GORDIAN®

Facilities Construction Costs with RSMeans data

Brian Adams, Senior Editor

2020

35th annual edition

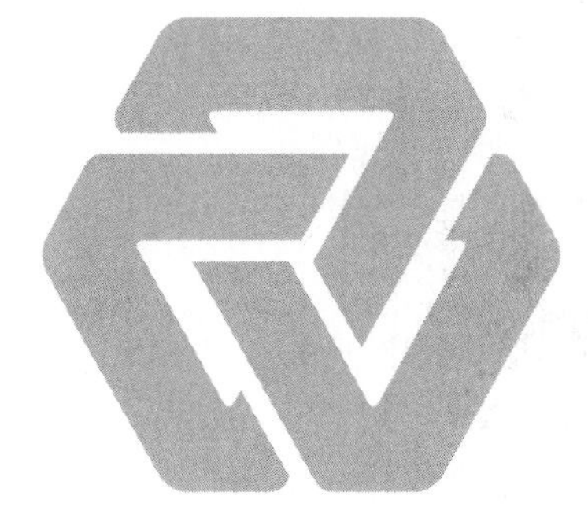

**Numbers in italics are the divisional responsibilities for each editor. Please contact the designated editor directly with any questions.*

RSMeans data from Gordian
Construction Publishers & Consultants
1099 Hingham Street, Suite 201
Rockland, MA 02370
United States of America
1.800.448.8182
RSMeans.com

Printed in the United States of America
ISSN 1540-6202
ISBN 978-1-950656-06-6

0200 $949.99 per copy (in United States)
Price is subject to change without prior notice.

Related Data and Services

Our engineers recommend the following products and services to complement *Facilities Construction Costs with RSMeans data:*

Annual Cost Data Books

2020 Assemblies Costs with RSMeans data
2020 Square Foot Costs with RSMeans data

Reference Books

Estimating Building Costs
RSMeans Estimating Handbook
Green Building: Project Planning & Estimating
How to Estimate with RSMeans data
Plan Reading & Material Takeoff
Project Scheduling & Management for Construction
Universal Design Ideas for Style, Comfort & Safety

Seminars and In-House Training

Unit Price Estimating
Training for our online estimating solution
Practical Project Management for Construction Professionals
Scheduling with MSProject for Construction Professionals
Mechanical & Electrical Estimating

RSMeans data

For access to the latest cost data, an intuitive search, and an easy-to-use estimate builder, take advantage of the time savings available from our online application.

To learn more visit: **RSMeans.com/online**

Enterprise Solutions

Building owners, facility managers, building product manufacturers, and attorneys across the public and private sectors engage with RSMeans data Enterprise to solve unique challenges where trusted construction cost data is critical.

To learn more visit: **RSMeans.com/Enterprise**

Custom Built Data Sets

Building and Space Models: Quickly plan construction costs across multiple locations based on geography, project size, building system component, product options, and other variables for precise budgeting and cost control.

Predictive Analytics: Accurately plan future builds with custom graphical interactive dashboards, negotiate future costs of tenant build-outs, and identify and compare national account pricing.

Consulting

Building Product Manufacturing Analytics: Validate your claims and assist with new product launches.

Third-Party Legal Resources: Used in cases of construction cost or estimate disputes, construction product failure vs. installation failure, eminent domain, class action construction product liability, and more.

API

For resellers or internal application integration, RSMeans data is offered via API. Deliver Unit, Assembly, and Square Foot Model data within your interface. To learn more about how you can provide your customers with the latest in localized construction cost data visit: **RSMeans.com/API**

Table of Contents

Foreword

The Value of RSMeans data from Gordian

Since 1942, RSMeans data has been the industry-standard materials, labor, and equipment cost information database for contractors, facility owners and managers, architects, engineers, and anyone else that requires the latest localized construction cost information. More than 75 years later, the objective remains the same: to provide facility and construction professionals with the most current and comprehensive construction cost database possible.

With the constant influx of new construction methods and materials, in addition to ever-changing labor and material costs, last year's cost data is not reliable for today's designs, estimates, or budgets. Gordian's cost engineers apply real-world construction experience to identify and quantify new building products and methodologies, adjust productivity rates, and adjust costs to local market conditions across the nation. This adds up to more than 22,000 hours in cost research annually. This unparalleled construction cost expertise is why so many facility and construction professionals rely on RSMeans data year over year.

About Gordian

Gordian originated in the spirit of innovation and a strong commitment to helping clients reach and exceed their construction goals. In 1982, Gordian's chairman and founder, Harry H. Mellon, created Job Order Contracting while serving as chief engineer at the Supreme Headquarters Allied Powers Europe. Job Order Contracting is a unique indefinite delivery/indefinite quantity (IDIQ) process, which enables facility owners to complete a substantial number of repair, maintenance, and construction projects with a single, competitively awarded contract. Realizing facility and infrastructure owners across various industries could greatly benefit from the time and cost saving advantages of this innovative construction procurement solution, he established Gordian in 1990.

Continuing the commitment to providing the most relevant and accurate facility and construction data, software, and expertise in the industry, Gordian enhanced the fortitude of its data with the acquisition of RSMeans in 2014. And in an effort to expand its facility management capabilities, Gordian acquired Sightlines, the leading provider of facilities benchmarking data and analysis, in 2015.

Our Offerings

Gordian is the leader in facility and construction cost data, software, and expertise for all phases of the building life cycle. From planning to design, procurement, construction, and operations, Gordian's solutions help clients maximize efficiency, optimize cost savings, and increase building quality with its highly specialized data engineers, software, and unique proprietary data sets.

Our Commitment

At Gordian, we do more than talk about the quality of our data and the usefulness of its application. We stand behind all of our RSMeans data—from historical cost indexes to construction materials and techniques—to craft current costs and predict future trends. If you have any questions about our products or services, please call us toll-free at 800.448.8182 or visit our website at gordian.com.

MasterFormat® 2016/ MasterFormat® 2018 Comparison Table

This table compares the 2016 edition of the Construction Specifications Institute's MasterFormat® to the expanded 2018 edition. For your convenience, all revised 2016 numbers and titles are listed along with the corresponding 2018 numbers and titles.

CSI 2016 MF ID	CSI 2016 MF Description	CSI 2018 MF ID	CSI 2018 MF Description
015632	Temporary Security	015733	Temporary Security
019308	Facility Maintenance Equipment	019308	Facilities Maintenance, Equipment
024200	Removal and Salvage of Construction Materials	024200	Removal and Diversion of Construction Materials
040130	Unit Masonry Cleaning	04012052	Cleaning Masonry
068010	Composite Decking	067300	Composite Decking
072127	Reflective Insulation	072153	Reflective Insulation
072610	Above-Grade Vapor Retarders	072613	Above-Grade Vapor Retarders
074473	Metal Faced Panels	074433	Metal Faced Panels
075430	Ketone Ethylene Ester Roofing	075416	Ketone Ethylene Ester Roofing
077280	Vents	077280	Vent Options
087125	Weatherstripping	087125	Door Weatherstripping
087530	Weatherstripping	087530	Window Weatherstripping
096223	Bamboo Flooring	096436	Bamboo Flooring
099103	Paint Restoration	090190	Maintenance of Painting and Coating
102833	Laundry Accessories	102823	Laundry Accessories
117610	Operating Room Equipment	117610	Equipment for Operating Rooms
117710	Radiology Equipment	117710	Equipment for Radiology
122310	Wood Interior Shutters	122313	Wood Interior Shutters
123580	Commercial Kitchen Casework	123539	Commercial Kitchen Casework
124636	Desk Accessories	124113	Desk Accessories
141210	Dumbwaiters	141000	Dumbwaiters
211113	Facility Water Distribution Piping	211113	Facility Fire Suppression Piping
233715	Louvers	233715	Air Outlets and Inlets, HVAC Louvers
260580	Wiring Connections	260583	Wiring Connections
270110	Operation and Maintenance of Communications Systems	270110	Operation and Maintenance of Communication Systems
272123	Data Communications Switches and Hubs	272129	Data Communications Switches and Hubs
283149	Carbon-Monoxide Detection Sensors	284611.21	Carbon-Monoxide Detection Sensors
284621	Fire Alarm	284620	Fire Alarm
316233	Drilled Micropiles	316333	Drilled Micropiles
323420	Fabricated Pedestrian Bridges	323413	Fabricated Pedestrian Bridges
337543	Shunt Reactors	337253	Shunt Reactors
350100	Operation and Maint. of Waterway & Marine Construction	350100	Operation and Maintenance of Waterway and Marine Construction

How the Cost Data Is Built: An Overview

Unit Prices*

All cost data have been divided into 50 divisions according to the MasterFormat® system of classification and numbering.

Assemblies*

The cost data in this section have been organized in an "Assemblies" format. These assemblies are the functional elements of a building and are arranged according to the 7 elements of the UNIFORMAT II classification system. For a complete explanation of a typical "Assembly", see "RSMeans data: Assemblies—How They Work."

*Residential Models**

Model buildings for four classes of construction—economy, average, custom, and luxury—are developed and shown with complete costs per square foot.

*Commercial/Industrial/ Institutional Models**

This section contains complete costs for 77 typical model buildings expressed as costs per square foot.

*Green Commercial/Industrial/ Institutional Models**

This section contains complete costs for 25 green model buildings expressed as costs per square foot.

*References**

This section includes information on Equipment Rental Costs, Crew Listings, Historical Cost Indexes, City Cost Indexes, Location Factors, Reference Tables, and Change Orders, as well as a listing of abbreviations.

- **Equipment Rental Costs:** Included are the average costs to rent and operate hundreds of pieces of construction equipment.
- **Crew Listings**: This section lists all the crews referenced in the cost data. A crew is composed of more than one trade classification and/or the addition of power equipment to any trade classification. Power equipment is included in the cost of the crew. Costs are shown both with bare labor rates and with the installing contractor's overhead and profit added. For each, the total crew cost per eight-hour day and the composite cost per labor-hour are listed.

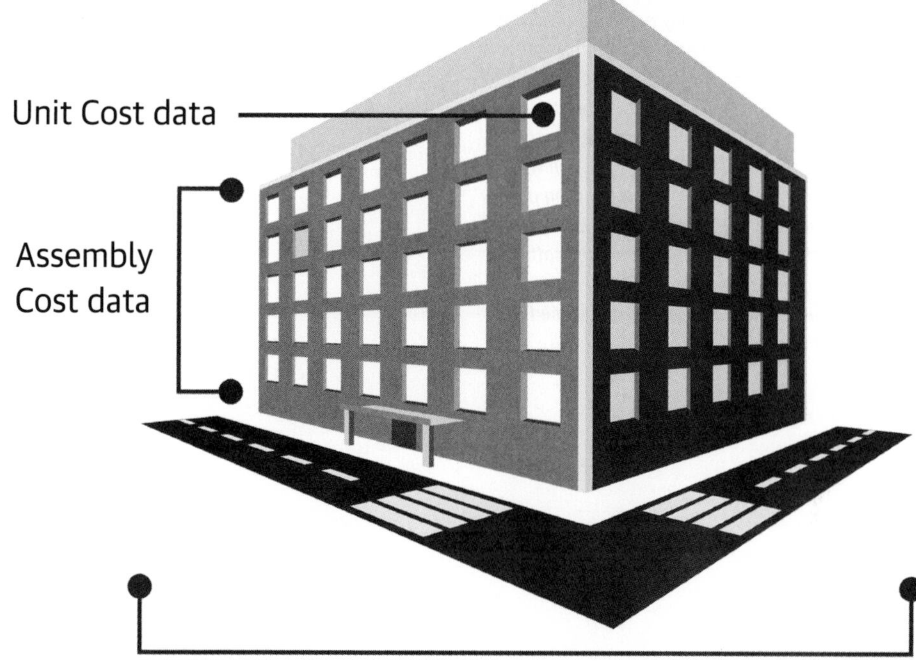

- **Historical Cost Indexes**: These indexes provide you with data to adjust construction costs over time.
- **City Cost Indexes**: All costs in this data set are U.S. national averages. Costs vary by region. You can adjust for this by CSI Division to over 730 cities in 900+ 3-digit zip codes throughout the U.S. and Canada by using this data.
- **Location Factors**: You can adjust total project costs to over 730 cities in 900+ 3-digit zip codes throughout the U.S. and Canada by using the weighted number, which applies across all divisions.
- **Reference Tables**: At the beginning of selected major classifications in the Unit Prices are reference numbers indicators. These numbers refer you to related information in the Reference Section. In this section, you'll find reference tables, explanations, and estimating information that support how we develop the unit price data, technical data, and estimating procedures.
- **Change Orders**: This section includes information on the factors that influence the pricing of change orders.
- **Abbreviations**: A listing of abbreviations used throughout this information, along with the terms they represent, is included.

Index (printed versions only)

A comprehensive listing of all terms and subjects will help you quickly find what you need when you are not sure where it occurs in MasterFormat®.

Conclusion

This information is designed to be as comprehensive and easy to use as possible.

The Construction Specifications Institute (CSI) and Construction Specifications Canada (CSC) have produced the 2018 edition of MasterFormat®, a system of titles and numbers used extensively to organize construction information.

All unit prices in the RSMeans cost data are now arranged in the 50-division MasterFormat® 2018 system.

* Not all information is available in all data sets

Note: The material prices in RSMeans cost data are "contractor's prices." They are the prices that contractors can expect to pay at the lumberyards, suppliers'/distributors' warehouses, etc. Small orders of specialty items would be higher than the costs shown, while very large orders, such as truckload lots, would be less. The variation would depend on the size, timing, and negotiating power of the contractor. The labor costs are primarily for new construction or major renovation rather than repairs or minor alterations. With reasonable exercise of judgment, the figures can be used for any building work.

Estimating with RSMeans data: Unit Prices

Following these steps will allow you to complete an accurate estimate using RSMeans data: Unit Prices.

1. Scope Out the Project

- Think through the project and identify the CSI divisions needed in your estimate.
- Identify the individual work tasks that will need to be covered in your estimate.
- The Unit Price data have been divided into 50 divisions according to CSI MasterFormat® 2018.
- In printed versions, the Unit Price Section Table of Contents on page 1 may also be helpful when scoping out your project.
- Experienced estimators find it helpful to begin with Division 2 and continue through completion. Division 1 can be estimated after the full project scope is known.

2. Quantify

- Determine the number of units required for each work task that you identified.
- Experienced estimators include an allowance for waste in their quantities. (Waste is not included in our Unit Price line items unless otherwise stated.)

3. Price the Quantities

- Use the search tools available to locate individual Unit Price line items for your estimate.
- Reference Numbers indicated within a Unit Price section refer to additional information that you may find useful.
- The crew indicates who is performing the work for that task. Crew codes are expanded in the Crew Listings in the Reference Section to include all trades and equipment that comprise the crew.
- The Daily Output is the amount of work the crew is expected to complete in one day.
- The Labor-Hours value is the amount of time it will take for the crew to install one unit of work.
- The abbreviated Unit designation indicates the unit of measure upon which the crew, productivity, and prices are based.
- Bare Costs are shown for materials, labor, and equipment needed to complete the Unit Price line item. Bare costs do not include waste, project overhead, payroll insurance, payroll taxes, main office overhead, or profit.
- The Total Incl O&P cost is the billing rate or invoice amount of the installing contractor or subcontractor who performs the work for the Unit Price line item.

4. Multiply

- Multiply the total number of units needed for your project by the Total Incl O&P cost for each Unit Price line item.
- Be careful that your take off unit of measure matches the unit of measure in the Unit column.
- The price you calculate is an estimate for a completed item of work.
- Keep scoping individual tasks, determining the number of units required for those tasks, matching each task with individual Unit Price line items, and multiplying quantities by Total Incl O&P costs.
- An estimate completed in this manner is priced as if a subcontractor, or set of subcontractors, is performing the work. The estimate does not yet include Project Overhead or Estimate Summary components such as general contractor markups on subcontracted work, general contractor office overhead and profit, contingency, and location factors.

5. Project Overhead

- Include project overhead items from Division 1-General Requirements.
- These items are needed to make the job run. They are typically, but not always, provided by the general contractor. Items include, but are not limited to, field personnel, insurance, performance bond, permits, testing, temporary utilities, field office and storage facilities, temporary scaffolding and platforms, equipment mobilization and demobilization, temporary roads and sidewalks, winter protection, temporary barricades and fencing, temporary security, temporary signs, field engineering and layout, final cleaning, and commissioning.
- Each item should be quantified and matched to individual Unit Price line items in Division 1, then priced and added to your estimate.
- An alternate method of estimating project overhead costs is to apply a percentage of the total project cost—usually 5% to 15% with an average of 10% (see General Conditions).
- Include other project related expenses in your estimate such as:
  - Rented equipment not itemized in the Crew Listings
  - Rubbish handling throughout the project (see section 02 41 19.19)

6. Estimate Summary

- Include sales tax as required by laws of your state or county.
- Include the general contractor's markup on self-performed work, usually 5% to 15% with an average of 10%.
- Include the general contractor's markup on subcontracted work, usually 5% to 15% with an average of 10%.
- Include the general contractor's main office overhead and profit:
  - RSMeans data provides general guidelines on the general contractor's main office overhead (see section 01 31 13.60 and Reference Number R013113-50).
  - Markups will depend on the size of the general contractor's operations, projected annual revenue, the level of risk, and the level of competition in the local area and for this project in particular.
- Include a contingency, usually 3% to 5%, if appropriate.
- Adjust your estimate to the project's location by using the City Cost Indexes or the Location Factors in the Reference Section:
  - Look at the rules in "How to Use the City Cost Indexes" to see how to apply the Indexes for your location.
  - When the proper Index or Factor has been identified for the project's location, convert it to a multiplier by dividing it by 100, then multiply that multiplier by your estimated total cost. The original estimated total cost will now be adjusted up or down from the national average to a total that is appropriate for your location.

Editors' Note:
We urge you to spend time reading and understanding the supporting material. An accurate estimate requires experience, knowledge, and careful calculation. The more you know about how we at RSMeans developed the data, the more accurate your estimate will be. In addition, it is important to take into consideration the reference material such as Equipment Listings, Crew Listings, City Cost Indexes, Location Factors, and Reference Tables.

How to Use the Cost Data: The Details

What's Behind the Numbers? The Development of Cost Data

RSMeans data engineers continually monitor developments in the construction industry in order to ensure reliable, thorough, and up-to-date cost information. While overall construction costs may vary relative to general economic conditions, price fluctuations within the industry are dependent upon many factors. Individual price variations may, in fact, be opposite to overall economic trends. Therefore, costs are constantly tracked and complete updates are performed yearly. Also, new items are frequently added in response to changes in materials and methods.

Costs in U.S. Dollars

All costs represent U.S. national averages and are given in U.S. dollars. The City Cost Index (CCI) with RSMeans data can be used to adjust costs to a particular location. The CCI for Canada can be used to adjust U.S. national averages to local costs in Canadian dollars. No exchange rate conversion is necessary because it has already been factored in.

G The processes or products identified by the green symbol in our publications have been determined to be environmentally responsible and/or resource-efficient solely by RSMeans data engineering staff. The inclusion of the green symbol does not represent compliance with any specific industry association or standard.

Material Costs

RSMeans data engineers contact manufacturers, dealers, distributors, and contractors all across the U.S. and Canada to determine national average material costs. If you have access to current material costs for your specific location, you may wish to make adjustments to reflect differences from the national average. Included within material costs are fasteners for a normal installation. RSMeans data engineers use manufacturers' recommendations, written specifications, and/or standard construction practices for the sizing and spacing of fasteners. Adjustments to material costs may be required for your specific application or location. The manufacturer's warranty is assumed. Extended warranties are not included in the material costs. **Material costs do not include sales tax.**

Labor Costs

Labor costs are based upon a mathematical average of trade-specific wages in 30 major U.S. cities. The type of wage (union, open shop, or residential) is identified on the inside back cover of printed publications or selected by the estimator when using the electronic products. Markups for the wages can also be found on the inside back cover of printed publications and/or under the labor references found in the electronic products.

- If wage rates in your area vary from those used, or if rate increases are expected within a given year, labor costs should be adjusted accordingly.

Labor costs reflect productivity based on actual working conditions. In addition to actual installation, these figures include time spent during a normal weekday on tasks, such as material receiving and handling, mobilization at the site, site movement, breaks, and cleanup.

Productivity data is developed over an extended period so as not to be influenced by abnormal variations and reflects a typical average.

Equipment Costs

Equipment costs include not only rental but also operating costs for equipment under normal use. The operating costs include parts and labor for routine servicing, such as the repair and replacement of pumps, filters, and worn lines. Normal operating expendables, such as fuel, lubricants, tires, and electricity (where applicable), are also included. Extraordinary operating expendables with highly variable wear patterns, such as diamond bits and blades, are excluded. These costs are included under materials. Equipment rental rates are obtained from industry sources throughout North America—contractors, suppliers, dealers, manufacturers, and distributors.

Rental rates can also be treated as reimbursement costs for contractor-owned equipment. Owned equipment costs include depreciation, loan payments, interest, taxes, insurance, storage, and major repairs.

Equipment costs do not include operators' wages.

Equipment Cost/Day—The cost of equipment required for each crew is included in the Crew Listings in the Reference Section (small tools that are considered essential everyday tools are not listed out separately). The Crew Listings itemize specialized tools and heavy equipment along with labor trades. The daily cost of itemized equipment included in a crew is based on dividing the weekly bare rental rate by 5 (number of working days per week), then adding the hourly operating cost times 8 (the number of hours per day). This Equipment Cost/Day is shown in the last column of the Equipment Rental Costs in the Reference Section.

Mobilization, Demobilization—The cost to move construction equipment from an equipment yard or rental company to the job site and back again is not included in equipment costs. Mobilization (to the site) and demobilization (from the site) costs can be found in the Unit Price Section. If a piece of equipment is already at the job site, it is not appropriate to utilize mobilization or demobilization costs again in an estimate.

Overhead and Profit

Total Cost including O&P for the installing contractor is shown in the last column of the Unit Price and/or Assemblies. This figure is the sum of the bare material cost plus 10% for profit, the bare labor cost plus total overhead and profit, and the bare equipment cost plus 10% for profit. Details for the calculation of overhead and profit on labor are shown on the inside back cover of the printed product and in the Reference Section of the electronic product.

General Conditions

Cost data in this data set are presented in two ways: Bare Costs and Total Cost including O&P (Overhead and Profit). General Conditions, or General Requirements, of the contract should also be added to the Total Cost including O&P when applicable. Costs for General Conditions are listed in Division 1 of the Unit Price Section and in the Reference Section.

General Conditions for the installing contractor may range from 0% to 10% of the Total Cost including O&P. For the general or prime contractor, costs for General Conditions may range from 5% to 15% of the Total Cost including O&P, with a figure of 10% as the most typical allowance. If applicable, the Assemblies and Models sections use costs that include the installing contractor's overhead and profit (O&P).

Factors Affecting Costs

Costs can vary depending upon a number of variables. Here's a listing of some factors that affect costs and points to consider.

Quality—The prices for materials and the workmanship upon which productivity is based represent sound construction work. They are also in line with industry standard and manufacturer specifications and are frequently used by federal, state, and local governments.

Overtime—We have made no allowance for overtime. If you anticipate premium time or work beyond normal working hours, be sure to make an appropriate adjustment to your labor costs.

Productivity—The productivity, daily output, and labor-hour figures for each line item are based on an eight-hour work day in daylight hours in moderate temperatures and up to a 14' working height unless otherwise indicated. For work that extends beyond normal work hours or is performed under adverse conditions, productivity may decrease.

Size of Project—The size, scope of work, and type of construction project will have a significant impact on cost. Economies of scale can reduce costs for large projects. Unit costs can often run higher for small projects.

Location—Material prices are for metropolitan areas. However, in dense urban areas, traffic and site storage limitations may increase costs. Beyond a 20-mile radius of metropolitan areas, extra trucking or transportation charges may also increase the material costs slightly. On the other hand, lower wage rates may be in effect. Be sure to consider both of these factors when preparing an estimate, particularly if the job site is located in a central city or remote rural location. In addition, highly specialized subcontract items may require travel and per-diem expenses for mechanics.

Other Factors—

- season of year
- contractor management
- weather conditions
- local union restrictions
- building code requirements
- availability of:
  - adequate energy
  - skilled labor
  - building materials
- owner's special requirements/restrictions
- safety requirements
- environmental considerations
- access

Unpredictable Factors—General business conditions influence "in-place" costs of all items. Substitute materials and construction methods may have to be employed. These may affect the installed cost and/or life cycle costs. Such factors may be difficult to evaluate and cannot necessarily be predicted on the basis of the job's location in a particular section of the country. Thus, where these factors apply, you may find significant but unavoidable cost variations for which you will have to apply a measure of judgment to your estimate.

Rounding of Costs

In printed publications only, all unit prices in excess of $5.00 have been rounded to make them easier to use and still maintain adequate precision of the results.

How Subcontracted Items Affect Costs

A considerable portion of all large construction jobs is usually subcontracted. In fact, the percentage done by subcontractors is constantly increasing and may run over 90%. Since the workers employed by these companies do nothing else but install their particular products, they soon become experts in that line. As a result, installation by these firms is accomplished so efficiently that the total in-place cost, even with the general contractor's overhead and profit, is no more, and often less, than if the principal contractor had handled the installation. Companies that deal with construction specialties are anxious to have their products perform well and, consequently, the installation will be the best possible.

Contingencies

The allowance for contingencies generally provides for unforeseen construction difficulties. On alterations or repair jobs, 20% is not too much. If drawings are final and only field contingencies are being considered, 2% or 3% is probably sufficient and often nothing needs to be added. Contractually, changes in plans will be covered by extras. The contractor should consider inflationary price trends and possible material shortages during the course of the job. These escalation factors are dependent upon both economic conditions and the anticipated time between the estimate and actual construction. If drawings are not complete or approved, or a budget cost is wanted, it is wise to add 5% to 10%. Contingencies, then, are a matter of judgment.

Important Estimating Considerations

The productivity, or daily output, of each craftsman or crew assumes a well-managed job where tradesmen with the proper tools and equipment, along with the appropriate construction materials, are present. Included are daily set-up and cleanup time, break time, and plan layout time. Unless otherwise indicated, time for material movement on site (for items that can be transported by hand) of up to 200' into the building and to the first or second floor is also included. If material has to be transported by other means, over greater distances, or to higher floors, an additional allowance should be considered by the estimator.

While horizontal movement is typically a sole function of distances, vertical transport introduces other variables that can significantly impact productivity. In an occupied building, the use of elevators (assuming access, size, and required protective measures are acceptable) must be understood at the time of the estimate. For new construction, hoist wait and cycle times can easily be 15 minutes and may result in scheduled access extending beyond the normal work day. Finally, all vertical transport will impose strict weight limits likely to preclude the use of any motorized material handling.

The productivity, or daily output, also assumes installation that meets manufacturer/designer/standard specifications. A time allowance for quality control checks, minor adjustments, and any task required to ensure proper function or operation is also included. For items that require connections to services, time is included for positioning, leveling, securing the unit, and making all the necessary connections (and start up where applicable) to ensure a complete installation. Estimating of the services themselves (electrical, plumbing, water, steam, hydraulics, dust collection, etc.) is separate.

In some cases, the estimator must consider the use of a crane and an appropriate crew for the installation of large or heavy items. For those situations where a crane is not included in the assigned crew and as part of the line item cost, then equipment rental costs, mobilization and demobilization costs, and operator and support personnel costs must be considered.

Labor-Hours

The labor-hours expressed in this publication are derived by dividing the total daily labor-hours for the crew by the daily output. Based on average installation time and the assumptions listed above, the labor-hours include: direct labor, indirect labor, and nonproductive time. A typical day for a craftsman might include but is not limited to:

- Direct Work
  - Measuring and layout
  - Preparing materials
  - Actual installation
  - Quality assurance/quality control
- Indirect Work
  - Reading plans or specifications
  - Preparing space
  - Receiving materials
  - Material movement
  - Giving or receiving instruction
  - Miscellaneous
- Non-Work
  - Chatting
  - Personal issues
  - Breaks
  - Interruptions (i.e., sickness, weather, material or equipment shortages, etc.)

If any of the items for a typical day do not apply to the particular work or project situation, the estimator should make any necessary adjustments.

Final Checklist

Estimating can be a straightforward process provided you remember the basics. Here's a checklist of some of the steps you should remember to complete before finalizing your estimate.

Did you remember to:

- factor in the City Cost Index for your locale?
- take into consideration which items have been marked up and by how much?
- mark up the entire estimate sufficiently for your purposes?
- read the background information on techniques and technical matters that could impact your project time span and cost?
- include all components of your project in the final estimate?
- double check your figures for accuracy?
- call RSMeans data engineers if you have any questions about your estimate or the data you've used? Remember, Gordian stands behind all of our products, including our extensive RSMeans data solutions. If you have any questions about your estimate, about the costs you've used from our data, or even about the technical aspects of the job that may affect your estimate, feel free to call the Gordian RSMeans editors at 1.800.448.8182.

Using Minimum Labor/Equipment Charges for Small Quantities

Estimating small construction or repair tasks often creates situations in which the quantity of work to be performed is very small. When this occurs, the labor and/or equipment costs to perform the work may be too low to allow for the crew to get to the job, receive instructions, find materials, get set up, perform the work, clean up, and get to the next job. In these situations, the estimator should compare the developed labor and/or equipment costs for performing the work (e.g., quantity × labor and/or equipment costs) with the *minimum labor/equipment charge* within the Unit Price section of the data set. (These minimum labor/equipment charge line items appear only in *Facilities Construction Costs with RSMeans data* and *Commercial Renovation Costs with RSMeans data.*)

If the labor and/or equipment costs developed by the estimator are LOWER THAN the *minimum labor/equipment charge* listed at the bottom of specific sections of Unit Price costs, the estimator should adjust the developed costs upward to the *minimum labor/equipment charge*. The proper use of a *minimum labor/equipment charge* results in having enough money in the estimate to cover the contractor's higher cost of performing a very small amount of work during a partial workday.

A *minimum labor/equipment charge* should be used only when the task being estimated is the only task the crew will perform at the job site that day. **If, however, the crew will be able to perform other tasks at the job site that day, the use of a *minimum labor/equipment charge* is not appropriate.**

08 52 Wood Windows

08 52 10 – Plain Wood Windows

08 52 10.40 Casement Window			Crew	Daily Output	Labor-Hours	Unit	Material	2020 Bare Costs Labor	Equipment	Total	Total Incl O&P
0010	**CASEMENT WINDOW**, including frame, screen and grilles										
0100	2'-0" x 3'-0" H, double insulated glass	G	1 Carp	10	.800	Ea.	275	42.50		317.50	370
0150	Low E glass	G		10	.800		285	42.50		327.50	385
0200	2'-0" x 4'-6" high, double insulated glass	G		9	.889		390	47		437	505
0250	Low E glass	G		9	.889		440	47		487	560
9000	Minimum labor/equipment charge		1 Carp	3	2.667	Job		142		142	227

Example:

Establish the bid price to install two casement windows. Assume installation of 2' × 4'-6" metal clad windows with insulating glass [Unit Price line number 08 52 10.40 0250], and that this is the only task this crew will perform at the job site that day.

Solution:

Step One – Develop the Bare Labor Cost for this task:
Bare Labor Cost = 2 windows @ $47 each = $94.00

Step Two – Evaluate the *minimum labor/equipment charge* for this Unit Price section against the developed Bare Labor Cost for this task:
Minimum labor/equipment charge = $142.00 (compare with $94.00)

Step Three – Choose to adjust the developed labor cost upward to the *minimum labor/equipment charge*.

Step Four – Develop the bid price for this task (including O&P):
Add together the marked-up Bare Material Cost for this task and the marked-up *minimum labor/equipment charge* for this Unit Price section.

2 × ($440.00 + 10%) + ($142.00 + 60.3%)

= 2 × ($440.00 + $44.00) + ($142.00 + $85.63)

= 2 × ($484.00) + $227.63

= $968.00 + $227.63

= $1,195.63

ANSWER: $1,195.63 is the correct bid price to use. This sum takes into consideration the material cost (with 10% for profit) for these two windows, plus the *minimum labor/equipment charge* (with O&P included) for this section of the Unit Price.

Unit Price Section

Table of Contents

Table of Contents (cont.)

Table of Contents (cont.)

RSMeans data: Unit Prices—How They Work

All RSMeans data: Unit Prices are organized in the same way.

03 30 Cast-In-Place Concrete

03 30 53 – Miscellaneous Cast-In-Place Concrete

03 30 53.40 Concrete In Place			Crew	Daily Output	Labor-Hours	Unit	2020 Bare Costs Material	Labor	Equipment	Total	Total Incl O&P
0010	CONCRETE IN PLACE	R033105-20									
0020	Including forms (4 uses), Grade 60 rebar, concrete (Portland cement	R033105-70									
0050	Type I), placement and finishing unless otherwise indicated	R033105-80									
0300	Beams (3500 psi), 5 kip/L.F., 10′ span		C-14A	15.62	12.804	C.Y.	370	675	28	1,073	1,500
0350	25′ span		"	18.55	10.782		390	570	23.50	983.50	1,375
0500	Chimney foundations (5000 psi), over 5 C.Y.	R033053-50	C-14C	32.22	3.476		178	175	.83	353.83	475
0510	(3500 psi), under 5 C.Y.		"	23.71	4.724		208	238	1.13	447.13	610
0700	Columns, square (4000 psi), 12″ x 12″, up to 1% reinforcing by area		C-14A	11.96	16.722		410	885	36.50	1,331.50	1,900
3540	Equipment pad (3000 psi), 3′ x 3′ x 6″ thick		C-14H	45	1.067	Ea.	50.50	55	.60	106.10	144
3550	4′ x 4′ x 6″ thick			30	1.600		78	82.50	.90	161.40	218
3560	5′ x 5′ x 8″ thick			18	2.667		138	138	1.49	277.49	375
3570	6′ x 6′ x 8″ thick			14	3.429		190	177	1.92	368.92	495
3580	8′ x 8′ x 10″ thick			8	6		395	310	3.36	708.36	935
3590	10′ x 10′ x 12″ thick		↓	5	9.600	↓	695	495	5.40	1,195.40	1,550

It is important to understand the structure of RSMeans data: Unit Prices so that you can find information easily and use it correctly.

1 Line Numbers

Line Numbers consist of 12 characters, which identify a unique location in the database for each task. The first 6 or 8 digits conform to the Construction Specifications Institute MasterFormat® 2018. The remainder of the digits are a further breakdown in order to arrange items in understandable groups of similar tasks. Line numbers are consistent across all of our publications, so a line number in any of our products will always refer to the same item of work.

2 Descriptions

Descriptions are shown in a hierarchical structure to make them readable. In order to read a complete description, read up through the indents to the top of the section. Include everything that is above and to the left that is not contradicted by information below. For instance, the complete description for line 03 30 53.40 3550 is "Concrete in place, including forms (4 uses), Grade 60 rebar, concrete (Portland cement Type 1), placement and finishing unless otherwise indicated; Equipment pad (3000 psi), 4' × 4' × 6" thick."

3 RSMeans data

When using **RSMeans data**, it is important to read through an entire section to ensure that you use the data that most closely matches your work. Note that sometimes there is additional information shown in the section that may improve your price. There are frequently lines that further describe, add to, or adjust data for specific situations.

4 Reference Information

Gordian's RSMeans engineers have created **reference** information to assist you in your estimate. **If** there is information that applies to a section, it will be indicated at the start of the section. The Reference Section is located in the back of the data set.

5 Crews

Crews include labor and/or equipment necessary to accomplish each task. In this case, Crew C-14H is used. Gordian's RSMeans staff selects a crew to represent the workers and equipment that are typically used for that task. In this case, Crew C-14H consists of one carpenter foreman (outside), two carpenters, one rodman, one laborer, one cement finisher, and one gas engine vibrator. Details of all crews can be found in the Reference Section.

Crews - Renovation

Crew No.	Bare Costs		Incl. Subs O & P		Cost Per Labor-Hour	
Crew C-14H	Hr.	Daily	Hr.	Daily	Bare Costs	Incl. O&P
1 Carpenter Foreman (outside)	$55.15	$441.20	$88.40	$707.20	$51.65	$82.24
2 Carpenters	53.15	850.40	85.15	1362.40		
1 Rodman (reinf.)	56.40	451.20	88.85	710.80		
1 Laborer	42.10	336.80	67.45	539.60		
1 Cement Finisher	49.95	399.60	78.45	627.60		
1 Gas Engine Vibrator		26.85		29.54	.56	.62
48 L.H., Daily Totals		$2506.05		$3977.14	$52.21	$82.86

6 Daily Output

The **Daily Output** is the amount of work that the crew can do in a normal 8-hour workday, including mobilization, layout, movement of materials, and cleanup. In this case, crew C-14H can install thirty 4' × 4' × 6" thick concrete pads in a day. Daily output is variable and based on many factors, including the size of the job, location, and environmental conditions. RSMeans data represents work done in daylight (or adequate lighting) and temperate conditions.

7 Labor-Hours

The figure in the **Labor-Hours** column is the amount of labor required to perform one unit of work–in this case the amount of labor required to construct one 4' × 4' equipment pad. This figure is calculated by dividing the number of hours of labor in the crew by the daily output (48 labor-hours divided by 30 pads = 1.6 hours of labor per pad). Multiply 1.6 times 60 to see the value in minutes: 60 × 1.6 = 96 minutes. Note: the labor-hour figure is not dependent on the crew size. A change in crew size will result in a corresponding change in daily output, but the labor-hours per unit of work will not change.

8 Unit of Measure

All RSMeans data: Unit Prices include the typical **Unit of Measure** used for estimating that item. For concrete-in-place the typical unit is cubic yards (C.Y.) or each (Ea.). For installing broadloom carpet it is square yard and for gypsum board it is square foot. The estimator needs to take special care that the unit in the data matches the unit in the take-off. Unit conversions may be found in the Reference Section.

9 Bare Costs

Bare Costs are the costs of materials, labor, and equipment that the installing contractor pays. They represent the cost, in U.S. dollars, for one unit of work. They do not include any markups for profit or labor burden.

10 Bare Total

The **Total column** represents the total bare cost for the installing contractor in U.S. dollars. In this case, the sum of $78 for material + $82.50 for labor + $.90 for equipment is $161.40.

11 Total Incl O&P

The **Total Incl O&P column** is the total cost, including overhead and profit, that the installing contractor will charge the customer. This represents the cost of materials plus 10% profit, the cost of labor plus labor burden and 10% profit, and the cost of equipment plus 10% profit. It does not include the general contractor's overhead and profit. Note: See the inside back cover of the printed product or the Reference Section of the electronic product for details on how the labor burden is calculated.

National Average

The RSMeans data in our print publications represent a "national average" cost. This data should be modified to the project location using the ***City Cost Indexes*** *or* ***Location Factors*** *tables found in the Reference Section. Use the Location Factors to adjust estimate totals if the project covers multiple trades. Use the City Cost Indexes (CCI) for single trade projects or projects where a more detailed analysis is required. All figures in the two tables are derived from the same research. The last row of data in the CCI—the weighted average—is the same as the numbers reported for each location in the location factor table.*

RSMeans data: Unit Prices—How They Work (Continued)

Project Name: Pre-Engineered Steel Building		Architect: As Shown						
Location:	**Anywhere, USA**						**01/01/20**	**R+R**
1 Line Number	Description	Qty	Unit	Material	Labor	Equipment	SubContract	Estimate Total
03 30 53.40 3940	Strip footing, 12" x 24", reinforced	15	C.Y.	$2,565.00	$1,770.00	$8.40	$0.00	
03 30 53.40 3950	Strip footing, 12" x 36", reinforced	34	C.Y.	$5,610.00	$3,196.00	$15.30	$0.00	
03 11 13.65 3000	Concrete slab edge forms	500	L.F.	$165.00	$1,345.00	$0.00	$0.00	
03 22 11.10 0200	Welded wire fabric reinforcing	150	C.S.F.	$2,872.50	$4,350.00	$0.00	$0.00	
03 31 13.35 0300	Ready mix concrete, 4000 psi for slab on grade	278	C.Y.	$35,306.00	$0.00	$0.00	$0.00	
03 31 13.70 4300	Place, strike off & consolidate concrete slab	278	C.Y.	$0.00	$5,309.80	$136.22	$0.00	
03 35 13.30 0250	Machine float & trowel concrete slab	15,000	S.F.	$0.00	$9,900.00	$750.00	$0.00	
03 15 16.20 0140	Cut control joints in concrete slab	950	L.F.	$47.50	$418.00	$57.00	$0.00	
03 39 23.13 0300	Sprayed concrete curing membrane	150	C.S.F.	$1,867.50	$1,065.00	$0.00	$0.00	
Division 03	**Subtotal**			**$48,433.50**	**$27,353.80**	**$966.92**	**$0.00**	**$76,754.22**
08 36 13.10 2650	Manual 10' x 10' steel sectional overhead door	8	Ea.	$11,000.00	$3,760.00	$0.00	$0.00	
08 36 13.10 2860	Insulation and steel back panel for OH door	800	S.F.	$4,000.00	$0.00	$0.00	$0.00	
Division 08	**Subtotal**			**$15,000.00**	**$3,760.00**	**$0.00**	**$0.00**	**$18,760.00**
13 34 19.50 1100	Pre-Engineered Steel Building, 100' x 150' x 24'	15,000	SF Flr.	$0.00	$0.00	$0.00	$375,000.00	
13 34 19.50 6050	Framing for PESB door opening, 3' x 7'	4	Opng.	$0.00	$0.00	$0.00	$2,400.00	
13 34 19.50 6100	Framing for PESB door opening, 10' x 10'	8	Opng.	$0.00	$0.00	$0.00	$9,800.00	
13 34 19.50 6200	Framing for PESB window opening, 4' x 3'	6	Opng.	$0.00	$0.00	$0.00	$3,510.00	
13 34 19.50 5750	PESB door, 3' x 7', single leaf	4	Opng.	$2,920.00	$736.00	$0.00	$0.00	
13 34 19.50 7750	PESB sliding window, 4' x 3' with screen	6	Opng.	$2,940.00	$630.00	$67.80	$0.00	
13 34 19.50 6550	PESB gutter, eave type, 26 ga., painted	300	L.F.	$2,415.00	$864.00	$0.00	$0.00	
13 34 19.50 8650	PESB roof vent, 12" wide x 10' long	15	Ea.	$570.00	$3,465.00	$0.00	$0.00	
13 34 19.50 6900	PESB insulation, vinyl faced, 4" thick	27,400	S.F.	$11,782.00	$10,138.00	$0.00	$0.00	
Division 13	**Subtotal**			**$20,627.00**	**$15,833.00**	**$67.80**	**$390,710.00**	**$427,237.80**
		Subtotal		**$84,060.50**	**$46,946.80**	**$1,034.72**	**$390,710.00**	**$522,752.02**
Division 01	2 **General Requirements @ 7%**			**5,884.24**	**3,286.28**	**72.43**	**27,349.70**	
		Estimate Subtotal		**$89,944.74**	**$50,233.08**	**$1,107.15**	**$418,059.70**	**$522,752.02**
	3	**Sales Tax @ 5%**		**4,497.24**		**55.36**	**10,451.49**	
		Subtotal A		**94,441.97**	**50,233.08**	**1,162.51**	**428,511.19**	
	4	**GC O & P**		**9,444.20**	**29,788.21**	**116.25**	**42,851.12**	
		Subtotal B		**103,886.17**	**80,021.29**	**1,278.76**	**471,362.31**	**$656,548.53**
	5	**Contingency @ 5%**						**32,827.43**
		Subtotal C						**$689,375.96**
	6	**Bond @ $12/1000 +10% O&P**						**9,099.76**
		Subtotal D						**$698,475.72**
	7	**Location Adjustment Factor**			**115.50**			**108,263.74**
		Grand Total						**$806,739.45**

This estimate is based on an interactive spreadsheet. You are free to download it and adjust it to your methodology. A copy of this spreadsheet is available at **RSMeans.com/2020books.**

Sample Estimate

This sample demonstrates the elements of an estimate, including a tally of the RSMeans data lines and a summary of the markups on a contractor's work to arrive at a total cost to the owner. The Location Factor with RSMeans data is added at the bottom of the estimate to adjust the cost of the work to a specific location.

---

1 Work Performed

The body of the estimate shows the RSMeans data selected, including the line number, a brief description of each item, its take-off unit and quantity, and the bare costs of materials, labor, and equipment. This estimate also includes a column titled "SubContract." This data is taken from the column "Total Incl O&P" and represents the total that a subcontractor would charge a general contractor for the work, including the sub's markup for overhead and profit.

2 Division 1, General Requirements

This is the first division numerically but the last division estimated. Division 1 includes project-wide needs provided by the general contractor. These requirements vary by project but may include temporary facilities and utilities, security, testing, project cleanup, etc. For small projects a percentage can be used—typically between 5% and 15% of project cost. For large projects the costs may be itemized and priced individually.

3 Sales Tax

If the work is subject to state or local sales taxes, the amount must be added to the estimate. Sales tax may be added to material costs, equipment costs, and subcontracted work. In this case, sales tax was added in all three categories. It was assumed that approximately half the subcontracted work would be material cost, so the tax was applied to 50% of the subcontract total.

4 GC O&P

This entry represents the general contractor's markup on material, labor, equipment, and subcontractor costs. Our standard markup on materials, equipment, and subcontracted work is 10%. In this estimate, the markup on the labor performed by the GC's workers uses "Skilled Workers Average" shown in Column F on the table "Installing Contractor's Overhead & Profit," which can be found on the inside back cover of the printed product or in the Reference Section of the electronic product.

5 Contingency

A factor for contingency may be added to any estimate to represent the cost of unknowns that may occur between the time that the estimate is performed and the time the project is constructed. The amount of the allowance will depend on the stage of design at which the estimate is done and the contractor's assessment of the risk involved. Refer to section 01 21 16.50 for contingency allowances.

6 Bonds

Bond costs should be added to the estimate. The figures here represent a typical performance bond, ensuring the owner that if the general contractor does not complete the obligations in the construction contract the bonding company will pay the cost for completion of the work.

7 Location Adjustment

Published prices are based on national average costs. If necessary, adjust the total cost of the project using a location factor from the "Location Factor" table or the "City Cost Index" table. Use location factors if the work is general, covering multiple trades. If the work is by a single trade (e.g., masonry) use the more specific data found in the "City Cost Indexes."

Estimating Tips

01 20 00 Price and Payment Procedures

- Allowances that should be added to estimates to cover contingencies and job conditions that are not included in the national average material and labor costs are shown in Section 01 21.
- When estimating historic preservation projects (depending on the condition of the existing structure and the owner's requirements), a 15–20% contingency or allowance is recommended, regardless of the stage of the drawings.

01 30 00 Administrative Requirements

- Before determining a final cost estimate, it is good practice to review all the items listed in Subdivisions 01 31 and 01 32 to make final adjustments for items that may need customizing to specific job conditions.
- Requirements for initial and periodic submittals can represent a significant cost to the General Requirements of a job. Thoroughly check the submittal specifications when estimating a project to determine any costs that should be included.

01 40 00 Quality Requirements

- All projects will require some degree of quality control. This cost is not included in the unit cost of construction listed in each division. Depending upon the terms of the contract, the various costs of inspection and testing can be the responsibility of either the owner or the contractor. Be sure to include the required costs in your estimate.

01 50 00 Temporary Facilities and Controls

- Barricades, access roads, safety nets, scaffolding, security, and many more requirements for the execution of a safe project are elements of direct cost. These costs can easily be overlooked when preparing an estimate. When looking through the major classifications of this subdivision, determine which items apply to each division in your estimate.
- Construction equipment rental costs can be found in the Reference Section in Section 01 54 33. Operators' wages are not included in equipment rental costs.
- Equipment mobilization and demobilization costs are not included in equipment rental costs and must be considered separately.
- The cost of small tools provided by the installing contractor for his workers is covered in the "Overhead" column on the "Installing Contractor's Overhead and Profit" table that lists labor trades, base rates, and markups. Therefore, it is included in the "Total Incl. O&P" cost of any unit price line item.

01 70 00 Execution and Closeout Requirements

- When preparing an estimate, thoroughly read the specifications to determine the requirements for Contract Closeout. Final cleaning, record documentation, operation and maintenance data, warranties and bonds, and spare parts and maintenance materials can all be elements of cost for the completion of a contract. Do not overlook these in your estimate.

Reference Numbers

Reference numbers are shown at the beginning of some major classifications. These numbers refer to related items in the Reference Section. The reference information may be an estimating procedure, an alternate pricing method, or technical information.

Note: Not all subdivisions listed here necessarily appear. ■

01 11 Summary of Work

01 11 31 – Professional Consultants

01 11 31.10 Architectural Fees		Crew	Daily Output	Labor-Hours	Unit	Material	2020 Bare Costs Labor	Equipment	Total	Total Incl O&P
0010	**ARCHITECTURAL FEES** R011110-10									
0020	For new construction									
0060	Minimum				Project				4.90%	4.90%
0090	Maximum								16%	16%
0100	For alteration work, to $500,000, add to new construction fee								50%	50%
0150	Over $500,000, add to new construction fee								25%	25%
2000	For "Greening" of building G				↓				3%	3%

01 11 31.20 Construction Management Fees		Crew	Daily Output	Labor-Hours	Unit	Material	Labor	Equipment	Total	Total Incl O&P
0010	**CONSTRUCTION MANAGEMENT FEES**									
0060	For work to $100,000				Project				10%	10%
0070	To $250,000								9%	9%
0090	To $1,000,000								6%	6%
0100	To $5,000,000								5%	5%
0110	To $10,000,000								4%	4%
0300	$50,000,000 job, minimum								2.50%	2.50%
0350	Maximum				↓				4%	4%

01 11 31.30 Engineering Fees		Crew	Daily Output	Labor-Hours	Unit	Material	Labor	Equipment	Total	Total Incl O&P
0010	**ENGINEERING FEES** R011110-30									
0020	Educational planning consultant, minimum				Project				.50%	.50%
0100	Maximum				"				2.50%	2.50%
0200	Electrical, minimum				Contrct				4.10%	4.10%
0300	Maximum								10.10%	10.10%
0400	Elevator & conveying systems, minimum								2.50%	2.50%
0500	Maximum								5%	5%
0600	Food service & kitchen equipment, minimum								8%	8%
0700	Maximum								12%	12%
0800	Landscaping & site development, minimum								2.50%	2.50%
0900	Maximum								6%	6%
1000	Mechanical (plumbing & HVAC), minimum								4.10%	4.10%
1100	Maximum				↓				10.10%	10.10%
1200	Structural, minimum				Project				1%	1%
1300	Maximum				"				2.50%	2.50%

01 11 31.50 Models		Crew	Daily Output	Labor-Hours	Unit	Material	Labor	Equipment	Total	Total Incl O&P
0010	**MODELS**									
0500	2 story building, scaled 100′ x 200′, simple materials and details				Ea.	4,750			4,750	5,225
0510	Elaborate materials and details				"	30,900			30,900	34,000

01 11 31.75 Renderings		Crew	Daily Output	Labor-Hours	Unit	Material	Labor	Equipment	Total	Total Incl O&P
0010	**RENDERINGS** Color, matted, 20″ x 30″, eye level,									
0020	1 building, minimum				Ea.	2,250			2,250	2,475
0050	Average					3,200			3,200	3,525
0100	Maximum					5,150			5,150	5,675
1000	5 buildings, minimum					4,250			4,250	4,675
1100	Maximum					8,525			8,525	9,375
2000	Aerial perspective, color, 1 building, minimum					3,100			3,100	3,425
2100	Maximum					8,450			8,450	9,300
3000	5 buildings, minimum					6,050			6,050	6,650
3100	Maximum				↓	12,100			12,100	13,400

01 21 Allowances

01 21 16 – Contingency Allowances

01 21 16.50 Contingencies

	01 21 16.50 Contingencies	Crew	Daily Output	Labor-Hours	Unit	Material	2020 Bare Costs Labor	Equipment	Total	Total Incl O&P
0010	**CONTINGENCIES**, Add to estimate									
0020	Conceptual stage				Project				20%	20%
0050	Schematic stage								15%	15%
0100	Preliminary working drawing stage (Design Dev.)								10%	10%
0150	Final working drawing stage				↓				3%	3%

01 21 53 – Factors Allowance

01 21 53.50 Factors

	01 21 53.50 Factors	Crew	Daily Output	Labor-Hours	Unit	Material	2020 Bare Costs Labor	Equipment	Total	Total Incl O&P
0010	**FACTORS** Cost adjustments R012153-10									
0100	Add to construction costs for particular job requirements									
0500	Cut & patch to match existing construction, add, minimum				Costs	2%	3%			
0550	Maximum					5%	9%			
0800	Dust protection, add, minimum					1%	2%			
0850	Maximum					4%	11%			
1100	Equipment usage curtailment, add, minimum					1%	1%			
1150	Maximum					3%	10%			
1400	Material handling & storage limitation, add, minimum					1%	1%			
1450	Maximum					6%	7%			
1700	Protection of existing work, add, minimum					2%	2%			
1750	Maximum					5%	7%			
2000	Shift work requirements, add, minimum						5%			
2050	Maximum						30%			
2300	Temporary shoring and bracing, add, minimum					2%	5%			
2350	Maximum				↓	5%	12%			

01 21 53.60 Security Factors

	01 21 53.60 Security Factors	Crew	Daily Output	Labor-Hours	Unit	Material	2020 Bare Costs Labor	Equipment	Total	Total Incl O&P
0010	**SECURITY FACTORS** R012153-60									
0100	Additional costs due to security requirements									
0110	Daily search of personnel, supplies, equipment and vehicles									
0120	Physical search, inventory and doc of assets, at entry				Costs		30%			
0130	At entry and exit						50%			
0140	Physical search, at entry						6.25%			
0150	At entry and exit						12.50%			
0160	Electronic scan search, at entry						2%			
0170	At entry and exit						4%			
0180	Visual inspection only, at entry						.25%			
0190	At entry and exit						.50%			
0200	ID card or display sticker only, at entry						.12%			
0210	At entry and exit				↓		.25%			
0220	Day 1 as described below, then visual only for up to 5 day job duration									
0230	Physical search, inventory and doc of assets, at entry				Costs		5%			
0240	At entry and exit						10%			
0250	Physical search, at entry						1.25%			
0260	At entry and exit						2.50%			
0270	Electronic scan search, at entry						.42%			
0280	At entry and exit				↓		.83%			
0290	Day 1 as described below, then visual only for 6-10 day job duration									
0300	Physical search, inventory and doc of assets, at entry				Costs		2.50%			
0310	At entry and exit						5%			
0320	Physical search, at entry						.63%			
0330	At entry and exit						1.25%			
0340	Electronic scan search, at entry						.21%			
0350	At entry and exit				↓		.42%			
0360	Day 1 as described below, then visual only for 11-20 day job duration									

01 21 Allowances

01 21 53 – Factors Allowance

01 21 53.60 Security Factors		Crew	Daily Output	Labor-Hours	Unit	2020 Bare Costs Material	Labor	Equipment	Total	Total Incl O&P
0370	Physical search, inventory and doc of assets, at entry				Costs		1.25%			
0380	At entry and exit						2.50%			
0390	Physical search, at entry						.31%			
0400	At entry and exit						.63%			
0410	Electronic scan search, at entry						.10%			
0420	At entry and exit						.21%			
0430	Beyond 20 days, costs are negligible									
0440	Escort required to be with tradesperson during work effort				Costs		6.25%			

01 21 55 – Job Conditions Allowance

01 21 55.50 Job Conditions		Crew	Daily Output	Labor-Hours	Unit	Material	Labor	Equipment	Total	Total Incl O&P
0010	**JOB CONDITIONS** Modifications to applicable									
0020	cost summaries									
0100	Economic conditions, favorable, deduct				Project				2%	2%
0200	Unfavorable, add								5%	5%
0300	Hoisting conditions, favorable, deduct								2%	2%
0400	Unfavorable, add								5%	5%
0500	General contractor management, experienced, deduct								2%	2%
0600	Inexperienced, add								10%	10%
0700	Labor availability, surplus, deduct								1%	1%
0800	Shortage, add								10%	10%
0900	Material storage area, available, deduct								1%	1%
1000	Not available, add								2%	2%
1100	Subcontractor availability, surplus, deduct								5%	5%
1200	Shortage, add								12%	12%
1300	Work space, available, deduct								2%	2%
1400	Not available, add								5%	5%

01 21 57 – Overtime Allowance

01 21 57.50 Overtime		Crew	Daily Output	Labor-Hours	Unit	Material	Labor	Equipment	Total	Total Incl O&P
0010	**OVERTIME** for early completion of projects or where R012909-90									
0020	labor shortages exist, add to usual labor, up to				Costs		100%			

01 21 63 – Taxes

01 21 63.10 Taxes		Crew	Daily Output	Labor-Hours	Unit	Material	Labor	Equipment	Total	Total Incl O&P
0010	**TAXES** R012909-80									
0020	Sales tax, State, average				%	5.08%				
0050	Maximum R012909-85					7.50%				
0200	Social Security, on first $118,500 of wages						7.65%			
0300	Unemployment, combined Federal and State, minimum R012909-86						.60%			
0350	Average						9.60%			
0400	Maximum						12%			

01 31 Project Management and Coordination

01 31 13 – Project Coordination

01 31 13.20 Field Personnel		Crew	Daily Output	Labor-Hours	Unit	Material	2020 Bare Costs Labor	Equipment	Total	Total Incl O&P
0010	**FIELD PERSONNEL**									
0020	Clerk, average				Week		495		495	790
0100	Field engineer, junior engineer						1,175		1,175	1,877
0120	Engineer						1,825		1,825	2,775
0140	Senior engineer						2,400		2,400	3,625
0160	General purpose laborer, average	1 Clab	.20	40			1,675		1,675	2,700
0180	Project manager, minimum						2,175		2,175	3,475
0200	Average						2,500		2,500	4,000
0220	Maximum						2,850		2,850	4,575
0240	Superintendent, minimum						2,125		2,125	3,400
0260	Average						2,325		2,325	3,725
0280	Maximum						2,650		2,650	4,225
0290	Timekeeper, average				↓		1,350		1,350	2,175

01 31 13.30 Insurance		Crew	Daily Output	Labor-Hours	Unit	Material	Labor	Equipment	Total	Total Incl O&P
0010	**INSURANCE** R013113-40									
0020	Builders risk, standard, minimum				Job				.24%	.24%
0050	Maximum R013113-50								.64%	.64%
0200	All-risk type, minimum								.25%	.25%
0250	Maximum R013113-60				↓				.62%	.62%
0400	Contractor's equipment floater, minimum				Value				.50%	.50%
0450	Maximum				"				1.50%	1.50%
0800	Workers' compensation & employer's liability, average									
0850	by trade, carpentry, general				Payroll		11.97%			
0900	Clerical						.38%			
0950	Concrete						10.84%			
1000	Electrical						4.91%			
1050	Excavation						7.81%			
1100	Glazing						11.29%			
1150	Insulation						10.07%			
1200	Lathing						7.58%			
1250	Masonry						13.40%			
1300	Painting & decorating						10.44%			
1350	Pile driving						12.06%			
1400	Plastering						10.24%			
1450	Plumbing						5.77%			
1500	Roofing						27.34%			
1550	Sheet metal work (HVAC)						7.56%			
1600	Steel erection, structural						17.21%			
1650	Tile work, interior ceramic						8.10%			
1700	Waterproofing, brush or hand caulking						6.05%			
1800	Wrecking						15.48%			
2000	Range of 35 trades in 50 states, excl. wrecking & clerical, min.						1.37%			
2100	Average						10.60%			
2200	Maximum				↓		120.29%			

01 31 13.40 Main Office Expense		Crew	Daily Output	Labor-Hours	Unit	Material	Labor	Equipment	Total	Total Incl O&P
0010	**MAIN OFFICE EXPENSE** Average for General Contractors R013113-50									
0020	As a percentage of their annual volume									
0030	Annual volume to $300,000, minimum				% Vol.				20%	
0040	Maximum								30%	
0060	To $500,000, minimum								17%	
0070	Maximum								22%	
0080	To $1,000,000, minimum				↓				16%	

01 31 Project Management and Coordination

01 31 13 – Project Coordination

01 31 13.40 Main Office Expense		Crew	Daily Output	Labor-Hours	Unit	Material	2020 Bare Costs Labor	Equipment	Total	Total Incl O&P
0090	Maximum				% Vol.				19%	
0110	To $3,000,000, minimum								14%	
0120	Maximum								16%	
0130	To $5,000,000, minimum								8%	
0140	Maximum				↓				10%	

01 31 13.50 General Contractor's Mark-Up		Crew	Daily Output	Labor-Hours	Unit	Material	Labor	Equipment	Total	Total Incl O&P
0010	**GENERAL CONTRACTOR'S MARK-UP** on Change Orders									
0200	Extra work, by subcontractors, add				%				10%	10%
0250	By General Contractor, add								15%	15%
0400	Omitted work, by subcontractors, deduct all but								5%	5%
0450	By General Contractor, deduct all but								7.50%	7.50%
0600	Overtime work, by subcontractors, add								15%	15%
0650	By General Contractor, add				↓				10%	10%

01 31 13.70 Overhead		Crew	Daily Output	Labor-Hours	Unit	Material	Labor	Equipment	Total	Total Incl O&P
0010	**OVERHEAD** As a percent of installing contractors direct costs R013113-50									
0040	Includes an allowance for home office expenses, FICA,									
0060	Risk & public liability insur. and unemploy.; minimum R013113-55				%				5%	
0080	Average								15%	
0120	With profit allowance, by size of project; under $50,000								40%	
0140	$50,000 to $100,000								35%	
0160	$100,000 to $500,000								25%	
0180	$500,000 to $1,000,000								20%	
0220	Handling subcontract; minimum								5%	5%
0260	Maximum				↓				15%	15%

01 31 13.90 Performance Bond		Crew	Daily Output	Labor-Hours	Unit	Material	Labor	Equipment	Total	Total Incl O&P
0010	**PERFORMANCE BOND** R013113-80									
0020	For buildings, minimum				Job				.60%	.60%
0100	Maximum				"				2.50%	2.50%

01 31 14 – Facilities Services Coordination

01 31 14.20 Lock Out/Tag Out		Crew	Daily Output	Labor-Hours	Unit	Material	Labor	Equipment	Total	Total Incl O&P
0010	**LOCK OUT / TAG OUT**									
0020	Miniature circuit breaker lock out device	1 Elec	220	.036	Ea.	24	2.23		26.23	30
0030	Miniature pin circuit breaker lock out device		220	.036		19.90	2.23		22.13	25.50
0040	Single circuit breaker lock out device		220	.036		21	2.23		23.23	26.50
0050	Multi-pole circuit breaker lock out device (15 to 225 Amp)		210	.038		19.90	2.34		22.24	25.50
0060	Large 3 pole circuit breaker lock out device (over 225 Amp)		210	.038		21	2.34		23.34	27
0080	Square D I-Line circuit breaker lock out device		210	.038		33	2.34		35.34	39.50
0090	Lock out disconnect switch, 30 to 100 Amp		330	.024		11	1.49		12.49	14.40
0100	100 to 400 Amp		330	.024		11	1.49		12.49	14.40
0110	Over 400 Amp		330	.024		11	1.49		12.49	14.40
0120	Lock out hasp for multiple lockout tags		200	.040		5.60	2.45		8.05	9.95
0130	Electrical cord plug lock out device		220	.036		10.50	2.23		12.73	15
0140	Electrical plug prong lock out device (3-wire grounding plug)		220	.036		6.95	2.23		9.18	11.10
0150	Wall switch lock out	↓	200	.040		25	2.45		27.45	32
0160	Fire alarm pull station lock out	1 Stpi	200	.040		18.75	2.62		21.37	24.50
0170	Sprinkler valve tamper and flow switch lock out device	1 Skwk	220	.036		16.60	1.99		18.59	21.50
0180	Lock out sign		330	.024		18.90	1.33		20.23	23
0190	Lock out tag	↓	440	.018	↓	5	1		6	7.10

01 32 Construction Progress Documentation

01 32 13 – Scheduling of Work

01 32 13.50 Scheduling of Work	Crew	Daily Output	Labor-Hours	Unit	Material	2020 Bare Costs Labor	Equipment	Total	Total Incl O&P
0010 **SCHEDULING**									
0025 Scheduling, critical path, $50 million project, initial schedule				Ea.				25,750	28,325
0030 Monthly updates				"				3,348	3,682

01 32 33 – Photographic Documentation

01 32 33.50 Photographs	Crew	Daily Output	Labor-Hours	Unit	Material	Labor	Equipment	Total	Total Incl O&P
0010 **PHOTOGRAPHS**									
0020 8" x 10", 4 shots, 2 prints ea., std. mounting				Set	545			545	600
0100 Hinged linen mounts					550			550	605
0200 8" x 10", 4 shots, 2 prints each, in color					535			535	590
0300 For I.D. slugs, add to all above					5.10			5.10	5.60
0500 Aerial photos, initial fly-over, 5 shots, digital images					455			455	500
0550 10 shots, digital images, 1 print					515			515	570
0600 For each additional print from fly-over					252			252	278
0700 For full color prints, add				▼	40%				
0750 Add for traffic control area				Ea.	355			355	390
0900 For over 30 miles from airport, add per				Mile	7.55			7.55	8.30
1500 Time lapse equipment, camera and projector, buy				Ea.	2,800			2,800	3,100
1550 Rent per month				"	1,325			1,325	1,450
1700 Cameraman and processing, black & white				Day	1,300			1,300	1,425
1720 Color				"	1,500			1,500	1,650

01 41 Regulatory Requirements

01 41 26 – Permit Requirements

01 41 26.50 Permits	Crew	Daily Output	Labor-Hours	Unit	Material	Labor	Equipment	Total	Total Incl O&P
0010 **PERMITS**									
0020 Rule of thumb, most cities, minimum				Job				.50%	.50%
0100 Maximum				"				2%	2%

01 45 Quality Control

01 45 23 – Testing and Inspecting Services

01 45 23.50 Testing	Crew	Daily Output	Labor-Hours	Unit	Material	Labor	Equipment	Total	Total Incl O&P
0010 **TESTING** and Inspecting Services									
0012 Testing, for concrete or steel building, minimum				Week				4,585	5,045
0025 Maximum				"				6,005	6,605
0200 Asphalt testing, compressive strength Marshall stability, set of 3				Ea.				545	600
0250 Extraction, individual tests on sample								236	260
0300 Penetration								91	100
0350 Mix design, 5 specimens								182	200
0360 Additional specimen								36	40
0400 Specific gravity								53	58
0420 Swell test								64	70
0450 Water effect and cohesion, set of 6								182	200
0470 Water effect and plastic flow								64	70
0600 Concrete testing, aggregates, abrasion, ASTM C 131								205	225
0650 Absorption, ASTM C 127								77	85
0800 Petrographic analysis, ASTM C 295								910	1,000
0900 Specific gravity, ASTM C 127								77	85
1000 Sieve analysis, washed, ASTM C 136								130	140
1050 Unwashed				▼				130	140

01 45 23 – Testing and Inspecting Services

01 45 23.50 Testing		Crew	Daily Output	Labor-Hours	Unit	2020 Bare Costs Material	Labor	Equipment	Total	Total Incl O&P
1200	Sulfate soundness				Ea.				182	200
1300	Weight per cubic foot								80	88
1500	Cement, physical tests, ASTM C 150								320	350
1600	Chemical tests, ASTM C 150								245	270
1800	Compressive test, cylinder, delivered to lab, ASTM C 39								36	40
1900	Picked up by lab, 30 minute round trip, add to above	1 Skwk	16	.500			27.50		27.50	43.50
1950	1 hour round trip, add to above		8	1			55		55	87.50
2000	2 hour round trip, add to above		4	2			110		110	175
2200	Compressive strength, cores (not incl. drilling), ASTM C 42								95	105
2250	Core drilling, 4" diameter (plus technician), up to 6" thick	B-89A	14	1.143		.65	55.50	8.15	64.30	98
2260	Technician for core drilling				Hr.				50	55
2300	Patching core holes, to 12" diameter	1 Cefi	22	.364	Ea.	36.50	18.15		54.65	68.50
2400	Drying shrinkage at 28 days								236	260
2500	Flexural test beams, ASTM C 78								136	150
2600	Mix design, one batch mix								259	285
2650	Added trial batches								120	132
2800	Modulus of elasticity, ASTM C 469								195	215
2900	Tensile test, cylinders, ASTM C 496								52	58
3000	Water-Cement ratio curve, 3 batches								141	155
3100	4 batches								186	205
3300	Masonry testing, absorption, per 5 brick, ASTM C 67								72	80
3350	Chemical resistance, per 2 brick								50	55
3400	Compressive strength, per 5 brick, ASTM C 67								95	105
3420	Efflorescence, per 5 brick, ASTM C 67								101	112
3440	Imperviousness, per 5 brick								87	96
3470	Modulus of rupture, per 5 brick								86	95
3500	Moisture, block only								68	75
3550	Mortar, compressive strength, set of 3								32	35
4100	Reinforcing steel, bend test								68	75
4200	Tensile test, up to #8 bar								68	75
4220	#9 to #11 bar								114	125
4240	#14 bar and larger								182	200
4400	Soil testing, Atterberg limits, liquid and plastic limits								91	100
4510	Hydrometer analysis								155	170
4600	Sieve analysis, washed, ASTM D 422								136	150
4710	Consolidation test (ASTM D 2435), minimum								455	500
4750	Moisture content, ASTM D 2216								18	20
4780	Permeability test, double ring infiltrometer								500	550
4800	Permeability, var. or constant head, undist., ASTM D 2434								264	290
4850	Recompacted								250	275
4900	Proctor compaction, 4" standard mold, ASTM D 698								230	253
4950	6" modified mold								110	120
5100	Shear tests, triaxial, minimum								410	450
5150	Maximum								545	600
5300	Direct shear, minimum, ASTM D 3080								320	350
5350	Maximum								410	450
5550	Technician for inspection, per day, earthwork								556	612
5570	Concrete								556	612
5650	Bolting								556	612
5750	Roofing								556	612
5790	Welding								556	612
5820	Non-destructive metal testing, dye penetrant				Day				556	612
5840	Magnetic particle								556	612

01 45 Quality Control

01 45 23 – Testing and Inspecting Services

01 45 23.50 Testing		Crew	Daily Output	Labor-Hours	Unit	Material	Labor	Equipment	Total	Total Incl O&P
						2020 Bare Costs				
5860	Radiography				Day				556	612
5880	Ultrasonic								556	612
6000	Welding certification, minimum				Ea.				91	100
6100	Maximum				"				364	400
7000	Underground storage tank									
7520	Volumetric tightness, <= 12,000 gal.				Ea.	470			470	520
7530	12,000 – 29,999 gal.					610			610	675
7540	>= 30,000 gal.					665			665	730
7600	Vadose zone (soil gas) sampling, 10-40 samples, min.				Day				1,375	1,500
7610	Maximum				"				2,275	2,500
7700	Ground water monitoring incl. drilling 3 wells, min.				Total				4,550	5,000
7710	Maximum				"				6,375	7,000
8000	X-ray concrete slabs				Ea.				182	200
9000	Thermographic testing, for bldg envelope heat loss, average 2,000 S.F. G				"				500	500

01 51 Temporary Utilities

01 51 13 – Temporary Electricity

01 51 13.80 Temporary Utilities		Crew	Daily Output	Labor-Hours	Unit	Material	Labor	Equipment	Total	Total Incl O&P
0010	**TEMPORARY UTILITIES**									
0350	Lighting, lamps, wiring, outlets, 40,000 S.F. building, 8 strings	1 Elec	34	.235	CSF Flr	5.85	14.45		20.30	28.50
0360	16 strings	"	17	.471		11.75	29		40.75	57.50
0400	Power for temp lighting only, 6.6 KWH, per month								.92	1.01
0430	11.8 KWH, per month								1.65	1.82
0450	23.6 KWH, per month								3.30	3.63
0600	Power for job duration incl. elevator, etc., minimum								53	58
0650	Maximum								110	121
0675	Temporary cooling				Ea.	1,025			1,025	1,125
0700	Temporary construction water bill per month, average				Month	77			77	84.50

01 52 Construction Facilities

01 52 13 – Field Offices and Sheds

01 52 13.20 Office and Storage Space		Crew	Daily Output	Labor-Hours	Unit	Material	Labor	Equipment	Total	Total Incl O&P
0010	**OFFICE AND STORAGE SPACE**									
0020	Office trailer, furnished, no hookups, 20′ x 8′, buy	2 Skwk	1	16	Ea.	9,525	880		10,405	11,900
0250	Rent per month					195			195	214
0300	32′ x 8′, buy	2 Skwk	.70	22.857		15,300	1,250		16,550	18,800
0350	Rent per month					245			245	269
0400	50′ x 10′, buy	2 Skwk	.60	26.667		30,300	1,475		31,775	35,600
0450	Rent per month					355			355	390
0500	50′ x 12′, buy	2 Skwk	.50	32		25,800	1,750		27,550	31,200
0550	Rent per month					460			460	505
0700	For air conditioning, rent per month, add					51.50			51.50	56.50
0800	For delivery, add per mile				Mile	12.20			12.20	13.40
0900	Bunk house trailer, 8′ x 40′ duplex dorm with kitchen, no hookups, buy	2 Carp	1	16	Ea.	89,000	850		89,850	99,000
0910	9 man with kitchen and bath, no hookups, buy		1	16		91,000	850		91,850	101,500
0920	18 man sleeper with bath, no hookups, buy		1	16		98,000	850		98,850	109,500
1000	Portable buildings, prefab, on skids, economy, 8′ x 8′		265	.060	S.F.	26.50	3.21		29.71	34
1100	Deluxe, 8′ x 12′		150	.107	"	29.50	5.65		35.15	41
1200	Storage boxes, 20′ x 8′, buy	2 Skwk	1.80	8.889	Ea.	3,150	490		3,640	4,250

01 52 Construction Facilities

01 52 13 – Field Offices and Sheds

01 52 13.20 Office and Storage Space		Crew	Daily Output	Labor-Hours	Unit	Material	2020 Bare Costs Labor	Equipment	Total	Total Incl O&P
1250	Rent per month				Ea.	86			86	94.50
1300	40′ x 8′, buy	2 Skwk	1.40	11.429		3,800	625		4,425	5,175
1350	Rent per month				↓	126			126	139
5000	Air supported structures, see Section 13 31 13.13									

01 52 13.40 Field Office Expense		Crew	Daily Output	Labor-Hours	Unit	Material	Labor	Equipment	Total	Total Incl O&P
0010	**FIELD OFFICE EXPENSE**									
0100	Office equipment rental average				Month	209			209	230
0120	Office supplies, average				"	87			87	96
0125	Office trailer rental, see Section 01 52 13.20									
0140	Telephone bill; avg. bill/month incl. long dist.				Month	87			87	96
0160	Lights & HVAC				"	163			163	179

01 54 Construction Aids

01 54 09 – Protection Equipment

01 54 09.50 Personnel Protective Equipment		Crew	Daily Output	Labor-Hours	Unit	Material	Labor	Equipment	Total	Total Incl O&P
0010	**PERSONNEL PROTECTIVE EQUIPMENT**									
0015	Hazardous waste protection									
0020	Respirator mask only, full face, silicone				Ea.	281			281	310
0030	Half face, silicone					52.50			52.50	58
0040	Respirator cartridges, 2 req'd/mask, dust or asbestos					4.11			4.11	4.52
0050	Chemical vapor					4.55			4.55	5
0060	Combination vapor and dust					14.15			14.15	15.55
0100	Emergency escape breathing apparatus, 5 minutes					695			695	765
0110	10 minutes					855			855	940
0150	Self contained breathing apparatus with full face piece, 30 minutes					2,175			2,175	2,400
0160	60 minutes					2,975			2,975	3,275
0200	Encapsulating suits, limited use, level A					2,000			2,000	2,200
0210	Level B				↓	405			405	445
0300	Over boots, latex				Pr.	8.15			8.15	8.95
0310	PVC					34.50			34.50	38
0320	Neoprene					37.50			37.50	41.50
0400	Gloves, nitrile/PVC					84			84	92.50
0410	Neoprene coated					50.50			50.50	55.50
0500	Fire protection equipment, shoes/boots				↓	197			197	216
0510	Hard hats				Ea.	330			330	360
0520	Gloves				Pr.	73			73	80

01 54 09.60 Safety Nets		Crew	Daily Output	Labor-Hours	Unit	Material	Labor	Equipment	Total	Total Incl O&P
0010	**SAFETY NETS**									
0020	No supports, stock sizes, nylon, 3-1/2″ mesh				S.F.	3.16			3.16	3.48
0100	Polypropylene, 6″ mesh					1.64			1.64	1.80
0200	Small mesh debris nets, 1/4″ mesh, stock sizes					.56			.56	.62
0220	Combined 3-1/2″ mesh and 1/4″ mesh, stock sizes					4.92			4.92	5.40
0300	Rental, 4″ mesh, stock sizes, 3 months					.84			.84	.92
0320	6 month rental					1.03			1.03	1.13
0340	12 months				↓	1.43			1.43	1.57

01 54 16 – Temporary Hoists

01 54 16.50 Weekly Forklift Crew		Crew	Daily Output	Labor-Hours	Unit	Material	2020 Bare Costs Labor	Equipment	Total	Total Incl O&P
0010	**WEEKLY FORKLIFT CREW**									
0100	All-terrain forklift, 45′ lift, 35′ reach, 9000 lb. capacity	A-3P	.20	40	Week		2,125	1,875	4,000	5,375

01 54 19 – Temporary Cranes

01 54 19.50 Daily Crane Crews		Crew	Daily Output	Labor-Hours	Unit	Material	2020 Bare Costs Labor	Equipment	Total	Total Incl O&P
0010	**DAILY CRANE CREWS** for small jobs, portal to portal R015433-15									
0100	12-ton truck-mounted hydraulic crane	A-3H	1	8	Day		475	725	1,200	1,525
0200	25-ton	A-3I	1	8			475	800	1,275	1,625
0300	40-ton	A-3J	1	8			475	1,275	1,750	2,150
0400	55-ton	A-3K	1	16			880	1,500	2,380	3,025
0500	80-ton	A-3L	1	16			880	2,250	3,130	3,850
0900	If crane is needed on a Saturday, Sunday or Holiday									
0910	At time-and-a-half, add				Day		50%			
0920	At double time, add				"		100%			

01 54 19.60 Monthly Tower Crane Crew		Crew	Daily Output	Labor-Hours	Unit	Material	2020 Bare Costs Labor	Equipment	Total	Total Incl O&P
0010	**MONTHLY TOWER CRANE CREW**, excludes concrete footing									
0100	Static tower crane, 130′ high, 106′ jib, 6200 lb. capacity	A-3N	.05	176	Month		10,400	37,300	47,700	57,500

01 54 23 – Temporary Scaffolding and Platforms

01 54 23.60 Pump Staging		Crew	Daily Output	Labor-Hours	Unit	Material	2020 Bare Costs Labor	Equipment	Total	Total Incl O&P
0010	**PUMP STAGING**, Aluminum R015423-20									
0200	24′ long pole section, buy				Ea.	380			380	420
0300	18′ long pole section, buy					310			310	345
0400	12′ long pole section, buy					199			199	219
0500	6′ long pole section, buy					125			125	138
0600	6′ long splice joint section, buy					78			78	86
0700	Pump jack, buy					150			150	165
0900	Foldable brace, buy					56			56	61.50
1000	Workbench/back safety rail support, buy					76.50			76.50	84
1100	Scaffolding planks/workbench, 14″ wide x 24′ long, buy					675			675	745
1200	Plank end safety rail, buy					360			360	395
1250	Safety net, 22′ long, buy					355			355	395
1300	System in place, 50′ working height, per use based on 50 uses	2 Carp	84.80	.189	C.S.F.	6.25	10.05		16.30	23
1400	100 uses		84.80	.189		3.13	10.05		13.18	19.50
1500	150 uses		84.80	.189		2.09	10.05		12.14	18.35

01 54 23.70 Scaffolding		Crew	Daily Output	Labor-Hours	Unit	Material	2020 Bare Costs Labor	Equipment	Total	Total Incl O&P
0010	**SCAFFOLDING** R015423-10									
0015	Steel tube, regular, no plank, labor only to erect & dismantle									
0090	Building exterior, wall face, 1 to 5 stories, 6′-4″ x 5′ frames	3 Carp	8	3	C.S.F.		159		159	255
0200	6 to 12 stories	4 Carp	8	4			213		213	340
0301	13 to 20 stories	5 Clab	8	5			211		211	335
0460	Building interior, wall face area, up to 16′ high	3 Carp	12	2			106		106	170
0560	16′ to 40′ high		10	2.400			128		128	204
0800	Building interior floor area, up to 30′ high		150	.160	C.C.F.		8.50		8.50	13.60
0900	Over 30′ high	4 Carp	160	.200	"		10.65		10.65	17.05
0906	Complete system for face of walls, no plank, material only rent/mo				C.S.F.	34			34	37.50
0908	Interior spaces, no plank, material only rent/mo				C.C.F.	3.88			3.88	4.27
0910	Steel tubular, heavy duty shoring, buy									
0920	Frames 5′ high 2′ wide				Ea.	89.50			89.50	98.50
0925	5′ high 4′ wide					106			106	116
0930	6′ high 2′ wide					107			107	118
0935	6′ high 4′ wide					115			115	126

01 54 23 – Temporary Scaffolding and Platforms

01 54 23.70 Scaffolding		Crew	Daily Output	Labor-Hours	Unit	2020 Bare Costs Material	Labor	Equipment	Total	Total Incl O&P
0940	Accessories									
0945	Cross braces				Ea.	17.05			17.05	18.80
0950	U-head, 8" x 8"					20.50			20.50	22.50
0955	J-head, 4" x 8"					14.90			14.90	16.40
0960	Base plate, 8" x 8"					16.05			16.05	17.65
0965	Leveling jack				↓	35			35	38
1000	Steel tubular, regular, buy									
1100	Frames 3' high 5' wide				Ea.	92.50			92.50	102
1150	5' high 5' wide					108			108	119
1200	6'-4" high 5' wide					90.50			90.50	99.50
1350	7'-6" high 6' wide					151			151	166
1500	Accessories, cross braces					18.85			18.85	20.50
1550	Guardrail post					23.50			23.50	25.50
1600	Guardrail 7' section					8.20			8.20	9.05
1650	Screw jacks & plates					26			26	28.50
1700	Sidearm brackets					23.50			23.50	26
1750	8" casters					37.50			37.50	41
1800	Plank 2" x 10" x 16'-0"					64.50			64.50	71
1900	Stairway section					292			292	320
1910	Stairway starter bar					32.50			32.50	35.50
1920	Stairway inside handrail					54.50			54.50	60
1930	Stairway outside handrail					87.50			87.50	96
1940	Walk-thru frame guardrail				↓	42			42	46.50
2000	Steel tubular, regular, rent/mo.									
2100	Frames 3' high 5' wide				Ea.	4.50			4.50	4.95
2150	5' high 5' wide					4.50			4.50	4.95
2200	6'-4" high 5' wide					5.70			5.70	6.30
2250	7'-6" high 6' wide					9.90			9.90	10.90
2500	Accessories, cross braces					.90			.90	.99
2550	Guardrail post					.90			.90	.99
2600	Guardrail 7' section					.90			.90	.99
2650	Screw jacks & plates					1.80			1.80	1.98
2700	Sidearm brackets					1.80			1.80	1.98
2750	8" casters					7.20			7.20	7.90
2800	Outrigger for rolling tower					2.70			2.70	2.97
2850	Plank 2" x 10" x 16'-0"					9.95			9.95	10.90
2900	Stairway section					31.50			31.50	35
2940	Walk-thru frame guardrail				↓	2.25			2.25	2.48
3000	Steel tubular, heavy duty shoring, rent/mo.									
3250	5' high 2' & 4' wide				Ea.	8.45			8.45	9.30
3300	6' high 2' & 4' wide					8.45			8.45	9.30
3500	Accessories, cross braces					.90			.90	.99
3600	U-head, 8" x 8"					2.50			2.50	2.75
3650	J-head, 4" x 8"					2.50			2.50	2.75
3700	Base plate, 8" x 8"					.90			.90	.99
3750	Leveling jack				↓	2.47			2.47	2.72
4000	Scaffolding, stl. tubular, reg., no plank, labor only to erect & dismantle									
4100	Building exterior 2 stories	3 Carp	8	3	C.S.F.		159		159	255
4150	4 stories	"	8	3			159		159	255
4200	6 stories	4 Carp	8	4			213		213	340
4250	8 stories		8	4			213		213	340
4300	10 stories		7.50	4.267			227		227	365
4350	12 stories	↓	7.50	4.267	↓		227		227	365

01 54 Construction Aids

01 54 23 – Temporary Scaffolding and Platforms

01 54 23.70 Scaffolding		Crew	Daily Output	Labor-Hours	Unit	Material	2020 Bare Costs Labor	Equipment	Total	Total Incl O&P
5700	Planks, 2″ x 10″ x 16′-0″, labor only to erect & remove to 50′ H	3 Carp	72	.333	Ea.		17.70		17.70	28.50
5800	Over 50′ high	4 Carp	80	.400	″		21.50		21.50	34
6000	Heavy duty shoring for elevated slab forms to 8′-2″ high, floor area									
6100	Labor only to erect & dismantle	4 Carp	16	2	C.S.F.		106		106	170
6110	Materials only, rent/mo.				″	43.50			43.50	48
6500	To 14′-8″ high									
6600	Labor only to erect & dismantle	4 Carp	10	3.200	C.S.F.		170		170	272
6610	Materials only, rent/mo				″	63.50			63.50	70

01 54 23.75 Scaffolding Specialties

		Crew	Daily Output	Labor-Hours	Unit	Material	Labor	Equipment	Total	Total Incl O&P
0010	**SCAFFOLDING SPECIALTIES**									
1200	Sidewalk bridge, heavy duty steel posts & beams, including									
1210	parapet protection & waterproofing (material cost is rent/month)									
1220	8′ to 10′ wide, 2 posts	3 Carp	15	1.600	L.F.	46	85		131	187
1230	3 posts	″	10	2.400	″	71	128		199	282
1500	Sidewalk bridge using tubular steel scaffold frames including									
1510	planking (material cost is rent/month)	3 Carp	45	.533	L.F.	8.40	28.50		36.90	55
1600	For 2 uses per month, deduct from all above					50%				
1700	For 1 use every 2 months, add to all above					100%				
1900	Catwalks, 20″ wide, no guardrails, 7′ span, buy				Ea.	153			153	168
2000	10′ span, buy					213			213	234
3720	Putlog, standard, 8′ span, with hangers, buy					80			80	88
3730	Rent per month					16.30			16.30	17.95
3750	12′ span, buy					101			101	111
3755	Rent per month					20.50			20.50	22.50
3760	Trussed type, 16′ span, buy					262			262	288
3770	Rent per month					24.50			24.50	27
3790	22′ span, buy					284			284	315
3795	Rent per month					32.50			32.50	35.50
3800	Rolling ladders with handrails, 30″ wide, buy, 2 step					320			320	355
4000	7 step					1,125			1,125	1,250
4050	10 step					1,625			1,625	1,775
4100	Rolling towers, buy, 5′ wide, 7′ long, 10′ high					1,375			1,375	1,500
4200	For additional 5′ high sections, to buy					255			255	280
4300	Complete incl. wheels, railings, outriggers,									
4350	21′ high, to buy				Ea.	2,325			2,325	2,550
4400	Rent/month = 5% of purchase cost				″	116			116	128
5000	Motorized work platform, mast climber									
5050	Base unit, 50′ W, less than 100′ tall, rent/mo				Ea.	3,625			3,625	3,975
5100	Less than 200′ tall, rent/mo					3,700			3,700	4,075
5150	Less than 300′ tall, rent/mo					4,475			4,475	4,925
5200	Less than 400′ tall, rent/mo					5,225			5,225	5,750
5250	Set up and demob, per unit, less than 100′ tall	B-68F	16.60	1.446	C.S.F.		80.50	16.85	97.35	147
5300	Less than 200′ tall		25	.960			53.50	11.20	64.70	97.50
5350	Less than 300′ tall		30	.800			44.50	9.35	53.85	81.50
5400	Less than 400′ tall		33	.727			40.50	8.50	49	74
5500	Mobilization (price includes freight in and out) per unit	B-34N	1	16	Ea.		845	680	1,525	2,075

01 54 23.80 Staging Aids

		Crew	Daily Output	Labor-Hours	Unit	Material	Labor	Equipment	Total	Total Incl O&P
0010	**STAGING AIDS** and fall protection equipment									
0100	Sidewall staging bracket, tubular, buy				Ea.	61.50			61.50	67.50
0110	Cost each per day, based on 250 days use				Day	.25			.25	.27
0200	Guard post, buy				Ea.	53.50			53.50	59
0210	Cost each per day, based on 250 days use				Day	.21			.21	.24

01 54 Construction Aids

01 54 23 – Temporary Scaffolding and Platforms

01 54 23.80 Staging Aids		Crew	Daily Output	Labor-Hours	Unit	2020 Bare Costs Material	Labor	Equipment	Total	Total Incl O&P
0300	End guard chains, buy per pair				Pair	47			47	51.50
0310	Cost per set per day, based on 250 days use				Day	.25			.25	.27
1000	Roof shingling bracket, steel, buy				Ea.	11.15			11.15	12.25
1010	Cost each per day, based on 250 days use				Day	.04			.04	.05
1100	Wood bracket, buy				Ea.	28			28	31
1110	Cost each per day, based on 250 days use				Day	.11			.11	.12
2000	Ladder jack, aluminum, buy per pair				Pair	139			139	153
2010	Cost per pair per day, based on 250 days use				Day	.56			.56	.61
3000	Laminated wood plank, 2" x 10" x 16', buy				Ea.	57.50			57.50	63
3010	Cost each per day, based on 250 days use				Day	.23			.23	.25
3100	Aluminum scaffolding plank, 20" wide x 24' long, buy				Ea.	635			635	695
3110	Cost each per day, based on 250 days use				Day	2.53			2.53	2.79
4000	Nylon full body harness, lanyard and rope grab				Ea.	186			186	205
4010	Cost each per day, based on 250 days use				Day	.74			.74	.82
4100	Rope for safety line, 5/8" x 100' nylon, buy				Ea.	61.50			61.50	67.50
4110	Cost each per day, based on 250 days use				Day	.25			.25	.27
4200	Permanent U-Bolt roof anchor, buy				Ea.	31			31	34
4300	Temporary (one use) roof ridge anchor, buy				"	6.40			6.40	7
5000	Installation (setup and removal) of staging aids									
5010	Sidewall staging bracket	2 Carp	64	.250	Ea.		13.30		13.30	21.50
5020	Guard post with 2 wood rails	"	64	.250			13.30		13.30	21.50
5030	End guard chains, set	1 Carp	64	.125			6.65		6.65	10.65
5100	Roof shingling bracket		96	.083			4.43		4.43	7.10
5200	Ladder jack	↓	64	.125			6.65		6.65	10.65
5300	Wood plank, 2" x 10" x 16'	2 Carp	80	.200			10.65		10.65	17.05
5310	Aluminum scaffold plank, 20" x 24'	"	40	.400			21.50		21.50	34
5410	Safety rope	1 Carp	40	.200			10.65		10.65	17.05
5420	Permanent U-Bolt roof anchor (install only)	2 Carp	40	.400			21.50		21.50	34
5430	Temporary roof ridge anchor (install only)	1 Carp	64	.125	↓		6.65		6.65	10.65

01 54 26 – Temporary Swing Staging

01 54 26.50 Swing Staging		Crew	Daily Output	Labor-Hours	Unit	Material	Labor	Equipment	Total	Total Incl O&P
0010	**SWING STAGING**, 500 lb. cap., 2' wide to 24' long, hand operated									
0020	Steel cable type, with 60' cables, buy				Ea.	6,325			6,325	6,950
0030	Rent per month				"	630			630	695
0600	Lightweight (not for masons) 24' long for 150' height,									
0610	manual type, buy				Ea.	12,000			12,000	13,200
0620	Rent per month					1,200			1,200	1,325
0700	Powered, electric or air, to 150' high, buy					32,100			32,100	35,300
0710	Rent per month					2,250			2,250	2,475
0780	To 300' high, buy					32,600			32,600	35,900
0800	Rent per month					2,275			2,275	2,525
1000	Bosun's chair or work basket 3' x 3.5', to 300' high, electric, buy					13,400			13,400	14,700
1010	Rent per month				↓	935			935	1,025
2200	Move swing staging (setup and remove)	E-4	2	16	Move		930	73.50	1,003.50	1,600

01 54 36 – Equipment Mobilization

01 54 36.50 Mobilization		Crew	Daily Output	Labor-Hours	Unit	Material	Labor	Equipment	Total	Total Incl O&P
0010	**MOBILIZATION** (Use line item again for demobilization) R015436-50									
0015	Up to 25 mi. haul dist. (50 mi. RT for mob/demob crew)									
1200	Small equipment, placed in rear of, or towed by pickup truck	A-3A	4	2	Ea.		106	44	150	215
1300	Equipment hauled on 3-ton capacity towed trailer	A-3Q	2.67	3			159	92.50	251.50	350
1400	20-ton capacity	B-34U	2	8			410	221	631	890
1500	40-ton capacity	B-34N	2	8	↓		425	340	765	1,050

01 54 Construction Aids

01 54 36 – Equipment Mobilization

01 54 36.50 Mobilization		Crew	Daily Output	Labor-Hours	Unit	Material	2020 Bare Costs Labor	Equipment	Total	Total Incl O&P
1600	50-ton capacity	B-34V	1	24	Ea.		1,300	985	2,285	3,100
1700	Crane, truck-mounted, up to 75 ton (driver only)	1 Eqhv	4	2			118		118	185
1800	Over 75 ton (with chase vehicle)	A-3E	2.50	6.400			345	70.50	415.50	625
2400	Crane, large lattice boom, requiring assembly	B-34W	.50	144	↓		7,425	7,025	14,450	19,500
2500	For each additional 5 miles haul distance, add						10%	10%		
3000	For large pieces of equipment, allow for assembly/knockdown									
3001	For mob/demob of vibroflotation equip, see Section 31 45 13.10									
3100	For mob/demob of micro-tunneling equip, see Section 33 05 07.36									
3200	For mob/demob of pile driving equip, see Section 31 62 19.10									
3300	For mob/demob of caisson drilling equip, see Section 31 63 26.13									

01 55 Vehicular Access and Parking

01 55 23 – Temporary Roads

01 55 23.50 Roads and Sidewalks		Crew	Daily Output	Labor-Hours	Unit	Material	Labor	Equipment	Total	Total Incl O&P
0010	**ROADS AND SIDEWALKS** Temporary									
0050	Roads, gravel fill, no surfacing, 4" gravel depth	B-14	715	.067	S.Y.	3.38	2.97	.30	6.65	8.80
0100	8" gravel depth	"	615	.078	"	6.75	3.45	.35	10.55	13.35
1000	Ramp, 3/4" plywood on 2" x 6" joists, 16" OC	2 Carp	300	.053	S.F.	1.62	2.83		4.45	6.30
1100	On 2" x 10" joists, 16" OC	"	275	.058	"	2.25	3.09		5.34	7.40

01 56 Temporary Barriers and Enclosures

01 56 13 – Temporary Air Barriers

01 56 13.60 Tarpaulins		Crew	Daily Output	Labor-Hours	Unit	Material	Labor	Equipment	Total	Total Incl O&P
0010	**TARPAULINS**									
0020	Cotton duck, 10-13.13 oz./S.Y., 6' x 8'				S.F.	.75			.75	.83
0050	30' x 30'					.63			.63	.69
0100	Polyvinyl coated nylon, 14-18 oz., minimum					1.49			1.49	1.64
0150	Maximum					1.49			1.49	1.64
0200	Reinforced polyethylene 3 mils thick, white					.05			.05	.06
0300	4 mils thick, white, clear or black					.16			.16	.18
0400	5.5 mils thick, clear					.19			.19	.21
0500	White, fire retardant					.60			.60	.66
0600	12 mils, oil resistant, fire retardant					.50			.50	.55
0700	8.5 mils, black					.26			.26	.29
0710	Woven polyethylene, 6 mils thick					.08			.08	.09
0720	Steel reinforced polyethylene, 20 mils thick					.54			.54	.59
0730	Polyester reinforced w/integral fastening system, 11 mils thick					.20			.20	.22
0740	Polyethylene, reflective, 23 mils thick				↓	1.34			1.34	1.47

01 56 13.90 Winter Protection		Crew	Daily Output	Labor-Hours	Unit	Material	Labor	Equipment	Total	Total Incl O&P
0010	**WINTER PROTECTION**									
0100	Framing to close openings	2 Clab	500	.032	S.F.	.51	1.35		1.86	2.72
0200	Tarpaulins hung over scaffolding, 8 uses, not incl. scaffolding		1500	.011		.25	.45		.70	1
0250	Tarpaulin polyester reinf. w/integral fastening system, 11 mils thick		1600	.010		.22	.42		.64	.91
0300	Prefab fiberglass panels, steel frame, 8 uses	↓	1200	.013	↓	2.78	.56		3.34	3.96

01 56 Temporary Barriers and Enclosures

01 56 16 – Temporary Dust Barriers

01 56 16.10 Dust Barriers, Temporary		Crew	Daily Output	Labor-Hours	Unit	Material	Labor	Equipment	Total	Total Incl O&P
						2020 Bare Costs				
0010	**DUST BARRIERS, TEMPORARY**, erect and dismantle									
0020	Spring loaded telescoping pole & head, to 12′, erect and dismantle	1 Clab	240	.033	Ea.		1.40		1.40	2.25
0025	Cost per day (based upon 250 days)				Day	.22			.22	.24
0030	To 21′, erect and dismantle	1 Clab	240	.033	Ea.		1.40		1.40	2.25
0035	Cost per day (based upon 250 days)				Day	.72			.72	.79
0040	Accessories, caution tape reel, erect and dismantle	1 Clab	480	.017	Ea.		.70		.70	1.12
0045	Cost per day (based upon 250 days)				Day	.33			.33	.37
0060	Foam rail and connector, erect and dismantle	1 Clab	240	.033	Ea.		1.40		1.40	2.25
0065	Cost per day (based upon 250 days)				Day	.10			.10	.11
0070	Caution tape	1 Clab	384	.021	C.L.F.	2.49	.88		3.37	4.15
0080	Zipper, standard duty		60	.133	Ea.	10.55	5.60		16.15	20.50
0090	Heavy duty		48	.167	″	10.65	7		17.65	23
0100	Polyethylene sheet, 4 mil		37	.216	Sq.	2.62	9.10		11.72	17.50
0110	6 mil		37	.216	″	4	9.10		13.10	19
1000	Dust partition, 6 mil polyethylene, 1″ x 3″ frame	2 Carp	2000	.008	S.F.	.33	.43		.76	1.04
1080	2″ x 4″ frame	″	2000	.008	″	.39	.43		.82	1.11
1085	Negative air machine, 1800 CFM				Ea.	890			890	975
1090	Adhesive strip application, 2″ width	1 Clab	192	.042	C.L.F.	6.75	1.75		8.50	10.25
4000	Dust & infectious control partition, adj. to 10′ high, obscured, 4′ panel	2 Carp	90	.178	Ea.	580	9.45		589.45	655
4010	3′ panel		90	.178		550	9.45		559.45	620
4020	2′ panel		90	.178		430	9.45		439.45	490
4030	1′ panel	1 Carp	90	.089		285	4.72		289.72	325
4040	6″ panel	″	90	.089		265	4.72		269.72	300
4050	2′ panel with HEPA filtered discharge port	2 Carp	90	.178		500	9.45		509.45	565
4060	3′ panel with 32″ door		90	.178		895	9.45		904.45	1,000
4070	4′ panel with 36″ door		90	.178		995	9.45		1,004.45	1,125
4080	4′-6″ panel with 44″ door		90	.178		1,200	9.45		1,209.45	1,350
4090	Hinged corner		80	.200		185	10.65		195.65	221
4100	Outside corner		80	.200		150	10.65		160.65	182
4110	T post		80	.200		150	10.65		160.65	182
4120	Accessories, ceiling grid clip		360	.044		7.45	2.36		9.81	12
4130	Panel locking clip		360	.044		5.15	2.36		7.51	9.45
4140	Panel joint closure strip		360	.044		8	2.36		10.36	12.60
4150	Screw jack		360	.044		6.65	2.36		9.01	11.10
4160	Digital pressure difference guage					275			275	305
4180	Combination lockset	1 Carp	13	.615		200	32.50		232.50	273
4185	Sealant tape, 2″ wide	1 Clab	192	.042	C.L.F.	6.75	1.75		8.50	10.25
4190	System in place, including door and accessories									
4200	Based upon 25 uses	2 Carp	51	.314	L.F.	9.85	16.65		26.50	37.50
4210	Based upon 50 uses		51	.314		4.94	16.65		21.59	32
4230	Based upon 100 uses		51	.314		2.47	16.65		19.12	29

01 56 23 – Temporary Barricades

01 56 23.10 Barricades		Crew	Daily Output	Labor-Hours	Unit	Material	Labor	Equipment	Total	Total Incl O&P
0010	**BARRICADES**									
0020	5′ high, 3 rail @ 2″ x 8″, fixed	2 Carp	20	.800	L.F.	6.75	42.50		49.25	75.50
0150	Movable	″	30	.533	″	5.60	28.50		34.10	51.50
0300	Stock units, 58′ high, 8′ wide, reflective, buy				Ea.	211			211	232
0350	With reflective tape, buy				″	425			425	465
0400	Break-a-way 3″ PVC pipe barricade									
0410	with 3 ea. 1′ x 4′ reflectorized panels, buy				Ea.	153			153	168
0500	Barricades, plastic, 8″ x 24″ wide, foldable					61			61	67
0800	Traffic cones, PVC, 18″ high					10.95			10.95	12.05

01 56 Temporary Barriers and Enclosures

01 56 23 – Temporary Barricades

01 56 23.10 Barricades		Crew	Daily Output	Labor-Hours	Unit	2020 Bare Costs Material	Labor	Equipment	Total	Total Incl O&P
0850	28″ high				Ea.	23.50			23.50	25.50
1000	Guardrail, wooden, 3′ high, 1″ x 6″ on 2″ x 4″ posts	2 Carp	200	.080	L.F.	1.44	4.25		5.69	8.40
1100	2″ x 6″ on 4″ x 4″ posts	″	165	.097		2.80	5.15		7.95	11.35
1200	Portable metal with base pads, buy					14.45			14.45	15.90
1250	Typical installation, assume 10 reuses	2 Carp	600	.027	↓	2.31	1.42		3.73	4.81
1300	Barricade tape, polyethylene, 7 mil, 3″ wide x 300′ long roll	1 Clab	128	.063	Ea.	7.50	2.63		10.13	12.45
3000	Detour signs, set up and remove									
3010	Reflective aluminum, MUTCD, 24″ x 24″, post mounted	1 Clab	20	.400	Ea.	2.80	16.85		19.65	30
4000	Roof edge portable barrier stands and warning flags, 50 uses	1 Rohe	9100	.001	L.F.	.07	.03		.10	.12
4010	100 uses	″	9100	.001	″	.03	.03		.06	.09

01 56 26 – Temporary Fencing

01 56 26.50 Temporary Fencing		Crew	Daily Output	Labor-Hours	Unit	Material	Labor	Equipment	Total	Total Incl O&P
0010	**TEMPORARY FENCING**									
0020	Chain link, 11 ga., 4′ high	2 Clab	400	.040	L.F.	2	1.68		3.68	4.90
0100	6′ high		300	.053		4.96	2.25		7.21	9.05
0200	Rented chain link, 6′ high, to 1000′ (up to 12 mo.)		400	.040		3.37	1.68		5.05	6.40
0250	Over 1000′ (up to 12 mo.)	↓	300	.053		4.14	2.25		6.39	8.15
0350	Plywood, painted, 2″ x 4″ frame, 4′ high	A-4	135	.178		6.90	8.95		15.85	22
0400	4″ x 4″ frame, 8′ high	″	110	.218		13.45	10.95		24.40	32.50
0500	Wire mesh on 4″ x 4″ posts, 4′ high	2 Carp	100	.160		11.15	8.50		19.65	26
0550	8′ high	″	80	.200		16.60	10.65		27.25	35.50
0600	Plastic safety fence, light duty, 4′ high, posts at 10′	B-1	500	.048		1.59	2.05		3.64	5.05
0610	Medium duty, 4′ high, posts at 10′		500	.048		1.90	2.05		3.95	5.40
0620	Heavy duty, 4′ high, posts at 10′		500	.048		2.51	2.05		4.56	6.05
0630	Reflective heavy duty, 4′ high, posts at 10′	↓	500	.048	↓	4.96	2.05		7.01	8.75

01 56 29 – Temporary Protective Walkways

01 56 29.50 Protection		Crew	Daily Output	Labor-Hours	Unit	Material	Labor	Equipment	Total	Total Incl O&P
0010	**PROTECTION**									
0020	Stair tread, 2″ x 12″ planks, 1 use	1 Carp	75	.107	Tread	5.15	5.65		10.80	14.80
0100	Exterior plywood, 1/2″ thick, 1 use		65	.123		2.05	6.55		8.60	12.75
0200	3/4″ thick, 1 use		60	.133	↓	2.84	7.10		9.94	14.45
2200	Sidewalks, 2″ x 12″ planks, 2 uses		350	.023	S.F.	.86	1.22		2.08	2.90
2300	Exterior plywood, 2 uses, 1/2″ thick		750	.011		.34	.57		.91	1.29
2500	3/4″ thick	↓	600	.013	↓	.47	.71		1.18	1.66

01 57 Temporary Controls

01 57 33 – Temporary Security

01 57 33.50 Watchman		Crew	Daily Output	Labor-Hours	Unit	Material	Labor	Equipment	Total	Total Incl O&P
0010	**WATCHMAN**									
0020	Service, monthly basis, uniformed person, minimum				Hr.				27.65	30.40
0100	Maximum								56	61.62
0200	Person and command dog, minimum								31	34
0300	Maximum				↓				60	65

01 58 Project Identification

01 58 13 – Temporary Project Signage

01 58 13.50 Signs		Crew	Daily Output	Labor-Hours	Unit	Material (2020 Bare Costs)	Labor (2020 Bare Costs)	Equipment (2020 Bare Costs)	Total (2020 Bare Costs)	Total Incl O&P
0010	**SIGNS**									
0020	High intensity reflectorized, no posts, buy				Ea.	27			27	29.50

01 66 Product Storage and Handling Requirements

01 66 19 – Material Handling

01 66 19.10 Material Handling		Crew	Daily Output	Labor-Hours	Unit	Material	Labor	Equipment	Total	Total Incl O&P
0010	**MATERIAL HANDLING**									
0020	Above 2nd story, via stairs, per C.Y. of material per floor	2 Clab	145	.110	C.Y.		4.65		4.65	7.45
0030	Via elevator, per C.Y. of material		240	.067			2.81		2.81	4.50
0050	Distances greater than 200′, per C.Y. of material per each addl 200′		300	.053			2.25		2.25	3.60

01 71 Examination and Preparation

01 71 23 – Field Engineering

01 71 23.13 Construction Layout		Crew	Daily Output	Labor-Hours	Unit	Material	Labor	Equipment	Total	Total Incl O&P
0010	**CONSTRUCTION LAYOUT**									
1100	Crew for layout of building, trenching or pipe laying, 2 person crew	A-6	1	16	Day		845	31	876	1,375
1200	3 person crew	A-7	1	24			1,375	30.50	1,405.50	2,200
1400	Crew for roadway layout, 4 person crew	A-8	1	32			1,775	30.50	1,805.50	2,850

01 71 23.19 Surveyor Stakes		Crew	Daily Output	Labor-Hours	Unit	Material	Labor	Equipment	Total	Total Incl O&P
0010	**SURVEYOR STAKES**									
0020	Hardwood, 1″ x 1″ x 48″ long				C	71			71	78
0100	2″ x 2″ x 18″ long					79			79	87
0150	2″ x 2″ x 24″ long					155			155	170

01 74 Cleaning and Waste Management

01 74 13 – Progress Cleaning

01 74 13.20 Cleaning Up		Crew	Daily Output	Labor-Hours	Unit	Material	Labor	Equipment	Total	Total Incl O&P
0010	**CLEANING UP**									
0020	After job completion, allow, minimum				Job				.30%	.30%
0040	Maximum				"				1%	1%
0042	Rubbish removal, see Section 02 41 19.19									
0052	Cleanup of floor area, continuous, per day, during const.	A-5	16	1.125	M.S.F.	2.47	48	3.06	53.53	83
0100	Final by GC at end of job	"	11.50	1.565	"	2.62	67	4.26	73.88	115

01 76 Protecting Installed Construction

01 76 13 – Temporary Protection of Installed Construction

01 76 13.20 Temporary Protection		Crew	Daily Output	Labor-Hours	Unit	Material	Labor	Equipment	Total	Total Incl O&P
0010	**TEMPORARY PROTECTION**									
0020	Flooring, 1/8″ tempered hardboard, taped seams	2 Carp	1500	.011	S.F.	.50	.57		1.07	1.46
0030	Peel away carpet protection	1 Clab	3200	.003	"	.13	.11		.24	.31

01 91 Commissioning

01 91 13 – General Commissioning Requirements

01 91 13.50 Building Commissioning		Crew	Daily Output	Labor-Hours	Unit	Material	2020 Bare Costs Labor	Equipment	Total	Total Incl O&P
0010	**BUILDING COMMISSIONING**									
0100	Systems operation and verification during turnover				%				.25%	.25%
0150	Including all systems subcontractors								.50%	.50%
0200	Systems design assistance, operation, verification and training								.50%	.50%
0250	Including all systems subcontractors				↓				1%	1%

01 93 Facility Maintenance

01 93 08 – Facilities Maintenance, Equipment

01 93 08.50 Equipment		Crew	Daily Output	Labor-Hours	Unit	Material	Labor	Equipment	Total	Total Incl O&P
0010	**EQUIPMENT**, Purchase									
0090	Carpet care equipment									
0110	Dual motor vac, 1 HP, 16" brush				Ea.	690			690	755
0120	Upright vacuum, 12" brush					227			227	250
0130	14" brush					455			455	505
0140	Soil extractor, hot water 5' wand, 12" head					3,050			3,050	3,375
0150	Dry foam, 13" brush					2,500			2,500	2,750
0160	24" brush				↓	4,250			4,250	4,675
0240	Floor care equipment									
0260	Polishing, buffing, waxing machine, 175 RPM									
0270	.33 HP, 20" diam. brush				Ea.	870			870	955
0280	1 HP, 16" diam. brush					1,225			1,225	1,325
0290	17" diam. brush					1,150			1,150	1,275
0300	1.5 HP, 20" diam. brush					1,400			1,400	1,525
0310	20" diam. brush, 1500 RPM				↓	1,625			1,625	1,800
0330	Scrubber, automatic, 2 stage 1 HP vacuum motor									
0340	20" diam. brush				Ea.	13,800			13,800	15,200
0350	28" diam. brush				"	18,900			18,900	20,800
1200	Plumbing maintenance equipment									
1220	Kinetic water ram				Ea.	223			223	245
1240	Cable pipe snake, 104 ft., self feed, electric, .5 HP				"	2,700			2,700	2,975
1500	Specialty equipment									
1550	Litterbuggy, 9 HP gasoline				Ea.	4,800			4,800	5,275
1560	Propane					4,075			4,075	4,475
1570	Pressure cleaner, hot water, 1000 psi					3,375			3,375	3,700
1580	3500 psi				↓	8,675			8,675	9,550
1590	Vacuum cleaners, steel canister									
1600	Dry only, two stage, 1 HP				Ea.	400			400	440
1610	Wet/dry, two stage, 2 HP					710			710	780
1620	4 HP					920			920	1,000
1630	Wet/dry, three stage, 1.5 HP					685			685	755
1670	Squeegee wet only, 2 stage, 1 HP					1,100			1,100	1,200
1680	2 HP					1,575			1,575	1,725
1690	Asbestos and hazardous waste dry vacs, 1 HP					925			925	1,025
1700	2 HP					790			790	870
1710	Three stage, 1 HP					630			630	695
1800	Upholstery soil extractor, dry-foam				↓	3,625			3,625	3,975
1900	Snow removal equipment									
1920	Thrower, 3 HP, 20", single stage, gas				Ea.	730			730	805
1930	Electric start					875			875	965
1940	5 HP, 24", 2 stage, gas					1,225			1,225	1,350
1950	11 HP, 36"				↓	2,400			2,400	2,650

01 93 Facility Maintenance

01 93 08 – Facilities Maintenance, Equipment

01 93 08.50 Equipment		Crew	Daily Output	Labor-Hours	Unit	2020 Bare Costs Material	Labor	Equipment	Total	Total Incl O&P
1970	Trash receptacles and closures; see Section 12 93 23.20									
2000	Tube cleaning equip., for heat exchangers, condensers, evap.									
2010	Tubes/pipes 1/4"-1", 115 volt, .5 HP drive				Ea.	1,675			1,675	1,825
2020	Air powered					1,725			1,725	1,875
2040	1" and up, 115 volt, 1 HP drive					2,125			2,125	2,350

01 93 09 – Facility Equipment

01 93 09.50 Moving Equipment		Crew	Daily Output	Labor-Hours	Unit	Material	Labor	Equipment	Total	Total Incl O&P
0010	**MOVING EQUIPMENT**, Remove and reset, 100' distance,									
0020	No obstructions, no assembly or leveling unless noted									
0100	Annealing furnace, 24' overall	B-67	4	4	Ea.		218	67.50	285.50	415
0200	Annealing oven, small		14	1.143			62	19.30	81.30	118
0240	Very large		1	16			870	270	1,140	1,650
0400	Band saw, small		12	1.333			72.50	22.50	95	138
0440	Large		8	2			109	33.50	142.50	207
0500	Blue print copy machine		7	2.286			124	38.50	162.50	237
0600	Bonding mill, 6"		7	2.286			124	38.50	162.50	237
0620	12"		6	2.667			145	45	190	276
0640	18"		4	4			218	67.50	285.50	415
0660	24"		2	8			435	135	570	830
0700	Boring machine (jig)	B-68	7	3.429			188	38.50	226.50	335
0800	Bridgeport mill, standard	B-67	14	1.143			62	19.30	81.30	118
1000	Calibrator, 6 unit	"	14	1.143			62	19.30	81.30	118
1100	Comparitor, bench top	2 Clab	14	1.143			48		48	77
1140	Floor mounted	B-67	7	2.286			124	38.50	162.50	237
1200	Computer, desk top	2 Clab	25	.640			27		27	43
1300	Copy machine	"	25	.640			27		27	43
1500	Deflasher	B-67	14	1.143			62	19.30	81.30	118
1600	Degreaser, small		14	1.143			62	19.30	81.30	118
1640	Large 24' overall		1	16			870	270	1,140	1,650
1700	Desk with chair	2 Clab	25	.640			27		27	43
1725	Desk only	"	22	.727			30.50		30.50	49
1750	Chair only	1 Clab	300	.027			1.12		1.12	1.80
1800	Dial press	B-67	7	2.286			124	38.50	162.50	237
1900	Drafting table	2 Clab	14	1.143			48		48	77
2000	Drill press, bench top	"	14	1.143			48		48	77
2040	Floor mounted	B-67	14	1.143			62	19.30	81.30	118
2080	Industrial radial	"	7	2.286			124	38.50	162.50	237
2100	Dust collector, portable	2 Clab	25	.640			27		27	43
2140	Stationary, small	B-67	7	2.286			124	38.50	162.50	237
2180	Stationary, large	"	2	8			435	135	570	830
2300	Electric discharge machine	B-68	7	3.429			188	38.50	226.50	335
2400	Environmental chamber walls, including assembly	4 Clab	18	1.778	L.F.		75		75	120
2600	File cabinet	2 Clab	25	.640	Ea.		27		27	43
2800	Grinder/sander, pedestal mount	B-67	14	1.143			62	19.30	81.30	118
3000	Hack saw, power	2 Clab	24	.667			28		28	45
3100	Hydraulic press	B-67	14	1.143			62	19.30	81.30	118
3500	Laminar flow tables	"	14	1.143			62	19.30	81.30	118
3600	Lathe, bench	2 Clab	14	1.143			48		48	77
3640	6"	B-67	14	1.143			62	19.30	81.30	118
3680	10"		13	1.231			67	21	88	127
3720	12"		12	1.333			72.50	22.50	95	138
4000	Milling machine		8	2			109	33.50	142.50	207

01 93 09 – Facility Equipment

01 93 09.50 Moving Equipment		Crew	Daily Output	Labor-Hours	Unit	Material	2020 Bare Costs Labor	Equipment	Total	Total Incl O&P
4100	Molding press, 25 ton	B-67	5	3.200	Ea.		174	54	228	330
4140	60 ton		4	4			218	67.50	285.50	415
4180	100 ton		2	8			435	135	570	830
4220	150 ton		1.50	10.667			580	180	760	1,100
4260	200 ton		1	16			870	270	1,140	1,650
4300	300 ton		.75	21.333			1,150	360	1,510	2,200
4700	Oil pot stand		14	1.143			62	19.30	81.30	118
5000	Press, 10 ton		14	1.143			62	19.30	81.30	118
5040	15 ton		12	1.333			72.50	22.50	95	138
5080	20 ton		10	1.600			87	27	114	166
5120	30 ton		8	2			109	33.50	142.50	207
5160	45 ton		6	2.667			145	45	190	276
5200	60 ton		4	4			218	67.50	285.50	415
5240	75 ton		2.50	6.400			350	108	458	665
5280	100 ton		2	8			435	135	570	830
5500	Raised floor, including assembly	2 Carp	250	.064	S.F.		3.40		3.40	5.45
5600	Rolling mill, 6″	B-67	7	2.286	Ea.		124	38.50	162.50	237
5640	9″		6	2.667			145	45	190	276
5680	12″		4	4			218	67.50	285.50	415
5720	13″		3.50	4.571			249	77	326	475
5760	18″		2	8			435	135	570	830
5800	25″		1	16			870	270	1,140	1,650
6000	Sander, floor stand		14	1.143			62	19.30	81.30	118
6100	Screw machine		7	2.286			124	38.50	162.50	237
6200	Shaper, 16″		14	1.143			62	19.30	81.30	118
6300	Shear, power assist		4	4			218	67.50	285.50	415
6400	Slitter, 6″		14	1.143			62	19.30	81.30	118
6440	8″		13	1.231			67	21	88	127
6480	10″		12	1.333			72.50	22.50	95	138
6520	12″		11	1.455			79	24.50	103.50	150
6560	16″		10	1.600			87	27	114	166
6600	20″		8	2			109	33.50	142.50	207
6640	24″		6	2.667			145	45	190	276
6800	Snag and tap machine		7	2.286			124	38.50	162.50	237
6900	Solder machine (auto)		7	2.286			124	38.50	162.50	237
7000	Storage cabinet metal, small	2 Clab	36	.444			18.70		18.70	30
7040	Large		25	.640			27		27	43
7100	Storage rack open, small		14	1.143			48		48	77
7140	Large		7	2.286			96		96	154
7200	Surface bench, small	B-67	14	1.143			62	19.30	81.30	118
7240	Large	″	5	3.200			174	54	228	330
7300	Surface grinder, large wet	B-68	5	4.800			264	54	318	470
7500	Time check machine	2 Clab	14	1.143			48		48	77
8000	Welder, 30 kVA (bench)		14	1.143			48		48	77
8100	Work bench with chair		25	.640			27		27	43
8200	Storage rack, relocate, disassemble, transport & reassemble, 8′ x 36″	B-68A	600	.040	S.F.		2.24	.43	2.67	3.94
8205	8′ x 42″		600	.040			2.24	.43	2.67	3.94
8210	8′ x 48″		600	.040			2.24	.43	2.67	3.94
8215	Excess or store, disassemble and deliver, 8′ x 36″		1200	.020			1.12	.21	1.33	1.97
8220	8′ x 42″		1200	.020			1.12	.21	1.33	1.97
8225	8′ x 48″		1200	.020			1.12	.21	1.33	1.97
8230	Work bench, remove, transport, reinstall	B-67B	4	4	Ea.		225		225	350
8235	Excess or store, remove and deliver	″	8	2			112		112	174

01 93 Facility Maintenance

01 93 09 – Facility Equipment

	01 93 09.50 Moving Equipment	Crew	Daily Output	Labor-Hours	Unit	Material	2020 Bare Costs Labor	Equipment	Total	Total Incl O&P
8245	Excess or store, remove and deliver	B-68A	8	3	Ea.		168	32	200	296
8250	Storage cabinet, relocate, remove, transport, and set in place	B-67B	8	2			112		112	174
8255	Excess or store, remove and deliver		10	1.600			90		90	139
8260	Bookcase cabinet, relocate, remove, transport, and set in place		16	1			56		56	87
8265	Excess or store, remove and deliver		20	.800			45		45	69.50
8270	Bench top equip, relocate, remove, transport, and set in place		8	2			112		112	174
8275	Excess or store, remove and deliver		12.80	1.250			70		70	109
8280	Floor mounted equipment									
8285	Small, relocate, generally less than 2000 lb.	B-68A	3	8	Ea.		450	85.50	535.50	790
8290	Excess or store, less than 2000 lb.		6	4			224	42.50	266.50	390
8295	Medium, relocate, 2000 to 6000 lb.		1.04	23.077			1,300	246	1,546	2,275
8300	Excess or store, 2000 to 6000 lb.		1.68	14.286			800	152	952	1,425
8305	Large, relocate, 6000 lb. or more		.40	60			3,375	640	4,015	5,900
8310	Excess or store, 6000 lb. or more		.64	37.500			2,100	400	2,500	3,700
8315	Bridge crane to 40′ length									
8320	1-5 ton	B-68E	1.36	29.412	Ea.		1,700	188	1,888	2,975
8325	5-10 ton		1.04	38.462	″		2,225	246	2,471	3,900
8330	Rails, both sides per L.F., removal, excludes support steel		8	5	L.F.		289	32	321	510
8335	Jib crane, remove and store, including support steel, rail track		2	20	Ea.		1,150	128	1,278	2,050
8340	Office desk, relocate, remove, transport, and set in place	B-67B	8	2			112		112	174
8345	Excess or store, remove and deliver	B-68A	10.64	2.256			126	24	150	223
8350	Office file cabinet, relocate, remove, transport, and set in place									
8355	2 drawer	B-67B	8	2	Ea.		112		112	174
8360	4 drawer	″	8	2	″		112		112	174
8365	Excess or store, remove and deliver									
8370	2 drawer	B-67B	16	1	Ea.		56		56	87
8375	4 drawer		10.64	1.504			84.50		84.50	131
8380	Office partition, excess or store, remove and deliver		16	1			56		56	87
8385	Gas bottle rack, relocate, remove, transport, and set in place	B-68A	8	3			168	32	200	296
8390	Excess or store, remove and deliver	″	10.64	2.256			126	24	150	223
8395	Fume hood & filter, bench mounted, relocate, remove, xport & set in place	B-67B	1.36	11.765			660		660	1,025
8400	Excess or store, remove and deliver	″	2	8			450		450	695
8405	Bench top microscope, relocate, remove, transport, and set in place	1 Mill	8	1			56		56	86.50
8410	Excess or store, remove and deliver	″	8	1			56		56	86.50
8415	Modular work station									
8420	Includes 2 desktops, 4 pedestal files	B-68A	1	24	Ea.		1,350	256	1,606	2,350
8425	Install, purchase new		1.60	15			840	160	1,000	1,475
8430	Relocate		1.60	15			840	160	1,000	1,475
8435	Excess or store		2.64	9.091			510	97	607	895
8440	Existing, install, includes 2 workstations		1.60	15			840	160	1,000	1,475
8445	Wall mounted shelf, relocate, remove, transport and install	B-67B	8	2			112		112	174
8450	Excess or store, remove and deliver		12	1.333			75		75	116
8455	Wall mounted coat rack, excess or store		16	1			56		56	87
8460	Office or work table, excess or store		10.64	1.504			84.50		84.50	131
8465	Floor stand testers									
8470	Small, relocate, footprint generally less than 12 S.F.	B-67B	8	2	Ea.		112		112	174
8475	Excess or store, footprint generally less than 12 S.F.		8	2			112		112	174
8480	Medium, relocate, footprint between 12 S.F. & 20 S.F.		2.64	6.061			340		340	525
8485	Excess or store, footprint between 12 S.F. & 20 S.F.		8	2			112		112	174
8490	Large, relocate, footprint generally greater than 20 S.F.	B-68A	1.04	23.077			1,300	246	1,546	2,275
8495	Excess or store, footprint greater than 20 S.F.		2	12			675	128	803	1,200
8500	Office safe, relocate, remove, transport, and set in place		8	3			168	32	200	296
8505	Excess or store, remove and deliver		10.64	2.256			126	24	150	223

01 93 Facility Maintenance

01 93 09 – Facility Equipment

01 93 09.50 Moving Equipment		Crew	Daily Output	Labor-Hours	Unit	Material	2020 Bare Costs Labor	Equipment	Total	Total Incl O&P
8510	Screen, relocate	1 Mill	16	.500	Ea.		28		28	43.50
8515	Excess or store, remove and deliver	″	16	.500			28		28	43.50
8520	Equipment pieces									
8525	Interconnect	L-1	8	2	Ea.		126		126	194
8530	Disconnect	″	8	2			126		126	194
8535	Vidmar cabinet, floor mounted, relocate, remove, xport and set in place	B-68A	4	6			335	64	399	590
8540	Excess or store, remove and deliver	″	8	3			168	32	200	296
8545	Work bench, relocate	B-67B	8	2			112		112	174
8550	Excess or store, remove and deliver		10.64	1.504			84.50		84.50	131
8555	Steel wire shelves, relocate, 18″ x 48″ x 5′ high		4	4			225		225	350
8560	Excess or store, remove and deliver, 18″ x 48″ x 5′ high		8	2			112		112	174
8565	Coat rack, relocate	1 Mill	16	.500			28		28	43.50
8570	Excess or store, remove and deliver	″	8	1			56		56	86.50
8575	Pallet racks, 8′ x 36″, disassemble & reassemble in another location	B-68A	504	.048	S.F.		2.67	.51	3.18	4.70
8580	Install clean room equipment	B-68B	1	56	Ea.		3,350	280	3,630	5,500
8585	Small	B-68C	2	16			950	140	1,090	1,625
8590	Large	B-68B	.24	233			14,000	1,175	15,175	22,900
8595	Install large 10′ x 25′ x 10′ equipment		.08	700			42,000	3,500	45,500	69,000
8600	Install large fume hood		.16	350			21,000	1,750	22,750	34,300
8605	Install equipment									
8610	Large	B-68B	.72	77.778	Ea.		4,675	390	5,065	7,625
8615	Medium		1.20	46.667			2,800	233	3,033	4,575
8620	Small		2.40	23.333			1,400	117	1,517	2,300
8900	For moving distance to 200′, add						25%			

01 93 13 – Facility Maintenance Procedures

01 93 13.03 Concrete Facilities Maintenance

		Crew	Daily Output	Labor-Hours	Unit	Material	Labor	Equipment	Total	Total Incl O&P
0010	**CONCRETE FACILITIES MAINTENANCE**									
0200	Resurface concrete floor with epoxy/silica finish									
0205	For new cured concrete floors with open surface texture:									
0210	Clean, roll base coat(s), broadcast silica, roll top coat									
0220	1/16″ thick (1 base coat) plus top coat	2 Cefi	600	.027	S.F.	1.56	1.33		2.89	3.80
0230	1/8″ thick (2 base coats) plus top coat, rolled	″	350	.046	″	2.14	2.28		4.42	5.95
0305	For old or coated concrete floors with open surface texture:									
0310	Medium shotblast, roll base coat(s), broadcast silica, roll top coat									
0320	1/16″ thick (1 base coat) plus top coat	2 Cefi	400	.040	S.F.	1.16	2		3.16	4.41
0330	1/8″ thick (2 base coats) plus top coat, rolled	″	300	.053	″	1.74	2.66		4.40	6.10
1000	Patching concrete, see Section 03 01 30.62									
1037	Stair cleaning, sweep	1 Clab	50	.160	Flight		6.75		6.75	10.80
1039	Mop or scrub	″	5	1.600	″	.20	67.50		67.70	108
1040	Sealing concrete floor, spray, oil or urethane base, 1 coat	1 Cefi	8000	.001	S.F.	.11	.05		.16	.20
1050	2 coats	″	4000	.002	″	.22	.10		.32	.40
1080	Sweep and damp mop concrete floor	1 Clab	20	.400	M.S.F.	4.51	16.85		21.36	32

01 93 13.04 Masonry Facilities Maintenance

		Crew	Daily Output	Labor-Hours	Unit	Material	Labor	Equipment	Total	Total Incl O&P
0010	**MASONRY FACILITIES MAINTENANCE**									
0040	Caulking masonry, no staging included									
0050	Re-caulk only, oil base, 1/2″ x 1/2″	1 Bric	225	.036	L.F.	1.04	1.85		2.89	4.13
0060	Butyl		205	.039		.57	2.03		2.60	3.90
0070	Polysulfide		200	.040		1.15	2.08		3.23	4.62
0080	Silicone		195	.041		.89	2.14		3.03	4.43
0100	Cut out and re-caulk, oil base		145	.055		1.04	2.87		3.91	5.80
0110	Butyl		130	.062		.57	3.20		3.77	5.80
0120	Polysulfide		125	.064		.57	3.33		3.90	6.05

01 93 Facility Maintenance

01 93 13 – Facility Maintenance Procedures

01 93 13.04 Masonry Facilities Maintenance		Crew	Daily Output	Labor-Hours	Unit	Material	Labor (2020 Bare Costs)	Equipment	Total	Total Incl O&P
0130	Silicone	1 Bric	125	.064	L.F.	.89	3.33		4.22	6.40
0800	Cleaning masonry, see Section 04 01 30.20									
3000	Pointing masonry, see Section 04 01 20.20									
01 93 13.05 Metals Facilities Maintenance										
0010	**METALS FACILITIES MAINTENANCE**									
2000	Steel surface treat., sand blast & paint, see Section 05 01 10.51									
2010	Steel siding painting, see Section 09 91 13									
01 93 13.07 Moisture-Thermal Control Facilities Maintenance										
0010	**MOISTURE-THERMAL CONTROL FACILITIES MAINTENANCE**									
0100	Caulking and sealants see Section 07 92									
0110	Caulking around exterior doors and windows, silicone	1 Carp	400	.020	L.F.	.22	1.06		1.28	1.95
2000	Roofing & siding demolition see Section 07 05 05.10									
2500	Single-ply or built-up roofing maintenance									
2510	Walk roof, clean drain trap, pick-up trash									
2520	Per inspection	1 Rofc	20000	.001	S.F.		.02		.02	.03
3000	Roof non-destructive testing, infrared analysis system									
3040	20,000 SF roof				S.F.				.07	.08
3050	Minimum fee				Ea.				2,500	2,750
01 93 13.08 Door and Window Facilities Maintenance										
0010	**DOOR & WINDOW FACILITIES MAINTENANCE**									
0250	Cylinder dead bolt lock set, repair	1 Skwk	11	.727	Ea.		40		40	63.50
0300	Demolition, windows see Section 08 05 05.20									
0500	Door closer, overhaul	1 Skwk	5	1.600	Ea.		88		88	140
0510	Repair concealed door check or closer		4	2			110		110	175
0520	Adjust		26	.308			16.90		16.90	27
0530	Replace, see Section 08 71 20.30									
1500	Remove and reset window, no interior/exterior trim or painting	1 Carp	4	2	Ea.		106		106	170
1510	Interior trim and painting only		2	4			213		213	340
1520	Interior/exterior trim and painting		1.33	6.015			320		320	510
1900	Washing door, 3′ x 7′, damp wipe	1 Clab	145	.055		.08	2.32		2.40	3.81
2000	Washing windows, at ground level w/ladder, sponge & squeegee									
2010	Both sides, 4′ x 7′, 2 pane	1 Clab	60	.133	Ea.	.20	5.60		5.80	9.20
2020	4′ x 6′, 4 pane		55	.145		.17	6.10		6.27	10
2030	3.5′ x 5.5′, 8 pane		45	.178		.14	7.50		7.64	12.15
2040	2.5′ x 5.7′, 12 pane		40	.200		.10	8.40		8.50	13.60
2050	4′ x 6′, 16 pane		35	.229		.17	9.60		9.77	15.60
2060	4′ x 7′, industrial, 20 pane		30	.267		.20	11.25		11.45	18.20
2070	6′ x 7′, austral casement, 6 pane		35	.229		.30	9.60		9.90	15.75
2080	Clear glass partition, 8 sq. ft. per unit		3	2.667	M.S.F.	3.55	112		115.55	184
2090	Opaque, 20 sq. ft. per unit		3	2.667	"	3.55	112		115.55	184
2100	Alternate pricing method by window area to be washed									
2110	Minimum productivity	1 Clab	1	8	M.S.F.	3.55	335		338.55	545
2120	Average productivity		2.50	3.200		3.55	135		138.55	220
2130	Maximum productivity		4	2		3.55	84		87.55	139
2200	Weatherstripping see Section 08 75 30.10									
01 93 13.09 Finishes, Facilities Maintenance										
0010	**FINISHES, FACILITIES MAINTENANCE** R019313-10									
0100	Ceiling maintenance									
0115	Acoustic tile cleaning, chemical spray, including masking									
0120	Unobstructed floor R019313-20	4 Clab	5400	.006	S.F.	.07	.25		.32	.48
0125	Obstructed floor	"	4000	.008	"	.07	.34		.41	.62
0128	Acoustic tile cleaning & coating, chemical, including masking									

01 93 13.09 Finishes, Facilities Maintenance		Crew	Daily Output	Labor-Hours	Unit	Material	2020 Bare Costs Labor	Equipment	Total	Total Incl O&P
0130	Unobstructed floor	4 Clab	4800	.007	S.F.	.14	.28		.42	.61
0135	Obstructed floor	"	3400	.009	"	.14	.40		.54	.79
0400	Floor maintenance R019313-30									
0500	Carpet fiber sealant, anti-soilant, commercial	1 Clab	64	.125	M.S.F.	10.65	5.25		15.90	20
0510	Residential	"	32	.250	"	10.65	10.55		21.20	28.50
0530	Carpet cleaning, vacuum, dry pick-up									
0540	Unobstructed	1 Clab	30	.267	M.S.F.		11.25		11.25	18
0550	Obstructed		22	.364			15.30		15.30	24.50
0560	Carpet cleaning, portable extractor with floor wand		20	.400		8.50	16.85		25.35	36.50
0570	Deep carpet cleaning, self-contained extractor									
0572	24" path at 100 fpm	1 Clab	64	.125	M.S.F.	10.65	5.25		15.90	20
0574	20" path		54	.148		10.65	6.25		16.90	21.50
0576	16" path		43	.186		10.65	7.85		18.50	24.50
0578	12" path		32	.250		10.65	10.55		21.20	28.50
0580	Carpet pile lifting, self-contained extractor									
0582	24" path at 90 fpm	1 Clab	60	.133	M.S.F.	11.70	5.60		17.30	22
0584	20" path		50	.160		11.70	6.75		18.45	23.50
0586	16" path		40	.200		11.70	8.40		20.10	26.50
0588	12" path		30	.267		11.70	11.25		22.95	31
0590	Carpet restoration cleaning, self-contained extractor									
0592	24" path at 50 fpm	1 Clab	32	.250	M.S.F.	12.75	10.55		23.30	31
0594	20" path		27	.296		12.75	12.45		25.20	34
0596	16" path		21	.381		12.75	16.05		28.80	39.50
0598	12" path		16	.500		12.75	21		33.75	47.50
0700	Composition, resilient or wood flooring									
0720	Dust mop, unobstructed	1 Clab	60	.133	M.S.F.		5.60		5.60	9
0730	Obstructed		35	.229			9.60		9.60	15.40
0740	Damp mop, unobstructed		26	.308		5.70	12.95		18.65	27.50
0750	Obstructed		16	.500		5.70	21		26.70	40
0780	Hand scrub, unobstructed		1.80	4.444		5.70	187		192.70	305
0790	Obstructed		1.30	6.154		5.70	259		264.70	420
0800	Machine scrub, unobstructed		17	.471		5.70	19.80		25.50	38
0810	Obstructed		12	.667		5.70	28		33.70	51.50
0820	Machine polish, unobstructed		28	.286			12.05		12.05	19.25
0830	Obstructed		14	.571			24		24	38.50
0850	Refinish old wood floors, minimum	1 Carp	400	.020	S.F.	.04	1.06		1.10	1.74
0860	Maximum cost	"	130	.062		.06	3.27		3.33	5.30
0862	Light sanding, clean and 1 coat of polyurethane	J-7	700	.023		.14	1.02	.11	1.27	1.88
0863	Heavy sanding, 1 coat sealer, 2 coats polyurethane	"	400	.040		.40	1.78	.19	2.37	3.48
0870	Strip & rewax/polish, unobstructed	1 Clab	4	2	M.S.F.	40.50	84		124.50	180
0880	Obstructed		3	2.667		40.50	112		152.50	225
0890	Sweeping, unobstructed		47	.170			7.15		7.15	11.50
0900	Obstructed		35	.229			9.60		9.60	15.40
0950	Painting concrete or wood floors, see Section 09 91 23.52									
1200	Paint removal, see Section 02 83 19.26									
1300	Scrape after fire damage, see Section 09 91 03.41									
2000	Wall maintenance									
2400	Washing enamel finish walls with mild cleanser	1 Clab	16	.500	M.S.F.	2.02	21		23.02	35.50

01 93 13 – Facility Maintenance Procedures

01 93 13.10 Specialties, Facilities Maintenance		Crew	Daily Output	Labor-Hours	Unit	Material	2020 Bare Costs Labor	Equipment	Total	Total Incl O&P
0010	**SPECIALTIES, FACILITIES MAINTENANCE**									
0100	Bathroom accessories									
0120	Accessory, replacement see Section 10 28 13.13									
0130	All purpose cleaner, concentrate				Gal.	8.10			8.10	8.90
0150	Clean mirror, 36″ x 24″	1 Clab	840	.010	Ea.	.01	.40		.41	.65
0160	48″ x 24″		630	.013		.02	.53		.55	.88
0170	72″ x 24″		420	.019		.02	.80		.82	1.31
0200	General cleaning of fixtures (basins, water closets, urinals)									
0210	Including shelves, partitions and dispenser servicing	1 Clab	80	.100	Fixture	.01	4.21		4.22	6.75
1500	Partition replacement, hospital, office, toilet see Section 10 21 13.13									

01 93 13.11 Architectural Equipment, Facilities Maintenance		Crew	Daily Output	Labor-Hours	Unit	Material	Labor	Equipment	Total	Total Incl O&P
0010	**ARCHITECTURAL EQUIPMENT, FACILITIES MAINTENANCE**									
0100	Bird control needle strips									
0110	Anchor clip mounted	2 Clab	500	.032	L.F.	5.50	1.35		6.85	8.20
0120	Adhesive mounted	″	550	.029	″	8.95	1.22		10.17	11.80
0500	Laundry equipment repair and maintenance									
0520	Dryer, domestic, remove or reinstall basket	1 Skwk	6	1.333	Ea.		73		73	117
0530	Commercial, disassemble, clean & reassemble		2	4			219		219	350
0550	Extractor, replace belts (2)		6	1.333			73		73	117
0560	Replace pressure pad		2	4			219		219	350
0570	Flatwork ironer, replace feed ribbons		2	4			219		219	350
0580	Press, overhaul air system		4	2			110		110	175
0590	Tumbler, extractor, replace bearings and seal		2	4			219		219	350
0600	Washer, domestic, replace bearings and seal		3	2.667			146		146	233
0610	Remove or reinstall tub		4	2			110		110	175
0700	Medical and ward equipment repair and maintenance									
0710	Aspirator-overhaul	1 Skwk	6	1.333	Ea.		73		73	117
0720	Autoclave-overhaul		4	2	″		110		110	175
0730	Bed rails, assemble or repair		10	.800	Pr.		44		44	70
0770	Cart, medication, ice, etc. - replace wheels		6	1.333	Carton		73		73	117
0780	Replace rubber bumper		6	1.333	″		73		73	117
0800	Centrifuge, repair or replace timer or switch		11	.727	Ea.		40		40	63.50
0850	Litter, attach safety belt		16	.500			27.50		27.50	43.50
0880	Operating table, overhaul and lubricate		6	1.333			73		73	117
0900	Oxygen tent apparatus, overhaul		2	4			219		219	350
0950	Still, clean		3	2.667			146		146	233
0970	Striker saw, overhaul		5	1.600			88		88	140
1100	Wheelchair, complete overhaul		3	2.667			146		146	233
1110	Replace wheel		11	.727			40		40	63.50
2000	Other equipment repair and maintenance									
2040	Cart or dolly, assemble or disassemble axles and wheels	1 Skwk	6	1.333	Carton		73		73	117
2090	Forge and harden tool		13	.615	Ea.		34		34	54
3100	Office furniture, repair		7	1.143			62.50		62.50	100
3120	Portable power tool, assemble or disassemble		6	1.333			73		73	117
3130	Repair or replace part		6	1.333			73		73	117
3180	Replace tool handle		16	.500			27.50		27.50	43.50
3200	Sharpen hand tool		16	.500			27.50		27.50	43.50

01 93 13.12 Furnishings Facilities Maintenance		Crew	Daily Output	Labor-Hours	Unit	Material	Labor	Equipment	Total	Total Incl O&P
0010	**FURNISHINGS FACILITIES MAINTENANCE**									
1000	Reupholster, side chair	1 Skwk	2.50	3.200	Ea.		176		176	280
1040	Tubular frame chair seat and back		3	2.667			146		146	233
1080	Sofa		2	4			219		219	350

01 93 13.12 Furnishings Facilities Maintenance		Crew	Daily Output	Labor-Hours	Unit	Material	2020 Bare Costs Labor	Equipment	Total	Total Incl O&P
1540	Vacuum, large divan	1 Clab	80	.100	Ea.		4.21		4.21	6.75

01 93 13.14 Conveying Systems Facilities Maintenance

Line	Description	Crew	Daily Output	Labor-Hours	Unit	Material	Labor	Equipment	Total	Total Incl O&P
0010	**CONVEYING SYSTEMS FACILITIES MAINTENANCE**									
0900	Elevator, general cleaning, vacuum, dust, mop, polish	1 Clab	16	.500	Ea.	.49	21		21.49	34
0920	Motor generator set, service	1 Elec	8	1			61.50		61.50	94.50
0930	Repair, minor		2	4			245		245	380
0940	Limit switch, repair		11	.727			44.50		44.50	68.50
0950	Troubleshoot, main control panel	2 Elec	2.50	6.400			395		395	605
1000	Escalator cleaning, dusting, polishing	1 Clab	20	.400	Flight	.40	16.85		17.25	27.50
1500	Vehicle maintenance, grease, light vehicle	1 Skwk	20	.400	Ea.		22		22	35
1510	Heavy vehicle		7	1.143			62.50		62.50	100
1520	Wash, light vehicle		8	1		.86	55		55.86	88.50
1530	Heavy vehicle		6	1.333		1.54	73		74.54	119
1540	Change oil and filter, light vehicle		10	.800		34.50	44		78.50	108
1570	Rotate tires, light vehicle		18	.444			24.50		24.50	39

01 93 13.15 Mechanical Facilities Maintenance

Line	Description	Crew	Daily Output	Labor-Hours	Unit	Material	Labor	Equipment	Total	Total Incl O&P
0010	**MECHANICAL FACILITIES MAINTENANCE**									
0100	Air conditioning system maintenance									
0130	Belt, replace	1 Stpi	15	.533	Ea.		35		35	54
0170	Fan, clean		16	.500			33		33	50.50
0180	Filter, remove, clean, replace		12	.667			43.50		43.50	67.50
0190	Flexible coupling alignment, inspect		40	.200			13.10		13.10	20.50
0200	Gas leak, locate and repair		4	2			131		131	203
0250	Pump packing gland, remove and replace		11	.727			47.50		47.50	73.50
0270	Tighten		32	.250			16.40		16.40	25.50
0290	Pump, disassemble and assemble		4	2			131		131	203
0300	Air pressure regulator, disassemble, clean, assemble	1 Skwk	4	2			110		110	175
0310	Repair or replace part	"	6	1.333			73		73	117
0320	Purging system	1 Stpi	16	.500			33		33	50.50
0400	Compressor, air, remove or install fan wheel	1 Skwk	20	.400			22		22	35
0410	Disassemble or assemble 2 cylinder, 2 stage		4	2			110		110	175
0420	4 cylinder, 4 stage		1	8			440		440	700
0430	Repair or replace part		2	4			219		219	350
0700	Demolition, for mech. demolition see Section 23 05 05.10 or 22 05 05.10									
0800	Ductwork, clean									
0810	Rectangular									
0820	6" G	1 Shee	187.50	.043	L.F.		2.66		2.66	4.16
0830	8" G		140.63	.057			3.54		3.54	5.55
0840	10" G		112.50	.071			4.43		4.43	6.95
0850	12" G		93.75	.085			5.30		5.30	8.35
0860	14" G		80.36	.100			6.20		6.20	9.70
0870	16" G		70.31	.114			7.10		7.10	11.10
0900	Round									
0910	4" G	1 Shee	358.10	.022	L.F.		1.39		1.39	2.18
0920	6" G		238.73	.034			2.09		2.09	3.27
0930	8" G		179.05	.045			2.78		2.78	4.36
0940	10" G		143.24	.056			3.48		3.48	5.45
0950	12" G		119.37	.067			4.18		4.18	6.55
0960	16" G		89.52	.089			5.55		5.55	8.70
1200	Fire protection equipment									
1220	Fire hydrant, replace	Q-1	3	5.333	Ea.		310		310	480
1230	Service, lubricate, inspect, flush, clean	1 Plum	7	1.143			73.50		73.50	114

01 93 Facility Maintenance

01 93 13 – Facility Maintenance Procedures

01 93 13.15 Mechanical Facilities Maintenance		Crew	Daily Output	Labor-Hours	Unit	Material	2020 Bare Costs Labor	Equipment	Total	Total Incl O&P
1240	Test	1 Plum	11	.727	Ea.		47		47	72.50
1310	Inspect valves, pressure, nozzle	1 Spri	4	2	System		127		127	196
1800	Plumbing fixtures, for installation see Section 22 41 00									
1850	Open drain with toilet auger	1 Plum	16	.500	Ea.		32		32	50
1870	Plaster trap, clean	"	6	1.333			86		86	133
1900	Relief valve, test and adjust	1 Stpi	20	.400			26		26	40.50
2000	Repair or replace, steam trap		8	1			65.50		65.50	101
2020	Y-type or bell strainer		6	1.333			87.50		87.50	135
2040	Water trap or vacuum breaker, screwed joints		13	.615			40.50		40.50	62.50
2100	Steam specialties, clean									
2120	Air separator with automatic trap, 1" fittings	1 Stpi	12	.667	Ea.		43.50		43.50	67.50
2130	Bucket trap, 2" pipe		7	1.143			75		75	116
2150	Drip leg, 2" fitting		45	.178			11.65		11.65	18
2200	Thermodynamic trap, 1" fittings		50	.160			10.50		10.50	16.20
2210	Thermostatic		65	.123			8.05		8.05	12.45
2240	Screen and seat in Y-type strainer, plug type		25	.320			21		21	32.50
2242	Screen and seat in Y-type strainer, flange type		12	.667			43.50		43.50	67.50
2500	Valve, replace broken handwheel		24	.333			22		22	34
3000	Valve, overhaul, regulator, relief, flushometer, mixing									
3040	Cold water, gas	1 Stpi	5	1.600	Ea.		105		105	162
3050	Hot water, steam		3	2.667			175		175	270
3080	Globe, gate, check up to 4" cold water, gas		10	.800			52.50		52.50	81
3090	Hot water, steam		5	1.600			105		105	162
3100	Over 4" ID hot or cold line		1.40	5.714			375		375	580
3120	Remove and replace, gate, globe or check up to 4"	Q-5	6	2.667			157		157	243
3130	Over 4"	"	2	8			470		470	730
3150	Repack up to 4"	1 Stpi	13	.615			40.50		40.50	62.50
3160	Over 4"	"	4	2			131		131	203

01 93 13.16 Electrical Facilities Maintenance		Crew	Daily Output	Labor-Hours	Unit	Material	Labor	Equipment	Total	Total Incl O&P
0010	**ELECTRICAL FACILITIES MAINTENANCE**									
0700	Cathodic protection systems									
0720	Check and adjust reading on rectifier	1 Elec	20	.400	Ea.		24.50		24.50	38
0730	Check pipe to soil potential		20	.400			24.50		24.50	38
0740	Replace lead connection		4	2			123		123	189
0800	Control device, install		5.70	1.404			86		86	133
0810	Disassemble, clean and reinstall		7	1.143			70		70	108
0820	Replace		10.70	.748			46		46	70.50
0830	Trouble shoot		10	.800			49		49	75.50
0900	Demolition, for electrical demolition see Section 26 05 05.10									
1000	Distribution systems and equipment install or repair a breaker									
1010	In power panels up to 200 amps	1 Elec	7	1.143	Ea.		70		70	108
1020	Over 200 amps		2	4			245		245	380
1030	Reset breaker or replace fuse		20	.400			24.50		24.50	38
1100	Megger test MCC (each stack)		4	2			123		123	189
1110	MCC vacuum and clean (each stack)		5.30	1.509			92.50		92.50	143
2500	Remove and replace or maint., road fixture & lamp		3	2.667		1,325	164		1,489	1,700
2510	Fluorescent fixture		7	1.143		87	70		157	204
2515	Relamp (fluor.) facility area each tube, spot	1 Clab	24	.333		2.15	14.05		16.20	25
2516	Group		100	.080		2.15	3.37		5.52	7.75
2518	Fluorescent fixture, clean (area)		44	.182		.05	7.65		7.70	12.30
2520	Incandescent fixture	1 Elec	11	.727		56.50	44.50		101	131
2530	Lamp (incandescent or fluorescent)	1 Clab	60	.133		2.15	5.60		7.75	11.35

01 93 13 – Facility Maintenance Procedures

01 93 13.16 Electrical Facilities Maintenance		Crew	Daily Output	Labor-Hours	Unit	Material	2020 Bare Costs Labor	Equipment	Total	Total Incl O&P
2535	Replace cord in socket lamp	1 Elec	13	.615	Ea.	1.79	38		39.79	60
2540	Ballast electronic type for two tubes		8	1		41.50	61.50		103	140
2541	Starter		30	.267		.87	16.35		17.22	26
2545	Replace other lighting parts		11	.727		20	44.50		64.50	90.50
2550	Switch		11	.727		3.46	44.50		47.96	72.50
2555	Receptacle		11	.727		10.65	44.50		55.15	80.50
2560	Floodlight		4	2		244	123		367	455
2570	Christmas lighting, indoor, per string	1 Clab	16	.500			21		21	33.50
2580	Outdoor		13	.615			26		26	41.50
2590	Test battery operated emergency lights		40	.200			8.40		8.40	13.50
2600	Repair and replace component in communication system	1 Elec	6	1.333		53.50	82		135.50	185
2700	Repair misc. appliances (incl. clocks, vent fan, blower, etc.)	"	6	1.333			82		82	126
2710	Reset clocks & timers	1 Clab	50	.160			6.75		6.75	10.80
2720	Adjust time delay relays	1 Elec	16	.500			30.50		30.50	47
2730	Test specific gravity of lead-acid batteries	1 Clab	80	.100			4.21		4.21	6.75
3000	Motors and generators									
3020	Disassemble, clean and reinstall motor, up to 1/4 HP	1 Elec	4	2	Ea.		123		123	189
3030	Up to 3/4 HP		3	2.667			164		164	252
3040	Up to 10 HP		2	4			245		245	380
3050	Replace part, up to 1/4 HP		6	1.333			82		82	126
3060	Up to 3/4 HP		4	2			123		123	189
3070	Up to 10 HP		3	2.667			164		164	252
3080	Megger test motor windings		5.33	1.501			92		92	142
3082	Motor vibration check		16	.500			30.50		30.50	47
3084	Oil motor bearings		25	.320			19.65		19.65	30
3086	Run test emergency generator for 30 minutes		11	.727			44.50		44.50	68.50
3090	Rewind motor, up to 1/4 HP		3	2.667			164		164	252
3100	Up to 3/4 HP		2	4			245		245	380
3110	Up to 10 HP		1.50	5.333			325		325	505
3150	Generator, repair or replace part		4	2			123		123	189
3160	Repair DC generator		2	4			245		245	380
4000	Stub pole, install or remove		3	2.667			164		164	252
4500	Transformer maintenance up to 15 kVA		2.70	2.963			182		182	280

Division Notes

I	CREW	DAILY OUTPUT	LABOR-HOURS	UNIT	BARE COSTS MAT.	BARE COSTS LABOR	BARE COSTS EQUIP.	BARE COSTS TOTAL	TOTAL INCL O&P

Estimating Tips

02 30 00 Subsurface Investigation

In preparing estimates on structures involving earthwork or foundations, all information concerning soil characteristics should be obtained. Look particularly for hazardous waste, evidence of prior dumping of debris, and previous stream beds.

02 40 00 Demolition and Structure Moving

The costs shown for selective demolition do not include rubbish handling or disposal. These items should be estimated separately using RSMeans data or other sources.

- Historic preservation often requires that the contractor remove materials from the existing structure, rehab them, and replace them. The estimator must be aware of any related measures and precautions that must be taken when doing selective demolition and cutting and patching. Requirements may include special handling and storage, as well as security.
- In addition to Subdivision 02 41 00, you can find selective demolition items in each division. Example: Roofing demolition is in Division 7.
- Absent of any other specific reference, an approximate demolish-in-place cost can be obtained by halving the new-install labor cost. To remove for reuse, allow the entire new-install labor figure.

02 40 00 Building Deconstruction

This section provides costs for the careful dismantling and recycling of most low-rise building materials.

02 50 00 Containment of Hazardous Waste

This section addresses on-site hazardous waste disposal costs.

02 80 00 Hazardous Material Disposal/Remediation

This subdivision includes information on hazardous waste handling, asbestos remediation, lead remediation, and mold remediation. See reference numbers R028213-20 and R028319-60 for further guidance in using these unit price lines.

02 90 00 Monitoring Chemical Sampling, Testing Analysis

This section provides costs for on-site sampling and testing hazardous waste.

Reference Numbers

Reference numbers are shown at the beginning of some major classifications. These numbers refer to related items in the Reference Section. The reference information may be an estimating procedure, an alternate pricing method, or technical information.

Note: Not all subdivisions listed here necessarily appear. ■

02 21 Surveys

02 21 13 – Site Surveys

02 21 13.09 Topographical Surveys		Crew	Daily Output	Labor-Hours	Unit	Material	2020 Bare Costs Labor	Equipment	Total	Total Incl O&P
0010	**TOPOGRAPHICAL SURVEYS**									
0020	Topographical surveying, conventional, minimum	A-7	3.30	7.273	Acre	23.50	415	9.30	447.80	690
0100	Maximum	A-8	.60	53.333	"	63	2,975	51	3,089	4,825

02 21 13.13 Boundary and Survey Markers		Crew	Daily Output	Labor-Hours	Unit	Material	Labor	Equipment	Total	Total Incl O&P
0010	**BOUNDARY AND SURVEY MARKERS**									
0300	Lot location and lines, large quantities, minimum	A-7	2	12	Acre	36	690	15.35	741.35	1,125
0320	Average	"	1.25	19.200		63.50	1,100	24.50	1,188	1,825
0400	Small quantities, maximum	A-8	1	32		76.50	1,775	30.50	1,882	2,950
0600	Monuments, 3′ long	A-7	10	2.400	Ea.	35.50	138	3.07	176.57	259
0800	Property lines, perimeter, cleared land	"	1000	.024	L.F.	.08	1.38	.03	1.49	2.29
0900	Wooded land	A-8	875	.037	"	.10	2.04	.04	2.18	3.37

02 21 13.16 Aerial Surveys		Crew	Daily Output	Labor-Hours	Unit	Material	Labor	Equipment	Total	Total Incl O&P
0010	**AERIAL SURVEYS**									
1500	Aerial surveying, including ground control, minimum fee, 10 acres				Total				4,700	4,700
1510	100 acres								9,400	9,400
1550	From existing photography, deduct								1,625	1,625
1600	2′ contours, 10 acres				Acre				470	470
1850	100 acres				"				94	94

02 32 Geotechnical Investigations

02 32 13 – Subsurface Drilling and Sampling

02 32 13.10 Boring and Exploratory Drilling		Crew	Daily Output	Labor-Hours	Unit	Material	Labor	Equipment	Total	Total Incl O&P
0010	**BORING AND EXPLORATORY DRILLING**									
0020	Borings, initial field stake out & determination of elevations	A-6	1	16	Day		845	31	876	1,375
0100	Drawings showing boring details				Total		335		335	425
0200	Report and recommendations from P.E.						775		775	970
0300	Mobilization and demobilization	B-55	4	6			263	305	568	755
0350	For over 100 miles, per added mile		450	.053	Mile		2.34	2.69	5.03	6.70
0600	Auger holes in earth, no samples, 2-1/2″ diameter		78.60	.305	L.F.		13.40	15.40	28.80	38.50
0650	4″ diameter		67.50	.356			15.60	17.95	33.55	45
0800	Cased borings in earth, with samples, 2-1/2″ diameter		55.50	.432		20.50	18.95	22	61.45	77
0850	4″ diameter		32.60	.736		29	32.50	37	98.50	125
1000	Drilling in rock, "BX" core, no sampling	B-56	34.90	.458			22	43.50	65.50	82.50
1050	With casing & sampling		31.70	.505		20.50	24	48	92.50	113
1200	"NX" core, no sampling		25.92	.617			29.50	58.50	88	111
1250	With casing and sampling		25	.640		21	30.50	60.50	112	138
1400	Borings, earth, drill rig and crew with truck mounted auger	B-55	1	24	Day		1,050	1,200	2,250	3,000
1450	Rock using crawler type drill	B-56	1	16	"		760	1,525	2,285	2,875
1500	For inner city borings, add, minimum								10%	10%
1510	Maximum								20%	20%

02 32 19 – Exploratory Excavations

02 32 19.10 Test Pits		Crew	Daily Output	Labor-Hours	Unit	Material	Labor	Equipment	Total	Total Incl O&P
0010	**TEST PITS**									
0020	Hand digging, light soil	1 Clab	4.50	1.778	C.Y.		75		75	120
0100	Heavy soil	"	2.50	3.200			135		135	216
0120	Loader-backhoe, light soil	B-11M	28	.571			28.50	8.30	36.80	53.50
0130	Heavy soil	"	20	.800			39.50	11.60	51.10	75.50
1000	Subsurface exploration, mobilization				Mile				6.75	8.40
1010	Difficult access for rig, add				Hr.				260	320
1020	Auger borings, drill rig, incl. samples				L.F.				26.50	33

02 32 Geotechnical Investigations

02 32 19 – Exploratory Excavations

02 32 19.10 Test Pits		Crew	Daily Output	Labor-Hours	Unit	Material	2020 Bare Costs Labor	Equipment	Total	Total Incl O&P
1030	Hand auger				L.F.				31.50	40
1050	Drill and sample every 5′, split spoon								31.50	40
1060	Extra samples				Ea.				36	45.50

02 41 Demolition

02 41 13 – Selective Site Demolition

02 41 13.15 Hydrodemolition

		Crew	Daily Output	Labor-Hours	Unit	Material	Labor	Equipment	Total	Total Incl O&P
0010	**HYDRODEMOLITION**									
0015	Hydrodemolition, concrete pavement									
0120	Includes removal but exlcudes disposal of concrete									
0130	2″ depth	B-5E	1000	.064	S.F.		3	2.75	5.75	7.80
0410	4″ depth		800	.080			3.75	3.44	7.19	9.75
0420	6″ depth		600	.107			5	4.59	9.59	13
0520	8″ depth		300	.213			10	9.20	19.20	26

02 41 13.16 Hydroexcavation

		Crew	Daily Output	Labor-Hours	Unit	Material	Labor	Equipment	Total	Total Incl O&P
0010	**HYDROEXCAVATION**									
0015	Hydroexcavation									
0110	Mobilization or Demobilization	B-6D	8	2.500	Ea.		121	158	279	365
0130	Normal Conditions		48	.417	B.C.Y.		20	26.50	46.50	61
0140	Adverse Conditions		30	.667	″		32	42	74	97.50
0160	Minimum labor/equipment charge		4	5	Ea.		242	315	557	730

02 41 13.17 Demolish, Remove Pavement and Curb

		Crew	Daily Output	Labor-Hours	Unit	Material	Labor	Equipment	Total	Total Incl O&P
0010	**DEMOLISH, REMOVE PAVEMENT AND CURB** R024119-10									
5010	Pavement removal, bituminous roads, up to 3″ thick	B-38	690	.058	S.Y.		2.76	1.80	4.56	6.35
5050	4″-6″ thick		420	.095			4.53	2.96	7.49	10.45
5100	Bituminous driveways		640	.063			2.98	1.94	4.92	6.85
5200	Concrete to 6″ thick, hydraulic hammer, mesh reinforced		255	.157			7.45	4.87	12.32	17.20
5300	Rod reinforced		200	.200			9.50	6.20	15.70	22
5400	Concrete, 7″-24″ thick, plain		33	1.212	C.Y.		57.50	37.50	95	133
5500	Reinforced		24	1.667	″		79.50	51.50	131	183
5590	Minimum labor/equipment charge		6	6.667	Job		315	207	522	735
5600	With hand held air equipment, bituminous, to 6″ thick	B-39	1900	.025	S.F.		1.12	.19	1.31	1.99
5700	Concrete to 6″ thick, no reinforcing		1600	.030			1.33	.22	1.55	2.37
5800	Mesh reinforced		1400	.034			1.52	.26	1.78	2.70
5900	Rod reinforced		765	.063			2.78	.47	3.25	4.94
5990	Minimum labor/equipment charge	B-38	6	6.667	Job		315	207	522	735
6000	Curbs, concrete, plain	B-6	360	.067	L.F.		3.05	.59	3.64	5.50
6100	Reinforced		275	.087			3.99	.78	4.77	7.20
6200	Granite		360	.067			3.05	.59	3.64	5.50
6300	Bituminous		528	.045			2.08	.41	2.49	3.75
6390	Minimum labor/equipment charge		6	4	Job		183	35.50	218.50	330
6500	Site demo, berms under 4″ in height, bituminous		528	.045	L.F.		2.08	.41	2.49	3.75
6600	4″ or over in height		300	.080	″		3.66	.71	4.37	6.60

02 41 13.20 Selective Demo, Highway Guard Rails & Barriers

		Crew	Daily Output	Labor-Hours	Unit	Material	Labor	Equipment	Total	Total Incl O&P
0010	**SELECTIVE DEMOLITION, HIGHWAY GUARD RAILS & BARRIERS**									
0100	Guard rail, corrugated steel	B-6	600	.040	L.F.		1.83	.36	2.19	3.29
0200	End sections		40	.600	Ea.		27.50	5.35	32.85	49.50
0300	Wrap around		40	.600	″		27.50	5.35	32.85	49.50
0400	Timber 4″ x 8″		600	.040	L.F.		1.83	.36	2.19	3.29
0500	Three 3/4″ cables		600	.040	″		1.83	.36	2.19	3.29

02 41 13 – Selective Site Demolition

02 41 13.20 Selective Demo, Highway Guard Rails & Barriers

	02 41 13.20 Selective Demo, Highway Guard Rails & Barriers	Crew	Daily Output	Labor-Hours	Unit	Material	2020 Bare Costs Labor	Equipment	Total	Total Incl O&P
0600	Wood posts	B-6	240	.100	Ea.		4.57	.89	5.46	8.25
0700	Guide rail, 6" x 6" box beam	B-80B	120	.267	L.F.		11.95	1.97	13.92	21
0800	Median barrier, 6" x 8" box beam		240	.133			6	.98	6.98	10.60
0850	Precast concrete 3'-6" high x 2' wide		300	.107			4.78	.79	5.57	8.45
0900	Impact barrier, MUTCD, barrel type	B-16	60	.533	Ea.		23.50	9.55	33.05	48.50
1000	Resilient guide fence and light shield 6' high	"	120	.267	L.F.		11.80	4.77	16.57	24
1100	Concrete posts, 6'-5" triangular	B-6	200	.120	Ea.		5.50	1.07	6.57	9.90
1200	Speed bumps 10-1/2" x 2-1/4" x 48"		300	.080	L.F.		3.66	.71	4.37	6.60
1300	Pavement marking channelizing		200	.120	Ea.		5.50	1.07	6.57	9.90
1400	Barrier and curb delineators		300	.080			3.66	.71	4.37	6.60
1500	Rumble strips 24" x 3-1/2" x 1/2"		150	.160			7.30	1.43	8.73	13.15

02 41 13.23 Utility Line Removal

	02 41 13.23 Utility Line Removal	Crew	Daily Output	Labor-Hours	Unit	Material	Labor	Equipment	Total	Total Incl O&P
0010	**UTILITY LINE REMOVAL**									
0015	No hauling, abandon catch basin or manhole	B-6	7	3.429	Ea.		157	30.50	187.50	283
0020	Remove existing catch basin or manhole, masonry		4	6			274	53.50	327.50	495
0030	Catch basin or manhole frames and covers, stored		13	1.846			84.50	16.45	100.95	152
0040	Remove and reset		7	3.429			157	30.50	187.50	283
0900	Hydrants, fire, remove only	B-21A	5	8			420	85.50	505.50	750
0950	Remove and reset	"	2	20			1,050	214	1,264	1,875
0990	Minimum labor/equipment charge	2 Plum	2	8	Job		515		515	795
2900	Pipe removal, sewer/water, no excavation, 12" diameter	B-6	175	.137	L.F.		6.25	1.22	7.47	11.30
2930	15"-18" diameter	B-12Z	150	.160			7.65	9.75	17.40	23
2960	21"-24" diameter		120	.200			9.55	12.20	21.75	28.50
3000	27"-36" diameter		90	.267			12.75	16.30	29.05	38
3200	Steel, welded connections, 4" diameter	B-6	160	.150			6.85	1.34	8.19	12.35
3300	10" diameter		80	.300			13.70	2.67	16.37	25
3390	Minimum labor/equipment charge		3	8	Job		365	71.50	436.50	660

02 41 13.30 Minor Site Demolition

	02 41 13.30 Minor Site Demolition	Crew	Daily Output	Labor-Hours	Unit	Material	Labor	Equipment	Total	Total Incl O&P
0010	**MINOR SITE DEMOLITION** R024119-10									
0100	Roadside delineators, remove only	B-80	175	.183	Ea.		8.50	5.85	14.35	19.95
0110	Remove and reset	"	100	.320	"		14.90	10.20	25.10	35
0400	Minimum labor/equipment charge	B-6	4	6	Job		274	53.50	327.50	495
0800	Guiderail, corrugated steel, remove only	B-80A	100	.240	L.F.		10.10	8.20	18.30	25
0850	Remove and reset	"	40	.600	"		25.50	20.50	46	63
0860	Guide posts, remove only	B-80B	120	.267	Ea.		11.95	1.97	13.92	21
0870	Remove and reset	B-55	50	.480	"		21	24	45	60
0890	Minimum labor/equipment charge	2 Clab	4	4	Job		168		168	270
1000	Masonry walls, block, solid	B-5	1800	.031	C.F.		1.45	.85	2.30	3.23
1200	Brick, solid		900	.062			2.90	1.70	4.60	6.45
1400	Stone, with mortar		900	.062			2.90	1.70	4.60	6.45
1500	Dry set		1500	.037			1.74	1.02	2.76	3.88
1600	Median barrier, precast concrete, remove and store	B-3	430	.112	L.F.		5.25	5.40	10.65	14.30
1610	Remove and reset	"	390	.123	"		5.80	5.95	11.75	15.80
1650	Minimum labor/equipment charge	A-1	4	2	Job		84	28	112	166
4000	Sidewalk removal, bituminous, 2" thick	B-6	350	.069	S.Y.		3.14	.61	3.75	5.65
4010	2-1/2" thick		325	.074			3.38	.66	4.04	6.05
4050	Brick, set in mortar		185	.130			5.95	1.16	7.11	10.65
4100	Concrete, plain, 4"		160	.150			6.85	1.34	8.19	12.35
4110	Plain, 5"		140	.171			7.85	1.53	9.38	14.15
4120	Plain, 6"		120	.200			9.15	1.78	10.93	16.45
4200	Mesh reinforced, concrete, 4"		150	.160			7.30	1.43	8.73	13.15
4210	5" thick		131	.183			8.40	1.63	10.03	15.10

02 41 13 – Selective Site Demolition

02 41 13.30 Minor Site Demolition		Crew	Daily Output	Labor-Hours	Unit	Material	2020 Bare Costs Labor	Equipment	Total	Total Incl O&P
4220	6" thick	B-6	112	.214	S.Y.		9.80	1.91	11.71	17.65
4290	Minimum labor/equipment charge	B-39	12	4	Job		177	30	207	315
6850	Runways, remove rubber skid marks, 4-6 passes	B-59A	35	.686	M.S.F.	75	30.50	13.10	118.60	145
6860	6-10 passes	"	35	.686	"	112	30.50	13.10	155.60	187
02 41 13.33 Railtrack Removal										
0010	**RAILTRACK REMOVAL**									
3500	Railroad track removal, ties and track	B-13	330	.170	L.F.		7.80	1.76	9.56	14.35
3600	Ballast	B-14	500	.096	C.Y.		4.25	.43	4.68	7.20
3700	Remove and re-install, ties & track using new bolts & spikes		50	.960	L.F.		42.50	4.27	46.77	72.50
3800	Turnouts using new bolts and spikes		1	48	Ea.		2,125	214	2,339	3,600
3890	Minimum labor/equipment charge	↓	5	9.600	Job		425	42.50	467.50	720
02 41 13.34 Selective Demolition, Utility Materials										
0010	**SELECTIVE DEMOLITION, UTILITY MATERIALS** R024119-10									
0015	Excludes excavation									
0020	See other utility items in Section 02 41 13.33									
0100	Fire hydrant extensions	B-20	14	1.714	Ea.		80.50		80.50	129
0200	Precast utility boxes up to 8' x 14' x 7'	B-13	2	28			1,300	290	1,590	2,375
0300	Handholes and meter pits	B-6	2	12			550	107	657	990
0400	Utility valves 4"-12"	B-20	4	6			282		282	450
0500	14"-24"	B-21	2	14	↓		685	94.50	779.50	1,175
02 41 13.36 Selective Demolition, Utility Valves and Accessories										
0010	**SELECTIVE DEMOLITION, UTILITY VALVES & ACCESSORIES**									
0015	Excludes excavation									
0100	Utility valves 4"-12" diam.	B-20	4	6	Ea.		282		282	450
0200	14"-24" diam.	B-21	2	14			685	94.50	779.50	1,175
0300	Crosses 4"-12" diam.	B-20	8	3			141		141	226
0400	14"-24" diam.	B-21	4	7			340	47.50	387.50	595
0500	Utility cut-in valves 4"-12" diam.	B-20	20	1.200			56.50		56.50	90
0600	Curb boxes	"	20	1.200	↓		56.50		56.50	90
02 41 13.38 Selective Demo., Water & Sewer Piping & Fittings										
0010	**SELECTIVE DEMOLITION, WATER & SEWER PIPING AND FITTINGS**									
0015	Excludes excavation									
0090	Concrete pipe 4"-10" diameter	B-6	250	.096	L.F.		4.39	.86	5.25	7.90
0100	42"-48" diameter	B-13B	96	.583			27	10.20	37.20	54
0200	60"-84" diameter	"	80	.700			32	12.25	44.25	64.50
0300	96" diameter	B-13C	80	.700			32	28.50	60.50	82.50
0400	108"-144" diameter	"	64	.875	↓		40.50	35.50	76	104
0450	Concrete fittings 12" diameter	B-6	24	1	Ea.		45.50	8.90	54.40	82.50
0480	Concrete end pieces 12" diameter		200	.120	L.F.		5.50	1.07	6.57	9.90
0485	15" diameter		150	.160			7.30	1.43	8.73	13.15
0490	18" diameter		150	.160			7.30	1.43	8.73	13.15
0500	24"-36" diameter		100	.240	↓		11	2.14	13.14	19.80
0600	Concrete fittings 24"-36" diameter	↓	12	2	Ea.		91.50	17.80	109.30	165
0700	48"-84" diameter	B-13B	12	4.667			215	81.50	296.50	430
0800	96" diameter	"	8	7			320	123	443	645
0900	108"-144" diameter	B-13C	4	14	↓		645	570	1,215	1,650
1000	Ductile iron pipe 4" diameter	B-21B	200	.200	L.F.		9.20	2.36	11.56	17.25
1100	6"-12" diameter		175	.229			10.50	2.69	13.19	19.65
1200	14"-24" diameter		120	.333	↓		15.30	3.93	19.23	29
1300	Ductile iron fittings 4"-12" diameter		24	1.667	Ea.		76.50	19.65	96.15	144
1400	14"-16" diameter		18	2.222			102	26	128	192
1500	18"-24" diameter	↓	12	3.333	↓		153	39.50	192.50	287

02 41 13.38 Selective Demo., Water & Sewer Piping & Fittings

	02 41 13.38 Selective Demo., Water & Sewer Piping & Fittings	Crew	Daily Output	Labor-Hours	Unit	Material	2020 Bare Costs Labor	Equipment	Total	Total Incl O&P
1600	Plastic pipe 3/4"-4" diameter	B-6	700	.034	L.F.		1.57	.31	1.88	2.83
1700	6"-8" diameter		500	.048			2.20	.43	2.63	3.96
1800	10"-18" diameter		300	.080			3.66	.71	4.37	6.60
1900	20"-36" diameter		200	.120			5.50	1.07	6.57	9.90
1910	42"-48" diameter		180	.133			6.10	1.19	7.29	11
1920	54"-60" diameter		160	.150			6.85	1.34	8.19	12.35
2000	Plastic fittings 4"-8" diameter		75	.320	Ea.		14.65	2.85	17.50	26
2100	10"-14" diameter		50	.480			22	4.28	26.28	39.50
2200	16"-24" diameter		20	1.200			55	10.70	65.70	99
2210	30"-36" diameter		15	1.600			73	14.25	87.25	132
2220	42"-48" diameter		12	2			91.50	17.80	109.30	165
2300	Copper pipe 3/4"-2" diameter	Q-1	500	.032	L.F.		1.86		1.86	2.87
2400	2-1/2"-3" diameter		300	.053			3.09		3.09	4.78
2500	4"-6" diameter		200	.080			4.64		4.64	7.15
2600	Copper fittings 3/4"-2" diameter		15	1.067	Ea.		62		62	95.50
2700	Cast iron pipe 4" diameter		200	.080	L.F.		4.64		4.64	7.15
2800	5"-6" diameter	Q-2	200	.120			7.20		7.20	11.15
2900	8"-12" diameter	Q-3	200	.160			9.80		9.80	15.20
3000	Cast iron fittings 4" diameter	Q-1	30	.533	Ea.		31		31	48
3100	5"-6" diameter	Q-2	30	.800			48		48	74.50
3200	8"-15" diameter	Q-3	30	1.067			65.50		65.50	101
3300	Vent cast iron pipe 4"-8" diameter	Q-1	200	.080	L.F.		4.64		4.64	7.15
3400	10"-15" diameter	Q-3	200	.160	"		9.80		9.80	15.20
3500	Vent cast iron fittings 4"-8" diameter	Q-2	30	.800	Ea.		48		48	74.50
3600	10"-15" diameter	"	20	1.200	"		72		72	112

02 41 13.40 Selective Demolition, Metal Drainage Piping

	02 41 13.40 Selective Demolition, Metal Drainage Piping	Crew	Daily Output	Labor-Hours	Unit	Material	Labor	Equipment	Total	Total Incl O&P
0010	**SELECTIVE DEMOLITION, METAL DRAINAGE PIPING**									
0015	Excludes excavation									
0100	CMP pipe, aluminum, 6"-10" diam.	B-11M	800	.020	L.F.		.99	.29	1.28	1.88
0110	12" diam.		600	.027			1.32	.39	1.71	2.51
0120	18" diam.		600	.027			1.32	.39	1.71	2.51
0140	Steel, 6"-10" diam.		800	.020			.99	.29	1.28	1.88
0150	12" diam.		600	.027			1.32	.39	1.71	2.51
0160	18" diam.		400	.040			1.98	.58	2.56	3.77
0170	24" diam.	B-13	300	.187			8.60	1.94	10.54	15.80
0180	30"-36" diam.		250	.224			10.30	2.32	12.62	18.95
0190	48"-60" diam.		200	.280			12.90	2.90	15.80	23.50
0200	72" diam.	B-13B	100	.560			26	9.80	35.80	52
0210	CMP end sections, steel, 10"-18" diam.	B-11M	40	.400	Ea.		19.75	5.80	25.55	38
0220	24"-36" diam.	B-13	30	1.867			86	19.35	105.35	159
0230	48" diam.		20	2.800			129	29	158	237
0240	60" diam.		10	5.600			258	58	316	475
0250	72" diam.	B-13B	10	5.600			258	98	356	520
0260	CMP fittings, 8"-12" diam.	B-11M	60	.267			13.20	3.87	17.07	25.50
0270	18" diam.	"	40	.400			19.75	5.80	25.55	38
0280	24"-48" diam.	B-13	30	1.867			86	19.35	105.35	159
0290	60" diam.	"	20	2.800			129	29	158	237
0300	72" diam.	B-13B	10	5.600			258	98	356	520
0310	Oval arch 17" x 13", 21" x 15", 15"-18" equivalent	B-11M	400	.040	L.F.		1.98	.58	2.56	3.77
0320	28" x 20", 24" equivalent	B-13	300	.187			8.60	1.94	10.54	15.80
0330	35" x 24", 42" x 29", 30"-36" equivalent		250	.224			10.30	2.32	12.62	18.95
0340	49" x 33", 57" x 38", 42"-48" equivalent		200	.280			12.90	2.90	15.80	23.50

02 41 13.40 Selective Demolition, Metal Drainage Piping		Crew	Daily Output	Labor-Hours	Unit	Material	2020 Bare Costs Labor	Equipment	Total	Total Incl O&P
0350	Oval arch 17″ x 13″ end piece, 15″ equivalent	B-11M	40	.400	Ea.		19.75	5.80	25.55	38
0360	42″ x 29″ end piece, 36″ equivalent	B-13	30	1.867	″		86	19.35	105.35	159

02 41 13.42 Selective Demolition, Manholes and Catch Basins

		Crew	Daily Output	Labor-Hours	Unit	Material	Labor	Equipment	Total	Total Incl O&P
0010	**SELECTIVE DEMOLITION, MANHOLES & CATCH BASINS**									
0015	Excludes excavation									
0100	Manholes, precast or brick over 8′ deep	B-6	8	3	V.L.F.		137	26.50	163.50	248
0200	Cast in place 4′-8′ deep	B-9	127	.315	SF Face		13.40	2.81	16.21	24.50
0300	Over 8′ deep	″	100	.400	″		17	3.57	20.57	31
0400	Top, precast, 8″ thick, 4′-6′ diam.	B-6	8	3	Ea.		137	26.50	163.50	248
0500	Steps	1 Clab	60	.133	″		5.60		5.60	9

02 41 13.43 Selective Demolition, Box Culvert

		Crew	Daily Output	Labor-Hours	Unit	Material	Labor	Equipment	Total	Total Incl O&P
0010	**SELECTIVE DEMOLITION, BOX CULVERT**									
0015	Excludes excavation									
0100	Box culvert 8′ x 6′ x 3′ to 8′ x 8′ x 8′	B-69	300	.160	L.F.		7.45	4.95	12.40	17.30
0200	8′ x 10′ x 3′ to 8′ x 12′ x 8′	″	200	.240	″		11.20	7.45	18.65	26

02 41 13.44 Selective Demolition, Septic Tanks and Related Components

		Crew	Daily Output	Labor-Hours	Unit	Material	Labor	Equipment	Total	Total Incl O&P
0010	**SELECTIVE DEMOLITION, SEPTIC TANKS & RELATED COMPONENTS**									
0020	Excludes excavation and disposal									
0100	Septic tanks, precast, 1000-1250 gal.	B-21	8	3.500	Ea.		171	23.50	194.50	298
0200	1500 gal.		7	4			195	27	222	340
0300	2000-2500 gal.		5	5.600			273	38	311	475
0400	4000 gal.		4	7			340	47.50	387.50	595
0500	Precast, 5000 gal., multiple sections	B-13	3	18.667			860	194	1,054	1,600
0600	15,000 gal.	B-13B	1.70	32.941			1,525	575	2,100	3,025
0800	40,000 gal.	″	.80	70			3,225	1,225	4,450	6,475
0900	Precast, 50,000 gal., 5 piece	B-13C	.60	93.333			4,300	3,800	8,100	11,000
1000	Cast-in-place, 75,000 gal.	B-13L	.50	32			1,900	4,150	6,050	7,550
1100	100,000 gal.	″	.30	53.333			3,150	6,925	10,075	12,600
1200	HDPE, 1000 gal.	B-21	9	3.111			152	21	173	265
1300	1500 gal.		8	3.500			171	23.50	194.50	298
1400	Galley, 4′ x 4′ x 4′		16	1.750			85.50	11.80	97.30	149
1500	Distribution boxes, concrete, 7 outlets	2 Clab	16	1			42		42	67.50
1600	9 outlets	″	8	2			84		84	135
1700	Leaching chambers 13′ x 3′-7″ x 1′-4″, standard	B-13	16	3.500			161	36.50	197.50	296
1800	8′ x 4′ x 1′-6″, heavy duty		14	4			184	41.50	225.50	340
1900	13′ x 3′-9″ x 1′-6″		12	4.667			215	48.50	263.50	395
2100	20′ x 4′ x 1′-6″		5	11.200			515	116	631	950
2200	Leaching pit 6′-6″ x 6′ deep	B-21	5	5.600			273	38	311	475
2300	6′-6″ x 8′ deep		4	7			340	47.50	387.50	595
2400	8′ x 6′ deep H20		4	7			340	47.50	387.50	595
2500	8′ x 8′ deep H20		3	9.333			455	63	518	795
2600	Velocity reducing pit, precast 6′ x 3′ deep		4.70	5.957			290	40	330	510

02 41 13.46 Selective Demolition, Steel Pipe With Insulation

		Crew	Daily Output	Labor-Hours	Unit	Material	Labor	Equipment	Total	Total Incl O&P
0010	**SELECTIVE DEMOLITION, STEEL PIPE WITH INSULATION**									
0020	Excludes excavation									
0100	Steel pipe, with insulation, 3/4″-4″	B-1A	400	.060	L.F.		2.57	.92	3.49	5.10
0200	5″-10″	B-1B	360	.089			4.17	2.33	6.50	9.20
0300	12″-16″		240	.133			6.25	3.50	9.75	13.80
0400	18″-24″		160	.200			9.40	5.25	14.65	20.50
0450	26″-36″		100	.320			15	8.40	23.40	33.50
0500	Steel gland seal, with insulation, 3/4″-4″	B-1A	100	.240	Ea.		10.25	3.69	13.94	20.50
0600	5″-10″	B-1B	75	.427			20	11.20	31.20	44.50

02 41 13 – Selective Site Demolition

02 41 13.46 Selective Demolition, Steel Pipe With Insulation		Crew	Daily Output	Labor-Hours	Unit	Material	2020 Bare Costs Labor	Equipment	Total	Total Incl O&P
0700	12"-16"	B-1B	60	.533	Ea.		25	14	39	55.50
0800	18"-24"		50	.640			30	16.80	46.80	66
0850	26"-36"		40	.800			37.50	21	58.50	82.50
0900	Demo steel fittings with insulation, 3/4"-4"	B-1A	60	.400			17.10	6.15	23.25	34.50
1000	5"-10"	B-1B	40	.800			37.50	21	58.50	82.50
1100	12"-16"		30	1.067			50	28	78	111
1200	18"-24"		20	1.600			75	42	117	165
1300	26"-36"		15	2.133			100	56	156	221
1400	Steel pipe anchors, 5"-10"		40	.800			37.50	21	58.50	82.50
1500	12"-16"		30	1.067			50	28	78	111
1600	18"-24"		20	1.600			75	42	117	165
1700	26"-36"		15	2.133			100	56	156	221
02 41 13.48 Selective Demolition, Gasoline Containment Piping										
0010	**SELECTIVE DEMOLITION, GASOLINE CONTAINMENT PIPING**									
0020	Excludes excavation									
0030	Excludes environmental site remediation									
0100	Gasoline plastic primary containment piping 2"-4"	Q-6	800	.030	L.F.		1.84		1.84	2.84
0200	Fittings 2"-4"		40	.600	Ea.		36.50		36.50	57
0300	Gasoline plastic secondary containment piping 3"-6"		800	.030	L.F.		1.84		1.84	2.84
0400	Fittings 3"-6"		40	.600	Ea.		36.50		36.50	57
02 41 13.50 Selective Demolition, Natural Gas, PE Pipe										
0010	**SELECTIVE DEMOLITION, NATURAL GAS, PE PIPE**									
0020	Excludes excavation									
0100	Natural gas coils, PE, 1-1/4"-3"	Q-6	800	.030	L.F.		1.84		1.84	2.84
0200	Joints, 40', PE, 3"-4"		800	.030			1.84		1.84	2.84
0300	6"-8"		600	.040			2.45		2.45	3.78
02 41 13.51 Selective Demolition, Natural Gas, Steel Pipe										
0010	**SELECTIVE DEMOLITION, NATURAL GAS, STEEL PIPE**									
0020	Excludes excavation									
0100	Natural gas steel pipe 1"-4"	B-1A	800	.030	L.F.		1.28	.46	1.74	2.57
0200	5"-10"	B-1B	360	.089			4.17	2.33	6.50	9.20
0300	12"-16"		240	.133			6.25	3.50	9.75	13.80
0400	18"-24"		160	.200			9.40	5.25	14.65	20.50
0500	Natural gas steel fittings 1"-4"	B-1A	160	.150	Ea.		6.40	2.30	8.70	12.85
0600	5"-10"	B-1B	160	.200			9.40	5.25	14.65	20.50
0700	12"-16"		108	.296			13.90	7.75	21.65	30.50
0800	18"-24"		70	.457			21.50	12	33.50	47
02 41 13.52 Selective Demo, Natural Gas, Valves, Fittings, Regulators										
0010	**SELECTIVE DEMO, NATURAL GAS, VALVES, FITTINGS, REGULATORS**									
0100	Gas stops 1-1/4"-2"	1 Plum	22	.364	Ea.		23.50		23.50	36
0200	Gas regulator 1-1/2"-2"	"	22	.364			23.50		23.50	36
0300	3"-4"	Q-1	22	.727			42		42	65
0400	Gas plug valve, 3/4"-2"	1 Plum	22	.364			23.50		23.50	36
0500	2-1/2"-3"	Q-1	10	1.600			93		93	143
02 41 13.54 Selective Demolition, Electric Ducts and Fittings										
0010	**SELECTIVE DEMOLITION, ELECTRIC DUCTS & FITTINGS**									
0020	Excludes excavation									
0100	Plastic conduit, 1/2"-2"	1 Elec	600	.013	L.F.		.82		.82	1.26
0200	3"-6"	2 Elec	400	.040	"		2.45		2.45	3.78
0300	Fittings, 1/2"-2"	1 Elec	50	.160	Ea.		9.80		9.80	15.10
0400	3"-6"	"	40	.200	"		12.25		12.25	18.90

02 41 13 – Selective Site Demolition

02 41 13.56 Selective Demolition, Electric Duct Banks		Crew	Daily Output	Labor-Hours	Unit	Material	2020 Bare Costs Labor	Equipment	Total	Total Incl O&P
0010	SELECTIVE DEMOLITION, ELECTRIC DUCT BANKS									
0020	Excludes excavation									
0100	Hand holes sized to 4' x 4' x 4'	R-3	7	2.857	Ea.		175	27	202	300
0200	Manholes sized to 6' x 10' x 7'	B-13	6	9.333	"		430	97	527	790
0300	Conduit 1 @ 2" diameter, EB plastic, no concrete	2 Elec	1000	.016	L.F.		.98		.98	1.51
0400	2 @ 2" diameter		500	.032			1.96		1.96	3.02
0500	4 @ 2" diameter		250	.064			3.93		3.93	6.05
0600	1 @ 3" diameter		800	.020			1.23		1.23	1.89
0700	2 @ 3" diameter		400	.040			2.45		2.45	3.78
0800	4 @ 3" diameter		200	.080			4.91		4.91	7.55
0900	1 @ 4" diameter		800	.020			1.23		1.23	1.89
1000	2 @ 4" diameter		400	.040			2.45		2.45	3.78
1100	4 @ 4" diameter		200	.080			4.91		4.91	7.55
1200	6 @ 4" diameter		100	.160			9.80		9.80	15.10
1300	1 @ 5" diameter		500	.032			1.96		1.96	3.02
1400	2 @ 5" diameter		250	.064			3.93		3.93	6.05
1500	4 @ 5" diameter		160	.100			6.15		6.15	9.45
1600	6 @ 5" diameter		100	.160			9.80		9.80	15.10
1700	1 @ 6" diameter		500	.032			1.96		1.96	3.02
1800	2 @ 6" diameter		250	.064			3.93		3.93	6.05
1900	4 @ 6" diameter		160	.100			6.15		6.15	9.45
2000	6 @ 6" diameter	↓	100	.160			9.80		9.80	15.10
2100	Conduit 1 EB plastic, with concrete, 0.92 C.F./L.F.	B-9	150	.267			11.35	2.38	13.73	21
2200	2 EB plastic 1.52 C.F./L.F.		100	.400			17	3.57	20.57	31
2300	2 x 2 EB plastic 2.51 C.F./L.F.		80	.500			21.50	4.47	25.97	39
2400	2 x 3 EB plastic 3.56 C.F./L.F.	↓	60	.667			28.50	5.95	34.45	52
2500	Conduit 2 @ 2" diameter, steel, no concrete	2 Elec	400	.040			2.45		2.45	3.78
2600	4 @ 2" diameter		200	.080			4.91		4.91	7.55
2700	2 @ 3" diameter		200	.080			4.91		4.91	7.55
2800	4 @ 3" diameter		100	.160			9.80		9.80	15.10
2900	2 @ 4" diameter		200	.080			4.91		4.91	7.55
3000	4 @ 4" diameter		100	.160			9.80		9.80	15.10
3100	6 @ 4" diameter		50	.320			19.65		19.65	30
3200	2 @ 5" diameter		120	.133			8.20		8.20	12.60
3300	4 @ 5" diameter		60	.267			16.35		16.35	25
3400	6 @ 5" diameter		40	.400			24.50		24.50	38
3500	2 @ 6" diameter		120	.133			8.20		8.20	12.60
3600	4 @ 6" diameter		60	.267			16.35		16.35	25
3700	6 @ 6" diameter	↓	40	.400			24.50		24.50	38
3800	Conduit 2 steel, with concrete, 1.52 C.F./L.F.	B-9	80	.500			21.50	4.47	25.97	39
3900	2 x 2 EB steel 2.51 C.F./L.F.		60	.667			28.50	5.95	34.45	52
4000	2 x 3 steel 3.56 C.F./L.F.	↓	50	.800	↓		34	7.15	41.15	62.50
4100	Conduit fittings, PVC type EB, 2"-3"	1 Elec	30	.267	Ea.		16.35		16.35	25
4200	4"-6"	"	20	.400	"		24.50		24.50	38

02 41 13.60 Selective Demolition Fencing		Crew	Daily Output	Labor-Hours	Unit	Material	Labor	Equipment	Total	Total Incl O&P
0010	SELECTIVE DEMOLITION FENCING R024119-10									
0700	Snow fence, 4' high	B-6	1000	.024	L.F.		1.10	.21	1.31	1.98
1600	Fencing, barbed wire, 3 strand	2 Clab	430	.037			1.57		1.57	2.51
1650	5 strand	"	280	.057			2.41		2.41	3.85
1700	Chain link, posts & fabric, 8'-10' high, remove only	B-6	445	.054	↓		2.47	.48	2.95	4.45

02 41 13 – Selective Site Demolition

02 41 13.62 Selective Demo., Chain Link Fences & Gates

		Crew	Daily Output	Labor-Hours	Unit	Material	2020 Bare Costs Labor	Equipment	Total	Total Incl O&P
0010	SELECTIVE DEMOLITION, CHAIN LINK FENCES & GATES									
0100	Chain link, gates, 3'-4' width	B-6	30	.800	Ea.		36.50	7.15	43.65	66
0200	10'-12' width		16	1.500			68.50	13.35	81.85	124
0300	14' width		15	1.600			73	14.25	87.25	132
0400	20' width		10	2.400			110	21.50	131.50	198
0500	18' width with overhead & cantilever		80	.300	L.F.		13.70	2.67	16.37	25
0510	Sliding		80	.300			13.70	2.67	16.37	25
0520	Cantilever to 40' wide		80	.300			13.70	2.67	16.37	25
0530	Motor operators	2 Skwk	1	16	Ea.		880		880	1,400
0540	Transmitter systems	"	15	1.067	"		58.50		58.50	93
0600	Chain link, fence, 5' high	B-6	890	.027	L.F.		1.23	.24	1.47	2.22
0650	3'-4' high		1000	.024			1.10	.21	1.31	1.98
0675	12' high		400	.060			2.74	.53	3.27	4.95
0800	Chain link, fence, braces	B-1	2000	.012	Ea.		.51		.51	.82
0900	Privacy slats	"	2000	.012			.51		.51	.82
1000	Fence posts, steel, in concrete	B-6	80	.300			13.70	2.67	16.37	25
1100	Fence fabric & accessories, fabric to 8' high		800	.030	L.F.		1.37	.27	1.64	2.47
1200	Barbed wire		5000	.005	"		.22	.04	.26	.40
1300	Extension arms & eye tops		300	.080	Ea.		3.66	.71	4.37	6.60
1400	Fence rails		2000	.012	L.F.		.55	.11	.66	.99
1500	Reinforcing wire		5000	.005	"		.22	.04	.26	.40

02 41 13.64 Selective Demolition, Vinyl Fences and Gates

		Crew	Daily Output	Labor-Hours	Unit	Material	Labor	Equipment	Total	Total Incl O&P
0010	SELECTIVE DEMOLITION, VINYL FENCES & GATES									
0100	Vinyl fence up to 6' high	B-6	1000	.024	L.F.		1.10	.21	1.31	1.98
0200	Gates up to 6' high	"	40	.600	Ea.		27.50	5.35	32.85	49.50

02 41 13.66 Selective Demolition, Misc Metal Fences and Gates

		Crew	Daily Output	Labor-Hours	Unit	Material	Labor	Equipment	Total	Total Incl O&P
0010	SELECTIVE DEMOLITION, MISC METAL FENCES & GATES									
0100	Misc steel mesh fences, 4'-6' high	B-6	600	.040	L.F.		1.83	.36	2.19	3.29
0200	Kennels, 6'-12' long	2 Clab	8	2	Ea.		84		84	135
0300	Tops, 6'-12' long	"	30	.533	"		22.50		22.50	36
0400	Security fences, 12'-16' high	B-6	100	.240	L.F.		11	2.14	13.14	19.80
0500	Metal tubular picket fences, 4'-6' high		500	.048	"		2.20	.43	2.63	3.96
0600	Gates, 3'-4' wide		20	1.200	Ea.		55	10.70	65.70	99

02 41 13.68 Selective Demolition, Wood Fences and Gates

		Crew	Daily Output	Labor-Hours	Unit	Material	Labor	Equipment	Total	Total Incl O&P
0010	SELECTIVE DEMOLITION, WOOD FENCES & GATES									
0100	Wood fence gates 3'-4' wide	2 Clab	20	.800	Ea.		33.50		33.50	54
0200	Wood fence, open rail, to 4' high		2560	.006	L.F.		.26		.26	.42
0300	to 8' high		368	.043	"		1.83		1.83	2.93
0400	Post, in concrete		50	.320	Ea.		13.45		13.45	21.50

02 41 13.70 Selective Demolition, Rip-Rap and Rock Lining

		Crew	Daily Output	Labor-Hours	Unit	Material	Labor	Equipment	Total	Total Incl O&P
0010	SELECTIVE DEMOLITION, RIP-RAP & ROCK LINING R024119-10									
0100	Slope protection, broken stone	B-13	62	.903	C.Y.		41.50	9.35	50.85	76.50
0200	3/8 to 1/4 C.Y. pieces		60	.933	S.Y.		43	9.70	52.70	79
0300	18" depth		60	.933	"		43	9.70	52.70	79
0400	Dumped stone		93	.602	Ton		27.50	6.25	33.75	51
0500	Gabions, 6"-12" deep		60	.933	S.Y.		43	9.70	52.70	79
0600	18"-36" deep		30	1.867	"		86	19.35	105.35	159

02 41 13.72 Selective Demo., Shore Protect/Mooring Struct.

		Crew	Daily Output	Labor-Hours	Unit	Material	Labor	Equipment	Total	Total Incl O&P
0010	SELECTIVE DEMOLITION, SHORE PROTECT/MOORING STRUCTURES									
0100	Breakwaters, bulkheads, concrete, maximum	B-9	12	3.333	L.F.		142	30	172	260
0200	Breakwaters, bulkheads, concrete, 12', minimum		10	4			170	35.50	205.50	310

02 41 Demolition

02 41 13 – Selective Site Demolition

02 41 13.72 Selective Demo., Shore Protect/Mooring Struct.		Crew	Daily Output	Labor-Hours	Unit	Material	2020 Bare Costs Labor	Equipment	Total	Total Incl O&P
0300	Maximum	B-9	9	4.444	L.F.		189	39.50	228.50	350
0400	Steel, from shore	B-40B	54	.889			41.50	37.50	79	108
0500	from barge	B-76A	30	2.133			97	79	176	242
0600	Jetties, docks, floating	B-21B	600	.067	S.F.		3.06	.79	3.85	5.75
0700	Pier supported, 3"-4" decking		300	.133			6.10	1.57	7.67	11.50
0800	Floating, prefab, small boat, minimum		600	.067			3.06	.79	3.85	5.75
0900	Maximum		300	.133			6.10	1.57	7.67	11.50
1000	Floating, prefab, per slip, minimum		3.20	12.500	Ea.		575	147	722	1,075
1010	Maximum		2.80	14.286	"		655	168	823	1,225

02 41 13.74 Selective Demolition, Piles

		Crew	Daily Output	Labor-Hours	Unit	Material	Labor	Equipment	Total	Total Incl O&P
0010	**SELECTIVE DEMOLITION, PILES**									
0100	Cast in place piles, corrugated, 8"-10"	B-19	600	.107	V.L.F.		5.90	3.35	9.25	13.05
0200	12"-14"		500	.128			7.05	4.02	11.07	15.65
0300	16"		400	.160			8.85	5	13.85	19.60
0400	fluted, 12"		600	.107			5.90	3.35	9.25	13.05
0500	14"-18"		500	.128			7.05	4.02	11.07	15.65
0600	end bearing, 12"		600	.107			5.90	3.35	9.25	13.05
0700	14"-18"		500	.128			7.05	4.02	11.07	15.65
0800	Precast prestressed piles, 12"-14" diam.		700	.091			5.05	2.87	7.92	11.15
0900	16"-24" diam.		600	.107			5.90	3.35	9.25	13.05
1000	36"-66" diam.		300	.213			11.80	6.70	18.50	26
1100	10"-14" thick		600	.107			5.90	3.35	9.25	13.05
1200	16"-24" thick	B-38	500	.080			3.81	2.48	6.29	8.80
1300	Pressure grouted pile, 5"	B-19	150	.427			23.50	13.40	36.90	52.50
1400	Steel piles, 8"-12" tip		600	.107			5.90	3.35	9.25	13.05
1500	H sections HP8 to HP12		600	.107			5.90	3.35	9.25	13.05
1600	HP14	B-19A	600	.107			5.90	4.91	10.81	14.75
1700	Steel pipe piles, 8"-12"	B-19	600	.107			5.90	3.35	9.25	13.05
1800	14"-18" plain		500	.128			7.05	4.02	11.07	15.65
1900	concrete filled, 14"-18"		400	.160			8.85	5	13.85	19.60
2000	Timber piles to 14" diam.		600	.107			5.90	3.35	9.25	13.05

02 41 13.76 Selective Demolition, Water Wells

		Crew	Daily Output	Labor-Hours	Unit	Material	Labor	Equipment	Total	Total Incl O&P
0010	**SELECTIVE DEMOLITION, WATER WELLS**									
0100	Well, 40' deep with casing & gravel pack, 24"-36" diam.	B-23	.25	160	Ea.		6,800	6,375	13,175	17,900
0200	Riser pipe, 1-1/4", for observation well	"	300	.133	L.F.		5.65	5.30	10.95	14.95
0300	Pump, 1/2 to 5 HP up to 100' depth	Q-1	3	5.333	Ea.		310		310	480
0400	Up to 150' well 25 HP pump	Q-22	1.50	10.667			620	315	935	1,300
0500	Up to 500' well 30 HP pump	"	1	16			930	470	1,400	1,950
0600	Well screen 2"-8"	B-23	300	.133	V.L.F.		5.65	5.30	10.95	14.95
0700	10"-16"		200	.200			8.50	7.95	16.45	22.50
0800	18"-26"		150	.267			11.35	10.60	21.95	30
0900	Slotted PVC for wells 1-1/4"-8"		600	.067			2.83	2.65	5.48	7.45
1000	Well screen and casing 6"-16"		300	.133			5.65	5.30	10.95	14.95
1100	18"-26"		150	.267			11.35	10.60	21.95	30
1200	30"-36"		100	.400			17	15.90	32.90	44.50

02 41 13.78 Selective Demolition, Radio Towers

		Crew	Daily Output	Labor-Hours	Unit	Material	Labor	Equipment	Total	Total Incl O&P
0010	**SELECTIVE DEMOLITION, RADIO TOWERS**									
0100	Radio tower, guyed, 50'	2 Skwk	16	1	Ea.		55		55	87.50
0200	190', 40 lb. section	K-2	.70	34.286			1,875	1,175	3,050	4,350
0300	200', 70 lb. section		.70	34.286			1,875	1,175	3,050	4,350
0400	300', 70 lb. section		.40	60			3,300	2,050	5,350	7,600
0500	270', 90 lb. section		.40	60			3,300	2,050	5,350	7,600

02 41 Demolition

02 41 13 – Selective Site Demolition

02 41 13.78 Selective Demolition, Radio Towers		Crew	Daily Output	Labor-Hours	Unit	Material	2020 Bare Costs Labor	Equipment	Total	Total Incl O&P
0600	400′	K-2	.30	80	Ea.		4,400	2,725	7,125	10,200
0700	Self supported, 60′		.90	26.667			1,475	910	2,385	3,375
0800	120′		.80	30			1,650	1,025	2,675	3,800
0900	190′		.40	60			3,300	2,050	5,350	7,600

02 41 13.80 Selective Demo., Utility Poles and Cross Arms

Line	Description	Crew	Daily Output	Labor-Hours	Unit	Material	Labor	Equipment	Total	Total Incl O&P
0010	**SELECTIVE DEMOLITION, UTILITY POLES & CROSS ARMS**									
0100	Utility poles, wood, 20′-30′ high	R-3	6	3.333	Ea.		204	31.50	235.50	350
0200	35′-45′ high	"	5	4			244	38	282	420
0300	Cross arms, wood, 4′-6′ long	1 Elec	5	1.600			98		98	151

02 41 13.82 Selective Removal, Pavement Lines and Markings

Line	Description	Crew	Daily Output	Labor-Hours	Unit	Material	Labor	Equipment	Total	Total Incl O&P
0010	**SELECTIVE REMOVAL, PAVEMENT LINES & MARKINGS**									
0015	Does not include traffic control costs									
0020	See other items in Section 32 17 23.13									
0100	Remove permanent painted traffic lines and markings	B-78A	500	.016	C.L.F.		.85	1.83	2.68	3.35
0200	Temporary traffic line tape	2 Clab	1500	.011	L.F.		.45		.45	.72
0300	Thermoplastic traffic lines and markings	B-79A	500	.024	C.L.F.		1.27	2.78	4.05	5.05
0400	Painted pavement markings	B-78B	500	.036	S.F.		1.56	.81	2.37	3.38

02 41 13.84 Selective Demolition, Walks, Steps and Pavers

Line	Description	Crew	Daily Output	Labor-Hours	Unit	Material	Labor	Equipment	Total	Total Incl O&P
0010	**SELECTIVE DEMOLITION, WALKS, STEPS AND PAVERS**									
0100	Splash blocks	1 Clab	300	.027	S.F.		1.12		1.12	1.80
0200	Tree grates	"	50	.160	Ea.		6.75		6.75	10.80
0300	Walks, limestone pavers	2 Clab	150	.107	S.F.		4.49		4.49	7.20
0400	Redwood sections		600	.027			1.12		1.12	1.80
0500	Redwood planks		480	.033			1.40		1.40	2.25
0600	Shale paver		300	.053			2.25		2.25	3.60
0700	Tile thinset paver		675	.024			1		1	1.60
0800	Wood round	B-1	350	.069	Ea.		2.93		2.93	4.70
0900	Asphalt block	2 Clab	450	.036	S.F.		1.50		1.50	2.40
1000	Bluestone		450	.036			1.50		1.50	2.40
1100	Slate, 1″ or thinner		675	.024			1		1	1.60
1200	Granite blocks		300	.053			2.25		2.25	3.60
1300	Precast patio blocks		450	.036			1.50		1.50	2.40
1400	Planter blocks		600	.027			1.12		1.12	1.80
1500	Brick paving, dry set		300	.053			2.25		2.25	3.60
1600	Mortar set		180	.089			3.74		3.74	6
1700	Dry set on edge		240	.067			2.81		2.81	4.50
1800	Steps, brick		200	.080	L.F.		3.37		3.37	5.40
1900	Railroad tie		150	.107			4.49		4.49	7.20
2000	Bluestone		180	.089			3.74		3.74	6
2100	Wood/steel edging for steps		1000	.016			.67		.67	1.08
2200	Timber or railroad tie edging for steps		400	.040			1.68		1.68	2.70

02 41 13.86 Selective Demolition, Athletic Surfaces

Line	Description	Crew	Daily Output	Labor-Hours	Unit	Material	Labor	Equipment	Total	Total Incl O&P
0010	**SELECTIVE DEMOLITION, ATHLETIC SURFACES**									
0100	Synthetic grass	2 Clab	2000	.008	S.F.		.34		.34	.54
0200	Surface coat latex rubber	"	2000	.008	"		.34		.34	.54
0300	Tennis court posts	B-11C	16	1	Ea.		49.50	13.35	62.85	92.50

02 41 13.88 Selective Demolition, Lawn Sprinkler Systems

Line	Description	Crew	Daily Output	Labor-Hours	Unit	Material	Labor	Equipment	Total	Total Incl O&P
0010	**SELECTIVE DEMOLITION, LAWN SPRINKLER SYSTEMS**									
0100	Golf course sprinkler system, 9 hole	4 Clab	.10	320	Ea.		13,500		13,500	21,600
0200	Sprinkler system, 24′ diam. @ 15′ OC, per head	B-20	110	.218	Head		10.25		10.25	16.40
0300	60′ diam. @ 24′ OC, per head	"	52	.462	"		21.50		21.50	34.50

02 41 13.88 Selective Demolition, Lawn Sprinkler Systems

	02 41 13.88 Selective Demolition, Lawn Sprinkler Systems	Crew	Daily Output	Labor-Hours	Unit	Material	2020 Bare Costs Labor	Equipment	Total	Total Incl O&P
0400	Sprinkler heads, plastic	2 Clab	150	.107	Ea.		4.49		4.49	7.20
0500	Impact circle pattern, 28′-76′ diam.		75	.213			9		9	14.40
0600	Pop-up, 42′-76′ diam.		50	.320			13.45		13.45	21.50
0700	39′-99′ diam.		50	.320			13.45		13.45	21.50
0800	Sprinkler valves		40	.400			16.85		16.85	27
0900	Valve boxes		40	.400			16.85		16.85	27
1000	Controls		2	8			335		335	540
1100	Backflow preventer		4	4			168		168	270
1200	Vacuum breaker		4	4			168		168	270

02 41 13.90 Selective Demolition, Retaining Walls

	02 41 13.90 Selective Demolition, Retaining Walls	Crew	Daily Output	Labor-Hours	Unit	Material	Labor	Equipment	Total	Total Incl O&P
0010	**SELECTIVE DEMOLITION, RETAINING WALLS**									
0100	Concrete retaining wall, 6′ high, no reinforcing	B-13K	200	.080	L.F.		4.74	10.15	14.89	18.55
0200	8′ high		150	.107			6.30	13.55	19.85	25
0300	10′ high		150	.107			6.30	13.55	19.85	25
0400	With reinforcing, 6′ high		200	.080			4.74	10.15	14.89	18.55
0500	8′ high		150	.107			6.30	13.55	19.85	25
0600	10′ high		120	.133			7.90	16.90	24.80	31
0700	20′ high		60	.267			15.80	34	49.80	61.50
0800	Concrete cribbing, 12′ high, open/closed face		150	.107	S.F.		6.30	13.55	19.85	25
0900	Interlocking segmental retaining wall	B-62	800	.030			1.37	.22	1.59	2.42
1000	Wall caps	"	600	.040			1.83	.30	2.13	3.23
1100	Metal bin retaining wall, 10′ wide, 4′-12′ high	B-13	1200	.047			2.15	.48	2.63	3.94
1200	10′ wide, 16′-28′ high		1000	.056			2.58	.58	3.16	4.74
1300	Stone filled gabions, 6′ x 3′ x 1′		170	.329	Ea.		15.15	3.42	18.57	28
1400	6′ x 3′ x 1′-6″		75	.747			34.50	7.75	42.25	63
1500	6′ x 3′ x 3′		25	2.240			103	23	126	190
1600	9′ x 3′ x 1′		75	.747			34.50	7.75	42.25	63
1700	9′ x 3′ x 1′-6″		33	1.697			78	17.60	95.60	143
1800	9′ x 3′ x 3′		12	4.667			215	48.50	263.50	395
1900	12′ x 3′ x 1′		42	1.333			61.50	13.85	75.35	113
2000	12′ x 3′ x 1′-6″		20	2.800			129	29	158	237
2100	12′ x 3′ x 3′		6	9.333			430	97	527	790

02 41 13.92 Selective Demolition, Parking Appurtenances

	02 41 13.92 Selective Demolition, Parking Appurtenances	Crew	Daily Output	Labor-Hours	Unit	Material	Labor	Equipment	Total	Total Incl O&P
0010	**SELECTIVE DEMOLITION, PARKING APPURTENANCES**									
0100	Bumper rails, garage, 6″ wide	B-6	300	.080	L.F.		3.66	.71	4.37	6.60
0200	12″ channel rail		300	.080			3.66	.71	4.37	6.60
0300	Parking bumper, timber		1000	.024			1.10	.21	1.31	1.98
0400	Folding, with locks	B-1	100	.240	Ea.		10.25		10.25	16.45
0500	Flexible fixed garage stanchion	B-6	150	.160			7.30	1.43	8.73	13.15
0600	Wheel stops, precast concrete		120	.200			9.15	1.78	10.93	16.45
0700	Thermoplastic		120	.200			9.15	1.78	10.93	16.45
0800	Pipe bollards, 6″-12″ diam.		80	.300			13.70	2.67	16.37	25

02 41 13.93 Selective Demolition, Site Furnishings

	02 41 13.93 Selective Demolition, Site Furnishings	Crew	Daily Output	Labor-Hours	Unit	Material	Labor	Equipment	Total	Total Incl O&P
0010	**SELECTIVE DEMOLITION, SITE FURNISHINGS**									
0100	Benches, all types	2 Clab	10	1.600	Ea.		67.50		67.50	108
0200	Trash receptacles, all types		80	.200			8.40		8.40	13.50
0300	Trash enclosures, steel or wood		10	1.600			67.50		67.50	108

02 41 13.94 Selective Demolition, Athletic Screening

	02 41 13.94 Selective Demolition, Athletic Screening	Crew	Daily Output	Labor-Hours	Unit	Material	Labor	Equipment	Total	Total Incl O&P
0010	**SELECTIVE DEMOLITION, ATHLETIC SCREENING**									
0020	See other items in Sections 02 41 13.60 & 02 41 13.62									
0100	Baseball backstops	B-6	2	12	Ea.		550	107	657	990
0200	Basketball goal, single		6	4			183	35.50	218.50	330

02 41 Demolition

02 41 13 – Selective Site Demolition

02 41 13.94 Selective Demolition, Athletic Screening

	02 41 13.94 Selective Demolition, Athletic Screening	Crew	Daily Output	Labor-Hours	Unit	Material	2020 Bare Costs Labor	Equipment	Total	Total Incl O&P
0300	Double	B-6	4	6	Ea.		274	53.50	327.50	495
0400	Tennis wire mesh, pair ends		5	4.800	Set		220	43	263	395
0500	Enclosed court		3	8	Ea.		365	71.50	436.50	660
0600	Wood or masonry handball court		1	24	"		1,100	214	1,314	1,975

02 41 13.95 Selective Demo., Athletic/Playground Equipment

	02 41 13.95 Selective Demo., Athletic/Playground Equipment	Crew	Daily Output	Labor-Hours	Unit	Material	2020 Bare Costs Labor	Equipment	Total	Total Incl O&P
0010	**SELECTIVE DEMO., ATHLETIC/PLAYGROUND EQUIPMENT**									
0020	See other items in Section 02 41 13.96									
0100	Bike rack	B-6	32	.750	Ea.		34.50	6.70	41.20	62
0200	Climber arch		16	1.500			68.50	13.35	81.85	124
0300	Fitness trail, 9 to 10 stations wood or metal		.50	48			2,200	430	2,630	3,950
0400	16 to 20 stations wood or metal		.25	96			4,400	855	5,255	7,925
0500	Monkey bars 14′ long		8	3			137	26.50	163.50	248
0600	Parallel bars 10′ long		8	3			137	26.50	163.50	248
0700	Post tether ball		32	.750			34.50	6.70	41.20	62
0800	Poles 10′-6″ long		32	.750			34.50	6.70	41.20	62
0900	Pole ground sockets		80	.300			13.70	2.67	16.37	25
1000	See-saw, 2 units		12	2			91.50	17.80	109.30	165
1100	4 units		10	2.400			110	21.50	131.50	198
1200	6 units		8	3			137	26.50	163.50	248
1300	Shelter, fiberglass, 3 person		10	2.400			110	21.50	131.50	198
1400	Slides, 12′ long		8	3			137	26.50	163.50	248
1500	20′ long		6	4			183	35.50	218.50	330
1600	Swings, 4 seats		6	4			183	35.50	218.50	330
1700	8 seats		4	6			274	53.50	327.50	495
1800	Whirlers, 8′-10′ diameter		6	4			183	35.50	218.50	330
1900	Football or football/soccer goal posts		3	8	Pair		365	71.50	436.50	660
2000	Soccer goal posts		4	6	"		274	53.50	327.50	495
2100	Playground surfacing, 4″ depth	B-62	2000	.012	S.F.		.55	.09	.64	.97
2200	2″ topping	"	3500	.007	"		.31	.05	.36	.56
2300	Platform/paddle tennis court, alum. deck & frame	B-1	.20	120	Court		5,125		5,125	8,225
2400	Aluminum deck and wood frame		.25	96	"		4,100		4,100	6,575
2500	Heater		2	12	Ea.		515		515	820
2600	Wood deck & frame		.25	96	Court		4,100		4,100	6,575
2700	Wood deck & steel frame		.25	96			4,100		4,100	6,575
2800	Steel deck & wood frame		.25	96			4,100		4,100	6,575

02 41 13.96 Selective Demo., Modular Playground Equipment

	02 41 13.96 Selective Demo., Modular Playground Equipment	Crew	Daily Output	Labor-Hours	Unit	Material	2020 Bare Costs Labor	Equipment	Total	Total Incl O&P
0010	**SELECTIVE DEMOLITION, MODULAR PLAYGROUND EQUIPMENT**									
0100	Modular playground, 48″ deck, square or triangle shape	B-1	2	12	Ea.		515		515	820
0200	Various posts		18	1.333			57		57	91.50
0300	Roof 54″ x 54″		18	1.333			57		57	91.50
0400	Wheel chair transfer module		6	4			171		171	274
0500	Guardrail pipe		90	.267	L.F.		11.40		11.40	18.25
0600	Steps, 3 risers		16	1.500	Ea.		64		64	103
0700	Activity panel crawl through		8	3			128		128	206
0800	Alphabet/spelling panel		8	3			128		128	206
0900	Crawl tunnel, each 4′ section		8	3			128		128	206
1000	Slide tunnel		16	1.500			64		64	103
1100	Spiral slide tunnel		12	2			85.50		85.50	137
1200	Ladder, horizontal or vertical		10	2.400			103		103	164
1300	Corkscrew climber		6	4			171		171	274
1400	Fire pole		12	2			85.50		85.50	137
1500	Bridge, ring chamber		8	3			128		128	206

02 41 Demolition

02 41 13 – Selective Site Demolition

02 41 13.96 Selective Demo., Modular Playground Equipment		Crew	Daily Output	Labor-Hours	Unit	2020 Bare Costs Material	Labor	Equipment	Total	Total Incl O&P
1600	Bridge, suspension	B-1	8	3	Ea.		128		128	206

02 41 13.97 Selective Demolition, Traffic Light Systems

		Crew	Daily Output	Labor-Hours	Unit	Material	Labor	Equipment	Total	Total Incl O&P
0010	**SELECTIVE DEMOLITION, TRAFFIC LIGHT SYSTEMS**									
0100	Traffic signal, pedestrian cross walk	R-2	.50	112	Total		6,450	755	7,205	10,800
0200	8 signal intersection, 2 each direction	"	.25	224			12,900	1,500	14,400	21,600
0300	Each traffic phase controller	L-9	3	12			575		575	920
0400	Each semi-actuated detector		1.50	24			1,150		1,150	1,850
0500	Each fully-actuated detector		1.50	24			1,150		1,150	1,850
0600	Pedestrian push button system		2	18			860		860	1,375
0650	Each optically programmed head		3	12			575		575	920
0700	School signal flashing system		1	36	Signal		1,725		1,725	2,750
0800	Complete traffic light intersection system, without lane control	R-2	.25	224	Ea.		12,900	1,500	14,400	21,600
0900	With lane control		.20	280			16,100	1,900	18,000	27,000
1000	Traffic light, each left turn protection system		.66	84.848			4,900	575	5,475	8,175

02 41 13.98 Select. Demo., Sod, Edging, Planters & Tree Guying

		Crew	Daily Output	Labor-Hours	Unit	Material	Labor	Equipment	Total	Total Incl O&P
0010	**SELECTIVE DEMOLITION, SOD, EDGING, PLANTERS & TREE GUYING**									
0100	Remove sod 1" deep, 3 M.S.F. or greater, level ground	B-63	40	1	M.S.F.		44.50	4.44	48.94	75.50
0200	Less than 3 M.S.F., level ground		27	1.481			65.50	6.60	72.10	112
0300	3 M.S.F. or greater, sloped ground		12	3.333			148	14.80	162.80	251
0400	Less than 3 M.S.F., sloped ground		8	5			221	22	243	380
0500	Edging, aluminum	B-1	800	.030	L.F.		1.28		1.28	2.06
0600	Brick horizontal		400	.060			2.57		2.57	4.11
0700	Brick vertical		200	.120			5.15		5.15	8.20
0800	Corrugated aluminum		1200	.020			.86		.86	1.37
0900	Granite 5" x 16"		600	.040			1.71		1.71	2.74
1000	Polyethylene grass barrier		1200	.020			.86		.86	1.37
1100	Precast 2" x 8" x 16"		800	.030			1.28		1.28	2.06
1200	Railroad ties		200	.120			5.15		5.15	8.20
1300	Wood		400	.060			2.57		2.57	4.11
1400	Steel strips including stakes		600	.040			1.71		1.71	2.74
1500	Planters, concrete, 48" diam.	2 Clab	30	.533	Ea.		22.50		22.50	36
1600	Concrete 7' diam.		20	.800			33.50		33.50	54
1700	Fiberglass 36" diam.		30	.533			22.50		22.50	36
1800	60" diam.		20	.800			33.50		33.50	54
1900	24" square		30	.533			22.50		22.50	36
2000	48" square		30	.533			22.50		22.50	36
2100	Wood 48" square		30	.533			22.50		22.50	36
2200	48" diam.		20	.800			33.50		33.50	54
2300	72" diam.		20	.800			33.50		33.50	54
2400	Planter/bench, fiberglass 72"-96" square		10	1.600			67.50		67.50	108
2500	Wood, 72" square		10	1.600			67.50		67.50	108
2600	Tree guying, stakes, 3"-4" caliper		40	.400			16.85		16.85	27
2700	anchors, 3" caliper		40	.400			16.85		16.85	27
2800	3"-6" caliper		30	.533			22.50		22.50	36
2900	6" caliper		20	.800			33.50		33.50	54
3000	8" caliper		15	1.067			45		45	72

02 41 16 – Structure Demolition

02 41 16.13 Building Demolition

		Crew	Daily Output	Labor-Hours	Unit	Material	Labor	Equipment	Total	Total Incl O&P
0010	**BUILDING DEMOLITION** Large urban projects, incl. 20 mi. haul R024119-10									
0011	No foundation or dump fees, C.F. is vol. of building standing									
0020	Steel	B-8	21500	.003	C.F.		.15	.13	.28	.38
0050	Concrete		15300	.004			.20	.19	.39	.53

02 41 16.13 Building Demolition		Crew	Daily Output	Labor-Hours	Unit	Material	2020 Bare Costs Labor	Equipment	Total	Total Incl O&P
0080	Masonry	B-8	20100	.003	C.F.		.16	.14	.30	.41
0100	Mixture of types		20100	.003			.16	.14	.30	.41
0500	Small bldgs, or single bldgs, no salvage included, steel	B-3	14800	.003			.15	.16	.31	.41
0600	Concrete		11300	.004			.20	.21	.41	.55
0650	Masonry		14800	.003			.15	.16	.31	.41
0700	Wood		14800	.003			.15	.16	.31	.41
0750	For buildings with no interior walls, deduct								30%	30%
1000	Demolition single family house, one story, wood 1600 S.F.	B-3	1	48	Ea.		2,275	2,325	4,600	6,150
1020	3200 S.F.		.50	96			4,525	4,625	9,150	12,300
1200	Demolition two family house, two story, wood 2400 S.F.		.67	71.964			3,400	3,475	6,875	9,225
1220	4200 S.F.		.38	128			6,025	6,175	12,200	16,400
1300	Demolition three family house, three story, wood 3200 S.F.		.50	96			4,525	4,625	9,150	12,300
1320	5400 S.F.		.30	160			7,550	7,725	15,275	20,500
5000	For buildings with no interior walls, deduct								30%	30%

02 41 16.17 Building Demolition Footings and Foundations

		Crew	Daily Output	Labor-Hours	Unit	Material	Labor	Equipment	Total	Total Incl O&P
0010	**BUILDING DEMOLITION FOOTINGS AND FOUNDATIONS** R024119-10									
0200	Floors, concrete slab on grade,									
0240	4" thick, plain concrete	B-13L	5000	.003	S.F.		.19	.42	.61	.76
0280	Reinforced, wire mesh		4000	.004			.24	.52	.76	.94
0300	Rods		4500	.004			.21	.46	.67	.84
0400	6" thick, plain concrete		4000	.004			.24	.52	.76	.94
0420	Reinforced, wire mesh		3200	.005			.30	.65	.95	1.17
0440	Rods		3600	.004			.26	.58	.84	1.04
1000	Footings, concrete, 1' thick, 2' wide		300	.053	L.F.		3.16	6.95	10.11	12.55
1080	1'-6" thick, 2' wide		250	.064			3.79	8.30	12.09	15.10
1120	3' wide		200	.080			4.74	10.40	15.14	18.85
1140	2' thick, 3' wide		175	.091			5.40	11.90	17.30	21.50
1200	Average reinforcing, add								10%	10%
1220	Heavy reinforcing, add								20%	20%
2000	Walls, block, 4" thick	B-13L	8000	.002	S.F.		.12	.26	.38	.48
2040	6" thick		6000	.003			.16	.35	.51	.63
2080	8" thick		4000	.004			.24	.52	.76	.94
2100	12" thick		3000	.005			.32	.69	1.01	1.25
2200	For horizontal reinforcing, add								10%	10%
2220	For vertical reinforcing, add								20%	20%
2400	Concrete, plain concrete, 6" thick	B-13L	4000	.004			.24	.52	.76	.94
2420	8" thick		3500	.005			.27	.59	.86	1.07
2440	10" thick		3000	.005			.32	.69	1.01	1.25
2500	12" thick		2500	.006			.38	.83	1.21	1.50
2600	For average reinforcing, add								10%	10%
2620	For heavy reinforcing, add								20%	20%
4300	Footings & foundation removal, with expansion grout, no hauling	B-9	5.68	7.042	C.Y.		299	63	362	550
9000	Minimum labor/equipment charge	A-1	2	4	Job		168	56	224	330

02 41 16.33 Bridge Demolition

		Crew	Daily Output	Labor-Hours	Unit	Material	Labor	Equipment	Total	Total Incl O&P
0010	**BRIDGE DEMOLITION**									
0100	Bridges, pedestrian, precast, 60'-150' long	B-21C	250	.224	S.F.		10.30	8.25	18.55	25.50
0200	Steel, 50'-160' long, 8'-10' wide	"	500	.112			5.15	4.13	9.28	12.75
0300	Laminated wood, 80'-130' long	C-12	300	.160			8.40	1.57	9.97	15.20

02 41 19.13 Selective Building Demolition		Crew	Daily Output	Labor-Hours	Unit	Material	2020 Bare Costs Labor	Equipment	Total	Total Incl O&P
0010	SELECTIVE BUILDING DEMOLITION									
0020	Costs related to selective demolition of specific building components									
0025	are included under Common Work Results (XX 05)									
0030	in the component's appropriate division.									

02 41 19.16 Selective Demolition, Cutout		Crew	Daily Output	Labor-Hours	Unit	Material	Labor	Equipment	Total	Total Incl O&P
0010	SELECTIVE DEMOLITION, CUTOUT R024119-10									
0020	Concrete, elev. slab, light reinforcement, under 6 C.F.	B-9	65	.615	C.F.		26	5.50	31.50	48
0050	Light reinforcing, over 6 C.F.		75	.533	"		22.50	4.76	27.26	42
0200	Slab on grade to 6" thick, not reinforced, under 8 S.F.		85	.471	S.F.		20	4.20	24.20	36.50
0250	8-16 S.F.	↓	175	.229	"		9.70	2.04	11.74	17.80
0255	For over 16 S.F. see Line 02 41 16.17 0400									
0600	Walls, not reinforced, under 6 C.F.	B-9	60	.667	C.F.		28.50	5.95	34.45	52
0650	6-12 C.F.	"	80	.500	"		21.50	4.47	25.97	39
0655	For over 12 C.F. see Line 02 41 16.17 2500									
1000	Concrete, elevated slab, bar reinforced, under 6 C.F.	B-9	45	.889	C.F.		38	7.95	45.95	69.50
1050	Bar reinforced, over 6 C.F.		50	.800	"		34	7.15	41.15	62.50
1200	Slab on grade to 6" thick, bar reinforced, under 8 S.F.		75	.533	S.F.		22.50	4.76	27.26	42
1250	8-16 S.F.	↓	150	.267	"		11.35	2.38	13.73	21
1255	For over 16 S.F. see Line 02 41 16.17 0440									
1400	Walls, bar reinforced, under 6 C.F.	B-9	50	.800	C.F.		34	7.15	41.15	62.50
1450	6-12 C.F.	"	70	.571	"		24.50	5.10	29.60	44.50
1455	For over 12 C.F. see Lines 02 41 16.17 2500 and 2600									
2000	Brick, to 4 S.F. opening, not including toothing									
2040	4" thick	B-9	30	1.333	Ea.		56.50	11.90	68.40	104
2060	8" thick		18	2.222			94.50	19.85	114.35	173
2080	12" thick		10	4			170	35.50	205.50	310
2400	Concrete block, to 4 S.F. opening, 2" thick		35	1.143			48.50	10.20	58.70	89
2420	4" thick		30	1.333			56.50	11.90	68.40	104
2440	8" thick		27	1.481			63	13.25	76.25	116
2460	12" thick		24	1.667			71	14.90	85.90	129
2600	Gypsum block, to 4 S.F. opening, 2" thick		80	.500			21.50	4.47	25.97	39
2620	4" thick		70	.571			24.50	5.10	29.60	44.50
2640	8" thick		55	.727			31	6.50	37.50	56.50
2800	Terra cotta, to 4 S.F. opening, 4" thick		70	.571			24.50	5.10	29.60	44.50
2840	8" thick		65	.615			26	5.50	31.50	48
2880	12" thick	↓	50	.800	↓		34	7.15	41.15	62.50
4000	For toothing masonry, see Section 04 01 20.50									
6000	Walls, interior, not including re-framing,									
6010	openings to 5 S.F.									
6100	Drywall to 5/8" thick	1 Clab	24	.333	Ea.		14.05		14.05	22.50
6200	Paneling to 3/4" thick		20	.400			16.85		16.85	27
6300	Plaster, on gypsum lath		20	.400			16.85		16.85	27
6340	On wire lath	↓	14	.571	↓		24		24	38.50
7000	Wood frame, not including re-framing, openings to 5 S.F.									
7200	Floors, sheathing and flooring to 2" thick	1 Clab	5	1.600	Ea.		67.50		67.50	108
7310	Roofs, sheathing to 1" thick, not including roofing		6	1.333			56		56	90
7410	Walls, sheathing to 1" thick, not including siding		7	1.143	↓		48		48	77
8500	Minimum labor/equipment charge	↓	4	2	Job		84		84	135

02 41 19.18 Selective Demolition, Disposal Only		Crew	Daily Output	Labor-Hours	Unit	Material	2020 Bare Costs Labor	Equipment	Total	Total Incl O&P
0010	**SELECTIVE DEMOLITION, DISPOSAL ONLY** R024119-10									
0015	Urban bldg w/salvage value allowed									
0020	Including loading and 5 mile haul to dump									
0200	Steel frame	B-3	430	.112	C.Y.		5.25	5.40	10.65	14.30
0300	Concrete frame		365	.132			6.20	6.35	12.55	16.90
0400	Masonry construction		445	.108			5.10	5.20	10.30	13.80
0500	Wood frame		247	.194			9.15	9.35	18.50	25

02 41 19.19 Selective Demolition

		Crew	Daily Output	Labor-Hours	Unit	Material	Labor	Equipment	Total	Total Incl O&P
0010	**SELECTIVE DEMOLITION**, Rubbish Handling R024119-10									
0020	The following are to be added to the demolition prices									
0050	The following are components for a complete chute system									
0100	Top chute circular steel, 4′ long, 18″ diameter R024119-30	B-1C	15	1.600	Ea.	285	68.50	20.50	374	450
0102	23″ diameter		15	1.600		310	68.50	20.50	399	475
0104	27″ diameter		15	1.600		330	68.50	20.50	419	500
0106	30″ diameter		15	1.600		355	68.50	20.50	444	525
0108	33″ diameter		15	1.600		380	68.50	20.50	469	555
0110	36″ diameter		15	1.600		405	68.50	20.50	494	580
0112	Regular chute, 18″ diameter		15	1.600		214	68.50	20.50	303	370
0114	23″ diameter		15	1.600		238	68.50	20.50	327	395
0116	27″ diameter		15	1.600		262	68.50	20.50	351	420
0118	30″ diameter		15	1.600		268	68.50	20.50	357	430
0120	33″ diameter		15	1.600		310	68.50	20.50	399	475
0122	36″ diameter		15	1.600		330	68.50	20.50	419	500
0124	Control door chute, 18″ diameter		15	1.600		405	68.50	20.50	494	580
0126	23″ diameter		15	1.600		430	68.50	20.50	519	605
0128	27″ diameter		15	1.600		450	68.50	20.50	539	630
0130	30″ diameter		15	1.600		475	68.50	20.50	564	660
0132	33″ diameter		15	1.600		500	68.50	20.50	589	685
0134	36″ diameter		15	1.600		520	68.50	20.50	609	710
0136	Chute liners, 14 ga., 18″-30″ diameter		15	1.600		228	68.50	20.50	317	385
0138	33″-36″ diameter		15	1.600		284	68.50	20.50	373	450
0140	17% thinner chute, 30″ diameter		15	1.600		233	68.50	20.50	322	390
0142	33% thinner chute, 30″ diameter		15	1.600		176	68.50	20.50	265	325
0144	Top chute cover	1 Clab	24	.333		155	14.05		169.05	194
0146	Door chute cover	"	24	.333		155	14.05		169.05	194
0148	Top chute trough	2 Clab	12	1.333		530	56		586	670
0150	Bolt down frame & counter weights, 250 lb.	B-1	4	6		4,325	257		4,582	5,150
0152	500 lb.		4	6		6,375	257		6,632	7,400
0154	750 lb.		4	6		11,100	257		11,357	12,600
0156	1000 lb.		2.67	8.989		12,000	385		12,385	13,800
0158	1500 lb.		2.67	8.989		14,000	385		14,385	16,000
0160	Chute warning light system, 5 stories	B-1C	4	6		8,525	257	77.50	8,859.50	9,875
0162	10 stories	"	2	12		14,800	515	155	15,470	17,300
0164	Dust control device for dumpsters	1 Clab	8	1		158	42		200	242
0166	Install or replace breakaway cord		8	1		28	42		70	98
0168	Install or replace warning sign		16	.500		11.10	21		32.10	45.50
0600	Dumpster, weekly rental, 1 dump/week, 6 C.Y. capacity (2 tons)				Week	415			415	455
0700	10 C.Y. capacity (3 tons)					480			480	530
0725	20 C.Y. capacity (5 tons) R024119-20					565			565	625
0800	30 C.Y. capacity (7 tons)					730			730	800
0840	40 C.Y. capacity (10 tons)					775			775	850
0900	Alternate pricing for dumpsters									

02 41 19.19 Selective Demolition		Crew	Daily Output	Labor-Hours	Unit	Material	Labor (2020 Bare Costs)	Equipment (2020 Bare Costs)	Total (2020 Bare Costs)	Total Incl O&P
0910	Delivery, average for all sizes				Ea.	75			75	82.50
0920	Haul, average for all sizes					235			235	259
0930	Rent per day, average for all sizes					20			20	22
0940	Rent per month, average for all sizes					80			80	88
0950	Disposal fee per ton, average for all sizes				Ton	88			88	97
2000	Load, haul, dump and return, 0′-50′ haul, hand carried	2 Clab	24	.667	C.Y.		28		28	45
2005	Wheeled		37	.432			18.20		18.20	29
2040	0′-100′ haul, hand carried		16.50	.970			41		41	65.50
2045	Wheeled		25	.640			27		27	43
2050	Forklift	A-3R	25	.320			16.95	10.25	27.20	38
2080	Haul and return, add per each extra 100′ haul, hand carried	2 Clab	35.50	.451			18.95		18.95	30.50
2085	Wheeled		54	.296			12.45		12.45	20
2120	For travel in elevators, up to 10 floors, add		140	.114			4.81		4.81	7.70
2130	0′-50′ haul, incl. up to 5 riser stairs, hand carried		23	.696			29.50		29.50	47
2135	Wheeled		35	.457			19.25		19.25	31
2140	6-10 riser stairs, hand carried		22	.727			30.50		30.50	49
2145	Wheeled		34	.471			19.80		19.80	31.50
2150	11-20 riser stairs, hand carried		20	.800			33.50		33.50	54
2155	Wheeled		31	.516			21.50		21.50	35
2160	21-40 riser stairs, hand carried		16	1			42		42	67.50
2165	Wheeled		24	.667			28		28	45
2170	0′-100′ haul, incl. 5 riser stairs, hand carried		15	1.067			45		45	72
2175	Wheeled		23	.696			29.50		29.50	47
2180	6-10 riser stairs, hand carried		14	1.143			48		48	77
2185	Wheeled		21	.762			32		32	51.50
2190	11-20 riser stairs, hand carried		12	1.333			56		56	90
2195	Wheeled		18	.889			37.50		37.50	60
2200	21-40 riser stairs, hand carried		8	2			84		84	135
2205	Wheeled		12	1.333			56		56	90
2210	Haul and return, add per each extra 100′ haul, hand carried		35.50	.451			18.95		18.95	30.50
2215	Wheeled		54	.296			12.45		12.45	20
2220	For each additional flight of stairs, up to 5 risers, add		550	.029	Flight		1.22		1.22	1.96
2225	6-10 risers, add		275	.058			2.45		2.45	3.92
2230	11-20 risers, add		138	.116			4.88		4.88	7.80
2235	21-40 risers, add		69	.232			9.75		9.75	15.65
3000	Loading & trucking, including 2 mile haul, chute loaded	B-16	45	.711	C.Y.		31.50	12.75	44.25	64.50
3040	Hand loading truck, 50′ haul	"	48	.667			29.50	11.95	41.45	60.50
3080	Machine loading truck	B-17	120	.267			12.40	5.35	17.75	25.50
5000	Haul, per mile, up to 8 C.Y. truck	B-34B	1165	.007			.34	.49	.83	1.08
5100	Over 8 C.Y. truck	"	1550	.005			.25	.37	.62	.81

02 41 19.20 Selective Demolition, Dump Charges		Crew	Daily Output	Labor-Hours	Unit	Material	Labor	Equipment	Total	Total Incl O&P
0010	**SELECTIVE DEMOLITION, DUMP CHARGES** R024119-10									
0020	Dump charges, typical urban city, tipping fees only									
0100	Building construction materials				Ton	74			74	81
0200	Trees, brush, lumber					63			63	69.50
0300	Rubbish only					63			63	69.50
0500	Reclamation station, usual charge					74			74	81

02 41 19.21 Selective Demolition, Gutting		Crew	Daily Output	Labor-Hours	Unit	Material	Labor	Equipment	Total	Total Incl O&P
0010	**SELECTIVE DEMOLITION, GUTTING** R024119-10									
0020	Building interior, including disposal, dumpster fees not included									
0500	Residential building									
0560	Minimum	B-16	400	.080	SF Flr.		3.54	1.43	4.97	7.30

02 41 Demolition

02 41 19 – Selective Demolition

02 41 19.21 Selective Demolition, Gutting		Crew	Daily Output	Labor-Hours	Unit	2020 Bare Costs Material	Labor	Equipment	Total	Total Incl O&P
0580	Maximum	B-16	360	.089	SF Flr.		3.94	1.59	5.53	8.05
0900	Commercial building									
1000	Minimum	B-16	350	.091	SF Flr.		4.05	1.64	5.69	8.30
1020	Maximum		250	.128	"		5.65	2.29	7.94	11.60
3000	Minimum labor/equipment charge		4	8	Job		355	143	498	730

02 41 19.25 Selective Demolition, Saw Cutting

Line	Description	Crew	Daily Output	Labor-Hours	Unit	Material	Labor	Equipment	Total	Total Incl O&P
0010	**SELECTIVE DEMOLITION, SAW CUTTING** R024119-10									
0015	Asphalt, up to 3" deep	B-89	1050	.015	L.F.	.20	.76	.98	1.94	2.51
0020	Each additional inch of depth	"	1800	.009		.07	.45	.57	1.09	1.40
1200	Masonry walls, hydraulic saw, brick, per inch of depth	B-89B	300	.053		.07	2.67	5	7.74	9.80
1220	Block walls, solid, per inch of depth	"	250	.064		.07	3.21	6	9.28	11.70
2000	Brick or masonry w/hand held saw, per inch of depth	A-1	125	.064		.05	2.69	.89	3.63	5.35
5000	Wood sheathing to 1" thick, on walls	1 Carp	200	.040			2.13		2.13	3.41
5020	On roof	"	250	.032			1.70		1.70	2.72
9000	Minimum labor/equipment charge	A-1	2	4	Job		168	56	224	330

02 41 19.27 Selective Demolition, Torch Cutting

Line	Description	Crew	Daily Output	Labor-Hours	Unit	Material	Labor	Equipment	Total	Total Incl O&P
0010	**SELECTIVE DEMOLITION, TORCH CUTTING** R024119-10									
0020	Steel, 1" thick plate	E-25	333	.024	L.F.	.90	1.43	.04	2.37	3.37
0040	1" diameter bar	"	600	.013	Ea.	.15	.79	.02	.96	1.49
1000	Oxygen lance cutting, reinforced concrete walls									
1040	12"-16" thick walls	1 Clab	10	.800	L.F.		33.50		33.50	54
1080	24" thick walls	"	6	1.333	"		56		56	90
1090	Minimum labor/equipment charge	E-25	2	4	Job		238	6.40	244.40	395
1100	See Section 05 05 21.10									

02 42 Removal and Diversion of Construction Materials

02 42 10 – Building Deconstruction

02 42 10.10 Estimated Salvage Value or Savings

Line	Description	Crew	Daily Output	Labor-Hours	Unit	Material	Labor	Equipment	Total	Total Incl O&P
0010	**ESTIMATED SALVAGE VALUE OR SAVINGS**									
0015	Excludes material handling, packaging, container costs and									
0020	transportation for salvage or disposal									
0050	All items in Section 02 42 10.10 are credit deducts and not costs									
0100	Copper wire salvage value G				Lb.	1.70			1.70	1.87
0200	Brass salvage value G					1.25			1.25	1.38
0250	Bronze					1.45			1.45	1.60
0300	Steel G					.05			.05	.06
0304	304 Stainless Steel					.30			.30	.33
0316	316 Stainless Steel					.45			.45	.50
0400	Cast Iron G					.06			.06	.07
0600	Aluminum siding G					.33			.33	.36
0610	Aluminum wire					.47			.47	.52

02 42 10.20 Deconstruction of Building Components

Line	Description	Crew	Daily Output	Labor-Hours	Unit	Material	Labor	Equipment	Total	Total Incl O&P
0010	**DECONSTRUCTION OF BUILDING COMPONENTS**									
0012	Buildings one or two stories only									
0015	Excludes material handling, packaging, container costs and									
0020	transportation for salvage or disposal									
0050	Deconstruction of plumbing fixtures									
0100	Wall hung or countertop lavatory G	2 Clab	16	1	Ea.		42		42	67.50
0110	Single or double compartment kitchen sink G		14	1.143			48		48	77
0120	Wall hung urinal G		14	1.143			48		48	77

02 42 10 – Building Deconstruction

02 42 10.20 Deconstruction of Building Components		Crew	Daily Output	Labor-Hours	Unit	Material	2020 Bare Costs Labor	Equipment	Total	Total Incl O&P
0130	Floor mounted G	2 Clab	8	2	Ea.		84		84	135
0140	Floor mounted water closet G		16	1			42		42	67.50
0150	Wall hung G		14	1.143			48		48	77
0160	Water fountain, free standing G		16	1			42		42	67.50
0170	Wall hung or deck mounted G		12	1.333			56		56	90
0180	Bathtub, steel or fiberglass G		10	1.600			67.50		67.50	108
0190	Cast iron G		8	2			84		84	135
0200	Shower, single G		6	2.667			112		112	180
0210	Group G		7	2.286			96		96	154
0300	Deconstruction of electrical fixtures									
0310	Surface mounted incandescent fixtures G	2 Clab	48	.333	Ea.		14.05		14.05	22.50
0320	Fluorescent, 2 lamp G		32	.500			21		21	33.50
0330	4 lamp G		24	.667			28		28	45
0340	Strip fluorescent, 1 lamp G		40	.400			16.85		16.85	27
0350	2 lamp G		32	.500			21		21	33.50
0400	Recessed drop-in fluorescent fixture, 2 lamp G		27	.593			25		25	40
0410	4 lamp G		18	.889			37.50		37.50	60
0500	Deconstruction of appliances									
0510	Cooking stoves G	2 Clab	26	.615	Ea.		26		26	41.50
0520	Dishwashers G	"	26	.615	"		26		26	41.50
0600	Deconstruction of millwork and trim									
0610	Cabinets, wood G	2 Carp	40	.400	L.F.		21.50		21.50	34
0620	Countertops G		100	.160	"		8.50		8.50	13.60
0630	Wall paneling, 1" thick G		500	.032	S.F.		1.70		1.70	2.72
0640	Ceiling trim G		500	.032	L.F.		1.70		1.70	2.72
0650	Wainscoting G		500	.032	S.F.		1.70		1.70	2.72
0660	Base, 3/4"-1" thick G		600	.027	L.F.		1.42		1.42	2.27
0700	Deconstruction of doors and windows									
0710	Doors, wrap, interior, wood, single, no closers G	2 Carp	21	.762	Ea.	4.75	40.50		45.25	70.50
0720	Double G		13	1.231		9.50	65.50		75	115
0730	Solid core, single, exterior or interior G		10	1.600		4.75	85		89.75	141
0740	Double G		8	2		9.50	106		115.50	180
0810	Windows, wrap, wood, single									
0812	with no casement or cladding G	2 Carp	21	.762	Ea.	4.75	40.50		45.25	70.50
0820	with casement and/or cladding G	"	18	.889	"	4.75	47		51.75	81
0900	Deconstruction of interior finishes									
0910	Drywall for recycling G	2 Clab	1775	.009	S.F.		.38		.38	.61
0920	Plaster wall, first floor G		1775	.009			.38		.38	.61
0930	Second floor G		1330	.012			.51		.51	.81
1000	Deconstruction of roofing and accessories									
1010	Built-up roofs G	2 Clab	570	.028	S.F.		1.18		1.18	1.89
1020	Gutters, fascia and rakes G	"	1140	.014	L.F.		.59		.59	.95
2000	Deconstruction of wood components									
2010	Roof sheeting G	2 Clab	570	.028	S.F.		1.18		1.18	1.89
2020	Main roof framing G		760	.021	L.F.		.89		.89	1.42
2030	Porch roof framing G		445	.036			1.51		1.51	2.43
2040	Beams 4" x 8" G	B-1	375	.064			2.74		2.74	4.39
2050	4" x 10" G		300	.080			3.42		3.42	5.50
2055	4" x 12" G		250	.096			4.11		4.11	6.60
2060	6" x 8" G		250	.096			4.11		4.11	6.60
2065	6" x 10" G		200	.120			5.15		5.15	8.20
2070	6" x 12" G		170	.141			6.05		6.05	9.65
2075	8" x 12" G		126	.190			8.15		8.15	13.05

02 42 Removal and Diversion of Construction Materials

02 42 10 – Building Deconstruction

02 42 10.20 Deconstruction of Building Components			Crew	Daily Output	Labor-Hours	Unit	Material	2020 Bare Costs Labor	Equipment	Total	Total Incl O&P
2080	10" x 12"	G	B-1	100	.240	L.F.		10.25		10.25	16.45
2100	Ceiling joists	G	2 Clab	800	.020			.84		.84	1.35
2150	Wall framing, interior	G		1230	.013			.55		.55	.88
2160	Sub-floor	G		2000	.008	S.F.		.34		.34	.54
2170	Floor joists	G		2000	.008	L.F.		.34		.34	.54
2200	Wood siding (no lead or asbestos)	G		1300	.012	S.F.		.52		.52	.83
2300	Wall framing, exterior	G		1600	.010	L.F.		.42		.42	.67
2400	Stair risers	G		53	.302	Ea.		12.70		12.70	20.50
2500	Posts	G		800	.020	L.F.		.84		.84	1.35
3000	Deconstruction of exterior brick walls										
3010	Exterior brick walls, first floor	G	2 Clab	200	.080	S.F.		3.37		3.37	5.40
3020	Second floor	G		64	.250	"		10.55		10.55	16.85
3030	Brick chimney	G		100	.160	C.F.		6.75		6.75	10.80
4000	Deconstruction of concrete										
4010	Slab on grade, 4" thick, plain concrete	G	B-9	500	.080	S.F.		3.40	.71	4.11	6.25
4020	Wire mesh reinforced	G		470	.085			3.62	.76	4.38	6.65
4030	Rod reinforced	G		400	.100			4.25	.89	5.14	7.80
4110	Foundation wall, 6" thick, plain concrete	G		160	.250			10.65	2.23	12.88	19.45
4120	8" thick	G		140	.286			12.15	2.55	14.70	22.50
4130	10" thick	G		120	.333			14.15	2.98	17.13	26
9000	Deconstruction process, support equipment as needed										
9010	Daily use, portal to portal, 12-ton truck-mounted hydraulic crane crew	G	A-3H	1	8	Day		475	725	1,200	1,525
9020	Daily use, skid steer and operator	G	A-3C	1	8			425	400	825	1,100
9030	Daily use, backhoe 48 HP, operator and labor	G	"	1	8			425	400	825	1,100

02 42 10.30 Deconstruction Material Handling			Crew	Daily Output	Labor-Hours	Unit	Material	Labor	Equipment	Total	Total Incl O&P
0010	**DECONSTRUCTION MATERIAL HANDLING**										
0012	Buildings one or two stories only										
0100	Clean and stack brick on pallet	G	2 Clab	1200	.013	Ea.		.56		.56	.90
0200	Haul 50' and load rough lumber up to 2" x 8" size	G		2000	.008	"		.34		.34	.54
0210	Lumber larger than 2" x 8"	G		3200	.005	B.F.		.21		.21	.34
0300	Finish wood for recycling stack and wrap per pallet	G		8	2	Ea.	38	84		122	177
0350	Light fixtures			6	2.667		68.50	112		180.50	255
0375	Windows			6	2.667		64.50	112		176.50	251
0400	Miscellaneous materials			8	2		19	84		103	156
1000	See Section 02 41 19.19 for bulk material handling										

02 43 Structure Moving

02 43 13 – Structure Relocation

02 43 13.13 Building Relocation			Crew	Daily Output	Labor-Hours	Unit	Material	Labor	Equipment	Total	Total Incl O&P
0010	**BUILDING RELOCATION**										
0011	One day move, up to 24' wide										
0020	Reset on existing foundation					Total				11,500	11,500
0040	Wood or steel frame bldg., based on ground floor area	G	B-4	185	.259	S.F.		11.30	2.67	13.97	21
0060	Masonry bldg., based on ground floor area	G	"	137	.350			15.25	3.60	18.85	28.50
0200	For 24'-42' wide, add									15%	15%

02 58 Snow Control

02 58 13 – Snow Fencing

02 58 13.10 Snow Fencing System		Crew	Daily Output	Labor-Hours	Unit	Material	2020 Bare Costs Labor	Equipment	Total	Total Incl O&P
0010	**SNOW FENCING SYSTEM**									
7001	Snow fence on steel posts 10′ OC, 4′ high	B-1	500	.048	L.F.	.39	2.05		2.44	3.72

02 65 Underground Storage Tank Removal

02 65 10 – Underground Tank and Contaminated Soil Removal

02 65 10.30 Removal of Underground Storage Tanks

Line	Description		Crew	Daily Output	Labor-Hours	Unit	Material	Labor	Equipment	Total	Total Incl O&P
0010	**REMOVAL OF UNDERGROUND STORAGE TANKS** R026510-20										
0011	Petroleum storage tanks, non-leaking										
0100	Excavate & load onto trailer										
0110	3,000 gal. to 5,000 gal. tank	G	B-14	4	12	Ea.		530	53.50	583.50	905
0120	6,000 gal. to 8,000 gal. tank	G	B-3A	3	13.333			600	229	829	1,200
0130	9,000 gal. to 12,000 gal. tank	G	″	2	20			900	345	1,245	1,800
0190	Known leaking tank, add					%				100%	100%
0200	Remove sludge, water and remaining product from bottom										
0201	of tank with vacuum truck										
0300	3,000 gal. to 5,000 gal. tank	G	A-13	5	1.600	Ea.		85	148	233	295
0310	6,000 gal. to 8,000 gal. tank	G		4	2			106	184	290	370
0320	9,000 gal. to 12,000 gal. tank	G		3	2.667			141	246	387	490
0390	Dispose of sludge off-site, average					Gal.				6.25	6.80
0400	Insert inert solid CO_2 "dry ice" into tank										
0401	For cleaning/transporting tanks (1.5 lb./100 gal. cap)	G	1 Clab	500	.016	Lb.	1.22	.67		1.89	2.42
0503	Disconnect and remove piping	G	1 Plum	160	.050	L.F.		3.22		3.22	4.98
0603	Transfer liquids, 10% of volume	G	″	1600	.005	Gal.		.32		.32	.50
0703	Cut accessway into underground storage tank	G	1 Clab	5.33	1.501	Ea.		63		63	101
0813	Remove sludge, wash and wipe tank, 500 gal.	G	1 Plum	8	1			64.50		64.50	99.50
0823	3,000 gal.	G		6.67	1.199			77.50		77.50	120
0833	5,000 gal.	G		6.15	1.301			84		84	130
0843	8,000 gal.	G		5.33	1.501			96.50		96.50	150
0853	10,000 gal.	G		4.57	1.751			113		113	174
0863	12,000 gal.	G		4.21	1.900			122		122	189
1020	Haul tank to certified salvage dump, 100 miles round trip										
1023	3,000 gal. to 5,000 gal. tank					Ea.				760	830
1026	6,000 gal. to 8,000 gal. tank									880	960
1029	9,000 gal. to 12,000 gal. tank									1,050	1,150
1100	Disposal of contaminated soil to landfill										
1110	Minimum					C.Y.				145	160
1111	Maximum					″				400	440
1120	Disposal of contaminated soil to										
1121	bituminous concrete batch plant										
1130	Minimum					C.Y.				80	88
1131	Maximum					″				115	125
1203	Excavate, pull, & load tank, backfill hole, 8,000 gal. +	G	B-12C	.50	32	Ea.		1,625	1,900	3,525	4,650
1213	Haul tank to certified dump, 100 miles rt, 8,000 gal. +	G	B-34K	1	8			390	855	1,245	1,575
1223	Excavate, pull, & load tank, backfill hole, 500 gal.	G	B-11C	1	16			790	214	1,004	1,475
1233	Excavate, pull, & load tank, backfill hole, 3,000-5,000 gal.	G	B-11M	.50	32			1,575	465	2,040	3,000
1243	Haul tank to certified dump, 100 miles rt, 500 gal.	G	B-34L	1	8			425	196	621	880
1253	Haul tank to certified dump, 100 miles rt, 3,000-5,000 gal.	G	B-34M	1	8			425	820	1,245	1,575
2010	Decontamination of soil on site incl poly tarp on top/bottom										
2011	Soil containment berm and chemical treatment										
2020	Minimum	G	B-11C	100	.160	C.Y.	7.65	7.90	2.14	17.69	23.50
2021	Maximum	G	″	100	.160		9.90	7.90	2.14	19.94	25.50

02 65 Underground Storage Tank Removal

02 65 10 – Underground Tank and Contaminated Soil Removal

02 65 10.30 Removal of Underground Storage Tanks		Crew	Daily Output	Labor-Hours	Unit	Material	Labor	Equipment	Total	Total Incl O&P
						2020 Bare Costs				
2050	Disposal of decontaminated soil, minimum				C.Y.				135	150
2055	Maximum				↓				400	440

02 81 Transportation and Disposal of Hazardous Materials

02 81 20 – Hazardous Waste Handling

02 81 20.10 Hazardous Waste Cleanup/Pickup/Disposal		Crew	Daily Output	Labor-Hours	Unit	Material	Labor	Equipment	Total	Total Incl O&P
0010	**HAZARDOUS WASTE CLEANUP/PICKUP/DISPOSAL**									
0100	For contractor rental equipment, i.e., dozer,									
0110	Front end loader, dump truck, etc., see 01 54 33 Reference Section									
1000	Solid pickup									
1100	55 gal. drums				Ea.				240	265
1120	Bulk material, minimum				Ton				190	210
1130	Maximum				"				595	655
1200	Transportation to disposal site									
1220	Truckload = 80 drums or 25 C.Y. or 18 tons									
1260	Minimum				Mile				3.95	4.45
1270	Maximum				"				7.25	7.98
3000	Liquid pickup, vacuum truck, stainless steel tank									
3100	Minimum charge, 4 hours									
3110	1 compartment, 2200 gallon				Hr.				140	155
3120	2 compartment, 5000 gallon				"				200	225
3400	Transportation in 6900 gallon bulk truck				Mile				7.95	8.75
3410	In teflon lined truck				"				10.20	11.25
5000	Heavy sludge or dry vacuumable material				Hr.				140	155
6000	Dumpsite disposal charge, minimum				Ton				140	155
6020	Maximum				"				415	455

02 82 Asbestos Remediation

02 82 13 – Asbestos Abatement

02 82 13.39 Asbestos Remediation Plans and Methods		Crew	Daily Output	Labor-Hours	Unit	Material	Labor	Equipment	Total	Total Incl O&P
0010	**ASBESTOS REMEDIATION PLANS AND METHODS**									
0100	Building Survey-Commercial Building				Ea.				2,200	2,400
0200	Asbestos Abatement Remediation Plan				"				1,350	1,475

02 82 13.41 Asbestos Abatement Equipment		Crew	Daily Output	Labor-Hours	Unit	Material	Labor	Equipment	Total	Total Incl O&P
0010	**ASBESTOS ABATEMENT EQUIPMENT** R028213-20									
0011	Equipment and supplies, buy									
0200	Air filtration device, 2000 CFM				Ea.	950			950	1,050
0250	Large volume air sampling pump, minimum					315			315	350
0260	Maximum					345			345	380
0300	Airless sprayer unit, 2 gun					2,625			2,625	2,875
0350	Light stand, 500 watt				↓	41			41	45
0400	Personal respirators									
0410	Negative pressure, 1/2 face, dual operation, minimum				Ea.	30			30	33
0420	Maximum					34			34	37.50
0450	P.A.P.R., full face, minimum					780			780	855
0460	Maximum					1,225			1,225	1,325
0470	Supplied air, full face, including air line, minimum					289			289	320
0480	Maximum					500			500	550
0500	Personnel sampling pump				↓	237			237	261

02 82 13 – Asbestos Abatement

02 82 13.41 Asbestos Abatement Equipment

	02 82 13.41 Asbestos Abatement Equipment	Crew	Daily Output	Labor-Hours	Unit	2020 Bare Costs Material	Labor	Equipment	Total	Total Incl O&P
1500	Power panel, 20 unit, including GFI				Ea.	475			475	525
1600	Shower unit, including pump and filters					1,125			1,125	1,250
1700	Supplied air system (type C)					3,700			3,700	4,075
1750	Vacuum cleaner, HEPA, 16 gal., stainless steel, wet/dry					1,275			1,275	1,400
1760	55 gallon					1,475			1,475	1,600
1800	Vacuum loader, 9-18 ton/hr.					99,000			99,000	109,000
1900	Water atomizer unit, including 55 gal. drum					305			305	340
2000	Worker protection, whole body, foot, head cover & gloves, plastic					7.45			7.45	8.20
2500	Respirator, single use					30			30	33
2550	Cartridge for respirator					5.15			5.15	5.65
2570	Glove bag, 7 mil, 50" x 64"					8.75			8.75	9.60
2580	10 mil, 44" x 60"					7.90			7.90	8.65
6000	Disposable polyethylene bags, 6 mil, 3 C.F.					.77			.77	.85
6300	Disposable fiber drums, 3 C.F.					18.90			18.90	21
6400	Pressure sensitive caution labels, 3" x 5"					3.08			3.08	3.39
6450	11" x 17"					7.75			7.75	8.55
6500	Negative air machine, 1800 CFM				↓	890			890	975

02 82 13.42 Preparation of Asbestos Containment Area

	02 82 13.42 Preparation of Asbestos Containment Area	Crew	Daily Output	Labor-Hours	Unit	Material	Labor	Equipment	Total	Total Incl O&P
0010	**PREPARATION OF ASBESTOS CONTAINMENT AREA**									
0100	Pre-cleaning, HEPA vacuum and wet wipe, flat surfaces	A-9	12000	.005	S.F.	.01	.31		.32	.51
0200	Protect carpeted area, 2 layers 6 mil poly on 3/4" plywood	"	1000	.064		2.10	3.76		5.86	8.25
0300	Separation barrier, 2" x 4" @ 16", 1/2" plywood ea. side, 8' high	2 Carp	400	.040		3.08	2.13		5.21	6.80
0310	12' high		320	.050		3.02	2.66		5.68	7.60
0320	16' high		200	.080		2.99	4.25		7.24	10.10
0400	Personnel decontam. chamber, 2" x 4" @ 16", 3/4" ply ea. side		280	.057		3.55	3.04		6.59	8.75
0450	Waste decontam. chamber, 2" x 4" studs @ 16", 3/4" ply ea. side	↓	360	.044	↓	3.55	2.36		5.91	7.70
0500	Cover surfaces with polyethylene sheeting									
0501	Including glue and tape									
0550	Floors, each layer, 6 mil	A-9	8000	.008	S.F.	.04	.47		.51	.79
0551	4 mil		9000	.007		.03	.42		.45	.69
0560	Walls, each layer, 6 mil		6000	.011		.04	.63		.67	1.04
0561	4 mil	↓	7000	.009	↓	.03	.54		.57	.88
0570	For heights above 14', add						20%			
0575	For heights above 20', add						30%			
0580	For fire retardant poly, add					100%				
0590	For large open areas, deduct					10%	20%			
0600	Seal floor penetrations with foam firestop to 36 sq. in.	2 Carp	200	.080	Ea.	13.90	4.25		18.15	22
0610	36 sq. in. to 72 sq. in.		125	.128		28	6.80		34.80	41.50
0615	72 sq. in. to 144 sq. in.		80	.200		55.50	10.65		66.15	78
0620	Wall penetrations, to 36 sq. in.		180	.089		13.90	4.72		18.62	23
0630	36 sq. in. to 72 sq. in.		100	.160		28	8.50		36.50	44
0640	72 sq. in. to 144 sq. in.	↓	60	.267	↓	55.50	14.15		69.65	83.50
0800	Caulk seams with latex	1 Carp	230	.035	L.F.	.17	1.85		2.02	3.15
0900	Set up neg. air machine, 1-2k CFM/25 M.C.F. volume	1 Asbe	4.30	1.860	Ea.		109		109	173
0950	Set up and remove portable shower unit	2 Asbe	4	4	"		235		235	370

02 82 13.43 Bulk Asbestos Removal

	02 82 13.43 Bulk Asbestos Removal	Crew	Daily Output	Labor-Hours	Unit	Material	Labor	Equipment	Total	Total Incl O&P
0010	**BULK ASBESTOS REMOVAL**									
0020	Includes disposable tools and 2 suits and 1 respirator filter/day/worker									
0100	Beams, W 10 x 19	A-9	235	.272	L.F.	.68	16		16.68	26.50
0110	W 12 x 22		210	.305		.76	17.90		18.66	29.50
0120	W 14 x 26		180	.356		.89	21		21.89	34
0130	W 16 x 31	↓	160	.400	↓	1	23.50		24.50	38

02 82 13.43 Bulk Asbestos Removal

02 82 13.43 Bulk Asbestos Removal		Crew	Daily Output	Labor-Hours	Unit	2020 Bare Costs Material	Labor	Equipment	Total	Total Incl O&P
0140	W 18 x 40	A-9	140	.457	L.F.	1.15	27		28.15	44
0150	W 24 x 55		110	.582		1.46	34		35.46	55.50
0160	W 30 x 108		85	.753		1.89	44		45.89	72
0170	W 36 x 150		72	.889		2.23	52		54.23	85.50
0200	Boiler insulation		480	.133	S.F.	.40	7.85		8.25	12.85
0210	With metal lath, add				%				50%	50%
0300	Boiler breeching or flue insulation	A-9	520	.123	S.F.	.31	7.25		7.56	11.80
0310	For active boiler, add				%				100%	100%
0400	Duct or AHU insulation	A-10B	440	.073	S.F.	.18	4.28		4.46	7
0500	Duct vibration isolation joints, up to 24 sq. in. duct	A-9	56	1.143	Ea.	2.87	67		69.87	109
0520	25 sq. in. to 48 sq. in. duct		48	1.333		3.35	78.50		81.85	128
0530	49 sq. in. to 76 sq. in. duct		40	1.600		4.01	94		98.01	153
0600	Pipe insulation, air cell type, up to 4" diameter pipe		900	.071	L.F.	.18	4.17		4.35	6.80
0610	4" to 8" diameter pipe		800	.080		.20	4.70		4.90	7.65
0620	10" to 12" diameter pipe		700	.091		.23	5.35		5.58	8.75
0630	14" to 16" diameter pipe		550	.116		.29	6.85		7.14	11.15
0650	Over 16" diameter pipe		650	.098	S.F.	.25	5.80		6.05	9.40
0700	With glove bag up to 3" diameter pipe		200	.320	L.F.	9.40	18.80		28.20	40.50
1000	Pipe fitting insulation up to 4" diameter pipe		320	.200	Ea.	.50	11.75		12.25	19.15
1100	6" to 8" diameter pipe		304	.211		.53	12.35		12.88	20
1110	10" to 12" diameter pipe		192	.333		.84	19.55		20.39	32
1120	14" to 16" diameter pipe		128	.500		1.25	29.50		30.75	48
1130	Over 16" diameter pipe		176	.364	S.F.	.91	21.50		22.41	35
1200	With glove bag, up to 8" diameter pipe		75	.853	L.F.	6.55	50		56.55	86.50
2000	Scrape foam fireproofing from flat surface		2400	.027	S.F.	.07	1.57		1.64	2.55
2100	Irregular surfaces		1200	.053		.13	3.13		3.26	5.10
3000	Remove cementitious material from flat surface		1800	.036		.09	2.09		2.18	3.41
3100	Irregular surface		1000	.064		.11	3.76		3.87	6.10
4000	Scrape acoustical coating/fireproofing, from ceiling		3200	.020		.05	1.17		1.22	1.92
5000	Remove VAT and mastic from floor by hand		2400	.027		.07	1.57		1.64	2.55
5100	By machine	A-11	4800	.013		.03	.78	.01	.82	1.29
5150	For 2 layers, add				%				50%	50%
6000	Remove contaminated soil from crawl space by hand	A-9	400	.160	C.F.	.40	9.40		9.80	15.35
6100	With large production vacuum loader	A-12	700	.091	"	.23	5.35	1.05	6.63	9.90
7000	Radiator backing, not including radiator removal	A-9	1200	.053	S.F.	.13	3.13		3.26	5.10
8000	Cement-asbestos transite board and cement wall board	2 Asbe	1000	.016		.16	.94		1.10	1.67
8100	Transite shingle siding	A-10B	750	.043		.21	2.51		2.72	4.21
8200	Shingle roofing	"	2000	.016		.08	.94		1.02	1.58
8250	Built-up, no gravel, non-friable	B-2	1400	.029		.08	1.21		1.29	2.04
8260	Bituminous flashing	1 Rofc	300	.027		.08	1.23		1.31	2.25
8300	Asbestos millboard, flat board and VAT contaminated plywood	2 Asbe	1000	.016		.08	.94		1.02	1.58
9000	For type B (supplied air) respirator equipment, add				%				10%	10%

02 82 13.44 Demolition In Asbestos Contaminated Area

02 82 13.44 Demolition In Asbestos Contaminated Area		Crew	Daily Output	Labor-Hours	Unit	Material	Labor	Equipment	Total	Total Incl O&P
0010	**DEMOLITION IN ASBESTOS CONTAMINATED AREA**									
0200	Ceiling, including suspension system, plaster and lath	A-9	2100	.030	S.F.	.08	1.79		1.87	2.92
0210	Finished plaster, leaving wire lath		585	.109		.27	6.40		6.67	10.50
0220	Suspended acoustical tile		3500	.018		.05	1.07		1.12	1.75
0230	Concealed tile grid system		3000	.021		.05	1.25		1.30	2.05
0240	Metal pan grid system		1500	.043		.11	2.51		2.62	4.09
0250	Gypsum board		2500	.026		.06	1.50		1.56	2.45
0260	Lighting fixtures up to 2' x 4'		72	.889	Ea.	2.23	52		54.23	85.50
0400	Partitions, non load bearing									

02 82 Asbestos Remediation

02 82 13 – Asbestos Abatement

02 82 13.44 Demolition In Asbestos Contaminated Area

	02 82 13.44 Demolition In Asbestos Contaminated Area	Crew	Daily Output	Labor-Hours	Unit	2020 Bare Costs Material	Labor	Equipment	Total	Total Incl O&P
0410	Plaster, lath, and studs	A-9	690	.093	S.F.	.90	5.45		6.35	9.65
0450	Gypsum board and studs	"	1390	.046	"	.12	2.70		2.82	4.42
9000	For type B (supplied air) respirator equipment, add				%				10%	10%

02 82 13.45 OSHA Testing

	02 82 13.45 OSHA Testing	Crew	Daily Output	Labor-Hours	Unit	Material	Labor	Equipment	Total	Total Incl O&P
0010	**OSHA TESTING**									
0100	Certified technician, minimum				Day		340		340	340
0110	Maximum						670		670	670
0121	Industrial hygienist, minimum						385		385	385
0130	Maximum						755		755	755
0200	Asbestos sampling and PCM analysis, NIOSH 7400, minimum	1 Asbe	8	1	Ea.	15.90	58.50		74.40	111
0210	Maximum		4	2		27	117		144	216
1000	Cleaned area samples		8	1		203	58.50		261.50	315
1100	PCM air sample analysis, NIOSH 7400, minimum		8	1		17.70	58.50		76.20	113
1110	Maximum		4	2		2.30	117		119.30	189
1200	TEM air sample analysis, NIOSH 7402, minimum								80	106
1210	Maximum								360	450

02 82 13.46 Decontamination of Asbestos Containment Area

	02 82 13.46 Decontamination of Asbestos Containment Area	Crew	Daily Output	Labor-Hours	Unit	Material	Labor	Equipment	Total	Total Incl O&P
0010	**DECONTAMINATION OF ASBESTOS CONTAINMENT AREA**									
0100	Spray exposed substrate with surfactant (bridging)									
0200	Flat surfaces	A-9	6000	.011	S.F.	.41	.63		1.04	1.44
0250	Irregular surfaces		4000	.016	"	.45	.94		1.39	1.99
0300	Pipes, beams, and columns		2000	.032	L.F.	.60	1.88		2.48	3.64
1000	Spray encapsulate polyethylene sheeting		8000	.008	S.F.	.41	.47		.88	1.19
1100	Roll down polyethylene sheeting		8000	.008	"		.47		.47	.74
1500	Bag polyethylene sheeting		400	.160	Ea.	1.01	9.40		10.41	16
2000	Fine clean exposed substrate, with nylon brush		2400	.027	S.F.		1.57		1.57	2.48
2500	Wet wipe substrate		4800	.013			.78		.78	1.24
2600	Vacuum surfaces, fine brush		6400	.010			.59		.59	.93
3000	Structural demolition									
3100	Wood stud walls	A-9	2800	.023	S.F.		1.34		1.34	2.13
3500	Window manifolds, not incl. window replacement		4200	.015			.89		.89	1.42
3600	Plywood carpet protection		2000	.032			1.88		1.88	2.98
4000	Remove custom decontamination facility	A-10A	8	3	Ea.	15.15	176		191.15	297
4100	Remove portable decontamination facility	3 Asbe	12	2	"	14.05	117		131.05	201
5000	HEPA vacuum, shampoo carpeting	A-9	4800	.013	S.F.	.11	.78		.89	1.36
9000	Final cleaning of protected surfaces	A-10A	8000	.003	"		.18		.18	.28

02 82 13.47 Asbestos Waste Pkg., Handling, and Disp.

	02 82 13.47 Asbestos Waste Pkg., Handling, and Disp.	Crew	Daily Output	Labor-Hours	Unit	Material	Labor	Equipment	Total	Total Incl O&P
0010	**ASBESTOS WASTE PACKAGING, HANDLING, AND DISPOSAL**									
0100	Collect and bag bulk material, 3 C.F. bags, by hand	A-9	400	.160	Ea.	.77	9.40		10.17	15.75
0200	Large production vacuum loader	A-12	880	.073		.99	4.27	.84	6.10	8.75
1000	Double bag and decontaminate	A-9	960	.067		.77	3.91		4.68	7.05
2000	Containerize bagged material in drums, per 3 C.F. drum	"	800	.080		18.90	4.70		23.60	28.50
3000	Cart bags 50′ to dumpster	2 Asbe	400	.040			2.35		2.35	3.72
5000	Disposal charges, not including haul, minimum				C.Y.				61	67
5020	Maximum				"				355	395
9000	For type B (supplied air) respirator equipment, add				%				10%	10%

02 82 13.48 Asbestos Encapsulation With Sealants

	02 82 13.48 Asbestos Encapsulation With Sealants	Crew	Daily Output	Labor-Hours	Unit	Material	Labor	Equipment	Total	Total Incl O&P
0010	**ASBESTOS ENCAPSULATION WITH SEALANTS**									
0100	Ceilings and walls, minimum	A-9	21000	.003	S.F.	.42	.18		.60	.74
0110	Maximum		10600	.006		.52	.35		.87	1.13
0200	Columns and beams, minimum		13300	.005		.42	.28		.70	.91
0210	Maximum		5325	.012		.63	.71		1.34	1.81

02 82 Asbestos Remediation

02 82 13 – Asbestos Abatement

02 82 13.48 Asbestos Encapsulation With Sealants		Crew	Daily Output	Labor-Hours	Unit	Material	Labor	Equipment	Total	Total Incl O&P
						2020 Bare Costs				
0300	Pipes to 12" diameter including minor repairs, minimum	A-9	800	.080	L.F.	.57	4.70		5.27	8.10
0310	Maximum	↓	400	.160	"	1.27	9.40		10.67	16.30

02 83 Lead Remediation

02 83 19 – Lead-Based Paint Remediation

		Crew	Daily Output	Labor-Hours	Unit	Material	Labor	Equipment	Total	Total Incl O&P
02 83 19.21 Lead Paint Remediation Plans and Methods										
0010	**LEAD PAINT REMEDIATION PLANS AND METHODS**									
0100	Building Survey-Commercial Building				Ea.				2,050	2,250
0200	Lead Abatement Remediation Plan								1,225	1,350
0300	Lead Paint Testing, AAS Analysis								51	56
0400	Lead Paint Testing, X-Ray Fluorescence				↓				51	56
02 83 19.22 Preparation of Lead Containment Area										
0010	**PREPARATION OF LEAD CONTAINMENT AREA**									
0020	Lead abatement work area, test kit, per swab	1 Skwk	16	.500	Ea.	3.40	27.50		30.90	47
0025	For dust barriers see Section 01 56 16.10									
0050	Caution sign	1 Skwk	48	.167	Ea.	9	9.15		18.15	24.50
0100	Pre-cleaning, HEPA vacuum and wet wipe, floor and wall surfaces	3 Skwk	5000	.005	S.F.	.01	.26		.27	.43
0105	Ceiling, 6'-11' high		4100	.006		.07	.32		.39	.59
0108	12'-15' high		3550	.007		.07	.37		.44	.67
0115	Over 15' high	↓	3000	.008	↓	.09	.44		.53	.80
0500	Cover surfaces with polyethylene sheeting									
0550	Floors, each layer, 6 mil	3 Skwk	8000	.003	S.F.	.04	.16		.20	.31
0560	Walls, each layer, 6 mil	"	6000	.004	"	.04	.22		.26	.40
0570	For heights above 14', add						20%			
2400	Post abatement cleaning of protective sheeting, HEPA vacuum & wet wipe	3 Skwk	5000	.005	S.F.	.01	.26		.27	.43
2450	Doff, bag and seal protective sheeting		12000	.002		.01	.11		.12	.18
2500	Post abatement cleaning, HEPA vacuum & wet wipe	↓	5000	.005	↓	.01	.26		.27	.43
02 83 19.23 Encapsulation of Lead-Based Paint										
0010	**ENCAPSULATION OF LEAD-BASED PAINT**									
0020	Interior, brushwork, trim, under 6"	1 Pord	240	.033	L.F.	2.45	1.48		3.93	5.05
0030	6"-12" wide		180	.044		3.19	1.97		5.16	6.65
0040	Balustrades		300	.027		2.07	1.18		3.25	4.16
0050	Pipe to 4" diameter		500	.016		1.27	.71		1.98	2.53
0060	To 8" diameter		375	.021		1.56	.95		2.51	3.22
0070	To 12" diameter		250	.032		2.36	1.42		3.78	4.86
0080	To 16" diameter		170	.047	↓	3.50	2.09		5.59	7.15
0090	Cabinets, ornate design		200	.040	S.F.	3.19	1.78		4.97	6.35
0100	Simple design	↓	250	.032	"	2.42	1.42		3.84	4.92
0110	Doors, 3' x 7', both sides, incl. frame & trim									
0120	Flush	1 Pord	6	1.333	Ea.	30	59		89	127
0130	French, 10-15 lite		3	2.667		6.45	118		124.45	195
0140	Panel		4	2		36.50	89		125.50	181
0150	Louvered	↓	2.75	2.909	↓	34.50	129		163.50	243
0160	Windows, per interior side, per 15 S.F.									
0170	1-6 lite	1 Pord	14	.571	Ea.	21.50	25.50		47	64
0180	7-10 lite		7.50	1.067		23.50	47.50		71	101
0190	12 lite		5.75	1.391		32	62		94	133
0200	Radiators		8	1	↓	74.50	44.50		119	153
0210	Grilles, vents		275	.029	S.F.	2.11	1.29		3.40	4.37
0220	Walls, roller, drywall or plaster	↓	1000	.008	↓	.63	.36		.99	1.25

02 83 Lead Remediation

02 83 19 – Lead-Based Paint Remediation

02 83 19.23 Encapsulation of Lead-Based Paint		Crew	Daily Output	Labor-Hours	Unit	2020 Bare Costs Material	Labor	Equipment	Total	Total Incl O&P
0230	With spunbonded reinforcing fabric	1 Pord	720	.011	S.F.	.73	.49		1.22	1.58
0240	Wood		800	.010		.72	.44		1.16	1.50
0250	Ceilings, roller, drywall or plaster		900	.009		.73	.39		1.12	1.43
0260	Wood		700	.011		.87	.51		1.38	1.77
0270	Exterior, brushwork, gutters and downspouts		300	.027	L.F.	2.04	1.18		3.22	4.12
0280	Columns		400	.020	S.F.	1.48	.89		2.37	3.04
0290	Spray, siding		600	.013	"	1.06	.59		1.65	2.11
0300	Miscellaneous									
0310	Electrical conduit, brushwork, to 2" diameter	1 Pord	500	.016	L.F.	1.27	.71		1.98	2.53
0320	Brick, block or concrete, spray		500	.016	S.F.	1.27	.71		1.98	2.53
0330	Steel, flat surfaces and tanks to 12"		500	.016		1.27	.71		1.98	2.53
0340	Beams, brushwork		400	.020		1.48	.89		2.37	3.04
0350	Trusses		400	.020		1.48	.89		2.37	3.04

02 83 19.26 Removal of Lead-Based Paint		Crew	Daily Output	Labor-Hours	Unit	Material	Labor	Equipment	Total	Total Incl O&P
0010	**REMOVAL OF LEAD-BASED PAINT** R028319-60									
0011	By chemicals, per application									
0050	Baseboard, to 6" wide	1 Pord	64	.125	L.F.	.84	5.55		6.39	9.70
0070	To 12" wide		32	.250	"	1.49	11.10		12.59	19.30
0200	Balustrades, one side		28	.286	S.F.	1.49	12.70		14.19	21.50
1400	Cabinets, simple design		32	.250		1.39	11.10		12.49	19.20
1420	Ornate design		25	.320		1.55	14.20		15.75	24
1600	Cornice, simple design		60	.133		1.59	5.90		7.49	11.15
1620	Ornate design		20	.400		5.50	17.75		23.25	34
2800	Doors, one side, flush		84	.095		1.88	4.23		6.11	8.75
2820	Two panel		80	.100		1.39	4.44		5.83	8.60
2840	Four panel		45	.178		1.40	7.90		9.30	14.10
2880	For trim, one side, add		64	.125	L.F.	.80	5.55		6.35	9.70
3000	Fence, picket, one side		30	.267	S.F.	1.39	11.85		13.24	20.50
3200	Grilles, one side, simple design		30	.267		1.40	11.85		13.25	20.50
3220	Ornate design		25	.320		1.46	14.20		15.66	24
3240	Handrails		90	.089	L.F.	1.40	3.95		5.35	7.80
4400	Pipes, to 4" diameter		90	.089		1.88	3.95		5.83	8.30
4420	To 8" diameter		50	.160		3.76	7.10		10.86	15.45
4440	To 12" diameter		36	.222		5.65	9.85		15.50	22
4460	To 16" diameter		20	.400		7.55	17.75		25.30	36.50
4500	For hangers, add		40	.200	Ea.	2.54	8.90		11.44	16.90
4800	Siding		90	.089	S.F.	1.36	3.95		5.31	7.75
5000	Trusses, open		55	.145	SF Face	2.05	6.45		8.50	12.50
6200	Windows, one side only, double-hung, 1/1 light, 24" x 48" high		4	2	Ea.	25.50	89		114.50	169
6220	30" x 60" high		3	2.667		33.50	118		151.50	225
6240	36" x 72" high		2.50	3.200		41	142		183	271
6280	40" x 80" high		2	4		51	178		229	340
6400	Colonial window, 6/6 light, 24" x 48" high		2	4		51	178		229	340
6420	30" x 60" high		1.50	5.333		67	237		304	450
6440	36" x 72" high		1	8		101	355		456	675
6480	40" x 80" high		1	8		101	355		456	675
6600	8/8 light, 24" x 48" high		2	4		51	178		229	340
6620	40" x 80" high		1	8		101	355		456	675
6800	12/12 light, 24" x 48" high		1	8		101	355		456	675
6820	40" x 80" high		.75	10.667		136	475		611	900
6840	Window frame & trim items, included in pricing above									
7000	Hand scraping and HEPA vacuum, less than 4 S.F.	1 Pord	8	1	Ea.	.77	44.50		45.27	71.50

02 83 Lead Remediation

02 83 19 – Lead-Based Paint Remediation

02 83 19.26 Removal of Lead-Based Paint		Crew	Daily Output	Labor-Hours	Unit	2020 Bare Costs Material	Labor	Equipment	Total	Total Incl O&P
8000	Collect and bag bulk material, 3 C.F. bags, by hand	1 Pord	30	.267	Ea.	.77	11.85		12.62	19.65
9000	Minimum labor/equipment charge	↓	3	2.667	Job		118		118	188

02 87 Biohazard Remediation

02 87 13 – Mold Remediation

02 87 13.16 Mold Remediation Preparation and Containment

		Crew	Daily Output	Labor-Hours	Unit	Material	Labor	Equipment	Total	Total Incl O&P
0010	**MOLD REMEDIATION PREPARATION AND CONTAINMENT**									
0015	Mold remediation plans and methods									
0020	Initial inspection, areas to 2,500 S.F.				Total				268	295
0030	Areas to 5,000 S.F.								440	485
0032	Areas to 10,000 S.F.								440	485
0040	Testing, air sample each								126	139
0050	Swab sample								115	127
0060	Tape sample								125	140
0070	Post remediation air test								126	139
0080	Mold abatement plan, area to 2,500 S.F.								1,225	1,325
0090	Areas to 5,000 S.F.								1,625	1,775
0095	Areas to 10,000 S.F.								2,550	2,800
0100	Packup & removal of contents, average 3 bedroom home, excl. storage								8,150	8,975
0110	Average 5 bedroom home, excl. storage				↓				15,300	16,800
0600	For demolition in mold contaminated areas, see Section 02 87 13.33									
0610	For personal protection equipment, see Section 02 82 13.41									
6010	Preparation of mold containment area									
6100	Pre-cleaning, HEPA vacuum and wet wipe, flat surfaces	A-9	12000	.005	S.F.	.01	.31		.32	.51
6300	Separation barrier, 2" x 4" @ 16", 1/2" plywood ea. side, 8' high	2 Carp	400	.040		3.70	2.13		5.83	7.50
6310	12' high		320	.050		3.52	2.66		6.18	8.15
6320	16' high		200	.080		2.40	4.25		6.65	9.45
6400	Personnel decontam. chamber, 2" x 4" @ 16", 3/4" ply ea. side		280	.057		4.47	3.04		7.51	9.80
6450	Waste decontam. chamber, 2" x 4" studs @ 16", 3/4" ply each side	↓	360	.044	↓	4.54	2.36		6.90	8.75
6500	Cover surfaces with polyethylene sheeting									
6501	Including glue and tape									
6550	Floors, each layer, 6 mil	A-9	8000	.008	S.F.	.04	.47		.51	.79
6551	4 mil		9000	.007		.03	.42		.45	.69
6560	Walls, each layer, 6 mil		6000	.011		.04	.63		.67	1.04
6561	4 mil	↓	7000	.009	↓	.03	.54		.57	.88
6570	For heights above 14', add						20%			
6575	For heights above 20', add						30%			
6580	For fire retardant poly, add					100%				
6590	For large open areas, deduct					10%	20%			
6600	Seal floor penetrations with foam firestop to 36 sq. in.	2 Carp	200	.080	Ea.	13.90	4.25		18.15	22
6610	36 sq. in. to 72 sq. in.		125	.128		28	6.80		34.80	41.50
6615	72 sq. in. to 144 sq. in.		80	.200		55.50	10.65		66.15	78
6620	Wall penetrations, to 36 sq. in.		180	.089		13.90	4.72		18.62	23
6630	36 sq. in. to 72 sq. in.		100	.160		28	8.50		36.50	44
6640	72 sq. in. to 144 sq. in.	↓	60	.267	↓	55.50	14.15		69.65	83.50
6800	Caulk seams with latex caulk	1 Carp	230	.035	L.F.	.17	1.85		2.02	3.15
6900	Set up neg. air machine, 1-2k CFM/25 M.C.F. volume	1 Asbe	4.30	1.860	Ea.		109		109	173

02 87 13 – Mold Remediation

02 87 13.33 Removal and Disposal of Materials With Mold		Crew	Daily Output	Labor-Hours	Unit	2020 Bare Costs Material	Labor	Equipment	Total	Total Incl O&P
0010	**REMOVAL AND DISPOSAL OF MATERIALS WITH MOLD**									
0015	Demolition in mold contaminated area									
0200	Ceiling, including suspension system, plaster and lath	A-9	2100	.030	S.F.	.08	1.79		1.87	2.92
0210	Finished plaster, leaving wire lath		585	.109		.27	6.40		6.67	10.50
0220	Suspended acoustical tile		3500	.018		.05	1.07		1.12	1.75
0230	Concealed tile grid system		3000	.021		.05	1.25		1.30	2.05
0240	Metal pan grid system		1500	.043		.11	2.51		2.62	4.09
0250	Gypsum board		2500	.026		.06	1.50		1.56	2.45
0255	Plywood		2500	.026		.06	1.50		1.56	2.45
0260	Lighting fixtures up to 2′ x 4′		72	.889	Ea.	2.23	52		54.23	85.50
0400	Partitions, non load bearing									
0410	Plaster, lath, and studs	A-9	690	.093	S.F.	.90	5.45		6.35	9.65
0450	Gypsum board and studs		1390	.046		.12	2.70		2.82	4.42
0465	Carpet & pad		1390	.046		.12	2.70		2.82	4.42
0600	Pipe insulation, air cell type, up to 4″ diameter pipe		900	.071	L.F.	.18	4.17		4.35	6.80
0610	4″ to 8″ diameter pipe		800	.080		.20	4.70		4.90	7.65
0620	10″ to 12″ diameter pipe		700	.091		.23	5.35		5.58	8.75
0630	14″ to 16″ diameter pipe		550	.116		.29	6.85		7.14	11.15
0650	Over 16″ diameter pipe		650	.098	S.F.	.25	5.80		6.05	9.40
9000	For type B (supplied air) respirator equipment, add				%				10%	10%

Division Notes

		CREW	DAILY OUTPUT	LABOR-HOURS	UNIT	BARE COSTS MAT.	BARE COSTS LABOR	BARE COSTS EQUIP.	BARE COSTS TOTAL	TOTAL INCL O&P

Estimating Tips

General

- Carefully check all the plans and specifications. Concrete often appears on drawings other than structural drawings, including mechanical and electrical drawings for equipment pads. The cost of cutting and patching is often difficult to estimate. See Subdivision 03 81 for Concrete Cutting, Subdivision 02 41 19.16 for Cutout Demolition, Subdivision 03 05 05.10 for Concrete Demolition, and Subdivision 02 41 19.19 for Rubbish Handling (handling, loading, and hauling of debris).
- Always obtain concrete prices from suppliers near the job site. A volume discount can often be negotiated, depending upon competition in the area. Remember to add for waste, particularly for slabs and footings on grade.

03 10 00 Concrete Forming and Accessories

- A primary cost for concrete construction is forming. Most jobs today are constructed with prefabricated forms. The selection of the forms best suited for the job and the total square feet of forms required for efficient concrete forming and placing are key elements in estimating concrete construction. Enough forms must be available for erection to make efficient use of the concrete placing equipment and crew.
- Concrete accessories for forming and placing depend upon the systems used. Study the plans and specifications to ensure that all special accessory requirements have been included in the cost estimate, such as anchor bolts, inserts, and hangers.
- Included within costs for forms-in-place are all necessary bracing and shoring.

03 20 00 Concrete Reinforcing

- Ascertain that the reinforcing steel supplier has included all accessories, cutting, bending, and an allowance for lapping, splicing, and waste. A good rule of thumb is 10% for lapping, splicing, and waste. Also, 10% waste should be allowed for welded wire fabric.
- The unit price items in the subdivisions for Reinforcing In Place, Glass Fiber Reinforcing, and Welded Wire Fabric include the labor to install accessories such as beam and slab bolsters, high chairs, and bar ties and tie wire. The material cost for these accessories is not included; they may be obtained from the Accessories Subdivisions.

03 30 00 Cast-In-Place Concrete

- When estimating structural concrete, pay particular attention to requirements for concrete additives, curing methods, and surface treatments. Special consideration for climate, hot or cold, must be included in your estimate. Be sure to include requirements for concrete placing equipment and concrete finishing.
- For accurate concrete estimating, the estimator must consider each of the following major components individually: forms, reinforcing steel, ready-mix concrete, placement of the concrete, and finishing of the top surface. For faster estimating, Subdivision 03 30 53.40 for Concrete-In-Place can be used; here, various items of concrete work are presented that include the costs of all five major components (unless specifically stated otherwise).

03 40 00 Precast Concrete
03 50 00 Cast Decks and Underlayment

- The cost of hauling precast concrete structural members is often an important factor. For this reason, it is important to get a quote from the nearest supplier. It may become economically feasible to set up precasting beds on the site if the hauling costs are prohibitive.

Reference Numbers

Reference numbers are shown at the beginning of some major classifications. These numbers refer to related items in the Reference Section. The reference information may be an estimating procedure, an alternate pricing method, or technical information.

Note: Not all subdivisions listed here necessarily appear. ■

03 01 30 – Maintenance of Cast-In-Place Concrete

03 01 30.62 Concrete Patching

03 01 30.62 Concrete Patching		Crew	Daily Output	Labor-Hours	Unit	2020 Bare Costs Material	Labor	Equipment	Total	Total Incl O&P
0010	CONCRETE PATCHING									
0100	Floors, 1/4" thick, small areas, regular grout	1 Cefi	170	.047	S.F.	1.61	2.35		3.96	5.45
0150	Epoxy grout	"	100	.080	"	9.90	4		13.90	17.20
2000	Walls, including chipping, cleaning and epoxy grout									
2100	1/4" deep	1 Cefi	65	.123	S.F.	7.80	6.15		13.95	18.25
2150	1/2" deep		50	.160		15.60	8		23.60	29.50
2200	3/4" deep		40	.200		23.50	10		33.50	41
9000	Minimum labor/equipment charge		4.50	1.778	Job		89		89	139

03 01 30.71 Concrete Crack Repair

03 01 30.71 Concrete Crack Repair		Crew	Daily Output	Labor-Hours	Unit	Material	Labor	Equipment	Total	Total Incl O&P
0010	CONCRETE CRACK REPAIR									
1000	Structural repair of concrete cracks by epoxy injection (ACI RAP-1)									
1001	suitable for horizontal, vertical and overhead repairs									
1010	Clean/grind concrete surface(s) free of contaminants	1 Cefi	400	.020	L.F.		1		1	1.57
1015	Rout crack with v-notch crack chaser, if needed	C-32	600	.027		.03	1.23	.25	1.51	2.26
1020	Blow out crack with oil-free dry compressed air (1 pass)	C-28	3000	.003			.13	.01	.14	.22
1030	Install surface-mounted entry ports (spacing = concrete depth)	1 Cefi	400	.020	Ea.	1.71	1		2.71	3.45
1040	Cap crack at surface with epoxy gel (per side/face)		400	.020	L.F.	.31	1		1.31	1.92
1050	Snap off ports, grind off epoxy cap residue after injection		200	.040	"		2		2	3.14
1100	Manual injection with 2-part epoxy cartridge, excludes prep									
1110	Up to 1/32" (0.03125") wide x 4" deep	1 Cefi	160	.050	L.F.	.16	2.50		2.66	4.09
1120	6" deep		120	.067		.24	3.33		3.57	5.50
1130	8" deep		107	.075		.31	3.73		4.04	6.20
1140	10" deep		100	.080		.39	4		4.39	6.75
1150	12" deep		96	.083		.47	4.16		4.63	7.05
1210	Up to 1/16" (0.0625") wide x 4" deep		160	.050		.31	2.50		2.81	4.27
1220	6" deep		120	.067		.47	3.33		3.80	5.75
1230	8" deep		107	.075		.63	3.73		4.36	6.55
1240	10" deep		100	.080		.78	4		4.78	7.15
1250	12" deep		96	.083		.94	4.16		5.10	7.60
1310	Up to 3/32" (0.09375") wide x 4" deep		160	.050		.47	2.50		2.97	4.44
1320	6" deep		120	.067		.71	3.33		4.04	6.05
1330	8" deep		107	.075		.94	3.73		4.67	6.90
1340	10" deep		100	.080		1.18	4		5.18	7.60
1350	12" deep		96	.083		1.41	4.16		5.57	8.10
1410	Up to 1/8" (0.125") wide x 4" deep		160	.050		.63	2.50		3.13	4.61
1420	6" deep		120	.067		.94	3.33		4.27	6.30
1430	8" deep		107	.075		1.26	3.73		4.99	7.25
1440	10" deep		100	.080		1.57	4		5.57	8.05
1450	12" deep		96	.083		1.88	4.16		6.04	8.60
1500	Pneumatic injection with 2-part bulk epoxy, excludes prep									
1510	Up to 5/32" (0.15625") wide x 4" deep	C-31	240	.033	L.F.	.27	1.66	1.32	3.25	4.36
1520	6" deep		180	.044		.40	2.22	1.77	4.39	5.85
1530	8" deep		160	.050		.54	2.50	1.99	5.03	6.70
1540	10" deep		150	.053		.67	2.66	2.12	5.45	7.25
1550	12" deep		144	.056		.80	2.78	2.21	5.79	7.65
1610	Up to 3/16" (0.1875") wide x 4" deep		240	.033		.32	1.66	1.32	3.30	4.42
1620	6" deep		180	.044		.48	2.22	1.77	4.47	5.95
1630	8" deep		160	.050		.64	2.50	1.99	5.13	6.80
1640	10" deep		150	.053		.80	2.66	2.12	5.58	7.40
1650	12" deep		144	.056		.96	2.78	2.21	5.95	7.85
1710	Up to 1/4" (0.25") wide x 4" deep		240	.033		.43	1.66	1.32	3.41	4.54
1720	6" deep		180	.044		.64	2.22	1.77	4.63	6.15

03 01 Maintenance of Concrete

03 01 30 – Maintenance of Cast-In-Place Concrete

03 01 30.71 Concrete Crack Repair		Crew	Daily Output	Labor-Hours	Unit	2020 Bare Costs Material	Labor	Equipment	Total	Total Incl O&P
1730	8" deep	C-31	160	.050	L.F.	.86	2.50	1.99	5.35	7.05
1740	10" deep		150	.053		1.07	2.66	2.12	5.85	7.70
1750	12" deep		144	.056		1.29	2.78	2.21	6.28	8.20
2000	Non-structural filling of concrete cracks by gravity-fed resin (ACI RAP-2)									
2001	suitable for individual cracks in stable horizontal surfaces only									
2010	Clean/grind concrete surface(s) free of contaminants	1 Cefi	400	.020	L.F.		1		1	1.57
2020	Rout crack with v-notch crack chaser, if needed	C-32	600	.027		.03	1.23	.25	1.51	2.26
2030	Blow out crack with oil-free dry compressed air (1 pass)	C-28	3000	.003			.13	.01	.14	.22
2040	Cap crack with epoxy gel at underside of elevated slabs, if needed	1 Cefi	400	.020		.31	1		1.31	1.92
2050	Insert backer rod into crack, if needed		400	.020		.03	1		1.03	1.60
2060	Partially fill crack with fine dry sand, if needed		800	.010		.02	.50		.52	.80
2070	Apply two beads of sealant alongside crack to form reservoir, if needed		400	.020		.28	1		1.28	1.88
2100	Manual filling with squeeze bottle of 2-part epoxy resin									
2110	Full depth crack up to 1/16" (0.0625") wide x 4" deep	1 Cefi	300	.027	L.F.	.10	1.33		1.43	2.20
2120	6" deep		240	.033		.15	1.66		1.81	2.77
2130	8" deep		200	.040		.19	2		2.19	3.35
2140	10" deep		185	.043		.24	2.16		2.40	3.66
2150	12" deep		175	.046		.29	2.28		2.57	3.91
2210	Full depth crack up to 1/8" (0.125") wide x 4" deep		270	.030		.19	1.48		1.67	2.53
2220	6" deep		215	.037		.29	1.86		2.15	3.24
2230	8" deep		180	.044		.39	2.22		2.61	3.92
2240	10" deep		165	.048		.49	2.42		2.91	4.33
2250	12" deep		155	.052		.58	2.58		3.16	4.69
2310	Partial depth crack up to 3/16" (0.1875") wide x 1" deep		300	.027		.07	1.33		1.40	2.17
2410	Up to 1/4" (0.250") wide x 1" deep		270	.030		.10	1.48		1.58	2.43
2510	Up to 5/16" (0.3125") wide x 1" deep		250	.032		.12	1.60		1.72	2.64
2610	Up to 3/8" (0.375") wide x 1" deep		240	.033		.15	1.66		1.81	2.77

03 01 30.72 Concrete Surface Repairs

Line	Description	Crew	Daily Output	Labor-Hours	Unit	Material	Labor	Equipment	Total	Total Incl O&P
0010	**CONCRETE SURFACE REPAIRS**									
1000	Spall repairs using low-pressure spraying (ACI RAP-3)									
1001	Suitable for vertical and overhead repairs up to 3" thick									
1010	Sound the concrete surface to locate delaminated areas	1 Cefi	2000	.004	S.F.		.20		.20	.31
1020	Remove concrete in repair areas to fully expose reinforcing bars	B-9E	160	.100			4.60	.20	4.80	7.50
1030	Mark the perimeter of each repair area	1 Cefi	2000	.004			.20		.20	.31
1040	Saw cut the perimeter of each repair area down to reinforcing bars	B-89C	160	.050	L.F.	.04	2.50	.38	2.92	4.38
1050	If corroded rebar is exposed, remove concrete to 3/4" below corroded rebars									
1051	Single layer of #4 rebar	B-9E	64	.250	S.F.		11.50	.51	12.01	18.80
1052	#5 rebar		53.30	.300			13.80	.61	14.41	22.50
1053	#6 rebar		45.70	.350			16.10	.71	16.81	26.50
1054	#7 rebar		40	.400			18.40	.81	19.21	30
1055	#8 rebar		35.60	.449			20.50	.91	21.41	34
1056	Double layer of #4 rebar		35.60	.449			20.50	.91	21.41	34
1057	#5 rebar		32	.500			23	1.02	24.02	37.50
1058	#6 rebar		29.10	.550			25.50	1.12	26.62	41
1059	#7 rebar		26.70	.599			27.50	1.22	28.72	45
1060	#8 rebar		24.60	.650			30	1.32	31.32	49
1100	See section 03 21 11.60 for placement of supplemental reinforcement bars									
1104	See section 02 41 19.19 for handling & removal of debris									
1106	See section 02 41 19.20 for disposal of debris									
1110	Final cleaning by sandblasting	E-11	1100	.029	S.F.	.51	1.36	.27	2.14	3.11
1120	By high pressure water	C-29	500	.016			.67	.19	.86	1.29
1130	Blow off dust and debris with oil-free dry compressed air	C-28	16000	.001			.03		.03	.04

03 01 Maintenance of Concrete

03 01 30 – Maintenance of Cast-In-Place Concrete

03 01 30.72 Concrete Surface Repairs		Crew	Daily Output	Labor-Hours	Unit	2020 Bare Costs Material	Labor	Equipment	Total	Total Incl O&P
1140	Mix up repair material with concrete mixer	C-30	200	.040	C.F.	7.65	1.68	.70	10.03	11.85
1150	Place repair material by low-pressure spray in lifts up to 3″ thick	C-9	200	.160	″		7.50	2.32	9.82	14.40
1160	Screed and initial float of repair material by hand	C-10	2000	.012	S.F.		.57		.57	.90
1170	Float and steel trowel finish repair material by hand	″	1265	.019			.90		.90	1.42
1180	Cure with sprayed membrane curing compound	2 Cefi	6200	.003		.12	.13		.25	.34
2000	Surface repairs using form-and-pour techniques (ACI RAP-4)									
2001	Suitable for walls, columns, beam sides/bottoms, and slab soffits									
2010	Sound the concrete surface to locate delaminated areas	1 Cefi	2000	.004	S.F.		.20		.20	.31
2020	Remove concrete in repair areas to fully expose reinforcing bars	B-9E	160	.100			4.60	.20	4.80	7.50
2030	Mark the perimeter of each repair area	1 Cefi	2000	.004			.20		.20	.31
2040	Saw cut the perimeter of each repair area down to reinforcing bars	B-89C	160	.050	L.F.	.04	2.50	.38	2.92	4.38
2050	If corroded rebar is exposed, remove concrete to 3/4″ below corroded rebars									
2051	Single layer of #4 rebar	B-9E	64	.250	S.F.		11.50	.51	12.01	18.80
2052	#5 rebar		53.30	.300			13.80	.61	14.41	22.50
2053	#6 rebar		45.70	.350			16.10	.71	16.81	26.50
2054	#7 rebar		40	.400			18.40	.81	19.21	30
2055	#8 rebar		35.60	.449			20.50	.91	21.41	34
2056	Double layer of #4 rebar		35.60	.449			20.50	.91	21.41	34
2057	#5 rebar		32	.500			23	1.02	24.02	37.50
2058	#6 rebar		29.10	.550			25.50	1.12	26.62	41
2059	#7 rebar		26.70	.599			27.50	1.22	28.72	45
2060	#8 rebar		24.60	.650			30	1.32	31.32	49
2100	See section 03 21 11.60 for placement of supplemental reinforcement bars									
2105	See section 02 41 19.19 for handling & removal of debris									
2110	See section 02 41 19.20 for disposal of debris									
2120	For slab soffit repairs, drill 2″ diam. holes to 6″ deep for conc placement	B-89A	16.50	.970	Ea.	.26	47	6.90	54.16	83
2130	For each additional inch of slab thickness in same hole, add	″	1080	.015		.04	.72	.11	.87	1.32
2140	Drill holes for expansion shields, 3/4″ diameter x 4″ deep	1 Carp	45	.178			9.45		9.45	15.15
2150	Final cleaning by sandblasting	E-11	1100	.029	S.F.	.51	1.36	.27	2.14	3.11
2160	By high pressure water	C-29	500	.016			.67	.19	.86	1.29
2170	Blow off dust and debris with oil-free dry compressed air	C-28	16000	.001			.03		.03	.04
2180	Build wood forms including chutes where needed for concrete placement	C-2	960	.050	SFCA	3.19	2.58		5.77	7.65
2190	Insert expansion shield, coil rod, coil tie and plastic cone in each hole	1 Carp	32	.250	Ea.	4.88	13.30		18.18	27
2200	Install wood forms and fasten with coil rod and coil nut	C-2	960	.050	SFCA	.48	2.58		3.06	4.67
2210	Mix up bagged repair material by hand with a mixing paddle	1 Cefi	30	.267	C.F.	7.65	13.30		20.95	29.50
2220	By concrete mixer	C-30	200	.040		7.65	1.68	.70	10.03	11.85
2230	Place repair material by hand and consolidate	C-6A	200	.080			4	.24	4.24	6.55
2240	Remove wood forms	C-2	1600	.030	SFCA		1.55		1.55	2.48
2250	Chip/grind off extra concrete at fill chutes	B-9E	16	1	Ea.	.12	46	2.03	48.15	75.50
2260	Break ties and patch voids	1 Cefi	270	.030	″	.09	1.48		1.57	2.42
2270	Cure with sprayed membrane curing compound	2 Cefi	6200	.003	S.F.	.12	.13		.25	.34
3000	Surface repairs using form-and-pump techniques (ACI RAP-5)									
3001	Suitable for walls, columns, beam sides/bottoms, and slab soffits									
3010	Sound the concrete surface to locate delaminated areas	1 Cefi	2000	.004	S.F.		.20		.20	.31
3020	Remove concrete in repair areas to fully expose reinforcing bars	B-9E	160	.100			4.60	.20	4.80	7.50
3030	Mark the perimeter of each repair area	1 Cefi	2000	.004			.20		.20	.31
3040	Saw cut the perimeter of each repair area down to reinforcing bars	B-89C	160	.050	L.F.	.04	2.50	.38	2.92	4.38
3050	If corroded rebar is exposed, remove concrete to 3/4″ below corroded rebars									
3051	Single layer of #4 rebar	B-9E	64	.250	S.F.		11.50	.51	12.01	18.80
3052	#5 rebar		53.30	.300			13.80	.61	14.41	22.50
3053	#6 rebar		45.70	.350			16.10	.71	16.81	26.50
3054	#7 rebar		40	.400			18.40	.81	19.21	30
3055	#8 rebar		35.60	.449			20.50	.91	21.41	34

03 01 30.72 Concrete Surface Repairs		Crew	Daily Output	Labor-Hours	Unit	Material	2020 Bare Costs Labor	Equipment	Total	Total Incl O&P
3056	Double layer of #4 rebar	B-9E	35.60	.449	S.F.		20.50	.91	21.41	34
3057	#5 rebar		32	.500			23	1.02	24.02	37.50
3058	#6 rebar		29.10	.550			25.50	1.12	26.62	41
3059	#7 rebar		26.70	.599			27.50	1.22	28.72	45
3060	#8 rebar		24.60	.650			30	1.32	31.32	49
3100	See section 03 21 11.60 for placement of supplemental reinforcement bars									
3104	See section 02 41 19.19 for handling & removal of debris									
3106	See section 02 41 19.20 for disposal of debris									
3110	Drill holes for expansion shields, 3/4" diameter x 4" deep	1 Carp	45	.178	Ea.		9.45		9.45	15.15
3120	Final cleaning by sandblasting	E-11	1100	.029	S.F.	.51	1.36	.27	2.14	3.11
3130	By high pressure water	C-29	500	.016			.67	.19	.86	1.29
3140	Blow off dust and debris with oil-free dry compressed air	C-28	16000	.001			.03		.03	.04
3150	Build wood forms and drill holes for injection ports	C-2	960	.050	SFCA	3.19	2.58		5.77	7.65
3160	Assemble PVC injection ports with ball valves and install in forms	1 Plum	8	1	Ea.	91.50	64.50		156	201
3170	Insert expansion shield, coil rod, coil tie and plastic cone in each hole	1 Carp	32	.250	"	4.88	13.30		18.18	27
3180	Install wood forms and fasten with coil rod and coil nut	C-2	960	.050	SFCA	.48	2.58		3.06	4.67
3190	Mix up self-consolidating repair material with concrete mixer	C-30	200	.040	C.F.	7.85	1.68	.70	10.23	12.10
3200	Place repair material by pump and pressurize	C-9	200	.160	"		7.50	2.32	9.82	14.40
3210	Remove wood forms	C-2	1600	.030	SFCA		1.55		1.55	2.48
3220	Chip/grind off extra concrete at injection ports	B-9E	64	.250	Ea.	.12	11.50	.51	12.13	18.95
3230	Break ties and patch voids	1 Cefi	270	.030	"	.09	1.48		1.57	2.42
3240	Cure with sprayed membrane curing compound	2 Cefi	6200	.003	S.F.	.12	.13		.25	.34
4000	Vertical & overhead spall repair by hand application (ACI RAP-6)									
4001	Generally recommended for thin repairs that are cosmetic in nature									
4010	Remove concrete in repair areas to fully expose reinforcing bars	B-9E	160	.100	S.F.		4.60	.20	4.80	7.50
4020	Saw cut the perimeter of each repair area to 1/2" depth	B-89C	160	.050	L.F.	.02	2.50	.38	2.90	4.36
4030	If corroded rebar is exposed, remove concrete to 3/4" below corroded rebars									
4031	Single layer of #4 rebar	B-9E	64	.250	S.F.		11.50	.51	12.01	18.80
4032	#5 rebar		53.30	.300			13.80	.61	14.41	22.50
4033	#6 rebar		45.70	.350			16.10	.71	16.81	26.50
4034	#7 rebar		40	.400			18.40	.81	19.21	30
4035	#8 rebar		35.60	.449			20.50	.91	21.41	34
4036	Double layer of #4 rebar		35.60	.449			20.50	.91	21.41	34
4037	#5 rebar		32	.500			23	1.02	24.02	37.50
4038	#6 rebar		29.10	.550			25.50	1.12	26.62	41
4039	#7 rebar		26.70	.599			27.50	1.22	28.72	45
4040	#8 rebar		24.60	.650			30	1.32	31.32	49
4080	See section 03 21 11.60 for placement of supplemental reinforcement bars									
4090	See section 02 41 19.19 for handling & removal of debris									
4095	See section 02 41 19.20 for disposal of debris									
4100	Final cleaning by sandblasting	E-11	1100	.029	S.F.	.51	1.36	.27	2.14	3.11
4110	By high pressure water	C-29	500	.016			.67	.19	.86	1.29
4120	Blow off dust and debris with oil-free dry compressed air	C-28	16000	.001			.03		.03	.04
4130	Brush on bonding material	1 Cefi	320	.025		.11	1.25		1.36	2.08
4140	Mix up bagged repair material by hand with a mixing paddle	"	30	.267	C.F.	77.50	13.30		90.80	106
4150	By concrete mixer	C-30	200	.040		77.50	1.68	.70	79.88	88.50
4160	Place repair material by hand in lifts up to 2" deep for vertical repairs	C-10	48	.500			23.50		23.50	37.50
4170	Up to 1" deep for overhead repairs		36	.667			31.50		31.50	50
4180	Screed and initial float of repair material by hand		2000	.012	S.F.		.57		.57	.90
4190	Float and steel trowel finish repair material by hand		1265	.019			.90		.90	1.42
4200	Cure with sprayed membrane curing compound	2 Cefi	6200	.003		.12	.13		.25	.34
5000	Spall repair of horizontal concrete surfaces (ACI RAP-7)									
5001	Suitable for structural slabs, slabs-on-grade, balconies, interior floors									

03 01 30.72 Concrete Surface Repairs		Crew	Daily Output	Labor-Hours	Unit	Material	Labor	Equipment	Total	Total Incl O&P
						2020 Bare Costs				
5010	Sound the concrete surface to locate delaminated areas	1 Cefi	2000	.004	SF Flr.		.20		.20	.31
5020	Mark the perimeter of each repair area	"	2000	.004	"		.20		.20	.31
5030	Saw cut the perimeter of each repair area down to reinforcing bars	B-89	1060	.015	L.F.	.04	.76	.98	1.78	2.31
5040	Remove concrete in repair areas to fully expose reinforcing bars	B-9E	160	.100	S.F.		4.60	.20	4.80	7.50
5050	If corroded rebar is exposed, remove concrete to 3/4" below corroded rebars									
5051	Single layer of #4 rebar	B-9E	64	.250	S.F.		11.50	.51	12.01	18.80
5052	#5 rebar		53.30	.300			13.80	.61	14.41	22.50
5053	#6 rebar		45.70	.350			16.10	.71	16.81	26.50
5054	#7 rebar		40	.400			18.40	.81	19.21	30
5055	#8 rebar		35.60	.449			20.50	.91	21.41	34
5056	Double layer of #4 rebar		35.60	.449			20.50	.91	21.41	34
5057	#5 rebar		32	.500			23	1.02	24.02	37.50
5058	#6 rebar		29.10	.550			25.50	1.12	26.62	41
5059	#7 rebar		26.70	.599			27.50	1.22	28.72	45
5060	#8 rebar		24.60	.650			30	1.32	31.32	49
5100	See section 03 21 11.60 for placement of supplemental reinforcement bars									
5110	See section 02 41 19.19 for handling & removal of debris									
5115	See section 02 41 19.20 for disposal of debris									
5120	Final cleaning by sandblasting	E-11	1100	.029	S.F.	.51	1.36	.27	2.14	3.11
5130	By high pressure water	C-29	500	.016			.67	.19	.86	1.29
5140	Blow off dust and debris with oil-free dry compressed air	C-28	16000	.001			.03		.03	.04
5160	Brush on bonding material	1 Cefi	320	.025		.11	1.25		1.36	2.08
5170	Mix up bagged repair material by hand with a mixing paddle	"	30	.267	C.F.	77.50	13.30		90.80	106
5180	By concrete mixer	C-30	200	.040		77.50	1.68	.70	79.88	88.50
5190	Place repair material by hand, consolidate and screed	C-6A	200	.080			4	.24	4.24	6.55
5200	Bull float, manual float & broom finish	C-10	1850	.013	S.F.		.61		.61	.97
5210	Bull float, manual float & trowel finish	"	1265	.019			.90		.90	1.42
5220	Cure with sprayed membrane curing compound	2 Cefi	6200	.003		.12	.13		.25	.34
6000	Spall repairs by the Preplaced Aggregate method (ACI RAP-9)									
6001	Suitable for vertical, overhead, and underwater repairs									
6010	Sound the concrete surface to locate delaminated areas	1 Cefi	2000	.004	S.F.		.20		.20	.31
6020	Remove concrete in repair areas to fully expose reinforcing bars	B-9E	160	.100			4.60	.20	4.80	7.50
6030	Mark the perimeter of each repair area	1 Cefi	2000	.004			.20		.20	.31
6040	Saw cut the perimeter of each repair area down to reinforcing bars	B-89C	160	.050	L.F.	.04	2.50	.38	2.92	4.38
6050	If corroded rebar is exposed, remove concrete to 3/4" below corroded rebars									
6051	Single layer of #4 rebar	B-9E	64	.250	S.F.		11.50	.51	12.01	18.80
6052	#5 rebar		53.30	.300			13.80	.61	14.41	22.50
6053	#6 rebar		45.70	.350			16.10	.71	16.81	26.50
6054	#7 rebar		40	.400			18.40	.81	19.21	30
6055	#8 rebar		35.60	.449			20.50	.91	21.41	34
6056	Double layer of #4 rebar		35.60	.449			20.50	.91	21.41	34
6057	#5 rebar		32	.500			23	1.02	24.02	37.50
6058	#6 rebar		29.10	.550			25.50	1.12	26.62	41
6059	#7 rebar		26.70	.599			27.50	1.22	28.72	45
6060	#8 rebar		24.60	.650			30	1.32	31.32	49
6100	See section 03 21 11.60 for placement of supplemental reinforcement bars									
6104	See section 02 41 19.19 for handling & removal of debris									
6106	See section 02 41 19.20 for disposal of debris									
6110	Drill holes for expansion shields, 3/4" diameter x 4" deep	1 Carp	45	.178	Ea.		9.45		9.45	15.15
6120	Final cleaning by sandblasting	E-11	1100	.029	S.F.	.51	1.36	.27	2.14	3.11
6130	By high pressure water	C-29	500	.016			.67	.19	.86	1.29
6140	Blow off dust and debris with oil-free dry compressed air	C-28	16000	.001			.03		.03	.04
6150	Build wood forms and drill holes for injection ports	C-2	960	.050	SFCA	3.19	2.58		5.77	7.65

03 01 30 – Maintenance of Cast-In-Place Concrete

03 01 30.72 Concrete Surface Repairs		Crew	Daily Output	Labor-Hours	Unit	2020 Bare Costs Material	Labor	Equipment	Total	Total Incl O&P
6160	Assemble PVC injection ports with ball valves and install in forms	1 Plum	8	1	Ea.	91.50	64.50		156	201
6170	Insert expansion shield, coil rod, coil tie and plastic cone in each hole	1 Carp	32	.250	"	4.88	13.30		18.18	27
6180	Install wood forms and fasten with coil rod and coil nut	C-2	960	.050	SFCA	.48	2.58		3.06	4.67
6185	Install pre-washed gap-graded aggregate as forms are installed	B-24	160	.150	C.F.	1.20	7.25		8.45	12.85
6190	Mix up pre-bagged grout material with concrete mixer	C-30	200	.040		16.25	1.68	.70	18.63	21.50
6200	Place grout by pump from bottom up	C-9	200	.160			7.50	2.32	9.82	14.40
6210	Remove wood forms	C-2	1600	.030	SFCA		1.55		1.55	2.48
6220	Chip/grind off extra concrete at injection ports	B-9E	64	.250	Ea.	.12	11.50	.51	12.13	18.95
6230	Break ties and patch voids	1 Cefi	270	.030	"	.09	1.48		1.57	2.42
6240	Cure with sprayed membrane curing compound	2 Cefi	6200	.003	S.F.	.12	.13		.25	.34
7000	Leveling & reprofiling of vertical & overhead surfaces (ACI RAP-10)									
7001	Used to provide an acceptable surface for aesthetic or protective coatings									
7010	Clean the surface by sandblasting, leaving an open profile	E-11	1100	.029	S.F.	.51	1.36	.27	2.14	3.11
7020	By high pressure water	C-29	500	.016			.67	.19	.86	1.29
7030	Blow off dust and debris with oil-free dry compressed air	C-28	16000	.001			.03		.03	.04
7040	Mix up bagged repair material by hand with a mixing paddle	1 Cefi	30	.267	C.F.	77.50	13.30		90.80	106
7050	Place repair material by hand with steel trowel to fill pores & honeycombs	C-10	1265	.019	S.F.		.90		.90	1.42
7060	Cure with sprayed membrane curing compound	2 Cefi	6200	.003	"	.12	.13		.25	.34
8000	Concrete repair by shotcrete application (ACI RAP-12)									
8001	Suitable for vertical and overhead surfaces									
8010	Sound the concrete surface to locate delaminated areas	1 Cefi	2000	.004	S.F.		.20		.20	.31
8020	Remove concrete in repair areas to fully expose reinforcing bars	B-9E	160	.100			4.60	.20	4.80	7.50
8030	Mark the perimeter of each repair area	1 Cefi	2000	.004			.20		.20	.31
8040	Saw cut the perimeter of each repair area down to reinforcing bars	B-89C	160	.050	L.F.	.04	2.50	.38	2.92	4.38
8050	If corroded rebar is exposed, remove concrete to 3/4" below corroded rebars									
8051	Single layer of #4 rebar	B-9E	64	.250	S.F.		11.50	.51	12.01	18.80
8052	#5 rebar		53.30	.300			13.80	.61	14.41	22.50
8053	#6 rebar		45.70	.350			16.10	.71	16.81	26.50
8054	#7 rebar		40	.400			18.40	.81	19.21	30
8055	#8 rebar		35.60	.449			20.50	.91	21.41	34
8056	Double layer of #4 rebar		35.60	.449			20.50	.91	21.41	34
8057	#5 rebar		32	.500			23	1.02	24.02	37.50
8058	#6 rebar		29.10	.550			25.50	1.12	26.62	41
8059	#7 rebar		26.70	.599			27.50	1.22	28.72	45
8060	#8 rebar		24.60	.650			30	1.32	31.32	49
8100	See section 03 21 11.60 for placement of supplemental reinforcement bars									
8110	Final cleaning by sandblasting	E-11	1100	.029	S.F.	.51	1.36	.27	2.14	3.11
8120	By high pressure water	C-29	500	.016			.67	.19	.86	1.29
8130	Blow off dust and debris with oil-free dry compressed air	C-28	16000	.001			.03		.03	.04
8140	Mix up repair material with concrete mixer	C-30	200	.040	C.F.	7.65	1.68	.70	10.03	11.85
8150	Place repair material by wet shotcrete process to full depth of repair	C-8C	200	.240	"		11.10	2.98	14.08	21
8160	Carefully trim excess material and steel trowel finish without overworking	C-10	1265	.019	S.F.		.90		.90	1.42
8170	Cure with sprayed membrane curing compound	2 Cefi	6200	.003	"	.12	.13		.25	.34
9000	Surface repair by Methacrylate flood coat (ACI RAP-13)									
9010	Suitable for healing and sealing horizontal surfaces only									
9020	Large cracks must previously have been repaired or filled									
9100	Shotblast entire surface to remove contaminants	A-1A	4000	.002	S.F.		.11	.05	.16	.23
9200	Blow off dust and debris with oil-free dry compressed air	C-28	16000	.001			.03		.03	.04
9300	Flood coat surface w/Methacrylate, distribute w/broom/squeegee, no prep.	3 Cefi	8000	.003		1.03	.15		1.18	1.37
9400	Lightly broadcast even coat of dry silica sand while sealer coat is tacky	1 Cefi	8000	.001		.02	.05		.07	.10

03 05 Common Work Results for Concrete

03 05 05 – Selective Demolition for Concrete

03 05 05.10 Selective Demolition, Concrete		Crew	Daily Output	Labor-Hours	Unit	Material	2020 Bare Costs Labor	Equipment	Total	Total Incl O&P
0010	**SELECTIVE DEMOLITION, CONCRETE** R024119-10									
0012	Excludes saw cutting, torch cutting, loading or hauling									
0050	Break into small pieces, reinf. less than 1% of cross-sectional area	B-9	24	1.667	C.Y.		71	14.90	85.90	129
0060	Reinforcing 1% to 2% of cross-sectional area		16	2.500			106	22.50	128.50	195
0070	Reinforcing more than 2% of cross-sectional area		8	5			213	44.50	257.50	390
0150	Remove whole pieces, up to 2 tons per piece	E-18	36	1.111	Ea.		64.50	41.50	106	150
0160	2-5 tons per piece		30	1.333			77	49.50	126.50	180
0170	5-10 tons per piece		24	1.667			96.50	62	158.50	225
0180	10-15 tons per piece		18	2.222			129	82.50	211.50	300
0250	Precast unit embedded in masonry, up to 1 C.F.	D-1	16	1			47		47	76
0260	1-2 C.F.		12	1.333			62.50		62.50	101
0270	2-5 C.F.		10	1.600			75		75	121
0280	5-10 C.F.		8	2			94		94	152
0990	For hydrodemolition see Section 02 41 13.15									
1910	Minimum labor/equipment charge	B-9	2	20	Job		850	179	1,029	1,550

03 05 13 – Basic Concrete Materials

03 05 13.20 Concrete Admixtures and Surface Treatments		Crew	Daily Output	Labor-Hours	Unit	Material	Labor	Equipment	Total	Total Incl O&P
0010	**CONCRETE ADMIXTURES AND SURFACE TREATMENTS**									
0040	Abrasives, aluminum oxide, over 20 tons				Lb.	2.51			2.51	2.76
0050	1 to 20 tons					2.61			2.61	2.87
0070	Under 1 ton					2.71			2.71	2.98
0100	Silicon carbide, black, over 20 tons					4.36			4.36	4.80
0110	1 to 20 tons					4.53			4.53	4.99
0120	Under 1 ton					4.71			4.71	5.20
0200	Air entraining agent, .7 to 1.5 oz. per bag, 55 gallon drum				Gal.	24.50			24.50	27
0220	5 gallon pail					36.50			36.50	40
0300	Bonding agent, acrylic latex, 250 S.F./gallon, 5 gallon pail					28			28	31
0320	Epoxy resin, 80 S.F./gallon, 4 gallon case					69			69	76
0400	Calcium chloride, 50 lb. bags, T.L. lots				Ton	1,525			1,525	1,675
0420	Less than truckload lots				Bag	24.50			24.50	27
0500	Carbon black, liquid, 2 to 8 lb. per bag of cement				Lb.	17			17	18.70
0600	Concrete admixture, integral colors, dry pigment, 5 lb. bag				Ea.	31			31	34.50
0610	10 lb. bag					38.50			38.50	42
0620	25 lb. bag					78.50			78.50	86
0920	Dustproofing compound, 250 S.F./gal., 5 gallon pail				Gal.	7.70			7.70	8.50
1010	Epoxy based, 125 S.F./gal., 5 gallon pail				"	63.50			63.50	69.50
1100	Hardeners, metallic, 55 lb. bags, natural (grey)				Lb.	.92			.92	1.01
1200	Colors					1.43			1.43	1.57
1300	Non-metallic, 55 lb. bags, natural grey					.74			.74	.82
1320	Colors					.91			.91	1
1550	Release agent, for tilt slabs, 5 gallon pail				Gal.	18.05			18.05	19.90
1570	For forms, 5 gallon pail					13.35			13.35	14.65
1590	Concrete release agent for forms, 100% biodegradable, zero VOC, 5 gal. pail G					23.50			23.50	26
1595	55 gallon drum G					19.30			19.30	21
1600	Sealer, hardener and dustproofer, epoxy-based, 125 S.F./gal., 5 gallon unit					63.50			63.50	69.50
1620	3 gallon unit					69.50			69.50	76.50
1630	Sealer, solvent-based, 250 S.F./gal., 55 gallon drum					23			23	25.50
1640	5 gallon pail					33			33	36
1650	Sealer, water based, 350 S.F., 55 gallon drum					23.50			23.50	26
1660	5 gallon pail					26			26	29
1900	Set retarder, 100 S.F./gal., 1 gallon pail					6.95			6.95	7.65
2000	Waterproofing, integral 1 lb. per bag of cement				Lb.	4.02			4.02	4.42

03 05 Common Work Results for Concrete

03 05 13 – Basic Concrete Materials

03 05 13.20 Concrete Admixtures and Surface Treatments		Crew	Daily Output	Labor-Hours	Unit	2020 Bare Costs Material	Labor	Equipment	Total	Total Incl O&P
2100	Powdered metallic, 40 lbs. per 100 S.F., standard colors				Lb.	4.77			4.77	5.25
2120	Premium colors					6.70			6.70	7.35
3000	For colored ready-mix concrete, add to prices in section 03 31 13.35									
3100	Subtle shades, 5 lb. dry pigment per C.Y., add				C.Y.	31			31	34.50
3400	Medium shades, 10 lb. dry pigment per C.Y., add					38.50			38.50	42
3700	Deep shades, 25 lb. dry pigment per C.Y., add					78.50			78.50	86
6000	Concrete ready mix additives, recycled coal fly ash, mixed at plant G				Ton	64			64	70
6010	Recycled blast furnace slag, mixed at plant G				″	90.50			90.50	99.50

03 05 13.25 Aggregate

		Crew	Daily Output	Labor-Hours	Unit	Material	Labor	Equipment	Total	Total Incl O&P
0010	**AGGREGATE**									
0100	Lightweight vermiculite or perlite, 4 C.F. bag, C.L. lots G				Bag	22			22	24.50
0150	L.C.L. lots G				″	24.50			24.50	27
0250	Sand & stone, loaded at pit, crushed bank gravel				Ton	19.85			19.85	22
0350	Sand, washed, for concrete					21.50			21.50	24
0400	For plaster or brick					21.50			21.50	24
0450	Stone, 3/4″ to 1-1/2″					20.50			20.50	22.50
0470	Round, river stone					41			41	45
0500	3/8″ roofing stone & 1/2″ pea stone					31.50			31.50	34.50
0550	For trucking 10-mile round trip, add to the above	B-34B	117	.068			3.35	4.90	8.25	10.75
0600	For trucking 30-mile round trip, add to the above	″	72	.111			5.45	7.95	13.40	17.45
0850	Sand & stone, loaded at pit, crushed bank gravel				C.Y.	28			28	30.50
0950	Sand, washed, for concrete					30			30	33
1000	For plaster or brick					30			30	33
1050	Stone, 3/4″ to 1-1/2″					39			39	43
1055	Round, river stone					44			44	48.50
1100	3/8″ roofing stone & 1/2″ pea stone					30			30	33
1150	For trucking 10-mile round trip, add to the above	B-34B	78	.103			5	7.35	12.35	16.15
1200	For trucking 30-mile round trip, add to the above	″	48	.167			8.15	11.95	20.10	26
1310	Onyx chips, 50 lb. bags				Cwt.	34			34	37
1330	Quartz chips, 50 lb. bags					34.50			34.50	37.50
1410	White marble, 3/8″ to 1/2″, 50 lb. bags					9.55			9.55	10.50
1430	3/4″, bulk				Ton	189			189	207

03 05 13.30 Cement

		Crew	Daily Output	Labor-Hours	Unit	Material	Labor	Equipment	Total	Total Incl O&P
0010	**CEMENT**									
0240	Portland, Type I/II, T.L. lots, 94 lb. bags				Bag	14.45			14.45	15.85
0250	L.T.L./L.C.L. lots				″	15.95			15.95	17.55
0300	Trucked in bulk, per cwt.				Cwt.	8.70			8.70	9.55
0400	Type III, high early strength, T.L. lots, 94 lb. bags				Bag	13.85			13.85	15.25
0420	L.T.L. or L.C.L. lots					19.30			19.30	21.50
0500	White, type III, high early strength, T.L. or C.L. lots, bags					27.50			27.50	30.50
0520	L.T.L. or L.C.L. lots					38.50			38.50	42.50
0600	White, type I, T.L. or C.L. lots, bags					30			30	33.50
0620	L.T.L. or L.C.L. lots					32			32	35

03 05 13.80 Waterproofing and Dampproofing

		Crew	Daily Output	Labor-Hours	Unit	Material	Labor	Equipment	Total	Total Incl O&P
0010	**WATERPROOFING AND DAMPPROOFING**									
0050	Integral waterproofing, add to cost of regular concrete				C.Y.	24			24	26.50

03 05 13.85 Winter Protection

		Crew	Daily Output	Labor-Hours	Unit	Material	Labor	Equipment	Total	Total Incl O&P
0010	**WINTER PROTECTION**									
0012	For heated ready mix, add				C.Y.	5.35			5.35	5.90
0100	Temporary heat to protect concrete, 24 hours	2 Clab	50	.320	M.S.F.	203	13.45		216.45	245
0200	Temporary shelter for slab on grade, wood frame/polyethylene sheeting									
0201	Build or remove, light framing for short spans	2 Carp	10	1.600	M.S.F.	345	85		430	515

03 05 Common Work Results for Concrete

03 05 13 – Basic Concrete Materials

03 05 13.85 Winter Protection		Crew	Daily Output	Labor-Hours	Unit	2020 Bare Costs Material	Labor	Equipment	Total	Total Incl O&P
0210	Large framing for long spans	2 Carp	3	5.333	M.S.F.	460	283		743	965
0500	Electrically heated pads, 110 volts, 15 watts/S.F., buy				S.F.	11.60			11.60	12.80
0600	20 watts/S.F., buy					15.45			15.45	17
0710	Electrically heated pads, 15 watts/S.F., 20 uses					.58			.58	.64

03 11 Concrete Forming

03 11 13 – Structural Cast-In-Place Concrete Forming

03 11 13.20 Forms In Place, Beams and Girders

		Crew	Daily Output	Labor-Hours	Unit	Material	Labor	Equipment	Total	Total Incl O&P
0010	**FORMS IN PLACE, BEAMS AND GIRDERS**									
0500	Exterior spandrel, job-built plywood, 12" wide, 1 use R031113-60	C-2	225	.213	SFCA	3.30	11		14.30	21.50
0550	2 use		275	.175		1.73	9		10.73	16.35
0600	3 use		295	.163		1.32	8.40		9.72	14.90
0650	4 use		310	.155		1.07	8		9.07	14
1000	18" wide, 1 use		250	.192		2.91	9.90		12.81	19.10
1050	2 use		275	.175		1.60	9		10.60	16.20
1100	3 use		305	.157		1.16	8.15		9.31	14.30
1150	4 use		315	.152		.95	7.85		8.80	13.65
1500	24" wide, 1 use		265	.181		2.65	9.35		12	17.90
1550	2 use		290	.166		1.49	8.55		10.04	15.35
1600	3 use		315	.152		1.06	7.85		8.91	13.75
1650	4 use		325	.148		.86	7.65		8.51	13.15
2000	Interior beam, job-built plywood, 12" wide, 1 use		300	.160		4.02	8.25		12.27	17.65
2050	2 use		340	.141		1.94	7.30		9.24	13.85
2100	3 use		364	.132		1.61	6.80		8.41	12.65
2150	4 use		377	.127		1.31	6.55		7.86	12
2500	24" wide, 1 use		320	.150		2.71	7.75		10.46	15.40
2550	2 use		365	.132		1.52	6.80		8.32	12.60
2600	3 use		385	.125		1.07	6.45		7.52	11.50
2650	4 use		395	.122		.87	6.30		7.17	11
3000	Encasing steel beam, hung, job-built plywood, 1 use		325	.148		3.24	7.65		10.89	15.75
3050	2 use		390	.123		1.78	6.35		8.13	12.15
3100	3 use		415	.116		1.30	5.95		7.25	11
3150	4 use		430	.112		1.05	5.75		6.80	10.40
3500	Bottoms only, to 30" wide, job-built plywood, 1 use		230	.209		4.49	10.80		15.29	22
3550	2 use		265	.181		2.51	9.35		11.86	17.75
3600	3 use		280	.171		1.79	8.85		10.64	16.15
3650	4 use		290	.166		1.46	8.55		10.01	15.30
4000	Sides only, vertical, 36" high, job-built plywood, 1 use		335	.143		5.95	7.40		13.35	18.40
4050	2 use		405	.119		3.28	6.10		9.38	13.40
4100	3 use		430	.112		2.39	5.75		8.14	11.90
4150	4 use		445	.108		1.94	5.55		7.49	11.10
4500	Sloped sides, 36" high, 1 use		305	.157		5.70	8.15		13.85	19.25
4550	2 use		370	.130		3.17	6.70		9.87	14.25
4600	3 use		405	.119		2.28	6.10		8.38	12.30
4650	4 use		425	.113		1.85	5.85		7.70	11.40
5000	Upstanding beams, 36" high, 1 use		225	.213		7.45	11		18.45	26
5050	2 use		255	.188		4.13	9.70		13.83	20
5100	3 use		275	.175		3	9		12	17.75
5150	4 use		280	.171		2.43	8.85		11.28	16.90
9000	Minimum labor/equipment charge	2 Carp	2	8	Job		425		425	680

03 11 13 – Structural Cast-In-Place Concrete Forming

03 11 13.25 Forms In Place, Columns		Crew	Daily Output	Labor-Hours	Unit	2020 Bare Costs Material	Labor	Equipment	Total	Total Incl O&P
0010	**FORMS IN PLACE, COLUMNS**									
0500	Round fiberglass, 4 use per mo., rent, 12" diameter	C-1	160	.200	L.F.	9.95	10.10		20.05	27
0550	16" diameter R031113-60		150	.213		11.85	10.75		22.60	30
0600	18" diameter		140	.229		13.25	11.50		24.75	33
0650	24" diameter		135	.237		16.50	11.95		28.45	37.50
0700	28" diameter		130	.246		18.45	12.40		30.85	40.50
0800	30" diameter		125	.256		19.25	12.90		32.15	41.50
0850	36" diameter		120	.267		25.50	13.45		38.95	49.50
1500	Round fiber tube, recycled paper, 1 use, 8" diameter G		155	.206		2.96	10.40		13.36	19.90
1550	10" diameter G		155	.206		3.91	10.40		14.31	21
1600	12" diameter G		150	.213		4.54	10.75		15.29	22
1650	14" diameter G		145	.221		4.54	11.10		15.64	23
1700	16" diameter G		140	.229		5.65	11.50		17.15	24.50
1720	18" diameter G		140	.229		6.60	11.50		18.10	25.50
1750	20" diameter G		135	.237		7.90	11.95		19.85	28
1800	24" diameter G		130	.246		11.40	12.40		23.80	32.50
1850	30" diameter G		125	.256		14.80	12.90		27.70	37
1900	36" diameter G		115	.278		20.50	14		34.50	45
1950	42" diameter G		100	.320		44.50	16.10		60.60	75
2000	48" diameter G		85	.376		55.50	18.95		74.45	91.50
2200	For seamless type, add					15%				
3000	Round, steel, 4 use per mo., rent, regular duty, 14" diameter G	C-1	145	.221	L.F.	18.40	11.10		29.50	38
3050	16" diameter G		125	.256		18.75	12.90		31.65	41
3100	Heavy duty, 20" diameter G		105	.305		20.50	15.35		35.85	47
3150	24" diameter G		85	.376		22	18.95		40.95	55
3200	30" diameter G		70	.457		25.50	23		48.50	65
3250	36" diameter G		60	.533		27.50	27		54.50	73
3300	48" diameter G		50	.640		39.50	32.50		72	95
3350	60" diameter G		45	.711		50	36		86	113
4500	For second and succeeding months, deduct					50%				
5000	Job-built plywood, 8" x 8" columns, 1 use	C-1	165	.194	SFCA	2.88	9.75		12.63	18.80
5050	2 use		195	.164		1.65	8.25		9.90	15.05
5100	3 use		210	.152		1.15	7.70		8.85	13.55
5150	4 use		215	.149		.95	7.50		8.45	13.05
5500	12" x 12" columns, 1 use		180	.178		2.80	8.95		11.75	17.45
5550	2 use		210	.152		1.54	7.70		9.24	14
5600	3 use		220	.145		1.12	7.35		8.47	13
5650	4 use		225	.142		.91	7.15		8.06	12.50
6000	16" x 16" columns, 1 use		185	.173		2.81	8.70		11.51	17.05
6050	2 use		215	.149		1.50	7.50		9	13.65
6100	3 use		230	.139		1.13	7		8.13	12.50
6150	4 use		235	.136		.92	6.85		7.77	12
6500	24" x 24" columns, 1 use		190	.168		3.19	8.50		11.69	17.10
6550	2 use		216	.148		1.75	7.45		9.20	13.90
6600	3 use		230	.139		1.27	7		8.27	12.65
6650	4 use		238	.134		1.04	6.75		7.79	12
7000	36" x 36" columns, 1 use		200	.160		2.33	8.05		10.38	15.45
7050	2 use		230	.139		1.32	7		8.32	12.70
7100	3 use		245	.131		.93	6.60		7.53	11.55
7150	4 use		250	.128		.76	6.45		7.21	11.20
7400	Steel framed plywood, based on 50 uses of purchased									
7420	forms, and 4 uses of bracing lumber									

03 11 13 – Structural Cast-In-Place Concrete Forming

03 11 13.25 Forms In Place, Columns		Crew	Daily Output	Labor-Hours	Unit	Material	Labor	Equipment	Total	Total Incl O&P
						2020 Bare Costs				
7500	8" x 8" column	C-1	340	.094	SFCA	2.22	4.74		6.96	10.05
7550	10" x 10"		350	.091		1.98	4.61		6.59	9.55
7600	12" x 12"		370	.086		1.68	4.36		6.04	8.85
7650	16" x 16"		400	.080		1.30	4.03		5.33	7.90
7700	20" x 20"		420	.076		1.16	3.84		5	7.40
7750	24" x 24"		440	.073		.81	3.66		4.47	6.75
7755	30" x 30"		440	.073		1.04	3.66		4.70	7
7760	36" x 36"		460	.070		.92	3.51		4.43	6.60
9000	Minimum labor/equipment charge	2 Carp	2	8	Job		425		425	680

03 11 13.30 Forms In Place, Culvert		Crew	Daily Output	Labor-Hours	Unit	Material	Labor	Equipment	Total	Total Incl O&P
0010	**FORMS IN PLACE, CULVERT**									
0015	5' to 8' square or rectangular, 1 use	C-1	170	.188	SFCA	4.25	9.50		13.75	19.90
0050	2 use R031113-60		180	.178		2.57	8.95		11.52	17.15
0100	3 use		190	.168		2.01	8.50		10.51	15.80
0150	4 use		200	.160		1.73	8.05		9.78	14.80

03 11 13.35 Forms In Place, Elevated Slabs		Crew	Daily Output	Labor-Hours	Unit	Material	Labor	Equipment	Total	Total Incl O&P
0010	**FORMS IN PLACE, ELEVATED SLABS**									
1000	Flat plate, job-built plywood, to 15' high, 1 use	C-2	470	.102	S.F.	4.09	5.25		9.34	12.95
1050	2 use R031113-60		520	.092		2.25	4.77		7.02	10.15
1100	3 use		545	.088		1.64	4.55		6.19	9.10
1150	4 use		560	.086		1.33	4.43		5.76	8.55
1500	15' to 20' high ceilings, 4 use		495	.097		1.32	5		6.32	9.45
1600	21' to 35' high ceilings, 4 use		450	.107		1.65	5.50		7.15	10.65
2000	Flat slab, drop panels, job-built plywood, to 15' high, 1 use		449	.107		4.85	5.50		10.35	14.20
2050	2 use		509	.094		2.67	4.87		7.54	10.75
2100	3 use		532	.090		1.94	4.66		6.60	9.60
2150	4 use		544	.088		1.58	4.56		6.14	9.05
2250	15' to 20' high ceilings, 4 use		480	.100		2.70	5.15		7.85	11.20
2350	20' to 35' high ceilings, 4 use		435	.110		3.03	5.70		8.73	12.50
3000	Floor slab hung from steel beams, 1 use		485	.099		2.66	5.10		7.76	11.15
3050	2 use		535	.090		2.08	4.63		6.71	9.70
3100	3 use		550	.087		1.89	4.51		6.40	9.25
3150	4 use		565	.085		1.79	4.39		6.18	9
3500	Floor slab, with 1-way joist pans, 1 use		415	.116		8.30	5.95		14.25	18.70
3550	2 use		445	.108		6.10	5.55		11.65	15.65
3600	3 use		475	.101		5.35	5.20		10.55	14.25
3650	4 use		500	.096		5	4.96		9.96	13.45
4500	With 2-way waffle domes, 1 use		405	.119		8.90	6.10		15	19.60
4520	2 use		450	.107		6.70	5.50		12.20	16.25
4530	3 use		460	.104		6	5.40		11.40	15.20
4550	4 use		470	.102		5.60	5.25		10.85	14.60
5000	Box out for slab openings, over 16" deep, 1 use		190	.253	SFCA	3.75	13.05		16.80	25
5050	2 use		240	.200	"	2.06	10.35		12.41	18.80
5500	Shallow slab box outs, to 10 S.F.		42	1.143	Ea.	12.60	59		71.60	108
5550	Over 10 S.F. (use perimeter)		600	.080	L.F.	1.69	4.13		5.82	8.45
6000	Bulkhead forms for slab, with keyway, 1 use, 2 piece		500	.096		1.80	4.96		6.76	9.95
6100	3 piece (see also edge forms)		460	.104		2.06	5.40		7.46	10.90
6200	Slab bulkhead form, 4-1/2" high, exp metal, w/keyway & stakes G	C-1	1200	.027		.94	1.34		2.28	3.18
6210	5-1/2" high G		1100	.029		1.19	1.47		2.66	3.66
6215	7-1/2" high G		960	.033		1.40	1.68		3.08	4.23
6220	9-1/2" high G		840	.038		1.56	1.92		3.48	4.80
6500	Curb forms, wood, 6" to 12" high, on elevated slabs, 1 use		180	.178	SFCA	1.80	8.95		10.75	16.35

03 11 Concrete Forming

03 11 13 – Structural Cast-In-Place Concrete Forming

03 11 13.35 Forms In Place, Elevated Slabs		Crew	Daily Output	Labor-Hours	Unit	Material	2020 Bare Costs Labor	Equipment	Total	Total Incl O&P
6550	2 use	C-1	205	.156	SFCA	.99	7.85		8.84	13.70
6600	3 use		220	.145		.72	7.35		8.07	12.55
6650	4 use		225	.142		.58	7.15		7.73	12.15
7000	Edge forms to 6" high, on elevated slab, 4 use		500	.064	L.F.	.22	3.23		3.45	5.40
7070	7" to 12" high, 1 use		162	.198	SFCA	1.33	9.95		11.28	17.40
7080	2 use		198	.162		.73	8.15		8.88	13.85
7090	3 use		222	.144		.53	7.25		7.78	12.25
7101	4 use		350	.091		.22	4.61		4.83	7.65
7500	Depressed area forms to 12" high, 4 use		300	.107	L.F.	.83	5.40		6.23	9.50
7550	12" to 24" high, 4 use		175	.183		1.13	9.20		10.33	16
8000	Perimeter deck and rail for elevated slabs, straight		90	.356		12.70	17.90		30.60	42.50
8050	Curved		65	.492		17.45	25		42.45	58.50
8500	Void forms, round plastic, 8" high x 3" diameter G		450	.071	Ea.	2.05	3.58		5.63	8
8550	4" diameter G		425	.075		2.32	3.79		6.11	8.65
8600	6" diameter G		400	.080		4.05	4.03		8.08	10.90
8650	8" diameter G		375	.085		7	4.30		11.30	14.60
9000	Minimum labor/equipment charge	2 Carp	2	8	Job		425		425	680

03 11 13.40 Forms In Place, Equipment Foundations		Crew	Daily Output	Labor-Hours	Unit	Material	Labor	Equipment	Total	Total Incl O&P
0010	**FORMS IN PLACE, EQUIPMENT FOUNDATIONS**									
0020	1 use	C-2	160	.300	SFCA	2.95	15.50		18.45	28.50
0050	2 use R031113-60		190	.253		1.62	13.05		14.67	23
0100	3 use		200	.240		1.18	12.40		13.58	21
0150	4 use		205	.234		.96	12.10		13.06	20.50
9000	Minimum labor/equipment charge	1 Carp	3	2.667	Job		142		142	227

03 11 13.45 Forms In Place, Footings		Crew	Daily Output	Labor-Hours	Unit	Material	Labor	Equipment	Total	Total Incl O&P
0010	**FORMS IN PLACE, FOOTINGS**									
0020	Continuous wall, plywood, 1 use	C-1	375	.085	SFCA	6.85	4.30		11.15	14.45
0050	2 use R031113-60		440	.073		3.77	3.66		7.43	10
0100	3 use		470	.068		2.74	3.43		6.17	8.50
0150	4 use		485	.066		2.23	3.32		5.55	7.80
0500	Dowel supports for footings or beams, 1 use		500	.064	L.F.	1.05	3.23		4.28	6.30
1000	Integral starter wall, to 4" high, 1 use		400	.080		1.01	4.03		5.04	7.55
1500	Keyway, 4 use, tapered wood, 2" x 4"	1 Carp	530	.015		.25	.80		1.05	1.56
1550	2" x 6"		500	.016		.34	.85		1.19	1.73
2000	Tapered plastic		530	.015		1.53	.80		2.33	2.96
2250	For keyway hung from supports, add		150	.053		1.05	2.83		3.88	5.70
3000	Pile cap, square or rectangular, job-built plywood, 1 use	C-1	290	.110	SFCA	3.16	5.55		8.71	12.40
3050	2 use		346	.092		1.74	4.66		6.40	9.35
3100	3 use		371	.086		1.26	4.35		5.61	8.35
3150	4 use		383	.084		1.03	4.21		5.24	7.90
4000	Triangular or hexagonal, 1 use		225	.142		3.70	7.15		10.85	15.55
4050	2 use		280	.114		2.03	5.75		7.78	11.50
4100	3 use		305	.105		1.48	5.30		6.78	10.10
4150	4 use		315	.102		1.20	5.10		6.30	9.50
5000	Spread footings, job-built lumber, 1 use		305	.105		2.28	5.30		7.58	10.95
5050	2 use		371	.086		1.27	4.35		5.62	8.35
5100	3 use		401	.080		.91	4.02		4.93	7.45
5150	4 use		414	.077		.74	3.89		4.63	7.05
6000	Supports for dowels, plinths or templates, 2' x 2' footing		25	1.280	Ea.	6.50	64.50		71	110
6050	4' x 4' footing		22	1.455		12.95	73.50		86.45	131
6100	8' x 8' footing		20	1.600		26	80.50		106.50	158
6150	12' x 12' footing		17	1.882		31	95		126	187

03 11 Concrete Forming

03 11 13 – Structural Cast-In-Place Concrete Forming

03 11 13.45 Forms In Place, Footings		Crew	Daily Output	Labor-Hours	Unit	2020 Bare Costs Material	Labor	Equipment	Total	Total Incl O&P
7000	Plinths, job-built plywood, 1 use	C-1	250	.128	SFCA	3.25	6.45		9.70	13.90
7100	4 use		270	.119	"	1.06	5.95		7.01	10.70
9000	Minimum labor/equipment charge	1 Carp	3	2.667	Job		142		142	227

03 11 13.47 Forms In Place, Gas Station Forms		Crew	Daily Output	Labor-Hours	Unit	Material	Labor	Equipment	Total	Total Incl O&P
0010	**FORMS IN PLACE, GAS STATION FORMS**									
0050	Curb fascia, with template, 12 ga. steel, left in place, 9" high G	1 Carp	50	.160	L.F.	15.05	8.50		23.55	30
1000	Sign or light bases, 18" diameter, 9" high G		9	.889	Ea.	95	47		142	180
1050	30" diameter, 13" high G		8	1	"	151	53		204	251
1990	Minimum labor/equipment charge		2	4	Job		213		213	340
2000	Island forms, 10' long, 9" high, 3'-6" wide G	C-1	10	3.200	Ea.	420	161		581	725
2050	4' wide G		9	3.556		435	179		614	765
2500	20' long, 9" high, 4' wide G		6	5.333		700	269		969	1,200
2550	5' wide G		5	6.400		725	325		1,050	1,325
9000	Minimum labor/equipment charge	1 Carp	3	2.667	Job		142		142	227

03 11 13.50 Forms In Place, Grade Beam		Crew	Daily Output	Labor-Hours	Unit	Material	Labor	Equipment	Total	Total Incl O&P
0010	**FORMS IN PLACE, GRADE BEAM**									
0020	Job-built plywood, 1 use	C-2	530	.091	SFCA	3.15	4.68		7.83	10.95
0050	2 use R031113-60		580	.083		1.73	4.27		6	8.75
0100	3 use		600	.080		1.26	4.13		5.39	8
0150	4 use		605	.079		1.02	4.10		5.12	7.70
9000	Minimum labor/equipment charge	2 Carp	2	8	Job		425		425	680

03 11 13.55 Forms In Place, Mat Foundation		Crew	Daily Output	Labor-Hours	Unit	Material	Labor	Equipment	Total	Total Incl O&P
0010	**FORMS IN PLACE, MAT FOUNDATION**									
0020	Job-built plywood, 1 use	C-2	290	.166	SFCA	3.16	8.55		11.71	17.20
0050	2 use R031113-60		310	.155		1.26	8		9.26	14.20
0100	3 use		330	.145		.81	7.50		8.31	12.95
0120	4 use		350	.137		.74	7.10		7.84	12.15

03 11 13.65 Forms In Place, Slab On Grade		Crew	Daily Output	Labor-Hours	Unit	Material	Labor	Equipment	Total	Total Incl O&P
0010	**FORMS IN PLACE, SLAB ON GRADE**									
1000	Bulkhead forms w/keyway, wood, 6" high, 1 use	C-1	510	.063	L.F.	1.10	3.16		4.26	6.25
1050	2 uses R031113-60		400	.080		.60	4.03		4.63	7.10
1100	4 uses		350	.091		.36	4.61		4.97	7.80
1400	Bulkhead form for slab, 4-1/2" high, exp metal, incl. keyway & stakes G		1200	.027		.94	1.34		2.28	3.18
1410	5-1/2" high G		1100	.029		1.19	1.47		2.66	3.66
1420	7-1/2" high G		960	.033		1.40	1.68		3.08	4.23
1430	9-1/2" high G		840	.038		1.56	1.92		3.48	4.80
2000	Curb forms, wood, 6" to 12" high, on grade, 1 use		215	.149	SFCA	2.19	7.50		9.69	14.40
2050	2 use		250	.128		1.21	6.45		7.66	11.70
2100	3 use		265	.121		.88	6.10		6.98	10.70
2150	4 use		275	.116		.71	5.85		6.56	10.20
3000	Edge forms, wood, 4 use, on grade, to 6" high		600	.053	L.F.	.33	2.69		3.02	4.67
3050	7" to 12" high		435	.074	SFCA	.75	3.71		4.46	6.80
3060	Over 12"		350	.091	"	.97	4.61		5.58	8.45
3500	For depressed slabs, 4 use, to 12" high		300	.107	L.F.	.81	5.40		6.21	9.50
3550	To 24" high		175	.183		1.04	9.20		10.24	15.90
4000	For slab blockouts, to 12" high, 1 use		200	.160		.82	8.05		8.87	13.80
4050	To 24" high, 1 use		120	.267		1.04	13.45		14.49	22.50
4100	Plastic (extruded), to 6" high, multiple use, on grade		800	.040		6.50	2.02		8.52	10.40
5000	Screed, 24 ga. metal key joint, see Section 03 15 16.30									
5020	Wood, incl. wood stakes, 1" x 3"	C-1	900	.036	L.F.	.87	1.79		2.66	3.83
5050	2" x 4"		900	.036	"	.90	1.79		2.69	3.86
6000	Trench forms in floor, wood, 1 use		160	.200	SFCA	2.05	10.10		12.15	18.40

03 11 Concrete Forming

03 11 13 – Structural Cast-In-Place Concrete Forming

03 11 13.65 Forms In Place, Slab On Grade		Crew	Daily Output	Labor-Hours	Unit	Material	Labor	Equipment	Total	Total Incl O&P
						2020 Bare Costs				
6050	2 use	C-1	175	.183	SFCA	1.13	9.20		10.33	16
6100	3 use		180	.178		.82	8.95		9.77	15.25
6150	4 use		185	.173		.67	8.70		9.37	14.70
8760	Void form, corrugated fiberboard, 4" x 12", 4' long G		3000	.011	S.F.	3.64	.54		4.18	4.86
8770	6" x 12", 4' long		3000	.011		4.47	.54		5.01	5.80
8780	1/4" thick hardboard protective cover for void form	2 Carp	1500	.011		.72	.57		1.29	1.70
9000	Minimum labor/equipment charge	1 Carp	2	4	Job		213		213	340

03 11 13.85 Forms In Place, Walls		Crew	Daily Output	Labor-Hours	Unit	Material	Labor	Equipment	Total	Total Incl O&P
0010	**FORMS IN PLACE, WALLS**									
0100	Box out for wall openings, to 16" thick, to 10 S.F.	C-2	24	2	Ea.	28	103		131	196
0150	Over 10 S.F. (use perimeter)	"	280	.171	L.F.	2.45	8.85		11.30	16.90
0250	Brick shelf, 4" w, add to wall forms, use wall area above shelf									
0260	1 use R031113-60	C-2	240	.200	SFCA	2.70	10.35		13.05	19.50
0300	2 use		275	.175		1.48	9		10.48	16.10
0350	4 use		300	.160		1.08	8.25		9.33	14.45
0500	Bulkhead, wood with keyway, 1 use, 2 piece		265	.181	L.F.	2.19	9.35		11.54	17.40
0600	Bulkhead forms with keyway, 1 piece expanded metal, 8" wall G	C-1	1000	.032		1.40	1.61		3.01	4.12
0610	10" wall G		800	.040		1.56	2.02		3.58	4.95
0620	12" wall G		525	.061		1.87	3.07		4.94	7
0700	Buttress, to 8' high, 1 use	C-2	350	.137	SFCA	4.85	7.10		11.95	16.70
0750	2 use		430	.112		2.67	5.75		8.42	12.20
0800	3 use		460	.104		1.95	5.40		7.35	10.80
0850	4 use		480	.100		1.60	5.15		6.75	10
1000	Corbel or haunch, to 12" wide, add to wall forms, 1 use		150	.320	L.F.	2.59	16.50		19.09	29.50
1050	2 use		170	.282		1.42	14.60		16.02	25
1100	3 use		175	.274		1.03	14.15		15.18	23.50
1150	4 use		180	.267		.84	13.75		14.59	23
2000	Wall, job-built plywood, to 8' high, 1 use		370	.130	SFCA	2.99	6.70		9.69	14.05
2050	2 use		435	.110		1.91	5.70		7.61	11.25
2100	3 use		495	.097		1.39	5		6.39	9.55
2150	4 use		505	.095		1.13	4.91		6.04	9.10
2400	Over 8' to 16' high, 1 use		280	.171		3.32	8.85		12.17	17.85
2450	2 use		345	.139		1.44	7.20		8.64	13.10
2500	3 use		375	.128		1.03	6.60		7.63	11.75
2550	4 use		395	.122		.84	6.30		7.14	11
2700	Over 16' high, 1 use		235	.204		2.94	10.55		13.49	20
2750	2 use		290	.166		1.62	8.55		10.17	15.50
2800	3 use		315	.152		1.18	7.85		9.03	13.90
2850	4 use		330	.145		.96	7.50		8.46	13.10
4000	Radial, smooth curved, job-built plywood, 1 use		245	.196		2.65	10.10		12.75	19.10
4050	2 use		300	.160		1.46	8.25		9.71	14.85
4100	3 use		325	.148		1.06	7.65		8.71	13.35
4150	4 use		335	.143		.86	7.40		8.26	12.80
4200	Below grade, job-built plywood, 1 use		225	.213		2.86	11		13.86	21
4210	2 use		225	.213		1.58	11		12.58	19.40
4220	3 use		225	.213		1.31	11		12.31	19.10
4230	4 use		225	.213		.93	11		11.93	18.70
4300	Curved, 2' chords, job-built plywood, to 8' high, 1 use		290	.166		2.32	8.55		10.87	16.25
4350	2 use		355	.135		1.27	7		8.27	12.60
4400	3 use		385	.125		.93	6.45		7.38	11.30
4450	4 use		400	.120		.75	6.20		6.95	10.80
4500	Over 8' to 16' high, 1 use		290	.166		.95	8.55		9.50	14.75

03 11 Concrete Forming

03 11 13 – Structural Cast-In-Place Concrete Forming

03 11 13.85 Forms In Place, Walls		Crew	Daily Output	Labor-Hours	Unit	Material	2020 Bare Costs Labor	Equipment	Total	Total Incl O&P
4525	2 use	C-2	355	.135	SFCA	.52	7		7.52	11.75
4550	3 use		385	.125		.38	6.45		6.83	10.70
4575	4 use		400	.120		.31	6.20		6.51	10.30
4600	Retaining wall, battered, job-built plyw'd, to 8' high, 1 use		300	.160		2.18	8.25		10.43	15.65
4650	2 use		355	.135		1.20	7		8.20	12.50
4700	3 use		375	.128		.87	6.60		7.47	11.55
4750	4 use		390	.123		.71	6.35		7.06	11
4900	Over 8' to 16' high, 1 use		240	.200		2.39	10.35		12.74	19.15
4950	2 use		295	.163		1.31	8.40		9.71	14.90
5000	3 use		305	.157		.95	8.15		9.10	14.05
5050	4 use		320	.150		.78	7.75		8.53	13.25
5500	For gang wall forming, 192 S.F. sections, deduct					10%	10%			
5550	384 S.F. sections, deduct					20%	20%			
7500	Lintel or sill forms, 1 use	1 Carp	30	.267		3.51	14.15		17.66	26.50
7520	2 use		34	.235		1.93	12.50		14.43	22
7540	3 use		36	.222		1.40	11.80		13.20	20.50
7560	4 use		37	.216		1.14	11.50		12.64	19.65
7800	Modular prefabricated plywood, based on 20 uses of purchased									
7820	forms, and 4 uses of bracing lumber									
7860	To 8' high	C-2	800	.060	SFCA	1.15	3.10		4.25	6.20
8060	Over 8' to 16' high		600	.080		1.20	4.13		5.33	7.90
8600	Pilasters, 1 use		270	.178		3.49	9.20		12.69	18.55
8620	2 use		330	.145		1.92	7.50		9.42	14.15
8640	3 use		370	.130		1.40	6.70		8.10	12.30
8660	4 use		385	.125		1.13	6.45		7.58	11.55
9010	Steel framed plywood, based on 50 uses of purchased									
9020	forms, and 4 uses of bracing lumber									
9060	To 8' high	C-2	600	.080	SFCA	.79	4.13		4.92	7.45
9260	Over 8' to 16' high		450	.107		.79	5.50		6.29	9.70
9460	Over 16' to 20' high		400	.120		.79	6.20		6.99	10.80
9475	For elevated walls, add						10%			
9480	For battered walls, 1 side battered, add					10%	10%			
9485	For battered walls, 2 sides battered, add					15%	15%			
9900	Minimum labor/equipment charge	2 Carp	2	8	Job		425		425	680

03 11 16 – Architectural Cast-in-Place Concrete Forming

03 11 16.13 Concrete Form Liners		Crew	Daily Output	Labor-Hours	Unit	Material	Labor	Equipment	Total	Total Incl O&P
0010	**CONCRETE FORM LINERS**									
5750	Liners for forms (add to wall forms), ABS plastic									
5800	Aged wood, 4" wide, 1 use	1 Carp	256	.031	SFCA	3.34	1.66		5	6.35
5820	2 use		256	.031		1.84	1.66		3.50	4.68
5830	3 use		256	.031		1.34	1.66		3	4.13
5840	4 use		256	.031		1.09	1.66		2.75	3.85
5900	Fractured rope rib, 1 use		192	.042		4.91	2.21		7.12	8.95
5925	2 use		192	.042		2.70	2.21		4.91	6.50
5950	3 use		192	.042		1.96	2.21		4.17	5.70
6000	4 use		192	.042		1.60	2.21		3.81	5.30
6100	Ribbed, 3/4" deep x 1-1/2" OC, 1 use		224	.036		5.05	1.90		6.95	8.65
6125	2 use		224	.036		2.79	1.90		4.69	6.10
6150	3 use		224	.036		2.03	1.90		3.93	5.25
6200	4 use		224	.036		1.65	1.90		3.55	4.85
6300	Rustic brick pattern, 1 use		224	.036		3.47	1.90		5.37	6.85
6325	2 use		224	.036		1.91	1.90		3.81	5.15

03 11 Concrete Forming

03 11 16 – Architectural Cast-in-Place Concrete Forming

03 11 16.13 Concrete Form Liners		Crew	Daily Output	Labor-Hours	Unit	Material	2020 Bare Costs Labor	Equipment	Total	Total Incl O&P
6350	3 use	1 Carp	224	.036	SFCA	1.39	1.90		3.29	4.57
6400	4 use		224	.036		1.13	1.90		3.03	4.28
6500	3/8" striated, random, 1 use		224	.036		3.54	1.90		5.44	6.95
6525	2 use		224	.036		1.95	1.90		3.85	5.20
6550	3 use		224	.036		1.42	1.90		3.32	4.60
6600	4 use		224	.036		1.15	1.90		3.05	4.31
6850	Random vertical rustication, 1 use		384	.021		6.60	1.11		7.71	9
6900	2 use		384	.021		3.62	1.11		4.73	5.75
6925	3 use		384	.021		2.63	1.11		3.74	4.67
6950	4 use		384	.021		2.14	1.11		3.25	4.12
7050	Wood, beveled edge, 3/4" deep, 1 use		384	.021	L.F.	.18	1.11		1.29	1.97
7100	1" deep, 1 use		384	.021	"	.29	1.11		1.40	2.09
7200	4" wide aged cedar, 1 use		256	.031	SFCA	3.47	1.66		5.13	6.50
7300	4" variable depth rough cedar		224	.036	"	4.85	1.90		6.75	8.40

03 11 19 – Insulating Concrete Forming

03 11 19.10 Insulating Forms, Left In Place		Crew	Daily Output	Labor-Hours	Unit	Material	Labor	Equipment	Total	Total Incl O&P
0010	**INSULATING FORMS, LEFT IN PLACE**									
0020	Forms include layout, exclude rebar, embedments, bucks for openings,									
0030	scaffolding, wall bracing, concrete, and concrete placing.									
0040	S.F. is for exterior face but includes forms for both faces of wall									
0100	Straight blocks or panels, molded, walls up to 4' high									
0110	4" core wall	4 Carp	1984	.016	S.F.	3.62	.86		4.48	5.35
0120	6" core wall		1808	.018		3.64	.94		4.58	5.50
0130	8" core wall		1536	.021		3.77	1.11		4.88	5.90
0140	10" core wall		1152	.028		4.30	1.48		5.78	7.10
0150	12" core wall		992	.032		4.89	1.71		6.60	8.15
0200	90 degree corner blocks or panels, molded, walls up to 4' high									
0210	4" core wall	4 Carp	1880	.017	S.F.	3.73	.90		4.63	5.55
0220	6" core wall		1708	.019		3.75	1		4.75	5.75
0230	8" core wall		1324	.024		3.88	1.28		5.16	6.35
0240	10" core wall		987	.032		4.49	1.72		6.21	7.70
0250	12" core wall		884	.036		4.81	1.92		6.73	8.40
0300	45 degree corner blocks or panels, molded, walls up to 4' high									
0310	4" core wall	4 Carp	1880	.017	S.F.	3.93	.90		4.83	5.75
0320	6" core wall		1712	.019		4.06	.99		5.05	6.05
0330	8" core wall		1324	.024		4.17	1.28		5.45	6.65
0400	T blocks or panels, molded, walls up to 4' high									
0420	6" core wall	4 Carp	1540	.021	S.F.	4.79	1.10		5.89	7
0430	8" core wall		1325	.024		4.84	1.28		6.12	7.35
0440	Non-standard corners or Ts requiring trimming & strapping		192	.167		4.89	8.85		13.74	19.60
0500	Radius blocks or panels, molded, walls up to 4' high, 6" core wall									
0520	5' to 10' diameter, molded blocks or panels	4 Carp	2400	.013	S.F.	5.90	.71		6.61	7.65
0530	10' to 15' diameter, requiring trimming and strapping, add		500	.064		4.44	3.40		7.84	10.35
0540	15'-1" to 30' diameter, requiring trimming and strapping, add		1200	.027		4.44	1.42		5.86	7.15
0550	30'-1" to 60' diameter, requiring trimming and strapping, add		1600	.020		4.44	1.06		5.50	6.60
0560	60'-1" to 100' diameter, requiring trimming and strapping, add		2800	.011		4.44	.61		5.05	5.85
0600	Additional labor for blocks/panels in higher walls (excludes scaffolding)									
0610	4'-1" to 9'-4" high, add						10%			
0620	9'-5" to 12'-0" high, add						20%			
0630	12'-1" to 20'-0" high, add						35%			
0640	Over 20'-0" high, add						55%			
0700	Taper block or panels, molded, single course									

03 11 19.10 Insulating Forms, Left In Place		Crew	Daily Output	Labor-Hours	Unit	2020 Bare Costs Material	Labor	Equipment	Total	Total Incl O&P
0720	6″ core wall	4 Carp	1600	.020	S.F.	3.91	1.06		4.97	6
0730	8″ core wall	″	1392	.023	″	3.96	1.22		5.18	6.30
0800	ICF brick ledge (corbel) block or panels, molded, single course									
0820	6″ core wall	4 Carp	1200	.027	S.F.	4.31	1.42		5.73	7
0830	8″ core wall	″	1152	.028	″	4.41	1.48		5.89	7.20
0900	ICF curb (shelf) block or panels, molded, single course									
0930	8″ core wall	4 Carp	688	.047	S.F.	3.72	2.47		6.19	8.05
0940	10″ core wall	″	544	.059	″	4.25	3.13		7.38	9.70
0950	Wood form to hold back concrete to form shelf, 8″ high	2 Carp	400	.040	L.F.	.87	2.13		3	4.36
1000	ICF half height block or panels, molded, single course									
1010	4″ core wall	4 Carp	1248	.026	S.F.	4.72	1.36		6.08	7.40
1020	6″ core wall		1152	.028		4.74	1.48		6.22	7.55
1030	8″ core wall		942	.034		4.79	1.81		6.60	8.15
1040	10″ core wall		752	.043		5.30	2.26		7.56	9.40
1050	12″ core wall		648	.049		5.20	2.62		7.82	9.95
1100	ICF half height block/panels, made by field sawing full height block/panels									
1110	4″ core wall	4 Carp	800	.040	S.F.	1.81	2.13		3.94	5.40
1120	6″ core wall		752	.043		1.82	2.26		4.08	5.60
1130	8″ core wall		600	.053		1.89	2.83		4.72	6.60
1140	10″ core wall		496	.065		2.15	3.43		5.58	7.85
1150	12″ core wall		400	.080		2.45	4.25		6.70	9.50
1200	Additional insulation inserted into forms between ties									
1210	1 layer (2″ thick)	4 Carp	14000	.002	S.F.	1.03	.12		1.15	1.33
1220	2 layers (4″ thick)		7000	.005		2.06	.24		2.30	2.66
1230	3 layers (6″ thick)		4622	.007		3.09	.37		3.46	3.99
1300	EPS window/door bucks, molded, permanent									
1310	4″ core wall (9″ wide)	2 Carp	200	.080	L.F.	3.47	4.25		7.72	10.60
1320	6″ core wall (11″ wide)		200	.080		3.56	4.25		7.81	10.70
1330	8″ core wall (13″ wide)		176	.091		3.73	4.83		8.56	11.85
1340	10″ core wall (15″ wide)		152	.105		4.86	5.60		10.46	14.30
1350	12″ core wall (17″ wide)		152	.105		5.45	5.60		11.05	14.90
1360	2″ x 6″ temporary buck bracing (includes installing and removing)		400	.040		.68	2.13		2.81	4.15
1400	Wood window/door bucks (instead of EPS bucks), permanent									
1410	4″ core wall (9″ wide)	2 Carp	400	.040	L.F.	1.31	2.13		3.44	4.85
1420	6″ core wall (11″ wide)		400	.040		1.72	2.13		3.85	5.30
1430	8″ core wall (13″ wide)		350	.046		7.35	2.43		9.78	11.95
1440	10″ core wall (15″ wide)		300	.053		10.15	2.83		12.98	15.75
1450	12″ core wall (17″ wide)		300	.053		10.15	2.83		12.98	15.75
1460	2″ x 6″ temporary buck bracing (includes installing and removing)		800	.020		.68	1.06		1.74	2.44
1500	ICF alignment brace (incl. stiff-back, diagonal kick-back, work platform									
1501	bracket & guard rail post), fastened to one face of wall forms @ 6′ O.C.									
1510	1st tier up to 10′ tall									
1520	Rental of ICF alignment brace set, per set				Week	10.05			10.05	11.05
1530	Labor (includes installing & removing)	2 Carp	30	.533	Ea.		28.50		28.50	45.50
1560	2nd tier from 10′ to 20′ tall (excludes mason's scaffolding up to 10′ high)									
1570	Rental of ICF alignment brace set, per set				Week	10.05			10.05	11.05
1580	Labor (includes installing & removing)	4 Carp	30	1.067	Ea.		56.50		56.50	91
1600	2″ x 10″ wood plank for work platform, 16′ long									
1610	Plank material cost pro-rated over 20 uses				Ea.	1.05			1.05	1.15
1620	Labor (includes installing & removing)	2 Carp	48	.333	″		17.70		17.70	28.50
1700	2″ x 4″ lumber for top & middle rails for work platform									
1710	Railing material cost pro-rated over 20 uses				Ea.	.02			.02	.03
1720	Labor (includes installing & removing)	2 Carp	2400	.007	L.F.		.35		.35	.57

03 11 Concrete Forming

03 11 19 – Insulating Concrete Forming

03 11 19.10 Insulating Forms, Left In Place		Crew	Daily Output	Labor-Hours	Unit	Material	Labor	Equipment	Total	Total Incl O&P
						2020 Bare Costs				
1800	ICF accessories									
1810	Wire clip to secure forms in place	2 Carp	2100	.008	Ea.	.39	.41		.80	1.08
1820	Masonry anchor embedment (excludes ties by mason)		1600	.010		4.24	.53		4.77	5.50
1830	Ledger anchor embedment (excludes timber hanger & screws)		128	.125		7.60	6.65		14.25	19
1900	See section 01 54 23.70 for mason's scaffolding components									
1910	See section 03 15 19.05 for anchor bolt sleeves									
1920	See section 03 15 19.10 for anchor bolts									
1930	See section 03 15 19.20 for dovetail anchor components									
1940	See section 03 15 19.30 for embedded inserts									
1950	See section 03 21 05.10 for rebar accessories									
1960	See section 03 21 11.60 for reinforcing bars in place									
1970	See section 03 31 13.35 for ready-mix concrete material									
1980	See section 03 31 13.70 for placement and consolidation of concrete									
1990	See section 06 05 23.60 for timber connectors									

03 11 19.60 Roof Deck Form Boards		Crew	Daily Output	Labor-Hours	Unit	Material	Labor	Equipment	Total	Total Incl O&P
0010	ROOF DECK FORM BOARDS R051223-50									
0050	Includes bulb tee sub-purlins @ 32-5/8" OC									
0070	Non-asbestos fiber cement, 5/16" thick	C-13	2950	.008	S.F.	3.28	.46	.05	3.79	4.40
0100	Fiberglass, 1" thick		2700	.009		3.89	.50	.05	4.44	5.15
0500	Wood fiber, 1" thick G		2700	.009		2.43	.50	.05	2.98	3.55

03 11 23 – Permanent Stair Forming

03 11 23.75 Forms In Place, Stairs		Crew	Daily Output	Labor-Hours	Unit	Material	Labor	Equipment	Total	Total Incl O&P
0010	FORMS IN PLACE, STAIRS									
0015	(Slant length x width), 1 use	C-2	165	.291	S.F.	6.15	15		21.15	31
0050	2 use R031113-60		170	.282		3.50	14.60		18.10	27.50
0100	3 use		180	.267		2.62	13.75		16.37	25
0150	4 use		190	.253		2.17	13.05		15.22	23.50
1000	Alternate pricing method (1.0 L.F./S.F.), 1 use		100	.480	LF Rsr	6.15	25		31.15	46.50
1050	2 use		105	.457		3.50	23.50		27	42
1100	3 use		110	.436		2.62	22.50		25.12	39
1150	4 use		115	.417		2.17	21.50		23.67	37
2000	Stairs, cast on sloping ground (length x width), 1 use		220	.218	S.F.	2.57	11.25		13.82	21
2025	2 use		232	.207		1.42	10.70		12.12	18.65
2050	3 use		244	.197		1.03	10.15		11.18	17.45
2100	4 use		256	.188		.84	9.70		10.54	16.40

03 15 Concrete Accessories

03 15 05 – Concrete Forming Accessories

03 15 05.12 Chamfer Strips		Crew	Daily Output	Labor-Hours	Unit	Material	Labor	Equipment	Total	Total Incl O&P
0010	CHAMFER STRIPS									
2000	Polyvinyl chloride, 1/2" wide with leg	1 Carp	535	.015	L.F.	.75	.79		1.54	2.10
2200	3/4" wide with leg		525	.015		.77	.81		1.58	2.15
2400	1" radius with leg		515	.016		.81	.83		1.64	2.21
2800	2" radius with leg		500	.016		1.79	.85		2.64	3.33
5000	Wood, 1/2" wide		535	.015		.14	.79		.93	1.42
5200	3/4" wide		525	.015		.18	.81		.99	1.50
5400	1" wide		515	.016		.29	.83		1.12	1.64

03 15 Concrete Accessories

03 15 05 – Concrete Forming Accessories

03 15 05.15 Column Form Accessories

Line	03 15 05.15 Column Form Accessories		Crew	Daily Output	Labor-Hours	Unit	2020 Bare Costs Material	Labor	Equipment	Total	Total Incl O&P
0010	**COLUMN FORM ACCESSORIES**										
1000	Column clamps, adjustable to 24" x 24", buy	G				Set	171			171	188
1100	Rent per month	G					13.65			13.65	15
1300	For sizes to 30" x 30", buy	G					220			220	242
1400	Rent per month	G					15.15			15.15	16.65
1600	For sizes to 36" x 36", buy	G					237			237	260
1700	Rent per month	G					17.15			17.15	18.85
2000	Bar type with wedges, 36" x 36", buy	G					190			190	209
2100	Rent per month	G					13.40			13.40	14.75
2300	48" x 48", buy	G					250			250	275
2400	Rent per month	G					18.40			18.40	20
3000	Scissor type with wedges, 36" x 36", buy	G					160			160	176
3100	Rent per month	G					15.25			15.25	16.80
3300	60" x 60", buy	G					228			228	250
3400	Rent per month	G					21.50			21.50	23.50
4000	Friction collars 2'-6" diam., buy	G					2,675			2,675	2,950
4100	Rent per month	G					205			205	226
4300	4'-0" diam., buy	G					3,275			3,275	3,600
4400	Rent per month	G					249			249	274

03 15 05.30 Hangers

Line	03 15 05.30 Hangers		Crew	Daily Output	Labor-Hours	Unit	Material	Labor	Equipment	Total	Total Incl O&P
0010	**HANGERS**										
0020	Slab and beam form										
0500	Banding iron										
0550	3/4" x 22 ga., 14 L.F./lb. or 1/2" x 14 ga., 7 L.F./lb.	G				Lb.	1.44			1.44	1.58
1000	Fascia ties, coil type, to 24" long	G				C	450			450	495
1500	Frame ties to 8-1/8"	G					520			520	570
1550	8-1/8" to 10-1/8"	G					545			545	600
5000	Snap tie hanger, to 30" overall length, 3000#	G					450			450	495
5050	To 36" overall length	G					500			500	550
5100	To 48" overall length	G					610			610	670
5500	Steel beam hanger										
5600	Flange to 8-1/8"	G				C	520			520	570
5650	8-1/8" to 10-1/8"	G				"	545			545	600
5900	Coil threaded rods, continuous, 1/2" diameter	G				L.F.	1.49			1.49	1.64
6000	Tie hangers to 30" overall length, 4000#	G				C	510			510	560
6100	To 36" overall length	G					555			555	610
6500	Tie back hanger, up to 12-1/8" flange	G					1,400			1,400	1,525
8500	Wire, black annealed, 15 gauge	G				Cwt.	187			187	206
8600	16 gauge					"	197			197	217

03 15 05.70 Shores

Line	03 15 05.70 Shores		Crew	Daily Output	Labor-Hours	Unit	Material	Labor	Equipment	Total	Total Incl O&P
0010	**SHORES**										
0020	Erect and strip, by hand, horizontal members										
0500	Aluminum joists and stringers	G	2 Carp	60	.267	Ea.		14.15		14.15	22.50
0600	Steel, adjustable beams	G		45	.356			18.90		18.90	30.50
0700	Wood joists			50	.320			17		17	27.50
0800	Wood stringers			30	.533			28.50		28.50	45.50
1000	Vertical members to 10' high	G		55	.291			15.45		15.45	25
1050	To 13' high	G		50	.320			17		17	27.50
1100	To 16' high	G		45	.356			18.90		18.90	30.50
1500	Reshoring	G		1400	.011	S.F.	.63	.61		1.24	1.66
1600	Flying truss system	G	C-17D	9600	.009	SFCA		.49	.08	.57	.86
1760	Horizontal, aluminum joists, 6-1/4" high x 5' to 21' span, buy	G				L.F.	16.40			16.40	18.05

03 15 05 – Concrete Forming Accessories

03 15 05.70 Shores		Crew	Daily Output	Labor-Hours	Unit	Material	Labor	Equipment	Total	Total Incl O&P
						2020 Bare Costs				
1770	Beams, 7-1/4" high x 4' to 30' span	G			L.F.	19.70			19.70	21.50
1810	Horizontal, steel beam, W8x10, 7' span, buy	G			Ea.	64.50			64.50	71
1830	10' span	G				75			75	82.50
1920	15' span	G				129			129	142
1940	20' span	G				181			181	199
1970	Steel stringer, W8x10, 4' to 16' span, buy	G			L.F.	7.45			7.45	8.20
3000	Rent for job duration, aluminum joist @ 2' OC, per mo.	G			SF Flr.	.41			.41	.45
3050	Steel W8x10	G				.19			.19	.21
3060	Steel adjustable	G				.19			.19	.21
3500	#1 post shore, steel, 5'-7" to 9'-6" high, 10,000# cap., buy	G			Ea.	161			161	178
3550	#2 post shore, 7'-3" to 12'-10" high, 7800# capacity	G				186			186	205
3600	#3 post shore, 8'-10" to 16'-1" high, 3800# capacity	G				204			204	224
5010	Frame shoring systems, steel, 12,000#/leg, buy									
5040	Frame, 2' wide x 6' high	G			Ea.	107			107	118
5250	X-brace	G				17.05			17.05	18.80
5550	Base plate	G				16.05			16.05	17.65
5600	Screw jack	G				35			35	38
5650	U-head, 8" x 8"	G				20.50			20.50	22.50

03 15 05.75 Sleeves and Chases

		Crew	Daily Output	Labor-Hours	Unit	Material	Labor	Equipment	Total	Total Incl O&P
0010	**SLEEVES AND CHASES**									
0100	Plastic, 1 use, 12" long, 2" diameter	1 Carp	100	.080	Ea.	2.32	4.25		6.57	9.35
0150	4" diameter		90	.089		4.97	4.72		9.69	13
0200	6" diameter		75	.107		10.85	5.65		16.50	21
0250	12" diameter		60	.133		28.50	7.10		35.60	42.50

03 15 05.80 Snap Ties

			Crew	Daily Output	Labor-Hours	Unit	Material	Labor	Equipment	Total	Total Incl O&P
0010	**SNAP TIES**, 8-1/4" L&W (Lumber and wedge)										
0100	2250 lb., w/flat washer, 8" wall	G				C	97.50			97.50	107
0150	10" wall	G					163			163	179
0200	12" wall	G					166			166	183
0250	16" wall	G					178			178	196
0300	18" wall	G					185			185	203
0500	With plastic cone, 8" wall	G					86			86	94.50
0550	10" wall	G					88			88	97
0600	12" wall	G					96.50			96.50	106
0650	16" wall	G					106			106	116
0700	18" wall	G					109			109	120
1000	3350 lb., w/flat washer, 8" wall	G					204			204	225
1100	10" wall	G					224			224	246
1150	12" wall	G					204			204	224
1200	16" wall	G					262			262	289
1250	18" wall	G					274			274	300
1500	With plastic cone, 8" wall	G					164			164	180
1550	10" wall	G					180			180	197
1600	12" wall	G					163			163	179
1650	16" wall	G					211			211	232
1700	18" wall	G					213			213	234

03 15 05.85 Stair Tread Inserts

		Crew	Daily Output	Labor-Hours	Unit	Material	Labor	Equipment	Total	Total Incl O&P
0010	**STAIR TREAD INSERTS**									
0105	Cast nosing insert, abrasive surface, pre-drilled, includes screws									
0110	Aluminum, 3" wide x 3' long	1 Cefi	32	.250	Ea.	58	12.50		70.50	83
0120	4' long		31	.258		75	12.90		87.90	103
0130	5' long		30	.267		90	13.30		103.30	120

03 15 Concrete Accessories

03 15 05 – Concrete Forming Accessories

03 15 05.85 Stair Tread Inserts		Crew	Daily Output	Labor-Hours	Unit	2020 Bare Costs Material	Labor	Equipment	Total	Total Incl O&P
0135	Extruded nosing insert, black abrasive strips, continuous anchor									
0140	Aluminum, 3" wide x 3' long	1 Cefi	64	.125	Ea.	35.50	6.25		41.75	49
0150	4' long		60	.133		48	6.65		54.65	63.50
0160	5' long	↓	56	.143	↓	62.50	7.15		69.65	80
0165	Extruded nosing insert, black abrasive strips, pre-drilled, incl. screws									
0170	Aluminum, 3" wide x 3' long	1 Cefi	32	.250	Ea.	45	12.50		57.50	69
0180	4' long		31	.258		59.50	12.90		72.40	85.50
0190	5' long	↓	30	.267	↓	83.50	13.30		96.80	113
9000	Minimum labor/equipment charge	1 Carp	4	2	Job		106		106	170

03 15 05.95 Wall and Foundation Form Accessories			Crew	Daily Output	Labor-Hours	Unit	Material	Labor	Equipment	Total	Total Incl O&P
0010	**WALL AND FOUNDATION FORM ACCESSORIES**										
2000	Footings, turnbuckle form aligner	G				Ea.	17.40			17.40	19.10
2050	Spreaders for footer, adjustable	G				"	25.50			25.50	28
3000	Form oil, up to 1200 S.F./gallon coverage					Gal.	15.95			15.95	17.55
3050	Up to 800 S.F./gallon					"	23.50			23.50	25.50
3500	Form patches, 1-3/4" diameter					C	27			27	29.50
3550	2-3/4" diameter					"	45.50			45.50	50.50
4000	Nail stakes, 3/4" diameter, 18" long	G				Ea.	2.70			2.70	2.97
4050	24" long	G					2.68			2.68	2.95
4200	30" long	G					3.38			3.38	3.72
4250	36" long	G				↓	4.25			4.25	4.68

03 15 13 – Waterstops

03 15 13.50 Waterstops		Crew	Daily Output	Labor-Hours	Unit	Material	Labor	Equipment	Total	Total Incl O&P
0010	**WATERSTOPS**, PVC and Rubber									
0020	PVC, ribbed 3/16" thick, 4" wide	1 Carp	155	.052	L.F.	1.55	2.74		4.29	6.10
0050	6" wide		145	.055		2.78	2.93		5.71	7.75
0500	With center bulb, 6" wide, 3/16" thick		135	.059		2.68	3.15		5.83	8
0550	3/8" thick		130	.062		4.85	3.27		8.12	10.60
0600	9" wide x 3/8" thick		125	.064		8.15	3.40		11.55	14.40
0800	Dumbbell type, 6" wide, 3/16" thick		150	.053		4.31	2.83		7.14	9.30
0850	3/8" thick		145	.055		4.13	2.93		7.06	9.25
1000	9" wide, 3/8" thick, plain		130	.062		6.60	3.27		9.87	12.50
1050	Center bulb		130	.062		9.15	3.27		12.42	15.30
1250	Ribbed type, split, 3/16" thick, 6" wide		145	.055		2.21	2.93		5.14	7.15
1300	3/8" thick		130	.062		5.10	3.27		8.37	10.85
2000	Rubber, flat dumbbell, 3/8" thick, 6" wide		145	.055		5.65	2.93		8.58	10.90
2050	9" wide		135	.059		10.40	3.15		13.55	16.45
2500	Flat dumbbell split, 3/8" thick, 6" wide		145	.055		2.21	2.93		5.14	7.15
2550	9" wide		135	.059		5.10	3.15		8.25	10.65
3000	Center bulb, 1/4" thick, 6" wide		145	.055		6.95	2.93		9.88	12.35
3050	9" wide		135	.059		14.85	3.15		18	21.50
3500	Center bulb split, 3/8" thick, 6" wide		145	.055		6.10	2.93		9.03	11.40
3550	9" wide	↓	135	.059	↓	10.25	3.15		13.40	16.30
5000	Waterstop fittings, rubber, flat									
5010	Dumbbell or center bulb, 3/8" thick,									
5200	Field union, 6" wide	1 Carp	50	.160	Ea.	31.50	8.50		40	48
5250	9" wide		50	.160		33.50	8.50		42	50.50
5500	Flat cross, 6" wide		30	.267		52.50	14.15		66.65	80.50
5550	9" wide		30	.267		78	14.15		92.15	109
6000	Flat tee, 6" wide		30	.267		50.50	14.15		64.65	78
6050	9" wide		30	.267		66	14.15		80.15	95.50
6500	Flat ell, 6" wide	↓	40	.200	↓	49.50	10.65		60.15	71.50

03 15 13 – Waterstops

03 15 13.50 Waterstops		Crew	Daily Output	Labor-Hours	Unit	Material	Labor	Equipment	Total	Total Incl O&P
						2020 Bare Costs				
6550	9" wide	1 Carp	40	.200	Ea.	61.50	10.65		72.15	84.50
7000	Vertical tee, 6" wide		25	.320		21	17		38	50.50
7050	9" wide		25	.320		28.50	17		45.50	59
7500	Vertical ell, 6" wide		35	.229		20.50	12.15		32.65	42
7550	9" wide		35	.229		39	12.15		51.15	62.50

03 15 16 – Concrete Construction Joints

03 15 16.20 Control Joints, Saw Cut

03 15 16.20 Control Joints, Saw Cut		Crew	Daily Output	Labor-Hours	Unit	Material	Labor	Equipment	Total	Total Incl O&P
0010	**CONTROL JOINTS, SAW CUT**									
0100	Sawcut control joints in green concrete									
0120	1" depth	C-27	2000	.008	L.F.	.03	.40	.06	.49	.73
0140	1-1/2" depth		1800	.009		.05	.44	.06	.55	.83
0160	2" depth		1600	.010		.07	.50	.07	.64	.93
0180	Sawcut joint reservoir in cured concrete									
0182	3/8" wide x 3/4" deep, with single saw blade	C-27	1000	.016	L.F.	.05	.80	.11	.96	1.44
0184	1/2" wide x 1" deep, with double saw blades		900	.018		.10	.89	.12	1.11	1.64
0186	3/4" wide x 1-1/2" deep, with double saw blades		800	.020		.20	1	.14	1.34	1.94
0190	Water blast joint to wash away laitance, 2 passes	C-29	2500	.003			.13	.04	.17	.26
0200	Air blast joint to blow out debris and air dry, 2 passes	C-28	2000	.004			.20	.02	.22	.33
0300	For backer rod, see Section 07 91 23.10									
0340	For joint sealant, see Section 03 15 16.30 or 07 92 13.20									
0900	For replacement of joint sealant, see Section 07 01 90.81									

03 15 16.30 Expansion Joints

03 15 16.30 Expansion Joints		Crew	Daily Output	Labor-Hours	Unit	Material	Labor	Equipment	Total	Total Incl O&P
0010	**EXPANSION JOINTS**									
0020	Keyed, cold, 24 ga., incl. stakes, 3-1/2" high G	1 Carp	200	.040	L.F.	.89	2.13		3.02	4.39
0050	4-1/2" high G		200	.040		.94	2.13		3.07	4.44
0100	5-1/2" high G		195	.041		1.19	2.18		3.37	4.80
0150	7-1/2" high G		190	.042		1.40	2.24		3.64	5.15
0160	9-1/2" high G		185	.043		1.56	2.30		3.86	5.40
0300	Poured asphalt, plain, 1/2" x 1"	1 Clab	450	.018		.24	.75		.99	1.46
0350	1" x 2"		400	.020		.97	.84		1.81	2.42
0500	Neoprene, liquid, cold applied, 1/2" x 1"		450	.018		2.53	.75		3.28	3.98
0550	1" x 2"		400	.020		9.95	.84		10.79	12.30
0700	Polyurethane, poured, 2 part, 1/2" x 1"		400	.020		1.40	.84		2.24	2.89
0750	1" x 2"		350	.023		5.60	.96		6.56	7.75
0900	Rubberized asphalt, hot or cold applied, 1/2" x 1"		450	.018		.32	.75		1.07	1.55
0950	1" x 2"		400	.020		1.27	.84		2.11	2.75
1100	Hot applied, fuel resistant, 1/2" x 1"		450	.018		.48	.75		1.23	1.73
1150	1" x 2"		400	.020		1.91	.84		2.75	3.45
2000	Premolded, bituminous fiber, 1/2" x 6"	1 Carp	375	.021		.44	1.13		1.57	2.30
2050	1" x 12"		300	.027		2.03	1.42		3.45	4.50
2140	Concrete expansion joint, recycled paper and fiber, 1/2" x 6" G		390	.021		.43	1.09		1.52	2.22
2150	1/2" x 12" G		360	.022		.86	1.18		2.04	2.84
2250	Cork with resin binder, 1/2" x 6"		375	.021		1.36	1.13		2.49	3.32
2300	1" x 12"		300	.027		3.50	1.42		4.92	6.10
2500	Neoprene sponge, closed cell, 1/2" x 6"		375	.021		2.43	1.13		3.56	4.49
2550	1" x 12"		300	.027		8.90	1.42		10.32	12
2750	Polyethylene foam, 1/2" x 6"		375	.021		.59	1.13		1.72	2.47
2800	1" x 12"		300	.027		3.11	1.42		4.53	5.70
3000	Polyethylene backer rod, 3/8" diameter		460	.017		.03	.92		.95	1.52
3050	3/4" diameter		460	.017		.07	.92		.99	1.55
3100	1" diameter		460	.017		.15	.92		1.07	1.64
3500	Polyurethane foam, with polybutylene, 1/2" x 1/2"		475	.017		1.25	.90		2.15	2.81

03 15 Concrete Accessories

03 15 16 – Concrete Construction Joints

03 15 16.30 Expansion Joints

	03 15 16.30 Expansion Joints	Crew	Daily Output	Labor-Hours	Unit	Material	Labor	Equipment	Total	Total Incl O&P
						2020 Bare Costs				
3550	1″ x 1″	1 Carp	450	.018	L.F.	3.19	.95		4.14	5
3750	Polyurethane foam, regular, closed cell, 1/2″ x 6″		375	.021		.93	1.13		2.06	2.84
3800	1″ x 12″		300	.027		3.32	1.42		4.74	5.90
4000	Polyvinyl chloride foam, closed cell, 1/2″ x 6″		375	.021		2.53	1.13		3.66	4.60
4050	1″ x 12″		300	.027		8.70	1.42		10.12	11.80
4250	Rubber, gray sponge, 1/2″ x 6″		375	.021		2.12	1.13		3.25	4.15
4300	1″ x 12″		300	.027		7.65	1.42		9.07	10.65
4400	Redwood heartwood, 1″ x 4″		400	.020		1.18	1.06		2.24	2.99
4450	1″ x 6″		375	.021		1.79	1.13		2.92	3.79
5000	For installation in walls, add						75%			
5250	For installation in boxouts, add						25%			

03 15 19 – Cast-In Concrete Anchors

03 15 19.05 Anchor Bolt Accessories

	03 15 19.05 Anchor Bolt Accessories	Crew	Daily Output	Labor-Hours	Unit	Material	Labor	Equipment	Total	Total Incl O&P
0010	**ANCHOR BOLT ACCESSORIES**									
0015	For anchor bolts set in fresh concrete, see Section 03 15 19.10									
8150	Anchor bolt sleeve, plastic, 1″ diameter bolts	1 Carp	60	.133	Ea.	11.60	7.10		18.70	24
8500	1-1/2″ diameter		28	.286		21.50	15.20		36.70	48
8600	2″ diameter		24	.333		19.10	17.70		36.80	49.50
8650	3″ diameter		20	.400		35	21.50		56.50	72.50

03 15 19.10 Anchor Bolts

	03 15 19.10 Anchor Bolts	Crew	Daily Output	Labor-Hours	Unit	Material	Labor	Equipment	Total	Total Incl O&P
0010	**ANCHOR BOLTS**									
0015	Made from recycled materials									
0025	Single bolts installed in fresh concrete, no templates									
0030	Hooked w/nut and washer, 1/2″ diameter, 8″ long G	1 Carp	132	.061	Ea.	1.62	3.22		4.84	6.95
0040	12″ long G		131	.061		1.80	3.25		5.05	7.20
0050	5/8″ diameter, 8″ long G		129	.062		3.39	3.30		6.69	9.05
0060	12″ long G		127	.063		4.18	3.35		7.53	9.95
0070	3/4″ diameter, 8″ long G		127	.063		4.18	3.35		7.53	9.95
0080	12″ long G		125	.064		5.20	3.40		8.60	11.20
0090	2-bolt pattern, including job-built 2-hole template, per set									
0100	J-type, incl. hex nut & washer, 1/2″ diameter x 6″ long G	1 Carp	21	.381	Set	5.60	20.50		26.10	38.50
0110	12″ long G		21	.381		6.35	20.50		26.85	39.50
0120	18″ long G		21	.381		7.40	20.50		27.90	40.50
0130	3/4″ diameter x 8″ long G		20	.400		11.10	21.50		32.60	46
0140	12″ long G		20	.400		13.15	21.50		34.65	48.50
0150	18″ long G		20	.400		16.30	21.50		37.80	52
0160	1″ diameter x 12″ long G		19	.421		26.50	22.50		49	65
0170	18″ long G		19	.421		31	22.50		53.50	70.50
0180	24″ long G		19	.421		37.50	22.50		60	77
0190	36″ long G		18	.444		50.50	23.50		74	93.50
0200	1-1/2″ diameter x 18″ long G		17	.471		47.50	25		72.50	92
0210	24″ long G		16	.500		56	26.50		82.50	104
0300	L-type, incl. hex nut & washer, 3/4″ diameter x 12″ long G		20	.400		18.20	21.50		39.70	54
0310	18″ long G		20	.400		22.50	21.50		44	59
0320	24″ long G		20	.400		27	21.50		48.50	63.50
0330	30″ long G		20	.400		33.50	21.50		55	71
0340	36″ long G		20	.400		38	21.50		59.50	75.50
0350	1″ diameter x 12″ long G		19	.421		24.50	22.50		47	63
0360	18″ long G		19	.421		29.50	22.50		52	68.50
0370	24″ long G		19	.421		36	22.50		58.50	75.50
0380	30″ long G		19	.421		42	22.50		64.50	82.50
0390	36″ long G		18	.444		47.50	23.50		71	90.50

03 15 19 – Cast-In Concrete Anchors

03 15 19.10 Anchor Bolts			Crew	Daily Output	Labor-Hours	Unit	2020 Bare Costs Material	Labor	Equipment	Total	Total Incl O&P
0400	42" long	G	1 Carp	18	.444	Set	57.50	23.50		81	101
0410	48" long	G		18	.444		64	23.50		87.50	109
0420	1-1/4" diameter x 18" long	G		18	.444		43.50	23.50		67	86
0430	24" long	G		18	.444		51.50	23.50		75	94.50
0440	30" long	G		17	.471		59	25		84	105
0450	36" long	G	↓	17	.471		66.50	25		91.50	114
0460	42" long	G	2 Carp	32	.500		75	26.50		101.50	125
0470	48" long	G		32	.500		85	26.50		111.50	136
0480	54" long	G		31	.516		100	27.50		127.50	154
0490	60" long	G		31	.516		109	27.50		136.50	164
0500	1-1/2" diameter x 18" long	G		33	.485		47	26		73	93
0510	24" long	G		32	.500		54	26.50		80.50	102
0520	30" long	G		31	.516		61	27.50		88.50	111
0530	36" long	G		30	.533		69.50	28.50		98	122
0540	42" long	G		30	.533		79	28.50		107.50	132
0550	48" long	G		29	.552		88	29.50		117.50	144
0560	54" long	G		28	.571		107	30.50		137.50	166
0570	60" long	G		28	.571		117	30.50		147.50	177
0580	1-3/4" diameter x 18" long	G		31	.516		71.50	27.50		99	123
0590	24" long	G		30	.533		83.50	28.50		112	138
0600	30" long	G		29	.552		96.50	29.50		126	153
0610	36" long	G		28	.571		110	30.50		140.50	170
0620	42" long	G		27	.593		123	31.50		154.50	187
0630	48" long	G		26	.615		135	32.50		167.50	202
0640	54" long	G		26	.615		167	32.50		199.50	237
0650	60" long	G		25	.640		180	34		214	253
0660	2" diameter x 24" long	G		27	.593		135	31.50		166.50	200
0670	30" long	G		27	.593		152	31.50		183.50	218
0680	36" long	G		26	.615		167	32.50		199.50	236
0690	42" long	G		25	.640		185	34		219	259
0700	48" long	G		24	.667		213	35.50		248.50	291
0710	54" long	G		23	.696		253	37		290	335
0720	60" long	G		23	.696		272	37		309	360
0730	66" long	G		22	.727		291	38.50		329.50	380
0740	72" long	G	↓	21	.762	↓	320	40.50		360.50	415
1000	4-bolt pattern, including job-built 4-hole template, per set										
1100	J-type, incl. hex nut & washer, 1/2" diameter x 6" long	G	1 Carp	19	.421	Set	8.50	22.50		31	45.50
1110	12" long	G		19	.421		9.95	22.50		32.45	47
1120	18" long	G		18	.444		12.10	23.50		35.60	51.50
1130	3/4" diameter x 8" long	G		17	.471		19.45	25		44.45	61.50
1140	12" long	G		17	.471		23.50	25		48.50	66
1150	18" long	G		17	.471		30	25		55	73
1160	1" diameter x 12" long	G		16	.500		50	26.50		76.50	97.50
1170	18" long	G		15	.533		59.50	28.50		88	111
1180	24" long	G		15	.533		72	28.50		100.50	125
1190	36" long	G		15	.533		98	28.50		126.50	154
1200	1-1/2" diameter x 18" long	G		13	.615		92	32.50		124.50	154
1210	24" long	G		12	.667		109	35.50		144.50	177
1300	L-type, incl. hex nut & washer, 3/4" diameter x 12" long	G		17	.471		33.50	25		58.50	77
1310	18" long	G		17	.471		42.50	25		67.50	86.50
1320	24" long	G		17	.471		51	25		76	96
1330	30" long	G		16	.500		64	26.50		90.50	113
1340	36" long	G	↓	16	.500	↓	73	26.50		99.50	123

03 15 19 – Cast-In Concrete Anchors

03 15 19.10 Anchor Bolts			Crew	Daily Output	Labor-Hours	Unit	Material	2020 Bare Costs Labor	Equipment	Total	Total Incl O&P
1350	1" diameter x 12" long	G	1 Carp	16	.500	Set	46	26.50		72.50	93
1360	18" long	G		15	.533		56.50	28.50		85	108
1370	24" long	G		15	.533		69.50	28.50		98	122
1380	30" long	G		15	.533		81.50	28.50		110	135
1390	36" long	G		15	.533		92.50	28.50		121	148
1400	42" long	G		14	.571		112	30.50		142.50	172
1410	48" long	G		14	.571		125	30.50		155.50	187
1420	1-1/4" diameter x 18" long	G		14	.571		84.50	30.50		115	142
1430	24" long	G		14	.571		100	30.50		130.50	159
1440	30" long	G		13	.615		115	32.50		147.50	180
1450	36" long	G		13	.615		131	32.50		163.50	197
1460	42" long	G	2 Carp	25	.640		147	34		181	217
1470	48" long	G		24	.667		168	35.50		203.50	241
1480	54" long	G		23	.696		197	37		234	276
1490	60" long	G		23	.696		216	37		253	297
1500	1-1/2" diameter x 18" long	G		25	.640		91	34		125	155
1510	24" long	G		24	.667		105	35.50		140.50	173
1520	30" long	G		23	.696		119	37		156	190
1530	36" long	G		22	.727		136	38.50		174.50	212
1540	42" long	G		22	.727		155	38.50		193.50	232
1550	48" long	G		21	.762		173	40.50		213.50	256
1560	54" long	G		20	.800		211	42.50		253.50	300
1570	60" long	G		20	.800		231	42.50		273.50	320
1580	1-3/4" diameter x 18" long	G		22	.727		140	38.50		178.50	216
1590	24" long	G		21	.762		164	40.50		204.50	246
1600	30" long	G		21	.762		191	40.50		231.50	275
1610	36" long	G		20	.800		217	42.50		259.50	305
1620	42" long	G		19	.842		244	45		289	340
1630	48" long	G		18	.889		267	47		314	370
1640	54" long	G		18	.889		330	47		377	440
1650	60" long	G		17	.941		360	50		410	475
1660	2" diameter x 24" long	G		19	.842		267	45		312	365
1670	30" long	G		18	.889		300	47		347	405
1680	36" long	G		18	.889		330	47		377	440
1690	42" long	G		17	.941		370	50		420	485
1700	48" long	G		16	1		425	53		478	550
1710	54" long	G		15	1.067		505	56.50		561.50	645
1720	60" long	G		15	1.067		540	56.50		596.50	685
1730	66" long	G		14	1.143		580	60.50		640.50	735
1740	72" long	G		14	1.143		635	60.50		695.50	795
1990	For galvanized, add					Ea.	75%				

03 15 19.20 Dovetail Anchor System			Crew	Daily Output	Labor-Hours	Unit	Material	Labor	Equipment	Total	Total Incl O&P
0010	**DOVETAIL ANCHOR SYSTEM**										
0500	Dovetail anchor slot, galvanized, foam-filled, 26 ga.	G	1 Carp	425	.019	L.F.	1.38	1		2.38	3.12
0600	24 ga.	G		400	.020		1.67	1.06		2.73	3.54
0625	22 ga.	G		400	.020		2.10	1.06		3.16	4.01
0900	Stainless steel, foam-filled, 26 ga.	G		375	.021		2.21	1.13		3.34	4.25
1200	Dovetail brick anchor, corrugated, galvanized, 3-1/2" long, 16 ga.	G	1 Bric	10.50	.762	C	44.50	39.50		84	113
1300	12 ga.	G		10.50	.762		67	39.50		106.50	138
1500	Seismic, galvanized, 3-1/2" long, 16 ga.	G		10.50	.762		75.50	39.50		115	147
1600	12 ga.	G		10.50	.762		86	39.50		125.50	159
2000	Dovetail cavity wall, corrugated, galvanized, 5-1/2" long, 16 ga.	G		10.50	.762		46.50	39.50		86	115

03 15 Concrete Accessories

03 15 19 – Cast-In Concrete Anchors

03 15 19.20 Dovetail Anchor System			Crew	Daily Output	Labor-Hours	Unit	Material	Labor	Equipment	Total	Total Incl O&P
							2020 Bare Costs				
2100	12 ga.	G	1 Bric	10.50	.762	C	58	39.50		97.50	128
3000	Dovetail furring anchors, corrugated, galvanized, 1-1/2" long, 16 ga.	G		10.50	.762		24.50	39.50		64	91
3100	12 ga.	G		10.50	.762		33.50	39.50		73	101
6000	Dovetail stone panel anchors, galvanized, 1/8" x 1" wide, 3-1/2" long	G		10.50	.762		103	39.50		142.50	177
6100	1/4" x 1" wide	G	↓	10.50	.762	↓	183	39.50		222.50	266

03 15 19.30 Inserts			Crew	Daily Output	Labor-Hours	Unit	Material	Labor	Equipment	Total	Total Incl O&P
0010	**INSERTS**										
1000	Inserts, slotted nut type for 3/4" bolts, 4" long	G	1 Carp	84	.095	Ea.	20.50	5.05		25.55	30.50
2100	6" long	G		84	.095		24	5.05		29.05	34.50
2150	8" long	G		84	.095		32	5.05		37.05	43.50
2200	Slotted, strap type, 4" long	G		84	.095		21.50	5.05		26.55	31.50
2250	6" long	G		84	.095		24	5.05		29.05	34.50
2300	8" long	G		84	.095		34	5.05		39.05	45.50
2350	Strap for slotted insert, 4" long	G		84	.095		12.05	5.05		17.10	21.50
4100	6" long	G		84	.095		13.10	5.05		18.15	22.50
4150	8" long	G		84	.095		16.25	5.05		21.30	26
4200	10" long	G	↓	84	.095	↓	19.75	5.05		24.80	29.50
7000	Loop ferrule type										
7100	1/4" diameter bolt	G	1 Carp	84	.095	Ea.	2.23	5.05		7.28	10.55
7350	7/8" diameter bolt	G	"	84	.095	"	10.35	5.05		15.40	19.50
9000	Wedge type										
9100	For 3/4" diameter bolt	G	1 Carp	60	.133	Ea.	9.10	7.10		16.20	21.50
9800	Cut washers, black										
9900	3/4" bolt	G				Ea.	1.30			1.30	1.43
9950	For galvanized inserts, add						30%				

03 15 19.45 Machinery Anchors			Crew	Daily Output	Labor-Hours	Unit	Material	Labor	Equipment	Total	Total Incl O&P
0010	**MACHINERY ANCHORS**, heavy duty, incl. sleeve, floating base nut,										
0020	lower stud & coupling nut, fiber plug, connecting stud, washer & nut.										
0030	For flush mounted embedment in poured concrete heavy equip. pads.										
0200	Stud & bolt, 1/2" diameter	G	E-16	40	.400	Ea.	61.50	23.50	3.68	88.68	111
0300	5/8" diameter	G		35	.457		67	26.50	4.20	97.70	122
0500	3/4" diameter	G		30	.533		79.50	31	4.90	115.40	144
0600	7/8" diameter	G		25	.640		93.50	37.50	5.90	136.90	171
0800	1" diameter	G		20	.800		102	47	7.35	156.35	197
0900	1-1/4" diameter	G	↓	15	1.067	↓	133	62.50	9.80	205.30	260

03 21 Reinforcement Bars

03 21 05 – Reinforcing Steel Accessories

03 21 05.10 Rebar Accessories			Crew	Daily Output	Labor-Hours	Unit	Material	Labor	Equipment	Total	Total Incl O&P
0010	**REBAR ACCESSORIES**										
0030	Steel & plastic made from recycled materials										
0100	Beam bolsters (BB), lower, 1-1/2" high, plain steel	G				C.L.F.	38.50			38.50	42
0102	Galvanized	G					41			41	45.50
0104	Stainless tipped legs	G					475			475	525
0106	Plastic tipped legs	G					36			36	39.50
0108	Epoxy dipped	G					98.50			98.50	108
0110	2" high, plain	G					49.50			49.50	54.50
0120	Galvanized	G					52			52	57.50
0140	Stainless tipped legs	G					535			535	590
0160	Plastic tipped legs	G				↓	66.50			66.50	73

03 21 05.10 Rebar Accessories			Crew	Daily Output	Labor-Hours	Unit	2020 Bare Costs Material	Labor	Equipment	Total	Total Incl O&P
0162	Epoxy dipped	G				C.L.F.	110			110	121
0200	Upper (BBU), 1-1/2" high, plain steel	G					78.50			78.50	86.50
0210	3" high	G					100			100	110
0500	Slab bolsters, continuous (SB), 1" high, plain steel	G					32.50			32.50	35.50
0502	Galvanized	G					38.50			38.50	42
0504	Stainless tipped legs	G					465			465	515
0506	Plastic tipped legs	G					46			46	51
0510	2" high, plain steel	G					40			40	44
0515	Galvanized	G					48.50			48.50	53.50
0520	Stainless tipped legs	G					555			555	610
0525	Plastic tipped legs	G					53.50			53.50	59
0530	For bolsters with wire runners (SBR), add	G					44			44	48
0540	For bolsters with plates (SBP), add	G				↓	99			99	109
0700	Bag ties, 16 ga., plain, 4" long	G				C	4.30			4.30	4.73
0710	5" long	G					5.55			5.55	6.10
0720	6" long	G					3.93			3.93	4.32
0730	7" long	G					5.85			5.85	6.45
1200	High chairs, individual (HC), 3" high, plain steel	G					66.50			66.50	73.50
1202	Galvanized	G					95.50			95.50	105
1204	Stainless tipped legs	G					530			530	585
1206	Plastic tipped legs	G					70.50			70.50	77.50
1210	5" high, plain	G					97			97	107
1212	Galvanized	G					124			124	136
1214	Stainless tipped legs	G					570			570	625
1216	Plastic tipped legs	G					105			105	116
1220	8" high, plain	G					141			141	155
1222	Galvanized	G					194			194	214
1224	Stainless tipped legs	G					645			645	705
1226	Plastic tipped legs	G					152			152	168
1230	12" high, plain	G					340			340	375
1232	Galvanized	G					385			385	425
1234	Stainless tipped legs	G					700			700	770
1236	Plastic tipped legs	G					360			360	400
1400	Individual high chairs, with plate (HCP), 5" high	G					192			192	211
1410	8" high	G					265			265	292
1500	Bar chair (BC), 1-1/2" high, plain steel	G					51			51	56
1520	Galvanized	G					45.50			45.50	50
1530	Stainless tipped legs	G					480			480	525
1540	Plastic tipped legs	G				↓	44			44	48.50
1700	Continuous high chairs (CHC), legs 8" OC, 4" high, plain steel	G				C.L.F.	57			57	63
1705	Galvanized	G					67.50			67.50	74.50
1710	Stainless tipped legs	G					540			540	590
1715	Plastic tipped legs	G					78.50			78.50	86
1718	Epoxy dipped	G					95			95	104
1720	6" high, plain	G					75			75	82.50
1725	Galvanized	G					109			109	119
1730	Stainless tipped legs	G					560			560	615
1735	Plastic tipped legs	G					101			101	111
1738	Epoxy dipped	G					129			129	142
1740	8" high, plain	G					132			132	145
1745	Galvanized	G					134			134	148
1750	Stainless tipped legs	G					630			630	695
1755	Plastic tipped legs	G				↓	132			132	145

03 21 05.10 Rebar Accessories

	03 21 05.10 Rebar Accessories		Crew	Daily Output	Labor-Hours	Unit	Material	Labor	Equipment	Total	Total Incl O&P
							2020 Bare Costs				
1758	Epoxy dipped	G				C.L.F.	168			168	185
1900	For continuous bottom wire runners, add	G					32			32	35.50
1940	For continuous bottom plate, add	G					191			191	210
2200	Screed chair base, 1/2" coil thread diam., 2-1/2" high, plain steel	G				C	345			345	380
2210	Galvanized	G					385			385	420
2220	5-1/2" high, plain	G					420			420	460
2250	Galvanized	G					460			460	510
2300	3/4" coil thread diam., 2-1/2" high, plain steel	G					420			420	460
2310	Galvanized	G					485			485	530
2320	5-1/2" high, plain steel	G					535			535	585
2350	Galvanized	G					665			665	730
2400	Screed holder, 1/2" coil thread diam. for pipe screed, plain steel, 6" long	G					445			445	490
2420	12" long	G					635			635	700
2500	3/4" coil thread diam. for pipe screed, plain steel, 6" long	G					590			590	645
2520	12" long	G					940			940	1,025
2700	Screw anchor for bolts, plain steel, 3/4" diameter x 4" long	G					535			535	585
2720	1" diameter x 6" long	G					1,000			1,000	1,100
2740	1-1/2" diameter x 8" long	G					1,225			1,225	1,350
2800	Screw anchor eye bolts, 3/4" x 3" long	G					3,125			3,125	3,450
2820	1" x 3-1/2" long	G					3,700			3,700	4,075
2840	1-1/2" x 6" long	G					12,400			12,400	13,600
2900	Screw anchor bolts, 3/4" x 9" long	G					1,775			1,775	1,950
2920	1" x 12" long	G					3,425			3,425	3,750
3001	Slab lifting inserts, single pickup, galv, 3/4" diam., 5" high	G					1,875			1,875	2,050
3010	6" high	G					1,900			1,900	2,075
3030	7" high	G					1,925			1,925	2,100
3100	1" diameter, 5-1/2" high	G					1,950			1,950	2,150
3120	7" high	G					2,025			2,025	2,225
3200	Double pickup lifting inserts, 1" diameter, 5-1/2" high	G					3,775			3,775	4,150
3220	7" high	G					4,175			4,175	4,575
3330	1-1/2" diameter, 8" high	G					5,075			5,075	5,575
3800	Subgrade chairs, #4 bar head, 3-1/2" high	G					44			44	48
3850	12" high	G					53			53	58
3900	#6 bar head, 3-1/2" high	G					44			44	48
3950	12" high	G					53			53	58
4200	Subgrade stakes, no nail holes, 3/4" diameter, 12" long	G					143			143	157
4250	24" long	G					274			274	300
4300	7/8" diameter, 12" long	G					420			420	465
4350	24" long	G					680			680	745
4500	Tie wire, 16 ga. annealed steel	G				Cwt.	197			197	217

03 21 05.75 Splicing Reinforcing Bars

	03 21 05.75 Splicing Reinforcing Bars		Crew	Daily Output	Labor-Hours	Unit	Material	Labor	Equipment	Total	Total Incl O&P
0010	**SPLICING REINFORCING BARS**										
0020	Including holding bars in place while splicing										
0100	Standard, self-aligning type, taper threaded, #4 bars	G	C-25	190	.168	Ea.	7.90	7.65		15.55	21.50
0105	#5 bars	G		170	.188		9.85	8.55		18.40	25
0110	#6 bars	G		150	.213		11.10	9.70		20.80	28
0120	#7 bars	G		130	.246		12.80	11.20		24	32.50
0300	#8 bars	G		115	.278		21	12.65		33.65	44.50
0305	#9 bars	G	C-5	105	.533		25	30	5.55	60.55	80.50
0310	#10 bars	G		95	.589		27.50	33	6.10	66.60	89.50
0320	#11 bars	G		85	.659		29.50	37	6.85	73.35	98
0330	#14 bars	G		65	.862		38	48.50	8.95	95.45	127

03 21 05 – Reinforcing Steel Accessories

03 21 05.75 Splicing Reinforcing Bars			Crew	Daily Output	Labor-Hours	Unit	2020 Bare Costs Material	Labor	Equipment	Total	Total Incl O&P
0340	#18 bars	G	C-5	45	1.244	Ea.	57.50	70	12.90	140.40	188
0500	Transition self-aligning, taper threaded, #18-14	G		45	1.244		77.50	70	12.90	160.40	210
0510	#18-11	G		45	1.244		79	70	12.90	161.90	211
0520	#14-11	G		65	.862		52.50	48.50	8.95	109.95	143
0540	#11-10	G		85	.659		34	37	6.85	77.85	104
0550	#10-9	G		95	.589		33	33	6.10	72.10	95.50
0560	#9-8	G	C-25	105	.305		29.50	13.85		43.35	55.50
0580	#8-7	G		115	.278		27.50	12.65		40.15	51.50
0590	#7-6	G		130	.246		21	11.20		32.20	41.50
0600	Position coupler for curved bars, taper threaded, #4 bars	G		160	.200		35.50	9.10		44.60	54
0610	#5 bars	G		145	.221		37	10.05		47.05	57.50
0620	#6 bars	G		130	.246		45	11.20		56.20	68
0630	#7 bars	G		110	.291		49.50	13.25		62.75	76
0640	#8 bars	G		100	.320		51.50	14.55		66.05	80.50
0650	#9 bars	G	C-5	90	.622		56	35	6.45	97.45	124
0660	#10 bars	G		80	.700		60.50	39.50	7.25	107.25	137
0670	#11 bars	G		70	.800		63	45	8.30	116.30	149
0680	#14 bars	G		55	1.018		78.50	57.50	10.55	146.55	188
0690	#18 bars	G		40	1.400		97.50	79	14.50	191	247
0700	Transition position coupler for curved bars, taper threaded, #18-14	G		40	1.400		141	79	14.50	234.50	295
0710	#18-11	G		40	1.400		143	79	14.50	236.50	297
0720	#14-11	G		55	1.018		106	57.50	10.55	174.05	219
0730	#11-10	G		70	.800		81.50	45	8.30	134.80	170
0740	#10-9	G		80	.700		77.50	39.50	7.25	124.25	156
0750	#9-8	G	C-25	90	.356		72	16.20		88.20	106
0760	#8-7	G		100	.320		66.50	14.55		81.05	97
0770	#7-6	G		110	.291		62.50	13.25		75.75	90
0800	Sleeve type w/grout filler, for precast concrete, #6 bars	G		72	.444		32.50	20		52.50	68.50
0802	#7 bars	G		64	.500		37.50	23		60.50	78.50
0805	#8 bars	G		56	.571		45	26		71	92
0807	#9 bars	G		48	.667		52.50	30.50		83	108
0810	#10 bars	G	C-5	40	1.400		62.50	79	14.50	156	208
0900	#11 bars	G		32	1.750		72	98.50	18.15	188.65	254
0920	#14 bars	G		24	2.333		110	131	24	265	355
1000	Sleeve type w/ferrous filler, for critical structures, #6 bars	G	C-25	72	.444		107	20		127	151
1210	#7 bars	G		64	.500		109	23		132	158
1220	#8 bars	G		56	.571		114	26		140	169
1230	#9 bars	G	C-5	48	1.167		119	65.50	12.10	196.60	247
1240	#10 bars	G		40	1.400		125	79	14.50	218.50	278
1250	#11 bars	G		32	1.750		152	98.50	18.15	268.65	340
1260	#14 bars	G		24	2.333		190	131	24	345	440
1270	#18 bars	G		16	3.500		236	197	36.50	469.50	610
2000	Weldable half coupler, taper threaded, #4 bars	G	E-16	120	.133		11.60	7.80	1.23	20.63	27
2100	#5 bars	G		112	.143		13.70	8.35	1.31	23.36	30
2200	#6 bars	G		104	.154		21.50	9	1.41	31.91	40.50
2300	#7 bars	G		96	.167		25.50	9.75	1.53	36.78	45.50
2400	#8 bars	G		88	.182		26.50	10.65	1.67	38.82	48
2500	#9 bars	G		80	.200		29	11.70	1.84	42.54	53
2600	#10 bars	G		72	.222		29.50	13	2.04	44.54	56.50
2700	#11 bars	G		64	.250		31.50	14.60	2.30	48.40	61.50
2800	#14 bars	G		56	.286		36.50	16.70	2.63	55.83	71
2900	#18 bars	G		48	.333		59.50	19.50	3.06	82.06	101

03 21 Reinforcement Bars

03 21 11 – Plain Steel Reinforcement Bars

03 21 11.60 Reinforcing In Place		Crew	Daily Output	Labor-Hours	Unit	2020 Bare Costs Material	Labor	Equipment	Total	Total Incl O&P
0010	**REINFORCING IN PLACE**, 50-60 ton lots, A615 Grade 60 R032110-10									
0020	Includes labor, but not material cost, to install accessories									
0030	Made from recycled materials									
0102	Beams & Girders, #3 to #7 G	4 Rodm	3200	.010	Lb.	.56	.56		1.12	1.51
0152	#8 to #18 G		5400	.006		.56	.33		.89	1.15
0202	Columns, #3 to #7 G		3000	.011		.56	.60		1.16	1.57
0252	#8 to #18 G		4600	.007		.56	.39		.95	1.24
0300	Spirals, hot rolled, 8″ to 15″ diameter G		2.20	14.545	Ton	1,675	820		2,495	3,125
0320	15″ to 24″ diameter G		2.20	14.545		1,600	820		2,420	3,050
0330	24″ to 36″ diameter G		2.30	13.913		1,525	785		2,310	2,900
0340	36″ to 48″ diameter G		2.40	13.333		1,450	750		2,200	2,750
0360	48″ to 64″ diameter G		2.50	12.800		1,600	720		2,320	2,875
0380	64″ to 84″ diameter G		2.60	12.308		1,675	695		2,370	2,925
0390	84″ to 96″ diameter G		2.70	11.852		1,750	670		2,420	2,975
0402	Elevated slabs, #4 to #7 R032110-80		5800	.006	Lb.	.56	.31		.87	1.11
0502	Footings, #4 to #7 G		4200	.008		.56	.43		.99	1.30
0552	#8 to #18 G		7200	.004		.56	.25		.81	1.01
0602	Slab on grade, #3 to #7 G		4200	.008		.56	.43		.99	1.30
0702	Walls, #3 to #7 G		6000	.005		.56	.30		.86	1.09
0752	#8 to #18 G		8000	.004		.56	.23		.79	.98
0900	For other than 50-60 ton lots									
1000	Under 10 ton job, #3 to #7, add					25%	10%			
1010	#8 to #18, add					20%	10%			
1050	10-50 ton job, #3 to #7, add					10%				
1060	#8 to #18, add					5%				
1100	60-100 ton job, #3 to #7, deduct					5%				
1110	#8 to #18, deduct					10%				
1150	Over 100 ton job, #3 to #7, deduct					10%				
1160	#8 to #18, deduct					15%				
1200	Reinforcing in place, A615 Grade 75, add G				Ton	102			102	112
1220	Grade 90, add					140			140	154
2000	Unloading & sorting, add to above	C-5	100	.560			31.50	5.80	37.30	56
2200	Crane cost for handling, 90 picks/day, up to 1.5 tons/bundle, add to above		135	.415			23.50	4.30	27.80	41
2210	1.0 ton/bundle		92	.609			34	6.30	40.30	61
2220	0.5 ton/bundle		35	1.600			90	16.60	106.60	160
2400	Dowels, 2 feet long, deformed, #3 G	2 Rodm	520	.031	Ea.	.47	1.74		2.21	3.24
2410	#4 G		480	.033		.83	1.88		2.71	3.87
2420	#5 G		435	.037		1.30	2.07		3.37	4.70
2430	#6 G		360	.044		1.87	2.51		4.38	6
2450	Longer and heavier dowels, add G		725	.022	Lb.	.62	1.24		1.86	2.64
2500	Smooth dowels, 12″ long, 1/4″ or 3/8″ diameter G		140	.114	Ea.	.87	6.45		7.32	11.10
2520	5/8″ diameter G		125	.128		1.53	7.20		8.73	13.05
2530	3/4″ diameter G		110	.145		1.89	8.20		10.09	15
2600	Dowel sleeves for CIP concrete, 2-part system									
2610	Sleeve base, plastic, for 5/8″ smooth dowel sleeve, fasten to edge form	1 Rodm	200	.040	Ea.	.54	2.26		2.80	4.14
2615	Sleeve, plastic, 12″ long, for 5/8″ smooth dowel, snap onto base		400	.020		1.54	1.13		2.67	3.47
2620	Sleeve base, for 3/4″ smooth dowel sleeve		175	.046		.54	2.58		3.12	4.65
2625	Sleeve, 12″ long, for 3/4″ smooth dowel		350	.023		1.22	1.29		2.51	3.37
2630	Sleeve base, for 1″ smooth dowel sleeve		150	.053		.71	3.01		3.72	5.50
2635	Sleeve, 12″ long, for 1″ smooth dowel		300	.027		1.67	1.50		3.17	4.21
2700	Dowel caps, visual warning only, plastic, #3 to #8	2 Rodm	800	.020		.37	1.13		1.50	2.19
2720	#8 to #18		750	.021		.89	1.20		2.09	2.88

03 21 Reinforcement Bars

03 21 11 – Plain Steel Reinforcement Bars

03 21 11.60 Reinforcing In Place		Crew	Daily Output	Labor-Hours	Unit	2020 Bare Costs Material	Labor	Equipment	Total	Total Incl O&P
2750	Impalement protective, plastic, #4 to #9	2 Rodm	800	.020	Ea.	1.18	1.13		2.31	3.08
9000	Minimum labor/equipment charge	1 Rodm	4	2	Job		113		113	178

03 21 13 – Galvanized Reinforcement Steel Bars

03 21 13.10 Galvanized Reinforcing		Crew	Daily Output	Labor-Hours	Unit	Material	Labor	Equipment	Total	Total Incl O&P
0010	**GALVANIZED REINFORCING**									
0150	Add to plain steel rebar pricing for galvanized rebar				Ton	510			510	560

03 21 16 – Epoxy-Coated Reinforcement Steel Bars

03 21 16.10 Epoxy-Coated Reinforcing		Crew	Daily Output	Labor-Hours	Unit	Material	Labor	Equipment	Total	Total Incl O&P
0010	**EPOXY-COATED REINFORCING**									
0100	Add to plain steel rebar pricing for epoxy-coated rebar				Ton	1,075			1,075	1,200

03 21 19 – Stainless Steel Reinforcement Bars

03 21 19.10 Stainless Steel Reinforcing		Crew	Daily Output	Labor-Hours	Unit	Material	Labor	Equipment	Total	Total Incl O&P
0010	**STAINLESS STEEL REINFORCING**									
0100	Add to plain steel rebar pricing for stainless steel rebar					300%				

03 21 21 – Composite Reinforcement Bars

03 21 21.11 Glass Fiber-Reinforced Polymer Reinf. Bars		Crew	Daily Output	Labor-Hours	Unit	Material	Labor	Equipment	Total	Total Incl O&P
0010	**GLASS FIBER-REINFORCED POLYMER REINFORCEMENT BARS**									
0020	Includes labor, but not material cost, to install accessories									
0050	#2 bar, .043 lb./L.F.	4 Rodm	9500	.003	L.F.	.41	.19		.60	.75
0100	#3 bar, .092 lb./L.F.		9300	.003		.65	.19		.84	1.03
0150	#4 bar, .160 lb./L.F.		9100	.004		.94	.20		1.14	1.34
0200	#5 bar, .258 lb./L.F.		8700	.004		1.14	.21		1.35	1.58
0250	#6 bar, .372 lb./L.F.		8300	.004		1.92	.22		2.14	2.45
0300	#7 bar, .497 lb./L.F.		7900	.004		2.59	.23		2.82	3.21
0350	#8 bar, .620 lb./L.F.		7400	.004		2.45	.24		2.69	3.08
0400	#9 bar, .800 lb./L.F.		6800	.005		4.22	.27		4.49	5.05
0450	#10 bar, 1.08 lb./L.F.	↓	5800	.006	↓	3.83	.31		4.14	4.70
0500	For bends, add per bend				Ea.	1.64			1.64	1.80

03 22 Fabric and Grid Reinforcing

03 22 11 – Plain Welded Wire Fabric Reinforcing

03 22 11.10 Plain Welded Wire Fabric		Crew	Daily Output	Labor-Hours	Unit	Material	Labor	Equipment	Total	Total Incl O&P
0010	**PLAIN WELDED WIRE FABRIC** ASTM A185									
0020	Includes labor, but not material cost, to install accessories									
0030	Made from recycled materials									
0050	Sheets									
0100	6 x 6 - W1.4 x W1.4 (10 x 10) 21 lb./C.S.F. G	2 Rodm	35	.457	C.S.F.	16.25	26		42.25	58.50
0200	6 x 6 - W2.1 x W2.1 (8 x 8) 30 lb./C.S.F. G		31	.516		19.15	29		48.15	67
0300	6 x 6 - W2.9 x W2.9 (6 x 6) 42 lb./C.S.F. G		29	.552		27.50	31		58.50	79.50
0400	6 x 6 - W4 x W4 (4 x 4) 58 lb./C.S.F. G		27	.593		38.50	33.50		72	95
0500	4 x 4 - W1.4 x W1.4 (10 x 10) 31 lb./C.S.F. G		31	.516		24.50	29		53.50	72.50
0600	4 x 4 - W2.1 x W2.1 (8 x 8) 44 lb./C.S.F. G		29	.552		30.50	31		61.50	82.50
0650	4 x 4 - W2.9 x W2.9 (6 x 6) 61 lb./C.S.F. G		27	.593		48.50	33.50		82	106
0700	4 x 4 - W4 x W4 (4 x 4) 85 lb./C.S.F. G	↓	25	.640	↓	60.50	36		96.50	124
0750	Rolls									
0800	2 x 2 - #14 galv., 21 lb./C.S.F., beam & column wrap G	2 Rodm	6.50	2.462	C.S.F.	48.50	139		187.50	273
0900	2 x 2 - #12 galv. for gunite reinforcing G	"	6.50	2.462	"	73	139		212	300
9000	Minimum labor/equipment charge	1 Rodm	4	2	Job		113		113	178

03 22 Fabric and Grid Reinforcing

03 22 13 – Galvanized Welded Wire Fabric Reinforcing

03 22 13.10 Galvanized Welded Wire Fabric			Crew	Daily Output	Labor-Hours	Unit	2020 Bare Costs Material	Labor	Equipment	Total	Total Incl O&P
0010	GALVANIZED WELDED WIRE FABRIC										
0100	Add to plain welded wire pricing for galvanized welded wire					Lb.	.26			.26	.28

03 22 16 – Epoxy-Coated Welded Wire Fabric Reinforcing

03 22 16.10 Epoxy-Coated Welded Wire Fabric			Crew	Daily Output	Labor-Hours	Unit	Material	Labor	Equipment	Total	Total Incl O&P
0010	EPOXY-COATED WELDED WIRE FABRIC										
0100	Add to plain welded wire pricing for epoxy-coated welded wire					Lb.	.54			.54	.60

03 23 Stressed Tendon Reinforcing

03 23 05 – Prestressing Tendons

03 23 05.50 Prestressing Steel			Crew	Daily Output	Labor-Hours	Unit	Material	Labor	Equipment	Total	Total Incl O&P
0010	PRESTRESSING STEEL										
0100	Grouted strand, in beams, post-tensioned in field, 50′ span, 100 kip	G	C-3	1200	.053	Lb.	3.15	2.81	.15	6.11	8.05
0150	300 kip	G		2700	.024		1.27	1.25	.07	2.59	3.44
0300	100′ span, 100 kip	G		1700	.038		3.15	1.98	.11	5.24	6.70
0350	300 kip	G		3200	.020		2.69	1.05	.06	3.80	4.68
0500	200′ span, 100 kip	G		2700	.024		3.14	1.25	.07	4.46	5.50
0550	300 kip	G		3500	.018		2.68	.96	.05	3.69	4.53
0800	Grouted bars, in beams, 50′ span, 42 kip	G		2600	.025		1.26	1.30	.07	2.63	3.52
0850	143 kip	G		3200	.020		1.21	1.05	.06	2.32	3.05
1000	75′ span, 42 kip	G		3200	.020		1.28	1.05	.06	2.39	3.13
1050	143 kip	G		4200	.015		1.07	.80	.04	1.91	2.50
1200	Ungrouted strand, in beams, 50′ span, 100 kip	G	C-4	1275	.025		.64	1.43	.04	2.11	3
1250	300 kip	G		1475	.022		.64	1.23	.04	1.91	2.68
1400	100′ span, 100 kip	G		1500	.021		.64	1.21	.04	1.89	2.65
1450	300 kip	G		1650	.019		.64	1.10	.03	1.77	2.48
1600	200′ span, 100 kip	G		1500	.021		.64	1.21	.04	1.89	2.65
1650	300 kip	G		1700	.019		.64	1.07	.03	1.74	2.43
1800	Ungrouted bars, in beams, 50′ span, 42 kip	G		1400	.023		.56	1.30	.04	1.90	2.71
1850	143 kip	G		1700	.019		.56	1.07	.03	1.66	2.35
2000	75′ span, 42 kip	G		1800	.018		.56	1.01	.03	1.60	2.24
2050	143 kip	G		2200	.015		.56	.83	.03	1.42	1.95
2220	Ungrouted single strand, 100′ elevated slab, 25 kip	G		1200	.027		.64	1.52	.05	2.21	3.14
2250	35 kip	G		1475	.022		.64	1.23	.04	1.91	2.68
3000	Slabs on grade, 0.5-inch diam. non-bonded strands, HDPE sheathed,										
3050	attached dead-end anchors, loose stressing-end anchors										
3100	25′ x 30′ slab, strands @ 36″ OC, placing		2 Rodm	2940	.005	S.F.	.68	.31		.99	1.23
3105	Stressing		C-4A	3750	.004			.24	.02	.26	.40
3110	42″ OC, placing		2 Rodm	3200	.005		.60	.28		.88	1.10
3115	Stressing		C-4A	4040	.004			.22	.02	.24	.37
3120	48″ OC, placing		2 Rodm	3510	.005		.53	.26		.79	.99
3125	Stressing		C-4A	4390	.004			.21	.02	.23	.34
3150	25′ x 40′ slab, strands @ 36″ OC, placing		2 Rodm	3370	.005		.66	.27		.93	1.14
3155	Stressing		C-4A	4360	.004			.21	.02	.23	.35
3160	42″ OC, placing		2 Rodm	3760	.004		.57	.24		.81	1.01
3165	Stressing		C-4A	4820	.003			.19	.02	.21	.32
3170	48″ OC, placing		2 Rodm	4090	.004		.51	.22		.73	.91
3175	Stressing		C-4A	5190	.003			.17	.01	.18	.29
3200	30′ x 30′ slab, strands @ 36″ OC, placing		2 Rodm	3260	.005		.66	.28		.94	1.16
3205	Stressing		C-4A	4190	.004			.22	.02	.24	.36
3210	42″ OC, placing		2 Rodm	3530	.005		.59	.26		.85	1.05

03 23 Stressed Tendon Reinforcing

03 23 05 – Prestressing Tendons

03 23 05.50 Prestressing Steel		Crew	Daily Output	Labor-Hours	Unit	Material	2020 Bare Costs Labor	Equipment	Total	Total Incl O&P
3215	Stressing	C-4A	4500	.004	S.F.		.20	.02	.22	.34
3220	48" OC, placing	2 Rodm	3840	.004		.52	.24		.76	.95
3225	Stressing	C-4A	4850	.003			.19	.02	.21	.31
3230	30' x 40' slab, strands @ 36" OC, placing	2 Rodm	3780	.004		.63	.24		.87	1.07
3235	Stressing	C-4A	4920	.003			.18	.02	.20	.31
3240	42" OC, placing	2 Rodm	4190	.004		.55	.22		.77	.95
3245	Stressing	C-4A	5410	.003			.17	.01	.18	.28
3250	48" OC, placing	2 Rodm	4520	.004		.50	.20		.70	.86
3255	Stressing	C-4A	5790	.003			.16	.01	.17	.26
3260	30' x 50' slab, strands @ 36" OC, placing	2 Rodm	4300	.004		.60	.21		.81	.98
3265	Stressing	C-4A	5650	.003			.16	.01	.17	.26
3270	42" OC, placing	2 Rodm	4720	.003		.53	.19		.72	.88
3275	Stressing	C-4A	6150	.003			.15	.01	.16	.24
3280	48" OC, placing	2 Rodm	5240	.003		.46	.17		.63	.78
3285	Stressing	C-4A	6760	.002	↓		.13	.01	.14	.22
3900	Minimum labor/equipment charge, placing	2 Rodm	2	8	Job		450		450	710
3905	Minimum labor/equipment charge, stressing	C-4A	2	8	"		450	38	488	750

03 24 Fibrous Reinforcing

03 24 05 – Reinforcing Fibers

03 24 05.30 Synthetic Fibers		Crew	Daily Output	Labor-Hours	Unit	Material	Labor	Equipment	Total	Total Incl O&P
0010	**SYNTHETIC FIBERS**									
0100	Synthetic fibers, add to concrete				Lb.	5.05			5.05	5.55
0110	1-1/2 lb./C.Y.				C.Y.	7.80			7.80	8.55

03 24 05.70 Steel Fibers		Crew	Daily Output	Labor-Hours	Unit	Material	Labor	Equipment	Total	Total Incl O&P
0010	**STEEL FIBERS**									
0140	ASTM A850, Type V, continuously deformed, 1-1/2" long x 0.045" diam.									
0150	Add to price of ready mix concrete G				Lb.	1.23			1.23	1.35
0205	Alternate pricing, dosing at 5 lb./C.Y., add to price of RMC G				C.Y.	6.15			6.15	6.75
0210	10 lb./C.Y. G					12.30			12.30	13.55
0215	15 lb./C.Y. G					18.45			18.45	20.50
0220	20 lb./C.Y. G					24.50			24.50	27
0225	25 lb./C.Y. G					31			31	34
0230	30 lb./C.Y. G					37			37	40.50
0235	35 lb./C.Y. G					43			43	47.50
0240	40 lb./C.Y. G					49			49	54
0250	50 lb./C.Y. G					61.50			61.50	67.50
0275	75 lb./C.Y. G					92.50			92.50	101
0300	100 lb./C.Y. G				↓	123			123	135

03 30 53.40 Concrete In Place		Crew	Daily Output	Labor-Hours	Unit	2020 Bare Costs Material	Labor	Equipment	Total	Total Incl O&P
0010	CONCRETE IN PLACE R033105-20									
0020	Including forms (4 uses), Grade 60 rebar, concrete (Portland cement R033105-70									
0050	Type I), placement and finishing unless otherwise indicated R033105-80									
0300	Beams (3500 psi), 5 kip/L.F., 10′ span	C-14A	15.62	12.804	C.Y.	370	675	28	1,073	1,500
0350	25′ span	″	18.55	10.782		390	570	23.50	983.50	1,375
0500	Chimney foundations (5000 psi), over 5 C.Y. R033053-50	C-14C	32.22	3.476		178	175	.83	353.83	475
0510	(3500 psi), under 5 C.Y.	″	23.71	4.724		208	238	1.13	447.13	610
0700	Columns, square (4000 psi), 12″ x 12″, up to 1% reinforcing by area	C-14A	11.96	16.722		410	885	36.50	1,331.50	1,900
0720	Up to 2% reinforcing by area		10.13	19.743		645	1,050	43	1,738	2,425
0740	Up to 3% reinforcing by area		9.03	22.148		965	1,175	48.50	2,188.50	2,975
0800	16″ x 16″, up to 1% reinforcing by area		16.22	12.330		325	650	27	1,002	1,450
0820	Up to 2% reinforcing by area		12.57	15.911		550	840	35	1,425	2,000
0840	Up to 3% reinforcing by area		10.25	19.512		850	1,025	42.50	1,917.50	2,625
0900	24″ x 24″, up to 1% reinforcing by area		23.66	8.453		278	445	18.50	741.50	1,050
0920	Up to 2% reinforcing by area		17.71	11.293		490	595	24.50	1,109.50	1,525
0940	Up to 3% reinforcing by area		14.15	14.134		780	745	31	1,556	2,100
1000	36″ x 36″, up to 1% reinforcing by area		33.69	5.936		247	315	13	575	785
1020	Up to 2% reinforcing by area		23.32	8.576		435	455	18.80	908.80	1,225
1040	Up to 3% reinforcing by area		17.82	11.223		725	595	24.50	1,344.50	1,775
1100	Columns, round (4000 psi), tied, 12″ diameter, up to 1% reinforcing by area		20.97	9.537		415	505	21	941	1,275
1120	Up to 2% reinforcing by area		15.27	13.098		645	695	28.50	1,368.50	1,850
1140	Up to 3% reinforcing by area		12.11	16.515		955	875	36	1,866	2,500
1200	16″ diameter, up to 1% reinforcing by area		31.49	6.351		340	335	13.90	688.90	920
1220	Up to 2% reinforcing by area		19.12	10.460		575	555	23	1,153	1,525
1240	Up to 3% reinforcing by area		13.77	14.524		855	770	32	1,657	2,200
1300	20″ diameter, up to 1% reinforcing by area		41.04	4.873		325	258	10.65	593.65	780
1320	Up to 2% reinforcing by area		24.05	8.316		540	440	18.20	998.20	1,325
1340	Up to 3% reinforcing by area		17.01	11.758		845	620	26	1,491	1,950
1400	24″ diameter, up to 1% reinforcing by area		51.85	3.857		315	204	8.45	527.45	685
1420	Up to 2% reinforcing by area		27.06	7.391		555	390	16.20	961.20	1,250
1440	Up to 3% reinforcing by area		18.29	10.935		840	580	24	1,444	1,875
1500	36″ diameter, up to 1% reinforcing by area		75.04	2.665		315	141	5.85	461.85	575
1520	Up to 2% reinforcing by area		37.49	5.335		525	282	11.70	818.70	1,050
1540	Up to 3% reinforcing by area		22.84	8.757		815	465	19.20	1,299.20	1,650
1900	Elevated slab (4000 psi), flat slab with drops, 125 psf Sup. Load, 20′ span	C-14B	38.45	5.410		300	285	11.35	596.35	800
1950	30′ span		50.99	4.079		320	215	8.55	543.55	710
2100	Flat plate, 125 psf Sup. Load, 15′ span		30.24	6.878		274	365	14.45	653.45	895
2150	25′ span		49.60	4.194		287	221	8.80	516.80	680
2300	Waffle const., 30″ domes, 125 psf Sup. Load, 20′ span		37.07	5.611		310	296	11.80	617.80	830
2350	30′ span		44.07	4.720		290	249	9.90	548.90	725
2500	One way joists, 30″ pans, 125 psf Sup. Load, 15′ span		27.38	7.597		405	400	15.95	820.95	1,100
2550	25′ span		31.15	6.677		375	350	14	739	990
2700	One way beam & slab, 125 psf Sup. Load, 15′ span		20.59	10.102		295	535	21	851	1,200
2750	25′ span		28.36	7.334		281	385	15.40	681.40	940
2900	Two way beam & slab, 125 psf Sup. Load, 15′ span		24.04	8.652		285	455	18.15	758.15	1,075
2950	25′ span		35.87	5.799		249	305	12.20	566.20	775
3100	Elevated slabs, flat plate, including finish, not									
3110	including forms or reinforcing									
3150	Regular concrete (4000 psi), 4″ slab	C-8	2613	.021	S.F.	1.69	1	.16	2.85	3.62
3200	6″ slab		2585	.022		2.48	1.01	.16	3.65	4.50
3250	2-1/2″ thick floor fill		2685	.021		1.10	.97	.15	2.22	2.94
3300	Lightweight, 110 #/C.F., 2-1/2″ thick floor fill		2585	.022		1.31	1.01	.16	2.48	3.23
3400	Cellular concrete, 1-5/8″ fill, under 5000 S.F.		2000	.028		.90	1.31	.21	2.42	3.29

03 30 53 – Miscellaneous Cast-In-Place Concrete

03 30 53.40 Concrete In Place		Crew	Daily Output	Labor-Hours	Unit	Material	Labor	Equipment	Total	Total Incl O&P
						2020 Bare Costs				
3450	Over 10,000 S.F.	C-8	2200	.025	S.F.	.86	1.19	.19	2.24	3.04
3500	Add per floor for 3 to 6 stories high		31800	.002			.08	.01	.09	.14
3520	For 7 to 20 stories high		21200	.003			.12	.02	.14	.22
3540	Equipment pad (3000 psi), 3′ x 3′ x 6″ thick	C-14H	45	1.067	Ea.	50.50	55	.60	106.10	144
3550	4′ x 4′ x 6″ thick		30	1.600		78	82.50	.90	161.40	218
3560	5′ x 5′ x 8″ thick		18	2.667		138	138	1.49	277.49	375
3570	6′ x 6′ x 8″ thick		14	3.429		190	177	1.92	368.92	495
3580	8′ x 8′ x 10″ thick		8	6		395	310	3.36	708.36	935
3590	10′ x 10′ x 12″ thick		5	9.600		695	495	5.40	1,195.40	1,550
3800	Footings (3000 psi), spread under 1 C.Y.	C-14C	28	4	C.Y.	203	201	.96	404.96	545
3813	Install new concrete (3000 psi) light pole base, 24″ diam. x 8′	C-1	2.66	12.030		325	605		930	1,325
3825	1 C.Y. to 5 C.Y.	C-14C	43	2.605		240	131	.63	371.63	475
3850	Over 5 C.Y.	"	75	1.493		220	75	.36	295.36	360
3900	Footings, strip (3000 psi), 18″ x 9″, unreinforced	C-14L	40	2.400		154	118	.67	272.67	360
3920	18″ x 9″, reinforced	C-14C	35	3.200		181	161	.77	342.77	455
3925	20″ x 10″, unreinforced	C-14L	45	2.133		150	105	.60	255.60	335
3930	20″ x 10″, reinforced	C-14C	40	2.800		172	141	.67	313.67	415
3935	24″ x 12″, unreinforced	C-14L	55	1.745		148	86	.49	234.49	300
3940	24″ x 12″, reinforced	C-14C	48	2.333		171	118	.56	289.56	375
3945	36″ x 12″, unreinforced	C-14L	70	1.371		144	67.50	.38	211.88	267
3950	36″ x 12″, reinforced	C-14C	60	1.867		165	94	.45	259.45	330
4000	Foundation mat (3000 psi), under 10 C.Y.		38.67	2.896		246	146	.70	392.70	505
4050	Over 20 C.Y.		56.40	1.986		217	100	.48	317.48	400
4200	Wall, free-standing (3000 psi), 8″ thick, 8′ high	C-14D	45.83	4.364		197	230	9.55	436.55	590
4250	14′ high		27.26	7.337		232	385	16.05	633.05	890
4260	12″ thick, 8′ high		64.32	3.109		179	164	6.80	349.80	465
4270	14′ high		40.01	4.999		190	263	10.95	463.95	640
4300	15″ thick, 8′ high		80.02	2.499		173	132	5.45	310.45	405
4350	12′ high		51.26	3.902		173	205	8.55	386.55	530
4500	18′ high		48.85	4.094		194	215	8.95	417.95	570
4520	Handicap access ramp (4000 psi), railing both sides, 3′ wide	C-14H	14.58	3.292	L.F.	390	170	1.84	561.84	705
4525	5′ wide		12.22	3.928		400	203	2.20	605.20	765
4530	With 6″ curb and rails both sides, 3′ wide		8.55	5.614		400	290	3.14	693.14	905
4535	5′ wide		7.31	6.566		405	340	3.68	748.68	995
4650	Slab on grade (3500 psi), not including finish, 4″ thick	C-14E	60.75	1.449	C.Y.	147	74	.45	221.45	280
4700	6″ thick	"	92	.957	"	141	49	.30	190.30	234
4701	Thickened slab edge (3500 psi), for slab on grade poured									
4702	monolithically with slab; depth is in addition to slab thickness;									
4703	formed vertical outside edge, earthen bottom and inside slope									
4705	8″ deep x 8″ wide bottom, unreinforced	C-14L	2190	.044	L.F.	4.07	2.16	.01	6.24	7.95
4710	8″ x 8″, reinforced	C-14C	1670	.067		6.70	3.38	.02	10.10	12.80
4715	12″ deep x 12″ wide bottom, unreinforced	C-14L	1800	.053		8.30	2.63	.01	10.94	13.40
4720	12″ x 12″, reinforced	C-14C	1310	.086		13.15	4.31	.02	17.48	21.50
4725	16″ deep x 16″ wide bottom, unreinforced	C-14L	1440	.067		14	3.29	.02	17.31	20.50
4730	16″ x 16″, reinforced	C-14C	1120	.100		19.75	5.05	.02	24.82	30
4735	20″ deep x 20″ wide bottom, unreinforced	C-14L	1150	.083		21	4.12	.02	25.14	30
4740	20″ x 20″, reinforced	C-14C	920	.122		28.50	6.15	.03	34.68	41.50
4745	24″ deep x 24″ wide bottom, unreinforced	C-14L	930	.103		30	5.10	.03	35.13	41
4750	24″ x 24″, reinforced	C-14C	740	.151		39.50	7.60	.04	47.14	55.50
4751	Slab on grade (3500 psi), incl. troweled finish, not incl. forms									
4760	or reinforcing, over 10,000 S.F., 4″ thick	C-14F	3425	.021	S.F.	1.62	1	.01	2.63	3.37
4820	6″ thick		3350	.021		2.36	1.02	.01	3.39	4.22
4840	8″ thick		3184	.023		3.24	1.08	.01	4.33	5.25

03 30 Cast-In-Place Concrete

03 30 53 – Miscellaneous Cast-In-Place Concrete

03 30 53.40 Concrete In Place		Crew	Daily Output	Labor-Hours	Unit	Material	Labor (2020 Bare Costs)	Equipment	Total	Total Incl O&P
4900	12" thick	C-14F	2734	.026	S.F.	4.85	1.25	.01	6.11	7.35
4950	15" thick		2505	.029		6.10	1.37	.01	7.48	8.85
5000	Slab on grade (3000 psi), incl. broom finish, not incl. forms									
5001	or reinforcing, 4" thick	C-14G	2873	.019	S.F.	1.67	.91	.01	2.59	3.30
5010	6" thick		2590	.022		2.62	1.01	.01	3.64	4.49
5020	8" thick		2320	.024		3.41	1.13	.01	4.55	5.55
5200	Lift slab in place above the foundation, incl. forms, reinforcing,									
5210	concrete (4000 psi) and columns, over 20,000 S.F./floor	C-14B	2113	.098	S.F.	7.75	5.20	.21	13.16	17.05
5250	10,000 S.F. to 20,000 S.F./floor		1650	.126		8.45	6.65	.26	15.36	20
5300	Under 10,000 S.F./floor		1500	.139		9.15	7.30	.29	16.74	22
5500	Lightweight, ready mix, including screed finish only,									
5510	not including forms or reinforcing									
5550	1:4 (2500 psi) for structural roof decks	C-14B	260	.800	C.Y.	125	42	1.68	168.68	206
5600	1:6 (3000 psi) for ground slab with radiant heat	C-14F	92	.783		134	37	.29	171.29	207
5650	1:3:2 (2000 psi) with sand aggregate, roof deck	C-14B	260	.800		114	42	1.68	157.68	194
5700	Ground slab (2000 psi)	C-14F	107	.673		114	32	.25	146.25	176
5900	Pile caps (3000 psi), incl. forms and reinf., sq. or rect., under 10 C.Y.	C-14C	54.14	2.069		200	104	.50	304.50	385
5950	Over 10 C.Y.		75	1.493		189	75	.36	264.36	330
6000	Triangular or hexagonal, under 10 C.Y.		53	2.113		153	106	.51	259.51	340
6050	Over 10 C.Y.		85	1.318		171	66.50	.32	237.82	294
6200	Retaining walls (3000 psi), gravity, 4' high see Section 32 32	C-14D	66.20	3.021		173	159	6.60	338.60	450
6250	10' high		125	1.600		166	84	3.50	253.50	320
6300	Cantilever, level backfill loading, 8' high		70	2.857		184	150	6.25	340.25	450
6350	16' high		91	2.198		179	116	4.81	299.81	385
6800	Stairs (3500 psi), not including safety treads, free standing, 3'-6" wide	C-14H	83	.578	LF Nose	6.40	30	.32	36.72	55
6850	Cast on ground		125	.384	"	5.40	19.85	.22	25.47	37.50
7000	Stair landings, free standing		200	.240	S.F.	5.10	12.40	.13	17.63	25.50
7050	Cast on ground		475	.101	"	4.15	5.20	.06	9.41	12.95
9000	Minimum labor/equipment charge	2 Carp	1	16	Job		850		850	1,350

03 31 Structural Concrete

03 31 13 – Heavyweight Structural Concrete

03 31 13.25 Concrete, Hand Mix		Crew	Daily Output	Labor-Hours	Unit	Material	Labor (2020 Bare Costs)	Equipment	Total	Total Incl O&P
0010	**CONCRETE, HAND MIX** for small quantities or remote areas									
0050	Includes bulk local aggregate, bulk sand, bagged Portland									
0060	cement (Type I) and water, using gas powered cement mixer									
0125	2500 psi	C-30	135	.059	C.F.	4.33	2.49	1.04	7.86	9.90
0130	3000 psi		135	.059		4.68	2.49	1.04	8.21	10.30
0135	3500 psi		135	.059		4.88	2.49	1.04	8.41	10.50
0140	4000 psi		135	.059		5.10	2.49	1.04	8.63	10.80
0145	4500 psi		135	.059		5.40	2.49	1.04	8.93	11.10
0150	5000 psi		135	.059		5.80	2.49	1.04	9.33	11.50
0300	Using pre-bagged dry mix and wheelbarrow (80-lb. bag = 0.6 C.F.)									
0340	4000 psi	1 Clab	48	.167	C.F.	8.50	7		15.50	20.50

03 31 13.30 Concrete, Volumetric Site-Mixed		Crew	Daily Output	Labor-Hours	Unit	Material	Labor (2020 Bare Costs)	Equipment	Total	Total Incl O&P
0010	**CONCRETE, VOLUMETRIC SITE-MIXED**									
0015	Mixed on-site in volumetric truck									
0020	Includes local aggregate, sand, Portland cement (Type I) and water									
0025	Excludes all additives and treatments									
0100	3000 psi, 1 C.Y. mixed and discharged				C.Y.	213			213	234
0110	2 C.Y.					164			164	180

03 31 Structural Concrete

03 31 13 – Heavyweight Structural Concrete

03 31 13.30 Concrete, Volumetric Site-Mixed		Crew	Daily Output	Labor-Hours	Unit	Material	Labor	Equipment	Total	Total Incl O&P
						2020 Bare Costs				
0120	3 C.Y.				C.Y.	143			143	157
0130	4 C.Y.					125			125	138
0140	5 C.Y.					113			113	125
0200	For truck holding/waiting time past first 2 on-site hours, add				Hr.	91.50			91.50	101
0210	For trip charge beyond first 20 miles, each way, add				Mile	3.65			3.65	4.02
0220	For each additional increase of 500 psi, add				Ea.	4.64			4.64	5.10

03 31 13.35 Heavyweight Concrete, Ready Mix		Crew	Daily Output	Labor-Hours	Unit	Material	Labor	Equipment	Total	Total Incl O&P
0010	**HEAVYWEIGHT CONCRETE, READY MIX**, delivered									
0012	Includes local aggregate, sand, Portland cement (Type I) and water									
0015	Excludes all additives and treatments									
0020	2000 psi				C.Y.	107			107	117
0100	2500 psi					110			110	121
0150	3000 psi					129			129	142
0200	3500 psi					124			124	137
0300	4000 psi					127			127	140
0350	4500 psi					131			131	144
0400	5000 psi					135			135	149
0411	6000 psi					139			139	153
0412	8000 psi					146			146	161
0413	10,000 psi					154			154	169
0414	12,000 psi					161			161	177
1000	For high early strength (Portland cement Type III), add					10%				
1010	For structural lightweight with regular sand, add					25%				
1300	For winter concrete (hot water), add					5.35			5.35	5.90
1410	For mid-range water reducer, add					4.18			4.18	4.60
1420	For high-range water reducer/superplasticizer, add					6.40			6.40	7
1430	For retarder, add					3.45			3.45	3.80
1440	For non-Chloride accelerator, add					6.80			6.80	7.45
1450	For Chloride accelerator, per 1%, add					4.10			4.10	4.51
1460	For fiber reinforcing, synthetic (1 lb./C.Y.), add					8.10			8.10	8.90
1500	For Saturday delivery, add					9.20			9.20	10.10
1510	For truck holding/waiting time past 1st hour per load, add				Hr.	109			109	120
1520	For short load (less than 4 C.Y.), add per load				Ea.	85			85	93.50
2000	For all lightweight aggregate, add				C.Y.	45%				
4000	Flowable fill: ash, cement, aggregate, water									
4100	40-80 psi				C.Y.	78.50			78.50	86.50
4150	Structural: ash, cement, aggregate, water & sand									
4200	50 psi				C.Y.	78.50			78.50	86.50
4250	140 psi					79.50			79.50	87
4300	500 psi					81.50			81.50	90
4350	1000 psi					85			85	93.50

03 31 13.70 Placing Concrete		Crew	Daily Output	Labor-Hours	Unit	Material	Labor	Equipment	Total	Total Incl O&P
0010	**PLACING CONCRETE** R033105-70									
0020	Includes labor and equipment to place, level (strike off) and consolidate									
0050	Beams, elevated, small beams, pumped	C-20	60	1.067	C.Y.		48	7.75	55.75	85
0100	With crane and bucket	C-7	45	1.600			73	24	97	143
0200	Large beams, pumped	C-20	90	.711			32	5.15	37.15	56.50
0250	With crane and bucket	C-7	65	1.108			50.50	16.60	67.10	99
0400	Columns, square or round, 12" thick, pumped	C-20	60	1.067			48	7.75	55.75	85
0450	With crane and bucket	C-7	40	1.800			82.50	27	109.50	161
0600	18" thick, pumped	C-20	90	.711			32	5.15	37.15	56.50
0650	With crane and bucket	C-7	55	1.309			60	19.65	79.65	117

03 31 13 – Heavyweight Structural Concrete

03 31 13.70 Placing Concrete		Crew	Daily Output	Labor-Hours	Unit	Material	2020 Bare Costs Labor	Equipment	Total	Total Incl O&P
0800	24" thick, pumped	C-20	92	.696	C.Y.		31.50	5.05	36.55	55.50
0850	With crane and bucket	C-7	70	1.029			47	15.45	62.45	92
1000	36" thick, pumped	C-20	140	.457			20.50	3.31	23.81	36.50
1050	With crane and bucket	C-7	100	.720			33	10.80	43.80	64.50
1200	Duct bank, direct chute	C-6	155	.310			13.55	.35	13.90	22
1400	Elevated slabs, less than 6" thick, pumped	C-20	140	.457			20.50	3.31	23.81	36.50
1450	With crane and bucket	C-7	95	.758			34.50	11.35	45.85	67.50
1500	6" to 10" thick, pumped	C-20	160	.400			18.05	2.90	20.95	32
1550	With crane and bucket	C-7	110	.655			30	9.80	39.80	58.50
1600	Slabs over 10" thick, pumped	C-20	180	.356			16.05	2.58	18.63	28.50
1650	With crane and bucket	C-7	130	.554			25.50	8.30	33.80	49.50
1900	Footings, continuous, shallow, direct chute	C-6	120	.400			17.50	.45	17.95	28.50
1950	Pumped	C-20	150	.427			19.25	3.09	22.34	34
2000	With crane and bucket	C-7	90	.800			36.50	12	48.50	71
2100	Footings, continuous, deep, direct chute	C-6	140	.343			15	.38	15.38	24.50
2150	Pumped	C-20	160	.400			18.05	2.90	20.95	32
2200	With crane and bucket	C-7	110	.655			30	9.80	39.80	58.50
2400	Footings, spread, under 1 C.Y., direct chute	C-6	55	.873			38	.98	38.98	62
2450	Pumped	C-20	65	.985			44.50	7.15	51.65	79
2500	With crane and bucket	C-7	45	1.600			73	24	97	143
2600	Over 5 C.Y., direct chute	C-6	120	.400			17.50	.45	17.95	28.50
2650	Pumped	C-20	150	.427			19.25	3.09	22.34	34
2700	With crane and bucket	C-7	100	.720			33	10.80	43.80	64.50
2900	Foundation mats, over 20 C.Y., direct chute	C-6	350	.137			6	.15	6.15	9.75
2950	Pumped	C-20	400	.160			7.25	1.16	8.41	12.80
3000	With crane and bucket	C-7	300	.240			11	3.60	14.60	21.50
3200	Grade beams, direct chute	C-6	150	.320			14	.36	14.36	23
3250	Pumped	C-20	180	.356			16.05	2.58	18.63	28.50
3300	With crane and bucket	C-7	120	.600			27.50	9	36.50	53.50
3500	High rise, for more than 5 stories, pumped, add per story	C-20	2100	.030			1.38	.22	1.60	2.43
3510	With crane and bucket, add per story	C-7	2100	.034			1.57	.51	2.08	3.06
3700	Pile caps, under 5 C.Y., direct chute	C-6	90	.533			23.50	.60	24.10	37.50
3750	Pumped	C-20	110	.582			26.50	4.22	30.72	46.50
3800	With crane and bucket	C-7	80	.900			41	13.50	54.50	80.50
3850	Pile cap, 5 C.Y. to 10 C.Y., direct chute	C-6	175	.274			12	.31	12.31	19.50
3900	Pumped	C-20	200	.320			14.45	2.32	16.77	25.50
3950	With crane and bucket	C-7	150	.480			22	7.20	29.20	43
4000	Over 10 C.Y., direct chute	C-6	215	.223			9.75	.25	10	15.85
4050	Pumped	C-20	240	.267			12.05	1.93	13.98	21.50
4100	With crane and bucket	C-7	185	.389			17.80	5.85	23.65	35
4300	Slab on grade, up to 6" thick, direct chute	C-6	110	.436			19.10	.49	19.59	31
4350	Pumped	C-20	130	.492			22	3.57	25.57	39.50
4400	With crane and bucket	C-7	110	.655			30	9.80	39.80	58.50
4600	Over 6" thick, direct chute	C-6	165	.291			12.70	.33	13.03	21
4650	Pumped	C-20	185	.346			15.60	2.51	18.11	28
4700	With crane and bucket	C-7	145	.497			22.50	7.45	29.95	44
4900	Walls, 8" thick, direct chute	C-6	90	.533			23.50	.60	24.10	37.50
4950	Pumped	C-20	100	.640			29	4.64	33.64	51
5000	With crane and bucket	C-7	80	.900			41	13.50	54.50	80.50
5050	12" thick, direct chute	C-6	100	.480			21	.54	21.54	34
5100	Pumped	C-20	110	.582			26.50	4.22	30.72	46.50
5200	With crane and bucket	C-7	90	.800			36.50	12	48.50	71
5300	15" thick, direct chute	C-6	105	.457			20	.51	20.51	32.50

03 31 Structural Concrete

03 31 13 – Heavyweight Structural Concrete

03 31 13.70 Placing Concrete		Crew	Daily Output	Labor-Hours	Unit	Material	2020 Bare Costs Labor	Equipment	Total	Total Incl O&P
5350	Pumped	C-20	120	.533	C.Y.		24	3.87	27.87	43
5400	With crane and bucket	C-7	95	.758	↓		34.50	11.35	45.85	67.50
5600	Wheeled concrete dumping, add to placing costs above									
5610	Walking cart, 50' haul, add	C-18	32	.281	C.Y.		11.90	4	15.90	23.50
5620	150' haul, add		24	.375			15.85	5.35	21.20	31.50
5700	250' haul, add	↓	18	.500			21	7.10	28.10	42
5800	Riding cart, 50' haul, add	C-19	80	.113			4.76	1.92	6.68	9.75
5810	150' haul, add		60	.150			6.35	2.56	8.91	12.95
5900	250' haul, add	↓	45	.200	↓		8.45	3.41	11.86	17.30
6000	Concrete in-fill for pan-type metal stairs and landings. Manual placement									
6010	includes up to 50' horizontal haul from point of concrete discharge.									
6100	Stair pan treads, 2" deep									
6110	Flights in 1st floor level up/down from discharge point	C-8A	3200	.015	S.F.		.68		.68	1.07
6120	2nd floor level		2500	.019			.87		.87	1.38
6130	3rd floor level		2000	.024			1.08		1.08	1.72
6140	4th floor level	↓	1800	.027	↓		1.20		1.20	1.91
6200	Intermediate stair landings, pan-type 4" deep									
6210	Flights in 1st floor level up/down from discharge point	C-8A	2000	.024	S.F.		1.08		1.08	1.72
6220	2nd floor level		1500	.032			1.44		1.44	2.29
6230	3rd floor level		1200	.040			1.80		1.80	2.87
6240	4th floor level	↓	1000	.048	↓		2.16		2.16	3.44
9000	Minimum labor/equipment charge	C-6	2	24	Job		1,050	27	1,077	1,700

03 35 Concrete Finishing

03 35 13 – High-Tolerance Concrete Floor Finishing

03 35 13.30 Finishing Floors, High Tolerance		Crew	Daily Output	Labor-Hours	Unit	Material	Labor	Equipment	Total	Total Incl O&P
0010	**FINISHING FLOORS, HIGH TOLERANCE**									
0012	Finishing of fresh concrete flatwork requires that concrete									
0013	first be placed, struck off & consolidated									
0015	Basic finishing for various unspecified flatwork									
0100	Bull float only	C-10	4000	.006	S.F.		.28		.28	.45
0125	Bull float & manual float		2000	.012			.57		.57	.90
0150	Bull float, manual float & broom finish, w/edging & joints		1850	.013			.61		.61	.97
0200	Bull float, manual float & manual steel trowel	↓	1265	.019	↓		.90		.90	1.42
0210	For specified Random Access Floors in ACI Classes 1, 2, 3 and 4 to achieve									
0215	Composite Overall Floor Flatness and Levelness values up to FF35/FL25									
0250	Bull float, machine float & machine trowel (walk-behind)	C-10C	1715	.014	S.F.		.66	.05	.71	1.11
0300	Power screed, bull float, machine float & trowel (walk-behind)	C-10D	2400	.010			.47	.07	.54	.82
0350	Power screed, bull float, machine float & trowel (ride-on)	C-10E	4000	.006	↓		.28	.06	.34	.52
0352	For specified Random Access Floors in ACI Classes 5, 6, 7 and 8 to achieve									
0354	Composite Overall Floor Flatness and Levelness values up to FF50/FL50									
0356	Add for two-dimensional restraightening after power float	C-10	6000	.004	S.F.		.19		.19	.30
0358	For specified Random or Defined Access Floors in ACI Class 9 to achieve									
0360	Composite Overall Floor Flatness and Levelness values up to FF100/FL100									
0362	Add for two-dimensional restraightening after bull float & power float	C-10	3000	.008	S.F.		.38		.38	.60
0364	For specified Superflat Defined Access Floors in ACI Class 9 to achieve									
0366	Minimum Floor Flatness and Levelness values of FF100/FL100									
0368	Add for 2-dim'l restraightening after bull float, power float, power trowel	C-10	2000	.012	S.F.		.57		.57	.90
9100	Minimum labor/equipment charge	"	2	12	Job		570		570	895

03 35 Concrete Finishing

03 35 16 – Heavy-Duty Concrete Floor Finishing

03 35 16.30 Finishing Floors, Heavy-Duty		Crew	Daily Output	Labor-Hours	Unit	2020 Bare Costs Material	Labor	Equipment	Total	Total Incl O&P
0010	FINISHING FLOORS, HEAVY-DUTY									
1800	Floor abrasives, dry shake on fresh concrete, .25 psf, aluminum oxide	1 Cefi	850	.009	S.F.	.60	.47		1.07	1.39
1850	Silicon carbide		850	.009		.84	.47		1.31	1.67
2000	Floor hardeners, dry shake, metallic, light service, .50 psf		850	.009		.50	.47		.97	1.29
2050	Medium service, .75 psf		750	.011		.75	.53		1.28	1.67
2100	Heavy service, 1.0 psf		650	.012		1	.61		1.61	2.07
2150	Extra heavy, 1.5 psf		575	.014		1.50	.69		2.19	2.74
2300	Non-metallic, light service, .50 psf		850	.009		.18	.47		.65	.93
2350	Medium service, .75 psf		750	.011		.26	.53		.79	1.13
2400	Heavy service, 1.0 psf		650	.012		.35	.61		.96	1.36
2450	Extra heavy, 1.5 psf		575	.014		.53	.69		1.22	1.67
2800	Trap rock wearing surface, dry shake, for monolithic floors									
2810	2.0 psf	C-10B	1250	.032	S.F.	.02	1.45	.25	1.72	2.61
3800	Dustproofing, liquid, for cured concrete, solvent-based, 1 coat	1 Cefi	1900	.004		.18	.21		.39	.53
3850	2 coats		1300	.006		.64	.31		.95	1.19
4000	Epoxy-based, 1 coat		1500	.005		.15	.27		.42	.59
4050	2 coats		1500	.005		.31	.27		.58	.76

03 35 19 – Colored Concrete Finishing

03 35 19.30 Finishing Floors, Colored		Crew	Daily Output	Labor-Hours	Unit	Material	Labor	Equipment	Total	Total Incl O&P
0010	FINISHING FLOORS, COLORED									
3000	Floor coloring, dry shake on fresh concrete (0.6 psf)	1 Cefi	1300	.006	S.F.	.45	.31		.76	.97
3050	(1.0 psf)	"	625	.013	"	.74	.64		1.38	1.82
3100	Colored dry shake powder only				Lb.	.74			.74	.82
3600	1/2" topping using 0.6 psf dry shake powdered color	C-10B	590	.068	S.F.	4.62	3.07	.53	8.22	10.55
3650	1.0 psf dry shake powdered color	"	590	.068	"	4.92	3.07	.53	8.52	10.85

03 35 23 – Exposed Aggregate Concrete Finishing

03 35 23.30 Finishing Floors, Exposed Aggregate		Crew	Daily Output	Labor-Hours	Unit	Material	Labor	Equipment	Total	Total Incl O&P
0010	FINISHING FLOORS, EXPOSED AGGREGATE									
1600	Exposed local aggregate finish, seeded on fresh concrete, 3 lb./S.F.	1 Cefi	625	.013	S.F.	.21	.64		.85	1.24
1650	4 lb./S.F.	"	465	.017	"	.36	.86		1.22	1.75

03 35 29 – Tooled Concrete Finishing

03 35 29.30 Finishing Floors, Tooled		Crew	Daily Output	Labor-Hours	Unit	Material	Labor	Equipment	Total	Total Incl O&P
0010	FINISHING FLOORS, TOOLED									
4400	Stair finish, fresh concrete, float finish	1 Cefi	275	.029	S.F.		1.45		1.45	2.28
4500	Steel trowel finish		200	.040			2		2	3.14
4600	Silicon carbide finish, dry shake on fresh concrete, .25 psf		150	.053		.60	2.66		3.26	4.83

03 35 29.60 Finishing Walls		Crew	Daily Output	Labor-Hours	Unit	Material	Labor	Equipment	Total	Total Incl O&P
0010	FINISHING WALLS									
0020	Break ties and patch voids	1 Cefi	540	.015	S.F.	.04	.74		.78	1.21
0050	Burlap rub with grout		450	.018		.04	.89		.93	1.44
0100	Carborundum rub, dry		270	.030			1.48		1.48	2.32
0150	Wet rub		175	.046			2.28		2.28	3.59
0300	Bush hammer, green concrete	B-39	1000	.048			2.12	.36	2.48	3.78
0350	Cured concrete	"	650	.074			3.27	.55	3.82	5.80
0500	Acid etch	1 Cefi	575	.014		.14	.69		.83	1.24
0600	Float finish, 1/16" thick	"	300	.027		.40	1.33		1.73	2.53
0700	Sandblast, light penetration	E-11	1100	.029		.51	1.36	.27	2.14	3.11
0750	Heavy penetration	"	375	.085		1.03	3.98	.80	5.81	8.60
0850	Grind form fins flush	1 Clab	700	.011	L.F.		.48		.48	.77
9000	Minimum labor/equipment charge	C-10	2	12	Job		570		570	895

03 35 Concrete Finishing

03 35 33 – Stamped Concrete Finishing

03 35 33.50 Slab Texture Stamping		Crew	Daily Output	Labor-Hours	Unit	Material	Labor	Equipment	Total	Total Incl O&P
						2020 Bare Costs				
0010	**SLAB TEXTURE STAMPING**									
0050	Stamping requires that concrete first be placed, struck off, consolidated,									
0060	bull floated and free of bleed water. Decorative stamping tasks include:									
0100	Step 1 - first application of dry shake colored hardener	1 Cefi	6400	.001	S.F.	.43	.06		.49	.57
0110	Step 2 - bull float		6400	.001			.06		.06	.10
0130	Step 3 - second application of dry shake colored hardener		6400	.001		.21	.06		.27	.33
0140	Step 4 - bull float, manual float & steel trowel	3 Cefi	1280	.019			.94		.94	1.47
0150	Step 5 - application of dry shake colored release agent	1 Cefi	6400	.001		.10	.06		.16	.21
0160	Step 6 - place, tamp & remove mats	3 Cefi	2400	.010		.87	.50		1.37	1.73
0170	Step 7 - touch up edges, mat joints & simulated grout lines	1 Cefi	1280	.006			.31		.31	.49
0300	Alternate stamping estimating method includes all tasks above	4 Cefi	800	.040		1.60	2		3.60	4.90
0400	Step 8 - pressure wash @ 3000 psi after 24 hours	1 Cefi	1600	.005			.25		.25	.39
0500	Step 9 - roll 2 coats cure/seal compound when dry	"	800	.010		.69	.50		1.19	1.54

03 35 43 – Polished Concrete Finishing

03 35 43.10 Polished Concrete Floors

		Crew	Daily Output	Labor-Hours	Unit	Material	Labor	Equipment	Total	Total Incl O&P
0010	**POLISHED CONCRETE FLOORS** R033543-10									
0015	Processing of cured concrete to include grinding, honing,									
0020	and polishing of interior floors with 22" segmented diamond									
0025	planetary floor grinder (2 passes in different directions per grit)									
0100	Removal of pre-existing coatings, dry, with carbide discs using									
0105	dry vacuum pick-up system, final hand sweeping									
0110	Glue, adhesive or tar	J-4	1.60	15	M.S.F.	22	710	139	871	1,300
0120	Paint, epoxy, 1 coat		3.60	6.667		22	315	62	399	595
0130	2 coats		1.80	13.333		22	630	124	776	1,150
0200	Grinding and edging, wet, including wet vac pick-up and auto									
0205	scrubbing between grit changes									
0210	40-grit diamond/metal matrix	J-4A	1.60	20	M.S.F.	41.50	920	251	1,212.50	1,775
0220	80-grit diamond/metal matrix		2	16		41.50	735	200	976.50	1,450
0230	120-grit diamond/metal matrix		2.40	13.333		41.50	615	167	823.50	1,200
0240	200-grit diamond/metal matrix		2.80	11.429		41.50	525	143	709.50	1,050
0300	Spray on dye or stain (1 coat)	1 Cefi	16	.500		209	25		234	269
0400	Spray on densifier/hardener (2 coats)	"	8	1		315	50		365	425
0410	Auto scrubbing after 2nd coat, when dry	J-4B	16	.500			21	11.15	32.15	46
0500	Honing and edging, wet, including wet vac pick-up and auto									
0505	scrubbing between grit changes									
0510	100-grit diamond/resin matrix	J-4A	2.80	11.429	M.S.F.	41.50	525	143	709.50	1,050
0520	200-grit diamond/resin matrix	"	2.80	11.429	"	41.50	525	143	709.50	1,050
0530	Dry, including dry vacuum pick-up system, final hand sweeping									
0540	400-grit diamond/resin matrix	J-4A	2.80	11.429	M.S.F.	41.50	525	143	709.50	1,050
0600	Polishing and edging, dry, including dry vac pick-up and hand									
0605	sweeping between grit changes									
0610	800-grit diamond/resin matrix	J-4A	2.80	11.429	M.S.F.	41.50	525	143	709.50	1,050
0620	1500-grit diamond/resin matrix		2.80	11.429		41.50	525	143	709.50	1,050
0630	3000-grit diamond/resin matrix		2.80	11.429		41.50	525	143	709.50	1,050
0700	Auto scrubbing after final polishing step	J-4B	16	.500			21	11.15	32.15	46

03 37 Specialty Placed Concrete

03 37 13 – Shotcrete

03 37 13.30 Gunite (Dry-Mix)		Crew	Daily Output	Labor-Hours	Unit	Material	Labor	Equipment	Total	Total Incl O&P
						2020 Bare Costs				
0010	**GUNITE (DRY-MIX)**									
0020	Typical in place, 1" layers, no mesh included	C-16	2000	.028	S.F.	.42	1.31	.20	1.93	2.75
0100	Mesh for gunite 2 x 2, #12	2 Rodm	800	.020		.73	1.13		1.86	2.58
0150	#4 reinforcing bars @ 6" each way	"	500	.032		1.88	1.80		3.68	4.91
0300	Typical in place, including mesh, 2" thick, flat surfaces	C-16	1000	.056		1.57	2.62	.39	4.58	6.30
0350	Curved surfaces		500	.112		1.57	5.25	.78	7.60	10.90
0500	4" thick, flat surfaces		750	.075		2.41	3.49	.52	6.42	8.75
0550	Curved surfaces		350	.160		2.41	7.50	1.12	11.03	15.75
0900	Prepare old walls, no scaffolding, good condition	C-10	1000	.024			1.14		1.14	1.79
0950	Poor condition	"	275	.087			4.13		4.13	6.55
1100	For high finish requirement or close tolerance, add						50%			
1150	Very high						110%			
9000	Minimum labor/equipment charge	C-10	1	24	Job		1,125		1,125	1,800

03 37 13.60 Shotcrete (Wet-Mix)		Crew	Daily Output	Labor-Hours	Unit	Material	Labor	Equipment	Total	Total Incl O&P
0010	**SHOTCRETE (WET-MIX)**									
0020	Wet mix, placed @ up to 12 C.Y./hour, 3000 psi	C-8C	80	.600	C.Y.	136	27.50	7.45	170.95	201
0100	Up to 35 C.Y./hour	C-8E	240	.200	"	122	9.10	2.70	133.80	151
1010	Fiber reinforced, 1" thick	C-8C	1740	.028	S.F.	1	1.27	.34	2.61	3.49
1020	2" thick		900	.053		1.99	2.46	.66	5.11	6.85
1030	3" thick		825	.058		2.99	2.69	.72	6.40	8.35
1040	4" thick		750	.064		3.98	2.96	.79	7.73	9.95

03 39 Concrete Curing

03 39 13 – Water Concrete Curing

03 39 13.50 Water Curing		Crew	Daily Output	Labor-Hours	Unit	Material	Labor	Equipment	Total	Total Incl O&P
0010	**WATER CURING**									
0015	With burlap, 4 uses assumed, 7.5 oz.	2 Clab	55	.291	C.S.F.	15.65	12.25		27.90	37
0100	10 oz.	"	55	.291	"	28	12.25		40.25	50.50
0400	Curing blankets, 1" to 2" thick, buy				S.F.	.26			.26	.29
9000	Minimum labor/equipment charge	1 Clab	5	1.600	Job		67.50		67.50	108

03 39 23 – Membrane Concrete Curing

03 39 23.13 Chemical Compound Membrane Concrete Curing		Crew	Daily Output	Labor-Hours	Unit	Material	Labor	Equipment	Total	Total Incl O&P
0010	**CHEMICAL COMPOUND MEMBRANE CONCRETE CURING**									
0300	Sprayed membrane curing compound	2 Clab	95	.168	C.S.F.	12.45	7.10		19.55	25
0700	Curing compound, solvent based, 400 S.F./gal., 55 gallon lots				Gal.	23			23	25.50
0720	5 gallon lots					33			33	36
0800	Curing compound, water based, 250 S.F./gal., 55 gallon lots					23.50			23.50	26
0820	5 gallon lots					26			26	29

03 39 23.23 Sheet Membrane Concrete Curing		Crew	Daily Output	Labor-Hours	Unit	Material	Labor	Equipment	Total	Total Incl O&P
0010	**SHEET MEMBRANE CONCRETE CURING**									
0200	Curing blanket, burlap/poly, 2-ply	2 Clab	70	.229	C.S.F.	20.50	9.60		30.10	38

03 41 Precast Structural Concrete

03 41 13 – Precast Concrete Hollow Core Planks

03 41 13.50 Precast Slab Planks		Crew	Daily Output	Labor-Hours	Unit	2020 Bare Costs Material	Labor	Equipment	Total	Total Incl O&P
0010	PRECAST SLAB PLANKS									
0020	Prestressed roof/floor members, grouted, solid, 4" thick	C-11	2400	.030	S.F.	8.55	1.72	.95	11.22	13.30
0050	6" thick		2800	.026		8.55	1.47	.81	10.83	12.70
0100	Hollow, 8" thick		3200	.023		11.90	1.29	.71	13.90	15.90
0150	10" thick		3600	.020		9.95	1.15	.63	11.73	13.45
0200	12" thick	↓	4000	.018	↓	11.60	1.03	.57	13.20	15.10

03 41 16 – Precast Concrete Slabs

03 41 16.20 Precast Concrete Channel Slabs		Crew	Daily Output	Labor-Hours	Unit	Material	Labor	Equipment	Total	Total Incl O&P
0010	PRECAST CONCRETE CHANNEL SLABS									
0335	Lightweight concrete channel slab, long runs, 2-3/4" thick	C-12	1575	.030	S.F.	12.85	1.60	.30	14.75	17
0375	3-3/4" thick		1550	.031		13.20	1.63	.30	15.13	17.50
0475	4-3/4" thick		1525	.031		13.30	1.66	.31	15.27	17.65
1275	Short pieces, 2-3/4" thick		785	.061		19.25	3.22	.60	23.07	27
1375	3-3/4" thick		770	.062		19.80	3.28	.61	23.69	28
1475	4-3/4" thick	↓	762	.063	↓	19.95	3.32	.62	23.89	28

03 41 16.50 Precast Lightweight Concrete Plank		Crew	Daily Output	Labor-Hours	Unit	Material	Labor	Equipment	Total	Total Incl O&P
0010	PRECAST LIGHTWEIGHT CONCRETE PLANK									
0015	Lightweight plank, nailable, T&G, 2" thick	C-12	1800	.027	S.F.	10.30	1.40	.26	11.96	13.90
0150	For premium ceiling finish, add				"	10%				
0200	For sloping roofs, slope over 4 in 12, add						25%			
0250	Slope over 6 in 12, add						150%			

03 41 23 – Precast Concrete Stairs

03 41 23.50 Precast Stairs		Crew	Daily Output	Labor-Hours	Unit	Material	Labor	Equipment	Total	Total Incl O&P
0010	PRECAST STAIRS									
0020	Precast concrete treads on steel stringers, 3' wide	C-12	75	.640	Riser	149	33.50	6.30	188.80	225
0300	Front entrance, 5' wide with 48" platform, 2 risers		16	3	Flight	675	158	29.50	862.50	1,025
0350	5 risers		12	4		1,125	211	39	1,375	1,600
0500	6' wide, 2 risers		15	3.200		750	168	31.50	949.50	1,125
0550	5 risers		11	4.364		1,250	230	43	1,523	1,750
0700	7' wide, 2 risers		14	3.429		1,050	181	33.50	1,264.50	1,500
0750	5 risers	↓	10	4.800		1,775	253	47	2,075	2,400
1200	Basement entrance stairwell, 6 steps, incl. steel bulkhead door	B-51	22	2.182		1,875	94.50	8.90	1,978.40	2,200
1250	14 steps	"	11	4.364	↓	3,350	189	17.85	3,556.85	4,000

03 41 33 – Precast Structural Pretensioned Concrete

03 41 33.10 Precast Beams		Crew	Daily Output	Labor-Hours	Unit	Material	Labor	Equipment	Total	Total Incl O&P
0010	PRECAST BEAMS									
0011	L-shaped, 20' span, 12" x 20"	C-11	32	2.250	Ea.	4,500	129	71	4,700	5,225
0060	18" x 36"		24	3		6,350	172	94.50	6,616.50	7,375
0100	24" x 44"		22	3.273		7,450	187	103	7,740	8,625
0150	30' span, 12" x 36"		24	3		8,800	172	94.50	9,066.50	10,100
0200	18" x 44"		20	3.600		10,500	206	113	10,819	12,000
0250	24" x 52"		16	4.500		12,800	258	142	13,200	14,700
0400	40' span, 12" x 52"		20	3.600		13,300	206	113	13,619	15,100
0450	18" x 52"		16	4.500		14,900	258	142	15,300	17,000
0500	24" x 52"		12	6		17,100	345	189	17,634	19,600
1200	Rectangular, 20' span, 12" x 20"		32	2.250		3,925	129	71	4,125	4,575
1250	18" x 36"		24	3		5,800	172	94.50	6,066.50	6,750
1300	24" x 44"		22	3.273		6,850	187	103	7,140	7,975
1400	30' span, 12" x 36"		24	3		6,475	172	94.50	6,741.50	7,500
1450	18" x 44"		20	3.600		9,000	206	113	9,319	10,400
1500	24" x 52"	↓	16	4.500	↓	11,200	258	142	11,600	12,900

03 41 Precast Structural Concrete

03 41 33 – Precast Structural Pretensioned Concrete

03 41 33.10 Precast Beams		Crew	Daily Output	Labor-Hours	Unit	2020 Bare Costs Material	Labor	Equipment	Total	Total Incl O&P
1600	40′ span, 12″ x 52″	C-11	20	3.600	Ea.	10,800	206	113	11,119	12,300
1650	18″ x 52″		16	4.500		13,000	258	142	13,400	14,900
1700	24″ x 52″		12	6		14,900	345	189	15,434	17,200
2000	"T" shaped, 20′ span, 12″ x 20″		32	2.250		5,575	129	71	5,775	6,400
2050	18″ x 36″		24	3		7,175	172	94.50	7,441.50	8,250
2100	24″ x 44″		22	3.273		8,075	187	103	8,365	9,300
2200	30′ span, 12″ x 36″		24	3		9,950	172	94.50	10,216.50	11,300
2250	18″ x 44″		20	3.600		12,500	206	113	12,819	14,300
2300	24″ x 52″		16	4.500		13,300	258	142	13,700	15,300
2500	40′ span, 12″ x 52″		20	3.600		16,200	206	113	16,519	18,300
2550	18″ x 52″		16	4.500		16,700	258	142	17,100	18,900
2600	24″ x 52″		12	6		17,800	345	189	18,334	20,400

03 41 33.15 Precast Columns		Crew	Daily Output	Labor-Hours	Unit	Material	Labor	Equipment	Total	Total Incl O&P
0010	**PRECAST COLUMNS**									
0020	Rectangular to 12′ high, 16″ x 16″	C-11	120	.600	L.F.	278	34.50	18.90	331.40	380
0050	24″ x 24″		96	.750		370	43	23.50	436.50	500
0300	24′ high, 28″ x 28″		192	.375		415	21.50	11.80	448.30	505
0350	36″ x 36″		144	.500		565	28.50	15.75	609.25	685
0700	24′ high, 1 haunch, 12″ x 12″		32	2.250	Ea.	5,025	129	71	5,225	5,800
0800	20″ x 20″		28	2.571	"	7,200	147	81	7,428	8,250

03 41 33.25 Precast Joists		Crew	Daily Output	Labor-Hours	Unit	Material	Labor	Equipment	Total	Total Incl O&P
0010	**PRECAST JOISTS**									
0015	40 psf L.L., 6″ deep for 12′ spans	C-12	600	.080	L.F.	33.50	4.21	.78	38.49	44
0050	8″ deep for 16′ spans		575	.083		55.50	4.40	.82	60.72	69
0100	10″ deep for 20′ spans		550	.087		97	4.59	.86	102.45	115
0150	12″ deep for 24′ spans		525	.091		133	4.81	.90	138.71	155

03 41 33.60 Precast Tees		Crew	Daily Output	Labor-Hours	Unit	Material	Labor	Equipment	Total	Total Incl O&P
0010	**PRECAST TEES**									
0020	Quad tee, short spans, roof	C-11	7200	.010	S.F.	11.20	.57	.32	12.09	13.60
0050	Floor		7200	.010		11.20	.57	.32	12.09	13.60
0200	Double tee, floor members, 60′ span		8400	.009		12.70	.49	.27	13.46	15.05
0250	80′ span		8000	.009		17.35	.52	.28	18.15	20
0300	Roof members, 30′ span		4800	.015		13.95	.86	.47	15.28	17.25
0350	50′ span		6400	.011		12.15	.64	.35	13.14	14.85
0400	Wall members, up to 55′ high		3600	.020		16.60	1.15	.63	18.38	21
0500	Single tee roof members, 40′ span		3200	.023		17.90	1.29	.71	19.90	22.50
0550	80′ span		5120	.014		19.80	.81	.44	21.05	24
0600	100′ span		6000	.012		29.50	.69	.38	30.57	34
0650	120′ span		6000	.012		31.50	.69	.38	32.57	36
1000	Double tees, floor members									
1100	Lightweight, 20″ x 8′ wide, 45′ span	C-11	20	3.600	Ea.	4,425	206	113	4,744	5,325
1150	24″ x 8′ wide, 50′ span		18	4		4,925	229	126	5,280	5,900
1200	32″ x 10′ wide, 60′ span		16	4.500		7,375	258	142	7,775	8,700
1250	Standard weight, 12″ x 8′ wide, 20′ span		22	3.273		1,800	187	103	2,090	2,400
1300	16″ x 8′ wide, 25′ span		20	3.600		2,225	206	113	2,544	2,900
1350	18″ x 8′ wide, 30′ span		20	3.600		2,675	206	113	2,994	3,400
1400	20″ x 8′ wide, 45′ span		18	4		4,025	229	126	4,380	4,925
1450	24″ x 8′ wide, 50′ span		16	4.500		4,475	258	142	4,875	5,500
1500	32″ x 10′ wide, 60′ span		14	5.143		6,700	294	162	7,156	8,025
2000	Roof members									
2050	Lightweight, 20″ x 8′ wide, 40′ span	C-11	20	3.600	Ea.	3,925	206	113	4,244	4,775
2100	24″ x 8′ wide, 50′ span		18	4		4,925	229	126	5,280	5,900

03 41 Precast Structural Concrete

03 41 33 – Precast Structural Pretensioned Concrete

03 41 33.60 Precast Tees		Crew	Daily Output	Labor-Hours	Unit	Material	2020 Bare Costs Labor	Equipment	Total	Total Incl O&P
2150	32" x 10' wide, 60' span	C-11	16	4.500	Ea.	7,375	258	142	7,775	8,700
2200	Standard weight, 12" x 8' wide, 30' span		22	3.273		2,675	187	103	2,965	3,375
2250	16" x 8' wide, 30' span		20	3.600		2,825	206	113	3,144	3,550
2300	18" x 8' wide, 30' span		20	3.600		2,950	206	113	3,269	3,700
2350	20" x 8' wide, 40' span		18	4		3,575	229	126	3,930	4,425
2400	24" x 8' wide, 50' span		16	4.500		4,475	258	142	4,875	5,500
2450	32" x 10' wide, 60' span	↓	14	5.143	↓	6,700	294	162	7,156	8,025

03 45 Precast Architectural Concrete

03 45 13 – Faced Architectural Precast Concrete

03 45 13.50 Precast Wall Panels		Crew	Daily Output	Labor-Hours	Unit	Material	Labor	Equipment	Total	Total Incl O&P
0010	**PRECAST WALL PANELS**									
0050	Uninsulated, smooth gray									
0150	Low rise, 4' x 8' x 4" thick	C-11	320	.225	S.F.	31.50	12.90	7.10	51.50	63.50
0210	8' x 8', 4" thick		576	.125		31	7.15	3.94	42.09	50.50
0250	8' x 16' x 4" thick		1024	.070		31	4.03	2.22	37.25	43
0600	High rise, 4' x 8' x 4" thick		288	.250		31.50	14.30	7.90	53.70	66
0650	8' x 8' x 4" thick		512	.141		31	8.05	4.43	43.48	52.50
0700	8' x 16' x 4" thick		768	.094		31	5.35	2.95	39.30	46
0750	10' x 20', 6" thick	↓	1400	.051		52.50	2.94	1.62	57.06	64.50
0800	Insulated panel, 2" polystyrene, add					1.22			1.22	1.34
0850	2" urethane, add					.74			.74	.81
1200	Finishes, white, add					3.54			3.54	3.89
1250	Exposed aggregate, add					.77			.77	.84
1300	Granite faced, domestic, add					30.50			30.50	33.50
1350	Brick faced, modular, red, add				↓	4.83			4.83	5.30
2200	Fiberglass reinforced cement with urethane core									
2210	R20, 8' x 8', 5" plain finish	E-2	750	.075	S.F.	29.50	4.27	2.26	36.03	42
2220	Exposed aggregate or brick finish	"	600	.093	"	48	5.35	2.83	56.18	64.50

03 47 Site-Cast Concrete

03 47 13 – Tilt-Up Concrete

03 47 13.50 Tilt-Up Wall Panels		Crew	Daily Output	Labor-Hours	Unit	Material	Labor	Equipment	Total	Total Incl O&P
0010	**TILT-UP WALL PANELS**									
0015	Wall panel construction, walls only, 5-1/2" thick	C-14	1600	.090	S.F.	6.40	4.62	.93	11.95	15.40
0100	7-1/2" thick		1550	.093		7.95	4.77	.96	13.68	17.40
0500	Walls and columns, 5-1/2" thick walls, 12" x 12" columns		1565	.092		9.45	4.73	.95	15.13	18.95
0550	7-1/2" thick wall, 12" x 12" columns		1370	.105	↓	11.60	5.40	1.08	18.08	22.50
0800	Columns only, site precast, 12" x 12"		200	.720	L.F.	22.50	37	7.45	66.95	91
0850	16" x 16"	↓	105	1.371	"	33	70.50	14.15	117.65	164

03 48 Precast Concrete Specialties

03 48 43 – Precast Concrete Trim

03 48 43.40 Precast Lintels		Crew	Daily Output	Labor-Hours	Unit	Material	Labor	Equipment	Total	Total Incl O&P
						2020 Bare Costs				
0010	**PRECAST LINTELS**, smooth gray, prestressed, stock units only									
0800	4″ wide x 8″ high x 4′ long	D-10	28	1.143	Ea.	35.50	59	15.30	109.80	150
0850	8′ long		24	1.333		81	69	17.85	167.85	220
1000	6″ wide x 8″ high x 4′ long		26	1.231		58.50	63.50	16.45	138.45	185
1050	10′ long		22	1.455		149	75.50	19.45	243.95	305
1200	8″ wide x 8″ high x 4′ long		24	1.333		61	69	17.85	147.85	198
1250	12′ long	↓	20	1.600		203	83	21.50	307.50	380
1275	For custom sizes, types, colors, or finishes of precast lintels, add				↓	150%				

03 48 43.90 Precast Window Sills		Crew	Daily Output	Labor-Hours	Unit	Material	Labor	Equipment	Total	Total Incl O&P
0010	**PRECAST WINDOW SILLS**									
0600	Precast concrete, 4″ tapers to 3″, 9″ wide	D-1	70	.229	L.F.	23	10.70		33.70	42.50
0650	11″ wide	″	60	.267	″	37	12.50		49.50	61

03 51 Cast Roof Decks

03 51 13 – Cementitious Wood Fiber Decks

03 51 13.50 Cementitious/Wood Fiber Planks		Crew	Daily Output	Labor-Hours	Unit	Material	Labor	Equipment	Total	Total Incl O&P
0010	**CEMENTITIOUS/WOOD FIBER PLANKS** R051223-50									
0050	Plank, beveled edge, 1″ thick	2 Carp	1000	.016	S.F.	3.68	.85		4.53	5.40
0100	1-1/2″ thick		975	.016		5.60	.87		6.47	7.55
0150	T&G, 2″ thick		950	.017		3.33	.90		4.23	5.10
0200	2-1/2″ thick		925	.017		3.71	.92		4.63	5.55
0250	3″ thick	↓	900	.018		4.15	.95		5.10	6.10
1000	Bulb tee, sub-purlin and grout, 6′ span, add	E-1	5000	.005		2.22	.27	.03	2.52	2.91
1100	8′ span	″	4200	.006	↓	2.22	.32	.04	2.58	3

03 51 16 – Gypsum Concrete Roof Decks

03 51 16.50 Gypsum Roof Deck		Crew	Daily Output	Labor-Hours	Unit	Material	Labor	Equipment	Total	Total Incl O&P
0010	**GYPSUM ROOF DECK**									
1000	Poured gypsum, 2″ thick	C-8	6000	.009	S.F.	1.77	.44	.07	2.28	2.71
1100	3″ thick	″	4800	.012	″	2.65	.55	.09	3.29	3.86

03 52 Lightweight Concrete Roof Insulation

03 52 16 – Lightweight Insulating Concrete

03 52 16.13 Lightweight Cellular Insulating Concrete		Crew	Daily Output	Labor-Hours	Unit	Material	Labor	Equipment	Total	Total Incl O&P
0010	**LIGHTWEIGHT CELLULAR INSULATING CONCRETE**									
0020	Portland cement and foaming agent G	C-8	50	1.120	C.Y.	136	52.50	8.20	196.70	241

03 52 16.16 Lightweight Aggregate Insulating Concrete		Crew	Daily Output	Labor-Hours	Unit	Material	Labor	Equipment	Total	Total Incl O&P
0010	**LIGHTWEIGHT AGGREGATE INSULATING CONCRETE**									
0100	Poured vermiculite or perlite, field mix,									
0110	1:6 field mix G	C-8	50	1.120	C.Y.	264	52.50	8.20	324.70	380
0200	Ready mix, 1:6 mix, roof fill, 2″ thick G		10000	.006	S.F.	1.47	.26	.04	1.77	2.08
0250	3″ thick G	↓	7700	.007		2.20	.34	.05	2.59	3.02
0400	Expanded volcanic glass rock, 1″ thick G	2 Carp	1500	.011		.62	.57		1.19	1.59
0450	3″ thick G	″	1200	.013	↓	1.85	.71		2.56	3.18

03 53 Concrete Topping

03 53 16 – Iron-Aggregate Concrete Topping

03 53 16.50 Floor Topping		Crew	Daily Output	Labor-Hours	Unit	Material	2020 Bare Costs Labor	Equipment	Total	Total Incl O&P
0010	FLOOR TOPPING									
0400	Integral topping/finish, on fresh concrete, using 1:1:2 mix, 3/16" thick	C-10B	1000	.040	S.F.	.14	1.81	.31	2.26	3.37
0450	1/2" thick		950	.042		.37	1.91	.33	2.61	3.80
0500	3/4" thick		850	.047		.56	2.13	.37	3.06	4.40
0600	1" thick		750	.053		.74	2.41	.42	3.57	5.10
0800	Granolithic topping, on fresh or cured concrete, 1:1:1-1/2 mix, 1/2" thick		590	.068		.41	3.07	.53	4.01	5.90
0820	3/4" thick		580	.069		.62	3.12	.54	4.28	6.25
0850	1" thick		575	.070		.82	3.15	.55	4.52	6.50
0950	2" thick		500	.080		1.65	3.62	.63	5.90	8.25
1200	Heavy duty, 1:1:2, 3/4" thick, preshrunk, gray, 20 M.S.F.		320	.125		.95	5.65	.98	7.58	11.15
1300	100 M.S.F.	↓	380	.105	↓	.56	4.76	.83	6.15	9.05

03 54 Cast Underlayment

03 54 13 – Gypsum Cement Underlayment

03 54 13.50 Poured Gypsum Underlayment

		Crew	Daily Output	Labor-Hours	Unit	Material	Labor	Equipment	Total	Total Incl O&P
0010	POURED GYPSUM UNDERLAYMENT									
0400	Underlayment, gypsum based, self-leveling 2500 psi, pumped, 1/2" thick	C-8	24000	.002	S.F.	.44	.11	.02	.57	.68
0500	3/4" thick		20000	.003		.66	.13	.02	.81	.96
0600	1" thick	↓	16000	.004		.88	.16	.03	1.07	1.26
1400	Hand placed, 1/2" thick	C-18	450	.020		.44	.85	.28	1.57	2.16
1500	3/4" thick	"	300	.030	↓	.66	1.27	.43	2.36	3.23

03 54 16 – Hydraulic Cement Underlayment

03 54 16.50 Cement Underlayment

		Crew	Daily Output	Labor-Hours	Unit	Material	Labor	Equipment	Total	Total Incl O&P
0010	CEMENT UNDERLAYMENT									
2510	Underlayment, P.C. based, self-leveling, 4100 psi, pumped, 1/4" thick	C-8	20000	.003	S.F.	1.75	.13	.02	1.90	2.16
2520	1/2" thick		19000	.003		3.50	.14	.02	3.66	4.10
2530	3/4" thick		18000	.003		5.25	.15	.02	5.42	6.05
2540	1" thick		17000	.003		7	.15	.02	7.17	7.95
2550	1-1/2" thick	↓	15000	.004		10.50	.17	.03	10.70	11.85
2560	Hand placed, 1/2" thick	C-18	450	.020		3.50	.85	.28	4.63	5.55
2610	Topping, P.C. based, self-leveling, 6100 psi, pumped, 1/4" thick	C-8	20000	.003		2.09	.13	.02	2.24	2.53
2620	1/2" thick		19000	.003		4.18	.14	.02	4.34	4.84
2630	3/4" thick		18000	.003		6.25	.15	.02	6.42	7.15
2660	1" thick		17000	.003		8.35	.15	.02	8.52	9.45
2670	1-1/2" thick	↓	15000	.004		12.55	.17	.03	12.75	14.10
2680	Hand placed, 1/2" thick	C-18	450	.020	↓	4.18	.85	.28	5.31	6.25

03 62 Non-Shrink Grouting

03 62 13 – Non-Metallic Non-Shrink Grouting

03 62 13.50 Grout, Non-Metallic Non-Shrink

		Crew	Daily Output	Labor-Hours	Unit	Material	Labor	Equipment	Total	Total Incl O&P
0010	GROUT, NON-METALLIC NON-SHRINK									
0300	Non-shrink, non-metallic, 1" deep	1 Cefi	35	.229	S.F.	7.60	11.40		19	26.50
0350	2" deep	"	25	.320	"	15.25	16		31.25	42

03 62 16 – Metallic Non-Shrink Grouting

03 62 16.50 Grout, Metallic Non-Shrink

		Crew	Daily Output	Labor-Hours	Unit	Material	Labor	Equipment	Total	Total Incl O&P
0010	GROUT, METALLIC NON-SHRINK									
0020	Column & machine bases, non-shrink, metallic, 1" deep	1 Cefi	35	.229	S.F.	12.05	11.40		23.45	31.50
0050	2" deep	"	25	.320	"	24	16		40	51.50

03 63 Epoxy Grouting

03 63 05 – Grouting of Dowels and Fasteners

03 63 05.10 Epoxy Only		Crew	Daily Output	Labor-Hours	Unit	2020 Bare Costs Material	Labor	Equipment	Total	Total Incl O&P
0010	**EPOXY ONLY**									
1500	Chemical anchoring, epoxy cartridge, excludes layout, drilling, fastener									
1530	For fastener 3/4" diam. x 6" embedment	2 Skwk	72	.222	Ea.	4.78	12.20		16.98	24.50
1535	1" diam. x 8" embedment		66	.242		7.15	13.30		20.45	29
1540	1-1/4" diam. x 10" embedment		60	.267		14.35	14.65		29	39.50
1545	1-3/4" diam. x 12" embedment		54	.296		24	16.25		40.25	52.50
1550	14" embedment		48	.333		28.50	18.30		46.80	60.50
1555	2" diam. x 12" embedment		42	.381		38	21		59	75.50
1560	18" embedment	↓	32	.500	↓	48	27.50		75.50	96

03 81 Concrete Cutting

03 81 13 – Flat Concrete Sawing

03 81 13.50 Concrete Floor/Slab Cutting		Crew	Daily Output	Labor-Hours	Unit	Material	Labor	Equipment	Total	Total Incl O&P
0010	**CONCRETE FLOOR/SLAB CUTTING**									
0050	Includes blade cost, layout and set-up time									
0300	Saw cut concrete slabs, plain, up to 3" deep	B-89	1060	.015	L.F.	.12	.76	.98	1.86	2.40
0320	Each additional inch of depth		3180	.005		.04	.25	.33	.62	.80
0400	Mesh reinforced, up to 3" deep		980	.016		.13	.82	1.06	2.01	2.60
0420	Each additional inch of depth		2940	.005		.04	.27	.35	.66	.87
0500	Rod reinforced, up to 3" deep		800	.020		.17	1	1.29	2.46	3.19
0520	Each additional inch of depth		2400	.007	↓	.06	.33	.43	.82	1.06
0590	Minimum labor/equipment charge	↓	2	8	Job		400	515	915	1,200

03 81 13.75 Concrete Saw Blades		Crew	Daily Output	Labor-Hours	Unit	Material	Labor	Equipment	Total	Total Incl O&P
0010	**CONCRETE SAW BLADES**									
3000	Blades for saw cutting, included in cutting line items									
3020	Diamond, 12" diameter				Ea.	249			249	274
3040	18" diameter					405			405	445
3080	24" diameter					585			585	645
3120	30" diameter					865			865	950
3160	36" diameter					1,150			1,150	1,250
3200	42" diameter				↓	2,450			2,450	2,700

03 81 16 – Track Mounted Concrete Wall Sawing

03 81 16.50 Concrete Wall Cutting		Crew	Daily Output	Labor-Hours	Unit	Material	Labor	Equipment	Total	Total Incl O&P
0010	**CONCRETE WALL CUTTING**									
0750	Includes blade cost, layout and set-up time									
0800	Concrete walls, hydraulic saw, plain, per inch of depth	B-89B	250	.064	L.F.	.04	3.21	6	9.25	11.70
0820	Rod reinforcing, per inch of depth		150	.107	"	.06	5.35	10.05	15.46	19.55
0890	Minimum labor/equipment charge	↓	2	8	Job		400	750	1,150	1,450

03 82 Concrete Boring

03 82 13 – Concrete Core Drilling

03 82 13.10 Core Drilling		Crew	Daily Output	Labor-Hours	Unit	Material	2020 Bare Costs Labor	Equipment	Total	Total Incl O&P
0010	**CORE DRILLING**									
0015	Includes bit cost, layout and set-up time									
0020	Reinforced concrete slab, up to 6″ thick									
0100	1″ diameter core	B-89A	17	.941	Ea.	.20	45.50	6.70	52.40	80.50
0150	For each additional inch of slab thickness in same hole, add		1440	.011		.03	.54	.08	.65	.99
0200	2″ diameter core		16.50	.970		.26	47	6.90	54.16	83
0250	For each additional inch of slab thickness in same hole, add		1080	.015		.04	.72	.11	.87	1.32
0300	3″ diameter core		16	1		.40	48.50	7.15	56.05	86
0350	For each additional inch of slab thickness in same hole, add		720	.022		.07	1.08	.16	1.31	1.96
0500	4″ diameter core		15	1.067		.45	51.50	7.60	59.55	91.50
0550	For each additional inch of slab thickness in same hole, add		480	.033		.07	1.62	.24	1.93	2.92
0700	6″ diameter core		14	1.143		.65	55.50	8.15	64.30	98
0750	For each additional inch of slab thickness in same hole, add		360	.044		.11	2.15	.32	2.58	3.91
0900	8″ diameter core		13	1.231		1.07	59.50	8.80	69.37	106
0950	For each additional inch of slab thickness in same hole, add		288	.056		.18	2.69	.40	3.27	4.94
1100	10″ diameter core		12	1.333		1.20	64.50	9.50	75.20	115
1150	For each additional inch of slab thickness in same hole, add		240	.067		.20	3.23	.48	3.91	5.90
1300	12″ diameter core		11	1.455		1.84	70.50	10.40	82.74	126
1350	For each additional inch of slab thickness in same hole, add		206	.078		.31	3.77	.55	4.63	6.95
1500	14″ diameter core		10	1.600		1.84	77.50	11.40	90.74	139
1550	For each additional inch of slab thickness in same hole, add		180	.089		.31	4.31	.63	5.25	7.95
1700	18″ diameter core		9	1.778		3.06	86	12.70	101.76	155
1750	For each additional inch of slab thickness in same hole, add		144	.111		.51	5.40	.79	6.70	10.05
1754	24″ diameter core		8	2		4.15	97	14.30	115.45	175
1756	For each additional inch of slab thickness in same hole, add		120	.133		.69	6.45	.95	8.09	12.10
1760	For horizontal holes, add to above						20%	20%		
1770	Prestressed hollow core plank, 8″ thick									
1780	1″ diameter core	B-89A	17.50	.914	Ea.	.27	44.50	6.55	51.32	78.50
1790	For each additional inch of plank thickness in same hole, add		3840	.004		.03	.20	.03	.26	.39
1794	2″ diameter core		17.25	.928		.35	45	6.60	51.95	79.50
1796	For each additional inch of plank thickness in same hole, add		2880	.006		.04	.27	.04	.35	.52
1800	3″ diameter core		17	.941		.53	45.50	6.70	52.73	81
1810	For each additional inch of plank thickness in same hole, add		1920	.008		.07	.40	.06	.53	.79
1820	4″ diameter core		16.50	.970		.60	47	6.90	54.50	83.50
1830	For each additional inch of plank thickness in same hole, add		1280	.013		.07	.61	.09	.77	1.15
1840	6″ diameter core		15.50	1.032		.87	50	7.35	58.22	89
1850	For each additional inch of plank thickness in same hole, add		960	.017		.11	.81	.12	1.04	1.54
1860	8″ diameter core		15	1.067		1.43	51.50	7.60	60.53	92.50
1870	For each additional inch of plank thickness in same hole, add		768	.021		.18	1.01	.15	1.34	1.97
1880	10″ diameter core		14	1.143		1.60	55.50	8.15	65.25	99.50
1890	For each additional inch of plank thickness in same hole, add		640	.025		.20	1.21	.18	1.59	2.36
1900	12″ diameter core		13.50	1.185		2.45	57.50	8.45	68.40	104
1910	For each additional inch of plank thickness in same hole, add		548	.029		.31	1.42	.21	1.94	2.83
1999	Drilling, core, minimum labor/equipment charge		2	8	Job		390	57	447	685
3000	Bits for core drilling, included in drilling line items									
3010	Diamond, premium, 1″ diameter				Ea.	80.50			80.50	89
3020	2″ diameter					104			104	114
3030	3″ diameter					159			159	175
3040	4″ diameter					180			180	198
3060	6″ diameter					262			262	288
3080	8″ diameter					430			430	470
3110	10″ diameter					480			480	530
3120	12″ diameter					735			735	810

03 82 13 – Concrete Core Drilling

03 82 13.10 Core Drilling		Crew	Daily Output	Labor-Hours	Unit	Material	Labor	Equipment	Total	Total Incl O&P
						2020 Bare Costs				
3140	14" diameter				Ea.	735			735	810
3180	18" diameter					1,225			1,225	1,350
3240	24" diameter					1,650			1,650	1,825

03 82 16 – Concrete Drilling

03 82 16.10 Concrete Impact Drilling

03 82 16.10 Concrete Impact Drilling		Crew	Daily Output	Labor-Hours	Unit	Material	Labor	Equipment	Total	Total Incl O&P
0010	**CONCRETE IMPACT DRILLING**									
0020	Includes bit cost, layout and set-up time, no anchors									
0050	Up to 4" deep in concrete/brick floors/walls									
0100	Holes, 1/4" diameter	1 Carp	75	.107	Ea.	.07	5.65		5.72	9.20
0150	For each additional inch of depth in same hole, add		430	.019		.02	.99		1.01	1.60
0200	3/8" diameter		63	.127		.05	6.75		6.80	10.85
0250	For each additional inch of depth in same hole, add		340	.024		.01	1.25		1.26	2.01
0300	1/2" diameter		50	.160		.05	8.50		8.55	13.65
0350	For each additional inch of depth in same hole, add		250	.032		.01	1.70		1.71	2.73
0400	5/8" diameter		48	.167		.10	8.85		8.95	14.30
0450	For each additional inch of depth in same hole, add		240	.033		.02	1.77		1.79	2.87
0500	3/4" diameter		45	.178		.11	9.45		9.56	15.25
0550	For each additional inch of depth in same hole, add		220	.036		.03	1.93		1.96	3.13
0600	7/8" diameter		43	.186		.17	9.90		10.07	16.05
0650	For each additional inch of depth in same hole, add		210	.038		.04	2.03		2.07	3.29
0700	1" diameter		40	.200		.18	10.65		10.83	17.25
0750	For each additional inch of depth in same hole, add		190	.042		.04	2.24		2.28	3.64
0800	1-1/4" diameter		38	.211		.33	11.20		11.53	18.30
0850	For each additional inch of depth in same hole, add		180	.044		.08	2.36		2.44	3.87
0900	1-1/2" diameter		35	.229		.43	12.15		12.58	19.90
0950	For each additional inch of depth in same hole, add		165	.048		.11	2.58		2.69	4.25
1000	For ceiling installations, add						40%			

Division Notes

I	CREW	DAILY OUTPUT	LABOR-HOURS	UNIT	BARE COSTS MAT.	LABOR	EQUIP.	TOTAL	TOTAL INCL O&P

Estimating Tips

04 05 00 Common Work Results for Masonry

- The terms mortar and grout are often used interchangeably—and incorrectly. Mortar is used to bed masonry units, seal the entry of air and moisture, provide architectural appearance, and allow for size variations in the units. Grout is used primarily in reinforced masonry construction and to bond the masonry to the reinforcing steel. Common mortar types are M (2500 psi), S (1800 psi), N (750 psi), and O (350 psi), and they conform to ASTM C270. Grout is either fine or coarse and conforms to ASTM C476, and in-place strengths generally exceed 2500 psi. Mortar and grout are different components of masonry construction and are placed by entirely different methods. An estimator should be aware of their unique uses and costs.
- Mortar is included in all assembled masonry line items. The mortar cost, part of the assembled masonry material cost, includes all ingredients, all labor, and all equipment required. Please see reference number R040513-10.
- Waste, specifically the loss/droppings of mortar and the breakage of brick and block, is included in all unit cost lines that include mortar and masonry units in this division. A factor of 25% is added for mortar and 3% for brick and concrete masonry units.
- Scaffolding or staging is not included in any of the Division 4 costs. Refer to Subdivision 01 54 23 for scaffolding and staging costs.

04 20 00 Unit Masonry

- The most common types of unit masonry are brick and concrete masonry. The major classifications of brick are building brick (ASTM C62), facing brick (ASTM C216), glazed brick, fire brick, and pavers. Many varieties of texture and appearance can exist within these classifications, and the estimator would be wise to check local custom and availability within the project area. For repair and remodeling jobs, matching the existing brick may be the most important criteria.
- Brick and concrete block are priced by the piece and then converted into a price per square foot of wall. Openings less than two square feet are generally ignored by the estimator because any savings in units used are offset by the cutting and trimming required.
- It is often difficult and expensive to find and purchase small lots of historic brick. Costs can vary widely. Many design issues affect costs, selection of mortar mix, and repairs or replacement of masonry materials. Cleaning techniques must be reflected in the estimate.
- All masonry walls, whether interior or exterior, require bracing. The cost of bracing walls during construction should be included by the estimator, and this bracing must remain in place until permanent bracing is complete. Permanent bracing of masonry walls is accomplished by masonry itself, in the form of pilasters or abutting wall corners, or by anchoring the walls to the structural frame. Accessories in the form of anchors, anchor slots, and ties are used, but their supply and installation can be by different trades. For instance, anchor slots on spandrel beams and columns are supplied and welded in place by the steel fabricator, but the ties from the slots into the masonry are installed by the bricklayer. Regardless of the installation method, the estimator must be certain that these accessories are accounted for in pricing.

Reference Numbers

Reference numbers are shown at the beginning of some major classifications. These numbers refer to related items in the Reference Section. The reference information may be an estimating procedure, an alternate pricing method, or technical information.

Note: Not all subdivisions listed here necessarily appear. ■

04 01 20 – Maintenance of Unit Masonry

04 01 20.20 Pointing Masonry		Crew	Daily Output	Labor-Hours	Unit	2020 Bare Costs Material	Labor	Equipment	Total	Total Incl O&P
0010	**POINTING MASONRY**									
0300	Cut and repoint brick, hard mortar, running bond	1 Bric	80	.100	S.F.	.57	5.20		5.77	9.05
0320	Common bond		77	.104		.57	5.40		5.97	9.40
0360	Flemish bond		70	.114		.60	5.95		6.55	10.25
0400	English bond		65	.123		.60	6.40		7	11
0600	Soft old mortar, running bond		100	.080		.57	4.16		4.73	7.40
0620	Common bond		96	.083		.57	4.34		4.91	7.65
0640	Flemish bond		90	.089		.60	4.63		5.23	8.15
0680	English bond		82	.098		.60	5.10		5.70	8.85
0700	Stonework, hard mortar		140	.057	L.F.	.76	2.97		3.73	5.65
0720	Soft old mortar		160	.050	"	.76	2.60		3.36	5.05
1000	Repoint, mask and grout method, running bond		95	.084	S.F.	.76	4.38		5.14	7.95
1020	Common bond		90	.089		.76	4.63		5.39	8.35
1040	Flemish bond		86	.093		.80	4.84		5.64	8.70
1060	English bond		77	.104		.80	5.40		6.20	9.65
2000	Scrub coat, sand grout on walls, thin mix, brushed		120	.067		3.59	3.47		7.06	9.55
2020	Troweled		98	.082		4.99	4.25		9.24	12.35
9000	Minimum labor/equipment charge		3	2.667	Job		139		139	224

04 01 20.30 Pointing CMU		Crew	Daily Output	Labor-Hours	Unit	Material	Labor	Equipment	Total	Total Incl O&P
0010	**POINTING CMU**									
0300	Cut and repoint block, hard mortar, running bond	1 Bric	190	.042	S.F.	.23	2.19		2.42	3.80
0310	Stacked bond		200	.040		.23	2.08		2.31	3.62
0600	Soft old mortar, running bond		230	.035		.23	1.81		2.04	3.19
0610	Stacked bond		245	.033		.23	1.70		1.93	3.01

04 01 20.40 Sawing Masonry		Crew	Daily Output	Labor-Hours	Unit	Material	Labor	Equipment	Total	Total Incl O&P
0010	**SAWING MASONRY**									
0050	Brick or block by hand, per inch depth	A-1	125	.064	L.F.	.05	2.69	.89	3.63	5.35

04 01 20.41 Unit Masonry Stabilization		Crew	Daily Output	Labor-Hours	Unit	Material	Labor	Equipment	Total	Total Incl O&P
0010	**UNIT MASONRY STABILIZATION**									
0100	Structural repointing method									
0110	Cut/grind mortar joint	1 Bric	240	.033	L.F.		1.73		1.73	2.80
0120	Clean and mask joint		2500	.003		.13	.17		.30	.41
0130	Epoxy paste and 1/4" FRP rod		240	.033		2.11	1.73		3.84	5.10
0132	3/8" FRP rod		160	.050		2.98	2.60		5.58	7.50
0140	Remove masking		14400	.001			.03		.03	.05
0300	Structural fabric method									
0310	Primer	1 Bric	600	.013	S.F.	1.08	.69		1.77	2.30
0320	Apply filling/leveling paste		720	.011		.89	.58		1.47	1.91
0330	Epoxy, glass fiber fabric		720	.011		9.85	.58		10.43	11.80
0340	Carbon fiber fabric		720	.011		23.50	.58		24.08	27

04 01 20.50 Toothing Masonry		Crew	Daily Output	Labor-Hours	Unit	Material	Labor	Equipment	Total	Total Incl O&P
0010	**TOOTHING MASONRY**									
0500	Brickwork, soft old mortar	1 Clab	40	.200	V.L.F.		8.40		8.40	13.50
0520	Hard mortar		30	.267			11.25		11.25	18
0700	Blockwork, soft old mortar		70	.114			4.81		4.81	7.70
0720	Hard mortar		50	.160			6.75		6.75	10.80
9000	Minimum labor/equipment charge		4	2	Job		84		84	135

04 01 20.52 Cleaning Masonry		Crew	Daily Output	Labor-Hours	Unit	Material	Labor	Equipment	Total	Total Incl O&P
0010	**CLEANING MASONRY**									
0200	By chemical, brush and rinse, new work, light construction dust	D-1	1000	.016	S.F.	.06	.75		.81	1.27
0220	Medium construction dust		800	.020		.08	.94		1.02	1.61

04 01 Maintenance of Masonry

04 01 20 – Maintenance of Unit Masonry

04 01 20.52 Cleaning Masonry		Crew	Daily Output	Labor-Hours	Unit	Material	2020 Bare Costs Labor	Equipment	Total	Total Incl O&P
0240	Heavy construction dust, drips or stains	D-1	600	.027	S.F.	.11	1.25		1.36	2.14
0260	Low pressure wash and rinse, light restoration, light soil		800	.020		.14	.94		1.08	1.68
0270	Average soil, biological staining		400	.040		.22	1.88		2.10	3.27
0280	Heavy soil, biological and mineral staining, paint		330	.048		.29	2.27		2.56	3.99
0300	High pressure wash and rinse, heavy restoration, light soil		600	.027		.10	1.25		1.35	2.13
0310	Average soil, biological staining		400	.040		.15	1.88		2.03	3.19
0320	Heavy soil, biological and mineral staining, paint		250	.064		.20	3		3.20	5.05
0400	High pressure wash, water only, light soil	C-29	500	.016			.67	.19	.86	1.29
0420	Average soil, biological staining		375	.021			.90	.26	1.16	1.72
0440	Heavy soil, biological and mineral staining, paint		250	.032			1.35	.39	1.74	2.59
0800	High pressure water and chemical, light soil		450	.018		.17	.75	.22	1.14	1.63
0820	Average soil, biological staining		300	.027		.26	1.12	.32	1.70	2.45
0840	Heavy soil, biological and mineral staining, paint		200	.040		.35	1.68	.48	2.51	3.61
1200	Sandblast, wet system, light soil	J-6	1750	.018		.34	.84	.17	1.35	1.90
1220	Average soil, biological staining		1100	.029		.51	1.34	.27	2.12	2.99
1240	Heavy soil, biological and mineral staining, paint		700	.046		.69	2.10	.43	3.22	4.55
1400	Dry system, light soil		2500	.013		.34	.59	.12	1.05	1.44
1420	Average soil, biological staining		1750	.018		.51	.84	.17	1.52	2.09
1440	Heavy soil, biological and mineral staining, paint		1000	.032		.69	1.47	.30	2.46	3.41
1800	For walnut shells, add					.92			.92	1.01
1820	For corn chips, add					.96			.96	1.06
2000	Steam cleaning, light soil	A-1H	750	.011			.45	.10	.55	.83
2020	Average soil, biological staining		625	.013			.54	.12	.66	.99
2040	Heavy soil, biological and mineral staining		375	.021			.90	.20	1.10	1.66
4000	Add for masking doors and windows	1 Clab	800	.010		.07	.42		.49	.75
4200	Add for pedestrian protection				Job				10%	10%
9000	Minimum labor/equipment charge	D-4	2	16	"		755	94	849	1,300

04 01 20.70 Brick Washing		Crew	Daily Output	Labor-Hours	Unit	Material	Labor	Equipment	Total	Total Incl O&P
0010	**BRICK WASHING**									
0012	Acid cleanser, smooth brick surface	1 Bric	560	.014	S.F.	.05	.74		.79	1.26
0050	Rough brick		400	.020		.07	1.04		1.11	1.76
0060	Stone, acid wash		600	.013		.08	.69		.77	1.21
1000	Muriatic acid, price per gallon in 5 gallon lots				Gal.	10.55			10.55	11.60

04 05 Common Work Results for Masonry

04 05 05 – Selective Demolition for Masonry

04 05 05.10 Selective Demolition		Crew	Daily Output	Labor-Hours	Unit	Material	Labor	Equipment	Total	Total Incl O&P
0010	**SELECTIVE DEMOLITION** R024119-10									
0200	Bond beams, 8" block with #4 bar	2 Clab	32	.500	L.F.		21		21	33.50
0300	Concrete block walls, unreinforced, 2" thick		1200	.013	S.F.		.56		.56	.90
0310	4" thick		1150	.014			.59		.59	.94
0320	6" thick		1100	.015			.61		.61	.98
0330	8" thick		1050	.015			.64		.64	1.03
0340	10" thick		1000	.016			.67		.67	1.08
0360	12" thick		950	.017			.71		.71	1.14
0380	Reinforced alternate courses, 2" thick		1130	.014			.60		.60	.96
0390	4" thick		1080	.015			.62		.62	1
0400	6" thick		1035	.015			.65		.65	1.04
0410	8" thick		990	.016			.68		.68	1.09
0420	10" thick		940	.017			.72		.72	1.15
0430	12" thick		890	.018			.76		.76	1.21

04 05 05.10 Selective Demolition		Crew	Daily Output	Labor-Hours	Unit	Material	2020 Bare Costs Labor	Equipment	Total	Total Incl O&P
0440	Reinforced alternate courses & vertically 48" OC, 4" thick	2 Clab	900	.018	S.F.		.75		.75	1.20
0450	6" thick		850	.019			.79		.79	1.27
0460	8" thick		800	.020			.84		.84	1.35
0480	10" thick		750	.021			.90		.90	1.44
0490	12" thick		700	.023			.96		.96	1.54
1000	Chimney, 16" x 16", soft old mortar	1 Clab	55	.145	C.F.		6.10		6.10	9.80
1020	Hard mortar		40	.200			8.40		8.40	13.50
1030	16" x 20", soft old mortar		55	.145			6.10		6.10	9.80
1040	Hard mortar		40	.200			8.40		8.40	13.50
1050	16" x 24", soft old mortar		55	.145			6.10		6.10	9.80
1060	Hard mortar		40	.200			8.40		8.40	13.50
1080	20" x 20", soft old mortar		55	.145			6.10		6.10	9.80
1100	Hard mortar		40	.200			8.40		8.40	13.50
1110	20" x 24", soft old mortar		55	.145			6.10		6.10	9.80
1120	Hard mortar		40	.200			8.40		8.40	13.50
1140	20" x 32", soft old mortar		55	.145			6.10		6.10	9.80
1160	Hard mortar		40	.200			8.40		8.40	13.50
1200	48" x 48", soft old mortar		55	.145			6.10		6.10	9.80
1220	Hard mortar		40	.200			8.40		8.40	13.50
1250	Metal, high temp steel jacket, 24" diameter	E-2	130	.431	V.L.F.		24.50	13.05	37.55	54.50
1260	60" diameter	"	60	.933			53.50	28.50	82	118
1280	Flue lining, up to 12" x 12"	1 Clab	200	.040			1.68		1.68	2.70
1282	Up to 24" x 24"		150	.053			2.25		2.25	3.60
2000	Columns, 8" x 8", soft old mortar		48	.167			7		7	11.25
2020	Hard mortar		40	.200			8.40		8.40	13.50
2060	16" x 16", soft old mortar		16	.500			21		21	33.50
2100	Hard mortar		14	.571			24		24	38.50
2140	24" x 24", soft old mortar		8	1			42		42	67.50
2160	Hard mortar		6	1.333			56		56	90
2200	36" x 36", soft old mortar		4	2			84		84	135
2220	Hard mortar		3	2.667			112		112	180
2230	Alternate pricing method, soft old mortar		30	.267	C.F.		11.25		11.25	18
2240	Hard mortar		23	.348	"		14.65		14.65	23.50
3000	Copings, precast or masonry, to 8" wide									
3020	Soft old mortar	1 Clab	180	.044	L.F.		1.87		1.87	3
3040	Hard mortar	"	160	.050	"		2.11		2.11	3.37
3100	To 12" wide									
3120	Soft old mortar	1 Clab	160	.050	L.F.		2.11		2.11	3.37
3140	Hard mortar	"	140	.057	"		2.41		2.41	3.85
4000	Fireplace, brick, 30" x 24" opening									
4020	Soft old mortar	1 Clab	2	4	Ea.		168		168	270
4040	Hard mortar		1.25	6.400			269		269	430
4100	Stone, soft old mortar		1.50	5.333			225		225	360
4120	Hard mortar		1	8			335		335	540
5000	Veneers, brick, soft old mortar		140	.057	S.F.		2.41		2.41	3.85
5020	Hard mortar		125	.064			2.69		2.69	4.32
5050	Glass block, up to 4" thick		500	.016			.67		.67	1.08
5100	Granite and marble, 2" thick		180	.044			1.87		1.87	3
5120	4" thick		170	.047			1.98		1.98	3.17
5140	Stone, 4" thick		180	.044			1.87		1.87	3
5160	8" thick		175	.046			1.92		1.92	3.08
5400	Alternate pricing method, stone, 4" thick		60	.133	C.F.		5.60		5.60	9
5420	8" thick		85	.094	"		3.96		3.96	6.35

04 05 05 – Selective Demolition for Masonry

04 05 05.10 Selective Demolition		Crew	Daily Output	Labor-Hours	Unit	2020 Bare Costs Material	Labor	Equipment	Total	Total Incl O&P
9000	Minimum labor/equipment charge	1 Clab	2	4	Job		168		168	270

04 05 13 – Masonry Mortaring

04 05 13.10 Cement

		Crew	Daily Output	Labor-Hours	Unit	Material	Labor	Equipment	Total	Total Incl O&P
0010	**CEMENT**									
0100	Masonry, 70 lb. bag, T.L. lots				Bag	11.40			11.40	12.50
0150	L.T.L. lots					12.05			12.05	13.25
0200	White, 70 lb. bag, T.L. lots					18.65			18.65	20.50
0250	L.T.L. lots					22.50			22.50	24.50

04 05 13.20 Lime

		Crew	Daily Output	Labor-Hours	Unit	Material	Labor	Equipment	Total	Total Incl O&P
0010	**LIME**									
0020	Masons, hydrated, 50 lb. bag, T.L. lots				Bag	11.25			11.25	12.40
0050	L.T.L. lots					12.40			12.40	13.65
0200	Finish, double hydrated, 50 lb. bag, T.L. lots					11.90			11.90	13.10
0250	L.T.L. lots					13.10			13.10	14.40

04 05 13.23 Surface Bonding Masonry Mortaring

		Crew	Daily Output	Labor-Hours	Unit	Material	Labor	Equipment	Total	Total Incl O&P
0010	**SURFACE BONDING MASONRY MORTARING**									
0020	Gray or white colors, not incl. block work	1 Bric	540	.015	S.F.	.20	.77		.97	1.47

04 05 13.30 Mortar

		Crew	Daily Output	Labor-Hours	Unit	Material	Labor	Equipment	Total	Total Incl O&P
0010	**MORTAR** R042110-50									
0020	With masonry cement									
0100	Type M, 1:1:6 mix	1 Brhe	143	.056	C.F.	6.15	2.34		8.49	10.50
0200	Type N, 1:3 mix		143	.056		5.35	2.34		7.69	9.65
0300	Type O, 1:3 mix		143	.056		5.20	2.34		7.54	9.45
0400	Type PM, 1:1:6 mix, 2500 psi		143	.056		6.20	2.34		8.54	10.55
0500	Type S, 1/2:1:4 mix		143	.056		6.55	2.34		8.89	10.95
2000	With Portland cement and lime									
2100	Type M, 1:1/4:3 mix	1 Brhe	143	.056	C.F.	9.10	2.34		11.44	13.75
2200	Type N, 1:1:6 mix, 750 psi		143	.056		7.30	2.34		9.64	11.80
2300	Type O, 1:2:9 mix (Pointing Mortar)		143	.056		8.75	2.34		11.09	13.40
2400	Type PL, 1:1/2:4 mix, 2500 psi		143	.056		6.35	2.34		8.69	10.75
2600	Type S, 1:1/2:4 mix, 1800 psi		143	.056		8.55	2.34		10.89	13.20
2650	Pre-mixed, type S or N					6.25			6.25	6.90
2700	Mortar for glass block	1 Brhe	143	.056		13.25	2.34		15.59	18.30
2900	Mortar for fire brick, dry mix, 10 lb. pail				Ea.	25			25	28

04 05 13.91 Masonry Restoration Mortaring

		Crew	Daily Output	Labor-Hours	Unit	Material	Labor	Equipment	Total	Total Incl O&P
0010	**MASONRY RESTORATION MORTARING**									
0020	Masonry restoration mix				Lb.	.85			.85	.94
0050	White				"	1.40			1.40	1.54

04 05 13.93 Mortar Pigments

		Crew	Daily Output	Labor-Hours	Unit	Material	Labor	Equipment	Total	Total Incl O&P
0010	**MORTAR PIGMENTS**, 50 lb. bags (2 bags per M bricks)									
0020	Color admixture, range 2 to 10 lb. per bag of cement, light colors				Lb.	6.30			6.30	6.95
0050	Medium colors					7.05			7.05	7.75
0100	Dark colors					15.95			15.95	17.55

04 05 13.95 Sand

		Crew	Daily Output	Labor-Hours	Unit	Material	Labor	Equipment	Total	Total Incl O&P
0010	**SAND**, screened and washed at pit									
0020	For mortar, per ton				Ton	21.50			21.50	24
0050	With 10 mile haul					42			42	46
0100	With 30 mile haul					67			67	74
0200	Screened and washed, at the pit				C.Y.	30			30	33
0250	With 10 mile haul					58.50			58.50	64

04 05 Common Work Results for Masonry

04 05 13 – Masonry Mortaring

04 05 13.95 Sand		Crew	Daily Output	Labor-Hours	Unit	2020 Bare Costs Material	Labor	Equipment	Total	Total Incl O&P
0300	With 30 mile haul				C.Y.	93.50			93.50	103

04 05 13.98 Mortar Admixtures

		Crew	Daily Output	Labor-Hours	Unit	Material	Labor	Equipment	Total	Total Incl O&P
0010	**MORTAR ADMIXTURES**									
0020	Waterproofing admixture, per quart (1 qt. to 2 bags of masonry cement)				Qt.	4.10			4.10	4.51

04 05 16 – Masonry Grouting

04 05 16.30 Grouting

		Crew	Daily Output	Labor-Hours	Unit	Material	Labor	Equipment	Total	Total Incl O&P
0010	**GROUTING**									
0011	Bond beams & lintels, 8" deep, 6" thick, 0.15 C.F./L.F.	D-4	1480	.022	L.F.	.80	1.02	.13	1.95	2.65
0020	8" thick, 0.2 C.F./L.F.		1400	.023		1.29	1.08	.13	2.50	3.30
0050	10" thick, 0.25 C.F./L.F.		1200	.027		1.34	1.26	.16	2.76	3.65
0060	12" thick, 0.3 C.F./L.F.		1040	.031		1.61	1.45	.18	3.24	4.29
0200	Concrete block cores, solid, 4" thk., by hand, 0.067 C.F./S.F. of wall	D-8	1100	.036	S.F.	.36	1.74		2.10	3.21
0210	6" thick, pumped, 0.175 C.F./S.F.	D-4	720	.044		.94	2.09	.26	3.29	4.67
0250	8" thick, pumped, 0.258 C.F./S.F.		680	.047		1.38	2.22	.28	3.88	5.35
0300	10" thick, pumped, 0.340 C.F./S.F.		660	.048		1.82	2.29	.29	4.40	5.95
0350	12" thick, pumped, 0.422 C.F./S.F.		640	.050		2.26	2.36	.29	4.91	6.60
0500	Cavity walls, 2" space, pumped, 0.167 C.F./S.F. of wall		1700	.019		.89	.89	.11	1.89	2.52
0550	3" space, 0.250 C.F./S.F.		1200	.027		1.34	1.26	.16	2.76	3.65
0600	4" space, 0.333 C.F./S.F.		1150	.028		1.78	1.31	.16	3.25	4.24
0700	6" space, 0.500 C.F./S.F.		800	.040		2.68	1.89	.24	4.81	6.25
0800	Door frames, 3' x 7' opening, 2.5 C.F. per opening		60	.533	Opng.	13.40	25	3.14	41.54	58.50
0850	6' x 7' opening, 3.5 C.F. per opening		45	.711	"	18.75	33.50	4.19	56.44	78.50
2000	Grout, C476, for bond beams, lintels and CMU cores		350	.091	C.F.	5.35	4.31	.54	10.20	13.40
9000	Minimum labor/equipment charge	1 Bric	2	4	Job		208		208	335

04 05 19 – Masonry Anchorage and Reinforcing

04 05 19.05 Anchor Bolts

		Crew	Daily Output	Labor-Hours	Unit	Material	Labor	Equipment	Total	Total Incl O&P
0010	**ANCHOR BOLTS**									
0015	Installed in fresh grout in CMU bond beams or filled cores, no templates									
0020	Hooked, with nut and washer, 1/2" diameter, 8" long	1 Bric	132	.061	Ea.	1.62	3.15		4.77	6.90
0030	12" long		131	.061		1.80	3.18		4.98	7.15
0040	5/8" diameter, 8" long		129	.062		3.39	3.23		6.62	8.95
0050	12" long		127	.063		4.18	3.28		7.46	9.90
0060	3/4" diameter, 8" long		127	.063		4.18	3.28		7.46	9.90
0070	12" long		125	.064		5.20	3.33		8.53	11.15

04 05 19.16 Masonry Anchors

		Crew	Daily Output	Labor-Hours	Unit	Material	Labor	Equipment	Total	Total Incl O&P
0010	**MASONRY ANCHORS**									
0020	For brick veneer, galv., corrugated, 7/8" x 7", 22 ga.	1 Bric	10.50	.762	C	16.70	39.50		56.20	82.50
0100	24 ga.		10.50	.762		10.35	39.50		49.85	75.50
0150	16 ga.		10.50	.762		31.50	39.50		71	99
0200	Buck anchors, galv., corrugated, 16 ga., 2" bend, 8" x 2"		10.50	.762		65	39.50		104.50	136
0250	8" x 3"		10.50	.762		69.50	39.50		109	141
0300	Adjustable, rectangular, 4-1/8" wide									
0350	Anchor and tie, 3/16" wire, mill galv.									
0400	2-3/4" eye, 3-1/4" tie	1 Bric	1.05	7.619	M	475	395		870	1,175
0500	4-3/4" tie		1.05	7.619		515	395		910	1,200
0520	5-1/2" tie		1.05	7.619		555	395		950	1,250
0550	4-3/4" eye, 3-1/4" tie		1.05	7.619		520	395		915	1,200
0570	4-3/4" tie		1.05	7.619		550	395		945	1,250
0580	5-1/2" tie		1.05	7.619		610	395		1,005	1,325
0660	Cavity wall, Z-type, galvanized, 6" long, 1/8" diam.		10.50	.762	C	25	39.50		64.50	91.50
0670	3/16" diameter		10.50	.762		31.50	39.50		71	98.50

04 05 19 – Masonry Anchorage and Reinforcing

04 05 19.16 Masonry Anchors		Crew	Daily Output	Labor-Hours	Unit	Material	Labor	Equipment	Total	Total Incl O&P
						2020 Bare Costs				
0680	1/4" diameter	1 Bric	10.50	.762	C	40.50	39.50		80	109
0850	8" long, 3/16" diameter		10.50	.762		27	39.50		66.50	94
0855	1/4" diameter		10.50	.762		49	39.50		88.50	118
1000	Rectangular type, galvanized, 1/4" diameter, 2" x 6"		10.50	.762		80	39.50		119.50	152
1050	4" x 6"		10.50	.762		91.50	39.50		131	164
1100	3/16" diameter, 2" x 6"		10.50	.762		54.50	39.50		94	124
1150	4" x 6"		10.50	.762		55	39.50		94.50	125
1200	Mesh wall tie, 1/2" mesh, hot dip galvanized									
1400	16 ga., 12" long, 3" wide	1 Bric	9	.889	C	92.50	46.50		139	177
1420	6" wide		9	.889		135	46.50		181.50	224
1440	12" wide		8.50	.941		214	49		263	315
1500	Rigid partition anchors, plain, 8" long, 1" x 1/8"		10.50	.762		241	39.50		280.50	330
1550	1" x 1/4"		10.50	.762		325	39.50		364.50	420
1580	1-1/2" x 1/8"		10.50	.762		261	39.50		300.50	350
1600	1-1/2" x 1/4"		10.50	.762		340	39.50		379.50	440
1650	2" x 1/8"		10.50	.762		330	39.50		369.50	430
1700	2" x 1/4"		10.50	.762		445	39.50		484.50	555
2000	Column flange ties, wire, galvanized									
2300	3/16" diameter, up to 3" wide	1 Bric	10.50	.762	C	88	39.50		127.50	161
2350	To 5" wide		10.50	.762		96	39.50		135.50	170
2400	To 7" wide		10.50	.762		104	39.50		143.50	178
2600	To 9" wide		10.50	.762		110	39.50		149.50	185
2650	1/4" diameter, up to 3" wide		10.50	.762		112	39.50		151.50	187
2700	To 5" wide		10.50	.762		133	39.50		172.50	211
2800	To 7" wide		10.50	.762		150	39.50		189.50	229
2850	To 9" wide		10.50	.762		162	39.50		201.50	243
2900	For hot dip galvanized, add					35%				
4000	Channel slots, 1-3/8" x 1/2" x 8"									
4100	12 ga., plain	1 Bric	10.50	.762	C	244	39.50		283.50	335
4150	16 ga., galvanized	"	10.50	.762	"	155	39.50		194.50	234
4200	Channel slot anchors									
4300	16 ga., galvanized, 1-1/4" x 3-1/2"				C	61.50			61.50	67.50
4350	1-1/4" x 5-1/2"					70.50			70.50	77.50
4400	1-1/4" x 7-1/2"					80			80	88
4500	1/8" plain, 1-1/4" x 3-1/2"					160			160	176
4550	1-1/4" x 5-1/2"					171			171	188
4600	1-1/4" x 7-1/2"					186			186	205
4700	For corrugation, add					86			86	94.50
4750	For hot dip galvanized, add					35%				
5000	Dowels									
5100	Plain, 1/4" diameter, 3" long				C	54.50			54.50	60
5150	4" long					60.50			60.50	66.50
5200	6" long					74			74	81
5300	3/8" diameter, 3" long					72			72	79
5350	4" long					89.50			89.50	98.50
5400	6" long					103			103	113
5500	1/2" diameter, 3" long					105			105	115
5550	4" long					124			124	137
5600	6" long					161			161	177
5700	5/8" diameter, 3" long					144			144	158
5750	4" long					177			177	195
5800	6" long					243			243	267
6000	3/4" diameter, 3" long					179			179	197

04 05 19 – Masonry Anchorage and Reinforcing

	04 05 19.16 Masonry Anchors	Crew	Daily Output	Labor-Hours	Unit	2020 Bare Costs Material	Labor	Equipment	Total	Total Incl O&P
6100	4" long				C	227			227	249
6150	6" long					325			325	355
6300	For hot dip galvanized, add					35%				

04 05 19.26 Masonry Reinforcing Bars

		Crew	Daily Output	Labor-Hours	Unit	Material	Labor	Equipment	Total	Total Incl O&P
0010	**MASONRY REINFORCING BARS** R040519-50									
0015	Steel bars A615, placed horiz., #3 & #4 bars	1 Bric	450	.018	Lb.	.56	.93		1.49	2.12
0020	#5 & #6 bars		800	.010		.56	.52		1.08	1.46
0050	Placed vertical, #3 & #4 bars		350	.023		.56	1.19		1.75	2.54
0060	#5 & #6 bars		650	.012		.56	.64		1.20	1.66
0200	Joint reinforcing, regular truss, to 6" wide, mill std galvanized		30	.267	C.L.F.	23	13.90		36.90	48
0250	12" wide		20	.400		29.50	21		50.50	66
0400	Cavity truss with drip section, to 6" wide		30	.267		23	13.90		36.90	48
0450	12" wide		20	.400		26.50	21		47.50	62.50
0500	Joint reinforcing, ladder type, mill std galvanized									
0600	9 ga. sides, 9 ga. ties, 4" wall	1 Bric	30	.267	C.L.F.	24	13.90		37.90	48.50
0650	6" wall		30	.267		22	13.90		35.90	46.50
0700	8" wall		25	.320		26.50	16.65		43.15	56
0750	10" wall		20	.400		24.50	21		45.50	60.50
0800	12" wall		20	.400		26	21		47	62
1000	Truss type									
1100	9 ga. sides, 9 ga. ties, 4" wall	1 Bric	30	.267	C.L.F.	24.50	13.90		38.40	49.50
1150	6" wall		30	.267		26	13.90		39.90	51
1200	8" wall		25	.320		28	16.65		44.65	58
1250	10" wall		20	.400		24.50	21		45.50	60
1300	12" wall		20	.400		25	21		46	61
1500	3/16" sides, 9 ga. ties, 4" wall		30	.267		28	13.90		41.90	53.50
1550	6" wall		30	.267		36	13.90		49.90	62
1600	8" wall		25	.320		32.50	16.65		49.15	62.50
1650	10" wall		20	.400		39	21		60	76.50
1700	12" wall		20	.400		38.50	21		59.50	75.50
2000	3/16" sides, 3/16" ties, 4" wall		30	.267		40	13.90		53.90	66.50
2050	6" wall		30	.267		41.50	13.90		55.40	68
2100	8" wall		25	.320		40	16.65		56.65	71
2150	10" wall		20	.400		45	21		66	83
2200	12" wall		20	.400		46.50	21		67.50	85
2500	Cavity truss type, galvanized									
2600	9 ga. sides, 9 ga. ties, 4" wall	1 Bric	25	.320	C.L.F.	37	16.65		53.65	68
2650	6" wall		25	.320		36	16.65		52.65	66.50
2700	8" wall		20	.400		51	21		72	89.50
2750	10" wall		15	.533		43.50	28		71.50	93
2800	12" wall		15	.533		49	28		77	99
3000	3/16" sides, 9 ga. ties, 4" wall		25	.320		36.50	16.65		53.15	67
3050	6" wall		25	.320		36.50	16.65		53.15	67.50
3100	8" wall		20	.400		56	21		77	95
3150	10" wall		15	.533		40	28		68	89
3200	12" wall		15	.533		41.50	28		69.50	90.50
3500	For hot dip galvanizing, add				Ton	510			510	560

04 05 23 – Masonry Accessories

04 05 23.13 Masonry Control and Expansion Joints

		Crew	Daily Output	Labor-Hours	Unit	Material	Labor	Equipment	Total	Total Incl O&P
0010	**MASONRY CONTROL AND EXPANSION JOINTS**									
0020	Rubber, for double wythe 8" minimum wall (Brick/CMU)	1 Bric	400	.020	L.F.	2.46	1.04		3.50	4.39
0025	"T" shaped		320	.025		1.22	1.30		2.52	3.44

04 05 Common Work Results for Masonry

04 05 23 – Masonry Accessories

04 05 23.13 Masonry Control and Expansion Joints		Crew	Daily Output	Labor-Hours	Unit	2020 Bare Costs Material	Labor	Equipment	Total	Total Incl O&P
0030	Cross-shaped for CMU units	1 Bric	280	.029	L.F.	1.66	1.49		3.15	4.23
0050	PVC, for double wythe 8" minimum wall (Brick/CMU)		400	.020		1.70	1.04		2.74	3.55
0120	"T" shaped		320	.025		.92	1.30		2.22	3.11
0160	Cross-shaped for CMU units		280	.029		1.14	1.49		2.63	3.65

04 05 23.19 Masonry Cavity Drainage, Weepholes, and Vents		Crew	Daily Output	Labor-Hours	Unit	Material	Labor	Equipment	Total	Total Incl O&P
0010	**MASONRY CAVITY DRAINAGE, WEEPHOLES, AND VENTS**									
0020	Extruded aluminum, 4" deep, 2-3/8" x 8-1/8"	1 Bric	30	.267	Ea.	41	13.90		54.90	67.50
0050	5" x 8-1/8"		25	.320		53	16.65		69.65	85.50
0100	2-1/4" x 25"		25	.320		89	16.65		105.65	125
0150	5" x 16-1/2"		22	.364		61.50	18.95		80.45	98.50
0175	5" x 24"		22	.364		93	18.95		111.95	133
0200	6" x 16-1/2"		22	.364		99.50	18.95		118.45	140
0250	7-3/4" x 16-1/2"		20	.400		92.50	21		113.50	136
0400	For baked enamel finish, add					35%				
0500	For cast aluminum, painted, add					60%				
1000	Stainless steel ventilators, 6" x 6"	1 Bric	25	.320		230	16.65		246.65	280
1050	8" x 8"		24	.333		257	17.35		274.35	310
1100	12" x 12"		23	.348		289	18.10		307.10	350
1150	12" x 6"		24	.333		279	17.35		296.35	335
1200	Foundation block vent, galv., 1-1/4" thk, 8" high, 16" long, no damper		30	.267		17.45	13.90		31.35	41.50
1250	For damper, add					3.34			3.34	3.67
1450	Drainage and ventilation fabric	2 Bric	1450	.011	S.F.	.82	.57		1.39	1.83

04 05 23.95 Wall Plugs		Crew	Daily Output	Labor-Hours	Unit	Material	Labor	Equipment	Total	Total Incl O&P
0010	**WALL PLUGS** (for nailing to brickwork)									
0020	25 ga., galvanized, plain	1 Bric	10.50	.762	C	26.50	39.50		66	93
0050	Wood filled	"	10.50	.762	"	73	39.50		112.50	145

04 21 Clay Unit Masonry

04 21 13 – Brick Masonry

04 21 13.13 Brick Veneer Masonry		Crew	Daily Output	Labor-Hours	Unit	Material	Labor	Equipment	Total	Total Incl O&P
0010	**BRICK VENEER MASONRY**, T.L. lots, excl. scaff., grout & reinforcing									
0015	Material costs incl. 3% brick and 25% mortar waste									
0020	Standard, select common, 4" x 2-2/3" x 8" (6.75/S.F.)	D-8	1.50	26.667	M	670	1,275		1,945	2,800
0050	Red, 4" x 2-2/3" x 8", running bond		1.50	26.667		720	1,275		1,995	2,875
0100	Full header every 6th course (7.88/S.F.)		1.45	27.586		720	1,325		2,045	2,925
0150	English, full header every 2nd course (10.13/S.F.)		1.40	28.571		715	1,375		2,090	3,025
0200	Flemish, alternate header every course (9.00/S.F.)		1.40	28.571		715	1,375		2,090	3,025
0250	Flemish, alt. header every 6th course (7.13/S.F.)		1.45	27.586		720	1,325		2,045	2,925
0300	Full headers throughout (13.50/S.F.)		1.40	28.571		715	1,375		2,090	3,000
0350	Rowlock course (13.50/S.F.)		1.35	29.630		715	1,425		2,140	3,075
0400	Rowlock stretcher (4.50/S.F.)		1.40	28.571		725	1,375		2,100	3,025
0450	Soldier course (6.75/S.F.)		1.40	28.571		720	1,375		2,095	3,025
0500	Sailor course (4.50/S.F.)		1.30	30.769		725	1,475		2,200	3,175
0601	Buff or gray face, running bond (6.75/S.F.)		1.50	26.667		720	1,275		1,995	2,875
0700	Glazed face, 4" x 2-2/3" x 8", running bond		1.40	28.571		2,175	1,375		3,550	4,625
0750	Full header every 6th course (7.88/S.F.)		1.35	29.630		2,075	1,425		3,500	4,575
1000	Jumbo, 6" x 4" x 12" (3.00/S.F.)		1.30	30.769		1,950	1,475		3,425	4,525
1051	Norman, 4" x 2-2/3" x 12" (4.50/S.F.)		1.45	27.586		1,225	1,325		2,550	3,475
1100	Norwegian, 4" x 3-1/5" x 12" (3.75/S.F.)		1.40	28.571		1,650	1,375		3,025	4,050
1150	Economy, 4" x 4" x 8" (4.50/S.F.)		1.40	28.571		1,050	1,375		2,425	3,375

04 21 Clay Unit Masonry

04 21 13 – Brick Masonry

04 21 13.13 Brick Veneer Masonry		Crew	Daily Output	Labor-Hours	Unit	2020 Bare Costs Material	Labor	Equipment	Total	Total Incl O&P
1201	Engineer, 4" x 3-1/5" x 8" (5.63/S.F.)	D-8	1.45	27.586	M	695	1,325		2,020	2,900
1251	Roman, 4" x 2" x 12" (6.00/S.F.)		1.50	26.667		1,300	1,275		2,575	3,500
1300	S.C.R., 6" x 2-2/3" x 12" (4.50/S.F.)		1.40	28.571		1,425	1,375		2,800	3,775
1350	Utility, 4" x 4" x 12" (3.00/S.F.)		1.08	37.037		1,850	1,775		3,625	4,900
1360	For less than truck load lots, add					15%				
1400	For battered walls, add						30%			
1450	For corbels, add						75%			
1500	For curved walls, add						30%			
1550	For pits and trenches, deduct						20%			
1999	Alternate method of figuring by square foot									
2000	Standard, sel. common, 4" x 2-2/3" x 8" (6.75/S.F.)	D-8	230	.174	S.F.	4.52	8.35		12.87	18.45
2020	Red, 4" x 2-2/3" x 8", running bond		220	.182		4.86	8.70		13.56	19.45
2050	Full header every 6th course (7.88/S.F.)		185	.216		5.65	10.35		16	23
2100	English, full header every 2nd course (10.13/S.F.)		140	.286		7.25	13.70		20.95	30
2150	Flemish, alternate header every course (9.00/S.F.)		150	.267		6.45	12.80		19.25	27.50
2200	Flemish, alt. header every 6th course (7.13/S.F.)		205	.195		5.15	9.35		14.50	21
2250	Full headers throughout (13.50/S.F.)		105	.381		9.65	18.25		27.90	40
2300	Rowlock course (13.50/S.F.)		100	.400		9.65	19.15		28.80	41.50
2350	Rowlock stretcher (4.50/S.F.)		310	.129		3.27	6.20		9.47	13.60
2400	Soldier course (6.75/S.F.)		200	.200		4.86	9.60		14.46	21
2450	Sailor course (4.50/S.F.)		290	.138		3.27	6.60		9.87	14.30
2600	Buff or gray face, running bond (6.75/S.F.)		220	.182		5.15	8.70		13.85	19.75
2700	Glazed face brick, running bond		210	.190		14	9.15		23.15	30
2750	Full header every 6th course (7.88/S.F.)		170	.235		16.30	11.30		27.60	36
3000	Jumbo, 6" x 4" x 12" running bond (3.00/S.F.)		435	.092		5.40	4.41		9.81	13
3050	Norman, 4" x 2-2/3" x 12" running bond (4.5/S.F.)		320	.125		6.65	6		12.65	17.05
3100	Norwegian, 4" x 3-1/5" x 12" (3.75/S.F.)		375	.107		6.05	5.10		11.15	14.90
3150	Economy, 4" x 4" x 8" (4.50/S.F.)		310	.129		4.63	6.20		10.83	15.10
3200	Engineer, 4" x 3-1/5" x 8" (5.63/S.F.)		260	.154		3.89	7.35		11.24	16.20
3250	Roman, 4" x 2" x 12" (6.00/S.F.)		250	.160		7.65	7.65		15.30	21
3300	S.C.R., 6" x 2-2/3" x 12" (4.50/S.F.)		310	.129		6.30	6.20		12.50	16.95
3350	Utility, 4" x 4" x 12" (3.00/S.F.)		360	.111		5.40	5.35		10.75	14.55
3360	For less than truck load lots, add					.10%				
3370	For battered walls, add						30%			
3380	For corbels, add						75%			
3400	For cavity wall construction, add						15%			
3450	For stacked bond, add						10%			
3500	For interior veneer construction, add						15%			
3510	For pits and trenches, deduct						20%			
3550	For curved walls, add						30%			
9000	Minimum labor/equipment charge	D-1	2	8	Job		375		375	605

04 21 13.14 Thin Brick Veneer		Crew	Daily Output	Labor-Hours	Unit	Material	Labor	Equipment	Total	Total Incl O&P
0010	**THIN BRICK VENEER**									
0015	Material costs incl. 3% brick and 25% mortar waste									
0020	On & incl. metal panel support sys, modular, 2-2/3" x 5/8" x 8", red	D-7	92	.174	S.F.	9.70	7.70		17.40	22.50
0100	Closure, 4" x 5/8" x 8"		110	.145		9.40	6.40		15.80	20.50
0110	Norman, 2-2/3" x 5/8" x 12"		110	.145		9.65	6.40		16.05	20.50
0120	Utility, 4" x 5/8" x 12"		125	.128		9.10	5.65		14.75	18.85
0130	Emperor, 4" x 3/4" x 16"		175	.091		10.30	4.04		14.34	17.65
0140	Super emperor, 8" x 3/4" x 16"		195	.082		11.15	3.62		14.77	17.95
0150	For L shaped corners with 4" return, add				L.F.	9.25			9.25	10.20
0200	On masonry/plaster back-up, modular, 2-2/3" x 5/8" x 8", red	D-7	137	.117	S.F.	4.66	5.15		9.81	13.25

04 21 13 – Brick Masonry

04 21 13.14 Thin Brick Veneer		Crew	Daily Output	Labor-Hours	Unit	Material	Labor	Equipment	Total	Total Incl O&P
						2020 Bare Costs				
0210	Closure, 4" x 5/8" x 8"	D-7	165	.097	S.F.	4.37	4.28		8.65	11.50
0220	Norman, 2-2/3" x 5/8" x 12"		165	.097		4.61	4.28		8.89	11.75
0230	Utility, 4" x 5/8" x 12"		185	.086		4.09	3.82		7.91	10.50
0240	Emperor, 4" x 3/4" x 16"		260	.062		5.25	2.72		7.97	10.05
0250	Super emperor, 8" x 3/4" x 16"		285	.056		6.15	2.48		8.63	10.65
0260	For L shaped corners with 4" return, add				L.F.	10			10	11
0270	For embedment into pre-cast concrete panels, add				S.F.	14.40			14.40	15.85

04 21 13.15 Chimney		Crew	Daily Output	Labor-Hours	Unit	Material	Labor	Equipment	Total	Total Incl O&P
0010	**CHIMNEY**, excludes foundation, scaffolding, grout and reinforcing									
0100	Brick, 16" x 16", 8" flue	D-1	18.20	.879	V.L.F.	26.50	41		67.50	95.50
0150	16" x 20" with one 8" x 12" flue		16	1		42.50	47		89.50	123
0200	16" x 24" with two 8" x 8" flues		14	1.143		63	53.50		116.50	156
0250	20" x 20" with one 12" x 12" flue		13.70	1.168		49.50	55		104.50	143
0300	20" x 24" with two 8" x 12" flues		12	1.333		71	62.50		133.50	179
0350	20" x 32" with two 12" x 12" flues		10	1.600		87	75		162	217

04 21 13.18 Columns		Crew	Daily Output	Labor-Hours	Unit	Material	Labor	Equipment	Total	Total Incl O&P
0010	**COLUMNS**, solid, excludes scaffolding, grout and reinforcing									
0050	Brick, 8" x 8", 9 brick/V.L.F.	D-1	56	.286	V.L.F.	6.30	13.40		19.70	28.50
0100	12" x 8", 13.5 brick/V.L.F.		37	.432		9.45	20.50		29.95	43.50
0200	12" x 12", 20 brick/V.L.F.		25	.640		14	30		44	64
0300	16" x 12", 27 brick/V.L.F.		19	.842		18.90	39.50		58.40	85
0400	16" x 16", 36 brick/V.L.F.		14	1.143		25	53.50		78.50	114
0500	20" x 16", 45 brick/V.L.F.		11	1.455		31.50	68		99.50	145
0600	20" x 20", 56 brick/V.L.F.		9	1.778		39	83.50		122.50	178
0700	24" x 20", 68 brick/V.L.F.		7	2.286		47.50	107		154.50	226
0800	24" x 24", 81 brick/V.L.F.		6	2.667		56.50	125		181.50	265
1000	36" x 36", 182 brick/V.L.F.		3	5.333		127	250		377	545
9000	Minimum labor/equipment charge		2	8	Job		375		375	605

04 21 13.30 Oversized Brick		Crew	Daily Output	Labor-Hours	Unit	Material	Labor	Equipment	Total	Total Incl O&P
0010	**OVERSIZED BRICK**, excludes scaffolding, grout and reinforcing									
0100	Veneer, 4" x 2.25" x 16"	D-8	387	.103	S.F.	5.05	4.95		10	13.55
0102	8" x 2.25" x 16", multicell		265	.151		17.55	7.25		24.80	31
0105	4" x 2.75" x 16"		412	.097		5.75	4.65		10.40	13.85
0107	8" x 2.75" x 16", multicell		295	.136		17.50	6.50		24	30
0110	4" x 4" x 16"		460	.087		3.80	4.17		7.97	10.95
0120	4" x 8" x 16"		533	.075		5.10	3.60		8.70	11.45
0122	4" x 8" x 16" multicell		327	.122		15.70	5.85		21.55	27
0125	Loadbearing, 6" x 4" x 16", grouted and reinforced		387	.103		11.60	4.95		16.55	21
0130	8" x 4" x 16", grouted and reinforced		327	.122		12.05	5.85		17.90	22.50
0132	10" x 4" x 16", grouted and reinforced		327	.122		27.50	5.85		33.35	40
0135	6" x 8" x 16", grouted and reinforced		440	.091		15.05	4.36		19.41	23.50
0140	8" x 8" x 16", grouted and reinforced		400	.100		16.05	4.79		20.84	25.50
0145	Curtainwall/reinforced veneer, 6" x 4" x 16"		387	.103		16.85	4.95		21.80	26.50
0150	8" x 4" x 16"		327	.122		19.95	5.85		25.80	31.50
0152	10" x 4" x 16"		327	.122		30.50	5.85		36.35	43
0155	6" x 8" x 16"		440	.091		21	4.36		25.36	30
0160	8" x 8" x 16"		400	.100		28.50	4.79		33.29	39.50
0200	For 1 to 3 slots in face, add					15%				
0210	For 4 to 7 slots in face, add					25%				
0220	For bond beams, add					20%				
0230	For bullnose shapes, add					20%				
0240	For open end knockout, add					10%				

04 21 Clay Unit Masonry

04 21 13 – Brick Masonry

04 21 13.30 Oversized Brick		Crew	Daily Output	Labor-Hours	Unit	Material	2020 Bare Costs Labor	Equipment	Total	Total Incl O&P
0250	For white or gray color group, add				S.F.	10%				
0260	For 135 degree corner, add				↓	250%				
04 21 13.35 Common Building Brick										
0010	**COMMON BUILDING BRICK**, C62, T.L. lots, material only R042110-20									
0020	Standard				M	590			590	650
0050	Select				"	555			555	615
04 21 13.40 Structural Brick										
0010	**STRUCTURAL BRICK** C652, Grade SW, incl. mortar, scaffolding not incl.									
0100	Standard unit, 4-5/8" x 2-3/4" x 9-5/8"	D-8	245	.163	S.F.	5.25	7.85		13.10	18.45
0120	Bond beam		225	.178		4.44	8.50		12.94	18.65
0140	V cut bond beam		225	.178		4.34	8.50		12.84	18.50
0160	Stretcher quoin, 5-5/8" x 2-3/4" x 9-5/8"		245	.163		8.30	7.85		16.15	22
0180	Corner quoin		245	.163		8.25	7.85		16.10	21.50
0200	Corner, 45 degree, 4-5/8" x 2-3/4" x 10-7/16"	↓	235	.170	↓	8.25	8.15		16.40	22.50
04 21 13.45 Face Brick										
0010	**FACE BRICK** Material Only, C216, T.L. lots R042110-20									
0300	Standard modular, 4" x 2-2/3" x 8"				M	605			605	665
0450	Economy, 4" x 4" x 8"					900			900	990
0510	Economy, 4" x 4" x 12"					1,425			1,425	1,550
0550	Jumbo, 6" x 4" x 12"					1,625			1,625	1,800
0610	Jumbo, 8" x 4" x 12"					1,625			1,625	1,800
0650	Norwegian, 4" x 3-1/5" x 12"					1,475			1,475	1,600
0710	Norwegian, 6" x 3-1/5" x 12"					1,725			1,725	1,900
0850	Standard glazed, plain colors, 4" x 2-2/3" x 8"					1,925			1,925	2,100
1000	Deep trim shades, 4" x 2-2/3" x 8"					2,350			2,350	2,600
1080	Jumbo utility, 4" x 4" x 12"					1,650			1,650	1,800
1120	4" x 8" x 8"					1,975			1,975	2,175
1140	4" x 8" x 16"					6,250			6,250	6,875
1260	Engineer, 4" x 3-1/5" x 8"					575			575	635
1350	King, 4" x 2-3/4" x 10"					600			600	660
1770	Standard modular, double glazed, 4" x 2-2/3" x 8"					2,950			2,950	3,250
1850	Jumbo, colored glazed ceramic, 6" x 4" x 12"					3,050			3,050	3,350
2050	Jumbo utility, glazed, 4" x 4" x 12"					6,125			6,125	6,725
2100	4" x 8" x 8"					7,200			7,200	7,925
2150	4" x 16" x 8"					7,875			7,875	8,675
3050	Used brick					480			480	530
3150	Add for brick to match existing work, minimum					5%				
3200	Maximum				↓	50%				

04 21 26 – Glazed Structural Clay Tile Masonry

04 21 26.10 Structural Facing Tile		Crew	Daily Output	Labor-Hours	Unit	Material	Labor	Equipment	Total	Total Incl O&P
0010	**STRUCTURAL FACING TILE**, std. colors, excl. scaffolding, grout, reinforcing									
0020	6T series, 5-1/3" x 12", 2.3 pieces per S.F., glazed 1 side, 2" thick	D-8	225	.178	S.F.	9	8.50		17.50	23.50
0100	4" thick		220	.182		13.50	8.70		22.20	29
0150	Glazed 2 sides		195	.205		16.40	9.85		26.25	34
0250	6" thick		210	.190		17.25	9.15		26.40	34
0300	Glazed 2 sides		185	.216		21	10.35		31.35	40
0400	8" thick		180	.222	↓	23	10.65		33.65	42.50
0500	Special shapes, group 1		400	.100	Ea.	7.80	4.79		12.59	16.30
0550	Group 2		375	.107		12.75	5.10		17.85	22.50
0600	Group 3		350	.114		17.50	5.50		23	28
0650	Group 4	↓	325	.123	↓	36	5.90		41.90	49.50

04 21 26 – Glazed Structural Clay Tile Masonry

04 21 26.10 Structural Facing Tile		Crew	Daily Output	Labor-Hours	Unit	Material	Labor	Equipment	Total	Total Incl O&P
						2020 Bare Costs				
0700	Group 5	D-8	300	.133	Ea.	43.50	6.40		49.90	58
0750	Group 6		275	.145		59	6.95		65.95	76.50
1000	Fire rated, 4" thick, 1 hr. rating		210	.190	S.F.	19	9.15		28.15	36
1300	Acoustic, 4" thick		210	.190	"	40	9.15		49.15	59
2000	8W series, 8" x 16", 1.125 pieces per S.F.									
2050	2" thick, glazed 1 side	D-8	360	.111	S.F.	9.95	5.35		15.30	19.55
2100	4" thick, glazed 1 side		345	.116		14.35	5.55		19.90	25
2150	Glazed 2 sides		325	.123		15.30	5.90		21.20	26.50
2200	6" thick, glazed 1 side		330	.121		28	5.80		33.80	40.50
2250	8" thick, glazed 1 side		310	.129		30.50	6.20		36.70	43.50
2500	Special shapes, group 1		300	.133	Ea.	18.35	6.40		24.75	30.50
2550	Group 2		280	.143		24.50	6.85		31.35	38
2600	Group 3		260	.154		26	7.35		33.35	40.50
2650	Group 4		250	.160		54	7.65		61.65	71.50
2700	Group 5		240	.167		47.50	8		55.50	65
2750	Group 6		230	.174		104	8.35		112.35	127
3000	4" thick, glazed 1 side		345	.116	S.F.	16.75	5.55		22.30	27.50
3100	Acoustic, 4" thick		345	.116	"	22.50	5.55		28.05	33.50
3120	4W series, 8" x 8", 2.25 pieces per S.F.									
3125	2" thick, glazed 1 side	D-8	360	.111	S.F.	9.75	5.35		15.10	19.35
3130	4" thick, glazed 1 side		345	.116		12.80	5.55		18.35	23
3135	Glazed 2 sides		325	.123		17.10	5.90		23	28.50
3140	6" thick, glazed 1 side		330	.121		17.35	5.80		23.15	28.50
3150	8" thick, glazed 1 side		310	.129		26.50	6.20		32.70	39
3155	Special shapes, group 1		300	.133	Ea.	8.25	6.40		14.65	19.45
3160	Group 2		280	.143	"	9.65	6.85		16.50	21.50
3200	For designer colors, add					25%				
3300	For epoxy mortar joints, add				S.F.	1.82			1.82	2
9000	Minimum labor/equipment charge	D-1	2	8	Job		375		375	605

04 21 29 – Terra Cotta Masonry

04 21 29.10 Terra Cotta Masonry Components		Crew	Daily Output	Labor-Hours	Unit	Material	Labor	Equipment	Total	Total Incl O&P
0010	**TERRA COTTA MASONRY COMPONENTS**									
0020	Coping, split type, not glazed, 9" wide	D-1	90	.178	L.F.	10.65	8.35		19	25
0100	13" wide		80	.200		16.30	9.40		25.70	33
0200	Coping, split type, glazed, 9" wide		90	.178		20.50	8.35		28.85	36
0250	13" wide		80	.200		24.50	9.40		33.90	42
0500	Partition or back-up blocks, scored, in C.L. lots									
0700	Non-load bearing 12" x 12", 3" thick, special order	D-8	550	.073	S.F.	19.40	3.49		22.89	27
0750	4" thick, standard		500	.080		6.80	3.83		10.63	13.65
0800	6" thick		450	.089		9.05	4.26		13.31	16.85
0850	8" thick		400	.100		11.25	4.79		16.04	20
1000	Load bearing, 12" x 12", 4" thick, in walls		500	.080		6.30	3.83		10.13	13.10
1050	In floors		750	.053		6.30	2.56		8.86	11.05
1200	6" thick, in walls		450	.089		9.75	4.26		14.01	17.65
1250	In floors		675	.059		9.75	2.84		12.59	15.35
1400	8" thick, in walls		400	.100		11.90	4.79		16.69	21
1450	In floors		575	.070		11.90	3.33		15.23	18.50
1600	10" thick, in walls, special order		350	.114		28	5.50		33.50	40
1650	In floors, special order		500	.080		28	3.83		31.83	37
1800	12" thick, in walls, special order		300	.133		28.50	6.40		34.90	42
1850	In floors, special order		450	.089		28.50	4.26		32.76	38.50
2000	For reinforcing with steel rods, add to above					15%	5%			

04 21 Clay Unit Masonry

04 21 29 – Terra Cotta Masonry

04 21 29.10 Terra Cotta Masonry Components		Crew	Daily Output	Labor-Hours	Unit	Material	Labor (2020 Bare Costs)	Equipment	Total	Total Incl O&P
2100	For smooth tile instead of scored, add				S.F.	2.94			2.94	3.23
2200	For L.C.L. quantities, add				S.F.	10%	10%			
9000	Minimum labor/equipment charge	D-1	2	8	Job		375		375	605

04 21 29.20 Terra Cotta Tile

		Crew	Daily Output	Labor-Hours	Unit	Material	Labor	Equipment	Total	Total Incl O&P
0010	**TERRA COTTA TILE**, on walls, dry set, 1/2" thick									
0100	Square, hexagonal or lattice shapes, unglazed	1 Tilf	135	.059	S.F.	5.20	2.93		8.13	10.30
0300	Glazed, plain colors	1 Tilf	130	.062	S.F.	7.90	3.05		10.95	13.45
0400	Intense colors	1 Tilf	125	.064	S.F.	9.25	3.17		12.42	15.10

04 22 Concrete Unit Masonry

04 22 10 – Concrete Masonry Units

04 22 10.10 Concrete Block

		Crew	Daily Output	Labor-Hours	Unit	Material	Labor	Equipment	Total	Total Incl O&P
0010	**CONCRETE BLOCK** Material Only R042210-20									
0020	2" x 8" x 16" solid, normal-weight, 2,000 psi				Ea.	1.19			1.19	1.31
0050	3,500 psi R042110-50				Ea.	1.31			1.31	1.44
0100	5,000 psi				Ea.	1.40			1.40	1.54
0150	Lightweight, std.				Ea.	1.43			1.43	1.57
0300	3" x 8" x 16" solid, normal-weight, 2,000 psi				Ea.	1.16			1.16	1.28
0350	3,500 psi				Ea.	1.45			1.45	1.60
0400	5,000 psi				Ea.	1.68			1.68	1.85
0450	Lightweight, std.				Ea.	1.48			1.48	1.63
0600	4" x 8" x 16" hollow, normal-weight, 2,000 psi				Ea.	1.50			1.50	1.65
0650	3,500 psi				Ea.	1.38			1.38	1.52
0700	5,000 psi				Ea.	1.73			1.73	1.90
0750	Lightweight, std.				Ea.	1.44			1.44	1.58
1300	Solid, normal-weight, 2,000 psi				Ea.	1.64			1.64	1.80
1350	3,500 psi				Ea.	1.77			1.77	1.95
1400	5,000 psi				Ea.	1.57			1.57	1.73
1450	Lightweight, std.				Ea.	1.44			1.44	1.58
1600	6" x 8" x 16" hollow, normal-weight, 2,000 psi				Ea.	1.87			1.87	2.06
1650	3,500 psi				Ea.	1.77			1.77	1.95
1700	5,000 psi				Ea.	2.08			2.08	2.29
1750	Lightweight, std.				Ea.	2			2	2.20
2300	Solid, normal-weight, 2,000 psi				Ea.	2.25			2.25	2.48
2350	3,500 psi				Ea.	1.67			1.67	1.84
2400	5,000 psi				Ea.	2.22			2.22	2.44
2450	Lightweight, std.				Ea.	2.61			2.61	2.87
2600	8" x 8" x 16" hollow, normal-weight, 2,000 psi				Ea.	1.75			1.75	1.93
2650	3,500 psi				Ea.	1.98			1.98	2.18
2700	5,000 psi				Ea.	2.54			2.54	2.79
2750	Lightweight, std.				Ea.	2.74			2.74	3.01
3200	Solid, normal-weight, 2,000 psi				Ea.	2.77			2.77	3.05
3250	3,500 psi				Ea.	2.93			2.93	3.22
3300	5,000 psi				Ea.	3.21			3.21	3.53
3350	Lightweight, std.				Ea.	2.80			2.80	3.08
3400	10" x 8" x 16" hollow, normal-weight, 2,000 psi				Ea.	1.75			1.75	1.93
3410	3,500 psi				Ea.	1.98			1.98	2.18
3420	5,000 psi				Ea.	2.54			2.54	2.79
3430	Lightweight, std.				Ea.	2.74			2.74	3.01
3480	Solid, normal-weight, 2,000 psi				Ea.	2.77			2.77	3.05
3490	3,500 psi				Ea.	2.93			2.93	3.22

04 22 Concrete Unit Masonry

04 22 10 – Concrete Masonry Units

04 22 10.10 Concrete Block		Crew	Daily Output	Labor-Hours	Unit	Material	2020 Bare Costs Labor	Equipment	Total	Total Incl O&P
3500	5,000 psi				Ea.	3.21			3.21	3.53
3510	Lightweight, std.					2.80			2.80	3.08
3600	12″ x 8″ x 16″ hollow, normal-weight, 2,000 psi					3.27			3.27	3.60
3650	3,500 psi					3.13			3.13	3.44
3700	5,000 psi					3.40			3.40	3.74
3750	Lightweight, std.					2.97			2.97	3.27
4300	Solid, normal-weight, 2,000 psi					4.37			4.37	4.81
4350	3,500 psi					3.72			3.72	4.09
4400	5,000 psi					3.81			3.81	4.19
4500	Lightweight, std.				↓	3.18			3.18	3.50

04 22 10.11 Autoclave Aerated Concrete Block		Crew	Daily Output	Labor-Hours	Unit	Material	Labor	Equipment	Total	Total Incl O&P
0010	**AUTOCLAVE AERATED CONCRETE BLOCK**, excl. scaffolding, grout & reinforcing									
0050	Solid, 4″ x 8″ x 24″, incl. mortar G	D-8	600	.067	S.F.	1.55	3.20		4.75	6.85
0060	6″ x 8″ x 24″ R042110-50 G		600	.067		2.50	3.20		5.70	7.90
0070	8″ x 8″ x 24″ G		575	.070		3.22	3.33		6.55	8.95
0080	10″ x 8″ x 24″ G		575	.070		4	3.33		7.33	9.80
0090	12″ x 8″ x 24″ G	↓	550	.073	↓	4.83	3.49		8.32	10.95

04 22 10.12 Chimney Block		Crew	Daily Output	Labor-Hours	Unit	Material	Labor	Equipment	Total	Total Incl O&P
0010	**CHIMNEY BLOCK**, excludes scaffolding, grout and reinforcing									
0220	1 piece, with 8″ x 8″ flue, 16″ x 16″	D-1	28	.571	V.L.F.	20.50	27		47.50	66
0230	2 piece, 16″ x 16″		26	.615		21.50	29		50.50	70
0240	2 piece, with 8″ x 12″ flue, 16″ x 20″	↓	24	.667	↓	35.50	31.50		67	89.50

04 22 10.14 Concrete Block, Back-Up		Crew	Daily Output	Labor-Hours	Unit	Material	Labor	Equipment	Total	Total Incl O&P
0010	**CONCRETE BLOCK, BACK-UP**, C90, 2000 psi R042210-20									
0020	Normal weight, 8″ x 16″ units, tooled joint 1 side									
0050	Not-reinforced, 2000 psi, 2″ thick R042110-50	D-8	475	.084	S.F.	1.68	4.04		5.72	8.35
0200	4″ thick		460	.087		2.16	4.17		6.33	9.15
0300	6″ thick		440	.091		2.72	4.36		7.08	10.05
0350	8″ thick		400	.100		2.73	4.79		7.52	10.75
0400	10″ thick	↓	330	.121		3.25	5.80		9.05	12.95
0450	12″ thick	D-9	310	.155		4.72	7.25		11.97	16.95
1000	Reinforced, alternate courses, 4″ thick	D-8	450	.089		2.34	4.26		6.60	9.50
1100	6″ thick		430	.093		2.91	4.46		7.37	10.40
1150	8″ thick		395	.101		2.94	4.85		7.79	11.10
1200	10″ thick	↓	320	.125		3.43	6		9.43	13.45
1250	12″ thick	D-9	300	.160		4.90	7.50		12.40	17.50
2000	Lightweight, not reinforced, 4″ thick	D-8	460	.087		2.09	4.17		6.26	9.05
2100	6″ thick		445	.090		2.86	4.31		7.17	10.10
2150	8″ thick		435	.092		3.84	4.41		8.25	11.35
2200	10″ thick	↓	410	.098		4.27	4.68		8.95	12.25
2250	12″ thick	D-9	390	.123		4.38	5.75		10.13	14.15
3000	Reinforced, alternate courses, 4″ thick	D-8	450	.089		2.28	4.26		6.54	9.40
3100	6″ thick		430	.093		3.06	4.46		7.52	10.55
3150	8″ thick		420	.095		4.05	4.56		8.61	11.85
3200	10″ thick	↓	400	.100		4.46	4.79		9.25	12.65
3250	12″ thick	D-9	380	.126	↓	4.56	5.90		10.46	14.55
9000	Minimum labor/equipment charge	D-1	2	8	Job		375		375	605

04 22 10.16 Concrete Block, Bond Beam		Crew	Daily Output	Labor-Hours	Unit	Material	Labor	Equipment	Total	Total Incl O&P
0010	**CONCRETE BLOCK, BOND BEAM**, C90, 2000 psi									
0020	Not including grout or reinforcing									
0125	Regular block, 6″ thick	D-8	584	.068	L.F.	2.77	3.28		6.05	8.35
0130	8″ high, 8″ thick	″	565	.071	↓	2.90	3.39		6.29	8.70

04 22 10 – Concrete Masonry Units

04 22 10.16 Concrete Block, Bond Beam		Crew	Daily Output	Labor-Hours	Unit	Material	Labor	Equipment	Total	Total Incl O&P
						2020 Bare Costs				
0150	12" thick	D-9	510	.094	L.F.	4.09	4.41		8.50	11.65
0525	Lightweight, 6" thick	D-8	592	.068		2.90	3.24		6.14	8.45
0530	8" high, 8" thick	"	575	.070		3.41	3.33		6.74	9.15
0550	12" thick	D-9	520	.092	↓	4.63	4.33		8.96	12.10
2000	Including grout and 2 #5 bars									
2100	Regular block, 8" high, 8" thick	D-8	300	.133	L.F.	5.35	6.40		11.75	16.25
2150	12" thick	D-9	250	.192		7.25	9		16.25	22.50
2500	Lightweight, 8" high, 8" thick	D-8	305	.131		5.90	6.30		12.20	16.60
2550	12" thick	D-9	255	.188	↓	7.80	8.85		16.65	23
9000	Minimum labor/equipment charge	D-1	2	8	Job		375		375	605

04 22 10.18 Concrete Block, Column		Crew	Daily Output	Labor-Hours	Unit	Material	Labor	Equipment	Total	Total Incl O&P
0010	**CONCRETE BLOCK, COLUMN** or pilaster									
0050	Including vertical reinforcing (4-#4 bars) and grout									
0160	1 piece unit, 16" x 16"	D-1	26	.615	V.L.F.	16.20	29		45.20	64.50
0170	2 piece units, 16" x 20"		24	.667		22	31.50		53.50	75
0180	20" x 20"		22	.727		30.50	34		64.50	88.50
0190	22" x 24"		18	.889		45.50	41.50		87	118
0200	20" x 32"	↓	14	1.143	↓	53.50	53.50		107	146

04 22 10.19 Concrete Block, Insulation Inserts		Crew	Daily Output	Labor-Hours	Unit	Material	Labor	Equipment	Total	Total Incl O&P
0010	**CONCRETE BLOCK, INSULATION INSERTS**									
0100	Styrofoam, plant installed, add to block prices									
0200	8" x 16" units, 6" thick				S.F.	1.25			1.25	1.38
0250	8" thick					1.57			1.57	1.73
0300	10" thick					1.47			1.47	1.62
0350	12" thick					1.85			1.85	2.04
0500	8" x 8" units, 8" thick					1.40			1.40	1.54
0550	12" thick				↓	1.47			1.47	1.62

04 22 10.23 Concrete Block, Decorative		Crew	Daily Output	Labor-Hours	Unit	Material	Labor	Equipment	Total	Total Incl O&P
0010	**CONCRETE BLOCK, DECORATIVE**, C90, 2000 psi									
0020	Embossed, simulated brick face									
0100	8" x 16" units, 4" thick	D-8	400	.100	S.F.	3.23	4.79		8.02	11.30
0200	8" thick		340	.118		3.25	5.65		8.90	12.70
0250	12" thick	↓	300	.133	↓	5.60	6.40		12	16.50
0400	Embossed both sides									
0500	8" thick	D-8	300	.133	S.F.	4.88	6.40		11.28	15.70
0550	12" thick	"	275	.145	"	6.10	6.95		13.05	17.95
1000	Fluted high strength									
1100	8" x 16" x 4" thick, flutes 1 side,	D-8	345	.116	S.F.	4.11	5.55		9.66	13.50
1150	Flutes 2 sides		335	.119		5.40	5.70		11.10	15.20
1200	8" thick	↓	300	.133		5.85	6.40		12.25	16.80
1250	For special colors, add				↓	.67			.67	.73
1400	Deep grooved, smooth face									
1450	8" x 16" x 4" thick	D-8	345	.116	S.F.	2.97	5.55		8.52	12.25
1500	8" thick	"	300	.133	"	4.48	6.40		10.88	15.30
2000	Formblock, incl. inserts & reinforcing									
2100	8" x 16" x 8" thick	D-8	345	.116	S.F.	4.22	5.55		9.77	13.65
2150	12" thick	"	310	.129	"	5.50	6.20		11.70	16.05
2500	Ground face									
2600	8" x 16" x 4" thick	D-8	345	.116	S.F.	4.86	5.55		10.41	14.35
2650	6" thick		325	.123		4.90	5.90		10.80	14.95
2700	8" thick	↓	300	.133		5.25	6.40		11.65	16.10
2750	12" thick	D-9	265	.181	↓	6.15	8.50		14.65	20.50

04 22 10.23 Concrete Block, Decorative		Crew	Daily Output	Labor-Hours	Unit	Material	Labor	Equipment	Total	Total Incl O&P
						2020 Bare Costs				
2900	For special colors, add, minimum					15%				
2950	For special colors, add, maximum					45%				
4000	Slump block									
4100	4" face height x 16" x 4" thick	D-1	165	.097	S.F.	4.76	4.55		9.31	12.60
4150	6" thick		160	.100		6.25	4.69		10.94	14.50
4200	8" thick		155	.103		8	4.84		12.84	16.60
4250	10" thick		140	.114		14.10	5.35		19.45	24
4300	12" thick		130	.123		12.55	5.75		18.30	23
4400	6" face height x 16" x 6" thick		155	.103		5.80	4.84		10.64	14.20
4450	8" thick		150	.107		8.75	5		13.75	17.75
4500	10" thick		130	.123		14.90	5.75		20.65	26
4550	12" thick		120	.133		14.50	6.25		20.75	26
5000	Split rib profile units, 1" deep ribs, 8 ribs									
5100	8" x 16" x 4" thick	D-8	345	.116	S.F.	4.61	5.55		10.16	14.10
5150	6" thick		325	.123		5.25	5.90		11.15	15.35
5200	8" thick		300	.133		6.55	6.40		12.95	17.60
5250	12" thick	D-9	275	.175		7.50	8.20		15.70	21.50
5400	For special deeper colors, 4" thick, add					1.32			1.32	1.45
5450	12" thick, add					1.41			1.41	1.55
5600	For white, 4" thick, add					1.32			1.32	1.45
5650	6" thick, add					1.39			1.39	1.53
5700	8" thick, add					1.48			1.48	1.63
5750	12" thick, add					1.47			1.47	1.62
6000	Split face									
6100	8" x 16" x 4" thick	D-8	350	.114	S.F.	3.90	5.50		9.40	13.15
6150	6" thick		325	.123		4.40	5.90		10.30	14.40
6200	8" thick		300	.133		5.60	6.40		12	16.55
6250	12" thick	D-9	270	.178		6.25	8.35		14.60	20.50
6300	For scored, add					.39			.39	.43
6400	For special deeper colors, 4" thick, add					.63			.63	.70
6450	6" thick, add					.77			.77	.85
6500	8" thick, add					.81			.81	.89
6550	12" thick, add					.86			.86	.94
6650	For white, 4" thick, add					1.32			1.32	1.45
6700	6" thick, add					1.29			1.29	1.42
6750	8" thick, add					1.41			1.41	1.55
6800	12" thick, add					1.41			1.41	1.55
7000	Scored ground face, 2 to 5 scores									
7100	8" x 16" x 4" thick	D-8	340	.118	S.F.	9	5.65		14.65	19.05
7150	6" thick		310	.129		10.10	6.20		16.30	21
7200	8" thick		290	.138		11.45	6.60		18.05	23.50
7250	12" thick	D-9	265	.181		15.65	8.50		24.15	31
8000	Hexagonal face profile units, 8" x 16" units									
8100	4" thick, hollow	D-8	340	.118	S.F.	4.06	5.65		9.71	13.55
8200	Solid		340	.118		5.15	5.65		10.80	14.80
8300	6" thick, hollow		310	.129		4.37	6.20		10.57	14.80
8350	8" thick, hollow		290	.138		4.74	6.60		11.34	15.90
8500	For stacked bond, add						26%			
8550	For high rise construction, add per story	D-8	67.80	.590	M.S.F.		28.50		28.50	45.50
8600	For scored block, add					10%				
8650	For honed or ground face, per face, add				Ea.	1.23			1.23	1.35
8700	For honed or ground end, per end, add				"	1.36			1.36	1.49
8750	For bullnose block, add					10%				

04 22 Concrete Unit Masonry

04 22 10 – Concrete Masonry Units

04 22 10.23 Concrete Block, Decorative		Crew	Daily Output	Labor-Hours	Unit	Material	2020 Bare Costs Labor	Equipment	Total	Total Incl O&P
8800	For special color, add					13%				
9000	Minimum labor/equipment charge	D-1	2	8	Job		375		375	605

04 22 10.24 Concrete Block, Exterior		Crew	Daily Output	Labor-Hours	Unit	Material	Labor	Equipment	Total	Total Incl O&P
0010	**CONCRETE BLOCK, EXTERIOR**, C90, 2000 psi									
0020	Reinforced alt courses, tooled joints 2 sides									
0100	Normal weight, 8″ x 16″ x 6″ thick	D-8	395	.101	S.F.	2.56	4.85		7.41	10.65
0200	8″ thick		360	.111		4.16	5.35		9.51	13.20
0250	10″ thick		290	.138		4.50	6.60		11.10	15.65
0300	12″ thick	D-9	250	.192		5.30	9		14.30	20.50
0500	Lightweight, 8″ x 16″ x 6″ thick	D-8	450	.089		3.62	4.26		7.88	10.90
0600	8″ thick		430	.093		4.01	4.46		8.47	11.60
0650	10″ thick		395	.101		4.50	4.85		9.35	12.80
0700	12″ thick	D-9	350	.137		4.71	6.45		11.16	15.60
9000	Minimum labor/equipment charge	D-1	2	8	Job		375		375	605

04 22 10.26 Concrete Block Foundation Wall		Crew	Daily Output	Labor-Hours	Unit	Material	Labor	Equipment	Total	Total Incl O&P
0010	**CONCRETE BLOCK FOUNDATION WALL**, C90/C145									
0050	Normal-weight, cut joints, horiz joint reinf, no vert reinf.									
0200	Hollow, 8″ x 16″ x 6″ thick	D-8	455	.088	S.F.	3.20	4.21		7.41	10.30
0250	8″ thick		425	.094		3.23	4.51		7.74	10.85
0300	10″ thick		350	.114		3.73	5.50		9.23	12.95
0350	12″ thick	D-9	300	.160		5.20	7.50		12.70	17.85
0500	Solid, 8″ x 16″ block, 6″ thick	D-8	440	.091		3.63	4.36		7.99	11.05
0550	8″ thick	″	415	.096		4.38	4.62		9	12.25
0600	12″ thick	D-9	350	.137		6.45	6.45		12.90	17.50
1000	Reinforced, #4 vert @ 48″									
1100	Hollow, 8″ x 16″ block, 4″ thick	D-8	455	.088	S.F.	3.29	4.21		7.50	10.40
1125	6″ thick		445	.090		4.25	4.31		8.56	11.65
1150	8″ thick		415	.096		4.72	4.62		9.34	12.65
1200	10″ thick		340	.118		5.65	5.65		11.30	15.35
1250	12″ thick	D-9	290	.166		7.60	7.75		15.35	21
1500	Solid, 8″ x 16″ block, 6″ thick	D-8	430	.093		3.64	4.46		8.10	11.20
1600	8″ thick	″	405	.099		4.38	4.73		9.11	12.45
1650	12″ thick	D-9	340	.141		6.45	6.60		13.05	17.80
9000	Minimum labor/equipment charge	D-1	2	8	Job		375		375	605

04 22 10.28 Concrete Block, High Strength		Crew	Daily Output	Labor-Hours	Unit	Material	Labor	Equipment	Total	Total Incl O&P
0010	**CONCRETE BLOCK, HIGH STRENGTH** R042210-20									
0050	Hollow, reinforced alternate courses, 8″ x 16″ units									
0200	3500 psi, 4″ thick	D-8	440	.091	S.F.	2.50	4.36		6.86	9.80
0250	6″ thick		395	.101		2.47	4.85		7.32	10.55
0300	8″ thick		360	.111		4.07	5.35		9.42	13.10
0350	12″ thick	D-9	250	.192		5.20	9		14.20	20.50
0500	5000 psi, 4″ thick	D-8	440	.091		2.32	4.36		6.68	9.60
0550	6″ thick		395	.101		3.18	4.85		8.03	11.35
0600	8″ thick		360	.111		4.48	5.35		9.83	13.50
0650	12″ thick	D-9	300	.160		5.40	7.50		12.90	18.05
1000	For 75% solid block, add					30%				
1050	For 100% solid block, add					50%				

04 22 10.30 Concrete Block, Interlocking		Crew	Daily Output	Labor-Hours	Unit	Material	Labor	Equipment	Total	Total Incl O&P
0010	**CONCRETE BLOCK, INTERLOCKING**									
0100	Not including grout or reinforcing									
0200	8″ x 16″ units, 2,000 psi, 8″ thick	D-1	245	.065	S.F.	3.08	3.06		6.14	8.35
0300	12″ thick		220	.073		4.57	3.41		7.98	10.55

04 22 10.30 Concrete Block, Interlocking

	04 22 10.30 Concrete Block, Interlocking	Crew	Daily Output	Labor-Hours	Unit	Material	2020 Bare Costs Labor	Equipment	Total	Total Incl O&P
0350	16" thick	D-1	185	.086	S.F.	6.70	4.06		10.76	13.90
0400	Including grout & reinforcing, 8" thick	D-4	245	.131		8.90	6.15	.77	15.82	20.50
0450	12" thick		220	.145		10.65	6.85	.86	18.36	23.50
0500	16" thick		185	.173		13	8.15	1.02	22.17	28.50
9000	Minimum labor/equipment charge	D-1	2	8	Job		375		375	605

04 22 10.32 Concrete Block, Lintels

	04 22 10.32 Concrete Block, Lintels	Crew	Daily Output	Labor-Hours	Unit	Material	Labor	Equipment	Total	Total Incl O&P
0010	**CONCRETE BLOCK, LINTELS**, C90, normal weight									
0100	Including grout and horizontal reinforcing									
0200	8" x 8" x 8", 1 #4 bar	D-4	300	.107	L.F.	4.54	5.05	.63	10.22	13.75
0250	2 #4 bars		295	.108		4.79	5.10	.64	10.53	14.15
0400	8" x 16" x 8", 1 #4 bar		275	.116		4.24	5.50	.69	10.43	14.20
0450	2 #4 bars		270	.119		4.49	5.60	.70	10.79	14.65
1000	12" x 8" x 8", 1 #4 bar		275	.116		6.05	5.50	.69	12.24	16.20
1100	2 #4 bars		270	.119		6.30	5.60	.70	12.60	16.60
1150	2 #5 bars		270	.119		6.55	5.60	.70	12.85	16.95
1200	2 #6 bars		265	.121		6.90	5.70	.71	13.31	17.50
1500	12" x 16" x 8", 1 #4 bar		250	.128		7.05	6.05	.75	13.85	18.25
1600	2 #3 bars		245	.131		7.10	6.15	.77	14.02	18.50
1650	2 #4 bars		245	.131		7.30	6.15	.77	14.22	18.75
1700	2 #5 bars		240	.133		7.60	6.30	.79	14.69	19.25
9000	Minimum labor/equipment charge	D-1	2	8	Job		375		375	605

04 22 10.33 Lintel Block

	04 22 10.33 Lintel Block	Crew	Daily Output	Labor-Hours	Unit	Material	Labor	Equipment	Total	Total Incl O&P
0010	**LINTEL BLOCK**									
3481	Lintel block 6" x 8" x 8"	D-1	300	.053	Ea.	1.37	2.50		3.87	5.55
3501	6" x 16" x 8"		275	.058		2.05	2.73		4.78	6.65
3521	8" x 8" x 8"		275	.058		1.29	2.73		4.02	5.85
3561	8" x 16" x 8"		250	.064		2.01	3		5.01	7.05

04 22 10.34 Concrete Block, Partitions

	04 22 10.34 Concrete Block, Partitions	Crew	Daily Output	Labor-Hours	Unit	Material	Labor	Equipment	Total	Total Incl O&P
0010	**CONCRETE BLOCK, PARTITIONS**, excludes scaffolding R042210-20									
0100	Acoustical slotted block									
0200	4" thick, type A-1	D-8	315	.127	S.F.	6.65	6.10		12.75	17.20
0210	8" thick		275	.145		7.60	6.95		14.55	19.60
0250	8" thick, type Q		275	.145		14.80	6.95		21.75	27.50
0260	4" thick, type RSC		315	.127		11.25	6.10		17.35	22.50
0270	6" thick		295	.136		11.80	6.50		18.30	23.50
0280	8" thick		275	.145		12.35	6.95		19.30	25
0290	12" thick		250	.160		12.85	7.65		20.50	26.50
0300	8" thick, type RSR		275	.145		12.35	6.95		19.30	25
0400	8" thick, type RSC/RF		275	.145		8.35	6.95		15.30	20.50
0410	10" thick		260	.154		10.45	7.35		17.80	23.50
0420	12" thick		250	.160		9.60	7.65		17.25	23
0430	12" thick, type RSC/RF-4		250	.160		16.40	7.65		24.05	30.50
0500	NRC .60 type R, 8" thick		265	.151		13.25	7.25		20.50	26.50
0600	NRC .65 type RR, 8" thick		265	.151		8.15	7.25		15.40	20.50
0700	NRC .65 type 4R-RF, 8" thick		265	.151		13.70	7.25		20.95	27
0710	NRC .70 type R, 12" thick		245	.163		14.15	7.85		22	28.50
1000	Lightweight block, tooled joints, 2 sides, hollow									
1100	Not reinforced, 8" x 16" x 4" thick	D-8	440	.091	S.F.	2	4.36		6.36	9.25
1150	6" thick		410	.098		2.77	4.68		7.45	10.60
1200	8" thick		385	.104		3.75	4.98		8.73	12.15
1250	10" thick		370	.108		4.18	5.20		9.38	12.95
1300	12" thick	D-9	350	.137		4.28	6.45		10.73	15.10

04 22 10.34 Concrete Block, Partitions		Crew	Daily Output	Labor-Hours	Unit	Material	2020 Bare Costs Labor	Equipment	Total	Total Incl O&P
1500	Reinforced alternate courses, 4" thick	D-8	435	.092	S.F.	2.18	4.41		6.59	9.50
1600	6" thick		405	.099		2.93	4.73		7.66	10.85
1650	8" thick		380	.105		3.95	5.05		9	12.50
1700	10" thick		365	.110		4.36	5.25		9.61	13.30
1750	12" thick	D-9	345	.139		4.48	6.55		11.03	15.45
2000	Not reinforced, 8" x 24" x 4" thick, hollow		460	.104		1.45	4.89		6.34	9.50
2100	6" thick		440	.109		1.98	5.10		7.08	10.45
2150	8" thick		415	.116		2.71	5.40		8.11	11.75
2200	10" thick		385	.125		3.04	5.85		8.89	12.80
2250	12" thick		365	.132		3.15	6.15		9.30	13.40
2400	Reinforced alternate courses, 4" thick		455	.105		1.63	4.95		6.58	9.80
2500	6" thick		435	.110		2.17	5.15		7.32	10.75
2550	8" thick		410	.117		2.90	5.50		8.40	12.05
2600	10" thick		380	.126		3.22	5.90		9.12	13.10
2650	12" thick		360	.133		3.35	6.25		9.60	13.80
2800	Solid, not reinforced, 8" x 16" x 2" thick	D-8	440	.091		1.86	4.36		6.22	9.10
2900	4" thick		420	.095		2	4.56		6.56	9.60
2950	6" thick		390	.103		3.46	4.92		8.38	11.75
3000	8" thick		365	.110		3.82	5.25		9.07	12.70
3050	10" thick		350	.114		4.31	5.50		9.81	13.60
3100	12" thick	D-9	330	.145		4.52	6.80		11.32	15.95
3300	Solid, reinforced alternate courses, 4" thick	D-8	415	.096		2.18	4.62		6.80	9.85
3400	6" thick		385	.104		3.84	4.98		8.82	12.25
3450	8" thick		360	.111		4.01	5.35		9.36	13
3500	10" thick		345	.116		4.50	5.55		10.05	13.95
3550	12" thick	D-9	325	.148		4.71	6.95		11.66	16.40
4000	Regular block, tooled joints, 2 sides, hollow									
4100	Not reinforced, 8" x 16" x 4" thick	D-8	430	.093	S.F.	2.06	4.46		6.52	9.45
4150	6" thick		400	.100		2.62	4.79		7.41	10.65
4200	8" thick		375	.107		2.63	5.10		7.73	11.15
4250	10" thick		360	.111		3.15	5.35		8.50	12.05
4300	12" thick	D-9	340	.141		4.62	6.60		11.22	15.80
4500	Reinforced alternate courses, 8" x 16" x 4" thick	D-8	425	.094		2.25	4.51		6.76	9.75
4550	6" thick		395	.101		2.81	4.85		7.66	10.95
4600	8" thick		370	.108		2.84	5.20		8.04	11.45
4650	10" thick		355	.113		4.67	5.40		10.07	13.90
4700	12" thick	D-9	335	.143		4.81	6.70		11.51	16.15
4900	Solid, not reinforced, 2" thick	D-8	435	.092		1.63	4.41		6.04	8.90
5000	3" thick		430	.093		1.63	4.46		6.09	9
5050	4" thick		415	.096		2.22	4.62		6.84	9.90
5100	6" thick		385	.104		3.05	4.98		8.03	11.40
5150	8" thick		360	.111		3.78	5.35		9.13	12.75
5200	12" thick	D-9	325	.148		5.85	6.95		12.80	17.65
5500	Solid, reinforced alternate courses, 4" thick	D-8	420	.095		2.40	4.56		6.96	10.05
5550	6" thick		380	.105		3.21	5.05		8.26	11.70
5600	8" thick		355	.113		3.98	5.40		9.38	13.15
5650	12" thick	D-9	320	.150		4.64	7.05		11.69	16.45
9000	Minimum labor/equipment charge	D-1	2	8	Job		375		375	605

04 22 10.38 Concrete Brick

		Crew	Daily Output	Labor-Hours	Unit	Material	Labor	Equipment	Total	Total Incl O&P
0010	**CONCRETE BRICK**, C55, grade N, type 1									
0100	Regular, 4" x 2-1/4" x 8"	D-8	660	.061	Ea.	.60	2.91		3.51	5.35
0125	Rusticated, 4" x 2-1/4" x 8"		660	.061		.67	2.91		3.58	5.45

04 22 Concrete Unit Masonry

04 22 10 – Concrete Masonry Units

04 22 10.38 Concrete Brick		Crew	Daily Output	Labor-Hours	Unit	Material	2020 Bare Costs Labor	Equipment	Total	Total Incl O&P
0150	Frog, 4" x 2-1/4" x 8"	D-8	660	.061	Ea.	.65	2.91		3.56	5.40
0200	Double, 4" x 4-7/8" x 8"	↓	535	.075	↓	1.05	3.58		4.63	6.95

04 22 10.42 Concrete Block, Screen Block		Crew	Daily Output	Labor-Hours	Unit	Material	Labor	Equipment	Total	Total Incl O&P
0010	**CONCRETE BLOCK, SCREEN BLOCK**									
0200	8" x 16", 4" thick	D-8	330	.121	S.F.	10.50	5.80		16.30	21
0300	8" thick		270	.148		11.75	7.10		18.85	24.50
0350	12" x 12", 4" thick		290	.138		7.35	6.60		13.95	18.80
0500	8" thick	↓	250	.160	↓	7.30	7.65		14.95	20.50

04 22 10.44 Glazed Concrete Block		Crew	Daily Output	Labor-Hours	Unit	Material	Labor	Equipment	Total	Total Incl O&P
0010	**GLAZED CONCRETE BLOCK** C744									
0100	Single face, 8" x 16" units, 2" thick	D-8	360	.111	S.F.	12.25	5.35		17.60	22
0200	4" thick		345	.116		12.55	5.55		18.10	23
0250	6" thick		330	.121		12.80	5.80		18.60	23.50
0300	8" thick		310	.129		15.35	6.20		21.55	27
0350	10" thick	↓	295	.136		16.90	6.50		23.40	29
0400	12" thick	D-9	280	.171		18.15	8.05		26.20	33
0700	Double face, 8" x 16" units, 4" thick	D-8	340	.118		16.80	5.65		22.45	27.50
0750	6" thick		320	.125		21.50	6		27.50	33
0800	8" thick		300	.133	↓	23	6.40		29.40	36
1000	Jambs, bullnose or square, single face, 8" x 16", 2" thick		315	.127	Ea.	20	6.10		26.10	32
1050	4" thick		285	.140	"	21	6.75		27.75	34
1200	Caps, bullnose or square, 8" x 16", 2" thick		420	.095	L.F.	19.90	4.56		24.46	29.50
1250	4" thick		380	.105	"	23	5.05		28.05	33
1256	Corner, bullnose or square, 2" thick		280	.143	Ea.	23.50	6.85		30.35	36.50
1258	4" thick		270	.148		26.50	7.10		33.60	41
1260	6" thick		260	.154		29.50	7.35		36.85	44.50
1270	8" thick		250	.160		39.50	7.65		47.15	56
1280	10" thick		240	.167		35.50	8		43.50	52
1290	12" thick		230	.174	↓	37.50	8.35		45.85	55
1500	Cove base, 8" x 16", 2" thick		315	.127	L.F.	11.25	6.10		17.35	22.50
1550	4" thick		285	.140		10.75	6.75		17.50	22.50
1600	6" thick		265	.151		11.45	7.25		18.70	24.50
1650	8" thick	↓	245	.163	↓	12	7.85		19.85	26
9000	Minimum labor/equipment charge	D-1	2	8	Job		375		375	605

04 23 Glass Unit Masonry

04 23 13 – Vertical Glass Unit Masonry

04 23 13.10 Glass Block		Crew	Daily Output	Labor-Hours	Unit	Material	Labor	Equipment	Total	Total Incl O&P
0010	**GLASS BLOCK**									
0100	Plain, 4" thick, under 1,000 S.F., 6" x 6"	D-8	115	.348	S.F.	23	16.65		39.65	52.50
0150	8" x 8"		160	.250		14.25	12		26.25	35
0160	end block		160	.250		64.50	12		76.50	90.50
0170	90 degree corner		160	.250		65	12		77	91
0180	45 degree corner		160	.250		67.50	12		79.50	93.50
0200	12" x 12"		175	.229		24	10.95		34.95	44
0210	4" x 8"		160	.250		36.50	12		48.50	59.50
0220	6" x 8"		160	.250		23	12		35	45
0300	1,000 to 5,000 S.F., 6" x 6"		135	.296		22.50	14.20		36.70	48
0350	8" x 8"		190	.211		14	10.10		24.10	31.50
0400	12" x 12"	↓	215	.186	↓	23.50	8.90		32.40	40.50

04 23 Glass Unit Masonry

04 23 13 – Vertical Glass Unit Masonry

04 23 13.10 Glass Block		Crew	Daily Output	Labor-Hours	Unit	Material	2020 Bare Costs Labor	Equipment	Total	Total Incl O&P
0410	4" x 8"	D-8	215	.186	S.F.	35.50	8.90		44.40	53.50
0420	6" x 8"		215	.186		22.50	8.90		31.40	39.50
0500	Over 5,000 S.F., 6" x 6"		145	.276		22	13.20		35.20	45.50
0550	8" x 8"		215	.186		13.55	8.90		22.45	29.50
0600	12" x 12"		240	.167		23	8		31	38
0610	4" x 8"		240	.167		34.50	8		42.50	51
0620	6" x 8"		240	.167		22	8		30	37
0700	For solar reflective blocks, add					100%				
1000	Thinline, plain, 3-1/8" thick, under 1,000 S.F., 6" x 6"	D-8	115	.348	S.F.	25	16.65		41.65	54
1050	8" x 8"		160	.250		15.35	12		27.35	36.50
1200	Over 5,000 S.F., 6" x 6"		145	.276		27.50	13.20		40.70	51.50
1250	8" x 8"		215	.186		15.30	8.90		24.20	31
1400	For cleaning block after installation (both sides), add		1000	.040		.16	1.92		2.08	3.28
4000	Accessories									
4100	Anchors, 20 ga. galv., 1-3/4" wide x 24" long				Ea.	5.95			5.95	6.55
4200	Emulsion asphalt				Gal.	11.60			11.60	12.80
4300	Expansion joint, fiberglass				L.F.	.74			.74	.81
4400	Steel mesh, double galvanized				"	1.05			1.05	1.16
9000	Minimum labor/equipment charge	D-1	2	8	Job		375		375	605

04 24 Adobe Unit Masonry

04 24 16 – Manufactured Adobe Unit Masonry

04 24 16.06 Adobe Brick		Crew	Daily Output	Labor-Hours	Unit	Material	Labor	Equipment	Total	Total Incl O&P
0010	**ADOBE BRICK**, Semi-stabilized, with cement mortar									
0060	Brick, 10" x 4" x 14", 2.6/S.F. G	D-8	560	.071	S.F.	5.05	3.42		8.47	11.10
0080	12" x 4" x 16", 2.3/S.F. G		580	.069		6.90	3.31		10.21	12.95
0100	10" x 4" x 16", 2.3/S.F. G		590	.068		7.20	3.25		10.45	13.15
0120	8" x 4" x 16", 2.3/S.F. G		560	.071		5.25	3.42		8.67	11.35
0140	4" x 4" x 16", 2.3/S.F. G		540	.074		4.75	3.55		8.30	10.95
0160	6" x 4" x 16", 2.3/S.F. G		540	.074		4.31	3.55		7.86	10.50
0180	4" x 4" x 12", 3.0/S.F. G		520	.077		6.20	3.69		9.89	12.75
0200	8" x 4" x 12", 3.0/S.F. G		520	.077		4.42	3.69		8.11	10.80

04 25 Unit Masonry Panels

04 25 20 – Pre-Fabricated Masonry Panels

04 25 20.10 Brick and Epoxy Mortar Panels		Crew	Daily Output	Labor-Hours	Unit	Material	Labor	Equipment	Total	Total Incl O&P
0010	**BRICK AND EPOXY MORTAR PANELS**									
0020	Prefabricated brick & epoxy mortar, 4" thick, minimum	C-11	775	.093	S.F.	8.45	5.30	2.93	16.68	21
0100	Maximum	"	500	.144		9.90	8.25	4.54	22.69	29.50
0200	For 2" concrete back-up, add					50%				
0300	For 1" urethane & 3" concrete back-up, add					70%				

04 27 Multiple-Wythe Unit Masonry

04 27 10 – Multiple-Wythe Masonry

04 27 10.10 Cornices		Crew	Daily Output	Labor-Hours	Unit	Material	Labor	Equipment	Total	Total Incl O&P
						2020 Bare Costs				
0010	**CORNICES**									
0110	Face bricks, 12 brick/S.F.	D-1	30	.533	SF Face	5.75	25		30.75	47
0150	15 brick/S.F.		23	.696	"	6.90	32.50		39.40	60
9000	Minimum labor/equipment charge	↓	1.50	10.667	Job		500		500	810

04 27 10.30 Brick Walls

Line	Description	Crew	Daily Output	Labor-Hours	Unit	Material	Labor	Equipment	Total	Total Incl O&P
0010	**BRICK WALLS**, including mortar, excludes scaffolding R042110-20									
0020	Estimating by number of brick									
0140	Face brick, 4" thick wall, 6.75 brick/S.F.	D-8	1.45	27.586	M	710	1,325		2,035	2,900
0150	Common brick, 4" thick wall, 6.75 brick/S.F. R042110-50		1.60	25		690	1,200		1,890	2,675
0204	8" thick, 13.50 brick/S.F.		1.80	22.222		710	1,075		1,785	2,500
0250	12" thick, 20.25 brick/S.F.		1.90	21.053		715	1,000		1,715	2,400
0304	16" thick, 27.00 brick/S.F.		2	20		720	960		1,680	2,350
0500	Reinforced, face brick, 4" thick wall, 6.75 brick/S.F.		1.40	28.571		735	1,375		2,110	3,025
0520	Common brick, 4" thick wall, 6.75 brick/S.F.		1.55	25.806		720	1,225		1,945	2,800
0550	8" thick, 13.50 brick/S.F.		1.75	22.857		740	1,100		1,840	2,575
0600	12" thick, 20.25 brick/S.F.		1.85	21.622		740	1,025		1,765	2,500
0650	16" thick, 27.00 brick/S.F.	↓	1.95	20.513	↓	750	985		1,735	2,425
0790	Alternate method of figuring by square foot									
0800	Face brick, 4" thick wall, 6.75 brick/S.F.	D-8	215	.186	S.F.	4.78	8.90		13.68	19.65
0850	Common brick, 4" thick wall, 6.75 brick/S.F.		240	.167		4.67	8		12.67	18.05
0900	8" thick, 13.50 brick/S.F.		135	.296		9.60	14.20		23.80	33.50
1000	12" thick, 20.25 brick/S.F.		95	.421		14.45	20		34.45	48.50
1050	16" thick, 27.00 brick/S.F.		75	.533		19.50	25.50		45	63
1200	Reinforced, face brick, 4" thick wall, 6.75 brick/S.F.		210	.190		4.96	9.15		14.11	20
1220	Common brick, 4" thick wall, 6.75 brick/S.F.		235	.170		4.86	8.15		13.01	18.55
1250	8" thick, 13.50 brick/S.F.		130	.308		9.95	14.75		24.70	35
1260	8" thick, 2.25 brick/S.F.		130	.308		1.79	14.75		16.54	26
1300	12" thick, 20.25 brick/S.F.		90	.444		15	21.50		36.50	51
1350	16" thick, 27.00 brick/S.F.	↓	70	.571	↓	20	27.50		47.50	66.50
9000	Minimum labor/equipment charge	D-1	2	8	Job		375		375	605

04 27 10.40 Steps

Line	Description	Crew	Daily Output	Labor-Hours	Unit	Material	Labor	Equipment	Total	Total Incl O&P
0010	**STEPS**									
0012	Entry steps, select common brick	D-1	.30	53.333	M	555	2,500		3,055	4,675

04 41 Dry-Placed Stone

04 41 10 – Dry Placed Stone

04 41 10.10 Rough Stone Wall

Line	Description	Crew	Daily Output	Labor-Hours	Unit	Material	Labor	Equipment	Total	Total Incl O&P
0011	**ROUGH STONE WALL**, Dry									
0012	Dry laid (no mortar), under 18" thick G	D-1	60	.267	C.F.	14.15	12.50		26.65	35.50
0100	Random fieldstone, under 18" thick G	D-12	60	.533		14.15	25.50		39.65	56.50
0150	Over 18" thick G	"	63	.508	↓	17	24		41	57.50
0500	Field stone veneer G	D-8	120	.333	S.F.	12.95	16		28.95	40.50
0510	Valley stone veneer G		120	.333		12.95	16		28.95	40.50
0520	River stone veneer G	↓	120	.333	↓	12.95	16		28.95	40.50
0600	Rubble stone walls, in mortar bed, up to 18" thick G	D-11	75	.320	C.F.	17.10	15.75		32.85	44.50
9000	Minimum labor/equipment charge	D-1	2	8	Job		375		375	605

04 43 10 – Masonry with Natural and Processed Stone

04 43 10.05 Ashlar Veneer

	04 43 10.05 Ashlar Veneer	Crew	Daily Output	Labor-Hours	Unit	Material	Labor	Equipment	Total	Total Incl O&P
						2020 Bare Costs				
0011	ASHLAR VENEER +/- 4" thk, random or random rectangular									
0150	Sawn face, split joints, low priced stone	D-8	140	.286	S.F.	12.15	13.70		25.85	35.50
0200	Medium priced stone		130	.308		15.40	14.75		30.15	41
0300	High priced stone		120	.333		18.45	16		34.45	46.50
0600	Seam face, split joints, medium price stone		125	.320		18.20	15.35		33.55	45
0700	High price stone		120	.333		18.10	16		34.10	46
1000	Split or rock face, split joints, medium price stone		125	.320		10.95	15.35		26.30	37
1100	High price stone	↓	120	.333	↓	17.25	16		33.25	45

04 43 10.10 Bluestone

	04 43 10.10 Bluestone	Crew	Daily Output	Labor-Hours	Unit	Material	Labor	Equipment	Total	Total Incl O&P
0010	BLUESTONE, cut to size									
0500	Sills, natural cleft, 10" wide to 6' long, 1-1/2" thick	D-11	70	.343	L.F.	13.50	16.90		30.40	42.50
0550	2" thick		63	.381		14.80	18.75		33.55	47
0600	Smooth finish, 1-1/2" thick		70	.343		14.45	16.90		31.35	43.50
0650	2" thick		63	.381		18.10	18.75		36.85	50.50
0800	Thermal finish, 1-1/2" thick		70	.343		15.35	16.90		32.25	44.50
0850	2" thick	↓	63	.381		12.20	18.75		30.95	44
1000	Stair treads, natural cleft, 12" wide, 6' long, 1-1/2" thick	D-10	115	.278		14.05	14.40	3.72	32.17	42.50
1050	2" thick		105	.305		15.30	15.75	4.08	35.13	47
1100	Smooth finish, 1-1/2" thick		115	.278		13.45	14.40	3.72	31.57	42
1150	2" thick		105	.305		14.55	15.75	4.08	34.38	46
1300	Thermal finish, 1-1/2" thick		115	.278		13.80	14.40	3.72	31.92	42.50
1350	2" thick	↓	105	.305	↓	13.50	15.75	4.08	33.33	45
2000	Coping, finished top & 2 sides, 12" to 6'									
2100	Natural cleft, 1-1/2" thick	D-10	115	.278	L.F.	15.05	14.40	3.72	33.17	43.50
2150	2" thick		105	.305		17.50	15.75	4.08	37.33	49
2200	Smooth finish, 1-1/2" thick		115	.278		15.05	14.40	3.72	33.17	43.50
2250	2" thick		105	.305		17.50	15.75	4.08	37.33	49
2300	Thermal finish, 1-1/2" thick		115	.278		15.20	14.40	3.72	33.32	44
2350	2" thick	↓	105	.305	↓	17.70	15.75	4.08	37.53	49.50
9000	Minimum labor/equipment charge	D-1	2.50	6.400	Job		300		300	485

04 43 10.45 Granite

	04 43 10.45 Granite	Crew	Daily Output	Labor-Hours	Unit	Material	Labor	Equipment	Total	Total Incl O&P
0010	GRANITE, cut to size									
0050	Veneer, polished face, 3/4" to 1-1/2" thick									
0150	Low price, gray, light gray, etc.	D-10	130	.246	S.F.	27	12.75	3.29	43.04	53.50
0180	Medium price, pink, brown, etc.		130	.246		30.50	12.75	3.29	46.54	57.50
0220	High price, red, black, etc.	↓	130	.246	↓	42.50	12.75	3.29	58.54	71
0300	1-1/2" to 2-1/2" thick, veneer									
0350	Low price, gray, light gray, etc.	D-10	130	.246	S.F.	29	12.75	3.29	45.04	56
0500	Medium price, pink, brown, etc.		130	.246		34	12.75	3.29	50.04	61
0550	High price, red, black, etc.	↓	130	.246	↓	52.50	12.75	3.29	68.54	82
0700	2-1/2" to 4" thick, veneer									
0750	Low price, gray, light gray, etc.	D-10	110	.291	S.F.	39	15.05	3.89	57.94	71.50
0850	Medium price, pink, brown, etc.		110	.291		44.50	15.05	3.89	63.44	77.50
0950	High price, red, black, etc.	↓	110	.291		64	15.05	3.89	82.94	99
1000	For bush hammered finish, deduct					5%				
1050	Coarse rubbed finish, deduct					10%				
1100	Honed finish, deduct					5%				
1150	Thermal finish, deduct				↓	18%				
1800	Carving or bas-relief, from templates or plaster molds									
1850	Low price, gray, light gray, etc.	D-10	80	.400	C.F.	181	20.50	5.35	206.85	238
1875	Medium price, pink, brown, etc.		80	.400		355	20.50	5.35	380.85	430
1900	High price, red, black, etc.	↓	80	.400	↓	525	20.50	5.35	550.85	620

04 43 10 – Masonry with Natural and Processed Stone

04 43 10.45 Granite

04 43 10.45 Granite		Crew	Daily Output	Labor-Hours	Unit	2020 Bare Costs Material	Labor	Equipment	Total	Total Incl O&P
2000	Intricate or hand finished pieces									
2010	Mouldings, radius cuts, bullnose edges, etc.									
2050	Add for low price gray, light gray, etc.					30%				
2075	Add for medium price, pink, brown, etc.					165%				
2100	Add for high price red, black, etc.					300%				
2450	For radius under 5′, add				L.F.	100%				
2500	Steps, copings, etc., finished on more than one surface									
2550	Low price, gray, light gray, etc.	D-10	50	.640	C.F.	92	33	8.55	133.55	163
2575	Medium price, pink, brown, etc.		50	.640		120	33	8.55	161.55	194
2600	High price, red, black, etc.		50	.640		147	33	8.55	188.55	224
2700	Pavers, Belgian block, 8″-13″ long, 4″-6″ wide, 4″-6″ deep	D-11	120	.200	S.F.	27	9.85		36.85	46
2800	Pavers, 4″ x 4″ x 4″ blocks, split face and joints									
2850	Low price, gray, light gray, etc.	D-11	80	.300	S.F.	13.30	14.80		28.10	38.50
2875	Medium price, pink, brown, etc.		80	.300		21.50	14.80		36.30	47.50
2900	High price, red, black, etc.		80	.300		29.50	14.80		44.30	56.50
3000	Pavers, 4″ x 4″ x 4″, thermal face, sawn joints									
3050	Low price, gray, light gray, etc.	D-11	65	.369	S.F.	24.50	18.20		42.70	56.50
3075	Medium price, pink, brown, etc.		65	.369		28.50	18.20		46.70	61
3100	High price, red, black, etc.		65	.369		32.50	18.20		50.70	65.50
4000	Soffits, 2″ thick, low price, gray, light gray	D-13	35	1.371		38	69	12.25	119.25	166
4050	Medium price, pink, brown, etc.		35	1.371		65.50	69	12.25	146.75	196
4100	High price, red, black, etc.		35	1.371		92.50	69	12.25	173.75	226
4200	Low price, gray, light gray, etc.		35	1.371		63.50	69	12.25	144.75	194
4250	Medium price, pink, brown, etc.		35	1.371		92	69	12.25	173.25	225
4300	High price, red, black, etc.		35	1.371		120	69	12.25	201.25	256
5000	Reclaimed or antique									
5010	Treads, up to 12″ wide	D-10	100	.320	L.F.	42	16.55	4.28	62.83	77
5020	Up to 18″ wide		100	.320		45.50	16.55	4.28	66.33	81
5030	Capstone, size varies		50	.640		26.50	33	8.55	68.05	91.50
5040	Posts		30	1.067	V.L.F.	31.50	55	14.25	100.75	139
9000	Minimum labor/equipment charge	D-1	2	8	Job		375		375	605

04 43 10.50 Lightweight Natural Stone

04 43 10.50 Lightweight Natural Stone		Crew	Daily Output	Labor-Hours	Unit	Material	Labor	Equipment	Total	Total Incl O&P
0011	**LIGHTWEIGHT NATURAL STONE** Lava type									
0100	Veneer, rubble face, sawed back, irregular shapes G	D-10	130	.246	S.F.	9.10	12.75	3.29	25.14	34
0200	Sawed face and back, irregular shapes G		130	.246	"	9.10	12.75	3.29	25.14	34
1000	Reclaimed or antique, barn or foundation stone		1	32	Ton	385	1,650	430	2,465	3,550

04 43 10.55 Limestone

04 43 10.55 Limestone		Crew	Daily Output	Labor-Hours	Unit	Material	Labor	Equipment	Total	Total Incl O&P
0010	**LIMESTONE**, cut to size									
0020	Veneer facing panels									
0500	Texture finish, light stick, 4-1/2″ thick, 5′ x 12′	D-4	300	.107	S.F.	22.50	5.05	.63	28.18	33.50
0750	5″ thick, 5′ x 14′ panels	D-10	275	.116		24	6	1.56	31.56	38
1000	Sugarcube finish, 2″ thick, 3′ x 5′ panels		275	.116		29.50	6	1.56	37.06	44
1050	3″ thick, 4′ x 9′ panels		275	.116		29	6	1.56	36.56	43
1200	4″ thick, 5′ x 11′ panels		275	.116		32.50	6	1.56	40.06	47
1400	Sugarcube, textured finish, 4-1/2″ thick, 5′ x 12′		275	.116		35.50	6	1.56	43.06	50.50
1450	5″ thick, 5′ x 14′ panels		275	.116		35.50	6	1.56	43.06	50.50
2000	Coping, sugarcube finish, top & 2 sides		30	1.067	C.F.	66.50	55	14.25	135.75	178
2100	Sills, lintels, jambs, trim, stops, sugarcube finish, simple		20	1.600		66.50	83	21.50	171	230
2150	Detailed		20	1.600		66.50	83	21.50	171	230
2300	Steps, extra hard, 14″ wide, 6″ rise		50	.640	L.F.	29	33	8.55	70.55	94.50
3000	Quoins, plain finish, 6″ x 12″ x 12″	D-12	25	1.280	Ea.	44.50	60.50		105	147
3050	6″ x 16″ x 24″	"	25	1.280	"	59.50	60.50		120	164

04 43 10.55 Limestone		Crew	Daily Output	Labor-Hours	Unit	Material	2020 Bare Costs Labor	Equipment	Total	Total Incl O&P
9000	Minimum labor/equipment charge	D-1	2	8	Job		375		375	605

04 43 10.60 Marble		Crew	Daily Output	Labor-Hours	Unit	Material	2020 Bare Costs Labor	Equipment	Total	Total Incl O&P
0011	**MARBLE**, ashlar, split face, +/- 4" thick, random									
0040	Lengths 1' to 4' & heights 2" to 7-1/2", average	D-8	175	.229	S.F.	17.60	10.95		28.55	37
0100	Base, polished, 3/4" or 7/8" thick, polished, 6" high	D-10	65	.492	L.F.	11.45	25.50	6.60	43.55	61
0300	Carvings or bas-relief, from templates, simple design		80	.400	S.F.	157	20.50	5.35	182.85	212
0350	Intricate design	↓	80	.400	"	355	20.50	5.35	380.85	430
0600	Columns, cornices, mouldings, etc.									
0650	Hand or special machine cut, simple design	D-10	35	.914	C.F.	61.50	47.50	12.25	121.25	157
0700	Intricate design	"	35	.914	"	325	47.50	12.25	384.75	445
1000	Facing, polished finish, cut to size, 3/4" to 7/8" thick									
1050	Carrara or equal	D-10	130	.246	S.F.	20.50	12.75	3.29	36.54	47
1100	Arabescato or equal		130	.246		38	12.75	3.29	54.04	65.50
1300	1-1/4" thick, Botticino Classico or equal		125	.256		24.50	13.25	3.43	41.18	52
1350	Statuarietto or equal		125	.256		39	13.25	3.43	55.68	68
1500	2" thick, Crema Marfil or equal		120	.267		48.50	13.80	3.57	65.87	79.50
1550	Cafe Pinta or equal	↓	120	.267	↓	65	13.80	3.57	82.37	97.50
1700	Rubbed finish, cut to size, 4" thick									
1740	Average	D-10	100	.320	S.F.	42.50	16.55	4.28	63.33	77.50
1780	Maximum	"	100	.320	"	71	16.55	4.28	91.83	110
2200	Window sills, 6" x 3/4" thick	D-1	85	.188	L.F.	13.10	8.85		21.95	28.50
2500	Flooring, polished tiles, 12" x 12" x 3/8" thick									
2510	Thin set, Giallo Solare or equal	D-11	90	.267	S.F.	15.65	13.15		28.80	38.50
2600	Sky Blue or equal		90	.267		15.05	13.15		28.20	37.50
2700	Mortar bed, Giallo Solare or equal		65	.369		15.75	18.20		33.95	47
2740	Sky Blue or equal	↓	65	.369		15.05	18.20		33.25	46
2780	Travertine, 3/8" thick, Sierra or equal	D-10	130	.246		10.80	12.75	3.29	26.84	36
2790	Silver or equal	"	130	.246		29	12.75	3.29	45.04	56
2800	Patio tile, non-slip, 1/2" thick, flame finish	D-11	75	.320	↓	12.40	15.75		28.15	39
2900	Shower or toilet partitions, 7/8" thick partitions									
3050	3/4" or 1-1/4" thick stiles, polished 2 sides, average	D-11	75	.320	S.F.	47.50	15.75		63.25	77.50
3201	Soffits, add to above prices				"	20%	100%			
3210	Stairs, risers, 7/8" thick x 6" high	D-10	115	.278	L.F.	15.15	14.40	3.72	33.27	44
3360	Treads, 12" wide x 1-1/4" thick	"	115	.278	"	50	14.40	3.72	68.12	82
3500	Thresholds, 3' long, 7/8" thick, 4" to 5" wide, plain	D-12	24	1.333	Ea.	36	63		99	142
3550	Beveled		24	1.333	"	76	63		139	186
3700	Window stools, polished, 7/8" thick, 5" wide	↓	85	.376	L.F.	24.50	17.85		42.35	56
9000	Minimum labor/equipment charge	D-1	2	8	Job		375		375	605

04 43 10.75 Sandstone or Brownstone		Crew	Daily Output	Labor-Hours	Unit	Material	2020 Bare Costs Labor	Equipment	Total	Total Incl O&P
0011	**SANDSTONE OR BROWNSTONE**									
0100	Sawed face veneer, 2-1/2" thick, to 2' x 4' panels	D-10	130	.246	S.F.	22	12.75	3.29	38.04	48
0150	4" thick, to 3'-6" x 8' panels		100	.320		22	16.55	4.28	42.83	55
0300	Split face, random sizes	↓	100	.320	↓	14.50	16.55	4.28	35.33	47
0350	Cut stone trim (limestone)									
0360	Ribbon stone, 4" thick, 5' pieces	D-8	120	.333	Ea.	162	16		178	204
0370	Cove stone, 4" thick, 5' pieces		105	.381		163	18.25		181.25	209
0380	Cornice stone, 10" to 12" wide		90	.444		201	21.50		222.50	256
0390	Band stone, 4" thick, 5' pieces		145	.276		107	13.20		120.20	139
0410	Window and door trim, 3" to 4" wide		160	.250		90.50	12		102.50	119
0420	Key stone, 18" long	↓	60	.667	↓	93	32		125	154
9000	Minimum labor/equipment charge	D-1	2.50	6.400	Job		300		300	485

04 43 10 – Masonry with Natural and Processed Stone

04 43 10.80 Slate		Crew	Daily Output	Labor-Hours	Unit	2020 Bare Costs Material	Labor	Equipment	Total	Total Incl O&P
0010	**SLATE**									
0040	Pennsylvania - blue gray to black									
0050	Vermont - unfading green, mottled green & purple, gray & purple									
0100	Virginia - blue black									
0200	Exterior paving, natural cleft, 1" thick									
0250	6" x 6" Pennsylvania	D-12	100	.320	S.F.	7.05	15.15		22.20	32.50
0300	Vermont		100	.320		11.20	15.15		26.35	37
0350	Virginia		100	.320		14.70	15.15		29.85	40.50
0500	24" x 24", Pennsylvania		120	.267		13.55	12.65		26.20	35.50
0550	Vermont		120	.267		28	12.65		40.65	51
0600	Virginia		120	.267		21.50	12.65		34.15	44
0700	18" x 30" Pennsylvania		120	.267		15.35	12.65		28	37.50
0750	Vermont		120	.267		28	12.65		40.65	51
0800	Virginia		120	.267		19	12.65		31.65	41.50
1000	Interior flooring, natural cleft, 1/2" thick									
1100	6" x 6" Pennsylvania	D-12	100	.320	S.F.	4.16	15.15		19.31	29
1150	Vermont		100	.320		9.85	15.15		25	35.50
1200	Virginia		100	.320		11.60	15.15		26.75	37.50
1300	24" x 24" Pennsylvania		120	.267		8.10	12.65		20.75	29.50
1350	Vermont		120	.267		22.50	12.65		35.15	45
1400	Virginia		120	.267		15.35	12.65		28	37.50
1500	18" x 24" Pennsylvania		120	.267		8.10	12.65		20.75	29.50
1550	Vermont		120	.267		16.95	12.65		29.60	39
1600	Virginia		120	.267		15.55	12.65		28.20	37.50
2000	Facing panels, 1-1/4" thick, to 4' x 4' panels									
2100	Natural cleft finish, Pennsylvania	D-10	180	.178	S.F.	35	9.20	2.38	46.58	56
2110	Vermont		180	.178		28.50	9.20	2.38	40.08	48.50
2120	Virginia		180	.178		34.50	9.20	2.38	46.08	55.50
2150	Sand rubbed finish, surface, add					10.60			10.60	11.65
2200	Honed finish, add					7.65			7.65	8.45
2500	Ribbon, natural cleft finish, 1" thick, to 9 S.F.	D-10	80	.400		13.65	20.50	5.35	39.50	54
2550	Sand rubbed finish		80	.400		18.50	20.50	5.35	44.35	59.50
2600	Honed finish		80	.400		17.20	20.50	5.35	43.05	58
2700	1-1/2" thick		78	.410		17.75	21	5.50	44.25	59.50
2750	Sand rubbed finish		78	.410		23.50	21	5.50	50	66
2800	Honed finish		78	.410		22	21	5.50	48.50	64.50
2850	2" thick		76	.421		21.50	22	5.65	49.15	64.50
2900	Sand rubbed finish		76	.421		29.50	22	5.65	57.15	73.50
2950	Honed finish		76	.421		27	22	5.65	54.65	71
3100	Stair landings, 1" thick, black, clear	D-1	65	.246		21	11.55		32.55	41.50
3200	Ribbon	"	65	.246		23	11.55		34.55	43.50
3500	Stair treads, sand finish, 1" thick x 12" wide									
3550	Under 3 L.F.	D-10	85	.376	L.F.	23	19.50	5.05	47.55	61.50
3600	3 L.F. to 6 L.F.	"	120	.267	"	25	13.80	3.57	42.37	53
3700	Ribbon, sand finish, 1" thick x 12" wide									
3750	To 6 L.F.	D-10	120	.267	L.F.	21	13.80	3.57	38.37	49
4000	Stools or sills, sand finish, 1" thick, 6" wide	D-12	160	.200		12	9.50		21.50	28.50
4100	Honed finish		160	.200		11.45	9.50		20.95	28
4200	10" wide		90	.356		18.50	16.85		35.35	47.50
4250	Honed finish		90	.356		17.20	16.85		34.05	46
4400	2" thick, 6" wide		140	.229		19.30	10.85		30.15	38.50
4450	Honed finish		140	.229		18.20	10.85		29.05	37.50

04 43 Stone Masonry

04 43 10 – Masonry with Natural and Processed Stone

04 43 10.80 Slate		Crew	Daily Output	Labor-Hours	Unit	Material	Labor	Equipment	Total	Total Incl O&P
						2020 Bare Costs				
4600	10" wide	D-12	90	.356	L.F.	30	16.85		46.85	60
4650	Honed finish	↓	90	.356		28.50	16.85		45.35	58.50
4800	For lengths over 3', add				↓	25%				
9000	Minimum labor/equipment charge	D-1	2.50	6.400	Job		300		300	485

04 43 10.85 Window Sill		Crew	Daily Output	Labor-Hours	Unit	Material	Labor	Equipment	Total	Total Incl O&P
0010	**WINDOW SILL**									
0020	Bluestone, thermal top, 10" wide, 1-1/2" thick	D-1	85	.188	S.F.	10.30	8.85		19.15	25.50
0050	2" thick		75	.213	"	10.95	10		20.95	28
0100	Cut stone, 5" x 8" plain		48	.333	L.F.	12.95	15.65		28.60	39.50
0200	Face brick on edge, brick, 8" wide		80	.200		3.42	9.40		12.82	18.90
0400	Marble, 9" wide, 1" thick		85	.188		9.25	8.85		18.10	24.50
0900	Slate, colored, unfading, honed, 12" wide, 1" thick		85	.188		10.45	8.85		19.30	26
0950	2" thick	↓	70	.229	↓	10.05	10.70		20.75	28.50
9000	Minimum labor/equipment charge	1 Bric	2	4	Job		208		208	335

04 51 Flue Liner Masonry

04 51 10 – Clay Flue Lining

04 51 10.10 Flue Lining		Crew	Daily Output	Labor-Hours	Unit	Material	Labor	Equipment	Total	Total Incl O&P
0010	**FLUE LINING**, including mortar									
0020	Clay, 8" x 8"	D-1	125	.128	V.L.F.	6.45	6		12.45	16.80
0100	8" x 12"		103	.155		9.60	7.30		16.90	22.50
0200	12" x 12"		93	.172		11.70	8.05		19.75	26
0300	12" x 18"		84	.190		26.50	8.95		35.45	43.50
0400	18" x 18"		75	.213		30	10		40	49
0500	20" x 20"		66	.242		48	11.35		59.35	71
0600	24" x 24"		56	.286		68.50	13.40		81.90	96.50
1000	Round, 18" diameter		66	.242		42.50	11.35		53.85	65.50
1100	24" diameter	↓	47	.340	↓	88.50	15.95		104.45	124

04 54 Refractory Brick Masonry

04 54 10 – Refractory Brick Work

04 54 10.10 Fire Brick		Crew	Daily Output	Labor-Hours	Unit	Material	Labor	Equipment	Total	Total Incl O&P
0010	**FIRE BRICK**									
0012	Low duty, 2000°F, 9" x 2-1/2" x 4-1/2"	D-1	.60	26.667	M	1,675	1,250		2,925	3,875
0050	High duty, 3000°F	"	.60	26.667	"	2,800	1,250		4,050	5,125

04 54 10.20 Fire Clay		Crew	Daily Output	Labor-Hours	Unit	Material	Labor	Equipment	Total	Total Incl O&P
0010	**FIRE CLAY**									
0020	Gray, high duty, 100 lb. bag				Bag	37			37	40.50
0050	100 lb. drum, premixed (400 brick per drum)				Drum	48.50			48.50	53.50

04 57 Masonry Fireplaces

04 57 10 – Brick or Stone Fireplaces

04 57 10.10 Fireplace		Crew	Daily Output	Labor-Hours	Unit	Material	Labor	Equipment	Total	Total Incl O&P
						2020 Bare Costs				
0010	**FIREPLACE**									
0100	Brick fireplace, not incl. foundations or chimneys									
0110	30" x 29" opening, incl. chamber, plain brickwork	D-1	.40	40	Ea.	625	1,875		2,500	3,700
0200	Fireplace box only (110 brick)	"	2	8	"	172	375		547	795
0300	For elaborate brickwork and details, add					35%	35%			
0400	For hearth, brick & stone, add	D-1	2	8	Ea.	215	375		590	840
0410	For steel, damper, cleanouts, add		4	4		18.25	188		206.25	325
0600	Plain brickwork, incl. metal circulator		.50	32		995	1,500		2,495	3,525
0800	Face brick only, standard size, 8" x 2-2/3" x 4"		.30	53.333	M	605	2,500		3,105	4,725
0900	Stone fireplace, fieldstone, add				SF Face	10			10	11
1000	Cut stone, add				"	8.50			8.50	9.35
9000	Minimum labor/equipment charge	D-1	2	8	Job		375		375	605

04 71 Manufactured Brick Masonry

04 71 10 – Simulated or Manufactured Brick

04 71 10.10 Simulated Brick		Crew	Daily Output	Labor-Hours	Unit	Material	Labor	Equipment	Total	Total Incl O&P
0010	**SIMULATED BRICK**									
0020	Aluminum, baked on colors	1 Carp	200	.040	S.F.	4.53	2.13		6.66	8.40
0050	Fiberglass panels		200	.040		9.90	2.13		12.03	14.30
0100	Urethane pieces cemented in mastic		150	.053		9	2.83		11.83	14.45
0150	Vinyl siding panels		200	.040		11.50	2.13		13.63	16.05
0160	Cement base, brick, incl. mastic	D-1	100	.160		9.75	7.50		17.25	23
0170	Corner		50	.320	V.L.F.	22	15		37	49
0180	Stone face, incl. mastic		100	.160	S.F.	10.05	7.50		17.55	23
0190	Corner		50	.320	V.L.F.	3.23	15		18.23	28

04 72 Cast Stone Masonry

04 72 10 – Cast Stone Masonry Features

04 72 10.10 Coping		Crew	Daily Output	Labor-Hours	Unit	Material	Labor	Equipment	Total	Total Incl O&P
0010	**COPING**, stock units									
0050	Precast concrete, 10" wide, 4" tapers to 3-1/2", 8" wall	D-1	75	.213	L.F.	25.50	10		35.50	44
0100	12" wide, 3-1/2" tapers to 3", 10" wall		70	.229		27.50	10.70		38.20	47.50
0110	14" wide, 4" tapers to 3-1/2", 12" wall		65	.246		31.50	11.55		43.05	53
0150	16" wide, 4" tapers to 3-1/2", 14" wall		60	.267		34	12.50		46.50	57
0250	Precast concrete corners		40	.400	Ea.	44	18.75		62.75	78.50
0300	Limestone for 12" wall, 4" thick		90	.178	L.F.	16.80	8.35		25.15	32
0350	6" thick		80	.200		24	9.40		33.40	41
0500	Marble, to 4" thick, no wash, 9" wide		90	.178		12.30	8.35		20.65	27
0550	12" wide		80	.200		19.90	9.40		29.30	37
0700	Terra cotta, 9" wide		90	.178		8.15	8.35		16.50	22.50
0750	12" wide		80	.200		8.60	9.40		18	24.50
0800	Aluminum, for 12" wall		80	.200		9.45	9.40		18.85	25.50
9000	Minimum labor/equipment charge		2	8	Job		375		375	605

04 72 20 – Cultured Stone Veneer

04 72 20.10 Cultured Stone Veneer Components		Crew	Daily Output	Labor-Hours	Unit	Material	Labor	Equipment	Total	Total Incl O&P
0010	**CULTURED STONE VENEER COMPONENTS**									
0110	On wood frame and sheathing substrate, random sized cobbles, corner stones	D-8	70	.571	V.L.F.	10.55	27.50		38.05	56
0120	Field stones		140	.286	S.F.	7.45	13.70		21.15	30
0130	Random sized flats, corner stones		70	.571	V.L.F.	10.55	27.50		38.05	56

04 72 Cast Stone Masonry

04 72 20 – Cultured Stone Veneer

04 72 20.10 Cultured Stone Veneer Components		Crew	Daily Output	Labor-Hours	Unit	2020 Bare Costs Material	Labor	Equipment	Total	Total Incl O&P
0140	Field stones	D-8	140	.286	S.F.	8.85	13.70		22.55	31.50
0150	Horizontal lined ledgestones, corner stones		75	.533	V.L.F.	10.35	25.50		35.85	53
0160	Field stones		150	.267	S.F.	7.80	12.80		20.60	29
0170	Random shaped flats, corner stones		65	.615	V.L.F.	10.35	29.50		39.85	59
0180	Field stones		150	.267	S.F.	7.75	12.80		20.55	29
0190	Random shaped/textured face, corner stones		65	.615	V.L.F.	10.55	29.50		40.05	59
0200	Field stones		130	.308	S.F.	7.70	14.75		22.45	32.50
0210	Random shaped river rock, corner stones		65	.615	V.L.F.	10.55	29.50		40.05	59
0220	Field stones		130	.308	S.F.	7.70	14.75		22.45	32.50
0240	On concrete or CMU substrate, random sized cobbles, corner stones		70	.571	V.L.F.	9.70	27.50		37.20	55
0250	Field stones		140	.286	S.F.	7.05	13.70		20.75	30
0260	Random sized flats, corner stones		70	.571	V.L.F.	9.70	27.50		37.20	55
0270	Field stones		140	.286	S.F.	8.40	13.70		22.10	31.50
0280	Horizontal lined ledgestones, corner stones		75	.533	V.L.F.	9.50	25.50		35	52
0290	Field stones		150	.267	S.F.	7.35	12.80		20.15	28.50
0300	Random shaped flats, corner stones		70	.571	V.L.F.	9.50	27.50		37	55
0310	Field stones		140	.286	S.F.	7.35	13.70		21.05	30
0320	Random shaped/textured face, corner stones		65	.615	V.L.F.	9.70	29.50		39.20	58
0330	Field stones		130	.308	S.F.	7.30	14.75		22.05	32
0340	Random shaped river rock, corner stones		65	.615	V.L.F.	9.70	29.50		39.20	58
0350	Field stones	↓	130	.308	S.F.	7.30	14.75		22.05	32
0360	Cultured stone veneer, #15 felt weather resistant barrier	1 Clab	3700	.002	Sq.	5.25	.09		5.34	5.95
0370	Expanded metal lath, diamond, 2.5 lb./S.Y., galvanized	1 Lath	85	.094	S.Y.	3.80	4.90		8.70	11.85
0390	Water table or window sill, 18" long	1 Bric	80	.100	Ea.	9.95	5.20		15.15	19.35

04 73 Manufactured Stone Masonry

04 73 20 – Simulated or Manufactured Stone

04 73 20.10 Simulated Stone		Crew	Daily Output	Labor-Hours	Unit	Material	Labor	Equipment	Total	Total Incl O&P
0010	**SIMULATED STONE**									
0100	Insulated fiberglass panels, 5/8" ply backer	L-4	200	.120	S.F.	10.85	6		16.85	21.50

Estimating Tips

05 05 00 Common Work Results for Metals

- Nuts, bolts, washers, connection angles, and plates can add a significant amount to both the tonnage of a structural steel job and the estimated cost. As a rule of thumb, add 10% to the total weight to account for these accessories.
- Type 2 steel construction, commonly referred to as "simple construction," consists generally of field-bolted connections with lateral bracing supplied by other elements of the building, such as masonry walls or x-bracing. The estimator should be aware, however, that shop connections may be accomplished by welding or bolting. The method may be particular to the fabrication shop and may have an impact on the estimated cost.

05 10 00 Structural Steel

- Steel items can be obtained from two sources: a fabrication shop or a metals service center. Fabrication shops can fabricate items under more controlled conditions than crews in the field can. They are also more efficient and can produce items more economically. Metal service centers serve as a source of long mill shapes to both fabrication shops and contractors.
- Most line items in this structural steel subdivision, and most items in 05 50 00 Metal Fabrications, are indicated as being shop fabricated. The bare material cost for these shop fabricated items is the "Invoice Cost" from the shop and includes the mill base price of steel plus mill extras, transportation to the shop, shop drawings and detailing where warranted, shop fabrication and handling, sandblasting and a shop coat of primer paint, all necessary structural bolts, and delivery to the job site. The bare labor cost and bare equipment cost for these shop fabricated items are for field installation or erection.
- Line items in Subdivision 05 12 23.40 Lightweight Framing, and other items scattered in Division 5, are indicated as being field fabricated. The bare material cost for these field fabricated items is the "Invoice Cost" from the metals service center and includes the mill base price of steel plus mill extras, transportation to the metals service center, material handling, and delivery of long lengths of mill shapes to the job site. Material costs for structural bolts and welding rods should be added to the estimate. The bare labor cost and bare equipment cost for these items are for both field fabrication and field installation or erection, and include time for cutting, welding, and drilling in the fabricated metal items. Drilling into concrete and fasteners to fasten field fabricated items to other work is not included and should be added to the estimate.

05 20 00 Steel Joist Framing

- In any given project the total weight of open web steel joists is determined by the loads to be supported and the design. However, economies can be realized in minimizing the amount of labor used to place the joists. This is done by maximizing the joist spacing and therefore minimizing the number of joists required to be installed on the job. Certain spacings and locations may be required by the design, but in other cases maximizing the spacing and keeping it as uniform as possible will keep the costs down.

05 30 00 Steel Decking

- The takeoff and estimating of a metal deck involve more than the area of the floor or roof and the type of deck specified or shown on the drawings. Many different sizes and types of openings may exist. Small openings for individual pipes or conduits may be drilled after the floor/roof is installed, but larger openings may require special deck lengths as well as reinforcing or structural support. The estimator should determine who will be supplying this reinforcing. Additionally, some deck terminations are part of the deck package, such as screed angles and pour stops, and others will be part of the steel contract, such as angles attached to structural members and cast-in-place angles and plates. The estimator must ensure that all pieces are accounted for in the complete estimate.

05 50 00 Metal Fabrications

- The most economical steel stairs are those that use common materials, standard details, and most importantly, a uniform and relatively simple method of field assembly. Commonly available A36/A992 channels and plates are very good choices for the main stringers of the stairs, as are angles and tees for the carrier members. Risers and treads are usually made by specialty shops, and it is most economical to use a typical detail in as many places as possible. The stairs should be pre-assembled and shipped directly to the site. The field connections should be simple and straightforward enough to be accomplished efficiently, and with minimum equipment and labor.

Reference Numbers

Reference numbers are shown at the beginning of some major classifications. These numbers refer to related items in the Reference Section. The reference information may be an estimating procedure, an alternate pricing method, or technical information.

Note: Not all subdivisions listed here necessarily appear. ■

05 01 Maintenance of Metals

05 01 10 – Maintenance of Structural Metal Framing

05 01 10.51 Cleaning of Structural Metal Framing		Crew	Daily Output	Labor-Hours	Unit	2020 Bare Costs Material	Labor	Equipment	Total	Total Incl O&P
0010	CLEANING OF STRUCTURAL METAL FRAMING									
6125	Steel surface treatments, PDCA guidelines									
6170	Wire brush, hand (SSPC-SP2)	1 Psst	400	.020	S.F.	.02	.92		.94	1.60
6180	Power tool (SSPC-SP3)	"	700	.011		.07	.52		.59	.98
6215	Pressure washing, up to 5,000 psi, 5,000-15,000 S.F./day	1 Pord	10000	.001			.04		.04	.06
6220	Steam cleaning, 600 psi @ 300 degree F, 1,250-2,500 S.F./day		2000	.004			.18		.18	.28
6225	Water blasting, up to 25,000 psi, 1,750-3,500 S.F./day		2500	.003			.14		.14	.23
6230	Brush-off blast (SSPC-SP7)	E-11	1750	.018		.17	.85	.17	1.19	1.79
6235	Com'l blast (SSPC-SP6), loose scale, fine pwder rust, 2.0#/S.F. sand		1200	.027		.34	1.25	.25	1.84	2.72
6240	Tight mill scale, little/no rust, 3.0#/S.F. sand		1000	.032		.51	1.49	.30	2.30	3.37
6245	Exist coat blistered/pitted, 4.0#/S.F. sand		875	.037		.69	1.71	.34	2.74	3.95
6250	Exist coat badly pitted/nodules, 6.7#/S.F. sand		825	.039		1.15	1.81	.36	3.32	4.65
6255	Near white blast (SSPC-SP10), loose scale, fine rust, 5.6#/S.F. sand		450	.071		.96	3.32	.67	4.95	7.30
6260	Tight mill scale, little/no rust, 6.9#/S.F. sand		325	.098		1.18	4.60	.93	6.71	9.90
6265	Exist coat blistered/pitted, 9.0#/S.F. sand		225	.142		1.54	6.65	1.34	9.53	14.10
6270	Exist coat badly pitted/nodules, 11.3#/S.F. sand		150	.213		1.94	9.95	2.01	13.90	21

05 05 Common Work Results for Metals

05 05 05 – Selective Demolition for Metals

05 05 05.10 Selective Demolition, Metals		Crew	Daily Output	Labor-Hours	Unit	Material	Labor	Equipment	Total	Total Incl O&P
0010	SELECTIVE DEMOLITION, METALS R024119-10									
0015	Excludes shores, bracing, cutting, loading, hauling, dumping									
0020	Remove nuts only up to 3/4" diameter	1 Sswk	480	.017	Ea.		.96		.96	1.57
0030	7/8" to 1-1/4" diameter		240	.033			1.92		1.92	3.15
0040	1-3/8" to 2" diameter		160	.050			2.88		2.88	4.72
0060	Unbolt and remove structural bolts up to 3/4" diameter		240	.033			1.92		1.92	3.15
0070	7/8" to 2" diameter		160	.050			2.88		2.88	4.72
0140	Light weight framing members, remove whole or cut up, up to 20 lb.		240	.033			1.92		1.92	3.15
0150	21-40 lb.	2 Sswk	210	.076			4.39		4.39	7.20
0160	41-80 lb.	3 Sswk	180	.133			7.70		7.70	12.60
0170	81-120 lb.	4 Sswk	150	.213			12.30		12.30	20
0230	Structural members, remove whole or cut up, up to 500 lb.	E-19	48	.500			28.50	31	59.50	80
0240	1/4-2 tons	E-18	36	1.111			64.50	41.50	106	150
0250	2-5 tons	E-24	30	1.067			61.50	19.35	80.85	121
0260	5-10 tons	E-20	24	2.667			153	84.50	237.50	340
0270	10-15 tons	E-2	18	3.111			178	94	272	390
0340	Fabricated item, remove whole or cut up, up to 20 lb.	1 Sswk	96	.083			4.80		4.80	7.85
0350	21-40 lb.	2 Sswk	84	.190			11		11	18
0360	41-80 lb.	3 Sswk	72	.333			19.20		19.20	31.50
0370	81-120 lb.	4 Sswk	60	.533			31		31	50.50
0380	121-500 lb.	E-19	48	.500			28.50	31	59.50	80
0390	501-1000 lb.	"	36	.667			38	41.50	79.50	107
0500	Steel roof decking, uncovered, bare	B-2	5000	.008	S.F.		.34		.34	.54
2290	Tube steel column, 8' high	B-1J	8	2	Ea.		84.50		84.50	136
2700	Floor grating, supports excluded	B-68D	800	.030	S.F.		1.38	.35	1.73	2.57
2705	Deckplate, supports excluded	"	800	.030			1.38	.35	1.73	2.57
2800	Structural steel platform, interior	B-68E	96	.417			24	2.67	26.67	42.50
2805	Exterior		48	.833			48	5.35	53.35	85
2810	Demolish, catwalk		32	1.250	L.F.		72	8	80	127
2815	Ladder with cage		100	.400	V.L.F.		23	2.56	25.56	41
2820	without cage		128	.313	"		18.05	2	20.05	31.50

05 05 Common Work Results for Metals

05 05 05 – Selective Demolition for Metals

05 05 05.10 Selective Demolition, Metals		Crew	Daily Output	Labor-Hours	Unit	Material	2020 Bare Costs Labor	Equipment	Total	Total Incl O&P
2825	Hand rail	B-68E	520	.077	L.F.		4.44	.49	4.93	7.80
2830	Guard rail		160	.250	"		14.45	1.60	16.05	25.50
2835	Guard post	2 Sswk	8	2	Ea.		115		115	189
2840	Steel bar grating	B-68E	600	.067	S.F.		3.85	.43	4.28	6.75
2845	Duct supports	"	80	.500	L.F.		29	3.20	32.20	51
2950	Minimum labor/equipment charge	E-19	2	12	Job		680	745	1,425	1,925

05 05 13 – Shop-Applied Coatings for Metal

05 05 13.50 Paints and Protective Coatings

		Crew	Daily Output	Labor-Hours	Unit	Material	Labor	Equipment	Total	Total Incl O&P
0010	**PAINTS AND PROTECTIVE COATINGS**									
5900	Galvanizing structural steel in shop, under 1 ton R050516-30				Ton	580			580	640
5950	1 ton to 20 tons					570			570	625
6000	Over 20 tons					535			535	590

05 05 19 – Post-Installed Concrete Anchors

05 05 19.10 Chemical Anchors

		Crew	Daily Output	Labor-Hours	Unit	Material	Labor	Equipment	Total	Total Incl O&P
0010	**CHEMICAL ANCHORS**									
0020	Includes layout & drilling									
1430	Chemical anchor, w/rod & epoxy cartridge, 3/4" diameter x 9-1/2" long	B-89A	27	.593	Ea.	10.15	28.50	4.23	42.88	62
1435	1" diameter x 11-3/4" long		24	.667		20.50	32.50	4.76	57.76	79.50
1440	1-1/4" diameter x 14" long		21	.762		39	37	5.45	81.45	108
1445	1-3/4" diameter x 15" long		20	.800		68.50	39	5.70	113.20	144
1450	18" long		17	.941		82.50	45.50	6.70	134.70	171
1455	2" diameter x 18" long		16	1		113	48.50	7.15	168.65	210
1460	24" long		15	1.067		148	51.50	7.60	207.10	254

05 05 19.20 Expansion Anchors

		Crew	Daily Output	Labor-Hours	Unit	Material	Labor	Equipment	Total	Total Incl O&P
0010	**EXPANSION ANCHORS**									
0100	Anchors for concrete, brick or stone, no layout and drilling									
0200	Expansion shields, zinc, 1/4" diameter, 1-5/16" long, single G	1 Carp	90	.089	Ea.	.46	4.72		5.18	8.05
0300	1-3/8" long, double G		85	.094		.66	5		5.66	8.75
0400	3/8" diameter, 1-1/2" long, single G		85	.094		.71	5		5.71	8.80
0500	2" long, double G		80	.100		1.20	5.30		6.50	9.80
0600	1/2" diameter, 2-1/16" long, single G		80	.100		1.28	5.30		6.58	9.90
0700	2-1/2" long, double G		75	.107		2.17	5.65		7.82	11.50
0800	5/8" diameter, 2-5/8" long, single G		75	.107		2.22	5.65		7.87	11.55
0900	2-3/4" long, double G		70	.114		3.64	6.05		9.69	13.75
1000	3/4" diameter, 2-3/4" long, single G		70	.114		3.66	6.05		9.71	13.80
1100	3-15/16" long, double G		65	.123		5.65	6.55		12.20	16.75
2100	Hollow wall anchors for gypsum wall board, plaster or tile									
2300	1/8" diameter, short G	1 Carp	160	.050	Ea.	.31	2.66		2.97	4.60
2400	Long G		150	.053		.32	2.83		3.15	4.89
2500	3/16" diameter, short G		150	.053		.58	2.83		3.41	5.20
2600	Long G		140	.057		.75	3.04		3.79	5.70
2700	1/4" diameter, short G		140	.057		.65	3.04		3.69	5.60
2800	Long G		130	.062		1.03	3.27		4.30	6.40
3000	Toggle bolts, bright steel, 1/8" diameter, 2" long G		85	.094		.21	5		5.21	8.25
3100	4" long G		80	.100		.26	5.30		5.56	8.80
3200	3/16" diameter, 3" long G		80	.100		.30	5.30		5.60	8.85
3300	6" long G		75	.107		.49	5.65		6.14	9.65
3400	1/4" diameter, 3" long G		75	.107		.44	5.65		6.09	9.60
3500	6" long G		70	.114		.57	6.05		6.62	10.40
3600	3/8" diameter, 3" long G		70	.114		.88	6.05		6.93	10.70
3700	6" long G		60	.133		1.58	7.10		8.68	13.10

05 05 19 – Post-Installed Concrete Anchors

05 05 19.20 Expansion Anchors			Crew	Daily Output	Labor-Hours	Unit	2020 Bare Costs Material	Labor	Equipment	Total	Total Incl O&P
3800	1/2" diameter, 4" long	G	1 Carp	60	.133	Ea.	1.86	7.10		8.96	13.40
3900	6" long	G		50	.160		2.37	8.50		10.87	16.20
4000	Nailing anchors										
4100	Nylon nailing anchor, 1/4" diameter, 1" long		1 Carp	3.20	2.500	C	19.75	133		152.75	235
4200	1-1/2" long			2.80	2.857		22	152		174	267
4300	2" long			2.40	3.333		24	177		201	310
4400	Metal nailing anchor, 1/4" diameter, 1" long	G		3.20	2.500		20	133		153	235
4500	1-1/2" long	G		2.80	2.857		25	152		177	271
4600	2" long	G		2.40	3.333		43	177		220	330
5000	Screw anchors for concrete, masonry,										
5100	stone & tile, no layout or drilling included										
5700	Lag screw shields, 1/4" diameter, short	G	1 Carp	90	.089	Ea.	.47	4.72		5.19	8.05
5800	Long	G		85	.094		.56	5		5.56	8.60
5900	3/8" diameter, short	G		85	.094		.74	5		5.74	8.80
6000	Long	G		80	.100		1.05	5.30		6.35	9.65
6100	1/2" diameter, short	G		80	.100		.98	5.30		6.28	9.60
6200	Long	G		75	.107		1.52	5.65		7.17	10.75
6300	5/8" diameter, short	G		70	.114		1.54	6.05		7.59	11.45
6400	Long	G		65	.123		2.21	6.55		8.76	12.95
6600	Lead, #6 & #8, 3/4" long	G		260	.031		.18	1.64		1.82	2.82
6700	#10 - #14, 1-1/2" long	G		200	.040		.48	2.13		2.61	3.94
6800	#16 & #18, 1-1/2" long	G		160	.050		.42	2.66		3.08	4.72
6900	Plastic, #6 & #8, 3/4" long			260	.031		.05	1.64		1.69	2.68
7000	#8 & #10, 7/8" long			240	.033		.05	1.77		1.82	2.90
7100	#10 & #12, 1" long			220	.036		.07	1.93		2	3.18
7200	#14 & #16, 1-1/2" long			160	.050		.07	2.66		2.73	4.34
8000	Wedge anchors, not including layout or drilling										
8050	Carbon steel, 1/4" diameter, 1-3/4" long	G	1 Carp	150	.053	Ea.	.62	2.83		3.45	5.20
8100	3-1/4" long	G		140	.057		.82	3.04		3.86	5.75
8150	3/8" diameter, 2-1/4" long	G		145	.055		.44	2.93		3.37	5.20
8200	5" long	G		140	.057		.78	3.04		3.82	5.75
8250	1/2" diameter, 2-3/4" long	G		140	.057		.99	3.04		4.03	5.95
8300	7" long	G		125	.064		1.70	3.40		5.10	7.30
8350	5/8" diameter, 3-1/2" long	G		130	.062		2.05	3.27		5.32	7.50
8400	8-1/2" long	G		115	.070		4.36	3.70		8.06	10.70
8450	3/4" diameter, 4-1/4" long	G		115	.070		2.97	3.70		6.67	9.15
8500	10" long	G		95	.084		6.75	4.48		11.23	14.60
8550	1" diameter, 6" long	G		100	.080		8.80	4.25		13.05	16.45
8575	9" long	G		85	.094		11.40	5		16.40	20.50
8600	12" long	G		75	.107		12.30	5.65		17.95	22.50
8650	1-1/4" diameter, 9" long	G		70	.114		23.50	6.05		29.55	35.50
8700	12" long	G		60	.133		30	7.10		37.10	44.50
8750	For type 303 stainless steel, add						350%				
8800	For type 316 stainless steel, add						450%				
8950	Self-drilling concrete screw, hex washer head, 3/16" diam. x 1-3/4" long	G	1 Carp	300	.027	Ea.	.18	1.42		1.60	2.47
8960	2-1/4" long	G		250	.032		.24	1.70		1.94	2.98
8970	Phillips flat head, 3/16" diam. x 1-3/4" long	G		300	.027		.19	1.42		1.61	2.48
8980	2-1/4" long	G		250	.032		.23	1.70		1.93	2.97
9000	Minimum labor/equipment charge			4	2	Job		106		106	170

05 05 21 – Fastening Methods for Metal

05 05 21.10 Cutting Steel		Crew	Daily Output	Labor-Hours	Unit	Material	Labor	Equipment	Total	Total Incl O&P
						2020 Bare Costs				
0010	CUTTING STEEL									
0020	Hand burning, incl. preparation, torch cutting & grinding, no staging									
0050	Steel to 1/4" thick	E-25	400	.020	L.F.	.21	1.19	.03	1.43	2.22
0100	1/2" thick		320	.025		.39	1.49	.04	1.92	2.91
0150	3/4" thick		260	.031		.63	1.83	.05	2.51	3.74
0200	1" thick		200	.040		.90	2.38	.06	3.34	4.96
9000	Minimum labor/equipment charge		2	4	Job		238	6.40	244.40	395

05 05 21.15 Drilling Steel		Crew	Daily Output	Labor-Hours	Unit	Material	Labor	Equipment	Total	Total Incl O&P
0010	DRILLING STEEL									
1910	Drilling & layout for steel, up to 1/4" deep, no anchor									
1920	Holes, 1/4" diameter	1 Sswk	112	.071	Ea.	.10	4.12		4.22	6.85
1925	For each additional 1/4" depth, add		336	.024		.10	1.37		1.47	2.36
1930	3/8" diameter		104	.077		.08	4.43		4.51	7.35
1935	For each additional 1/4" depth, add		312	.026		.08	1.48		1.56	2.51
1940	1/2" diameter		96	.083		.08	4.80		4.88	7.95
1945	For each additional 1/4" depth, add		288	.028		.08	1.60		1.68	2.71
1950	5/8" diameter		88	.091		.16	5.25		5.41	8.75
1955	For each additional 1/4" depth, add		264	.030		.16	1.75		1.91	3.03
1960	3/4" diameter		80	.100		.17	5.75		5.92	9.65
1965	For each additional 1/4" depth, add		240	.033		.17	1.92		2.09	3.33
1970	7/8" diameter		72	.111		.24	6.40		6.64	10.75
1975	For each additional 1/4" depth, add		216	.037		.24	2.14		2.38	3.77
1980	1" diameter		64	.125		.26	7.20		7.46	12.10
1985	For each additional 1/4" depth, add		192	.042		.26	2.40		2.66	4.22
1990	For drilling up, add						40%			
2000	Minimum labor/equipment charge	1 Carp	4	2	Job		106		106	170

05 05 21.90 Welding Steel		Crew	Daily Output	Labor-Hours	Unit	Material	Labor	Equipment	Total	Total Incl O&P
0010	WELDING STEEL, Structural R050521-20									
0020	Field welding, 1/8" E6011, cost per welder, no operating engineer	E-14	8	1	Hr.	5.55	59.50	18.40	83.45	124
0200	With 1/2 operating engineer	E-13	8	1.500		5.55	86	18.40	109.95	165
0300	With 1 operating engineer	E-12	8	2		5.55	112	18.40	135.95	206
0500	With no operating engineer, 2# weld rod per ton	E-14	8	1	Ton	5.55	59.50	18.40	83.45	124
0600	8# E6011 per ton	"	2	4		22	238	73.50	333.50	495
0800	With one operating engineer per welder, 2# E6011 per ton	E-12	8	2		5.55	112	18.40	135.95	206
0900	8# E6011 per ton	"	2	8		22	450	73.50	545.50	825
1200	Continuous fillet, down welding									
1300	Single pass, 1/8" thick, 0.1#/L.F.	E-14	150	.053	L.F.	.28	3.17	.98	4.43	6.60
1400	3/16" thick, 0.2#/L.F.		75	.107		.55	6.35	1.96	8.86	13.15
1500	1/4" thick, 0.3#/L.F.		50	.160		.83	9.50	2.94	13.27	19.75
1610	5/16" thick, 0.4#/L.F.		38	.211		1.11	12.50	3.87	17.48	26
1800	3 passes, 3/8" thick, 0.5#/L.F.		30	.267		1.38	15.85	4.90	22.13	33
2010	4 passes, 1/2" thick, 0.7#/L.F.		22	.364		1.93	21.50	6.70	30.13	45
2200	5 to 6 passes, 3/4" thick, 1.3#/L.F.		12	.667		3.59	39.50	12.25	55.34	82.50
2400	8 to 11 passes, 1" thick, 2.4#/L.F.		6	1.333		6.65	79.50	24.50	110.65	164
2600	For vertical joint welding, add						20%			
2700	Overhead joint welding, add						300%			
2900	For semi-automatic welding, obstructed joints, deduct						5%			
3000	Exposed joints, deduct						15%			
4000	Cleaning and welding plates, bars, or rods									
4010	to existing beams, columns, or trusses	E-14	12	.667	L.F.	1.38	39.50	12.25	53.13	80
9000	Minimum labor/equipment charge	"	4	2	Job		119	37	156	236

05 05 23.10 Bolts and Hex Nuts

Line	05 05 23.10 Bolts and Hex Nuts		Crew	Daily Output	Labor-Hours	Unit	Material	Labor	Equipment	Total	Total Incl O&P
							2020 Bare Costs				
0010	**BOLTS & HEX NUTS**, Steel, A307										
0100	1/4" diameter, 1/2" long	G	1 Sswk	140	.057	Ea.	.06	3.29		3.35	5.45
0200	1" long	G		140	.057		.07	3.29		3.36	5.50
0300	2" long	G		130	.062		.10	3.55		3.65	5.90
0400	3" long	G		130	.062		.15	3.55		3.70	5.95
0500	4" long	G		120	.067		.17	3.84		4.01	6.50
0600	3/8" diameter, 1" long	G		130	.062		.14	3.55		3.69	5.95
0700	2" long	G		130	.062		.17	3.55		3.72	6
0800	3" long	G		120	.067		.23	3.84		4.07	6.55
0900	4" long	G		120	.067		.29	3.84		4.13	6.60
1000	5" long	G		115	.070		.36	4.01		4.37	6.95
1100	1/2" diameter, 1-1/2" long	G		120	.067		.41	3.84		4.25	6.75
1200	2" long	G		120	.067		.48	3.84		4.32	6.80
1300	4" long	G		115	.070		.77	4.01		4.78	7.40
1400	6" long	G		110	.073		1.08	4.19		5.27	8.05
1500	8" long	G		105	.076		1.42	4.39		5.81	8.75
1600	5/8" diameter, 1-1/2" long	G		120	.067		.88	3.84		4.72	7.25
1700	2" long	G		120	.067		.96	3.84		4.80	7.35
1800	4" long	G		115	.070		1.37	4.01		5.38	8.05
1900	6" long	G		110	.073		1.74	4.19		5.93	8.75
2000	8" long	G		105	.076		2.56	4.39		6.95	10
2100	10" long	G		100	.080		3.21	4.61		7.82	11.10
2200	3/4" diameter, 2" long	G		120	.067		1.22	3.84		5.06	7.65
2300	4" long	G		110	.073		1.78	4.19		5.97	8.80
2400	6" long	G		105	.076		2.30	4.39		6.69	9.75
2500	8" long	G		95	.084		3.51	4.85		8.36	11.80
2600	10" long	G		85	.094		4.62	5.45		10.07	14
2700	12" long	G		80	.100		5.45	5.75		11.20	15.40
2800	1" diameter, 3" long	G		105	.076		3.02	4.39		7.41	10.50
2900	6" long	G		90	.089		4.49	5.10		9.59	13.35
3000	12" long	G		75	.107		8.20	6.15		14.35	19.15
3100	For galvanized, add						75%				
3200	For stainless, add						350%				

05 05 23.25 High Strength Bolts

Line	05 05 23.25 High Strength Bolts		Crew	Daily Output	Labor-Hours	Unit	Material	Labor	Equipment	Total	Total Incl O&P
0010	**HIGH STRENGTH BOLTS** R050523-10										
0020	A325 Type 1, structural steel, bolt-nut-washer set										
0100	1/2" diameter x 1-1/2" long	G	1 Sswk	130	.062	Ea.	1.02	3.55		4.57	6.90
0120	2" long	G		125	.064		1.10	3.69		4.79	7.25
0150	3" long	G		120	.067		1.52	3.84		5.36	7.95
0170	5/8" diameter x 1-1/2" long	G		125	.064		2	3.69		5.69	8.25
0180	2" long	G		120	.067		2.16	3.84		6	8.70
0190	3" long	G		115	.070		2.70	4.01		6.71	9.50
0200	3/4" diameter x 2" long	G		120	.067		3.30	3.84		7.14	9.95
0220	3" long	G		115	.070		3.99	4.01		8	10.95
0250	4" long	G		110	.073		4.92	4.19		9.11	12.25
0300	6" long	G		105	.076		6.45	4.39		10.84	14.30
0350	8" long	G		95	.084		12.85	4.85		17.70	22
0360	7/8" diameter x 2" long	G		115	.070		4.08	4.01		8.09	11.05
0365	3" long	G		110	.073		4.82	4.19		9.01	12.15
0370	4" long	G		105	.076		5.80	4.39		10.19	13.60
0380	6" long	G		100	.080		7.35	4.61		11.96	15.65
0390	8" long	G		90	.089		11.70	5.10		16.80	21.50

05 05 Common Work Results for Metals

05 05 23 – Metal Fastenings

05 05 23.25 High Strength Bolts			Crew	Daily Output	Labor-Hours	Unit	Material (2020 Bare Costs)	Labor (2020 Bare Costs)	Equipment (2020 Bare Costs)	Total (2020 Bare Costs)	Total Incl O&P
0400	1" diameter x 2" long	G	1 Sswk	105	.076	Ea.	4.61	4.39		9	12.25
0420	3" long	G		100	.080		5.20	4.61		9.81	13.25
0450	4" long	G		95	.084		5.85	4.85		10.70	14.40
0500	6" long	G		90	.089		7.75	5.10		12.85	16.95
0550	8" long	G		85	.094		13.40	5.45		18.85	23.50
0600	1-1/4" diameter x 3" long	G		85	.094		10.25	5.45		15.70	20
0650	4" long	G		80	.100		11.15	5.75		16.90	21.50
0700	6" long	G		75	.107		14.50	6.15		20.65	26
0750	8" long	G		70	.114		18.45	6.60		25.05	31.50
1020	A490, bolt-nut-washer set										
1170	5/8" diameter x 1-1/2" long	G	1 Sswk	125	.064	Ea.	5.10	3.69		8.79	11.65
1180	2" long	G		120	.067		6	3.84		9.84	12.90
1190	3" long	G		115	.070		7.30	4.01		11.31	14.60
1200	3/4" diameter x 2" long	G		120	.067		5.45	3.84		9.29	12.30
1220	3" long	G		115	.070		6.40	4.01		10.41	13.60
1250	4" long	G		110	.073		7.45	4.19		11.64	15.05
1300	6" long	G		105	.076		10.85	4.39		15.24	19.15
1350	8" long	G		95	.084		18.30	4.85		23.15	28
1360	7/8" diameter x 2" long	G		115	.070		7.75	4.01		11.76	15.10
1365	3" long	G		110	.073		9.05	4.19		13.24	16.80
1370	4" long	G		105	.076		11.10	4.39		15.49	19.40
1380	6" long	G		100	.080		15.45	4.61		20.06	24.50
1390	8" long	G		90	.089		22	5.10		27.10	33
1400	1" diameter x 2" long	G		105	.076		9.40	4.39		13.79	17.55
1420	3" long	G		100	.080		11.15	4.61		15.76	19.85
1450	4" long	G		95	.084		12.70	4.85		17.55	22
1500	6" long	G		90	.089		16.70	5.10		21.80	27
1550	8" long	G		85	.094		26	5.45		31.45	37.50
1600	1-1/4" diameter x 3" long	G		85	.094		50.50	5.45		55.95	64.50
1650	4" long	G		80	.100		58.50	5.75		64.25	73.50
1700	6" long	G		75	.107		80.50	6.15		86.65	99
1750	8" long	G		70	.114		105	6.60		111.60	127

05 05 23.30 Lag Screws			Crew	Daily Output	Labor-Hours	Unit	Material	Labor	Equipment	Total	Total Incl O&P
0010	**LAG SCREWS**										
0020	Steel, 1/4" diameter, 2" long	G	1 Carp	200	.040	Ea.	.10	2.13		2.23	3.52
0100	3/8" diameter, 3" long	G		150	.053		.31	2.83		3.14	4.88
0200	1/2" diameter, 3" long	G		130	.062		.72	3.27		3.99	6.05
0300	5/8" diameter, 3" long	G		120	.067		1.26	3.54		4.80	7.10

05 05 23.35 Machine Screws			Crew	Daily Output	Labor-Hours	Unit	Material	Labor	Equipment	Total	Total Incl O&P
0010	**MACHINE SCREWS**										
0020	Steel, round head, #8 x 1" long	G	1 Carp	4.80	1.667	C	3.98	88.50		92.48	146
0110	#8 x 2" long	G		2.40	3.333		8.80	177		185.80	294
0200	#10 x 1" long	G		4	2		4.89	106		110.89	175
0300	#10 x 2" long	G		2	4		9.80	213		222.80	350

05 05 23.50 Powder Actuated Tools and Fasteners			Crew	Daily Output	Labor-Hours	Unit	Material	Labor	Equipment	Total	Total Incl O&P
0010	**POWDER ACTUATED TOOLS & FASTENERS**										
0020	Stud driver, .22 caliber, single shot					Ea.	159			159	175
0100	.27 caliber, semi automatic, strip					"	460			460	505
0300	Powder load, single shot, .22 cal, power level 2, brown					C	6.25			6.25	6.85
0400	Strip, .27 cal, power level 4, red						10.30			10.30	11.35
0600	Drive pin, .300 x 3/4" long	G	1 Carp	4.80	1.667		4.48	88.50		92.98	147
0700	.300 x 3" long with washer	G	"	4	2		11.95	106		117.95	183

05 05 23 – Metal Fastenings

05 05 23.55 Rivets		Crew	Daily Output	Labor-Hours	Unit	Material	Labor	Equipment	Total	Total Incl O&P
0010	RIVETS									
0100	Aluminum rivet & mandrel, 1/2" grip length x 1/8" diameter G	1 Carp	4.80	1.667	C	8.15	88.50		96.65	151
0200	3/16" diameter G		4	2		11.65	106		117.65	183
0300	Aluminum rivet, steel mandrel, 1/8" diameter G		4.80	1.667		10.30	88.50		98.80	153
0400	3/16" diameter G		4	2		18.35	106		124.35	190
0500	Copper rivet, steel mandrel, 1/8" diameter G		4.80	1.667		10.30	88.50		98.80	153
0800	Stainless rivet & mandrel, 1/8" diameter G		4.80	1.667		25	88.50		113.50	170
0900	3/16" diameter G		4	2		41	106		147	215
1000	Stainless rivet, steel mandrel, 1/8" diameter G		4.80	1.667		16.70	88.50		105.20	160
1100	3/16" diameter G		4	2		26	106		132	199
1200	Steel rivet and mandrel, 1/8" diameter G		4.80	1.667		8.15	88.50		96.65	151
1300	3/16" diameter G		4	2		11	106		117	182
1400	Hand riveting tool, standard				Ea.	78			78	86
1500	Deluxe					415			415	460
1600	Power riveting tool, standard					545			545	600
1700	Deluxe					1,575			1,575	1,725

Column group "2020 Bare Costs" spans Material, Labor, Equipment, Total.

05 05 23.70 Structural Blind Bolts

		Crew	Daily Output	Labor-Hours	Unit	Material	Labor	Equipment	Total	Total Incl O&P
0010	STRUCTURAL BLIND BOLTS									
0100	1/4" diameter x 1/4" grip G	1 Sswk	240	.033	Ea.	1.84	1.92		3.76	5.15
0150	1/2" grip G		216	.037		1.89	2.14		4.03	5.60
0200	3/8" diameter x 1/2" grip G		232	.034		3.50	1.99		5.49	7.10
0250	3/4" grip G		208	.038		3.14	2.22		5.36	7.10
0300	1/2" diameter x 1/2" grip G		224	.036		6.25	2.06		8.31	10.25
0350	3/4" grip G		200	.040		9.10	2.31		11.41	13.80
0400	5/8" diameter x 3/4" grip G		216	.037		11.25	2.14		13.39	15.90
0450	1" grip G		192	.042		15.70	2.40		18.10	21

05 05 23.80 Vibration and Bearing Pads

		Crew	Daily Output	Labor-Hours	Unit	Material	Labor	Equipment	Total	Total Incl O&P
0010	VIBRATION & BEARING PADS									
0300	Laminated synthetic rubber impregnated cotton duck, 1/2" thick	2 Sswk	24	.667	S.F.	74.50	38.50		113	145
0400	1" thick		20	.800		156	46		202	248
0600	Neoprene bearing pads, 1/2" thick		24	.667		30	38.50		68.50	96
0700	1" thick		20	.800		60.50	46		106.50	142
0900	Fabric reinforced neoprene, 5000 psi, 1/2" thick		24	.667		12.30	38.50		50.80	76.50
1000	1" thick		20	.800		24.50	46		70.50	103
1200	Felt surfaced vinyl pads, cork and sisal, 5/8" thick		24	.667		6.35	38.50		44.85	70
1300	1" thick		20	.800		11.40	46		57.40	88
1500	Teflon bonded to 10 ga. carbon steel, 1/32" layer		24	.667		59	38.50		97.50	128
1600	3/32" layer		24	.667		88.50	38.50		127	161
1800	Bonded to 10 ga. stainless steel, 1/32" layer		24	.667		105	38.50		143.50	178
1900	3/32" layer		24	.667		124	38.50		162.50	200
2100	Circular machine leveling pad & stud				Kip	7.55			7.55	8.30

05 05 23.85 Weld Shear Connectors

		Crew	Daily Output	Labor-Hours	Unit	Material	Labor	Equipment	Total	Total Incl O&P
0010	WELD SHEAR CONNECTORS									
0020	3/4" diameter, 3-3/16" long G	E-10	960	.017	Ea.	.56	.97	1.01	2.54	3.32
0030	3-3/8" long G		950	.017		.58	.98	1.02	2.58	3.37
0200	3-7/8" long G		945	.017		.62	.99	1.02	2.63	3.44
0300	4-3/16" long G		935	.017		.65	1	1.03	2.68	3.50
0500	4-7/8" long G		930	.017		.73	1.01	1.04	2.78	3.59
0600	5-3/16" long G		920	.017		.76	1.02	1.05	2.83	3.67
0800	5-3/8" long G		910	.018		.77	1.03	1.06	2.86	3.69
0900	6-3/16" long G		905	.018		.84	1.03	1.07	2.94	3.79

05 05 23 – Metal Fastenings

05 05 23.85 Weld Shear Connectors		Crew	Daily Output	Labor-Hours	Unit	Material	Labor	Equipment	Total	Total Incl O&P
1000	7-3/16" long	G	E-10	895	.018	Ea.	1.05	1.05	1.08	3.18	4.05
1100	8-3/16" long	G		890	.018		1.14	1.05	1.09	3.28	4.18
1500	7/8" diameter, 3-11/16" long	G		920	.017		.91	1.02	1.05	2.98	3.83
1600	4-3/16" long	G		910	.018		.97	1.03	1.06	3.06	3.92
1700	5-3/16" long	G		905	.018		1.10	1.03	1.07	3.20	4.08
1800	6-3/16" long	G		895	.018		1.23	1.05	1.08	3.36	4.25
1900	7-3/16" long	G		890	.018		1.37	1.05	1.09	3.51	4.42
2000	8-3/16" long	G	↓	880	.018	↓	1.49	1.06	1.10	3.65	4.59
9000	Minimum labor/equipment charge		1 Sswk	2	4	Job		231		231	380

05 05 23.87 Weld Studs			Crew	Daily Output	Labor-Hours	Unit	Material	Labor	Equipment	Total	Total Incl O&P
0010	WELD STUDS										
0020	1/4" diameter, 2-11/16" long	G	E-10	1120	.014	Ea.	.37	.84	.86	2.07	2.73
0100	4-1/8" long	G		1080	.015		.35	.87	.90	2.12	2.80
0200	3/8" diameter, 4-1/8" long	G		1080	.015		.40	.87	.90	2.17	2.85
0300	6-1/8" long	G		1040	.015		.52	.90	.93	2.35	3.06
0400	1/2" diameter, 2-1/8" long	G		1040	.015		.37	.90	.93	2.20	2.90
0500	3-1/8" long	G		1025	.016		.45	.91	.94	2.30	3.04
0600	4-1/8" long	G		1010	.016		.53	.93	.96	2.42	3.15
0700	5-5/16" long	G		990	.016		.65	.94	.98	2.57	3.34
0800	6-1/8" long	G		975	.016		.70	.96	.99	2.65	3.44
0900	8-1/8" long	G		960	.017		.99	.97	1.01	2.97	3.80
1000	5/8" diameter, 2-11/16" long	G		1000	.016		.64	.94	.97	2.55	3.30
1010	4-3/16" long	G		990	.016		.80	.94	.98	2.72	3.50
1100	6-9/16" long	G		975	.016		1.04	.96	.99	2.99	3.80
1200	8-3/16" long	G	↓	960	.017	↓	1.39	.97	1.01	3.37	4.24
9000	Minimum labor/equipment charge		1 Sswk	2	4	Job		231		231	380

05 05 23.90 Welding Rod		Crew	Daily Output	Labor-Hours	Unit	Material	Labor	Equipment	Total	Total Incl O&P
0010	WELDING ROD									
0020	Steel, type 6011, 1/8" diam., less than 500#				Lb.	2.76			2.76	3.04
0100	500# to 2,000#					2.49			2.49	2.74
0200	2,000# to 5,000#					2.34			2.34	2.57
0300	5/32" diam., less than 500#					2.46			2.46	2.71
0310	500# to 2,000#					2.22			2.22	2.44
0320	2,000# to 5,000#					2.09			2.09	2.30
0400	3/16" diam., less than 500#					2.75			2.75	3.03
0500	500# to 2,000#					2.48			2.48	2.73
0600	2,000# to 5,000#					2.33			2.33	2.56
0620	Steel, type 6010, 1/8" diam., less than 500#					2.49			2.49	2.74
0630	500# to 2,000#					2.24			2.24	2.46
0640	2,000# to 5,000#					2.11			2.11	2.32
0650	Steel, type 7018 Low Hydrogen, 1/8" diam., less than 500#					2.94			2.94	3.24
0660	500# to 2,000#					2.65			2.65	2.92
0670	2,000# to 5,000#					2.49			2.49	2.74
0700	Steel, type 7024 Jet Weld, 1/8" diam., less than 500#					2.55			2.55	2.81
0710	500# to 2,000#					2.30			2.30	2.53
0720	2,000# to 5,000#					2.16			2.16	2.38
1550	Aluminum, type 4043 TIG, 1/8" diam., less than 10#					5.20			5.20	5.75
1560	10# to 60#					4.70			4.70	5.15
1570	Over 60#					4.42			4.42	4.86
1600	Aluminum, type 5356 TIG, 1/8" diam., less than 10#					5.55			5.55	6.10
1610	10# to 60#					4.99			4.99	5.50
1620	Over 60#				↓	4.69			4.69	5.15

(Material, Labor, Equipment and Total columns: 2020 Bare Costs)

05 05 Common Work Results for Metals

05 05 23 – Metal Fastenings

05 05 23.90 Welding Rod		Crew	Daily Output	Labor-Hours	Unit	2020 Bare Costs Material	Labor	Equipment	Total	Total Incl O&P
1900	Cast iron, type 8 Nickel, 1/8" diam., less than 500#				Lb.	22.50			22.50	24.50
1910	500# to 1,000#					20			20	22
1920	Over 1,000#					18.90			18.90	21
2000	Stainless steel, type 316/316L, 1/8" diam., less than 500#					7.15			7.15	7.85
2100	500# to 1,000#					6.45			6.45	7.10
2220	Over 1,000#					6.05			6.05	6.65

05 12 Structural Steel Framing

05 12 23 – Structural Steel for Buildings

05 12 23.05 Canopy Framing

Line	Description		Crew	Daily Output	Labor-Hours	Unit	Material	Labor	Equipment	Total	Total Incl O&P
0010	**CANOPY FRAMING**										
0020	6" and 8" members, shop fabricated	G	E-4	3000	.011	Lb.	1.72	.62	.05	2.39	2.97
9000	Minimum labor/equipment charge		1 Sswk	1	8	Job		460		460	755

05 12 23.10 Ceiling Supports

Line	Description		Crew	Daily Output	Labor-Hours	Unit	Material	Labor	Equipment	Total	Total Incl O&P
0010	**CEILING SUPPORTS**										
1000	Entrance door/folding partition supports, shop fabricated	G	E-4	60	.533	L.F.	29	31	2.45	62.45	85
1100	Linear accelerator door supports	G		14	2.286		131	133	10.50	274.50	375
1200	Lintels or shelf angles, hung, exterior hot dipped galv.	G		267	.120		19.60	6.95	.55	27.10	33.50
1250	Two coats primer paint instead of galv.	G		267	.120		17	6.95	.55	24.50	30.50
1400	Monitor support, ceiling hung, expansion bolted	G		4	8	Ea.	455	465	36.50	956.50	1,300
1450	Hung from pre-set inserts	G		6	5.333		490	310	24.50	824.50	1,075
1600	Motor supports for overhead doors	G		4	8		232	465	36.50	733.50	1,050
1700	Partition support for heavy folding partitions, without pocket	G		24	1.333	L.F.	65.50	77.50	6.10	149.10	206
1750	Supports at pocket only	G		12	2.667		131	155	12.25	298.25	410
2000	Rolling grilles & fire door supports	G		34	.941		56	54.50	4.32	114.82	156
2100	Spider-leg light supports, expansion bolted to ceiling slab	G		8	4	Ea.	187	233	18.35	438.35	605
2150	Hung from pre-set inserts	G		12	2.667	"	201	155	12.25	368.25	490
2400	Toilet partition support	G		36	.889	L.F.	65.50	51.50	4.08	121.08	161
2500	X-ray travel gantry support	G		12	2.667	"	224	155	12.25	391.25	515

05 12 23.15 Columns, Lightweight

Line	Description		Crew	Daily Output	Labor-Hours	Unit	Material	Labor	Equipment	Total	Total Incl O&P
0010	**COLUMNS, LIGHTWEIGHT**										
1000	Lightweight units (lally), 3-1/2" diameter		E-2	780	.072	L.F.	6.30	4.10	2.17	12.57	16
1050	4" diameter		"	900	.062	"	9.40	3.56	1.88	14.84	18.10
5800	Adjustable jack post, 8' maximum height, 2-3/4" diameter	G				Ea.	52			52	57
5850	4" diameter	G				"	83			83	91.50

05 12 23.17 Columns, Structural

Line	Description		Crew	Daily Output	Labor-Hours	Unit	Material	Labor	Equipment	Total	Total Incl O&P
0010	**COLUMNS, STRUCTURAL**	R051223-20									
0015	Made from recycled materials										
0020	Shop fab'd for 100-ton, 1-2 story project, bolted connections										
0800	Steel, concrete filled, extra strong pipe, 3-1/2" diameter		E-2	660	.085	L.F.	47.50	4.85	2.57	54.92	62.50
0830	4" diameter			780	.072		53	4.10	2.17	59.27	67
0890	5" diameter			1020	.055		63	3.14	1.66	67.80	76
0930	6" diameter			1200	.047		83.50	2.67	1.41	87.58	98
0940	8" diameter			1100	.051		83.50	2.91	1.54	87.95	98.50
1100	For galvanizing, add					Lb.	.29			.29	.31
1300	For web ties, angles, etc., add per added lb.		1 Sswk	945	.008		1.44	.49		1.93	2.38
1500	Steel pipe, extra strong, no concrete, 3" to 5" diameter	G	E-2	16000	.004		1.44	.20	.11	1.75	2.02
1600	6" to 12" diameter	G		14000	.004		1.44	.23	.12	1.79	2.08
1700	Steel pipe, extra strong, no concrete, 3" diameter x 12'-0"	G		60	.933	Ea.	177	53.50	28.50	259	310
1750	4" diameter x 12'-0"	G		58	.966		258	55	29.50	342.50	405

05 12 23.17 Columns, Structural		Crew	Daily Output	Labor-Hours	Unit	Material	Labor	Equipment	Total	Total Incl O&P
						2020 Bare Costs				
1800	6″ diameter x 12′-0″ G	E-2	54	1.037	Ea.	495	59.50	31.50	586	670
1850	8″ diameter x 14′-0″ G		50	1.120		875	64	34	973	1,100
1900	10″ diameter x 16′-0″ G		48	1.167		1,250	66.50	35.50	1,352	1,525
1950	12″ diameter x 18′-0″ G		45	1.244		1,700	71	37.50	1,808.50	2,000
3300	Structural tubing, square, A500GrB, 4″ to 6″ square, light section G		11270	.005	Lb.	1.44	.28	.15	1.87	2.21
3600	Heavy section G		32000	.002	″	1.44	.10	.05	1.59	1.80
4000	Concrete filled, add				L.F.	5.10			5.10	5.60
4500	Structural tubing, square, 4″ x 4″ x 1/4″ x 12′-0″ G	E-2	58	.966	Ea.	237	55	29.50	321.50	385
4550	6″ x 6″ x 1/4″ x 12′-0″ G		54	1.037		390	59.50	31.50	481	555
4600	8″ x 8″ x 3/8″ x 14′-0″ G		50	1.120		840	64	34	938	1,075
4650	10″ x 10″ x 1/2″ x 16′-0″ G		48	1.167		1,550	66.50	35.50	1,652	1,875
5100	Structural tubing, rect., 5″ to 6″ wide, light section G		8000	.007	Lb.	1.44	.40	.21	2.05	2.46
5200	Heavy section G		12000	.005		1.44	.27	.14	1.85	2.17
5300	7″ to 10″ wide, light section G		15000	.004		1.44	.21	.11	1.76	2.04
5400	Heavy section G		18000	.003		1.44	.18	.09	1.71	1.97
5500	Structural tubing, rect., 5″ x 3″ x 1/4″ x 12′-0″ G		58	.966	Ea.	230	55	29.50	314.50	375
5550	6″ x 4″ x 5/16″ x 12′-0″ G		54	1.037		360	59.50	31.50	451	525
5600	8″ x 4″ x 3/8″ x 12′-0″ G		54	1.037		525	59.50	31.50	616	705
5650	10″ x 6″ x 3/8″ x 14′-0″ G		50	1.120		840	64	34	938	1,075
5700	12″ x 8″ x 1/2″ x 16′-0″ G		48	1.167		1,550	66.50	35.50	1,652	1,850
6800	W Shape, A992 steel, 2 tier, W8 x 24 G		1080	.052	L.F.	38	2.96	1.57	42.53	48
6850	W8 x 31 G		1080	.052		49	2.96	1.57	53.53	60.50
6900	W8 x 48 G		1032	.054		76	3.10	1.64	80.74	90.50
6950	W8 x 67 G		984	.057		106	3.25	1.72	110.97	123
7000	W10 x 45 G		1032	.054		71	3.10	1.64	75.74	85
7050	W10 x 68 G		984	.057		107	3.25	1.72	111.97	125
7100	W10 x 112 G		960	.058		177	3.33	1.77	182.10	202
7150	W12 x 50 G		1032	.054		79	3.10	1.64	83.74	94
7200	W12 x 87 G		984	.057		138	3.25	1.72	142.97	158
7250	W12 x 120 G		960	.058		190	3.33	1.77	195.10	216
7300	W12 x 190 G		912	.061		300	3.51	1.86	305.37	340
7350	W14 x 74 G		984	.057		117	3.25	1.72	121.97	136
7400	W14 x 120 G		960	.058		190	3.33	1.77	195.10	216
7450	W14 x 176 G		912	.061		278	3.51	1.86	283.37	315
8090	For projects 75 to 99 tons, add				%	10%				
8092	50 to 74 tons, add					20%				
8094	25 to 49 tons, add					30%	10%			
8096	10 to 24 tons, add					50%	25%			
8098	2 to 9 tons, add					75%	50%			
8099	Less than 2 tons, add					100%	100%			
9000	Minimum labor/equipment charge	1 Sswk	1	8	Job		460		460	755

05 12 23.18 Corner Guards		Crew	Daily Output	Labor-Hours	Unit	Material	Labor	Equipment	Total	Total Incl O&P
0010	**CORNER GUARDS**									
0020	Steel angle w/anchors, 1″ x 1″ x 1/4″, 1.5#/L.F.	2 Carp	160	.100	L.F.	7.95	5.30		13.25	17.20
0100	2″ x 2″ x 1/4″ angles, 3.2#/L.F.		150	.107		10.60	5.65		16.25	21
0200	3″ x 3″ x 5/16″ angles, 6.1#/L.F.		140	.114		17.95	6.05		24	29.50
0300	4″ x 4″ x 5/16″ angles, 8.2#/L.F.		120	.133		18.10	7.10		25.20	31.50
0350	For angles drilled and anchored to masonry, add					15%	120%			
0370	Drilled and anchored to concrete, add					20%	170%			
0400	For galvanized angles, add					35%				
0450	For stainless steel angles, add					100%				
9000	Minimum labor/equipment charge	1 Carp	2	4	Job		213		213	340

05 12 23.20 Curb Edging

	05 12 23.20 Curb Edging		Crew	Daily Output	Labor-Hours	Unit	Material	2020 Bare Costs Labor	Equipment	Total	Total Incl O&P
0010	**CURB EDGING**										
0020	Steel angle w/anchors, shop fabricated, on forms, 1" x 1", 0.8#/L.F.	G	E-4	350	.091	L.F.	1.74	5.30	.42	7.46	11.05
0100	2" x 2" angles, 3.92#/L.F.	G		330	.097		7.10	5.65	.45	13.20	17.60
0200	3" x 3" angles, 6.1#/L.F.	G		300	.107		11.70	6.20	.49	18.39	23.50
0300	4" x 4" angles, 8.2#/L.F.	G		275	.116		15.30	6.75	.53	22.58	28.50
1000	6" x 4" angles, 12.3#/L.F.	G		250	.128		22.50	7.45	.59	30.54	37.50
1050	Steel channels with anchors, on forms, 3" channel, 5#/L.F.	G		290	.110		9	6.40	.51	15.91	21
1100	4" channel, 5.4#/L.F.	G		270	.119		9.65	6.90	.54	17.09	22.50
1200	6" channel, 8.2#/L.F.	G		255	.125		15.30	7.30	.58	23.18	29.50
1300	8" channel, 11.5#/L.F.	G		225	.142		21	8.25	.65	29.90	37.50
1400	10" channel, 15.3#/L.F.	G		180	.178		27.50	10.35	.82	38.67	48.50
1500	12" channel, 20.7#/L.F.	G		140	.229		37	13.30	1.05	51.35	63.50
2000	For curved edging, add						35%	10%			
9000	Minimum labor/equipment charge		E-4	4	8	Job		465	36.50	501.50	800

05 12 23.40 Lightweight Framing

	05 12 23.40 Lightweight Framing		Crew	Daily Output	Labor-Hours	Unit	Material	Labor	Equipment	Total	Total Incl O&P
0010	**LIGHTWEIGHT FRAMING** R051223-35										
0015	Made from recycled materials										
0200	For load-bearing steel studs see Section 05 41 13.30										
0400	Angle framing, field fabricated, 4" and larger R051223-45	G	E-3	440	.055	Lb.	.83	3.18	.33	4.34	6.50
0450	Less than 4" angles	G		265	.091	"	.86	5.30	.56	6.72	10.20
0460	1/2" x 1/2" x 1/8"	G		200	.120	L.F.	.17	7	.74	7.91	12.45
0462	3/4" x 3/4" x 1/8"	G		160	.150		.48	8.75	.92	10.15	15.85
0464	1" x 1" x 1/8"	G		135	.178		.69	10.35	1.09	12.13	18.90
0466	1-1/4" x 1-1/4" x 3/16"	G		115	.209		1.28	12.15	1.28	14.71	22.50
0468	1-1/2" x 1-1/2" x 3/16"	G		100	.240		1.55	14	1.47	17.02	26.50
0470	2" x 2" x 1/4"	G		90	.267		2.75	15.55	1.63	19.93	30.50
0472	2-1/2" x 2-1/2" x 1/4"	G		72	.333		3.53	19.40	2.04	24.97	38
0474	3" x 2" x 3/8"	G		65	.369		5.10	21.50	2.26	28.86	43
0476	3" x 3" x 3/8"	G		57	.421		6.20	24.50	2.58	33.28	49.50
0600	Channel framing, field fabricated, 8" and larger	G		500	.048	Lb.	.86	2.80	.29	3.95	5.85
0650	Less than 8" channels	G		335	.072	"	.86	4.17	.44	5.47	8.30
0660	C2 x 1.78	G		115	.209	L.F.	1.53	12.15	1.28	14.96	23
0662	C3 x 4.1	G		80	.300		3.53	17.50	1.84	22.87	34.50
0664	C4 x 5.4	G		66	.364		4.66	21	2.23	27.89	42
0666	C5 x 6.7	G		57	.421		5.80	24.50	2.58	32.88	49
0668	C6 x 8.2	G		55	.436		6.85	25.50	2.67	35.02	52
0670	C7 x 9.8	G		40	.600		8.45	35	3.68	47.13	71
0672	C8 x 11.5	G		36	.667		9.90	39	4.09	52.99	79
0710	Structural bar tee, field fabricated, 3/4" x 3/4" x 1/8"	G		160	.150		.48	8.75	.92	10.15	15.85
0712	1" x 1" x 1/8"	G		135	.178		.69	10.35	1.09	12.13	18.90
0714	1-1/2" x 1-1/2" x 1/4"	G		114	.211		2.02	12.25	1.29	15.56	23.50
0716	2" x 2" x 1/4"	G		89	.270		2.75	15.70	1.65	20.10	30.50
0718	2-1/2" x 2-1/2" x 3/8"	G		72	.333		5.10	19.40	2.04	26.54	40
0720	3" x 3" x 3/8"	G		57	.421		6.20	24.50	2.58	33.28	49.50
0730	Structural zee, field fabricated, 1-1/4" x 1-3/4" x 1-3/4"	G		114	.211		.66	12.25	1.29	14.20	22
0732	2-11/16" x 3" x 2-11/16"	G		114	.211		1.53	12.25	1.29	15.07	23
0734	3-1/16" x 4" x 3-1/16"	G		133	.180		2.32	10.50	1.11	13.93	21
0736	3-1/4" x 5" x 3-1/4"	G		133	.180		3.16	10.50	1.11	14.77	22
0738	3-1/2" x 6" x 3-1/2"	G		160	.150		4.77	8.75	.92	14.44	20.50
0740	Junior beam, field fabricated, 3"	G		80	.300		4.91	17.50	1.84	24.25	36
0742	4"	G		72	.333		6.65	19.40	2.04	28.09	41.50
0744	5"	G		67	.358		8.60	21	2.20	31.80	46

05 12 23 – Structural Steel for Buildings

05 12 23.40 Lightweight Framing

	05 12 23.40 Lightweight Framing		Crew	Daily Output	Labor-Hours	Unit	Material	Labor	Equipment	Total	Total Incl O&P
							2020 Bare Costs				
0746	6"	G	E-3	62	.387	L.F.	10.80	22.50	2.37	35.67	51.50
0748	7"	G		57	.421		13.20	24.50	2.58	40.28	57.50
0750	8"	G		53	.453		15.85	26.50	2.78	45.13	63.50
1000	Continuous slotted channel framing system, shop fab, simple framing	G	2 Sswk	2400	.007	Lb.	4.45	.38		4.83	5.55
1200	Complex framing	G	"	1600	.010		5.05	.58		5.63	6.50
1250	Plate & bar stock for reinforcing beams and trusses	G					1.58			1.58	1.74
1300	Cross bracing, rods, shop fabricated, 3/4" diameter	G	E-3	700	.034		1.72	2	.21	3.93	5.40
1310	7/8" diameter	G		850	.028		1.72	1.65	.17	3.54	4.79
1320	1" diameter	G		1000	.024		1.72	1.40	.15	3.27	4.35
1330	Angle, 5" x 5" x 3/8"	G		2800	.009		1.72	.50	.05	2.27	2.78
1350	Hanging lintels, shop fabricated	G		850	.028		1.72	1.65	.17	3.54	4.79
1380	Roof frames, shop fabricated, 3'-0" square, 5' span	G	E-2	4200	.013		1.72	.76	.40	2.88	3.57
1400	Tie rod, not upset, 1-1/2" to 4" diameter, with turnbuckle	G	2 Sswk	800	.020		1.87	1.15		3.02	3.94
1420	No turnbuckle	G		700	.023		1.80	1.32		3.12	4.14
1500	Upset, 1-3/4" to 4" diameter, with turnbuckle	G		800	.020		1.87	1.15		3.02	3.94
1520	No turnbuckle	G		700	.023		1.80	1.32		3.12	4.14
9000	Minimum labor/equipment charge			2	8	Job		460		460	755

05 12 23.45 Lintels

	05 12 23.45 Lintels		Crew	Daily Output	Labor-Hours	Unit	Material	Labor	Equipment	Total	Total Incl O&P
0010	**LINTELS**										
0015	Made from recycled materials										
0020	Plain steel angles, shop fabricated, under 500 lb.	G	1 Bric	550	.015	Lb.	1.11	.76		1.87	2.44
0100	500 to 1,000 lb.	G		640	.013		1.08	.65		1.73	2.24
0200	1,000 to 2,000 lb.	G		640	.013		1.05	.65		1.70	2.20
0300	2,000 to 4,000 lb.	G		640	.013		1.02	.65		1.67	2.17
0500	For built-up angles and plates, add to above	G					1.44			1.44	1.58
0700	For engineering, add to above						.14			.14	.16
0900	For galvanizing, add to above, under 500 lb.						.32			.32	.35
0950	500 to 2,000 lb.						.29			.29	.32
1000	Over 2,000 lb.						.29			.29	.31
2000	Steel angles, 3-1/2" x 3", 1/4" thick, 2'-6" long	G	1 Bric	47	.170	Ea.	15.50	8.85		24.35	31.50
2100	4'-6" long	G		26	.308		28	16		44	56.50
2500	3-1/2" x 3-1/2" x 5/16", 5'-0" long	G		18	.444		41.50	23		64.50	83
2600	4" x 3-1/2", 1/4" thick, 5'-0" long	G		21	.381		35.50	19.85		55.35	71
2700	9'-0" long	G		12	.667		64	34.50		98.50	127
2800	4" x 3-1/2" x 5/16", 7'-0" long	G		12	.667		62	34.50		96.50	124
2900	5" x 3-1/2" x 5/16", 10'-0" long	G		8	1		100	52		152	194
9000	Minimum labor/equipment charge			4	2	Job		104		104	168

05 12 23.60 Pipe Support Framing

	05 12 23.60 Pipe Support Framing		Crew	Daily Output	Labor-Hours	Unit	Material	Labor	Equipment	Total	Total Incl O&P
0010	**PIPE SUPPORT FRAMING**										
0020	Under 10#/L.F., shop fabricated	G	E-4	3900	.008	Lb.	1.93	.48	.04	2.45	2.94
0200	10.1 to 15#/L.F.	G		4300	.007		1.90	.43	.03	2.36	2.84
0400	15.1 to 20#/L.F.	G		4800	.007		1.87	.39	.03	2.29	2.72
0600	Over 20#/L.F.	G		5400	.006		1.84	.34	.03	2.21	2.62

05 12 23.65 Plates

	05 12 23.65 Plates		Crew	Daily Output	Labor-Hours	Unit	Material	Labor	Equipment	Total	Total Incl O&P
0010	**PLATES**	R051223-80									
0015	Made from recycled materials										
0020	For connections & stiffener plates, shop fabricated										
0050	1/8" thick (5.1 lb./S.F.)	G				S.F.	7.35			7.35	8.05
0100	1/4" thick (10.2 lb./S.F.)	G					14.65			14.65	16.10
0300	3/8" thick (15.3 lb./S.F.)	G					22			22	24
0400	1/2" thick (20.4 lb./S.F.)	G					29.50			29.50	32
0450	3/4" thick (30.6 lb./S.F.)	G					44			44	48.50

05 12 23.65 Plates			Crew	Daily Output	Labor-Hours	Unit	2020 Bare Costs Material	Labor	Equipment	Total	Total Incl O&P
0500	1" thick (40.8 lb./S.F.)	G				S.F.	58.50			58.50	64.50
2000	Steel plate, warehouse prices, no shop fabrication										
2100	1/4" thick (10.2 lb./S.F.)	G				S.F.	7.20			7.20	7.90

05 12 23.70 Stressed Skin Steel Roof and Ceiling System

			Crew	Daily Output	Labor-Hours	Unit	Material	Labor	Equipment	Total	Total Incl O&P
0010	**STRESSED SKIN STEEL ROOF & CEILING SYSTEM**										
0020	Double panel flat roof, spans to 100'	G	E-2	1150	.049	S.F.	11.50	2.78	1.48	15.76	18.75
0100	Double panel convex roof, spans to 200'	G		960	.058		18.70	3.33	1.77	23.80	28
0200	Double panel arched roof, spans to 300'	G		760	.074		28.50	4.21	2.23	34.94	41

05 12 23.75 Structural Steel Members

			Crew	Daily Output	Labor-Hours	Unit	Material	Labor	Equipment	Total	Total Incl O&P
0010	**STRUCTURAL STEEL MEMBERS** R051223-10										
0015	Made from recycled materials										
0020	Shop fab'd for 100-ton, 1-2 story project, bolted connections										
0102	Beam or girder, W 6 x 9 R051223-15	G	E-2	600	.093	L.F.	14.25	5.35	2.83	22.43	27.50
0302	W 8 x 10	G		600	.093		15.80	5.35	2.83	23.98	29
0502	x 31	G		550	.102		49	5.80	3.08	57.88	67
0702	W 10 x 22	G		600	.093		35	5.35	2.83	43.18	50.50
0902	x 49	G		550	.102		77.50	5.80	3.08	86.38	98
1102	W 12 x 16	G		880	.064		25.50	3.64	1.93	31.07	36
1302	x 22	G		880	.064		35	3.64	1.93	40.57	46.50
1502	x 26	G		880	.064		41	3.64	1.93	46.57	53
1702	x 72	G		640	.088		114	5	2.65	121.65	136
1902	W 14 x 26	G		990	.057		41	3.23	1.71	45.94	52
2102	x 30	G		900	.062		47.50	3.56	1.88	52.94	60
2302	x 34	G		810	.069		53.50	3.95	2.09	59.54	67.50
2502	x 120	G		720	.078		190	4.44	2.36	196.80	219
2702	W 16 x 26	G		1000	.056		41	3.20	1.70	45.90	52
2902	x 31	G		900	.062		49	3.56	1.88	54.44	62
3102	x 40	G		800	.070		63	4	2.12	69.12	78.50
3302	W 18 x 35	G	E-5	960	.083		55.50	4.79	1.92	62.21	71
3502	x 40	G		960	.083		63	4.79	1.92	69.71	79.50
3702	x 50	G		912	.088		79	5.05	2.02	86.07	97.50
3902	x 55	G		912	.088		87	5.05	2.02	94.07	106
4102	W 21 x 44	G		1064	.075		69.50	4.32	1.73	75.55	85.50
4302	x 50	G		1064	.075		79	4.32	1.73	85.05	96
4502	x 62	G		1036	.077		98	4.44	1.78	104.22	117
4702	x 68	G		1036	.077		107	4.44	1.78	113.22	127
4902	W 24 x 55	G		1110	.072		87	4.14	1.66	92.80	104
5102	x 62	G		1110	.072		98	4.14	1.66	103.80	117
5302	x 68	G		1110	.072		107	4.14	1.66	112.80	127
5502	x 76	G		1110	.072		120	4.14	1.66	125.80	141
5702	x 84	G		1080	.074		133	4.26	1.71	138.97	155
5902	W 27 x 94	G		1190	.067		149	3.86	1.55	154.41	171
6102	W 30 x 99	G		1200	.067		156	3.83	1.54	161.37	180
6302	x 108	G		1200	.067		171	3.83	1.54	176.37	196
6502	x 116	G		1160	.069		183	3.96	1.59	188.55	210
6702	W 33 x 118	G		1176	.068		187	3.91	1.57	192.48	213
6902	x 130	G		1134	.071		205	4.06	1.63	210.69	234
7102	x 141	G		1134	.071		223	4.06	1.63	228.69	253
7302	W 36 x 135	G		1170	.068		213	3.93	1.58	218.51	243
7502	x 150	G		1170	.068		237	3.93	1.58	242.51	269
7702	x 194	G		1125	.071		305	4.09	1.64	310.73	345
7902	x 231	G		1125	.071		365	4.09	1.64	370.73	410

05 12 23.75 Structural Steel Members		Crew	Daily Output	Labor-Hours	Unit	Material	2020 Bare Costs Labor	Equipment	Total	Total Incl O&P
8102	x 302 G	E-5	1035	.077	L.F.	475	4.44	1.78	481.22	535
8490	For projects 75 to 99 tons, add					10%				
8492	50 to 74 tons, add					20%				
8494	25 to 49 tons, add					30%	10%			
8496	10 to 24 tons, add					50%	25%			
8498	2 to 9 tons, add					75%	50%			
8499	Less than 2 tons, add					100%	100%			
9000	Minimum labor/equipment charge	E-2	2	28	Job		1,600	850	2,450	3,525

05 12 23.77 Structural Steel Projects		Crew	Daily Output	Labor-Hours	Unit	Material	Labor	Equipment	Total	Total Incl O&P
0010	STRUCTURAL STEEL PROJECTS R050516-30									
0015	Made from recycled materials									
0020	Shop fab'd for 100-ton, 1-2 story project, bolted connections									
0200	Apartments, nursing homes, etc., 1 to 2 stories R050523-10 G	E-5	10.30	7.767	Ton	2,875	445	179	3,499	4,075
0300	3 to 6 stories G	"	10.10	7.921		2,925	455	183	3,563	4,175
0400	7 to 15 stories R051223-10 G	E-6	14.20	9.014		3,000	515	151	3,666	4,300
0500	Over 15 stories G	"	13.90	9.209		3,100	530	155	3,785	4,450
0700	Offices, hospitals, etc., steel bearing, 1 to 2 stories R051223-15 G	E-5	10.30	7.767		2,875	445	179	3,499	4,075
0800	3 to 6 stories G	E-6	14.40	8.889		2,925	510	149	3,584	4,225
0900	7 to 15 stories R051223-20 G		14.20	9.014		3,000	515	151	3,666	4,300
1000	Over 15 stories G		13.90	9.209		3,100	530	155	3,785	4,450
1100	For multi-story masonry wall bearing construction, add R051223-25						30%			
1300	Industrial bldgs., 1 story, beams & girders, steel bearing G	E-5	12.90	6.202		2,875	355	143	3,373	3,875
1400	Masonry bearing G	"	10	8		2,875	460	184	3,519	4,100
1500	Industrial bldgs., 1 story, under 10 tons,									
1510	steel from warehouse, trucked G	E-2	7.50	7.467	Ton	3,450	425	226	4,101	4,750
1600	1 story with roof trusses, steel bearing G	E-5	10.60	7.547		3,400	435	174	4,009	4,625
1700	Masonry bearing G	"	8.30	9.639		3,400	555	222	4,177	4,875
1900	Monumental structures, banks, stores, etc., simple connections G	E-6	13	9.846		2,875	565	165	3,605	4,250
2000	Moment/composite connections G	"	9	14.222		4,775	815	239	5,829	6,850
2200	Churches, simple connections G	E-5	11.60	6.897		2,675	395	159	3,229	3,775
2300	Moment/composite connections G	"	5.20	15.385		3,575	885	355	4,815	5,750
2800	Power stations, fossil fuels, simple connections G	E-6	11	11.636		2,875	670	195	3,740	4,450
2900	Moment/composite connections G		5.70	22.456		4,300	1,300	375	5,975	7,275
2950	Nuclear fuels, non-safety steel, simple connections G		7	18.286		2,875	1,050	305	4,230	5,200
3000	Moment/composite connections G		5.50	23.273		4,300	1,325	390	6,015	7,350
3040	Safety steel, simple connections G		2.50	51.200		4,200	2,950	860	8,010	10,300
3070	Moment/composite connections G		1.50	85.333		5,525	4,900	1,425	11,850	15,600
3100	Roof trusses, simple connections G	E-5	13	6.154		4,025	355	142	4,522	5,150
3200	Moment/composite connections G		8.30	9.639		4,875	555	222	5,652	6,525
3210	Schools, simple connections G		14.50	5.517		2,875	315	127	3,317	3,800
3220	Moment/composite connections G		8.30	9.639		4,200	555	222	4,977	5,775
3400	Welded construction, simple commercial bldgs., 1 to 2 stories G	E-7	7.60	10.526		2,925	605	262	3,792	4,500
3500	7 to 15 stories G	E-9	8.30	15.422		3,400	885	293	4,578	5,475
3700	Welded rigid frame, 1 story, simple connections G	E-7	15.80	5.063		3,000	291	126	3,417	3,900
3800	Moment/composite connections G	"	5.50	14.545		3,875	835	360	5,070	6,025
3900	High strength steel mill spec extras:									
3950	A529, A572 (50 ksi) and A36: same as A992 steel (no extra)									
4000	Add to A992 price for A572 (60, 65 ksi) G				Ton	87.50			87.50	96.50
4300	Column base plates, light, up to 150 lb. G	2 Sswk	2000	.008	Lb.	1.58	.46		2.04	2.50
4400	Heavy, over 150 lb. G	E-2	7500	.007	"	1.65	.43	.23	2.31	2.76
4600	Castellated beams, light sections, to 50#/L.F., simple connections G		10.70	5.234	Ton	3,025	299	159	3,483	3,975
4700	Moment/composite connections G		7	8		3,300	455	242	3,997	4,625

05 12 23 – Structural Steel for Buildings

05 12 23.77 Structural Steel Projects		Crew	Daily Output	Labor-Hours	Unit	Material	2020 Bare Costs Labor	Equipment	Total	Total Incl O&P
4900	Heavy sections, over 50 plf, simple connections G	E-2	11.70	4.786	Ton	3,150	273	145	3,568	4,075
5000	Moment/composite connections G	↓	7.80	7.179		3,450	410	217	4,077	4,700
5390	For projects 75 to 99 tons, add					10%				
5392	50 to 74 tons, add					20%				
5394	25 to 49 tons, add					30%	10%			
5396	10 to 24 tons, add					50%	25%			
5398	2 to 9 tons, add					75%	50%			
5399	Less than 2 tons, add				↓	100%	100%			

05 12 23.78 Structural Steel Secondary Members		Crew	Daily Output	Labor-Hours	Unit	Material	Labor	Equipment	Total	Total Incl O&P
0010	**STRUCTURAL STEEL SECONDARY MEMBERS**									
0015	Made from recycled materials									
0020	Shop fabricated for 20-ton girt/purlin framing package, materials only									
0100	Girts/purlins, C/Z-shapes, includes clips and bolts									
0110	6" x 2-1/2" x 2-1/2", 16 ga., 3.0 lb./L.F.				L.F.	3.88			3.88	4.27
0115	14 ga., 3.5 lb./L.F.					4.53			4.53	4.98
0120	8" x 2-3/4" x 2-3/4", 16 ga., 3.4 lb./L.F.					4.40			4.40	4.84
0125	14 ga., 4.1 lb./L.F.					5.30			5.30	5.85
0130	12 ga., 5.6 lb./L.F.					7.25			7.25	7.95
0135	10" x 3-1/2" x 3-1/2", 14 ga., 4.7 lb./L.F.					6.10			6.10	6.70
0140	12 ga., 6.7 lb./L.F.					8.65			8.65	9.55
0145	12" x 3-1/2" x 3-1/2", 14 ga., 5.3 lb./L.F.					6.85			6.85	7.55
0150	12 ga., 7.4 lb./L.F.				↓	9.55			9.55	10.55
0200	Eave struts, C-shape, includes clips and bolts									
0210	6" x 4" x 3", 16 ga., 3.1 lb./L.F.				L.F.	4.01			4.01	4.41
0215	14 ga., 3.9 lb./L.F.					5.05			5.05	5.55
0220	8" x 4" x 3", 16 ga., 3.5 lb./L.F.					4.53			4.53	4.98
0225	14 ga., 4.4 lb./L.F.					5.70			5.70	6.25
0230	12 ga., 6.2 lb./L.F.					8			8	8.80
0235	10" x 5" x 3", 14 ga., 5.2 lb./L.F.					6.70			6.70	7.40
0240	12 ga., 7.3 lb./L.F.					9.45			9.45	10.40
0245	12" x 5" x 4", 14 ga., 6.0 lb./L.F.					7.75			7.75	8.55
0250	12 ga., 8.4 lb./L.F.				↓	10.85			10.85	11.95
0300	Rake/base angle, excludes concrete drilling and expansion anchors									
0310	2" x 2", 14 ga., 1.0 lb./L.F.	2 Sswk	640	.025	L.F.	1.29	1.44		2.73	3.78
0315	3" x 2", 14 ga., 1.3 lb./L.F.		535	.030		1.68	1.72		3.40	4.68
0320	3" x 3", 14 ga., 1.6 lb./L.F.		500	.032		2.07	1.84		3.91	5.30
0325	4" x 3", 14 ga., 1.8 lb./L.F.	↓	480	.033	↓	2.33	1.92		4.25	5.70
0600	Installation of secondary members, erection only									
0610	Girts, purlins, eave struts, 16 ga., 6" deep	E-18	100	.400	Ea.		23	14.90	37.90	54
0615	8" deep		80	.500			29	18.60	47.60	67.50
0620	14 ga., 6" deep		80	.500			29	18.60	47.60	67.50
0625	8" deep		65	.615			35.50	23	58.50	83
0630	10" deep		55	.727			42	27	69	98.50
0635	12" deep		50	.800			46.50	30	76.50	108
0640	12 ga., 8" deep		50	.800			46.50	30	76.50	108
0645	10" deep		45	.889			51.50	33	84.50	120
0650	12" deep	↓	40	1	↓		58	37	95	135
0900	For less than 20-ton job lots									
0905	For 15 to 19 tons, add				%	10%				
0910	For 10 to 14 tons, add					25%				
0915	For 5 to 9 tons, add					50%	50%	50%		
0920	For 1 to 4 tons, add				↓	75%	75%	75%		

05 12 Structural Steel Framing

05 12 23 – Structural Steel for Buildings

05 12 23.78 Structural Steel Secondary Members		Crew	Daily Output	Labor-Hours	Unit	2020 Bare Costs Material	Labor	Equipment	Total	Total Incl O&P
0925	For less than 1 ton, add				%	100%	100%	100%		
0990	Minimum labor/equipment charge	E-18	1	40	Job		2,325	1,500	3,825	5,375

05 12 23.80 Subpurlins

05 12 23.80 Subpurlins		Crew	Daily Output	Labor-Hours	Unit	Material	Labor	Equipment	Total	Total Incl O&P
0010	SUBPURLINS R051223-50									
0015	Made from recycled materials									
0020	Bulb tees, shop fabricated, painted, 32-5/8" OC, 40 psf L.L.									
0200	Type 218, max 10'-2" span, 3.19 plf, 2-1/8" high x 2-1/8" wide G	E-1	3100	.008	S.F.	1.79	.44	.05	2.28	2.73
1420	For 24-5/8" spacing, add					33%	33%			
1430	For 48-5/8" spacing, deduct					33%	33%			

05 14 Structural Aluminum Framing

05 14 23 – Non-Exposed Structural Aluminum Framing

05 14 23.05 Aluminum Shapes

05 14 23.05 Aluminum Shapes		Crew	Daily Output	Labor-Hours	Unit	Material	Labor	Equipment	Total	Total Incl O&P
0010	ALUMINUM SHAPES									
0015	Made from recycled materials									
0020	Structural shapes, 1" to 10" members, under 1 ton G	E-2	4000	.014	Lb.	4.46	.80	.42	5.68	6.65
0050	1 to 5 tons G		4300	.013		3.77	.74	.39	4.90	5.80
0100	Over 5 tons G		4600	.012		3.58	.70	.37	4.65	5.50
0300	Extrusions, over 5 tons, stock shapes G		1330	.042		3.39	2.41	1.28	7.08	9
0400	Custom shapes G		1330	.042		4.35	2.41	1.28	8.04	10.10

05 15 Wire Rope Assemblies

05 15 16 – Steel Wire Rope Assemblies

05 15 16.05 Accessories for Steel Wire Rope

05 15 16.05 Accessories for Steel Wire Rope		Crew	Daily Output	Labor-Hours	Unit	Material	Labor	Equipment	Total	Total Incl O&P
0010	ACCESSORIES FOR STEEL WIRE ROPE									
0015	Made from recycled materials									
1500	Thimbles, heavy duty, 1/4" G	E-17	160	.100	Ea.	.51	5.85		6.36	10.15
1510	1/2" G		160	.100		2.24	5.85		8.09	12.05
1520	3/4" G		105	.152		5.10	8.95		14.05	20.50
1530	1" G		52	.308		10.15	18.05		28.20	40.50
1540	1-1/4" G		38	.421		15.65	24.50		40.15	58
1550	1-1/2" G		13	1.231		44	72		116	167
1560	1-3/4" G		8	2		91	117		208	292
1570	2" G		6	2.667		132	156		288	400
1580	2-1/4" G		4	4		179	235		414	580
1600	Clips, 1/4" diameter G		160	.100		1.81	5.85		7.66	11.60
1610	3/8" diameter G		160	.100		1.98	5.85		7.83	11.80
1620	1/2" diameter G		160	.100		3.19	5.85		9.04	13.10
1630	3/4" diameter G		102	.157		5.20	9.20		14.40	21
1640	1" diameter G		64	.250		8.65	14.65		23.30	33.50
1650	1-1/4" diameter G		35	.457		14.15	27		41.15	59.50
1670	1-1/2" diameter G		26	.615		19.05	36		55.05	80
1680	1-3/4" diameter G		16	1		44.50	58.50		103	145
1690	2" diameter G		12	1.333		49.50	78		127.50	183
1700	2-1/4" diameter G		10	1.600		73	94		167	234
1800	Sockets, open swage, 1/4" diameter G		160	.100		31.50	5.85		37.35	44.50
1810	1/2" diameter G		77	.208		45.50	12.20		57.70	70
1820	3/4" diameter G		19	.842		71	49.50		120.50	159
1830	1" diameter G		9	1.778		127	104		231	310

05 15 Wire Rope Assemblies

05 15 16 – Steel Wire Rope Assemblies

05 15 16.05 Accessories for Steel Wire Rope			Crew	Daily Output	Labor-Hours	Unit	2020 Bare Costs Material	Labor	Equipment	Total	Total Incl O&P
1840	1-1/4" diameter	G	E-17	5	3.200	Ea.	176	188		364	505
1850	1-1/2" diameter	G		3	5.333		385	315		700	940
1860	1-3/4" diameter	G		3	5.333		685	315		1,000	1,275
1870	2" diameter	G		1.50	10.667		1,025	625		1,650	2,175
1900	Closed swage, 1/4" diameter	G		160	.100		19	5.85		24.85	30.50
1910	1/2" diameter	G		104	.154		33	9		42	51
1920	3/4" diameter	G		32	.500		49	29.50		78.50	102
1930	1" diameter	G		15	1.067		86.50	62.50		149	198
1940	1-1/4" diameter	G		7	2.286		129	134		263	360
1950	1-1/2" diameter	G		4	4		235	235		470	645
1960	1-3/4" diameter	G		3	5.333		345	315		660	895
1970	2" diameter	G		2	8		675	470		1,145	1,500
2000	Open spelter, galv., 1/4" diameter	G		160	.100		42	5.85		47.85	55.50
2010	1/2" diameter	G		70	.229		43.50	13.40		56.90	70
2020	3/4" diameter	G		26	.615		65.50	36		101.50	131
2030	1" diameter	G		10	1.600		182	94		276	355
2040	1-1/4" diameter	G		5	3.200		260	188		448	595
2050	1-1/2" diameter	G		4	4		550	235		785	990
2060	1-3/4" diameter	G		2	8		960	470		1,430	1,825
2070	2" diameter	G		1.20	13.333		1,100	780		1,880	2,500
2080	2-1/2" diameter	G		1	16		2,025	940		2,965	3,800
2100	Closed spelter, galv., 1/4" diameter	G		160	.100		33	5.85		38.85	45.50
2110	1/2" diameter	G		88	.182		35	10.65		45.65	56
2120	3/4" diameter	G		30	.533		53	31.50		84.50	110
2130	1" diameter	G		13	1.231		113	72		185	243
2140	1-1/4" diameter	G		7	2.286		181	134		315	420
2150	1-1/2" diameter	G		6	2.667		390	156		546	685
2160	1-3/4" diameter	G		2.80	5.714		520	335		855	1,125
2170	2" diameter	G		2	8		640	470		1,110	1,475
2200	Jaw & jaw turnbuckles, 1/4" x 4"	G		160	.100		9.65	5.85		15.50	20
2250	1/2" x 6"	G		96	.167		12.20	9.80		22	29.50
2260	1/2" x 9"	G		77	.208		16.25	12.20		28.45	38
2270	1/2" x 12"	G		66	.242		18.30	14.20		32.50	43.50
2300	3/4" x 6"	G		38	.421		24	24.50		48.50	67
2310	3/4" x 9"	G		30	.533		26.50	31.50		58	80.50
2320	3/4" x 12"	G		28	.571		34	33.50		67.50	92.50
2330	3/4" x 18"	G		23	.696		40.50	41		81.50	112
2350	1" x 6"	G		17	.941		46	55		101	142
2360	1" x 12"	G		13	1.231		51	72		123	174
2370	1" x 18"	G		10	1.600		76	94		170	238
2380	1" x 24"	G		9	1.778		84	104		188	263
2400	1-1/4" x 12"	G		7	2.286		85.50	134		219.50	315
2410	1-1/4" x 18"	G		6.50	2.462		106	144		250	355
2420	1-1/4" x 24"	G		5.60	2.857		142	168		310	430
2450	1-1/2" x 12"	G		5.20	3.077		238	180		418	560
2460	1-1/2" x 18"	G		4	4		254	235		489	665
2470	1-1/2" x 24"	G		3.20	5		340	293		633	855
2500	1-3/4" x 18"	G		3.20	5		515	293		808	1,050
2510	1-3/4" x 24"	G		2.80	5.714		585	335		920	1,200
2550	2" x 24"	G		1.60	10		795	585		1,380	1,825

05 15 Wire Rope Assemblies

05 15 16 – Steel Wire Rope Assemblies

05 15 16.50 Steel Wire Rope			Crew	Daily Output	Labor-Hours	Unit	2020 Bare Costs Material	Labor	Equipment	Total	Total Incl O&P
0010	**STEEL WIRE ROPE**										
0015	Made from recycled materials										
0020	6 x 19, bright, fiber core, 5000′ rolls, 1/2″ diameter	G				L.F.	.70			.70	.77
0050	Steel core	G					.92			.92	1.01
0100	Fiber core, 1″ diameter	G					2.36			2.36	2.59
0150	Steel core	G					2.69			2.69	2.96
0300	6 x 19, galvanized, fiber core, 1/2″ diameter	G					1.03			1.03	1.13
0350	Steel core	G					1.18			1.18	1.30
0400	Fiber core, 1″ diameter	G					3.02			3.02	3.32
0450	Steel core	G					3.16			3.16	3.48
0500	6 x 7, bright, IPS, fiber core, <500 L.F. w/acc., 1/4″ diameter	G	E-17	6400	.003		1.20	.15		1.35	1.56
0510	1/2″ diameter	G		2100	.008		2.93	.45		3.38	3.95
0520	3/4″ diameter	G		960	.017		5.30	.98		6.28	7.45
0550	6 x 19, bright, IPS, IWRC, <500 L.F. w/acc., 1/4″ diameter	G		5760	.003		.77	.16		.93	1.12
0560	1/2″ diameter	G		1730	.009		1.25	.54		1.79	2.27
0570	3/4″ diameter	G		770	.021		2.17	1.22		3.39	4.39
0580	1″ diameter	G		420	.038		3.68	2.23		5.91	7.70
0590	1-1/4″ diameter	G		290	.055		6.10	3.24		9.34	12
0600	1-1/2″ diameter	G		192	.083		7.50	4.89		12.39	16.25
0610	1-3/4″ diameter	G	E-18	240	.167		11.95	9.65	6.20	27.80	35.50
0620	2″ diameter	G		160	.250		15.35	14.45	9.30	39.10	50.50
0630	2-1/4″ diameter	G		160	.250		20.50	14.45	9.30	44.25	56.50
0650	6 x 37, bright, IPS, IWRC, <500 L.F. w/acc., 1/4″ diameter	G	E-17	6400	.003		1.12	.15		1.27	1.47
0660	1/2″ diameter	G		1730	.009		1.90	.54		2.44	2.98
0670	3/4″ diameter	G		770	.021		3.07	1.22		4.29	5.40
0680	1″ diameter	G		430	.037		4.88	2.18		7.06	8.95
0690	1-1/4″ diameter	G		290	.055		7.35	3.24		10.59	13.40
0700	1-1/2″ diameter	G		190	.084		10.55	4.94		15.49	19.70
0710	1-3/4″ diameter	G	E-18	260	.154		16.75	8.90	5.70	31.35	39
0720	2″ diameter	G		200	.200		21.50	11.55	7.45	40.50	51
0730	2-1/4″ diameter	G		160	.250		28.50	14.45	9.30	52.25	65.50
0800	6 x 19 & 6 x 37, swaged, 1/2″ diameter	G	E-17	1220	.013		3.91	.77		4.68	5.55
0810	9/16″ diameter	G		1120	.014		4.55	.84		5.39	6.35
0820	5/8″ diameter	G		930	.017		5.40	1.01		6.41	7.60
0830	3/4″ diameter	G		640	.025		6.85	1.47		8.32	9.95
0840	7/8″ diameter	G		480	.033		8.65	1.95		10.60	12.75
0850	1″ diameter	G		350	.046		10.55	2.68		13.23	16.05
0860	1-1/8″ diameter	G		288	.056		13	3.26		16.26	19.65
0870	1-1/4″ diameter	G		230	.070		15.75	4.08		19.83	24
0880	1-3/8″ diameter	G		192	.083		18.20	4.89		23.09	28
0890	1-1/2″ diameter	G	E-18	300	.133		22	7.70	4.96	34.66	42.50

05 15 16.60 Galvanized Steel Wire Rope and Accessories			Crew	Daily Output	Labor-Hours	Unit	Material	Labor	Equipment	Total	Total Incl O&P
0010	**GALVANIZED STEEL WIRE ROPE & ACCESSORIES**										
0015	Made from recycled materials										
3000	Aircraft cable, galvanized, 7 x 7 x 1/8″	G	E-17	5000	.003	L.F.	.18	.19		.37	.51
3100	Clamps, 1/8″	G	″	125	.128	Ea.	1.29	7.50		8.79	13.70

05 15 16.70 Temporary Cable Safety Railing			Crew	Daily Output	Labor-Hours	Unit	Material	Labor	Equipment	Total	Total Incl O&P
0010	**TEMPORARY CABLE SAFETY RAILING**, Each 100′ strand incl.										
0020	2 eyebolts, 1 turnbuckle, 100′ cable, 2 thimbles, 6 clips										
0025	Made from recycled materials										
0100	One strand using 1/4″ cable & accessories	G	2 Sswk	4	4	C.L.F.	158	231		389	555
0200	1/2″ cable & accessories	G	″	2	8	″	345	460		805	1,125

05 21 Steel Joist Framing

05 21 13 – Deep Longspan Steel Joist Framing

05 21 13.50 Deep Longspan Joists			Crew	Daily Output	Labor-Hours	Unit	2020 Bare Costs Material	Labor	Equipment	Total	Total Incl O&P
0010	**DEEP LONGSPAN JOISTS**										
3010	DLH series, 40-ton job lots, bolted cross bridging, shop primer										
3015	Made from recycled materials										
3040	Spans to 144′ (shipped in 2 pieces)	G	E-7	13	6.154	Ton	2,275	355	153	2,783	3,250
3200	52DLH11, 26 lb./L.F.	G		2000	.040	L.F.	28.50	2.30	1	31.80	36.50
3220	52DLH16, 45 lb./L.F.	G		2000	.040		51	2.30	1	54.30	61.50
3240	56DLH11, 26 lb./L.F.	G		2000	.040		29.50	2.30	1	32.80	37.50
3260	56DLH16, 46 lb./L.F.	G		2000	.040		52.50	2.30	1	55.80	62.50
3280	60DLH12, 29 lb./L.F.	G		2000	.040		33	2.30	1	36.30	41.50
3300	60DLH17, 52 lb./L.F.	G		2000	.040		59	2.30	1	62.30	70
3320	64DLH12, 31 lb./L.F.	G		2200	.036		35.50	2.09	.90	38.49	43.50
3340	64DLH17, 52 lb./L.F.	G		2200	.036		59	2.09	.90	61.99	69.50
3360	68DLH13, 37 lb./L.F.	G		2200	.036		42	2.09	.90	44.99	51
3380	68DLH18, 61 lb./L.F.	G		2200	.036		69.50	2.09	.90	72.49	81
3400	72DLH14, 41 lb./L.F.	G		2200	.036		46.50	2.09	.90	49.49	56
3420	72DLH19, 70 lb./L.F.	G		2200	.036		79.50	2.09	.90	82.49	92
3500	For less than 40-ton job lots										
3502	For 30 to 39 tons, add					%	10%				
3504	20 to 29 tons, add						20%				
3506	10 to 19 tons, add						30%				
3507	5 to 9 tons, add						50%	25%			
3508	1 to 4 tons, add						75%	50%			
3509	Less than 1 ton, add						100%	100%			
4010	SLH series, 40-ton job lots, bolted cross bridging, shop primer										
4040	Spans to 200′ (shipped in 3 pieces)	G	E-7	13	6.154	Ton	2,350	355	153	2,858	3,350
4200	80SLH15, 40 lb./L.F.	G		1500	.053	L.F.	47	3.06	1.33	51.39	58.50
4220	80SLH20, 75 lb./L.F.	G		1500	.053		88	3.06	1.33	92.39	103
4240	88SLH16, 46 lb./L.F.	G		1500	.053		54	3.06	1.33	58.39	66
4260	88SLH21, 89 lb./L.F.	G		1500	.053		105	3.06	1.33	109.39	121
4280	96SLH17, 52 lb./L.F.	G		1500	.053		61	3.06	1.33	65.39	74
4300	96SLH22, 102 lb./L.F.	G		1500	.053		120	3.06	1.33	124.39	138
4320	104SLH18, 59 lb./L.F.	G		1800	.044		69.50	2.55	1.11	73.16	82
4340	104SLH23, 109 lb./L.F.	G		1800	.044		128	2.55	1.11	131.66	146
4360	112SLH19, 67 lb./L.F.	G		1800	.044		79	2.55	1.11	82.66	92
4380	112SLH24, 131 lb./L.F.	G		1800	.044		154	2.55	1.11	157.66	175
4400	120SLH20, 77 lb./L.F.	G		1800	.044		90.50	2.55	1.11	94.16	105
4420	120SLH25, 152 lb./L.F.	G		1800	.044		179	2.55	1.11	182.66	202
6100	For less than 40-ton job lots										
6102	For 30 to 39 tons, add					%	10%				
6104	20 to 29 tons, add						20%				
6106	10 to 19 tons, add						30%				
6107	5 to 9 tons, add						50%	25%			
6108	1 to 4 tons, add						75%	50%			
6109	Less than 1 ton, add						100%	100%			

05 21 16 – Longspan Steel Joist Framing

05 21 16.50 Longspan Joists			Crew	Daily Output	Labor-Hours	Unit	2020 Bare Costs Material	Labor	Equipment	Total	Total Incl O&P
0010	**LONGSPAN JOISTS**										
2000	LH series, 40-ton job lots, bolted cross bridging, shop primer										
2015	Made from recycled materials										
2040	Longspan joists, LH series, up to 96′	G	E-7	13	6.154	Ton	2,175	355	153	2,683	3,150
2200	18LH04, 12 lb./L.F.	G		1400	.057	L.F.	13.05	3.28	1.42	17.75	21.50
2220	18LH08, 19 lb./L.F.	G		1400	.057		20.50	3.28	1.42	25.20	30

05 21 16 – Longspan Steel Joist Framing

05 21 16.50 Longspan Joists

	05 21 16.50 Longspan Joists		Crew	Daily Output	Labor-Hours	Unit	2020 Bare Costs Material	Labor	Equipment	Total	Total Incl O&P
2240	20LH04, 12 lb./L.F.	G	E-7	1400	.057	L.F.	13.05	3.28	1.42	17.75	21.50
2260	20LH08, 19 lb./L.F.	G		1400	.057		20.50	3.28	1.42	25.20	30
2280	24LH05, 13 lb./L.F.	G		1400	.057		14.15	3.28	1.42	18.85	22.50
2300	24LH10, 23 lb./L.F.	G		1400	.057		25	3.28	1.42	29.70	34.50
2320	28LH06, 16 lb./L.F.	G		1800	.044		17.40	2.55	1.11	21.06	24.50
2340	28LH11, 25 lb./L.F.	G		1800	.044		27	2.55	1.11	30.66	35.50
2360	32LH08, 17 lb./L.F.	G		1800	.044		18.50	2.55	1.11	22.16	26
2380	32LH13, 30 lb./L.F.	G		1800	.044		32.50	2.55	1.11	36.16	41.50
2400	36LH09, 21 lb./L.F.	G		1800	.044		23	2.55	1.11	26.66	30.50
2420	36LH14, 36 lb./L.F.	G		1800	.044		39	2.55	1.11	42.66	48.50
2440	40LH10, 21 lb./L.F.	G		2200	.036		23	2.09	.90	25.99	29.50
2460	40LH15, 36 lb./L.F.	G		2200	.036		39	2.09	.90	41.99	47.50
2480	44LH11, 22 lb./L.F.	G		2200	.036		24	2.09	.90	26.99	31
2500	44LH16, 42 lb./L.F.	G		2200	.036		45.50	2.09	.90	48.49	55
2520	48LH11, 22 lb./L.F.	G		2200	.036		24	2.09	.90	26.99	31
2540	48LH16, 42 lb./L.F.	G	↓	2200	.036	↓	45.50	2.09	.90	48.49	55
2600	For less than 40-ton job lots										
2602	For 30 to 39 tons, add					%	10%				
2604	20 to 29 tons, add						20%				
2606	10 to 19 tons, add						30%				
2607	5 to 9 tons, add						50%	25%			
2608	1 to 4 tons, add						75%	50%			
2609	Less than 1 ton, add					↓	100%	100%			
6000	For welded cross bridging, add							30%			

05 21 19 – Open Web Steel Joist Framing

05 21 19.10 Open Web Joists

	05 21 19.10 Open Web Joists		Crew	Daily Output	Labor-Hours	Unit	Material	Labor	Equipment	Total	Total Incl O&P
0010	**OPEN WEB JOISTS**										
0015	Made from recycled materials										
0050	K series, 40-ton lots, horiz. bridging, spans to 30′, shop primer	G	E-7	12	6.667	Ton	1,925	385	166	2,476	2,925
0130	8K1, 5.1 lb./L.F.	G		1200	.067	L.F.	4.91	3.83	1.66	10.40	13.40
0140	10K1, 5.0 lb./L.F.	G		1200	.067		4.81	3.83	1.66	10.30	13.30
0160	12K3, 5.7 lb./L.F.	G		1500	.053		5.50	3.06	1.33	9.89	12.50
0180	14K3, 6.0 lb./L.F.	G		1500	.053		5.80	3.06	1.33	10.19	12.80
0200	16K3, 6.3 lb./L.F.	G		1800	.044		6.05	2.55	1.11	9.71	12
0220	16K6, 8.1 lb./L.F.	G		1800	.044		7.80	2.55	1.11	11.46	13.95
0240	18K5, 7.7 lb./L.F.	G		2000	.040		7.40	2.30	1	10.70	12.95
0260	18K9, 10.2 lb./L.F.	G		2000	.040	↓	9.80	2.30	1	13.10	15.60
0440	K series, 30′ to 50′ spans	G		17	4.706	Ton	1,950	270	117	2,337	2,725
0500	20K5, 8.2 lb./L.F.	G		2000	.040	L.F.	8.05	2.30	1	11.35	13.65
0520	20K9, 10.8 lb./L.F.	G		2000	.040		10.55	2.30	1	13.85	16.45
0540	22K5, 8.8 lb./L.F.	G		2000	.040		8.60	2.30	1	11.90	14.30
0560	22K9, 11.3 lb./L.F.	G		2000	.040		11.05	2.30	1	14.35	16.95
0580	24K6, 9.7 lb./L.F.	G		2200	.036		9.50	2.09	.90	12.49	14.85
0600	24K10, 13.1 lb./L.F.	G		2200	.036		12.85	2.09	.90	15.84	18.50
0620	26K6, 10.6 lb./L.F.	G		2200	.036		10.40	2.09	.90	13.39	15.80
0640	26K10, 13.8 lb./L.F.	G		2200	.036		13.50	2.09	.90	16.49	19.25
0660	28K8, 12.7 lb./L.F.	G		2400	.033		12.45	1.91	.83	15.19	17.70
0680	28K12, 17.1 lb./L.F.	G		2400	.033		16.75	1.91	.83	19.49	22.50
0700	30K8, 13.2 lb./L.F.	G		2400	.033		12.90	1.91	.83	15.64	18.20
0720	30K12, 17.6 lb./L.F.	G	↓	2400	.033	↓	17.25	1.91	.83	19.99	23
0800	For less than 40-ton job lots										
0802	For 30 to 39 tons, add					%	10%				

05 21 Steel Joist Framing

05 21 19 – Open Web Steel Joist Framing

05 21 19.10 Open Web Joists		Crew	Daily Output	Labor-Hours	Unit	Material	2020 Bare Costs Labor	Equipment	Total	Total Incl O&P
0804	20 to 29 tons, add				%	20%				
0806	10 to 19 tons, add					30%				
0807	5 to 9 tons, add					50%	25%			
0808	1 to 4 tons, add					75%	50%			
0809	Less than 1 ton, add					100%	100%			
1010	CS series, 40-ton job lots, horizontal bridging, shop primer									
1040	Spans to 30′ G	E-7	12	6.667	Ton	2,000	385	166	2,551	3,000
1100	10CS2, 7.5 lb./L.F. G		1200	.067	L.F.	7.45	3.83	1.66	12.94	16.20
1120	12CS2, 8.0 lb./L.F. G		1500	.053		7.95	3.06	1.33	12.34	15.20
1140	14CS2, 8.0 lb./L.F. G		1500	.053		7.95	3.06	1.33	12.34	15.20
1160	16CS2, 8.5 lb./L.F. G		1800	.044		8.45	2.55	1.11	12.11	14.65
1180	16CS4, 14.5 lb./L.F. G		1800	.044		14.45	2.55	1.11	18.11	21
1200	18CS2, 9.0 lb./L.F. G		2000	.040		8.95	2.30	1	12.25	14.65
1220	18CS4, 15.0 lb./L.F. G		2000	.040		14.95	2.30	1	18.25	21
1240	20CS2, 9.5 lb./L.F. G		2000	.040		9.45	2.30	1	12.75	15.20
1260	20CS4, 16.5 lb./L.F. G		2000	.040		16.40	2.30	1	19.70	23
1280	22CS2, 10.0 lb./L.F. G		2000	.040		9.95	2.30	1	13.25	15.75
1300	22CS4, 16.5 lb./L.F. G		2000	.040		16.40	2.30	1	19.70	23
1320	24CS2, 10.0 lb./L.F. G		2200	.036		9.95	2.09	.90	12.94	15.35
1340	24CS4, 16.5 lb./L.F. G		2200	.036		16.40	2.09	.90	19.39	22.50
1360	26CS2, 10.0 lb./L.F. G		2200	.036		9.95	2.09	.90	12.94	15.35
1380	26CS4, 16.5 lb./L.F. G		2200	.036		16.40	2.09	.90	19.39	22.50
1400	28CS2, 10.5 lb./L.F. G		2400	.033		10.45	1.91	.83	13.19	15.50
1420	28CS4, 16.5 lb./L.F. G		2400	.033		16.40	1.91	.83	19.14	22
1440	30CS2, 11.0 lb./L.F. G		2400	.033		10.95	1.91	.83	13.69	16.05
1460	30CS4, 16.5 lb./L.F. G		2400	.033		16.40	1.91	.83	19.14	22
1500	For less than 40-ton job lots									
1502	For 30 to 39 tons, add				%	10%				
1504	20 to 29 tons, add					20%				
1506	10 to 19 tons, add					30%				
1507	5 to 9 tons, add					50%	25%			
1508	1 to 4 tons, add					75%	50%			
1509	Less than 1 ton, add					100%	100%			
6200	For shop prime paint other than mfrs. standard, add					20%				
6300	For bottom chord extensions, add per chord G				Ea.	42.50			42.50	47
6400	Individual steel bearing plate, 6″ x 6″ x 1/4″ with J-hook G	1 Bric	160	.050	″	8.60	2.60		11.20	13.70
9000	Minimum labor and equipment, bar joists	F-6	1	40	Job		2,000	470	2,470	3,700

05 21 23 – Steel Joist Girder Framing

05 21 23.50 Joist Girders		Crew	Daily Output	Labor-Hours	Unit	Material	Labor	Equipment	Total	Total Incl O&P
0010	JOIST GIRDERS									
0015	Made from recycled materials									
7020	Joist girders, 40-ton job lots, shop primer G	E-5	13	6.154	Ton	1,925	355	142	2,422	2,850
7100	For less than 40-ton job lots									
7102	For 30 to 39 tons, add				Ton	10%				
7104	20 to 29 tons, add					20%				
7106	10 to 19 tons, add					30%				
7107	5 to 9 tons, add					50%	25%			
7108	1 to 4 tons, add					75%	50%			
7109	Less than 1 ton, add					100%	100%			
8000	Trusses, 40-ton job lots, shop fabricated WT chords, shop primer G	E-5	11	7.273		6,350	420	168	6,938	7,850
8100	For less than 40-ton job lots									
8102	For 30 to 39 tons, add				Ton	10%				

05 21 Steel Joist Framing

05 21 23 – Steel Joist Girder Framing

05 21 23.50 Joist Girders		Crew	Daily Output	Labor-Hours	Unit	Material	2020 Bare Costs Labor	Equipment	Total	Total Incl O&P
8104	20 to 29 tons, add				Ton	20%				
8106	10 to 19 tons, add					30%				
8107	5 to 9 tons, add					50%	25%			
8108	1 to 4 tons, add					75%	50%			
8109	Less than 1 ton, add					100%	100%			

05 31 Steel Decking

05 31 13 – Steel Floor Decking

05 31 13.50 Floor Decking			Crew	Daily Output	Labor-Hours	Unit	Material	Labor	Equipment	Total	Total Incl O&P
0010	FLOOR DECKING	R053100-10									
0015	Made from recycled materials										
5100	Non-cellular composite decking, galvanized, 1-1/2" deep, 16 ga.	G	E-4	3500	.009	S.F.	4.82	.53	.04	5.39	6.20
5120	18 ga.	G		3650	.009		3	.51	.04	3.55	4.17
5140	20 ga.	G		3800	.008		3.52	.49	.04	4.05	4.71
5200	2" deep, 22 ga.	G		3860	.008		2.38	.48	.04	2.90	3.45
5300	20 ga.	G		3600	.009		3.38	.52	.04	3.94	4.61
5400	18 ga.	G		3380	.009		3.12	.55	.04	3.71	4.38
5500	16 ga.	G		3200	.010		4.77	.58	.05	5.40	6.25
5700	3" deep, 22 ga.	G		3200	.010		2.59	.58	.05	3.22	3.85
5800	20 ga.	G		3000	.011		3.72	.62	.05	4.39	5.15
5900	18 ga.	G		2850	.011		3.12	.65	.05	3.82	4.56
6000	16 ga.	G		2700	.012		5.40	.69	.05	6.14	7.15
9000	Minimum labor/equipment charge		1 Sswk	1	8	Job		460		460	755

05 31 23 – Steel Roof Decking

05 31 23.50 Roof Decking			Crew	Daily Output	Labor-Hours	Unit	Material	Labor	Equipment	Total	Total Incl O&P
0010	ROOF DECKING										
0015	Made from recycled materials										
2100	Open type, 1-1/2" deep, Type B, wide rib, galv., 22 ga., under 50 sq.	G	E-4	4500	.007	S.F.	2.68	.41	.03	3.12	3.67
2200	50-500 squares	G		4900	.007		2.08	.38	.03	2.49	2.94
2400	Over 500 squares	G		5100	.006		1.93	.36	.03	2.32	2.75
2600	20 ga., under 50 squares	G		3865	.008		2.99	.48	.04	3.51	4.12
2650	50-500 squares	G		4170	.008		2.39	.45	.04	2.88	3.40
2700	Over 500 squares	G		4300	.007		2.15	.43	.03	2.61	3.12
2900	18 ga., under 50 squares	G		3800	.008		3.84	.49	.04	4.37	5.05
2950	50-500 squares	G		4100	.008		3.07	.45	.04	3.56	4.16
3000	Over 500 squares	G		4300	.007		2.76	.43	.03	3.22	3.79
3050	16 ga., under 50 squares	G		3700	.009		5.20	.50	.04	5.74	6.55
3060	50-500 squares	G		4000	.008		4.15	.47	.04	4.66	5.35
3100	Over 500 squares	G		4200	.008		3.74	.44	.04	4.22	4.88
3200	3" deep, Type N, 22 ga., under 50 squares	G		3600	.009		3.72	.52	.04	4.28	4.98
3250	50-500 squares	G		3800	.008		2.98	.49	.04	3.51	4.11
3260	over 500 squares	G		4000	.008		2.68	.47	.04	3.19	3.75
3300	20 ga., under 50 squares	G		3400	.009		4.04	.55	.04	4.63	5.40
3350	50-500 squares	G		3600	.009		3.23	.52	.04	3.79	4.45
3360	over 500 squares	G		3800	.008		2.91	.49	.04	3.44	4.04
3400	18 ga., under 50 squares	G		3200	.010		5.25	.58	.05	5.88	6.75
3450	50-500 squares	G		3400	.009		4.19	.55	.04	4.78	5.55
3460	over 500 squares	G		3600	.009		3.77	.52	.04	4.33	5.05
3500	16 ga., under 50 squares	G		3000	.011		6.95	.62	.05	7.62	8.65
3550	50-500 squares	G		3200	.010		5.55	.58	.05	6.18	7.10

05 31 Steel Decking

05 31 23 – Steel Roof Decking

05 31 23.50 Roof Decking

05 31 23.50 Roof Decking			Crew	Daily Output	Labor-Hours	Unit	Material	2020 Bare Costs Labor	Equipment	Total	Total Incl O&P
3560	over 500 squares	G	E-4	3400	.009	S.F.	4.99	.55	.04	5.58	6.45
3700	4-1/2" deep, Type J, 20 ga., over 50 squares	G		2700	.012		4.47	.69	.05	5.21	6.10
3800	18 ga.	G		2460	.013		5.85	.76	.06	6.67	7.75
3900	16 ga.	G		2350	.014		7.10	.79	.06	7.95	9.15
4100	6" deep, Type H, 18 ga., over 50 squares	G		2000	.016		6.55	.93	.07	7.55	8.80
4200	16 ga.	G		1930	.017		8.95	.96	.08	9.99	11.50
4300	14 ga.	G		1860	.017		11.50	1	.08	12.58	14.40
4500	7-1/2" deep, Type H, 18 ga., over 50 squares	G		1690	.019		7.75	1.10	.09	8.94	10.45
4600	16 ga.	G		1590	.020		9.65	1.17	.09	10.91	12.65
4700	14 ga.	G	↓	1490	.021	↓	12.05	1.25	.10	13.40	15.40
4800	For painted instead of galvanized, deduct						5%				
5000	For acoustical perforated with fiberglass insulation, add					S.F.	25%				
5100	For type F intermediate rib instead of type B wide rib, add	G					25%				
5150	For type A narrow rib instead of type B wide rib, add	G				↓	25%				

05 31 33 – Steel Form Decking

05 31 33.50 Form Decking

05 31 33.50 Form Decking			Crew	Daily Output	Labor-Hours	Unit	Material	Labor	Equipment	Total	Total Incl O&P
0010	**FORM DECKING**										
0015	Made from recycled materials										
6100	Slab form, steel, 28 ga., 9/16" deep, Type UFS, uncoated	G	E-4	4000	.008	S.F.	1.90	.47	.04	2.41	2.89
6200	Galvanized	G		4000	.008		1.68	.47	.04	2.19	2.65
6220	24 ga., 1" deep, Type UF1X, uncoated	G		3900	.008		1.84	.48	.04	2.36	2.85
6240	Galvanized	G		3900	.008		2.17	.48	.04	2.69	3.21
6300	24 ga., 1-5/16" deep, Type UFX, uncoated	G		3800	.008		1.96	.49	.04	2.49	3
6400	Galvanized	G		3800	.008		2.31	.49	.04	2.84	3.38
6500	22 ga., 1-5/16" deep, uncoated	G		3700	.009		2.48	.50	.04	3.02	3.59
6600	Galvanized	G		3700	.009		2.53	.50	.04	3.07	3.64
6700	22 ga., 2" deep, uncoated	G		3600	.009		3.23	.52	.04	3.79	4.45
6800	Galvanized	G	↓	3600	.009	↓	3.17	.52	.04	3.73	4.38
7000	Sheet metal edge closure form, 12" wide with 2 bends, galvanized										
7100	18 ga.	G	E-14	360	.022	L.F.	5.25	1.32	.41	6.98	8.35
7200	16 ga.	G	"	360	.022	"	7.10	1.32	.41	8.83	10.40

05 35 Raceway Decking Assemblies

05 35 13 – Steel Cellular Decking

05 35 13.50 Cellular Decking

05 35 13.50 Cellular Decking			Crew	Daily Output	Labor-Hours	Unit	Material	Labor	Equipment	Total	Total Incl O&P
0010	**CELLULAR DECKING**										
0015	Made from recycled materials										
0200	Cellular units, galv, 1-1/2" deep, Type BC, 20-20 ga., over 15 squares	G	E-4	1460	.022	S.F.	11.55	1.27	.10	12.92	14.90
0250	18-20 ga.	G		1420	.023		10.25	1.31	.10	11.66	13.55
0300	18-18 ga.	G		1390	.023		10.55	1.34	.11	12	13.90
0320	16-18 ga.	G		1360	.024		16.05	1.37	.11	17.53	20
0340	16-16 ga.	G		1330	.024		17.85	1.40	.11	19.36	22
0400	3" deep, Type NC, galvanized, 20-20 ga.	G		1375	.023		12.70	1.35	.11	14.16	16.35
0500	18-20 ga.	G		1350	.024		12	1.38	.11	13.49	15.60
0600	18-18 ga.	G		1290	.025		11.95	1.44	.11	13.50	15.65
0700	16-18 ga.	G		1230	.026		17.25	1.51	.12	18.88	21.50
0800	16-16 ga.	G		1150	.028		18.80	1.62	.13	20.55	23.50
1000	4-1/2" deep, Type JC, galvanized, 18-20 ga.	G		1100	.029		13.85	1.69	.13	15.67	18.15
1100	18-18 ga.	G		1040	.031		13.75	1.79	.14	15.68	18.25
1200	16-18 ga.	G	↓	980	.033	↓	19.85	1.90	.15	21.90	25.50

05 35 Raceway Decking Assemblies

05 35 13 – Steel Cellular Decking

05 35 13.50 Cellular Decking		Crew	Daily Output	Labor-Hours	Unit	Material	Labor	Equipment	Total	Total Incl O&P
						2020 Bare Costs				
1300	16-16 ga. G	E-4	935	.034	S.F.	21.50	1.99	.16	23.65	27.50
1500	For acoustical deck, add					15%				
1700	For cells used for ventilation, add					15%				
1900	For multi-story or congested site, add						50%			
8000	Metal deck and trench, 2" thick, 20 ga., combination									
8010	60% cellular, 40% non-cellular, inserts and trench G	R-4	1100	.036	S.F.	23	2.14	.13	25.27	29

05 41 Structural Metal Stud Framing

05 41 13 – Load-Bearing Metal Stud Framing

05 41 13.05 Bracing

Line	Description	Crew	Daily Output	Labor-Hours	Unit	Material	Labor	Equipment	Total	Total Incl O&P
0010	**BRACING**, shear wall X-bracing, per 10' x 10' bay, one face									
0015	Made of recycled materials									
0120	Metal strap, 20 ga. x 4" wide G	2 Carp	18	.889	Ea.	19.35	47		66.35	97
0130	6" wide G		18	.889		32	47		79	111
0160	18 ga. x 4" wide G		16	1		32	53		85	120
0170	6" wide G	↓	16	1	↓	47	53		100	137
0410	Continuous strap bracing, per horizontal row on both faces									
0420	Metal strap, 20 ga. x 2" wide, studs 12" OC G	1 Carp	7	1.143	C.L.F.	57	60.50		117.50	161
0430	16" OC G		8	1		57	53		110	148
0440	24" OC G		10	.800		57	42.50		99.50	131
0450	18 ga. x 2" wide, studs 12" OC G		6	1.333		77	71		148	199
0460	16" OC G		7	1.143		77	60.50		137.50	182
0470	24" OC G	↓	8	1	↓	77	53		130	170

05 41 13.10 Bridging

Line	Description	Crew	Daily Output	Labor-Hours	Unit	Material	Labor	Equipment	Total	Total Incl O&P
0010	**BRIDGING**, solid between studs w/1-1/4" leg track, per stud bay									
0015	Made from recycled materials									
0200	Studs 12" OC, 18 ga. x 2-1/2" wide G	1 Carp	125	.064	Ea.	.93	3.40		4.33	6.45
0210	3-5/8" wide G		120	.067		1.11	3.54		4.65	6.90
0220	4" wide G		120	.067		1.11	3.54		4.65	6.90
0230	6" wide G		115	.070		1.53	3.70		5.23	7.60
0240	8" wide G		110	.073		1.88	3.87		5.75	8.25
0300	16 ga. x 2-1/2" wide G		115	.070		1.15	3.70		4.85	7.15
0310	3-5/8" wide G		110	.073		1.39	3.87		5.26	7.75
0320	4" wide G		110	.073		1.54	3.87		5.41	7.90
0330	6" wide G		105	.076		1.94	4.05		5.99	8.65
0340	8" wide G		100	.080		2.39	4.25		6.64	9.45
1200	Studs 16" OC, 18 ga. x 2-1/2" wide G		125	.064		1.19	3.40		4.59	6.75
1210	3-5/8" wide G		120	.067		1.42	3.54		4.96	7.25
1220	4" wide G		120	.067		1.42	3.54		4.96	7.25
1230	6" wide G		115	.070		1.96	3.70		5.66	8.05
1240	8" wide G		110	.073		2.41	3.87		6.28	8.85
1300	16 ga. x 2-1/2" wide G		115	.070		1.47	3.70		5.17	7.50
1310	3-5/8" wide G		110	.073		1.78	3.87		5.65	8.15
1320	4" wide G		110	.073		1.98	3.87		5.85	8.40
1330	6" wide G		105	.076		2.49	4.05		6.54	9.25
1340	8" wide G		100	.080		3.07	4.25		7.32	10.20
2200	Studs 24" OC, 18 ga. x 2-1/2" wide G		125	.064		1.72	3.40		5.12	7.35
2210	3-5/8" wide G		120	.067		2.05	3.54		5.59	7.95
2220	4" wide G		120	.067		2.05	3.54		5.59	7.95
2230	6" wide G		115	.070		2.84	3.70		6.54	9
2240	8" wide G	↓	110	.073	↓	3.49	3.87		7.36	10.05

05 41 13 – Load-Bearing Metal Stud Framing

05 41 13.10 Bridging

05 41 13.10 Bridging			Crew	Daily Output	Labor-Hours	Unit	2020 Bare Costs Material	Labor	Equipment	Total	Total Incl O&P
2300	16 ga. x 2-1/2" wide	G	1 Carp	115	.070	Ea.	2.12	3.70		5.82	8.25
2310	3-5/8" wide	G		110	.073		2.58	3.87		6.45	9.05
2320	4" wide	G		110	.073		2.86	3.87		6.73	9.35
2330	6" wide	G		105	.076		3.60	4.05		7.65	10.45
2340	8" wide	G		100	.080		4.44	4.25		8.69	11.70
3000	Continuous bridging, per row										
3100	16 ga. x 1-1/2" channel thru studs 12" OC	G	1 Carp	6	1.333	C.L.F.	51.50	71		122.50	171
3110	16" OC	G		7	1.143		51.50	60.50		112	155
3120	24" OC	G		8.80	.909		51.50	48.50		100	135
4100	2" x 2" angle x 18 ga., studs 12" OC	G		7	1.143		78	60.50		138.50	184
4110	16" OC	G		9	.889		78	47		125	162
4120	24" OC	G		12	.667		78	35.50		113.50	143
4200	16 ga., studs 12" OC	G		5	1.600		98	85		183	244
4210	16" OC	G		7	1.143		98	60.50		158.50	206
4220	24" OC	G		10	.800		98	42.50		140.50	176

05 41 13.25 Framing, Boxed Headers/Beams

05 41 13.25 Framing, Boxed Headers/Beams			Crew	Daily Output	Labor-Hours	Unit	Material	Labor	Equipment	Total	Total Incl O&P
0010	**FRAMING, BOXED HEADERS/BEAMS**										
0015	Made from recycled materials										
0200	Double, 18 ga. x 6" deep	G	2 Carp	220	.073	L.F.	5.55	3.87		9.42	12.30
0210	8" deep	G		210	.076		5.80	4.05		9.85	12.85
0220	10" deep	G		200	.080		7.30	4.25		11.55	14.85
0230	12" deep	G		190	.084		7.95	4.48		12.43	15.90
0300	16 ga. x 8" deep	G		180	.089		6.70	4.72		11.42	14.90
0310	10" deep	G		170	.094		8.30	5		13.30	17.15
0320	12" deep	G		160	.100		9	5.30		14.30	18.40
0400	14 ga. x 10" deep	G		140	.114		9.60	6.05		15.65	20.50
0410	12" deep	G		130	.123		10.50	6.55		17.05	22
1210	Triple, 18 ga. x 8" deep	G		170	.094		8.55	5		13.55	17.40
1220	10" deep	G		165	.097		10.40	5.15		15.55	19.70
1230	12" deep	G		160	.100		11.40	5.30		16.70	21
1300	16 ga. x 8" deep	G		145	.110		9.95	5.85		15.80	20.50
1310	10" deep	G		140	.114		11.90	6.05		17.95	23
1320	12" deep	G		135	.119		13	6.30		19.30	24.50
1400	14 ga. x 10" deep	G		115	.139		13.10	7.40		20.50	26.50
1410	12" deep	G		110	.145		14.45	7.75		22.20	28.50

05 41 13.30 Framing, Stud Walls

05 41 13.30 Framing, Stud Walls			Crew	Daily Output	Labor-Hours	Unit	Material	Labor	Equipment	Total	Total Incl O&P
0010	**FRAMING, STUD WALLS** w/top & bottom track, no openings,										
0020	Headers, beams, bridging or bracing										
0025	Made from recycled materials										
4100	8' high walls, 18 ga. x 2-1/2" wide, studs 12" OC	G	2 Carp	54	.296	L.F.	8.40	15.75		24.15	34.50
4110	16" OC	G		77	.208		6.75	11.05		17.80	25
4120	24" OC	G		107	.150		5.10	7.95		13.05	18.35
4130	3-5/8" wide, studs 12" OC	G		53	.302		9.95	16.05		26	36.50
4140	16" OC	G		76	.211		8	11.20		19.20	27
4150	24" OC	G		105	.152		6.05	8.10		14.15	19.65
4160	4" wide, studs 12" OC	G		52	.308		10.25	16.35		26.60	37.50
4170	16" OC	G		74	.216		8.25	11.50		19.75	27.50
4180	24" OC	G		103	.155		6.20	8.25		14.45	20
4190	6" wide, studs 12" OC	G		51	.314		13.25	16.65		29.90	41
4200	16" OC	G		73	.219		10.65	11.65		22.30	30.50
4210	24" OC	G		101	.158		8.10	8.40		16.50	22.50
4220	8" wide, studs 12" OC	G		50	.320		16.95	17		33.95	46

05 41 13.30 Framing, Stud Walls			Crew	Daily Output	Labor-Hours	Unit	2020 Bare Costs Material	Labor	Equipment	Total	Total Incl O&P
4230	16" OC	G	2 Carp	72	.222	L.F.	13.60	11.80		25.40	34
4240	24" OC	G		100	.160		10.25	8.50		18.75	25
4300	16 ga. x 2-1/2" wide, studs 12" OC	G		47	.340		10.50	18.10		28.60	40.50
4310	16" OC	G		68	.235		8.30	12.50		20.80	29
4320	24" OC	G		94	.170		6.10	9.05		15.15	21.50
4330	3-5/8" wide, studs 12" OC	G		46	.348		12.35	18.50		30.85	43
4340	16" OC	G		66	.242		9.80	12.90		22.70	31.50
4350	24" OC	G		92	.174		7.25	9.25		16.50	23
4360	4" wide, studs 12" OC	G		45	.356		13.05	18.90		31.95	45
4370	16" OC	G		65	.246		10.35	13.10		23.45	32.50
4380	24" OC	G		90	.178		7.60	9.45		17.05	23.50
4390	6" wide, studs 12" OC	G		44	.364		16.30	19.35		35.65	49
4400	16" OC	G		64	.250		12.95	13.30		26.25	36
4410	24" OC	G		88	.182		9.60	9.65		19.25	26
4420	8" wide, studs 12" OC	G		43	.372		20	19.80		39.80	53.50
4430	16" OC	G		63	.254		16	13.50		29.50	39
4440	24" OC	G		86	.186		11.85	9.90		21.75	29
5100	10' high walls, 18 ga. x 2-1/2" wide, studs 12" OC	G		54	.296		10.05	15.75		25.80	36
5110	16" OC	G		77	.208		8	11.05		19.05	26.50
5120	24" OC	G		107	.150		5.90	7.95		13.85	19.25
5130	3-5/8" wide, studs 12" OC	G		53	.302		11.90	16.05		27.95	38.50
5140	16" OC	G		76	.211		9.45	11.20		20.65	28.50
5150	24" OC	G		105	.152		7	8.10		15.10	20.50
5160	4" wide, studs 12" OC	G		52	.308		12.30	16.35		28.65	39.50
5170	16" OC	G		74	.216		9.75	11.50		21.25	29
5180	24" OC	G		103	.155		7.20	8.25		15.45	21
5190	6" wide, studs 12" OC	G		51	.314		15.80	16.65		32.45	44
5200	16" OC	G		73	.219		12.60	11.65		24.25	32.50
5210	24" OC	G		101	.158		9.35	8.40		17.75	24
5220	8" wide, studs 12" OC	G		50	.320		20.50	17		37.50	50
5230	16" OC	G		72	.222		16.10	11.80		27.90	36.50
5240	24" OC	G		100	.160		11.95	8.50		20.45	26.50
5300	16 ga. x 2-1/2" wide, studs 12" OC	G		47	.340		12.65	18.10		30.75	43
5310	16" OC	G		68	.235		9.95	12.50		22.45	31
5320	24" OC	G		94	.170		7.20	9.05		16.25	22.50
5330	3-5/8" wide, studs 12" OC	G		46	.348		14.90	18.50		33.40	46
5340	16" OC	G		66	.242		11.70	12.90		24.60	33.50
5350	24" OC	G		92	.174		8.50	9.25		17.75	24
5360	4" wide, studs 12" OC	G		45	.356		15.80	18.90		34.70	48
5370	16" OC	G		65	.246		12.40	13.10		25.50	34.50
5380	24" OC	G		90	.178		8.95	9.45		18.40	25
5390	6" wide, studs 12" OC	G		44	.364		19.60	19.35		38.95	52.50
5400	16" OC	G		64	.250		15.45	13.30		28.75	38.50
5410	24" OC	G		88	.182		11.25	9.65		20.90	28
5420	8" wide, studs 12" OC	G		43	.372		24.50	19.80		44.30	58
5430	16" OC	G		63	.254		19.10	13.50		32.60	42.50
5440	24" OC	G		86	.186		13.95	9.90		23.85	31
6190	12' high walls, 18 ga. x 6" wide, studs 12" OC	G		41	.390		18.40	20.50		38.90	53
6200	16" OC	G		58	.276		14.55	14.65		29.20	39.50
6210	24" OC	G		81	.198		10.65	10.50		21.15	28.50
6220	8" wide, studs 12" OC	G		40	.400		23.50	21.50		45	60
6230	16" OC	G		57	.281		18.60	14.90		33.50	44.50
6240	24" OC	G		80	.200		13.60	10.65		24.25	32

05 41 13 – Load-Bearing Metal Stud Framing

05 41 13.30 Framing, Stud Walls			Crew	Daily Output	Labor-Hours	Unit	2020 Bare Costs Material	Labor	Equipment	Total	Total Incl O&P
6390	16 ga. x 6" wide, studs 12" OC	G	2 Carp	35	.457	L.F.	23	24.50		47.50	64.50
6400	16" OC	G		51	.314		17.95	16.65		34.60	46.50
6410	24" OC	G		70	.229		12.95	12.15		25.10	33.50
6420	8" wide, studs 12" OC	G		34	.471		28.50	25		53.50	71.50
6430	16" OC	G		50	.320		22	17		39	52
6440	24" OC	G		69	.232		16	12.30		28.30	37.50
6530	14 ga. x 3-5/8" wide, studs 12" OC	G		34	.471		22	25		47	64.50
6540	16" OC	G		48	.333		17.30	17.70		35	47.50
6550	24" OC	G		65	.246		12.45	13.10		25.55	34.50
6560	4" wide, studs 12" OC	G		33	.485		23.50	26		49.50	67.50
6570	16" OC	G		47	.340		18.35	18.10		36.45	49
6580	24" OC	G		64	.250		13.20	13.30		26.50	36
6730	12 ga. x 3-5/8" wide, studs 12" OC	G		31	.516		30.50	27.50		58	78
6740	16" OC	G		43	.372		23.50	19.80		43.30	57.50
6750	24" OC	G		59	.271		16.70	14.40		31.10	41.50
6760	4" wide, studs 12" OC	G		30	.533		33	28.50		61.50	82
6770	16" OC	G		42	.381		25.50	20.50		46	60.50
6780	24" OC	G		58	.276		18	14.65		32.65	43.50
7390	16' high walls, 16 ga. x 6" wide, studs 12" OC	G		33	.485		29.50	26		55.50	74
7400	16" OC	G		48	.333		23	17.70		40.70	54
7410	24" OC	G		67	.239		16.30	12.70		29	38.50
7420	8" wide, studs 12" OC	G		32	.500		36.50	26.50		63	83
7430	16" OC	G		47	.340		28.50	18.10		46.60	60.50
7440	24" OC	G		66	.242		20	12.90		32.90	42.50
7560	14 ga. x 4" wide, studs 12" OC	G		31	.516		30.50	27.50		58	77.50
7570	16" OC	G		45	.356		23.50	18.90		42.40	56.50
7580	24" OC	G		61	.262		16.60	13.95		30.55	41
7590	6" wide, studs 12" OC	G		30	.533		38.50	28.50		67	87.50
7600	16" OC	G		44	.364		29.50	19.35		48.85	63.50
7610	24" OC	G		60	.267		21	14.15		35.15	45.50
7760	12 ga. x 4" wide, studs 12" OC	G		29	.552		43	29.50		72.50	94.50
7770	16" OC	G		40	.400		33	21.50		54.50	70.50
7780	24" OC	G		55	.291		23	15.45		38.45	50.50
7790	6" wide, studs 12" OC	G		28	.571		54	30.50		84.50	108
7800	16" OC	G		39	.410		41.50	22		63.50	80.50
7810	24" OC	G		54	.296		29	15.75		44.75	57
8590	20' high walls, 14 ga. x 6" wide, studs 12" OC	G		29	.552		47	29.50		76.50	98.50
8600	16" OC	G		42	.381		36	20.50		56.50	72
8610	24" OC	G		57	.281		25.50	14.90		40.40	52
8620	8" wide, studs 12" OC	G		28	.571		53	30.50		83.50	107
8630	16" OC	G		41	.390		40.50	20.50		61	78
8640	24" OC	G		56	.286		28.50	15.20		43.70	56
8790	12 ga. x 6" wide, studs 12" OC	G		27	.593		66.50	31.50		98	124
8800	16" OC	G		37	.432		51	23		74	93
8810	24" OC	G		51	.314		35	16.65		51.65	65
8820	8" wide, studs 12" OC	G		26	.615		81	32.50		113.50	142
8830	16" OC	G		36	.444		62	23.50		85.50	106
8840	24" OC	G		50	.320		43	17		60	74.50
9000	Minimum labor/equipment charge			4	4	Job		213		213	340

05 42 Cold-Formed Metal Joist Framing

05 42 13 – Cold-Formed Metal Floor Joist Framing

05 42 13.05 Bracing		Crew	Daily Output	Labor-Hours	Unit	Material	2020 Bare Costs Labor	Equipment	Total	Total Incl O&P
0010	**BRACING**, continuous, per row, top & bottom									
0015	Made from recycled materials									
0120	Flat strap, 20 ga. x 2" wide, joists at 12" OC G	1 Carp	4.67	1.713	C.L.F.	60	91		151	212
0130	16" OC G		5.33	1.501		57.50	80		137.50	192
0140	24" OC G		6.66	1.201		55.50	64		119.50	163
0150	18 ga. x 2" wide, joists at 12" OC G		4	2		76.50	106		182.50	254
0160	16" OC G		4.67	1.713		75	91		166	229
0170	24" OC G		5.33	1.501		73.50	80		153.50	209

05 42 13.10 Bridging

		Crew	Daily Output	Labor-Hours	Unit	Material	Labor	Equipment	Total	Total Incl O&P
0010	**BRIDGING**, solid between joists w/1-1/4" leg track, per joist bay									
0015	Made from recycled materials									
0230	Joists 12" OC, 18 ga. track x 6" wide G	1 Carp	80	.100	Ea.	1.53	5.30		6.83	10.20
0240	8" wide G		75	.107		1.88	5.65		7.53	11.15
0250	10" wide G		70	.114		2.39	6.05		8.44	12.40
0260	12" wide G		65	.123		2.74	6.55		9.29	13.50
0330	16 ga. track x 6" wide G		70	.114		1.94	6.05		7.99	11.90
0340	8" wide G		65	.123		2.39	6.55		8.94	13.15
0350	10" wide G		60	.133		3.06	7.10		10.16	14.70
0360	12" wide G		55	.145		3.53	7.75		11.28	16.30
0440	14 ga. track x 8" wide G		60	.133		3.05	7.10		10.15	14.70
0450	10" wide G		55	.145		3.78	7.75		11.53	16.55
0460	12" wide G		50	.160		4.38	8.50		12.88	18.40
0550	12 ga. track x 10" wide G		45	.178		5.55	9.45		15	21.50
0560	12" wide G		40	.200		5.90	10.65		16.55	23.50
1230	16" OC, 18 ga. track x 6" wide G		80	.100		1.96	5.30		7.26	10.65
1240	8" wide G		75	.107		2.41	5.65		8.06	11.75
1250	10" wide G		70	.114		3.07	6.05		9.12	13.15
1260	12" wide G		65	.123		3.51	6.55		10.06	14.35
1330	16 ga. track x 6" wide G		70	.114		2.49	6.05		8.54	12.50
1340	8" wide G		65	.123		3.07	6.55		9.62	13.90
1350	10" wide G		60	.133		3.93	7.10		11.03	15.65
1360	12" wide G		55	.145		4.52	7.75		12.27	17.35
1440	14 ga. track x 8" wide G		60	.133		3.91	7.10		11.01	15.65
1450	10" wide G		55	.145		4.85	7.75		12.60	17.75
1460	12" wide G		50	.160		5.60	8.50		14.10	19.75
1550	12 ga. track x 10" wide G		45	.178		7.10	9.45		16.55	23
1560	12" wide G		40	.200		7.55	10.65		18.20	25.50
2230	24" OC, 18 ga. track x 6" wide G		80	.100		2.84	5.30		8.14	11.60
2240	8" wide G		75	.107		3.49	5.65		9.14	12.95
2250	10" wide G		70	.114		4.44	6.05		10.49	14.65
2260	12" wide G		65	.123		5.10	6.55		11.65	16.10
2330	16 ga. track x 6" wide G		70	.114		3.60	6.05		9.65	13.70
2340	8" wide G		65	.123		4.44	6.55		10.99	15.40
2350	10" wide G		60	.133		5.70	7.10		12.80	17.60
2360	12" wide G		55	.145		6.55	7.75		14.30	19.60
2440	14 ga. track x 8" wide G		60	.133		5.65	7.10		12.75	17.55
2450	10" wide G		55	.145		7	7.75		14.75	20
2460	12" wide G		50	.160		8.10	8.50		16.60	22.50
2550	12 ga. track x 10" wide G		45	.178		10.30	9.45		19.75	26.50
2560	12" wide G		40	.200		10.95	10.65		21.60	29

05 42 13 – Cold-Formed Metal Floor Joist Framing

05 42 13.25 Framing, Band Joist		Crew	Daily Output	Labor-Hours	Unit	Material	Labor	Equipment	Total	Total Incl O&P
						2020 Bare Costs				
0010	FRAMING, BAND JOIST (track) fastened to bearing wall									
0015	Made from recycled materials									
0220	18 ga. track x 6" deep G	2 Carp	1000	.016	L.F.	1.25	.85		2.10	2.73
0230	8" deep G		920	.017		1.53	.92		2.45	3.17
0240	10" deep G		860	.019		1.95	.99		2.94	3.73
0320	16 ga. track x 6" deep G		900	.018		1.59	.95		2.54	3.25
0330	8" deep G		840	.019		1.95	1.01		2.96	3.77
0340	10" deep G		780	.021		2.50	1.09		3.59	4.50
0350	12" deep G		740	.022		2.88	1.15		4.03	5
0430	14 ga. track x 8" deep G		750	.021		2.49	1.13		3.62	4.56
0440	10" deep G		720	.022		3.09	1.18		4.27	5.30
0450	12" deep G		700	.023		3.57	1.22		4.79	5.90
0540	12 ga. track x 10" deep G		670	.024		4.53	1.27		5.80	7
0550	12" deep G	↓	650	.025	↓	4.81	1.31		6.12	7.40

05 42 13.30 Framing, Boxed Headers/Beams		Crew	Daily Output	Labor-Hours	Unit	Material	Labor	Equipment	Total	Total Incl O&P
0010	FRAMING, BOXED HEADERS/BEAMS									
0015	Made from recycled materials									
0200	Double, 18 ga. x 6" deep G	2 Carp	220	.073	L.F.	5.55	3.87		9.42	12.30
0210	8" deep G		210	.076		5.80	4.05		9.85	12.85
0220	10" deep G		200	.080		7.30	4.25		11.55	14.85
0230	12" deep G		190	.084		7.95	4.48		12.43	15.90
0300	16 ga. x 8" deep G		180	.089		6.70	4.72		11.42	14.90
0310	10" deep G		170	.094		8.30	5		13.30	17.15
0320	12" deep G		160	.100		9	5.30		14.30	18.40
0400	14 ga. x 10" deep G		140	.114		9.60	6.05		15.65	20.50
0410	12" deep G		130	.123		10.50	6.55		17.05	22
0500	12 ga. x 10" deep G		110	.145		12.65	7.75		20.40	26.50
0510	12" deep G		100	.160		14	8.50		22.50	29
1210	Triple, 18 ga. x 8" deep G		170	.094		8.55	5		13.55	17.40
1220	10" deep G		165	.097		10.40	5.15		15.55	19.70
1230	12" deep G		160	.100		11.40	5.30		16.70	21
1300	16 ga. x 8" deep G		145	.110		9.95	5.85		15.80	20.50
1310	10" deep G		140	.114		11.90	6.05		17.95	23
1320	12" deep G		135	.119		13	6.30		19.30	24.50
1400	14 ga. x 10" deep G		115	.139		13.85	7.40		21.25	27
1410	12" deep G		110	.145		15.25	7.75		23	29
1500	12 ga. x 10" deep G		90	.178		18.45	9.45		27.90	35.50
1510	12" deep G	↓	85	.188	↓	20.50	10		30.50	38.50

05 42 13.40 Framing, Joists		Crew	Daily Output	Labor-Hours	Unit	Material	Labor	Equipment	Total	Total Incl O&P
0010	FRAMING, JOISTS, no band joists (track), web stiffeners, headers,									
0020	Beams, bridging or bracing									
0025	Made from recycled materials									
0030	Joists (2" flange) and fasteners, materials only									
0220	18 ga. x 6" deep G				L.F.	1.84			1.84	2.02
0230	8" deep G					1.95			1.95	2.15
0240	10" deep G					2.32			2.32	2.55
0320	16 ga. x 6" deep G					2.04			2.04	2.24
0330	8" deep G					2.44			2.44	2.68
0340	10" deep G					2.85			2.85	3.13
0350	12" deep G					3.22			3.22	3.55
0430	14 ga. x 8" deep G					3.07			3.07	3.37
0440	10" deep G				↓	3.53			3.53	3.88

05 42 13 – Cold-Formed Metal Floor Joist Framing

05 42 13.40 Framing, Joists		Crew	Daily Output	Labor-Hours	Unit	Material	Labor	Equipment	Total	Total Incl O&P
0450	12″ deep G				L.F.	4.01			4.01	4.41
0540	12 ga. x 10″ deep G					5.15			5.15	5.65
0550	12″ deep G					5.85			5.85	6.40
1010	Installation of joists to band joists, beams & headers, labor only									
1220	18 ga. x 6″ deep	2 Carp	110	.145	Ea.		7.75		7.75	12.40
1230	8″ deep		90	.178			9.45		9.45	15.15
1240	10″ deep		80	.200			10.65		10.65	17.05
1320	16 ga. x 6″ deep		95	.168			8.95		8.95	14.35
1330	8″ deep		70	.229			12.15		12.15	19.45
1340	10″ deep		60	.267			14.15		14.15	22.50
1350	12″ deep		55	.291			15.45		15.45	25
1430	14 ga. x 8″ deep		65	.246			13.10		13.10	21
1440	10″ deep		45	.356			18.90		18.90	30.50
1450	12″ deep		35	.457			24.50		24.50	39
1540	12 ga. x 10″ deep		40	.400			21.50		21.50	34
1550	12″ deep		30	.533			28.50		28.50	45.50
9000	Minimum labor/equipment charge		4	4	Job		213		213	340

05 42 13.45 Framing, Web Stiffeners		Crew	Daily Output	Labor-Hours	Unit	Material	Labor	Equipment	Total	Total Incl O&P
0010	**FRAMING, WEB STIFFENERS** at joist bearing, fabricated from									
0020	Stud piece (1-5/8″ flange) to stiffen joist (2″ flange)									
0025	Made from recycled materials									
2120	For 6″ deep joist, with 18 ga. x 2-1/2″ stud G	1 Carp	120	.067	Ea.	.83	3.54		4.37	6.60
2130	3-5/8″ stud G		110	.073		.98	3.87		4.85	7.30
2140	4″ stud G		105	.076		1.02	4.05		5.07	7.60
2150	6″ stud G		100	.080		1.29	4.25		5.54	8.20
2160	8″ stud G		95	.084		1.67	4.48		6.15	9
2220	8″ deep joist, with 2-1/2″ stud G		120	.067		1.11	3.54		4.65	6.90
2230	3-5/8″ stud G		110	.073		1.31	3.87		5.18	7.65
2240	4″ stud G		105	.076		1.37	4.05		5.42	8
2250	6″ stud G		100	.080		1.73	4.25		5.98	8.70
2260	8″ stud G		95	.084		2.24	4.48		6.72	9.60
2320	10″ deep joist, with 2-1/2″ stud G		110	.073		1.38	3.87		5.25	7.70
2330	3-5/8″ stud G		100	.080		1.63	4.25		5.88	8.60
2340	4″ stud G		95	.084		1.69	4.48		6.17	9
2350	6″ stud G		90	.089		2.14	4.72		6.86	9.90
2360	8″ stud G		85	.094		2.77	5		7.77	11.05
2420	12″ deep joist, with 2-1/2″ stud G		110	.073		1.66	3.87		5.53	8.05
2430	3-5/8″ stud G		100	.080		1.96	4.25		6.21	8.95
2440	4″ stud G		95	.084		2.04	4.48		6.52	9.40
2450	6″ stud G		90	.089		2.58	4.72		7.30	10.40
2460	8″ stud G		85	.094		3.34	5		8.34	11.65
3130	For 6″ deep joist, with 16 ga. x 3-5/8″ stud G		100	.080		1.28	4.25		5.53	8.20
3140	4″ stud G		95	.084		1.37	4.48		5.85	8.65
3150	6″ stud G		90	.089		1.67	4.72		6.39	9.40
3160	8″ stud G		85	.094		2.07	5		7.07	10.30
3230	8″ deep joist, with 3-5/8″ stud G		100	.080		1.72	4.25		5.97	8.70
3240	4″ stud G		95	.084		1.84	4.48		6.32	9.15
3250	6″ stud G		90	.089		2.24	4.72		6.96	10
3260	8″ stud G		85	.094		2.77	5		7.77	11.05
3330	10″ deep joist, with 3-5/8″ stud G		85	.094		2.12	5		7.12	10.35
3340	4″ stud G		80	.100		2.27	5.30		7.57	11
3350	6″ stud G		75	.107		2.77	5.65		8.42	12.15

05 42 Cold-Formed Metal Joist Framing

05 42 13 – Cold-Formed Metal Floor Joist Framing

05 42 13.45 Framing, Web Stiffeners			Crew	Daily Output	Labor-Hours	Unit	Material	Labor	Equipment	Total	Total Incl O&P
							2020 Bare Costs				
3360	8" stud	G	1 Carp	70	.114	Ea.	3.44	6.05		9.49	13.55
3430	12" deep joist, with 3-5/8" stud	G		85	.094		2.56	5		7.56	10.80
3440	4" stud	G		80	.100		2.74	5.30		8.04	11.50
3450	6" stud	G		75	.107		3.34	5.65		8.99	12.75
3460	8" stud	G		70	.114		4.14	6.05		10.19	14.30
4230	For 8" deep joist, with 14 ga. x 3-5/8" stud	G		90	.089		2.18	4.72		6.90	9.95
4240	4" stud	G		85	.094		2.29	5		7.29	10.50
4250	6" stud	G		80	.100		2.89	5.30		8.19	11.70
4260	8" stud	G		75	.107		3.23	5.65		8.88	12.65
4330	10" deep joist, with 3-5/8" stud	G		75	.107		2.71	5.65		8.36	12.10
4340	4" stud	G		70	.114		2.84	6.05		8.89	12.85
4350	6" stud	G		65	.123		3.59	6.55		10.14	14.45
4360	8" stud	G		60	.133		4	7.10		11.10	15.75
4430	12" deep joist, with 3-5/8" stud	G		75	.107		3.26	5.65		8.91	12.70
4440	4" stud	G		70	.114		3.42	6.05		9.47	13.50
4450	6" stud	G		65	.123		4.32	6.55		10.87	15.25
4460	8" stud	G		60	.133		4.82	7.10		11.92	16.65
5330	For 10" deep joist, with 12 ga. x 3-5/8" stud	G		65	.123		3.88	6.55		10.43	14.75
5340	4" stud	G		60	.133		4.17	7.10		11.27	15.95
5350	6" stud	G		55	.145		5.25	7.75		13	18.15
5360	8" stud	G		50	.160		6.35	8.50		14.85	20.50
5430	12" deep joist, with 3-5/8" stud	G		65	.123		4.68	6.55		11.23	15.65
5440	4" stud	G		60	.133		5	7.10		12.10	16.85
5450	6" stud	G		55	.145		6.30	7.75		14.05	19.35
5460	8" stud	G		50	.160		7.65	8.50		16.15	22

05 42 23 – Cold-Formed Metal Roof Joist Framing

05 42 23.05 Framing, Bracing			Crew	Daily Output	Labor-Hours	Unit	Material	Labor	Equipment	Total	Total Incl O&P
0010	**FRAMING, BRACING**										
0015	Made from recycled materials										
0020	Continuous bracing, per row										
0100	16 ga. x 1-1/2" channel thru rafters/trusses @ 16" OC	G	1 Carp	4.50	1.778	C.L.F.	51.50	94.50		146	208
0120	24" OC	G		6	1.333		51.50	71		122.50	171
0300	2" x 2" angle x 18 ga., rafters/trusses @ 16" OC	G		6	1.333		78	71		149	200
0320	24" OC	G		8	1		78	53		131	171
0400	16 ga., rafters/trusses @ 16" OC	G		4.50	1.778		98	94.50		192.50	259
0420	24" OC	G		6.50	1.231		98	65.50		163.50	213

05 42 23.10 Framing, Bridging			Crew	Daily Output	Labor-Hours	Unit	Material	Labor	Equipment	Total	Total Incl O&P
0010	**FRAMING, BRIDGING**										
0015	Made from recycled materials										
0020	Solid, between rafters w/1-1/4" leg track, per rafter bay										
1200	Rafters 16" OC, 18 ga. x 4" deep	G	1 Carp	60	.133	Ea.	1.42	7.10		8.52	12.90
1210	6" deep	G		57	.140		1.96	7.45		9.41	14.10
1220	8" deep	G		55	.145		2.41	7.75		10.16	15.05
1230	10" deep	G		52	.154		3.07	8.20		11.27	16.50
1240	12" deep	G		50	.160		3.51	8.50		12.01	17.45
2200	24" OC, 18 ga. x 4" deep	G		60	.133		2.05	7.10		9.15	13.60
2210	6" deep	G		57	.140		2.84	7.45		10.29	15.05
2220	8" deep	G		55	.145		3.49	7.75		11.24	16.25
2230	10" deep	G		52	.154		4.44	8.20		12.64	18
2240	12" deep	G		50	.160		5.10	8.50		13.60	19.20

05 42 Cold-Formed Metal Joist Framing

05 42 23 – Cold-Formed Metal Roof Joist Framing

05 42 23.50 Framing, Parapets

	05 42 23.50 Framing, Parapets		Crew	Daily Output	Labor-Hours	Unit	2020 Bare Costs Material	Labor	Equipment	Total	Total Incl O&P
0010	**FRAMING, PARAPETS**										
0015	Made from recycled materials										
0100	3′ high installed on 1st story, 18 ga. x 4″ wide studs, 12″ OC	G	2 Carp	100	.160	L.F.	5.15	8.50		13.65	19.30
0110	16″ OC	G		150	.107		4.40	5.65		10.05	13.95
0120	24″ OC	G		200	.080		3.64	4.25		7.89	10.80
0200	6″ wide studs, 12″ OC	G		100	.160		6.80	8.50		15.30	21
0210	16″ OC	G		150	.107		5.80	5.65		11.45	15.50
0220	24″ OC	G		200	.080		4.85	4.25		9.10	12.15
1100	Installed on 2nd story, 18 ga. x 4″ wide studs, 12″ OC	G		95	.168		5.15	8.95		14.10	20
1110	16″ OC	G		145	.110		4.40	5.85		10.25	14.25
1120	24″ OC	G		190	.084		3.64	4.48		8.12	11.15
1200	6″ wide studs, 12″ OC	G		95	.168		6.80	8.95		15.75	22
1210	16″ OC	G		145	.110		5.80	5.85		11.65	15.80
1220	24″ OC	G		190	.084		4.85	4.48		9.33	12.50
2100	Installed on gable, 18 ga. x 4″ wide studs, 12″ OC	G		85	.188		5.15	10		15.15	22
2110	16″ OC	G		130	.123		4.40	6.55		10.95	15.35
2120	24″ OC	G		170	.094		3.64	5		8.64	12
2200	6″ wide studs, 12″ OC	G		85	.188		6.80	10		16.80	23.50
2210	16″ OC	G		130	.123		5.80	6.55		12.35	16.90
2220	24″ OC	G		170	.094		4.85	5		9.85	13.35

05 42 23.60 Framing, Roof Rafters

	05 42 23.60 Framing, Roof Rafters		Crew	Daily Output	Labor-Hours	Unit	Material	Labor	Equipment	Total	Total Incl O&P
0010	**FRAMING, ROOF RAFTERS**										
0015	Made from recycled materials										
0100	Boxed ridge beam, double, 18 ga. x 6″ deep	G	2 Carp	160	.100	L.F.	5.55	5.30		10.85	14.60
0110	8″ deep	G		150	.107		5.80	5.65		11.45	15.45
0120	10″ deep	G		140	.114		7.30	6.05		13.35	17.80
0130	12″ deep	G		130	.123		7.95	6.55		14.50	19.25
0200	16 ga. x 6″ deep	G		150	.107		5.95	5.65		11.60	15.65
0210	8″ deep	G		140	.114		6.70	6.05		12.75	17.10
0220	10″ deep	G		130	.123		8.30	6.55		14.85	19.65
0230	12″ deep	G		120	.133		9	7.10		16.10	21.50
1100	Rafters, 2″ flange, material only, 18 ga. x 6″ deep	G					1.84			1.84	2.02
1110	8″ deep	G					1.95			1.95	2.15
1120	10″ deep	G					2.32			2.32	2.55
1130	12″ deep	G					2.67			2.67	2.93
1200	16 ga. x 6″ deep	G					2.04			2.04	2.24
1210	8″ deep	G					2.44			2.44	2.68
1220	10″ deep	G					2.85			2.85	3.13
1230	12″ deep	G					3.22			3.22	3.55
2100	Installation only, ordinary rafter to 4:12 pitch, 18 ga. x 6″ deep		2 Carp	35	.457	Ea.		24.50		24.50	39
2110	8″ deep			30	.533			28.50		28.50	45.50
2120	10″ deep			25	.640			34		34	54.50
2130	12″ deep			20	.800			42.50		42.50	68
2200	16 ga. x 6″ deep			30	.533			28.50		28.50	45.50
2210	8″ deep			25	.640			34		34	54.50
2220	10″ deep			20	.800			42.50		42.50	68
2230	12″ deep			15	1.067			56.50		56.50	91
8100	Add to labor, ordinary rafters on steep roofs							25%			
8110	Dormers & complex roofs							50%			
8200	Hip & valley rafters to 4:12 pitch							25%			
8210	Steep roofs							50%			
8220	Dormers & complex roofs							75%			

05 42 Cold-Formed Metal Joist Framing

05 42 23 – Cold-Formed Metal Roof Joist Framing

05 42 23.60 Framing, Roof Rafters			Crew	Daily Output	Labor-Hours	Unit	2020 Bare Costs Material	Labor	Equipment	Total	Total Incl O&P
8300	Hip & valley jack rafters to 4:12 pitch							50%			
8310	Steep roofs							75%			
8320	Dormers & complex roofs							100%			
9000	Minimum labor/equipment charge		2 Carp	4	4	Job		213		213	340

05 42 23.70 Framing, Soffits and Canopies

			Crew	Daily Output	Labor-Hours	Unit	Material	Labor	Equipment	Total	Total Incl O&P
0010	**FRAMING, SOFFITS & CANOPIES**										
0015	Made from recycled materials										
0130	Continuous ledger track @ wall, studs @ 16" OC, 18 ga. x 4" wide	G	2 Carp	535	.030	L.F.	.95	1.59		2.54	3.59
0140	6" wide	G		500	.032		1.31	1.70		3.01	4.16
0150	8" wide	G		465	.034		1.61	1.83		3.44	4.70
0160	10" wide	G		430	.037		2.05	1.98		4.03	5.40
0230	Studs @ 24" OC, 18 ga. x 4" wide	G		800	.020		.90	1.06		1.96	2.69
0240	6" wide	G		750	.021		1.25	1.13		2.38	3.19
0250	8" wide	G		700	.023		1.53	1.22		2.75	3.64
0260	10" wide	G		650	.025		1.95	1.31		3.26	4.25
1000	Horizontal soffit and canopy members, material only										
1030	1-5/8" flange studs, 18 ga. x 4" deep	G				L.F.	1.22			1.22	1.35
1040	6" deep	G					1.55			1.55	1.70
1050	8" deep	G					2			2	2.20
1140	2" flange joists, 18 ga. x 6" deep	G					2.10			2.10	2.31
1150	8" deep	G					2.23			2.23	2.46
1160	10" deep	G					2.65			2.65	2.92
4030	Installation only, 18 ga., 1-5/8" flange x 4" deep		2 Carp	130	.123	Ea.		6.55		6.55	10.50
4040	6" deep			110	.145			7.75		7.75	12.40
4050	8" deep			90	.178			9.45		9.45	15.15
4140	2" flange, 18 ga. x 6" deep			110	.145			7.75		7.75	12.40
4150	8" deep			90	.178			9.45		9.45	15.15
4160	10" deep			80	.200			10.65		10.65	17.05
6010	Clips to attach fascia to rafter tails, 2" x 2" x 18 ga. angle	G	1 Carp	120	.067		.92	3.54		4.46	6.70
6020	16 ga. angle	G	"	100	.080		1.16	4.25		5.41	8.05
9000	Minimum labor/equipment charge		2 Carp	4	4	Job		213		213	340

05 44 Cold-Formed Metal Trusses

05 44 13 – Cold-Formed Metal Roof Trusses

05 44 13.60 Framing, Roof Trusses

			Crew	Daily Output	Labor-Hours	Unit	Material	Labor	Equipment	Total	Total Incl O&P
0010	**FRAMING, ROOF TRUSSES**										
0015	Made from recycled materials										
0020	Fabrication of trusses on ground, Fink (W) or King Post, to 4:12 pitch										
0120	18 ga. x 4" chords, 16' span	G	2 Carp	12	1.333	Ea.	57	71		128	177
0130	20' span	G		11	1.455		71.50	77.50		149	203
0140	24' span	G		11	1.455		85.50	77.50		163	219
0150	28' span	G		10	1.600		100	85		185	246
0160	32' span	G		10	1.600		114	85		199	262
0250	6" chords, 28' span	G		9	1.778		126	94.50		220.50	290
0260	32' span	G		9	1.778		144	94.50		238.50	310
0270	36' span	G		8	2		163	106		269	350
0280	40' span	G		8	2		181	106		287	370
1120	5:12 to 8:12 pitch, 18 ga. x 4" chords, 16' span	G		10	1.600		65.50	85		150.50	208
1130	20' span	G		9	1.778		81.50	94.50		176	241
1140	24' span	G		9	1.778		98	94.50		192.50	259
1150	28' span	G		8	2		114	106		220	296

05 44 Cold-Formed Metal Trusses

05 44 13 – Cold-Formed Metal Roof Trusses

05 44 13.60 Framing, Roof Trusses		Crew	Daily Output	Labor-Hours	Unit	Material	2020 Bare Costs Labor	Equipment	Total	Total Incl O&P
1160	32′ span	G	2 Carp	8	2	Ea.	131	106		237	315
1250	6″ chords, 28′ span	G		7	2.286		144	121		265	355
1260	32′ span	G		7	2.286		165	121		286	375
1270	36′ span	G		6	2.667		186	142		328	430
1280	40′ span	G		6	2.667		206	142		348	455
2120	9:12 to 12:12 pitch, 18 ga. x 4″ chords, 16′ span	G		8	2		81.50	106		187.50	260
2130	20′ span	G		7	2.286		102	121		223	305
2140	24′ span	G		7	2.286		122	121		243	330
2150	28′ span	G		6	2.667		143	142		285	385
2160	32′ span	G		6	2.667		163	142		305	405
2250	6″ chords, 28′ span	G		5	3.200		181	170		351	470
2260	32′ span	G		5	3.200		206	170		376	500
2270	36′ span	G		4	4		232	213		445	595
2280	40′ span	G		4	4		258	213		471	625
4900	Minimum labor/equipment charge			4	4	Job		213		213	340
5120	Erection only of roof trusses, to 4:12 pitch, 16′ span		F-6	48	.833	Ea.		41.50	9.80	51.30	77.50
5130	20′ span			46	.870			43.50	10.25	53.75	80.50
5140	24′ span			44	.909			45.50	10.70	56.20	84.50
5150	28′ span			42	.952			47.50	11.20	58.70	88.50
5160	32′ span			40	1			50	11.80	61.80	92.50
5170	36′ span			38	1.053			52.50	12.40	64.90	97.50
5180	40′ span			36	1.111			55.50	13.10	68.60	103
5220	5:12 to 8:12 pitch, 16′ span			42	.952			47.50	11.20	58.70	88.50
5230	20′ span			40	1			50	11.80	61.80	92.50
5240	24′ span			38	1.053			52.50	12.40	64.90	97.50
5250	28′ span			36	1.111			55.50	13.10	68.60	103
5260	32′ span			34	1.176			59	13.85	72.85	109
5270	36′ span			32	1.250			62.50	14.75	77.25	116
5280	40′ span			30	1.333			66.50	15.70	82.20	123
5320	9:12 to 12:12 pitch, 16′ span			36	1.111			55.50	13.10	68.60	103
5330	20′ span			34	1.176			59	13.85	72.85	109
5340	24′ span			32	1.250			62.50	14.75	77.25	116
5350	28′ span			30	1.333			66.50	15.70	82.20	123
5360	32′ span			28	1.429			71.50	16.85	88.35	133
5370	36′ span			26	1.538			77	18.10	95.10	142
5380	40′ span			24	1.667			83	19.65	102.65	155
9000	Minimum labor/equipment charge			2	20	Job		1,000	236	1,236	1,850

05 51 Metal Stairs

05 51 13 – Metal Pan Stairs

05 51 13.50 Pan Stairs			Crew	Daily Output	Labor-Hours	Unit	Material	Labor	Equipment	Total	Total Incl O&P
0010	**PAN STAIRS**, shop fabricated, steel stringers										
0015	Made from recycled materials										
0200	Metal pan tread for concrete in-fill, picket rail, 3′-6″ wide	G	E-4	35	.914	Riser	600	53	4.20	657.20	750
0300	4′-0″ wide	G		30	1.067		590	62	4.90	656.90	755
0350	Wall rail, both sides, 3′-6″ wide	G		53	.604		460	35	2.77	497.77	565
1500	Landing, steel pan, conventional	G		160	.200	S.F.	79.50	11.65	.92	92.07	108
1600	Pre-erected	G		255	.125	"	138	7.30	.58	145.88	165
1700	Pre-erected, steel pan tread, 3′-6″ wide, 2 line pipe rail	G	E-2	87	.644	Riser	585	37	19.50	641.50	725

05 51 Metal Stairs

05 51 16 – Metal Floor Plate Stairs

05 51 16.50 Floor Plate Stairs			Crew	Daily Output	Labor-Hours	Unit	2020 Bare Costs Material	Labor	Equipment	Total	Total Incl O&P
0010	**FLOOR PLATE STAIRS**, shop fabricated, steel stringers										
0015	Made from recycled materials										
0400	Cast iron tread and pipe rail, 3'-6" wide	G	E-4	35	.914	Riser	570	53	4.20	627.20	715
0500	Checkered plate tread, industrial, 3'-6" wide	G		28	1.143		385	66.50	5.25	456.75	540
0550	Circular, for tanks, 3'-0" wide	G		33	.970		435	56.50	4.45	495.95	575
0600	For isolated stairs, add							100%			
0800	Custom steel stairs, 3'-6" wide, economy	G	E-4	35	.914		585	53	4.20	642.20	735
0810	Medium priced	G		30	1.067		775	62	4.90	841.90	960
0900	Deluxe	G		20	1.600		970	93	7.35	1,070.35	1,225
1100	For 4' wide stairs, add						5%	5%			
1300	For 5' wide stairs, add						10%	10%			

05 51 19 – Metal Grating Stairs

05 51 19.50 Grating Stairs			Crew	Daily Output	Labor-Hours	Unit	Material	Labor	Equipment	Total	Total Incl O&P
0010	**GRATING STAIRS**, shop fabricated, steel stringers, safety nosing on treads										
0015	Made from recycled materials										
0020	Grating tread and pipe railing, 3'-6" wide	G	E-4	35	.914	Riser	390	53	4.20	447.20	520
0100	4'-0" wide	G		30	1.067	"	455	62	4.90	521.90	605
9000	Minimum labor/equipment charge			2	16	Job		930	73.50	1,003.50	1,600

05 51 23 – Metal Fire Escapes

05 51 23.25 Fire Escapes			Crew	Daily Output	Labor-Hours	Unit	Material	Labor	Equipment	Total	Total Incl O&P
0010	**FIRE ESCAPES**, shop fabricated										
0200	2' wide balcony, 1" x 1/4" bars 1-1/2" OC, with railing	G	2 Sswk	10	1.600	L.F.	63.50	92		155.50	221
0400	1st story cantilevered stair, standard, with railing	G		.50	32	Ea.	2,550	1,850		4,400	5,825
0500	Cable counterweighted, with railing	G		.40	40	"	2,425	2,300		4,725	6,450
0700	36" x 40" platform & fixed stair, with railing	G		.40	40	Flight	1,125	2,300		3,425	5,025
0900	For 3'-6" wide escapes, add to above						100%	150%			

05 51 23.50 Fire Escape Stairs			Crew	Daily Output	Labor-Hours	Unit	Material	Labor	Equipment	Total	Total Incl O&P
0010	**FIRE ESCAPE STAIRS**, portable										
0100	Portable ladder					Ea.	125			125	137

05 51 33 – Metal Ladders

05 51 33.13 Vertical Metal Ladders			Crew	Daily Output	Labor-Hours	Unit	Material	Labor	Equipment	Total	Total Incl O&P
0010	**VERTICAL METAL LADDERS**, shop fabricated										
0015	Made from recycled materials										
0020	Steel, 20" wide, bolted to concrete, with cage	G	E-4	50	.640	V.L.F.	92	37	2.94	131.94	165
0100	Without cage	G		85	.376		42	22	1.73	65.73	84.50
0300	Aluminum, bolted to concrete, with cage	G		50	.640		130	37	2.94	169.94	207
0400	Without cage	G		85	.376		51	22	1.73	74.73	94
9000	Minimum labor/equipment charge			2	16	Job		930	73.50	1,003.50	1,600

05 51 33.16 Inclined Metal Ladders			Crew	Daily Output	Labor-Hours	Unit	Material	Labor	Equipment	Total	Total Incl O&P
0010	**INCLINED METAL LADDERS**, shop fabricated										
0015	Made from recycled materials										
3900	Industrial ships ladder, steel, 24" W, grating treads, 2 line pipe rail	G	E-4	30	1.067	Riser	205	62	4.90	271.90	330
4000	Aluminum	G	"	30	1.067	"	300	62	4.90	366.90	435

05 51 33.23 Alternating Tread Ladders			Crew	Daily Output	Labor-Hours	Unit	Material	Labor	Equipment	Total	Total Incl O&P
0010	**ALTERNATING TREAD LADDERS**, shop fabricated										
0015	Made from recycled materials										
0800	Alternating tread ladders, 68-degree angle of incline										
0810	8' vertical rise, steel, 149 lb., standard paint color		B-68G	3	5.333	Ea.	1,950	305	93.50	2,348.50	2,725
0820	Non-standard paint color			3	5.333		2,400	305	93.50	2,798.50	3,250
0830	Galvanized			3	5.333		2,425	305	93.50	2,823.50	3,250

05 51 Metal Stairs

05 51 33 – Metal Ladders

05 51 33.23 Alternating Tread Ladders		Crew	Daily Output	Labor-Hours	Unit	Material	2020 Bare Costs Labor	Equipment	Total	Total Incl O&P
0840	Stainless	B-68G	3	5.333	Ea.	3,700	305	93.50	4,098.50	4,650
0850	Aluminum, 87 lb.		3	5.333		2,700	305	93.50	3,098.50	3,575
1010	10′ vertical rise, steel, 181 lb., standard paint color		2.75	5.818		2,775	335	102	3,212	3,725
1020	Non-standard paint color		2.75	5.818		3,025	335	102	3,462	3,975
1030	Galvanized		2.75	5.818		3,075	335	102	3,512	4,025
1040	Stainless		2.75	5.818		4,450	335	102	4,887	5,550
1050	Aluminum, 103 lb.		2.75	5.818		3,250	335	102	3,687	4,225
1210	12′ vertical rise, steel, 245 lb., standard paint color		2.50	6.400		3,100	370	112	3,582	4,125
1220	Non-standard paint color		2.50	6.400		3,525	370	112	4,007	4,600
1230	Galvanized		2.50	6.400		3,600	370	112	4,082	4,700
1240	Stainless		2.50	6.400		5,200	370	112	5,682	6,450
1250	Aluminum, 103 lb.		2.50	6.400		3,775	370	112	4,257	4,875
1410	14′ vertical rise, steel, 281 lb., standard paint color		2.25	7.111		3,550	410	124	4,084	4,700
1420	Non-standard paint color		2.25	7.111		4,025	410	124	4,559	5,225
1430	Galvanized		2.25	7.111		4,150	410	124	4,684	5,350
1440	Stainless		2.25	7.111		5,950	410	124	6,484	7,350
1450	Aluminum, 136 lb.		2.25	7.111		4,375	410	124	4,909	5,600
1610	16′ vertical rise, steel, 317 lb., standard paint color		2	8		4,025	460	140	4,625	5,325
1620	Non-standard paint color		2	8		4,525	460	140	5,125	5,875
1630	Galvanized		2	8		4,675	460	140	5,275	6,050
1640	Stainless		2	8		6,700	460	140	7,300	8,275
1650	Aluminum, 153 lb.	↓	2	8	↓	4,675	460	140	5,275	6,025

05 52 Metal Railings

05 52 13 – Pipe and Tube Railings

05 52 13.50 Railings, Pipe			Crew	Daily Output	Labor-Hours	Unit	Material	Labor	Equipment	Total	Total Incl O&P
0010	**RAILINGS, PIPE**, shop fab'd, 3′-6″ high, posts @ 5′ OC										
0015	Made from recycled materials										
0020	Aluminum, 2 rail, satin finish, 1-1/4″ diameter	G	E-4	160	.200	L.F.	51.50	11.65	.92	64.07	76.50
0030	Clear anodized	G		160	.200		61.50	11.65	.92	74.07	87.50
0040	Dark anodized	G		160	.200		68	11.65	.92	80.57	95
0080	1-1/2″ diameter, satin finish	G		160	.200		58.50	11.65	.92	71.07	84.50
0090	Clear anodized	G		160	.200		66	11.65	.92	78.57	92.50
0100	Dark anodized	G		160	.200		72.50	11.65	.92	85.07	100
0140	Aluminum, 3 rail, 1-1/4″ diam., satin finish	G		137	.234		67	13.60	1.07	81.67	97.50
0150	Clear anodized	G		137	.234		83.50	13.60	1.07	98.17	116
0160	Dark anodized	G		137	.234		92	13.60	1.07	106.67	125
0200	1-1/2″ diameter, satin finish	G		137	.234		79.50	13.60	1.07	94.17	111
0210	Clear anodized	G		137	.234		90.50	13.60	1.07	105.17	123
0220	Dark anodized	G		137	.234		99.50	13.60	1.07	114.17	133
0500	Steel, 2 rail, on stairs, primed, 1-1/4″ diameter	G		160	.200		32	11.65	.92	44.57	55.50
0520	1-1/2″ diameter	G		160	.200		31	11.65	.92	43.57	54
0540	Galvanized, 1-1/4″ diameter	G		160	.200		41.50	11.65	.92	54.07	66
0560	1-1/2″ diameter	G		160	.200		49	11.65	.92	61.57	74
0580	Steel, 3 rail, primed, 1-1/4″ diameter	G		137	.234		46	13.60	1.07	60.67	74
0600	1-1/2″ diameter	G		137	.234		50.50	13.60	1.07	65.17	79
0620	Galvanized, 1-1/4″ diameter	G		137	.234		64.50	13.60	1.07	79.17	94.50
0640	1-1/2″ diameter	G		137	.234		77.50	13.60	1.07	92.17	109
0700	Stainless steel, 2 rail, 1-1/4″ diam., #4 finish	G		137	.234		141	13.60	1.07	155.67	179
0720	High polish	G		137	.234		228	13.60	1.07	242.67	275
0740	Mirror polish	G	↓	137	.234	↓	285	13.60	1.07	299.67	340

05 52 Metal Railings

05 52 13 – Pipe and Tube Railings

05 52 13.50 Railings, Pipe			Crew	Daily Output	Labor-Hours	Unit	Material	2020 Bare Costs Labor	Equipment	Total	Total Incl O&P
0760	Stainless steel, 3 rail, 1-1/2" diam., #4 finish	G	E-4	120	.267	L.F.	212	15.50	1.22	228.72	260
0770	High polish	G		120	.267		350	15.50	1.22	366.72	410
0780	Mirror finish	G		120	.267		430	15.50	1.22	446.72	495
0900	Wall rail, alum. pipe, 1-1/4" diam., satin finish	G		213	.150		25	8.75	.69	34.44	42.50
0905	Clear anodized	G		213	.150		31	8.75	.69	40.44	49
0910	Dark anodized	G		213	.150		36.50	8.75	.69	45.94	55.50
0915	1-1/2" diameter, satin finish	G		213	.150		27.50	8.75	.69	36.94	45.50
0920	Clear anodized	G		213	.150		34.50	8.75	.69	43.94	53
0925	Dark anodized	G		213	.150		43.50	8.75	.69	52.94	63
0930	Steel pipe, 1-1/4" diameter, primed	G		213	.150		18.50	8.75	.69	27.94	35.50
0935	Galvanized	G		213	.150		27	8.75	.69	36.44	44.50
0940	1-1/2" diameter	G		176	.182		17.75	10.55	.83	29.13	37.50
0945	Galvanized	G		213	.150		27	8.75	.69	36.44	44.50
0955	Stainless steel pipe, 1-1/2" diam., #4 finish	G		107	.299		113	17.40	1.37	131.77	154
0960	High polish	G		107	.299		229	17.40	1.37	247.77	282
0965	Mirror polish	G	↓	107	.299	↓	271	17.40	1.37	289.77	330
2000	2-line pipe rail (1-1/2" T&B) with 1/2" pickets @ 4-1/2" OC,										
2005	attached handrail on brackets										
2010	42" high aluminum, satin finish, straight & level	G	E-4	120	.267	L.F.	355	15.50	1.22	371.72	415
2050	42" high steel, primed, straight & level	G	"	120	.267		149	15.50	1.22	165.72	191
4000	For curved and level rails, add						10%	10%			
4100	For sloped rails for stairs, add					↓	30%	30%			
9000	Minimum labor/equipment charge		1 Sswk	2	4	Job		231		231	380

05 52 16 – Industrial Railings

05 52 16.50 Railings, Industrial			Crew	Daily Output	Labor-Hours	Unit	Material	Labor	Equipment	Total	Total Incl O&P
0010	**RAILINGS, INDUSTRIAL**, shop fab'd, 3'-6" high, posts @ 5' OC										
0020	2 rail, 3'-6" high, 1-1/2" pipe	G	E-4	255	.125	L.F.	39	7.30	.58	46.88	55.50
0100	2" angle rail	G	"	255	.125		35	7.30	.58	42.88	51
0200	For 4" high kick plate, 10 ga., add	G					6.20			6.20	6.80
0300	1/4" thick, add	G					8.15			8.15	8.95
0500	For curved level rails, add						10%	10%			
0550	For sloped rails for stairs, add					↓	30%	30%			
9000	Minimum labor/equipment charge		1 Sswk	2	4	Job		231		231	380

05 53 Metal Gratings

05 53 13 – Bar Gratings

05 53 13.10 Floor Grating, Aluminum			Crew	Daily Output	Labor-Hours	Unit	Material	Labor	Equipment	Total	Total Incl O&P
0010	**FLOOR GRATING, ALUMINUM**, field fabricated from panels										
0015	Made from recycled materials										
0110	Bearing bars @ 1-3/16" OC, cross bars @ 4" OC,										
0111	Up to 300 S.F., 1" x 1/8" bar	G	E-4	900	.036	S.F.	22	2.07	.16	24.23	28
0112	Over 300 S.F.	G		850	.038		20	2.19	.17	22.36	26
0113	1-1/4" x 1/8" bar, up to 300 S.F.	G		800	.040		19.10	2.33	.18	21.61	25
0114	Over 300 S.F.	G		1000	.032		17.35	1.86	.15	19.36	22.50
0122	1-1/4" x 3/16" bar, up to 300 S.F.	G		750	.043		28	2.48	.20	30.68	35.50
0124	Over 300 S.F.	G		1000	.032		25.50	1.86	.15	27.51	31
0132	1-1/2" x 3/16" bar, up to 300 S.F.	G		700	.046		41	2.66	.21	43.87	49.50
0134	Over 300 S.F.	G		1000	.032		37.50	1.86	.15	39.51	44
0136	1-3/4" x 3/16" bar, up to 300 S.F.	G		500	.064		43.50	3.72	.29	47.51	54
0138	Over 300 S.F.	G	↓	1000	.032	↓	39.50	1.86	.15	41.51	46.50

05 53 Metal Gratings

05 53 13 – Bar Gratings

05 53 13.10 Floor Grating, Aluminum

	05 53 13.10 Floor Grating, Aluminum		Crew	Daily Output	Labor-Hours	Unit	Material	Labor	Equipment	Total	Total Incl O&P
							2020 Bare Costs				
0146	2-1/4" x 3/16" bar, up to 300 S.F.	G	E-4	600	.053	S.F.	59	3.10	.24	62.34	70
0148	Over 300 S.F.	G		1000	.032		53.50	1.86	.15	55.51	62
0162	Cross bars @ 2" OC, 1" x 1/8", up to 300 S.F.	G		600	.053		31.50	3.10	.24	34.84	40
0164	Over 300 S.F.	G		1000	.032		28.50	1.86	.15	30.51	34.50
0172	1-1/4" x 3/16" bar, up to 300 S.F.	G		600	.053		50.50	3.10	.24	53.84	61
0174	Over 300 S.F.	G		1000	.032		46	1.86	.15	48.01	53.50
0182	1-1/2" x 3/16" bar, up to 300 S.F.	G		600	.053		61.50	3.10	.24	64.84	73.50
0184	Over 300 S.F.	G		1000	.032		56	1.86	.15	58.01	64.50
0186	1-3/4" x 3/16" bar, up to 300 S.F.	G		600	.053		63	3.10	.24	66.34	75
0188	Over 300 S.F.	G		1000	.032		57.50	1.86	.15	59.51	66
0200	For straight cuts, add					L.F.	4.53			4.53	4.98
0212	Bearing bars @ 15/16" OC, 1" x 1/8", up to 300 S.F.	G	E-4	520	.062	S.F.	32	3.58	.28	35.86	41.50
0214	Over 300 S.F.	G		920	.035		29	2.02	.16	31.18	35.50
0222	1-1/4" x 3/16", up to 300 S.F.	G		520	.062		51	3.58	.28	54.86	62
0224	Over 300 S.F.	G		920	.035		46.50	2.02	.16	48.68	54.50
0232	1-1/2" x 3/16", up to 300 S.F.	G		520	.062		57	3.58	.28	60.86	68.50
0234	Over 300 S.F.	G		920	.035		51.50	2.02	.16	53.68	60.50
0300	For curved cuts, add					L.F.	5.55			5.55	6.15
0400	For straight banding, add	G					5.80			5.80	6.35
0500	For curved banding, add	G					7.10			7.10	7.85
0600	For aluminum checkered plate nosings, add	G					7.50			7.50	8.25
0700	For straight toe plate, add	G					11.35			11.35	12.50
0800	For curved toe plate, add	G					13.25			13.25	14.55
1000	For cast aluminum abrasive nosings, add	G					11.05			11.05	12.15
1400	Extruded I bars are 10% less than 3/16" bars										
1600	Heavy duty, all extruded plank, 3/4" deep, 1.8 #/S.F.	G	E-4	1100	.029	S.F.	27.50	1.69	.13	29.32	33
1700	1-1/4" deep, 2.9 #/S.F.	G		1000	.032		34.50	1.86	.15	36.51	41
1800	1-3/4" deep, 4.2 #/S.F.	G		925	.035		43.50	2.01	.16	45.67	51.50
1900	2-1/4" deep, 5.0 #/S.F.	G		875	.037		67.50	2.13	.17	69.80	78
2100	For safety serrated surface, add						15%				

05 53 13.70 Floor Grating, Steel

	05 53 13.70 Floor Grating, Steel		Crew	Daily Output	Labor-Hours	Unit	Material	Labor	Equipment	Total	Total Incl O&P
0010	**FLOOR GRATING, STEEL**, field fabricated from panels										
0015	Made from recycled materials										
0300	Platforms, to 12' high, rectangular	G	E-4	3150	.010	Lb.	3.34	.59	.05	3.98	4.69
0400	Circular	G	"	2300	.014	"	4.17	.81	.06	5.04	6
0410	Painted bearing bars @ 1-3/16"										
0412	Cross bars @ 4" OC, 3/4" x 1/8" bar, up to 300 S.F.	G	E-2	500	.112	S.F.	8.35	6.40	3.39	18.14	23.50
0414	Over 300 S.F.	G		750	.075		7.60	4.27	2.26	14.13	17.75
0422	1-1/4" x 3/16", up to 300 S.F.	G		400	.140		13.30	8	4.24	25.54	32.50
0424	Over 300 S.F.	G		600	.093		12.10	5.35	2.83	20.28	25
0432	1-1/2" x 3/16", up to 300 S.F.	G		400	.140		15.45	8	4.24	27.69	34.50
0434	Over 300 S.F.	G		600	.093		14.05	5.35	2.83	22.23	27
0436	1-3/4" x 3/16", up to 300 S.F.	G		400	.140		20.50	8	4.24	32.74	40
0438	Over 300 S.F.	G		600	.093		18.50	5.35	2.83	26.68	32.50
0452	2-1/4" x 3/16", up to 300 S.F.	G		300	.187		24	10.65	5.65	40.30	49.50
0454	Over 300 S.F.	G		450	.124		21.50	7.10	3.77	32.37	39.50
0462	Cross bars @ 2" OC, 3/4" x 1/8", up to 300 S.F.	G		500	.112		15.20	6.40	3.39	24.99	31
0464	Over 300 S.F.	G		750	.075		12.65	4.27	2.26	19.18	23.50
0472	1-1/4" x 3/16", up to 300 S.F.	G		400	.140		19.90	8	4.24	32.14	39.50
0474	Over 300 S.F.	G		600	.093		16.55	5.35	2.83	24.73	30
0482	1-1/2" x 3/16", up to 300 S.F.	G		400	.140		22.50	8	4.24	34.74	42
0484	Over 300 S.F.	G		600	.093		18.65	5.35	2.83	26.83	32.50

05 53 Metal Gratings

05 53 13 – Bar Gratings

05 53 13.70 Floor Grating, Steel			Crew	Daily Output	Labor-Hours	Unit	Material	2020 Bare Costs Labor	Equipment	Total	Total Incl O&P
0486	1-3/4" x 3/16", up to 300 S.F.	G	E-2	400	.140	S.F.	33.50	8	4.24	45.74	54.50
0488	Over 300 S.F.	G		600	.093		28	5.35	2.83	36.18	43
0502	2-1/4" x 3/16", up to 300 S.F.	G		300	.187		31	10.65	5.65	47.30	57.50
0504	Over 300 S.F.	G		450	.124		26	7.10	3.77	36.87	44
0601	Painted bearing bars @ 15/16" OC, cross bars @ 4" OC,										
0612	Up to 300 S.F., 3/4" x 3/16" bars	G	E-4	850	.038	S.F.	12.45	2.19	.17	14.81	17.50
0622	1-1/4" x 3/16" bars	G		600	.053		16.90	3.10	.24	20.24	24
0632	1-1/2" x 3/16" bars	G		550	.058		19.25	3.38	.27	22.90	27
0636	1-3/4" x 3/16" bars	G		450	.071		24.50	4.14	.33	28.97	34
0652	2-1/4" x 3/16" bars	G	E-2	300	.187		26	10.65	5.65	42.30	52
0662	Cross bars @ 2" OC, up to 300 S.F., 3/4" x 3/16"	G		500	.112		15.80	6.40	3.39	25.59	31.50
0672	1-1/4" x 3/16" bars	G		400	.140		20.50	8	4.24	32.74	40
0682	1-1/2" x 3/16" bars	G		400	.140		23	8	4.24	35.24	42.50
0686	1-3/4" x 3/16" bars	G		300	.187		29	10.65	5.65	45.30	55.50
0690	For galvanized grating, add						25%				
0800	For straight cuts, add					L.F.	6.45			6.45	7.10
0900	For curved cuts, add						8.20			8.20	9.05
1000	For straight banding, add	G					7.10			7.10	7.85
1100	For curved banding, add	G					9.15			9.15	10.05
1200	For checkered plate nosings, add	G					8.05			8.05	8.85
1300	For straight toe or kick plate, add	G					14.40			14.40	15.85
1400	For curved toe or kick plate, add	G					16.30			16.30	17.95
1500	For abrasive nosings, add	G					10.65			10.65	11.75
1510	For stair treads, see Section 05 55 13.50										
1600	For safety serrated surface, bearing bars @ 1-3/16" OC, add						15%				
1700	Bearing bars @ 15/16" OC, add						25%				
2000	Stainless steel gratings, close spaced, 1" x 1/8" bars, up to 300 S.F.	G	E-4	450	.071	S.F.	76.50	4.14	.33	80.97	91
2100	Standard spacing, 3/4" x 1/8" bars	G		500	.064		96.50	3.72	.29	100.51	112
2200	1-1/4" x 3/16" bars	G		400	.080		118	4.65	.37	123.02	137

05 53 16 – Plank Gratings

05 53 16.50 Grating Planks			Crew	Daily Output	Labor-Hours	Unit	Material	Labor	Equipment	Total	Total Incl O&P
0010	**GRATING PLANKS**, field fabricated from planks										
0020	Aluminum, 9-1/2" wide, 14 ga., 2" rib	G	E-4	950	.034	L.F.	36	1.96	.15	38.11	43
0200	Galvanized steel, 9-1/2" wide, 14 ga., 2-1/2" rib	G		950	.034		16.95	1.96	.15	19.06	22
0300	4" rib	G		950	.034		18.20	1.96	.15	20.31	23.50
0500	12 ga., 2-1/2" rib	G		950	.034		19.80	1.96	.15	21.91	25.50
0600	3" rib	G		950	.034		21.50	1.96	.15	23.61	27.50
0800	Stainless steel, type 304, 16 ga., 2" rib	G		950	.034		42	1.96	.15	44.11	49.50
0900	Type 316	G		950	.034		56.50	1.96	.15	58.61	65.50

05 53 19 – Expanded Metal Gratings

05 53 19.10 Expanded Grating, Aluminum			Crew	Daily Output	Labor-Hours	Unit	Material	Labor	Equipment	Total	Total Incl O&P
0010	**EXPANDED GRATING, ALUMINUM**										
1200	Expanded aluminum, .65 #/S.F.	G	E-4	1050	.030	S.F.	22	1.77	.14	23.91	27

05 53 19.20 Expanded Grating, Steel			Crew	Daily Output	Labor-Hours	Unit	Material	Labor	Equipment	Total	Total Incl O&P
0010	**EXPANDED GRATING, STEEL**										
2400	Expanded steel grating, at ground, 3.0 #/S.F.	G	E-4	900	.036	S.F.	8.10	2.07	.16	10.33	12.45
2500	3.14 #/S.F.	G		900	.036		7.35	2.07	.16	9.58	11.65
2600	4.0 #/S.F.	G		850	.038		8.85	2.19	.17	11.21	13.55
2650	4.27 #/S.F.	G		850	.038		9.70	2.19	.17	12.06	14.50
2700	5.0 #/S.F.	G		800	.040		13.80	2.33	.18	16.31	19.20
2800	6.25 #/S.F.	G		750	.043		17.85	2.48	.20	20.53	24

05 53 Metal Gratings

05 53 19 – Expanded Metal Gratings

05 53 19.20 Expanded Grating, Steel			Crew	Daily Output	Labor-Hours	Unit	Material	2020 Bare Costs Labor	Equipment	Total	Total Incl O&P
2900	7.0 #/S.F.	G	E-4	700	.046	S.F.	19.70	2.66	.21	22.57	26
3100	For flattened expanded steel grating, add						8%				
3300	For elevated installation above 15', add					↓		15%			

05 53 19.30 Grating Frame

			Crew	Daily Output	Labor-Hours	Unit	Material	Labor	Equipment	Total	Total Incl O&P
0010	**GRATING FRAME**, field fabricated										
0020	Aluminum, for gratings 1" to 1-1/2" deep	G	1 Sswk	70	.114	L.F.	3.82	6.60		10.42	15
0100	For each corner, add	G				Ea.	5.70			5.70	6.25

05 54 Metal Floor Plates

05 54 13 – Floor Plates

05 54 13.20 Checkered Plates

			Crew	Daily Output	Labor-Hours	Unit	Material	Labor	Equipment	Total	Total Incl O&P
0010	**CHECKERED PLATES**, steel, field fabricated										
0015	Made from recycled materials										
0020	1/4" & 3/8", 2000 to 5000 S.F., bolted	G	E-4	2900	.011	Lb.	.80	.64	.05	1.49	1.99
0100	Welded	G		4400	.007	"	.91	.42	.03	1.36	1.73
0300	Pit or trench cover and frame, 1/4" plate, 2' to 3' wide	G	↓	100	.320	S.F.	10.95	18.60	1.47	31.02	44
0400	For galvanizing, add	G				Lb.	.32			.32	.35
0500	Platforms, 1/4" plate, no handrails included, rectangular	G	E-4	4200	.008		3.55	.44	.04	4.03	4.68
0600	Circular	G	"	2500	.013	↓	4.44	.74	.06	5.24	6.15

05 54 13.70 Trench Covers

			Crew	Daily Output	Labor-Hours	Unit	Material	Labor	Equipment	Total	Total Incl O&P
0010	**TRENCH COVERS**, field fabricated										
0020	Cast iron grating with bar stops and angle frame, to 18" wide	G	1 Sswk	20	.400	L.F.	212	23		235	271
0100	Frame only (both sides of trench), 1" grating	G		45	.178		1.41	10.25		11.66	18.35
0150	2" grating	G	↓	35	.229	↓	3.23	13.20		16.43	25
0200	Aluminum, stock units, including frames and										
0210	3/8" plain cover plate, 4" opening	G	E-4	205	.156	L.F.	15.85	9.10	.72	25.67	33
0300	6" opening	G		185	.173		19.95	10.05	.79	30.79	39.50
0400	10" opening	G		170	.188		30	10.95	.86	41.81	52
0500	16" opening	G	↓	155	.206		43.50	12	.95	56.45	68
0700	Add per inch for additional widths to 24"	G					1.87			1.87	2.05
0900	For custom fabrication, add						50%				
1100	For 1/4" plain cover plate, deduct						12%				
1500	For cover recessed for tile, 1/4" thick, deduct						12%				
1600	3/8" thick, add						5%				
1800	For checkered plate cover, 1/4" thick, deduct						12%				
1900	3/8" thick, add						2%				
2100	For slotted or round holes in cover, 1/4" thick, add						3%				
2200	3/8" thick, add						4%				
2300	For abrasive cover, add					↓	12%				

05 55 Metal Stair Treads and Nosings

05 55 13 – Metal Stair Treads

05 55 13.50 Stair Treads		Crew	Daily Output	Labor-Hours	Unit	Material	2020 Bare Costs Labor	Equipment	Total	Total Incl O&P
0010	**STAIR TREADS**, stringers and bolts not included									
3000	Diamond plate treads, steel, 1/8" thick									
3005	Open riser, black enamel									
3010	9" deep x 36" long G	2 Sswk	48	.333	Ea.	112	19.20		131.20	155
3020	42" long G		48	.333		118	19.20		137.20	162
3030	48" long G		48	.333		123	19.20		142.20	167
3040	11" deep x 36" long G		44	.364		118	21		139	165
3050	42" long G		44	.364		125	21		146	172
3060	48" long G		44	.364		133	21		154	181
3110	Galvanized, 9" deep x 36" long G		48	.333		177	19.20		196.20	226
3120	42" long G		48	.333		188	19.20		207.20	239
3130	48" long G		48	.333		199	19.20		218.20	251
3140	11" deep x 36" long G		44	.364		183	21		204	236
3150	42" long G		44	.364		199	21		220	254
3160	48" long G	↓	44	.364	↓	207	21		228	263
3200	Closed riser, black enamel									
3210	12" deep x 36" long G	2 Sswk	40	.400	Ea.	135	23		158	187
3220	42" long G		40	.400		147	23		170	199
3230	48" long G		40	.400		155	23		178	208
3240	Galvanized, 12" deep x 36" long G		40	.400		213	23		236	273
3250	42" long G		40	.400		233	23		256	294
3260	48" long G	↓	40	.400	↓	244	23		267	305
4000	Bar grating treads									
4005	Steel, 1-1/4" x 3/16" bars, anti-skid nosing, black enamel									
4010	8-5/8" deep x 30" long G	2 Sswk	48	.333	Ea.	56	19.20		75.20	93
4020	36" long G		48	.333		65.50	19.20		84.70	104
4030	48" long G		48	.333		101	19.20		120.20	143
4040	10-15/16" deep x 36" long G		44	.364		72.50	21		93.50	115
4050	48" long G		44	.364		104	21		125	149
4060	Galvanized, 8-5/8" deep x 30" long G		48	.333		64	19.20		83.20	102
4070	36" long G		48	.333		76	19.20		95.20	115
4080	48" long G		48	.333		112	19.20		131.20	155
4090	10-15/16" deep x 36" long G		44	.364		89	21		110	133
4100	48" long G	↓	44	.364	↓	115	21		136	162
4200	Aluminum, 1-1/4" x 3/16" bars, serrated, with nosing									
4210	7-5/8" deep x 18" long G	2 Sswk	52	.308	Ea.	53	17.75		70.75	87.50
4220	24" long G		52	.308		63	17.75		80.75	98
4230	30" long G		52	.308		73	17.75		90.75	109
4240	36" long G		52	.308		188	17.75		205.75	236
4250	8-13/16" deep x 18" long G		48	.333		71.50	19.20		90.70	110
4260	24" long G		48	.333		106	19.20		125.20	148
4270	30" long G		48	.333		122	19.20		141.20	166
4280	36" long G		48	.333		207	19.20		226.20	260
4290	10" deep x 18" long G		44	.364		139	21		160	187
4300	30" long G		44	.364		184	21		205	238
4310	36" long G	↓	44	.364	↓	224	21		245	281
5000	Channel grating treads									
5005	Steel, 14 ga., 2-1/2" thick, galvanized									
5010	9" deep x 36" long G	2 Sswk	48	.333	Ea.	106	19.20		125.20	149
5020	48" long G		48	.333	"	143	19.20		162.20	190
9000	Minimum labor/equipment charge	↓	4	4	Job		231		231	380

05 55 Metal Stair Treads and Nosings

05 55 19 – Metal Stair Tread Covers

05 55 19.50 Stair Tread Covers for Renovation		Crew	Daily Output	Labor-Hours	Unit	Material	Labor	Equipment	Total	Total Incl O&P
						2020 Bare Costs				
0010	STAIR TREAD COVERS FOR RENOVATION									
0205	Extruded tread cover with nosing, pre-drilled, includes screws									
0210	Aluminum with black abrasive strips, 9" wide x 3' long	1 Carp	24	.333	Ea.	104	17.70		121.70	144
0220	4' long		22	.364		136	19.35		155.35	181
0230	5' long		20	.400		186	21.50		207.50	238
0240	11" wide x 3' long		24	.333		147	17.70		164.70	190
0250	4' long		22	.364		199	19.35		218.35	249
0260	5' long		20	.400		238	21.50		259.50	295
0305	Black abrasive strips with yellow front strips									
0310	Aluminum, 9" wide x 3' long	1 Carp	24	.333	Ea.	124	17.70		141.70	166
0320	4' long		22	.364		166	19.35		185.35	213
0330	5' long		20	.400		213	21.50		234.50	268
0340	11" wide x 3' long		24	.333		161	17.70		178.70	206
0350	4' long		22	.364		204	19.35		223.35	256
0360	5' long		20	.400		267	21.50		288.50	325
0405	Black abrasive strips with photoluminescent front strips									
0410	Aluminum, 9" wide x 3' long	1 Carp	24	.333	Ea.	157	17.70		174.70	202
0420	4' long		22	.364		178	19.35		197.35	227
0430	5' long		20	.400		223	21.50		244.50	279
0440	11" wide x 3' long		24	.333		162	17.70		179.70	207
0450	4' long		22	.364		216	19.35		235.35	268
0460	5' long		20	.400		270	21.50		291.50	330
9000	Minimum labor/equipment charge		4	2	Job		106		106	170

05 56 Metal Castings

05 56 13 – Metal Construction Castings

05 56 13.50 Construction Castings		Crew	Daily Output	Labor-Hours	Unit	Material	Labor	Equipment	Total	Total Incl O&P
0010	CONSTRUCTION CASTINGS									
0020	Manhole covers and frames, see Section 33 44 13.13									
0100	Column bases, cast iron, 16" x 16", approx. 65 lb. G	E-4	46	.696	Ea.	140	40.50	3.19	183.69	224
0200	32" x 32", approx. 256 lb. G	"	23	1.391		525	81	6.40	612.40	715
0400	Cast aluminum for wood columns, 8" x 8" G	1 Carp	32	.250		43.50	13.30		56.80	69
0500	12" x 12" G	"	32	.250		70	13.30		83.30	98.50
0600	Miscellaneous C.I. castings, light sections, less than 150 lb. G	E-4	3200	.010	Lb.	9.10	.58	.05	9.73	11.05
1100	Heavy sections, more than 150 lb. G		4200	.008		5.30	.44	.04	5.78	6.60
1300	Special low volume items G		3200	.010		11.05	.58	.05	11.68	13.15
1500	For ductile iron, add					100%				

05 58 Formed Metal Fabrications

05 58 13 – Column Covers

05 58 13.05 Column Covers		Crew	Daily Output	Labor-Hours	Unit	Material	Labor	Equipment	Total	Total Incl O&P
0010	COLUMN COVERS									
0015	Made from recycled materials									
0020	Excludes structural steel, light ga. metal framing, misc. metals, sealants									
0100	Round covers, 2 halves with 2 vertical joints for backer rod and sealant									
0110	Up to 12' high, no horizontal joints									
0120	12" diameter, 0.125" aluminum, anodized/painted finish G	2 Sswk	32	.500	V.L.F.	33	29		62	83
0130	Type 304 stainless steel, 16 gauge, #4 brushed finish G		32	.500		47	29		76	99
0140	Type 316 stainless steel, 16 gauge, #4 brushed finish G		32	.500		57.50	29		86.50	111

05 58 Formed Metal Fabrications

05 58 13 – Column Covers

05 58 13.05 Column Covers

	05 58 13.05 Column Covers		Crew	Daily Output	Labor-Hours	Unit	Material	2020 Bare Costs Labor	Equipment	Total	Total Incl O&P
0150	18″ diameter, aluminum	G	2 Sswk	32	.500	V.L.F.	49	29		78	101
0160	Type 304 stainless steel	G		32	.500		70.50	29		99.50	125
0170	Type 316 stainless steel	G		32	.500		86.50	29		115.50	142
0180	24″ diameter, aluminum	G		32	.500		65.50	29		94.50	119
0190	Type 304 stainless steel	G		32	.500		94.50	29		123.50	151
0200	Type 316 stainless steel	G		32	.500		115	29		144	174
0210	30″ diameter, aluminum	G		30	.533		82	31		113	141
0220	Type 304 stainless steel	G		30	.533		118	31		149	181
0230	Type 316 stainless steel	G		30	.533		144	31		175	210
0240	36″ diameter, aluminum			30	.533		98.50	31		129.50	159
0250	Type 304 stainless steel	G		30	.533		141	31		172	207
0260	Type 316 stainless steel	G		30	.533		173	31		204	241
0400	Up to 24′ high, 2 stacked sections with 1 horizontal joint										
0410	18″ diameter, aluminum	G	2 Sswk	28	.571	V.L.F.	51.50	33		84.50	111
0450	Type 304 stainless steel	G		28	.571		74.50	33		107.50	136
0460	Type 316 stainless steel	G		28	.571		91	33		124	154
0470	24″ diameter, aluminum	G		28	.571		69	33		102	130
0480	Type 304 stainless steel	G		28	.571		99	33		132	163
0490	Type 316 stainless steel	G		28	.571		121	33		154	187
0500	30″ diameter, aluminum	G		24	.667		86	38.50		124.50	158
0510	Type 304 stainless steel	G		24	.667		124	38.50		162.50	199
0520	Type 316 stainless steel	G		24	.667		151	38.50		189.50	230
0530	36″ diameter, aluminum	G		24	.667		103	38.50		141.50	177
0540	Type 304 stainless steel	G		24	.667		149	38.50		187.50	226
0550	Type 316 stainless steel	G		24	.667		182	38.50		220.50	263

05 58 21 – Formed Chain

05 58 21.05 Alloy Steel Chain

	05 58 21.05 Alloy Steel Chain		Crew	Daily Output	Labor-Hours	Unit	Material	Labor	Equipment	Total	Total Incl O&P
0010	**ALLOY STEEL CHAIN**, Grade 80, for lifting										
0015	Self-colored, cut lengths, 1/4″	G	E-17	4	4	C.L.F.	920	235		1,155	1,400
0020	3/8″	G		2	8		1,400	470		1,870	2,325
0030	1/2″	G		1.20	13.333		2,300	780		3,080	3,800
0040	5/8″	G		.72	22.222		3,800	1,300		5,100	6,300
0050	3/4″	G	E-18	.48	83.333		3,750	4,825	3,100	11,675	15,400
0060	7/8″	G		.40	100		7,425	5,775	3,725	16,925	21,700
0070	1″	G		.35	114		9,200	6,625	4,250	20,075	25,500
0080	1-1/4″	G		.24	167		13,600	9,650	6,200	29,450	37,400
0110	Hook, Grade 80, Clevis slip, 1/4″	G				Ea.	32.50			32.50	35.50
0120	3/8″	G					51.50			51.50	56.50
0130	1/2″	G					75			75	82.50
0140	5/8″	G					107			107	117
0150	3/4″	G					123			123	135
0160	Hook, Grade 80, eye/sling w/hammerlock coupling, 15 ton	G					425			425	470
0170	22 ton	G					945			945	1,050
0180	37 ton	G					3,050			3,050	3,350

05 58 23 – Formed Metal Guards

05 58 23.90 Window Guards

	05 58 23.90 Window Guards		Crew	Daily Output	Labor-Hours	Unit	Material	Labor	Equipment	Total	Total Incl O&P
0010	**WINDOW GUARDS**, shop fabricated										
0015	Expanded metal, steel angle frame, permanent	G	E-4	350	.091	S.F.	25	5.30	.42	30.72	36.50
0025	Steel bars, 1/2″ x 1/2″, spaced 5″ OC	G	″	290	.110	″	17.30	6.40	.51	24.21	30
0030	Hinge mounted, add	G				Opng.	50			50	55
0040	Removable type, add	G				″	31.50			31.50	35
0050	For galvanized guards, add					S.F.	35%				

05 58 Formed Metal Fabrications

05 58 23 – Formed Metal Guards

05 58 23.90 Window Guards			Crew	Daily Output	Labor-Hours	Unit	Material	2020 Bare Costs Labor	Equipment	Total	Total Incl O&P
0070	For pivoted or projected type, add					S.F.	105%	40%			
0100	Mild steel, stock units, economy	G	E-4	405	.079		6.80	4.59	.36	11.75	15.40
0200	Deluxe	G		405	.079		13.85	4.59	.36	18.80	23
0400	Woven wire, stock units, 3/8" channel frame, 3' x 5' opening	G		40	.800	Opng.	184	46.50	3.67	234.17	282
0500	4' x 6' opening	G		38	.842		293	49	3.87	345.87	410
0800	Basket guards for above, add	G					263			263	290
1000	Swinging guards for above, add	G					90			90	99
9000	Minimum labor/equipment charge		1 Sswk	2	4	Job		231		231	380

05 58 25 – Formed Lamp Posts

05 58 25.40 Lamp Posts			Crew	Daily Output	Labor-Hours	Unit	Material	Labor	Equipment	Total	Total Incl O&P
0010	**LAMP POSTS**										
0020	Aluminum, 7' high, stock units, post only	G	1 Carp	16	.500	Ea.	84	26.50		110.50	135
0100	Mild steel, plain	G		16	.500	"	86	26.50		112.50	137
9000	Minimum labor/equipment charge			4	2	Job		106		106	170

05 71 Decorative Metal Stairs

05 71 13 – Fabricated Metal Spiral Stairs

05 71 13.50 Spiral Stairs			Crew	Daily Output	Labor-Hours	Unit	Material	Labor	Equipment	Total	Total Incl O&P
0010	**SPIRAL STAIRS**										
1805	Shop fabricated, custom ordered										
1810	Aluminum, 5'-0" diameter, plain units	G	E-4	45	.711	Riser	720	41.50	3.26	764.76	860
1820	Fancy units	G		45	.711		1,200	41.50	3.26	1,244.76	1,400
1900	Cast iron, 4'-0" diameter, plain units	G		45	.711		715	41.50	3.26	759.76	855
1920	Fancy units	G		25	1.280		1,400	74.50	5.90	1,480.40	1,675
2000	Steel, industrial checkered plate, 4' diameter	G		45	.711		700	41.50	3.26	744.76	835
2200	6' diameter	G		40	.800		715	46.50	3.67	765.17	865
3100	Spiral stair kits, 12 stacking risers to fit exact floor height										
3110	Steel, flat metal treads, primed, 3'-6" diameter	G	2 Carp	1.60	10	Flight	1,300	530		1,830	2,275
3120	4'-0" diameter	G		1.45	11.034		1,500	585		2,085	2,600
3130	4'-6" diameter	G		1.35	11.852		1,650	630		2,280	2,825
3140	5'-0" diameter	G		1.25	12.800		1,850	680		2,530	3,125
3210	Galvanized, 3'-6" diameter	G		1.60	10		1,875	530		2,405	2,900
3220	4'-0" diameter	G		1.45	11.034		2,150	585		2,735	3,300
3230	4'-6" diameter	G		1.35	11.852		2,325	630		2,955	3,550
3240	5'-0" diameter	G		1.25	12.800		2,575	680		3,255	3,925
3310	Checkered plate tread, primed, 3'-6" diameter	G		1.45	11.034		1,575	585		2,160	2,700
3320	4'-0" diameter	G		1.35	11.852		1,725	630		2,355	2,900
3330	4'-6" diameter	G		1.25	12.800		1,875	680		2,555	3,150
3340	5'-0" diameter	G		1.15	13.913		1,950	740		2,690	3,325
3410	Galvanized, 3'-6" diameter	G		1.45	11.034		2,150	585		2,735	3,325
3420	4'-0" diameter	G		1.35	11.852		2,425	630		3,055	3,650
3430	4'-6" diameter	G		1.25	12.800		2,625	680		3,305	3,975
3440	5'-0" diameter	G		1.15	13.913		2,925	740		3,665	4,375
3510	Red oak covers on flat metal treads, 3'-6" diameter			1.35	11.852		2,375	630		3,005	3,625
3520	4'-0" diameter			1.25	12.800		2,850	680		3,530	4,250
3530	4'-6" diameter			1.15	13.913		3,075	740		3,815	4,575
3540	5'-0" diameter			1.05	15.238		3,300	810		4,110	4,925

05 73 Decorative Metal Railings

05 73 16 – Wire Rope Decorative Metal Railings

05 73 16.10 Cable Railings		Crew	Daily Output	Labor-Hours	Unit	Material	2020 Bare Costs Labor	Equipment	Total	Total Incl O&P
0010 CABLE RAILINGS, with 316 stainless steel 1 x 19 cable, 3/16" diameter										
0015 Made from recycled materials										
0100 1-3/4" diameter stainless steel posts x 42" high, cables 4" OC	G	2 Sswk	25	.640	L.F.	43.50	37		80.50	109

05 73 23 – Ornamental Railings

05 73 23.50 Railings, Ornamental		Crew	Daily Output	Labor-Hours	Unit	Material	Labor	Equipment	Total	Total Incl O&P
0010 RAILINGS, ORNAMENTAL, 3'-6" high, posts @ 6' OC										
0020 Bronze or stainless, hand forged, plain	G	2 Sswk	24	.667	L.F.	265	38.50		303.50	355
0100 Fancy	G		18	.889		525	51		576	665
0200 Aluminum, panelized, plain	G		24	.667		13.35	38.50		51.85	77.50
0300 Fancy	G		18	.889		25.50	51		76.50	112
0400 Wrought iron, hand forged, plain	G		24	.667		96.50	38.50		135	169
0500 Fancy	G		18	.889		231	51		282	340
0550 Steel, panelized, plain	G		24	.667		21	38.50		59.50	86.50
0560 Fancy	G		18	.889		33	51		84	121
0600 Composite metal/wood/glass, plain			18	.889		145	51		196	243
0700 Fancy			12	1.333		290	77		367	445
9000 Minimum labor/equipment charge		1 Sswk	2	4	Job		231		231	380

05 75 Decorative Formed Metal

05 75 13 – Columns

05 75 13.10 Aluminum Columns		Crew	Daily Output	Labor-Hours	Unit	Material	Labor	Equipment	Total	Total Incl O&P
0010 ALUMINUM COLUMNS										
0015 Made from recycled materials										
0020 Aluminum, extruded, stock units, no cap or base, 6" diameter	G	E-4	240	.133	L.F.	16.85	7.75	.61	25.21	32
0100 8" diameter	G		170	.188		18.60	10.95	.86	30.41	39.50
0200 10" diameter	G		150	.213		25.50	12.40	.98	38.88	49.50
0300 12" diameter	G		140	.229		40	13.30	1.05	54.35	67
0400 15" diameter	G		120	.267		56.50	15.50	1.22	73.22	89.50
0410 Caps and bases, plain, 6" diameter	G				Set	25.50			25.50	28
0420 8" diameter	G					31.50			31.50	34.50
0430 10" diameter	G					50.50			50.50	56
0440 12" diameter	G					79.50			79.50	87.50
0450 15" diameter	G					143			143	158
0460 Caps, ornamental, plain	G					315			315	345
0470 Fancy	G					1,750			1,750	1,925
0500 For square columns, add to column prices above					L.F.	50%				
0700 Residential, flat, 8' high, plain	G	E-4	20	1.600	Ea.	103	93	7.35	203.35	273
0720 Fancy	G		20	1.600		200	93	7.35	300.35	380
0740 Corner type, plain	G		20	1.600		188	93	7.35	288.35	365
0760 Fancy	G		20	1.600		310	93	7.35	410.35	500

05 75 13.20 Columns, Ornamental		Crew	Daily Output	Labor-Hours	Unit	Material	Labor	Equipment	Total	Total Incl O&P
0010 COLUMNS, ORNAMENTAL, shop fabricated										
6400 Mild steel, flat, 9" wide, stock units, painted, plain	G	E-4	160	.200	V.L.F.	9.65	11.65	.92	22.22	30.50
6450 Fancy	G		160	.200		18.70	11.65	.92	31.27	40.50
6500 Corner columns, painted, plain	G		160	.200		17.65	11.65	.92	30.22	39.50
6550 Fancy	G		160	.200		29	11.65	.92	41.57	52

Estimating Tips

06 05 00 Common Work Results for Wood, Plastics, and Composites

- Common to any wood-framed structure are the accessory connector items such as screws, nails, adhesives, hangers, connector plates, straps, angles, and hold-downs. For typical wood-framed buildings, such as residential projects, the aggregate total for these items can be significant, especially in areas where seismic loading is a concern. For floor and wall framing, the material cost is based on 10 to 25 lbs. of accessory connectors per MBF. Hold-downs, hangers, and other connectors should be taken off by the piece.

  Included with material costs are fasteners for a normal installation. Gordian's RSMeans engineers use manufacturers' recommendations, written specifications, and/or standard construction practice for the sizing and spacing of fasteners. Prices for various fasteners are shown for informational purposes only. Adjustments should be made if unusual fastening conditions exist.

06 10 00 Carpentry

- Lumber is a traded commodity and therefore sensitive to supply and demand in the marketplace. Even with "budgetary" estimating of wood-framed projects, it is advisable to call local suppliers for the latest market pricing.
- The common quantity unit for wood-framed projects is "thousand board feet" (MBF). A board foot is a volume of wood—1" x 1' x 1' or 144 cubic inches. Board-foot quantities are generally calculated using nominal material dimensions—dressed sizes are ignored. Board foot per lineal foot of any stick of lumber can be calculated by dividing the nominal cross-sectional area by 12. As an example, 2,000 lineal feet of 2 x 12 equates to 4 MBF by dividing the nominal area, 2 x 12, by 12, which equals 2, and multiplying that by 2,000 to give 4,000 board feet. This simple rule applies to all nominal dimensioned lumber.
- Waste is an issue of concern at the quantity takeoff for any area of construction. Framing lumber is sold in even foot lengths, i.e., 8', 10', 12', 14', 16', and depending on spans, wall heights, and the grade of lumber, waste is inevitable. A rule of thumb for lumber waste is 5–10% depending on material quality and the complexity of the framing.
- Wood in various forms and shapes is used in many projects, even where the main structural framing is steel, concrete, or masonry. Plywood as a back-up partition material and 2x boards used as blocking and cant strips around roof edges are two common examples. The estimator should ensure that the costs of all wood materials are included in the final estimate.

06 20 00 Finish Carpentry

- It is necessary to consider the grade of workmanship when estimating labor costs for erecting millwork and an interior finish. In practice, there are three grades: premium, custom, and economy. The RSMeans daily output for base and case moldings is in the range of 200 to 250 L.F. per carpenter per day. This is appropriate for most average custom-grade projects. For premium projects, an adjustment to productivity of 25–50% should be made, depending on the complexity of the job.

Reference Numbers

Reference numbers are shown at the beginning of some major classifications. These numbers refer to related items in the Reference Section. The reference information may be an estimating procedure, an alternate pricing method, or technical information.

Note: Not all subdivisions listed here necessarily appear. ■

06 05 05 – Selective Demolition for Wood, Plastics, and Composites

06 05 05.10 Selective Demolition Wood Framing		Crew	Daily Output	Labor-Hours	Unit	Material	2020 Bare Costs Labor	Equipment	Total	Total Incl O&P
0010	**SELECTIVE DEMOLITION WOOD FRAMING** R024119-10									
0100	Timber connector, nailed, small	1 Clab	96	.083	Ea.		3.51		3.51	5.60
0110	Medium		60	.133			5.60		5.60	9
0120	Large		48	.167			7		7	11.25
0130	Bolted, small		48	.167			7		7	11.25
0140	Medium		32	.250			10.55		10.55	16.85
0150	Large		24	.333			14.05		14.05	22.50
2958	Beams, 2" x 6"	2 Clab	1100	.015	L.F.		.61		.61	.98
2960	2" x 8"		825	.019			.82		.82	1.31
2965	2" x 10"		665	.024			1.01		1.01	1.62
2970	2" x 12"		550	.029			1.22		1.22	1.96
2972	2" x 14"		470	.034			1.43		1.43	2.30
2975	4" x 8"	B-1	413	.058			2.49		2.49	3.98
2980	4" x 10"		330	.073			3.11		3.11	4.98
2985	4" x 12"		275	.087			3.73		3.73	6
3000	6" x 8"		275	.087			3.73		3.73	6
3040	6" x 10"		220	.109			4.67		4.67	7.45
3080	6" x 12"		185	.130			5.55		5.55	8.90
3120	8" x 12"		140	.171			7.35		7.35	11.75
3160	10" x 12"		110	.218			9.35		9.35	14.95
3162	Alternate pricing method		1.10	21.818	M.B.F.		935		935	1,500
3170	Blocking, in 16" OC wall framing, 2" x 4"	1 Clab	600	.013	L.F.		.56		.56	.90
3172	2" x 6"		400	.020			.84		.84	1.35
3174	In 24" OC wall framing, 2" x 4"		600	.013			.56		.56	.90
3176	2" x 6"		400	.020			.84		.84	1.35
3178	Alt method, wood blocking removal from wood framing		.40	20	M.B.F.		840		840	1,350
3179	Wood blocking removal from steel framing		.36	22.222	"		935		935	1,500
3180	Bracing, let in, 1" x 3", studs 16" OC		1050	.008	L.F.		.32		.32	.51
3181	Studs 24" OC		1080	.007			.31		.31	.50
3182	1" x 4", studs 16" OC		1050	.008			.32		.32	.51
3183	Studs 24" OC		1080	.007			.31		.31	.50
3184	1" x 6", studs 16" OC		1050	.008			.32		.32	.51
3185	Studs 24" OC		1080	.007			.31		.31	.50
3186	2" x 3", studs 16" OC		800	.010			.42		.42	.67
3187	Studs 24" OC		830	.010			.41		.41	.65
3188	2" x 4", studs 16" OC		800	.010			.42		.42	.67
3189	Studs 24" OC		830	.010			.41		.41	.65
3190	2" x 6", studs 16" OC		800	.010			.42		.42	.67
3191	Studs 24" OC		830	.010			.41		.41	.65
3192	2" x 8", studs 16" OC		800	.010			.42		.42	.67
3193	Studs 24" OC		830	.010			.41		.41	.65
3194	"T" shaped metal bracing, studs at 16" OC		1060	.008			.32		.32	.51
3195	Studs at 24" OC		1200	.007			.28		.28	.45
3196	Metal straps, studs at 16" OC		1200	.007			.28		.28	.45
3197	Studs at 24" OC		1240	.006			.27		.27	.44
3200	Columns, round, 8' to 14' tall		40	.200	Ea.		8.40		8.40	13.50
3202	Dimensional lumber sizes	2 Clab	1.10	14.545	M.B.F.		610		610	980
3250	Blocking, between joists	1 Clab	320	.025	Ea.		1.05		1.05	1.69
3252	Bridging, metal strap, between joists		320	.025	Pr.		1.05		1.05	1.69
3254	Wood, between joists		320	.025	"		1.05		1.05	1.69
3260	Door buck, studs, header & access., 8' high 2" x 4" wall, 3' wide		32	.250	Ea.		10.55		10.55	16.85
3261	4' wide		32	.250			10.55		10.55	16.85
3262	5' wide		32	.250			10.55		10.55	16.85

06 05 05 – Selective Demolition for Wood, Plastics, and Composites

06 05 05.10 Selective Demolition Wood Framing		Crew	Daily Output	Labor-Hours	Unit	Material	2020 Bare Costs Labor	Equipment	Total	Total Incl O&P
3263	6′ wide	1 Clab	32	.250	Ea.		10.55		10.55	16.85
3264	8′ wide		30	.267			11.25		11.25	18
3265	10′ wide		30	.267			11.25		11.25	18
3266	12′ wide		30	.267			11.25		11.25	18
3267	2″ x 6″ wall, 3′ wide		32	.250			10.55		10.55	16.85
3268	4′ wide		32	.250			10.55		10.55	16.85
3269	5′ wide		32	.250			10.55		10.55	16.85
3270	6′ wide		32	.250			10.55		10.55	16.85
3271	8′ wide		30	.267			11.25		11.25	18
3272	10′ wide		30	.267			11.25		11.25	18
3273	12′ wide		30	.267			11.25		11.25	18
3274	Window buck, studs, header & access., 8′ high 2″ x 4″ wall, 2′ wide		24	.333			14.05		14.05	22.50
3275	3′ wide		24	.333			14.05		14.05	22.50
3276	4′ wide		24	.333			14.05		14.05	22.50
3277	5′ wide		24	.333			14.05		14.05	22.50
3278	6′ wide		24	.333			14.05		14.05	22.50
3279	7′ wide		24	.333			14.05		14.05	22.50
3280	8′ wide		22	.364			15.30		15.30	24.50
3281	10′ wide		22	.364			15.30		15.30	24.50
3282	12′ wide		22	.364			15.30		15.30	24.50
3283	2″ x 6″ wall, 2′ wide		24	.333			14.05		14.05	22.50
3284	3′ wide		24	.333			14.05		14.05	22.50
3285	4′ wide		24	.333			14.05		14.05	22.50
3286	5′ wide		24	.333			14.05		14.05	22.50
3287	6′ wide		24	.333			14.05		14.05	22.50
3288	7′ wide		24	.333			14.05		14.05	22.50
3289	8′ wide		22	.364			15.30		15.30	24.50
3290	10′ wide		22	.364			15.30		15.30	24.50
3291	12′ wide		22	.364			15.30		15.30	24.50
3360	Deck or porch decking		825	.010	L.F.		.41		.41	.65
3400	Fascia boards, 1″ x 6″		500	.016			.67		.67	1.08
3440	1″ x 8″		450	.018			.75		.75	1.20
3480	1″ x 10″		400	.020			.84		.84	1.35
3490	2″ x 6″		450	.018			.75		.75	1.20
3500	2″ x 8″		400	.020			.84		.84	1.35
3510	2″ x 10″		350	.023			.96		.96	1.54
3610	Furring, on wood walls or ceiling		4000	.002	S.F.		.08		.08	.13
3620	On masonry or concrete walls or ceiling		1200	.007	″		.28		.28	.45
3800	Headers over openings, 2 @ 2″ x 6″		110	.073	L.F.		3.06		3.06	4.91
3840	2 @ 2″ x 8″		100	.080			3.37		3.37	5.40
3880	2 @ 2″ x 10″		90	.089			3.74		3.74	6
3885	Alternate pricing method		.26	30.651	M.B.F.		1,300		1,300	2,075
3920	Joists, 1″ x 4″		1250	.006	L.F.		.27		.27	.43
3930	1″ x 6″		1135	.007			.30		.30	.48
3940	1″ x 8″		1000	.008			.34		.34	.54
3950	1″ x 10″		895	.009			.38		.38	.60
3960	1″ x 12″		765	.010			.44		.44	.71
4200	2″ x 4″	2 Clab	1000	.016			.67		.67	1.08
4230	2″ x 6″		970	.016			.69		.69	1.11
4240	2″ x 8″		940	.017			.72		.72	1.15
4250	2″ x 10″		910	.018			.74		.74	1.19
4280	2″ x 12″		880	.018			.77		.77	1.23
4281	2″ x 14″		850	.019			.79		.79	1.27

06 05 05 – Selective Demolition for Wood, Plastics, and Composites

06 05 05.10 Selective Demolition Wood Framing		Crew	Daily Output	Labor-Hours	Unit	Material	2020 Bare Costs Labor	Equipment	Total	Total Incl O&P
4282	Composite joists, 9-1/2"	2 Clab	960	.017	L.F.		.70		.70	1.12
4283	11-7/8"		930	.017			.72		.72	1.16
4284	14"		897	.018			.75		.75	1.20
4285	16"		865	.019			.78		.78	1.25
4290	Wood joists, alternate pricing method		1.50	10.667	M.B.F.		450		450	720
4500	Open web joist, 12" deep		500	.032	L.F.		1.35		1.35	2.16
4505	14" deep		475	.034			1.42		1.42	2.27
4510	16" deep		450	.036			1.50		1.50	2.40
4520	18" deep		425	.038			1.59		1.59	2.54
4530	24" deep		400	.040			1.68		1.68	2.70
4550	Ledger strips, 1" x 2"	1 Clab	1200	.007			.28		.28	.45
4560	1" x 3"		1200	.007			.28		.28	.45
4570	1" x 4"		1200	.007			.28		.28	.45
4580	2" x 2"		1100	.007			.31		.31	.49
4590	2" x 4"		1000	.008			.34		.34	.54
4600	2" x 6"		1000	.008			.34		.34	.54
4601	2" x 8 or 2" x 10"		800	.010			.42		.42	.67
4602	4" x 6"		600	.013			.56		.56	.90
4604	4" x 8"		450	.018			.75		.75	1.20
5400	Posts, 4" x 4"	2 Clab	800	.020			.84		.84	1.35
5405	4" x 6"		550	.029			1.22		1.22	1.96
5410	4" x 8"		440	.036			1.53		1.53	2.45
5425	4" x 10"		390	.041			1.73		1.73	2.77
5430	4" x 12"		350	.046			1.92		1.92	3.08
5440	6" x 6"		400	.040			1.68		1.68	2.70
5445	6" x 8"		350	.046			1.92		1.92	3.08
5450	6" x 10"		320	.050			2.11		2.11	3.37
5455	6" x 12"		290	.055			2.32		2.32	3.72
5480	8" x 8"		300	.053			2.25		2.25	3.60
5500	10" x 10"		240	.067			2.81		2.81	4.50
5660	T&G floor planks		2	8	M.B.F.		335		335	540
5682	Rafters, ordinary, 16" OC, 2" x 4"		880	.018	S.F.		.77		.77	1.23
5683	2" x 6"		840	.019			.80		.80	1.28
5684	2" x 8"		820	.020			.82		.82	1.32
5685	2" x 10"		820	.020			.82		.82	1.32
5686	2" x 12"		810	.020			.83		.83	1.33
5687	24" OC, 2" x 4"		1170	.014			.58		.58	.92
5688	2" x 6"		1117	.014			.60		.60	.97
5689	2" x 8"		1091	.015			.62		.62	.99
5690	2" x 10"		1091	.015			.62		.62	.99
5691	2" x 12"		1077	.015			.63		.63	1
5795	Rafters, ordinary, 2" x 4" (alternate method)		862	.019	L.F.		.78		.78	1.25
5800	2" x 6" (alternate method)		850	.019			.79		.79	1.27
5840	2" x 8" (alternate method)		837	.019			.81		.81	1.29
5855	2" x 10" (alternate method)		825	.019			.82		.82	1.31
5865	2" x 12" (alternate method)		812	.020			.83		.83	1.33
5870	Sill plate, 2" x 4"	1 Clab	1170	.007			.29		.29	.46
5871	2" x 6"		780	.010			.43		.43	.69
5872	2" x 8"		586	.014			.57		.57	.92
5873	Alternate pricing method		.78	10.256	M.B.F.		430		430	690
5885	Ridge board, 1" x 4"	2 Clab	900	.018	L.F.		.75		.75	1.20
5886	1" x 6"		875	.018			.77		.77	1.23
5887	1" x 8"		850	.019			.79		.79	1.27

06 05 05 – Selective Demolition for Wood, Plastics, and Composites

06 05 05.10 Selective Demolition Wood Framing		Crew	Daily Output	Labor-Hours	Unit	Material	2020 Bare Costs Labor	Equipment	Total	Total Incl O&P
5888	1" x 10"	2 Clab	825	.019	L.F.		.82		.82	1.31
5889	1" x 12"		800	.020			.84		.84	1.35
5890	2" x 4"		900	.018			.75		.75	1.20
5892	2" x 6"		875	.018			.77		.77	1.23
5894	2" x 8"		850	.019			.79		.79	1.27
5896	2" x 10"		825	.019			.82		.82	1.31
5898	2" x 12"		800	.020			.84		.84	1.35
5900	Hip & valley rafters, 2" x 6"		500	.032			1.35		1.35	2.16
5940	2" x 8"		420	.038			1.60		1.60	2.57
6050	Rafter tie, 1" x 4"		1250	.013			.54		.54	.86
6052	1" x 6"		1135	.014			.59		.59	.95
6054	2" x 4"		1000	.016			.67		.67	1.08
6056	2" x 6"	↓	970	.016			.69		.69	1.11
6070	Sleepers, on concrete, 1" x 2"	1 Clab	4700	.002			.07		.07	.11
6075	1" x 3"		4000	.002			.08		.08	.13
6080	2" x 4"		3000	.003			.11		.11	.18
6085	2" x 6"	↓	2600	.003	↓		.13		.13	.21
6086	Sheathing from roof, 5/16"	2 Clab	1600	.010	S.F.		.42		.42	.67
6088	3/8"		1525	.010			.44		.44	.71
6090	1/2"		1400	.011			.48		.48	.77
6092	5/8"		1300	.012			.52		.52	.83
6094	3/4"		1200	.013			.56		.56	.90
6096	Board sheathing from roof		1400	.011			.48		.48	.77
6100	Sheathing, from walls, 1/4"		1200	.013			.56		.56	.90
6110	5/16"		1175	.014			.57		.57	.92
6120	3/8"		1150	.014			.59		.59	.94
6130	1/2"		1125	.014			.60		.60	.96
6140	5/8"		1100	.015			.61		.61	.98
6150	3/4"		1075	.015			.63		.63	1
6152	Board sheathing from walls		1500	.011			.45		.45	.72
6158	Subfloor/roof deck, with boards		2200	.007			.31		.31	.49
6159	Subfloor/roof deck, with tongue & groove boards		2000	.008			.34		.34	.54
6160	Plywood, 1/2" thick		768	.021			.88		.88	1.41
6162	5/8" thick		760	.021			.89		.89	1.42
6164	3/4" thick		750	.021			.90		.90	1.44
6165	1-1/8" thick	↓	720	.022			.94		.94	1.50
6166	Underlayment, particle board, 3/8" thick	1 Clab	780	.010			.43		.43	.69
6168	1/2" thick		768	.010			.44		.44	.70
6170	5/8" thick		760	.011			.44		.44	.71
6172	3/4" thick	↓	750	.011	↓		.45		.45	.72
6200	Stairs and stringers, straight run	2 Clab	40	.400	Riser		16.85		16.85	27
6240	With platforms, winders or curves	"	26	.615	"		26		26	41.50
6300	Components, tread	1 Clab	110	.073	Ea.		3.06		3.06	4.91
6320	Riser		80	.100	"		4.21		4.21	6.75
6390	Stringer, 2" x 10"		260	.031	L.F.		1.30		1.30	2.08
6400	2" x 12"		260	.031			1.30		1.30	2.08
6410	3" x 10"		250	.032			1.35		1.35	2.16
6420	3" x 12"	↓	250	.032			1.35		1.35	2.16
6590	Wood studs, 2" x 3"	2 Clab	3076	.005			.22		.22	.35
6600	2" x 4"		2000	.008			.34		.34	.54
6640	2" x 6"	↓	1600	.010	↓		.42		.42	.67
6720	Wall framing, including studs, plates and blocking, 2" x 4"	1 Clab	600	.013	S.F.		.56		.56	.90
6740	2" x 6"	↓	480	.017	"		.70		.70	1.12

06 05 05 – Selective Demolition for Wood, Plastics, and Composites

06 05 05.10 Selective Demolition Wood Framing

		Crew	Daily Output	Labor-Hours	Unit	Material	2020 Bare Costs Labor	Equipment	Total	Total Incl O&P
6750	Headers, 2" x 4"	1 Clab	1125	.007	L.F.		.30		.30	.48
6755	2" x 6"		1125	.007			.30		.30	.48
6760	2" x 8"		1050	.008			.32		.32	.51
6765	2" x 10"		1050	.008			.32		.32	.51
6770	2" x 12"		1000	.008			.34		.34	.54
6780	4" x 10"		525	.015			.64		.64	1.03
6785	4" x 12"		500	.016			.67		.67	1.08
6790	6" x 8"		560	.014			.60		.60	.96
6795	6" x 10"		525	.015			.64		.64	1.03
6797	6" x 12"		500	.016			.67		.67	1.08
7000	Trusses									
7050	12' span	2 Clab	74	.216	Ea.		9.10		9.10	14.60
7150	24' span	F-3	66	.606			33	7.15	40.15	60.50
7200	26' span		64	.625			34	7.35	41.35	62
7250	28' span		62	.645			35	7.60	42.60	64.50
7300	30' span		58	.690			37.50	8.10	45.60	69
7350	32' span		56	.714			39	8.40	47.40	71.50
7400	34' span		54	.741			40.50	8.75	49.25	73.50
7450	36' span		52	.769			42	9.05	51.05	76.50
8000	Soffit, T&G wood	1 Clab	520	.015	S.F.		.65		.65	1.04
8010	Hardboard, vinyl or aluminum	"	640	.013			.53		.53	.84
8030	Plywood	2 Carp	315	.051			2.70		2.70	4.32
9000	Minimum labor/equipment charge	1 Clab	4	2	Job		84		84	135
9500	See Section 02 41 19.19 for rubbish handling									

06 05 05.20 Selective Demolition Millwork and Trim

		Crew	Daily Output	Labor-Hours	Unit	Material	2020 Bare Costs Labor	Equipment	Total	Total Incl O&P
0010	**SELECTIVE DEMOLITION MILLWORK AND TRIM** R024119-10									
1000	Cabinets, wood, base cabinets, per L.F.	2 Clab	80	.200	L.F.		8.40		8.40	13.50
1020	Wall cabinets, per L.F.	"	80	.200	"		8.40		8.40	13.50
1060	Remove and reset, base cabinets	2 Carp	18	.889	Ea.		47		47	75.50
1070	Wall cabinets		20	.800			42.50		42.50	68
1072	Oven cabinet, 7' high		11	1.455			77.50		77.50	124
1074	Cabinet door, up to 2' high	1 Clab	66	.121			5.10		5.10	8.20
1076	2' - 4' high	"	46	.174			7.30		7.30	11.75
1100	Steel, painted, base cabinets	2 Clab	60	.267	L.F.		11.25		11.25	18
1120	Wall cabinets		60	.267	"		11.25		11.25	18
1200	Casework, large area		320	.050	S.F.		2.11		2.11	3.37
1220	Selective		200	.080	"		3.37		3.37	5.40
1500	Counter top, straight runs		200	.080	L.F.		3.37		3.37	5.40
1510	L, U or C shapes		120	.133			5.60		5.60	9
1550	Remove and reset, straight runs	2 Carp	50	.320			17		17	27.50
1560	L, U or C shape	"	40	.400			21.50		21.50	34
2000	Paneling, 4' x 8' sheets	2 Clab	2000	.008	S.F.		.34		.34	.54
2100	Boards, 1" x 4"		700	.023			.96		.96	1.54
2120	1" x 6"		750	.021			.90		.90	1.44
2140	1" x 8"		800	.020			.84		.84	1.35
3000	Trim, baseboard, to 6" wide		1200	.013	L.F.		.56		.56	.90
3040	Greater than 6" and up to 12" wide		1000	.016			.67		.67	1.08
3080	Remove and reset, minimum	2 Carp	400	.040			2.13		2.13	3.41
3090	Maximum	"	300	.053			2.83		2.83	4.54
3100	Ceiling trim	2 Clab	1000	.016			.67		.67	1.08
3120	Chair rail		1200	.013			.56		.56	.90
3140	Railings with balusters		240	.067			2.81		2.81	4.50

06 05 05 – Selective Demolition for Wood, Plastics, and Composites

06 05 05.20 Selective Demolition Millwork and Trim		Crew	Daily Output	Labor-Hours	Unit	Material	2020 Bare Costs Labor	Equipment	Total	Total Incl O&P
3160	Wainscoting	2 Clab	700	.023	S.F.		.96		.96	1.54
9000	Minimum labor/equipment charge	1 Clab	4	2	Job		84		84	135

06 05 23 – Wood, Plastic, and Composite Fastenings

06 05 23.10 Nails

		Crew	Daily Output	Labor-Hours	Unit	Material	Labor	Equipment	Total	Total Incl O&P
0010	**NAILS**, material only, based upon 50# box purchase									
0020	Copper nails, plain				Lb.	11.65			11.65	12.85
0400	Stainless steel, plain					10.10			10.10	11.10
0500	Box, 3d to 20d, bright					1.52			1.52	1.67
0520	Galvanized					2.55			2.55	2.81
0600	Common, 3d to 60d, plain					1.13			1.13	1.24
0700	Galvanized					2.48			2.48	2.73
0800	Aluminum					11.05			11.05	12.15
1000	Annular or spiral thread, 4d to 60d, plain					2.85			2.85	3.14
1200	Galvanized					3.49			3.49	3.84
1400	Drywall nails, plain					2.01			2.01	2.21
1600	Galvanized					1.89			1.89	2.08
1800	Finish nails, 4d to 10d, plain					1.57			1.57	1.73
2000	Galvanized					1.93			1.93	2.12
2100	Aluminum					8.75			8.75	9.65
2300	Flooring nails, hardened steel, 2d to 10d, plain					3.88			3.88	4.27
2400	Galvanized					4.09			4.09	4.50
2500	Gypsum lath nails, 1-1/8", 13 ga. flathead, blued					3.64			3.64	4
2600	Masonry nails, hardened steel, 3/4" to 3" long, plain					2.34			2.34	2.57
2700	Galvanized					4.14			4.14	4.55
2900	Roofing nails, threaded, galvanized					2.93			2.93	3.22
3100	Aluminum					7.60			7.60	8.40
3300	Compressed lead head, threaded, galvanized					3.06			3.06	3.37
3600	Siding nails, plain shank, galvanized					2.59			2.59	2.85
3800	Aluminum					5.95			5.95	6.55
5000	Add to prices above for cement coating					.16			.16	.18
5200	Zinc or tin plating					.27			.27	.30
5500	Vinyl coated sinkers, 8d to 16d					2.51			2.51	2.76

06 05 23.40 Sheet Metal Screws

		Crew	Daily Output	Labor-Hours	Unit	Material	Labor	Equipment	Total	Total Incl O&P
0010	**SHEET METAL SCREWS**									
0020	Steel, standard, #8 x 3/4", plain				C	2.04			2.04	2.24
0100	Galvanized					2.49			2.49	2.74
0300	#10 x 1", plain					3.77			3.77	4.15
0400	Galvanized					4.34			4.34	4.77
0600	With washers, #14 x 1", plain					7.60			7.60	8.35
0700	Galvanized					8			8	8.80
0900	#14 x 2", plain					24			24	26.50
1000	Galvanized					11.45			11.45	12.60
1500	Self-drilling, with washers (pinch point), #8 x 3/4", plain					10.35			10.35	11.40
1600	Galvanized					3.13			3.13	3.44
1800	#10 x 3/4", plain					3.62			3.62	3.98
1900	Galvanized					3.99			3.99	4.39
3000	Stainless steel w/aluminum or neoprene washers, #14 x 1", plain					14			14	15.40
3100	#14 x 2", plain					23			23	25.50

06 05 23.50 Wood Screws

		Crew	Daily Output	Labor-Hours	Unit	Material	Labor	Equipment	Total	Total Incl O&P
0010	**WOOD SCREWS**									
0020	#8, 1" long, steel				C	4.03			4.03	4.43
0100	Brass					12.30			12.30	13.50

06 05 23 – Wood, Plastic, and Composite Fastenings

06 05 23.50 Wood Screws		Crew	Daily Output	Labor-Hours	Unit	Material	Labor	Equipment	Total	Total Incl O&P
0200	#8, 2" long, steel				C	6.90			6.90	7.60
0300	Brass					27.50			27.50	30
0400	#10, 1" long, steel					3.55			3.55	3.91
0500	Brass					15.65			15.65	17.25
0600	#10, 2" long, steel					5.15			5.15	5.65
0700	Brass					26			26	28.50
0800	#10, 3" long, steel					9.20			9.20	10.15
1000	#12, 2" long, steel					7.70			7.70	8.45
1100	Brass					36.50			36.50	40
1500	#12, 3" long, steel					11.55			11.55	12.70
2000	#12, 4" long, steel				▼	26.50			26.50	29
2050	Composite trim screws									
2100	Stainless steel, trim head, #7 x 1-5/8" long				C	27.50			27.50	30
2200	#7 x 2" long					27			27	30
2300	#7 x 2-1/2" long					30.50			30.50	33.50
2400	#7 x 3" long					35.50			35.50	39
2500	#9 x 3-1/2" long					76.50			76.50	84.50
2600	#9 x 4" long					90			90	99
3000	Stainless steel, flat head, #6 x 1-5/8" long					25.50			25.50	28
3100	#8 x 2" long					18.05			18.05	19.85
3200	#8 x 2-1/2" long					21.50			21.50	23.50
3300	#8 x 3" long					25			25	27.50
3400	#10 x 2" long					20.50			20.50	22.50
3500	#10 x 3-1/2" long					41			41	45.50
3600	#12 x 3" long					47.50			47.50	52
3700	#12 x 4" long					72.50			72.50	80
3800	#12 x 6" long				▼	124			124	137

06 05 23.60 Timber Connectors

		Crew	Daily Output	Labor-Hours	Unit	Material	Labor	Equipment	Total	Total Incl O&P
0010	**TIMBER CONNECTORS**									
0020	Add up cost of each part for total cost of connection									
0100	Connector plates, steel, with bolts, straight	2 Carp	75	.213	Ea.	27.50	11.35		38.85	48.50
0110	Tee, 7 ga.		50	.320		41.50	17		58.50	73
0120	T- Strap, 14 ga., 12" x 8" x 2"		50	.320		41.50	17		58.50	73
0150	Anchor plates, 7 ga., 9" x 7"	▼	75	.213		27.50	11.35		38.85	48.50
0200	Bolts, machine, sq. hd. with nut & washer, 1/2" diameter, 4" long	1 Carp	140	.057		.77	3.04		3.81	5.70
0300	7-1/2" long		130	.062		1.41	3.27		4.68	6.80
0500	3/4" diameter, 7-1/2" long		130	.062		3.51	3.27		6.78	9.10
0610	Machine bolts, w/nut, washer, 3/4" diameter, 15" L, HD's & beam hangers		95	.084	▼	6.60	4.48		11.08	14.40
0800	Drilling bolt holes in timber, 1/2" diameter		450	.018	Inch		.95		.95	1.51
0900	1" diameter		350	.023	"		1.22		1.22	1.95
1100	Framing anchor, angle, 3" x 3" x 1-1/2", 12 ga.		175	.046	Ea.	2.53	2.43		4.96	6.65
1150	Framing anchors, 18 ga., 4-1/2" x 2-3/4"		175	.046		2.53	2.43		4.96	6.65
1160	Framing anchors, 18 ga., 4-1/2" x 3"		175	.046		2.53	2.43		4.96	6.65
1170	Clip anchors plates, 18 ga., 12" x 1-1/8"		175	.046		2.53	2.43		4.96	6.65
1250	Holdowns, 3 ga. base, 10 ga. body		8	1		46.50	53		99.50	136
1260	Holdowns, 7 ga. 11-1/16" x 3-1/4"		8	1		46.50	53		99.50	136
1270	Holdowns, 7 ga. 14-3/8" x 3-1/8"		8	1		46.50	53		99.50	136
1275	Holdowns, 12 ga. 8" x 2-1/2"		8	1		46.50	53		99.50	136
1300	Joist and beam hangers, 18 ga. galv., for 2" x 4" joist		175	.046		.85	2.43		3.28	4.83
1400	2" x 6" to 2" x 10" joist	▼	165	.048	▼	1.56	2.58		4.14	5.85

06 05 23 – Wood, Plastic, and Composite Fastenings

06 05 23.60 Timber Connectors		Crew	Daily Output	Labor-Hours	Unit	Material	Labor	Equipment	Total	Total Incl O&P
						2020 Bare Costs				
1600	16 ga. galv., 3" x 6" to 3" x 10" joist	1 Carp	160	.050	Ea.	3.03	2.66		5.69	7.60
1700	3" x 10" to 3" x 14" joist		160	.050		4.64	2.66		7.30	9.35
1800	4" x 6" to 4" x 10" joist		155	.052		3.59	2.74		6.33	8.35
1900	4" x 10" to 4" x 14" joist		155	.052		5	2.74		7.74	9.90
2000	Two-2" x 6" to two-2" x 10" joists		150	.053		3.98	2.83		6.81	8.90
2100	Two-2" x 10" to two-2" x 14" joists		150	.053		4.55	2.83		7.38	9.55
2300	3/16" thick, 6" x 8" joist		145	.055		75	2.93		77.93	87
2400	6" x 10" joist		140	.057		77.50	3.04		80.54	90.50
2500	6" x 12" joist		135	.059		79.50	3.15		82.65	92.50
2700	1/4" thick, 6" x 14" joist	↓	130	.062		82	3.27		85.27	95.50
2900	Plywood clips, extruded aluminum H clip, for 3/4" panels					.24			.24	.26
3000	Galvanized 18 ga. back-up clip					.19			.19	.21
3200	Post framing, 16 ga. galv. for 4" x 4" base, 2 piece	1 Carp	130	.062		15.25	3.27		18.52	22
3300	Cap		130	.062		25	3.27		28.27	33
3500	Rafter anchors, 18 ga. galv., 1-1/2" wide, 5-1/4" long		145	.055		.41	2.93		3.34	5.15
3600	10-3/4" long		145	.055		1.15	2.93		4.08	5.95
3800	Shear plates, 2-5/8" diameter		120	.067		3.20	3.54		6.74	9.20
3900	4" diameter		115	.070		2.69	3.70		6.39	8.85
4000	Sill anchors, embedded in concrete or block, 25-1/2" long		115	.070		16.10	3.70		19.80	23.50
4100	Spike grids, 3" x 6"		120	.067		1.34	3.54		4.88	7.15
4400	Split rings, 2-1/2" diameter		120	.067		3.31	3.54		6.85	9.35
4500	4" diameter		110	.073		3.51	3.87		7.38	10.05
4550	Tie plate, 20 ga., 7" x 3-1/8"		110	.073		3.51	3.87		7.38	10.05
4560	5" x 4-1/8"		110	.073		3.51	3.87		7.38	10.05
4575	Twist straps, 18 ga., 12" x 1-1/4"		110	.073		3.51	3.87		7.38	10.05
4580	16" x 1-1/4"		110	.073		3.51	3.87		7.38	10.05
4600	Strap ties, 20 ga., 2-1/16" wide, 12-13/16" long		180	.044		1.01	2.36		3.37	4.89
4700	16 ga., 1-3/8" wide, 12" long		180	.044		1.01	2.36		3.37	4.89
4800	1-1/4" wide, 21-5/8" long		160	.050		2.81	2.66		5.47	7.35
5000	Toothed rings, 2-5/8" or 4" diameter		90	.089	↓	2.33	4.72		7.05	10.10
5200	Truss plates, nailed, 20 ga., up to 32' span	↓	17	.471	Truss	15.15	25		40.15	56.50
5400	Washers, 2" x 2" x 1/8"				Ea.	.48			.48	.53
5500	3" x 3" x 3/16"				"	1.26			1.26	1.39
9000	Minimum labor/equipment charge	1 Carp	4	2	Job		106		106	170

06 05 23.80 Metal Bracing		Crew	Daily Output	Labor-Hours	Unit	Material	Labor	Equipment	Total	Total Incl O&P
0010	**METAL BRACING**									
0302	Let-in, "T" shaped, 22 ga. galv. steel, studs at 16" OC	1 Carp	580	.014	L.F.	.85	.73		1.58	2.11
0402	Studs at 24" OC		600	.013		.85	.71		1.56	2.08
0502	Steel straps, 16 ga. galv. steel, studs at 16" OC		600	.013		1.10	.71		1.81	2.35
0602	Studs at 24" OC	↓	620	.013	↓	1.10	.69		1.79	2.31

06 11 Wood Framing

06 11 10 – Framing with Dimensional, Engineered or Composite Lumber

06 11 10.01 Forest Stewardship Council Certification

	06 11 10.01 Forest Stewardship Council Certification	Crew	Daily Output	Labor-Hours	Unit	Material	2020 Bare Costs Labor	Equipment	Total	Total Incl O&P
0010	**FOREST STEWARDSHIP COUNCIL CERTIFICATION**									
0020	For Forest Stewardship Council (FSC) cert dimension lumber, add [G]					65%				

06 11 10.02 Blocking

	06 11 10.02 Blocking	Crew	Daily Output	Labor-Hours	Unit	Material	Labor	Equipment	Total	Total Incl O&P
0010	**BLOCKING**									
2600	Miscellaneous, to wood construction									
2620	2″ x 4″	1 Carp	.17	47.059	M.B.F.	710	2,500		3,210	4,775
2625	Pneumatic nailed		.21	38.095		720	2,025		2,745	4,050
2660	2″ x 8″		.27	29.630		750	1,575		2,325	3,350
2665	Pneumatic nailed		.33	24.242		760	1,300		2,060	2,900
2720	To steel construction									
2740	2″ x 4″	1 Carp	.14	57.143	M.B.F.	710	3,025		3,735	5,650
2780	2″ x 8″		.21	38.095	″	750	2,025		2,775	4,075
9000	Minimum labor/equipment charge		4	2	Job		106		106	170

06 11 10.04 Wood Bracing

	06 11 10.04 Wood Bracing	Crew	Daily Output	Labor-Hours	Unit	Material	Labor	Equipment	Total	Total Incl O&P
0010	**WOOD BRACING**									
0012	Let-in, with 1″ x 6″ boards, studs @ 16″ OC	1 Carp	150	.053	L.F.	.83	2.83		3.66	5.45
0202	Studs @ 24″ OC	″	230	.035	″	.83	1.85		2.68	3.87

06 11 10.06 Bridging

	06 11 10.06 Bridging	Crew	Daily Output	Labor-Hours	Unit	Material	Labor	Equipment	Total	Total Incl O&P
0010	**BRIDGING**									
0012	Wood, for joists 16″ OC, 1″ x 3″	1 Carp	130	.062	Pr.	.70	3.27		3.97	6
0017	Pneumatic nailed		170	.047		.79	2.50		3.29	4.88
0102	2″ x 3″ bridging		130	.062		.72	3.27		3.99	6.05
0107	Pneumatic nailed		170	.047		.77	2.50		3.27	4.85
0302	Steel, galvanized, 18 ga., for 2″ x 10″ joists at 12″ OC		130	.062		1.71	3.27		4.98	7.15
0352	16″ OC		135	.059		1.72	3.15		4.87	6.95
0402	24″ OC		140	.057		2.63	3.04		5.67	7.75
0602	For 2″ x 14″ joists at 16″ OC		130	.062		1.98	3.27		5.25	7.45
0902	Compression type, 16″ OC, 2″ x 8″ joists		200	.040		1.43	2.13		3.56	4.98
1002	2″ x 12″ joists		200	.040		1.43	2.13		3.56	4.98

06 11 10.10 Beam and Girder Framing

	06 11 10.10 Beam and Girder Framing	Crew	Daily Output	Labor-Hours	Unit	Material	Labor	Equipment	Total	Total Incl O&P
0010	**BEAM AND GIRDER FRAMING** R061110-30									
1000	Single, 2″ x 6″	2 Carp	700	.023	L.F.	.69	1.22		1.91	2.70
1005	Pneumatic nailed		812	.020		.69	1.05		1.74	2.44
1020	2″ x 8″		650	.025		1	1.31		2.31	3.20
1025	Pneumatic nailed		754	.021		1.01	1.13		2.14	2.92
1040	2″ x 10″		600	.027		1.33	1.42		2.75	3.73
1045	Pneumatic nailed		696	.023		1.34	1.22		2.56	3.44
1060	2″ x 12″		550	.029		1.74	1.55		3.29	4.39
1065	Pneumatic nailed		638	.025		1.76	1.33		3.09	4.07
1080	2″ x 14″		500	.032		2.38	1.70		4.08	5.35
1085	Pneumatic nailed		580	.028		2.40	1.47		3.87	4.99
1100	3″ x 8″		550	.029		2.95	1.55		4.50	5.70
1120	3″ x 10″		500	.032		3.57	1.70		5.27	6.65
1140	3″ x 12″		450	.036		4.83	1.89		6.72	8.35
1160	3″ x 14″		400	.040		5.30	2.13		7.43	9.25
1170	4″ x 6″	F-3	1100	.036		3.06	1.98	.43	5.47	7
1180	4″ x 8″		1000	.040		4.33	2.17	.47	6.97	8.75
1200	4″ x 10″		950	.042		5.45	2.29	.50	8.24	10.20
1220	4″ x 12″		900	.044		5.80	2.42	.52	8.74	10.85
1240	4″ x 14″		850	.047		6.70	2.56	.55	9.81	12.05
1250	6″ x 8″		525	.076		8.35	4.14	.90	13.39	16.75
1260	6″ x 10″		500	.080		7.10	4.35	.94	12.39	15.80

06 11 10 – Framing with Dimensional, Engineered or Composite Lumber

06 11 10.10 Beam and Girder Framing		Crew	Daily Output	Labor-Hours	Unit	Material	Labor	Equipment	Total	Total Incl O&P
						2020 Bare Costs				
1290	8" x 12"	F-3	300	.133	L.F.	17.45	7.25	1.57	26.27	32.50
2000	Double, 2" x 6"	2 Carp	625	.026		1.37	1.36		2.73	3.69
2005	Pneumatic nailed		725	.022		1.39	1.17		2.56	3.41
2020	2" x 8"		575	.028		2	1.48		3.48	4.57
2025	Pneumatic nailed		667	.024		2.02	1.28		3.30	4.26
2040	2" x 10"		550	.029		2.65	1.55		4.20	5.40
2045	Pneumatic nailed		638	.025		2.68	1.33		4.01	5.10
2060	2" x 12"		525	.030		3.48	1.62		5.10	6.45
2065	Pneumatic nailed		610	.026		3.51	1.39		4.90	6.10
2080	2" x 14"		475	.034		4.76	1.79		6.55	8.10
2085	Pneumatic nailed		551	.029		4.80	1.54		6.34	7.75
3000	Triple, 2" x 6"		550	.029		2.06	1.55		3.61	4.74
3005	Pneumatic nailed		638	.025		2.08	1.33		3.41	4.43
3020	2" x 8"		525	.030		3	1.62		4.62	5.90
3025	Pneumatic nailed		609	.026		3.03	1.40		4.43	5.55
3040	2" x 10"		500	.032		3.98	1.70		5.68	7.10
3045	Pneumatic nailed		580	.028		4.03	1.47		5.50	6.80
3060	2" x 12"		475	.034		5.20	1.79		6.99	8.60
3065	Pneumatic nailed		551	.029		5.25	1.54		6.79	8.25
3080	2" x 14"		450	.036		7.70	1.89		9.59	11.55
3085	Pneumatic nailed		522	.031		7.20	1.63		8.83	10.55
9000	Minimum labor/equipment charge	1 Carp	2	4	Job		213		213	340

06 11 10.12 Ceiling Framing		Crew	Daily Output	Labor-Hours	Unit	Material	Labor	Equipment	Total	Total Incl O&P
0010	**CEILING FRAMING**									
6000	Suspended, 2" x 3"	2 Carp	1000	.016	L.F.	.42	.85		1.27	1.83
6050	2" x 4"		900	.018		.47	.95		1.42	2.03
6100	2" x 6"		800	.020		.69	1.06		1.75	2.45
6150	2" x 8"		650	.025		1	1.31		2.31	3.20
9000	Minimum labor/equipment charge	1 Carp	4	2	Job		106		106	170

06 11 10.14 Posts and Columns		Crew	Daily Output	Labor-Hours	Unit	Material	Labor	Equipment	Total	Total Incl O&P
0010	**POSTS AND COLUMNS**									
0100	4" x 4"	2 Carp	390	.041	L.F.	2.04	2.18		4.22	5.75
0150	4" x 6"		275	.058		3.06	3.09		6.15	8.30
0200	4" x 8"		220	.073		4.33	3.87		8.20	10.95
0250	6" x 6"		215	.074		5.35	3.96		9.31	12.25
0300	6" x 8"		175	.091		8.35	4.86		13.21	16.95
0350	6" x 10"		150	.107		7.10	5.65		12.75	16.90
9000	Minimum labor/equipment charge	1 Carp	2	4	Job		213		213	340

06 11 10.18 Joist Framing		Crew	Daily Output	Labor-Hours	Unit	Material	Labor	Equipment	Total	Total Incl O&P
0010	**JOIST FRAMING** R061110-30									
2000	Joists, 2" x 4"	2 Carp	1250	.013	L.F.	.47	.68		1.15	1.61
2005	Pneumatic nailed		1438	.011		.48	.59		1.07	1.48
2100	2" x 6"		1250	.013		.69	.68		1.37	1.84
2105	Pneumatic nailed		1438	.011		.69	.59		1.28	1.71
2150	2" x 8"		1100	.015		1	.77		1.77	2.34
2155	Pneumatic nailed		1265	.013		1.01	.67		1.68	2.19
2200	2" x 10"		900	.018		1.33	.95		2.28	2.97
2205	Pneumatic nailed		1035	.015		1.34	.82		2.16	2.80
2250	2" x 12"		875	.018		1.74	.97		2.71	3.47
2255	Pneumatic nailed		1006	.016		1.76	.85		2.61	3.28
2300	2" x 14"		770	.021		2.38	1.10		3.48	4.39
2305	Pneumatic nailed		886	.018		2.40	.96		3.36	4.18

06 11 10 – Framing with Dimensional, Engineered or Composite Lumber

06 11 10.18 Joist Framing		Crew	Daily Output	Labor-Hours	Unit	Material (2020 Bare Costs)	Labor (2020 Bare Costs)	Equipment (2020 Bare Costs)	Total (2020 Bare Costs)	Total Incl O&P
2350	3" x 6"	2 Carp	925	.017	L.F.	2.21	.92		3.13	3.90
2400	3" x 10"		780	.021		3.57	1.09		4.66	5.70
2450	3" x 12"		600	.027		4.83	1.42		6.25	7.55
2500	4" x 6"		800	.020		3.06	1.06		4.12	5.05
2550	4" x 10"		600	.027		5.45	1.42		6.87	8.25
2600	4" x 12"		450	.036		5.80	1.89		7.69	9.45
2605	Sister joist, 2" x 6"		800	.020		.69	1.06		1.75	2.45
2606	Pneumatic nailed		960	.017		.69	.89		1.58	2.18
2610	2" x 8"		640	.025		1	1.33		2.33	3.23
2611	Pneumatic nailed		768	.021		1.01	1.11		2.12	2.88
2615	2" x 10"		535	.030		1.33	1.59		2.92	4.01
2616	Pneumatic nailed		642	.025		1.34	1.32		2.66	3.60
2620	2" x 12"		455	.035		1.74	1.87		3.61	4.90
2625	Pneumatic nailed		546	.029		1.76	1.56		3.32	4.42
3000	Composite wood joist 9-1/2" deep		.90	17.778	M.L.F.	1,600	945		2,545	3,275
3010	11-1/2" deep		.88	18.182		2,000	965		2,965	3,750
3020	14" deep		.82	19.512		2,375	1,025		3,400	4,250
3030	16" deep		.78	20.513		4,225	1,100		5,325	6,375
4000	Open web joist 12" deep		.88	18.182		3,850	965		4,815	5,800
4002	Per linear foot		880	.018	L.F.	3.86	.97		4.83	5.80
4004	Treated, per linear foot		880	.018	"	4.85	.97		5.82	6.90
4010	14" deep		.82	19.512	M.L.F.	3,900	1,025		4,925	5,950
4012	Per linear foot		820	.020	L.F.	3.91	1.04		4.95	5.95
4014	Treated, per linear foot		820	.020	"	5.05	1.04		6.09	7.25
4020	16" deep		.78	20.513	M.L.F.	4,200	1,100		5,300	6,375
4022	Per linear foot		780	.021	L.F.	4.21	1.09		5.30	6.40
4024	Treated, per linear foot		780	.021	"	5.55	1.09		6.64	7.85
4030	18" deep		.74	21.622	M.L.F.	4,400	1,150		5,550	6,675
4032	Per linear foot		740	.022	L.F.	4.39	1.15		5.54	6.65
4034	Treated, per linear foot		740	.022	"	5.90	1.15		7.05	8.30
6000	Composite rim joist, 1-1/4" x 9-1/2"		.90	17.778	M.L.F.	2,075	945		3,020	3,825
6010	1-1/4" x 11-1/2"		.88	18.182		2,250	965		3,215	4,025
6020	1-1/4" x 14-1/2"		.82	19.512		3,200	1,025		4,225	5,175
6030	1-1/4" x 16-1/2"		.78	20.513		2,850	1,100		3,950	4,900
9000	Minimum labor/equipment charge	1 Carp	4	2	Job		106		106	170

06 11 10.24 Miscellaneous Framing		Crew	Daily Output	Labor-Hours	Unit	Material	Labor	Equipment	Total	Total Incl O&P
0010	**MISCELLANEOUS FRAMING**									
2000	Firestops, 2" x 4"	2 Carp	780	.021	L.F.	.47	1.09		1.56	2.27
2005	Pneumatic nailed		952	.017		.48	.89		1.37	1.96
2100	2" x 6"		600	.027		.69	1.42		2.11	3.02
2105	Pneumatic nailed		732	.022		.69	1.16		1.85	2.62
5000	Nailers, treated, wood construction, 2" x 4"		800	.020		.57	1.06		1.63	2.33
5005	Pneumatic nailed		960	.017		.58	.89		1.47	2.06
5100	2" x 6"		750	.021		.76	1.13		1.89	2.66
5105	Pneumatic nailed		900	.018		.77	.95		1.72	2.36
5120	2" x 8"		700	.023		1.23	1.22		2.45	3.30
5125	Pneumatic nailed		840	.019		1.24	1.01		2.25	2.99
5200	Steel construction, 2" x 4"		750	.021		.57	1.13		1.70	2.45
5220	2" x 6"		700	.023		.76	1.22		1.98	2.79
5240	2" x 8"		650	.025		1.23	1.31		2.54	3.46
7000	Rough bucks, treated, for doors or windows, 2" x 6"		400	.040		.76	2.13		2.89	4.25
7005	Pneumatic nailed		480	.033		.77	1.77		2.54	3.69

06 11 10 – Framing with Dimensional, Engineered or Composite Lumber

06 11 10.24 Miscellaneous Framing		Crew	Daily Output	Labor-Hours	Unit	2020 Bare Costs Material	Labor	Equipment	Total	Total Incl O&P
7100	2″ x 8″	2 Carp	380	.042	L.F.	1.23	2.24		3.47	4.94
7105	Pneumatic nailed		456	.035		1.24	1.87		3.11	4.36
8000	Stair stringers, 2″ x 10″		130	.123		1.33	6.55		7.88	11.95
8100	2″ x 12″		130	.123		1.74	6.55		8.29	12.40
8150	3″ x 10″		125	.128		3.57	6.80		10.37	14.85
8200	3″ x 12″		125	.128		4.83	6.80		11.63	16.20
9000	Minimum labor/equipment charge	1 Carp	4	2	Job		106		106	170

06 11 10.26 Partitions		Crew	Daily Output	Labor-Hours	Unit	Material	Labor	Equipment	Total	Total Incl O&P
0010	**PARTITIONS**									
0020	Single bottom and double top plate, no waste, std. & better lumber									
0180	2″ x 4″ studs, 8′ high, studs 12″ OC	2 Carp	80	.200	L.F.	5.20	10.65		15.85	23
0185	12″ OC, pneumatic nailed		96	.167		5.25	8.85		14.10	20
0200	16″ OC		100	.160		4.26	8.50		12.76	18.30
0205	16″ OC, pneumatic nailed		120	.133		4.31	7.10		11.41	16.10
0300	24″ OC		125	.128		3.31	6.80		10.11	14.55
0305	24″ OC, pneumatic nailed		150	.107		3.35	5.65		9	12.80
0380	10′ high, studs 12″ OC		80	.200		6.15	10.65		16.80	24
0385	12″ OC, pneumatic nailed		96	.167		6.25	8.85		15.10	21
0400	16″ OC		100	.160		4.97	8.50		13.47	19.05
0405	16″ OC, pneumatic nailed		120	.133		5.05	7.10		12.15	16.90
0500	24″ OC		125	.128		3.79	6.80		10.59	15.05
0505	24″ OC, pneumatic nailed		150	.107		3.83	5.65		9.48	13.30
0580	12′ high, studs 12″ OC		65	.246		7.10	13.10		20.20	29
0585	12″ OC, pneumatic nailed		78	.205		7.20	10.90		18.10	25.50
0600	16″ OC		80	.200		5.70	10.65		16.35	23.50
0605	16″ OC, pneumatic nailed		96	.167		5.75	8.85		14.60	20.50
0700	24″ OC		100	.160		4.26	8.50		12.76	18.30
0705	24″ OC, pneumatic nailed		120	.133		4.31	7.10		11.41	16.10
0780	2″ x 6″ studs, 8′ high, studs 12″ OC		70	.229		7.55	12.15		19.70	28
0785	12″ OC, pneumatic nailed		84	.190		7.65	10.10		17.75	24.50
0800	16″ OC		90	.178		6.15	9.45		15.60	22
0805	16″ OC, pneumatic nailed		108	.148		6.25	7.85		14.10	19.45
0900	24″ OC		115	.139		4.80	7.40		12.20	17.15
0905	24″ OC, pneumatic nailed		138	.116		4.86	6.15		11.01	15.20
0980	10′ high, studs 12″ OC		70	.229		8.90	12.15		21.05	29.50
0985	12″ OC, pneumatic nailed		84	.190		9	10.10		19.10	26
1000	16″ OC		90	.178		7.20	9.45		16.65	23
1005	16″ OC, pneumatic nailed		108	.148		7.30	7.85		15.15	20.50
1100	24″ OC		115	.139		5.50	7.40		12.90	17.90
1105	24″ OC, pneumatic nailed		138	.116		5.55	6.15		11.70	15.95
1180	12′ high, studs 12″ OC		55	.291		10.30	15.45		25.75	36.50
1185	12″ OC, pneumatic nailed		66	.242		10.40	12.90		23.30	32
1200	16″ OC		70	.229		8.25	12.15		20.40	28.50
1205	16″ OC, pneumatic nailed		84	.190		8.35	10.10		18.45	25.50
1300	24″ OC		90	.178		6.15	9.45		15.60	22
1305	24″ OC, pneumatic nailed		108	.148		6.25	7.85		14.10	19.45
1400	For horizontal blocking, 2″ x 4″, add		600	.027		.47	1.42		1.89	2.79
1500	2″ x 6″, add		600	.027		.69	1.42		2.11	3.02
1600	For openings, add		250	.064			3.40		3.40	5.45
1702	Headers for above openings, material only, add				B.F.	.83			.83	.91
9000	Minimum labor/equipment charge	1 Carp	4	2	Job		106		106	170

06 11 10 – Framing with Dimensional, Engineered or Composite Lumber

06 11 10.28 Porch or Deck Framing		Crew	Daily Output	Labor-Hours	Unit	2020 Bare Costs Material	Labor	Equipment	Total	Total Incl O&P
0010	**PORCH OR DECK FRAMING**									
0100	Treated lumber, posts or columns, 4" x 4"	2 Carp	390	.041	L.F.	1.41	2.18		3.59	5.05
0110	4" x 6"		275	.058		2.24	3.09		5.33	7.40
0120	4" x 8"		220	.073		4.28	3.87		8.15	10.90
0130	Girder, single, 4" x 4"		675	.024		1.41	1.26		2.67	3.57
0140	4" x 6"		600	.027		2.24	1.42		3.66	4.73
0150	4" x 8"		525	.030		4.28	1.62		5.90	7.30
0160	Double, 2" x 4"		625	.026		1.18	1.36		2.54	3.48
0170	2" x 6"		600	.027		1.58	1.42		3	4.01
0180	2" x 8"		575	.028		2.54	1.48		4.02	5.15
0190	2" x 10"		550	.029		2.94	1.55		4.49	5.70
0200	2" x 12"		525	.030		4.75	1.62		6.37	7.85
0210	Triple, 2" x 4"		575	.028		1.77	1.48		3.25	4.32
0220	2" x 6"		550	.029		2.37	1.55		3.92	5.10
0230	2" x 8"		525	.030		3.81	1.62		5.43	6.80
0240	2" x 10"		500	.032		4.41	1.70		6.11	7.55
0250	2" x 12"		475	.034		7.15	1.79		8.94	10.70
0260	Ledger, bolted 4' OC, 2" x 4"		400	.040		.75	2.13		2.88	4.23
0270	2" x 6"		395	.041		.93	2.15		3.08	4.48
0280	2" x 8"		390	.041		1.40	2.18		3.58	5.05
0290	2" x 10"		385	.042		1.59	2.21		3.80	5.30
0300	2" x 12"		380	.042		2.48	2.24		4.72	6.30
0310	Joists, 2" x 4"		1250	.013		.59	.68		1.27	1.74
0320	2" x 6"		1250	.013		.79	.68		1.47	1.96
0330	2" x 8"		1100	.015		1.27	.77		2.04	2.64
0340	2" x 10"		900	.018		1.47	.95		2.42	3.13
0350	2" x 12"		875	.018		1.77	.97		2.74	3.50
0360	Railings and trim, 1" x 4"	1 Carp	300	.027		.54	1.42		1.96	2.86
0370	2" x 2"		300	.027		.48	1.42		1.90	2.79
0380	2" x 4"		300	.027		.58	1.42		2	2.91
0390	2" x 6"		300	.027		.77	1.42		2.19	3.12
0400	Decking, 1" x 4"		275	.029	S.F.	3.09	1.55		4.64	5.90
0410	2" x 4"		300	.027		1.95	1.42		3.37	4.42
0420	2" x 6"		320	.025		1.66	1.33		2.99	3.96
0430	5/4" x 6"		320	.025		2.17	1.33		3.50	4.51
0440	Balusters, square, 2" x 2"	2 Carp	660	.024	L.F.	.48	1.29		1.77	2.59
0450	Turned, 2" x 2"		420	.038		.64	2.03		2.67	3.94
0460	Stair stringer, 2" x 10"		130	.123		1.47	6.55		8.02	12.10
0470	2" x 12"		130	.123		1.77	6.55		8.32	12.45
0480	Stair treads, 1" x 4"		140	.114		3.09	6.05		9.14	13.15
0490	2" x 4"		140	.114		.59	6.05		6.64	10.40
0500	2" x 6"		160	.100		.89	5.30		6.19	9.50
0510	5/4" x 6"		160	.100		1.01	5.30		6.31	9.60
0520	Turned handrail post, 4" x 4"		64	.250	Ea.	36	13.30		49.30	61.50
0530	Lattice panel, 4' x 8', 1/2"		1600	.010	S.F.	.68	.53		1.21	1.60
0535	3/4"		1600	.010	"	1.01	.53		1.54	1.96
0540	Cedar, posts or columns, 4" x 4"		390	.041	L.F.	4.14	2.18		6.32	8.05
0550	4" x 6"		275	.058		6.65	3.09		9.74	12.25
0560	4" x 8"		220	.073		12.50	3.87		16.37	19.95
0800	Decking, 1" x 4"		550	.029		2.85	1.55		4.40	5.60
0810	2" x 4"		600	.027		5.45	1.42		6.87	8.25
0820	2" x 6"		640	.025		9.90	1.33		11.23	13

06 11 10 – Framing with Dimensional, Engineered or Composite Lumber

06 11 10.28 Porch or Deck Framing		Crew	Daily Output	Labor-Hours	Unit	Material	Labor	Equipment	Total	Total Incl O&P
0830	5/4" x 6"	2 Carp	640	.025	L.F.	7.55	1.33		8.88	10.45
0840	Railings and trim, 1" x 4"		600	.027		2.85	1.42		4.27	5.40
0860	2" x 4"		600	.027		5.45	1.42		6.87	8.25
0870	2" x 6"		600	.027		9.90	1.42		11.32	13.10
0920	Stair treads, 1" x 4"		140	.114		2.85	6.05		8.90	12.90
0930	2" x 4"		140	.114		5.45	6.05		11.50	15.75
0940	2" x 6"		160	.100		9.90	5.30		15.20	19.35
0950	5/4" x 6"		160	.100		7.55	5.30		12.85	16.80
0980	Redwood, posts or columns, 4" x 4"		390	.041		6.60	2.18		8.78	10.80
0990	4" x 6"		275	.058		12.85	3.09		15.94	19.10
1000	4" x 8"		220	.073		24	3.87		27.87	32.50
1240	Decking, 1" x 4"	1 Carp	275	.029	S.F.	4.03	1.55		5.58	6.90
1260	2" x 6"		340	.024		6.95	1.25		8.20	9.60
1270	5/4" x 6"		320	.025		4.89	1.33		6.22	7.55
1280	Railings and trim, 1" x 4"	2 Carp	600	.027	L.F.	1.19	1.42		2.61	3.58
1310	2" x 6"		600	.027		6.95	1.42		8.37	9.90
1420	Alternative decking, wood/plastic composite, 5/4" x 6" G		640	.025		3.44	1.33		4.77	5.90
1440	1" x 4" square edge fir		550	.029		3.10	1.55		4.65	5.90
1450	1" x 4" tongue and groove fir		450	.036		1.53	1.89		3.42	4.71
1460	1" x 4" mahogany		550	.029		2.13	1.55		3.68	4.82
1462	5/4" x 6" PVC		550	.029		3.45	1.55		5	6.25
1465	Framing, porch or deck, alt deck fastening, screws, add	1 Carp	240	.033	S.F.		1.77		1.77	2.84
1470	Accessories, joist hangers, 2" x 4"		160	.050	Ea.	.85	2.66		3.51	5.20
1480	2" x 6" through 2" x 12"		150	.053		1.56	2.83		4.39	6.25
1530	Post footing, incl excav, backfill, tube form & concrete, 4' deep, 8" diam.	F-7	12	2.667		18.30	127		145.30	223
1540	10" diameter		11	2.909		26	139		165	251
1550	12" diameter		10	3.200		33.50	152		185.50	281

06 11 10.30 Roof Framing		Crew	Daily Output	Labor-Hours	Unit	Material	Labor	Equipment	Total	Total Incl O&P
0010	**ROOF FRAMING**									
1900	Rough fascia, 2" x 6"	2 Carp	250	.064	L.F.	.69	3.40		4.09	6.20
2000	2" x 8"		225	.071		1	3.78		4.78	7.15
2100	2" x 10"		180	.089		1.33	4.72		6.05	9
2200	2" x 12"		180	.089		1.74	4.72		6.46	9.45
5000	Rafters, to 4 in 12 pitch, 2" x 6", ordinary		1000	.016		.69	.85		1.54	2.11
5060	2" x 8", ordinary		950	.017		1	.90		1.90	2.53
5120	2" x 10", ordinary		630	.025		1.33	1.35		2.68	3.62
5180	2" x 12", ordinary		575	.028		1.74	1.48		3.22	4.28
5300	Hip and valley rafters, 2" x 6", ordinary		760	.021		.69	1.12		1.81	2.54
5360	2" x 8", ordinary		720	.022		1	1.18		2.18	2.99
5420	2" x 10", ordinary		570	.028		1.33	1.49		2.82	3.85
5480	2" x 12", ordinary		525	.030		1.74	1.62		3.36	4.51
5540	Hip and valley jacks, 2" x 6", ordinary		600	.027		.69	1.42		2.11	3.02
5600	2" x 8", ordinary		490	.033		1	1.74		2.74	3.88
5660	2" x 10", ordinary		450	.036		1.33	1.89		3.22	4.49
5720	2" x 12", ordinary		375	.043		1.74	2.27		4.01	5.55
5761	For slopes steeper than 4 in 12, add						30%			
5780	Rafter tie, 1" x 4", #3	2 Carp	800	.020	L.F.	.53	1.06		1.59	2.29
5790	2" x 4", #3		800	.020		.47	1.06		1.53	2.22
5800	Ridge board, #2 or better, 1" x 6"		600	.027		.83	1.42		2.25	3.18
5820	1" x 8"		550	.029		1.29	1.55		2.84	3.90
5840	1" x 10"		500	.032		1.98	1.70		3.68	4.90
5860	2" x 6"		500	.032		.69	1.70		2.39	3.47

06 11 10 – Framing with Dimensional, Engineered or Composite Lumber

06 11 10.30 Roof Framing		Crew	Daily Output	Labor-Hours	Unit	Material	Labor (2020 Bare Costs)	Equipment (2020 Bare Costs)	Total (2020 Bare Costs)	Total Incl O&P
5880	2" x 8"	2 Carp	450	.036	L.F.	1	1.89		2.89	4.13
5900	2" x 10"		400	.040		1.33	2.13		3.46	4.87
5920	Roof cants, split, 4" x 4"		650	.025		2.04	1.31		3.35	4.34
5940	6" x 6"		600	.027		5.35	1.42		6.77	8.15
5960	Roof curbs, untreated, 2" x 6"		520	.031		.69	1.64		2.33	3.37
5980	2" x 12"		400	.040		1.74	2.13		3.87	5.30
6000	Sister rafters, 2" x 6"		800	.020		.69	1.06		1.75	2.45
6020	2" x 8"		640	.025		1	1.33		2.33	3.23
6040	2" x 10"		535	.030		1.33	1.59		2.92	4.01
6060	2" x 12"		455	.035		1.74	1.87		3.61	4.90
9000	Minimum labor/equipment charge	1 Carp	4	2	Job		106		106	170

06 11 10.32 Sill and Ledger Framing		Crew	Daily Output	Labor-Hours	Unit	Material	Labor (2020 Bare Costs)	Equipment (2020 Bare Costs)	Total (2020 Bare Costs)	Total Incl O&P
0010	**SILL AND LEDGER FRAMING**									
0020	Extruded polystyrene sill sealer, 5-1/2" wide	1 Carp	1600	.005	L.F.	.14	.27		.41	.58
2002	Ledgers, nailed, 2" x 4"	2 Carp	755	.021		.47	1.13		1.60	2.32
2052	2" x 6"		600	.027		.69	1.42		2.11	3.02
2102	Bolted, not including bolts, 3" x 6"		325	.049		2.19	2.62		4.81	6.60
2152	3" x 12"		233	.069		4.81	3.65		8.46	11.15
2602	Mud sills, redwood, construction grade, 2" x 4"		895	.018		2.31	.95		3.26	4.06
2622	2" x 6"		780	.021		3.14	1.09		4.23	5.20
4002	Sills, 2" x 4"		600	.027		.47	1.42		1.89	2.78
4052	2" x 6"		550	.029		.68	1.55		2.23	3.22
4082	2" x 8"		500	.032		.99	1.70		2.69	3.81
4101	2" x 10"		450	.036		1.31	1.89		3.20	4.47
4121	2" x 12"		400	.040		1.72	2.13		3.85	5.30
4202	Treated, 2" x 4"		550	.029		.57	1.55		2.12	3.10
4222	2" x 6"		500	.032		.75	1.70		2.45	3.55
4242	2" x 8"		450	.036		1.22	1.89		3.11	4.37
4261	2" x 10"		400	.040		1.41	2.13		3.54	4.96
4281	2" x 12"		350	.046		2.30	2.43		4.73	6.40
4402	4" x 4"		450	.036		1.36	1.89		3.25	4.52
4422	4" x 6"		350	.046		2.16	2.43		4.59	6.25
4462	4" x 8"		300	.053		4.18	2.83		7.01	9.15
4481	4" x 10"		260	.062		5.75	3.27		9.02	11.55
9000	Minimum labor/equipment charge	1 Carp	4	2	Job		106		106	170

06 11 10.34 Sleepers		Crew	Daily Output	Labor-Hours	Unit	Material	Labor (2020 Bare Costs)	Equipment (2020 Bare Costs)	Total (2020 Bare Costs)	Total Incl O&P
0010	**SLEEPERS**									
0100	On concrete, treated, 1" x 2"	2 Carp	2350	.007	L.F.	.31	.36		.67	.92
0150	1" x 3"		2000	.008		.51	.43		.94	1.24
0200	2" x 4"		1500	.011		.61	.57		1.18	1.58
0250	2" x 6"		1300	.012		.84	.65		1.49	1.98
9000	Minimum labor/equipment charge	1 Carp	4	2	Job		106		106	170

06 11 10.36 Soffit and Canopy Framing		Crew	Daily Output	Labor-Hours	Unit	Material	Labor (2020 Bare Costs)	Equipment (2020 Bare Costs)	Total (2020 Bare Costs)	Total Incl O&P
0010	**SOFFIT AND CANOPY FRAMING**									
1002	Canopy or soffit framing, 1" x 4"	2 Carp	900	.018	L.F.	.53	.95		1.48	2.10
1021	1" x 6"		850	.019		.83	1		1.83	2.51
1042	1" x 8"		750	.021		1.29	1.13		2.42	3.24
1102	2" x 4"		620	.026		.47	1.37		1.84	2.72
1121	2" x 6"		560	.029		.69	1.52		2.21	3.18
1142	2" x 8"		500	.032		1	1.70		2.70	3.82
1202	3" x 4"		500	.032		1.38	1.70		3.08	4.23
1221	3" x 6"		400	.040		2.21	2.13		4.34	5.85

06 11 10 – Framing with Dimensional, Engineered or Composite Lumber

06 11 10.36 Soffit and Canopy Framing		Crew	Daily Output	Labor-Hours	Unit	Material	2020 Bare Costs Labor	Equipment	Total	Total Incl O&P
1242	3″ x 10″	2 Carp	300	.053	L.F.	3.57	2.83		6.40	8.45
9000	Minimum labor/equipment charge	1 Carp	4	2	Job		106		106	170
06 11 10.38 Treated Lumber Framing Material										
0010	**TREATED LUMBER FRAMING MATERIAL**									
0100	2″ x 4″				M.B.F.	850			850	935
0110	2″ x 6″					755			755	830
0120	2″ x 8″					915			915	1,000
0130	2″ x 10″					845			845	930
0140	2″ x 12″					1,150			1,150	1,275
0200	4″ x 4″					1,025			1,025	1,125
0210	4″ x 6″					1,075			1,075	1,200
0220	4″ x 8″					1,575			1,575	1,725
06 11 10.40 Wall Framing										
0010	**WALL FRAMING** R061110-30									
0100	Door buck, studs, header, access, 8′ high, 2″ x 4″ wall, 3′ wide	1 Carp	32	.250	Ea.	19.60	13.30		32.90	43
0110	4′ wide		32	.250		21	13.30		34.30	44.50
0120	5′ wide		32	.250		25.50	13.30		38.80	49.50
0130	6′ wide		32	.250		27.50	13.30		40.80	52
0140	8′ wide		30	.267		37	14.15		51.15	63
0150	10′ wide		30	.267		50.50	14.15		64.65	78
0160	12′ wide		30	.267		73	14.15		87.15	103
0170	2″ x 6″ wall, 3′ wide		32	.250		26.50	13.30		39.80	50.50
0180	4′ wide		32	.250		28	13.30		41.30	52
0190	5′ wide		32	.250		32.50	13.30		45.80	57
0200	6′ wide		32	.250		34.50	13.30		47.80	59.50
0210	8′ wide		30	.267		43.50	14.15		57.65	70.50
0220	10′ wide		30	.267		57.50	14.15		71.65	85.50
0230	12′ wide		30	.267		80	14.15		94.15	111
0240	Window buck, studs, header & access, 8′ high 2″ x 4″ wall, 2′ wide		24	.333		21	17.70		38.70	51.50
0250	3′ wide		24	.333		24.50	17.70		42.20	55.50
0260	4′ wide		24	.333		26.50	17.70		44.20	58
0270	5′ wide		24	.333		31.50	17.70		49.20	63
0280	6′ wide		24	.333		34.50	17.70		52.20	66.50
0290	7′ wide		24	.333		42.50	17.70		60.20	75
0300	8′ wide		22	.364		47	19.35		66.35	82.50
0310	10′ wide		22	.364		62	19.35		81.35	99
0320	12′ wide		22	.364		87.50	19.35		106.85	127
0330	2″ x 6″ wall, 2′ wide		24	.333		29.50	17.70		47.20	61
0340	3′ wide		24	.333		33	17.70		50.70	64.50
0350	4′ wide		24	.333		35.50	17.70		53.20	67.50
0360	5′ wide		24	.333		40.50	17.70		58.20	73
0370	6′ wide		24	.333		44.50	17.70		62.20	77.50
0380	7′ wide		24	.333		53.50	17.70		71.20	87.50
0390	8′ wide		22	.364		58	19.35		77.35	95
0400	10′ wide		22	.364		74	19.35		93.35	113
0410	12′ wide		22	.364		101	19.35		120.35	142
2002	Headers over openings, 2″ x 6″	2 Carp	360	.044	L.F.	.69	2.36		3.05	4.53
2007	2″ x 6″, pneumatic nailed		432	.037		.69	1.97		2.66	3.91
2052	2″ x 8″		340	.047		1	2.50		3.50	5.10
2057	2″ x 8″, pneumatic nailed		408	.039		1.01	2.08		3.09	4.45
2101	2″ x 10″		320	.050		1.33	2.66		3.99	5.70
2106	2″ x 10″, pneumatic nailed		384	.042		1.34	2.21		3.55	5.05

06 11 10 – Framing with Dimensional, Engineered or Composite Lumber

06 11 10.40 Wall Framing		Crew	Daily Output	Labor-Hours	Unit	Material	2020 Bare Costs Labor	Equipment	Total	Total Incl O&P
2152	2" x 12"	2 Carp	300	.053	L.F.	1.74	2.83		4.57	6.45
2157	2" x 12", pneumatic nailed		360	.044		1.76	2.36		4.12	5.70
2180	4" x 8"		260	.062		4.33	3.27		7.60	10
2185	4" x 8", pneumatic nailed		312	.051		4.35	2.73		7.08	9.15
2191	4" x 10"		240	.067		5.45	3.54		8.99	11.70
2196	4" x 10", pneumatic nailed		288	.056		5.45	2.95		8.40	10.75
2202	4" x 12"		190	.084		5.80	4.48		10.28	13.55
2207	4" x 12", pneumatic nailed		228	.070		5.85	3.73		9.58	12.45
2241	6" x 10"		165	.097		7.05	5.15		12.20	16.05
2246	6" x 10", pneumatic nailed		198	.081		7.10	4.30		11.40	14.75
2251	6" x 12"		140	.114		9	6.05		15.05	19.65
2256	6" x 12", pneumatic nailed		168	.095		9.05	5.05		14.10	18.05
5002	Plates, untreated, 2" x 3"		850	.019		.42	1		1.42	2.07
5007	2" x 3", pneumatic nailed		1020	.016		.43	.83		1.26	1.81
5022	2" x 4"		800	.020		.47	1.06		1.53	2.22
5027	2" x 4", pneumatic nailed		960	.017		.48	.89		1.37	1.95
5041	2" x 6"		750	.021		.69	1.13		1.82	2.57
5046	2" x 6", pneumatic nailed		900	.018		.69	.95		1.64	2.27
5122	Studs, 8' high wall, 2" x 3"		1200	.013		.42	.71		1.13	1.61
5127	2" x 3", pneumatic nailed		1440	.011		.43	.59		1.02	1.42
5142	2" x 4"		1100	.015		.47	.77		1.24	1.75
5147	2" x 4", pneumatic nailed		1320	.012		.48	.64		1.12	1.56
5162	2" x 6"		1000	.016		.69	.85		1.54	2.11
5167	2" x 6", pneumatic nailed		1200	.013		.69	.71		1.40	1.90
5182	3" x 4"		800	.020		1.38	1.06		2.44	3.21
5187	3" x 4", pneumatic nailed		960	.017		1.38	.89		2.27	2.94
8200	For 12' high walls, deduct						5%			
8220	For stub wall, 6' high, add						20%			
8240	3' high, add						40%			
8250	For second story & above, add						5%			
8300	For dormer & gable, add						15%			
9000	Minimum labor/equipment charge	1 Carp	4	2	Job		106		106	170

06 11 10.42 Furring		Crew	Daily Output	Labor-Hours	Unit	Material	Labor	Equipment	Total	Total Incl O&P
0010	**FURRING**									
0012	Wood strips, 1" x 2", on walls, on wood	1 Carp	550	.015	L.F.	.29	.77		1.06	1.56
0015	On wood, pneumatic nailed		710	.011		.29	.60		.89	1.28
0300	On masonry		495	.016		.31	.86		1.17	1.73
0400	On concrete		260	.031		.31	1.64		1.95	2.97
0600	1" x 3", on walls, on wood		550	.015		.44	.77		1.21	1.72
0605	On wood, pneumatic nailed		710	.011		.44	.60		1.04	1.44
0700	On masonry		495	.016		.47	.86		1.33	1.90
0800	On concrete		260	.031		.47	1.64		2.11	3.14
0850	On ceilings, on wood		350	.023		.44	1.22		1.66	2.43
0855	On wood, pneumatic nailed		450	.018		.44	.95		1.39	1.99
0900	On masonry		320	.025		.47	1.33		1.80	2.65
0950	On concrete		210	.038		.47	2.03		2.50	3.76
9000	Minimum labor/equipment charge		4	2	Job		106		106	170

06 11 10.44 Grounds		Crew	Daily Output	Labor-Hours	Unit	Material	Labor	Equipment	Total	Total Incl O&P
0010	**GROUNDS**									
0020	For casework, 1" x 2" wood strips, on wood	1 Carp	330	.024	L.F.	.29	1.29		1.58	2.38
0100	On masonry		285	.028		.31	1.49		1.80	2.74
0200	On concrete		250	.032		.31	1.70		2.01	3.07

06 11 Wood Framing

06 11 10 – Framing with Dimensional, Engineered or Composite Lumber

06 11 10.44 Grounds		Crew	Daily Output	Labor-Hours	Unit	2020 Bare Costs Material	Labor	Equipment	Total	Total Incl O&P
0400	For plaster, 3/4" deep, on wood	1 Carp	450	.018	L.F.	.29	.95		1.24	1.83
0500	On masonry		225	.036		.31	1.89		2.20	3.38
0600	On concrete		175	.046		.31	2.43		2.74	4.24
0700	On metal lath		200	.040		.31	2.13		2.44	3.76
9000	Minimum labor/equipment charge		4	2	Job		106		106	170

06 12 Structural Panels

06 12 10 – Structural Insulated Panels

06 12 10.10 OSB Faced Panels

Line	Description		Crew	Daily Output	Labor-Hours	Unit	Material	Labor	Equipment	Total	Total Incl O&P
0010	**OSB FACED PANELS**										
0100	Structural insul. panels, 7/16" OSB both faces, EPS insul., 3-5/8" T	G	F-3	2075	.019	S.F.	4.18	1.05	.23	5.46	6.50
0110	5-5/8" thick	G		1725	.023		4.41	1.26	.27	5.94	7.15
0120	7-3/8" thick	G		1425	.028		5.50	1.53	.33	7.36	8.85
0130	9-3/8" thick	G		1125	.036		5.95	1.93	.42	8.30	10.10
0140	7/16" OSB one face, EPS insul., 3-5/8" thick	G		2175	.018		4.27	1	.22	5.49	6.55
0150	5-5/8" thick	G		1825	.022		4.62	1.19	.26	6.07	7.30
0160	7-3/8" thick	G		1525	.026		5.40	1.43	.31	7.14	8.50
0170	9-3/8" thick	G		1225	.033		5.85	1.77	.38	8	9.70
0190	7/16" OSB - 1/2" GWB faces, EPS insul., 3-5/8" T	G		2075	.019		3.61	1.05	.23	4.89	5.90
0200	5-5/8" thick	G		1725	.023		4.74	1.26	.27	6.27	7.50
0210	7-3/8" thick	G		1425	.028		5	1.53	.33	6.86	8.30
0220	9-3/8" thick	G		1125	.036		5.60	1.93	.42	7.95	9.70
0240	7/16" OSB - 1/2" MRGWB faces, EPS insul., 3-5/8" T	G		2075	.019		4.45	1.05	.23	5.73	6.80
0250	5-5/8" thick	G		1725	.023		4.56	1.26	.27	6.09	7.30
0260	7-3/8" thick	G		1425	.028		5.25	1.53	.33	7.11	8.60
0270	9-3/8" thick	G		1125	.036		5.85	1.93	.42	8.20	9.95
0300	For 1/2" GWB added to OSB skin, add	G					1.52			1.52	1.67
0310	For 1/2" MRGWB added to OSB skin, add	G					1.54			1.54	1.69
0320	For one T1-11 skin, add to OSB-OSB	G					2.22			2.22	2.44
0330	For one 19/32" CDX skin, add to OSB-OSB	G					1.63			1.63	1.79
0500	Structural insulated panel, 7/16" OSB both sides, straw core										
0510	4-3/8" T, walls (w/sill, splines, plates)	G	F-6	2400	.017	S.F.	7.30	.83	.20	8.33	9.55
0520	Floors (w/splines)	G		2400	.017		7.30	.83	.20	8.33	9.55
0530	Roof (w/splines)	G		2400	.017		7.30	.83	.20	8.33	9.55
0550	7-7/8" T, walls (w/sill, splines, plates)	G		2400	.017		11.45	.83	.20	12.48	14.15
0560	Floors (w/splines)	G		2400	.017		11.45	.83	.20	12.48	14.15
0570	Roof (w/splines)	G		2400	.017		11.45	.83	.20	12.48	14.15

06 12 19 – Composite Shearwall Panels

06 12 19.10 Steel and Wood Composite Shearwall Panels

Line	Description	Crew	Daily Output	Labor-Hours	Unit	Material	Labor	Equipment	Total	Total Incl O&P
0010	**STEEL & WOOD COMPOSITE SHEARWALL PANELS**									
0020	Anchor bolts, 36" long (must be placed in wet concrete)	1 Carp	150	.053	Ea.	43.50	2.83		46.33	52.50
0030	On concrete, 2" x 4" & 2" x 6" walls, 7'-10' high, 360 lb. shear, 12" wide	2 Carp	8	2		495	106		601	715
0040	715 lb. shear, 15" wide		8	2		650	106		756	885
0050	1860 lb. shear, 18" wide		8	2		700	106		806	940
0060	2780 lb. shear, 21" wide		8	2		830	106		936	1,075
0070	3790 lb. shear, 24" wide		8	2		780	106		886	1,025
0080	2" x 6" walls, 11'-13' high, 1180 lb. shear, 18" wide		6	2.667		735	142		877	1,025
0090	1555 lb. shear, 21" wide		6	2.667		755	142		897	1,050
0100	2280 lb. shear, 24" wide		6	2.667		895	142		1,037	1,200
0110	For installing above on wood floor frame, add									

06 12 Structural Panels

06 12 19 – Composite Shearwall Panels

06 12 19.10 Steel and Wood Composite Shearwall Panels		Crew	Daily Output	Labor-Hours	Unit	Material	2020 Bare Costs Labor	Equipment	Total	Total Incl O&P
0120	Coupler nuts, threaded rods, bolts, shear transfer plate kit	1 Carp	16	.500	Ea.	63	26.50		89.50	112
0130	Framing anchors, angle (2 required)	"	96	.083	"	2.53	4.43		6.96	9.90
0140	For blocking see Section 06 11 10.02									
0150	For installing above, first floor to second floor, wood floor frame, add									
0160	Add stack option to first floor wall panel				Ea.	69.50			69.50	76.50
0170	Threaded rods, bolts, shear transfer plate kit	1 Carp	16	.500		78	26.50		104.50	128
0180	Framing anchors, angle (2 required)	"	96	.083	↓	2.53	4.43		6.96	9.90
0190	For blocking see section 06 11 10.02									
0200	For installing stacked panels, balloon framing									
0210	Add stack option to first floor wall panel				Ea.	69.50			69.50	76.50
0220	Threaded rods, bolts kit	1 Carp	16	.500	"	44.50	26.50		71	91.50

06 13 Heavy Timber Construction

06 13 23 – Heavy Timber Framing

06 13 23.10 Heavy Framing

		Crew	Daily Output	Labor-Hours	Unit	Material	Labor	Equipment	Total	Total Incl O&P
0010	**HEAVY FRAMING**									
0020	Beams, single 6" x 10"	2 Carp	1.10	14.545	M.B.F.	1,575	775		2,350	2,975
0100	Single 8" x 16"		1.20	13.333	"	2,000	710		2,710	3,325
0202	Built from 2" lumber, multiple 2" x 14"		900	.018	B.F.	1.01	.95		1.96	2.62
0212	Built from 3" lumber, multiple 3" x 6"		700	.023		1.46	1.22		2.68	3.56
0222	Multiple 3" x 8"		800	.020		1.46	1.06		2.52	3.31
0232	Multiple 3" x 10"		900	.018		1.42	.95		2.37	3.07
0242	Multiple 3" x 12"		1000	.016		1.60	.85		2.45	3.12
0252	Built from 4" lumber, multiple 4" x 6"		800	.020		1.52	1.06		2.58	3.37
0262	Multiple 4" x 8"		900	.018		1.61	.95		2.56	3.28
0272	Multiple 4" x 10"		1000	.016		1.62	.85		2.47	3.15
0282	Multiple 4" x 12"		1100	.015	↓	1.45	.77		2.22	2.83
0292	Columns, structural grade, 1500Fb, 4" x 4"		450	.036	L.F.	2.32	1.89		4.21	5.60
0302	6" x 6"		225	.071		4.07	3.78		7.85	10.55
0402	8" x 8"		240	.067		8.25	3.54		11.79	14.75
0502	10" x 10"		90	.178		14.60	9.45		24.05	31
0602	12" x 12"		70	.229	↓	18.30	12.15		30.45	39.50
0802	Floor planks, 2" thick, T&G, 2" x 6"		1050	.015	B.F.	1.45	.81		2.26	2.89
0902	2" x 10"		1100	.015		1.46	.77		2.23	2.85
1102	3" thick, 3" x 6"		1050	.015		1.62	.81		2.43	3.08
1202	3" x 10"		1100	.015		1.65	.77		2.42	3.05
1402	Girders, structural grade, 12" x 12"		800	.020		1.52	1.06		2.58	3.38
1502	10" x 16"		1000	.016		2.56	.85		3.41	4.18
2302	Roof purlins, 4" thick, structural grade	↓	1050	.015	↓	1.61	.81		2.42	3.07
9000	Minimum labor/equipment charge	1 Carp	2	4	Job		213		213	340

06 15 Wood Decking

06 15 16 – Wood Roof Decking

06 15 16.10 Solid Wood Roof Decking		Crew	Daily Output	Labor-Hours	Unit	Material	Labor	Equipment	Total	Total Incl O&P
0010	SOLID WOOD ROOF DECKING									
0350	Cedar planks, 2" thick	2 Carp	350	.046	S.F.	7.05	2.43		9.48	11.65
0400	3" thick		320	.050		10.60	2.66		13.26	15.90
0500	4" thick		250	.064		14.15	3.40		17.55	21
0550	6" thick		200	.080		21	4.25		25.25	30.50
0650	Douglas fir, 2" thick		350	.046		3.10	2.43		5.53	7.30
0700	3" thick		320	.050		4.65	2.66		7.31	9.35
0800	4" thick		250	.064		6.20	3.40		9.60	12.30
0850	6" thick		200	.080		9.30	4.25		13.55	17.05
0950	Hemlock, 2" thick		350	.046		3.16	2.43		5.59	7.35
1000	3" thick		320	.050		4.74	2.66		7.40	9.45
1100	4" thick		250	.064		6.30	3.40		9.70	12.40
1150	6" thick		200	.080		9.50	4.25		13.75	17.20
1250	Western white spruce, 2" thick		350	.046		2.02	2.43		4.45	6.10
1300	3" thick		320	.050		3.03	2.66		5.69	7.60
1400	4" thick		250	.064		4.03	3.40		7.43	9.90
1450	6" thick	▼	200	.080	▼	6.05	4.25		10.30	13.45
9000	Minimum labor/equipment charge	1 Carp	2	4	Job		213		213	340

06 15 23 – Laminated Wood Decking

06 15 23.10 Laminated Roof Deck		Crew	Daily Output	Labor-Hours	Unit	Material	Labor	Equipment	Total	Total Incl O&P
0010	LAMINATED ROOF DECK									
0020	Pine or hemlock, 3" thick	2 Carp	425	.038	S.F.	5.70	2		7.70	9.45
0100	4" thick		325	.049		7.40	2.62		10.02	12.35
0300	Cedar, 3" thick		425	.038		8.20	2		10.20	12.20
0400	4" thick		325	.049		11.05	2.62		13.67	16.35
0600	Fir, 3" thick		425	.038		6.60	2		8.60	10.45
0700	4" thick	▼	325	.049	▼	8.40	2.62		11.02	13.45
9000	Minimum labor/equipment charge	1 Carp	3	2.667	Job		142		142	227

06 16 Sheathing

06 16 13 – Insulating Sheathing

06 16 13.10 Insulating Sheathing		Crew	Daily Output	Labor-Hours	Unit	Material	Labor	Equipment	Total	Total Incl O&P
0010	INSULATING SHEATHING									
0020	Expanded polystyrene, 1#/C.F. density, 3/4" thick, R2.89 G	2 Carp	1400	.011	S.F.	.37	.61		.98	1.38
0030	1" thick, R3.85 G		1300	.012		.45	.65		1.10	1.54
0040	2" thick, R7.69 G		1200	.013		.74	.71		1.45	1.95
0050	Extruded polystyrene, 15 psi compressive strength, 1" thick, R5 G		1300	.012		.68	.65		1.33	1.80
0060	2" thick, R10 G		1200	.013		.83	.71		1.54	2.06
0070	Polyisocyanurate, 2#/C.F. density, 3/4" thick G		1400	.011		.68	.61		1.29	1.72
0080	1" thick G		1300	.012		.61	.65		1.26	1.72
0090	1-1/2" thick G		1250	.013		.76	.68		1.44	1.92
0100	2" thick G	▼	1200	.013	▼	1.08	.71		1.79	2.33

06 16 23 – Subflooring

06 16 23.10 Subfloor		Crew	Daily Output	Labor-Hours	Unit	Material	Labor	Equipment	Total	Total Incl O&P
0010	SUBFLOOR									
0011	Plywood, CDX, 1/2" thick	2 Carp	1500	.011	SF Flr.	.68	.57		1.25	1.66
0017	Pneumatic nailed		1860	.009		.68	.46		1.14	1.48
0102	5/8" thick		1350	.012		.77	.63		1.40	1.86
0107	Pneumatic nailed		1674	.010		.77	.51		1.28	1.66
0202	3/4" thick	▼	1250	.013	▼	.95	.68		1.63	2.13

06 16 Sheathing

06 16 23 – Subflooring

06 16 23.10 Subfloor		Crew	Daily Output	Labor-Hours	Unit	Material	2020 Bare Costs Labor	Equipment	Total	Total Incl O&P
0207	Pneumatic nailed	2 Carp	1550	.010	SF Flr.	.95	.55		1.50	1.92
0302	1-1/8" thick, 2-4-1 including underlayment		1050	.015		2.12	.81		2.93	3.63
0440	With boards, 1" x 6", S4S, laid regular		900	.018		1.79	.95		2.74	3.48
0452	1" x 8", laid regular		1000	.016		2.06	.85		2.91	3.62
0462	Laid diagonal		850	.019		2.06	1		3.06	3.86
0502	1" x 10", laid regular		1100	.015		2.49	.77		3.26	3.98
0602	Laid diagonal		900	.018		2.49	.95		3.44	4.25
8990	Subfloor adhesive, 3/8" bead	1 Carp	2300	.003	L.F.	.10	.19		.29	.41
9000	Minimum labor/equipment charge	"	4	2	Job		106		106	170

06 16 26 – Underlayment

06 16 26.10 Wood Product Underlayment		Crew	Daily Output	Labor-Hours	Unit	Material	Labor	Equipment	Total	Total Incl O&P
0010	**WOOD PRODUCT UNDERLAYMENT** R061636-20									
0015	Plywood, underlayment grade, 1/4" thick	2 Carp	1500	.011	S.F.	.95	.57		1.52	1.95
0018	Pneumatic nailed		1860	.009		.95	.46		1.41	1.77
0030	3/8" thick		1500	.011		1.05	.57		1.62	2.07
0070	Pneumatic nailed		1860	.009		1.05	.46		1.51	1.89
0102	1/2" thick		1450	.011		1.18	.59		1.77	2.24
0107	Pneumatic nailed		1798	.009		1.18	.47		1.65	2.06
0202	5/8" thick		1400	.011		1.54	.61		2.15	2.66
0207	Pneumatic nailed		1736	.009		1.54	.49		2.03	2.48
0302	3/4" thick		1300	.012		1.49	.65		2.14	2.69
0306	Pneumatic nailed		1612	.010		1.49	.53		2.02	2.49
0502	Particle board, 3/8" thick		1500	.011		.39	.57		.96	1.34
0507	Pneumatic nailed		1860	.009		.39	.46		.85	1.16
0602	1/2" thick		1450	.011		.41	.59		1	1.39
0607	Pneumatic nailed		1798	.009		.41	.47		.88	1.21
0802	5/8" thick		1400	.011		.61	.61		1.22	1.64
0807	Pneumatic nailed		1736	.009		.61	.49		1.10	1.46
0902	3/4" thick		1300	.012		.74	.65		1.39	1.86
0907	Pneumatic nailed		1612	.010		.74	.53		1.27	1.66
1100	Hardboard, underlayment grade, 4' x 4', .215" thick G		1500	.011		.72	.57		1.29	1.70
9000	Minimum labor/equipment charge	1 Carp	4	2	Job		106		106	170

06 16 33 – Wood Board Sheathing

06 16 33.10 Board Sheathing		Crew	Daily Output	Labor-Hours	Unit	Material	Labor	Equipment	Total	Total Incl O&P
0009	**BOARD SHEATHING**									
0010	Roof, 1" x 6" boards, laid horizontal	2 Carp	725	.022	S.F.	1.79	1.17		2.96	3.85
0020	On steep roof		520	.031		1.79	1.64		3.43	4.59
0040	On dormers, hips, & valleys		480	.033		1.79	1.77		3.56	4.81
0050	Laid diagonal		650	.025		1.79	1.31		3.10	4.07
0070	1" x 8" boards, laid horizontal		875	.018		2.06	.97		3.03	3.82
0080	On steep roof		635	.025		2.06	1.34		3.40	4.41
0090	On dormers, hips, & valleys		580	.028		2.10	1.47		3.57	4.66
0100	Laid diagonal		725	.022		2.06	1.17		3.23	4.14
0110	Skip sheathing, 1" x 4", 7" OC	1 Carp	1200	.007		.66	.35		1.01	1.30
0120	1" x 6", 9" OC		1450	.006		.82	.29		1.11	1.37
0180	T&G sheathing/decking, 1" x 6"		1000	.008		1.78	.43		2.21	2.64
0190	2" x 6"		1000	.008		3.95	.43		4.38	5.05
0200	Walls, 1" x 6" boards, laid regular	2 Carp	650	.025		1.79	1.31		3.10	4.07
0210	Laid diagonal		585	.027		1.79	1.45		3.24	4.30
0220	1" x 8" boards, laid regular		765	.021		2.06	1.11		3.17	4.04
0230	Laid diagonal		650	.025		2.06	1.31		3.37	4.36

06 16 36.10 Sheathing		Crew	Daily Output	Labor-Hours	Unit	Material	2020 Bare Costs Labor	Equipment	Total	Total Incl O&P
0010	**SHEATHING** R061636-20									
0012	Plywood on roofs, CDX									
0032	5/16" thick	2 Carp	1600	.010	S.F.	.60	.53		1.13	1.50
0037	Pneumatic nailed R061110-30		1952	.008		.60	.44		1.04	1.35
0052	3/8" thick		1525	.010		.53	.56		1.09	1.47
0057	Pneumatic nailed		1860	.009		.53	.46		.99	1.31
0102	1/2" thick		1400	.011		.68	.61		1.29	1.72
0103	Pneumatic nailed		1708	.009		.68	.50		1.18	1.55
0202	5/8" thick		1300	.012		.77	.65		1.42	1.90
0207	Pneumatic nailed		1586	.010		.77	.54		1.31	1.71
0302	3/4" thick		1200	.013		.95	.71		1.66	2.18
0307	Pneumatic nailed		1464	.011		.95	.58		1.53	1.97
0502	Plywood on walls, with exterior CDX, 3/8" thick		1200	.013		.53	.71		1.24	1.72
0507	Pneumatic nailed		1488	.011		.53	.57		1.10	1.50
0602	1/2" thick		1125	.014		.68	.76		1.44	1.96
0607	Pneumatic nailed		1395	.011		.68	.61		1.29	1.73
0702	5/8" thick		1050	.015		.77	.81		1.58	2.15
0707	Pneumatic nailed		1302	.012		.77	.65		1.42	1.90
0802	3/4" thick		975	.016		.95	.87		1.82	2.44
0807	Pneumatic nailed		1209	.013		.95	.70		1.65	2.17
1000	For shear wall construction, add						20%			
1200	For structural 1 exterior plywood, add				S.F.	10%				
3000	Wood fiber, regular, no vapor barrier, 1/2" thick	2 Carp	1200	.013		.64	.71		1.35	1.84
3100	5/8" thick		1200	.013		.70	.71		1.41	1.91
3300	No vapor barrier, in colors, 1/2" thick		1200	.013		.82	.71		1.53	2.04
3400	5/8" thick		1200	.013		.86	.71		1.57	2.09
3600	With vapor barrier one side, white, 1/2" thick		1200	.013		.63	.71		1.34	1.83
3700	Vapor barrier 2 sides, 1/2" thick		1200	.013		.86	.71		1.57	2.09
3800	Asphalt impregnated, 25/32" thick		1200	.013		.32	.71		1.03	1.49
3850	Intermediate, 1/2" thick		1200	.013		.34	.71		1.05	1.51
4500	Oriented strand board, on roof, 7/16" thick G		1460	.011		.48	.58		1.06	1.46
4505	Pneumatic nailed G		1780	.009		.48	.48		.96	1.30
4550	1/2" thick G		1400	.011		.48	.61		1.09	1.50
4555	Pneumatic nailed G		1736	.009		.48	.49		.97	1.32
4600	5/8" thick G		1300	.012		.66	.65		1.31	1.78
4605	Pneumatic nailed G		1586	.010		.66	.54		1.20	1.59
4610	On walls, 7/16" thick G		1200	.013		.48	.71		1.19	1.67
4615	Pneumatic nailed G		1488	.011		.48	.57		1.05	1.45
4620	1/2" thick G		1195	.013		.48	.71		1.19	1.67
4625	Pneumatic nailed G		1325	.012		.48	.64		1.12	1.56
4630	5/8" thick G		1050	.015		.66	.81		1.47	2.03
4635	Pneumatic nailed G		1302	.012		.66	.65		1.31	1.78
4700	Oriented strand board, factory laminated W.R. barrier, on roof, 1/2" thick G		1400	.011		.79	.61		1.40	1.84
4705	Pneumatic nailed G		1736	.009		.79	.49		1.28	1.66
4720	5/8" thick G		1300	.012		.98	.65		1.63	2.13
4725	Pneumatic nailed G		1586	.010		.98	.54		1.52	1.94
4730	5/8" thick, T&G G		1150	.014		1.06	.74		1.80	2.35
4735	Pneumatic nailed, T&G G		1400	.011		1.06	.61		1.67	2.14
4740	On walls, 7/16" thick G		1200	.013		.73	.71		1.44	1.94
4745	Pneumatic nailed G		1488	.011		.73	.57		1.30	1.72
4750	1/2" thick G		1195	.013		.79	.71		1.50	2.01
4755	Pneumatic nailed G		1325	.012		.79	.64		1.43	1.90

06 16 Sheathing

06 16 36 – Wood Panel Product Sheathing

06 16 36.10 Sheathing		Crew	Daily Output	Labor-Hours	Unit	2020 Bare Costs Material	Labor	Equipment	Total	Total Incl O&P
4800	Joint sealant tape, 3-1/2"	2 Carp	7600	.002	L.F.	.36	.11		.47	.58
4810	Joint sealant tape, 6"	↓	7600	.002	"	.41	.11		.52	.63
9000	Minimum labor/equipment charge	1 Carp	2	4	Job		213		213	340

06 16 43 – Gypsum Sheathing

06 16 43.10 Gypsum Sheathing		Crew	Daily Output	Labor-Hours	Unit	Material	Labor	Equipment	Total	Total Incl O&P
0010	GYPSUM SHEATHING									
0020	Gypsum, weatherproof, 1/2" thick	2 Carp	1125	.014	S.F.	.53	.76		1.29	1.79
0040	With embedded glass mats	"	1100	.015	"	.71	.77		1.48	2.02

06 17 Shop-Fabricated Structural Wood

06 17 33 – Wood I-Joists

06 17 33.10 Wood and Composite I-Joists		Crew	Daily Output	Labor-Hours	Unit	Material	Labor	Equipment	Total	Total Incl O&P
0010	WOOD AND COMPOSITE I-JOISTS									
0100	Plywood webs, incl. bridging & blocking, panels 24" OC									
1200	15' to 24' span, 50 psf live load	F-5	2400	.013	SF Flr.	1.81	.72		2.53	3.14
1300	55 psf live load		2250	.014		2.26	.76		3.02	3.71
1400	24' to 30' span, 45 psf live load		2600	.012		2.68	.66		3.34	4
1500	55 psf live load	↓	2400	.013	↓	4.77	.72		5.49	6.40

06 17 53 – Shop-Fabricated Wood Trusses

06 17 53.10 Roof Trusses		Crew	Daily Output	Labor-Hours	Unit	Material	Labor	Equipment	Total	Total Incl O&P
0010	ROOF TRUSSES									
0100	Fink (W) or King post type, 2'-0" OC									
0200	Metal plate connected, 4 in 12 slope									
0210	24' to 29' span	F-3	3000	.013	SF Flr.	1.82	.72	.16	2.70	3.33
0300	30' to 43' span		3000	.013		2.40	.72	.16	3.28	3.97
0400	44' to 60' span	↓	3000	.013		2.41	.72	.16	3.29	3.98
0700	Glued and nailed, add				↓	50%				

06 18 Glued-Laminated Construction

06 18 13 – Glued-Laminated Beams

06 18 13.20 Laminated Framing		Crew	Daily Output	Labor-Hours	Unit	Material	Labor	Equipment	Total	Total Incl O&P
0010	LAMINATED FRAMING									
0020	30 lb., short term live load, 15 lb. dead load									
0200	Straight roof beams, 20' clear span, beams 8' OC	F-3	2560	.016	SF Flr.	2.27	.85	.18	3.30	4.05
0300	Beams 16' OC		3200	.013		1.66	.68	.15	2.49	3.07
0500	40' clear span, beams 8' OC		3200	.013		4.31	.68	.15	5.14	6
0600	Beams 16' OC	↓	3840	.010		3.57	.57	.12	4.26	4.96
0800	60' clear span, beams 8' OC	F-4	2880	.017		7.40	.90	.34	8.64	9.90
0900	Beams 16' OC	"	3840	.013		5.55	.67	.26	6.48	7.45
1100	Tudor arches, 30' to 40' clear span, frames 8' OC	F-3	1680	.024		9.60	1.29	.28	11.17	12.95
1200	Frames 16' OC	"	2240	.018		7.50	.97	.21	8.68	10.05
1400	50' to 60' clear span, frames 8' OC	F-4	2200	.022		10.35	1.17	.45	11.97	13.75
1500	Frames 16' OC		2640	.018		8.85	.98	.37	10.20	11.65
1700	Radial arches, 60' clear span, frames 8' OC		1920	.025		9.70	1.34	.51	11.55	13.35
1800	Frames 16' OC		2880	.017		7.75	.90	.34	8.99	10.30
2000	100' clear span, frames 8' OC		1600	.030		10	1.61	.61	12.22	14.25
2100	Frames 16' OC		2400	.020		8.80	1.07	.41	10.28	11.85
2300	120' clear span, frames 8' OC	↓	1440	.033	↓	13.25	1.79	.68	15.72	18.20

06 18 Glued-Laminated Construction

06 18 13 – Glued-Laminated Beams

06 18 13.20 Laminated Framing		Crew	Daily Output	Labor-Hours	Unit	2020 Bare Costs Material	Labor	Equipment	Total	Total Incl O&P
2400	Frames 16′ OC	F-4	1920	.025	SF Flr.	12.10	1.34	.51	13.95	16.05
2600	Bowstring trusses, 20′ OC, 40′ clear span	F-3	2400	.017		6	.91	.20	7.11	8.25
2700	60′ clear span	F-4	3600	.013		5.40	.72	.27	6.39	7.40
2800	100′ clear span		4000	.012		7.65	.64	.25	8.54	9.70
2900	120′ clear span		3600	.013		8.10	.72	.27	9.09	10.40
3000	For less than 1000 B.F., add					20%				
3100	For premium appearance, add to S.F. prices					5%				
3300	For industrial type, deduct					15%				
3500	For stain and varnish, add					5%				
3900	For 3/4″ laminations, add to straight					25%				
4100	Add to curved					15%				
4300	Alternate pricing method: (use nominal footage of									
4310	components). Straight beams, camber less than 6″	F-3	3.50	11.429	M.B.F.	3,075	620	135	3,830	4,525
4400	Columns, including hardware		2	20		3,300	1,075	236	4,611	5,600
4600	Curved members, radius over 32′		2.50	16		3,375	870	188	4,433	5,275
4700	Radius 10′ to 32′		3	13.333		3,350	725	157	4,232	5,000
4900	For complicated shapes, add maximum					100%				
5100	For pressure treating, add to straight					35%				
5200	Add to curved					45%				
6000	Laminated veneer members, southern pine or western species									
6050	1-3/4″ wide x 5-1/2″ deep	2 Carp	480	.033	L.F.	3.36	1.77		5.13	6.55
6100	9-1/2″ deep		480	.033		4.56	1.77		6.33	7.85
6150	14″ deep		450	.036		7.35	1.89		9.24	11.10
6200	18″ deep		450	.036		10.15	1.89		12.04	14.25
6300	Parallel strand members, southern pine or western species									
6350	1-3/4″ wide x 9-1/4″ deep	2 Carp	480	.033	L.F.	5.05	1.77		6.82	8.40
6400	11-1/4″ deep		450	.036		5.65	1.89		7.54	9.30
6450	14″ deep		400	.040		8.35	2.13		10.48	12.60
6500	3-1/2″ wide x 9-1/4″ deep		480	.033		18.10	1.77		19.87	22.50
6550	11-1/4″ deep		450	.036		19.70	1.89		21.59	24.50
6600	14″ deep		400	.040		22.50	2.13		24.63	28.50
6650	7″ wide x 9-1/4″ deep		450	.036		33	1.89		34.89	39.50
6700	11-1/4″ deep		420	.038		42	2.03		44.03	49
6750	14″ deep		400	.040		50	2.13		52.13	58.50
9000	Minimum labor/equipment charge	F-3	2.50	16	Job		870	188	1,058	1,575

06 22 Millwork

06 22 13 – Standard Pattern Wood Trim

06 22 13.15 Moldings, Base		Crew	Daily Output	Labor-Hours	Unit	Material	Labor	Equipment	Total	Total Incl O&P
0010	**MOLDINGS, BASE**									
5100	Classic profile, 5/8″ x 5-1/2″, finger jointed and primed	1 Carp	250	.032	L.F.	1.76	1.70		3.46	4.66
5105	Poplar		240	.033		2.18	1.77		3.95	5.25
5110	Red oak		220	.036		2.54	1.93		4.47	5.90
5115	Maple		220	.036		4.15	1.93		6.08	7.65
5120	Cherry		220	.036		4.55	1.93		6.48	8.10
5125	3/4″ x 7-1/2″, finger jointed and primed		250	.032		2.05	1.70		3.75	4.98
5130	Poplar		240	.033		2.66	1.77		4.43	5.75
5135	Red oak		220	.036		3.70	1.93		5.63	7.15
5140	Maple		220	.036		5.15	1.93		7.08	8.75
5145	Cherry		220	.036		6.10	1.93		8.03	9.85
5150	Modern profile, 5/8″ x 3-1/2″, finger jointed and primed		250	.032		.95	1.70		2.65	3.77

06 22 13 – Standard Pattern Wood Trim

06 22 13.15 Moldings, Base		Crew	Daily Output	Labor-Hours	Unit	Material	Labor	Equipment	Total	Total Incl O&P
						2020 Bare Costs				
5155	Poplar	1 Carp	240	.033	L.F.	1.06	1.77		2.83	4.01
5160	Red oak		220	.036		1.74	1.93		3.67	5
5165	Maple		220	.036		2.70	1.93		4.63	6.05
5170	Cherry		220	.036		2.94	1.93		4.87	6.35
5175	Ogee profile, 7/16" x 3", finger jointed and primed		250	.032		.68	1.70		2.38	3.47
5180	Poplar		240	.033		.76	1.77		2.53	3.68
5185	Red oak		220	.036		.99	1.93		2.92	4.19
5200	9/16" x 3-1/2", finger jointed and primed		250	.032		.68	1.70		2.38	3.47
5205	Pine		240	.033		1.28	1.77		3.05	4.25
5210	Red oak		220	.036		2.92	1.93		4.85	6.30
5215	9/16" x 4-1/2", red oak		220	.036		4.38	1.93		6.31	7.90
5220	5/8" x 3-1/2", finger jointed and primed		250	.032		1.04	1.70		2.74	3.86
5225	Poplar		240	.033		1.06	1.77		2.83	4.01
5230	Red oak		220	.036		1.74	1.93		3.67	5
5235	Maple		220	.036		2.70	1.93		4.63	6.05
5240	Cherry		220	.036		2.92	1.93		4.85	6.30
5245	5/8" x 4", finger jointed and primed		250	.032		1.28	1.70		2.98	4.13
5250	Poplar		240	.033		1.37	1.77		3.14	4.35
5255	Red oak		220	.036		1.96	1.93		3.89	5.25
5260	Maple		220	.036		3.02	1.93		4.95	6.40
5265	Cherry		220	.036		3.17	1.93		5.10	6.60
5270	Rectangular profile, oak, 3/8" x 1-1/4"		260	.031		1.58	1.64		3.22	4.35
5275	1/2" x 2-1/2"		255	.031		2.27	1.67		3.94	5.15
5280	1/2" x 3-1/2"		250	.032		2.85	1.70		4.55	5.85
5285	1" x 6"		240	.033		4.06	1.77		5.83	7.30
5290	1" x 8"		240	.033		5.05	1.77		6.82	8.45
5295	Pine, 3/8" x 1-3/4"		260	.031		.50	1.64		2.14	3.17
5300	7/16" x 2-1/2"		255	.031		.82	1.67		2.49	3.57
5305	1" x 6"		240	.033		.84	1.77		2.61	3.76
5310	1" x 8"		240	.033		.95	1.77		2.72	3.89
5315	Shoe, 1/2" x 3/4", primed		260	.031		.59	1.64		2.23	3.27
5320	Pine		240	.033		.43	1.77		2.20	3.31
5325	Poplar		240	.033		.43	1.77		2.20	3.31
5330	Red oak		220	.036		.57	1.93		2.50	3.73
5335	Maple		220	.036		.82	1.93		2.75	4
5340	Cherry		220	.036		.84	1.93		2.77	4.02
5345	11/16" x 1-1/2", pine		240	.033		.87	1.77		2.64	3.80
5350	Caps, 11/16" x 1-3/8", pine		240	.033		.69	1.77		2.46	3.60
5355	3/4" x 1-3/4", finger jointed and primed		260	.031		.93	1.64		2.57	3.64
5360	Poplar		240	.033		1.07	1.77		2.84	4.02
5365	Red oak		220	.036		1.33	1.93		3.26	4.56
5370	Maple		220	.036		1.68	1.93		3.61	4.95
5375	Cherry		220	.036		3.69	1.93		5.62	7.15
5380	Combination base & shoe, 9/16" x 3-1/2" & 1/2" x 3/4", pine		125	.064		1.71	3.40		5.11	7.35
5385	Three piece oak, 6" high		80	.100		5.95	5.30		11.25	15.05
5390	Including 3/4" x 1" base shoe		70	.114		6.60	6.05		12.65	17
5395	Flooring cant strip, 3/4" x 3/4", pre-finished pine		260	.031		.48	1.64		2.12	3.15
5400	For pre-finished, stain and clear coat, add					.59			.59	.65
5405	Clear coat only, add					.45			.45	.50
9000	Minimum labor/equipment charge	1 Carp	4	2	Job		106		106	170

06 22 13 – Standard Pattern Wood Trim

06 22 13.30 Moldings, Casings		Crew	Daily Output	Labor-Hours	Unit	Material	2020 Bare Costs Labor	Equipment	Total	Total Incl O&P
0010	**MOLDINGS, CASINGS**									
0085	Apron, 9/16" x 2-1/2", pine	1 Carp	250	.032	L.F.	1.51	1.70		3.21	4.38
0090	5/8" x 2-1/2", pine		250	.032		1.81	1.70		3.51	4.71
0110	5/8" x 3-1/2", pine		220	.036		2.27	1.93		4.20	5.60
0300	Band, 11/16" x 1-1/8", pine		270	.030		.77	1.57		2.34	3.37
0310	11/16" x 1-1/2", finger jointed and primed		270	.030		.76	1.57		2.33	3.36
0320	Pine		270	.030		1.02	1.57		2.59	3.64
0330	11/16" x 1-3/4", finger jointed and primed		270	.030		.96	1.57		2.53	3.58
0350	Pine		270	.030		1.09	1.57		2.66	3.72
0355	Beaded, 3/4" x 3-1/2", finger jointed and primed		220	.036		1.03	1.93		2.96	4.23
0360	Poplar		220	.036		1.20	1.93		3.13	4.42
0365	Red oak		220	.036		1.74	1.93		3.67	5
0370	Maple		220	.036		2.69	1.93		4.62	6.05
0375	Cherry		220	.036		3.25	1.93		5.18	6.70
0380	3/4" x 4", finger jointed and primed		220	.036		1.27	1.93		3.20	4.50
0385	Poplar		220	.036		1.84	1.93		3.77	5.10
0390	Red oak		220	.036		2.47	1.93		4.40	5.80
0395	Maple		220	.036		2.71	1.93		4.64	6.10
0400	Cherry		220	.036		4.03	1.93		5.96	7.55
0405	3/4" x 5-1/2", finger jointed and primed		200	.040		1.59	2.13		3.72	5.15
0410	Poplar		200	.040		2.04	2.13		4.17	5.65
0415	Red oak		200	.040		2.99	2.13		5.12	6.70
0420	Maple		200	.040		3.95	2.13		6.08	7.75
0425	Cherry		200	.040		4.53	2.13		6.66	8.40
0430	Classic profile, 3/4" x 2-3/4", finger jointed and primed		250	.032		.87	1.70		2.57	3.68
0435	Poplar		250	.032		1.05	1.70		2.75	3.88
0440	Red oak		250	.032		1.57	1.70		3.27	4.45
0445	Maple		250	.032		2.21	1.70		3.91	5.15
0450	Cherry		250	.032		2.54	1.70		4.24	5.50
0455	Fluted, 3/4" x 3-1/2", poplar		220	.036		1.20	1.93		3.13	4.42
0460	Red oak		220	.036		1.74	1.93		3.67	5
0465	Maple		220	.036		3.04	1.93		4.97	6.45
0470	Cherry		220	.036		3.25	1.93		5.18	6.70
0475	3/4" x 4", poplar		220	.036		1.47	1.93		3.40	4.72
0480	Red oak		220	.036		2.31	1.93		4.24	5.65
0485	Maple		220	.036		2.71	1.93		4.64	6.10
0490	Cherry		220	.036		4.03	1.93		5.96	7.55
0495	3/4" x 5-1/2", poplar		200	.040		2.04	2.13		4.17	5.65
0500	Red oak		200	.040		3.08	2.13		5.21	6.80
0505	Maple		200	.040		3.95	2.13		6.08	7.75
0510	Cherry		200	.040		4.52	2.13		6.65	8.40
0515	3/4" x 7-1/2", poplar		190	.042		1.59	2.24		3.83	5.35
0520	Red oak		190	.042		4	2.24		6.24	8
0525	Maple		190	.042		7.20	2.24		9.44	11.50
0530	Cherry		190	.042		8.60	2.24		10.84	13.05
0535	3/4" x 9-1/2", poplar		180	.044		4.21	2.36		6.57	8.40
0540	Red oak		180	.044		6.60	2.36		8.96	11.10
0545	Maple		180	.044		11.45	2.36		13.81	16.40
0550	Cherry		180	.044		12.50	2.36		14.86	17.55
0555	Modern profile, 9/16" x 2-1/4", poplar		250	.032		.88	1.70		2.58	3.69
0560	Red oak		250	.032		.99	1.70		2.69	3.81
0565	11/16" x 2-1/2", finger jointed & primed		250	.032		.77	1.70		2.47	3.57

06 22 13 – Standard Pattern Wood Trim

06 22 13.30 Moldings, Casings		Crew	Daily Output	Labor-Hours	Unit	2020 Bare Costs Material	Labor	Equipment	Total	Total Incl O&P
0570	Pine	1 Carp	250	.032	L.F.	1.29	1.70		2.99	4.14
0575	3/4" x 2-1/2", poplar		250	.032		.94	1.70		2.64	3.75
0580	Red oak		250	.032		1.26	1.70		2.96	4.11
0585	Maple		250	.032		1.94	1.70		3.64	4.85
0590	Cherry		250	.032		2.75	1.70		4.45	5.75
0595	Mullion, 5/16" x 2", pine		270	.030		.98	1.57		2.55	3.60
0600	9/16" x 2-1/2", finger jointed and primed		250	.032		.98	1.70		2.68	3.80
0605	Pine		250	.032		1.35	1.70		3.05	4.21
0610	Red oak		250	.032		3.40	1.70		5.10	6.45
0615	1-1/16" x 3-3/4", red oak		220	.036		7.05	1.93		8.98	10.85
0620	Ogee, 7/16" x 2-1/2", poplar		250	.032		.79	1.70		2.49	3.59
0625	Red oak		250	.032		.93	1.70		2.63	3.74
0630	9/16" x 2-1/4", finger jointed and primed		250	.032		.61	1.70		2.31	3.39
0635	Poplar		250	.032		.62	1.70		2.32	3.40
0640	Red oak		250	.032		.85	1.70		2.55	3.66
0645	11/16" x 2-1/2", finger jointed and primed		250	.032		.73	1.70		2.43	3.52
0700	Pine		250	.032		1.41	1.70		3.11	4.27
0701	Red oak		250	.032		3.29	1.70		4.99	6.35
0730	11/16" x 3-1/2", finger jointed and primed		220	.036		1.36	1.93		3.29	4.60
0750	Pine		220	.036		1.95	1.93		3.88	5.25
0755	3/4" x 2-1/2", finger jointed and primed		250	.032		.77	1.70		2.47	3.57
0760	Poplar		250	.032		.96	1.70		2.66	3.78
0765	Red oak		250	.032		1.32	1.70		3.02	4.17
0770	Maple		250	.032		1.91	1.70		3.61	4.82
0775	Cherry		250	.032		2.40	1.70		4.10	5.35
0780	3/4" x 3-1/2", finger jointed and primed		220	.036		.86	1.93		2.79	4.05
0785	Poplar		220	.036		1.20	1.93		3.13	4.42
0790	Red oak		220	.036		1.74	1.93		3.67	5
0795	Maple		220	.036		2.68	1.93		4.61	6.05
0800	Cherry		220	.036		3.30	1.93		5.23	6.75
4700	Square profile, 1" x 1", teak		215	.037		2.23	1.98		4.21	5.60
4800	Rectangular profile, 1" x 3", teak		200	.040		6.60	2.13		8.73	10.65
9000	Minimum labor/equipment charge		4	2	Job		106		106	170

06 22 13.35 Moldings, Ceilings		Crew	Daily Output	Labor-Hours	Unit	Material	Labor	Equipment	Total	Total Incl O&P
0010	**MOLDINGS, CEILINGS**									
0600	Bed, 9/16" x 1-3/4", pine	1 Carp	270	.030	L.F.	1.13	1.57		2.70	3.76
0650	9/16" x 2", pine		270	.030		1.30	1.57		2.87	3.95
0710	9/16" x 1-3/4", oak		270	.030		2.44	1.57		4.01	5.20
1200	Cornice, 9/16" x 1-3/4", pine		270	.030		1.02	1.57		2.59	3.64
1300	9/16" x 2-1/4", pine		265	.030		1.37	1.60		2.97	4.08
1350	Cove, 1/2" x 2-1/4", poplar		265	.030		1.26	1.60		2.86	3.96
1360	Red oak		265	.030		1.73	1.60		3.33	4.47
1370	Hard maple		265	.030		1.71	1.60		3.31	4.45
1380	Cherry		265	.030		2.31	1.60		3.91	5.10
2400	9/16" x 1-3/4", pine		270	.030		1.10	1.57		2.67	3.73
2500	11/16" x 2-3/4", pine		265	.030		2.21	1.60		3.81	5
2510	Crown, 5/8" x 5/8", poplar		300	.027		.55	1.42		1.97	2.88
2520	Red oak		300	.027		.56	1.42		1.98	2.89
2530	Hard maple		300	.027		.76	1.42		2.18	3.11
2540	Cherry		300	.027		.80	1.42		2.22	3.15
2600	9/16" x 3-5/8", pine		250	.032		1.83	1.70		3.53	4.73
2700	11/16" x 4-1/4", pine		250	.032		3.05	1.70		4.75	6.10

06 22 13.35 Moldings, Ceilings		Crew	Daily Output	Labor-Hours	Unit	2020 Bare Costs Material	Labor	Equipment	Total	Total Incl O&P
2705	Oak	1 Carp	250	.032	L.F.	6.50	1.70		8.20	9.85
2710	3/4" x 1-3/4", poplar		270	.030		.76	1.57		2.33	3.36
2720	Red oak		270	.030		1.12	1.57		2.69	3.75
2730	Hard maple		270	.030		1.47	1.57		3.04	4.14
2740	Cherry		270	.030		1.81	1.57		3.38	4.51
2750	3/4" x 2", poplar		270	.030		1	1.57		2.57	3.62
2760	Red oak		270	.030		1.50	1.57		3.07	4.17
2770	Hard maple		270	.030		2.05	1.57		3.62	4.78
2780	Cherry		270	.030		2.28	1.57		3.85	5.05
2790	3/4" x 2-3/4", poplar		265	.030		1.08	1.60		2.68	3.76
2800	Red oak		265	.030		1.67	1.60		3.27	4.41
2810	Hard maple		265	.030		2.28	1.60		3.88	5.10
2820	Cherry		265	.030		2.59	1.60		4.19	5.40
2830	3/4" x 3-1/2", poplar		250	.032		1.40	1.70		3.10	4.26
2840	Red oak		250	.032		2.08	1.70		3.78	5
2850	Hard maple		250	.032		2.95	1.70		4.65	5.95
2860	Cherry		250	.032		3.16	1.70		4.86	6.20
2870	FJP poplar		250	.032		1.06	1.70		2.76	3.89
2880	3/4" x 5", poplar		245	.033		2.04	1.74		3.78	5.05
2890	Red oak		245	.033		3.02	1.74		4.76	6.10
2900	Hard maple		245	.033		3.99	1.74		5.73	7.15
2910	Cherry		245	.033		4.76	1.74		6.50	8.05
2920	FJP poplar		245	.033		1.48	1.74		3.22	4.41
2930	3/4" x 6-1/4", poplar		240	.033		2.45	1.77		4.22	5.55
2940	Red oak		240	.033		3.69	1.77		5.46	6.90
2950	Hard maple		240	.033		4.96	1.77		6.73	8.30
2960	Cherry		240	.033		5.90	1.77		7.67	9.35
2970	7/8" x 8-3/4", poplar		220	.036		4.97	1.93		6.90	8.55
2980	Red oak		220	.036		5.20	1.93		7.13	8.80
2990	Hard maple		220	.036		6.85	1.93		8.78	10.65
3000	Cherry		220	.036		7.85	1.93		9.78	11.75
3010	1" x 7-1/4", poplar		220	.036		4.45	1.93		6.38	8
3020	Red oak		220	.036		6.25	1.93		8.18	9.95
3030	Hard maple		220	.036		9.05	1.93		10.98	13.05
3040	Cherry		220	.036		10.30	1.93		12.23	14.40
3050	1-1/16" x 4-1/4", poplar		250	.032		2.54	1.70		4.24	5.50
3060	Red oak		250	.032		2.86	1.70		4.56	5.85
3070	Hard maple		250	.032		4.20	1.70		5.90	7.35
3080	Cherry		250	.032		5.45	1.70		7.15	8.65
3090	Dentil crown, 3/4" x 5", poplar		250	.032		2.04	1.70		3.74	4.97
3100	Red oak		250	.032		3.02	1.70		4.72	6.05
3110	Hard maple		250	.032		3.98	1.70		5.68	7.10
3120	Cherry		250	.032		4.59	1.70		6.29	7.75
3130	Dentil piece for above, 1/2" x 1/2", poplar		300	.027		3.22	1.42		4.64	5.80
3140	Red oak		300	.027		3.33	1.42		4.75	5.95
3150	Hard maple		300	.027		4.17	1.42		5.59	6.85
3160	Cherry		300	.027		4.14	1.42		5.56	6.80
9000	Minimum labor/equipment charge		4	2	Job		106		106	170

06 22 13.40 Moldings, Exterior		Crew	Daily Output	Labor-Hours	Unit	Material	Labor	Equipment	Total	Total Incl O&P
0010	**MOLDINGS, EXTERIOR**									
0100	Band board, cedar, rough sawn, 1" x 2"	1 Carp	300	.027	L.F.	.72	1.42		2.14	3.06
0110	1" x 3"		300	.027		1.07	1.42		2.49	3.45

06 22 13 – Standard Pattern Wood Trim

06 22 13.40 Moldings, Exterior		Crew	Daily Output	Labor-Hours	Unit	2020 Bare Costs Material	Labor	Equipment	Total	Total Incl O&P
0120	1" x 4"	1 Carp	250	.032	L.F.	1.43	1.70		3.13	4.29
0130	1" x 6"		250	.032		2.15	1.70		3.85	5.10
0140	1" x 8"		225	.036		2.86	1.89		4.75	6.20
0150	1" x 10"		225	.036		3.56	1.89		5.45	6.95
0160	1" x 12"		200	.040		4.27	2.13		6.40	8.10
0240	STK, 1" x 2"		300	.027		.44	1.42		1.86	2.75
0250	1" x 3"		300	.027		.48	1.42		1.90	2.80
0260	1" x 4"		250	.032		.85	1.70		2.55	3.66
0270	1" x 6"		250	.032		1.39	1.70		3.09	4.25
0280	1" x 8"		225	.036		2.35	1.89		4.24	5.60
0290	1" x 10"		225	.036		2.86	1.89		4.75	6.20
0300	1" x 12"		200	.040		4.44	2.13		6.57	8.30
0310	Pine, #2, 1" x 2"		300	.027		.31	1.42		1.73	2.61
0320	1" x 3"		300	.027		.45	1.42		1.87	2.77
0330	1" x 4"		250	.032		.57	1.70		2.27	3.35
0340	1" x 6"		250	.032		.86	1.70		2.56	3.67
0350	1" x 8"		225	.036		1.33	1.89		3.22	4.49
0360	1" x 10"		225	.036		2.04	1.89		3.93	5.25
0370	1" x 12"		200	.040		2.66	2.13		4.79	6.35
0380	D & better, 1" x 2"		300	.027		.51	1.42		1.93	2.83
0390	1" x 3"		300	.027		.73	1.42		2.15	3.07
0400	1" x 4"		250	.032		.95	1.70		2.65	3.77
0410	1" x 6"		250	.032		1.10	1.70		2.80	3.93
0420	1" x 8"		225	.036		1.64	1.89		3.53	4.83
0430	1" x 10"		225	.036		2.16	1.89		4.05	5.40
0440	1" x 12"		200	.040		2.73	2.13		4.86	6.40
0450	Redwood, clear all heart, 1" x 2"		300	.027		.66	1.42		2.08	3
0460	1" x 3"		300	.027		.98	1.42		2.40	3.35
0470	1" x 4"		250	.032		1.22	1.70		2.92	4.06
0480	1" x 6"		252	.032		1.83	1.69		3.52	4.72
0490	1" x 8"		225	.036		2.43	1.89		4.32	5.70
0500	1" x 10"		225	.036		4.03	1.89		5.92	7.45
0510	1" x 12"		200	.040		4.40	2.13		6.53	8.25
0530	Corner board, cedar, rough sawn, 1" x 2"		225	.036		.72	1.89		2.61	3.82
0540	1" x 3"		225	.036		1.07	1.89		2.96	4.21
0550	1" x 4"		200	.040		1.43	2.13		3.56	4.98
0560	1" x 6"		200	.040		2.15	2.13		4.28	5.75
0570	1" x 8"		200	.040		2.86	2.13		4.99	6.55
0580	1" x 10"		175	.046		3.56	2.43		5.99	7.80
0590	1" x 12"		175	.046		4.27	2.43		6.70	8.60
0670	STK, 1" x 2"		225	.036		.44	1.89		2.33	3.51
0680	1" x 3"		225	.036		.48	1.89		2.37	3.56
0690	1" x 4"		200	.040		.83	2.13		2.96	4.32
0700	1" x 6"		200	.040		1.39	2.13		3.52	4.94
0710	1" x 8"		200	.040		2.35	2.13		4.48	6
0720	1" x 10"		175	.046		2.86	2.43		5.29	7.05
0730	1" x 12"		175	.046		4.44	2.43		6.87	8.80
0740	Pine, #2, 1" x 2"		225	.036		.31	1.89		2.20	3.37
0750	1" x 3"		225	.036		.45	1.89		2.34	3.53
0760	1" x 4"		200	.040		.57	2.13		2.70	4.04
0770	1" x 6"		200	.040		.86	2.13		2.99	4.36
0780	1" x 8"		200	.040		1.33	2.13		3.46	4.87
0790	1" x 10"		175	.046		2.04	2.43		4.47	6.15

06 22 13 – Standard Pattern Wood Trim

06 22 13.40 Moldings, Exterior		Crew	Daily Output	Labor-Hours	Unit	Material	Labor	Equipment	Total	Total Incl O&P
						2020 Bare Costs				
0800	1" x 12"	1 Carp	175	.046	L.F.	2.66	2.43		5.09	6.80
0810	D & better, 1" x 2"		225	.036		.51	1.89		2.40	3.59
0820	1" x 3"		225	.036		.73	1.89		2.62	3.83
0830	1" x 4"		200	.040		.95	2.13		3.08	4.46
0840	1" x 6"		200	.040		1.10	2.13		3.23	4.62
0850	1" x 8"		200	.040		1.64	2.13		3.77	5.20
0860	1" x 10"		175	.046		2.16	2.43		4.59	6.25
0870	1" x 12"		175	.046		2.73	2.43		5.16	6.90
0880	Redwood, clear all heart, 1" x 2"		225	.036		.66	1.89		2.55	3.76
0890	1" x 3"		225	.036		.98	1.89		2.87	4.11
0900	1" x 4"		200	.040		1.22	2.13		3.35	4.75
0910	1" x 6"		200	.040		1.83	2.13		3.96	5.45
0920	1" x 8"		200	.040		2.43	2.13		4.56	6.10
0930	1" x 10"		175	.046		4.03	2.43		6.46	8.35
0940	1" x 12"		175	.046		4.40	2.43		6.83	8.75
0950	Cornice board, cedar, rough sawn, 1" x 2"		330	.024		.72	1.29		2.01	2.85
0960	1" x 3"		290	.028		1.07	1.47		2.54	3.53
0970	1" x 4"		250	.032		1.43	1.70		3.13	4.29
0980	1" x 6"		250	.032		2.15	1.70		3.85	5.10
0990	1" x 8"		200	.040		2.86	2.13		4.99	6.55
1000	1" x 10"		180	.044		3.56	2.36		5.92	7.70
1010	1" x 12"		180	.044		4.27	2.36		6.63	8.50
1020	STK, 1" x 2"		330	.024		.44	1.29		1.73	2.54
1030	1" x 3"		290	.028		.48	1.47		1.95	2.88
1040	1" x 4"		250	.032		.85	1.70		2.55	3.66
1050	1" x 6"		250	.032		1.39	1.70		3.09	4.25
1060	1" x 8"		200	.040		2.35	2.13		4.48	6
1070	1" x 10"		180	.044		2.86	2.36		5.22	6.95
1080	1" x 12"		180	.044		4.44	2.36		6.80	8.65
1500	Pine, #2, 1" x 2"		330	.024		.31	1.29		1.60	2.40
1510	1" x 3"		290	.028		.31	1.47		1.78	2.69
1600	1" x 4"		250	.032		.57	1.70		2.27	3.35
1700	1" x 6"		250	.032		.86	1.70		2.56	3.67
1800	1" x 8"		200	.040		1.33	2.13		3.46	4.87
1900	1" x 10"		180	.044		2.04	2.36		4.40	6
2000	1" x 12"		180	.044		2.66	2.36		5.02	6.70
2020	D & better, 1" x 2"		330	.024		.51	1.29		1.80	2.62
2030	1" x 3"		290	.028		.73	1.47		2.20	3.15
2040	1" x 4"		250	.032		.95	1.70		2.65	3.77
2050	1" x 6"		250	.032		1.10	1.70		2.80	3.93
2060	1" x 8"		200	.040		1.64	2.13		3.77	5.20
2070	1" x 10"		180	.044		2.16	2.36		4.52	6.15
2080	1" x 12"		180	.044		2.73	2.36		5.09	6.80
2090	Redwood, clear all heart, 1" x 2"		330	.024		.66	1.29		1.95	2.79
2100	1" x 3"		290	.028		.98	1.47		2.45	3.43
2110	1" x 4"		250	.032		1.22	1.70		2.92	4.06
2120	1" x 6"		250	.032		1.83	1.70		3.53	4.74
2130	1" x 8"		200	.040		2.43	2.13		4.56	6.10
2140	1" x 10"		180	.044		4.03	2.36		6.39	8.20
2150	1" x 12"		180	.044		4.40	2.36		6.76	8.60
2160	3 piece, 1" x 2", 1" x 4", 1" x 6", rough sawn cedar		80	.100		4.31	5.30		9.61	13.25
2180	STK cedar		80	.100		2.68	5.30		7.98	11.45
2200	#2 pine		80	.100		1.75	5.30		7.05	10.40

06 22 13.40 Moldings, Exterior		Crew	Daily Output	Labor-Hours	Unit	Material	Labor	Equipment	Total	Total Incl O&P
						2020 Bare Costs				
2210	D & better pine	1 Carp	80	.100	L.F.	2.56	5.30		7.86	11.30
2220	Clear all heart redwood		80	.100		3.71	5.30		9.01	12.60
2230	1" x 8", 1" x 10", 1" x 12", rough sawn cedar		65	.123		10.65	6.55		17.20	22
2240	STK cedar		65	.123		9.60	6.55		16.15	21
2300	#2 pine		65	.123		6	6.55		12.55	17.10
2320	D & better pine		65	.123		6.50	6.55		13.05	17.65
2330	Clear all heart redwood		65	.123		10.80	6.55		17.35	22.50
2340	Door/window casing, cedar, rough sawn, 1" x 2"		275	.029		.72	1.55		2.27	3.27
2350	1" x 3"		275	.029		1.07	1.55		2.62	3.66
2360	1" x 4"		250	.032		1.43	1.70		3.13	4.29
2370	1" x 6"		250	.032		2.15	1.70		3.85	5.10
2380	1" x 8"		230	.035		2.86	1.85		4.71	6.10
2390	1" x 10"		230	.035		3.56	1.85		5.41	6.85
2395	1" x 12"		210	.038		4.27	2.03		6.30	7.95
2410	STK, 1" x 2"		275	.029		.44	1.55		1.99	2.96
2420	1" x 3"		275	.029		.48	1.55		2.03	3.01
2430	1" x 4"		250	.032		.85	1.70		2.55	3.66
2440	1" x 6"		250	.032		1.39	1.70		3.09	4.25
2450	1" x 8"		230	.035		2.35	1.85		4.20	5.55
2460	1" x 10"		230	.035		2.86	1.85		4.71	6.10
2470	1" x 12"		210	.038		4.44	2.03		6.47	8.15
2550	Pine, #2, 1" x 2"		275	.029		.31	1.55		1.86	2.82
2560	1" x 3"		275	.029		.45	1.55		2	2.98
2570	1" x 4"		250	.032		.57	1.70		2.27	3.35
2580	1" x 6"		250	.032		.86	1.70		2.56	3.67
2590	1" x 8"		230	.035		1.33	1.85		3.18	4.42
2600	1" x 10"		230	.035		2.04	1.85		3.89	5.20
2610	1" x 12"		210	.038		2.66	2.03		4.69	6.15
2620	Pine, D & better, 1" x 2"		275	.029		.51	1.55		2.06	3.04
2630	1" x 3"		275	.029		.73	1.55		2.28	3.28
2640	1" x 4"		250	.032		.95	1.70		2.65	3.77
2650	1" x 6"		250	.032		1.10	1.70		2.80	3.93
2660	1" x 8"		230	.035		1.64	1.85		3.49	4.76
2670	1" x 10"		230	.035		2.16	1.85		4.01	5.35
2680	1" x 12"		210	.038		2.73	2.03		4.76	6.25
2690	Redwood, clear all heart, 1" x 2"		275	.029		.66	1.55		2.21	3.21
2695	1" x 3"		275	.029		.98	1.55		2.53	3.56
2710	1" x 4"		250	.032		1.22	1.70		2.92	4.06
2715	1" x 6"		250	.032		1.83	1.70		3.53	4.74
2730	1" x 8"		230	.035		2.43	1.85		4.28	5.65
2740	1" x 10"		230	.035		4.03	1.85		5.88	7.40
2750	1" x 12"		210	.038		4.40	2.03		6.43	8.10
3500	Bellyband, pine, 11/16" x 4-1/4"		250	.032		2.82	1.70		4.52	5.80
3610	Brickmold, pine, 1-1/4" x 2"		200	.040		2.13	2.13		4.26	5.75
3620	FJP, 1-1/4" x 2"		200	.040		1.03	2.13		3.16	4.54
5100	Fascia, cedar, rough sawn, 1" x 2"		275	.029		.72	1.55		2.27	3.27
5110	1" x 3"		275	.029		1.07	1.55		2.62	3.66
5120	1" x 4"		250	.032		1.43	1.70		3.13	4.29
5200	1" x 6"		250	.032		2.15	1.70		3.85	5.10
5300	1" x 8"		230	.035		2.86	1.85		4.71	6.10
5310	1" x 10"		230	.035		3.56	1.85		5.41	6.85
5320	1" x 12"		210	.038		4.27	2.03		6.30	7.95
5400	2" x 4"		220	.036		1.26	1.93		3.19	4.48

06 22 13.40 Moldings, Exterior		Crew	Daily Output	Labor-Hours	Unit	Material	Labor	Equipment	Total	Total Incl O&P
						2020 Bare Costs				
5500	2" x 6"	1 Carp	220	.036	L.F.	1.88	1.93		3.81	5.15
5600	2" x 8"		200	.040		2.51	2.13		4.64	6.15
5700	2" x 10"		180	.044		3.12	2.36		5.48	7.20
5800	2" x 12"		170	.047		7.45	2.50		9.95	12.15
6120	STK, 1" x 2"		275	.029		.44	1.55		1.99	2.96
6130	1" x 3"		275	.029		.48	1.55		2.03	3.01
6140	1" x 4"		250	.032		.85	1.70		2.55	3.66
6150	1" x 6"		250	.032		1.39	1.70		3.09	4.25
6160	1" x 8"		230	.035		2.35	1.85		4.20	5.55
6170	1" x 10"		230	.035		2.86	1.85		4.71	6.10
6180	1" x 12"		210	.038		4.44	2.03		6.47	8.15
6185	2" x 2"		260	.031		.76	1.64		2.40	3.46
6190	Pine, #2, 1" x 2"		275	.029		.31	1.55		1.86	2.82
6200	1" x 3"		275	.029		.45	1.55		2	2.98
6210	1" x 4"		250	.032		.57	1.70		2.27	3.35
6220	1" x 6"		250	.032		.86	1.70		2.56	3.67
6230	1" x 8"		230	.035		1.33	1.85		3.18	4.42
6240	1" x 10"		230	.035		2.04	1.85		3.89	5.20
6250	1" x 12"		210	.038		2.66	2.03		4.69	6.15
6260	D & better, 1" x 2"		275	.029		.51	1.55		2.06	3.04
6270	1" x 3"		275	.029		.73	1.55		2.28	3.28
6280	1" x 4"		250	.032		.95	1.70		2.65	3.77
6290	1" x 6"		250	.032		1.10	1.70		2.80	3.93
6300	1" x 8"		230	.035		1.64	1.85		3.49	4.76
6310	1" x 10"		230	.035		2.16	1.85		4.01	5.35
6312	1" x 12"		210	.038		2.73	2.03		4.76	6.25
6330	Southern yellow, 1-1/4" x 5"		240	.033		3.12	1.77		4.89	6.25
6340	1-1/4" x 6"		240	.033		2.55	1.77		4.32	5.65
6350	1-1/4" x 8"		215	.037		3.83	1.98		5.81	7.40
6360	1-1/4" x 12"		190	.042		6.15	2.24		8.39	10.40
6370	Redwood, clear all heart, 1" x 2"		275	.029		.66	1.55		2.21	3.21
6380	1" x 3"		275	.029		1.22	1.55		2.77	3.82
6390	1" x 4"		250	.032		1.22	1.70		2.92	4.06
6400	1" x 6"		250	.032		1.83	1.70		3.53	4.74
6410	1" x 8"		230	.035		2.43	1.85		4.28	5.65
6420	1" x 10"		230	.035		4.03	1.85		5.88	7.40
6430	1" x 12"		210	.038		4.40	2.03		6.43	8.10
6440	1-1/4" x 5"		240	.033		1.92	1.77		3.69	4.95
6450	1-1/4" x 6"		240	.033		2.29	1.77		4.06	5.35
6460	1-1/4" x 8"		215	.037		3.63	1.98		5.61	7.15
6470	1-1/4" x 12"		190	.042		6.60	2.24		8.84	10.85
6580	Frieze, cedar, rough sawn, 1" x 2"		275	.029		.72	1.55		2.27	3.27
6590	1" x 3"		275	.029		1.07	1.55		2.62	3.66
6600	1" x 4"		250	.032		1.43	1.70		3.13	4.29
6610	1" x 6"		250	.032		2.15	1.70		3.85	5.10
6620	1" x 8"		250	.032		2.86	1.70		4.56	5.85
6630	1" x 10"		225	.036		3.56	1.89		5.45	6.95
6640	1" x 12"		200	.040		4.23	2.13		6.36	8.05
6650	STK, 1" x 2"		275	.029		.44	1.55		1.99	2.96
6660	1" x 3"		275	.029		.48	1.55		2.03	3.01
6670	1" x 4"		250	.032		.85	1.70		2.55	3.66
6680	1" x 6"		250	.032		1.39	1.70		3.09	4.25
6690	1" x 8"		250	.032		2.35	1.70		4.05	5.30

06 22 13 – Standard Pattern Wood Trim

06 22 13.40 Moldings, Exterior		Crew	Daily Output	Labor-Hours	Unit	Material	2020 Bare Costs Labor	Equipment	Total	Total Incl O&P
6700	1" x 10"	1 Carp	225	.036	L.F.	2.86	1.89		4.75	6.20
6710	1" x 12"		200	.040		4.44	2.13		6.57	8.30
6790	Pine, #2, 1" x 2"		275	.029		.31	1.55		1.86	2.82
6800	1" x 3"		275	.029		.45	1.55		2	2.98
6810	1" x 4"		250	.032		.57	1.70		2.27	3.35
6820	1" x 6"		250	.032		.86	1.70		2.56	3.67
6830	1" x 8"		250	.032		1.33	1.70		3.03	4.18
6840	1" x 10"		225	.036		2.04	1.89		3.93	5.25
6850	1" x 12"		200	.040		2.66	2.13		4.79	6.35
6860	D & better, 1" x 2"		275	.029		.51	1.55		2.06	3.04
6870	1" x 3"		275	.029		.73	1.55		2.28	3.28
6880	1" x 4"		250	.032		.95	1.70		2.65	3.77
6890	1" x 6"		250	.032		1.10	1.70		2.80	3.93
6900	1" x 8"		250	.032		1.64	1.70		3.34	4.52
6910	1" x 10"		225	.036		2.16	1.89		4.05	5.40
6920	1" x 12"		200	.040		2.73	2.13		4.86	6.40
6930	Redwood, clear all heart, 1" x 2"		275	.029		.66	1.55		2.21	3.21
6940	1" x 3"		275	.029		.98	1.55		2.53	3.56
6950	1" x 4"		250	.032		1.22	1.70		2.92	4.06
6960	1" x 6"		250	.032		1.83	1.70		3.53	4.74
6970	1" x 8"		250	.032		2.43	1.70		4.13	5.40
6980	1" x 10"		225	.036		4.03	1.89		5.92	7.45
6990	1" x 12"		200	.040		4.40	2.13		6.53	8.25
7000	Grounds, 1" x 1", cedar, rough sawn		300	.027		.37	1.42		1.79	2.68
7010	STK		300	.027		.27	1.42		1.69	2.57
7020	Pine, #2		300	.027		.20	1.42		1.62	2.48
7030	D & better		300	.027		.31	1.42		1.73	2.62
7050	Redwood		300	.027		.40	1.42		1.82	2.71
7060	Rake/verge board, cedar, rough sawn, 1" x 2"		225	.036		.72	1.89		2.61	3.82
7070	1" x 3"		225	.036		1.07	1.89		2.96	4.21
7080	1" x 4"		200	.040		1.43	2.13		3.56	4.98
7090	1" x 6"		200	.040		2.15	2.13		4.28	5.75
7100	1" x 8"		190	.042		2.86	2.24		5.10	6.75
7110	1" x 10"		190	.042		3.56	2.24		5.80	7.50
7120	1" x 12"		180	.044		4.27	2.36		6.63	8.50
7130	STK, 1" x 2"		225	.036		.44	1.89		2.33	3.51
7140	1" x 3"		225	.036		.48	1.89		2.37	3.56
7150	1" x 4"		200	.040		.85	2.13		2.98	4.35
7160	1" x 6"		200	.040		1.39	2.13		3.52	4.94
7170	1" x 8"		190	.042		2.35	2.24		4.59	6.20
7180	1" x 10"		190	.042		2.86	2.24		5.10	6.75
7190	1" x 12"		180	.044		4.44	2.36		6.80	8.65
7200	Pine, #2, 1" x 2"		225	.036		.31	1.89		2.20	3.37
7210	1" x 3"		225	.036		.45	1.89		2.34	3.53
7220	1" x 4"		200	.040		.57	2.13		2.70	4.04
7230	1" x 6"		200	.040		.86	2.13		2.99	4.36
7240	1" x 8"		190	.042		1.33	2.24		3.57	5.05
7250	1" x 10"		190	.042		2.04	2.24		4.28	5.85
7260	1" x 12"		180	.044		2.66	2.36		5.02	6.70
7340	D & better, 1" x 2"		225	.036		.51	1.89		2.40	3.59
7350	1" x 3"		225	.036		.73	1.89		2.62	3.83
7360	1" x 4"		200	.040		.95	2.13		3.08	4.46
7370	1" x 6"		200	.040		1.10	2.13		3.23	4.62

06 22 13 – Standard Pattern Wood Trim

06 22 13.40 Moldings, Exterior		Crew	Daily Output	Labor-Hours	Unit	Material	Labor	Equipment	Total	Total Incl O&P
						2020 Bare Costs				
7380	1" x 8"	1 Carp	190	.042	L.F.	1.64	2.24		3.88	5.40
7390	1" x 10"		190	.042		2.16	2.24		4.40	5.95
7400	1" x 12"		180	.044		2.73	2.36		5.09	6.80
7410	Redwood, clear all heart, 1" x 2"		225	.036		.66	1.89		2.55	3.76
7420	1" x 3"		225	.036		.98	1.89		2.87	4.11
7430	1" x 4"		200	.040		1.22	2.13		3.35	4.75
7440	1" x 6"		200	.040		1.83	2.13		3.96	5.45
7450	1" x 8"		190	.042		2.43	2.24		4.67	6.25
7460	1" x 10"		190	.042		4.03	2.24		6.27	8.05
7470	1" x 12"		180	.044		4.40	2.36		6.76	8.60
7480	2" x 4"		200	.040		2.35	2.13		4.48	6
7490	2" x 6"		182	.044		3.19	2.34		5.53	7.25
7500	2" x 8"		165	.048		5.10	2.58		7.68	9.80
7630	Soffit, cedar, rough sawn, 1" x 2"	2 Carp	440	.036		.72	1.93		2.65	3.89
7640	1" x 3"		440	.036		1.07	1.93		3	4.28
7650	1" x 4"		420	.038		1.43	2.03		3.46	4.81
7660	1" x 6"		420	.038		2.15	2.03		4.18	5.60
7670	1" x 8"		420	.038		2.86	2.03		4.89	6.40
7680	1" x 10"		400	.040		3.56	2.13		5.69	7.30
7690	1" x 12"		400	.040		4.27	2.13		6.40	8.10
7700	STK, 1" x 2"		440	.036		.44	1.93		2.37	3.58
7710	1" x 3"		440	.036		.48	1.93		2.41	3.63
7720	1" x 4"		420	.038		.85	2.03		2.88	4.18
7730	1" x 6"		420	.038		1.39	2.03		3.42	4.77
7740	1" x 8"		420	.038		2.35	2.03		4.38	5.85
7750	1" x 10"		400	.040		2.86	2.13		4.99	6.55
7760	1" x 12"		400	.040		4.44	2.13		6.57	8.30
7770	Pine, #2, 1" x 2"		440	.036		.31	1.93		2.24	3.44
7780	1" x 3"		440	.036		.45	1.93		2.38	3.60
7790	1" x 4"		420	.038		.57	2.03		2.60	3.87
7800	1" x 6"		420	.038		.86	2.03		2.89	4.19
7810	1" x 8"		420	.038		1.33	2.03		3.36	4.70
7820	1" x 10"		400	.040		2.04	2.13		4.17	5.65
7830	1" x 12"		400	.040		2.66	2.13		4.79	6.35
7840	D & better, 1" x 2"		440	.036		.51	1.93		2.44	3.66
7850	1" x 3"		440	.036		.73	1.93		2.66	3.90
7860	1" x 4"		420	.038		.95	2.03		2.98	4.29
7870	1" x 6"		420	.038		1.10	2.03		3.13	4.45
7880	1" x 8"		420	.038		1.64	2.03		3.67	5.05
7890	1" x 10"		400	.040		2.16	2.13		4.29	5.80
7900	1" x 12"		400	.040		2.73	2.13		4.86	6.40
7910	Redwood, clear all heart, 1" x 2"		440	.036		.66	1.93		2.59	3.83
7920	1" x 3"		440	.036		.98	1.93		2.91	4.18
7930	1" x 4"		420	.038		1.22	2.03		3.25	4.58
7940	1" x 6"		420	.038		1.83	2.03		3.86	5.25
7950	1" x 8"		420	.038		2.43	2.03		4.46	5.90
7960	1" x 10"		400	.040		4.03	2.13		6.16	7.85
7970	1" x 12"		400	.040		4.40	2.13		6.53	8.25
8050	Trim, crown molding, pine, 11/16" x 4-1/4"	1 Carp	250	.032		4.10	1.70		5.80	7.25
8060	Back band, 11/16" x 1-1/16"		250	.032		.92	1.70		2.62	3.73
8070	Insect screen frame stock, 1-1/16" x 1-3/4"		395	.020		2.24	1.08		3.32	4.18
8080	Dentils, 2-1/2" x 2-1/2" x 4", 6" OC		30	.267		1.27	14.15		15.42	24
8100	Fluted, 5-1/2"		165	.048		5.15	2.58		7.73	9.80

06 22 13 – Standard Pattern Wood Trim

06 22 13.40 Moldings, Exterior

	06 22 13.40 Moldings, Exterior	Crew	Daily Output	Labor-Hours	Unit	Material	Labor	Equipment	Total	Total Incl O&P
						2020 Bare Costs				
8110	Stucco bead, 1-3/8" x 1-5/8"	1 Carp	250	.032	L.F.	2.53	1.70		4.23	5.50
9000	Minimum labor/equipment charge	↓	4	2	Job		106		106	170

06 22 13.45 Moldings, Trim

	06 22 13.45 Moldings, Trim	Crew	Daily Output	Labor-Hours	Unit	Material	Labor	Equipment	Total	Total Incl O&P
0010	**MOLDINGS, TRIM**									
0200	Astragal, stock pine, 11/16" x 1-3/4"	1 Carp	255	.031	L.F.	1.26	1.67		2.93	4.06
0250	1-5/16" x 2-3/16"		240	.033		2.65	1.77		4.42	5.75
0800	Chair rail, stock pine, 5/8" x 2-1/2"		270	.030		1.54	1.57		3.11	4.21
0900	5/8" x 3-1/2"		240	.033		2.36	1.77		4.13	5.45
1000	Closet pole, stock pine, 1-1/8" diameter		200	.040		1.19	2.13		3.32	4.72
1100	Fir, 1-5/8" diameter		200	.040		2.28	2.13		4.41	5.90
3300	Half round, stock pine, 1/4" x 1/2"		270	.030		.25	1.57		1.82	2.80
3350	1/2" x 1"	↓	255	.031	↓	.64	1.67		2.31	3.37
3400	Handrail, fir, single piece, stock, hardware not included									
3450	1-1/2" x 1-3/4"	1 Carp	80	.100	L.F.	2.24	5.30		7.54	10.95
3470	Pine, 1-1/2" x 1-3/4"		80	.100		2.07	5.30		7.37	10.80
3500	1-1/2" x 2-1/2"		76	.105		2.59	5.60		8.19	11.80
3600	Lattice, stock pine, 1/4" x 1-1/8"		270	.030		.40	1.57		1.97	2.96
3700	1/4" x 1-3/4"		250	.032		.72	1.70		2.42	3.51
3800	Miscellaneous, custom, pine, 1" x 1"		270	.030		.42	1.57		1.99	2.98
3850	1" x 2"		265	.030		.84	1.60		2.44	3.49
3900	1" x 3"		240	.033		1.26	1.77		3.03	4.22
4100	Birch or oak, nominal 1" x 1"		240	.033		.40	1.77		2.17	3.28
4200	Nominal 1" x 3"		215	.037		1.20	1.98		3.18	4.49
4400	Walnut, nominal 1" x 1"		215	.037		.65	1.98		2.63	3.89
4500	Nominal 1" x 3"		200	.040		1.96	2.13		4.09	5.55
4700	Teak, nominal 1" x 1"		215	.037		2.91	1.98		4.89	6.35
4800	Nominal 1" x 3"		200	.040		8.75	2.13		10.88	13
4900	Quarter round, stock pine, 1/4" x 1/4"		275	.029		.29	1.55		1.84	2.80
4950	3/4" x 3/4"		255	.031	↓	.54	1.67		2.21	3.26
5600	Wainscot moldings, 1-1/8" x 9/16", 2' high, minimum		76	.105	S.F.	12.70	5.60		18.30	23
5700	Maximum		65	.123	"	16.80	6.55		23.35	29
9000	Minimum labor/equipment charge	↓	4	2	Job		106		106	170

06 22 13.50 Moldings, Window and Door

	06 22 13.50 Moldings, Window and Door	Crew	Daily Output	Labor-Hours	Unit	Material	Labor	Equipment	Total	Total Incl O&P
0010	**MOLDINGS, WINDOW AND DOOR**									
2800	Door moldings, stock, decorative, 1-1/8" wide, plain	1 Carp	17	.471	Set	47.50	25		72.50	92.50
2900	Detailed		17	.471	"	112	25		137	163
2960	Clear pine door jamb, no stops, 11/16" x 4-9/16"		240	.033	L.F.	5.45	1.77		7.22	8.85
3150	Door trim set, 1 head and 2 sides, pine, 2-1/2" wide		12	.667	Opng.	24	35.50		59.50	83
3170	3-1/2" wide		11	.727	"	33	38.50		71.50	98.50
3250	Glass beads, stock pine, 3/8" x 1/2"		275	.029	L.F.	.35	1.55		1.90	2.87
3270	3/8" x 7/8"		270	.030		.44	1.57		2.01	3
4850	Parting bead, stock pine, 3/8" x 3/4"		275	.029		.47	1.55		2.02	3
4870	1/2" x 3/4"		255	.031		.46	1.67		2.13	3.18
5000	Stool caps, stock pine, 11/16" x 3-1/2"		200	.040		2.56	2.13		4.69	6.25
5100	1-1/16" x 3-1/4"		150	.053	↓	3.80	2.83		6.63	8.70
5300	Threshold, oak, 3' long, inside, 5/8" x 3-5/8"		32	.250	Ea.	12.70	13.30		26	35.50
5400	Outside, 1-1/2" x 7-5/8"	↓	16	.500	"	48	26.50		74.50	95.50
5900	Window trim sets, including casings, header, stops,									
5910	stool and apron, 2-1/2" wide, FJP	1 Carp	13	.615	Opng.	33	32.50		65.50	89
5950	Pine		10	.800		39	42.50		81.50	111
6000	Oak		6	1.333	↓	79.50	71		150.50	202
9000	Minimum labor/equipment charge	↓	4	2	Job		106		106	170

06 22 Millwork

06 22 13 – Standard Pattern Wood Trim

06 22 13.60 Moldings, Soffits		Crew	Daily Output	Labor-Hours	Unit	2020 Bare Costs Material	Labor	Equipment	Total	Total Incl O&P
0010	**MOLDINGS, SOFFITS**									
0200	Soffits, pine, 1″ x 4″	2 Carp	420	.038	L.F.	.53	2.03		2.56	3.82
0210	1″ x 6″		420	.038		.82	2.03		2.85	4.14
0220	1″ x 8″		420	.038		1.29	2.03		3.32	4.65
0230	1″ x 10″		400	.040		1.97	2.13		4.10	5.60
0240	1″ x 12″		400	.040		2.60	2.13		4.73	6.25
0250	STK cedar, 1″ x 4″		420	.038		.81	2.03		2.84	4.13
0260	1″ x 6″		420	.038		1.35	2.03		3.38	4.73
0270	1″ x 8″		420	.038		2.31	2.03		4.34	5.80
0280	1″ x 10″		400	.040		2.80	2.13		4.93	6.50
0290	1″ x 12″		400	.040		4.38	2.13		6.51	8.25
1000	Exterior AC plywood, 1/4″ thick		400	.040	S.F.	.99	2.13		3.12	4.50
1050	3/8″ thick		400	.040		1.05	2.13		3.18	4.57
1100	1/2″ thick		400	.040		1.18	2.13		3.31	4.71
1150	Polyvinyl chloride, white, solid	1 Carp	230	.035		2.24	1.85		4.09	5.40
1160	Perforated	″	230	.035		2.24	1.85		4.09	5.40
1170	Accessories, "J" channel 5/8″	2 Carp	700	.023	L.F.	.56	1.22		1.78	2.57
9000	Minimum labor/equipment charge	″	5	3.200	Job		170		170	272

06 25 Prefinished Paneling

06 25 13 – Prefinished Hardboard Paneling

06 25 13.10 Paneling, Hardboard			Crew	Daily Output	Labor-Hours	Unit	Material	Labor	Equipment	Total	Total Incl O&P
0010	**PANELING, HARDBOARD**										
0050	Not incl. furring or trim, hardboard, tempered, 1/8″ thick	G	2 Carp	500	.032	S.F.	.50	1.70		2.20	3.27
0100	1/4″ thick	G		500	.032		.58	1.70		2.28	3.36
0300	Tempered pegboard, 1/8″ thick	G		500	.032		.43	1.70		2.13	3.19
0400	1/4″ thick	G		500	.032		.76	1.70		2.46	3.56
0600	Untempered hardboard, natural finish, 1/8″ thick	G		500	.032		.44	1.70		2.14	3.20
0700	1/4″ thick	G		500	.032		.53	1.70		2.23	3.30
0900	Untempered pegboard, 1/8″ thick	G		500	.032		.49	1.70		2.19	3.26
1000	1/4″ thick	G		500	.032		.55	1.70		2.25	3.33
1200	Plastic faced hardboard, 1/8″ thick	G		500	.032		.63	1.70		2.33	3.41
1300	1/4″ thick	G		500	.032		.87	1.70		2.57	3.68
1500	Plastic faced pegboard, 1/8″ thick	G		500	.032		.68	1.70		2.38	3.47
1600	1/4″ thick	G		500	.032		.83	1.70		2.53	3.63
1800	Wood grained, plain or grooved, 1/8″ thick	G		500	.032		.68	1.70		2.38	3.47
1900	1/4″ thick	G		425	.038		1.43	2		3.43	4.78
2100	Moldings, wood grained MDF			500	.032	L.F.	.41	1.70		2.11	3.17
2200	Pine			425	.038	″	1.40	2		3.40	4.75
9000	Minimum labor/equipment charge		1 Carp	2	4	Job		213		213	340

06 25 16 – Prefinished Plywood Paneling

06 25 16.10 Paneling, Plywood		Crew	Daily Output	Labor-Hours	Unit	Material	Labor	Equipment	Total	Total Incl O&P
0010	**PANELING, PLYWOOD**									
2400	Plywood, prefinished, 1/4″ thick, 4′ x 8′ sheets									
2410	with vertical grooves. Birch faced, economy	2 Carp	500	.032	S.F.	1.69	1.70		3.39	4.58
2420	Average		420	.038		1.24	2.03		3.27	4.60
2430	Custom		350	.046		1.08	2.43		3.51	5.10
2600	Mahogany, African		400	.040		2.43	2.13		4.56	6.10
2700	Philippine (Lauan)		500	.032		.57	1.70		2.27	3.35
2900	Oak		500	.032		1.36	1.70		3.06	4.22

06 25 Prefinished Paneling

06 25 16 – Prefinished Plywood Paneling

06 25 16.10 Paneling, Plywood		Crew	Daily Output	Labor-Hours	Unit	Material	2020 Bare Costs Labor	Equipment	Total	Total Incl O&P
3000	Cherry	2 Carp	400	.040	S.F.	2.03	2.13		4.16	5.65
3200	Rosewood		320	.050		3.08	2.66		5.74	7.65
3400	Teak		400	.040		3.26	2.13		5.39	7
3600	Chestnut		375	.043		5.45	2.27		7.72	9.65
3800	Pecan		400	.040		2.57	2.13		4.70	6.25
3900	Walnut, average		500	.032		2.64	1.70		4.34	5.60
3950	Custom		400	.040		2.38	2.13		4.51	6.05
4000	Plywood, prefinished, 3/4" thick, stock grades, economy		320	.050		1.38	2.66		4.04	5.80
4100	Average		224	.071		5.20	3.80		9	11.80
4300	Architectural grade, custom		224	.071		5.30	3.80		9.10	11.95
4400	Luxury		160	.100		5.55	5.30		10.85	14.65
4600	Plywood, "A" face, birch, VC, 1/2" thick, natural		450	.036		1.36	1.89		3.25	4.53
4700	Select		450	.036		1.95	1.89		3.84	5.20
4900	Veneer core, 3/4" thick, natural		320	.050		2.29	2.66		4.95	6.80
5000	Select		320	.050		2.68	2.66		5.34	7.20
5200	Lumber core, 3/4" thick, natural		320	.050		3.05	2.66		5.71	7.60
5500	Plywood, knotty pine, 1/4" thick, A2 grade		450	.036		1.62	1.89		3.51	4.81
5600	A3 grade		450	.036		2.25	1.89		4.14	5.50
5800	3/4" thick, veneer core, A2 grade		320	.050		2.42	2.66		5.08	6.90
5900	A3 grade		320	.050		2.49	2.66		5.15	7
6100	Aromatic cedar, 1/4" thick, plywood		400	.040		2.26	2.13		4.39	5.90
6200	1/4" thick, particle board	↓	400	.040	↓	1.16	2.13		3.29	4.69
9000	Minimum labor/equipment charge	1 Carp	2	4	Job		213		213	340

06 25 26 – Panel System

06 25 26.10 Panel Systems		Crew	Daily Output	Labor-Hours	Unit	Material	Labor	Equipment	Total	Total Incl O&P
0010	**PANEL SYSTEMS**									
0100	Raised panel, eng. wood core w/wood veneer, std., paint grade	2 Carp	300	.053	S.F.	13.10	2.83		15.93	19
0110	Oak veneer		300	.053		26.50	2.83		29.33	33.50
0120	Maple veneer		300	.053		30	2.83		32.83	37.50
0130	Cherry veneer		300	.053		40	2.83		42.83	49
0300	Class I fire rated, paint grade		300	.053		14	2.83		16.83	19.95
0310	Oak veneer		300	.053		30.50	2.83		33.33	38.50
0320	Maple veneer		300	.053		34.50	2.83		37.33	42.50
0330	Cherry veneer		300	.053		50	2.83		52.83	59.50
0510	Beadboard, 5/8" MDF, standard, primed		300	.053		9.70	2.83		12.53	15.20
0520	Oak veneer, unfinished		300	.053		16	2.83		18.83	22
0530	Maple veneer, unfinished		300	.053		16.50	2.83		19.33	22.50
0610	Rustic paneling, 5/8" MDF, standard, maple veneer, unfinished	↓	300	.053	↓	18.20	2.83		21.03	24.50

06 26 Board Paneling

06 26 13 – Profile Board Paneling

06 26 13.10 Paneling, Boards		Crew	Daily Output	Labor-Hours	Unit	Material	Labor	Equipment	Total	Total Incl O&P
0010	**PANELING, BOARDS**									
6400	Wood board paneling, 3/4" thick, knotty pine	2 Carp	300	.053	S.F.	2.07	2.83		4.90	6.80
6500	Rough sawn cedar		300	.053		3.48	2.83		6.31	8.35
6700	Redwood, clear, 1" x 4" boards		300	.053		5.05	2.83		7.88	10.10
6900	Aromatic cedar, closet lining, boards	↓	275	.058	↓	2.28	3.09		5.37	7.45
9000	Minimum labor/equipment charge	1 Carp	2	4	Job		213		213	340

06 43 13.20 Prefabricated Wood Stairs		Crew	Daily Output	Labor-Hours	Unit	Material	2020 Bare Costs Labor	Equipment	Total	Total Incl O&P
0010	**PREFABRICATED WOOD STAIRS**									
0100	Box stairs, prefabricated, 3'-0" wide									
0110	Oak treads, up to 14 risers	2 Carp	39	.410	Riser	98	22		120	143
0600	With pine treads for carpet, up to 14 risers	"	39	.410	"	63.50	22		85.50	105
1100	For 4' wide stairs, add				Flight	25%				
1550	Stairs, prefabricated stair handrail with balusters	1 Carp	30	.267	L.F.	83	14.15		97.15	114
1700	Basement stairs, prefabricated, pine treads									
1710	Pine risers, 3' wide, up to 14 risers	2 Carp	52	.308	Riser	63.50	16.35		79.85	96
4000	Residential, wood, oak treads, prefabricated		1.50	10.667	Flight	1,275	565		1,840	2,300
4200	Built in place		.44	36.364	"	2,250	1,925		4,175	5,575
4400	Spiral, oak, 4'-6" diameter, unfinished, prefabricated,									
4500	incl. railing, 9' high	2 Carp	1.50	10.667	Flight	2,625	565		3,190	3,775
9000	Minimum labor/equipment charge	"	3	5.333	Job		283		283	455

06 43 13.40 Wood Stair Parts		Crew	Daily Output	Labor-Hours	Unit	Material	Labor	Equipment	Total	Total Incl O&P
0010	**WOOD STAIR PARTS**									
0020	Pin top balusters, 1-1/4", oak, 34"	1 Carp	96	.083	Ea.	5.65	4.43		10.08	13.30
0030	38"		96	.083		5.75	4.43		10.18	13.45
0040	42"		96	.083		6.85	4.43		11.28	14.60
0050	Poplar, 34"		96	.083		3.34	4.43		7.77	10.75
0060	38"		96	.083		5.60	4.43		10.03	13.25
0070	42"		96	.083		5.95	4.43		10.38	13.65
0080	Maple, 34"		96	.083		6	4.43		10.43	13.70
0090	38"		96	.083		5.55	4.43		9.98	13.20
0100	42"		96	.083		7.85	4.43		12.28	15.75
0130	Primed, 34"		96	.083		3.67	4.43		8.10	11.15
0140	38"		96	.083		4.20	4.43		8.63	11.70
0150	42"		96	.083		4.84	4.43		9.27	12.40
0180	Box top balusters, 1-1/4", oak, 34"		60	.133		7.45	7.10		14.55	19.55
0190	38"		60	.133		11.75	7.10		18.85	24.50
0200	42"		60	.133		12.85	7.10		19.95	25.50
0210	Poplar, 34"		60	.133		7.65	7.10		14.75	19.80
0220	38"		60	.133		8.25	7.10		15.35	20.50
0230	42"		60	.133		9.10	7.10		16.20	21.50
0240	Maple, 34"		60	.133		9.75	7.10		16.85	22
0250	38"		60	.133		10.65	7.10		17.75	23
0260	42"		60	.133		11.75	7.10		18.85	24.50
0290	Primed, 34"		60	.133		8.20	7.10		15.30	20.50
0300	38"		60	.133		12.05	7.10		19.15	24.50
0310	42"		60	.133		11.30	7.10		18.40	24
0340	Square balusters, cut from lineal stock, pine, 1-1/16" x 1-1/16"		180	.044	L.F.	1.29	2.36		3.65	5.20
0350	1-5/16" x 1-5/16"		180	.044		1.99	2.36		4.35	5.95
0360	1-5/8" x 1-5/8"		180	.044		2.48	2.36		4.84	6.50
0370	Turned newel, oak, 3-1/2" square, 48" high		8	1	Ea.	94	53		147	188
0380	62" high		8	1		99.50	53		152.50	194
0390	Poplar, 3-1/2" square, 48" high		8	1		46	53		99	136
0400	62" high		8	1		56.50	53		109.50	148
0410	Maple, 3-1/2" square, 48" high		8	1		79	53		132	172
0420	62" high		8	1		79	53		132	172
0430	Square newel, oak, 3-1/2" square, 48" high		8	1		56.50	53		109.50	148
0440	58" high		8	1		65	53		118	157
0450	Poplar, 3-1/2" square, 48" high		8	1		42	53		95	131
0460	58" high		8	1		62	53		115	153

06 43 13 – Wood Stairs

06 43 13.40 Wood Stair Parts		Crew	Daily Output	Labor-Hours	Unit	Material	2020 Bare Costs Labor	Equipment	Total	Total Incl O&P
0470	Maple, 3" square, 48" high	1 Carp	8	1	Ea.	53	53		106	144
0480	58" high		8	1		67	53		120	159
0490	Railings, oak, economy		96	.083	L.F.	9	4.43		13.43	17
0500	Average		96	.083		15.10	4.43		19.53	24
0510	Custom		96	.083		18.75	4.43		23.18	27.50
0520	Maple, economy		96	.083		11.70	4.43		16.13	20
0530	Average		96	.083		13.25	4.43		17.68	21.50
0540	Custom		96	.083		22	4.43		26.43	31.50
0550	Oak, for bending rail, economy		48	.167		26.50	8.85		35.35	43
0560	Average		48	.167		29.50	8.85		38.35	46.50
0570	Custom		48	.167		33	8.85		41.85	50
0580	Maple, for bending rail, economy		48	.167		29	8.85		37.85	46
0590	Average		48	.167		31.50	8.85		40.35	48.50
0600	Custom		48	.167		36.50	8.85		45.35	54
0610	Risers, oak, 3/4" x 8", 36" long		80	.100	Ea.	12.85	5.30		18.15	22.50
0620	42" long		70	.114		15	6.05		21.05	26.50
0630	48" long		63	.127		17.10	6.75		23.85	29.50
0640	54" long		56	.143		19.25	7.60		26.85	33
0650	60" long		50	.160		21.50	8.50		30	37
0660	72" long		42	.190		25.50	10.10		35.60	44.50
0670	Poplar, 3/4" x 8", 36" long		80	.100		13.10	5.30		18.40	23
0680	42" long		71	.113		15.25	6		21.25	26.50
0690	48" long		63	.127		17.45	6.75		24.20	30
0700	54" long		56	.143		19.60	7.60		27.20	33.50
0710	60" long		50	.160		22	8.50		30.50	37.50
0720	72" long		42	.190		26	10.10		36.10	45
0730	Pine, 1" x 8", 36" long		80	.100		3.86	5.30		9.16	12.75
0740	42" long		70	.114		4.50	6.05		10.55	14.70
0750	48" long		63	.127		5.15	6.75		11.90	16.45
0760	54" long		56	.143		5.80	7.60		13.40	18.50
0770	60" long		50	.160		6.45	8.50		14.95	20.50
0780	72" long		42	.190		7.70	10.10		17.80	24.50
0790	Treads, oak, no returns, 1-1/32" x 11-1/2" x 36" long		32	.250		31.50	13.30		44.80	56
0800	42" long		32	.250		36.50	13.30		49.80	62
0810	48" long		32	.250		42	13.30		55.30	67.50
0820	54" long		32	.250		47	13.30		60.30	73.50
0830	60" long		32	.250		52.50	13.30		65.80	79
0840	72" long		32	.250		63	13.30		76.30	90.50
0850	Mitred return one end, 1-1/32" x 11-1/2" x 36" long		24	.333		38.50	17.70		56.20	70.50
0860	42" long		24	.333		45	17.70		62.70	77.50
0870	48" long		24	.333		51	17.70		68.70	85
0880	54" long		24	.333		57.50	17.70		75.20	92
0890	60" long		24	.333		64	17.70		81.70	99
0900	72" long		24	.333		76.50	17.70		94.20	113
0910	Mitred return two ends, 1-1/32" x 11-1/2" x 36" long		12	.667		45.50	35.50		81	107
0920	42" long		12	.667		53	35.50		88.50	116
0930	48" long		12	.667		61	35.50		96.50	124
0940	54" long		12	.667		68.50	35.50		104	133
0950	60" long		12	.667		76	35.50		111.50	141
0960	72" long		12	.667		91.50	35.50		127	157
0970	Starting step, oak, 48", bullnose		8	1		176	53		229	279
0980	Double end bullnose		8	1		268	53		321	380
1030	Skirt board, pine, 1" x 10"		55	.145	L.F.	1.97	7.75		9.72	14.55

06 43 Wood Stairs and Railings

06 43 13 – Wood Stairs

06 43 13.40 Wood Stair Parts		Crew	Daily Output	Labor-Hours	Unit	Material	Labor	Equipment	Total	Total Incl O&P
						2020 Bare Costs				
1040	1" x 12"	1 Carp	52	.154	L.F.	2.60	8.20		10.80	15.95
1050	Oak landing tread, 1-1/16" thick		54	.148	S.F.	7.90	7.85		15.75	21.50
1060	Oak cove molding		96	.083	L.F.	.95	4.43		5.38	8.15
1070	Oak stringer molding		96	.083	"	3.68	4.43		8.11	11.15
1090	Rail bolt, 5/16" x 3-1/2"		48	.167	Ea.	3.55	8.85		12.40	18.10
1100	5/16" x 4-1/2"		48	.167		3.04	8.85		11.89	17.55
1120	Newel post anchor		16	.500		12.85	26.50		39.35	56.50
1130	Tapered plug, 1/2"		240	.033		.95	1.77		2.72	3.89
1140	1"		240	.033		.98	1.77		2.75	3.92
9000	Minimum labor/equipment charge		3	2.667	Job		142		142	227

06 43 16 – Wood Railings

06 43 16.10 Wood Handrails and Railings		Crew	Daily Output	Labor-Hours	Unit	Material	Labor	Equipment	Total	Total Incl O&P
0010	**WOOD HANDRAILS AND RAILINGS**									
0020	Custom design, architectural grade, hardwood, plain	1 Carp	38	.211	L.F.	12.60	11.20		23.80	32
0100	Shaped		30	.267		56	14.15		70.15	84
0300	Stock interior railing with spindles 4" OC, 4' long		40	.200		46.50	10.65		57.15	68
0400	8' long		48	.167		46.50	8.85		55.35	65
9000	Minimum labor/equipment charge		3	2.667	Job		142		142	227

06 44 Ornamental Woodwork

06 44 19 – Wood Grilles

06 44 19.10 Grilles		Crew	Daily Output	Labor-Hours	Unit	Material	Labor	Equipment	Total	Total Incl O&P
0010	**GRILLES** and panels, hardwood, sanded									
0020	2' x 4' to 4' x 8', custom designs, unfinished, economy	1 Carp	38	.211	S.F.	61	11.20		72.20	85
0050	Average		30	.267		73.50	14.15		87.65	104
0100	Custom		19	.421		70.50	22.50		93	114
9000	Minimum labor/equipment charge		2	4	Job		213		213	340

06 44 33 – Wood Mantels

06 44 33.10 Fireplace Mantels		Crew	Daily Output	Labor-Hours	Unit	Material	Labor	Equipment	Total	Total Incl O&P
0010	**FIREPLACE MANTELS**									
0015	6" molding, 6' x 3'-6" opening, plain, paint grade	1 Carp	5	1.600	Opng.	490	85		575	675
0100	Ornate, oak		5	1.600		640	85		725	840
0300	Prefabricated pine, colonial type, stock, deluxe		2	4		1,900	213		2,113	2,450
0400	Economy		3	2.667		815	142		957	1,125
9000	Minimum labor/equipment charge		3	2.667	Job		142		142	227

06 44 33.20 Fireplace Mantel Beam		Crew	Daily Output	Labor-Hours	Unit	Material	Labor	Equipment	Total	Total Incl O&P
0010	**FIREPLACE MANTEL BEAM**									
0020	Rough texture wood, 4" x 8"	1 Carp	36	.222	L.F.	8.80	11.80		20.60	28.50
0100	4" x 10"		35	.229	"	12.45	12.15		24.60	33
0300	Laminated hardwood, 2-1/4" x 10-1/2" wide, 6' long		5	1.600	Ea.	109	85		194	256
0400	8' long		5	1.600	"	162	85		247	315
0600	Brackets for above, rough sawn		12	.667	Pr.	11.50	35.50		47	69.50
0700	Laminated		12	.667	"	22.50	35.50		58	81.50
9000	Minimum labor/equipment charge		4	2	Job		106		106	170

06 44 39 – Wood Posts and Columns

06 44 39.10 Decorative Beams		Crew	Daily Output	Labor-Hours	Unit	Material	Labor	Equipment	Total	Total Incl O&P
0010	**DECORATIVE BEAMS**									
0020	Rough sawn cedar, non-load bearing, 4" x 4"	2 Carp	180	.089	L.F.	1.61	4.72		6.33	9.30
0100	4" x 6"		170	.094		1.84	5		6.84	10

06 44 Ornamental Woodwork

06 44 39 – Wood Posts and Columns

06 44 39.10 Decorative Beams		Crew	Daily Output	Labor-Hours	Unit	Material	2020 Bare Costs Labor	Equipment	Total	Total Incl O&P
0200	4" x 8"	2 Carp	160	.100	L.F.	2.44	5.30		7.74	11.20
0300	4" x 10"		150	.107		4.66	5.65		10.31	14.25
0400	4" x 12"		140	.114		5.50	6.05		11.55	15.80
0500	8" x 8"	↓	130	.123	↓	5	6.55		11.55	16
9000	Minimum labor/equipment charge	1 Carp	3	2.667	Job		142		142	227

06 44 39.20 Columns		Crew	Daily Output	Labor-Hours	Unit	Material	Labor	Equipment	Total	Total Incl O&P
0010	**COLUMNS**									
0050	Aluminum, round colonial, 6" diameter	2 Carp	80	.200	V.L.F.	21.50	10.65		32.15	41
0100	8" diameter		62.25	.257		20.50	13.65		34.15	44.50
0200	10" diameter		55	.291		24.50	15.45		39.95	52
0250	Fir, stock units, hollow round, 6" diameter		80	.200		31	10.65		41.65	51
0300	8" diameter		80	.200		36.50	10.65		47.15	57.50
0350	10" diameter		70	.229		45.50	12.15		57.65	69.50
0360	12" diameter		65	.246		65	13.10		78.10	92.50
0400	Solid turned, to 8' high, 3-1/2" diameter		80	.200		11.30	10.65		21.95	29.50
0500	4-1/2" diameter		75	.213		14.30	11.35		25.65	34
0600	5-1/2" diameter		70	.229		19.75	12.15		31.90	41
0800	Square columns, built-up, 5" x 5"		65	.246		34.50	13.10		47.60	59
0900	Solid, 3-1/2" x 3-1/2"		130	.123		11.95	6.55		18.50	23.50
1600	Hemlock, tapered, T&G, 12" diam., 10' high		100	.160		48	8.50		56.50	66
1700	16' high		65	.246		82.50	13.10		95.60	112
1900	14" diameter, 10' high		100	.160		102	8.50		110.50	126
2000	18' high		65	.246		102	13.10		115.10	133
2200	18" diameter, 12' high		65	.246		204	13.10		217.10	246
2300	20' high		50	.320		125	17		142	166
2500	20" diameter, 14' high		40	.400		201	21.50		222.50	255
2600	20' high	↓	35	.457		188	24.50		212.50	245
2800	For flat pilasters, deduct				↓	33%				
3000	For splitting into halves, add				Ea.	130			130	143
4000	Rough sawn cedar posts, 4" x 4"	2 Carp	250	.064	V.L.F.	4.61	3.40		8.01	10.50
4100	4" x 6"		235	.068		8.55	3.62		12.17	15.20
4200	6" x 6"		220	.073		13.10	3.87		16.97	20.50
4300	8" x 8"	↓	200	.080	↓	21	4.25		25.25	30
9000	Minimum labor/equipment charge	1 Carp	3	2.667	Job		142		142	227

06 48 Wood Frames

06 48 13 – Exterior Wood Door Frames

06 48 13.10 Exterior Wood Door Frames and Accessories		Crew	Daily Output	Labor-Hours	Unit	Material	Labor	Equipment	Total	Total Incl O&P
0010	**EXTERIOR WOOD DOOR FRAMES AND ACCESSORIES**									
0400	Exterior frame, incl. ext. trim, pine, 5/4 x 4-9/16" deep	2 Carp	375	.043	L.F.	6.90	2.27		9.17	11.25
0420	5-3/16" deep		375	.043		8.05	2.27		10.32	12.50
0440	6-9/16" deep		375	.043		10	2.27		12.27	14.65
0600	Oak, 5/4 x 4-9/16" deep		350	.046		12.30	2.43		14.73	17.40
0620	5-3/16" deep		350	.046		13.50	2.43		15.93	18.75
0640	6-9/16" deep		350	.046		21	2.43		23.43	27
1000	Sills, 8/4 x 8" deep, oak, no horns		100	.160		8.35	8.50		16.85	23
1020	2" horns		100	.160		21.50	8.50		30	37
1040	3" horns		100	.160		21.50	8.50		30	37
1100	8/4 x 10" deep, oak, no horns		90	.178		6.85	9.45		16.30	22.50
1120	2" horns		90	.178		27.50	9.45		36.95	45.50
1140	3" horns	↓	90	.178	↓	27.50	9.45		36.95	45.50

06 48 Wood Frames

06 48 13 – Exterior Wood Door Frames

06 48 13.10 Exterior Wood Door Frames and Accessories		Crew	Daily Output	Labor-Hours	Unit	Material	2020 Bare Costs Labor	Equipment	Total	Total Incl O&P
2000	Wood frame & trim, ext., colonial, 3′ opng., fluted pilasters, flat head	2 Carp	22	.727	Ea.	510	38.50		548.50	620
2010	Dentil head		21	.762		620	40.50		660.50	750
2020	Ram's head		20	.800		735	42.50		777.50	875
2100	5′-4″ opening, in-swing, fluted pilasters, flat head		17	.941		485	50		535	615
2120	Ram's head		15	1.067		1,325	56.50		1,381.50	1,575
2140	Out-swing, fluted pilasters, flat head		17	.941		480	50		530	610
2160	Ram's head		15	1.067		1,550	56.50		1,606.50	1,800
2400	6′-0″ opening, in-swing, fluted pilasters, flat head		16	1		490	53		543	620
2420	Ram's head		10	1.600		1,525	85		1,610	1,800
2460	Out-swing, fluted pilasters, flat head		16	1		485	53		538	620
2480	Ram's head		10	1.600		1,525	85		1,610	1,800
2600	For two sidelights, flat head, add		30	.533	Opng.	330	28.50		358.50	410
2620	Ram's head, add		20	.800	″	865	42.50		907.50	1,025
2700	Custom birch frame, 3′-0″ opening		16	1	Ea.	246	53		299	355
2750	6′-0″ opng.		16	1		395	53		448	520
2900	Exterior, modern, plain trim, 3′ opng., in-swing, FJP		26	.615		55	32.50		87.50	113
2920	Fir		24	.667		59	35.50		94.50	122
2940	Oak		22	.727		68.50	38.50		107	138

06 48 16 – Interior Wood Door Frames

06 48 16.10 Interior Wood Door Jamb and Frames

Line	Description	Crew	Daily Output	Labor-Hours	Unit	Material	Labor	Equipment	Total	Total Incl O&P
0010	**INTERIOR WOOD DOOR JAMB AND FRAMES**									
3000	Interior frame, pine, 11/16″ x 3-5/8″ deep	2 Carp	375	.043	L.F.	4.27	2.27		6.54	8.35
3020	4-9/16″ deep		375	.043		5.40	2.27		7.67	9.60
3200	Oak, 11/16″ x 3-5/8″ deep		350	.046		2.53	2.43		4.96	6.65
3220	4-9/16″ deep		350	.046		10.45	2.43		12.88	15.35
3240	5-3/16″ deep		350	.046		18.60	2.43		21.03	24.50
3400	Walnut, 11/16″ x 3-5/8″ deep		350	.046		9.75	2.43		12.18	14.60
3420	4-9/16″ deep		350	.046		12.10	2.43		14.53	17.20
3440	5-3/16″ deep		350	.046		9.95	2.43		12.38	14.85
3600	Pocket door frame		16	1	Ea.	113	53		166	209
3800	Threshold, oak, 5/8″ x 3-5/8″ deep		200	.080	L.F.	3.73	4.25		7.98	10.90
3820	4-5/8″ deep		190	.084		4.25	4.48		8.73	11.85
3840	5-5/8″ deep		180	.089		6.80	4.72		11.52	15.05
9000	Minimum labor/equipment charge	1 Carp	4	2	Job		106		106	170

06 49 Wood Screens and Exterior Wood Shutters

06 49 19 – Exterior Wood Shutters

06 49 19.10 Shutters, Exterior

Line	Description	Crew	Daily Output	Labor-Hours	Unit	Material	Labor	Equipment	Total	Total Incl O&P
0010	**SHUTTERS, EXTERIOR**									
0012	Aluminum, louvered, 1′-4″ wide, 3′-0″ long	1 Carp	10	.800	Pr.	219	42.50		261.50	310
0200	4′-0″ long		10	.800		265	42.50		307.50	360
0300	5′-4″ long		10	.800		315	42.50		357.50	415
0400	6′-8″ long		9	.889		390	47		437	505
1000	Pine, louvered, primed, each 1′-2″ wide, 3′-3″ long		10	.800		263	42.50		305.50	355
1100	4′-7″ long		10	.800		305	42.50		347.50	405
1250	Each 1′-4″ wide, 3′-0″ long		10	.800		297	42.50		339.50	395
1350	5′-3″ long		10	.800		410	42.50		452.50	520
1500	Each 1′-6″ wide, 3′-3″ long		10	.800		294	42.50		336.50	395
1600	4′-7″ long		10	.800		405	42.50		447.50	515
1620	Cedar, louvered, 1′-2″ wide, 5′-7″ long		10	.800		360	42.50		402.50	465

06 49 Wood Screens and Exterior Wood Shutters

06 49 19 – Exterior Wood Shutters

06 49 19.10 Shutters, Exterior		Crew	Daily Output	Labor-Hours	Unit	Material	2020 Bare Costs Labor	Equipment	Total	Total Incl O&P
1630	Each 1′-4″ wide, 2′-2″ long	1 Carp	10	.800	Pr.	195	42.50		237.50	282
1640	3′-0″ long		10	.800		257	42.50		299.50	350
1650	3′-3″ long		10	.800		270	42.50		312.50	365
1660	3′-11″ long		10	.800		310	42.50		352.50	410
1670	4′-3″ long		10	.800		345	42.50		387.50	450
1680	5′-3″ long		10	.800		410	42.50		452.50	525
1690	5′-11″ long		10	.800		455	42.50		497.50	570
1700	Door blinds, 6′-9″ long, each 1′-3″ wide		9	.889		500	47		547	625
1710	1′-6″ wide		9	.889		455	47		502	580
1720	Cedar, solid raised panel, each 1′-4″ wide, 3′-3″ long		10	.800		375	42.50		417.50	480
1730	3′-11″ long		10	.800		375	42.50		417.50	485
1740	4′-3″ long		10	.800		370	42.50		412.50	475
1750	4′-7″ long		10	.800		385	42.50		427.50	495
1760	4′-11″ long		10	.800		420	42.50		462.50	530
1770	5′-11″ long		10	.800		590	42.50		632.50	720
1800	Door blinds, 6′-9″ long, each 1′-3″ wide		9	.889		515	47		562	640
1900	1′-6″ wide		9	.889		770	47		817	925
2500	Polystyrene, solid raised panel, each 1′-4″ wide, 3′-3″ long		10	.800		89	42.50		131.50	166
2600	3′-11″ long		10	.800		93.50	42.50		136	171
2700	4′-7″ long		10	.800		106	42.50		148.50	185
2800	5′-3″ long		10	.800		120	42.50		162.50	200
2900	6′-8″ long		9	.889		153	47		200	245
4500	Polystyrene, louvered, each 1′-2″ wide, 3′-3″ long		10	.800		37.50	42.50		80	109
4600	4′-7″ long		10	.800		50	42.50		92.50	123
4750	5′-3″ long		10	.800		58	42.50		100.50	132
4850	6′-8″ long		9	.889		72.50	47		119.50	156
6000	Vinyl, louvered, each 1′-2″ x 4′-7″ long		10	.800		67	42.50		109.50	142
6200	Each 1′-4″ x 6′-8″ long		9	.889		89	47		136	174
9000	Minimum labor/equipment charge		4	2	Job		106		106	170

06 51 Structural Plastic Shapes and Plates

06 51 13 – Plastic Lumber

06 51 13.10 Recycled Plastic Lumber			Crew	Daily Output	Labor-Hours	Unit	Material	Labor	Equipment	Total	Total Incl O&P
0010	RECYCLED PLASTIC LUMBER										
4000	Sheeting, recycled plastic, black or white, 4′ x 8′ x 1/8″	G	2 Carp	1100	.015	S.F.	1.05	.77		1.82	2.40
4010	4′ x 8′ x 3/16″	G		1100	.015		1.46	.77		2.23	2.85
4020	4′ x 8′ x 1/4″	G		950	.017		1.62	.90		2.52	3.21
4030	4′ x 8′ x 3/8″	G		950	.017		2.80	.90		3.70	4.51
4040	4′ x 8′ x 1/2″	G		900	.018		3.74	.95		4.69	5.60
4050	4′ x 8′ x 5/8″	G		900	.018		6.95	.95		7.90	9.15
4060	4′ x 8′ x 3/4″	G		850	.019		8.10	1		9.10	10.50
4070	Add for colors	G				Ea.	5%				
8500	100% recycled plastic, var colors, NLB, 2″ x 2″	G				L.F.	1.89			1.89	2.08
8510	2″ x 4″	G					3.96			3.96	4.36
8520	2″ x 6″	G					6.20			6.20	6.85
8530	2″ x 8″	G					8.80			8.80	9.65
8540	2″ x 10″	G					12.45			12.45	13.70
8550	5/4″ x 4″	G					4.32			4.32	4.75
8560	5/4″ x 6″	G					5.90			5.90	6.50
8570	1″ x 6″	G					2.90			2.90	3.19
8580	1/2″ x 8″	G					3.41			3.41	3.75

06 51 Structural Plastic Shapes and Plates

06 51 13 – Plastic Lumber

06 51 13.10 Recycled Plastic Lumber		Crew	Daily Output	Labor-Hours	Unit	Material	2020 Bare Costs Labor	Equipment	Total	Total Incl O&P
8590	2" x 10" T&G G				L.F.	12.60			12.60	13.85
8600	3" x 10" T&G G					19.80			19.80	22
8610	Add for premium colors G				↓	20%				

06 51 13.12 Structural Plastic Lumber		Crew	Daily Output	Labor-Hours	Unit	Material	Labor	Equipment	Total	Total Incl O&P
0010	STRUCTURAL PLASTIC LUMBER									
1320	Plastic lumber, posts or columns, 4" x 4"	2 Carp	390	.041	L.F.	10.95	2.18		13.13	15.55
1325	4" x 6"		275	.058		14.45	3.09		17.54	21
1330	4" x 8"		220	.073		21	3.87		24.87	29.50
1340	Girder, single, 4" x 4"		675	.024		10.95	1.26		12.21	14.05
1345	4" x 6"		600	.027		14.45	1.42		15.87	18.15
1350	4" x 8"		525	.030		21	1.62		22.62	26
1352	Double, 2" x 4"		625	.026		9.55	1.36		10.91	12.75
1354	2" x 6"		600	.027		10.45	1.42		11.87	13.75
1356	2" x 8"		575	.028		18.30	1.48		19.78	22.50
1358	2" x 10"		550	.029		22.50	1.55		24.05	27.50
1360	2" x 12"		525	.030		26	1.62		27.62	31
1362	Triple, 2" x 4"		575	.028		14.35	1.48		15.83	18.15
1364	2" x 6"		550	.029		15.70	1.55		17.25	19.75
1366	2" x 8"		525	.030		27.50	1.62		29.12	32.50
1368	2" x 10"		500	.032		34	1.70		35.70	40
1370	2" x 12"		475	.034		38.50	1.79		40.29	45.50
1372	Ledger, bolted 4' OC, 2" x 4"		400	.040		4.94	2.13		7.07	8.85
1374	2" x 6"		550	.029		5.30	1.55		6.85	8.35
1376	2" x 8"		390	.041		9.30	2.18		11.48	13.70
1378	2" x 10"		385	.042		11.40	2.21		13.61	16.10
1380	2" x 12"		380	.042		13	2.24		15.24	17.90
1382	Joists, 2" x 4"		1250	.013		4.79	.68		5.47	6.35
1384	2" x 6"		1250	.013		5.25	.68		5.93	6.85
1386	2" x 8"		1100	.015		9.15	.77		9.92	11.30
1388	2" x 10"		500	.032		11.45	1.70		13.15	15.25
1390	2" x 12"	↓	875	.018		12.90	.97		13.87	15.75
1392	Railings and trim, 5/4" x 4"	1 Carp	300	.027		4.90	1.42		6.32	7.65
1394	2" x 2"		300	.027		2.20	1.42		3.62	4.69
1396	2" x 4"		300	.027		4.76	1.42		6.18	7.50
1398	2" x 6"	↓	300	.027	↓	5.20	1.42		6.62	7.95

06 52 Plastic Structural Assemblies

06 52 10 – Fiberglass Structural Assemblies

06 52 10.20 Fiberglass Stair Treads		Crew	Daily Output	Labor-Hours	Unit	Material	Labor	Equipment	Total	Total Incl O&P
0010	FIBERGLASS STAIR TREADS									
0100	24" wide	2 Sswk	52	.308	Ea.	37.50	17.75		55.25	70
0140	30" wide		52	.308		46.50	17.75		64.25	80
0180	36" wide		52	.308		56	17.75		73.75	90.50
0220	42" wide	↓	52	.308	↓	65	17.75		82.75	101

06 52 10.30 Fiberglass Grating		Crew	Daily Output	Labor-Hours	Unit	Material	Labor	Equipment	Total	Total Incl O&P
0010	FIBERGLASS GRATING									
0100	Molded, green (for mod. corrosive environment)									
0140	1" x 4" mesh, 1" thick	2 Sswk	400	.040	S.F.	7.30	2.31		9.61	11.85
0180	1-1/2" square mesh, 1" thick		400	.040		16.40	2.31		18.71	22
0220	1-1/4" thick	↓	400	.040	↓	9.10	2.31		11.41	13.85

06 52 Plastic Structural Assemblies

06 52 10 – Fiberglass Structural Assemblies

06 52 10.30 Fiberglass Grating		Crew	Daily Output	Labor-Hours	Unit	Material	2020 Bare Costs Labor	Equipment	Total	Total Incl O&P
0260	1-1/2" thick	2 Sswk	400	.040	S.F.	24	2.31		26.31	30.50
0300	2" square mesh, 2" thick		320	.050		27	2.88		29.88	34.50
1000	Orange (for highly corrosive environment)									
1040	1" x 4" mesh, 1" thick	2 Sswk	400	.040	S.F.	17.85	2.31		20.16	23.50
1080	1-1/2" square mesh, 1" thick		400	.040		19.70	2.31		22.01	25.50
1120	1-1/4" thick		400	.040		18.55	2.31		20.86	24.50
1160	1-1/2" thick		400	.040		21	2.31		23.31	27
1200	2" square mesh, 2" thick		320	.050		21	2.88		23.88	27.50
3000	Pultruded, green (for mod. corrosive environment)									
3040	1" OC bar spacing, 1" thick	2 Sswk	400	.040	S.F.	14.35	2.31		16.66	19.55
3080	1-1/2" thick		320	.050		16.10	2.88		18.98	22.50
3120	1-1/2" OC bar spacing, 1" thick		400	.040		14.95	2.31		17.26	20
3160	1-1/2" thick		400	.040		21.50	2.31		23.81	27.50
4000	Grating support legs, fixed height, no base				Ea.	45.50			45.50	50
4040	With base					38.50			38.50	42.50
4080	Adjustable to 60"					58			58	63.50

06 52 10.40 Fiberglass Floor Grating		Crew	Daily Output	Labor-Hours	Unit	Material	Labor	Equipment	Total	Total Incl O&P
0010	FIBERGLASS FLOOR GRATING									
0100	Reinforced polyester, fire retardant, 1" x 4" grid, 1" thick	E-4	510	.063	S.F.	14.55	3.65	.29	18.49	22.50
0200	1-1/2" x 6" mesh, 1-1/2" thick		500	.064		16.80	3.72	.29	20.81	25
0300	With grit surface, 1-1/2" x 6" grid, 1-1/2" thick		500	.064		17.15	3.72	.29	21.16	25.50

06 63 Plastic Railings

06 63 10 – Plastic (PVC) Railings

06 63 10.10 Plastic Railings		Crew	Daily Output	Labor-Hours	Unit	Material	Labor	Equipment	Total	Total Incl O&P
0010	PLASTIC RAILINGS									
0100	Horizontal PVC handrail with balusters, 3-1/2" wide, 36" high	1 Carp	96	.083	L.F.	30	4.43		34.43	40
0150	42" high		96	.083		29	4.43		33.43	38.50
0200	Angled PVC handrail with balusters, 3-1/2" wide, 36" high		72	.111		20.50	5.90		26.40	32
0250	42" high		72	.111		30.50	5.90		36.40	43
0300	Post sleeve for 4 x 4 post		96	.083		16.45	4.43		20.88	25
0400	Post cap for 4 x 4 post, flat profile		48	.167	Ea.	15.55	8.85		24.40	31.50
0450	Newel post style profile		48	.167		28.50	8.85		37.35	45.50
0500	Raised corbeled profile		48	.167		35	8.85		43.85	53
0550	Post base trim for 4 x 4 post		96	.083		20.50	4.43		24.93	29.50

06 65 Plastic Trim

06 65 10 – PVC Trim

06 65 10.10 PVC Trim, Exterior		Crew	Daily Output	Labor-Hours	Unit	Material	Labor	Equipment	Total	Total Incl O&P
0010	PVC TRIM, EXTERIOR									
0100	Cornerboards, 5/4" x 6" x 6"	1 Carp	240	.033	L.F.	8.50	1.77		10.27	12.20
0110	Door/window casing, 1" x 4"		200	.040		2.09	2.13		4.22	5.70
0120	1" x 6"		200	.040		2.60	2.13		4.73	6.25
0130	1" x 8"		195	.041		3.43	2.18		5.61	7.25
0140	1" x 10"		195	.041		4.50	2.18		6.68	8.45
0150	1" x 12"		190	.042		5.10	2.24		7.34	9.20
0160	5/4" x 4"		195	.041		2.48	2.18		4.66	6.20
0170	5/4" x 6"		195	.041		3.74	2.18		5.92	7.60
0180	5/4" x 8"		190	.042		4.75	2.24		6.99	8.80

06 65 Plastic Trim

06 65 10 – PVC Trim

06 65 10.10 PVC Trim, Exterior		Crew	Daily Output	Labor-Hours	Unit	Material	Labor	Equipment	Total	Total Incl O&P
						2020 Bare Costs				
0190	5/4" x 10"	1 Carp	190	.042	L.F.	5.95	2.24		8.19	10.15
0200	5/4" x 12"		185	.043		6.75	2.30		9.05	11.10
0210	Fascia, 1" x 4"		250	.032		2.09	1.70		3.79	5
0220	1" x 6"		250	.032		2.60	1.70		4.30	5.60
0230	1" x 8"		225	.036		3.43	1.89		5.32	6.80
0240	1" x 10"		225	.036		4.50	1.89		6.39	8
0250	1" x 12"		200	.040		5.10	2.13		7.23	9
0260	5/4" x 4"		240	.033		2.48	1.77		4.25	5.55
0270	5/4" x 6"		240	.033		3.74	1.77		5.51	6.95
0280	5/4" x 8"		215	.037		4.75	1.98		6.73	8.35
0290	5/4" x 10"		215	.037		5.95	1.98		7.93	9.70
0300	5/4" x 12"		190	.042		6.75	2.24		8.99	11
0310	Frieze, 1" x 4"		250	.032		2.09	1.70		3.79	5
0320	1" x 6"		250	.032		2.60	1.70		4.30	5.60
0330	1" x 8"		225	.036		3.43	1.89		5.32	6.80
0340	1" x 10"		225	.036		4.50	1.89		6.39	8
0350	1" x 12"		200	.040		5.10	2.13		7.23	9
0360	5/4" x 4"		240	.033		2.48	1.77		4.25	5.55
0370	5/4" x 6"		240	.033		3.74	1.77		5.51	6.95
0380	5/4" x 8"		215	.037		4.75	1.98		6.73	8.35
0390	5/4" x 10"		215	.037		5.95	1.98		7.93	9.70
0400	5/4" x 12"		190	.042		6.75	2.24		8.99	11
0410	Rake, 1" x 4"		200	.040		2.09	2.13		4.22	5.70
0420	1" x 6"		200	.040		2.60	2.13		4.73	6.25
0430	1" x 8"		190	.042		3.43	2.24		5.67	7.35
0440	1" x 10"		190	.042		4.50	2.24		6.74	8.55
0450	1" x 12"		180	.044		5.10	2.36		7.46	9.40
0460	5/4" x 4"		195	.041		2.48	2.18		4.66	6.20
0470	5/4" x 6"		195	.041		3.74	2.18		5.92	7.60
0480	5/4" x 8"		185	.043		4.75	2.30		7.05	8.90
0490	5/4" x 10"		185	.043		5.95	2.30		8.25	10.25
0500	5/4" x 12"		175	.046		6.75	2.43		9.18	11.30
0510	Rake trim, 1" x 4"		225	.036		2.09	1.89		3.98	5.35
0520	1" x 6"		225	.036		2.60	1.89		4.49	5.90
0560	5/4" x 4"		220	.036		2.48	1.93		4.41	5.85
0570	5/4" x 6"	↓	220	.036		3.74	1.93		5.67	7.20
0610	Soffit, 1" x 4"	2 Carp	420	.038		2.09	2.03		4.12	5.55
0620	1" x 6"		420	.038		2.60	2.03		4.63	6.10
0630	1" x 8"		420	.038		3.43	2.03		5.46	7
0640	1" x 10"		400	.040		4.50	2.13		6.63	8.35
0650	1" x 12"		400	.040		5.10	2.13		7.23	9
0660	5/4" x 4"		410	.039		2.48	2.07		4.55	6.05
0670	5/4" x 6"		410	.039		3.74	2.07		5.81	7.45
0680	5/4" x 8"		410	.039		4.75	2.07		6.82	8.50
0690	5/4" x 10"		390	.041		5.95	2.18		8.13	10.05
0700	5/4" x 12"	↓	390	.041	↓	6.75	2.18		8.93	10.90

06 73 Composite Decking

06 73 13 – Composite Decking

06 73 13.10 Woodgrained Composite Decking		Crew	Daily Output	Labor-Hours	Unit	Material	Labor	Equipment	Total	Total Incl O&P
0010	WOODGRAINED COMPOSITE DECKING									
0100	Woodgrained composite decking, 1" x 6"	2 Carp	640	.025	L.F.	4.20	1.33		5.53	6.75
0110	Grooved edge		660	.024		4.39	1.29		5.68	6.90
0120	2" x 6"		640	.025		4.02	1.33		5.35	6.55
0130	Encased, 1" x 6"		640	.025		4.71	1.33		6.04	7.35
0140	Grooved edge		660	.024		4.66	1.29		5.95	7.15
0150	2" x 6"		640	.025		5.45	1.33		6.78	8.15

06 81 Composite Railings

06 81 10 – Encased Railings

06 81 10.10 Encased Composite Railings		Crew	Daily Output	Labor-Hours	Unit	Material	Labor	Equipment	Total	Total Incl O&P
0010	ENCASED COMPOSITE RAILINGS									
0100	Encased composite railing, 6' long, 36" high, incl. balusters	1 Carp	16	.500	Ea.	146	26.50		172.50	203
0110	42" high, incl. balusters		16	.500		225	26.50		251.50	290
0120	8' long, 36" high, incl. balusters		12	.667		170	35.50		205.50	243
0130	42" high, incl. balusters		12	.667		178	35.50		213.50	253
0140	Accessories, post sleeve, 4" x 4", 39" long		32	.250		30	13.30		43.30	54.50
0150	96" long		24	.333		76	17.70		93.70	113
0160	6" x 6", 39" long		32	.250		53.50	13.30		66.80	80
0170	96" long		24	.333		159	17.70		176.70	204
0180	Accessories, post skirt, 4" x 4"		96	.083		5.75	4.43		10.18	13.40
0190	6" x 6"		96	.083		8.05	4.43		12.48	15.95
0200	Post cap, 4" x 4", flat		48	.167		9.15	8.85		18	24.50
0210	Pyramid		48	.167		7.95	8.85		16.80	23
0220	Post cap, 6" x 6", flat		48	.167		14.25	8.85		23.10	30
0230	Pyramid		48	.167		10.85	8.85		19.70	26

Estimating Tips

07 10 00 Dampproofing and Waterproofing

- Be sure of the job specifications before pricing this subdivision. The difference in cost between waterproofing and dampproofing can be great. Waterproofing will hold back standing water. Dampproofing prevents the transmission of water vapor. Also included in this section are vapor retarding membranes.

07 20 00 Thermal Protection

- Insulation and fireproofing products are measured by area, thickness, volume, or R-value. Specifications may give only what the specific R-value should be in a certain situation. The estimator may need to choose the type of insulation to meet that R-value.

07 30 00 Steep Slope Roofing 07 40 00 Roofing and Siding Panels

- Many roofing and siding products are bought and sold by the square. One square is equal to an area that measures 100 square feet.

  This simple change in unit of measure could create a large error if the estimator is not observant. Accessories necessary for a complete installation must be figured into any calculations for both material and labor.

07 50 00 Membrane Roofing 07 60 00 Flashing and Sheet Metal 07 70 00 Roofing and Wall Specialties and Accessories

- The items in these subdivisions compose a roofing system. No one component completes the installation, and all must be estimated. Built-up or single-ply membrane roofing systems are made up of many products and installation trades. Wood blocking at roof perimeters or penetrations, parapet coverings, reglets, roof drains, gutters, downspouts, sheet metal flashing, skylights, smoke vents, and roof hatches all need to be considered along with the roofing material. Several different installation trades will need to work together on the roofing system. Inherent difficulties in the scheduling and coordination of various trades must be accounted for when estimating labor costs.

07 90 00 Joint Protection

- To complete the weather-tight shell, the sealants and caulkings must be estimated. Where different materials meet—at expansion joints, at flashing penetrations, and at hundreds of other locations throughout a construction project—caulking and sealants provide another line of defense against water penetration. Often, an entire system is based on the proper location and placement of caulking or sealants. The detailed drawings that are included as part of a set of architectural plans show typical locations for these materials. When caulking or sealants are shown at typical locations, this means the estimator must include them for all the locations where this detail is applicable. Be careful to keep different types of sealants separate, and remember to consider backer rods and primers if necessary.

Reference Numbers

Reference numbers are shown at the beginning of some major classifications. These numbers refer to related items in the Reference Section. The reference information may be an estimating procedure, an alternate pricing method, or technical information.

Note: Not all subdivisions listed here necessarily appear. ■

07 01 Operation and Maint. of Thermal and Moisture Protection

07 01 50 – Maintenance of Membrane Roofing

07 01 50.10 Roof Coatings		Crew	Daily Output	Labor-Hours	Unit	2020 Bare Costs Material	Labor	Equipment	Total	Total Incl O&P
0010	**ROOF COATINGS**									
0012	Asphalt, brush grade, material only				Gal.	9.50			9.50	10.45
0200	Asphalt base, fibered aluminum coating G					9.40			9.40	10.35
0300	Asphalt primer, 5 gal.				↓	7.15			7.15	7.85
0600	Coal tar pitch, 200 lb. barrels				Ton	1,225			1,225	1,350
0700	Tar roof cement, 5 gal. lots				Gal.	16.15			16.15	17.75
0800	Glass fibered roof & patching cement, 5 gal.				"	9.40			9.40	10.35
0900	Reinforcing glass membrane, 450 S.F./roll				Ea.	61			61	67
1000	Neoprene roof coating, 5 gal., 2 gal./sq.				Gal.	32.50			32.50	35.50
1100	Roof patch & flashing cement, 5 gal.					8.90			8.90	9.75
1200	Roof resaturant, glass fibered, 3 gal./sq.					9.35			9.35	10.25
1600	Reflective roof coating, white, elastomeric, approx. 50 S.F./gal. G				↓	17.35			17.35	19.10

07 01 90 – Maintenance of Joint Protection

07 01 90.81 Joint Sealant Replacement		Crew	Daily Output	Labor-Hours	Unit	Material	Labor	Equipment	Total	Total Incl O&P
0010	**JOINT SEALANT REPLACEMENT**									
0050	Control joints in concrete floors/slabs									
0100	Option 1 for joints with hard dry sealant									
0110	Step 1: Sawcut to remove 95% of old sealant									
0112	1/4" wide x 1/2" deep, with single saw blade	C-27	4800	.003	L.F.	.02	.17	.02	.21	.31
0114	3/8" wide x 3/4" deep, with single saw blade		4000	.004		.03	.20	.03	.26	.37
0116	1/2" wide x 1" deep, with double saw blades		3600	.004		.06	.22	.03	.31	.44
0118	3/4" wide x 1-1/2" deep, with double saw blades	↓	3200	.005		.11	.25	.03	.39	.55
0120	Step 2: Water blast joint faces and edges	C-29	2500	.003			.13	.04	.17	.26
0130	Step 3: Air blast joint faces and edges	C-28	2000	.004			.20	.02	.22	.33
0140	Step 4: Sand blast joint faces and edges	E-11	2000	.016			.75	.15	.90	1.40
0150	Step 5: Air blast joint faces and edges	C-28	2000	.004	↓		.20	.02	.22	.33
0200	Option 2 for joints with soft pliable sealant									
0210	Step 1: Plow joint with rectangular blade	B-62	2600	.009	L.F.		.42	.07	.49	.75
0220	Step 2: Sawcut to re-face joint faces									
0222	1/4" wide x 1/2" deep, with single saw blade	C-27	2400	.007	L.F.	.02	.33	.05	.40	.59
0224	3/8" wide x 3/4" deep, with single saw blade		2000	.008		.04	.40	.06	.50	.73
0226	1/2" wide x 1" deep, with double saw blades		1800	.009		.07	.44	.06	.57	.85
0228	3/4" wide x 1-1/2" deep, with double saw blades	↓	1600	.010		.15	.50	.07	.72	1.03
0230	Step 3: Water blast joint faces and edges	C-29	2500	.003			.13	.04	.17	.26
0240	Step 4: Air blast joint faces and edges	C-28	2000	.004			.20	.02	.22	.33
0250	Step 5: Sand blast joint faces and edges	E-11	2000	.016			.75	.15	.90	1.40
0260	Step 6: Air blast joint faces and edges	C-28	2000	.004	↓		.20	.02	.22	.33
0290	For saw cutting new control joints, see Section 03 15 16.20									
8910	For backer rod, see Section 07 91 23.10									
8920	For joint sealant, see Section 03 15 16.30 or 07 92 13.20									

07 05 05 – Selective Demolition for Thermal and Moisture Protection

07 05 05.10 Selective Demo., Thermal and Moist. Protection		Crew	Daily Output	Labor-Hours	Unit	Material	2020 Bare Costs Labor	Equipment	Total	Total Incl O&P
0010	**SELECTIVE DEMO., THERMAL AND MOISTURE PROTECTION**									
0020	Caulking/sealant, to 1″ x 1″ joint R024119-10	1 Clab	600	.013	L.F.		.56		.56	.90
0120	Downspouts, including hangers		350	.023	″		.96		.96	1.54
0220	Flashing, sheet metal		290	.028	S.F.		1.16		1.16	1.86
0420	Gutters, aluminum or wood, edge hung		240	.033	L.F.		1.40		1.40	2.25
0520	Built-in		100	.080	″		3.37		3.37	5.40
0620	Insulation, air/vapor barrier		3500	.002	S.F.		.10		.10	.15
0670	Batts or blankets		1400	.006	C.F.		.24		.24	.39
0720	Foamed or sprayed in place	2 Clab	1000	.016	B.F.		.67		.67	1.08
0770	Loose fitting	1 Clab	3000	.003	C.F.		.11		.11	.18
0870	Rigid board		3450	.002	B.F.		.10		.10	.16
1120	Roll roofing, cold adhesive		12	.667	Sq.		28		28	45
1170	Roof accessories, adjustable metal chimney flashing		9	.889	Ea.		37.50		37.50	60
1325	Plumbing vent flashing		32	.250	″		10.55		10.55	16.85
1375	Ridge vent strip, aluminum		310	.026	L.F.		1.09		1.09	1.74
1620	Skylight to 10 S.F.		8	1	Ea.		42		42	67.50
2120	Roof edge, aluminum soffit and fascia		570	.014	L.F.		.59		.59	.95
2170	Concrete coping, up to 12″ wide	2 Clab	160	.100			4.21		4.21	6.75
2220	Drip edge	1 Clab	1000	.008			.34		.34	.54
2270	Gravel stop		950	.008			.35		.35	.57
2370	Sheet metal coping, up to 12″ wide		240	.033			1.40		1.40	2.25
2470	Roof insulation board, over 2″ thick	B-2	7800	.005	B.F.		.22		.22	.35
2520	Up to 2″ thick	″	3900	.010	S.F.		.44		.44	.70
2620	Roof ventilation, louvered gable vent	1 Clab	16	.500	Ea.		21		21	33.50
2670	Remove, roof hatch	G-3	15	2.133			111		111	176
2675	Rafter vents	1 Clab	960	.008			.35		.35	.56
2720	Soffit vent and/or fascia vent		575	.014	L.F.		.59		.59	.94
2775	Soffit vent strip, aluminum, 3″ to 4″ wide		160	.050			2.11		2.11	3.37
2820	Roofing accessories, shingle moulding, to 1″ x 4″		1600	.005			.21		.21	.34
2870	Cant strip	B-2	2000	.020			.85		.85	1.36
2920	Concrete block walkway	1 Clab	230	.035			1.46		1.46	2.35
3070	Roofing, felt paper, #15		70	.114	Sq.		4.81		4.81	7.70
3125	#30 felt		30	.267	″		11.25		11.25	18
3170	Asphalt shingles, 1 layer	B-2	3500	.011	S.F.		.49		.49	.78
3180	2 layers		1750	.023	″		.97		.97	1.56
3370	Modified bitumen		26	1.538	Sq.		65.50		65.50	105
3420	Built-up, no gravel, 3 ply		25	1.600			68		68	109
3470	4 ply		21	1.905			81		81	130
3620	5 ply		1600	.025	S.F.		1.06		1.06	1.70
3720	5 ply, with gravel		890	.045			1.91		1.91	3.06
3725	Loose gravel removal		5000	.008			.34		.34	.54
3730	Embedded gravel removal		2000	.020			.85		.85	1.36
3870	Fiberglass sheet		1200	.033			1.42		1.42	2.27
4120	Slate shingles		1900	.021			.89		.89	1.43
4170	Ridge shingles, clay or slate		2000	.020	L.F.		.85		.85	1.36
4320	Single ply membrane, attached at seams		52	.769	Sq.		32.50		32.50	52.50
4370	Ballasted		75	.533			22.50		22.50	36.50
4420	Fully adhered		39	1.026			43.50		43.50	70
4550	Roof hatch, 2′-6″ x 3′-0″	1 Clab	10	.800	Ea.		33.50		33.50	54
4670	Wood shingles	B-2	2200	.018	S.F.		.77		.77	1.24
4820	Sheet metal roofing	″	2150	.019			.79		.79	1.27
4970	Siding, horizontal wood clapboards	1 Clab	380	.021			.89		.89	1.42
5025	Exterior insulation finish system	″	120	.067			2.81		2.81	4.50

07 05 Common Work Results for Thermal and Moisture Protection

07 05 05 – Selective Demolition for Thermal and Moisture Protection

07 05 05.10 Selective Demo., Thermal and Moist. Protection		Crew	Daily Output	Labor-Hours	Unit	Material	2020 Bare Costs Labor	Equipment	Total	Total Incl O&P
5070	Tempered hardboard, remove and reset	1 Carp	380	.021	S.F.		1.12		1.12	1.79
5120	Tempered hardboard sheet siding	"	375	.021	↓		1.13		1.13	1.82
5170	Metal, corner strips	1 Clab	850	.009	L.F.		.40		.40	.63
5225	Horizontal strips		444	.018	S.F.		.76		.76	1.22
5320	Vertical strips		400	.020			.84		.84	1.35
5520	Wood shingles		350	.023			.96		.96	1.54
5620	Stucco siding		360	.022			.94		.94	1.50
5670	Textured plywood		725	.011			.46		.46	.74
5720	Vinyl siding		510	.016	↓		.66		.66	1.06
5770	Corner strips		900	.009	L.F.		.37		.37	.60
5870	Wood, boards, vertical		400	.020	S.F.		.84		.84	1.35
5880	Steel siding, corrugated/ribbed	↓	402.50	.020	"		.84		.84	1.34
5920	Waterproofing, protection/drain board	2 Clab	3900	.004	B.F.		.17		.17	.28
5970	Over 1/2" thick		1750	.009	S.F.		.38		.38	.62
6020	To 1/2" thick	↓	2000	.008	"		.34		.34	.54
9000	Minimum labor/equipment charge	1 Clab	2	4	Job		168		168	270

07 11 Dampproofing

07 11 13 – Bituminous Dampproofing

07 11 13.10 Bituminous Asphalt Coating

		Crew	Daily Output	Labor-Hours	Unit	Material	Labor	Equipment	Total	Total Incl O&P
0010	**BITUMINOUS ASPHALT COATING**									
0030	Brushed on, below grade, 1 coat	1 Rofc	665	.012	S.F.	.24	.56		.80	1.23
0100	2 coat		500	.016		.48	.74		1.22	1.81
0300	Sprayed on, below grade, 1 coat		830	.010		.24	.45		.69	1.04
0400	2 coat	↓	500	.016	↓	.46	.74		1.20	1.80
0500	Asphalt coating, with fibers				Gal.	9.40			9.40	10.35
0600	Troweled on, asphalt with fibers, 1/16" thick	1 Rofc	500	.016	S.F.	.41	.74		1.15	1.74
0700	1/8" thick		400	.020		.72	.92		1.64	2.41
1000	1/2" thick		350	.023	↓	2.35	1.06		3.41	4.43
9000	Minimum labor/equipment charge	↓	3	2.667	Job		123		123	216

07 11 16 – Cementitious Dampproofing

07 11 16.20 Cementitious Parging

		Crew	Daily Output	Labor-Hours	Unit	Material	Labor	Equipment	Total	Total Incl O&P
0010	**CEMENTITIOUS PARGING**									
0020	Portland cement, 2 coats, 1/2" thick	D-1	250	.064	S.F.	.42	3		3.42	5.30
0100	Waterproofed Portland cement, 1/2" thick, 2 coats	"	250	.064	"	7.25	3		10.25	12.85

07 12 Built-up Bituminous Waterproofing

07 12 13 – Built-Up Asphalt Waterproofing

07 12 13.20 Membrane Waterproofing

		Crew	Daily Output	Labor-Hours	Unit	Material	Labor	Equipment	Total	Total Incl O&P
0010	**MEMBRANE WATERPROOFING**									
0012	On slabs, 1 ply, felt, mopped	G-1	3000	.019	S.F.	.40	.81	.19	1.40	2.06
0015	On walls, 1 ply, felt, mopped		3000	.019		.40	.81	.19	1.40	2.06
0100	On slabs, 1 ply, glass fiber fabric, mopped		2100	.027		.45	1.15	.27	1.87	2.80
0105	On walls, 1 ply, glass fiber fabric, mopped		2100	.027		.45	1.15	.27	1.87	2.80
0300	On slabs, 2 ply, felt, mopped		2500	.022		.80	.97	.23	2	2.82
0305	On walls, 2 ply, felt, mopped		2500	.022		.80	.97	.23	2	2.82
0400	On slabs, 2 ply, glass fiber fabric, mopped		1650	.034		.99	1.47	.34	2.80	4.03
0405	On walls, 2 ply, glass fiber fabric, mopped		1650	.034		.99	1.47	.34	2.80	4.03
0600	On slabs, 3 ply, felt, mopped	↓	2100	.027	↓	1.20	1.15	.27	2.62	3.63

07 12 Built-up Bituminous Waterproofing

07 12 13 – Built-Up Asphalt Waterproofing

07 12 13.20 Membrane Waterproofing		Crew	Daily Output	Labor-Hours	Unit	Material	2020 Bare Costs Labor	Equipment	Total	Total Incl O&P
0605	On walls, 3 ply, felt, mopped	G-1	2100	.027	S.F.	1.20	1.15	.27	2.62	3.63
0700	On slabs, 3 ply, glass fiber fabric, mopped		1550	.036		1.35	1.56	.37	3.28	4.61
0705	On walls, 3 ply, glass fiber fabric, mopped		1550	.036		1.35	1.56	.37	3.28	4.61
0710	Asphaltic hardboard protection board, 1/8″ thick	2 Rofc	500	.032		.76	1.48		2.24	3.43
0715	1/4″ thick		450	.036		1.51	1.64		3.15	4.54
1000	EPS membrane protection board, 1/4″ thick		3500	.005		.33	.21		.54	.73
1050	3/8″ thick		3500	.005		.36	.21		.57	.77
1060	1/2″ thick		3500	.005		.40	.21		.61	.81
1070	Fiberglass fabric, black, 20/10 mesh		116	.138	Sq.	15.10	6.35		21.45	28
9000	Minimum labor/equipment charge		2	8	Job		370		370	645

07 13 Sheet Waterproofing

07 13 53 – Elastomeric Sheet Waterproofing

07 13 53.10 Elastomeric Sheet Waterproofing and Access.		Crew	Daily Output	Labor-Hours	Unit	Material	Labor	Equipment	Total	Total Incl O&P
0010	**ELASTOMERIC SHEET WATERPROOFING AND ACCESS.**									
0090	EPDM, plain, 45 mils thick	2 Rofc	580	.028	S.F.	1.50	1.27		2.77	3.88
0100	60 mils thick		570	.028		1.72	1.30		3.02	4.16
0300	Nylon reinforced sheets, 45 mils thick		580	.028		1.62	1.27		2.89	4.01
0400	60 mils thick		570	.028		1.69	1.30		2.99	4.13
0600	Vulcanizing splicing tape for above, 2″ wide				C.L.F.	61.50			61.50	67.50
0700	4″ wide				″	124			124	136
0900	Adhesive, bonding, 60 S.F./gal.				Gal.	28			28	31
1000	Splicing, 75 S.F./gal.				″	37.50			37.50	41.50
1200	Neoprene sheets, plain, 45 mils thick	2 Rofc	580	.028	S.F.	2.17	1.27		3.44	4.62
1300	60 mils thick		570	.028		2.26	1.30		3.56	4.76
1500	Nylon reinforced, 45 mils thick		580	.028		2.40	1.27		3.67	4.87
1600	60 mils thick		570	.028		2.56	1.30		3.86	5.10
1800	120 mils thick		500	.032		5.25	1.48		6.73	8.40
1900	Adhesive, splicing, 150 S.F./gal. per coat				Gal.	37.50			37.50	41.50
2100	Fiberglass reinforced, fluid applied, 1/8″ thick	2 Rofc	500	.032	S.F.	1.72	1.48		3.20	4.48
2200	Polyethylene and rubberized asphalt sheets, 60 mils thick		550	.029		.96	1.34		2.30	3.41
2400	Polyvinyl chloride sheets, plain, 10 mils thick		580	.028		.15	1.27		1.42	2.40
2500	20 mils thick		570	.028		.22	1.30		1.52	2.51
2700	30 mils thick		560	.029		.24	1.32		1.56	2.57
3000	Adhesives, trowel grade, 40-100 S.F./gal.				Gal.	23.50			23.50	26
3100	Brush grade, 100-250 S.F./gal.				″	21			21	23.50
3300	Bitumen modified polyurethane, fluid applied, 55 mils thick	2 Rofc	665	.024	S.F.	1.12	1.11		2.23	3.18
9000	Minimum labor/equipment charge	″	2	8	Job		370		370	645

07 16 Cementitious and Reactive Waterproofing

07 16 16 – Crystalline Waterproofing

07 16 16.20 Cementitious Waterproofing		Crew	Daily Output	Labor-Hours	Unit	Material	Labor	Equipment	Total	Total Incl O&P
0010	**CEMENTITIOUS WATERPROOFING**									
0020	1/8″ application, sprayed on	G-2A	1000	.024	S.F.	.71	.98	.63	2.32	3.15
0050	4 coat cementitious metallic slurry	1 Cefi	1.20	6.667	C.S.F.	75	335		410	610

07 17 Bentonite Waterproofing

07 17 13 – Bentonite Panel Waterproofing

07 17 13.10 Bentonite		Crew	Daily Output	Labor-Hours	Unit	Material	2020 Bare Costs Labor	Equipment	Total	Total Incl O&P
0010	**BENTONITE**									
0020	Panels, 4′ x 4′, 3/16″ thick	1 Rofc	625	.013	S.F.	1.83	.59		2.42	3.04
0100	Rolls, 3/8″ thick, with geotextile fabric both sides	″	550	.015	″	1.78	.67		2.45	3.14
0300	Granular bentonite, 50 lb. bags (.625 C.F.)				Bag	17.65			17.65	19.40
0400	3/8″ thick, troweled on	1 Rofc	475	.017	S.F.	.88	.78		1.66	2.33
0500	Drain board, expanded polystyrene, 1-1/2″ thick	1 Rohe	1600	.005		.44	.17		.61	.78
0510	2″ thick		1600	.005		.58	.17		.75	.94
0520	3″ thick		1600	.005		.87	.17		1.04	1.26
0530	4″ thick		1600	.005		1.16	.17		1.33	1.58
0600	With filter fabric, 1-1/2″ thick		1600	.005		.50	.17		.67	.85
0625	2″ thick		1600	.005		.65	.17		.82	1.01
0650	3″ thick		1600	.005		.94	.17		1.11	1.33
0675	4″ thick	↓	1600	.005	↓	1.23	.17		1.40	1.65

07 19 Water Repellents

07 19 19 – Silicone Water Repellents

07 19 19.10 Silicone Based Water Repellents		Crew	Daily Output	Labor-Hours	Unit	Material	Labor	Equipment	Total	Total Incl O&P
0010	**SILICONE BASED WATER REPELLENTS**									
0020	Water base liquid, roller applied	2 Rofc	7000	.002	S.F.	.43	.11		.54	.66
0200	Silicone or stearate, sprayed on CMU, 1 coat	1 Rofc	4000	.002		.40	.09		.49	.60
0300	2 coats		3000	.003	↓	.79	.12		.91	1.09
9000	Minimum labor/equipment charge	↓	3	2.667	Job		123		123	216

07 21 Thermal Insulation

07 21 13 – Board Insulation

07 21 13.10 Rigid Insulation			Crew	Daily Output	Labor-Hours	Unit	Material	Labor	Equipment	Total	Total Incl O&P
0010	**RIGID INSULATION**, for walls										
0040	Fiberglass, 1.5#/C.F., unfaced, 1″ thick, R4.1	G	1 Carp	1000	.008	S.F.	.33	.43		.76	1.04
0060	1-1/2″ thick, R6.2	G		1000	.008		.43	.43		.86	1.15
0080	2″ thick, R8.3	G		1000	.008		.49	.43		.92	1.22
0120	3″ thick, R12.4	G		800	.010		.56	.53		1.09	1.47
0370	3#/C.F., unfaced, 1″ thick, R4.3	G		1000	.008		.57	.43		1	1.31
0390	1-1/2″ thick, R6.5	G		1000	.008		.78	.43		1.21	1.54
0400	2″ thick, R8.7	G		890	.009		1.09	.48		1.57	1.97
0420	2-1/2″ thick, R10.9	G		800	.010		1.12	.53		1.65	2.08
0440	3″ thick, R13	G		800	.010		1.65	.53		2.18	2.67
0520	Foil faced, 1″ thick, R4.3	G		1000	.008		.85	.43		1.28	1.62
0540	1-1/2″ thick, R6.5	G		1000	.008		1.28	.43		1.71	2.09
0560	2″ thick, R8.7	G		890	.009		1.62	.48		2.10	2.55
0580	2-1/2″ thick, R10.9	G		800	.010		1.91	.53		2.44	2.95
0600	3″ thick, R13	G	↓	800	.010	↓	2.01	.53		2.54	3.06
1600	Isocyanurate, 4′ x 8′ sheet, foil faced, both sides										
1610	1/2″ thick	G	1 Carp	800	.010	S.F.	.31	.53		.84	1.19
1620	5/8″ thick	G		800	.010		.62	.53		1.15	1.53
1630	3/4″ thick	G		800	.010		.43	.53		.96	1.32
1640	1″ thick	G		800	.010		.62	.53		1.15	1.53
1650	1-1/2″ thick	G		730	.011		.76	.58		1.34	1.77
1660	2″ thick	G		730	.011		.74	.58		1.32	1.74
1670	3″ thick	G	↓	730	.011	↓	3.11	.58		3.69	4.35

07 21 Thermal Insulation

07 21 13 – Board Insulation

07 21 13.10 Rigid Insulation			Crew	Daily Output	Labor-Hours	Unit	Material	Labor	Equipment	Total	Total Incl O&P
1680	4" thick	G	1 Carp	730	.011	S.F.	2.58	.58		3.16	3.77
1700	Perlite, 1" thick, R2.77	G		800	.010		.47	.53		1	1.37
1750	2" thick, R5.55	G		730	.011		.86	.58		1.44	1.88
1900	Extruded polystyrene, 25 psi compressive strength, 1" thick, R5	G		800	.010		.62	.53		1.15	1.53
1940	2" thick, R10	G		730	.011		1.22	.58		1.80	2.27
1960	3" thick, R15	G		730	.011		1.67	.58		2.25	2.77
2100	Expanded polystyrene, 1" thick, R3.85	G		800	.010		.29	.53		.82	1.17
2120	2" thick, R7.69	G		730	.011		.58	.58		1.16	1.57
2140	3" thick, R11.49	G		730	.011		.87	.58		1.45	1.89
9000	Minimum labor/equipment charge			4	2	Job		106		106	170

07 21 13.13 Foam Board Insulation			Crew	Daily Output	Labor-Hours	Unit	Material	Labor	Equipment	Total	Total Incl O&P
0010	**FOAM BOARD INSULATION**										
0600	Polystyrene, expanded, 1" thick, R4	G	1 Carp	680	.012	S.F.	.29	.63		.92	1.32
0700	2" thick, R8	G		675	.012	"	.58	.63		1.21	1.65
9000	Minimum labor/equipment charge			4	2	Job		106		106	170

07 21 16 – Blanket Insulation

07 21 16.10 Blanket Insulation for Floors/Ceilings			Crew	Daily Output	Labor-Hours	Unit	Material	Labor	Equipment	Total	Total Incl O&P
0010	**BLANKET INSULATION FOR FLOORS/CEILINGS**										
0020	Including spring type wire fasteners										
2000	Fiberglass, blankets or batts, paper or foil backing										
2100	3-1/2" thick, R13	G	1 Carp	700	.011	S.F.	.41	.61		1.02	1.42
2150	6-1/4" thick, R19	G		600	.013		.51	.71		1.22	1.70
2210	9-1/2" thick, R30	G		500	.016		.86	.85		1.71	2.31
2220	12" thick, R38	G		475	.017		1.04	.90		1.94	2.57
3000	Unfaced, 3-1/2" thick, R13	G		600	.013		.33	.71		1.04	1.50
3010	6-1/4" thick, R19	G		500	.016		.39	.85		1.24	1.79
3020	9-1/2" thick, R30	G		450	.018		.68	.95		1.63	2.26
3030	12" thick, R38	G		425	.019		.77	1		1.77	2.45
9000	Minimum labor/equipment charge			4	2	Job		106		106	170

07 21 16.20 Blanket Insulation for Walls			Crew	Daily Output	Labor-Hours	Unit	Material	Labor	Equipment	Total	Total Incl O&P
0010	**BLANKET INSULATION FOR WALLS**										
0020	Kraft faced fiberglass, 3-1/2" thick, R11, 15" wide	G	1 Carp	1350	.006	S.F.	.30	.32		.62	.83
0030	23" wide	G		1600	.005		.30	.27		.57	.76
0060	R13, 11" wide	G		1150	.007		.34	.37		.71	.96
0080	15" wide	G		1350	.006		.34	.32		.66	.87
0100	23" wide	G		1600	.005		.34	.27		.61	.80
0110	R15, 11" wide	G		1150	.007		.55	.37		.92	1.20
0120	15" wide	G		1350	.006		.55	.32		.87	1.11
0130	23" wide	G		1600	.005		.55	.27		.82	1.04
0140	6" thick, R19, 11" wide	G		1150	.007		.44	.37		.81	1.07
0160	15" wide	G		1350	.006		.44	.32		.76	.98
0180	23" wide	G		1600	.005		.44	.27		.71	.91
0182	R21, 11" wide	G		1150	.007		.70	.37		1.07	1.36
0184	15" wide	G		1350	.006		.70	.32		1.02	1.27
0186	23" wide	G		1600	.005		.70	.27		.97	1.20
0188	9" thick, R30, 11" wide	G		985	.008		.86	.43		1.29	1.64
0200	15" wide	G		1150	.007		.86	.37		1.23	1.54
0220	23" wide	G		1350	.006		.86	.32		1.18	1.45
0230	12" thick, R38, 11" wide	G		985	.008		1.04	.43		1.47	1.83
0240	15" wide	G		1150	.007		1.04	.37		1.41	1.73
0260	23" wide	G		1350	.006		1.04	.32		1.36	1.64
0410	Foil faced fiberglass, 3-1/2" thick, R13, 11" wide	G		1150	.007		.47	.37		.84	1.11

07 21 16 – Blanket Insulation

07 21 16.20 Blanket Insulation for Walls			Crew	Daily Output	Labor-Hours	Unit	2020 Bare Costs Material	Labor	Equipment	Total	Total Incl O&P
0420	15" wide	G	1 Carp	1350	.006	S.F.	.47	.32		.79	1.02
0440	23" wide	G		1600	.005		.47	.27		.74	.95
0442	R15, 11" wide	G		1150	.007		.46	.37		.83	1.10
0444	15" wide	G		1350	.006		.46	.32		.78	1.01
0446	23" wide	G		1600	.005		.46	.27		.73	.94
0448	6" thick, R19, 11" wide	G		1150	.007		.63	.37		1	1.28
0460	15" wide	G		1350	.006		.63	.32		.95	1.19
0480	23" wide	G		1600	.005		.63	.27		.90	1.12
0482	R21, 11" wide	G		1150	.007		.65	.37		1.02	1.31
0484	15" wide	G		1350	.006		.65	.32		.97	1.22
0486	23" wide	G		1600	.005		.65	.27		.92	1.15
0488	9" thick, R30, 11" wide	G		985	.008		1.05	.43		1.48	1.85
0500	15" wide	G		1150	.007		1.05	.37		1.42	1.75
0550	23" wide	G		1350	.006		1.05	.32		1.37	1.66
0560	12" thick, R38, 11" wide	G		985	.008		1.14	.43		1.57	1.94
0570	15" wide	G		1150	.007		1.14	.37		1.51	1.84
0580	23" wide	G		1350	.006		1.14	.32		1.46	1.75
0620	Unfaced fiberglass, 3-1/2" thick, R13, 11" wide	G		1150	.007		.33	.37		.70	.95
0820	15" wide	G		1350	.006		.33	.32		.65	.86
0830	23" wide	G		1600	.005		.33	.27		.60	.79
0832	R15, 11" wide	G		1150	.007		.46	.37		.83	1.10
0836	23" wide	G		1600	.005		.46	.27		.73	.94
0838	6" thick, R19, 11" wide	G		1150	.007		.39	.37		.76	1.02
0860	15" wide	G		1150	.007		.39	.37		.76	1.02
0880	23" wide	G		1350	.006		.39	.32		.71	.93
0882	R21, 11" wide	G		1150	.007		.68	.37		1.05	1.34
0886	15" wide	G		1350	.006		.68	.32		1	1.25
0888	23" wide	G		1600	.005		.68	.27		.95	1.18
0890	9" thick, R30, 11" wide	G		985	.008		.68	.43		1.11	1.44
0900	15" wide	G		1150	.007		.68	.37		1.05	1.34
0920	23" wide	G		1350	.006		.68	.32		1	1.25
0930	12" thick, R38, 11" wide	G		985	.008		.77	.43		1.20	1.54
0940	15" wide	G		1000	.008		.77	.43		1.20	1.53
0960	23" wide	G	↓	1150	.007	↓	.77	.37		1.14	1.44
1300	Wall or ceiling insulation, mineral wool batts										
1320	3-1/2" thick, R15	G	1 Carp	1600	.005	S.F.	.82	.27		1.09	1.33
1340	5-1/2" thick, R23	G		1600	.005		1.29	.27		1.56	1.85
1380	7-1/4" thick, R30	G		1350	.006		1.70	.32		2.02	2.37
1700	Non-rigid insul., recycled blue cotton fiber, unfaced batts, R13, 16" wide	G		1600	.005		1.18	.27		1.45	1.73
1710	R19, 16" wide	G		1600	.005	↓	1.45	.27		1.72	2.03
1850	Friction fit wire insulation supports, 16" OC			960	.008	Ea.	.07	.44		.51	.79
9000	Minimum labor/equipment charge		↓	4	2	Job		106		106	170

07 21 19 – Foamed In Place Insulation

07 21 19.10 Masonry Foamed In Place Insulation			Crew	Daily Output	Labor-Hours	Unit	Material	Labor	Equipment	Total	Total Incl O&P
0010	**MASONRY FOAMED IN PLACE INSULATION**										
0100	Amino-plast foam, injected into block core, 6" block	G	G-2A	6000	.004	Ea.	.17	.16	.11	.44	.59
0110	8" block	G		5000	.005		.20	.20	.13	.53	.69
0120	10" block	G		4000	.006		.26	.25	.16	.67	.87
0130	12" block	G		3000	.008	↓	.34	.33	.21	.88	1.17
0140	Injected into cavity wall	G	↓	13000	.002	B.F.	.06	.08	.05	.19	.25
0150	Preparation, drill holes into mortar joint every 4 V.L.F., 5/8" diameter		1 Clab	960	.008	Ea.		.35		.35	.56
0160	7/8" diameter		↓	680	.012	↓		.50		.50	.79

07 21 Thermal Insulation

07 21 19 – Foamed In Place Insulation

07 21 19.10 Masonry Foamed In Place Insulation		Crew	Daily Output	Labor-Hours	Unit	Material	2020 Bare Costs Labor	Equipment	Total	Total Incl O&P
0170	Patch drilled holes, 5/8" diameter	1 Clab	1800	.004	Ea.	.04	.19		.23	.34
0180	7/8" diameter	↓	1200	.007	↓	.05	.28		.33	.51
9000	Minimum labor/equipment charge	G-2A	2	12	Job		490	315	805	1,175
9010	Minimum labor/equipment charge	1 Clab	4	2	"		84		84	135

07 21 23 – Loose-Fill Insulation

07 21 23.10 Poured Loose-Fill Insulation

			Crew	Daily Output	Labor-Hours	Unit	Material	Labor	Equipment	Total	Total Incl O&P
0010	**POURED LOOSE-FILL INSULATION**										
0020	Cellulose fiber, R3.8 per inch	G	1 Carp	200	.040	C.F.	.70	2.13		2.83	4.18
0021	4" thick	G		1000	.008	S.F.	.17	.43		.60	.86
0022	6" thick	G		800	.010	"	.28	.53		.81	1.16
0080	Fiberglass wool, R4 per inch	G		200	.040	C.F.	.70	2.13		2.83	4.18
0081	4" thick	G		600	.013	S.F.	.24	.71		.95	1.40
0082	6" thick	G		400	.020	"	.34	1.06		1.40	2.07
0100	Mineral wool, R3 per inch	G		200	.040	C.F.	.56	2.13		2.69	4.03
0101	4" thick	G		600	.013	S.F.	.18	.71		.89	1.34
0102	6" thick	G		400	.020	"	.28	1.06		1.34	2.01
0300	Polystyrene, R4 per inch	G		200	.040	C.F.	1.56	2.13		3.69	5.15
0301	4" thick	G		600	.013	S.F.	.51	.71		1.22	1.71
0302	6" thick	G		400	.020	"	.78	1.06		1.84	2.56
0400	Perlite, R2.78 per inch	G		200	.040	C.F.	5.30	2.13		7.43	9.20
0401	4" thick	G		1000	.008	S.F.	1.76	.43		2.19	2.62
0402	6" thick	G		800	.010	"	2.65	.53		3.18	3.76
9000	Minimum labor/equipment charge		↓	4	2	Job		106		106	170

07 21 23.20 Masonry Loose-Fill Insulation

			Crew	Daily Output	Labor-Hours	Unit	Material	Labor	Equipment	Total	Total Incl O&P
0010	**MASONRY LOOSE-FILL INSULATION**, vermiculite or perlite										
0100	In cores of concrete block, 4" thick wall, .115 C.F./S.F.	G	D-1	4800	.003	S.F.	.61	.16		.77	.92
0200	6" thick wall, .175 C.F./S.F.	G		3000	.005		.93	.25		1.18	1.42
0300	8" thick wall, .258 C.F./S.F.	G		2400	.007		1.36	.31		1.67	2.01
0400	10" thick wall, .340 C.F./S.F.	G		1850	.009		1.80	.41		2.21	2.64
0500	12" thick wall, .422 C.F./S.F.	G		1200	.013	↓	2.23	.63		2.86	3.47
0600	Poured cavity wall, vermiculite or perlite, water repellent	G	↓	250	.064	C.F.	5.30	3		8.30	10.65
0700	Foamed in place, urethane in 2-5/8" cavity	G	G-2A	1035	.023	S.F.	1.39	.95	.61	2.95	3.81
0800	For each 1" added thickness, add	G	"	2372	.010	"	.53	.41	.27	1.21	1.57

07 21 26 – Blown Insulation

07 21 26.10 Blown Insulation

			Crew	Daily Output	Labor-Hours	Unit	Material	Labor	Equipment	Total	Total Incl O&P
0010	**BLOWN INSULATION** Ceilings, with open access										
0020	Cellulose, 3-1/2" thick, R13	G	G-4	5000	.005	S.F.	.24	.21	.08	.53	.68
0030	5-3/16" thick, R19	G		3800	.006		.35	.27	.11	.73	.94
0050	6-1/2" thick, R22	G		3000	.008		.45	.34	.13	.92	1.20
0100	8-11/16" thick, R30	G		2600	.009		.61	.39	.15	1.15	1.47
0120	10-7/8" thick, R38	G		1800	.013		.78	.57	.22	1.57	2.01
1000	Fiberglass, 5.5" thick, R11	G		3800	.006		.24	.27	.11	.62	.81
1050	6" thick, R12	G		3000	.008		.34	.34	.13	.81	1.07
1100	8.8" thick, R19	G		2200	.011		.42	.47	.18	1.07	1.41
1200	10" thick, R22	G		1800	.013		.49	.57	.22	1.28	1.69
1300	11.5" thick, R26	G		1500	.016		.58	.68	.27	1.53	2.03
1350	13" thick, R30	G		1400	.017		.68	.73	.28	1.69	2.22
1450	16" thick, R38	G		1145	.021		.86	.90	.35	2.11	2.77
1500	20" thick, R49	G		920	.026	↓	1.14	1.12	.43	2.69	3.53
9000	Minimum labor/equipment charge		↓	4	6	Job		257	99.50	356.50	520

07 21 Thermal Insulation

07 21 29 – Sprayed Insulation

07 21 29.10 Sprayed-On Insulation			Crew	Daily Output	Labor-Hours	Unit	Material	Labor (2020 Bare Costs)	Equipment (2020 Bare Costs)	Total (2020 Bare Costs)	Total Incl O&P
0010	**SPRAYED-ON INSULATION**										
0020	Fibrous/cementitious, finished wall, 1" thick, R3.7	G	G-2	2050	.012	S.F.	.36	.52	.09	.97	1.33
0100	Attic, 5.2" thick, R19	G		1550	.015	"	.43	.69	.12	1.24	1.70
0200	Fiberglass, R4 per inch, vertical	G		1600	.015	B.F.	.20	.67	.12	.99	1.41
0210	Horizontal	G		1200	.020	"	.20	.89	.16	1.25	1.81
0300	Closed cell, spray polyurethane foam, 2 lb./C.F. density										
0310	1" thick	G	G-2A	6000	.004	S.F.	.53	.16	.11	.80	.98
0320	2" thick	G		3000	.008		1.06	.33	.21	1.60	1.96
0330	3" thick	G		2000	.012		1.59	.49	.32	2.40	2.94
0335	3-1/2" thick	G		1715	.014		1.86	.57	.37	2.80	3.42
0340	4" thick	G		1500	.016		2.12	.66	.42	3.20	3.92
0350	5" thick	G		1200	.020		2.65	.82	.53	4	4.89
0355	5-1/2" thick	G		1090	.022		2.92	.90	.58	4.40	5.40
0360	6" thick	G		1000	.024		3.18	.98	.63	4.79	5.85
9000	Minimum labor/equipment charge		G-2	2	12	Job		530	94	624	955

07 21 53 – Reflective Insulation

07 21 53.10 Reflective Insulation Options			Crew	Daily Output	Labor-Hours	Unit	Material	Labor	Equipment	Total	Total Incl O&P
0010	**REFLECTIVE INSULATION OPTIONS**										
0020	Aluminum foil on reinforced scrim	G	1 Carp	19	.421	C.S.F.	15	22.50		37.50	52.50
0100	Reinforced with woven polyolefin	G		19	.421		22.50	22.50		45	61
0500	With single bubble air space, R8.8	G		15	.533		26.50	28.50		55	74.50
0600	With double bubble air space, R9.8	G		15	.533		32.50	28.50		61	81.50
9000	Minimum labor/equipment charge			4	2	Job		106		106	170

07 22 Roof and Deck Insulation

07 22 16 – Roof Board Insulation

07 22 16.10 Roof Deck Insulation			Crew	Daily Output	Labor-Hours	Unit	Material	Labor	Equipment	Total	Total Incl O&P
0010	**ROOF DECK INSULATION**, fastening excluded										
0016	Asphaltic cover board, fiberglass lined, 1/8" thick		1 Rofc	1400	.006	S.F.	.49	.26		.75	1
0018	1/4" thick			1400	.006		.99	.26		1.25	1.55
0020	Fiberboard low density, 1/2" thick, R1.39	G		1300	.006		.36	.28		.64	.90
0030	1" thick, R2.78	G		1040	.008		.69	.36		1.05	1.38
0080	1-1/2" thick, R4.17	G		1040	.008		1.04	.36		1.40	1.76
0100	2" thick, R5.56	G		1040	.008		1.19	.36		1.55	1.93
0110	Fiberboard high density, 1/2" thick, R1.3	G		1300	.006		.28	.28		.56	.81
0120	1" thick, R2.5	G		1040	.008		.64	.36		1	1.32
0130	1-1/2" thick, R3.8	G		1040	.008		1.24	.36		1.60	1.98
0200	Fiberglass, 3/4" thick, R2.78	G		1300	.006		.63	.28		.91	1.19
0400	15/16" thick, R3.70	G		1300	.006		.84	.28		1.12	1.42
0460	1-1/16" thick, R4.17	G		1300	.006		1.17	.28		1.45	1.79
0600	1-5/16" thick, R5.26	G		1300	.006		1.45	.28		1.73	2.10
0650	2-1/16" thick, R8.33	G		1040	.008		1.63	.36		1.99	2.41
0700	2-7/16" thick, R10	G		1040	.008		1.75	.36		2.11	2.55
0800	Gypsum cover board, fiberglass mat facer, 1/4" thick			1400	.006		.51	.26		.77	1.02
0810	1/2" thick			1300	.006		.67	.28		.95	1.24
0820	5/8" thick			1200	.007		.66	.31		.97	1.27
0830	Primed fiberglass mat facer, 1/4" thick			1400	.006		.50	.26		.76	1.01
0840	1/2" thick			1300	.006		.58	.28		.86	1.14
0850	5/8" thick			1200	.007		.61	.31		.92	1.21
1650	Perlite, 1/2" thick, R1.32	G		1365	.006		.29	.27		.56	.79

07 22 16 – Roof Board Insulation

07 22 16.10 Roof Deck Insulation			Crew	Daily Output	Labor-Hours	Unit	Material	2020 Bare Costs Labor	Equipment	Total	Total Incl O&P
1655	3/4" thick, R2.08	G	1 Rofc	1040	.008	S.F.	.39	.36		.75	1.05
1660	1" thick, R2.78	G		1040	.008		.57	.36		.93	1.25
1670	1-1/2" thick, R4.17	G		1040	.008		.83	.36		1.19	1.53
1680	2" thick, R5.56	G		910	.009		1.09	.41		1.50	1.91
1685	2-1/2" thick, R6.67	G		910	.009		1.42	.41		1.83	2.27
1690	Tapered for drainage	G		1040	.008	B.F.	1.09	.36		1.45	1.82
1700	Polyisocyanurate, 2#/C.F. density, 3/4" thick	G		1950	.004	S.F.	.52	.19		.71	.90
1705	1" thick	G		1820	.004		.45	.20		.65	.86
1715	1-1/2" thick	G		1625	.005		.60	.23		.83	1.06
1725	2" thick	G		1430	.006		.92	.26		1.18	1.46
1735	2-1/2" thick	G		1365	.006		1	.27		1.27	1.57
1745	3" thick	G		1300	.006		1.13	.28		1.41	1.74
1755	3-1/2" thick	G		1300	.006		1.67	.28		1.95	2.34
1765	Tapered for drainage	G		1820	.004	B.F.	.53	.20		.73	.94
1900	Extruded polystyrene										
1910	15 psi compressive strength, 1" thick, R5	G	1 Rofc	1950	.004	S.F.	.52	.19		.71	.90
1920	2" thick, R10	G		1625	.005		.68	.23		.91	1.14
1930	3" thick, R15	G		1300	.006		1.35	.28		1.63	1.99
1932	4" thick, R20	G		1300	.006		1.82	.28		2.10	2.50
1934	Tapered for drainage	G		1950	.004	B.F.	.51	.19		.70	.89
1940	25 psi compressive strength, 1" thick, R5	G		1950	.004	S.F.	1.06	.19		1.25	1.50
1942	2" thick, R10	G		1625	.005		2.01	.23		2.24	2.62
1944	3" thick, R15	G		1300	.006		3.07	.28		3.35	3.88
1946	4" thick, R20	G		1300	.006		4.24	.28		4.52	5.15
1948	Tapered for drainage	G		1950	.004	B.F.	.57	.19		.76	.96
1950	40 psi compressive strength, 1" thick, R5	G		1950	.004	S.F.	.87	.19		1.06	1.29
1952	2" thick, R10	G		1625	.005		1.65	.23		1.88	2.22
1954	3" thick, R15	G		1300	.006		2.39	.28		2.67	3.13
1956	4" thick, R20	G		1300	.006		3.13	.28		3.41	3.95
1958	Tapered for drainage	G		1820	.004	B.F.	.88	.20		1.08	1.33
1960	60 psi compressive strength, 1" thick, R5	G		1885	.004	S.F.	1.09	.20		1.29	1.54
1962	2" thick, R10	G		1560	.005		2.07	.24		2.31	2.69
1964	3" thick, R15	G		1270	.006		3.38	.29		3.67	4.23
1966	4" thick, R20	G		1235	.006		4.20	.30		4.50	5.15
1968	Tapered for drainage	G		1820	.004	B.F.	.99	.20		1.19	1.45
2010	Expanded polystyrene, 1#/C.F. density, 3/4" thick, R2.89	G		1950	.004	S.F.	.22	.19		.41	.57
2020	1" thick, R3.85	G		1950	.004		.29	.19		.48	.65
2100	2" thick, R7.69	G		1625	.005		.58	.23		.81	1.04
2110	3" thick, R11.49	G		1625	.005		.87	.23		1.10	1.36
2120	4" thick, R15.38	G		1625	.005		1.16	.23		1.39	1.68
2130	5" thick, R19.23	G		1495	.005		1.45	.25		1.70	2.03
2140	6" thick, R23.26	G		1495	.005		1.74	.25		1.99	2.34
2150	Tapered for drainage	G		1950	.004	B.F.	.51	.19		.70	.89
2400	Composites with 2" EPS										
2410	1" fiberboard	G	1 Rofc	1325	.006	S.F.	1.48	.28		1.76	2.12
2420	7/16" oriented strand board	G		1040	.008		1.25	.36		1.61	2
2430	1/2" plywood	G		1040	.008		1.47	.36		1.83	2.24
2440	1" perlite	G		1040	.008		1.16	.36		1.52	1.90
2450	Composites with 1-1/2" polyisocyanurate										
2460	1" fiberboard	G	1 Rofc	1040	.008	S.F.	1.21	.36		1.57	1.95
2470	1" perlite	G		1105	.007		.98	.33		1.31	1.67
2480	7/16" oriented strand board	G		1040	.008		.91	.36		1.27	1.62
3000	Fastening alternatives, coated screws, 2" long			3744	.002	Ea.	.06	.10		.16	.24

07 22 Roof and Deck Insulation

07 22 16 – Roof Board Insulation

07 22 16.10 Roof Deck Insulation		Crew	Daily Output	Labor-Hours	Unit	Material	2020 Bare Costs Labor	Equipment	Total	Total Incl O&P
3010	4″ long	1 Rofc	3120	.003	Ea.	.11	.12		.23	.33
3020	6″ long		2675	.003		.20	.14		.34	.46
3030	8″ long		2340	.003		.28	.16		.44	.59
3040	10″ long		1872	.004		.52	.20		.72	.92
3050	Pre-drill and drive wedge spike, 2-1/2″		1248	.006		.40	.30		.70	.96
3060	3-1/2″		1101	.007		.59	.34		.93	1.24
3070	4-1/2″		936	.009		.65	.40		1.05	1.41
3075	3″ galvanized deck plates		7488	.001		.10	.05		.15	.20
3080	Spot mop asphalt	G-1	295	.190	Sq.	5.30	8.20	1.92	15.42	22.50
3090	Full mop asphalt	"	192	.292		10.65	12.60	2.95	26.20	37
3110	Low-rise polyurethane adhesive, 12″ OC beads	1 Rofc	45	.178		41.50	8.20		49.70	60
3120	6″ OC beads		32	.250		83	11.55		94.55	111
9000	Minimum labor/equipment charge		3.25	2.462	Job		114		114	199

07 24 Exterior Insulation and Finish Systems

07 24 13 – Polymer-Based Exterior Insulation and Finish System

07 24 13.10 Exterior Insulation and Finish Systems		Crew	Daily Output	Labor-Hours	Unit	Material	Labor	Equipment	Total	Total Incl O&P
0010	**EXTERIOR INSULATION AND FINISH SYSTEMS**									
0095	Field applied, 1″ EPS insulation G	J-1	390	.103	S.F.	1.80	4.73	.30	6.83	9.85
0100	With 1/2″ cement board sheathing G		268	.149		2.58	6.90	.44	9.92	14.25
0105	2″ EPS insulation G		390	.103		2.09	4.73	.30	7.12	10.20
0110	With 1/2″ cement board sheathing G		268	.149		2.87	6.90	.44	10.21	14.60
0115	3″ EPS insulation G		390	.103		2.38	4.73	.30	7.41	10.50
0120	With 1/2″ cement board sheathing G		268	.149		3.16	6.90	.44	10.50	14.90
0125	4″ EPS insulation G		390	.103		2.67	4.73	.30	7.70	10.80
0130	With 1/2″ cement board sheathing G		268	.149		4.22	6.90	.44	11.56	16.10
0140	Premium finish add		1265	.032		.45	1.46	.09	2	2.92
0145	Drainage and ventilation cavity, add	2 Plas	1450	.011		.82	.54		1.36	1.75
0150	Heavy duty reinforcement add	J-1	914	.044		.59	2.02	.13	2.74	4
0160	2.5#/S.Y. metal lath substrate add	1 Lath	75	.107	S.Y.	3.80	5.55		9.35	12.90
0170	3.4#/S.Y. metal lath substrate add	"	75	.107	"	4.72	5.55		10.27	13.90
0180	Color or texture change	J-1	1265	.032	S.F.	.74	1.46	.09	2.29	3.23
0190	With substrate leveling base coat	1 Plas	530	.015		.84	.73		1.57	2.09
0210	With substrate sealing base coat	1 Pord	1224	.007		.14	.29		.43	.61
0370	V groove shape in panel face				L.F.	.70			.70	.77
0380	U groove shape in panel face				"	.86			.86	.95
0433	Crack repair, acrylic rubber, fluid applied, 20 mils thick	1 Plas	350	.023	S.F.	1.55	1.11		2.66	3.48
0437	50 mils thick, reinforced	"	200	.040	"	2.77	1.94		4.71	6.15
0440	For higher than one story, add						25%			

07 25 Weather Barriers

07 25 10 – Weather Barriers or Wraps

07 25 10.10 Weather Barriers		Crew	Daily Output	Labor-Hours	Unit	Material	2020 Bare Costs Labor	Equipment	Total	Total Incl O&P
0010	**WEATHER BARRIERS**									
0400	Asphalt felt paper, #15	1 Carp	37	.216	Sq.	5.25	11.50		16.75	24
0401	Per square foot	"	3700	.002	S.F.	.05	.11		.16	.24
0450	Housewrap, exterior, spun bonded polypropylene									
0470	Small roll	1 Carp	3800	.002	S.F.	.15	.11		.26	.35
0480	Large roll	"	4000	.002	"	.16	.11		.27	.35
2100	Asphalt felt roof deck vapor barrier, class 1 metal decks	1 Rofc	37	.216	Sq.	21	10		31	41
2200	For all other decks	"	37	.216		15.90	10		25.90	35
2800	Asphalt felt, 50% recycled content, 15 lb., 4 sq./roll	1 Carp	36	.222		6.10	11.80		17.90	25.50
2810	30 lb., 2 sq./roll	"	36	.222	↓	10.45	11.80		22.25	30.50
3000	Building wrap, spun bonded polyethylene	2 Carp	8000	.002	S.F.	.14	.11		.25	.32
9960	Minimum labor/equipment charge	1 Carp	2	4	Job		213		213	340

07 26 Vapor Retarders

07 26 13 – Above-Grade Vapor Retarders

07 26 13.10 Vapor Retarders		Crew	Daily Output	Labor-Hours	Unit	Material	Labor	Equipment	Total	Total Incl O&P
0010	**VAPOR RETARDERS**									
0020	Aluminum and kraft laminated, foil 1 side G	1 Carp	37	.216	Sq.	16.05	11.50		27.55	36
0100	Foil 2 sides G		37	.216		15.50	11.50		27	35.50
0600	Polyethylene vapor barrier, standard, 2 mil G		37	.216		1.63	11.50		13.13	20
0700	4 mil G		37	.216		2.62	11.50		14.12	21.50
0900	6 mil G		37	.216		4	11.50		15.50	23
1200	10 mil G		37	.216		9.70	11.50		21.20	29
1300	Clear reinforced, fire retardant, 8 mil G		37	.216		28	11.50		39.50	49.50
1350	Cross laminated type, 3 mil G		37	.216		12.90	11.50		24.40	32.50
1400	4 mil G		37	.216		14.15	11.50		25.65	34
1800	Reinf. waterproof, 2 mil polyethylene backing, 1 side		37	.216		11.35	11.50		22.85	31
1900	2 sides		37	.216		14.35	11.50		25.85	34
2400	Waterproofed kraft with sisal or fiberglass fibers		37	.216	↓	21	11.50		32.50	42
9950	Minimum labor/equipment charge	↓	4	2	Job		106		106	170

07 27 Air Barriers

07 27 13 – Modified Bituminous Sheet Air Barriers

07 27 13.10 Modified Bituminous Sheet Air Barrier		Crew	Daily Output	Labor-Hours	Unit	Material	Labor	Equipment	Total	Total Incl O&P
0010	**MODIFIED BITUMINOUS SHEET AIR BARRIER**									
0100	SBS modified sheet laminated to polyethylene sheet, 40 mils, 4" wide	1 Carp	1200	.007	L.F.	.32	.35		.67	.93
0120	6" wide		1100	.007		.44	.39		.83	1.11
0140	9" wide		1000	.008		.62	.43		1.05	1.36
0160	12" wide	↓	900	.009	↓	.79	.47		1.26	1.63
0180	18" wide	2 Carp	1700	.009	S.F.	.74	.50		1.24	1.61
0200	36" wide	"	1800	.009		.72	.47		1.19	1.55
0220	Adhesive for above	1 Carp	1400	.006	↓	.32	.30		.62	.84

07 27 26 – Fluid-Applied Membrane Air Barriers

07 27 26.10 Fluid Applied Membrane Air Barrier		Crew	Daily Output	Labor-Hours	Unit	Material	Labor	Equipment	Total	Total Incl O&P
0010	**FLUID APPLIED MEMBRANE AIR BARRIER**									
0100	Spray applied vapor barrier, 25 S.F./gallon	1 Pord	1375	.006	S.F.	.02	.26		.28	.43

07 31 Shingles and Shakes

07 31 13 – Asphalt Shingles

07 31 13.10 Asphalt Roof Shingles		Crew	Daily Output	Labor-Hours	Unit	2020 Bare Costs Material	Labor	Equipment	Total	Total Incl O&P
0010	**ASPHALT ROOF SHINGLES**									
0100	Standard strip shingles									
0150	Inorganic, class A, 25 year	1 Rofc	5.50	1.455	Sq.	75	67		142	201
0155	Pneumatic nailed		7	1.143		75	53		128	175
0200	30 year		5	1.600		111	74		185	251
0205	Pneumatic nailed		6.25	1.280		111	59		170	225
0250	Standard laminated multi-layered shingles									
0300	Class A, 240-260 lb./square	1 Rofc	4.50	1.778	Sq.	103	82		185	257
0305	Pneumatic nailed		5.63	1.422		103	65.50		168.50	228
0350	Class A, 250-270 lb./square		4	2		103	92.50		195.50	275
0355	Pneumatic nailed		5	1.600		103	74		177	242
0400	Premium, laminated multi-layered shingles									
0450	Class A, 260-300 lb./square	1 Rofc	3.50	2.286	Sq.	148	106		254	350
0455	Pneumatic nailed		4.37	1.831		148	84.50		232.50	310
0500	Class A, 300-385 lb./square		3	2.667		310	123		433	555
0505	Pneumatic nailed		3.75	2.133		310	98.50		408.50	510
0800	#15 felt underlayment		64	.125		5.25	5.80		11.05	15.90
0825	#30 felt underlayment		58	.138		10.20	6.35		16.55	22.50
0850	Self adhering polyethylene and rubberized asphalt underlayment		22	.364		76	16.80		92.80	113
0900	Ridge shingles		330	.024	L.F.	2.31	1.12		3.43	4.50
0905	Pneumatic nailed		412.50	.019	"	2.31	.90		3.21	4.11
1000	For steep roofs (7 to 12 pitch or greater), add						50%			
9000	Minimum labor/equipment charge	1 Rofc	3	2.667	Job		123		123	216

07 31 16 – Metal Shingles

07 31 16.10 Aluminum Shingles		Crew	Daily Output	Labor-Hours	Unit	Material	Labor	Equipment	Total	Total Incl O&P
0010	**ALUMINUM SHINGLES**									
0020	Mill finish, .019" thick	1 Carp	5	1.600	Sq.	232	85		317	390
0100	.020" thick	"	5	1.600		256	85		341	420
0300	For colors, add					21.50			21.50	23.50
0600	Ridge cap, .024" thick	1 Carp	170	.047	L.F.	4.02	2.50		6.52	8.45
0700	End wall flashing, .024" thick		170	.047		2.48	2.50		4.98	6.75
0900	Valley section, .024" thick		170	.047		4.08	2.50		6.58	8.50
1000	Starter strip, .024" thick		400	.020		2.01	1.06		3.07	3.91
1200	Side wall flashing, .024" thick		170	.047		2.41	2.50		4.91	6.65
1500	Gable flashing, .024" thick		400	.020		1.89	1.06		2.95	3.78
9000	Minimum labor/equipment charge		3	2.667	Job		142		142	227

07 31 16.20 Steel Shingles		Crew	Daily Output	Labor-Hours	Unit	Material	Labor	Equipment	Total	Total Incl O&P
0010	**STEEL SHINGLES**									
0012	Galvanized, 26 ga.	1 Rots	2.20	3.636	Sq.	390	168		558	725
0200	24 ga.	"	2.20	3.636		380	168		548	715
0300	For colored galvanized shingles, add					59.50			59.50	65.50
9000	Minimum labor/equipment charge	1 Rots	3	2.667	Job		123		123	216

07 31 26 – Slate Shingles

07 31 26.10 Slate Roof Shingles			Crew	Daily Output	Labor-Hours	Unit	Material	Labor	Equipment	Total	Total Incl O&P
0010	**SLATE ROOF SHINGLES**										
0100	Buckingham Virginia black, 3/16" - 1/4" thick	G	1 Rots	1.75	4.571	Sq.	560	211		771	985
0200	1/4" thick	G		1.75	4.571		560	211		771	985
0900	Pennsylvania black, Bangor, #1 clear	G		1.75	4.571		500	211		711	920
1200	Vermont, unfading, green, mottled green	G		1.75	4.571		505	211		716	925
1300	Semi-weathering green & gray	G		1.75	4.571		405	211		616	815
1400	Purple	G		1.75	4.571		440	211		651	855

07 31 Shingles and Shakes

07 31 26 – Slate Shingles

07 31 26.10 Slate Roof Shingles		Crew	Daily Output	Labor-Hours	Unit	Material	2020 Bare Costs Labor	Equipment	Total	Total Incl O&P
1500	Black or gray G	1 Rots	1.75	4.571	Sq.	485	211		696	905
2500	Slate roof repair, extensive replacement		1	8		605	370		975	1,325
2600	Repair individual pieces, scattered		19	.421	Ea.	7.15	19.45		26.60	42
2700	Ridge shingles, slate		200	.040	L.F.	10.20	1.85		12.05	14.45
9000	Minimum labor/equipment charge		3	2.667	Job		123		123	216

07 31 29 – Wood Shingles and Shakes

07 31 29.13 Wood Shingles

		Crew	Daily Output	Labor-Hours	Unit	Material	Labor	Equipment	Total	Total Incl O&P
0010	**WOOD SHINGLES**									
0012	16" No. 1 red cedar shingles, 5" exposure, on roof	1 Carp	2.50	3.200	Sq.	320	170		490	620
0015	Pneumatic nailed		3.25	2.462		320	131		451	560
0200	7-1/2" exposure, on walls		2.05	3.902		213	207		420	565
0205	Pneumatic nailed		2.67	2.996		213	159		372	490
0300	18" No. 1 red cedar perfections, 5-1/2" exposure, on roof		2.75	2.909		286	155		441	565
0305	Pneumatic nailed		3.57	2.241		286	119		405	505
0500	7-1/2" exposure, on walls		2.25	3.556		210	189		399	535
0505	Pneumatic nailed		2.92	2.740		210	146		356	465
0600	Resquared and rebutted, 5-1/2" exposure, on roof		3	2.667		280	142		422	535
0605	Pneumatic nailed		3.90	2.051		280	109		389	485
0900	7-1/2" exposure, on walls		2.45	3.265		206	174		380	505
0905	Pneumatic nailed		3.18	2.516		206	134		340	440
1000	Add to above for fire retardant shingles					60			60	66
1060	Preformed ridge shingles	1 Carp	400	.020	L.F.	4.06	1.06		5.12	6.15
2000	White cedar shingles, 16" long, extras, 5" exposure, on roof		2.40	3.333	Sq.	194	177		371	495
2005	Pneumatic nailed		3.12	2.564		194	136		330	430
2050	5" exposure on walls		2	4		194	213		407	555
2055	Pneumatic nailed		2.60	3.077		194	164		358	475
2100	7-1/2" exposure, on walls		2	4		138	213		351	490
2105	Pneumatic nailed		2.60	3.077		138	164		302	415
2150	"B" grade, 5" exposure on walls		2	4		168	213		381	525
2155	Pneumatic nailed		2.60	3.077		168	164		332	445
2300	For #15 organic felt underlayment on roof, 1 layer, add		64	.125		5.25	6.65		11.90	16.45
2400	2 layers, add		32	.250		10.50	13.30		23.80	33
2600	For steep roofs (7/12 pitch or greater), add to above						50%			
2700	Panelized systems, No.1 cedar shingles on 5/16" CDX plywood									
2800	On walls, 8' strips, 7" or 14" exposure	2 Carp	700	.023	S.F.	6.85	1.22		8.07	9.50
3500	On roofs, 8' strips, 7" or 14" exposure	1 Carp	3	2.667	Sq.	665	142		807	955
3505	Pneumatic nailed		4	2	"	665	106		771	900
9000	Minimum labor/equipment charge		3	2.667	Job		142		142	227

07 31 29.16 Wood Shakes

		Crew	Daily Output	Labor-Hours	Unit	Material	Labor	Equipment	Total	Total Incl O&P
0010	**WOOD SHAKES**									
1100	Hand-split red cedar shakes, 1/2" thick x 24" long, 10" exp. on roof	1 Carp	2.50	3.200	Sq.	340	170		510	645
1105	Pneumatic nailed		3.25	2.462		340	131		471	585
1110	3/4" thick x 24" long, 10" exp. on roof		2.25	3.556		340	189		529	680
1115	Pneumatic nailed		2.92	2.740		340	146		486	610
1200	1/2" thick, 18" long, 8-1/2" exp. on roof		2	4		294	213		507	665
1205	Pneumatic nailed		2.60	3.077		294	164		458	585
1210	3/4" thick x 18" long, 8-1/2" exp. on roof		1.80	4.444		294	236		530	705
1215	Pneumatic nailed		2.34	3.419		294	182		476	615
1255	10" exposure on walls		2	4		283	213		496	650
1260	10" exposure on walls, pneumatic nailed		2.60	3.077		283	164		447	570
1700	Add to above for fire retardant shakes, 24" long					60			60	66
1800	18" long					60			60	66

07 31 Shingles and Shakes

07 31 29 – Wood Shingles and Shakes

07 31 29.16 Wood Shakes		Crew	Daily Output	Labor-Hours	Unit	2020 Bare Costs Material	Labor	Equipment	Total	Total Incl O&P
1810	Ridge shakes	1 Carp	350	.023	L.F.	5.75	1.22		6.97	8.30

07 32 Roof Tiles

07 32 13 – Clay Roof Tiles

07 32 13.10 Clay Tiles

		Crew	Daily Output	Labor-Hours	Unit	Material	Labor	Equipment	Total	Total Incl O&P
0010	**CLAY TILES**, including accessories									
0300	Flat shingle, interlocking, 15", 166 pcs./sq., fireflashed blend	3 Rots	6	4	Sq.	535	185		720	915
0500	Terra cotta red		6	4		545	185		730	925
0600	Roman pan and top, 18", 102 pcs./sq., fireflashed blend		5.50	4.364		495	202		697	900
0640	Terra cotta red	1 Rots	2.40	3.333		585	154		739	915
1100	Barrel mission tile, 18", 166 pcs./sq., fireflashed blend	3 Rots	5.50	4.364		460	202		662	860
1140	Terra cotta red		5.50	4.364		465	202		667	865
1700	Scalloped edge flat shingle, 14", 145 pcs./sq., fireflashed blend		6	4		1,125	185		1,310	1,550
1800	Terra cotta red		6	4		1,050	185		1,235	1,475
3010	#15 felt underlayment	1 Rofc	64	.125		5.25	5.80		11.05	15.90
3020	#30 felt underlayment		58	.138		10.20	6.35		16.55	22.50
3040	Polyethylene and rubberized asph. underlayment		22	.364		76	16.80		92.80	113
9010	Minimum labor/equipment charge	1 Rots	2	4	Job		185		185	325

07 32 16 – Concrete Roof Tiles

07 32 16.10 Concrete Tiles

		Crew	Daily Output	Labor-Hours	Unit	Material	Labor	Equipment	Total	Total Incl O&P
0010	**CONCRETE TILES**									
0020	Corrugated, 13" x 16-1/2", 90 per sq., 950 lb./sq.									
0050	Earthtone colors, nailed to wood deck	1 Rots	1.35	5.926	Sq.	107	274		381	595
0150	Blues		1.35	5.926		106	274		380	595
0200	Greens		1.35	5.926		117	274		391	610
0250	Premium colors		1.35	5.926		117	274		391	610
0500	Shakes, 13" x 16-1/2", 90 per sq., 950 lb./sq.									
0600	All colors, nailed to wood deck	1 Rots	1.50	5.333	Sq.	153	247		400	600
1500	Accessory pieces, ridge & hip, 10" x 16-1/2", 8 lb. each	"	120	.067	Ea.	3.72	3.08		6.80	9.50
1700	Rake, 6-1/2" x 16-3/4", 9 lb. each					3.72			3.72	4.09
1800	Mansard hip, 10" x 16-1/2", 9.2 lb. each					3.72			3.72	4.09
1900	Hip starter, 10" x 16-1/2", 10.5 lb. each					10.65			10.65	11.75
2000	3 or 4 way apex, 10" each side, 11.5 lb. each					12.20			12.20	13.40
9000	Minimum labor/equipment charge	1 Rots	3	2.667	Job		123		123	216

07 32 19 – Metal Roof Tiles

07 32 19.10 Metal Roof Tiles

		Crew	Daily Output	Labor-Hours	Unit	Material	Labor	Equipment	Total	Total Incl O&P
0010	**METAL ROOF TILES**									
0020	Accessories included, .032" thick aluminum, mission tile	1 Carp	2.50	3.200	Sq.	830	170		1,000	1,175
0200	Spanish tiles		3	2.667	"	550	142		692	830
9000	Minimum labor/equipment charge		3	2.667	Job		142		142	227

07 33 Natural Roof Coverings

07 33 63 – Vegetated Roofing

07 33 63.10 Green Roof Systems			Crew	Daily Output	Labor-Hours	Unit	2020 Bare Costs Material	Labor	Equipment	Total	Total Incl O&P
0010	**GREEN ROOF SYSTEMS**										
0020	Soil mixture for green roof 30% sand, 55% gravel, 15% soil										
0100	Hoist and spread soil mixture 4″ depth up to 5 stories tall roof	G	B-13B	4000	.014	S.F.	.23	.64	.25	1.12	1.55
0150	6″ depth	G		2667	.021		.35	.97	.37	1.69	2.33
0200	8″ depth	G		2000	.028		.47	1.29	.49	2.25	3.11
0250	10″ depth	G		1600	.035		.59	1.61	.61	2.81	3.87
0300	12″ depth	G		1335	.042		.70	1.93	.73	3.36	4.65
0310	Alt. man-made soil mix, hoist & spread, 4″ deep up to 5 stories tall roof	G		4000	.014		1.92	.64	.25	2.81	3.40
0350	Mobilization 55 ton crane to site	G	1 Eqhv	3.60	2.222	Ea.		132		132	206
0355	Hoisting cost to 5 stories per day (Avg. 28 picks per day)	G	B-13B	1	56	Day		2,575	980	3,555	5,175
0360	Mobilization or demobilization, 100 ton crane to site driver & escort	G	A-3E	2.50	6.400	Ea.		345	70.50	415.50	625
0365	Hoisting cost 6-10 stories per day (Avg. 21 picks per day)	G	B-13C	1	56	Day		2,575	2,275	4,850	6,625
0370	Hoist and spread soil mixture 4″ depth 6-10 stories tall roof	G		4000	.014	S.F.	.23	.64	.57	1.44	1.91
0375	6″ depth	G		2667	.021		.35	.97	.86	2.18	2.87
0380	8″ depth	G		2000	.028		.47	1.29	1.14	2.90	3.83
0385	10″ depth	G		1600	.035		.59	1.61	1.43	3.63	4.77
0390	12″ depth	G		1335	.042		.70	1.93	1.71	4.34	5.70
0400	Green roof edging treated lumber 4″ x 4″, no hoisting included	G	2 Carp	400	.040	L.F.	1.41	2.13		3.54	4.96
0410	4″ x 6″	G		400	.040		2.24	2.13		4.37	5.85
0420	4″ x 8″	G		360	.044		4.28	2.36		6.64	8.50
0430	4″ x 6″ double stacked	G		300	.053		4.48	2.83		7.31	9.45
0500	Green roof edging redwood lumber 4″ x 4″, no hoisting included	G		400	.040		6.60	2.13		8.73	10.70
0510	4″ x 6″	G		400	.040		12.85	2.13		14.98	17.55
0520	4″ x 8″	G		360	.044		24	2.36		26.36	30.50
0530	4″ x 6″ double stacked	G		300	.053		25.50	2.83		28.33	33
0550	Components, not including membrane or insulation:										
0560	Fluid applied rubber membrane, reinforced, 215 mil thick	G	G-5	350	.114	S.F.	.29	4.80	.55	5.64	9.35
0570	Root barrier	G	2 Rofc	775	.021		.70	.95		1.65	2.44
0580	Moisture retention barrier and reservoir	G	″	900	.018		2.66	.82		3.48	4.37
0600	Planting sedum, light soil, potted, 2-1/4″ diameter, 2 per S.F.	G	1 Clab	420	.019		6.35	.80		7.15	8.30
0610	1 per S.F.	G	″	840	.010		3.18	.40		3.58	4.14
0630	Planting sedum mat per S.F. including shipping (4000 S.F. min)	G	4 Clab	4000	.008		7.65	.34		7.99	8.95
0640	Installation sedum mat system (no soil required) per S.F. (4000 S.F. min)	G	″	4000	.008		10.60	.34		10.94	12.25
0645	Note: pricing of sedum mats shipped in full truck loads (4000-5000 S.F.)										

07 41 Roof Panels

07 41 13 – Metal Roof Panels

07 41 13.10 Aluminum Roof Panels		Crew	Daily Output	Labor-Hours	Unit	Material	Labor	Equipment	Total	Total Incl O&P
0010	**ALUMINUM ROOF PANELS**									
0020	Corrugated or ribbed, .0155″ thick, natural	G-3	1200	.027	S.F.	1.02	1.39		2.41	3.32
0300	Painted		1200	.027		1.50	1.39		2.89	3.85
0400	Corrugated, .018″ thick, on steel frame, natural finish		1200	.027		1.25	1.39		2.64	3.58
0600	Painted		1200	.027		1.57	1.39		2.96	3.93
0700	Corrugated, on steel frame, natural, .024″ thick		1200	.027		1.82	1.39		3.21	4.20
0800	Painted		1200	.027		2.20	1.39		3.59	4.62
0900	.032″ thick, natural		1200	.027		3.15	1.39		4.54	5.65
1200	Painted		1200	.027		3.41	1.39		4.80	5.95
1300	V-Beam, on steel frame construction, .032″ thick, natural		1200	.027		2.73	1.39		4.12	5.20
1500	Painted		1200	.027		4.03	1.39		5.42	6.65
1600	.040″ thick, natural		1200	.027		3.98	1.39		5.37	6.60
1800	Painted		1200	.027		4.79	1.39		6.18	7.45

07 41 Roof Panels

07 41 13 – Metal Roof Panels

07 41 13.10 Aluminum Roof Panels		Crew	Daily Output	Labor-Hours	Unit	Material	2020 Bare Costs Labor	Equipment	Total	Total Incl O&P
1900	.050" thick, natural	G-3	1200	.027	S.F.	4.11	1.39		5.50	6.70
2100	Painted		1200	.027		4.94	1.39		6.33	7.65
2200	For roofing on wood frame, deduct		4600	.007		.08	.36		.44	.66
2400	Ridge cap, .032" thick, natural		800	.040	L.F.	3.36	2.09		5.45	7
9000	Minimum labor/equipment charge	1 Rofc	3	2.667	Job		123		123	216

07 41 13.20 Steel Roofing Panels		Crew	Daily Output	Labor-Hours	Unit	Material	Labor	Equipment	Total	Total Incl O&P
0010	**STEEL ROOFING PANELS**									
0012	Corrugated or ribbed, on steel framing, 30 ga. galv	G-3	1100	.029	S.F.	1.80	1.52		3.32	4.38
0100	28 ga.		1050	.030		1.63	1.59		3.22	4.31
0300	26 ga.		1000	.032		2	1.67		3.67	4.84
0400	24 ga.		950	.034		3.06	1.76		4.82	6.15
0600	Colored, 28 ga.		1050	.030		1.90	1.59		3.49	4.61
0700	26 ga.		1000	.032		2.15	1.67		3.82	5
0710	Flat profile, 1-3/4" standing seams, 10" wide, standard finish, 26 ga.		1000	.032		4.16	1.67		5.83	7.20
0715	24 ga.		950	.034		4.85	1.76		6.61	8.15
0720	22 ga.		900	.036		6	1.86		7.86	9.55
0725	Zinc aluminum alloy finish, 26 ga.		1000	.032		3.35	1.67		5.02	6.35
0730	24 ga.		950	.034		3.92	1.76		5.68	7.10
0735	22 ga.		900	.036		4.47	1.86		6.33	7.85
0740	12" wide, standard finish, 26 ga.		1000	.032		4.19	1.67		5.86	7.25
0745	24 ga.		950	.034		5.45	1.76		7.21	8.80
0750	Zinc aluminum alloy finish, 26 ga.		1000	.032		4.71	1.67		6.38	7.85
0755	24 ga.		950	.034		3.92	1.76		5.68	7.10
0840	Flat profile, 1" x 3/8" batten, 12" wide, standard finish, 26 ga.		1000	.032		3.66	1.67		5.33	6.65
0845	24 ga.		950	.034		4.31	1.76		6.07	7.50
0850	22 ga.		900	.036		5.15	1.86		7.01	8.65
0855	Zinc aluminum alloy finish, 26 ga.		1000	.032		3.53	1.67		5.20	6.50
0860	24 ga.		950	.034		4.02	1.76		5.78	7.20
0865	22 ga.		900	.036		4.56	1.86		6.42	7.95
0870	16-1/2" wide, standard finish, 24 ga.		950	.034		4.25	1.76		6.01	7.45
0875	22 ga.		900	.036		4.78	1.86		6.64	8.20
0880	Zinc aluminum alloy finish, 24 ga.		950	.034		3.71	1.76		5.47	6.85
0885	22 ga.		900	.036		4.20	1.86		6.06	7.55
0890	Flat profile, 2" x 2" batten, 12" wide, standard finish, 26 ga.		1000	.032		4.24	1.67		5.91	7.30
0895	24 ga.		950	.034		5.10	1.76		6.86	8.40
0900	22 ga.		900	.036		6.20	1.86		8.06	9.80
0905	Zinc aluminum alloy finish, 26 ga.		1000	.032		4.02	1.67		5.69	7.05
0910	24 ga.		950	.034		4.51	1.76		6.27	7.75
0915	22 ga.		900	.036		5.25	1.86		7.11	8.70
0920	16-1/2" wide, standard finish, 24 ga.		950	.034		4.67	1.76		6.43	7.95
0925	22 ga.		900	.036		5.40	1.86		7.26	8.90
0930	Zinc aluminum alloy finish, 24 ga.		950	.034		4.20	1.76		5.96	7.40
0935	22 ga.		900	.036		4.80	1.86		6.66	8.25
1200	Ridge, galvanized, 10" wide		800	.040	L.F.	3.29	2.09		5.38	6.90
1210	20" wide		750	.043	"	4.23	2.23		6.46	8.15
9000	Minimum labor/equipment charge	1 Rofc	2	4	Job		185		185	325

07 41 33 – Plastic Roof Panels

07 41 33.10 Fiberglass Panels		Crew	Daily Output	Labor-Hours	Unit	Material	Labor	Equipment	Total	Total Incl O&P
0010	**FIBERGLASS PANELS**									
0012	Corrugated panels, roofing, 8 oz./S.F.	G-3	1000	.032	S.F.	2.55	1.67		4.22	5.45
0100	12 oz./S.F.		1000	.032		4.68	1.67		6.35	7.80
0300	Corrugated siding, 6 oz./S.F.		880	.036		2.02	1.90		3.92	5.20

07 41 Roof Panels

07 41 33 – Plastic Roof Panels

07 41 33.10 Fiberglass Panels		Crew	Daily Output	Labor-Hours	Unit	2020 Bare Costs Material	Labor	Equipment	Total	Total Incl O&P
0400	8 oz./S.F.	G-3	880	.036	S.F.	2.55	1.90		4.45	5.80
0500	Fire retardant		880	.036		4.02	1.90		5.92	7.40
0600	12 oz. siding, textured		880	.036		3.78	1.90		5.68	7.15
0700	Fire retardant		880	.036		4.74	1.90		6.64	8.20
0900	Flat panels, 6 oz./S.F., clear or colors		880	.036		2.78	1.90		4.68	6.05
1100	Fire retardant, class A		880	.036		3.64	1.90		5.54	7
1300	8 oz./S.F., clear or colors		880	.036		2.41	1.90		4.31	5.65
1700	Sandwich panels, fiberglass, 1-9/16" thick, panels to 20 S.F.		180	.178		36.50	9.30		45.80	54.50
1900	As above, but 2-3/4" thick, panels to 100 S.F.		265	.121		25.50	6.30		31.80	38
9000	Minimum labor/equipment charge	1 Rofc	2	4	Job		185		185	325

07 42 Wall Panels

07 42 13 – Metal Wall Panels

07 42 13.10 Mansard Panels		Crew	Daily Output	Labor-Hours	Unit	Material	Labor	Equipment	Total	Total Incl O&P
0010	**MANSARD PANELS**									
0600	Aluminum, stock units, straight surfaces	1 Shee	115	.070	S.F.	4.48	4.33		8.81	11.75
0700	Concave or convex surfaces, add		75	.107	"	2.16	6.65		8.81	12.80
0800	For framing, to 5' high, add		115	.070	L.F.	3.91	4.33		8.24	11.10
0900	Soffits, to 1' wide		125	.064	S.F.	2.94	3.99		6.93	9.50
9000	Minimum labor/equipment charge		2.50	3.200	Job		199		199	310

07 42 13.20 Aluminum Siding Panels		Crew	Daily Output	Labor-Hours	Unit	Material	Labor	Equipment	Total	Total Incl O&P
0010	**ALUMINUM SIDING PANELS**									
0012	Corrugated, on steel framing, .019" thick, natural finish	G-3	775	.041	S.F.	1.72	2.16		3.88	5.30
0100	Painted		775	.041		1.86	2.16		4.02	5.45
0400	Farm type, .021" thick on steel frame, natural		775	.041		1.76	2.16		3.92	5.35
0600	Painted		775	.041		1.86	2.16		4.02	5.45
0700	Industrial type, corrugated, on steel, .024" thick, mill		775	.041		2.42	2.16		4.58	6.05
0900	Painted		775	.041		2.57	2.16		4.73	6.25
1000	.032" thick, mill		775	.041		2.65	2.16		4.81	6.35
1200	Painted		775	.041		3.28	2.16		5.44	7
1300	V-Beam, on steel frame, .032" thick, mill		775	.041		2.94	2.16		5.10	6.65
1500	Painted		775	.041		3.28	2.16		5.44	7
1600	.040" thick, mill		775	.041		3.71	2.16		5.87	7.50
1800	Painted		775	.041		4.31	2.16		6.47	8.15
1900	.050" thick, mill		775	.041		4.37	2.16		6.53	8.20
2100	Painted		775	.041		5.05	2.16		7.21	8.95
2200	Ribbed, 3" profile, on steel frame, .032" thick, natural		775	.041		2.67	2.16		4.83	6.35
2400	Painted		775	.041		3.33	2.16		5.49	7.05
2500	.040" thick, natural		775	.041		3.09	2.16		5.25	6.80
2700	Painted		775	.041		3.57	2.16		5.73	7.35
2750	.050" thick, natural		775	.041		3.52	2.16		5.68	7.30
2760	Painted		775	.041		4.06	2.16		6.22	7.90
3300	For siding on wood frame, deduct from above		2800	.011		.09	.60		.69	1.04
3400	Screw fasteners, aluminum, self tapping, neoprene washer, 1"				M	219			219	241
3600	Stitch screws, self tapping, with neoprene washer, 5/8"				"	165			165	181
3630	Flashing, sidewall, .032" thick	G-3	800	.040	L.F.	3.11	2.09		5.20	6.70
3650	End wall, .040" thick		800	.040		3.64	2.09		5.73	7.30
3670	Closure strips, corrugated, .032" thick		800	.040		.94	2.09		3.03	4.33
3680	Ribbed, 4" or 8", .032" thick		800	.040		.99	2.09		3.08	4.39
3690	V-beam, .040" thick		800	.040		1.45	2.09		3.54	4.90
3800	Horizontal, colored clapboard, 8" wide, plain	2 Carp	515	.031	S.F.	2.72	1.65		4.37	5.65

07 42 Wall Panels

07 42 13 – Metal Wall Panels

07 42 13.20 Aluminum Siding Panels		Crew	Daily Output	Labor-Hours	Unit	Material	Labor	Equipment	Total	Total Incl O&P
						2020 Bare Costs				
3810	Insulated	2 Carp	515	.031	S.F.	3.17	1.65		4.82	6.15
3830	8″ embossed, painted		515	.031		3.14	1.65		4.79	6.10
3840	Insulated		515	.031		3.22	1.65		4.87	6.20
3860	12″ painted, smooth		600	.027		2.94	1.42		4.36	5.50
3870	Insulated		600	.027		3.16	1.42		4.58	5.75
3890	12″ embossed, painted		600	.027		3.19	1.42		4.61	5.80
3900	Insulated		515	.031		2.86	1.65		4.51	5.80
4000	Vertical board & batten, colored, non-insulated		515	.031		2.36	1.65		4.01	5.25
4200	For simulated wood design, add					.16			.16	.18
4300	Corners for above, outside	2 Carp	515	.031	V.L.F.	3.92	1.65		5.57	6.95
4500	Inside corners	″	515	.031	″	1.81	1.65		3.46	4.64
9000	Minimum labor/equipment charge	1 Carp	3	2.667	Job		142		142	227

07 42 13.30 Steel Siding		Crew	Daily Output	Labor-Hours	Unit	Material	Labor	Equipment	Total	Total Incl O&P
0010	**STEEL SIDING**									
0020	Beveled, vinyl coated, 8″ wide	1 Carp	265	.030	S.F.	1.87	1.60		3.47	4.63
0050	10″ wide	″	275	.029		1.95	1.55		3.50	4.63
0080	Galv, corrugated or ribbed, on steel frame, 30 ga.	G-3	800	.040		1.20	2.09		3.29	4.62
0100	28 ga.		795	.040		1.30	2.10		3.40	4.75
0300	26 ga.		790	.041		1.69	2.11		3.80	5.20
0400	24 ga.		785	.041		2.22	2.13		4.35	5.80
0600	22 ga.		770	.042		2.31	2.17		4.48	5.95
0700	Colored, corrugated/ribbed, on steel frame, 10 yr. finish, 28 ga.		800	.040		2.33	2.09		4.42	5.85
0900	26 ga.		795	.040		2.01	2.10		4.11	5.55
1000	24 ga.		790	.041		2.24	2.11		4.35	5.80
1020	20 ga.		785	.041		2.85	2.13		4.98	6.50
1200	Factory sandwich panel, 26 ga., 1″ insulation, galvanized		380	.084		5.20	4.40		9.60	12.65
1300	Colored 1 side		380	.084		7	4.40		11.40	14.65
1500	Galvanized 2 sides		380	.084		7.95	4.40		12.35	15.70
1600	Colored 2 sides		380	.084		8.20	4.40		12.60	15.95
1800	Acrylic paint face, regular paint liner		380	.084		6.05	4.40		10.45	13.60
1900	For 2″ thick polystyrene, add					1.01			1.01	1.11
2000	22 ga., galv, 2″ insulation, baked enamel exterior	G-3	360	.089		11.80	4.64		16.44	20.50
2100	Polyvinylidene exterior finish	″	360	.089		12.55	4.64		17.19	21
9000	Minimum labor/equipment charge	1 Carp	3	2.667	Job		142		142	227

07 44 Faced Panels

07 44 33 – Metal Faced Panels

07 44 33.10 Metal Faced Panels and Accessories		Crew	Daily Output	Labor-Hours	Unit	Material	Labor	Equipment	Total	Total Incl O&P
0010	**METAL FACED PANELS AND ACCESSORIES**									
0400	Textured aluminum, 4′ x 8′ x 5/16″ plywood backing, single face	2 Shee	375	.043	S.F.	4.22	2.66		6.88	8.80
0600	Double face		375	.043		5.30	2.66		7.96	10
0700	4′ x 10′ x 5/16″ plywood backing, single face		375	.043		4.28	2.66		6.94	8.85
0900	Double face		375	.043		5.80	2.66		8.46	10.50
1000	4′ x 12′ x 5/16″ plywood backing, single face		375	.043		4.52	2.66		7.18	9.15
1300	Smooth aluminum, 1/4″ plywood panel, fluoropolymer finish, double face		375	.043		6.05	2.66		8.71	10.80
1350	Clear anodized finish, double face		375	.043		10.25	2.66		12.91	15.40
1400	Double face textured aluminum, structural panel, 1″ EPS insulation		375	.043		6	2.66		8.66	10.75
1500	Accessories, outside corner	1 Shee	175	.046	L.F.	1.73	2.85		4.58	6.35
1600	Inside corner		175	.046		1.38	2.85		4.23	6
1800	Batten mounting clip		200	.040		.50	2.49		2.99	4.45
1900	Low profile batten		480	.017		.63	1.04		1.67	2.32

07 44 Faced Panels

07 44 33 – Metal Faced Panels

07 44 33.10 Metal Faced Panels and Accessories		Crew	Daily Output	Labor-Hours	Unit	Material	2020 Bare Costs Labor	Equipment	Total	Total Incl O&P
2100	High profile batten	1 Shee	480	.017	L.F.	1.43	1.04		2.47	3.20
2200	Water table		200	.040		2.16	2.49		4.65	6.30
2400	Horizontal joint connector		200	.040		1.66	2.49		4.15	5.75
2500	Corner cap		200	.040		1.88	2.49		4.37	5.95
2700	H - moulding		480	.017		1.27	1.04		2.31	3.03

07 46 Siding

07 46 23 – Wood Siding

07 46 23.10 Wood Board Siding

		Crew	Daily Output	Labor-Hours	Unit	Material	Labor	Equipment	Total	Total Incl O&P
0010	**WOOD BOARD SIDING**									
3200	Wood, cedar bevel, A grade, 1/2" x 6"	1 Carp	295	.027	S.F.	4.46	1.44		5.90	7.20
3300	1/2" x 8"		330	.024		8.50	1.29		9.79	11.40
3500	3/4" x 10", clear grade		375	.021		8.45	1.13		9.58	11.10
3600	"B" grade		375	.021		4.16	1.13		5.29	6.40
3800	Cedar, rough sawn, 1" x 4", A grade, natural		220	.036		7.90	1.93		9.83	11.80
3900	Stained		220	.036		8	1.93		9.93	11.90
4100	1" x 12", board & batten, #3 & Btr., natural		420	.019		4.77	1.01		5.78	6.85
4200	Stained		420	.019		5.20	1.01		6.21	7.30
4400	1" x 8" channel siding, #3 & Btr., natural		330	.024		5.15	1.29		6.44	7.70
4500	Stained		330	.024		5.05	1.29		6.34	7.60
4700	Redwood, clear, beveled, vertical grain, 1/2" x 4"		220	.036		5	1.93		6.93	8.60
4750	1/2" x 6"		295	.027		5.35	1.44		6.79	8.15
4800	1/2" x 8"		330	.024		5.45	1.29		6.74	8
5000	3/4" x 10"		375	.021		5.35	1.13		6.48	7.65
5200	Channel siding, 1" x 10", B grade		375	.021		4.87	1.13		6	7.15
5250	Redwood, T&G boards, B grade, 1" x 4"		220	.036		7.10	1.93		9.03	10.90
5270	1" x 8"		330	.024		8.35	1.29		9.64	11.20
5400	White pine, rough sawn, 1" x 8", natural		330	.024		2.55	1.29		3.84	4.87
5500	Stained		330	.024		2.41	1.29		3.70	4.71
9000	Minimum labor/equipment charge		2	4	Job		213		213	340

07 46 29 – Plywood Siding

07 46 29.10 Plywood Siding Options

		Crew	Daily Output	Labor-Hours	Unit	Material	Labor	Equipment	Total	Total Incl O&P
0010	**PLYWOOD SIDING OPTIONS**									
0900	Plywood, medium density overlaid, 3/8" thick	2 Carp	750	.021	S.F.	1.35	1.13		2.48	3.31
1000	1/2" thick		700	.023		1.58	1.22		2.80	3.69
1100	3/4" thick		650	.025		2.29	1.31		3.60	4.62
1600	Texture 1-11, cedar, 5/8" thick, natural		675	.024		2.59	1.26		3.85	4.87
1700	Factory stained		675	.024		2.93	1.26		4.19	5.25
1900	Texture 1-11, fir, 5/8" thick, natural		675	.024		1.21	1.26		2.47	3.35
2000	Factory stained		675	.024		2.05	1.26		3.31	4.28
2050	Texture 1-11, S.Y.P., 5/8" thick, natural		675	.024		1.44	1.26		2.70	3.60
2100	Factory stained		675	.024		1.54	1.26		2.80	3.71
2200	Rough sawn cedar, 3/8" thick, natural		675	.024		1.28	1.26		2.54	3.43
2300	Factory stained		675	.024		1.59	1.26		2.85	3.77
2500	Rough sawn fir, 3/8" thick, natural		675	.024		.97	1.26		2.23	3.09
2600	Factory stained		675	.024		1.10	1.26		2.36	3.23
2800	Redwood, textured siding, 5/8" thick		675	.024		2.05	1.26		3.31	4.28
3000	Polyvinyl chloride coated, 3/8" thick		750	.021		1.18	1.13		2.31	3.12
9000	Minimum labor/equipment charge	1 Carp	2	4	Job		213		213	340

07 46 33 – Plastic Siding

07 46 33.10 Vinyl Siding		Crew	Daily Output	Labor-Hours	Unit	Material	Labor	Equipment	Total	Total Incl O&P
						2020 Bare Costs				
0010	**VINYL SIDING**									
3995	Clapboard profile, woodgrain texture, .048 thick, double 4	2 Carp	495	.032	S.F.	1.09	1.72		2.81	3.95
4000	Double 5		550	.029		1.09	1.55		2.64	3.68
4005	Single 8		495	.032		1.65	1.72		3.37	4.57
4010	Single 10		550	.029		1.98	1.55		3.53	4.66
4015	.044 thick, double 4		495	.032		1.08	1.72		2.80	3.94
4020	Double 5		550	.029		1.10	1.55		2.65	3.70
4025	.042 thick, double 4		495	.032		1.08	1.72		2.80	3.94
4030	Double 5		550	.029		1.08	1.55		2.63	3.67
4035	Cross sawn texture, .040 thick, double 4		495	.032		.72	1.72		2.44	3.55
4040	Double 5		550	.029		.65	1.55		2.20	3.20
4045	Smooth texture, .042 thick, double 4		495	.032		.91	1.72		2.63	3.76
4050	Double 5		550	.029		.80	1.55		2.35	3.37
4055	Single 8		495	.032		.80	1.72		2.52	3.64
4060	Cedar texture, .044 thick, double 4		495	.032		1.18	1.72		2.90	4.05
4065	Double 6		600	.027		1.31	1.42		2.73	3.72
4070	Dutch lap profile, woodgrain texture, .048 thick, double 5		550	.029		1.15	1.55		2.70	3.75
4075	.044 thick, double 4.5		525	.030		1.09	1.62		2.71	3.80
4080	.042 thick, double 4.5		525	.030		.92	1.62		2.54	3.62
4085	.040 thick, double 4.5		525	.030		.72	1.62		2.34	3.40
4100	Shake profile, 10" wide		400	.040		4.04	2.13		6.17	7.85
4105	Vertical pattern, .046 thick, double 5		550	.029		1.77	1.55		3.32	4.43
4110	.044 thick, triple 3		550	.029		1.77	1.55		3.32	4.43
4115	.040 thick, triple 4		550	.029		1.72	1.55		3.27	4.38
4120	.040 thick, triple 2.66		550	.029		1.88	1.55		3.43	4.55
4125	Insulation, fan folded extruded polystyrene, 1/4"		2000	.008		.29	.43		.72	1
4130	3/8"		2000	.008		.32	.43		.75	1.04
4135	Accessories, J channel, 5/8" pocket		700	.023	L.F.	.57	1.22		1.79	2.57
4140	3/4" pocket		695	.023		.63	1.22		1.85	2.65
4145	1-1/4" pocket		680	.024		1.15	1.25		2.40	3.26
4150	Flexible, 3/4" pocket		600	.027		2.85	1.42		4.27	5.40
4155	Under sill finish trim		500	.032		.59	1.70		2.29	3.37
4160	Vinyl starter strip		700	.023		.73	1.22		1.95	2.75
4165	Aluminum starter strip		700	.023		.31	1.22		1.53	2.29
4170	Window casing, 2-1/2" wide, 3/4" pocket		510	.031		1.78	1.67		3.45	4.63
4175	Outside corner, woodgrain finish, 4" face, 3/4" pocket		700	.023		2.35	1.22		3.57	4.54
4180	5/8" pocket		700	.023		2.29	1.22		3.51	4.47
4185	Smooth finish, 4" face, 3/4" pocket		700	.023		2.36	1.22		3.58	4.55
4190	7/8" pocket		690	.023		1.91	1.23		3.14	4.08
4195	1-1/4" pocket		700	.023		1.35	1.22		2.57	3.44
4200	Soffit and fascia, 1' overhang, solid		120	.133		5.15	7.10		12.25	17
4205	Vented		120	.133		5.15	7.10		12.25	17
4207	18" overhang, solid		110	.145		6.05	7.75		13.80	19.05
4208	Vented		110	.145		6.05	7.75		13.80	19.05
4210	2' overhang, solid		100	.160		6.90	8.50		15.40	21
4215	Vented		100	.160		6.90	8.50		15.40	21
4217	3' overhang, solid		100	.160		8.70	8.50		17.20	23
4218	Vented		100	.160		8.70	8.50		17.20	23
4220	Colors for siding and soffits, add				S.F.	.15			.15	.17
4225	Colors for accessories and trim, add				L.F.	.31			.31	.34
9000	Minimum labor/equipment charge	1 Carp	3	2.667	Job		142		142	227

07 46 Siding

07 46 33 – Plastic Siding

07 46 33.20 Polypropylene Siding		Crew	Daily Output	Labor-Hours	Unit	Material	Labor	Equipment	Total	Total Incl O&P
						2020 Bare Costs				
0010	POLYPROPYLENE SIDING									
4090	Shingle profile, random grooves, double 7	2 Carp	400	.040	S.F.	3.51	2.13		5.64	7.30
4092	Cornerpost for above	1 Carp	365	.022	L.F.	12.85	1.17		14.02	16
4095	Triple 5	2 Carp	400	.040	S.F.	3.72	2.13		5.85	7.50
4097	Cornerpost for above	1 Carp	365	.022	L.F.	12.40	1.17		13.57	15.50
5000	Staggered butt, double 7"	2 Carp	400	.040	S.F.	3.64	2.13		5.77	7.40
5002	Cornerpost for above	1 Carp	365	.022	L.F.	13.35	1.17		14.52	16.55
5010	Half round, double 6-1/4"	2 Carp	360	.044	S.F.	4.27	2.36		6.63	8.50
5020	Shake profile, staggered butt, double 9"	"	510	.031	"	3.80	1.67		5.47	6.85
5022	Cornerpost for above	1 Carp	365	.022	L.F.	10	1.17		11.17	12.85
5030	Straight butt, double 7"	2 Carp	400	.040	S.F.	4.27	2.13		6.40	8.10
5032	Cornerpost for above	1 Carp	365	.022	L.F.	14.55	1.17		15.72	17.85
6000	Accessories, J channel, 5/8" pocket	2 Carp	700	.023		.57	1.22		1.79	2.57
6010	3/4" pocket		695	.023		.63	1.22		1.85	2.65
6020	1-1/4" pocket		680	.024		1.15	1.25		2.40	3.26
6030	Aluminum starter strip	↓	700	.023	↓	.31	1.22		1.53	2.29

07 46 46 – Fiber Cement Siding

07 46 46.10 Fiber Cement Siding		Crew	Daily Output	Labor-Hours	Unit	Material	Labor	Equipment	Total	Total Incl O&P
0010	FIBER CEMENT SIDING									
0020	Lap siding, 5/16" thick, 6" wide, 4-3/4" exposure, smooth texture	2 Carp	415	.039	S.F.	1.41	2.05		3.46	4.83
0025	Woodgrain texture		415	.039		1.41	2.05		3.46	4.83
0030	7-1/2" wide, 6-1/4" exposure, smooth texture		425	.038		2	2		4	5.40
0035	Woodgrain texture		425	.038		2	2		4	5.40
0040	8" wide, 6-3/4" exposure, smooth texture		425	.038		1.31	2		3.31	4.66
0045	Rough sawn texture		425	.038		1.31	2		3.31	4.66
0050	9-1/2" wide, 8-1/4" exposure, smooth texture		440	.036		1.19	1.93		3.12	4.41
0055	Woodgrain texture		440	.036		1.19	1.93		3.12	4.41
0060	12" wide, 10-3/8" exposure, smooth texture		455	.035		2.18	1.87		4.05	5.40
0065	Woodgrain texture		455	.035		2.18	1.87		4.05	5.40
0070	Panel siding, 5/16" thick, smooth texture		750	.021		1.22	1.13		2.35	3.16
0075	Stucco texture		750	.021		1.22	1.13		2.35	3.16
0080	Grooved woodgrain texture		750	.021		1.22	1.13		2.35	3.16
0088	Shingle siding, 48" x 15-1/4" panels, 7" exposure		700	.023	↓	4.31	1.22		5.53	6.70
0090	Wood starter strip	↓	400	.040	L.F.	.47	2.13		2.60	3.93

07 46 73 – Soffit

07 46 73.10 Soffit Options		Crew	Daily Output	Labor-Hours	Unit	Material	Labor	Equipment	Total	Total Incl O&P
0010	SOFFIT OPTIONS									
0012	Aluminum, residential, .020" thick	1 Carp	210	.038	S.F.	2.25	2.03		4.28	5.70
0100	Baked enamel on steel, 16 or 18 ga.		105	.076		6.45	4.05		10.50	13.60
0300	Polyvinyl chloride, white, solid		230	.035		2.24	1.85		4.09	5.40
0400	Perforated	↓	230	.035		2.24	1.85		4.09	5.40
0500	For colors, add				↓	.15			.15	.17
9000	Minimum labor/equipment charge	1 Carp	3	2.667	Job		142		142	227

07 51 Built-Up Bituminous Roofing

07 51 13 – Built-Up Asphalt Roofing

07 51 13.10 Built-Up Roofing Components		Crew	Daily Output	Labor-Hours	Unit	2020 Bare Costs Material	Labor	Equipment	Total	Total Incl O&P
0010	**BUILT-UP ROOFING COMPONENTS**									
0012	Asphalt saturated felt, #30, 2 sq./roll	1 Rofc	58	.138	Sq.	10.20	6.35		16.55	22.50
0200	#15, 4 sq./roll, plain or perforated, not mopped		58	.138		5.25	6.35		11.60	16.95
0300	Roll roofing, smooth, #65		15	.533		10.30	24.50		34.80	54.50
0500	#90		12	.667		37.50	31		68.50	95
0520	Mineralized		12	.667		36	31		67	93.50
0540	D.C. (double coverage), 19" selvage edge		10	.800		48	37		85	117
0580	Adhesive (lap cement)				Gal.	8.50			8.50	9.35
0800	Steep, flat or dead level asphalt, 10 ton lots, packaged				Ton	885			885	975
9000	Minimum labor/equipment charge	1 Rofc	4	2	Job		92.50		92.50	162

07 51 13.13 Cold-Applied Built-Up Asphalt Roofing		Crew	Daily Output	Labor-Hours	Unit	Material	Labor	Equipment	Total	Total Incl O&P
0010	**COLD-APPLIED BUILT-UP ASPHALT ROOFING**									
0020	3 ply system, installation only (components listed below)	G-5	50	.800	Sq.		33.50	3.86	37.36	62.50
0100	Spunbond poly. fabric, 1.35 oz./S.Y., 36"W, 10.8 sq./roll				Ea.	133			133	146
0500	Base & finish coat, 3 gal./sq., 5 gal./can				Gal.	8.10			8.10	8.95
0600	Coating, ceramic granules, 1/2 sq./bag				Ea.	23			23	25.50
0700	Aluminum, 2 gal./sq.				Gal.	12.45			12.45	13.70
0800	Emulsion, fibered or non-fibered, 4 gal./sq.				"	6.65			6.65	7.30

07 51 13.20 Built-Up Roofing Systems		Crew	Daily Output	Labor-Hours	Unit	Material	Labor	Equipment	Total	Total Incl O&P
0010	**BUILT-UP ROOFING SYSTEMS**									
0120	Asphalt flood coat with gravel/slag surfacing, not including									
0140	Insulation, flashing or wood nailers									
0200	Asphalt base sheet, 3 plies #15 asphalt felt, mopped	G-1	22	2.545	Sq.	108	110	26	244	340
0350	On nailable decks		21	2.667		100	115	27	242	340
0500	4 plies #15 asphalt felt, mopped		20	2.800		133	121	28.50	282.50	390
0550	On nailable decks		19	2.947		118	127	30	275	385
0700	Coated glass base sheet, 2 plies glass (type IV), mopped		22	2.545		108	110	26	244	340
0850	3 plies glass, mopped		20	2.800		130	121	28.50	279.50	385
0950	On nailable decks		19	2.947		123	127	30	280	390
1100	4 plies glass fiber felt (type IV), mopped		20	2.800		161	121	28.50	310.50	420
1150	On nailable decks		19	2.947		146	127	30	303	415
1200	Coated & saturated base sheet, 3 plies #15 asph. felt, mopped		20	2.800		109	121	28.50	258.50	365
1250	On nailable decks		19	2.947		102	127	30	259	370
1300	4 plies #15 asphalt felt, mopped		22	2.545		127	110	26	263	360
2000	Asphalt flood coat, smooth surface									
2200	Asphalt base sheet & 3 plies #15 asphalt felt, mopped	G-1	24	2.333	Sq.	102	101	23.50	226.50	315
2400	On nailable decks		23	2.435		95	105	24.50	224.50	315
2600	4 plies #15 asphalt felt, mopped		24	2.333		120	101	23.50	244.50	335
2700	On nailable decks		23	2.435		113	105	24.50	242.50	335
2900	Coated glass fiber base sheet, mopped, and 2 plies of									
2910	glass fiber felt (type IV)	G-1	25	2.240	Sq.	102	96.50	22.50	221	305
3100	On nailable decks		24	2.333		97	101	23.50	221.50	310
3200	3 plies, mopped		23	2.435		125	105	24.50	254.50	350
3300	On nailable decks		22	2.545		118	110	26	254	350
3800	4 plies glass fiber felt (type IV), mopped		23	2.435		148	105	24.50	277.50	375
3900	On nailable decks		22	2.545		140	110	26	276	375
4000	Coated & saturated base sheet, 3 plies #15 asph. felt, mopped		24	2.333		104	101	23.50	228.50	315
4200	On nailable decks		23	2.435		96	105	24.50	225.50	315
4300	4 plies #15 organic felt, mopped		22	2.545		121	110	26	257	355
4500	Coal tar pitch with gravel/slag surfacing									
4600	4 plies #15 tarred felt, mopped	G-1	21	2.667	Sq.	204	115	27	346	455
4800	3 plies glass fiber felt (type IV), mopped	"	19	2.947	"	168	127	30	325	440

07 51 Built-Up Bituminous Roofing

07 51 13 – Built-Up Asphalt Roofing

07 51 13.20 Built-Up Roofing Systems		Crew	Daily Output	Labor-Hours	Unit	Material	Labor	Equipment	Total	Total Incl O&P
						2020 Bare Costs				
5000	Coated glass fiber base sheet, and 2 plies of									
5010	glass fiber felt (type IV), mopped	G-1	19	2.947	Sq.	172	127	30	329	445
5300	On nailable decks		18	3.111		150	134	31.50	315.50	435
5600	4 plies glass fiber felt (type IV), mopped		21	2.667		233	115	27	375	485
5800	On nailable decks		20	2.800		212	121	28.50	361.50	475

07 51 13.30 Cants		Crew	Daily Output	Labor-Hours	Unit	Material	Labor	Equipment	Total	Total Incl O&P
0010	**CANTS**									
0012	Lumber, treated, 4″ x 4″ cut diagonally	1 Rofc	325	.025	L.F.	1.89	1.14		3.03	4.07
0300	Mineral or fiber, trapezoidal, 1″ x 4″ x 48″		325	.025		.31	1.14		1.45	2.33
0400	1-1/2″ x 5-5/8″ x 48″		325	.025		.48	1.14		1.62	2.52
9000	Minimum labor/equipment charge		4	2	Job		92.50		92.50	162

07 51 13.40 Felts		Crew	Daily Output	Labor-Hours	Unit	Material	Labor	Equipment	Total	Total Incl O&P
0010	**FELTS**									
0012	Glass fibered roofing felt, #15, not mopped	1 Rofc	58	.138	Sq.	10.20	6.35		16.55	22.50
0300	Base sheet, #80, channel vented		58	.138		48.50	6.35		54.85	64.50
0400	#70, coated		58	.138		18.70	6.35		25.05	31.50
0500	Cap, #87, mineral surfaced		58	.138		84.50	6.35		90.85	104
0600	Flashing membrane, #65		16	.500		10.30	23		33.30	52
0800	Coal tar fibered, #15, no mopping		58	.138		17.55	6.35		23.90	30.50
0900	Asphalt felt, #15, 4 sq./roll, no mopping		58	.138		5.25	6.35		11.60	16.95
1100	#30, 2 sq./roll		58	.138		10.20	6.35		16.55	22.50
1200	Double coated, #33		58	.138		11.90	6.35		18.25	24.50
1400	#40, base sheet		58	.138		10.95	6.35		17.30	23
1450	Coated and saturated		58	.138		12.20	6.35		18.55	24.50
1500	Tarred felt, organic, #15, 4 sq. rolls		58	.138		14.70	6.35		21.05	27.50
1550	#30, 2 sq. roll		58	.138		25	6.35		31.35	38.50
1700	Add for mopping above felts, per ply, asphalt, 24 lb./sq.	G-1	192	.292		10.65	12.60	2.95	26.20	37
1800	Coal tar mopping, 30 lb./sq.		186	.301		18.50	13	3.05	34.55	47
1900	Flood coat, with asphalt, 60 lb./sq.		60	.933		26.50	40.50	9.45	76.45	110
2000	With coal tar, 75 lb./sq.		56	1		46	43	10.10	99.10	138
9000	Minimum labor/equipment charge	1 Rofc	4	2	Job		92.50		92.50	162

07 51 13.50 Walkways for Built-Up Roofs		Crew	Daily Output	Labor-Hours	Unit	Material	Labor	Equipment	Total	Total Incl O&P
0010	**WALKWAYS FOR BUILT-UP ROOFS**									
0020	Asphalt impregnated, 3′ x 6′ x 1/2″ thick	1 Rofc	400	.020	S.F.	1.98	.92		2.90	3.80
0100	3′ x 3′ x 3/4″ thick	″	400	.020		5.70	.92		6.62	7.90
0300	Concrete patio blocks, 2″ thick, natural	1 Clab	115	.070		3.58	2.93		6.51	8.65
0400	Colors	″	115	.070		3.71	2.93		6.64	8.75
0600	100% recycled rubber, 3′ x 4′ x 3/8″ G	1 Rofc	400	.020	L.F.	7.35	.92		8.27	9.70
0610	3′ x 4′ x 1/2″ G		400	.020		8.20	.92		9.12	10.65
0620	3′ x 4′ x 3/4″ G		400	.020		8.60	.92		9.52	11.05
9000	Minimum labor/equipment charge		2.75	2.909	Job		134		134	235

07 52 Modified Bituminous Membrane Roofing

07 52 13 – Atactic-Polypropylene-Modified Bituminous Membrane Roofing

07 52 13.10 APP Modified Bituminous Membrane		Crew	Daily Output	Labor-Hours	Unit	2020 Bare Costs Material	Labor	Equipment	Total	Total Incl O&P
0010	**APP MODIFIED BITUMINOUS MEMBRANE** R075213-30									
0020	Base sheet, #15 glass fiber felt, nailed to deck	1 Rofc	58	.138	Sq.	11.85	6.35		18.20	24
0030	Spot mopped to deck	G-1	295	.190		15.55	8.20	1.92	25.67	33.50
0040	Fully mopped to deck	"	192	.292		21	12.60	2.95	36.55	48.50
0050	#15 organic felt, nailed to deck	1 Rofc	58	.138		6.90	6.35		13.25	18.75
0060	Spot mopped to deck	G-1	295	.190		10.60	8.20	1.92	20.72	28
0070	Fully mopped to deck	"	192	.292	↓	15.90	12.60	2.95	31.45	43
2100	APP mod., smooth surf. cap sheet, poly. reinf., torched, 160 mils	G-5	2100	.019	S.F.	.84	.80	.09	1.73	2.42
2150	170 mils		2100	.019		.76	.80	.09	1.65	2.34
2200	Granule surface cap sheet, poly. reinf., torched, 180 mils		2000	.020		.96	.84	.10	1.90	2.64
2250	Smooth surface flashing, torched, 160 mils		1260	.032		.84	1.33	.15	2.32	3.42
2300	170 mils		1260	.032		.76	1.33	.15	2.24	3.34
2350	Granule surface flashing, torched, 180 mils	↓	1260	.032		.96	1.33	.15	2.44	3.56
2400	Fibrated aluminum coating	1 Rofc	3800	.002	↓	.09	.10		.19	.27
2450	Seam heat welding	"	205	.039	L.F.	.09	1.80		1.89	3.25

07 52 16 – Styrene-Butadiene-Styrene Modified Bituminous Membrane Roofing

07 52 16.10 SBS Modified Bituminous Membrane		Crew	Daily Output	Labor-Hours	Unit	Material	Labor	Equipment	Total	Total Incl O&P
0010	**SBS MODIFIED BITUMINOUS MEMBRANE**									
0080	Mod. bit. rfng., SBS mod, gran surf. cap sheet, poly. reinf.									
0650	120 to 149 mils thick	G-1	2000	.028	S.F.	1.36	1.21	.28	2.85	3.93
0750	150 to 160 mils		2000	.028		1.80	1.21	.28	3.29	4.41
1600	Smooth surface cap sheet, mopped, 145 mils		2100	.027		.83	1.15	.27	2.25	3.22
1620	Lightweight base sheet, fiberglass reinforced, 35 to 47 mil		2100	.027		.29	1.15	.27	1.71	2.63
1625	Heavyweight base/ply sheet, reinforced, 87 to 120 mil thick	↓	2100	.027		.93	1.15	.27	2.35	3.33
1650	Granulated walkpad, 180 to 220 mils	1 Rofc	400	.020		1.86	.92		2.78	3.67
1700	Smooth surface flashing, 145 mils	G-1	1260	.044		.83	1.92	.45	3.20	4.77
1800	150 mils		1260	.044		.51	1.92	.45	2.88	4.42
1900	Granular surface flashing, 150 mils		1260	.044		.69	1.92	.45	3.06	4.62
2000	160 mils	↓	1260	.044		.78	1.92	.45	3.15	4.72
2010	Elastomeric asphalt primer	1 Rofc	2600	.003		.17	.14		.31	.44
2015	Roofing asphalt, 30 lb./square	G-1	19000	.003		.13	.13	.03	.29	.40
2020	Cold process adhesive, 20 to 30 mils thick	1 Rofc	750	.011		.26	.49		.75	1.15
2025	Self adhering vapor retarder, 30 to 45 mils thick	G-5	2150	.019	↓	1.07	.78	.09	1.94	2.65
2050	Seam heat welding	1 Rofc	205	.039	L.F.	.09	1.80		1.89	3.25

07 53 Elastomeric Membrane Roofing

07 53 16 – Chlorosulfonate-Polyethylene Roofing

07 53 16.10 Chlorosulfonated Polyethylene Roofing		Crew	Daily Output	Labor-Hours	Unit	Material	Labor	Equipment	Total	Total Incl O&P
0010	**CHLOROSULFONATED POLYETHYLENE ROOFING**									
0800	Chlorosulfonated polyethylene (CSPE)									
0900	45 mils, heat welded seams, plate attachment	G-5	35	1.143	Sq.	249	48	5.50	302.50	365
1100	Heat welded seams, plate attachment and ballasted		26	1.538		260	64.50	7.40	331.90	405
1200	60 mils, heat welded seams, plate attachment		35	1.143		335	48	5.50	388.50	455
1300	Heat welded seams, plate attachment and ballasted	↓	26	1.538	↓	345	64.50	7.40	416.90	500

07 53 23 – Ethylene-Propylene-Diene-Monomer Roofing

07 53 23.20 Ethylene-Propylene-Diene-Monomer Roofing		Crew	Daily Output	Labor-Hours	Unit	Material	Labor	Equipment	Total	Total Incl O&P
0010	**ETHYLENE-PROPYLENE-DIENE-MONOMER ROOFING (EPDM)**									
3500	Ethylene-propylene-diene-monomer (EPDM), 45 mils, 0.28 psf									
3600	Loose-laid & ballasted with stone (10 psf)	G-5	51	.784	Sq.	92	33	3.78	128.78	163
3700	Mechanically attached	↓	35	1.143	↓	85.50	48	5.50	139	184

07 53 Elastomeric Membrane Roofing

07 53 23 – Ethylene-Propylene-Diene-Monomer Roofing

07 53 23.20 Ethylene-Propylene-Diene-Monomer Roofing		Crew	Daily Output	Labor-Hours	Unit	2020 Bare Costs Material	Labor	Equipment	Total	Total Incl O&P
3800	Fully adhered with adhesive	G-5	26	1.538	Sq.	120	64.50	7.40	191.90	253
4500	60 mils, 0.40 psf									
4600	Loose-laid & ballasted with stone (10 psf)	G-5	51	.784	Sq.	112	33	3.78	148.78	186
4700	Mechanically attached		35	1.143		105	48	5.50	158.50	205
4800	Fully adhered with adhesive		26	1.538		139	64.50	7.40	210.90	274
4810	45 mil, 0.28 psf, membrane only					55			55	60.50
4820	60 mil, 0.40 psf, membrane only					72			72	79.50
4850	Seam tape for membrane, 3" x 100' roll				Ea.	49.50			49.50	54.50
4900	Batten strips, 10' sections					4.05			4.05	4.46
4910	Cover tape for batten strips, 6" x 100' roll					181			181	199
4930	Plate anchors				M	82.50			82.50	91
4970	Adhesive for fully adhered systems, 60 S.F./gal.				Gal.	21			21	23

07 53 29 – Polyisobutylene Roofing

07 53 29.10 Polyisobutylene Roofing		Crew	Daily Output	Labor-Hours	Unit	Material	Labor	Equipment	Total	Total Incl O&P
0010	**POLYISOBUTYLENE ROOFING**									
7500	Polyisobutylene (PIB), 100 mils, 0.57 psf									
7600	Loose-laid & ballasted with stone/gravel (10 psf)	G-5	51	.784	Sq.	215	33	3.78	251.78	298
7700	Partially adhered with adhesive		35	1.143		253	48	5.50	306.50	370
7800	Hot asphalt attachment		35	1.143		241	48	5.50	294.50	355
7900	Fully adhered with contact cement		26	1.538		266	64.50	7.40	337.90	415

07 54 Thermoplastic Membrane Roofing

07 54 16 – Ketone Ethylene Ester Roofing

07 54 16.10 Ketone Ethylene Ester Roofing		Crew	Daily Output	Labor-Hours	Unit	Material	Labor	Equipment	Total	Total Incl O&P
0010	**KETONE ETHYLENE ESTER ROOFING**									
0100	Ketone ethylene ester roofing, 50 mil, fully adhered	G-5	26	1.538	Sq.	216	64.50	7.40	287.90	360
0120	Mechanically attached		35	1.143		140	48	5.50	193.50	244
0140	Ballasted with stone		51	.784		148	33	3.78	184.78	225
0160	50 mil, fleece backed, adhered w/hot asphalt	G-1	26	2.154		160	93	22	275	360
0180	Accessories, pipe boot	1 Rofc	32	.250	Ea.	26	11.55		37.55	48.50
0200	Pre-formed corners		32	.250	"	9.75	11.55		21.30	31
0220	Ketone clad metal, including up to 4 bends		330	.024	S.F.	4.02	1.12		5.14	6.40
0240	Walkway pad	2 Rofc	800	.020	"	4.27	.92		5.19	6.30
0260	Stripping material	1 Rofc	310	.026	L.F.	.99	1.19		2.18	3.18

07 54 19 – Polyvinyl-Chloride Roofing

07 54 19.10 Polyvinyl-Chloride Roofing (PVC)		Crew	Daily Output	Labor-Hours	Unit	Material	Labor	Equipment	Total	Total Incl O&P
0010	**POLYVINYL-CHLORIDE ROOFING (PVC)**									
8200	Heat welded seams									
8700	Reinforced, 48 mils, 0.33 psf									
8750	Loose-laid & ballasted with stone/gravel (12 psf)	G-5	51	.784	Sq.	127	33	3.78	163.78	202
8800	Mechanically attached		35	1.143		120	48	5.50	173.50	222
8850	Fully adhered with adhesive		26	1.538		166	64.50	7.40	237.90	305
8860	Reinforced, 60 mils, 0.40 psf									
8870	Loose-laid & ballasted with stone/gravel (12 psf)	G-5	51	.784	Sq.	129	33	3.78	165.78	204
8880	Mechanically attached		35	1.143		121	48	5.50	174.50	224
8890	Fully adhered with adhesive		26	1.538		167	64.50	7.40	238.90	305

07 54 Thermoplastic Membrane Roofing

07 54 23 – Thermoplastic-Polyolefin Roofing

07 54 23.10 Thermoplastic Polyolefin Roofing (T.P.O.)		Crew	Daily Output	Labor-Hours	Unit	2020 Bare Costs Material	Labor	Equipment	Total	Total Incl O&P
0010	**THERMOPLASTIC POLYOLEFIN ROOFING (T.P.O.)**									
0100	45 mil, loose laid & ballasted with stone (1/2 ton/sq.)	G-5	51	.784	Sq.	86.50	33	3.78	123.28	157
0120	Fully adhered		25	1.600		76	67	7.70	150.70	209
0140	Mechanically attached		34	1.176		75.50	49.50	5.65	130.65	176
0160	Self adhered		35	1.143		79	48	5.50	132.50	177
0180	60 mil membrane, heat welded seams, ballasted		50	.800		100	33.50	3.86	137.36	173
0200	Fully adhered		25	1.600		89	67	7.70	163.70	224
0220	Mechanically attached		34	1.176		93	49.50	5.65	148.15	195
0240	Self adhered		35	1.143		106	48	5.50	159.50	207

07 55 Protected Membrane Roofing

07 55 10 – Protected Membrane Roofing Components

07 55 10.10 Protected Membrane Roofing Components		Crew	Daily Output	Labor-Hours	Unit	Material	Labor	Equipment	Total	Total Incl O&P
0010	**PROTECTED MEMBRANE ROOFING COMPONENTS**									
0100	Choose roofing membrane from 07 50									
0120	Then choose roof deck insulation from 07 22									
0130	Filter fabric	2 Rofc	10000	.002	S.F.	.10	.07		.17	.24
0140	Ballast, 3/8″ - 1/2″ in place	G-1	36	1.556	Ton	22	67	15.75	104.75	159
0150	3/4″ - 1-1/2″ in place	″	36	1.556	″	22	67	15.75	104.75	159
0200	2″ concrete blocks, natural	1 Clab	115	.070	S.F.	3.58	2.93		6.51	8.65
0210	Colors	″	115	.070	″	3.71	2.93		6.64	8.75

07 56 Fluid-Applied Roofing

07 56 10 – Fluid-Applied Roofing Elastomers

07 56 10.10 Elastomeric Roofing		Crew	Daily Output	Labor-Hours	Unit	Material	Labor	Equipment	Total	Total Incl O&P
0010	**ELASTOMERIC ROOFING**									
0020	Acrylic, 44% solids, 2 coats, on corrugated metal	2 Rofc	2400	.007	S.F.	.56	.31		.87	1.16
0025	On smooth metal		3000	.005		.45	.25		.70	.92
0030	On foam or modified bitumen		1500	.011		.90	.49		1.39	1.84
0035	On concrete		1500	.011		.90	.49		1.39	1.84
0040	On tar and gravel		1500	.011		.90	.49		1.39	1.84
0045	36% solids, 2 coats, on corrugated metal		2400	.007		.51	.31		.82	1.10
0050	On smooth metal		3000	.005		.41	.25		.66	.88
0055	On foam or modified bitumen		1500	.011		.82	.49		1.31	1.76
0060	On concrete		1500	.011		.82	.49		1.31	1.76
0065	On tar and gravel		1500	.011		.82	.49		1.31	1.76
0070	Primer if required, 2 coats on corrugated metal		2400	.007		.51	.31		.82	1.10
0075	On smooth metal		3000	.005		.51	.25		.76	.99
0080	On foam or modified bitumen		1500	.011		.69	.49		1.18	1.62
0085	On concrete		1500	.011		.52	.49		1.01	1.43
0090	On tar & gravel/rolled roof		1500	.011		.69	.49		1.18	1.62
0110	Acrylic rubber, fluid applied, 20 mils thick	G-5	2000	.020		2.20	.84	.10	3.14	4
0120	50 mils, reinforced		1200	.033		3.23	1.40	.16	4.79	6.20
0130	For walking surface, add		900	.044		1.34	1.86	.21	3.41	4.97
0300	Neoprene, fluid applied, 20 mil thick, not reinforced	G-1	1135	.049		1.31	2.13	.50	3.94	5.70
0600	Non-woven polyester, reinforced		960	.058		1.43	2.52	.59	4.54	6.65
0700	5 coat neoprene deck, 60 mil thick, under 10,000 S.F.		325	.172		4.37	7.45	1.74	13.56	19.70
0900	Over 10,000 S.F.		625	.090		4.37	3.87	.91	9.15	12.55
9000	Minimum labor/equipment charge	1 Rofc	2	4	Job		185		185	325

07 57 Coated Foamed Roofing

07 57 13 – Sprayed Polyurethane Foam Roofing

07 57 13.10 Sprayed Polyurethane Foam Roofing (S.P.F.)		Crew	Daily Output	Labor-Hours	Unit	2020 Bare Costs Material	Labor	Equipment	Total	Total Incl O&P
0010	SPRAYED POLYURETHANE FOAM ROOFING (S.P.F.)									
0100	Primer for metal substrate (when required)	G-2A	3000	.008	S.F.	.51	.33	.21	1.05	1.35
0200	Primer for non-metal substrate (when required)		3000	.008		.19	.33	.21	.73	.99
0300	Closed cell spray, polyurethane foam, 3 lb./C.F. density, 1", R6.7		15000	.002		.64	.07	.04	.75	.87
0400	2", R13.4		13125	.002		1.29	.08	.05	1.42	1.60
0500	3", R18.6		11485	.002		1.93	.09	.06	2.08	2.34
0550	4", R24.8		10080	.002		2.58	.10	.06	2.74	3.07
0700	Spray-on silicone coating	↓	2500	.010	↓	1.28	.39	.25	1.92	2.36

07 58 Roll Roofing

07 58 10 – Asphalt Roll Roofing

07 58 10.10 Roll Roofing		Crew	Daily Output	Labor-Hours	Unit	Material	Labor	Equipment	Total	Total Incl O&P
0010	ROLL ROOFING									
0100	Asphalt, mineral surface									
0200	1 ply #15 organic felt, 1 ply mineral surfaced									
0300	Selvage roofing, lap 19", nailed & mopped	G-1	27	2.074	Sq.	69	89.50	21	179.50	256
0400	3 plies glass fiber felt (type IV), 1 ply mineral surfaced									
0500	Selvage roofing, lapped 19", mopped	G-1	25	2.240	Sq.	121	96.50	22.50	240	325
0600	Coated glass fiber base sheet, 2 plies of glass fiber									
0700	Felt (type IV), 1 ply mineral surfaced selvage									
0800	Roofing, lapped 19", mopped	G-1	25	2.240	Sq.	130	96.50	22.50	249	335
0900	On nailable decks	"	24	2.333	"	119	101	23.50	243.50	335
1000	3 plies glass fiber felt (type III), 1 ply mineral surfaced									
1100	Selvage roofing, lapped 19", mopped	G-1	25	2.240	Sq.	121	96.50	22.50	240	325

07 61 Sheet Metal Roofing

07 61 13 – Standing Seam Sheet Metal Roofing

07 61 13.10 Standing Seam Sheet Metal Roofing, Field Fab.		Crew	Daily Output	Labor-Hours	Unit	Material	Labor	Equipment	Total	Total Incl O&P
0010	STANDING SEAM SHEET METAL ROOFING, FIELD FABRICATED									
0400	Copper standing seam roofing, over 10 squares, 16 oz., 125 lb./sq.	1 Shee	1.30	6.154	Sq.	1,025	385		1,410	1,725
0600	18 oz., 140 lb./sq.		1.20	6.667		1,150	415		1,565	1,900
0700	20 oz., 150 lb./sq.	↓	1.10	7.273		1,300	455		1,755	2,150
1200	For abnormal conditions or small areas, add					25%	100%			
1300	For lead-coated copper, add				↓	25%				
9000	Minimum labor/equipment charge	1 Shee	2	4	Job		249		249	390

07 61 16 – Batten Seam Sheet Metal Roofing

07 61 16.10 Batten Seam Sheet Metal Roofing, Field Fab.		Crew	Daily Output	Labor-Hours	Unit	Material	Labor	Equipment	Total	Total Incl O&P
0010	BATTEN SEAM SHEET METAL ROOFING, FIELD FABRICATED									
0012	Copper batten seam roofing, over 10 sq., 16 oz., 130 lb./sq.	1 Shee	1.10	7.273	Sq.	1,300	455		1,755	2,125
0020	Lead batten seam roofing, 5 lb./S.F.		1.20	6.667		1,700	415		2,115	2,525
0100	Zinc/copper alloy batten seam roofing, .020" thick		1.20	6.667		1,550	415		1,965	2,350
0200	Copper roofing, batten seam, over 10 sq., 18 oz., 145 lb./sq.		1	8		1,450	500		1,950	2,375
0300	20 oz., 160 lb./sq.		1	8		1,650	500		2,150	2,600
0500	Stainless steel batten seam roofing, type 304, 28 ga.		1.20	6.667		680	415		1,095	1,400
0600	26 ga.		1.15	6.957		770	435		1,205	1,525
0800	Zinc, copper alloy roofing, batten seam, .027" thick		1.15	6.957		1,900	435		2,335	2,775
0900	.032" thick		1.10	7.273		2,025	455		2,480	2,925
1000	.040" thick		1.05	7.619	↓	2,675	475		3,150	3,700
9000	Minimum labor/equipment charge	↓	2	4	Job		249		249	390

07 61 Sheet Metal Roofing

07 61 19 – Flat Seam Sheet Metal Roofing

07 61 19.10 Flat Seam Sheet Metal Roofing, Field Fabricated		Crew	Daily Output	Labor-Hours	Unit	Material	2020 Bare Costs Labor	Equipment	Total	Total Incl O&P
0010	**FLAT SEAM SHEET METAL ROOFING, FIELD FABRICATED**									
0900	Copper flat seam roofing, over 10 squares, 16 oz., 115 lb./sq.	1 Shee	1.20	6.667	Sq.	945	415		1,360	1,700
0950	18 oz., 130 lb./sq.		1.15	6.957		1,050	435		1,485	1,850
1000	20 oz., 145 lb./sq.		1.10	7.273		1,225	455		1,680	2,025
1008	Zinc flat seam roofing, .020" thick		1.20	6.667		1,325	415		1,740	2,100
1010	.027" thick		1.15	6.957		1,625	435		2,060	2,450
1020	.032" thick		1.12	7.143		1,725	445		2,170	2,600
1030	.040" thick		1.05	7.619		2,300	475		2,775	3,275
1100	Lead flat seam roofing, 5 lb./S.F.		1.30	6.154		1,450	385		1,835	2,200
9000	Minimum labor/equipment charge		2.75	2.909	Job		181		181	284

07 62 Sheet Metal Flashing and Trim

07 62 10 – Sheet Metal Trim

07 62 10.10 Sheet Metal Cladding

		Crew	Daily Output	Labor-Hours	Unit	Material	Labor	Equipment	Total	Total Incl O&P
0010	**SHEET METAL CLADDING**									
0100	Aluminum, up to 6 bends, .032" thick, window casing	1 Carp	180	.044	S.F.	1.95	2.36		4.31	5.95
0200	Window sill		72	.111	L.F.	1.95	5.90		7.85	11.60
0300	Door casing		180	.044	S.F.	1.95	2.36		4.31	5.95
0400	Fascia		250	.032		1.95	1.70		3.65	4.87
0500	Rake trim		225	.036		1.95	1.89		3.84	5.20
0700	.024" thick, window casing		180	.044		1.47	2.36		3.83	5.40
0800	Window sill		72	.111	L.F.	1.47	5.90		7.37	11.05
0900	Door casing		180	.044	S.F.	1.47	2.36		3.83	5.40
1000	Fascia		250	.032		1.47	1.70		3.17	4.34
1100	Rake trim		225	.036		1.47	1.89		3.36	4.65
1200	Vinyl coated aluminum, up to 6 bends, window casing		180	.044		1.94	2.36		4.30	5.90
1300	Window sill		72	.111	L.F.	1.94	5.90		7.84	11.60
1400	Door casing		180	.044	S.F.	1.94	2.36		4.30	5.90
1500	Fascia		250	.032		1.94	1.70		3.64	4.85
1600	Rake trim		225	.036		1.94	1.89		3.83	5.15

07 65 Flexible Flashing

07 65 10 – Sheet Metal Flashing

07 65 10.10 Sheet Metal Flashing and Counter Flashing

		Crew	Daily Output	Labor-Hours	Unit	Material	Labor	Equipment	Total	Total Incl O&P
0010	**SHEET METAL FLASHING AND COUNTER FLASHING**									
0011	Including up to 4 bends									
0020	Aluminum, mill finish, .013" thick	1 Rofc	145	.055	S.F.	1.01	2.55		3.56	5.55
0030	.016" thick		145	.055		1.15	2.55		3.70	5.75
0060	.019" thick		145	.055		1.50	2.55		4.05	6.10
0100	.032" thick		145	.055		1.38	2.55		3.93	6
0200	.040" thick		145	.055		2.33	2.55		4.88	7
0300	.050" thick		145	.055		3.22	2.55		5.77	8
0325	Mill finish 5" x 7" step flashing, .016" thick		1920	.004	Ea.	.15	.19		.34	.51
0350	Mill finish 12" x 12" step flashing, .016" thick		1600	.005	"	.60	.23		.83	1.06
0400	Painted finish, add				S.F.	.34			.34	.37
1000	Mastic-coated 2 sides, .005" thick	1 Rofc	330	.024		1.89	1.12		3.01	4.04
1100	.016" thick		330	.024		2.09	1.12		3.21	4.26
1600	Copper, 16 oz. sheets, under 1000 lb.		115	.070		8.15	3.21		11.36	14.60
1700	Over 4000 lb.		155	.052		8.15	2.38		10.53	13.15

07 65 Flexible Flashing

07 65 10 – Sheet Metal Flashing

07 65 10.10 Sheet Metal Flashing and Counter Flashing		Crew	Daily Output	Labor-Hours	Unit	2020 Bare Costs Material	Labor	Equipment	Total	Total Incl O&P
1900	20 oz. sheets, under 1000 lb.	1 Rofc	110	.073	S.F.	10.45	3.36		13.81	17.40
2000	Over 4000 lb.		145	.055		9.95	2.55		12.50	15.40
2200	24 oz. sheets, under 1000 lb.		105	.076		15.30	3.52		18.82	23
2300	Over 4000 lb.		135	.059		14.55	2.74		17.29	21
2500	32 oz. sheets, under 1000 lb.		100	.080		21	3.70		24.70	29.50
2600	Over 4000 lb.		130	.062		19.75	2.84		22.59	26.50
2700	W shape for valleys, 16 oz., 24" wide		100	.080	L.F.	16.65	3.70		20.35	25
5800	Lead, 2.5 lb./S.F., up to 12" wide		135	.059	S.F.	6.10	2.74		8.84	11.50
5900	Over 12" wide		135	.059		4.05	2.74		6.79	9.25
8900	Stainless steel sheets, 32 ga.		155	.052		3.51	2.38		5.89	8.05
9000	28 ga.		155	.052		5	2.38		7.38	9.65
9100	26 ga.		155	.052		5.10	2.38		7.48	9.75
9200	24 ga.		155	.052		5.85	2.38		8.23	10.55
9290	For mechanically keyed flashing, add					40%				
9320	Steel sheets, galvanized, 20 ga.	1 Rofc	130	.062	S.F.	1.22	2.84		4.06	6.30
9322	22 ga.		135	.059		1.27	2.74		4.01	6.20
9324	24 ga.		140	.057		.97	2.64		3.61	5.70
9326	26 ga.		148	.054		.85	2.50		3.35	5.30
9328	28 ga.		155	.052		.73	2.38		3.11	4.98
9340	30 ga.		160	.050		.62	2.31		2.93	4.72
9400	Terne coated stainless steel, .015" thick, 28 ga.		155	.052		8.20	2.38		10.58	13.15
9500	.018" thick, 26 ga.		155	.052		9.15	2.38		11.53	14.20
9600	Zinc and copper alloy (brass), .020" thick		155	.052		10.10	2.38		12.48	15.25
9700	.027" thick		155	.052		12.25	2.38		14.63	17.65
9800	.032" thick		155	.052		15.45	2.38		17.83	21
9900	.040" thick		155	.052		20.50	2.38		22.88	26.50
9950	Minimum labor/equipment charge		3	2.667	Job		123		123	216

07 65 12 – Fabric and Mastic Flashings

07 65 12.10 Fabric and Mastic Flashing and Counter Flashing		Crew	Daily Output	Labor-Hours	Unit	Material	Labor	Equipment	Total	Total Incl O&P
0010	**FABRIC AND MASTIC FLASHING AND COUNTER FLASHING**									
1300	Asphalt flashing cement, 5 gallon				Gal.	11.35			11.35	12.50
4900	Fabric, asphalt-saturated cotton, specification grade	1 Rofc	35	.229	S.Y.	3.26	10.55		13.81	22
5000	Utility grade		35	.229		1.55	10.55		12.10	20
5300	Close-mesh fabric, saturated, 17 oz./S.Y.		35	.229		2.21	10.55		12.76	21
5500	Fiberglass, resin-coated		35	.229		1.24	10.55		11.79	19.85
8500	Shower pan, bituminous membrane, 7 oz.		155	.052	S.F.	1.95	2.38		4.33	6.30

07 65 13 – Laminated Sheet Flashing

07 65 13.10 Laminated Sheet Flashing		Crew	Daily Output	Labor-Hours	Unit	Material	Labor	Equipment	Total	Total Incl O&P
0010	**LAMINATED SHEET FLASHING**, Including up to 4 bends									
0500	Aluminum, fabric-backed 2 sides, mill finish, .004" thick	1 Rofc	330	.024	S.F.	1.58	1.12		2.70	3.70
0700	.005" thick		330	.024		1.88	1.12		3	4.03
0750	Mastic-backed, self adhesive		460	.017		3.41	.80		4.21	5.15
0800	Mastic-coated 2 sides, .004" thick		330	.024		1.58	1.12		2.70	3.70
2800	Copper, paperbacked 1 side, 2 oz.		330	.024		2.16	1.12		3.28	4.34
2900	3 oz.		330	.024		3.15	1.12		4.27	5.45
3100	Paperbacked 2 sides, 2 oz.		330	.024		2.39	1.12		3.51	4.59
3150	3 oz.		330	.024		2.24	1.12		3.36	4.42
3200	5 oz.		330	.024		3.65	1.12		4.77	6
3250	7 oz.		330	.024		6.85	1.12		7.97	9.45
3400	Mastic-backed 2 sides, copper, 2 oz.		330	.024		2.15	1.12		3.27	4.33
3500	3 oz.		330	.024		2.57	1.12		3.69	4.79
3700	5 oz.		330	.024		3.90	1.12		5.02	6.25

07 65 Flexible Flashing

07 65 13 – Laminated Sheet Flashing

07 65 13.10 Laminated Sheet Flashing		Crew	Daily Output	Labor-Hours	Unit	Material	2020 Bare Costs Labor	Equipment	Total	Total Incl O&P
3800	Fabric-backed 2 sides, copper, 2 oz.	1 Rofc	330	.024	S.F.	2.05	1.12		3.17	4.22
4000	3 oz.		330	.024		2.88	1.12		4	5.15
4100	5 oz.		330	.024		4.03	1.12		5.15	6.40
4300	Copper-clad stainless steel, .015" thick, under 500 lb.		115	.070		7.05	3.21		10.26	13.35
4400	Over 2000 lb.		155	.052		7.05	2.38		9.43	11.90
4600	.018" thick, under 500 lb.		100	.080		7.95	3.70		11.65	15.20
4700	Over 2000 lb.		145	.055		8.10	2.55		10.65	13.35
8550	Shower pan, 3 ply copper and fabric, 3 oz.		155	.052		4.12	2.38		6.50	8.70
8600	7 oz.		155	.052		5.45	2.38		7.83	10.10
9300	Stainless steel, paperbacked 2 sides, .005" thick	↓	330	.024	↓	4.04	1.12		5.16	6.40

07 65 19 – Plastic Sheet Flashing

07 65 19.10 Plastic Sheet Flashing and Counter Flashing		Crew	Daily Output	Labor-Hours	Unit	Material	Labor	Equipment	Total	Total Incl O&P
0010	PLASTIC SHEET FLASHING AND COUNTER FLASHING									
7300	Polyvinyl chloride, black, 10 mil	1 Rofc	285	.028	S.F.	.27	1.30		1.57	2.57
7400	20 mil		285	.028		.26	1.30		1.56	2.56
7600	30 mil		285	.028		.33	1.30		1.63	2.63
7700	60 mil		285	.028		.88	1.30		2.18	3.24
7900	Black or white for exposed roofs, 60 mil	↓	285	.028	↓	1.11	1.30		2.41	3.49
8060	PVC tape, 5" x 45 mils, for joint covers, 100 L.F./roll				Ea.	183			183	201
8850	Polyvinyl chloride, 30 mil	1 Rofc	160	.050	S.F.	1.47	2.31		3.78	5.65

07 65 23 – Rubber Sheet Flashing

07 65 23.10 Rubber Sheet Flashing and Counter Flashing		Crew	Daily Output	Labor-Hours	Unit	Material	Labor	Equipment	Total	Total Incl O&P
0010	RUBBER SHEET FLASHING AND COUNTER FLASHING									
4810	EPDM 90 mils, 1" diameter pipe flashing	1 Rofc	32	.250	Ea.	21	11.55		32.55	43
4820	2" diameter		30	.267		21	12.30		33.30	44.50
4830	3" diameter		28	.286		24	13.20		37.20	49.50
4840	4" diameter		24	.333		29.50	15.40		44.90	59.50
4850	6" diameter		22	.364	↓	29.50	16.80		46.30	61.50
8100	Rubber, butyl, 1/32" thick		285	.028	S.F.	2.56	1.30		3.86	5.10
8200	1/16" thick		285	.028		3.83	1.30		5.13	6.50
8300	Neoprene, cured, 1/16" thick		285	.028		2.77	1.30		4.07	5.30
8400	1/8" thick	↓	285	.028	↓	6.15	1.30		7.45	9.05

07 65 26 – Self-Adhering Sheet Flashing

07 65 26.10 Self-Adhering Sheet or Roll Flashing		Crew	Daily Output	Labor-Hours	Unit	Material	Labor	Equipment	Total	Total Incl O&P
0010	SELF-ADHERING SHEET OR ROLL FLASHING									
0020	Self-adhered flashing, 25 mil cross laminated HDPE, 4" wide	1 Rofc	960	.008	L.F.	.20	.38		.58	.89
0040	6" wide		896	.009		.31	.41		.72	1.06
0060	9" wide		832	.010		.46	.44		.90	1.28
0080	12" wide	↓	768	.010	↓	.61	.48		1.09	1.51

07 71 Roof Specialties

07 71 16 – Manufactured Counterflashing Systems

07 71 16.10 Roof Drain Boot		Crew	Daily Output	Labor-Hours	Unit	Material	Labor	Equipment	Total	Total Incl O&P
						2020 Bare Costs				
0010	**ROOF DRAIN BOOT**									
0100	Cast iron, 4" diameter	1 Shee	125	.064	L.F.	46	3.99		49.99	57
0300	4" x 3"		125	.064		72.50	3.99		76.49	86.50
0400	5" x 4"		125	.064		135	3.99		138.99	155

07 71 16.20 Pitch Pockets, Variable Sizes		Crew	Daily Output	Labor-Hours	Unit	Material	Labor	Equipment	Total	Total Incl O&P
0010	**PITCH POCKETS, VARIABLE SIZES**									
0100	Adjustable, 4" to 7", welded corners, 4" deep	1 Rofc	48	.167	Ea.	21.50	7.70		29.20	37.50
0200	Side extenders, 6"	"	240	.033	"	3.34	1.54		4.88	6.35

07 71 19 – Manufactured Gravel Stops and Fasciae

07 71 19.10 Gravel Stop		Crew	Daily Output	Labor-Hours	Unit	Material	Labor	Equipment	Total	Total Incl O&P
0010	**GRAVEL STOP**									
0020	Aluminum, .050" thick, 4" face height, mill finish	1 Shee	145	.055	L.F.	6.95	3.44		10.39	13.05
0080	Duranodic finish		145	.055		8.25	3.44		11.69	14.45
0100	Painted		145	.055		7.60	3.44		11.04	13.75
0300	6" face height		135	.059		7.45	3.69		11.14	13.95
0350	Duranodic finish		135	.059		8.25	3.69		11.94	14.85
0400	Painted		135	.059		8.85	3.69		12.54	15.55
0600	8" face height		125	.064		7.95	3.99		11.94	14.95
0650	Duranodic finish		125	.064		9.40	3.99		13.39	16.60
0700	Painted		125	.064		9.60	3.99		13.59	16.80
0900	12" face height, .080" thick, 2 piece		100	.080		11	4.98		15.98	19.90
0950	Duranodic finish		100	.080		10.95	4.98		15.93	19.85
1000	Painted		100	.080		14.45	4.98		19.43	23.50
1200	Copper, 16 oz., 3" face height		145	.055		28.50	3.44		31.94	37
1300	6" face height		135	.059		37	3.69		40.69	46.50
1350	Galv steel, 24 ga., 4" leg, plain, with continuous cleat, 4" face		145	.055		6.60	3.44		10.04	12.70
1360	6" face height		145	.055		6.95	3.44		10.39	13.05
1500	Polyvinyl chloride, 6" face height		135	.059		5.85	3.69		9.54	12.20
1600	9" face height		125	.064		7.05	3.99		11.04	14
1800	Stainless steel, 24 ga., 6" face height		135	.059		17	3.69		20.69	24.50
1900	12" face height		100	.080		22.50	4.98		27.48	33
2100	20 ga., 6" face height		135	.059		22.50	3.69		26.19	30.50
2200	12" face height		100	.080		27	4.98		31.98	37.50
9000	Minimum labor/equipment charge		3.50	2.286	Job		142		142	223

07 71 19.30 Fascia		Crew	Daily Output	Labor-Hours	Unit	Material	Labor	Equipment	Total	Total Incl O&P
0010	**FASCIA**									
0100	Aluminum, reverse board and batten, .032" thick, colored, no furring incl.	1 Shee	145	.055	S.F.	7.20	3.44		10.64	13.35
0300	Steel, galv and enameled, stock, no furring, long panels		145	.055		5.65	3.44		9.09	11.60
0600	Short panels		115	.070		5.35	4.33		9.68	12.70
9000	Minimum labor/equipment charge		4	2	Job		125		125	195

07 71 23 – Manufactured Gutters and Downspouts

07 71 23.10 Downspouts		Crew	Daily Output	Labor-Hours	Unit	Material	Labor	Equipment	Total	Total Incl O&P
0010	**DOWNSPOUTS**									
0020	Aluminum, embossed, .020" thick, 2" x 3"	1 Shee	190	.042	L.F.	.95	2.62		3.57	5.15
0100	Enameled		190	.042		1.37	2.62		3.99	5.60
0300	.024" thick, 2" x 3"		180	.044		2.17	2.77		4.94	6.75
0400	3" x 4"		140	.057		2.29	3.56		5.85	8.10
0600	Round, corrugated aluminum, 3" diameter, .020" thick		190	.042		2.13	2.62		4.75	6.45
0700	4" diameter, .025" thick		140	.057		3.21	3.56		6.77	9.15
0900	Wire strainer, round, 2" diameter		155	.052	Ea.	2.05	3.22		5.27	7.30
1000	4" diameter		155	.052		2.66	3.22		5.88	8

07 71 Roof Specialties

07 71 23 – Manufactured Gutters and Downspouts

07 71 23.10 Downspouts		Crew	Daily Output	Labor-Hours	Unit	Material	Labor	Equipment	Total	Total Incl O&P
						2020 Bare Costs				
1200	Rectangular, perforated, 2″ x 3″	1 Shee	145	.055	Ea.	2.36	3.44		5.80	8
1300	3″ x 4″		145	.055		3.42	3.44		6.86	9.15
1500	Copper, round, 16 oz., stock, 2″ diameter		190	.042	L.F.	8	2.62		10.62	12.90
1600	3″ diameter		190	.042		9	2.62		11.62	14
1800	4″ diameter		145	.055		10.65	3.44		14.09	17.15
1900	5″ diameter		130	.062		14.85	3.83		18.68	22.50
2100	Rectangular, corrugated copper, stock, 2″ x 3″		190	.042		9.20	2.62		11.82	14.25
2200	3″ x 4″		145	.055		9.40	3.44		12.84	15.70
2400	Rectangular, plain copper, stock, 2″ x 3″		190	.042		13.25	2.62		15.87	18.70
2500	3″ x 4″		145	.055		13.85	3.44		17.29	20.50
2700	Wire strainers, rectangular, 2″ x 3″		145	.055	Ea.	17.15	3.44		20.59	24.50
2800	3″ x 4″		145	.055		20	3.44		23.44	27.50
3000	Round, 2″ diameter		145	.055		7.65	3.44		11.09	13.85
3100	3″ diameter		145	.055		10.40	3.44		13.84	16.85
3300	4″ diameter		145	.055		12	3.44		15.44	18.60
3400	5″ diameter		115	.070		23	4.33		27.33	32
3600	Lead-coated copper, round, stock, 2″ diameter		190	.042	L.F.	25.50	2.62		28.12	32
3700	3″ diameter		190	.042		24.50	2.62		27.12	31
3900	4″ diameter		145	.055		27.50	3.44		30.94	36
4000	5″ diameter, corrugated		130	.062		26	3.83		29.83	34.50
4200	6″ diameter, corrugated		105	.076		32	4.75		36.75	42.50
4300	Rectangular, corrugated, stock, 2″ x 3″		190	.042		15.25	2.62		17.87	21
4500	Plain, stock, 2″ x 3″		190	.042		27.50	2.62		30.12	34.50
4600	3″ x 4″		145	.055		40.50	3.44		43.94	50
4800	Steel, galvanized, round, corrugated, 2″ or 3″ diameter, 28 ga.		190	.042		2.30	2.62		4.92	6.65
4900	4″ diameter, 28 ga.		145	.055		2.19	3.44		5.63	7.80
5100	5″ diameter, 26 ga.		130	.062		3.86	3.83		7.69	10.25
5400	6″ diameter, 28 ga.		105	.076		3.72	4.75		8.47	11.55
5500	26 ga.		105	.076		4.08	4.75		8.83	11.95
5700	Rectangular, corrugated, 28 ga., 2″ x 3″		190	.042		2.38	2.62		5	6.75
5800	3″ x 4″		145	.055		1.96	3.44		5.40	7.55
6000	Rectangular, plain, 28 ga., galvanized, 2″ x 3″		190	.042		3.98	2.62		6.60	8.50
6100	3″ x 4″		145	.055		4.83	3.44		8.27	10.70
6300	Epoxy painted, 24 ga., corrugated, 2″ x 3″		190	.042		2.53	2.62		5.15	6.90
6400	3″ x 4″		145	.055		2.99	3.44		6.43	8.70
6600	Wire strainers, rectangular, 2″ x 3″		145	.055	Ea.	20.50	3.44		23.94	28
6700	3″ x 4″		145	.055		19.30	3.44		22.74	26.50
6900	Round strainers, 2″ or 3″ diameter		145	.055		4.50	3.44		7.94	10.35
7000	4″ diameter		145	.055		6.55	3.44		9.99	12.60
7800	Stainless steel tubing, schedule 5, 2″ x 3″ or 3″ diameter		190	.042	L.F.	69	2.62		71.62	79.50
7900	3″ x 4″ or 4″ diameter		145	.055		125	3.44		128.44	142
8100	4″ x 5″ or 5″ diameter		135	.059		152	3.69		155.69	173
8200	Vinyl, rectangular, 2″ x 3″		210	.038		2.17	2.37		4.54	6.10
8300	Round, 2-1/2″		220	.036		1.41	2.27		3.68	5.10
9000	Minimum labor/equipment charge		4	2	Job		125		125	195

07 71 23.20 Downspout Elbows		Crew	Daily Output	Labor-Hours	Unit	Material	Labor	Equipment	Total	Total Incl O&P
0010	**DOWNSPOUT ELBOWS**									
0020	Aluminum, embossed, 2″ x 3″, .020″ thick	1 Shee	100	.080	Ea.	1	4.98		5.98	8.90
0100	Enameled		100	.080		2.45	4.98		7.43	10.50
0200	Embossed, 3″ x 4″, .025″ thick		100	.080		2.96	4.98		7.94	11.05
0300	Enameled		100	.080		3.81	4.98		8.79	12
0400	Embossed, corrugated, 3″ diameter, .020″ thick		100	.080		3.15	4.98		8.13	11.25

07 71 23.20 Downspout Elbows		Crew	Daily Output	Labor-Hours	Unit	Material	Labor	Equipment	Total	Total Incl O&P
						2020 Bare Costs				
0500	4" diameter, .025" thick	1 Shee	100	.080	Ea.	7	4.98		11.98	15.50
0600	Copper, 16 oz., 2" diameter		100	.080		10.90	4.98		15.88	19.80
0700	3" diameter		100	.080		10.30	4.98		15.28	19.15
0800	4" diameter		100	.080		12.95	4.98		17.93	22
1000	Rectangular, 2" x 3" corrugated		100	.080		10.20	4.98		15.18	19.05
1100	3" x 4" corrugated		100	.080		15.15	4.98		20.13	24.50
1300	Vinyl, 2-1/2" diameter, 45 or 75 degree bend		100	.080		4.04	4.98		9.02	12.25
1400	Tee Y junction		75	.107		13.10	6.65		19.75	25
9000	Minimum labor/equipment charge		4	2	Job		125		125	195

07 71 23.30 Gutters		Crew	Daily Output	Labor-Hours	Unit	Material	Labor	Equipment	Total	Total Incl O&P
0010	**GUTTERS**									
0012	Aluminum, stock units, 5" K type, .027" thick, plain	1 Shee	125	.064	L.F.	2.85	3.99		6.84	9.40
0100	Enameled		125	.064		2.94	3.99		6.93	9.50
0300	5" K type, .032" thick, plain		125	.064		3.79	3.99		7.78	10.40
0400	Enameled		125	.064		3.58	3.99		7.57	10.20
0700	Copper, half round, 16 oz., stock units, 4" wide		125	.064		8.60	3.99		12.59	15.75
0900	5" wide		125	.064		7.65	3.99		11.64	14.70
1000	6" wide		118	.068		10.90	4.22		15.12	18.55
1200	K type, 16 oz., stock, 5" wide		125	.064		8.75	3.99		12.74	15.85
1300	6" wide		125	.064		8.45	3.99		12.44	15.55
1500	Lead coated copper, 16 oz., half round, stock, 4" wide		125	.064		17.55	3.99		21.54	25.50
1600	6" wide		118	.068		20	4.22		24.22	28.50
1800	K type, stock, 5" wide		125	.064		18.55	3.99		22.54	27
1900	6" wide		125	.064		21	3.99		24.99	29.50
2100	Copper clad stainless steel, K type, 5" wide		125	.064		7.60	3.99		11.59	14.60
2200	6" wide		125	.064		10.30	3.99		14.29	17.60
2400	Steel, galv, half round or box, 28 ga., 5" wide, plain		125	.064		2.40	3.99		6.39	8.90
2500	Enameled		125	.064		2.24	3.99		6.23	8.70
2700	26 ga., stock, 5" wide		125	.064		2.66	3.99		6.65	9.20
2800	6" wide		125	.064		2.64	3.99		6.63	9.15
3000	Vinyl, O.G., 4" wide	1 Carp	115	.070		1.39	3.70		5.09	7.45
3100	5" wide		115	.070		1.51	3.70		5.21	7.55
3200	4" half round, stock units		115	.070		1.42	3.70		5.12	7.45
3250	Joint connectors				Ea.	3.10			3.10	3.41
3300	Wood, clear treated cedar, fir or hemlock, 3" x 4"	1 Carp	100	.080	L.F.	12.70	4.25		16.95	21
3400	4" x 5"	"	100	.080	"	24.50	4.25		28.75	34
5000	Accessories, end cap, K type, aluminum 5"	1 Shee	625	.013	Ea.	.74	.80		1.54	2.06
5010	6"		625	.013		1.49	.80		2.29	2.89
5020	Copper, 5"		625	.013		3.41	.80		4.21	5
5030	6"		625	.013		3.59	.80		4.39	5.20
5040	Lead coated copper, 5"		625	.013		13.20	.80		14	15.75
5050	6"		625	.013		14.10	.80		14.90	16.75
5060	Copper clad stainless steel, 5"		625	.013		3.58	.80		4.38	5.20
5070	6"		625	.013		3.58	.80		4.38	5.20
5080	Galvanized steel, 5"		625	.013		1.57	.80		2.37	2.98
5090	6"		625	.013		2.72	.80		3.52	4.24
5100	Vinyl, 4"	1 Carp	625	.013		6.50	.68		7.18	8.25
5110	5"	"	625	.013		6.85	.68		7.53	8.65
5120	Half round, copper, 4"	1 Shee	625	.013		4.65	.80		5.45	6.35
5130	5"		625	.013		5.25	.80		6.05	7
5140	6"		625	.013		8.50	.80		9.30	10.60
5150	Lead coated copper, 5"		625	.013		14.85	.80		15.65	17.60

07 71 Roof Specialties

07 71 23 – Manufactured Gutters and Downspouts

07 71 23.30 Gutters		Crew	Daily Output	Labor-Hours	Unit	Material	2020 Bare Costs Labor	Equipment	Total	Total Incl O&P
5160	6"	1 Shee	625	.013	Ea.	22	.80		22.80	26
5170	Copper clad stainless steel, 5"		625	.013		5	.80		5.80	6.75
5180	6"		625	.013		4.62	.80		5.42	6.35
5190	Galvanized steel, 5"		625	.013		2.80	.80		3.60	4.33
5200	6"		625	.013		3.47	.80		4.27	5.05
5210	Outlet, aluminum, 2" x 3"		420	.019		.72	1.19		1.91	2.65
5220	3" x 4"		420	.019		1.18	1.19		2.37	3.16
5230	2-3/8" round		420	.019		.66	1.19		1.85	2.59
5240	Copper, 2" x 3"		420	.019		7.80	1.19		8.99	10.45
5250	3" x 4"		420	.019		8.80	1.19		9.99	11.55
5260	2-3/8" round		420	.019		4.87	1.19		6.06	7.20
5270	Lead coated copper, 2" x 3"		420	.019		26	1.19		27.19	31
5280	3" x 4"		420	.019		30	1.19		31.19	35
5290	2-3/8" round		420	.019		26.50	1.19		27.69	31
5300	Copper clad stainless steel, 2" x 3"		420	.019		7.30	1.19		8.49	9.90
5310	3" x 4"		420	.019		8.35	1.19		9.54	11
5320	2-3/8" round		420	.019		4.87	1.19		6.06	7.20
5330	Galvanized steel, 2" x 3"		420	.019		3.81	1.19		5	6.05
5340	3" x 4"		420	.019		6.10	1.19		7.29	8.55
5350	2-3/8" round		420	.019		4.95	1.19		6.14	7.30
5360	K type mitres, aluminum		65	.123		4.94	7.65		12.59	17.45
5370	Copper		65	.123		17.45	7.65		25.10	31
5380	Lead coated copper		65	.123		56	7.65		63.65	73.50
5390	Copper clad stainless steel		65	.123		28	7.65		35.65	42.50
5400	Galvanized steel		65	.123		28	7.65		35.65	43
5420	Half round mitres, copper		65	.123		70.50	7.65		78.15	89.50
5430	Lead coated copper		65	.123		91.50	7.65		99.15	113
5440	Copper clad stainless steel		65	.123		57.50	7.65		65.15	75
5450	Galvanized steel		65	.123		33.50	7.65		41.15	49
5460	Vinyl mitres and outlets		65	.123		11	7.65		18.65	24
5470	Sealant		940	.009	L.F.	.01	.53		.54	.84
5480	Soldering		96	.083	"	.29	5.20		5.49	8.45
9000	Minimum labor/equipment charge		3.75	2.133	Job		133		133	208

07 71 23.35 Gutter Guard		Crew	Daily Output	Labor-Hours	Unit	Material	Labor	Equipment	Total	Total Incl O&P
0010	**GUTTER GUARD**									
0020	6" wide strip, aluminum mesh	1 Carp	500	.016	L.F.	2.46	.85		3.31	4.07
0100	Vinyl mesh		500	.016	"	2.89	.85		3.74	4.54
9000	Minimum labor/equipment charge		4	2	Job		106		106	170

07 71 26 – Reglets

07 71 26.10 Reglets and Accessories		Crew	Daily Output	Labor-Hours	Unit	Material	Labor	Equipment	Total	Total Incl O&P
0010	**REGLETS AND ACCESSORIES**									
0020	Reglet, aluminum, .025" thick, in parapet	1 Carp	225	.036	L.F.	1.84	1.89		3.73	5.05
0300	16 oz. copper		225	.036		6.55	1.89		8.44	10.25
0400	Galvanized steel, 24 ga.		225	.036		1.37	1.89		3.26	4.54
0600	Stainless steel, .020" thick		225	.036		4.21	1.89		6.10	7.65
0900	Counter flashing for above, 12" wide, .032" aluminum	1 Shee	150	.053		2.09	3.32		5.41	7.50
1200	16 oz. copper		150	.053		6.35	3.32		9.67	12.20
1300	Galvanized steel, 26 ga.		150	.053		1.33	3.32		4.65	6.65
1500	Stainless steel, .020" thick		150	.053		6.80	3.32		10.12	12.70
9000	Minimum labor/equipment charge	1 Carp	3	2.667	Job		142		142	227

07 71 Roof Specialties

07 71 29 – Manufactured Roof Expansion Joints

07 71 29.10 Expansion Joints		Crew	Daily Output	Labor-Hours	Unit	Material	Labor	Equipment	Total	Total Incl O&P
						2020 Bare Costs				
0010	**EXPANSION JOINTS**									
0300	Butyl or neoprene center with foam insulation, metal flanges									
0400	Aluminum, .032″ thick for openings to 2-1/2″	1 Rofc	165	.048	L.F.	12.85	2.24		15.09	18.05
0600	For joint openings to 3-1/2″		165	.048		12.40	2.24		14.64	17.55
0610	For joint openings to 5″		165	.048		14.75	2.24		16.99	20
0620	For joint openings to 8″		165	.048		16.70	2.24		18.94	22.50
0700	Copper, 16 oz. for openings to 2-1/2″		165	.048		21.50	2.24		23.74	28
0900	For joint openings to 3-1/2″		165	.048		19.70	2.24		21.94	25.50
0910	For joint openings to 5″		165	.048		22.50	2.24		24.74	29
0920	For joint openings to 8″		165	.048		25	2.24		27.24	31
1000	Galvanized steel, 26 ga. for openings to 2-1/2″		165	.048		10.80	2.24		13.04	15.80
1200	For joint openings to 3-1/2″		165	.048		10.55	2.24		12.79	15.50
1210	For joint openings to 5″		165	.048		14.10	2.24		16.34	19.45
1220	For joint openings to 8″		165	.048		15.20	2.24		17.44	20.50
1300	Lead-coated copper, 16 oz. for openings to 2-1/2″		165	.048		35.50	2.24		37.74	43
1500	For joint openings to 3-1/2″		165	.048		37.50	2.24		39.74	45
1600	Stainless steel, .018″, for openings to 2-1/2″		165	.048		13.35	2.24		15.59	18.55
1800	For joint openings to 3-1/2″		165	.048		14.15	2.24		16.39	19.45
1810	For joint openings to 5″		165	.048		15.25	2.24		17.49	20.50
1820	For joint openings to 8″		165	.048		23.50	2.24		25.74	29.50
1900	Neoprene, double-seal type with thick center, 4-1/2″ wide		125	.064		14.95	2.96		17.91	21.50
1950	Polyethylene bellows, with galv steel flat flanges		100	.080		7.20	3.70		10.90	14.35
1960	With galvanized angle flanges		100	.080		7.75	3.70		11.45	15
2000	Roof joint with extruded aluminum cover, 2″	1 Shee	115	.070		31.50	4.33		35.83	41.50
2100	Expansion, stainless flange, foam center, standard	1 Rofc	100	.080		13.80	3.70		17.50	21.50
2200	Large	″	100	.080		18.35	3.70		22.05	26.50
2500	Roof to wall joint with extruded aluminum cover	1 Shee	115	.070		30.50	4.33		34.83	40.50
2700	Wall joint, closed cell foam on PVC cover, 9″ wide	1 Rofc	125	.064		6.10	2.96		9.06	11.90
2800	12″ wide	″	115	.070		7	3.21		10.21	13.30
9000	Minimum labor/equipment charge	1 Shee	3	2.667	Job		166		166	260

07 71 43 – Drip Edge

07 71 43.10 Drip Edge, Rake Edge, Ice Belts		Crew	Daily Output	Labor-Hours	Unit	Material	Labor	Equipment	Total	Total Incl O&P
0010	**DRIP EDGE, RAKE EDGE, ICE BELTS**									
0020	Aluminum, .016″ thick, 5″ wide, mill finish	1 Carp	400	.020	L.F.	.59	1.06		1.65	2.35
0100	White finish		400	.020		.65	1.06		1.71	2.42
0200	8″ wide, mill finish		400	.020		1.48	1.06		2.54	3.33
0300	Ice belt, 28″ wide, mill finish		100	.080		8	4.25		12.25	15.60
0310	Vented, mill finish		400	.020		2.29	1.06		3.35	4.22
0320	Painted finish		400	.020		2.54	1.06		3.60	4.49
0400	Galvanized, 5″ wide		400	.020		.62	1.06		1.68	2.38
0500	8″ wide, mill finish		400	.020		.84	1.06		1.90	2.62
0510	Rake edge, aluminum, 1-1/2″ x 1-1/2″		400	.020		.35	1.06		1.41	2.09
0520	3-1/2″ x 1-1/2″		400	.020		.51	1.06		1.57	2.26
9000	Minimum labor/equipment charge		4	2	Job		106		106	170

07 72 Roof Accessories

07 72 23 – Relief Vents

07 72 23.10 Roof Vents		Crew	Daily Output	Labor-Hours	Unit	Material	2020 Bare Costs Labor	Equipment	Total	Total Incl O&P
0010	ROOF VENTS									
0020	Mushroom shape, for built-up roofs, aluminum	1 Rofc	30	.267	Ea.	74.50	12.30		86.80	104
0100	PVC, 6" high		30	.267	"	26	12.30		38.30	50
9000	Minimum labor/equipment charge		2.75	2.909	Job		134		134	235

07 72 23.20 Vents		Crew	Daily Output	Labor-Hours	Unit	Material	Labor	Equipment	Total	Total Incl O&P
0010	VENTS									
0100	Soffit or eave, aluminum, mill finish, strips, 2-1/2" wide	1 Carp	200	.040	L.F.	.49	2.13		2.62	3.95
0200	3" wide		200	.040		.47	2.13		2.60	3.93
0300	Enamel finish, 3" wide		200	.040		.54	2.13		2.67	4
0400	Mill finish, rectangular, 4" x 16"		72	.111	Ea.	1.53	5.90		7.43	11.15
0500	8" x 16"		72	.111		2.60	5.90		8.50	12.30
2420	Roof ventilator	Q-9	16	1		54	56		110	148
2500	Vent, roof vent	1 Rofc	24	.333		27	15.40		42.40	56.50

07 72 26 – Ridge Vents

07 72 26.10 Ridge Vents and Accessories		Crew	Daily Output	Labor-Hours	Unit	Material	Labor	Equipment	Total	Total Incl O&P
0010	RIDGE VENTS AND ACCESSORIES									
0100	Aluminum strips, mill finish	1 Rofc	160	.050	L.F.	2.67	2.31		4.98	7
0150	Painted finish		160	.050	"	4.23	2.31		6.54	8.70
0200	Connectors		48	.167	Ea.	5.20	7.70		12.90	19.20
0300	End caps		48	.167	"	2.49	7.70		10.19	16.25
0400	Galvanized strips		160	.050	L.F.	3.81	2.31		6.12	8.25
0430	Molded polyethylene, shingles not included		160	.050	"	2.85	2.31		5.16	7.20
0440	End plugs		48	.167	Ea.	2.49	7.70		10.19	16.25
0450	Flexible roll, shingles not included		160	.050	L.F.	2.54	2.31		4.85	6.85
2300	Ridge vent strip, mill finish	1 Shee	155	.052	"	3.95	3.22		7.17	9.40

07 72 33 – Roof Hatches

07 72 33.10 Roof Hatch Options		Crew	Daily Output	Labor-Hours	Unit	Material	Labor	Equipment	Total	Total Incl O&P
0010	ROOF HATCH OPTIONS									
0500	2'-6" x 3', aluminum curb and cover	G-3	10	3.200	Ea.	900	167		1,067	1,250
0520	Galvanized steel curb and aluminum cover		10	3.200		850	167		1,017	1,200
0540	Galvanized steel curb and cover		10	3.200		610	167		777	935
0600	2'-6" x 4'-6", aluminum curb and cover		9	3.556		1,150	186		1,336	1,575
0800	Galvanized steel curb and aluminum cover		9	3.556		1,025	186		1,211	1,425
0900	Galvanized steel curb and cover		9	3.556		995	186		1,181	1,400
1100	4' x 4' aluminum curb and cover		8	4		1,700	209		1,909	2,200
1120	Galvanized steel curb and aluminum cover		8	4		1,800	209		2,009	2,300
1140	Galvanized steel curb and cover		8	4		1,150	209		1,359	1,575
1200	2'-6" x 8'-0", aluminum curb and cover		6.60	4.848		2,175	253		2,428	2,800
1400	Galvanized steel curb and aluminum cover		6.60	4.848		1,900	253		2,153	2,500
1500	Galvanized steel curb and cover		6.60	4.848		1,200	253		1,453	1,725
1800	For plexiglass panels, 2'-6" x 3'-0", add to above					485			485	535
9000	Minimum labor/equipment charge	2 Carp	2	8	Job		425		425	680

07 72 36 – Smoke Vents

07 72 36.10 Smoke Hatches		Crew	Daily Output	Labor-Hours	Unit	Material	Labor	Equipment	Total	Total Incl O&P
0010	SMOKE HATCHES									
0200	For 3'-0" long, add to roof hatches from Section 07 72 33.10				Ea.	25%	5%			
0250	For 4'-0" long, add to roof hatches from Section 07 72 33.10					20%	5%			
0300	For 8'-0" long, add to roof hatches from Section 07 72 33.10					10%	5%			

07 72 Roof Accessories

07 72 36 – Smoke Vents

	07 72 36.20 Smoke Vent Options	Crew	Daily Output	Labor-Hours	Unit	Material	Labor	Equipment	Total	Total Incl O&P
						2020 Bare Costs				
0010	**SMOKE VENT OPTIONS**									
0100	4′ x 4′ aluminum cover and frame	G-3	13	2.462	Ea.	2,200	128		2,328	2,625
0200	Galvanized steel cover and frame		13	2.462		1,725	128		1,853	2,100
0300	4′ x 8′ aluminum cover and frame		8	4		3,275	209		3,484	3,925
0400	Galvanized steel cover and frame		8	4		2,750	209		2,959	3,350
9000	Minimum labor/equipment charge	2 Carp	2	8	Job		425		425	680

07 72 53 – Snow Guards

	07 72 53.10 Snow Guard Options	Crew	Daily Output	Labor-Hours	Unit	Material	Labor	Equipment	Total	Total Incl O&P
0010	**SNOW GUARD OPTIONS**									
0100	Slate & asphalt shingle roofs, fastened with nails	1 Rofc	160	.050	Ea.	12.45	2.31		14.76	17.75
0200	Standing seam metal roofs, fastened with set screws		48	.167		17.75	7.70		25.45	33
0300	Surface mount for metal roofs, fastened with solder		48	.167		7.75	7.70		15.45	22
0400	Double rail pipe type, including pipe		130	.062	L.F.	35	2.84		37.84	43.50

07 72 80 – Vent Options

	07 72 80.30 Vent Options	Crew	Daily Output	Labor-Hours	Unit	Material	Labor	Equipment	Total	Total Incl O&P
0010	**VENT OPTIONS**									
0020	Plastic, for insulated decks, 1 per M.S.F.	1 Rofc	40	.200	Ea.	24	9.25		33.25	42.50
0100	Heavy duty		20	.400		56.50	18.50		75	95
0300	Aluminum		30	.267		21.50	12.30		33.80	45
0800	Polystyrene baffles, 12″ wide for 16″ OC rafter spacing	1 Carp	90	.089		.53	4.72		5.25	8.15
0900	For 24″ OC rafter spacing		110	.073		.88	3.87		4.75	7.15
9000	Minimum labor/equipment charge		3	2.667	Job		142		142	227

07 76 Roof Pavers

07 76 16 – Roof Decking Pavers

	07 76 16.10 Roof Pavers and Supports	Crew	Daily Output	Labor-Hours	Unit	Material	Labor	Equipment	Total	Total Incl O&P
0010	**ROOF PAVERS AND SUPPORTS**									
1000	Roof decking pavers, concrete blocks, 2″ thick, natural	1 Clab	115	.070	S.F.	3.58	2.93		6.51	8.65
1100	Colors		115	.070	″	3.71	2.93		6.64	8.75
1200	Support pedestal, bottom cap		960	.008	Ea.	2.66	.35		3.01	3.49
1300	Top cap		960	.008		4.88	.35		5.23	5.90
1400	Leveling shims, 1/16″		1920	.004		1.22	.18		1.40	1.62
1500	1/8″		1920	.004		1.22	.18		1.40	1.62
1600	Buffer pad		960	.008		2.54	.35		2.89	3.35
1700	PVC legs (4″ SDR 35)		2880	.003	Inch	.14	.12		.26	.34
2000	Alternate pricing method, system in place		101	.079	S.F.	7.20	3.33		10.53	13.25

07 81 Applied Fireproofing

07 81 16 – Cementitious Fireproofing

	07 81 16.10 Sprayed Cementitious Fireproofing	Crew	Daily Output	Labor-Hours	Unit	Material	Labor	Equipment	Total	Total Incl O&P
0010	**SPRAYED CEMENTITIOUS FIREPROOFING**									
0050	Not including canvas protection, normal density									
0100	Per 1″ thick, on flat plate steel	G-2	3000	.008	S.F.	.59	.36	.06	1.01	1.29
0200	Flat decking		2400	.010		.59	.44	.08	1.11	1.45
0400	Beams		1500	.016		.59	.71	.13	1.43	1.92
0500	Corrugated or fluted decks		1250	.019		.88	.85	.15	1.88	2.50
0700	Columns, 1-1/8″ thick		1100	.022		.66	.97	.17	1.80	2.46
0800	2-3/16″ thick		700	.034		1.53	1.52	.27	3.32	4.41

07 81 Applied Fireproofing

07 81 16 – Cementitious Fireproofing

07 81 16.10 Sprayed Cementitious Fireproofing		Crew	Daily Output	Labor-Hours	Unit	Material	2020 Bare Costs Labor	Equipment	Total	Total Incl O&P
0900	For canvas protection, add	G-2	5000	.005	S.F.	.12	.21	.04	.37	.51
1000	Not including canvas protection, high density									
1100	Per 1" thick, on flat plate steel	G-2	3000	.008	S.F.	2.28	.36	.06	2.70	3.15
1110	On flat decking		2400	.010		2.28	.44	.08	2.80	3.31
1120	On beams		1500	.016		2.28	.71	.13	3.12	3.78
1130	Corrugated or fluted decks		1250	.019		2.28	.85	.15	3.28	4.04
1140	Columns, 1-1/8" thick		1100	.022		2.56	.97	.17	3.70	4.55
1150	2-3/16" thick		1100	.022		5.15	.97	.17	6.29	7.40
1170	For canvas protection, add		5000	.005		.12	.21	.04	.37	.51
1200	Not including canvas protection, retrofitting									
1210	Per 1" thick, on flat plate steel	G-2	1500	.016	S.F.	.52	.71	.13	1.36	1.84
1220	On flat decking		1200	.020		.52	.89	.16	1.57	2.16
1230	On beams		750	.032		.52	1.42	.25	2.19	3.12
1240	Corrugated or fluted decks		625	.038		.77	1.70	.30	2.77	3.90
1250	Columns, 1-1/8" thick		550	.044		.58	1.94	.34	2.86	4.11
1260	2-3/16" thick		500	.048		1.16	2.13	.38	3.67	5.10
1400	Accessories, preliminary spattered texture coat		4500	.005		.05	.24	.04	.33	.48
1410	Bonding agent	1 Plas	1000	.008		.10	.39		.49	.73
1500	Intumescent epoxy fireproofing on wire mesh, 3/16" thick									
1550	1 hour rating, exterior use	G-2	136	.176	S.F.	7.75	7.85	1.39	16.99	22.50
1551	2 hour rating, exterior use		136	.176		7.75	7.85	1.39	16.99	22.50
1600	Magnesium oxychloride, 35# to 40# density, 1/4" thick		3000	.008		1.65	.36	.06	2.07	2.46
1650	1/2" thick		2000	.012		3.36	.53	.09	3.98	4.65
1700	60# to 70# density, 1/4" thick		3000	.008		2.17	.36	.06	2.59	3.03
1750	1/2" thick		2000	.012		4.34	.53	.09	4.96	5.70
2000	Vermiculite cement, troweled or sprayed, 1/4" thick		3000	.008		1.49	.36	.06	1.91	2.28
2050	1/2" thick		2000	.012		2.99	.53	.09	3.61	4.24
9000	Minimum labor/equipment charge		3	8	Job		355	63	418	635

07 84 Firestopping

07 84 13 – Penetration Firestopping

07 84 13.10 Firestopping		Crew	Daily Output	Labor-Hours	Unit	Material	Labor	Equipment	Total	Total Incl O&P
0010	**FIRESTOPPING** R078413-30									
0100	Metallic piping, non insulated									
0110	Through walls, 2" diameter	1 Carp	16	.500	Ea.	3.67	26.50		30.17	46.50
0120	4" diameter		14	.571		8.20	30.50		38.70	57.50
0130	6" diameter		12	.667		12.60	35.50		48.10	71
0140	12" diameter		10	.800		25	42.50		67.50	95.50
0150	Through floors, 2" diameter		32	.250		1.86	13.30		15.16	23.50
0160	4" diameter		28	.286		3.28	15.20		18.48	28
0170	6" diameter		24	.333		6.45	17.70		24.15	35.50
0180	12" diameter		20	.400		12.80	21.50		34.30	48
0190	Metallic piping, insulated									
0200	Through walls, 2" diameter	1 Carp	16	.500	Ea.	8.60	26.50		35.10	52
0210	4" diameter		14	.571		12.55	30.50		43.05	62.50
0220	6" diameter		12	.667		16.75	35.50		52.25	75.50
0230	12" diameter		10	.800		28.50	42.50		71	99.50
0240	Through floors, 2" diameter		32	.250		3.27	13.30		16.57	25
0250	4" diameter		28	.286		6.30	15.20		21.50	31.50
0260	6" diameter		24	.333		8.70	17.70		26.40	38
0270	12" diameter		20	.400		12.95	21.50		34.45	48

07 84 Firestopping

07 84 13 – Penetration Firestopping

07 84 13.10 Firestopping		Crew	Daily Output	Labor-Hours	Unit	2020 Bare Costs Material	Labor	Equipment	Total	Total Incl O&P
0280	Non metallic piping, non insulated									
0290	Through walls, 2" diameter	1 Carp	12	.667	Ea.	52.50	35.50		88	115
0300	4" diameter		10	.800		101	42.50		143.50	179
0310	6" diameter		8	1		186	53		239	290
0330	Through floors, 2" diameter		16	.500		36.50	26.50		63	82.50
0340	4" diameter		6	1.333		62	71		133	183
0350	6" diameter		6	1.333		135	71		206	263
0370	Ductwork, insulated & non insulated, round									
0380	Through walls, 6" diameter	1 Carp	12	.667	Ea.	16.55	35.50		52.05	75
0390	12" diameter		10	.800		28.50	42.50		71	99.50
0400	18" diameter		8	1		33.50	53		86.50	122
0410	Through floors, 6" diameter		16	.500		8.50	26.50		35	52
0420	12" diameter		14	.571		14.60	30.50		45.10	64.50
0430	18" diameter		12	.667		17.60	35.50		53.10	76.50
0440	Ductwork, insulated & non insulated, rectangular									
0450	With stiffener/closure angle, through walls, 6" x 12"	1 Carp	8	1	Ea.	21	53		74	108
0460	12" x 24"		6	1.333		36	71		107	154
0470	24" x 48"		4	2		84	106		190	262
0480	With stiffener/closure angle, through floors, 6" x 12"		10	.800		10.40	42.50		52.90	79.50
0490	12" x 24"		8	1		20	53		73	107
0500	24" x 48"		6	1.333		42	71		113	160
0510	Multi trade openings									
0520	Through walls, 6" x 12"	1 Carp	2	4	Ea.	52	213		265	400
0530	12" x 24"	"	1	8		189	425		614	890
0540	24" x 48"	2 Carp	1	16		670	850		1,520	2,100
0550	48" x 96"	"	.75	21.333		2,550	1,125		3,675	4,625
0560	Through floors, 6" x 12"	1 Carp	2	4		44.50	213		257.50	390
0570	12" x 24"	"	1	8		69.50	425		494.50	755
0580	24" x 48"	2 Carp	.75	21.333		154	1,125		1,279	2,000
0590	48" x 96"	"	.50	32		365	1,700		2,065	3,125
0600	Structural penetrations, through walls									
0610	Steel beams, W8 x 10	1 Carp	8	1	Ea.	75	53		128	168
0620	W12 x 14		6	1.333		109	71		180	233
0630	W21 x 44		5	1.600		154	85		239	305
0640	W36 x 135		3	2.667		229	142		371	480
0650	Bar joists, 18" deep		6	1.333		42	71		113	161
0660	24" deep		6	1.333		59	71		130	179
0670	36" deep		5	1.600		88.50	85		173.50	234
0680	48" deep		4	2		120	106		226	300
0690	Construction joints, floor slab at exterior wall									
0700	Precast, brick, block or drywall exterior									
0710	2" wide joint	1 Carp	125	.064	L.F.	10.30	3.40		13.70	16.80
0720	4" wide joint	"	75	.107	"	16.60	5.65		22.25	27.50
0730	Metal panel, glass or curtain wall exterior									
0740	2" wide joint	1 Carp	40	.200	L.F.	12.45	10.65		23.10	31
0750	4" wide joint	"	25	.320	"	18.65	17		35.65	48
0760	Floor slab to drywall partition									
0770	Flat joint	1 Carp	100	.080	L.F.	14.55	4.25		18.80	23
0780	Fluted joint		50	.160		16.35	8.50		24.85	31.50
0790	Etched fluted joint		75	.107		20.50	5.65		26.15	31.50
0800	Floor slab to concrete/masonry partition									
0810	Flat joint	1 Carp	75	.107	L.F.	15.45	5.65		21.10	26
0820	Fluted joint	"	50	.160	"	17.65	8.50		26.15	33

07 84 Firestopping

07 84 13 – Penetration Firestopping

07 84 13.10 Firestopping		Crew	Daily Output	Labor-Hours	Unit	Material	2020 Bare Costs Labor	Equipment	Total	Total Incl O&P
0830	Concrete/CMU wall joints									
0840	1" wide	1 Carp	100	.080	L.F.	20.50	4.25		24.75	29.50
0850	2" wide		75	.107		30	5.65		35.65	42
0860	4" wide		50	.160		36	8.50		44.50	53
0870	Concrete/CMU floor joints									
0880	1" wide	1 Carp	200	.040	L.F.	20.50	2.13		22.63	26.50
0890	2" wide		150	.053		25	2.83		27.83	32
0900	4" wide		100	.080		36.50	4.25		40.75	47

07 91 Preformed Joint Seals

07 91 13 – Compression Seals

07 91 13.10 Compression Seals

		Crew	Daily Output	Labor-Hours	Unit	Material	Labor	Equipment	Total	Total Incl O&P
0010	**COMPRESSION SEALS**									
4900	O-ring type cord, 1/4"	1 Bric	472	.017	L.F.	.40	.88		1.28	1.87
4910	1/2"		440	.018		1.01	.95		1.96	2.64
4920	3/4"		424	.019		2	.98		2.98	3.79
4930	1"		408	.020		3.93	1.02		4.95	5.95
4940	1-1/4"		384	.021		9.40	1.08		10.48	12.10
4950	1-1/2"		368	.022		11.50	1.13		12.63	14.50
4960	1-3/4"		352	.023		13.50	1.18		14.68	16.75
4970	2"		344	.023		22.50	1.21		23.71	27

07 91 16 – Joint Gaskets

07 91 16.10 Joint Gaskets

		Crew	Daily Output	Labor-Hours	Unit	Material	Labor	Equipment	Total	Total Incl O&P
0010	**JOINT GASKETS**									
4400	Joint gaskets, neoprene, closed cell w/adh, 1/8" x 3/8"	1 Bric	240	.033	L.F.	.36	1.73		2.09	3.20
4500	1/4" x 3/4"		215	.037		.66	1.94		2.60	3.86
4700	1/2" x 1"		200	.040		1.71	2.08		3.79	5.25
4800	3/4" x 1-1/2"		165	.048		1.85	2.52		4.37	6.10

07 91 23 – Backer Rods

07 91 23.10 Backer Rods

		Crew	Daily Output	Labor-Hours	Unit	Material	Labor	Equipment	Total	Total Incl O&P
0010	**BACKER RODS**									
0032	Backer rod, polyethylene, 1/4" diameter	1 Bric	460	.017	L.F.	.03	.91		.94	1.49
0052	1/2" diameter		460	.017		.04	.91		.95	1.51
0072	3/4" diameter		460	.017		.07	.91		.98	1.53
0092	1" diameter		460	.017		.15	.91		1.06	1.62

07 91 26 – Joint Fillers

07 91 26.10 Joint Fillers

		Crew	Daily Output	Labor-Hours	Unit	Material	Labor	Equipment	Total	Total Incl O&P
0010	**JOINT FILLERS**									
4360	Butyl rubber filler, 1/4" x 1/4"	1 Bric	290	.028	L.F.	.23	1.44		1.67	2.57
4365	1/2" x 1/2"		250	.032		.92	1.67		2.59	3.70
4370	1/2" x 3/4"		210	.038		1.38	1.98		3.36	4.71
4375	3/4" x 3/4"		230	.035		2.06	1.81		3.87	5.20
4380	1" x 1"		180	.044		2.75	2.31		5.06	6.75
4390	For coloring, add					12%				
4980	Polyethylene joint backing, 1/4" x 2"	1 Bric	2.08	3.846	C.L.F.	14.25	200		214.25	340
4990	1/4" x 6"		1.28	6.250	"	32.50	325		357.50	560
5600	Silicone, room temp vulcanizing foam seal, 1/4" x 1/2"		1312	.006	L.F.	.46	.32		.78	1.01
5610	1/2" x 1/2"		656	.012		.92	.64		1.56	2.04
5620	1/2" x 3/4"		442	.018		1.37	.94		2.31	3.03

07 91 Preformed Joint Seals

07 91 26 – Joint Fillers

07 91 26.10 Joint Fillers		Crew	Daily Output	Labor-Hours	Unit	2020 Bare Costs Material	Labor	Equipment	Total	Total Incl O&P
5630	3/4" x 3/4"	1 Bric	328	.024	L.F.	2.06	1.27		3.33	4.32
5640	1/8" x 1"		1312	.006		.46	.32		.78	1.01
5650	1/8" x 3"		442	.018		1.37	.94		2.31	3.03
5670	1/4" x 3"		295	.027		2.75	1.41		4.16	5.30
5680	1/4" x 6"		148	.054		5.50	2.81		8.31	10.60
5690	1/2" x 6"		82	.098		11	5.10		16.10	20.50
5700	1/2" x 9"		52.50	.152		16.50	7.95		24.45	31
5710	1/2" x 12"		33	.242		22	12.60		34.60	44.50

07 92 Joint Sealants

07 92 13 – Elastomeric Joint Sealants

07 92 13.20 Caulking and Sealant Options

		Crew	Daily Output	Labor-Hours	Unit	Material	Labor	Equipment	Total	Total Incl O&P
0010	**CAULKING AND SEALANT OPTIONS**									
0050	Latex acrylic based, bulk				Gal.	31			31	34
0055	Bulk in place 1/4" x 1/4" bead	1 Bric	300	.027	L.F.	.10	1.39		1.49	2.35
0060	1/4" x 3/8"		294	.027		.16	1.42		1.58	2.47
0065	1/4" x 1/2"		288	.028		.22	1.45		1.67	2.58
0075	3/8" x 3/8"		284	.028		.24	1.47		1.71	2.64
0080	3/8" x 1/2"		280	.029		.32	1.49		1.81	2.76
0085	3/8" x 5/8"		276	.029		.40	1.51		1.91	2.89
0095	3/8" x 3/4"		272	.029		.48	1.53		2.01	3
0100	1/2" x 1/2"		275	.029		.43	1.51		1.94	2.93
0105	1/2" x 5/8"		269	.030		.54	1.55		2.09	3.09
0110	1/2" x 3/4"		263	.030		.65	1.58		2.23	3.27
0115	1/2" x 7/8"		256	.031		.76	1.63		2.39	3.46
0120	1/2" x 1"		250	.032		.87	1.67		2.54	3.64
0125	3/4" x 3/4"		244	.033		.97	1.71		2.68	3.83
0130	3/4" x 1"		225	.036		1.30	1.85		3.15	4.42
0135	1" x 1"		200	.040		1.73	2.08		3.81	5.25
0190	Cartridges				Gal.	33.50			33.50	37
0200	11 fl. oz. cartridge				Ea.	2.89			2.89	3.18
0500	1/4" x 1/2"	1 Bric	288	.028	L.F.	.24	1.45		1.69	2.60
0600	1/2" x 1/2"		275	.029		.47	1.51		1.98	2.97
0800	3/4" x 3/4"		244	.033		1.06	1.71		2.77	3.93
0900	3/4" x 1"		225	.036		1.42	1.85		3.27	4.55
1000	1" x 1"		200	.040		1.77	2.08		3.85	5.30
1400	Butyl based, bulk				Gal.	43.50			43.50	48
1500	Cartridges				"	43.50			43.50	48
1700	1/4" x 1/2", 154 L.F./gal.	1 Bric	288	.028	L.F.	.28	1.45		1.73	2.65
1800	1/2" x 1/2", 77 L.F./gal.	"	275	.029	"	.57	1.51		2.08	3.07
2300	Polysulfide compounds, 1 component, bulk				Gal.	88.50			88.50	97.50
2600	1 or 2 component, in place, 1/4" x 1/4", 308 L.F./gal.	1 Bric	300	.027	L.F.	.29	1.39		1.68	2.56
2700	1/2" x 1/4", 154 L.F./gal.		288	.028		.57	1.45		2.02	2.97
2900	3/4" x 3/8", 68 L.F./gal.		272	.029		1.30	1.53		2.83	3.90
3000	1" x 1/2", 38 L.F./gal.		250	.032		2.33	1.67		4	5.25
3200	Polyurethane, 1 or 2 component				Gal.	54			54	59.50
3500	Bulk, in place, 1/4" x 1/4"	1 Bric	300	.027	L.F.	.18	1.39		1.57	2.43
3655	1/2" x 1/4"		288	.028		.35	1.45		1.80	2.73
3800	3/4" x 3/8"		272	.029		.80	1.53		2.33	3.34
3900	1" x 1/2"		250	.032		1.40	1.67		3.07	4.23
4100	Silicone rubber, bulk				Gal.	68.50			68.50	75.50

07 92 Joint Sealants

07 92 13 – Elastomeric Joint Sealants

07 92 13.20 Caulking and Sealant Options		Crew	Daily Output	Labor-Hours	Unit	Material	2020 Bare Costs Labor	Equipment	Total	Total Incl O&P
4200	Cartridges				Gal.	51.50			51.50	56.50

07 92 16 – Rigid Joint Sealants

07 92 16.10 Rigid Joint Sealants		Crew	Daily Output	Labor-Hours	Unit	Material	Labor	Equipment	Total	Total Incl O&P
0010	RIGID JOINT SEALANTS									
5802	Tapes, sealant, PVC foam adhesive, 1/16" x 1/4"				L.F.	.09			.09	.10
5902	1/16" x 1/2"					.09			.09	.10
5952	1/16" x 1"					.16			.16	.18
6002	1/8" x 1/2"					.09			.09	.10

07 92 19 – Acoustical Joint Sealants

07 92 19.10 Acoustical Sealant		Crew	Daily Output	Labor-Hours	Unit	Material	Labor	Equipment	Total	Total Incl O&P
0010	ACOUSTICAL SEALANT									
0020	Acoustical sealant, elastomeric, cartridges				Ea.	8.65			8.65	9.55
0025	In place, 1/4" x 1/4"	1 Bric	300	.027	L.F.	.35	1.39		1.74	2.63
0030	1/4" x 1/2"		288	.028		.71	1.45		2.16	3.12
0035	1/2" x 1/2"		275	.029		1.41	1.51		2.92	4.01
0040	1/2" x 3/4"		263	.030		2.12	1.58		3.70	4.89
0045	3/4" x 3/4"		244	.033		3.18	1.71		4.89	6.25
0050	1" x 1"		200	.040		5.65	2.08		7.73	9.60

07 95 Expansion Control

07 95 13 – Expansion Joint Cover Assemblies

07 95 13.50 Expansion Joint Assemblies		Crew	Daily Output	Labor-Hours	Unit	Material	Labor	Equipment	Total	Total Incl O&P
0010	EXPANSION JOINT ASSEMBLIES									
0200	Floor cover assemblies, 1" space, aluminum	1 Sswk	38	.211	L.F.	26	12.15		38.15	48.50
0300	Bronze		38	.211		52.50	12.15		64.65	78
0500	2" space, aluminum		38	.211		18.55	12.15		30.70	40.50
0600	Bronze		38	.211		61	12.15		73.15	87
0800	Wall and ceiling assemblies, 1" space, aluminum		38	.211		17.20	12.15		29.35	39
0900	Bronze		38	.211		54	12.15		66.15	79.50
1100	2" space, aluminum		38	.211		22	12.15		34.15	44
1200	Bronze		38	.211		52.50	12.15		64.65	77.50
1400	Floor to wall assemblies, 1" space, aluminum		38	.211		20	12.15		32.15	42
1500	Bronze or stainless		38	.211		65	12.15		77.15	91.50
1700	Gym floor angle covers, aluminum, 3" x 3" angle		46	.174		19.70	10.05		29.75	38
1800	3" x 4" angle		46	.174		22	10.05		32.05	40.50
2000	Roof closures, aluminum, flat roof, low profile, 1" space		57	.140		24	8.10		32.10	40
2100	High profile		57	.140		34	8.10		42.10	51
2300	Roof to wall, low profile, 1" space		57	.140		23.50	8.10		31.60	39.50
2400	High profile		57	.140		34	8.10		42.10	51
9000	Minimum labor/equipment charge		2	4	Job		231		231	380

Estimating Tips

08 10 00 Doors and Frames

All exterior doors should be addressed for their energy conservation (insulation and seals).

- Most metal doors and frames look alike, but there may be significant differences among them. When estimating these items, be sure to choose the line item that most closely compares to the specification or door schedule requirements regarding:
  - ☐ type of metal
  - ☐ metal gauge
  - ☐ door core material
  - ☐ fire rating
  - ☐ finish
- Wood and plastic doors vary considerably in price. The primary determinant is the veneer material. Lauan, birch, and oak are the most common veneers. Other variables include the following:
  - ☐ hollow or solid core
  - ☐ fire rating
  - ☐ flush or raised panel
  - ☐ finish
- Door pricing includes bore for cylindrical locksets and mortise for hinges.

08 30 00 Specialty Doors and Frames

- There are many varieties of special doors, and they are usually priced per each. Add frames, hardware, or operators required for a complete installation.

08 40 00 Entrances, Storefronts, and Curtain Walls

- Glazed curtain walls consist of the metal tube framing and the glazing material. The cost data in this subdivision is presented for the metal tube framing alone or the composite wall. If your estimate requires a detailed takeoff of the framing, be sure to add the glazing cost and any tints.

08 50 00 Windows

- Steel windows are unglazed and aluminum can be glazed or unglazed. Some metal windows are priced without glass. Refer to 08 80 00 Glazing for glass pricing. The grade C indicates commercial grade windows, usually ASTM C-35.
- All wood windows and vinyl are priced preglazed. The glazing is insulating glass. Add the cost of screens and grills if required and not already included.

08 70 00 Hardware

- Hardware costs add considerably to the cost of a door. The most efficient method to determine the hardware requirements for a project is to review the door and hardware schedule together. One type of door may have different hardware, depending on the door usage.
- Door hinges are priced by the pair, with most doors requiring 1-1/2 pairs per door. The hinge prices do not include installation labor because it is included in door installation. Hinges are classified according to the frequency of use, base material, and finish.

08 80 00 Glazing

- Different openings require different types of glass. The most common types are:
  - ☐ float
  - ☐ tempered
  - ☐ insulating
  - ☐ impact-resistant
  - ☐ ballistic-resistant
- Most exterior windows are glazed with insulating glass. Entrance doors and window walls, where the glass is less than 18" from the floor, are generally glazed with tempered glass. Interior windows and some residential windows are glazed with float glass.
- Coastal communities require the use of impact-resistant glass, dependent on wind speed.
- The insulation or 'u' value is a strong consideration, along with solar heat gain, to determine total energy efficiency.

Reference Numbers

Reference numbers are shown at the beginning of some major classifications. These numbers refer to related items in the Reference Section. The reference information may be an estimating procedure, an alternate pricing method, or technical information.

Note: Not all subdivisions listed here necessarily appear. ■

08 01 Operation and Maintenance of Openings

08 01 11 – Operation and Maintenance of Metal Doors and Frames

08 01 11.10 Door and Window Maintenance		Crew	Daily Output	Labor-Hours	Unit	Material	2020 Bare Costs Labor	Equipment	Total	Total Incl O&P
0010	DOOR & WINDOW MAINTENANCE									
0012	Remove weatherstripping from door or window	1 Carp	45	.178	Ea.		9.45		9.45	15.15
0060	Remove lockset		20	.400			21.50		21.50	34
0200	Remove damaged louver		58.46	.137			7.25		7.25	11.65
1130	Install door	2 Carp	13.07	1.224			65		65	104
1140	Remove deadbolt	1 Carp	13.85	.578			30.50		30.50	49
1150	Remove panic bar		7.69	1.040			55.50		55.50	88.50
1420	Remove door, plane to fit, rail, reinstall door		9.60	.833			44.50		44.50	71

08 01 14 – Operation and Maintenance of Wood Doors

08 01 14.15 Door and Window Maintenance		Crew	Daily Output	Labor-Hours	Unit	Material	Labor	Equipment	Total	Total Incl O&P
0010	DOOR & WINDOW MAINTENANCE									
0050	Rehang single door	1 Carp	32	.250	Ea.		13.30		13.30	21.50
0100	Rehang double door		16	.500	Pr.		26.50		26.50	42.50
0350	Install window sash cord		16	.500	Ea.	5	26.50		31.50	48

08 01 53 – Operation and Maintenance of Plastic Windows

08 01 53.81 Solid Vinyl Replacement Windows		Crew	Daily Output	Labor-Hours	Unit	Material	Labor	Equipment	Total	Total Incl O&P
0010	SOLID VINYL REPLACEMENT WINDOWS									
0020	Double-hung, insulated glass, up to 83 united inches G	2 Carp	8	2	Ea.	183	106		289	370
0040	84 to 93 G		8	2		195	106		301	385
0060	94 to 101 G		6	2.667		195	142		337	440
0080	102 to 111 G		6	2.667		207	142		349	455
0100	112 to 120 G		6	2.667		223	142		365	470
0120	For each united inch over 120, add G		800	.020	Inch	2.92	1.06		3.98	4.91
0140	Casement windows, one operating sash, 42 to 60 united inches G		8	2	Ea.	282	106		388	480
0160	61 to 70 G		8	2		305	106		411	510
0180	71 to 80 G		8	2		340	106		446	540
0200	81 to 96 G		8	2		370	106		476	575
0220	Two operating sash, 58 to 78 united inches G		8	2		655	106		761	890
0240	79 to 88 G		8	2		755	106		861	1,000
0260	89 to 98 G		8	2		950	106		1,056	1,225
0280	99 to 108 G		6	2.667		780	142		922	1,075
0300	109 to 121 G		6	2.667		855	142		997	1,175
0320	Two operating, one fixed sash, 73 to 108 united inches G		8	2		620	106		726	850
0340	109 to 118 G		8	2		675	106		781	910
0360	119 to 128 G		6	2.667		770	142		912	1,075
0380	129 to 138 G		6	2.667		800	142		942	1,100
0400	139 to 156 G		6	2.667		990	142		1,132	1,300
0420	Four operating sash, 98 to 118 united inches G		8	2		1,325	106		1,431	1,625
0440	119 to 128 G		8	2		1,375	106		1,481	1,700
0460	129 to 138 G		6	2.667		1,425	142		1,567	1,775
0480	139 to 148 G		6	2.667		1,475	142		1,617	1,850
0500	149 to 168 G		6	2.667		1,575	142		1,717	1,975
0520	169 to 178 G		6	2.667		1,625	142		1,767	2,025
0560	Fixed picture window, up to 63 united inches G		8	2		181	106		287	370
0580	64 to 83 G		8	2		219	106		325	410
0600	84 to 101 G		8	2		273	106		379	470
0620	For each united inch over 101, add G		900	.018	Inch	3.39	.95		4.34	5.25
0800	Cellulose fiber insulation, poured into sash balance cavity G	1 Carp	36	.222	C.F.	.70	11.80		12.50	19.65
0820	Silicone caulking at perimeter G	"	800	.010	L.F.	.17	.53		.70	1.03
2000	Impact resistant replacement windows									
2005	Laminated glass, 120 MPH rating, measure in united inches									
2010	Installation labor does not cover any rework of the window opening									

08 01 Operation and Maintenance of Openings

08 01 53 – Operation and Maintenance of Plastic Windows

08 01 53.81 Solid Vinyl Replacement Windows		Crew	Daily Output	Labor-Hours	Unit	Material	2020 Bare Costs Labor	Equipment	Total	Total Incl O&P
2020	Double-hung, insulated glass, up to 101 united inches	2 Carp	8	2	Ea.	555	106		661	780
2025	For each united inch over 101, add		80	.200	Inch	3.48	10.65		14.13	21
2100	Casement windows, impact resistant, up to 60 united inches		8	2	Ea.	515	106		621	735
2120	61 to 70		8	2		545	106		651	765
2130	71 to 80		8	2		570	106		676	800
2140	81 to 100		8	2		590	106		696	815
2150	For each united inch over 100, add		80	.200	Inch	5.30	10.65		15.95	23
2200	Awning windows, impact resistant, up to 60 united inches		8	2	Ea.	525	106		631	750
2220	61 to 70		8	2		550	106		656	775
2230	71 to 80		8	2		560	106		666	790
2240	For each united inch over 80, add		80	.200	Inch	4.99	10.65		15.64	22.50
2300	Picture windows, impact resistant, up to 63 united inches		8	2	Ea.	380	106		486	590
2320	63 to 83		8	2		410	106		516	620
2330	84 to 101		8	2		450	106		556	665
2340	For each united inch over 101, add		80	.200	Inch	3.35	10.65		14	20.50

08 05 Common Work Results for Openings

08 05 05 – Selective Demolition for Openings

08 05 05.10 Selective Demolition Doors		Crew	Daily Output	Labor-Hours	Unit	Material	Labor	Equipment	Total	Total Incl O&P
0010	**SELECTIVE DEMOLITION DOORS** R024119-10									
0200	Doors, exterior, 1-3/4" thick, single, 3' x 7' high	1 Clab	16	.500	Ea.		21		21	33.50
0202	Doors, exterior, 3' - 6' wide x 7' high		12	.667			28		28	45
0210	3' x 8' high		10	.800			33.50		33.50	54
0215	Double, 3' x 8' high		6	1.333			56		56	90
0220	Double, 6' x 7' high		12	.667			28		28	45
0500	Interior, 1-3/8" thick, single, 3' x 7' high		20	.400			16.85		16.85	27
0520	Double, 6' x 7' high		16	.500			21		21	33.50
0700	Bi-folding, 3' x 6'-8" high		20	.400			16.85		16.85	27
0720	6' x 6'-8" high		18	.444			18.70		18.70	30
0900	Bi-passing, 3' x 6'-8" high		16	.500			21		21	33.50
0940	6' x 6'-8" high		14	.571			24		24	38.50
0960	Interior metal door 1-3/4" thick, 3'-0" x 6'-8"		18	.444			18.70		18.70	30
0980	Interior metal door 1-3/4" thick, 3'-0" x 7'-0"		18	.444			18.70		18.70	30
1000	Interior wood door 1-3/4" thick, 3'-0" x 6'-8"		20	.400			16.85		16.85	27
1020	Interior wood door 1-3/4" thick, 3'-0" x 7'-0"		20	.400			16.85		16.85	27
1100	Door demo, floor door	2 Sswk	5	3.200			184		184	300
1500	Remove and reset, hollow core	1 Carp	8	1			53		53	85
1520	Solid		6	1.333			71		71	114
2000	Frames, including trim, metal		8	1			53		53	85
2200	Wood	2 Carp	32	.500			26.50		26.50	42.50
2201	Alternate pricing method	1 Carp	200	.040	L.F.		2.13		2.13	3.41
2205	Remove door hardware	"	45.70	.175	Ea.		9.30		9.30	14.90
3000	Special doors, counter doors	2 Carp	6	2.667			142		142	227
3001	Special doors, counter doors	2 Clab	6	2.667			112		112	180
3100	Double acting	2 Carp	10	1.600			85		85	136
3101	Double acting	2 Clab	10	1.600			67.50		67.50	108
3200	Floor door (trap type), or access type	2 Carp	8	2			106		106	170
3300	Glass, sliding, including frames		12	1.333			71		71	114
3400	Overhead, commercial, 12' x 12' high		4	4			213		213	340
3440	up to 20' x 16' high		3	5.333			283		283	455
3445	up to 35' x 30' high		1	16			850		850	1,350

08 05 Common Work Results for Openings

08 05 05 – Selective Demolition for Openings

	08 05 05.10 Selective Demolition Doors	Crew	Daily Output	Labor-Hours	Unit	Material	2020 Bare Costs Labor	Equipment	Total	Total Incl O&P
3500	Residential, 9′ x 7′ high	2 Carp	8	2	Ea.		106		106	170
3540	16′ x 7′ high		7	2.286			121		121	195
3600	Remove and reset, small		4	4			213		213	340
3620	Large		2.50	6.400			340		340	545
3700	Roll-up grille		5	3.200			170		170	272
3800	Revolving door, no wire or connections		2	8			425		425	680
3900	Storefront swing door		3	5.333			283		283	455
3902	Cafe/bar swing door	2 Clab	8	2			84		84	135
4000	Residential lockset, exterior	1 Carp	28	.286			15.20		15.20	24.50
4224	Pocket door, no frame		8	1			53		53	85
4300	Remove and reset deadbolt		8	1			53		53	85
5500	Remove door closer	1 Clab	10	.800			33.50		33.50	54
5520	Remove push/pull plate		20	.400			16.85		16.85	27
5540	Remove door bolt		30	.267			11.25		11.25	18
5560	Remove kick/mop plate		20	.400			16.85		16.85	27
5585	Remove panic device		10	.800			33.50		33.50	54
5590	Remove mail slot		45	.178			7.50		7.50	12
5595	Remove peep hole		45	.178			7.50		7.50	12
5600	Remove door sidelight	1 Carp	6	1.333			71		71	114
5604	Remove 9′ x 7′ garage door	2 Carp	9.50	1.684			89.50		89.50	143
5624	16′ x 7′ garage door		8	2			106		106	170
5626	9′ x 7′ swing up garage door		7	2.286			121		121	195
5628	16′ x 7′ swing up garage door		7	2.286			121		121	195
5630	Garage door demolition, remove garage door track	1 Carp	6	1.333			71		71	114
5644	Remove overhead door opener	1 Clab	5	1.600			67.50		67.50	108
5774	Remove heavy gage sectional door, 20 ga., 8′ x 8′	2 Carp	7	2.286			121		121	195
5784	10′ x 10′		6	2.667			142		142	227
5794	12′ x 12′		5	3.200			170		170	272
5805	14′ x 14′		4	4			213		213	340
5850	8′ x 8′ sliding door & frame	B-68D	1.60	15			690	175	865	1,300
5860	12′ x 22′ wand fire door and frame	"	.64	37.500			1,725	435	2,160	3,200
6334	Remove shower door unit	1 Clab	16	.500			21		21	33.50
6384	Remove French door unit	"	8	1			42		42	67.50
6600	Demo flexible transparent strip entrance	3 Shee	115	.209	SF Surf		13		13	20.50
7100	Remove double swing pneumatic doors, openers and sensors	2 Skwk	.50	32	Opng.		1,750		1,750	2,800
7110	Remove automatic operators, industrial, sliding doors, to 12′ wide	"	.40	40	"		2,200		2,200	3,500
7550	Hangar door demo	2 Sswk	330	.048	S.F.		2.79		2.79	4.58
7570	Remove shock absorbing door	"	1.90	8.421	Opng.		485		485	795
9000	Minimum labor/equipment charge	1 Carp	4	2	Job		106		106	170

	08 05 05.20 Selective Demolition of Windows	Crew	Daily Output	Labor-Hours	Unit	Material	Labor	Equipment	Total	Total Incl O&P
0010	**SELECTIVE DEMOLITION OF WINDOWS** R024119-10									
0200	Aluminum, including trim, to 12 S.F.	1 Clab	16	.500	Ea.		21		21	33.50
0240	To 25 S.F.		11	.727			30.50		30.50	49
0280	To 50 S.F.		5	1.600			67.50		67.50	108
0320	Storm windows/screens, to 12 S.F.		27	.296			12.45		12.45	20
0360	To 25 S.F.		21	.381			16.05		16.05	25.50
0400	To 50 S.F.		16	.500			21		21	33.50
0600	Glass, up to 10 S.F./window		200	.040	S.F.		1.68		1.68	2.70
0620	Over 10 S.F./window		150	.053	"		2.25		2.25	3.60
1000	Steel, including trim, to 12 S.F.		13	.615	Ea.		26		26	41.50
1020	To 25 S.F.		9	.889			37.50		37.50	60
1040	To 50 S.F.		4	2			84		84	135

08 05 Common Work Results for Openings

08 05 05 – Selective Demolition for Openings

08 05 05.20 Selective Demolition of Windows		Crew	Daily Output	Labor-Hours	Unit	Material	2020 Bare Costs Labor	Equipment	Total	Total Incl O&P
2000	Wood, including trim, to 12 S.F.	1 Clab	22	.364	Ea.		15.30		15.30	24.50
2020	To 25 S.F.		18	.444			18.70		18.70	30
2060	To 50 S.F.		13	.615			26		26	41.50
2065	To 180 S.F.		8	1			42		42	67.50
4300	Remove bay/bow window	2 Carp	6	2.667			142		142	227
4400	Remove skylight, prefabricated glass block with metal frame	G-3	180	.178	S.F.		9.30		9.30	14.65
4410	Remove skylight, plstc domes, flush/curb mtd	2 Carp	395	.041	"		2.15		2.15	3.45
4420	Remove skylight, plstc/glass up to 2′ x 3′	1 Carp	15	.533	Ea.		28.50		28.50	45.50
4440	Remove skylight, plstc/glass up to 4′ x 6′	2 Carp	10	1.600			85		85	136
4480	Remove roof window up to 3′ x 4′	1 Carp	8	1			53		53	85
4500	Remove roof window up to 4′ x 6′	2 Carp	6	2.667			142		142	227
5020	Remove and reset window, up to a 2′ x 2′ window	1 Carp	6	1.333			71		71	114
5040	Up to a 3′ x 3′ window		4	2			106		106	170
5080	Up to a 4′ x 5′ window		2	4			213		213	340
6000	Screening only	1 Clab	4000	.002	S.F.		.08		.08	.13
9000	Minimum labor/equipment charge	"	4	2	Job		84		84	135

08 11 Metal Doors and Frames

08 11 16 – Aluminum Doors and Frames

08 11 16.10 Entrance Doors		Crew	Daily Output	Labor-Hours	Unit	Material	Labor	Equipment	Total	Total Incl O&P
0010	**ENTRANCE DOORS** and frame, aluminum, narrow stile									
0011	Including standard hardware, clear finish, no glass									
0012	Top and bottom offset pivots, 1/4″ beveled glass stops, threshold									
0013	Dead bolt lock with inside thumb screw, standard push pull									
0020	3′-0″ x 7′-0″ opening	2 Sswk	2	8	Ea.	1,075	460		1,535	1,925
0025	Anodizing aluminum entr. door & frame, add					115			115	126
0030	3′-6″ x 7′-0″ opening	2 Sswk	2	8		865	460		1,325	1,700
0100	3′-0″ x 10′-0″ opening, 3′ high transom		1.80	8.889		1,325	510		1,835	2,325
0200	3′-6″ x 10′-0″ opening, 3′ high transom		1.80	8.889		1,400	510		1,910	2,400
0280	5′-0″ x 7′-0″ opening		2	8		1,575	460		2,035	2,475
0300	6′-0″ x 7′-0″ opening		1.30	12.308		1,050	710		1,760	2,325
0301	6′-0″ x 7′-0″ opening		1.30	12.308	Pr.	1,050	710		1,760	2,325
0400	6′-0″ x 10′-0″ opening, 3′ high transom		1.10	14.545	"	1,500	840		2,340	3,025
0520	3′-0″ x 7′-0″ opening, wide stile		2	8	Ea.	905	460		1,365	1,750
0540	3′-6″ x 7′-0″ opening		2	8		990	460		1,450	1,850
0560	5′-0″ x 7′-0″ opening		2	8		1,500	460		1,960	2,400
0580	6′-0″ x 7′-0″ opening		1.30	12.308	Pr.	1,600	710		2,310	2,925
0600	7′-0″ x 7′-0″ opening		1	16	"	2,275	920		3,195	4,000
1200	For non-standard size, add				Leaf	80%				
1250	For installation of non-standard size, add						20%			
1300	Light bronze finish, add				Leaf	36%				
1400	Dark bronze finish, add					25%				
1500	For black finish, add					40%				
1600	Concealed panic device, add					1,075			1,075	1,175
1700	Electric striker release, add				Opng.	325			325	360
1800	Floor check, add				Leaf	705			705	780
1900	Concealed closer, add				"	575			575	630
9000	Minimum labor/equipment charge	2 Carp	4	4	Job		213		213	340

08 11 Metal Doors and Frames

08 11 63 – Metal Screen and Storm Doors and Frames

08 11 63.23 Aluminum Screen and Storm Doors and Frames		Crew	Daily Output	Labor-Hours	Unit	2020 Bare Costs Material	Labor	Equipment	Total	Total Incl O&P
0010	**ALUMINUM SCREEN AND STORM DOORS AND FRAMES**									
0020	Combination storm and screen									
0420	Clear anodic coating, 2'-8" wide	2 Carp	14	1.143	Ea.	238	60.50		298.50	360
0440	3'-0" wide	"	14	1.143	"	192	60.50		252.50	310
0500	For 7'-0" door height, add					8%				
1020	Mill finish, 2'-8" wide	2 Carp	14	1.143	Ea.	249	60.50		309.50	370
1040	3'-0" wide	"	14	1.143		268	60.50		328.50	395
1100	For 7'-0" door, add					8%				
1520	White painted, 2'-8" wide	2 Carp	14	1.143		269	60.50		329.50	395
1540	3'-0" wide	"	14	1.143		315	60.50		375.50	445
1600	For 7'-0" door, add					8%				
2000	Wood door & screen, see Section 08 14 33.20									
9000	Minimum labor/equipment charge	1 Carp	4	2	Job		106		106	170

08 12 Metal Frames

08 12 13 – Hollow Metal Frames

08 12 13.13 Standard Hollow Metal Frames			Crew	Daily Output	Labor-Hours	Unit	Material	Labor	Equipment	Total	Total Incl O&P
0010	**STANDARD HOLLOW METAL FRAMES**										
0020	16 ga., up to 5-3/4" jamb depth										
0025	3'-0" x 6'-8" single	G	2 Carp	16	1	Ea.	217	53		270	325
0028	3'-6" wide, single	G		16	1		270	53		323	380
0030	4'-0" wide, single	G		16	1		305	53		358	420
0040	6'-0" wide, double	G		14	1.143		256	60.50		316.50	380
0045	8'-0" wide, double	G		14	1.143		248	60.50		308.50	370
0100	3'-0" x 7'-0" single			16	1		214	53		267	320
0110	3'-6" wide, single			16	1		221	53		274	330
0112	4'-0" wide, single			16	1		252	53		305	360
0140	6'-0" wide, double			14	1.143		268	60.50		328.50	395
0145	8'-0" wide, double			14	1.143		236	60.50		296.50	360
1000	16 ga., up to 4-7/8" deep, 3'-0" x 7'-0" single	G		16	1		202	53		255	305
1140	6'-0" wide, double	G		14	1.143		215	60.50		275.50	335
1200	16 ga., 8-3/4" deep, 3'-0" x 7'-0" single	G		16	1		206	53		259	310
1240	6'-0" wide, double	G		14	1.143		310	60.50		370.50	440
2800	14 ga., up to 3-7/8" deep, 3'-0" x 7'-0" single	G		16	1		251	53		304	360
2840	6'-0" wide, double	G		14	1.143		360	60.50		420.50	495
2900	14 ga., 4-3/4" deep, 2'-0" x 7'-0" high, single	G		15	1.067		214	56.50		270.50	325
2910	2'-4" wide	G		15	1.067		266	56.50		322.50	385
2920	2'-6" wide	G		15	1.067		266	56.50		322.50	385
2930	2'-8" wide	G		15	1.067		266	56.50		322.50	385
2940	3'-0" wide	G		15	1.067		265	56.50		321.50	385
2950	3'-4" wide	G		15	1.067		276	56.50		332.50	395
2960	6'-0" wide, double	G		12	1.333		310	71		381	455
3000	14 ga., up to 5-3/4" deep, 3'-0" x 6'-8" single	G		16	1		159	53		212	260
3002	3'-6" wide, single	G		16	1		267	53		320	380
3005	4'-0" wide, single	G		16	1		158	53		211	259
3600	up to 5-3/4" jamb depth, 4'-0" x 7'-0" single			15	1.067		256	56.50		312.50	375
3620	6'-0" wide, double			12	1.333		201	71		272	335
3640	8'-0" wide, double			12	1.333		300	71		371	445
3700	8'-0" high, 4'-0" wide, single			15	1.067		305	56.50		361.50	425
3740	8'-0" wide, double			12	1.333		345	71		416	495

08 12 Metal Frames

08 12 13 – Hollow Metal Frames

08 12 13.13 Standard Hollow Metal Frames		Crew	Daily Output	Labor-Hours	Unit	Material	2020 Bare Costs Labor	Equipment	Total	Total Incl O&P
4000	6-3/4" deep, 4'-0" x 7'-0" single G	2 Carp	15	1.067	Ea.	310	56.50		366.50	430
4020	6'-0" wide, double G		12	1.333		360	71		431	510
4040	8'-0" wide, double G		12	1.333		258	71		329	400
4100	8'-0" high, 4'-0" wide, single G		15	1.067		188	56.50		244.50	297
4140	8'-0" wide, double G		12	1.333		450	71		521	610
4400	8-3/4" deep, 4'-0" x 7'-0", single		15	1.067		430	56.50		486.50	565
4440	8'-0" wide, double		12	1.333		455	71		526	615
4500	4'-0" x 8'-0", single		15	1.067		480	56.50		536.50	615
4540	8'-0" wide, double		12	1.333		490	71		561	650
4900	For welded frames, add					59.50			59.50	65.50
5380	Steel frames, KD, 14 ga., "B" label, to 5-3/4" throat, to 3'-0" x 7'-0"	2 Carp	15	1.067		248	56.50		304.50	365
5400	14 ga., "B" label, up to 5-3/4" deep, 4'-0" x 7'-0" single G		15	1.067		190	56.50		246.50	300
5440	8'-0" wide, double G		12	1.333		243	71		314	380
5800	6-3/4" deep, 7'-0" high, 4'-0" wide, single G		15	1.067		169	56.50		225.50	277
5840	8'-0" wide, double G		12	1.333		335	71		406	485
6200	8-3/4" deep, 4'-0" x 7'-0" single G		15	1.067		269	56.50		325.50	385
6240	8'-0" wide, double G		12	1.333		370	71		441	525
6300	For "A" label use same price as "B" label									
6400	For baked enamel finish, add					30%	15%			
6500	For galvanizing, add					20%				
6600	For hospital stop, add				Ea.	251			251	276
6620	For hospital stop, stainless steel, add				"	118			118	130
7900	Transom lite frames, fixed, add	2 Carp	155	.103	S.F.	49	5.50		54.50	63
8000	Movable, add	"	130	.123	"	64	6.55		70.55	81
9000	Minimum labor/equipment charge	1 Carp	4	2	Job		106		106	170

08 12 13.15 Borrowed Lites		Crew	Daily Output	Labor-Hours	Unit	Material	Labor	Equipment	Total	Total Incl O&P
0010	**BORROWED LITES**									
0100	Hollow metal section 20 ga., 3-1/2" with glass stop	2 Carp	100	.160	L.F.	19.90	8.50		28.40	35.50
0110	3-3/4" with glass stop		100	.160		9.15	8.50		17.65	23.50
0120	4" with glass stop		100	.160		9.30	8.50		17.80	24
0130	4-5/8" with glass stop		100	.160		9.60	8.50		18.10	24
0140	4-7/8" with glass stop		100	.160		9.80	8.50		18.30	24.50
0150	5" with glass stop		100	.160		10.65	8.50		19.15	25.50
0160	5-3/8" with glass stop		100	.160		12.55	8.50		21.05	27.50
0300	Hollow metal section 18 ga., 3-1/2" with glass stop		80	.200		9.90	10.65		20.55	28
0310	3-3/4" with glass stop		80	.200		10.10	10.65		20.75	28
0320	4" with glass stop		80	.200		14	10.65		24.65	32.50
0330	4-5/8" with glass stop		80	.200		10.75	10.65		21.40	29
0340	4-7/8" with glass stop		80	.200		11	10.65		21.65	29
0350	5" with glass stop		80	.200		11.10	10.65		21.75	29.50
0360	5-3/8" with glass stop		80	.200		11.10	10.65		21.75	29.50
0370	6-5/8" with glass stop		80	.200		12.35	10.65		23	30.50
0380	6-7/8" with glass stop		80	.200		12.55	10.65		23.20	31
0390	7-1/4" with glass stop		80	.200		13.10	10.65		23.75	31.50
0500	Mullion section 18 ga., 3-1/2" with double stop		50	.320		23	17		40	53
0510	3-3/4" with double stop		50	.320		23.50	17		40.50	53.50
0530	4-5/8" with double stop		50	.320		25	17		42	55
0540	4-7/8" with double stop		50	.320		25.50	17		42.50	55.50
0550	5" with double stop		50	.320		26	17		43	56
0560	5-3/8" with double stop		50	.320		26.50	17		43.50	56.50
0570	6-5/8" with double stop		50	.320		29	17		46	59
0580	6-7/8" with double stop		50	.320		29.50	17		46.50	60

08 12 Metal Frames

08 12 13 – Hollow Metal Frames

08 12 13.15 Borrowed Lites		Crew	Daily Output	Labor-Hours	Unit	Material	Labor	Equipment	Total	Total Incl O&P
						2020 Bare Costs				
0590	7-1/4" with double stop	2 Carp	50	.320	L.F.	30	17		47	60.50
1000	Assembled frame 2'-0" x 2'-0" 16 ga., 3-3/4" with glass stop		10	1.600	Ea.	217	85		302	375
1010	3'-0" x 2'-0"		10	1.600		217	85		302	375
1020	4'-0" x 2'-0"		10	1.600		218	85		303	375
1030	5'-0" x 2'-0"		10	1.600		224	85		309	385
1040	6'-0" x 2'-0"		8	2		325	106		431	530
1050	7'-0" x 2'-0"		8	2		325	106		431	530
1060	8'-0" x 2'-0"		8	2		325	106		431	530
1100	2'-0" x 7'-0"		8	2		325	106		431	530
1110	3'-0" x 7'-0"		8	2		325	106		431	530
1120	4'-0" x 7'-0"		8	2		325	106		431	530
1130	5'-0" x 7'-0"		8	2		325	106		431	530
1140	6'-0" x 7'-0"		6	2.667		350	142		492	610
1150	7'-0" x 7'-0"		6	2.667		350	142		492	610
1160	8'-0" x 7'-0"		6	2.667		350	142		492	610
1200	Assembled frame 2'-0" x 2'-0" 16 ga., 4-7/8" with glass stop		10	1.600		218	85		303	375
1210	3'-0" x 2'-0"		10	1.600		218	85		303	375
1220	4'-0" x 2'-0"		8	2		218	106		324	410
1230	5'-0" x 2'-0"		8	2		325	106		431	530
1240	6'-0" x 2'-0"		8	2		325	106		431	530
1250	7'-0" x 2'-0"		8	2		325	106		431	530
1260	8'-0" x 2'-0"		8	2		325	106		431	530
1300	2'-0" x 7'-0"		8	2		325	106		431	530
1310	3'-0" x 7'-0"		8	2		325	106		431	530
1320	4'-0" x 7'-0"		8	2		325	106		431	530
1330	5'-0" x 7'-0"		6	2.667		325	142		467	585
1340	6'-0" x 7'-0"		6	2.667		350	142		492	610
1350	7'-0" x 7'-0"		6	2.667		360	142		502	625
1360	8'-0" x 7'-0"		6	2.667		360	142		502	625
1400	Assembled frame 2'-0" x 3'-0" 16 ga., 6-1/8" with glass stop		6	2.667		219	142		361	465
1410	3'-0" x 3'-0"		8	2		218	106		324	410
1420	4'-0" x 3'-0"		8	2		218	106		324	410
1430	5'-0" x 3'-0"		6	2.667		325	142		467	585
1440	6'-0" x 3'-0"		6	2.667		300	142		442	555
1450	7'-0" x 3'-0"		6	2.667		325	142		467	585
1460	8'-0" x 3'-0"	↓	10	1.600	↓	325	85		410	495

08 12 13.18 Transoms and Sidelights		Crew	Daily Output	Labor-Hours	Unit	Material	Labor	Equipment	Total	Total Incl O&P
0010	**TRANSOMS & SIDELIGHTS**									
0160	Transom sash, 6-3/4" thick, 3'-0" x 1'-4"	2 Carp	11	1.455	Ea.	194	77.50		271.50	340
0180	3'-4" x 1'-4"		11	1.455		198	77.50		275.50	340
0200	6'-0" x 1'-4"		6	2.667		226	142		368	475
0220	4-3/4" thick, 3'-0" x 1'-4"		11	1.455		199	77.50		276.50	345
0240	3'-4" x 1'-4"		11	1.455		181	77.50		258.50	325
0260	6'-0" x 1'-4"	↓	6	2.667	↓	206	142		348	455

08 12 13.20 Wrap Around Drywall Frames		Crew	Daily Output	Labor-Hours	Unit	Material	Labor	Equipment	Total	Total Incl O&P
0010	**WRAP AROUND DRYWALL FRAMES**									
0400	Wrap-around drywall frame, 16 ga., 6-1/4" x 7'-0" x 2'-6" wide	2 Carp	16	1	Ea.	186	53		239	289
0430	2'-8" wide		16	1		186	53		239	289
0460	3'-0" wide		15	1.067		186	56.50		242.50	295
0490	3'-6" wide		15	1.067		199	56.50		255.50	310
0520	5'-0" wide		14	1.143		233	60.50		293.50	355
0580	6'-0" wide	↓	14	1.143	↓	250	60.50		310.50	375

08 12 Metal Frames

08 12 13 – Hollow Metal Frames

08 12 13.20 Wrap Around Drywall Frames		Crew	Daily Output	Labor-Hours	Unit	2020 Bare Costs Material	Labor	Equipment	Total	Total Incl O&P
0610	7'-0" wide	2 Carp	13	1.231	Ea.	249	65.50		314.50	380
2000	18 ga., 4-3/4" x 7'-0 x 2'-6" wide		16	1		173	53		226	275
2030	2'-8" wide		16	1		173	53		226	275
2060	3'-0" wide		15	1.067		173	56.50		229.50	281
2090	3'-6" wide		15	1.067		175	56.50		231.50	284
2120	5'-0" wide		14	1.143		197	60.50		257.50	315
2150	5'-4" wide		14	1.143		209	60.50		269.50	330
2200	6'-0" wide		14	1.143		209	60.50		269.50	330
2250	7'-0" wide		13	1.231		215	65.50		280.50	340
3700	Fire door frames, 6-1/4" x 7'-0" x 2'-6" wide		16	1		204	53		257	310
3730	2'-8" wide		16	1		132	53		185	230
3760	3'-0" wide		15	1.067		204	56.50		260.50	315
3820	5'-0" wide		14	1.143		220	60.50		280.50	340
3850	5'-4" wide		14	1.143		158	60.50		218.50	272
3880	6'-0" wide		14	1.143		236	60.50		296.50	360
9000	Minimum labor/equipment charge	1 Carp	2	4	Job		213		213	340

08 12 13.25 Channel Metal Frames		Crew	Daily Output	Labor-Hours	Unit	Material	Labor	Equipment	Total	Total Incl O&P
0010	**CHANNEL METAL FRAMES**									
0020	Steel channels with anchors and bar stops									
0100	6" channel @ 8.2#/L.F., 3' x 7' door, weighs 150# G	E-4	13	2.462	Ea.	259	143	11.30	413.30	530
0200	8" channel @ 11.5#/L.F., 6' x 8' door, weighs 275# G		9	3.556		475	207	16.30	698.30	880
0300	8' x 12' door, weighs 400# G		6.50	4.923		690	286	22.50	998.50	1,250
0400	10" channel @ 15.3#/L.F., 10' x 10' door, weighs 500# G		6	5.333		860	310	24.50	1,194.50	1,475
0500	12' x 12' door, weighs 600# G		5.50	5.818		1,025	340	26.50	1,391.50	1,725
0600	12" channel @ 20.7#/L.F., 12' x 12' door, weighs 825# G		4.50	7.111		1,425	415	32.50	1,872.50	2,300
0700	12' x 16' door, weighs 1000# G		4	8		1,725	465	36.50	2,226.50	2,700
0800	For frames without bar stops, light sections, deduct					15%				
0900	Heavy sections, deduct					10%				
9000	Minimum labor/equipment charge	E-4	4	8	Job		465	36.50	501.50	800

08 13 Metal Doors

08 13 13 – Hollow Metal Doors

08 13 13.13 Standard Hollow Metal Doors		Crew	Daily Output	Labor-Hours	Unit	Material	Labor	Equipment	Total	Total Incl O&P
0010	**STANDARD HOLLOW METAL DOORS** R081313-20									
0015	Flush, full panel, hollow core									
0017	When noted doors are prepared but do not include glass or louvers									
0020	1-3/8" thick, 20 ga., 2'-0" x 6'-8" G	2 Carp	20	.800	Ea.	345	42.50		387.50	450
0040	2'-8" x 6'-8" G		18	.889		355	47		402	470
0060	3'-0" x 6'-8" G		17	.941		360	50		410	475
0100	3'-0" x 7'-0"		17	.941		375	50		425	490
0120	For vision lite, add					111			111	122
0140	For narrow lite, add					115			115	126
0320	Half glass, 20 ga., 2'-0" x 6'-8" G	2 Carp	20	.800		695	42.50		737.50	835
0340	2'-8" x 6'-8" G		18	.889		610	47		657	745
0360	3'-0" x 6'-8"		17	.941		605	50		655	745
0400	3'-0" x 7'-0" G		17	.941		475	50		525	605
0410	1-3/8" thick, 18 ga., 2'-0" x 6'-8" G		20	.800		465	42.50		507.50	585
0420	3'-0" x 6'-8" G		17	.941		470	50		520	595
0425	3'-0" x 7'-0"		17	.941		460	50		510	585
0450	For vision lite, add					111			111	122
0452	For narrow lite, add					115			115	126

08 13 Metal Doors

08 13 13 – Hollow Metal Doors

08 13 13.13 Standard Hollow Metal Doors			Crew	Daily Output	Labor-Hours	Unit	2020 Bare Costs Material	Labor	Equipment	Total	Total Incl O&P
0460	Half glass, 18 ga., 2'-0" x 6'-8"	G	2 Carp	20	.800	Ea.	635	42.50		677.50	770
0465	2'-8" x 6'-8"	G		18	.889		660	47		707	800
0470	3'-0" x 6'-8"	G		17	.941		645	50		695	790
0475	3'-0" x 7'-0"	G		17	.941		605	50		655	745
0500	Hollow core, 1-3/4" thick, full panel, 20 ga., 2'-8" x 6'-8"	G		18	.889		495	47		542	620
0520	3'-0" x 6'-8"	G		17	.941		495	50		545	625
0640	3'-0" x 7'-0"	G		17	.941		515	50		565	645
0680	4'-0" x 7'-0"	G		15	1.067		650	56.50		706.50	805
0700	4'-0" x 8'-0"	G		13	1.231		755	65.50		820.50	935
1000	18 ga., 2'-8" x 6'-8"	G		17	.941		525	50		575	655
1020	3'-0" x 6'-8"	G		16	1		510	53		563	645
1120	3'-0" x 7'-0"	G		17	.941		505	50		555	635
1150	3'-6" x 7'-0"		1 Carp	14	.571		555	30.50		585.50	660
1180	4'-0" x 7'-0"	G	2 Carp	14	1.143		740	60.50		800.50	915
1200	4'-0" x 8'-0"	G	"	17	.941		850	50		900	1,025
1212	For vision lite, add						111			111	122
1214	For narrow lite, add						115			115	126
1230	Half glass, 20 ga., 2'-8" x 6'-8"	G	2 Carp	20	.800		575	42.50		617.50	700
1240	3'-0" x 6'-8"	G		18	.889		570	47		617	700
1260	3'-0" x 7'-0"	G		18	.889		600	47		647	735
1280	Embossed panel, 1-3/4" thick, poly core, 20 ga., 3'-0" x 7'-0"	G		18	.889		425	47		472	545
1290	Half glass, 1-3/4" thick, poly core, 20 ga., 3'-0" x 7'-0"	G		18	.889		625	47		672	760
1320	18 ga., 2'-8" x 6'-8"	G		18	.889		645	47		692	785
1340	3'-0" x 6'-8"	G		17	.941		640	50		690	785
1360	3'-0" x 7'-0"	G		17	.941		655	50		705	800
1380	4'-0" x 7'-0"			15	1.067		785	56.50		841.50	955
1400	4'-0" x 8'-0"	G		14	1.143		880	60.50		940.50	1,075
1500	Flush full panel, 16 ga., steel hollow core										
1520	2'-0" x 6'-8"	G	2 Carp	20	.800	Ea.	580	42.50		622.50	705
1530	2'-8" x 6'-8"	G		20	.800		580	42.50		622.50	705
1540	3'-0" x 6'-8"	G		20	.800		495	42.50		537.50	615
1560	2'-8" x 7'-0"	G		18	.889		590	47		637	725
1570	3'-0" x 7'-0"	G		18	.889		550	47		597	680
1580	3'-6" x 7'-0"	G		18	.889		675	47		722	820
1590	4'-0" x 7'-0"	G		18	.889		745	47		792	890
1600	2'-8" x 8'-0"	G		18	.889		725	47		772	875
1620	3'-0" x 8'-0"	G		18	.889		740	47		787	890
1630	3'-6" x 8'-0"	G		18	.889		820	47		867	980
1640	4'-0" x 8'-0"	G		18	.889		865	47		912	1,025
1650	1-13/16", 14 ga., 2'-8" x 7'-0"	G		10	1.600		1,150	85		1,235	1,400
1670	3'-0" x 7'-0"	G		10	1.600		1,125	85		1,210	1,350
1690	3'-6" x 7'-0"	G		10	1.600		1,225	85		1,310	1,475
1700	4'-0" x 7'-0"	G		10	1.600		1,300	85		1,385	1,550
1720	Insulated, 1-3/4" thick, full panel, 18 ga., 3'-0" x 6'-8"	G		15	1.067		500	56.50		556.50	645
1740	2'-8" x 7'-0"	G		16	1		505	53		558	640
1760	3'-0" x 7'-0"	G		15	1.067		520	56.50		576.50	665
1800	4'-0" x 8'-0"	G		13	1.231		810	65.50		875.50	1,000
1820	Half glass, 18 ga., 3'-0" x 6'-8"	G		16	1		630	53		683	780
1840	2'-8" x 7'-0"	G		17	.941		660	50		710	805
1860	3'-0" x 7'-0"	G		16	1		690	53		743	845
1900	4'-0" x 8'-0"	G		14	1.143		590	60.50		650.50	750
2000	For vision lite, add						111			111	122
2010	For narrow lite, add						115			115	126

08 13 13 – Hollow Metal Doors

08 13 13.13 Standard Hollow Metal Doors		Crew	Daily Output	Labor-Hours	Unit	Material	Labor	Equipment	Total	Total Incl O&P
						2020 Bare Costs				
8100	For bottom louver, add				Ea.	243			243	267
8110	For baked enamel finish, add					30%	15%			
8120	For galvanizing, add					20%				
8190	Commercial door, flush steel, fire door see section 08 13 13 15									
8270	For dutch door shelf, add to standard door				Ea.	300			300	330
9000	Minimum labor/equipment charge	1 Carp	4	2	Job		106		106	170

08 13 13.15 Metal Fire Doors		Crew	Daily Output	Labor-Hours	Unit	Material	Labor	Equipment	Total	Total Incl O&P
0010	**METAL FIRE DOORS** R081313-20									
0015	Steel, flush, "B" label, 90 minutes									
0020	Full panel, 20 ga., 2'-0" x 6'-8"	2 Carp	20	.800	Ea.	420	42.50		462.50	530
0040	2'-8" x 6'-8"		18	.889		435	47		482	555
0060	3'-0" x 6'-8"		17	.941		435	50		485	560
0080	3'-0" x 7'-0"		17	.941		450	50		500	575
0140	18 ga., 3'-0" x 6'-8"		16	1		530	53		583	670
0160	2'-8" x 7'-0"		17	.941		560	50		610	700
0180	3'-0" x 7'-0"		16	1		510	53		563	645
0200	4'-0" x 7'-0"		15	1.067		645	56.50		701.50	800
0220	For "A" label, 3 hour, 18 ga., use same price as "B" label									
0240	For vision lite, add				Ea.	183			183	201
0300	Full panel, 16 ga., 2'-0" x 6'-8"	2 Carp	20	.800		320	42.50		362.50	420
0310	2'-8" x 6'-8"		18	.889		560	47		607	690
0320	3'-0" x 6'-8"		17	.941		450	50		500	575
0350	2'-8" x 7'-0"		17	.941		580	50		630	715
0360	3'-0" x 7'-0"		16	1		560	53		613	705
0370	4'-0" x 7'-0"		15	1.067		725	56.50		781.50	890
0520	Flush, "B" label, 90 minutes, egress core, 20 ga., 2'-0" x 6'-8"		18	.889		680	47		727	825
0540	2'-8" x 6'-8"		17	.941		685	50		735	830
0560	3'-0" x 6'-8"		16	1		685	53		738	840
0580	3'-0" x 7'-0"		16	1		715	53		768	870
0640	Flush, "A" label, 3 hour, egress core, 18 ga., 3'-0" x 6'-8"		15	1.067		740	56.50		796.50	905
0660	2'-8" x 7'-0"		16	1		780	53		833	940
0680	3'-0" x 7'-0"		15	1.067		410	56.50		466.50	545
0700	4'-0" x 7'-0"		14	1.143		945	60.50		1,005.50	1,150
9000	Minimum labor/equipment charge	1 Carp	4	2	Job		106		106	170

08 13 13.20 Residential Steel Doors		Crew	Daily Output	Labor-Hours	Unit	Material	Labor	Equipment	Total	Total Incl O&P
0010	**RESIDENTIAL STEEL DOORS**									
0020	Prehung, insulated, exterior									
0030	Embossed, full panel, 2'-8" x 6'-8" G	2 Carp	17	.941	Ea.	410	50		460	530
0040	3'-0" x 6'-8" G		15	1.067		300	56.50		356.50	420
0070	5'-4" x 6'-8", double G		8	2		590	106		696	820
0220	Half glass, 2'-8" x 6'-8" G		17	.941		330	50		380	445
0240	3'-0" x 6'-8" G		16	1		330	53		383	450
0260	3'-0" x 7'-0" G		16	1		400	53		453	525
0270	5'-4" x 6'-8", double G		8	2		985	106		1,091	1,250
1320	Flush face, full panel, 2'-8" x 6'-8" G		16	1		335	53		388	455
1340	3'-0" x 6'-8" G		15	1.067		335	56.50		391.50	460
1360	3'-0" x 7'-0" G		15	1.067		315	56.50		371.50	435
1380	5'-4" x 6'-8", double G		8	2		810	106		916	1,075
1420	Half glass, 2'-8" x 6'-8" G		17	.941		350	50		400	465
1440	3'-0" x 6'-8" G		16	1		350	53		403	470
1460	3'-0" x 7'-0" G		16	1		455	53		508	590
1480	5'-4" x 6'-8", double G		8	2		680	106		786	915

08 13 Metal Doors

08 13 13 – Hollow Metal Doors

08 13 13.20 Residential Steel Doors		Crew	Daily Output	Labor-Hours	Unit	Material	Labor	Equipment	Total	Total Incl O&P
						2020 Bare Costs				
1500	Sidelight, full lite, 1'-0" x 6'-8" with grille G				Ea.	395			395	435
1510	1'-0" x 6'-8", low E G					410			410	450
1520	1'-0" x 6'-8", half lite G					299			299	330
1530	1'-0" x 6'-8", half lite, low E G					300			300	330
2300	Interior, residential, closet, bi-fold, 2'-0" x 6'-8" G	2 Carp	16	1		251	53		304	360
2330	3'-0" wide G		16	1		284	53		337	400
2360	4'-0" wide G		15	1.067		293	56.50		349.50	410
2400	5'-0" wide G		14	1.143		370	60.50		430.50	505
2420	6'-0" wide G		13	1.231		300	65.50		365.50	435
9000	Minimum labor/equipment charge	1 Carp	4	2	Job		106		106	170

08 13 13.25 Doors Hollow Metal		Crew	Daily Output	Labor-Hours	Unit	Material	Labor	Equipment	Total	Total Incl O&P
0010	**DOORS HOLLOW METAL**									
0500	Exterior, commercial, flush, 20 ga., 1-3/4" x 7'-0" x 2'-6" wide G	2 Carp	15	1.067	Ea.	287	56.50		343.50	405
0530	2'-8" wide G		15	1.067		345	56.50		401.50	470
0560	3'-0" wide G		14	1.143		345	60.50		405.50	480
0590	3'-6" wide G		14	1.143		375	60.50		435.50	515
1000	18 ga., 1-3/4" x 7'-0" x 2'-6" wide G		15	1.067		300	56.50		356.50	420
1030	2'-8" wide G		15	1.067		305	56.50		361.50	425
1060	3'-0" wide G		14	1.143		296	60.50		356.50	425
1500	16 ga., 1-3/4" x 7'-0" x 2'-6" wide G		15	1.067		385	56.50		441.50	515
1530	2'-8" wide G		15	1.067		385	56.50		441.50	510
1560	3'-0" wide G		14	1.143		455	60.50		515.50	600
1590	3'-6" wide G		14	1.143		735	60.50		795.50	910
2900	Fire door, "A" label, 18 gauge, 1-3/4" x 2'-6" x 7'-0" G		15	1.067		275	56.50		331.50	395
2930	2'-8" wide G		15	1.067		275	56.50		331.50	395
2960	3'-0" wide G		14	1.143		275	60.50		335.50	405
2990	3'-6" wide G		14	1.143		495	60.50		555.50	645
3100	"B" label, 2'-6" wide G		15	1.067		655	56.50		711.50	810
3130	2'-8" wide G		15	1.067		675	56.50		731.50	830
3160	3'-0" wide G		14	1.143		670	60.50		730.50	835

08 13 16 – Aluminum Doors

08 13 16.10 Commercial Aluminum Doors		Crew	Daily Output	Labor-Hours	Unit	Material	Labor	Equipment	Total	Total Incl O&P
0010	**COMMERCIAL ALUMINUM DOORS**, flush, no glazing									
5000	Flush panel doors, pair of 2'-6" x 7'-0"	2 Sswk	2	8	Pr.	1,250	460		1,710	2,100
5050	3'-0" x 7'-0", single		2.50	6.400	Ea.	730	370		1,100	1,400
5100	Pair of 3'-0" x 7'-0"		2	8	Pr.	1,500	460		1,960	2,400
5150	3'-6" x 7'-0", single		2.50	6.400	Ea.	955	370		1,325	1,650

08 14 Wood Doors

08 14 13 – Carved Wood Doors

08 14 13.10 Types of Wood Doors, Carved		Crew	Daily Output	Labor-Hours	Unit	Material	Labor	Equipment	Total	Total Incl O&P
0010	**TYPES OF WOOD DOORS, CARVED**									
3000	Solid wood, 1-3/4" thick stile and rail									
3020	Mahogany, 3'-0" x 7'-0", six panel	2 Carp	14	1.143	Ea.	1,125	60.50		1,185.50	1,350
3030	With two lites		10	1.600		3,475	85		3,560	3,950
3040	3'-6" x 8'-0", six panel		10	1.600		1,800	85		1,885	2,100
3050	With two lites		8	2		3,050	106		3,156	3,550
3100	Pine, 3'-0" x 7'-0", six panel		14	1.143		655	60.50		715.50	820
3110	With two lites		10	1.600		880	85		965	1,100
3120	3'-6" x 8'-0", six panel		10	1.600		1,125	85		1,210	1,375

08 14 Wood Doors

08 14 13 – Carved Wood Doors

08 14 13.10 Types of Wood Doors, Carved		Crew	Daily Output	Labor-Hours	Unit	Material	Labor	Equipment	Total	Total Incl O&P
						2020 Bare Costs				
3130	With two lites	2 Carp	8	2	Ea.	2,000	106		2,106	2,400
3200	Red oak, 3′-0″ x 7′-0″, six panel		14	1.143		1,325	60.50		1,385.50	1,575
3210	With two lites		10	1.600		2,675	85		2,760	3,075
3220	3′-6″ x 8′-0″, six panel		10	1.600		2,825	85		2,910	3,225
3230	With two lites		8	2		3,575	106		3,681	4,125
4000	Hand carved door, mahogany									
4020	3′-0″ x 7′-0″, simple design	2 Carp	14	1.143	Ea.	1,875	60.50		1,935.50	2,175
4030	Intricate design		11	1.455		3,825	77.50		3,902.50	4,325
4040	3′-6″ x 8′-0″, simple design		10	1.600		2,875	85		2,960	3,300
4050	Intricate design		8	2		3,875	106		3,981	4,450
4400	For custom finish, add					640			640	705
4600	Side light, mahogany, 7′-0″ x 1′-6″ wide, 4 lites	2 Carp	18	.889		1,200	47		1,247	1,375
4610	6 lites		14	1.143		2,800	60.50		2,860.50	3,175
4620	8′-0″ x 1′-6″ wide, 4 lites		14	1.143		2,225	60.50		2,285.50	2,550
4630	6 lites		10	1.600		2,375	85		2,460	2,750
4640	Side light, oak, 7′-0″ x 1′-6″ wide, 4 lites		18	.889		1,425	47		1,472	1,625
4650	6 lites		14	1.143		2,475	60.50		2,535.50	2,825
4660	8′-0″ x 1′-6″ wide, 4 lites		14	1.143		1,325	60.50		1,385.50	1,575
4670	6 lites		10	1.600		2,475	85		2,560	2,850

08 14 16 – Flush Wood Doors

08 14 16.09 Smooth Wood Doors

		Crew	Daily Output	Labor-Hours	Unit	Material	Labor	Equipment	Total	Total Incl O&P
0010	**SMOOTH WOOD DOORS**									
0015	Flush, interior, hollow core									
0025	Lauan face, 1-3/8″, 3′-0″ x 6′-8″	2 Carp	17	.941	Ea.	53.50	50		103.50	139
0030	4′-0″ x 6′-8″		16	1		134	53		187	232
0080	1-3/4″, 2′-0″ x 6′-8″		17	.941		60	50		110	146
0085	3′-0″ x 6′-8″	1 Carp	16	.500		53.50	26.50		80	101
0108	3′-0″ x 7′-0″	2 Carp	16	1		226	53		279	335
0112	Pair of 3′-0″ x 7′-0″		9	1.778	Pr.	153	94.50		247.50	320
0140	Birch face, 1-3/8″, 2′-6″ x 6′-8″		17	.941	Ea.	122	50		172	214
0180	3′-0″ x 6′-8″		17	.941		101	50		151	191
0200	4′-0″ x 6′-8″		16	1		167	53		220	269
0202	1-3/4″, 2′-0″ x 6′-8″		17	.941		69	50		119	156
0204	2′-4″ x 7′-0″		16	1		133	53		186	232
0206	2′-6″ x 7′-0″		16	1		137	53		190	236
0210	3′-0″ x 7′-0″		16	1		174	53		227	277
0214	Pair of 3′-0″ x 7′-0″		9	1.778	Pr.	274	94.50		368.50	450
0218	Birch face, 1-3/4″, 3′-0″ x 8′-0″		7	2.286	Ea.	255	121		376	475
0220	Oak face, 1-3/8″, 2′-0″ x 6′-8″		17	.941		117	50		167	208
0280	3′-0″ x 6′-8″		17	.941		112	50		162	203
0300	4′-0″ x 6′-8″		16	1		148	53		201	248
0305	1-3/4″, 2′-6″ x 6′-8″		17	.941		137	50		187	230
0310	3′-0″ x 7′-0″		16	1		272	53		325	385
0320	Walnut face, 1-3/8″, 2′-0″ x 6′-8″		17	.941		206	50		256	305
0340	2′-6″ x 6′-8″		17	.941		201	50		251	300
0380	3′-0″ x 6′-8″		17	.941		210	50		260	310
0400	4′-0″ x 6′-8″		16	1		231	53		284	340
0430	For 7′-0″ high, add					31.50			31.50	35
0440	For 8′-0″ high, add					46			46	50.50
0480	For prefinishing, clear, add					57			57	63
0500	For prefinishing, stain, add					67.50			67.50	74
1320	M.D. overlay on hardboard, 1-3/8″, 2′-0″ x 6′-8″	2 Carp	17	.941		134	50		184	227

08 14 16.09 Smooth Wood Doors		Crew	Daily Output	Labor-Hours	Unit	Material	Labor	Equipment	Total	Total Incl O&P
						2020 Bare Costs				
1340	2′-6″ x 6′-8″	2 Carp	17	.941	Ea.	135	50		185	229
1380	3′-0″ x 6′-8″		17	.941		136	50		186	230
1400	4′-0″ x 6′-8″		16	1		231	53		284	340
1420	For 7′-0″ high, add					18.95			18.95	21
1440	For 8′-0″ high, add					42			42	46
1720	H.P. plastic laminate, 1-3/8″, 2′-0″ x 6′-8″	2 Carp	16	1		286	53		339	400
1740	2′-6″ x 6′-8″		16	1		282	53		335	395
1780	3′-0″ x 6′-8″		15	1.067		305	56.50		361.50	425
1800	4′-0″ x 6′-8″		14	1.143		405	60.50		465.50	545
1820	For 7′-0″ high, add					22.50			22.50	24.50
1840	For 8′-0″ high, add					39			39	43
2020	Particle core, lauan face, 1-3/8″, 2′-6″ x 6′-8″	2 Carp	15	1.067		105	56.50		161.50	207
2040	3′-0″ x 6′-8″		14	1.143		108	60.50		168.50	217
2080	3′-0″ x 7′-0″		13	1.231		128	65.50		193.50	246
2085	4′-0″ x 7′-0″		12	1.333		140	71		211	268
2110	1-3/4″, 3′-0″ x 7′-0″		13	1.231		226	65.50		291.50	355
2120	Birch face, 1-3/8″, 2′-6″ x 6′-8″		15	1.067		120	56.50		176.50	223
2140	3′-0″ x 6′-8″		14	1.143		130	60.50		190.50	241
2180	3′-0″ x 7′-0″		13	1.231		143	65.50		208.50	262
2200	4′-0″ x 7′-0″		12	1.333		156	71		227	286
2205	1-3/4″, 3′-0″ x 7′-0″		13	1.231		211	65.50		276.50	335
2220	Oak face, 1-3/8″, 2′-6″ x 6′-8″		15	1.067		132	56.50		188.50	237
2240	3′-0″ x 6′-8″		14	1.143		145	60.50		205.50	258
2280	3′-0″ x 7′-0″		13	1.231		152	65.50		217.50	272
2300	4′-0″ x 7′-0″		12	1.333		176	71		247	310
2305	1-3/4″, 3′-0″ x 7′-0″		13	1.231		256	65.50		321.50	385
2320	Walnut face, 1-3/8″, 2′-0″ x 6′-8″		15	1.067		138	56.50		194.50	243
2340	2′-6″ x 6′-8″		14	1.143		155	60.50		215.50	268
2380	3′-0″ x 6′-8″		13	1.231		175	65.50		240.50	297
2400	4′-0″ x 6′-8″		12	1.333		229	71		300	365
2440	For 8′-0″ high, add					46.50			46.50	51.50
2460	For 8′-0″ high walnut, add					42.50			42.50	46.50
2720	For prefinishing, clear, add					41.50			41.50	45.50
2740	For prefinishing, stain, add					59.50			59.50	65.50
3320	M.D. overlay on hardboard, 1-3/8″, 2′-6″ x 6′-8″	2 Carp	14	1.143		189	60.50		249.50	305
3340	3′-0″ x 6′-8″		13	1.231		213	65.50		278.50	340
3380	3′-0″ x 7′-0″		12	1.333		216	71		287	350
3400	4′-0″ x 7′-0″		10	1.600		298	85		383	465
3440	For 8′-0″ height, add					42			42	46
3460	For solid wood core, add					75.50			75.50	83
3720	H.P. plastic laminate, 1-3/8″, 2′-6″ x 6′-8″	2 Carp	13	1.231		172	65.50		237.50	294
3740	3′-0″ x 6′-8″		12	1.333		194	71		265	330
3780	3′-0″ x 7′-0″		11	1.455		208	77.50		285.50	350
3800	4′-0″ x 7′-0″		8	2		415	106		521	630
3840	For 8′-0″ height, add					42			42	46
3860	For solid wood core, add					44			44	48.50
4000	Exterior, flush, solid core, birch, 1-3/4″ x 2′-6″ x 7′-0″	2 Carp	15	1.067		165	56.50		221.50	273
4020	2′-8″ wide		15	1.067		239	56.50		295.50	355
4040	3′-0″ wide		14	1.143		182	60.50		242.50	298
4045	3′-0″ x 8′-0″	1 Carp	8	1		530	53		583	665
4100	Oak faced 1-3/4″ x 2′-6″ x 7′-0″	2 Carp	15	1.067		238	56.50		294.50	355
4120	2′-8″ wide		15	1.067		264	56.50		320.50	380
4140	3′-0″ wide		14	1.143		255	60.50		315.50	380

08 14 16.09 Smooth Wood Doors		Crew	Daily Output	Labor-Hours	Unit	2020 Bare Costs Material	Labor	Equipment	Total	Total Incl O&P
4200	Walnut faced, 1-3/4" x 2'-6" x 7'-0"	2 Carp	15	1.067	Ea.	340	56.50		396.50	465
4220	2'-8" wide		15	1.067		375	56.50		431.50	500
4240	3'-0" wide		14	1.143		315	60.50		375.50	445
4300	For 6'-8" high door, deduct from 7'-0" door					18.70			18.70	20.50
5000	Wood doors, for vision lite, add					111			111	122
5010	Wood doors, for narrow lite, add					115			115	126
5015	Wood doors, for bottom (or top) louver, add					243			243	267
9000	Minimum labor/equipment charge	1 Carp	4	2	Job		106		106	170

08 14 16.10 Wood Doors Decorator		Crew	Daily Output	Labor-Hours	Unit	Material	Labor	Equipment	Total	Total Incl O&P
0010	**WOOD DOORS DECORATOR**									
1800	Exterior, flush, solid wood core, birch 1-3/4" x 2'-6" x 7'-0"	2 Carp	15	1.067	Ea.	370	56.50		426.50	495
1820	2'-8" wide		15	1.067		360	56.50		416.50	485
1840	3'-0" wide		14	1.143		360	60.50		420.50	500
1900	Oak faced, 1-3/4" x 2'-6" x 7'-0"		15	1.067		189	56.50		245.50	299
1920	2'-8" wide		15	1.067		495	56.50		551.50	635
1940	3'-0" wide		14	1.143		465	60.50		525.50	610
2100	Walnut faced, 1-3/4" x 2'-6" x 7'-0"		15	1.067		475	56.50		531.50	610
2120	2'-8" wide		15	1.067		510	56.50		566.50	650
2140	3'-0" wide		14	1.143		475	60.50		535.50	625

08 14 16.20 Wood Fire Doors		Crew	Daily Output	Labor-Hours	Unit	Material	Labor	Equipment	Total	Total Incl O&P
0010	**WOOD FIRE DOORS**									
0020	Particle core, 7 face plys, "B" label,									
0040	1 hour, birch face, 1-3/4" x 2'-6" x 6'-8"	2 Carp	14	1.143	Ea.	465	60.50		525.50	615
0080	3'-0" x 6'-8"		13	1.231		540	65.50		605.50	700
0090	3'-0" x 7'-0"		12	1.333		500	71		571	665
0100	4'-0" x 7'-0"		12	1.333		730	71		801	920
0140	Oak face, 2'-6" x 6'-8"		14	1.143		585	60.50		645.50	745
0180	3'-0" x 6'-8"		13	1.231		525	65.50		590.50	680
0190	3'-0" x 7'-0"		12	1.333		585	71		656	755
0200	4'-0" x 7'-0"		12	1.333		685	71		756	870
0240	Walnut face, 2'-6" x 6'-8"		14	1.143		545	60.50		605.50	700
0280	3'-0" x 6'-8"		13	1.231		555	65.50		620.50	715
0290	3'-0" x 7'-0"		12	1.333		620	71		691	800
0300	4'-0" x 7'-0"		12	1.333		865	71		936	1,075
0440	M.D. overlay on hardboard, 2'-6" x 6'-8"		15	1.067		370	56.50		426.50	500
0480	3'-0" x 6'-8"		14	1.143		460	60.50		520.50	605
0490	3'-0" x 7'-0"		13	1.231		490	65.50		555.50	640
0500	4'-0" x 7'-0"		12	1.333		475	71		546	635
0740	90 minutes, birch face, 1-3/4" x 2'-6" x 6'-8"		14	1.143		345	60.50		405.50	475
0780	3'-0" x 6'-8"		13	1.231		410	65.50		475.50	555
0790	3'-0" x 7'-0"		12	1.333		460	71		531	620
0800	4'-0" x 7'-0"		12	1.333		555	71		626	725
0840	Oak face, 2'-6" x 6'-8"		14	1.143		470	60.50		530.50	615
0880	3'-0" x 6'-8"		13	1.231		520	65.50		585.50	680
0890	3'-0" x 7'-0"		12	1.333		555	71		626	725
0900	4'-0" x 7'-0"		12	1.333		680	71		751	865
0940	Walnut face, 2'-6" x 6'-8"		14	1.143		545	60.50		605.50	700
0980	3'-0" x 6'-8"		13	1.231		465	65.50		530.50	615
0990	3'-0" x 7'-0"		12	1.333		570	71		641	740
1000	4'-0" x 7'-0"		12	1.333		695	71		766	880
1140	M.D. overlay on hardboard, 2'-6" x 6'-8"		15	1.067		455	56.50		511.50	590
1180	3'-0" x 6'-8"		14	1.143		435	60.50		495.50	580

08 14 Wood Doors

08 14 16 – Flush Wood Doors

08 14 16.20 Wood Fire Doors		Crew	Daily Output	Labor-Hours	Unit	Material	2020 Bare Costs Labor	Equipment	Total	Total Incl O&P
1190	3'-0" x 7'-0"	2 Carp	13	1.231	Ea.	515	65.50		580.50	675
1200	4'-0" x 7'-0"		12	1.333		595	71		666	765
1240	For 8'-0" height, add					86			86	94.50
1260	For 8'-0" height walnut, add					105			105	115
2200	Custom architectural "B" label, flush, 1-3/4" thick, birch,									
2210	Solid core									
2220	2'-6" x 7'-0"	2 Carp	15	1.067	Ea.	425	56.50		481.50	560
2260	3'-0" x 7'-0"		14	1.143		370	60.50		430.50	505
2300	4'-0" x 7'-0"		13	1.231		565	65.50		630.50	725
2420	4'-0" x 8'-0"		11	1.455		565	77.50		642.50	750
2480	For oak veneer, add					50%				
2500	For walnut veneer, add					75%				
9000	Minimum labor/equipment charge	1 Carp	4	2	Job		106		106	170

08 14 33 – Stile and Rail Wood Doors

08 14 33.10 Wood Doors Paneled		Crew	Daily Output	Labor-Hours	Unit	Material	Labor	Equipment	Total	Total Incl O&P
0010	**WOOD DOORS PANELED**									
0020	Interior, six panel, hollow core, 1-3/8" thick									
0040	Molded hardboard, 2'-0" x 6'-8"	2 Carp	17	.941	Ea.	65.50	50		115.50	153
0060	2'-6" x 6'-8"		17	.941		69	50		119	156
0070	2'-8" x 6'-8"		17	.941		72	50		122	159
0080	3'-0" x 6'-8"		17	.941		89.50	50		139.50	179
0140	Embossed print, molded hardboard, 2'-0" x 6'-8"		17	.941		69	50		119	156
0160	2'-6" x 6'-8"		17	.941		69	50		119	156
0180	3'-0" x 6'-8"		17	.941		89.50	50		139.50	179
0540	Six panel, solid, 1-3/8" thick, pine, 2'-0" x 6'-8"		15	1.067		166	56.50		222.50	274
0560	2'-6" x 6'-8"		14	1.143		164	60.50		224.50	279
0580	3'-0" x 6'-8"		13	1.231		150	65.50		215.50	270
1020	Two panel, bored rail, solid, 1-3/8" thick, pine, 1'-6" x 6'-8"		16	1		257	53		310	370
1040	2'-0" x 6'-8"		15	1.067		335	56.50		391.50	460
1060	2'-6" x 6'-8"		14	1.143		380	60.50		440.50	520
1340	Two panel, solid, 1-3/8" thick, fir, 2'-0" x 6'-8"		15	1.067		173	56.50		229.50	281
1360	2'-6" x 6'-8"		14	1.143		230	60.50		290.50	350
1380	3'-0" x 6'-8"		13	1.231		420	65.50		485.50	565
1740	Five panel, solid, 1-3/8" thick, fir, 2'-0" x 6'-8"		15	1.067		310	56.50		366.50	435
1760	2'-6" x 6'-8"		14	1.143		400	60.50		460.50	540
1780	3'-0" x 6'-8"		13	1.231		400	65.50		465.50	545
4000	Ext., hollow core, 1-3/4" thick, hdbd., with 10" x 10" wdw., 2'-8" x 6'-8"		16	1		183	53		236	286
4050	3'-0" x 6'-8"		16	1		195	53		248	299
4100	Solid core, 2'-8" x 6'-8"		16	1		178	53		231	281
4150	3'-0" x 6'-8"		16	1		193	53		246	298
4190	Exterior, Knotty pine, paneled, 1-3/4", 3'-0" x 6'-8"		16	1		880	53		933	1,050
4195	Double 1-3/4", 3'-0" x 6'-8"		16	1		1,750	53		1,803	2,000
4200	Ash, paneled, 1-3/4", 3'-0" x 6'-8"		16	1		1,000	53		1,053	1,175
4205	Double 1-3/4", 3'-0" x 6'-8"		16	1		2,000	53		2,053	2,275
4210	Cherry, paneled, 1-3/4", 3'-0" x 6'-8"		16	1		1,150	53		1,203	1,350
4215	Double 1-3/4", 3'-0" x 6'-8"		16	1		2,300	53		2,353	2,600
4230	Ash, paneled, 1-3/4", 3'-0" x 8'-0"		16	1		1,350	53		1,403	1,550
4235	Double 1-3/4", 3'-0" x 8'-0"		16	1		2,675	53		2,728	3,025
4240	Hard maple, paneled, 1-3/4", 3'-0" x 8'-0"		16	1		1,350	53		1,403	1,550
4245	Double 1-3/4", 3'-0" x 8'-0"		16	1		2,675	53		2,728	3,025
4250	Cherry, paneled, 1-3/4", 3'-0" x 8'-0"		16	1		1,350	53		1,403	1,550
4255	Double 1-3/4", 3'-0" x 8'-0"		16	1		2,675	53		2,728	3,025

08 14 33 – Stile and Rail Wood Doors

08 14 33.10 Wood Doors Paneled		Crew	Daily Output	Labor-Hours	Unit	Material	Labor	Equipment	Total	Total Incl O&P
						2020 Bare Costs				
9000	Minimum labor/equipment charge	1 Carp	4	2	Job		106		106	170

08 14 33.20 Wood Doors Residential		Crew	Daily Output	Labor-Hours	Unit	Material	Labor	Equipment	Total	Total Incl O&P
0010	**WOOD DOORS RESIDENTIAL**									
0200	Exterior, combination storm & screen, pine									
0220	Cross buck, 6'-9" x 2'-6" wide	2 Carp	11	1.455	Ea.	360	77.50		437.50	520
0260	2'-8" wide		10	1.600		330	85		415	495
0280	3'-0" wide		9	1.778		345	94.50		439.50	530
0300	7'-1" x 3'-0" wide		9	1.778		365	94.50		459.50	550
0400	Full lite, 6'-9" x 2'-6" wide		11	1.455		355	77.50		432.50	515
0420	2'-8" wide		10	1.600		345	85		430	515
0440	3'-0" wide		9	1.778		355	94.50		449.50	540
0500	7'-1" x 3'-0" wide		9	1.778		385	94.50		479.50	570
0700	Dutch door, pine, 1-3/4" x 2'-8" x 6'-8", 6 panel		12	1.333		740	71		811	930
0720	Half glass		10	1.600		1,025	85		1,110	1,250
0800	3'-0" wide, 6 panel		12	1.333		660	71		731	845
0820	Half glass		10	1.600		1,050	85		1,135	1,275
1000	Entrance door, colonial, 1-3/4" x 6'-8" x 2'-8" wide		16	1		625	53		678	775
1020	6 panel pine, 3'-0" wide		15	1.067		610	56.50		666.50	760
1100	8 panel pine, 2'-8" wide		16	1		735	53		788	895
1120	3'-0" wide		15	1.067		670	56.50		726.50	830
1200	For tempered safety glass lites (min. of 2), add					92.50			92.50	102
1300	Flush, birch, solid core, 1-3/4" x 6'-8" x 2'-8" wide	2 Carp	16	1		169	53		222	271
1320	3'-0" wide		15	1.067		153	56.50		209.50	260
1350	7'-0" x 2'-8" wide		16	1		141	53		194	240
1360	3'-0" wide		15	1.067		149	56.50		205.50	255
1380	For tempered safety glass lites, add					120			120	132
1420	6'-8" x 3'-0" wide, fir	2 Carp	16	1		540	53		593	680
1720	Carved mahogany 3'-0" x 6'-8"		15	1.067		1,600	56.50		1,656.50	1,850
1800	Lauan, solid core, 1-3/4" x 7'-0" x 2'-4" wide		16	1		143	53		196	243
1810	2'-6" wide		15	1.067		150	56.50		206.50	256
1820	2'-8" wide		9	1.778		187	94.50		281.50	355
1830	3'-0" wide		16	1		204	53		257	310
1840	3'-4" wide		16	1		276	53		329	390
1850	Pair of 3'-0" wide		15	1.067	Pr.	405	56.50		461.50	540
2700	Interior, closet, bi-fold, w/hardware, no frame or trim incl.									
2720	Flush, birch, 2'-6" x 6'-8"	2 Carp	13	1.231	Ea.	80.50	65.50		146	194
2740	3'-0" wide		13	1.231		78	65.50		143.50	191
2760	4'-0" wide		12	1.333		123	71		194	249
2780	5'-0" wide		11	1.455		121	77.50		198.50	257
2800	6'-0" wide		10	1.600		137	85		222	287
2804	Flush lauan 2'-0" x 6'-8"		14	1.143		57	60.50		117.50	161
2810	8'-0" wide		9	1.778		214	94.50		308.50	385
2817	6'-0" wide		9	1.778		180	94.50		274.50	350
2820	Flush, hardboard, primed, 6'-8" x 2'-6" wide		13	1.231		77.50	65.50		143	191
2840	3'-0" wide		13	1.231		95.50	65.50		161	210
2860	4'-0" wide		12	1.333		156	71		227	286
2880	5'-0" wide		11	1.455		206	77.50		283.50	350
2900	6'-0" wide		10	1.600		193	85		278	350
3000	Raised panel pine, 6'-6" or 6'-8" x 2'-6" wide		13	1.231		207	65.50		272.50	335
3020	3'-0" wide		13	1.231		305	65.50		370.50	440
3040	4'-0" wide		12	1.333		330	71		401	480
3060	5'-0" wide		11	1.455		425	77.50		502.50	595

08 14 33 – Stile and Rail Wood Doors

08 14 33.20 Wood Doors Residential		Crew	Daily Output	Labor-Hours	Unit	Material	Labor	Equipment	Total	Total Incl O&P
						2020 Bare Costs				
3080	6'-0" wide	2 Carp	10	1.600	Ea.	470	85		555	650
3200	Louvered, pine, 6'-6" or 6'-8" x 2'-6" wide		13	1.231		170	65.50		235.50	292
3220	3'-0" wide		13	1.231		284	65.50		349.50	420
3240	4'-0" wide		12	1.333		256	71		327	395
3260	5'-0" wide		11	1.455		276	77.50		353.50	430
3280	6'-0" wide		10	1.600		335	85		420	500
4400	Bi-passing closet, incl. hardware and frame, no trim incl.									
4420	Flush, lauan, 6'-8" x 4'-0" wide	2 Carp	12	1.333	Opng.	178	71		249	310
4440	5'-0" wide		11	1.455		196	77.50		273.50	340
4460	6'-0" wide		10	1.600		177	85		262	330
4600	Flush, birch, 6'-8" x 4'-0" wide		12	1.333		270	71		341	410
4620	5'-0" wide		11	1.455		227	77.50		304.50	375
4640	6'-0" wide		10	1.600		360	85		445	530
4800	Louvered, pine, 6'-8" x 4'-0" wide		12	1.333		525	71		596	690
4820	5'-0" wide		11	1.455		740	77.50		817.50	940
4840	6'-0" wide		10	1.600		770	85		855	980
5000	Paneled, pine, 6'-8" x 4'-0" wide		12	1.333		505	71		576	670
5020	5'-0" wide		11	1.455		760	77.50		837.50	960
5040	6'-0" wide		10	1.600		900	85		985	1,125
5042	8'-0" wide		12	1.333		1,000	71		1,071	1,225
6100	Folding accordion, closet, including track and frame									
6120	Vinyl, 2 layer, stock	2 Carp	10	1.600	Ea.	81	85		166	225
6140	Woven mahogany and vinyl, stock		10	1.600		65	85		150	208
6160	Wood slats with vinyl overlay, stock		10	1.600		174	85		259	330
6180	Economy vinyl, stock		10	1.600		42	85		127	183
6200	Rigid PVC		10	1.600		124	85		209	273
7310	Passage doors, flush, no frame included									
7320	Hardboard, hollow core, 1-3/8" x 6'-8" x 1'-6" wide	2 Carp	18	.889	Ea.	43	47		90	123
7330	2'-0" wide		18	.889		48	47		95	128
7340	2'-6" wide		18	.889		55	47		102	136
7350	2'-8" wide		18	.889		57	47		104	139
7360	3'-0" wide		17	.941		58	50		108	144
7420	Lauan, hollow core, 1-3/8" x 6'-8" x 1'-6" wide		18	.889		59.50	47		106.50	141
7440	2'-0" wide		18	.889		63	47		110	145
7450	2'-4" wide		18	.889		71.50	47		118.50	155
7460	2'-6" wide		18	.889		71.50	47		118.50	155
7480	2'-8" wide		18	.889		74	47		121	157
7500	3'-0" wide		17	.941		90.50	50		140.50	180
7700	Birch, hollow core, 1-3/8" x 6'-8" x 1'-6" wide		18	.889		78	47		125	161
7720	2'-0" wide		18	.889		86	47		133	170
7740	2'-6" wide		18	.889		87.50	47		134.50	172
7760	2'-8" wide		18	.889		91	47		138	176
7780	3'-0" wide		17	.941		94.50	50		144.50	184
8000	Pine louvered, 1-3/8" x 6'-8" x 1'-6" wide		19	.842		160	45		205	248
8020	2'-0" wide		18	.889		177	47		224	270
8040	2'-6" wide		18	.889		200	47		247	296
8060	2'-8" wide		18	.889		220	47		267	320
8080	3'-0" wide		17	.941		233	50		283	335
8300	Pine paneled, 1-3/8" x 6'-8" x 1'-6" wide		19	.842		201	45		246	293
8320	2'-0" wide		18	.889		251	47		298	350
8330	2'-4" wide		18	.889		252	47		299	355
8340	2'-6" wide		18	.889		265	47		312	365
8360	2'-8" wide		18	.889		271	47		318	375

08 14 Wood Doors

08 14 33 – Stile and Rail Wood Doors

08 14 33.20 Wood Doors Residential		Crew	Daily Output	Labor-Hours	Unit	Material	2020 Bare Costs Labor	Equipment	Total	Total Incl O&P
8380	3'-0" wide	2 Carp	17	.941	Ea.	291	50		341	400
9000	Passage doors, flush, no frame, birch, solid core, 1-3/8" x 2'-4" x 7'-0"		16	1		133	53		186	231
9020	2'-8" wide		16	1		135	53		188	233
9040	3'-0" wide		16	1		167	53		220	269
9060	3'-4" wide		15	1.067		320	56.50		376.50	440
9080	Pair of 3'-0" wide		9	1.778	Pr.	296	94.50		390.50	475
9100	Lauan, solid core, 1-3/8" x 7'-0" x 2'-4" wide		16	1	Ea.	171	53		224	273
9120	2'-8" wide		16	1		152	53		205	253
9140	3'-0" wide		16	1		201	53		254	305
9160	3'-4" wide		15	1.067		218	56.50		274.50	330
9180	Pair of 3'-0" wide		9	1.778	Pr.	425	94.50		519.50	615
9200	Hardboard, solid core, 1-3/8" x 7'-0" x 2'-4" wide		16	1	Ea.	177	53		230	279
9220	2'-8" wide		16	1		180	53		233	283
9240	3'-0" wide		16	1		186	53		239	290
9260	3'-4" wide		15	1.067		375	56.50		431.50	500
9900	Minimum labor/equipment charge	1 Carp	4	2	Job		106		106	170

08 14 35 – Torrified Doors

08 14 35.10 Torrified Exterior Doors		Crew	Daily Output	Labor-Hours	Unit	Material	Labor	Equipment	Total	Total Incl O&P
0010	**TORRIFIED EXTERIOR DOORS**									
0020	Wood doors made from torrified wood, exterior									
0030	All doors require a finish be applied, all glass is insulated									
0040	All doors require pilot holes for all fasteners									
0100	6 panel, paint grade poplar, 1-3/4" x 3'-0" x 6'-8"	2 Carp	12	1.333	Ea.	1,425	71		1,496	1,700
0120	Half glass 3'-0" x 6'-8"	"	12	1.333		1,550	71		1,621	1,825
0200	Side lite, full glass, 1-3/4" x 1'-2" x 6'-8"					1,025			1,025	1,125
0220	Side lite, half glass, 1-3/4" x 1'-2" x 6'-8"					995			995	1,100
0300	Raised face, 2 panel, paint grade poplar, 1-3/4" x 3'-0" x 7'-0"	2 Carp	12	1.333		1,425	71		1,496	1,700
0320	Side lite, raised face, half glass, 1-3/4" x 1'-2" x 7'-0"					1,150			1,150	1,275
0500	6 panel, Fir, 1-3/4" x 3'-0" x 6'-8"	2 Carp	12	1.333		2,050	71		2,121	2,375
0520	Half glass 3'-0" x 6'-8"	"	12	1.333		2,025	71		2,096	2,350
0600	Side lite, full glass, 1-3/4" x 1'-2" x 6'-8"					1,025			1,025	1,125
0620	Side lite, half glass, 1-3/4" x 1'-2" x 6'-8"					1,050			1,050	1,175
0700	6 panel, Mahogany, 1-3/4" x 3'-0" x 6'-8"	2 Carp	12	1.333		2,525	71		2,596	2,900
0800	Side lite, full glass, 1-3/4" x 1'-2" x 6'-8"					1,150			1,150	1,250
0820	Side lite, half glass, 1-3/4" x 1'-2" x 6'-8"					1,150			1,150	1,275

08 14 40 – Interior Cafe Doors

08 14 40.10 Cafe Style Doors		Crew	Daily Output	Labor-Hours	Unit	Material	Labor	Equipment	Total	Total Incl O&P
0010	**CAFE STYLE DOORS**									
6520	Interior cafe doors, 2'-6" opening, stock, panel pine	2 Carp	16	1	Ea.	430	53		483	560
6540	3'-0" opening	"	16	1	"	455	53		508	585
6550	Louvered pine									
6560	2'-6" opening	2 Carp	16	1	Ea.	345	53		398	460
8000	3'-0" opening		16	1		370	53		423	495
8010	2'-6" opening, hardwood		16	1		375	53		428	500
8020	3'-0" opening		16	1		425	53		478	550
9000	Minimum labor/equipment charge	1 Carp	4	2	Job		106		106	170

08 16 Composite Doors

08 16 13 – Fiberglass Doors

08 16 13.10 Entrance Doors, Fibrous Glass		Crew	Daily Output	Labor-Hours	Unit	Material	Labor	Equipment	Total	Total Incl O&P
0010	**ENTRANCE DOORS, FIBROUS GLASS**									
0020	Exterior, fiberglass, door, 2'-8" wide x 6'-8" high G	2 Carp	15	1.067	Ea.	276	56.50		332.50	395
0040	3'-0" wide x 6'-8" high G		15	1.067		279	56.50		335.50	395
0060	3'-0" wide x 7'-0" high G		15	1.067		520	56.50		576.50	660
0080	3'-0" wide x 6'-8" high, with two lites G		15	1.067		345	56.50		401.50	470
0100	3'-0" wide x 8'-0" high, with two lites G		15	1.067		535	56.50		591.50	680
0110	Half glass, 3'-0" wide x 6'-8" high G		15	1.067		485	56.50		541.50	625
0120	3'-0" wide x 6'-8" high, low E G		15	1.067		525	56.50		581.50	670
0130	3'-0" wide x 8'-0" high G		15	1.067		605	56.50		661.50	755
0140	3'-0" wide x 8'-0" high, low E G		15	1.067		685	56.50		741.50	845
0150	Side lights, 1'-0" wide x 6'-8" high G					284			284	310
0160	1'-0" wide x 6'-8" high, low E G					305			305	335
0180	1'-0" wide x 6'-8" high, full glass G					340			340	375
0190	1'-0" wide x 6'-8" high, low E G					385			385	420

08 16 14 – French Doors

08 16 14.10 Exterior Doors With Glass Lites		Crew	Daily Output	Labor-Hours	Unit	Material	Labor	Equipment	Total	Total Incl O&P
0010	**EXTERIOR DOORS WITH GLASS LITES**									
0020	French, Fir, 1-3/4", 3'-0" wide x 6'-8" high	2 Carp	12	1.333	Ea.	650	71		721	830
0025	Double		12	1.333		1,300	71		1,371	1,550
0030	Maple, 1-3/4", 3'-0" wide x 6'-8" high		12	1.333		725	71		796	915
0035	Double		12	1.333		1,450	71		1,521	1,725
0040	Cherry, 1-3/4", 3'-0" wide x 6'-8" high		12	1.333		850	71		921	1,050
0045	Double		12	1.333		1,700	71		1,771	2,000
0100	Mahogany, 1-3/4", 3'-0" wide x 8'-0" high		10	1.600		875	85		960	1,100
0105	Double		10	1.600		1,750	85		1,835	2,050
0110	Fir, 1-3/4", 3'-0" wide x 8'-0" high		10	1.600		1,300	85		1,385	1,550
0115	Double		10	1.600		2,625	85		2,710	3,000
0120	Oak, 1-3/4", 3'-0" wide x 8'-0" high		10	1.600		1,975	85		2,060	2,275
0125	Double		10	1.600		3,925	85		4,010	4,450

08 17 Integrated Door Opening Assemblies

08 17 13 – Integrated Metal Door Opening Assemblies

08 17 13.10 Hollow Metal Doors and Frames		Crew	Daily Output	Labor-Hours	Unit	Material	Labor	Equipment	Total	Total Incl O&P
0010	**HOLLOW METAL DOORS AND FRAMES**									
0100	Prehung, flush, 18 ga., 1-3/4" x 6'-8" x 2'-8" wide	1 Carp	16	.500	Ea.	640	26.50		666.50	750
0120	3'-0" wide		15	.533		640	28.50		668.50	750
0130	3'-6" wide		13	.615		740	32.50		772.50	865
0140	4'-0" wide		10	.800		740	42.50		782.50	885
0300	Double, 1-3/4" x 3'-0" x 6'-8"		5	1.600		1,175	85		1,260	1,425
0350	3'-0" x 8'-0"		4	2		1,425	106		1,531	1,750

08 17 13.20 Stainless Steel Doors and Frames		Crew	Daily Output	Labor-Hours	Unit	Material	Labor	Equipment	Total	Total Incl O&P
0010	**STAINLESS STEEL DOORS AND FRAMES**									
0020	Stainless steel (304) prehung 24 ga. 2'-6" x 6'-8" door w/16 g frame G	2 Carp	6	2.667	Ea.	1,900	142		2,042	2,300
0025	2'-8" x 6'-8" G		6	2.667		1,950	142		2,092	2,375
0030	3'-0" x 6'-8" G		6	2.667		1,925	142		2,067	2,350
0040	3'-0" x 7'-0" G		6	2.667		2,000	142		2,142	2,425
0050	4'-0" x 7'-0" G		5	3.200		2,200	170		2,370	2,675
0100	Stainless steel (316) prehung 24 ga. 2'-6" x 6'-8" door w/16 g frame G		6	2.667		2,500	142		2,642	2,975
0110	2'-8" x 6'-8" G		6	2.667		2,300	142		2,442	2,750
0120	3'-0" x 6'-8" G		6	2.667		2,400	142		2,542	2,875

08 17 Integrated Door Opening Assemblies

08 17 13 – Integrated Metal Door Opening Assemblies

08 17 13.20 Stainless Steel Doors and Frames			Crew	Daily Output	Labor-Hours	Unit	Material	2020 Bare Costs Labor	Equipment	Total	Total Incl O&P
0150	3'-0" x 7'-0"	G	2 Carp	6	2.667	Ea.	2,550	142		2,692	3,025
0160	4'-0" x 7'-0"	G		6	2.667		3,175	142		3,317	3,700
0300	Stainless steel (304) prehung 18 ga. 2'-6" x 6'-8" door w/16 g frame	G		6	2.667		2,850	142		2,992	3,350
0310	2'-8" x 6'-8"	G		6	2.667		2,900	142		3,042	3,400
0320	3'-0" x 6'-8"	G		6	2.667		2,950	142		3,092	3,450
0350	3'-0" x 7'-0"	G		6	2.667		2,975	142		3,117	3,500
0360	4'-0" x 7'-0"	G		5	3.200		3,125	170		3,295	3,700
0500	Stainless steel, prehung door, foam core, 14 ga., 3'-0" x 7'-0"	G		5	3.200		3,125	170		3,295	3,700
0600	Stainless steel, prehung double door, foam core, 14 ga., 3'-0" x 7'-0"	G	↓	4	4	↓	6,150	213		6,363	7,125

08 17 23 – Integrated Wood Door Opening Assemblies

08 17 23.10 Pre-Hung Doors		Crew	Daily Output	Labor-Hours	Unit	Material	Labor	Equipment	Total	Total Incl O&P
0010	**PRE-HUNG DOORS**									
0300	Exterior, wood, comb. storm & screen, 6'-9" x 2'-6" wide	2 Carp	15	1.067	Ea.	262	56.50		318.50	380
0320	2'-8" wide		15	1.067		350	56.50		406.50	475
0340	3'-0" wide	↓	15	1.067		330	56.50		386.50	455
0360	For 7'-0" high door, add				↓	45.50			45.50	50
1600	Entrance door, flush, birch, solid core									
1620	4-5/8" solid jamb, 1-3/4" x 6'-8" x 2'-8" wide	2 Carp	16	1	Ea.	299	53		352	415
1640	3'-0" wide	"	16	1		405	53		458	535
1680	For 7'-0" high door, add				↓	25.50			25.50	28
2000	Entrance door, colonial, 6 panel pine									
2020	4-5/8" solid jamb, 1-3/4" x 6'-8" x 2'-8" wide	2 Carp	16	1	Ea.	670	53		723	825
2040	3'-0" wide	"	16	1		705	53		758	860
2060	For 7'-0" high door, add					57			57	63
2200	For 5-5/8" solid jamb, add					44			44	48.50
2230	French style, exterior, 1 lite, 1-3/4" x 3'-0" x 6'-8"	1 Carp	14	.571		755	30.50		785.50	880
2235	9 lites	"	14	.571		755	30.50		785.50	880
2245	15 lites	2 Carp	14	1.143	↓	815	60.50		875.50	995
2250	Double, 15 lites, 2'-0" x 6'-8", 4'-0" opening		7	2.286	Pr.	1,300	121		1,421	1,625
2260	2'-6" x 6'-8", 5'-0" opening		7	2.286		1,425	121		1,546	1,775
2280	3'-0" x 6'-8", 6'-0" opening		7	2.286	↓	1,700	121		1,821	2,075
2430	3'-0" x 7'-0", 15 lites		14	1.143	Ea.	1,000	60.50		1,060.50	1,200
2432	Two 3'-0" x 7'-0"		7	2.286	Pr.	2,075	121		2,196	2,475
2435	3'-0" x 8'-0"		14	1.143	Ea.	1,075	60.50		1,135.50	1,275
2437	Two, 3'-0" x 8'-0"	↓	7	2.286	Pr.	2,200	121		2,321	2,625
4000	Interior, passage door, 4-5/8" solid jamb									
4350	Paneled, primed, hollow core, 2'-8" wide	2 Carp	17	.941	Ea.	140	50		190	234
4360	3'-0" wide		17	.941		191	50		241	291
4370	Pine, louvered, 2'-8" x 6'-8"		17	.941		209	50		259	310
4380	3'-0"		17	.941		222	50		272	325
4400	Lauan, flush, solid core, 1-3/8" x 6'-8" x 2'-6" wide		17	.941		194	50		244	294
4420	2'-8" wide		17	.941		194	50		244	294
4440	3'-0" wide		16	1		212	53		265	320
4600	Hollow core, 1-3/8" x 6'-8" x 2'-6" wide		17	.941		136	50		186	230
4620	2'-8" wide		17	.941		136	50		186	229
4640	3'-0" wide	↓	16	1		145	53		198	245
4700	For 7'-0" high door, add					41			41	45
5000	Birch, flush, solid core, 1-3/8" x 6'-8" x 2'-6" wide	2 Carp	17	.941		315	50		365	425
5020	2'-8" wide		17	.941		210	50		260	310
5040	3'-0" wide		16	1		315	53		368	430
5200	Hollow core, 1-3/8" x 6'-8" x 2'-6" wide		17	.941		260	50		310	365
5220	2'-8" wide	↓	17	.941	↓	273	50		323	380

08 17 Integrated Door Opening Assemblies

08 17 23 – Integrated Wood Door Opening Assemblies

08 17 23.10 Pre-Hung Doors		Crew	Daily Output	Labor-Hours	Unit	Material	2020 Bare Costs Labor	Equipment	Total	Total Incl O&P
5240	3'-0" wide	2 Carp	16	1	Ea.	277	53		330	390
5280	For 7'-0" high door, add					35			35	38.50
5500	Hardboard paneled, 1-3/8" x 6'-8" x 2'-6" wide	2 Carp	17	.941		156	50		206	252
5520	2'-8" wide		17	.941		163	50		213	259
5540	3'-0" wide		16	1		165	53		218	266
6000	Pine paneled, 1-3/8" x 6'-8" x 2'-6" wide		17	.941		278	50		328	385
6020	2'-8" wide		17	.941		294	50		344	405
6040	3'-0" wide		16	1		305	53		358	420
7600	Oak, 6 panel, 1-3/4" x 6'-8" x 3'-0"	1 Carp	17	.471		900	25		925	1,025
8200	Birch, flush, solid core, 1-3/4" x 6'-8" x 2'-4" wide		17	.471		237	25		262	300
8220	2'-6" wide		17	.471		219	25		244	281
8240	2'-8" wide		17	.471		231	25		256	294
8260	3'-0" wide		16	.500		235	26.50		261.50	300
8280	3'-6" wide		15	.533		370	28.50		398.50	455
8500	Pocket door frame with lauan, flush, hollow core, 1-3/8" x 3'-0" x 6'-8"		17	.471		320	25		345	395
9000	Minimum labor/equipment charge		4	2	Job		106		106	170

08 31 Access Doors and Panels

08 31 13 – Access Doors and Frames

08 31 13.10 Types of Framed Access Doors		Crew	Daily Output	Labor-Hours	Unit	Material	Labor	Equipment	Total	Total Incl O&P
0010	**TYPES OF FRAMED ACCESS DOORS**									
1000	Fire rated door with lock									
1100	Metal, 12" x 12"	1 Carp	10	.800	Ea.	164	42.50		206.50	248
1150	18" x 18"		9	.889		221	47		268	320
1200	24" x 24"		9	.889		345	47		392	455
1250	24" x 36"		8	1		350	53		403	470
1300	24" x 48"		8	1		415	53		468	540
1350	36" x 36"		7.50	1.067		515	56.50		571.50	655
1400	48" x 48"		7.50	1.067		670	56.50		726.50	825
1600	Stainless steel, 12" x 12"		10	.800		340	42.50		382.50	445
1650	18" x 18"		9	.889		375	47		422	490
1700	24" x 24"		9	.889		550	47		597	680
1750	24" x 36"		8	1		655	53		708	810
2000	Flush door for finishing									
2100	Metal 8" x 8"	1 Carp	10	.800	Ea.	37.50	42.50		80	109
2150	12" x 12"	"	10	.800	"	37.50	42.50		80	110
3000	Recessed door for acoustic tile									
3100	Metal, 12" x 12"	1 Carp	4.50	1.778	Ea.	57	94.50		151.50	214
3150	12" x 24"		4.50	1.778		111	94.50		205.50	273
3200	24" x 24"		4	2		107	106		213	288
3250	24" x 36"		4	2		165	106		271	350
4000	Recessed door for drywall									
4100	Metal 12" x 12"	1 Carp	6	1.333	Ea.	76.50	71		147.50	198
4150	12" x 24"		5.50	1.455		119	77.50		196.50	254
4200	24" x 36"		5	1.600		177	85		262	330
6000	Standard door									
6100	Metal, 8" x 8"	1 Carp	10	.800	Ea.	30	42.50		72.50	101
6150	12" x 12"		10	.800		30.50	42.50		73	102
6200	18" x 18"		9	.889		35.50	47		82.50	115
6250	24" x 24"		9	.889		50	47		97	131
6300	24" x 36"		8	1		88	53		141	182

08 31 13 – Access Doors and Frames

08 31 13.10 Types of Framed Access Doors		Crew	Daily Output	Labor-Hours	Unit	Material	Labor	Equipment	Total	Total Incl O&P
6350	36" x 36"	1 Carp	8	1	Ea.	103	53		156	198
6500	Stainless steel, 8" x 8"		10	.800		80.50	42.50		123	157
6550	12" x 12"		10	.800		97	42.50		139.50	174
6600	18" x 18"		9	.889		177	47		224	271
6650	24" x 24"		9	.889		276	47		323	380
9000	Minimum labor/equipment charge		4	2	Job		106		106	170

08 31 13.20 Bulkhead/Cellar Doors		Crew	Daily Output	Labor-Hours	Unit	Material	Labor	Equipment	Total	Total Incl O&P
0010	**BULKHEAD/CELLAR DOORS**									
0020	Steel, not incl. sides, 44" x 62"	1 Carp	5.50	1.455	Ea.	705	77.50		782.50	900
0100	52" x 73"		5.10	1.569		865	83.50		948.50	1,075
0500	With sides and foundation plates, 57" x 45" x 24"		4.70	1.702		895	90.50		985.50	1,125
0600	42" x 49" x 51"		4.30	1.860		600	99		699	820
9000	Minimum labor/equipment charge		2	4	Job		213		213	340

08 31 13.30 Commercial Floor Doors		Crew	Daily Output	Labor-Hours	Unit	Material	Labor	Equipment	Total	Total Incl O&P
0010	**COMMERCIAL FLOOR DOORS**									
0020	Aluminum tile, steel frame, one leaf, 2' x 2' opng.	2 Sswk	3.50	4.571	Opng.	725	264		989	1,225
0021	Aluminum tile, steel frame, one leaf, 2' x 2' opng.	L-4	3.50	6.857		725	340		1,065	1,350
0050	3'-6" x 3'-6" opening	2 Sswk	3.50	4.571		1,325	264		1,589	1,900
0051	3'-6" x 3'-6" opening	L-4	3.50	6.857		1,325	340		1,665	2,025
0500	Double leaf, 4' x 4' opening	2 Sswk	3	5.333		1,650	305		1,955	2,300
0501	Double leaf, 4' x 4' opening	L-4	3	8		1,650	400		2,050	2,450
0550	5' x 5' opening	2 Sswk	3	5.333		2,700	305		3,005	3,475
0551	5' x 5' opening	L-4	3	8		2,700	400		3,100	3,625
9000	Minimum labor/equipment charge	2 Sswk	2	8	Job		460		460	755

08 31 13.35 Industrial Floor Doors		Crew	Daily Output	Labor-Hours	Unit	Material	Labor	Equipment	Total	Total Incl O&P
0010	**INDUSTRIAL FLOOR DOORS**									
0020	Steel 300 psf L.L., single leaf, 2' x 2', 175#	2 Sswk	6	2.667	Opng.	810	154		964	1,150
0050	3' x 3' opening, 300#		5.50	2.909		1,300	168		1,468	1,700
0300	Double leaf, 4' x 4' opening, 455#		5	3.200		2,450	184		2,634	3,000
0350	5' x 5' opening, 645#		4.50	3.556		2,925	205		3,130	3,525
1000	Aluminum, 300 psf L.L., single leaf, 2' x 2', 60#		6	2.667		720	154		874	1,050
1050	3' x 3' opening, 100#		5.50	2.909		995	168		1,163	1,375
1500	Double leaf, 4' x 4' opening, 160#		5	3.200		2,825	184		3,009	3,400
1550	5' x 5' opening, 235#		4.50	3.556		3,250	205		3,455	3,900
2000	Aluminum, 150 psf L.L., single leaf, 2' x 2', 60#		6	2.667		715	154		869	1,025
2050	3' x 3' opening, 95#		5.50	2.909		1,050	168		1,218	1,425
2500	Double leaf, 4' x 4' opening, 150#		5	3.200		1,475	184		1,659	1,925
2550	5' x 5' opening, 230#		4.50	3.556		1,975	205		2,180	2,500
9000	Minimum labor/equipment charge		2	8	Job		460		460	755

08 31 13.40 Kennel Doors		Crew	Daily Output	Labor-Hours	Unit	Material	Labor	Equipment	Total	Total Incl O&P
0010	**KENNEL DOORS**									
0020	2 way, swinging type, 13" x 19" opening	2 Carp	11	1.455	Opng.	88.50	77.50		166	222
0100	17" x 29" opening		11	1.455		132	77.50		209.50	269
0200	9" x 9" opening, electronic with accessories		11	1.455		153	77.50		230.50	292

Material, Labor, Equipment and Total columns: 2020 Bare Costs

08 32 Sliding Glass Doors

08 32 13 – Sliding Aluminum-Framed Glass Doors

08 32 13.10 Sliding Aluminum Doors		Crew	Daily Output	Labor-Hours	Unit	2020 Bare Costs Material	Labor	Equipment	Total	Total Incl O&P
0010	**SLIDING ALUMINUM DOORS**									
0350	Aluminum, 5/8" tempered insulated glass, 6' wide									
0400	Premium	2 Carp	4	4	Ea.	1,600	213		1,813	2,125
0450	Economy		4	4		890	213		1,103	1,325
0500	8' wide, premium		3	5.333		1,775	283		2,058	2,400
0550	Economy		3	5.333		1,575	283		1,858	2,175
0600	12' wide, premium		2.50	6.400		3,100	340		3,440	3,975
0650	Economy		2.50	6.400		1,675	340		2,015	2,375
4000	Aluminum, baked on enamel, temp glass, 6'-8" x 10'-0" wide		4	4		1,200	213		1,413	1,650
4020	Insulating glass, 6'-8" x 6'-0" wide		4	4		975	213		1,188	1,425
4040	8'-0" wide		3	5.333		1,175	283		1,458	1,750
4060	10'-0" wide		2	8		1,450	425		1,875	2,275
4080	Anodized, temp glass, 6'-8" x 6'-0" wide		4	4		495	213		708	885
4100	8'-0" wide		3	5.333		605	283		888	1,125
4120	10'-0" wide		2	8		710	425		1,135	1,450
5000	Aluminum sliding glass door system									
5010	Sliding door 4' wide opening single side	2 Carp	2	8	Ea.	6,475	425		6,900	7,775
5015	8' wide opening single side		2	8		9,700	425		10,125	11,400
5020	Telescoping glass door system, 4' wide opening biparting		2	8		5,525	425		5,950	6,750
5025	8' wide opening biparting		2	8		6,475	425		6,900	7,775
5030	Folding glass door, 4' wide opening biparting		2	8		8,625	425		9,050	10,200
5035	8' wide opening biparting		2	8		10,800	425		11,225	12,600
5040	ICU-CCU sliding telescoping glass door, 4' x 7', single side opening		2	8		3,300	425		3,725	4,300
5045	8' x 7', single side opening		2	8		5,000	425		5,425	6,175
7000	Electric swing door operator and control, single door w/sensors	1 Carp	4	2		2,950	106		3,056	3,400
7005	Double door w/sensors		2	4		5,925	213		6,138	6,850
7010	Electric folding door operator and control, single door w/sensors		4	2		6,950	106		7,056	7,825
7015	Bi-folding door		4	2		8,175	106		8,281	9,175
7020	Electric swing door operator and control, single door		4	2		2,950	106		3,056	3,425

08 32 19 – Sliding Wood-Framed Glass Doors

08 32 19.15 Sliding Glass Vinyl-Clad Wood Doors		Crew	Daily Output	Labor-Hours	Unit	Material	Labor	Equipment	Total	Total Incl O&P
0010	**SLIDING GLASS VINYL-CLAD WOOD DOORS**									
0020	Glass, sliding vinyl-clad, insul. glass, 6'-0" x 6'-8" G	2 Carp	4	4	Opng.	1,575	213		1,788	2,100
0025	6'-0" x 6'-10" high G		4	4		1,775	213		1,988	2,300
0030	6'-0" x 8'-0" high G		4	4		2,175	213		2,388	2,750
0050	5'-0" x 6'-8" high G		4	4		1,625	213		1,838	2,150
0100	8'-0" x 6'-10" high G		4	4		2,125	213		2,338	2,675
0150	8'-0" x 8'-0" high G		4	4		2,475	213		2,688	3,075
0500	4 leaf, 9'-0" x 6'-10" high G		3	5.333		3,675	283		3,958	4,500
0550	9'-0" x 8'-0" high G		3	5.333		4,050	283		4,333	4,900
0600	12'-0" x 6'-10" high G		3	5.333		4,350	283		4,633	5,225
0650	12'-0" x 8'-0" high G		3	5.333		4,350	283		4,633	5,225
9000	Minimum labor/equipment charge	1 Carp	4	2	Job		106		106	170

08 33 Coiling Doors and Grilles

08 33 13 – Coiling Counter Doors

08 33 13.10 Counter Doors, Coiling Type		Crew	Daily Output	Labor-Hours	Unit	Material	Labor	Equipment	Total	Total Incl O&P
						2020 Bare Costs				
0010	**COUNTER DOORS, COILING TYPE**									
0020	Manual, incl. frame and hardware, galv. stl., 4′ roll-up, 6′ long	2 Carp	2	8	Opng.	1,325	425		1,750	2,125
0300	Galvanized steel, UL label		1.80	8.889		1,325	470		1,795	2,200
0600	Stainless steel, 4′ high roll-up, 6′ long		2	8		2,200	425		2,625	3,100
0700	10′ long		1.80	8.889		2,625	470		3,095	3,650
2000	Aluminum, 4′ high, 4′ long		2.20	7.273		1,725	385		2,110	2,500
2020	6′ long		2	8		2,000	425		2,425	2,875
2040	8′ long		1.90	8.421		2,300	450		2,750	3,250
2060	10′ long		1.80	8.889		2,300	470		2,770	3,275
2080	14′ long		1.40	11.429		2,975	605		3,580	4,250
2100	6′ high, 4′ long		2	8		2,025	425		2,450	2,900
2120	6′ long		1.60	10		1,825	530		2,355	2,850
2140	10′ long		1.40	11.429		2,400	605		3,005	3,625
9000	Minimum labor/equipment charge	1 Carp	2	4	Job		213		213	340

08 33 16 – Coiling Counter Grilles

08 33 16.10 Coiling Grilles		Crew	Daily Output	Labor-Hours	Unit	Material	Labor	Equipment	Total	Total Incl O&P
0010	**COILING GRILLES**									
0015	Aluminum, manual, incl. frame, mill finish									
0020	Top coiling, 4′ high, 4′ long	2 Sswk	3.20	5	Opng.	1,500	288		1,788	2,125
0030	6′ long		3.20	5		1,725	288		2,013	2,375
0040	8′ long		2.40	6.667		1,925	385		2,310	2,750
0050	12′ long		2.40	6.667		2,375	385		2,760	3,250
0060	16′ long		1.60	10		2,550	575		3,125	3,775
0070	6′ high, 4′ long		3.20	5		1,775	288		2,063	2,425
0080	6′ long		3.20	5		1,775	288		2,063	2,425
0090	8′ long		2.40	6.667		1,725	385		2,110	2,525
0100	12′ long		1.60	10		2,300	575		2,875	3,475
0110	16′ long		1.20	13.333		2,800	770		3,570	4,325
0200	Side coiling, 8′ high, 12′ long		.60	26.667		2,250	1,525		3,775	5,000
0220	18′ long		.50	32		5,050	1,850		6,900	8,575
0240	24′ long		.40	40		3,675	2,300		5,975	7,825
0260	12′ high, 12′ long		.50	32		3,275	1,850		5,125	6,625
0280	18′ long		.40	40		4,950	2,300		7,250	9,200
0300	24′ long		.28	57.143		6,500	3,300		9,800	12,500
9000	Minimum labor/equipment charge		1	16	Job		920		920	1,500

08 33 23 – Overhead Coiling Doors

08 33 23.10 Coiling Service Doors		Crew	Daily Output	Labor-Hours	Unit	Material	Labor	Equipment	Total	Total Incl O&P
0010	**COILING SERVICE DOORS** Steel, manual, 20 ga., incl. hardware									
0050	8′ x 8′ high	2 Sswk	1.60	10	Ea.	710	575		1,285	1,725
0100	10′ x 10′ high	"	1.40	11.429		1,050	660		1,710	2,250
0101	10′ x 10′ high	L-4	1.40	17.143		1,050	855		1,905	2,550
0130	12′ x 12′ high, standard	2 Sswk	1.20	13.333		2,175	770		2,945	3,650
0160	10′ x 20′ high, standard		.50	32		2,600	1,850		4,450	5,900
0200	20′ x 10′ high		1	16		3,050	920		3,970	4,850
0300	12′ x 12′ high		1.20	13.333		1,250	770		2,020	2,625
0303	Doors, rolling service, steel, manual, 20ga., 12′ x 12′, incl. hardware		.92	17.391		1,250	1,000		2,250	3,025
0400	20′ x 12′ high		.90	17.778		2,050	1,025		3,075	3,925
0420	14′ x 12′ high		.80	20		2,925	1,150		4,075	5,125
0430	14′ x 13′ high		.80	20		3,100	1,150		4,250	5,300
0500	14′ x 14′ high		.80	20		2,925	1,150		4,075	5,125
0520	14′ x 16′ high		.80	20	Opng.	2,500	1,150		3,650	4,650
0540	14′ x 20′ high		.56	28.571		3,700	1,650		5,350	6,775

08 33 Coiling Doors and Grilles

08 33 23 – Overhead Coiling Doors

08 33 23.10 Coiling Service Doors		Crew	Daily Output	Labor-Hours	Unit	Material	2020 Bare Costs Labor	Equipment	Total	Total Incl O&P
0550	18′ x 18′ high	2 Sswk	.60	26.667	Opng.	2,900	1,525		4,425	5,700
0600	20′ x 16′ high		.60	26.667	Ea.	3,700	1,525		5,225	6,600
0700	10′ x 20′ high		.50	32	″	2,600	1,850		4,450	5,900
0750	16′ x 24′ high		.48	33.333	Opng.	5,975	1,925		7,900	9,725
1000	12′ x 12′, crank operated, crank on door side		.80	20	Ea.	1,800	1,150		2,950	3,900
1100	Crank thru wall		.70	22.857		2,075	1,325		3,400	4,425
1300	For vision panel, add					360			360	395
1600	3′ x 7′ pass door within rolling steel door, new construction					1,725			1,725	1,900
1700	Existing construction	2 Sswk	2	8		2,050	460		2,510	3,000
2000	Class A fire doors, manual, 20 ga., 8′ x 8′ high		1.40	11.429		1,575	660		2,235	2,825
2100	10′ x 10′ high		1.10	14.545		2,150	840		2,990	3,750
2200	20′ x 10′ high		.80	20		4,400	1,150		5,550	6,750
2300	12′ x 12′ high		1	16		3,400	920		4,320	5,250
2304	Overhead door, roll up, fire rated 12′ x 14′		.90	17.778		4,000	1,025		5,025	6,075
2400	20′ x 12′ high		.80	20		4,775	1,150		5,925	7,150
2500	14′ x 14′ high		.60	26.667		3,700	1,525		5,225	6,600
2600	20′ x 16′ high		.50	32		5,950	1,850		7,800	9,550
2700	10′ x 20′ high		.40	40		4,725	2,300		7,025	8,975
2730	Manual steel rollup fire doors		160	.100	S.F.	34.50	5.75		40.25	47.50
2740	For motor operated, add		1	16	Ea.	800	920		1,720	2,375
3000	For 18 ga. doors, add				S.F.	1.52			1.52	1.67
3300	For enamel finish, add				″	2.84			2.84	3.12
3600	For safety edge bottom bar, pneumatic, add				L.F.	24.50			24.50	27
3700	Electric, add					42.50			42.50	47
4000	For weatherstripping, extruded rubber, jambs, add					19.15			19.15	21
4100	Hood, add					8.65			8.65	9.50
4200	Sill, add					8.85			8.85	9.75
4500	Motor operators, to 14′ x 14′ opening	2 Sswk	5	3.200	Ea.	1,200	184		1,384	1,625
4600	Over 14′ x 14′, jack shaft type	″	5	3.200		1,325	184		1,509	1,750
4700	For fire door, additional fusible link, add					77.50			77.50	85.50
5100	Radio control operator to 12′ x 12′ door	2 Sswk	3	5.333		820	305		1,125	1,400
5120	Receiver to 12′ x 12′ door		13	1.231		243	71		314	385
5140	Transmitter to 12′ x 12′ door		49	.327		54	18.80		72.80	90.50
9000	Minimum labor/equipment charge		1	16	Job		920		920	1,500

08 34 Special Function Doors

08 34 13 – Cold Storage Doors

08 34 13.10 Doors for Cold Area Storage		Crew	Daily Output	Labor-Hours	Unit	Material	Labor	Equipment	Total	Total Incl O&P
0010	**DOORS FOR COLD AREA STORAGE**									
0020	Single, 20 ga. galvanized steel									
0300	Horizontal sliding, 5′ x 7′, manual operation, 3.5″ thick	2 Carp	2	8	Ea.	3,150	425		3,575	4,125
0400	4″ thick		2	8		3,650	425		4,075	4,675
0500	6″ thick		2	8		3,400	425		3,825	4,425
0800	5′ x 7′, power operation, 2″ thick		1.90	8.421		5,675	450		6,125	6,975
0900	4″ thick		1.90	8.421		5,675	450		6,125	6,975
1000	6″ thick		1.90	8.421		6,475	450		6,925	7,825
1300	9′ x 10′, manual operation, 2″ insulation		1.70	9.412		4,450	500		4,950	5,700
1400	4″ insulation		1.70	9.412		4,825	500		5,325	6,100
1500	6″ insulation		1.70	9.412		5,675	500		6,175	7,050
1800	Power operation, 2″ insulation		1.60	10		7,825	530		8,355	9,450
1900	4″ insulation		1.60	10		7,975	530		8,505	9,625

08 34 Special Function Doors

08 34 13 – Cold Storage Doors

08 34 13.10 Doors for Cold Area Storage		Crew	Daily Output	Labor-Hours	Unit	Material	2020 Bare Costs Labor	Equipment	Total	Total Incl O&P
2000	6" insulation	2 Carp	1.70	9.412	Ea.	8,225	500		8,725	9,825
2300	For stainless steel face, add					25%				
3000	Hinged, lightweight, 3' x 7'-0", 2" thick	2 Carp	2	8	Ea.	1,300	425		1,725	2,100
3050	4" thick		1.90	8.421		1,625	450		2,075	2,500
3300	Polymer doors, 3' x 7'-0"		1.90	8.421		1,350	450		1,800	2,200
3350	6" thick		1.40	11.429		2,375	605		2,980	3,600
3600	Stainless steel, 3' x 7'-0", 4" thick		1.90	8.421		1,725	450		2,175	2,625
3650	6" thick		1.40	11.429		2,875	605		3,480	4,125
3900	Painted, 3' x 7'-0", 4" thick		1.90	8.421		1,350	450		1,800	2,200
3950	6" thick		1.40	11.429		2,325	605		2,930	3,550
5000	Bi-parting, electric operated									
5010	6' x 8' opening, galv. faces, 4" thick for cooler	2 Carp	.80	20	Opng.	7,350	1,075		8,425	9,800
5050	For freezer, 4" thick		.80	20		8,100	1,075		9,175	10,600
5300	For door buck framing and door protection, add		2.50	6.400		680	340		1,020	1,300
6000	Galvanized batten door, galvanized hinges, 4' x 7'		2	8		1,875	425		2,300	2,725
6050	6' x 8'		1.80	8.889		2,475	470		2,945	3,475
6500	Fire door, 3 hr., 6' x 8', single slide		.80	20		8,725	1,075		9,800	11,300
6550	Double, bi-parting		.70	22.857		11,300	1,225		12,525	14,500
9000	Minimum labor/equipment charge	1 Carp	2	4	Job		213		213	340

08 34 36 – Darkroom Doors

08 34 36.10 Various Types of Darkroom Doors

		Crew	Daily Output	Labor-Hours	Unit	Material	Labor	Equipment	Total	Total Incl O&P
0010	**VARIOUS TYPES OF DARKROOM DOORS**									
0015	Revolving, standard, 2 way, 36" diameter	2 Carp	3.10	5.161	Opng.	3,350	274		3,624	4,150
0020	41" diameter		3.10	5.161		3,850	274		4,124	4,700
0050	3 way, 51" diameter		1.40	11.429		4,125	605		4,730	5,500
1000	4 way, 49" diameter		1.40	11.429		4,075	605		4,680	5,450
2000	Hinged safety, 2 way, 41" diameter		2.30	6.957		4,050	370		4,420	5,075
2500	3 way, 51" diameter		1.40	11.429		5,025	605		5,630	6,500
3000	Pop out safety, 2 way, 41" diameter		3.10	5.161		4,875	274		5,149	5,800
4000	3 way, 51" diameter		1.40	11.429		5,375	605		5,980	6,900
5000	Wheelchair-type, pop out, 51" diameter		1.40	11.429		5,925	605		6,530	7,500
5020	72" diameter		.90	17.778		10,400	945		11,345	13,000
9300	For complete darkrooms, see Section 13 21 53.50									

08 34 53 – Security Doors and Frames

08 34 53.20 Steel Door

		Crew	Daily Output	Labor-Hours	Unit	Material	Labor	Equipment	Total	Total Incl O&P
0010	**STEEL DOOR** flush with ballistic core and welded frame both 14 ga.									
0120	UL 752 Level 3, 1-3/4", 3'-0" x 7'-0"	2 Carp	1.50	10.667	Opng.	2,700	565		3,265	3,850
0125	1-3/4", 3'-6" x 7'-0"		1.50	10.667		2,950	565		3,515	4,150
0130	1-3/4", 4'-0" x 7'-0"		1.20	13.333		3,175	710		3,885	4,600
0150	UL 752 Level 8, 1-3/4", 3'-0" x 7'-0"		1.50	10.667		14,100	565		14,665	16,400
0155	1-3/4", 3'-6" x 7'-0"		1.50	10.667		15,400	565		15,965	17,800
0160	1-3/4", 4'-0" x 7'-0"		1.20	13.333		15,300	710		16,010	18,000
1000	Safe room sliding door and hardware, 1-3/4", 3'-0" x 7'-0" UL 752 Level 3		.50	32		24,300	1,700		26,000	29,400
1050	Safe room swinging door and hardware, 1-3/4", 3'-0" x 7'-0" UL 752 Level 3		.50	32		29,100	1,700		30,800	34,700

08 34 53.30 Wood Ballistic Doors

		Crew	Daily Output	Labor-Hours	Unit	Material	Labor	Equipment	Total	Total Incl O&P
0010	**WOOD BALLISTIC DOORS** with frames and hardware									
0050	Wood, 1-3/4", 3'-0" x 7'-0" UL 752 Level 3	2 Carp	1.50	10.667	Opng.	2,700	565		3,265	3,850

08 34 56 – Security Gates

08 34 56.10 Gates		Crew	Daily Output	Labor-Hours	Unit	2020 Bare Costs Material	Labor	Equipment	Total	Total Incl O&P
0010	**GATES**									
0015	Driveway gates include mounting hardware									
0500	Wood, security gate, driveway, dual, 10′ wide	H-4	.80	25	Opng.	3,825	1,250		5,075	6,200
0505	12′ wide		.80	25		3,700	1,250		4,950	6,025
0510	15′ wide		.80	25		4,400	1,250		5,650	6,800
0600	Steel, security gate, driveway, single, 10′ wide		.80	25		1,825	1,250		3,075	3,975
0605	12′ wide		.80	25		2,150	1,250		3,400	4,350
0620	Steel, security gate, driveway, dual, 12′ wide		.80	25		2,200	1,250		3,450	4,375
0625	14′ wide		.80	25		2,525	1,250		3,775	4,750
0630	16′ wide		.80	25		2,500	1,250		3,750	4,725
0700	Aluminum, security gate, driveway, dual, 10′ wide		.80	25		3,275	1,250		4,525	5,575
0705	12′ wide		.80	25		4,050	1,250		5,300	6,450
0710	16′ wide		.80	25		4,625	1,250		5,875	7,050
1000	Security gate, driveway, opener 12 VDC				Ea.	840			840	925
1010	Wireless					1,950			1,950	2,150
1020	Security gate, driveway, opener 24 VDC					1,500			1,500	1,625
1030	Wireless					1,650			1,650	1,825
1040	Security gate, driveway, opener 12 VDC, solar panel 10 watt	1 Elec	2	4		495	245		740	925
1050	20 watt	"	2	4		635	245		880	1,075

08 34 59 – Vault Doors and Day Gates

08 34 59.10 Secure Storage Doors		Crew	Daily Output	Labor-Hours	Unit	Material	Labor	Equipment	Total	Total Incl O&P
0010	**SECURE STORAGE DOORS**									
0020	Door and frame, 32″ x 78″, clear opening									
0100	1 hour test, 32″ door, weighs 750 lb.	2 Sswk	1.50	10.667	Opng.	7,150	615		7,765	8,850
0200	2 hour test, 32″ door, weighs 950 lb.		1.30	12.308		8,250	710		8,960	10,200
0250	40″ door, weighs 1130 lb.		1	16		8,975	920		9,895	11,400
0300	4 hour test, 32″ door, weighs 1025 lb.		1.20	13.333		10,300	770		11,070	12,600
0350	40″ door, weighs 1140 lb.		.90	17.778		11,400	1,025		12,425	14,200
0600	For time lock, two movement, add	1 Elec	2	4	Ea.	1,775	245		2,020	2,325
0800	Day gate, painted, steel, 32″ wide	2 Sswk	1.50	10.667		2,025	615		2,640	3,250
0850	40″ wide		1.40	11.429		2,300	660		2,960	3,600
0900	Aluminum, 32″ wide		1.50	10.667		2,900	615		3,515	4,200
0950	40″ wide		1.40	11.429		3,150	660		3,810	4,550
2050	Security vault door, class I, 3′ wide, 3-1/2″ thick	E-24	.19	167	Opng.	14,600	9,575	3,025	27,200	34,900
2100	Class II, 3′ wide, 7″ thick		.19	167		17,400	9,575	3,025	30,000	37,900
2150	Class III, 9R, 3′ wide, 10″ thick		.13	250		21,800	14,400	4,550	40,750	52,500
2160	Class V, type 1, 40″ door		2.48	12.903	Ea.	6,500	740	234	7,474	8,600
2170	Class V, type 2, 40″ door		2.48	12.903		6,400	740	234	7,374	8,500
2180	Day gate for class V vault	2 Sswk	2	8		1,675	460		2,135	2,600

08 34 73 – Sound Control Door Assemblies

08 34 73.10 Acoustical Doors		Crew	Daily Output	Labor-Hours	Unit	Material	Labor	Equipment	Total	Total Incl O&P
0010	**ACOUSTICAL DOORS**									
0020	Including framed seals, 3′ x 7′, wood, 40 STC rating	2 Carp	1.50	10.667	Ea.	3,875	565		4,440	5,175
0100	Steel, 41 STC rating		1.50	10.667		3,850	565		4,415	5,125
0200	45 STC rating		1.50	10.667		4,175	565		4,740	5,500
0300	48 STC rating		1.50	10.667		4,925	565		5,490	6,300
0400	52 STC rating		1.50	10.667		5,400	565		5,965	6,850
9000	Minimum labor/equipment charge	1 Carp	4	2	Job		106		106	170

08 36 Panel Doors

08 36 13 – Sectional Doors

08 36 13.10 Overhead Commercial Doors		Crew	Daily Output	Labor-Hours	Unit	2020 Bare Costs Material	Labor	Equipment	Total	Total Incl O&P
0010	**OVERHEAD COMMERCIAL DOORS**									
1000	Stock, sectional, heavy duty, wood, 1-3/4" thick, 8' x 8' high	2 Carp	2	8	Ea.	1,250	425		1,675	2,050
1100	10' x 10' high		1.80	8.889		1,725	470		2,195	2,650
1200	12' x 12' high		1.50	10.667		2,775	565		3,340	3,950
1300	Chain hoist, 14' x 14' high		1.30	12.308		3,825	655		4,480	5,250
1400	12' x 16' high		1	16		3,875	850		4,725	5,600
1500	20' x 8' high		1.30	12.270		2,950	650		3,600	4,300
1600	20' x 16' high		.65	24.615		5,975	1,300		7,275	8,675
1800	Center mullion openings, 8' high		4	4		1,300	213		1,513	1,775
1900	20' high		2	8		2,225	425		2,650	3,125
2100	For medium duty custom door, deduct					5%	5%			
2150	For medium duty stock doors, deduct					10%	5%			
2300	Fiberglass and aluminum, heavy duty, sectional, 12' x 12' high	2 Carp	1.50	10.667	Ea.	3,425	565		3,990	4,650
2450	Chain hoist, 20' x 20' high		.50	32		7,775	1,700		9,475	11,300
2600	Steel, 24 ga. sectional, manual, 8' x 8' high		2	8		1,125	425		1,550	1,925
2650	10' x 10' high		1.80	8.889		1,375	470		1,845	2,275
2700	12' x 12' high		1.50	10.667		1,650	565		2,215	2,700
2800	Chain hoist, 20' x 14' high		.70	22.857		3,725	1,225		4,950	6,050
2850	For 1-1/4" rigid insulation and 26 ga. galv.									
2860	back panel, add				S.F.	5			5	5.50
2900	For electric trolley operator, 1/3 HP, to 12' x 12', add	1 Carp	2	4	Ea.	1,125	213		1,338	1,575
2950	Over 12' x 12', 1/2 HP, add		1	8		1,250	425		1,675	2,050
2980	Overhead, for row of clear lites, add		1	8		169	425		594	865
9000	Minimum labor/equipment charge	2 Carp	1.50	10.667	Job		565		565	910

08 36 13.20 Residential Garage Doors		Crew	Daily Output	Labor-Hours	Unit	Material	Labor	Equipment	Total	Total Incl O&P
0010	**RESIDENTIAL GARAGE DOORS**									
0050	Hinged, wood, custom, double door, 9' x 7'	2 Carp	4	4	Ea.	935	213		1,148	1,375
0070	16' x 7'		3	5.333		1,300	283		1,583	1,875
0200	Overhead, sectional, incl. hardware, fiberglass, 9' x 7', standard		5	3.200		975	170		1,145	1,350
0220	Deluxe		5	3.200		1,200	170		1,370	1,600
0300	16' x 7', standard		6	2.667		1,950	142		2,092	2,375
0320	Deluxe		6	2.667		2,325	142		2,467	2,800
0500	Hardboard, 9' x 7', standard		8	2		750	106		856	995
0520	Deluxe		8	2		905	106		1,011	1,175
0600	16' x 7', standard		6	2.667		1,325	142		1,467	1,675
0620	Deluxe		6	2.667		1,600	142		1,742	2,000
0700	Metal, 9' x 7', standard		8	2		1,000	106		1,106	1,275
0720	Deluxe		6	2.667		1,025	142		1,167	1,350
0800	16' x 7', standard		6	2.667		1,200	142		1,342	1,550
0820	Deluxe		5	3.200		1,450	170		1,620	1,875
0900	Wood, 9' x 7', standard		8	2		1,075	106		1,181	1,350
0920	Deluxe		8	2		2,325	106		2,431	2,725
1000	16' x 7', standard		6	2.667		1,750	142		1,892	2,150
1020	Deluxe		6	2.667		3,150	142		3,292	3,700
1800	Door hardware, sectional	1 Carp	4	2		375	106		481	585
1810	Door tracks only		4	2		173	106		279	360
1820	One side only		7	1.143		129	60.50		189.50	239
4000	For electric operator, economy, add		8	1		460	53		513	590
4100	Deluxe, including remote control		8	1		640	53		693	785
4500	For transmitter/receiver control, add to operator				Total	119			119	131
4600	Transmitters, additional				"	67			67	74
7010	Garage doors, row of lites				Ea.	133			133	146

08 36 Panel Doors

08 36 13 – Sectional Doors

08 36 13.20 Residential Garage Doors		Crew	Daily Output	Labor-Hours	Unit	2020 Bare Costs Material	Labor	Equipment	Total	Total Incl O&P
9000	Minimum labor/equipment charge	1 Carp	2.50	3.200	Job		170		170	272

08 36 19 – Multi-Leaf Vertical Lift Doors

08 36 19.10 Sectional Vertical Lift Doors		Crew	Daily Output	Labor-Hours	Unit	Material	Labor	Equipment	Total	Total Incl O&P
0010	SECTIONAL VERTICAL LIFT DOORS									
0020	Motorized, 14 ga. steel, incl. frame and control panel									
0050	16' x 16' high	L-10	.50	48	Ea.	22,600	2,825	940	26,365	30,500
0100	10' x 20' high		1.30	18.462		42,800	1,075	360	44,235	49,300
0120	15' x 20' high		1.30	18.462		51,500	1,075	360	52,935	59,000
0140	20' x 20' high		1	24		61,500	1,400	470	63,370	70,500
0160	25' x 20' high		1	24		59,000	1,400	470	60,870	67,500
0170	32' x 24' high		.75	32		54,500	1,875	630	57,005	63,500
0180	20' x 25' high		1	24		60,500	1,400	470	62,370	69,500
0200	25' x 25' high		.70	34.286		78,000	2,025	675	80,700	90,000
0220	25' x 30' high		.70	34.286		83,500	2,025	675	86,200	95,500
0240	30' x 30' high		.70	34.286		99,000	2,025	675	101,700	113,000
0260	35' x 30' high	↓	.70	34.286	↓	98,500	2,025	675	101,200	112,500

08 38 Traffic Doors

08 38 13 – Flexible Strip Doors

08 38 13.10 Flexible Transparent Strip Doors		Crew	Daily Output	Labor-Hours	Unit	Material	Labor	Equipment	Total	Total Incl O&P
0010	FLEXIBLE TRANSPARENT STRIP DOORS									
0100	12" strip width, 2/3 overlap	3 Shee	135	.178	SF Surf	7.75	11.10		18.85	26
0200	Full overlap		115	.209		9.85	13		22.85	31.50
0220	8" strip width, 1/2 overlap		140	.171		6.35	10.70		17.05	24
0240	Full overlap	↓	120	.200	↓	7.90	12.45		20.35	28
0300	Add for suspension system, header mount				L.F.	9.20			9.20	10.10
0400	Wall mount				"	9.75			9.75	10.70

08 38 16 – Flexible Traffic Doors

08 38 16.10 Rubber Doors		Crew	Daily Output	Labor-Hours	Unit	Material	Labor	Equipment	Total	Total Incl O&P
0010	RUBBER DOORS									
0015	Incl. frame and hardware, 14" x 16" vision panel,									
0020	48" wear panel, pair of 2'-6" x 7'	2 Sswk	1.60	10	Ea.	2,275	575		2,850	3,450
0040	Pair of 3' x 7'		1.50	10.667		2,650	615		3,265	3,900
0060	Pair of 4' x 7'	↓	1.40	11.429	↓	3,100	660		3,760	4,475

08 38 19 – Rigid Traffic Doors

08 38 19.20 Double Acting Swing Doors		Crew	Daily Output	Labor-Hours	Unit	Material	Labor	Equipment	Total	Total Incl O&P
0010	DOUBLE ACTING SWING DOORS									
0020	Including frame, closer, hardware and vision panel									
1000	Polymer, 7'-0" high, 4'-0" wide	2 Carp	4.20	3.810	Pr.	2,100	202		2,302	2,625
1025	6'-0" wide		4	4		2,225	213		2,438	2,800
1050	6'-8" wide	↓	4	4	↓	2,425	213		2,638	3,025
2000	3/4" thick, stainless steel									
2010	Stainless steel, 7' high opening, 4' wide	2 Carp	4	4	Pr.	2,625	213		2,838	3,225
2050	7' wide		3.80	4.211	"	2,800	224		3,024	3,425
9000	Minimum labor/equipment charge	↓	2	8	Job		425		425	680

08 38 19.30 Shock Absorbing Doors		Crew	Daily Output	Labor-Hours	Unit	Material	Labor	Equipment	Total	Total Incl O&P
0010	SHOCK ABSORBING DOORS									
0020	Rigid, no frame, 1-1/2" thick, 5' x 7'	2 Sswk	1.90	8.421	Opng.	1,550	485		2,035	2,500
0100	8' x 8'	↓	1.80	8.889	↓	2,025	510		2,535	3,075

08 38 Traffic Doors

08 38 19 – Rigid Traffic Doors

08 38 19.30 Shock Absorbing Doors		Crew	Daily Output	Labor-Hours	Unit	Material	Labor	Equipment	Total	Total Incl O&P
						2020 Bare Costs				
0500	Flexible, no frame, insulated, .16″ thick, economy, 5′ x 7′	2 Sswk	2	8	Opng.	1,775	460		2,235	2,725
0600	Deluxe		1.90	8.421		2,675	485		3,160	3,750
1000	8′ x 8′ opening, economy		2	8		2,825	460		3,285	3,850
1100	Deluxe		1.90	8.421		3,575	485		4,060	4,725
9000	Minimum labor/equipment charge		2	8	Job		460		460	755

08 41 Entrances and Storefronts

08 41 13 – Aluminum-Framed Entrances and Storefronts

08 41 13.20 Tube Framing

		Crew	Daily Output	Labor-Hours	Unit	Material	Labor	Equipment	Total	Total Incl O&P
0010	**TUBE FRAMING**, For window walls and storefronts, aluminum stock									
0050	Plain tube frame, mill finish, 1-3/4″ x 1-3/4″	2 Glaz	103	.155	L.F.	11.15	7.90		19.05	25
0150	1-3/4″ x 4″		98	.163		14.95	8.35		23.30	30
0200	1-3/4″ x 4-1/2″		95	.168		17	8.60		25.60	32.50
0250	2″ x 6″		89	.180		26	9.15		35.15	43
0350	4″ x 4″		87	.184		27.50	9.40		36.90	45.50
0400	4-1/2″ x 4-1/2″		85	.188		31	9.60		40.60	49.50
0450	Glass bead		240	.067		3.36	3.40		6.76	9.15
1000	Flush tube frame, mill finish, 1/4″ glass, 1-3/4″ x 4″, open header		80	.200		14.90	10.20		25.10	32.50
1050	Open sill		82	.195		12	9.95		21.95	29
1100	Closed back header		83	.193		21	9.85		30.85	39
1150	Closed back sill		85	.188		20	9.60		29.60	37.50
1160	Tube fmg., spandrel cover both sides, alum 1″ wide	1 Sswk	85	.094	S.F.	98	5.45		103.45	117
1170	Tube fmg., spandrel cover both sides, alum 2″ wide	″	85	.094	″	40	5.45		45.45	53
1200	Vertical mullion, one piece	2 Glaz	75	.213	L.F.	21.50	10.90		32.40	41
1250	Two piece		73	.219		23	11.20		34.20	43
1300	90° or 180° vertical corner post		75	.213		34.50	10.90		45.40	55.50
1400	1-3/4″ x 4-1/2″, open header		80	.200		17.45	10.20		27.65	35.50
1450	Open sill		82	.195		14.40	9.95		24.35	32
1500	Closed back header		83	.193		21	9.85		30.85	38.50
1550	Closed back sill		85	.188		20	9.60		29.60	37.50
1600	Vertical mullion, one piece		75	.213		22.50	10.90		33.40	42.50
1650	Two piece		73	.219		24	11.20		35.20	44.50
1700	90° or 180° vertical corner post		75	.213		24.50	10.90		35.40	44.50
2000	Flush tube frame, mil fin.,ins. glass w/thml brk, 2″ x 4-1/2″, open header		75	.213		15.95	10.90		26.85	35
2050	Open sill		77	.208		13.45	10.60		24.05	31.50
2100	Closed back header		78	.205		14.85	10.45		25.30	33
2150	Closed back sill		80	.200		15.75	10.20		25.95	33.50
2200	Vertical mullion, one piece		70	.229		16.95	11.65		28.60	37.50
2250	Two piece		68	.235		18.30	12		30.30	39
2300	90° or 180° vertical corner post		70	.229		19.05	11.65		30.70	39.50
5000	Flush tube frame, mill fin., thermal brk., 2-1/4″ x 4-1/2″, open header		74	.216		16.50	11.05		27.55	36
5050	Open sill		75	.213		14.25	10.90		25.15	33
5100	Vertical mullion, one piece		69	.232		19.35	11.85		31.20	40.50
5150	Two piece		67	.239		21	12.20		33.20	42.50
5200	90° or 180° vertical corner post		69	.232		18.80	11.85		30.65	39.50
5295	Flush tube frame, mill finish, 4-1/2″ x 4-1/2″		85	.188		31	9.60		40.60	49.50
5300	Plain tube frame, mill finish, 2″ x 3″ jamb		115	.139		12.15	7.10		19.25	24.50
5310	2″ x 4″		115	.139		16.45	7.10		23.55	29.50
5320	3″ x 5-1/2″		107	.150		29	7.65		36.65	44
5330	Head, 2″ x 3″		200	.080		21	4.08		25.08	29.50
5340	Mullion, 2″ x 3″		200	.080		21.50	4.08		25.58	30

08 41 13 – Aluminum-Framed Entrances and Storefronts

08 41 13.20 Tube Framing		Crew	Daily Output	Labor-Hours	Unit	Material	Labor	Equipment	Total	Total Incl O&P
						2020 Bare Costs				
5350	2" x 4"	2 Glaz	115	.139	L.F.	27.50	7.10		34.60	41.50
5360	3" x 5-1/2"		107	.150		36	7.65		43.65	51.50
5370	4" corner mullion		90	.178		30.50	9.05		39.55	48
5380	Horizontal, 2" x 3"		115	.139		25	7.10		32.10	39
5390	3" x 5-1/2"		107	.150		29	7.65		36.65	43.50
5430	Sill section, 1/8" x 6"		133	.120		35.50	6.15		41.65	49.50
5440	1/8" x 7"		133	.120		43	6.15		49.15	57
5450	1/8" x 8-1/2"		133	.120		50	6.15		56.15	65
5460	Column covers, aluminum, 1/8" x 26"		57	.281		69	14.30		83.30	99
5470	1/8" x 34"		53	.302		87	15.40		102.40	120
5480	1/8" x 38"		53	.302		99.50	15.40		114.90	134
5700	Vertical mullions, clear finish, 1/4" thick glass		528	.030		15.25	1.55		16.80	19.25
5720	3/8" thick		445	.036		13.70	1.83		15.53	18.05
5740	1/2" thick		400	.040		14.25	2.04		16.29	18.95
5760	3/4" thick		344	.047		16.30	2.37		18.67	21.50
5780	1" thick		304	.053		15.10	2.68		17.78	21
6980	Door stop (snap in)		380	.042		3.61	2.15		5.76	7.40
7000	For joints, 90°, clip type, add				Ea.	26.50			26.50	29
7050	Screw spline joint, add					24.50			24.50	27
7100	For joint other than 90°, add					51			51	56
8000	For bronze anodized aluminum, add					15%				
8020	For black finish, add					30%				
8050	For stainless steel materials, add					350%				
8100	For monumental grade, add					53%				
8150	For steel stiffener, add	2 Glaz	200	.080	L.F.	11.55	4.08		15.63	19.20
8200	For 2 to 5 stories, add per story				Story		8%			
9000	Minimum labor/equipment charge	2 Glaz	2	8	Job		410		410	650

08 41 19 – Stainless-Steel-Framed Entrances and Storefronts

08 41 19.10 Stainless-Steel and Glass Entrance Unit		Crew	Daily Output	Labor-Hours	Unit	Material	Labor	Equipment	Total	Total Incl O&P
0010	**STAINLESS-STEEL AND GLASS ENTRANCE UNIT**, narrow stiles									
0020	3' x 7' opening, including hardware, minimum	2 Sswk	1.60	10	Opng.	7,650	575		8,225	9,375
0050	Average		1.40	11.429		8,075	660		8,735	9,950
0100	Maximum		1.20	13.333		8,700	770		9,470	10,800
1000	For solid bronze entrance units, statuary finish, add					64%				
1100	Without statuary finish, add					45%				
2000	Balanced doors, 3' x 7', economy	2 Sswk	.90	17.778	Ea.	10,300	1,025		11,325	13,100
2100	Premium		.70	22.857	"	16,800	1,325		18,125	20,700
9000	Minimum labor/equipment charge		2	8	Job		460		460	755

08 41 26 – All-Glass Entrances and Storefronts

08 41 26.10 Window Walls Aluminum, Stock		Crew	Daily Output	Labor-Hours	Unit	Material	Labor	Equipment	Total	Total Incl O&P
0010	**WINDOW WALLS ALUMINUM, STOCK**, including glazing									
0020	Minimum	H-2	160	.150	S.F.	51.50	7.20		58.70	68
0050	Average		140	.171		71.50	8.25		79.75	91.50
0100	Maximum		110	.218		188	10.50		198.50	224
0500	For translucent sandwich wall systems, see Section 07 41 33.10									
0850	Cost of the above walls depends on material,									
0860	finish, repetition, and size of units.									
0870	The larger the opening, the lower the S.F. cost									

08 41 Entrances and Storefronts

08 41 26 – All-Glass Entrances and Storefronts

08 41 26.20 All-Glass Entrance Doors		Crew	Daily Output	Labor-Hours	Unit	Material	Labor	Equipment	Total	Total Incl O&P
						2020 Bare Costs				
0010	**ALL-GLASS ENTRANCE DOORS**									
0015	Hardware and sst trim, temp glass, 1/2" thk, 3' x 7', single	2 Sswk	1.85	8.649	Ea.	3,050	500		3,550	4,175
0020	Pair of 3' x 7'		1.05	15.238	Pr.	5,400	880		6,280	7,400
0040	3/4" thick, 3' x 7', single		1.85	8.649	Ea.	6,250	500		6,750	7,700
0060	Pair of 3' x 7'	↓	1.05	15.238	Pr.	12,300	880		13,180	15,000

08 42 Entrances

08 42 26 – All-Glass Entrances

08 42 26.10 Swinging Glass Doors		Crew	Daily Output	Labor-Hours	Unit	Material	Labor	Equipment	Total	Total Incl O&P
0010	**SWINGING GLASS DOORS**									
0020	Including hardware, 1/2" thick, tempered, 3' x 7' opening	2 Glaz	2	8	Opng.	2,400	410		2,810	3,275
0100	6' x 7' opening		1.40	11.429	"	4,650	585		5,235	6,050
9000	Minimum labor/equipment charge	↓	2	8	Job		410		410	650

08 42 33 – Revolving Door Entrances

08 42 33.10 Circular Rotating Entrance Doors		Crew	Daily Output	Labor-Hours	Unit	Material	Labor	Equipment	Total	Total Incl O&P
0010	**CIRCULAR ROTATING ENTRANCE DOORS**, Aluminum									
0020	6'-10" to 7' high, stock units, minimum	4 Sswk	.75	42.667	Opng.	38,000	2,450		40,450	45,900
0050	Average		.60	53.333		43,400	3,075		46,475	53,000
0100	Maximum		.45	71.111		44,300	4,100		48,400	55,500
1000	Stainless steel		.30	106		50,500	6,100		56,600	65,500
1100	Solid bronze	↓	.15	213		50,500	12,300		62,800	75,500
1500	For automatic controls, add	2 Elec	2	8	↓	15,400	490		15,890	17,800

08 42 36 – Balanced Door Entrances

08 42 36.10 Balanced Entrance Doors		Crew	Daily Output	Labor-Hours	Unit	Material	Labor	Equipment	Total	Total Incl O&P
0010	**BALANCED ENTRANCE DOORS**									
0020	Hardware & frame, alum. & glass, 3' x 7', econ.	2 Sswk	.90	17.778	Ea.	7,200	1,025		8,225	9,600
0150	Premium		.70	22.857	"	8,725	1,325		10,050	11,800
9000	Minimum labor/equipment charge	↓	1	16	Job		920		920	1,500

08 43 Storefronts

08 43 13 – Aluminum-Framed Storefronts

08 43 13.10 Aluminum-Framed Entrance Doors and Frames		Crew	Daily Output	Labor-Hours	Unit	Material	Labor	Equipment	Total	Total Incl O&P
0010	**ALUMINUM-FRAMED ENTRANCE DOORS AND FRAMES**									
0015	Standard hardware and glass stops but no glass									
0020	Entrance door, 3' x 7' opening, clear anodized finish	2 Sswk	7	2.286	Opng.	680	132		812	965
0040	Bronze finish		7	2.286		835	132		967	1,125
0060	Black finish		7	2.286		815	132		947	1,125
0200	3'-6" x 7'-0", mill finish		7	2.286		835	132		967	1,125
0220	Bronze finish		7	2.286		960	132		1,092	1,275
0240	Black finish		7	2.286		985	132		1,117	1,300
0500	6' x 7' opening, clear finish		6	2.667		1,075	154		1,229	1,425
0520	Bronze finish		6	2.667		1,175	154		1,329	1,550
0600	Door frame for above doors 3'-0" x 7'-0", mill finish		6	2.667		560	154		714	865
0620	Bronze finish		6	2.667		595	154		749	905
0640	Black finish		6	2.667		670	154		824	990
0700	3'-6" x 7'-0", mill finish		6	2.667		410	154		564	700
0720	Bronze finish		6	2.667		410	154		564	700
0740	Black finish	↓	6	2.667	↓	410	154		564	700

08 43 Storefronts

08 43 13 – Aluminum-Framed Storefronts

08 43 13.10 Aluminum-Framed Entrance Doors and Frames		Crew	Daily Output	Labor-Hours	Unit	2020 Bare Costs Material	Labor	Equipment	Total	Total Incl O&P
0800	6'-0" x 7'-0", mill finish	2 Sswk	6	2.667	Opng.	410	154		564	700
0820	Bronze finish		6	2.667		415	154		569	705
0840	Black finish		6	2.667		445	154		599	740
1000	With 3' high transom above, 3' x 7' opening, clear finish		5.50	2.909		635	168		803	975
1050	Bronze finish		5.50	2.909		630	168		798	970
1100	Black finish		5.50	2.909		695	168		863	1,025
1300	3'-6" x 7'-0" opening, clear finish		5.50	2.909		420	168		588	735
1320	Bronze finish		5.50	2.909		440	168		608	760
1340	Black finish		5.50	2.909		450	168		618	770
1500	6' x 7' opening, clear finish		5.50	2.909		695	168		863	1,050
1550	Bronze finish		5.50	2.909		785	168		953	1,150
1600	Black finish	↓	5.50	2.909	↓	870	168		1,038	1,225
8000	For 8' high doors, add				Ea.	174			174	174
9000	Minimum labor/equipment charge	2 Sswk	4	4	Job		231		231	380

08 43 13.20 Storefront Systems		Crew	Daily Output	Labor-Hours	Unit	Material	Labor	Equipment	Total	Total Incl O&P
0010	**STOREFRONT SYSTEMS**, aluminum frame clear 3/8" plate glass									
0020	incl. 3' x 7' door with hardware (400 sq. ft. max. wall)									
0500	Wall height to 12' high, commercial grade	2 Glaz	150	.107	S.F.	27.50	5.45		32.95	38.50
0600	Institutional grade		130	.123		32	6.30		38.30	45
0700	Monumental grade		115	.139		84.50	7.10		91.60	104
1000	6' x 7' door with hardware, commercial grade		135	.119		76	6.05		82.05	93
1100	Institutional grade		115	.139		50	7.10		57.10	66.50
1200	Monumental grade	↓	100	.160		117	8.15		125.15	141
1500	For bronze anodized finish, add					15%				
1600	For black anodized finish, add					36%				
1700	For stainless steel framing, add to monumental				↓	78%				
9000	Minimum labor/equipment charge	2 Glaz	1	16	Job		815		815	1,300

08 43 29 – Sliding Storefronts

08 43 29.10 Sliding Panels		Crew	Daily Output	Labor-Hours	Unit	Material	Labor	Equipment	Total	Total Incl O&P
0010	**SLIDING PANELS**									
0020	Mall fronts, aluminum & glass, 15' x 9' high	2 Glaz	1.30	12.308	Opng.	4,300	630		4,930	5,750
0100	24' x 9' high		.70	22.857		6,125	1,175		7,300	8,625
0200	48' x 9' high, with fixed panels	↓	.90	17.778		11,000	905		11,905	13,600
0500	For bronze finish, add				↓	17%				
9000	Minimum labor/equipment charge	2 Glaz	1	16	Job		815		815	1,300

08 44 Curtain Wall and Glazed Assemblies

08 44 13 – Glazed Aluminum Curtain Walls

08 44 13.10 Glazed Curtain Walls		Crew	Daily Output	Labor-Hours	Unit	Material	Labor	Equipment	Total	Total Incl O&P
0010	**GLAZED CURTAIN WALLS**, aluminum, stock, including glazing									
0020	Minimum	H-1	205	.156	S.F.	43.50	8.50		52	62
0050	Average, single glazed		195	.164		62.50	8.90		71.40	83.50
0150	Average, double glazed		180	.178		81	9.65		90.65	105
0200	Maximum	↓	160	.200	↓	206	10.85		216.85	244

08 45 Translucent Wall and Roof Assemblies

08 45 10 – Translucent Roof Assemblies

08 45 10.10 Skyroofs		Crew	Daily Output	Labor-Hours	Unit	Material	2020 Bare Costs Labor	Equipment	Total	Total Incl O&P
0010	**SKYROOFS**									
1200	Skylights, circular, clear, double glazed acrylic									
1230	30" diameter	2 Carp	3	5.333	Ea.	3,075	283		3,358	3,825
1250	60" diameter		3	5.333		4,075	283		4,358	4,950
1290	96" diameter		2	8		5,100	425		5,525	6,300
1300	Skylight Barrel Vault, clear, double glazed, acrylic									
1330	3'-0" x 12'-0"	G-3	3	10.667	Ea.	5,100	555		5,655	6,500
1350	4'-0" x 12'-0"		3	10.667		5,625	555		6,180	7,050
1390	5'-0" x 12'-0"		2	16		6,125	835		6,960	8,075
1400	Skylight Pyramid, aluminum frame, clear low E laminated glass									
1410	The glass is installed in the frame except where noted									
1430	Square, 3' x 3'	G-3	3	10.667	Ea.	6,125	555		6,680	7,625
1440	4' x 4'		3	10.667		7,150	555		7,705	8,725
1450	5' x 5', glass must be field installed		3	10.667		8,175	555		8,730	9,850
1460	6' x 6', glass must be field installed		2	16		10,200	835		11,035	12,500
1550	Install pre-cut laminated glass in aluminum frame on a flat roof	2 Glaz	55	.291	SF Surf		14.85		14.85	23.50
1560	Install pre-cut laminated glass in aluminum frame on a sloped roof	"	40	.400	"		20.50		20.50	32.50
9000	Minimum labor/equipment charge	2 Carp	8	2	Job		106		106	170

08 51 Metal Windows

08 51 13 – Aluminum Windows

08 51 13.10 Aluminum Sash

		Crew	Daily Output	Labor-Hours	Unit	Material	Labor	Equipment	Total	Total Incl O&P
0010	**ALUMINUM SASH**									
0020	Stock, grade C, glaze & trim not incl., casement	2 Sswk	200	.080	S.F.	43	4.61		47.61	54.50
0050	Double-hung		200	.080		42	4.61		46.61	54
0100	Fixed casement		200	.080		18.45	4.61		23.06	28
0150	Picture window		200	.080		19.75	4.61		24.36	29
0200	Projected window		200	.080		38.50	4.61		43.11	50
0250	Single-hung		200	.080		17.40	4.61		22.01	26.50
0300	Sliding		200	.080		23	4.61		27.61	33
1000	Mullions for above, tubular		240	.067	L.F.	6.55	3.84		10.39	13.50
2000	Custom aluminum sash, grade HC, glazing not included		140	.114	S.F.	41.50	6.60		48.10	56.50
9000	Minimum labor/equipment charge	1 Sswk	2	4	Job		231		231	380

08 51 13.20 Aluminum Windows

		Crew	Daily Output	Labor-Hours	Unit	Material	Labor	Equipment	Total	Total Incl O&P
0010	**ALUMINUM WINDOWS**, incl. frame and glazing, commercial grade									
1000	Stock units, casement, 3'-1" x 3'-2" opening	2 Sswk	10	1.600	Ea.	395	92		487	580
1050	Add for storms					126			126	138
1600	Projected, with screen, 3'-1" x 3'-2" opening	2 Sswk	10	1.600		370	92		462	560
1700	Add for storms					123			123	135
2000	4'-5" x 5'-3" opening	2 Sswk	8	2		410	115		525	645
2100	Add for storms					132			132	145
2500	Enamel finish windows, 3'-1" x 3'-2"	2 Sswk	10	1.600		380	92		472	565
2600	4'-5" x 5'-3"		8	2		425	115		540	660
3000	Single-hung, 2' x 3' opening, enameled, standard glazed		10	1.600		219	92		311	390
3100	Insulating glass		10	1.600		265	92		357	445
3300	2'-8" x 6'-8" opening, standard glazed		8	2		375	115		490	600
3400	Insulating glass		8	2		480	115		595	720
3700	3'-4" x 5'-0" opening, standard glazed		9	1.778		315	102		417	520
3800	Insulating glass		9	1.778		345	102		447	545
3890	Awning type, 3' x 3' opening, standard glass		14	1.143		435	66		501	590
3900	Insulating glass		14	1.143		475	66		541	630

08 51 Metal Windows

08 51 13 – Aluminum Windows

08 51 13.20 Aluminum Windows

	08 51 13.20 Aluminum Windows	Crew	Daily Output	Labor-Hours	Unit	2020 Bare Costs Material	Labor	Equipment	Total	Total Incl O&P
3910	3′ x 4′ opening, standard glass	2 Sswk	10	1.600	Ea.	510	92		602	710
3920	Insulating glass		10	1.600		590	92		682	800
3930	3′ x 5′-4″ opening, standard glass		10	1.600		615	92		707	825
3940	Insulating glass		10	1.600		725	92		817	950
3950	4′ x 5′-4″ opening, standard glass		9	1.778		675	102		777	915
3960	Insulating glass		9	1.778		780	102		882	1,025
4000	Sliding aluminum, 3′ x 2′ opening, standard glazed		10	1.600		218	92		310	390
4100	Insulating glass		10	1.600		241	92		333	415
4300	5′ x 3′ opening, standard glazed		9	1.778		345	102		447	550
4400	Insulating glass		9	1.778		385	102		487	595
4600	8′ x 4′ opening, standard glazed		6	2.667		365	154		519	650
4700	Insulating glass		6	2.667		590	154		744	900
5000	9′ x 5′ opening, standard glazed		4	4		550	231		781	985
5100	Insulating glass		4	4		850	231		1,081	1,325
5500	Sliding, with thermal barrier and screen, 6′ x 4′, 2 track		8	2		725	115		840	990
5700	4 track		8	2		915	115		1,030	1,200
6000	For above units with bronze finish, add					15%				
6200	For installation in concrete openings, add					8%				
7000	Double-hung, insulating glass, 2′-0″ x 2′-0″	2 Sswk	11	1.455		118	84		202	267
7020	2′-0″ x 2′-6″		11	1.455		134	84		218	284
7040	3′-0″ x 1′-6″		10	1.600		271	92		363	450
7060	3′-0″ x 2′-0″		10	1.600		315	92		407	500
7080	3′-0″ x 2′-6″		10	1.600		385	92		477	570
7100	3′-0″ x 3′-0″		10	1.600		425	92		517	620
7120	3′-0″ x 3′-6″		10	1.600		450	92		542	645
7140	3′-0″ x 4′-0″		10	1.600		485	92		577	685
7160	3′-0″ x 5′-0″		10	1.600		530	92		622	735
7180	3′-0″ x 6′-0″		9	1.778		565	102		667	790

08 51 13.30 Impact Resistant Aluminum Windows

	08 51 13.30 Impact Resistant Aluminum Windows	Crew	Daily Output	Labor-Hours	Unit	Material	Labor	Equipment	Total	Total Incl O&P
0010	**IMPACT RESISTANT ALUMINUM WINDOWS**, incl. frame and glazing									
0100	Single-hung, impact resistant, 2′-8″ x 5′-0″	2 Carp	9	1.778	Ea.	1,250	94.50		1,344.50	1,525
0120	3′-0″ x 5′-0″		9	1.778		1,350	94.50		1,444.50	1,650
0130	4′-0″ x 5′-0″		9	1.778		1,450	94.50		1,544.50	1,750
0250	Horizontal slider, impact resistant, 5′-5″ x 5′-2″		9	1.778		1,625	94.50		1,719.50	1,925

08 51 23 – Steel Windows

08 51 23.10 Steel Sash

	08 51 23.10 Steel Sash	Crew	Daily Output	Labor-Hours	Unit	Material	Labor	Equipment	Total	Total Incl O&P
0010	**STEEL SASH** Custom units, glazing and trim not included									
0100	Casement, 100% vented	2 Sswk	200	.080	S.F.	69.50	4.61		74.11	83.50
0200	50% vented		200	.080		56	4.61		60.61	69
0300	Fixed		200	.080		29	4.61		33.61	39
1000	Projected, commercial, 40% vented		200	.080		53	4.61		57.61	66
1100	Intermediate, 50% vented		200	.080		62.50	4.61		67.11	76
1500	Industrial, horizontally pivoted		200	.080		56	4.61		60.61	69.50
1600	Fixed		200	.080		32.50	4.61		37.11	43
2000	Industrial security sash, 50% vented		200	.080		61.50	4.61		66.11	75
2100	Fixed		200	.080		49	4.61		53.61	61.50
2500	Picture window		200	.080		32	4.61		36.61	42.50
3000	Double-hung		200	.080		62	4.61		66.61	76
5000	Mullions for above, open interior face		240	.067	L.F.	10.95	3.84		14.79	18.35
5100	With interior cover		240	.067	″	18.10	3.84		21.94	26
6100	Triple glazing for above, add	2 Glaz	85	.188	S.F.	12.65	9.60		22.25	29.50
9000	Minimum labor/equipment charge	1 Sswk	2	4	Job		231		231	380

08 51 Metal Windows

08 51 23 – Steel Windows

08 51 23.20 Steel Windows		Crew	Daily Output	Labor-Hours	Unit	Material	2020 Bare Costs Labor	Equipment	Total	Total Incl O&P
0010	**STEEL WINDOWS** Stock, including frame, trim and insul. glass									
0020	See Section 13 34 19.50									
1000	Custom units, double-hung, 2'-8" x 4'-6" opening	2 Sswk	12	1.333	Ea.	730	77		807	925
1100	2'-4" x 3'-9" opening		12	1.333		600	77		677	785
1500	Commercial projected, 3'-9" x 5'-5" opening		10	1.600		1,275	92		1,367	1,550
1600	6'-9" x 4'-1" opening		7	2.286		1,675	132		1,807	2,075
2000	Intermediate projected, 2'-9" x 4'-1" opening		12	1.333		710	77		787	905
2100	4'-1" x 5'-5" opening		10	1.600		1,450	92		1,542	1,750
9000	Minimum labor/equipment charge	1 Sswk	3	2.667	Job		154		154	252

08 51 23.40 Basement Utility Windows		Crew	Daily Output	Labor-Hours	Unit	Material	Labor	Equipment	Total	Total Incl O&P
0010	**BASEMENT UTILITY WINDOWS**									
0015	1'-3" x 2'-8"	1 Carp	16	.500	Ea.	156	26.50		182.50	215
1100	1'-7" x 2'-8"	"	16	.500	"	151	26.50		177.50	209

08 51 66 – Metal Window Screens

08 51 66.10 Screens		Crew	Daily Output	Labor-Hours	Unit	Material	Labor	Equipment	Total	Total Incl O&P
0010	**SCREENS**									
0020	For metal sash, aluminum or bronze mesh, flat screen	2 Sswk	1200	.013	S.F.	4.63	.77		5.40	6.35
0500	Wicket screen, inside window		1000	.016		6.95	.92		7.87	9.15
0800	Security screen, aluminum frame with stainless steel cloth		1200	.013		24.50	.77		25.27	28.50
0900	Steel grate, painted, on steel frame		1600	.010		13.95	.58		14.53	16.25
1000	Screens for solar louvers		160	.100		25.50	5.75		31.25	37.50
4000	See Section 05 58 23.90 for window guards									

08 52 Wood Windows

08 52 10 – Plain Wood Windows

08 52 10.20 Awning Window		Crew	Daily Output	Labor-Hours	Unit	Material	Labor	Equipment	Total	Total Incl O&P
0010	**AWNING WINDOW**, Including frame, screens and grilles									
0100	34" x 22", insulated glass	1 Carp	10	.800	Ea.	300	42.50		342.50	400
0200	Low E glass		10	.800		330	42.50		372.50	430
0300	40" x 28", insulated glass		9	.889		315	47		362	425
0400	Low E glass		9	.889		350	47		397	460
0500	48" x 36", insulated glass		8	1		480	53		533	610
0600	Low E glass		8	1		505	53		558	640
4000	Impact windows, minimum, add					60%				
4010	Impact windows, maximum, add					160%				
9000	Minimum labor/equipment charge	1 Carp	4	2	Job		106		106	170

08 52 10.30 Wood Windows		Crew	Daily Output	Labor-Hours	Unit	Material	Labor	Equipment	Total	Total Incl O&P
0010	**WOOD WINDOWS**, double-hung									
0020	Including frame, double insulated glass, screens and grilles									
0040	Double-hung, 2'-2" x 3'-4" high	2 Carp	15	1.067	Ea.	221	56.50		277.50	335
0060	2'-2" x 4'-4"		14	1.143		236	60.50		296.50	360
0080	2'-6" x 3'-4"		13	1.231		233	65.50		298.50	360
0100	2'-6" x 4'-0"		12	1.333		238	71		309	375
0120	2'-6" x 4'-8"		12	1.333		254	71		325	395
0140	2'-10" x 3'-4"		10	1.600		229	85		314	390
0160	2'-10" x 4'-0"		10	1.600		262	85		347	425
0180	3'-7" x 3'-4"		9	1.778		262	94.50		356.50	440
0200	3'-7" x 5'-4"		9	1.778		300	94.50		394.50	480
0220	3'-10" x 5'-4"		8	2		525	106		631	745

08 52 10 – Plain Wood Windows

08 52 10.30 Wood Windows

	08 52 10.30 Wood Windows		Crew	Daily Output	Labor-Hours	Unit	Material	2020 Bare Costs Labor	Equipment	Total	Total Incl O&P
3800	Triple glazing for above, add					Ea.	25%				

08 52 10.40 Casement Window

	08 52 10.40 Casement Window		Crew	Daily Output	Labor-Hours	Unit	Material	2020 Bare Costs Labor	Equipment	Total	Total Incl O&P
0010	**CASEMENT WINDOW**, including frame, screen and grilles										
0100	2'-0" x 3'-0" H, double insulated glass	G	1 Carp	10	.800	Ea.	275	42.50		317.50	370
0150	Low E glass	G		10	.800		285	42.50		327.50	385
0200	2'-0" x 4'-6" high, double insulated glass	G		9	.889		390	47		437	505
0250	Low E glass	G		9	.889		440	47		487	560
0260	Casement 4'-2" x 4'-2" double insulated glass	G		11	.727		940	38.50		978.50	1,075
0270	4'-0" x 4'-0" Low E glass	G		11	.727		575	38.50		613.50	690
0290	6'-4" x 5'-7" Low E glass	G		9	.889		1,200	47		1,247	1,400
0300	2'-4" x 6'-0" high, double insulated glass	G		8	1		470	53		523	605
0350	Low E glass	G		8	1		450	53		503	580
0522	Vinyl-clad, premium, double insulated glass, 2'-0" x 3'-0"	G		10	.800		281	42.50		323.50	380
0524	2'-0" x 4'-0"	G		9	.889		335	47		382	440
0525	2'-0" x 5'-0"	G		8	1		365	53		418	485
0528	2'-0" x 6'-0"	G		8	1		410	53		463	535
0600	3'-0" x 5'-0"	G		8	1		685	53		738	840
0700	4'-0" x 3'-0"	G		8	1		770	53		823	930
0710	4'-0" x 4'-0"	G		8	1		645	53		698	795
0720	4'-8" x 4'-0"	G		8	1		710	53		763	865
0730	4'-8" x 5'-0"	G		6	1.333		830	71		901	1,025
0740	4'-8" x 6'-0"	G		6	1.333		920	71		991	1,125
0750	6'-0" x 4'-0"	G		6	1.333		835	71		906	1,025
0800	6'-0" x 5'-0"	G		6	1.333		1,025	71		1,096	1,250
0900	5'-6" x 5'-6"	G	2 Carp	15	1.067		1,500	56.50		1,556.50	1,750
2000	Bay, casement units, 8' x 5', w/screens, double insulated glass			2.50	6.400	Opng.	1,675	340		2,015	2,400
2100	Low E glass			2.50	6.400	"	1,750	340		2,090	2,475
8190	For installation, add per leaf					Ea.		15%			
8200	For multiple leaf units, deduct for stationary sash										
8220	2' high					Ea.	24.50			24.50	27
8240	4'-6" high						28			28	30.50
8260	6' high						37			37	41
8300	Impact windows, minimum, add						60%				
8310	Impact windows, maximum, add						160%				
9000	Minimum labor/equipment charge		1 Carp	3	2.667	Job		142		142	227

08 52 10.50 Double-Hung

	08 52 10.50 Double-Hung		Crew	Daily Output	Labor-Hours	Unit	Material	2020 Bare Costs Labor	Equipment	Total	Total Incl O&P
0010	**DOUBLE-HUNG**, Including frame, screens and grilles										
0100	2'-0" x 3'-0" high, low E insul. glass	G	1 Carp	10	.800	Ea.	193	42.50		235.50	280
0200	3'-0" x 4'-0" high, double insulated glass	G		9	.889		290	47		337	395
0300	4'-0" x 4'-6" high, low E insulated glass	G		8	1		325	53		378	445
8000	Impact windows, minimum, add						60%				
8010	Impact windows, maximum, add						160%				
9000	Minimum labor/equipment charge		1 Carp	3	2.667	Job		142		142	227

08 52 10.55 Picture Window

	08 52 10.55 Picture Window		Crew	Daily Output	Labor-Hours	Unit	Material	2020 Bare Costs Labor	Equipment	Total	Total Incl O&P
0010	**PICTURE WINDOW**, Including frame and grilles										
0100	3'-6" x 4'-0" high, double insulated glass		2 Carp	12	1.333	Ea.	440	71		511	600
0150	Low E glass			12	1.333		450	71		521	610
0200	4'-0" x 4'-6" high, double insulated glass			11	1.455		545	77.50		622.50	725
0250	Low E glass			11	1.455		560	77.50		637.50	740
0300	5'-0" x 4'-0" high, double insulated glass			11	1.455		610	77.50		687.50	800
0350	Low E glass			11	1.455		620	77.50		697.50	805
0400	6'-0" x 4'-6" high, double insulated glass			10	1.600		655	85		740	855

08 52 Wood Windows

08 52 10 – Plain Wood Windows

08 52 10.55 Picture Window		Crew	Daily Output	Labor-Hours	Unit	Material	2020 Bare Costs Labor	Equipment	Total	Total Incl O&P
0450	Low E glass	2 Carp	10	1.600	Ea.	645	85		730	845
08 52 10.65 Wood Sash										
0010	**WOOD SASH**, Including glazing but not trim									
0050	Custom, 5'-0" x 4'-0", 1" double glazed, 3/16" thick lites	2 Carp	3.20	5	Ea.	244	266		510	695
0100	1/4" thick lites		5	3.200		293	170		463	590
0200	1" thick, triple glazed		5	3.200		430	170		600	745
0300	7'-0" x 4'-6" high, 1" double glazed, 3/16" thick lites		4.30	3.721		430	198		628	790
0400	1/4" thick lites		4.30	3.721		495	198		693	855
0500	1" thick, triple glazed		4.30	3.721		560	198		758	935
0600	8'-6" x 5'-0" high, 1" double glazed, 3/16" thick lites		3.50	4.571		585	243		828	1,025
0700	1/4" thick lites		3.50	4.571		640	243		883	1,100
0800	1" thick, triple glazed		3.50	4.571		725	243		968	1,175
0900	Window frames only, based on perimeter length				L.F.	4.10			4.10	4.51
1200	Window sill, stock, per lineal foot					9.05			9.05	9.95
1250	Casing, stock					3.36			3.36	3.70
08 52 10.70 Sliding Windows										
0010	**SLIDING WINDOWS**									
0100	3'-0" x 3'-0" high, double insulated G	1 Carp	10	.800	Ea.	286	42.50		328.50	385
0120	Low E glass G		10	.800		310	42.50		352.50	415
0200	4'-0" x 3'-6" high, double insulated G		9	.889		425	47		472	540
0220	Low E glass G		9	.889		420	47		467	540
0300	6'-0" x 5'-0" high, double insulated G		8	1		495	53		548	625
0320	Low E glass G		8	1		605	53		658	750
9000	Minimum labor/equipment charge		3	2.667	Job		142		142	227

08 52 13 – Metal-Clad Wood Windows

08 52 13.10 Awning Windows, Metal-Clad		Crew	Daily Output	Labor-Hours	Unit	Material	Labor	Equipment	Total	Total Incl O&P
0010	**AWNING WINDOWS, METAL-CLAD**									
2000	Metal-clad, awning deluxe, double insulated glass, 34" x 22"	1 Carp	9	.889	Ea.	251	47		298	355
2050	36" x 25"		9	.889		280	47		327	385
2100	40" x 22"		9	.889		297	47		344	400
2150	40" x 30"		9	.889		345	47		392	455
2200	48" x 28"		8	1		355	53		408	475
2250	60" x 36"		8	1		380	53		433	505
08 52 13.20 Casement Windows, Metal-Clad										
0010	**CASEMENT WINDOWS, METAL-CLAD**									
0100	Metal-clad, deluxe, dbl. insul. glass, 2'-0" x 3'-0" high G	1 Carp	10	.800	Ea.	300	42.50		342.50	400
0120	2'-0" x 4'-0" high G		9	.889		325	47		372	435
0130	2'-0" x 5'-0" high G		8	1		360	53		413	480
0140	2'-0" x 6'-0" high G		8	1		390	53		443	515
0300	Metal-clad, casement, bldrs mdl, 6'-0" x 4'-0", dbl. insul. glass, 3 panels	2 Carp	10	1.600		1,250	85		1,335	1,525
0310	9'-0" x 4'-0", 4 panels		8	2		1,600	106		1,706	1,925
0320	10'-0" x 5'-0", 5 panels		7	2.286		2,175	121		2,296	2,600
0330	12'-0" x 6'-0", 6 panels		6	2.667		2,750	142		2,892	3,250
08 52 13.30 Double-Hung Windows, Metal-Clad										
0010	**DOUBLE-HUNG WINDOWS, METAL-CLAD**									
0100	Metal-clad, deluxe, dbl. insul. glass, 2'-6" x 3'-0" high G	1 Carp	10	.800	Ea.	294	42.50		336.50	395
0120	3'-0" x 3'-6" high G		10	.800		330	42.50		372.50	435
0140	3'-0" x 4'-0" high G		9	.889		340	47		387	450
0160	3'-0" x 4'-6" high G		9	.889		365	47		412	475
0180	3'-0" x 5'-0" high G		8	1		395	53		448	520
0200	3'-6" x 6'-0" high G		8	1		475	53		528	610

08 52 Wood Windows

08 52 13 – Metal-Clad Wood Windows

08 52 13.35 Picture and Sliding Windows Metal-Clad			Crew	Daily Output	Labor-Hours	Unit	Material	2020 Bare Costs Labor	Equipment	Total	Total Incl O&P
0010	PICTURE AND SLIDING WINDOWS METAL-CLAD										
2000	Metal-clad, dlx picture, dbl. insul. glass, 4'-0" x 4'-0" high		2 Carp	12	1.333	Ea.	385	71		456	540
2100	4'-0" x 6'-0" high			11	1.455		575	77.50		652.50	755
2200	5'-0" x 6'-0" high			10	1.600		635	85		720	835
2300	6'-0" x 6'-0" high			10	1.600		725	85		810	935
2400	Metal-clad, dlx sliding, dbl. insul. glass, 3'-0" x 3'-0" high	G	1 Carp	10	.800		330	42.50		372.50	430
2420	4'-0" x 3'-6" high	G		9	.889		400	47		447	515
2440	5'-0" x 4'-0" high	G		9	.889		480	47		527	605
2460	6'-0" x 5'-0" high	G		8	1		770	53		823	930
9000	Minimum labor/equipment charge		2 Carp	2.75	5.818	Job		310		310	495

08 52 13.40 Bow and Bay Windows, Metal-Clad		Crew	Daily Output	Labor-Hours	Unit	Material	Labor	Equipment	Total	Total Incl O&P
0010	BOW AND BAY WINDOWS, METAL-CLAD									
0100	Metal-clad, deluxe, dbl. insul. glass, 8'-0" x 5'-0" high, 4 panels	2 Carp	10	1.600	Ea.	1,725	85		1,810	2,025
0120	10'-0" x 5'-0" high, 5 panels		8	2		1,875	106		1,981	2,225
0140	10'-0" x 6'-0" high, 5 panels		7	2.286		2,200	121		2,321	2,625
0160	12'-0" x 6'-0" high, 6 panels		6	2.667		2,950	142		3,092	3,475
0400	Double-hung, bldrs. model, bay, 8' x 4' high, dbl. insul. glass		10	1.600		1,375	85		1,460	1,650
0440	Low E glass		10	1.600		1,475	85		1,560	1,750
0460	9'-0" x 5'-0" high, dbl. insul. glass		6	2.667		1,475	142		1,617	1,850
0480	Low E glass		6	2.667		1,550	142		1,692	1,950
0500	Metal-clad, deluxe, dbl. insul. glass, 7'-0" x 4'-0" high		10	1.600		1,325	85		1,410	1,575
0520	8'-0" x 4'-0" high		8	2		1,375	106		1,481	1,675
0540	8'-0" x 5'-0" high		7	2.286		1,400	121		1,521	1,750
0560	9'-0" x 5'-0" high		6	2.667		1,500	142		1,642	1,875

08 52 16 – Plastic-Clad Wood Windows

08 52 16.10 Bow Window		Crew	Daily Output	Labor-Hours	Unit	Material	Labor	Equipment	Total	Total Incl O&P
0010	BOW WINDOW including frames, screens, and grilles									
0020	End panels operable									
1000	Bow type, casement, wood, bldrs. mdl., 8' x 5' dbl. insul. glass, 4 panel	2 Carp	10	1.600	Ea.	1,600	85		1,685	1,875
1050	Low E glass		10	1.600		1,325	85		1,410	1,575
1100	10'-0" x 5'-0", dbl. insul. glass, 6 panels		6	2.667		1,375	142		1,517	1,750
1200	Low E glass, 6 panels		6	2.667		1,475	142		1,617	1,850
1300	Vinyl-clad, bldrs. model, dbl. insul. glass, 6'-0" x 4'-0", 3 panel		10	1.600		1,050	85		1,135	1,275
1340	9'-0" x 4'-0", 4 panel		8	2		1,450	106		1,556	1,775
1380	10'-0" x 6'-0", 5 panels		7	2.286		2,375	121		2,496	2,800
1420	12'-0" x 6'-0", 6 panels		6	2.667		3,100	142		3,242	3,650
2000	Bay window, 8' x 5', dbl. insul. glass		10	1.600		1,975	85		2,060	2,275
2050	Low E glass		10	1.600		2,375	85		2,460	2,725
2100	12'-0" x 6'-0", dbl. insul. glass, 6 panels		6	2.667		2,450	142		2,592	2,925
2200	Low E glass		6	2.667		3,175	142		3,317	3,725
2280	6'-0" x 4'-0"		11	1.455		1,300	77.50		1,377.50	1,575
2300	Vinyl-clad, premium, dbl. insul. glass, 8'-0" x 5'-0"		10	1.600		1,825	85		1,910	2,125
2340	10'-0" x 5'-0"		8	2		2,500	106		2,606	2,925
2380	10'-0" x 6'-0"		7	2.286		3,075	121		3,196	3,575
2420	12'-0" x 6'-0"		6	2.667		3,325	142		3,467	3,875
3300	Vinyl-clad, premium, dbl. insul. glass, 7'-0" x 4'-6"		10	1.600		1,400	85		1,485	1,650
3340	8'-0" x 4'-6"		8	2		1,450	106		1,556	1,775
3380	8'-0" x 5'-0"		7	2.286		1,525	121		1,646	1,875
3420	9'-0" x 5'-0"		6	2.667		1,550	142		1,692	1,950
9000	Minimum labor/equipment charge		2.50	6.400	Job		340		340	545

08 52 16 – Plastic-Clad Wood Windows

08 52 16.15 Awning Window Vinyl-Clad		Crew	Daily Output	Labor-Hours	Unit	Material	Labor	Equipment	Total	Total Incl O&P
						2020 Bare Costs				
0010	**AWNING WINDOW VINYL-CLAD** including frames, screens, and grilles									
0240	Vinyl-clad, 34" x 22"	1 Carp	10	.800	Ea.	272	42.50		314.50	365
0280	36" x 28"		9	.889		310	47		357	415
0300	36" x 36"		9	.889		350	47		397	460
0340	40" x 22"		10	.800		298	42.50		340.50	395
0360	48" x 28"		8	1		375	53		428	495
0380	60" x 36"		8	1		515	53		568	650

08 52 16.30 Palladian Windows

		Crew	Daily Output	Labor-Hours	Unit	Material	Labor	Equipment	Total	Total Incl O&P
0010	**PALLADIAN WINDOWS**									
0020	Vinyl-clad, double insulated glass, including frame and grilles									
0040	3'-2" x 2'-6" high	2 Carp	11	1.455	Ea.	1,275	77.50		1,352.50	1,525
0060	3'-2" x 4'-10"		11	1.455		1,775	77.50		1,852.50	2,075
0080	3'-2" x 6'-4"		10	1.600		1,775	85		1,860	2,075
0100	4'-0" x 4'-0"		10	1.600		1,575	85		1,660	1,875
0120	4'-0" x 5'-4"	3 Carp	10	2.400		1,900	128		2,028	2,300
0140	4'-0" x 6'-0"		9	2.667		2,000	142		2,142	2,450
0160	4'-0" x 7'-4"		9	2.667		2,150	142		2,292	2,575
0180	5'-5" x 4'-10"		9	2.667		2,325	142		2,467	2,775
0200	5'-5" x 6'-10"		9	2.667		2,650	142		2,792	3,150
0220	5'-5" x 7'-9"		9	2.667		2,825	142		2,967	3,350
0240	6'-0" x 7'-11"		8	3		3,625	159		3,784	4,225
0260	8'-0" x 6'-0"		8	3		3,225	159		3,384	3,800

08 52 16.35 Double-Hung Window

		Crew	Daily Output	Labor-Hours	Unit	Material	Labor	Equipment	Total	Total Incl O&P
0010	**DOUBLE-HUNG WINDOW** including frames, screens, and grilles									
0300	Vinyl-clad, premium, double insulated glass, 2'-6" x 3'-0" G	1 Carp	10	.800	Ea.	365	42.50		407.50	470
0305	2'-6" x 4'-0" G		10	.800		385	42.50		427.50	490
0400	3'-0" x 3'-6" G		10	.800		355	42.50		397.50	460
0500	3'-0" x 4'-0" G		9	.889		405	47		452	520
0600	3'-0" x 4'-6" G		9	.889		445	47		492	565
0700	3'-0" x 5'-0" G		8	1		465	53		518	600
0790	3'-4" x 5'-0" G		8	1		470	53		523	605
0800	3'-6" x 6'-0" G		8	1		525	53		578	665
0820	4'-0" x 5'-0" G		7	1.143		570	60.50		630.50	730
0830	4'-0" x 6'-0" G		7	1.143		715	60.50		775.50	885

08 52 16.40 Transom Windows

		Crew	Daily Output	Labor-Hours	Unit	Material	Labor	Equipment	Total	Total Incl O&P
0010	**TRANSOM WINDOWS**									
0050	Vinyl-clad, premium, dbl. insul. glass, 32" x 8"	1 Carp	16	.500	Ea.	189	26.50		215.50	251
0100	36" x 8"		16	.500		209	26.50		235.50	273
0110	36" x 12"		16	.500		225	26.50		251.50	290
0200	44" x 48"		12	.667		590	35.50		625.50	705
1000	Vinyl-clad, premium, dbl. insul. glass, 4'-0" x 4'-0"	2 Carp	12	1.333		535	71		606	705
1100	4'-0" x 6'-0"		11	1.455		995	77.50		1,072.50	1,225
1200	5'-0" x 6'-0"		10	1.600		1,100	85		1,185	1,325
1300	6'-0" x 6'-0"		10	1.600		1,125	85		1,210	1,350

08 52 16.70 Vinyl-Clad, Premium, DBL. Insul. Glass

		Crew	Daily Output	Labor-Hours	Unit	Material	Labor	Equipment	Total	Total Incl O&P
0010	**VINYL-CLAD, PREMIUM, DBL. INSUL. GLASS**									
1000	Sliding, 3'-0" x 3'-0" G	1 Carp	10	.800	Ea.	645	42.50		687.50	780
1050	4'-0" x 3'-6" G		9	.889		695	47		742	840
1100	5'-0" x 4'-0" G		9	.889		940	47		987	1,100
1150	6'-0" x 5'-0" G		8	1		1,200	53		1,253	1,400

08 52 Wood Windows

08 52 50 – Window Accessories

08 52 50.10 Window Grille or Muntin		Crew	Daily Output	Labor-Hours	Unit	Material	Labor	Equipment	Total	Total Incl O&P
						2020 Bare Costs				
0010	**WINDOW GRILLE OR MUNTIN**, snap in type									
0020	Standard pattern interior grilles									
2000	Wood, awning window, glass size, 28" x 16" high	1 Carp	30	.267	Ea.	30.50	14.15		44.65	56
2060	44" x 24" high		32	.250		44	13.30		57.30	70
2100	Casement, glass size, 20" x 36" high		30	.267		34.50	14.15		48.65	60.50
2180	20" x 56" high		32	.250		46	13.30		59.30	72.50
2200	Double-hung, glass size, 16" x 24" high		24	.333	Set	61	17.70		78.70	95.50
2280	32" x 32" high		34	.235	"	141	12.50		153.50	175
2500	Picture, glass size, 48" x 48" high		30	.267	Ea.	128	14.15		142.15	164
2580	60" x 68" high		28	.286	"	206	15.20		221.20	251
2600	Sliding, glass size, 14" x 36" high		24	.333	Set	41	17.70		58.70	73.50
2680	36" x 36" high		22	.364	"	47	19.35		66.35	83
9000	Minimum labor/equipment charge		5	1.600	Job		85		85	136

08 52 66 – Wood Window Screens

08 52 66.10 Wood Screens		Crew	Daily Output	Labor-Hours	Unit	Material	Labor	Equipment	Total	Total Incl O&P
0010	**WOOD SCREENS**									
0020	Over 3 S.F., 3/4" frames	2 Carp	375	.043	S.F.	4.79	2.27		7.06	8.90
0100	1-1/8" frames	"	375	.043	"	8.45	2.27		10.72	12.95
9000	Minimum labor/equipment charge	1 Carp	4	2	Job		106		106	170

08 52 69 – Wood Storm Windows

08 52 69.10 Storm Windows		Crew	Daily Output	Labor-Hours	Unit	Material	Labor	Equipment	Total	Total Incl O&P
0010	**STORM WINDOWS**, aluminum residential									
0300	Basement, mill finish, incl. fiberglass screen									
0320	1'-10" x 1'-0" high G	2 Carp	30	.533	Ea.	37.50	28.50		66	87
0340	2'-9" x 1'-6" high G		30	.533		41.50	28.50		70	91
0360	3'-4" x 2'-0" high G		30	.533		44	28.50		72.50	94
1600	Double-hung, combination, storm & screen									
2000	Clear anodic coating, 2'-0" x 3'-5" high G	2 Carp	30	.533	Ea.	98	28.50		126.50	154
2020	2'-6" x 5'-0" high G		28	.571		126	30.50		156.50	187
2040	4'-0" x 6'-0" high G		25	.640		133	34		167	201
2400	White painted, 2'-0" x 3'-5" high G		30	.533		104	28.50		132.50	160
2420	2'-6" x 5'-0" high G		28	.571		102	30.50		132.50	161
2440	4'-0" x 6'-0" high G		25	.640		120	34		154	187
2600	Mill finish, 2'-0" x 3'-5" high G		30	.533		86.50	28.50		115	141
2620	2'-6" x 5'-0" high G		28	.571		103	30.50		133.50	162
2640	4'-0" x 6'-8" high G		25	.640		130	34		164	199
9410	Minimum labor/equipment charge	1 Carp	4	2	Job		106		106	170

08 53 Plastic Windows

08 53 13 – Vinyl Windows

08 53 13.20 Vinyl Single-Hung Windows		Crew	Daily Output	Labor-Hours	Unit	Material	Labor	Equipment	Total	Total Incl O&P
0010	**VINYL SINGLE-HUNG WINDOWS**, insulated glass									
0020	Grids, low E, J fin, extension jambs									
0130	25" x 41" G	2 Carp	20	.800	Ea.	201	42.50		243.50	289
0140	25" x 49" G		18	.889		208	47		255	305
0150	25" x 57" G		17	.941		225	50		275	325
0160	25" x 65" G		16	1		239	53		292	350
0170	29" x 41" G		18	.889		202	47		249	298
0180	29" x 53" G		18	.889		217	47		264	315

08 53 13 – Vinyl Windows

08 53 13.20 Vinyl Single-Hung Windows			Crew	Daily Output	Labor-Hours	Unit	Material	Labor	Equipment	Total	Total Incl O&P
							2020 Bare Costs				
0190	29" x 57"	G	2 Carp	17	.941	Ea.	235	50		285	340
0200	29" x 65"	G		16	1		279	53		332	390
0210	33" x 41"	G		20	.800		216	42.50		258.50	305
0220	33" x 53"	G		18	.889		231	47		278	330
0230	33" x 57"	G		17	.941		279	50		329	385
0240	33" x 65"	G		16	1		284	53		337	400
0250	37" x 41"	G		20	.800		236	42.50		278.50	330
0260	37" x 53"	G		18	.889		271	47		318	375
0270	37" x 57"	G		17	.941		282	50		332	390
0280	37" x 65"	G		16	1		320	53		373	440

08 53 13.30 Vinyl Double-Hung Windows

Code	Description		Crew	Daily Output	Labor-Hours	Unit	Material	Labor	Equipment	Total	Total Incl O&P
0010	**VINYL DOUBLE-HUNG WINDOWS**, insulated glass										
0100	Grids, low E, J fin, ext. jambs, 21" x 53"	G	2 Carp	18	.889	Ea.	213	47		260	310
0102	21" x 37"	G		18	.889		249	47		296	350
0104	21" x 41"	G		18	.889		259	47		306	360
0106	21" x 49"	G		18	.889		263	47		310	365
0110	21" x 57"	G		17	.941		283	50		333	390
0120	21" x 65"	G		16	1		300	53		353	415
0128	25" x 37"	G		20	.800		249	42.50		291.50	340
0130	25" x 41"	G		20	.800		258	42.50		300.50	350
0140	25" x 49"	G		18	.889		260	47		307	360
0145	25" x 53"	G		18	.889		285	47		332	390
0150	25" x 57"	G		17	.941		293	50		343	400
0160	25" x 65"	G		16	1		269	53		322	380
0162	25" x 69"	G		16	1		315	53		368	435
0164	25" x 77"	G		16	1		360	53		413	480
0168	29" x 37"	G		18	.889		258	47		305	360
0170	29" x 41"	G		18	.889		278	47		325	380
0172	29" x 49"	G		18	.889		283	47		330	385
0180	29" x 53"	G		18	.889		292	47		339	395
0190	29" x 57"	G		17	.941		305	50		355	415
0200	29" x 65"	G		16	1		320	53		373	440
0202	29" x 69"	G		16	1		345	53		398	465
0205	29" x 77"	G		16	1		340	53		393	460
0208	33" x 37"	G		20	.800		273	42.50		315.50	370
0210	33" x 41"	G		20	.800		278	42.50		320.50	375
0215	33" x 49"	G		20	.800		295	42.50		337.50	395
0220	33" x 53"	G		18	.889		310	47		357	415
0230	33" x 57"	G		17	.941		287	50		337	395
0240	33" x 65"	G		16	1		330	53		383	450
0242	33" x 69"	G		16	1		360	53		413	480
0246	33" x 77"	G		16	1		340	53		393	460
0250	37" x 41"	G		20	.800		305	42.50		347.50	405
0255	37" x 49"	G		20	.800		320	42.50		362.50	420
0260	37" x 53"	G		18	.889		330	47		377	435
0270	37" x 57"	G		17	.941		350	50		400	465
0280	37" x 65"	G		16	1		380	53		433	505
0282	37" x 69"	G		16	1		395	53		448	520
0286	37" x 77"	G		16	1		390	53		443	515
0300	Solid vinyl, average quality, double insulated glass, 2'-0" x 3'-0"	G	1 Carp	10	.800		295	42.50		337.50	395
0310	3'-0" x 4'-0"	G		9	.889		247	47		294	345
0330	Premium, double insulated glass, 2'-6" x 3'-0"	G		10	.800		305	42.50		347.50	405

08 53 13.30 Vinyl Double-Hung Windows		Crew	Daily Output	Labor-Hours	Unit	Material	Labor	Equipment	Total	Total Incl O&P
0340	3'-0" x 3'-6" G	1 Carp	9	.889	Ea.	335	47		382	440
0350	3'-0" x 4'-0" G		9	.889		345	47		392	455
0360	3'-0" x 4'-6" G		9	.889		360	47		407	470
0370	3'-0" x 5'-0" G		8	1		400	53		453	525
0380	3'-6" x 6'-0" G		8	1		420	53		473	545

(Material, Labor, Equipment and Total are 2020 Bare Costs.)

08 53 13.40 Vinyl Casement Windows

		Crew	Daily Output	Labor-Hours	Unit	Material	Labor	Equipment	Total	Total Incl O&P
0010	**VINYL CASEMENT WINDOWS**, insulated glass									
0015	Grids, low E, J fin, extension jambs, screens									
0100	One lite, 21" x 41" G	2 Carp	20	.800	Ea.	350	42.50		392.50	455
0110	21" x 47" G		20	.800		360	42.50		402.50	465
0120	21" x 53" G		20	.800		390	42.50		432.50	500
0128	24" x 35" G		19	.842		305	45		350	405
0130	24" x 41" G		19	.842		355	45		400	460
0140	24" x 47" G		19	.842		380	45		425	485
0150	24" x 53" G		19	.842		400	45		445	510
0158	28" x 35" G		19	.842		350	45		395	455
0160	28" x 41" G		19	.842		310	45		355	415
0170	28" x 47" G		19	.842		400	45		445	510
0180	28" x 53" G		19	.842		390	45		435	500
0184	28" x 59" G		19	.842		440	45		485	550
0188	Two lites, 33" x 35" G		18	.889		525	47		572	650
0190	33" x 41" G		18	.889		510	47		557	635
0200	33" x 47" G		18	.889		540	47		587	665
0210	33" x 53" G		18	.889		550	47		597	680
0212	33" x 59" G		18	.889		605	47		652	745
0215	33" x 72" G		18	.889		620	47		667	760
0220	41" x 41" G		18	.889		580	47		627	710
0230	41" x 47" G		18	.889		580	47		627	715
0240	41" x 53" G		17	.941		625	50		675	765
0242	41" x 59" G		17	.941		660	50		710	805
0246	41" x 72" G		17	.941		685	50		735	835
0250	47" x 41" G		17	.941		560	50		610	695
0260	47" x 47" G		17	.941		585	50		635	720
0270	47" x 53" G		17	.941		620	50		670	765
0272	47" x 59" G		17	.941		695	50		745	845
0280	56" x 41" G		15	1.067		600	56.50		656.50	750
0290	56" x 47" G		15	1.067		620	56.50		676.50	775
0300	56" x 53" G		15	1.067		695	56.50		751.50	855
0302	56" x 59" G		15	1.067		720	56.50		776.50	880
0310	56" x 72" G		15	1.067		880	56.50		936.50	1,050
0340	Solid vinyl, premium, double insulated glass, 2'-0" x 3'-0" high G	1 Carp	10	.800		276	42.50		318.50	375
0360	2'-0" x 4'-0" high G		9	.889		300	47		347	405
0380	2'-0" x 5'-0" high G		8	1		345	53		398	460

08 53 13.50 Vinyl Picture Windows

		Crew	Daily Output	Labor-Hours	Unit	Material	Labor	Equipment	Total	Total Incl O&P
0010	**VINYL PICTURE WINDOWS**, insulated glass									
0120	Grids, low E, J fin, ext. jambs, 47" x 35"	2 Carp	12	1.333	Ea.	290	71		361	435
0130	47" x 41"		12	1.333		430	71		501	585
0140	47" x 47"		12	1.333		355	71		426	505
0150	47" x 53"		11	1.455		410	77.50		487.50	580
0160	71" x 35"		11	1.455		420	77.50		497.50	585
0170	71" x 41"		11	1.455		440	77.50		517.50	610
0180	71" x 47"		11	1.455		470	77.50		547.50	645

08 54 Composite Windows

08 54 13 – Fiberglass Windows

08 54 13.10 Fiberglass Single-Hung Windows		Crew	Daily Output	Labor-Hours	Unit	Material	2020 Bare Costs Labor	Equipment	Total	Total Incl O&P
0010	**FIBERGLASS SINGLE-HUNG WINDOWS**									
0100	Grids, low E, 18" x 24" G	2 Carp	18	.889	Ea.	390	47		437	505
0110	18" x 40" G		17	.941		400	50		450	520
0130	24" x 40" G		20	.800		400	42.50		442.50	510
0230	36" x 36" G		17	.941		420	50		470	545
0250	36" x 48" G		20	.800		455	42.50		497.50	570
0260	36" x 60" G		18	.889		505	47		552	630
0280	36" x 72" G		16	1		550	53		603	690
0290	48" x 40" G		16	1		550	53		603	690

08 54 13.30 Fiberglass Slider Windows		Crew	Daily Output	Labor-Hours	Unit	Material	Labor	Equipment	Total	Total Incl O&P
0010	**FIBERGLASS SLIDER WINDOWS**									
0100	Grids, low E, 36" x 24" G	2 Carp	20	.800	Ea.	390	42.50		432.50	500
0110	36" x 36" G	"	20	.800	"	395	42.50		437.50	505

08 54 13.50 Fiberglass Bay Windows		Crew	Daily Output	Labor-Hours	Unit	Material	Labor	Equipment	Total	Total Incl O&P
0010	**FIBERGLASS BAY WINDOWS**									
0150	48" x 36" G	2 Carp	11	1.455	Ea.	1,125	77.50		1,202.50	1,375

08 56 Special Function Windows

08 56 46 – Radio-Frequency-Interference Shielding Windows

08 56 46.10 Radio-Frequency-Interference Mesh		Crew	Daily Output	Labor-Hours	Unit	Material	Labor	Equipment	Total	Total Incl O&P
0010	**RADIO-FREQUENCY-INTERFERENCE MESH**									
0100	16 mesh copper 0.011" wire				S.F.	4.88			4.88	5.35
0150	22 mesh copper 0.015" wire					6			6	6.60
0200	100 mesh copper 0.022" wire					8.60			8.60	9.50
0250	100 mesh stainless steel 0.0012" wire					14.90			14.90	16.40

08 61 Roof Windows

08 61 13 – Metal Roof Windows

08 61 13.10 Roof Windows		Crew	Daily Output	Labor-Hours	Unit	Material	Labor	Equipment	Total	Total Incl O&P
0010	**ROOF WINDOWS**, fixed high perf tmpd glazing, metallic framed									
0020	46" x 21-1/2", Flashed for shingled roof	1 Carp	8	1	Ea.	298	53		351	415
0100	46" x 28"		8	1		335	53		388	450
0125	57" x 44"		6	1.333		390	71		461	545
0130	72" x 28"		7	1.143		390	60.50		450.50	530
0150	Fixed, laminated tempered glazing, 46" x 21-1/2"		8	1		480	53		533	615
0175	46" x 28"		8	1		535	53		588	670
0200	57" x 44"		6	1.333		495	71		566	660
0500	Vented flashing set for shingled roof, 46" x 21-1/2"		7	1.143		540	60.50		600.50	695
0525	46" x 28"		6	1.333		600	71		671	770
0550	57" x 44"		5	1.600		750	85		835	960
0560	72" x 28"		5	1.600		750	85		835	960
0575	Flashing set for low pitched roof, 46" x 21-1/2"		7	1.143		580	60.50		640.50	735
0600	46" x 28"		7	1.143		635	60.50		695.50	800
0625	57" x 44"		5	1.600		790	85		875	1,000
0650	Flashing set for curb, 46" x 21-1/2"		7	1.143		680	60.50		740.50	850
0675	46" x 28"		7	1.143		745	60.50		805.50	915
0700	57" x 44"		5	1.600		920	85		1,005	1,125

08 62 Unit Skylights

08 62 13 – Domed Unit Skylights

08 62 13.10 Domed Skylights			Crew	Daily Output	Labor-Hours	Unit	2020 Bare Costs Material	Labor	Equipment	Total	Total Incl O&P
0010	**DOMED SKYLIGHTS**										
0020	Skylight, fixed dome type, 22" x 22"	G	G-3	12	2.667	Ea.	214	139		353	455
0030	22" x 46"	G		10	3.200		273	167		440	565
0040	30" x 30"	G		12	2.667		288	139		427	535
0050	30" x 46"	G		10	3.200		375	167		542	680
0110	Fixed, double glazed, 22" x 27"	G		12	2.667		274	139		413	520
0120	22" x 46"	G		10	3.200		320	167		487	615
0130	44" x 46"	G		10	3.200		430	167		597	735
0210	Operable, double glazed, 22" x 27"	G		12	2.667		410	139		549	670
0220	22" x 46"	G		10	3.200		485	167		652	800
0230	44" x 46"	G		10	3.200		965	167		1,132	1,325
9000	Minimum labor/equipment charge			2	16	Job		835		835	1,325

08 62 13.20 Skylights			Crew	Daily Output	Labor-Hours	Unit	Material	Labor	Equipment	Total	Total Incl O&P
0010	**SKYLIGHTS**, flush or curb mounted										
2120	Ventilating insulated plexiglass dome with										
2130	curb mounting, 36" x 36"	G	G-3	12	2.667	Ea.	490	139		629	755
2150	52" x 52"	G		12	2.667		665	139		804	955
2160	28" x 52"	G		10	3.200		525	167		692	840
2170	36" x 52"	G		10	3.200		630	167		797	955
2180	For electric opening system, add	G					335			335	370
2210	Operating skylight, with thermopane glass, 24" x 48"	G	G-3	10	3.200		595	167		762	920
2220	32" x 48"	G		9	3.556		620	186		806	980
2300	Insulated safety glass with aluminum frame	G		160	.200	S.F.	95	10.45		105.45	122
2386	Skylight,non venting,non insul. plexiglass dome with curb mount, 46"x46"			13.91	2.301	Ea.	330	120		450	555
4000	Skylight, solar tube kit, incl. dome, flashing, diffuser, 1 pipe, 10" diam.	G	1 Carp	2	4		300	213		513	675
4010	14" diam.	G		2	4		365	213		578	745
4020	21" diam.	G		2	4		440	213		653	825
4030	Accessories for, 1' long x 9" diam. pipe	G		24	.333		53	17.70		70.70	86.50
4040	2' long x 9" diam. pipe	G		24	.333		41.50	17.70		59.20	74
4050	4' long x 9" diam. pipe	G		20	.400		73.50	21.50		95	115
4060	1' long x 13" diam. pipe	G		24	.333		73.50	17.70		91.20	110
4070	2' long x 13" diam. pipe	G		24	.333		54	17.70		71.70	87.50
4080	4' long x 13" diam. pipe	G		20	.400		101	21.50		122.50	145
4090	6.5" turret ext. for 21" diam. pipe	G		16	.500		105	26.50		131.50	158
4100	12' long x 21" diam. flexible pipe	G		12	.667		91.50	35.50		127	158
4110	45 degree elbow, 10"	G		16	.500		168	26.50		194.50	227
4120	14"	G		16	.500		83.50	26.50		110	135
4130	Interior decorative ring, 9"	G		20	.400		48.50	21.50		70	87.50
4140	13"	G		20	.400		70	21.50		91.50	111

08 63 Metal-Framed Skylights

08 63 13 – Domed Metal-Framed Skylights

08 63 13.20 Skylight Rigid Metal-Framed		Crew	Daily Output	Labor-Hours	Unit	Material	Labor	Equipment	Total	Total Incl O&P
0010	**SKYLIGHT RIGID METAL-FRAMED** skylight framing is aluminum									
0050	Fixed acrylic double domes, curb mount, 25-1/2" x 25-1/2"	G-3	10	3.200	Ea.	222	167		389	510
0060	25-1/2" x 33-1/2"		10	3.200		250	167		417	540
0070	25-1/2" x 49-1/2"		6	5.333		294	278		572	765
0080	33-1/2" x 33-1/2"		8	4		320	209		529	680
0090	37-1/2" x 25-1/2"		8	4		258	209		467	615
0100	37-1/2" x 37-1/2"		8	4		253	209		462	610
0110	37-1/2" x 49-1/2"		6	5.333		435	278		713	915

08 63 Metal-Framed Skylights

08 63 13 – Domed Metal-Framed Skylights

08 63 13.20 Skylight Rigid Metal-Framed		Crew	Daily Output	Labor-Hours	Unit	Material	Labor	Equipment	Total	Total Incl O&P
						2020 Bare Costs				
0120	49-1/2" x 33-1/2"	G-3	6	5.333	Ea.	385	278		663	865
0130	49-1/2" x 49-1/2"		6	5.333		495	278		773	980
1000	Fixed tempered glass, curb mount, 17-1/2" x 33-1/2"		10	3.200		159	167		326	440
1020	17-1/2" x 49-1/2"		6	5.333		180	278		458	640
1030	25-1/2" x 25-1/2"		10	3.200		159	167		326	440
1040	25-1/2" x 33-1/2"		10	3.200		189	167		356	470
1050	25-1/2" x 37-1/2"		8	4		201	209		410	550
1060	25-1/2" x 49-1/2"		6	5.333		229	278		507	690
1070	25-1/2" x 73-1/2"		6	5.333		365	278		643	845
1080	33-1/2" x 33-1/2"		8	4		220	209		429	570
2000	Manual vent tempered glass & screen, curb, 25-1/2" x 25-1/2"		10	3.200		420	167		587	725
2020	25-1/2" x 37-1/2"		10	3.200		480	167		647	795
2030	25-1/2" x 49-1/2"		8	4		520	209		729	900
2040	33-1/2" x 33-1/2"		8	4		555	209		764	940
2050	33-1/2" x 49-1/2"		6	5.333		700	278		978	1,200
2060	37-1/2" x 37-1/2"		6	5.333		635	278		913	1,150
2070	49-1/2" x 49-1/2"		6	5.333		820	278		1,098	1,350
3000	Electric vent tempered glass, curb mount, 25-1/2" x 25-1/2"		10	3.200		1,075	167		1,242	1,450
3020	25-1/2" x 37-1/2"		10	3.200		1,150	167		1,317	1,550
3030	25-1/2" x 49-1/2"		8	4		1,225	209		1,434	1,675
3040	33-1/2" x 33-1/2"		8	4		1,250	209		1,459	1,700
3050	33-1/2" x 49-1/2"		6	5.333		1,325	278		1,603	1,900
3060	37-1/2" x 37-1/2"		6	5.333		1,300	278		1,578	1,875
3070	49-1/2" x 49-1/2"		6	5.333		1,425	278		1,703	2,000

08 71 Door Hardware

08 71 13 – Automatic Door Operators

08 71 13.10 Automatic Openers Commercial

Line	Description	Crew	Daily Output	Labor-Hours	Unit	Material	Labor	Equipment	Total	Total Incl O&P
0010	**AUTOMATIC OPENERS COMMERCIAL**									
0020	Pneumatic door opener incl. motion sens, control box, tubing, compressor									
0050	For single swing door, per opening	2 Skwk	.80	20	Ea.	4,900	1,100		6,000	7,125
0100	Pair, per opening		.50	32	Opng.	7,875	1,750		9,625	11,500
1000	For single sliding door, per opening		.60	26.667		5,275	1,475		6,750	8,125
1300	Bi-parting pair		.50	32		7,800	1,750		9,550	11,400
1420	Electronic door opener incl. motion sens, 12 V control box, motor									
1450	For single swing door, per opening	2 Skwk	.80	20	Opng.	3,925	1,100		5,025	6,075
1500	Pair, per opening		.50	32		7,825	1,750		9,575	11,400
1600	For single sliding door, per opening		.60	26.667		5,150	1,475		6,625	7,975
1700	Bi-parting pair		.50	32		5,725	1,750		7,475	9,100
1750	Handicap actuator buttons, 2, incl. 12 V DC wiring, add	1 Carp	1.50	5.333	Pr.	495	283		778	1,000
2000	Electric panic button for door	2 Skwk	.50	32	Opng.	232	1,750		1,982	3,050

08 71 13.20 Automatic Openers Industrial

Line	Description	Crew	Daily Output	Labor-Hours	Unit	Material	Labor	Equipment	Total	Total Incl O&P
0010	**AUTOMATIC OPENERS INDUSTRIAL**									
0015	Sliding doors up to 6' wide	2 Skwk	.60	26.667	Opng.	6,250	1,475		7,725	9,200
0200	To 12' wide	"	.40	40	"	7,500	2,200		9,700	11,700
0400	Over 12' wide, add per L.F. of excess				L.F.	840			840	925
1000	Swing doors, to 5' wide	2 Skwk	.80	20	Ea.	4,025	1,100		5,125	6,175
1860	Add for controls, wall pushbutton, 3 button		4	4		241	219		460	615
1870	Control pull cord		4.30	3.721		206	204		410	550
1880	For addl elec eye for sliding door operation, one side, add					9%				

08 71 Door Hardware

08 71 20 – Hardware

08 71 20.10 Bolts, Flush		Crew	Daily Output	Labor-Hours	Unit	2020 Bare Costs Material	Labor	Equipment	Total	Total Incl O&P
0010	**BOLTS, FLUSH**									
0020	Standard, concealed	1 Carp	7	1.143	Ea.	25.50	60.50		86	126
0800	Automatic fire exit	"	5	1.600		189	85		274	345
1600	Electrified dead bolt	1 Elec	3	2.667		183	164		347	455
3000	Barrel, brass, 2" long	1 Carp	40	.200		5.25	10.65		15.90	23
3020	4" long		40	.200		13.75	10.65		24.40	32
3060	6" long		40	.200		24.50	10.65		35.15	44

08 71 20.15 Hardware		Crew	Daily Output	Labor-Hours	Unit	Material	Labor	Equipment	Total	Total Incl O&P
0010	**HARDWARE**									
0020	Average hardware cost									
1000	Door hardware, apartment, interior	1 Carp	4	2	Door	570	106		676	795
1300	Average, door hardware, motel/hotel interior, with access card		4	2		700	106		806	940
1500	Hospital bedroom, average quality		4	2		720	106		826	960
2000	High quality		3	2.667		725	142		867	1,025
2100	Pocket door		6	1.333	Ea.	118	71		189	244
2250	School, single exterior, incl. lever, incl. panic device		3	2.667	Door	1,600	142		1,742	2,000
2500	Single interior, regular use, lever included		3	2.667		585	142		727	870
2550	Avg., door hdwe., school, classroom, ANSI F84, lever handle		3	2.667		845	142		987	1,150
2600	Avg., door hdwe. set, school, classroom, ANSI F88, incl. lever		3	2.667		915	142		1,057	1,225
2850	Stairway, single interior		3	2.667		645	142		787	935
3100	Double exterior, with panic device		2	4	Pr.	3,025	213		3,238	3,700
6020	Add for fire alarm door holder, electro-magnetic	1 Elec	4	2	Ea.	92	123		215	290

08 71 20.20 Door Protectors		Crew	Daily Output	Labor-Hours	Unit	Material	Labor	Equipment	Total	Total Incl O&P
0010	**DOOR PROTECTORS**									
0020	1-3/4" x 3/4" U channel	2 Carp	80	.200	L.F.	25	10.65		35.65	44.50
0021	1-3/4" x 1-1/4" U channel		80	.200	"	22.50	10.65		33.15	41.50
1000	Tear drop, spring stl., 8" high x 19" long		15	1.067	Ea.	107	56.50		163.50	209
1010	8" high x 32" long		15	1.067		133	56.50		189.50	237
1100	Tear drop, stainless stl., 8" high x 19" long		15	1.067		315	56.50		371.50	435
1200	8" high x 32" long		15	1.067		375	56.50		431.50	500

08 71 20.30 Door Closers		Crew	Daily Output	Labor-Hours	Unit	Material	Labor	Equipment	Total	Total Incl O&P
0010	**DOOR CLOSERS** adjustable backcheck, multiple mounting									
0015	and rack and pinion									
0020	Standard Regular Arm	1 Carp	6	1.333	Ea.	211	71		282	345
0040	Hold open arm		6	1.333		99.50	71		170.50	223
0100	Fusible link		6.50	1.231		205	65.50		270.50	330
0210	Light duty, regular arm		6	1.333		122	71		193	248
0220	Parallel arm		6	1.333		147	71		218	276
0230	Hold open arm		6	1.333		147	71		218	276
0240	Fusible link arm		6	1.333		164	71		235	294
0250	Medium duty, regular arm		6	1.333		146	71		217	274
0500	Surface mount regular arm		6.50	1.231		160	65.50		225.50	281
0550	Fusible link		6.50	1.231		156	65.50		221.50	277
1500	Concealed closers, normal use, head, pivot hung, interior		5.50	1.455		390	77.50		467.50	555
1510	Exterior		5.50	1.455		460	77.50		537.50	630
1520	Overhead concealed, all sizes, regular arm		5.50	1.455		246	77.50		323.50	395
1525	Concealed arm		5	1.600		365	85		450	535
1530	Concealed in door, all sizes, regular arm		5.50	1.455		385	77.50		462.50	545
1535	Concealed arm		5	1.600		272	85		357	435
1560	Floor concealed, all sizes, single acting		2.20	3.636		570	193		763	935
1565	Double acting		2.20	3.636		675	193		868	1,050

08 71 20.30 Door Closers		Crew	Daily Output	Labor-Hours	Unit	Material	2020 Bare Costs Labor	Equipment	Total	Total Incl O&P
1570	Interior, floor, offset pivot, single acting	1 Carp	3.50	2.286	Ea.	955	121		1,076	1,250
1590	Exterior		3.50	2.286		1,050	121		1,171	1,350
1610	Hold open arm		6	1.333		455	71		526	615
1620	Double acting, standard arm		6	1.333		905	71		976	1,100
1630	Hold open arm		6	1.333		935	71		1,006	1,150
1640	Floor, center hung, single acting, bottom arm		6	1.333		480	71		551	645
1650	Double acting		6	1.333		540	71		611	710
1660	Offset hung, single acting, bottom arm		6	1.333		645	71		716	825
2000	Backcheck and adjustable power, hinge face mount									
5000	For cast aluminum cylinder, deduct				Ea.	41.50			41.50	45.50
5040	For delayed action, add					55			55	60.50
5080	For fusible link arm, add					54			54	59.50
5120	For shock absorbing arm, add					55			55	60.50
5160	For spring power adjustment, add					41.50			41.50	46
6000	Closer-holder, hinge face mount, all sizes, exposed arm	1 Carp	6.50	1.231		235	65.50		300.50	365
6500	Electro magnetic closer/holder									
6510	Single point, no detector	1 Carp	4	2	Ea.	595	106		701	825
6515	Including detector		4	2		795	106		901	1,050
6520	Multi-point, no detector		4	2		995	106		1,101	1,275
6524	Including detector		4	2		1,550	106		1,656	1,875
6550	Electric automatic operators									
6555	Operator	1 Carp	4	2	Ea.	2,150	106		2,256	2,550
6570	Wall plate actuator		4	2		269	106		375	465
7000	Electronic closer-holder, hinge facemount, concealed arm		5	1.600		490	85		575	670
7400	With built-in detector		5	1.600		620	85		705	815
8000	Surface mounted, standard duty, parallel arm, primed, traditional		6	1.333		210	71		281	345
8030	Light duty		6	1.333		170	71		241	300
8050	Heavy duty		6	1.333		255	71		326	395
8100	Standard duty, parallel arm, modern		6	1.333		275	71		346	420
8150	Heavy duty		6	1.333		340	71		411	490
9000	Minimum labor/equipment charge		4	2	Job		106		106	170

08 71 20.36 Panic Devices

		Crew	Daily Output	Labor-Hours	Unit	Material	Labor	Equipment	Total	Total Incl O&P
0010	**PANIC DEVICES** R087110-10									
0015	Touch bars various styles									
0040	Single door exit only, rim device, wide stile									
0050	Economy US28	1 Carp	5	1.600	Ea.	287	85		372	450
0060	Standard duty US28		5	1.600		820	85		905	1,050
0065	US26D		5	1.600		810	85		895	1,025
0070	US10		5	1.600		880	85		965	1,100
0075	US3		5	1.600		1,000	85		1,085	1,225
0080	Night latch, economy US28		5	1.600		365	85		450	535
0085	Standard duty, night latch US28		5	1.600		910	85		995	1,125
0090	US26D		5	1.600		990	85		1,075	1,225
0095	US10		5	1.600		1,050	85		1,135	1,275
0100	US3		5	1.600		1,225	85		1,310	1,475
0500	Single door exit only, rim device, narrow stile									
0540	Economy US28	1 Carp	5	1.600	Ea.	470	85		555	650
0550	Standard duty US28		5	1.600		895	85		980	1,125
0565	US26D		5	1.600		815	85		900	1,025
0570	US10		5	1.600		900	85		985	1,125
0575	US3		5	1.600		1,125	85		1,210	1,350
0580	Economy with night latch US28		5	1.600		355	85		440	525

08 71 20 – Hardware

08 71 20.36 Panic Devices		Crew	Daily Output	Labor-Hours	Unit	Material	2020 Bare Costs Labor	Equipment	Total	Total Incl O&P
0585	Standard duty US28	1 Carp	5	1.600	Ea.	985	85		1,070	1,200
0590	US26D		5	1.600		1,100	85		1,185	1,325
0595	US10		5	1.600		990	85		1,075	1,225
0600	US32D		5	1.600		1,275	85		1,360	1,525
1040	Single door exit only, surface vertical rod									
1050	Surface vertical rod, economy US28	1 Carp	4	2	Ea.	540	106		646	760
1060	Surface vertical rod, standard duty US28		4	2		1,075	106		1,181	1,350
1065	US26D		4	2		1,150	106		1,256	1,450
1070	US10		4	2		1,175	106		1,281	1,475
1075	US3		4	2		1,200	106		1,306	1,475
1080	US32D		4	2		1,425	106		1,531	1,750
1500	Single door exit, rim, narrow stile, economy, surface vertical rod									
1540	Surface vertical rod, narrow, economy US28	1 Carp	4	2	Ea.	540	106		646	765
2040	Single door exit only, concealed vertical rod									
2042	These devices will require separate and extra door preparation									
2060	Concealed rod, standard duty US28	1 Carp	4	2	Ea.	1,125	106		1,231	1,400
2065	US26D		4	2		1,225	106		1,331	1,525
2070	US10		4	2		1,350	106		1,456	1,650
2560	Concealed rod, narrow, standard duty US28		4	2		1,275	106		1,381	1,575
2565	US26D		4	2		1,500	106		1,606	1,825
2567	US26		4	2		1,575	106		1,681	1,900
2570	US10		4	2		1,325	106		1,431	1,650
2575	US3		4	2		1,550	106		1,656	1,875
3040	Single door exit only, mortise device, wide stile									
3042	These devices must be combined with a mortise lock									
3060	Mortise, standard duty US28	1 Carp	4	2	Ea.	1,175	106		1,281	1,450
3065	US26D		4	2		1,150	106		1,256	1,450
3067	US26		4	2		1,400	106		1,506	1,725
3070	US10		4	2		1,275	106		1,381	1,575
3075	US3		4	2		1,175	106		1,281	1,475
3080	US32D		4	2		1,225	106		1,331	1,525
3500	Single door exit only, mortise device, narrow stile									
3510	These devices must be combined with a mortise lock									
3540	Mortise, economy US28	1 Carp	4	2	Ea.	425	106		531	635

08 71 20.40 Lockset		Crew	Daily Output	Labor-Hours	Unit	Material	Labor	Equipment	Total	Total Incl O&P
0010	**LOCKSET**, Standard duty									
0020	Non-keyed, passage, w/sect. trim	1 Carp	12	.667	Ea.	81	35.50		116.50	146
0100	Privacy		12	.667		77.50	35.50		113	143
0400	Keyed, single cylinder function		10	.800		163	42.50		205.50	247
0420	Hotel (see also Section 08 71 20.15)		8	1		210	53		263	315
0500	Lever handled, keyed, single cylinder function		10	.800		159	42.50		201.50	243
1000	Heavy duty with sectional trim, non-keyed, passages		12	.667		129	35.50		164.50	199
1100	Privacy		12	.667		160	35.50		195.50	232
1400	Keyed, single cylinder function		10	.800		201	42.50		243.50	289
1420	Hotel		8	1		580	53		633	725
1600	Communicating		10	.800		325	42.50		367.50	425
1690	For re-core cylinder, add					62.50			62.50	68.50
1800	Average quality	1 Carp	14	.571		50.50	30.50		81	104
1820	Heavy duty		8	1		244	53		297	355
3800	Cipher lockset w/key pad (security item)		13	.615		1,050	32.50		1,082.50	1,225
3900	Cipher lockset with dial for swinging doors (security item)		13	.615		1,950	32.50		1,982.50	2,200
3920	with dial for swinging doors & drill resistant plate (security item)		12	.667		2,375	35.50		2,410.50	2,650

08 71 20.40 Lockset

	08 71 20.40 Lockset	Crew	Daily Output	Labor-Hours	Unit	2020 Bare Costs Material	Labor	Equipment	Total	Total Incl O&P
3950	Cipher lockset with dial for safe/vault door (security item)	1 Carp	12	.667	Ea.	1,675	35.50		1,710.50	1,875
3980	Keyless, pushbutton type									
4000	Residential/light commercial, deadbolt, standard	1 Carp	9	.889	Ea.	156	47		203	247
4010	Heavy duty		9	.889		258	47		305	360
4020	Industrial, heavy duty, with deadbolt		9	.889		475	47		522	600
4030	Key override		9	.889		470	47		517	595
4040	Lever activated handle		9	.889		480	47		527	605
4050	Key override		9	.889		495	47		542	615
4060	Double sided pushbutton type		8	1		960	53		1,013	1,125
4070	Key override		8	1		875	53		928	1,050
9000	Minimum labor/equipment charge		6	1.333	Job		71		71	114

08 71 20.41 Dead Locks

	08 71 20.41 Dead Locks	Crew	Daily Output	Labor-Hours	Unit	Material	Labor	Equipment	Total	Total Incl O&P
0010	**DEAD LOCKS**									
0011	Mortise heavy duty outside key (security item)	1 Carp	9	.889	Ea.	193	47		240	289
0020	Double cylinder		9	.889		193	47		240	288
0100	Medium duty, outside key		10	.800		117	42.50		159.50	196
0110	Double cylinder		10	.800		143	42.50		185.50	225
1000	Tubular, standard duty, outside key		10	.800		54	42.50		96.50	128
1010	Double cylinder		10	.800		71.50	42.50		114	147
1200	Night latch, outside key		10	.800		58	42.50		100.50	132
1203	Deadlock night latch		7.70	1.039		58	55		113	152
1420	Deadbolt lock, single cylinder		10	.800		48.50	42.50		91	122
1440	Double cylinder		10	.800		67.50	42.50		110	142

08 71 20.42 Mortise Locksets

	08 71 20.42 Mortise Locksets	Crew	Daily Output	Labor-Hours	Unit	Material	Labor	Equipment	Total	Total Incl O&P
0010	**MORTISE LOCKSETS**, Comm., wrought knobs & full escutcheon trim									
0015	Assumes mortise is cut									
0020	Non-keyed, passage, Grade 3	1 Carp	9	.889	Ea.	164	47		211	256
0030	Grade 1		8	1		440	53		493	570
0040	Privacy set, Grade 3		9	.889		176	47		223	269
0050	Grade 1		8	1		485	53		538	620
0100	Keyed, office/entrance/apartment, Grade 2		8	1		205	53		258	310
0110	Grade 1		7	1.143		510	60.50		570.50	660
0120	Single cylinder, typical, Grade 3		8	1		199	53		252	305
0130	Grade 1		7	1.143		540	60.50		600.50	695
0200	Hotel, room, Grade 3		7	1.143		234	60.50		294.50	355
0210	Grade 1 (see also Section 08 71 20.15)		6	1.333		570	71		641	740
0300	Double cylinder, Grade 3		8	1		241	53		294	350
0310	Grade 1		7	1.143		560	60.50		620.50	715
1000	Wrought knobs and sectional trim, non-keyed, passage, Grade 3		10	.800		138	42.50		180.50	220
1010	Grade 1		9	.889		440	47		487	560
1040	Privacy, Grade 3		10	.800		161	42.50		203.50	245
1050	Grade 1		9	.889		495	47		542	620
1100	Keyed, entrance, office/apartment, Grade 3		9	.889		239	47		286	340
1103	Install lockset		6.92	1.156		239	61.50		300.50	360
1110	Grade 1		8	1		565	53		618	705
1120	Single cylinder, Grade 3		9	.889		236	47		283	335
1130	Grade 1		8	1		520	53		573	660
2000	Cast knobs and full escutcheon trim									
2010	Non-keyed, passage, Grade 3	1 Carp	9	.889	Ea.	287	47		334	390
2020	Grade 1		8	1		425	53		478	555
2040	Privacy, Grade 3		9	.889		340	47		387	445
2050	Grade 1		8	1		450	53		503	580

08 71 20 – Hardware

08 71 20.42 Mortise Locksets		Crew	Daily Output	Labor-Hours	Unit	Material	2020 Bare Costs Labor	Equipment	Total	Total Incl O&P
2120	Keyed, single cylinder, Grade 3	1 Carp	8	1	Ea.	345	53		398	465
2123	Mortise lock		6.15	1.301		345	69		414	490
2130	Grade 1		7	1.143		550	60.50		610.50	700
3000	Cast knob and sectional trim, non-keyed, passage, Grade 3		10	.800		221	42.50		263.50	310
3010	Grade 1		10	.800		395	42.50		437.50	505
3040	Privacy, Grade 3		10	.800		231	42.50		273.50	320
3050	Grade 1		10	.800		480	42.50		522.50	595
3100	Keyed, office/entrance/apartment, Grade 3		9	.889		261	47		308	365
3110	Grade 1		9	.889		605	47		652	740
3120	Single cylinder, Grade 3		9	.889		273	47		320	375
3130	Grade 1		9	.889		550	47		597	685
3190	For re-core cylinder, add					90.50			90.50	99.50
5000	Wrought steel case, brass base, knob US26D									
5020	Closet, non-keyed passage	1 Carp	8	1	Ea.	297	53		350	410
5040	Bath/bedroom, keyed		8	1		330	53		383	445
5060	Entrance, keyed		8	1		415	53		468	540
5080	Classroom, outside keyed		8	1		440	53		493	570
5100	Storeroom, keyed		8	1		440	53		493	570
5120	Front door, keyed		8	1		440	53		493	565
5140	Dormitory/exit, keyed		8	1		435	53		488	565

08 71 20.44 Anti-Ligature Locksets Grade 1		Crew	Daily Output	Labor-Hours	Unit	Material	Labor	Equipment	Total	Total Incl O&P
0010	**ANTI-LIGATURE LOCKSETS GRADE 1**									
0100	Anti-ligature cylindrical locksets									
0500	Arch handle passage set, US32D	1 Carp	8	1	Ea.	535	53		588	675
0510	privacy set		8	1		575	53		628	715
0520	classroom set		8	1		590	53		643	735
0530	storeroom set		8	1		590	53		643	735
0550	Lever handle passage set US32D		8	1		405	53		458	530
0560	Hospital/privacy set		8	1		420	53		473	545
0570	Office set		8	1		425	53		478	550
0580	Classroom set		8	1		435	53		488	565
0590	Storeroom set		8	1		420	53		473	550
0600	Exit set		8	1		515	53		568	650
0610	Entry set		8	1		445	53		498	575
0620	Asylum set		8	1		485	53		538	620
0630	Back to back dummy set		8	1		310	53		363	425
0690	Anti-ligature mortise locksets									
0700	Knob handle privacy set, US32D	1 Carp	8	1	Ea.	605	53		658	750
0710	Office set		8	1		590	53		643	730
0720	Classroom set		8	1		590	53		643	730
0730	Hotel set		8	1		605	53		658	750
0740	Apartment set		8	1		605	53		658	750
0750	Institutional privacy set		8	1		650	53		703	800
0760	Lever handle office set		8	1		585	53		638	730
0770	Classroom set sectional trim		8	1		585	53		638	730
0780	Hotel set		8	1		600	53		653	745
0790	Apartment set		8	1		600	53		653	745
0800	Institutional privacy set		8	1		650	53		703	800
0810	Office set US32D		8	1		630	53		683	775
0820	Classroom set escutcheon trim		8	1		745	53		798	905
0830	Hotel set escutcheon trim		8	1		765	53		818	925
0840	Apartment set escutcheon trim		8	1		765	53		818	925

08 71 20.44 Anti-Ligature Locksets Grade 1		Crew	Daily Output	Labor-Hours	Unit	Material	Labor	Equipment	Total	Total Incl O&P
						2020 Bare Costs				
0850	Institutional privacy set escutcheon trim	1 Carp	8	1	Ea.	765	53		818	925

08 71 20.45 Peepholes

		Crew	Daily Output	Labor-Hours	Unit	Material	Labor	Equipment	Total	Total Incl O&P
0010	**PEEPHOLES**									
2010	Peephole	1 Carp	32	.250	Ea.	9.45	13.30		22.75	32
2020	Peephole, wide view	"	32	.250	"	11.25	13.30		24.55	34

08 71 20.50 Door Stops

		Crew	Daily Output	Labor-Hours	Unit	Material	Labor	Equipment	Total	Total Incl O&P
0010	**DOOR STOPS**									
0020	Holder & bumper, floor or wall	1 Carp	32	.250	Ea.	36.50	13.30		49.80	61.50
1300	Wall bumper, 4" diameter, with rubber pad, aluminum		32	.250		13.55	13.30		26.85	36.50
1600	Door bumper, floor type, aluminum		32	.250		3.94	13.30		17.24	26
1620	Brass		32	.250		12.95	13.30		26.25	36
1630	Bronze		32	.250		22	13.30		35.30	45.50
1900	Plunger type, door mounted		32	.250		32	13.30		45.30	57
2500	Holder, floor type, aluminum		32	.250		21.50	13.30		34.80	45.50
2520	Wall type, aluminum		32	.250		35.50	13.30		48.80	60.50
2530	Overhead type, bronze		32	.250		116	13.30		129.30	150
2540	Plunger type, aluminum		32	.250		32	13.30		45.30	57
2560	Brass		32	.250		47	13.30		60.30	73
3000	Electromagnetic, wall mounted, US3		3	2.667		298	142		440	555
3020	Floor mounted, US3		3	2.667		345	142		487	605
4000	Doorstop, ceiling mounted		3	2.667		19.35	142		161.35	249
4030	Doorstop, header		3	2.667		45.50	142		187.50	277
9000	Minimum labor/equipment charge		6	1.333	Job		71		71	114

08 71 20.55 Push-Pull Plates

		Crew	Daily Output	Labor-Hours	Unit	Material	Labor	Equipment	Total	Total Incl O&P
0010	**PUSH-PULL PLATES**									
0090	Push plate, 0.050 thick, 3" x 12", aluminum	1 Carp	12	.667	Ea.	6.20	35.50		41.70	64
0100	4" x 16"		12	.667		15.50	35.50		51	74
0110	6" x 16"		12	.667		8.30	35.50		43.80	66
0120	8" x 16"		12	.667		9.75	35.50		45.25	68
0200	Push plate, 0.050 thick, 3" x 12", brass		12	.667		14.20	35.50		49.70	72.50
0210	4" x 16"		12	.667		17.75	35.50		53.25	76.50
0220	6" x 16"		12	.667		27.50	35.50		63	87
0230	8" x 16"		12	.667		35.50	35.50		71	96.50
0250	Push plate, 0.050 thick, 3" x 12", satin brass		12	.667		14.45	35.50		49.95	73
0260	4" x 16"		12	.667		17.95	35.50		53.45	77
0270	6" x 16"		12	.667		27.50	35.50		63	87
0280	8" x 16"		12	.667		36	35.50		71.50	96.50
0490	Push plate, 0.050 thick, 3" x 12", bronze		12	.667		17.15	35.50		52.65	76
0500	4" x 16"		12	.667		27	35.50		62.50	86.50
0510	6" x 16"		12	.667		32.50	35.50		68	93
0520	8" x 16"		12	.667		39.50	35.50		75	101
0600	Push plate, antimicrobial copper alloy finish, 3.5" x 15"		13	.615		54	32.50		86.50	112
0610	4" x 16"		13	.615		60.50	32.50		93	119
0620	6" x 16"		13	.615		23	32.50		55.50	78
0630	6" x 20"		13	.615		42	32.50		74.50	99
0740	Push plate, 0.050 thick, 3" x 12", stainless steel		12	.667		12.10	35.50		47.60	70.50
0750	4" x 16"		12	.667		30	35.50		65.50	90
0760	6" x 16"		12	.667		19	35.50		54.50	78
0780	8" x 16"		12	.667		24	35.50		59.50	83.50
0790	Push plate, 0.050 thick, 3" x 12", satin stainless steel		12	.667		6.85	35.50		42.35	64.50
0810	4" x 16"		12	.667		8.40	35.50		43.90	66.50
0820	6" x 16"		12	.667		11.95	35.50		47.45	70

08 71 20.55 Push-Pull Plates		Crew	Daily Output	Labor-Hours	Unit	Material	2020 Bare Costs Labor	Equipment	Total	Total Incl O&P
0830	8" x 16"	1 Carp	12	.667	Ea.	16.15	35.50		51.65	75
0980	Pull plate, 0.050 thick, 3" x 12", aluminum		12	.667		26.50	35.50		62	86
1000	4" x 16"		12	.667		29.50	35.50		65	89.50
1050	Pull plate, 0.050 thick, 3" x 12", brass		12	.667		41	35.50		76.50	102
1060	4" x 16"		12	.667		36	35.50		71.50	96.50
1080	Pull plate, 0.050 thick, 3" x 12", bronze		12	.667		51.50	35.50		87	114
1100	4" x 16"		12	.667		59.50	35.50		95	123
1180	Pull plate, 0.050 thick, 3" x 12", stainless steel		12	.667		47	35.50		82.50	109
1200	4" x 16"		12	.667		62	35.50		97.50	125
1250	Pull plate, 0.050 thick, 3" x 12", chrome		12	.667		47	35.50		82.50	109
1270	4" x 16"		12	.667		49	35.50		84.50	111
1500	Pull handle and push bar, aluminum		11	.727		125	38.50		163.50	199
2000	Bronze		10	.800		169	42.50		211.50	254
9800	Minimum labor/equipment charge		5	1.600	Job		85		85	136

08 71 20.60 Entrance Locks		Crew	Daily Output	Labor-Hours	Unit	Material	Labor	Equipment	Total	Total Incl O&P
0010	**ENTRANCE LOCKS**									
0015	Cylinder, grip handle deadlocking latch	1 Carp	9	.889	Ea.	190	47		237	285
0020	Deadbolt		8	1		195	53		248	299
0100	Push and pull plate, dead bolt		8	1		241	53		294	350
0200	Push bar and pull, dead bolt, bronze		7	1.143		276	60.50		336.50	405
0240	Push bar and pull bar, dead bolt, bronze		7	1.143		415	60.50		475.50	555
0900	For handicapped lever, add					158			158	173

08 71 20.65 Thresholds		Crew	Daily Output	Labor-Hours	Unit	Material	Labor	Equipment	Total	Total Incl O&P
0010	**THRESHOLDS**									
0011	Threshold 3' long saddles aluminum	1 Carp	48	.167	L.F.	11.10	8.85		19.95	26.50
0100	Aluminum, 8" wide, 1/2" thick		12	.667	Ea.	56.50	35.50		92	119
0500	Bronze		60	.133	L.F.	40.50	7.10		47.60	56.50
0600	Bronze, panic threshold, 5" wide, 1/2" thick		12	.667	Ea.	165	35.50		200.50	239
0700	Rubber, 1/2" thick, 5-1/2" wide		20	.400		43	21.50		64.50	81.50
0800	2-3/4" wide		20	.400		54.50	21.50		76	93.50
1950	ADA compliant thresholds									
2000	Threshold, wood oak 3-1/2" wide x 24" long	1 Carp	12	.667	Ea.	12.55	35.50		48.05	71
2010	3-1/2" wide x 36" long		12	.667		16.90	35.50		52.40	75.50
2020	3-1/2" wide x 48" long		12	.667		22.50	35.50		58	82
2030	4-1/2" wide x 24" long		12	.667		14.55	35.50		50.05	73
2040	4-1/2" wide x 36" long		12	.667		21.50	35.50		57	80.50
2050	4-1/2" wide x 48" long		12	.667		28	35.50		63.50	88
2060	6-1/2" wide x 24" long		12	.667		20	35.50		55.50	79
2070	6-1/2" wide x 36" long		12	.667		30.50	35.50		66	90.50
2080	6-1/2" wide x 48" long		12	.667		40.50	35.50		76	102
2090	Threshold, wood cherry 3-1/2" wide x 24" long		12	.667		15.05	35.50		50.55	73.50
2100	3-1/2" wide x 36" long		12	.667		36	35.50		71.50	96.50
2110	3-1/2" wide x 48" long		12	.667		40	35.50		75.50	101
2120	4-1/2" wide x 24" long		12	.667		19.20	35.50		54.70	78
2130	4-1/2" wide x 36" long		12	.667		38.50	35.50		74	99.50
2140	4-1/2" wide x 48" long		12	.667		47	35.50		82.50	109
2150	6-1/2" wide x 24" long		12	.667		29	35.50		64.50	89
2160	6-1/2" wide x 36" long		12	.667		52.50	35.50		88	115
2170	6-1/2" wide x 48" long		12	.667		62.50	35.50		98	126
2180	Threshold, wood walnut 3-1/2" wide x 24" long		12	.667		18.35	35.50		53.85	77
2190	3-1/2" wide x 36" long		12	.667		36	35.50		71.50	96.50
2200	3-1/2" wide x 48" long		12	.667		41.50	35.50		77	103

08 71 Door Hardware

08 71 20 – Hardware

08 71 20.65 Thresholds		Crew	Daily Output	Labor-Hours	Unit	Material	Labor	Equipment	Total	Total Incl O&P
						2020 Bare Costs				
2210	4-1/2" wide x 24" long	1 Carp	12	.667	Ea.	29.50	35.50		65	89
2220	4-1/2" wide x 36" long		12	.667		46.50	35.50		82	108
2230	4-1/2" wide x 48" long		12	.667		59	35.50		94.50	122
2240	6-1/2" wide x 24" long		12	.667		45	35.50		80.50	107
2250	6-1/2" wide x 36" long		12	.667		72.50	35.50		108	137
2260	6-1/2" wide x 48" long		12	.667		94.50	35.50		130	161
2300	Threshold, aluminum 4" wide x 36" long		12	.667		26	35.50		61.50	86
2310	4" wide x 48" long		12	.667		31.50	35.50		67	91.50
2320	4" wide x 72" long		12	.667		47	35.50		82.50	109
2330	5" wide x 36" long		12	.667		34.50	35.50		70	95
2340	5" wide x 48" long		12	.667		41	35.50		76.50	102
2350	5" wide x 72" long		12	.667		76	35.50		111.50	141
2360	6" wide x 36" long		12	.667		47	35.50		82.50	109
2370	6" wide x 48" long		12	.667		50.50	35.50		86	113
2380	6" wide x 72" long		12	.667		73.50	35.50		109	138
2390	7" wide x 36" long		12	.667		51.50	35.50		87	114
2400	7" wide x 48" long		12	.667		69	35.50		104.50	133
2410	7" wide x 72" long		12	.667		98.50	35.50		134	165
2500	Threshold, ramp, aluminum or rubber 24" x 24"		12	.667		173	35.50		208.50	247
9000	Minimum labor/equipment charge		4	2	Job		106		106	170

08 71 20.70 Floor Checks		Crew	Daily Output	Labor-Hours	Unit	Material	Labor	Equipment	Total	Total Incl O&P
0010	**FLOOR CHECKS**									
0020	For over 3' wide doors single acting	1 Carp	2.50	3.200	Ea.	800	170		970	1,150
0500	Double acting	"	2.50	3.200	"	915	170		1,085	1,275

08 71 20.75 Door Hardware Accessories		Crew	Daily Output	Labor-Hours	Unit	Material	Labor	Equipment	Total	Total Incl O&P
0010	**DOOR HARDWARE ACCESSORIES**									
0050	Door closing coordinator, 36" (for paired openings up to 56")	1 Carp	8	1	Ea.	131	53		184	229
0060	48" (for paired openings up to 84")		8	1		139	53		192	238
0070	56" (for paired openings up to 96")		8	1		148	53		201	248
2000	Torsion springs for overhead doors									
2050	1-3/4" diam., 32" long, 0.243" spring	1 Carp	8	1	Ea.	60	53		113	151
2060	1-3/4" diam., 32" long, 0.250" spring		8	1		43	53		96	132
2070	2" diam., 32" long, 0.243" spring		8	1		44.50	53		97.50	134
2080	2" diam., 32" long, 0.253" spring		8	1		53	53		106	144

08 71 20.80 Hasps		Crew	Daily Output	Labor-Hours	Unit	Material	Labor	Equipment	Total	Total Incl O&P
0010	**HASPS**, steel assembly									
0015	3"	1 Carp	26	.308	Ea.	5.95	16.35		22.30	32.50
0020	4-1/2"		13	.615		7.10	32.50		39.60	60.50
0040	6"		12.50	.640		10.45	34		44.45	66

08 71 20.90 Hinges		Crew	Daily Output	Labor-Hours	Unit	Material	Labor	Equipment	Total	Total Incl O&P
0010	**HINGES**									
0012	Full mortise, avg. freq., steel base, USP, 4-1/2" x 4-1/2"				Pr.	47.50			47.50	52.50
0040	US26D					71			71	78
0080	US10A					60			60	66
0100	5" x 5", USP					68.50			68.50	75
0200	6" x 6", USP					144			144	158
0400	Brass base, 4-1/2" x 4-1/2", US10					73.50			73.50	81
0440	US26D					69			69	76
0480	US10B					73			73	80.50
0500	5" x 5", US10					78			78	85.50
0600	6" x 6", US10					171			171	188
0800	Stainless steel base, 4-1/2" x 4-1/2", US32					93			93	102

08 71 20 – Hardware

08 71 20.90 Hinges		Crew	Daily Output	Labor-Hours	Unit	Material	Labor	Equipment	Total	Total Incl O&P
						2020 Bare Costs				
0900	For non removable pin, add (security item)				Ea.	5.80			5.80	6.40
0910	For floating pin, driven tips, add					3.29			3.29	3.62
0930	For hospital type tip on pin, add					14.85			14.85	16.35
0940	For steeple type tip on pin, add					20.50			20.50	23
0950	Full mortise, high frequency, steel base, 3-1/2" x 3-1/2", US26D				Pr.	31.50			31.50	34.50
1000	4-1/2" x 4-1/2", USP					67.50			67.50	74
1040	US26D					59			59	65
1080	US26					74.50			74.50	81.50
1100	5" x 5", USP					54			54	59.50
1200	6" x 6", USP					143			143	157
1300	8" x 8", USP					320			320	350
1400	Brass base, 3-1/2" x 3-1/2", US4					53.50			53.50	58.50
1430	4-1/2" x 4-1/2", US10					74.50			74.50	82
1440	US26D					118			118	130
1480	US10B					127			127	140
1500	5" x 5", US10					139			139	152
1600	6" x 6", US10					175			175	192
1700	8" x 8", US10					284			284	315
1800	Stainless steel base, 4-1/2" x 4-1/2", US32					107			107	117
1810	5" x 4-1/2", US32					141			141	155
1930	For hospital type tip on pin, add				Ea.	14.65			14.65	16.15
1950	Full mortise, low frequency, steel base, 3-1/2" x 3-1/2", US26D				Pr.	33			33	36.50
2000	4-1/2" x 4-1/2", USP					25			25	27.50
2040	US26D					26			26	29
2080	US10A					51			51	56
2100	5" x 5", USP					54			54	59.50
2200	6" x 6", USP					97			97	107
2300	4-1/2" x 4-1/2", US3					18.75			18.75	20.50
2310	5" x 5", US3					43			43	47.50
2400	Brass bass, 4-1/2" x 4-1/2", US10					58.50			58.50	64.50
2440	US26D					74.50			74.50	82
2480	US10A					67.50			67.50	74
2500	5" x 5", US10					93			93	102
2800	Stainless steel base, 4-1/2" x 4-1/2", US32					80			80	88
3000	Half surface, half mortise, or full surface, average frequency									
3010	Steel base, 4-1/2" x 4-1/2", USP				Pr.	54			54	59.50
3040	US26D					74.50			74.50	82
3080	US10A					104			104	114
3100	5" x 5", USP					95			95	104
3400	Brass base, 4-1/2" x 4-1/2", US10					185			185	203
3440	US26D					179			179	196
3480	US10B					212			212	233
3500	5" x 5", US10					184			184	202
3800	Stainless steel base, 4-1/2" x 4-1/2", US32					248			248	273
4000	Half surface, half mortise or full surface, high frequency									
4010	Steel base, 4-1/2" x 4-1/2", USP				Pr.	107			107	117
4040	US26D					142			142	156
4080	US10A					165			165	182
4100	5" x 5", USP					151			151	166
4400	Brass base, 4-1/2" x 4-1/2", US10					194			194	213
4440	US26D					202			202	222
4480	US10B					256			256	281
4500	5" x 5", US10					246			246	271

08 71 20.90 Hinges		Crew	Daily Output	Labor-Hours	Unit	2020 Bare Costs Material	Labor	Equipment	Total	Total Incl O&P
4800	Stainless steel base, 4-1/2" x 4-1/2", US32				Pr.	198			198	218
5000	Half surface, half mortise, or full surface, low frequency									
5010	Steel base, 4-1/2" x 4-1/2", USP				Pr.	44			44	48.50
5040	US26D					60.50			60.50	66.50
5080	US10A					88.50			88.50	97
8000	Install hinge	1 Carp	34	.235	↓		12.50		12.50	20

08 71 20.91 Special Hinges		Crew	Daily Output	Labor-Hours	Unit	Material	Labor	Equipment	Total	Total Incl O&P
0010	**SPECIAL HINGES**									
0015	Paumelle, high frequency									
0020	Steel base, 6" x 4-1/2", US10				Pr.	138			138	152
0040	US26D					172			172	189
0080	US10A				↓	181			181	199
0100	Brass base, 5" x 4-1/2", US10				Ea.	226			226	248
0140	US26D					229			229	252
0180	US3				↓	230			230	253
0200	Paumelle, average frequency, steel base, 4-1/2" x 3-1/2", US10				Pr.	93.50			93.50	103
0240	US26D					97.50			97.50	107
0280	US10A				↓	101			101	111
0400	Olive knuckle, low frequency, brass base, 6" x 4-1/2", US10				Ea.	151			151	167
0440	US26					158			158	174
0480	US3				↓	156			156	172
0800	Emergency door pivot, average frequency									
0810	Brass base, 4-7/8" jamb plate, USP				Pr.	72			72	79
0840	US26D					79.50			79.50	87
0880	For emergency door stop and hold back, add				↓	64.50			64.50	71
1000	Electric hinge with concealed conductor, average frequency									
1010	Steel base, 4-1/2" x 4-1/2", US26D				Pr.	345			345	380
1100	Bronze base, 4-1/2" x 4-1/2", US26D				"	350			350	385
1200	Electric hinge with concealed conductor, high frequency									
1210	Steel base, 4-1/2" x 4-1/2", US26D				Pr.	261			261	287
1400	Non template, full mortise, low frequency									
1410	Steel base, 4" x 4", USP				Pr.	29			29	32
1440	US26D					33.50			33.50	36.50
1480	US10A				↓	24			24	26.50
1500	Non template, full mortise, average frequency									
1510	Steel base, 4" x 4", USP				Pr.	32			32	35
1540	US26D					36			36	39.50
1580	US10A					43			43	47
1600	Double weight, 800 lb., steel base, removable pin, 5" x 6", USP					535			535	590
1700	Steel base-welded pin, 5" x 6", USP					172			172	190
1800	Triple weight, 2000 lb., steel base, welded pin, 5" x 6", USP					700			700	770
2000	Pivot reinf., high frequency, steel base, 7-3/4" door plate, USP					182			182	200
2040	US26D					253			253	279
2080	US10A					249			249	274
2200	Bronze base, 7-3/4" door plate, US10					248			248	273
2240	US26D					325			325	360
2280	US10A				↓	355			355	390
3000	Swing clear, full mortise, full or half surface, high frequency,									
3010	Steel base, 5" high, USP				Pr.	150			150	165
3040	US26D					156			156	172
3080	US10A				↓	197			197	217
3200	Swing clear, full mortise, average frequency									

08 71 20 – Hardware

08 71 20.91 Special Hinges		Crew	Daily Output	Labor-Hours	Unit	2020 Bare Costs Material	Labor	Equipment	Total	Total Incl O&P
3210	Steel base, 4-1/2" high, USP				Pr.	133			133	146
3280	US10A				"	151			151	166
3400	Swing clear, half mortise, high frequency									
3410	Steel base, 4-1/2" high, USP				Pr.	157			157	172
3440	US26D					157			157	173
3480	US10A					253			253	278
4000	Wide throw, average frequency, steel base, 4-1/2" x 6", USP					94.50			94.50	104
4040	US26D					109			109	120
4080	US10A					113			113	125
4100	5" x 7", USP					109			109	120
4200	High frequency, steel base, 4-1/2" x 6", USP					121			121	133
4240	US26D					161			161	177
4280	US10A					188			188	207
4300	5" x 7", USP					129			129	142
4400	Wide throw, low frequency, steel base, 4-1/2" x 6", USP					81			81	89
4440	US26D					93			93	102
4480	US10A					91.50			91.50	101
4500	5" x 7", USP				↓	92.50			92.50	102
4600	Spring hinge, single acting, 6" flange, steel				Ea.	49.50			49.50	54.50
4700	Brass					96			96	106
4900	Double acting, 6" flange, steel					89			89	98
4950	Brass				↓	133			133	146
8000	Continuous hinges									
8010	Steel, piano, 2" x 72"	1 Carp	20	.400	Ea.	24	21.50		45.50	60.50
8020	Brass, piano, 1-1/16" x 30"		30	.267		9.45	14.15		23.60	33
8030	Acrylic, piano, 1-3/4" x 12"		40	.200		15.65	10.65		26.30	34.50
8040	Aluminum, door, standard duty, 7'		3	2.667		120	142		262	360
8050	Heavy duty, 7'		3	2.667		152	142		294	395
8060	8'		3	2.667		183	142		325	430
8070	Steel, door, heavy duty, 7'		3	2.667		210	142		352	460
8080	8'		3	2.667		263	142		405	515
8090	Stainless steel, door, heavy duty, 7'		3	2.667		268	142		410	520
8100	8'		3	2.667		300	142		442	555
9000	Continuous hinge, steel, full mortise, heavy duty, 96"		2	4		725	213		938	1,150
9200	Continuous geared hinge, aluminum, full mortise, standard duty, 83"		3	2.667		157	142		299	400
9250	Continuous geared hinge, aluminum, full mortise, heavy duty, 83"	↓	3	2.667	↓	225	142		367	475

08 71 20.92 Mortised Hinges		Crew	Daily Output	Labor-Hours	Unit	Material	Labor	Equipment	Total	Total Incl O&P
0010	**MORTISED HINGES**									
0200	Average frequency, steel plated, ball bearing, 3-1/2" x 3-1/2"				Pr.	31			31	34
0300	Bronze, ball bearing					36.50			36.50	40
0900	High frequency, steel plated, ball bearing					102			102	113
1100	Bronze, ball bearing					110			110	121
1300	Average frequency, steel plated, ball bearing, 4-1/2" x 4-1/2"					37			37	40.50
1500	Bronze, ball bearing, to 36" wide					41.50			41.50	45.50
1700	Low frequency, steel, plated, plain bearing					23.50			23.50	25.50
1900	Bronze, plain bearing				↓	29			29	32

08 71 20.95 Kick Plates		Crew	Daily Output	Labor-Hours	Unit	Material	Labor	Equipment	Total	Total Incl O&P
0010	**KICK PLATES**									
0020	Stainless steel, .050", 16 ga., 8" x 28", US32	1 Carp	15	.533	Ea.	43	28.50		71.50	92.50
0030	8" x 30"		15	.533		46.50	28.50		75	96.50
0040	8" x 34"		15	.533		52.50	28.50		81	103
0050	10" x 28"	↓	15	.533	↓	86	28.50		114.50	140

08 71 Door Hardware

08 71 20 – Hardware

08 71 20.95 Kick Plates		Crew	Daily Output	Labor-Hours	Unit	Material	2020 Bare Costs Labor	Equipment	Total	Total Incl O&P
0060	10″ x 30″	1 Carp	15	.533	Ea.	93	28.50		121.50	148
0070	10″ x 34″		15	.533		104	28.50		132.50	161
0080	Mop/Kick, 4″ x 28″		15	.533		38.50	28.50		67	88
0090	4″ x 30″		15	.533		41	28.50		69.50	90.50
0100	4″ x 34″		15	.533		46.50	28.50		75	96.50
0110	6″ x 28″		15	.533		44	28.50		72.50	94
0120	6″ x 30″		15	.533		53.50	28.50		82	104
0130	6″ x 34″		15	.533		60	28.50		88.50	112
0500	Bronze, .050″, 8″ x 28″		15	.533		77.50	28.50		106	131
0510	8″ x 30″		15	.533		73	28.50		101.50	126
0520	8″ x 34″		15	.533		75.50	28.50		104	129
0530	10″ x 28″		15	.533		79	28.50		107.50	133
0540	10″ x 30″		15	.533		86	28.50		114.50	140
0550	10″ x 34″		15	.533		96	28.50		124.50	152
0560	Mop/Kick, 4″ x 28″		15	.533		37	28.50		65.50	86
0570	4″ x 30″		15	.533		41	28.50		69.50	90.50
0580	4″ x 34″		15	.533		41.50	28.50		70	91
0590	6″ x 28″		15	.533		51.50	28.50		80	102
0600	6″ x 30″		15	.533		52	28.50		80.50	103
0610	6″ x 34″		15	.533		64.50	28.50		93	117
1000	Acrylic, .125″, 8″ x 26″		15	.533		28	28.50		56.50	76.50
1010	8″ x 36″		15	.533		39.50	28.50		68	89
1020	8″ x 42″		15	.533		45.50	28.50		74	95.50
1030	10″ x 26″		15	.533		35.50	28.50		64	84.50
1040	10″ x 36″		15	.533		49.50	28.50		78	100
1050	10″ x 42″		15	.533		57.50	28.50		86	109
1060	Mop/Kick, 4″ x 26″		15	.533		16.70	28.50		45.20	64
1070	4″ x 36″		15	.533		23.50	28.50		52	71.50
1080	4″ x 42″		15	.533		27	28.50		55.50	75.50
1090	6″ x 26″		15	.533		25	28.50		53.50	73
1100	6″ x 36″		15	.533		34.50	28.50		63	83.50
1110	6″ x 42″		15	.533		40.50	28.50		69	90
1220	Brass, .050″, 8″ x 26″		15	.533		72.50	28.50		101	126
1230	8″ x 36″		15	.533		72	28.50		100.50	125
1240	8″ x 42″		15	.533		88.50	28.50		117	143
1250	10″ x 26″		15	.533		73.50	28.50		102	126
1260	10″ x 36″		15	.533		94	28.50		122.50	150
1270	10″ x 42″		15	.533		108	28.50		136.50	165
1320	Mop/Kick, 4″ x 26″		15	.533		27	28.50		55.50	75
1330	4″ x 36″		15	.533		37.50	28.50		66	86.50
1340	4″ x 42″		15	.533		43.50	28.50		72	93.50
1350	6″ x 26″		15	.533		40.50	28.50		69	90
1360	6″ x 36″		15	.533		48.50	28.50		77	99
1370	6″ x 42″		15	.533		57.50	28.50		86	109
1800	Aluminum, .050″, 8″ x 26″		15	.533		40.50	28.50		69	90
1810	8″ x 36″		15	.533		37.50	28.50		66	87
1820	8″ x 42″		15	.533		44.50	28.50		73	94.50
1830	10″ x 26″		15	.533		34	28.50		62.50	83
1840	10″ x 36″		15	.533		47	28.50		75.50	97
1850	10″ x 42″		15	.533		65.50	28.50		94	118
1860	Mop/Kick, 4″ x 26″		15	.533		16.80	28.50		45.30	64
1870	4″ x 36″		15	.533		22	28.50		50.50	70
1880	4″ x 42″		15	.533		27	28.50		55.50	75

08 71 Door Hardware

08 71 20 – Hardware

08 71 20.95 Kick Plates		Crew	Daily Output	Labor-Hours	Unit	Material	2020 Bare Costs Labor	Equipment	Total	Total Incl O&P
1890	6" x 26"	1 Carp	15	.533	Ea.	24.50	28.50		53	72.50
1900	6" x 36"		15	.533		33	28.50		61.50	82
1910	6" x 42"		15	.533	↓	38.50	28.50		67	88
9000	Minimum labor/equipment charge	↓	6	1.333	Job		71		71	114

08 71 21 – Astragals

08 71 21.10 Exterior Mouldings, Astragals

		Crew	Daily Output	Labor-Hours	Unit	Material	Labor	Equipment	Total	Total Incl O&P
0010	**EXTERIOR MOULDINGS, ASTRAGALS**									
0400	One piece, overlapping cadmium plated steel, flat, 3/16" x 2"	1 Carp	90	.089	L.F.	4.12	4.72		8.84	12.10
0600	Prime coated steel, flat, 1/8" x 3"		90	.089		6.25	4.72		10.97	14.40
0800	Stainless steel, flat, 3/32" x 1-5/8"		90	.089		18.10	4.72		22.82	27.50
1000	Aluminum, flat, 1/8" x 2"		90	.089		4.16	4.72		8.88	12.15
1200	Nail on, "T" extrusion		120	.067		2.21	3.54		5.75	8.15
1300	Vinyl bulb insert		105	.076		2.78	4.05		6.83	9.55
1600	Screw on, "T" extrusion		90	.089		4.16	4.72		8.88	12.15
1700	Vinyl insert		75	.107		4.65	5.65		10.30	14.20
2000	"L" extrusion, neoprene bulbs		75	.107		4.51	5.65		10.16	14.05
2100	Neoprene sponge insert		75	.107		6.75	5.65		12.40	16.55
2200	Magnetic		75	.107		10.50	5.65		16.15	20.50
2400	Spring hinged security seal, with cam		75	.107		6.75	5.65		12.40	16.50
2600	Spring loaded locking bolt, vinyl insert		45	.178		9.15	9.45		18.60	25
2800	Neoprene sponge strip, "Z" shaped, aluminum		60	.133		8.25	7.10		15.35	20.50
2900	Solid neoprene strip, nail on aluminum strip	↓	90	.089	↓	3.82	4.72		8.54	11.75
3000	One piece stile protection									
3020	Neoprene fabric loop, nail on aluminum strips	1 Carp	60	.133	L.F.	1.25	7.10		8.35	12.75
3110	Flush mounted aluminum extrusion, 1/2" x 1-1/4"		60	.133		6.85	7.10		13.95	18.85
3140	3/4" x 1-3/8"		60	.133		4.13	7.10		11.23	15.90
3160	1-1/8" x 1-3/4"		60	.133		5	7.10		12.10	16.85
3300	Mortise, 9/16" x 3/4"		60	.133		4.13	7.10		11.23	15.90
3320	13/16" x 1-3/8"		60	.133		4.20	7.10		11.30	15.95
3600	Spring bronze strip, nail on type		105	.076		2.16	4.05		6.21	8.90
3620	Screw on, with retainer		75	.107		2.69	5.65		8.34	12.05
3800	Flexible stainless steel housing, pile insert, 1/2" door		105	.076		7.25	4.05		11.30	14.50
3820	3/4" door		105	.076		8.05	4.05		12.10	15.35
4000	Extruded aluminum retainer, flush mount, pile insert		105	.076		2.49	4.05		6.54	9.25
4080	Mortise, felt insert		90	.089		4.50	4.72		9.22	12.50
4160	Mortise with spring, pile insert		90	.089		3.62	4.72		8.34	11.55
4400	Rigid vinyl retainer, mortise, pile insert		105	.076		2.85	4.05		6.90	9.65
4600	Wool pile filler strip, aluminum backing	↓	105	.076	↓	2.94	4.05		6.99	9.75
5000	Two piece overlapping astragal, extruded aluminum retainer									
5010	Pile insert	1 Carp	60	.133	L.F.	3.26	7.10		10.36	14.95
5020	Vinyl bulb insert		60	.133		1.91	7.10		9.01	13.45
5040	Vinyl flap insert		60	.133		3.97	7.10		11.07	15.70
5060	Solid neoprene flap insert		60	.133		6.50	7.10		13.60	18.50
5080	Hypalon rubber flap insert		60	.133		6.65	7.10		13.75	18.65
5090	Snap on cover, pile insert		60	.133		9.75	7.10		16.85	22
5400	Magnetic aluminum, surface mounted		60	.133		25.50	7.10		32.60	39.50
5500	Interlocking aluminum, 5/8" x 1" neoprene bulb insert		45	.178		5.30	9.45		14.75	21
5600	Adjustable aluminum, 9/16" x 21/32", pile insert		45	.178		17.65	9.45		27.10	34.50
5800	Magnetic, adjustable, 9/16" x 21/32"	↓	45	.178	↓	23	9.45		32.45	40.50
6000	Two piece stile protection									
6010	Cloth backed rubber loop, 1" gap, nail on aluminum strips	1 Carp	45	.178	L.F.	4.33	9.45		13.78	19.90
6040	Screw on aluminum strips	↓	45	.178	↓	6.50	9.45		15.95	22.50

08 71 Door Hardware

08 71 21 – Astragals

08 71 21.10 Exterior Mouldings, Astragals		Crew	Daily Output	Labor-Hours	Unit	Material	Labor	Equipment	Total	Total Incl O&P
						2020 Bare Costs				
6100	1-1/2" gap, screw on aluminum extrusion	1 Carp	45	.178	L.F.	5.80	9.45		15.25	21.50
6240	Vinyl fabric loop, slotted aluminum extrusion, 1" gap		45	.178		2.28	9.45		11.73	17.65
6300	1-1/4" gap	↓	45	.178	↓	6.15	9.45		15.60	22

08 71 25 – Door Weatherstripping

08 71 25.10 Mechanical Seals, Weatherstripping

		Crew	Daily Output	Labor-Hours	Unit	Material	Labor	Equipment	Total	Total Incl O&P
0010	**MECHANICAL SEALS, WEATHERSTRIPPING**									
1000	Doors, wood frame, interlocking, for 3' x 7' door, zinc	1 Carp	3	2.667	Opng.	58.50	142		200.50	292
1100	Bronze		3	2.667		71	142		213	305
1300	6' x 7' opening, zinc		2	4		60.50	213		273.50	405
1400	Bronze	↓	2	4	↓	71	213		284	420
1700	Wood frame, spring type, bronze									
1800	3' x 7' door	1 Carp	7.60	1.053	Opng.	24.50	56		80.50	116
1900	6' x 7' door	"	7	1.143	"	32	60.50		92.50	133
2200	Metal frame, spring type, bronze									
2300	3' x 7' door	1 Carp	3	2.667	Opng.	52	142		194	284
2400	6' x 7' door	"	2.50	3.200	"	53.50	170		223.50	330
2500	For stainless steel, spring type, add					133%				
2700	Metal frame, extruded sections, 3' x 7' door, aluminum	1 Carp	3	2.667	Opng.	31	142		173	261
2800	Bronze		3	2.667		90	142		232	325
3100	6' x 7' door, aluminum		1.50	5.333		38	283		321	495
3200	Bronze	↓	1.50	5.333	↓	148	283		431	615
3500	Threshold weatherstripping									
3650	Door sweep, flush mounted, aluminum	1 Carp	25	.320	Ea.	22	17		39	51.50
3700	Vinyl		25	.320		19.90	17		36.90	49.50
5000	Garage door bottom weatherstrip, 12' aluminum, clear		14	.571		26	30.50		56.50	77
5010	Bronze		14	.571		94.50	30.50		125	153
5050	Bottom protection, rubber		14	.571		54.50	30.50		85	109
5100	Threshold		14	.571	↓	70.50	30.50		101	126
9000	Minimum labor/equipment charge	↓	3	2.667	Job		142		142	227

08 75 Window Hardware

08 75 10 – Window Handles and Latches

08 75 10.10 Handles and Latches

		Crew	Daily Output	Labor-Hours	Unit	Material	Labor	Equipment	Total	Total Incl O&P
0010	**HANDLES AND LATCHES**									
1000	Handles, surface mounted, aluminum	1 Carp	24	.333	Ea.	6.15	17.70		23.85	35.50
1020	Brass		24	.333		5.35	17.70		23.05	34.50
1040	Chrome		24	.333		8.65	17.70		26.35	38
1200	Window handles window crank ADA		24	.333		17.45	17.70		35.15	47.50
1500	Recessed, aluminum		12	.667		3.90	35.50		39.40	61.50
1520	Brass		12	.667		5.60	35.50		41.10	63
1540	Chrome		12	.667		4.63	35.50		40.13	62
2000	Latches, aluminum		20	.400		4.41	21.50		25.91	39
2020	Brass		20	.400		6.70	21.50		28.20	41.50
2040	Chrome		20	.400		4.39	21.50		25.89	39
3000	Window locks, keyed sash locks, brass		32	.250	↓	19.10	13.30		32.40	42.50
9000	Minimum labor/equipment charge	↓	6	1.333	Job		71		71	114

08 75 Window Hardware

08 75 10 – Window Handles and Latches

08 75 10.15 Window Opening Control		Crew	Daily Output	Labor-Hours	Unit	Material	Labor	Equipment	Total	Total Incl O&P
						2020 Bare Costs				
0010	**WINDOW OPENING CONTROL**									
0015	Window stops									
0020	For double-hung window	1 Carp	35	.229	Ea.	40.50	12.15		52.65	64
0030	Cam action for sliding window		35	.229		4.55	12.15		16.70	24.50
0040	Thumb screw for sliding window	↓	35	.229	↓	4.91	12.15		17.06	25
0100	Window guards, child safety bars for single or double-hung									
0110	14" to 17" wide max vert opening 26"	1 Carp	16	.500	Ea.	68.50	26.50		95	118
0120	17" to 23" wide max vert opening 26"		16	.500		78	26.50		104.50	129
0130	23" to 36" wide max vert opening 26"		16	.500		89	26.50		115.50	141
0140	35" to 58" wide max vert opening 26"		16	.500		103	26.50		129.50	157
0150	58" to 90" wide max vert opening 26"		16	.500		139	26.50		165.50	196
0160	73" to 120" wide max vert opening 26"	↓	16	.500	↓	243	26.50		269.50	310

08 75 30 – Window Weatherstripping

08 75 30.10 Mechanical Weather Seals

		Crew	Daily Output	Labor-Hours	Unit	Material	Labor	Equipment	Total	Total Incl O&P
0010	**MECHANICAL WEATHER SEALS**, Window, double-hung, 3' x 5'									
0020	Zinc	1 Carp	7.20	1.111	Opng.	22.50	59		81.50	120
0100	Bronze		7.20	1.111		42.50	59		101.50	142
0500	As above but heavy duty, zinc		4.60	1.739		21	92.50		113.50	171
0600	Bronze	↓	4.60	1.739	↓	79	92.50		171.50	235
9000	Minimum labor/equipment charge	1 Clab	4.60	1.739	Job		73		73	117

08 79 Hardware Accessories

08 79 13 – Key Storage Equipment

08 79 13.10 Key Cabinets

		Crew	Daily Output	Labor-Hours	Unit	Material	Labor	Equipment	Total	Total Incl O&P
0010	**KEY CABINETS**									
0020	Wall mounted, 60 key capacity	1 Carp	20	.400	Ea.	107	21.50		128.50	152
0400	Wall mounted, 30 key capacity	1 Clab	50	.160		75.50	6.75		82.25	94.50
0500	Wall mounted, 80 key capacity	"	40	.200	↓	142	8.40		150.40	170

08 79 20 – Door Accessories

08 79 20.10 Door Hardware Accessories

		Crew	Daily Output	Labor-Hours	Unit	Material	Labor	Equipment	Total	Total Incl O&P
0010	**DOOR HARDWARE ACCESSORIES**									
0140	Door bolt, surface, 4"	1 Carp	32	.250	Ea.	18.80	13.30		32.10	42
0160	Door latch	"	12	.667	"	8.80	35.50		44.30	66.50
0200	Sliding closet door									
0220	Track and hanger, single	1 Carp	10	.800	Ea.	67.50	42.50		110	143
0240	Double		8	1		85	53		138	179
0260	Door guide, single		48	.167		32.50	8.85		41.35	49.50
0280	Double		48	.167		41.50	8.85		50.35	59.50
0600	Deadbolt and lock cover plate, brass or stainless steel		30	.267		33	14.15		47.15	58.50
0620	Hole cover plate, brass or chrome		35	.229		8.70	12.15		20.85	29
2240	Mortise lockset, passage, lever handle		9	.889		173	47		220	266
4000	Security chain, standard	↓	18	.444	↓	15	23.50		38.50	54.50

08 81 Glass Glazing

08 81 10 – Float Glass

08 81 10.10 Various Types and Thickness of Float Glass		Crew	Daily Output	Labor-Hours	Unit	2020 Bare Costs Material	Labor	Equipment	Total	Total Incl O&P
0010	**VARIOUS TYPES AND THICKNESS OF FLOAT GLASS**									
0020	3/16" plain	2 Glaz	130	.123	S.F.	7.10	6.30		13.40	17.90
0200	Tempered, clear		130	.123		7.50	6.30		13.80	18.30
0300	Tinted		130	.123		7.60	6.30		13.90	18.40
0600	1/4" thick, clear, plain		120	.133		8.40	6.80		15.20	20
0700	Tinted		120	.133		9.75	6.80		16.55	21.50
0800	Tempered, clear		120	.133		7.95	6.80		14.75	19.60
0900	Tinted		120	.133		12.15	6.80		18.95	24.50
1600	3/8" thick, clear, plain		75	.213		11.70	10.90		22.60	30.50
1700	Tinted		75	.213		16.85	10.90		27.75	36
1800	Tempered, clear		75	.213		18.45	10.90		29.35	38
1900	Tinted		75	.213		21	10.90		31.90	41
2200	1/2" thick, clear, plain		55	.291		18.95	14.85		33.80	44.50
2300	Tinted		55	.291		29.50	14.85		44.35	56
2400	Tempered, clear		55	.291		27.50	14.85		42.35	54
2500	Tinted		55	.291		28	14.85		42.85	54
2800	5/8" thick, clear, plain		45	.356		29.50	18.15		47.65	61.50
2900	Tempered, clear		45	.356		33.50	18.15		51.65	66
3200	3/4" thick, clear, plain		35	.457		38	23.50		61.50	79.50
3300	Tempered, clear		35	.457		45	23.50		68.50	86.50
3600	1" thick, clear, plain	↓	30	.533		63	27		90	113
8900	For low emissivity coating for 3/16" & 1/4" only, add to above				↓	18%				
9000	Minimum labor/equipment charge	1 Glaz	2	4	Job		204		204	325

08 81 13 – Decorative Glass Glazing

08 81 13.10 Beveled Glass		Crew	Daily Output	Labor-Hours	Unit	Material	Labor	Equipment	Total	Total Incl O&P
0010	**BEVELED GLASS**, with design patterns									
0020	Simple pattern	2 Glaz	150	.107	S.F.	65.50	5.45		70.95	80.50
0050	Intricate pattern	"	125	.128	"	144	6.55		150.55	168

08 81 13.30 Sandblasted Glass		Crew	Daily Output	Labor-Hours	Unit	Material	Labor	Equipment	Total	Total Incl O&P
0010	**SANDBLASTED GLASS**, float glass									
0020	1/8" thick	2 Glaz	160	.100	S.F.	12.20	5.10		17.30	21.50
0100	3/16" thick		130	.123		13.50	6.30		19.80	25
0500	1/4" thick		120	.133		14.05	6.80		20.85	26.50
0600	3/8" thick	↓	75	.213	↓	14.80	10.90		25.70	33.50

08 81 17 – Fire Glass

08 81 17.10 Fire Resistant Glass		Crew	Daily Output	Labor-Hours	Unit	Material	Labor	Equipment	Total	Total Incl O&P
0010	**FIRE RESISTANT GLASS**									
0020	Fire glass minimum	2 Glaz	40	.400	S.F.	42.50	20.50		63	79
0030	Mid range		40	.400		85.50	20.50		106	127
0050	High end	↓	40	.400	↓	375	20.50		395.50	445

08 81 20 – Vision Panels

08 81 20.10 Full Vision		Crew	Daily Output	Labor-Hours	Unit	Material	Labor	Equipment	Total	Total Incl O&P
0010	**FULL VISION**, window system with 3/4" glass mullions									
0020	Up to 10' high	H-2	130	.185	S.F.	71	8.85		79.85	92
0100	10' to 20' high, minimum		110	.218		75.50	10.50		86	100
0150	Average		100	.240		81	11.55		92.55	107
0200	Maximum	↓	80	.300	↓	91	14.40		105.40	123
9000	Minimum labor/equipment charge	1 Glaz	2	4	Job		204		204	325

08 81 Glass Glazing

08 81 25 – Glazing Variables

08 81 25.10 Applications of Glazing		Crew	Daily Output	Labor-Hours	Unit	Material	2020 Bare Costs Labor	Equipment	Total	Total Incl O&P
0010	**APPLICATIONS OF GLAZING**									
0600	For glass replacement, add				S.F.		100%			
0700	For gasket settings, add				L.F.	6.40			6.40	7
0900	For sloped glazing, add				S.F.		26%			
2000	Fabrication, polished edges, 1/4" thick				Inch	.61			.61	.67
2100	1/2" thick					1.44			1.44	1.58
2500	Mitered edges, 1/4" thick					1.44			1.44	1.58
2600	1/2" thick					2.38			2.38	2.62

08 81 30 – Insulating Glass

08 81 30.10 Reduce Heat Transfer Glass

Line	Description	Crew	Daily Output	Labor-Hours	Unit	Material	Labor	Equipment	Total	Total Incl O&P
0010	**REDUCE HEAT TRANSFER GLASS**									
0015	2 lites 1/8" float, 1/2" thk under 15 S.F.									
0020	Clear G	2 Glaz	95	.168	S.F.	10.60	8.60		19.20	25.50
0100	Tinted G		95	.168		14.75	8.60		23.35	30
0200	2 lites 3/16" float, for 5/8" thk unit, 15 to 30 S.F., clear G		90	.178		14.25	9.05		23.30	30
0300	Tinted G		90	.178		14.70	9.05		23.75	30.50
0400	1" thk, dbl. glazed, 1/4" float, 30 to 70 S.F., clear G		75	.213		17.10	10.90		28	36
0500	Tinted G		75	.213		24	10.90		34.90	44
0600	1" thk, dbl. glazed, 1/4" float, 1/4" wire		75	.213		24.50	10.90		35.40	44.50
0700	1/4" float, 1/4" tempered		75	.213		34	10.90		44.90	55
0800	1/4" wire, 1/4" tempered		75	.213		35	10.90		45.90	56
2000	Both lites, light & heat reflective G		85	.188		33	9.60		42.60	51.50
2500	Heat reflective, film inside, 1" thick unit, clear G		85	.188		28.50	9.60		38.10	47
2600	Tinted G		85	.188		31	9.60		40.60	49.50
3000	Film on weatherside, clear, 1/2" thick unit G		95	.168		20	8.60		28.60	36
3100	5/8" thick unit G		90	.178		21	9.05		30.05	37.50
3200	1" thick unit G		85	.188		28.50	9.60		38.10	47
3350	Clear heat reflective film on inside G	1 Glaz	50	.160		13.65	8.15		21.80	28
3360	Heat reflective film on inside, tinted G		25	.320		14.30	16.30		30.60	41.50
3370	Heat reflective film on the inside, metalized G		20	.400		19.40	20.50		39.90	54
5000	Spectrally selective film, on ext., blocks solar gain/allows 70% of light G	2 Glaz	95	.168		15.80	8.60		24.40	31
9000	Minimum labor/equipment charge	1 Glaz	2	4	Job		204		204	325

08 81 35 – Translucent Glass

08 81 35.10 Obscure Glass

Line	Description	Crew	Daily Output	Labor-Hours	Unit	Material	Labor	Equipment	Total	Total Incl O&P
0010	**OBSCURE GLASS**									
0020	1/8" thick, textured	2 Glaz	140	.114	S.F.	12.65	5.85		18.50	23
0100	Color		125	.128		14.95	6.55		21.50	27
0300	7/32" thick, textured		120	.133		13.85	6.80		20.65	26
0400	Color		105	.152		17.55	7.75		25.30	31.50

08 81 35.20 Patterned Glass

Line	Description	Crew	Daily Output	Labor-Hours	Unit	Material	Labor	Equipment	Total	Total Incl O&P
0010	**PATTERNED GLASS**, colored									
0020	1/8" thick	2 Glaz	140	.114	S.F.	10.30	5.85		16.15	20.50
0300	7/32" thick	"	120	.133	"	13.75	6.80		20.55	26

08 81 45 – Sheet Glass

08 81 45.10 Window Glass, Sheet

Line	Description	Crew	Daily Output	Labor-Hours	Unit	Material	Labor	Equipment	Total	Total Incl O&P
0010	**WINDOW GLASS, SHEET** gray									
0020	1/8" thick	2 Glaz	160	.100	S.F.	6.80	5.10		11.90	15.60
0200	1/4" thick	"	130	.123	"	8	6.30		14.30	18.85

08 81 Glass Glazing

08 81 50 – Spandrel Glass

08 81 50.10 Glass for Non Vision Areas		Crew	Daily Output	Labor-Hours	Unit	2020 Bare Costs Material	Labor	Equipment	Total	Total Incl O&P
0010	**GLASS FOR NON VISION AREAS**, 1/4" thick standard colors									
0020	Up to 1,000 S.F.	2 Glaz	110	.145	S.F.	19.50	7.40		26.90	33.50
0200	1,000 to 2,000 S.F.	"	120	.133	"	16.80	6.80		23.60	29.50
0300	For custom colors, add				Total	10%				
0500	For 3/8" thick, add				S.F.	13.15			13.15	14.45
1000	For double coated, 1/4" thick, add					4.70			4.70	5.15
1200	For insulation on panels, add					7.70			7.70	8.45
2000	Panels, insulated, with aluminum backed fiberglass, 1" thick	2 Glaz	120	.133		17.50	6.80		24.30	30
2100	2" thick	"	120	.133		22.50	6.80		29.30	36

08 81 55 – Window Glass

08 81 55.10 Sheet Glass		Crew	Daily Output	Labor-Hours	Unit	Material	Labor	Equipment	Total	Total Incl O&P
0010	**SHEET GLASS** (window), clear float, stops, putty bed									
0015	1/8" thick, clear float	2 Glaz	480	.033	S.F.	4.27	1.70		5.97	7.40
0500	3/16" thick, clear		480	.033		6.65	1.70		8.35	10
0600	Tinted		480	.033		8.15	1.70		9.85	11.65
0700	Tempered		480	.033		10	1.70		11.70	13.70
9000	Minimum labor/equipment charge		5	3.200	Job		163		163	261

08 81 65 – Wire Glass

08 81 65.10 Glass Reinforced With Wire		Crew	Daily Output	Labor-Hours	Unit	Material	Labor	Equipment	Total	Total Incl O&P
0010	**GLASS REINFORCED WITH WIRE**									
0012	1/4" thick rough obscure	2 Glaz	135	.119	S.F.	27.50	6.05		33.55	39.50
1000	Polished wire, 1/4" thick, diamond, clear		135	.119		33	6.05		39.05	45.50
1500	Pinstripe, obscure		135	.119		55	6.05		61.05	70

08 83 Mirrors

08 83 13 – Mirrored Glass Glazing

08 83 13.10 Mirrors		Crew	Daily Output	Labor-Hours	Unit	Material	Labor	Equipment	Total	Total Incl O&P
0010	**MIRRORS**, No frames, wall type, 1/4" plate glass, polished edge									
0100	Up to 5 S.F.	2 Glaz	125	.128	S.F.	9.85	6.55		16.40	21.50
0200	Over 5 S.F.		160	.100		9.45	5.10		14.55	18.55
0500	Door type, 1/4" plate glass, up to 12 S.F.		160	.100		9.55	5.10		14.65	18.65
1000	Float glass, up to 10 S.F., 1/8" thick		160	.100		6.75	5.10		11.85	15.55
1100	3/16" thick		150	.107		7.85	5.45		13.30	17.35
1500	12" x 12" wall tiles, square edge, clear		195	.082		2.80	4.18		6.98	9.80
1600	Veined		195	.082		6.60	4.18		10.78	13.95
2000	1/4" thick, stock sizes, one way transparent		125	.128		21	6.55		27.55	34
2010	Bathroom, unframed, laminated		160	.100		16	5.10		21.10	26
2500	Tempered		160	.100		18.75	5.10		23.85	28.50

08 83 13.15 Reflective Glass		Crew	Daily Output	Labor-Hours	Unit	Material	Labor	Equipment	Total	Total Incl O&P
0010	**REFLECTIVE GLASS**									
0100	1/4" float with fused metallic oxide fixed G	2 Glaz	115	.139	S.F.	18.30	7.10		25.40	31.50
0500	1/4" float glass with reflective applied coating G	"	115	.139	"	14.95	7.10		22.05	28

08 84 Plastic Glazing

08 84 10 – Plexiglass Glazing

08 84 10.10 Plexiglass Acrylic		Crew	Daily Output	Labor-Hours	Unit	Material	2020 Bare Costs Labor	Equipment	Total	Total Incl O&P
0010	**PLEXIGLASS ACRYLIC**, clear, masked,									
0020	1/8" thick, cut sheets	2 Glaz	170	.094	S.F.	13.90	4.80		18.70	23
0200	Full sheets		195	.082		6.20	4.18		10.38	13.55
0500	1/4" thick, cut sheets		165	.097		14.65	4.95		19.60	24
0600	Full sheets		185	.086		10.10	4.41		14.51	18.15
0900	3/8" thick, cut sheets		155	.103		23	5.25		28.25	33.50
1000	Full sheets		180	.089		17	4.53		21.53	26
1300	1/2" thick, cut sheets		135	.119		34	6.05		40.05	47
1400	Full sheets		150	.107		22	5.45		27.45	33
1700	3/4" thick, cut sheets		115	.139		32	7.10		39.10	46.50
1800	Full sheets		130	.123		18.35	6.30		24.65	30
2100	1" thick, cut sheets		105	.152		36	7.75		43.75	52
2200	Full sheets		125	.128		22	6.55		28.55	35
3000	Colored, 1/8" thick, cut sheets		170	.094		20	4.80		24.80	29.50
3200	Full sheets		195	.082		11.95	4.18		16.13	19.85
3500	1/4" thick, cut sheets		165	.097		22.50	4.95		27.45	32.50
3600	Full sheets		185	.086		16.35	4.41		20.76	25
4000	Mirrors, untinted, cut sheets, 1/8" thick		185	.086		16.15	4.41		20.56	25
4200	1/4" thick	↓	180	.089	↓	18.70	4.53		23.23	28

08 84 20 – Polycarbonate

08 84 20.10 Thermoplastic		Crew	Daily Output	Labor-Hours	Unit	Material	Labor	Equipment	Total	Total Incl O&P
0010	**THERMOPLASTIC**, clear, masked, cut sheets									
0020	1/8" thick	2 Glaz	170	.094	S.F.	16	4.80		20.80	25.50
0500	3/16" thick		165	.097		20	4.95		24.95	30
1000	1/4" thick		155	.103		19.20	5.25		24.45	29.50
1500	3/8" thick	↓	150	.107	↓	30	5.45		35.45	41.50
9000	Minimum labor/equipment charge	1 Glaz	2	4	Job		204		204	325

08 85 Glazing Accessories

08 85 10 – Miscellaneous Glazing Accessories

08 85 10.10 Glazing Gaskets		Crew	Daily Output	Labor-Hours	Unit	Material	Labor	Equipment	Total	Total Incl O&P
0010	**GLAZING GASKETS**, Neoprene for glass tongued mullion									
0015	1/4" glass	2 Glaz	200	.080	L.F.	.98	4.08		5.06	7.60
0020	3/8"		200	.080		1.18	4.08		5.26	7.80
0040	1/2"		200	.080		1.29	4.08		5.37	7.90
0060	3/4"		180	.089		1.98	4.53		6.51	9.45
0080	1"	↓	180	.089		1.64	4.53		6.17	9.05
1000	Glazing compound, wood	1 Glaz	58	.138		1.31	7.05		8.36	12.70
1005	Glazing compound, per window, up to 30 L.F.				↓	1.31			1.31	1.45
1006	Glazing compound, per window, up to 30 L.F.				Ea.	39.50			39.50	43.50
1020	Metal	1 Glaz	58	.138	L.F.	1.31	7.05		8.36	12.70

08 85 20 – Structural Glazing Sealants

08 85 20.10 Structural Glazing Adhesives		Crew	Daily Output	Labor-Hours	Unit	Material	Labor	Equipment	Total	Total Incl O&P
0010	**STRUCTURAL GLAZING ADHESIVES**									
0050	Structural glazing adhesive, 1/4" x 1/4" joint seal	1 Glaz	200	.040	L.F.	14.65	2.04		16.69	19.35

08 87 Glazing Surface Films

08 87 13 – Solar Control Films

08 87 13.10 Solar Films On Glass		Crew	Daily Output	Labor-Hours	Unit	2020 Bare Costs Material	Labor	Equipment	Total	Total Incl O&P
0010	**SOLAR FILMS ON GLASS** (glass not included)									
1000	Bronze, 20% VLT	H-2	180	.133	S.F.	1.67	6.40		8.07	12.10
1020	50% VLT		180	.133		1.87	6.40		8.27	12.30
1050	Neutral, 20% VLT		180	.133		1.87	6.40		8.27	12.30
1100	Silver, 15% VLT		180	.133		1.08	6.40		7.48	11.45
1120	35% VLT		180	.133		3.14	6.40		9.54	13.70
1150	68% VLT		180	.133		.42	6.40		6.82	10.70
3000	One way mirror with night vision 5% in and 15% out VLT		180	.133		3.15	6.40		9.55	13.70

08 87 16 – Glass Safety Films

08 87 16.10 Safety Films On Glass		Crew	Daily Output	Labor-Hours	Unit	Material	Labor	Equipment	Total	Total Incl O&P
0010	**SAFETY FILMS ON GLASS** (glass not included)									
0015	Safety film helps hold glass together when broken									
0050	Safety film, clear, 2 mil	H-2	180	.133	S.F.	1.26	6.40		7.66	11.65
0100	Safety film, clear, 4 mil		180	.133		1.88	6.40		8.28	12.30
0110	Safety film, tinted, 4 mil		180	.133		3.60	6.40		10	14.20
0150	Safety film, black tint, 8 mil		180	.133		2.32	6.40		8.72	12.80

08 87 26 – Bird Control Film

08 87 26.10 Bird Control Film		Crew	Daily Output	Labor-Hours	Unit	Material	Labor	Equipment	Total	Total Incl O&P
0010	**BIRD CONTROL FILM**									
0050	Patterned, adhered to glass	2 Glaz	180	.089	S.F.	7	4.53		11.53	14.95
0200	Decals small, adhered to glass	1 Glaz	50	.160	Ea.	2.57	8.15		10.72	15.90
0250	Decals large, adhered to glass		50	.160	"	8.20	8.15		16.35	22
0300	Decals set of 4, adhered to glass		25	.320	Set	7.55	16.30		23.85	34.50
0400	Bird control film tape		10	.800	Roll	20.50	41		61.50	88

08 87 33 – Electrically Tinted Window Film

08 87 33.20 Window Film		Crew	Daily Output	Labor-Hours	Unit	Material	Labor	Equipment	Total	Total Incl O&P
0010	**WINDOW FILM** adhered on glass (glass not included)									
0015	Film is pre-wired and can be trimmed									
0100	Window film 4 S.F. adhered on glass	1 Glaz	200	.040	S.F.	224	2.04		226.04	250
0120	6 S.F.		180	.044		335	2.27		337.27	375
0160	9 S.F.		170	.047		505	2.40		507.40	560
0180	10 S.F.		150	.053		585	2.72		587.72	645
0200	12 S.F.		144	.056		670	2.83		672.83	745
0220	14 S.F.		140	.057		785	2.91		787.91	865
0240	16 S.F.		128	.063		895	3.19		898.19	990
0260	18 S.F.		126	.063		1,000	3.24		1,003.24	1,100
0280	20 S.F.		120	.067		1,125	3.40		1,128.40	1,225

08 87 53 – Security Films On Glass

08 87 53.10 Security Film Adhered On Glass		Crew	Daily Output	Labor-Hours	Unit	Material	Labor	Equipment	Total	Total Incl O&P
0010	**SECURITY FILM ADHERED ON GLASS** (glass not included)									
0020	Security film, clear, 7 mil	1 Glaz	200	.040	S.F.	1.13	2.04		3.17	4.50
0030	8 mil		200	.040		2.04	2.04		4.08	5.50
0040	9 mil		200	.040		1.12	2.04		3.16	4.49
0050	10 mil		200	.040		1.93	2.04		3.97	5.40
0060	12 mil		200	.040		1.91	2.04		3.95	5.35
0075	15 mil		200	.040		2.41	2.04		4.45	5.90
0100	Security film, sealed with structural adhesive, 7 mil		180	.044		5.75	2.27		8.02	9.95
0110	8 mil		180	.044		6.70	2.27		8.97	10.95
0140	14 mil		180	.044		7.35	2.27		9.62	11.65

08 88 40 – Acoustical Glass Units

08 88 40.10 Sound Reduction Units		Crew	Daily Output	Labor-Hours	Unit	Material	2020 Bare Costs Labor	Equipment	Total	Total Incl O&P
0010	**SOUND REDUCTION UNITS**, 1 lite at 3/8", 1 lite at 3/16"									
0020	For 1" thick	2 Glaz	100	.160	S.F.	37	8.15		45.15	53.50
0100	For 4" thick	"	80	.200	"	62	10.20		72.20	84.50

08 88 52 – Prefabricated Glass Block Windows

08 88 52.10 Prefabricated Glass Block Windows		Crew	Daily Output	Labor-Hours	Unit	Material	Labor	Equipment	Total	Total Incl O&P
0010	**PREFABRICATED GLASS BLOCK WINDOWS**									
0015	Includes frame and silicone seal									
0020	Glass Block Window 16" x 8"	2 Glaz	6	2.667	Ea.	168	136		304	400
0025	16" x 16"		6	2.667		192	136		328	430
0030	16" x 24"		6	2.667		223	136		359	460
0050	16" x 48"		4	4		360	204		564	720
0100	Glass Block Window 24" x 8"		6	2.667		181	136		317	415
0110	24" x 16"		6	2.667		232	136		368	470
0120	24" x 24"		6	2.667		292	136		428	535
0150	24" x 48"		6	2.667		430	136		566	690
0200	Glass Block Window 32" x 8"		6	2.667		182	136		318	415
0210	32" x 16"		6	2.667		266	136		402	510
0220	32" x 24"		4	4		330	204		534	690
0250	32" x 48"		4	4		525	204		729	905
0300	Glass Block Window 40" x 8"		6	2.667		229	136		365	470
0310	40" x 16"		6	2.667		305	136		441	550
0320	40" x 24"		6	2.667		375	136		511	625
0350	40" x 48"		6	2.667		615	136		751	895
0400	Glass Block Window 48" x 8"		6	2.667		274	136		410	515
0410	48" x 16"		6	2.667		360	136		496	610
0420	48" x 24"		6	2.667		430	136		566	690
0450	48" x 48"		6	2.667		700	136		836	985
0500	Glass Block Window 56" x 8"		6	2.667		272	136		408	515
0510	56" x 16"		4	4		405	204		609	770
0520	56" x 24"		4	4		500	204		704	875
0550	56" x 48"		4	4		785	204		989	1,200

08 88 52.20 Prefab. Glass Block Windows Hurricane Resistant		Crew	Daily Output	Labor-Hours	Unit	Material	Labor	Equipment	Total	Total Incl O&P
0010	**PREFABRICATED GLASS BLOCK WINDOWS HURRICANE RESISTANT**									
0015	Includes frame and silicone seal									
3020	Glass Block Window 16" x 16"	2 Glaz	6	2.667	Ea.	430	136		566	690
3030	16" x 24"		6	2.667		515	136		651	785
3040	16" x 32"		4	4		590	204		794	975
3050	16" x 40"		4	4		645	204		849	1,025
3060	16" x 48"		4	4		680	204		884	1,075
3070	16" x 56"		4	4		845	204		1,049	1,250
3080	Glass Block Window 24" x 24"		6	2.667		600	136		736	875
3090	24" x 32"		4	4		680	204		884	1,075
3100	24" x 40"		4	4		835	204		1,039	1,250
3110	24" x 48"		4	4		890	204		1,094	1,300
3120	24" x 56"		4	4		920	204		1,124	1,325
3130	Glass Block Window 32" x 32"		4	4		760	204		964	1,175
3140	32" x 40"		4	4		565	204		769	945
3150	32" x 48"		4	4		1,000	204		1,204	1,425
3160	32" x 56"		4	4		865	204		1,069	1,275
3200	Glass Block Window 40" x 40"		4	4		1,000	204		1,204	1,425
3210	40" x 48"		3	5.333		1,225	272		1,497	1,750
3220	40" x 56"		3	5.333		1,350	272		1,622	1,925

08 88 52 – Prefabricated Glass Block Windows

08 88 52.20 Prefab. Glass Block Windows Hurricane Resistant		Crew	Daily Output	Labor-Hours	Unit	Material	Labor	Equipment	Total	Total Incl O&P
						2020 Bare Costs				
3260	Glass Block Window 48" x 48"	2 Glaz	3	5.333	Ea.	1,425	272		1,697	1,975
3270	48" x 56"		3	5.333		1,575	272		1,847	2,150
3310	Glass Block Window 56" x 56"		3	5.333		1,750	272		2,022	2,350

08 88 52.30 Prefabricated Acrylic Block Casement Windows

		Crew	Daily Output	Labor-Hours	Unit	Material	Labor	Equipment	Total	Total Incl O&P
0010	PREFABRICATED ACRYLIC BLOCK CASEMENT WINDOWS									
0015	Windows include frame and nail fin.									
3050	6" x 6" x 1-3/4" Block Casement Window 2 x 2 Pattern	2 Glaz	6	2.667	Ea.	450	136		586	710
3055	2 x 3		6	2.667		465	136		601	730
3060	2 x 4		6	2.667		510	136		646	775
3065	2 x 5		6	2.667		540	136		676	810
3070	2 x 6		6	2.667		595	136		731	870
3075	2 x 7		4	4		710	204		914	1,100
3080	2 x 8		4	4		740	204		944	1,150
3085	2 x 9		4	4		810	204		1,014	1,225
3090	3 x 3		4	4		490	204		694	865
3095	3 x 4		4	4		540	204		744	920
3100	3 x 5		4	4		610	204		814	995
3110	3 x 6		4	4		645	204		849	1,025
3115	3 x 7		4	4		740	204		944	1,150
3120	3 x 8		4	4		815	204		1,019	1,225
3125	3 x 9		3	5.333		880	272		1,152	1,400
3130	4 x 4		3	5.333		635	272		907	1,125
3135	4 x 5		3	5.333		675	272		947	1,175
3140	4 x 6		3	5.333		725	272		997	1,225
3145	4 x 7		3	5.333		765	272		1,037	1,275
3150	4 x 8		3	5.333		880	272		1,152	1,400
3500	8" x 8" x 1-3/4" Block Casement Window 2 x 2 Pattern		6	2.667		415	136		551	670
3510	2 x 3		6	2.667		460	136		596	720
3515	2 x 4		6	2.667		515	136		651	780
3520	2 x 5		6	2.667		595	136		731	870
3525	2 x 6		4	4		660	204		864	1,050
3530	2 x 7		4	4		695	204		899	1,100
3535	2 x 8		3	5.333		755	272		1,027	1,275
3540	2 x 9		3	5.333		795	272		1,067	1,300
3545	2 x 10		3	5.333		800	272		1,072	1,325
3550	2 x 11		3	5.333		815	272		1,087	1,325
3555	3 x 3		4	4		515	204		719	890
3560	3 x 4		4	4		570	204		774	950
3565	3 x 5		4	4		685	204		889	1,075
3570	3 x 6		4	4		750	204		954	1,150
3575	3 x 7		4	4		855	204		1,059	1,275
3580	3 x 8		3	5.333		950	272		1,222	1,475
3585	3 x 9		3	5.333		875	272		1,147	1,400
3590	4 x 4		3	5.333		700	272		972	1,200
3595	4 x 5		3	5.333		855	272		1,127	1,375
3600	4 x 6		3	5.333		920	272		1,192	1,450

08 88 56 – Ballistics-Resistant Glazing

08 88 56.10 Laminated Glass

		Crew	Daily Output	Labor-Hours	Unit	Material	Labor	Equipment	Total	Total Incl O&P
0010	LAMINATED GLASS									
0020	Clear float .03" vinyl 1/4" thick	2 Glaz	90	.178	S.F.	13.30	9.05		22.35	29
0100	3/8" thick		78	.205		31	10.45		41.45	50.50
0200	.06" vinyl, 1/2" thick		65	.246		28	12.55		40.55	51

08 88 Special Function Glazing

08 88 56 – Ballistics-Resistant Glazing

08 88 56.10 Laminated Glass		Crew	Daily Output	Labor-Hours	Unit	Material	2020 Bare Costs Labor	Equipment	Total	Total Incl O&P
1000	5/8" thick	2 Glaz	90	.178	S.F.	32.50	9.05		41.55	50.50
2000	Bullet-resisting, 1-3/16" thick, to 15 S.F.		16	1		126	51		177	220
2100	Over 15 S.F.		16	1		157	51		208	255
2200	2" thick, to 15 S.F.		10	1.600		152	81.50		233.50	297
2300	Over 15 S.F.		10	1.600		133	81.50		214.50	276
2500	2-1/4" thick, to 15 S.F.		12	1.333		216	68		284	345
2600	Over 15 S.F.		12	1.333		205	68		273	335
2700	Level 2 (.357 magnum), NIJ and UL		12	1.333		56	68		124	171
2750	Level 3A (.44 magnum) NIJ, UL 3		12	1.333		60	68		128	175
2800	Level 4 (AK-47) NIJ, UL 7 & 8		12	1.333		82.50	68		150.50	200
2850	Level 5 (M-16) UL		12	1.333		81.50	68		149.50	199
2900	Level 3 (7.62 Armor Piercing) NIJ, UL 4 & 5	▼	12	1.333	▼	96	68		164	215

08 91 Louvers

08 91 16 – Operable Wall Louvers

08 91 16.10 Movable Blade Louvers		Crew	Daily Output	Labor-Hours	Unit	Material	Labor	Equipment	Total	Total Incl O&P
0010	**MOVABLE BLADE LOUVERS**									
0100	PVC, commercial grade, 12" x 12"	1 Shee	20	.400	Ea.	82.50	25		107.50	130
0110	16" x 16"		14	.571		96	35.50		131.50	162
0120	18" x 18"		14	.571		98.50	35.50		134	165
0130	24" x 24"		14	.571		141	35.50		176.50	211
0140	30" x 30"		12	.667		170	41.50		211.50	251
0150	36" x 36"		12	.667		200	41.50		241.50	285
0300	Stainless steel, commercial grade, 12" x 12"		20	.400		230	25		255	292
0310	16" x 16"		14	.571		310	35.50		345.50	395
0320	18" x 18"		14	.571		310	35.50		345.50	395
0330	20" x 20"		14	.571		360	35.50		395.50	450
0340	24" x 24"		14	.571		425	35.50		460.50	525
0350	30" x 30"		12	.667		475	41.50		516.50	590
0360	36" x 36"	▼	12	.667	▼	580	41.50		621.50	700

08 91 19 – Fixed Louvers

08 91 19.10 Aluminum Louvers		Crew	Daily Output	Labor-Hours	Unit	Material	Labor	Equipment	Total	Total Incl O&P
0010	**ALUMINUM LOUVERS**									
0020	Aluminum with screen, residential, 8" x 8"	1 Carp	38	.211	Ea.	21.50	11.20		32.70	41.50
0100	12" x 12"		38	.211		17.85	11.20		29.05	37.50
0200	12" x 18"		35	.229		22.50	12.15		34.65	44
0250	14" x 24"		30	.267		32.50	14.15		46.65	58
0300	18" x 24"		27	.296		33.50	15.75		49.25	62
0500	24" x 30"		24	.333		70.50	17.70		88.20	106
0700	Triangle, adjustable, small		20	.400		73.50	21.50		95	115
0800	Large	▼	15	.533	▼	94	28.50		122.50	149
1200	Extruded aluminum, see Section 23 37 15.40									
2100	Midget, aluminum, 3/4" deep, 1" diameter	1 Carp	85	.094	Ea.	.98	5		5.98	9.10
2150	3" diameter		60	.133		3.22	7.10		10.32	14.90
2200	4" diameter		50	.160		6	8.50		14.50	20
2250	6" diameter	▼	30	.267		3.87	14.15		18.02	27
3000	PVC, commercial grade, 12" x 12"	1 Shee	20	.400		97	25		122	146
3010	12" x 18"		20	.400		109	25		134	159
3020	12" x 24"		20	.400		147	25		172	200
3030	14" x 24"	▼	20	.400	▼	153	25		178	208

08 91 Louvers

08 91 19 – Fixed Louvers

08 91 19.10 Aluminum Louvers		Crew	Daily Output	Labor-Hours	Unit	Material	Labor	Equipment	Total	Total Incl O&P
						2020 Bare Costs				
3100	Aluminum, commercial grade, 12″ x 12″	1 Shee	20	.400	Ea.	170	25		195	226
3110	24″ x 24″		16	.500		234	31		265	305
3120	24″ x 36″		16	.500		335	31		366	415
3130	36″ x 24″		16	.500		345	31		376	430
3140	36″ x 36″		14	.571		430	35.50		465.50	530
3150	36″ x 48″		10	.800		465	50		515	590
3160	48″ x 36″		10	.800		500	50		550	630
3170	48″ x 48″		10	.800		455	50		505	580
3180	60″ x 48″		8	1		515	62.50		577.50	670
3190	60″ x 60″		8	1		750	62.50		812.50	925

08 91 19.20 Steel Louvers		Crew	Daily Output	Labor-Hours	Unit	Material	Labor	Equipment	Total	Total Incl O&P
0010	**STEEL LOUVERS**									
3300	Galvanized steel, fixed blades, commercial grade, 18″ x 18″	1 Shee	20	.400	Ea.	163	25		188	218
3310	24″ x 24″		20	.400		218	25		243	279
3320	24″ x 36″		16	.500		284	31		315	360
3330	36″ x 24″		16	.500		291	31		322	370
3340	36″ x 36″		14	.571		355	35.50		390.50	450
3350	36″ x 48″		10	.800		420	50		470	545
3360	48″ x 36″		10	.800		410	50		460	530
3370	48″ x 48″		10	.800		510	50		560	640
3380	60″ x 48″		10	.800		560	50		610	695
3390	60″ x 60″		10	.800		670	50		720	820

08 91 26 – Door Louvers

08 91 26.10 Steel Louvers, 18 Gauge, Fixed Blade		Crew	Daily Output	Labor-Hours	Unit	Material	Labor	Equipment	Total	Total Incl O&P
0010	**STEEL LOUVERS, 18 GAUGE, FIXED BLADE**									
0050	12″ x 12″, with powder coat	1 Carp	20	.400	Ea.	101	21.50		122.50	145
0055	18″ x 12″		20	.400		106	21.50		127.50	151
0060	18″ x 18″		20	.400		109	21.50		130.50	154
0065	24″ x 12″		20	.400		123	21.50		144.50	169
0070	24″ x 18″		20	.400		130	21.50		151.50	177
0075	24″ x 24″		20	.400		164	21.50		185.50	215
0100	12″ x 12″, galvanized		20	.400		82	21.50		103.50	124
0105	18″ x 12″		20	.400		94	21.50		115.50	137
0115	24″ x 12″		20	.400		110	21.50		131.50	155
0125	24″ x 24″		20	.400		150	21.50		171.50	199
0300	6″ x 12″, painted		20	.400		45.50	21.50		67	84
0320	10″ x 16″		20	.400		71	21.50		92.50	113
0340	12″ x 22″		20	.400		108	21.50		129.50	152
0360	14″ x 14″		20	.400		109	21.50		130.50	153
0400	18″ x 18″		20	.400		85	21.50		106.50	128
0420	20″ x 26″		20	.400		147	21.50		168.50	196
0440	22″ x 22″		20	.400		139	21.50		160.50	187
0460	26″ x 26″		20	.400		226	21.50		247.50	282
3000	Fire rated									
3010	12″ x 12″	1 Carp	10	.800	Ea.	183	42.50		225.50	269
3050	18″ x 18″		10	.800		268	42.50		310.50	365
3100	24″ x 12″		10	.800		274	42.50		316.50	370
3120	24″ x 18″		10	.800		300	42.50		342.50	400
3130	24″ x 24″		10	.800		300	42.50		342.50	400

08 91 Louvers

08 91 26 – Door Louvers

08 91 26.20 Aluminum Louvers		Crew	Daily Output	Labor-Hours	Unit	Material	Labor	Equipment	Total	Total Incl O&P
0010	**ALUMINUM LOUVERS**, 18 ga., fixed blade, clear anodized									
0050	6" x 12"	1 Carp	20	.400	Ea.	65.50	21.50		87	106
0060	10" x 10"		20	.400		61.50	21.50		83	102
0065	10" x 14"		20	.400		70	21.50		91.50	111
0080	12" x 22"		20	.400		98	21.50		119.50	142
0090	14" x 14"		20	.400		88	21.50		109.50	131
0100	18" x 10"		20	.400		73	21.50		94.50	114
0110	18" x 18"		20	.400		131	21.50		152.50	178
0120	22" x 22"		20	.400		159	21.50		180.50	209
0130	26" x 26"		20	.400		203	21.50		224.50	258

(2020 Bare Costs: Material, Labor, Equipment, Total)

08 95 Vents

08 95 13 – Soffit Vents

08 95 13.10 Wall Louvers

		Crew	Daily Output	Labor-Hours	Unit	Material	Labor	Equipment	Total	Total Incl O&P
0010	**WALL LOUVERS**									
2340	Baked enamel finish	1 Carp	200	.040	L.F.	4.81	2.13		6.94	8.70
2400	Under eaves vent, aluminum, mill finish, 16" x 4"		48	.167	Ea.	2.01	8.85		10.86	16.40
2500	16" x 8"		48	.167	"	2.85	8.85		11.70	17.35

08 95 16 – Wall Vents

08 95 16.10 Louvers

		Crew	Daily Output	Labor-Hours	Unit	Material	Labor	Equipment	Total	Total Incl O&P
0010	**LOUVERS**									
0020	Redwood, 2'-0" diameter, full circle	1 Carp	16	.500	Ea.	310	26.50		336.50	390
0100	Half circle		16	.500		210	26.50		236.50	274
0200	Octagonal		16	.500		214	26.50		240.50	278
0300	Triangular, 5/12 pitch, 5'-0" at base		16	.500		655	26.50		681.50	765
1000	Rectangular, 1'-4" x 1'-3"		16	.500		28.50	26.50		55	74
1100	Rectangular, 1'-4" x 1'-8"		16	.500		43	26.50		69.50	90
1200	1'-4" x 2'-2"		15	.533		50.50	28.50		79	102
1300	1'-9" x 2'-2"		15	.533		62	28.50		90.50	114
1400	2'-3" x 2'-2"		14	.571		75.50	30.50		106	132
1700	2'-4" x 2'-11"		13	.615		75.50	32.50		108	136
2000	Aluminum, 12" x 16"		25	.320		24.50	17		41.50	54.50
2010	16" x 20"		25	.320		35	17		52	66
2020	24" x 30"		25	.320		69	17		86	104
2100	6' triangle		12	.667		193	35.50		228.50	270
3100	Round, 2'-2" diameter		16	.500		200	26.50		226.50	263
7000	Vinyl gable vent, 8" x 8"		38	.211		16.55	11.20		27.75	36
7020	12" x 12"		38	.211		32	11.20		43.20	53
7080	12" x 18"		35	.229		41	12.15		53.15	65
7200	18" x 24"		30	.267		58.50	14.15		72.65	87
9000	Minimum labor/equipment charge		3.50	2.286	Job		121		121	195

Estimating Tips

General

- Room Finish Schedule: A complete set of plans should contain a room finish schedule. If one is not available, it would be well worth the time and effort to obtain one.

09 20 00 Plaster and Gypsum Board

- Lath is estimated by the square yard plus a 5% allowance for waste. Furring, channels, and accessories are measured by the linear foot. An extra foot should be allowed for each accessory miter or stop.
- Plaster is also estimated by the square yard. Deductions for openings vary by preference, from zero deduction to 50% of all openings over 2 feet in width. The estimator should allow one extra square foot for each linear foot of horizontal interior or exterior angle located below the ceiling level. Also, double the areas of small radius work.
- Drywall accessories, studs, track, and acoustical caulking are all measured by the linear foot. Drywall taping is figured by the square foot. Gypsum wallboard is estimated by the square foot. No material deductions should be made for door or window openings under 32 S.F.

09 60 00 Flooring

- Tile and terrazzo areas are taken off on a square foot basis. Trim and base materials are measured by the linear foot. Accent tiles are listed per each. Two basic methods of installation are used. Mud set is approximately 30% more expensive than thin set. The cost of grout is included with tile unit price lines unless otherwise noted. In terrazzo work, be sure to include the linear footage of embedded decorative strips, grounds, machine rubbing, and power cleanup.
- Wood flooring is available in strip, parquet, or block configuration. The latter two types are set in adhesives with quantities estimated by the square foot. The laying pattern will influence labor costs and material waste. In addition to the material and labor for laying wood floors, the estimator must make allowances for sanding and finishing these areas, unless the flooring is prefinished.
- Sheet flooring is measured by the square yard. Roll widths vary, so consideration should be given to use the most economical width, as waste must be figured into the total quantity. Consider also the installation methods available—direct glue down or stretched. Direct glue-down installation is assumed with sheet carpet unit price lines unless otherwise noted.

09 70 00 Wall Finishes

- Wall coverings are estimated by the square foot. The area to be covered is measured—length by height of the wall above the baseboards—to calculate the square footage of each wall. This figure is divided by the number of square feet in the single roll which is being used. Deduct, in full, the areas of openings such as doors and windows. Where a pattern match is required allow 25–30% waste.

09 80 00 Acoustic Treatment

- Acoustical systems fall into several categories. The takeoff of these materials should be by the square foot of area with a 5% allowance for waste. Do not forget about scaffolding, if applicable, when estimating these systems.

09 90 00 Painting and Coating

- New line items created for cut-ins with reference diagram.
- A major portion of the work in painting involves surface preparation. Be sure to include cleaning, sanding, filling, and masking costs in the estimate.
- Protection of adjacent surfaces is not included in painting costs. When considering the method of paint application, an important factor is the amount of protection and masking required. These must be estimated separately and may be the determining factor in choosing the method of application.

Reference Numbers

Reference numbers are shown at the beginning of some major classifications. These numbers refer to related items in the Reference Section. The reference information may be an estimating procedure, an alternate pricing method, or technical information.

Note: Not all subdivisions listed here necessarily appear. ■

09 01 Maintenance of Finishes

09 01 60 – Maintenance of Flooring

09 01 60.10 Carpet Maintenance		Crew	Daily Output	Labor-Hours	Unit	Material	2020 Bare Costs Labor	Equipment	Total	Total Incl O&P
0011	**CARPET MAINTENANCE** See Section 01 93 13.09									

09 01 70 – Maintenance of Wall Finishes

09 01 70.10 Gypsum Wallboard Repairs

		Crew	Daily Output	Labor-Hours	Unit	Material	Labor	Equipment	Total	Total Incl O&P
0010	**GYPSUM WALLBOARD REPAIRS**									
0100	Fill and sand, pin/nail holes	1 Carp	960	.008	Ea.		.44		.44	.71
0110	Screw head pops		480	.017			.89		.89	1.42
0120	Dents, up to 2" square		48	.167		.01	8.85		8.86	14.20
0130	2" to 4" square		24	.333		.04	17.70		17.74	28.50
0140	Cut square, patch, sand and finish, holes, up to 2" square		12	.667		.04	35.50		35.54	57
0150	2" to 4" square		11	.727		.09	38.50		38.59	62
0160	4" to 8" square		10	.800		.24	42.50		42.74	68.50
0170	8" to 12" square		8	1		.47	53		53.47	85.50
0180	12" to 32" square		6	1.333		1.58	71		72.58	116
0210	16" by 48"		5	1.600		2.68	85		87.68	139
0220	32" by 48"		4	2		4.45	106		110.45	175
0230	48" square		3.50	2.286		6.30	121		127.30	202
0240	60" square		3.20	2.500		9.55	133		142.55	224
0500	Skim coat surface with joint compound		1600	.005	S.F.	.04	.27		.31	.47
0510	Prepare, retape and refinish joints		60	.133	L.F.	.70	7.10		7.80	12.10
9000	Minimum labor/equipment charge		2	4	Job		213		213	340

09 01 90 – Maintenance of Painting and Coating

09 01 90.92 Sanding

		Crew	Daily Output	Labor-Hours	Unit	Material	Labor	Equipment	Total	Total Incl O&P
0010	**SANDING** and puttying interior trim, compared to R099100-10									
0100	Painting 1 coat, on quality work				L.F.		100%			
0300	Medium work						50%			
0400	Industrial grade						25%			
0500	Surface protection, placement and removal									
0510	Surface protection, placement and removal, basic drop cloths	1 Pord	6400	.001	S.F.		.06		.06	.09
0520	Masking with paper		800	.010		.07	.44		.51	.79
0530	Volume cover up (using plastic sheathing or building paper)		16000	.001			.02		.02	.04

09 01 90.93 Exterior Surface Preparation

		Crew	Daily Output	Labor-Hours	Unit	Material	Labor	Equipment	Total	Total Incl O&P
0010	**EXTERIOR SURFACE PREPARATION** R099100-10									
0015	Doors, per side, not incl. frames or trim									
0020	Scrape & sand									
0030	Wood, flush	1 Pord	616	.013	S.F.		.58		.58	.92
0040	Wood, detail		496	.016			.72		.72	1.14
0050	Wood, louvered		280	.029			1.27		1.27	2.01
0060	Wood, overhead		616	.013			.58		.58	.92
0070	Wire brush									
0080	Metal, flush	1 Pord	640	.013	S.F.		.56		.56	.88
0090	Metal, detail		520	.015			.68		.68	1.08
0100	Metal, louvered		360	.022			.99		.99	1.57
0110	Metal or fibr., overhead		640	.013			.56		.56	.88
0120	Metal, roll up		560	.014			.63		.63	1.01
0130	Metal, bulkhead		640	.013			.56		.56	.88
0140	Power wash, based on 2500 lb. operating pressure									
0150	Metal, flush	A-1H	2240	.004	S.F.		.15	.03	.18	.28
0160	Metal, detail		2120	.004			.16	.04	.20	.29
0170	Metal, louvered		2000	.004			.17	.04	.21	.31
0180	Metal or fibr., overhead		2400	.003			.14	.03	.17	.25
0190	Metal, roll up		2400	.003			.14	.03	.17	.25

09 01 90 – Maintenance of Painting and Coating

09 01 90.93 Exterior Surface Preparation		Crew	Daily Output	Labor-Hours	Unit	Material	2020 Bare Costs Labor	Equipment	Total	Total Incl O&P
0200	Metal, bulkhead	A-1H	2200	.004	S.F.		.15	.03	.18	.29
0400	Windows, per side, not incl. trim									
0410	Scrape & sand									
0420	Wood, 1-2 lite	1 Pord	320	.025	S.F.		1.11		1.11	1.76
0430	Wood, 3-6 lite		280	.029			1.27		1.27	2.01
0440	Wood, 7-10 lite		240	.033			1.48		1.48	2.35
0450	Wood, 12 lite		200	.040			1.78		1.78	2.82
0460	Wood, Bay/Bow		320	.025			1.11		1.11	1.76
0470	Wire brush									
0480	Metal, 1-2 lite	1 Pord	480	.017	S.F.		.74		.74	1.18
0490	Metal, 3-6 lite		400	.020			.89		.89	1.41
0500	Metal, Bay/Bow		480	.017			.74		.74	1.18
0510	Power wash, based on 2500 lb. operating pressure									
0520	1-2 lite	A-1H	4400	.002	S.F.		.08	.02	.10	.14
0530	3-6 lite		4320	.002			.08	.02	.10	.14
0540	7-10 lite		4240	.002			.08	.02	.10	.15
0550	12 lite		4160	.002			.08	.02	.10	.15
0560	Bay/Bow		4400	.002			.08	.02	.10	.14
0600	Siding, scrape and sand, light=10-30%, med.=30-70%									
0610	Heavy=70-100% of surface to sand									
0650	Texture 1-11, light	1 Pord	480	.017	S.F.		.74		.74	1.18
0660	Med.		440	.018			.81		.81	1.28
0670	Heavy		360	.022			.99		.99	1.57
0680	Wood shingles, shakes, light		440	.018			.81		.81	1.28
0690	Med.		360	.022			.99		.99	1.57
0700	Heavy		280	.029			1.27		1.27	2.01
0710	Clapboard, light		520	.015			.68		.68	1.08
0720	Med.		480	.017			.74		.74	1.18
0730	Heavy		400	.020			.89		.89	1.41
0740	Wire brush									
0750	Aluminum, light	1 Pord	600	.013	S.F.		.59		.59	.94
0760	Med.		520	.015			.68		.68	1.08
0770	Heavy		440	.018			.81		.81	1.28
0780	Pressure wash, based on 2500 lb. operating pressure									
0790	Stucco	A-1H	3080	.003	S.F.		.11	.02	.13	.21
0800	Aluminum or vinyl		3200	.003			.11	.02	.13	.20
0810	Siding, masonry, brick & block		2400	.003			.14	.03	.17	.25
1300	Miscellaneous, wire brush									
1310	Metal, pedestrian gate	1 Pord	100	.080	S.F.		3.55		3.55	5.65
1320	Aluminum chain link, both sides		250	.032			1.42		1.42	2.26
1400	Existing galvanized surface, clean and prime, prep for painting		380	.021		.13	.93		1.06	1.62
8000	For chemical washing, see Section 04 01 30									
8010	For steam cleaning, see Section 04 01 30.20									
8020	For sand blasting, see Sections 03 35 29.60 and 05 01 10.51									

09 01 90.94 Interior Surface Preparation		Crew	Daily Output	Labor-Hours	Unit	Material	Labor	Equipment	Total	Total Incl O&P
0010	**INTERIOR SURFACE PREPARATION** R099100-10									
0020	Doors, per side, not incl. frames or trim									
0030	Scrape & sand									
0040	Wood, flush	1 Pord	616	.013	S.F.		.58		.58	.92
0050	Wood, detail		496	.016			.72		.72	1.14
0060	Wood, louvered		280	.029			1.27		1.27	2.01
0070	Wire brush									

09 01 90 – Maintenance of Painting and Coating

09 01 90.94 Interior Surface Preparation		Crew	Daily Output	Labor-Hours	Unit	Material	2020 Bare Costs Labor	Equipment	Total	Total Incl O&P
0080	Metal, flush	1 Pord	640	.013	S.F.		.56		.56	.88
0090	Metal, detail		520	.015			.68		.68	1.08
0100	Metal, louvered		360	.022			.99		.99	1.57
0110	Hand wash									
0120	Wood, flush	1 Pord	2160	.004	S.F.		.16		.16	.26
0130	Wood, detail		2000	.004			.18		.18	.28
0140	Wood, louvered		1360	.006			.26		.26	.41
0150	Metal, flush		2160	.004			.16		.16	.26
0160	Metal, detail		2000	.004			.18		.18	.28
0170	Metal, louvered		1360	.006			.26		.26	.41
0400	Windows, per side, not incl. trim									
0410	Scrape & sand									
0420	Wood, 1-2 lite	1 Pord	360	.022	S.F.		.99		.99	1.57
0430	Wood, 3-6 lite		320	.025			1.11		1.11	1.76
0440	Wood, 7-10 lite		280	.029			1.27		1.27	2.01
0450	Wood, 12 lite		240	.033			1.48		1.48	2.35
0460	Wood, Bay/Bow		360	.022			.99		.99	1.57
0470	Wire brush									
0480	Metal, 1-2 lite	1 Pord	520	.015	S.F.		.68		.68	1.08
0490	Metal, 3-6 lite		440	.018			.81		.81	1.28
0500	Metal, Bay/Bow		520	.015			.68		.68	1.08
0600	Walls, sanding, light=10-30%, medium=30-70%,									
0610	heavy=70-100% of surface to sand									
0650	Walls, sand									
0660	Gypsum board or plaster, light	1 Pord	3077	.003	S.F.		.12		.12	.18
0670	Gypsum board or plaster, medium		2160	.004			.16		.16	.26
0680	Gypsum board or plaster, heavy		923	.009			.38		.38	.61
0690	Wood, T&G, light		2400	.003			.15		.15	.23
0700	Wood, T&G, medium		1600	.005			.22		.22	.35
0710	Wood, T&G, heavy		800	.010			.44		.44	.71
0720	Walls, wash									
0730	Gypsum board or plaster	1 Pord	3200	.003	S.F.		.11		.11	.18
0740	Wood, T&G		3200	.003			.11		.11	.18
0750	Masonry, brick & block, smooth		2800	.003			.13		.13	.20
0760	Masonry, brick & block, coarse		2000	.004			.18		.18	.28
8000	For chemical washing, see Section 04 01 30									
8010	For steam cleaning, see Section 04 01 30.20									
8020	For sand blasting, see Sections 03 35 29.60 and 05 01 10.51									
9010	Minimum labor/equipment charge	1 Pord	3	2.667	Job		118		118	188

09 01 90.95 Scrape After Fire Damage		Crew	Daily Output	Labor-Hours	Unit	Material	Labor	Equipment	Total	Total Incl O&P
0010	**SCRAPE AFTER FIRE DAMAGE**									
0050	Boards, 1" x 4"	1 Pord	336	.024	L.F.		1.06		1.06	1.68
0060	1" x 6"		260	.031			1.37		1.37	2.17
0070	1" x 8"		207	.039			1.72		1.72	2.72
0080	1" x 10"		174	.046			2.04		2.04	3.24
0500	Framing, 2" x 4"		265	.030			1.34		1.34	2.13
0510	2" x 6"		221	.036			1.61		1.61	2.55
0520	2" x 8"		190	.042			1.87		1.87	2.97
0530	2" x 10"		165	.048			2.15		2.15	3.42
0540	2" x 12"		144	.056			2.47		2.47	3.92
1000	Heavy framing, 3" x 4"		226	.035			1.57		1.57	2.50
1010	4" x 4"		210	.038			1.69		1.69	2.69

09 01 Maintenance of Finishes

09 01 90 – Maintenance of Painting and Coating

09 01 90.95 Scrape After Fire Damage		Crew	Daily Output	Labor-Hours	Unit	2020 Bare Costs Material	Labor	Equipment	Total	Total Incl O&P
1020	4″ x 6″	1 Pord	191	.042	L.F.		1.86		1.86	2.95
1030	4″ x 8″		165	.048			2.15		2.15	3.42
1040	4″ x 10″		144	.056			2.47		2.47	3.92
1060	4″ x 12″		131	.061			2.71		2.71	4.31
2900	For sealing, light damage		825	.010	S.F.	.15	.43		.58	.85
2920	Heavy damage		460	.017	″	.35	.77		1.12	1.62
9000	Minimum labor/equipment charge		3	2.667	Job		118		118	188

09 05 Common Work Results for Finishes

09 05 05 – Selective Demolition for Finishes

09 05 05.10 Selective Demolition, Ceilings

		Crew	Daily Output	Labor-Hours	Unit	Material	Labor	Equipment	Total	Total Incl O&P
0010	**SELECTIVE DEMOLITION, CEILINGS** R024119-10									
0200	Ceiling, gypsum wall board, furred and nailed or screwed	2 Clab	800	.020	S.F.		.84		.84	1.35
0220	On metal frame		760	.021			.89		.89	1.42
0240	On suspension system, including system		720	.022			.94		.94	1.50
1000	Plaster, lime and horse hair, on wood lath, incl. lath		700	.023			.96		.96	1.54
1020	On metal lath		570	.028			1.18		1.18	1.89
1100	Gypsum, on gypsum lath		720	.022			.94		.94	1.50
1120	On metal lath		500	.032			1.35		1.35	2.16
1200	Suspended ceiling, mineral fiber, 2′ x 2′ or 2′ x 4′		1500	.011			.45		.45	.72
1250	On suspension system, incl. system		1200	.013			.56		.56	.90
1500	Tile, wood fiber, 12″ x 12″, glued		900	.018			.75		.75	1.20
1540	Stapled		1500	.011			.45		.45	.72
1580	On suspension system, incl. system		760	.021			.89		.89	1.42
2000	Wood, tongue and groove, 1″ x 4″		1000	.016			.67		.67	1.08
2040	1″ x 8″		1100	.015			.61		.61	.98
2400	Plywood or wood fiberboard, 4′ x 8′ sheets		1200	.013			.56		.56	.90
2500	Remove & refinish textured ceiling	1 Plas	222	.036		.04	1.75		1.79	2.83
9000	Minimum labor/equipment charge	1 Clab	2	4	Job		168		168	270

09 05 05.20 Selective Demolition, Flooring

		Crew	Daily Output	Labor-Hours	Unit	Material	Labor	Equipment	Total	Total Incl O&P
0010	**SELECTIVE DEMOLITION, FLOORING** R024119-10									
0200	Brick with mortar	2 Clab	475	.034	S.F.		1.42		1.42	2.27
0400	Carpet, bonded, including surface scraping		2000	.008			.34		.34	.54
0480	Tackless		9000	.002			.07		.07	.12
0550	Carpet tile, releasable adhesive		5000	.003			.13		.13	.22
0560	Permanent adhesive		1850	.009			.36		.36	.58
0600	Composition, acrylic or epoxy		400	.040			1.68		1.68	2.70
0700	Concrete, scarify skin	A-1A	225	.036			1.95	.92	2.87	4.12
0800	Resilient, sheet goods	2 Clab	1400	.011			.48		.48	.77
0820	For gym floors	″	900	.018			.75		.75	1.20
0850	Vinyl or rubber cove base	1 Clab	1000	.008	L.F.		.34		.34	.54
0860	Vinyl or rubber cove base, molded corner	″	1000	.008	Ea.		.34		.34	.54
0870	For glued and caulked installation, add to labor						50%			
0900	Vinyl composition tile, 12″ x 12″	2 Clab	1000	.016	S.F.		.67		.67	1.08
2000	Tile, ceramic, thin set		675	.024			1		1	1.60
2020	Mud set		625	.026			1.08		1.08	1.73
2200	Marble, slate, thin set		675	.024			1		1	1.60
2220	Mud set		625	.026			1.08		1.08	1.73
2600	Terrazzo, thin set		450	.036			1.50		1.50	2.40
2620	Mud set		425	.038			1.59		1.59	2.54
2640	Terrazzo, cast in place		300	.053			2.25		2.25	3.60

09 05 Common Work Results for Finishes

09 05 05 – Selective Demolition for Finishes

09 05 05.20 Selective Demolition, Flooring

	09 05 05.20 Selective Demolition, Flooring	Crew	Daily Output	Labor-Hours	Unit	2020 Bare Costs Material	Labor	Equipment	Total	Total Incl O&P
3000	Wood, block, on end	1 Carp	400	.020	S.F.		1.06		1.06	1.70
3010	Wood blk floor, includes shot blasting	B-63B	103.28	.310			13.95	3.87	17.82	26.50
3200	Parquet	1 Carp	450	.018			.95		.95	1.51
3400	Strip flooring, interior, 2-1/4" x 25/32" thick		325	.025			1.31		1.31	2.10
3500	Exterior, porch flooring, 1" x 4"		220	.036			1.93		1.93	3.10
3800	Subfloor, tongue and groove, 1" x 6"		325	.025			1.31		1.31	2.10
3820	1" x 8"		430	.019			.99		.99	1.58
3840	1" x 10"		520	.015			.82		.82	1.31
4000	Plywood, nailed		600	.013			.71		.71	1.14
4100	Glued and nailed		400	.020			1.06		1.06	1.70
4200	Hardboard, 1/4" thick		760	.011			.56		.56	.90
8000	Remove flooring, bead blast, simple floor plan	A-1A	1000	.008		.05	.44	.21	.70	.99
8100	Complex floor plan		400	.020		.05	1.10	.52	1.67	2.38
8150	Mastic only		1500	.005		.05	.29	.14	.48	.68
8200	Floor demolition, raised access floor	B-1J	400	.040			1.69		1.69	2.71
9000	Minimum labor/equipment charge	1 Clab	4	2	Job		84		84	135
9050	For grinding concrete floors, see Section 03 35 43.10									

09 05 05.30 Selective Demolition, Walls and Partitions

	09 05 05.30 Selective Demolition, Walls and Partitions	Crew	Daily Output	Labor-Hours	Unit	Material	Labor	Equipment	Total	Total Incl O&P
0010	**SELECTIVE DEMOLITION, WALLS AND PARTITIONS** R024119-10									
0020	Walls, concrete, reinforced	B-39	120	.400	C.F.		17.70	2.98	20.68	31.50
0025	Plain	"	160	.300			13.30	2.23	15.53	23.50
0100	Brick, 4" to 12" thick	B-9	220	.182			7.75	1.62	9.37	14.20
0200	Concrete block, 4" thick		1150	.035	S.F.		1.48	.31	1.79	2.71
0280	8" thick		1050	.038			1.62	.34	1.96	2.96
0300	Exterior stucco 1" thick over mesh		3200	.013			.53	.11	.64	.97
1000	Gypsum wallboard, nailed or screwed	1 Clab	1000	.008			.34		.34	.54
1010	2 layers		400	.020			.84		.84	1.35
1020	Glued and nailed		900	.009			.37		.37	.60
1500	Fiberboard, nailed		900	.009			.37		.37	.60
1520	Glued and nailed		800	.010			.42		.42	.67
1568	Plenum barrier, sheet lead		300	.027			1.12		1.12	1.80
1600	Glass block		65	.123			5.20		5.20	8.30
2000	Movable walls, metal, 5' high		300	.027			1.12		1.12	1.80
2020	8' high		400	.020			.84		.84	1.35
2200	Metal or wood studs, finish 2 sides, fiberboard	B-1	520	.046			1.97		1.97	3.16
2250	Lath and plaster		260	.092			3.95		3.95	6.35
2300	Gypsum wallboard		520	.046			1.97		1.97	3.16
2350	Plywood		450	.053			2.28		2.28	3.65
2800	Paneling, 4' x 8' sheets	1 Clab	475	.017			.71		.71	1.14
3000	Plaster, lime and horsehair, on wood lath		400	.020			.84		.84	1.35
3020	On metal lath		335	.024			1.01		1.01	1.61
3400	Gypsum or perlite, on gypsum lath		410	.020			.82		.82	1.32
3420	On metal lath		300	.027			1.12		1.12	1.80
3450	Plaster, interior gypsum, acoustic, or cement		60	.133	S.Y.		5.60		5.60	9
3500	Stucco, on masonry		145	.055			2.32		2.32	3.72
3510	Commercial 3-coat		80	.100			4.21		4.21	6.75
3520	Interior stucco		25	.320			13.45		13.45	21.50
3750	Terra cotta block and plaster, to 6" thick	B-1	175	.137	S.F.		5.85		5.85	9.40
3753	Remove damaged glass block	"	134.62	.178			7.65		7.65	12.20
3760	Tile, ceramic, on walls, thin set	1 Clab	300	.027			1.12		1.12	1.80
3765	Mud set		250	.032			1.35		1.35	2.16
3800	Toilet partitions, slate or marble		5	1.600	Ea.		67.50		67.50	108

09 05 Common Work Results for Finishes

09 05 05 – Selective Demolition for Finishes

09 05 05.30 Selective Demolition, Walls and Partitions		Crew	Daily Output	Labor-Hours	Unit	Material	2020 Bare Costs Labor	Equipment	Total	Total Incl O&P
3820	Metal or plastic	1 Clab	8	1	Ea.		42		42	67.50
3950	Demolish coreboard partitions	B-1J	1548.64	.010	S.F.		.44		.44	.70
3955	w/lift included	B-68D	900	.027			1.22	.31	1.53	2.28
3960	Demolish metal wall panels	B-1J	774.32	.021			.88		.88	1.40
3965	w/lift included	B-1K	400	.040			2.14		2.14	3.42
3970	Demolish weld curtain with frame		600	.027			1.42		1.42	2.28
3975	Demolish wire mesh panels		400	.040			2.14		2.14	3.42
3980	Demolish gyp & mtl stud wall, forktruck included	B-68D	520	.046			2.12	.54	2.66	3.95
3990	Demolish dust curtain, lift included (curtain furn by ASI)	B-1K	2000	.008			.43		.43	.68
3995	Demolish dust curtain, lift included	"	2000	.008			.43		.43	.68
5000	Wallcovering, vinyl	1 Pape	700	.011			.51		.51	.81
5010	With release agent		1500	.005			.24		.24	.38
5025	Wallpaper, 2 layers or less, by hand		250	.032			1.43		1.43	2.26
5035	3 layers or more		165	.048			2.16		2.16	3.43
9000	Minimum labor/equipment charge	1 Clab	4	2	Job		84		84	135

09 05 71 – Acoustic Underlayment

09 05 71.10 Acoustical Underlayment

		Crew	Daily Output	Labor-Hours	Unit	Material	Labor	Equipment	Total	Total Incl O&P
0010	**ACOUSTICAL UNDERLAYMENT**									
0100	Rubber underlayment, 5/64"	1 Tilf	275	.029	S.F.	.32	1.44		1.76	2.61
0110	1/8" thk.		275	.029		.52	1.44		1.96	2.83
0120	13/64" thk.		270	.030		.71	1.47		2.18	3.08
0130	15/64" thk.		270	.030		.94	1.47		2.41	3.33
0140	23/64" thk.		265	.030		1.29	1.49		2.78	3.76
0150	15/32" thk.		265	.030		1.50	1.49		2.99	3.99
0400	Cork underlayment, 3/32"		275	.029		.40	1.44		1.84	2.70
0410	1/8" thk.		275	.029		.41	1.44		1.85	2.71
0420	5/32" thk.		275	.029		.48	1.44		1.92	2.79
0430	15/64" thk.		270	.030		.62	1.47		2.09	2.98
0440	15/32" thk.		265	.030		1.30	1.49		2.79	3.77
0600	Rubber cork underlayment, 3/64"		275	.029		.31	1.44		1.75	2.60
0610	13/64" thk.		270	.030		1.60	1.47		3.07	4.06
0620	3/8" thk.		270	.030		2.70	1.47		4.17	5.25
0630	25/64" thk.		270	.030		2.79	1.47		4.26	5.35
0800	Foam underlayment, 5/64"		275	.029		.55	1.44		1.99	2.87
0810	1/8" thk.		275	.029		.67	1.44		2.11	3
0820	15/32" thk.		265	.030		.54	1.49		2.03	2.93
4000	Nylon matting 0.4" thick, with carbon black spinerette									
4010	plus polyester fabric, on floor	D-7	1600	.010	S.F.	1.44	.44		1.88	2.27

09 21 Plaster and Gypsum Board Assemblies

09 21 13 – Plaster Assemblies

09 21 13.10 Plaster Partition Wall

		Crew	Daily Output	Labor-Hours	Unit	Material	Labor	Equipment	Total	Total Incl O&P
0010	**PLASTER PARTITION WALL**									
0400	Stud walls, 3.4 lb. metal lath, 3 coat gypsum plaster, 2 sides									
0600	2" x 4" wood studs, 16" OC	J-2	315	.152	S.F.	3.01	7.20	.37	10.58	15.10
0700	2-1/2" metal studs, 25 ga., 12" OC		325	.148		2.71	6.95	.36	10.02	14.45
0800	3-5/8" metal studs, 25 ga., 16" OC		320	.150		2.68	7.05	.37	10.10	14.55
0900	Gypsum lath, 2 coat vermiculite plaster, 2 sides									
1000	2" x 4" wood studs, 16" OC	J-2	355	.135	S.F.	2.63	6.35	.33	9.31	13.35
1200	2-1/2" metal studs, 25 ga., 12" OC		365	.132		2.54	6.20	.32	9.06	13

09 21 Plaster and Gypsum Board Assemblies

09 21 13 – Plaster Assemblies

09 21 13.10 Plaster Partition Wall		Crew	Daily Output	Labor-Hours	Unit	Material	2020 Bare Costs Labor	Equipment	Total	Total Incl O&P
1300	3-5/8" metal studs, 25 ga., 16" OC	J-2	360	.133	S.F.	2.51	6.30	.33	9.14	13.05

09 21 16 – Gypsum Board Assemblies

09 21 16.23 Gypsum Board Shaft Wall Assemblies

Line	Description	Crew	Daily Output	Labor-Hours	Unit	Material	Labor	Equipment	Total	Total Incl O&P
0010	**GYPSUM BOARD SHAFT WALL ASSEMBLIES**									
0020	Cavity type on 25 ga. J-track & C-H studs, 24" OC									
0030	1" thick coreboard wall liner on shaft side									
0040	2-hour assembly with double layer									
0060	5/8" f.r. gyp bd on rm side, 2-1/2" J-track & C-H studs	2 Carp	220	.073	S.F.	2.99	3.87		6.86	9.50
0065	4" J-track & C-H studs		220	.073		3.15	3.87		7.02	9.65
0070	6" J-track & C-H studs		220	.073		3.36	3.87		7.23	9.90
0100	3-hour assembly with triple layer									
0300	5/8" f.r. gyp bd on rm side, 2-1/2" J-track & C-H studs	2 Carp	180	.089	S.F.	2.66	4.72		7.38	10.45
0305	4" J-track & C-H studs		180	.089		2.82	4.72		7.54	10.65
0310	6" J-track & C-H studs		180	.089		3.03	4.72		7.75	10.90
0400	4-hour assembly, 1" coreboard, 5/8" fire rated gypsum board									
0600	and 3/4" galv. metal furring channels, 24" OC, with									
0700	Dbl. layer 5/8" f.r. gyp bd on rm side, 2-1/2" trk. & C-H studs	2 Carp	110	.145	S.F.	3.36	7.75		11.11	16.10
0705	4" J-track & C-H studs		110	.145		3.15	7.75		10.90	15.85
0710	6" J-track & C-H studs		110	.145		3.36	7.75		11.11	16.10
0900	For taping & finishing, add per side	1 Carp	1050	.008		.05	.41		.46	.71
1000	For insulation, see Section 07 21									
5200	For work over 8' high, add	2 Carp	3060	.005	S.F.		.28		.28	.45
5300	For distribution cost 3 stories and above, add per story	"	6100	.003	"		.14		.14	.22

09 21 16.33 Partition Wall

Line	Description	Crew	Daily Output	Labor-Hours	Unit	Material	Labor	Equipment	Total	Total Incl O&P
0010	**PARTITION WALL** Stud wall, 8' to 12' high									
0050	1/2", interior, gypsum board, std, tape & finish 2 sides									
0500	Installed on and incl., 2" x 4" wood studs, 16" OC	2 Carp	310	.052	S.F.	1.29	2.74		4.03	5.80
1000	Metal studs, NLB, 25 ga., 16" OC, 3-5/8" wide		350	.046		1.17	2.43		3.60	5.20
1200	6" wide		330	.048		1.30	2.58		3.88	5.55
1400	Water resistant, on 2" x 4" wood studs, 16" OC		310	.052		1.45	2.74		4.19	6
1600	Metal studs, NLB, 25 ga., 16" OC, 3-5/8" wide		350	.046		1.33	2.43		3.76	5.35
1800	6" wide		330	.048		1.46	2.58		4.04	5.75
2000	Fire res., 2 layers, 1-1/2 hr., on 2" x 4" wood studs, 16" OC		210	.076		2.15	4.05		6.20	8.85
2200	Metal studs, NLB, 25 ga., 16" OC, 3-5/8" wide		250	.064		2.03	3.40		5.43	7.70
2400	6" wide		230	.070		2.16	3.70		5.86	8.30
2600	Fire & water res., 2 layers, 1-1/2 hr., 2" x 4" studs, 16" OC		210	.076		2.15	4.05		6.20	8.85
2800	Metal studs, NLB, 25 ga., 16" OC, 3-5/8" wide		250	.064		2.03	3.40		5.43	7.70
3000	6" wide		230	.070		2.16	3.70		5.86	8.30
3200	5/8", interior, gypsum board, standard, tape & finish 2 sides									
3400	Installed on and including 2" x 4" wood studs, 16" OC	2 Carp	300	.053	S.F.	1.29	2.83		4.12	5.95
3600	24" OC		330	.048		1.17	2.58		3.75	5.40
3800	Metal studs, NLB, 25 ga., 16" OC, 3-5/8" wide		340	.047		1.17	2.50		3.67	5.30
4000	6" wide		320	.050		1.30	2.66		3.96	5.70
4200	24" OC, 3-5/8" wide		360	.044		1.06	2.36		3.42	4.95
4400	6" wide		340	.047		1.16	2.50		3.66	5.30
4800	Water resistant, on 2" x 4" wood studs, 16" OC		300	.053		1.51	2.83		4.34	6.20
5000	24" OC		330	.048		1.39	2.58		3.97	5.65
5200	Metal studs, NLB, 25 ga. 16" OC, 3-5/8" wide		340	.047		1.39	2.50		3.89	5.55
5400	6" wide		320	.050		1.52	2.66		4.18	5.95
5600	24" OC, 3-5/8" wide		360	.044		1.28	2.36		3.64	5.20
5800	6" wide		340	.047		1.38	2.50		3.88	5.55
6000	Fire resistant, 2 layers, 2 hr., on 2" x 4" wood studs, 16" OC		205	.078		2.15	4.15		6.30	9

09 21 Plaster and Gypsum Board Assemblies

09 21 16 – Gypsum Board Assemblies

09 21 16.33 Partition Wall		Crew	Daily Output	Labor-Hours	Unit	Material	2020 Bare Costs Labor	Equipment	Total	Total Incl O&P
6200	24" OC	2 Carp	235	.068	S.F.	2.03	3.62		5.65	8.05
6400	Metal studs, NLB, 25 ga., 16" OC, 3-5/8" wide		245	.065		2.06	3.47		5.53	7.80
6600	6" wide		225	.071		2.16	3.78		5.94	8.45
6800	24" OC, 3-5/8" wide		265	.060		1.92	3.21		5.13	7.25
7000	6" wide		245	.065		2.02	3.47		5.49	7.75
7200	Fire & water resistant, 2 layers, 2 hr., 2" x 4" studs, 16" OC		205	.078		2.15	4.15		6.30	9
7400	24" OC		235	.068		2.03	3.62		5.65	8.05
7600	Metal studs, NLB, 25 ga., 16" OC, 3-5/8" wide		245	.065		2.03	3.47		5.50	7.80
7800	6" wide		225	.071		2.16	3.78		5.94	8.45
8000	24" OC, 3-5/8" wide		265	.060		1.92	3.21		5.13	7.25
8200	6" wide	↓	245	.065	↓	2.02	3.47		5.49	7.75
8600	1/2" blueboard, mesh tape both sides									
8620	Installed on and including 2" x 4" wood studs, 16" OC	2 Carp	300	.053	S.F.	1.45	2.83		4.28	6.15
8640	Metal studs, NLB, 25 ga., 16" OC, 3-5/8" wide		340	.047		1.33	2.50		3.83	5.50
8660	6" wide	↓	320	.050	↓	1.46	2.66		4.12	5.85
8800	Hospital security partition, 5/8" fiber reinf. high abuse gyp. bd.									
8810	Mtl. studs, NLB, 20 ga., 16" OC, 3-5/8" wide, w/sec. mesh, gyp. bd.	2 Carp	208	.077	S.F.	4.67	4.09		8.76	11.70
9000	Exterior, 1/2" gypsum sheathing, 1/2" gypsum finished, interior,									
9100	including foil faced insulation, metal studs, 20 ga.									
9200	16" OC, 3-5/8" wide	2 Carp	290	.055	S.F.	1.90	2.93		4.83	6.80
9400	6" wide		270	.059		2.15	3.15		5.30	7.40
9600	Partitions, for work over 8' high, add	↓	1530	.010	↓		.56		.56	.89

09 22 Supports for Plaster and Gypsum Board

09 22 03 – Fastening Methods for Finishes

09 22 03.20 Drilling Plaster/Drywall

		Crew	Daily Output	Labor-Hours	Unit	Material	Labor	Equipment	Total	Total Incl O&P
0010	**DRILLING PLASTER/DRYWALL**									
1100	Drilling & layout for drywall/plaster walls, up to 1" deep, no anchor									
1200	Holes, 1/4" diameter	1 Carp	150	.053	Ea.	.01	2.83		2.84	4.55
1300	3/8" diameter		140	.057		.01	3.04		3.05	4.88
1400	1/2" diameter		130	.062		.01	3.27		3.28	5.25
1500	3/4" diameter		120	.067		.01	3.54		3.55	5.70
1600	1" diameter		110	.073		.02	3.87		3.89	6.20
1700	1-1/4" diameter		100	.080		.04	4.25		4.29	6.85
1800	1-1/2" diameter	↓	90	.089		.05	4.72		4.77	7.60
1900	For ceiling installations, add				↓		40%			

09 22 13 – Metal Furring

09 22 13.13 Metal Channel Furring

		Crew	Daily Output	Labor-Hours	Unit	Material	Labor	Equipment	Total	Total Incl O&P
0010	**METAL CHANNEL FURRING**									
0030	Beams and columns, 7/8" hat channels, galvanized, 12" OC	1 Lath	155	.052	S.F.	.46	2.69		3.15	4.71
0050	16" OC		170	.047		.37	2.45		2.82	4.25
0070	24" OC		185	.043		.25	2.25		2.50	3.79
0100	Ceilings, on steel, 7/8" hat channels, galvanized, 12" OC		210	.038		.42	1.98		2.40	3.57
0300	16" OC		290	.028		.37	1.44		1.81	2.66
0400	24" OC		420	.019		.25	.99		1.24	1.82
0600	1-5/8" hat channels, galvanized, 12" OC		190	.042		.56	2.19		2.75	4.05
0700	16" OC		260	.031		.50	1.60		2.10	3.06
0900	24" OC		390	.021		.34	1.07		1.41	2.04
0930	7/8" hat channels with sound isolation clips, 12" OC		120	.067		1.73	3.47		5.20	7.35
0940	16" OC	↓	100	.080	↓	1.30	4.16		5.46	7.95

09 22 Supports for Plaster and Gypsum Board

09 22 13 – Metal Furring

09 22 13.13 Metal Channel Furring		Crew	Daily Output	Labor-Hours	Unit	Material	2020 Bare Costs Labor	Equipment	Total	Total Incl O&P
0950	24" OC	1 Lath	165	.048	S.F.	.87	2.52		3.39	4.90
0960	1-5/8" hat channels, galvanized, 12" OC		110	.073		1.88	3.79		5.67	8
0970	16" OC		100	.080		1.41	4.16		5.57	8.05
0980	24" OC		155	.052		.94	2.69		3.63	5.25
1000	Walls, 7/8" hat channels, galvanized, 12" OC		235	.034		.42	1.77		2.19	3.23
1200	16" OC		265	.030		.37	1.57		1.94	2.87
1300	24" OC		350	.023		.25	1.19		1.44	2.13
1500	1-5/8" hat channels, galvanized, 12" OC		210	.038		.56	1.98		2.54	3.73
1600	16" OC		240	.033		.50	1.73		2.23	3.27
1800	24" OC		305	.026		.34	1.37		1.71	2.51
1920	7/8" hat channels with sound isolation clips, 12" OC		125	.064		1.73	3.33		5.06	7.10
1940	16" OC		100	.080		1.30	4.16		5.46	7.95
1950	24" OC		150	.053		.87	2.78		3.65	5.30
1960	1-5/8" hat channels, galvanized, 12" OC		115	.070		1.88	3.62		5.50	7.70
1970	16" OC		95	.084		1.41	4.38		5.79	8.40
1980	24" OC		140	.057		.94	2.97		3.91	5.70
3000	Z Furring, walls, 1" deep, 25 ga., 24" OC		350	.023		1.37	1.19		2.56	3.36
3010	48" OC		700	.011		.68	.59		1.27	1.68
3020	1-1/2" deep, 24" OC		345	.023		1.59	1.21		2.80	3.64
3030	48" OC		695	.012		.80	.60		1.40	1.81
3040	2" deep, 24" OC		340	.024		1.91	1.22		3.13	4.02
3050	48" OC		690	.012		.96	.60		1.56	1.99
3060	1" deep, 20 ga., 24" OC		350	.023		2.26	1.19		3.45	4.34
3070	48" OC		700	.011		1.13	.59		1.72	2.17
3080	1-1/2" deep, 24" OC		345	.023		2.60	1.21		3.81	4.74
3090	48" OC		695	.012		1.30	.60		1.90	2.37
4000	2" deep, 24" OC		340	.024		3.19	1.22		4.41	5.40
4010	48" OC		690	.012		1.59	.60		2.19	2.69
9000	Minimum labor/equipment charge		4	2	Job		104		104	163

09 22 16 – Non-Structural Metal Framing

09 22 16.13 Non-Structural Metal Stud Framing		Crew	Daily Output	Labor-Hours	Unit	Material	Labor	Equipment	Total	Total Incl O&P
0010	**NON-STRUCTURAL METAL STUD FRAMING**									
1600	Non-load bearing, galv., 8' high, 25 ga. 1-5/8" wide, 16" OC	1 Carp	619	.013	S.F.	.28	.69		.97	1.41
1610	24" OC		950	.008		.21	.45		.66	.95
1620	2-1/2" wide, 16" OC		613	.013		.36	.69		1.05	1.51
1630	24" OC		938	.009		.27	.45		.72	1.03
1640	3-5/8" wide, 16" OC		600	.013		.43	.71		1.14	1.61
1650	24" OC		925	.009		.32	.46		.78	1.09
1660	4" wide, 16" OC		594	.013		.48	.72		1.20	1.68
1670	24" OC		925	.009		.36	.46		.82	1.14
1680	6" wide, 16" OC		588	.014		.57	.72		1.29	1.79
1690	24" OC		906	.009		.43	.47		.90	1.22
1700	20 ga. studs, 1-5/8" wide, 16" OC		494	.016		.38	.86		1.24	1.80
1710	24" OC		763	.010		.28	.56		.84	1.20
1720	2-1/2" wide, 16" OC		488	.016		.48	.87		1.35	1.92
1730	24" OC		750	.011		.36	.57		.93	1.31
1740	3-5/8" wide, 16" OC		481	.017		.55	.88		1.43	2.02
1750	24" OC		738	.011		.41	.58		.99	1.37
1760	4" wide, 16" OC		475	.017		.67	.90		1.57	2.17
1770	24" OC		738	.011		.50	.58		1.08	1.47
1780	6" wide, 16" OC		469	.017		.81	.91		1.72	2.34
1790	24" OC		725	.011		.61	.59		1.20	1.61

09 22 16.13 Non-Structural Metal Stud Framing		Crew	Daily Output	Labor-Hours	Unit	Material	2020 Bare Costs Labor	Equipment	Total	Total Incl O&P
2000	Non-load bearing, galv., 10′ high, 25 ga. 1-5/8″ wide, 16″ OC	1 Carp	495	.016	S.F.	.26	.86		1.12	1.67
2100	24″ OC		760	.011		.19	.56		.75	1.11
2200	2-1/2″ wide, 16″ OC		490	.016		.35	.87		1.22	1.77
2250	24″ OC		750	.011		.25	.57		.82	1.19
2300	3-5/8″ wide, 16″ OC		480	.017		.41	.89		1.30	1.87
2350	24″ OC		740	.011		.30	.57		.87	1.25
2400	4″ wide, 16″ OC		475	.017		.46	.90		1.36	1.93
2450	24″ OC		740	.011		.34	.57		.91	1.29
2500	6″ wide, 16″ OC		470	.017		.54	.90		1.44	2.05
2550	24″ OC		725	.011		.40	.59		.99	1.38
2600	20 ga. studs, 1-5/8″ wide, 16″ OC		395	.020		.36	1.08		1.44	2.11
2650	24″ OC		610	.013		.26	.70		.96	1.41
2700	2-1/2″ wide, 16″ OC		390	.021		.45	1.09		1.54	2.25
2750	24″ OC		600	.013		.33	.71		1.04	1.50
2800	3-5/8″ wide, 16″ OC		385	.021		.52	1.10		1.62	2.34
2850	24″ OC		590	.014		.38	.72		1.10	1.57
2900	4″ wide, 16″ OC		380	.021		.63	1.12		1.75	2.49
2950	24″ OC		590	.014		.47	.72		1.19	1.66
3000	6″ wide, 16″ OC		375	.021		.77	1.13		1.90	2.66
3050	24″ OC		580	.014		.56	.73		1.29	1.79
3060	Non-load bearing, galv., 12′ high, 25 ga. 1-5/8″ wide, 16″ OC		413	.019		.25	1.03		1.28	1.92
3070	24″ OC		633	.013		.18	.67		.85	1.28
3080	2-1/2″ wide, 16″ OC		408	.020		.33	1.04		1.37	2.03
3090	24″ OC		625	.013		.24	.68		.92	1.35
3100	3-5/8″ wide, 16″ OC		400	.020		.39	1.06		1.45	2.13
3110	24″ OC		617	.013		.28	.69		.97	1.41
3120	4″ wide, 16″ OC		396	.020		.44	1.07		1.51	2.20
3130	24″ OC		617	.013		.32	.69		1.01	1.45
3140	6″ wide, 16″ OC		392	.020		.52	1.08		1.60	2.31
3150	24″ OC		604	.013		.38	.70		1.08	1.54
3160	20 ga. studs, 1-5/8″ wide, 16″ OC		329	.024		.34	1.29		1.63	2.45
3170	24″ OC		508	.016		.25	.84		1.09	1.61
3180	2-1/2″ wide, 16″ OC		325	.025		.43	1.31		1.74	2.57
3190	24″ OC		500	.016		.31	.85		1.16	1.71
3200	3-5/8″ wide, 16″ OC		321	.025		.50	1.32		1.82	2.67
3210	24″ OC		492	.016		.36	.86		1.22	1.78
3220	4″ wide, 16″ OC		317	.025		.61	1.34		1.95	2.82
3230	24″ OC		492	.016		.44	.86		1.30	1.87
3240	6″ wide, 16″ OC		313	.026		.73	1.36		2.09	2.99
3250	24″ OC		483	.017		.53	.88		1.41	2
3260	Non-load bearing, galv., 16′ high, 25 ga. 4″ wide, 12″ OC		195	.041		.56	2.18		2.74	4.11
3270	16″ OC		275	.029		.44	1.55		1.99	2.97
3280	24″ OC		400	.020		.32	1.06		1.38	2.05
3290	6″ wide, 12″ OC		190	.042		.67	2.24		2.91	4.32
3300	16″ OC		280	.029		.52	1.52		2.04	3.01
3310	24″ OC		400	.020		.38	1.06		1.44	2.12
3320	20 ga. studs, 2-1/2″ wide, 12″ OC		180	.044		.55	2.36		2.91	4.39
3330	16″ OC		254	.032		.43	1.67		2.10	3.16
3340	24″ OC		390	.021		.32	1.09		1.41	2.10
3350	3-5/8″ wide, 12″ OC		170	.047		.65	2.50		3.15	4.72
3360	16″ OC		251	.032		.51	1.69		2.20	3.27
3370	24″ OC		384	.021		.37	1.11		1.48	2.18
3380	4″ wide, 12″ OC		170	.047		.78	2.50		3.28	4.86

09 22 Supports for Plaster and Gypsum Board

09 22 16 – Non-Structural Metal Framing

09 22 16.13 Non-Structural Metal Stud Framing		Crew	Daily Output	Labor-Hours	Unit	Material	2020 Bare Costs Labor	Equipment	Total	Total Incl O&P
3390	16" OC	1 Carp	247	.032	S.F.	.61	1.72		2.33	3.43
3400	24" OC		384	.021		.45	1.11		1.56	2.26
3410	6" wide, 12" OC		175	.046		.94	2.43		3.37	4.92
3420	16" OC		245	.033		.74	1.74		2.48	3.59
3430	24" OC		400	.020		.54	1.06		1.60	2.29
3440	Non-load bearing, galv., 20' high, 25 ga. 6" wide, 12" OC		125	.064		.65	3.40		4.05	6.15
3450	16" OC		220	.036		.51	1.93		2.44	3.66
3460	24" OC		360	.022		.37	1.18		1.55	2.30
3470	20 ga. studs, 4" wide, 12" OC		120	.067		.76	3.54		4.30	6.55
3480	16" OC		215	.037		.61	1.98		2.59	3.84
3490	6" wide, 12" OC		115	.070		.92	3.70		4.62	6.90
3500	16" OC		215	.037		.72	1.98		2.70	3.96
3510	24" OC		331	.024		.52	1.28		1.80	2.63
5000	For load bearing studs, see Section 05 41 13.30									
9000	Minimum labor/equipment charge	1 Carp	4	2	Job		106		106	170

09 22 26 – Suspension Systems

09 22 26.13 Ceiling Suspension Systems		Crew	Daily Output	Labor-Hours	Unit	Material	Labor	Equipment	Total	Total Incl O&P
0010	**CEILING SUSPENSION SYSTEMS** for gypsum board or plaster									
8000	Suspended ceilings, including carriers									
8200	1-1/2" carriers, 24" OC with:									
8300	7/8" channels, 16" OC	1 Lath	275	.029	S.F.	.59	1.51		2.10	3.02
8320	24" OC		310	.026		.47	1.34		1.81	2.62
8400	1-5/8" channels, 16" OC		205	.039		.72	2.03		2.75	3.98
8420	24" OC		250	.032		.56	1.67		2.23	3.22
8600	2" carriers, 24" OC with:									
8700	7/8" channels, 16" OC	1 Lath	250	.032	S.F.	.69	1.67		2.36	3.37
8720	24" OC		285	.028		.57	1.46		2.03	2.92
8800	1-5/8" channels, 16" OC		190	.042		.82	2.19		3.01	4.34
8820	24" OC		225	.036		.66	1.85		2.51	3.62

09 22 36 – Lath

09 22 36.13 Gypsum Lath		Crew	Daily Output	Labor-Hours	Unit	Material	Labor	Equipment	Total	Total Incl O&P
0010	**GYPSUM LATH**									
0020	Plain or perforated, nailed, 3/8" thick	1 Lath	85	.094	S.Y.	3.51	4.90		8.41	11.50
0100	1/2" thick		80	.100		2.70	5.20		7.90	11.10
0300	Clipped to steel studs, 3/8" thick		75	.107		3.51	5.55		9.06	12.55
0400	1/2" thick		70	.114		2.70	5.95		8.65	12.25
1500	For ceiling installations, add		216	.037			1.93		1.93	3.02
1600	For columns and beams, add		170	.047			2.45		2.45	3.84
9000	Minimum labor/equipment charge		4.25	1.882	Job		98		98	153

09 22 36.23 Metal Lath		Crew	Daily Output	Labor-Hours	Unit	Material	Labor	Equipment	Total	Total Incl O&P
0010	**METAL LATH**									
0020	Diamond, expanded, 2.5 lb./S.Y., painted				S.Y.	4.04			4.04	4.44
0100	Galvanized					3.80			3.80	4.18
0300	3.4 lb./S.Y., painted					4.23			4.23	4.65
0400	Galvanized					4.72			4.72	5.20
0600	For #15 asphalt sheathing paper, add					.47			.47	.52
0900	Flat rib, 1/8" high, 2.75 lb., painted					3.51			3.51	3.86
1000	Foil backed					3.69			3.69	4.06
1200	3.4 lb./S.Y., painted					4.29			4.29	4.72
1300	Galvanized					4.60			4.60	5.05
1500	For #15 asphalt sheathing paper, add					.47			.47	.52

09 22 36 – Lath

09 22 36.23 Metal Lath		Crew	Daily Output	Labor-Hours	Unit	Material	2020 Bare Costs Labor	Equipment	Total	Total Incl O&P
1800	High rib, 3/8" high, 3.4 lb./S.Y., painted				S.Y.	4.83			4.83	5.30
1900	Galvanized				↓	4.01			4.01	4.41
2400	3/4" high, painted, .60 lb./S.F.				S.F.	.77			.77	.85
2500	.75 lb./S.F.				"	1.61			1.61	1.77
2800	Stucco mesh, painted, 3.6 lb.				S.Y.	4.17			4.17	4.59
3000	K-lath, perforated, absorbent paper, regular					4.36			4.36	4.80
3100	Heavy duty					5.15			5.15	5.65
3300	Waterproof, heavy duty, grade B backing					5.05			5.05	5.55
3400	Fire resistant backing					5.55			5.55	6.15
3600	2.5 lb. diamond painted, on wood framing, on walls	1 Lath	85	.094		4.04	4.90		8.94	12.10
3700	On ceilings		75	.107		4.04	5.55		9.59	13.15
3900	3.4 lb. diamond painted, on wood framing, on walls		80	.100		4.29	5.20		9.49	12.85
4000	On ceilings		70	.114		4.29	5.95		10.24	14
4200	3.4 lb. diamond painted, wired to steel framing		75	.107		4.29	5.55		9.84	13.40
4300	On ceilings		60	.133		4.29	6.95		11.24	15.55
4500	Columns and beams, wired to steel		40	.200		4.29	10.40		14.69	21
4600	Cornices, wired to steel		35	.229		4.29	11.90		16.19	23.50
4800	Screwed to steel studs, 2.5 lb.		80	.100		4.04	5.20		9.24	12.60
4900	3.4 lb.		75	.107		4.23	5.55		9.78	13.35
5100	Rib lath, painted, wired to steel, on walls, 2.5 lb.		75	.107		3.51	5.55		9.06	12.55
5200	3.4 lb.		70	.114		4.83	5.95		10.78	14.60
5400	4.0 lb.	↓	65	.123		5.70	6.40		12.10	16.30
5500	For self-furring lath, add					.12			.12	.13
5700	Suspended ceiling system, incl. 3.4 lb. diamond lath, painted	1 Lath	15	.533		4.30	28		32.30	48
5800	Galvanized	"	15	.533	↓	4.64	28		32.64	48.50
6000	Hollow metal stud partitions, 3.4 lb. painted lath both sides									
6010	Non-load bearing, 25 ga., w/rib lath, 2-1/2" studs, 12" OC	1 Lath	20.30	.394	S.Y.	13.60	20.50		34.10	47
6300	16" OC		21.10	.379		12.75	19.75		32.50	45
6350	24" OC		22.70	.352		11.95	18.35		30.30	41.50
6400	3-5/8" studs, 16" OC		19.50	.410		13.35	21.50		34.85	48
6600	24" OC		20.40	.392		12.35	20.50		32.85	45.50
6700	4" studs, 16" OC		20.40	.392		13.80	20.50		34.30	47
6900	24" OC		21.60	.370		12.70	19.30		32	44
7000	6" studs, 16" OC		19.50	.410		14.55	21.50		36.05	49.50
7100	24" OC		21.10	.379		13.25	19.75		33	45.50
7200	L.B. partitions, 16 ga., w/rib lath, 2-1/2" studs, 16" OC		20	.400		13.05	21		34.05	47
7300	3-5/8" studs, 16 ga.		19.70	.406		14.70	21		35.70	49
7500	4" studs, 16 ga.		19.50	.410		15.50	21.50		37	50.50
7600	6" studs, 16 ga.		18.70	.428	↓	18.10	22.50		40.60	55
9000	Minimum labor/equipment charge	↓	4.25	1.882	Job		98		98	153

09 22 36.43 Security Mesh		Crew	Daily Output	Labor-Hours	Unit	Material	Labor	Equipment	Total	Total Incl O&P
0010	**SECURITY MESH**, expanded metal, flat, screwed to framing									
0100	On walls, 3/4", 1.76 lb./S.F.	2 Carp	1500	.011	S.F.	2.31	.57		2.88	3.45
0110	1-1/2", 1.14 lb./S.F.		1600	.010		1.74	.53		2.27	2.76
0200	On ceilings, 3/4", 1.76 lb./S.F.		1350	.012		2.31	.63		2.94	3.55
0210	1-1/2", 1.14 lb./S.F.	↓	1450	.011	↓	1.74	.59		2.33	2.85

09 22 36.83 Accessories, Plaster		Crew	Daily Output	Labor-Hours	Unit	Material	Labor	Equipment	Total	Total Incl O&P
0010	**ACCESSORIES, PLASTER**									
0020	Casing bead, expanded flange, galvanized	1 Lath	2.70	2.963	C.L.F.	52.50	154		206.50	299
0200	Foundation weep screed, galvanized	"	2.70	2.963		53	154		207	300
0900	Channels, cold rolled, 16 ga., 3/4" deep, galvanized					41.50			41.50	46
1200	1-1/2" deep, 16 ga., galvanized				↓	56			56	61.50

09 22 Supports for Plaster and Gypsum Board

09 22 36 – Lath

09 22 36.83 Accessories, Plaster		Crew	Daily Output	Labor-Hours	Unit	Material	2020 Bare Costs Labor	Equipment	Total	Total Incl O&P
1620	Corner bead, expanded bullnose, 3/4" radius, #10, galvanized	1 Lath	2.60	3.077	C.L.F.	26	160		186	280
1650	#1, galvanized		2.55	3.137		47.50	163		210.50	310
1670	Expanded wing, 2-3/4" wide, #1, galvanized		2.65	3.019		41	157		198	291
1700	Inside corner (corner rite), 3" x 3", painted		2.60	3.077		21	160		181	275
1750	Strip-ex, 4" wide, painted		2.55	3.137		27.50	163		190.50	287
1800	Expansion joint, 3/4" grounds, limited expansion, galv., 1 piece		2.70	2.963		71	154		225	320
2100	Extreme expansion, galvanized, 2 piece	↓	2.60	3.077	↓	137	160		297	400

09 23 Gypsum Plastering

09 23 13 – Acoustical Gypsum Plastering

09 23 13.10 Perlite or Vermiculite Plaster

		Crew	Daily Output	Labor-Hours	Unit	Material	Labor	Equipment	Total	Total Incl O&P
0010	**PERLITE OR VERMICULITE PLASTER**									
0020	In 100 lb. bags, under 200 bags				Bag	19.10			19.10	21
0100	Over 200 bags				"	17.60			17.60	19.35
0300	2 coats, no lath included, on walls	J-1	92	.435	S.Y.	5.95	20	1.27	27.22	40
0400	On ceilings	"	79	.506		5.95	23.50	1.48	30.93	45
0600	On and incl. 3/8" gypsum lath, on metal studs	J-2	84	.571		9.95	27	1.39	38.34	55
0700	On ceilings	"	70	.686		9.95	32.50	1.67	44.12	64.50
0900	3 coats, no lath included, on walls	J-1	74	.541		6.65	25	1.58	33.23	48.50
1000	On ceilings	"	63	.635		6.65	29.50	1.86	38.01	56
1200	On and incl. painted metal lath, on metal studs	J-2	72	.667		11.50	31.50	1.63	44.63	64.50
1300	On ceilings		61	.787		11.50	37	1.92	50.42	74
1500	On and incl. suspended metal lath ceiling	↓	37	1.297		10.95	61	3.17	75.12	113
1700	For irregular or curved surfaces, add to above						30%			
1800	For columns and beams, add to above						50%			
1900	For soffits, add to ceiling prices				↓		40%			
9000	Minimum labor/equipment charge	1 Plas	1	8	Job		390		390	620

09 23 20 – Gypsum Plaster

09 23 20.10 Gypsum Plaster On Walls and Ceilings

		Crew	Daily Output	Labor-Hours	Unit	Material	Labor	Equipment	Total	Total Incl O&P
0010	**GYPSUM PLASTER ON WALLS AND CEILINGS**									
0020	80# bag, less than 1 ton				Bag	15.90			15.90	17.50
0100	Over 1 ton				"	13.70			13.70	15.05
0300	2 coats, no lath included, on walls	J-1	105	.381	S.Y.	3.90	17.55	1.12	22.57	33.50
0400	On ceilings	"	92	.435		3.90	20	1.27	25.17	37.50
0600	On and incl. 3/8" gypsum lath on steel, on walls	J-2	97	.495		7.40	23.50	1.21	32.11	46.50
0700	On ceilings	"	83	.578		7.40	27	1.41	35.81	53
0900	3 coats, no lath included, on walls	J-1	87	.460		5.55	21	1.35	27.90	41.50
1000	On ceilings	"	78	.513		5.55	23.50	1.50	30.55	45.50
1200	On and including painted metal lath, on wood studs	J-2	86	.558		10.70	26.50	1.36	38.56	55
1300	On ceilings	"	76.50	.627	↓	10.70	29.50	1.53	41.73	60.50
1600	For irregular or curved surfaces, add						30%			
1800	For columns & beams, add						50%			
9000	Minimum labor/equipment charge	1 Plas	1	8	Job		390		390	620

09 23 20.20 Gauging Plaster

		Crew	Daily Output	Labor-Hours	Unit	Material	Labor	Equipment	Total	Total Incl O&P
0010	**GAUGING PLASTER**									
0020	100 lb. bags, less than 1 ton				Bag	20.50			20.50	22.50
0100	Over 1 ton				"	19.45			19.45	21.50

09 23 Gypsum Plastering

09 23 20 – Gypsum Plaster

09 23 20.30 Keenes Cement		Crew	Daily Output	Labor-Hours	Unit	2020 Bare Costs Material	Labor	Equipment	Total	Total Incl O&P
0010	**KEENES CEMENT**									
0020	In 100 lb. bags, less than 1 ton				Bag	24			24	26.50
0100	Over 1 ton				"	24			24	26
0300	Finish only, add to plaster prices, standard	J-1	215	.186	S.Y.	2.29	8.60	.55	11.44	16.75
0400	High quality	"	144	.278	"	2.32	12.80	.81	15.93	24

09 24 Cement Plastering

09 24 23 – Cement Stucco

09 24 23.40 Stucco

		Crew	Daily Output	Labor-Hours	Unit	Material	Labor	Equipment	Total	Total Incl O&P
0010	**STUCCO**									
0015	3 coats 7/8" thick, float finish, with mesh, on wood frame	J-2	63	.762	S.Y.	7.65	36	1.86	45.51	67.50
0100	On masonry construction, no mesh incl.	J-1	67	.597		3.02	27.50	1.75	32.27	49.50
0300	For trowel finish, add	1 Plas	170	.047			2.29		2.29	3.64
0600	For coloring, add	J-1	685	.058		.43	2.69	.17	3.29	4.95
0700	For special texture, add	"	200	.200		1.50	9.20	.59	11.29	17
0900	For soffits, add	J-2	155	.310		2.29	14.60	.76	17.65	26.50
1000	Stucco, with bonding agent, 3 coats, on walls, no mesh incl.	J-1	200	.200		4.23	9.20	.59	14.02	20
1200	Ceilings		180	.222		3.83	10.25	.65	14.73	21
1300	Beams		80	.500		3.83	23	1.47	28.30	42.50
1500	Columns		100	.400		3.83	18.45	1.17	23.45	35
1550	Minimum labor/equipment charge	1 Plas	1	8	Job		390		390	620
1600	Mesh, galvanized, nailed to wood, 1.8 lb.	1 Lath	60	.133	S.Y.	7.50	6.95		14.45	19.10
1800	3.6 lb.		55	.145		4.17	7.55		11.72	16.45
1900	Wired to steel, galvanized, 1.8 lb.		53	.151		7.50	7.85		15.35	20.50
2100	3.6 lb.		50	.160		4.17	8.35		12.52	17.65
9000	Minimum labor/equipment charge		4	2	Job		104		104	163

09 25 Other Plastering

09 25 23 – Lime Based Plastering

09 25 23.10 Venetian Plaster

		Crew	Daily Output	Labor-Hours	Unit	Material	Labor	Equipment	Total	Total Incl O&P
0010	**VENETIAN PLASTER**									
0100	Walls, 1 coat primer, roller applied	1 Plas	950	.008	S.F.	.17	.41		.58	.84
0200	Plaster, 3 coats, incl. sanding	2 Plas	700	.023		.41	1.11		1.52	2.22
0210	For pigment, light colors add per S.F. plaster					.02			.02	.02
0220	For pigment, dark colors add per S.F. plaster					.04			.04	.04
0300	For sealer/wax coat incl. burnishing, add	1 Plas	300	.027		.38	1.30		1.68	2.48

09 26 Veneer Plastering

09 26 13 – Gypsum Veneer Plastering

09 26 13.20 Blueboard		Crew	Daily Output	Labor-Hours	Unit	Material	Labor	Equipment	Total	Total Incl O&P
						2020 Bare Costs				
0010	**BLUEBOARD** For use with thin coat									
0100	plaster application see Section 09 26 13.80									
1000	3/8" thick, on walls or ceilings, standard, no finish included	2 Carp	1900	.008	S.F.	.33	.45		.78	1.08
1100	With thin coat plaster finish		875	.018		.43	.97		1.40	2.04
1400	On beams, columns, or soffits, standard, no finish included		675	.024		.38	1.26		1.64	2.44
1450	With thin coat plaster finish		475	.034		.48	1.79		2.27	3.40
3000	1/2" thick, on walls or ceilings, standard, no finish included		1900	.008		.33	.45		.78	1.08
3100	With thin coat plaster finish		875	.018		.43	.97		1.40	2.04
3300	Fire resistant, no finish included		1900	.008		.33	.45		.78	1.08
3400	With thin coat plaster finish		875	.018		.43	.97		1.40	2.04
3450	On beams, columns, or soffits, standard, no finish included		675	.024		.38	1.26		1.64	2.44
3500	With thin coat plaster finish		475	.034		.48	1.79		2.27	3.40
3700	Fire resistant, no finish included		675	.024		.38	1.26		1.64	2.44
3800	With thin coat plaster finish		475	.034		.48	1.79		2.27	3.40
5000	5/8" thick, on walls or ceilings, fire resistant, no finish included		1900	.008		.35	.45		.80	1.11
5100	With thin coat plaster finish		875	.018		.45	.97		1.42	2.06
5500	On beams, columns, or soffits, no finish included		675	.024		.40	1.26		1.66	2.46
5600	With thin coat plaster finish		475	.034		.51	1.79		2.30	3.43
6000	For high ceilings, over 8' high, add		3060	.005			.28		.28	.45
6500	For distribution costs 3 stories and above, add per story		6100	.003			.14		.14	.22
9000	Minimum labor/equipment charge	1 Carp	2	4	Job		213		213	340

09 26 13.80 Thin Coat Plaster

		Crew	Daily Output	Labor-Hours	Unit	Material	Labor	Equipment	Total	Total Incl O&P
0010	**THIN COAT PLASTER**									
0012	1 coat veneer, not incl. lath	J-1	3600	.011	S.F.	.10	.51	.03	.64	.97
1000	In 50 lb. bags				Bag	14			14	15.40

09 28 Backing Boards and Underlayments

09 28 13 – Cementitious Backing Boards

09 28 13.10 Cementitious Backerboard

		Crew	Daily Output	Labor-Hours	Unit	Material	Labor	Equipment	Total	Total Incl O&P
0010	**CEMENTITIOUS BACKERBOARD**									
0070	Cementitious backerboard, on floor, 3' x 4' x 1/2" sheets	2 Carp	525	.030	S.F.	.92	1.62		2.54	3.61
0080	3' x 5' x 1/2" sheets		525	.030		.80	1.62		2.42	3.47
0090	3' x 6' x 1/2" sheets		525	.030		.78	1.62		2.40	3.46
0100	3' x 4' x 5/8" sheets		525	.030		.98	1.62		2.60	3.68
0110	3' x 5' x 5/8" sheets		525	.030		1	1.62		2.62	3.70
0120	3' x 6' x 5/8" sheets		525	.030		.98	1.62		2.60	3.68
0150	On wall, 3' x 4' x 1/2" sheets		350	.046		.92	2.43		3.35	4.90
0160	3' x 5' x 1/2" sheets		350	.046		.80	2.43		3.23	4.76
0170	3' x 6' x 1/2" sheets		350	.046		.78	2.43		3.21	4.75
0180	3' x 4' x 5/8" sheets		350	.046		.98	2.43		3.41	4.97
0190	3' x 5' x 5/8" sheets		350	.046		1	2.43		3.43	4.99
0200	3' x 6' x 5/8" sheets		350	.046		.98	2.43		3.41	4.97
0250	On counter, 3' x 4' x 1/2" sheets		180	.089		.92	4.72		5.64	8.55
0260	3' x 5' x 1/2" sheets		180	.089		.80	4.72		5.52	8.40
0270	3' x 6' x 1/2" sheets		180	.089		.78	4.72		5.50	8.40
0300	3' x 4' x 5/8" sheets		180	.089		.98	4.72		5.70	8.65
0310	3' x 5' x 5/8" sheets		180	.089		1	4.72		5.72	8.65
0320	3' x 6' x 5/8" sheets		180	.089		.98	4.72		5.70	8.65

09 29 Gypsum Board

09 29 10 – Gypsum Board Panels

09 29 10.30 Gypsum Board		Crew	Daily Output	Labor-Hours	Unit	Material	2020 Bare Costs Labor	Equipment	Total	Total Incl O&P
0010	**GYPSUM BOARD** on walls & ceilings R092910-10									
0100	Nailed or screwed to studs unless otherwise noted									
0110	1/4" thick, on walls or ceilings, standard, no finish included	2 Carp	1330	.012	S.F.	.38	.64		1.02	1.44
0115	1/4" thick, on walls or ceilings, flexible, no finish included		1050	.015		.54	.81		1.35	1.89
0117	1/4" thick, on columns or soffits, flexible, no finish included		1050	.015		.54	.81		1.35	1.89
0130	1/4" thick, standard, no finish included, less than 800 S.F.		510	.031		.38	1.67		2.05	3.09
0150	3/8" thick, on walls, standard, no finish included		2000	.008		.35	.43		.78	1.07
0200	On ceilings, standard, no finish included		1800	.009		.35	.47		.82	1.15
0250	On beams, columns, or soffits, no finish included		675	.024		.35	1.26		1.61	2.41
0300	1/2" thick, on walls, standard, no finish included		2000	.008		.33	.43		.76	1.04
0350	Taped and finished (level 4 finish)		965	.017		.38	.88		1.26	1.83
0390	With compound skim coat (level 5 finish)		775	.021		.44	1.10		1.54	2.24
0400	Fire resistant, no finish included		2000	.008		.38	.43		.81	1.10
0450	Taped and finished (level 4 finish)		965	.017		.43	.88		1.31	1.88
0490	With compound skim coat (level 5 finish)		775	.021		.49	1.10		1.59	2.30
0500	Water resistant, no finish included		2000	.008		.41	.43		.84	1.13
0550	Taped and finished (level 4 finish)		965	.017		.46	.88		1.34	1.92
0590	With compound skim coat (level 5 finish)		775	.021		.52	1.10		1.62	2.33
0600	Prefinished, vinyl, clipped to studs		900	.018		.52	.95		1.47	2.08
0700	Mold resistant, no finish included		2000	.008		.44	.43		.87	1.16
0710	Taped and finished (level 4 finish)		965	.017		.49	.88		1.37	1.95
0720	With compound skim coat (level 5 finish)		775	.021		.55	1.10		1.65	2.36
1000	On ceilings, standard, no finish included		1800	.009		.33	.47		.80	1.12
1050	Taped and finished (level 4 finish)		765	.021		.38	1.11		1.49	2.20
1090	With compound skim coat (level 5 finish)		610	.026		.44	1.39		1.83	2.71
1100	Fire resistant, no finish included		1800	.009		.38	.47		.85	1.18
1150	Taped and finished (level 4 finish)		765	.021		.43	1.11		1.54	2.25
1195	With compound skim coat (level 5 finish)		610	.026		.49	1.39		1.88	2.77
1200	Water resistant, no finish included		1800	.009		.41	.47		.88	1.21
1250	Taped and finished (level 4 finish)		765	.021		.46	1.11		1.57	2.29
1290	With compound skim coat (level 5 finish)		610	.026		.52	1.39		1.91	2.80
1310	Mold resistant, no finish included		1800	.009		.44	.47		.91	1.24
1320	Taped and finished (level 4 finish)		765	.021		.49	1.11		1.60	2.32
1330	With compound skim coat (level 5 finish)		610	.026		.55	1.39		1.94	2.83
1350	Sag resistant, no finish included		1600	.010		.33	.53		.86	1.21
1360	Taped and finished (level 4 finish)		765	.021		.38	1.11		1.49	2.20
1370	With compound skim coat (level 5 finish)		610	.026		.44	1.39		1.83	2.71
1500	On beams, columns, or soffits, standard, no finish included		675	.024		.38	1.26		1.64	2.44
1550	Taped and finished (level 4 finish)		540	.030		.38	1.57		1.95	2.94
1590	With compound skim coat (level 5 finish)		475	.034		.44	1.79		2.23	3.35
1600	Fire resistant, no finish included		675	.024		.38	1.26		1.64	2.44
1650	Taped and finished (level 4 finish)		540	.030		.43	1.57		2	2.99
1690	With compound skim coat (level 5 finish)		475	.034		.49	1.79		2.28	3.41
1700	Water resistant, no finish included		675	.024		.47	1.26		1.73	2.54
1750	Taped and finished (level 4 finish)		540	.030		.46	1.57		2.03	3.03
1790	With compound skim coat (level 5 finish)		475	.034		.52	1.79		2.31	3.44
1800	Mold resistant, no finish included		675	.024		.51	1.26		1.77	2.58
1810	Taped and finished (level 4 finish)		540	.030		.49	1.57		2.06	3.06
1820	With compound skim coat (level 5 finish)		475	.034		.55	1.79		2.34	3.47
1850	Sag resistant, no finish included		675	.024		.38	1.26		1.64	2.44
1860	Taped and finished (level 4 finish)		540	.030		.38	1.57		1.95	2.94
1870	With compound skim coat (level 5 finish)		475	.034		.44	1.79		2.23	3.35
2000	5/8" thick, on walls, standard, no finish included		2000	.008		.33	.43		.76	1.04

09 29 Gypsum Board

09 29 10 – Gypsum Board Panels

09 29 10.30 Gypsum Board		Crew	Daily Output	Labor-Hours	Unit	Material	Labor	Equipment	Total	Total Incl O&P
						2020 Bare Costs				
2050	Taped and finished (level 4 finish)	2 Carp	965	.017	S.F.	.38	.88		1.26	1.83
2090	With compound skim coat (level 5 finish)		775	.021		.44	1.10		1.54	2.24
2100	Fire resistant, no finish included		2000	.008		.38	.43		.81	1.10
2150	Taped and finished (level 4 finish)		965	.017		.43	.88		1.31	1.88
2195	With compound skim coat (level 5 finish)		775	.021		.49	1.10		1.59	2.30
2200	Water resistant, no finish included		2000	.008		.44	.43		.87	1.16
2250	Taped and finished (level 4 finish)		965	.017		.49	.88		1.37	1.95
2290	With compound skim coat (level 5 finish)		775	.021		.55	1.10		1.65	2.36
2300	Prefinished, vinyl, clipped to studs		900	.018		.95	.95		1.90	2.56
2510	Mold resistant, no finish included		2000	.008		.54	.43		.97	1.27
2520	Taped and finished (level 4 finish)		965	.017		.59	.88		1.47	2.06
2530	With compound skim coat (level 5 finish)		775	.021		.65	1.10		1.75	2.47
3000	On ceilings, standard, no finish included		1800	.009		.33	.47		.80	1.12
3050	Taped and finished (level 4 finish)		765	.021		.38	1.11		1.49	2.20
3090	With compound skim coat (level 5 finish)		615	.026		.44	1.38		1.82	2.70
3100	Fire resistant, no finish included		1800	.009		.38	.47		.85	1.18
3150	Taped and finished (level 4 finish)		765	.021		.43	1.11		1.54	2.25
3190	With compound skim coat (level 5 finish)		615	.026		.49	1.38		1.87	2.76
3200	Water resistant, no finish included		1800	.009		.44	.47		.91	1.24
3250	Taped and finished (level 4 finish)		765	.021		.49	1.11		1.60	2.32
3290	With compound skim coat (level 5 finish)		615	.026		.55	1.38		1.93	2.82
3300	Mold resistant, no finish included		1800	.009		.54	.47		1.01	1.35
3310	Taped and finished (level 4 finish)		765	.021		.59	1.11		1.70	2.43
3320	With compound skim coat (level 5 finish)		615	.026		.65	1.38		2.03	2.93
3500	On beams, columns, or soffits, no finish included		675	.024		.38	1.26		1.64	2.44
3550	Taped and finished (level 4 finish)		475	.034		.44	1.79		2.23	3.35
3590	With compound skim coat (level 5 finish)		380	.042		.51	2.24		2.75	4.15
3600	Fire resistant, no finish included		675	.024		.44	1.26		1.70	2.50
3650	Taped and finished (level 4 finish)		475	.034		.50	1.79		2.29	3.42
3690	With compound skim coat (level 5 finish)		380	.042		.49	2.24		2.73	4.13
3700	Water resistant, no finish included		675	.024		.51	1.26		1.77	2.58
3750	Taped and finished (level 4 finish)		475	.034		.55	1.79		2.34	3.47
3790	With compound skim coat (level 5 finish)		380	.042		.56	2.24		2.80	4.21
3800	Mold resistant, no finish included		675	.024		.62	1.26		1.88	2.70
3810	Taped and finished (level 4 finish)		475	.034		.65	1.79		2.44	3.58
3820	With compound skim coat (level 5 finish)		380	.042		.68	2.24		2.92	4.34
4000	Fireproofing, beams or columns, 2 layers, 1/2" thick, incl finish		330	.048		.86	2.58		3.44	5.10
4010	Mold resistant		330	.048		.98	2.58		3.56	5.20
4050	5/8" thick		300	.053		.86	2.83		3.69	5.50
4060	Mold resistant		300	.053		1.18	2.83		4.01	5.85
4100	3 layers, 1/2" thick		225	.071		1.29	3.78		5.07	7.45
4110	Mold resistant		225	.071		1.47	3.78		5.25	7.65
4150	5/8" thick		210	.076		1.29	4.05		5.34	7.90
4160	Mold resistant		210	.076		1.77	4.05		5.82	8.45
5050	For 1" thick coreboard on columns		480	.033		.89	1.77		2.66	3.82
5100	For foil-backed board, add					.21			.21	.23
5200	For work over 8' high, add	2 Carp	3060	.005			.28		.28	.45
5270	For textured spray, add	2 Lath	1600	.010		.04	.52		.56	.86
5300	For distribution cost 3 stories and above, add per story	2 Carp	6100	.003			.14		.14	.22
5350	For finishing inner corners, add		950	.017	L.F.	.11	.90		1.01	1.55
5355	For finishing outer corners, add		1250	.013		.24	.68		.92	1.36
5500	For acoustical sealant, add per bead	1 Carp	500	.016		.04	.85		.89	1.41
5550	Sealant, 1 quart tube				Ea.	7.15			7.15	7.90

09 29 10.30 Gypsum Board

	09 29 10.30 Gypsum Board	Crew	Daily Output	Labor-Hours	Unit	Material	2020 Bare Costs Labor	Equipment	Total	Total Incl O&P
6000	Gypsum sound dampening panels									
6010	1/2" thick on walls, multi-layer, lightweight, no finish included	2 Carp	1500	.011	S.F.	2.25	.57		2.82	3.39
6015	Taped and finished (level 4 finish)		725	.022		2.30	1.17		3.47	4.41
6020	With compound skim coat (level 5 finish)		580	.028		2.36	1.47		3.83	4.95
6025	5/8" thick on walls, for wood studs, no finish included		1500	.011		2.46	.57		3.03	3.62
6030	Taped and finished (level 4 finish)		725	.022		2.51	1.17		3.68	4.64
6035	With compound skim coat (level 5 finish)		580	.028		2.57	1.47		4.04	5.20
6040	For metal stud, no finish included		1500	.011		2.22	.57		2.79	3.35
6045	Taped and finished (level 4 finish)		725	.022		2.27	1.17		3.44	4.38
6050	With compound skim coat (level 5 finish)		580	.028		2.33	1.47		3.80	4.91
6055	Abuse resist, no finish included		1500	.011		4.23	.57		4.80	5.55
6060	Taped and finished (level 4 finish)		725	.022		4.28	1.17		5.45	6.60
6065	With compound skim coat (level 5 finish)		580	.028		4.34	1.47		5.81	7.10
6070	Shear rated, no finish included		1500	.011		5.60	.57		6.17	7.10
6075	Taped and finished (level 4 finish)		725	.022		5.65	1.17		6.82	8.15
6080	With compound skim coat (level 5 finish)		580	.028		5.75	1.47		7.22	8.65
6085	For SCIF applications, no finish included		1500	.011		5.30	.57		5.87	6.75
6090	Taped and finished (level 4 finish)		725	.022		5.35	1.17		6.52	7.80
6095	With compound skim coat (level 5 finish)		580	.028		5.40	1.47		6.87	8.30
6100	1-3/8" thick on walls, THX certified, no finish included		1500	.011		9.35	.57		9.92	11.20
6105	Taped and finished (level 4 finish)		725	.022		9.40	1.17		10.57	12.25
6110	With compound skim coat (level 5 finish)		580	.028		9.50	1.47		10.97	12.80
6115	5/8" thick on walls, score & snap installation, no finish included		2000	.008		1.99	.43		2.42	2.87
6120	Taped and finished (level 4 finish)		965	.017		2.04	.88		2.92	3.66
6125	With compound skim coat (level 5 finish)		775	.021		2.10	1.10		3.20	4.07
7020	5/8" thick on ceilings, for wood joists, no finish included		1200	.013		2.46	.71		3.17	3.85
7025	Taped and finished (level 4 finish)		510	.031		2.51	1.67		4.18	5.45
7030	With compound skim coat (level 5 finish)		410	.039		2.57	2.07		4.64	6.15
7035	For metal joists, no finish included		1200	.013		2.22	.71		2.93	3.58
7040	Taped and finished (level 4 finish)		510	.031		2.27	1.67		3.94	5.15
7045	With compound skim coat (level 5 finish)		410	.039		2.33	2.07		4.40	5.90
7050	Abuse resist, no finish included		1200	.013		4.23	.71		4.94	5.80
7055	Taped and finished (level 4 finish)		510	.031		4.28	1.67		5.95	7.40
7060	With compound skim coat (level 5 finish)		410	.039		4.34	2.07		6.41	8.10
7065	Shear rated, no finish included		1200	.013		5.60	.71		6.31	7.35
7070	Taped and finished (level 4 finish)		510	.031		5.65	1.67		7.32	8.90
7075	With compound skim coat (level 5 finish)		410	.039		5.75	2.07		7.82	9.60
7080	For SCIF applications, no finish included		1200	.013		5.30	.71		6.01	7
7085	Taped and finished (level 4 finish)		510	.031		5.35	1.67		7.02	8.55
7090	With compound skim coat (level 5 finish)		410	.039		5.40	2.07		7.47	9.25
8010	5/8" thick on ceilings, score & snap installation, no finish included		1600	.010		1.99	.53		2.52	3.04
8015	Taped and finished (level 4 finish)		680	.024		2.04	1.25		3.29	4.25
8020	With compound skim coat (level 5 finish)	↓	545	.029	↓	2.10	1.56		3.66	4.81
9000	Minimum labor/equipment charge	1 Carp	2	4	Job		213		213	340

09 29 10.50 High Abuse Gypsum Board

	09 29 10.50 High Abuse Gypsum Board	Crew	Daily Output	Labor-Hours	Unit	Material	Labor	Equipment	Total	Total Incl O&P
0010	**HIGH ABUSE GYPSUM BOARD**, fiber reinforced, nailed or									
0100	screwed to studs unless otherwise noted									
0110	1/2" thick, on walls, no finish included	2 Carp	1800	.009	S.F.	.82	.47		1.29	1.66
0120	Taped and finished (level 4 finish)		870	.018		.87	.98		1.85	2.53
0130	With compound skim coat (level 5 finish)		700	.023		.93	1.22		2.15	2.97
0150	On ceilings, no finish included		1620	.010		.82	.53		1.35	1.74
0160	Taped and finished (level 4 finish)	↓	690	.023	↓	.87	1.23		2.10	2.93

09 29 Gypsum Board

09 29 10 – Gypsum Board Panels

09 29 10.50 High Abuse Gypsum Board		Crew	Daily Output	Labor-Hours	Unit	Material	Labor	Equipment	Total	Total Incl O&P
						2020 Bare Costs				
0170	With compound skim coat (level 5 finish)	2 Carp	550	.029	S.F.	.93	1.55		2.48	3.50
0210	5/8" thick, on walls, no finish included		1800	.009		.87	.47		1.34	1.72
0220	Taped and finished (level 4 finish)		870	.018		.92	.98		1.90	2.58
0230	With compound skim coat (level 5 finish)		700	.023		.98	1.22		2.20	3.03
0250	On ceilings, no finish included		1620	.010		.87	.53		1.40	1.80
0260	Taped and finished (level 4 finish)		690	.023		.92	1.23		2.15	2.98
0270	With compound skim coat (level 5 finish)		550	.029		.98	1.55		2.53	3.56
0310	5/8" thick, on walls, very high impact, no finish included		1800	.009		.95	.47		1.42	1.81
0320	Taped and finished (level 4 finish)		870	.018		1	.98		1.98	2.67
0330	With compound skim coat (level 5 finish)		700	.023		1.06	1.22		2.28	3.12
0350	On ceilings, no finish included		1620	.010		.95	.53		1.48	1.89
0360	Taped and finished (level 4 finish)		690	.023		1	1.23		2.23	3.07
0370	With compound skim coat (level 5 finish)	↓	550	.029	↓	1.06	1.55		2.61	3.65
0400	High abuse, gypsum core, paper face									
0410	1/2" thick, on walls, no finish included	2 Carp	1800	.009	S.F.	.74	.47		1.21	1.57
0420	Taped and finished (level 4 finish)		870	.018		.79	.98		1.77	2.44
0430	With compound skim coat (level 5 finish)		700	.023		.85	1.22		2.07	2.88
0450	On ceilings, no finish included		1620	.010		.74	.53		1.27	1.65
0460	Taped and finished (level 4 finish)		690	.023		.79	1.23		2.02	2.84
0470	With compound skim coat (level 5 finish)		550	.029		.85	1.55		2.40	3.41
0510	5/8" thick, on walls, no finish included		1800	.009		.82	.47		1.29	1.66
0520	Taped and finished (level 4 finish)		870	.018		.87	.98		1.85	2.53
0530	With compound skim coat (level 5 finish)		700	.023		.93	1.22		2.15	2.97
0550	On ceilings, no finish included		1620	.010		.82	.53		1.35	1.74
0560	Taped and finished (level 4 finish)		690	.023		.87	1.23		2.10	2.93
0570	With compound skim coat (level 5 finish)		550	.029		.93	1.55		2.48	3.50
1000	For high ceilings, over 8' high, add		2750	.006			.31		.31	.50
1010	For distribution cost 3 stories and above, add per story	↓	5500	.003	↓		.15		.15	.25

09 29 15 – Gypsum Board Accessories

09 29 15.10 Accessories, Gypsum Board		Crew	Daily Output	Labor-Hours	Unit	Material	Labor	Equipment	Total	Total Incl O&P
0010	**ACCESSORIES, GYPSUM BOARD**									
0020	Casing bead, galvanized steel	1 Carp	2.90	2.759	C.L.F.	24.50	147		171.50	262
0100	Vinyl		3	2.667		23.50	142		165.50	253
0300	Corner bead, galvanized steel, 1" x 1"		4	2		15.90	106		121.90	188
0400	1-1/4" x 1-1/4"		3.50	2.286		17.35	121		138.35	214
0600	Vinyl		4	2		19.95	106		125.95	192
0900	Furring channel, galv. steel, 7/8" deep, standard		2.60	3.077		35	164		199	300
1000	Resilient		2.55	3.137		27	167		194	297
1100	J trim, galvanized steel, 1/2" wide		3	2.667		23	142		165	253
1120	5/8" wide		2.95	2.712		31	144		175	265
1140	L trim, galvanized		3	2.667		22	142		164	252
1150	U trim, galvanized	↓	2.95	2.712	↓	24	144		168	258
1160	Screws #6 x 1" A				M	12.35			12.35	13.60
1170	#6 x 1-5/8" A				"	14.60			14.60	16.05
1200	For stud partitions, see Sections 05 41 13.30 and 09 22 16.13									
1500	Z stud, galvanized steel, 1-1/2" wide	1 Carp	2.60	3.077	C.L.F.	43.50	164		207.50	310
1600	2" wide		2.55	3.137	"	68.50	167		235.50	345
9000	Minimum labor/equipment charge	↓	3	2.667	Job		142		142	227

09 30 Tiling

09 30 13 – Ceramic Tiling

09 30 13.20 Ceramic Tile Repairs		Crew	Daily Output	Labor-Hours	Unit	Material	Labor	Equipment	Total	Total Incl O&P
0010	CERAMIC TILE REPAIRS									
1000	Grout removal, carbide tipped, rotary grinder	1 Clab	240	.033	L.F.		1.40		1.40	2.25
1100	Regrout tile 4-1/2" x 4-1/2" or larger, wall	1 Tilf	100	.080	S.F.	.15	3.96		4.11	6.35
1150	Floor		125	.064		.16	3.17		3.33	5.15
1200	Seal tile and grout		360	.022			1.10		1.10	1.72

09 30 13.45 Ceramic Tile Accessories		Crew	Daily Output	Labor-Hours	Unit	Material	Labor	Equipment	Total	Total Incl O&P
0010	CERAMIC TILE ACCESSORIES									
0100	Spacers, 1/8"				C	2.27			2.27	2.50
1310	Sealer for natural stone tile, installed	1 Tilf	650	.012	S.F.	.05	.61		.66	1.01

09 30 29 – Metal Tiling

09 30 29.10 Metal Tile		Crew	Daily Output	Labor-Hours	Unit	Material	Labor	Equipment	Total	Total Incl O&P
0010	**METAL TILE** 4' x 4' sheet, 24 ga., tile pattern, nailed									
0200	Stainless steel	2 Carp	512	.031	S.F.	28	1.66		29.66	33.50
0400	Aluminized steel	"	512	.031	"	20.50	1.66		22.16	25
9000	Minimum labor/equipment charge	1 Carp	4	2	Job		106		106	170

09 30 95 – Tile & Stone Setting Materials and Specialties

09 30 95.10 Moisture Resistant, Anti-Fracture Membrane		Crew	Daily Output	Labor-Hours	Unit	Material	Labor	Equipment	Total	Total Incl O&P
0010	MOISTURE RESISTANT, ANTI-FRACTURE MEMBRANE									
0200	Elastomeric membrane, 1/16" thick	D-7	275	.058	S.F.	1.14	2.57		3.71	5.25

09 31 Thin-Set Tiling

09 31 13 – Thin-Set Ceramic Tiling

09 31 13.10 Thin-Set Ceramic Tile		Crew	Daily Output	Labor-Hours	Unit	Material	Labor	Equipment	Total	Total Incl O&P
0010	THIN-SET CERAMIC TILE									
0020	Backsplash, average grade tiles	1 Tilf	50	.160	S.F.	3.30	7.90		11.20	16.05
0022	Custom grade tiles		50	.160		6.60	7.90		14.50	19.65
0024	Luxury grade tiles		50	.160		13.20	7.90		21.10	27
0026	Economy grade tiles		50	.160		3.02	7.90		10.92	15.70
0100	Base, using 1' x 4" high piece with 1" x 1" tiles	D-7	128	.125	L.F.	5	5.50		10.50	14.15
0300	For 6" high base, 1" x 1" tile face, add					1.29			1.29	1.42
0400	For 2" x 2" tile face, add to above					.70			.70	.77
0700	Cove base, 4-1/4" x 4-1/4"	D-7	128	.125		4.11	5.50		9.61	13.15
1000	6" x 4-1/4" high		137	.117		5	5.15		10.15	13.60
1300	Sanitary cove base, 6" x 4-1/4" high		124	.129		4.84	5.70		10.54	14.25
1600	6" x 6" high		117	.137		5	6.05		11.05	14.95
1800	Bathroom accessories, average (soap dish, toothbrush holder)		82	.195	Ea.	10.20	8.60		18.80	24.50
1900	Bathtub, 5', rec. 4-1/4" x 4-1/4" tile wainscot, adhesive set 6' high		2.90	5.517		204	244		448	605
2100	7' high wainscot		2.50	6.400		238	283		521	705
2200	8' high wainscot		2.20	7.273		272	320		592	805
2500	Bullnose trim, 4-1/4" x 4-1/4"		128	.125	L.F.	4.35	5.50		9.85	13.45
2800	2" x 6"		124	.129	"	4.75	5.70		10.45	14.20
3300	Ceramic tile, porcelain type, 1 color, color group 2, 1" x 1"		183	.087	S.F.	6.25	3.86		10.11	12.90
3310	2" x 2" or 2" x 1"		190	.084		6.35	3.72		10.07	12.85
3350	For random blend, 2 colors, add					1.02			1.02	1.12
3360	4 colors, add					1.43			1.43	1.57
3370	For color group 3, add					.65			.65	.72
3380	For abrasive non-slip tile, add					.45			.45	.50
4300	Specialty tile, 4-1/4" x 4-1/4" x 1/2", decorator finish	D-7	183	.087		12.85	3.86		16.71	20
4500	Add for epoxy grout, 1/16" joint, 1" x 1" tile		800	.020		.67	.88		1.55	2.12
4600	2" x 2" tile		820	.020		.63	.86		1.49	2.04

09 31 Thin-Set Tiling

09 31 13 – Thin-Set Ceramic Tiling

09 31 13.10 Thin-Set Ceramic Tile		Crew	Daily Output	Labor-Hours	Unit	Material	Labor (2020 Bare Costs)	Equipment	Total	Total Incl O&P
4610	Add for epoxy grout, 1/8" joint, 8" x 8" x 3/8" tile, add	D-7	900	.018	S.F.	1.43	.79		2.22	2.80
4800	Pregrouted sheets, walls, 4-1/4" x 4-1/4", 6" x 4-1/4"									
4810	and 8-1/2" x 4-1/4", 4 S.F. sheets, silicone grout	D-7	240	.067	S.F.	5.70	2.94		8.64	10.90
5100	Floors, unglazed, 2 S.F. sheets,									
5110	urethane adhesive	D-7	180	.089	S.F.	2.17	3.92		6.09	8.55
5400	Walls, interior, 4-1/4" x 4-1/4" tile		190	.084		3.10	3.72		6.82	9.25
5500	6" x 4-1/4" tile		190	.084		3.25	3.72		6.97	9.40
5700	8-1/2" x 4-1/4" tile		190	.084		5.95	3.72		9.67	12.40
5800	6" x 6" tile		175	.091		4.08	4.04		8.12	10.85
5810	8" x 8" tile		170	.094		5.20	4.16		9.36	12.25
5820	12" x 12" tile		160	.100		5	4.42		9.42	12.40
5830	16" x 16" tile		150	.107		5.45	4.71		10.16	13.40
6000	Decorated wall tile, 4-1/4" x 4-1/4", color group 1		270	.059		3.52	2.62		6.14	7.95
6100	Color group 4		180	.089		54.50	3.92		58.42	66
9300	Ceramic tiles, recycled glass, standard colors, 2" x 2" thru 6" x 6" G		190	.084		23	3.72		26.72	31
9310	6" x 6" G		175	.091		23	4.04		27.04	31.50
9320	8" x 8" G		170	.094		23.50	4.16		27.66	32.50
9330	12" x 12" G		160	.100		23.50	4.42		27.92	33
9340	Earthtones, 2" x 2" to 4" x 8" G		190	.084		27	3.72		30.72	35.50
9350	6" x 6" G		175	.091		27	4.04		31.04	36
9360	8" x 8" G		170	.094		27	4.16		31.16	36.50
9370	12" x 12" G		160	.100		27	4.42		31.42	37
9380	Deep colors, 2" x 2" to 4" x 8" G		190	.084		30.50	3.72		34.22	40
9390	6" x 6" G		175	.091		30.50	4.04		34.54	40
9400	8" x 8" G		170	.094		32	4.16		36.16	42
9410	12" x 12" G		160	.100		32	4.42		36.42	42.50

09 31 33 – Thin-Set Stone Tiling

09 31 33.10 Tiling, Thin-Set Stone		Crew	Daily Output	Labor-Hours	Unit	Material	Labor	Equipment	Total	Total Incl O&P
0010	**TILING, THIN-SET STONE**									
3000	Floors, natural clay, random or uniform, color group 1	D-7	183	.087	S.F.	4.55	3.86		8.41	11.05
3100	Color group 2		183	.087		5.95	3.86		9.81	12.55
3255	Floors, glazed, 6" x 6", color group 1		300	.053		5.60	2.35		7.95	9.85
3260	8" x 8" tile		300	.053		5.55	2.35		7.90	9.80
3270	12" x 12" tile		290	.055		6.30	2.44		8.74	10.75
3280	16" x 16" tile		280	.057		7.75	2.52		10.27	12.50
3281	18" x 18" tile		270	.059		7.40	2.62		10.02	12.20
3282	20" x 20" tile		260	.062		3.94	2.72		6.66	8.60
3283	24" x 24" tile		250	.064		9.65	2.83		12.48	15.05
3285	Border, 6" x 12" tile		200	.080		12.20	3.53		15.73	19
3290	3" x 12" tile		200	.080		12.70	3.53		16.23	19.50

09 32 Mortar-Bed Tiling

09 32 13 – Mortar-Bed Ceramic Tiling

09 32 13.10 Ceramic Tile		Crew	Daily Output	Labor-Hours	Unit	2020 Bare Costs Material	Labor	Equipment	Total	Total Incl O&P
0010	**CERAMIC TILE**									
0050	Base, using 1' x 4" high pc. with 1" x 1" tiles	D-7	82	.195	L.F.	5.35	8.60		13.95	19.40
0600	Cove base, 4-1/4" x 4-1/4" high		91	.176		4.22	7.75		11.97	16.80
0900	6" x 4-1/4" high		100	.160		5.10	7.05		12.15	16.70
1200	Sanitary cove base, 6" x 4-1/4" high		93	.172		4.95	7.60		12.55	17.35
1500	6" x 6" high		84	.190		5.10	8.40		13.50	18.85
2400	Bullnose trim, 4-1/4" x 4-1/4"		82	.195		4.43	8.60		13.03	18.35
2700	2" x 6" bullnose trim		84	.190		4.82	8.40		13.22	18.50
6210	Wall tile, 4-1/4" x 4-1/4", better grade	1 Tilf	50	.160	S.F.	9.55	7.90		17.45	23
6240	2" x 2"		50	.160		7.75	7.90		15.65	21
6250	6" x 6"		55	.145		9.75	7.20		16.95	22
6260	8" x 8"		60	.133		9.75	6.60		16.35	21
6300	Exterior walls, frostproof, 4-1/4" x 4-1/4"	D-7	102	.157		7.25	6.95		14.20	18.85
6400	1-3/8" x 1-3/8"		93	.172		6.60	7.60		14.20	19.15
6600	Crystalline glazed, 4-1/4" x 4-1/4", plain		100	.160		4.77	7.05		11.82	16.30
6700	4-1/4" x 4-1/4", scored tile		100	.160		6.60	7.05		13.65	18.30
6900	6" x 6" plain		93	.172		5.70	7.60		13.30	18.15
7000	For epoxy grout, 1/16" joints, 4-1/4" tile, add		800	.020		.42	.88		1.30	1.84
7200	For tile set in dry mortar, add		1735	.009			.41		.41	.64
7300	For tile set in Portland cement mortar, add		290	.055		.17	2.44		2.61	4.01
9500	Minimum labor/equipment charge		3.25	4.923	Job		217		217	340

09 32 16 – Mortar-Bed Quarry Tiling

09 32 16.10 Quarry Tile		Crew	Daily Output	Labor-Hours	Unit	Material	Labor	Equipment	Total	Total Incl O&P
0010	**QUARRY TILE**									
0100	Base, cove or sanitary, to 5" high, 1/2" thick	D-7	110	.145	L.F.	6.35	6.40		12.75	17
0300	Bullnose trim, red, 6" x 6" x 1/2" thick		120	.133		5.10	5.90		11	14.80
0400	4" x 4" x 1/2" thick		110	.145		5.15	6.40		11.55	15.70
0600	4" x 8" x 1/2" thick, using 8" as edge		130	.123		5.40	5.45		10.85	14.45
0700	Floors, 1,000 S.F. lots, red, 4" x 4" x 1/2" thick		120	.133	S.F.	8.55	5.90		14.45	18.60
0900	6" x 6" x 1/2" thick		140	.114		8.70	5.05		13.75	17.50
1000	4" x 8" x 1/2" thick		130	.123		6.65	5.45		12.10	15.80
1300	For waxed coating, add					.77			.77	.85
1500	For non-standard colors, add					.49			.49	.54
1600	For abrasive surface, add					.53			.53	.58
1800	Brown tile, imported, 6" x 6" x 3/4"	D-7	120	.133		7.20	5.90		13.10	17.10
1900	8" x 8" x 1"		110	.145		9.60	6.40		16	20.50
2100	For thin set mortar application, deduct		700	.023			1.01		1.01	1.58
2200	For epoxy grout & mortar, 6" x 6" x 1/2", add		350	.046		2.36	2.02		4.38	5.75
2700	Stair tread, 6" x 6" x 3/4", plain		50	.320		7.80	14.15		21.95	30.50
2800	Abrasive		47	.340		9.10	15.05		24.15	33.50
3000	Wainscot, 6" x 6" x 1/2", thin set, red		105	.152		6.55	6.75		13.30	17.75
3100	Non-standard colors		105	.152		6.55	6.75		13.30	17.75
3300	Window sill, 6" wide, 3/4" thick		90	.178	L.F.	8.85	7.85		16.70	22
3400	Corners		80	.200	Ea.	6.10	8.85		14.95	20.50
9000	Minimum labor/equipment charge		3.25	4.923	Job		217		217	340

09 32 23 – Mortar-Bed Glass Mosaic Tiling

09 32 23.10 Glass Mosaics		Crew	Daily Output	Labor-Hours	Unit	Material	Labor	Equipment	Total	Total Incl O&P
0010	**GLASS MOSAICS** 3/4" tile on 12" sheets, standard grout									
0300	Color group 1 & 2	D-7	73	.219	S.F.	21	9.70		30.70	38
0350	Color group 3		73	.219		22	9.70		31.70	39
0400	Color group 4		73	.219		29.50	9.70		39.20	47.50
0450	Color group 5		73	.219		31	9.70		40.70	49

09 32 Mortar-Bed Tiling

09 32 23 – Mortar-Bed Glass Mosaic Tiling

09 32 23.10 Glass Mosaics		Crew	Daily Output	Labor-Hours	Unit	Material	2020 Bare Costs Labor	Equipment	Total	Total Incl O&P
0500	Color group 6	D-7	73	.219	S.F.	39.50	9.70		49.20	58.50
0600	Color group 7		73	.219		41.50	9.70		51.20	60.50
0700	Color group 8, golds, silvers & specialties		64	.250		42	11.05		53.05	64
1020	1" tile on 12" sheets, opalescent finish		73	.219		18.40	9.70		28.10	35
1040	1" x 2" tile on 12" sheet, blend		73	.219		21	9.70		30.70	38
1060	2" tile on 12" sheet, blend		73	.219		17.15	9.70		26.85	34
1080	5/8" x random tile, linear, on 12" sheet, blend		73	.219		25	9.70		34.70	42.50
1600	Dots on 12" sheet		73	.219		25	9.70		34.70	42.50
1700	For glass mosaic tiles set in dry mortar, add		290	.055		.45	2.44		2.89	4.32
1720	For glass mosaic tiles set in Portland cement mortar, add		290	.055		.01	2.44		2.45	3.83
1730	For polyblend sanded tile grout		96.15	.166	Lb.	2.19	7.35		9.54	13.90

09 34 Waterproofing-Membrane Tiling

09 34 13 – Waterproofing-Membrane Ceramic Tiling

09 34 13.10 Ceramic Tile Waterproofing Membrane		Crew	Daily Output	Labor-Hours	Unit	Material	Labor	Equipment	Total	Total Incl O&P
0010	CERAMIC TILE WATERPROOFING MEMBRANE									
0020	On floors, including thinset									
0030	Fleece laminated polyethylene grid, 1/8" thick	D-7	250	.064	S.F.	2.13	2.83		4.96	6.75
0040	5/16" thick	"	250	.064	"	2.61	2.83		5.44	7.30
0050	On walls, including thinset									
0060	Fleece laminated polyethylene sheet, 8 mil thick	D-7	480	.033	S.F.	2.30	1.47		3.77	4.84
0070	Accessories, including thinset									
0080	Joint and corner sheet, 4 mils thick, 5" wide	1 Tilf	240	.033	L.F.	1.35	1.65		3	4.06
0090	7-1/4" wide		180	.044		1.72	2.20		3.92	5.35
0100	10" wide		120	.067		2.09	3.30		5.39	7.45
0110	Pre-formed corners, inside		32	.250	Ea.	7.85	12.40		20.25	28
0120	Outside		32	.250		7.70	12.40		20.10	28
0130	2" flanged floor drain with 6" stainless steel grate		16	.500		390	25		415	470
0140	EPS, sloped shower floor		480	.017	S.F.	5.55	.83		6.38	7.40
0150	Curb		32	.250	L.F.	14.10	12.40		26.50	35

09 35 Chemical-Resistant Tiling

09 35 13 – Chemical-Resistant Ceramic Tiling

09 35 13.10 Chemical-Resistant Ceramic Tiling		Crew	Daily Output	Labor-Hours	Unit	Material	Labor	Equipment	Total	Total Incl O&P
0010	CHEMICAL-RESISTANT CERAMIC TILING									
0100	4-1/4" x 4-1/4" x 1/4", 1/8" joint	D-7	130	.123	S.F.	12.25	5.45		17.70	22
0200	6" x 6" x 1/2" thick		120	.133		9.95	5.90		15.85	20
0300	8" x 8" x 1/2" thick		110	.145		11.35	6.40		17.75	22.50
0400	4-1/4" x 4-1/4" x 1/4", 1/4" joint		130	.123		13.05	5.45		18.50	23
0500	6" x 6" x 1/2" thick		120	.133		11.15	5.90		17.05	21.50
0600	8" x 8" x 1/2" thick		110	.145		12	6.40		18.40	23.50
0700	4-1/4" x 4-1/4" x 1/4", 3/8" joint		130	.123		13.80	5.45		19.25	23.50
0800	6" x 6" x 1/2" thick		120	.133		12.15	5.90		18.05	22.50
0900	8" x 8" x 1/2" thick		110	.145		13.25	6.40		19.65	24.50

09 35 16 – Chemical-Resistant Quarry Tiling

09 35 16.10 Chemical-Resistant Quarry Tiling		Crew	Daily Output	Labor-Hours	Unit	Material	Labor	Equipment	Total	Total Incl O&P
0010	CHEMICAL-RESISTANT QUARRY TILING									
0100	4" x 8" x 1/2" thick, 1/8" joint	D-7	130	.123	S.F.	11.45	5.45		16.90	21
0200	6" x 6" x 1/2" thick		120	.133		11.50	5.90		17.40	22

09 35 Chemical-Resistant Tiling

09 35 16 – Chemical-Resistant Quarry Tiling

09 35 16.10 Chemical-Resistant Quarry Tiling		Crew	Daily Output	Labor-Hours	Unit	2020 Bare Costs Material	Labor	Equipment	Total	Total Incl O&P
0300	8" x 8" x 1/2" thick	D-7	110	.145	S.F.	10.55	6.40		16.95	21.50
0400	4" x 8" x 1/2" thick, 1/4" joint		130	.123		12.75	5.45		18.20	22.50
0500	6" x 6" x 1/2" thick		120	.133		12.65	5.90		18.55	23
0600	8" x 8" x 1/2" thick		110	.145		11.25	6.40		17.65	22.50
0700	4" x 8" x 1/2" thick, 3/8" joint		130	.123		13.90	5.45		19.35	24
0800	6" x 6" x 1/2" thick		120	.133		13.65	5.90		19.55	24.50
0900	8" x 8" x 1/2" thick	↓	110	.145	↓	12.50	6.40		18.90	24

09 51 Acoustical Ceilings

09 51 13 – Acoustical Panel Ceilings

09 51 13.10 Ceiling, Acoustical Panel		Crew	Daily Output	Labor-Hours	Unit	Material	Labor	Equipment	Total	Total Incl O&P
0010	**CEILING, ACOUSTICAL PANEL**									
0100	Fiberglass boards, film faced, 2' x 2' or 2' x 4', 5/8" thick	1 Carp	625	.013	S.F.	1.34	.68		2.02	2.56
0120	3/4" thick		600	.013		3.09	.71		3.80	4.54
0130	3" thick, thermal, R11	↓	450	.018	↓	3.87	.95		4.82	5.75

09 51 14 – Acoustical Fabric-Faced Panel Ceilings

09 51 14.10 Ceiling, Acoustical Fabric-Faced Panel		Crew	Daily Output	Labor-Hours	Unit	Material	Labor	Equipment	Total	Total Incl O&P
0010	**CEILING, ACOUSTICAL FABRIC-FACED PANEL**									
0100	Glass cloth faced fiberglass, 3/4" thick	1 Carp	500	.016	S.F.	3.57	.85		4.42	5.30
0120	1" thick		485	.016		3.48	.88		4.36	5.25
0130	1-1/2" thick, nubby face	↓	475	.017	↓	2.80	.90		3.70	4.51

09 51 23 – Acoustical Tile Ceilings

09 51 23.10 Suspended Acoustic Ceiling Tiles		Crew	Daily Output	Labor-Hours	Unit	Material	Labor	Equipment	Total	Total Incl O&P
0010	**SUSPENDED ACOUSTIC CEILING TILES**, not including									
0100	suspension system									
1110	Mineral fiber tile, lay-in, 2' x 2' or 2' x 4', 5/8" thick, fine texture	1 Carp	625	.013	S.F.	.85	.68		1.53	2.03
1115	Rough textured		625	.013		.64	.68		1.32	1.79
1125	3/4" thick, fine textured		600	.013		2.16	.71		2.87	3.52
1130	Rough textured		600	.013		1.99	.71		2.70	3.33
1135	Fissured		600	.013		2.16	.71		2.87	3.52
1150	Tegular, 5/8" thick, fine textured		470	.017		1.28	.90		2.18	2.86
1155	Rough textured		470	.017		1.40	.90		2.30	2.99
1165	3/4" thick, fine textured		450	.018		2.38	.95		3.33	4.13
1170	Rough textured		450	.018		1.51	.95		2.46	3.17
1175	Fissured	↓	450	.018		2.83	.95		3.78	4.62
1185	For plastic film face, add					.86			.86	.95
1190	For fire rating, add					.54			.54	.59
3720	Mineral fiber, 24" x 24" or 48", reveal edge, painted, 5/8" thick	1 Carp	600	.013		1.36	.71		2.07	2.64
3740	3/4" thick		575	.014		1.72	.74		2.46	3.07
5020	66-78% recycled content, 3/4" thick G		600	.013		2.08	.71		2.79	3.43
5040	Mylar, 42% recycled content, 3/4" thick G		600	.013		4.66	.71		5.37	6.30
6000	Remove & install new ceiling tiles, min fiber, 2' x 2' or 2' x 4', 5/8"thk.		335	.024	↓	.85	1.27		2.12	2.97
9000	Minimum labor/equipment charge	↓	4	2	Job		106		106	170

09 51 23.30 Suspended Ceilings, Complete		Crew	Daily Output	Labor-Hours	Unit	Material	Labor	Equipment	Total	Total Incl O&P
0010	**SUSPENDED CEILINGS, COMPLETE**, incl. standard									
0100	suspension system but not incl. 1-1/2" carrier channels									
0600	Fiberglass ceiling board, 2' x 4' x 5/8", plain faced	1 Carp	500	.016	S.F.	2.27	.85		3.12	3.86
0700	Offices, 2' x 4' x 3/4"		380	.021		4.02	1.12		5.14	6.20
0800	Mineral fiber, on 15/16" T bar susp. 2' x 2' x 3/4" lay-in board		345	.023		3.32	1.23		4.55	5.65
0810	2' x 4' x 5/8" tile	↓	380	.021	↓	2.44	1.12		3.56	4.47

09 51 Acoustical Ceilings

09 51 23 – Acoustical Tile Ceilings

09 51 23.30 Suspended Ceilings, Complete		Crew	Daily Output	Labor-Hours	Unit	Material	2020 Bare Costs Labor	Equipment	Total	Total Incl O&P
0820	Tegular, 2′ x 2′ x 5/8″ tile on 9/16″ grid	1 Carp	250	.032	S.F.	2.92	1.70		4.62	5.95
0830	2′ x 4′ x 3/4″ tile		275	.029		3	1.55		4.55	5.80
0900	Luminous panels, prismatic, acrylic		255	.031		4.08	1.67		5.75	7.15
1200	Metal pan with acoustic pad, steel		75	.107		5.15	5.65		10.80	14.75
1300	Painted aluminum		75	.107		3.83	5.65		9.48	13.30
1500	Aluminum, degreased finish		75	.107		5.10	5.65		10.75	14.70
1600	Stainless steel		75	.107		9.65	5.65		15.30	19.75
1800	Tile, Z bar suspension, 5/8″ mineral fiber tile		150	.053		2.15	2.83		4.98	6.90
1900	3/4″ mineral fiber tile		150	.053		2.44	2.83		5.27	7.20
2402	For strip lighting, see Section 26 51 13.50									
2500	For rooms under 500 S.F., add				S.F.		25%			
9000	Minimum labor/equipment charge	1 Carp	2	4	Job		213		213	340

09 51 33 – Acoustical Metal Pan Ceilings

09 51 33.10 Ceiling, Acoustical Metal Pan		Crew	Daily Output	Labor-Hours	Unit	Material	Labor	Equipment	Total	Total Incl O&P
0010	**CEILING, ACOUSTICAL METAL PAN**									
0100	Metal panel, lay-in, 2′ x 2′, sq. edge	1 Carp	500	.016	S.F.	11.10	.85		11.95	13.55
0110	Tegular edge		500	.016		14.20	.85		15.05	16.95
0120	2′ x 4′, sq. edge		500	.016		13.95	.85		14.80	16.70
0130	Tegular edge		500	.016		14.05	.85		14.90	16.80
0140	Perforated alum. clip-in, 2′ x 2′		500	.016		14.20	.85		15.05	16.95
0150	2′ x 4′		500	.016		11.60	.85		12.45	14.10

09 51 53 – Direct-Applied Acoustical Ceilings

09 51 53.10 Ceiling Tile		Crew	Daily Output	Labor-Hours	Unit	Material	Labor	Equipment	Total	Total Incl O&P
0010	**CEILING TILE**, stapled or cemented									
0100	12″ x 12″ or 12″ x 24″, not including furring									
0600	Mineral fiber, vinyl coated, 5/8″ thick	1 Carp	300	.027	S.F.	2.23	1.42		3.65	4.72
0700	3/4″ thick		300	.027		3.15	1.42		4.57	5.75
0900	Fire rated, 3/4″ thick, plain faced		300	.027		1.47	1.42		2.89	3.89
1000	Plastic coated face		300	.027		2.26	1.42		3.68	4.76
1200	Aluminum faced, 5/8″ thick, plain		300	.027		1.92	1.42		3.34	4.38
3700	Wall application of above, add		1000	.008			.43		.43	.68
3900	For ceiling primer, add					.12			.12	.13
4000	For ceiling cement, add					.42			.42	.46
9000	Minimum labor/equipment charge	1 Carp	4	2	Job		106		106	170

09 53 Acoustical Ceiling Suspension Assemblies

09 53 23 – Metal Acoustical Ceiling Suspension Assemblies

09 53 23.30 Ceiling Suspension Systems		Crew	Daily Output	Labor-Hours	Unit	Material	Labor	Equipment	Total	Total Incl O&P
0010	**CEILING SUSPENSION SYSTEMS** for boards and tile									
0050	Class A suspension system, 15/16″ T bar, 2′ x 4′ grid	1 Carp	800	.010	S.F.	.93	.53		1.46	1.88
0300	2′ x 2′ grid		650	.012		1.16	.65		1.81	2.33
0310	25% recycled steel, 2′ x 4′ grid G		800	.010		.91	.53		1.44	1.85
0320	2′ x 2′ grid G		650	.012		1.13	.65		1.78	2.29
0350	For 9/16″ grid, add					.16			.16	.18
0360	For fire rated grid, add					.09			.09	.10
0370	For colored grid, add					.22			.22	.24
0400	Concealed Z bar suspension system, 12″ module	1 Carp	520	.015		.86	.82		1.68	2.26
0600	1-1/2″ carrier channels, 4′ OC, add	″	470	.017		.12	.90		1.02	1.58
0700	Carrier channels for ceilings with									
0900	recessed lighting fixtures, add	1 Carp	460	.017	S.F.	.22	.92		1.14	1.72

09 53 Acoustical Ceiling Suspension Assemblies

09 53 23 – Metal Acoustical Ceiling Suspension Assemblies

	09 53 23.30 Ceiling Suspension Systems	Crew	Daily Output	Labor-Hours	Unit	Material	Labor	Equipment	Total	Total Incl O&P
						2020 Bare Costs				
1040	Hanging wire, 12 ga., 4' long	1 Carp	65	.123	C.S.F.	3.10	6.55		9.65	13.90
1080	8' long	↓	65	.123	"	6.20	6.55		12.75	17.30
3000	Seismic ceiling bracing, IBC Site Class D, Occupancy Category II									
3050	For ceilings less than 2500 S.F.									
3060	Seismic clips at attached walls	1 Carp	180	.044	Ea.	.98	2.36		3.34	4.86
3100	For ceilings greater than 2500 S.F., add									
3120	Seismic clips, joints at cross tees	1 Carp	120	.067	Ea.	2	3.54		5.54	7.90
3140	At cross tees and mains, mains field cut	"	60	.133	"	2	7.10		9.10	13.55
3200	Compression posts, telescopic, attached to structure above									
3210	To 30" high	1 Carp	26	.308	Ea.	40	16.35		56.35	70
3220	30" to 48" high		25.50	.314		42	16.65		58.65	72.50
3230	48" to 84" high		25	.320		50.50	17		67.50	83
3240	84" to 102" high		24.50	.327		57.50	17.35		74.85	91
3250	102" to 120" high		24	.333		83	17.70		100.70	120
3260	120" to 144" high	↓	24	.333	↓	92	17.70		109.70	130
3300	Stabilizer bars									
3310	12" long	1 Carp	240	.033	Ea.	1.11	1.77		2.88	4.06
3320	24" long		235	.034		1.04	1.81		2.85	4.04
3330	36" long		230	.035		1.02	1.85		2.87	4.08
3340	48" long	↓	220	.036	↓	.82	1.93		2.75	4.01
3400	Wire support for light fixtures, per L.F. height to structure above									
3410	Less than 10 lb.	1 Carp	400	.020	L.F.	.30	1.06		1.36	2.03
3420	10 lb. to 56 lb.	"	240	.033	"	.59	1.77		2.36	3.49
3500	Retrofit existing suspended ceiling grid to current code									
3510	Less than 2500 S.F. (using clips @ perimeter)	1 Carp	1455	.006	S.F.	.06	.29		.35	.54
3520	Greater than 2500 S.F. (using compression posts and clips)		550	.015		.42	.77		1.19	1.70
4000	Remove and replace ceiling grid, class A, 15/16" T bar, 2' x 2' grid		435	.018		1.16	.98		2.14	2.85
4010	2' x 4' grid	↓	535	.015	↓	.93	.79		1.72	2.30

09 54 Specialty Ceilings

09 54 16 – Luminous Ceilings

	09 54 16.10 Ceiling, Luminous	Crew	Daily Output	Labor-Hours	Unit	Material	Labor	Equipment	Total	Total Incl O&P
0010	**CEILING, LUMINOUS**									
0020	Translucent lay-in panels, 2' x 2'	1 Carp	500	.016	S.F.	23.50	.85		24.35	27
0030	2' x 6'	"	500	.016	"	18.05	.85		18.90	21

09 54 23 – Linear Metal Ceilings

	09 54 23.10 Metal Ceilings	Crew	Daily Output	Labor-Hours	Unit	Material	Labor	Equipment	Total	Total Incl O&P
0010	**METAL CEILINGS**									
0015	Solid alum. planks, 3-1/4" x 12', open reveal	1 Carp	500	.016	S.F.	2.38	.85		3.23	3.98
0020	Closed reveal		500	.016		3.05	.85		3.90	4.72
0030	7-1/4" x 12', open reveal		500	.016		4.07	.85		4.92	5.85
0040	Closed reveal		500	.016		5.20	.85		6.05	7.05
0050	Metal, open cell, 2' x 2', 6" cell		500	.016		8.80	.85		9.65	11
0060	8" cell		500	.016		9.70	.85		10.55	12.05
0070	2' x 4', 6" cell		500	.016		5.80	.85		6.65	7.70
0080	8" cell	↓	500	.016	↓	5.80	.85		6.65	7.70

09 54 26 – Suspended Wood Ceilings

	09 54 26.10 Wood Ceilings	Crew	Daily Output	Labor-Hours	Unit	Material	Labor	Equipment	Total	Total Incl O&P
0010	**WOOD CEILINGS**									
1000	4"-6" wood slats on heavy duty 15/16" T-bar grid	2 Carp	250	.064	S.F.	26	3.40		29.40	34

09 54 Specialty Ceilings

09 54 33 – Decorative Panel Ceilings

09 54 33.20 Metal Panel Ceilings		Crew	Daily Output	Labor-Hours	Unit	2020 Bare Costs Material	Labor	Equipment	Total	Total Incl O&P
0010	**METAL PANEL CEILINGS**									
0020	Lay-in or screwed to furring, not including grid									
0100	Tin ceilings, 2′ x 2′ or 2′ x 4′, bare steel finish	2 Carp	300	.053	S.F.	2.22	2.83		5.05	7
0120	Painted white finish		300	.053	"	3.53	2.83		6.36	8.40
0140	Copper, chrome or brass finish		300	.053	L.F.	6.30	2.83		9.13	11.50
0200	Cornice molding, 2-1/2″ to 3-1/2″ wide, 4′ long, bare steel finish		200	.080	S.F.	2.17	4.25		6.42	9.20
0220	Painted white finish		200	.080		2.87	4.25		7.12	9.95
0240	Copper, chrome or brass finish		200	.080		3.96	4.25		8.21	11.15
0320	5″ to 6-1/2″ wide, 4′ long, bare steel finish		150	.107		3.37	5.65		9.02	12.80
0340	Painted white finish		150	.107		3.89	5.65		9.54	13.40
0360	Copper, chrome or brass finish		150	.107		5.70	5.65		11.35	15.35
0420	Flat molding, 3-1/2″ to 5″ wide, 4′ long, bare steel finish		250	.064		3.89	3.40		7.29	9.70
0440	Painted white finish		250	.064		3.94	3.40		7.34	9.80
0460	Copper, chrome or brass finish		250	.064		7.15	3.40		10.55	13.30

09 57 Special Function Ceilings

09 57 53 – Security Ceiling Assemblies

09 57 53.10 Ceiling Assem., Security, Radio Freq. Shielding		Crew	Daily Output	Labor-Hours	Unit	Material	Labor	Equipment	Total	Total Incl O&P
0010	**CEILING ASSEMBLY, SECURITY, RADIO FREQUENCY SHIELDING**									
0020	Prefabricated, galvanized steel	2 Carp	375	.043	SF Surf	5.65	2.27		7.92	9.90
0050	5 oz., copper ceiling panel		155	.103		4.72	5.50		10.22	14
0110	12 oz., copper ceiling panel		140	.114		9.65	6.05		15.70	20.50
0250	Ceiling hangers	E-1	45	.533	Ea.	40	30	3.27	73.27	96.50
0300	Shielding transition, 11 ga. preformed angles	"	1365	.018	L.F.	40	1	.11	41.11	45.50

09 61 Flooring Treatment

09 61 19 – Concrete Floor Staining

09 61 19.40 Floors, Interior		Crew	Daily Output	Labor-Hours	Unit	Material	Labor	Equipment	Total	Total Incl O&P
0010	**FLOORS, INTERIOR**									
0300	Acid stain and sealer									
0310	Stain, one coat	1 Pord	650	.012	S.F.	.15	.55		.70	1.03
0320	Two coats		570	.014		.29	.62		.91	1.31
0330	Acrylic sealer, one coat		2600	.003		.25	.14		.39	.50
0340	Two coats		1400	.006		.51	.25		.76	.96

09 62 Specialty Flooring

09 62 19 – Laminate Flooring

09 62 19.10 Floating Floor		Crew	Daily Output	Labor-Hours	Unit	Material	Labor	Equipment	Total	Total Incl O&P
0010	**FLOATING FLOOR**									
8300	Floating floor, laminate, wood pattern strip, complete	1 Clab	133	.060	S.F.	4.95	2.53		7.48	9.50
8310	Components, T&G wood composite strips					4.36			4.36	4.79
8320	Film					.21			.21	.23
8330	Foam					.30			.30	.33
8340	Adhesive					.47			.47	.52
8350	Installation kit					.24			.24	.26
8360	Trim, 2″ wide x 3′ long				L.F.	5.20			5.20	5.70
8370	Reducer moulding				"	6.15			6.15	6.80

09 62 Specialty Flooring

09 62 29 – Cork Flooring

09 62 29.10 Cork Tile Flooring			Crew	Daily Output	Labor-Hours	Unit	2020 Bare Costs Material	Labor	Equipment	Total	Total Incl O&P
0010	**CORK TILE FLOORING**										
2200	Cork tile, standard finish, 1/8" thick	G	1 Tilf	315	.025	S.F.	5.25	1.26		6.51	7.70
2250	3/16" thick	G		315	.025		6.20	1.26		7.46	8.75
2300	5/16" thick	G		315	.025		6.65	1.26		7.91	9.30
2350	1/2" thick	G		315	.025		6.95	1.26		8.21	9.60
2500	Urethane finish, 1/8" thick	G		315	.025		5.10	1.26		6.36	7.55
2550	3/16" thick	G		315	.025		8.20	1.26		9.46	10.95
2600	5/16" thick	G		315	.025		7.50	1.26		8.76	10.20
2650	1/2" thick	G		315	.025		7.80	1.26		9.06	10.50

09 63 Masonry Flooring

09 63 13 – Brick Flooring

09 63 13.10 Miscellaneous Brick Flooring			Crew	Daily Output	Labor-Hours	Unit	Material	Labor	Equipment	Total	Total Incl O&P
0010	**MISCELLANEOUS BRICK FLOORING**										
0020	Acid-proof shales, red, 8" x 3-3/4" x 1-1/4" thick		D-7	.43	37.209	M	725	1,650		2,375	3,375
0050	2-1/4" thick		D-1	.40	40		1,075	1,875		2,950	4,225
0200	Acid-proof clay brick, 8" x 3-3/4" x 2-1/4" thick	G		.40	40		1,025	1,875		2,900	4,150
0250	9" x 4-1/2" x 3"	G		95	.168	S.F.	4.42	7.90		12.32	17.60
0260	Cast ceramic, pressed, 4" x 8" x 1/2", unglazed		D-7	100	.160		7.55	7.05		14.60	19.40
0270	Glazed			100	.160		10.10	7.05		17.15	22
0280	Hand molded flooring, 4" x 8" x 3/4", unglazed			95	.168		10	7.45		17.45	22.50
0290	Glazed			95	.168		11.40	7.45		18.85	24
0300	8" hexagonal, 3/4" thick, unglazed			85	.188		9.95	8.30		18.25	24
0310	Glazed			85	.188		18	8.30		26.30	33
0400	Heavy duty industrial, cement mortar bed, 2" thick, not incl. brick		D-1	80	.200		1.18	9.40		10.58	16.45
0450	Acid-proof joints, 1/4" wide		"	65	.246		1.74	11.55		13.29	20.50
0500	Pavers, 8" x 4", 1" to 1-1/4" thick, red		D-7	95	.168		4.39	7.45		11.84	16.50
0510	Ironspot		"	95	.168		5.65	7.45		13.10	17.85
0540	1-3/8" to 1-3/4" thick, red		D-1	95	.168		4.24	7.90		12.14	17.40
0560	Ironspot			95	.168		5.60	7.90		13.50	18.90
0580	2-1/4" thick, red			90	.178		4.32	8.35		12.67	18.20
0590	Ironspot			90	.178		6.70	8.35		15.05	21
0700	Paver, adobe brick, 6" x 12", 1/2" joint	G		42	.381		1.47	17.85		19.32	30.50
0710	Mexican red, 12" x 12"	G	1 Tilf	48	.167		2.08	8.25		10.33	15.25
0720	Saltillo, 12" x 12"	G	"	48	.167		1.66	8.25		9.91	14.80
0800	For sidewalks and patios with pavers, see Section 32 14 16.10										
0870	For epoxy joints, add		D-1	600	.027	S.F.	3.15	1.25		4.40	5.50
0880	For Furan underlayment, add		"	600	.027		2.60	1.25		3.85	4.88
0890	For waxed surface, steam cleaned, add		A-1H	1000	.008		.21	.34	.08	.63	.85
9000	Minimum labor/equipment charge		1 Bric	2	4	Job		208		208	335

09 63 40 – Stone Flooring

09 63 40.10 Marble		Crew	Daily Output	Labor-Hours	Unit	Material	Labor	Equipment	Total	Total Incl O&P
0010	**MARBLE**									
0020	Thin gauge tile, 12" x 6", 3/8", white Carara	D-7	60	.267	S.F.	17.05	11.75		28.80	37.50
0100	Travertine		60	.267		9.50	11.75		21.25	29
0200	12" x 12" x 3/8", thin set, floors		60	.267		11.65	11.75		23.40	31.50
0300	On walls		52	.308		10	13.60		23.60	32.50
1000	Marble threshold, 4" wide x 36" long x 5/8" thick, white		60	.267	Ea.	11.65	11.75		23.40	31.50
9000	Minimum labor/equipment charge		3	5.333	Job		235		235	370

09 63 Masonry Flooring

09 63 40 – Stone Flooring

09 63 40.20 Slate Tile		Crew	Daily Output	Labor-Hours	Unit	Material	Labor	Equipment	Total	Total Incl O&P
						2020 Bare Costs				
0010	**SLATE TILE**									
0020	Vermont, 6" x 6" x 1/4" thick, thin set	D-7	180	.089	S.F.	7.85	3.92		11.77	14.75
0200	See also Section 32 14 40.10									
9000	Minimum labor/equipment charge	D-7	3	5.333	Job		235		235	370

09 63 40.95 Concrete Floors

0010	**CONCRETE FLOORS** And toppings, see Section 03 35 29.30									

09 64 Wood Flooring

09 64 16 – Wood Block Flooring

09 64 16.10 End Grain Block Flooring

		Crew	Daily Output	Labor-Hours	Unit	Material	Labor	Equipment	Total	Total Incl O&P
0010	**END GRAIN BLOCK FLOORING**									
0020	End grain flooring, coated, 2" thick	1 Carp	295	.027	S.F.	3.78	1.44		5.22	6.45
0400	Natural finish, 1" thick, fir		125	.064		3.91	3.40		7.31	9.75
0600	1-1/2" thick, pine		125	.064		3.83	3.40		7.23	9.65
0700	2" thick, pine		125	.064		5.70	3.40		9.10	11.70
9000	Minimum labor/equipment charge		2	4	Job		213		213	340

09 64 19 – Wood Composition Flooring

09 64 19.10 Wood Composition

		Crew	Daily Output	Labor-Hours	Unit	Material	Labor	Equipment	Total	Total Incl O&P
0010	**WOOD COMPOSITION** Gym floors									
0100	2-1/4" x 6-7/8" x 3/8", on adh, corkbd & bond coat	D-7	150	.107	S.F.	8.20	4.71		12.91	16.45
0200	Thin set, on concrete	"	250	.064		5.80	2.83		8.63	10.85
0300	Sanding and finishing, add	1 Carp	200	.040		.90	2.13		3.03	4.40

09 64 23 – Wood Parquet Flooring

09 64 23.10 Wood Parquet

		Crew	Daily Output	Labor-Hours	Unit	Material	Labor	Equipment	Total	Total Incl O&P
0010	**WOOD PARQUET** flooring									
5200	Parquetry, 5/16" thk, no finish, oak, plain pattern	1 Carp	160	.050	S.F.	5.50	2.66		8.16	10.30
5300	Intricate pattern		100	.080		10.55	4.25		14.80	18.40
5500	Teak, plain pattern		160	.050		6.05	2.66		8.71	10.95
5600	Intricate pattern		100	.080		10.35	4.25		14.60	18.20
5650	13/16" thick, select grade oak, plain pattern		160	.050		11.35	2.66		14.01	16.75
5700	Intricate pattern		100	.080		17.05	4.25		21.30	25.50
5800	Custom parquetry, including finish, plain pattern		100	.080		17.75	4.25		22	26.50
5900	Intricate pattern		50	.160		24.50	8.50		33	40.50
6700	Parquetry, prefinished white oak, 5/16" thick, plain pattern		160	.050		8.80	2.66		11.46	13.95
6800	Intricate pattern		100	.080		8.60	4.25		12.85	16.30
7000	Walnut or teak, parquetry, plain pattern		160	.050		9.20	2.66		11.86	14.40
7100	Intricate pattern		100	.080		16.40	4.25		20.65	25
7200	Acrylic wood parquet blocks, 12" x 12" x 5/16",									
7210	Irradiated, set in epoxy	1 Carp	160	.050	S.F.	10.90	2.66		13.56	16.25

09 64 29 – Wood Strip and Plank Flooring

09 64 29.10 Wood

		Crew	Daily Output	Labor-Hours	Unit	Material	Labor	Equipment	Total	Total Incl O&P
0010	**WOOD**									
0020	Fir, vertical grain, 1" x 4", not incl. finish, grade B & better	1 Carp	255	.031	S.F.	3.65	1.67		5.32	6.70
0100	Grade C & better		255	.031		3.43	1.67		5.10	6.45
4000	Maple, strip, 25/32" x 2-1/4", not incl. finish, select		170	.047		5.10	2.50		7.60	9.60
4100	#2 & better		170	.047		5.15	2.50		7.65	9.65
4300	33/32" x 3-1/4", not incl. finish, #1 grade		170	.047		6.05	2.50		8.55	10.65
4400	#2 & better		170	.047		5.25	2.50		7.75	9.75

09 64 Wood Flooring

09 64 29 – Wood Strip and Plank Flooring

09 64 29.10 Wood		Crew	Daily Output	Labor-Hours	Unit	Material	2020 Bare Costs Labor	Equipment	Total	Total Incl O&P
4600	Oak, white or red, 25/32" x 2-1/4", not incl. finish									
4700	#1 common	1 Carp	170	.047	S.F.	3.62	2.50		6.12	8
4900	Select quartered, 2-1/4" wide		170	.047		4.48	2.50		6.98	8.95
5000	Clear		170	.047		4.43	2.50		6.93	8.90
6100	Prefinished, white oak, prime grade, 2-1/4" wide		170	.047		6.60	2.50		9.10	11.25
6200	3-1/4" wide		185	.043		5.85	2.30		8.15	10.10
6400	Ranch plank		145	.055		6.90	2.93		9.83	12.30
6500	Hardwood blocks, 9" x 9", 25/32" thick		160	.050		7.90	2.66		10.56	12.95
7400	Yellow pine, 3/4" x 3-1/8", T&G, C & better, not incl. finish		200	.040		1.93	2.13		4.06	5.55
7550	Refinish old floors, see Section 01 93 13.09									
7800	Sanding and finishing, 2 coats polyurethane	1 Clab	295	.027	S.F.	.23	1.14		1.37	2.08
7900	Subfloor and underlayment, see Section 06 16									
8015	Transition molding, 2-1/4" wide, 5' long	1 Carp	19.20	.417	Ea.	21	22		43	59
9000	Minimum labor/equipment charge	"	2	4	Job		213		213	340

09 64 36 – Bamboo Flooring

09 64 36.10 Flooring, Bamboo		Crew	Daily Output	Labor-Hours	Unit	Material	Labor	Equipment	Total	Total Incl O&P
0010	**FLOORING, BAMBOO**									
8600	Flooring, wood, bamboo strips, unfinished, 5/8" x 4" x 3' G	1 Carp	255	.031	S.F.	6	1.67		7.67	9.25
8610	5/8" x 4" x 4' G		275	.029		6.25	1.55		7.80	9.35
8620	5/8" x 4" x 6' G		295	.027		6.85	1.44		8.29	9.85
8630	Finished, 5/8" x 4" x 3' G		255	.031		6.60	1.67		8.27	9.90
8640	5/8" x 4" x 4' G		275	.029		6.95	1.55		8.50	10.15
8650	5/8" x 4" x 6' G		295	.027		5.10	1.44		6.54	7.95
8660	Stair treads, unfinished, 1-1/16" x 11-1/2" x 4' G		18	.444	Ea.	57.50	23.50		81	102
8670	Finished, 1-1/16" x 11-1/2" x 4' G		18	.444		91	23.50		114.50	138
8680	Stair risers, unfinished, 5/8" x 7-1/2" x 4' G		18	.444		21.50	23.50		45	61.50
8690	Finished, 5/8" x 7-1/2" x 4' G		18	.444		40.50	23.50		64	82.50
8700	Stair nosing, unfinished, 6' long G		16	.500		47	26.50		73.50	94.50
8710	Finished, 6' long G		16	.500		45	26.50		71.50	92

09 64 66 – Wood Athletic Flooring

09 64 66.10 Gymnasium Flooring		Crew	Daily Output	Labor-Hours	Unit	Material	Labor	Equipment	Total	Total Incl O&P
0010	**GYMNASIUM FLOORING**									
0600	Gym floor, in mastic, over 2 ply felt, #2 & better									
0700	25/32" thick maple	1 Carp	100	.080	S.F.	4.30	4.25		8.55	11.55
0900	33/32" thick maple		98	.082		7.15	4.34		11.49	14.80
1000	For 1/2" corkboard underlayment, add		750	.011		1.25	.57		1.82	2.29
1300	For #1 grade maple, add					.64			.64	.70
1600	Maple flooring, over sleepers, #2 & better									
1700	25/32" thick	1 Carp	85	.094	S.F.	6.25	5		11.25	14.85
1900	33/32" thick	"	83	.096		7.05	5.10		12.15	15.95
2000	For #1 grade, add					.71			.71	.78
2200	For 3/4" subfloor, add	1 Carp	350	.023		1.60	1.22		2.82	3.71
2300	With two 1/2" subfloors, 25/32" thick	"	69	.116		7.80	6.15		13.95	18.40
2500	Maple, incl. finish, #2 & btr., 25/32" thick, on rubber									
2600	Sleepers, with two 1/2" subfloors	1 Carp	76	.105	S.F.	8.10	5.60		13.70	17.85
2800	With steel spline, double connection to channels	"	73	.110		8.90	5.80		14.70	19.15
2900	For 33/32" maple, add					.92			.92	1.01
3100	For #1 grade maple, add					.71			.71	.78
3500	For termite proofing all of the above, add					.38			.38	.42
3700	Portable hardwood, prefinished panels	1 Carp	83	.096		10.80	5.10		15.90	20
3720	Insulated with polystyrene, 1" thick, add		165	.048		.92	2.58		3.50	5.15
3750	Running tracks, Sitka spruce surface, 25/32" x 2-1/4"		62	.129		20.50	6.85		27.35	33.50

09 64 Wood Flooring

09 64 66 – Wood Athletic Flooring

09 64 66.10 Gymnasium Flooring		Crew	Daily Output	Labor-Hours	Unit	Material	2020 Bare Costs Labor	Equipment	Total	Total Incl O&P
3770	3/4" plywood surface, finished	1 Carp	100	.080	S.F.	4.70	4.25		8.95	11.95

09 65 Resilient Flooring

09 65 10 – Resilient Tile Underlayment

09 65 10.10 Latex Underlayment

		Crew	Daily Output	Labor-Hours	Unit	Material	Labor	Equipment	Total	Total Incl O&P
0010	LATEX UNDERLAYMENT									
3600	Latex underlayment, 1/8" thk., cementitious for resilient flooring	1 Tilf	160	.050	S.F.	1.29	2.48		3.77	5.30
4000	Liquid, fortified				Gal.	30.50			30.50	33.50

09 65 13 – Resilient Base and Accessories

09 65 13.13 Resilient Base

		Crew	Daily Output	Labor-Hours	Unit	Material	Labor	Equipment	Total	Total Incl O&P
0010	RESILIENT BASE									
0690	1/8" vinyl base, 2-1/2" H, straight or cove, standard colors	1 Tilf	315	.025	L.F.	.78	1.26		2.04	2.83
0700	4" high		315	.025		1.17	1.26		2.43	3.26
0710	6" high		315	.025		1.50	1.26		2.76	3.62
0720	Corners, 2-1/2" high		315	.025	Ea.	2.27	1.26		3.53	4.47
0730	4" high		315	.025		3.08	1.26		4.34	5.35
0740	6" high		315	.025		2.91	1.26		4.17	5.15
0800	1/8" rubber base, 2-1/2" H, straight or cove, standard colors		315	.025	L.F.	1.20	1.26		2.46	3.29
1100	4" high		315	.025		1.43	1.26		2.69	3.54
1110	6" high		315	.025		2.08	1.26		3.34	4.26
1150	Corners, 2-1/2" high		315	.025	Ea.	2.52	1.26		3.78	4.74
1153	4" high		315	.025		2.60	1.26		3.86	4.83
1155	6" high		315	.025		3.58	1.26		4.84	5.90
1450	For premium color/finish add					50%				
1500	Millwork profile	1 Tilf	315	.025	L.F.	6.45	1.26		7.71	9

09 65 13.23 Resilient Stair Treads and Risers

		Crew	Daily Output	Labor-Hours	Unit	Material	Labor	Equipment	Total	Total Incl O&P
0010	RESILIENT STAIR TREADS AND RISERS									
0300	Rubber, molded tread, 12" wide, 5/16" thick, black	1 Tilf	115	.070	L.F.	14.35	3.44		17.79	21
0400	Colors		115	.070		15	3.44		18.44	22
0600	1/4" thick, black		115	.070		13.45	3.44		16.89	20
0700	Colors		115	.070		16.25	3.44		19.69	23.50
0900	Grip strip safety tread, colors, 5/16" thick		115	.070		20.50	3.44		23.94	28
1000	3/16" thick		120	.067		16.70	3.30		20	23.50
1200	Landings, smooth sheet rubber, 1/8" thick		120	.067	S.F.	8.60	3.30		11.90	14.60
1300	3/16" thick		120	.067	"	9.35	3.30		12.65	15.40
1500	Nosings, 3" wide, 3/16" thick, black		140	.057	L.F.	4.79	2.83		7.62	9.70
1600	Colors		140	.057		6.50	2.83		9.33	11.60
1800	Risers, 7" high, 1/8" thick, flat		250	.032		8.55	1.58		10.13	11.90
1900	Coved		250	.032		9.55	1.58		11.13	13
2100	Vinyl, molded tread, 12" wide, colors, 1/8" thick		115	.070		6.05	3.44		9.49	12.05
2200	1/4" thick		115	.070		8.15	3.44		11.59	14.35
2300	Landing material, 1/8" thick		200	.040	S.F.	6.35	1.98		8.33	10.10
2400	Riser, 7" high, 1/8" thick, coved		175	.046	L.F.	3.39	2.26		5.65	7.25
2500	Tread and riser combined, 1/8" thick		80	.100	"	9.90	4.95		14.85	18.65
9000	Minimum labor/equipment charge		3	2.667	Job		132		132	207

09 65 13.37 Vinyl Transition Strips

		Crew	Daily Output	Labor-Hours	Unit	Material	Labor	Equipment	Total	Total Incl O&P
0010	VINYL TRANSITION STRIPS									
0100	Various mats. to various mats., adhesive applied, 1/4" to 1/8"	1 Tilf	315	.025	L.F.	1.51	1.26		2.77	3.63
0105	0.08" to 1/8"		315	.025		1.36	1.26		2.62	3.47
0110	0.08" to 1/4"		315	.025		1.49	1.26		2.75	3.61

09 65 Resilient Flooring

09 65 13 – Resilient Base and Accessories

09 65 13.37 Vinyl Transition Strips		Crew	Daily Output	Labor-Hours	Unit	2020 Bare Costs Material	Labor	Equipment	Total	Total Incl O&P
0115	1/4" to 3/8"	1 Tilf	315	.025	L.F.	1.38	1.26		2.64	3.49
0120	1/4" to 1/2"		315	.025		1.38	1.26		2.64	3.49
0125	1/4" to 0.08"		315	.025		1.49	1.26		2.75	3.61
0200	Vinyl wheeled trans. strips, carpet to var. mats., 1/4" to 1/8" x 2-1/2"		315	.025		5.15	1.26		6.41	7.60
0205	1/4" to 1/8" x 4"		315	.025		6.25	1.26		7.51	8.85
0210	Various mats. to various mats. 1/4" to 0.08" x 2-1/2"		315	.025		5.15	1.26		6.41	7.65
0215	Carpet to various materials, 1/4" to flush x 2-1/2"		315	.025		4.29	1.26		5.55	6.70
0220	1/4" to flush x 4"		315	.025		6.60	1.26		7.86	9.20
0225	Various materials to resilient, 3/8" to 1/8" x 2-1/2"		315	.025		4.29	1.26		5.55	6.70
0230	Carpet to various materials, 3/8" to 1/4" x 2-1/2"		315	.025		5.90	1.26		7.16	8.45
0235	1/4" to 1/4" x 2-1/2"		315	.025		6.60	1.26		7.86	9.20
0240	Various materials to resilient, 1/8" to 1/8" x 2-1/2"		315	.025		5	1.26		6.26	7.45
0245	Various materials to var. mats., 1/8" to flush x 2-1/2"		315	.025		3.35	1.26		4.61	5.65
0250	3/8" to flush x 4"		315	.025		6.75	1.26		8.01	9.35
0255	1/2" to flush x 4"		315	.025		9.05	1.26		10.31	11.90
0260	Various materials to resilient, 1/8" to 0.08" x 2-1/2"		315	.025		3.88	1.26		5.14	6.25
0265	0.08" to 0.08" x 2-1/2"		315	.025		3.68	1.26		4.94	6
0270	3/8" to 0.08" x 2-1/2"	↓	315	.025	↓	3.68	1.26		4.94	6

09 65 16 – Resilient Sheet Flooring

09 65 16.10 Rubber and Vinyl Sheet Flooring		Crew	Daily Output	Labor-Hours	Unit	Material	Labor	Equipment	Total	Total Incl O&P
0010	**RUBBER AND VINYL SHEET FLOORING**									
5500	Linoleum, sheet goods G	1 Tilf	360	.022	S.F.	3.63	1.10		4.73	5.70
5900	Rubber, sheet goods, 36" wide, 1/8" thick		120	.067		9.25	3.30		12.55	15.30
5950	3/16" thick		100	.080		10	3.96		13.96	17.20
6000	1/4" thick		90	.089		11.75	4.40		16.15	19.85
8000	Vinyl sheet goods, backed, .065" thick, plain pattern/colors		250	.032		4.52	1.58		6.10	7.45
8050	Intricate pattern/colors		200	.040		3.82	1.98		5.80	7.30
8100	.080" thick, plain pattern/colors		230	.035		4.15	1.72		5.87	7.25
8150	Intricate pattern/colors		200	.040		6.40	1.98		8.38	10.15
8200	.125" thick, plain pattern/colors		230	.035		4.19	1.72		5.91	7.30
8250	Intricate pattern/colors		200	.040	↓	7.35	1.98		9.33	11.15
8400	For welding seams, add		100	.080	L.F.	.24	3.96		4.20	6.45
8450	For integral cove base, add	↓	175	.046	"	.82	2.26		3.08	4.44
8700	Adhesive cement, 1 gallon per 200 to 300 S.F.				Gal.	33			33	36
8800	Asphalt primer, 1 gallon per 300 S.F.					15.10			15.10	16.60
8900	Emulsion, 1 gallon per 140 S.F.				↓	19.95			19.95	22

09 65 19 – Resilient Tile Flooring

09 65 19.19 Vinyl Composition Tile Flooring		Crew	Daily Output	Labor-Hours	Unit	Material	Labor	Equipment	Total	Total Incl O&P
0010	**VINYL COMPOSITION TILE FLOORING**									
7000	Vinyl composition tile, 12" x 12", 1/16" thick	1 Tilf	500	.016	S.F.	1.23	.79		2.02	2.59
7050	Embossed		500	.016		2.72	.79		3.51	4.23
7100	Marbleized		500	.016		2.72	.79		3.51	4.23
7150	Solid		500	.016		3.51	.79		4.30	5.10
7200	3/32" thick, embossed		500	.016		1.53	.79		2.32	2.92
7250	Marbleized		500	.016		3.13	.79		3.92	4.68
7300	Solid		500	.016		2.89	.79		3.68	4.42
7350	1/8" thick, marbleized		500	.016		2.47	.79		3.26	3.96
7400	Solid		500	.016		1.75	.79		2.54	3.17
7450	Conductive	↓	500	.016	↓	5.90	.79		6.69	7.70

09 65 Resilient Flooring

09 65 19 – Resilient Tile Flooring

09 65 19.23 Vinyl Tile Flooring		Crew	Daily Output	Labor-Hours	Unit	Material	2020 Bare Costs Labor	Equipment	Total	Total Incl O&P
0010	**VINYL TILE FLOORING**									
7500	Vinyl tile, 12" x 12", 3/32" thick, standard colors/patterns	1 Tilf	500	.016	S.F.	3.58	.79		4.37	5.20
7550	1/8" thick, standard colors/patterns		500	.016		5.25	.79		6.04	7
7600	1/8" thick, premium colors/patterns		500	.016		6.90	.79		7.69	8.85
7650	Solid colors		500	.016		3.09	.79		3.88	4.64
7700	Marbleized or Travertine pattern		500	.016		6.35	.79		7.14	8.20
7750	Florentine pattern		500	.016		7.20	.79		7.99	9.15
7800	Premium colors/patterns		500	.016		7.25	.79		8.04	9.20
9500	Minimum labor/equipment charge		4	2	Job		99		99	155

09 65 19.33 Rubber Tile Flooring		Crew	Daily Output	Labor-Hours	Unit	Material	Labor	Equipment	Total	Total Incl O&P
0010	**RUBBER TILE FLOORING**									
6050	Rubber tile, marbleized colors, 12" x 12", 1/8" thick	1 Tilf	400	.020	S.F.	6	.99		6.99	8.15
6100	3/16" thick		400	.020		8.40	.99		9.39	10.80
6300	Special tile, plain colors, 1/8" thick		400	.020		8.30	.99		9.29	10.65
6350	3/16" thick		400	.020		11.15	.99		12.14	13.80
6410	Raised, radial or square, .5 mm black		400	.020		8.05	.99		9.04	10.40
6430	.5 mm colored		400	.020		7.60	.99		8.59	9.90
6450	For golf course, skating rink, etc., 1/4" thick		275	.029		11	1.44		12.44	14.35

09 65 33 – Conductive Resilient Flooring

09 65 33.10 Conductive Rubber and Vinyl Flooring		Crew	Daily Output	Labor-Hours	Unit	Material	Labor	Equipment	Total	Total Incl O&P
0010	**CONDUCTIVE RUBBER AND VINYL FLOORING**									
1700	Conductive flooring, rubber tile, 1/8" thick	1 Tilf	315	.025	S.F.	7.35	1.26		8.61	10.05
1800	Homogeneous vinyl tile, 1/8" thick	"	315	.025	"	7.35	1.26		8.61	10.05

09 65 66 – Resilient Athletic Flooring

09 65 66.10 Resilient Athletic Flooring		Crew	Daily Output	Labor-Hours	Unit	Material	Labor	Equipment	Total	Total Incl O&P
0010	**RESILIENT ATHLETIC FLOORING**									
1000	Recycled rubber rolled goods, for weight rooms, 3/8" thk.	1 Tilf	315	.025	S.F.	2.86	1.26		4.12	5.10
1050	Interlocking 2' x 2' squares, rubber, 1/4" thk.		310	.026		1.93	1.28		3.21	4.12
1055	5/16" thk.		310	.026		2.59	1.28		3.87	4.85
1060	3/8" thk.		300	.027		2.94	1.32		4.26	5.30
1065	1/2" thk.		310	.026		3.73	1.28		5.01	6.10
2000	Vinyl sheet flooring, 1/4" thk.		315	.025		4.93	1.26		6.19	7.35

09 66 Terrazzo Flooring

09 66 13 – Portland Cement Terrazzo Flooring

09 66 13.10 Portland Cement Terrazzo		Crew	Daily Output	Labor-Hours	Unit	Material	Labor	Equipment	Total	Total Incl O&P
0010	**PORTLAND CEMENT TERRAZZO**, cast-in-place									
0020	Cove base, 6" high, 16 ga. zinc	1 Mstz	20	.400	L.F.	3.75	19.75		23.50	35
0100	Curb, 6" high and 6" wide		6	1.333		6.60	66		72.60	110
0300	Divider strip for floors, 14 ga., 1-1/4" deep, zinc		375	.021		1.43	1.05		2.48	3.22
0400	Brass		375	.021		2.83	1.05		3.88	4.76
0600	Heavy top strip 1/4" thick, 1-1/4" deep, zinc		300	.027		2.26	1.32		3.58	4.55
0900	Galv. bottoms, brass		300	.027		2.78	1.32		4.10	5.10
1200	For thin set floors, 16 ga., 1/2" x 1/2", zinc		350	.023		1.40	1.13		2.53	3.31
1300	Brass		350	.023		2.87	1.13		4	4.93
1500	Floor, bonded to concrete, 1-3/4" thick, gray cement	J-3	75	.213	S.F.	3.84	9.60	3.54	16.98	23
1600	White cement, mud set		75	.213		4.53	9.60	3.54	17.67	24
1800	Not bonded, 3" total thickness, gray cement		70	.229		4.73	10.25	3.80	18.78	25.50
1900	White cement, mud set		70	.229		5.55	10.25	3.80	19.60	26.50

09 66 Terrazzo Flooring

09 66 13 – Portland Cement Terrazzo Flooring

09 66 13.10 Portland Cement Terrazzo		Crew	Daily Output	Labor-Hours	Unit	2020 Bare Costs Material	Labor	Equipment	Total	Total Incl O&P
2100	For Venetian terrazzo, 1" topping, add					50%	50%			
2200	For heavy duty abrasive terrazzo, add					50%	50%			
2700	Monolithic terrazzo, 1/2" thick									
2710	10' panels	J-3	125	.128	S.F.	3.45	5.75	2.13	11.33	15.15
3000	Stairs, cast-in-place, pan filled treads		30	.533	L.F.	4.07	24	8.85	36.92	51.50
3100	Treads and risers		14	1.143	"	6.25	51.50	19	76.75	108
3300	For stair landings, add to floor prices						50%			
3400	Stair stringers and fascia	J-3	30	.533	S.F.	5.40	24	8.85	38.25	53
3600	For abrasive metal nosings on stairs, add		150	.107	L.F.	9.90	4.79	1.77	16.46	20.50
3700	For abrasive surface finish, add		600	.027	S.F.	1.84	1.20	.44	3.48	4.39
3900	For raised abrasive strips, add		150	.107	L.F.	1.39	4.79	1.77	7.95	11
4000	Wainscot, bonded, 1-1/2" thick		30	.533	S.F.	4.06	24	8.85	36.91	51.50
4200	1/4" thick		40	.400	"	6.50	18	6.65	31.15	42.50
4300	Stone chips, onyx gemstone, per 50 lb. bag				Bag	19.25			19.25	21
9000	Minimum labor/equipment charge	1 Mstz	1	8	Job		395		395	620

09 66 16 – Terrazzo Floor Tile

09 66 16.10 Tile or Terrazzo Base

		Crew	Daily Output	Labor-Hours	Unit	Material	Labor	Equipment	Total	Total Incl O&P
0010	**TILE OR TERRAZZO BASE**									
0020	Scratch coat only	1 Mstz	150	.053	S.F.	.58	2.63		3.21	4.76
0500	Scratch and brown coat only		75	.107	"	1.15	5.25		6.40	9.50
9000	Minimum labor/equipment charge		1	8	Job		395		395	620

09 66 16.13 Portland Cement Terrazzo Floor Tile

		Crew	Daily Output	Labor-Hours	Unit	Material	Labor	Equipment	Total	Total Incl O&P
0010	**PORTLAND CEMENT TERRAZZO FLOOR TILE**									
1200	Floor tiles, non-slip, 1" thick, 12" x 12"	D-1	60	.267	S.F.	25.50	12.50		38	48
1300	1-1/4" thick, 12" x 12"		60	.267		26	12.50		38.50	49
1500	16" x 16"		50	.320		28	15		43	55.50
1600	1-1/2" thick, 16" x 16"		45	.356		26	16.70		42.70	55.50
1800	For Venetian terrazzo, add					7.75			7.75	8.55
1900	For white cement, add					.73			.73	.80

09 66 16.16 Plastic Matrix Terrazzo Floor Tile

		Crew	Daily Output	Labor-Hours	Unit	Material	Labor	Equipment	Total	Total Incl O&P
0010	**PLASTIC MATRIX TERRAZZO FLOOR TILE**									
0100	12" x 12", 3/16" thick, floor tiles w/marble chips	1 Tilf	500	.016	S.F.	7.65	.79		8.44	9.65
0200	12" x 12", 3/16" thick, floor tiles w/glass chips		500	.016		8.25	.79		9.04	10.30
0300	12" x 12", 3/16" thick, floor tiles w/recycled content		500	.016		6.50	.79		7.29	8.40

09 66 16.30 Terrazzo, Precast

		Crew	Daily Output	Labor-Hours	Unit	Material	Labor	Equipment	Total	Total Incl O&P
0010	**TERRAZZO, PRECAST**									
0020	Base, 6" high, straight	1 Mstz	70	.114	L.F.	12.65	5.65		18.30	23
0100	Cove		60	.133		16.95	6.60		23.55	29
0300	8" high, straight		60	.133		16.45	6.60		23.05	28.50
0400	Cove		50	.160		24.50	7.90		32.40	39.50
0600	For white cement, add					.60			.60	.66
0700	For 16 ga. zinc toe strip, add					2.33			2.33	2.56
0900	Curbs, 4" x 4" high	1 Mstz	40	.200		43.50	9.90		53.40	63.50
1000	8" x 8" high	"	30	.267		47	13.15		60.15	72
2400	Stair treads, 1-1/2" thick, non-slip, three line pattern	2 Mstz	70	.229		54	11.30		65.30	77
2500	Nosing and two lines		70	.229		54	11.30		65.30	77
2700	2" thick treads, straight		60	.267		64.50	13.15		77.65	91.50
2800	Curved		50	.320		84.50	15.80		100.30	118
3000	Stair risers, 1" thick, to 6" high, straight sections		60	.267		15.30	13.15		28.45	37.50
3100	Cove		50	.320		19.50	15.80		35.30	46.50
3300	Curved, 1" thick, to 6" high, vertical		48	.333		27	16.45		43.45	56

09 66 Terrazzo Flooring

09 66 16 – Terrazzo Floor Tile

09 66 16.30 Terrazzo, Precast		Crew	Daily Output	Labor-Hours	Unit	2020 Bare Costs Material	Labor	Equipment	Total	Total Incl O&P
3400	Cove	2 Mstz	38	.421	L.F.	43	21		64	80
3600	Stair tread and riser, single piece, straight, smooth surface		60	.267		66.50	13.15		79.65	94
3700	Non skid surface		40	.400		86	19.75		105.75	126
3900	Curved tread and riser, smooth surface		40	.400		96	19.75		115.75	136
4000	Non skid surface		32	.500		118	24.50		142.50	168
4200	Stair stringers, notched, 1" thick		25	.640		38	31.50		69.50	91.50
4300	2" thick		22	.727		45	36		81	106
4500	Stair landings, structural, non-slip, 1-1/2" thick		85	.188	S.F.	42.50	9.30		51.80	61.50
4600	3" thick		75	.213		59	10.55		69.55	81.50
4800	Wainscot, 12" x 12" x 1" tiles	1 Mstz	12	.667		9.75	33		42.75	62.50
4900	16" x 16" x 1-1/2" tiles	"	8	1		18.75	49.50		68.25	98
9500	Minimum labor/equipment charge	1 Tilf	2	4	Job		198		198	310

09 66 23 – Resinous Matrix Terrazzo Flooring

09 66 23.13 Polyacrylate Mod. Cementitious Terrazzo Flr.

		Crew	Daily Output	Labor-Hours	Unit	Material	Labor	Equipment	Total	Total Incl O&P
0010	**POLYACRYLATE MODIFIED CEMENTITIOUS TERRAZZO FLOORING**									
3150	Polyacrylate, 1/4" thick, granite chips	C-6	735	.065	S.F.	4.20	2.86	.07	7.13	9.25
3170	Recycled porcelain		480	.100		5	4.37	.11	9.48	12.60
3200	3/8" thick, granite chips		620	.077		5.25	3.39	.09	8.73	11.30
3220	Recycled porcelain		480	.100		7.15	4.37	.11	11.63	15

09 66 23.16 Epoxy-Resin Terrazzo Flooring

		Crew	Daily Output	Labor-Hours	Unit	Material	Labor	Equipment	Total	Total Incl O&P
0010	**EPOXY-RESIN TERRAZZO FLOORING**									
1800	Epoxy terrazzo, 1/4" thick, chemical resistant, granite chips	J-3	200	.080	S.F.	6.60	3.60	1.33	11.53	14.35
1900	Recycled porcelain		150	.107		10.40	4.79	1.77	16.96	21
2500	Epoxy terrazzo, 1/4" thick, granite chips		200	.080		6	3.60	1.33	10.93	13.70
2550	Average		175	.091		5.95	4.11	1.52	11.58	14.65
2600	Recycled aggregate		150	.107		5.75	4.79	1.77	12.31	15.80
2650	Epoxy terrazzo, 3/8" thick, marble chips		200	.080		5.75	3.60	1.33	10.68	13.45
2675	Glass or mother of pearl		200	.080		7.20	3.60	1.33	12.13	15

09 66 33 – Conductive Terrazzo Flooring

09 66 33.10 Conductive Terrazzo

		Crew	Daily Output	Labor-Hours	Unit	Material	Labor	Equipment	Total	Total Incl O&P
0010	**CONDUCTIVE TERRAZZO**									
2400	Bonded conductive floor for hospitals	J-3	90	.178	S.F.	5.60	8	2.95	16.55	22

09 66 33.13 Conductive Epoxy-Resin Terrazzo

		Crew	Daily Output	Labor-Hours	Unit	Material	Labor	Equipment	Total	Total Incl O&P
0010	**CONDUCTIVE EPOXY-RESIN TERRAZZO**									
2100	Epoxy terrazzo, 1/4" thick, conductive, granite chips	J-3	100	.160	S.F.	9.50	7.20	2.66	19.36	24.50
2200	Recycled porcelain	"	90	.178	"	11.95	8	2.95	22.90	29

09 66 33.19 Conductive Plastic-Matrix Terrazzo Flooring

		Crew	Daily Output	Labor-Hours	Unit	Material	Labor	Equipment	Total	Total Incl O&P
0010	**CONDUCTIVE PLASTIC-MATRIX TERRAZZO FLOORING**									
3300	Conductive, 1/4" thick, granite chips	C-6	450	.107	S.F.	8.65	4.67	.12	13.44	17.10
3330	Recycled porcelain		305	.157		11.10	6.90	.18	18.18	23.50
3350	3/8" thick, granite chips		365	.132		12.20	5.75	.15	18.10	23
3370	Recycled porcelain		255	.188		14.50	8.25	.21	22.96	29.50
3450	Granite, conductive, 1/4" thick, 20% chip		695	.069		10.10	3.02	.08	13.20	16
3470	50% chip		420	.114		13	5	.13	18.13	22.50
3500	3/8" thick, 20% chip		695	.069		14.75	3.02	.08	17.85	21
3520	50% chip		380	.126		17.85	5.55	.14	23.54	28.50

09 67 Fluid-Applied Flooring

09 67 13 – Elastomeric Liquid Flooring

09 67 13.13 Elastomeric Liquid Flooring		Crew	Daily Output	Labor-Hours	Unit	Material	Labor	Equipment	Total	Total Incl O&P
						2020 Bare Costs				
0010	**ELASTOMERIC LIQUID FLOORING**									
0020	Cementitious acrylic, 1/4" thick	C-6	520	.092	S.F.	1.81	4.04	.10	5.95	8.55
0100	3/8" thick	"	450	.107		2.39	4.67	.12	7.18	10.20
0200	Methyl methachrylate, 1/4" thick	C-8A	3000	.016		7.20	.72		7.92	9.10
0210	1/8" thick	"	3000	.016		6.05	.72		6.77	7.80
0300	Cupric oxychloride, on bond coat, simple configs and patterns	C-6	480	.100		4.10	4.37	.11	8.58	11.65
0400	Complex configurations and patterns		420	.114		6.85	5	.13	11.98	15.65
2400	Mastic, hot laid, 2 coat, 1-1/2" thick, std., simple configs and patterns		690	.070		4.74	3.04	.08	7.86	10.15
2500	Maximum		520	.092		6.10	4.04	.10	10.24	13.25
2700	Acid-proof, minimum		605	.079		6.10	3.47	.09	9.66	12.35
2800	Maximum		350	.137		8.45	6	.15	14.60	19
3000	Neoprene, troweled on, 1/4" thick, minimum		545	.088		4.67	3.85	.10	8.62	11.40
3100	Maximum		430	.112		6.45	4.88	.13	11.46	15
4300	Polyurethane, with suspended vinyl chips, clear		1065	.045		7.80	1.97	.05	9.82	11.75
4500	Pigmented	↓	860	.056	↓	11.45	2.44	.06	13.95	16.55

09 67 23 – Resinous Flooring

09 67 23.23 Resinous Flooring		Crew	Daily Output	Labor-Hours	Unit	Material	Labor	Equipment	Total	Total Incl O&P
0010	**RESINOUS FLOORING**									
1200	Heavy duty epoxy topping, 1/4" thick,									
1300	500 to 1,000 S.F.	C-6	420	.114	S.F.	6.20	5	.13	11.33	14.95
1500	1,000 to 2,000 S.F.		450	.107		5.25	4.67	.12	10.04	13.40
1600	Over 10,000 S.F.	↓	480	.100	↓	5	4.37	.11	9.48	12.60

09 67 26 – Quartz Flooring

09 67 26.26 Quartz Flooring		Crew	Daily Output	Labor-Hours	Unit	Material	Labor	Equipment	Total	Total Incl O&P
0010	**QUARTZ FLOORING**									
0600	Epoxy, with colored quartz chips, broadcast, 3/8" thick	C-6	675	.071	S.F.	3.22	3.11	.08	6.41	8.60
0700	1/2" thick		490	.098		4.52	4.28	.11	8.91	11.95
0900	Troweled, minimum		560	.086		3.64	3.75	.10	7.49	10.10
1000	Maximum		480	.100		6.15	4.37	.11	10.63	13.85
3600	Polyester, with colored quartz chips, 1/16" thick, minimum		1065	.045		3.63	1.97	.05	5.65	7.20
3700	Maximum		560	.086		4.95	3.75	.10	8.80	11.55
3900	1/8" thick, minimum		810	.059		4.20	2.59	.07	6.86	8.85
4000	Maximum		675	.071		5.60	3.11	.08	8.79	11.25
4200	Polyester, heavy duty, compared to epoxy, add	↓	2590	.019	↓	1.69	.81	.02	2.52	3.17

09 67 66 – Fluid-Applied Athletic Flooring

09 67 66.10 Polyurethane		Crew	Daily Output	Labor-Hours	Unit	Material	Labor	Equipment	Total	Total Incl O&P
0010	**POLYURETHANE**									
4400	Thermoset, prefabricated in place, indoor									
4500	3/8" thick for basketball, gyms, etc.	1 Tilf	100	.080	S.F.	5.95	3.96		9.91	12.75
4600	1/2" thick for professional sports		95	.084		8	4.17		12.17	15.35
4700	Outdoor, 1/4" thick, smooth, for tennis		100	.080		6.20	3.96		10.16	13
5000	Poured in place, indoor, with finish, 1/4" thick		80	.100		4.41	4.95		9.36	12.60
5050	3/8" thick		65	.123		5.35	6.10		11.45	15.45
5100	1/2" thick	↓	50	.160	↓	6.40	7.90		14.30	19.45

09 68 Carpeting

09 68 05 – Carpet Accessories

09 68 05.11 Flooring Transition Strip		Crew	Daily Output	Labor-Hours	Unit	Material	Labor	Equipment	Total	Total Incl O&P
						2020 Bare Costs				
0010	**FLOORING TRANSITION STRIP**									
0107	Clamp down brass divider, 12′ strip, vinyl to carpet	1 Tilf	31.25	.256	Ea.	15	12.65		27.65	36.50
0117	Vinyl to hard surface	"	31.25	.256	"	15	12.65		27.65	36.50

09 68 10 – Carpet Pad

09 68 10.10 Commercial Grade Carpet Pad

		Crew	Daily Output	Labor-Hours	Unit	Material	Labor	Equipment	Total	Total Incl O&P
0010	**COMMERCIAL GRADE CARPET PAD**									
9000	Sponge rubber pad, 20 oz./sq. yd.	1 Tilf	150	.053	S.Y.	4.85	2.64		7.49	9.50
9100	40 to 62 oz./sq. yd.		150	.053		8.85	2.64		11.49	13.85
9200	Felt pad, 20 oz./sq. yd.		150	.053		6.40	2.64		9.04	11.20
9300	32 to 56 oz./sq. yd.		150	.053		12.10	2.64		14.74	17.50
9400	Bonded urethane pad, 2.7 density		150	.053		5.95	2.64		8.59	10.70
9500	13.0 density		150	.053		8.05	2.64		10.69	13
9600	Prime urethane pad, 2.7 density		150	.053		3.69	2.64		6.33	8.20
9700	13.0 density		150	.053		7.95	2.64		10.59	12.90
9800	Carpet pad, for 'double stick' installation, add					.90	4.10		5	7

09 68 13 – Tile Carpeting

09 68 13.10 Carpet Tile

		Crew	Daily Output	Labor-Hours	Unit	Material	Labor	Equipment	Total	Total Incl O&P
0010	**CARPET TILE**									
0100	Tufted nylon, 18″ x 18″, hard back, 20 oz.	1 Tilf	80	.100	S.Y.	28	4.95		32.95	39
0110	26 oz.		80	.100		26.50	4.95		31.45	37
0200	Cushion back, 20 oz.		80	.100		26	4.95		30.95	36.50
0210	26 oz.		80	.100		31	4.95		35.95	42.50
1100	Tufted, 24″ x 24″, hard back, 24 oz. nylon		80	.100		32	4.95		36.95	43
1180	35 oz.		80	.100		36.50	4.95		41.45	48
5060	42 oz.		80	.100		46.50	4.95		51.45	59
6000	Electrostatic dissapative carpet tile, 24″ x 24″, 24 oz.		80	.100		38.50	4.95		43.45	50.50
6100	Electrostatic dissapative carpet tile for access floors, 24″ x 24″, 24 oz.		80	.100		48	4.95		52.95	61

09 68 16 – Sheet Carpeting

09 68 16.10 Sheet Carpet

		Crew	Daily Output	Labor-Hours	Unit	Material	Labor	Equipment	Total	Total Incl O&P
0010	**SHEET CARPET**									
0700	Nylon, level loop, 26 oz., light to medium traffic	1 Tilf	75	.107	S.Y.	24.50	5.30		29.80	35.50
0720	28 oz., light to medium traffic		75	.107		32.50	5.30		37.80	44.50
0900	32 oz., medium traffic		75	.107		41	5.30		46.30	53.50
1100	40 oz., medium to heavy traffic		75	.107		48	5.30		53.30	61
2920	Nylon plush, 30 oz., medium traffic		75	.107		28	5.30		33.30	39
3000	36 oz., medium traffic		75	.107		37.50	5.30		42.80	49.50
3100	42 oz., medium to heavy traffic		70	.114		46.50	5.65		52.15	60
3200	46 oz., medium to heavy traffic		70	.114		53.50	5.65		59.15	68
3300	54 oz., heavy traffic		70	.114		62	5.65		67.65	77
3340	60 oz., heavy traffic		70	.114		68	5.65		73.65	83.50
3665	Olefin, 24 oz., light to medium traffic		75	.107		19.50	5.30		24.80	30
3670	26 oz., medium traffic		75	.107		13.20	5.30		18.50	23
3680	28 oz., medium to heavy traffic		75	.107		25	5.30		30.30	36
3700	32 oz., medium to heavy traffic		75	.107		30.50	5.30		35.80	42
3730	42 oz., heavy traffic		70	.114		28	5.65		33.65	39.50
4110	Wool, level loop, 40 oz., medium traffic		70	.114		115	5.65		120.65	136
4500	50 oz., medium to heavy traffic		70	.114		111	5.65		116.65	131
4700	Patterned, 32 oz., medium to heavy traffic		70	.114		102	5.65		107.65	121
4900	48 oz., heavy traffic		70	.114		112	5.65		117.65	132
5000	For less than full roll (approx. 1500 S.F.), add					25%				
5100	For small rooms, less than 12′ wide, add						25%			

09 68 Carpeting

09 68 16 – Sheet Carpeting

09 68 16.10 Sheet Carpet		Crew	Daily Output	Labor-Hours	Unit	Material	2020 Bare Costs Labor	Equipment	Total	Total Incl O&P
5200	For large open areas (no cuts), deduct						25%			
5600	For bound carpet baseboard, add	1 Tilf	300	.027	L.F.	1.79	1.32		3.11	4.04
5610	For stairs, not incl. price of carpet, add	"	30	.267	Riser		13.20		13.20	20.50
5620	For borders and patterns, add to labor						18%			
8950	For tackless, stretched installation, add padding from 09 68 10.10 to above									
9850	For brand-named specific fiber, add				S.Y.	25%				
9910	Minimum labor/equipment charge	1 Tilf	3	2.667	Job		132		132	207

09 68 20 – Athletic Carpet

09 68 20.10 Indoor Athletic Carpet		Crew	Daily Output	Labor-Hours	Unit	Material	Labor	Equipment	Total	Total Incl O&P
0010	**INDOOR ATHLETIC CARPET**									
3700	Polyethylene, in rolls, no base incl., landscape surfaces	1 Tilf	275	.029	S.F.	4.47	1.44		5.91	7.20
3800	Nylon action surface, 1/8" thick		275	.029		4.17	1.44		5.61	6.85
3900	1/4" thick		275	.029		6	1.44		7.44	8.85
4000	3/8" thick		275	.029		7.55	1.44		8.99	10.55
4100	Golf tee surface with foam back		235	.034		7.50	1.69		9.19	10.90
4200	Practice putting, knitted nylon surface		235	.034		6.30	1.69		7.99	9.60
4300	Synthetic turf, 1/2" ht.		90	.089		4.67	4.40		9.07	12.05
4350	3/4" ht.		210	.038		4.61	1.89		6.50	8
4400	1" ht.		190	.042		5.75	2.08		7.83	9.55
5500	Polyvinyl chloride, sheet goods for gyms, 1/4" thick		80	.100		7.25	4.95		12.20	15.75
5600	3/8" thick	↓	60	.133	↓	9.95	6.60		16.55	21.50

09 69 Access Flooring

09 69 13 – Rigid-Grid Access Flooring

09 69 13.10 Access Floors		Crew	Daily Output	Labor-Hours	Unit	Material	Labor	Equipment	Total	Total Incl O&P
0010	**ACCESS FLOORS**									
0015	Access floor pkg. including panel, pedestal, & stringers 1500 lbs. load									
0100	Package pricing, conc. fill panels, no fin., 6" ht.	4 Carp	750	.043	S.F.	18.45	2.27		20.72	24
0105	12" ht.		750	.043		18.55	2.27		20.82	24
0110	18" ht.		750	.043		19.20	2.27		21.47	24.50
0115	24" ht.		750	.043		19.05	2.27		21.32	24.50
0120	Package pricing, steel panels, no fin., 6" ht.		750	.043		15.15	2.27		17.42	20.50
0125	12" ht.		750	.043		20	2.27		22.27	25.50
0130	18" ht.		750	.043		20.50	2.27		22.77	26
0135	24" ht.		750	.043		20.50	2.27		22.77	26
0140	Package pricing, wood core panels, no fin., 6" ht.		750	.043		13.65	2.27		15.92	18.65
0145	12" ht.		750	.043		13.90	2.27		16.17	18.90
0150	18" ht.		750	.043		13.90	2.27		16.17	18.90
0155	24" ht.		750	.043		14.15	2.27		16.42	19.20
0160	Pkg. pricing, conc. fill pnls, no stringers, no fin., 6" ht, 1250 lbs. load		700	.046		11.05	2.43		13.48	16.05
0165	12" ht.		700	.046		10.25	2.43		12.68	15.15
0170	18" ht.		700	.046		11.20	2.43		13.63	16.20
0175	24" ht.	↓	700	.046		12.15	2.43		14.58	17.30
0250	Panels, 2' x 2' conc. fill, no fin.	2 Carp	500	.032		23	1.70		24.70	27.50
0255	With 1/8" high pressure laminate		500	.032		57.50	1.70		59.20	66
0260	Metal panels with 1/8" high pressure laminate		500	.032		68	1.70		69.70	77.50
0265	Wood core panels with 1/8" high pressure laminate		500	.032		44.50	1.70		46.20	51.50
0400	Aluminum panels, no fin.	↓	500	.032		35	1.70		36.70	41
0600	For carpet covering, add					9.25			9.25	10.15
0700	For vinyl floor covering, add				↓	9.90			9.90	10.90

09 69 Access Flooring

09 69 13 – Rigid-Grid Access Flooring

09 69 13.10 Access Floors		Crew	Daily Output	Labor-Hours	Unit	Material	2020 Bare Costs Labor	Equipment	Total	Total Incl O&P
0900	For high pressure laminate covering, add				S.F.	8.15			8.15	8.95
0910	For snap on stringer system, add	2 Carp	1000	.016	↓	1.60	.85		2.45	3.12
1000	Machine cutouts after initial installation	1 Carp	50	.160	Ea.	20.50	8.50		29	36
1050	Pedestals, 6" to 12"	2 Carp	85	.188		9.10	10		19.10	26
1100	Air conditioning grilles, 4" x 12"	1 Carp	17	.471		72.50	25		97.50	120
1150	4" x 18"	"	14	.571	↓	99	30.50		129.50	158
1200	Approach ramps, steel	2 Carp	60	.267	S.F.	28	14.15		42.15	53.50
1300	Aluminum	"	40	.400	"	34.50	21.50		56	72
1500	Handrail, 2 rail, aluminum	1 Carp	15	.533	L.F.	126	28.50		154.50	184

09 72 Wall Coverings

09 72 13 – Cork Wall Coverings

09 72 13.10 Covering, Cork Wall

		Crew	Daily Output	Labor-Hours	Unit	Material	Labor	Equipment	Total	Total Incl O&P
0010	**COVERING, CORK WALL**									
0600	Cork tiles, light or dark, 12" x 12" x 3/16"	1 Pape	240	.033	S.F.	4.26	1.48		5.74	7.05
0700	5/16" thick		235	.034		3.40	1.52		4.92	6.15
0900	1/4" basket weave		240	.033		3.29	1.48		4.77	6
1000	1/2" natural, non-directional pattern		240	.033		6.80	1.48		8.28	9.85
1100	3/4" natural, non-directional pattern		240	.033		11.85	1.48		13.33	15.35
1200	Granular surface, 12" x 36", 1/2" thick		385	.021		1.38	.93		2.31	2.99
1300	1" thick		370	.022		1.72	.96		2.68	3.42
1500	Polyurethane coated, 12" x 12" x 3/16" thick		240	.033		4.17	1.48		5.65	6.95
1600	5/16" thick		235	.034		6	1.52		7.52	9
1800	Cork wallpaper, paperbacked, natural		480	.017		1.68	.74		2.42	3.03
1900	Colors	↓	480	.017	↓	2.89	.74		3.63	4.36

09 72 16 – Vinyl-Coated Fabric Wall Coverings

09 72 16.13 Flexible Vinyl Wall Coverings

		Crew	Daily Output	Labor-Hours	Unit	Material	Labor	Equipment	Total	Total Incl O&P
0010	**FLEXIBLE VINYL WALL COVERINGS**									
3000	Vinyl wall covering, fabric-backed, lightweight, type 1 (12-15 oz./S.Y.)	1 Pape	640	.013	S.F.	1.45	.56		2.01	2.48
3300	Medium weight, type 2 (20-24 oz./S.Y.)		480	.017		1.06	.74		1.80	2.35
3400	Heavy weight, type 3 (28 oz./S.Y.)	↓	435	.018	↓	1.59	.82		2.41	3.05
3600	Adhesive, 5 gal. lots (18 S.Y./gal.)				Gal.	12.60			12.60	13.85

09 72 16.16 Rigid-Sheet Vinyl Wall Coverings

		Crew	Daily Output	Labor-Hours	Unit	Material	Labor	Equipment	Total	Total Incl O&P
0010	**RIGID-SHEET VINYL WALL COVERINGS**									
0100	Acrylic, modified, semi-rigid PVC, .028" thick	2 Carp	330	.048	S.F.	1.32	2.58		3.90	5.60
0110	.040" thick	"	320	.050	"	2.05	2.66		4.71	6.50

09 72 19 – Textile Wall Coverings

09 72 19.10 Textile Wall Covering

		Crew	Daily Output	Labor-Hours	Unit	Material	Labor	Equipment	Total	Total Incl O&P
0010	**TEXTILE WALL COVERING**, including sizing; add 10-30% waste @ takeoff									
0020	Silk	1 Pape	640	.013	S.F.	4.53	.56		5.09	5.85
0030	Cotton		640	.013		6.90	.56		7.46	8.50
0040	Linen		640	.013		1.88	.56		2.44	2.95
0050	Blend	↓	640	.013	↓	3.23	.56		3.79	4.43
0060	Linen wall covering, paper backed									
0070	Flame treatment				S.F.	1.03			1.03	1.13
0080	Stain resistance treatment					1.92			1.92	2.11
0090	Grass cloth, natural fabric G	1 Pape	400	.020		1.94	.89		2.83	3.55
0100	Grass cloths with lining paper G		400	.020		1.30	.89		2.19	2.85
0110	Premium texture/color G	↓	350	.023	↓	3.29	1.02		4.31	5.25

09 72 Wall Coverings

09 72 20 – Natural Fiber Wall Covering

09 72 20.10 Natural Fiber Wall Covering		Crew	Daily Output	Labor-Hours	Unit	Material	Labor	Equipment	Total	Total Incl O&P
						2020 Bare Costs				
0010	**NATURAL FIBER WALL COVERING**, including sizing; add 10-30% waste @ takeoff									
0015	Bamboo	1 Pape	640	.013	S.F.	2.20	.56		2.76	3.30
0030	Burlap		640	.013		1.90	.56		2.46	2.97
0045	Jute		640	.013		1.31	.56		1.87	2.32
0060	Sisal		640	.013		1.72	.56		2.28	2.77

09 72 23 – Wallpapering

09 72 23.10 Wallpaper		Crew	Daily Output	Labor-Hours	Unit	Material	Labor	Equipment	Total	Total Incl O&P
0010	**WALLPAPER** including sizing; add 10-30% waste @ takeoff									
0050	Aluminum foil	1 Pape	275	.029	S.F.	1.05	1.30		2.35	3.22
0100	Copper sheets, .025" thick, vinyl backing		240	.033		5.60	1.48		7.08	8.50
0300	Phenolic backing		240	.033		7.25	1.48		8.73	10.30
2400	Gypsum-based, fabric-backed, fire resistant									
2500	for masonry walls, 21 oz./S.Y.	1 Pape	800	.010	S.F.	.85	.45		1.30	1.65
2600	Average		720	.011		1.34	.50		1.84	2.26
2700	Small quantities		640	.013		.85	.56		1.41	1.82
3700	Wallpaper, average workmanship, solid pattern, low cost paper		640	.013		.63	.56		1.19	1.57
3900	Basic patterns (matching required), avg. cost paper		535	.015		1.28	.67		1.95	2.47
4000	Paper at $85 per double roll, quality workmanship		435	.018		2.20	.82		3.02	3.72
5990	Wallpaper removal, 1 layer		800	.010		.02	.45		.47	.73
6000	Wallpaper removal, 3 layer		400	.020		.07	.89		.96	1.49
9000	Minimum labor/equipment charge		2	4	Job		178		178	283

09 74 Flexible Wood Sheets

09 74 16 – Flexible Wood Veneers

09 74 16.10 Veneer, Flexible Wood		Crew	Daily Output	Labor-Hours	Unit	Material	Labor	Equipment	Total	Total Incl O&P
0010	**VENEER, FLEXIBLE WOOD**									
0100	Flexible wood veneer, 1/32" thick, plain woods	1 Pape	100	.080	S.F.	2.50	3.56		6.06	8.40
0110	Exotic woods	"	95	.084	"	3.77	3.75		7.52	10.10

09 77 Special Wall Surfacing

09 77 30 – Fiberglass Reinforced Panels

09 77 30.10 Fiberglass Reinforced Plastic Panels		Crew	Daily Output	Labor-Hours	Unit	Material	Labor	Equipment	Total	Total Incl O&P
0010	**FIBERGLASS REINFORCED PLASTIC PANELS**, .090" thick									
0020	On walls, adhesive mounted, embossed surface	2 Carp	640	.025	S.F.	1.19	1.33		2.52	3.44
0030	Smooth surface		640	.025		1.47	1.33		2.80	3.75
0040	Fire rated, embossed surface		640	.025		2.12	1.33		3.45	4.46
0050	Nylon rivet mounted, on drywall, embossed surface		480	.033		1.18	1.77		2.95	4.14
0060	Smooth surface		480	.033		1.47	1.77		3.24	4.46
0070	Fire rated, embossed surface		480	.033		2.12	1.77		3.89	5.15
0080	On masonry, embossed surface		320	.050		1.18	2.66		3.84	5.55
0090	Smooth surface		320	.050		1.47	2.66		4.13	5.90
0100	Fire rated, embossed surface		320	.050		2.12	2.66		4.78	6.60
0110	Nylon rivet and adhesive mounted, on drywall, embossed surface		240	.067		1.35	3.54		4.89	7.20
0120	Smooth surface		240	.067		1.43	3.54		4.97	7.25
0130	Fire rated, embossed surface		240	.067		2.34	3.54		5.88	8.25
0140	On masonry, embossed surface		190	.084		1.35	4.48		5.83	8.65
0150	Smooth surface		190	.084		1.43	4.48		5.91	8.70
0160	Fire rated, embossed surface		190	.084		2.34	4.48		6.82	9.70

09 77 Special Wall Surfacing

09 77 30 – Fiberglass Reinforced Panels

09 77 30.10 Fiberglass Reinforced Plastic Panels		Crew	Daily Output	Labor-Hours	Unit	Material	Labor	Equipment	Total	Total Incl O&P
						2020 Bare Costs				
0170	For moldings, add	1 Carp	250	.032	L.F.	.29	1.70		1.99	3.04
0180	On ceilings, for lay in grid system, embossed surface		400	.020	S.F.	1.19	1.06		2.25	3.01
0190	Smooth surface		400	.020		1.47	1.06		2.53	3.32
0200	Fire rated, embossed surface	↓	400	.020	↓	2.12	1.06		3.18	4.03

09 77 43 – Panel Systems

09 77 43.20 Slatwall Panels and Accessories		Crew	Daily Output	Labor-Hours	Unit	Material	Labor	Equipment	Total	Total Incl O&P
0010	**SLATWALL PANELS AND ACCESSORIES**									
0100	Slatwall panel, 4′ x 8′ x 3/4″ T, MDF, paint grade	1 Carp	500	.016	S.F.	1.58	.85		2.43	3.10
0110	Melamine finish		500	.016		2.13	.85		2.98	3.70
0120	High pressure plastic laminate finish		500	.016		3.38	.85		4.23	5.10
0125	Wood veneer	↓	500	.016		4.23	.85		5.08	6
0130	Aluminum channel inserts, add				↓	2.66			2.66	2.93
0200	Accessories, corner forms, 8′ L				L.F.	5.70			5.70	6.30
0210	T-connector, 8′ L					7.45			7.45	8.20
0220	J-mold, 8′ L					1.25			1.25	1.38
0230	Edge cap, 8′ L					1.74			1.74	1.91
0240	Finish end cap, 8′ L				↓	3.39			3.39	3.73
0300	Display hook, metal, 4″ L				Ea.	.47			.47	.52
0310	6″ L					.52			.52	.57
0320	8″ L					.56			.56	.62
0330	10″ L					.54			.54	.59
0340	12″ L					.51			.51	.56
0350	Acrylic, 4″ L					1.27			1.27	1.40
0360	6″ L					.97			.97	1.07
0400	Waterfall hanger, metal, 12″-16″					3.92			3.92	4.31
0410	Acrylic					11.35			11.35	12.50
0500	Shelf bracket, metal, 8″					2.06			2.06	2.27
0510	10″					2.04			2.04	2.24
0520	12″					2.26			2.26	2.49
0530	14″					2.49			2.49	2.74
0540	16″					3.07			3.07	3.38
0550	Acrylic, 8″					3.94			3.94	4.33
0560	10″					4.10			4.10	4.51
0570	12″					4.64			4.64	5.10
0580	14″					6			6	6.60
0600	Shelf, acrylic, 12″ x 16″ x 1/4″					23			23	25
0610	12″ x 24″ x 1/4″				↓	25			25	27.50

09 81 Acoustic Insulation

09 81 13 – Acoustic Board Insulation

09 81 13.10 Acoustic Board Insulation		Crew	Daily Output	Labor-Hours	Unit	Material	Labor	Equipment	Total	Total Incl O&P
0010	**ACOUSTIC BOARD INSULATION**									
0020	Cellulose fiber board, 1/2″ thk.	1 Carp	800	.010	S.F.	.91	.53		1.44	1.85

09 81 16 – Acoustic Blanket Insulation

09 81 16.10 Sound Attenuation Blanket		Crew	Daily Output	Labor-Hours	Unit	Material	Labor	Equipment	Total	Total Incl O&P
0010	**SOUND ATTENUATION BLANKET**									
0020	Blanket, 1″ thick	1 Carp	925	.009	S.F.	.28	.46		.74	1.05
0500	1-1/2″ thick		920	.009		.33	.46		.79	1.10
1000	2″ thick		915	.009		.40	.46		.86	1.18
1500	3″ thick	↓	910	.009	↓	.59	.47		1.06	1.40

09 81 Acoustic Insulation

09 81 16 – Acoustic Blanket Insulation

	09 81 16.10 Sound Attenuation Blanket	Crew	Daily Output	Labor-Hours	Unit	2020 Bare Costs Material	Labor	Equipment	Total	Total Incl O&P
2000	Wall hung, STC 18-21, 1″ thick, 4′ x 20′	2 Carp	22	.727	Ea.	550	38.50		588.50	665
2010	10′ x 20′	″	19	.842		1,375	45		1,420	1,600
2020	Wall hung, STC 27-28, 3″ thick, 4′ x 20′	3 Carp	12	2		590	106		696	820
2030	10′ x 20′	″	9	2.667		1,475	142		1,617	1,850
3000	Thermal or acoustical batt above ceiling, 2″ thick	1 Carp	900	.009	S.F.	.52	.47		.99	1.33
3100	3″ thick		900	.009		.80	.47		1.27	1.64
3200	4″ thick		900	.009		1	.47		1.47	1.86
3400	Urethane plastic foam, open cell, on wall, 2″ thick	2 Carp	2050	.008		3.29	.41		3.70	4.28
3500	3″ thick		1550	.010		4.37	.55		4.92	5.70
3600	4″ thick		1050	.015		6.15	.81		6.96	8.05
3700	On ceiling, 2″ thick		1700	.009		3.28	.50		3.78	4.41
3800	3″ thick		1300	.012		4.37	.65		5.02	5.85
3900	4″ thick		900	.018		6.15	.95		7.10	8.25
9000	Minimum labor/equipment charge	1 Carp	5	1.600	Job		85		85	136

09 84 Acoustic Room Components

09 84 13 – Fixed Sound-Absorptive Panels

	09 84 13.10 Fixed Panels	Crew	Daily Output	Labor-Hours	Unit	Material	Labor	Equipment	Total	Total Incl O&P
0010	**FIXED PANELS** Perforated steel facing, painted with									
0100	Fiberglass or mineral filler, no backs, 2-1/4″ thick, modular									
0200	space units, ceiling or wall hung, white or colored	1 Carp	100	.080	S.F.	9.95	4.25		14.20	17.70
0300	Fiberboard sound deadening panels, 1/2″ thick	″	600	.013	″	.33	.71		1.04	1.50
0500	Fiberglass panels, 4′ x 8′ x 1″ thick, with									
0600	glass cloth face for walls, cemented	1 Carp	155	.052	S.F.	9.15	2.74		11.89	14.50
0700	1-1/2″ thick, dacron covered, inner aluminum frame,									
0710	wall mounted	1 Carp	300	.027	S.F.	8.85	1.42		10.27	11.95
0900	Mineral fiberboard panels, fabric covered, 30″ x 108″,									
1000	3/4″ thick, concealed spline, wall mounted	1 Carp	150	.053	S.F.	6.35	2.83		9.18	11.55
9000	Minimum labor/equipment charge	″	4	2	Job		106		106	170

09 84 36 – Sound-Absorbing Ceiling Units

	09 84 36.10 Barriers	Crew	Daily Output	Labor-Hours	Unit	Material	Labor	Equipment	Total	Total Incl O&P
0010	**BARRIERS** Plenum									
0600	Aluminum foil, fiberglass reinf., parallel with joists	1 Carp	275	.029	S.F.	1.14	1.55		2.69	3.73
0700	Perpendicular to joists		180	.044		1.14	2.36		3.50	5.05
0900	Aluminum mesh, kraft paperbacked		275	.029		.78	1.55		2.33	3.34
0970	Fiberglass batts, kraft faced, 3-1/2″ thick		1400	.006		.37	.30		.67	.90
0980	6″ thick		1300	.006		.65	.33		.98	1.24
1000	Sheet lead, 1 lb., 1/64″ thick, perpendicular to joists		150	.053		7.90	2.83		10.73	13.25
1100	Vinyl foam reinforced, 1/8″ thick, 1.0 lb./S.F.		150	.053		5.95	2.83		8.78	11.10

09 91 Painting

09 91 13 – Exterior Painting

09 91 13.30 Fences		Crew	Daily Output	Labor-Hours	Unit	Material	2020 Bare Costs Labor	Equipment	Total	Total Incl O&P
0010	**FENCES** R099100-20									
0100	Chain link or wire metal, one side, water base									
0110	Roll & brush, first coat	1 Pord	960	.008	S.F.	.08	.37		.45	.68
0120	Second coat		1280	.006		.08	.28		.36	.52
0130	Spray, first coat		2275	.004		.08	.16		.24	.34
0140	Second coat		2600	.003		.08	.14		.22	.31
0150	Picket, water base									
0160	Roll & brush, first coat	1 Pord	865	.009	S.F.	.09	.41		.50	.75
0170	Second coat		1050	.008		.09	.34		.43	.64
0180	Spray, first coat		2275	.004		.09	.16		.25	.35
0190	Second coat		2600	.003		.09	.14		.23	.32
0200	Stockade, water base									
0210	Roll & brush, first coat	1 Pord	1040	.008	S.F.	.09	.34		.43	.64
0220	Second coat		1200	.007		.09	.30		.39	.57
0230	Spray, first coat		2275	.004		.09	.16		.25	.35
0240	Second coat		2600	.003		.09	.14		.23	.32
9000	Minimum labor/equipment charge		2	4	Job		178		178	282

09 91 13.42 Miscellaneous, Exterior		Crew	Daily Output	Labor-Hours	Unit	Material	Labor	Equipment	Total	Total Incl O&P
0010	**MISCELLANEOUS, EXTERIOR** R099100-20									
0015	For painting metals, see Section 09 97 13.23									
0100	Railing, ext., decorative wood, incl. cap & baluster									
0110	Newels & spindles @ 12" OC									
0120	Brushwork, stain, sand, seal & varnish									
0130	First coat	1 Pord	90	.089	L.F.	.88	3.95		4.83	7.20
0140	Second coat	"	120	.067	"	.88	2.96		3.84	5.65
0150	Rough sawn wood, 42" high, 2" x 2" verticals, 6" OC									
0160	Brushwork, stain, each coat	1 Pord	90	.089	L.F.	.29	3.95		4.24	6.55
0170	Wrought iron, 1" rail, 1/2" sq. verticals									
0180	Brushwork, zinc chromate, 60" high, bars 6" OC									
0190	Primer	1 Pord	130	.062	L.F.	.85	2.73		3.58	5.30
0200	Finish coat		130	.062		1.13	2.73		3.86	5.60
0210	Additional coat		190	.042		1.31	1.87		3.18	4.42
0220	Shutters or blinds, single panel, 2' x 4', paint all sides									
0230	Brushwork, primer	1 Pord	20	.400	Ea.	.72	17.75		18.47	29
0240	Finish coat, exterior latex		20	.400		.65	17.75		18.40	28.50
0250	Primer & 1 coat, exterior latex		13	.615		1.21	27.50		28.71	45
0260	Spray, primer		35	.229		1.04	10.15		11.19	17.25
0270	Finish coat, exterior latex		35	.229		1.38	10.15		11.53	17.60
0280	Primer & 1 coat, exterior latex		20	.400		1.13	17.75		18.88	29
0290	For louvered shutters, add				S.F.	10%				
0300	Stair stringers, exterior, metal									
0310	Roll & brush, zinc chromate, to 14", each coat	1 Pord	320	.025	L.F.	.38	1.11		1.49	2.17
0320	Rough sawn wood, 4" x 12"									
0330	Roll & brush, exterior latex, each coat	1 Pord	215	.037	L.F.	.10	1.65		1.75	2.73
0340	Trellis/lattice, 2" x 2" @ 3" OC with 2" x 8" supports									
0350	Spray, latex, per side, each coat	1 Pord	475	.017	S.F.	.10	.75		.85	1.30
0450	Decking, ext., sealer, alkyd, brushwork, sealer coat		1140	.007		.10	.31		.41	.60
0460	1st coat		1140	.007		.11	.31		.42	.61
0470	2nd coat		1300	.006		.08	.27		.35	.52
0500	Paint, alkyd, brushwork, primer coat		1140	.007		.11	.31		.42	.61
0510	1st coat		1140	.007		.14	.31		.45	.64
0520	2nd coat		1300	.006		.10	.27		.37	.54

09 91 Painting

09 91 13 – Exterior Painting

09 91 13.42 Miscellaneous, Exterior		Crew	Daily Output	Labor-Hours	Unit	Material	2020 Bare Costs Labor	Equipment	Total	Total Incl O&P
0600	Sand paint, alkyd, brushwork, 1 coat	1 Pord	150	.053	S.F.	.14	2.37		2.51	3.92
9000	Minimum labor/equipment charge		2	4	Job		178		178	282

09 91 13.60 Siding Exterior		Crew	Daily Output	Labor-Hours	Unit	Material	2020 Bare Costs Labor	Equipment	Total	Total Incl O&P
0010	**SIDING EXTERIOR**, Alkyd (oil base) R099100-10									
0450	Steel siding, oil base, paint 1 coat, brushwork	2 Pord	2015	.008	S.F.	.11	.35		.46	.68
0500	Spray R099100-20		4550	.004		.17	.16		.33	.43
0800	Paint 2 coats, brushwork		1300	.012		.22	.55		.77	1.11
1000	Spray		2750	.006		.15	.26		.41	.58
1200	Stucco, rough, oil base, paint 2 coats, brushwork		1300	.012		.22	.55		.77	1.11
1400	Roller		1625	.010		.23	.44		.67	.94
1600	Spray		2925	.005		.24	.24		.48	.66
1800	Texture 1-11 or clapboard, oil base, primer coat, brushwork		1300	.012		.14	.55		.69	1.02
2000	Spray		4550	.004		.14	.16		.30	.40
2100	Paint 1 coat, brushwork		1300	.012		.16	.55		.71	1.05
2200	Spray		4550	.004		.16	.16		.32	.43
2400	Paint 2 coats, brushwork		810	.020		.32	.88		1.20	1.74
2600	Spray		2600	.006		.36	.27		.63	.82
3000	Stain 1 coat, brushwork		1520	.011		.10	.47		.57	.85
3200	Spray		5320	.003		.11	.13		.24	.33
3400	Stain 2 coats, brushwork		950	.017		.20	.75		.95	1.41
4000	Spray		3050	.005		.22	.23		.45	.61
4200	Wood shingles, oil base primer coat, brushwork		1300	.012		.13	.55		.68	1.01
4400	Spray		3900	.004		.12	.18		.30	.42
4600	Paint 1 coat, brushwork		1300	.012		.13	.55		.68	1.02
4800	Spray		3900	.004		.17	.18		.35	.47
5000	Paint 2 coats, brushwork		810	.020		.27	.88		1.15	1.68
5200	Spray		2275	.007		.25	.31		.56	.78
5800	Stain 1 coat, brushwork		1500	.011		.10	.47		.57	.86
6000	Spray		3900	.004		.10	.18		.28	.40
6500	Stain 2 coats, brushwork		950	.017		.20	.75		.95	1.41
7000	Spray		2660	.006		.27	.27		.54	.72
8000	For latex paint, deduct					10%				
8100	For work over 12′ H, from pipe scaffolding, add						15%			
8200	For work over 12′ H, from extension ladder, add						25%			
8300	For work over 12′ H, from swing staging, add						35%			
9000	Minimum labor/equipment charge	1 Pord	2	4	Job		178		178	282

09 91 13.62 Siding, Misc.		Crew	Daily Output	Labor-Hours	Unit	Material	2020 Bare Costs Labor	Equipment	Total	Total Incl O&P
0010	**SIDING, MISC.**, latex paint R099100-10									
0100	Aluminum siding									
0110	Brushwork, primer R099100-20	2 Pord	2275	.007	S.F.	.07	.31		.38	.57
0120	Finish coat, exterior latex		2275	.007		.06	.31		.37	.57
0130	Primer & 1 coat exterior latex		1300	.012		.14	.55		.69	1.02
0140	Primer & 2 coats exterior latex		975	.016		.20	.73		.93	1.38
0150	Mineral fiber shingles									
0160	Brushwork, primer	2 Pord	1495	.011	S.F.	.14	.48		.62	.90
0170	Finish coat, industrial enamel		1495	.011		.20	.48		.68	.97
0180	Primer & 1 coat enamel		810	.020		.34	.88		1.22	1.76
0190	Primer & 2 coats enamel		540	.030		.54	1.32		1.86	2.68
0200	Roll, primer		1625	.010		.16	.44		.60	.86
0210	Finish coat, industrial enamel		1625	.010		.22	.44		.66	.93
0220	Primer & 1 coat enamel		975	.016		.37	.73		1.10	1.57
0230	Primer & 2 coats enamel		650	.025		.59	1.09		1.68	2.38

09 91 Painting

09 91 13 – Exterior Painting

09 91 13.62 Siding, Misc.		Crew	Daily Output	Labor-Hours	Unit	Material	Labor	Equipment	Total	Total Incl O&P
						2020 Bare Costs				
0240	Spray, primer	2 Pord	3900	.004	S.F.	.12	.18		.30	.42
0250	Finish coat, industrial enamel		3900	.004		.18	.18		.36	.49
0260	Primer & 1 coat enamel		2275	.007		.30	.31		.61	.83
0270	Primer & 2 coats enamel		1625	.010		.48	.44		.92	1.22
0280	Waterproof sealer, first coat		4485	.004		.12	.16		.28	.39
0290	Second coat	↓	5235	.003	↓	.12	.14		.26	.35
0300	Rough wood incl. shingles, shakes or rough sawn siding									
0310	Brushwork, primer	2 Pord	1280	.013	S.F.	.14	.56		.70	1.04
0320	Finish coat, exterior latex		1280	.013		.11	.56		.67	1
0330	Primer & 1 coat exterior latex		960	.017		.25	.74		.99	1.46
0340	Primer & 2 coats exterior latex		700	.023		.36	1.02		1.38	2
0350	Roll, primer		2925	.005		.19	.24		.43	.60
0360	Finish coat, exterior latex		2925	.005		.13	.24		.37	.53
0370	Primer & 1 coat exterior latex		1790	.009		.32	.40		.72	.98
0380	Primer & 2 coats exterior latex		1300	.012		.45	.55		1	1.36
0390	Spray, primer		3900	.004		.16	.18		.34	.47
0400	Finish coat, exterior latex		3900	.004		.10	.18		.28	.40
0410	Primer & 1 coat exterior latex		2600	.006		.26	.27		.53	.72
0420	Primer & 2 coats exterior latex		2080	.008		.36	.34		.70	.94
0430	Waterproof sealer, first coat		4485	.004		.22	.16		.38	.50
0440	Second coat	↓	4485	.004	↓	.12	.16		.28	.39
0450	Smooth wood incl. butt, T&G, beveled, drop or B&B siding									
0460	Brushwork, primer	2 Pord	2325	.007	S.F.	.10	.31		.41	.60
0470	Finish coat, exterior latex		1280	.013		.11	.56		.67	1
0480	Primer & 1 coat exterior latex		800	.020		.21	.89		1.10	1.64
0490	Primer & 2 coats exterior latex		630	.025		.32	1.13		1.45	2.14
0500	Roll, primer		2275	.007		.12	.31		.43	.63
0510	Finish coat, exterior latex		2275	.007		.11	.31		.42	.63
0520	Primer & 1 coat exterior latex		1300	.012		.23	.55		.78	1.12
0530	Primer & 2 coats exterior latex		975	.016		.35	.73		1.08	1.54
0540	Spray, primer		4550	.004		.09	.16		.25	.35
0550	Finish coat, exterior latex		4550	.004		.10	.16		.26	.36
0560	Primer & 1 coat exterior latex		2600	.006		.19	.27		.46	.64
0570	Primer & 2 coats exterior latex		1950	.008		.29	.36		.65	.90
0580	Waterproof sealer, first coat		5230	.003		.12	.14		.26	.36
0590	Second coat	↓	5980	.003		.12	.12		.24	.33
0600	For oil base paint, add				↓	10%				
9000	Minimum labor/equipment charge	1 Pord	2	4	Job		178		178	282

09 91 13.70 Doors and Windows, Exterior		Crew	Daily Output	Labor-Hours	Unit	Material	Labor	Equipment	Total	Total Incl O&P
0010	**DOORS AND WINDOWS, EXTERIOR** R099100-10									
0100	Door frames & trim, only									
0110	Brushwork, primer R099100-20	1 Pord	512	.016	L.F.	.07	.69		.76	1.17
0120	Finish coat, exterior latex		512	.016		.08	.69		.77	1.19
0130	Primer & 1 coat, exterior latex		300	.027	↓	.15	1.18		1.33	2.04
0135	2 coats, exterior latex, both sides		15	.533	Ea.	7.25	23.50		30.75	45.50
0140	Primer & 2 coats, exterior latex	↓	265	.030	L.F.	.23	1.34		1.57	2.38
0150	Doors, flush, both sides, incl. frame & trim									
0160	Roll & brush, primer	1 Pord	10	.800	Ea.	4.96	35.50		40.46	62
0170	Finish coat, exterior latex		10	.800		6.20	35.50		41.70	63.50
0180	Primer & 1 coat, exterior latex		7	1.143		11.15	50.50		61.65	93
0190	Primer & 2 coats, exterior latex		5	1.600		17.30	71		88.30	132
0200	Brushwork, stain, sealer & 2 coats polyurethane	↓	4	2	↓	31	89		120	175

09 91 Painting

09 91 13 – Exterior Painting

09 91 13.70 Doors and Windows, Exterior		Crew	Daily Output	Labor-Hours	Unit	Material	Labor	Equipment	Total	Total Incl O&P
						2020 Bare Costs				
0210	Doors, French, both sides, 10-15 lite, incl. frame & trim									
0220	Brushwork, primer	1 Pord	6	1.333	Ea.	2.48	59		61.48	96.50
0230	Finish coat, exterior latex		6	1.333		3.09	59		62.09	97.50
0240	Primer & 1 coat, exterior latex		3	2.667		5.55	118		123.55	194
0250	Primer & 2 coats, exterior latex		2	4		8.50	178		186.50	291
0260	Brushwork, stain, sealer & 2 coats polyurethane		2.50	3.200		11.15	142		153.15	238
0270	Doors, louvered, both sides, incl. frame & trim									
0280	Brushwork, primer	1 Pord	7	1.143	Ea.	4.96	50.50		55.46	86
0290	Finish coat, exterior latex		7	1.143		6.20	50.50		56.70	87.50
0300	Primer & 1 coat, exterior latex		4	2		11.15	89		100.15	153
0310	Primer & 2 coats, exterior latex		3	2.667		16.95	118		134.95	207
0320	Brushwork, stain, sealer & 2 coats polyurethane		4.50	1.778		31	79		110	159
0330	Doors, panel, both sides, incl. frame & trim									
0340	Roll & brush, primer	1 Pord	6	1.333	Ea.	4.96	59		63.96	99.50
0350	Finish coat, exterior latex		6	1.333		6.20	59		65.20	101
0360	Primer & 1 coat, exterior latex		3	2.667		11.15	118		129.15	200
0370	Primer & 2 coats, exterior latex		2.50	3.200		16.95	142		158.95	245
0380	Brushwork, stain, sealer & 2 coats polyurethane		3	2.667		31	118		149	222
0400	Windows, per ext. side, based on 15 S.F.									
0410	1 to 6 lite									
0420	Brushwork, primer	1 Pord	13	.615	Ea.	.98	27.50		28.48	44.50
0430	Finish coat, exterior latex		13	.615		1.22	27.50		28.72	45
0440	Primer & 1 coat, exterior latex		8	1		2.20	44.50		46.70	73
0450	Primer & 2 coats, exterior latex		6	1.333		3.35	59		62.35	97.50
0460	Stain, sealer & 1 coat varnish		7	1.143		4.40	50.50		54.90	85.50
0470	7 to 10 lite									
0480	Brushwork, primer	1 Pord	11	.727	Ea.	.98	32.50		33.48	52.50
0490	Finish coat, exterior latex		11	.727		1.22	32.50		33.72	53
0500	Primer & 1 coat, exterior latex		7	1.143		2.20	50.50		52.70	83
0510	Primer & 2 coats, exterior latex		5	1.600		3.35	71		74.35	117
0520	Stain, sealer & 1 coat varnish		6	1.333		4.40	59		63.40	99
0530	12 lite									
0540	Brushwork, primer	1 Pord	10	.800	Ea.	.98	35.50		36.48	57.50
0550	Finish coat, exterior latex		10	.800		1.22	35.50		36.72	58
0560	Primer & 1 coat, exterior latex		6	1.333		2.20	59		61.20	96.50
0570	Primer & 2 coats, exterior latex		5	1.600		3.35	71		74.35	117
0580	Stain, sealer & 1 coat varnish		6	1.333		4.45	59		63.45	99
0590	For oil base paint, add					10%				
9000	Minimum labor/equipment charge	1 Pord	2	4	Job		178		178	282

09 91 13.80 Trim, Exterior		Crew	Daily Output	Labor-Hours	Unit	Material	Labor	Equipment	Total	Total Incl O&P
0010	**TRIM, EXTERIOR** R099100-10									
0100	Door frames & trim (see Doors, interior or exterior)									
0110	Fascia, latex paint, one coat coverage R099100-20									
0120	1" x 4", brushwork	1 Pord	640	.013	L.F.	.02	.56		.58	.91
0130	Roll		1280	.006		.03	.28		.31	.47
0140	Spray		2080	.004		.02	.17		.19	.29
0150	1" x 6" to 1" x 10", brushwork		640	.013		.09	.56		.65	.98
0160	Roll		1230	.007		.09	.29		.38	.56
0170	Spray		2100	.004		.07	.17		.24	.35
0180	1" x 12", brushwork		640	.013		.09	.56		.65	.98
0190	Roll		1050	.008		.09	.34		.43	.64
0200	Spray		2200	.004		.07	.16		.23	.34

09 91 13 – Exterior Painting

09 91 13.80 Trim, Exterior		Crew	Daily Output	Labor-Hours	Unit	2020 Bare Costs Material	Labor	Equipment	Total	Total Incl O&P
0210	Gutters & downspouts, metal, zinc chromate paint									
0220	Brushwork, gutters, 5", first coat	1 Pord	640	.013	L.F.	.40	.56		.96	1.31
0230	Second coat		960	.008		.38	.37		.75	1
0240	Third coat		1280	.006		.30	.28		.58	.77
0250	Downspouts, 4", first coat		640	.013		.40	.56		.96	1.31
0260	Second coat		960	.008		.38	.37		.75	1
0270	Third coat		1280	.006		.30	.28		.58	.77
0280	Gutters & downspouts, wood									
0290	Brushwork, gutters, 5", primer	1 Pord	640	.013	L.F.	.07	.56		.63	.95
0300	Finish coat, exterior latex		640	.013		.07	.56		.63	.96
0310	Primer & 1 coat exterior latex		400	.020		.15	.89		1.04	1.57
0320	Primer & 2 coats exterior latex		325	.025		.23	1.09		1.32	1.99
0330	Downspouts, 4", primer		640	.013		.07	.56		.63	.95
0340	Finish coat, exterior latex		640	.013		.07	.56		.63	.96
0350	Primer & 1 coat exterior latex		400	.020		.15	.89		1.04	1.57
0360	Primer & 2 coats exterior latex		325	.025		.11	1.09		1.20	1.87
0370	Molding, exterior, up to 14" wide									
0380	Brushwork, primer	1 Pord	640	.013	L.F.	.08	.56		.64	.97
0390	Finish coat, exterior latex		640	.013		.09	.56		.65	.98
0400	Primer & 1 coat exterior latex		400	.020		.18	.89		1.07	1.60
0410	Primer & 2 coats exterior latex		315	.025		.18	1.13		1.31	1.98
0420	Stain & fill		1050	.008		.12	.34		.46	.67
0430	Shellac		1850	.004		.16	.19		.35	.47
0440	Varnish		1275	.006		.12	.28		.40	.57
9000	Minimum labor/equipment charge		2	4	Job		178		178	282

09 91 13.90 Walls, Masonry (CMU), Exterior

Line	Description	Crew	Daily Output	Labor-Hours	Unit	Material	Labor	Equipment	Total	Total Incl O&P
0010	**WALLS, MASONRY (CMU), EXTERIOR**									
0360	Concrete masonry units (CMU), smooth surface									
0370	Brushwork, latex, first coat	1 Pord	640	.013	S.F.	.06	.56		.62	.95
0380	Second coat		960	.008		.05	.37		.42	.64
0390	Waterproof sealer, first coat		736	.011		.28	.48		.76	1.08
0400	Second coat		1104	.007		.28	.32		.60	.82
0410	Roll, latex, paint, first coat		1465	.005		.07	.24		.31	.46
0420	Second coat		1790	.004		.06	.20		.26	.38
0430	Waterproof sealer, first coat		1680	.005		.28	.21		.49	.65
0440	Second coat		2060	.004		.28	.17		.45	.58
0450	Spray, latex, paint, first coat		1950	.004		.06	.18		.24	.35
0460	Second coat		2600	.003		.05	.14		.19	.27
0470	Waterproof sealer, first coat		2245	.004		.28	.16		.44	.56
0480	Second coat		2990	.003		.28	.12		.40	.50
0490	Concrete masonry unit (CMU), porous									
0500	Brushwork, latex, first coat	1 Pord	640	.013	S.F.	.12	.56		.68	1.01
0510	Second coat		960	.008		.06	.37		.43	.66
0520	Waterproof sealer, first coat		736	.011		.28	.48		.76	1.08
0530	Second coat		1104	.007		.28	.32		.60	.82
0540	Roll latex, first coat		1465	.005		.09	.24		.33	.48
0550	Second coat		1790	.004		.06	.20		.26	.38
0560	Waterproof sealer, first coat		1680	.005		.28	.21		.49	.65
0570	Second coat		2060	.004		.28	.17		.45	.58
0580	Spray latex, first coat		1950	.004		.07	.18		.25	.36
0590	Second coat		2600	.003		.05	.14		.19	.27
0600	Waterproof sealer, first coat		2245	.004		.28	.16		.44	.56

09 91 Painting

09 91 13 – Exterior Painting

09 91 13.90 Walls, Masonry (CMU), Exterior		Crew	Daily Output	Labor-Hours	Unit	Material	2020 Bare Costs Labor	Equipment	Total	Total Incl O&P
0610	Second coat	1 Pord	2990	.003	S.F.	.28	.12		.40	.50
9000	Minimum labor/equipment charge		2	4	Job		178		178	282

09 91 23 – Interior Painting

09 91 23.20 Cabinets and Casework

Line	Description	Crew	Daily Output	Labor-Hours	Unit	Material	Labor	Equipment	Total	Total Incl O&P
0010	**CABINETS AND CASEWORK** R099100-10									
1000	Primer coat, oil base, brushwork	1 Pord	650	.012	S.F.	.07	.55		.62	.95
2000	Paint, oil base, brushwork, 1 coat R099100-20		650	.012		.12	.55		.67	1
2500	2 coats		400	.020		.23	.89		1.12	1.67
3000	Stain, brushwork, wipe off		650	.012		.10	.55		.65	.98
4000	Shellac, 1 coat, brushwork		650	.012		.13	.55		.68	1.02
4500	Varnish, 3 coats, brushwork, sand after 1st coat		325	.025		.29	1.09		1.38	2.06
5000	For latex paint, deduct					10%				
6300	Strip, prep and refinish wood furniture									
6310	Remove paint using chemicals, wood furniture	1 Pord	28	.286	S.F.	1.55	12.70		14.25	21.50
6320	Prep for painting, sanding		75	.107		.25	4.74		4.99	7.75
6350	Stain and wipe, brushwork		600	.013		.10	.59		.69	1.05
6355	Spray applied		900	.009		.09	.39		.48	.73
6360	Sealer or varnish, brushwork		1080	.007		.10	.33		.43	.63
6365	Spray applied		2100	.004		.10	.17		.27	.38
6370	Paint, primer, brushwork		720	.011		.07	.49		.56	.86
6375	Spray applied		2100	.004		.07	.17		.24	.34
6380	Finish coat, brushwork		810	.010		.12	.44		.56	.83
6385	Spray applied		2100	.004		.11	.17		.28	.40
9010	Minimum labor/equipment charge		2	4	Job		178		178	282

09 91 23.33 Doors and Windows, Interior Alkyd (Oil Base)

Line	Description	Crew	Daily Output	Labor-Hours	Unit	Material	Labor	Equipment	Total	Total Incl O&P
0010	**DOORS AND WINDOWS, INTERIOR ALKYD (OIL BASE)** R099100-10									
0500	Flush door & frame, 3′ x 7′, oil, primer, brushwork	1 Pord	10	.800	Ea.	4.21	35.50		39.71	61
1000	Paint, 1 coat R099100-20		10	.800		4.82	35.50		40.32	62
1200	2 coats		6	1.333		4.98	59		63.98	99.50
1220	3 coats		5	1.600		17.10	71		88.10	132
1400	Stain, brushwork, wipe off		18	.444		2.06	19.75		21.81	34
1600	Shellac, 1 coat, brushwork		25	.320		2.77	14.20		16.97	25.50
1800	Varnish, 3 coats, brushwork, sand after 1st coat		9	.889		6.05	39.50		45.55	69
2000	Panel door & frame, 3′ x 7′, oil, primer, brushwork		6	1.333		2.64	59		61.64	97
2200	Paint, 1 coat		6	1.333		4.82	59		63.82	99.50
2400	2 coats		3	2.667		12.30	118		130.30	202
2420	3 coats		2	4		16.55	178		194.55	300
2600	Stain, brushwork, panel door, 3′ x 7′, not incl. frame		16	.500		2.06	22		24.06	38
2800	Shellac, 1 coat, brushwork		22	.364		2.77	16.15		18.92	28.50
3000	Varnish, 3 coats, brushwork, sand after 1st coat		7.50	1.067		6.05	47.50		53.55	81.50
4400	Windows, including frame and trim, per side									
4600	Colonial type, 6/6 lites, 2′ x 3′, oil, primer, brushwork	1 Pord	14	.571	Ea.	.42	25.50		25.92	41
5800	Paint, 1 coat		14	.571		.76	25.50		26.26	41.50
6000	2 coats		9	.889		1.48	39.50		40.98	64
6010	3 coats		7	1.143		2.19	50.50		52.69	83
6200	3′ x 5′ opening, 6/6 lites, primer coat, brushwork		12	.667		1.04	29.50		30.54	48
6400	Paint, 1 coat		12	.667		1.90	29.50		31.40	49
6600	2 coats		7	1.143		3.70	50.50		54.20	84.50
6610	3 coats		6	1.333		5.50	59		64.50	100
6800	4′ x 8′ opening, 6/6 lites, primer coat, brushwork		8	1		2.23	44.50		46.73	73
7000	Paint, 1 coat		8	1		4.06	44.50		48.56	75
7200	2 coats		5	1.600		7.90	71		78.90	122

09 91 23.33 Doors and Windows, Interior Alkyd (Oil Base)		Crew	Daily Output	Labor-Hours	Unit	Material	2020 Bare Costs Labor	Equipment	Total	Total Incl O&P
7210	3 coats	1 Pord	4	2	Ea.	11.70	89		100.70	154
7500	Standard, 6/6 lites, 2′ x 3′, primer coat, brushwork		14	.571		.42	25.50		25.92	41
7520	Paint 1 coat		14	.571		.76	25.50		26.26	41.50
7540	2 coats		9	.889		1.48	39.50		40.98	64
7560	3 coats		7	1.143		2.19	50.50		52.69	83
7580	3′ x 5′, 6/6 lites, primer coat, brushwork		12	.667		1.04	29.50		30.54	48
7600	Paint 1 coat		12	.667		1.90	29.50		31.40	49
7620	2 coats		7	1.143		3.70	50.50		54.20	84.50
7640	3 coats		6	1.333		5.50	59		64.50	100
7660	4′ x 8′, 6/6 lites, primer coat, brushwork		8	1		2.23	44.50		46.73	73
7680	Paint 1 coat		8	1		4.06	44.50		48.56	75
7700	2 coats		5	1.600		7.90	71		78.90	122
7720	3 coats		4	2		11.70	89		100.70	154
8000	Single lite type, 2′ x 3′, oil base, primer coat, brushwork		33	.242		.42	10.75		11.17	17.55
8200	Paint, 1 coat		33	.242		.76	10.75		11.51	17.95
8400	2 coats		20	.400		1.48	17.75		19.23	29.50
8410	3 coats		16	.500		2.19	22		24.19	38
8600	3′ x 5′ opening, primer coat, brushwork		20	.400		1.04	17.75		18.79	29
8800	Paint, 1 coat		20	.400		1.90	17.75		19.65	30
8900	2 coats		13	.615		3.70	27.50		31.20	47.50
9010	3 coats		10	.800		5.50	35.50		41	62.50
9200	4′ x 8′ opening, primer coat, brushwork		14	.571		2.23	25.50		27.73	43
9400	Paint, 1 coat		14	.571		4.06	25.50		29.56	45
9600	2 coats		8	1		7.90	44.50		52.40	79
9610	3 coats		7	1.143		11.70	50.50		62.20	93.50
9900	Minimum labor/equipment charge		2	4	Job		178		178	282

09 91 23.35 Doors and Windows, Interior Latex		Crew	Daily Output	Labor-Hours	Unit	Material	Labor	Equipment	Total	Total Incl O&P
0010	**DOORS & WINDOWS, INTERIOR LATEX** R099100-10									
0100	Doors, flush, both sides, incl. frame & trim									
0110	Roll & brush, primer	1 Pord	10	.800	Ea.	4.17	35.50		39.67	61
0120	Finish coat, latex		10	.800		5.95	35.50		41.45	63
0130	Primer & 1 coat latex		7	1.143		10.10	50.50		60.60	91.50
0140	Primer & 2 coats latex		5	1.600		15.70	71		86.70	130
0160	Spray, both sides, primer		20	.400		4.39	17.75		22.14	33
0170	Finish coat, latex		20	.400		6.20	17.75		23.95	35
0180	Primer & 1 coat latex		11	.727		10.70	32.50		43.20	63.50
0190	Primer & 2 coats latex		8	1		16.60	44.50		61.10	89
0200	Doors, French, both sides, 10-15 lite, incl. frame & trim									
0210	Roll & brush, primer	1 Pord	6	1.333	Ea.	2.08	59		61.08	96.50
0220	Finish coat, latex		6	1.333		2.96	59		61.96	97.50
0230	Primer & 1 coat latex		3	2.667		5.05	118		123.05	194
0240	Primer & 2 coats latex		2	4		7.85	178		185.85	291
0260	Doors, louvered, both sides, incl. frame & trim									
0270	Roll & brush, primer	1 Pord	7	1.143	Ea.	4.17	50.50		54.67	85
0280	Finish coat, latex		7	1.143		5.95	50.50		56.45	87
0290	Primer & 1 coat, latex		4	2		9.85	89		98.85	152
0300	Primer & 2 coats, latex		3	2.667		16.05	118		134.05	206
0320	Spray, both sides, primer		20	.400		4.39	17.75		22.14	33
0330	Finish coat, latex		20	.400		6.20	17.75		23.95	35
0340	Primer & 1 coat, latex		11	.727		10.70	32.50		43.20	63.50
0350	Primer & 2 coats, latex		8	1		17	44.50		61.50	89
0360	Doors, panel, both sides, incl. frame & trim									

09 91 23.35 Doors and Windows, Interior Latex		Crew	Daily Output	Labor-Hours	Unit	Material	2020 Bare Costs Labor	Equipment	Total	Total Incl O&P
0370	Roll & brush, primer	1 Pord	6	1.333	Ea.	4.39	59		63.39	99
0380	Finish coat, latex		6	1.333		5.95	59		64.95	101
0390	Primer & 1 coat, latex		3	2.667		10.10	118		128.10	199
0400	Primer & 2 coats, latex		2.50	3.200		16.05	142		158.05	244
0420	Spray, both sides, primer		10	.800		4.39	35.50		39.89	61.50
0430	Finish coat, latex		10	.800		6.20	35.50		41.70	63.50
0440	Primer & 1 coat, latex		5	1.600		10.70	71		81.70	125
0450	Primer & 2 coats, latex		4	2		17	89		106	160
0460	Windows, per interior side, based on 15 S.F.									
0470	1 to 6 lite									
0480	Brushwork, primer	1 Pord	13	.615	Ea.	.82	27.50		28.32	44.50
0490	Finish coat, enamel		13	.615		1.17	27.50		28.67	45
0500	Primer & 1 coat enamel		8	1		1.99	44.50		46.49	72.50
0510	Primer & 2 coats enamel		6	1.333		3.16	59		62.16	97.50
0530	7 to 10 lite									
0540	Brushwork, primer	1 Pord	11	.727	Ea.	.82	32.50		33.32	52.50
0550	Finish coat, enamel		11	.727		1.17	32.50		33.67	53
0560	Primer & 1 coat enamel		7	1.143		1.99	50.50		52.49	82.50
0570	Primer & 2 coats enamel		5	1.600		3.16	71		74.16	116
0590	12 lite									
0600	Brushwork, primer	1 Pord	10	.800	Ea.	.82	35.50		36.32	57.50
0610	Finish coat, enamel		10	.800		1.17	35.50		36.67	58
0620	Primer & 1 coat enamel		6	1.333		1.99	59		60.99	96
0630	Primer & 2 coats enamel		5	1.600		3.16	71		74.16	116
0650	For oil base paint, add					10%				
9000	Minimum labor/equipment charge	1 Pord	2	4	Job		178		178	282

09 91 23.39 Doors and Windows, Interior Latex, Zero Voc			Crew	Daily Output	Labor-Hours	Unit	Material	Labor	Equipment	Total	Total Incl O&P
0010	**DOORS & WINDOWS, INTERIOR LATEX, ZERO VOC**										
0100	Doors flush, both sides, incl. frame & trim										
0110	Roll & brush, primer	G	1 Pord	10	.800	Ea.	5.70	35.50		41.20	63
0120	Finish coat, latex	G		10	.800		6.60	35.50		42.10	64
0130	Primer & 1 coat latex	G		7	1.143		12.25	50.50		62.75	94
0140	Primer & 2 coats latex	G		5	1.600		18.50	71		89.50	134
0160	Spray, both sides, primer	G		20	.400		6	17.75		23.75	34.50
0170	Finish coat, latex	G		20	.400		6.90	17.75		24.65	35.50
0180	Primer & 1 coat latex	G		11	.727		13	32.50		45.50	66
0190	Primer & 2 coats latex	G		8	1		19.55	44.50		64.05	92
0200	Doors, French, both sides, 10-15 lite, incl. frame & trim										
0210	Roll & brush, primer	G	1 Pord	6	1.333	Ea.	2.84	59		61.84	97
0220	Finish coat, latex	G		6	1.333		3.29	59		62.29	97.50
0230	Primer & 1 coat latex	G		3	2.667		6.15	118		124.15	195
0240	Primer & 2 coats latex	G		2	4		9.25	178		187.25	292
0360	Doors, panel, both sides, incl. frame & trim										
0370	Roll & brush, primer	G	1 Pord	6	1.333	Ea.	6	59		65	101
0380	Finish coat, latex	G		6	1.333		6.60	59		65.60	101
0390	Primer & 1 coat, latex	G		3	2.667		12.25	118		130.25	202
0400	Primer & 2 coats, latex	G		2.50	3.200		18.85	142		160.85	247
0420	Spray, both sides, primer	G		10	.800		6	35.50		41.50	63
0430	Finish coat, latex	G		10	.800		6.90	35.50		42.40	64
0440	Primer & 1 coat, latex	G		5	1.600		13	71		84	127
0450	Primer & 2 coats, latex	G		4	2		20	89		109	163
0460	Windows, per interior side, based on 15 S.F.										

09 91 Painting

09 91 23 – Interior Painting

09 91 23.39 Doors and Windows, Interior Latex, Zero Voc

	09 91 23.39 Doors and Windows, Interior Latex, Zero Voc	Crew	Daily Output	Labor-Hours	Unit	Material	Labor	Equipment	Total	Total Incl O&P
0470	1 to 6 lite									
0480	Brushwork, primer G	1 Pord	13	.615	Ea.	1.12	27.50		28.62	44.50
0490	Finish coat, enamel G		13	.615		1.30	27.50		28.80	45
0500	Primer & 1 coat enamel G		8	1		2.42	44.50		46.92	73
0510	Primer & 2 coats enamel G		6	1.333		3.72	59		62.72	98
9000	Minimum labor/equipment charge		2	4	Job		178		178	282

09 91 23.40 Floors, Interior

	09 91 23.40 Floors, Interior	Crew	Daily Output	Labor-Hours	Unit	Material	Labor	Equipment	Total	Total Incl O&P
0010	**FLOORS, INTERIOR** R099100-10									
0100	Concrete paint, latex									
0110	Brushwork									
0120	1st coat	1 Pord	975	.008	S.F.	.15	.36		.51	.75
0130	2nd coat		1150	.007		.10	.31		.41	.60
0140	3rd coat		1300	.006		.08	.27		.35	.52
0150	Roll									
0160	1st coat	1 Pord	2600	.003	S.F.	.20	.14		.34	.44
0170	2nd coat		3250	.002		.12	.11		.23	.30
0180	3rd coat		3900	.002		.09	.09		.18	.24
0190	Spray									
0200	1st coat	1 Pord	2600	.003	S.F.	.17	.14		.31	.41
0210	2nd coat		3250	.002		.10	.11		.21	.27
0220	3rd coat		3900	.002		.08	.09		.17	.22

09 91 23.44 Anti-Slip Floor Treatments

	09 91 23.44 Anti-Slip Floor Treatments	Crew	Daily Output	Labor-Hours	Unit	Material	Labor	Equipment	Total	Total Incl O&P
0010	**ANTI-SLIP FLOOR TREATMENTS**									
1000	Walking surface treatment, ADA compliant, mop on and rinse									
1100	For tile, terrazzo, stone or smooth concrete	1 Pord	4000	.002	S.F.	.34	.09		.43	.52
1110	For marble		4000	.002		.25	.09		.34	.41
1120	For wood		4000	.002		.23	.09		.32	.39
1130	For baths and showers		500	.016		.39	.71		1.10	1.56
2000	Granular additive for paint or sealer, add to paint cost					.02			.02	.02

09 91 23.52 Miscellaneous, Interior

	09 91 23.52 Miscellaneous, Interior	Crew	Daily Output	Labor-Hours	Unit	Material	Labor	Equipment	Total	Total Incl O&P
0010	**MISCELLANEOUS, INTERIOR** R099100-10									
2400	Floors, conc./wood, oil base, primer/sealer coat, brushwork	2 Pord	1950	.008	S.F.	.09	.36		.45	.68
2450	Roller		5200	.003		.09	.14		.23	.32
2600	Spray		6000	.003		.09	.12		.21	.29
2650	Paint 1 coat, brushwork		1950	.008		.10	.36		.46	.69
2800	Roller		5200	.003		.10	.14		.24	.33
2850	Spray		6000	.003		.11	.12		.23	.31
3000	Stain, wood floor, brushwork, 1 coat		4550	.004		.10	.16		.26	.36
3200	Roller		5200	.003		.10	.14		.24	.33
3250	Spray		6000	.003		.10	.12		.22	.30
3400	Varnish, wood floor, brushwork		4550	.004		.10	.16		.26	.36
3450	Roller		5200	.003		.10	.14		.24	.33
3600	Spray		6000	.003		.11	.12		.23	.31
3650	For anti-skid, see Section 09 91 23.44									
3800	Grilles, per side, oil base, primer coat, brushwork	1 Pord	520	.015	S.F.	.14	.68		.82	1.23
3850	Spray		1140	.007		.15	.31		.46	.65
3880	Paint 1 coat, brushwork		520	.015		.25	.68		.93	1.36
3900	Spray		1140	.007		.28	.31		.59	.80
3920	Paint 2 coats, brushwork		325	.025		.49	1.09		1.58	2.28
3940	Spray		650	.012		.56	.55		1.11	1.49
3950	Prime & paint 1 coat		325	.025		.39	1.09		1.48	2.17
3960	Prime & paint 2 coats		270	.030		.39	1.32		1.71	2.51

09 91 23 – Interior Painting

09 91 23.52 Miscellaneous, Interior		Crew	Daily Output	Labor-Hours	Unit	Material	Labor	Equipment	Total	Total Incl O&P
						2020 Bare Costs				
4500	Louvers, 1 side, primer, brushwork	1 Pord	524	.015	S.F.	.09	.68		.77	1.18
4520	Paint 1 coat, brushwork		520	.015		.11	.68		.79	1.20
4530	Spray		1140	.007		.12	.31		.43	.63
4540	Paint 2 coats, brushwork		325	.025		.22	1.09		1.31	1.98
4550	Spray		650	.012		.24	.55		.79	1.13
4560	Paint 3 coats, brushwork		270	.030		.32	1.32		1.64	2.44
4570	Spray		500	.016		.36	.71		1.07	1.52
4600	Miscellaneous surfaces, metallic paint, spray applied									
4610	Water based, non-tintable, warm silver	1 Pord	1140	.007	S.F.	.59	.31		.90	1.14
4620	Rusted iron		1140	.007		1.04	.31		1.35	1.63
4630	Low VOC, tintable		1140	.007		.40	.31		.71	.93
5000	Pipe, 1"-4" diameter, primer or sealer coat, oil base, brushwork	2 Pord	1250	.013	L.F.	.09	.57		.66	1
5100	Spray		2165	.007		.09	.33		.42	.62
5200	Paint 1 coat, brushwork		1250	.013		.13	.57		.70	1.04
5300	Spray		2165	.007		.11	.33		.44	.65
5350	Paint 2 coats, brushwork		775	.021		.23	.92		1.15	1.71
5400	Spray		1240	.013		.25	.57		.82	1.19
5420	Paint 3 coats, brushwork		775	.021		.34	.92		1.26	1.83
5450	5"-8" diameter, primer or sealer coat, brushwork		620	.026		.18	1.15		1.33	2.02
5500	Spray		1085	.015		.30	.65		.95	1.37
5550	Paint 1 coat, brushwork		620	.026		.35	1.15		1.50	2.20
5600	Spray		1085	.015		.38	.65		1.03	1.46
5650	Paint 2 coats, brushwork		385	.042		.45	1.85		2.30	3.43
5700	Spray		620	.026		.51	1.15		1.66	2.38
5720	Paint 3 coats, brushwork		385	.042		.68	1.85		2.53	3.67
5750	9"-12" diameter, primer or sealer coat, brushwork		415	.039		.28	1.71		1.99	3.02
5800	Spray		725	.022		.41	.98		1.39	2.01
5850	Paint 1 coat, brushwork		415	.039		.35	1.71		2.06	3.11
6000	Spray		725	.022		.39	.98		1.37	1.99
6200	Paint 2 coats, brushwork		260	.062		.68	2.73		3.41	5.10
6250	Spray		415	.039		.76	1.71		2.47	3.55
6270	Paint 3 coats, brushwork		260	.062		1.01	2.73		3.74	5.45
6300	13"-16" diameter, primer or sealer coat, brushwork		310	.052		.37	2.29		2.66	4.04
6350	Spray		540	.030		.41	1.32		1.73	2.54
6400	Paint 1 coat, brushwork		310	.052		.47	2.29		2.76	4.16
6450	Spray		540	.030		.52	1.32		1.84	2.66
6500	Paint 2 coats, brushwork		195	.082		.91	3.64		4.55	6.80
6550	Spray		310	.052		1.01	2.29		3.30	4.75
6600	Radiators, per side, primer, brushwork	1 Pord	520	.015	S.F.	.09	.68		.77	1.18
6620	Paint, 1 coat		520	.015		.08	.68		.76	1.17
6640	2 coats		340	.024		.22	1.04		1.26	1.90
6660	3 coats		283	.028		.32	1.26		1.58	2.34
7000	Trim, wood, incl. puttying, under 6" wide									
7200	Primer coat, oil base, brushwork	1 Pord	650	.012	L.F.	.03	.55		.58	.91
7250	Paint, 1 coat, brushwork		650	.012		.06	.55		.61	.94
7400	2 coats		400	.020		.12	.89		1.01	1.55
7450	3 coats		325	.025		.18	1.09		1.27	1.94
7500	Over 6" wide, primer coat, brushwork		650	.012		.07	.55		.62	.95
7550	Paint, 1 coat, brushwork		650	.012		.13	.55		.68	1.01
7600	2 coats		400	.020		.25	.89		1.14	1.68
7650	3 coats		325	.025		.37	1.09		1.46	2.14
8000	Cornice, simple design, primer coat, oil base, brushwork		650	.012	S.F.	.07	.55		.62	.95
8250	Paint, 1 coat		650	.012		.13	.55		.68	1.01

09 91 23.52 Miscellaneous, Interior		Crew	Daily Output	Labor-Hours	Unit	Material	2020 Bare Costs Labor	Equipment	Total	Total Incl O&P
8300	2 coats	1 Pord	400	.020	S.F.	.25	.89		1.14	1.68
8350	Ornate design, primer coat		350	.023		.07	1.02		1.09	1.69
8400	Paint, 1 coat		350	.023		.13	1.02		1.15	1.75
8450	2 coats		400	.020		.25	.89		1.14	1.68
8600	Balustrades, primer coat, oil base, brushwork		520	.015		.07	.68		.75	1.16
8650	Paint, 1 coat		520	.015		.13	.68		.81	1.22
8700	2 coats		325	.025		.25	1.09		1.34	2.01
8900	Trusses and wood frames, primer coat, oil base, brushwork		800	.010		.07	.44		.51	.79
8950	Spray		1200	.007		.07	.30		.37	.55
9000	Paint 1 coat, brushwork		750	.011		.13	.47		.60	.89
9200	Spray		1200	.007		.14	.30		.44	.62
9220	Paint 2 coats, brushwork		500	.016		.25	.71		.96	1.40
9240	Spray		600	.013		.27	.59		.86	1.24
9260	Stain, brushwork, wipe off		600	.013		.10	.59		.69	1.05
9280	Varnish, 3 coats, brushwork		275	.029		.29	1.29		1.58	2.37
9350	For latex paint, deduct					10%				
9900	Minimum labor/equipment charge	1 Pord	2	4	Job		178		178	282

09 91 23.62 Electrostatic Painting		Crew	Daily Output	Labor-Hours	Unit	Material	Labor	Equipment	Total	Total Incl O&P
0010	**ELECTROSTATIC PAINTING**									
0100	In shop									
0200	Flat surfaces (lockers, casework, elevator doors, etc.)									
0300	One coat	1 Pord	200	.040	S.F.	.65	1.78		2.43	3.53
0400	Two coats	"	120	.067	"	.94	2.96		3.90	5.75
0500	Irregular surfaces (furniture, door frames, etc.)									
0600	One coat	1 Pord	150	.053	S.F.	.65	2.37		3.02	4.47
0700	Two coats	"	100	.080	"	.94	3.55		4.49	6.70
0800	On site									
0900	Flat surfaces (lockers, casework, elevator doors, etc.)									
1000	One coat	1 Pord	150	.053	S.F.	.65	2.37		3.02	4.47
1100	Two coats	"	100	.080	"	.94	3.55		4.49	6.70
1200	Irregular surfaces (furniture, door frames, etc.)									
1300	One coat	1 Pord	115	.070	S.F.	.65	3.09		3.74	5.60
1400	Two coats		70	.114		.94	5.05		5.99	9.10
2000	Anti-microbial coating, hospital application		150	.053		.06	2.37		2.43	3.82

09 91 23.72 Walls and Ceilings, Interior		Crew	Daily Output	Labor-Hours	Unit	Material	Labor	Equipment	Total	Total Incl O&P
0010	**WALLS AND CEILINGS, INTERIOR** R099100-10									
0100	Concrete, drywall or plaster, latex, primer or sealer coat R099100-20									
0150	Smooth finish, cut-in by brush	1 Pord	1150	.007	L.F.	.02	.31		.33	.51
0200	Brushwork		1150	.007	S.F.	.06	.31		.37	.56
0240	Roller		1350	.006		.06	.26		.32	.49
0280	Spray		2750	.003		.05	.13		.18	.27
0290	Sand finish, cut-in by brush		975	.008	L.F.	.02	.36		.38	.60
0300	Brushwork		975	.008	S.F.	.06	.36		.42	.65
0340	Roller		1150	.007		.06	.31		.37	.56
0380	Spray		2275	.004		.05	.16		.21	.31
0390	Paint 1 coat, smooth finish, cut-in by brush		1200	.007	L.F.	.02	.30		.32	.49
0400	Brushwork		1200	.007	S.F.	.08	.30		.38	.55
0440	Roller		1300	.006		.08	.27		.35	.51
0480	Spray		2275	.004		.06	.16		.22	.32
0490	Sand finish, cut-in by brush		1050	.008	L.F.	.02	.34		.36	.56
0500	Brushwork		1050	.008	S.F.	.07	.34		.41	.62
0540	Roller		1600	.005		.08	.22		.30	.43

09 91 23.72 Walls and Ceilings, Interior		Crew	Daily Output	Labor-Hours	Unit	Material	Labor	Equipment	Total	Total Incl O&P
						2020 Bare Costs				
0580	Spray	1 Pord	2100	.004	S.F.	.03	.17		.20	.30
0590	Paint 2 coats, smooth finish, cut-in by brush		680	.012		.04	.52		.56	.87
0800	Brushwork		680	.012		.15	.52		.67	1
0840	Roller		800	.010		.15	.44		.59	.88
0880	Spray		1625	.005		.14	.22		.36	.51
0890	Sand finish, cut-in by brush		605	.013	L.F.	.04	.59		.63	.97
0900	Brushwork		605	.013	S.F.	.15	.59		.74	1.10
0940	Roller		1020	.008		.15	.35		.50	.72
0980	Spray		1700	.005		.14	.21		.35	.49
1190	Paint 3 coats, smooth finish, cut-in by brush		510	.016	L.F.	.06	.70		.76	1.17
1200	Brushwork		510	.016	S.F.	.23	.70		.93	1.36
1240	Roller		650	.012		.23	.55		.78	1.12
1280	Spray		850	.009		.21	.42		.63	.90
1290	Sand finish, cut-in by brush		454	.018	L.F.	.10	.78		.88	1.35
1300	Brushwork		454	.018	S.F.	.39	.78		1.17	1.67
1340	Roller		680	.012		.42	.52		.94	1.29
1380	Spray		1133	.007		.36	.31		.67	.89
1600	Glaze coating, 2 coats, spray, clear		1200	.007		.56	.30		.86	1.09
1640	Multicolor		1200	.007		.87	.30		1.17	1.43
1660	Painting walls, complete, including surface prep, primer &									
1670	2 coats finish, on drywall or plaster, with roller	1 Pord	325	.025	S.F.	.22	1.09		1.31	1.99
1700	For oil base paint, add					10%				
1800	For ceiling installations, add						25%			
2000	Masonry or concrete block, primer/sealer, latex paint									
2090	Primer, smooth finish, cut-in by brush	1 Pord	1000	.008	L.F.	.04	.36		.40	.61
2100	Brushwork		1000	.008	S.F.	.17	.36		.53	.75
2110	Roller		1150	.007		.11	.31		.42	.61
2180	Spray		2400	.003		.10	.15		.25	.34
2190	Sand finish, cut-in by brush		850	.009	L.F.	.03	.42		.45	.69
2200	Brushwork		850	.009	S.F.	.11	.42		.53	.78
2210	Roller		975	.008		.11	.36		.47	.70
2280	Spray		2050	.004		.10	.17		.27	.39
2290	Finish coat, smooth finish, cut-in by brush		1100	.007	L.F.	.02	.32		.34	.53
2400	Brushwork		1100	.007	S.F.	.08	.32		.40	.60
2410	Roller		1300	.006		.08	.27		.35	.52
2480	Spray		2400	.003		.07	.15		.22	.31
2490	Sand finish, cut-in by brush		950	.008	L.F.	.02	.37		.39	.61
2500	Brushwork		950	.008	S.F.	.08	.37		.45	.68
2510	Roller		1090	.007		.08	.33		.41	.61
2580	Spray		2040	.004		.07	.17		.24	.36
2590	Primer plus one finish coat, smooth cut-in by brush		525	.015	L.F.	.08	.68		.76	1.15
2800	Brushwork		525	.015	S.F.	.30	.68		.98	1.40
2810	Roller		615	.013		.19	.58		.77	1.13
2880	Spray		1200	.007		.17	.30		.47	.66
2890	Sand finish, cut-in by brush		450	.018	L.F.	.05	.79		.84	1.30
2900	Brushwork		450	.018	S.F.	.19	.79		.98	1.46
2910	Roller		515	.016		.19	.69		.88	1.30
2980	Spray		1025	.008		.17	.35		.52	.74
3190	Primer plus 2 finish coats, smooth, cut-in by brush		355	.023	L.F.	.07	1		1.07	1.67
3200	Brushwork		355	.023	S.F.	.28	1		1.28	1.90
3210	Roller		415	.019		.28	.86		1.14	1.67
3280	Spray		800	.010		.24	.44		.68	.98
3290	Sand finish, cut-in by brush		305	.026	L.F.	.07	1.16		1.23	1.93

09 91 Painting

09 91 23 – Interior Painting

09 91 23.72 Walls and Ceilings, Interior			Crew	Daily Output	Labor-Hours	Unit	Material	2020 Bare Costs Labor	Equipment	Total	Total Incl O&P
3300	Brushwork		1 Pord	305	.026	S.F.	.28	1.16		1.44	2.16
3310	Roller			350	.023		.28	1.02		1.30	1.92
3380	Spray			675	.012		.24	.53		.77	1.11
3600	Glaze coating, 3 coats, spray, clear			900	.009		.80	.39		1.19	1.51
3620	Multicolor			900	.009		1.08	.39		1.47	1.82
4000	Block filler, 1 coat, brushwork			425	.019		.13	.84		.97	1.48
4100	Silicone, water repellent, 2 coats, spray			2000	.004		.48	.18		.66	.80
4120	For oil base paint, add						10%				
8200	For work 8'-15' H, add							10%			
8300	For work over 15' H, add							20%			
8400	For light textured surfaces, add							10%			
8410	Heavy textured, add							25%			
9900	Minimum labor/equipment charge		1 Pord	2	4	Job		178		178	282

09 91 23.74 Walls and Ceilings, Interior, Zero VOC Latex			Crew	Daily Output	Labor-Hours	Unit	Material	Labor	Equipment	Total	Total Incl O&P
0010	**WALLS AND CEILINGS, INTERIOR, ZERO VOC LATEX**										
0100	Concrete, dry wall or plaster, latex, primer or sealer coat										
0190	Smooth finish, cut-in by brush		1 Pord	1150	.007	L.F.	.02	.31		.33	.51
0200	Brushwork	G		1150	.007	S.F.	.07	.31		.38	.57
0240	Roller	G		1350	.006		.07	.26		.33	.50
0280	Spray	G		2750	.003		.05	.13		.18	.27
0290	Sand finish, cut-in by brush			975	.008	L.F.	.02	.36		.38	.60
0300	Brushwork	G		975	.008	S.F.	.07	.36		.43	.66
0340	Roller	G		1150	.007		.08	.31		.39	.58
0380	Spray	G		2275	.004		.06	.16		.22	.32
0390	Paint 1 coat, smooth finish, cut-in by brush			1200	.007	L.F.	.02	.30		.32	.49
0400	Brushwork	G		1200	.007	S.F.	.08	.30		.38	.56
0440	Roller	G		1300	.006		.08	.27		.35	.52
0480	Spray	G		2275	.004		.07	.16		.23	.33
0490	Sand finish, cut-in by brush			1050	.008	L.F.	.02	.34		.36	.56
0500	Brushwork	G		1050	.008	S.F.	.02	.34		.36	.56
0540	Roller	G		1600	.005		.08	.22		.30	.44
0580	Spray	G		2100	.004		.07	.17		.24	.35
0790	Paint 2 coats, smooth finish, cut-in by brush			680	.012	L.F.	.04	.52		.56	.87
0800	Brushwork	G		680	.012	S.F.	.04	.52		.56	.87
0840	Roller	G		800	.010		.17	.44		.61	.89
0880	Spray	G		1625	.005		.14	.22		.36	.51
0890	Sand finish, cut-in by brush			605	.013	L.F.	.04	.59		.63	.97
0900	Brushwork	G		605	.013	S.F.	.16	.59		.75	1.10
0940	Roller	G		1020	.008		.17	.35		.52	.73
0980	Spray	G		1700	.005		.14	.21		.35	.49
1190	Paint 3 coats, smooth finish, cut-in my brush			510	.016	L.F.	.06	.70		.76	1.17
1200	Brushwork	G		510	.016	S.F.	.24	.70		.94	1.37
1240	Roller	G		650	.012		.25	.55		.80	1.14
1280	Spray	G		850	.009		.21	.42		.63	.90
1800	For ceiling installations, add	G						25%			
8200	For work 8' - 15' H, add							10%			
8300	For work over 15' H, add							20%			
9900	Minimum labor/equipment charge		1 Pord	2	4	Job		178		178	282

09 91 Painting

09 91 23 – Interior Painting

09 91 23.75 Dry Fall Painting		Crew	Daily Output	Labor-Hours	Unit	Material	Labor	Equipment	Total	Total Incl O&P
						2020 Bare Costs				
0010	**DRY FALL PAINTING** R099100-10									
0100	Sprayed on walls, gypsum board or plaster									
0220	One coat R099100-20	1 Pord	2600	.003	S.F.	.08	.14		.22	.31
0250	Two coats		1560	.005		.16	.23		.39	.54
0280	Concrete or textured plaster, one coat		1560	.005		.08	.23		.31	.45
0310	Two coats		1300	.006		.16	.27		.43	.61
0340	Concrete block, one coat		1560	.005		.08	.23		.31	.45
0370	Two coats		1300	.006		.16	.27		.43	.61
0400	Wood, one coat		877	.009		.08	.40		.48	.73
0430	Two coats	↓	650	.012	↓	.16	.55		.71	1.05
0440	On ceilings, gypsum board or plaster									
0470	One coat	1 Pord	1560	.005	S.F.	.08	.23		.31	.45
0500	Two coats		1300	.006		.16	.27		.43	.61
0530	Concrete or textured plaster, one coat		1560	.005		.08	.23		.31	.45
0560	Two coats		1300	.006		.16	.27		.43	.61
0570	Structural steel, bar joists or metal deck, one coat		1560	.005		.08	.23		.31	.45
0580	Two coats		1040	.008	↓	.16	.34		.50	.72
9900	Minimum labor/equipment charge	↓	2	4	Job		178		178	282

09 93 Staining and Transparent Finishing

09 93 23 – Interior Staining and Finishing

09 93 23.10 Varnish		Crew	Daily Output	Labor-Hours	Unit	Material	Labor	Equipment	Total	Total Incl O&P
0010	**VARNISH**									
0012	1 coat + sealer, on wood trim, brush, no sanding included	1 Pord	400	.020	S.F.	.07	.89		.96	1.49
0020	1 coat + sealer, on wood trim, brush, no sanding included, no VOC		400	.020		.22	.89		1.11	1.65
0100	Hardwood floors, 2 coats, no sanding included, roller		1890	.004	↓	.15	.19		.34	.47
9000	Minimum labor/equipment charge	↓	4	2	Job		89		89	141

09 96 High-Performance Coatings

09 96 23 – Graffiti-Resistant Coatings

09 96 23.10 Graffiti-Resistant Treatments		Crew	Daily Output	Labor-Hours	Unit	Material	Labor	Equipment	Total	Total Incl O&P
0010	**GRAFFITI-RESISTANT TREATMENTS**, sprayed on walls									
0100	Non-sacrificial, permanent non-stick coating, clear, on metals	1 Pord	2000	.004	S.F.	2.05	.18		2.23	2.54
0200	Concrete		2000	.004		2.33	.18		2.51	2.84
0300	Concrete block		2000	.004		3.02	.18		3.20	3.60
0400	Brick		2000	.004		3.42	.18		3.60	4.04
0500	Stone		2000	.004		3.42	.18		3.60	4.04
0600	Unpainted wood		2000	.004		3.94	.18		4.12	4.62
2000	Semi-permanent cross linking polymer primer, on metals		2000	.004		.62	.18		.80	.96
2100	Concrete		2000	.004		.74	.18		.92	1.09
2200	Concrete block		2000	.004		.92	.18		1.10	1.30
2300	Brick		2000	.004		.74	.18		.92	1.09
2400	Stone		2000	.004		.74	.18		.92	1.09
2500	Unpainted wood		2000	.004		1.03	.18		1.21	1.41
3000	Top coat, on metals		2000	.004		.56	.18		.74	.89
3100	Concrete		2000	.004		.64	.18		.82	.98
3200	Concrete block		2000	.004		.89	.18		1.07	1.26
3300	Brick		2000	.004		.74	.18		.92	1.10
3400	Stone	↓	2000	.004	↓	.74	.18		.92	1.10

09 96 High-Performance Coatings

09 96 23 – Graffiti-Resistant Coatings

09 96 23.10 Graffiti-Resistant Treatments		Crew	Daily Output	Labor-Hours	Unit	2020 Bare Costs Material	Labor	Equipment	Total	Total Incl O&P
3500	Unpainted wood	1 Pord	2000	.004	S.F.	.89	.18		1.07	1.26
5000	Sacrificial, water based, on metal		2000	.004		.34	.18		.52	.65
5100	Concrete		2000	.004		.34	.18		.52	.65
5200	Concrete block		2000	.004		.34	.18		.52	.65
5300	Brick		2000	.004		.34	.18		.52	.65
5400	Stone		2000	.004		.34	.18		.52	.65
5500	Unpainted wood		2000	.004		.34	.18		.52	.65
8000	Cleaner for use after treatment									
8100	Towels or wipes, per package of 30				Ea.	.72			.72	.79
8200	Aerosol spray, 24 oz. can				"	19			19	21
8500	Graffiti removal with chemicals									
8510	Brush on, spray rinse off, on brick, masonry or stone	A-1H	200	.040	S.F.	.34	1.68	.38	2.40	3.49
8520	On smooth concrete	"	400	.020		.27	.84	.19	1.30	1.86
8530	Wipe on, wipe off, on plastic or painted metal	1 Clab	1500	.005		.72	.22		.94	1.15
8540	Brush on, wipe off, on painted wood	"	500	.016		.36	.67		1.03	1.47

09 96 46 – Intumescent Painting

09 96 46.10 Coatings, Intumescent		Crew	Daily Output	Labor-Hours	Unit	Material	Labor	Equipment	Total	Total Incl O&P
0010	**COATINGS, INTUMESCENT**, spray applied									
0100	On exterior structural steel, 0.25" d.f.t.	1 Pord	475	.017	S.F.	.50	.75		1.25	1.74
0150	0.51" d.f.t.		350	.023		.50	1.02		1.52	2.16
0200	0.98" d.f.t.		280	.029		.50	1.27		1.77	2.56
0300	On interior structural steel, 0.108" d.f.t.		300	.027		.49	1.18		1.67	2.42
0350	0.310" d.f.t.		150	.053		.49	2.37		2.86	4.30
0400	0.670" d.f.t.		100	.080		.49	3.55		4.04	6.20

09 96 53 – Elastomeric Coatings

09 96 53.10 Coatings, Elastomeric		Crew	Daily Output	Labor-Hours	Unit	Material	Labor	Equipment	Total	Total Incl O&P
0010	**COATINGS, ELASTOMERIC**									
0020	High build, water proof, one coat system									
0100	Concrete, brush	1 Pord	650	.012	S.F.	.29	.55		.84	1.19

09 96 56 – Epoxy Coatings

09 96 56.20 Wall Coatings		Crew	Daily Output	Labor-Hours	Unit	Material	Labor	Equipment	Total	Total Incl O&P
0010	**WALL COATINGS**									
0100	Acrylic glazed coatings, matte	1 Pord	525	.015	S.F.	.38	.68		1.06	1.49
0200	Gloss		305	.026		.78	1.16		1.94	2.71
0300	Epoxy coatings, solvent based		525	.015		.47	.68		1.15	1.59
0400	Water based		170	.047		.38	2.09		2.47	3.74
0600	Exposed aggregate, troweled on, 1/16" to 1/4", solvent based		235	.034		.76	1.51		2.27	3.24
0700	Water based (epoxy or polyacrylate)		130	.062		1.63	2.73		4.36	6.15
0900	1/2" to 5/8" aggregate, solvent based		130	.062		1.42	2.73		4.15	5.90
1000	Water based		80	.100		2.47	4.44		6.91	9.75
1200	1" aggregate size, solvent based		90	.089		2.52	3.95		6.47	9
1300	Water based		55	.145		3.83	6.45		10.28	14.45
1500	Exposed aggregate, sprayed on, 1/8" aggregate, solvent based		295	.027		.58	1.20		1.78	2.55
1600	Water based		145	.055		1.24	2.45		3.69	5.25
1800	High build epoxy, 50 mil, solvent based		390	.021		.76	.91		1.67	2.29
1900	Water based		95	.084		1.36	3.74		5.10	7.45
2100	Laminated epoxy with fiberglass, solvent based		295	.027		.86	1.20		2.06	2.86
2200	Water based		145	.055		1.62	2.45		4.07	5.65
2400	Sprayed perlite or vermiculite, 1/16" thick, solvent based		2935	.003		.28	.12		.40	.50
2500	Water based		640	.013		.87	.56		1.43	1.84
2700	Vinyl plastic wall coating, solvent based		735	.011		.41	.48		.89	1.22

09 96 High-Performance Coatings

09 96 56 – Epoxy Coatings

09 96 56.20 Wall Coatings		Crew	Daily Output	Labor-Hours	Unit	Material	Labor	Equipment	Total	Total Incl O&P
						2020 Bare Costs				
2800	Water based	1 Pord	240	.033	S.F.	1	1.48		2.48	3.45
3000	Urethane on smooth surface, 2 coats, solvent based		1135	.007		.35	.31		.66	.89
3100	Water based		665	.012		.62	.53		1.15	1.53
3300	3 coat, solvent based		840	.010		.44	.42		.86	1.15
3400	Water based		470	.017		.98	.76		1.74	2.28
3600	Ceramic-like glazed coating, cementitious, solvent based		440	.018		.49	.81		1.30	1.82
3700	Water based		345	.023		.99	1.03		2.02	2.72
3900	Resin base, solvent based		640	.013		.34	.56		.90	1.25
4000	Water based	↓	330	.024	↓	.64	1.08		1.72	2.41

09 97 Special Coatings

09 97 10 – Coatings and Paints

09 97 10.10 Coatings and Paints

		Crew	Daily Output	Labor-Hours	Unit	Material	Labor	Equipment	Total	Total Incl O&P
0010	**COATINGS & PAINTS** in 5 gallon lots R099100-20									
0050	For 100 gallons or more, deduct					10%				
0100	Paint, exterior alkyd (oil base)									
0200	Flat				Gal.	48			48	53
0300	Gloss					47.50			47.50	52
0400	Primer				↓	35			35	38.50
0500	Latex (water base)									
0600	Acrylic stain				Gal.	41			41	45
0700	Gloss enamel					34.50			34.50	38
0800	Flat					28.50			28.50	31.50
0900	Primer					26			26	28.50
1000	Semi-gloss				↓	35			35	38.50
1054	Low VOC									
1055	Primer				Gal.	53			53	58.50
1060	Satin					18.05			18.05	19.85
1065	Door and trim				↓	75			75	82.50
1100	Interior, alkyd (oil base)									
1200	Enamel undercoat				Gal.	44.50			44.50	49
1300	Flat					59			59	65
1400	Gloss					51			51	56.50
1500	Primer sealer					28			28	30.50
1600	Semi-gloss				↓	54			54	59.50
1700	Latex (water base)									
1800	Enamel undercoat				Gal.	27			27	29.50
1900	Flat					25			25	27.50
2000	Floor and deck					38			38	41.50
2100	Gloss					32			32	35
2200	Primer sealer					22			22	24
2300	Semi-gloss				↓	33			33	36.50
2320	Low VOC									
2330	Wallboard primer				Gal.	50.50			50.50	55.50
2335	Flat					52			52	57
2340	Semi-gloss					54.50			54.50	60
2345	Eggshell					55			55	60.50
2350	Stain				↓	57.50			57.50	63
2400	Masonry, exterior									
2500	Alkali-resistant primer				Gal.	24.50			24.50	27
2600	Block filler, epoxy				↓	30.50			30.50	33.50

09 97 Special Coatings

09 97 10 – Coatings and Paints

09 97 10.10 Coatings and Paints		Crew	Daily Output	Labor-Hours	Unit	2020 Bare Costs Material	Labor	Equipment	Total	Total Incl O&P
2700	Latex				Gal.	18.25			18.25	20
2800	Latex, flat					25.50			25.50	28.50
2900	Semi-gloss				▼	33			33	36.50
3000	Masonry, interior									
3100	Alkali-resistant primer				Gal.	28			28	31
3200	Block filler, epoxy					26.50			26.50	29.50
3300	Latex					27			27	29.50
3400	Floor, alkyd					42			42	46
3500	Latex					30.50			30.50	33.50
3600	Latex, flat acrylic					25			25	27.50
3700	Flat emulsion					27.50			27.50	30.50
3800	Sealer					26			26	28.50
3900	Semi-gloss				▼	29.50			29.50	32.50
4000	Metal									
4100	Galvanizing paint				Gal.	92			92	101
4200	High heat					67			67	73.50
4300	Heat-resistant					40.50			40.50	44.50
4400	Machinery enamel, alkyd					54			54	59
4500	Metal pretreatment (polyvinyl butyral)					117			117	128
4600	Rust inhibitor, ferrous metal					40.50			40.50	44.50
4700	Zinc chromate					150			150	165
4800	Zinc rich primer				▼	170			170	187
4900	Varnish and stain									
5000	Alkyd, clear				Gal.	38.50			38.50	42.50
5100	Polyurethane, clear					45.50			45.50	50.50
5200	Primer sealer					32.50			32.50	35.50
5300	Stain, semi-transparent					39.50			39.50	43
5400	Solid color				▼	40.50			40.50	44.50
5500	Coatings									
5600	Heavy duty									
5700	Acrylic urethane				Gal.	59.50			59.50	65.50
5800	Chlorinated rubber					70.50			70.50	77.50
5900	Coal tar epoxy					75			75	82.50
6000	Polyamide epoxy, finish					72.50			72.50	80
6100	Primer					52			52	57.50
6200	Silicone alkyd					62.50			62.50	68.50
6300	2 component solvent based acrylic epoxy					113			113	124
6400	Polyester epoxy					100			100	110
6500	Vinyl				▼	37			37	40.50
6600	Special/Miscellaneous									
6700	Aluminum				Gal.	47.50			47.50	52
6900	Dry fall out, flat					24.50			24.50	27
7000	Fire retardant, intumescent					50			50	55
7100	Linseed oil					26			26	28.50
7200	Shellac					52.50			52.50	58
7300	Swimming pool, epoxy or urethane base					54			54	59
7400	Rubber base					74			74	81.50
7500	Texture paint					27.50			27.50	30.50
7600	Turpentine					37.50			37.50	41.50
7700	Water repellent, 5% silicone					36			36	39.50
7800	Insulating additive				▼	19.90			19.90	22

09 97 Special Coatings

09 97 13 – Steel Coatings

09 97 13.23 Exterior Steel Coatings		Crew	Daily Output	Labor-Hours	Unit	Material	Labor	Equipment	Total	Total Incl O&P
						2020 Bare Costs				
0010	EXTERIOR STEEL COATINGS R050516-30									
6100	Cold galvanizing, brush in field	1 Psst	1100	.007	S.F.	.23	.33		.56	.82
6510	Paints & protective coatings, sprayed in field									
6520	Alkyds, primer	2 Psst	3600	.004	S.F.	.09	.20		.29	.45
6540	Gloss topcoats		3200	.005		.09	.23		.32	.50
6560	Silicone alkyd		3200	.005		.18	.23		.41	.60
6610	Epoxy, primer		3000	.005		.26	.24		.50	.71
6630	Intermediate or topcoat		2800	.006		.29	.26		.55	.77
6650	Enamel coat		2800	.006		.36	.26		.62	.85
6700	Epoxy ester, primer		2800	.006		.50	.26		.76	1
6720	Topcoats		2800	.006		.24	.26		.50	.72
6810	Latex primer		3600	.004		.07	.20		.27	.42
6830	Topcoats		3200	.005		.08	.23		.31	.49
6910	Universal primers, one part, phenolic, modified alkyd		2000	.008		.39	.37		.76	1.06
6940	Two part, epoxy spray		2000	.008		.45	.37		.82	1.13
7000	Zinc rich primers, self cure, spray, inorganic		1800	.009		.85	.41		1.26	1.64
7010	Epoxy, spray, organic	▼	1800	.009	▼	.29	.41		.70	1.02
7020	Above one story, spray painting simple structures, add						25%			
7030	Intricate structures, add						50%			

09 97 35 – Dry Erase Coatings

09 97 35.10 Dry Erase Coatings		Crew	Daily Output	Labor-Hours	Unit	Material	Labor	Equipment	Total	Total Incl O&P
0010	DRY ERASE COATINGS									
0020	Dry erase coatings, clear, roller applied	1 Pord	1325	.006	S.F.	2.14	.27		2.41	2.78

Division Notes

	CREW	DAILY OUTPUT	LABOR-HOURS	UNIT	BARE COSTS MAT.	BARE COSTS LABOR	BARE COSTS EQUIP.	BARE COSTS TOTAL	TOTAL INCL O&P

Estimating Tips

General

- The items in this division are usually priced per square foot or each.
- Many items in Division 10 require some type of support system or special anchors that are not usually furnished with the item. The required anchors must be added to the estimate in the appropriate division.
- Some items in Division 10, such as lockers, may require assembly before installation. Verify the amount of assembly required. Assembly can often exceed installation time.

10 20 00 Interior Specialties

- Support angles and blocking are not included in the installation of toilet compartments, shower/dressing compartments, or cubicles. Appropriate line items from Division 5 or 6 may need to be added to support the installations.
- Toilet partitions are priced by the stall. A stall consists of a side wall, pilaster, and door with hardware. Toilet tissue holders and grab bars are extra.
- The required acoustical rating of a folding partition can have a significant impact on costs. Verify the sound transmission coefficient rating of the panel priced against the specification requirements.
- Grab bar installation does not include supplemental blocking or backing to support the required load. When grab bars are installed at an existing facility, provisions must be made to attach the grab bars to a solid structure.

Reference Numbers

Reference numbers are shown at the beginning of some major classifications. These numbers refer to related items in the Reference Section. The reference information may be an estimating procedure, an alternate pricing method, or technical information.

Note: Not all subdivisions listed here necessarily appear. ■

10 05 Common Work Results for Specialties

10 05 05 – Selective Demolition for Specialties

10 05 05.10 Selective Demolition, Specialties		Crew	Daily Output	Labor-Hours	Unit	Material	2020 Bare Costs Labor	Equipment	Total	Total Incl O&P
0010	**SELECTIVE DEMOLITION, SPECIALTIES** R024119-10									
1100	Boards and panels, wall mounted	2 Clab	15	1.067	Ea.		45		45	72
1105	Demolition, mirror, wall mounted	1 Clab	30	.267			11.25		11.25	18
1200	Cases, for directory and/or bulletin boards, including doors	2 Clab	24	.667			28		28	45
1850	Shower partitions, cabinet or stall, including base and door	″	8	2			84		84	135
1855	Shower receptor, terrazzo or concrete	1 Clab	14	.571			24		24	38.50
1900	Curtain track or rod, hospital type, ceiling mounted or suspended	″	220	.036	L.F.		1.53		1.53	2.45
1910	Toilet cubicles, remove	2 Clab	8	2	Ea.		84		84	135
1930	Urinal screen, remove	1 Clab	12	.667	″		28		28	45
2650	Wall guard, misc. wall or corner protection	″	320	.025	L.F.		1.05		1.05	1.69
2750	Access floor, metal panel system, including pedestals, covering	2 Clab	850	.019	S.F.		.79		.79	1.27
3050	Fireplace, prefab, freestanding or wall hung, including hood and screen	1 Clab	2	4	Ea.		168		168	270
3054	Chimney top, simulated brick, 4′ high	″	15	.533			22.50		22.50	36
3200	Stove, woodburning, cast iron	2 Clab	2	8			335		335	540
3440	Weathervane, residential	1 Clab	12	.667			28		28	45
3500	Flagpole, groundset, to 70′ high, excluding base/foundation	K-1	1	16			805	820	1,625	2,175
3555	To 30′ high	″	2.50	6.400			320	330	650	875
4000	Removal of traffic signs, including supports									
4020	To 10 S.F.	B-80B	16	2	Ea.		89.50	14.75	104.25	159
4030	11 S.F. to 20 S.F.	″	5	6.400			287	47	334	505
4040	21 S.F. to 40 S.F.	B-14	1.80	26.667			1,175	119	1,294	2,000
4050	41 S.F. to 100 S.F.	B-13	1.30	43.077			1,975	445	2,420	3,650
4070	Remove traffic posts to 12′-0″ high	B-6	100	.240			11	2.14	13.14	19.80
4300	Letter, signs or plaques, exterior on wall	1 Clab	20	.400			16.85		16.85	27
4310	Signs, street, reflective aluminum, including post and bracket		60	.133			5.60		5.60	9
4320	Door signs interior on door 6″ x 6″, selective demolition		20	.400			16.85		16.85	27
4550	Turnstiles, manual or electric	2 Clab	2	8			335		335	540
5050	Lockers	1 Clab	15	.533	Opng.		22.50		22.50	36
5250	Cabinets, recessed	Q-12	12	1.333	Ea.		76		76	118
5260	Mail boxes, horiz., key lock, front loading, remove	1 Carp	34	.235	″		12.50		12.50	20
5350	Awning, fabric, including frame	2 Clab	100	.160	S.F.		6.75		6.75	10.80
6050	Partition, woven wire		1400	.011			.48		.48	.77
6100	Folding gate, security, door or window		500	.032			1.35		1.35	2.16
6580	Acoustic air wall		650	.025			1.04		1.04	1.66
7550	Telephone enclosure, exterior, post mounted		3	5.333	Ea.		225		225	360
8850	Scale, platform, excludes foundation or pit		.25	64	″		2,700		2,700	4,325

10 11 Visual Display Units

10 11 13 – Chalkboards

10 11 13.13 Fixed Chalkboards		Crew	Daily Output	Labor-Hours	Unit	Material	Labor	Equipment	Total	Total Incl O&P
0010	**FIXED CHALKBOARDS** Porcelain enamel steel									
3900	Wall hung									
4000	Aluminum frame and chalktrough									
4200	3′ x 4′	2 Carp	16	1	Ea.	250	53		303	360
4300	3′ x 5′		15	1.067		315	56.50		371.50	440
4500	4′ x 8′		14	1.143		410	60.50		470.50	555
4600	4′ x 12′		13	1.231		565	65.50		630.50	725
4700	Wood frame and chalktrough									
4800	3′ x 4′	2 Carp	16	1	Ea.	194	53		247	298
5000	3′ x 5′		15	1.067		261	56.50		317.50	380
5100	4′ x 5′		14	1.143		274	60.50		334.50	400

10 11 13.13 Fixed Chalkboards		Crew	Daily Output	Labor-Hours	Unit	2020 Bare Costs Material	Labor	Equipment	Total	Total Incl O&P
5300	4′ x 8′	2 Carp	13	1.231	Ea.	375	65.50		440.50	515
5400	Liquid chalk, white porcelain enamel, wall hung									
5420	Deluxe units, aluminum trim and chalktrough									
5450	4′ x 4′	2 Carp	16	1	Ea.	315	53		368	430
5500	4′ x 8′		14	1.143		530	60.50		590.50	685
5550	4′ x 12′		12	1.333		635	71		706	810
5700	Wood trim and chalktrough									
5900	4′ x 4′	2 Carp	16	1	Ea.	765	53		818	925
6000	4′ x 6′		15	1.067		850	56.50		906.50	1,025
6200	4′ x 8′		14	1.143		1,050	60.50		1,110.50	1,250
6250	Economy dry erase board, melamine, 36″ high		104	.154	S.F.	18.10	8.20		26.30	33
6300	Liquid chalk, felt tip markers				Ea.	2.18			2.18	2.40
6500	Erasers					2.29			2.29	2.52
6600	Board cleaner, 8 oz. bottle					6.55			6.55	7.20
9000	Minimum labor/equipment charge	2 Carp	3	5.333	Job		283		283	455

10 11 13.23 Modular-Support-Mounted Chalkboards

Line	Description	Crew	Daily Output	Labor-Hours	Unit	Material	Labor	Equipment	Total	Total Incl O&P
0010	**MODULAR-SUPPORT-MOUNTED CHALKBOARDS**									
0400	Sliding chalkboards									
0450	Vertical, one sliding board with back panel, wall mounted									
0500	8′ x 4′	2 Carp	8	2	Ea.	3,300	106		3,406	3,800
0520	8′ x 8′		7.50	2.133		3,900	113		4,013	4,450
0540	8′ x 12′		7	2.286		4,725	121		4,846	5,375
0600	Two sliding boards, with back panel									
0620	8′ x 4′	2 Carp	8	2	Ea.	5,275	106		5,381	6,000
0640	8′ x 8′		7.50	2.133		5,175	113		5,288	5,850
0660	8′ x 12′		7	2.286		9,525	121		9,646	10,700
0700	Horizontal, two track									
0800	4′ x 8′, 2 sliding panels	2 Carp	8	2	Ea.	2,075	106		2,181	2,450
0820	4′ x 12′, 2 sliding panels		7.50	2.133		3,100	113		3,213	3,600
0840	4′ x 16′, 4 sliding panels		7	2.286		4,200	121		4,321	4,800
0900	Four track, four sliding panels									
0920	4′ x 8′	2 Carp	8	2	Ea.	3,775	106		3,881	4,325
0940	4′ x 12′		7.50	2.133		4,950	113		5,063	5,625
0960	4′ x 16′		7	2.286		6,525	121		6,646	7,375
1200	Vertical, motor operated									
1400	One sliding panel with back panel									
1450	10′ x 4′	2 Carp	4	4	Ea.	5,950	213		6,163	6,875
1500	10′ x 10′		3.75	4.267		7,050	227		7,277	8,125
1550	10′ x 16′		3.50	4.571		8,225	243		8,468	9,450
1700	Two sliding panels with back panel									
1750	10′ x 4′	2 Carp	4	4	Ea.	9,550	213		9,763	10,800
1800	10′ x 10′		3.75	4.267		11,500	227		11,727	13,100
1850	10′ x 16′		3.50	4.571		12,700	243		12,943	14,400
2000	Three sliding panels with back panel									
2100	10′ x 4′	2 Carp	4	4	Ea.	16,300	213		16,513	18,200
2150	10′ x 10′		3.75	4.267		16,500	227		16,727	18,500
2200	10′ x 16′		3.50	4.571		19,500	243		19,743	21,800
2400	For projection screen, glass beaded, add				S.F.	5.35			5.35	5.90
2500	For remote control, 1 panel control, add				Ea.	365			365	400
2600	2 panel control, add				″	625			625	690
2800	For units without back panels, deduct				S.F.	5			5	5.50
2850	For liquid chalk porcelain panels, add				″	5.25			5.25	5.75

10 11 Visual Display Units

10 11 13 – Chalkboards

10 11 13.23 Modular-Support-Mounted Chalkboards		Crew	Daily Output	Labor-Hours	Unit	Material	2020 Bare Costs Labor	Equipment	Total	Total Incl O&P
3000	Swing leaf, any comb. of chalkboard & cork, aluminum frame									
3100	Floor style, 6 panels									
3150	30" x 40" panels				Ea.	1,550			1,550	1,700
3200	48" x 40" panels				"	2,675			2,675	2,950
3300	Wall mounted, 6 panels									
3400	30" x 40" panels	2 Carp	16	1	Ea.	1,500	53		1,553	1,725
3450	48" x 40" panels	"	16	1	"	1,825	53		1,878	2,100
3600	Extra panels for swing leaf units									
3700	30" x 40" panels				Ea.	298			298	325
3750	48" x 40" panels				"	380			380	420

10 11 13.43 Portable Chalkboards

		Crew	Daily Output	Labor-Hours	Unit	Material	Labor	Equipment	Total	Total Incl O&P
0010	**PORTABLE CHALKBOARDS**									
0100	Freestanding, reversible									
0120	Economy, wood frame, 4' x 6'									
0140	Chalkboard both sides				Ea.	630			630	695
0160	Chalkboard one side, cork other side				"	625			625	685
0200	Standard, lightweight satin finished aluminum, 4' x 6'									
0220	Chalkboard both sides				Ea.	565			565	625
0240	Chalkboard one side, cork other side				"	610			610	675
0300	Deluxe, heavy duty extruded aluminum, 4' x 6'									
0320	Chalkboard both sides				Ea.	1,150			1,150	1,250
0340	Chalkboard one side, cork other side				"	1,175			1,175	1,300

10 11 16 – Markerboards

10 11 16.53 Electronic Markerboards

		Crew	Daily Output	Labor-Hours	Unit	Material	Labor	Equipment	Total	Total Incl O&P
0010	**ELECTRONIC MARKERBOARDS**									
0100	Wall hung or free standing, 3' x 4' to 4' x 6'	2 Carp	8	2	S.F.	86.50	106		192.50	265
0150	5' x 6' to 4' x 8'		8	2	"	62.50	106		168.50	239
0500	Interactive projection module for existing whiteboards		8	2	Ea.	1,300	106		1,406	1,625

10 11 23 – Tackboards

10 11 23.10 Fixed Tackboards

		Crew	Daily Output	Labor-Hours	Unit	Material	Labor	Equipment	Total	Total Incl O&P
0010	**FIXED TACKBOARDS**									
0020	Cork sheets, unbacked, no frame, 1/4" thick	2 Carp	290	.055	S.F.	1.68	2.93		4.61	6.55
0100	1/2" thick		290	.055		4.46	2.93		7.39	9.60
0300	Fabric-face, no frame, on 7/32" cork underlay		290	.055		7.35	2.93		10.28	12.75
0400	On 1/4" cork on 1/4" hardboard		290	.055		9.60	2.93		12.53	15.25
0600	With edges wrapped		290	.055		11.35	2.93		14.28	17.20
0700	On 7/16" fire retardant core		290	.055		6.45	2.93		9.38	11.75
0900	With edges wrapped		290	.055		10.05	2.93		12.98	15.75
1000	Designer fabric only, cut to size					2.92			2.92	3.21
1200	1/4" vinyl cork, on 1/4" hardboard, no frame	2 Carp	290	.055		9.95	2.93		12.88	15.60
1300	On 1/4" coreboard		290	.055		5.25	2.93		8.18	10.50
2000	For map and display rail, economy, add		385	.042	L.F.	3.29	2.21		5.50	7.15
2100	Deluxe, add		350	.046	"	5.80	2.43		8.23	10.25
2120	Prefabricated, 1/4" cork, 3' x 5' with aluminum frame		16	1	Ea.	146	53		199	246
2140	Wood frame		16	1		168	53		221	269
2160	4' x 4' with aluminum frame		16	1		174	53		227	276
2180	Wood frame		16	1		218	53		271	325
2200	4' x 8' with aluminum frame		14	1.143		310	60.50		370.50	445
2210	Wood frame		14	1.143		340	60.50		400.50	475
2220	4' x 12' with aluminum frame		12	1.333		375	71		446	530
2230	Bulletin board case, single glass door, with lock									

10 11 Visual Display Units

10 11 23 – Tackboards

10 11 23.10 Fixed Tackboards		Crew	Daily Output	Labor-Hours	Unit	Material	Labor	Equipment	Total	Total Incl O&P
						2020 Bare Costs				
2240	36″ x 24″, economy	2 Carp	12	1.333	Ea.	360	71		431	515
2250	Deluxe		12	1.333		445	71		516	605
2260	42″ x 30″, economy		12	1.333		405	71		476	560
2270	Deluxe		12	1.333		700	71		771	885
2300	Glass enclosed cabinets, alum., cork panel, hinged doors									
2400	3′ x 3′, 1 door	2 Carp	12	1.333	Ea.	645	71		716	825
2500	4′ x 4′, 2 door		11	1.455		1,075	77.50		1,152.50	1,300
2600	4′ x 7′, 3 door		10	1.600		2,125	85		2,210	2,450
2800	4′ x 10′, 4 door		8	2		2,550	106		2,656	2,975
2900	For lights, add per door opening	1 Elec	13	.615		180	38		218	256
3100	Horizontal sliding units, 4 doors, 4′ x 8′, 8′ x 4′	2 Carp	9	1.778		2,075	94.50		2,169.50	2,425
3200	4′ x 12′		7	2.286		3,100	121		3,221	3,625
3400	8 doors, 4′ x 16′		5	3.200		4,175	170		4,345	4,875
3500	4′ x 24′		4	4		5,650	213		5,863	6,575
9000	Minimum labor/equipment charge		4	4	Job		213		213	340

10 11 23.20 Control Boards		Crew	Daily Output	Labor-Hours	Unit	Material	Labor	Equipment	Total	Total Incl O&P
0010	**CONTROL BOARDS**									
0020	Magnetic, porcelain finish, 18″ x 24″, framed	2 Carp	8	2	Ea.	199	106		305	390
0100	24″ x 36″		7.50	2.133		278	113		391	485
0200	36″ x 48″		7	2.286		490	121		611	735
0300	48″ x 72″		6	2.667		660	142		802	955
0400	48″ x 96″		5	3.200		1,075	170		1,245	1,475
1000	Hospital patient display board, 4-color custom design									
1010	Porcelain steel dry erase board, 36″ x 24″	2 Carp	7.50	2.133	Ea.	238	113		351	445

10 13 Directories

10 13 10 – Building Directories

10 13 10.10 Directory Boards		Crew	Daily Output	Labor-Hours	Unit	Material	Labor	Equipment	Total	Total Incl O&P
0010	**DIRECTORY BOARDS**									
0050	Plastic, glass covered, 30″ x 20″	2 Carp	3	5.333	Ea.	179	283		462	650
0100	36″ x 48″		2	8		880	425		1,305	1,650
0300	Grooved cork, 30″ x 20″		3	5.333		420	283		703	915
0400	36″ x 48″		2	8		575	425		1,000	1,325
0600	Black felt, 30″ x 20″		3	5.333		223	283		506	700
0700	36″ x 48″		2	8		460	425		885	1,175
0900	Outdoor, weatherproof, black plastic, 36″ x 24″		2	8		700	425		1,125	1,450
1000	36″ x 36″		1.50	10.667		815	565		1,380	1,800
1800	Indoor, economy, open face, 18″ x 24″		7	2.286		185	121		306	400
1900	24″ x 36″		7	2.286		158	121		279	370
2000	36″ x 24″		6	2.667		160	142		302	405
2100	36″ x 48″		6	2.667		265	142		407	520
2400	Building directory, alum., black felt panels, 1 door, 24″ x 18″		4	4		296	213		509	665
2500	36″ x 24″		3.50	4.571		390	243		633	820
2600	48″ x 32″		3	5.333		555	283		838	1,075
2700	2 door, 36″ x 48″		2.50	6.400		680	340		1,020	1,300
2800	36″ x 60″		2	8		790	425		1,215	1,550
2900	48″ x 60″		1	16		885	850		1,735	2,325
3100	For bronze enamel finish, add					15%				
3200	For bronze anodized finish, add					25%				
3400	For illuminated directory, single door unit, add					147			147	161
3500	For 6″ header panel, 6 letters per foot, add				L.F.	20.50			20.50	22.50

10 13 Directories

10 13 10 – Building Directories

10 13 10.10 Directory Boards		Crew	Daily Output	Labor-Hours	Unit	Material	Labor	Equipment	Total	Total Incl O&P
						2020 Bare Costs				
5000	Building directory, illuminated, 7" sections, bronze finish									
5100	19" x 36", 2 sections, 80 name capacity				Ea.	2,425			2,425	2,650
5200	26-1/4" x 36", 3 sections, 120 name capacity					2,875			2,875	3,175
5300	33-5/8" x 36", 4 sections, 160 name capacity					3,300			3,300	3,625
5400	48-3/4" x 36", 6 sections, 240 name capacity					4,275			4,275	4,700
5500	63-1/2" x 36", 8 sections, 320 name capacity					5,175			5,175	5,700
5600	Engraving charge per namestrip					12.90			12.90	14.20
6050	Building directory, electronic display, alum. frame, wall mounted	2 Carp	32	.500		4,100	26.50		4,126.50	4,550
6100	Free standing	"	60	.267		3,600	14.15		3,614.15	3,975
9000	Minimum labor/equipment charge	1 Carp	1	8	Job		425		425	680

10 14 Signage

10 14 19 – Dimensional Letter Signage

10 14 19.10 Exterior Signs		Crew	Daily Output	Labor-Hours	Unit	Material	Labor	Equipment	Total	Total Incl O&P
0010	**EXTERIOR SIGNS**									
0020	Letters, 2" high, 3/8" deep, cast bronze	1 Carp	24	.333	Ea.	33	17.70		50.70	65
0140	1/2" deep, cast aluminum		18	.444		33	23.50		56.50	74.50
0160	Cast bronze		32	.250		34.50	13.30		47.80	59.50
0300	6" high, 5/8" deep, cast aluminum		24	.333		30	17.70		47.70	61.50
0400	Cast bronze		24	.333		60.50	17.70		78.20	95
0600	8" high, 3/4" deep, cast aluminum		14	.571		33	30.50		63.50	85
0700	Cast bronze		20	.400		100	21.50		121.50	144
0900	10" high, 1" deep, cast aluminum		18	.444		64.50	23.50		88	109
1000	Cast bronze		18	.444		102	23.50		125.50	150
1200	12" high, 1-1/4" deep, cast aluminum		12	.667		59.50	35.50		95	123
1500	Cast bronze		18	.444		163	23.50		186.50	217
1600	14" high, 2-5/16" deep, cast aluminum		12	.667		110	35.50		145.50	178
1800	Fabricated stainless steel, 6" high, 2" deep		20	.400		45.50	21.50		67	84
1900	12" high, 3" deep		18	.444		85	23.50		108.50	132
2100	18" high, 3" deep		12	.667		101	35.50		136.50	168
2200	24" high, 4" deep		10	.800		198	42.50		240.50	286
2700	Acrylic, on high density foam, 12" high, 2" deep		20	.400		18.65	21.50		40.15	54.50
2800	18" high, 2" deep		18	.444		38	23.50		61.50	80
3900	Plaques, custom, 20" x 30", for up to 450 letters, cast aluminum	2 Carp	4	4		1,875	213		2,088	2,400
4000	Cast bronze		4	4		1,950	213		2,163	2,500
4200	30" x 36", up to 900 letters, cast aluminum		3	5.333		2,925	283		3,208	3,675
4300	Cast bronze		3	5.333		4,200	283		4,483	5,075
4500	36" x 48", for up to 1300 letters, cast bronze		2	8		4,500	425		4,925	5,625
4800	Signs, reflective alum. directional signs, dbl. face, 2-way, w/bracket		30	.533		107	28.50		135.50	164
4900	4-way		30	.533		197	28.50		225.50	263
5100	Exit signs, 24 ga. alum., 14" x 12" surface mounted	1 Carp	30	.267		52	14.15		66.15	79.50
5200	10" x 7"		20	.400		32.50	21.50		54	70
5400	Bracket mounted, double face, 12" x 10"		30	.267		59.50	14.15		73.65	88
5500	Sticky back, stock decals, 14" x 10"	1 Clab	50	.160		24.50	6.75		31.25	37.50
6400	Replacement sign faces, 6" or 8"	"	50	.160		66.50	6.75		73.25	84.50
8000	Internally illuminated, custom									
8100	On pedestal, 84" x 30" x 12"	L-7	.50	56	Ea.	11,500	2,875		14,375	17,300
9000	Minimum labor/equipment charge	1 Carp	4	2	Job		106		106	170

10 14 23 – Panel Signage

10 14 23.13 Engraved Panel Signage		Crew	Daily Output	Labor-Hours	Unit	2020 Bare Costs Material	Labor	Equipment	Total	Total Incl O&P
0010	**ENGRAVED PANEL SIGNAGE**, interior									
1010	Flexible door sign, adhesive back, w/Braille, 5/8″ letters, 4″ x 4″	1 Clab	32	.250	Ea.	35	10.55		45.55	55
1050	6″ x 6″		32	.250		50	10.55		60.55	72
1100	8″ x 2″		32	.250		35.50	10.55		46.05	56
1150	8″ x 4″		32	.250		44.50	10.55		55.05	66
1200	8″ x 8″		32	.250		74	10.55		84.55	98
1250	12″ x 2″		32	.250		36	10.55		46.55	56.50
1300	12″ x 6″		32	.250		42.50	10.55		53.05	64
1350	12″ x 12″		32	.250		158	10.55		168.55	191
1500	Graphic symbols, 2″ x 2″		32	.250		12.05	10.55		22.60	30
1550	6″ x 6″		32	.250		31	10.55		41.55	51
1600	8″ x 8″		32	.250		39	10.55		49.55	59.50
2010	Corridor, stock acrylic, 2-sided, with mounting bracket, 2″ x 8″	1 Carp	24	.333		29.50	17.70		47.20	61
2020	2″ x 10″		24	.333		34	17.70		51.70	66
2050	3″ x 8″		24	.333		27	17.70		44.70	58
2060	3″ x 10″		24	.333		39	17.70		56.70	71.50
2070	3″ x 12″		24	.333		46.50	17.70		64.20	79.50
2100	4″ x 8″		24	.333		23.50	17.70		41.20	54.50
2110	4″ x 10″		24	.333		40	17.70		57.70	72.50
2120	4″ x 12″		24	.333		49.50	17.70		67.20	83
7000	Wayfinding signage, custom									
7010	Plastic, flexible, 1/8″ thick, incl. mounting									
7020	6″ x 6″	1 Carp	32	.250	Ea.	16.95	13.30		30.25	40
7030	6″ x 12″		32	.250		35	13.30		48.30	60
7040	6″ x 24″		28	.286		66.50	15.20		81.70	97.50
7050	8″ x 8″		32	.250		28.50	13.30		41.80	53
7060	8″ x 16″		28	.286		60.50	15.20		75.70	91
7070	8″ x 24″		26	.308		92	16.35		108.35	127
7080	12″ x 12″		28	.286		70.50	15.20		85.70	102
7090	12″ x 24″		24	.333		136	17.70		153.70	179
7100	12″ x 36″		20	.400		205	21.50		226.50	260
7210	Weather resistant, engraved and color filled									
7220	8″ x 8″	1 Carp	32	.250	Ea.	38.50	13.30		51.80	64
7230	8″ x 16″		28	.286		64	15.20		79.20	95
7240	8″ x 24″		26	.308		93	16.35		109.35	128
7250	12″ x 12″		28	.286		67.50	15.20		82.70	98.50
7260	12″ x 24″		24	.333		120	17.70		137.70	161
7270	12″ x 36″		20	.400		210	21.50		231.50	265
7280	16″ x 32″	2 Carp	60	.267		218	14.15		232.15	263
7290	16″ x 48″	″	60	.267		355	14.15		369.15	415
7320	For engraved letters, 1/2″ high, add					.70			.70	.77
7330	1″ high, add					.75			.75	.83
7340	2″ high, add					1.77			1.77	1.95
7350	3″ high, add					2.31			2.31	2.54
7360	For engraved graphic symbols, 3″ high, add					14.60			14.60	16.05
7370	7″ high, add					17.75			17.75	19.50
7380	10″ high, add					24			24	26.50
9990	Replace interior labels	1 Carp	12	.667		26.50	35.50		62	86
9991	Replace interior plaques	″	12	.667		84.50	35.50		120	150

10 14 26 – Post and Panel/Pylon Signage

10 14 26.10 Post and Panel Signage		Crew	Daily Output	Labor-Hours	Unit	2020 Bare Costs Material	Labor	Equipment	Total	Total Incl O&P
0010	**POST AND PANEL SIGNAGE**, Alum., incl. 2 posts,									
0011	2-sided panel (blank), concrete footings per post									
0025	24″ W x 18″ H	B-1	25	.960	Ea.	905	41		946	1,075
0050	36″ W x 24″ H		25	.960		950	41		991	1,125
0100	57″ W x 36″ H		25	.960		2,150	41		2,191	2,425
0150	81″ W x 46″ H		25	.960		2,225	41		2,266	2,525

10 14 43 – Photoluminescent Signage

10 14 43.10 Photoluminescent Signage		Crew	Daily Output	Labor-Hours	Unit	Material	Labor	Equipment	Total	Total Incl O&P
0010	**PHOTOLUMINESCENT SIGNAGE**									
0011	Photoluminescent exit sign, plastic, 7″x10″	1 Carp	32	.250	Ea.	24	13.30		37.30	48
0022	Polyester		32	.250		15.20	13.30		28.50	38
0033	Aluminum		32	.250		53.50	13.30		66.80	80
0044	Plastic, 10″x14″		32	.250		59.50	13.30		72.80	87
0055	Polyester		32	.250		35.50	13.30		48.80	60.50
0066	Aluminum		32	.250		70.50	13.30		83.80	99
0111	Directional exit sign, plastic, 10″x14″		32	.250		58.50	13.30		71.80	85.50
0122	Polyester		32	.250		46	13.30		59.30	72
0133	Aluminum		32	.250		68.50	13.30		81.80	96.50
0153	Kick plate exit sign, 10″x34″		28	.286		278	15.20		293.20	330
0173	Photoluminescent tape, 1″x60′		300	.027	L.F.	.50	1.42		1.92	2.82
0183	2″x60′		290	.028	″	1.01	1.47		2.48	3.46
0211	Photoluminescent directional arrows		42	.190	Ea.	1.30	10.10		11.40	17.65

10 14 53 – Traffic Signage

10 14 53.20 Traffic Signs		Crew	Daily Output	Labor-Hours	Unit	Material	Labor	Equipment	Total	Total Incl O&P
0010	**TRAFFIC SIGNS**									
0012	Stock, 24″ x 24″, no posts, .080″ alum. reflectorized	B-80	70	.457	Ea.	93.50	21.50	14.60	129.60	153
0100	High intensity		70	.457		93.50	21.50	14.60	129.60	153
0300	30″ x 30″, reflectorized		70	.457		65	21.50	14.60	101.10	122
0400	High intensity		70	.457		172	21.50	14.60	208.10	239
0600	Guide and directional signs, 12″ x 18″, reflectorized		70	.457		34	21.50	14.60	70.10	87
0700	High intensity		70	.457		65.50	21.50	14.60	101.60	122
0900	18″ x 24″, stock signs, reflectorized		70	.457		48.50	21.50	14.60	84.60	103
1000	High intensity		70	.457		53.50	21.50	14.60	89.60	109
1200	24″ x 24″, stock signs, reflectorized		70	.457		59	21.50	14.60	95.10	115
1300	High intensity		70	.457		63.50	21.50	14.60	99.60	120
1500	Add to above for steel posts, galvanized, 10′-0″ upright, bolted		200	.160		44	7.45	5.10	56.55	66
1600	12′-0″ upright, bolted		140	.229		38.50	10.65	7.30	56.45	67.50
1800	Highway road signs, aluminum, over 20 S.F., reflectorized		350	.091	S.F.	16.45	4.26	2.92	23.63	28
2000	High intensity		350	.091		16.45	4.26	2.92	23.63	28
2200	Highway, suspended over road, 80 S.F. min., reflectorized		165	.194		19.35	9.05	6.20	34.60	42.50
2300	High intensity		165	.194		19.20	9.05	6.20	34.45	42
9000	Replace directional sign		6	5.333	Ea.	172	249	170	591	770

10 17 Telephone Specialties

10 17 16 – Telephone Enclosures

10 17 16.10 Commercial Telephone Enclosures		Crew	Daily Output	Labor-Hours	Unit	2020 Bare Costs Material	Labor	Equipment	Total	Total Incl O&P
0010	**COMMERCIAL TELEPHONE ENCLOSURES**									
0300	Shelf type, wall hung, recessed	2 Carp	5	3.200	Ea.	885	170		1,055	1,250
0400	Surface mount		5	3.200	″	1,700	170		1,870	2,150
9000	Minimum labor/equipment charge		4	4	Job		213		213	340

10 21 Compartments and Cubicles

10 21 13 – Toilet Compartments

10 21 13.13 Metal Toilet Compartments

Line	Description	Crew	Daily Output	Labor-Hours	Unit	Material	Labor	Equipment	Total	Total Incl O&P
0010	**METAL TOILET COMPARTMENTS**									
0110	Cubicles, ceiling hung									
0200	Powder coated steel	2 Carp	4	4	Ea.	575	213		788	975
0500	Stainless steel	″	4	4		1,100	213		1,313	1,550
0600	For handicap units, add					465			465	510
0900	Floor and ceiling anchored									
1000	Powder coated steel	2 Carp	5	3.200	Ea.	580	170		750	910
1300	Stainless steel	″	5	3.200		1,225	170		1,395	1,625
1400	For handicap units, add					370			370	405
1610	Floor anchored									
1700	Powder coated steel	2 Carp	7	2.286	Ea.	615	121		736	870
2000	Stainless steel	″	7	2.286		1,350	121		1,471	1,675
2100	For handicap units, add					360			360	395
2200	For juvenile units, deduct					44.50			44.50	49
2450	Floor anchored, headrail braced									
2500	Powder coated steel	2 Carp	6	2.667	Ea.	385	142		527	650
2804	Stainless steel	″	6	2.667		1,050	142		1,192	1,400
2900	For handicap units, add					325			325	360
3000	Wall hung partitions, powder coated steel	2 Carp	7	2.286		675	121		796	940
3300	Stainless steel	″	7	2.286		1,775	121		1,896	2,150
3400	For handicap units, add					325			325	360
4000	Screens, entrance, floor mounted, 58″ high, 48″ wide									
4200	Powder coated steel	2 Carp	15	1.067	Ea.	240	56.50		296.50	355
4500	Stainless steel	″	15	1.067	″	970	56.50		1,026.50	1,175
4650	Urinal screen, 18″ wide									
4704	Powder coated steel	2 Carp	6.15	2.602	Ea.	226	138		364	470
5004	Stainless steel	″	6.15	2.602	″	535	138		673	805
5100	Floor mounted, headrail braced									
5300	Powder coated steel	2 Carp	8	2	Ea.	234	106		340	425
5600	Stainless steel	″	8	2	″	580	106		686	810
5750	Pilaster, flush									
5800	Powder coated steel	2 Carp	10	1.600	Ea.	275	85		360	435
6100	Stainless steel		10	1.600		610	85		695	810
6300	Post braced, powder coated steel		10	1.600		167	85		252	320
6800	Powder coated steel		10	1.600		176	85		261	330
7800	Wedge type, powder coated steel		10	1.600		138	85		223	288
8100	Stainless steel		10	1.600		615	85		700	810
9000	Minimum labor/equipment charge	1 Carp	2.50	3.200	Job		170		170	272

10 21 13.16 Plastic-Laminate-Clad Toilet Compartments

Line	Description	Crew	Daily Output	Labor-Hours	Unit	Material	Labor	Equipment	Total	Total Incl O&P
0010	**PLASTIC-LAMINATE-CLAD TOILET COMPARTMENTS**									
0110	Cubicles, ceiling hung									
0300	Plastic laminate on particle board	2 Carp	4	4	Ea.	550	213		763	945
0600	For handicap units, add				″	465			465	510

10 21 Compartments and Cubicles

10 21 13 – Toilet Compartments

10 21 13.16 Plastic-Laminate-Clad Toilet Compartments		Crew	Daily Output	Labor-Hours	Unit	Material	2020 Bare Costs Labor	Equipment	Total	Total Incl O&P
0900	Floor and ceiling anchored									
1100	Plastic laminate on particle board	2 Carp	5	3.200	Ea.	800	170		970	1,150
1400	For handicap units, add				"	370			370	405
1610	Floor mounted									
1800	Plastic laminate on particle board	2 Carp	7	2.286	Ea.	540	121		661	790
2450	Floor mounted, headrail braced									
2600	Plastic laminate on particle board	2 Carp	6	2.667	Ea.	810	142		952	1,125
3400	For handicap units, add					325			325	360
4300	Entrance screen, floor mtd., plas. lam., 58" high, 48" wide	2 Carp	15	1.067		520	56.50		576.50	660
4800	Urinal screen, 18" wide, ceiling braced, plastic laminate		8	2		210	106		316	400
5400	Floor mounted, headrail braced		8	2		208	106		314	400
5900	Pilaster, flush, plastic laminate		10	1.600		430	85		515	610
6400	Post braced, plastic laminate		10	1.600		237	85		322	395
6700	Wall hung, bracket supported									
6900	Plastic laminate on particle board	2 Carp	10	1.600	Ea.	97	85		182	242
7450	Flange supported									
7500	Plastic laminate on particle board	2 Carp	10	1.600	Ea.	251	85		336	410
9000	Minimum labor/equipment charge	1 Carp	2.50	3.200	Job		170		170	272

10 21 13.19 Plastic Toilet Compartments		Crew	Daily Output	Labor-Hours	Unit	Material	Labor	Equipment	Total	Total Incl O&P
0010	PLASTIC TOILET COMPARTMENTS									
0110	Cubicles, ceiling hung									
0250	Phenolic	2 Carp	4	4	Ea.	885	213		1,098	1,325
0260	Polymer plastic	"	4	4		1,025	213		1,238	1,475
0600	For handicap units, add					465			465	510
0900	Floor and ceiling anchored									
1050	Phenolic	2 Carp	5	3.200	Ea.	940	170		1,110	1,300
1060	Polymer plastic	"	5	3.200		1,200	170		1,370	1,600
1400	For handicap units, add					370			370	405
1610	Floor mounted									
1750	Phenolic	2 Carp	7	2.286	Ea.	880	121		1,001	1,175
1760	Polymer plastic	"	7	2.286		855	121		976	1,125
2100	For handicap units, add					360			360	395
2200	For juvenile units, deduct					44.50			44.50	49
2450	Floor mounted, headrail braced									
2550	Phenolic	2 Carp	6	2.667	Ea.	740	142		882	1,050
3600	Polymer plastic		6	2.667		870	142		1,012	1,175
3810	Entrance screen, polymer plastic, flr. mtd., 48"x58"		6	2.667		495	142		637	770
3820	Entrance screen, polymer plastic, flr. to clg pilaster, 48"x58"		6	2.667		590	142		732	875
6110	Urinal screen, polymer plastic, pilaster flush, 18" w		6	2.667		430	142		572	700
7110	Wall hung		6	2.667		167	142		309	410
7710	Flange mounted		6	2.667		920	142		1,062	1,250
9000	Minimum labor/equipment charge	1 Carp	2.50	3.200	Job		170		170	272

10 21 13.40 Stone Toilet Compartments		Crew	Daily Output	Labor-Hours	Unit	Material	Labor	Equipment	Total	Total Incl O&P
0010	STONE TOILET COMPARTMENTS									
0100	Cubicles, ceiling hung, marble	2 Marb	2	8	Ea.	1,650	405		2,055	2,475
0600	For handicap units, add					465			465	510
0800	Floor & ceiling anchored, marble	2 Marb	2.50	6.400		1,775	325		2,100	2,475
1400	For handicap units, add					370			370	405
1600	Floor mounted, marble	2 Marb	3	5.333		1,050	271		1,321	1,600
2400	Floor mounted, headrail braced, marble	"	3	5.333		1,225	271		1,496	1,800
2900	For handicap units, add					325			325	360
4100	Entrance screen, floor mounted marble, 58" high, 48" wide	2 Marb	9	1.778		735	90.50		825.50	955

10 21 13 – Toilet Compartments

10 21 13.40 Stone Toilet Compartments		Crew	Daily Output	Labor-Hours	Unit	Material	Labor	Equipment	Total	Total Incl O&P
						2020 Bare Costs				
4600	Urinal screen, 18" wide, ceiling braced, marble	D-1	6	2.667	Ea.	770	125		895	1,050
5100	Floor mounted, headrail braced									
5200	Marble	D-1	6	2.667	Ea.	660	125		785	925
5700	Pilaster, flush, marble		9	1.778		860	83.50		943.50	1,075
6200	Post braced, marble	↓	9	1.778	↓	845	83.50		928.50	1,075
9000	Minimum labor/equipment charge	1 Carp	2.50	3.200	Job		170		170	272

10 21 14 – Toilet Compartment Components

10 21 14.13 Metal Toilet Compartment Components

		Crew	Daily Output	Labor-Hours	Unit	Material	Labor	Equipment	Total	Total Incl O&P
0010	**METAL TOILET COMPARTMENT COMPONENTS**									
0100	Pilasters									
0110	Overhead braced, powder coated steel, 7" wide x 82" high	2 Carp	22.20	.721	Ea.	72.50	38.50		111	142
0120	Stainless steel		22.20	.721		132	38.50		170.50	207
0130	Floor braced, powder coated steel, 7" wide x 70" high		23.30	.687		106	36.50		142.50	176
0140	Stainless steel		23.30	.687		227	36.50		263.50	310
0150	Ceiling hung, powder coated steel, 7" wide x 83" high		13.30	1.203		130	64		194	245
0160	Stainless steel		13.30	1.203		320	64		384	450
0170	Wall hung, powder coated steel, 3" wide x 58" high		18.90	.847		142	45		187	228
0180	Stainless steel	↓	18.90	.847	↓	234	45		279	330
0200	Panels									
0210	Powder coated steel, 31" wide x 58" high	2 Carp	18.90	.847	Ea.	139	45		184	225
0220	Stainless steel		18.90	.847		340	45		385	445
0230	Powder coated steel, 53" wide x 58" high		18.90	.847		177	45		222	266
0240	Stainless steel		18.90	.847		405	45		450	515
0250	Powder coated steel, 63" wide x 58" high		18.90	.847		176	45		221	265
0260	Stainless steel	↓	18.90	.847	↓	485	45		530	600
0300	Doors									
0310	Powder coated steel, 24" wide x 58" high	2 Carp	14.10	1.135	Ea.	125	60.50		185.50	234
0320	Stainless steel		14.10	1.135		325	60.50		385.50	455
0330	Powder coated steel, 26" wide x 58" high		14.10	1.135		161	60.50		221.50	275
0340	Stainless steel		14.10	1.135		335	60.50		395.50	465
0350	Powder coated steel, 28" wide x 58" high		14.10	1.135		178	60.50		238.50	292
0360	Stainless steel		14.10	1.135		330	60.50		390.50	460
0370	Powder coated steel, 36" wide x 58" high		14.10	1.135		182	60.50		242.50	297
0380	Stainless steel	↓	14.10	1.135	↓	395	60.50		455.50	530
0400	Headrails									
0410	For powder coated steel, 62" long	2 Carp	65	.246	Ea.	20	13.10		33.10	43
0420	Stainless steel		65	.246		20.50	13.10		33.60	43.50
0430	For powder coated steel, 84" long		50	.320		30.50	17		47.50	61
0440	Stainless steel		50	.320		30.50	17		47.50	61
0450	For powder coated steel, 120" long		30	.533		41	28.50		69.50	90.50
0460	Stainless steel	↓	30	.533	↓	41	28.50		69.50	90.50
9000	Minimum labor/equipment charge	1 Carp	4	2	Job		106		106	170

10 21 14.16 Plastic-Laminate Clad Toilet Compartment Components

		Crew	Daily Output	Labor-Hours	Unit	Material	Labor	Equipment	Total	Total Incl O&P
0010	**PLASTIC-LAMINATE CLAD TOILET COMPARTMENT COMPONENTS**									
0100	Pilasters									
0110	Overhead braced, 7" wide x 82" high	2 Carp	22.20	.721	Ea.	102	38.50		140.50	175
0130	Floor anchored, 7" wide x 70" high		23.30	.687		108	36.50		144.50	177
0150	Ceiling hung, 7" wide x 83" high		13.30	1.203		130	64		194	245
0180	Wall hung, 3" wide x 58" high	↓	18.90	.847	↓	98	45		143	180
0200	Panels									
0210	31" wide x 58" high	2 Carp	18.90	.847	Ea.	169	45		214	258
0230	51" wide x 58" high	↓	18.90	.847	↓	208	45		253	300

10 21 Compartments and Cubicles

10 21 14 – Toilet Compartment Components

10 21 14.16 Plastic-Laminate Clad Toilet Compartment Components		Crew	Daily Output	Labor-Hours	Unit	2020 Bare Costs Material	Labor	Equipment	Total	Total Incl O&P
0250	63″ wide x 58″ high	2 Carp	18.90	.847	Ea.	243	45		288	340
0300	Doors									
0310	24″ wide x 58″ high	2 Carp	14.10	1.135	Ea.	154	60.50		214.50	266
0330	26″ wide x 58″ high		14.10	1.135		159	60.50		219.50	271
0350	28″ wide x 58″ high		14.10	1.135		164	60.50		224.50	277
0370	36″ wide x 58″ high		14.10	1.135		174	60.50		234.50	289
0400	Headrails									
0410	62″ long	2 Carp	65	.246	Ea.	26	13.10		39.10	49.50
0430	84″ long		60	.267		31.50	14.15		45.65	57
0450	120″ long		30	.533		42	28.50		70.50	92
9000	Minimum labor/equipment charge	1 Carp	4	2	Job		106		106	170

10 21 14.19 Plastic Toilet Compartment Components		Crew	Daily Output	Labor-Hours	Unit	Material	Labor	Equipment	Total	Total Incl O&P
0010	**PLASTIC TOILET COMPARTMENT COMPONENTS**									
0100	Pilasters									
0110	Overhead braced, polymer plastic, 7″ wide x 82″ high	2 Carp	22.20	.721	Ea.	116	38.50		154.50	190
0120	Phenolic		22.20	.721		140	38.50		178.50	216
0130	Floor braced, polymer plastic, 7″ wide x 70″ high		23.30	.687		185	36.50		221.50	262
0140	Phenolic		23.30	.687		135	36.50		171.50	207
0150	Ceiling hung, polymer plastic, 7″ wide x 83″ high		13.30	1.203		179	64		243	299
0160	Phenolic		13.30	1.203		168	64		232	286
0180	Wall hung, phenolic, 3″ wide x 58″ high		18.90	.847		105	45		150	188
0200	Panels									
0203	Polymer plastic, 18″ wide x 55″ high	2 Carp	18.90	.847	Ea.	310	45		355	415
0206	Phenolic, 18″ wide x 58″ high		18.90	.847		256	45		301	355
0210	Polymer plastic, 31″ wide x 55″ high		18.90	.847		340	45		385	445
0220	Phenolic, 31″ wide x 58″ high		18.90	.847		274	45		319	370
0223	Polymer plastic, 48″ wide x 55″ high		18.90	.847		520	45		565	645
0226	Phenolic, 48″ wide x 58″ high		18.90	.847		460	45		505	580
0230	Polymer plastic, 51″ wide x 55″ high		18.90	.847		490	45		535	610
0240	Phenolic, 51″ wide x 58″ high		18.90	.847		435	45		480	550
0250	Polymer plastic, 63″ wide x 55″ high		18.90	.847		690	45		735	830
0260	Phenolic, 63″ wide x 58″ high		18.90	.847		475	45		520	595
0300	Doors									
0310	Polymer plastic, 24″ wide x 55″ high	2 Carp	14.10	1.135	Ea.	252	60.50		312.50	375
0320	Phenolic, 24″ wide x 58″ high		14.10	1.135		340	60.50		400.50	465
0330	Polymer plastic, 26″ wide x 55″ high		14.10	1.135		262	60.50		322.50	385
0340	Phenolic, 26″ wide x 58″ high		14.10	1.135		360	60.50		420.50	490
0350	Polymer plastic, 28″ wide x 55″ high		14.10	1.135		284	60.50		344.50	405
0360	Phenolic, 28″ wide x 58″ high		14.10	1.135		375	60.50		435.50	510
0370	Polymer plastic, 36″ wide x 55″ high		14.10	1.135		335	60.50		395.50	465
0380	Phenolic, 36″ wide x 58″ high		14.10	1.135		520	60.50		580.50	665
0400	Headrails									
0410	For polymer plastic, 62″ long	2 Carp	65	.246	Ea.	22.50	13.10		35.60	46
0420	Phenolic		65	.246		22.50	13.10		35.60	45.50
0430	For polymer plastic, 84″ long		50	.320		31	17		48	61.50
0440	Phenolic		50	.320		39.50	17		56.50	71
0450	For polymer plastic, 120″ long		30	.533		44	28.50		72.50	93.50
0460	Phenolic		30	.533		42	28.50		70.50	91.50
9000	Minimum labor/equipment charge	1 Carp	4	2	Job		106		106	170

10 21 Compartments and Cubicles

10 21 23 – Cubicle Curtains and Track

10 21 23.16 Cubicle Track and Hardware		Crew	Daily Output	Labor-Hours	Unit	Material	Labor	Equipment	Total	Total Incl O&P
						2020 Bare Costs				
0010	CUBICLE TRACK AND HARDWARE									
0020	Curtain track, box channel, ceiling mounted	1 Carp	135	.059	L.F.	6.75	3.15		9.90	12.45
0100	Suspended	"	100	.080	"	8.35	4.25		12.60	16
0300	Curtains, nylon mesh tops, fire resistant, 11 oz. per lineal yard									
0310	Polyester oxford cloth, 9' ceiling height	1 Carp	425	.019	L.F.	17.25	1		18.25	20.50
0500	8' ceiling height		425	.019		15.75	1		16.75	18.95
0550	Polyester, antimicrobial, 9' ceiling height		425	.019		19.40	1		20.40	23
0560	8' ceiling height		425	.019		17.30	1		18.30	20.50
0700	Designer oxford cloth	↓	425	.019	↓	6.90	1		7.90	9.15
0800	I.V. track systems									
0820	I.V. track, oval	1 Carp	135	.059	L.F.	8.30	3.15		11.45	14.20
0830	I.V. trolley		32	.250	Ea.	40.50	13.30		53.80	66
0840	I.V. pendant (tree, 5 hook)	↓	32	.250	"	176	13.30		189.30	216

10 22 Partitions

10 22 13 – Wire Mesh Partitions

10 22 13.10 Partitions, Woven Wire		Crew	Daily Output	Labor-Hours	Unit	Material	Labor	Equipment	Total	Total Incl O&P
0010	**PARTITIONS, WOVEN WIRE** for tool or stockroom enclosures									
0100	Channel frame, 1-1/2" diamond mesh, 10 ga. wire, painted									
0300	Wall panels, 4'-0" wide, 7' high	2 Carp	25	.640	Ea.	142	34		176	212
0400	8' high		23	.696		142	37		179	215
0600	10' high	↓	18	.889		199	47		246	295
0700	For 5' wide panels, add					5%				
0900	Ceiling panels, 10' long, 2' wide	2 Carp	25	.640		140	34		174	209
1000	4' wide		15	1.067		159	56.50		215.50	265
1200	Panel with service window & shelf, 5' wide, 7' high		20	.800		335	42.50		377.50	440
1300	8' high		15	1.067		510	56.50		566.50	650
1500	Sliding doors, full height, 3' wide, 7' high		6	2.667		440	142		582	710
1600	10' high		5	3.200		560	170		730	885
1800	6' wide sliding door, 7' full height		5	3.200		605	170		775	940
1900	10' high		4	4		1,125	213		1,338	1,575
2100	Swinging doors, 3' wide, 7' high, no transom		6	2.667		294	142		436	550
2200	7' high, 3' transom	↓	5	3.200	↓	535	170		705	860

10 22 16 – Folding Gates

10 22 16.10 Security Gates		Crew	Daily Output	Labor-Hours	Unit	Material	Labor	Equipment	Total	Total Incl O&P
0010	**SECURITY GATES**									
0015	For roll up type, see Section 08 33 13.10									
0300	Scissors type folding gate, ptd. steel, single, 6-1/2' high, 5-1/2'wide	2 Sswk	4	4	Opng.	235	231		466	640
0350	6-1/2' wide		4	4		240	231		471	645
0400	7-1/2' wide		4	4		223	231		454	625
0600	Double gate, 8' high, 8' wide		2.50	6.400		385	370		755	1,025
0650	10' wide		2.50	6.400		455	370		825	1,100
0700	12' wide		2	8		605	460		1,065	1,425
0750	14' wide		2	8		625	460		1,085	1,450
0900	Door gate, folding steel, 4' wide, 61" high		4	4		144	231		375	540
1000	71" high		4	4		179	231		410	575
1200	81" high		4	4		196	231		427	595
1300	Window gates, 2' to 4' wide, 31" high		4	4		90	231		321	480
1500	55" high		3.75	4.267		125	246		371	545
1600	79" high	↓	3.50	4.571	↓	148	264		412	595

10 22 Partitions

10 22 19 – Demountable Partitions

10 22 19.43 Demountable Composite Partitions		Crew	Daily Output	Labor-Hours	Unit	2020 Bare Costs Material	Labor	Equipment	Total	Total Incl O&P
0010	**DEMOUNTABLE COMPOSITE PARTITIONS**, add for doors									
0100	Do not deduct door openings from total L.F.									
0900	Demountable gypsum system on 2" to 2-1/2"									
1000	Steel studs, 9' high, 3" to 3-3/4" thick									
1200	Vinyl-clad gypsum	2 Carp	48	.333	L.F.	61	17.70		78.70	95.50
1300	Fabric clad gypsum		44	.364		152	19.35		171.35	198
1500	Steel clad gypsum		40	.400		170	21.50		191.50	221
1600	1.75 system, aluminum framing, vinyl-clad hardboard,									
1800	Paper honeycomb core panel, 1-3/4" to 2-1/2" thick									
1900	9' high	2 Carp	48	.333	L.F.	102	17.70		119.70	141
2100	7' high		60	.267		91.50	14.15		105.65	124
2200	5' high		80	.200		77.50	10.65		88.15	103
2250	Unitized gypsum system									
2300	Unitized panel, 9' high, 2" to 2-1/2" thick									
2350	Vinyl-clad gypsum	2 Carp	48	.333	L.F.	132	17.70		149.70	174
2400	Fabric clad gypsum	"	44	.364	"	217	19.35		236.35	269
2500	Unitized mineral fiber system									
2510	Unitized panel, 9' high, 2-1/4" thick, aluminum frame									
2550	Vinyl-clad mineral fiber	2 Carp	48	.333	L.F.	131	17.70		148.70	173
2600	Fabric clad mineral fiber	"	44	.364	"	195	19.35		214.35	246
2800	Movable steel walls, modular system									
2900	Unitized panels, 9' high, 48" wide									
3100	Baked enamel, pre-finished	2 Carp	60	.267	L.F.	148	14.15		162.15	186
3200	Fabric clad steel		56	.286	"	214	15.20		229.20	261
5310	Trackless wall, cork finish, semi-acoustic, 1-5/8" thick, unsealed		325	.049	S.F.	39.50	2.62		42.12	47.50
5320	Sealed		190	.084		49.50	4.48		53.98	61.50
5330	Acoustic, 2" thick, unsealed		305	.052		42.50	2.79		45.29	51
5340	Sealed		225	.071		59	3.78		62.78	71
5500	For acoustical partitions, add, unsealed					2.38			2.38	2.62
5550	Sealed					11.10			11.10	12.20
5700	For doors, see Section 08 16									
5800	For door hardware, see Section 08 71									
6100	In-plant modular office system, w/prehung hollow core door									
6200	3" thick polystyrene core panels									
6250	12' x 12', 2 wall	2 Clab	3.80	4.211	Ea.	4,975	177		5,152	5,750
6300	4 wall		1.90	8.421		7,125	355		7,480	8,425
6350	16' x 16', 2 wall		3.60	4.444		8,175	187		8,362	9,300
6400	4 wall		1.80	8.889		11,200	375		11,575	12,900
9000	Minimum labor/equipment charge	2 Carp	3	5.333	Job		283		283	455

10 22 23 – Portable Partitions, Screens, and Panels

10 22 23.13 Wall Screens

Code	Description	Crew	Daily Output	Labor-Hours	Unit	Material	Labor	Equipment	Total	Total Incl O&P
0010	**WALL SCREENS**, divider panels, free standing, fiber core									
0020	Fabric face straight									
0100	3'-0" long, 4'-0" high	2 Carp	100	.160	L.F.	136	8.50		144.50	164
0200	5'-0" high		90	.178		126	9.45		135.45	154
0500	6'-0" high		75	.213		127	11.35		138.35	157
0900	5'-0" long, 4'-0" high		175	.091		61.50	4.86		66.36	76
1000	5'-0" high		150	.107		84.50	5.65		90.15	102
1500	6'-0" high		125	.128		113	6.80		119.80	135
1600	6'-0" long, 5'-0" high		162	.099		113	5.25		118.25	132
3200	Economical panels, fabric face, 4'-0" long, 5'-0" high		132	.121		59.50	6.45		65.95	75.50
3250	6'-0" high		112	.143		61.50	7.60		69.10	79.50

10 22 23 – Portable Partitions, Screens, and Panels

10 22 23.13 Wall Screens		Crew	Daily Output	Labor-Hours	Unit	Material	2020 Bare Costs Labor	Equipment	Total	Total Incl O&P
3300	5′-0″ long, 5′-0″ high	2 Carp	150	.107	L.F.	34.50	5.65		40.15	47
3350	6′-0″ high		125	.128		52.50	6.80		59.30	68.50
3450	Acoustical panels, 60 to 90 NRC, 3′-0″ long, 5′-0″ high		90	.178		87	9.45		96.45	111
3550	6′-0″ high		75	.213		88	11.35		99.35	115
3600	5′-0″ long, 5′-0″ high		150	.107		61	5.65		66.65	76
3650	6′-0″ high		125	.128		59	6.80		65.80	76
3700	6′-0″ long, 5′-0″ high		162	.099		53	5.25		58.25	67
3750	6′-0″ high		138	.116		83.50	6.15		89.65	102
3800	Economy acoustical panels, 40 NRC, 4′-0″ long, 5′-0″ high		132	.121		59.50	6.45		65.95	75.50
3850	6′-0″ high		112	.143		61.50	7.60		69.10	79.50
3900	5′-0″ long, 6′-0″ high		125	.128		52.50	6.80		59.30	68.50
3950	6′-0″ long, 5′-0″ high		162	.099		46	5.25		51.25	59
4000	Metal chalkboard, 6′-6″ high, chalkboard, 1 side		125	.128		126	6.80		132.80	150
4100	Metal chalkboard, 2 sides		120	.133		144	7.10		151.10	170
4300	Tackboard, both sides		123	.130		114	6.90		120.90	136
9000	Minimum labor/equipment charge		3	5.333	Job		283		283	455

10 22 23.23 Movable Panel Systems		Crew	Daily Output	Labor-Hours	Unit	Material	Labor	Equipment	Total	Total Incl O&P
0010	**MOVABLE PANEL SYSTEMS**									
0030	Fabric panel, class A fire rated									
0040	Minimum N.R.C. 0.95, STC 28, fabric edged									
0200	42″ high, 24″ wide	2 Clab	30	.533	Ea.	155	22.50		177.50	206
0220	36″ wide		28	.571		181	24		205	238
0240	48″ wide		26	.615		187	26		213	248
0260	60″ wide		25	.640		232	27		259	299
0400	54″ high, 24″ wide		30	.533		186	22.50		208.50	240
0420	36″ wide		28	.571		191	24		215	250
0440	48″ wide		26	.615		195	26		221	257
0460	60″ wide		25	.640		198	27		225	260
0600	60″ high, 24″ wide		29	.552		181	23		204	236
0620	36″ wide		27	.593		202	25		227	262
0640	48″ wide		25	.640		208	27		235	272
0660	60″ wide		24	.667		228	28		256	296
0800	72″ high, 24″ wide		29	.552		206	23		229	264
1000	36″ wide		27	.593		228	25		253	291
1200	48″ wide		25	.640		243	27		270	310
1400	60″ wide		24	.667		282	28		310	355
2000	Hardwood edged, 42″ high, 24″ wide		30	.533		161	22.50		183.50	213
2100	36″ wide		28	.571		184	24		208	241
2120	48″ wide		26	.615		200	26		226	262
2180	60″ wide		25	.640		222	27		249	287
2200	54″ high, 24″ wide		30	.533		165	22.50		187.50	218
2220	36″ wide		28	.571		206	24		230	266
2240	48″ wide		26	.615		220	26		246	284
2300	60″ wide		25	.640		257	27		284	325
2400	60″ high, 24″ wide		29	.552		185	23		208	240
2420	36″ wide		27	.593		199	25		224	259
2440	48″ wide		25	.640		223	27		250	289
2460	60″ wide		24	.667		243	28		271	315
2800	72″ high, 24″ wide		29	.552		226	23		249	286
2820	36″ wide		27	.593		228	25		253	291
2840	48″ wide		25	.640		250	27		277	320
2860	60″ wide		24	.667		290	28		318	365

10 22 23 – Portable Partitions, Screens, and Panels

10 22 23.23 Movable Panel Systems		Crew	Daily Output	Labor-Hours	Unit	Material	Labor	Equipment	Total	Total Incl O&P
						2020 Bare Costs				
3000	Fabric panel, straight, N.R.C. less than or equal to 0.50									
3110	60" high, 24" wide	2 Clab	29	.552	Ea.	195	23		218	251
3120	30" wide		28	.571		199	24		223	258
3130	36" wide		27	.593		204	25		229	264
3140	48" wide		26	.615		237	26		263	305
3150	60" wide		25	.640		172	27		199	232
3160	72" wide		24	.667		276	28		304	350
3210	72" high, 24" wide		29	.552		136	23		159	186
3220	30" wide		28	.571		289	24		313	360
3230	36" wide		27	.593		206	25		231	267
3240	48" wide		26	.615		246	26		272	310
3250	60" wide		25	.640		262	27		289	330
3260	72" wide		24	.667		325	28		353	400
3310	N.R.C. greater than or equal to 0.075, 42" high, 24" wide		30	.533		198	22.50		220.50	254
3320	30" wide		29	.552		195	23		218	251
3330	36" wide		28	.571		232	24		256	294
3340	48" wide		27	.593		258	25		283	325
3350	60" wide		26	.615		241	26		267	305
3360	72" wide		25	.640		355	27		382	435
3410	60" high, 24" wide		29	.552		206	23		229	264
3420	30" wide		28	.571		270	24		294	335
3430	36" wide		27	.593		261	25		286	330
3440	48" wide		26	.615		285	26		311	355
3450	60" wide		25	.640		305	27		332	380
3460	72" wide		24	.667		320	28		348	395
3510	72" high, 24" wide		29	.552		272	23		295	335
3520	30" wide		28	.571		281	24		305	350
3530	36" wide		27	.593		264	25		289	330
3540	48" wide		26	.615		305	26		331	375
3550	60" wide		25	.640		296	27		323	370
3560	72" wide		24	.667		385	28		413	470
3910	Connector kit, straight					34			34	37.50
3920	Corner, 2-way					32			32	35
3930	3-way					31			31	34.50
3940	4-way					15.50			15.50	17.05
3950	T-leg					27			27	30

10 22 33 – Accordion Folding Partitions

10 22 33.10 Partitions, Accordion Folding		Crew	Daily Output	Labor-Hours	Unit	Material	Labor	Equipment	Total	Total Incl O&P
0010	**PARTITIONS, ACCORDION FOLDING**									
0100	Vinyl covered, over 150 S.F., frame not included									
0300	Residential, 1.25 lb./S.F., 8' maximum height	2 Carp	300	.053	S.F.	22	2.83		24.83	29
0400	Commercial, 1.75 lb./S.F., 8' maximum height		225	.071		25.50	3.78		29.28	34
0600	2 lb./S.F., 17' maximum height		150	.107		26	5.65		31.65	38
0700	Industrial, 4 lb./S.F., 20' maximum height		75	.213		39	11.35		50.35	61
0900	Acoustical, 3 lb./S.F., 17' maximum height		100	.160		27.50	8.50		36	44
1200	5 lb./S.F., 20' maximum height		95	.168		44.50	8.95		53.45	63
1300	5.5 lb./S.F., 17' maximum height		90	.178		52	9.45		61.45	72
1400	Fire rated, 4.5 psf, 20' maximum height		160	.100		52	5.30		57.30	65.50
1500	Vinyl-clad wood or steel, electric operation, 5.0 psf		160	.100		55	5.30		60.30	69.50
1900	Wood, non-acoustic, birch or mahogany, to 10' high		300	.053		30	2.83		32.83	37
9000	Minimum labor/equipment charge		4	4	Job		213		213	340

10 22 Partitions

10 22 39 – Folding Panel Partitions

10 22 39.10 Partitions, Folding Panel		Crew	Daily Output	Labor-Hours	Unit	2020 Bare Costs Material	Labor	Equipment	Total	Total Incl O&P
0010	**PARTITIONS, FOLDING PANEL**, acoustic, wood									
0100	Vinyl faced, to 18′ high, 6 psf, economy trim	2 Carp	60	.267	S.F.	70.50	14.15		84.65	100
0150	Standard trim		45	.356		84	18.90		102.90	123
0200	Premium trim		30	.533		108	28.50		136.50	165
0400	Plastic laminate or hardwood finish, standard trim		60	.267		72.50	14.15		86.65	102
0500	Premium trim		30	.533		77	28.50		105.50	131
0600	Wood, low acoustical type, 4.5 psf, to 14′ high		50	.320		52.50	17		69.50	85.50
1100	Steel, acoustical, 9 to 12 lb./S.F., vinyl faced, standard trim		60	.267		75	14.15		89.15	105
1200	Premium trim		30	.533		91.50	28.50		120	147
1700	Aluminum framed, acoustical, to 12′ high, 5.5 psf, standard trim		60	.267		50.50	14.15		64.65	78
1800	Premium trim		30	.533		61	28.50		89.50	113
2000	6.5 lb./S.F., standard trim		60	.267		53	14.15		67.15	81
2100	Premium trim		30	.533		65.50	28.50		94	118
9000	Minimum labor/equipment charge		4	4	Job		213		213	340

10 22 43 – Sliding Partitions

10 22 43.10 Partitions, Sliding		Crew	Daily Output	Labor-Hours	Unit	Material	Labor	Equipment	Total	Total Incl O&P
0010	**PARTITIONS, SLIDING**									
0020	Acoustic air wall, 1-5/8″ thick, standard trim	2 Carp	375	.043	S.F.	34.50	2.27		36.77	41.50
0100	Premium trim		365	.044		59	2.33		61.33	68
0300	2-1/4″ thick, standard trim		360	.044		38.50	2.36		40.86	46.50
0400	Premium trim		330	.048		67.50	2.58		70.08	78.50
0600	For track type, add to above				L.F.	126			126	138
0700	Overhead track type, acoustical, 3″ thick, 11 psf, standard trim	2 Carp	350	.046	S.F.	86.50	2.43		88.93	99.50
0800	Premium trim	″	300	.053	″	104	2.83		106.83	120

10 26 Wall and Door Protection

10 26 13 – Corner Guards

10 26 13.20 Corner Protection		Crew	Daily Output	Labor-Hours	Unit	Material	Labor	Equipment	Total	Total Incl O&P
0010	**CORNER PROTECTION**									
0100	Stainless steel, 16 ga., adhesive mount, 3-1/2″ leg	1 Carp	80	.100	L.F.	22	5.30		27.30	32.50
0200	12 ga. stainless, adhesive mount	″	80	.100		20.50	5.30		25.80	31
0300	For screw mount, add						10%			
0500	Vinyl acrylic, adhesive mount, 3″ leg	1 Carp	128	.063		11.40	3.32		14.72	17.85
0550	1-1/2″ leg		160	.050		5.25	2.66		7.91	10.05
0600	Screw mounted, 3″ leg		80	.100		9.65	5.30		14.95	19.10
0650	1-1/2″ leg		100	.080		4.64	4.25		8.89	11.90
0700	Clear plastic, screw mounted, 2-1/2″		60	.133		4.73	7.10		11.83	16.55
1000	Vinyl cover, alum. retainer, surface mount, 3″ x 3″		48	.167		11.10	8.85		19.95	26.50
1050	2″ x 2″		48	.167		9.95	8.85		18.80	25
1100	Flush mounted, 3″ x 3″		32	.250		21	13.30		34.30	44.50
1150	2″ x 2″		32	.250		16.80	13.30		30.10	40

10 26 16 – Bumper Guards

10 26 16.10 Guard, Bumper		Crew	Daily Output	Labor-Hours	Unit	Material	Labor	Equipment	Total	Total Incl O&P
0010	**GUARD, BUMPER**									
1200	Bed bumper, vinyl acrylic, alum. retainer, 21″ long	1 Carp	10	.800	Ea.	42.50	42.50		85	115
1300	53″ long with aligner		9	.889	″	107	47		154	194
1400	Bumper, vinyl cover, alum. retain., cush. mnt., 1-1/2″ x 2-3/4″		80	.100	L.F.	14.35	5.30		19.65	24.50
1500	2″ x 4-1/4″		80	.100		21	5.30		26.30	31.50
1600	Surface mounted, 1-3/4″ x 3-5/8″		80	.100		12.05	5.30		17.35	22

10 26 Wall and Door Protection

10 26 16 – Bumper Guards

10 26 16.16 Protective Corridor Handrails		Crew	Daily Output	Labor-Hours	Unit	Material	Labor	Equipment	Total	Total Incl O&P
						2020 Bare Costs				
0010	PROTECTIVE CORRIDOR HANDRAILS									
3000	Handrail/bumper, vinyl cover, alum. retainer									
3010	Bracket mounted, flat rail, 5-1/2"	1 Carp	80	.100	L.F.	18.45	5.30		23.75	29
3100	6-1/2"		80	.100		23	5.30		28.30	34
3200	Bronze bracket, 1-3/4" diam. rail	↓	80	.100	↓	16.65	5.30		21.95	27

10 26 23 – Protective Wall Covering

10 26 23.10 Wall Covering, Protective		Crew	Daily Output	Labor-Hours	Unit	Material	Labor	Equipment	Total	Total Incl O&P
0010	WALL COVERING, PROTECTIVE									
0400	Rub rail, vinyl, adhesive mounted	1 Carp	185	.043	L.F.	9.65	2.30		11.95	14.35
0500	Neoprene, aluminum backing, 1-1/2" x 2"		110	.073		9.25	3.87		13.12	16.40
1000	Trolley rail, PVC, clipped to wall, 5" high		185	.043		9	2.30		11.30	13.60
1050	8" high	↓	180	.044		15.50	2.36		17.86	21
1700	Bumper rail, stainless steel, flat bar on brackets, 4" x 1/4"	2 Skwk	120	.133	↓	37.50	7.30		44.80	52.50
1775	Wall end guard, stainless steel, 16 ga., 36" tall, screwed to studs	1 Skwk	30	.267	Ea.	26	14.65		40.65	52
2000	Crash rail, vinyl cover, alum. retainer, 1" x 4"	1 Carp	110	.073	L.F.	11.15	3.87		15.02	18.45
2100	1" x 8"		90	.089		19	4.72		23.72	28.50
2150	Vinyl inserts, aluminum plate, 1" x 2-1/2"		110	.073		15.05	3.87		18.92	23
2200	1" x 5"	↓	90	.089	↓	24.50	4.72		29.22	34

10 26 23.13 Impact Resistant Wall Protection		Crew	Daily Output	Labor-Hours	Unit	Material	Labor	Equipment	Total	Total Incl O&P
0010	IMPACT RESISTANT WALL PROTECTION									
0100	Vinyl wall protection, complete instl. incl. panels, trim, adhesive									
0110	.040" thk., std. colors	2 Carp	320	.050	S.F.	15.60	2.66		18.26	21.50
0120	.040" thk., element patterns		300	.053		17.35	2.83		20.18	23.50
0130	.060" thk., std. colors		300	.053		17.90	2.83		20.73	24
0140	.060" thk., element patterns	↓	280	.057	↓	18.90	3.04		21.94	26
1750	Wallguard stainless steel baseboard, 12" tall, adhesive applied	2 Skwk	260	.062	L.F.	33.50	3.38		36.88	42
1775	Wall protection, stainless steel, 16 ga., 48" x 36" tall, screwed to studs	"	500	.032	S.F.	8	1.76		9.76	11.60

10 26 33 – Door and Frame Protection

10 26 33.10 Protection, Door and Frame		Crew	Daily Output	Labor-Hours	Unit	Material	Labor	Equipment	Total	Total Incl O&P
0010	PROTECTION, DOOR AND FRAME									
0100	Door frame guard, vinyl, 3" x 3" x 4'	1 Carp	30	.267	Ea.	88.50	14.15		102.65	120
0110	Door frame guard, vinyl, 3" x 3" x 8'		26	.308		134	16.35		150.35	173
0120	Door frame guard, stainless steel, 3" x 3" x 4'	↓	30	.267		440	14.15		454.15	510
0500	Steel door track/wheel guards, 4'-0" high	E-4	22	1.455	↓	116	84.50	6.70	207.20	274

10 28 Toilet, Bath, and Laundry Accessories

10 28 13 – Toilet Accessories

10 28 13.13 Commercial Toilet Accessories		Crew	Daily Output	Labor-Hours	Unit	Material	Labor	Equipment	Total	Total Incl O&P
0010	COMMERCIAL TOILET ACCESSORIES									
0200	Curtain rod, stainless steel, 5' long, 1" diameter	1 Carp	13	.615	Ea.	26.50	32.50		59	82
0300	1-1/4" diameter		13	.615		29	32.50		61.50	84.50
0350	Chrome, 1" diameter		13	.615	↓	26	32.50		58.50	81
0360	For vinyl curtain, add		1950	.004	S.F.	.97	.22		1.19	1.42
0400	Diaper changing station, horizontal, wall mounted, plastic		10	.800	Ea.	246	42.50		288.50	340
0420	Vertical		10	.800		221	42.50		263.50	310
0430	Oval shaped		10	.800		305	42.50		347.50	405
0440	Recessed, with stainless steel flange	↓	6	1.333	↓	515	71		586	680
0500	Dispenser units, combined soap & towel dispensers,									
0510	Mirror and shelf, flush mounted	1 Carp	10	.800	Ea.	345	42.50		387.50	450

10 28 13.13 Commercial Toilet Accessories		Crew	Daily Output	Labor-Hours	Unit	Material	Labor	Equipment	Total	Total Incl O&P
						2020 Bare Costs				
0600	Towel dispenser and waste receptacle,									
0610	18 gallon capacity	1 Carp	10	.800	Ea.	365	42.50		407.50	475
0800	Grab bar, straight, 1-1/4" diameter, stainless steel, 18" long		24	.333		30	17.70		47.70	61
0900	24" long		23	.348		29	18.50		47.50	61.50
1000	30" long		22	.364		26.50	19.35		45.85	60
1100	36" long		20	.400		34.50	21.50		56	72
1105	42" long		20	.400		40	21.50		61.50	78
1120	Corner, 36" long		20	.400		95	21.50		116.50	138
1200	1-1/2" diameter, 24" long		23	.348		33.50	18.50		52	66.50
1300	36" long		20	.400		37	21.50		58.50	74.50
1310	42" long		18	.444		33.50	23.50		57	75
1500	Tub bar, 1-1/4" diameter, 24" x 36"		14	.571		90	30.50		120.50	148
1600	Plus vertical arm		12	.667		100	35.50		135.50	167
1900	End tub bar, 1" diameter, 90° angle, 16" x 32"		12	.667		111	35.50		146.50	179
2010	Tub/shower/toilet, 2-wall, 36" x 24"		12	.667		90.50	35.50		126	157
2300	Hand dryer, surface mounted, electric, 115 volt, 20 amp		4	2		425	106		531	635
2400	230 volt, 10 amp		4	2		645	106		751	880
2450	Hand dryer, touch free, 1400 watt, 81,000 rpm		4	2		1,350	106		1,456	1,650
2600	Hat and coat strip, stainless steel, 4 hook, 36" long		24	.333		68	17.70		85.70	104
2700	6 hook, 60" long		20	.400		118	21.50		139.50	164
3000	Mirror, with stainless steel 3/4" square frame, 18" x 24"		20	.400		51.50	21.50		73	90.50
3100	36" x 24"		15	.533		104	28.50		132.50	160
3200	48" x 24"		10	.800		186	42.50		228.50	272
3300	72" x 24"		6	1.333		305	71		376	450
3500	With 5" stainless steel shelf, 18" x 24"		20	.400		185	21.50		206.50	237
3600	36" x 24"		15	.533		236	28.50		264.50	305
3700	48" x 24"		10	.800		246	42.50		288.50	340
3800	72" x 24"		6	1.333		277	71		348	420
4100	Mop holder strip, stainless steel, 5 holders, 48" long		20	.400		76.50	21.50		98	119
4200	Napkin/tampon dispenser, recessed		15	.533		560	28.50		588.50	660
4220	Semi-recessed		6.50	1.231		370	65.50		435.50	510
4250	Napkin receptacle, recessed		6.50	1.231		171	65.50		236.50	293
4300	Robe hook, single, regular		96	.083		20.50	4.43		24.93	29.50
4400	Heavy duty, concealed mounting		56	.143		25.50	7.60		33.10	40
4600	Soap dispenser, chrome, surface mounted, liquid		20	.400		58.50	21.50		80	98
5000	Recessed stainless steel, liquid		10	.800		163	42.50		205.50	247
5600	Shelf, stainless steel, 5" wide, 18 ga., 24" long		24	.333		82	17.70		99.70	119
5700	48" long		16	.500		168	26.50		194.50	228
5800	8" wide shelf, 18 ga., 24" long		22	.364		63	19.35		82.35	101
5900	48" long		14	.571		145	30.50		175.50	208
6000	Toilet seat cover dispenser, stainless steel, recessed		20	.400		181	21.50		202.50	233
6050	Surface mounted		15	.533		32.50	28.50		61	81.50
6100	Toilet tissue dispenser, surface mounted, SS, single roll		30	.267		16	14.15		30.15	40
6200	Double roll		24	.333		23	17.70		40.70	53.50
6240	Plastic, twin/jumbo dbl. roll		24	.333		30.50	17.70		48.20	62
6400	Towel bar, stainless steel, 18" long		23	.348		40	18.50		58.50	73.50
6500	30" long		21	.381		48.50	20.50		69	85.50
6700	Towel dispenser, stainless steel, surface mounted		16	.500		41.50	26.50		68	88.50
6800	Flush mounted, recessed		10	.800		67.50	42.50		110	143
6900	Plastic, touchless, battery operated		16	.500		107	26.50		133.50	161
7000	Towel holder, hotel type, 2 guest size		20	.400		58	21.50		79.50	97.50
7200	Towel shelf, stainless steel, 24" long, 8" wide		20	.400		65	21.50		86.50	106
7400	Tumbler holder, for tumbler only		30	.267		37	14.15		51.15	63.50

10 28 Toilet, Bath, and Laundry Accessories

10 28 13 – Toilet Accessories

10 28 13.13 Commercial Toilet Accessories		Crew	Daily Output	Labor-Hours	Unit	2020 Bare Costs Material	Labor	Equipment	Total	Total Incl O&P
7410	Tumbler holder, recessed	1 Carp	20	.400	Ea.	9.40	21.50		30.90	44.50
7500	Soap, tumbler & toothbrush		30	.267		20	14.15		34.15	44.50
7510	Tumbler & toothbrush holder		20	.400		13.45	21.50		34.95	49
7590	Air freshener, stainless steel, recessed		16	.500		49.50	26.50		76	97
8000	Waste receptacles, stainless steel, with top, 13 gallon		10	.800		335	42.50		377.50	435
8100	36 gallon		8	1		460	53		513	590
8200	Shower seat, wall mounted		30	.267		370	14.15		384.15	435
9000	Minimum labor/equipment charge		5	1.600	Job		85		85	136
9996	Bathroom access., grab bar, straight, 1-1/2" diam., SS, 42" L install only		18	.444	Ea.		23.50		23.50	38

10 28 16 – Bath Accessories

10 28 16.20 Medicine Cabinets		Crew	Daily Output	Labor-Hours	Unit	Material	Labor	Equipment	Total	Total Incl O&P
0010	**MEDICINE CABINETS**									
0020	With mirror, sst frame, 16" x 22", unlighted	1 Carp	14	.571	Ea.	98.50	30.50		129	157
0100	Wood frame		14	.571		128	30.50		158.50	190
0300	Sliding mirror doors, 20" x 16" x 4-3/4", unlighted		7	1.143		130	60.50		190.50	241
0400	24" x 19" x 8-1/2", lighted		5	1.600		219	85		304	375
0600	Triple door, 30" x 32", unlighted, plywood body		7	1.143		355	60.50		415.50	490
0700	Steel body		7	1.143		355	60.50		415.50	490
0900	Oak door, wood body, beveled mirror, single door		7	1.143		170	60.50		230.50	284
1000	Double door		6	1.333		370	71		441	520
1200	Hotel cabinets, stainless, with lower shelf, unlighted		10	.800		218	42.50		260.50	310
1300	Lighted		5	1.600		330	85		415	495
9000	Minimum labor/equipment charge		4	2	Job		106		106	170

10 28 19 – Tub and Shower Enclosures

10 28 19.10 Partitions, Shower		Crew	Daily Output	Labor-Hours	Unit	Material	Labor	Equipment	Total	Total Incl O&P
0010	**PARTITIONS, SHOWER** floor mounted, no plumbing									
0400	Cabinet, one piece, fiberglass, 32" x 32"	2 Carp	5	3.200	Ea.	625	170		795	960
0420	36" x 36"		5	3.200		550	170		720	875
0440	36" x 48"		5	3.200		1,375	170		1,545	1,800
0460	Acrylic, 32" x 32"		5	3.200		335	170		505	640
0480	36" x 36"		5	3.200		1,050	170		1,220	1,425
0500	36" x 48"		5	3.200		1,300	170		1,470	1,700
0520	Shower door for above, clear plastic, 24" wide	1 Carp	8	1		185	53		238	289
0540	28" wide		8	1		241	53		294	350
0560	Tempered glass, 24" wide		8	1		260	53		313	370
0580	28" wide		8	1		293	53		346	405
2400	Glass stalls, with doors, no receptors, chrome on brass	2 Shee	3	5.333		1,625	330		1,955	2,325
2700	Anodized aluminum	"	4	4		1,300	249		1,549	1,850
2900	Marble shower stall, stock design, with shower door	2 Marb	1.20	13.333		2,275	680		2,955	3,600
3000	With curtain		1.30	12.308		2,000	625		2,625	3,200
3200	Receptors, precast terrazzo, 32" x 32"		14	1.143		300	58		358	425
3300	48" x 34"		9.50	1.684		440	85.50		525.50	620
3500	Plastic, simulated terrazzo receptor, 32" x 32"		14	1.143		172	58		230	283
3600	32" x 48"		12	1.333		315	68		383	455
3800	Precast concrete, colors, 32" x 32"		14	1.143		220	58		278	335
3900	48" x 48"		8	2		310	102		412	505
4100	Shower doors, economy plastic, 24" wide	1 Shee	9	.889		134	55.50		189.50	234
4200	Tempered glass door, economy		8	1		279	62.50		341.50	405
4400	Folding, tempered glass, aluminum frame		6	1.333		430	83		513	605
4500	Sliding, tempered glass, 48" opening		6	1.333		575	83		658	760
4700	Deluxe, tempered glass, chrome on brass frame, 42" to 44"		8	1		485	62.50		547.50	635
4800	39" to 48" wide		1	8		700	500		1,200	1,550

10 28 Toilet, Bath, and Laundry Accessories

10 28 19 – Tub and Shower Enclosures

10 28 19.10 Partitions, Shower		Crew	Daily Output	Labor-Hours	Unit	Material	Labor	Equipment	Total	Total Incl O&P
						2020 Bare Costs				
4850	On anodized aluminum frame, obscure glass	1 Shee	2	4	Ea.	570	249		819	1,025
4900	Clear glass		1	8		690	500		1,190	1,550
5100	Shower enclosure, tempered glass, anodized alum. frame									
5120	2 panel & door, corner unit, 32" x 32"	1 Shee	2	4	Ea.	965	249		1,214	1,450
5140	Neo-angle corner unit, 16" x 24" x 16"	"	2	4		1,025	249		1,274	1,525
5200	Shower surround, 3 wall, polypropylene, 32" x 32"	1 Carp	4	2		635	106		741	870
5220	PVC, 32" x 32"		4	2		405	106		511	615
5240	Fiberglass		4	2		390	106		496	600
5250	2 wall, polypropylene, 32" x 32"		4	2		310	106		416	510
5270	PVC		4	2		370	106		476	575
5290	Fiberglass		4	2		375	106		481	580
5300	Tub doors, tempered glass & frame, obscure glass	1 Shee	8	1		219	62.50		281.50	340
5400	Clear glass		6	1.333		510	83		593	690
5600	Chrome plated, brass frame, obscure glass		8	1		289	62.50		351.50	420
5700	Clear glass		6	1.333		710	83		793	910
5900	Tub/shower enclosure, temp. glass, alum. frame, obscure glass		2	4		395	249		644	820
6200	Clear glass		1.50	5.333		820	330		1,150	1,425
6500	On chrome-plated brass frame, obscure glass		2	4		540	249		789	985
6600	Clear glass		1.50	5.333		1,150	330		1,480	1,800
6800	Tub surround, 3 wall, polypropylene	1 Carp	4	2		251	106		357	445
6900	PVC		4	2		365	106		471	575
7000	Fiberglass, obscure glass		4	2		385	106		491	595
7100	Clear glass		3	2.667		650	142		792	940
9990	Minimum labor/equipment charge	1 Shee	2.50	3.200	Job		199		199	310

10 28 23 – Laundry Accessories

10 28 23.13 Built-In Ironing Boards		Crew	Daily Output	Labor-Hours	Unit	Material	Labor	Equipment	Total	Total Incl O&P
0010	**BUILT-IN IRONING BOARDS**									
0020	Including cabinet, board & light, 42"	1 Carp	2	4	Ea.	245	213		458	610
0100	46"	"	1.50	5.333	"	580	283		863	1,100

10 31 Manufactured Fireplaces

10 31 13 – Manufactured Fireplace Chimneys

10 31 13.10 Fireplace Chimneys		Crew	Daily Output	Labor-Hours	Unit	Material	Labor	Equipment	Total	Total Incl O&P
0010	**FIREPLACE CHIMNEYS**									
0500	Chimney dbl. wall, all stainless, over 8'-6", 7" diam., add to fireplace	1 Carp	33	.242	V.L.F.	94	12.90		106.90	124
0600	10" diameter, add to fireplace		32	.250		103	13.30		116.30	135
0700	12" diameter, add to fireplace		31	.258		181	13.70		194.70	221
0800	14" diameter, add to fireplace		30	.267		216	14.15		230.15	261
1000	Simulated brick chimney top, 4' high, 16" x 16"		10	.800	Ea.	515	42.50		557.50	635
1100	24" x 24"		7	1.143	"	575	60.50		635.50	730

10 31 13.20 Chimney Accessories		Crew	Daily Output	Labor-Hours	Unit	Material	Labor	Equipment	Total	Total Incl O&P
0010	**CHIMNEY ACCESSORIES**									
0020	Chimney screens, galv., 13" x 13" flue	1 Bric	8	1	Ea.	56.50	52		108.50	146
0050	24" x 24" flue		5	1.600		128	83.50		211.50	275
0200	Stainless steel, 13" x 13" flue		8	1		101	52		153	196
0250	20" x 20" flue		5	1.600		147	83.50		230.50	297
2400	Squirrel and bird screens, galvanized, 8" x 8" flue		16	.500		56	26		82	104
2450	13" x 13" flue		12	.667		66	34.50		100.50	129
9000	Minimum labor/equipment charge		3.50	2.286	Job		119		119	192

10 31 Manufactured Fireplaces

10 31 16 – Manufactured Fireplace Forms

10 31 16.10 Fireplace Forms		Crew	Daily Output	Labor-Hours	Unit	Material	2020 Bare Costs Labor	Equipment	Total	Total Incl O&P
0010	**FIREPLACE FORMS**									
1800	Fireplace forms, no accessories, 32" opening	1 Bric	3	2.667	Ea.	735	139		874	1,025
1900	36" opening		2.50	3.200		940	167		1,107	1,300
2000	40" opening		2	4		1,250	208		1,458	1,700
2100	78" opening		1.50	5.333		1,825	278		2,103	2,450

10 31 23 – Prefabricated Fireplaces

10 31 23.10 Fireplace, Prefabricated		Crew	Daily Output	Labor-Hours	Unit	Material	Labor	Equipment	Total	Total Incl O&P
0010	**FIREPLACE, PREFABRICATED**, free standing or wall hung									
0100	With hood & screen, painted	1 Carp	1.30	6.154	Ea.	1,650	325		1,975	2,325
0150	Average		1	8		1,775	425		2,200	2,625
0200	Stainless steel		.90	8.889		3,300	470		3,770	4,375
1500	Simulated logs, gas fired, 40,000 BTU, 2' long, manual safety pilot		7	1.143	Set	560	60.50		620.50	715
1600	Adjustable flame remote pilot		6	1.333		1,325	71		1,396	1,575
1700	Electric, 1,500 BTU, 1'-6" long, incandescent flame		7	1.143		265	60.50		325.50	390
1800	1,500 BTU, LED flame		6	1.333		345	71		416	490

10 32 Fireplace Specialties

10 32 13 – Fireplace Dampers

10 32 13.10 Dampers		Crew	Daily Output	Labor-Hours	Unit	Material	Labor	Equipment	Total	Total Incl O&P
0010	**DAMPERS**									
0800	Damper, rotary control, steel, 30" opening	1 Bric	6	1.333	Ea.	123	69.50		192.50	248
0850	Cast iron, 30" opening		6	1.333		124	69.50		193.50	248
0880	36" opening		6	1.333		126	69.50		195.50	250
0900	48" opening		6	1.333		175	69.50		244.50	305
0920	60" opening		6	1.333		355	69.50		424.50	505
0950	72" opening		5	1.600		430	83.50		513.50	610
1000	84" opening, special order		5	1.600		920	83.50		1,003.50	1,150
1050	96" opening, special order		4	2		935	104		1,039	1,200
1200	Steel plate, poker control, 60" opening		8	1		325	52		377	445
1250	84" opening, special order		5	1.600		595	83.50		678.50	790
1400	"Universal" type, chain operated, 32" x 20" opening		8	1		254	52		306	365
1450	48" x 24" opening		5	1.600		380	83.50		463.50	550

10 32 23 – Fireplace Doors

10 32 23.10 Doors		Crew	Daily Output	Labor-Hours	Unit	Material	Labor	Equipment	Total	Total Incl O&P
0010	**DOORS**									
0400	Cleanout doors and frames, cast iron, 8" x 8"	1 Bric	12	.667	Ea.	57	34.50		91.50	119
0450	12" x 12"		10	.800		90.50	41.50		132	167
0500	18" x 24"		8	1		152	52		204	251
0550	Cast iron frame, steel door, 24" x 30"		5	1.600		325	83.50		408.50	495
1600	Dutch oven door and frame, cast iron, 12" x 15" opening		13	.615		133	32		165	198
1650	Copper plated, 12" x 15" opening		13	.615		260	32		292	340

10 35 Stoves

10 35 13 – Heating Stoves

10 35 13.10 Wood Burning Stoves		Crew	Daily Output	Labor-Hours	Unit	Material	Labor	Equipment	Total	Total Incl O&P
						2020 Bare Costs				
0010	**WOOD BURNING STOVES**									
0015	Cast iron, less than 1,500 S.F.	2 Carp	1.30	12.308	Ea.	1,500	655		2,155	2,700
0020	1,500 to 2,000 S.F.		1	16		2,550	850		3,400	4,150
0030	greater than 2,000 S.F.		.80	20		2,925	1,075		4,000	4,925
0050	For gas log lighter, add					49.50			49.50	54.50

10 43 Emergency Aid Specialties

10 43 13 – Defibrillator Cabinets

10 43 13.05 Defibrillator Cabinets		Crew	Daily Output	Labor-Hours	Unit	Material	Labor	Equipment	Total	Total Incl O&P
0010	**DEFIBRILLATOR CABINETS**, not equipped, stainless steel									
0050	Defibrillator cabinet, stainless steel with strobe & alarm 12" x 27"	1 Carp	10	.800	Ea.	430	42.50		472.50	540
0100	Automatic External Defibrillator	"	30	.267	"	1,400	14.15		1,414.15	1,550

10 44 Fire Protection Specialties

10 44 13 – Fire Protection Cabinets

10 44 13.53 Fire Equipment Cabinets		Crew	Daily Output	Labor-Hours	Unit	Material	Labor	Equipment	Total	Total Incl O&P
0010	**FIRE EQUIPMENT CABINETS**, not equipped, 20 ga. steel box									
0040	Recessed, D.S. glass in door, box size given									
1000	Portable extinguisher, single, 8" x 12" x 27", alum. door & frame	Q-12	8	2	Ea.	142	114		256	330
1100	Steel door and frame		8	2		164	114		278	355
1200	Stainless steel door and frame		8	2		227	114		341	425
2000	Portable extinguisher, large, 8" x 12" x 36", alum. door & frame		8	2		281	114		395	485
2100	Steel door and frame		8	2		248	114		362	450
2200	Stainless steel door and frame		8	2		410	114		524	630
2500	8" x 16" x 38", aluminum door & frame		8	2		251	114		365	450
2600	Steel door and frame		8	2		246	114		360	445
2700	Fire blanket & extinguisher cab, inc blanket, rec stl., 14" x 40" x 8"		7	2.286		219	130		349	445
2800	Fire blanket cab, inc blanket, surf mtd, stl, 15"x10"x5", w/pwdr coat fin		8	2		103	114		217	289
3000	Hose rack assy., 1-1/2" valve & 100' hose, 24" x 40" x 5-1/2"									
3100	Aluminum door and frame	Q-12	6	2.667	Ea.	565	152		717	855
3200	Steel door and frame		6	2.667		259	152		411	520
3300	Stainless steel door and frame		6	2.667		590	152		742	885
4000	Hose rack assy., 2-1/2" x 1-1/2" valve, 100' hose, 24" x 40" x 8"									
4100	Aluminum door and frame	Q-12	6	2.667	Ea.	545	152		697	835
4200	Steel door and frame		6	2.667		385	152		537	660
4300	Stainless steel door and frame		6	2.667		760	152		912	1,075
5000	Hose rack assy., 2-1/2" x 1-1/2" valve, 100' hose									
5010	and extinguisher, 30" x 40" x 8"									
5100	Aluminum door and frame	Q-12	5	3.200	Ea.	725	182		907	1,075
5200	Steel door and frame		5	3.200		299	182		481	610
5300	Stainless steel door and frame		5	3.200		630	182		812	975
6000	Hose rack assy., 1-1/2" valve, 100' hose									
6010	and 2-1/2" FD valve, 24" x 44" x 8"									
6100	Aluminum door and frame	Q-12	5	3.200	Ea.	590	182		772	930
6200	Steel door and frame		5	3.200		375	182		557	695
6300	Stainless steel door and frame		5	3.200		855	182		1,037	1,225
7000	Hose rack assy., 1-1/2" valve & 100' hose, 2-1/2" FD valve									
7010	and extinguisher, 30" x 44" x 8"									
7100	Aluminum door and frame	Q-12	5	3.200	Ea.	765	182		947	1,125

10 44 Fire Protection Specialties

10 44 13 – Fire Protection Cabinets

10 44 13.53 Fire Equipment Cabinets		Crew	Daily Output	Labor-Hours	Unit	2020 Bare Costs Material	Labor	Equipment	Total	Total Incl O&P
7200	Steel door and frame	Q-12	5	3.200	Ea.	465	182		647	790
7300	Stainless steel door and frame		5	3.200		720	182		902	1,075
8000	Valve cabinet for 2-1/2" FD angle valve, 18" x 18" x 8"									
8100	Aluminum door and frame	Q-12	12	1.333	Ea.	251	76		327	395
8200	Steel door and frame		12	1.333		207	76		283	345
8300	Stainless steel door and frame		12	1.333		335	76		411	490

10 44 16 – Fire Extinguishers

10 44 16.13 Portable Fire Extinguishers		Crew	Daily Output	Labor-Hours	Unit	Material	Labor	Equipment	Total	Total Incl O&P
0010	**PORTABLE FIRE EXTINGUISHERS**									
0140	CO_2, with hose and "H" horn, 10 lb.				Ea.	300			300	330
0160	15 lb.					385			385	425
0180	20 lb.					425			425	470
1000	Dry chemical, pressurized									
1040	Standard type, portable, painted, 2-1/2 lb.				Ea.	44			44	48.50
1060	5 lb.					62			62	68
1080	10 lb.					96			96	106
1100	20 lb.					138			138	152
1120	30 lb.					430			430	470
1300	Standard type, wheeled, 150 lb.					1,500			1,500	1,650
2000	ABC all purpose type, portable, 2-1/2 lb.					23.50			23.50	26
2060	5 lb.					28.50			28.50	31.50
2080	9-1/2 lb.					52			52	57
2100	20 lb.					91			91	100
3500	Halotron 1, 2-1/2 lb.					133			133	146
3600	5 lb.					207			207	228
3700	60 lb.					425			425	470
5000	Pressurized water, 2-1/2 gallon, stainless steel					105			105	115
5060	With anti-freeze					118			118	130
9400	Installation of extinguishers, 12 or more, on nailable surface	1 Carp	30	.267			14.15		14.15	22.50
9420	On masonry or concrete	"	15	.533			28.50		28.50	45.50

10 44 16.16 Wheeled Fire Extinguisher Units		Crew	Daily Output	Labor-Hours	Unit	Material	Labor	Equipment	Total	Total Incl O&P
0010	**WHEELED FIRE EXTINGUISHER UNITS**									
0350	CO_2, portable, with swivel horn									
0360	Wheeled type, cart mounted, 50 lb.				Ea.	1,375			1,375	1,525
0400	100 lb.				"	4,250			4,250	4,675
2200	ABC all purpose type									
2300	Wheeled, 45 lb.				Ea.	775			775	850
2360	150 lb.				"	1,975			1,975	2,150

10 51 Lockers

10 51 13 – Metal Lockers

10 51 13.10 Lockers		Crew	Daily Output	Labor-Hours	Unit	Material	Labor	Equipment	Total	Total Incl O&P
0011	**LOCKERS** steel, baked enamel, knock down construction									
0012	Body- 24ga, door- 16 or 18ga									
0110	1-tier locker, 12" x 15" x 72"	1 Shee	20	.400	Ea.	271	25		296	335
0120	18" x 15" x 72"		20	.400		335	25		360	410
0130	12" x 18" x 72"		20	.400		230	25		255	292
0140	18" x 18" x 72"		20	.400		299	25		324	370
0410	2- tier, 12" x 15" x 36"		30	.267		290	16.60		306.60	345
0420	18" x 15" x 36"		30	.267		239	16.60		255.60	289

10 51 13 – Metal Lockers

10 51 13.10 Lockers		Crew	Daily Output	Labor-Hours	Unit	Material	Labor	Equipment	Total	Total Incl O&P
						2020 Bare Costs				
0430	12" x 18" x 36"	1 Shee	30	.267	Ea.	320	16.60		336.60	380
0440	18" x 18" x 36"		30	.267		255	16.60		271.60	305
0450	3- tier 12" x 15" x 24"		30	.267		293	16.60		309.60	345
0455	12" x 18" x 24"		30	.267		300	16.60		316.60	355
0480	4- tier 12" x 15" x 18"		30	.267		375	16.60		391.60	435
0485	12" x 18" x 18"		30	.267		390	16.60		406.60	455
0500	Two person, 18" x 15" x 72"		20	.400		325	25		350	400
0510	18" x 18" x 72"		20	.400		305	25		330	375
0520	Duplex, 15" x 15" x 72"		20	.400		345	25		370	420
0530	15" x 21" x 72"		20	.400		385	25		410	465
0600	5 tier box lockers, unassembled		30	.267	Opng.	45.50	16.60		62.10	76
0700	Set up		24	.333		52	21		73	90
0900	6 tier box lockers, unassembled		36	.222		39	13.85		52.85	64.50
1000	Set up		30	.267		45.50	16.60		62.10	76
1098	All welded, ventilated athletic lockers, body- 16 ga., door- 14 ga.									
1100	1-tier, 12" x 18" x 72"	1 Shee	7.50	1.067	Ea.	350	66.50		416.50	490
1110	18" x 15" x 72"		7.50	1.067		555	66.50		621.50	720
1115	18" x 18" x 72"		7.50	1.067		570	66.50		636.50	730
1130	2-tier, 12" x 15" x 36"		7.50	1.067		580	66.50		646.50	745
1135	12" x 18" x 36"		7.50	1.067		620	66.50		686.50	790
1140	18" x 15" x 36"		7.50	1.067		675	66.50		741.50	850
1145	18" x 18" x 36"		7.50	1.067		690	66.50		756.50	860
1160	3-tier, 12" x 18" x 24"		7.50	1.067		795	66.50		861.50	980
1180	4-tier, 12" x 18" x 18"		7.50	1.067		605	66.50		671.50	770
1190	5H box, 12" x 18" x 14.4"		7.50	1.067		620	66.50		686.50	790
1195	18" x 18" x 14.4"		7.50	1.067		795	66.50		861.50	980
1196	6H box, 12" x 15" x 12"		7.50	1.067		535	66.50		601.50	690
1197	12" x 18" x 12"		7.50	1.067		605	66.50		671.50	770
1198	18" x 18" x 12"		7.50	1.067		755	66.50		821.50	935
2399	Standard duty lockers, body- 24 ga., doors- 16-18 ga.									
2400	16-person locker unit with clothing rack									
2500	72" wide x 15" deep x 72" high	1 Shee	15	.533	Ea.	610	33		643	725
2550	18" deep	"	15	.533	"	590	33		623	700
3000	Wall mounted lockers, 4 person, with coat bar									
3100	48" wide x 18" deep x 12" high	1 Shee	20	.400	Ea.	325	25		350	400
3250	Rack w/24 wire mesh baskets		1.50	5.333	Set	460	330		790	1,025
3260	30 baskets		1.25	6.400		420	400		820	1,075
3270	36 baskets		.95	8.421		450	525		975	1,325
3280	42 baskets		.80	10		465	625		1,090	1,475
3300	For built-in lock with 2 keys, add				Ea.	12.80			12.80	14.10
3600	For hanger rods, add					2.53			2.53	2.78
3650	For number plate kit, 100 plates #1 - #100, add	1 Shee	4	2		85.50	125		210.50	289
3700	For locker base, closed front panel		90	.089		5.40	5.55		10.95	14.65
3710	End panel, bolted		36	.222		8.35	13.85		22.20	30.50
3800	For sloping top, 12" wide		24	.333		31.50	21		52.50	67
3810	15" wide		24	.333		32	21		53	67.50
3820	18" wide		24	.333		33	21		54	69
3850	Sloping top end panel, 12" deep		72	.111		15.95	6.90		22.85	28.50
3860	15" deep		72	.111		17.25	6.90		24.15	30
3870	18" deep		72	.111		19.75	6.90		26.65	32.50
3900	For finish end panels, steel, 60" high, 15" deep		12	.667		30.50	41.50		72	98.50
3910	72" high, 12" deep		12	.667		27.50	41.50		69	95
3920	18" deep		12	.667		44	41.50		85.50	113

10 51 Lockers

10 51 13 – Metal Lockers

10 51 13.10 Lockers		Crew	Daily Output	Labor-Hours	Unit	Material	2020 Bare Costs Labor	Equipment	Total	Total Incl O&P
5000	For "ready to assemble" lockers,									
5010	Add to labor						75%			
5020	Deduct from material					20%				
6000	Heavy duty for detention facility, tamper proof, 14 ga. welded steel, solid									
6100	24" W x 24" D x 74" H, single tier	1 Shee	18	.444	Ea.	610	27.50		637.50	715
6110	Double tier		18	.444		560	27.50		587.50	660
6120	Triple tier		18	.444		610	27.50		637.50	715
9000	Minimum labor/equipment charge		2.50	3.200	Job		199		199	310

10 51 26 – Plastic Lockers

10 51 26.13 Recycled Plastic Lockers		Crew	Daily Output	Labor-Hours	Unit	Material	Labor	Equipment	Total	Total Incl O&P
0011	**RECYCLED PLASTIC LOCKERS**, 30% recycled									
0110	Single tier box locker, 12" x 12" x 72" G	1 Shee	8	1	Ea.	430	62.50		492.50	570
0120	12" x 15" x 72" G		8	1		440	62.50		502.50	580
0130	12" x 18" x 72" G		8	1		520	62.50		582.50	675
0410	Double tier, 12" x 12" x 72" G		21	.381		460	23.50		483.50	540
0420	12" x 15" x 72" G		21	.381		515	23.50		538.50	605
0430	12" x 18" x 72" G		21	.381		525	23.50		548.50	610

10 51 53 – Locker Room Benches

10 51 53.10 Benches		Crew	Daily Output	Labor-Hours	Unit	Material	Labor	Equipment	Total	Total Incl O&P
0010	**BENCHES**									
2100	Locker bench, laminated maple, top only	1 Shee	100	.080	L.F.	27	4.98		31.98	37.50
2200	Pedestals, steel pipe		25	.320	Ea.	48	19.95		67.95	83.50
2250	Plastic, 9.5" top with PVC pedestals		80	.100	L.F.	73	6.25		79.25	90.50

10 55 Postal Specialties

10 55 23 – Mail Boxes

10 55 23.10 Commercial Mail Boxes		Crew	Daily Output	Labor-Hours	Unit	Material	Labor	Equipment	Total	Total Incl O&P
0010	**COMMERCIAL MAIL BOXES**									
0020	Horiz., key lock, 5"H x 6"W x 15"D, alum., rear load	1 Carp	34	.235	Ea.	40.50	12.50		53	64.50
0100	Front loading		34	.235		40.50	12.50		53	64.50
0200	Double, 5"H x 12"W x 15"D, rear loading		26	.308		70.50	16.35		86.85	104
0300	Front loading		26	.308		77.50	16.35		93.85	112
0500	Quadruple, 10"H x 12"W x 15"D, rear loading		20	.400		145	21.50		166.50	193
0600	Front loading		20	.400		94.50	21.50		116	138
0800	Vertical, front load, 15"H x 5"W x 6"D, alum., per compartment		34	.235		47	12.50		59.50	71.50
0900	Bronze, duranodic finish		34	.235		51.50	12.50		64	76.50
1000	Steel, enameled		34	.235		46.50	12.50		59	71
1700	Alphabetical directories, 120 names		10	.800		103	42.50		145.50	182
1800	Letter collection box		6	1.333		203	71		274	335
1830	Lobby collection boxes, aluminum	2 Shee	5	3.200		1,750	199		1,949	2,225
1840	Bronze or stainless	"	4.50	3.556		1,975	222		2,197	2,525
1900	Letter slot, residential	1 Carp	20	.400		80	21.50		101.50	122
2000	Post office type		8	1		33.50	53		86.50	122
2250	Key keeper, single key, aluminum		26	.308		42.50	16.35		58.85	72.50
2300	Steel, enameled		26	.308		78	16.35		94.35	112
9000	Minimum labor/equipment charge		5	1.600	Job		85		85	136

10 56 Storage Assemblies

10 56 13 – Metal Storage Shelving

10 56 13.10 Shelving		Crew	Daily Output	Labor-Hours	Unit	Material	Labor	Equipment	Total	Total Incl O&P
						2020 Bare Costs				
0010	**SHELVING**									
0020	Metal, industrial, cross-braced, 3′ W, 12″ D	1 Sswk	175	.046	SF Shlf	7.60	2.64		10.24	12.65
0100	24″ D		330	.024		5.65	1.40		7.05	8.55
0300	4′ W, 12″ D		185	.043		7.85	2.49		10.34	12.70
0400	24″ D		380	.021		5.65	1.21		6.86	8.20
1200	Enclosed sides, cross-braced back, 3′ W, 12″ D		175	.046		14.10	2.64		16.74	19.80
1300	24″ D		290	.028		8.10	1.59		9.69	11.50
1500	Fully enclosed, sides and back, 3′ W, 12″ D		150	.053		17.25	3.07		20.32	24
1600	24″ D		255	.031		11.35	1.81		13.16	15.45
1800	4′ W, 12″ D		150	.053		10.30	3.07		13.37	16.40
1900	24″ D		290	.028		8.65	1.59		10.24	12.10
2200	Wide span, 1600 lb. capacity per shelf, 6′ W, 24″ D		380	.021		7.25	1.21		8.46	10
2400	36″ D		440	.018		6.55	1.05		7.60	8.95
2600	8′ W, 24″ D		440	.018		7.10	1.05		8.15	9.55
2800	36″ D	↓	520	.015		7.40	.89		8.29	9.60
4000	Pallet racks, steel frame 5,000 lb. capacity, 8′ long, 36″ D	2 Sswk	450	.036		9.70	2.05		11.75	14
4200	42″ D		500	.032		8.30	1.84		10.14	12.15
4400	48″ D	↓	520	.031	↓	7.50	1.77		9.27	11.15
9000	Minimum labor/equipment charge	1 Carp	4	2	Job		106		106	170

10 56 13.20 Parts Bins		Crew	Daily Output	Labor-Hours	Unit	Material	Labor	Equipment	Total	Total Incl O&P
0010	**PARTS BINS** metal, gray baked enamel finish									
0100	6′-3″ high, 3′ wide									
0300	12 bins, 18″ wide x 12″ high, 12″ deep	2 Clab	10	1.600	Ea.	365	67.50		432.50	510
0400	24″ deep		10	1.600		485	67.50		552.50	640
0600	72 bins, 6″ wide x 6″ high, 12″ deep		8	2		545	84		629	735
0700	18″ deep	↓	8	2	↓	875	84		959	1,100
1000	7′-3″ high, 3′ wide									
1200	14 bins, 18″ wide x 12″ high, 12″ deep	2 Clab	10	1.600	Ea.	420	67.50		487.50	570
1300	24″ deep		10	1.600		440	67.50		507.50	595
1500	84 bins, 6″ wide x 6″ high, 12″ deep		8	2		1,100	84		1,184	1,350
1600	24″ deep	↓	8	2	↓	1,350	84		1,434	1,600

10 57 Wardrobe and Closet Specialties

10 57 13 – Hat and Coat Racks

10 57 13.10 Coat Racks and Wardrobes		Crew	Daily Output	Labor-Hours	Unit	Material	Labor	Equipment	Total	Total Incl O&P
0010	**COAT RACKS AND WARDROBES**									
0020	Hat & coat rack, floor model, 6 hangers									
0050	Standing, beech wood, 21″ x 21″ x 72″, chrome				Ea.	267			267	293
0100	18 ga. tubular steel, 21″ x 21″ x 69″, wood walnut				″	370			370	405
0500	16 ga. steel frame, 22 ga. steel shelves									
0650	Single pedestal, 30″ x 18″ x 63″				Ea.	375			375	410
0800	Single face rack, 29″ x 18-1/2″ x 62″					345			345	380
0900	51″ x 18-1/2″ x 70″					445			445	490
0910	Double face rack, 39″ x 26″ x 70″					450			450	495
0920	63″ x 26″ x 70″				↓	565			565	620
0940	For 2″ ball casters, add				Set	82.50			82.50	90.50
1400	Utility hook strips, 3/8″ x 2-1/2″ x 18″, 6 hooks	1 Carp	48	.167	Ea.	58	8.85		66.85	78
1500	34″ long, 12 hooks	″	48	.167	″	71	8.85		79.85	92
1650	Wall mounted racks, 16 ga. steel frame, 22 ga. steel shelves									
1850	12″ x 15″ x 26″, 6 hangers	1 Carp	32	.250	Ea.	141	13.30		154.30	177
2000	12″ x 15″ x 50″, 12 hangers	″	32	.250	″	164	13.30		177.30	203

10 57 Wardrobe and Closet Specialties

10 57 13 – Hat and Coat Racks

10 57 13.10 Coat Racks and Wardrobes		Crew	Daily Output	Labor-Hours	Unit	2020 Bare Costs Material	Labor	Equipment	Total	Total Incl O&P
2150	Wardrobe cabinet, steel, baked enamel finish									
2300	36" x 21" x 78", incl. top shelf & hanger rod				Ea.	335			335	365
2400	Wardrobe, 24" x 24" x 76", KD, w/door, hospital, baked enamel steel	1 Carp	2	4		770	213		983	1,175
2500	Hardwood	"	2	4		1,250	213		1,463	1,725

10 57 23 – Closet and Utility Shelving

10 57 23.19 Wood Closet and Utility Shelving

		Crew	Daily Output	Labor-Hours	Unit	Material	Labor	Equipment	Total	Total Incl O&P
0010	**WOOD CLOSET AND UTILITY SHELVING**									
0020	Pine, clear grade, no edge band, 1" x 8"	1 Carp	115	.070	L.F.	3.35	3.70		7.05	9.60
0100	1" x 10"		110	.073		4.17	3.87		8.04	10.80
0200	1" x 12"		105	.076		5.05	4.05		9.10	12.05
0600	Plywood, 3/4" thick with lumber edge, 12" wide		75	.107		1.91	5.65		7.56	11.20
0700	24" wide		70	.114		3.40	6.05		9.45	13.50
0900	Bookcase, clear grade pine, shelves 12" OC, 8" deep, per S.F. shelf		70	.114	S.F.	10.90	6.05		16.95	22
1000	12" deep shelves		65	.123	"	16.35	6.55		22.90	28.50
1200	Adjustable closet rod and shelf, 12" wide, 3' long		20	.400	Ea.	13.80	21.50		35.30	49
1300	8' long		15	.533	"	25.50	28.50		54	73.50
1500	Prefinished shelves with supports, stock, 8" wide		75	.107	L.F.	5.85	5.65		11.50	15.55
1600	10" wide		70	.114	"	5.65	6.05		11.70	16
9000	Minimum labor/equipment charge		4	2	Job		106		106	170

10 73 Protective Covers

10 73 13 – Awnings

10 73 13.10 Awnings, Fabric

		Crew	Daily Output	Labor-Hours	Unit	Material	Labor	Equipment	Total	Total Incl O&P
0010	**AWNINGS, FABRIC**									
0020	Including acrylic canvas and frame, standard design									
0100	Door and window, slope, 3' high, 4' wide	1 Carp	4.50	1.778	Ea.	755	94.50		849.50	980
0110	6' wide		3.50	2.286		970	121		1,091	1,275
0120	8' wide		3	2.667		1,175	142		1,317	1,525
0200	Quarter round convex, 4' wide		3	2.667		1,175	142		1,317	1,525
0210	6' wide		2.25	3.556		1,525	189		1,714	1,975
0220	8' wide		1.80	4.444		1,875	236		2,111	2,425
0300	Dome, 4' wide		7.50	1.067		455	56.50		511.50	590
0310	6' wide		3.50	2.286		1,025	121		1,146	1,325
0320	8' wide		2	4		1,800	213		2,013	2,350
0350	Elongated dome, 4' wide		1.33	6.015		1,700	320		2,020	2,375
0360	6' wide		1.11	7.207		2,025	385		2,410	2,850
0370	8' wide		1	8		2,375	425		2,800	3,275
1000	Entry or walkway, peak, 12' long, 4' wide	2 Carp	.90	17.778		5,375	945		6,320	7,450
1010	6' wide		.60	26.667		8,300	1,425		9,725	11,400
1020	8' wide		.40	40		11,500	2,125		13,625	16,000
1100	Radius with dome end, 4' wide		1.10	14.545		4,075	775		4,850	5,750
1110	6' wide		.70	22.857		6,575	1,225		7,800	9,175
1120	8' wide		.50	32		9,325	1,700		11,025	13,000
2000	Retractable lateral arm awning, manual									
2010	To 12' wide, 8'-6" projection	2 Carp	1.70	9.412	Ea.	1,225	500		1,725	2,150
2020	To 14' wide, 8'-6" projection		1.10	14.545		1,425	775		2,200	2,825
2030	To 19' wide, 8'-6" projection		.85	18.824		1,925	1,000		2,925	3,725
2040	To 24' wide, 8'-6" projection		.67	23.881		2,425	1,275		3,700	4,700
2050	Motor for above, add	1 Carp	2.67	3		1,050	159		1,209	1,400
3000	Patio/deck canopy with frame									

10 73 Protective Covers

10 73 13 – Awnings

10 73 13.10 Awnings, Fabric		Crew	Daily Output	Labor-Hours	Unit	2020 Bare Costs Material	Labor	Equipment	Total	Total Incl O&P
3010	12′ wide, 12′ projection	2 Carp	2	8	Ea.	1,725	425		2,150	2,575
3020	16′ wide, 14′ projection	"	1.20	13.333		2,675	710		3,385	4,075
9000	For fire retardant canvas, add					7%				
9010	For lettering or graphics, add					35%				
9020	For painted or coated acrylic canvas, deduct					8%				
9030	For translucent or opaque vinyl canvas, add					10%				
9040	For 6 or more units, deduct					20%	15%			

10 73 16 – Canopies

10 73 16.20 Metal Canopies		Crew	Daily Output	Labor-Hours	Unit	Material	Labor	Equipment	Total	Total Incl O&P
0010	**METAL CANOPIES**									
0020	Wall hung, .032″, aluminum, prefinished, 8′ x 10′	K-2	1.30	18.462	Ea.	2,500	1,025	630	4,155	5,100
0300	8′ x 20′		1.10	21.818		4,200	1,200	745	6,145	7,375
0500	10′ x 10′		1.30	18.462		3,075	1,025	630	4,730	5,725
0700	10′ x 20′		1.10	21.818		4,725	1,200	745	6,670	7,975
1000	12′ x 20′		1	24		5,825	1,325	820	7,970	9,450
1360	12′ x 30′		.80	30		8,775	1,650	1,025	11,450	13,500
1700	12′ x 40′		.60	40		11,700	2,200	1,375	15,275	18,000
1900	For free standing units, add					20%	10%			
2300	Aluminum entrance canopies, flat soffit, .032″									
2500	3′-6″ x 4′-0″, clear anodized	2 Carp	4	4	Ea.	1,050	213		1,263	1,500
2700	Bronze anodized		4	4		1,850	213		2,063	2,400
3000	Polyurethane painted		4	4		1,500	213		1,713	2,000
3300	4′-6″ x 10′-0″, clear anodized		2	8		2,900	425		3,325	3,875
3500	Bronze anodized		2	8		3,700	425		4,125	4,750
3700	Polyurethane painted		2	8		3,100	425		3,525	4,100
4000	Wall downspout, 10 L.F., clear anodized	1 Carp	7	1.143		180	60.50		240.50	296
4300	Bronze anodized		7	1.143		315	60.50		375.50	445
4500	Polyurethane painted		7	1.143		270	60.50		330.50	395
7000	Carport, baked vinyl finish, .032″, 20′ x 10′, no foundations, flat panel	K-2	4	6	Car	4,125	330	205	4,660	5,300
7250	Insulated flat panel		2	12	"	4,950	660	410	6,020	6,975
7500	Walkway cover, to 12′ wide, stl., vinyl finish, .032″, no fndtns., flat		250	.096	S.F.	22	5.25	3.28	30.53	36
7750	Arched		200	.120	"	53.50	6.60	4.10	64.20	74
9000	Minimum labor/equipment charge	2 Carp	2	8	Job		425		425	680

10 74 Manufactured Exterior Specialties

10 74 29 – Steeples

10 74 29.10 Prefabricated Steeples		Crew	Daily Output	Labor-Hours	Unit	Material	Labor	Equipment	Total	Total Incl O&P
0010	**PREFABRICATED STEEPLES**									
4000	Steeples, translucent fiberglass, 30″ square, 15′ high	F-3	2	20	Ea.	9,925	1,075	236	11,236	12,900
4150	25′ high		1.80	22.222		11,500	1,200	262	12,962	14,800
4350	Opaque fiberglass, 24″ square, 14′ high		2	20		8,225	1,075	236	9,536	11,000
4500	28′ high		1.80	22.222		6,675	1,200	262	8,137	9,550
4600	Aluminum, baked finish, 16″ square, 14′ high					6,250			6,250	6,850
4620	20′ high, 3′-6″ base					10,900			10,900	12,000
4640	35′ high, 8′ base					38,700			38,700	42,600
4660	60′ high, 14′ base					87,000			87,000	95,500
4680	152′ high, custom					660,500			660,500	727,000
4700	Porcelain enamel steeples, custom, 40′ high	F-3	.50	80		15,100	4,350	940	20,390	24,600
4800	60′ high	"	.30	133		18,600	7,250	1,575	27,425	33,700

10 74 Manufactured Exterior Specialties

10 74 33 – Weathervanes

10 74 33.10 Residential Weathervanes		Crew	Daily Output	Labor-Hours	Unit	Material	2020 Bare Costs Labor	Equipment	Total	Total Incl O&P
0010	**RESIDENTIAL WEATHERVANES**									
0020	Residential types, 18" to 24"	1 Carp	8	1	Ea.	138	53		191	237
0100	24" to 48"		2	4	"	1,900	213		2,113	2,425
9000	Minimum labor/equipment charge		4	2	Job		106		106	170

10 74 46 – Window Wells

10 74 46.10 Area Window Wells

		Crew	Daily Output	Labor-Hours	Unit	Material	Labor	Equipment	Total	Total Incl O&P
0010	**AREA WINDOW WELLS**, Galvanized steel									
0020	20 ga., 3'-2" wide, 1' deep	1 Sswk	29	.276	Ea.	18.70	15.90		34.60	46.50
0100	2' deep		23	.348		30	20		50	66
0300	16 ga., 3'-2" wide, 1' deep		29	.276		25.50	15.90		41.40	54
0400	3' deep		23	.348		48.50	20		68.50	86.50
0600	Welded grating for above, 15 lb., painted		45	.178		103	10.25		113.25	130
0700	Galvanized		45	.178		126	10.25		136.25	155
0900	Translucent plastic cap for above		60	.133		21.50	7.70		29.20	36

10 75 Flagpoles

10 75 16 – Ground-Set Flagpoles

10 75 16.10 Flagpoles

		Crew	Daily Output	Labor-Hours	Unit	Material	Labor	Equipment	Total	Total Incl O&P
0010	**FLAGPOLES**, ground set									
0050	Not including base or foundation									
0100	Aluminum, tapered, ground set 20' high	K-1	2	8	Ea.	1,175	400	410	1,985	2,375
0200	25' high		1.70	9.412		1,275	470	485	2,230	2,675
0300	30' high		1.50	10.667		1,250	535	545	2,330	2,825
0400	35' high		1.40	11.429		1,800	575	585	2,960	3,550
0500	40' high		1.20	13.333		3,100	670	685	4,455	5,250
0600	50' high		1	16		3,950	805	820	5,575	6,500
0700	60' high		.90	17.778		4,900	890	910	6,700	7,825
0800	70' high		.80	20		7,725	1,000	1,025	9,750	11,200
1100	Counterbalanced, internal halyard, 20' high		1.80	8.889		3,125	445	455	4,025	4,650
1200	30' high		1.50	10.667		2,900	535	545	3,980	4,625
1300	40' high		1.30	12.308		7,350	620	630	8,600	9,775
1400	50' high		1	16		10,200	805	820	11,825	13,500
2820	Aluminum, electronically operated, 30' high		1.40	11.429		4,525	575	585	5,685	6,550
2840	35' high		1.30	12.308		5,450	620	630	6,700	7,650
2860	39' high		1.10	14.545		6,775	730	745	8,250	9,450
2880	45' high		1	16		6,250	805	820	7,875	9,050
2900	50' high		.90	17.778		9,100	890	910	10,900	12,400
3000	Fiberglass, tapered, ground set, 23' high		2	8		785	400	410	1,595	1,950
3100	29'-7" high		1.50	10.667		1,300	535	545	2,380	2,875
3200	36'-1" high		1.40	11.429		1,825	575	585	2,985	3,575
3300	39'-5" high		1.20	13.333		2,050	670	685	3,405	4,075
3400	49'-2" high		1	16		4,825	805	820	6,450	7,475
3500	59' high		.90	17.778		5,075	890	910	6,875	8,025
4300	Steel, direct imbedded installation									
4400	Internal halyard, 20' high	K-1	2.50	6.400	Ea.	1,575	320	330	2,225	2,600
4500	25' high		2.50	6.400		2,150	320	330	2,800	3,225
4600	30' high		2.30	6.957		2,475	350	355	3,180	3,675
4700	40' high		2.10	7.619		3,525	380	390	4,295	4,925
4800	50' high		1.90	8.421		4,500	425	430	5,355	6,100
5000	60' high		1.80	8.889		7,150	445	455	8,050	9,075

10 75 Flagpoles

10 75 16 – Ground-Set Flagpoles

10 75 16.10 Flagpoles		Crew	Daily Output	Labor-Hours	Unit	Material	Labor	Equipment	Total	Total Incl O&P
						2020 Bare Costs				
5100	70′ high	K-1	1.60	10	Ea.	8,500	500	515	9,515	10,700
5200	80′ high		1.40	11.429		11,100	575	585	12,260	13,800
5300	90′ high		1.20	13.333		17,100	670	685	18,455	20,600
5500	100′ high	↓	1	16	↓	18,600	805	820	20,225	22,700
6400	Wood poles, tapered, clear vertical grain fir with tilting									
6410	base, not incl. foundation, 4″ butt, 25′ high	K-1	1.90	8.421	Ea.	1,875	425	430	2,730	3,200
6800	6″ butt, 30′ high	″	1.30	12.308	″	3,275	620	630	4,525	5,275
7300	Foundations for flagpoles, including									
7400	excavation and concrete, to 35′ high poles	C-1	10	3.200	Ea.	730	161		891	1,050
7600	40′ to 50′ high		3.50	9.143		1,350	460		1,810	2,225
7700	Over 60′ high	↓	2	16	↓	1,675	805		2,480	3,150

10 75 23 – Wall-Mounted Flagpoles

10 75 23.10 Flagpoles		Crew	Daily Output	Labor-Hours	Unit	Material	Labor	Equipment	Total	Total Incl O&P
0010	**FLAGPOLES**, structure mounted									
0100	Fiberglass, vertical wall set, 19′-8″ long	K-1	1.50	10.667	Ea.	1,125	535	545	2,205	2,675
0200	23′ long		1.40	11.429		1,750	575	585	2,910	3,500
0300	26′-3″ long		1.30	12.308		1,625	620	630	2,875	3,450
0800	19′-8″ long outrigger		1.30	12.308		1,500	620	630	2,750	3,325
1300	Aluminum, vertical wall set, tapered, with base, 20′ high		1.20	13.333		1,300	670	685	2,655	3,250
1400	29′-6″ high		1	16		3,350	805	820	4,975	5,850
2400	Outrigger poles with base, 12′ long		1.30	12.308		1,375	620	630	2,625	3,200
2500	14′ long	↓	1	16	↓	1,575	805	820	3,200	3,900

10 86 Security Mirrors and Domes

10 86 10 – Security Mirrors

10 86 10.10 Exterior Traffic Control Mirrors		Crew	Daily Output	Labor-Hours	Unit	Material	Labor	Equipment	Total	Total Incl O&P
0010	**EXTERIOR TRAFFIC CONTROL MIRRORS**									
0100	Convex, stainless steel, 20 ga., 26″ diameter	1 Carp	12	.667	Ea.	188	35.50		223.50	264

10 86 20 – Security Domes

10 86 20.10 Domes		Crew	Daily Output	Labor-Hours	Unit	Material	Labor	Equipment	Total	Total Incl O&P
0010	**DOMES** for security cameras (CCTV)									
0100	Ceiling mounted, 10″ diameter	1 Carp	30	.267	Ea.	12.30	14.15		26.45	36
0110	12″ diameter	″	30	.267	″	8.85	14.15		23	32

10 88 Scales

10 88 05 – Commercial Scales

10 88 05.10 Scales		Crew	Daily Output	Labor-Hours	Unit	Material	Labor	Equipment	Total	Total Incl O&P
0010	**SCALES**									
0700	Truck scales, incl. steel weigh bridge,									
0800	not including foundation, pits									
1550	Digital, electronic, 100 ton capacity, steel deck 12′ x 10′ platform	3 Carp	.20	120	Ea.	14,900	6,375		21,275	26,600
1600	40′ x 10′ platform		.14	171		29,700	9,100		38,800	47,300
1640	60′ x 10′ platform		.13	185		43,700	9,800		53,500	64,000
1680	70′ x 10′ platform	↓	.12	200		41,800	10,600		52,400	63,000
2000	For standard automatic printing device, add					1,550			1,550	1,700
2100	For remote reading electronic system, add					2,925			2,925	3,225
2300	Concrete foundation pits for above, 8′ x 6′, 5 C.Y. required	C-1	.50	64		1,125	3,225		4,350	6,425
2400	14′ x 6′ platform, 10 C.Y. required	↓	.35	91.429	↓	1,675	4,600		6,275	9,200

10 88 Scales

10 88 05 – Commercial Scales

10 88 05.10 Scales		Crew	Daily Output	Labor-Hours	Unit	2020 Bare Costs Material	Labor	Equipment	Total	Total Incl O&P
2600	50′ x 10′ platform, 30 C.Y. required	C-1	.25	128	Ea.	2,250	6,450		8,700	12,800
2700	70′ x 10′ platform, 40 C.Y. required	↓	.15	213		4,900	10,700		15,600	22,600
2750	Crane scales, dial, 1 ton capacity					1,175			1,175	1,275
2780	5 ton capacity					1,625			1,625	1,775
2800	Digital, 1 ton capacity					1,925			1,925	2,100
2850	10 ton capacity				↓	4,775			4,775	5,250
2900	Low profile electronic warehouse scale,									
3000	not incl. printer, 4′ x 4′ platform, 10,000 lb. capacity	2 Carp	.30	53.333	Ea.	1,375	2,825		4,200	6,050
3300	5′ x 7′ platform, 10,000 lb. capacity		.25	64		5,525	3,400		8,925	11,500
3400	20,000 lb. capacity	↓	.20	80		6,725	4,250		10,975	14,200
3500	For printers, incl. time, date & numbering, add					700			700	770
3800	Portable, beam type, capacity 1000 lb., platform 18″ x 24″					875			875	965
3900	Dial type, capacity 2000 lb., platform 24″ x 24″					1,625			1,625	1,800
4000	Digital type, capacity 1000 lb., platform 24″ x 30″					2,600			2,600	2,875
4100	Portable contractor truck scales, 50 ton cap., 40′ x 10′ platform					29,700			29,700	32,700
4200	60′ x 10′ platform					36,300			36,300	39,900
4400	Heavy-Duty Steel Deck Truck Scales 20′ x 10′	3 Carp	.20	120		18,200	6,375		24,575	30,300
4500	Heavy-Duty Steel Deck Truck Scales 80′ x 10′		.10	240		48,100	12,800		60,900	73,500
4600	Heavy-Duty Steel Deck Truck Scales 160′ x 10′		.04	600		105,500	31,900		137,400	167,000
4700	Heavy-Duty Steel Deck Truck Scales 20′ x 12′		.18	137		21,300	7,300		28,600	35,200
4800	Heavy-Duty Steel Deck Truck Scales 80′ x 12′		.09	267		61,000	14,200		75,200	89,500
4900	Heavy-Duty Steel Deck Truck Scales 160′ x 12′		.04	600		123,000	31,900		154,900	186,500
5000	Heavy-Duty Steel Deck Truck Scales 20′ x 14′		.16	150		28,300	7,975		36,275	44,000
5100	Heavy-Duty Steel Deck Truck Scales 80′ x 14′		.06	400		67,000	21,300		88,300	107,500
5200	Heavy-Duty Steel Deck Truck Scales 160′ x 14′	↓	.03	800	↓	131,000	42,500		173,500	212,000

Estimating Tips

General

- The items in this division are usually priced per square foot or each. Many of these items are purchased by the owner for installation by the contractor. Check the specifications for responsibilities and include time for receiving, storage, installation, and mechanical and electrical hookups in the appropriate divisions.
- Many items in Division 11 require some type of support system that is not usually furnished with the item. Examples of these systems include blocking for the attachment of casework and support angles for ceiling-hung projection screens. The required blocking or supports must be added to the estimate in the appropriate division.
- Some items in Division 11 may require assembly or electrical hookups. Verify the amount of assembly required or the need for a hard electrical connection and add the appropriate costs.

Reference Numbers

Reference numbers are shown at the beginning of some major classifications. These numbers refer to related items in the Reference Section. The reference information may be an estimating procedure, an alternate pricing method, or technical information.

11 05 Common Work Results for Equipment

11 05 05 – Selective Demolition for Equipment

11 05 05.10 Selective Demolition		Crew	Daily Output	Labor-Hours	Unit	Material	2020 Bare Costs Labor	Equipment	Total	Total Incl O&P
0010	**SELECTIVE DEMOLITION** R024119-10									
0130	Central vacuum, motor unit, residential or commercial	1 Clab	2	4	Ea.		168		168	270
0210	Vault door and frame	2 Skwk	2	8			440		440	700
0215	Day gate, for vault	"	3	5.333			293		293	465
0380	Bank equipment, teller window, bullet resistant	1 Clab	1.20	6.667			281		281	450
0381	Counter	2 Clab	1.50	10.667	Station		450		450	720
0382	Drive-up window, including drawer and glass		1.50	10.667	"		450		450	720
0383	Thru-wall boxes and chests, selective demolition		2.50	6.400	Ea.		269		269	430
0384	Bullet resistant partitions		20	.800	L.F.		33.50		33.50	54
0385	Pneumatic tube system, 2 lane drive-up	L-3	.45	35.556	Ea.		2,050		2,050	3,225
0386	Safety deposit box	1 Clab	50	.160	Opng.		6.75		6.75	10.80
0387	Surveillance system, video, complete	2 Elec	2	8	Ea.		490		490	755
0410	Church equipment, misc movable fixtures	2 Clab	1	16			675		675	1,075
0412	Steeple, to 28′ high	F-3	3	13.333			725	157	882	1,325
0414	40′ to 60′ high	"	.80	50			2,725	590	3,315	4,975
0510	Library equipment, bookshelves, wood, to 90″ high	1 Clab	20	.400	L.F.		16.85		16.85	27
0515	Carrels, hardwood, 36″ x 24″	"	9	.889	Ea.		37.50		37.50	60
0630	Stage equipment, light control panel	1 Elec	1	8	"		490		490	755
0632	Border lights		40	.200	L.F.		12.25		12.25	18.90
0634	Spotlights		8	1	Ea.		61.50		61.50	94.50
0636	Telescoping platforms and risers	2 Clab	175	.091	SF Stg.		3.85		3.85	6.15
1020	Barber equipment, hydraulic chair	1 Clab	40	.200	Ea.		8.40		8.40	13.50
1030	Checkout counter, supermarket or warehouse conveyor	2 Clab	18	.889			37.50		37.50	60
1040	Food cases, refrigerated or frozen	Q-5	6	2.667			157		157	243
1190	Laundry equipment, commercial	L-6	3	4			254		254	390
1360	Movie equipment, lamphouse, to 4000 watt, incl. rectifier	1 Elec	4	2			123		123	189
1365	Sound system, incl. amplifier	"	1.25	6.400			395		395	605
1410	Air compressor, to 5 HP	2 Clab	2.50	6.400			269		269	430
1412	Lubrication equipment, automotive, 3 reel type, incl. pump, excl. piping	L-4	1	24	Set		1,200		1,200	1,925
1414	Booth, spray paint, complete, to 26′ long	"	.80	30	"		1,500		1,500	2,400
1560	Parking equipment, cashier booth	B-22	2	15	Ea.		740	142	882	1,325
1600	Loading dock equipment, dock bumpers, rubber	1 Clab	50	.160	"		6.75		6.75	10.80
1610	Door seal for door perimeter	"	50	.160	L.F.		6.75		6.75	10.80
1611	Loading dock equipment, dock seal for perimeter, selective demolition	2 Clab	13	1.231	"		52		52	83
1620	Platform lifter, fixed, 6′ x 8′, 5000 lb. capacity	E-16	1.50	10.667	Ea.		625	98	723	1,125
1630	Dock leveller	"	2	8			470	73.50	543.50	845
1640	Lights, single or double arm	1 Elec	8	1			61.50		61.50	94.50
1650	Shelter, fabric, truck or train	1 Clab	1.50	5.333			225		225	360
1790	Waste handling equipment, commercial compactor	L-4	2	12			600		600	960
1792	Commercial or municipal incinerator, gas	"	2	12			600		600	960
1795	Crematory, excluding building	Q-3	.25	128			7,850		7,850	12,100
1910	Detection equipment, cell bar front	E-4	4	8			465	36.50	501.50	800
1912	Cell door and frame		8	4			233	18.35	251.35	400
1914	Prefab cell, 4′ to 5′ wide, 7′ to 8′ high, 7′ deep		8	4			233	18.35	251.35	400
1916	Cot, bolted, single		40	.800			46.50	3.67	50.17	80
1918	Visitor cubicle		4	8			465	36.50	501.50	800
2850	Hydraulic gates, canal, flap, knife, slide or sluice, to 18″ diameter	L-5A	8	4			234	143	377	535
2852	19″ to 36″ diameter		6	5.333			310	191	501	715
2854	37″ to 48″ diameter		2	16			935	570	1,505	2,150
2856	49″ to 60″ diameter		1	32			1,875	1,150	3,025	4,275
2858	Over 60″ diameter		.30	107			6,250	3,825	10,075	14,300
3100	Sewage pumping system, prefabricated, to 1000 GPM	C-17D	.20	420			23,300	3,725	27,025	41,200
3110	Sewage treatment, holding tank for recirc chemical water closet	1 Plum	8	1			64.50		64.50	99.50

11 05 Common Work Results for Equipment

11 05 05 – Selective Demolition for Equipment

11 05 05.10 Selective Demolition		Crew	Daily Output	Labor-Hours	Unit	Material	2020 Bare Costs Labor	Equipment	Total	Total Incl O&P
3900	Wastewater treatment system, to 1500 gal.	B-21	2	14	Ea.		685	94.50	779.50	1,175
4050	Food storage equipment, walk-in refrigerator/freezer	2 Clab	64	.250	S.F.		10.55		10.55	16.85
4052	Shelving, stainless steel, 4 tier or dunnage rack	1 Clab	12	.667	Ea.		28		28	45
4100	Food preparation equipment, small countertop		18	.444			18.70		18.70	30
4150	Food delivery carts, heated cabinets		18	.444			18.70		18.70	30
4200	Cooking equipment, commercial range	Q-1	12	1.333			77.50		77.50	120
4250	Hood and ventilation equipment, kitchen exhaust hood, excl. fire prot	1 Clab	3	2.667			112		112	180
4255	Fire protection system	Q-1	3	5.333			310		310	480
4300	Food dispensing equipment, countertop items	1 Clab	15	.533			22.50		22.50	36
4310	Serving counter	"	65	.123	L.F.		5.20		5.20	8.30
4350	Ice machine, ice cube maker, flakers and storage bins, to 2000 lb./day	Q-1	1.60	10	Ea.		580		580	895
4400	Cleaning and disposal, commercial dishwasher, to 50 racks/hour	L-6	1	12			760		760	1,175
4405	To 275 racks/hour	L-4	1	24			1,200		1,200	1,925
4410	Dishwasher hood	2 Clab	5	3.200			135		135	216
4420	Garbage disposal, commercial, to 5 HP	L-1	8	2			126		126	194
4540	Water heater, residential, to 80 gal./day	"	5	3.200			201		201	310
4542	Water softener, automatic	2 Plum	10	1.600			103		103	159
4544	Disappearing stairway, to 15' floor height	2 Clab	6	2.667			112		112	180
4710	Darkroom equipment, light	L-7	10	2.800			143		143	228
4712	Heavy	"	1.50	18.667			955		955	1,525
4720	Doors	2 Clab	3.50	4.571	Opng.		192		192	310
4830	Bowling alley, complete, incl. pinsetter, scorer, counters, misc supplies	4 Clab	.40	80	Lane		3,375		3,375	5,400
4840	Health club equipment, circuit training apparatus	2 Clab	2	8	Set		335		335	540
4842	Squat racks	"	10	1.600	Ea.		67.50		67.50	108
4860	School equipment, basketball backstop	L-2	2	8			370		370	600
4862	Table and benches, folding, in wall, 14' long	L-4	4	6			299		299	480
4864	Bleachers, telescoping, to 30 tier	F-5	120	.267	Seat		14.30		14.30	23
4865	to 15 tier	"	160	.200	"		10.75		10.75	17.20
4866	Boxing ring, elevated	L-4	.20	120	Ea.		5,975		5,975	9,600
4867	Boxing ring, floor level	"	2	12			600		600	960
4868	Exercise equipment	1 Clab	6	1.333			56		56	90
4870	Gym divider	L-4	1000	.024	S.F.		1.20		1.20	1.92
4875	Scoreboard	R-3	2	10	Ea.		610	94.50	704.50	1,050
4880	Shooting range, incl. bullet traps, targets, excl. structure	L-9	1	36	Point		1,725		1,725	2,750
5200	Vocational shop equipment	2 Clab	8	2	Ea.		84		84	135
6200	Fume hood, incl. countertop, excl. HVAC	"	6	2.667	L.F.		112		112	180
7100	Medical sterilizing, distiller, water, steam heated, 50 gal. capacity	1 Plum	2.80	2.857	Ea.		184		184	285
7200	Medical equipment, surgery table, minor	1 Clab	1	8			335		335	540
7210	Surgical lights, doctor's office, single or double arm	2 Elec	3	5.333			325		325	505
7300	Physical therapy, table	2 Clab	4	4			168		168	270
7310	Whirlpool bath, fixed, incl. mixing valves	1 Plum	4	2			129		129	199
7400	Dental equipment, chair, electric or hydraulic	1 Clab	.75	10.667			450		450	720
7410	Central suction system	1 Plum	2	4			258		258	400
7420	Drill console with accessories	1 Clab	3.20	2.500			105		105	169
7430	X-ray unit	"	4	2			84		84	135
7440	X-ray developer	1 Plum	10	.800			51.50		51.50	79.50

11 05 10 – Equipment Installation

11 05 10.10 Industrial Equipment Installation		Crew	Daily Output	Labor-Hours	Unit	Material	Labor	Equipment	Total	Total Incl O&P
0010	**INDUSTRIAL EQUIPMENT INSTALLATION**									
0020	Industrial equipment, minimum	E-2	12	4.667	Ton		267	141	408	585
0200	Maximum	"	2	28	"		1,600	850	2,450	3,525

11 11 Vehicle Service Equipment

11 11 13 – Compressed-Air Vehicle Service Equipment

11 11 13.10 Compressed Air Equipment		Crew	Daily Output	Labor-Hours	Unit	Material	Labor (2020 Bare Costs)	Equipment	Total	Total Incl O&P
0010	**COMPRESSED AIR EQUIPMENT**									
0030	Compressors, electric, 1-1/2 HP, standard controls	L-4	1.50	16	Ea.	505	800		1,305	1,825
0550	Dual controls		1.50	16		1,050	800		1,850	2,425
0600	5 HP, 115/230 volt, standard controls		1	24		1,725	1,200		2,925	3,825
0650	Dual controls		1	24		3,225	1,200		4,425	5,475

11 11 19 – Vehicle Lubrication Equipment

11 11 19.10 Lubrication Equipment		Crew	Daily Output	Labor-Hours	Unit	Material	Labor	Equipment	Total	Total Incl O&P
0010	**LUBRICATION EQUIPMENT**									
3000	Lube equipment, 3 reel type, with pumps, not including piping	L-4	.50	48	Set	10,400	2,400		12,800	15,300
3100	Hose reel, including hose, oil/lube, 1000 psi	2 Sswk	2	8	Ea.	480	460		940	1,275
3200	Grease, 5000 psi		2	8		585	460		1,045	1,400
3300	Air, 50′, 160 psi		2	8		605	460		1,065	1,425
3350	25′, 160 psi		2	8		365	460		825	1,150

11 11 33 – Vehicle Spray Painting Equipment

11 11 33.10 Spray Painting Equipment		Crew	Daily Output	Labor-Hours	Unit	Material	Labor	Equipment	Total	Total Incl O&P
0010	**SPRAY PAINTING EQUIPMENT**									
4000	Spray painting booth, 26′ long, complete	L-4	.40	60	Ea.	11,600	3,000		14,600	17,500

11 12 Parking Control Equipment

11 12 13 – Parking Key and Card Control Units

11 12 13.10 Parking Control Units		Crew	Daily Output	Labor-Hours	Unit	Material	Labor	Equipment	Total	Total Incl O&P
0010	**PARKING CONTROL UNITS**									
5100	Card reader	1 Elec	2	4	Ea.	1,475	245		1,720	2,000
5120	Proximity with customer display	2 Elec	1	16		6,025	980		7,005	8,125
6000	Parking control software, basic functionality	1 Elec	.50	16		24,000	980		24,980	27,900
6020	Multi-function	″	.20	40		96,000	2,450		98,450	109,500

11 12 16 – Parking Ticket Dispensers

11 12 16.10 Ticket Dispensers		Crew	Daily Output	Labor-Hours	Unit	Material	Labor	Equipment	Total	Total Incl O&P
0010	**TICKET DISPENSERS**									
5900	Ticket spitter with time/date stamp, standard	2 Elec	2	8	Ea.	6,625	490		7,115	8,025
5920	Mag stripe encoding	″	2	8	″	18,900	490		19,390	21,600

11 12 26 – Parking Fee Collection Equipment

11 12 26.13 Parking Fee Coin Collection Equipment		Crew	Daily Output	Labor-Hours	Unit	Material	Labor	Equipment	Total	Total Incl O&P
0010	**PARKING FEE COIN COLLECTION EQUIPMENT**									
5200	Cashier booth, average	B-22	1	30	Ea.	10,600	1,475	284	12,359	14,400
5300	Collector station, pay on foot	2 Elec	.20	80		111,500	4,900		116,400	130,500
5320	Credit card only	″	.50	32		20,600	1,975		22,575	25,700

11 12 26.23 Fee Equipment		Crew	Daily Output	Labor-Hours	Unit	Material	Labor	Equipment	Total	Total Incl O&P
0010	**FEE EQUIPMENT**									
5600	Fee computer	1 Elec	1.50	5.333	Ea.	13,000	325		13,325	14,800

11 12 33 – Parking Gates

11 12 33.13 Lift Arm Parking Gates		Crew	Daily Output	Labor-Hours	Unit	Material	Labor	Equipment	Total	Total Incl O&P
0010	**LIFT ARM PARKING GATES**									
5000	Barrier gate with programmable controller	2 Elec	3	5.333	Ea.	3,300	325		3,625	4,125
5020	Industrial		3	5.333		5,225	325		5,550	6,250
5050	Non-programmable, with reader and 12′ arm		3	5.333		1,700	325		2,025	2,375
5500	Exit verifier		1	16		19,600	980		20,580	23,100
5700	Full sign, 4″ letters	1 Elec	2	4		1,250	245		1,495	1,725

11 12 Parking Control Equipment

11 12 33 – Parking Gates

11 12 33.13 Lift Arm Parking Gates		Crew	Daily Output	Labor-Hours	Unit	2020 Bare Costs Material	Labor	Equipment	Total	Total Incl O&P
5800	Inductive loop	2 Elec	4	4	Ea.	209	245		454	610
5950	Vehicle detector, microprocessor based	1 Elec	3	2.667		470	164		634	765
7100	Traffic spike unit, flush mount, spring loaded, 72″ L	B-89	4	4		1,275	201	258	1,734	2,000
7200	Surface mount, 72″ L	2 Skwk	10	1.600		2,425	88		2,513	2,800

11 13 Loading Dock Equipment

11 13 13 – Loading Dock Bumpers

11 13 13.10 Dock Bumpers		Crew	Daily Output	Labor-Hours	Unit	Material	Labor	Equipment	Total	Total Incl O&P
0010	**DOCK BUMPERS** Bolts not included									
0012	2″ x 6″ to 4″ x 8″, average	1 Carp	300	.027	B.F.	.74	1.42		2.16	3.08
0050	Bumpers, lam. rubber blocks 4-1/2″ thick, 10″ high, 14″ long		26	.308	Ea.	60.50	16.35		76.85	92.50
0200	24″ long		22	.364		103	19.35		122.35	145
0300	36″ long		17	.471		340	25		365	415
0500	12″ high, 14″ long		25	.320		81	17		98	117
0550	24″ long		20	.400		85.50	21.50		107	128
0600	36″ long		15	.533		127	28.50		155.50	186
0800	Laminated rubber blocks 6″ thick, 10″ high, 14″ long		22	.364		72.50	19.35		91.85	111
0850	24″ long		18	.444		127	23.50		150.50	177
0900	36″ long		13	.615		212	32.50		244.50	286
0910	20″ high, 11″ long		13	.615		136	32.50		168.50	203
0920	Extruded rubber bumpers, T section, 22″ x 22″ x 3″ thick		41	.195		73.50	10.35		83.85	97.50
0940	Molded rubber bumpers, 24″ x 12″ x 3″ thick		20	.400		66.50	21.50		88	107
1000	Welded installation of above bumpers	E-14	8	1		4	59.50	18.40	81.90	122
1100	For drilled anchors, add per anchor	1 Carp	36	.222		24	11.80		35.80	45
1300	Steel bumpers, see Section 10 26 13.10									

11 13 16 – Loading Dock Seals and Shelters

11 13 16.10 Dock Seals and Shelters		Crew	Daily Output	Labor-Hours	Unit	Material	Labor	Equipment	Total	Total Incl O&P
0010	**DOCK SEALS AND SHELTERS**									
3600	Door seal for door perimeter, 12″ x 12″, vinyl covered	1 Carp	26	.308	L.F.	53.50	16.35		69.85	85
3700	Loading dock, seal for perimeter, 9′ x 8′, with 12″ vinyl	2 Carp	6	2.667	Ea.	1,450	142		1,592	1,825
3900	Folding gates, see Section 10 22 16.10									
6200	Shelters, fabric, for truck or train, scissor arms, minimum	1 Carp	1	8	Ea.	2,250	425		2,675	3,150
6300	Maximum	″	.50	16	″	2,550	850		3,400	4,150

11 13 19 – Stationary Loading Dock Equipment

11 13 19.10 Dock Equipment		Crew	Daily Output	Labor-Hours	Unit	Material	Labor	Equipment	Total	Total Incl O&P
0010	**DOCK EQUIPMENT**									
2200	Dock boards, heavy duty, 60″ x 60″, aluminum, 5,000 lb. capacity				Ea.	1,375			1,375	1,525
2700	9,000 lb. capacity					1,325			1,325	1,475
3200	15,000 lb. capacity					1,450			1,450	1,575
4200	Platform lifter, 6′ x 6′, portable, 3,000 lb. capacity					9,800			9,800	10,800
4250	4,000 lb. capacity					12,100			12,100	13,300
4400	Fixed, 6′ x 8′, 5,000 lb. capacity	E-16	.70	22.857		8,725	1,325	210	10,260	12,000
4500	Levelers, hinged for trucks, 10 ton capacity, 6′ x 8′		1.08	14.815		4,575	865	136	5,576	6,600
4650	7′ x 8′		1.08	14.815		6,975	865	136	7,976	9,250
4670	Air bag power operated, 10 ton cap., 6′ x 8′		1.08	14.815		5,050	865	136	6,051	7,125
4680	7′ x 8′		1.08	14.815		5,475	865	136	6,476	7,600
4700	Hydraulic, 10 ton capacity, 6′ x 8′		1.08	14.815		6,275	865	136	7,276	8,475
4800	7′ x 8′		1.08	14.815		6,650	865	136	7,651	8,900
6000	Dock leveler, 15 ton capacity									
6100	I-beam construction, mechanical, 6′ x 8′	E-16	.50	32	Ea.	5,425	1,875	294	7,594	9,375

11 13 Loading Dock Equipment

11 13 19 – Stationary Loading Dock Equipment

11 13 19.10 Dock Equipment		Crew	Daily Output	Labor-Hours	Unit	Material	2020 Bare Costs Labor	Equipment	Total	Total Incl O&P
6150	Hydraulic	E-16	.50	32	Ea.	6,375	1,875	294	8,544	10,400
6200	Formed beam deck construction, mechanical, 6' x 8'		.50	32		4,175	1,875	294	6,344	8,000
6250	Hydraulic		.50	32		5,725	1,875	294	7,894	9,700
6300	Edge of dock leveler, mechanical, 15 ton capacity		2	8		1,250	470	73.50	1,793.50	2,225
6301	Edge of dock leveler, mechanical, 10 ton capacity	↓	2	8	↓	1,250	470	73.50	1,793.50	2,225
7000	22.5 ton capacity									
7100	Vertical storing dock leveler, hydraulic, 6' x 6'	E-16	.40	40	Ea.	8,300	2,350	370	11,020	13,400

11 13 26 – Loading Dock Lights

11 13 26.10 Dock Lights		Crew	Daily Output	Labor-Hours	Unit	Material	Labor	Equipment	Total	Total Incl O&P
0010	**DOCK LIGHTS**									
5000	Lights for loading docks, single arm, 24" long	1 Elec	3.80	2.105	Ea.	201	129		330	420
5700	Double arm, 60" long	"	3.80	2.105	"	229	129		358	450

11 14 Pedestrian Control Equipment

11 14 13 – Pedestrian Gates

11 14 13.19 Turnstiles		Crew	Daily Output	Labor-Hours	Unit	Material	Labor	Equipment	Total	Total Incl O&P
0010	**TURNSTILES**									
0020	One way, 4 arm, 46" diameter, economy, manual	2 Carp	5	3.200	Ea.	2,250	170		2,420	2,750
0100	Electric		1.20	13.333		2,375	710		3,085	3,750
0300	High security, galv., 5'-5" diameter, 7' high, manual		1	16		5,725	850		6,575	7,650
0350	Electric		.60	26.667		9,000	1,425		10,425	12,200
0420	Three arm, 24" opening, light duty, manual		2	8		3,100	425		3,525	4,075
0450	Heavy duty		1.50	10.667		4,925	565		5,490	6,300
0460	Manual, with registering & controls, light duty		2	8		4,600	425		5,025	5,725
0470	Heavy duty		1.50	10.667		4,900	565		5,465	6,300
0480	Electric, heavy duty	↓	1.10	14.545		6,300	775		7,075	8,175
0500	For coin or token operating, add				↓	795			795	875
1200	One way gate with horizontal bars, 5'-5" diameter									
1300	7' high, recreation or transit type	2 Carp	.80	20	Ea.	6,950	1,075		8,025	9,350
1500	For electronic counter, add				"	287			287	315

11 14 19 – Portable Posts and Railings

11 14 19.13 Portable Posts and Railings		Crew	Daily Output	Labor-Hours	Unit	Material	Labor	Equipment	Total	Total Incl O&P
0010	**PORTABLE POSTS AND RAILINGS**									
0020	Portable for pedestrian traffic control, standard				Ea.	166			166	182
0300	Deluxe posts				"	239			239	263
0600	Ropes for above posts, plastic covered, 1-1/2" diameter				L.F.	20			20	22
0700	Chain core				"	14.65			14.65	16.10
1500	Portable security or safety barrier, black with 7' yellow strap				Ea.	206			206	227
1510	12' yellow strap					251			251	276
1550	Sign holder, standard design				↓	86			86	95

11 21 Retail and Service Equipment

11 21 13 – Cash Registers and Checking Equipment

11 21 13.10 Checkout Counter		Crew	Daily Output	Labor-Hours	Unit	Material	2020 Bare Costs Labor	Equipment	Total	Total Incl O&P
0010	CHECKOUT COUNTER									
0020	Supermarket conveyor, single belt	2 Clab	10	1.600	Ea.	3,775	67.50		3,842.50	4,275
0100	Double belt, power take-away		9	1.778		5,175	75		5,250	5,825
0400	Double belt, power take-away, incl. side scanning		7	2.286		5,975	96		6,071	6,725
0800	Warehouse or bulk type		6	2.667		6,950	112		7,062	7,800
1000	Scanning system, 2 lanes, w/registers, scan gun & memory				System	18,800			18,800	20,700
1100	10 lanes, single processor, full scan, with scales				"	178,500			178,500	196,500
2000	Register, restaurant, minimum				Ea.	800			800	880
2100	Maximum					3,575			3,575	3,925
2150	Store, minimum					800			800	880
2200	Maximum					3,575			3,575	3,925

11 21 33 – Checkroom Equipment

11 21 33.10 Clothes Check Equipment		Crew	Daily Output	Labor-Hours	Unit	Material	Labor	Equipment	Total	Total Incl O&P
0010	CLOTHES CHECK EQUIPMENT									
0030	Clothes check rack, free standing, st. stl., 2-tier, 90 bag capacity				Ea.	1,575			1,575	1,725
0050	Wall mounted, 45 bag capacity	L-2	8	2		905	93		998	1,150
0100	Garment checking bag, green mesh fabric, 21" H x 17" W with 4.5" hook					28			28	31

11 21 53 – Barber and Beauty Shop Equipment

11 21 53.10 Barber Equipment		Crew	Daily Output	Labor-Hours	Unit	Material	Labor	Equipment	Total	Total Incl O&P
0010	BARBER EQUIPMENT									
0020	Chair, hydraulic, movable, minimum	1 Carp	24	.333	Ea.	640	17.70		657.70	735
0050	Maximum	"	16	.500		4,150	26.50		4,176.50	4,625
0200	Wall hung styling station with mirrors, minimum	L-2	8	2		600	93		693	810
0300	Maximum	"	4	4		2,775	186		2,961	3,350
0500	Sink, hair washing basin, rough plumbing not incl.	1 Plum	8	1		530	64.50		594.50	685
1000	Sterilizer, liquid solution for tools					171			171	189
1100	Total equipment, rule of thumb, per chair, minimum	L-8	1	20		2,225	1,100		3,325	4,200
1150	Maximum	"	1	20		6,175	1,100		7,275	8,525

11 21 73 – Commercial Laundry and Dry Cleaning Equipment

11 21 73.13 Dry Cleaning Equipment		Crew	Daily Output	Labor-Hours	Unit	Material	Labor	Equipment	Total	Total Incl O&P
0010	DRY CLEANING EQUIPMENT									
2000	Dry cleaners, electric, 20 lb. capacity, not incl. rough-in	L-1	.20	80	Ea.	35,100	5,025		40,125	46,400
2050	25 lb. capacity		.17	94.118		52,000	5,925		57,925	66,000
2100	30 lb. capacity		.15	107		55,000	6,700		61,700	71,000
2150	60 lb. capacity		.09	178		79,000	11,200		90,200	104,500

11 21 73.16 Drying and Conditioning Equipment		Crew	Daily Output	Labor-Hours	Unit	Material	Labor	Equipment	Total	Total Incl O&P
0010	DRYING AND CONDITIONING EQUIPMENT									
0100	Dryers, not including rough-in									
1500	Industrial, 30 lb. capacity	1 Plum	2	4	Ea.	3,350	258		3,608	4,075
1600	50 lb. capacity	"	1.70	4.706		4,550	305		4,855	5,500
4700	Lint collector, ductwork not included, 8,000 to 10,000 CFM	Q-10	.30	80		9,900	4,650		14,550	18,200

11 21 73.19 Finishing Equipment		Crew	Daily Output	Labor-Hours	Unit	Material	Labor	Equipment	Total	Total Incl O&P
0010	FINISHING EQUIPMENT									
3500	Folders, blankets & sheets, minimum	1 Elec	.17	47.059	Ea.	35,600	2,875		38,475	43,600
3700	King size with automatic stacker		.10	80		64,500	4,900		69,400	78,000
3800	For conveyor delivery, add		.45	17.778		16,600	1,100		17,700	20,000
4900	Spreader feeders, 240V, 2 station	L-6	.70	17.143		60,500	1,075		61,575	68,000
4920	4 station	"	.35	34.286		73,000	2,175		75,175	84,000

11 21 Retail and Service Equipment

11 21 73 – Commercial Laundry and Dry Cleaning Equipment

11 21 73.23 Commercial Ironing Equipment		Crew	Daily Output	Labor-Hours	Unit	Material	Labor	Equipment	Total	Total Incl O&P
						2020 Bare Costs				
0010	**COMMERCIAL IRONING EQUIPMENT**									
4500	Ironers, institutional, 110", single roll	1 Elec	.20	40	Ea.	34,400	2,450		36,850	41,700
4800	Pressers, low capacity air operated	L-6	1.75	6.857		10,400	435		10,835	12,100
4820	Hand operated		1.75	6.857		9,650	435		10,085	11,300
4840	Ironer 48", 240V		3.50	3.429		119,000	217		119,217	131,500
6600	Hand operated presser		.70	17.143		6,400	1,075		7,475	8,725
6620	Mushroom press 115V		.70	17.143		8,175	1,075		9,250	10,700

11 21 73.26 Commercial Washers and Extractors

		Crew	Daily Output	Labor-Hours	Unit	Material	Labor	Equipment	Total	Total Incl O&P
0010	**COMMERCIAL WASHERS AND EXTRACTORS**, not including rough-in									
6000	Combination washer/extractor, 20 lb. capacity	L-6	1.50	8	Ea.	6,725	505		7,230	8,175
6100	30 lb. capacity		.80	15		10,300	950		11,250	12,800
6200	50 lb. capacity		.68	17.647		12,900	1,125		14,025	15,900
6300	75 lb. capacity		.30	40		19,200	2,525		21,725	25,000
6350	125 lb. capacity		.16	75		29,900	4,750		34,650	40,300
6380	Washer extractor/dryer, 110 lb., 240V		1	12		19,900	760		20,660	23,100
6400	Washer extractor, 135 lb., 240V		1	12		33,200	760		33,960	37,700
6450	Pass through		1	12		57,500	760		58,260	64,500
6500	200 lb. washer extractor		1	12		75,000	760		75,760	83,500
6550	Pass through		1	12		79,000	760		79,760	88,000
6600	Extractor, low capacity		1.75	6.857		7,700	435		8,135	9,150

11 21 73.33 Coin-Operated Laundry Equipment

		Crew	Daily Output	Labor-Hours	Unit	Material	Labor	Equipment	Total	Total Incl O&P
0010	**COIN-OPERATED LAUNDRY EQUIPMENT**									
0990	Dryer, gas fired									
1000	Commercial, 30 lb. capacity, coin operated, single	1 Plum	3	2.667	Ea.	3,475	172		3,647	4,100
1100	Double stacked	"	2	4		8,275	258		8,533	9,500
4860	Coin dry cleaner 20 lb.	L-6	1.75	6.857		29,900	435		30,335	33,600
5290	Clothes washer									
5300	Commercial, coin operated, average	1 Plum	3	2.667	Ea.	1,350	172		1,522	1,775

11 21 83 – Photo Processing Equipment

11 21 83.13 Darkroom Equipment

		Crew	Daily Output	Labor-Hours	Unit	Material	Labor	Equipment	Total	Total Incl O&P
0010	**DARKROOM EQUIPMENT**									
0020	Developing sink, 5" deep, 24" x 48"	Q-1	2	8	Ea.	500	465		965	1,275
0050	48" x 52"		1.70	9.412		1,250	545		1,795	2,225
0200	10" deep, 24" x 48"		1.70	9.412		1,625	545		2,170	2,650
0250	24" x 108"		1.50	10.667		3,875	620		4,495	5,200
0500	Dryers, dehumidified filtered air, 36" x 25" x 68" high	L-7	6	4.667		2,175	239		2,414	2,775
3000	Viewing lights, 20" x 24"		6	4.667		276	239		515	685
3100	20" x 24" with color correction		6	4.667		410	239		649	830

11 22 Banking Equipment

11 22 13 – Vault Equipment

11 22 13.16 Safes

		Crew	Daily Output	Labor-Hours	Unit	Material	Labor	Equipment	Total	Total Incl O&P
0010	**SAFES**									
0200	Office, 1 hr. rating, 30" x 18" x 18"				Ea.	2,175			2,175	2,400
0250	40" x 18" x 18"					4,975			4,975	5,475
0300	60" x 36" x 18", double door					9,800			9,800	10,800
0600	Data, 1 hr. rating, 27" x 19" x 16"					5,875			5,875	6,450
0700	63" x 34" x 16"					16,600			16,600	18,300
0750	Diskette, 1 hr., 14" x 12" x 11", inside					4,875			4,875	5,350

11 22 13 – Vault Equipment

11 22 13.16 Safes		Crew	Daily Output	Labor-Hours	Unit	Material	2020 Bare Costs Labor	Equipment	Total	Total Incl O&P
0800	Money, "B" label, 9" x 14" x 14"				Ea.	605			605	670
0900	Tool resistive, 24" x 24" x 20"					4,175			4,175	4,600
1050	Tool and torch resistive, 24" x 24" x 20"					9,525			9,525	10,500
1150	Jewelers, 23" x 20" x 18"					9,900			9,900	10,900
1200	63" x 25" x 18"					14,300			14,300	15,700
1300	For handling into building, add, minimum	A-2	8.50	2.824			124	23	147	224
1400	Maximum	"	.78	30.769			1,350	251	1,601	2,425

11 22 16 – Teller and Service Equipment

11 22 16.13 Teller Equipment Systems		Crew	Daily Output	Labor-Hours	Unit	Material	2020 Bare Costs Labor	Equipment	Total	Total Incl O&P
0010	**TELLER EQUIPMENT SYSTEMS**									
0020	Alarm system, police	2 Elec	1.60	10	Ea.	5,075	615		5,690	6,525
0100	With vault alarm	"	.40	40		20,100	2,450		22,550	25,900
0400	Bullet resistant teller window, 44" x 60"	1 Glaz	.60	13.333		4,625	680		5,305	6,150
0500	48" x 60"	"	.60	13.333		6,475	680		7,155	8,200
3000	Counters for banks, frontal only	2 Carp	1	16	Station	1,925	850		2,775	3,450
3100	Complete with steel undercounter	"	.50	32	"	3,700	1,700		5,400	6,800
4600	Door and frame, bullet-resistant, with vision panel, minimum	2 Sswk	1.10	14.545	Ea.	5,600	840		6,440	7,525
4700	Maximum		1.10	14.545		7,600	840		8,440	9,725
4800	Drive-up window, drawer & mike, not incl. glass, minimum		1	16		8,075	920		8,995	10,400
4900	Maximum		.50	32		10,200	1,850		12,050	14,200
5000	Night depository, with chest, minimum		1	16		8,400	920		9,320	10,800
5100	Maximum		.50	32		12,100	1,850		13,950	16,300
5200	Package receiver, painted		3.20	5		1,400	288		1,688	2,025
5300	Stainless steel		3.20	5		2,400	288		2,688	3,125
5400	Partitions, bullet-resistant, 1-3/16" glass, 8' high	2 Carp	10	1.600	L.F.	205	85		290	360
5450	Acrylic	"	10	1.600	"	390	85		475	560
5500	Pneumatic tube systems, 2 lane drive-up, complete	L-3	.25	64	Total	26,700	3,675		30,375	35,100
5550	With T.V. viewer	"	.20	80	"	51,500	4,600		56,100	64,500
5570	Safety deposit boxes, minimum	1 Sswk	44	.182	Opng.	59.50	10.50		70	82
5580	Maximum, 10" x 15" opening		19	.421		126	24.50		150.50	178
5590	Teller locker, average		15	.533		1,625	31		1,656	1,825
5600	Pass thru, bullet-res. window, painted steel, 24" x 36"	2 Sswk	1.60	10	Ea.	3,025	575		3,600	4,275
5700	48" x 48"		1.20	13.333		2,750	770		3,520	4,275
5800	72" x 40"		.80	20		4,500	1,150		5,650	6,850
5900	For stainless steel frames, add					20%				
6100	Surveillance system, video camera, complete	2 Elec	1	16	Ea.	10,000	980		10,980	12,500
6110	For each additional camera, add				"	1,025			1,025	1,150
6120	CCTV system, see Section 28 23 13.10									
6200	24 hour teller, single unit,									
6300	automated deposit, cash and memo	L-3	.25	64	Ea.	47,500	3,675		51,175	58,500
7000	Vault front, see Section 08 34 59.10									

11 30 13 – Residential Appliances

11 30 13.15 Cooking Equipment		Crew	Daily Output	Labor-Hours	Unit	Material	Labor	Equipment	Total	Total Incl O&P
0010	COOKING EQUIPMENT									
0020	Cooking range, 30" free standing, 1 oven, minimum	2 Clab	10	1.600	Ea.	465	67.50		532.50	620
0050	Maximum		4	4		2,200	168		2,368	2,700
0150	2 oven, minimum		10	1.600		1,050	67.50		1,117.50	1,275
0200	Maximum		10	1.600		3,425	67.50		3,492.50	3,850
0350	Built-in, 30" wide, 1 oven, minimum	1 Elec	6	1.333		870	82		952	1,075
0400	Maximum	2 Carp	2	8		2,000	425		2,425	2,875
0500	2 oven, conventional, minimum		4	4		1,250	213		1,463	1,725
0550	1 conventional, 1 microwave, maximum		2	8		2,750	425		3,175	3,700
0700	Free standing, 1 oven, 21" wide range, minimum	2 Clab	10	1.600		465	67.50		532.50	625
0750	21" wide, maximum	"	4	4		755	168		923	1,100
0900	Countertop cooktops, 4 burner, standard, minimum	1 Elec	6	1.333		340	82		422	495
0950	Maximum		3	2.667		2,025	164		2,189	2,475
1050	As above, but with grill and griddle attachment, minimum		6	1.333		1,375	82		1,457	1,650
1100	Maximum		3	2.667		3,875	164		4,039	4,525
1200	Induction cooktop, 30" wide		3	2.667		1,450	164		1,614	1,850
1250	Microwave oven, minimum		4	2		98.50	123		221.50	297
1300	Maximum		2	4		470	245		715	900

11 30 13.16 Refrigeration Equipment		Crew	Daily Output	Labor-Hours	Unit	Material	Labor	Equipment	Total	Total Incl O&P
0010	REFRIGERATION EQUIPMENT									
2000	Deep freeze, 15 to 23 C.F., minimum	2 Clab	10	1.600	Ea.	680	67.50		747.50	860
2050	Maximum		5	3.200		785	135		920	1,075
2200	30 C.F., minimum		8	2		895	84		979	1,125
2250	Maximum		3	5.333		905	225		1,130	1,350
5200	Icemaker, automatic, 20 lbs./day	1 Plum	7	1.143		1,375	73.50		1,448.50	1,625
5350	51 lbs./day	"	2	4		1,400	258		1,658	1,950
5450	Refrigerator, no frost, 6 C.F.	2 Clab	15	1.067		365	45		410	470
5500	Refrigerator, no frost, 10 C.F. to 12 C.F., minimum		10	1.600		455	67.50		522.50	610
5600	Maximum		6	2.667		560	112		672	800
5750	14 C.F. to 16 C.F., minimum		9	1.778		605	75		680	790
5800	Maximum		5	3.200		1,050	135		1,185	1,375
5950	18 C.F. to 20 C.F., minimum		8	2		775	84		859	990
6000	Maximum		4	4		1,925	168		2,093	2,400
6150	21 C.F. to 29 C.F., minimum		7	2.286		1,100	96		1,196	1,375
6200	Maximum		3	5.333		2,425	225		2,650	3,025
6790	Energy-star qualified, 18 C.F., minimum G	2 Carp	4	4		510	213		723	905
6795	Maximum G		2	8		1,950	425		2,375	2,800
6797	21.7 C.F., minimum G		4	4		1,025	213		1,238	1,475
6799	Maximum G		4	4		1,800	213		2,013	2,325

11 30 13.17 Kitchen Cleaning Equipment		Crew	Daily Output	Labor-Hours	Unit	Material	Labor	Equipment	Total	Total Incl O&P
0010	KITCHEN CLEANING EQUIPMENT									
2750	Dishwasher, built-in, 2 cycles, minimum	L-1	4	4	Ea.	310	252		562	730
2800	Maximum		2	8		445	505		950	1,275
2950	4 or more cycles, minimum		4	4		400	252		652	830
2960	Average		4	4		530	252		782	975
3000	Maximum		2	8		1,850	505		2,355	2,800
3100	Energy-star qualified, minimum G		4	4		405	252		657	840
3110	Maximum G		2	8		1,975	505		2,480	2,950

11 30 13.18 Waste Disposal Equipment		Crew	Daily Output	Labor-Hours	Unit	Material	Labor	Equipment	Total	Total Incl O&P
0010	WASTE DISPOSAL EQUIPMENT									
1750	Compactor, residential size, 4 to 1 compaction, minimum	1 Carp	5	1.600	Ea.	705	85		790	910
1800	Maximum	"	3	2.667		1,150	142		1,292	1,500

11 30 13 – Residential Appliances

11 30 13.18 Waste Disposal Equipment		Crew	Daily Output	Labor-Hours	Unit	Material	Labor	Equipment	Total	Total Incl O&P
3300	Garbage disposal, sink type, minimum	L-1	10	1.600	Ea.	111	101		212	277
3350	Maximum	"	10	1.600	↓	216	101		317	390
11 30 13.19 Kitchen Ventilation Equipment										
0010	**KITCHEN VENTILATION EQUIPMENT**									
4150	Hood for range, 2 speed, vented, 30" wide, minimum	L-3	5	3.200	Ea.	107	184		291	410
4200	Maximum		3	5.333		1,000	305		1,305	1,575
4300	42" wide, minimum		5	3.200		151	184		335	455
4330	Custom		5	3.200		1,800	184		1,984	2,275
4350	Maximum	↓	3	5.333		2,175	305		2,480	2,875
4500	For ventless hood, 2 speed, add					19.05			19.05	21
4650	For vented 1 speed, deduct from maximum				↓	74			74	81.50
11 30 13.24 Washers										
0010	**WASHERS**									
5000	Residential, 4 cycle, average	1 Plum	3	2.667	Ea.	990	172		1,162	1,375
6650	Washing machine, automatic, minimum		3	2.667		560	172		732	880
6700	Maximum		1	8		1,400	515		1,915	2,350
6750	Energy star qualified, front loading, minimum [G]		3	2.667		880	172		1,052	1,225
6760	Maximum [G]		1	8		1,925	515		2,440	2,900
6764	Top loading, minimum [G]		3	2.667		735	172		907	1,075
6766	Maximum [G]	↓	3	2.667	↓	1,125	172		1,297	1,525
11 30 13.25 Dryers										
0010	**DRYERS**									
0500	Gas fired residential, 16 lb. capacity, average	1 Plum	3	2.667	Ea.	720	172		892	1,050
6770	Electric, front loading, energy-star qualified, minimum [G]	L-2	3	5.333		480	248		728	925
6780	Maximum [G]	"	2	8		1,500	370		1,870	2,225
7450	Vent kits for dryers	1 Carp	10	.800	↓	48	42.50		90.50	121

11 30 15 – Miscellaneous Residential Appliances

11 30 15.13 Sump Pumps		Crew	Daily Output	Labor-Hours	Unit	Material	Labor	Equipment	Total	Total Incl O&P
0010	**SUMP PUMPS**									
6400	Cellar drainer, pedestal, 1/3 HP, molded PVC base	1 Plum	3	2.667	Ea.	141	172		313	420
6450	Solid brass	"	2	4	"	243	258		501	665
6460	Sump pump, see also Section 22 14 29.16									
11 30 15.23 Water Heaters										
0010	**WATER HEATERS**									
6900	Electric, glass lined, 30 gallon, minimum	L-1	5	3.200	Ea.	970	201		1,171	1,375
6950	Maximum		3	5.333		1,350	335		1,685	2,000
7100	80 gallon, minimum		2	8		1,950	505		2,455	2,925
7150	Maximum	↓	1	16		2,725	1,000		3,725	4,550
7180	Gas, glass lined, 30 gallon, minimum	2 Plum	5	3.200		1,900	206		2,106	2,400
7220	Maximum		3	5.333		2,625	345		2,970	3,425
7260	50 gallon, minimum		2.50	6.400		1,750	410		2,160	2,575
7300	Maximum	↓	1.50	10.667	↓	2,450	685		3,135	3,775
7310	Water heater, see also Section 22 33 30.13									
11 30 15.43 Air Quality										
0010	**AIR QUALITY**									
2450	Dehumidifier, portable, automatic, 15 pint	1 Elec	4	2	Ea.	208	123		331	420
2550	40 pint		3.75	2.133		247	131		378	470
3550	Heater, electric, built-in, 1250 watt, ceiling type, minimum		4	2		125	123		248	325
3600	Maximum		3	2.667		184	164		348	455
3700	Wall type, minimum		4	2		221	123		344	430
3750	Maximum	↓	3	2.667	↓	198	164		362	470

Material, Labor, Equipment and Total columns: 2020 Bare Costs.

11 30 Residential Equipment

11 30 15 – Miscellaneous Residential Appliances

11 30 15.43 Air Quality		Crew	Daily Output	Labor-Hours	Unit	Material	2020 Bare Costs Labor	Equipment	Total	Total Incl O&P
3900	1500 watt wall type, with blower	1 Elec	4	2	Ea.	186	123		309	395
3950	3000 watt		3	2.667		515	164		679	815
4850	Humidifier, portable, 8 gallons/day					158			158	174
5000	15 gallons/day					207			207	227

11 30 33 – Retractable Stairs

11 30 33.10 Disappearing Stairway		Crew	Daily Output	Labor-Hours	Unit	Material	Labor	Equipment	Total	Total Incl O&P
0010	**DISAPPEARING STAIRWAY** No trim included									
0100	Custom grade, pine, 8'-6" ceiling, minimum	1 Carp	4	2	Ea.	129	106		235	310
0150	Average		3.50	2.286		292	121		413	515
0200	Maximum		3	2.667		279	142		421	530
0500	Heavy duty, pivoted, from 7'-7" to 12'-10" floor to floor		3	2.667		1,375	142		1,517	1,725
0600	16'-0" ceiling		2	4		1,600	213		1,813	2,100
0800	Economy folding, pine, 8'-6" ceiling		4	2		178	106		284	365
0900	9'-6" ceiling		4	2		232	106		338	425
1100	Automatic electric, aluminum, floor to floor height, 8' to 9'	2 Carp	1	16		9,200	850		10,050	11,500
1400	11' to 12'		.90	17.778		9,725	945		10,670	12,200
1700	14' to 15'		.70	22.857		10,400	1,225		11,625	13,400
9000	Minimum labor/equipment charge	1 Carp	2	4	Job		213		213	340

11 32 Unit Kitchens

11 32 13 – Metal Unit Kitchens

11 32 13.10 Commercial Unit Kitchens		Crew	Daily Output	Labor-Hours	Unit	Material	Labor	Equipment	Total	Total Incl O&P
0010	**COMMERCIAL UNIT KITCHENS**									
1500	Combination range, refrigerator and sink, 30" wide, minimum	L-1	2	8	Ea.	710	505		1,215	1,550
1550	Maximum		1	16		1,100	1,000		2,100	2,750
1570	60" wide, average		1.40	11.429		1,225	720		1,945	2,450
1590	72" wide, average		1.20	13.333		2,000	840		2,840	3,500
1600	Office model, 48" wide		2	8		1,425	505		1,930	2,350
1620	Refrigerator and sink only		2.40	6.667		2,625	420		3,045	3,550
1640	Combination range, refrigerator, sink, microwave									
1660	Oven and ice maker	L-1	.80	20	Ea.	4,900	1,250		6,150	7,325

11 41 Foodservice Storage Equipment

11 41 13 – Refrigerated Food Storage Cases

11 41 13.10 Refrigerated Food Cases		Crew	Daily Output	Labor-Hours	Unit	Material	Labor	Equipment	Total	Total Incl O&P
0010	**REFRIGERATED FOOD CASES**									
0030	Dairy, multi-deck, 12' long	Q-5	3	5.333	Ea.	11,500	315		11,815	13,100
0100	For rear sliding doors, add					2,050			2,050	2,250
0200	Delicatessen case, service deli, 12' long, single deck	Q-5	3.90	4.103		8,750	242		8,992	10,000
0300	Multi-deck, 18 S.F. shelf display		3	5.333		8,375	315		8,690	9,700
0400	Freezer, self-contained, chest-type, 30 C.F.		3.90	4.103		4,800	242		5,042	5,650
0500	Glass door, upright, 78 C.F.		3.30	4.848		9,375	286		9,661	10,700
0600	Frozen food, chest type, 12' long		3.30	4.848		7,575	286		7,861	8,775
0700	Glass door, reach-in, 5 door		3	5.333		10,400	315		10,715	11,900
0800	Island case, 12' long, single deck		3.30	4.848		7,900	286		8,186	9,125
0900	Multi-deck		3	5.333		10,100	315		10,415	11,600
1000	Meat case, 12' long, single deck		3.30	4.848		7,875	286		8,161	9,100
1050	Multi-deck		3.10	5.161		10,200	305		10,505	11,700
1100	Produce, 12' long, single deck		3.30	4.848		6,475	286		6,761	7,575

11 41 Foodservice Storage Equipment

11 41 13 – Refrigerated Food Storage Cases

11 41 13.10 Refrigerated Food Cases			Crew	Daily Output	Labor-Hours	Unit	Material	Labor	Equipment	Total	Total Incl O&P
							2020 Bare Costs				
1200	Multi-deck		Q-5	3.10	5.161	Ea.	8,550	305		8,855	9,875

11 41 13.20 Refrigerated Food Storage Equipment

			Crew	Daily Output	Labor-Hours	Unit	Material	Labor	Equipment	Total	Total Incl O&P
0010	**REFRIGERATED FOOD STORAGE EQUIPMENT**										
2350	Cooler, reach-in, beverage, 6′ long		Q-1	6	2.667	Ea.	3,125	155		3,280	3,675
4300	Freezers, reach-in, 44 C.F.			4	4		4,425	232		4,657	5,225
4500	68 C.F.			3	5.333		5,575	310		5,885	6,600
4600	Freezer, pre-fab, 8′ x 8′ w/refrigeration		2 Carp	.45	35.556		9,200	1,900		11,100	13,100
4620	8′ x 12′			.35	45.714		10,900	2,425		13,325	15,900
4640	8′ x 16′			.25	64		13,200	3,400		16,600	20,000
4660	8′ x 20′			.17	94.118		20,500	5,000		25,500	30,500
4680	Reach-in, 1 compartment		Q-1	4	4		3,350	232		3,582	4,025
4685	Energy star rated	G	R-18	7.80	3.333		3,575	180		3,755	4,200
4700	2 compartment		Q-1	3	5.333		5,000	310		5,310	5,975
4705	Energy star rated	G	R-18	6.20	4.194		3,750	226		3,976	4,475
4710	3 compartment		Q-1	3	5.333		6,225	310		6,535	7,325
4715	Energy star rated	G	R-18	5.60	4.643		5,250	250		5,500	6,150
8320	Refrigerator, reach-in, 1 compartment			7.80	3.333		3,600	180		3,780	4,225
8325	Energy star rated	G		7.80	3.333		3,575	180		3,755	4,200
8330	2 compartment			6.20	4.194		3,425	226		3,651	4,125
8335	Energy star rated	G		6.20	4.194		3,750	226		3,976	4,475
8340	3 compartment			5.60	4.643		4,825	250		5,075	5,675
8345	Energy star rated	G		5.60	4.643		5,250	250		5,500	6,150
8350	Pre-fab, with refrigeration, 8′ x 8′		2 Carp	.45	35.556		6,950	1,900		8,850	10,700
8360	8′ x 12′			.35	45.714		8,900	2,425		11,325	13,700
8370	8′ x 16′			.25	64		11,500	3,400		14,900	18,200
8380	8′ x 20′			.17	94.118		14,600	5,000		19,600	24,100
8390	Pass-thru/roll-in, 1 compartment		R-18	7.80	3.333		4,100	180		4,280	4,800
8400	2 compartment			6.24	4.167		5,375	224		5,599	6,275
8410	3 compartment			5.60	4.643		9,625	250		9,875	11,000
8420	Walk-in, alum, door & floor only, no refrig, 6′ x 6′ x 7′-6″		2 Carp	1.40	11.429		6,800	605		7,405	8,450
8430	10′ x 6′ x 7′-6″			.55	29.091		9,475	1,550		11,025	12,900
8440	12′ x 14′ x 7′-6″			.25	64		13,200	3,400		16,600	20,000
8450	12′ x 20′ x 7′-6″			.17	94.118		13,100	5,000		18,100	22,400
8451	12′ x 20′ x 7′-6″			.17	94.118		6,525	5,000		11,525	15,200
8460	Refrigerated cabinets, mobile						6,350			6,350	6,975
8470	Refrigerator/freezer, reach-in, 1 compartment		R-18	5.60	4.643		6,950	250		7,200	8,000
8480	2 compartment		"	4.80	5.417		7,575	292		7,867	8,800

11 41 13.30 Wine Cellar

		Crew	Daily Output	Labor-Hours	Unit	Material	Labor	Equipment	Total	Total Incl O&P
0010	**WINE CELLAR**, refrigerated, Redwood interior, carpeted, walk-in type									
0020	6′-8″ high, including racks									
0200	80″ W x 48″ D for 900 bottles	2 Carp	1.50	10.667	Ea.	4,200	565		4,765	5,525
0250	80″ W x 72″ D for 1300 bottles		1.33	12.030		5,725	640		6,365	7,325
0300	80″ W x 94″ D for 1900 bottles		1.17	13.675		6,725	725		7,450	8,575
0400	80″ W x 124″ D for 2500 bottles		1	16		7,800	850		8,650	9,950
0600	Portable cabinets, red oak, reach-in temp. & humidity controlled									
0650	26-5/8″ W x 26-1/2″ D x 68″ H for 235 bottles				Ea.	4,050			4,050	4,450
0660	32″ W x 21-1/2″ D x 73-1/2″ H for 144 bottles					2,875			2,875	3,175
0670	32″ W x 29-1/2″ D x 73-1/2″ H for 288 bottles					4,800			4,800	5,275
0680	39-1/2″ W x 29-1/2″ D x 86-1/2″ H for 440 bottles					4,650			4,650	5,100
0690	52-1/2″ W x 29-1/2″ D x 73-1/2″ H for 468 bottles					5,200			5,200	5,725
0700	52-1/2″ W x 29-1/2″ D x 86-1/2″ H for 572 bottles					5,575			5,575	6,125
0730	Portable, red oak, can be built-in with glass door									

11 41 Foodservice Storage Equipment

11 41 13 – Refrigerated Food Storage Cases

11 41 13.30 Wine Cellar		Crew	Daily Output	Labor-Hours	Unit	Material	2020 Bare Costs Labor	Equipment	Total	Total Incl O&P
0750	23-7/8" W x 24" D x 34-1/2" H for 50 bottles				Ea.	1,050			1,050	1,150

11 41 33 – Foodservice Shelving

11 41 33.20 Metal Food Storage Shelving		Crew	Daily Output	Labor-Hours	Unit	Material	Labor	Equipment	Total	Total Incl O&P
0010	METAL FOOD STORAGE SHELVING									
8600	Stainless steel shelving, louvered 4-tier, 20" x 3'	1 Clab	6	1.333	Ea.	1,500	56		1,556	1,750
8605	20" x 4'		6	1.333		1,650	56		1,706	1,925
8610	20" x 6'		6	1.333		1,750	56		1,806	2,025
8615	24" x 3'		6	1.333		2,050	56		2,106	2,350
8620	24" x 4'		6	1.333		2,525	56		2,581	2,875
8625	24" x 6'		6	1.333		3,450	56		3,506	3,900
8630	Flat 4-tier, 20" x 3'		6	1.333		1,300	56		1,356	1,525
8635	20" x 4'		6	1.333		1,400	56		1,456	1,650
8640	20" x 5'		6	1.333		1,675	56		1,731	1,950
8645	24" x 3'		6	1.333		885	56		941	1,050
8650	24" x 4'		6	1.333		2,300	56		2,356	2,650
8655	24" x 6'		6	1.333		2,825	56		2,881	3,200
8700	Galvanized shelving, louvered 4-tier, 20" x 3'		6	1.333		810	56		866	980
8705	20" x 4'		6	1.333		900	56		956	1,075
8710	20" x 6'		6	1.333		940	56		996	1,125
8715	24" x 3'		6	1.333		790	56		846	960
8720	24" x 4'		6	1.333		945	56		1,001	1,125
8725	24" x 6'		6	1.333		1,350	56		1,406	1,600
8730	Flat 4-tier, 20" x 3'		6	1.333		750	56		806	915
8735	20" x 4'		6	1.333		685	56		741	840
8740	20" x 6'		6	1.333		985	56		1,041	1,175
8745	24" x 3'		6	1.333		770	56		826	935
8750	24" x 4'		6	1.333		570	56		626	715
8755	24" x 6'		6	1.333		955	56		1,011	1,150
8760	Stainless steel dunnage rack, 24" x 3'		8	1		325	42		367	430
8765	24" x 4'		8	1		400	42		442	510
8770	Galvanized dunnage rack, 24" x 3'		8	1		146	42		188	228
8775	24" x 4'	↓	8	1	↓	191	42		233	279

11 42 Food Preparation Equipment

11 42 10 – Commercial Food Preparation Equipment

11 42 10.10 Choppers, Mixers and Misc. Equipment		Crew	Daily Output	Labor-Hours	Unit	Material	Labor	Equipment	Total	Total Incl O&P
0010	CHOPPERS, MIXERS AND MISC. EQUIPMENT									
1700	Choppers, 5 pounds	R-18	7	3.714	Ea.	3,375	200		3,575	4,000
1720	16 pounds		5	5.200		2,550	280		2,830	3,225
1740	35 to 40 pounds	↓	4	6.500		3,675	350		4,025	4,600
1840	Coffee brewer, 5 burners	1 Plum	3	2.667		960	172		1,132	1,325
1850	Coffee urn, twin 6 gallon urns		2	4		2,525	258		2,783	3,175
1860	Single, 3 gallon	↓	3	2.667		1,825	172		1,997	2,275
3000	Fast food equipment, total package, minimum	6 Skwk	.08	600		220,000	32,900		252,900	294,500
3100	Maximum	"	.07	686		286,500	37,600		324,100	375,000
3800	Food mixers, bench type, 20 quarts	L-7	7	4		3,150	205		3,355	3,775
3850	40 quarts		5.40	5.185		8,325	265		8,590	9,575
3900	60 quarts		5	5.600		11,500	287		11,787	13,200
4040	80 quarts		3.90	7.179		17,300	365		17,665	19,600
4100	Floor type, 20 quarts	↓	15	1.867	↓	3,725	95.50		3,820.50	4,250

11 42 Food Preparation Equipment

11 42 10 – Commercial Food Preparation Equipment

11 42 10.10 Choppers, Mixers and Misc. Equipment		Crew	Daily Output	Labor-Hours	Unit	Material	Labor	Equipment	Total	Total Incl O&P
4120	60 quarts	L-7	14	2	Ea.	11,000	102		11,102	12,300
4140	80 quarts		12	2.333		19,200	119		19,319	21,400
4160	140 quarts		8.60	3.256		27,100	167		27,267	30,100
6700	Peelers, small	R-18	8	3.250		2,100	175		2,275	2,575
6720	Large	"	6	4.333		3,175	233		3,408	3,850
6800	Pulper/extractor, close coupled, 5 HP	1 Plum	1.90	4.211		2,475	271		2,746	3,150
8580	Slicer with table	R-18	9	2.889		4,325	156		4,481	5,025

(2020 Bare Costs: Material, Labor, Equipment, Total)

11 43 Food Delivery Carts and Conveyors

11 43 13 – Food Delivery Carts

11 43 13.10 Mobile Carts, Racks and Trays

		Crew	Daily Output	Labor-Hours	Unit	Material	Labor	Equipment	Total	Total Incl O&P
0010	**MOBILE CARTS, RACKS AND TRAYS**									
1650	Cabinet, heated, 1 compartment, reach-in	R-18	5.60	4.643	Ea.	4,375	250		4,625	5,175
1655	Pass-thru roll-in		5.60	4.643		3,475	250		3,725	4,200
1660	2 compartment, reach-in		4.80	5.417		5,975	292		6,267	7,025
1670	Mobile					3,925			3,925	4,325
2000	Hospital food cart, hot and cold service, 20 tray capacity					15,500			15,500	17,000
6850	Mobile rack w/pan slide					1,525			1,525	1,675
9180	Tray and silver dispenser, mobile	1 Clab	16	.500		675	21		696	780

11 44 Food Cooking Equipment

11 44 13 – Commercial Ranges

11 44 13.10 Cooking Equipment

		Crew	Daily Output	Labor-Hours	Unit	Material	Labor	Equipment	Total	Total Incl O&P
0010	**COOKING EQUIPMENT**									
0020	Bake oven, gas, one section	Q-1	8	2	Ea.	5,675	116		5,791	6,425
0300	Two sections		7	2.286		9,375	133		9,508	10,500
0600	Three sections		6	2.667		12,400	155		12,555	13,800
0900	Electric convection, single deck	L-7	4	7		5,575	360		5,935	6,700
1300	Broiler, without oven, standard	Q-1	8	2		3,650	116		3,766	4,200
1550	Infrared	L-7	4	7		7,500	360		7,860	8,825
4750	Fryer, with twin baskets, modular model	Q-1	7	2.286		1,525	133		1,658	1,875
5000	Floor model, on 6" legs	"	5	3.200		2,550	186		2,736	3,075
5100	Extra single basket, large					51			51	56
5170	Energy star rated, 50 lb. capacity [G]	R-18	4	6.500		5,025	350		5,375	6,075
5175	85 lb. capacity [G]	"	4	6.500		7,725	350		8,075	9,025
5300	Griddle, SS, 24" plate, w/4" legs, elec, 208 V, 3 phase, 3' long	Q-1	7	2.286		2,300	133		2,433	2,750
5550	4' long	"	6	2.667		2,450	155		2,605	2,925
6200	Iced tea brewer	1 Plum	3.44	2.326		705	150		855	1,000
6350	Kettle, w/steam jacket, tilting, w/positive lock, SS, 20 gallons	L-7	7	4		8,650	205		8,855	9,850
6600	60 gallons	"	6	4.667		17,100	239		17,339	19,200
6900	Range, restaurant type, 6 burners and 1 standard oven, 36" wide	Q-1	7	2.286		2,675	133		2,808	3,150
6950	Convection		7	2.286		4,800	133		4,933	5,500
7150	2 standard ovens, 24" griddle, 60" wide		6	2.667		5,225	155		5,380	6,000
7200	1 standard, 1 convection oven		6	2.667		9,750	155		9,905	10,900
7450	Heavy duty, single 34" standard oven, open top		5	3.200		5,400	186		5,586	6,225
7500	Convection oven		5	3.200		6,300	186		6,486	7,200
7700	Griddle top		6	2.667		2,400	155		2,555	2,900
7750	Convection oven		6	2.667		3,125	155		3,280	3,675
7760	Induction cooker, electric	L-7	7	4		1,850	205		2,055	2,350

11 44 Food Cooking Equipment

11 44 13 – Commercial Ranges

11 44 13.10 Cooking Equipment		Crew	Daily Output	Labor-Hours	Unit	Material	Labor	Equipment	Total	Total Incl O&P
						2020 Bare Costs				
8850	Steamer, electric 27 KW	L-7	7	4	Ea.	13,300	205		13,505	14,900
9100	Electric, 10 KW or gas 100,000 BTU		5	5.600		7,625	287		7,912	8,825
9150	Toaster, conveyor type, 16-22 slices/minute					1,200			1,200	1,325
9160	Pop-up, 2 slot					99			99	109
9200	For deluxe models of above equipment, add					75%				
9400	Rule of thumb: Equipment cost based									
9410	on kitchen work area									
9420	Office buildings, minimum	L-7	77	.364	S.F.	97	18.60		115.60	137
9450	Maximum		58	.483		158	24.50		182.50	213
9550	Public eating facilities, minimum		77	.364		127	18.60		145.60	170
9600	Maximum		46	.609		207	31		238	278
9750	Hospitals, minimum		58	.483		126	24.50		150.50	178
9800	Maximum		39	.718		231	36.50		267.50	315

11 46 Food Dispensing Equipment

11 46 13 – Bar Equipment

11 46 13.10 Bar Equipment

		Crew	Daily Output	Labor-Hours	Unit	Material	Labor	Equipment	Total	Total Incl O&P
0010	**BAR EQUIPMENT**									
0100	Bar die, flat wall type 41" high	Q-1	15	1.067	L.F.	197	62		259	315
0150	Blender station with sink, 14"		6	2.667	Ea.	650	155		805	955
0200	Cocktail station 36"		6	2.667		1,350	155		1,505	1,750
0250	Drainboard with glass rack		6	2.667		425	155		580	710
0300	Glass storage cabinet	1 Clab	6	1.333		670	56		726	825
0350	Glass and plate chiller, 4.4 C.F.	Q-1	6	2.667		2,425	155		2,580	2,900

11 46 16 – Service Line Equipment

11 46 16.10 Commercial Food Dispensing Equipment

		Crew	Daily Output	Labor-Hours	Unit	Material	Labor	Equipment	Total	Total Incl O&P
0010	**COMMERCIAL FOOD DISPENSING EQUIPMENT**									
1050	Butter pat dispenser	1 Clab	13	.615	Ea.	865	26		891	990
1100	Bread dispenser, counter top		13	.615		825	26		851	950
1900	Cup and glass dispenser, drop in		4	2		560	84		644	750
1920	Disposable cup, drop in		16	.500		750	21		771	855
2650	Dish dispenser, drop in, 12"		11	.727		2,850	30.50		2,880.50	3,200
2660	Mobile		10	.800		3,275	33.50		3,308.50	3,650
3300	Food warmer, counter, 1.2 KW					765			765	840
3550	1.6 KW					2,200			2,200	2,425
3600	Well, hot food, built-in, rectangular, 12" x 20"	R-30	10	2.600		330	129		459	570
3610	Circular, 7 qt.		10	2.600		400	129		529	645
3620	Refrigerated, 2 compartments		10	2.600		2,825	129		2,954	3,300
3630	3 compartments		9	2.889		3,975	143		4,118	4,600
3640	4 compartments		8	3.250		4,750	161		4,911	5,475
4720	Frost cold plate		9	2.889		22,700	143		22,843	25,100
5700	Hot chocolate dispenser	1 Plum	4	2		1,175	129		1,304	1,500
5750	Ice dispenser 567 pound	Q-1	6	2.667		6,075	155		6,230	6,950
6250	Jet spray dispenser	R-18	4.50	5.778		2,125	310		2,435	2,825
6300	Juice dispenser, concentrate	"	4.50	5.778		1,975	310		2,285	2,650
6690	Milk dispenser, bulk, 2 flavor	R-30	8	3.250		2,100	161		2,261	2,575
6695	3 flavor	"	8	3.250		2,650	161		2,811	3,175
8800	Serving counter, straight	1 Carp	40	.200	L.F.	2,000	10.65		2,010.65	2,225
8820	Curved section	"	30	.267	"	2,225	14.15		2,239.15	2,475
8825	Solid surface, see Section 12 36 61.16									

11 46 Food Dispensing Equipment

11 46 16 – Service Line Equipment

11 46 16.10 Commercial Food Dispensing Equipment		Crew	Daily Output	Labor-Hours	Unit	Material	2020 Bare Costs Labor	Equipment	Total	Total Incl O&P
8860	Sneeze guard with lights, 60" L	1 Clab	16	.500	Ea.	325	21		346	390
8900	Sneeze guard, stainless steel and glass, single sided									
8910	Portable, 48" W				Ea.	395			395	435
8920	Portable, 72" W					445			445	490
8930	Adjustable, 36" W	1 Carp	24	.333		360	17.70		377.70	425
8940	Adjustable, 48" W	"	20	.400		320	21.50		341.50	385
9100	Soft serve ice cream machine, medium	R-18	11	2.364		8,125	127		8,252	9,125
9110	Large	"	9	2.889		22,200	156		22,356	24,600

11 46 83 – Ice Machines

11 46 83.10 Commercial Ice Equipment		Crew	Daily Output	Labor-Hours	Unit	Material	Labor	Equipment	Total	Total Incl O&P
0010	**COMMERCIAL ICE EQUIPMENT**									
5800	Ice cube maker, 50 lbs./day	Q-1	6	2.667	Ea.	1,800	155		1,955	2,225
5810	65 lbs./day, energy star rated		6	2.667		1,675	155		1,830	2,075
5900	250 lbs./day		1.20	13.333		2,525	775		3,300	3,975
5950	300 lbs./day, remote condensing		1.20	13.333		2,500	775		3,275	3,950
6050	500 lbs./day		4	4		2,600	232		2,832	3,200
6060	With bin		1.20	13.333		3,950	775		4,725	5,550
6070	Modular, with bin and condenser		1.20	13.333		4,125	775		4,900	5,725
6090	1000 lbs./day, with bin		1	16		5,300	930		6,230	7,250
6100	Ice flakers, 300 lbs./day		1.60	10		3,750	580		4,330	5,025
6120	600 lbs./day		.95	16.842		4,300	975		5,275	6,250
6130	1000 lbs./day		.75	21.333		4,925	1,225		6,150	7,325
6140	2000 lbs./day		.65	24.615		22,200	1,425		23,625	26,600
6160	Ice storage bin, 500 pound capacity	Q-5	1	16		1,025	945		1,970	2,575
6180	1000 pound	"	.56	28.571		3,050	1,675		4,725	5,975

11 48 Foodservice Cleaning and Disposal Equipment

11 48 13 – Commercial Dishwashers

11 48 13.10 Dishwashers		Crew	Daily Output	Labor-Hours	Unit	Material	Labor	Equipment	Total	Total Incl O&P
0010	**DISHWASHERS**									
2700	Dishwasher, commercial, rack type									
2720	10 to 12 racks/hour	Q-1	3.20	5	Ea.	3,450	290		3,740	4,250
2730	Energy star rated, 35 to 40 racks/hour G		1.30	12.308		5,500	715		6,215	7,150
2740	50 to 60 racks/hour G		1.30	12.308		10,600	715		11,315	12,700
2800	Automatic, 190 to 230 racks/hour	L-6	.35	34.286		14,600	2,175		16,775	19,400
2820	235 to 275 racks/hour		.25	48		28,000	3,050		31,050	35,500
2840	8,750 to 12,500 dishes/hour		.10	120		47,500	7,600		55,100	64,500
2950	Dishwasher hood, canopy type	L-3A	10	1.200	L.F.	1,250	69		1,319	1,475
2960	Pant leg type	"	2.50	4.800	Ea.	7,825	276		8,101	9,075
5200	Garbage disposal 1.5 HP, 100 GPH	L-1	4.80	3.333		1,875	210		2,085	2,375
5210	3 HP, 120 GPH		4.60	3.478		2,400	219		2,619	3,000
5220	5 HP, 250 GPH		4.50	3.556		2,975	224		3,199	3,625
6750	Pot sink, 3 compartment	1 Plum	7.25	1.103	L.F.	910	71		981	1,100
6760	Pot washer, low temp wash/rinse		1.60	5	Ea.	5,375	320		5,695	6,425
6770	High pressure wash, high temperature rinse		1.20	6.667		37,200	430		37,630	41,600
9170	Trash compactor, small, up to 125 lb. compacted weight	L-4	4	6		21,500	299		21,799	24,200
9175	Large, up to 175 lb. compacted weight	"	3	8		28,500	400		28,900	31,900

11 52 Audio-Visual Equipment

11 52 13 – Projection Screens

11 52 13.10 Projection Screens, Wall or Ceiling Hung		Crew	Daily Output	Labor-Hours	Unit	2020 Bare Costs Material	Labor	Equipment	Total	Total Incl O&P
0010	**PROJECTION SCREENS, WALL OR CEILING HUNG**, matte white									
0100	Manually operated, economy	2 Carp	500	.032	S.F.	6.55	1.70		8.25	9.90
0300	Intermediate		450	.036		7.90	1.89		9.79	11.75
0400	Deluxe		400	.040		10	2.13		12.13	14.40
0600	Electric operated, matte white, 25 S.F., economy		5	3.200	Ea.	1,125	170		1,295	1,525
0700	Deluxe		4	4		1,325	213		1,538	1,800
0900	50 S.F., economy		3	5.333		1,050	283		1,333	1,600
1000	Deluxe		2	8		2,975	425		3,400	3,950
1200	Heavy duty, electric operated, 200 S.F.		1.50	10.667		4,250	565		4,815	5,575
1300	400 S.F.		1	16		5,175	850		6,025	7,050
1500	Rigid acrylic in wall, for rear projection, 1/4" thick	2 Glaz	30	.533	S.F.	81	27		108	133
1600	1/2" thick (maximum size 10' x 20')	"	25	.640	"	82	32.50		114.50	142
9000	Minimum labor/equipment charge	2 Carp	3	5.333	Job		283		283	455

11 52 16 – Projectors

11 52 16.10 Movie Equipment		Crew	Daily Output	Labor-Hours	Unit	Material	Labor	Equipment	Total	Total Incl O&P
0010	**MOVIE EQUIPMENT**									
0020	Changeover, minimum				Ea.	520			520	570
0100	Maximum					1,000			1,000	1,100
0400	Film transport, incl. platters and autowind, minimum					5,575			5,575	6,125
0500	Maximum					15,800			15,800	17,400
0800	Lamphouses, incl. rectifiers, xenon, 1,000 watt	1 Elec	2	4		7,350	245		7,595	8,475
0900	1,600 watt		2	4		7,850	245		8,095	9,025
1000	2,000 watt		1.50	5.333		8,675	325		9,000	10,100
1100	4,000 watt		1.50	5.333		11,700	325		12,025	13,400
1400	Lenses, anamorphic, minimum					1,400			1,400	1,550
1500	Maximum					3,175			3,175	3,475
1800	Flat 35 mm, minimum					1,225			1,225	1,350
1900	Maximum					1,900			1,900	2,075
2200	Pedestals, for projectors					1,700			1,700	1,875
2300	Console type					12,200			12,200	13,500
2600	Projector mechanisms, incl. soundhead, 35 mm, minimum					12,600			12,600	13,900
2700	Maximum					17,400			17,400	19,100
3000	Projection screens, rigid, in wall, acrylic, 1/4" thick	2 Glaz	195	.082	S.F.	48.50	4.18		52.68	59.50
3100	1/2" thick	"	130	.123	"	55.50	6.30		61.80	71
3300	Electric operated, heavy duty, 400 S.F.	2 Carp	1	16	Ea.	3,225	850		4,075	4,900
3320	Theater projection screens, matte white, including frames	"	200	.080	S.F.	7.70	4.25		11.95	15.25
3400	Also see Section 11 52 13.10									
3700	Sound systems, incl. amplifier, mono, minimum	1 Elec	.90	8.889	Ea.	3,825	545		4,370	5,050
3800	Dolby/Super Sound, maximum		.40	20		20,600	1,225		21,825	24,600
4100	Dual system, 2 channel, front surround, minimum		.70	11.429		5,275	700		5,975	6,875
4200	Dolby/Super Sound, 4 channel, maximum		.40	20		19,700	1,225		20,925	23,600
4500	Sound heads, 35 mm					6,125			6,125	6,750
4900	Splicer, tape system, minimum					855			855	940
5000	Tape type, maximum					1,525			1,525	1,675
5300	Speakers, recessed behind screen, minimum	1 Elec	2	4		1,225	245		1,470	1,725
5400	Maximum	"	1	8		3,525	490		4,015	4,625
5700	Seating, painted steel, upholstered, minimum	2 Carp	35	.457		151	24.50		175.50	205
5800	Maximum	"	28	.571		550	30.50		580.50	655
6100	Rewind tables, minimum					3,050			3,050	3,350
6200	Maximum					5,425			5,425	5,975
7000	For automation, varying sophistication, minimum	1 Elec	1	8	System	2,725	490		3,215	3,750
7100	Maximum	2 Elec	.30	53.333	"	6,350	3,275		9,625	12,000

11 52 Audio-Visual Equipment

11 52 16 – Projectors

11 52 16.20 Movie Equipment- Digital		Crew	Daily Output	Labor-Hours	Unit	2020 Bare Costs Material	Labor	Equipment	Total	Total Incl O&P
0010	**MOVIE EQUIPMENT- DIGITAL**									
1000	Digital 2K projection system, 98" DMD	1 Elec	2	4	Ea.	53,000	245		53,245	58,500
1100	OEM lens		2	4		6,050	245		6,295	7,050
2000	Pedestal with power distribution		2	4		2,325	245		2,570	2,950
3000	Software		2	4		1,975	245		2,220	2,550

11 53 Laboratory Equipment

11 53 03 – Laboratory Test Equipment

11 53 03.13 Test Equipment

		Crew	Daily Output	Labor-Hours	Unit	Material	Labor	Equipment	Total	Total Incl O&P
0010	**TEST EQUIPMENT**									
1700	Thermometer, electric, portable				Ea.	340			340	375
1800	Titration unit, four 2000 ml reservoirs				"	6,275			6,275	6,900

11 53 13 – Laboratory Fume Hoods

11 53 13.13 Recirculating Laboratory Fume Hoods

		Crew	Daily Output	Labor-Hours	Unit	Material	Labor	Equipment	Total	Total Incl O&P
0010	**RECIRCULATING LABORATORY FUME HOODS**									
0600	Fume hood, with countertop & base, not including HVAC									
0610	Simple, minimum	2 Carp	5.40	2.963	L.F.	580	157		737	890
0620	Complex, including fixtures		2.40	6.667		1,125	355		1,480	1,825
0630	Special, maximum		1.70	9.412		1,325	500		1,825	2,250
0670	Service fixtures, average				Ea.	365			365	405
0680	For sink assembly with hot and cold water, add	1 Plum	1.40	5.714		780	370		1,150	1,425
0750	Glove box, fiberglass, bacteriological					16,700			16,700	18,400
0760	Controlled atmosphere					21,300			21,300	23,400
0770	Radioisotope					16,700			16,700	18,400
0780	Carcinogenic					16,700			16,700	18,400

11 53 16 – Laboratory Incubators

11 53 16.13 Incubators

		Crew	Daily Output	Labor-Hours	Unit	Material	Labor	Equipment	Total	Total Incl O&P
0010	**INCUBATORS**									
1000	Incubators, minimum				Ea.	3,225			3,225	3,550
1010	Maximum				"	15,700			15,700	17,200

11 53 19 – Laboratory Sterilizers

11 53 19.13 Sterilizers

		Crew	Daily Output	Labor-Hours	Unit	Material	Labor	Equipment	Total	Total Incl O&P
0010	**STERILIZERS**									
0700	Glassware washer, undercounter, minimum	L-1	1.80	8.889	Ea.	6,850	560		7,410	8,400
0710	Maximum	"	1	16		15,200	1,000		16,200	18,300
1850	Utensil washer-sanitizer	1 Plum	2	4		8,725	258		8,983	10,000

11 53 23 – Laboratory Refrigerators

11 53 23.13 Refrigerators

		Crew	Daily Output	Labor-Hours	Unit	Material	Labor	Equipment	Total	Total Incl O&P
0010	**REFRIGERATORS**									
1200	Blood bank, 28.6 C.F. emergency signal				Ea.	14,400			14,400	15,900
1210	Reach-in, 16.9 C.F.				"	9,400			9,400	10,300

11 53 33 – Emergency Safety Appliances

11 53 33.13 Emergency Equipment

		Crew	Daily Output	Labor-Hours	Unit	Material	Labor	Equipment	Total	Total Incl O&P
0010	**EMERGENCY EQUIPMENT**									
1400	Safety equipment, eye wash, hand held				Ea.	400			400	440
1450	Deluge shower				"	840			840	920

11 53 Laboratory Equipment

11 53 43 – Service Fittings and Accessories

11 53 43.13 Fittings		Crew	Daily Output	Labor-Hours	Unit	2020 Bare Costs Material	Labor	Equipment	Total	Total Incl O&P
0010	**FITTINGS**									
1600	Sink, one piece plastic, flask wash, hose, free standing	1 Plum	1.60	5	Ea.	2,075	320		2,395	2,775
1610	Epoxy resin sink, 25" x 16" x 10"	"	2	4	"	227	258		485	650
1950	Utility table, acid resistant top with drawers	2 Carp	30	.533	L.F.	179	28.50		207.50	243
8000	Alternate pricing method: as percent of lab furniture									
8050	Installation, not incl. plumbing & duct work				% Furn.				22%	22%
8100	Plumbing, final connections, simple system								10%	10%
8110	Moderately complex system								15%	15%
8120	Complex system								20%	20%
8150	Electrical, simple system								10%	10%
8160	Moderately complex system								20%	20%
8170	Complex system								35%	35%

11 53 53 – Biological Safety Cabinets

11 53 53.10 Pharmacy Cabinets		Crew	Daily Output	Labor-Hours	Unit	Material	Labor	Equipment	Total	Total Incl O&P
0010	**PHARMACY CABINETS**, vertical flow									
0100	Class II, type B2, 6' L	2 Carp	1.50	10.667	Ea.	15,400	565		15,965	17,800

11 57 Vocational Shop Equipment

11 57 10 – Shop Equipment

11 57 10.10 Vocational School Shop Equipment		Crew	Daily Output	Labor-Hours	Unit	Material	Labor	Equipment	Total	Total Incl O&P
0010	**VOCATIONAL SCHOOL SHOP EQUIPMENT**									
0020	Benches, work, wood, average	2 Carp	5	3.200	Ea.	755	170		925	1,100
0100	Metal, average		5	3.200		440	170		610	755
0400	Combination belt & disc sander, 6"		4	4		1,700	213		1,913	2,225
0700	Drill press, floor mounted, 12", 1/2 HP		4	4		470	213		683	855
0800	Dust collector, not incl. ductwork, 6" diameter	1 Shee	1.10	7.273		5,925	455		6,380	7,225
0810	Dust collector bag, 20" diameter	"	5	1.600		600	99.50		699.50	820
1000	Grinders, double wheel, 1/2 HP	2 Carp	5	3.200		239	170		409	535
1300	Jointer, 4", 3/4 HP		4	4		1,500	213		1,713	2,000
1600	Kilns, 16 C.F., to 2000°		4	4		1,875	213		2,088	2,425
1900	Lathe, woodworking, 10", 1/2 HP		4	4		580	213		793	975
2200	Planer, 13" x 6"		4	4		1,325	213		1,538	1,825
2500	Potter's wheel, motorized		4	4		1,150	213		1,363	1,625
2800	Saws, band, 14", 3/4 HP		4	4		1,150	213		1,363	1,600
3100	Metal cutting band saw, 14"		4	4		2,275	213		2,488	2,850
3400	Radial arm saw, 10", 2 HP		4	4		1,350	213		1,563	1,850
3700	Scroll saw, 24"		4	4		625	213		838	1,025
4000	Table saw, 10", 3 HP		4	4		3,150	213		3,363	3,825
4300	Welder AC arc, 30 amp capacity		4	4		3,500	213		3,713	4,175

11 61 Broadcast, Theater, and Stage Equipment

11 61 23 – Folding and Portable Stages

11 61 23.10 Portable Stages		Crew	Daily Output	Labor-Hours	Unit	Material	2020 Bare Costs Labor	Equipment	Total	Total Incl O&P
0010	PORTABLE STAGES									
1500	Flooring, portable oak parquet, 3' x 3' sections				S.F.	14.55			14.55	16
1600	Cart to carry 225 S.F. of flooring				Ea.	445			445	490
5000	Stages, portable with steps, folding legs, stock, 8" high				SF Stg.	46.50			46.50	51
5100	16" high					61			61	67
5200	32" high					62			62	68.50
5300	40" high					64.50			64.50	71
6000	Telescoping platforms, extruded alum., straight, minimum	4 Carp	157	.204		38	10.85		48.85	59
6100	Maximum		77	.416		55	22		77	96
6500	Pie-shaped, minimum		150	.213		82	11.35		93.35	108
6600	Maximum		70	.457		91.50	24.50		116	140
6800	For 3/4" plywood covered deck, deduct					5.40			5.40	5.95
7000	Band risers, steel frame, plywood deck, minimum	4 Carp	275	.116		33	6.20		39.20	46.50
7100	Maximum	"	138	.232		83	12.30		95.30	111
7500	Chairs for above, self-storing, minimum	2 Carp	43	.372	Ea.	122	19.80		141.80	166
7600	Maximum	"	40	.400	"	215	21.50		236.50	270

11 61 33 – Rigging Systems and Controls

11 61 33.10 Controls		Crew	Daily Output	Labor-Hours	Unit	Material	Labor	Equipment	Total	Total Incl O&P
0010	CONTROLS									
0050	Control boards with dimmers and breakers, minimum	1 Elec	1	8	Ea.	19,700	490		20,190	22,500
0100	Average		.50	16		55,500	980		56,480	62,500
0150	Maximum		.20	40		148,500	2,450		150,950	167,000
8000	Rule of thumb: total stage equipment, minimum	4 Carp	100	.320	SF Stg.	106	17		123	144
8100	Maximum	"	25	1.280	"	595	68		663	765

11 61 43 – Stage Curtains

11 61 43.10 Curtains		Crew	Daily Output	Labor-Hours	Unit	Material	Labor	Equipment	Total	Total Incl O&P
0010	CURTAINS									
0500	Curtain track, straight, light duty	2 Carp	20	.800	L.F.	31	42.50		73.50	103
0600	Heavy duty		18	.889		67.50	47		114.50	150
0700	Curved sections		12	1.333		201	71		272	335
1000	Curtains, velour, medium weight		600	.027	S.F.	8.75	1.42		10.17	11.90
1150	Silica based yarn, inherently fire retardant		50	.320	"	16.50	17		33.50	45.50

11 62 Musical Equipment

11 62 16 – Carillons

11 62 16.10 Bell Tower Equipment		Crew	Daily Output	Labor-Hours	Unit	Material	Labor	Equipment	Total	Total Incl O&P
0010	BELL TOWER EQUIPMENT									
0300	Carillon, 4 octave (48 bells), with keyboard				System	1,204,500			1,204,500	1,325,000
0320	2 octave (24 bells)					567,000			567,000	623,500
0340	3 to 4 bell peal, minimum					141,500			141,500	156,000
0360	Maximum					850,500			850,500	935,500
0380	Cast bronze bell, average				Ea.	127,500			127,500	140,500
0400	Electronic, digital, minimum					21,300			21,300	23,400
0410	With keyboard, maximum					106,500			106,500	117,000

11 66 Athletic Equipment

11 66 13 – Exercise Equipment

11 66 13.10 Physical Training Equipment		Crew	Daily Output	Labor-Hours	Unit	Material	2020 Bare Costs Labor	Equipment	Total	Total Incl O&P
0010	**PHYSICAL TRAINING EQUIPMENT**									
0020	Abdominal rack, 2 board capacity				Ea.	525			525	575
0050	Abdominal board, upholstered					775			775	850
0200	Bicycle trainer, minimum					610			610	670
0300	Deluxe, electric					4,850			4,850	5,350
0400	Barbell set, chrome plated steel, 25 lb.					305			305	335
0420	100 lb.					620			620	680
0450	200 lb.					740			740	815
0500	Weight plates, cast iron, per lb.				Lb.	3.70			3.70	4.07
0520	Storage rack, 10 station				Ea.	1,225			1,225	1,325
0600	Circuit training apparatus, 12 machines minimum	2 Clab	1.25	12.800	Set	36,800	540		37,340	41,400
0700	Average		1	16		41,700	675		42,375	47,000
0800	Maximum		.75	21.333		50,000	900		50,900	56,500
0820	Dumbbell set, cast iron, with rack and 5 pair					475			475	520
0900	Squat racks	2 Clab	5	3.200	Ea.	825	135		960	1,125
1200	Multi-station gym machine, 5 station					4,900			4,900	5,400
1250	9 station					11,400			11,400	12,600
1280	Rowing machine, hydraulic					1,675			1,675	1,825
1300	Treadmill, manual					1,075			1,075	1,200
1320	Motorized					4,050			4,050	4,450
1340	Electronic					4,100			4,100	4,500
1360	Cardio-testing					5,300			5,300	5,850
1400	Treatment/massage tables, minimum					625			625	690
1420	Deluxe, with accessories					870			870	955
4150	Exercise equipment, bicycle trainer					1,125			1,125	1,250
4180	Chinning bar, adjustable, wall mounted	1 Carp	5	1.600		237	85		322	395
4200	Exercise ladder, 16′ x 1′-7″, suspended	L-2	3	5.333		1,375	248		1,623	1,925
4210	High bar, floor plate attached	1 Carp	4	2		2,725	106		2,831	3,150
4240	Parallel bars, adjustable		4	2		1,625	106		1,731	1,975
4270	Uneven parallel bars, adjustable		4	2		3,950	106		4,056	4,525
4280	Wall mounted, adjustable	L-2	1.50	10.667	Set	955	495		1,450	1,850
4300	Rope, ceiling mounted, 18′ long	1 Carp	3.66	2.186	Ea.	208	116		324	415
4330	Side horse, vaulting		5	1.600		1,600	85		1,685	1,900
4360	Treadmill, motorized, deluxe, training type		5	1.600		3,975	85		4,060	4,475
4390	Weight lifting multi-station, minimum	2 Clab	1	16		305	675		980	1,400
4450	Maximum	″	.50	32		15,000	1,350		16,350	18,700

11 66 23 – Gymnasium Equipment

11 66 23.13 Basketball Equipment

		Crew	Daily Output	Labor-Hours	Unit	Material	Labor	Equipment	Total	Total Incl O&P
0010	**BASKETBALL EQUIPMENT**									
1000	Backstops, wall mtd., 6′ extended, fixed, minimum	L-2	1	16	Ea.	1,750	745		2,495	3,125
1100	Maximum		1	16		2,075	745		2,820	3,475
1200	Swing up, minimum		1	16		1,750	745		2,495	3,125
1250	Maximum		1	16		3,300	745		4,045	4,825
1300	Portable, manual, heavy duty, spring operated		1.90	8.421		13,200	390		13,590	15,200
1400	Ceiling suspended, stationary, minimum		.78	20.513		4,350	955		5,305	6,350
1450	Fold up, with accessories, maximum		.40	40		6,325	1,850		8,175	9,950
1600	For electrically operated, add	1 Elec	1	8		2,600	490		3,090	3,600
5800	Wall pads, 1-1/2″ thick, standard (not fire rated)	2 Carp	640	.025	S.F.	6.85	1.33		8.18	9.70

11 66 23.19 Boxing Ring

		Crew	Daily Output	Labor-Hours	Unit	Material	Labor	Equipment	Total	Total Incl O&P
0010	**BOXING RING**									
4100	Elevated, 22′ x 22′	L-4	.10	240	Ea.	6,150	12,000		18,150	26,000
4110	For cellular plastic foam padding, add		.10	240		1,250	12,000		13,250	20,600

11 66 Athletic Equipment

11 66 23 – Gymnasium Equipment

11 66 23.19 Boxing Ring		Crew	Daily Output	Labor-Hours	Unit	Material	Labor	Equipment	Total	Total Incl O&P
						2020 Bare Costs				
4120	Floor level, including posts and ropes only, 20′ x 20′	L-4	.80	30	Ea.	4,825	1,500		6,325	7,700
4130	Canvas, 30′ x 30′	↓	5	4.800	↓	1,500	239		1,739	2,025

11 66 23.47 Gym Mats		Crew	Daily Output	Labor-Hours	Unit	Material	Labor	Equipment	Total	Total Incl O&P
0010	**GYM MATS**									
5500	2″ thick, naugahyde covered				S.F.	5.35			5.35	5.90
5600	Vinyl/nylon covered					8.55			8.55	9.45
6000	Wrestling mats, 1″ thick, heavy duty				↓	5.05			5.05	5.55

11 66 43 – Interior Scoreboards

11 66 43.10 Scoreboards		Crew	Daily Output	Labor-Hours	Unit	Material	Labor	Equipment	Total	Total Incl O&P
0010	**SCOREBOARDS**									
7000	Baseball, minimum	R-3	1.30	15.385	Ea.	4,125	940	145	5,210	6,150
7200	Maximum		.05	400		19,800	24,400	3,775	47,975	64,000
7300	Football, minimum		.86	23.256		5,225	1,425	220	6,870	8,200
7400	Maximum		.20	100		20,200	6,100	945	27,245	32,700
7500	Basketball (one side), minimum		2.07	9.662		2,475	590	91.50	3,156.50	3,725
7600	Maximum		.30	66.667		8,900	4,075	630	13,605	16,800
7700	Hockey-basketball (four sides), minimum		.25	80		10,100	4,900	755	15,755	19,500
7800	Maximum	↓	.15	133	↓	16,300	8,150	1,250	25,700	31,900

11 66 53 – Gymnasium Dividers

11 66 53.10 Divider Curtains		Crew	Daily Output	Labor-Hours	Unit	Material	Labor	Equipment	Total	Total Incl O&P
0010	**DIVIDER CURTAINS**									
4500	Gym divider curtain, mesh top, vinyl bottom, manual	L-4	500	.048	S.F.	9.90	2.39		12.29	14.75
4700	Electric roll up	L-7	400	.070	″	13.30	3.58		16.88	20.50

11 67 Recreational Equipment

11 67 13 – Bowling Alley Equipment

11 67 13.10 Bowling Alleys		Crew	Daily Output	Labor-Hours	Unit	Material	Labor	Equipment	Total	Total Incl O&P
0010	**BOWLING ALLEYS** Including alley, pinsetter, scorer,									
0020	Counters and misc. supplies, minimum	4 Carp	.20	160	Lane	46,900	8,500		55,400	65,000
0150	Average		.19	168		52,000	8,950		60,950	71,500
0300	Maximum	↓	.18	178		59,500	9,450		68,950	80,500
0400	Combo table ball rack, add					1,400			1,400	1,550
0600	For automatic scorer, add, minimum					6,275			6,275	6,900
0700	Maximum				↓	10,500			10,500	11,500

11 67 23 – Shooting Range Equipment

11 67 23.10 Shooting Range		Crew	Daily Output	Labor-Hours	Unit	Material	Labor	Equipment	Total	Total Incl O&P
0010	**SHOOTING RANGE** Incl. bullet traps, target provisions, controls,									
0100	Separators, ceiling system, etc. Not incl. structural shell									
0200	Commercial	L-9	.64	56.250	Point	36,500	2,700		39,200	44,400
0300	Law enforcement		.28	129		47,600	6,150		53,750	62,500
0400	National Guard armories		.71	50.704		26,800	2,425		29,225	33,400
0500	Reserve training centers		.71	50.704		19,200	2,425		21,625	25,000
0600	Schools and colleges		.32	113		47,800	5,375		53,175	61,000
0700	Major academies	↓	.19	189		63,500	9,050		72,550	84,500
0800	For acoustical treatment, add					10%	10%			
0900	For lighting, add					28%	25%			
1000	For plumbing, add					5%	5%			
1100	For ventilating system, add, minimum					40%	40%			
1200	Add, average				↓	25%	25%			

11 67 Recreational Equipment

11 67 23 – Shooting Range Equipment

11 67 23.10 Shooting Range		Crew	Daily Output	Labor-Hours	Unit	Material	Labor	Equipment	Total	Total Incl O&P
						2020 Bare Costs				
1300	Add, maximum				Point	35%	35%			

11 68 Play Field Equipment and Structures

11 68 13 – Playground Equipment

11 68 13.10 Free-Standing Playground Equipment

		Crew	Daily Output	Labor-Hours	Unit	Material	Labor	Equipment	Total	Total Incl O&P
0010	**FREE-STANDING PLAYGROUND EQUIPMENT** See also individual items									
0200	Bike rack, 10′ long, permanent G	B-1	12	2	Ea.	540	85.50		625.50	730
0240	Climber, arch, 6′ high, 12′ long, 5′ wide		4	6		810	257		1,067	1,300
0260	Fitness trail, with signs, 9 to 10 stations, treated pine, minimum		.25	96		7,725	4,100		11,825	15,100
0270	Maximum		.17	141		23,100	6,050		29,150	35,100
0280	Metal, minimum		.25	96		11,700	4,100		15,800	19,500
0285	Maximum		.17	141		24,600	6,050		30,650	36,700
0300	Redwood, minimum		.25	96		11,800	4,100		15,900	19,600
0310	Maximum		.17	141		12,600	6,050		18,650	23,500
0320	16 to 20 stations, treated pine, minimum		.17	141		15,300	6,050		21,350	26,600
0330	Maximum		.13	185		16,300	7,900		24,200	30,600
0340	Metal, minimum		.17	141		13,200	6,050		19,250	24,300
0350	Maximum		.13	185		20,300	7,900		28,200	34,900
0360	Redwood, minimum		.17	141		17,800	6,050		23,850	29,300
0370	Maximum		.13	185		32,300	7,900		40,200	48,100
0392	Upper body warm-up station		2.60	9.231		3,475	395		3,870	4,425
0394	Bench stepper station		2.60	9.231		1,750	395		2,145	2,550
0396	Standing push up station		2.60	9.231		1,200	395		1,595	1,950
0398	Upper body stretch station		2.60	9.231		2,450	395		2,845	3,325
0400	Horizontal monkey ladder, 14′ long, 6′ high		4	6		1,200	257		1,457	1,725
0590	Parallel bars, 10′ long		4	6		635	257		892	1,100
0600	Posts, tether ball set, 2-3/8″ OD		12	2		430	85.50		515.50	610
0800	Poles, multiple purpose, 10′-6″ long		12	2	Pr.	259	85.50		344.50	420
1000	Ground socket for movable posts, 2-3/8″ post		10	2.400		114	103		217	290
1100	3-1/2″ post		10	2.400		225	103		328	410
1300	See-saw, spring, steel, 2 units		6	4	Ea.	730	171		901	1,075
1400	4 units		4	6		1,275	257		1,532	1,800
1500	6 units		3	8		2,000	340		2,340	2,750
1700	Shelter, fiberglass golf tee, 3 person		4.60	5.217		4,375	223		4,598	5,175
1900	Slides, stainless steel bed, 12′ long, 6′ high		3	8		4,400	340		4,740	5,400
2000	20′ long, 10′ high		2	12		8,250	515		8,765	9,900
2200	Swings, plain seats, 8′ high, 4 seats		2	12		1,675	515		2,190	2,675
2300	8 seats		1.30	18.462		2,925	790		3,715	4,475
2500	12′ high, 4 seats		2	12		2,150	515		2,665	3,175
2600	8 seats		1.30	18.462		4,825	790		5,615	6,600
2800	Whirlers, 8′ diameter		3	8		3,175	340		3,515	4,025
2900	10′ diameter		3	8		6,625	340		6,965	7,850

11 68 13.20 Modular Playground

		Crew	Daily Output	Labor-Hours	Unit	Material	Labor	Equipment	Total	Total Incl O&P
0010	**MODULAR PLAYGROUND** Basic components									
0100	Deck, square, steel, 48″ x 48″	B-1	1	24	Ea.	695	1,025		1,720	2,425
0110	Recycled polyurethane		1	24		690	1,025		1,715	2,400
0120	Triangular, steel, 48″ side		1	24		815	1,025		1,840	2,550
0130	Post, steel, 5″ square		18	1.333	L.F.	51.50	57		108.50	148
0140	Aluminum, 2-3/8″ square		20	1.200		59.50	51.50		111	147
0150	5″ square		18	1.333		50.50	57		107.50	147
0160	Roof, square poly, 54″ side		18	1.333	Ea.	1,975	57		2,032	2,275

11 68 Play Field Equipment and Structures

11 68 13 – Playground Equipment

11 68 13.20 Modular Playground		Crew	Daily Output	Labor-Hours	Unit	2020 Bare Costs Material	Labor	Equipment	Total	Total Incl O&P
0170	Wheelchair transfer module, for 3′ high deck	B-1	3	8	Ea.	3,425	340		3,765	4,325
0180	Guardrail, pipe, 36″ high		60	.400	L.F.	294	17.10		311.10	355
0190	Steps, deck-to-deck, three 8″ steps		8	3	Ea.	1,375	128		1,503	1,700
0200	Activity panel, crawl through panel		2	12		580	515		1,095	1,450
0210	Alphabet/spelling panel		2	12		660	515		1,175	1,550
0360	With guardrails		3	8		2,000	340		2,340	2,750
0370	Crawl tunnel, straight, 56″ long		4	6		1,375	257		1,632	1,925
0380	90°, 4′ long		4	6		1,575	257		1,832	2,150
1200	Slide, tunnel, for 56″ high deck		8	3		2,300	128		2,428	2,750
1210	Straight, poly		8	3		520	128		648	775
1220	Stainless steel, 54″ high deck		6	4		995	171		1,166	1,375
1230	Curved, poly, 40″ high deck		6	4		940	171		1,111	1,300
1240	Spiral slide, 56″-72″ high		5	4.800		4,900	205		5,105	5,725
1300	Ladder, vertical, for 24″-72″ high deck		5	4.800		545	205		750	930
1310	Horizontal, 8′ long		5	4.800		1,050	205		1,255	1,500
1320	Corkscrew climber, 6′ high		3	8		1,200	340		1,540	1,875
1330	Fire pole for 72″ high deck		6	4		360	171		531	670
1340	Bridge, ring climber, 8′ long		4	6		2,775	257		3,032	3,450
1350	Suspension		4	6	L.F.	375	257		632	820

11 68 16 – Play Structures

11 68 16.10 Handball/Squash Court		Crew	Daily Output	Labor-Hours	Unit	Material	Labor	Equipment	Total	Total Incl O&P
0010	**HANDBALL/SQUASH COURT**, outdoor									
0900	Handball or squash court, outdoor, wood	2 Carp	.50	32	Ea.	5,075	1,700		6,775	8,325
1000	Masonry handball/squash court	D-1	.30	53.333	"	26,500	2,500		29,000	33,300

11 68 16.30 Platform/Paddle Tennis Court		Crew	Daily Output	Labor-Hours	Unit	Material	Labor	Equipment	Total	Total Incl O&P
0010	**PLATFORM/PADDLE TENNIS COURT** Complete with lighting, etc.									
0100	Aluminum slat deck with aluminum frame	B-1	.08	300	Court	68,500	12,800		81,300	96,000
0500	Aluminum slat deck with wood frame	C-1	.12	267		80,500	13,400		93,900	110,000
0800	Aluminum deck heater, add	B-1	1.18	20.339		2,925	870		3,795	4,625
0900	Douglas fir planking with wood frame 2″ x 6″ x 30′	C-1	.12	267		75,500	13,400		88,900	104,500
1000	Plywood deck with steel frame		.12	267		75,500	13,400		88,900	104,500
1100	Steel slat deck with wood frame		.12	267		45,600	13,400		59,000	71,500

11 68 33 – Athletic Field Equipment

11 68 33.13 Football Field Equipment		Crew	Daily Output	Labor-Hours	Unit	Material	Labor	Equipment	Total	Total Incl O&P
0010	**FOOTBALL FIELD EQUIPMENT**									
0020	Goal posts, steel, football, double post	B-1	1.50	16	Pr.	4,600	685		5,285	6,175
0100	Deluxe, single post		1.50	16		3,525	685		4,210	5,000
0300	Football, convertible to soccer		1.50	16		3,150	685		3,835	4,575
0500	Soccer, regulation		2	12		2,075	515		2,590	3,100

11 71 Medical Sterilizing Equipment

11 71 10 – Medical Sterilizers & Distillers

11 71 10.10 Sterilizers and Distillers		Crew	Daily Output	Labor-Hours	Unit	Material	2020 Bare Costs Labor	Equipment	Total	Total Incl O&P
0010	**STERILIZERS AND DISTILLERS**									
0700	Distiller, water, steam heated, 50 gal. capacity	1 Plum	1.40	5.714	Ea.	25,500	370		25,870	28,700
3010	Portable, top loading, 105-135 degree C, 3 to 30 psi, 50 L chamber					9,775			9,775	10,800
3020	Stainless steel basket, 10.7" diam. x 11.8" H					214			214	236
3025	Stainless steel pail, 10.7" diam. x 10.7" H					320			320	350
3050	85 L chamber					17,800			17,800	19,600
3060	Stainless steel basket, 15.3" diam. x 11.5" H					455			455	505
3065	Stainless steel pail, 15.3" diam. x 11" H					675			675	745
5600	Sterilizers, floor loading, 26" x 62" x 42", single door, steam					155,000			155,000	170,500
5650	Double door, steam					239,000			239,000	262,500
5800	General purpose, 20" x 20" x 38", single door					12,500			12,500	13,700
6000	Portable, counter top, steam, minimum					2,900			2,900	3,175
6020	Maximum					5,075			5,075	5,575
6050	Portable, counter top, gas, 17" x 15" x 32-1/2"					45,400			45,400	49,900
6150	Manual washer/sterilizer, 16" x 16" x 26"	1 Plum	2	4	↓	62,000	258		62,258	69,000
6200	Steam generators, electric 10 kW to 180 kW, freestanding									
6250	Minimum	1 Elec	3	2.667	Ea.	12,200	164		12,364	13,700
6300	Maximum	"	.70	11.429		22,700	700		23,400	26,100
8200	Bed pan washer-sanitizer	1 Plum	2	4	↓	9,775	258		10,033	11,100

11 72 Examination and Treatment Equipment

11 72 13 – Examination Equipment

11 72 13.13 Examination Equipment		Crew	Daily Output	Labor-Hours	Unit	Material	Labor	Equipment	Total	Total Incl O&P
0010	**EXAMINATION EQUIPMENT**									
0300	Blood pressure unit, mercurial, wall				Ea.	158			158	174
0400	Diagnostic set, wall					675			675	740
4400	Scale, physician's, with height rod				↓	465			465	510

11 72 53 – Treatment Equipment

11 72 53.13 Medical Treatment Equipment		Crew	Daily Output	Labor-Hours	Unit	Material	Labor	Equipment	Total	Total Incl O&P
0010	**MEDICAL TREATMENT EQUIPMENT**									
6300	Exam light, portable, 14" flexible arm				Ea.	181			181	199
6500	Surgery table, minor minimum	1 Sswk	.70	11.429		14,900	660		15,560	17,500
6520	Maximum	"	.50	16		21,400	920		22,320	25,000
6700	Surgical lights, doctor's office, single arm	2 Elec	2	8		3,350	490		3,840	4,425
6750	Dual arm	"	1	16	↓	6,325	980		7,305	8,450

11 73 Patient Care Equipment

11 73 10 – Patient Treatment Equipment

11 73 10.10 Treatment Equipment		Crew	Daily Output	Labor-Hours	Unit	Material	Labor	Equipment	Total	Total Incl O&P
0010	**TREATMENT EQUIPMENT**									
0750	Exam room furnishings, average per room				Ea.	5,900			5,900	6,500
1800	Heat therapy unit, humidified, 26" x 78" x 28"				"	4,250			4,250	4,675
2100	Hubbard tank with accessories, stainless steel,									
2110	125 GPM at 45 psi water pressure				Ea.	15,100			15,100	16,600
2150	For electric overhead hoist, add					3,275			3,275	3,600
2900	K-Module for heat therapy, 20 oz. capacity, 75°F to 110°F					600			600	660
3600	Paraffin bath, 126°F, auto controlled					280			280	305
3900	Parallel bars for walking training, 12'-0"					2,050			2,050	2,275
4600	Station, dietary, medium, with ice				↓	22,200			22,200	24,400

11 73 Patient Care Equipment

11 73 10 – Patient Treatment Equipment

11 73 10.10 Treatment Equipment		Crew	Daily Output	Labor-Hours	Unit	Material	2020 Bare Costs Labor	Equipment	Total	Total Incl O&P
4700	Medicine				Ea.	10,000			10,000	11,000
7000	Tables, physical therapy, walk off, electric	2 Carp	3	5.333		2,350	283		2,633	3,050
7150	Standard, vinyl top with base cabinets, minimum		3	5.333		1,125	283		1,408	1,675
7200	Maximum		2	8		5,650	425		6,075	6,875
7250	Table, hospital, adjustable height					1,650			1,650	1,800
8400	Whirlpool bath, mobile, sst, 18" x 24" x 60"					6,550			6,550	7,200
8450	Fixed, incl. mixing valves	1 Plum	2	4		4,825	258		5,083	5,725

11 73 10.20 Bariatric Equipment		Crew	Daily Output	Labor-Hours	Unit	Material	Labor	Equipment	Total	Total Incl O&P
0010	**BARIATRIC EQUIPMENT**									
5000	Patient lift, electric operated, arm style									
5110	400 lb. capacity				Ea.	1,225			1,225	1,325
5120	450 lb. capacity					2,550			2,550	2,825
5130	600 lb. capacity					3,125			3,125	3,450
5140	700 lb. capacity					4,800			4,800	5,275
5150	1,000 lb. capacity					7,925			7,925	8,725
5200	Overhead, 4-post, 1,000 lb. capacity					10,200			10,200	11,200
5300	Overhead, track type, 450 lb. capacity, not including track					3,275			3,275	3,600
5500	For fabric sling, add					340			340	375
5550	For digital scale, add					815			815	895

11 74 Dental Equipment

11 74 10 – Dental Office Equipment

11 74 10.10 Diagnostic and Treatment Equipment		Crew	Daily Output	Labor-Hours	Unit	Material	Labor	Equipment	Total	Total Incl O&P
0010	**DIAGNOSTIC AND TREATMENT EQUIPMENT**									
0020	Central suction system, minimum	1 Plum	1.20	6.667	Ea.	1,175	430		1,605	1,950
0100	Maximum	"	.90	8.889		6,425	575		7,000	7,950
0300	Air compressor, minimum	1 Skwk	.80	10		2,850	550		3,400	4,000
0400	Maximum		.50	16		9,075	880		9,955	11,400
0600	Chair, electric or hydraulic, minimum		.50	16		2,625	880		3,505	4,275
0700	Maximum		.25	32		3,975	1,750		5,725	7,175
0800	Doctor's/assistant's stool, minimum					273			273	300
0850	Maximum					760			760	835
1000	Drill console with accessories, minimum	1 Skwk	1.60	5		3,175	274		3,449	3,925
1100	Maximum		1.60	5		4,800	274		5,074	5,700
2000	Light, ceiling mounted, minimum		8	1		875	55		930	1,050
2100	Maximum		8	1		2,050	55		2,105	2,350
2200	Unit light, minimum	2 Skwk	5.33	3.002		775	165		940	1,100
2210	Maximum		5.33	3.002		1,475	165		1,640	1,850
2220	Track light, minimum		3.20	5		1,725	274		1,999	2,325
2230	Maximum		3.20	5		2,300	274		2,574	2,950
2300	Sterilizers, steam portable, minimum					1,875			1,875	2,050
2350	Maximum					6,350			6,350	7,000
2600	Steam, institutional					2,650			2,650	2,900
2650	Dry heat, electric, portable, 3 trays					1,375			1,375	1,500
2700	Ultra-sonic cleaner, portable, minimum					435			435	475
2750	Maximum (institutional)					1,325			1,325	1,450
3000	X-ray unit, wall, minimum	1 Skwk	4	2		2,925	110		3,035	3,400
3010	Maximum		4	2		5,300	110		5,410	6,000
3100	Panoramic unit		.60	13.333		16,900	730		17,630	19,800
3105	Deluxe, minimum	2 Skwk	1.60	10		11,700	550		12,250	13,700
3110	Maximum	"	1.60	10		35,500	550		36,050	39,900

11 74 Dental Equipment

11 74 10 – Dental Office Equipment

11 74 10.10 Diagnostic and Treatment Equipment		Crew	Daily Output	Labor-Hours	Unit	Material	2020 Bare Costs Labor	Equipment	Total	Total Incl O&P
3500	Developers, X-ray, average	1 Plum	5.33	1.501	Ea.	5,375	96.50		5,471.50	6,075
3600	Maximum	"	5.33	1.501	↓	11,000	96.50		11,096.50	12,300

11 76 Operating Room Equipment

11 76 10 – Equipment for Operating Rooms

11 76 10.10 Surgical Equipment

		Crew	Daily Output	Labor-Hours	Unit	Material	Labor	Equipment	Total	Total Incl O&P
0010	SURGICAL EQUIPMENT									
5000	Scrub, surgical, stainless steel, single station, minimum	1 Plum	3	2.667	Ea.	5,875	172		6,047	6,750
5100	Maximum					9,000			9,000	9,900
6550	Major surgery table, minimum	1 Sswk	.50	16		25,900	920		26,820	30,000
6570	Maximum		.50	16		39,300	920		40,220	44,800
6600	Hydraulic, hand-held control, general surgery		.60	13.333		41,100	770		41,870	46,500
6650	Stationary, universal	↓	.50	16		53,500	920		54,420	60,000
6800	Surgical lights, major operating room, dual head, minimum	2 Elec	1	16		5,975	980		6,955	8,075
6850	Maximum		1	16		40,100	980		41,080	45,600
6900	Ceiling mount articulation, single arm	↓	1	16	↓	4,875	980		5,855	6,850

11 77 Radiology Equipment

11 77 10 – Equipment for Radiology

11 77 10.10 X-Ray Equipment

		Crew	Daily Output	Labor-Hours	Unit	Material	Labor	Equipment	Total	Total Incl O&P
0010	X-RAY EQUIPMENT									
8700	X-ray, mobile, minimum				Ea.	17,400			17,400	19,100
8750	Maximum					97,000			97,000	106,500
8900	Stationary, minimum					37,200			37,200	40,900
8950	Maximum					280,500			280,500	308,500
9150	Developing processors, minimum					5,175			5,175	5,700
9200	Maximum				↓	13,900			13,900	15,300

11 78 Mortuary Equipment

11 78 13 – Mortuary Refrigerators

11 78 13.10 Mortuary and Autopsy Equipment

		Crew	Daily Output	Labor-Hours	Unit	Material	Labor	Equipment	Total	Total Incl O&P
0010	MORTUARY AND AUTOPSY EQUIPMENT									
0015	Autopsy table, standard	1 Plum	1	8	Ea.	10,000	515		10,515	11,800
0020	Deluxe	"	.60	13.333		16,400	860		17,260	19,300
3200	Mortuary refrigerator, end operated, 2 capacity					9,625			9,625	10,600
3300	6 capacity				↓	16,300			16,300	17,900

11 78 16 – Crematorium Equipment

11 78 16.10 Crematory

		Crew	Daily Output	Labor-Hours	Unit	Material	Labor	Equipment	Total	Total Incl O&P
0010	CREMATORY									
1500	Crematory, not including building, 1 place	Q-3	.20	160	Ea.	80,500	9,825		90,325	103,500
1750	2 place	"	.10	320	"	115,000	19,600		134,600	157,000

11 81 Facility Maintenance Equipment

11 81 19 – Vacuum Cleaning Systems

11 81 19.10 Vacuum Cleaning		Crew	Daily Output	Labor-Hours	Unit	2020 Bare Costs Material	Labor	Equipment	Total	Total Incl O&P
0010	**VACUUM CLEANING**									
0020	Central, 3 inlet, residential	1 Skwk	.90	8.889	Total	1,325	490		1,815	2,250
0200	Commercial		.70	11.429		1,350	625		1,975	2,500
0400	5 inlet system, residential		.50	16		2,025	880		2,905	3,625
0600	7 inlet system, commercial		.40	20		2,600	1,100		3,700	4,600
0800	9 inlet system, residential	↓	.30	26.667	↓	4,225	1,475		5,700	6,975

11 82 Facility Solid Waste Handling Equipment

11 82 19 – Packaged Incinerators

11 82 19.10 Packaged Gas Fired Incinerators		Crew	Daily Output	Labor-Hours	Unit	Material	Labor	Equipment	Total	Total Incl O&P
0010	**PACKAGED GAS FIRED INCINERATORS**									
4400	Incinerator, gas, not incl. chimney, elec. or pipe, 50 lbs./hr., minimum	Q-3	.80	40	Ea.	42,000	2,450		44,450	50,000
4420	Maximum		.70	45.714		43,500	2,800		46,300	52,000
4440	200 lbs./hr., minimum (batch type)		.60	53.333		72,000	3,275		75,275	84,000
4460	Maximum (with feeder)		.50	64		84,500	3,925		88,425	99,000
4480	400 lbs./hr., minimum (batch type)		.30	107		88,000	6,550		94,550	106,500
4500	Maximum (with feeder)		.25	128		95,500	7,850		103,350	117,000
4520	800 lbs./hr., with feeder, minimum		.20	160		122,000	9,825		131,825	149,500
4540	Maximum		.17	188		211,000	11,500		222,500	250,000
4560	1,200 lbs./hr., with feeder, minimum		.15	213		161,500	13,100		174,600	197,500
4580	Maximum		.11	291		204,500	17,800		222,300	252,000
4600	2,000 lbs./hr., with feeder, minimum		.10	320		417,500	19,600		437,100	489,500
4620	Maximum		.05	640		653,500	39,300		692,800	779,000
4700	For heat recovery system, add, minimum		.25	128		101,500	7,850		109,350	124,000
4710	Add, maximum		.11	291		266,500	17,800		284,300	320,500
4720	For automatic ash conveyer, add		.50	64	↓	41,900	3,925		45,825	52,000
4750	Large municipal incinerators, incl. stack, minimum		.25	128	Ton/day	21,700	7,850		29,550	36,000
4850	Maximum	↓	.10	320	"	58,000	19,600		77,600	94,000

11 82 26 – Facility Waste Compactors

11 82 26.10 Compactors		Crew	Daily Output	Labor-Hours	Unit	Material	Labor	Equipment	Total	Total Incl O&P
0010	**COMPACTORS**									
0020	Compactors, 115 volt, 250 lbs./hr., chute fed	L-4	1	24	Ea.	12,600	1,200		13,800	15,700
0100	Hand fed		2.40	10		17,200	500		17,700	19,700
0300	Multi-bag, 230 volt, 600 lbs./hr., chute fed		1	24		16,800	1,200		18,000	20,400
0400	Hand fed		1	24		17,200	1,200		18,400	20,800
0500	Containerized, hand fed, 2 to 6 C.Y. containers, 250 lbs./hr.		1	24		16,700	1,200		17,900	20,300
0550	For chute fed, add per floor		1	24		1,500	1,200		2,700	3,575
1000	Heavy duty industrial compactor, 0.5 C.Y. capacity		1	24		11,200	1,200		12,400	14,200
1050	1.0 C.Y. capacity		1	24		17,100	1,200		18,300	20,700
1100	3.0 C.Y. capacity		.50	48		29,200	2,400		31,600	35,900
1150	5.0 C.Y. capacity		.50	48		36,900	2,400		39,300	44,400
1200	Combination shredder/compactor (5,000 lbs./hr.)	↓	.50	48		71,500	2,400		73,900	82,500
1400	For handling hazardous waste materials, 55 gallon drum packer, std.					21,700			21,700	23,900
1410	55 gallon drum packer w/HEPA filter					26,400			26,400	29,100
1420	55 gallon drum packer w/charcoal & HEPA filter					35,200			35,200	38,800
1430	All of the above made explosion proof, add					1,725			1,725	1,900
5500	Shredder, municipal use, 35 tons/hour					378,500			378,500	416,500
5600	60 tons/hour					806,500			806,500	887,500
5750	Shredder & baler, 50 tons/day					756,500			756,500	832,000
5800	Shredder, industrial, minimum				↓	29,900			29,900	32,900

11 82 Facility Solid Waste Handling Equipment

11 82 26 – Facility Waste Compactors

11 82 26.10 Compactors		Crew	Daily Output	Labor-Hours	Unit	Material (2020 Bare Costs)	Labor (2020 Bare Costs)	Equipment (2020 Bare Costs)	Total (2020 Bare Costs)	Total Incl O&P
5850	Maximum				Ea.	160,000			160,000	175,500
5900	Baler, industrial, minimum					12,000			12,000	13,200
5950	Maximum				↓	697,500			697,500	767,000
6000	Transfer station compactor, with power unit									
6050	and pedestal, not including pit, 50 tons/hour				Ea.	239,000			239,000	263,000

11 82 39 – Medical Waste Disposal Systems

11 82 39.10 Off-Site Disposal		Crew	Daily Output	Labor-Hours	Unit	Material	Labor	Equipment	Total	Total Incl O&P
0010	**OFF-SITE DISPOSAL**									
0100	Medical waste disposal, Red Bag system, pick up & treat, 200 lbs./week				Week	208			208	228
0110	Per month				Month	745			745	820
0150	Red bags, 7-10 gal., 1.2 mil, pkg of 500				Ea.	66			66	72.50
0200	15 gal., package of 250					64			64	70
0250	33 gal., package of 250					73			73	80
0300	45 gal., package of 100				↓	59.50			59.50	65.50

11 82 39.20 Disposal Carts		Crew	Daily Output	Labor-Hours	Unit	Material	Labor	Equipment	Total	Total Incl O&P
0010	**DISPOSAL CARTS**									
2010	Medical waste disposal cart, HDPE, w/lid, 28 gal. capacity				Ea.	289			289	320
2020	96 gal. capacity					335			335	370
2030	150 gal. capacity, low profile					500			500	555
2040	200 gal. capacity				↓	1,000			1,000	1,100

11 82 39.30 Medical Waste Sanitizers		Crew	Daily Output	Labor-Hours	Unit	Material	Labor	Equipment	Total	Total Incl O&P
0010	**MEDICAL WASTE SANITIZERS**									
2010	Small, hand loaded, 1.5 C.Y., 225 lb. capacity				Ea.	84,000			84,000	92,500
2020	Medium, cart loaded, 6.25 C.Y., 938 lb. capacity					112,000			112,000	123,500
2030	Large, cart loaded, 15 C.Y., 2250 lb. capacity					134,500			134,500	148,000
3010	Cart, aluminum, 75 lb. capacity					2,225			2,225	2,450
3020	95 lb. capacity					2,425			2,425	2,675
4010	Stainless steel, 173 lb. capacity					3,175			3,175	3,500
4020	232 lb. capacity					3,550			3,550	3,900
4030	Cart lift, hydraulic scissor type					6,300			6,300	6,925
4040	Portable aluminum ramp					1,775			1,775	1,950
4050	Fold-down steel tracks					1,450			1,450	1,600
4060	Pull-out drawer, small					6,775			6,775	7,475
4070	Medium					9,800			9,800	10,800
4080	Large				↓	13,800			13,800	15,200
5000	Medical waste treatment, sanitize, on-site									
5010	Less than 15,000 lbs./month				Lb.	.20			.20	.22
5020	Over 15,000 lbs./month				"	.16			.16	.18

11 91 Religious Equipment

11 91 13 – Baptisteries

11 91 13.10 Baptistry		Crew	Daily Output	Labor-Hours	Unit	Material	Labor	Equipment	Total	Total Incl O&P
0010	**BAPTISTRY**									
0150	Fiberglass, 3'-6" deep, x 13'-7" long,									
0160	steps at both ends, incl. plumbing, minimum	L-8	1	20	Ea.	6,300	1,100		7,400	8,700
0200	Maximum	"	.70	28.571		10,500	1,575		12,075	14,000
0250	Add for filter, heater and lights				↓	1,850			1,850	2,025

11 91 Religious Equipment

11 91 23 – Sanctuary Equipment

11 91 23.10 Sanctuary Furnishings		Crew	Daily Output	Labor-Hours	Unit	Material	2020 Bare Costs Labor	Equipment	Total	Total Incl O&P
0010	**SANCTUARY FURNISHINGS**									
0020	Altar, wood, custom design, plain	1 Carp	1.40	5.714	Ea.	2,725	305		3,030	3,475
0050	Deluxe	″	.20	40		13,100	2,125		15,225	17,800
0070	Granite or marble, average	2 Marb	.50	32		13,200	1,625		14,825	17,200
0090	Deluxe	″	.20	80		36,000	4,075		40,075	46,200
0100	Arks, prefabricated, plain	2 Carp	.80	20		9,250	1,075		10,325	11,900
0130	Deluxe, maximum	″	.20	80		132,000	4,250		136,250	152,000
0500	Reconciliation room, wood, prefabricated, single, plain	1 Carp	.60	13.333		3,450	710		4,160	4,925
0550	Deluxe		.40	20		8,600	1,075		9,675	11,200
0650	Double, plain		.40	20		6,300	1,075		7,375	8,625
0700	Deluxe		.20	40		18,600	2,125		20,725	23,900
1000	Lecterns, wood, plain		5	1.600		745	85		830	955
1100	Deluxe		2	4		6,125	213		6,338	7,100
2000	Pulpits, hardwood, prefabricated, plain		2	4		1,500	213		1,713	2,000
2100	Deluxe		1.60	5		13,700	266		13,966	15,500
2500	Railing, hardwood, average		25	.320	L.F.	208	17		225	257
3000	Seating, individual, oak, contour, laminated		21	.381	Person	241	20.50		261.50	298
3100	Cushion seat		21	.381		219	20.50		239.50	274
3200	Fully upholstered		21	.381		197	20.50		217.50	250
3300	Combination, self-rising		21	.381		365	20.50		385.50	435
3500	For cherry, add					30%				
5000	Wall cross, aluminum, extruded, 2″ x 2″ section	1 Carp	34	.235	L.F.	246	12.50		258.50	291
5150	4″ x 4″ section		29	.276		355	14.65		369.65	415
5300	Bronze, extruded, 1″ x 2″ section		31	.258		485	13.70		498.70	555
5350	2-1/2″ x 2-1/2″ section		34	.235		735	12.50		747.50	830
5450	Solid bar stock, 1/2″ x 3″ section		29	.276		965	14.65		979.65	1,100
5600	Fiberglass, stock		34	.235		220	12.50		232.50	262
5700	Stainless steel, 4″ deep, channel section		29	.276		780	14.65		794.65	880
5800	4″ deep box section		29	.276		1,125	14.65		1,139.65	1,275

11 92 Agricultural Equipment

11 92 16 – Stock Feeders

11 92 16.16 Barns		Crew	Daily Output	Labor-Hours	Unit	Material	Labor	Equipment	Total	Total Incl O&P
0010	**BARNS**									
0015	Swine barn, farrowing pens and equipment	B-1	1000	.024	S.F.	13.90	1.03		14.93	16.95
0020	Gestation pens and equipment		800	.030		12.20	1.28		13.48	15.50
0030	Nursery pens and equipment		1150	.021		10.70	.89		11.59	13.20
0040	Finishing pens and equipment		1400	.017		8.15	.73		8.88	10.10
0120	Poultry barn, cages and equipment		700	.034		17.90	1.47		19.37	22
0220	Animal barn stall, feed and water equipment		1000	.024		15.90	1.03		16.93	19.15
0230	Manure floor scraper system		1500	.016		2.98	.68		3.66	4.38
0240	Below slab manure gutter and shuttle stroker		400	.060		10.20	2.57		12.77	15.30
0250	Exhaust system		1400	.017		1.23	.73		1.96	2.52
0260	Milking barn, milking equipment		220	.109		89.50	4.67		94.17	106
0270	Milk storage equipment		450	.053		27	2.28		29.28	33
0280	Sheep barn, equipment, sheep shear street gates		1000	.024		1,025	1.03		1,026.03	1,125
0290	Gate in frame		1000	.024		105	1.03		106.03	118
0295	Maternity fence with drinking trough		1000	.024		61.50	1.03		62.53	69.50
0300	Tobacco barn, fruit & tobacco auto. dryer machine/heat pump dryer	Q-20	2.90	6.897	Ea.	6,550	395		6,945	7,850
0310	Tobacco curing generator		4.90	4.082		4,600	233		4,833	5,450

11 92 Agricultural Equipment

11 92 16 – Stock Feeders

11 92 16.16 Barns		Crew	Daily Output	Labor-Hours	Unit	Material	Labor	Equipment	Total	Total Incl O&P
						2020 Bare Costs				
0320	Automatic system controller	Q-20	6.90	2.899	Ea.	1,675	166		1,841	2,100
0330	Humidity & temp. transmitters for humidity measurement		6.90	2.899		930	166		1,096	1,275
0340	Handling system for leaf tobacco	↓	5.90	3.390	↓	580	194		774	935

11 97 Security Equipment

11 97 30 – Security Drawers

11 97 30.10 Pass Through Drawer

		Crew	Daily Output	Labor-Hours	Unit	Material	Labor	Equipment	Total	Total Incl O&P
0010	**PASS THROUGH DRAWER**									
0100	Pass-thru drawer for personal items, 18" x 15" x 24"	1 Skwk	2	4	Ea.	2,725	219		2,944	3,350
0110	Including speakers	"	1.50	5.333	"	3,050	293		3,343	3,850

11 98 Detention Equipment

11 98 21 – Detention Windows

11 98 21.13 Visitor Cubicle Windows

		Crew	Daily Output	Labor-Hours	Unit	Material	Labor	Equipment	Total	Total Incl O&P
0010	**VISITOR CUBICLE WINDOWS**									
4000	Visitor cubicle, vision panel, no intercom	E-4	2	16	Ea.	3,550	930	73.50	4,553.50	5,500

11 98 30 – Detention Cell Equipment

11 98 30.10 Cell Equipment

		Crew	Daily Output	Labor-Hours	Unit	Material	Labor	Equipment	Total	Total Incl O&P
0010	**CELL EQUIPMENT**									
3000	Toilet apparatus including wash basin, average	L-8	1.50	13.333	Ea.	3,550	740		4,290	5,075

Estimating Tips

General

- The items in this division are usually priced per square foot or each. Most of these items are purchased by the owner and installed by the contractor. Do not assume the items in Division 12 will be purchased and installed by the contractor. Check the specifications for responsibilities and include receiving, storage, installation, and mechanical and electrical hookups in the appropriate divisions.
- Some items in this division require some type of support system that is not usually furnished with the item. Examples of these systems include blocking for the attachment of casework and heavy drapery rods. The required blocking must be added to the estimate in the appropriate division.

Reference Numbers

Reference numbers are shown at the beginning of some major classifications. These numbers refer to related items in the Reference Section. The reference information may be an estimating procedure, an alternate pricing method, or technical information.

12 05 Common Work Results for Furnishings

12 05 05 – Selective Demolition for Furnishings

12 05 05.10 Selective Demolition, Interiors		Crew	Daily Output	Labor-Hours	Unit	Material	2020 Bare Costs Labor	Equipment	Total	Total Incl O&P
0010	SELECTIVE DEMOLITION, INTERIORS R024119-10									
1100	Casework, wood base cabinets	2 Clab	24	.667	L.F.		28		28	45
1200	Countertop		96	.167			7		7	11.25
3100	Casework, metal base cabinets		20	.800			33.50		33.50	54
3110	Cabinet base trim		400	.040			1.68		1.68	2.70
3120	Countertop, stainless steel or acid proof		80	.200			8.40		8.40	13.50
3122	custom	1 Carp	48	.167	S.F.		8.85		8.85	14.20
3125	Ceramic tile	1 Clab	120	.067	"		2.81		2.81	4.50
3127	Trim	1 Tilf	72	.111	L.F.		5.50		5.50	8.60
3130	Wall cabinets, wood, 84" high	2 Clab	30	.533			22.50		22.50	36
3500	Laboratory casework, tall storage cabinets, 84" high		40	.400			16.85		16.85	27
3510	Wall cabinets, metal		40	.400			16.85		16.85	27
4830	Floor mats, recessed or link mats	1 Clab	300	.027	S.F.		1.12		1.12	1.80
4832	Skate lock tile		200	.040			1.68		1.68	2.70
4834	Duckboard		300	.027			1.12		1.12	1.80
4920	Blinds, interior, horizontal or vertical		150	.053	L.F.		2.25		2.25	3.60
4922	Wood folding panels		35	.229	Pr.		9.60		9.60	15.40
4924	Shades, interior		700	.011	S.F.		.48		.48	.77
4930	Drapery hardware, traverse rods		35	.229	Ea.		9.60		9.60	15.40
4950	Blast curtains, including hardware		25	.320			13.45		13.45	21.50
5200	Fixed seating, per seat	2 Carp	44	.364			19.35		19.35	31
6400	Booth, restaurant	2 Clab	80	.200	L.F.		8.40		8.40	13.50
7400	Office systems, furniture, cubicle	"	3000	.005	S.F.		.22		.22	.36

12 05 13 – Fabrics

12 05 13.10 Upholstery Materials		Crew	Daily Output	Labor-Hours	Unit	Material	Labor	Equipment	Total	Total Incl O&P
0010	UPHOLSTERY MATERIALS									
1000	Fabrics, fabric blends, minimum				S.Y.	45			45	49.50
1020	Maximum					111			111	122
1200	Nylons, minimum					43			43	47
1220	Maximum					75			75	82.50
1400	Polyester, minimum					32.50			32.50	36
1420	Maximum					68.50			68.50	75.50
1600	Silk, minimum					87.50			87.50	96.50
1620	Maximum					150			150	164
1800	Wool, minimum					65.50			65.50	72.50
1820	Maximum					147			147	162
5000	Leather, minimum				S.F.	9.45			9.45	10.40
5020	Maximum				"	10.70			10.70	11.75
6000	Vinyl, minimum				S.Y.	26			26	28.50
6020	Maximum				"	42.50			42.50	46.50

12 12 Wall Decorations

12 12 19 – Framed Prints

12 12 19.10 Art Work		Crew	Daily Output	Labor-Hours	Unit	Material	2020 Bare Costs Labor	Equipment	Total	Total Incl O&P
0010	**ART WORK** framed									
1000	Photography, minimum	1 Carp	36	.222	Ea.	94.50	11.80		106.30	123
1050	Maximum		30	.267		550	14.15		564.15	630
2000	Posters, minimum		36	.222		44	11.80		55.80	67.50
2050	Maximum		30	.267		795	14.15		809.15	900
3000	Reproductions, minimum		36	.222		100	11.80		111.80	130
3050	Maximum		30	.267		820	14.15		834.15	930

12 21 Window Blinds

12 21 13 – Horizontal Louver Blinds

12 21 13.13 Metal Horizontal Louver Blinds

		Crew	Daily Output	Labor-Hours	Unit	Material	Labor	Equipment	Total	Total Incl O&P
0010	**METAL HORIZONTAL LOUVER BLINDS**									
0020	Horizontal, 1" aluminum slats, solid color, stock	1 Carp	590	.014	S.F.	5.80	.72		6.52	7.55
0250	2" aluminum slats, solid color, stock	"	590	.014	"	5.05	.72		5.77	6.70
1000	Alternate method of figuring:									
1300	1" aluminum slats, 48" wide, 48" high	1 Carp	30	.267	Ea.	56	14.15		70.15	84
1320	72" high		29	.276		70	14.65		84.65	101
1340	96" high		28	.286		81	15.20		96.20	114
1400	72" wide, 72" high		25	.320		98.50	17		115.50	136
1420	96" high		23	.348		153	18.50		171.50	198
1480	96" wide, 96" high		20	.400		126	21.50		147.50	173

12 21 13.33 Vinyl Horizontal Louver Blinds

		Crew	Daily Output	Labor-Hours	Unit	Material	Labor	Equipment	Total	Total Incl O&P
0010	**VINYL HORIZONTAL LOUVER BLINDS**									
0100	2" composite, 48" wide, 48" high	1 Carp	30	.267	Ea.	93	14.15		107.15	125
0120	72" high		29	.276		125	14.65		139.65	162
0140	96" high		28	.286		171	15.20		186.20	213
0200	60" wide, 60" high		27	.296		135	15.75		150.75	174
0220	72" high		25	.320		156	17		173	199
0240	96" high		24	.333		233	17.70		250.70	285
0300	72" wide, 72" high		25	.320		206	17		223	254
0320	96" high		23	.348		284	18.50		302.50	340
0400	96" wide, 96" high		20	.400		505	21.50		526.50	590
1000	2" faux wood, 48" wide, 48" high		30	.267		101	14.15		115.15	134
1020	72" high		29	.276		131	14.65		145.65	168
1040	96" high		28	.286		156	15.20		171.20	196
1300	72" wide, 72" high		25	.320		129	17		146	170
1320	96" high		23	.348		194	18.50		212.50	244
1400	96" wide, 96" high		20	.400		265	21.50		286.50	325

12 21 16 – Vertical Louver Blinds

12 21 16.13 Metal Vertical Louver Blinds

		Crew	Daily Output	Labor-Hours	Unit	Material	Labor	Equipment	Total	Total Incl O&P
0010	**METAL VERTICAL LOUVER BLINDS**									
1500	Vertical, 3" PVC strips, minimum	1 Carp	460	.017	S.F.	9.40	.92		10.32	11.85
1600	Maximum		400	.020		21	1.06		22.06	24.50
1800	4" aluminum slats, minimum		460	.017		9.70	.92		10.62	12.15
1900	Maximum		400	.020		18.80	1.06		19.86	22
1990	Alternate method of figuring:									
2000	2" aluminum slats, 48" wide, 48" high	1 Carp	30	.267	Ea.	88.50	14.15		102.65	120
2050	72" high		29	.276		121	14.65		135.65	157
2100	96" high		28	.286		131	15.20		146.20	169
2200	72" wide, 72" high		25	.320		147	17		164	189

12 21 Window Blinds

12 21 16 – Vertical Louver Blinds

12 21 16.13 Metal Vertical Louver Blinds		Crew	Daily Output	Labor-Hours	Unit	Material	2020 Bare Costs Labor	Equipment	Total	Total Incl O&P
2250	96″ high	1 Carp	23	.348	Ea.	176	18.50		194.50	224
2300	96″ wide, 96″ high	↓	20	.400	↓	293	21.50		314.50	355

12 22 Curtains and Drapes

12 22 13 – Draperies

12 22 13.10 Custom Draperies		Crew	Daily Output	Labor-Hours	Unit	Material	Labor	Equipment	Total	Total Incl O&P
0010	**CUSTOM DRAPERIES** (Material only)									
0050	Lined, minimum				S.Y.	68			68	75
0100	Maximum					163			163	179
0200	Unlined, minimum					32.50			32.50	35.50
0300	Maximum					94			94	104
0400	Lightproof type, add, minimum					1.58			1.58	1.74
0500	Maximum				↓	66			66	72.50
0800	Valances, pleated, 10″ to 18″ depth, minimum				L.F.	1.98			1.98	2.18
0900	Maximum				″	9.65			9.65	10.65
2000	Alternate method, lined overlaps and returns, average									
2020	32″ to 48″ wide, 26″ to 39″ long				Ea.	193			193	212
2040	40″ to 63″ long					195			195	214
2060	64″ to 72″ long					201			201	221
2080	73″ to 81″ long					227			227	249
2100	82″ to 90″ long					233			233	257
2200	91″ to 99″ long					251			251	276
2400	100″ to 108″ long					261			261	287
2600	109″ to 120″ long					276			276	305
2800	121″ to 130″ long					281			281	310
3000	48″ to 72″ wide, 26″ to 39″ long					305			305	340
3050	40″ to 63″ long					330			330	360
3100	64″ to 72″ long					345			345	375
3150	73″ to 81″ long					360			360	395
3200	82″ to 90″ long					395			395	435
3250	91″ to 99″ long					410			410	450
3300	100″ to 108″ long					425			425	465
3350	109″ to 120″ long					450			450	495
3400	121″ to 130″ long					490			490	535
3450	64″ to 96″ wide, 26″ to 39″ long					370			370	405
3500	40″ to 63″ long					400			400	440
3550	64″ to 72″ long					410			410	450
3600	73″ to 81″ long					440			440	485
3650	82″ to 90″ long					465			465	510
3700	91″ to 99″ long					495			495	545
3750	100″ to 108″ long					510			510	560
3800	109″ to 120″ long					540			540	595
3820	121″ to 130″ long					580			580	635
3840	80″ to 120″ wide, 26″ to 39″ long					705			705	775
3860	40″ to 63″ long					750			750	825
3880	64″to 72″ long					770			770	845
4000	73″ to 81″ long					835			835	915
4020	82″ to 90″ long					875			875	965
4080	91″ to 99″ long					925			925	1,025
4100	100″ to 108″ long					980			980	1,075
4150	109″ to 120″ long				↓	1,025			1,025	1,125

12 22 Curtains and Drapes

12 22 13 – Draperies

12 22 13.10 Custom Draperies		Crew	Daily Output	Labor-Hours	Unit	Material	2020 Bare Costs Labor	Equipment	Total	Total Incl O&P
4200	121" to 130" long				Ea.	1,075			1,075	1,175
4250	96" to 144" wide, 26" to 39" long					710			710	780
4300	40" to 63" long					765			765	845
4400	64" to 72" long					780			780	860
4500	73" to 81" long					825			825	905
4600	82" to 90" long					895			895	985
4700	91" to 99" long					950			950	1,050
4800	100" to 108" long					995			995	1,100
4900	109" to 120" long					1,050			1,050	1,150
5000	121" to 130" long					1,125			1,125	1,250
5100	112" to 168" wide, 26" to 39" long					775			775	855
5200	40" to 63" long					830			830	915
5300	64" to 72" long					860			860	950
5400	73" to 81" long					910			910	1,000
5500	82" to 90" long					965			965	1,050
5600	91" to 99" long					1,025			1,025	1,125
5700	100" to 108" long					1,075			1,075	1,200
5800	109" to 120" long					1,225			1,225	1,350
6000	121" to 130" long					1,275			1,275	1,425
6100	128" to 192" wide, 26" to 39" long					805			805	885
6200	40" to 63" long					865			865	955
6300	64" to 72" long					900			900	990
6400	73" to 81" long					960			960	1,050
6450	82" to 90" long					1,000			1,000	1,100
6500	91" to 99" long					1,075			1,075	1,175
6550	100" to 108" long					1,125			1,125	1,250
6600	109" to 120" long					1,200			1,200	1,325
6650	121" to 130" long					1,275			1,275	1,400
6700	144" to 216" wide, 26" to 39" long					875			875	965
6750	40" to 63" long					965			965	1,050
6800	64" to 72" long					1,025			1,025	1,125
6850	73" to 81" long					1,050			1,050	1,150
6880	82" to 90" long					1,100			1,100	1,200
6900	91" to 99" long					1,150			1,150	1,275
6920	100" to 108" long					1,225			1,225	1,350
6980	109" to 120" long					1,300			1,300	1,425
7000	121" to 130" long					1,375			1,375	1,525
7100	160" to 240" wide, 26" to 39" long					840			840	925
7150	40" to 63" long					910			910	1,000
7200	64" to 72" long					935			935	1,025
7250	73" to 81" long					990			990	1,100
7300	82" to 90" long					1,050			1,050	1,175
7350	91" to 99" long					1,100			1,100	1,225
7400	100" to 108" long					1,175			1,175	1,300
7500	109" to 120" long					1,225			1,225	1,350
7600	121" to 130" long				↓	1,325			1,325	1,450
8800	Drapery installation, hardware & drapes,									
9000	Labor cost only, minimum	1 Clab	75	.107	L.F.		4.49		4.49	7.20
9100	Maximum	"	20	.400	"		16.85		16.85	27

12 22 Curtains and Drapes

12 22 16 – Drapery Track and Accessories

12 22 16.10 Drapery Hardware		Crew	Daily Output	Labor-Hours	Unit	Material	2020 Bare Costs Labor	Equipment	Total	Total Incl O&P
0010	**DRAPERY HARDWARE**									
0030	Standard traverse, per foot, minimum	1 Carp	59	.136	L.F.	6.85	7.20		14.05	19.05
0100	Maximum		51	.157	"	15.10	8.35		23.45	30
0200	Decorative traverse, 28" to 48", minimum		22	.364	Ea.	24.50	19.35		43.85	58
0220	Maximum		21	.381		55	20.50		75.50	93
0300	48" to 84", minimum		20	.400		27	21.50		48.50	63.50
0320	Maximum		19	.421		70	22.50		92.50	113
0400	66" to 120", minimum		18	.444		37	23.50		60.50	78.50
0420	Maximum		17	.471		107	25		132	158
0500	84" to 156", minimum		16	.500		47	26.50		73.50	94.50
0520	Maximum		15	.533		146	28.50		174.50	207
0600	130" to 240", minimum		14	.571		68.50	30.50		99	124
0620	Maximum	↓	13	.615		202	32.50		234.50	275
0700	Slide rings, each, minimum					1.30			1.30	1.43
0720	Maximum					2.17			2.17	2.39
4000	Traverse rods, adjustable, 28" to 48"	1 Carp	22	.364		37	19.35		56.35	72
4020	48" to 84"		20	.400		47.50	21.50		69	86
4040	66" to 120"		18	.444		57.50	23.50		81	101
4060	84" to 156"		16	.500		65	26.50		91.50	114
4080	100" to 180"		14	.571		76	30.50		106.50	132
4090	156" to 228"		13	.615		91	32.50		123.50	153
4100	228" to 312"		13	.615		105	32.50		137.50	169
4200	Double rods, adjustable, 30" to 48"		9	.889		56.50	47		103.50	138
4220	48" to 86"		9	.889		79	47		126	162
4240	86" to 150"		8	1		87.50	53		140.50	181
4260	100" to 180"		7	1.143		92	60.50		152.50	199
4300	Curtain rod & brackets, adjustable, 30" to 48"		9	.889		32.50	47		79.50	111
4320	48" to 86"		9	.889		46.50	47		93.50	127
4340	86" to 150"		8	1		57	53		110	148
4360	100" to 180"	↓	7	1.143		70.50	60.50		131	175
5000	Stationary rods, first 2'				↓	8.30			8.30	9.15
5020	Each additional foot, add				L.F.	4.54			4.54	4.99

12 22 16.20 Blast Curtains		Crew	Daily Output	Labor-Hours	Unit	Material	Labor	Equipment	Total	Total Incl O&P
0010	**BLAST CURTAINS** per L.F. horizontal opening width, off-white or gray fabric									
0100	Blast curtains, drapery system, complete, including hardware, minimum	1 Carp	10.25	.780	L.F.	279	41.50		320.50	370
0120	Average		10.25	.780		269	41.50		310.50	360
0140	Maximum	↓	10.25	.780	↓	288	41.50		329.50	380

12 23 Interior Shutters

12 23 13 – Wood Interior Shutters

12 23 13.10 Wood Interior Shutters		Crew	Daily Output	Labor-Hours	Unit	Material	Labor	Equipment	Total	Total Incl O&P
0010	**WOOD INTERIOR SHUTTERS**, louvered									
0200	Two panel, 27" wide, 36" high	1 Carp	5	1.600	Set	173	85		258	325
0300	33" wide, 36" high		5	1.600		223	85		308	380
0500	47" wide, 36" high		5	1.600		299	85		384	465
1000	Four panel, 27" wide, 36" high		5	1.600		161	85		246	315
1100	33" wide, 36" high		5	1.600		206	85		291	360
1300	47" wide, 36" high		5	1.600	↓	275	85		360	440
1400	Plantation shutters, 16" x 48"		5	1.600	Ea.	146	85		231	297
1450	16" x 96"	↓	4	2	↓	240	106		346	435

12 23 Interior Shutters

12 23 13 – Wood Interior Shutters

12 23 13.10 Wood Interior Shutters		Crew	Daily Output	Labor-Hours	Unit	Material	2020 Bare Costs Labor	Equipment	Total	Total Incl O&P
1460	36″ x 96″	1 Carp	3	2.667	Ea.	530	142		672	805

12 23 13.13 Wood Panels

		Crew	Daily Output	Labor-Hours	Unit	Material	Labor	Equipment	Total	Total Incl O&P
0010	**WOOD PANELS**									
3000	Wood folding panels with movable louvers, 7″ x 20″ each	1 Carp	17	.471	Pr.	95.50	25		120.50	145
3300	8″ x 28″ each		17	.471		95.50	25		120.50	145
3450	9″ x 36″ each		17	.471		110	25		135	161
3600	10″ x 40″ each		17	.471		120	25		145	172
4000	Fixed louver type, stock units, 8″ x 20″ each		17	.471		107	25		132	158
4150	10″ x 28″ each		17	.471		90.50	25		115.50	140
4300	12″ x 36″ each		17	.471		107	25		132	158
4450	18″ x 40″ each		17	.471		153	25		178	208
5000	Insert panel type, stock, 7″ x 20″ each		17	.471		28	25		53	70.50
5150	8″ x 28″ each		17	.471		51	25		76	96
5300	9″ x 36″ each		17	.471		64.50	25		89.50	111
5450	10″ x 40″ each		17	.471		69.50	25		94.50	117
5600	Raised panel type, stock, 10″ x 24″ each		17	.471		118	25		143	170
5650	12″ x 26″ each		17	.471		118	25		143	170
5700	14″ x 30″ each		17	.471		131	25		156	184
5750	16″ x 36″ each	↓	17	.471		144	25		169	199
6000	For custom built pine, add					22%				
6500	For custom built hardwood blinds, add				↓	42%				

12 24 Window Shades

12 24 13 – Roller Window Shades

12 24 13.10 Shades

		Crew	Daily Output	Labor-Hours	Unit	Material	Labor	Equipment	Total	Total Incl O&P
0010	**SHADES**									
0020	Basswood, roll-up, stain finish, 3/8″ slats	1 Carp	300	.027	S.F.	23	1.42		24.42	27.50
0200	7/8″ slats		300	.027		21	1.42		22.42	25.50
0300	Vertical side slide, stain finish, 3/8″ slats		300	.027		24.50	1.42		25.92	29.50
0400	7/8″ slats	↓	300	.027		26.50	1.42		27.92	31.50
0500	For fire retardant finishes, add					16%				
0600	For ″B″ rated finishes, add					20%				
0900	Mylar, single layer, non-heat reflective	1 Carp	685	.012		3.14	.62		3.76	4.44
0910	Mylar, single layer, heat reflective		685	.012		2.83	.62		3.45	4.10
1000	Double layered, heat reflective		685	.012		6.65	.62		7.27	8.30
1100	Triple layered, heat reflective	↓	685	.012	↓	8.35	.62		8.97	10.15
1200	For metal roller instead of wood, add per				Shade	6.90			6.90	7.60
1300	Vinyl coated cotton, standard	1 Carp	685	.012	S.F.	4.55	.62		5.17	6
1400	Lightproof decorator shades		685	.012		7	.62		7.62	8.70
1500	Vinyl, lightweight, 4 ga.		685	.012		.92	.62		1.54	2
1600	Heavyweight, 6 ga.		685	.012		2.88	.62		3.50	4.16
1700	Vinyl laminated fiberglass, 6 ga., translucent		685	.012		5.55	.62		6.17	7.15
1800	Lightproof	↓	685	.012	↓	6.15	.62		6.77	7.75
2000	Polyester, room darkening, with continuous cord, GEI									
2010	36″ x 72″	1 Carp	38	.211	Ea.	122	11.20		133.20	152
2020	48″ x 72″		28	.286		166	15.20		181.20	208
2030	60″ x 72″		23	.348		188	18.50		206.50	237
2040	72″ x 72″		19	.421	↓	213	22.50		235.50	270
5011	Insulative shades [G]		125	.064	S.F.	17.15	3.40		20.55	24.50
6011	Solar screening, fiberglass [G]	↓	85	.094	″	8.25	5		13.25	17.10
8011	Interior insulative shutter									

12 24 Window Shades

12 24 13 – Roller Window Shades

12 24 13.10 Shades		Crew	Daily Output	Labor-Hours	Unit	Material	2020 Bare Costs Labor	Equipment	Total	Total Incl O&P
8111	Stock unit, 15" x 60" G	1 Carp	17	.471	Pr.	18.35	25		43.35	60

12 32 Manufactured Wood Casework

12 32 23 – Hardwood Casework

12 32 23.10 Manufactured Wood Casework, Stock Units

		Crew	Daily Output	Labor-Hours	Unit	Material	Labor	Equipment	Total	Total Incl O&P
0010	**MANUFACTURED WOOD CASEWORK, STOCK UNITS**									
0700	Kitchen base cabinets, hardwood, not incl. counter tops,									
0710	24" deep, 35" high, prefinished									
0800	One top drawer, one door below, 12" wide	2 Carp	24.80	.645	Ea.	325	34.50		359.50	410
0820	15" wide		24	.667		385	35.50		420.50	480
0840	18" wide		23.30	.687		299	36.50		335.50	390
0860	21" wide		22.70	.705		315	37.50		352.50	405
0880	24" wide		22.30	.717		435	38		473	540
1000	Four drawers, 12" wide		24.80	.645		330	34.50		364.50	420
1020	15" wide		24	.667		335	35.50		370.50	425
1040	18" wide		23.30	.687		370	36.50		406.50	465
1060	24" wide		22.30	.717		415	38		453	520
1200	Two top drawers, two doors below, 27" wide		22	.727		455	38.50		493.50	560
1220	30" wide		21.40	.748		515	39.50		554.50	635
1240	33" wide		20.90	.766		545	40.50		585.50	665
1260	36" wide		20.30	.788		570	42		612	695
1280	42" wide		19.80	.808		595	43		638	725
1300	48" wide		18.90	.847		695	45		740	835
1500	Range or sink base, two doors below, 30" wide		21.40	.748		460	39.50		499.50	570
1520	33" wide		20.90	.766		485	40.50		525.50	600
1540	36" wide		20.30	.788		495	42		537	610
1560	42" wide		19.80	.808		540	43		583	665
1580	48" wide		18.90	.847		545	45		590	670
1800	For sink front units, deduct					188			188	207
2000	Corner base cabinets, 36" wide, standard	2 Carp	18	.889		790	47		837	945
2100	Lazy Susan with revolving door	"	16.50	.970		1,025	51.50		1,076.50	1,200
4000	Kitchen wall cabinets, hardwood, 12" deep with two doors									
4050	12" high, 30" wide	2 Carp	24.80	.645	Ea.	279	34.50		313.50	360
4400	15" high, 30" wide		24	.667		285	35.50		320.50	370
4420	33" wide		23.30	.687		350	36.50		386.50	445
4440	36" wide		22.70	.705		355	37.50		392.50	450
4450	42" wide		22.70	.705		410	37.50		447.50	510
4700	24" high, 30" wide		23.30	.687		415	36.50		451.50	520
4720	36" wide		22.70	.705		485	37.50		522.50	595
4740	42" wide		22.30	.717		296	38		334	385
5000	30" high, one door, 12" wide		22	.727		280	38.50		318.50	370
5020	15" wide		21.40	.748		289	39.50		328.50	385
5040	18" wide		20.90	.766		315	40.50		355.50	415
5060	24" wide		20.30	.788		375	42		417	480
5300	Two doors, 27" wide		19.80	.808		430	43		473	545
5320	30" wide		19.30	.829		485	44		529	605
5340	36" wide		18.80	.851		465	45		510	585
5360	42" wide		18.50	.865		535	46		581	660
5380	48" wide		18.40	.870		625	46		671	760
6000	Corner wall, 30" high, 24" wide		18	.889		435	47		482	555
6050	30" wide		17.20	.930		445	49.50		494.50	570

12 32 23 – Hardwood Casework

12 32 23.10 Manufactured Wood Casework, Stock Units

	12 32 23.10 Manufactured Wood Casework, Stock Units	Crew	Daily Output	Labor-Hours	Unit	2020 Bare Costs Material	Labor	Equipment	Total	Total Incl O&P
6100	36″ wide	2 Carp	16.50	.970	Ea.	475	51.50		526.50	610
6500	Revolving Lazy Susan		15.20	1.053		139	56		195	243
7000	Broom cabinet, 84″ high, 24″ deep, 18″ wide		10	1.600		785	85		870	1,000
7500	Oven cabinets, 84″ high, 24″ deep, 27″ wide		8	2		1,075	106		1,181	1,350
7750	Valance board trim		396	.040	L.F.	20.50	2.15		22.65	26.50
7780	Toe kick trim	1 Carp	256	.031	″	3.61	1.66		5.27	6.65
7790	Base cabinet corner filler		16	.500	Ea.	50	26.50		76.50	97.50
7800	Cabinet filler, 3″ x 24″		20	.400		19.95	21.50		41.45	56
7810	3″ x 30″		20	.400		25	21.50		46.50	61.50
7820	3″ x 42″		18	.444		35	23.50		58.50	76.50
7830	3″ x 80″		16	.500		66.50	26.50		93	116
7850	Cabinet panel		50	.160	S.F.	11.20	8.50		19.70	26
9000	For deluxe models of all cabinets, add					40%				
9500	For custom built in place, add					25%	10%			
9558	Rule of thumb, kitchen cabinets not including									
9560	appliances & counter top, minimum	2 Carp	30	.533	L.F.	215	28.50		243.50	282
9600	Maximum	″	25	.640	″	460	34		494	560
9610	For metal cabinets, see Section 12 35 70.13									
9700	Minimum labor/equipment charge	1 Carp	3	2.667	Job		142		142	227

12 32 23.30 Manufactured Wood Casework Vanities

		Crew	Daily Output	Labor-Hours	Unit	Material	Labor	Equipment	Total	Total Incl O&P
0010	**MANUFACTURED WOOD CASEWORK VANITIES**									
8000	Vanity bases, 2 doors, 30″ high, 21″ deep, 24″ wide	2 Carp	20	.800	Ea.	420	42.50		462.50	530
8050	30″ wide		16	1		460	53		513	590
8100	36″ wide		13.33	1.200		400	64		464	540
8150	48″ wide		11.43	1.400		610	74.50		684.50	790
9000	For deluxe models of all vanities, add to above					40%				
9500	For custom built in place, add to above					25%	10%			

12 32 23.35 Manufactured Wood Casework Hardware

		Crew	Daily Output	Labor-Hours	Unit	Material	Labor	Equipment	Total	Total Incl O&P
0010	**MANUFACTURED WOOD CASEWORK HARDWARE**									
1000	Catches, minimum	1 Carp	235	.034	Ea.	1.56	1.81		3.37	4.62
1020	Average		119.40	.067		4.37	3.56		7.93	10.50
1040	Maximum		80	.100		9.25	5.30		14.55	18.70
2000	Door/drawer pulls, handles									
2200	Handles and pulls, projecting, metal, minimum	1 Carp	48	.167	Ea.	5.75	8.85		14.60	20.50
2220	Average		42	.190		8.75	10.10		18.85	26
2240	Maximum		36	.222		11.45	11.80		23.25	31.50
2300	Wood, minimum		48	.167		5.85	8.85		14.70	20.50
2320	Average		42	.190		7.85	10.10		17.95	25
2340	Maximum		36	.222		10.80	11.80		22.60	31
2400	Drawer pulls, antimicrobial copper alloy finish		50	.160		20.50	8.50		29	36
2600	Flush, metal, minimum		48	.167		5.85	8.85		14.70	20.50
2620	Average		42	.190		7.95	10.10		18.05	25
2640	Maximum		36	.222		10.95	11.80		22.75	31
2900	Drawer knobs, antimicrobial copper alloy finish		50	.160		13.45	8.50		21.95	28.50
3000	Drawer tracks/glides, minimum		48	.167	Pr.	9.60	8.85		18.45	25
3020	Average		32	.250		17	13.30		30.30	40
3040	Maximum		24	.333		28	17.70		45.70	59
4000	Cabinet hinges, minimum		160	.050		3.40	2.66		6.06	8
4020	Average		95.24	.084		5.55	4.46		10.01	13.25
4040	Maximum		68	.118		14.20	6.25		20.45	25.50
5000	Cabinet locks, minimum		47.90	.167	Ea.	5.85	8.90		14.75	20.50
5020	Average		23.95	.334		6.15	17.75		23.90	35.50

12 32 Manufactured Wood Casework

12 32 23 – Hardwood Casework

12 32 23.35 Manufactured Wood Casework Hardware		Crew	Daily Output	Labor-Hours	Unit	Material	Labor	Equipment	Total	Total Incl O&P
						2020 Bare Costs				
5040	Maximum	1 Carp	16	.500	Ea.	29.50	26.50		56	75
7000	Appliance pulls, antimicrobial copper alloy finish	↓	50	.160	L.F.	73.50	8.50		82	94.50

12 35 Specialty Casework

12 35 39 – Commercial Kitchen Casework

12 35 39.13 Metal Kitchen Casework

		Crew	Daily Output	Labor-Hours	Unit	Material	Labor	Equipment	Total	Total Incl O&P
0010	**METAL KITCHEN CASEWORK**									
3500	Base cabinets, metal, minimum	2 Carp	30	.533	L.F.	106	28.50		134.50	162
3600	Maximum		25	.640		269	34		303	350
3700	Wall cabinets, metal, minimum		30	.533		106	28.50		134.50	162
3800	Maximum	↓	25	.640	↓	243	34		277	320

12 35 50 – Educational/Library Casework

12 35 50.13 Educational Casework

		Crew	Daily Output	Labor-Hours	Unit	Material	Labor	Equipment	Total	Total Incl O&P
0010	**EDUCATIONAL CASEWORK**									
5000	School, 24" deep, metal, 84" high units	2 Carp	15	1.067	L.F.	610	56.50		666.50	765
5150	Counter height units		20	.800		460	42.50		502.50	575
5450	Wood, custom fabricated, 32" high counter		20	.800		263	42.50		305.50	355
5600	Add for counter top		56	.286		28	15.20		43.20	55.50
5800	84" high wall units	↓	15	1.067	↓	670	56.50		726.50	825
6000	Laminated plastic finish is same price as wood									

12 35 53 – Laboratory Casework

12 35 53.13 Metal Laboratory Casework

		Crew	Daily Output	Labor-Hours	Unit	Material	Labor	Equipment	Total	Total Incl O&P
0010	**METAL LABORATORY CASEWORK**									
0020	Cabinets, base, door units, metal	2 Carp	18	.889	L.F.	270	47		317	375
0300	Drawer units		18	.889		595	47		642	730
0700	Tall storage cabinets, open, 7' high		20	.800		575	42.50		617.50	700
0900	With glazed doors		20	.800		1,025	42.50		1,067.50	1,200
1300	Wall cabinets, metal, 12-1/2" deep, open		20	.800		227	42.50		269.50	320
1500	With doors	↓	20	.800	↓	445	42.50		487.50	560
6300	Rule of thumb: lab furniture including installation & connection									

12 35 59 – Display Casework

12 35 59.10 Display Cases

		Crew	Daily Output	Labor-Hours	Unit	Material	Labor	Equipment	Total	Total Incl O&P
0010	**DISPLAY CASES** Free standing, all glass									
0020	Aluminum frame, 42" high x 36" wide x 12" deep	2 Carp	8	2	Ea.	1,375	106		1,481	1,700
0100	70" high x 48" wide x 18" deep	"	6	2.667		4,425	142		4,567	5,100
0500	For wood bases, add					9%				
0600	For hardwood frames, deduct					8%				
0700	For bronze, baked enamel finish, add				↓	10%				
2000	Wall mounted, glass front, aluminum frame									
2010	Non-illuminated, one section 3' x 4' x 1'-4"	2 Carp	5	3.200	Ea.	2,475	170		2,645	3,000
2100	5' x 4' x 1'-4"		5	3.200		2,975	170		3,145	3,550
2200	6' x 4' x 1'-4"		4	4		3,475	213		3,688	4,175
2500	Two sections, 8' x 4' x 1'-4"		2	8		2,775	425		3,200	3,725
2600	10' x 4' x 1'-4"		2	8		5,300	425		5,725	6,500
3000	Three sections, 16' x 4' x 1'-4"	↓	1.50	10.667	↓	5,125	565		5,690	6,525
3500	For fluorescent lights, add				Section	460			460	505
4000	Table exhibit cases, 2' wide, 3' high, 4' long, flat top	2 Carp	5	3.200	Ea.	1,225	170		1,395	1,625
4100	3' wide, 3' high, 4' long, sloping top	"	3	5.333	"	835	283		1,118	1,375

12 35 Specialty Casework

12 35 70 – Healthcare Casework

12 35 70.13 Hospital Casework		Crew	Daily Output	Labor-Hours	Unit	2020 Bare Costs Material	Labor	Equipment	Total	Total Incl O&P
0010	**HOSPITAL CASEWORK**									
0500	Base cabinets, laminated plastic	2 Carp	10	1.600	L.F.	390	85		475	565
0700	Enameled steel		10	1.600		360	85		445	530
1000	Stainless steel		10	1.600		720	85		805	930
1200	For all drawers, add					41			41	45
1300	Cabinet base trim, 4″ high, enameled steel	2 Carp	200	.080		66	4.25		70.25	79.50
1400	Stainless steel		200	.080		132	4.25		136.25	152
1450	Countertop, laminated plastic, no backsplash		40	.400		55.50	21.50		77	95
1650	With backsplash		40	.400		69	21.50		90.50	110
1800	For sink cutout, add		12.20	1.311	Ea.		69.50		69.50	112
1900	Stainless steel counter top		40	.400	L.F.	179	21.50		200.50	231
2000	For drop-in stainless 43″ x 21″ sink, add				Ea.	1,450			1,450	1,575
2050	Laminate with antimicrobial finish #4	2 Carp	40	.400	L.F.	36.50	21.50		58	74
2500	Wall cabinets, laminated plastic		15	1.067		294	56.50		350.50	415
2600	Enameled steel		15	1.067		360	56.50		416.50	485
2700	Stainless steel		15	1.067		720	56.50		776.50	880
3000	Hospital cabinets, stainless steel with glass door(s), lockable									
3010	One door, 24″ W x 18″ D x 60″ H	2 Clab	18	.889	Ea.	3,175	37.50		3,212.50	3,550
3020	Two doors, 36″ W x 18″ D x 60″ H		15	1.067		4,325	45		4,370	4,825
3030	36″ W x 24″ D x 67″ H		15	1.067		5,800	45		5,845	6,450
3040	48″ W x 24″ D x 66″ H		12	1.333		4,925	56		4,981	5,500
3050	48″ W x 24″ D x 72″ H		12	1.333		6,125	56		6,181	6,850
3060	60″ W x 24″ D x 72″ H		9	1.778		7,075	75		7,150	7,900

12 35 70.16 Nurse Station Casework

		Crew	Daily Output	Labor-Hours	Unit	Material	Labor	Equipment	Total	Total Incl O&P
0010	**NURSE STATION CASEWORK**									
2100	Door type, laminated plastic	2 Carp	10	1.600	L.F.	455	85		540	635
2200	Enameled steel		10	1.600		435	85		520	615
2300	Stainless steel		10	1.600		870	85		955	1,100
2400	For drawer type, add					370			370	405

12 36 Countertops

12 36 16 – Metal Countertops

12 36 16.10 Stainless Steel Countertops

		Crew	Daily Output	Labor-Hours	Unit	Material	Labor	Equipment	Total	Total Incl O&P
0010	**STAINLESS STEEL COUNTERTOPS**									
3200	Stainless steel, custom	1 Carp	24	.333	S.F.	184	17.70		201.70	232
3210	Stainless steel	"	12	.667	L.F.	370	35.50		405.50	460

12 36 23 – Plastic Countertops

12 36 23.13 Plastic-Laminate-Clad Countertops

		Crew	Daily Output	Labor-Hours	Unit	Material	Labor	Equipment	Total	Total Incl O&P
0010	**PLASTIC-LAMINATE-CLAD COUNTERTOPS**									
0020	Stock, 24″ wide w/backsplash, minimum	1 Carp	30	.267	L.F.	17.75	14.15		31.90	42
0100	Maximum	"	25	.320	"	42.50	17		59.50	74
1700	For end splash, add				Ea.	22			22	24
9000	Minimum labor/equipment charge	1 Carp	3.75	2.133	Job		113		113	182

12 36 23.30 Plastic Laminate Countertop Components

		Crew	Daily Output	Labor-Hours	Unit	Material	Labor	Equipment	Total	Total Incl O&P
0010	**PLASTIC LAMINATE COUNTERTOP COMPONENTS**									
1000	Edging, 24″ wide, 1-1/2″ thick									
1500	Plastic laminate, minimum	1 Carp	25	.320	L.F.	3.96	17		20.96	32
1520	Average		24	.333		7.25	17.70		24.95	36.50
1540	Maximum		22	.364		11.35	19.35		30.70	43.50

12 36 Countertops

12 36 23 – Plastic Countertops

12 36 23.30 Plastic Laminate Countertop Components		Crew	Daily Output	Labor-Hours	Unit	2020 Bare Costs Material	Labor	Equipment	Total	Total Incl O&P
2500	Hardwood, minimum	1 Carp	20	.400	L.F.	6.40	21.50		27.90	41
2520	Average		18	.444		12.95	23.50		36.45	52.50
2540	Maximum		16	.500		19.65	26.50		46.15	64
2600	Backsplash, add to above, minimum		36	.222		1.59	11.80		13.39	20.50
2620	Average		35	.229		2.19	12.15		14.34	22
2640	Maximum		34	.235		4.41	12.50		16.91	25
2700	For metal cove, add					2.69			2.69	2.96
2900	Postformed backsplash, add to above									
2920	Minimum	1 Carp	96	.083	L.F.	7.90	4.43		12.33	15.80
2940	Average		96	.083		9	4.43		13.43	17
2960	Maximum		96	.083		11.75	4.43		16.18	20
3500	Well openings (for computers etc.)		2.50	3.200	Ea.		170		170	272
3900	Cutouts for sinks, lavatories		12	.667	″		35.50		35.50	57

12 36 40 – Stone Countertops

12 36 40.10 Natural Stone Countertops		Crew	Daily Output	Labor-Hours	Unit	Material	Labor	Equipment	Total	Total Incl O&P
0010	**NATURAL STONE COUNTERTOPS**									
2500	Marble, stock, with splash, 1/2″ thick, minimum	1 Bric	17	.471	L.F.	52.50	24.50		77	97.50
2700	3/4″ thick, maximum		13	.615		129	32		161	194
2720	Marble, 24″ wide, no splash		10	.800		56	41.50		97.50	129
2740	4″ backsplash		10	.800		64.50	41.50		106	139
2800	Granite, average, 1-1/4″ thick, 24″ wide, no splash		13.01	.615		171	32		203	240

12 36 53 – Laboratory Countertops

12 36 53.10 Laboratory Countertops and Sinks		Crew	Daily Output	Labor-Hours	Unit	Material	Labor	Equipment	Total	Total Incl O&P
0010	**LABORATORY COUNTERTOPS AND SINKS**									
0020	Countertops, epoxy resin, not incl. base cabinets, acid-proof, minimum	2 Carp	82	.195	S.F.	50.50	10.35		60.85	72
0030	Maximum		70	.229		48	12.15		60.15	72.50
0040	Stainless steel		82	.195		218	10.35		228.35	257

12 36 61 – Simulated Stone Countertops

12 36 61.16 Solid Surface Countertops		Crew	Daily Output	Labor-Hours	Unit	Material	Labor	Equipment	Total	Total Incl O&P
0010	**SOLID SURFACE COUNTERTOPS**, Acrylic polymer									
0020	Pricing for orders of 100 L.F. or greater									
0100	25″ wide, solid colors	2 Carp	28	.571	L.F.	52	30.50		82.50	106
0200	Patterned colors		28	.571		66	30.50		96.50	122
0300	Premium patterned colors		28	.571		88	30.50		118.50	146
0400	With silicone attached 4″ backsplash, solid colors		27	.593		61	31.50		92.50	118
0500	Patterned colors		27	.593		77	31.50		108.50	136
0600	Premium patterned colors		27	.593		96	31.50		127.50	157
0700	With hard seam attached 4″ backsplash, solid colors		23	.696		61	37		98	126
0800	Patterned colors		23	.696		77	37		114	144
0900	Premium patterned colors		23	.696		96	37		133	165
1000	Pricing for order of 51-99 L.F.									
1100	25″ wide, solid colors	2 Carp	24	.667	L.F.	60	35.50		95.50	123
1200	Patterned colors		24	.667		76	35.50		111.50	141
1300	Premium patterned colors		24	.667		101	35.50		136.50	168
1400	With silicone attached 4″ backsplash, solid colors		23	.696		70	37		107	136
1500	Patterned colors		23	.696		89	37		126	157
1600	Premium patterned colors		23	.696		111	37		148	181
1700	With hard seam attached 4″ backsplash, solid colors		20	.800		70	42.50		112.50	145
1800	Patterned colors		20	.800		89	42.50		131.50	166
1900	Premium patterned colors		20	.800		111	42.50		153.50	190
2000	Pricing for order of 1-50 L.F.									

12 36 61 – Simulated Stone Countertops

12 36 61.16 Solid Surface Countertops		Crew	Daily Output	Labor-Hours	Unit	Material	2020 Bare Costs Labor	Equipment	Total	Total Incl O&P
2100	25″ wide, solid colors	2 Carp	20	.800	L.F.	70	42.50		112.50	146
2200	Patterned colors		20	.800		89.50	42.50		132	167
2300	Premium patterned colors		20	.800		119	42.50		161.50	199
2400	With silicone attached 4″ backsplash, solid colors		19	.842		82.50	45		127.50	162
2500	Patterned colors		19	.842		104	45		149	187
2600	Premium patterned colors		19	.842		130	45		175	215
2700	With hard seam attached 4″ backsplash, solid colors		15	1.067		82.50	56.50		139	182
2800	Patterned colors		15	1.067		104	56.50		160.50	206
2900	Premium patterned colors		15	1.067		130	56.50		186.50	234
3000	Sinks, pricing for order of 100 or greater units									
3100	Single bowl, hard seamed, solid colors, 13″ x 17″	1 Carp	3	2.667	Ea.	375	142		517	640
3200	10″ x 15″		7	1.143		174	60.50		234.50	289
3300	Cutouts for sinks		8	1			53		53	85
3400	Sinks, pricing for order of 51-99 units									
3500	Single bowl, hard seamed, solid colors, 13″ x 17″	1 Carp	2.55	3.137	Ea.	430	167		597	740
3600	10″ x 15″		6	1.333		200	71		271	335
3700	Cutouts for sinks		7	1.143			60.50		60.50	97.50
3800	Sinks, pricing for order of 1-50 units									
3900	Single bowl, hard seamed, solid colors, 13″ x 17″	1 Carp	2	4	Ea.	505	213		718	895
4000	10″ x 15″		4.55	1.758		234	93.50		327.50	410
4100	Cutouts for sinks		5.25	1.524			81		81	130
4200	Cooktop cutouts, pricing for 100 or greater units		4	2		25.50	106		131.50	198
4300	51-99 units		3.40	2.353		29.50	125		154.50	232
4400	1-50 units		3	2.667		34.50	142		176.50	265

12 36 61.17 Solid Surface Vanity Tops		Crew	Daily Output	Labor-Hours	Unit	Material	Labor	Equipment	Total	Total Incl O&P
0010	**SOLID SURFACE VANITY TOPS**									
0015	Solid surface, center bowl, 17″ x 19″	1 Carp	12	.667	Ea.	171	35.50		206.50	245
0020	19″ x 25″		12	.667		216	35.50		251.50	294
0030	19″ x 31″		12	.667		243	35.50		278.50	325
0040	19″ x 37″		12	.667		283	35.50		318.50	365
0050	22″ x 25″		10	.800		335	42.50		377.50	440
0060	22″ x 31″		10	.800		395	42.50		437.50	505
0070	22″ x 37″		10	.800		460	42.50		502.50	575
0080	22″ x 43″		10	.800		525	42.50		567.50	650
0090	22″ x 49″		10	.800		575	42.50		617.50	700
0110	22″ x 55″		8	1		425	53		478	550
0120	22″ x 61″		8	1		480	53		533	615
0220	Double bowl, 22″ x 61″		8	1		535	53		588	675
0230	Double bowl, 22″ x 73″		8	1		1,075	53		1,128	1,275
0240	For aggregate colors, add					35%				
0250	For faucets and fittings, see Section 22 41 39.10									

12 36 61.19 Quartz Agglomerate Countertops		Crew	Daily Output	Labor-Hours	Unit	Material	Labor	Equipment	Total	Total Incl O&P
0010	**QUARTZ AGGLOMERATE COUNTERTOPS**									
0100	25″ wide, 4″ backsplash, color group A, minimum	2 Carp	15	1.067	L.F.	68	56.50		124.50	166
0110	Maximum		15	1.067		95	56.50		151.50	196
0120	Color group B, minimum		15	1.067		77	56.50		133.50	176
0130	Maximum		15	1.067		108	56.50		164.50	210
0140	Color group C, minimum		15	1.067		82	56.50		138.50	181
0150	Maximum		15	1.067		125	56.50		181.50	229
0160	Color group D, minimum		15	1.067		89	56.50		145.50	189
0170	Maximum		15	1.067		124	56.50		180.50	227

12 41 Desk Accessories

12 41 13 – Desk Accessories

12 41 13.10 Commercial Desk Accessories		Crew	Daily Output	Labor-Hours	Unit	Material	2020 Bare Costs Labor	Equipment	Total	Total Incl O&P
0010	**COMMERCIAL DESK ACCESSORIES**									
0300	Bookends, minimum				Ea.	13.30			13.30	14.60
0320	Maximum					25			25	27.50
1000	Calendar with pad, minimum					25.50			25.50	28
1020	Maximum					40			40	44
1200	Carafe, tray, minimum					33.50			33.50	37
1220	Maximum					107			107	117
1300	Desk pad, minimum					27.50			27.50	30
1320	Maximum					175			175	193
1400	Double pen set with pens, minimum					56			56	62
1420	Maximum					135			135	148
1500	Letter tray, minimum					16.80			16.80	18.50
1520	Maximum					59			59	64.50
1600	Memo box, minimum					8.30			8.30	9.10
1620	Maximum				↓	34.50			34.50	38

12 43 Portable Lamps

12 43 13 – Lamps

12 43 13.23 Miscellaneous Lamps		Crew	Daily Output	Labor-Hours	Unit	Material	Labor	Equipment	Total	Total Incl O&P
0010	**MISCELLANEOUS LAMPS**									
1000	Ceramic, desk, minimum				Ea.	60			60	66
1020	Maximum					138			138	152
1200	End table, minimum					66			66	72.50
1220	Maximum					116			116	128
1300	Night stand, minimum					49			49	54
1320	Maximum					113			113	124
1400	Wall, minimum					90.50			90.50	99.50
1420	Maximum					289			289	320
2000	Glass, desk, minimum					141			141	155
2020	Maximum					485			485	535
2100	End table, minimum					121			121	133
2120	Maximum					450			450	495
2200	Floor, minimum					475			475	525
2250	Maximum					1,525			1,525	1,675
2280	Night stand, minimum					192			192	211
2300	Maximum					395			395	435
2350	Pendant, minimum					120			120	132
2400	Maximum					900			900	990
3000	Metal, desk, minimum					56			56	61.50
3100	Maximum					131			131	145
3200	End table, minimum					63.50			63.50	70
3250	Maximum					91.50			91.50	101
3300	Wall, minimum					63			63	69
3350	Maximum					139			139	152
3400	Floor, minimum					103			103	113
3450	Maximum					455			455	500
3500	Night stand, minimum					49			49	54
3550	Maximum					126			126	139
3600	Pendant, minimum					52.50			52.50	58
3650	Maximum					905			905	995
4000	Stone, desk, minimum				↓	52.50			52.50	57.50

12 43 Portable Lamps

12 43 13 – Lamps

12 43 13.23 Miscellaneous Lamps		Crew	Daily Output	Labor-Hours	Unit	Material	2020 Bare Costs Labor	Equipment	Total	Total Incl O&P
4100	Maximum				Ea.	205			205	225
4200	End table, minimum					61			61	67
4300	Maximum					135			135	148
4400	Night stand, minimum					48.50			48.50	53.50
4500	Maximum					95			95	105
5000	Wall, minimum					77			77	85
6000	Maximum					150			150	165
8000	Replacement shades, lamp, desk, minimum					11.25			11.25	12.40
8020	Maximum					45			45	49.50
8040	End table, minimum					18.30			18.30	20
8060	Maximum					45			45	49.50
8100	Night stand, minimum					15.45			15.45	17
8120	Maximum					38			38	41.50
8140	Wall, minimum					10.10			10.10	11.10
8160	Maximum				↓	18.60			18.60	20.50

12 45 Bedroom Furnishings

12 45 13 – Bed Linens

12 45 13.13 Blankets		Crew	Daily Output	Labor-Hours	Unit	Material	Labor	Equipment	Total	Total Incl O&P
0010	**BLANKETS**									
7700	Bedspreads, unquilted, twin, minimum				Ea.	63.50			63.50	69.50
7800	Maximum					104			104	114
7850	Full, minimum					87.50			87.50	96.50
7900	Maximum					164			164	180
8000	Queen, minimum					97			97	107
8100	Maximum					176			176	194
8200	Quilted, twin, minimum					81.50			81.50	89.50
8300	Maximum					161			161	177
8400	Full, minimum					113			113	124
8500	Maximum					157			157	173
8600	Queen, minimum					142			142	156
8700	Maximum				↓	217			217	238

12 46 Furnishing Accessories

12 46 13 – Ash Receptacles

12 46 13.10 Ash/Trash Receivers		Crew	Daily Output	Labor-Hours	Unit	Material	Labor	Equipment	Total	Total Incl O&P
0010	**ASH/TRASH RECEIVERS**									
1000	Ash urn, cylindrical metal									
1020	8″ diameter, 20″ high	1 Clab	60	.133	Ea.	204	5.60		209.60	234
1040	8″ diameter, 25″ high		60	.133		298	5.60		303.60	340
1060	10″ diameter, 26″ high		60	.133		163	5.60		168.60	189
1080	12″ diameter, 30″ high	↓	60	.133	↓	231	5.60		236.60	263
2000	Combination ash/trash urn, metal									
2020	8″ diameter, 20″ high	1 Clab	60	.133	Ea.	300	5.60		305.60	340
2040	8″ diameter, 25″ high		60	.133		360	5.60		365.60	410
2050	10″ diameter, 26″ high		60	.133		241	5.60		246.60	275
2060	12″ diameter, 30″ high	↓	60	.133	↓	580	5.60		585.60	650

12 46 Furnishing Accessories

12 46 19 – Clocks

12 46 19.50 Wall Clocks		Crew	Daily Output	Labor-Hours	Unit	Material	2020 Bare Costs Labor	Equipment	Total	Total Incl O&P
0010	**WALL CLOCKS**									
0080	12" diameter, single face	1 Elec	8	1	Ea.	109	61.50		170.50	215
0100	Double face	"	6.20	1.290	"	161	79		240	299

12 46 33 – Waste Receptacles

12 46 33.13 Trash Receptacles		Crew	Daily Output	Labor-Hours	Unit	Material	Labor	Equipment	Total	Total Incl O&P
0010	**TRASH RECEPTACLES**									
4000	Trash receptacle, metal									
4020	8" diameter, 15" high	1 Clab	60	.133	Ea.	79	5.60		84.60	96
4040	10" diameter, 18" high		60	.133		148	5.60		153.60	172
4060	16" diameter, 16" high		60	.133		158	5.60		163.60	183
4100	18" diameter, 32" high	↓	60	.133	↓	221	5.60		226.60	252
5000	Plastic, fire resistant									
5020	Rectangular 11" x 8" x 12" high	1 Clab	60	.133	Ea.	33	5.60		38.60	45.50
5040	16" x 8" x 14" high	"	60	.133	"	46.50	5.60		52.10	60
5500	Plastic, with lid									
5520	35 gal.	1 Clab	60	.133	Ea.	216	5.60		221.60	247
5540	45 gal.		60	.133		350	5.60		355.60	395
5550	Plastic recycling barrel, w/lid & wheels, 32 gal. G		60	.133		108	5.60		113.60	128
5560	65 gal. G		60	.133		695	5.60		700.60	770
5570	95 gal. G	↓	60	.133	↓	1,300	5.60		1,305.60	1,425

12 48 Rugs and Mats

12 48 13 – Entrance Floor Mats and Frames

12 48 13.13 Entrance Floor Mats		Crew	Daily Output	Labor-Hours	Unit	Material	Labor	Equipment	Total	Total Incl O&P
0010	**ENTRANCE FLOOR MATS**									
0020	Recessed, black rubber, 3/8" thick, solid	1 Clab	155	.052	S.F.	27.50	2.17		29.67	34
0050	Perforated		155	.052		23.50	2.17		25.67	29.50
0100	1/2" thick, solid		155	.052		28	2.17		30.17	34.50
0150	Perforated		155	.052		30	2.17		32.17	36.50
0200	In colors, 3/8" thick, solid		155	.052		33.50	2.17		35.67	40.50
0250	Perforated		155	.052		30.50	2.17		32.67	37
0300	1/2" thick, solid		155	.052		39	2.17		41.17	46.50
0350	Perforated	↓	155	.052	↓	40	2.17		42.17	47.50
1225	Recessed, alum. rail, hinged mat, 7/16" thick									
1250	Carpet insert	1 Clab	360	.022	S.F.	70	.94		70.94	78.50
1275	Vinyl insert		360	.022		70	.94		70.94	78.50
1300	Abrasive insert	↓	360	.022	↓	70	.94		70.94	78.50
1325	Recessed, vinyl rail, hinged mat, 7/16" thick									
1350	Carpet insert	1 Clab	360	.022	S.F.	77.50	.94		78.44	87
1375	Vinyl insert		360	.022		77.50	.94		78.44	87
1400	Abrasive insert		360	.022		77.50	.94		78.44	87
2000	Recycled rubber tire tile, 12" x 12" x 3/8" thick G		125	.064		11.50	2.69		14.19	16.95
2510	Natural cocoa fiber, 1/2" thick G		125	.064		8	2.69		10.69	13.10
2520	3/4" thick G		125	.064		7.15	2.69		9.84	12.15
2530	1" thick G	↓	125	.064	↓	13.40	2.69		16.09	19.05
3000	Hospital tacky mats, package of 30 with frame				Ea.	60			60	66
3010	4 packages of 30				"	89			89	98

12 51 Office Furniture

12 51 16 – Case Goods

12 51 16.13 Metal Case Goods		Crew	Daily Output	Labor-Hours	Unit	Material	2020 Bare Costs Labor	Equipment	Total	Total Incl O&P
0010	**METAL CASE GOODS**									
0020	Desks, 29" high, double pedestal, 30" x 60", metal, minimum				Ea.	590			590	645
0030	Maximum					915			915	1,000
0400	36" x 72", metal, minimum					840			840	925
0420	Maximum					1,825			1,825	2,000
0600	Desks, single pedestal, 30" x 60", metal, minimum					515			515	570
0620	Maximum					870			870	955
0720	Desks, secretarial, 30" x 60", metal, minimum					495			495	545
0730	Maximum					855			855	940
0740	Return, 20" x 42", minimum					510			510	560
0750	Maximum					870			870	955
0940	59" x 12" x 23" high, steel, minimum					405			405	445
0960	Maximum					505			505	555
0970	Keyboard shelf, standard					104			104	114
0980	Articulating					650			650	715
0990	Center drawer, 18" wide					145			145	160
1020	Credenza, 18" to 22" x 60" to 72" metal, minimum					635			635	695
1040	Maximum					1,750			1,750	1,900
1240	Bookcase, 36" x 12" x 29" high					195			195	215
1260	52" high					315			315	345
1320	Computer stand, mobile, 25" x 24" x 38" high					310			310	340
1370	Computer desk with hutch, 47" x 24" x 46" high					525			525	580
1400	Printer stand, 22" x 24" x 34" high				↓	251			251	276

12 51 16.16 Wood Case Goods		Crew	Daily Output	Labor-Hours	Unit	Material	Labor	Equipment	Total	Total Incl O&P
0010	**WOOD CASE GOODS**									
0150	Desk, 29" high, double pedestal, 30" x 60"									
0160	Wood, minimum				Ea.	750			750	825
0180	Maximum				"	1,050			1,050	1,175
0550	Desk, 29" high, double pedestal, 36" x 72"									
0560	Wood, minimum				Ea.	830			830	910
0580	Maximum				"	4,025			4,025	4,425
0630	Single pedestal, 30" x 60"									
0640	Wood, minimum				Ea.	670			670	735
0650	Maximum					995			995	1,100
0670	Executive return, 24" x 42", with box, file, wood, minimum					440			440	485
0680	Maximum				↓	990			990	1,100
0790	Desk, 29" high, secretarial, 30" x 60"									
0800	Wood, minimum				Ea.	465			465	510
0810	Maximum					1,100			1,100	1,200
0820	Return, 20" x 42", minimum					430			430	470
0830	Maximum					715			715	785
0900	Desktop organizer, 72" x 14" x 36" high, wood, minimum					190			190	209
0920	Maximum				↓	465			465	510
1110	Furniture, credenza, 29" high, 18" to 22" x 60" to 72"									
1120	Wood, minimum				Ea.	885			885	975
1140	Maximum					2,825			2,825	3,125
1150	Hutch/bookcase, minimum					600			600	660
1160	Maximum					2,550			2,550	2,800
1200	Bookcase, 36" x 12" x 30" high, wood					291			291	320
1220	48" high					283			283	310
1300	Computer stand, mobile, wood, 25" x 24" x 38" high					595			595	655
1360	Computer desk with hutch, wood, 47" x 24" x 46" high				↓	715			715	785

12 51 Office Furniture

12 51 16 – Case Goods

12 51 16.26 Resinite Case Goods		Crew	Daily Output	Labor-Hours	Unit	Material	2020 Bare Costs Labor	Equipment	Total	Total Incl O&P
0010	**RESINITE CASE GOODS**									
1340	Computer stand, resinite				Ea.	330			330	365
1380	Desk with hutch				"	520			520	570

12 51 19 – Filing Cabinets

12 51 19.13 Lateral Filing Cabinets		Crew	Daily Output	Labor-Hours	Unit	Material	Labor	Equipment	Total	Total Incl O&P
0010	**LATERAL FILING CABINETS**, metal, baked enamel finish									
1060	Lateral, 36″ wide, minimum				Ea.	315			315	350
1080	Maximum					535			535	585
1160	Lateral, 36″ wide, minimum					485			485	535
1180	Maximum					1,050			1,050	1,150
1200	Wood, 2 drawer, lateral, minimum					550			550	605
1210	Maximum				↓	1,100			1,100	1,225

12 51 19.16 Vertical Filing Cabinets		Crew	Daily Output	Labor-Hours	Unit	Material	Labor	Equipment	Total	Total Incl O&P
0010	**VERTICAL FILING CABINETS**, metal, baked enamel finish									
1000	2 drawer, vertical, minimum				Ea.	222			222	244
1020	Maximum					440			440	485
1100	4 drawer, vertical, minimum					355			355	390
1120	Maximum				↓	565			565	625

12 51 19.23 Flat Files		Crew	Daily Output	Labor-Hours	Unit	Material	Labor	Equipment	Total	Total Incl O&P
0010	**FLAT FILES**, metal, baked enamel finish									
2010	Steel, 5 drawer, 40″ wide x 27″ deep				Ea.	1,150			1,150	1,275
2020	46″ wide x 33″ deep					1,175			1,175	1,300
2030	8 drawer, 40″ wide x 27″ deep					1,825			1,825	2,025
2040	46″ wide x 33″ deep					1,875			1,875	2,050
2050	Base, 40″ wide x 27″ deep x 5″ high					325			325	355
2060	46″ wide x 33″ deep				↓	355			355	395

12 51 19.26 Hanging Files		Crew	Daily Output	Labor-Hours	Unit	Material	Labor	Equipment	Total	Total Incl O&P
0010	**HANGING FILES**, metal, baked enamel finish									
2100	File stand				Ea.	335			335	370
2110	Clamps, package of 6				"	167			167	184

12 51 23 – Office Tables

12 51 23.13 Wood Tables		Crew	Daily Output	Labor-Hours	Unit	Material	Labor	Equipment	Total	Total Incl O&P
0010	**WOOD TABLES**									
5000	Sled base, laminate top, coffee, minimum				Ea.	216			216	238
5100	Maximum					550			550	605
5150	End, minimum					213			213	234
5200	Maximum					265			265	292
5250	Wood cube, coffee					900			900	990
5350	End					730			730	805
5700	Designer table, Mies Barcelona, 40″ square, glass top, s.s. legs					1,200			1,200	1,325
5750	Saarinen, 42″ round, plastic top, metal pedestal base					2,000			2,000	2,225
5800	Aulenti coffee table, 45″ square, marble top and leg base					12,600			12,600	13,800
5840	Bruer, Laccio, 21-1/2″ x 19″, plastic top, tubular legs					585			585	640
5860	F. knoll, 24″ square, glass top, solid steel leg base, chrome				↓	2,400			2,400	2,625

12 51 23.23 Metal Tables		Crew	Daily Output	Labor-Hours	Unit	Material	Labor	Equipment	Total	Total Incl O&P
0010	**METAL TABLES**									
5400	All metal drum, 14″ diameter, 14″ high				Ea.	470			470	515
5420	21″ high					545			545	600
5480	18″ diameter, 18″ high					705			705	775
5500	22″ diameter, 15″ high				↓	910			910	1,000

12 51 23.23 Metal Tables		Crew	Daily Output	Labor-Hours	Unit	Material	Labor	Equipment	Total	Total Incl O&P
5550	20" high				Ea.	1,000			1,000	1,100
5600	For 3/4" glass top, add				S.F.	89			89	98
5650	For 1" marble top, add				"	89			89	98
7500	Table, training, with modesty panel, 30" x 60"				Ea.	630			630	690
8000	Tables, modular									
8010	Rectangular, 24" x 48"				Ea.	315			315	345
8020	24" x 60"					350			350	385
8030	24" x 72"					440			440	485
8040	30" x 60"					1,225			1,225	1,350
8050	30" x 72"					1,325			1,325	1,450
8100	Corner triangle, 24" deep					440			440	485
8110	30" deep					495			495	545
8200	Half round, 30" x 60"					1,000			1,000	1,100
8300	Trapezoid, 30" x 60"					560			560	615
8400	Crescent, 30" x 60"				↓	1,225			1,225	1,350
12 51 23.33 Conference Tables										
0010	**CONFERENCE TABLES**									
6010	Segmented, oval, 72" x 144"				Ea.	5,925			5,925	6,525
6050	Boat, 96" x 42", minimum					940			940	1,025
6150	Maximum					3,650			3,650	4,025
6200	120" x 48", minimum					1,575			1,575	1,750
6250	Maximum					5,625			5,625	6,175
6300	144" x 48", minimum					1,950			1,950	2,150
6350	Maximum					6,575			6,575	7,225
6400	168" x 60", minimum					3,350			3,350	3,700
6450	Maximum					9,725			9,725	10,700
6500	192" x 60", minimum					3,900			3,900	4,300
6550	Maximum					10,600			10,600	11,700
6680	240" x 60", minimum					5,675			5,675	6,250
6700	Maximum					17,800			17,800	19,600
6720	Rectangle, 96" x 42", minimum					1,475			1,475	1,625
6740	Maximum					3,450			3,450	3,800
6760	120" x 48", minimum					3,250			3,250	3,600
6780	Maximum					4,250			4,250	4,675
6800	144" x 60", minimum					4,900			4,900	5,400
6820	Maximum					5,950			5,950	6,525
6840	168" x 60", minimum					5,100			5,100	5,600
6860	Maximum					7,600			7,600	8,350
6880	192" x 60", minimum					6,900			6,900	7,600
6900	Maximum					8,400			8,400	9,250
6960	240" x 60", minimum					8,225			8,225	9,050
6980	Maximum				↓	18,000			18,000	19,800

Labor, Equipment, Material and Total columns are under "2020 Bare Costs".

12 52 Seating

12 52 13 – Chairs

12 52 13.10 Chairs, Folding and Stack		Crew	Daily Output	Labor-Hours	Unit	Material	2020 Bare Costs Labor	Equipment	Total	Total Incl O&P
0010	**CHAIRS, FOLDING & STACK**									
2000	Folding, all steel, baked enamel finish,									
2100	Form fitting seat and backrests				Ea.	44			44	48.50
2200	Upholstered seat and back					71			71	78
2300	Polypropylene seat and back w/frame					58.50			58.50	64.50
2500	Chair caddy for above				↓	375			375	415
5000	Stack chair									
5300	Hardwood frame, seat and back									
5320	Minimum				Ea.	66			66	72.50
5340	Maximum				"	255			255	280
5400	Upholstered seat and back									
5420	Minimum				Ea.	66.50			66.50	73
5440	Maximum				"	370			370	410
5600	Metal frame, upholstered seat and back									
5620	Minimum				Ea.	56.50			56.50	62
5640	Maximum				"	415			415	455
5700	Plastic shell, metal legs									
5720	Minimum				Ea.	53.50			53.50	58.50
5740	Maximum					148			148	163
5800	Chair caddy for above					273			273	300
5900	Tablet arms, minimum					175			175	192
5920	Maximum				↓	249			249	274

12 52 19 – Upholstered Seating

12 52 19.13 Upholstered Office Seating		Crew	Daily Output	Labor-Hours	Unit	Material	Labor	Equipment	Total	Total Incl O&P
0010	**UPHOLSTERED OFFICE SEATING**									
3800	Lounge chair, upholstered, minimum				Ea.	540			540	595
3850	Maximum					2,950			2,950	3,250
4000	Sofa, two seat, upholstered, minimum					1,050			1,050	1,150
4100	Maximum					4,300			4,300	4,725
4650	Three seat, minimum					1,525			1,525	1,675
4680	Maximum					6,025			6,025	6,625
4700	Modular seating, lounge chair unit, upholstered, minimum					2,425			2,425	2,650
4750	Maximum					3,500			3,500	3,875
4780	Corner unit, minimum					1,650			1,650	1,825
4800	Maximum				↓	3,875			3,875	4,250

12 52 23 – Office Seating

12 52 23.13 Office Chairs		Crew	Daily Output	Labor-Hours	Unit	Material	Labor	Equipment	Total	Total Incl O&P
0010	**OFFICE CHAIRS**									
2000	Standard office chair, executive, minimum				Ea.	340			340	370
2150	Maximum					1,950			1,950	2,150
2200	Management, minimum					241			241	265
2250	Maximum					1,925			1,925	2,125
2280	Task, minimum					228			228	251
2290	Maximum					640			640	700
2300	Arm kit, minimum					86			86	94.50
2320	Maximum					137			137	150
2340	Ergonomic, executive, minimum					745			745	815
2380	Maximum					830			830	910
2390	Management, minimum					560			560	615
2400	Maximum					2,000			2,000	2,175
2450	Task, minimum					168			168	185
2500	Maximum				↓	670			670	735

12 52 Seating

12 52 23 – Office Seating

12 52 23.13 Office Chairs		Crew	Daily Output	Labor-Hours	Unit	2020 Bare Costs Material	Labor	Equipment	Total	Total Incl O&P
2550	Arm kit, minimum				Ea.	52.50			52.50	57.50
2600	Maximum					124			124	137
3000	Side/guest chairs, upholstered, sled base, metal, min.					147			147	162
3100	Maximum					725			725	795
3200	Wood, minimum					250			250	275
3250	Maximum					330			330	365
3400	Traditional, wood leg, minimum					201			201	221
3500	Maximum					1,425			1,425	1,575
3600	Conference chair, upholstered, metal leg, minimum					233			233	257
3700	Maximum				↓	1,175			1,175	1,275

12 52 23.23 Multiple Office Seating Units		Crew	Daily Output	Labor-Hours	Unit	Material	Labor	Equipment	Total	Total Incl O&P
0010	**MULTIPLE OFFICE SEATING UNITS**									
1000	Area seating, full upholstered, 3 seat straight unit, minimum				Ea.	1,350			1,350	1,475
1020	Maximum					2,925			2,925	3,225
1100	Four seat with corner table, minimum					4,000			4,000	4,400
1120	Maximum					3,000			3,000	3,300
1200	2 seat with in-line table, minimum					1,850			1,850	2,025
1220	Maximum					1,550			1,550	1,700
2000	Individual seat with table, minimum					890			890	980
2020	Maximum				↓	1,100			1,100	1,200

12 54 Hospitality Furniture

12 54 13 – Hotel and Motel Furniture

12 54 13.10 Hotel Furniture		Crew	Daily Output	Labor-Hours	Unit	Material	Labor	Equipment	Total	Total Incl O&P
0010	**HOTEL FURNITURE**									
0020	Standard quality set, minimum				Room	2,650			2,650	2,900
0200	Maximum				"	8,625			8,625	9,475
0300	Bed frame				Ea.	74			74	81
0400	Bench, upholstered, 42" x 18" x 18"					226			226	249
0420	18" x 18" x 23"					440			440	485
0500	Desk section, one drawer, 34" x 20" x 30"					267			267	293
0600	Free standing, 42" x 22" x 30"					335			335	370
0700	Desk chair, upholstered, minimum					141			141	155
0720	Maximum					178			178	196
1000	Dressers, uniplex, 2 drawer					390			390	430
1100	3 drawer					470			470	520
2000	Guest tables, 30" diameter					139			139	153
2100	34" diameter					256			256	282
3000	Headboards, free standing, twin					185			185	203
3050	Full					214			214	236
3100	Queen					227			227	249
3200	Wall mounted, twin					133			133	146
3250	Full					243			243	268
3300	Queen					315			315	350
4000	Lounge chair, full upholstered, minimum					480			480	530
4050	Maximum					435			435	480
4200	Open arms, minimum					121			121	133
4250	Maximum					455			455	500
5000	Mattress/box springs, twin					375			375	410
5050	Full					530			530	580
5100	Queen				↓	635			635	700

12 54 Hospitality Furniture

12 54 13 – Hotel and Motel Furniture

12 54 13.10 Hotel Furniture		Crew	Daily Output	Labor-Hours	Unit	Material	2020 Bare Costs Labor	Equipment	Total	Total Incl O&P
5150	Mirror, framed, 29″ x 45″				Ea.	160			160	176
6000	Sleep sofas, twin, minimum					390			390	430
6050	Maximum					830			830	915
6100	Full, minimum					765			765	845
6150	Maximum					1,350			1,350	1,475
6200	Queen, minimum					980			980	1,075
6250	Maximum					1,475			1,475	1,625
7000	Table, wood top, cocktail, 54″ x 24″ x 16″ high					310			310	340
7020	Corner, 30″ x 30″ x 21″ high					335			335	365
7040	End, 22″ x 28″ x 21″ high				↓	296			296	325

12 54 13.20 Mattress and Box Springs		Crew	Daily Output	Labor-Hours	Unit	Material	Labor	Equipment	Total	Total Incl O&P
0010	**MATTRESS & BOX SPRINGS** per set									
1000	Hospital, 34″ x 84″				Ea.	590			590	645
2000	Hotel/motel, twin, minimum					460			460	505
2020	Maximum					785			785	865
2200	Full, minimum					595			595	655
2220	Maximum					885			885	975
2400	Queen, minimum					605			605	665
2420	Maximum					935			935	1,025
3000	Stow away bed with head board, twin size				↓	755			755	830

12 54 16 – Restaurant Furniture

12 54 16.10 Tables, Folding		Crew	Daily Output	Labor-Hours	Unit	Material	Labor	Equipment	Total	Total Incl O&P
0010	**TABLES, FOLDING** Laminated plastic tops									
1000	Tubular steel legs with glides									
1020	18″ x 60″, minimum				Ea.	325			325	360
1040	Maximum					1,400			1,400	1,525
1100	18″ x 72″, minimum					330			330	365
1120	Maximum					1,450			1,450	1,575
1200	18″ x 96″, minimum					365			365	400
1220	Maximum					1,725			1,725	1,900
1400	30″ x 48″, minimum					189			189	208
1420	Maximum					315			315	345
1500	30″ x 60″, minimum					240			240	264
1520	Maximum					1,450			1,450	1,600
1600	30″ x 72″, minimum					375			375	415
1620	Maximum					1,625			1,625	1,800
1700	30″ x 96″, minimum					289			289	320
1720	Maximum					2,050			2,050	2,250
1800	36″ x 72″, minimum					305			305	335
1820	Maximum					1,800			1,800	2,000
1840	36″ x 96″, minimum					340			340	375
1860	Maximum					2,350			2,350	2,600
2000	Round, wood stained, plywood top, 60″ diameter, minimum					228			228	251
2020	Maximum					350			350	385
2200	72″ diameter, minimum					365			365	405
2220	Maximum				↓	465			465	510
4000	Mobile storage carts									
4020	For 72″ tables, flat, maximum 10 tables				Ea.	910			910	1,000
4040	96″ tables, flat, maximum 10 tables					840			840	920
4060	Rounds, on edge, maximum 10 tables				↓	1,025			1,025	1,150

12 54 Hospitality Furniture

12 54 16 – Restaurant Furniture

12 54 16.20 Furniture, Restaurant		Crew	Daily Output	Labor-Hours	Unit	Material	2020 Bare Costs Labor	Equipment	Total	Total Incl O&P
0010	**FURNITURE, RESTAURANT**									
0020	Bars, built-in, front bar	1 Carp	5	1.600	L.F.	315	85		400	480
0200	Back bar	"	5	1.600	"	230	85		315	390
0300	Booth seating, see Section 12 54 16.70									
2000	Chair, bentwood side chair, metal, minimum				Ea.	106			106	116
2020	Maximum					124			124	136
2100	Wood, minimum					220			220	242
2120	Maximum					258			258	284
2400	Bruer Cesca, cane seat & back, arms, minimum					665			665	730
2420	Maximum					725			725	795
2500	Side, minimum					650			650	715
2520	Maximum					715			715	790
2600	Upholstered seat & back, arms, minimum					170			170	187
2620	Maximum					520			520	570
2640	Side, minimum					475			475	525
2660	Maximum					595			595	655
2700	Corbusier, arm chair, cane seat and back, minimum					238			238	261
2720	Maximum					259			259	284
2740	Fledermaus, fabric seat, ash frame, minimum					219			219	241
2760	Maximum					274			274	300
2780	Hoffman arm, fabric seat, cane back, minimum					250			250	275
2800	Maximum					273			273	300
2820	Side, minimum					195			195	215
2840	Maximum					210			210	231
2860	Lombard, minimum					134			134	148
2880	Maximum					175			175	193
2900	Mies, side chair, leather seat and back, minimum					580			580	635
2920	Maximum					850			850	940
2940	Napoleon, upholstered seat, wood back, minimum					96.50			96.50	106
2960	Maximum					130			130	143
3000	Prague, arm, upholstered, minimum					190			190	209
3020	Maximum					385			385	425
3040	Side, upholstered, minimum					189			189	208
3060	Maximum					219			219	241
3080	Contemporary leather seat & back, chrome frame, min.					360			360	395
3100	Maximum					520			520	575
3120	Foam padded seat & back, s. steel barstock, min.					1,025			1,025	1,125
3140	Maximum					1,075			1,075	1,175
3160	Chrome tube cantilever frame, padded arms, min.					147			147	162
3180	Maximum					400			400	440
4200	"Tub" style, minimum					252			252	278
4220	Maximum					610			610	670
4280	Bent wood arms and legs with back bow, minimum					70.50			70.50	77.50
4300	Maximum					465			465	510
4320	Sled base, minimum					147			147	162
4340	Maximum					465			465	510
4360	Armchair foam padded seat & back, carved wood frame min.					425			425	470
4380	Maximum					815			815	895
4400	Ornate wood frame and legs, minimum					455			455	500
4420	Maximum					540			540	595
4440	Heavy wood frame and legs, tailored back, minimum					725			725	800
4460	Maximum				▼	845			845	925

12 54 16 – Restaurant Furniture

12 54 16.20 Furniture, Restaurant		Crew	Daily Output	Labor-Hours	Unit	Material	Labor	Equipment	Total	Total Incl O&P
						2020 Bare Costs				
4480	Button tufted back, minimum				Ea.	500			500	550
4500	Maximum					1,075			1,075	1,200
4520	"Dining" wood frame and legs, minimum					345			345	380
4550	Maximum					430			430	475
4600	"Queen Ann" wood frame and legs, minimum					475			475	520
4620	Maximum					525			525	575
4660	Upholstered arms and wood legs, minimum					670			670	740
4680	Maximum					790			790	865
4700	Wicker, foam padded seat and back, barrel design, minimum					590			590	650
4720	Maximum					670			670	735
4740	Couch design, minimum					440			440	485
4760	Maximum					535			535	590
4780	Chrome tube round cantilever base, minimum					268			268	295
4800	Maximum					555			555	610
4820	Chrome tube square cantilever base, minimum					243			243	267
4840	Maximum					540			540	595
4860	Misc. chair, upholst. seat, wood arms, bow back & legs, min.					198			198	217
4880	Maximum					275			275	300
4900	Open curved wood back, minimum					445			445	490
4950	Maximum					495			495	545
5000	Upholstered seat and open back, wood frame arms, min.					415			415	455
5100	Maximum					475			475	525
5150	Wood frame sled base, minimum					243			243	267
5200	Maximum					380			380	415
5250	Wood saddle seat, "Windsor", wood frame & legs, min.					234			234	257
5300	Maximum					273			273	300
5350	Brass nail trim leather, minimum					275			275	300
5400	Maximum					315			315	345
5450	Wood frame, fabric, minimum					250			250	275
5500	Maximum					465			465	515
6000	Stools, upholst. seat & back, chrome tube cantilever frame, min.					390			390	430
6150	Swivel on chrome post mount with spread base, minimum					231			231	254
6250	"Vienna" wood back and legs, minimum					289			289	320
6350	"Napoleon" wood back and legs, minimum					99			99	109
6450	Wood swivel seat and curved spindle back, leg base					172			172	189
6600	Upholstered seat no back, bent wood legs					345			345	380
7000	With back & straight wood frame and legs					345			345	375
7200	Wood swivel seat, "Captain" wood back and legs					219			219	240
7400	Veneer seat and curved wood back with arms, leg base					208			208	228
7600	Upholstered swivel seat no back, wood legs					82.50			82.50	90.50
8000	With back & wood legs					640			640	705
8200	Upholstered seat & wood back, wood legs					570			570	630
8400	Upholstered swivel seat and back, wood legs					247			247	272
8600	Square tube frame and legs					59.50			59.50	65.50
8800	Metal frame and legs					59.50			59.50	65.50

12 54 16.30 Table Bases		Crew	Daily Output	Labor-Hours	Unit	Material	Labor	Equipment	Total	Total Incl O&P
0010	**TABLE BASES**									
0040	Dining height									
1000	Metal disk design, minimum				Ea.	112			112	123
1200	Maximum					278			278	305
1400	Wood, minimum					505			505	555
1600	Maximum					555			555	615

12 54 16 – Restaurant Furniture

12 54 16.30 Table Bases		Crew	Daily Output	Labor-Hours	Unit	Material	Labor	Equipment	Total	Total Incl O&P
							2020 Bare Costs			
2000	Fluted wheel, minimum				Ea.	131			131	144
2200	Maximum					200			200	220
2400	Heavy cast iron, minimum					101			101	111
2600	Maximum					290			290	320
2800	Hobnail design, minimum					112			112	123
3000	Maximum					340			340	375
3200	Manhole design, minimum					144			144	158
3400	Maximum					292			292	320
3600	Ring design, minimum					214			214	236
3800	Maximum					320			320	350
4000	Trumpet design, minimum					360			360	395
4200	Maximum					400			400	440
4400	Tubular, round or rectangular shape, minimum					142			142	157
4600	Maximum					292			292	320
5000	Cocktail height bases, minimum					148			148	163
5200	Maximum					210			210	231
5400	For foot ring, add, minimum					37.50			37.50	41.50
5600	Maximum				↓	50.50			50.50	55.50

12 54 16.40 Table Tops

12 54 16.40 Table Tops		Crew	Daily Output	Labor-Hours	Unit	Material	Labor	Equipment	Total	Total Incl O&P
0010	**TABLE TOPS** Laminated plastic top and edge									
0040	24" wide, 24" long, minimum				Ea.	143			143	158
1000	Maximum					271			271	298
1200	30" long, minimum					179			179	197
1400	Maximum					400			400	440
1600	36" long, minimum					215			215	236
1800	Maximum					425			425	465
2000	48" long, minimum					287			287	315
2200	Maximum					495			495	545
2400	60" long, minimum					360			360	395
2600	Maximum					775			775	855
2800	72" long, minimum					430			430	475
3000	Maximum					820			820	900
3050	30" wide, 30" long, minimum					174			174	191
3150	Maximum					415			415	460
3200	36" long, minimum					208			208	229
3250	Maximum					495			495	545
3400	42" long, minimum					243			243	267
3600	Maximum					540			540	595
3800	48" long, minimum					278			278	305
4000	Maximum					590			590	650
4200	60" long, minimum					345			345	380
4400	Maximum					700			700	770
4600	72" long, minimum					415			415	460
4800	Maximum					825			825	905
5000	36" wide, 36" long, minimum					199			199	219
5200	Maximum					425			425	465
5400	48" long, minimum					265			265	292
5600	Maximum					550			550	605
5800	60" long, minimum					330			330	365
6000	Maximum					710			710	780
6200	72" long, minimum					400			400	440
6400	Maximum				↓	835			835	920

12 54 Hospitality Furniture

12 54 16 – Restaurant Furniture

12 54 16.40 Table Tops		Crew	Daily Output	Labor-Hours	Unit	Material	Labor	Equipment	Total	Total Incl O&P
						2020 Bare Costs				
6500	42" wide, 42" long, minimum				Ea.	268			268	295
6600	Maximum					515			515	570
6700	Round, 24" diameter, minimum					95.50			95.50	105
6800	Maximum					375			375	410
6900	30" diameter, minimum					129			129	141
7000	Maximum					560			560	615
7200	36" diameter, minimum					164			164	181
7400	Maximum					645			645	705
7600	42" diameter, minimum					263			263	289
8000	Maximum					985			985	1,075
8200	48" diameter, minimum					299			299	330
8400	Maximum					1,050			1,050	1,150
8800	54" diameter, minimum					480			480	525
9000	Maximum					1,350			1,350	1,500
9200	60" diameter, minimum					510			510	560
9400	Maximum				▼	1,700			1,700	1,850

12 54 16.50 Table Tops		Crew	Daily Output	Labor-Hours	Unit	Material	Labor	Equipment	Total	Total Incl O&P
0010	**TABLE TOPS** laminate top and hardwood edge									
0040	24" wide, 24" long, minimum				Ea.	298			298	330
1000	Maximum					750			750	830
1200	30" long, minimum					330			330	365
1400	Maximum					805			805	885
1600	36" long, minimum					260			260	286
1800	Maximum					845			845	930
2000	48" long, minimum					385			385	425
2100	Maximum					930			930	1,025
2200	60" long, minimum					620			620	685
2400	Maximum					1,025			1,025	1,125
2600	72" long, minimum					670			670	740
2800	Maximum					1,125			1,125	1,250
3000	30" wide, 30" long, minimum					320			320	350
3100	Maximum					860			860	945
3200	36" long, minimum					370			370	405
3400	Maximum					920			920	1,025
3600	42" long, minimum					380			380	415
3800	Maximum					1,175			1,175	1,300
4000	48" long, minimum					475			475	525
4100	Maximum					990			990	1,100
4200	60" long, minimum					565			565	625
4300	Maximum					1,075			1,075	1,175
4400	72" long, minimum					655			655	720
4600	Maximum					1,125			1,125	1,225
4800	36" wide, 36" long, minimum					390			390	430
5000	Maximum					945			945	1,025
5200	48" long, minimum					485			485	535
5400	Maximum					1,075			1,075	1,175
5600	60" long, minimum					555			555	610
5800	Maximum					1,200			1,200	1,325
6000	72" long, minimum					805			805	885
6200	Maximum					1,325			1,325	1,450
6400	42" wide, 42" long, minimum					530			530	585
6600	Maximum				▼	1,150			1,150	1,250

12 54 16.50 Table Tops		Crew	Daily Output	Labor-Hours	Unit	Material	2020 Bare Costs Labor	Equipment	Total	Total Incl O&P
6800	Round, 24" diameter, minimum				Ea.	515			515	565
6900	Maximum					840			840	925
7000	30" diameter, minimum					565			565	625
7200	Maximum					950			950	1,050
7400	36" diameter, minimum					630			630	690
7600	Maximum					1,125			1,125	1,225
7800	42" diameter, minimum					725			725	800
7900	Maximum					1,375			1,375	1,500
8000	48" diameter, minimum					775			775	855
8150	Maximum					1,450			1,450	1,600
8200	54" diameter, minimum					1,175			1,175	1,300
8250	Maximum					1,900			1,900	2,100
8300	60" diameter, minimum					1,175			1,175	1,300
8350	Maximum					1,900			1,900	2,100

12 54 16.60 Table Tops

Code	Description	Crew	Daily Output	Labor-Hours	Unit	Material	Labor	Equipment	Total	Total Incl O&P
0010	**TABLE TOPS** Polyester resin top & edge									
0050	Add inlay top material to cost									
0100	24" wide, 24" long				Ea.	450			450	495
1200	30" long					450			450	495
1600	36" long					470			470	515
2000	48" long					485			485	530
2400	60" long					575			575	630
2800	72" long					675			675	740
3200	30" wide, 30" long					495			495	545
3600	36" long					585			585	645
4000	42" long					660			660	725
4400	48" long					665			665	730
4800	60" long					735			735	810
5200	72" long					900			900	990
5600	36" wide, 36" long					750			750	825
6000	48" long					905			905	1,000
6400	60" long					1,100			1,100	1,200
6800	72" long					1,325			1,325	1,475
7200	42" wide, 42" long					930			930	1,025
7600	Round, 24" diameter					660			660	725
8000	30" diameter					685			685	755
8400	36" diameter					815			815	895
8800	42" diameter					1,100			1,100	1,200
9150	48" diameter					1,200			1,200	1,325
9250	54" diameter					1,625			1,625	1,800
9350	60" diameter					1,700			1,700	1,875

12 54 16.70 Booths

Code	Description	Crew	Daily Output	Labor-Hours	Unit	Material	Labor	Equipment	Total	Total Incl O&P
0010	**BOOTHS**									
1000	Banquet, upholstered seat and back, custom									
1500	Straight, minimum	2 Carp	40	.400	L.F.	209	21.50		230.50	264
1520	Maximum		36	.444		405	23.50		428.50	485
1600	"L" or "U" shape, minimum		35	.457		213	24.50		237.50	273
1620	Maximum		30	.533		375	28.50		403.50	460
1800	Upholstered outside finished backs for									
1810	single booths and custom banquets									
1820	Minimum	2 Carp	44	.364	L.F.	29	19.35		48.35	63
1840	Maximum	"	40	.400	"	78.50	21.50		100	120

12 54 Hospitality Furniture

12 54 16 – Restaurant Furniture

12 54 16.70 Booths		Crew	Daily Output	Labor-Hours	Unit	Material	2020 Bare Costs Labor	Equipment	Total	Total Incl O&P
3000	Fixed seating, one piece plastic chair and									
3010	plastic laminate table top									
3100	Two seat, 24" x 24" table, minimum	F-7	30	1.067	Ea.	835	51		886	1,000
3120	Maximum		26	1.231		1,200	58.50		1,258.50	1,425
3200	Four seat, 24" x 48" table, minimum		28	1.143		830	54.50		884.50	1,000
3220	Maximum		24	1.333		1,425	63.50		1,488.50	1,650
3300	Six seat, 24" x 76" table, minimum		26	1.231		1,500	58.50		1,558.50	1,750
3320	Maximum		22	1.455		2,075	69.50		2,144.50	2,375
3400	Eight seat, 24" x 102" table, minimum		20	1.600		2,000	76		2,076	2,325
3420	Maximum		18	1.778		2,625	84.50		2,709.50	3,000
4000	Free standing, wood fiber core with									
4010	plastic laminate face, single booth									
4100	24" wide	2 Carp	38	.421	Ea.	395	22.50		417.50	470
4150	48" wide		34	.471		480	25		505	570
4200	60" wide		30	.533		585	28.50		613.50	690
4300	Double booth, 24" wide		32	.500		585	26.50		611.50	690
4350	48" wide		28	.571		785	30.50		815.50	915
4400	60" wide		26	.615		985	32.50		1,017.50	1,125
4600	Upholstered seat and back									
4650	Foursome, single booth, minimum	2 Carp	38	.421	Ea.	450	22.50		472.50	530
4700	Maximum		30	.533		1,700	28.50		1,728.50	1,925
4800	Double booth, minimum		32	.500		985	26.50		1,011.50	1,125
4850	Maximum		26	.615		2,625	32.50		2,657.50	2,925
5000	Mount in floor, wood fiber core with									
5010	plastic laminate face, single booth									
5050	24" wide	F-7	30	1.067	Ea.	320	51		371	435
5100	48" wide		28	1.143		410	54.50		464.50	535
5150	60" wide		26	1.231		585	58.50		643.50	740
5200	Double booth, 24" wide		26	1.231		520	58.50		578.50	665
5250	48" wide		24	1.333		660	63.50		723.50	830
5300	60" wide		22	1.455		940	69.50		1,009.50	1,125

12 55 Detention Furniture

12 55 13 – Detention Bunks

12 55 13.13 Cots

		Crew	Daily Output	Labor-Hours	Unit	Material	Labor	Equipment	Total	Total Incl O&P
0010	**COTS**									
2500	Bolted, single, painted steel	E-4	20	1.600	Ea.	350	93	7.35	450.35	545
2700	Stainless steel	"	20	1.600	"	1,025	93	7.35	1,125.35	1,275

12 56 Institutional Furniture

12 56 33 – Classroom Furniture

12 56 33.10 Furniture, School

		Crew	Daily Output	Labor-Hours	Unit	Material	Labor	Equipment	Total	Total Incl O&P
0010	**FURNITURE, SCHOOL**									
1000	Chair, molded plastic									
1100	Integral tablet arm, minimum				Ea.	92			92	101
1150	Maximum					224			224	246
2000	Desk, single pedestal, top book compartment, minimum					110			110	121
2020	Maximum					202			202	222
2100	Side book compartment, minimum					142			142	156

12 56 Institutional Furniture

12 56 33 – Classroom Furniture

12 56 33.10 Furniture, School		Crew	Daily Output	Labor-Hours	Unit	2020 Bare Costs Material	Labor	Equipment	Total	Total Incl O&P
2120	Maximum				Ea.	159			159	175
2200	Flip top, minimum					234			234	257
2220	Maximum					270			270	297
3000	Preschool, moulded plastic chairs, minimum					24.50			24.50	27
3020	Maximum					61.50			61.50	67.50
3800	Tables, plastic laminate top, 24" wide, 36" long					70			70	77
3820	48" long					81			81	89
3840	60" long					125			125	137
3900	30" wide, 48" long					103			103	113
3920	60" long					147			147	162
3940	72" long					170			170	187
4000	36" wide, 36" long					148			148	163
4020	60" long					204			204	224
4040	72" long					256			256	282
4200	Round, 36" diam.					102			102	113
4220	42" diam.					147			147	162
4230	48" diam.					161			161	177
4240	60" diam.					175			175	193

12 56 43 – Dormitory Furniture

12 56 43.10 Dormitory Furnishings

		Crew	Daily Output	Labor-Hours	Unit	Material	Labor	Equipment	Total	Total Incl O&P
0010	**DORMITORY FURNISHINGS**									
0200	Bookcase, two shelf				Ea.	286			286	315
0300	Bunkable bed, twin, minimum					390			390	430
0320	Maximum					600			600	660
1000	Chest, four drawer, minimum					390			390	430
1020	Maximum					775			775	850
1050	Built-in, minimum	2 Carp	13	1.231	L.F.	140	65.50		205.50	259
1150	Maximum		10	1.600		259	85		344	420
1200	Desk top, built-in, laminated plastic, 24" deep, minimum		50	.320		49.50	17		66.50	81.50
1300	Maximum		40	.400		148	21.50		169.50	197
1450	30" deep, minimum		50	.320		63	17		80	97
1550	Maximum		40	.400		276	21.50		297.50	340
1750	Dressing unit, built-in, minimum		12	1.333		210	71		281	345
1850	Maximum		8	2		630	106		736	865
2000	Desk, single pedestal				Ea.	580			580	640
2100	Hutch/bookcase, with light					282			282	310
3000	Ladder					121			121	134
4000	Mirror, with frame					115			115	126
5000	Nightstand					285			285	315
6000	Wall unit, open shelving					645			645	710
7000	Wardrobe					565			565	620

12 56 51 – Library Furniture

12 56 51.10 Library Furnishings

		Crew	Daily Output	Labor-Hours	Unit	Material	Labor	Equipment	Total	Total Incl O&P
0010	**LIBRARY FURNISHINGS**									
0100	Attendant desk, 36" x 62" x 29" high	1 Carp	16	.500	Ea.	2,225	26.50		2,251.50	2,500
0200	Book display, "A" frame display, both sides, 42" x 42" x 60" high		16	.500		1,250	26.50		1,276.50	1,425
0220	Table with bulletin board, 42" x 24" x 49" high		16	.500		735	26.50		761.50	850
0300	Book trucks, descending platform,									
0320	Small, 14" x 30" x 35" high	1 Carp	16	.500	Ea.	815	26.50		841.50	940
0340	Large, 14" x 40" x 42" high		16	.500		835	26.50		861.50	965
0800	Card catalog, 30 tray unit		16	.500		3,600	26.50		3,626.50	4,025
0840	60 tray unit		16	.500		7,500	26.50		7,526.50	8,325

12 56 Institutional Furniture

12 56 51 – Library Furniture

12 56 51.10 Library Furnishings		Crew	Daily Output	Labor-Hours	Unit	Material	2020 Bare Costs Labor	Equipment	Total	Total Incl O&P
0880	72 tray unit	2 Carp	16	1	Ea.	8,475	53		8,528	9,400
0960	120 tray unit	"	16	1		14,800	53		14,853	16,400
1000	Carrels, single face, initial unit	1 Carp	16	.500		990	26.50		1,016.50	1,150
1050	Additional unit	"	16	.500		765	26.50		791.50	885
1500	Double face, initial unit	2 Carp	16	1		1,450	53		1,503	1,675
1550	Additional unit		16	1		850	53		903	1,025
1600	Cloverleaf		11	1.455		2,525	77.50		2,602.50	2,900
1710	Carrels, hardwood, 36" x 24", minimum	1 Carp	5	1.600		580	85		665	775
1720	Maximum		4	2		1,825	106		1,931	2,175
2000	Chairs, sled base, arms, minimum		24	.333		202	17.70		219.70	251
2050	Maximum		16	.500		208	26.50		234.50	272
2100	No arms, minimum		24	.333		142	17.70		159.70	185
2150	Maximum		16	.500		204	26.50		230.50	267
2500	Standard leg base, arms, minimum		24	.333		232	17.70		249.70	284
2520	Maximum		16	.500		440	26.50		466.50	525
2600	No arms, minimum		24	.333		163	17.70		180.70	208
2620	Maximum		16	.500		305	26.50		331.50	380
2700	Card catalog file, 60 trays, complete					10,200			10,200	11,200
2720	Alternate method: each tray					169			169	186
3000	Charge desk, modular unit, 35" x 27" x 39" high									
3020	Wood front and edges, plastic laminate tops									
3100	Book return	1 Carp	16	.500	Ea.	1,050	26.50		1,076.50	1,225
3150	Book truck port		16	.500		730	26.50		756.50	845
3400	Corner		16	.500		1,275	26.50		1,301.50	1,475
3450	Cupboard		16	.500		1,500	26.50		1,526.50	1,700
3500	Detachable end panel		16	.500		440	26.50		466.50	530
3550	Gate		16	.500		525	26.50		551.50	620
3600	Knee space		16	.500		1,325	26.50		1,351.50	1,500
3650	Open storage		16	.500		955	26.50		981.50	1,100
3700	Station charge		16	.500		1,575	26.50		1,601.50	1,800
3750	Work station		16	.500		1,600	26.50		1,626.50	1,825
3800	Charging desk, built-in, with counter, plastic laminated top		7	1.143	L.F.	500	60.50		560.50	645
4000	Dictionary stand, stationary		16	.500	Ea.	725	26.50		751.50	840
4020	Revolving		16	.500		255	26.50		281.50	325
4200	Exhibit case, table style, 60" x 28" x 36"		11	.727		2,950	38.50		2,988.50	3,300
4500	Globe stand		16	.500		980	26.50		1,006.50	1,125
4800	Magazine rack		16	.500		880	26.50		906.50	1,025
5000	Newspaper rack		16	.500		850	26.50		876.50	980
6010	Bookshelf, metal, 90" high, 10" shelf, double face		11.50	.696	L.F.	206	37		243	285
6020	Single face		12	.667	"	117	35.50		152.50	185
6050	For 8" shelving, subtract from above					10%				
6060	For 12" shelving, add to above					10%				
6070	For 42" high with countertop, subtract from above					20%				
6100	Mobile compacted shelving, hand crank, 9'-0" high									
6110	Double face, including track, 3' section				Ea.	2,225			2,225	2,450
6150	For electrical operation, add					25%				
6200	Magazine shelving, 82" high, 12" deep, single face	1 Carp	11.50	.696	L.F.	169	37		206	245
6210	Double face		11.50	.696	"	290	37		327	380
7100	Index, single tier, 48" x 72"		16	.500	Ea.	3,075	26.50		3,101.50	3,425
7150	Double tier, 48" x 72"		16	.500		2,250	26.50		2,276.50	2,525
7200	Reading table, laminated top, 60" x 36"					450			450	495
8000	Parsons table, 29" high, plastic lam. top, wood legs & edges									
8010	36" x 36"	2 Carp	16	1	Ea.	500	53		553	630

12 56 Institutional Furniture

12 56 51 – Library Furniture

12 56 51.10 Library Furnishings		Crew	Daily Output	Labor-Hours	Unit	Material	Labor	Equipment	Total	Total Incl O&P
						2020 Bare Costs				
8020	36″ x 60″	2 Carp	16	1	Ea.	610	53		663	755
8030	36″ x 72″		16	1		795	53		848	960
8040	36″ x 84″		16	1		1,150	53		1,203	1,325
8050	42″ x 90″		16	1		1,100	53		1,153	1,275
8060	48″ x 72″		16	1		685	53		738	840
8070	48″ x 120″		16	1		1,850	53		1,903	2,100
8110	42″ diameter		16	1		640	53		693	790
8120	48″ diameter		16	1		700	53		753	855
8130	60″ diameter		16	1		865	53		918	1,025
8500	Study, panel ends, plastic laminate surfaces 29″ high, 36″ x 60″		16	1		600	53		653	745
8510	36″ x 72″		16	1		635	53		688	785
8515	36″ x 90″		16	1		660	53		713	810
8525	48″ x 72″		16	1		1,100	53		1,153	1,300

12 56 70 – Healthcare Furniture

12 56 70.10 Furniture, Hospital		Crew	Daily Output	Labor-Hours	Unit	Material	Labor	Equipment	Total	Total Incl O&P
0010	**FURNITURE, HOSPITAL**									
0020	Beds, manual, minimum				Ea.	800			800	880
0100	Maximum					2,950			2,950	3,250
0600	All electric hospital beds, minimum					1,975			1,975	2,175
0700	Maximum					3,675			3,675	4,050
0900	Manual, nursing home beds, minimum					690			690	760
1000	Maximum					1,925			1,925	2,125
1020	Overbed table, laminated top, minimum					365			365	400
1040	Maximum					770			770	850
1100	Patient wall systems, not incl. plumbing, minimum				Room	1,375			1,375	1,500
1200	Maximum				″	2,275			2,275	2,500
2000	Geriatric chairs, minimum				Ea.	475			475	525
2020	Maximum				″	770			770	850

12 59 Systems Furniture

12 59 13 – Panel-Hung Component System Furniture

12 59 13.10 Furniture, Office Systems		Crew	Daily Output	Labor-Hours	Unit	Material	Labor	Equipment	Total	Total Incl O&P
0010	**FURNITURE, OFFICE SYSTEMS**, Panel hung									
0100	Acoustic panel, 43″ high, 24″ wide, NRC .85				Ea.	400			400	440
0110	30″ wide					445			445	490
0120	36″ wide					495			495	545
0130	48″ wide					590			590	645
0200	64″ high, 30″ wide					530			530	580
0210	36″ wide					570			570	625
0220	42″ wide					660			660	725
0230	48″ wide					695			695	765
0240	60″ wide					765			765	845
0400	Bookshelf, 36″ wide					184			184	203
0410	42″ wide					194			194	213
0420	48″ wide					200			200	220
1000	Connectors, brackets and supports									
1200	Bracket, cantilever, 20″ deep, 24″ deep				Ea.	62.50			62.50	69
1220	Countertop, per pair					26			26	28.50
1240	Flat, 20″ deep, 24″ wide					32			32	35.50
1260	Worksurface kit, per pair					26			26	28.50

12 59 Systems Furniture

12 59 13 – Panel-Hung Component System Furniture

12 59 13.10 Furniture, Office Systems		Crew	Daily Output	Labor-Hours	Unit	Material	2020 Bare Costs Labor	Equipment	Total	Total Incl O&P
1300	Connector kit, ell, 43" high, 90" deep				Ea.	111			111	122
1320	64" high					65.50			65.50	72
1340	Straight, 43" high					45.50			45.50	50.50
1360	64" high					45.50			45.50	50.50
1400	Support column for peninsula					160			160	176
2000	Countertop, 15" deep, 36" wide					211			211	232
2020	48" wide					235			235	258
3000	Duplex receptacle circuit 1					23			23	25
3010	Circuit 2					23			23	25
3100	Electric power harness, 36" wide					149			149	164
3120	42" wide					315			315	345
3140	60" wide					157			157	173
4000	End cover, panel, 43" high					45.50			45.50	50.50
4020	64" high					45.50			45.50	50.50
4100	Finish, variable height, 2-way					68.50			68.50	75.50
4600	Overhead cabinet with door, 42" wide					425			425	465
4620	60" wide					655			655	720
5000	Pedestal spacer, 22" deep, 15" wide					74.50			74.50	82
5100	28" deep, 15" wide					90			90	99
5200	Pedestal box, box, file, 22" deep, 26" high					525			525	575
5300	28" deep					545			545	600
5400	Pedestal file, file, 22" deep, 26" high					485			485	535
5600	Lateral file, 2 drawer, 30" wide					690			690	760
6000	Power, base in-feed cable					166			166	183
7400	Task light, recessed, 30"-36" wide					203			203	223
7420	42"-48" wide					220			220	242
7600	60" wide					238			238	262
8000	Worksurface, radius edge, 24" deep, 36" wide					218			218	240
8020	42" wide					269			269	296
8040	48" wide					287			287	315
8060	60" wide					360			360	395
8080	72" wide					405			405	445
8200	30" deep, 42" wide					345			345	380
8220	60" wide					400			400	440
8240	72" wide					460			460	505
8400	Corner, 24" deep, 36" wide					525			525	580
8420	42" wide					595			595	655
8500	Peninsula, 36" wide, 66" long					690			690	755
9000	For installation of systems furniture, add 5%									

12 59 16 – Free-Standing Component System Furniture

12 59 16.10 Furniture, Office Systems		Crew	Daily Output	Labor-Hours	Unit	Material	Labor	Equipment	Total	Total Incl O&P
0010	**FURNITURE, OFFICE SYSTEMS**, Freestanding									
0100	Desk table, 24" deep x 48" wide				Ea.	223			223	246
0110	30" deep x 48" wide					410			410	450
0120	30" deep x 60" wide					570			570	625
0130	30" deep x 72" wide					800			800	880
0200	File, mobile, 2 drawer, 22" deep					565			565	620
0210	Suspended, 2 file					565			565	620
0220	Undercounter					680			680	750
0230	2 box, 1 file					660			660	725
0400	Keyboard platform with articulating arm					217			217	239
0500	End support leg					206			206	227

12 59 Systems Furniture

12 59 16 – Free-Standing Component System Furniture

12 59 16.10 Furniture, Office Systems		Crew	Daily Output	Labor-Hours	Unit	Material (2020 Bare Costs)	Labor (2020 Bare Costs)	Equipment (2020 Bare Costs)	Total (2020 Bare Costs)	Total Incl O&P
0800	Modesty panel, 48" wide				Ea.	127			127	139
0810	Return					138			138	151
0820	Pencil/utility drawer					161			161	177
0830	Privacy screen, 48" wide x 17"-20" high					365			365	405
0840	72" wide					475			475	525
0910	Printer stand, 30" deep x 36" wide					780			780	860
0920	Universal					760			760	840
0930	Paper basket					91.50			91.50	101
1010	Storage unit with doors, 60" wide					840			840	920
1020	72" wide					1,250			1,250	1,375
1210	Table, peninsula, 30" x 72"					1,050			1,050	1,150
1220	Return, 24" x 48"					820			820	900
1230	Round conference return, 42" diameter					690			690	755
1510	Work surface, 24" deep x 48" wide					385			385	420
1520	30" deep x 30" wide					400			400	440
1530	60" wide					495			495	545
1540	72" wide					450			450	495
1550	Corner bridge, 30" x 42"					227			227	250
1560	Peninsula, 30" x 60"					625			625	685
1570	Connector plate				↓	24			24	26.50

12 59 23 – Desk System Furniture

12 59 23.10 Work Stations		Crew	Daily Output	Labor-Hours	Unit	Material	Labor	Equipment	Total	Total Incl O&P
0010	**WORK STATIONS**									
1000	Secretarial work station, minimum				Ea.	5,150			5,150	5,650
1020	Maximum					17,600			17,600	19,300
1400	Management work station, minimum					5,850			5,850	6,450
1420	Maximum					18,900			18,900	20,800
1800	Executive work station, minimum					13,000			13,000	14,300
1820	Maximum				↓	33,400			33,400	36,800
1840	For installation of systems furniture, add 15%									

12 61 Fixed Audience Seating

12 61 13 – Upholstered Audience Seating

12 61 13.13 Auditorium Chairs		Crew	Daily Output	Labor-Hours	Unit	Material	Labor	Equipment	Total	Total Incl O&P
0010	**AUDITORIUM CHAIRS**									
2000	All veneer construction	2 Carp	22	.727	Ea.	300	38.50		338.50	395
2200	Veneer back, padded seat		22	.727		305	38.50		343.50	395
2350	Fully upholstered, spring seat	↓	22	.727		255	38.50		293.50	340
2450	For tablet arms, add					85			85	93.50
2500	For fire retardancy, CATB-133, add				↓	30.50			30.50	33.50

12 61 13.23 Lecture Hall Seating		Crew	Daily Output	Labor-Hours	Unit	Material	Labor	Equipment	Total	Total Incl O&P
0010	**LECTURE HALL SEATING**									
1000	Pedestal type, minimum	2 Carp	22	.727	Ea.	272	38.50		310.50	360
1200	Maximum	"	14.50	1.103	"	550	58.50		608.50	700

12 63 Stadium and Arena Seating

12 63 13 – Stadium and Arena Bench Seating

12 63 13.13 Bleachers		Crew	Daily Output	Labor-Hours	Unit	Material	2020 Bare Costs Labor	Equipment	Total	Total Incl O&P
0010	**BLEACHERS**									
3000	Telescoping, manual to 15 tier, minimum	F-5	65	.492	Seat	114	26.50		140.50	169
3100	Maximum		60	.533		172	28.50		200.50	235
3300	16 to 20 tier, minimum		60	.533		250	28.50		278.50	320
3400	Maximum		55	.582		340	31		371	425
3600	21 to 30 tier, minimum		50	.640		272	34.50		306.50	355
3700	Maximum		40	.800		435	43		478	550
3900	For integral power operation, add, minimum	2 Elec	300	.053		55.50	3.27		58.77	66.50
4000	Maximum	"	250	.064		91	3.93		94.93	106
5000	Benches, folding, in wall, 14' table, 2 benches	L-4	2	12	Set	880	600		1,480	1,925

12 67 Pews and Benches

12 67 13 – Pews

12 67 13.13 Sanctuary Pews

		Crew	Daily Output	Labor-Hours	Unit	Material	Labor	Equipment	Total	Total Incl O&P
0010	**SANCTUARY PEWS**									
1500	Bench type, hardwood, minimum	1 Carp	20	.400	L.F.	111	21.50		132.50	156
1550	Maximum	"	15	.533		172	28.50		200.50	235
1570	For kneeler, add					22			22	24.50

12 92 Interior Planters and Artificial Plants

12 92 13 – Interior Artificial Plants

12 92 13.10 Plants

		Crew	Daily Output	Labor-Hours	Unit	Material	Labor	Equipment	Total	Total Incl O&P
0010	**PLANTS** Permanent only, weighted									
0020	Preserved or polyester leaf, natural wood									
0030	Or molded trunk. For pots see Section 12 92 33.10									
0100	Plants, acuba, 5' high				Ea.	360			360	395
0200	Apidistra, 4' high					315			315	345
0300	Beech, variegated, 5' high					310			310	340
0400	Birds nest fern, 6' high					550			550	610
0500	Croton, 4' high					135			135	148
0600	Diffenbachia, 3' high					124			124	136
0700	Ficus Benjamina, 3' high					147			147	162
0720	6' high					370			370	410
0760	Nitida, 3' high					330			330	365
0780	6' high					580			580	635
0800	Helicona					310			310	340
1200	Palm, green date fan, 5' high					320			320	350
1220	7' high					405			405	445
1280	Green chamdora date, 5' high					290			290	320
1300	7' high					440			440	485
1400	Green giant, 7' high					685			685	755
1600	Rubber plant, 5' high					233			233	257
1800	Schefflera, 3' high					160			160	176
1820	4' high					192			192	211
1840	5' high					305			305	335
1860	6' high					415			415	460
1880	7' high					625			625	685
1900	Spathiphyllum, 3' high					222			222	244
4000	Trees, polyester or preserved, with									

12 92 Interior Planters and Artificial Plants

12 92 13 – Interior Artificial Plants

12 92 13.10 Plants		Crew	Daily Output	Labor-Hours	Unit	Material	2020 Bare Costs Labor	Equipment	Total	Total Incl O&P
4020	Natural trunks									
4100	Acuba, 10' high				Ea.	1,750			1,750	1,925
4120	12' high					2,450			2,450	2,675
4140	14' high					3,400			3,400	3,750
4400	Bamboo, 10' high					760			760	835
4420	12' high					880			880	965
4800	Beech, 10' high					2,325			2,325	2,575
4820	12' high					2,675			2,675	2,950
4840	14' high					3,625			3,625	3,975
5000	Birch, 10' high					2,325			2,325	2,575
5020	12' high					2,675			2,675	2,950
5040	14' high					2,925			2,925	3,200
5060	16' high					3,625			3,625	3,975
5080	18' high					4,075			4,075	4,500
5500	Ficus Benjamina, 10' high					1,725			1,725	1,900
5520	12' high					2,200			2,200	2,425
5540	14' high					2,450			2,450	2,700
5560	16' high					3,425			3,425	3,775
5580	18' high					5,150			5,150	5,675
6000	Magnolia, 10' high					2,525			2,525	2,775
6020	12' high					3,575			3,575	3,925
6040	14' high					4,875			4,875	5,375
6500	Maple, 10' high					270			270	297
6520	12' high					405			405	445
6540	14' high					500			500	550
7000	Palms, chamadora, 10' high					490			490	540
7020	12' high					625			625	690
7040	Date, 10' high					675			675	745
7060	12' high				↓	775			775	850

12 92 33 – Interior Planters

12 92 33.10 Planters		Crew	Daily Output	Labor-Hours	Unit	Material	Labor	Equipment	Total	Total Incl O&P
0010	**PLANTERS**									
1000	Fiberglass, hanging, 12" diameter, 7" high				Ea.	109			109	120
1100	15" diameter, 7" high					158			158	174
1200	36" diameter, 8" high					218			218	240
1500	Rectangular, 48" long, 16" high, 15" wide					615			615	675
1550	16" high, 24" wide					670			670	735
1600	24" high, 24" wide					980			980	1,075
1650	60" long, 30" high, 28" wide					1,025			1,025	1,150
1700	72" long, 16" high, 15" wide					845			845	930
1750	21" high, 24" wide					1,075			1,075	1,200
1800	30" high, 24" wide					1,200			1,200	1,300
2000	Round, 12" diameter, 13" high					203			203	223
2050	25" high					241			241	265
2150	14" diameter, 15" high					188			188	207
2200	16" diameter, 16" high					198			198	217
2250	18" diameter, 19" high					239			239	262
2300	23" high					370			370	410
2350	20" diameter, 16" high					223			223	245
2400	18" high					245			245	270
2450	21" high					315			315	350
2500	22" diameter, 10" high				↓	240			240	264

12 92 33 – Interior Planters

12 92 33.10 Planters		Crew	Daily Output	Labor-Hours	Unit	Material	2020 Bare Costs Labor	Equipment	Total	Total Incl O&P
2550	24" diameter, 16" high				Ea.	288			288	315
2600	19" high					345			345	380
2650	25" high					395			395	435
2700	36" high					620			620	680
2750	48" high					865			865	950
2800	30" diameter, 16" high					340			340	375
2850	18" high					345			345	375
2900	21" high					385			385	420
3000	24" high					395			395	435
3350	27" high					435			435	480
3400	36" diameter, 16" high					425			425	470
3450	18" high					445			445	485
3500	21" high					475			475	525
3550	24" high					510			510	560
3600	27" high					555			555	610
3650	30" high					600			600	660
3700	48" diameter, 16" high					745			745	820
3750	21" high					805			805	885
3800	24" high					900			900	990
3850	27" high					945			945	1,050
3900	30" high					1,025			1,025	1,125
3950	36" high					1,075			1,075	1,175
4000	60" diameter, 16" high					985			985	1,075
4100	21" high					1,125			1,125	1,225
4150	27" high					1,250			1,250	1,375
4200	30" high					1,375			1,375	1,525
4250	33" high					1,575			1,575	1,725
4300	36" high					1,925			1,925	2,100
4400	39" high					2,075			2,075	2,275
5000	Square, 10" side, 20" high					206			206	227
5100	14" side, 15" high					238			238	262
5200	18" side, 19" high					250			250	274
5300	20" side, 16" high					305			305	335
5320	18" high					475			475	525
5340	21" high					395			395	435
5400	24" side, 16" high					385			385	425
5420	21" high					580			580	640
5440	25" high					540			540	595
5460	30" side, 16" high					515			515	565
5480	24" high					815			815	895
5490	27" high					600			600	660
5500	Round, 36" diameter, 16" high					580			580	640
5510	18" high					805			805	890
5520	21" high					890			890	975
5530	24" high					985			985	1,075
5540	27" high					1,125			1,125	1,225
5550	30" high					1,550			1,550	1,700
5800	48" diameter, 16" high					820			820	900
5820	21" high					915			915	1,000
5840	24" high					1,125			1,125	1,250
5860	27" high					1,225			1,225	1,350
5880	30" high					325			325	360
5900	60" diameter, 16" high				↓	985			985	1,075

12 92 33.10 Planters		Crew	Daily Output	Labor-Hours	Unit	Material	Labor	Equipment	Total	Total Incl O&P
						2020 Bare Costs				
5920	21″ high				Ea.	1,150			1,150	1,250
5940	27″ high					1,250			1,250	1,375
5960	30″ high					1,400			1,400	1,550
5980	36″ high					1,525			1,525	1,675
6000	Metal bowl, 32″ diameter, 8″ high, minimum					580			580	640
6050	Maximum					860			860	945
6100	Rectangle, 30″ long x 12″ wide, 6″ high, minimum					435			435	475
6200	Maximum					610			610	670
6300	36″ long x 12″ wide, 6″ high, minimum					855			855	940
6400	Maximum					560			560	620
6500	Square, 15″ side, minimum					635			635	700
6600	Maximum					940			940	1,025
6700	20″ side, minimum					1,350			1,350	1,475
6800	Maximum					650			650	715
6900	Round, 6″ diameter x 6″ high, minimum					62.50			62.50	68.50
7000	Maximum					74.50			74.50	81.50
7100	8″ diameter x 8″ high, minimum					83			83	91.50
7200	Maximum					89.50			89.50	98
7300	10″ diameter x 11″ high, minimum					107			107	117
7400	Maximum					157			157	173
7420	12″ diameter x 13″ high, minimum					98.50			98.50	109
7440	Maximum					227			227	250
7500	14″ diameter x 15″ high, minimum					133			133	146
7550	Maximum					233			233	257
7580	16″ diameter x 17″ high, minimum					144			144	158
7600	Maximum					255			255	281
7620	18″ diameter x 19″ high, minimum					174			174	191
7640	Maximum					310			310	345
7680	22″ diameter x 20″ high, minimum					219			219	241
7700	Maximum					400			400	440
7750	24″ diameter x 21″ high, minimum					285			285	315
7800	Maximum					535			535	585
7850	31″ diameter x 18″ high, minimum					620			620	680
7900	Maximum					1,575			1,575	1,750
7950	38″ diameter x 24″ high, minimum					1,050			1,050	1,150
8000	Maximum					2,650			2,650	2,900
8050	48″ diameter x 24″ high, minimum					1,375			1,375	1,525
8150	Maximum					2,725			2,725	3,000
8750	Wood, fiberglass liner, square									
8780	14″ square, 15″ high, minimum				Ea.	430			430	470
8800	Maximum					525			525	580
8820	24″ square, 16″ high, minimum					525			525	580
8840	Maximum					695			695	765
8860	36″ square, 21″ high, minimum					675			675	745
8880	Maximum					1,025			1,025	1,125
9000	Rectangle, 36″ long x 12″ wide, 10″ high, minimum					480			480	525
9050	Maximum					620			620	685
9100	48″ long x 12″ wide, 10″ high, minimum					515			515	570
9120	Maximum					650			650	715
9200	48″ long x 12″ wide, 24″ high, minimum					620			620	685
9300	Maximum					915			915	1,000
9400	Plastic cylinder, molded, 10″ diameter, 10″ high					47.50			47.50	52.50
9500	11″ diameter, 11″ high					90.50			90.50	99.50

12 92 Interior Planters and Artificial Plants

12 92 33 – Interior Planters

12 92 33.10 Planters		Crew	Daily Output	Labor-Hours	Unit	Material	Labor	Equipment	Total	Total Incl O&P
						2020 Bare Costs				
9600	13″ diameter, 12″ high				Ea.	148			148	162
9700	16″ diameter, 14″ high				↓	172			172	189

12 93 Interior Public Space Furnishings

12 93 13 – Bicycle Racks

12 93 13.10 Bicycle Racks

		Crew	Daily Output	Labor-Hours	Unit	Material	Labor	Equipment	Total	Total Incl O&P
0010	**BICYCLE RACKS**									
0020	Single side, grid, 1-5/8″ OD stl. pipe, w/1/2″ bars, galv, 5 bike cap	2 Clab	10	1.600	Ea.	282	67.50		349.50	420
0025	Powder coat finish		10	1.600		288	67.50		355.50	425
0030	Single side, grid, 1-5/8″ OD stl. pipe, w/1/2″ bars, galv, 9 bike cap		8	2		400	84		484	575
0035	Powder coat finish		8	2		430	84		514	605
0040	Single side, grid, 1-5/8″ OD stl. pipe, w/1/2″ bars, galv, 18 bike cap		4	4		520	168		688	845
0045	Powder coat finish		4	4		535	168		703	860
0050	S curve, 1-7/8″ OD stl. pipe, 11 ga., galv, 5 bike cap		10	1.600		160	67.50		227.50	284
0055	Powder coat finish		10	1.600		160	67.50		227.50	284
0060	S curve, 1-7/8″ OD stl. pipe, 11 ga., galv, 7 bike cap		9	1.778		203	75		278	345
0065	Powder coat finish		9	1.778		219	75		294	360
0070	S curve, 1-7/8″ OD stl. pipe, 11 ga., galv, 9 bike cap		8	2		320	84		404	485
0075	Powder coat finish		8	2		310	84		394	475
0080	S curve, 1-7/8″ OD stl. pipe, 11 ga., galv, 11 bike cap		6	2.667		425	112		537	645
0085	Powder coat finish	↓	6	2.667	↓	395	112		507	615

12 93 23 – Trash and Litter Receptacles

12 93 23.10 Trash Receptacles

		Crew	Daily Output	Labor-Hours	Unit	Material	Labor	Equipment	Total	Total Incl O&P
0010	**TRASH RECEPTACLES**									
0020	Fiberglass, 2′ square, 18″ high	2 Clab	30	.533	Ea.	610	22.50		632.50	710
0100	2′ square, 2′-6″ high		30	.533		975	22.50		997.50	1,100
0300	Circular, 2′ diameter, 18″ high		30	.533		575	22.50		597.50	670
0400	2′ diameter, 2′-6″ high		30	.533		615	22.50		637.50	710
0500	Recycled plastic, var colors, round, 32 gal., 28″ x 38″ high G		5	3.200		710	135		845	995
0510	32 gal., 31″ x 32″ high G	↓	5	3.200	↓	770	135		905	1,050
1000	Alum. frame, hardboard panels, steel drum base,									
1020	30 gal. capacity, silk screen on plastic finish	2 Clab	25	.640	Ea.	515	27		542	610
1040	Aggregate finish		25	.640		655	27		682	765
1100	50 gal. capacity, silk screen on plastic finish		20	.800		770	33.50		803.50	905
1140	Aggregate finish		20	.800		825	33.50		858.50	965
1200	Formed plastic liner, 14 gal., silk screen on plastic finish		40	.400		430	16.85		446.85	500
1240	Aggregate finish		40	.400		470	16.85		486.85	540
1300	30 gal. capacity, silk screen on plastic finish		35	.457		555	19.25		574.25	640
1340	Aggregate finish	↓	35	.457	↓	605	19.25		624.25	695
1400	Redwood slats, plastic liner, leg base, 14 gal. capacity,									
1420	Varnish w/routed message	2 Clab	40	.400	Ea.	450	16.85		466.85	520
2000	Concrete, precast, 2′ to 2-1/2′ wide, 3′ high, sandblasted	″	15	1.067	″	590	45		635	720
3000	Galv. steel frame and panels, leg base, poly bag retainer,									
3020	40 gal. capacity, silk screen on enamel finish	2 Clab	25	.640	Ea.	630	27		657	740
3040	Aggregate finish		25	.640		725	27		752	840
3200	Formed plastic liner, 50 gal., silk screen on enamel finish		20	.800		1,050	33.50		1,083.50	1,200
3240	Aggregate finish		20	.800		1,125	33.50		1,158.50	1,275
4000	Perforated steel, pole mounted, 12″ diam., 10 gal., painted		25	.640		166	27		193	226
4040	Redwood slats		25	.640		330	27		357	410
4100	22 gal. capacity, painted	↓	25	.640	↓	217	27		244	282

12 93 Interior Public Space Furnishings

12 93 23 – Trash and Litter Receptacles

12 93 23.10 Trash Receptacles		Crew	Daily Output	Labor-Hours	Unit	2020 Bare Costs Material	Labor	Equipment	Total	Total Incl O&P
4140	Redwood slats	2 Clab	25	.640	Ea.	620	27		647	725
4500	Galvanized steel street basket, 52 gal. capacity, unpainted		40	.400		390	16.85		406.85	450
9110	Plastic, with dome lid, 32 gal. capacity		35	.457		61	19.25		80.25	98
9120	Recycled plastic slats, plastic dome lid, 32 gal. capacity	↓	35	.457	↓	297	19.25		316.25	355
12 93 23.20 Trash Closure										
0010	**TRASH CLOSURE**									
0020	Steel with pullover cover, 2′-3″ wide, 4′-7″ high, 6′-2″ long	2 Clab	5	3.200	Ea.	2,050	135		2,185	2,475
0100	10′-1″ long		4	4		2,600	168		2,768	3,125
0300	Wood, 10′ wide, 6′ high, 10′ long	↓	1.20	13.333	↓	2,325	560		2,885	3,450

Division Notes

	CREW	DAILY OUTPUT	LABOR-HOURS	UNIT	BARE COSTS MAT.	LABOR	EQUIP.	TOTAL	TOTAL INCL O&P

Estimating Tips

General

- The items and systems in this division are usually estimated, purchased, supplied, and installed as a unit by one or more subcontractors. The estimator must ensure that all parties are operating from the same set of specifications and assumptions, and that all necessary items are estimated and will be provided. Many times the complex items and systems are covered, but the more common ones, such as excavation or a crane, are overlooked for the very reason that everyone assumes nobody could miss them. The estimator should be the central focus and be able to ensure that all systems are complete.
- It is important to consider factors such as site conditions, weather, shape and size of building, as well as labor availability as they may impact the overall cost of erecting special structures and systems included in this division.
- Another area where problems can develop in this division is at the interface between systems. The estimator must ensure, for instance, that anchor bolts, nuts, and washers are estimated and included for the air-supported structures and pre-engineered buildings to be bolted to their foundations. Utility supply is a common area where essential items or pieces of equipment can be missed or overlooked because each subcontractor may feel it is another's responsibility. The estimator should also be aware of certain items which may be supplied as part of a package but installed by others, and ensure that the installing contractor's estimate includes the cost of installation. Conversely, the estimator must also ensure that items are not costed by two different subcontractors, resulting in an inflated overall estimate.

13 30 00 Special Structures

- The foundations and floor slab, as well as rough mechanical and electrical, should be estimated, as this work is required for the assembly and erection of the structure. Generally, as noted in the data set, the pre-engineered building comes as a shell. Pricing is based on the size and structural design parameters stated in the reference section. Additional features, such as windows and doors with their related structural framing, must also be included by the estimator. Here again, the estimator must have a clear understanding of the scope of each portion of the work and all the necessary interfaces.

Reference Numbers

Reference numbers are shown at the beginning of some major classifications. These numbers refer to related items in the Reference Section. The reference information may be an estimating procedure, an alternate pricing method, or technical information.

Note: Not all subdivisions listed here necessarily appear. ■

13 05 Common Work Results for Special Construction

13 05 05 – Selective Demolition for Special Construction

13 05 05.10 Selective Demolition, Air Supported Structures		Crew	Daily Output	Labor-Hours	Unit	Material	2020 Bare Costs Labor	Equipment	Total	Total Incl O&P
0010	SELECTIVE DEMOLITION, AIR SUPPORTED STRUCTURES									
0020	Tank covers, scrim, dbl. layer, vinyl poly w/hdwe., blower & controls									
0050	Round and rectangular R024119-10	B-2	9000	.004	S.F.		.19		.19	.30
0100	Warehouse structures									
0120	Poly/vinyl fabric, 28 oz., incl. tension cables & inflation system	4 Clab	9000	.004	SF Flr.		.15		.15	.24
0150	Reinforced vinyl, 12 oz., 3,000 S.F.	"	5000	.006			.27		.27	.43
0200	12,000 to 24,000 S.F.	8 Clab	20000	.003			.13		.13	.22
0250	Tedlar vinyl fabric, 28 oz. w/liner, to 3,000 S.F.	4 Clab	5000	.006			.27		.27	.43
0300	12,000 to 24,000 S.F.	8 Clab	20000	.003	↓		.13		.13	.22
0350	Greenhouse/shelter, woven polyethylene with liner									
0400	3,000 S.F.	4 Clab	5000	.006	SF Flr.		.27		.27	.43
0450	12,000 to 24,000 S.F.	8 Clab	20000	.003			.13		.13	.22
0500	Tennis/gymnasium, poly/vinyl fabric, 28 oz., incl. thermal liner	4 Clab	9000	.004			.15		.15	.24
0600	Stadium/convention center, teflon coated fiberglass, incl. thermal liner	9 Clab	40000	.002	↓		.08		.08	.12
0700	Doors, air lock, 15' long, 10' x 10'	2 Carp	1.50	10.667	Ea.		565		565	910
0720	15' x 15'		.80	20			1,075		1,075	1,700
0750	Revolving personnel door, 6' diam. x 6'-6" high	↓	1.50	10.667	↓		565		565	910

13 05 05.20 Selective Demolition, Garden Houses		Crew	Daily Output	Labor-Hours	Unit	Material	Labor	Equipment	Total	Total Incl O&P
0010	SELECTIVE DEMOLITION, GARDEN HOUSES R024119-10									
0020	Prefab, wood, excl. foundation, average	2 Clab	400	.040	SF Flr.		1.68		1.68	2.70

13 05 05.25 Selective Demolition, Geodesic Domes		Crew	Daily Output	Labor-Hours	Unit	Material	Labor	Equipment	Total	Total Incl O&P
0010	SELECTIVE DEMOLITION, GEODESIC DOMES									
0050	Shell only, interlocking plywood panels, 30' diameter	F-5	3.20	10	Ea.		535		535	860
0060	34' diameter		2.30	13.913			745		745	1,200
0070	39' diameter	↓	2	16			860		860	1,375
0080	45' diameter	F-3	2.20	18.182			990	214	1,204	1,800
0090	55' diameter		2	20			1,075	236	1,311	1,975
0100	60' diameter		2	20			1,075	236	1,311	1,975
0110	65' diameter	↓	1.60	25	↓		1,350	295	1,645	2,500

13 05 05.30 Selective Demolition, Greenhouses		Crew	Daily Output	Labor-Hours	Unit	Material	Labor	Equipment	Total	Total Incl O&P
0010	SELECTIVE DEMOLITION, GREENHOUSES R024119-10									
0020	Resi-type, free standing, excl. foundations, 9' long x 8' wide	2 Clab	160	.100	SF Flr.		4.21		4.21	6.75
0030	9' long x 11' wide		170	.094			3.96		3.96	6.35
0040	9' long x 14' wide		220	.073			3.06		3.06	4.91
0050	9' long x 17' wide		320	.050			2.11		2.11	3.37
0060	Lean-to type, 4' wide		64	.250			10.55		10.55	16.85
0070	7' wide		120	.133	↓		5.60		5.60	9
0080	Geodesic hemisphere, 1/8" plexiglass glazing, 8' diam.		4	4	Ea.		168		168	270
0090	24' diam.		.80	20			840		840	1,350
0100	48' diam.	↓	.40	40	↓		1,675		1,675	2,700

13 05 05.35 Selective Demolition, Hangars		Crew	Daily Output	Labor-Hours	Unit	Material	Labor	Equipment	Total	Total Incl O&P
0010	SELECTIVE DEMOLITION, HANGARS									
0020	T type hangars, prefab, steel, galv roof & walls, incl doors, excl fndtn	E-2	2550	.022	SF Flr.		1.25	.67	1.92	2.76
0030	Circular type, prefab, steel frame, plastic skin, incl foundation, 80' diam	"	.50	112	Total		6,400	3,400	9,800	14,100

13 05 05.45 Selective Demolition, Lightning Protection		Crew	Daily Output	Labor-Hours	Unit	Material	Labor	Equipment	Total	Total Incl O&P
0010	SELECTIVE DEMOLITION, LIGHTNING PROTECTION									
0020	Air terminal & base, copper, 3/8" diam. x 10", to 75' H	1 Clab	16	.500	Ea.		21		21	33.50
0030	1/2" diam. x 12", over 75' H		16	.500			21		21	33.50
0050	Aluminum, 1/2" diam. x 12", to 75' H		16	.500			21		21	33.50
0060	5/8" diam. x 12", over 75' H		16	.500	↓		21		21	33.50
0070	Cable, copper, 220 lb. per thousand feet, to 75' H	↓	640	.013	L.F.		.53		.53	.84

13 05 05 – Selective Demolition for Special Construction

13 05 05.45 Selective Demolition, Lightning Protection		Crew	Daily Output	Labor-Hours	Unit	Material	2020 Bare Costs Labor	Equipment	Total	Total Incl O&P
0080	375 lb. per thousand feet, over 75′ H	1 Clab	460	.017	L.F.		.73		.73	1.17
0090	Aluminum, 101 lb. per thousand feet, to 75′ H		560	.014			.60		.60	.96
0100	199 lb. per thousand feet, over 75′ H		480	.017			.70		.70	1.12
0110	Arrester, 175 V AC, to ground		16	.500	Ea.		21		21	33.50
0120	650 V AC, to ground		13	.615	"		26		26	41.50

13 05 05.50 Selective Demolition, Pre-Engineered Steel Buildings		Crew	Daily Output	Labor-Hours	Unit	Material	Labor	Equipment	Total	Total Incl O&P
0010	**SELECTIVE DEMOLITION, PRE-ENGINEERED STEEL BUILDINGS**									
0500	Pre-engd. steel bldgs., rigid frame, clear span & multi post, excl. salvage									
0550	3,500 to 7,500 S.F.	L-10	1000	.024	SF Flr.		1.41	.47	1.88	2.80
0600	7,501 to 12,500 S.F.		1500	.016			.94	.31	1.25	1.87
0650	12,501 S.F. or greater		1650	.015			.86	.29	1.15	1.69
0700	Pre-engd. steel building components									
0710	Entrance canopy, including frame 4′ x 4′	E-24	8	4	Ea.		230	72.50	302.50	450
0720	4′ x 8′	"	7	4.571			263	83	346	515
0730	HM doors, self framing, single leaf	2 Skwk	8	2			110		110	175
0740	Double leaf		5	3.200			176		176	280
0760	Gutter, eave type		600	.027	L.F.		1.46		1.46	2.33
0770	Sash, single slide, double slide or fixed		24	.667	Ea.		36.50		36.50	58.50
0780	Skylight, fiberglass, to 30 S.F.		16	1			55		55	87.50
0785	Roof vents, circular, 12″ to 24″ diameter		12	1.333			73		73	117
0790	Continuous, 10′ long		8	2			110		110	175
0900	Shelters, aluminum frame									
0910	Acrylic glazing, 3′ x 9′ x 8′ high	2 Skwk	2	8	Ea.		440		440	700
0920	9′ x 12′ x 8′ high	"	1.50	10.667	"		585		585	930

13 05 05.55 Selective Demolition, Residential Garages		Crew	Daily Output	Labor-Hours	Unit	Material	Labor	Equipment	Total	Total Incl O&P
0010	**SELECTIVE DEMOLITION, RESIDENTIAL GARAGES** R024119-10									
0020	Garage, residential, prefab shell, stock, wood, single car	2 Clab	1.50	10.667	Ea.		450		450	720
0030	Two car	"	1.10	14.545	"		610		610	980

13 05 05.60 Selective Demolition, Silos		Crew	Daily Output	Labor-Hours	Unit	Material	Labor	Equipment	Total	Total Incl O&P
0010	**SELECTIVE DEMOLITION, SILOS**									
0020	Conc stave, indstrl, conical/sloping bott, excl fndtn, 12′ diam., 35′ H	E-24	.18	178	Ea.		10,200	3,225	13,425	20,100
0030	16′ diam., 45′ H		.12	267			15,300	4,850	20,150	30,100
0040	25′ diam., 75′ H		.08	400			23,000	7,250	30,250	45,200
0050	Steel, factory fabricated, 30,000 gal. cap, painted or epoxy lined	L-5	2	28			1,625	290	1,915	2,975

13 05 05.65 Selective Demolition, Sound Control		Crew	Daily Output	Labor-Hours	Unit	Material	Labor	Equipment	Total	Total Incl O&P
0010	**SELECTIVE DEMOLITION, SOUND CONTROL** R024119-10									
0120	Acoustical enclosure, 4″ thick walls & ceiling panels, 8 lb./S.F.	3 Carp	144	.167	SF Surf		8.85		8.85	14.20
0130	10.5 lb./S.F.		128	.188			9.95		9.95	15.95
0140	Reverb chamber, parallel walls, 4″ thick		120	.200			10.65		10.65	17.05
0150	Skewed walls, parallel roof, 4″ thick		110	.218			11.60		11.60	18.60
0160	Skewed walls/roof, 4″ layer/air space		96	.250			13.30		13.30	21.50
0170	Sound-absorbing panels, painted metal, 2′-6″ x 8′, under 1,000 S.F.		430	.056			2.97		2.97	4.75
0180	Over 1,000 S.F.		480	.050			2.66		2.66	4.26
0190	Flexible transparent curtain, clear	3 Shee	430	.056			3.48		3.48	5.45
0192	50% clear, 50% foam		430	.056			3.48		3.48	5.45
0194	25% clear, 75% foam		430	.056			3.48		3.48	5.45
0196	100% foam		430	.056			3.48		3.48	5.45
0200	Audio-masking sys., incl. speakers, amplfr., signal gnrtr.									
0205	Ceiling mounted, 5,000 S.F.	2 Elec	4800	.003	S.F.		.20		.20	.31
0210	10,000 S.F.		5600	.003			.18		.18	.27
0220	Plenum mounted, 5,000 S.F.		7600	.002			.13		.13	.20
0230	10,000 S.F.		8800	.002			.11		.11	.17

13 05 Common Work Results for Special Construction

13 05 05 – Selective Demolition for Special Construction

13 05 05.70 Selective Demolition, Special Purpose Rooms		Crew	Daily Output	Labor-Hours	Unit	Material	2020 Bare Costs Labor	Equipment	Total	Total Incl O&P
0010	**SELECTIVE DEMOLITION, SPECIAL PURPOSE ROOMS** R024119-10									
0100	Audiometric rooms, under 500 S.F. surface	4 Carp	200	.160	SF Surf		8.50		8.50	13.60
0110	Over 500 S.F. surface	"	240	.133	"		7.10		7.10	11.35
0200	Clean rooms, 12′ x 12′ soft wall, class 100	1 Carp	.30	26.667	Ea.		1,425		1,425	2,275
0210	Class 1,000		.30	26.667			1,425		1,425	2,275
0220	Class 10,000		.35	22.857			1,225		1,225	1,950
0230	Class 100,000		.35	22.857			1,225		1,225	1,950
0300	Darkrooms, shell complete, 8′ high	2 Carp	220	.073	SF Flr.		3.87		3.87	6.20
0310	12′ high		110	.145	"		7.75		7.75	12.40
0350	Darkrooms doors, mini-cylindrical, revolving		4	4	Ea.		213		213	340
0400	Music room, practice modular		140	.114	SF Surf		6.05		6.05	9.75
0500	Refrigeration structures and finishes									
0510	Wall finish, 2 coat Portland cement plaster, 1/2″ thick	1 Clab	200	.040	S.F.		1.68		1.68	2.70
0520	Fiberglass panels, 1/8″ thick		400	.020			.84		.84	1.35
0530	Ceiling finish, polystyrene plastic, 1″ to 2″ thick		500	.016			.67		.67	1.08
0540	4″ thick		450	.018			.75		.75	1.20
0550	Refrigerator, prefab aluminum walk-in, 7′-6″ high, 6′ x 6′ OD	2 Carp	100	.160	SF Flr.		8.50		8.50	13.60
0560	10′ x 10′ OD		160	.100			5.30		5.30	8.50
0570	Over 150 S.F.		200	.080			4.25		4.25	6.80
0600	Sauna, prefabricated, including heater & controls, 7′ high, to 30 S.F.		120	.133			7.10		7.10	11.35
0610	To 40 S.F.		140	.114			6.05		6.05	9.75
0620	To 60 S.F.		175	.091			4.86		4.86	7.80
0630	To 100 S.F.		220	.073			3.87		3.87	6.20
0640	To 130 S.F.		250	.064			3.40		3.40	5.45
0650	Steam bath, heater, timer, head, single, to 140 C.F.	1 Plum	2.20	3.636	Ea.		234		234	360
0660	To 300 C.F.		2.20	3.636			234		234	360
0670	Steam bath, comm. size, w/blow-down assembly, to 800 C.F.		1.80	4.444			286		286	445
0680	To 2,500 C.F.		1.60	5			320		320	500
0690	Steam bath, comm. size, multiple, for motels, apts, 500 C.F., 2 baths		2	4			258		258	400
0700	1,000 C.F., 4 baths		1.40	5.714			370		370	570

13 05 05.75 Selective Demolition, Storage Tanks		Crew	Daily Output	Labor-Hours	Unit	Material	Labor	Equipment	Total	Total Incl O&P
0010	**SELECTIVE DEMOLITION, STORAGE TANKS**									
0500	Steel tank, single wall, above ground, not incl. fdn., pumps or piping									
0510	Single wall, 275 gallon R024119-10	Q-1	3	5.333	Ea.		310		310	480
0520	550 thru 2,000 gallon	B-34P	2	12			680	515	1,195	1,650
0530	5,000 thru 10,000 gallon	B-34Q	2	12			690	570	1,260	1,700
0540	15,000 thru 30,000 gallon	B-34S	2	16			955	1,550	2,505	3,225
0600	Steel tank, double wall, above ground not incl. fdn., pumps & piping									
0620	500 thru 2,000 gallon	B-34P	2	12	Ea.		680	515	1,195	1,650

13 05 05.85 Selective Demolition, Swimming Pool Equip		Crew	Daily Output	Labor-Hours	Unit	Material	Labor	Equipment	Total	Total Incl O&P
0010	**SELECTIVE DEMOLITION, SWIMMING POOL EQUIP**									
0020	Diving stand, stainless steel, 3 meter	2 Clab	3	5.333	Ea.		225		225	360
0030	1 meter		5	3.200			135		135	216
0040	Diving board, 16′ long, aluminum		5.40	2.963			125		125	200
0050	Fiberglass		5.40	2.963			125		125	200
0070	Ladders, heavy duty, stainless steel, 2 tread		14	1.143			48		48	77
0080	4 tread		12	1.333			56		56	90
0090	Lifeguard chair, stainless steel, fixed		5	3.200			135		135	216
0100	Slide, tubular, fiberglass, aluminum handrails & ladder, 5′, straight		4	4			168		168	270
0110	8′, curved		6	2.667			112		112	180
0120	10′, curved		3	5.333			225		225	360
0130	12′ straight, with platform		2.50	6.400			269		269	430

13 05 Common Work Results for Special Construction

13 05 05 – Selective Demolition for Special Construction

13 05 05.85 Selective Demolition, Swimming Pool Equip		Crew	Daily Output	Labor-Hours	Unit	Material	2020 Bare Costs Labor	Equipment	Total	Total Incl O&P
0140	Removable access ramp, stainless steel	2 Clab	4	4	Ea.		168		168	270
0150	Removable stairs, stainless steel, collapsible	↓	4	4	↓		168		168	270

13 05 05.90 Selective Demolition, Tension Structures

		Crew	Daily Output	Labor-Hours	Unit	Material	Labor	Equipment	Total	Total Incl O&P
0010	**SELECTIVE DEMOLITION, TENSION STRUCTURES**									
0020	Steel/alum. frame, fabric shell, 60′ clear span, 6,000 S.F.	B-41	2000	.022	SF Flr.		.96	.15	1.11	1.70
0030	12,000 S.F.		2200	.020			.87	.14	1.01	1.54
0040	80′ clear span, 20,800 S.F.	↓	2440	.018			.79	.12	.91	1.40
0050	100′ clear span, 10,000 S.F.	L-5	4350	.013			.75	.13	.88	1.37
0060	26,000 S.F.		4600	.012			.71	.13	.84	1.29
0070	36,000 S.F.		5000	.011			.65	.12	.77	1.19
0080	120′ clear span, 24,000 S.F.		6000	.009			.54	.10	.64	.99
0090	150′ clear span, 30,000 S.F.	↓	7000	.008			.47	.08	.55	.85
0100	200′ clear span, 40,000 S.F.	E-6	16000	.008	↓		.46	.13	.59	.90
0110	For roll-up door, 12′ x 14′	L-2	2	8	Ea.		370		370	600

13 05 05.95 Selective Demo, X-Ray/Radio Freq Protection

		Crew	Daily Output	Labor-Hours	Unit	Material	Labor	Equipment	Total	Total Incl O&P
0010	**SELECTIVE DEMO, X-RAY/RADIO FREQ PROTECTION**									
0020	Shielding lead, lined door frame, excl. hdwe., 1/16″ thick	1 Clab	4.80	1.667	Ea.		70		70	112
0030	Lead sheets, 1/16″ thick R024119-10	2 Clab	270	.059	S.F.		2.49		2.49	4
0040	1/8″ thick		240	.067			2.81		2.81	4.50
0050	Lead shielding, 1/4″ thick		270	.059			2.49		2.49	4
0060	1/2″ thick	↓	240	.067	↓		2.81		2.81	4.50
0070	Lead glass, 1/4″ thick, 2.0 mm LE, 12″ x 16″	2 Glaz	16	1	Ea.		51		51	81.50
0080	24″ x 36″		8	2			102		102	163
0090	36″ x 60″		4	4			204		204	325
0100	Lead glass window frame, with 1/16″ lead & voice passage, 36″ x 60″		4	4			204		204	325
0110	Lead glass window frame, 24″ x 36″	↓	8	2	↓		102		102	163
0120	Lead gypsum board, 5/8″ thick with 1/16″ lead	2 Clab	320	.050	S.F.		2.11		2.11	3.37
0130	1/8″ lead		280	.057			2.41		2.41	3.85
0140	1/32″ lead		400	.040	↓		1.68		1.68	2.70
0150	Butt joints, 1/8″ lead or thicker, 2″ x 7′ long batten strip		480	.033	Ea.		1.40		1.40	2.25
0160	X-ray protection, average radiography room, up to 300 S.F., 1/16″ lead, min		.50	32	Total		1,350		1,350	2,150
0170	Maximum		.30	53.333			2,250		2,250	3,600
0180	Deep therapy X-ray room, 250 kV cap, up to 300 S.F., 1/4″ lead, min		.20	80			3,375		3,375	5,400
0190	Maximum		.12	133	↓		5,625		5,625	9,000
0880	Radio frequency shielding, prefab or screen-type copper or steel, minimum		360	.044	SF Surf		1.87		1.87	3
0890	Average		310	.052			2.17		2.17	3.48
0895	Maximum	↓	290	.055	↓		2.32		2.32	3.72

13 11 Swimming Pools

13 11 13 – Below-Grade Swimming Pools

13 11 13.50 Swimming Pools

		Crew	Daily Output	Labor-Hours	Unit	Material	Labor	Equipment	Total	Total Incl O&P
0010	**SWIMMING POOLS** Residential in-ground, vinyl lined									
0020	Concrete sides, w/equip, sand bottom	B-52	300	.187	SF Surf	28.50	9.10	1.95	39.55	48
0100	Metal or polystyrene sides	B-14	410	.117		24	5.20	.52	29.72	35.50
0200	Add for vermiculite bottom				↓	1.82			1.82	2
0500	Gunite bottom and sides, white plaster finish									
0600	12′ x 30′ pool	B-52	145	.386	SF Surf	53	18.80	4.03	75.83	93
0720	16′ x 32′ pool		155	.361		48	17.60	3.77	69.37	84.50
0750	20′ x 40′ pool	↓	250	.224	↓	42.50	10.90	2.34	55.74	67
0810	Concrete bottom and sides, tile finish									

13 11 Swimming Pools

13 11 13 – Below-Grade Swimming Pools

13 11 13.50 Swimming Pools		Crew	Daily Output	Labor-Hours	Unit	2020 Bare Costs Material	Labor	Equipment	Total	Total Incl O&P
0820	12' x 30' pool	B-52	80	.700	SF Surf	53.50	34	7.30	94.80	122
0830	16' x 32' pool		95	.589		44.50	28.50	6.15	79.15	101
0840	20' x 40' pool		130	.431		35.50	21	4.50	61	77.50
1100	Motel, gunite with plaster finish, incl. medium									
1150	capacity filtration & chlorination	B-52	115	.487	SF Surf	65.50	23.50	5.10	94.10	116
1200	Municipal, gunite with plaster finish, incl. high									
1250	capacity filtration & chlorination	B-52	100	.560	SF Surf	84.50	27.50	5.85	117.85	143
1350	Add for formed gutters				L.F.	124			124	137
1360	Add for stainless steel gutters				"	370			370	405
1600	For water heating system, see Section 23 52 28.10									
1700	Filtration and deck equipment only, as % of total				Total				20%	20%
1800	Automatic vacuum, hand tools, etc., 20' x 40' pool				SF Pool				.56	.62
1900	5,000 S.F. pool				"				.11	.12
3000	Painting pools, preparation + 3 coats, 20' x 40' pool, epoxy	2 Pord	.33	48.485	Total	1,450	2,150		3,600	5,025
3100	Rubber base paint, 18 gallons	"	.33	48.485		1,250	2,150		3,400	4,800
3500	42' x 82' pool, 75 gallons, epoxy paint	3 Pord	.14	171		6,150	7,600		13,750	18,900
3600	Rubber base paint	"	.14	171		5,150	7,600		12,750	17,800
4000	Replaster pool, 2 coat Portland cement plaster, 1/2" thick	2 Plas	75	.213	SF Surf	.80	10.35		11.15	17.40

13 11 23 – On-Grade Swimming Pools

13 11 23.50 Swimming Pools		Crew	Daily Output	Labor-Hours	Unit	Material	Labor	Equipment	Total	Total Incl O&P
0010	**SWIMMING POOLS** Residential above ground, steel construction									
0100	Round, 15' diam.	B-80A	3	8	Ea.	895	335	273	1,503	1,825
0120	18' diam.		2.50	9.600		1,025	405	330	1,760	2,125
0140	21' diam.		2	12		1,175	505	410	2,090	2,550
0160	24' diam.		1.80	13.333		1,350	560	455	2,365	2,900
0180	27' diam.		1.50	16		1,550	675	545	2,770	3,400
0200	30' diam.		1	24		1,575	1,000	820	3,395	4,275
0220	Oval, 12' x 24'		2.30	10.435		1,700	440	355	2,495	2,950
0240	15' x 30'		1.80	13.333		1,925	560	455	2,940	3,525
0260	18' x 33'		1	24		2,150	1,000	820	3,970	4,875

13 11 46 – Swimming Pool Accessories

13 11 46.50 Swimming Pool Equipment		Crew	Daily Output	Labor-Hours	Unit	Material	Labor	Equipment	Total	Total Incl O&P
0010	**SWIMMING POOL EQUIPMENT**									
0020	Diving stand, stainless steel, 3 meter	2 Carp	.40	40	Ea.	18,000	2,125		20,125	23,200
0300	1 meter		2.70	5.926		10,900	315		11,215	12,500
0600	Diving boards, 16' long, aluminum		2.70	5.926		4,275	315		4,590	5,200
0700	Fiberglass		2.70	5.926		3,475	315		3,790	4,300
0800	14' long, aluminum		2.70	5.926		3,925	315		4,240	4,800
0850	Fiberglass		2.70	5.926		3,675	315		3,990	4,550
1100	Bulkhead, movable, PVC, 8'-2" wide	2 Clab	8	2		3,075	84		3,159	3,525
1120	7'-9" wide		8	2		2,975	84		3,059	3,400
1140	7'-3" wide		8	2		2,300	84		2,384	2,650
1160	6'-9" wide		8	2		2,300	84		2,384	2,650
1200	Ladders, heavy duty, stainless steel, 2 tread	2 Carp	7	2.286		960	121		1,081	1,250
1500	4 tread		6	2.667		910	142		1,052	1,225
1800	Lifeguard chair, stainless steel, fixed		2.70	5.926		3,150	315		3,465	3,975
1900	Portable					3,050			3,050	3,350
2100	Lights, underwater, 12 volt, with transformer, 300 watt	1 Elec	1	8		365	490		855	1,150
2200	110 volt, 500 watt, standard		1	8		291	490		781	1,075
2400	Low water cutoff type		1	8		310	490		800	1,100
2800	Heaters, see Section 23 52 28.10									
3000	Pool covers, reinforced vinyl	3 Clab	1800	.013	S.F.	1.28	.56		1.84	2.31

13 11 Swimming Pools

13 11 46 – Swimming Pool Accessories

13 11 46.50 Swimming Pool Equipment		Crew	Daily Output	Labor-Hours	Unit	Material	2020 Bare Costs Labor	Equipment	Total	Total Incl O&P
3050	Automatic, electric				S.F.				9.15	10.10
3100	Vinyl, for winter, 400 S.F. max pool surface	3 Clab	3200	.008		.23	.32		.55	.76
3200	With water tubes, 400 S.F. max pool surface	"	3000	.008		.34	.34		.68	.91
3250	Sealed air bubble polyethylene solar blanket, 16 mils				↓	.35			.35	.39
3300	Slides, tubular, fiberglass, aluminum handrails & ladder, 5'-0", straight	2 Carp	1.60	10	Ea.	3,850	530		4,380	5,075
3320	8'-0", curved		3	5.333		8,000	283		8,283	9,250
3400	10'-0", curved		1	16		23,800	850		24,650	27,600
3420	12'-0", straight with platform	↓	1.20	13.333	↓	16,600	710		17,310	19,300
4500	Hydraulic lift, movable pool bottom, single ram									
4520	Under 1,000 S.F. area	L-9	72	.500	S.F.	195	24		219	253
4600	Four ram lift, over 1,000 S.F.	"	109	.330	"	162	15.80		177.80	204
5000	Removable access ramp, stainless steel	2 Clab	2	8	Ea.	6,375	335		6,710	7,575

13 17 Tubs and Pools

13 17 13 – Hot Tubs

13 17 13.10 Redwood Hot Tub System

		Crew	Daily Output	Labor-Hours	Unit	Material	Labor	Equipment	Total	Total Incl O&P
0010	**REDWOOD HOT TUB SYSTEM**									
7050	4' diameter x 4' deep	Q-1	1	16	Ea.	3,250	930		4,180	5,000
7100	5' diameter x 4' deep		1	16		4,125	930		5,055	5,950
7150	6' diameter x 4' deep		.80	20		4,975	1,150		6,125	7,275
7200	8' diameter x 4' deep	↓	.80	20	↓	7,300	1,150		8,450	9,825

13 17 33 – Whirlpool Tubs

13 17 33.10 Whirlpool Bath

		Crew	Daily Output	Labor-Hours	Unit	Material	Labor	Equipment	Total	Total Incl O&P
0010	**WHIRLPOOL BATH**									
6000	Whirlpool, bath with vented overflow, molded fiberglass									
6100	66" x 36" x 24"	Q-1	1	16	Ea.	910	930		1,840	2,425
6400	72" x 36" x 21"		1	16		1,450	930		2,380	3,025
6500	60" x 34" x 21"		1	16		1,450	930		2,380	3,025
6600	72" x 42" x 23"	↓	1	16	↓	1,350	930		2,280	2,925

13 18 Ice Rinks

13 18 13 – Ice Rink Floor Systems

13 18 13.50 Ice Skating

		Crew	Daily Output	Labor-Hours	Unit	Material	Labor	Equipment	Total	Total Incl O&P
0010	**ICE SKATING** Equipment incl. refrigeration, plumbing & cooling									
0020	coils & concrete slab, 85' x 200' rink									
0300	55° system, 5 mos., 100 ton				Total	591,000			591,000	650,000
0700	90° system, 12 mos., 135 ton				"	683,000			683,000	751,500
1200	Subsoil heating system (recycled from compressor), 85' x 200'	Q-7	.27	119	Ea.	41,400	7,400		48,800	57,000
1300	Subsoil insulation, 2 lb. polystyrene with vapor barrier, 85' x 200'	2 Carp	.14	114	"	30,900	6,075		36,975	43,700

13 18 16 – Ice Rink Dasher Boards

13 18 16.50 Ice Rink Dasher Boards

		Crew	Daily Output	Labor-Hours	Unit	Material	Labor	Equipment	Total	Total Incl O&P
0010	**ICE RINK DASHER BOARDS**									
1000	Dasher boards, 1/2" H.D. polyethylene faced steel frame, 3' acrylic									
1020	screen at sides, 5' acrylic ends, 85' x 200'	F-5	.06	533	Ea.	144,500	28,600		173,100	205,000
1100	Fiberglass & aluminum construction, same sides and ends	"	.06	533	"	162,000	28,600		190,600	224,000

13 21 Controlled Environment Rooms

13 21 13 – Clean Rooms

13 21 13.50 Clean Room Components		Crew	Daily Output	Labor-Hours	Unit	2020 Bare Costs Material	Labor	Equipment	Total	Total Incl O&P
0010	**CLEAN ROOM COMPONENTS**									
1100	Clean room, soft wall, 12′ x 12′, Class 100	1 Carp	.18	44.444	Ea.	16,500	2,350		18,850	21,900
1110	Class 1,000		.18	44.444		16,800	2,350		19,150	22,300
1120	Class 10,000		.21	38.095		15,100	2,025		17,125	19,900
1130	Class 100,000		.21	38.095		12,800	2,025		14,825	17,400
2800	Ceiling grid support, slotted channel struts 4′-0″ OC, ea. way				S.F.				5.90	6.50
3000	Ceiling panel, vinyl coated foil on mineral substrate									
3020	Sealed, non-perforated				S.F.				3.72	4.13
4000	Ceiling panel seal, silicone sealant, 150 L.F./gal.	1 Carp	150	.053	L.F.	.30	2.83		3.13	4.87
4100	Two sided adhesive tape	″	240	.033	″	.13	1.77		1.90	2.98
4200	Clips, one per panel				Ea.	1			1	1.10
6000	HEPA filter, 2′ x 4′, 99.97% eff., 3″ dp beveled frame (silicone seal)					360			360	395
6040	6″ deep skirted frame (channel seal)					500			500	550
6100	99.99% efficient, 3″ deep beveled frame (silicone seal)					440			440	485
6140	6″ deep skirted frame (channel seal)					485			485	530
6200	99.999% efficient, 3″ deep beveled frame (silicone seal)					525			525	580
6240	6″ deep skirted frame (channel seal)					565			565	620
7000	Wall panel systems, including channel strut framing									
7020	Polyester coated aluminum, particle board				S.F.				18.20	18.20
7100	Porcelain coated aluminum, particle board								32	32
7400	Wall panel support, slotted channel struts, to 12′ high								16.35	16.35

13 21 26 – Cold Storage Rooms

13 21 26.50 Refrigeration		Crew	Daily Output	Labor-Hours	Unit	Material	Labor	Equipment	Total	Total Incl O&P
0010	**REFRIGERATION**									
0020	Curbs, 12″ high, 4″ thick, concrete	2 Carp	58	.276	L.F.	3.30	14.65		17.95	27
1000	Doors, see Section 08 34 13.10									
2400	Finishes, 2 coat Portland cement plaster, 1/2″ thick	1 Plas	48	.167	S.F.	.80	8.10		8.90	13.80
2500	For galvanized reinforcing mesh, add	1 Lath	335	.024		.61	1.24		1.85	2.62
2700	3/16″ thick latex cement	1 Plas	88	.091		1.50	4.42		5.92	8.70
2900	For glass cloth reinforced ceilings, add	″	450	.018		.64	.86		1.50	2.08
3100	Fiberglass panels, 1/8″ thick	1 Carp	149.45	.054		2.08	2.85		4.93	6.85
3200	Polystyrene, plastic finish ceiling, 1″ thick		274	.029		2.84	1.55		4.39	5.60
3400	2″ thick		274	.029		3.19	1.55		4.74	6
3500	4″ thick		219	.037		4.05	1.94		5.99	7.55
3800	Floors, concrete, 4″ thick	1 Cefi	93	.086		1.67	4.30		5.97	8.60
3900	6″ thick	″	85	.094		2.62	4.70		7.32	10.30
4000	Insulation, 1″ to 6″ thick, cork				B.F.	1.50			1.50	1.65
4100	Urethane					.62			.62	.68
4300	Polystyrene, regular					.62			.62	.68
4400	Bead board					.29			.29	.32
4600	Installation of above, add per layer	2 Carp	657.60	.024	S.F.	.44	1.29		1.73	2.55
4700	Wall and ceiling juncture		298.90	.054	L.F.	1.41	2.85		4.26	6.10
4900	Partitions, galvanized sandwich panels, 4″ thick, stock		219.20	.073	S.F.	10.10	3.88		13.98	17.30
5000	Aluminum or fiberglass		219.20	.073	″	6.45	3.88		10.33	13.30
5200	Prefab walk-in, 7′-6″ high, aluminum, incl. refrigeration, door & floor									
5210	not incl. partitions, 6′ x 6′	2 Carp	54.80	.292	SF Flr.	105	15.50		120.50	141
5500	10′ x 10′		82.20	.195		85	10.35		95.35	110
5700	12′ x 14′		109.60	.146		76	7.75		83.75	96
5800	12′ x 20′		109.60	.146		115	7.75		122.75	139
6100	For 8′-6″ high, add					5%				
6300	Rule of thumb for complete units, w/o doors & refrigeration, cooler	2 Carp	146	.110		166	5.80		171.80	192
6400	Freezer		109.60	.146		113	7.75		120.75	136

13 21 Controlled Environment Rooms

13 21 26 – Cold Storage Rooms

13 21 26.50 Refrigeration		Crew	Daily Output	Labor-Hours	Unit	Material	2020 Bare Costs Labor	Equipment	Total	Total Incl O&P
6600	Shelving, plated or galvanized, steel wire type	2 Carp	360	.044	SF Hor.	8.35	2.36		10.71	12.95
6700	Slat shelf type	↓	375	.043		17.65	2.27		19.92	23
6900	For stainless steel shelving, add				↓	300%				
7000	Vapor barrier, on wood walls	2 Carp	1644	.010	S.F.	.19	.52		.71	1.04
7200	On masonry walls	"	1315	.012	"	.33	.65		.98	1.40
7500	For air curtain doors, see Section 23 34 33.10									

13 21 48 – Sound-Conditioned Rooms

13 21 48.10 Anechoic Chambers		Crew	Daily Output	Labor-Hours	Unit	Material	Labor	Equipment	Total	Total Incl O&P
0010	**ANECHOIC CHAMBERS** Standard units, 7′ ceiling heights									
0100	Area for pricing is net inside dimensions									
0300	200 cycles per second cutoff, 25 S.F. floor area				SF Flr.	1,875			1,875	2,050
0700	100 S.F.					1,400			1,400	1,550
0900	For 150 cycles per second cutoff, add to 100 S.F. room								30%	30%
1000	For 100 cycles per second cutoff, add to 100 S.F. room				↓				45%	45%

13 21 48.15 Audiometric Rooms		Crew	Daily Output	Labor-Hours	Unit	Material	Labor	Equipment	Total	Total Incl O&P
0010	**AUDIOMETRIC ROOMS**									
0020	Under 500 S.F. surface	4 Carp	98	.327	SF Surf	60	17.35		77.35	94
0100	Over 500 S.F. surface	"	120	.267	"	58	14.15		72.15	86

13 21 53 – Darkrooms

13 21 53.50 Darkrooms		Crew	Daily Output	Labor-Hours	Unit	Material	Labor	Equipment	Total	Total Incl O&P
0010	**DARKROOMS**									
0020	Shell, complete except for door, 64 S.F., 8′ high	2 Carp	128	.125	SF Flr.	54	6.65		60.65	70
0100	12′ high		64	.250		70	13.30		83.30	98.50
0500	120 S.F. floor, 8′ high		120	.133		39	7.10		46.10	54.50
0600	12′ high		60	.267		53	14.15		67.15	81
0800	240 S.F. floor, 8′ high		120	.133		27.50	7.10		34.60	41.50
0900	12′ high		60	.267	↓	39	14.15		53.15	65.50
1200	Mini-cylindrical, revolving, unlined, 4′ diameter		3.50	4.571	Ea.	3,125	243		3,368	3,850
1400	5′-6″ diameter	↓	2.50	6.400		5,050	340		5,390	6,100
1600	Add for lead lining, inner cylinder, 1/32″ thick					1,650			1,650	1,825
1700	1/16″ thick					4,425			4,425	4,850
1800	Add for lead lining, inner and outer cylinder, 1/32″ thick					3,050			3,050	3,350
1900	1/16″ thick				↓	6,800			6,800	7,475
2000	For darkroom door, see Section 08 34 36.10									

13 21 56 – Music Rooms

13 21 56.50 Music Rooms		Crew	Daily Output	Labor-Hours	Unit	Material	Labor	Equipment	Total	Total Incl O&P
0010	**MUSIC ROOMS**									
0020	Practice room, modular, perforated steel, under 500 S.F.	2 Carp	70	.229	SF Surf	38	12.15		50.15	61
0100	Over 500 S.F.	"	80	.200	"	32	10.65		42.65	52

13 24 Special Activity Rooms

13 24 16 – Saunas

13 24 16.50 Saunas and Heaters		Crew	Daily Output	Labor-Hours	Unit	Material	Labor	Equipment	Total	Total Incl O&P
						2020 Bare Costs				
0010	**SAUNAS AND HEATERS**									
0020	Prefabricated, incl. heater & controls, 7′ high, 6′ x 4′, C/C	L-7	2.20	12.727	Ea.	5,250	650		5,900	6,800
0050	6′ x 4′, C/P		2	14		4,325	715		5,040	5,925
0400	6′ x 5′, C/C		2	14		5,300	715		6,015	6,975
0450	6′ x 5′, C/P		2	14		5,225	715		5,940	6,900
0600	6′ x 6′, C/C		1.80	15.556		6,950	795		7,745	8,925
0650	6′ x 6′, C/P		1.80	15.556		5,525	795		6,320	7,350
0800	6′ x 9′, C/C		1.60	17.500		7,325	895		8,220	9,475
0850	6′ x 9′, C/P		1.60	17.500		6,500	895		7,395	8,575
1000	8′ x 12′, C/C		1.10	25.455		11,500	1,300		12,800	14,800
1050	8′ x 12′, C/P		1.10	25.455		8,575	1,300		9,875	11,500
1200	8′ x 8′, C/C		1.40	20		9,600	1,025		10,625	12,200
1250	8′ x 8′, C/P		1.40	20		7,200	1,025		8,225	9,550
1400	8′ x 10′, C/C		1.20	23.333		8,925	1,200		10,125	11,700
1450	8′ x 10′, C/P		1.20	23.333		8,375	1,200		9,575	11,100
1600	10′ x 12′, C/C		1	28		13,500	1,425		14,925	17,100
1650	10′ x 12′, C/P	↓	1	28		13,500	1,425		14,925	17,200
1700	Door only, cedar, 2′ x 6′, w/ 1′ x 4′ tempered insulated glass window	2 Carp	3.40	4.706		610	250		860	1,075
1800	Prehung, incl. jambs, pulls & hardware	″	12	1.333		620	71		691	800
2500	Heaters only (incl. above), wall mounted, to 200 C.F.					1,025			1,025	1,125
2750	To 300 C.F.					1,150			1,150	1,250
3000	Floor standing, to 720 C.F., 10,000 watts, w/controls	1 Elec	3	2.667		2,575	164		2,739	3,100
3250	To 1,000 C.F., 16,000 watts	″	3	2.667	↓	3,725	164		3,889	4,350

13 24 26 – Steam Baths

13 24 26.50 Steam Baths and Components		Crew	Daily Output	Labor-Hours	Unit	Material	Labor	Equipment	Total	Total Incl O&P
0010	**STEAM BATHS AND COMPONENTS**									
0020	Heater, timer & head, single, to 140 C.F.	1 Plum	1.20	6.667	Ea.	2,350	430		2,780	3,275
0500	To 300 C.F.		1.10	7.273		2,700	470		3,170	3,675
1000	Commercial size, with blow-down assembly, to 800 C.F.		.90	8.889		5,800	575		6,375	7,275
1500	To 2,500 C.F.	↓	.80	10		8,200	645		8,845	10,000
2000	Multiple, motels, apts., 2 baths, w/blow-down assm., 500 C.F.	Q-1	1.30	12.308		6,650	715		7,365	8,425
2500	4 baths	″	.70	22.857		10,500	1,325		11,825	13,600
2700	Conversion unit for residential tub, including door				↓	4,050			4,050	4,450

13 28 Athletic and Recreational Special Construction

13 28 33 – Athletic and Recreational Court Walls

13 28 33.50 Sport Court		Crew	Daily Output	Labor-Hours	Unit	Material	Labor	Equipment	Total	Total Incl O&P
0010	**SPORT COURT**									
0020	Sport court floors, no. 2 & better maple, 25/32″ thick, see 09 64 66.10				SF Flr.					
0300	Squash, regulation court in existing building, minimum				Court	41,900			41,900	46,000
0400	Maximum				″	46,600			46,600	51,500
0450	Rule of thumb for components:									
0470	Walls	3 Carp	.15	160	Court	12,700	8,500		21,200	27,600
0500	Floor	″	.25	96		10,000	5,100		15,100	19,200
0550	Lighting	2 Elec	.60	26.667		2,400	1,625		4,025	5,150
0600	Handball, racquetball court in existing building, minimum	C-1	.20	160		45,800	8,050		53,850	63,500
0800	Maximum	″	.10	320		49,000	16,100		65,100	80,000
0900	Rule of thumb for components: walls	3 Carp	.12	200		14,300	10,600		24,900	32,800
1000	Floor		.25	96		10,000	5,100		15,100	19,200
1100	Ceiling	↓	.33	72.727	↓	4,775	3,875		8,650	11,500

13 28 Athletic and Recreational Special Construction

13 28 33 – Athletic and Recreational Court Walls

	13 28 33.50 Sport Court	Crew	Daily Output	Labor-Hours	Unit	Material	2020 Bare Costs Labor	Equipment	Total	Total Incl O&P
1200	Lighting	2 Elec	.60	26.667	Court	2,550	1,625		4,175	5,325

13 31 Fabric Structures

13 31 13 – Air-Supported Fabric Structures

13 31 13.09 Air Supported Tank Covers

		Crew	Daily Output	Labor-Hours	Unit	Material	Labor	Equipment	Total	Total Incl O&P
0010	**AIR SUPPORTED TANK COVERS**, vinyl polyester									
0100	Scrim, double layer, with hardware, blower, standby & controls									
0200	Round, 75′ diameter	B-2	4500	.009	S.F.	12.70	.38		13.08	14.60
0300	100′ diameter		5000	.008		11.55	.34		11.89	13.25
0400	150′ diameter		5000	.008		9.15	.34		9.49	10.60
0500	Rectangular, 20′ x 20′		4500	.009		24.50	.38		24.88	27.50
0600	30′ x 40′		4500	.009		24.50	.38		24.88	27.50
0700	50′ x 60′		4500	.009		24.50	.38		24.88	27.50
0800	For single wall construction, deduct, minimum					.86			.86	.95
0900	Maximum					2.48			2.48	2.73
1000	For maximum resistance to atmosphere or cold, add					1.24			1.24	1.36
1100	For average shipping charges, add				Total	2,125			2,125	2,350

13 31 13.13 Single-Walled Air-Supported Structures

		Crew	Daily Output	Labor-Hours	Unit	Material	Labor	Equipment	Total	Total Incl O&P
0010	**SINGLE-WALLED AIR-SUPPORTED STRUCTURES**									
0020	Site preparation, incl. anchor placement and utilities	B-11B	1000	.016	SF Flr.	1.25	.76	.40	2.41	3.02
0030	For concrete, see Section 03 30 53.40									
0050	Warehouse, polyester/vinyl fabric, 28 oz., over 10 yr. life, welded									
0060	Seams, tension cables, primary & auxiliary inflation system,									
0070	airlock, personnel doors and liner									
0100	5,000 S.F.	4 Clab	5000	.006	SF Flr.	29.50	.27		29.77	33
0250	12,000 S.F.	″	6000	.005		20	.22		20.22	22.50
0400	24,000 S.F.	8 Clab	12000	.005		15	.22		15.22	16.85
0500	50,000 S.F.	″	12500	.005		13.20	.22		13.42	14.85
0700	12 oz. reinforced vinyl fabric, 5 yr. life, sewn seams,									
0710	accordion door, including liner									
0750	3,000 S.F.	4 Clab	3000	.011	SF Flr.	13.75	.45		14.20	15.85
0800	12,000 S.F.	″	6000	.005		11.70	.22		11.92	13.25
0850	24,000 S.F.	8 Clab	12000	.005		9.90	.22		10.12	11.25
0950	Deduct for single layer					1.12			1.12	1.23
1000	Add for welded seams					1.56			1.56	1.72
1050	Add for double layer, welded seams included					3.13			3.13	3.44
1250	Tedlar/vinyl fabric, 28 oz., with liner, over 10 yr. life,									
1260	incl. overhead and personnel doors									
1300	3,000 S.F.	4 Clab	3000	.011	SF Flr.	26	.45		26.45	29
1450	12,000 S.F.	″	6000	.005		17.50	.22		17.72	19.60
1550	24,000 S.F.	8 Clab	12000	.005		14.05	.22		14.27	15.80
1700	Deduct for single layer					2.08			2.08	2.29
2250	Greenhouse/shelter, woven polyethylene with liner, 2 yr. life,									
2260	sewn seams, including doors									
2300	3,000 S.F.	4 Clab	3000	.011	SF Flr.	16.80	.45		17.25	19.20
2350	12,000 S.F.	″	6000	.005		14.70	.22		14.92	16.50
2450	24,000 S.F.	8 Clab	12000	.005		12.60	.22		12.82	14.20
2550	Deduct for single layer					1.03			1.03	1.13
2600	Tennis/gymnasium, polyester/vinyl fabric, 28 oz., over 10 yr. life,									
2610	including thermal liner, heat and lights									
2650	7,200 S.F.	4 Clab	6000	.005	SF Flr.	24	.22		24.22	26.50

13 31 Fabric Structures

13 31 13 – Air-Supported Fabric Structures

13 31 13.13 Single-Walled Air-Supported Structures		Crew	Daily Output	Labor-Hours	Unit	2020 Bare Costs Material	Labor	Equipment	Total	Total Incl O&P
2750	13,000 S.F.	4 Clab	6500	.005	SF Flr.	19	.21		19.21	21.50
2850	Over 24,000 S.F.	8 Clab	12000	.005		17.35	.22		17.57	19.45
2860	For low temperature conditions, add					1.23			1.23	1.35
2870	For average shipping charges, add				Total	5,825			5,825	6,425
2900	Thermal liner, translucent reinforced vinyl				SF Flr.	1.23			1.23	1.35
2950	Metalized mylar fabric and mesh, double liner				"	2.47			2.47	2.72
3050	Stadium/convention center, teflon coated fiberglass, heavy weight,									
3060	over 20 yr. life, incl. thermal liner and heating system									
3100	Minimum	9 Clab	26000	.003	SF Flr.	60.50	.12		60.62	66.50
3110	Maximum	"	19000	.004	"	74.50	.16		74.66	82.50
3400	Doors, air lock, 15′ long, 10′ x 10′	2 Carp	.80	20	Ea.	21,600	1,075		22,675	25,500
3600	15′ x 15′	"	.50	32		33,200	1,700		34,900	39,200
3700	For each added 5′ length, add					5,925			5,925	6,525
3900	Revolving personnel door, 6′ diameter, 6′-6″ high	2 Carp	.80	20		16,600	1,075		17,675	19,900

13 31 23 – Tensioned Fabric Structures

13 31 23.50 Tension Structures		Crew	Daily Output	Labor-Hours	Unit	Material	Labor	Equipment	Total	Total Incl O&P
0010	**TENSION STRUCTURES** Rigid steel/alum. frame, vinyl coated poly									
0100	Fabric shell, 60′ clear span, not incl. foundations or floors									
0200	6,000 S.F.	B-41	1000	.044	SF Flr.	17.10	1.92	.30	19.32	22
0300	12,000 S.F.		1100	.040		19.35	1.75	.27	21.37	24.50
0400	80′ to 99′ clear span, 20,800 S.F.		1220	.036		18.60	1.57	.25	20.42	23.50
0410	100′ to 119′ clear span, 10,000 S.F.	L-5	2175	.026		16.80	1.50	.27	18.57	21
0430	26,000 S.F.		2300	.024		17.20	1.42	.25	18.87	21.50
0450	36,000 S.F.		2500	.022		17.25	1.30	.23	18.78	21.50
0460	120′ to 149′ clear span, 24,000 S.F.		3000	.019		17.05	1.09	.19	18.33	20.50
0470	150′ to 199′ clear span, 30,000 S.F.		6000	.009		19.75	.54	.10	20.39	22.50
0480	200′ clear span, 40,000 S.F.	E-6	8000	.016		25	.92	.27	26.19	29.50
0500	For roll-up door, 12′ x 14′, add	L-2	1	16	Ea.	6,925	745		7,670	8,825

13 33 Geodesic Structures

13 33 13 – Geodesic Domes

13 33 13.35 Geodesic Domes		Crew	Daily Output	Labor-Hours	Unit	Material	Labor	Equipment	Total	Total Incl O&P
0010	**GEODESIC DOMES** Shell only, interlocking plywood panels									
0400	30′ diameter	F-5	1.60	20	Ea.	27,000	1,075		28,075	31,400
0500	33′ diameter		1.14	28.070		26,700	1,500		28,200	31,800
0600	40′ diameter		1	32		30,500	1,725		32,225	36,300
0700	45′ diameter	F-3	1.13	35.556		32,100	1,925	420	34,445	38,800
0750	56′ diameter		1	40		58,500	2,175	470	61,145	68,000
0800	60′ diameter		1	40		61,000	2,175	470	63,645	71,500
0850	67′ diameter		.80	50		94,000	2,725	590	97,315	108,500
1100	Aluminum panel, with 6″ insulation									
1200	100′ diameter				SF Flr.	25			25	27.50
1300	500′ diameter				"	24.50			24.50	27
1600	Aluminum framed, plexiglass closure panels									
1700	40′ diameter				SF Flr.	61.50			61.50	67.50
1800	200′ diameter				"	58			58	63.50
2100	Aluminum framed, aluminum closure panels									
2200	40′ diameter				SF Flr.	20			20	22
2300	100′ diameter					24			24	26.50
2400	200′ diameter					20.50			20.50	22.50

13 33 Geodesic Structures

13 33 13 – Geodesic Domes

13 33 13.35 Geodesic Domes		Crew	Daily Output	Labor-Hours	Unit	2020 Bare Costs Material	Labor	Equipment	Total	Total Incl O&P
2700	Aluminum framed, fiberglass sandwich panel closure									
2800	6′ diameter	2 Carp	150	.107	SF Flr.	27	5.65		32.65	39
2900	28′ diameter	"	350	.046	"	25	2.43		27.43	31

13 34 Fabricated Engineered Structures

13 34 13 – Glazed Structures

13 34 13.13 Greenhouses		Crew	Daily Output	Labor-Hours	Unit	Material	Labor	Equipment	Total	Total Incl O&P
0010	**GREENHOUSES**, Shell only, stock units, not incl. 2′ stub walls,									
0020	foundation, floors, heat or compartments									
0300	Residential type, free standing, 8′-6″ long x 7′-6″ wide	2 Carp	59	.271	SF Flr.	24	14.40		38.40	49
0400	10′-6″ wide		85	.188		42.50	10		52.50	63
0600	13′-6″ wide		108	.148		45	7.85		52.85	62
0700	17′-0″ wide		160	.100		43.50	5.30		48.80	56
0900	Lean-to type, 3′-10″ wide		34	.471		43.50	25		68.50	87.50
1000	6′-10″ wide		58	.276		28	14.65		42.65	54.50
1500	Commercial, custom, truss frame, incl. equip., plumbing, elec.,									
1550	benches and controls, under 2,000 S.F.				SF Flr.	13.50			13.50	14.85
1700	Over 5,000 S.F.				"	11.95			11.95	13.15
2000	Institutional, custom, rigid frame, including compartments and									
2050	multi-controls, under 500 S.F.				SF Flr.	29.50			29.50	32.50
2150	Over 2,000 S.F.				"	10.50			10.50	11.55
3700	For 1/4″ tempered glass, add				SF Surf	1.75			1.75	1.93
3900	Cooling, 1,200 CFM exhaust fan, add				Ea.	325			325	355
4000	7,850 CFM					1,025			1,025	1,150
4200	For heaters, 10 MBH, add					246			246	270
4300	60 MBH, add					845			845	925
4500	For benches, 2′ x 8′, add					188			188	207
4600	4′ x 10′, add					227			227	249
4800	For ventilation & humidity control w/4 integrated outlets, add				Total	273			273	300
4900	For environmental controls and automation, 8 outputs, 9 stages, add				"	1,200			1,200	1,325
5100	For humidification equipment, add				Ea.	325			325	355
5200	For vinyl shading, add				S.F.	.44			.44	.48
6000	Geodesic hemisphere, 1/8″ plexiglass glazing									
6050	8′ diameter	2 Carp	2	8	Ea.	5,950	425		6,375	7,225
6150	24′ diameter		.35	45.714		16,000	2,425		18,425	21,500
6250	48′ diameter		.20	80		34,700	4,250		38,950	45,000

13 34 13.19 Swimming Pool Enclosures		Crew	Daily Output	Labor-Hours	Unit	Material	Labor	Equipment	Total	Total Incl O&P
0010	**SWIMMING POOL ENCLOSURES** Translucent, free standing									
0020	not including foundations, heat or light									
0200	Economy	2 Carp	200	.080	SF Hor.	46.50	4.25		50.75	58
0600	Deluxe	"	70	.229		103	12.15		115.15	132
0700	For motorized roof, 40% opening, solid roof, add					23			23	25.50
0800	Skylight type roof, add					12.95			12.95	14.25

13 34 16 – Grandstands and Bleachers

13 34 16.13 Grandstands		Crew	Daily Output	Labor-Hours	Unit	Material	Labor	Equipment	Total	Total Incl O&P
0010	**GRANDSTANDS** Permanent, municipal, including foundation									
0300	Steel, economy				Seat	27			27	30
0400	Steel, deluxe					30.50			30.50	34
0900	Composite, steel, wood and plastic, stock design, economy					38			38	42
1000	Deluxe					91.50			91.50	101

13 34 Fabricated Engineered Structures

13 34 16 – Grandstands and Bleachers

13 34 16.53 Bleachers		Crew	Daily Output	Labor-Hours	Unit	2020 Bare Costs Material	Labor	Equipment	Total	Total Incl O&P
0010	BLEACHERS									
0020	Bleachers, outdoor, portable, 5 tiers, 42 seats	2 Sswk	120	.133	Seat	71.50	7.70		79.20	91
0100	5 tiers, 54 seats		80	.200		70	11.55		81.55	96
0200	10 tiers, 104 seats		120	.133		87.50	7.70		95.20	109
0300	10 tiers, 144 seats		80	.200		64.50	11.55		76.05	90
0500	Permanent bleachers, aluminum seat, steel frame, 24" row									
0600	8 tiers, 80 seats	2 Sswk	60	.267	Seat	74	15.35		89.35	106
0700	8 tiers, 160 seats		48	.333		66.50	19.20		85.70	105
0925	15 tiers, 154 to 165 seats		60	.267		74	15.35		89.35	106
0975	15 tiers, 214 to 225 seats		60	.267		66	15.35		81.35	97.50
1050	15 tiers, 274 to 285 seats		60	.267		68.50	15.35		83.85	100
1200	Seat backs only, 30" row, fiberglass		160	.100		25	5.75		30.75	37
1300	Steel and wood		160	.100		29	5.75		34.75	41.50
1400	NOTE: average seating is 1.5′ in width									

13 34 19 – Metal Building Systems

13 34 19.50 Pre-Engineered Steel Buildings

Line	Description	Crew	Daily Output	Labor-Hours	Unit	Material	Labor	Equipment	Total	Total Incl O&P
0010	PRE-ENGINEERED STEEL BUILDINGS R133419-10									
0100	Clear span rigid frame, 26 ga. colored roofing and siding									
0150	20′ to 29′ wide, 10′ eave height	E-2	425	.132	SF Flr.	9.50	7.55	3.99	21.04	27
0160	14′ eave height		350	.160		10.25	9.15	4.85	24.25	31.50
0170	16′ eave height		320	.175		10.85	10	5.30	26.15	34
0180	20′ eave height		275	.204		11.95	11.65	6.15	29.75	39
0190	24′ eave height		240	.233		13.30	13.35	7.05	33.70	44
0200	30′ to 49′ wide, 10′ eave height		535	.105		7	6	3.17	16.17	21
0300	14′ eave height		450	.124		7.55	7.10	3.77	18.42	24
0400	16′ eave height		415	.135		8.05	7.70	4.09	19.84	26
0500	20′ eave height		360	.156		8.70	8.90	4.71	22.31	29
0600	24′ eave height		320	.175		9.60	10	5.30	24.90	32.50
0700	50′ to 100′ wide, 10′ eave height		770	.073		5.95	4.16	2.20	12.31	15.65
0900	16′ eave height		600	.093		6.85	5.35	2.83	15.03	19.25
1000	20′ eave height		490	.114		7.40	6.55	3.46	17.41	22.50
1100	24′ eave height		435	.129		8.15	7.35	3.90	19.40	25
1200	Clear span tapered beam frame, 26 ga. colored roofing/siding									
1300	30′ to 39′ wide, 10′ eave height	E-2	535	.105	SF Flr.	7.90	6	3.17	17.07	22
1400	14′ eave height		450	.124		8.75	7.10	3.77	19.62	25.50
1500	16′ eave height		415	.135		9.15	7.70	4.09	20.94	27
1600	20′ eave height		360	.156		10.15	8.90	4.71	23.76	31
1700	40′ wide, 10′ eave height		600	.093		7.05	5.35	2.83	15.23	19.50
1800	14′ eave height		510	.110		7.80	6.25	3.33	17.38	22.50
1900	16′ eave height		475	.118		8.15	6.75	3.57	18.47	24
2000	20′ eave height		415	.135		8.95	7.70	4.09	20.74	27
2100	50′ to 79′ wide, 10′ eave height		770	.073		6.55	4.16	2.20	12.91	16.40
2200	14′ eave height		675	.083		7.10	4.74	2.51	14.35	18.25
2300	16′ eave height		635	.088		7.40	5.05	2.67	15.12	19.25
2400	20′ eave height		490	.114		8.70	6.55	3.46	18.71	24
2410	80′ to 100′ wide, 10′ eave height		935	.060		5.85	3.42	1.81	11.08	13.95
2420	14′ eave height		750	.075		6.40	4.27	2.26	12.93	16.45
2430	16′ eave height		685	.082		6.70	4.67	2.48	13.85	17.65
2440	20′ eave height		560	.100		7.15	5.70	3.03	15.88	20.50
2460	101′ to 120′ wide, 10′ eave height		950	.059		5.35	3.37	1.79	10.51	13.30
2470	14′ eave height		770	.073		5.95	4.16	2.20	12.31	15.70
2480	16′ eave height		675	.083		6.30	4.74	2.51	13.55	17.35

13 34 19 – Metal Building Systems

13 34 19.50 Pre-Engineered Steel Buildings		Crew	Daily Output	Labor-Hours	Unit	Material	Labor	Equipment	Total	Total Incl O&P
						2020 Bare Costs				
2490	20′ eave height	E-2	560	.100	SF Flr.	6.75	5.70	3.03	15.48	20
2500	Single post 2-span frame, 26 ga. colored roofing and siding									
2600	80′ wide, 14′ eave height	E-2	740	.076	SF Flr.	5.90	4.32	2.29	12.51	16
2700	16′ eave height		695	.081		6.30	4.60	2.44	13.34	17.10
2800	20′ eave height		625	.090		6.85	5.10	2.71	14.66	18.80
2900	24′ eave height		570	.098		7.45	5.60	2.98	16.03	20.50
3000	100′ wide, 14′ eave height		835	.067		5.75	3.83	2.03	11.61	14.80
3100	16′ eave height		795	.070		5.35	4.02	2.13	11.50	14.75
3200	20′ eave height		730	.077		6.50	4.38	2.32	13.20	16.80
3300	24′ eave height		670	.084		7.20	4.78	2.53	14.51	18.45
3400	120′ wide, 14′ eave height		870	.064		7.60	3.68	1.95	13.23	16.45
3500	16′ eave height		830	.067		5.95	3.86	2.04	11.85	15.05
3600	20′ eave height		765	.073		6.90	4.18	2.22	13.30	16.80
3700	24′ eave height		705	.079		7.55	4.54	2.41	14.50	18.30
3800	Double post 3-span frame, 26 ga. colored roofing and siding									
3900	150′ wide, 14′ eave height	E-2	925	.061	SF Flr.	4.71	3.46	1.83	10	12.80
4000	16′ eave height		890	.063		4.90	3.60	1.91	10.41	13.30
4100	20′ eave height		820	.068		5.35	3.90	2.07	11.32	14.40
4200	24′ eave height		765	.073		5.90	4.18	2.22	12.30	15.70
4300	Triple post 4-span frame, 26 ga. colored roofing and siding									
4400	160′ wide, 14′ eave height	E-2	970	.058	SF Flr.	4.58	3.30	1.75	9.63	12.30
4500	16′ eave height		930	.060		4.81	3.44	1.82	10.07	12.85
4600	20′ eave height		870	.064		4.73	3.68	1.95	10.36	13.30
4700	24′ eave height		815	.069		5.35	3.93	2.08	11.36	14.55
4800	200′ wide, 14′ eave height		1030	.054		4.23	3.11	1.65	8.99	11.50
4900	16′ eave height		995	.056		4.38	3.22	1.70	9.30	11.90
5000	20′ eave height		935	.060		4.82	3.42	1.81	10.05	12.85
5100	24′ eave height		885	.063		5.40	3.62	1.92	10.94	13.90
5200	Accessory items: add to the basic building cost above									
5250	Eave overhang, 2′ wide, 26 ga., with soffit	E-2	360	.156	L.F.	36	8.90	4.71	49.61	59.50
5300	4′ wide, without soffit		300	.187		32	10.65	5.65	48.30	59
5350	With soffit		250	.224		42	12.80	6.80	61.60	74
5400	6′ wide, without soffit		250	.224		37	12.80	6.80	56.60	69
5450	With soffit		200	.280		49.50	16	8.50	74	90
5500	Entrance canopy, incl. frame, 4′ x 4′		25	2.240	Ea.	550	128	68	746	885
5550	4′ x 8′		19	2.947	"	615	168	89.50	872.50	1,050
5600	End wall roof overhang, 4′ wide, without soffit		850	.066	L.F.	18.25	3.76	2	24.01	28.50
5650	With soffit		500	.112	"	33.50	6.40	3.39	43.29	51
5700	Doors, HM self-framing, incl. butts, lockset and trim									
5750	Single leaf, 3070 (3′ x 7′), economy	2 Sswk	5	3.200	Opng.	730	184		914	1,100
5800	Deluxe		4	4		735	231		966	1,200
5825	Glazed		4	4		815	231		1,046	1,275
5850	3670 (3′-6″ x 7′)		4	4		910	231		1,141	1,375
5900	4070 (4′ x 7′)		3	5.333		1,175	305		1,480	1,800
5950	Double leaf, 6070 (6′ x 7′)		2	8		1,375	460		1,835	2,275
6000	Glazed		2	8		1,550	460		2,010	2,475
6050	Framing only, for openings, 3′ x 7′		4	4		200	231		431	600
6100	10′ x 10′		3	5.333		660	305		965	1,225
6150	For windows below, 2020 (2′ x 2′)		6	2.667		212	154		366	485
6200	4030 (4′ x 3′)		5	3.200		259	184		443	585
6250	Flashings, 26 ga., corner or eave, painted		240	.067	L.F.	4.72	3.84		8.56	11.50
6300	Galvanized		240	.067		4.59	3.84		8.43	11.35
6350	Rake flashing, painted		240	.067		5.10	3.84		8.94	11.95

13 34 19 – Metal Building Systems

13 34 19.50 Pre-Engineered Steel Buildings		Crew	Daily Output	Labor-Hours	Unit	Material	2020 Bare Costs Labor	Equipment	Total	Total Incl O&P
6400	Galvanized	2 Sswk	240	.067	L.F.	4.91	3.84		8.75	11.70
6450	Ridge flashing, 18" wide, painted		240	.067		6.80	3.84		10.64	13.80
6500	Galvanized		240	.067		7.65	3.84		11.49	14.75
6550	Gutter, eave type, 26 ga., painted		320	.050		8.05	2.88		10.93	13.55
6650	Valley type, between buildings, painted		120	.133		14.95	7.70		22.65	29
6710	Insulation, rated .6 lb. density, unfaced 4" thick, R13	2 Carp	2300	.007	S.F.	.49	.37		.86	1.13
6720	6" thick, R19		2300	.007		.64	.37		1.01	1.29
6730	10" thick, R30		2300	.007		1.22	.37		1.59	1.93
6750	Insulation, rated .6 lb. density, poly/scrim/foil (PSF) faced									
6760	4" thick, R13	2 Carp	2300	.007	S.F.	.70	.37		1.07	1.36
6770	6" thick, R19		2300	.007		.93	.37		1.30	1.61
6780	9-1/2" thick, R30		2300	.007		1.03	.37		1.40	1.72
6800	Insulation, rated .6 lb. density, vinyl faced 1-1/2" thick, R5		2300	.007		.41	.37		.78	1.04
6850	3" thick, R10		2300	.007		.40	.37		.77	1.03
6900	4" thick, R13		2300	.007		.43	.37		.80	1.06
6920	6" thick, R19		2300	.007		.60	.37		.97	1.25
6930	10" thick, R30		2300	.007		1.80	.37		2.17	2.57
6950	Foil/scrim/kraft (FSK) faced, 1-1/2" thick, R5		2300	.007		.44	.37		.81	1.07
7000	2" thick, R6		2300	.007		.56	.37		.93	1.21
7050	3" thick, R10		2300	.007		.54	.37		.91	1.18
7100	4" thick, R13		2300	.007		.49	.37		.86	1.13
7110	6" thick, R19		2300	.007		.67	.37		1.04	1.33
7120	10" thick, R30		2300	.007		.93	.37		1.30	1.61
7150	Metalized polyester/scrim/kraft (PSK) facing,1-1/2" thk, R5		2300	.007		.68	.37		1.05	1.34
7200	2" thick, R6		2300	.007		.81	.37		1.18	1.48
7250	3" thick, R11		2300	.007		1.02	.37		1.39	1.71
7300	4" thick, R13		2300	.007		1.03	.37		1.40	1.72
7310	6" thick, R19		2300	.007		1.28	.37		1.65	2
7320	10" thick, R30		2300	.007		1.45	.37		1.82	2.19
7350	Vinyl/scrim/foil (VSF), 1-1/2" thick, R5		2300	.007		.59	.37		.96	1.24
7400	2" thick, R6		2300	.007		.78	.37		1.15	1.45
7450	3" thick, R10		2300	.007		.80	.37		1.17	1.47
7500	4" thick, R13		2300	.007		1.03	.37		1.40	1.72
7510	Vinyl/scrim/vinyl (VSV), 4" thick, R13		2300	.007		.57	.37		.94	1.22
7520	6" thick, R19		2300	.007		.71	.37		1.08	1.37
7530	9-1/2" thick, R30		2300	.007		.95	.37		1.32	1.64
7580	10" thick, R19		2300	.007		1.75	.37		2.12	2.52
7585	Vinyl/scrim/polyester (VSP), 4" thick, R13		2300	.007		.63	.37		1	1.28
7590	6" thick, R19		2300	.007		.84	.37		1.21	1.51
7600	10" thick, R30		2300	.007		1.31	.37		1.68	2.03
7635	Insulation installation, over the purlin, second layer, up to 4" thick, add						90%			
7640	Insulation installation, between the purlins, up to 4" thick, add						100%			
7650	Sash, single slide, glazed, with screens, 2020 (2' x 2')	E-1	22	1.091	Opng.	163	62	6.70	231.70	286
7700	3030 (3' x 3')		14	1.714		365	97	10.50	472.50	575
7750	4030 (4' x 3')		13	1.846		490	105	11.30	606.30	715
7800	6040 (6' x 4')		12	2		975	113	12.25	1,100.25	1,275
7850	Double slide sash, 3030 (3' x 3')		14	1.714		237	97	10.50	344.50	430
7900	6040 (6' x 4')		12	2		635	113	12.25	760.25	890
7950	Fixed glass, no screens, 3030 (3' x 3')		14	1.714		221	97	10.50	328.50	410
8000	6040 (6' x 4')		12	2		590	113	12.25	715.25	845
8050	Prefinished storm sash, 3030 (3' x 3')		70	.343		86.50	19.40	2.10	108	129
8100	Siding and roofing, see Sections 07 41 13 & 07 42 13									
8200	Skylight, fiberglass panels, to 30 S.F.	E-1	10	2.400	Ea.	140	136	14.70	290.70	390

13 34 19 – Metal Building Systems

13 34 19.50 Pre-Engineered Steel Buildings		Crew	Daily Output	Labor-Hours	Unit	2020 Bare Costs Material	Labor	Equipment	Total	Total Incl O&P
8250	Larger sizes, add for excess over 30 S.F.	E-1	300	.080	S.F.	4.68	4.53	.49	9.70	13
8300	Roof vents, turbine ventilator, wind driven									
8350	No damper, includes base, galvanized									
8400	12" diameter	Q-9	10	1.600	Ea.	73	89.50		162.50	221
8450	20" diameter		8	2		209	112		321	405
8500	24" diameter		8	2		320	112		432	530
8600	Continuous, 26 ga., 10' long, 9" wide	2 Sswk	4	4		38	231		269	420
8650	12" wide	"	4	4		38	231		269	420

13 34 23 – Fabricated Structures

13 34 23.10 Comfort Stations

Line	Description	Crew	Daily Output	Labor-Hours	Unit	Material	Labor	Equipment	Total	Total Incl O&P
0010	**COMFORT STATIONS** Prefab., stock, w/doors, windows & fixt.									
0100	Not incl. interior finish or electrical									
0300	Mobile, on steel frame, 2 unit				S.F.	190			190	209
0350	7 unit					320			320	350
0400	Permanent, including concrete slab, 2 unit	B-12J	50	.320		251	16.20	16.95	284.15	320
0500	6 unit	"	43	.372		201	18.85	19.70	239.55	273
0600	Alternate pricing method, mobile, 2 fixture				Fixture	6,675			6,675	7,350
0650	7 fixture					11,600			11,600	12,800
0700	Permanent, 2 unit	B-12J	.70	22.857		20,700	1,150	1,200	23,050	26,000
0750	6 unit	"	.50	32		17,800	1,625	1,700	21,125	24,000

13 34 23.15 Domes

Line	Description	Crew	Daily Output	Labor-Hours	Unit	Material	Labor	Equipment	Total	Total Incl O&P
0010	**DOMES**									
1500	Domes, bulk storage, shell only, dual radius hemisphere, arch, steel									
1600	framing, corrugated steel covering, 150' diameter	E-2	550	.102	SF Flr.	34	5.80	3.08	42.88	50
1700	400' diameter	"	720	.078		27.50	4.44	2.36	34.30	40.50
1800	Wood framing, wood decking, to 400' diameter	F-4	400	.120		35	6.45	2.45	43.90	51.50
1900	Radial framed wood (2" x 6"), 1/2" thick									
2000	plywood, asphalt shingles, 50' diameter	F-3	2000	.020	SF Flr.	69.50	1.09	.24	70.83	78.50
2100	60' diameter		1900	.021		59.50	1.14	.25	60.89	67.50
2200	72' diameter		1800	.022		49.50	1.21	.26	50.97	56.50
2300	116' diameter		1730	.023		33.50	1.26	.27	35.03	39.50
2400	150' diameter		1500	.027		36	1.45	.31	37.76	42.50

13 34 23.16 Fabricated Control Booths

Line	Description	Crew	Daily Output	Labor-Hours	Unit	Material	Labor	Equipment	Total	Total Incl O&P
0010	**FABRICATED CONTROL BOOTHS**									
0100	Guard House, prefab conc. w/bullet resistant doors & windows, roof & wiring									
0110	8' x 8', Level III	L-10	1	24	Ea.	54,500	1,400	470	56,370	63,000
0120	8' x 8', Level IV	"	1	24	"	72,000	1,400	470	73,870	82,000

13 34 23.25 Garage Costs

Line	Description	Crew	Daily Output	Labor-Hours	Unit	Material	Labor	Equipment	Total	Total Incl O&P
0010	**GARAGE COSTS**									
0020	Public parking, average				Car				15,200	16,700
0300	Residential, wood, 12' x 20', one car prefab shell, stock, economy	2 Carp	1	16	Total	6,350	850		7,200	8,350
0350	Custom		.67	23.881		7,125	1,275		8,400	9,850
0400	Two car, 24' x 20', economy		.67	23.881		12,900	1,275		14,175	16,200
0450	Custom		.50	32		14,100	1,700		15,800	18,200

13 34 23.30 Garden House

Line	Description	Crew	Daily Output	Labor-Hours	Unit	Material	Labor	Equipment	Total	Total Incl O&P
0010	**GARDEN HOUSE** Prefab wood, no floors or foundations									
0100	6' x 6'	2 Carp	200	.080	SF Flr.	52.50	4.25		56.75	65
0300	8' x 12'	"	48	.333	"	30	17.70		47.70	61.50

13 34 Fabricated Engineered Structures

13 34 23 – Fabricated Structures

13 34 23.45 Kiosks

	13 34 23.45 Kiosks	Crew	Daily Output	Labor-Hours	Unit	Material	Labor	Equipment	Total	Total Incl O&P
						2020 Bare Costs				
0010	**KIOSKS**									
0020	Round, advertising type, 5′ diameter, 7′ high, aluminum wall, illuminated				Ea.	23,500			23,500	25,900
0100	Aluminum wall, non-illuminated					22,500			22,500	24,800
0500	Rectangular, 5′ x 9′, 7′-6″ high, aluminum wall, illuminated					25,500			25,500	28,100
0600	Aluminum wall, non-illuminated					23,800			23,800	26,200

13 34 23.60 Portable Booths

	13 34 23.60 Portable Booths	Crew	Daily Output	Labor-Hours	Unit	Material	Labor	Equipment	Total	Total Incl O&P
0010	**PORTABLE BOOTHS** Prefab. aluminum with doors, windows, ext. roof									
0100	lights wiring & insulation, 15 S.F. building, OD, painted				S.F.	282			282	310
0300	30 S.F. building					259			259	285
0400	50 S.F. building					189			189	208
0600	80 S.F. building					161			161	177
0700	100 S.F. building					147			147	162
0900	Acoustical booth, 27 Db @ 1,000 Hz, 15 S.F. floor				Ea.	4,250			4,250	4,700
1000	7′ x 7′-6″, including light & ventilation					8,775			8,775	9,650
1200	Ticket booth, galv. steel, not incl. foundations., 4′ x 4′					5,275			5,275	5,800
1300	4′ x 6′					8,200			8,200	9,025

13 34 23.70 Shelters

	13 34 23.70 Shelters	Crew	Daily Output	Labor-Hours	Unit	Material	Labor	Equipment	Total	Total Incl O&P
0010	**SHELTERS**									
0020	Aluminum frame, acrylic glazing, 3′ x 9′ x 8′ high	2 Sswk	1.14	14.035	Ea.	3,450	810		4,260	5,125
0100	9′ x 12′ x 8′ high		.73	21.918		7,575	1,275		8,850	10,400
1000	Gable end shelter, 16′ x 32′, 8′ eave ht, excl. footings, slab, anchors		.68	23.529		13,300	1,350		14,650	16,800
1050	Gable end shelter, 30′ x 64′, 8′ eave ht, excl. footings, slab, anchors		.34	47.059		41,300	2,725		44,025	49,900

13 34 43 – Aircraft Hangars

13 34 43.50 Hangars

	13 34 43.50 Hangars	Crew	Daily Output	Labor-Hours	Unit	Material	Labor	Equipment	Total	Total Incl O&P
0010	**HANGARS** Prefabricated steel T hangars, galv. steel roof &									
0100	walls, incl. electric bi-folding doors									
0110	not including floors or foundations, 4 unit	E-2	1275	.044	SF Flr.	14.85	2.51	1.33	18.69	22
0130	8 unit		1063	.053		13.05	3.01	1.60	17.66	21
0900	With bottom rolling doors, 4 unit		1386	.040		13.75	2.31	1.22	17.28	20
1000	8 unit		966	.058		11.50	3.31	1.76	16.57	19.95
1200	Alternate pricing method:									
1300	Galv. roof and walls, electric bi-folding doors, 4 plane	E-2	1.06	52.830	Plane	19,600	3,025	1,600	24,225	28,100
1500	8 plane		.91	61.538		13,900	3,525	1,875	19,300	23,100
1600	With bottom rolling doors, 4 plane		1.25	44.800		19,200	2,550	1,350	23,100	26,800
1800	8 plane		.97	57.732		14,300	3,300	1,750	19,350	23,100
2000	Circular type, prefab., steel frame, plastic skin, electric									
2010	door, including foundations, 80′ diameter									

13 34 53 – Agricultural Structures

13 34 53.50 Silos

	13 34 53.50 Silos	Crew	Daily Output	Labor-Hours	Unit	Material	Labor	Equipment	Total	Total Incl O&P
0010	**SILOS**									
0500	Steel, factory fab., 30,000 gallon cap., painted, economy	L-5	1	56	Ea.	26,300	3,250	580	30,130	34,800
0700	Deluxe		.50	112		41,800	6,525	1,150	49,475	58,000
0800	Epoxy lined, economy		1	56		43,000	3,250	580	46,830	53,000
1000	Deluxe		.50	112		54,500	6,525	1,150	62,175	72,000

13 34 56 – Observatories

13 34 56.15 Domes

	13 34 56.15 Domes	Crew	Daily Output	Labor-Hours	Unit	Material	Labor	Equipment	Total	Total Incl O&P
0010	**DOMES**									
0020	Domes, rev. alum., elec. drive, for astronomy obsv. shell only, stock units									
0600	10′-6″ diameter	2 Carp	.25	64	Ea.	47,000	3,400		50,400	57,000
0900	18′-6″ diameter		.17	94.118		78,000	5,000		83,000	94,000

13 34 Fabricated Engineered Structures

13 34 56 – Observatories

13 34 56.15 Domes		Crew	Daily Output	Labor-Hours	Unit	Material	Labor	Equipment	Total	Total Incl O&P
						2020 Bare Costs				
1200	24'-6" diameter	2 Carp	.08	200	Ea.	134,000	10,600		144,600	164,500

13 34 63 – Natural Fiber Construction

13 34 63.50 Straw Bale Construction		Crew	Daily Output	Labor-Hours	Unit	Material	Labor	Equipment	Total	Total Incl O&P
0010	STRAW BALE CONSTRUCTION									
2020	Straw bales in walls w/modified post and beam frame G	2 Carp	320	.050	S.F.	6.85	2.66		9.51	11.80

13 36 Towers

13 36 13 – Metal Towers

13 36 13.50 Control Towers		Crew	Daily Output	Labor-Hours	Unit	Material	Labor	Equipment	Total	Total Incl O&P
0010	CONTROL TOWERS									
0020	Modular 12' x 10', incl. instruments				Ea.	810,500			810,500	891,500
0500	With standard 40' tower				"	1,250,500			1,250,500	1,375,500
1000	Temporary portable control towers, 8' x 12',									

13 42 Building Modules

13 42 63 – Detention Cell Modules

13 42 63.16 Steel Detention Cell Modules		Crew	Daily Output	Labor-Hours	Unit	Material	Labor	Equipment	Total	Total Incl O&P
0010	STEEL DETENTION CELL MODULES									
2000	Cells, prefab., 5' to 6' wide, 7' to 8' high, 7' to 8' deep,									
2010	bar front, cot, not incl. plumbing	E-4	1.50	21.333	Ea.	9,650	1,250	98	10,998	12,700

13 47 Facility Protection

13 47 13 – Cathodic Protection

13 47 13.16 Cathodic Prot. for Underground Storage Tanks		Crew	Daily Output	Labor-Hours	Unit	Material	Labor	Equipment	Total	Total Incl O&P
0010	CATHODIC PROTECTION FOR UNDERGROUND STORAGE TANKS									
1000	Anodes, magnesium type, 9 #	R-15	18.50	2.595	Ea.	38.50	156	15.05	209.55	299
1010	17 #		13	3.692		81.50	222	21.50	325	455
1020	32 #		10	4.800		123	288	28	439	610
1030	48 #		7.20	6.667		160	400	38.50	598.50	840
1100	Graphite type w/epoxy cap, 3" x 60" (32 #)	R-22	8.40	4.438		148	249		397	550
1110	4" x 80" (68 #)		6	6.213		246	350		596	805
1120	6" x 72" (80 #)		5.20	7.169		1,525	405		1,930	2,300
1130	6" x 36" (45 #)		9.60	3.883		770	218		988	1,175
2000	Rectifiers, silicon type, air cooled, 28 V/10 A	R-19	3.50	5.714		2,325	350		2,675	3,100
2010	20 V/20 A		3.50	5.714		2,325	350		2,675	3,125
2100	Oil immersed, 28 V/10 A		3	6.667		2,875	410		3,285	3,800
2110	20 V/20 A		3	6.667		3,250	410		3,660	4,200
3000	Anode backfill, coke breeze	R-22	3850	.010	Lb.	.30	.54		.84	1.17
4000	Cable, HMWPE, No. 8		2.40	15.533	M.L.F.	605	870		1,475	2,025
4010	No. 6		2.40	15.533		820	870		1,690	2,250
4020	No. 4		2.40	15.533		1,225	870		2,095	2,700
4030	No. 2		2.40	15.533		1,900	870		2,770	3,450
4040	No. 1		2.20	16.945		2,575	950		3,525	4,325
4050	No. 1/0		2.20	16.945		3,350	950		4,300	5,150
4060	No. 2/0		2.20	16.945		5,250	950		6,200	7,250
4070	No. 4/0		2	18.640		7,225	1,050		8,275	9,550
5000	Test station, 7 terminal box, flush curb type w/lockable cover	R-19	12	1.667	Ea.	79.50	102		181.50	246

13 47 Facility Protection

13 47 13 – Cathodic Protection

13 47 13.16 Cathodic Prot. for Underground Storage Tanks		Crew	Daily Output	Labor-Hours	Unit	Material	2020 Bare Costs Labor	Equipment	Total	Total Incl O&P
5010	Reference cell, 2″ diam. PVC conduit, cplg., plug, set flush	R-19	4.80	4.167	Ea.	158	256		414	570

13 48 Sound, Vibration, and Seismic Control

13 48 13 – Manufactured Sound and Vibration Control Components

13 48 13.50 Audio Masking

		Crew	Daily Output	Labor-Hours	Unit	Material	Labor	Equipment	Total	Total Incl O&P
0010	**AUDIO MASKING**, acoustical enclosure, 4″ thick wall and ceiling									
0020	8 lb./S.F., up to 12′ span	3 Carp	72	.333	SF Surf	31.50	17.70		49.20	63
0300	Better quality panels, 10.5 #/S.F		64	.375		36	19.95		55.95	71.50
0400	Reverb-chamber, 4″ thick, parallel walls		60	.400		44.50	21.50		66	83
0600	Skewed wall, parallel roof, 4″ thick panels		55	.436		51	23		74	93
0700	Skewed walls, skewed roof, 4″ layers, 4″ air space		48	.500		57.50	26.50		84	106
0900	Sound-absorbing panels, pntd. mtl., 2′-6″ x 8′, under 1,000 S.F.		215	.112		11.90	5.95		17.85	22.50
1100	Over 1,000 S.F.		240	.100		11.45	5.30		16.75	21
1200	Fabric faced		240	.100		9.25	5.30		14.55	18.70
1500	Flexible transparent curtain, clear	3 Shee	215	.112		7.15	6.95		14.10	18.80
1600	50% foam		215	.112		10	6.95		16.95	22
1700	75% foam		215	.112		10	6.95		16.95	22
1800	100% foam		215	.112		10	6.95		16.95	22
3100	Audio masking system, including speakers, amplification									
3110	and signal generator									
3200	Ceiling mounted, 5,000 S.F.	2 Elec	2400	.007	S.F.	1.19	.41		1.60	1.94
3300	10,000 S.F.		2800	.006		.97	.35		1.32	1.61
3400	Plenum mounted, 5,000 S.F.		3800	.004		1.04	.26		1.30	1.54
3500	10,000 S.F.		4400	.004		.70	.22		.92	1.11

13 49 Radiation Protection

13 49 13 – Integrated X-Ray Shielding Assemblies

13 49 13.50 Lead Sheets

		Crew	Daily Output	Labor-Hours	Unit	Material	Labor	Equipment	Total	Total Incl O&P
0010	**LEAD SHEETS**									
0300	Lead sheets, 1/16″ thick	2 Lath	135	.119	S.F.	11.40	6.15		17.55	22
0400	1/8″ thick		120	.133		24	6.95		30.95	37.50
0500	Lead shielding, 1/4″ thick		135	.119		40	6.15		46.15	53.50
0550	1/2″ thick		120	.133		91.50	6.95		98.45	112
0950	Lead headed nails (average 1 lb. per sheet)				Lb.	8.75			8.75	9.65
1000	Butt joints in 1/8″ lead or thicker, 2″ batten strip x 7′ long	2 Lath	240	.067	Ea.	32	3.47		35.47	41
1200	X-ray protection, average radiography or fluoroscopy									
1210	room, up to 300 S.F. floor, 1/16″ lead, economy	2 Lath	.25	64	Total	12,100	3,325		15,425	18,600
1500	7′-0″ walls, deluxe	″	.15	107	″	14,600	5,550		20,150	24,700
1600	Deep therapy X-ray room, 250 kV capacity,									
1800	up to 300 S.F. floor, 1/4″ lead, economy	2 Lath	.08	200	Total	33,900	10,400		44,300	53,500
1900	7′-0″ walls, deluxe		.06	267	″	41,800	13,900		55,700	67,500
1999	Minimum labor/equipment charge		4.50	3.556	Job		185		185	290

13 49 19 – Lead-Lined Materials

13 49 19.50 Shielding Lead

		Crew	Daily Output	Labor-Hours	Unit	Material	Labor	Equipment	Total	Total Incl O&P
0010	**SHIELDING LEAD**									
0100	Laminated lead in wood doors, 1/16″ thick, no hardware				S.F.	56.50			56.50	62.50
0200	Lead lined door frame, not incl. hardware,									
0210	1/16″ thick lead, butt prepared for hardware	1 Lath	2.40	3.333	Ea.	925	174		1,099	1,300
0850	Window frame with 1/16″ lead and voice passage, 36″ x 60″	2 Glaz	2	8		4,675	410		5,085	5,800

13 49 Radiation Protection

13 49 19 – Lead-Lined Materials

13 49 19.50 Shielding Lead		Crew	Daily Output	Labor-Hours	Unit	Material	2020 Bare Costs Labor	Equipment	Total	Total Incl O&P
0870	24" x 36" frame	2 Glaz	4	4	Ea.	2,550	204		2,754	3,125
0900	Lead gypsum board, 5/8" thick with 1/16" lead		160	.100	S.F.	16.40	5.10		21.50	26
0910	1/8" lead		140	.114		28	5.85		33.85	40.50
0930	1/32" lead	2 Lath	200	.080		10.10	4.16		14.26	17.60

13 49 21 – Lead Glazing

13 49 21.50 Lead Glazing		Crew	Daily Output	Labor-Hours	Unit	Material	Labor	Equipment	Total	Total Incl O&P
0010	**LEAD GLAZING**									
0600	Lead glass, 1/4" thick, 2.0 mm LE, 12" x 16"	2 Glaz	13	1.231	Ea.	435	63		498	580
0700	24" x 36"		8	2		1,500	102		1,602	1,825
0800	36" x 60"		2	8		4,000	410		4,410	5,025
2000	X-ray viewing panels, clear lead plastic									
2010	7 mm thick, 0.3 mm LE, 2.3 lb./S.F.	H-3	139	.115	S.F.	277	5.25		282.25	315
2020	12 mm thick, 0.5 mm LE, 3.9 lb./S.F.		82	.195		385	8.85		393.85	440
2030	18 mm thick, 0.8 mm LE, 5.9 lb./S.F.		54	.296		470	13.50		483.50	535
2040	22 mm thick, 1.0 mm LE, 7.2 lb./S.F.		44	.364		665	16.55		681.55	755
2050	35 mm thick, 1.5 mm LE, 11.5 lb./S.F.		28	.571		1,000	26		1,026	1,150
2060	46 mm thick, 2.0 mm LE, 15.0 lb./S.F.		21	.762		1,300	34.50		1,334.50	1,475

13 49 23 – Integrated RFI/EMI Shielding Assemblies

13 49 23.50 Modular Shielding Partitions		Crew	Daily Output	Labor-Hours	Unit	Material	Labor	Equipment	Total	Total Incl O&P
0010	**MODULAR SHIELDING PARTITIONS**									
4000	X-ray barriers, modular, panels mounted within framework for									
4002	attaching to floor, wall or ceiling, upper portion is clear lead									
4005	plastic window panels 48"H, lower portion is opaque leaded									
4008	steel panels 36"H, structural supports not incl.									
4010	1-section barrier, 36"W x 84"H overall									
4020	0.5 mm LE panels	H-3	6.40	2.500	Ea.	9,025	114		9,139	10,100
4030	0.8 mm LE panels		6.40	2.500		9,725	114		9,839	10,900
4040	1.0 mm LE panels		5.33	3.002		13,500	137		13,637	15,100
4050	1.5 mm LE panels		5.33	3.002		15,800	137		15,937	17,600
4060	2-section barrier, 72"W x 84"H overall									
4070	0.5 mm LE panels	H-3	4	4	Ea.	14,400	182		14,582	16,100
4080	0.8 mm LE panels		4	4		14,800	182		14,982	16,600
4090	1.0 mm LE panels		3.56	4.494		16,300	204		16,504	18,200
5000	1.5 mm LE panels		3.20	5		27,400	227		27,627	30,600
5010	3-section barrier, 108"W x 84"H overall									
5020	0.5 mm LE panels	H-3	3.20	5	Ea.	21,500	227		21,727	24,100
5030	0.8 mm LE panels		3.20	5		22,200	227		22,427	24,800
5040	1.0 mm LE panels		2.67	5.993		24,400	273		24,673	27,300
5050	1.5 mm LE panels		2.46	6.504		41,500	296		41,796	46,100
7000	X-ray barriers, mobile, mounted within framework w/casters on									
7005	bottom, clear lead plastic window panels on upper portion,									
7010	opaque on lower, 30"W x 75"H overall, incl. framework									
7020	24"H upper w/0.5 mm LE, 48"H lower w/0.8 mm LE	1 Carp	16	.500	Ea.	3,975	26.50		4,001.50	4,425
7030	48"W x 75"H overall, incl. framework									
7040	36"H upper w/0.5 mm LE, 36"H lower w/0.8 mm LE	1 Carp	16	.500	Ea.	7,075	26.50		7,101.50	7,850
7050	36"H upper w/1.0 mm LE, 36"H lower w/1.5 mm LE	"	16	.500	"	8,400	26.50		8,426.50	9,275
7060	72"W x 75"H overall, incl. framework									
7070	36"H upper w/0.5 mm LE, 36"H lower w/0.8 mm LE	1 Carp	16	.500	Ea.	8,400	26.50		8,426.50	9,275
7080	36"H upper w/1.0 mm LE, 36"H lower w/1.5 mm LE	"	16	.500	"	10,500	26.50		10,526.50	11,600

13 49 Radiation Protection

13 49 33 – Radio Frequency Shielding

13 49 33.50 Shielding, Radio Frequency		Crew	Daily Output	Labor-Hours	Unit	2020 Bare Costs Material	Labor	Equipment	Total	Total Incl O&P
0010	SHIELDING, RADIO FREQUENCY									
0020	Prefabricated, galvanized steel	2 Carp	375	.043	SF Surf	5.65	2.27		7.92	9.90
0040	5 oz., copper floor panel		480	.033		4.72	1.77		6.49	8.05
0050	5 oz., copper wall/ceiling panel		155	.103		4.72	5.50		10.22	14
0100	12 oz., copper floor panel		470	.034		9.65	1.81		11.46	13.50
0110	12 oz., copper wall/ceiling panel		140	.114		9.65	6.05		15.70	20.50
0150	Door, copper/wood laminate, 4' x 7'		1.50	10.667	Ea.	8,825	565		9,390	10,600
0200	RF modular shielding panels, walls & ceilings	E-1	180	.133	S.F.	24	7.55	.82	32.37	39.50
0210	Floor liner, steel sheet, 14 ga.		430	.056		2.55	3.16	.34	6.05	8.30
0215	11 ga.		140	.171		3.75	9.70	1.05	14.50	21
0220	Wall liner, steel sheet, 14 ga.		180	.133		2.58	7.55	.82	10.95	15.95
0225	11 ga.		140	.171		3.53	9.70	1.05	14.28	20.50
0230	Steel plate, 1/4"		90	.267		7.30	15.10	1.63	24.03	34.50
0235	Ceiling liner, steel sheet, 14 ga.		180	.133		2.76	7.55	.82	11.13	16.15
0250	Ceiling hangers		45	.533	Ea.	40	30	3.27	73.27	96.50
0275	Wall supports		45	.533	"	40	30	3.27	73.27	96.50
0300	Shielding transition, 11 ga. preformed angles		1365	.018	L.F.	40	1	.11	41.11	45.50
0500	Protection, door, 3' W x 7' H, to 120 Db elec/plane wave	Q-11	9	3.556	Ea.	10,800	211		11,011	12,200
0550	Double door, 6' W x 7' H		6	5.333		17,000	315		17,315	19,200
0600	Wave guide vents, 2" diameter		67	.478		233	28.50		261.50	300
0610	12" diameter		17	1.882		1,100	112		1,212	1,375
0620	12" x 6"		17	1.882		375	112		487	585
0630	12" x 12"		17	1.882		495	112		607	715
0640	15" x 15"		17	1.882		580	112		692	810
0650	30" x 14"		11	2.909		1,075	173		1,248	1,475

13 53 Meteorological Instrumentation

13 53 09 – Weather Instrumentation

13 53 09.50 Weather Station		Crew	Daily Output	Labor-Hours	Unit	Material	Labor	Equipment	Total	Total Incl O&P
0010	WEATHER STATION									
0020	Remote recording, solar powered, with rain gauge & display, 400' range				Ea.	850			850	935
0100	1 mile range				"	1,950			1,950	2,150

Division 14 Conveying Equipment

Estimating Tips

General

- Many products in Division 14 will require some type of support or blocking for installation not included with the item itself. Examples are supports for conveyors or tube systems, attachment points for lifts, and footings for hoists or cranes. Add these supports in the appropriate division.

14 10 00 Dumbwaiters
14 20 00 Elevators

- Dumbwaiters and elevators are estimated and purchased in a method similar to buying a car. The manufacturer has a base unit with standard features. Added to this base unit price will be whatever options the owner or specifications require. Increased load capacity, additional vertical travel, additional stops, higher speed, and cab finish options are items to be considered. When developing an estimate for dumbwaiters and elevators, remember that some items needed by the installers may have to be included as part of the general contract.

  Examples are:
  - ☐ shaftway
  - ☐ rail support brackets
  - ☐ machine room
  - ☐ electrical supply
  - ☐ sill angles
  - ☐ electrical connections
  - ☐ pits
  - ☐ roof penthouses
  - ☐ pit ladders

  Check the job specifications and drawings before pricing.
- Installation of elevators and handicapped lifts in historic structures can require significant additional costs. The associated structural requirements may involve cutting into and repairing finishes, moldings, flooring, etc. The estimator must account for these special conditions.

14 30 00 Escalators and Moving Walks

- Escalators and moving walks are specialty items installed by specialty contractors. There are numerous options associated with these items. For specific options, contact a manufacturer or contractor. In a method similar to estimating dumbwaiters and elevators, you should verify the extent of general contract work and add items as necessary.

14 40 00 Lifts
14 90 00 Other Conveying Equipment

- Products such as correspondence lifts, chutes, and pneumatic tube systems, as well as other items specified in this subdivision, may require trained installers. The general contractor might not have any choice as to who will perform the installation or when it will be performed. Long lead times are often required for these products, making early decisions in scheduling necessary.

Reference Numbers

Reference numbers are shown at the beginning of some major classifications. These numbers refer to related items in the Reference Section. The reference information may be an estimating procedure, an alternate pricing method, or technical information.

Note: Not all subdivisions listed here necessarily appear. ■

14 05 Common Work Results for Conveying Equipment

14 05 05 – Selective Demolition for Conveying Equipment

14 05 05.10 Dumbwaiter Removal		Crew	Daily Output	Labor-Hours	Unit	Material	2020 Bare Costs Labor	Equipment	Total	Total Incl O&P
0010	**DUMBWAITER REMOVAL**									
0050	Dumbwaiter removal, cab, track and equipment	2 Elev	4	4	Stop		340		340	525

14 05 05.20 Elevator Removal

		Crew	Daily Output	Labor-Hours	Unit	Material	Labor	Equipment	Total	Total Incl O&P
0010	**ELEVATOR REMOVAL**									
0050	Elevator removal, cab, track and equipment	2 Elev	2	8	Stop		685		685	1,050

14 05 05.30 Escalator/Moving Walk Removal

		Crew	Daily Output	Labor-Hours	Unit	Material	Labor	Equipment	Total	Total Incl O&P
0010	**ESCALATOR/MOVING WALK REMOVAL**									
0050	Escalator removal, 10′ floor to floor	M-1	.14	229	Ea.		18,600	355	18,955	28,900
0100	Escalator removal, 15′ floor to floor		.12	267			21,700	415	22,115	33,800
0200	Escalator removal, 20′ floor to floor		.10	320			26,000	500	26,500	40,500
0300	Escalator removal, 25′ floor to floor		.08	400			32,500	625	33,125	50,500
0400	Moving ramp/walk removal		10	3.200	L.F.		260	4.99	264.99	405

14 05 05.40 Lift Removal

		Crew	Daily Output	Labor-Hours	Unit	Material	Labor	Equipment	Total	Total Incl O&P
0010	**LIFT REMOVAL**									
0050	Chair lift removal, minimum	2 Elev	2	8	Ea.		685		685	1,050
0060	Maximum	"	1	16			1,375		1,375	2,100
0100	Single post lift removal	L-4	7.50	3.200			160		160	256
0200	Double post lift removal		5	4.800			239		239	385
0300	Four post lift removal		3	8			400		400	640

14 05 05.50 Distribution System Removal

		Crew	Daily Output	Labor-Hours	Unit	Material	Labor	Equipment	Total	Total Incl O&P
0010	**DISTRIBUTION SYSTEM REMOVAL**									
0050	Distribution system removal, motorized cart	4 Mill	.30	107	Station		5,975		5,975	9,225
0100	Distribution system removal, conveyor belt	2 Mill	80	.200	L.F.		11.20		11.20	17.30
0200	Distribution system removal, overhead monorail	1 Elev	50	.160			13.70		13.70	21
0300	Distribution system removal, pneumatic tube	2 Stpi	150	.107			7		7	10.80
0400	Remove linen/trash chute	2 Shee	8	2	Floor		125		125	195

14 05 05.60 Crane Removal

		Crew	Daily Output	Labor-Hours	Unit	Material	Labor	Equipment	Total	Total Incl O&P
0010	**CRANE REMOVAL**									
0050	Remove crane rail	E-4	300	.107	L.F.		6.20	.49	6.69	10.70
0100	Remove overhead bridge crane	M-4	1	36	Ea.		2,300	189	2,489	3,750

14 11 Manual Dumbwaiters

14 11 10 – Hand Operated Dumbwaiters

14 11 10.20 Manual Dumbwaiters		Crew	Daily Output	Labor-Hours	Unit	Material	Labor	Equipment	Total	Total Incl O&P
0010	**MANUAL DUMBWAITERS**									
0020	2 stop, hand powered, up to 75 lb. capacity	2 Elev	.75	21.333	Ea.	3,450	1,825		5,275	6,600
0100	76 lb. capacity and up		.50	32	"	7,075	2,750		9,825	12,000
0300	For each additional stop, add		.75	21.333	Stop	1,125	1,825		2,950	4,050

14 12 Electric Dumbwaiters

14 12 10 – Dumbwaiters

14 12 10.10 Electric Dumbwaiters		Crew	Daily Output	Labor-Hours	Unit	2020 Bare Costs Material	Labor	Equipment	Total	Total Incl O&P
0010	ELECTRIC DUMBWAITERS									
0020	2 stop, up to 75 lb. capacity	2 Elev	.13	123	Ea.	8,950	10,500		19,450	26,100
0100	76 lb. capacity and up		.11	145	"	23,600	12,400		36,000	45,000
0600	For each additional stop, add	↓	.54	29.630	Stop	3,575	2,525		6,100	7,825

14 21 Electric Traction Elevators

14 21 13 – Electric Traction Freight Elevators

14 21 13.10 Electric Traction Freight Elevators and Options			Crew	Daily Output	Labor-Hours	Unit	Material	Labor	Equipment	Total	Total Incl O&P
0010	ELECTRIC TRACTION FREIGHT ELEVATORS AND OPTIONS	R142000-10									
0425	Electric freight, base unit, 4000 lb., 200 fpm, 4 stop, std. fin.	R142000-20	2 Elev	.05	320	Ea.	124,500	27,400		151,900	178,500
0450	For 5000 lb. capacity, add	R142000-30					6,350			6,350	6,975
0500	For 6000 lb. capacity, add	R142000-40					14,900			14,900	16,400
0525	For 7000 lb. capacity, add						20,400			20,400	22,400
0550	For 8000 lb. capacity, add						23,800			23,800	26,200
0575	For 10000 lb. capacity, add						33,200			33,200	36,500
0600	For 12000 lb. capacity, add						40,100			40,100	44,100
0625	For 16000 lb. capacity, add						48,900			48,900	53,500
0650	For 20000 lb. capacity, add						55,500			55,500	61,000
0675	For increased speed, 250 fpm, add						14,800			14,800	16,300
0700	300 fpm, geared electric, add						18,500			18,500	20,300
0725	350 fpm, geared electric, add						22,600			22,600	24,800
0750	400 fpm, geared electric, add						26,600			26,600	29,300
0775	500 fpm, gearless electric, add						36,900			36,900	40,500
0800	600 fpm, gearless electric, add						40,900			40,900	45,000
0825	700 fpm, gearless electric, add						48,400			48,400	53,500
0850	800 fpm, gearless electric, add						54,000			54,000	59,000
0875	For class "B" loading, add						3,000			3,000	3,300
0900	For class "C-1" loading, add						7,375			7,375	8,125
0925	For class "C-2" loading, add						8,675			8,675	9,550
0950	For class "C-3" loading, add					↓	12,100			12,100	13,400
0975	For travel over 40 V.L.F., add		2 Elev	7.25	2.207	V.L.F.	620	189		809	970
1000	For number of stops over 4, add		"	.27	59.259	Stop	4,675	5,075		9,750	12,900

14 21 23 – Electric Traction Passenger Elevators

14 21 23.10 Electric Traction Passenger Elevators and Options			Crew	Daily Output	Labor-Hours	Unit	Material	Labor	Equipment	Total	Total Incl O&P
0010	ELECTRIC TRACTION PASSENGER ELEVATORS AND OPTIONS	R142000-10									
1625	Electric pass., base unit, 2,000 lb., 200 fpm, 4 stop, std. fin.	R142000-20	2 Elev	.05	320	Ea.	104,500	27,400		131,900	157,000
1650	For 2,500 lb. capacity, add	R142000-30					4,475			4,475	4,925
1675	For 3,000 lb. capacity, add	R142000-40					4,900			4,900	5,400
1700	For 3,500 lb. capacity, add						6,225			6,225	6,850
1725	For 4,000 lb. capacity, add						7,325			7,325	8,050
1750	For 4,500 lb. capacity, add						10,100			10,100	11,100
1775	For 5000 lb. capacity, add						13,000			13,000	14,300
1800	For increased speed, 250 fpm, geared electric, add						3,550			3,550	3,900
1825	300 fpm, geared electric, add						6,850			6,850	7,525
1850	350 fpm, geared electric, add						8,800			8,800	9,675
1875	400 fpm, geared electric, add						12,800			12,800	14,100
1900	500 fpm, gearless electric, add						30,900			30,900	34,000
1925	600 fpm, gearless electric, add						50,000			50,000	55,000
1950	700 fpm, gearless electric, add						56,500			56,500	62,000
1975	800 fpm, gearless electric, add					↓	62,500			62,500	69,000

14 21 Electric Traction Elevators

14 21 23 – Electric Traction Passenger Elevators

14 21 23.10 Electric Traction Passenger Elevators and Options		Crew	Daily Output	Labor-Hours	Unit	2020 Bare Costs Material	Labor	Equipment	Total	Total Incl O&P
2000	For travel over 40 V.L.F., add	2 Elev	7.25	2.207	V.L.F.	740	189		929	1,100
2025	For number of stops over 4, add		.27	59.259	Stop	3,225	5,075		8,300	11,300
2400	Electric hospital, base unit, 4,000 lb., 200 fpm, 4 stop, std fin.		.05	320	Ea.	97,000	27,400		124,400	148,500
2425	For 4,500 lb. capacity, add					6,200			6,200	6,825
2450	For 5,000 lb. capacity, add					8,125			8,125	8,925
2475	For increased speed, 250 fpm, geared electric, add					3,950			3,950	4,350
2500	300 fpm, geared electric, add					7,575			7,575	8,325
2525	350 fpm, geared electric, add					9,125			9,125	10,000
2550	400 fpm, geared electric, add					13,200			13,200	14,600
2575	500 fpm, gearless electric, add					36,300			36,300	40,000
2600	600 fpm, gearless electric, add					54,500			54,500	60,000
2625	700 fpm, gearless electric, add					60,500			60,500	66,500
2650	800 fpm, gearless electric, add					68,000			68,000	75,000
2675	For travel over 40 V.L.F., add	2 Elev	7.25	2.207	V.L.F.	196	189		385	505
2700	For number of stops over 4, add	"	.27	59.259	Stop	4,775	5,075		9,850	13,000

14 21 33 – Electric Traction Residential Elevators

14 21 33.20 Residential Elevators

		Crew	Daily Output	Labor-Hours	Unit	Material	Labor	Equipment	Total	Total Incl O&P
0010	**RESIDENTIAL ELEVATORS**									
7000	Residential, cab type, 1 floor, 2 stop, economy model	2 Elev	.20	80	Ea.	10,700	6,850		17,550	22,300
7100	Custom model		.10	160		18,100	13,700		31,800	40,900
7200	2 floor, 3 stop, economy model		.12	133		15,900	11,400		27,300	35,000
7300	Custom model		.06	267		26,000	22,800		48,800	63,500

14 24 Hydraulic Elevators

14 24 13 – Hydraulic Freight Elevators

14 24 13.10 Hydraulic Freight Elevators and Options

			Crew	Daily Output	Labor-Hours	Unit	Material	Labor	Equipment	Total	Total Incl O&P
0010	**HYDRAULIC FREIGHT ELEVATORS AND OPTIONS**	R142000-10									
1025	Hydraulic freight, base unit, 2,000 lb., 50 fpm, 2 stop, std. fin.	R142000-20	2 Elev	.10	160	Ea.	83,500	13,700		97,200	113,000
1050	For 2,500 lb. capacity, add	R142000-30					4,225			4,225	4,625
1075	For 3,000 lb. capacity, add	R142000-40					5,575			5,575	6,125
1100	For 3,500 lb. capacity, add						9,100			9,100	10,000
1125	For 4,000 lb. capacity, add						10,400			10,400	11,500
1150	For 4,500 lb. capacity, add						12,100			12,100	13,300
1175	For 5,000 lb. capacity, add						16,400			16,400	18,000
1200	For 6,000 lb. capacity, add						16,700			16,700	18,400
1225	For 7,000 lb. capacity, add						25,700			25,700	28,200
1250	For 8,000 lb. capacity, add						31,100			31,100	34,200
1275	For 10,000 lb. capacity, add						32,900			32,900	36,200
1300	For 12,000 lb. capacity, add						40,000			40,000	44,000
1325	For 16,000 lb. capacity, add						53,500			53,500	59,000
1350	For 20,000 lb. capacity, add						59,500			59,500	65,500
1375	For increased speed, 100 fpm, add						1,200			1,200	1,325
1400	125 fpm, add						3,150			3,150	3,450
1425	150 fpm, add						4,950			4,950	5,450
1450	175 fpm, add						7,100			7,100	7,800
1475	For class "B" loading, add						2,825			2,825	3,100
1500	For class "C-1" loading, add						7,125			7,125	7,825
1525	For class "C-2" loading, add						8,550			8,550	9,400
1550	For class "C-3" loading, add						11,700			11,700	12,900
1575	For travel over 20 V.L.F., add		2 Elev	7.25	2.207	V.L.F.	895	189		1,084	1,275

14 24 Hydraulic Elevators

14 24 13 – Hydraulic Freight Elevators

14 24 13.10 Hydraulic Freight Elevators and Options		Crew	Daily Output	Labor-Hours	Unit	Material	2020 Bare Costs Labor	Equipment	Total	Total Incl O&P
1600	For number of stops over 2, add	2 Elev	.27	59.259	Stop	2,250	5,075		7,325	10,300

14 24 23 – Hydraulic Passenger Elevators

14 24 23.10 Hydraulic Passenger Elevators and Options		Crew	Daily Output	Labor-Hours	Unit	Material	Labor	Equipment	Total	Total Incl O&P
0010	HYDRAULIC PASSENGER ELEVATORS AND OPTIONS R142000-10									
2050	Hyd. pass., base unit, 1,500 lb., 100 fpm, 2 stop, std. fin. R142000-20	2 Elev	.10	160	Ea.	40,800	13,700		54,500	66,000
2075	For 2,000 lb. capacity, add R142000-30					845			845	930
2100	For 2,500 lb. capacity, add R142000-40					3,175			3,175	3,500
2125	For 3,000 lb. capacity, add					4,550			4,550	5,025
2150	For 3,500 lb. capacity, add					7,625			7,625	8,375
2175	For 4,000 lb. capacity, add					9,175			9,175	10,100
2200	For 4,500 lb. capacity, add					12,100			12,100	13,300
2225	For 5,000 lb. capacity, add					17,000			17,000	18,700
2250	For increased speed, 125 fpm, add					1,225			1,225	1,350
2275	150 fpm, add					2,750			2,750	3,025
2300	175 fpm, add					5,350			5,350	5,875
2325	200 fpm, add				↓	10,300			10,300	11,400
2350	For travel over 12 V.L.F., add	2 Elev	7.25	2.207	V.L.F.	730	189		919	1,100
2375	For number of stops over 2, add		.27	59.259	Stop	1,000	5,075		6,075	8,900
2725	Hydraulic hospital, base unit, 4,000 lb., 100 fpm, 2 stop, std. fin.	↓	.10	160	Ea.	67,000	13,700		80,700	94,500
2775	For 4,500 lb. capacity, add					7,525			7,525	8,275
2800	For 5,000 lb. capacity, add					10,900			10,900	12,000
2825	For increased speed, 125 fpm, add					2,250			2,250	2,475
2850	150 fpm, add					3,750			3,750	4,125
2875	175 fpm, add					6,325			6,325	6,950
2900	200 fpm, add				↓	9,300			9,300	10,200
2925	For travel over 12 V.L.F., add	2 Elev	7.25	2.207	V.L.F.	410	189		599	740
2950	For number of stops over 2, add	"	.27	59.259	Stop	4,675	5,075		9,750	12,900

14 27 Custom Elevator Cabs and Doors

14 27 13 – Custom Elevator Cab Finishes

14 27 13.10 Cab Finishes		Crew	Daily Output	Labor-Hours	Unit	Material	Labor	Equipment	Total	Total Incl O&P
0010	CAB FINISHES									
3325	Passenger elevator cab finishes (based on 3,500 lb. cab size)									
3350	Acrylic panel ceiling				Ea.	795			795	875
3375	Aluminum eggcrate ceiling					715			715	790
3400	Stainless steel doors					4,150			4,150	4,550
3425	Carpet flooring					635			635	695
3450	Epoxy flooring					480			480	525
3475	Quarry tile flooring					920			920	1,025
3500	Slate flooring					1,650			1,650	1,825
3525	Textured rubber flooring					665			665	735
3550	Stainless steel walls					4,150			4,150	4,575
3575	Stainless steel returns at door				↓	1,200			1,200	1,300
4450	Hospital elevator cab finishes (based on 3,500 lb. cab size)									
4475	Aluminum eggcrate ceiling				Ea.	725			725	795
4500	Stainless steel doors					4,125			4,125	4,550
4525	Epoxy flooring					475			475	525
4550	Quarry tile flooring					920			920	1,025
4575	Textured rubber flooring					660			660	730
4600	Stainless steel walls				↓	4,700			4,700	5,175

14 27 Custom Elevator Cabs and Doors

14 27 13 – Custom Elevator Cab Finishes

14 27 13.10 Cab Finishes		Crew	Daily Output	Labor-Hours	Unit	Material (2020 Bare Costs)	Labor (2020 Bare Costs)	Equipment (2020 Bare Costs)	Total (2020 Bare Costs)	Total Incl O&P
4625	Stainless steel returns at door				Ea.	930			930	1,025

14 28 Elevator Equipment and Controls

14 28 10 – Elevator Equipment and Control Options

14 28 10.10 Elevator Controls and Doors

		Crew	Daily Output	Labor-Hours	Unit	Material	Labor	Equipment	Total	Total Incl O&P
0010	**ELEVATOR CONTROLS AND DOORS**									
2975	Passenger elevator options									
3000	2 car group automatic controls	2 Elev	.66	24.242	Ea.	3,625	2,075		5,700	7,150
3025	3 car group automatic controls		.44	36.364		5,525	3,100		8,625	10,900
3050	4 car group automatic controls		.33	48.485		9,650	4,150		13,800	17,000
3075	5 car group automatic controls		.26	61.538		14,500	5,275		19,775	24,100
3100	6 car group automatic controls		.22	72.727		22,100	6,225		28,325	33,900
3125	Intercom service		3	5.333		580	455		1,035	1,350
3150	Duplex car selective collective		.66	24.242		4,175	2,075		6,250	7,775
3175	Center opening 1 speed doors		2	8		2,075	685		2,760	3,350
3200	Center opening 2 speed doors		2	8		2,925	685		3,610	4,275
3225	Rear opening doors (opposite front)		2	8		4,475	685		5,160	5,975
3250	Side opening 2 speed doors		2	8		7,650	685		8,335	9,475
3275	Automatic emergency power switching		.66	24.242		1,275	2,075		3,350	4,575
3300	Manual emergency power switching		8	2		540	171		711	860
3625	Hall finishes, stainless steel doors					1,475			1,475	1,625
3650	Stainless steel frames					1,550			1,550	1,700
3675	12 month maintenance contract								4,425	4,425
3700	Signal devices, hall lanterns	2 Elev	8	2		545	171		716	865
3725	Position indicators, up to 3		9.40	1.702		350	146		496	610
3750	Position indicators, per each over 3		32	.500		94.50	43		137.50	170
3775	High speed heavy duty door opener					2,775			2,775	3,050
3800	Variable voltage, O.H. gearless machine, min.	2 Elev	.16	100		34,600	8,550		43,150	51,000
3815	Maximum		.07	229		77,500	19,600		97,100	115,500
3825	Basement installed geared machine		.33	48.485		13,600	4,150		17,750	21,400
3850	Freight elevator options									
3875	Doors, bi-parting	2 Elev	.66	24.242	Ea.	6,250	2,075		8,325	10,100
3900	Power operated door and gate	″	.66	24.242		25,100	2,075		27,175	30,800
3925	Finishes, steel plate floor					1,125			1,125	1,250
3950	14 ga. 1/4″ x 4′ steel plate walls					2,075			2,075	2,275
3975	12 month maintenance contract								3,400	3,400
4000	Signal devices, hall lanterns	2 Elev	8	2		530	171		701	850
4025	Position indicators, up to 3		9.40	1.702		375	146		521	640
4050	Position indicators, per each over 3		32	.500		103	43		146	179
4075	Variable voltage basement installed geared machine		.66	24.242		21,000	2,075		23,075	26,300
4100	Hospital elevator options									
4125	2 car group automatic controls	2 Elev	.66	24.242	Ea.	3,625	2,075		5,700	7,150
4150	3 car group automatic controls		.44	36.364		5,500	3,100		8,600	10,800
4175	4 car group automatic controls		.33	48.485		13,100	4,150		17,250	20,800
4200	5 car group automatic controls		.26	61.538		13,100	5,275		18,375	22,500
4225	6 car group automatic controls		.22	72.727		20,000	6,225		26,225	31,600
4250	Intercom service		3	5.333		550	455		1,005	1,300
4275	Duplex car selective collective		.66	24.242		4,075	2,075		6,150	7,675
4300	Center opening 1 speed doors		2	8		2,075	685		2,760	3,325
4325	Center opening 2 speed doors		2	8		2,750	685		3,435	4,050
4350	Rear opening doors (opposite front)		2	8		4,475	685		5,160	5,975

14 28 Elevator Equipment and Controls

14 28 10 – Elevator Equipment and Control Options

14 28 10.10 Elevator Controls and Doors		Crew	Daily Output	Labor-Hours	Unit	Material	Labor	Equipment	Total	Total Incl O&P
						2020 Bare Costs				
4375	Side opening 2 speed doors	2 Elev	2	8	Ea.	6,650	685		7,335	8,350
4400	Automatic emergency power switching		.66	24.242		1,250	2,075		3,325	4,550
4425	Manual emergency power switching		8	2		525	171		696	840
4675	Hall finishes, stainless steel doors					1,550			1,550	1,700
4700	Stainless steel frames					1,550			1,550	1,725
4725	12 month maintenance contract								6,125	6,125
4750	Signal devices, hall lanterns	2 Elev	8	2		520	171		691	835
4775	Position indicators, up to 3		9.40	1.702		355	146		501	615
4800	Position indicators, per each over 3		32	.500		95	43		138	171
4825	High speed heavy duty door opener					2,800			2,800	3,075
4850	Variable voltage, O.H. gearless machine, min.	2 Elev	.16	100		36,400	8,550		44,950	53,000
4865	Maximum		.07	229		79,000	19,600		98,600	117,000
4875	Basement installed geared machine		.33	48.485		18,200	4,150		22,350	26,500
5000	Drilling for piston, casing included, 18" diameter	B-48	80	.700	V.L.F.	65.50	33.50	13.10	112.10	139

14 31 Escalators

14 31 10 – Glass and Steel Escalators

14 31 10.10 Escalators		Crew	Daily Output	Labor-Hours	Unit	Material	Labor	Equipment	Total	Total Incl O&P
0010	**ESCALATORS**									
1000	Glass, 32" wide x 10' floor to floor height	M-1	.07	457	Ea.	90,500	37,200	715	128,415	157,500
1010	48" wide x 10' floor to floor height		.07	457		97,500	37,200	715	135,415	165,000
1020	32" wide x 15' floor to floor height		.06	533		95,500	43,300	830	139,630	172,500
1030	48" wide x 15' floor to floor height		.06	533		101,000	43,300	830	145,130	179,000
1040	32" wide x 20' floor to floor height		.05	653		101,500	53,000	1,025	155,525	194,000
1050	48" wide x 20' floor to floor height		.05	653		110,000	53,000	1,025	164,025	203,500
1060	32" wide x 25' floor to floor height		.04	800		112,500	65,000	1,250	178,750	225,000
1070	48" wide x 25' floor to floor height		.04	800		130,000	65,000	1,250	196,250	244,500
1080	Enameled steel, 32" wide x 10' floor to floor height		.07	457		98,000	37,200	715	135,915	166,000
1090	48" wide x 10' floor to floor height		.07	457		107,000	37,200	715	144,915	176,000
1110	32" wide x 15' floor to floor height		.06	533		105,000	43,300	830	149,130	183,000
1120	48" wide x 15' floor to floor height		.06	533		109,000	43,300	830	153,130	187,500
1130	32" wide x 20' floor to floor height		.05	653		111,000	53,000	1,025	165,025	205,000
1140	48" wide x 20' floor to floor height		.05	653		120,500	53,000	1,025	174,525	215,000
1150	32" wide x 25' floor to floor height		.04	800		123,000	65,000	1,250	189,250	236,500
1160	48" wide x 25' floor to floor height		.04	800		140,000	65,000	1,250	206,250	255,500
1170	Stainless steel, 32" wide x 10' floor to floor height		.07	457		104,500	37,200	715	142,415	173,000
1180	48" wide x 10' floor to floor height		.07	457		112,500	37,200	715	150,415	182,000
1500	32" wide x 15' floor to floor height		.06	533		111,500	43,300	830	155,630	190,000
1700	48" wide x 15' floor to floor height		.06	533		116,000	43,300	830	160,130	195,500
1750	32" wide x 18' floor to floor height		.05	615		140,500	50,000	960	191,460	232,000
1775	48" wide x 18' floor to floor height		.05	615		152,500	50,000	960	203,460	245,000
2300	32" wide x 25' floor to floor height		.04	800		139,500	65,000	1,250	205,750	255,000
2500	48" wide x 25' floor to floor height		.04	800		147,500	65,000	1,250	213,750	264,000

14 32 Moving Walks

14 32 10 – Moving Walkways

14 32 10.10 Moving Walks		Crew	Daily Output	Labor-Hours	Unit	Material	2020 Bare Costs Labor	Equipment	Total	Total Incl O&P
0010	MOVING WALKS R143210-20									
0020	Walk, 27" tread width, minimum	M-1	6.50	4.923	L.F.	945	400	7.70	1,352.70	1,650
0100	300' to 500', maximum		4.43	7.223		1,300	585	11.25	1,896.25	2,325
0300	48" tread width walk, minimum		4.43	7.223		2,150	585	11.25	2,746.25	3,275
0400	100' to 350', maximum		3.82	8.377		2,475	680	13.05	3,168.05	3,800
0600	Ramp, 12° incline, 36" tread width, minimum		5.27	6.072		1,725	495	9.45	2,229.45	2,675
0700	70' to 90' maximum		3.82	8.377		2,700	680	13.05	3,393.05	4,050
0900	48" tread width, minimum		3.57	8.964		2,575	730	14	3,319	3,975
1000	40' to 70', maximum	↓	2.91	10.997	↓	3,175	895	17.15	4,087.15	4,900

14 42 Wheelchair Lifts

14 42 13 – Inclined Wheelchair Lifts

14 42 13.10 Inclined Wheelchair Lifts and Stairclimbers		Crew	Daily Output	Labor-Hours	Unit	Material	Labor	Equipment	Total	Total Incl O&P
0010	INCLINED WHEELCHAIR LIFTS AND STAIRCLIMBERS									
7700	Stair climber (chair lift), single seat, minimum	2 Elev	1	16	Ea.	4,975	1,375		6,350	7,575
7800	Maximum		.20	80		6,825	6,850		13,675	18,000
8700	Stair lift, minimum		1	16		13,500	1,375		14,875	17,000
8900	Maximum	↓	.20	80	↓	21,400	6,850		28,250	34,000

14 42 16 – Vertical Wheelchair Lifts

14 42 16.10 Wheelchair Lifts		Crew	Daily Output	Labor-Hours	Unit	Material	Labor	Equipment	Total	Total Incl O&P
0010	WHEELCHAIR LIFTS									
8000	Wheelchair lift, minimum	2 Elev	1	16	Ea.	6,825	1,375		8,200	9,600
8500	Maximum	"	.50	32	"	16,100	2,750		18,850	22,000

14 45 Vehicle Lifts

14 45 10 – Hydraulic Vehicle Lifts

14 45 10.10 Hydraulic Lifts		Crew	Daily Output	Labor-Hours	Unit	Material	Labor	Equipment	Total	Total Incl O&P
0010	HYDRAULIC LIFTS									
2200	Single post, 8,000 lb. capacity	L-4	.40	60	Ea.	7,100	3,000		10,100	12,600
2810	Double post, 6,000 lb. capacity		2.67	8.989		9,775	450		10,225	11,500
2815	9,000 lb. capacity		2.29	10.480		23,200	525		23,725	26,300
2820	15,000 lb. capacity		2	12		26,200	600		26,800	29,900
2822	Four post, 26,000 lb. capacity		1.80	13.333		15,100	665		15,765	17,700
2825	30,000 lb. capacity		1.60	15		58,000	750		58,750	65,000
2830	Ramp style, 4 post, 25,000 lb. capacity		2	12		22,700	600		23,300	26,000
2835	35,000 lb. capacity		1	24		106,000	1,200		107,200	118,500
2840	50,000 lb. capacity		1	24		121,000	1,200		122,200	135,500
2845	75,000 lb. capacity	↓	1	24		141,000	1,200		142,200	157,000
2850	For drive thru tracks, add, minimum					1,150			1,150	1,275
2855	Maximum					1,975			1,975	2,175
2860	Ramp extensions, 3' (set of 2)					1,200			1,200	1,325
2865	Rolling jack platform					4,175			4,175	4,575
2870	Electric/hydraulic jacking beam					11,400			11,400	12,600
2880	Scissor lift, portable, 6,000 lb. capacity				↓	10,900			10,900	12,000

14 91 Facility Chutes

14 91 33 – Laundry and Linen Chutes

14 91 33.10 Chutes		Crew	Daily Output	Labor-Hours	Unit	Material	2020 Bare Costs Labor	Equipment	Total	Total Incl O&P
0011	**CHUTES**, linen, trash or refuse									
0050	Aluminized steel, 16 ga., 18" diameter	2 Shee	3.50	4.571	Floor	1,800	285		2,085	2,450
0100	24" diameter		3.20	5		2,000	310		2,310	2,700
0200	30" diameter		3	5.333		2,375	330		2,705	3,125
0300	36" diameter		2.80	5.714		2,725	355		3,080	3,550
0400	Galvanized steel, 16 ga., 18" diameter		3.50	4.571		1,075	285		1,360	1,650
0500	24" diameter		3.20	5		1,225	310		1,535	1,850
0600	30" diameter		3	5.333		1,375	330		1,705	2,025
0700	36" diameter		2.80	5.714		1,625	355		1,980	2,325
0800	Stainless steel, 18" diameter		3.50	4.571		3,375	285		3,660	4,175
0900	24" diameter		3.20	5		3,250	310		3,560	4,075
1000	30" diameter		3	5.333		3,975	330		4,305	4,900
1005	36" diameter		2.80	5.714		4,200	355		4,555	5,175
1200	Linen chute bottom collector, aluminized steel		4	4	Ea.	1,525	249		1,774	2,075
1300	Stainless steel		4	4		1,950	249		2,199	2,550
1500	Refuse, bottom hopper, aluminized steel, 18" diameter		3	5.333		1,150	330		1,480	1,775
1600	24" diameter		3	5.333		1,350	330		1,680	2,000
1800	36" diameter		3	5.333		2,600	330		2,930	3,400

14 91 82 – Trash Chutes

14 91 82.10 Trash Chutes and Accessories		Crew	Daily Output	Labor-Hours	Unit	Material	Labor	Equipment	Total	Total Incl O&P
0010	**TRASH CHUTES AND ACCESSORIES**									
2900	Package chutes, spiral type, minimum	2 Shee	4.50	3.556	Floor	2,575	222		2,797	3,200
3000	Maximum	"	1.50	10.667	"	6,750	665		7,415	8,475
9000	Minimum labor/equipment charge	1 Shee	1	8	Job		500		500	780

14 92 Pneumatic Tube Systems

14 92 10 – Conventional, Automatic and Computer Controlled Pneumatic Tube Systems

14 92 10.10 Pneumatic Tube Systems		Crew	Daily Output	Labor-Hours	Unit	Material	Labor	Equipment	Total	Total Incl O&P
0010	**PNEUMATIC TUBE SYSTEMS**									
0020	100' long, single tube, 2 stations, stock									
0100	3" diameter	2 Stpi	.12	133	Total	3,675	8,750		12,425	17,600
0300	4" diameter	"	.09	178	"	4,600	11,700		16,300	23,100
0400	Twin tube, two stations or more, conventional system									
0600	2-1/2" round	2 Stpi	62.50	.256	L.F.	43	16.80		59.80	73
0700	3" round		46	.348		40	23		63	79.50
0900	4" round		49.60	.323		50.50	21		71.50	88
1000	4" x 7" oval		37.60	.426		99.50	28		127.50	152
1050	Add for blower		2	8	System	5,200	525		5,725	6,525
1110	Plus for each round station, add		7.50	2.133	Ea.	1,375	140		1,515	1,725
1150	Plus for each oval station, add		7.50	2.133	"	1,375	140		1,515	1,725
1200	Alternate pricing method: base cost, economy model		.75	21.333	Total	5,850	1,400		7,250	8,575
1300	Custom model		.25	64	"	11,700	4,200		15,900	19,300
1500	Plus total system length, add, for economy model		93.40	.171	L.F.	8.25	11.25		19.50	26.50
1600	For custom model		37.60	.426	"	25	28		53	70.50
1800	Completely automatic system, 4" round, 15 to 50 stations		.29	55.172	Station	22,600	3,625		26,225	30,400
2200	51 to 144 stations		.32	50		15,600	3,275		18,875	22,300
2400	6" round or 4" x 7" oval, 15 to 50 stations		.24	66.667		27,600	4,375		31,975	37,200
2800	51 to 144 stations		.23	69.565		21,100	4,550		25,650	30,300

Division Notes

I	CREW	DAILY OUTPUT	LABOR-HOURS	UNIT	BARE COSTS MAT.	BARE COSTS LABOR	BARE COSTS EQUIP.	BARE COSTS TOTAL	TOTAL INCL O&P

Estimating Tips

Pipe for fire protection and all uses is located in Subdivisions 21 11 13 and 22 11 13.

The labor adjustment factors listed in Subdivision 22 01 02.20 also apply to Division 21.

Many, but not all, areas in the U.S. require backflow protection in the fire system. Insurance underwriters may have specific requirements for the type of materials to be installed or design requirements based on the hazard to be protected. Local jurisdictions may have requirements not covered by code. It is advisable to be aware of any special conditions.

For your reference, the following is a list of the most applicable Fire Codes and Standards, which may be purchased from the NFPA, 1 Batterymarch Park, Quincy, MA 02169-7471.

- NFPA 1: Uniform Fire Code
- NFPA 10: Portable Fire Extinguishers
- NFPA 11: Low-, Medium-, and High-Expansion Foam
- NFPA 12: Carbon Dioxide Extinguishing Systems (Also companion 12A)
- NFPA 13: Installation of Sprinkler Systems (Also companion 13D, 13E, and 13R)
- NFPA 14: Installation of Standpipe and Hose Systems
- NFPA 15: Water Spray Fixed Systems for Fire Protection
- NFPA 16: Installation of Foam-Water Sprinkler and Foam-Water Spray Systems
- NFPA 17: Dry Chemical Extinguishing Systems (Also companion 17A)
- NFPA 18: Wetting Agents
- NFPA 20: Installation of Stationary Pumps for Fire Protection
- NFPA 22: Water Tanks for Private Fire Protection
- NFPA 24: Installation of Private Fire Service Mains and their Appurtenances
- NFPA 25: Inspection, Testing and Maintenance of Water-Based Fire Protection

Reference Numbers

Reference numbers are shown at the beginning of some major classifications. These numbers refer to related items in the Reference Section. The reference information may be an estimating procedure, an alternate pricing method, or technical information.

Note: "Powered in part by CINX™, based on licensed proprietary information of Harrison Publishing House, Inc."

Note: Trade Service, in part, has been used as a reference source for some of the material prices used in Division 21.

21 05 23 – General-Duty Valves for Water-Based Fire-Suppression Piping

21 05 23.50 General-Duty Valves		Crew	Daily Output	Labor-Hours	Unit	2020 Bare Costs Material	Labor	Equipment	Total	Total Incl O&P
0010	**GENERAL-DUTY VALVES,** for water-based fire suppression									
6200	Valves and components									
6210	Wet alarm, includes									
6220	retard chamber, trim, gauges, alarm line strainer									
6260	3″ size	Q-12	3	5.333	Ea.	1,725	305		2,030	2,375
6280	4″ size	″	2	8		2,325	455		2,780	3,250
6300	6″ size	Q-13	4	8		2,000	480		2,480	2,950
6320	8″ size	″	3	10.667		2,425	640		3,065	3,675
6400	Dry alarm, includes									
6405	retard chamber, trim, gauges, alarm line strainer									
6410	1-1/2″ size	Q-12	3	5.333	Ea.	5,100	305		5,405	6,075
6420	2″ size		3	5.333		5,100	305		5,405	6,075
6430	3″ size		3	5.333		5,175	305		5,480	6,175
6440	4″ size		2	8		5,525	455		5,980	6,775
6450	6″ size	Q-13	3	10.667		6,375	640		7,015	8,000
6460	8″ size	″	3	10.667		9,350	640		9,990	11,300
6500	Check, swing, C.I. body, brass fittings, auto. ball drip									
6520	4″ size	Q-12	3	5.333	Ea.	420	305		725	930
6540	6″ size	Q-13	4	8		780	480		1,260	1,600
6580	8″ size	″	3	10.667		1,225	640		1,865	2,350
6800	Check, wafer, butterfly type, C.I. body, bronze fittings									
6820	4″ size	Q-12	4	4	Ea.	1,375	228		1,603	1,850
6840	6″ size	Q-13	5.50	5.818		1,975	350		2,325	2,725
6860	8″ size		5	6.400		3,425	385		3,810	4,375
6880	10″ size		4.50	7.111		2,650	430		3,080	3,600
8700	Floor control valve, includes trim and gauges, 2″ size	Q-12	6	2.667		1,000	152		1,152	1,325
8710	2-1/2″ size		6	2.667		1,150	152		1,302	1,475
8720	3″ size		6	2.667		1,150	152		1,302	1,475
8730	4″ size		6	2.667		1,150	152		1,302	1,475
8740	6″ size		5	3.200		1,150	182		1,332	1,525
8800	Flow control valve, includes trim and gauges, 2″ size		2	8		5,475	455		5,930	6,725
8820	3″ size		1.50	10.667		6,275	605		6,880	7,850
8840	4″ size	Q-13	2.80	11.429		6,650	690		7,340	8,400
8860	6″ size	″	2	16		8,325	965		9,290	10,700
9200	Pressure operated relief valve, brass body	1 Spri	18	.444		695	28		723	810
9600	Waterflow indicator, vane type, with recycling retard and									
9610	two single pole retard switches, 2″ thru 6″ pipe size	1 Spri	8	1	Ea.	171	63.50		234.50	287
9990	Minimum labor/equipment charge	″	3	2.667	Job		169		169	261

21 05 53 – Identification For Fire-Suppression Piping and Equipment

21 05 53.50 Identification		Crew	Daily Output	Labor-Hours	Unit	Material	Labor	Equipment	Total	Total Incl O&P
0010	**IDENTIFICATION,** for fire suppression piping and equipment									
3010	Plates and escutcheons for identification of fire dept. service/connections									
3100	Wall mount, round, aluminum									
3110	4″	1 Plum	96	.083	Ea.	25.50	5.35		30.85	37
3120	6″	″	96	.083	″	67	5.35		72.35	82
3200	Wall mount, round, cast brass									
3210	2-1/2″	1 Plum	70	.114	Ea.	61.50	7.35		68.85	79.50
3220	3″		70	.114		72	7.35		79.35	90.50
3230	4″		70	.114		123	7.35		130.35	147
3240	6″		70	.114		144	7.35		151.35	169
3250	For polished brass, add					25%				
3260	For rough chrome, add					33%				

21 05 Common Work Results for Fire Suppression

21 05 53 – Identification For Fire-Suppression Piping and Equipment

21 05 53.50 Identification		Crew	Daily Output	Labor-Hours	Unit	2020 Bare Costs Material	Labor	Equipment	Total	Total Incl O&P
3270	For polished chrome, add					55%				
3300	Wall mount, square, cast brass									
3310	2-1/2"	1 Plum	70	.114	Ea.	164	7.35		171.35	192
3320	3"	"	70	.114	"	175	7.35		182.35	203
3330	For polished brass, add					15%				
3340	For rough chrome, add					25%				
3350	For polished chrome, add					33%				
3400	Wall mount, cast brass, multiple outlets									
3410	rect. 2 way	Q-1	5	3.200	Ea.	227	186		413	535
3420	rect. 3 way		4	4		535	232		767	950
3430	rect. 4 way		4	4		670	232		902	1,100
3440	square 4 way		4	4		665	232		897	1,100
3450	rect. 6 way		3	5.333		915	310		1,225	1,475
3460	For polished brass, add					10%				
3470	For rough chrome, add					20%				
3480	For polished chrome, add					25%				
3500	Base mount, free standing fdc, cast brass									
3510	4"	1 Plum	60	.133	Ea.	123	8.60		131.60	149
3520	6"	"	60	.133	"	164	8.60		172.60	194
3530	For polished brass, add					25%				
3540	For rough chrome, add					30%				
3550	For polished chrome, add					45%				

21 11 Facility Fire-Suppression Water-Service Piping

21 11 11 – Fire-Suppression, Pipe Fittings, Grooved Joint

21 11 11.05 Corrosion Monitors		Crew	Daily Output	Labor-Hours	Unit	Material	Labor	Equipment	Total	Total Incl O&P
0010	**CORROSION MONITORS**, pipe, in line spool									
1100	Powder coated, schedule 10									
1140	2"	1 Plum	17	.471	Ea.	180	30.50		210.50	245
1150	2-1/2"	Q-1	27	.593		188	34.50		222.50	260
1160	3"		22	.727		196	42		238	281
1170	4"		17	.941		216	54.50		270.50	320
1180	6"	Q-2	17	1.412		227	85		312	380
1190	8"	"	14	1.714		255	103		358	440
1200	Powder coated, schedule 40									
1240	2"	1 Plum	17	.471	Ea.	180	30.50		210.50	245
1250	2-1/2"	Q-1	27	.593		188	34.50		222.50	260
1260	3"		22	.727		196	42		238	281
1270	4"		17	.941		216	54.50		270.50	320
1280	6"	Q-2	17	1.412		227	85		312	380
1290	8"	"	14	1.714		255	103		358	440
1300	Galvanized, schedule 10									
1340	2"	1 Plum	17	.471	Ea.	188	30.50		218.50	254
1350	2-1/2"	Q-1	27	.593		200	34.50		234.50	273
1360	3"		22	.727		208	42		250	294
1370	4"		17	.941		227	54.50		281.50	335
1380	6"	Q-2	17	1.412		247	85		332	405
1390	8"	"	14	1.714		274	103		377	460
1400	Galvanized, schedule 40									
1440	2"	1 Plum	17	.471	Ea.	180	30.50		210.50	245
1450	2-1/2"	Q-1	27	.593		193	34.50		227.50	265

21 11 11 – Fire-Suppression, Pipe Fittings, Grooved Joint

21 11 11.05 Corrosion Monitors		Crew	Daily Output	Labor-Hours	Unit	2020 Bare Costs Material	Labor	Equipment	Total	Total Incl O&P
1460	3"	Q-1	22	.727	Ea.	196	42		238	281
1470	4"	↓	17	.941		216	54.50		270.50	320
1480	6"	Q-2	17	1.412		227	85		312	380
1490	8"	"	14	1.714	↓	255	103		358	440
1500	Mechanical tee, painted									
1540	2"	1 Plum	50	.160	Ea.	153	10.30		163.30	184
1550	2-1/2"	Q-1	80	.200		161	11.60		172.60	195
1560	3"		67	.239		165	13.85		178.85	203
1570	4"	↓	50	.320		169	18.55		187.55	214
1580	6"	Q-2	50	.480		172	29		201	235
1590	8"	"	42	.571	↓	196	34.50		230.50	269
1600	Mechanical tee, galvanized									
1640	2"	1 Plum	50	.160	Ea.	165	10.30		175.30	197
1650	2-1/2"	Q-1	80	.200		172	11.60		183.60	208
1660	3"		67	.239		192	13.85		205.85	233
1670	4"	↓	50	.320		180	18.55		198.55	227
1680	6"	Q-2	50	.480		192	29		221	256
1690	8"	"	42	.571	↓	216	34.50		250.50	290

21 11 11.16 Pipe Fittings, Grooved Joint		Crew	Daily Output	Labor-Hours	Unit	Material	Labor	Equipment	Total	Total Incl O&P
0010	**PIPE FITTINGS, GROOVED JOINT,** For fire-suppression									
0020	Fittings, ductile iron									
0030	Coupling required at joints not incl. in fitting price.									
0034	Add 1 coupling, material only, per joint for installed price.									
0038	For standard grooved joint materials see Div. 22 11 13.48									
0040	90° elbow									
0110	2"	1 Plum	25	.320	Ea.	22	20.50		42.50	56.50
0120	2-1/2"	Q-1	40	.400		29.50	23		52.50	68.50
0130	3"		33	.485		46.50	28		74.50	95
0140	4"		25	.640		54.50	37		91.50	118
0150	5"	↓	20	.800		128	46.50		174.50	212
0160	6"	Q-2	25	.960		150	57.50		207.50	255
0170	8"	"	21	1.143	↓	292	68.50		360.50	425
0200	45° elbow									
0210	2"	1 Plum	25	.320	Ea.	22	20.50		42.50	56.50
0220	2-1/2"	Q-1	40	.400		29.50	23		52.50	68.50
0230	3"		33	.485		38	28		66	85.50
0240	4"		25	.640		54.50	37		91.50	118
0250	5"	↓	20	.800		128	46.50		174.50	212
0260	6"	Q-2	25	.960		150	57.50		207.50	255
0270	8"	"	21	1.143	↓	296	68.50		364.50	430
0300	Tee									
0310	2"	1 Plum	17	.471	Ea.	34	30.50		64.50	84
0320	2-1/2"	Q-1	27	.593		44	34.50		78.50	102
0330	3"		22	.727		61	42		103	132
0340	4"		17	.941		92	54.50		146.50	186
0350	5"	↓	13	1.231		212	71.50		283.50	345
0360	6"	Q-2	17	1.412		244	85		329	400
0370	8"	"	14	1.714	↓	520	103		623	730
0400	Cap									
0410	1-1/4"	1 Plum	76	.105	Ea.	12.15	6.80		18.95	24
0420	1-1/2"		63	.127		12.80	8.20		21	27
0430	2"	↓	47	.170	↓	15.20	10.95		26.15	33.50

21 11 11 – Fire-Suppression, Pipe Fittings, Grooved Joint

21 11 11.16 Pipe Fittings, Grooved Joint		Crew	Daily Output	Labor-Hours	Unit	2020 Bare Costs Material	Labor	Equipment	Total	Total Incl O&P
0440	2-1/2"	Q-1	76	.211	Ea.	17.55	12.20		29.75	38
0450	3"		63	.254		21	14.75		35.75	46
0460	4"		47	.340		30.50	19.75		50.25	64
0470	5"		37	.432		61	25		86	106
0480	6"	Q-2	47	.511		68	30.50		98.50	122
0490	8"	"	39	.615		127	37		164	197
0500	Coupling, rigid									
0510	1-1/4"	1 Plum	100	.080	Ea.	27.50	5.15		32.65	38
0520	1-1/2"		67	.119		36.50	7.70		44.20	52
0530	2"		50	.160		39.50	10.30		49.80	59
0540	2-1/2"	Q-1	80	.200		46.50	11.60		58.10	69.50
0550	3"		67	.239		50.50	13.85		64.35	77
0560	4"		50	.320		70.50	18.55		89.05	107
0570	5"		40	.400		74.50	23		97.50	118
0580	6"	Q-2	50	.480		89.50	29		118.50	143
0590	8"	"	42	.571		150	34.50		184.50	218
0700	End of run fitting									
0710	1-1/4" x 1/2" NPT	1 Plum	76	.105	Ea.	41.50	6.80		48.30	56
0714	1-1/4" x 3/4" NPT		76	.105		41.50	6.80		48.30	56
0718	1-1/4" x 1" NPT		76	.105		41.50	6.80		48.30	56
0722	1-1/2" x 1/2" NPT		63	.127		41.50	8.20		49.70	58
0726	1-1/2" x 3/4" NPT		63	.127		41.50	8.20		49.70	58
0730	1-1/2" x 1" NPT		63	.127		41.50	8.20		49.70	58
0734	2" x 1/2" NPT		47	.170		44	10.95		54.95	65.50
0738	2" x 3/4" NPT		47	.170		44	10.95		54.95	65.50
0742	2" x 1" NPT		47	.170		44	10.95		54.95	65.50
0746	2-1/2" x 1/2" NPT		40	.200		52.50	12.90		65.40	77.50
0750	2-1/2" x 3/4" NPT		40	.200		52.50	12.90		65.40	77.50
0754	2-1/2" x 1" NPT		40	.200		52.50	12.90		65.40	77.50
0800	Drain elbow									
0820	2-1/2"	Q-1	40	.400	Ea.	103	23		126	150
0830	3"		33	.485		132	28		160	190
0840	4"		25	.640		134	37		171	206
0850	6"	Q-2	25	.960		235	57.50		292.50	350
1000	Valves, grooved joint									
1002	Coupling required at joints not incl. in fitting price.									
1004	Add 1 coupling, material only, per joint for installed price.									
1010	Ball valve with weatherproof actuator									
1020	1-1/4"	1 Plum	39	.205	Ea.	156	13.20		169.20	193
1030	1-1/2"		31	.258		185	16.65		201.65	229
1040	2"		24	.333		223	21.50		244.50	278
1100	Butterfly valve, high pressure, with actuator									
1110	Supervised open									
1120	2"	1 Plum	23	.348	Ea.	415	22.50		437.50	495
1130	2-1/2"	Q-1	38	.421		430	24.50		454.50	515
1140	3"		30	.533		465	31		496	560
1150	4"		22	.727		490	42		532	600
1160	5"		19	.842		730	49		779	880
1170	6"	Q-2	23	1.043		695	63		758	855
1180	8"		18	1.333		1,050	80		1,130	1,300
1190	10"		15	1.600		2,575	96		2,671	2,975
1200	12"		12	2		3,675	120		3,795	4,200
1300	Gate valve, OS&Y									

21 11 11 – Fire-Suppression, Pipe Fittings, Grooved Joint

21 11 11.16 Pipe Fittings, Grooved Joint		Crew	Daily Output	Labor-Hours	Unit	Material	2020 Bare Costs Labor	Equipment	Total	Total Incl O&P
1310	2-1/2"	Q-1	35	.457	Ea.	470	26.50		496.50	555
1320	3"		28	.571		645	33		678	760
1330	4"		20	.800		760	46.50		806.50	905
1340	6"	Q-2	21	1.143		965	68.50		1,033.50	1,150
1350	8"		16	1.500		1,500	90		1,590	1,800
1360	10"		13	1.846		2,025	111		2,136	2,400
1370	12"		10	2.400		3,175	144		3,319	3,700
1400	Gate valve, non-rising stem									
1410	2-1/2"	Q-1	36	.444	Ea.	435	26		461	520
1420	3"		29	.552		585	32		617	695
1430	4"		21	.762		695	44		739	835
1440	6"	Q-2	22	1.091		895	65.50		960.50	1,075
1450	8"		17	1.412		1,325	85		1,410	1,575
1460	10"		14	1.714		1,925	103		2,028	2,250
1470	12"		11	2.182		2,675	131		2,806	3,150
2000	Alarm check valve, pre-trimmed									
2010	1-1/2"	Q-1	6	2.667	Ea.	1,925	155		2,080	2,375
2020	2"	"	5	3.200		1,925	186		2,111	2,400
2030	2-1/2"	Q-2	7	3.429		1,925	206		2,131	2,450
2040	3"		6	4		1,925	241		2,166	2,500
2050	4"		5	4.800		2,125	289		2,414	2,775
2060	6"		4	6		2,450	360		2,810	3,250
2070	8"		2	12		3,400	720		4,120	4,875
2200	Dry valve, pre-trimmed									
2210	1-1/2"	Q-1	5	3.200	Ea.	2,875	186		3,061	3,450
2220	2"	"	4	4		2,875	232		3,107	3,525
2230	2-1/2"	Q-2	6	4		2,925	241		3,166	3,600
2240	3"		5	4.800		2,925	289		3,214	3,675
2250	4"		4	6		3,125	360		3,485	4,000
2260	6"		2	12		3,600	720		4,320	5,100
2270	8"		1	24		5,300	1,450		6,750	8,050
2300	Deluge valve, pre-trimmed, with electric solenoid									
2310	1-1/2"	Q-1	5	3.200	Ea.	4,125	186		4,311	4,800
2320	2"	"	4	4		4,125	232		4,357	4,875
2330	2-1/2"	Q-2	6	4		4,175	241		4,416	4,975
2340	3"		5	4.800		4,175	289		4,464	5,050
2350	4"		4	6		4,975	360		5,335	6,025
2360	6"		2	12		5,950	720		6,670	7,675
2370	8"		1	24		6,725	1,450		8,175	9,625
2400	Preaction valve									
2410	Valve has double interlock,									
2420	pneumatic/electric actuation and trim.									
2430	1-1/2"	Q-1	5	3.200	Ea.	4,225	186		4,411	4,925
2440	2"	"	4	4		4,225	232		4,457	5,000
2450	2-1/2"	Q-2	6	4		4,225	241		4,466	5,025
2460	3"		5	4.800		4,225	289		4,514	5,100
2470	4"		4	6		4,650	360		5,010	5,650
2480	6"		2	12		4,650	720		5,370	6,225
2490	8"		1	24		5,100	1,450		6,550	7,850
3000	Fittings, ductile iron, ready to install									
3010	Includes bolts & grade "E" gaskets									
3012	Add 1 coupling, material only, per joint for installed price.									
3200	90° elbow									

21 11 11 – Fire-Suppression, Pipe Fittings, Grooved Joint

21 11 11.16 Pipe Fittings, Grooved Joint		Crew	Daily Output	Labor-Hours	Unit	Material	2020 Bare Costs Labor	Equipment	Total	Total Incl O&P
3210	1-1/4"	1 Plum	88	.091	Ea.	113	5.85		118.85	133
3220	1-1/2"		72	.111		116	7.15		123.15	138
3230	2"		60	.133		117	8.60		125.60	141
3240	2-1/2"	Q-1	85	.188		141	10.90		151.90	172
3300	45° elbow									
3310	1-1/4"	1 Plum	88	.091	Ea.	113	5.85		118.85	133
3320	1-1/2"		72	.111		116	7.15		123.15	138
3330	2"		60	.133		117	8.60		125.60	141
3340	2-1/2"	Q-1	85	.188		141	10.90		151.90	172
3400	Tee									
3410	1-1/4"	1 Plum	60	.133	Ea.	174	8.60		182.60	204
3420	1-1/2"		48	.167		180	10.75		190.75	215
3430	2"		38	.211		182	13.55		195.55	221
3440	2-1/2"	Q-1	60	.267		210	15.45		225.45	255
3500	Coupling									
3510	1-1/4"	1 Plum	200	.040	Ea.	42.50	2.58		45.08	50.50
3520	1-1/2"		134	.060		43	3.85		46.85	53.50
3530	2"		100	.080		49	5.15		54.15	61.50
3540	2-1/2"	Q-1	160	.100		56	5.80		61.80	70.50
3550	3"		134	.119		63	6.95		69.95	80
3560	4"		100	.160		88	9.30		97.30	111
3570	5"		80	.200		117	11.60		128.60	147
3580	6"	Q-2	100	.240		140	14.45		154.45	177
3590	8"	"	84	.286		140	17.20		157.20	181
4000	For seismic bracing, see Section 22 05 48.40									
5000	For hangers and supports, see Section 22 05 29.10									

21 11 13 – Facility Fire Suppression Piping

21 11 13.16 Pipe, Plastic

		Crew	Daily Output	Labor-Hours	Unit	Material	Labor	Equipment	Total	Total Incl O&P
0010	**PIPE, PLASTIC**									
0020	CPVC, fire suppression (C-UL-S, FM, NFPA 13, 13D & 13R)									
0030	Socket joint, no couplings or hangers									
0100	SDR 13.5 (ASTM F442)									
0120	3/4" diameter	Q-12	420	.038	L.F.	1.46	2.17		3.63	4.97
0130	1" diameter		340	.047		2.26	2.68		4.94	6.65
0140	1-1/4" diameter		260	.062		3.58	3.50		7.08	9.40
0150	1-1/2" diameter		190	.084		4.94	4.79		9.73	12.90
0160	2" diameter		140	.114		7.70	6.50		14.20	18.60
0170	2-1/2" diameter		130	.123		13.55	7		20.55	26
0180	3" diameter		120	.133		20.50	7.60		28.10	34.50

21 11 13.18 Pipe Fittings, Plastic

		Crew	Daily Output	Labor-Hours	Unit	Material	Labor	Equipment	Total	Total Incl O&P
0010	**PIPE FITTINGS, PLASTIC**									
0020	CPVC, fire suppression (C-UL-S, FM, NFPA 13, 13D & 13R)									
0030	Socket joint									
0100	90° elbow									
0120	3/4"	1 Plum	26	.308	Ea.	1.88	19.85		21.73	32.50
0130	1"		22.70	.352		4.12	22.50		26.62	39.50
0140	1-1/4"		20.20	.396		5.20	25.50		30.70	45.50
0150	1-1/2"		18.20	.440		7.40	28.50		35.90	52
0160	2"	Q-1	33.10	.483		9.20	28		37.20	53.50
0170	2-1/2"		24.20	.661		17.70	38.50		56.20	79
0180	3"		20.80	.769		24	44.50		68.50	95.50
0200	45° elbow									

21 11 13.18 Pipe Fittings, Plastic		Crew	Daily Output	Labor-Hours	Unit	Material	Labor	Equipment	Total	Total Incl O&P
						2020 Bare Costs				
0210	3/4"	1 Plum	26	.308	Ea.	2.58	19.85		22.43	33.50
0220	1"		22.70	.352		3.03	22.50		25.53	38.50
0230	1-1/4"		20.20	.396		4.38	25.50		29.88	44.50
0240	1-1/2"		18.20	.440		6.10	28.50		34.60	51
0250	2"	Q-1	33.10	.483		7.60	28		35.60	52
0260	2-1/2"		24.20	.661		13.65	38.50		52.15	74.50
0270	3"		20.80	.769		19.55	44.50		64.05	90.50
0300	Tee									
0310	3/4"	1 Plum	17.30	.462	Ea.	2.58	30		32.58	49
0320	1"		15.20	.526		5.10	34		39.10	58
0330	1-1/4"		13.50	.593		7.65	38		45.65	67.50
0340	1-1/2"		12.10	.661		11.25	42.50		53.75	78.50
0350	2"	Q-1	20	.800		16.65	46.50		63.15	90
0360	2-1/2"		16.20	.988		27	57.50		84.50	119
0370	3"		13.90	1.151		42	67		109	149
0400	Tee, reducing x any size									
0420	1"	1 Plum	15.20	.526	Ea.	4.32	34		38.32	57.50
0430	1-1/4"		13.50	.593		7.90	38		45.90	67.50
0440	1-1/2"		12.10	.661		9.60	42.50		52.10	76.50
0450	2"	Q-1	20	.800		18.60	46.50		65.10	92
0460	2-1/2"		16.20	.988		21	57.50		78.50	112
0470	3"		13.90	1.151		24.50	67		91.50	130
0500	Coupling									
0510	3/4"	1 Plum	26	.308	Ea.	1.81	19.85		21.66	32.50
0520	1"		22.70	.352		2.39	22.50		24.89	37.50
0530	1-1/4"		20.20	.396		3.48	25.50		28.98	43.50
0540	1-1/2"		18.20	.440		4.96	28.50		33.46	49.50
0550	2"	Q-1	33.10	.483		6.70	28		34.70	51
0560	2-1/2"		24.20	.661		10.25	38.50		48.75	71
0570	3"		20.80	.769		13.30	44.50		57.80	83.50
0600	Coupling, reducing									
0610	1" x 3/4"	1 Plum	22.70	.352	Ea.	2.39	22.50		24.89	37.50
0620	1-1/4" x 1"		20.20	.396		3.61	25.50		29.11	43.50
0630	1-1/2" x 3/4"		18.20	.440		5.40	28.50		33.90	50
0640	1-1/2" x 1"		18.20	.440		5.20	28.50		33.70	50
0650	1-1/2" x 1-1/4"		18.20	.440		4.96	28.50		33.46	49.50
0660	2" x 1"	Q-1	33.10	.483		6.95	28		34.95	51
0670	2" x 1-1/2"	"	33.10	.483		6.70	28		34.70	51
0700	Cross									
0720	3/4"	1 Plum	13	.615	Ea.	4.06	39.50		43.56	66
0730	1"		11.30	.708		5.10	45.50		50.60	76
0740	1-1/4"		10.10	.792		7	51		58	86.50
0750	1-1/2"		9.10	.879		9.65	56.50		66.15	98
0760	2"	Q-1	16.60	.964		15.85	56		71.85	104
0770	2-1/2"	"	12.10	1.322		35	76.50		111.50	157
0800	Cap									
0820	3/4"	1 Plum	52	.154	Ea.	1.09	9.90		10.99	16.55
0830	1"		45	.178		1.55	11.45		13	19.40
0840	1-1/4"		40	.200		2.52	12.90		15.42	22.50
0850	1-1/2"		36.40	.220		3.48	14.15		17.63	26
0860	2"	Q-1	66	.242		5.20	14.05		19.25	27.50
0870	2-1/2"		48.40	.331		7.55	19.15		26.70	38
0880	3"		41.60	.385		12.15	22.50		34.65	48

21 11 Facility Fire-Suppression Water-Service Piping

21 11 13 – Facility Fire Suppression Piping

21 11 13.18 Pipe Fittings, Plastic		Crew	Daily Output	Labor-Hours	Unit	Material	Labor	Equipment	Total	Total Incl O&P
						2020 Bare Costs				
0900	Adapter, sprinkler head, female w/metal thd. insert (s x FNPT)									
0920	3/4" x 1/2"	1 Plum	52	.154	Ea.	4.96	9.90		14.86	21
0930	1" x 1/2"		45	.178		5.25	11.45		16.70	23.50
0940	1" x 3/4"	↓	45	.178	↓	8.25	11.45		19.70	27

21 11 16 – Facility Fire Hydrants

21 11 16.50 Fire Hydrants for Buildings

		Crew	Daily Output	Labor-Hours	Unit	Material	Labor	Equipment	Total	Total Incl O&P
0010	FIRE HYDRANTS FOR BUILDINGS									
3750	Hydrants, wall, w/caps, single, flush, polished brass									
3800	2-1/2" x 2-1/2"	Q-12	5	3.200	Ea.	273	182		455	580
3840	2-1/2" x 3"		5	3.200		495	182		677	825
3860	3" x 3"	↓	4.80	3.333		400	190		590	735
3900	For polished chrome, add				↓	20%				
3950	Double, flush, polished brass									
4000	2-1/2" x 2-1/2" x 4"	Q-12	5	3.200	Ea.	775	182		957	1,125
4040	2-1/2" x 2-1/2" x 6"		4.60	3.478		1,300	198		1,498	1,725
4080	3" x 3" x 4"		4.90	3.265		1,175	186		1,361	1,575
4120	3" x 3" x 6"	↓	4.50	3.556		1,575	202		1,777	2,050
4200	For polished chrome, add				↓	10%				
4350	Double, projecting, polished brass									
4400	2-1/2" x 2-1/2" x 4"	Q-12	5	3.200	Ea.	290	182		472	600
4450	2-1/2" x 2-1/2" x 6"	"	4.60	3.478	"	595	198		793	955
4460	Valve control, dbl. flush/projecting hydrant, cap &									
4470	chain, extension rod & cplg., escutcheon, polished brass	Q-12	8	2	Ea.	268	114		382	470
4480	Four-way square, flush, polished brass									
4540	2-1/2" (4) x 6"	Q-12	3.60	4.444	Ea.	3,900	253		4,153	4,675

21 11 19 – Fire-Department Connections

21 11 19.50 Connections for the Fire-Department

		Crew	Daily Output	Labor-Hours	Unit	Material	Labor	Equipment	Total	Total Incl O&P
0010	CONNECTIONS FOR THE FIRE-DEPARTMENT									
0020	For fire pro. cabinets, see Section 10 44 13.53									
4000	Storz type, with cap and chain									
6000	Roof manifold, horiz., brass, without valves & caps									
6040	2-1/2" x 2-1/2" x 4"	Q-12	4.80	3.333	Ea.	228	190		418	545
6060	2-1/2" x 2-1/2" x 6"		4.60	3.478		236	198		434	565
6080	2-1/2" x 2-1/2" x 2-1/2" x 4"		4.60	3.478		380	198		578	725
6090	2-1/2" x 2-1/2" x 2-1/2" x 6"	↓	4.60	3.478		405	198		603	750
7000	Sprinkler line tester, cast brass				↓	38.50			38.50	42.50
7140	Standpipe connections, wall, w/plugs & chains									
7160	Single, flush, brass, 2-1/2" x 2-1/2", Fire Dept Conn.	Q-12	5	3.200	Ea.	186	182		368	485
7180	2-1/2" x 3"	"	5	3.200	"	191	182		373	490
7240	For polished chrome, add					15%				
7280	Double, flush, polished brass									
7300	2-1/2" x 2-1/2" x 4"	Q-12	5	3.200	Ea.	775	182		957	1,125
7330	2-1/2" x 2-1/2" x 6"		4.60	3.478		855	198		1,053	1,250
7340	3" x 3" x 4"		4.90	3.265		1,150	186		1,336	1,575
7370	3" x 3" x 6"	↓	4.50	3.556	↓	1,300	202		1,502	1,775
7400	For polished chrome, add					15%				
7440	For sill cock combination, add				Ea.	101			101	112
7580	Double projecting, polished brass									
7600	2-1/2" x 2-1/2" x 4"	Q-12	5	3.200	Ea.	610	182		792	950
7630	2-1/2" x 2-1/2" x 6"	"	4.60	3.478	"	1,025	198		1,223	1,425
7680	For polished chrome, add					15%				
7900	Three way, flush, polished brass									

21 11 Facility Fire-Suppression Water-Service Piping

21 11 19 – Fire-Department Connections

21 11 19.50 Connections for the Fire-Department		Crew	Daily Output	Labor-Hours	Unit	Material	Labor	Equipment	Total	Total Incl O&P
						2020 Bare Costs				
7920	2-1/2" (3) x 4"	Q-12	4.80	3.333	Ea.	2,000	190		2,190	2,500
7930	2-1/2" (3) x 6"	"	4.60	3.478		2,150	198		2,348	2,675
8000	For polished chrome, add				↓	9%				
8020	Three way, projecting, polished brass									
8040	2-1/2" (3) x 4"	Q-12	4.80	3.333	Ea.	910	190		1,100	1,300
8070	2-1/2" (3) x 6"	"	4.60	3.478		1,800	198		1,998	2,275
8100	For polished chrome, add				↓	12%				
8200	Four way, square, flush, polished brass,									
8240	2-1/2" (4) x 6"	Q-12	3.60	4.444	Ea.	1,750	253		2,003	2,325
8300	For polished chrome, add				"	10%				
8550	Wall, vertical, flush, cast brass									
8600	Two way, 2-1/2" x 2-1/2" x 4"	Q-12	5	3.200	Ea.	400	182		582	720
8660	Four way, 2-1/2" (4) x 6"		3.80	4.211		1,350	240		1,590	1,850
8680	Six way, 2-1/2" (6) x 6"	↓	3.40	4.706		1,600	268		1,868	2,175
8700	For polished chrome, add				↓	10%				
8800	Free standing siamese unit, polished brass, two way									
8820	2-1/2" x 2-1/2" x 4"	Q-12	2.50	6.400	Ea.	750	365		1,115	1,400
8850	2-1/2" x 2-1/2" x 6"		2	8		830	455		1,285	1,625
8860	3" x 3" x 4"		2.50	6.400		570	365		935	1,200
8890	3" x 3" x 6"	↓	2	8	↓	1,475	455		1,930	2,325
8940	For polished chrome, add					12%				
9100	Free standing siamese unit, polished brass, three way									
9120	2-1/2" x 2-1/2" x 2-1/2" x 6"	Q-12	2	8	Ea.	980	455		1,435	1,775
9160	For polished chrome, add				"	15%				
9990	Minimum labor/equipment charge	1 Spri	4	2	Job		127		127	196

21 12 Fire-Suppression Standpipes

21 12 13 – Fire-Suppression Hoses and Nozzles

21 12 13.50 Fire Hoses and Nozzles		Crew	Daily Output	Labor-Hours	Unit	Material	Labor	Equipment	Total	Total Incl O&P
0010	**FIRE HOSES AND NOZZLES** R211226-10									
0200	Adapters, rough brass, straight hose threads									
0220	One piece, female to male, rocker lugs R211226-20									
0240	1" x 1"				Ea.	56.50			56.50	62
0260	1-1/2" x 1"					45.50			45.50	50
0280	1-1/2" x 1-1/2"					10.15			10.15	11.20
0300	2" x 1-1/2"					78.50			78.50	86.50
0320	2" x 2"					42.50			42.50	46.50
0340	2-1/2" x 1-1/2"					32			32	35
0360	3" x 1-1/2"					56.50			56.50	62.50
0380	2-1/2" x 2-1/2"					19.55			19.55	21.50
0400	3" x 2-1/2"					141			141	156
0420	3" x 3"					85			85	93.50
0500	For polished brass, add					50%				
0520	For polished chrome, add				↓	75%				
0700	One piece, female to male, hexagon									
0740	1-1/2" x 3/4"				Ea.	50			50	55
0760	2" x 1-1/2"					114			114	125
0780	2-1/2" x 1"					196			196	215
0800	2-1/2" x 1-1/2"					64.50			64.50	71
0820	2-1/2" x 2"					45.50			45.50	50
0840	3" x 2-1/2"				↓	85.50			85.50	94

21 12 13 – Fire-Suppression Hoses and Nozzles

21 12 13.50 Fire Hoses and Nozzles		Crew	Daily Output	Labor-Hours	Unit	2020 Bare Costs Material	Labor	Equipment	Total	Total Incl O&P
0900	For polished chrome, add				Ea.	75%				
1100	Swivel, female to female, pin lugs									
1120	1-1/2" x 1-1/2"				Ea.	74.50			74.50	82
1200	2-1/2" x 2-1/2"					147			147	162
1260	For polished brass, add					50%				
1280	For polished chrome, add					75%				
1400	Couplings, sngl. & dbl. jacket, pin lug or rocker lug, cast brass									
1410	1-1/2"				Ea.	63			63	69.50
1420	2-1/2"				"	51.50			51.50	57
1500	For polished brass, add					20%				
1520	For polished chrome, add					40%				
1580	Reducing, F x M, interior installation, cast brass									
1590	2" x 1-1/2"				Ea.	83			83	91.50
1600	2-1/2" x 1-1/2"					19.50			19.50	21.50
1680	For polished brass, add					50%				
1720	For polished chrome, add					75%				
2200	Hose, less couplings									
2260	Synthetic jacket, lined, 300 lb. test, 1-1/2" diameter	Q-12	2600	.006	L.F.	3.31	.35		3.66	4.18
2270	2" diameter		2200	.007		2.59	.41		3	3.49
2280	2-1/2" diameter		2200	.007		5.90	.41		6.31	7.15
2290	3" diameter		2200	.007		3.27	.41		3.68	4.24
2360	High strength, 500 lb. test, 1-1/2" diameter		2600	.006		2.59	.35		2.94	3.39
2380	2-1/2" diameter		2200	.007		6.05	.41		6.46	7.30
5000	Nipples, straight hose to tapered iron pipe, brass									
5060	Female to female, 1-1/2" x 1-1/2"				Ea.	26			26	28.50
5100	2-1/2" x 2-1/2"					45			45	49.50
5190	For polished chrome, add					75%				
5200	Double male or male to female, 1" x 1"					56.50			56.50	62
5220	1-1/2" x 1"					72			72	79.50
5230	1-1/2" x 1-1/2"					21			21	23
5260	2" x 1-1/2"					82			82	90
5270	2" x 2"					115			115	127
5280	2-1/2" x 1-1/2"					70.50			70.50	77.50
5300	2-1/2" x 2"					56			56	62
5310	2-1/2" x 2-1/2"					32			32	35
5340	For polished chrome, add					75%				
5600	Nozzles, brass									
5620	Adjustable fog, 3/4" booster line				Ea.	150			150	165
5630	1" booster line					145			145	159
5640	1-1/2" leader line					117			117	128
5660	2-1/2" direct connection					199			199	219
5680	2-1/2" playpipe nozzle					238			238	262
5780	For chrome plated, add					8%				
5850	Electrical fire, adjustable fog, no shock									
5900	1-1/2"				Ea.	455			455	500
5920	2-1/2"					900			900	990
5980	For polished chrome, add					6%				
6200	Heavy duty, comb. adj. fog and str. stream, with handle									
6210	1" booster line				Ea.	208			208	229
6240	1-1/2"					440			440	485
6260	2-1/2", for playpipe					675			675	745
6280	2-1/2" direct connection					470			470	520
6300	2-1/2" playpipe combination					605			605	665

21 12 Fire-Suppression Standpipes

21 12 13 – Fire-Suppression Hoses and Nozzles

21 12 13.50 Fire Hoses and Nozzles		Crew	Daily Output	Labor-Hours	Unit	Material	Labor	Equipment	Total	Total Incl O&P
						2020 Bare Costs				
6480	For polished chrome, add				Ea.	7%				
6500	Plain fog, polished brass, 1-1/2"					154			154	170
6540	Chrome plated, 1-1/2"					128			128	141
6700	Plain stream, polished brass, 1-1/2" x 10"					58			58	63.50
6760	2-1/2" x 15" x 7/8" or 1-1/2"					109			109	119
6860	For polished chrome, add					20%				
7000	Underwriters playpipe, 2-1/2" x 30" with 1-1/8" tip				Ea.	640			640	705
9200	Storage house, hose only, primed steel					1,175			1,175	1,275
9220	Aluminum					2,550			2,550	2,800
9280	Hose and hydrant house, primed steel					1,350			1,350	1,475
9300	Aluminum					1,625			1,625	1,800
9340	Tools, crowbar and brackets	1 Carp	12	.667		97.50	35.50		133	164
9360	Combination hydrant wrench and spanner					39.50			39.50	43.50
9380	Fire axe and brackets									
9400	6 lb.	1 Carp	12	.667	Ea.	137	35.50		172.50	208
9500	For fire equipment cabinets, Section 10 44 13.53									
9900	Minimum labor/equipment charge	1 Plum	2	4	Job		258		258	400

21 12 16 – Fire-Suppression Hose Reels

21 12 16.50 Fire-Suppression Hose Reels		Crew	Daily Output	Labor-Hours	Unit	Material	Labor	Equipment	Total	Total Incl O&P
0010	**FIRE-SUPPRESSION HOSE REELS**									
2990	Hose reel, swinging, for 1-1/2" polyester neoprene lined hose									
3000	50' long	Q-12	14	1.143	Ea.	158	65		223	275
3020	100' long		14	1.143		248	65		313	375
3060	For 2-1/2" cotton rubber hose, 75' long		14	1.143		296	65		361	425
3100	150' long		14	1.143		296	65		361	425

21 12 19 – Fire-Suppression Hose Racks

21 12 19.50 Fire Hose Racks		Crew	Daily Output	Labor-Hours	Unit	Material	Labor	Equipment	Total	Total Incl O&P
0010	**FIRE HOSE RACKS**									
2600	Hose rack, swinging, for 1-1/2" diameter hose,									
2620	Enameled steel, 50' and 75' lengths of hose	Q-12	20	.800	Ea.	72	45.50		117.50	150
2640	100' and 125' lengths of hose		20	.800		89	45.50		134.50	168
2680	Chrome plated, 50' and 75' lengths of hose		20	.800		73.50	45.50		119	151
2700	100' and 125' lengths of hose		20	.800		153	45.50		198.50	239
2780	For hose rack nipple, 1-1/2" polished brass, add					32.50			32.50	35.50
2820	2-1/2" polished brass, add					55.50			55.50	61
2840	1-1/2" polished chrome, add					38.50			38.50	42
2860	2-1/2" polished chrome, add					75.50			75.50	83

21 12 23 – Fire-Suppression Hose Valves

21 12 23.70 Fire Hose Valves		Crew	Daily Output	Labor-Hours	Unit	Material	Labor	Equipment	Total	Total Incl O&P
0010	**FIRE HOSE VALVES**									
0020	Angle, combination pressure adjust/restricting, rough brass									
0030	1-1/2" R211226-20	1 Spri	12	.667	Ea.	110	42		152	187
0040	2-1/2"	"	7	1.143	"	202	72.50		274.50	335
0042	Nonpressure adjustable/restricting, rough brass									
0044	1-1/2"	1 Spri	12	.667	Ea.	48	42		90	118
0046	2-1/2"	"	7	1.143	"	143	72.50		215.50	269
0050	For polished brass, add					30%				
0060	For polished chrome, add					40%				
0080	Wheel handle, 300 lb., 1-1/2"	1 Spri	12	.667	Ea.	105	42		147	182
0090	2-1/2"	"	7	1.143	"	198	72.50		270.50	330
0100	For polished brass, add					35%				

21 12 23.70 Fire Hose Valves		Crew	Daily Output	Labor-Hours	Unit	Material	2020 Bare Costs Labor	Equipment	Total	Total Incl O&P
0110	For polished chrome, add					50%				
1000	Ball drip, automatic, rough brass, 1/2"	1 Spri	20	.400	Ea.	20.50	25.50		46	61.50
1010	3/4"	"	20	.400	"	26	25.50		51.50	68
1100	Ball, 175 lb., sprinkler system, FM/UL, threaded, bronze									
1120	Slow close									
1150	1" size	1 Spri	19	.421	Ea.	305	26.50		331.50	375
1160	1-1/4" size		15	.533		330	33.50		363.50	410
1170	1-1/2" size		13	.615		425	39		464	530
1180	2" size	↓	11	.727		525	46		571	650
1190	2-1/2" size	Q-12	15	1.067	↓	710	60.50		770.50	875
1230	For supervisory switch kit, all sizes									
1240	One circuit, add	1 Spri	48	.167	Ea.	157	10.55		167.55	189
1280	Quarter turn for trim									
1300	1/2" size	1 Spri	22	.364	Ea.	42	23		65	81.50
1310	3/4" size		20	.400		45	25.50		70.50	88.50
1320	1" size		19	.421		50	26.50		76.50	96
1330	1-1/4" size		15	.533		81.50	33.50		115	142
1340	1-1/2" size		13	.615		102	39		141	174
1350	2" size	↓	11	.727		122	46		168	205
1400	Caps, polished brass with chain, 3/4"					68			68	74.50
1420	1"					85			85	93.50
1440	1-1/2"					20.50			20.50	22.50
1460	2-1/2"					30			30	33
1480	3"					40			40	44
1900	Escutcheon plate, for angle valves, polished brass, 1-1/2"					16.60			16.60	18.25
1920	2-1/2"					24			24	26.50
1940	3"					31.50			31.50	34.50
1980	For polished chrome, add					15%				
2000	Foam, control valve, 3"	1 Spri	6	1.333		2,250	84.50		2,334.50	2,600
2020	Supply valve, 2-1/2"		7	1.143		158	72.50		230.50	285
2040	Proportioner, 8"	↓	2	4		3,600	253		3,853	4,375
2060	Oscillating foam monitor with electric remote control	Q-12	5.33	3.002		17,900	171		18,071	20,000
3000	Gate, hose, wheel handle, N.R.S., rough brass, 1-1/2"	1 Spri	12	.667		166	42		208	248
3040	2-1/2", 300 lb.	"	7	1.143		217	72.50		289.50	350
3080	For polished brass, add					40%				
3090	For polished chrome, add				↓	50%				
3800	Hydrant, screw type, crank handle, brass									
3840	2-1/2" size	Q-12	11	1.455	Ea.	375	83		458	540
3880	For chrome, same price									
4200	Hydrolator, vent and draining, rough brass, 1-1/2"	1 Spri	12	.667	Ea.	115	42		157	192
4280	For polished brass, add					50%				
4290	For polished chrome, add					90%				
5000	Pressure reducing rough brass, 1-1/2"	1 Spri	12	.667		325	42		367	420
5020	2-1/2"	"	7	1.143		445	72.50		517.50	600
5080	For polished brass, add					105%				
5090	For polished chrome, add				↓	140%				
8000	Wye, leader line, ball type, swivel female x male x male									
8040	2-1/2" x 1-1/2" x 1-1/2" polished brass				Ea.	360			360	395
8060	2-1/2" x 1-1/2" x 1-1/2" polished chrome				"	310			310	340

21 13 13.50 Wet-Pipe Sprinkler System Components		Crew	Daily Output	Labor-Hours	Unit	Material	2020 Bare Costs Labor	Equipment	Total	Total Incl O&P
0010	WET-PIPE SPRINKLER SYSTEM COMPONENTS									
1100	Alarm, electric pressure switch (circuit closer)	1 Spri	26	.308	Ea.	112	19.45		131.45	153
1140	For explosion proof, max 20 psi, contacts close or open		26	.308		740	19.45		759.45	845
1220	Water motor gong		4	2		480	127		607	725
1900	Flexible sprinkler head connectors									
1910	Braided stainless steel hose with mounting bracket									
1920	1/2" and 3/4" outlet size									
1940	40" length	1 Spri	30	.267	Ea.	73	16.85		89.85	106
1960	60" length	"	22	.364	"	85	23		108	129
1982	May replace hard-pipe armovers									
1984	for wet and pre-action systems.									
2000	Release, emergency, manual, for hydraulic or pneumatic system	1 Spri	12	.667	Ea.	206	42		248	293
2060	Release, thermostatic, for hydraulic or pneumatic release line		20	.400		800	25.50		825.50	920
2200	Sprinkler cabinets, 6 head capacity		16	.500		77.50	31.50		109	134
2260	12 head capacity		16	.500		88	31.50		119.50	146
2340	Sprinkler head escutcheons, standard, brass tone, 1" size		40	.200		3.56	12.65		16.21	23.50
2360	Chrome, 1" size		40	.200		3.50	12.65		16.15	23.50
2400	Recessed type, bright brass		40	.200		11	12.65		23.65	31.50
2440	Chrome or white enamel		40	.200		4.22	12.65		16.87	24
2600	Sprinkler heads, not including supply piping									
3700	Standard spray, pendent or upright, brass, 135°F to 286°F									
3720	1/2" NPT, K5.6	1 Spri	16	.500	Ea.	16.75	31.50		48.25	67.50
3730	1/2" NPT, 7/16" orifice		16	.500		16.80	31.50		48.30	67.50
3740	1/2" NPT,K5.6		16	.500		11.20	31.50		42.70	61.50
3760	1/2" NPT, 17/32" orifice		16	.500		13.75	31.50		45.25	64
3780	3/4" NPT, 17/32" orifice		16	.500		13.45	31.50		44.95	64
3800	For open sprinklers, deduct					15%				
3840	For chrome, add				Ea.	4.06			4.06	4.47
3920	For 360°F, same cost									
3930	For 400°F	1 Spri	16	.500	Ea.	110	31.50		141.50	170
3940	For 500°F	"	16	.500	"	110	31.50		141.50	170
4200	Sidewall, vertical brass, 135°F to 286°F									
4240	1/2" NPT, 1/2" orifice	1 Spri	16	.500	Ea.	29.50	31.50		61	81.50
4280	3/4" NPT, 17/32" orifice	"	16	.500		81.50	31.50		113	139
4360	For satin chrome, add					4.66			4.66	5.15
4400	For 360°F, same cost									
4500	Sidewall, horizontal, brass, 135°F to 286°F									
4520	1/2" NPT, 1/2" orifice	1 Spri	16	.500	Ea.	29.50	31.50		61	81.50
4540	For 360°F, same cost									
4800	Recessed pendent, brass, 135°F to 286°F									
4820	1/2" NPT, K5.6	1 Spri	10	.800	Ea.	49.50	50.50		100	133
4830	1/2" NPT, 7/16" orifice		10	.800		22	50.50		72.50	103
4840	1/2" NPT, K5.6		10	.800		17.10	50.50		67.60	97.50
4860	1/2" NPT, 17/32" orifice		10	.800		50	50.50		100.50	133
4900	For satin chrome, add					6.30			6.30	6.90
5000	Recessed-vertical sidewall, brass, 135°F to 286°F									
5020	1/2" NPT, K5.6	1 Spri	10	.800	Ea.	36	50.50		86.50	118
5030	1/2" NPT, 7/16" orifice		10	.800		36	50.50		86.50	118
5040	1/2" NPT, K5.6		10	.800		36	50.50		86.50	118
5100	For bright nickel, same cost									
5600	Concealed, complete with cover plate									
5620	1/2" NPT, 1/2" orifice, 135°F to 212°F	1 Spri	9	.889	Ea.	27.50	56		83.50	118
5800	Window, brass, 1/2" NPT, 1/4" orifice		16	.500		40	31.50		71.50	93

21 13 13 – Wet-Pipe Sprinkler Systems

21 13 13.50 Wet-Pipe Sprinkler System Components		Crew	Daily Output	Labor-Hours	Unit	Material	2020 Bare Costs Labor	Equipment	Total	Total Incl O&P
5810	1/2" NPT, 5/16" orifice	1 Spri	16	.500	Ea.	40	31.50		71.50	93
5820	1/2" NPT, 3/8" orifice		16	.500		40	31.50		71.50	93
5830	1/2" NPT, 7/16" orifice		16	.500		40	31.50		71.50	93
5840	1/2" NPT, 1/2" orifice		16	.500		42	31.50		73.50	95.50
5860	For polished chrome, add					5.10			5.10	5.60
5880	3/4" NPT, 5/8" orifice	1 Spri	16	.500		44	31.50		75.50	97
5890	3/4" NPT, 3/4" orifice	"	16	.500		44	31.50		75.50	97
6000	Sprinkler head guards, bright zinc, 1/2" NPT					5.70			5.70	6.30
6020	Bright zinc, 3/4" NPT					5.75			5.75	6.35
6100	Sprinkler head wrenches, standard head					28			28	30.50
6120	Recessed head					42			42	46
6160	Tamper switch (valve supervisory switch)	1 Spri	16	.500		265	31.50		296.50	340
6165	Flow switch (valve supervisory switch)	"	16	.500		265	31.50		296.50	340

21 13 16 – Dry-Pipe Sprinkler Systems

21 13 16.50 Dry-Pipe Sprinkler System Components		Crew	Daily Output	Labor-Hours	Unit	Material	Labor	Equipment	Total	Total Incl O&P
0010	**DRY-PIPE SPRINKLER SYSTEM COMPONENTS**									
0600	Accelerator	1 Spri	8	1	Ea.	900	63.50		963.50	1,100
0800	Air compressor for dry pipe system, automatic, complete R211313-20									
0820	30 gal. system capacity, 3/4 HP	1 Spri	1.30	6.154	Ea.	1,200	390		1,590	1,925
0860	30 gal. system capacity, 1 HP		1.30	6.154		1,575	390		1,965	2,350
0910	30 gal. system capacity, 1-1/2 HP		1.30	6.154		1,000	390		1,390	1,700
0920	30 gal. system capacity, 2 HP		1.30	6.154		905	390		1,295	1,600
0960	Air pressure maintenance control		24	.333		375	21		396	445
1600	Dehydrator package, incl. valves and nipples		12	.667		845	42		887	995
2600	Sprinkler heads, not including supply piping									
2640	Dry, pendent, 1/2" orifice, 3/4" or 1" NPT									
2660	3" to 6" length	1 Spri	14	.571	Ea.	145	36		181	216
2670	6-1/4" to 8" length		14	.571		147	36		183	218
2680	8-1/4" to 12" length		14	.571		158	36		194	230
2690	12-1/4" to 15" length		14	.571		164	36		200	236
2700	15-1/4" to 18" length		14	.571		170	36		206	243
2710	18-1/4" to 21" length		13	.615		177	39		216	255
2720	21-1/4" to 24" length		13	.615		183	39		222	262
2730	24-1/4" to 27" length		13	.615		189	39		228	269
2740	27-1/4" to 30" length		13	.615		196	39		235	277
2750	30-1/4" to 33" length		13	.615		203	39		242	284
2760	33-1/4" to 36" length		13	.615		209	39		248	291
2780	36-1/4" to 39" length		12	.667		216	42		258	305
2790	39-1/4" to 42" length		12	.667		222	42		264	310
2800	For each inch or fraction, add					4.07			4.07	4.48
6330	Valves and components									
6340	Alarm test/shut off valve, 1/2"	1 Spri	20	.400	Ea.	26	25.50		51.50	67.50
8000	Dry pipe air check valve, 3" size	Q-12	2	8		2,100	455		2,555	3,025
8200	Dry pipe valve, incl. trim and gauges, 3" size		2	8		3,075	455		3,530	4,075
8220	4" size		1	16		3,300	910		4,210	5,025
8240	6" size	Q-13	2	16		4,000	965		4,965	5,900
8280	For accelerator trim with gauges, add	1 Spri	8	1		284	63.50		347.50	410

21 13 Fire-Suppression Sprinkler Systems

21 13 19 – Preaction Sprinkler Systems

21 13 19.50 Preaction Sprinkler System Components		Crew	Daily Output	Labor-Hours	Unit	Material (2020 Bare Costs)	Labor	Equipment	Total	Total Incl O&P
0010	**PREACTION SPRINKLER SYSTEM COMPONENTS**									
3000	Preaction valve cabinet									
3100	Single interlock, pneum. release, panel, 1/2 HP comp. regul. air trim									
3110	1-1/2"	Q-12	3	5.333	Ea.	49,600	305		49,905	55,000
3120	2"		3	5.333		49,600	305		49,905	55,000
3130	2-1/2"		3	5.333		49,600	305		49,905	55,000
3140	3"		3	5.333		49,700	305		50,005	55,000
3150	4"		2	8		51,500	455		51,955	57,000
3160	6"	Q-13	4	8		54,000	480		54,480	60,000
3200	Double interlock, pneum. release, panel, 1/2 HP comp. regul. air trim									
3210	1-1/2"	Q-12	3	5.333	Ea.	48,200	305		48,505	53,500
3220	2"		3	5.333		48,200	305		48,505	53,500
3230	2-1/2"		3	5.333		48,200	305		48,505	53,500
3240	3"		3	5.333		48,300	305		48,605	53,500
3250	4"		2	8		50,500	455		50,955	56,500
3260	6"	Q-13	4	8		53,000	480		53,480	58,500

21 13 20 – On-Off Multicycle Sprinkler System

21 13 20.50 On-Off Multicycle Fire-Suppression Sprinkler Systems		Crew	Daily Output	Labor-Hours	Unit	Material	Labor	Equipment	Total	Total Incl O&P
0010	**ON-OFF MULTICYCLE FIRE-SUPPRESSION SPRINKLER SYSTEMS**									
8400	On-off multicycle package, includes swing check									
8420	and flow control valves with required trim									
8440	2" size	Q-12	2	8	Ea.	5,500	455		5,955	6,750
8460	3" size		1.50	10.667		6,025	605		6,630	7,575
8480	4" size		1	16		6,725	910		7,635	8,800
8500	6" size	Q-13	1.40	22.857		7,750	1,375		9,125	10,700

21 13 26 – Deluge Fire-Suppression Sprinkler Systems

21 13 26.50 Deluge Fire-Suppression Sprinkler Sys. Comp.		Crew	Daily Output	Labor-Hours	Unit	Material	Labor	Equipment	Total	Total Incl O&P
0010	**DELUGE FIRE-SUPPRESSION SPRINKLER SYSTEM COMPONENTS**									
1400	Deluge system, monitoring panel w/deluge valve & trim	1 Spri	18	.444	Ea.	7,475	28		7,503	8,275
6200	Valves and components									
7000	Deluge, assembly, incl. trim, pressure									
7020	operated relief, emergency release, gauges									
7040	2" size	Q-12	2	8	Ea.	4,400	455		4,855	5,525
7060	3" size		1.50	10.667		4,925	605		5,530	6,375
7080	4" size		1	16		5,625	910		6,535	7,600
7100	6" size	Q-13	1.80	17.778		6,650	1,075		7,725	8,975
7800	Pneumatic actuator, bronze, required on all									
7820	pneumatic release systems, any size deluge	1 Spri	18	.444	Ea.	480	28		508	575

21 13 39 – Foam-Water Systems

21 13 39.50 Foam-Water System Components		Crew	Daily Output	Labor-Hours	Unit	Material	Labor	Equipment	Total	Total Incl O&P
0010	**FOAM-WATER SYSTEM COMPONENTS**									
2600	Sprinkler heads, not including supply piping									
3600	Foam-water, pendent or upright, 1/2" NPT	1 Spri	12	.667	Ea.	237	42		279	325

21 21 Carbon-Dioxide Fire-Extinguishing Systems

21 21 16 – Carbon-Dioxide Fire-Extinguishing Equipment

21 21 16.50 CO2 Fire Extinguishing System		Crew	Daily Output	Labor-Hours	Unit	Material	Labor	Equipment	Total	Total Incl O&P
						2020 Bare Costs				
0010	**CO_2 FIRE EXTINGUISHING SYSTEM**									
0042	For detectors and control stations, see Section 28 31 23.50									
0100	Control panel, single zone with batteries (2 zones det., 1 suppr.)	1 Elec	1	8	Ea.	1,050	490		1,540	1,900
0150	Multizone (4) with batteries (8 zones det., 4 suppr.)	"	.50	16		3,075	980		4,055	4,875
1000	Dispersion nozzle, CO_2, 3" x 5"	1 Plum	18	.444		166	28.50		194.50	227
2000	Extinguisher, CO_2 system, high pressure, 75 lb. cylinder	Q-1	6	2.667		1,475	155		1,630	1,875
2100	100 lb. cylinder	"	5	3.200		1,950	186		2,136	2,425
3000	Electro/mechanical release	L-1	4	4		1,050	252		1,302	1,550
3400	Manual pull station	1 Plum	6	1.333		92.50	86		178.50	235
4000	Pneumatic damper release	"	8	1	↓	182	64.50		246.50	300

21 22 Clean-Agent Fire-Extinguishing Systems

21 22 16 – Clean-Agent Fire-Extinguishing Equipment

21 22 16.50 Clean-Agent Extinguishing Systems		Crew	Daily Output	Labor-Hours	Unit	Material	Labor	Equipment	Total	Total Incl O&P
0010	**CLEAN-AGENT EXTINGUISHING SYSTEMS**									
0020	FM200 fire extinguishing system									
1100	Dispersion nozzle FM200, 1-1/2"	1 Plum	14	.571	Ea.	236	37		273	315
2400	Extinguisher, FM200 system, filled, with mounting bracket									
2460	26 lb. container	Q-1	8	2	Ea.	2,075	116		2,191	2,450
2480	44 lb. container		7	2.286		2,825	133		2,958	3,300
2500	63 lb. container		6	2.667		3,675	155		3,830	4,275
2520	101 lb. container		5	3.200		5,375	186		5,561	6,175
2540	196 lb. container	↓	4	4	↓	7,000	232		7,232	8,050
6000	FM200 system, simple nozzle layout, with broad dispersion				C.F.	1.90			1.90	2.09
6010	Extinguisher, FM200 system, filled, with mounting bracket									
6020	Complex nozzle layout and/or including underfloor dispersion				C.F.	3.78			3.78	4.16
6100	20,000 C.F. 2 exits, 8' clng					2.08			2.08	2.29
6200	100,000 C.F. 4 exits, 8' clng					1.88			1.88	2.07
6300	250,000 C.F. 6 exits, 8' clng				↓	1.59			1.59	1.75
7010	HFC-227ea fire extinguishing system									
7100	Cylinders with clean-agent									
7110	Does not include pallete jack/fork lift rental fees									
7120	70 lb. cyl, w/35 lb. agent, no solenoid	Q-12	14	1.143	Ea.	2,500	65		2,565	2,850
7130	70 lb. cyl w/70 lb. agent, no solenoid		10	1.600		3,275	91		3,366	3,750
7140	70 lb. cyl w/35 lb. agent, w/solenoid		14	1.143		3,475	65		3,540	3,925
7150	70 lb. cyl w/70 lb. agent, w/solenoid		10	1.600		4,325	91		4,416	4,925
7220	250 lb. cyl, w/125 lb. agent, no solenoid		8	2		5,750	114		5,864	6,500
7230	250 lb. cyl, w/250 lb. agent, no solenoid		5	3.200		5,750	182		5,932	6,600
7240	250 lb. cyl, w/125 lb. agent, w/solenoid		8	2		7,425	114		7,539	8,350
7250	250 lb. cyl, w/250 lb. agent, w/solenoid		5	3.200		10,700	182		10,882	12,000
7320	560 lb. cyl, w/300 lb. agent, no solenoid		4	4		10,800	228		11,028	12,300
7330	560 lb. cyl, w/560 lb. agent, no solenoid		2.50	6.400		16,400	365		16,765	18,600
7340	560 lb. cyl, w/300 lb. agent, w/solenoid		4	4		13,700	228		13,928	15,500
7350	560 lb. cyl, w/560 lb. agent, w/solenoid	↓	2.50	6.400		20,700	365		21,065	23,300
7420	1,200 lb. cyl, w/600 lb. agent, no solenoid	Q-13	4	8		20,500	480		20,980	23,200
7430	1,200 lb. cyl, w/1,200 lb. agent, no solenoid		3	10.667		33,600	640		34,240	38,000
7440	1,200 lb. cyl, w/600 lb. agent, w/solenoid		4	8		48,300	480		48,780	53,500
7450	1,200 lb. cyl, w/1,200 lb. agent, w/solenoid	↓	3	10.667	↓	75,000	640		75,640	83,500
7500	Accessories									
7510	Dispersion nozzle	1 Spri	16	.500	Ea.	98.50	31.50		130	157
7520	Agent release panel	1 Elec	4	2	↓	61	123		184	256

21 22 Clean-Agent Fire-Extinguishing Systems

21 22 16 – Clean-Agent Fire-Extinguishing Equipment

21 22 16.50 Clean-Agent Extinguishing Systems		Crew	Daily Output	Labor-Hours	Unit	Material	2020 Bare Costs Labor	Equipment	Total	Total Incl O&P
7530	Maintenance switch	1 Elec	6	1.333	Ea.	39	82		121	169
7540	Solenoid valve, 12v dc	1 Spri	8	1		233	63.50		296.50	355
7550	12v ac		8	1		500	63.50		563.50	650
7560	12v dc, explosion proof		8	1		395	63.50		458.50	535

21 31 Centrifugal Fire Pumps

21 31 13 – Electric-Drive, Centrifugal Fire Pumps

21 31 13.50 Electric-Drive Fire Pumps

		Crew	Daily Output	Labor-Hours	Unit	Material	Labor	Equipment	Total	Total Incl O&P
0010	**ELECTRIC-DRIVE FIRE PUMPS** Including controller, fittings and relief valve									
3100	250 GPM, 55 psi, 15 HP, 3550 RPM, 2″ pump	Q-13	.70	45.714	Ea.	15,500	2,750		18,250	21,300
3200	500 GPM, 50 psi, 27 HP, 1770 RPM, 4″ pump		.68	47.059		15,400	2,825		18,225	21,300
3250	500 GPM, 100 psi, 47 HP, 3550 RPM, 3″ pump		.66	48.485		16,300	2,925		19,225	22,400
3300	500 GPM, 125 psi, 64 HP, 3550 RPM, 3″ pump		.62	51.613		18,300	3,100		21,400	25,000
3350	750 GPM, 50 psi, 44 HP, 1770 RPM, 5″ pump		.64	50		19,000	3,000		22,000	25,600
3400	750 GPM, 100 psi, 66 HP, 3550 RPM, 4″ pump		.58	55.172		18,500	3,325		21,825	25,600
3450	750 GPM, 165 psi, 120 HP, 3550 RPM, 4″ pump		.56	57.143		25,000	3,450		28,450	32,800
3500	1000 GPM, 50 psi, 48 HP, 1770 RPM, 5″ pump		.60	53.333		20,300	3,200		23,500	27,300
3550	1000 GPM, 100 psi, 86 HP, 3550 RPM, 5″ pump		.54	59.259		25,000	3,575		28,575	33,000
3600	1000 GPM, 150 psi, 142 HP, 3550 RPM, 5″ pump		.50	64		29,200	3,850		33,050	38,100
3650	1000 GPM, 200 psi, 245 HP, 1770 RPM, 6″ pump		.36	88.889		47,900	5,350		53,250	61,000
3660	1250 GPM, 75 psi, 75 HP, 1770 RPM, 5″ pump		.55	58.182		23,600	3,500		27,100	31,400
3700	1500 GPM, 50 psi, 66 HP, 1770 RPM, 6″ pump		.50	64		22,600	3,850		26,450	30,800
3750	1500 GPM, 100 psi, 139 HP, 1770 RPM, 6″ pump		.46	69.565		28,500	4,200		32,700	37,800
3800	1500 GPM, 150 psi, 200 HP, 1770 RPM, 6″ pump		.36	88.889		48,800	5,350		54,150	62,000
3850	1500 GPM, 200 psi, 279 HP, 1770 RPM, 6″ pump		.32	100		52,500	6,025		58,525	67,000
3900	2000 GPM, 100 psi, 167 HP, 1770 RPM, 6″ pump		.34	94.118		34,400	5,675		40,075	46,600
3950	2000 GPM, 150 psi, 292 HP, 1770 RPM, 6″ pump		.28	114		47,000	6,875		53,875	62,000
4000	2500 GPM, 100 psi, 213 HP, 1770 RPM, 8″ pump		.30	107		40,600	6,425		47,025	54,500
4040	2500 GPM, 135 psi, 339 HP, 1770 RPM, 8″ pump		.26	123		61,000	7,400		68,400	78,500
4100	3000 GPM, 100 psi, 250 HP, 1770 RPM, 8″ pump		.28	114		62,000	6,875		68,875	78,500
4150	3000 GPM, 140 psi, 428 HP, 1770 RPM, 10″ pump		.24	133		74,000	8,025		82,025	94,000
4200	3500 GPM, 100 psi, 300 HP, 1770 RPM, 10″ pump		.26	123		66,500	7,400		73,900	84,500
4250	3500 GPM, 140 psi, 450 HP, 1770 RPM, 10″ pump		.24	133		85,000	8,025		93,025	106,000
5000	For jockey pump 1″, 3 HP, with control, add	Q-12	2	8		3,025	455		3,480	4,050

21 31 16 – Diesel-Drive, Centrifugal Fire Pumps

21 31 16.50 Diesel-Drive Fire Pumps

		Crew	Daily Output	Labor-Hours	Unit	Material	Labor	Equipment	Total	Total Incl O&P
0010	**DIESEL-DRIVE FIRE PUMPS** Including controller, fittings and relief valve									
0050	500 GPM, 50 psi, 27 HP, 4″ pump	Q-13	.64	50	Ea.	43,600	3,000		46,600	52,500
0100	500 GPM, 100 psi, 62 HP, 4″ pump		.60	53.333		53,500	3,200		56,700	64,000
0150	500 GPM, 125 psi, 78 HP, 4″ pump		.56	57.143		57,000	3,450		60,450	68,000
0200	750 GPM, 50 psi, 44 HP, 5″ pump		.60	53.333		45,000	3,200		48,200	54,500
0250	750 GPM, 100 psi, 80 HP, 4″ pump		.56	57.143		49,700	3,450		53,150	60,000
0300	750 GPM, 165 psi, 203 HP, 5″ pump		.52	61.538		56,500	3,700		60,200	68,000
0350	1000 GPM, 50 psi, 48 HP, 5″ pump		.58	55.172		48,600	3,325		51,925	58,500
0400	1000 GPM, 100 psi, 89 HP, 4″ pump		.56	57.143		50,000	3,450		53,450	60,500
0450	1000 GPM, 150 psi, 148 HP, 4″ pump		.48	66.667		54,500	4,025		58,525	66,000
0470	1000 GPM, 200 psi, 280 HP, 5″ pump		.40	80		71,500	4,825		76,325	86,000
0480	1250 GPM, 75 psi, 75 HP, 5″ pump		.54	59.259		52,500	3,575		56,075	63,000
0500	1500 GPM, 50 psi, 66 HP, 6″ pump		.50	64		50,500	3,850		54,350	61,500
0550	1500 GPM, 100 psi, 140 HP, 6″ pump		.46	69.565		54,000	4,200		58,200	66,000

21 31 Centrifugal Fire Pumps

21 31 16 – Diesel-Drive, Centrifugal Fire Pumps

21 31 16.50 Diesel-Drive Fire Pumps		Crew	Daily Output	Labor-Hours	Unit	2020 Bare Costs Material	Labor	Equipment	Total	Total Incl O&P
0600	1500 GPM, 150 psi, 228 HP, 6" pump	Q-13	.42	76.190	Ea.	69,500	4,575		74,075	83,500
0650	1500 GPM, 200 psi, 279 HP, 6" pump		.38	84.211		97,000	5,075		102,075	114,500
0700	2000 GPM, 100 psi, 167 HP, 6" pump		.34	94.118		65,000	5,675		70,675	80,500
0750	2000 GPM, 150 psi, 284 HP, 6" pump		.30	107		82,500	6,425		88,925	101,000
0800	2500 GPM, 100 psi, 213 HP, 8" pump		.32	100		67,500	6,025		73,525	83,500
0820	2500 GPM, 150 psi, 365 HP, 8" pump		.26	123		88,500	7,400		95,900	108,500
0850	3000 GPM, 100 psi, 250 HP, 8" pump		.28	114		96,000	6,875		102,875	116,000
0900	3000 GPM, 150 psi, 384 HP, 10" pump		.20	160		118,000	9,625		127,625	144,500
0950	3500 GPM, 100 psi, 300 HP, 10" pump		.24	133		89,000	8,025		97,025	110,500
1000	3500 GPM, 150 psi, 518 HP, 10" pump	▼	.20	160	▼	119,000	9,625		128,625	146,000

Division Notes

		CREW	DAILY OUTPUT	LABOR-HOURS	UNIT	BARE COSTS MAT.	BARE COSTS LABOR	BARE COSTS EQUIP.	BARE COSTS TOTAL	TOTAL INCL O&P

Estimating Tips

22 10 00 Plumbing Piping and Pumps

This subdivision is primarily basic pipe and related materials. The pipe may be used by any of the mechanical disciplines, i.e., plumbing, fire protection, heating, and air conditioning.

Note: CPVC plastic piping approved for fire protection is located in 21 11 13.

- The labor adjustment factors listed in Subdivision 22 01 02.20 apply throughout Divisions 21, 22, and 23. CAUTION: the correct percentage may vary for the same items. For example, the percentage add for the basic pipe installation should be based on the maximum height that the installer must install for that particular section. If the pipe is to be located 14' above the floor but it is suspended on threaded rod from beams, the bottom flange of which is 18' high (4' rods), then the height is actually 18' and the add is 20%. The pipe cover, however, does not have to go above the 14' and so the add should be 10%.
- Most pipe is priced first as straight pipe with a joint (coupling, weld, etc.) every 10' and a hanger usually every 10'. There are exceptions with hanger spacing such as for cast iron pipe (5') and plastic pipe (3 per 10'). Following each type of pipe there are several lines listing sizes and the amount to be subtracted to delete couplings and hangers. This is for pipe that is to be buried or supported together on trapeze hangers. The reason that the couplings are deleted is that these runs are usually long, and frequently longer lengths of pipe are used. By deleting the couplings, the estimator is expected to look up and add back the correct reduced number of couplings.
- When preparing an estimate, it may be necessary to approximate the fittings. Fittings usually run between 25% and 50% of the cost of the pipe. The lower percentage is for simpler runs, and the higher number is for complex areas, such as mechanical rooms.
- For historic restoration projects, the systems must be as invisible as possible, and pathways must be sought for pipes, conduit, and ductwork. While installations in accessible spaces (such as basements and attics) are relatively straightforward to estimate, labor costs may be more difficult to determine when delivery systems must be concealed.

22 40 00 Plumbing Fixtures

- Plumbing fixture costs usually require two lines: the fixture itself and its "rough-in, supply, and waste."
- In the Assemblies Section (Plumbing D2010) for the desired fixture, the System Components Group at the center of the page shows the fixture on the first line. The rest of the list (fittings, pipe, tubing, etc.) will total up to what we refer to in the Unit Price section as "Rough-in, supply, waste, and vent." Note that for most fixtures we allow a nominal 5' of tubing to reach from the fixture to a main or riser.
- Remember that gas- and oil-fired units need venting.

Reference Numbers

Reference numbers are shown at the beginning of some major classifications. These numbers refer to related items in the Reference Section. The reference information may be an estimating procedure, an alternate pricing method, or technical information.

Note: Not all subdivisions listed here necessarily appear. ■

Note: "Powered in part by CINX™, based on liscensed proprietary information of Harrison Publishing House, Inc."

Note: Trade Service, in part, has been used as a reference source for some of the material prices used in Division 22.

22 01 Operation and Maintenance of Plumbing

22 01 02 – Labor Adjustments

22 01 02.10 Boilers, General		Crew	Daily Output	Labor-Hours	Unit	2020 Bare Costs Material	Labor	Equipment	Total	Total Incl O&P
0010	**BOILERS, GENERAL**, Prices do not include flue piping, elec. wiring,									
0020	gas or oil piping, boiler base, pad, or tankless unless noted									
0100	Boiler H.P.: 10 KW = 34 lb./steam/hr. = 33,475 BTU/hr.									
0150	To convert SFR to BTU rating: Hot water, 150 x SFR;									
0160	Forced hot water, 180 x SFR; steam, 240 x SFR									

22 01 02.20 Labor Adjustment Factors

		Crew	Daily Output	Labor-Hours	Unit	Material	Labor	Equipment	Total	Total Incl O&P
0010	**LABOR ADJUSTMENT FACTORS** (For Div. 21, 22 and 23) R220102-20									
0100	Labor factors: The below are reasonable suggestions, but									
0110	each project must be evaluated for its own peculiarities, and									
0120	the adjustments be increased or decreased depending on the									
0130	severity of the special conditions.									
1000	Add to labor for elevated installation (Above floor level)									
1080	10′ to 14.5′ high						10%			
1100	15′ to 19.5′ high						20%			
1120	20′ to 24.5′ high						25%			
1140	25′ to 29.5′ high						35%			
1160	30′ to 34.5′ high						40%			
1180	35′ to 39.5′ high						50%			
1200	40′ and higher						55%			
2000	Add to labor for crawl space									
2100	3′ high						40%			
2140	4′ high						30%			
3000	Add to labor for multi-story building									
3010	For new construction (No elevator available)									
3100	Add for floors 3 thru 10						5%			
3110	Add for floors 11 thru 15						10%			
3120	Add for floors 16 thru 20						15%			
3130	Add for floors 21 thru 30						20%			
3140	Add for floors 31 and up						30%			
3170	For existing structure (Elevator available)									
3180	Add for work on floor 3 and above						2%			
4000	Add to labor for working in existing occupied buildings									
4100	Hospital						35%			
4140	Office building						25%			
4180	School						20%			
4220	Factory or warehouse						15%			
4260	Multi dwelling						15%			
5000	Add to labor, miscellaneous									
5100	Cramped shaft						35%			
5140	Congested area						15%			
5180	Excessive heat or cold						30%			
9000	Labor factors: The above are reasonable suggestions, but									
9010	each project should be evaluated for its own peculiarities.									
9100	Other factors to be considered are:									
9140	Movement of material and equipment through finished areas									
9180	Equipment room									
9220	Attic space									
9260	No service road									
9300	Poor unloading/storage area									
9340	Congested site area/heavy traffic									

22 05 05 – Selective Demolition for Plumbing

22 05 05.10 Plumbing Demolition		Crew	Daily Output	Labor-Hours	Unit	Material	2020 Bare Costs Labor	Equipment	Total	Total Incl O&P
0010	**PLUMBING DEMOLITION** R220105-10									
0400	Air compressor, up thru 2 HP	Q-1	10	1.600	Ea.		93		93	143
0410	3 HP thru 7-1/2 HP R024119-10		5.60	2.857			166		166	256
0420	10 HP thru 15 HP		1.40	11.429			665		665	1,025
0430	20 HP thru 30 HP	Q-2	1.30	18.462			1,100		1,100	1,725
0500	Backflow preventer, up thru 2" diameter	1 Plum	17	.471			30.50		30.50	47
0510	2-1/2" thru 3" diameter	Q-1	10	1.600			93		93	143
0520	4" thru 6" diameter	"	5	3.200			186		186	287
0530	8" thru 10" diameter	Q-2	3	8			480		480	745
0700	Carriers and supports									
0710	Fountains, sinks, lavatories and urinals	1 Plum	14	.571	Ea.		37		37	57
0720	Water closets	"	12	.667	"		43		43	66.50
0730	Grinder pump or sewage ejector system									
0732	Simplex	Q-1	7	2.286	Ea.		133		133	205
0734	Duplex	"	2.80	5.714			330		330	510
0738	Hot water dispenser	1 Plum	36	.222			14.30		14.30	22
0740	Hydrant, wall		26	.308			19.85		19.85	30.50
0744	Ground		12	.667			43		43	66.50
0760	Cleanouts and drains, up thru 4" pipe diameter		10	.800			51.50		51.50	79.50
0764	5" thru 8" pipe diameter	Q-1	10	1.600			93		93	143
0780	Industrial safety fixtures	1 Plum	8	1			64.50		64.50	99.50
1020	Fixtures, including 10' piping									
1100	Bathtubs, cast iron	1 Plum	4	2	Ea.		129		129	199
1120	Fiberglass		6	1.333			86		86	133
1140	Steel		5	1.600			103		103	159
1150	Bidet	Q-1	7	2.286			133		133	205
1200	Lavatory, wall hung	1 Plum	10	.800			51.50		51.50	79.50
1220	Counter top		8	1			64.50		64.50	99.50
1300	Sink, single compartment		8	1			64.50		64.50	99.50
1320	Double compartment		7	1.143			73.50		73.50	114
1340	Shower, stall and receptor	Q-1	6	2.667			155		155	239
1350	Group	"	7	2.286			133		133	205
1400	Water closet, floor mounted	1 Plum	8	1			64.50		64.50	99.50
1420	Wall mounted	"	7	1.143			73.50		73.50	114
1440	Wash fountain, 36" diameter	Q-2	8	3			180		180	279
1442	54" diameter	"	7	3.429			206		206	320
1500	Urinal, floor mounted	1 Plum	4	2			129		129	199
1520	Wall mounted	"	7	1.143			73.50		73.50	114
1590	Whirl pool or hot tub	Q-1	2.60	6.154			355		355	550
1600	Water fountains, free standing	1 Plum	8	1			64.50		64.50	99.50
1620	Wall or deck mounted		6	1.333			86		86	133
1800	Medical gas specialties		8	1			64.50		64.50	99.50
1900	Piping fittings, single connection, up thru 1-1/2" diameter		30	.267			17.20		17.20	26.50
1910	2" thru 4" diameter		14	.571			37		37	57
1980	Pipe hanger/support removal		80	.100			6.45		6.45	9.95
1990	Glass pipe with fittings, 1" thru 3" diameter		200	.040	L.F.		2.58		2.58	3.99
1992	4" thru 6" diameter		150	.053			3.44		3.44	5.30
2000	Piping, metal, up thru 1-1/2" diameter		200	.040			2.58		2.58	3.99
2050	2" thru 3-1/2" diameter		150	.053			3.44		3.44	5.30
2100	4" thru 6" diameter	2 Plum	100	.160			10.30		10.30	15.95
2150	8" thru 14" diameter	"	60	.267			17.20		17.20	26.50
2153	16" thru 20" diameter	Q-18	70	.343			21	1.52	22.52	34
2155	24" thru 26" diameter		55	.436			26.50	1.93	28.43	43.50

22 05 05 – Selective Demolition for Plumbing

22 05 05.10 Plumbing Demolition		Crew	Daily Output	Labor-Hours	Unit	Material	2020 Bare Costs Labor	Equipment	Total	Total Incl O&P
2156	30" thru 36" diameter	Q-18	40	.600	L.F.		36.50	2.66	39.16	60
2160	Plastic pipe with fittings, up thru 1-1/2" diameter	1 Plum	250	.032			2.06		2.06	3.19
2162	2" thru 3" diameter	"	200	.040			2.58		2.58	3.99
2164	4" thru 6" diameter	Q-1	200	.080			4.64		4.64	7.15
2166	8" thru 14" diameter		150	.107			6.20		6.20	9.55
2168	16" diameter		100	.160			9.30		9.30	14.35
2170	Prison fixtures, lavatory or sink		18	.889	Ea.		51.50		51.50	79.50
2172	Shower		5.60	2.857			166		166	256
2174	Urinal or water closet		13	1.231			71.50		71.50	110
2180	Pumps, all fractional horse-power		12	1.333			77.50		77.50	120
2184	1 HP thru 5 HP		6	2.667			155		155	239
2186	7-1/2 HP thru 15 HP		2.50	6.400			370		370	575
2188	20 HP thru 25 HP	Q-2	4	6			360		360	560
2190	30 HP thru 60 HP		.80	30			1,800		1,800	2,800
2192	75 HP thru 100 HP		.60	40			2,400		2,400	3,725
2194	150 HP		.50	48			2,875		2,875	4,475
2198	Pump, sump or submersible	1 Plum	12	.667			43		43	66.50
2200	Receptors and interceptors, up thru 20 GPM	"	8	1			64.50		64.50	99.50
2204	25 thru 100 GPM	Q-1	6	2.667			155		155	239
2208	125 thru 300 GPM	"	2.40	6.667			385		385	600
2211	325 thru 500 GPM	Q-2	2.60	9.231			555		555	860
2212	Deduct for salvage, aluminum scrap				Ton				635	695
2214	Brass scrap								2,700	2,975
2216	Copper scrap								4,475	4,925
2218	Lead scrap								850	935
2220	Steel scrap								204	224
2230	Temperature maintenance cable	1 Plum	1200	.007	L.F.		.43		.43	.66
2250	Water heater, 40 gal.	"	6	1.333	Ea.		86		86	133
3100	Tanks, water heaters and liquid containers									
3110	Up thru 45 gallons	Q-1	22	.727	Ea.		42		42	65
3120	50 thru 120 gallons		14	1.143			66.50		66.50	102
3130	130 thru 240 gallons		7.60	2.105			122		122	189
3140	250 thru 500 gallons		5.40	2.963			172		172	266
3150	600 thru 1,000 gallons	Q-2	1.60	15			900		900	1,400
3160	1,100 thru 2,000 gallons		.70	34.286			2,050		2,050	3,200
3170	2,100 thru 4,000 gallons		.50	48			2,875		2,875	4,475
6000	Remove and reset fixtures, easy access	1 Plum	6	1.333			86		86	133
6100	Difficult access		4	2			129		129	199
9000	Minimum labor/equipment charge		2	4	Job		258		258	400
9100	Valve, metal valves or strainers and similar, up thru 1-1/2" diameter	1 Stpi	28	.286	Ea.		18.75		18.75	29
9110	2" thru 3" diameter	Q-1	11	1.455			84.50		84.50	130
9120	4" thru 6" diameter	"	8	2			116		116	179
9130	8" thru 14" diameter	Q-2	8	3			180		180	279
9140	16" thru 20" diameter		2	12			720		720	1,125
9150	24" diameter		1.20	20			1,200		1,200	1,850
9200	Valve, plastic, up thru 1-1/2" diameter	1 Plum	42	.190			12.30		12.30	19
9210	2" thru 3" diameter		15	.533			34.50		34.50	53
9220	4" thru 6" diameter		12	.667			43		43	66.50
9300	Vent flashing and caps		55	.145			9.35		9.35	14.50
9350	Water filter, commercial, 1" thru 1-1/2"	Q-1	2	8			465		465	715
9360	2" thru 2-1/2"	"	1.60	10			580		580	895
9400	Water heaters									
9410	Up thru 245 GPH	Q-1	2.40	6.667	Ea.		385		385	600

22 05 05 – Selective Demolition for Plumbing

22 05 05.10 Plumbing Demolition		Crew	Daily Output	Labor-Hours	Unit	Material	2020 Bare Costs Labor	Equipment	Total	Total Incl O&P
9420	250 thru 756 GPH	Q-1	1.60	10	Ea.		580		580	895
9430	775 thru 1,640 GPH		.80	20			1,150		1,150	1,800
9440	1,650 thru 4,000 GPH	Q-2	.50	48			2,875		2,875	4,475
9470	Water softener	Q-1	2	8			465		465	715

22 05 23 – General-Duty Valves for Plumbing Piping

22 05 23.10 Valves, Brass

		Crew	Daily Output	Labor-Hours	Unit	Material	Labor	Equipment	Total	Total Incl O&P
0010	**VALVES, BRASS**									
0032	For motorized valves, see Section 23 09 53.10									
0500	Gas cocks, threaded									
0510	1/4"	1 Plum	26	.308	Ea.	15.80	19.85		35.65	48
0520	3/8"		24	.333		15.80	21.50		37.30	50.50
0530	1/2"		24	.333		13.40	21.50		34.90	48
0540	3/4"		22	.364		18.10	23.50		41.60	56
0550	1"		19	.421		34	27		61	79.50
0560	1-1/4"		15	.533		57	34.50		91.50	116
0570	1-1/2"		13	.615		85.50	39.50		125	156
0580	2"		11	.727		111	47		158	195
0672	For larger sizes use lubricated plug valve, Section 23 05 23.70									

22 05 23.20 Valves, Bronze

		Crew	Daily Output	Labor-Hours	Unit	Material	Labor	Equipment	Total	Total Incl O&P
0010	**VALVES, BRONZE** R220523-90									
1020	Angle, 150 lb., rising stem, threaded									
1030	1/8"	1 Plum	24	.333	Ea.	142	21.50		163.50	189
1040	1/4"		24	.333		162	21.50		183.50	211
1050	3/8"		24	.333		165	21.50		186.50	214
1060	1/2"		22	.364		169	23.50		192.50	222
1070	3/4"		20	.400		230	26		256	292
1080	1"		19	.421		315	27		342	385
1090	1-1/4"		15	.533		355	34.50		389.50	450
1100	1-1/2"		13	.615		525	39.50		564.50	635
1102	Soldered same price as threaded									
1110	2"	1 Plum	11	.727	Ea.	845	47		892	1,000
1300	Ball									
1304	Soldered									
1312	3/8"	1 Plum	21	.381	Ea.	19.70	24.50		44.20	59.50
1316	1/2"		18	.444		18.70	28.50		47.20	65
1320	3/4"		17	.471		33	30.50		63.50	83.50
1324	1"		15	.533		44	34.50		78.50	102
1328	1-1/4"		13	.615		46	39.50		85.50	112
1332	1-1/2"		11	.727		86.50	47		133.50	168
1336	2"		9	.889		92	57.50		149.50	190
1340	2-1/2"		7	1.143		485	73.50		558.50	650
1344	3"		5	1.600		570	103		673	785
1350	Single union end									
1358	3/8"	1 Plum	21	.381	Ea.	26.50	24.50		51	67
1362	1/2"		18	.444		29.50	28.50		58	77
1366	3/4"		17	.471		56	30.50		86.50	109
1370	1"		15	.533		65.50	34.50		100	125
1374	1-1/4"		13	.615		102	39.50		141.50	174
1378	1-1/2"		11	.727		145	47		192	233
1382	2"		9	.889		194	57.50		251.50	300
1398	Threaded, 150 psi									
1400	1/4"	1 Plum	24	.333	Ea.	19.55	21.50		41.05	54.50

22 05 23.20 Valves, Bronze		Crew	Daily Output	Labor-Hours	Unit	Material	2020 Bare Costs Labor	Equipment	Total	Total Incl O&P
1430	3/8"	1 Plum	24	.333	Ea.	18.25	21.50		39.75	53
1450	1/2"		22	.364		18	23.50		41.50	56
1460	3/4"		20	.400		35.50	26		61.50	79
1470	1"		19	.421		32	27		59	77
1480	1-1/4"		15	.533		50	34.50		84.50	108
1490	1-1/2"		13	.615		75.50	39.50		115	145
1500	2"		11	.727		83.50	47		130.50	165
1510	2-1/2"		9	.889		264	57.50		321.50	380
1520	3"		8	1		425	64.50		489.50	565
1522	Solder the same price as threaded									
1600	Butterfly, 175 psi, full port, solder or threaded ends									
1610	Stainless steel disc and stem									
1620	1/4"	1 Plum	24	.333	Ea.	24	21.50		45.50	59.50
1630	3/8"		24	.333		16.85	21.50		38.35	51.50
1640	1/2"		22	.364		21	23.50		44.50	59.50
1650	3/4"		20	.400		29	26		55	72
1660	1"		19	.421		36	27		63	81.50
1670	1-1/4"		15	.533		69	34.50		103.50	129
1680	1-1/2"		13	.615		73.50	39.50		113	143
1690	2"		11	.727		93	47		140	176
1750	Check, swing, class 150, regrinding disc, threaded									
1800	1/8"	1 Plum	24	.333	Ea.	82.50	21.50		104	124
1830	1/4"		24	.333		76.50	21.50		98	118
1840	3/8"		24	.333		71	21.50		92.50	112
1850	1/2"		24	.333		77	21.50		98.50	118
1860	3/4"		20	.400		94	26		120	143
1870	1"		19	.421		178	27		205	238
1880	1-1/4"		15	.533		210	34.50		244.50	284
1890	1-1/2"		13	.615		300	39.50		339.50	390
1900	2"		11	.727		380	47		427	495
1910	2-1/2"	Q-1	15	1.067		925	62		987	1,125
1920	3"	"	13	1.231		1,275	71.50		1,346.50	1,500
2000	For 200 lb., add					5%	10%			
2040	For 300 lb., add					15%	15%			
2060	Check swing, 300 lb., lead free unless noted, sweat, 3/8" size	1 Plum	24	.333	Ea.	96	21.50		117.50	139
2070	1/2"		24	.333		96	21.50		117.50	139
2080	3/4"		20	.400		130	26		156	183
2090	1"		19	.421		191	27		218	252
2100	1-1/4"		15	.533		269	34.50		303.50	350
2110	1-1/2"		13	.615		315	39.50		354.50	410
2120	2"		11	.727		465	47		512	585
2130	2-1/2", not lead free	Q-1	15	1.067		750	62		812	920
2140	3", not lead free	"	13	1.231		950	71.50		1,021.50	1,150
2350	Check, lift, class 150, horizontal composition disc, threaded									
2430	1/4"	1 Plum	24	.333	Ea.	162	21.50		183.50	212
2440	3/8"		24	.333		206	21.50		227.50	260
2450	1/2"		24	.333		179	21.50		200.50	230
2460	3/4"		20	.400		219	26		245	281
2470	1"		19	.421		320	27		347	390
2480	1-1/4"		15	.533		420	34.50		454.50	515
2490	1-1/2"		13	.615		500	39.50		539.50	610
2500	2"		11	.727		845	47		892	1,000
2850	Gate, N.R.S., soldered, 125 psi									

22 05 23.20 Valves, Bronze		Crew	Daily Output	Labor-Hours	Unit	Material	Labor	Equipment	Total	Total Incl O&P
						2020 Bare Costs				
2900	3/8"	1 Plum	24	.333	Ea.	75.50	21.50		97	116
2920	1/2"		24	.333		60.50	21.50		82	99.50
2940	3/4"		20	.400		74	26		100	122
2950	1"		19	.421		78.50	27		105.50	129
2960	1-1/4"		15	.533		152	34.50		186.50	220
2970	1-1/2"		13	.615		184	39.50		223.50	265
2980	2"		11	.727		213	47		260	305
2990	2-1/2"	Q-1	15	1.067		515	62		577	660
3000	3"	"	13	1.231		570	71.50		641.50	740
3350	Threaded, class 150									
3410	1/4"	1 Plum	24	.333	Ea.	81	21.50		102.50	122
3420	3/8"		24	.333		98.50	21.50		120	141
3430	1/2"		24	.333		78.50	21.50		100	120
3440	3/4"		20	.400		104	26		130	154
3450	1"		19	.421		134	27		161	190
3460	1-1/4"		15	.533		137	34.50		171.50	204
3470	1-1/2"		13	.615		269	39.50		308.50	360
3480	2"		11	.727		234	47		281	330
3490	2-1/2"	Q-1	15	1.067		700	62		762	865
3500	3"	"	13	1.231		865	71.50		936.50	1,050
3850	Rising stem, soldered, 300 psi									
3900	3/8"	1 Plum	24	.333	Ea.	155	21.50		176.50	204
3920	1/2"		24	.333		157	21.50		178.50	206
3940	3/4"		20	.400		169	26		195	226
3950	1"		19	.421		229	27		256	294
3960	1-1/4"		15	.533		315	34.50		349.50	405
3970	1-1/2"		13	.615		385	39.50		424.50	485
3980	2"		11	.727		615	47		662	755
3990	2-1/2"	Q-1	15	1.067		1,325	62		1,387	1,550
4000	3"	"	13	1.231		2,025	71.50		2,096.50	2,350
4250	Threaded, class 150									
4310	1/4"	1 Plum	24	.333	Ea.	90	21.50		111.50	132
4320	3/8"		24	.333		90	21.50		111.50	132
4330	1/2"		24	.333		69.50	21.50		91	110
4340	3/4"		20	.400		81.50	26		107.50	130
4350	1"		19	.421		125	27		152	180
4360	1-1/4"		15	.533		132	34.50		166.50	198
4370	1-1/2"		13	.615		167	39.50		206.50	245
4380	2"		11	.727		224	47		271	320
4390	2-1/2"	Q-1	15	1.067		670	62		732	835
4400	3"	"	13	1.231		935	71.50		1,006.50	1,125
4500	For 300 psi, threaded, add					100%	15%			
4540	For chain operated type, add					15%				
4850	Globe, class 150, rising stem, threaded									
4920	1/4"	1 Plum	24	.333	Ea.	123	21.50		144.50	169
4940	3/8"		24	.333		121	21.50		142.50	167
4950	1/2"		24	.333		106	21.50		127.50	150
4960	3/4"		20	.400		172	26		198	229
4970	1"		19	.421		239	27		266	305
4980	1-1/4"		15	.533		292	34.50		326.50	375
4990	1-1/2"		13	.615		460	39.50		499.50	565
5000	2"		11	.727		690	47		737	835
5010	2-1/2"	Q-1	15	1.067		1,475	62		1,537	1,725

22 05 23.20 Valves, Bronze		Crew	Daily Output	Labor-Hours	Unit	Material	2020 Bare Costs Labor	Equipment	Total	Total Incl O&P
5020	3"	Q-1	13	1.231	Ea.	2,125	71.50		2,196.50	2,425
5120	For 300 lb. threaded, add					50%	15%			
5130	Globe, 300 lb., sweat, 3/8" size	1 Plum	24	.333	Ea.	134	21.50		155.50	181
5140	1/2"		24	.333		130	21.50		151.50	176
5150	3/4"		20	.400		187	26		213	246
5160	1"		19	.421		297	27		324	365
5170	1-1/4"		15	.533		365	34.50		399.50	455
5180	1-1/2"		13	.615		445	39.50		484.50	550
5190	2"		11	.727		1,050	47		1,097	1,225
5200	2-1/2"	Q-1	15	1.067		1,625	62		1,687	1,900
5210	3"	"	13	1.231		2,100	71.50		2,171.50	2,400
5600	Relief, pressure & temperature, self-closing, ASME, threaded									
5640	3/4"	1 Plum	28	.286	Ea.	257	18.40		275.40	310
5650	1"		24	.333		425	21.50		446.50	500
5660	1-1/4"		20	.400		805	26		831	925
5670	1-1/2"		18	.444		1,550	28.50		1,578.50	1,750
5680	2"		16	.500		1,675	32		1,707	1,900
5950	Pressure, poppet type, threaded									
6000	1/2"	1 Plum	30	.267	Ea.	90	17.20		107.20	126
6040	3/4"	"	28	.286	"	104	18.40		122.40	144
6400	Pressure, water, ASME, threaded									
6440	3/4"	1 Plum	28	.286	Ea.	119	18.40		137.40	160
6450	1"		24	.333		340	21.50		361.50	410
6460	1-1/4"		20	.400		485	26		511	575
6470	1-1/2"		18	.444		750	28.50		778.50	865
6480	2"		16	.500		1,075	32		1,107	1,250
6490	2-1/2"		15	.533		4,400	34.50		4,434.50	4,900
6900	Reducing, water pressure									
6920	300 psi to 25-75 psi, threaded or sweat									
6940	1/2"	1 Plum	24	.333	Ea.	470	21.50		491.50	555
6950	3/4"		20	.400		550	26		576	645
6960	1"		19	.421		850	27		877	975
6970	1-1/4"		15	.533		1,475	34.50		1,509.50	1,675
6980	1-1/2"		13	.615		2,225	39.50		2,264.50	2,500
6990	2"		11	.727		3,350	47		3,397	3,750
7100	For built-in by-pass or 10-35 psi, add					53			53	58.50
7700	High capacity, 250 psi to 25-75 psi, threaded									
7740	1/2"	1 Plum	24	.333	Ea.	890	21.50		911.50	1,025
7780	3/4"		20	.400		780	26		806	895
7790	1"		19	.421		1,050	27		1,077	1,225
7800	1-1/4"		15	.533		1,825	34.50		1,859.50	2,075
7810	1-1/2"		13	.615		2,675	39.50		2,714.50	3,000
7820	2"		11	.727		3,900	47		3,947	4,375
7830	2-1/2"		9	.889		5,575	57.50		5,632.50	6,225
7840	3"		8	1		6,575	64.50		6,639.50	7,350
7850	3" flanged (iron body)	Q-1	10	1.600		6,575	93		6,668	7,375
7860	4" flanged (iron body)	"	8	2		8,450	116		8,566	9,475
7920	For higher pressure, add					25%				
8350	Tempering, water, sweat connections									
8400	1/2"	1 Plum	24	.333	Ea.	118	21.50		139.50	163
8440	3/4"	"	20	.400	"	167	26		193	223
8650	Threaded connections									
8700	1/2"	1 Plum	24	.333	Ea.	162	21.50		183.50	211

22 05 23 – General-Duty Valves for Plumbing Piping

22 05 23.20 Valves, Bronze		Crew	Daily Output	Labor-Hours	Unit	Material	Labor	Equipment	Total	Total Incl O&P
8740	3/4"	1 Plum	20	.400	Ea.	1,050	26		1,076	1,225
8750	1"		19	.421		1,200	27		1,227	1,350
8760	1-1/4"		15	.533		1,975	34.50		2,009.50	2,225
8770	1-1/2"		13	.615		2,025	39.50		2,064.50	2,275
8780	2"		11	.727		3,000	47		3,047	3,375
8800	Water heater water & gas safety shut off									
8810	Protection against a leaking water heater									
8814	Shut off valve	1 Plum	16	.500	Ea.	196	32		228	266
8818	Water heater dam		32	.250		32.50	16.10		48.60	60.50
8822	Gas control wiring harness		32	.250		24.50	16.10		40.60	52
8830	Whole house flood safety shut off									
8834	Connections									
8838	3/4" NPT	1 Plum	12	.667	Ea.	1,025	43		1,068	1,200
8842	1" NPT		11	.727		1,100	47		1,147	1,275
8846	1-1/4" NPT		10	.800		1,100	51.50		1,151.50	1,275
9000	Minimum labor/equipment charge		4	2	Job		129		129	199

22 05 23.40 Valves, Lined, Corrosion Resistant/High Purity		Crew	Daily Output	Labor-Hours	Unit	Material	Labor	Equipment	Total	Total Incl O&P
0010	**VALVES, LINED, CORROSION RESISTANT/HIGH PURITY** R220523-90									
3500	Check lift, 125 lb., cast iron flanged									
3510	Horizontal PPL or SL lined									
3530	1"	1 Plum	14	.571	Ea.	660	37		697	780
3540	1-1/2"		11	.727		795	47		842	950
3550	2"		8	1		930	64.50		994.50	1,125
3560	2-1/2"	Q-1	5	3.200		1,200	186		1,386	1,600
3570	3"		4.50	3.556		1,525	206		1,731	2,000
3590	4"		3	5.333		2,000	310		2,310	2,675
3610	6"	Q-2	3	8		3,400	480		3,880	4,475
3620	8"	"	2.50	9.600		7,475	575		8,050	9,125
4250	Vertical PPL or SL lined									
4270	1"	1 Plum	14	.571	Ea.	665	37		702	785
4290	1-1/2"		11	.727		770	47		817	925
4300	2"		8	1		910	64.50		974.50	1,100
4310	2-1/2"	Q-1	5	3.200		1,325	186		1,511	1,725
4320	3"		4.50	3.556		1,375	206		1,581	1,825
4340	4"		3	5.333		1,875	310		2,185	2,550
4360	6"	Q-2	3	8		2,825	480		3,305	3,875
4370	8"	"	2.50	9.600		5,625	575		6,200	7,100

22 05 23.60 Valves, Plastic		Crew	Daily Output	Labor-Hours	Unit	Material	Labor	Equipment	Total	Total Incl O&P
0010	**VALVES, PLASTIC** R220523-90									
1100	Angle, PVC, threaded									
1110	1/4"	1 Plum	26	.308	Ea.	60	19.85		79.85	96.50
1120	1/2"		26	.308		81.50	19.85		101.35	120
1130	3/4"		25	.320		96.50	20.50		117	138
1140	1"		23	.348		117	22.50		139.50	164
1150	Ball, PVC, socket or threaded, true union									
1230	1/2"	1 Plum	26	.308	Ea.	39	19.85		58.85	73.50
1240	3/4"		25	.320		50	20.50		70.50	87
1250	1"		23	.348		55.50	22.50		78	95.50
1260	1-1/4"		21	.381		94	24.50		118.50	141
1270	1-1/2"		20	.400		86	26		112	135
1280	2"		17	.471		136	30.50		166.50	196
1290	2-1/2"	Q-1	26	.615		236	35.50		271.50	315

Material, Labor and Equipment are 2020 Bare Costs.

22 05 23.60 Valves, Plastic		Crew	Daily Output	Labor-Hours	Unit	Material	2020 Bare Costs Labor	Equipment	Total	Total Incl O&P
1300	3"	Q-1	24	.667	Ea.	272	38.50		310.50	360
1310	4"		20	.800		455	46.50		501.50	575
1360	For PVC, flanged, add					100%	15%			
1450	Double union 1/2"	1 Plum	26	.308		37.50	19.85		57.35	71.50
1460	3/4"		25	.320		43.50	20.50		64	80
1470	1"		23	.348		57	22.50		79.50	97.50
1480	1-1/4"		21	.381		72.50	24.50		97	118
1490	1-1/2"		20	.400		95	26		121	144
1500	2"		17	.471		109	30.50		139.50	167
1650	CPVC, socket or threaded, single union									
1700	1/2"	1 Plum	26	.308	Ea.	61.50	19.85		81.35	98
1720	3/4"		25	.320		81	20.50		101.50	121
1730	1"		23	.348		92.50	22.50		115	137
1750	1-1/4"		21	.381		144	24.50		168.50	197
1760	1-1/2"		20	.400		146	26		172	201
1770	2"		17	.471		202	30.50		232.50	269
1780	3"	Q-1	24	.667		755	38.50		793.50	890
1840	For CPVC, flanged, add					65%	15%			
1880	For true union, socket or threaded, add					50%	5%			
2050	Polypropylene, threaded									
2100	1/4"	1 Plum	26	.308	Ea.	44	19.85		63.85	79
2120	3/8"		26	.308		45.50	19.85		65.35	80.50
2130	1/2"		26	.308		39.50	19.85		59.35	74
2140	3/4"		25	.320		49	20.50		69.50	86
2150	1"		23	.348		55.50	22.50		78	95.50
2160	1-1/4"		21	.381		74	24.50		98.50	120
2170	1-1/2"		20	.400		91.50	26		117.50	141
2180	2"		17	.471		123	30.50		153.50	183
2190	3"	Q-1	24	.667		300	38.50		338.50	390
2200	4"	"	20	.800		500	46.50		546.50	620
2550	PVC, three way, socket or threaded									
2600	1/2"	1 Plum	26	.308	Ea.	57	19.85		76.85	93
2640	3/4"		25	.320		67.50	20.50		88	107
2650	1"		23	.348		81.50	22.50		104	124
2660	1-1/2"		20	.400		117	26		143	169
2670	2"		17	.471		155	30.50		185.50	218
2680	3"	Q-1	24	.667		465	38.50		503.50	570
2740	For flanged, add					60%	15%			
3150	Ball check, PVC, socket or threaded									
3200	1/4"	1 Plum	26	.308	Ea.	48	19.85		67.85	83.50
3220	3/8"		26	.308		48	19.85		67.85	83.50
3240	1/2"		26	.308		46.50	19.85		66.35	81.50
3250	3/4"		25	.320		52	20.50		72.50	89
3260	1"		23	.348		65	22.50		87.50	106
3270	1-1/4"		21	.381		124	24.50		148.50	174
3280	1-1/2"		20	.400		109	26		135	160
3290	2"		17	.471		149	30.50		179.50	210
3310	3"	Q-1	24	.667		415	38.50		453.50	515
3320	4"	"	20	.800		585	46.50		631.50	710
3360	For PVC, flanged, add					50%	15%			
3750	CPVC, socket or threaded									
3800	1/2"	1 Plum	26	.308	Ea.	66.50	19.85		86.35	104
3840	3/4"		25	.320		85.50	20.50		106	127

22 05 23.60 Valves, Plastic		Crew	Daily Output	Labor-Hours	Unit	Material	Labor	Equipment	Total	Total Incl O&P
						2020 Bare Costs				
3850	1"	1 Plum	23	.348	Ea.	104	22.50		126.50	149
3860	1-1/2"		20	.400		175	26		201	232
3870	2"		17	.471		227	30.50		257.50	296
3880	3"	Q-1	24	.667		700	38.50		738.50	830
3920	4"	"	20	.800		1,000	46.50		1,046.50	1,175
3930	For CPVC, flanged, add					40%	15%			
4340	Polypropylene, threaded									
4360	1/2"	1 Plum	26	.308	Ea.	49.50	19.85		69.35	85
4400	3/4"		25	.320		69	20.50		89.50	108
4440	1"		23	.348		73	22.50		95.50	115
4450	1-1/2"		20	.400		141	26		167	195
4460	2"		17	.471		179	30.50		209.50	244
4500	For polypropylene flanged, add					200%	15%			
4850	Foot valve, PVC, socket or threaded									
4900	1/2"	1 Plum	34	.235	Ea.	62	15.15		77.15	92
4930	3/4"		32	.250		70	16.10		86.10	102
4940	1"		28	.286		91.50	18.40		109.90	129
4950	1-1/4"		27	.296		173	19.10		192.10	220
4960	1-1/2"		26	.308		175	19.85		194.85	224
4970	2"		24	.333		202	21.50		223.50	255
4980	3"		20	.400		480	26		506	570
4990	4"		18	.444		845	28.50		873.50	975
5000	For flanged, add					25%	10%			
5050	CPVC, socket or threaded									
5060	1/2"	1 Plum	34	.235	Ea.	77.50	15.15		92.65	109
5070	3/4"		32	.250		99.50	16.10		115.60	135
5080	1"		28	.286		122	18.40		140.40	163
5090	1-1/4"		27	.296		194	19.10		213.10	243
5100	1-1/2"		26	.308		194	19.85		213.85	244
5110	2"		24	.333		243	21.50		264.50	300
5120	3"		20	.400		495	26		521	585
5130	4"		18	.444		895	28.50		923.50	1,025
5140	For flanged, add					25%	10%			
5280	Needle valve, PVC, threaded									
5300	1/4"	1 Plum	26	.308	Ea.	66	19.85		85.85	103
5340	3/8"		26	.308		76.50	19.85		96.35	115
5360	1/2"		26	.308		76	19.85		95.85	114
5380	For polypropylene, add					10%				
5800	Y check, PVC, socket or threaded									
5820	1/2"	1 Plum	26	.308	Ea.	68	19.85		87.85	106
5840	3/4"		25	.320		74	20.50		94.50	113
5850	1"		23	.348		82	22.50		104.50	125
5860	1-1/4"		21	.381		129	24.50		153.50	180
5870	1-1/2"		20	.400		141	26		167	195
5880	2"		17	.471		175	30.50		205.50	239
5890	2-1/2"		15	.533		370	34.50		404.50	465
5900	3"	Q-1	24	.667		365	38.50		403.50	460
5910	4"	"	20	.800		610	46.50		656.50	740
5960	For PVC flanged, add					45%	15%			
6350	Y sediment strainer, PVC, socket or threaded									
6400	1/2"	1 Plum	26	.308	Ea.	56.50	19.85		76.35	92.50
6440	3/4"		24	.333		65.50	21.50		87	106
6450	1"		23	.348		67.50	22.50		90	109

22 05 23 – General-Duty Valves for Plumbing Piping

22 05 23.60 Valves, Plastic		Crew	Daily Output	Labor-Hours	Unit	Material	2020 Bare Costs Labor	Equipment	Total	Total Incl O&P
6460	1-1/4"	1 Plum	21	.381	Ea.	113	24.50		137.50	163
6470	1-1/2"		20	.400		133	26		159	186
6480	2"		17	.471		142	30.50		172.50	203
6490	2-1/2"		15	.533		345	34.50		379.50	435
6500	3"	Q-1	24	.667		340	38.50		378.50	435
6510	4"	"	20	.800		580	46.50		626.50	705
6560	For PVC, flanged, add					55%	15%			
9000	Minimum labor/equipment charge	1 Plum	3.50	2.286	Job		147		147	228

22 05 29 – Hangers and Supports for Plumbing Piping and Equipment

22 05 29.10 Hangers & Supp. for Plumb'g/HVAC Pipe/Equip.		Crew	Daily Output	Labor-Hours	Unit	Material	Labor	Equipment	Total	Total Incl O&P
0010	**HANGERS AND SUPPORTS FOR PLUMB'G/HVAC PIPE/EQUIP.**									
0011	TYPE numbers per MSS-SP58									
0050	Brackets									
0060	Beam side or wall, malleable iron, TYPE 34									
0070	3/8" threaded rod size	1 Plum	48	.167	Ea.	5.10	10.75		15.85	22
0080	1/2" threaded rod size		48	.167		4	10.75		14.75	21
0090	5/8" threaded rod size		48	.167		11.70	10.75		22.45	29.50
0100	3/4" threaded rod size		48	.167		21.50	10.75		32.25	40
0110	7/8" threaded rod size		48	.167		13.95	10.75		24.70	32
0120	For concrete installation, add						30%			
0150	Wall, welded steel, medium, TYPE 32									
0160	0 size, 12" wide, 18" deep	1 Plum	34	.235	Ea.	183	15.15		198.15	226
0170	1 size, 18" wide, 24" deep		34	.235		218	15.15		233.15	264
0180	2 size, 24" wide, 30" deep		34	.235		289	15.15		304.15	340
0300	Clamps									
0310	C-clamp, for mounting on steel beam flange, w/locknut, TYPE 23									
0320	3/8" threaded rod size	1 Plum	160	.050	Ea.	3.38	3.22		6.60	8.70
0330	1/2" threaded rod size		160	.050		4.76	3.22		7.98	10.25
0340	5/8" threaded rod size		160	.050		5.30	3.22		8.52	10.80
0350	3/4" threaded rod size		160	.050		6.85	3.22		10.07	12.50
0400	High temperature to 1050°F, alloy steel									
0410	4" pipe size	Q-1	106	.151	Ea.	28	8.75		36.75	44
0420	6" pipe size		106	.151		43	8.75		51.75	60.50
0430	8" pipe size		97	.165		48	9.55		57.55	68
0440	10" pipe size		84	.190		81.50	11.05		92.55	107
0450	12" pipe size		72	.222		82	12.90		94.90	110
0460	14" pipe size		64	.250		253	14.50		267.50	300
0470	16" pipe size		56	.286		269	16.55		285.55	320
0500	I-beam, for mounting on bottom flange, strap iron, TYPE 21									
0530	4" flange size	1 Plum	93	.086	Ea.	6.75	5.55		12.30	15.95
0540	5" flange size		92	.087		7.40	5.60		13	16.80
0550	6" flange size		90	.089		8.40	5.75		14.15	18.10
0560	7" flange size		88	.091		9.35	5.85		15.20	19.35
0570	8" flange size		86	.093		10.05	6		16.05	20.50
0600	One hole, vertical mounting, malleable iron									
0610	1/2" pipe size	1 Plum	160	.050	Ea.	1.27	3.22		4.49	6.40
0620	3/4" pipe size		145	.055		1.35	3.56		4.91	7
0630	1" pipe size		136	.059		1.46	3.79		5.25	7.45
0640	1-1/4" pipe size		128	.063		2.56	4.03		6.59	9.05
0650	1-1/2" pipe size		120	.067		2.96	4.30		7.26	9.90
0660	2" pipe size		112	.071		4.27	4.60		8.87	11.80
0670	2-1/2" pipe size		104	.077		8.45	4.96		13.41	16.90

22 05 29.10 Hangers & Supp. for Plumb'g/HVAC Pipe/Equip.	Crew	Daily Output	Labor-Hours	Unit	Material	2020 Bare Costs Labor	Equipment	Total	Total Incl O&P
0680 3" pipe size	1 Plum	96	.083	Ea.	9.85	5.35		15.20	19.10
0690 3-1/2" pipe size		90	.089		9.55	5.75		15.30	19.35
0700 4" pipe size		84	.095		13.50	6.15		19.65	24.50
0750 Riser or extension pipe, carbon steel, TYPE 8									
0760 3/4" pipe size	1 Plum	48	.167	Ea.	3.48	10.75		14.23	20.50
0770 1" pipe size		47	.170		4.04	10.95		14.99	21.50
0780 1-1/4" pipe size		46	.174		5.45	11.20		16.65	23.50
0790 1-1/2" pipe size		45	.178		5.65	11.45		17.10	24
0800 2" pipe size		43	.186		5.65	12		17.65	25
0810 2-1/2" pipe size		41	.195		6.65	12.60		19.25	27
0820 3" pipe size		40	.200		6.70	12.90		19.60	27.50
0830 3-1/2" pipe size		39	.205		6.05	13.20		19.25	27
0840 4" pipe size		38	.211		9.20	13.55		22.75	31
0850 5" pipe size		37	.216		14.30	13.95		28.25	37.50
0860 6" pipe size		36	.222		16.30	14.30		30.60	40
0870 8" pipe size		34	.235		30	15.15		45.15	56.50
0880 10" pipe size		32	.250		30.50	16.10		46.60	58.50
0890 12" pipe size		28	.286		43.50	18.40		61.90	76.50
0900 For plastic coating 3/4" to 4", add					190%				
0910 For copper plating 3/4" to 4", add					58%				
0950 Two piece, complete, carbon steel, medium weight, TYPE 4									
0960 1/2" pipe size	Q-1	137	.117	Ea.	2.67	6.75		9.42	13.40
0970 3/4" pipe size		134	.119		2.73	6.95		9.68	13.70
0980 1" pipe size		132	.121		2.51	7.05		9.56	13.60
0990 1-1/4" pipe size		130	.123		3.30	7.15		10.45	14.70
1000 1-1/2" pipe size		126	.127		4.40	7.35		11.75	16.25
1010 2" pipe size		124	.129		5.20	7.50		12.70	17.25
1020 2-1/2" pipe size		120	.133		5.60	7.75		13.35	18.10
1030 3" pipe size		117	.137		5.25	7.95		13.20	18.05
1040 3-1/2" pipe size		114	.140		11.80	8.15		19.95	25.50
1050 4" pipe size		110	.145		9	8.45		17.45	23
1060 5" pipe size		106	.151		17	8.75		25.75	32.50
1070 6" pipe size		104	.154		22	8.90		30.90	38
1080 8" pipe size		100	.160		21.50	9.30		30.80	38.50
1090 10" pipe size		96	.167		51.50	9.65		61.15	71.50
1100 12" pipe size		89	.180		77.50	10.45		87.95	102
1110 14" pipe size		82	.195		87	11.30		98.30	113
1120 16" pipe size		68	.235		136	13.65		149.65	171
1130 For galvanized, add					45%				
1150 Insert, concrete									
1160 Wedge type, carbon steel body, malleable iron nut, galvanized									
1170 1/4" threaded rod size	1 Plum	96	.083	Ea.	8.40	5.35		13.75	17.55
1180 3/8" threaded rod size		96	.083		20.50	5.35		25.85	31
1190 1/2" threaded rod size		96	.083		39	5.35		44.35	51.50
1200 5/8" threaded rod size		96	.083		6.55	5.35		11.90	15.55
1210 3/4" threaded rod size		96	.083		12.30	5.35		17.65	22
1220 7/8" threaded rod size		96	.083		12.30	5.35		17.65	22
1250 Pipe guide sized for insulation									
1260 No. 1, 1" pipe size, 1" thick insulation	1 Stpi	26	.308	Ea.	143	20		163	188
1270 No. 2, 1-1/4"-2" pipe size, 1" thick insulation		23	.348		169	23		192	222
1280 No. 3, 1-1/4"-2" pipe size, 1-1/2" thick insulation		21	.381		169	25		194	225
1290 No. 4, 2-1/2"-3-1/2" pipe size, 1-1/2" thick insulation		18	.444		183	29		212	246
1300 No. 5, 4"-5" pipe size, 1-1/2" thick insulation		16	.500		197	33		230	267

22 05 29.10 Hangers & Supp. for Plumb'g/HVAC Pipe/Equip.		Crew	Daily Output	Labor-Hours	Unit	2020 Bare Costs Material	Labor	Equipment	Total	Total Incl O&P
1310	No. 6, 5"-6" pipe size, 2" thick insulation	Q-5	21	.762	Ea.	226	45		271	320
1320	No. 7, 8" pipe size, 2" thick insulation		16	1		292	59		351	410
1330	No. 8, 10" pipe size, 2" thick insulation		12	1.333		435	78.50		513.50	600
1340	No. 9, 12" pipe size, 2" thick insulation	Q-6	17	1.412		435	86.50		521.50	615
1350	No. 10, 12"-14" pipe size, 2-1/2" thick insulation		16	1.500		520	92		612	710
1360	No. 11, 16" pipe size, 2-1/2" thick insulation		10.50	2.286		520	140		660	785
1370	No. 12, 16"-18" pipe size, 3" thick insulation		9	2.667		740	163		903	1,075
1380	No. 13, 20" pipe size, 3" thick insulation		7.50	3.200		740	196		936	1,125
1390	No. 14, 24" pipe size, 3" thick insulation		7	3.429		1,025	210		1,235	1,475
1400	Bands									
1410	Adjustable band, carbon steel, for non-insulated pipe, TYPE 7									
1420	1/2" pipe size	Q-1	142	.113	Ea.	.38	6.55		6.93	10.50
1430	3/4" pipe size		140	.114		.38	6.65		7.03	10.65
1440	1" pipe size		137	.117		.38	6.75		7.13	10.85
1450	1-1/4" pipe size		134	.119		.46	6.95		7.41	11.20
1460	1-1/2" pipe size		131	.122		.41	7.10		7.51	11.40
1470	2" pipe size		129	.124		.41	7.20		7.61	11.55
1480	2-1/2" pipe size		125	.128		.74	7.40		8.14	12.30
1490	3" pipe size		122	.131		.79	7.60		8.39	12.60
1500	3-1/2" pipe size		119	.134		.95	7.80		8.75	13.10
1510	4" pipe size		114	.140		1.29	8.15		9.44	14
1520	5" pipe size		110	.145		2.23	8.45		10.68	15.50
1530	6" pipe size		108	.148		2.68	8.60		11.28	16.25
1540	8" pipe size		104	.154		4.08	8.90		12.98	18.30
1550	For copper plated, add					50%				
1560	For galvanized, add					30%				
1570	For plastic coating, add					30%				
1600	Adjusting nut malleable iron, steel band, TYPE 9									
1610	1/2" pipe size, galvanized band	Q-1	137	.117	Ea.	5.75	6.75		12.50	16.80
1620	3/4" pipe size, galvanized band		135	.119		5.70	6.85		12.55	16.95
1630	1" pipe size, galvanized band		132	.121		5.40	7.05		12.45	16.80
1640	1-1/4" pipe size, galvanized band		129	.124		5.85	7.20		13.05	17.50
1650	1-1/2" pipe size, galvanized band		126	.127		5.60	7.35		12.95	17.55
1660	2" pipe size, galvanized band		124	.129		6.35	7.50		13.85	18.50
1670	2-1/2" pipe size, galvanized band		120	.133		9.95	7.75		17.70	23
1680	3" pipe size, galvanized band		117	.137		10.40	7.95		18.35	23.50
1690	3-1/2" pipe size, galvanized band		114	.140		32	8.15		40.15	47.50
1700	4" pipe size, cadmium plated band		110	.145		28	8.45		36.45	44
1740	For plastic coated band, add					35%				
1750	For completely copper coated, add					45%				
1800	Clevis, adjustable, carbon steel, for non-insulated pipe, TYPE 1									
1810	1/2" pipe size	Q-1	137	.117	Ea.	1.12	6.75		7.87	11.70
1820	3/4" pipe size		135	.119		1.19	6.85		8.04	11.95
1830	1" pipe size		132	.121		1.51	7.05		8.56	12.50
1840	1-1/4" pipe size		129	.124		1.68	7.20		8.88	12.95
1850	1-1/2" pipe size		126	.127		1.15	7.35		8.50	12.65
1860	2" pipe size		124	.129		2.15	7.50		9.65	13.90
1870	2-1/2" pipe size		120	.133		3.55	7.75		11.30	15.85
1880	3" pipe size		117	.137		4.22	7.95		12.17	16.90
1890	3-1/2" pipe size		114	.140		4.50	8.15		12.65	17.55
1900	4" pipe size		110	.145		3.15	8.45		11.60	16.50
1910	5" pipe size		106	.151		4.27	8.75		13.02	18.25
1920	6" pipe size		104	.154		5.40	8.90		14.30	19.75

22 05 29.10 Hangers & Supp. for Plumb'g/HVAC Pipe/Equip.		Crew	Daily Output	Labor-Hours	Unit	2020 Bare Costs Material	Labor	Equipment	Total	Total Incl O&P
1930	8" pipe size	Q-1	100	.160	Ea.	10.60	9.30		19.90	26
1940	10" pipe size		96	.167		17.05	9.65		26.70	33.50
1950	12" pipe size		89	.180		22	10.45		32.45	40.50
1960	14" pipe size		82	.195		36	11.30		47.30	57.50
1970	16" pipe size		68	.235		70.50	13.65		84.15	98.50
1980	For galvanized, add					66%				
1990	For copper plated 1/2" to 4", add					77%				
2000	For light weight 1/2" to 4", deduct					13%				
2010	Insulated pipe type, 3/4" to 12" pipe, add					180%				
2020	Insulated pipe type, chrome-moly U-strap, add					530%				
2250	Split ring, malleable iron, for non-insulated pipe, TYPE 11									
2260	1/2" pipe size	Q-1	137	.117	Ea.	5.05	6.75		11.80	16
2270	3/4" pipe size		135	.119		5.15	6.85		12	16.35
2280	1" pipe size		132	.121		5.45	7.05		12.50	16.85
2290	1-1/4" pipe size		129	.124		6.80	7.20		14	18.55
2300	1-1/2" pipe size		126	.127		8.25	7.35		15.60	20.50
2310	2" pipe size		124	.129		9.35	7.50		16.85	22
2320	2-1/2" pipe size		120	.133		13.45	7.75		21.20	27
2330	3" pipe size		117	.137		16.85	7.95		24.80	31
2340	3-1/2" pipe size		114	.140		17.25	8.15		25.40	31.50
2350	4" pipe size		110	.145		17.55	8.45		26	32.50
2360	5" pipe size		106	.151		23.50	8.75		32.25	39
2370	6" pipe size		104	.154		52	8.90		60.90	71.50
2380	8" pipe size		100	.160		78	9.30		87.30	100
2390	For copper plated, add					8%				
2532	Turnbuckle, TYPE 13									
2534	3/8"	1 Plum	80	.100	Ea.	5.10	6.45		11.55	15.55
2535	1/2"		72	.111		5.65	7.15		12.80	17.30
2536	5/8"		64	.125		10.05	8.05		18.10	23.50
2537	3/4"		56	.143		12.80	9.20		22	28.50
2538	7/8"		48	.167		33.50	10.75		44.25	53.50
2539	1"		40	.200		45.50	12.90		58.40	70
2540	1-1/4"		32	.250		84	16.10		100.10	118
2650	Rods, carbon steel									
2660	Continuous thread									
2670	1/4" thread size	1 Plum	144	.056	L.F.	2.43	3.58		6.01	8.20
2680	3/8" thread size		144	.056		2.59	3.58		6.17	8.40
2690	1/2" thread size		144	.056		4.07	3.58		7.65	10.05
2700	5/8" thread size		144	.056		5.75	3.58		9.33	11.90
2710	3/4" thread size		144	.056		10.60	3.58		14.18	17.25
2720	7/8" thread size		144	.056		13.35	3.58		16.93	20
2725	1/4" thread size, bright finish		144	.056		1.57	3.58		5.15	7.30
2726	1/2" thread size, bright finish		144	.056		5.10	3.58		8.68	11.15
2730	For galvanized, add					40%				
2860	Pipe hanger assy, adj. clevis, saddle, rod, clamp, insul. allowance									
2864	1/2" pipe size	Q-5	35	.457	Ea.	26	27		53	70
2866	3/4" pipe size		34.80	.460		27.50	27		54.50	72
2868	1" pipe size		34.60	.462		27	27.50		54.50	71.50
2869	1-1/4" pipe size		34.30	.466		27	27.50		54.50	72.50
2870	1-1/2" pipe size		33.90	.472		27.50	28		55.50	73
2872	2" pipe size		33.30	.480		30	28.50		58.50	77
2874	2-1/2" pipe size		32.30	.495		33.50	29		62.50	81.50
2876	3" pipe size		31.20	.513		36	30.50		66.50	86.50

22 05 29 – Hangers and Supports for Plumbing Piping and Equipment

22 05 29.10 Hangers & Supp. for Plumb'g/HVAC Pipe/Equip.		Crew	Daily Output	Labor-Hours	Unit	Material	Labor	Equipment	Total	Total Incl O&P
						2020 Bare Costs				
2880	4" pipe size	Q-5	30.70	.521	Ea.	38.50	31		69.50	90
2884	6" pipe size		29.80	.537		49.50	31.50		81	104
2888	8" pipe size		28	.571		67.50	33.50		101	127
2892	10" pipe size		25.20	.635		74.50	37.50		112	140
2896	12" pipe size	↓	23.20	.690	↓	117	40.50		157.50	191
2900	Rolls									
2910	Adjustable yoke, carbon steel with CI roll, TYPE 43									
2918	2" pipe size	Q-1	140	.114	Ea.	11.70	6.65		18.35	23
2920	2-1/2" pipe size		137	.117		12.40	6.75		19.15	24
2930	3" pipe size		131	.122		13.25	7.10		20.35	25.50
2940	3-1/2" pipe size		124	.129		18.45	7.50		25.95	32
2950	4" pipe size		117	.137		18.15	7.95		26.10	32
2960	5" pipe size		110	.145		30	8.45		38.45	46
2970	6" pipe size		104	.154		40.50	8.90		49.40	58.50
2980	8" pipe size		96	.167		58	9.65		67.65	79
2990	10" pipe size		80	.200		74.50	11.60		86.10	100
3000	12" pipe size		68	.235		108	13.65		121.65	140
3010	14" pipe size		56	.286		185	16.55		201.55	230
3020	16" pipe size	↓	48	.333	↓	224	19.35		243.35	276
3050	Chair, carbon steel with CI roll									
3060	2" pipe size	1 Plum	68	.118	Ea.	14.65	7.60		22.25	28
3070	2-1/2" pipe size		65	.123		15.75	7.95		23.70	29.50
3080	3" pipe size		62	.129		16.80	8.30		25.10	31.50
3090	3-1/2" pipe size		60	.133		20.50	8.60		29.10	36
3100	4" pipe size		58	.138		19.90	8.90		28.80	36
3110	5" pipe size		56	.143		23	9.20		32.20	40
3120	6" pipe size		53	.151		31	9.75		40.75	49
3130	8" pipe size		50	.160		42	10.30		52.30	62
3140	10" pipe size		48	.167		56.50	10.75		67.25	78.50
3150	12" pipe size	↓	46	.174		86	11.20		97.20	112
3170	Single pipe roll (see line 2650 for rods), TYPE 41, 1" pipe size	Q-1	137	.117		9.85	6.75		16.60	21.50
3180	1-1/4" pipe size		131	.122		10.45	7.10		17.55	22.50
3190	1-1/2" pipe size		129	.124		10.45	7.20		17.65	22.50
3200	2" pipe size		124	.129		10.45	7.50		17.95	23
3210	2-1/2" pipe size		118	.136		11.15	7.85		19	24.50
3220	3" pipe size		115	.139		11.75	8.05		19.80	25.50
3230	3-1/2" pipe size		113	.142		12.80	8.20		21	27
3240	4" pipe size		112	.143		13	8.30		21.30	27
3250	5" pipe size		110	.145		15.10	8.45		23.55	29.50
3260	6" pipe size		101	.158		30	9.20		39.20	47
3270	8" pipe size		90	.178		32.50	10.30		42.80	51.50
3280	10" pipe size		80	.200		42	11.60		53.60	64
3290	12" pipe size		68	.235		68.50	13.65		82.15	96
3291	14" pipe size		56	.286		104	16.55		120.55	140
3292	16" pipe size		48	.333		135	19.35		154.35	178
3293	18" pipe size		40	.400		229	23		252	288
3294	20" pipe size		35	.457		189	26.50		215.50	249
3296	24" pipe size		30	.533		320	31		351	400
3297	30" pipe size		25	.640		585	37		622	700
3298	36" pipe size	↓	20	.800	↓	620	46.50		666.50	750
3300	Saddles (add vertical pipe riser, usually 3" diameter)									
3310	Pipe support, complete, adjust., CI saddle, TYPE 36									
3320	2-1/2" pipe size	1 Plum	96	.083	Ea.	111	5.35		116.35	130

22 05 29 – Hangers and Supports for Plumbing Piping and Equipment

22 05 29.10 Hangers & Supp. for Plumb'g/HVAC Pipe/Equip.		Crew	Daily Output	Labor-Hours	Unit	Material	2020 Bare Costs Labor	Equipment	Total	Total Incl O&P
3330	3" pipe size	1 Plum	88	.091	Ea.	113	5.85		118.85	133
3340	3-1/2" pipe size		79	.101		111	6.55		117.55	132
3350	4" pipe size		68	.118		160	7.60		167.60	188
3360	5" pipe size		64	.125		169	8.05		177.05	198
3370	6" pipe size		59	.136		165	8.75		173.75	195
3380	8" pipe size		53	.151		170	9.75		179.75	202
3390	10" pipe size		50	.160		198	10.30		208.30	233
3400	12" pipe size		48	.167		209	10.75		219.75	247
3450	For standard pipe support, one piece, CI, deduct					34%				
3460	For stanchion support, CI with steel yoke, deduct					60%				
3550	Insulation shield 1" thick, 1/2" pipe size, TYPE 40	1 Asbe	100	.080	Ea.	3.37	4.69		8.06	11.15
3560	3/4" pipe size		100	.080		4.06	4.69		8.75	11.90
3570	1" pipe size		98	.082		4.39	4.79		9.18	12.45
3580	1-1/4" pipe size		98	.082		4.68	4.79		9.47	12.75
3590	1-1/2" pipe size		96	.083		4.35	4.89		9.24	12.55
3600	2" pipe size		96	.083		4.53	4.89		9.42	12.75
3610	2-1/2" pipe size		94	.085		4.67	4.99		9.66	13.05
3620	3" pipe size		94	.085		5.55	4.99		10.54	14
3630	2" thick, 3-1/2" pipe size		92	.087		5.95	5.10		11.05	14.60
3640	4" pipe size		92	.087		9.85	5.10		14.95	18.95
3650	5" pipe size		90	.089		12.05	5.20		17.25	21.50
3660	6" pipe size		90	.089		13.85	5.20		19.05	23.50
3670	8" pipe size		88	.091		19.65	5.35		25	30
3680	10" pipe size		88	.091		27.50	5.35		32.85	39
3690	12" pipe size		86	.093		29.50	5.45		34.95	41
3700	14" pipe size		86	.093		57.50	5.45		62.95	71.50
3710	16" pipe size		84	.095		56.50	5.60		62.10	71.50
3720	18" pipe size		84	.095		59	5.60		64.60	73.50
3730	20" pipe size		82	.098		64.50	5.70		70.20	80
3732	24" pipe size		80	.100		76	5.85		81.85	93.50
3750	Covering protection saddle, TYPE 39									
3760	1" covering size									
3770	3/4" pipe size	1 Plum	68	.118	Ea.	5.60	7.60		13.20	17.85
3780	1" pipe size		68	.118		5.60	7.60		13.20	17.85
3790	1-1/4" pipe size		68	.118		5.60	7.60		13.20	17.85
3800	1-1/2" pipe size		66	.121		6	7.80		13.80	18.70
3810	2" pipe size		66	.121		6	7.80		13.80	18.70
3820	2-1/2" pipe size		64	.125		6.05	8.05		14.10	19.15
3830	3" pipe size		64	.125		8	8.05		16.05	21.50
3840	3-1/2" pipe size		62	.129		9.55	8.30		17.85	23.50
3850	4" pipe size		62	.129		8.85	8.30		17.15	22.50
3860	5" pipe size		60	.133		9.65	8.60		18.25	24
3870	6" pipe size		60	.133		11.55	8.60		20.15	26
3900	1-1/2" covering size									
3910	3/4" pipe size	1 Plum	68	.118	Ea.	7.20	7.60		14.80	19.60
3920	1" pipe size		68	.118		7.15	7.60		14.75	19.60
3930	1-1/4" pipe size		68	.118		7.20	7.60		14.80	19.60
3940	1-1/2" pipe size		66	.121		6.70	7.80		14.50	19.50
3950	2" pipe size		66	.121		7.25	7.80		15.05	20
3960	2-1/2" pipe size		64	.125		8.70	8.05		16.75	22
3970	3" pipe size		64	.125		8.70	8.05		16.75	22
3980	3-1/2" pipe size		62	.129		9	8.30		17.30	23
3990	4" pipe size		62	.129		9.15	8.30		17.45	23

22 05 29.10 Hangers & Supp. for Plumb'g/HVAC Pipe/Equip.		Crew	Daily Output	Labor-Hours	Unit	Material	2020 Bare Costs Labor	Equipment	Total	Total Incl O&P
4000	5" pipe size	1 Plum	60	.133	Ea.	9.90	8.60		18.50	24
4010	6" pipe size		60	.133		12.45	8.60		21.05	27
4020	8" pipe size		58	.138		15.60	8.90		24.50	31
4028	2" covering size									
4029	2-1/2" pipe size	1 Plum	62	.129	Ea.	9.50	8.30		17.80	23.50
4032	3" pipe size		60	.133		10.40	8.60		19	25
4033	4" pipe size		58	.138		10.40	8.90		19.30	25
4034	6" pipe size		56	.143		14.90	9.20		24.10	30.50
4035	8" pipe size		54	.148		17.70	9.55		27.25	34
4080	10" pipe size		58	.138		20	8.90		28.90	36
4090	12" pipe size		56	.143		42	9.20		51.20	61
4100	14" pipe size		56	.143		43	9.20		52.20	62
4110	16" pipe size		54	.148		57	9.55		66.55	78
4120	18" pipe size		54	.148		61	9.55		70.55	82.50
4130	20" pipe size		52	.154		68	9.90		77.90	90.50
4150	24" pipe size		50	.160		79.50	10.30		89.80	103
4160	30" pipe size		48	.167		87.50	10.75		98.25	113
4180	36" pipe size		45	.178		101	11.45		112.45	129
4200	Sockets									
4210	Rod end, malleable iron, TYPE 16									
4220	1/4" thread size	1 Plum	240	.033	Ea.	1.64	2.15		3.79	5.10
4230	3/8" thread size		240	.033		1.67	2.15		3.82	5.15
4240	1/2" thread size		230	.035		2.20	2.24		4.44	5.90
4250	5/8" thread size		225	.036		4.51	2.29		6.80	8.50
4260	3/4" thread size		220	.036		5.90	2.34		8.24	10.10
4270	7/8" thread size		210	.038		8.65	2.46		11.11	13.35
4290	Strap, 1/2" pipe size, TYPE 26	Q-1	142	.113		1.94	6.55		8.49	12.25
4300	3/4" pipe size		140	.114		1.97	6.65		8.62	12.40
4310	1" pipe size		137	.117		2.77	6.75		9.52	13.50
4320	1-1/4" pipe size		134	.119		2.84	6.95		9.79	13.80
4330	1-1/2" pipe size		131	.122		3.08	7.10		10.18	14.35
4340	2" pipe size		129	.124		3.13	7.20		10.33	14.55
4350	2-1/2" pipe size		125	.128		4.91	7.40		12.31	16.90
4360	3" pipe size		122	.131		6.25	7.60		13.85	18.60
4370	3-1/2" pipe size		119	.134		7	7.80		14.80	19.75
4380	4" pipe size		114	.140		7.35	8.15		15.50	20.50
4400	U-bolt, carbon steel									
4410	Standard, with nuts, TYPE 42									
4420	1/2" pipe size	1 Plum	160	.050	Ea.	1.41	3.22		4.63	6.55
4430	3/4" pipe size		158	.051		1.22	3.26		4.48	6.40
4450	1" pipe size		152	.053		1.25	3.39		4.64	6.65
4460	1-1/4" pipe size		148	.054		1.57	3.48		5.05	7.15
4470	1-1/2" pipe size		143	.056		1.67	3.61		5.28	7.40
4480	2" pipe size		139	.058		1.81	3.71		5.52	7.75
4490	2-1/2" pipe size		134	.060		2.96	3.85		6.81	9.20
4500	3" pipe size		128	.063		3.31	4.03		7.34	9.90
4510	3-1/2" pipe size		122	.066		4.29	4.23		8.52	11.25
4520	4" pipe size		117	.068		4.42	4.41		8.83	11.65
4530	5" pipe size		114	.070		4.39	4.52		8.91	11.85
4540	6" pipe size		111	.072		7.35	4.64		11.99	15.30
4550	8" pipe size		109	.073		8.85	4.73		13.58	17
4560	10" pipe size		107	.075		15.15	4.82		19.97	24
4570	12" pipe size		104	.077		21	4.96		25.96	30.50

22 05 29 – Hangers and Supports for Plumbing Piping and Equipment

22 05 29.10 Hangers & Supp. for Plumb'g/HVAC Pipe/Equip.		Crew	Daily Output	Labor-Hours	Unit	2020 Bare Costs Material	Labor	Equipment	Total	Total Incl O&P
4580	For plastic coating on 1/2" thru 6" size, add					150%				
4700	U-hook, carbon steel, requires mounting screws or bolts									
4710	3/4" thru 2" pipe size									
4720	6" long	1 Plum	96	.083	Ea.	1.16	5.35		6.51	9.60
4730	8" long		96	.083		1.46	5.35		6.81	9.90
4740	10" long		96	.083		1.59	5.35		6.94	10.05
4750	12" long		96	.083		1.75	5.35		7.10	10.25
4760	For copper plated, add					50%				
7000	Roof supports									
7006	Duct									
7010	Rectangular, open, 12" off roof									
7020	To 18" wide	Q-9	26	.615	Ea.	163	34.50		197.50	233
7030	To 24" wide	"	22	.727		190	41		231	273
7040	To 36" wide	Q-10	30	.800		217	46.50		263.50	310
7050	To 48" wide		28	.857		244	50		294	345
7060	To 60" wide		24	1		271	58		329	390
7100	Equipment									
7120	Equipment support	Q-5	20	.800	Ea.	71	47		118	151
7300	Pipe									
7310	Roller type									
7320	Up to 2-1/2" diam. pipe									
7324	3-1/2" off roof	Q-5	24	.667	Ea.	22.50	39.50		62	85.50
7326	Up to 10" off roof	"	20	.800	"	28	47		75	104
7340	2-1/2" to 3-1/2" diam. pipe									
7342	Up to 16" off roof	Q-5	18	.889	Ea.	49.50	52.50		102	136
7360	4" to 5" diam. pipe									
7362	Up to 12" off roof	Q-5	16	1	Ea.	70.50	59		129.50	169
7400	Strut/channel type									
7410	Up to 2-1/2" diam. pipe									
7424	3-1/2" off roof	Q-5	24	.667	Ea.	20	39.50		59.50	83
7426	Up to 10" off roof	"	20	.800	"	25.50	47		72.50	101
7440	Strut and roller type									
7452	2-1/2" to 3-1/2" diam. pipe									
7454	Up to 16" off roof	Q-5	18	.889	Ea.	71	52.50		123.50	160
7460	Strut and hanger type									
7470	Up to 3" diam. pipe									
7474	Up to 8" off roof	Q-5	19	.842	Ea.	45	49.50		94.50	127
8000	Pipe clamp, plastic, 1/2" CTS	1 Plum	80	.100		.20	6.45		6.65	10.15
8010	3/4" CTS		73	.110		.24	7.05		7.29	11.15
8020	1" CTS		68	.118		.54	7.60		8.14	12.30
8080	Economy clamp, 1/4" CTS		175	.046		.05	2.95		3	4.62
8090	3/8" CTS		168	.048		.05	3.07		3.12	4.81
8100	1/2" CTS		160	.050		.05	3.22		3.27	5.05
8110	3/4" CTS		145	.055		.05	3.56		3.61	5.55
8200	Half clamp, 1/2" CTS		80	.100		.07	6.45		6.52	10.05
8210	3/4" CTS		73	.110		.10	7.05		7.15	11
8300	Suspension clamp, 1/2" CTS		80	.100		.24	6.45		6.69	10.20
8310	3/4" CTS		73	.110		.22	7.05		7.27	11.15
8320	1" CTS		68	.118		.53	7.60		8.13	12.30
8400	Insulator, 1/2" CTS		80	.100		.36	6.45		6.81	10.35
8410	3/4" CTS		73	.110		.37	7.05		7.42	11.30
8420	1" CTS		68	.118		.39	7.60		7.99	12.15
8800	Wire cable support system									

22 05 29 – Hangers and Supports for Plumbing Piping and Equipment

22 05 29.10 Hangers & Supp. for Plumb'g/HVAC Pipe/Equip.		Crew	Daily Output	Labor-Hours	Unit	Material	2020 Bare Costs Labor	Equipment	Total	Total Incl O&P
8810	Cable with hook terminal and locking device									
8830	2 mm (.079") diam. cable (100 lb. cap.)									
8840	1 m (3.3') length, with hook	1 Shee	96	.083	Ea.	3.90	5.20		9.10	12.45
8850	2 m (6.6') length, with hook		84	.095		4.53	5.95		10.48	14.30
8860	3 m (9.9') length, with hook		72	.111		4.47	6.90		11.37	15.75
8870	5 m (16.4') length, with hook	Q-9	60	.267		5.55	14.95		20.50	29.50
8880	10 m (32.8') length, with hook	"	30	.533		8.05	30		38.05	56
8900	3 mm (.118") diam. cable (200 lb. cap.)									
8910	1 m (3.3') length, with hook	1 Shee	96	.083	Ea.	9.80	5.20		15	18.95
8920	2 m (6.6') length, with hook		84	.095		10.85	5.95		16.80	21.50
8930	3 m (9.9') length, with hook		72	.111		11.80	6.90		18.70	24
8940	5 m (16.4') length, with hook	Q-9	60	.267		10.40	14.95		25.35	35
8950	10 m (32.8') length, with hook	"	30	.533		13.90	30		43.90	62.50
9000	Cable system accessories									
9010	Anchor bolt, 3/8", with nut	1 Shee	140	.057	Ea.	1.13	3.56		4.69	6.85
9020	Air duct corner protector		160	.050		1.51	3.12		4.63	6.55
9030	Air duct support attachment		140	.057		1.82	3.56		5.38	7.60
9040	Flange clip, hammer-on style									
9044	For flange thickness 3/32"-9/64", 160 lb. cap.	1 Shee	180	.044	Ea.	.44	2.77		3.21	4.82
9048	For flange thickness 1/8"-1/4", 200 lb. cap.		160	.050		.32	3.12		3.44	5.25
9052	For flange thickness 5/16"-1/2", 200 lb. cap.		150	.053		.65	3.32		3.97	5.90
9056	For flange thickness 9/16"-3/4", 200 lb. cap.		140	.057		.89	3.56		4.45	6.60
9060	Wire insulation protection tube		180	.044	L.F.	.54	2.77		3.31	4.93
9070	Wire cutter				Ea.	44			44	48.50

22 05 33 – Heat Tracing for Plumbing Piping

22 05 33.20 Temperature Maintenance Cable		Crew	Daily Output	Labor-Hours	Unit	Material	Labor	Equipment	Total	Total Incl O&P
0010	**TEMPERATURE MAINTENANCE CABLE**									
0040	Components									
0080	Heating cable									
0100	208 V									
0150	140°F	Q-1	1060.80	.015	L.F.	10.15	.87		11.02	12.55
0200	120 V									
0220	125°F	Q-1	1060.80	.015	L.F.	9.50	.87		10.37	11.80
0300	Power kit w/1 end seal	1 Elec	48.80	.164	Ea.	136	10.05		146.05	166
0310	Splice kit		35.50	.225		161	13.85		174.85	199
0320	End seal		160	.050		12.75	3.07		15.82	18.70
0330	Tee kit w/1 end seal		26.80	.299		166	18.30		184.30	210
0340	Powered splice w/2 end seals		20	.400		118	24.50		142.50	168
0350	Powered tee kit w/3 end seals		18	.444		152	27.50		179.50	209
0360	Cross kit w/2 end seals		18.60	.430		156	26.50		182.50	213
0500	Recommended thickness of fiberglass insulation									
0510	Pipe size									
0520	1/2" to 1" use 1" insulation									
0530	1-1/4" to 2" use 1-1/2" insulation									
0540	2-1/2" to 6" use 2" insulation									
0560	NOTE: For pipe sizes 1-1/4" and smaller use 1/4" larger diameter									
0570	insulation to allow room for installation over cable.									

22 05 48 – Vibration and Seismic Controls for Plumbing Piping and Equipment

	22 05 48.40 Vibration Absorbers	Crew	Daily Output	Labor-Hours	Unit	2020 Bare Costs Material	Labor	Equipment	Total	Total Incl O&P
0010	VIBRATION ABSORBERS									
0100	Hangers, neoprene flex									
0200	10-120 lb. capacity				Ea.	24			24	26.50
0220	75-550 lb. capacity					42			42	46
0240	250-1,100 lb. capacity					78			78	86
0260	1,000-4,000 lb. capacity					144			144	158
0500	Spring flex, 60 lb. capacity					35.50			35.50	39.50
0520	450 lb. capacity					64.50			64.50	71
0540	900 lb. capacity					81			81	89
0560	1,100-1,300 lb. capacity					78			78	86
0600	Rubber in shear									
0610	45-340 lb., up to 1/2" rod size	1 Stpi	22	.364	Ea.	23	24		47	62.50
0620	130-700 lb., up to 3/4" rod size		20	.400		50	26		76	95.50
0630	50-1,000 lb., up to 3/4" rod size		18	.444		66	29		95	118
1000	Mounts, neoprene, 45-380 lb. capacity					17.35			17.35	19.10
1020	250-1,100 lb. capacity					63			63	69
1040	1,000-4,000 lb. capacity					116			116	127
1100	Spring flex, 60 lb. capacity					79			79	86.50
1120	165 lb. capacity					82			82	90
1140	260 lb. capacity					80.50			80.50	89
1160	450 lb. capacity					115			115	127
1180	600 lb. capacity					100			100	110
1200	750 lb. capacity					157			157	173
1220	900 lb. capacity					152			152	167
1240	1,100 lb. capacity					175			175	192
1260	1,300 lb. capacity					168			168	184
1280	1,500 lb. capacity					178			178	195
1300	1,800 lb. capacity					238			238	262
1320	2,200 lb. capacity					299			299	330
1340	2,600 lb. capacity					330			330	365
2000	Pads, cork rib, 18" x 18" x 1", 10-50 psi					196			196	215
2020	18" x 36" x 1", 10-50 psi					390			390	430
2100	Shear flexible pads, 18" x 18" x 3/8", 20-70 psi					78.50			78.50	86.50
2120	18" x 36" x 3/8", 20-70 psi					195			195	215
3000	Note overlap in capacities due to deflections									

22 05 53 – Identification for Plumbing Piping and Equipment

	22 05 53.10 Piping System Identification Labels	Crew	Daily Output	Labor-Hours	Unit	Material	Labor	Equipment	Total	Total Incl O&P
0010	PIPING SYSTEM IDENTIFICATION LABELS									
0100	Indicate contents and flow direction									
0106	Pipe markers									
0110	Plastic snap around									
0114	1/2" pipe	1 Plum	80	.100	Ea.	6.90	6.45		13.35	17.55
0116	3/4" pipe		80	.100		7.25	6.45		13.70	17.90
0118	1" pipe		80	.100		7.25	6.45		13.70	17.90
0120	2" pipe		75	.107		8.60	6.85		15.45	20
0122	3" pipe		70	.114		11.25	7.35		18.60	24
0124	4" pipe		60	.133		11.55	8.60		20.15	26
0126	6" pipe		60	.133		12.15	8.60		20.75	26.50
0128	8" pipe		56	.143		17.10	9.20		26.30	33
0130	10" pipe		56	.143		18.80	9.20		28	35
0200	Over 10" pipe size		50	.160		22	10.30		32.30	40
1110	Self adhesive									

22 05 53 – Identification for Plumbing Piping and Equipment

22 05 53.10 Piping System Identification Labels		Crew	Daily Output	Labor-Hours	Unit	Material	2020 Bare Costs Labor	Equipment	Total	Total Incl O&P
1114	1″ pipe	1 Plum	80	.100	Ea.	3.29	6.45		9.74	13.55
1116	2″ pipe		75	.107		3.46	6.85		10.31	14.45
1118	3″ pipe		70	.114		5.30	7.35		12.65	17.25
1120	4″ pipe		60	.133		5.30	8.60		13.90	19.10
1122	6″ pipe		60	.133		5.30	8.60		13.90	19.15
1124	8″ pipe		56	.143		7.60	9.20		16.80	22.50
1126	10″ pipe		56	.143		7.60	9.20		16.80	22.50
1200	Over 10″ pipe size		50	.160		10.65	10.30		20.95	27.50
2000	Valve tags									
2010	Numbered plus identifying legend									
2100	Brass, 2″ diameter	1 Plum	40	.200	Ea.	3.16	12.90		16.06	23.50
2200	Plastic, 1-1/2″ diameter	″	40	.200	″	2.99	12.90		15.89	23

22 05 76 – Facility Drainage Piping Cleanouts

22 05 76.10 Cleanouts		Crew	Daily Output	Labor-Hours	Unit	Material	Labor	Equipment	Total	Total Incl O&P
0010	**CLEANOUTS**									
0060	Floor type									
0080	Round or square, scoriated nickel bronze top									
0100	2″ pipe size	1 Plum	10	.800	Ea.	183	51.50		234.50	282
0120	3″ pipe size		8	1		264	64.50		328.50	390
0140	4″ pipe size		6	1.333		284	86		370	445
0160	5″ pipe size		4	2		485	129		614	735
0180	6″ pipe size	Q-1	6	2.667		530	155		685	820
0200	8″ pipe size	″	4	4		845	232		1,077	1,300
0340	Recessed for tile, same price									
0980	Round top, recessed for terrazzo									
1000	2″ pipe size	1 Plum	9	.889	Ea.	545	57.50		602.50	690
1080	3″ pipe size		6	1.333		620	86		706	815
1100	4″ pipe size		4	2		710	129		839	980
1120	5″ pipe size	Q-1	6	2.667		1,175	155		1,330	1,550
1140	6″ pipe size		5	3.200		1,175	186		1,361	1,575
1160	8″ pipe size		4	4		1,100	232		1,332	1,575
2000	Round scoriated nickel bronze top, extra heavy duty									
2060	2″ pipe size	1 Plum	9	.889	Ea.	335	57.50		392.50	460
2080	3″ pipe size		6	1.333		355	86		441	525
2100	4″ pipe size		4	2		500	129		629	750
2120	5″ pipe size	Q-1	6	2.667		730	155		885	1,050
2140	6″ pipe size		5	3.200		730	186		916	1,100
2160	8″ pipe size		4	4		905	232		1,137	1,350
4000	Wall type, square smooth cover, over wall frame									
4060	2″ pipe size	1 Plum	14	.571	Ea.	355	37		392	445
4080	3″ pipe size		12	.667		345	43		388	445
4100	4″ pipe size		10	.800		360	51.50		411.50	480
4120	5″ pipe size		9	.889		585	57.50		642.50	730
4140	6″ pipe size		8	1		655	64.50		719.50	825
4160	8″ pipe size	Q-1	11	1.455		885	84.50		969.50	1,100
5000	Extension, CI; bronze countersunk plug, 8″ long									
5040	2″ pipe size	1 Plum	16	.500	Ea.	200	32		232	270
5060	3″ pipe size		14	.571		273	37		310	355
5080	4″ pipe size		13	.615		234	39.50		273.50	320
5100	5″ pipe size		12	.667		380	43		423	480
5120	6″ pipe size		11	.727		710	47		757	855
9000	Minimum labor/equipment charge		3	2.667	Job		172		172	266

22 05 76 – Facility Drainage Piping Cleanouts

22 05 76.20 Cleanout Tees		Crew	Daily Output	Labor-Hours	Unit	2020 Bare Costs Material	Labor	Equipment	Total	Total Incl O&P
0010	**CLEANOUT TEES**									
0100	Cast iron, B&S, with countersunk plug									
0200	2" pipe size	1 Plum	4	2	Ea.	143	129		272	355
0220	3" pipe size		3.60	2.222		181	143		324	420
0240	4" pipe size		3.30	2.424		275	156		431	540
0260	5" pipe size	Q-1	5.50	2.909		545	169		714	860
0280	6" pipe size	"	5	3.200		790	186		976	1,150
0300	8" pipe size	Q-3	5	6.400		1,100	395		1,495	1,825
0500	For round smooth access cover, same price									
0600	For round scoriated access cover, same price									
0700	For square smooth access cover, add				Ea.	60%				
4000	Plastic, tees and adapters. Add plugs									
4010	ABS, DWV									
4020	Cleanout tee, 1-1/2" pipe size	1 Plum	15	.533	Ea.	14.15	34.50		48.65	68.50
4030	2" pipe size	Q-1	27	.593		12.65	34.50		47.15	67
4040	3" pipe size		21	.762		25.50	44		69.50	96.50
4050	4" pipe size		16	1		64.50	58		122.50	161
4100	Cleanout plug, 1-1/2" pipe size	1 Plum	32	.250		2.72	16.10		18.82	28
4110	2" pipe size	Q-1	56	.286		3.28	16.55		19.83	29
4120	3" pipe size		36	.444		5.35	26		31.35	46
4130	4" pipe size		30	.533		9.15	31		40.15	58
4180	Cleanout adapter fitting, 1-1/2" pipe size	1 Plum	32	.250		4.18	16.10		20.28	29.50
4190	2" pipe size	Q-1	56	.286		6.10	16.55		22.65	32
4200	3" pipe size		36	.444		15.60	26		41.60	57
4210	4" pipe size		30	.533		29	31		60	80
5000	PVC, DWV									
5010	Cleanout tee, 1-1/2" pipe size	1 Plum	15	.533	Ea.	11.90	34.50		46.40	66
5020	2" pipe size	Q-1	27	.593		9.15	34.50		43.65	63
5030	3" pipe size		21	.762		27	44		71	98.50
5040	4" pipe size		16	1		36.50	58		94.50	130
5090	Cleanout plug, 1-1/2" pipe size	1 Plum	32	.250		1.87	16.10		17.97	27
5100	2" pipe size	Q-1	56	.286		2.09	16.55		18.64	28
5110	3" pipe size		36	.444		3.73	26		29.73	44
5120	4" pipe size		30	.533		5.50	31		36.50	54
5130	6" pipe size		24	.667		17.75	38.50		56.25	79.50
5170	Cleanout adapter fitting, 1-1/2" pipe size	1 Plum	32	.250		2.51	16.10		18.61	28
5180	2" pipe size	Q-1	56	.286		3.08	16.55		19.63	29
5190	3" pipe size		36	.444		8.45	26		34.45	49.50
5200	4" pipe size		30	.533		13.90	31		44.90	63.50
5210	6" pipe size		24	.667		40	38.50		78.50	104
9000	Minimum labor/equipment charge	1 Plum	2.75	2.909	Job		187		187	290

22 07 Plumbing Insulation

22 07 16 – Plumbing Equipment Insulation

22 07 16.10 Insulation for Plumbing Equipment		Crew	Daily Output	Labor-Hours	Unit	Material	2020 Bare Costs Labor	Equipment	Total	Total Incl O&P
0010	INSULATION FOR PLUMBING EQUIPMENT									
2900	Domestic water heater wrap kit									
2920	1-1/2" with vinyl jacket, 20 to 60 gal. G	1 Plum	8	1	Ea.	16.70	64.50		81.20	118
2925	50 to 80 gal. G	"	8	1	"	28	64.50		92.50	131
9000	Minimum labor/equipment charge	1 Stpi	4	2	Job		131		131	203

22 07 19 – Plumbing Piping Insulation

22 07 19.10 Piping Insulation		Crew	Daily Output	Labor-Hours	Unit	Material	Labor	Equipment	Total	Total Incl O&P
0010	PIPING INSULATION									
0110	Insulation req'd. is based on the surface size/area to be covered									
0230	Insulated protectors (ADA)									
0235	For exposed piping under sinks or lavatories									
0240	Vinyl coated foam, velcro tabs									
0245	P Trap, 1-1/4" or 1-1/2"	1 Plum	32	.250	Ea.	16.95	16.10		33.05	43.50
0260	Valve and supply cover									
0265	1/2", 3/8", and 7/16" pipe size	1 Plum	32	.250	Ea.	16.40	16.10		32.50	43
0280	Tailpiece offset (wheelchair)									
0285	1-1/4" pipe size	1 Plum	32	.250	Ea.	13.15	16.10		29.25	39.50
0600	Pipe covering (price copper tube one size less than IPS)									
4280	Cellular glass, closed cell foam, all service jacket, sealant,									
4281	working temp. (-450°F to +900°F), 0 water vapor transmission									
4284	1" wall									
4286	1/2" iron pipe size G	Q-14	120	.133	L.F.	9.40	7.05		16.45	21.50
4300	1-1/2" wall									
4301	1" iron pipe size G	Q-14	105	.152	L.F.	10.40	8.05		18.45	24
4304	2-1/2" iron pipe size G		90	.178		13.70	9.40		23.10	30
4306	3" iron pipe size G		85	.188		17.50	9.95		27.45	35
4308	4" iron pipe size G		70	.229		24.50	12.05		36.55	46
4310	5" iron pipe size G	↓	65	.246	↓	25.50	13		38.50	48.50
4320	2" wall									
4322	1" iron pipe size G	Q-14	100	.160	L.F.	16.45	8.45		24.90	31.50
4324	2-1/2" iron pipe size G		85	.188		21	9.95		30.95	39.50
4326	3" iron pipe size G		80	.200		21	10.55		31.55	40.50
4328	4" iron pipe size G		65	.246		29	13		42	52.50
4330	5" iron pipe size G		60	.267		31.50	14.05		45.55	57
4332	6" iron pipe size G		50	.320		36.50	16.90		53.40	67
4336	8" iron pipe size G		40	.400		42	21		63	79.50
4338	10" iron pipe size G	↓	35	.457	↓	45	24		69	88
4350	2-1/2" wall									
4360	12" iron pipe size G	Q-14	32	.500	L.F.	75	26.50		101.50	125
4362	14" iron pipe size G	"	28	.571	"	82.50	30		112.50	139
4370	3" wall									
4378	6" iron pipe size G	Q-14	48	.333	L.F.	55	17.60		72.60	88.50
4380	8" iron pipe size G		38	.421		60	22		82	101
4382	10" iron pipe size G		33	.485		69	25.50		94.50	116
4384	16" iron pipe size G		25	.640		98.50	34		132.50	163
4386	18" iron pipe size G		22	.727		127	38.50		165.50	201
4388	20" iron pipe size G	↓	20	.800	↓	135	42		177	216
4400	3-1/2" wall									
4412	12" iron pipe size G	Q-14	27	.593	L.F.	79.50	31.50		111	137
4414	14" iron pipe size G	"	25	.640	"	96.50	34		130.50	160
4430	4" wall									
4446	16" iron pipe size G	Q-14	22	.727	L.F.	92.50	38.50		131	163

22 07 19.10 Piping Insulation			Crew	Daily Output	Labor-Hours	Unit	2020 Bare Costs Material	Labor	Equipment	Total	Total Incl O&P
4448	18" iron pipe size	G	Q-14	20	.800	L.F.	118	42		160	197
4450	20" iron pipe size	G	↓	18	.889	↓	135	47		182	223
4480	Fittings, average with fabric and mastic										
4484	1" wall										
4486	1/2" iron pipe size	G	1 Asbe	40	.200	Ea.	7.40	11.75		19.15	27
4500	1-1/2" wall										
4502	1" iron pipe size	G	1 Asbe	38	.211	Ea.	8.15	12.35		20.50	28.50
4504	2-1/2" iron pipe size	G		32	.250		18.40	14.65		33.05	43.50
4506	3" iron pipe size	G		30	.267		18.20	15.65		33.85	45
4508	4" iron pipe size	G		28	.286		19.30	16.75		36.05	47.50
4510	5" iron pipe size	G	↓	24	.333	↓	25	19.55		44.55	58.50
4520	2" wall										
4522	1" iron pipe size	G	1 Asbe	36	.222	Ea.	13.10	13.05		26.15	35
4524	2-1/2" iron pipe size	G		30	.267		20	15.65		35.65	47
4526	3" iron pipe size	G		28	.286		22.50	16.75		39.25	51.50
4528	4" iron pipe size	G		24	.333		25.50	19.55		45.05	59
4530	5" iron pipe size	G		22	.364		42.50	21.50		64	81
4532	6" iron pipe size	G		20	.400		51.50	23.50		75	94
4536	8" iron pipe size	G		12	.667		89	39		128	160
4538	10" iron pipe size	G	↓	8	1	↓	102	58.50		160.50	205
4550	2-1/2" wall										
4560	12" iron pipe size	G	1 Asbe	6	1.333	Ea.	189	78		267	330
4562	14" iron pipe size	G	"	4	2	"	207	117		324	415
4570	3" wall										
4578	6" iron pipe size	G	1 Asbe	16	.500	Ea.	63.50	29.50		93	117
4580	8" iron pipe size	G		10	.800		94.50	47		141.50	179
4582	10" iron pipe size	G	↓	6	1.333	↓	128	78		206	265
4900	Calcium silicate, with 8 oz. canvas cover										
5100	1" wall, 1/2" iron pipe size	G	Q-14	170	.094	L.F.	4.29	4.97		9.26	12.60
5130	3/4" iron pipe size	G		170	.094		4.32	4.97		9.29	12.65
5140	1" iron pipe size	G		170	.094		4.23	4.97		9.20	12.55
5150	1-1/4" iron pipe size	G		165	.097		4.29	5.10		9.39	12.80
5160	1-1/2" iron pipe size	G		165	.097		4.34	5.10		9.44	12.85
5170	2" iron pipe size	G		160	.100		5	5.30		10.30	13.85
5180	2-1/2" iron pipe size	G		160	.100		5.30	5.30		10.60	14.20
5190	3" iron pipe size	G		150	.107		5.80	5.65		11.45	15.30
5200	4" iron pipe size	G		140	.114		7.15	6.05		13.20	17.40
5210	5" iron pipe size	G		135	.119		7.55	6.25		13.80	18.20
5220	6" iron pipe size	G		130	.123		8	6.50		14.50	19.10
5280	1-1/2" wall, 1/2" iron pipe size	G		150	.107		4.65	5.65		10.30	14.05
5310	3/4" iron pipe size	G		150	.107		4.74	5.65		10.39	14.15
5320	1" iron pipe size	G		150	.107		5.15	5.65		10.80	14.60
5330	1-1/4" iron pipe size	G		145	.110		5.55	5.80		11.35	15.35
5340	1-1/2" iron pipe size	G		145	.110		5.95	5.80		11.75	15.80
5350	2" iron pipe size	G		140	.114		6.55	6.05		12.60	16.75
5360	2-1/2" iron pipe size	G		140	.114		7.15	6.05		13.20	17.40
5370	3" iron pipe size	G		135	.119		7.50	6.25		13.75	18.15
5380	4" iron pipe size	G		125	.128		8.70	6.75		15.45	20.50
5390	5" iron pipe size	G		120	.133		9.80	7.05		16.85	22
5400	6" iron pipe size	G		110	.145		10.10	7.70		17.80	23.50
5460	2" wall, 1/2" iron pipe size	G		135	.119		7.20	6.25		13.45	17.80
5490	3/4" iron pipe size	G		135	.119		7.55	6.25		13.80	18.20
5500	1" iron pipe size	G	↓	135	.119	↓	7.95	6.25		14.20	18.65

22 07 19.10 Piping Insulation			Crew	Daily Output	Labor-Hours	Unit	Material	Labor	Equipment	Total	Total Incl O&P
							2020 Bare Costs				
5510	1-1/4" iron pipe size	G	Q-14	130	.123	L.F.	8.50	6.50		15	19.65
5520	1-1/2" iron pipe size	G		130	.123		8.90	6.50		15.40	20
5530	2" iron pipe size	G		125	.128		9.40	6.75		16.15	21
5540	2-1/2" iron pipe size	G		125	.128		11.10	6.75		17.85	23
5550	3" iron pipe size	G		120	.133		11.20	7.05		18.25	23.50
5560	4" iron pipe size	G		115	.139		12.95	7.35		20.30	26
5570	5" iron pipe size	G		110	.145		14.75	7.70		22.45	28.50
5580	6" iron pipe size	G	↓	105	.152	↓	16.15	8.05		24.20	30.50
5600	Calcium silicate, no cover										
5720	1" wall, 1/2" iron pipe size	G	Q-14	180	.089	L.F.	3.91	4.69		8.60	11.75
5740	3/4" iron pipe size	G		180	.089		3.91	4.69		8.60	11.75
5750	1" iron pipe size	G		180	.089		3.74	4.69		8.43	11.55
5760	1-1/4" iron pipe size	G		175	.091		3.82	4.83		8.65	11.85
5770	1-1/2" iron pipe size	G		175	.091		3.85	4.83		8.68	11.90
5780	2" iron pipe size	G		170	.094		4.42	4.97		9.39	12.75
5790	2-1/2" iron pipe size	G		170	.094		4.69	4.97		9.66	13.05
5800	3" iron pipe size	G		160	.100		5.05	5.30		10.35	13.90
5810	4" iron pipe size	G		150	.107		6.30	5.65		11.95	15.85
5820	5" iron pipe size	G		145	.110		6.55	5.80		12.35	16.45
5830	6" iron pipe size	G		140	.114		6.90	6.05		12.95	17.10
5900	1-1/2" wall, 1/2" iron pipe size	G		160	.100		4.14	5.30		9.44	12.90
5920	3/4" iron pipe size	G		160	.100		4.20	5.30		9.50	12.95
5930	1" iron pipe size	G		160	.100		4.57	5.30		9.87	13.40
5940	1-1/4" iron pipe size	G		155	.103		4.94	5.45		10.39	14.10
5950	1-1/2" iron pipe size	G		155	.103		5.30	5.45		10.75	14.50
5960	2" iron pipe size	G		150	.107		5.85	5.65		11.50	15.40
5970	2-1/2" iron pipe size	G		150	.107		6.40	5.65		12.05	15.95
5980	3" iron pipe size	G		145	.110		6.65	5.80		12.45	16.60
5990	4" iron pipe size	G		135	.119		7.70	6.25		13.95	18.40
6000	5" iron pipe size	G		130	.123		8.70	6.50		15.20	19.85
6010	6" iron pipe size	G		120	.133		8.90	7.05		15.95	21
6020	7" iron pipe size	G		115	.139		10.45	7.35		17.80	23
6030	8" iron pipe size	G		105	.152		11.75	8.05		19.80	25.50
6040	9" iron pipe size	G		100	.160		14.10	8.45		22.55	29
6050	10" iron pipe size	G		95	.168		15.55	8.90		24.45	31.50
6060	12" iron pipe size	G		90	.178		18.50	9.40		27.90	35.50
6070	14" iron pipe size	G		85	.188		21	9.95		30.95	39
6080	16" iron pipe size	G		80	.200		23.50	10.55		34.05	42.50
6090	18" iron pipe size	G		75	.213		26	11.25		37.25	46.50
6120	2" wall, 1/2" iron pipe size	G		145	.110		6.55	5.80		12.35	16.45
6140	3/4" iron pipe size	G		145	.110		6.90	5.80		12.70	16.80
6150	1" iron pipe size	G		145	.110		7.25	5.80		13.05	17.25
6160	1-1/4" iron pipe size	G		140	.114		7.75	6.05		13.80	18.10
6170	1-1/2" iron pipe size	G		140	.114		8.10	6.05		14.15	18.45
6180	2" iron pipe size	G		135	.119		8.50	6.25		14.75	19.25
6190	2-1/2" iron pipe size	G		135	.119		10.25	6.25		16.50	21
6200	3" iron pipe size	G		130	.123		10.30	6.50		16.80	21.50
6210	4" iron pipe size	G		125	.128		11.90	6.75		18.65	24
6220	5" iron pipe size	G		120	.133		13.55	7.05		20.60	26
6230	6" iron pipe size	G		115	.139		14.80	7.35		22.15	28
6240	7" iron pipe size	G		110	.145		16.05	7.70		23.75	30
6250	8" iron pipe size	G		105	.152		18.05	8.05		26.10	32.50
6260	9" iron pipe size	G	↓	100	.160	↓	20	8.45		28.45	35.50

22 07 19.10 Piping Insulation			Crew	Daily Output	Labor-Hours	Unit	2020 Bare Costs Material	Labor	Equipment	Total	Total Incl O&P
6270	10" iron pipe size	G	Q-14	95	.168	L.F.	22	8.90		30.90	38.50
6280	12" iron pipe size	G		90	.178		24.50	9.40		33.90	42
6290	14" iron pipe size	G		85	.188		27	9.95		36.95	46
6300	16" iron pipe size	G		80	.200		30	10.55		40.55	50
6310	18" iron pipe size	G		75	.213		32.50	11.25		43.75	54
6320	20" iron pipe size	G		65	.246		40	13		53	64.50
6330	22" iron pipe size	G		60	.267		44.50	14.05		58.55	71.50
6340	24" iron pipe size	G		55	.291		45.50	15.35		60.85	74.50
6360	3" wall, 1/2" iron pipe size	G		115	.139		11.85	7.35		19.20	24.50
6380	3/4" iron pipe size	G		115	.139		11.90	7.35		19.25	25
6390	1" iron pipe size	G		115	.139		12	7.35		19.35	25
6400	1-1/4" iron pipe size	G		110	.145		12.20	7.70		19.90	25.50
6410	1-1/2" iron pipe size	G		110	.145		12.20	7.70		19.90	25.50
6420	2" iron pipe size	G		105	.152		12.65	8.05		20.70	26.50
6430	2-1/2" iron pipe size	G		105	.152		14.90	8.05		22.95	29
6440	3" iron pipe size	G		100	.160		15	8.45		23.45	30
6450	4" iron pipe size	G		95	.168		19.15	8.90		28.05	35
6460	5" iron pipe size	G		90	.178		21	9.40		30.40	38
6470	6" iron pipe size	G		90	.178		23.50	9.40		32.90	40.50
6480	7" iron pipe size	G		85	.188		26	9.95		35.95	44.50
6490	8" iron pipe size	G		85	.188		28	9.95		37.95	46.50
6500	9" iron pipe size	G		80	.200		31.50	10.55		42.05	51.50
6510	10" iron pipe size	G		75	.213		33.50	11.25		44.75	55
6520	12" iron pipe size	G		70	.229		37	12.05		49.05	60
6530	14" iron pipe size	G		65	.246		41.50	13		54.50	66.50
6540	16" iron pipe size	G		60	.267		46	14.05		60.05	73
6550	18" iron pipe size	G		55	.291		50.50	15.35		65.85	80
6560	20" iron pipe size	G		50	.320		60	16.90		76.90	93
6570	22" iron pipe size	G		45	.356		65	18.75		83.75	102
6580	24" iron pipe size	G	▼	40	.400	▼	70	21		91	111
6600	Fiberglass, with all service jacket										
6640	1/2" wall, 1/2" iron pipe size	G	Q-14	250	.064	L.F.	.75	3.38		4.13	6.20
6660	3/4" iron pipe size	G		240	.067		.86	3.52		4.38	6.55
6670	1" iron pipe size	G		230	.070		.89	3.67		4.56	6.80
6680	1-1/4" iron pipe size	G		220	.073		.95	3.84		4.79	7.15
6690	1-1/2" iron pipe size	G		220	.073		1.08	3.84		4.92	7.30
6700	2" iron pipe size	G		210	.076		1.16	4.02		5.18	7.70
6710	2-1/2" iron pipe size	G		200	.080		1.21	4.22		5.43	8.05
6840	1" wall, 1/2" iron pipe size	G		240	.067		.92	3.52		4.44	6.60
6860	3/4" iron pipe size	G		230	.070		.98	3.67		4.65	6.90
6870	1" iron pipe size	G		220	.073		1.08	3.84		4.92	7.30
6880	1-1/4" iron pipe size	G		210	.076		1.16	4.02		5.18	7.70
6890	1-1/2" iron pipe size	G		210	.076		1.25	4.02		5.27	7.80
6900	2" iron pipe size	G		200	.080		1.78	4.22		6	8.65
6910	2-1/2" iron pipe size	G		190	.084		1.79	4.44		6.23	9
6920	3" iron pipe size	G		180	.089		1.94	4.69		6.63	9.60
6930	3-1/2" iron pipe size	G		170	.094		2.32	4.97		7.29	10.45
6940	4" iron pipe size	G		150	.107		2.58	5.65		8.23	11.80
6950	5" iron pipe size	G		140	.114		2.84	6.05		8.89	12.65
6960	6" iron pipe size	G		120	.133		3.07	7.05		10.12	14.55
6970	7" iron pipe size	G		110	.145		3.65	7.70		11.35	16.15
6980	8" iron pipe size	G		100	.160		5	8.45		13.45	18.90
6990	9" iron pipe size	G	▼	90	.178	▼	5.30	9.40		14.70	20.50

22 07 19 – Plumbing Piping Insulation

22 07 19.10 Piping Insulation		Crew	Daily Output	Labor-Hours	Unit	Material	Labor	Equipment	Total	Total Incl O&P
						2020 Bare Costs				
7000	10" iron pipe size G	Q-14	90	.178	L.F.	5.35	9.40		14.75	21
7010	12" iron pipe size G		80	.200		5.85	10.55		16.40	23
7020	14" iron pipe size G		80	.200		7.05	10.55		17.60	24.50
7030	16" iron pipe size G		70	.229		9	12.05		21.05	29
7040	18" iron pipe size G		70	.229		10	12.05		22.05	30
7050	20" iron pipe size G		60	.267		11.15	14.05		25.20	35
7060	24" iron pipe size G		60	.267		13.60	14.05		27.65	37.50
7080	1-1/2" wall, 1/2" iron pipe size G		230	.070		2.06	3.67		5.73	8.05
7100	3/4" iron pipe size G		220	.073		2.06	3.84		5.90	8.35
7110	1" iron pipe size G		210	.076		2.21	4.02		6.23	8.85
7120	1-1/4" iron pipe size G		200	.080		2.40	4.22		6.62	9.35
7130	1-1/2" iron pipe size G		200	.080		2.81	4.22		7.03	9.80
7140	2" iron pipe size G		190	.084		2.79	4.44		7.23	10.10
7150	2-1/2" iron pipe size G		180	.089		2.98	4.69		7.67	10.75
7160	3" iron pipe size G		170	.094		3.12	4.97		8.09	11.35
7170	3-1/2" iron pipe size G		160	.100		3.42	5.30		8.72	12.10
7180	4" iron pipe size G		140	.114		3.95	6.05		10	13.90
7190	5" iron pipe size G		130	.123		3.97	6.50		10.47	14.65
7200	6" iron pipe size G		110	.145		4.20	7.70		11.90	16.75
7210	7" iron pipe size G		100	.160		4.66	8.45		13.11	18.55
7220	8" iron pipe size G		90	.178		5.90	9.40		15.30	21.50
7230	9" iron pipe size G		85	.188		6.10	9.95		16.05	22.50
7240	10" iron pipe size G		80	.200		6.35	10.55		16.90	23.50
7250	12" iron pipe size G		75	.213		7.20	11.25		18.45	26
7260	14" iron pipe size G		70	.229		8.65	12.05		20.70	28.50
7270	16" iron pipe size G		65	.246		11.30	13		24.30	33
7280	18" iron pipe size G		60	.267		12.65	14.05		26.70	36.50
7290	20" iron pipe size G		55	.291		12.95	15.35		28.30	38.50
7300	24" iron pipe size G		50	.320		15.85	16.90		32.75	44.50
7320	2" wall, 1/2" iron pipe size G		220	.073		3.20	3.84		7.04	9.60
7340	3/4" iron pipe size G		210	.076		3.30	4.02		7.32	10.05
7350	1" iron pipe size G		200	.080		3.52	4.22		7.74	10.55
7360	1-1/4" iron pipe size G		190	.084		3.71	4.44		8.15	11.15
7370	1-1/2" iron pipe size G		190	.084		4.47	4.44		8.91	11.95
7380	2" iron pipe size G		180	.089		4.08	4.69		8.77	11.95
7390	2-1/2" iron pipe size G		170	.094		4.39	4.97		9.36	12.75
7400	3" iron pipe size G		160	.100		4.67	5.30		9.97	13.50
7410	3-1/2" iron pipe size G		150	.107		5.05	5.65		10.70	14.50
7420	4" iron pipe size G		130	.123		6.25	6.50		12.75	17.20
7430	5" iron pipe size G		120	.133		6.20	7.05		13.25	18
7440	6" iron pipe size G		100	.160		6.90	8.45		15.35	21
7450	7" iron pipe size G		90	.178		7.30	9.40		16.70	23
7460	8" iron pipe size G		80	.200		7.80	10.55		18.35	25.50
7470	9" iron pipe size G		75	.213		8.55	11.25		19.80	27.50
7480	10" iron pipe size G		70	.229		9.30	12.05		21.35	29.50
7490	12" iron pipe size G		65	.246		10.45	13		23.45	32
7500	14" iron pipe size G		60	.267		13.60	14.05		27.65	37.50
7510	16" iron pipe size G		55	.291		14.90	15.35		30.25	41
7520	18" iron pipe size G		50	.320		16.60	16.90		33.50	45.50
7530	20" iron pipe size G		45	.356		18.90	18.75		37.65	51
7540	24" iron pipe size G		40	.400		20	21		41	55.50
7560	2-1/2" wall, 1/2" iron pipe size G	Q-14	210	.076		3.79	4.02		7.81	10.55
7562	3/4" iron pipe size G		200	.080		3.96	4.22		8.18	11.05

22 07 19 – Plumbing Piping Insulation

22 07 19.10 Piping Insulation		Crew	Daily Output	Labor-Hours	Unit	2020 Bare Costs Material	Labor	Equipment	Total	Total Incl O&P
7564	1" iron pipe size	G	Q-14	190	.084	L.F.	4.13	4.44		8.57	11.60
7566	1-1/4" iron pipe size	G		185	.086		4.28	4.56		8.84	11.95
7568	1-1/2" iron pipe size	G		180	.089		4.50	4.69		9.19	12.40
7570	2" iron pipe size	G		170	.094		4.72	4.97		9.69	13.10
7572	2-1/2" iron pipe size	G		160	.100		5.45	5.30		10.75	14.35
7574	3" iron pipe size	G		150	.107		5.70	5.65		11.35	15.25
7576	3-1/2" iron pipe size	G		140	.114		6.25	6.05		12.30	16.40
7578	4" iron pipe size	G		120	.133		6.55	7.05		13.60	18.40
7580	5" iron pipe size	G		110	.145		5.65	7.70		13.35	18.40
7582	6" iron pipe size	G		90	.178		9.45	9.40		18.85	25.50
7584	7" iron pipe size	G		80	.200		9.50	10.55		20.05	27
7586	8" iron pipe size	G		70	.229		10.05	12.05		22.10	30
7588	9" iron pipe size	G		65	.246		11	13		24	32.50
7590	10" iron pipe size	G		60	.267		12	14.05		26.05	35.50
7592	12" iron pipe size	G		55	.291		14.55	15.35		29.90	40.50
7594	14" iron pipe size	G		50	.320		17.15	16.90		34.05	46
7596	16" iron pipe size	G		45	.356		19.70	18.75		38.45	51.50
7598	18" iron pipe size	G		40	.400		21.50	21		42.50	57
7602	24" iron pipe size	G		30	.533		28	28		56	75.50
7620	3" wall, 1/2" iron pipe size	G		200	.080		4.85	4.22		9.07	12.05
7622	3/4" iron pipe size	G		190	.084		5.15	4.44		9.59	12.70
7624	1" iron pipe size	G		180	.089		5.45	4.69		10.14	13.45
7626	1-1/4" iron pipe size	G		175	.091		5.55	4.83		10.38	13.75
7628	1-1/2" iron pipe size	G		170	.094		5.80	4.97		10.77	14.30
7630	2" iron pipe size	G		160	.100		6.25	5.30		11.55	15.25
7632	2-1/2" iron pipe size	G		150	.107		6.50	5.65		12.15	16.10
7634	3" iron pipe size	G		140	.114		6.95	6.05		13	17.15
7636	3-1/2" iron pipe size	G		130	.123		7.65	6.50		14.15	18.75
7638	4" iron pipe size	G		110	.145		8.20	7.70		15.90	21
7640	5" iron pipe size	G		100	.160		9.30	8.45		17.75	23.50
7642	6" iron pipe size	G		80	.200		9.95	10.55		20.50	27.50
7644	7" iron pipe size	G		70	.229		11.40	12.05		23.45	31.50
7646	8" iron pipe size	G		60	.267		12.40	14.05		26.45	36
7648	9" iron pipe size	G		55	.291		13.35	15.35		28.70	39
7650	10" iron pipe size	G		50	.320		14.35	16.90		31.25	43
7652	12" iron pipe size	G		45	.356		17.90	18.75		36.65	49.50
7654	14" iron pipe size	G		40	.400		21	21		42	56.50
7656	16" iron pipe size	G		35	.457		23.50	24		47.50	64.50
7658	18" iron pipe size	G		32	.500		25	26.50		51.50	69.50
7660	20" iron pipe size	G		30	.533		27.50	28		55.50	74.50
7662	24" iron pipe size	G		28	.571		34.50	30		64.50	86
7664	26" iron pipe size	G		26	.615		38.50	32.50		71	93.50
7666	30" iron pipe size	G		24	.667		48	35		83	109
7800	For fiberglass with standard canvas jacket, deduct						5%				
7802	For fittings, add 3 L.F. for each fitting										
7804	plus 4 L.F. for each flange of the fitting										
7879	Rubber tubing, flexible closed cell foam										
7880	3/8" wall, 1/4" iron pipe size	G	1 Asbe	120	.067	L.F.	.33	3.91		4.24	6.55
7900	3/8" iron pipe size	G		120	.067		.46	3.91		4.37	6.70
7910	1/2" iron pipe size	G		115	.070		.50	4.08		4.58	7
7920	3/4" iron pipe size	G		115	.070		.57	4.08		4.65	7.10
7930	1" iron pipe size	G		110	.073		.60	4.27		4.87	7.40
7940	1-1/4" iron pipe size	G		110	.073		.68	4.27		4.95	7.50

22 07 19.10 Piping Insulation		Crew	Daily Output	Labor-Hours	Unit	Material	2020 Bare Costs Labor	Equipment	Total	Total Incl O&P
7950	1-1/2" iron pipe size	G	1 Asbe	110	.073	L.F.	.79	4.27		5.06	7.60
8100	1/2" wall, 1/4" iron pipe size	G		90	.089		.99	5.20		6.19	9.35
8120	3/8" iron pipe size	G		90	.089		1.05	5.20		6.25	9.40
8130	1/2" iron pipe size	G		89	.090		1.06	5.25		6.31	9.50
8140	3/4" iron pipe size	G		89	.090		1.22	5.25		6.47	9.70
8150	1" iron pipe size	G		88	.091		.86	5.35		6.21	9.40
8160	1-1/4" iron pipe size	G		87	.092		1.21	5.40		6.61	9.90
8170	1-1/2" iron pipe size	G		87	.092		1.89	5.40		7.29	10.65
8180	2" iron pipe size	G		86	.093		2.38	5.45		7.83	11.25
8190	2-1/2" iron pipe size	G		86	.093		2.60	5.45		8.05	11.50
8200	3" iron pipe size	G		85	.094		2.78	5.50		8.28	11.80
8210	3-1/2" iron pipe size	G		85	.094		3.81	5.50		9.31	12.95
8220	4" iron pipe size	G		80	.100		4.12	5.85		9.97	13.85
8230	5" iron pipe size	G		80	.100		5.60	5.85		11.45	15.45
8240	6" iron pipe size	G		75	.107		5.60	6.25		11.85	16.05
8300	3/4" wall, 1/4" iron pipe size	G		90	.089		.94	5.20		6.14	9.30
8320	3/8" iron pipe size	G		90	.089		1.10	5.20		6.30	9.45
8330	1/2" iron pipe size	G		89	.090		1.13	5.25		6.38	9.60
8340	3/4" iron pipe size	G		89	.090		2.13	5.25		7.38	10.70
8350	1" iron pipe size	G		88	.091		2.10	5.35		7.45	10.75
8360	1-1/4" iron pipe size	G		87	.092		2.46	5.40		7.86	11.25
8370	1-1/2" iron pipe size	G		87	.092		3.07	5.40		8.47	11.95
8380	2" iron pipe size	G		86	.093		4.27	5.45		9.72	13.35
8390	2-1/2" iron pipe size	G		86	.093		4.89	5.45		10.34	14.05
8400	3" iron pipe size	G		85	.094		5.50	5.50		11	14.80
8410	3-1/2" iron pipe size	G		85	.094		5.20	5.50		10.70	14.50
8420	4" iron pipe size	G		80	.100		6.80	5.85		12.65	16.80
8430	5" iron pipe size	G		80	.100		8.15	5.85		14	18.25
8440	6" iron pipe size	G		80	.100		9.90	5.85		15.75	20
8444	1" wall, 1/2" iron pipe size	G		86	.093		3.18	5.45		8.63	12.15
8445	3/4" iron pipe size	G		84	.095		3.90	5.60		9.50	13.15
8446	1" iron pipe size	G		84	.095		3.49	5.60		9.09	12.70
8447	1-1/4" iron pipe size	G		82	.098		3.83	5.70		9.53	13.25
8448	1-1/2" iron pipe size	G		82	.098		6.15	5.70		11.85	15.80
8449	2" iron pipe size	G		80	.100		7.55	5.85		13.40	17.60
8450	2-1/2" iron pipe size	G		80	.100		8.45	5.85		14.30	18.60
8456	Rubber insulation tape, 1/8" x 2" x 30'	G				Ea.	23			23	25
9600	Minimum labor/equipment charge		1 Plum	4	2	Job		129		129	199

22 07 19.30 Piping Insulation Protective Jacketing, PVC		Crew	Daily Output	Labor-Hours	Unit	Material	Labor	Equipment	Total	Total Incl O&P
0010	**PIPING INSULATION PROTECTIVE JACKETING, PVC**									
0100	PVC, white, 48" lengths cut from roll goods									
0120	20 mil thick									
0140	Size based on OD of insulation									
0150	1-1/2" ID	Q-14	270	.059	L.F.	.29	3.13		3.42	5.30
0152	2" ID		260	.062		.39	3.25		3.64	5.60
0154	2-1/2" ID		250	.064		.45	3.38		3.83	5.85
0156	3" ID		240	.067		.53	3.52		4.05	6.20
0158	3-1/2" ID		230	.070		.61	3.67		4.28	6.45
0160	4" ID		220	.073		.68	3.84		4.52	6.85
0162	4-1/2" ID		210	.076		.77	4.02		4.79	7.25
0164	5" ID		200	.080		.84	4.22		5.06	7.60
0166	5-1/2" ID		190	.084		.92	4.44		5.36	8.05

22 07 19.30 Piping Insulation Protective Jacketing, PVC		Crew	Daily Output	Labor-Hours	Unit	Material	2020 Bare Costs Labor	Equipment	Total	Total Incl O&P
0168	6" ID	Q-14	180	.089	L.F.	1	4.69		5.69	8.55
0170	6-1/2" ID		175	.091		1.08	4.83		5.91	8.85
0172	7" ID		170	.094		1.15	4.97		6.12	9.15
0174	7-1/2" ID		164	.098		1.25	5.15		6.40	9.55
0176	8" ID		161	.099		1.33	5.25		6.58	9.75
0178	8-1/2" ID		158	.101		1.40	5.35		6.75	10
0180	9" ID		155	.103		1.48	5.45		6.93	10.30
0182	9-1/2" ID		152	.105		1.56	5.55		7.11	10.50
0184	10" ID		149	.107		1.63	5.65		7.28	10.80
0186	10-1/2" ID		146	.110		1.72	5.80		7.52	11.05
0188	11" ID		143	.112		1.80	5.90		7.70	11.35
0190	11-1/2" ID		140	.114		1.87	6.05		7.92	11.60
0192	12" ID		137	.117		1.95	6.15		8.10	11.90
0194	12-1/2" ID		134	.119		2.03	6.30		8.33	12.25
0195	13" ID		132	.121		2.11	6.40		8.51	12.45
0196	13-1/2" ID		132	.121		2.19	6.40		8.59	12.55
0198	14" ID		130	.123		2.27	6.50		8.77	12.80
0200	15" ID		128	.125		2.42	6.60		9.02	13.10
0202	16" ID		126	.127		2.58	6.70		9.28	13.50
0204	17" ID		124	.129		2.74	6.80		9.54	13.80
0206	18" ID		122	.131		2.90	6.90		9.80	14.15
0208	19" ID		120	.133		3.05	7.05		10.10	14.50
0210	20" ID		118	.136		3.24	7.15		10.39	14.90
0212	21" ID		116	.138		3.37	7.30		10.67	15.25
0214	22" ID		114	.140		3.54	7.40		10.94	15.65
0216	23" ID		112	.143		3.70	7.55		11.25	16
0218	24" ID		110	.145		3.86	7.70		11.56	16.40
0220	25" ID		108	.148		4.02	7.80		11.82	16.80
0222	26" ID		106	.151		4.18	7.95		12.13	17.25
0224	27" ID		104	.154		4.34	8.10		12.44	17.60
0226	28" ID		102	.157		4.49	8.30		12.79	18.10
0228	29" ID		100	.160		4.66	8.45		13.11	18.55
0230	30" ID		98	.163		4.82	8.60		13.42	18.95
0300	For colors, add				Ea.	10%				
1000	30 mil thick									
1010	Size based on OD of insulation									
1020	2" ID	Q-14	260	.062	L.F.	.55	3.25		3.80	5.75
1022	2-1/2" ID		250	.064		.67	3.38		4.05	6.10
1024	3" ID		240	.067		.80	3.52		4.32	6.50
1026	3-1/2" ID		230	.070		.91	3.67		4.58	6.80
1028	4" ID		220	.073		1.02	3.84		4.86	7.20
1030	4-1/2" ID		210	.076		1.14	4.02		5.16	7.65
1032	5" ID		200	.080		1.26	4.22		5.48	8.10
1034	5-1/2" ID		190	.084		1.38	4.44		5.82	8.55
1036	6" ID		180	.089		1.49	4.69		6.18	9.10
1038	6-1/2" ID		175	.091		1.61	4.83		6.44	9.40
1040	7" ID		170	.094		1.73	4.97		6.70	9.80
1042	7-1/2" ID		164	.098		1.86	5.15		7.01	10.20
1044	8" ID		161	.099		1.97	5.25		7.22	10.45
1046	8-1/2" ID		158	.101		2.09	5.35		7.44	10.75
1048	9" ID		155	.103		2.21	5.45		7.66	11.10
1050	9-1/2" ID		152	.105		2.33	5.55		7.88	11.35
1052	10" ID		149	.107		2.44	5.65		8.09	11.70

22 07 19 – Plumbing Piping Insulation

22 07 19.30 Piping Insulation Protective Jacketing, PVC		Crew	Daily Output	Labor-Hours	Unit	Material	2020 Bare Costs Labor	Equipment	Total	Total Incl O&P
1054	10-1/2" ID	Q-14	146	.110	L.F.	2.56	5.80		8.36	11.95
1056	11" ID		143	.112		2.69	5.90		8.59	12.30
1058	11-1/2" ID		140	.114		2.80	6.05		8.85	12.65
1060	12" ID		137	.117		2.91	6.15		9.06	12.95
1062	12-1/2" ID		134	.119		3.04	6.30		9.34	13.35
1063	13" ID		132	.121		3.16	6.40		9.56	13.65
1064	13-1/2" ID		132	.121		3.28	6.40		9.68	13.75
1066	14" ID		130	.123		3.39	6.50		9.89	14.05
1068	15" ID		128	.125		3.63	6.60		10.23	14.45
1070	16" ID		126	.127		3.87	6.70		10.57	14.90
1072	17" ID		124	.129		4.11	6.80		10.91	15.30
1074	18" ID		122	.131		4.34	6.90		11.24	15.70
1076	19" ID		120	.133		4.58	7.05		11.63	16.20
1078	20" ID		118	.136		4.80	7.15		11.95	16.65
1080	21" ID		116	.138		5.05	7.30		12.35	17.10
1082	22" ID		114	.140		5.25	7.40		12.65	17.55
1084	23" ID		112	.143		5.55	7.55		13.10	18.05
1086	24" ID		110	.145		5.75	7.70		13.45	18.50
1088	25" ID		108	.148		6	7.80		13.80	19
1090	26" ID		106	.151		6.25	7.95		14.20	19.50
1092	27" ID		104	.154		6.50	8.10		14.60	20
1094	28" ID		102	.157		6.70	8.30		15	20.50
1096	29" ID		100	.160		6.95	8.45		15.40	21
1098	30" ID	↓	98	.163	↓	7.20	8.60		15.80	21.50
1300	For colors, add				Ea.	10%				
2000	PVC, white, fitting covers									
2020	Fiberglass insulation inserts included with sizes 1-3/4" thru 9-3/4"									
2030	Size is based on OD of insulation									
2040	90° elbow fitting									
2060	1-3/4"	Q-14	135	.119	Ea.	.55	6.25		6.80	10.50
2062	2"		130	.123		.68	6.50		7.18	11.05
2064	2-1/4"		128	.125		.79	6.60		7.39	11.30
2068	2-1/2"		126	.127		.84	6.70		7.54	11.55
2070	2-3/4"		123	.130		.99	6.85		7.84	12
2072	3"		120	.133		.96	7.05		8.01	12.20
2074	3-3/8"		116	.138		1.13	7.30		8.43	12.80
2076	3-3/4"		113	.142		1.21	7.45		8.66	13.20
2078	4-1/8"		110	.145		1.60	7.70		9.30	13.90
2080	4-3/4"		105	.152		1.93	8.05		9.98	14.85
2082	5-1/4"		100	.160		2.25	8.45		10.70	15.90
2084	5-3/4"		95	.168		2.84	8.90		11.74	17.20
2086	6-1/4"		90	.178		4.73	9.40		14.13	20
2088	6-3/4"		87	.184		4.99	9.70		14.69	21
2090	7-1/4"		85	.188		6.25	9.95		16.20	22.50
2092	7-3/4"		83	.193		6.60	10.15		16.75	23.50
2094	8-3/4"		80	.200		8.45	10.55		19	26
2096	9-3/4"		77	.208		11.30	10.95		22.25	30
2098	10-7/8"		74	.216		12.60	11.40		24	32
2100	11-7/8"		71	.225		14.45	11.90		26.35	34.50
2102	12-7/8"		68	.235		19.95	12.40		32.35	41.50
2104	14-1/8"		66	.242		21	12.80		33.80	43.50
2106	15-1/8"		64	.250		22.50	13.20		35.70	45.50
2108	16-1/8"	↓	63	.254	↓	24.50	13.40		37.90	48.50

22 07 19.30 Piping Insulation Protective Jacketing, PVC		Crew	Daily Output	Labor-Hours	Unit	2020 Bare Costs Material	Labor	Equipment	Total	Total Incl O&P
2110	17-1/8"	Q-14	62	.258	Ea.	27	13.60		40.60	51.50
2112	18-1/8"		61	.262		36.50	13.85		50.35	62
2114	19-1/8"		60	.267		47.50	14.05		61.55	74.50
2116	20-1/8"		59	.271		61	14.30		75.30	89.50
2200	45° elbow fitting									
2220	1-3/4" thru 9-3/4" same price as 90° elbow fitting									
2320	10-7/8"	Q-14	74	.216	Ea.	12.60	11.40		24	32
2322	11-7/8"		71	.225		13.85	11.90		25.75	34
2324	12-7/8"		68	.235		15.50	12.40		27.90	37
2326	14-1/8"		66	.242		17.65	12.80		30.45	40
2328	15-1/8"		64	.250		19	13.20		32.20	42
2330	16-1/8"		63	.254		21.50	13.40		34.90	45.50
2332	17-1/8"		62	.258		24.50	13.60		38.10	48.50
2334	18-1/8"		61	.262		30	13.85		43.85	55
2336	19-1/8"		60	.267		40.50	14.05		54.55	67.50
2338	20-1/8"		59	.271		46	14.30		60.30	73
2400	Tee fitting									
2410	1-3/4"	Q-14	96	.167	Ea.	1.04	8.80		9.84	15.10
2412	2"		94	.170		1.18	9		10.18	15.55
2414	2-1/4"		91	.176		1.27	9.30		10.57	16.10
2416	2-1/2"		88	.182		1.39	9.60		10.99	16.75
2418	2-3/4"		85	.188		1.53	9.95		11.48	17.45
2420	3"		82	.195		1.66	10.30		11.96	18.20
2422	3-3/8"		79	.203		1.91	10.70		12.61	19.05
2424	3-3/4"		76	.211		2.37	11.10		13.47	20
2426	4-1/8"		73	.219		2.57	11.55		14.12	21
2428	4-3/4"		70	.229		3.19	12.05		15.24	22.50
2430	5-1/4"		67	.239		3.84	12.60		16.44	24
2432	5-3/4"		63	.254		5.10	13.40		18.50	27
2434	6-1/4"		60	.267		6.70	14.05		20.75	30
2436	6-3/4"		59	.271		8.30	14.30		22.60	31.50
2438	7-1/4"		57	.281		13.40	14.80		28.20	38.50
2440	7-3/4"		54	.296		14.70	15.65		30.35	41
2442	8-3/4"		52	.308		17.85	16.25		34.10	45.50
2444	9-3/4"		50	.320		21	16.90		37.90	50
2446	10-7/8"		48	.333		21.50	17.60		39.10	51.50
2448	11-7/8"		47	.340		23.50	17.95		41.45	54.50
2450	12-7/8"		46	.348		26	18.35		44.35	57.50
2452	14-1/8"		45	.356		28.50	18.75		47.25	61.50
2454	15-1/8"		44	.364		31	19.20		50.20	64.50
2456	16-1/8"		43	.372		33	19.65		52.65	67.50
2458	17-1/8"		42	.381		35.50	20		55.50	71
2460	18-1/8"		41	.390		39	20.50		59.50	75.50
2462	19-1/8"		40	.400		42.50	21		63.50	80.50
2464	20-1/8"		39	.410		47	21.50		68.50	86.50
4000	Mechanical grooved fitting cover, including insert									
4020	90° elbow fitting									
4030	3/4" & 1"	Q-14	140	.114	Ea.	5.05	6.05		11.10	15.10
4040	1-1/4" & 1-1/2"		135	.119		6.20	6.25		12.45	16.70
4042	2"		130	.123		8.60	6.50		15.10	19.80
4044	2-1/2"		125	.128		9.60	6.75		16.35	21.50
4046	3"		120	.133		10.75	7.05		17.80	23
4048	3-1/2"		115	.139		12.45	7.35		19.80	25.50

22 07 19.30 Piping Insulation Protective Jacketing, PVC

22 07 19.30 Piping Insulation Protective Jacketing, PVC		Crew	Daily Output	Labor-Hours	Unit	Material	2020 Bare Costs Labor	Equipment	Total	Total Incl O&P
4050	4"	Q-14	110	.145	Ea.	13.85	7.70		21.55	27.50
4052	5"		100	.160		17.25	8.45		25.70	32.50
4054	6"		90	.178		25.50	9.40		34.90	43
4056	8"		80	.200		27.50	10.55		38.05	47.50
4058	10"		75	.213		35.50	11.25		46.75	57
4060	12"		68	.235		51	12.40		63.40	75.50
4062	14"		65	.246		61	13		74	88
4064	16"		63	.254		83	13.40		96.40	113
4066	18"	↓	61	.262	↓	114	13.85		127.85	148
4100	45° elbow fitting									
4120	3/4" & 1"	Q-14	140	.114	Ea.	4.53	6.05		10.58	14.55
4130	1-1/4" & 1-1/2"		135	.119		5.60	6.25		11.85	16.05
4140	2"		130	.123		8.40	6.50		14.90	19.55
4142	2-1/2"		125	.128		9.35	6.75		16.10	21
4144	3"		120	.133		9.60	7.05		16.65	21.50
4146	3-1/2"		115	.139		12.15	7.35		19.50	25
4148	4"		110	.145		15.45	7.70		23.15	29
4150	5"		100	.160		16.75	8.45		25.20	32
4152	6"		90	.178		23.50	9.40		32.90	41
4154	8"		80	.200		24.50	10.55		35.05	44
4156	10"		75	.213		31.50	11.25		42.75	53
4158	12"		68	.235		47.50	12.40		59.90	71.50
4160	14"		65	.246		64	13		77	91
4162	16"		63	.254		78.50	13.40		91.90	108
4164	18"	↓	61	.262	↓	101	13.85		114.85	133
4200	Tee fitting									
4220	3/4" & 1"	Q-14	93	.172	Ea.	6.60	9.10		15.70	21.50
4230	1-1/4" & 1-1/2"		90	.178		8.10	9.40		17.50	24
4240	2"		87	.184		13.55	9.70		23.25	30.50
4242	2-1/2"		84	.190		15.10	10.05		25.15	32.50
4244	3"		80	.200		16.20	10.55		26.75	34.50
4246	3-1/2"		77	.208		17.35	10.95		28.30	36.50
4248	4"		73	.219		21.50	11.55		33.05	42.50
4250	5"		67	.239		23.50	12.60		36.10	46
4252	6"		60	.267		37	14.05		51.05	63
4254	8"		54	.296		38.50	15.65		54.15	67.50
4256	10"		50	.320		46.50	16.90		63.40	78
4258	12"		46	.348		72	18.35		90.35	109
4260	14"		43	.372		101	19.65		120.65	142
4262	16"		42	.381		105	20		125	148
4264	18"	↓	41	.390	↓	127	20.50		147.50	172

22 07 19.40 Pipe Insulation Protective Jacketing, Aluminum

22 07 19.40 Pipe Insulation Protective Jacketing, Aluminum		Crew	Daily Output	Labor-Hours	Unit	Material	Labor	Equipment	Total	Total Incl O&P
0010	**PIPE INSULATION PROTECTIVE JACKETING, ALUMINUM**									
0100	Metal roll jacketing									
0120	Aluminum with polykraft moisture barrier									
0140	Smooth, based on OD of insulation, .016" thick									
0180	1/2" ID	Q-14	220	.073	L.F.	.29	3.84		4.13	6.40
0190	3/4" ID		215	.074		.37	3.93		4.30	6.65
0200	1" ID		210	.076		.45	4.02		4.47	6.90
0210	1-1/4" ID		205	.078		.54	4.12		4.66	7.15
0220	1-1/2" ID		202	.079		.62	4.18		4.80	7.35
0230	1-3/4" ID	↓	199	.080	↓	.79	4.24		5.03	7.60

22 07 19.40 Pipe Insulation Protective Jacketing, Aluminum		Crew	Daily Output	Labor-Hours	Unit	Material	2020 Bare Costs Labor	Equipment	Total	Total Incl O&P
0240	2" ID	Q-14	195	.082	L.F.	.88	4.33		5.21	7.80
0250	2-1/4" ID		191	.084		.98	4.42		5.40	8.10
0260	2-1/2" ID		187	.086		.97	4.52		5.49	8.20
0270	2-3/4" ID		184	.087		1.06	4.59		5.65	8.45
0280	3" ID		180	.089		1.27	4.69		5.96	8.85
0290	3-1/4" ID		176	.091		1.37	4.80		6.17	9.10
0300	3-1/2" ID		172	.093		1.33	4.91		6.24	9.25
0310	3-3/4" ID		169	.095		1.41	5		6.41	9.45
0320	4" ID		165	.097		1.50	5.10		6.60	9.75
0330	4-1/4" ID		161	.099		1.74	5.25		6.99	10.20
0340	4-1/2" ID		157	.102		1.68	5.40		7.08	10.40
0350	4-3/4" ID		154	.104		1.93	5.50		7.43	10.80
0360	5" ID		150	.107		2.03	5.65		7.68	11.20
0370	5-1/4" ID		146	.110		2.12	5.80		7.92	11.50
0380	5-1/2" ID		143	.112		2.02	5.90		7.92	11.55
0390	5-3/4" ID		139	.115		2.33	6.10		8.43	12.20
0400	6" ID		135	.119		2.42	6.25		8.67	12.55
0410	6-1/4" ID		133	.120		2.51	6.35		8.86	12.80
0420	6-1/2" ID		131	.122		2.37	6.45		8.82	12.80
0430	7" ID		128	.125		2.52	6.60		9.12	13.20
0440	7-1/4" ID		125	.128		2.61	6.75		9.36	13.55
0450	7-1/2" ID		123	.130		2.69	6.85		9.54	13.85
0460	8" ID		121	.132		2.86	7		9.86	14.20
0470	8-1/2" ID		119	.134		3.03	7.10		10.13	14.60
0480	9" ID		116	.138		3.20	7.30		10.50	15.05
0490	9-1/2" ID		114	.140		3.38	7.40		10.78	15.45
0500	10" ID		112	.143		3.55	7.55		11.10	15.85
0510	10-1/2" ID		110	.145		4.09	7.70		11.79	16.65
0520	11" ID		107	.150		3.88	7.90		11.78	16.75
0530	11-1/2" ID		105	.152		4.06	8.05		12.11	17.20
0540	12" ID		103	.155		4.24	8.20		12.44	17.65
0550	12-1/2" ID		100	.160		4.41	8.45		12.86	18.25
0560	13" ID		99	.162		4.58	8.55		13.13	18.55
0570	14" ID		98	.163		4.93	8.60		13.53	19.05
0580	15" ID		96	.167		5.25	8.80		14.05	19.75
0590	16" ID		95	.168		5.60	8.90		14.50	20.50
0600	17" ID		93	.172		5.95	9.10		15.05	21
0610	18" ID		92	.174		6.20	9.20		15.40	21.50
0620	19" ID		90	.178		6.65	9.40		16.05	22
0630	20" ID		89	.180		7	9.50		16.50	23
0640	21" ID		87	.184		7.35	9.70		17.05	23.50
0650	22" ID		86	.186		7.65	9.80		17.45	24
0660	23" ID		84	.190		8	10.05		18.05	25
0670	24" ID	▼	83	.193		8.35	10.15		18.50	25.50
0710	For smooth .020" thick, add					27%	10%			
0720	For smooth .024" thick, add					52%	20%			
0730	For smooth .032" thick, add					104%	33%			
0800	For stucco embossed, add					1%				
0820	For corrugated, add				▼	2.50%				
0900	White aluminum with polysurlyn moisture barrier									
0910	Smooth, % is an add to polykraft lines of same thickness									
0940	For smooth .016" thick, add				L.F.	35%				
0960	For smooth .024" thick, add				"	22%				

22 07 19.40 Pipe Insulation Protective Jacketing, Aluminum		Crew	Daily Output	Labor-Hours	Unit	Material	2020 Bare Costs Labor	Equipment	Total	Total Incl O&P
1000	Aluminum fitting covers									
1010	Size is based on OD of insulation									
1020	90° LR elbow, 2 piece									
1100	1-1/2"	Q-14	140	.114	Ea.	4.37	6.05		10.42	14.35
1110	1-3/4"		135	.119		4.37	6.25		10.62	14.70
1120	2"		130	.123		5.50	6.50		12	16.35
1130	2-1/4"		128	.125		5.50	6.60		12.10	16.50
1140	2-1/2"		126	.127		5.50	6.70		12.20	16.70
1150	2-3/4"		123	.130		5.50	6.85		12.35	16.95
1160	3"		120	.133		5.95	7.05		13	17.70
1170	3-1/4"		117	.137		5.95	7.20		13.15	18
1180	3-1/2"		115	.139		7.05	7.35		14.40	19.40
1190	3-3/4"		113	.142		7.25	7.45		14.70	19.80
1200	4"		110	.145		7.55	7.70		15.25	20.50
1210	4-1/4"		108	.148		7.80	7.80		15.60	21
1220	4-1/2"		106	.151		7.80	7.95		15.75	21
1230	4-3/4"		104	.154		7.80	8.10		15.90	21.50
1240	5"		102	.157		8.75	8.30		17.05	23
1250	5-1/4"		100	.160		10.10	8.45		18.55	24.50
1260	5-1/2"		97	.165		12.60	8.70		21.30	27.50
1270	5-3/4"		95	.168		12.60	8.90		21.50	28
1280	6"		92	.174		10.75	9.20		19.95	26.50
1290	6-1/4"		90	.178		10.75	9.40		20.15	26.50
1300	6-1/2"		87	.184		12.95	9.70		22.65	29.50
1310	7"		85	.188		15.70	9.95		25.65	33
1320	7-1/4"		84	.190		15.70	10.05		25.75	33.50
1330	7-1/2"		83	.193		23	10.15		33.15	41.50
1340	8"		82	.195		17.20	10.30		27.50	35.50
1350	8-1/2"		80	.200		32	10.55		42.55	52
1360	9"		78	.205		32	10.85		42.85	52
1370	9-1/2"		77	.208		23.50	10.95		34.45	43.50
1380	10"		76	.211		23.50	11.10		34.60	43.50
1390	10-1/2"		75	.213		24	11.25		35.25	44.50
1400	11"		74	.216		24	11.40		35.40	44.50
1410	11-1/2"		72	.222		27	11.75		38.75	48.50
1420	12"		71	.225		27	11.90		38.90	49
1430	12-1/2"		69	.232		48.50	12.25		60.75	72.50
1440	13"		68	.235		48.50	12.40		60.90	72.50
1450	14"		66	.242		65	12.80		77.80	92
1460	15"		64	.250		68	13.20		81.20	95.50
1470	16"		63	.254		74	13.40		87.40	103
2000	45° elbow, 2 piece									
2010	2-1/2"	Q-14	126	.127	Ea.	4.54	6.70		11.24	15.65
2020	2-3/4"		123	.130		4.54	6.85		11.39	15.90
2030	3"		120	.133		5.15	7.05		12.20	16.80
2040	3-1/4"		117	.137		5.15	7.20		12.35	17.10
2050	3-1/2"		115	.139		5.85	7.35		13.20	18.10
2060	3-3/4"		113	.142		5.85	7.45		13.30	18.30
2070	4"		110	.145		5.95	7.70		13.65	18.70
2080	4-1/4"		108	.148		6.80	7.80		14.60	19.85
2090	4-1/2"		106	.151		6.80	7.95		14.75	20
2100	4-3/4"		104	.154		6.80	8.10		14.90	20.50
2110	5"		102	.157		7.95	8.30		16.25	22

22 07 19.40 Pipe Insulation Protective Jacketing, Aluminum		Crew	Daily Output	Labor-Hours	Unit	2020 Bare Costs Material	Labor	Equipment	Total	Total Incl O&P
2120	5-1/4"	Q-14	100	.160	Ea.	7.95	8.45		16.40	22
2130	5-1/2"		97	.165		8.35	8.70		17.05	23
2140	6"		92	.174		8.35	9.20		17.55	24
2150	6-1/2"		87	.184		11.55	9.70		21.25	28
2160	7"		85	.188		11.55	9.95		21.50	28.50
2170	7-1/2"		83	.193		11.65	10.15		21.80	29
2180	8"		82	.195		11.65	10.30		21.95	29
2190	8-1/2"		80	.200		14.50	10.55		25.05	32.50
2200	9"		78	.205		14.50	10.85		25.35	33
2210	9-1/2"		77	.208		20	10.95		30.95	39.50
2220	10"		76	.211		20	11.10		31.10	39.50
2230	10-1/2"		75	.213		18.80	11.25		30.05	38.50
2240	11"		74	.216		18.80	11.40		30.20	38.50
2250	11-1/2"		72	.222		22.50	11.75		34.25	43.50
2260	12"		71	.225		22.50	11.90		34.40	44
2270	13"		68	.235		26.50	12.40		38.90	48.50
2280	14"		66	.242		32.50	12.80		45.30	56.50
2290	15"		64	.250		50	13.20		63.20	76
2300	16"		63	.254		54.50	13.40		67.90	81.50
2310	17"		62	.258		53	13.60		66.60	80
2320	18"		61	.262		60.50	13.85		74.35	88.50
2330	19"		60	.267		74	14.05		88.05	104
2340	20"		59	.271		71.50	14.30		85.80	101
2350	21"		58	.276		77	14.55		91.55	108
3000	Tee, 4 piece									
3010	2-1/2"	Q-14	88	.182	Ea.	29.50	9.60		39.10	47.50
3020	2-3/4"		86	.186		29.50	9.80		39.30	48
3030	3"		84	.190		33.50	10.05		43.55	53
3040	3-1/4"		82	.195		33.50	10.30		43.80	53.50
3050	3-1/2"		80	.200		35	10.55		45.55	55.50
3060	4"		78	.205		35.50	10.85		46.35	56
3070	4-1/4"		76	.211		37	11.10		48.10	58.50
3080	4-1/2"		74	.216		37	11.40		48.40	59
3090	4-3/4"		72	.222		37	11.75		48.75	59.50
3100	5"		70	.229		38.50	12.05		50.55	61.50
3110	5-1/4"		68	.235		38.50	12.40		50.90	62
3120	5-1/2"		66	.242		40.50	12.80		53.30	65
3130	6"		64	.250		40.50	13.20		53.70	65.50
3140	6-1/2"		60	.267		44.50	14.05		58.55	71
3150	7"		58	.276		44.50	14.55		59.05	71.50
3160	7-1/2"		56	.286		49	15.10		64.10	78
3170	8"		54	.296		49	15.65		64.65	79
3180	8-1/2"		52	.308		50.50	16.25		66.75	81.50
3190	9"		50	.320		50.50	16.90		67.40	82.50
3200	9-1/2"		49	.327		37.50	17.25		54.75	68.50
3210	10"		48	.333		37.50	17.60		55.10	69
3220	10-1/2"		47	.340		40	17.95		57.95	72.50
3230	11"		46	.348		40	18.35		58.35	73
3240	11-1/2"		45	.356		42	18.75		60.75	76.50
3250	12"		44	.364		42	19.20		61.20	77
3260	13"		43	.372		44	19.65		63.65	79.50
3270	14"		42	.381		46.50	20		66.50	83
3280	15"		41	.390		50	20.50		70.50	87.50

22 07 19.40 Pipe Insulation Protective Jacketing, Aluminum		Crew	Daily Output	Labor-Hours	Unit	Material	Labor	Equipment	Total	Total Incl O&P
						2020 Bare Costs				
3290	16"	Q-14	40	.400	Ea.	51.50	21		72.50	90
3300	17"		39	.410		59	21.50		80.50	99.50
3310	18"		38	.421		61.50	22		83.50	103
3320	19"		37	.432		68	23		91	111
3330	20"		36	.444		69.50	23.50		93	114
3340	22"		35	.457		88.50	24		112.50	136
3350	23"		34	.471		90.50	25		115.50	140
3360	24"		31	.516		93	27		120	145

22 07 19.50 Pipe Insulation Protective Jacketing, St. Stl.

Line	Description	Crew	Daily Output	Labor-Hours	Unit	Material	Labor	Equipment	Total	Total Incl O&P
0010	**PIPE INSULATION PROTECTIVE JACKETING, STAINLESS STEEL**									
0100	Metal roll jacketing									
0120	Type 304 with moisture barrier									
0140	Smooth, based on OD of insulation, .010" thick									
0260	2-1/2" ID	Q-14	250	.064	L.F.	3.10	3.38		6.48	8.75
0270	2-3/4" ID		245	.065		3.37	3.45		6.82	9.15
0280	3" ID		240	.067		3.64	3.52		7.16	9.60
0290	3-1/4" ID		235	.068		3.95	3.59		7.54	10.05
0300	3-1/2" ID		230	.070		4.19	3.67		7.86	10.40
0310	3-3/4" ID		225	.071		4.50	3.75		8.25	10.90
0320	4" ID		220	.073		4.75	3.84		8.59	11.35
0330	4-1/4" ID		215	.074		5	3.93		8.93	11.75
0340	4-1/2" ID		210	.076		5.30	4.02		9.32	12.25
0350	5" ID		200	.080		5.85	4.22		10.07	13.15
0360	5-1/2" ID		190	.084		6.40	4.44		10.84	14.10
0370	6" ID		180	.089		6.95	4.69		11.64	15.10
0380	6-1/2" ID		175	.091		7.50	4.83		12.33	15.90
0390	7" ID		170	.094		8.05	4.97		13.02	16.75
0400	7-1/2" ID		164	.098		8.60	5.15		13.75	17.60
0410	8" ID		161	.099		9.15	5.25		14.40	18.35
0420	8-1/2" ID		158	.101		9.70	5.35		15.05	19.10
0430	9" ID		155	.103		10.25	5.45		15.70	19.95
0440	9-1/2" ID		152	.105		10.80	5.55		16.35	20.50
0450	10" ID		149	.107		11.35	5.65		17	21.50
0460	10-1/2" ID		146	.110		11.90	5.80		17.70	22.50
0470	11" ID		143	.112		12.45	5.90		18.35	23
0480	12" ID		137	.117		13.55	6.15		19.70	24.50
0490	13" ID		132	.121		14.65	6.40		21.05	26.50
0500	14" ID		130	.123		15.75	6.50		22.25	27.50
0700	For smooth .016" thick, add					45%	33%			
1000	Stainless steel, Type 316, fitting covers									
1010	Size is based on OD of insulation									
1020	90° LR elbow, 2 piece									
1100	1-1/2"	Q-14	126	.127	Ea.	12.90	6.70		19.60	25
1110	2-3/4"		123	.130		13.20	6.85		20.05	25.50
1120	3"		120	.133		13.80	7.05		20.85	26.50
1130	3-1/4"		117	.137		13.80	7.20		21	26.50
1140	3-1/2"		115	.139		14.50	7.35		21.85	27.50
1150	3-3/4"		113	.142		15.35	7.45		22.80	28.50
1160	4"		110	.145		16.85	7.70		24.55	30.50
1170	4-1/4"		108	.148		22.50	7.80		30.30	37
1180	4-1/2"		106	.151		22.50	7.95		30.45	37
1190	5"		102	.157		23	8.30		31.30	38

22 07 19.50 Pipe Insulation Protective Jacketing, St. Stl.		Crew	Daily Output	Labor-Hours	Unit	2020 Bare Costs Material	Labor	Equipment	Total	Total Incl O&P
1200	5-1/2"	Q-14	97	.165	Ea.	34	8.70		42.70	51.50
1210	6"		92	.174		37.50	9.20		46.70	55.50
1220	6-1/2"		87	.184		51.50	9.70		61.20	72.50
1230	7"		85	.188		51.50	9.95		61.45	73
1240	7-1/2"		83	.193		60	10.15		70.15	82.50
1250	8"		80	.200		60	10.55		70.55	83.50
1260	8-1/2"		80	.200		62.50	10.55		73.05	85.50
1270	9"		78	.205		95	10.85		105.85	121
1280	9-1/2"		77	.208		93	10.95		103.95	119
1290	10"		76	.211		93	11.10		104.10	120
1300	10-1/2"		75	.213		111	11.25		122.25	141
1310	11"		74	.216		106	11.40		117.40	135
1320	12"		71	.225		121	11.90		132.90	152
1330	13"		68	.235		167	12.40		179.40	204
1340	14"		66	.242		168	12.80		180.80	206
2000	45° elbow, 2 piece									
2010	2-1/2"	Q-14	126	.127	Ea.	11	6.70		17.70	23
2020	2-3/4"		123	.130		11	6.85		17.85	23
2030	3"		120	.133		11.85	7.05		18.90	24
2040	3-1/4"		117	.137		11.85	7.20		19.05	24.50
2050	3-1/2"		115	.139		12	7.35		19.35	25
2060	3-3/4"		113	.142		12	7.45		19.45	25
2070	4"		110	.145		15.40	7.70		23.10	29
2080	4-1/4"		108	.148		21.50	7.80		29.30	36.50
2090	4-1/2"		106	.151		21.50	7.95		29.45	36.50
2100	4-3/4"		104	.154		21.50	8.10		29.60	37
2110	5"		102	.157		22	8.30		30.30	37
2120	5-1/2"		97	.165		22	8.70		30.70	38
2130	6"		92	.174		26	9.20		35.20	43
2140	6-1/2"		87	.184		26	9.70		35.70	44
2150	7"		85	.188		44	9.95		53.95	64.50
2160	7-1/2"		83	.193		45	10.15		55.15	65.50
2170	8"		82	.195		45	10.30		55.30	66
2180	8-1/2"		80	.200		52.50	10.55		63.05	75
2190	9"		78	.205		52.50	10.85		63.35	75
2200	9-1/2"		77	.208		62	10.95		72.95	85.50
2210	10"		76	.211		62	11.10		73.10	85.50
2220	10-1/2"		75	.213		74.50	11.25		85.75	100
2230	11"		74	.216		74.50	11.40		85.90	100
2240	12"		71	.225		81	11.90		92.90	108
2250	13"		68	.235		95	12.40		107.40	125

22 11 13 – Facility Water Distribution Piping

22 11 13.23 Pipe/Tube, Copper		Crew	Daily Output	Labor-Hours	Unit	Material	Labor	Equipment	Total	Total Incl O&P
						2020 Bare Costs				
0010	**PIPE/TUBE, COPPER**, Solder joints R221113-50									
1000	Type K tubing, couplings & clevis hanger assemblies 10' OC									
1100	1/4" diameter	1 Plum	84	.095	L.F.	4.25	6.15		10.40	14.15
1120	3/8" diameter		82	.098		4.76	6.30		11.06	14.95
1140	1/2" diameter		78	.103		5.30	6.60		11.90	16
1160	5/8" diameter		77	.104		6.75	6.70		13.45	17.75
1180	3/4" diameter		74	.108		8.65	6.95		15.60	20.50
1200	1" diameter		66	.121		12.80	7.80		20.60	26
1220	1-1/4" diameter		56	.143		15.30	9.20		24.50	31
1240	1-1/2" diameter		50	.160		18.60	10.30		28.90	36.50
1260	2" diameter		40	.200		27	12.90		39.90	50
1280	2-1/2" diameter	Q-1	60	.267		44	15.45		59.45	72.50
1300	3" diameter		54	.296		58	17.20		75.20	90
1320	3-1/2" diameter		42	.381		80	22		102	122
1330	4" diameter		38	.421		97.50	24.50		122	145
1340	5" diameter		32	.500		140	29		169	199
1360	6" diameter	Q-2	38	.632		207	38		245	286
1380	8" diameter	"	34	.706		360	42.50		402.50	465
1390	For other than full hard temper, add					13%				
1440	For silver solder, add						15%			
1800	For medical clean (oxygen class), add					12%				
1950	To delete cplgs. & hngrs., 1/4"-1" pipe, subtract					27%	60%			
1960	1-1/4" -3" pipe, subtract					14%	52%			
1970	3-1/2"-5" pipe, subtract					10%	60%			
1980	6"-8" pipe, subtract					19%	53%			
2000	Type L tubing, couplings & clevis hanger assemblies 10' OC									
2100	1/4" diameter	1 Plum	88	.091	L.F.	2.67	5.85		8.52	12
2120	3/8" diameter		84	.095		3.34	6.15		9.49	13.15
2140	1/2" diameter		81	.099		3.68	6.35		10.03	13.90
2160	5/8" diameter		79	.101		5.75	6.55		12.30	16.40
2180	3/4" diameter		76	.105		4.71	6.80		11.51	15.70
2200	1" diameter		68	.118		7.60	7.60		15.20	20
2220	1-1/4" diameter		58	.138		12	8.90		20.90	27
2240	1-1/2" diameter		52	.154		11.60	9.90		21.50	28
2260	2" diameter		42	.190		18.85	12.30		31.15	40
2280	2-1/2" diameter	Q-1	62	.258		30	14.95		44.95	56
2300	3" diameter		56	.286		46.50	16.55		63.05	76.50
2320	3-1/2" diameter		43	.372		60	21.50		81.50	99.50
2340	4" diameter		39	.410		65.50	24		89.50	109
2360	5" diameter		34	.471		148	27.50		175.50	204
2380	6" diameter	Q-2	40	.600		151	36		187	222
2400	8" diameter	"	36	.667		250	40		290	335
2410	For other than full hard temper, add					21%				
2590	For silver solder, add						15%			
2900	For medical clean (oxygen class), add					12%				
2940	To delete cplgs. & hngrs., 1/4"-1" pipe, subtract					37%	63%			
2960	1-1/4"-3" pipe, subtract					12%	53%			
2970	3-1/2"-5" pipe, subtract					12%	63%			
2980	6"-8" pipe, subtract					24%	55%			
3000	Type M tubing, couplings & clevis hanger assemblies 10' OC									
3100	1/4" diameter	1 Plum	90	.089	L.F.	3.72	5.75		9.47	12.95
3120	3/8" diameter		87	.092		4.09	5.95		10.04	13.65
3140	1/2" diameter		84	.095		3.56	6.15		9.71	13.40

22 11 13 – Facility Water Distribution Piping

22 11 13.23 Pipe/Tube, Copper		Crew	Daily Output	Labor-Hours	Unit	Material	Labor	Equipment	Total	Total Incl O&P
						2020 Bare Costs				
3160	5/8″ diameter	1 Plum	81	.099	L.F.	4.94	6.35		11.29	15.30
3180	3/4″ diameter		78	.103		5.05	6.60		11.65	15.75
3200	1″ diameter		70	.114		8.50	7.35		15.85	21
3220	1-1/4″ diameter		60	.133		12.30	8.60		20.90	27
3240	1-1/2″ diameter		54	.148		14.45	9.55		24	30.50
3260	2″ diameter		44	.182		21.50	11.70		33.20	41.50
3280	2-1/2″ diameter	Q-1	64	.250		31.50	14.50		46	57.50
3300	3″ diameter		58	.276		39.50	16		55.50	68
3320	3-1/2″ diameter		45	.356		58.50	20.50		79	96
3340	4″ diameter		40	.400		75	23		98	119
3360	5″ diameter		36	.444		138	26		164	192
3370	6″ diameter	Q-2	42	.571		197	34.50		231.50	270
3380	8″ diameter	″	38	.632		325	38		363	420
3440	For silver solder, add						15%			
3960	To delete cplgs. & hngrs., 1/4″-1″ pipe, subtract					35%	65%			
3970	1-1/4″-3″ pipe, subtract					19%	56%			
3980	3-1/2″-5″ pipe, subtract					13%	65%			
3990	6″-8″ pipe, subtract					28%	58%			
4000	Type DWV tubing, couplings & clevis hanger assemblies 10′ OC									
4100	1-1/4″ diameter	1 Plum	60	.133	L.F.	12.45	8.60		21.05	27
4120	1-1/2″ diameter		54	.148		12.60	9.55		22.15	28.50
4140	2″ diameter		44	.182		18.30	11.70		30	38
4160	3″ diameter	Q-1	58	.276		29	16		45	56.50
4180	4″ diameter		40	.400		60.50	23		83.50	103
4200	5″ diameter		36	.444		110	26		136	161
4220	6″ diameter	Q-2	42	.571		161	34.50		195.50	230
4240	8″ diameter	″	38	.632		510	38		548	625
4730	To delete cplgs. & hngrs., 1-1/4″-2″ pipe, subtract					16%	53%			
4740	3″-4″ pipe, subtract					13%	60%			
4750	5″-8″ pipe, subtract					23%	58%			
5200	ACR tubing, type L, hard temper, cleaned and									
5220	capped, no couplings or hangers									
5240	3/8″ OD				L.F.	1.76			1.76	1.94
5250	1/2″ OD					2.64			2.64	2.90
5260	5/8″ OD					3.30			3.30	3.63
5270	3/4″ OD					4.48			4.48	4.93
5280	7/8″ OD					5.05			5.05	5.55
5290	1-1/8″ OD					7.20			7.20	7.90
5300	1-3/8″ OD					9.65			9.65	10.60
5310	1-5/8″ OD					12.60			12.60	13.85
5320	2-1/8″ OD					20.50			20.50	22.50
5330	2-5/8″ OD					27.50			27.50	30.50
5340	3-1/8″ OD					34			34	37.50
5350	3-5/8″ OD					63.50			63.50	70
5360	4-1/8″ OD					65			65	72
5380	ACR tubing, type L, hard, cleaned and capped									
5381	No couplings or hangers									
5384	3/8″	1 Stpi	160	.050	L.F.	1.76	3.28		5.04	7
5385	1/2″		160	.050		2.64	3.28		5.92	7.95
5386	5/8″		160	.050		3.30	3.28		6.58	8.70
5387	3/4″		130	.062		4.48	4.03		8.51	11.20
5388	7/8″		130	.062		5.05	4.03		9.08	11.80
5389	1-1/8″		115	.070		7.20	4.56		11.76	14.95

22 11 13 – Facility Water Distribution Piping

22 11 13.23 Pipe/Tube, Copper		Crew	Daily Output	Labor-Hours	Unit	Material	2020 Bare Costs Labor	Equipment	Total	Total Incl O&P
5390	1-3/8"	1 Stpi	100	.080	L.F.	9.65	5.25		14.90	18.70
5391	1-5/8"		90	.089		12.60	5.85		18.45	23
5392	2-1/8"		80	.100		20.50	6.55		27.05	32.50
5393	2-5/8"	Q-5	125	.128		27.50	7.55		35.05	42
5394	3-1/8"		105	.152		34	9		43	51.50
5395	4-1/8"		95	.168		65	9.95		74.95	87.50
5800	Refrigeration tubing, dryseal, 50' coils									
5840	1/8" OD				Coil	34			34	37.50
5850	3/16" OD					46			46	50.50
5860	1/4" OD					46.50			46.50	51
5870	5/16" OD					71			71	78
5880	3/8" OD					64.50			64.50	71
5890	1/2" OD					93			93	102
5900	5/8" OD					124			124	136
5910	3/4" OD					145			145	159
5920	7/8" OD					147			147	162
5930	1-1/8" OD					330			330	365
5940	1-3/8" OD					555			555	615
5950	1-5/8" OD					710			710	780
9000	Minimum labor/equipment charge	1 Plum	4	2	Job		129		129	199
9400	Sub assemblies used in assembly systems									
9410	Chilled water unit, coil connections per unit under 10 ton	Q-5	.80	20	System	1,200	1,175		2,375	3,150
9420	Chilled water unit, coil connections per unit 10 ton and up		1	16		2,175	945		3,120	3,850
9430	Chilled water dist. piping per ton, less than 61 ton systems		26	.615		23	36.50		59.50	81
9440	Chilled water dist. piping per ton, 61 through 120 ton systems	Q-6	31	.774		53	47.50		100.50	132
9450	Chilled water dist. piping/ton, 135 ton systems and up	Q-8	25.40	1.260		71	78.50	4.20	153.70	205
9510	Refrigerant piping/ton of cooling for remote condensers	Q-5	2	8		375	470		845	1,150
9520	Refrigerant piping per ton up to 10 ton w/remote condensing unit		2.40	6.667		201	395		596	830
9530	Refrigerant piping per ton, 20 ton w/remote condensing unit		2	8		245	470		715	1,000
9540	Refrigerant piping per ton, 40 ton w/remote condensing unit		1.90	8.421		330	495		825	1,125
9550	Refrigerant piping per ton, 75-80 ton w/remote condensing unit	Q-6	2.40	10		515	610		1,125	1,500
9560	Refrigerant piping per ton, 100 ton w/remote condensing unit	"	2.20	10.909		710	665		1,375	1,800

22 11 13.25 Pipe/Tube Fittings, Copper		Crew	Daily Output	Labor-Hours	Unit	Material	Labor	Equipment	Total	Total Incl O&P
0010	**PIPE/TUBE FITTINGS, COPPER**, Wrought unless otherwise noted									
0020	For silver solder, add						15%			
0040	Solder joints, copper x copper									
0070	90° elbow, 1/4"	1 Plum	22	.364	Ea.	3.88	23.50		27.38	40.50
0090	3/8"		22	.364		4.15	23.50		27.65	40.50
0100	1/2"		20	.400		1.26	26		27.26	41.50
0110	5/8"		19	.421		2.77	27		29.77	45
0120	3/4"		19	.421		2.68	27		29.68	45
0130	1"		16	.500		7.05	32		39.05	58
0140	1-1/4"		15	.533		11.65	34.50		46.15	66
0150	1-1/2"		13	.615		16.10	39.50		55.60	79
0160	2"		11	.727		29.50	47		76.50	105
0170	2-1/2"	Q-1	13	1.231		51	71.50		122.50	166
0180	3"		11	1.455		86.50	84.50		171	225
0190	3-1/2"		10	1.600		281	93		374	455
0200	4"		9	1.778		236	103		339	420
0210	5"		6	2.667		775	155		930	1,100
0220	6"	Q-2	9	2.667		1,025	160		1,185	1,375
0230	8"	"	8	3		3,800	180		3,980	4,475

22 11 13.25 Pipe/Tube Fittings, Copper		Crew	Daily Output	Labor-Hours	Unit	Material	Labor (2020 Bare Costs)	Equipment	Total	Total Incl O&P
0250	45° elbow, 1/4"	1 Plum	22	.364	Ea.	7.75	23.50		31.25	44.50
0270	3/8"		22	.364		6.80	23.50		30.30	43.50
0280	1/2"		20	.400		2.75	26		28.75	43
0290	5/8"		19	.421		11.45	27		38.45	54.50
0300	3/4"		19	.421		4.46	27		31.46	47
0310	1"		16	.500		11.20	32		43.20	62.50
0320	1-1/4"		15	.533		16	34.50		50.50	70.50
0330	1-1/2"		13	.615		18.20	39.50		57.70	81.50
0340	2"		11	.727		30.50	47		77.50	106
0350	2-1/2"	Q-1	13	1.231		49.50	71.50		121	165
0360	3"		13	1.231		88	71.50		159.50	207
0370	3-1/2"		10	1.600		122	93		215	277
0380	4"		9	1.778		173	103		276	350
0390	5"		6	2.667		635	155		790	940
0400	6"	Q-2	9	2.667		995	160		1,155	1,350
0410	8"	"	8	3		4,300	180		4,480	5,000
0450	Tee, 1/4"	1 Plum	14	.571		8.45	37		45.45	66.50
0470	3/8"		14	.571		6.85	37		43.85	64.50
0480	1/2"		13	.615		2.44	39.50		41.94	64
0490	5/8"		12	.667		16.25	43		59.25	84.50
0500	3/4"		12	.667		6.05	43		49.05	73
0510	1"		10	.800		17.05	51.50		68.55	98.50
0520	1-1/4"		9	.889		24.50	57.50		82	116
0530	1-1/2"		8	1		36.50	64.50		101	140
0540	2"		7	1.143		58	73.50		131.50	178
0550	2-1/2"	Q-1	8	2		111	116		227	300
0560	3"		7	2.286		158	133		291	380
0570	3-1/2"		6	2.667		475	155		630	760
0580	4"		5	3.200		365	186		551	685
0590	5"		4	4		1,075	232		1,307	1,550
0600	6"	Q-2	6	4		1,475	241		1,716	2,000
0610	8"	"	5	4.800		5,875	289		6,164	6,925
0612	Tee, reducing on the outlet, 1/4"	1 Plum	15	.533		17.05	34.50		51.55	72
0613	3/8"		15	.533		15.55	34.50		50.05	70
0614	1/2"		14	.571		15.60	37		52.60	74
0615	5/8"		13	.615		30.50	39.50		70	95
0616	3/4"		12	.667		8	43		51	75.50
0617	1"		11	.727		29	47		76	105
0618	1-1/4"		10	.800		32.50	51.50		84	115
0619	1-1/2"		9	.889		32	57.50		89.50	124
0620	2"		8	1		64.50	64.50		129	171
0621	2-1/2"	Q-1	9	1.778		137	103		240	310
0622	3"		8	2		163	116		279	360
0623	4"		6	2.667		299	155		454	570
0624	5"		5	3.200		1,725	186		1,911	2,175
0625	6"	Q-2	7	3.429		2,150	206		2,356	2,700
0626	8"	"	6	4		10,800	241		11,041	12,300
0630	Tee, reducing on the run, 1/4"	1 Plum	15	.533		23.50	34.50		58	79
0631	3/8"		15	.533		31	34.50		65.50	87
0632	1/2"		14	.571		20.50	37		57.50	79.50
0633	5/8"		13	.615		29	39.50		68.50	93.50
0634	3/4"		12	.667		21	43		64	89.50
0635	1"		11	.727		26	47		73	101

22 11 13.25 Pipe/Tube Fittings, Copper		Crew	Daily Output	Labor-Hours	Unit	2020 Bare Costs Material	Labor	Equipment	Total	Total Incl O&P
0636	1-1/4"	1 Plum	10	.800	Ea.	41	51.50		92.50	125
0637	1-1/2"		9	.889		72	57.50		129.50	168
0638	2"		8	1		82.50	64.50		147	191
0639	2-1/2"	Q-1	9	1.778		177	103		280	355
0640	3"		8	2		320	116		436	535
0641	4"		6	2.667		585	155		740	885
0642	5"		5	3.200		1,950	186		2,136	2,425
0643	6"	Q-2	7	3.429		2,950	206		3,156	3,575
0644	8"	"	6	4		10,200	241		10,441	11,600
0650	Coupling, 1/4"	1 Plum	24	.333		1.07	21.50		22.57	34
0670	3/8"		24	.333		1.53	21.50		23.03	34.50
0680	1/2"		22	.364		.96	23.50		24.46	37
0690	5/8"		21	.381		4.23	24.50		28.73	42.50
0700	3/4"		21	.381		2.71	24.50		27.21	41
0710	1"		18	.444		5.30	28.50		33.80	50.50
0715	1-1/4"		17	.471		7.80	30.50		38.30	55.50
0716	1-1/2"		15	.533		10.80	34.50		45.30	65
0718	2"		13	.615		15.30	39.50		54.80	78.50
0721	2-1/2"	Q-1	15	1.067		40.50	62		102.50	140
0722	3"		13	1.231		47	71.50		118.50	162
0724	3-1/2"		8	2		124	116		240	315
0726	4"		7	2.286		154	133		287	375
0728	5"		6	2.667		228	155		383	490
0731	6"	Q-2	8	3		355	180		535	670
0732	8"	"	7	3.429		1,275	206		1,481	1,725
0741	Coupling, reducing, concentric									
0743	1/2"	1 Plum	23	.348	Ea.	2.85	22.50		25.35	37.50
0745	3/4"		21.50	.372		5.65	24		29.65	43
0747	1"		19.50	.410		8.80	26.50		35.30	50.50
0748	1-1/4"		18	.444		12.15	28.50		40.65	58
0749	1-1/2"		16	.500		17.05	32		49.05	69
0751	2"		14	.571		25.50	37		62.50	85
0752	2-1/2"		13	.615		56	39.50		95.50	123
0753	3"	Q-1	14	1.143		44.50	66.50		111	151
0755	4"	"	8	2		133	116		249	325
0757	5"	Q-2	7.50	3.200		755	192		947	1,125
0759	6"		7	3.429		1,200	206		1,406	1,650
0761	8"		6.50	3.692		3,825	222		4,047	4,550
0771	Cap, sweat									
0773	1/2"	1 Plum	40	.200	Ea.	1.27	12.90		14.17	21.50
0775	3/4"		38	.211		2.36	13.55		15.91	23.50
0777	1"		32	.250		5.50	16.10		21.60	31
0778	1-1/4"		29	.276		6.95	17.80		24.75	35
0779	1-1/2"		26	.308		12.80	19.85		32.65	44.50
0781	2"		22	.364		19.70	23.50		43.20	57.50
0791	Flange, sweat									
0793	3"	Q-1	22	.727	Ea.	275	42		317	370
0795	4"		18	.889		415	51.50		466.50	535
0797	5"		12	1.333		720	77.50		797.50	910
0799	6"	Q-2	18	1.333		750	80		830	950
0801	8"	"	16	1.500		1,275	90		1,365	1,550
0850	Unions, 1/4"	1 Plum	21	.381		41	24.50		65.50	83.50
0870	3/8"		21	.381		41.50	24.50		66	83.50

22 11 13.25 Pipe/Tube Fittings, Copper		Crew	Daily Output	Labor-Hours	Unit	Material	Labor	Equipment	Total	Total Incl O&P
						2020 Bare Costs				
0880	1/2"	1 Plum	19	.421	Ea.	22.50	27		49.50	66.50
0890	5/8"		18	.444		89	28.50		117.50	143
0900	3/4"		18	.444		23	28.50		51.50	69.50
0910	1"		15	.533		47.50	34.50		82	106
0920	1-1/4"		14	.571		78.50	37		115.50	144
0930	1-1/2"		12	.667		118	43		161	197
0940	2"		10	.800		152	51.50		203.50	248
0950	2-1/2"	Q-1	12	1.333		385	77.50		462.50	545
0960	3"	"	10	1.600		1,000	93		1,093	1,250
0980	Adapter, copper x male IPS, 1/4"	1 Plum	20	.400		13.55	26		39.55	55
0990	3/8"		20	.400		8.50	26		34.50	49.50
1000	1/2"		18	.444		3.64	28.50		32.14	48.50
1010	3/4"		17	.471		6.10	30.50		36.60	53.50
1020	1"		15	.533		15.55	34.50		50.05	70
1030	1-1/4"		13	.615		26.50	39.50		66	90.50
1040	1-1/2"		12	.667		26.50	43		69.50	95.50
1050	2"		11	.727		44.50	47		91.50	122
1060	2-1/2"	Q-1	10.50	1.524		182	88.50		270.50	335
1070	3"		10	1.600		226	93		319	390
1080	3-1/2"		9	1.778		219	103		322	400
1090	4"		8	2		247	116		363	450
1200	5", cast		6	2.667		2,200	155		2,355	2,675
1210	6", cast	Q-2	8.50	2.824		2,475	170		2,645	3,000
1250	Cross, 1/2"	1 Plum	10	.800		28.50	51.50		80	111
1260	3/4"		9.50	.842		55	54.50		109.50	145
1270	1"		8	1		93.50	64.50		158	203
1280	1-1/4"		7.50	1.067		135	69		204	254
1290	1-1/2"		6.50	1.231		192	79.50		271.50	335
1300	2"		5.50	1.455		365	94		459	545
1310	2-1/2"	Q-1	6.50	2.462		835	143		978	1,150
1320	3"	"	5.50	2.909		710	169		879	1,050
1500	Tee fitting, mechanically formed (Type 1, 'branch sizes up to 2 in.')									
1520	1/2" run size, 3/8" to 1/2" branch size	1 Plum	80	.100	Ea.		6.45		6.45	9.95
1530	3/4" run size, 3/8" to 3/4" branch size		60	.133			8.60		8.60	13.30
1540	1" run size, 3/8" to 1" branch size		54	.148			9.55		9.55	14.75
1550	1-1/4" run size, 3/8" to 1-1/4" branch size		48	.167			10.75		10.75	16.60
1560	1-1/2" run size, 3/8" to 1-1/2" branch size		40	.200			12.90		12.90	19.95
1570	2" run size, 3/8" to 2" branch size		35	.229			14.75		14.75	23
1580	2-1/2" run size, 1/2" to 2" branch size		32	.250			16.10		16.10	25
1590	3" run size, 1" to 2" branch size		26	.308			19.85		19.85	30.50
1600	4" run size, 1" to 2" branch size		24	.333			21.50		21.50	33
1640	Tee fitting, mechanically formed (Type 2, branches 2-1/2" thru 4")									
1650	2-1/2" run size, 2-1/2" branch size	1 Plum	12.50	.640	Ea.		41.50		41.50	64
1660	3" run size, 2-1/2" to 3" branch size		12	.667			43		43	66.50
1670	3-1/2" run size, 2-1/2" to 3-1/2" branch size		11	.727			47		47	72.50
1680	4" run size, 2-1/2" to 4" branch size		10.50	.762			49		49	76
1698	5" run size, 2" to 4" branch size		9.50	.842			54.50		54.50	84
1700	6" run size, 2" to 4" branch size		8.50	.941			60.50		60.50	94
1710	8" run size, 2" to 4" branch size		7	1.143			73.50		73.50	114
1800	ACR fittings, OD size									
1802	Tee, straight									
1808	5/8"	1 Stpi	12	.667	Ea.	2.96	43.50		46.46	71
1810	3/4"		12	.667		22	43.50		65.50	91.50

22 11 13.25 Pipe/Tube Fittings, Copper		Crew	Daily Output	Labor-Hours	Unit	Material	Labor	Equipment	Total	Total Incl O&P
						2020 Bare Costs				
1812	7/8"	1 Stpi	10	.800	Ea.	7.15	52.50		59.65	89
1813	1"		10	.800		60	52.50		112.50	147
1814	1-1/8"		10	.800		25.50	52.50		78	109
1816	1-3/8"		9	.889		29.50	58.50		88	123
1818	1-5/8"		8	1		53	65.50		118.50	159
1820	2-1/8"		7	1.143		71	75		146	195
1822	2-5/8"	Q-5	8	2		153	118		271	350
1824	3-1/8"		7	2.286		220	135		355	450
1826	4-1/8"		5	3.200		415	189		604	750
1830	90° elbow									
1836	5/8"	1 Stpi	19	.421	Ea.	6.10	27.50		33.60	49
1838	3/4"		19	.421		11.05	27.50		38.55	54.50
1840	7/8"		16	.500		11	33		44	62.50
1842	1-1/8"		16	.500		14.70	33		47.70	66.50
1844	1-3/8"		15	.533		14.20	35		49.20	69.50
1846	1-5/8"		13	.615		22	40.50		62.50	87
1848	2-1/8"		11	.727		40.50	47.50		88	118
1850	2-5/8"	Q-5	13	1.231		86	72.50		158.50	207
1852	3-1/8"		11	1.455		104	86		190	248
1854	4-1/8"		9	1.778		228	105		333	415
1860	Coupling									
1866	5/8"	1 Stpi	21	.381	Ea.	1.27	25		26.27	40
1868	3/4"		21	.381		3.46	25		28.46	42.50
1870	7/8"		18	.444		2.49	29		31.49	47.50
1871	1"		18	.444		8.55	29		37.55	54.50
1872	1-1/8"		18	.444		5.25	29		34.25	51
1874	1-3/8"		17	.471		9.20	31		40.20	57.50
1876	1-5/8"		15	.533		12.15	35		47.15	67.50
1878	2-1/8"		13	.615		20.50	40.50		61	85
1880	2-5/8"	Q-5	15	1.067		39.50	63		102.50	141
1882	3-1/8"		13	1.231		53.50	72.50		126	171
1884	4-1/8"		7	2.286		118	135		253	340
2000	DWV, solder joints, copper x copper									
2030	90° elbow, 1-1/4"	1 Plum	13	.615	Ea.	18.50	39.50		58	82
2050	1-1/2"		12	.667		24.50	43		67.50	93.50
2070	2"		10	.800		46	51.50		97.50	130
2090	3"	Q-1	10	1.600		85.50	93		178.50	237
2100	4"	"	9	1.778		545	103		648	760
2150	45° elbow, 1-1/4"	1 Plum	13	.615		15.15	39.50		54.65	78
2170	1-1/2"		12	.667		14.05	43		57.05	82
2180	2"		10	.800		29	51.50		80.50	112
2190	3"	Q-1	10	1.600		59	93		152	208
2200	4"	"	9	1.778		94	103		197	263
2250	Tee, sanitary, 1-1/4"	1 Plum	9	.889		28.50	57.50		86	120
2270	1-1/2"		8	1		35.50	64.50		100	139
2290	2"		7	1.143		55	73.50		128.50	175
2310	3"	Q-1	7	2.286		213	133		346	440
2330	4"	"	6	2.667		520	155		675	815
2400	Coupling, 1-1/4"	1 Plum	14	.571		7.75	37		44.75	65.50
2420	1-1/2"		13	.615		9.60	39.50		49.10	72
2440	2"		11	.727		13.35	47		60.35	87
2460	3"	Q-1	11	1.455		30.50	84.50		115	164
2480	4"	"	10	1.600		68	93		161	218

22 11 13 – Facility Water Distribution Piping

22 11 13.25 Pipe/Tube Fittings, Copper		Crew	Daily Output	Labor-Hours	Unit	Material	Labor (2020 Bare Costs)	Equipment	Total	Total Incl O&P
2602	Traps, see Section 22 13 16.60									
6992	Tube connector fittings, See Section 22 11 13.76 for plastic ftng.									
7000	Insert type brass/copper, 100 psi @ 180°F, CTS									
7010	Adapter MPT 3/8" x 1/2" CTS	1 Plum	29	.276	Ea.	2.90	17.80		20.70	30.50
7020	1/2" x 1/2"		26	.308		3	19.85		22.85	34
7030	3/4" x 1/2"		26	.308		3.82	19.85		23.67	34.50
7040	3/4" x 3/4"		25	.320		4.52	20.50		25.02	37
7050	Adapter CTS 1/2" x 1/2" sweat		24	.333		3.82	21.50		25.32	37
7060	3/4" x 3/4" sweat		22	.364		1.43	23.50		24.93	37.50
7070	Coupler center set 3/8" CTS		25	.320		1.43	20.50		21.93	33.50
7080	1/2" CTS		23	.348		3.67	22.50		26.17	38.50
7090	3/4" CTS		22	.364		1.47	23.50		24.97	37.50
7100	Elbow 90°, copper 3/8"		25	.320		2.98	20.50		23.48	35.50
7110	1/2" CTS		23	.348		2	22.50		24.50	36.50
7120	3/4" CTS		22	.364		2.63	23.50		26.13	39
7130	Tee copper 3/8" CTS		17	.471		3.79	30.50		34.29	51
7140	1/2" CTS		15	.533		2.67	34.50		37.17	56
7150	3/4" CTS		14	.571		4.16	37		41.16	61.50
7160	3/8" x 3/8" x 1/2"		16	.500		4.21	32		36.21	54.50
7170	1/2" x 3/8" x 1/2"		15	.533		2.84	34.50		37.34	56
7180	3/4" x 1/2" x 3/4"		14	.571		3.91	37		40.91	61.50
9000	Minimum labor/equipment charge		4	2	Job		129		129	199

22 11 13.29 Pipe, Fittings and Valves, Copper, Pressed-Joint		Crew	Daily Output	Labor-Hours	Unit	Material	Labor	Equipment	Total	Total Incl O&P
0010	**PIPE, FITTINGS AND VALVES, COPPER, PRESSED-JOINT**									
0040	Pipe/tube includes coupling & clevis type hanger assy's, 10' OC									
0120	Type K									
0130	1/2" diameter	1 Plum	78	.103	L.F.	5.50	6.60		12.10	16.25
0134	3/4" diameter		74	.108		8.85	6.95		15.80	20.50
0138	1" diameter		66	.121		13.25	7.80		21.05	26.50
0142	1-1/4" diameter		56	.143		15.75	9.20		24.95	31.50
0146	1-1/2" diameter		50	.160		19.75	10.30		30.05	38
0150	2" diameter		40	.200		28.50	12.90		41.40	51.50
0154	2-1/2" diameter	Q-1	60	.267		49	15.45		64.45	77.50
0158	3" diameter		54	.296		64.50	17.20		81.70	97.50
0162	4" diameter		38	.421		98	24.50		122.50	146
0180	To delete cplgs. & hngrs., 1/2" pipe, subtract					19%	48%			
0184	3/4" -2" pipe, subtract					14%	46%			
0186	2-1/2"-4" pipe, subtract					24%	34%			
0220	Type L									
0230	1/2" diameter	1 Plum	81	.099	L.F.	3.90	6.35		10.25	14.15
0234	3/4" diameter		76	.105		4.93	6.80		11.73	15.90
0238	1" diameter		68	.118		8.05	7.60		15.65	20.50
0242	1-1/4" diameter		58	.138		12.45	8.90		21.35	27.50
0246	1-1/2" diameter		52	.154		12.80	9.90		22.70	29.50
0250	2" diameter		42	.190		20	12.30		32.30	41
0254	2-1/2" diameter	Q-1	62	.258		35	14.95		49.95	61.50
0258	3" diameter		56	.286		53	16.55		69.55	84
0262	4" diameter		39	.410		66	24		90	110
0280	To delete cplgs. & hngrs., 1/2" pipe, subtract					21%	52%			
0284	3/4"-2" pipe, subtract					17%	46%			
0286	2-1/2"-4" pipe, subtract					23%	35%			
0320	Type M									

	22 11 13.29 Pipe, Fittings and Valves, Copper, Pressed-Joint	Crew	Daily Output	Labor-Hours	Unit	Material	2020 Bare Costs Labor	Equipment	Total	Total Incl O&P
0330	1/2" diameter	1 Plum	84	.095	L.F.	3.78	6.15		9.93	13.65
0334	3/4" diameter		78	.103		5.25	6.60		11.85	16
0338	1" diameter		70	.114		8.95	7.35		16.30	21.50
0342	1-1/4" diameter		60	.133		12.75	8.60		21.35	27.50
0346	1-1/2" diameter		54	.148		15.65	9.55		25.20	32
0350	2" diameter		44	.182		22.50	11.70		34.20	43
0354	2-1/2" diameter	Q-1	64	.250		36.50	14.50		51	62.50
0358	3" diameter		58	.276		46.50	16		62.50	75.50
0362	4" diameter		40	.400		75.50	23		98.50	119
0380	To delete cplgs. & hngrs., 1/2" pipe, subtract					32%	49%			
0384	3/4"-2" pipe, subtract					21%	46%			
0386	2-1/2"-4" pipe, subtract					25%	36%			
1600	Fittings									
1610	Press joints, copper x copper									
1620	Note: Reducing fittings show most expensive size combination.									
1800	90° elbow, 1/2"	1 Plum	36.60	.219	Ea.	3.58	14.10		17.68	26
1810	3/4"		27.50	.291		5.80	18.75		24.55	35.50
1820	1"		25.90	.309		11.60	19.90		31.50	44
1830	1-1/4"		20.90	.383		23	24.50		47.50	63.50
1840	1-1/2"		18.30	.437		44	28		72	92
1850	2"		15.70	.510		62	33		95	119
1860	2-1/2"	Q-1	25.90	.618		179	36		215	252
1870	3"		22	.727		225	42		267	310
1880	4"		16.30	.982		278	57		335	395
2000	45° elbow, 1/2"	1 Plum	36.60	.219		4.20	14.10		18.30	26.50
2010	3/4"		27.50	.291		4.95	18.75		23.70	34.50
2020	1"		25.90	.309		15.80	19.90		35.70	48.50
2030	1-1/4"		20.90	.383		23	24.50		47.50	63
2040	1-1/2"		18.30	.437		36.50	28		64.50	84
2050	2"		15.70	.510		51	33		84	107
2060	2-1/2"	Q-1	25.90	.618		120	36		156	188
2070	3"		22	.727		169	42		211	250
2080	4"		16.30	.982		238	57		295	350
2200	Tee, 1/2"	1 Plum	27.50	.291		5.45	18.75		24.20	35
2210	3/4"		20.70	.386		9.40	25		34.40	49
2220	1"		19.40	.412		17.15	26.50		43.65	60
2230	1-1/4"		15.70	.510		29.50	33		62.50	83.50
2240	1-1/2"		13.80	.580		56.50	37.50		94	120
2250	2"		11.80	.678		69.50	43.50		113	144
2260	2-1/2"	Q-1	19.40	.825		224	48		272	320
2270	3"		16.50	.970		276	56		332	390
2280	4"		12.20	1.311		395	76		471	550
2400	Tee, reducing on the outlet									
2410	3/4"	1 Plum	20.70	.386	Ea.	8.15	25		33.15	47.50
2420	1"		19.40	.412		19.95	26.50		46.45	63
2430	1-1/4"		15.70	.510		29	33		62	83
2440	1-1/2"		13.80	.580		61	37.50		98.50	125
2450	2"		11.80	.678		96	43.50		139.50	174
2460	2-1/2"	Q-1	19.40	.825		300	48		348	405
2470	3"		16.50	.970		360	56		416	480
2480	4"		12.20	1.311		450	76		526	615
2600	Tee, reducing on the run									
2610	3/4"	1 Plum	20.70	.386	Ea.	16	25		41	56

22 11 13.29 Pipe, Fittings and Valves, Copper, Pressed-Joint		Crew	Daily Output	Labor-Hours	Unit	Material	Labor	Equipment	Total	Total Incl O&P
						2020 Bare Costs				
2620	1"	1 Plum	19.40	.412	Ea.	30.50	26.50		57	74.50
2630	1-1/4"		15.70	.510		61	33		94	118
2640	1-1/2"		13.80	.580		93.50	37.50		131	161
2650	2"		11.80	.678		102	43.50		145.50	181
2660	2-1/2"	Q-1	19.40	.825		335	48		383	445
2670	3"		16.50	.970		400	56		456	525
2680	4"		12.20	1.311		525	76		601	695
2800	Coupling, 1/2"	1 Plum	36.60	.219		3.16	14.10		17.26	25.50
2810	3/4"		27.50	.291		4.88	18.75		23.63	34.50
2820	1"		25.90	.309		9.85	19.90		29.75	42
2830	1-1/4"		20.90	.383		12.30	24.50		36.80	51.50
2840	1-1/2"		18.30	.437		22.50	28		50.50	68.50
2850	2"		15.70	.510		29	33		62	82.50
2860	2-1/2"	Q-1	25.90	.618		89	36		125	154
2870	3"		22	.727		113	42		155	190
2880	4"		16.30	.982		160	57		217	264
3000	Union, 1/2"	1 Plum	36.60	.219		28	14.10		42.10	52.50
3010	3/4"		27.50	.291		35	18.75		53.75	67.50
3020	1"		25.90	.309		56.50	19.90		76.40	93
3030	1-1/4"		20.90	.383		82.50	24.50		107	129
3040	1-1/2"		18.30	.437		108	28		136	163
3050	2"		15.70	.510		174	33		207	242
3200	Adapter, tube to MPT									
3210	1/2"	1 Plum	15.10	.530	Ea.	4.10	34		38.10	57.50
3220	3/4"		13.60	.588		7.60	38		45.60	67
3230	1"		11.60	.690		13.70	44.50		58.20	83.50
3240	1-1/4"		9.90	.808		29.50	52		81.50	113
3250	1-1/2"		9	.889		41.50	57.50		99	134
3260	2"		7.90	1.013		80	65.50		145.50	189
3270	2-1/2"	Q-1	13.40	1.194		181	69.50		250.50	305
3280	3"		10.60	1.509		228	87.50		315.50	385
3290	4"		7.70	2.078		277	121		398	490
3400	Adapter, tube to FPT									
3410	1/2"	1 Plum	15.10	.530	Ea.	5.05	34		39.05	58.50
3420	3/4"		13.60	.588		8	38		46	67.50
3430	1"		11.60	.690		15.15	44.50		59.65	85
3440	1-1/4"		9.90	.808		34.50	52		86.50	119
3450	1-1/2"		9	.889		49	57.50		106.50	143
3460	2"		7.90	1.013		82.50	65.50		148	192
3470	2-1/2"	Q-1	13.40	1.194		203	69.50		272.50	330
3480	3"		10.60	1.509		335	87.50		422.50	500
3490	4"		7.70	2.078		430	121		551	655
3600	Flange									
3620	1"	1 Plum	36.20	.221	Ea.	147	14.25		161.25	183
3630	1-1/4"		29.30	.273		210	17.60		227.60	258
3640	1-1/2"		25.60	.313		232	20		252	286
3650	2"		22	.364		261	23.50		284.50	325
3660	2-1/2"	Q-1	36.20	.442		273	25.50		298.50	340
3670	3"		30.80	.519		330	30		360	410
3680	4"		22.80	.702		370	40.50		410.50	470
3800	Cap, 1/2"	1 Plum	53.10	.151		6.85	9.70		16.55	22.50
3810	3/4"		39.80	.201		11.50	12.95		24.45	32.50
3820	1"		37.50	.213		17.75	13.75		31.50	41

22 11 13.29 Pipe, Fittings and Valves, Copper, Pressed-Joint		Crew	Daily Output	Labor-Hours	Unit	Material	2020 Bare Costs Labor	Equipment	Total	Total Incl O&P
3830	1-1/4"	1 Plum	30.40	.263	Ea.	21	16.95		37.95	49
3840	1-1/2"		26.60	.301		32.50	19.40		51.90	66
3850	2"		22.80	.351		39.50	22.50		62	78.50
3860	2-1/2"	Q-1	37.50	.427		124	25		149	176
3870	3"		31.90	.502		157	29		186	218
3880	4"		23.60	.678		192	39.50		231.50	272
4000	Reducer									
4010	3/4"	1 Plum	27.50	.291	Ea.	14.45	18.75		33.20	45
4020	1"		25.90	.309		26.50	19.90		46.40	60
4030	1-1/4"		20.90	.383		39	24.50		63.50	81
4040	1-1/2"		18.30	.437		46.50	28		74.50	95
4050	2"		15.70	.510		64.50	33		97.50	122
4060	2-1/2"	Q-1	25.90	.618		179	36		215	253
4070	3"		22	.727		242	42		284	330
4080	4"		16.30	.982		310	57		367	435
4100	Stub out, 1/2"	1 Plum	50	.160		9.65	10.30		19.95	26.50
4110	3/4"		35	.229		14.90	14.75		29.65	39.50
4120	1"		30	.267		20.50	17.20		37.70	49
6000	Valves									
6200	Ball valve									
6210	1/2"	1 Plum	25.60	.313	Ea.	40	20		60	74.50
6220	3/4"		19.20	.417		53	27		80	99.50
6230	1"		18.10	.442		63.50	28.50		92	114
6240	1-1/4"		14.70	.544		103	35		138	168
6250	1-1/2"		12.80	.625		254	40.50		294.50	340
6260	2"		11	.727		365	47		412	480
6400	Check valve									
6410	1/2"	1 Plum	30	.267	Ea.	34.50	17.20		51.70	64.50
6420	3/4"		22.50	.356		44.50	23		67.50	84.50
6430	1"		21.20	.377		51	24.50		75.50	94
6440	1-1/4"		17.20	.465		74.50	30		104.50	129
6450	1-1/2"		15	.533		105	34.50		139.50	169
6460	2"		12.90	.620		191	40		231	272
6600	Butterfly valve, lug type									
6660	2-1/2"	Q-1	9	1.778	Ea.	163	103		266	340
6670	3"		8	2		200	116		316	400
6680	4"		5	3.200		250	186		436	560

22 11 13.44 Pipe, Steel

		Crew	Daily Output	Labor-Hours	Unit	Material	Labor	Equipment	Total	Total Incl O&P
0010	**PIPE, STEEL** R221113-50									
0020	All pipe sizes are to Spec. A-53 unless noted otherwise									
0032	Schedule 10, see Line 22 11 13.48 0500									
0050	Schedule 40, threaded, with couplings, and clevis hanger									
0060	assemblies sized for covering, 10' OC									
0540	Black, 1/4" diameter	1 Plum	66	.121	L.F.	6.15	7.80		13.95	18.90
0550	3/8" diameter		65	.123		6.85	7.95		14.80	19.80
0560	1/2" diameter		63	.127		4.03	8.20		12.23	17.10
0570	3/4" diameter		61	.131		4.40	8.45		12.85	17.90
0580	1" diameter		53	.151		7.80	9.75		17.55	23.50
0590	1-1/4" diameter	Q-1	89	.180		8.20	10.45		18.65	25
0600	1-1/2" diameter		80	.200		8.75	11.60		20.35	27.50
0610	2" diameter		64	.250		6.90	14.50		21.40	30
0620	2-1/2" diameter		50	.320		11.40	18.55		29.95	41

22 11 13.44 Pipe, Steel		Crew	Daily Output	Labor-Hours	Unit	Material	Labor (2020 Bare Costs)	Equipment	Total	Total Incl O&P
0630	3" diameter	Q-1	43	.372	L.F.	14.35	21.50		35.85	49.50
0640	3-1/2" diameter		40	.400		21	23		44	59.50
0650	4" diameter		36	.444		23	26		49	65.50
0809	A-106, gr. A/B, seamless w/cplgs. & clevis hanger assemblies									
0811	1/4" diameter	1 Plum	66	.121	L.F.	9.15	7.80		16.95	22
0812	3/8" diameter		65	.123		9.20	7.95		17.15	22.50
0813	1/2" diameter		63	.127		9.40	8.20		17.60	23
0814	3/4" diameter		61	.131		12.30	8.45		20.75	26.50
0815	1" diameter		53	.151		11.30	9.75		21.05	27.50
0816	1-1/4" diameter	Q-1	89	.180		13.20	10.45		23.65	30.50
0817	1-1/2" diameter		80	.200		19.55	11.60		31.15	39.50
0819	2" diameter		64	.250		15.70	14.50		30.20	40
0821	2-1/2" diameter		50	.320		21	18.55		39.55	51.50
0822	3" diameter		43	.372		26.50	21.50		48	62.50
0823	4" diameter		36	.444		41	26		67	85
1220	To delete coupling & hanger, subtract									
1230	1/4" diam. to 3/4" diam.					31%	56%			
1240	1" diam. to 1-1/2" diam.					23%	51%			
1250	2" diam. to 4" diam.					23%	41%			
1280	All pipe sizes are to Spec. A-53 unless noted otherwise									
1281	Schedule 40, threaded, with couplings and clevis hanger									
1282	assemblies sized for covering, 10' OC									
1290	Galvanized, 1/4" diameter	1 Plum	66	.121	L.F.	8.10	7.80		15.90	21
1300	3/8" diameter		65	.123		8.35	7.95		16.30	21.50
1310	1/2" diameter		63	.127		4.10	8.20		12.30	17.15
1320	3/4" diameter		61	.131		5	8.45		13.45	18.55
1330	1" diameter		53	.151		7.90	9.75		17.65	24
1340	1-1/4" diameter	Q-1	89	.180		8.85	10.45		19.30	26
1350	1-1/2" diameter		80	.200		9.45	11.60		21.05	28.50
1360	2" diameter		64	.250		7.90	14.50		22.40	31
1370	2-1/2" diameter		50	.320		13.15	18.55		31.70	43
1380	3" diameter		43	.372		16	21.50		37.50	51
1390	3-1/2" diameter		40	.400		23	23		46	61
1400	4" diameter		36	.444		25.50	26		51.50	68
1750	To delete coupling & hanger, subtract									
1760	1/4" diam. to 3/4" diam.					31%	56%			
1770	1" diam. to 1-1/2" diam.					23%	51%			
1780	2" diam. to 4" diam.					23%	41%			
2000	Welded, sch. 40, on yoke & roll hanger assy's, sized for covering, 10' OC									
2040	Black, 1" diameter	Q-15	93	.172	L.F.	7.15	10	1.14	18.29	24.50
2050	1-1/4" diameter		84	.190		8.20	11.05	1.27	20.52	27.50
2060	1-1/2" diameter		76	.211		8.65	12.20	1.40	22.25	30
2070	2" diameter		61	.262		6.60	15.20	1.74	23.54	32.50
2080	2-1/2" diameter		47	.340		11.80	19.75	2.26	33.81	46
2090	3" diameter		43	.372		16.65	21.50	2.47	40.62	54.50
2100	3-1/2" diameter		39	.410		17.45	24	2.73	44.18	59
2110	4" diameter		37	.432		24	25	2.88	51.88	68.50
2120	5" diameter		32	.500		37	29	3.33	69.33	89
2130	6" diameter	Q-16	36	.667		45	40	2.95	87.95	115
2140	8" diameter		29	.828		71.50	50	3.67	125.17	160
2150	10" diameter		24	1		90	60	4.43	154.43	197
2160	12" diameter		19	1.263		107	76	5.60	188.60	240
2170	14" diameter (two rod roll type hanger for 14" diam. and up)		15	1.600		104	96	7.10	207.10	272

22 11 13.44 Pipe, Steel		Crew	Daily Output	Labor-Hours	Unit	2020 Bare Costs Material	Labor	Equipment	Total	Total Incl O&P
2180	16″ diameter (two rod roll type hanger)	Q-16	13	1.846	L.F.	172	111	8.20	291.20	370
2190	18″ diameter (two rod roll type hanger)		11	2.182		147	131	9.65	287.65	375
2200	20″ diameter (two rod roll type hanger)		9	2.667		156	160	11.80	327.80	430
2220	24″ diameter (two rod roll type hanger)		8	3		203	180	13.30	396.30	515
2345	Sch. 40, A-53, gr. A/B, ERW, welded w/hngrs.									
2346	2″ diameter	Q-15	61	.262	L.F.	10.20	15.20	1.74	27.14	36.50
2347	2-1/2″ diameter		47	.340		12.40	19.75	2.26	34.41	46.50
2348	3″ diameter		43	.372		14.15	21.50	2.47	38.12	52
2349	4″ diameter		38	.421		20.50	24.50	2.80	47.80	63.50
2350	6″ diameter	Q-16	37	.649		30.50	39	2.87	72.37	97.50
2351	8″ diameter		29	.828		73	50	3.67	126.67	162
2352	10″ diameter		24	1		93.50	60	4.43	157.93	201
2560	To delete hanger, subtract									
2570	1″ diam. to 1-1/2″ diam.					15%	34%			
2580	2″ diam. to 3-1/2″ diam.					9%	21%			
2590	4″ diam. to 12″ diam.					5%	12%			
2596	14″ diam. to 24″ diam.					3%	10%			
3250	Flanged, 150 lb. weld neck, on yoke & roll hangers									
3260	sized for covering, 10′ OC									
3290	Black, 1″ diameter	Q-15	70	.229	L.F.	10.60	13.25	1.52	25.37	34
3300	1-1/4″ diameter		64	.250		11.65	14.50	1.66	27.81	37
3310	1-1/2″ diameter		58	.276		12.10	16	1.83	29.93	40
3320	2″ diameter		45	.356		15.25	20.50	2.36	38.11	51.50
3330	2-1/2″ diameter		36	.444		23	26	2.96	51.96	69
3340	3″ diameter		32	.500		28.50	29	3.33	60.83	80
3350	3-1/2″ diameter		29	.552		31	32	3.67	66.67	87.50
3360	4″ diameter		26	.615		33	35.50	4.09	72.59	96
3370	5″ diameter		21	.762		51.50	44	5.05	100.55	131
3380	6″ diameter	Q-16	25	.960		62.50	57.50	4.25	124.25	163
3390	8″ diameter		19	1.263		101	76	5.60	182.60	234
3400	10″ diameter		16	1.500		141	90	6.65	237.65	300
3410	12″ diameter		14	1.714		175	103	7.60	285.60	360
3470	For 300 lb. flanges, add					63%				
3480	For 600 lb. flanges, add					310%				
3960	To delete flanges & hanger, subtract									
3970	1″ diam. to 2″ diam.					76%	65%			
3980	2-1/2″ diam. to 4″ diam.					62%	59%			
3990	5″ diam. to 12″ diam.					60%	46%			
4750	Schedule 80, threaded, with couplings, and clevis hanger assemblies									
4760	sized for covering, 10′ OC									
4790	Black, 1/4″ diameter	1 Plum	54	.148	L.F.	7.60	9.55		17.15	23
4800	3/8″ diameter		53	.151		9.70	9.75		19.45	25.50
4810	1/2″ diameter		52	.154		4.87	9.90		14.77	20.50
4820	3/4″ diameter		50	.160		6.45	10.30		16.75	23
4830	1″ diameter		45	.178		7.35	11.45		18.80	26
4840	1-1/4″ diameter	Q-1	75	.213		9.40	12.35		21.75	29.50
4850	1-1/2″ diameter		69	.232		10.35	13.45		23.80	32.50
4860	2″ diameter		56	.286		13.20	16.55		29.75	40
4870	2-1/2″ diameter		44	.364		22	21		43	56.50
4880	3″ diameter		38	.421		27	24.50		51.50	68
4890	3-1/2″ diameter		35	.457		32.50	26.50		59	77
4900	4″ diameter		32	.500		36	29		65	85
5061	A-106, gr. A/B seamless with cplgs. & clevis hanger assemblies, 1/4″ diam.	1 Plum	63	.127		7	8.20		15.20	20.50

22 11 13.44 Pipe, Steel		Crew	Daily Output	Labor-Hours	Unit	Material	Labor	Equipment	Total	Total Incl O&P
						2020 Bare Costs				
5062	3/8" diameter	1 Plum	62	.129	L.F.	8	8.30		16.30	21.50
5063	1/2" diameter		61	.131		6.65	8.45		15.10	20.50
5064	3/4" diameter		57	.140		7.65	9.05		16.70	22.50
5065	1" diameter		51	.157		8.90	10.10		19	25.50
5066	1-1/4" diameter	Q-1	85	.188		12.40	10.90		23.30	30.50
5067	1-1/2" diameter		77	.208		16.95	12.05		29	37.50
5071	2" diameter		61	.262		15.60	15.20		30.80	40.50
5072	2-1/2" diameter		48	.333		18.35	19.35		37.70	50
5073	3" diameter		41	.390		24	22.50		46.50	61.50
5074	4" diameter		35	.457		34.50	26.50		61	79
5430	To delete coupling & hanger, subtract									
5440	1/4" diam. to 1/2" diam.					31%	54%			
5450	3/4" diam. to 1-1/2" diam.					28%	49%			
5460	2" diam. to 4" diam.					21%	40%			
5510	Galvanized, 1/4" diameter	1 Plum	54	.148	L.F.	9.90	9.55		19.45	25.50
5520	3/8" diameter		53	.151		10.70	9.75		20.45	27
5530	1/2" diameter		52	.154		5.65	9.90		15.55	21.50
5540	3/4" diameter		50	.160		7.10	10.30		17.40	24
5550	1" diameter		45	.178		8.25	11.45		19.70	27
5560	1-1/4" diameter	Q-1	75	.213		10.70	12.35		23.05	31
5570	1-1/2" diameter		69	.232		11.90	13.45		25.35	34
5580	2" diameter		56	.286		15.45	16.55		32	42.50
5590	2-1/2" diameter		44	.364		25	21		46	60
5600	3" diameter		38	.421		31	24.50		55.50	72
5610	3-1/2" diameter		35	.457		38	26.50		64.50	83
5620	4" diameter		32	.500		39.50	29		68.50	88.50
5930	To delete coupling & hanger, subtract									
5940	1/4" diam. to 1/2" diam.					31%	54%			
5950	3/4" diam. to 1-1/2" diam.					28%	49%			
5960	2" diam. to 4" diam.					21%	40%			
6000	Welded, on yoke & roller hangers									
6010	sized for covering, 10' OC									
6040	Black, 1" diameter	Q-15	85	.188	L.F.	7.15	10.90	1.25	19.30	26
6050	1-1/4" diameter		79	.203		9	11.75	1.35	22.10	29.50
6060	1-1/2" diameter		72	.222		9.95	12.90	1.48	24.33	32.50
6070	2" diameter		57	.281		12	16.30	1.87	30.17	40.50
6080	2-1/2" diameter		44	.364		20.50	21	2.42	43.92	57.50
6090	3" diameter		40	.400		25	23	2.66	50.66	66.50
6100	3-1/2" diameter		34	.471		30	27.50	3.13	60.63	78.50
6110	4" diameter		33	.485		32	28	3.22	63.22	82.50
6120	5" diameter, A-106B		26	.615		57.50	35.50	4.09	97.09	123
6130	6" diameter, A-106B	Q-16	30	.800		67.50	48	3.54	119.04	153
6140	8" diameter, A-106B		25	.960		103	57.50	4.25	164.75	208
6150	10" diameter, A-106B		20	1.200		153	72	5.30	230.30	286
6160	12" diameter, A-106B		15	1.600		345	96	7.10	448.10	535
6540	To delete hanger, subtract									
6550	1" diam. to 1-1/2" diam.					30%	14%			
6560	2" diam. to 3" diam.					23%	9%			
6570	3-1/2" diam. to 5" diam.					12%	6%			
6580	6" diam. to 12" diam.					10%	4%			
7250	Flanged, 300 lb. weld neck, on yoke & roll hangers									
7260	sized for covering, 10' OC									
7290	Black, 1" diameter	Q-15	66	.242	L.F.	14.40	14.05	1.61	30.06	39

22 11 13 – Facility Water Distribution Piping

22 11 13.44 Pipe, Steel		Crew	Daily Output	Labor-Hours	Unit	Material	Labor	Equipment	Total	Total Incl O&P
						2020 Bare Costs				
7300	1-1/4" diameter	Q-15	61	.262	L.F.	16.25	15.20	1.74	33.19	43.50
7310	1-1/2" diameter		54	.296		17.20	17.20	1.97	36.37	47.50
7320	2" diameter		42	.381		22	22	2.53	46.53	61
7330	2-1/2" diameter		33	.485		31.50	28	3.22	62.72	81.50
7340	3" diameter		29	.552		34.50	32	3.67	70.17	91.50
7350	3-1/2" diameter		24	.667		47	38.50	4.43	89.93	116
7360	4" diameter		23	.696		49	40.50	4.63	94.13	122
7370	5" diameter		19	.842		82.50	49	5.60	137.10	172
7380	6" diameter	Q-16	23	1.043		92.50	63	4.62	160.12	204
7390	8" diameter		17	1.412		149	85	6.25	240.25	300
7400	10" diameter		14	1.714		241	103	7.60	351.60	430
7410	12" diameter		12	2		440	120	8.85	568.85	680
7470	For 600 lb. flanges, add					100%				
7940	To delete flanges & hanger, subtract									
7950	1" diam. to 1-1/2" diam.					75%	66%			
7960	2" diam. to 3" diam.					62%	60%			
7970	3-1/2" diam. to 5" diam.					54%	66%			
7980	6" diam. to 12" diam.					55%	62%			
8040	Galvanized, 1" diameter	Q-15	66	.242	L.F.	15.30	14.05	1.61	30.96	40
8050	1-1/4" diameter		61	.262		17.55	15.20	1.74	34.49	44.50
8060	1-1/2" diameter		54	.296		18.75	17.20	1.97	37.92	49
8070	2" diameter		42	.381		24	22	2.53	48.53	63.50
8080	2-1/2" diameter		33	.485		34.50	28	3.22	65.72	85
8090	3" diameter		29	.552		38.50	32	3.67	74.17	96
8100	3-1/2" diameter		24	.667		52.50	38.50	4.43	95.43	122
8110	4" diameter		23	.696		52.50	40.50	4.63	97.63	125
8120	5" diameter, A-106B		19	.842		67	49	5.60	121.60	155
8130	6" diameter, A-106B	Q-16	23	1.043		84	63	4.62	151.62	195
8140	8" diameter, A-106B		17	1.412		155	85	6.25	246.25	310
8150	10" diameter, A-106B		14	1.714		335	103	7.60	445.60	535
8160	12" diameter, A-106B		12	2		660	120	8.85	788.85	925
8240	For 600 lb. flanges, add					100%				
8300	To delete flanges & hangers, subtract									
8310	1" diam. to 1-1/2" diam.					72%	66%			
8320	2" diam. to 3" diam.					59%	60%			
8330	3-1/2" diam. to 5" diam.					51%	66%			
8340	6" diam. to 12" diam.					49%	62%			
9000	Threading pipe labor, one end, all schedules through 80									
9010	1/4" through 3/4" pipe size	1 Plum	80	.100	Ea.		6.45		6.45	9.95
9020	1" through 2" pipe size		73	.110			7.05		7.05	10.90
9030	2-1/2" pipe size		53	.151			9.75		9.75	15.05
9040	3" pipe size		50	.160			10.30		10.30	15.95
9050	3-1/2" pipe size	Q-1	89	.180			10.45		10.45	16.10
9060	4" pipe size		73	.219			12.70		12.70	19.65
9070	5" pipe size		53	.302			17.50		17.50	27
9080	6" pipe size		46	.348			20		20	31
9090	8" pipe size		29	.552			32		32	49.50
9100	10" pipe size		21	.762			44		44	68.50
9110	12" pipe size		13	1.231			71.50		71.50	110
9120	Cutting pipe labor, one cut									
9124	Shop fabrication, machine cut									
9126	Schedule 40, straight pipe									
9128	2" pipe size or less	1 Stpi	62	.129	Ea.		8.45		8.45	13.10

22 11 13.44 Pipe, Steel		Crew	Daily Output	Labor-Hours	Unit	Material	2020 Bare Costs Labor	Equipment	Total	Total Incl O&P
9130	2-1/2" pipe size	1 Stpi	56	.143	Ea.		9.35		9.35	14.50
9132	3" pipe size		42	.190			12.50		12.50	19.30
9134	4" pipe size		31	.258			16.90		16.90	26
9136	5" pipe size		26	.308			20		20	31
9138	6" pipe size		19	.421			27.50		27.50	42.50
9140	8" pipe size		14	.571			37.50		37.50	58
9142	10" pipe size		10	.800			52.50		52.50	81
9144	12" pipe size		7	1.143			75		75	116
9146	14" pipe size	Q-5	10.50	1.524			90		90	139
9148	16" pipe size		8.60	1.860			110		110	170
9150	18" pipe size		7	2.286			135		135	209
9152	20" pipe size		5.80	2.759			163		163	252
9154	24" pipe size		4	4			236		236	365
9160	Schedule 80, straight pipe									
9164	2" pipe size or less	1 Stpi	42	.190	Ea.		12.50		12.50	19.30
9166	2-1/2" pipe size		37	.216			14.15		14.15	22
9168	3" pipe size		31	.258			16.90		16.90	26
9170	4" pipe size		23	.348			23		23	35.50
9172	5" pipe size		18	.444			29		29	45
9174	6" pipe size		14.60	.548			36		36	55.50
9176	8" pipe size		10	.800			52.50		52.50	81
9178	10" pipe size	Q-5	14	1.143			67.50		67.50	104
9180	12" pipe size	"	10	1.600			94.50		94.50	146
9200	Welding labor per joint									
9210	Schedule 40									
9230	1/2" pipe size	Q-15	32	.500	Ea.		29	3.33	32.33	48.50
9240	3/4" pipe size		27	.593			34.50	3.94	38.44	57.50
9250	1" pipe size		23	.696			40.50	4.63	45.13	67.50
9260	1-1/4" pipe size		20	.800			46.50	5.30	51.80	77.50
9270	1-1/2" pipe size		19	.842			49	5.60	54.60	81.50
9280	2" pipe size		16	1			58	6.65	64.65	97
9290	2-1/2" pipe size		13	1.231			71.50	8.20	79.70	119
9300	3" pipe size		12	1.333			77.50	8.85	86.35	130
9310	4" pipe size		10	1.600			93	10.65	103.65	155
9320	5" pipe size		9	1.778			103	11.80	114.80	172
9330	6" pipe size		8	2			116	13.30	129.30	194
9340	8" pipe size		5	3.200			186	21.50	207.50	310
9350	10" pipe size		4	4			232	26.50	258.50	390
9360	12" pipe size		3	5.333			310	35.50	345.50	520
9370	14" pipe size		2.60	6.154			355	41	396	595
9380	16" pipe size		2.20	7.273			420	48.50	468.50	705
9390	18" pipe size		2	8			465	53	518	775
9400	20" pipe size		1.80	8.889			515	59	574	860
9410	22" pipe size		1.70	9.412			545	62.50	607.50	915
9420	24" pipe size		1.50	10.667			620	71	691	1,025
9450	Schedule 80									
9460	1/2" pipe size	Q-15	27	.593	Ea.		34.50	3.94	38.44	57.50
9470	3/4" pipe size		23	.696			40.50	4.63	45.13	67.50
9480	1" pipe size		20	.800			46.50	5.30	51.80	77.50
9490	1-1/4" pipe size		19	.842			49	5.60	54.60	81.50
9500	1-1/2" pipe size		18	.889			51.50	5.90	57.40	86
9510	2" pipe size		15	1.067			62	7.10	69.10	103
9520	2-1/2" pipe size		12	1.333			77.50	8.85	86.35	130

22 11 13.44 Pipe, Steel		Crew	Daily Output	Labor-Hours	Unit	Material	2020 Bare Costs Labor	Equipment	Total	Total Incl O&P
9530	3" pipe size	Q-15	11	1.455	Ea.		84.50	9.65	94.15	141
9540	4" pipe size		8	2			116	13.30	129.30	194
9550	5" pipe size		6	2.667			155	17.75	172.75	259
9560	6" pipe size		5	3.200			186	21.50	207.50	310
9570	8" pipe size		4	4			232	26.50	258.50	390
9580	10" pipe size		3	5.333			310	35.50	345.50	520
9590	12" pipe size		2	8			465	53	518	775
9600	14" pipe size	Q-16	2.60	9.231			555	41	596	905
9610	16" pipe size		2.30	10.435			630	46	676	1,025
9620	18" pipe size		2	12			720	53	773	1,175
9630	20" pipe size		1.80	13.333			800	59	859	1,325
9640	22" pipe size		1.60	15			900	66.50	966.50	1,475
9650	24" pipe size		1.50	16			960	71	1,031	1,575
9990	Minimum labor/equipment charge	1 Plum	3	2.667	Job		172		172	266

22 11 13.45 Pipe Fittings, Steel, Threaded

		Crew	Daily Output	Labor-Hours	Unit	Material	Labor	Equipment	Total	Total Incl O&P
0010	**PIPE FITTINGS, STEEL, THREADED** R221113-50									
0020	Cast iron									
0040	Standard weight, black									
0060	90° elbow, straight									
0070	1/4"	1 Plum	16	.500	Ea.	11.95	32		43.95	63
0080	3/8"		16	.500		17.30	32		49.30	69
0090	1/2"		15	.533		7.60	34.50		42.10	61.50
0100	3/4"		14	.571		7.90	37		44.90	65.50
0110	1"		13	.615		9.35	39.50		48.85	72
0120	1-1/4"	Q-1	22	.727		13.25	42		55.25	79.50
0130	1-1/2"		20	.800		18.35	46.50		64.85	91.50
0140	2"		18	.889		28.50	51.50		80	111
0150	2-1/2"		14	1.143		68.50	66.50		135	178
0160	3"		10	1.600		113	93		206	267
0170	3-1/2"		8	2		305	116		421	515
0180	4"		6	2.667		209	155		364	470
0250	45° elbow, straight									
0260	1/4"	1 Plum	16	.500	Ea.	14.75	32		46.75	66.50
0270	3/8"		16	.500		15.85	32		47.85	67.50
0280	1/2"		15	.533		11.60	34.50		46.10	66
0300	3/4"		14	.571		11.65	37		48.65	70
0320	1"		13	.615		13.70	39.50		53.20	76.50
0330	1-1/4"	Q-1	22	.727		18.40	42		60.40	85
0340	1-1/2"		20	.800		30.50	46.50		77	105
0350	2"		18	.889		35	51.50		86.50	118
0360	2-1/2"		14	1.143		91.50	66.50		158	203
0370	3"		10	1.600		145	93		238	300
0380	3-1/2"		8	2		340	116		456	555
0400	4"		6	2.667		300	155		455	570
0500	Tee, straight									
0510	1/4"	1 Plum	10	.800	Ea.	18.70	51.50		70.20	100
0520	3/8"		10	.800		18.20	51.50		69.70	99.50
0530	1/2"		9	.889		11.80	57.50		69.30	102
0540	3/4"		9	.889		13.75	57.50		71.25	104
0550	1"		8	1		12.25	64.50		76.75	113
0560	1-1/4"	Q-1	14	1.143		22.50	66.50		89	127
0570	1-1/2"		13	1.231		29	71.50		100.50	142

22 11 13.45 Pipe Fittings, Steel, Threaded		Crew	Daily Output	Labor-Hours	Unit	2020 Bare Costs Material	Labor	Equipment	Total	Total Incl O&P
0580	2"	Q-1	11	1.455	Ea.	40.50	84.50		125	175
0590	2-1/2"		9	1.778		105	103		208	275
0600	3"		6	2.667		161	155		316	415
0610	3-1/2"		5	3.200		325	186		511	645
0620	4"		4	4		315	232		547	705
0700	Standard weight, galvanized cast iron									
0720	90° elbow, straight									
0730	1/4"	1 Plum	16	.500	Ea.	21.50	32		53.50	73.50
0740	3/8"		16	.500		21.50	32		53.50	73.50
0750	1/2"		15	.533		22	34.50		56.50	77
0760	3/4"		14	.571		21	37		58	80
0770	1"		13	.615		28	39.50		67.50	92
0780	1-1/4"	Q-1	22	.727		37.50	42		79.50	106
0790	1-1/2"		20	.800		59.50	46.50		106	137
0800	2"		18	.889		76	51.50		127.50	163
0810	2-1/2"		14	1.143		156	66.50		222.50	273
0820	3"		10	1.600		237	93		330	405
0830	3-1/2"		8	2		380	116		496	595
0840	4"		6	2.667		435	155		590	720
0900	45° elbow, straight									
0910	1/4"	1 Plum	16	.500	Ea.	23	32		55	75.50
0920	3/8"		16	.500		20	32		52	72
0930	1/2"		15	.533		21	34.50		55.50	76.50
0940	3/4"		14	.571		24.50	37		61.50	83.50
0950	1"		13	.615		34	39.50		73.50	99
0960	1-1/4"	Q-1	22	.727		52	42		94	122
0970	1-1/2"		20	.800		69	46.50		115.50	148
0980	2"		18	.889		96.50	51.50		148	186
0990	2-1/2"		14	1.143		136	66.50		202.50	252
1000	3"		10	1.600		305	93		398	480
1010	3-1/2"		8	2		525	116		641	760
1020	4"		6	2.667		475	155		630	765
1100	Tee, straight									
1110	1/4"	1 Plum	10	.800	Ea.	20	51.50		71.50	102
1120	3/8"		10	.800		22.50	51.50		74	105
1130	1/2"		9	.889		26.50	57.50		84	118
1140	3/4"		9	.889		34.50	57.50		92	127
1150	1"		8	1		32.50	64.50		97	136
1160	1-1/4"	Q-1	14	1.143		57	66.50		123.50	165
1170	1-1/2"		13	1.231		75.50	71.50		147	193
1180	2"		11	1.455		94	84.50		178.50	233
1190	2-1/2"		9	1.778		191	103		294	370
1200	3"		6	2.667		455	155		610	740
1210	3-1/2"		5	3.200		535	186		721	875
1220	4"		4	4		590	232		822	1,000
1300	Extra heavy weight, black									
1310	Couplings, steel straight									
1320	1/4"	1 Plum	19	.421	Ea.	5.55	27		32.55	48
1330	3/8"		19	.421		6.05	27		33.05	48.50
1340	1/2"		19	.421		8.20	27		35.20	51
1350	3/4"		18	.444		8.75	28.50		37.25	54
1360	1"		15	.533		11.15	34.50		45.65	65.50
1370	1-1/4"	Q-1	26	.615		17.85	35.50		53.35	74.50

22 11 13.45 Pipe Fittings, Steel, Threaded		Crew	Daily Output	Labor-Hours	Unit	2020 Bare Costs Material	Labor	Equipment	Total	Total Incl O&P
1380	1-1/2"	Q-1	24	.667	Ea.	17.85	38.50		56.35	79.50
1390	2"		21	.762		27	44		71	98.50
1400	2-1/2"		18	.889		40.50	51.50		92	124
1410	3"		14	1.143		48	66.50		114.50	155
1420	3-1/2"		12	1.333		64.50	77.50		142	191
1430	4"		10	1.600		76	93		169	227
1510	90° elbow, straight									
1520	1/2"	1 Plum	15	.533	Ea.	39	34.50		73.50	96
1530	3/4"		14	.571		39.50	37		76.50	101
1540	1"		13	.615		48	39.50		87.50	115
1550	1-1/4"	Q-1	22	.727		71.50	42		113.50	144
1560	1-1/2"		20	.800		88.50	46.50		135	169
1580	2"		18	.889		109	51.50		160.50	200
1590	2-1/2"		14	1.143		266	66.50		332.50	395
1600	3"		10	1.600		360	93		453	540
1610	4"		6	2.667		785	155		940	1,100
1650	45° elbow, straight									
1660	1/2"	1 Plum	15	.533	Ea.	55.50	34.50		90	114
1670	3/4"		14	.571		53.50	37		90.50	116
1680	1"		13	.615		64.50	39.50		104	133
1690	1-1/4"	Q-1	22	.727		106	42		148	182
1700	1-1/2"		20	.800		117	46.50		163.50	201
1710	2"		18	.889		166	51.50		217.50	263
1720	2-1/2"		14	1.143		287	66.50		353.50	415
1800	Tee, straight									
1810	1/2"	1 Plum	9	.889	Ea.	61	57.50		118.50	156
1820	3/4"		9	.889		61	57.50		118.50	156
1830	1"		8	1		73.50	64.50		138	181
1840	1-1/4"	Q-1	14	1.143		110	66.50		176.50	223
1850	1-1/2"		13	1.231		141	71.50		212.50	265
1860	2"		11	1.455		175	84.50		259.50	325
1870	2-1/2"		9	1.778		370	103		473	565
1880	3"		6	2.667		525	155		680	815
1890	4"		4	4		1,025	232		1,257	1,475
4000	Standard weight, black									
4010	Couplings, steel straight, merchants									
4030	1/4"	1 Plum	19	.421	Ea.	1.48	27		28.48	43.50
4040	3/8"		19	.421		1.79	27		28.79	44
4050	1/2"		19	.421		1.91	27		28.91	44
4060	3/4"		18	.444		2.42	28.50		30.92	47
4070	1"		15	.533		3.39	34.50		37.89	56.50
4080	1-1/4"	Q-1	26	.615		4.32	35.50		39.82	60
4090	1-1/2"		24	.667		5.50	38.50		44	66
4100	2"		21	.762		7.85	44		51.85	77
4110	2-1/2"		18	.889		25.50	51.50		77	108
4120	3"		14	1.143		36	66.50		102.50	142
4130	3-1/2"		12	1.333		64	77.50		141.50	190
4140	4"		10	1.600		64	93		157	213
4166	Plug, 1/4"	1 Plum	38	.211		3.37	13.55		16.92	24.50
4167	3/8"		38	.211		3.03	13.55		16.58	24.50
4168	1/2"		38	.211		3.37	13.55		16.92	24.50
4169	3/4"		32	.250		8.90	16.10		25	35
4170	1"		30	.267		9.50	17.20		26.70	37

22 11 13.45 Pipe Fittings, Steel, Threaded		Crew	Daily Output	Labor-Hours	Unit	2020 Bare Costs Material	Labor	Equipment	Total	Total Incl O&P
4171	1-1/4"	Q-1	52	.308	Ea.	10.90	17.85		28.75	39.50
4172	1-1/2"		48	.333		15.50	19.35		34.85	47
4173	2"		42	.381		20	22		42	56
4176	2-1/2"		36	.444		29.50	26		55.50	72.50
4180	4"		20	.800		62.50	46.50		109	140
4700	Nipple, black									
4710	1/2" x 4" long	1 Plum	19	.421	Ea.	3.40	27		30.40	45.50
4712	3/4" x 4" long		18	.444		4.11	28.50		32.61	49
4714	1" x 4" long		15	.533		5.70	34.50		40.20	59.50
4716	1-1/4" x 4" long	Q-1	26	.615		7.15	35.50		42.65	63
4718	1-1/2" x 4" long		24	.667		8.40	38.50		46.90	69.50
4720	2" x 4" long		21	.762		11.75	44		55.75	81.50
4722	2-1/2" x 4" long		18	.889		32	51.50		83.50	115
4724	3" x 4" long		14	1.143		40.50	66.50		107	147
4726	4" x 4" long		10	1.600		48.50	93		141.50	197
4800	Nipple, galvanized									
4810	1/2" x 4" long	1 Plum	19	.421	Ea.	4.17	27		31.17	46.50
4812	3/4" x 4" long		18	.444		5.15	28.50		33.65	50
4814	1" x 4" long		15	.533		6.90	34.50		41.40	60.50
4816	1-1/4" x 4" long	Q-1	26	.615		8.50	35.50		44	64.50
4818	1-1/2" x 4" long		24	.667		10.75	38.50		49.25	72
4820	2" x 4" long		21	.762		13.65	44		57.65	83.50
4822	2-1/2" x 4" long		18	.889		36	51.50		87.50	119
4824	3" x 4" long		14	1.143		47.50	66.50		114	155
4826	4" x 4" long		10	1.600		64	93		157	214
5000	Malleable iron, 150 lb.									
5020	Black									
5040	90° elbow, straight									
5060	1/4"	1 Plum	16	.500	Ea.	5.90	32		37.90	56.50
5070	3/8"		16	.500		5.90	32		37.90	56.50
5080	1/2"		15	.533		4	34.50		38.50	57.50
5090	3/4"		14	.571		4.12	37		41.12	61.50
5100	1"		13	.615		8.35	39.50		47.85	70.50
5110	1-1/4"	Q-1	22	.727		13.85	42		55.85	80.50
5120	1-1/2"		20	.800		14.60	46.50		61.10	87.50
5130	2"		18	.889		25.50	51.50		77	108
5140	2-1/2"		14	1.143		71	66.50		137.50	181
5150	3"		10	1.600		104	93		197	258
5160	3-1/2"		8	2		287	116		403	495
5170	4"		6	2.667		196	155		351	455
5250	45° elbow, straight									
5270	1/4"	1 Plum	16	.500	Ea.	8.85	32		40.85	60
5280	3/8"		16	.500		8.85	32		40.85	60
5290	1/2"		15	.533		6.75	34.50		41.25	60.50
5300	3/4"		14	.571		8.35	37		45.35	66
5310	1"		13	.615		10.50	39.50		50	73
5320	1-1/4"	Q-1	22	.727		18.55	42		60.55	85.50
5330	1-1/2"		20	.800		23	46.50		69.50	97
5340	2"		18	.889		34.50	51.50		86	118
5350	2-1/2"		14	1.143		100	66.50		166.50	212
5360	3"		10	1.600		131	93		224	287
5370	3-1/2"		8	2		261	116		377	465
5380	4"		6	2.667		256	155		411	520

22 11 Facility Water Distribution

22 11 13 – Facility Water Distribution Piping

22 11 13.45 Pipe Fittings, Steel, Threaded		Crew	Daily Output	Labor-Hours	Unit	2020 Bare Costs Material	Labor	Equipment	Total	Total Incl O&P
5450	Tee, straight									
5470	1/4"	1 Plum	10	.800	Ea.	8.55	51.50		60.05	89
5480	3/8"		10	.800		8.55	51.50		60.05	89
5490	1/2"		9	.889		5.45	57.50		62.95	94.50
5500	3/4"		9	.889		7.85	57.50		65.35	97
5510	1"		8	1		13.40	64.50		77.90	114
5520	1-1/4"	Q-1	14	1.143		21.50	66.50		88	126
5530	1-1/2"		13	1.231		27	71.50		98.50	140
5540	2"		11	1.455		46	84.50		130.50	181
5550	2-1/2"		9	1.778		99	103		202	268
5560	3"		6	2.667		146	155		301	400
5570	3-1/2"		5	3.200		340	186		526	660
5580	4"		4	4		355	232		587	750
5650	Coupling									
5670	1/4"	1 Plum	19	.421	Ea.	7.30	27		34.30	50
5680	3/8"		19	.421		7.30	27		34.30	50
5690	1/2"		19	.421		5.65	27		32.65	48
5700	3/4"		18	.444		6.60	28.50		35.10	52
5710	1"		15	.533		9.90	34.50		44.40	64
5720	1-1/4"	Q-1	26	.615		11.95	35.50		47.45	68
5730	1-1/2"		24	.667		15.80	38.50		54.30	77.50
5740	2"		21	.762		23.50	44		67.50	94
5750	2-1/2"		18	.889		64.50	51.50		116	151
5760	3"		14	1.143		87.50	66.50		154	198
5770	3-1/2"		12	1.333		161	77.50		238.50	297
5780	4"		10	1.600		176	93		269	335
5840	Reducer, concentric, 1/4"	1 Plum	19	.421		7.70	27		34.70	50.50
5850	3/8"		19	.421		8.40	27		35.40	51.50
5860	1/2"		19	.421		7	27		34	49.50
5870	3/4"		16	.500		8	32		40	59
5880	1"		15	.533		12.65	34.50		47.15	67
5890	1-1/4"	Q-1	26	.615		15.65	35.50		51.15	72.50
5900	1-1/2"		24	.667		21	38.50		59.50	83
5910	2"		21	.762		30	44		74	102
5911	2-1/2"		18	.889		68.50	51.50		120	155
5981	Bushing, 1/4"	1 Plum	19	.421		1.09	27		28.09	43
5982	3/8"		19	.421		5.70	27		32.70	48.50
5983	1/2"		19	.421		6	27		33	48.50
5984	3/4"		16	.500		6.95	32		38.95	57.50
5985	1"		15	.533		10.70	34.50		45.20	65
5986	1-1/4"	Q-1	26	.615		13.25	35.50		48.75	69.50
5987	1-1/2"		24	.667		11.45	38.50		49.95	72.50
5988	2"		21	.762		14.35	44		58.35	84.50
5989	Cap, 1/4"	1 Plum	38	.211		6.85	13.55		20.40	28.50
5991	3/8"		38	.211		6.45	13.55		20	28
5992	1/2"		38	.211		4.13	13.55		17.68	25.50
5993	3/4"		32	.250		5.60	16.10		21.70	31
5994	1"		30	.267		6.75	17.20		23.95	34
5995	1-1/4"	Q-1	52	.308		8.90	17.85		26.75	37.50
5996	1-1/2"		48	.333		12.25	19.35		31.60	43.50
5997	2"		42	.381		17.85	22		39.85	53.50
6000	For galvanized elbows, tees, and couplings, add					20%				
6058	For galvanized reducers, caps and bushings, add					20%				

22 11 13.45 Pipe Fittings, Steel, Threaded		Crew	Daily Output	Labor-Hours	Unit	Material	Labor	Equipment	Total	Total Incl O&P
						2020 Bare Costs				
7000	Union, with brass seat									
7010	1/4"	1 Plum	15	.533	Ea.	26.50	34.50		61	82
7020	3/8"		15	.533		21	34.50		55.50	76.50
7030	1/2"		14	.571		19.15	37		56.15	78
7040	3/4"		13	.615		18.95	39.50		58.45	82.50
7050	1"		12	.667		24.50	43		67.50	93.50
7060	1-1/4"	Q-1	21	.762		41	44		85	114
7070	1-1/2"		19	.842		51.50	49		100.50	132
7080	2"		17	.941		59.50	54.50		114	150
7090	2-1/2"		13	1.231		176	71.50		247.50	305
7100	3"		9	1.778		184	103		287	360
7250	For galvanized unions, add					15%				
9757	Forged steel, 3000 lb.									
9758	Black									
9760	90° elbow, 1/4"	1 Plum	16	.500	Ea.	19.25	32		51.25	71
9761	3/8"		16	.500		19.25	32		51.25	71
9762	1/2"		15	.533		15	34.50		49.50	69.50
9763	3/4"		14	.571		18.45	37		55.45	77.50
9764	1"		13	.615		29.50	39.50		69	94
9765	1-1/4"	Q-1	22	.727		52.50	42		94.50	123
9766	1-1/2"		20	.800		67.50	46.50		114	146
9767	2"		18	.889		82	51.50		133.50	170
9780	45° elbow, 1/4"	1 Plum	16	.500		24.50	32		56.50	77
9781	3/8"		16	.500		24.50	32		56.50	77
9782	1/2"		15	.533		24	34.50		58.50	79.50
9783	3/4"		14	.571		28	37		65	87.50
9784	1"		13	.615		38	39.50		77.50	104
9785	1-1/4"	Q-1	22	.727		52.50	42		94.50	123
9786	1-1/2"		20	.800		75.50	46.50		122	155
9787	2"		18	.889		104	51.50		155.50	194
9800	Tee, 1/4"	1 Plum	10	.800		23.50	51.50		75	105
9801	3/8"		10	.800		23.50	51.50		75	105
9802	1/2"		9	.889		21	57.50		78.50	112
9803	3/4"		9	.889		28.50	57.50		86	120
9804	1"		8	1		37	64.50		101.50	141
9805	1-1/4"	Q-1	14	1.143		73	66.50		139.50	182
9806	1-1/2"		13	1.231		85	71.50		156.50	204
9807	2"		11	1.455		99.50	84.50		184	239
9820	Reducer, concentric, 1/4"	1 Plum	19	.421		12.85	27		39.85	56
9821	3/8"		19	.421		13.40	27		40.40	57
9822	1/2"		17	.471		13.40	30.50		43.90	62
9823	3/4"		16	.500		15.90	32		47.90	67.50
9824	1"		15	.533		20.50	34.50		55	76
9825	1-1/4"	Q-1	26	.615		36	35.50		71.50	94.50
9826	1-1/2"		24	.667		39	38.50		77.50	103
9827	2"		21	.762		57	44		101	131
9840	Cap, 1/4"	1 Plum	38	.211		8.10	13.55		21.65	30
9841	3/8"		38	.211		7.75	13.55		21.30	29.50
9842	1/2"		34	.235		7.55	15.15		22.70	32
9843	3/4"		32	.250		10.50	16.10		26.60	36.50
9844	1"		30	.267		16.10	17.20		33.30	44
9845	1-1/4"	Q-1	52	.308		24.50	17.85		42.35	54
9846	1-1/2"		48	.333		29	19.35		48.35	61.50

22 11 13 – Facility Water Distribution Piping

22 11 13.45 Pipe Fittings, Steel, Threaded		Crew	Daily Output	Labor-Hours	Unit	Material	Labor	Equipment	Total	Total Incl O&P
						2020 Bare Costs				
9847	2"	Q-1	42	.381	Ea.	42	22		64	80
9860	Plug, 1/4"	1 Plum	38	.211		4.09	13.55		17.64	25.50
9861	3/8"		38	.211		3.72	13.55		17.27	25
9862	1/2"		34	.235		3.58	15.15		18.73	27.50
9863	3/4"		32	.250		5.55	16.10		21.65	31
9864	1"		30	.267		8.50	17.20		25.70	36
9865	1-1/4"	Q-1	52	.308		16.40	17.85		34.25	45.50
9866	1-1/2"		48	.333		18.65	19.35		38	50.50
9867	2"		42	.381		29.50	22		51.50	66.50
9880	Union, bronze seat, 1/4"	1 Plum	15	.533		65	34.50		99.50	125
9881	3/8"		15	.533		65	34.50		99.50	125
9882	1/2"		14	.571		62.50	37		99.50	126
9883	3/4"		13	.615		84.50	39.50		124	155
9884	1"		12	.667		94.50	43		137.50	171
9885	1-1/4"	Q-1	21	.762		173	44		217	259
9886	1-1/2"		19	.842		189	49		238	284
9887	2"		17	.941		224	54.50		278.50	330
9900	Coupling, 1/4"	1 Plum	19	.421		7.60	27		34.60	50.50
9901	3/8"		19	.421		7.60	27		34.60	50.50
9902	1/2"		17	.471		6.20	30.50		36.70	54
9903	3/4"		16	.500		8	32		40	59
9904	1"		15	.533		13.90	34.50		48.40	68.50
9905	1-1/4"	Q-1	26	.615		23.50	35.50		59	80.50
9906	1-1/2"		24	.667		30	38.50		68.50	93
9907	2"		21	.762		40	44		84	113
9990	Minimum labor/equipment charge	1 Plum	4	2	Job		129		129	199

22 11 13.47 Pipe Fittings, Steel		Crew	Daily Output	Labor-Hours	Unit	Material	Labor	Equipment	Total	Total Incl O&P
0010	**PIPE FITTINGS, STEEL**, flanged, welded & special									
0020	Flanged joints, CI, standard weight, black. One gasket & bolt									
0040	set, mat'l only, required at each joint, not included (see line 0620)									
0060	90° elbow, straight, 1-1/2" pipe size	Q-1	14	1.143	Ea.	690	66.50		756.50	860
0080	2" pipe size		13	1.231		400	71.50		471.50	550
0090	2-1/2" pipe size		12	1.333		430	77.50		507.50	595
0100	3" pipe size		11	1.455		360	84.50		444.50	525
0110	4" pipe size		8	2		445	116		561	670
0120	5" pipe size		7	2.286		1,050	133		1,183	1,350
0130	6" pipe size	Q-2	9	2.667		695	160		855	1,025
0140	8" pipe size		8	3		1,200	180		1,380	1,600
0150	10" pipe size		7	3.429		2,625	206		2,831	3,225
0160	12" pipe size		6	4		5,300	241		5,541	6,200
0200	45° elbow, straight, 1-1/2" pipe size	Q-1	14	1.143		840	66.50		906.50	1,025
0220	2" pipe size		13	1.231		575	71.50		646.50	745
0230	2-1/2" pipe size		12	1.333		615	77.50		692.50	795
0240	3" pipe size		11	1.455		600	84.50		684.50	790
0250	4" pipe size		8	2		675	116		791	920
0260	5" pipe size		7	2.286		1,625	133		1,758	1,975
0270	6" pipe size	Q-2	9	2.667		1,100	160		1,260	1,450
0280	8" pipe size		8	3		1,600	180		1,780	2,050
0290	10" pipe size		7	3.429		3,400	206		3,606	4,075
0300	12" pipe size		6	4		5,200	241		5,441	6,100
0350	Tee, straight, 1-1/2" pipe size	Q-1	10	1.600		795	93		888	1,025
0370	2" pipe size		9	1.778		435	103		538	640

22 11 13.47 Pipe Fittings, Steel		Crew	Daily Output	Labor-Hours	Unit	Material	Labor	Equipment	Total	Total Incl O&P
						2020 Bare Costs				
0380	2-1/2" pipe size	Q-1	8	2	Ea.	635	116		751	880
0390	3" pipe size		7	2.286		445	133		578	695
0400	4" pipe size		5	3.200		675	186		861	1,025
0410	5" pipe size		4	4		1,825	232		2,057	2,350
0420	6" pipe size	Q-2	6	4		985	241		1,226	1,450
0430	8" pipe size		5	4.800		1,675	289		1,964	2,300
0440	10" pipe size		4	6		4,525	360		4,885	5,525
0450	12" pipe size		3	8		7,100	480		7,580	8,550
0500	For galvanized elbows and tees, add					100%				
0520	For extra heavy weight elbows and tees, add					140%				
0620	Gasket and bolt set, 150 lb., 1/2" pipe size	1 Plum	20	.400		3.83	26		29.83	44
0622	3/4" pipe size		19	.421		4.13	27		31.13	46.50
0624	1" pipe size		18	.444		4.62	28.50		33.12	49.50
0626	1-1/4" pipe size		17	.471		4.36	30.50		34.86	52
0628	1-1/2" pipe size		15	.533		4.38	34.50		38.88	58
0630	2" pipe size		13	.615		8.90	39.50		48.40	71.50
0640	2-1/2" pipe size		12	.667		8.45	43		51.45	76
0650	3" pipe size		11	.727		9.10	47		56.10	82.50
0660	3-1/2" pipe size		9	.889		20	57.50		77.50	111
0670	4" pipe size		8	1		17.90	64.50		82.40	119
0680	5" pipe size		7	1.143		27	73.50		100.50	144
0690	6" pipe size		6	1.333		28.50	86		114.50	164
0700	8" pipe size		5	1.600		31	103		134	193
0710	10" pipe size		4.50	1.778		57	115		172	240
0720	12" pipe size		4.20	1.905		59	123		182	255
0730	14" pipe size		4	2		54.50	129		183.50	259
0740	16" pipe size		3	2.667		65.50	172		237.50	340
0750	18" pipe size		2.70	2.963		127	191		318	435
0760	20" pipe size		2.30	3.478		206	224		430	570
0780	24" pipe size		1.90	4.211		257	271		528	705
0790	26" pipe size		1.60	5		350	320		670	885
0810	30" pipe size		1.40	5.714		680	370		1,050	1,325
0830	36" pipe size		1.10	7.273		1,275	470		1,745	2,125
0850	For 300 lb. gasket set, add					40%				
2000	Flanged unions, 125 lb., black, 1/2" pipe size	1 Plum	17	.471	Ea.	103	30.50		133.50	160
2040	3/4" pipe size		17	.471		143	30.50		173.50	204
2050	1" pipe size		16	.500		138	32		170	202
2060	1-1/4" pipe size	Q-1	28	.571		164	33		197	232
2070	1-1/2" pipe size		27	.593		151	34.50		185.50	219
2080	2" pipe size		26	.615		178	35.50		213.50	251
2090	2-1/2" pipe size		24	.667		241	38.50		279.50	325
2100	3" pipe size		22	.727		273	42		315	365
2110	3-1/2" pipe size		18	.889		460	51.50		511.50	590
2120	4" pipe size		16	1		370	58		428	500
2130	5" pipe size		14	1.143		865	66.50		931.50	1,050
2140	6" pipe size	Q-2	19	1.263		810	76		886	1,000
2150	8" pipe size	"	16	1.500		1,875	90		1,965	2,200
2200	For galvanized unions, add					150%				
2290	Threaded flange									
2300	Cast iron									
2310	Black, 125 lb., per flange									
2320	1" pipe size	1 Plum	27	.296	Ea.	53	19.10		72.10	88
2330	1-1/4" pipe size	Q-1	44	.364		69.50	21		90.50	109

22 11 13 – Facility Water Distribution Piping

22 11 13.47 Pipe Fittings, Steel		Crew	Daily Output	Labor-Hours	Unit	2020 Bare Costs Material	Labor	Equipment	Total	Total Incl O&P
2340	1-1/2" pipe size	Q-1	40	.400	Ea.	59	23		82	101
2350	2" pipe size		36	.444		64.50	26		90.50	111
2360	2-1/2" pipe size		28	.571		75	33		108	134
2370	3" pipe size		20	.800		97	46.50		143.50	179
2380	3-1/2" pipe size		16	1		138	58		196	242
2390	4" pipe size		12	1.333		119	77.50		196.50	251
2400	5" pipe size		10	1.600		184	93		277	345
2410	6" pipe size	Q-2	14	1.714		209	103		312	390
2420	8" pipe size		12	2		330	120		450	545
2430	10" pipe size		10	2.400		585	144		729	865
2440	12" pipe size		8	3		1,325	180		1,505	1,725
2460	For galvanized flanges, add					95%				
2490	Blind flange									
2492	Cast iron									
2494	Black, 125 lb., per flange									
2496	1" pipe size	1 Plum	27	.296	Ea.	92.50	19.10		111.60	132
2500	1-1/2" pipe size	Q-1	40	.400		100	23		123	146
2502	2" pipe size		36	.444		117	26		143	169
2504	2-1/2" pipe size		28	.571		125	33		158	189
2506	3" pipe size		20	.800		155	46.50		201.50	242
2508	4" pipe size		12	1.333		198	77.50		275.50	340
2510	5" pipe size		10	1.600		320	93		413	500
2512	6" pipe size	Q-2	14	1.714		350	103		453	545
2514	8" pipe size		12	2		550	120		670	790
2516	10" pipe size		10	2.400		815	144		959	1,125
2518	12" pipe size		8	3		1,575	180		1,755	2,000
2520	For galvanized flanges, add					80%				
2570	Threaded flange									
2580	Forged steel									
2590	Black, 150 lb., per flange									
2600	1/2" pipe size	1 Plum	30	.267	Ea.	27.50	17.20		44.70	57
2610	3/4" pipe size		28	.286		27.50	18.40		45.90	59
2620	1" pipe size		27	.296		27.50	19.10		46.60	60
2630	1-1/4" pipe size	Q-1	44	.364		27.50	21		48.50	63
2640	1-1/2" pipe size		40	.400		27.50	23		50.50	66.50
2650	2" pipe size		36	.444		31	26		57	74
2660	2-1/2" pipe size		28	.571		42.50	33		75.50	98.50
2670	3" pipe size		20	.800		43	46.50		89.50	119
2690	4" pipe size		12	1.333		48.50	77.50		126	173
2700	5" pipe size		10	1.600		79.50	93		172.50	231
2710	6" pipe size	Q-2	14	1.714		85	103		188	253
2720	8" pipe size		12	2		149	120		269	350
2730	10" pipe size		10	2.400		266	144		410	515
2860	Black, 300 lb., per flange									
2870	1/2" pipe size	1 Plum	30	.267	Ea.	30.50	17.20		47.70	60
2880	3/4" pipe size		28	.286		30.50	18.40		48.90	62
2890	1" pipe size		27	.296		30.50	19.10		49.60	63
2900	1-1/4" pipe size	Q-1	44	.364		30.50	21		51.50	66
2910	1-1/2" pipe size		40	.400		30.50	23		53.50	69.50
2920	2" pipe size		36	.444		40	26		66	84
2930	2-1/2" pipe size		28	.571		49.50	33		82.50	106
2940	3" pipe size		20	.800		50.50	46.50		97	127
2960	4" pipe size		12	1.333		72.50	77.50		150	200

22 11 13.47 Pipe Fittings, Steel		Crew	Daily Output	Labor-Hours	Unit	Material	2020 Bare Costs Labor	Equipment	Total	Total Incl O&P
2970	6" pipe size	Q-2	14	1.714	Ea.	157	103		260	330
3000	Weld joint, butt, carbon steel, standard weight									
3040	90° elbow, long radius									
3050	1/2" pipe size	Q-15	16	1	Ea.	44	58	6.65	108.65	145
3060	3/4" pipe size		16	1		44	58	6.65	108.65	145
3070	1" pipe size		16	1		26.50	58	6.65	91.15	126
3080	1-1/4" pipe size		14	1.143		26.50	66.50	7.60	100.60	139
3090	1-1/2" pipe size		13	1.231		26.50	71.50	8.20	106.20	148
3100	2" pipe size		10	1.600		22.50	93	10.65	126.15	179
3110	2-1/2" pipe size		8	2		35	116	13.30	164.30	232
3120	3" pipe size		7	2.286		30	133	15.20	178.20	255
3130	4" pipe size		5	3.200		56.50	186	21.50	264	375
3136	5" pipe size		4	4		112	232	26.50	370.50	515
3140	6" pipe size	Q-16	5	4.800		126	289	21.50	436.50	605
3150	8" pipe size		3.75	6.400		237	385	28.50	650.50	885
3160	10" pipe size		3	8		470	480	35.50	985.50	1,300
3170	12" pipe size		2.50	9.600		665	575	42.50	1,282.50	1,675
3180	14" pipe size		2	12		1,025	720	53	1,798	2,300
3190	16" pipe size		1.50	16		1,400	960	71	2,431	3,125
3191	18" pipe size		1.25	19.200		1,750	1,150	85	2,985	3,800
3192	20" pipe size		1.15	20.870		2,425	1,250	92.50	3,767.50	4,700
3194	24" pipe size		1.02	23.529		3,625	1,425	104	5,154	6,325
3200	45° elbow, long									
3210	1/2" pipe size	Q-15	16	1	Ea.	65.50	58	6.65	130.15	169
3220	3/4" pipe size		16	1		65.50	58	6.65	130.15	169
3230	1" pipe size		16	1		23	58	6.65	87.65	122
3240	1-1/4" pipe size		14	1.143		23	66.50	7.60	97.10	136
3250	1-1/2" pipe size		13	1.231		23	71.50	8.20	102.70	145
3260	2" pipe size		10	1.600		23	93	10.65	126.65	180
3270	2-1/2" pipe size		8	2		28	116	13.30	157.30	225
3280	3" pipe size		7	2.286		29	133	15.20	177.20	254
3290	4" pipe size		5	3.200		52	186	21.50	259.50	370
3296	5" pipe size		4	4		79	232	26.50	337.50	475
3300	6" pipe size	Q-16	5	4.800		103	289	21.50	413.50	580
3310	8" pipe size		3.75	6.400		171	385	28.50	584.50	815
3320	10" pipe size		3	8		340	480	35.50	855.50	1,150
3330	12" pipe size		2.50	9.600		480	575	42.50	1,097.50	1,475
3340	14" pipe size		2	12		650	720	53	1,423	1,900
3341	16" pipe size		1.50	16		1,150	960	71	2,181	2,850
3342	18" pipe size		1.25	19.200		1,650	1,150	85	2,885	3,675
3343	20" pipe size		1.15	20.870		1,700	1,250	92.50	3,042.50	3,925
3345	24" pipe size		1.05	22.857		2,525	1,375	101	4,001	5,000
3346	26" pipe size		.85	28.235		2,825	1,700	125	4,650	5,900
3347	30" pipe size		.45	53.333		3,125	3,200	236	6,561	8,625
3349	36" pipe size		.38	63.158		3,425	3,800	280	7,505	9,950
3350	Tee, straight									
3360	1/2" pipe size	Q-15	10	1.600	Ea.	116	93	10.65	219.65	283
3370	3/4" pipe size		10	1.600		116	93	10.65	219.65	283
3380	1" pipe size		10	1.600		57.50	93	10.65	161.15	218
3390	1-1/4" pipe size		9	1.778		72	103	11.80	186.80	251
3400	1-1/2" pipe size		8	2		72	116	13.30	201.30	273
3410	2" pipe size		6	2.667		57.50	155	17.75	230.25	320
3420	2-1/2" pipe size		5	3.200		79.50	186	21.50	287	400

22 11 13.47 Pipe Fittings, Steel		Crew	Daily Output	Labor-Hours	Unit	Material	Labor	Equipment	Total	Total Incl O&P
						2020 Bare Costs				
3430	3" pipe size	Q-15	4	4	Ea.	88.50	232	26.50	347	485
3440	4" pipe size		3	5.333		124	310	35.50	469.50	655
3446	5" pipe size		2.50	6.400		205	370	42.50	617.50	845
3450	6" pipe size	Q-16	3	8		213	480	35.50	728.50	1,025
3460	8" pipe size		2.50	9.600		370	575	42.50	987.50	1,350
3470	10" pipe size		2	12		730	720	53	1,503	2,000
3480	12" pipe size		1.60	15		1,025	900	66.50	1,991.50	2,600
3481	14" pipe size		1.30	18.462		1,800	1,100	82	2,982	3,800
3482	16" pipe size		1	24		2,000	1,450	106	3,556	4,550
3483	18" pipe size		.80	30		3,175	1,800	133	5,108	6,450
3484	20" pipe size		.75	32		5,000	1,925	142	7,067	8,625
3486	24" pipe size		.70	34.286		6,450	2,050	152	8,652	10,500
3487	26" pipe size		.55	43.636		7,075	2,625	193	9,893	12,000
3488	30" pipe size		.30	80		7,750	4,800	355	12,905	16,400
3490	36" pipe size		.25	96		8,550	5,775	425	14,750	18,800
3491	Eccentric reducer, 1-1/2" pipe size	Q-15	14	1.143		44.50	66.50	7.60	118.60	159
3492	2" pipe size		11	1.455		58	84.50	9.65	152.15	204
3493	2-1/2" pipe size		9	1.778		63.50	103	11.80	178.30	242
3494	3" pipe size		8	2		83	116	13.30	212.30	285
3495	4" pipe size		6	2.667		166	155	17.75	338.75	440
3496	6" pipe size	Q-16	5	4.800		223	289	21.50	533.50	715
3497	8" pipe size		4	6		258	360	26.50	644.50	875
3498	10" pipe size		3	8		440	480	35.50	955.50	1,275
3499	12" pipe size		2.50	9.600		680	575	42.50	1,297.50	1,675
3501	Cap, 1-1/2" pipe size	Q-15	28	.571		20.50	33	3.80	57.30	78
3502	2" pipe size		22	.727		22.50	42	4.84	69.34	95
3503	2-1/2" pipe size		18	.889		23.50	51.50	5.90	80.90	112
3504	3" pipe size		16	1		23.50	58	6.65	88.15	122
3505	4" pipe size		12	1.333		34	77.50	8.85	120.35	167
3506	6" pipe size	Q-16	10	2.400		58.50	144	10.65	213.15	299
3507	8" pipe size		8	3		89	180	13.30	282.30	390
3508	10" pipe size		6	4		160	241	17.70	418.70	565
3509	12" pipe size		5	4.800		247	289	21.50	557.50	740
3511	14" pipe size		4	6		480	360	26.50	866.50	1,125
3512	16" pipe size		4	6		535	360	26.50	921.50	1,175
3513	18" pipe size		3	8		515	480	35.50	1,030.50	1,350
3517	Weld joint, butt, carbon steel, extra strong									
3519	90° elbow, long									
3520	1/2" pipe size	Q-15	13	1.231	Ea.	58.50	71.50	8.20	138.20	184
3530	3/4" pipe size		12	1.333		58.50	77.50	8.85	144.85	194
3540	1" pipe size		11	1.455		28.50	84.50	9.65	122.65	172
3550	1-1/4" pipe size		10	1.600		28.50	93	10.65	132.15	186
3560	1-1/2" pipe size		9	1.778		28.50	103	11.80	143.30	204
3570	2" pipe size		8	2		29	116	13.30	158.30	226
3580	2-1/2" pipe size		7	2.286		40.50	133	15.20	188.70	267
3590	3" pipe size		6	2.667		52	155	17.75	224.75	315
3600	4" pipe size		4	4		86	232	26.50	344.50	485
3606	5" pipe size		3.50	4.571		206	265	30.50	501.50	670
3610	6" pipe size	Q-16	4.50	5.333		217	320	23.50	560.50	760
3620	8" pipe size		3.50	6.857		410	410	30.50	850.50	1,125
3630	10" pipe size		2.50	9.600		870	575	42.50	1,487.50	1,900
3640	12" pipe size		2.25	10.667		1,075	640	47.50	1,762.50	2,225
3650	45° elbow, long									

22 11 13.47 Pipe Fittings, Steel		Crew	Daily Output	Labor-Hours	Unit	2020 Bare Costs Material	Labor	Equipment	Total	Total Incl O&P
3660	1/2" pipe size	Q-15	13	1.231	Ea.	65	71.50	8.20	144.70	191
3670	3/4" pipe size		12	1.333		65	77.50	8.85	151.35	201
3680	1" pipe size		11	1.455		30.50	84.50	9.65	124.65	174
3690	1-1/4" pipe size		10	1.600		30.50	93	10.65	134.15	188
3700	1-1/2" pipe size		9	1.778		30.50	103	11.80	145.30	206
3710	2" pipe size		8	2		30.50	116	13.30	159.80	227
3720	2-1/2" pipe size		7	2.286		67.50	133	15.20	215.70	296
3730	3" pipe size		6	2.667		39	155	17.75	211.75	300
3740	4" pipe size		4	4		61.50	232	26.50	320	455
3746	5" pipe size		3.50	4.571		146	265	30.50	441.50	605
3750	6" pipe size	Q-16	4.50	5.333		169	320	23.50	512.50	705
3760	8" pipe size		3.50	6.857		291	410	30.50	731.50	995
3770	10" pipe size		2.50	9.600		560	575	42.50	1,177.50	1,550
3780	12" pipe size		2.25	10.667		830	640	47.50	1,517.50	1,950
3800	Tee, straight									
3810	1/2" pipe size	Q-15	9	1.778	Ea.	143	103	11.80	257.80	330
3820	3/4" pipe size		8.50	1.882		137	109	12.50	258.50	335
3830	1" pipe size		8	2		54.50	116	13.30	183.80	253
3840	1-1/4" pipe size		7	2.286		54	133	15.20	202.20	281
3850	1-1/2" pipe size		6	2.667		54.50	155	17.75	227.25	320
3860	2" pipe size		5	3.200		61	186	21.50	268.50	380
3870	2-1/2" pipe size		4	4		102	232	26.50	360.50	500
3880	3" pipe size		3.50	4.571		128	265	30.50	423.50	585
3890	4" pipe size		2.50	6.400		153	370	42.50	565.50	790
3896	5" pipe size		2.25	7.111		390	410	47.50	847.50	1,125
3900	6" pipe size	Q-16	2.25	10.667		300	640	47.50	987.50	1,375
3910	8" pipe size		2	12		575	720	53	1,348	1,825
3920	10" pipe size		1.75	13.714		865	825	61	1,751	2,300
3930	12" pipe size		1.50	16		1,275	960	71	2,306	2,975
4000	Eccentric reducer, 1-1/2" pipe size	Q-15	10	1.600		20.50	93	10.65	124.15	178
4010	2" pipe size		9	1.778		56.50	103	11.80	171.30	235
4020	2-1/2" pipe size		8	2		82.50	116	13.30	211.80	284
4030	3" pipe size		7	2.286		69	133	15.20	217.20	298
4040	4" pipe size		5	3.200		111	186	21.50	318.50	435
4046	5" pipe size		4.70	3.404		300	197	22.50	519.50	660
4050	6" pipe size	Q-16	4.50	5.333		305	320	23.50	648.50	855
4060	8" pipe size		3.50	6.857		405	410	30.50	845.50	1,125
4070	10" pipe size		2.50	9.600		690	575	42.50	1,307.50	1,700
4080	12" pipe size		2.25	10.667		880	640	47.50	1,567.50	2,000
4090	14" pipe size		2.10	11.429		1,650	685	50.50	2,385.50	2,925
4100	16" pipe size		1.90	12.632		2,125	760	56	2,941	3,550
4151	Cap, 1-1/2" pipe size	Q-15	24	.667		25.50	38.50	4.43	68.43	93
4152	2" pipe size		18	.889		22.50	51.50	5.90	79.90	111
4153	2-1/2" pipe size		16	1		30.50	58	6.65	95.15	130
4154	3" pipe size		14	1.143		34	66.50	7.60	108.10	148
4155	4" pipe size		10	1.600		45.50	93	10.65	149.15	205
4156	6" pipe size	Q-16	9	2.667		95.50	160	11.80	267.30	365
4157	8" pipe size		7	3.429		143	206	15.20	364.20	495
4158	10" pipe size		5	4.800		217	289	21.50	527.50	710
4159	12" pipe size		4	6		294	360	26.50	680.50	915
4190	Weld fittings, reducing, standard weight									
4200	Welding ring w/spacer pins, 2" pipe size				Ea.	2.23			2.23	2.45
4210	2-1/2" pipe size					2.16			2.16	2.38

22 11 13 – Facility Water Distribution Piping

22 11 13.47 Pipe Fittings, Steel		Crew	Daily Output	Labor-Hours	Unit	Material	Labor	Equipment	Total	Total Incl O&P
						2020 Bare Costs				
4220	3" pipe size				Ea.	2.25			2.25	2.48
4230	4" pipe size					2.81			2.81	3.09
4236	5" pipe size					3.46			3.46	3.81
4240	6" pipe size					3.29			3.29	3.62
4250	8" pipe size					3.65			3.65	4.02
4260	10" pipe size					4.22			4.22	4.64
4270	12" pipe size					4.93			4.93	5.40
4280	14" pipe size					5.65			5.65	6.25
4290	16" pipe size					6.65			6.65	7.30
4300	18" pipe size					7.35			7.35	8.05
4310	20" pipe size					9.25			9.25	10.20
4330	24" pipe size					12.85			12.85	14.10
4340	26" pipe size					15.85			15.85	17.40
4350	30" pipe size					18.40			18.40	20.50
4370	36" pipe size				↓	23			23	25.50
5000	Weld joint, socket, forged steel, 3000 lb., schedule 40 pipe									
5010	90° elbow, straight									
5020	1/4" pipe size	Q-15	22	.727	Ea.	29.50	42	4.84	76.34	103
5030	3/8" pipe size		22	.727		29.50	42	4.84	76.34	103
5040	1/2" pipe size		20	.800		16.45	46.50	5.30	68.25	95.50
5050	3/4" pipe size		20	.800		13.65	46.50	5.30	65.45	92.50
5060	1" pipe size		20	.800		22	46.50	5.30	73.80	101
5070	1-1/4" pipe size		18	.889		42.50	51.50	5.90	99.90	133
5080	1-1/2" pipe size		16	1		41.50	58	6.65	106.15	142
5090	2" pipe size		12	1.333		62	77.50	8.85	148.35	198
5100	2-1/2" pipe size		10	1.600		206	93	10.65	309.65	380
5110	3" pipe size		8	2		355	116	13.30	484.30	585
5120	4" pipe size	↓	6	2.667	↓	925	155	17.75	1,097.75	1,275
5130	45° elbow, straight									
5134	1/4" pipe size	Q-15	22	.727	Ea.	29.50	42	4.84	76.34	103
5135	3/8" pipe size		22	.727		29.50	42	4.84	76.34	103
5136	1/2" pipe size		20	.800		22	46.50	5.30	73.80	101
5137	3/4" pipe size		20	.800		25.50	46.50	5.30	77.30	105
5140	1" pipe size		20	.800		33.50	46.50	5.30	85.30	114
5150	1-1/4" pipe size		18	.889		46	51.50	5.90	103.40	137
5160	1-1/2" pipe size		16	1		56	58	6.65	120.65	158
5170	2" pipe size		12	1.333		90.50	77.50	8.85	176.85	229
5180	2-1/2" pipe size		10	1.600		238	93	10.65	341.65	415
5190	3" pipe size		8	2		395	116	13.30	524.30	630
5200	4" pipe size	↓	6	2.667	↓	775	155	17.75	947.75	1,125
5250	Tee, straight									
5254	1/4" pipe size	Q-15	15	1.067	Ea.	32.50	62	7.10	101.60	139
5255	3/8" pipe size		15	1.067		32.50	62	7.10	101.60	139
5256	1/2" pipe size		13	1.231		20.50	71.50	8.20	100.20	142
5257	3/4" pipe size		13	1.231		25	71.50	8.20	104.70	147
5260	1" pipe size		13	1.231		34	71.50	8.20	113.70	157
5270	1-1/4" pipe size		12	1.333		52.50	77.50	8.85	138.85	188
5280	1-1/2" pipe size		11	1.455		68.50	84.50	9.65	162.65	216
5290	2" pipe size		8	2		100	116	13.30	229.30	305
5300	2-1/2" pipe size		6	2.667		246	155	17.75	418.75	530
5310	3" pipe size		5	3.200		590	186	21.50	797.50	955
5320	4" pipe size	↓	4	4	↓	1,100	232	26.50	1,358.50	1,625
5350	For reducing sizes, add					60%				

22 11 13.47 Pipe Fittings, Steel		Crew	Daily Output	Labor-Hours	Unit	Material	Labor	Equipment	Total	Total Incl O&P
						2020 Bare Costs				
5450	Couplings									
5451	1/4" pipe size	Q-15	23	.696	Ea.	19.25	40.50	4.63	64.38	88.50
5452	3/8" pipe size		23	.696		19.25	40.50	4.63	64.38	88.50
5453	1/2" pipe size		21	.762		8.75	44	5.05	57.80	83.50
5454	3/4" pipe size		21	.762		11.30	44	5.05	60.35	86.50
5460	1" pipe size		20	.800		12.45	46.50	5.30	64.25	91
5470	1-1/4" pipe size		20	.800		22	46.50	5.30	73.80	102
5480	1-1/2" pipe size		18	.889		24.50	51.50	5.90	81.90	113
5490	2" pipe size		14	1.143		39	66.50	7.60	113.10	153
5500	2-1/2" pipe size		12	1.333		87.50	77.50	8.85	173.85	226
5510	3" pipe size		9	1.778		196	103	11.80	310.80	390
5520	4" pipe size		7	2.286		295	133	15.20	443.20	545
5570	Union, 1/4" pipe size		21	.762		40.50	44	5.05	89.55	119
5571	3/8" pipe size		21	.762		40.50	44	5.05	89.55	119
5572	1/2" pipe size		19	.842		34	49	5.60	88.60	119
5573	3/4" pipe size		19	.842		39.50	49	5.60	94.10	125
5574	1" pipe size		19	.842		51	49	5.60	105.60	138
5575	1-1/4" pipe size		17	.941		82	54.50	6.25	142.75	182
5576	1-1/2" pipe size		15	1.067		89.50	62	7.10	158.60	202
5577	2" pipe size		11	1.455		109	84.50	9.65	203.15	261
5600	Reducer, 1/4" pipe size		23	.696		45.50	40.50	4.63	90.63	118
5601	3/8" pipe size		23	.696		48.50	40.50	4.63	93.63	121
5602	1/2" pipe size		21	.762		31	44	5.05	80.05	108
5603	3/4" pipe size		21	.762		31	44	5.05	80.05	108
5604	1" pipe size		21	.762		36.50	44	5.05	85.55	115
5605	1-1/4" pipe size		19	.842		55	49	5.60	109.60	142
5607	1-1/2" pipe size		17	.941		53	54.50	6.25	113.75	149
5608	2" pipe size		13	1.231		58.50	71.50	8.20	138.20	184
5612	Cap, 1/4" pipe size		46	.348		18.10	20	2.31	40.41	53.50
5613	3/8" pipe size		46	.348		18.10	20	2.31	40.41	53.50
5614	1/2" pipe size		42	.381		10.95	22	2.53	35.48	49
5615	3/4" pipe size		42	.381		12.75	22	2.53	37.28	51
5616	1" pipe size		42	.381		19.35	22	2.53	43.88	58.50
5617	1-1/4" pipe size		38	.421		23	24.50	2.80	50.30	66
5618	1-1/2" pipe size		34	.471		32	27.50	3.13	62.63	80.50
5619	2" pipe size		26	.615		49	35.50	4.09	88.59	114
5630	T-O-L, 1/4" pipe size, nozzle		23	.696		8.25	40.50	4.63	53.38	76.50
5631	3/8" pipe size, nozzle		23	.696		8.40	40.50	4.63	53.53	77
5632	1/2" pipe size, nozzle		22	.727		8.25	42	4.84	55.09	79.50
5633	3/4" pipe size, nozzle		21	.762		9.50	44	5.05	58.55	84.50
5634	1" pipe size, nozzle		20	.800		11.10	46.50	5.30	62.90	89.50
5635	1-1/4" pipe size, nozzle		18	.889		15.80	51.50	5.90	73.20	103
5636	1-1/2" pipe size, nozzle		16	1		17.55	58	6.65	82.20	116
5637	2" pipe size, nozzle		12	1.333		19.15	77.50	8.85	105.50	151
5638	2-1/2" pipe size, nozzle		10	1.600		67	93	10.65	170.65	229
5639	4" pipe size, nozzle		6	2.667		151	155	17.75	323.75	425
5640	W-O-L, 1/4" pipe size, nozzle		23	.696		17.60	40.50	4.63	62.73	87
5641	3/8" pipe size, nozzle		23	.696		16.65	40.50	4.63	61.78	86
5642	1/2" pipe size, nozzle		22	.727		17.10	42	4.84	63.94	89
5643	3/4" pipe size, nozzle		21	.762		18.05	44	5.05	67.10	94
5644	1" pipe size, nozzle		20	.800		18.85	46.50	5.30	70.65	98
5645	1-1/4" pipe size, nozzle		18	.889		21.50	51.50	5.90	78.90	110
5646	1-1/2" pipe size, nozzle	↓	16	1	↓	22.50	58	6.65	87.15	121

22 11 13.47 Pipe Fittings, Steel		Crew	Daily Output	Labor-Hours	Unit	Material	Labor	Equipment	Total	Total Incl O&P
						2020 Bare Costs				
5647	2" pipe size, nozzle	Q-15	12	1.333	Ea.	22.50	77.50	8.85	108.85	155
5648	2-1/2" pipe size, nozzle		10	1.600		51.50	93	10.65	155.15	212
5649	3" pipe size, nozzle		8	2		56.50	116	13.30	185.80	256
5650	4" pipe size, nozzle		6	2.667		71.50	155	17.75	244.25	335
5651	5" pipe size, nozzle		5	3.200		176	186	21.50	383.50	505
5652	6" pipe size, nozzle		4	4		199	232	26.50	457.50	610
5653	8" pipe size, nozzle		3	5.333		380	310	35.50	725.50	935
5654	10" pipe size, nozzle		2.60	6.154		525	355	41	921	1,175
5655	12" pipe size, nozzle		2.20	7.273		1,000	420	48.50	1,468.50	1,800
5674	S-O-L, 1/4" pipe size, outlet		23	.696		9.80	40.50	4.63	54.93	78.50
5675	3/8" pipe size, outlet		23	.696		9.80	40.50	4.63	54.93	78.50
5676	1/2" pipe size, outlet		22	.727		10.25	42	4.84	57.09	81.50
5677	3/4" pipe size, outlet		21	.762		10.40	44	5.05	59.45	85.50
5678	1" pipe size, outlet		20	.800		11.50	46.50	5.30	63.30	90
5679	1-1/4" pipe size, outlet		18	.889		19.25	51.50	5.90	76.65	107
5680	1-1/2" pipe size, outlet		16	1		19.25	58	6.65	83.90	118
5681	2" pipe size, outlet	↓	12	1.333	↓	19.75	77.50	8.85	106.10	151
6000	Weld-on flange, forged steel									
6020	Slip-on, 150 lb. flange (welded front and back)									
6050	1/2" pipe size	Q-15	18	.889	Ea.	19.85	51.50	5.90	77.25	108
6060	3/4" pipe size		18	.889		19.85	51.50	5.90	77.25	108
6070	1" pipe size		17	.941		19.85	54.50	6.25	80.60	113
6080	1-1/4" pipe size		16	1		19.85	58	6.65	84.50	119
6090	1-1/2" pipe size		15	1.067		19.85	62	7.10	88.95	125
6100	2" pipe size		12	1.333		26	77.50	8.85	112.35	159
6110	2-1/2" pipe size		10	1.600		38.50	93	10.65	142.15	197
6120	3" pipe size		9	1.778		32	103	11.80	146.80	208
6130	3-1/2" pipe size		7	2.286		34	133	15.20	182.20	259
6140	4" pipe size		6	2.667		33	155	17.75	205.75	295
6150	5" pipe size	↓	5	3.200		67.50	186	21.50	275	385
6160	6" pipe size	Q-16	6	4		54	241	17.70	312.70	450
6170	8" pipe size		5	4.800		94.50	289	21.50	405	575
6180	10" pipe size		4	6		166	360	26.50	552.50	775
6190	12" pipe size		3	8		249	480	35.50	764.50	1,050
6191	14" pipe size		2.50	9.600		280	575	42.50	897.50	1,250
6192	16" pipe size	↓	1.80	13.333	↓	495	800	59	1,354	1,850
6200	300 lb. flange									
6210	1/2" pipe size	Q-15	17	.941	Ea.	24.50	54.50	6.25	85.25	118
6220	3/4" pipe size		17	.941		24.50	54.50	6.25	85.25	118
6230	1" pipe size		16	1		24.50	58	6.65	89.15	124
6240	1-1/4" pipe size		13	1.231		24.50	71.50	8.20	104.20	146
6250	1-1/2" pipe size		12	1.333		24.50	77.50	8.85	110.85	157
6260	2" pipe size		11	1.455		36.50	84.50	9.65	130.65	181
6270	2-1/2" pipe size		9	1.778		36.50	103	11.80	151.30	212
6280	3" pipe size		7	2.286		38	133	15.20	186.20	264
6290	4" pipe size		6	2.667		57.50	155	17.75	230.25	320
6300	5" pipe size	↓	4	4		111	232	26.50	369.50	510
6310	6" pipe size	Q-16	5	4.800		108	289	21.50	418.50	590
6320	8" pipe size		4	6		184	360	26.50	570.50	790
6330	10" pipe size		3.40	7.059		325	425	31.50	781.50	1,050
6340	12" pipe size	↓	2.80	8.571	↓	340	515	38	893	1,200
6400	Welding neck, 150 lb. flange									
6410	1/2" pipe size	Q-15	40	.400	Ea.	25.50	23	2.66	51.16	67

22 11 13 – Facility Water Distribution Piping

22 11 13.47 Pipe Fittings, Steel		Crew	Daily Output	Labor-Hours	Unit	Material	2020 Bare Costs Labor	Equipment	Total	Total Incl O&P
6420	3/4" pipe size	Q-15	36	.444	Ea.	25.50	26	2.96	54.46	71.50
6430	1" pipe size		32	.500		25.50	29	3.33	57.83	76.50
6440	1-1/4" pipe size		29	.552		25.50	32	3.67	61.17	81.50
6450	1-1/2" pipe size		26	.615		25.50	35.50	4.09	65.09	87.50
6460	2" pipe size		20	.800		34	46.50	5.30	85.80	115
6470	2-1/2" pipe size		16	1		37	58	6.65	101.65	138
6480	3" pipe size		14	1.143		41	66.50	7.60	115.10	155
6500	4" pipe size		10	1.600		42	93	10.65	145.65	201
6510	5" pipe size		8	2		66.50	116	13.30	195.80	267
6520	6" pipe size	Q-16	10	2.400		63.50	144	10.65	218.15	305
6530	8" pipe size		7	3.429		118	206	15.20	339.20	465
6540	10" pipe size		6	4		188	241	17.70	446.70	595
6550	12" pipe size		5	4.800		296	289	21.50	606.50	795
6551	14" pipe size		4.50	5.333		440	320	23.50	783.50	1,000
6552	16" pipe size		3	8		595	480	35.50	1,110.50	1,450
6553	18" pipe size		2.50	9.600		800	575	42.50	1,417.50	1,825
6554	20" pipe size		2.30	10.435		970	630	46	1,646	2,100
6556	24" pipe size		2	12		1,300	720	53	2,073	2,625
6557	26" pipe size		1.70	14.118		1,425	850	62.50	2,337.50	2,975
6558	30" pipe size		.90	26.667		1,625	1,600	118	3,343	4,400
6559	36" pipe size		.75	32		1,875	1,925	142	3,942	5,200
6560	300 lb. flange									
6570	1/2" pipe size	Q-15	36	.444	Ea.	34.50	26	2.96	63.46	81.50
6580	3/4" pipe size		34	.471		34.50	27.50	3.13	65.13	83.50
6590	1" pipe size		30	.533		34.50	31	3.55	69.05	90
6600	1-1/4" pipe size		28	.571		34.50	33	3.80	71.30	93.50
6610	1-1/2" pipe size		24	.667		34.50	38.50	4.43	77.43	103
6620	2" pipe size		18	.889		45.50	51.50	5.90	102.90	136
6630	2-1/2" pipe size		14	1.143		52	66.50	7.60	126.10	167
6640	3" pipe size		12	1.333		44.50	77.50	8.85	130.85	179
6650	4" pipe size		8	2		76	116	13.30	205.30	277
6660	5" pipe size		7	2.286		111	133	15.20	259.20	345
6670	6" pipe size	Q-16	9	2.667		113	160	11.80	284.80	385
6680	8" pipe size		6	4		216	241	17.70	474.70	630
6690	10" pipe size		5	4.800		415	289	21.50	725.50	925
6700	12" pipe size		4	6		450	360	26.50	836.50	1,075
6710	14" pipe size		3.50	6.857		975	410	30.50	1,415.50	1,750
6720	16" pipe size		2	12		1,150	720	53	1,923	2,450
7740	Plain ends for plain end pipe, mechanically coupled									
7750	Cplg. & labor required at joints not included, add 1 per									
7760	joint for installed price, see line 9180									
7770	Malleable iron, painted, unless noted otherwise									
7800	90° elbow, 1"				Ea.	154			154	170
7810	1-1/2"					182			182	201
7820	2"					271			271	299
7830	2-1/2"					320			320	350
7840	3"					330			330	365
7860	4"					375			375	410
7870	5" welded steel					430			430	470
7880	6"					540			540	595
7890	8" welded steel					1,025			1,025	1,125
7900	10" welded steel					1,300			1,300	1,425
7910	12" welded steel					1,450			1,450	1,600

22 11 13.47 Pipe Fittings, Steel		Crew	Daily Output	Labor-Hours	Unit	2020 Bare Costs Material	Labor	Equipment	Total	Total Incl O&P
7970	45° elbow, 1"				Ea.	92.50			92.50	102
7980	1-1/2"					137			137	151
7990	2"					288			288	315
8000	2-1/2"					288			288	315
8010	3"					330			330	365
8030	4"					345			345	380
8040	5" welded steel					430			430	470
8050	6"					490			490	540
8060	8"					575			575	635
8070	10" welded steel					595			595	655
8080	12" welded steel					1,075			1,075	1,175
8140	Tee, straight 1"					172			172	189
8150	1-1/2"					222			222	244
8160	2"					222			222	244
8170	2-1/2"					289			289	320
8180	3"					445			445	490
8200	4"					635			635	700
8210	5" welded steel					870			870	955
8220	6"					755			755	830
8230	8" welded steel					1,100			1,100	1,200
8240	10" welded steel					1,725			1,725	1,900
8250	12" welded steel					2,000			2,000	2,200
8340	Segmentally welded steel, painted									
8390	Wye 2"				Ea.	251			251	276
8400	2-1/2"					251			251	276
8410	3"					300			300	330
8430	4"					390			390	430
8440	5"					570			570	625
8450	6"					695			695	760
8460	8"					1,025			1,025	1,125
8470	10"					1,675			1,675	1,850
8480	12"					1,725			1,725	1,900
8540	Wye, lateral 2"					299			299	330
8550	2-1/2"					345			345	380
8560	3"					410			410	450
8580	4"					560			560	620
8590	5"					955			955	1,050
8600	6"					985			985	1,075
8610	8"					1,650			1,650	1,825
8620	10"					2,425			2,425	2,675
8630	12"					3,025			3,025	3,325
8690	Cross, 2"					181			181	199
8700	2-1/2"					181			181	199
8710	3"					216			216	237
8730	4"					490			490	540
8740	5"					560			560	615
8750	6"					710			710	780
8760	8"					1,100			1,100	1,225
8770	10"					1,425			1,425	1,575
8780	12"					1,825			1,825	2,025
8800	Tees, reducing 2" x 1"					235			235	259
8810	2" x 1-1/2"					235			235	259
8820	3" x 1"					217			217	238

22 11 13 – Facility Water Distribution Piping

22 11 13.47 Pipe Fittings, Steel		Crew	Daily Output	Labor-Hours	Unit	Material	2020 Bare Costs Labor	Equipment	Total	Total Incl O&P
8830	3" x 1-1/2"				Ea.	298			298	330
8840	3" x 2"					216			216	237
8850	4" x 1"					315			315	350
8860	4" x 1-1/2"					315			315	350
8870	4" x 2"					470			470	515
8880	4" x 2-1/2"					485			485	530
8890	4" x 3"					485			485	530
8900	6" x 2"					400			400	440
8910	6" x 3"					298			298	330
8920	6" x 4"					380			380	420
8930	8" x 2"					575			575	635
8940	8" x 3"					625			625	690
8950	8" x 4"					670			670	735
8960	8" x 5"					470			470	515
8970	8" x 6"					640			640	705
8980	10" x 4"					530			530	585
8990	10" x 6"					545			545	600
9000	10" x 8"					850			850	940
9010	12" x 6"					1,325			1,325	1,450
9020	12" x 8"					1,050			1,050	1,150
9030	12" x 10"				↓	1,025			1,025	1,150
9080	Adapter nipples 3" long									
9090	1"				Ea.	27			27	29.50
9100	1-1/2"					27			27	29.50
9110	2"					27			27	29.50
9120	2-1/2"					31			31	34.50
9130	3"					38			38	41.50
9140	4"					61.50			61.50	67.50
9150	6"				↓	157			157	172
9180	Coupling, mechanical, plain end pipe to plain end pipe or fitting									
9190	1"	Q-1	29	.552	Ea.	93.50	32		125.50	153
9200	1-1/2"		28	.571		93.50	33		126.50	155
9210	2"		27	.593		93.50	34.50		128	156
9220	2-1/2"		26	.615		93.50	35.50		129	158
9230	3"		25	.640		136	37		173	208
9240	3-1/2"		24	.667		157	38.50		195.50	233
9250	4"	↓	22	.727		157	42		199	238
9260	5"	Q-2	28	.857		223	51.50		274.50	325
9270	6"		24	1		273	60		333	395
9280	8"		19	1.263		480	76		556	645
9290	10"		16	1.500		625	90		715	825
9300	12"	↓	12	2		785	120		905	1,050
9805	2" x 1"	1 Plum	50	.160		101	10.30		111.30	127
9810	2" x 1-1/4"		50	.160		117	10.30		127.30	144
9815	2" x 1-1/2"	↓	50	.160		125	10.30		135.30	154
9820	2-1/2" x 1"	Q-1	80	.200		106	11.60		117.60	134
9825	2-1/2" x 1-1/4"		80	.200		156	11.60		167.60	190
9830	2-1/2" x 1-1/2"		80	.200		156	11.60		167.60	190
9835	2-1/2" x 2"		80	.200		106	11.60		117.60	134
9840	3" x 1"		80	.200		115	11.60		126.60	144
9845	3" x 1-1/4"		80	.200		153	11.60		164.60	187
9850	3" x 1-1/2"		80	.200		153	11.60		164.60	187
9852	3" x 2"	↓	80	.200	↓	184	11.60		195.60	220

22 11 13.47 Pipe Fittings, Steel		Crew	Daily Output	Labor-Hours	Unit	Material	2020 Bare Costs Labor	Equipment	Total	Total Incl O&P
9856	4" x 1"	Q-1	67	.239	Ea.	142	13.85		155.85	179
9858	4" x 1-1/4"		67	.239		187	13.85		200.85	228
9860	4" x 1-1/2"		67	.239		191	13.85		204.85	232
9862	4" x 2"		67	.239		196	13.85		209.85	237
9864	4" x 2-1/2"		67	.239		203	13.85		216.85	245
9866	4" x 3"		67	.239		207	13.85		220.85	250
9870	6" x 1-1/2"	Q-2	50	.480		239	29		268	310
9872	6" x 2"		50	.480		242	29		271	310
9874	6" x 2-1/2"		50	.480		246	29		275	315
9876	6" x 3"		50	.480		274	29		303	345
9878	6" x 4"		50	.480		305	29		334	380
9880	8" x 2"		42	.571		415	34.50		449.50	510
9882	8" x 2-1/2"		42	.571		415	34.50		449.50	510
9884	8" x 3"		42	.571		435	34.50		469.50	535
9886	8" x 4"		42	.571		890	34.50		924.50	1,025
9940	For galvanized fittings for plain end pipe, add					20%				
9990	Minimum labor/equipment charge	Q-15	3	5.333	Job		310	35.50	345.50	520

22 11 13.48 Pipe, Fittings and Valves, Steel, Grooved-Joint		Crew	Daily Output	Labor-Hours	Unit	Material	Labor	Equipment	Total	Total Incl O&P
0010	**PIPE, FITTINGS AND VALVES, STEEL, GROOVED-JOINT**									
0012	Fittings are ductile iron. Steel fittings noted.									
0020	Pipe includes coupling & clevis type hanger assemblies, 10' OC									
0500	Schedule 10, black									
0550	2" diameter	1 Plum	43	.186	L.F.	9.40	12		21.40	29
0560	2-1/2" diameter	Q-1	61	.262		13.25	15.20		28.45	38
0570	3" diameter		55	.291		14.65	16.85		31.50	42
0580	3-1/2" diameter		53	.302		18.40	17.50		35.90	47.50
0590	4" diameter		49	.327		15.15	18.95		34.10	46
0600	5" diameter		40	.400		18.75	23		41.75	56.50
0610	6" diameter	Q-2	46	.522		24.50	31.50		56	75.50
0620	8" diameter	"	41	.585		36.50	35		71.50	94.50
0700	To delete couplings & hangers, subtract									
0710	2" diam. to 5" diam.					25%	20%			
0720	6" diam. to 8" diam.					27%	15%			
1000	Schedule 40, black									
1040	3/4" diameter	1 Plum	71	.113	L.F.	6.55	7.25		13.80	18.45
1050	1" diameter		63	.127		6.45	8.20		14.65	19.75
1060	1-1/4" diameter		58	.138		7.70	8.90		16.60	22.50
1070	1-1/2" diameter		51	.157		8.40	10.10		18.50	25
1080	2" diameter		40	.200		9.70	12.90		22.60	30.50
1090	2-1/2" diameter	Q-1	57	.281		16.50	16.30		32.80	43
1100	3" diameter		50	.320		22	18.55		40.55	52.50
1110	4" diameter		45	.356		26	20.50		46.50	60.50
1120	5" diameter		37	.432		41	25		66	84
1130	6" diameter	Q-2	42	.571		52.50	34.50		87	111
1140	8" diameter		37	.649		84.50	39		123.50	154
1150	10" diameter		31	.774		114	46.50		160.50	197
1160	12" diameter		27	.889		128	53.50		181.50	223
1170	14" diameter		20	1.200		138	72		210	264
1180	16" diameter		17	1.412		215	85		300	365
1190	18" diameter		14	1.714		207	103		310	385
1200	20" diameter		12	2		252	120		372	465
1210	24" diameter		10	2.400		284	144		428	535

22 11 13.48 Pipe, Fittings and Valves, Steel, Grooved-Joint		Crew	Daily Output	Labor-Hours	Unit	2020 Bare Costs Material	Labor	Equipment	Total	Total Incl O&P
1740	To delete coupling & hanger, subtract									
1750	3/4" diam. to 2" diam.					65%	27%			
1760	2-1/2" diam. to 5" diam.					41%	18%			
1770	6" diam. to 12" diam.					31%	13%			
1780	14" diam. to 24" diam.					35%	10%			
1800	Galvanized									
1840	3/4" diameter	1 Plum	71	.113	L.F.	7.15	7.25		14.40	19.10
1850	1" diameter		63	.127		7.10	8.20		15.30	20.50
1860	1-1/4" diameter		58	.138		8.50	8.90		17.40	23
1870	1-1/2" diameter		51	.157		9.25	10.10		19.35	26
1880	2" diameter		40	.200		10.80	12.90		23.70	32
1890	2-1/2" diameter	Q-1	57	.281		16.70	16.30		33	43.50
1900	3" diameter		50	.320		19.65	18.55		38.20	50
1910	4" diameter		45	.356		23	20.50		43.50	57.50
1920	5" diameter		37	.432		30.50	25		55.50	72.50
1930	6" diameter	Q-2	42	.571		39.50	34.50		74	96.50
1940	8" diameter		37	.649		58.50	39		97.50	125
1950	10" diameter		31	.774		114	46.50		160.50	197
1960	12" diameter		27	.889		137	53.50		190.50	234
2540	To delete coupling & hanger, subtract									
2550	3/4" diam. to 2" diam.					36%	27%			
2560	2-1/2" diam. to 5" diam.					19%	18%			
2570	6" diam. to 12" diam.					14%	13%			
2600	Schedule 80, black									
2610	3/4" diameter	1 Plum	65	.123	L.F.	8	7.95		15.95	21
2650	1" diameter		61	.131		8.50	8.45		16.95	22.50
2660	1-1/4" diameter		55	.145		10.55	9.35		19.90	26
2670	1-1/2" diameter		49	.163		11.75	10.50		22.25	29
2680	2" diameter		38	.211		14.20	13.55		27.75	36.50
2690	2-1/2" diameter	Q-1	54	.296		22	17.20		39.20	50.50
2700	3" diameter		48	.333		27	19.35		46.35	60
2710	4" diameter		44	.364		35.50	21		56.50	71.50
2720	5" diameter		35	.457		63	26.50		89.50	110
2730	6" diameter	Q-2	40	.600		72.50	36		108.50	136
2740	8" diameter		35	.686		113	41.50		154.50	189
2750	10" diameter		29	.828		169	50		219	262
2760	12" diameter		24	1		365	60		425	495
3240	To delete coupling & hanger, subtract									
3250	3/4" diam. to 2" diam.					30%	25%			
3260	2-1/2" diam. to 5" diam.					14%	17%			
3270	6" diam. to 12" diam.					12%	12%			
3300	Galvanized									
3310	3/4" diameter	1 Plum	65	.123	L.F.	8.60	7.95		16.55	21.50
3350	1" diameter		61	.131		9.40	8.45		17.85	23.50
3360	1-1/4" diameter		55	.145		11.85	9.35		21.20	27.50
3370	1-1/2" diameter		46	.174		13.30	11.20		24.50	32
3380	2" diameter		38	.211		17	13.55		30.55	39.50
3390	2-1/2" diameter	Q-1	54	.296		26	17.20		43.20	55
3400	3" diameter		48	.333		32	19.35		51.35	65
3410	4" diameter		44	.364		39.50	21		60.50	76
3420	5" diameter		35	.457		47	26.50		73.50	92.50
3430	6" diameter	Q-2	40	.600		65	36		101	128
3440	8" diameter		35	.686		120	41.50		161.50	196

22 11 13 – Facility Water Distribution Piping

22 11 13.48 Pipe, Fittings and Valves, Steel, Grooved-Joint		Crew	Daily Output	Labor-Hours	Unit	Material	2020 Bare Costs Labor	Equipment	Total	Total Incl O&P
3450	10″ diameter	Q-2	29	.828	L.F.	273	50		323	375
3460	12″ diameter	↓	24	1	↓	590	60		650	745
3920	To delete coupling & hanger, subtract									
3930	3/4″ diam. to 2″ diam.					30%	25%			
3940	2-1/2″ diam. to 5″ diam.					15%	17%			
3950	6″ diam. to 12″ diam.					11%	12%			
3990	Fittings: coupling material required at joints not incl. in fitting price.									
3994	Add 1 selected coupling, material only, per joint for installed price.									
4000	Elbow, 90° or 45°, painted									
4030	3/4″ diameter	1 Plum	50	.160	Ea.	82.50	10.30		92.80	107
4040	1″ diameter		50	.160		44.50	10.30		54.80	65
4050	1-1/4″ diameter		40	.200		44.50	12.90		57.40	69
4060	1-1/2″ diameter		33	.242		44.50	15.60		60.10	73
4070	2″ diameter	↓	25	.320		44.50	20.50		65	81
4080	2-1/2″ diameter	Q-1	40	.400		44.50	23		67.50	85
4090	3″ diameter		33	.485		78	28		106	129
4100	4″ diameter		25	.640		84.50	37		121.50	151
4110	5″ diameter	↓	20	.800		201	46.50		247.50	293
4120	6″ diameter	Q-2	25	.960		236	57.50		293.50	350
4130	8″ diameter		21	1.143		490	68.50		558.50	645
4140	10″ diameter		18	1.333		900	80		980	1,125
4150	12″ diameter		15	1.600		975	96		1,071	1,225
4170	14″ diameter		12	2		1,025	120		1,145	1,300
4180	16″ diameter	↓	11	2.182		1,125	131		1,256	1,425
4190	18″ diameter	Q-3	14	2.286		1,700	140		1,840	2,075
4200	20″ diameter		12	2.667		2,225	164		2,389	2,700
4210	24″ diameter	↓	10	3.200		3,225	196		3,421	3,850
4250	For galvanized elbows, add				↓	26%				
4690	Tee, painted									
4700	3/4″ diameter	1 Plum	38	.211	Ea.	88.50	13.55		102.05	119
4740	1″ diameter		33	.242		68.50	15.60		84.10	99
4750	1-1/4″ diameter		27	.296		68.50	19.10		87.60	105
4760	1-1/2″ diameter		22	.364		68.50	23.50		92	111
4770	2″ diameter	↓	17	.471		68.50	30.50		99	122
4780	2-1/2″ diameter	Q-1	27	.593		68.50	34.50		103	128
4790	3″ diameter		22	.727		93.50	42		135.50	168
4800	4″ diameter		17	.941		142	54.50		196.50	241
4810	5″ diameter	↓	13	1.231		330	71.50		401.50	475
4820	6″ diameter	Q-2	17	1.412		380	85		465	550
4830	8″ diameter		14	1.714		835	103		938	1,075
4840	10″ diameter		12	2		1,050	120		1,170	1,350
4850	12″ diameter		10	2.400		1,375	144		1,519	1,750
4851	14″ diameter		9	2.667		1,375	160		1,535	1,750
4852	16″ diameter	↓	8	3		1,550	180		1,730	1,975
4853	18″ diameter	Q-3	10	3.200		1,925	196		2,121	2,425
4854	20″ diameter		9	3.556		2,775	218		2,993	3,375
4855	24″ diameter	↓	8	4		4,225	245		4,470	5,025
4900	For galvanized tees, add				↓	24%				
4939	Couplings									
4940	Flexible, standard, painted									
4950	3/4″ diameter	1 Plum	100	.080	Ea.	25	5.15		30.15	35.50
4960	1″ diameter		100	.080		25	5.15		30.15	35.50
4970	1-1/4″ diameter	↓	80	.100	↓	32.50	6.45		38.95	45.50

22 11 13.48 Pipe, Fittings and Valves, Steel, Grooved-Joint		Crew	Daily Output	Labor-Hours	Unit	Material	Labor	Equipment	Total	Total Incl O&P
						2020 Bare Costs				
4980	1-1/2" diameter	1 Plum	67	.119	Ea.	35	7.70		42.70	50.50
4990	2" diameter		50	.160		38	10.30		48.30	57.50
5000	2-1/2" diameter	Q-1	80	.200		44	11.60		55.60	66.50
5010	3" diameter		67	.239		48.50	13.85		62.35	75
5020	3-1/2" diameter		57	.281		69.50	16.30		85.80	102
5030	4" diameter		50	.320		70	18.55		88.55	106
5040	5" diameter		40	.400		105	23		128	152
5050	6" diameter	Q-2	50	.480		125	29		154	182
5070	8" diameter		42	.571		202	34.50		236.50	275
5090	10" diameter		35	.686		330	41.50		371.50	425
5110	12" diameter		32	.750		375	45		420	480
5120	14" diameter		24	1		530	60		590	675
5130	16" diameter		20	1.200		695	72		767	870
5140	18" diameter		18	1.333		810	80		890	1,025
5150	20" diameter		16	1.500		1,275	90		1,365	1,550
5160	24" diameter		13	1.846		1,400	111		1,511	1,700
5200	For galvanized couplings, add					33%				
5750	Flange, w/groove gasket, black steel									
5754	See Line 22 11 13.47 0620 for gasket & bolt set									
5760	ANSI class 125 and 150, painted									
5780	2" pipe size	1 Plum	23	.348	Ea.	145	22.50		167.50	195
5790	2-1/2" pipe size	Q-1	37	.432		181	25		206	238
5800	3" pipe size		31	.516		195	30		225	261
5820	4" pipe size		23	.696		260	40.50		300.50	350
5830	5" pipe size		19	.842		300	49		349	405
5840	6" pipe size	Q-2	23	1.043		330	63		393	455
5850	8" pipe size		17	1.412		370	85		455	540
5860	10" pipe size		14	1.714		585	103		688	805
5870	12" pipe size		12	2		765	120		885	1,025
5880	14" pipe size		10	2.400		1,400	144		1,544	1,775
5890	16" pipe size		9	2.667		1,625	160		1,785	2,050
5900	18" pipe size		6	4		2,000	241		2,241	2,575
5910	20" pipe size		5	4.800		2,425	289		2,714	3,100
5920	24" pipe size		4.50	5.333		3,100	320		3,420	3,900
7790	Valves: coupling material required at joints not incl. in valve price.									
7794	Add 1 selected coupling, material only, per joint for installed price.									
8000	Butterfly valve, 2 position handle, with standard trim									
8010	1-1/2" pipe size	1 Plum	30	.267	Ea.	350	17.20		367.20	410
8020	2" pipe size	"	23	.348		350	22.50		372.50	420
8030	3" pipe size	Q-1	30	.533		505	31		536	605
8050	4" pipe size	"	22	.727		555	42		597	675
8070	6" pipe size	Q-2	23	1.043		1,125	63		1,188	1,325
8080	8" pipe size		18	1.333		1,450	80		1,530	1,725
8090	10" pipe size		15	1.600		2,000	96		2,096	2,350
8200	With stainless steel trim									
8240	1-1/2" pipe size	1 Plum	30	.267	Ea.	445	17.20		462.20	515
8250	2" pipe size	"	23	.348		445	22.50		467.50	525
8270	3" pipe size	Q-1	30	.533		600	31		631	710
8280	4" pipe size	"	22	.727		650	42		692	780
8300	6" pipe size	Q-2	23	1.043		1,200	63		1,263	1,425
8310	8" pipe size		18	1.333		3,000	80		3,080	3,425
8320	10" pipe size		15	1.600		4,800	96		4,896	5,425
9000	Cut one groove, labor									

22 11 13.48 Pipe, Fittings and Valves, Steel, Grooved-Joint		Crew	Daily Output	Labor-Hours	Unit	Material	2020 Bare Costs Labor	Equipment	Total	Total Incl O&P
9010	3/4" pipe size	Q-1	152	.105	Ea.		6.10		6.10	9.45
9020	1" pipe size		140	.114			6.65		6.65	10.25
9030	1-1/4" pipe size		124	.129			7.50		7.50	11.55
9040	1-1/2" pipe size		114	.140			8.15		8.15	12.60
9050	2" pipe size		104	.154			8.90		8.90	13.80
9060	2-1/2" pipe size		96	.167			9.65		9.65	14.95
9070	3" pipe size		88	.182			10.55		10.55	16.30
9080	3-1/2" pipe size		83	.193			11.20		11.20	17.30
9090	4" pipe size		78	.205			11.90		11.90	18.40
9100	5" pipe size		72	.222			12.90		12.90	19.95
9110	6" pipe size		70	.229			13.25		13.25	20.50
9120	8" pipe size		54	.296			17.20		17.20	26.50
9130	10" pipe size		38	.421			24.50		24.50	38
9140	12" pipe size		30	.533			31		31	48
9150	14" pipe size		20	.800			46.50		46.50	71.50
9160	16" pipe size		19	.842			49		49	75.50
9170	18" pipe size		18	.889			51.50		51.50	79.50
9180	20" pipe size		17	.941			54.50		54.50	84.50
9190	24" pipe size		15	1.067			62		62	95.50
9210	Roll one groove									
9220	3/4" pipe size	Q-1	266	.060	Ea.		3.49		3.49	5.40
9230	1" pipe size		228	.070			4.07		4.07	6.30
9240	1-1/4" pipe size		200	.080			4.64		4.64	7.15
9250	1-1/2" pipe size		178	.090			5.20		5.20	8.05
9260	2" pipe size		116	.138			8		8	12.35
9270	2-1/2" pipe size		110	.145			8.45		8.45	13.05
9280	3" pipe size		100	.160			9.30		9.30	14.35
9290	3-1/2" pipe size		94	.170			9.85		9.85	15.25
9300	4" pipe size		86	.186			10.80		10.80	16.70
9310	5" pipe size		84	.190			11.05		11.05	17.10
9320	6" pipe size		80	.200			11.60		11.60	17.95
9330	8" pipe size		66	.242			14.05		14.05	21.50
9340	10" pipe size		58	.276			16		16	24.50
9350	12" pipe size		46	.348			20		20	31
9360	14" pipe size		30	.533			31		31	48
9370	16" pipe size		28	.571			33		33	51.50
9380	18" pipe size		27	.593			34.50		34.50	53
9390	20" pipe size		25	.640			37		37	57.50
9400	24" pipe size		23	.696			40.50		40.50	62.50
9990	Minimum labor/equipment charge	1 Plum	4	2	Job		129		129	199

22 11 13.60 Tubing, Stainless Steel		Crew	Daily Output	Labor-Hours	Unit	Material	Labor	Equipment	Total	Total Incl O&P
0010	**TUBING, STAINLESS STEEL**									
5000	Tubing									
5010	Type 304, no joints, no hangers									
5020	.035 wall									
5021	1/4"	1 Plum	160	.050	L.F.	6.70	3.22		9.92	12.40
5022	3/8"		160	.050		6.25	3.22		9.47	11.90
5023	1/2"		160	.050		10.60	3.22		13.82	16.65
5024	5/8"		160	.050		10.25	3.22		13.47	16.25
5025	3/4"		133	.060		10.80	3.88		14.68	17.85
5026	7/8"		133	.060		11.05	3.88		14.93	18.15
5027	1"		114	.070		14.65	4.52		19.17	23

22 11 13 – Facility Water Distribution Piping

22 11 13.60 Tubing, Stainless Steel		Crew	Daily Output	Labor-Hours	Unit	2020 Bare Costs Material	Labor	Equipment	Total	Total Incl O&P
5040	.049 wall									
5041	1/4"	1 Plum	160	.050	L.F.	7.05	3.22		10.27	12.80
5042	3/8"		160	.050		10.60	3.22		13.82	16.65
5043	1/2"		160	.050		10.45	3.22		13.67	16.45
5044	5/8"		160	.050		11.70	3.22		14.92	17.85
5045	3/4"		133	.060		12.40	3.88		16.28	19.65
5046	7/8"		133	.060		10.80	3.88		14.68	17.85
5047	1"		114	.070		12	4.52		16.52	20
5060	.065 wall									
5061	1/4"	1 Plum	160	.050	L.F.	7.60	3.22		10.82	13.40
5062	3/8"		160	.050		10.60	3.22		13.82	16.65
5063	1/2"		160	.050		10.15	3.22		13.37	16.15
5064	5/8"		160	.050		13	3.22		16.22	19.30
5065	3/4"		133	.060		14.55	3.88		18.43	22
5066	7/8"		133	.060		13.20	3.88		17.08	20.50
5067	1"		114	.070		12.55	4.52		17.07	21
5210	Type 316									
5220	.035 wall									
5221	1/4"	1 Plum	160	.050	L.F.	6.10	3.22		9.32	11.70
5222	3/8"		160	.050		6.30	3.22		9.52	11.95
5223	1/2"		160	.050		9.45	3.22		12.67	15.40
5224	5/8"		160	.050		9.80	3.22		13.02	15.80
5225	3/4"		133	.060		12.15	3.88		16.03	19.35
5226	7/8"		133	.060		18.10	3.88		21.98	26
5227	1"		114	.070		16.30	4.52		20.82	25
5240	.049 wall									
5241	1/4"	1 Plum	160	.050	L.F.	5.30	3.22		8.52	10.85
5242	3/8"		160	.050		9.80	3.22		13.02	15.75
5243	1/2"		160	.050		10.35	3.22		13.57	16.35
5244	5/8"		160	.050		11.25	3.22		14.47	17.35
5245	3/4"		133	.060		12.20	3.88		16.08	19.40
5246	7/8"		133	.060		20.50	3.88		24.38	28.50
5247	1"		114	.070		14.55	4.52		19.07	23
5260	.065 wall									
5261	1/4"	1 Plum	160	.050	L.F.	7.20	3.22		10.42	12.95
5262	3/8"		160	.050		10.35	3.22		13.57	16.40
5263	1/2"		160	.050		10.05	3.22		13.27	16.05
5264	5/8"		160	.050		12.10	3.22		15.32	18.30
5265	3/4"		133	.060		14.05	3.88		17.93	21.50
5266	7/8"		133	.060		21	3.88		24.88	29
5267	1"		114	.070		18.25	4.52		22.77	27

22 11 13.61 Tubing Fittings, Stainless Steel

		Crew	Daily Output	Labor-Hours	Unit	Material	Labor	Equipment	Total	Total Incl O&P
0010	**TUBING FITTINGS, STAINLESS STEEL**									
8200	Tube fittings, compression type									
8202	Type 316									
8204	90° elbow									
8206	1/4"	1 Plum	24	.333	Ea.	17.30	21.50		38.80	52
8207	3/8"		22	.364		21.50	23.50		45	59.50
8208	1/2"		22	.364		35	23.50		58.50	74.50
8209	5/8"		21	.381		39.50	24.50		64	81.50
8210	3/4"		21	.381		62	24.50		86.50	106
8211	7/8"		20	.400		95	26		121	145

22 11 13.61 Tubing Fittings, Stainless Steel		Crew	Daily Output	Labor-Hours	Unit	2020 Bare Costs Material	Labor	Equipment	Total	Total Incl O&P
8212	1"	1 Plum	20	.400	Ea.	118	26		144	170
8220	Union tee									
8222	1/4"	1 Plum	15	.533	Ea.	24.50	34.50		59	80
8224	3/8"		15	.533		31.50	34.50		66	87.50
8225	1/2"		15	.533		48.50	34.50		83	107
8226	5/8"		14	.571		54.50	37		91.50	117
8227	3/4"		14	.571		73	37		110	137
8228	7/8"		13	.615		141	39.50		180.50	217
8229	1"		13	.615		156	39.50		195.50	234
8234	Union									
8236	1/4"	1 Plum	24	.333	Ea.	12	21.50		33.50	46
8237	3/8"		22	.364		17.10	23.50		40.60	55
8238	1/2"		22	.364		25.50	23.50		49	64
8239	5/8"		21	.381		33	24.50		57.50	74.50
8240	3/4"		21	.381		41.50	24.50		66	83.50
8241	7/8"		20	.400		67.50	26		93.50	115
8242	1"		20	.400		71.50	26		97.50	119
8250	Male connector									
8252	1/4" x 1/4"	1 Plum	24	.333	Ea.	7.70	21.50		29.20	41.50
8253	3/8" x 3/8"		22	.364		12	23.50		35.50	49
8254	1/2" x 1/2"		22	.364		17.75	23.50		41.25	55.50
8256	3/4" x 3/4"		21	.381		27	24.50		51.50	68
8258	1" x 1"		20	.400		47.50	26		73.50	92

22 11 13.64 Pipe, Stainless Steel		Crew	Daily Output	Labor-Hours	Unit	Material	Labor	Equipment	Total	Total Incl O&P
0010	**PIPE, STAINLESS STEEL**									
0020	Welded, with clevis type hanger assemblies, 10' OC									
0500	Schedule 5, type 304									
0540	1/2" diameter	Q-15	128	.125	L.F.	13.70	7.25	.83	21.78	27
0550	3/4" diameter		116	.138		18.70	8	.92	27.62	34
0560	1" diameter		103	.155		17.05	9	1.03	27.08	34
0570	1-1/4" diameter		93	.172		20.50	10	1.14	31.64	39.50
0580	1-1/2" diameter		85	.188		23	10.90	1.25	35.15	44
0590	2" diameter		69	.232		33.50	13.45	1.54	48.49	59.50
0600	2-1/2" diameter		53	.302		47.50	17.50	2.01	67.01	81.50
0610	3" diameter		48	.333		73.50	19.35	2.22	95.07	113
0620	4" diameter		44	.364		127	21	2.42	150.42	174
0630	5" diameter		36	.444		154	26	2.96	182.96	212
0640	6" diameter	Q-16	42	.571		320	34.50	2.53	357.03	405
0650	8" diameter		34	.706		490	42.50	3.13	535.63	610
0660	10" diameter		26	.923		300	55.50	4.09	359.59	420
0670	12" diameter		21	1.143		395	68.50	5.05	468.55	545
0700	To delete hangers, subtract									
0710	1/2" diam. to 1-1/2" diam.					8%	19%			
0720	2" diam. to 5" diam.					4%	9%			
0730	6" diam. to 12" diam.					3%	4%			
0750	For small quantities, add				L.F.	10%				
1250	Schedule 5, type 316									
1290	1/2" diameter	Q-15	128	.125	L.F.	13.05	7.25	.83	21.13	26.50
1300	3/4" diameter		116	.138		21	8	.92	29.92	37
1310	1" diameter		103	.155		21.50	9	1.03	31.53	39
1320	1-1/4" diameter		93	.172		25	10	1.14	36.14	44
1330	1-1/2" diameter		85	.188		44	10.90	1.25	56.15	67

22 11 13 – Facility Water Distribution Piping

22 11 13.64 Pipe, Stainless Steel		Crew	Daily Output	Labor-Hours	Unit	2020 Bare Costs Material	Labor	Equipment	Total	Total Incl O&P
1340	2" diameter	Q-15	69	.232	L.F.	59	13.45	1.54	73.99	87.50
1350	2-1/2" diameter		53	.302		91.50	17.50	2.01	111.01	130
1360	3" diameter		48	.333		115	19.35	2.22	136.57	159
1370	4" diameter		44	.364		139	21	2.42	162.42	188
1380	5" diameter		36	.444		159	26	2.96	187.96	218
1390	6" diameter	Q-16	42	.571		190	34.50	2.53	227.03	265
1400	8" diameter		34	.706		286	42.50	3.13	331.63	385
1410	10" diameter		26	.923		335	55.50	4.09	394.59	455
1420	12" diameter		21	1.143		430	68.50	5.05	503.55	585
1490	For small quantities, add					10%				
1940	To delete hanger, subtract									
1950	1/2" diam. to 1-1/2" diam.					5%	19%			
1960	2" diam. to 5" diam.					3%	9%			
1970	6" diam. to 12" diam.					2%	4%			
2000	Schedule 10, type 304									
2040	1/4" diameter	Q-15	131	.122	L.F.	9.10	7.10	.81	17.01	22
2050	3/8" diameter		128	.125		8.15	7.25	.83	16.23	21
2060	1/2" diameter		125	.128		13.95	7.40	.85	22.20	28
2070	3/4" diameter		113	.142		16.40	8.20	.94	25.54	32
2080	1" diameter		100	.160		15.30	9.30	1.06	25.66	32.50
2090	1-1/4" diameter		91	.176		25.50	10.20	1.17	36.87	45.50
2100	1-1/2" diameter		83	.193		22	11.20	1.28	34.48	42.50
2110	2" diameter		67	.239		23.50	13.85	1.59	38.94	49.50
2120	2-1/2" diameter		51	.314		42	18.20	2.09	62.29	76.50
2130	3" diameter		46	.348		50.50	20	2.31	72.81	89
2140	4" diameter		42	.381		44	22	2.53	68.53	85.50
2150	5" diameter		35	.457		66.50	26.50	3.04	96.04	117
2160	6" diameter	Q-16	40	.600		60	36	2.66	98.66	125
2170	8" diameter		33	.727		102	44	3.22	149.22	183
2180	10" diameter		25	.960		140	57.50	4.25	201.75	248
2190	12" diameter		21	1.143		174	68.50	5.05	247.55	305
2250	For small quantities, add					10%				
2650	To delete hanger, subtract									
2660	1/4" diam. to 3/4" diam.					9%	22%			
2670	1" diam. to 2" diam.					4%	15%			
2680	2-1/2" diam. to 5" diam.					3%	8%			
2690	6" diam. to 12" diam.					3%	4%			
2750	Schedule 10, type 316									
2790	1/4" diameter	Q-15	131	.122	L.F.	7.15	7.10	.81	15.06	19.70
2800	3/8" diameter		128	.125		8.70	7.25	.83	16.78	21.50
2810	1/2" diameter		125	.128		11.10	7.40	.85	19.35	24.50
2820	3/4" diameter		113	.142		20.50	8.20	.94	29.64	36
2830	1" diameter		100	.160		23.50	9.30	1.06	33.86	41.50
2840	1-1/4" diameter		91	.176		22.50	10.20	1.17	33.87	42
2850	1-1/2" diameter		83	.193		30.50	11.20	1.28	42.98	52
2860	2" diameter		67	.239		32	13.85	1.59	47.44	58.50
2870	2-1/2" diameter		51	.314		39	18.20	2.09	59.29	73.50
2880	3" diameter		46	.348		59	20	2.31	81.31	98.50
2890	4" diameter		42	.381		59	22	2.53	83.53	102
2900	5" diameter		35	.457		56.50	26.50	3.04	86.04	106
2910	6" diameter	Q-16	40	.600		91	36	2.66	129.66	159
2920	8" diameter		33	.727		133	44	3.22	180.22	217
2930	10" diameter		25	.960		174	57.50	4.25	235.75	286

22 11 13.64 Pipe, Stainless Steel		Crew	Daily Output	Labor-Hours	Unit	2020 Bare Costs Material	Labor	Equipment	Total	Total Incl O&P
2940	12" diameter	Q-16	21	1.143	L.F.	218	68.50	5.05	291.55	350
2990	For small quantities, add					10%				
3430	To delete hanger, subtract									
3440	1/4" diam. to 3/4" diam.					6%	22%			
3450	1" diam. to 2" diam.					3%	15%			
3460	2-1/2" diam. to 5" diam.					2%	8%			
3470	6" diam. to 12" diam.					2%	4%			
3500	Threaded, couplings and clevis hanger assemblies, 10' OC									
3520	Schedule 40, type 304									
3540	1/4" diameter	1 Plum	54	.148	L.F.	10.20	9.55		19.75	26
3550	3/8" diameter		53	.151		11.15	9.75		20.90	27.50
3560	1/2" diameter		52	.154		12.40	9.90		22.30	29
3570	3/4" diameter		51	.157		20.50	10.10		30.60	38
3580	1" diameter		45	.178		24.50	11.45		35.95	44.50
3590	1-1/4" diameter	Q-1	76	.211		39	12.20		51.20	62
3600	1-1/2" diameter		69	.232		40	13.45		53.45	65.50
3610	2" diameter		57	.281		52	16.30		68.30	82.50
3620	2-1/2" diameter		44	.364		98	21		119	141
3630	3" diameter		38	.421		113	24.50		137.50	163
3640	4" diameter	Q-2	51	.471		131	28.50		159.50	188
3740	For small quantities, add					10%				
4200	To delete couplings & hangers, subtract									
4210	1/4" diam. to 3/4" diam.					15%	56%			
4220	1" diam. to 2" diam.					18%	49%			
4230	2-1/2" diam. to 4" diam.					34%	40%			
4250	Schedule 40, type 316									
4290	1/4" diameter	1 Plum	54	.148	L.F.	23.50	9.55		33.05	40.50
4300	3/8" diameter		53	.151		25.50	9.75		35.25	43
4310	1/2" diameter		52	.154		28.50	9.90		38.40	47
4320	3/4" diameter		51	.157		32.50	10.10		42.60	51
4330	1" diameter		45	.178		45	11.45		56.45	67
4340	1-1/4" diameter	Q-1	76	.211		74	12.20		86.20	100
4350	1-1/2" diameter		69	.232		67.50	13.45		80.95	95
4360	2" diameter		57	.281		90.50	16.30		106.80	125
4370	2-1/2" diameter		44	.364		130	21		151	176
4380	3" diameter		38	.421		152	24.50		176.50	205
4390	4" diameter	Q-2	51	.471		187	28.50		215.50	250
4490	For small quantities, add					10%				
4900	To delete couplings & hangers, subtract									
4910	1/4" diam. to 3/4" diam.					12%	56%			
4920	1" diam. to 2" diam.					14%	49%			
4930	2-1/2" diam. to 4" diam.					27%	40%			
5000	Schedule 80, type 304									
5040	1/4" diameter	1 Plum	53	.151	L.F.	21	9.75		30.75	38
5050	3/8" diameter		52	.154		37	9.90		46.90	56.50
5060	1/2" diameter		51	.157		31.50	10.10		41.60	50
5070	3/4" diameter		48	.167		38	10.75		48.75	58
5080	1" diameter		43	.186		55	12		67	79
5090	1-1/4" diameter	Q-1	73	.219		59	12.70		71.70	84
5100	1-1/2" diameter		67	.239		67.50	13.85		81.35	96
5110	2" diameter		54	.296		108	17.20		125.20	146
5190	For small quantities, add					10%				
5700	To delete couplings & hangers, subtract									

22 11 13.64 Pipe, Stainless Steel		Crew	Daily Output	Labor-Hours	Unit	Material	2020 Bare Costs Labor	Equipment	Total	Total Incl O&P
5710	1/4" diam. to 3/4" diam.					10%	53%			
5720	1" diam. to 2" diam.					14%	47%			
5750	Schedule 80, type 316									
5790	1/4" diameter	1 Plum	53	.151	L.F.	42.50	9.75		52.25	62
5800	3/8" diameter		52	.154		42.50	9.90		52.40	62
5810	1/2" diameter		51	.157		54.50	10.10		64.60	75.50
5820	3/4" diameter		48	.167		54	10.75		64.75	76
5830	1" diameter		43	.186		51.50	12		63.50	75.50
5840	1-1/4" diameter	Q-1	73	.219		77.50	12.70		90.20	105
5850	1-1/2" diameter		67	.239		104	13.85		117.85	136
5860	2" diameter		54	.296		117	17.20		134.20	155
5950	For small quantities, add					10%				
7000	To delete couplings & hangers, subtract									
7010	1/4" diam. to 3/4" diam.					9%	53%			
7020	1" diam. to 2" diam.					14%	47%			
8000	Weld joints with clevis type hanger assemblies, 10' OC									
8010	Schedule 40, type 304									
8050	1/8" pipe size	Q-15	126	.127	L.F.	7.55	7.35	.84	15.74	20.50
8060	1/4" pipe size		125	.128		8.85	7.40	.85	17.10	22
8070	3/8" pipe size		122	.131		9.55	7.60	.87	18.02	23
8080	1/2" pipe size		118	.136		10.25	7.85	.90	19	24.50
8090	3/4" pipe size		109	.147		17.70	8.50	.98	27.18	33.50
8100	1" pipe size		95	.168		19.80	9.75	1.12	30.67	38.50
8110	1-1/4" pipe size		86	.186		30.50	10.80	1.24	42.54	51.50
8120	1-1/2" pipe size		78	.205		32	11.90	1.36	45.26	55
8130	2" pipe size		62	.258		38.50	14.95	1.72	55.17	67.50
8140	2-1/2" pipe size		49	.327		66	18.95	2.17	87.12	104
8150	3" pipe size		44	.364		69.50	21	2.42	92.92	112
8160	3-1/2" pipe size		44	.364		88	21	2.42	111.42	132
8170	4" pipe size		39	.410		71	24	2.73	97.73	118
8180	5" pipe size		32	.500		121	29	3.33	153.33	182
8190	6" pipe size	Q-16	37	.649		160	39	2.87	201.87	240
8191	6" pipe size		37	.649		157	39	2.87	198.87	236
8200	8" pipe size		29	.828		164	50	3.67	217.67	262
8210	10" pipe size		24	1		273	60	4.43	337.43	400
8220	12" pipe size		20	1.200		515	72	5.30	592.30	685
8300	Schedule 40, type 316									
8310	1/8" pipe size	Q-15	126	.127	L.F.	20	7.35	.84	28.19	34.50
8320	1/4" pipe size		125	.128		22	7.40	.85	30.25	36.50
8330	3/8" pipe size		122	.131		23.50	7.60	.87	31.97	38.50
8340	1/2" pipe size		118	.136		26	7.85	.90	34.75	41.50
8350	3/4" pipe size		109	.147		29	8.50	.98	38.48	46
8360	1" pipe size		95	.168		39.50	9.75	1.12	50.37	60
8370	1-1/4" pipe size		86	.186		63	10.80	1.24	75.04	87.50
8380	1-1/2" pipe size		78	.205		57.50	11.90	1.36	70.76	83.50
8390	2" pipe size		62	.258		74	14.95	1.72	90.67	106
8400	2-1/2" pipe size		49	.327		91.50	18.95	2.17	112.62	133
8410	3" pipe size		44	.364		99.50	21	2.42	122.92	144
8420	3-1/2" pipe size		44	.364		109	21	2.42	132.42	155
8430	4" pipe size		39	.410		114	24	2.73	140.73	166
8440	5" pipe size		32	.500		180	29	3.33	212.33	247
8450	6" pipe size	Q-16	37	.649		263	39	2.87	304.87	355
8460	8" pipe size		29	.828		271	50	3.67	324.67	380

22 11 13.64 Pipe, Stainless Steel		Crew	Daily Output	Labor-Hours	Unit	Material	2020 Bare Costs Labor	Equipment	Total	Total Incl O&P
8470	10" pipe size	Q-16	24	1	L.F.	335	60	4.43	399.43	470
8480	12" pipe size		20	1.200		455	72	5.30	532.30	620
8500	Schedule 80, type 304									
8510	1/4" pipe size	Q-15	110	.145	L.F.	19.40	8.45	.97	28.82	35.50
8520	3/8" pipe size		109	.147		35	8.50	.98	44.48	53
8530	1/2" pipe size		106	.151		29.50	8.75	1	39.25	47
8540	3/4" pipe size		96	.167		35	9.65	1.11	45.76	54.50
8550	1" pipe size		87	.184		50.50	10.65	1.22	62.37	73.50
8560	1-1/4" pipe size		81	.198		49.50	11.45	1.31	62.26	73.50
8570	1-1/2" pipe size		74	.216		55	12.55	1.44	68.99	81
8580	2" pipe size		58	.276		93.50	16	1.83	111.33	130
8590	2-1/2" pipe size		46	.348		104	20	2.31	126.31	148
8600	3" pipe size		41	.390		123	22.50	2.60	148.10	173
8610	4" pipe size		33	.485		154	28	3.22	185.22	216
8630	6" pipe size	Q-16	30	.800		279	48	3.54	330.54	385
8640	Schedule 80, type 316									
8650	1/4" pipe size	Q-15	110	.145	L.F.	40.50	8.45	.97	49.92	58.50
8660	3/8" pipe size		109	.147		40.50	8.50	.98	49.98	58.50
8670	1/2" pipe size		106	.151		52	8.75	1	61.75	72
8680	3/4" pipe size		96	.167		51	9.65	1.11	61.76	72
8690	1" pipe size		87	.184		45.50	10.65	1.22	57.37	68
8700	1-1/4" pipe size		81	.198		63.50	11.45	1.31	76.26	88.50
8710	1-1/2" pipe size		74	.216		87	12.55	1.44	100.99	117
8720	2" pipe size		58	.276		94	16	1.83	111.83	130
8730	2-1/2" pipe size		46	.348		95.50	20	2.31	117.81	139
8740	3" pipe size		41	.390		148	22.50	2.60	173.10	201
8760	4" pipe size		33	.485		195	28	3.22	226.22	261
8770	6" pipe size	Q-16	30	.800		340	48	3.54	391.54	455
9100	Threading pipe labor, sst, one end, schedules 40 & 80									
9110	1/4" through 3/4" pipe size	1 Plum	61.50	.130	Ea.		8.40		8.40	12.95
9120	1" through 2" pipe size		55.90	.143			9.20		9.20	14.25
9130	2-1/2" pipe size		41.50	.193			12.40		12.40	19.20
9140	3" pipe size		38.50	.208			13.40		13.40	20.50
9150	3-1/2" pipe size	Q-1	68.40	.234			13.55		13.55	21
9160	4" pipe size		73	.219			12.70		12.70	19.65
9170	5" pipe size		40.70	.393			23		23	35.50
9180	6" pipe size		35.40	.452			26		26	40.50
9190	8" pipe size		22.30	.717			41.50		41.50	64.50
9200	10" pipe size		16.10	.994			57.50		57.50	89
9210	12" pipe size		12.30	1.301			75.50		75.50	117
9250	Welding labor per joint for stainless steel									
9260	Schedule 5 and 10									
9270	1/4" pipe size	Q-15	36	.444	Ea.		26	2.96	28.96	43.50
9280	3/8" pipe size		35	.457			26.50	3.04	29.54	44.50
9290	1/2" pipe size		35	.457			26.50	3.04	29.54	44.50
9300	3/4" pipe size		28	.571			33	3.80	36.80	55.50
9310	1" pipe size		25	.640			37	4.26	41.26	62
9320	1-1/4" pipe size		22	.727			42	4.84	46.84	70.50
9330	1-1/2" pipe size		21	.762			44	5.05	49.05	74
9340	2" pipe size		18	.889			51.50	5.90	57.40	86
9350	2-1/2" pipe size		12	1.333			77.50	8.85	86.35	130
9360	3" pipe size		9.73	1.644			95.50	10.95	106.45	159
9370	4" pipe size		7.37	2.171			126	14.45	140.45	211

22 11 13 – Facility Water Distribution Piping

22 11 13.64 Pipe, Stainless Steel		Crew	Daily Output	Labor-Hours	Unit	Material	2020 Bare Costs Labor	Equipment	Total	Total Incl O&P
9380	5" pipe size	Q-15	6.15	2.602	Ea.		151	17.30	168.30	252
9390	6" pipe size		5.71	2.802			163	18.65	181.65	272
9400	8" pipe size		3.69	4.336			251	29	280	420
9410	10" pipe size		2.91	5.498			320	36.50	356.50	535
9420	12" pipe size	↓	2.31	6.926	↓		400	46	446	670
9500	Schedule 40									
9510	1/4" pipe size	Q-15	28	.571	Ea.		33	3.80	36.80	55.50
9520	3/8" pipe size		27	.593			34.50	3.94	38.44	57.50
9530	1/2" pipe size		25.40	.630			36.50	4.19	40.69	61
9540	3/4" pipe size		22.22	.720			42	4.79	46.79	70
9550	1" pipe size		20.25	.790			46	5.25	51.25	77
9560	1-1/4" pipe size		18.82	.850			49.50	5.65	55.15	82
9570	1-1/2" pipe size		17.78	.900			52	6	58	87
9580	2" pipe size		15.09	1.060			61.50	7.05	68.55	103
9590	2-1/2" pipe size		7.96	2.010			117	13.35	130.35	195
9600	3" pipe size		6.43	2.488			144	16.55	160.55	241
9610	4" pipe size		4.88	3.279			190	22	212	320
9620	5" pipe size		4.26	3.756			218	25	243	365
9630	6" pipe size		3.77	4.244			246	28	274	410
9640	8" pipe size		2.44	6.557			380	43.50	423.50	640
9650	10" pipe size		1.92	8.333			485	55.50	540.50	805
9660	12" pipe size	↓	1.52	10.526	↓		610	70	680	1,025
9750	Schedule 80									
9760	1/4" pipe size	Q-15	21.55	.742	Ea.		43	4.94	47.94	72
9770	3/8" pipe size		20.75	.771			44.50	5.15	49.65	74.50
9780	1/2" pipe size		19.54	.819			47.50	5.45	52.95	79.50
9790	3/4" pipe size		17.09	.936			54.50	6.25	60.75	91
9800	1" pipe size		15.58	1.027			59.50	6.85	66.35	99.50
9810	1-1/4" pipe size		14.48	1.105			64	7.35	71.35	107
9820	1-1/2" pipe size		13.68	1.170			68	7.80	75.80	114
9830	2" pipe size		11.61	1.378			80	9.15	89.15	134
9840	2-1/2" pipe size		6.12	2.614			152	17.40	169.40	253
9850	3" pipe size		4.94	3.239			188	21.50	209.50	315
9860	4" pipe size		3.75	4.267			247	28.50	275.50	415
9870	5" pipe size		3.27	4.893			284	32.50	316.50	475
9880	6" pipe size		2.90	5.517			320	36.50	356.50	535
9890	8" pipe size		1.87	8.556			495	57	552	830
9900	10" pipe size		1.48	10.811			625	72	697	1,050
9910	12" pipe size		1.17	13.675			795	91	886	1,325
9920	Schedule 160, 1/2" pipe size		17	.941			54.50	6.25	60.75	91.50
9930	3/4" pipe size		14.81	1.080			62.50	7.20	69.70	105
9940	1" pipe size		13.50	1.185			68.50	7.90	76.40	115
9950	1-1/4" pipe size		12.55	1.275			74	8.50	82.50	123
9960	1-1/2" pipe size		11.85	1.350			78.50	9	87.50	131
9970	2" pipe size		10	1.600			93	10.65	103.65	155
9980	3" pipe size		4.28	3.738			217	25	242	365
9990	4" pipe size	↓	3.25	4.923	↓		286	32.50	318.50	475

22 11 13 – Facility Water Distribution Piping

22 11 13.66 Pipe Fittings, Stainless Steel		Crew	Daily Output	Labor-Hours	Unit	Material	Labor	Equipment	Total	Total Incl O&P
						2020 Bare Costs				
0010	**PIPE FITTINGS, STAINLESS STEEL**									
0100	Butt weld joint, schedule 5, type 304									
0120	90° elbow, long									
0140	1/2"	Q-15	17.50	.914	Ea.	18.05	53	6.10	77.15	109
0150	3/4"		14	1.143		18.05	66.50	7.60	92.15	130
0160	1"		12.50	1.280		19.05	74	8.50	101.55	145
0170	1-1/4"		11	1.455		26	84.50	9.65	120.15	169
0180	1-1/2"		10.50	1.524		22	88.50	10.15	120.65	173
0190	2"		9	1.778		26	103	11.80	140.80	201
0200	2-1/2"		6	2.667		60	155	17.75	232.75	325
0210	3"		4.86	3.292		56	191	22	269	380
0220	3-1/2"		4.27	3.747		165	217	25	407	545
0230	4"		3.69	4.336		94	251	29	374	525
0240	5"		3.08	5.195		350	300	34.50	684.50	890
0250	6"	Q-16	4.29	5.594		271	335	25	631	845
0260	8"		2.76	8.696		570	525	38.50	1,133.50	1,475
0270	10"		2.18	11.009		880	660	49	1,589	2,050
0280	12"		1.73	13.873		1,250	835	61.50	2,146.50	2,750
0320	For schedule 5, type 316, add					30%				
0600	45° elbow, long									
0620	1/2"	Q-15	17.50	.914	Ea.	18.05	53	6.10	77.15	109
0630	3/4"		14	1.143		18.05	66.50	7.60	92.15	130
0640	1"		12.50	1.280		19.05	74	8.50	101.55	145
0650	1-1/4"		11	1.455		26	84.50	9.65	120.15	169
0660	1-1/2"		10.50	1.524		22	88.50	10.15	120.65	173
0670	2"		9	1.778		26	103	11.80	140.80	201
0680	2-1/2"		6	2.667		60	155	17.75	232.75	325
0690	3"		4.86	3.292		45	191	22	258	370
0700	3-1/2"		4.27	3.747		165	217	25	407	545
0710	4"		3.69	4.336		76	251	29	356	505
0720	5"		3.08	5.195		281	300	34.50	615.50	815
0730	6"	Q-16	4.29	5.594		190	335	25	550	760
0740	8"		2.76	8.696		400	525	38.50	963.50	1,300
0750	10"		2.18	11.009		700	660	49	1,409	1,850
0760	12"		1.73	13.873		875	835	61.50	1,771.50	2,325
0800	For schedule 5, type 316, add					25%				
1100	Tee, straight									
1130	1/2"	Q-15	11.66	1.372	Ea.	54	79.50	9.15	142.65	193
1140	3/4"		9.33	1.715		54	99.50	11.40	164.90	226
1150	1"		8.33	1.921		57	111	12.75	180.75	249
1160	1-1/4"		7.33	2.183		46	127	14.50	187.50	263
1170	1-1/2"		7	2.286		45	133	15.20	193.20	271
1180	2"		6	2.667		47	155	17.75	219.75	310
1190	2-1/2"		4	4		113	232	26.50	371.50	515
1200	3"		3.24	4.938		86	286	33	405	575
1210	3-1/2"		2.85	5.614		237	325	37.50	599.50	805
1220	4"		2.46	6.504		128	375	43.50	546.50	775
1230	5"		2	8		405	465	53	923	1,225
1240	6"	Q-16	2.85	8.421		320	505	37.50	862.50	1,175
1250	8"		1.84	13.043		680	785	58	1,523	2,050
1260	10"		1.45	16.552		1,100	995	73.50	2,168.50	2,850
1270	12"		1.15	20.870		1,525	1,250	92.50	2,867.50	3,725

22 11 13.66 Pipe Fittings, Stainless Steel		Crew	Daily Output	Labor-Hours	Unit	Material	Labor	Equipment	Total	Total Incl O&P
						2020 Bare Costs				
1320	For schedule 5, type 316, add				Ea.	25%				
2000	Butt weld joint, schedule 10, type 304									
2020	90° elbow, long									
2040	1/2"	Q-15	17	.941	Ea.	16.55	54.50	6.25	77.30	110
2050	3/4"		14	1.143		12.75	66.50	7.60	86.85	124
2060	1"		12.50	1.280		14.65	74	8.50	97.15	141
2070	1-1/4"		11	1.455		24	84.50	9.65	118.15	167
2080	1-1/2"		10.50	1.524		17	88.50	10.15	115.65	167
2090	2"		9	1.778		20	103	11.80	134.80	194
2100	2-1/2"		6	2.667		55	155	17.75	227.75	320
2110	3"		4.86	3.292		51.50	191	22	264.50	375
2120	3-1/2"		4.27	3.747		152	217	25	394	530
2130	4"		3.69	4.336		72.50	251	29	352.50	500
2140	5"		3.08	5.195		320	300	34.50	654.50	860
2150	6"	Q-16	4.29	5.594		208	335	25	568	775
2160	8"		2.76	8.696		440	525	38.50	1,003.50	1,350
2170	10"		2.18	11.009		680	660	49	1,389	1,825
2180	12"		1.73	13.873		965	835	61.50	1,861.50	2,425
2500	45° elbow, long									
2520	1/2"	Q-15	17.50	.914	Ea.	16.55	53	6.10	75.65	107
2530	3/4"		14	1.143		16.55	66.50	7.60	90.65	129
2540	1"		12.50	1.280		14.65	74	8.50	97.15	141
2550	1-1/4"		11	1.455		24	84.50	9.65	118.15	167
2560	1-1/2"		10.50	1.524		17	88.50	10.15	115.65	167
2570	2"		9	1.778		20	103	11.80	134.80	194
2580	2-1/2"		6	2.667		55	155	17.75	227.75	320
2590	3"		4.86	3.292		41.50	191	22	254.50	365
2600	3-1/2"		4.27	3.747		152	217	25	394	530
2610	4"		3.69	4.336		58.50	251	29	338.50	485
2620	5"		3.08	5.195		257	300	34.50	591.50	785
2630	6"	Q-16	4.29	5.594		147	335	25	507	710
2640	8"		2.76	8.696		310	525	38.50	873.50	1,200
2650	10"		2.18	11.009		540	660	49	1,249	1,675
2660	12"		1.73	13.873		675	835	61.50	1,571.50	2,125
3000	Tee, straight									
3030	1/2"	Q-15	11.66	1.372	Ea.	49.50	79.50	9.15	138.15	188
3040	3/4"		9.33	1.715		49.50	99.50	11.40	160.40	221
3050	1"		8.33	1.921		44	111	12.75	167.75	235
3060	1-1/4"		7.33	2.183		55.50	127	14.50	197	273
3070	1-1/2"		7	2.286		47	133	15.20	195.20	274
3080	2"		6	2.667		49.50	155	17.75	222.25	315
3090	2-1/2"		4	4		63	232	26.50	321.50	460
3100	3"		3.24	4.938		82	286	33	401	570
3110	3-1/2"		2.85	5.614		121	325	37.50	483.50	680
3120	4"		2.46	6.504		99	375	43.50	517.50	740
3130	5"		2	8		375	465	53	893	1,175
3140	6"	Q-16	2.85	8.421		247	505	37.50	789.50	1,100
3150	8"		1.84	13.043		525	785	58	1,368	1,875
3151	10"		1.45	16.552		850	995	73.50	1,918.50	2,575
3152	12"		1.15	20.870		1,175	1,250	92.50	2,517.50	3,350
3154	For schedule 10, type 316, add					25%				
3281	Butt weld joint, schedule 40, type 304									
3284	90° elbow, long, 1/2"	Q-15	12.70	1.260	Ea.	19.30	73	8.40	100.70	143

22 11 13.66 Pipe Fittings, Stainless Steel		Crew	Daily Output	Labor-Hours	Unit	2020 Bare Costs Material	Labor	Equipment	Total	Total Incl O&P
3288	3/4"	Q-15	11.10	1.441	Ea.	19.30	83.50	9.60	112.40	161
3289	1"		10.13	1.579		20	91.50	10.50	122	176
3290	1-1/4"		9.40	1.702		26.50	98.50	11.30	136.30	195
3300	1-1/2"		8.89	1.800		21	104	11.95	136.95	197
3310	2"		7.55	2.119		30.50	123	14.10	167.60	239
3320	2-1/2"		3.98	4.020		58	233	26.50	317.50	455
3330	3"		3.21	4.984		74.50	289	33	396.50	565
3340	3-1/2"		2.83	5.654		276	330	37.50	643.50	850
3350	4"		2.44	6.557		129	380	43.50	552.50	780
3360	5"		2.13	7.512		430	435	50	915	1,200
3370	6"	Q-16	2.83	8.481		375	510	37.50	922.50	1,250
3380	8"		1.83	13.115		635	790	58	1,483	1,975
3390	10"		1.44	16.667		1,325	1,000	74	2,399	3,100
3400	12"		1.14	21.053		1,700	1,275	93.50	3,068.50	3,925
3410	For schedule 40, type 316, add					25%				
3460	45° elbow, long, 1/2"	Q-15	12.70	1.260	Ea.	19.30	73	8.40	100.70	143
3470	3/4"		11.10	1.441		19.30	83.50	9.60	112.40	161
3480	1"		10.13	1.579		20	91.50	10.50	122	176
3490	1-1/4"		9.40	1.702		26.50	98.50	11.30	136.30	195
3500	1-1/2"		8.89	1.800		21	104	11.95	136.95	197
3510	2"		7.55	2.119		30.50	123	14.10	167.60	239
3520	2-1/2"		3.98	4.020		58	233	26.50	317.50	455
3530	3"		3.21	4.984		59	289	33	381	545
3540	3-1/2"		2.83	5.654		276	330	37.50	643.50	850
3550	4"		2.44	6.557		92	380	43.50	515.50	740
3560	5"		2.13	7.512		305	435	50	790	1,075
3570	6"	Q-16	2.83	8.481		266	510	37.50	813.50	1,125
3580	8"		1.83	13.115		445	790	58	1,293	1,775
3590	10"		1.44	16.667		935	1,000	74	2,009	2,650
3600	12"		1.14	21.053		1,200	1,275	93.50	2,568.50	3,350
3610	For schedule 40, type 316, add					25%				
3660	Tee, straight 1/2"	Q-15	8.46	1.891	Ea.	49.50	110	12.60	172.10	238
3670	3/4"		7.40	2.162		49.50	125	14.40	188.90	264
3680	1"		6.74	2.374		52.50	138	15.80	206.30	288
3690	1-1/4"		6.27	2.552		122	148	16.95	286.95	380
3700	1-1/2"		5.92	2.703		56	157	17.95	230.95	325
3710	2"		5.03	3.181		83.50	184	21	288.50	400
3720	2-1/2"		2.65	6.038		106	350	40	496	700
3730	3"		2.14	7.477		107	435	49.50	591.50	840
3740	3-1/2"		1.88	8.511		201	495	56.50	752.50	1,050
3750	4"		1.62	9.877		201	575	65.50	841.50	1,175
3760	5"		1.42	11.268		465	655	75	1,195	1,600
3770	6"	Q-16	1.88	12.766		375	770	56.50	1,201.50	1,650
3780	8"		1.22	19.672		755	1,175	87	2,017	2,750
3790	10"		.96	25		1,475	1,500	111	3,086	4,075
3800	12"		.76	31.579		1,950	1,900	140	3,990	5,225
3810	For schedule 40, type 316, add					25%				
3820	Tee, reducing on outlet, 3/4" x 1/2"	Q-15	7.73	2.070	Ea.	57	120	13.75	190.75	264
3822	1" x 1/2"		7.24	2.210		71.50	128	14.70	214.20	293
3824	1" x 3/4"		6.96	2.299		66	133	15.30	214.30	295
3826	1-1/4" x 1"		6.43	2.488		162	144	16.55	322.55	420
3828	1-1/2" x 1/2"		6.58	2.432		97.50	141	16.15	254.65	345
3830	1-1/2" x 3/4"		6.35	2.520		91	146	16.75	253.75	345

22 11 Facility Water Distribution

22 11 13 – Facility Water Distribution Piping

22 11 13.66 Pipe Fittings, Stainless Steel		Crew	Daily Output	Labor-Hours	Unit	Material	Labor (2020 Bare Costs)	Equipment	Total	Total Incl O&P
3832	1-1/2" x 1"	Q-15	6.18	2.589	Ea.	70	150	17.20	237.20	330
3834	2" x 1"		5.50	2.909		125	169	19.35	313.35	420
3836	2" x 1-1/2"		5.30	3.019		105	175	20	300	410
3838	2-1/2" x 2"		3.15	5.079		158	295	34	487	665
3840	3" x 1-1/2"		2.72	5.882		160	340	39	539	750
3842	3" x 2"		2.65	6.038		133	350	40	523	730
3844	4" x 2"		2.10	7.619		335	440	50.50	825.50	1,100
3846	4" x 3"		1.77	9.040		242	525	60	827	1,150
3848	5" x 4"		1.48	10.811		560	625	72	1,257	1,675
3850	6" x 3"	Q-16	2.19	10.959		615	660	48.50	1,323.50	1,750
3852	6" x 4"		2.04	11.765		535	710	52	1,297	1,750
3854	8" x 4"		1.46	16.438		1,250	990	73	2,313	2,975
3856	10" x 8"		.69	34.783		2,100	2,100	154	4,354	5,700
3858	12" x 10"		.55	43.636		2,800	2,625	193	5,618	7,350
3950	Reducer, concentric, 3/4" x 1/2"	Q-15	11.85	1.350		36.50	78.50	9	124	171
3952	1" x 3/4"		10.60	1.509		38.50	87.50	10.05	136.05	189
3954	1-1/4" x 3/4"		10.19	1.570		94.50	91	10.45	195.95	257
3956	1-1/4" x 1"		9.76	1.639		48	95	10.90	153.90	212
3958	1-1/2" x 3/4"		9.88	1.619		80	94	10.75	184.75	245
3960	1-1/2" x 1"		9.47	1.690		64.50	98	11.25	173.75	235
3962	2" x 1"		8.65	1.850		32.50	107	12.30	151.80	216
3964	2" x 1-1/2"		8.16	1.961		32	114	13.05	159.05	226
3966	2-1/2" x 1"		5.71	2.802		139	163	18.65	320.65	425
3968	2-1/2" x 2"		5.21	3.071		70.50	178	20.50	269	375
3970	3" x 1"		4.88	3.279		95.50	190	22	307.50	425
3972	3" x 1-1/2"		4.72	3.390		48.50	197	22.50	268	385
3974	3" x 2"		4.51	3.548		46	206	23.50	275.50	395
3976	4" x 2"		3.69	4.336		65	251	29	345	495
3978	4" x 3"		2.77	5.776		48.50	335	38.50	422	615
3980	5" x 3"		2.56	6.250		290	365	41.50	696.50	925
3982	5" x 4"		2.27	7.048		234	410	47	691	940
3984	6" x 3"	Q-16	3.57	6.723		150	405	30	585	825
3986	6" x 4"		3.19	7.524		123	455	33.50	611.50	870
3988	8" x 4"		2.44	9.836		325	590	43.50	958.50	1,325
3990	8" x 6"		2.22	10.811		242	650	48	940	1,325
3992	10" x 6"		1.91	12.565		450	755	55.50	1,260.50	1,725
3994	10" x 8"		1.61	14.907		375	895	66	1,336	1,875
3995	12" x 6"		1.63	14.724		825	885	65	1,775	2,350
3996	12" x 8"		1.41	17.021		640	1,025	75.50	1,740.50	2,350
3997	12" x 10"		1.27	18.898		450	1,125	83.50	1,658.50	2,325
4000	Socket weld joint, 3,000 lb., type 304									
4100	90° elbow									
4140	1/4"	Q-15	13.47	1.188	Ea.	45	69	7.90	121.90	165
4150	3/8"		12.97	1.234		59	71.50	8.20	138.70	185
4160	1/2"		12.21	1.310		63	76	8.70	147.70	197
4170	3/4"		10.68	1.498		74	87	9.95	170.95	226
4180	1"		9.74	1.643		111	95.50	10.90	217.40	282
4190	1-1/4"		9.05	1.768		194	103	11.75	308.75	385
4200	1-1/2"		8.55	1.871		236	109	12.45	357.45	440
4210	2"		7.26	2.204		380	128	14.65	522.65	635
4300	45° elbow									
4340	1/4"	Q-15	13.47	1.188	Ea.	85	69	7.90	161.90	209
4350	3/8"		12.97	1.234		85	71.50	8.20	164.70	214

22 11 13.66 Pipe Fittings, Stainless Steel		Crew	Daily Output	Labor-Hours	Unit	Material	2020 Bare Costs Labor	Equipment	Total	Total Incl O&P
4360	1/2"	Q-15	12.21	1.310	Ea.	85	76	8.70	169.70	221
4370	3/4"		10.68	1.498		96.50	87	9.95	193.45	251
4380	1"		9.74	1.643		140	95.50	10.90	246.40	315
4390	1-1/4"		9.05	1.768		230	103	11.75	344.75	425
4400	1-1/2"		8.55	1.871		231	109	12.45	352.45	435
4410	2"		7.26	2.204		420	128	14.65	562.65	675
4500	Tee									
4540	1/4"	Q-15	8.97	1.784	Ea.	60	103	11.85	174.85	239
4550	3/8"		8.64	1.852		72	107	12.30	191.30	259
4560	1/2"		8.13	1.968		88	114	13.10	215.10	287
4570	3/4"		7.12	2.247		102	130	14.95	246.95	330
4580	1"		6.48	2.469		137	143	16.40	296.40	390
4590	1-1/4"		6.03	2.653		243	154	17.65	414.65	525
4600	1-1/2"		5.69	2.812		350	163	18.70	531.70	660
4610	2"		4.83	3.313		530	192	22	744	900
5000	Socket weld joint, 3,000 lb., type 316									
5100	90° elbow									
5140	1/4"	Q-15	13.47	1.188	Ea.	55.50	69	7.90	132.40	177
5150	3/8"		12.97	1.234		64.50	71.50	8.20	144.20	191
5160	1/2"		12.21	1.310		78	76	8.70	162.70	213
5170	3/4"		10.68	1.498		103	87	9.95	199.95	258
5180	1"		9.74	1.643		146	95.50	10.90	252.40	320
5190	1-1/4"		9.05	1.768		260	103	11.75	374.75	460
5200	1-1/2"		8.55	1.871		293	109	12.45	414.45	505
5210	2"		7.26	2.204		500	128	14.65	642.65	765
5300	45° elbow									
5340	1/4"	Q-15	13.47	1.188	Ea.	110	69	7.90	186.90	237
5350	3/8"		12.97	1.234		110	71.50	8.20	189.70	241
5360	1/2"		12.21	1.310		112	76	8.70	196.70	251
5370	3/4"		10.68	1.498		124	87	9.95	220.95	281
5380	1"		9.74	1.643		187	95.50	10.90	293.40	365
5390	1-1/4"		9.05	1.768		267	103	11.75	381.75	465
5400	1-1/2"		8.55	1.871		300	109	12.45	421.45	510
5410	2"		7.26	2.204		450	128	14.65	592.65	705
5500	Tee									
5540	1/4"	Q-15	8.97	1.784	Ea.	75	103	11.85	189.85	256
5550	3/8"		8.64	1.852		91.50	107	12.30	210.80	281
5560	1/2"		8.13	1.968		102	114	13.10	229.10	300
5570	3/4"		7.12	2.247		126	130	14.95	270.95	355
5580	1"		6.48	2.469		193	143	16.40	352.40	450
5590	1-1/4"		6.03	2.653		305	154	17.65	476.65	595
5600	1-1/2"		5.69	2.812		430	163	18.70	611.70	750
5610	2"		4.83	3.313		690	192	22	904	1,075
5700	For socket weld joint, 6,000 lb., type 304 and 316, add					100%				
6000	Threaded companion flange									
6010	Stainless steel, 150 lb., type 304									
6020	1/2" diam.	1 Plum	30	.267	Ea.	46	17.20		63.20	77
6030	3/4" diam.		28	.286		51	18.40		69.40	84.50
6040	1" diam.		27	.296		56	19.10		75.10	91
6050	1-1/4" diam.	Q-1	44	.364		72.50	21		93.50	112
6060	1-1/2" diam.		40	.400		72.50	23		95.50	116
6070	2" diam.		36	.444		94.50	26		120.50	144
6080	2-1/2" diam.		28	.571		132	33		165	198

22 11 13.66 Pipe Fittings, Stainless Steel		Crew	Daily Output	Labor-Hours	Unit	2020 Bare Costs Material	Labor	Equipment	Total	Total Incl O&P
6090	3" diam.	Q-1	20	.800	Ea.	139	46.50		185.50	224
6110	4" diam.		12	1.333		189	77.50		266.50	330
6130	6" diam.	Q-2	14	1.714		340	103		443	535
6140	8" diam.	"	12	2		655	120		775	905
6150	For type 316, add					40%				
6260	Weld flanges, stainless steel, type 304									
6270	Slip on, 150 lb. (welded, front and back)									
6280	1/2" diam.	Q-15	12.70	1.260	Ea.	41	73	8.40	122.40	167
6290	3/4" diam.		11.11	1.440		42	83.50	9.60	135.10	186
6300	1" diam.		10.13	1.579		46.50	91.50	10.50	148.50	205
6310	1-1/4" diam.		9.41	1.700		62.50	98.50	11.30	172.30	233
6320	1-1/2" diam.		8.89	1.800		62.50	104	11.95	178.45	243
6330	2" diam.		7.55	2.119		81	123	14.10	218.10	295
6340	2-1/2" diam.		3.98	4.020		113	233	26.50	372.50	515
6350	3" diam.		3.21	4.984		122	289	33	444	615
6370	4" diam.		2.44	6.557		166	380	43.50	589.50	820
6390	6" diam.	Q-16	1.89	12.698		252	765	56.50	1,073.50	1,525
6400	8" diam.	"	1.22	19.672		475	1,175	87	1,737	2,450
6410	For type 316, add					40%				
6530	Weld neck 150 lb.									
6540	1/2" diam.	Q-15	25.40	.630	Ea.	21.50	36.50	4.19	62.19	85
6550	3/4" diam.		22.22	.720		26	42	4.79	72.79	98.50
6560	1" diam.		20.25	.790		29.50	46	5.25	80.75	109
6570	1-1/4" diam.		18.82	.850		41.50	49.50	5.65	96.65	128
6580	1-1/2" diam.		17.78	.900		41.50	52	6	99.50	133
6590	2" diam.		15.09	1.060		46.50	61.50	7.05	115.05	154
6600	2-1/2" diam.		7.96	2.010		75.50	117	13.35	205.85	278
6610	3" diam.		6.43	2.488		76	144	16.55	236.55	325
6630	4" diam.		4.88	3.279		110	190	22	322	440
6640	5" diam.		4.26	3.756		139	218	25	382	515
6650	6" diam.	Q-16	5.66	4.240		163	255	18.80	436.80	595
6652	8" diam.		3.65	6.575		286	395	29	710	955
6654	10" diam.		2.88	8.333		400	500	37	937	1,250
6656	12" diam.		2.28	10.526		740	635	46.50	1,421.50	1,850
6670	For type 316, add					23%				
7000	Threaded joint, 150 lb., type 304									
7030	90° elbow									
7040	1/8"	1 Plum	13	.615	Ea.	29.50	39.50		69	94
7050	1/4"		13	.615		29.50	39.50		69	94
7070	3/8"		13	.615		34.50	39.50		74	99.50
7080	1/2"		12	.667		30.50	43		73.50	100
7090	3/4"		11	.727		36.50	47		83.50	113
7100	1"		10	.800		50.50	51.50		102	135
7110	1-1/4"	Q-1	17	.941		78.50	54.50		133	171
7120	1-1/2"		16	1		90	58		148	189
7130	2"		14	1.143		130	66.50		196.50	245
7140	2-1/2"		11	1.455		315	84.50		399.50	480
7150	3"		8	2		460	116		576	685
7160	4"	Q-2	11	2.182		775	131		906	1,050
7180	45° elbow									
7190	1/8"	1 Plum	13	.615	Ea.	43	39.50		82.50	109
7200	1/4"		13	.615		43	39.50		82.50	109
7210	3/8"		13	.615		43.50	39.50		83	110

22 11 13.66 Pipe Fittings, Stainless Steel		Crew	Daily Output	Labor-Hours	Unit	Material	Labor	Equipment	Total	Total Incl O&P
						2020 Bare Costs				
7220	1/2"	1 Plum	12	.667	Ea.	44	43		87	115
7230	3/4"		11	.727		48	47		95	125
7240	1"		10	.800		55.50	51.50		107	141
7250	1-1/4"	Q-1	17	.941		77.50	54.50		132	170
7260	1-1/2"		16	1		100	58		158	200
7270	2"		14	1.143		143	66.50		209.50	260
7280	2-1/2"		11	1.455		445	84.50		529.50	620
7290	3"		8	2		655	116		771	900
7300	4"	Q-2	11	2.182		1,175	131		1,306	1,500
7320	Tee, straight									
7330	1/8"	1 Plum	9	.889	Ea.	45.50	57.50		103	139
7340	1/4"		9	.889		45.50	57.50		103	139
7350	3/8"		9	.889		48.50	57.50		106	142
7360	1/2"		8	1		46	64.50		110.50	151
7370	3/4"		7	1.143		49.50	73.50		123	169
7380	1"		6.50	1.231		64	79.50		143.50	194
7390	1-1/4"	Q-1	11	1.455		107	84.50		191.50	248
7400	1-1/2"		10	1.600		138	93		231	295
7410	2"		9	1.778		171	103		274	345
7420	2-1/2"		7	2.286		455	133		588	705
7430	3"		5	3.200		690	186		876	1,050
7440	4"	Q-2	7	3.429		1,700	206		1,906	2,200
7460	Coupling, straight									
7470	1/8"	1 Plum	19	.421	Ea.	10.85	27		37.85	54
7480	1/4"		19	.421		13.35	27		40.35	56.50
7490	3/8"		19	.421		16	27		43	59.50
7500	1/2"		19	.421		21.50	27		48.50	65.50
7510	3/4"		18	.444		28.50	28.50		57	75.50
7520	1"		15	.533		45.50	34.50		80	104
7530	1-1/4"	Q-1	26	.615		88.50	35.50		124	153
7540	1-1/2"		24	.667		82	38.50		120.50	151
7550	2"		21	.762		137	44		181	220
7560	2-1/2"		18	.889		320	51.50		371.50	430
7570	3"		14	1.143		435	66.50		501.50	580
7580	4"	Q-2	16	1.500		605	90		695	805
7600	Reducer, concentric, 1/2"	1 Plum	12	.667		23	43		66	91.50
7610	3/4"		11	.727		29.50	47		76.50	105
7612	1"		10	.800		49	51.50		100.50	133
7614	1-1/4"	Q-1	17	.941		104	54.50		158.50	199
7616	1-1/2"		16	1		114	58		172	215
7618	2"		14	1.143		177	66.50		243.50	297
7620	2-1/2"		11	1.455		470	84.50		554.50	645
7622	3"		8	2		520	116		636	755
7624	4"	Q-2	11	2.182		865	131		996	1,150
7710	Union									
7720	1/8"	1 Plum	12	.667	Ea.	53.50	43		96.50	126
7730	1/4"		12	.667		53.50	43		96.50	126
7740	3/8"		12	.667		63	43		106	136
7750	1/2"		11	.727		78	47		125	158
7760	3/4"		10	.800		107	51.50		158.50	197
7770	1"		9	.889		155	57.50		212.50	259
7780	1-1/4"	Q-1	16	1		385	58		443	515
7790	1-1/2"		15	1.067		415	62		477	550

22 11 13.66 Pipe Fittings, Stainless Steel		Crew	Daily Output	Labor-Hours	Unit	Material	2020 Bare Costs Labor	Equipment	Total	Total Incl O&P
7800	2"	Q-1	13	1.231	Ea.	525	71.50		596.50	685
7810	2-1/2"		10	1.600		1,000	93		1,093	1,250
7820	3"		7	2.286		1,325	133		1,458	1,675
7830	4"	Q-2	10	2.400		1,825	144		1,969	2,250
7850	For 150 lb., type 316, add					25%				
8750	Threaded joint, 2,000 lb., type 304									
8770	90° elbow									
8780	1/8"	1 Plum	13	.615	Ea.	33.50	39.50		73	98
8790	1/4"		13	.615		33.50	39.50		73	98
8800	3/8"		13	.615		41	39.50		80.50	107
8810	1/2"		12	.667		56	43		99	128
8820	3/4"		11	.727		64.50	47		111.50	144
8830	1"		10	.800		86	51.50		137.50	174
8840	1-1/4"	Q-1	17	.941		140	54.50		194.50	239
8850	1-1/2"		16	1		224	58		282	335
8860	2"		14	1.143		315	66.50		381.50	445
8880	45° elbow									
8890	1/8"	1 Plum	13	.615	Ea.	69.50	39.50		109	138
8900	1/4"		13	.615		69.50	39.50		109	138
8910	3/8"		13	.615		87	39.50		126.50	157
8920	1/2"		12	.667		87	43		130	162
8930	3/4"		11	.727		96	47		143	178
8940	1"		10	.800		114	51.50		165.50	206
8950	1-1/4"	Q-1	17	.941		199	54.50		253.50	305
8960	1-1/2"		16	1		267	58		325	385
8970	2"		14	1.143		310	66.50		376.50	440
8990	Tee, straight									
9000	1/8"	1 Plum	9	.889	Ea.	43	57.50		100.50	136
9010	1/4"		9	.889		43	57.50		100.50	136
9020	3/8"		9	.889		55.50	57.50		113	150
9030	1/2"		8	1		71	64.50		135.50	178
9040	3/4"		7	1.143		90.50	73.50		164	214
9050	1"		6.50	1.231		113	79.50		192.50	247
9060	1-1/4"	Q-1	11	1.455		195	84.50		279.50	345
9070	1-1/2"		10	1.600		282	93		375	455
9080	2"		9	1.778		435	103		538	635
9100	For couplings and unions use 3,000 lb., type 304									
9120	2,000 lb., type 316									
9130	90° elbow									
9140	1/8"	1 Plum	13	.615	Ea.	41.50	39.50		81	107
9150	1/4"		13	.615		41.50	39.50		81	107
9160	3/8"		13	.615		48.50	39.50		88	115
9170	1/2"		12	.667		63	43		106	136
9180	3/4"		11	.727		77	47		124	157
9190	1"		10	.800		116	51.50		167.50	208
9200	1-1/4"	Q-1	17	.941		208	54.50		262.50	315
9210	1-1/2"		16	1		239	58		297	355
9220	2"		14	1.143		405	66.50		471.50	545
9240	45° elbow									
9250	1/8"	1 Plum	13	.615	Ea.	88	39.50		127.50	158
9260	1/4"		13	.615		88	39.50		127.50	158
9270	3/8"		13	.615		88	39.50		127.50	158
9280	1/2"		12	.667		88	43		131	163

22 11 13.66 Pipe Fittings, Stainless Steel		Crew	Daily Output	Labor-Hours	Unit	Material	2020 Bare Costs Labor	Equipment	Total	Total Incl O&P
9300	3/4"	1 Plum	11	.727	Ea.	99	47		146	182
9310	1"		10	.800		150	51.50		201.50	245
9320	1-1/4"	Q-1	17	.941		214	54.50		268.50	320
9330	1-1/2"		16	1		293	58		351	410
9340	2"		14	1.143		360	66.50		426.50	495
9360	Tee, straight									
9370	1/8"	1 Plum	9	.889	Ea.	53	57.50		110.50	147
9380	1/4"		9	.889		53	57.50		110.50	147
9390	3/8"		9	.889		64.50	57.50		122	159
9400	1/2"		8	1		80.50	64.50		145	188
9410	3/4"		7	1.143		99.50	73.50		173	223
9420	1"		6.50	1.231		152	79.50		231.50	290
9430	1-1/4"	Q-1	11	1.455		246	84.50		330.50	400
9440	1-1/2"		10	1.600		355	93		448	535
9450	2"		9	1.778		565	103		668	780
9470	For couplings and unions use 3,000 lb., type 316									
9490	3,000 lb., type 304									
9510	Coupling									
9520	1/8"	1 Plum	19	.421	Ea.	13.15	27		40.15	56.50
9530	1/4"		19	.421		14.30	27		41.30	58
9540	3/8"		19	.421		18.40	27		45.40	62.50
9550	1/2"		19	.421		21.50	27		48.50	65.50
9560	3/4"		18	.444		30	28.50		58.50	77
9570	1"		15	.533		44	34.50		78.50	101
9580	1-1/4"	Q-1	26	.615		93	35.50		128.50	157
9590	1-1/2"		24	.667		129	38.50		167.50	202
9600	2"		21	.762		147	44		191	231
9620	Union									
9630	1/8"	1 Plum	12	.667	Ea.	101	43		144	178
9640	1/4"		12	.667		110	43		153	188
9650	3/8"		12	.667		112	43		155	190
9660	1/2"		11	.727		89.50	47		136.50	171
9670	3/4"		10	.800		138	51.50		189.50	231
9680	1"		9	.889		212	57.50		269.50	320
9690	1-1/4"	Q-1	16	1		390	58		448	520
9700	1-1/2"		15	1.067		440	62		502	580
9710	2"		13	1.231		590	71.50		661.50	760
9730	3,000 lb., type 316									
9750	Coupling									
9770	1/8"	1 Plum	19	.421	Ea.	14.80	27		41.80	58.50
9780	1/4"		19	.421		19.55	27		46.55	63.50
9790	3/8"		19	.421		21	27		48	65
9800	1/2"		19	.421		23.50	27		50.50	67.50
9810	3/4"		18	.444		33.50	28.50		62	81
9820	1"		15	.533		62.50	34.50		97	122
9830	1-1/4"	Q-1	26	.615		140	35.50		175.50	209
9840	1-1/2"		24	.667		163	38.50		201.50	240
9850	2"		21	.762		226	44		270	320
9870	Union									
9880	1/8"	1 Plum	12	.667	Ea.	115	43		158	194
9890	1/4"		12	.667		115	43		158	194
9900	3/8"		12	.667		133	43		176	214
9910	1/2"		11	.727		134	47		181	220

22 11 13.66 Pipe Fittings, Stainless Steel		Crew	Daily Output	Labor-Hours	Unit	Material	2020 Bare Costs Labor	Equipment	Total	Total Incl O&P
9920	3/4"	1 Plum	10	.800	Ea.	174	51.50		225.50	272
9930	1"		9	.889		275	57.50		332.50	390
9940	1-1/4"	Q-1	16	1		480	58		538	620
9950	1-1/2"		15	1.067		595	62		657	750
9960	2"		13	1.231		755	71.50		826.50	940

22 11 13.74 Pipe, Plastic		Crew	Daily Output	Labor-Hours	Unit	Material	Labor	Equipment	Total	Total Incl O&P
0010	**PIPE, PLASTIC**									
0020	Fiberglass reinforced, couplings 10′ OC, clevis hanger assy's, 3 per 10′									
0080	General service									
0120	2" diameter	Q-1	59	.271	L.F.	17.50	15.75		33.25	44
0140	3" diameter		52	.308		27.50	17.85		45.35	58
0150	4" diameter		48	.333		23	19.35		42.35	55.50
0160	6" diameter		39	.410		43	24		67	84.50
0170	8" diameter	Q-2	49	.490		67.50	29.50		97	120
0180	10" diameter		41	.585		95.50	35		130.50	160
0190	12" diameter		36	.667		125	40		165	200
0600	PVC, high impact/pressure, cplgs. 10′ OC, clevis hanger assy's, 3 per 10′									
1020	Schedule 80									
1070	1/2" diameter	1 Plum	50	.160	L.F.	6.85	10.30		17.15	23.50
1080	3/4" diameter		47	.170		7.95	10.95		18.90	25.50
1090	1" diameter		43	.186		13.25	12		25.25	33
1100	1-1/4" diameter		39	.205		15.40	13.20		28.60	37.50
1110	1-1/2" diameter		34	.235		16.75	15.15		31.90	42
1120	2" diameter	Q-1	55	.291		20.50	16.85		37.35	48.50
1140	3" diameter		50	.320		40.50	18.55		59.05	73.50
1150	4" diameter		46	.348		43	20		63	78
1170	6" diameter		38	.421		66.50	24.50		91	111
1730	To delete coupling & hangers, subtract									
1740	1/2" diam.					62%	80%			
1750	3/4" diam. to 1-1/4" diam.					58%	73%			
1760	1-1/2" diam. to 6" diam.					40%	57%			
1800	PVC, couplings 10′ OC, clevis hanger assemblies, 3 per 10′									
1820	Schedule 40									
1860	1/2" diameter	1 Plum	54	.148	L.F.	4.90	9.55		14.45	20
1870	3/4" diameter		51	.157		5.40	10.10		15.50	21.50
1880	1" diameter		46	.174		9.45	11.20		20.65	28
1890	1-1/4" diameter		42	.190		9.95	12.30		22.25	30
1900	1-1/2" diameter		36	.222		10.25	14.30		24.55	33.50
1910	2" diameter	Q-1	59	.271		12.15	15.75		27.90	38
1920	2-1/2" diameter		56	.286		20	16.55		36.55	48
1930	3" diameter		53	.302		22	17.50		39.50	51.50
1940	4" diameter		48	.333		14.80	19.35		34.15	46.50
1950	5" diameter		43	.372		23.50	21.50		45	59.50
1960	6" diameter		39	.410		30.50	24		54.50	70.50
1970	8" diameter	Q-2	48	.500		37	30		67	87.50
1980	10" diameter		43	.558		77	33.50		110.50	137
1990	12" diameter		42	.571		93	34.50		127.50	155
2000	14" diameter		31	.774		151	46.50		197.50	238
2010	16" diameter		23	1.043		220	63		283	340
2340	To delete coupling & hangers, subtract									
2360	1/2" diam. to 1-1/4" diam.					65%	74%			
2370	1-1/2" diam. to 6" diam.					44%	57%			

22 11 13 – Facility Water Distribution Piping

22 11 13.74 Pipe, Plastic		Crew	Daily Output	Labor-Hours	Unit	Material	Labor	Equipment	Total	Total Incl O&P
						2020 Bare Costs				
2380	8″ diam. to 12″ diam.					41%	53%			
2390	14″ diam. to 16″ diam.					48%	45%			
2420	Schedule 80									
2440	1/4″ diameter	1 Plum	58	.138	L.F.	4.46	8.90		13.36	18.65
2450	3/8″ diameter		55	.145		4.46	9.35		13.81	19.40
2460	1/2″ diameter		50	.160		5.15	10.30		15.45	21.50
2470	3/4″ diameter		47	.170		5.55	10.95		16.50	23
2480	1″ diameter		43	.186		9.90	12		21.90	29.50
2490	1-1/4″ diameter		39	.205		10.45	13.20		23.65	32
2500	1-1/2″ diameter		34	.235		10.65	15.15		25.80	35.50
2510	2″ diameter	Q-1	55	.291		11.55	16.85		28.40	38.50
2520	2-1/2″ diameter		52	.308		19.55	17.85		37.40	49
2530	3″ diameter		50	.320		24	18.55		42.55	55
2540	4″ diameter		46	.348		19.30	20		39.30	52
2550	5″ diameter		42	.381		29.50	22		51.50	66.50
2560	6″ diameter		38	.421		37	24.50		61.50	78.50
2570	8″ diameter	Q-2	47	.511		47	30.50		77.50	99
2580	10″ diameter		42	.571		110	34.50		144.50	174
2590	12″ diameter		38	.632		134	38		172	206
2830	To delete coupling & hangers, subtract									
2840	1/4″ diam. to 1/2″ diam.					66%	80%			
2850	3/4″ diam. to 1-1/4″ diam.					61%	73%			
2860	1-1/2″ diam. to 6″ diam.					41%	57%			
2870	8″ diam. to 12″ diam.					31%	50%			
2900	Schedule 120									
2910	1/2″ diameter	1 Plum	50	.160	L.F.	5.60	10.30		15.90	22
2950	3/4″ diameter		47	.170		6.20	10.95		17.15	24
2960	1″ diameter		43	.186		10.60	12		22.60	30
2970	1-1/4″ diameter		39	.205		11.90	13.20		25.10	33.50
2980	1-1/2″ diameter		33	.242		12.25	15.60		27.85	37.50
2990	2″ diameter	Q-1	54	.296		14.35	17.20		31.55	42.50
3000	2-1/2″ diameter		52	.308		25	17.85		42.85	55.50
3010	3″ diameter		49	.327		36	18.95		54.95	69
3020	4″ diameter		45	.356		25.50	20.50		46	60
3030	6″ diameter		37	.432		63	25		88	108
3240	To delete coupling & hangers, subtract									
3250	1/2″ diam. to 1-1/4″ diam.					52%	74%			
3260	1-1/2″ diam. to 4″ diam.					30%	57%			
3270	6″ diam.					17%	50%			
3300	PVC, pressure, couplings 10′ OC, clevis hanger assy's, 3 per 10′									
3310	SDR 26, 160 psi									
3350	1-1/4″ diameter	1 Plum	42	.190	L.F.	9.85	12.30		22.15	30
3360	1-1/2″ diameter	″	36	.222		10.60	14.30		24.90	33.50
3370	2″ diameter	Q-1	59	.271		11.05	15.75		26.80	36.50
3380	2-1/2″ diameter		56	.286		19.35	16.55		35.90	47
3390	3″ diameter		53	.302		22	17.50		39.50	51.50
3400	4″ diameter		48	.333		16.35	19.35		35.70	48
3420	6″ diameter		39	.410		33.50	24		57.50	74
3430	8″ diameter	Q-2	48	.500		49	30		79	100
3660	To delete coupling & clevis hanger assy's, subtract									
3670	1-1/4″ diam.					63%	68%			
3680	1-1/2″ diam. to 4″ diam.					48%	57%			
3690	6″ diam. to 8″ diam.					60%	54%			

22 11 13 – Facility Water Distribution Piping

22 11 13.74 Pipe, Plastic		Crew	Daily Output	Labor-Hours	Unit	Material	2020 Bare Costs Labor	Equipment	Total	Total Incl O&P
3720	SDR 21, 200 psi, 1/2" diameter	1 Plum	54	.148	L.F.	4.80	9.55		14.35	20
3740	3/4" diameter		51	.157		5.05	10.10		15.15	21
3750	1" diameter		46	.174		8.90	11.20		20.10	27
3760	1-1/4" diameter		42	.190		10	12.30		22.30	30
3770	1-1/2" diameter		36	.222		10.55	14.30		24.85	33.50
3780	2" diameter	Q-1	59	.271		10.45	15.75		26.20	36
3790	2-1/2" diameter		56	.286		19.85	16.55		36.40	47.50
3800	3" diameter		53	.302		21	17.50		38.50	50
3810	4" diameter		48	.333		16.95	19.35		36.30	48.50
3830	6" diameter		39	.410		29.50	24		53.50	69.50
3840	8" diameter	Q-2	48	.500		48	30		78	99
4000	To delete coupling & hangers, subtract									
4010	1/2" diam. to 3/4" diam.					71%	77%			
4020	1" diam. to 1-1/4" diam.					63%	70%			
4030	1-1/2" diam. to 6" diam.					44%	57%			
4040	8" diam.					46%	54%			
4100	DWV type, schedule 40, couplings 10' OC, clevis hanger assy's, 3 per 10'									
4210	ABS, schedule 40, foam core type									
4212	Plain end black									
4214	1-1/2" diameter	1 Plum	39	.205	L.F.	8.60	13.20		21.80	30
4216	2" diameter	Q-1	62	.258		9.25	14.95		24.20	33
4218	3" diameter		56	.286		17.60	16.55		34.15	45
4220	4" diameter		51	.314		9.80	18.20		28	39
4222	6" diameter		42	.381		21.50	22		43.50	57.50
4240	To delete coupling & hangers, subtract									
4244	1-1/2" diam. to 6" diam.					43%	48%			
4400	PVC									
4410	1-1/4" diameter	1 Plum	42	.190	L.F.	9.10	12.30		21.40	29
4420	1-1/2" diameter	"	36	.222		8.25	14.30		22.55	31
4460	2" diameter	Q-1	59	.271		9.15	15.75		24.90	34.50
4470	3" diameter		53	.302		17.45	17.50		34.95	46
4480	4" diameter		48	.333		19.20	19.35		38.55	51
4490	6" diameter		39	.410		18.60	24		42.60	57.50
4500	8" diameter	Q-2	48	.500		25.50	30		55.50	74.50
4510	To delete coupling & hangers, subtract									
4520	1-1/4" diam. to 1-1/2" diam.					48%	60%			
4530	2" diam. to 8" diam.					42%	54%			
4532	to delete hangers, 2" diam. to 8" diam.	Q-1	50	.320	L.F.	5.50	18.55		24.05	34.50
4550	PVC, schedule 40, foam core type									
4552	Plain end, white									
4554	1-1/2" diameter	1 Plum	39	.205	L.F.	8.40	13.20		21.60	29.50
4556	2" diameter	Q-1	62	.258		8.95	14.95		23.90	33
4558	3" diameter		56	.286		17.05	16.55		33.60	44.50
4560	4" diameter		51	.314		18.55	18.20		36.75	48.50
4562	6" diameter		42	.381		17.25	22		39.25	53
4564	8" diameter	Q-2	51	.471		23.50	28.50		52	69.50
4568	10" diameter		48	.500		27.50	30		57.50	76.50
4570	12" diameter		46	.522		30.50	31.50		62	82
4580	To delete coupling & hangers, subtract									
4582	1-1/2" diam. to 2" diam.					58%	54%			
4584	3" diam. to 12" diam.					46%	42%			
4800	PVC, clear pipe, cplgs. 10' OC, clevis hanger assy's 3 per 10', Sched. 40									
4840	1/4" diameter	1 Plum	59	.136	L.F.	4.93	8.75		13.68	18.90

22 11 13.74 Pipe, Plastic		Crew	Daily Output	Labor-Hours	Unit	Material	Labor	Equipment	Total	Total Incl O&P
						2020 Bare Costs				
4850	3/8" diameter	1 Plum	56	.143	L.F.	5.05	9.20		14.25	19.80
4860	1/2" diameter		54	.148		5.90	9.55		15.45	21.50
4870	3/4" diameter		51	.157		6.60	10.10		16.70	23
4880	1" diameter		46	.174		11.60	11.20		22.80	30
4890	1-1/4" diameter		42	.190		13.10	12.30		25.40	33.50
4900	1-1/2" diameter		36	.222		13.95	14.30		28.25	37.50
4910	2" diameter	Q-1	59	.271		16.45	15.75		32.20	42.50
4920	2-1/2" diameter		56	.286		28	16.55		44.55	56.50
4930	3" diameter		53	.302		32.50	17.50		50	63
4940	3-1/2" diameter		50	.320		40	18.55		58.55	72.50
4950	4" diameter		48	.333		28.50	19.35		47.85	61
5250	To delete coupling & hangers, subtract									
5260	1/4" diam. to 3/8" diam.					60%	81%			
5270	1/2" diam. to 3/4" diam.					41%	77%			
5280	1" diam. to 1-1/2" diam.					26%	67%			
5290	2" diam. to 4" diam.					16%	58%			
5300	CPVC, socket joint, couplings 10' OC, clevis hanger assemblies, 3 per 10'									
5302	Schedule 40									
5304	1/2" diameter	1 Plum	54	.148	L.F.	5.80	9.55		15.35	21
5305	3/4" diameter		51	.157		6.95	10.10		17.05	23.50
5306	1" diameter		46	.174		11.35	11.20		22.55	30
5307	1-1/4" diameter		42	.190		12.85	12.30		25.15	33
5308	1-1/2" diameter		36	.222		12.45	14.30		26.75	35.50
5309	2" diameter	Q-1	59	.271		15.70	15.75		31.45	42
5310	2-1/2" diameter		56	.286		28	16.55		44.55	56.50
5311	3" diameter		53	.302		32	17.50		49.50	62
5312	4" diameter		48	.333		32.50	19.35		51.85	66
5314	6" diameter		43	.372		55.50	21.50		77	94.50
5318	To delete coupling & hangers, subtract									
5319	1/2" diam. to 3/4" diam.					37%	77%			
5320	1" diam. to 1-1/4" diam.					27%	70%			
5321	1-1/2" diam. to 3" diam.					21%	57%			
5322	4" diam. to 6" diam.					16%	57%			
5324	Schedule 80									
5325	1/2" diameter	1 Plum	50	.160	L.F.	6.10	10.30		16.40	22.50
5326	3/4" diameter		47	.170		6.95	10.95		17.90	24.50
5327	1" diameter		43	.186		12	12		24	32
5328	1-1/4" diameter		39	.205		13.55	13.20		26.75	35.50
5329	1-1/2" diameter		34	.235		14.70	15.15		29.85	39.50
5330	2" diameter	Q-1	55	.291		17.30	16.85		34.15	45
5331	2-1/2" diameter		52	.308		30.50	17.85		48.35	61
5332	3" diameter		50	.320		32.50	18.55		51.05	64
5333	4" diameter		46	.348		31.50	20		51.50	65.50
5334	6" diameter		38	.421		61.50	24.50		86	106
5335	8" diameter	Q-2	47	.511		170	30.50		200.50	235
5339	To delete couplings & hangers, subtract									
5340	1/2" diam. to 3/4" diam.					44%	77%			
5341	1" diam. to 1-1/4" diam.					32%	71%			
5342	1-1/2" diam. to 4" diam.					25%	58%			
5343	6" diam. to 8" diam.					20%	53%			
5360	CPVC, threaded, couplings 10' OC, clevis hanger assemblies, 3 per 10'									
5380	Schedule 40									
5460	1/2" diameter	1 Plum	54	.148	L.F.	6.65	9.55		16.20	22

22 11 13 – Facility Water Distribution Piping

22 11 13.74 Pipe, Plastic		Crew	Daily Output	Labor-Hours	Unit	Material	Labor	Equipment	Total	Total Incl O&P
						2020 Bare Costs				
5470	3/4" diameter	1 Plum	51	.157	L.F.	8.45	10.10		18.55	25
5480	1" diameter		46	.174		12.90	11.20		24.10	31.50
5490	1-1/4" diameter		42	.190		14.05	12.30		26.35	34.50
5500	1-1/2" diameter		36	.222		13.45	14.30		27.75	37
5510	2" diameter	Q-1	59	.271		16.90	15.75		32.65	43
5520	2-1/2" diameter		56	.286		29.50	16.55		46.05	58
5530	3" diameter		53	.302		33.50	17.50		51	64
5540	4" diameter		48	.333		40	19.35		59.35	74.50
5550	6" diameter		43	.372		59	21.50		80.50	98.50
5730	To delete coupling & hangers, subtract									
5740	1/2" diam. to 3/4" diam.					37%	77%			
5750	1" diam. to 1-1/4" diam.					27%	70%			
5760	1-1/2" diam. to 3" diam.					21%	57%			
5770	4" diam. to 6" diam.					16%	57%			
5800	Schedule 80									
5860	1/2" diameter	1 Plum	50	.160	L.F.	6.95	10.30		17.25	23.50
5870	3/4" diameter		47	.170		8.45	10.95		19.40	26.50
5880	1" diameter		43	.186		13.55	12		25.55	33.50
5890	1-1/4" diameter		39	.205		14.75	13.20		27.95	37
5900	1-1/2" diameter		34	.235		15.70	15.15		30.85	41
5910	2" diameter	Q-1	55	.291		18.55	16.85		35.40	46.50
5920	2-1/2" diameter		52	.308		31.50	17.85		49.35	62.50
5930	3" diameter		50	.320		34.50	18.55		53.05	66.50
5940	4" diameter		46	.348		39	20		59	74
5950	6" diameter		38	.421		65	24.50		89.50	110
5960	8" diameter	Q-2	47	.511		168	30.50		198.50	233
6060	To delete couplings & hangers, subtract									
6070	1/2" diam. to 3/4" diam.					44%	77%			
6080	1" diam. to 1-1/4" diam.					32%	71%			
6090	1-1/2" diam. to 4" diam.					25%	58%			
6100	6" diam. to 8" diam.					20%	53%			
6240	CTS, 1/2" diameter	1 Plum	54	.148	L.F.	4.80	9.55		14.35	20
6250	3/4" diameter		51	.157		10.05	10.10		20.15	26.50
6260	1" diameter		46	.174		13.85	11.20		25.05	32.50
6270	1-1/4" diameter		42	.190		18.65	12.30		30.95	39.50
6280	1-1/2" diameter		36	.222		22	14.30		36.30	46.50
6290	2" diameter	Q-1	59	.271		33	15.75		48.75	60.50
6370	To delete coupling & hangers, subtract									
6380	1/2" diam.					51%	79%			
6390	3/4" diam.					40%	76%			
6392	1" thru 2" diam.					72%	68%			
6500	Residential installation, plastic pipe									
6510	Couplings 10' OC, strap hangers 3 per 10'									
6520	PVC, Schedule 40									
6530	1/2" diameter	1 Plum	138	.058	L.F.	1.06	3.74		4.80	6.95
6540	3/4" diameter		128	.063		1.30	4.03		5.33	7.70
6550	1" diameter		119	.067		1.87	4.33		6.20	8.75
6560	1-1/4" diameter		111	.072		2.13	4.64		6.77	9.55
6570	1-1/2" diameter		104	.077		2.53	4.96		7.49	10.45
6580	2" diameter	Q-1	197	.081		3.45	4.71		8.16	11.10
6590	2-1/2" diameter		162	.099		5.70	5.75		11.45	15.15
6600	4" diameter		123	.130		8.15	7.55		15.70	20.50
6700	PVC, DWV, Schedule 40									

22 11 13.74 Pipe, Plastic		Crew	Daily Output	Labor-Hours	Unit	Material	Labor	Equipment	Total	Total Incl O&P
						2020 Bare Costs				
6720	1-1/4" diameter	1 Plum	100	.080	L.F.	2.14	5.15		7.29	10.30
6730	1-1/2" diameter	"	94	.085		1.54	5.50		7.04	10.20
6740	2" diameter	Q-1	178	.090		2.16	5.20		7.36	10.45
6760	4" diameter	"	110	.145		6.20	8.45		14.65	19.85
7280	PEX, flexible, no couplings or hangers									
7282	Note: For labor costs add 25% to the couplings and fittings labor total.									
7285	For fittings see section 23 83 16.10 7000									
7300	Non-barrier type, hot/cold tubing rolls									
7310	1/4" diameter x 100'				L.F.	.56			.56	.62
7350	3/8" diameter x 100'					.55			.55	.61
7360	1/2" diameter x 100'					.76			.76	.84
7370	1/2" diameter x 500'					.74			.74	.81
7380	1/2" diameter x 1000'					.72			.72	.79
7400	3/4" diameter x 100'					1.04			1.04	1.14
7410	3/4" diameter x 500'					1.17			1.17	1.29
7420	3/4" diameter x 1000'					1.17			1.17	1.29
7460	1" diameter x 100'					2.02			2.02	2.22
7470	1" diameter x 300'					2.02			2.02	2.22
7480	1" diameter x 500'					2.03			2.03	2.23
7500	1-1/4" diameter x 100'					3.44			3.44	3.78
7510	1-1/4" diameter x 300'					3.44			3.44	3.78
7540	1-1/2" diameter x 100'					4.71			4.71	5.20
7550	1-1/2" diameter x 300'					4.72			4.72	5.20
7596	Most sizes available in red or blue									
7700	Non-barrier type, hot/cold tubing straight lengths									
7710	1/2" diameter x 20'				L.F.	.68			.68	.75
7750	3/4" diameter x 20'					1.20			1.20	1.32
7760	1" diameter x 20'					2.04			2.04	2.24
7770	1-1/4" diameter x 20'					3.65			3.65	4.02
7780	1-1/2" diameter x 20'					4.73			4.73	5.20
7790	2" diameter					9.30			9.30	10.25
7796	Most sizes available in red or blue									
9000	Polypropylene pipe									
9002	For fusion weld fittings and accessories see line 22 11 13.76 9400									
9004	Note: sizes 1/2" thru 4" use socket fusion									
9005	Sizes 6" thru 10" use butt fusion									
9010	SDR 7.4 (domestic hot water piping)									
9011	Enhanced to minimize thermal expansion and high temperature life									
9016	13' lengths, size is ID, includes joints 13' OC and hangers 3 per 10'									
9020	3/8" diameter	1 Plum	53	.151	L.F.	3.20	9.75		12.95	18.55
9022	1/2" diameter		52	.154		3.67	9.90		13.57	19.40
9024	3/4" diameter		50	.160		4.25	10.30		14.55	20.50
9026	1" diameter		45	.178		5.40	11.45		16.85	23.50
9028	1-1/4" diameter		40	.200		7.90	12.90		20.80	28.50
9030	1-1/2" diameter		35	.229		10.45	14.75		25.20	34.50
9032	2" diameter	Q-1	58	.276		14.50	16		30.50	40.50
9034	2-1/2" diameter		55	.291		19.55	16.85		36.40	47.50
9036	3" diameter		52	.308		25.50	17.85		43.35	55.50
9038	3-1/2" diameter		49	.327		35	18.95		53.95	67.50
9040	4" diameter		46	.348		37	20		57	72
9042	6" diameter		39	.410		50	24		74	92
9044	8" diameter	Q-2	48	.500		75	30		105	129
9046	10" diameter	"	43	.558		118	33.50		151.50	182

22 11 13.74 Pipe, Plastic		Crew	Daily Output	Labor-Hours	Unit	Material	2020 Bare Costs Labor	Equipment	Total	Total Incl O&P
9050	To delete joint & hangers, subtract									
9052	3/8" diam. to 1" diam.					45%	65%			
9054	1-1/4" diam. to 4" diam.					15%	45%			
9056	6" diam. to 10" diam.					5%	24%			
9060	SDR 11 (domestic cold water piping)									
9062	13' lengths, size is ID, includes joints 13' OC and hangers 3 per 10'									
9064	1/2" diameter	1 Plum	57	.140	L.F.	3.33	9.05		12.38	17.65
9066	3/4" diameter		54	.148		3.70	9.55		13.25	18.80
9068	1" diameter		49	.163		4.43	10.50		14.93	21
9070	1-1/4" diameter		45	.178		6.45	11.45		17.90	25
9072	1-1/2" diameter		40	.200		7.95	12.90		20.85	28.50
9074	2" diameter	Q-1	62	.258		10.95	14.95		25.90	35
9076	2-1/2" diameter		59	.271		14.05	15.75		29.80	40
9078	3" diameter		56	.286		18.65	16.55		35.20	46
9080	3-1/2" diameter		53	.302		26.50	17.50		44	56
9082	4" diameter		50	.320		29.50	18.55		48.05	61
9084	6" diameter		47	.340		34.50	19.75		54.25	68
9086	8" diameter	Q-2	51	.471		52.50	28.50		81	102
9088	10" diameter	"	46	.522		78.50	31.50		110	135
9090	To delete joint & hangers, subtract									
9092	1/2" diam. to 1" diam.					45%	65%			
9094	1-1/4" diam. to 4" diam.					15%	45%			
9096	6" diam. to 10" diam.					5%	24%			
9900	Minimum labor/equipment charge	1 Plum	4	2	Job		129		129	199

22 11 13.76 Pipe Fittings, Plastic		Crew	Daily Output	Labor-Hours	Unit	Material	Labor	Equipment	Total	Total Incl O&P
0010	**PIPE FITTINGS, PLASTIC**									
0030	Epoxy resin, fiberglass reinforced, general service									
0100	3"	Q-1	20.80	.769	Ea.	118	44.50		162.50	198
0110	4"		16.50	.970		121	56		177	220
0120	6"		10.10	1.584		234	92		326	400
0130	8"	Q-2	9.30	2.581		430	155		585	715
0140	10"		8.50	2.824		540	170		710	860
0150	12"		7.60	3.158		775	190		965	1,150
0380	Couplings									
0410	2"	Q-1	33.10	.483	Ea.	27	28		55	73
0420	3"		20.80	.769		31.50	44.50		76	104
0430	4"		16.50	.970		43.50	56		99.50	135
0440	6"		10.10	1.584		101	92		193	253
0450	8"	Q-2	9.30	2.581		171	155		326	430
0460	10"		8.50	2.824		233	170		403	520
0470	12"		7.60	3.158		335	190		525	665
0473	High corrosion resistant couplings, add					30%				
2100	PVC schedule 80, socket joint									
2110	90° elbow, 1/2"	1 Plum	30.30	.264	Ea.	2.67	17		19.67	29.50
2130	3/4"		26	.308		3.42	19.85		23.27	34.50
2140	1"		22.70	.352		5.55	22.50		28.05	41
2150	1-1/4"		20.20	.396		7.35	25.50		32.85	47.50
2160	1-1/2"		18.20	.440		7.90	28.50		36.40	52.50
2170	2"	Q-1	33.10	.483		9.55	28		37.55	54
2180	3"		20.80	.769		25	44.50		69.50	96.50
2190	4"		16.50	.970		38	56		94	129
2200	6"		10.10	1.584		109	92		201	261

22 11 13.76 Pipe Fittings, Plastic		Crew	Daily Output	Labor-Hours	Unit	2020 Bare Costs Material	Labor	Equipment	Total	Total Incl O&P
2210	8"	Q-2	9.30	2.581	Ea.	299	155		454	570
2250	45° elbow, 1/2"	1 Plum	30.30	.264		5.05	17		22.05	32
2270	3/4"		26	.308		7.70	19.85		27.55	39
2280	1"		22.70	.352		11.55	22.50		34.05	47.50
2290	1-1/4"		20.20	.396		14.60	25.50		40.10	55.50
2300	1-1/2"		18.20	.440		17.35	28.50		45.85	63
2310	2"	Q-1	33.10	.483		22.50	28		50.50	68.50
2320	3"		20.80	.769		57	44.50		101.50	132
2330	4"		16.50	.970		104	56		160	201
2340	6"		10.10	1.584		130	92		222	286
2350	8"	Q-2	9.30	2.581		283	155		438	550
2400	Tee, 1/2"	1 Plum	20.20	.396		7.55	25.50		33.05	48
2420	3/4"		17.30	.462		7.90	30		37.90	54.50
2430	1"		15.20	.526		9.90	34		43.90	63.50
2440	1-1/4"		13.50	.593		27	38		65	88.50
2450	1-1/2"		12.10	.661		27	42.50		69.50	96
2460	2"	Q-1	20	.800		34	46.50		80.50	109
2470	3"		13.90	1.151		46	67		113	154
2480	4"		11	1.455		53.50	84.50		138	189
2490	6"		6.70	2.388		183	139		322	415
2500	8"	Q-2	6.20	3.871		420	233		653	820
2510	Flange, socket, 150 lb., 1/2"	1 Plum	55.60	.144		14.55	9.25		23.80	30.50
2514	3/4"		47.60	.168		15.55	10.85		26.40	34
2518	1"		41.70	.192		17.35	12.35		29.70	38
2522	1-1/2"		33.30	.240		18.30	15.50		33.80	44
2526	2"	Q-1	60.60	.264		24.50	15.30		39.80	50
2530	4"		30.30	.528		52.50	30.50		83	105
2534	6"		18.50	.865		82.50	50		132.50	168
2538	8"	Q-2	17.10	1.404		147	84.50		231.50	293
2550	Coupling, 1/2"	1 Plum	30.30	.264		5.10	17		22.10	32
2570	3/4"		26	.308		7.55	19.85		27.40	39
2580	1"		22.70	.352		7.45	22.50		29.95	43
2590	1-1/4"		20.20	.396		10.15	25.50		35.65	50.50
2600	1-1/2"		18.20	.440		11.40	28.50		39.90	56.50
2610	2"	Q-1	33.10	.483		12.25	28		40.25	57
2620	3"		20.80	.769		33.50	44.50		78	106
2630	4"		16.50	.970		43.50	56		99.50	135
2640	6"		10.10	1.584		93	92		185	245
2650	8"	Q-2	9.30	2.581		122	155		277	375
2660	10"		8.50	2.824		420	170		590	725
2670	12"		7.60	3.158		485	190		675	825
2700	PVC (white), schedule 40, socket joints									
2760	90° elbow, 1/2"	1 Plum	33.30	.240	Ea.	.52	15.50		16.02	24.50
2770	3/4"		28.60	.280		.59	18.05		18.64	28.50
2780	1"		25	.320		1.05	20.50		21.55	33
2790	1-1/4"		22.20	.360		1.83	23		24.83	38
2800	1-1/2"		20	.400		1.99	26		27.99	42
2810	2"	Q-1	36.40	.440		3.11	25.50		28.61	43
2820	2-1/2"		26.70	.599		9.60	35		44.60	64
2830	3"		22.90	.699		11.35	40.50		51.85	75
2840	4"		18.20	.879		20.50	51		71.50	102
2850	5"		12.10	1.322		52	76.50		128.50	177
2860	6"		11.10	1.441		64.50	83.50		148	200

22 11 13.76 Pipe Fittings, Plastic		Crew	Daily Output	Labor-Hours	Unit	2020 Bare Costs Material	Labor	Equipment	Total	Total Incl O&P
2870	8″	Q-2	10.30	2.330	Ea.	166	140		306	400
2980	45° elbow, 1/2″	1 Plum	33.30	.240		.86	15.50		16.36	25
2990	3/4″		28.60	.280		1.34	18.05		19.39	29.50
3000	1″		25	.320		1.60	20.50		22.10	34
3010	1-1/4″		22.20	.360		2.23	23		25.23	38.50
3020	1-1/2″		20	.400		2.80	26		28.80	43
3030	2″	Q-1	36.40	.440		3.65	25.50		29.15	43.50
3040	2-1/2″		26.70	.599		9.50	35		44.50	64
3050	3″		22.90	.699		14.65	40.50		55.15	78.50
3060	4″		18.20	.879		26.50	51		77.50	108
3070	5″		12.10	1.322		52	76.50		128.50	177
3080	6″		11.10	1.441		65.50	83.50		149	201
3090	8″	Q-2	10.30	2.330		157	140		297	390
3180	Tee, 1/2″	1 Plum	22.20	.360		.65	23		23.65	36.50
3190	3/4″		19	.421		.75	27		27.75	43
3200	1″		16.70	.479		1.39	31		32.39	49
3210	1-1/4″		14.80	.541		2.16	35		37.16	56.50
3220	1-1/2″		13.30	.602		2.64	39		41.64	63
3230	2″	Q-1	24.20	.661		3.84	38.50		42.34	63.50
3240	2-1/2″		17.80	.899		12.60	52		64.60	94.50
3250	3″		15.20	1.053		17.15	61		78.15	113
3260	4″		12.10	1.322		30	76.50		106.50	152
3270	5″		8.10	1.975		72.50	115		187.50	257
3280	6″		7.40	2.162		101	125		226	305
3290	8″	Q-2	6.80	3.529		235	212		447	590
3380	Coupling, 1/2″	1 Plum	33.30	.240		.34	15.50		15.84	24.50
3390	3/4″		28.60	.280		.47	18.05		18.52	28.50
3400	1″		25	.320		.83	20.50		21.33	33
3410	1-1/4″		22.20	.360		1.13	23		24.13	37
3420	1-1/2″		20	.400		1.22	26		27.22	41.50
3430	2″	Q-1	36.40	.440		1.86	25.50		27.36	41.50
3440	2-1/2″		26.70	.599		4.11	35		39.11	58
3450	3″		22.90	.699		7.70	40.50		48.20	71
3460	4″		18.20	.879		9.60	51		60.60	89.50
3470	5″		12.10	1.322		16.95	76.50		93.45	138
3480	6″		11.10	1.441		30.50	83.50		114	163
3490	8″	Q-2	10.30	2.330		54.50	140		194.50	277
4500	DWV, ABS, non pressure, socket joints									
4540	1/4 bend, 1-1/4″	1 Plum	20.20	.396	Ea.	4.48	25.50		29.98	44.50
4560	1-1/2″	″	18.20	.440		3.44	28.50		31.94	48
4570	2″	Q-1	33.10	.483		5.30	28		33.30	49.50
4580	3″		20.80	.769		13.45	44.50		57.95	84
4590	4″		16.50	.970		27.50	56		83.50	118
4600	6″		10.10	1.584		118	92		210	271
4650	1/8 bend, same as 1/4 bend									
4800	Tee, sanitary									
4820	1-1/4″	1 Plum	13.50	.593	Ea.	5.95	38		43.95	65.50
4830	1-1/2″	″	12.10	.661		5.15	42.50		47.65	71.50
4840	2″	Q-1	20	.800		7.90	46.50		54.40	80
4850	3″		13.90	1.151		21.50	67		88.50	127
4860	4″		11	1.455		38.50	84.50		123	173
4862	Tee, sanitary, reducing, 2″ x 1-1/2″		22	.727		7	42		49	72.50
4864	3″ x 2″		15.30	1.046		13.40	60.50		73.90	109

22 11 13.76 Pipe Fittings, Plastic		Crew	Daily Output	Labor-Hours	Unit	Material	Labor	Equipment	Total	Total Incl O&P
						2020 Bare Costs				
4868	4" x 3"	Q-1	12.10	1.322	Ea.	37.50	76.50		114	161
4870	Combination Y and 1/8 bend									
4872	1-1/2"	1 Plum	12.10	.661	Ea.	12.25	42.50		54.75	79.50
4874	2"	Q-1	20	.800		13.70	46.50		60.20	86.50
4876	3"		13.90	1.151		32	67		99	139
4878	4"		11	1.455		62	84.50		146.50	198
4880	3" x 1-1/2"		15.50	1.032		32.50	60		92.50	129
4882	4" x 3"		12.10	1.322		49.50	76.50		126	174
4900	Wye, 1-1/4"	1 Plum	13.50	.593		6.85	38		44.85	66.50
4902	1-1/2"	"	12.10	.661		7.85	42.50		50.35	74.50
4904	2"	Q-1	20	.800		10.20	46.50		56.70	82.50
4906	3"		13.90	1.151		24	67		91	130
4908	4"		11	1.455		49	84.50		133.50	184
4910	6"		6.70	2.388		146	139		285	375
4918	3" x 1-1/2"		15.50	1.032		19.40	60		79.40	114
4920	4" x 3"		12.10	1.322		38.50	76.50		115	162
4922	6" x 4"		6.90	2.319		116	134		250	335
4930	Double wye, 1-1/2"	1 Plum	9.10	.879		25	56.50		81.50	115
4932	2"	Q-1	16.60	.964		30	56		86	120
4934	3"		10.40	1.538		70.50	89		159.50	216
4936	4"		8.25	1.939		138	112		250	325
4940	2" x 1-1/2"		16.80	.952		27.50	55		82.50	116
4942	3" x 2"		10.60	1.509		49	87.50		136.50	189
4944	4" x 3"		8.45	1.893		109	110		219	290
4946	6" x 4"		7.25	2.207		156	128		284	370
4950	Reducer bushing, 2" x 1-1/2"		36.40	.440		2.72	25.50		28.22	42.50
4952	3" x 1-1/2"		27.30	.586		11.95	34		45.95	65.50
4954	4" x 2"		18.20	.879		22.50	51		73.50	104
4956	6" x 4"		11.10	1.441		62.50	83.50		146	198
4960	Couplings, 1-1/2"	1 Plum	18.20	.440		1.66	28.50		30.16	46
4962	2"	Q-1	33.10	.483		2.22	28		30.22	46
4963	3"		20.80	.769		6.30	44.50		50.80	76
4964	4"		16.50	.970		11.35	56		67.35	99.50
4966	6"		10.10	1.584		47.50	92		139.50	194
4970	2" x 1-1/2"		33.30	.480		4.80	28		32.80	48.50
4972	3" x 1-1/2"		21	.762		13.85	44		57.85	84
4974	4" x 3"		16.70	.958		21	55.50		76.50	110
4978	Closet flange, 4"	1 Plum	32	.250		11.60	16.10		27.70	38
4980	4" x 3"	"	34	.235		14.10	15.15		29.25	39
5000	DWV, PVC, schedule 40, socket joints									
5040	1/4 bend, 1-1/4"	1 Plum	20.20	.396	Ea.	8.80	25.50		34.30	49
5060	1-1/2"	"	18.20	.440		2.48	28.50		30.98	46.50
5070	2"	Q-1	33.10	.483		3.91	28		31.91	48
5080	3"		20.80	.769		11.65	44.50		56.15	82
5090	4"		16.50	.970		23	56		79	112
5100	6"		10.10	1.584		79.50	92		171.50	230
5105	8"	Q-2	9.30	2.581		122	155		277	375
5110	1/4 bend, long sweep, 1-1/2"	1 Plum	18.20	.440		5.85	28.50		34.35	50.50
5112	2"	Q-1	33.10	.483		6.50	28		34.50	50.50
5114	3"		20.80	.769		15.05	44.50		59.55	85.50
5116	4"		16.50	.970		28.50	56		84.50	119
5150	1/8 bend, 1-1/4"	1 Plum	20.20	.396		6.05	25.50		31.55	46
5170	1-1/2"	"	18.20	.440		2.46	28.50		30.96	46.50

22 11 13 – Facility Water Distribution Piping

22 11 13.76 Pipe Fittings, Plastic		Crew	Daily Output	Labor-Hours	Unit	Material	Labor	Equipment	Total	Total Incl O&P
						2020 Bare Costs				
5180	2"	Q-1	33.10	.483	Ea.	3.64	28		31.64	47.50
5190	3"		20.80	.769		10.45	44.50		54.95	80.50
5200	4"		16.50	.970		18.95	56		74.95	108
5210	6"		10.10	1.584		71.50	92		163.50	221
5215	8"	Q-2	9.30	2.581		117	155		272	370
5250	Tee, sanitary 1-1/4"	1 Plum	13.50	.593		9.45	38		47.45	69.50
5254	1-1/2"	"	12.10	.661		4.41	42.50		46.91	71
5255	2"	Q-1	20	.800		6.50	46.50		53	78.50
5256	3"		13.90	1.151		17.05	67		84.05	122
5257	4"		11	1.455		31	84.50		115.50	165
5259	6"		6.70	2.388		126	139		265	355
5261	8"	Q-2	6.20	3.871		293	233		526	685
5264	2" x 1-1/2"	Q-1	22	.727		5.75	42		47.75	71.50
5266	3" x 1-1/2"		15.50	1.032		12.45	60		72.45	106
5268	4" x 3"		12.10	1.322		37	76.50		113.50	160
5271	6" x 4"		6.90	2.319		122	134		256	340
5314	Combination Y & 1/8 bend, 1-1/2"	1 Plum	12.10	.661		10.75	42.50		53.25	78
5315	2"	Q-1	20	.800		12.90	46.50		59.40	85.50
5317	3"		13.90	1.151		29	67		96	135
5318	4"		11	1.455		58	84.50		142.50	194
5324	Combination Y & 1/8 bend, reducing									
5325	2" x 2" x 1-1/2"	Q-1	22	.727	Ea.	15.15	42		57.15	81.50
5327	3" x 3" x 1-1/2"		15.50	1.032		27	60		87	122
5328	3" x 3" x 2"		15.30	1.046		19.95	60.50		80.45	116
5329	4" x 4" x 2"		12.20	1.311		30.50	76		106.50	152
5331	Wye, 1-1/4"	1 Plum	13.50	.593		12.10	38		50.10	72.50
5332	1-1/2"	"	12.10	.661		8.20	42.50		50.70	75
5333	2"	Q-1	20	.800		7.90	46.50		54.40	80
5334	3"		13.90	1.151		21.50	67		88.50	127
5335	4"		11	1.455		39	84.50		123.50	173
5336	6"		6.70	2.388		114	139		253	340
5337	8"	Q-2	6.20	3.871		217	233		450	600
5341	2" x 1-1/2"	Q-1	22	.727		9.85	42		51.85	76
5342	3" x 1-1/2"		15.50	1.032		14.60	60		74.60	109
5343	4" x 3"		12.10	1.322		32	76.50		108.50	154
5344	6" x 4"		6.90	2.319		87	134		221	305
5345	8" x 6"	Q-2	6.40	3.750		188	226		414	555
5347	Double wye, 1-1/2"	1 Plum	9.10	.879		18.25	56.50		74.75	108
5348	2"	Q-1	16.60	.964		20.50	56		76.50	109
5349	3"		10.40	1.538		42.50	89		131.50	185
5350	4"		8.25	1.939		85.50	112		197.50	269
5353	Double wye, reducing									
5354	2" x 2" x 1-1/2" x 1-1/2"	Q-1	16.80	.952	Ea.	18.60	55		73.60	106
5355	3" x 3" x 2" x 2"		10.60	1.509		31.50	87.50		119	170
5356	4" x 4" x 3" x 3"		8.45	1.893		68	110		178	245
5357	6" x 6" x 4" x 4"		7.25	2.207		239	128		367	460
5374	Coupling, 1-1/4"	1 Plum	20.20	.396		5.70	25.50		31.20	46
5376	1-1/2"	"	18.20	.440		1.19	28.50		29.69	45.50
5378	2"	Q-1	33.10	.483		1.62	28		29.62	45.50
5380	3"		20.80	.769		5.65	44.50		50.15	75
5390	4"		16.50	.970		9.65	56		65.65	97.50
5400	6"		10.10	1.584		31.50	92		123.50	177
5402	8"	Q-2	9.30	2.581		57	155		212	305

22 11 13 – Facility Water Distribution Piping

22 11 13.76 Pipe Fittings, Plastic		Crew	Daily Output	Labor-Hours	Unit	Material	Labor	Equipment	Total	Total Incl O&P
						2020 Bare Costs				
5404	2″ x 1-1/2″	Q-1	33.30	.480	Ea.	3.63	28		31.63	47
5406	3″ x 1-1/2″		21	.762		10.65	44		54.65	80.50
5408	4″ x 3″		16.70	.958		17.35	55.50		72.85	105
5410	Reducer bushing, 2″ x 1-1/4″		36.50	.438		3.38	25.50		28.88	43
5412	3″ x 1-1/2″		27.30	.586		10.10	34		44.10	63.50
5414	4″ x 2″		18.20	.879		17.35	51		68.35	98
5416	6″ x 4″		11.10	1.441		45	83.50		128.50	179
5418	8″ x 6″	Q-2	10.20	2.353		89	142		231	315
5425	Closet flange 4″	Q-1	32	.500		13.70	29		42.70	60
5426	4″ x 3″	″	34	.471		13.50	27.50		41	57
5450	Solvent cement for PVC, industrial grade, per quart				Qt.	31			31	34
5500	CPVC, Schedule 80, threaded joints									
5540	90° elbow, 1/4″	1 Plum	32	.250	Ea.	13	16.10		29.10	39.50
5560	1/2″		30.30	.264		7.55	17		24.55	35
5570	3/4″		26	.308		11.30	19.85		31.15	43
5580	1″		22.70	.352		15.85	22.50		38.35	52.50
5590	1-1/4″		20.20	.396		30.50	25.50		56	73
5600	1-1/2″		18.20	.440		33	28.50		61.50	80
5610	2″	Q-1	33.10	.483		44	28		72	92
5620	2-1/2″		24.20	.661		137	38.50		175.50	210
5630	3″		20.80	.769		147	44.50		191.50	230
5640	4″		16.50	.970		230	56		286	340
5650	6″		10.10	1.584		267	92		359	435
5660	45° elbow same as 90° elbow									
5700	Tee, 1/4″	1 Plum	22	.364	Ea.	25.50	23.50		49	64
5702	1/2″		20.20	.396		25.50	25.50		51	67.50
5704	3/4″		17.30	.462		36.50	30		66.50	86
5706	1″		15.20	.526		39.50	34		73.50	95.50
5708	1-1/4″		13.50	.593		39.50	38		77.50	103
5710	1-1/2″		12.10	.661		41.50	42.50		84	112
5712	2″	Q-1	20	.800		46	46.50		92.50	122
5714	2-1/2″		16.20	.988		225	57.50		282.50	335
5716	3″		13.90	1.151		263	67		330	390
5718	4″		11	1.455		625	84.50		709.50	815
5720	6″		6.70	2.388		710	139		849	995
5730	Coupling, 1/4″	1 Plum	32	.250		16.65	16.10		32.75	43.50
5732	1/2″		30.30	.264		13.75	17		30.75	41.50
5734	3/4″		26	.308		22	19.85		41.85	55
5736	1″		22.70	.352		25	22.50		47.50	62.50
5738	1-1/4″		20.20	.396		26.50	25.50		52	68.50
5740	1-1/2″		18.20	.440		28.50	28.50		57	75.50
5742	2″	Q-1	33.10	.483		33.50	28		61.50	80.50
5744	2-1/2″		24.20	.661		60.50	38.50		99	126
5746	3″		20.80	.769		70	44.50		114.50	147
5748	4″		16.50	.970		143	56		199	244
5750	6″		10.10	1.584		196	92		288	360
5752	8″	Q-2	9.30	2.581		410	155		565	690
5900	CPVC, Schedule 80, socket joints									
5904	90° elbow, 1/4″	1 Plum	32	.250	Ea.	12.45	16.10		28.55	38.50
5906	1/2″		30.30	.264		4.88	17		21.88	32
5908	3/4″		26	.308		6.25	19.85		26.10	37.50
5910	1″		22.70	.352		9.90	22.50		32.40	46
5912	1-1/4″		20.20	.396		21.50	25.50		47	63

22 11 13.76 Pipe Fittings, Plastic		Crew	Daily Output	Labor-Hours	Unit	Material	2020 Bare Costs Labor	Equipment	Total	Total Incl O&P
5914	1-1/2"	1 Plum	18.20	.440	Ea.	24	28.50		52.50	70
5916	2"	Q-1	33.10	.483		29	28		57	75
5918	2-1/2"		24.20	.661		66.50	38.50		105	133
5920	3"		20.80	.769		75	44.50		119.50	152
5922	4"		16.50	.970		135	56		191	236
5924	6"		10.10	1.584		272	92		364	440
5926	8"		9.30	1.720		665	100		765	885
5930	45° elbow, 1/4"	1 Plum	32	.250		18.50	16.10		34.60	45.50
5932	1/2"		30.30	.264		5.95	17		22.95	33
5934	3/4"		26	.308		7.80	19.85		27.65	39
5936	1"		22.70	.352		13.70	22.50		36.20	50
5938	1-1/4"		20.20	.396		27	25.50		52.50	69
5940	1-1/2"		18.20	.440		27.50	28.50		56	74.50
5942	2"	Q-1	33.10	.483		27.50	28		55.50	73.50
5944	2-1/2"		24.20	.661		63.50	38.50		102	129
5946	3"		20.80	.769		81.50	44.50		126	159
5948	4"		16.50	.970		98.50	56		154.50	195
5950	6"		10.10	1.584		310	92		402	480
5952	8"		9.30	1.720		715	100		815	940
5960	Tee, 1/4"	1 Plum	22	.364		11.45	23.50		34.95	48.50
5962	1/2"		20.20	.396		11.45	25.50		36.95	52
5964	3/4"		17.30	.462		11.65	30		41.65	59
5966	1"		15.20	.526		14.25	34		48.25	68
5968	1-1/4"		13.50	.593		30	38		68	92
5970	1-1/2"		12.10	.661		34.50	42.50		77	104
5972	2"	Q-1	20	.800		34	46.50		80.50	109
5974	2-1/2"		16.20	.988		97.50	57.50		155	196
5976	3"		13.90	1.151		97.50	67		164.50	210
5978	4"		11	1.455		130	84.50		214.50	273
5980	6"		6.70	2.388		340	139		479	585
5982	8"	Q-2	6.20	3.871		965	233		1,198	1,400
5990	Coupling, 1/4"	1 Plum	32	.250		13.25	16.10		29.35	39.50
5992	1/2"		30.30	.264		5.15	17		22.15	32
5994	3/4"		26	.308		7.20	19.85		27.05	38.50
5996	1"		22.70	.352		9.70	22.50		32.20	45.50
5998	1-1/4"		20.20	.396		14.55	25.50		40.05	55.50
6000	1-1/2"		18.20	.440		18.25	28.50		46.75	64
6002	2"	Q-1	33.10	.483		21.50	28		49.50	67
6004	2-1/2"		24.20	.661		47.50	38.50		86	112
6006	3"		20.80	.769		51.50	44.50		96	126
6008	4"		16.50	.970		67.50	56		123.50	161
6010	6"		10.10	1.584		159	92		251	315
6012	8"	Q-2	9.30	2.581		430	155		585	710
6200	CTS, 100 psi at 180°F, hot and cold water									
6230	90° elbow, 1/2"	1 Plum	20	.400	Ea.	.32	26		26.32	40.50
6250	3/4"		19	.421		.52	27		27.52	42.50
6251	1"		16	.500		1.65	32		33.65	52
6252	1-1/4"		15	.533		3.04	34.50		37.54	56.50
6253	1-1/2"		14	.571		5.50	37		42.50	63
6254	2"	Q-1	23	.696		10.55	40.50		51.05	74
6260	45° elbow, 1/2"	1 Plum	20	.400		.40	26		26.40	40.50
6280	3/4"		19	.421		.69	27		27.69	43
6281	1"		16	.500		1.87	32		33.87	52

22 11 13 – Facility Water Distribution Piping

22 11 13.76 Pipe Fittings, Plastic		Crew	Daily Output	Labor-Hours	Unit	Material	Labor	Equipment	Total	Total Incl O&P
						2020 Bare Costs				
6282	1-1/4"	1 Plum	15	.533	Ea.	3.75	34.50		38.25	57
6283	1-1/2"		14	.571		5.45	37		42.45	63
6284	2"	Q-1	23	.696		12.40	40.50		52.90	76
6290	Tee, 1/2"	1 Plum	13	.615		.40	39.50		39.90	62
6310	3/4"		12	.667		.75	43		43.75	67.50
6311	1"		11	.727		3.80	47		50.80	76.50
6312	1-1/4"		10	.800		5.85	51.50		57.35	86
6313	1-1/2"		10	.800		7.65	51.50		59.15	88
6314	2"	Q-1	17	.941		12.35	54.50		66.85	98
6320	Coupling, 1/2"	1 Plum	22	.364		.26	23.50		23.76	36.50
6340	3/4"		21	.381		.34	24.50		24.84	38.50
6341	1"		18	.444		1.55	28.50		30.05	46
6342	1-1/4"		17	.471		2.01	30.50		32.51	49
6343	1-1/2"		16	.500		3.07	32		35.07	53.50
6344	2"	Q-1	28	.571		5.40	33		38.40	57.50
6360	Solvent cement for CPVC, commercial grade, per quart				Qt.	51.50			51.50	56.50
7340	PVC flange, slip-on, Sch 80 std., 1/2"	1 Plum	22	.364	Ea.	14.65	23.50		38.15	52
7350	3/4"		21	.381		15.65	24.50		40.15	55.50
7360	1"		18	.444		17.45	28.50		45.95	63.50
7370	1-1/4"		17	.471		17.95	30.50		48.45	67
7380	1-1/2"		16	.500		18.35	32		50.35	70
7390	2"	Q-1	26	.615		24.50	35.50		60	82
7400	2-1/2"		24	.667		37.50	38.50		76	102
7410	3"		18	.889		41.50	51.50		93	126
7420	4"		15	1.067		52.50	62		114.50	154
7430	6"		10	1.600		83	93		176	234
7440	8"	Q-2	11	2.182		148	131		279	365
7550	Union, schedule 40, socket joints, 1/2"	1 Plum	19	.421		5.20	27		32.20	48
7560	3/4"		18	.444		5.80	28.50		34.30	51
7570	1"		15	.533		6	34.50		40.50	59.50
7580	1-1/4"		14	.571		17.85	37		54.85	76.50
7590	1-1/2"		13	.615		20	39.50		59.50	83.50
7600	2"	Q-1	20	.800		27	46.50		73.50	102
7992	Polybutyl/polyethyl pipe, for copper fittings see Line 22 11 13.25 7000									
8000	Compression type, PVC, 160 psi cold water									
8010	Coupling, 3/4" CTS	1 Plum	21	.381	Ea.	4.27	24.50		28.77	42.50
8020	1" CTS		18	.444		6	28.50		34.50	51
8030	1-1/4" CTS		17	.471		8.35	30.50		38.85	56
8040	1-1/2" CTS		16	.500		10	32		42	61
8050	2" CTS		15	.533		15.85	34.50		50.35	70.50
8060	Female adapter, 3/4" FPT x 3/4" CTS		23	.348		6.35	22.50		28.85	41.50
8070	3/4" FPT x 1" CTS		21	.381		7.35	24.50		31.85	46
8080	1" FPT x 1" CTS		20	.400		7.35	26		33.35	48
8090	1-1/4" FPT x 1-1/4" CTS		18	.444		9.60	28.50		38.10	55
8100	1-1/2" FPT x 1-1/2" CTS		16	.500		10.95	32		42.95	62
8110	2" FPT x 2" CTS		13	.615		16.05	39.50		55.55	79
8130	Male adapter, 3/4" MPT x 3/4" CTS		23	.348		5.30	22.50		27.80	40.50
8140	3/4" MPT x 1" CTS		21	.381		6.25	24.50		30.75	45
8150	1" MPT x 1" CTS		20	.400		6.25	26		32.25	47
8160	1-1/4" MPT x 1-1/4" CTS		18	.444		8.45	28.50		36.95	54
8170	1-1/2" MPT x 1-1/2" CTS		16	.500		10.15	32		42.15	61
8180	2" MPT x 2" CTS		13	.615		13.15	39.50		52.65	76
8200	Spigot adapter, 3/4" IPS x 3/4" CTS		23	.348		2.22	22.50		24.72	37

22 11 13 – Facility Water Distribution Piping

22 11 13.76 Pipe Fittings, Plastic		Crew	Daily Output	Labor-Hours	Unit	Material	Labor	Equipment	Total	Total Incl O&P
						2020 Bare Costs				
8210	3/4" IPS x 1" CTS	1 Plum	21	.381	Ea.	2.71	24.50		27.21	41
8220	1" IPS x 1" CTS		20	.400		2.70	26		28.70	43
8230	1-1/4" IPS x 1-1/4" CTS		18	.444		4.07	28.50		32.57	49
8240	1-1/2" IPS x 1-1/2" CTS		16	.500		4.26	32		36.26	54.50
8250	2" IPS x 2" CTS		13	.615		5.30	39.50		44.80	67.50
8270	Price includes insert stiffeners									
8280	250 psi is same price as 160 psi									
8300	Insert type, nylon, 160 & 250 psi, cold water									
8310	Clamp ring stainless steel, 3/4" IPS	1 Plum	115	.070	Ea.	2.78	4.48		7.26	10
8320	1" IPS		107	.075		3.07	4.82		7.89	10.85
8330	1-1/4" IPS		101	.079		2.85	5.10		7.95	11.05
8340	1-1/2" IPS		95	.084		3.95	5.45		9.40	12.75
8350	2" IPS		85	.094		4.74	6.05		10.79	14.60
8370	Coupling, 3/4" IPS		22	.364		.95	23.50		24.45	37
8380	1" IPS		19	.421		.99	27		27.99	43
8390	1-1/4" IPS		18	.444		1.47	28.50		29.97	46
8400	1-1/2" IPS		17	.471		1.73	30.50		32.23	49
8410	2" IPS		16	.500		3.36	32		35.36	53.50
8430	Elbow, 90°, 3/4" IPS		22	.364		1.89	23.50		25.39	38
8440	1" IPS		19	.421		2.10	27		29.10	44.50
8450	1-1/4" IPS		18	.444		2.35	28.50		30.85	47
8460	1-1/2" IPS		17	.471		2.77	30.50		33.27	50
8470	2" IPS		16	.500		3.86	32		35.86	54.50
8490	Male adapter, 3/4" IPS x 3/4" MPT		25	.320		.95	20.50		21.45	33
8500	1" IPS x 1" MPT		21	.381		.98	24.50		25.48	39
8510	1-1/4" IPS x 1-1/4" MPT		20	.400		1.55	26		27.55	41.50
8520	1-1/2" IPS x 1-1/2" MPT		18	.444		1.73	28.50		30.23	46.50
8530	2" IPS x 2" MPT		15	.533		3.33	34.50		37.83	56.50
8550	Tee, 3/4" IPS		14	.571		1.83	37		38.83	59
8560	1" IPS		13	.615		2.40	39.50		41.90	64
8570	1-1/4" IPS		12	.667		3.73	43		46.73	70.50
8580	1-1/2" IPS		11	.727		4.24	47		51.24	77
8590	2" IPS		10	.800		8.35	51.50		59.85	88.50
8610	Insert type, PVC, 100 psi @ 180°F, hot & cold water									
8620	Coupler, male, 3/8" CTS x 3/8" MPT	1 Plum	29	.276	Ea.	.76	17.80		18.56	28.50
8630	3/8" CTS x 1/2" MPT		28	.286		.76	18.40		19.16	29.50
8640	1/2" CTS x 1/2" MPT		27	.296		.78	19.10		19.88	30.50
8650	1/2" CTS x 3/4" MPT		26	.308		2.46	19.85		22.31	33
8660	3/4" CTS x 1/2" MPT		25	.320		2.22	20.50		22.72	34.50
8670	3/4" CTS x 3/4" MPT		25	.320		.93	20.50		21.43	33
8700	Coupling, 3/8" CTS x 1/2" CTS		25	.320		4.44	20.50		24.94	37
8710	1/2" CTS		23	.348		5.65	22.50		28.15	41
8730	3/4" CTS		22	.364		12.35	23.50		35.85	49.50
8750	Elbow 90°, 3/8" CTS		25	.320		4.15	20.50		24.65	36.50
8760	1/2" CTS		23	.348		5.10	22.50		27.60	40
8770	3/4" CTS		22	.364		7.65	23.50		31.15	44.50
8800	Rings, crimp, copper, 3/8" CTS		120	.067		.18	4.30		4.48	6.85
8810	1/2" CTS		117	.068		.20	4.41		4.61	7
8820	3/4" CTS		115	.070		.26	4.48		4.74	7.25
8850	Reducer tee, bronze, 3/8" x 3/8" x 1/2" CTS		17	.471		5.50	30.50		36	53
8860	1/2" x 1/2" x 3/4" CTS		15	.533		7.40	34.50		41.90	61
8870	3/4" x 1/2" x 1/2" CTS		14	.571		7.40	37		44.40	65
8890	3/4" x 3/4" x 1/2" CTS		14	.571		7.50	37		44.50	65.50

22 11 13.76 Pipe Fittings, Plastic		Crew	Daily Output	Labor-Hours	Unit	Material	Labor	Equipment	Total	Total Incl O&P
						2020 Bare Costs				
8900	1" x 1/2" x 1/2" CTS	1 Plum	14	.571	Ea.	12.40	37		49.40	70.50
8930	Tee, 3/8" CTS		17	.471		3.24	30.50		33.74	50.50
8940	1/2" CTS		15	.533		2.92	34.50		37.42	56
8950	3/4" CTS		14	.571		3.61	37		40.61	61
8960	Copper rings included in fitting price									
9000	Flare type, assembled, acetal, hot & cold water									
9010	Coupling, 1/4" & 3/8" CTS	1 Plum	24	.333	Ea.	3.55	21.50		25.05	37
9020	1/2" CTS		22	.364		4.07	23.50		27.57	40.50
9030	3/4" CTS		21	.381		6	24.50		30.50	44.50
9040	1" CTS		18	.444		7.65	28.50		36.15	53
9050	Elbow 90°, 1/4" CTS		26	.308		3.94	19.85		23.79	35
9060	3/8" CTS		24	.333		4.23	21.50		25.73	37.50
9070	1/2" CTS		22	.364		5	23.50		28.50	41.50
9080	3/4" CTS		21	.381		7.65	24.50		32.15	46.50
9090	1" CTS		18	.444		9.90	28.50		38.40	55.50
9110	Tee, 1/4" CTS		16	.500		4.33	32		36.33	55
9114	3/8" CTS		15	.533		4.42	34.50		38.92	58
9120	1/2" CTS		14	.571		5.55	37		42.55	63
9130	3/4" CTS		13	.615		8.70	39.50		48.20	71
9140	1" CTS		12	.667		11.70	43		54.70	79.50
9400	Polypropylene, fittings and accessories									
9404	Fittings fusion welded, sizes are ID									
9408	Note: sizes 1/2" thru 4" use socket fusion									
9410	Sizes 6" thru 10" use butt fusion									
9416	Coupling									
9420	3/8"	1 Plum	39	.205	Ea.	.92	13.20		14.12	21.50
9422	1/2"		37.40	.214		1.23	13.80		15.03	23
9424	3/4"		35.40	.226		1.36	14.55		15.91	24
9426	1"		29.70	.269		1.79	17.35		19.14	29
9428	1-1/4"		27.60	.290		2.14	18.70		20.84	31.50
9430	1-1/2"		24.80	.323		4.51	21		25.51	37
9432	2"	Q-1	43	.372		9	21.50		30.50	43.50
9434	2-1/2"		35.60	.449		10.10	26		36.10	51.50
9436	3"		30.90	.518		22	30		52	71
9438	3-1/2"		27.80	.576		36	33.50		69.50	91
9440	4"		25	.640		47.50	37		84.50	110
9442	Reducing coupling, female to female									
9446	2" to 1-1/2"	Q-1	49	.327	Ea.	13	18.95		31.95	44
9448	2-1/2" to 2"		41.20	.388		14.25	22.50		36.75	50.50
9450	3" to 2-1/2"		33.10	.483		19.25	28		47.25	64.50
9470	Reducing bushing, female to female									
9472	1/2" to 3/8"	1 Plum	38.20	.209	Ea.	1.23	13.50		14.73	22.50
9474	3/4" to 3/8" or 1/2"		36.50	.219		1.36	14.15		15.51	23.50
9476	1" to 3/4" or 1/2"		33.10	.242		1.81	15.60		17.41	26
9478	1-1/4" to 3/4" or 1"		28.70	.279		2.79	17.95		20.74	31
9480	1-1/2" to 1/2" thru 1-1/4"		26.20	.305		4.62	19.70		24.32	35.50
9482	2" to 1/2" thru 1-1/2"	Q-1	43	.372		9.25	21.50		30.75	43.50
9484	2-1/2" to 1/2" thru 2"		41.20	.388		10.35	22.50		32.85	46.50
9486	3" to 1-1/2" thru 2-1/2"		33.10	.483		23	28		51	69
9488	3-1/2" to 2" thru 3"		29.20	.548		37	32		69	89.50
9490	4" to 2-1/2" thru 3-1/2"		26.20	.611		57.50	35.50		93	119
9491	6" to 4" SDR 7.4		16.50	.970		69	56		125	163
9492	6" to 4" SDR 11		16.50	.970		69	56		125	163

22 11 13.76	Pipe Fittings, Plastic	Crew	Daily Output	Labor-Hours	Unit	2020 Bare Costs Material	Labor	Equipment	Total	Total Incl O&P
9493	8″ to 6″ SDR 7.4	Q-1	10.10	1.584	Ea.	102	92		194	254
9494	8″ to 6″ SDR 11		10.10	1.584		69.50	92		161.50	219
9495	10″ to 8″ SDR 7.4		7.80	2.051		139	119		258	335
9496	10″ to 8″ SDR 11		7.80	2.051		100	119		219	294
9500	90° elbow									
9504	3/8″	1 Plum	39	.205	Ea.	1.26	13.20		14.46	22
9506	1/2″		37.40	.214		1.31	13.80		15.11	23
9508	3/4″		35.40	.226		1.68	14.55		16.23	24.50
9510	1″		29.70	.269		2.42	17.35		19.77	29.50
9512	1-1/4″		27.60	.290		3.73	18.70		22.43	33
9514	1-1/2″		24.80	.323		8	21		29	41
9516	2″	Q-1	43	.372		12.30	21.50		33.80	47
9518	2-1/2″		35.60	.449		27.50	26		53.50	70.50
9520	3″		30.90	.518		45.50	30		75.50	96.50
9522	3-1/2″		27.80	.576		64.50	33.50		98	123
9524	4″		25	.640		99.50	37		136.50	167
9526	6″ SDR 7.4		5.55	2.883		113	167		280	385
9528	6″ SDR 11		5.55	2.883		89	167		256	355
9530	8″ SDR 7.4	Q-2	8.10	2.963		295	178		473	600
9532	8″ SDR 11		8.10	2.963		229	178		407	530
9534	10″ SDR 7.4		7.50	3.200		390	192		582	730
9536	10″ SDR 11		7.50	3.200		365	192		557	705
9551	45° elbow									
9554	3/8″	1 Plum	39	.205	Ea.	1.26	13.20		14.46	22
9556	1/2″		37.40	.214		1.31	13.80		15.11	23
9558	3/4″		35.40	.226		1.68	14.55		16.23	24.50
9564	1″		29.70	.269		2.42	17.35		19.77	29.50
9566	1-1/4″		27.60	.290		3.73	18.70		22.43	33
9568	1-1/2″		24.80	.323		8	21		29	41
9570	2″	Q-1	43	.372		12.15	21.50		33.65	47
9572	2-1/2″		35.60	.449		27	26		53	70
9574	3″		30.90	.518		50	30		80	102
9576	3-1/2″		27.80	.576		71	33.50		104.50	130
9578	4″		25	.640		109	37		146	178
9580	6″ SDR 7.4		5.55	2.883		126	167		293	400
9582	6″ SDR 11		5.55	2.883		104	167		271	375
9584	8″ SDR 7.4	Q-2	8.10	2.963		274	178		452	575
9586	8″ SDR 11		8.10	2.963		239	178		417	540
9588	10″ SDR 7.4		7.50	3.200		440	192		632	785
9590	10″ SDR 11		7.50	3.200		365	192		557	705
9600	Tee									
9604	3/8″	1 Plum	26	.308	Ea.	1.68	19.85		21.53	32.50
9606	1/2″		24.90	.321		2.31	20.50		22.81	34.50
9608	3/4″		23.70	.338		2.42	22		24.42	36
9610	1″		19.90	.402		3.05	26		29.05	43.50
9612	1-1/4″		18.50	.432		4.69	28		32.69	48
9614	1-1/2″		16.60	.482		13.40	31		44.40	63
9616	2″	Q-1	26.80	.597		18.25	34.50		52.75	73.50
9618	2-1/2″		23.80	.672		30	39		69	93.50
9620	3″		20.60	.777		60	45		105	136
9622	3-1/2″		18.40	.870		94	50.50		144.50	181
9624	4″		16.70	.958		110	55.50		165.50	207
9626	6″ SDR 7.4		3.70	4.324		137	251		388	540

22 11 13.76 Pipe Fittings, Plastic		Crew	Daily Output	Labor-Hours	Unit	2020 Bare Costs Material	Labor	Equipment	Total	Total Incl O&P
9628	6" SDR 11	Q-1	3.70	4.324	Ea.	167	251		418	575
9630	8" SDR 7.4	Q-2	5.40	4.444		375	267		642	830
9632	8" SDR 11		5.40	4.444		330	267		597	780
9634	10" SDR 7.4		5	4.800		645	289		934	1,150
9636	10" SDR 11		5	4.800		560	289		849	1,075
9638	For reducing tee use same tee price									
9660	End cap									
9662	3/8"	1 Plum	78	.103	Ea.	1.80	6.60		8.40	12.20
9664	1/2"		74.60	.107		1.90	6.90		8.80	12.80
9666	3/4"		71.40	.112		2.42	7.20		9.62	13.80
9668	1"		59.50	.134		2.92	8.65		11.57	16.60
9670	1-1/4"		55.60	.144		4.62	9.25		13.87	19.45
9672	1-1/2"		49.50	.162		6.35	10.40		16.75	23
9674	2"	Q-1	80	.200		10.70	11.60		22.30	29.50
9676	2-1/2"		71.40	.224		15.45	13		28.45	37
9678	3"		61.70	.259		35	15.05		50.05	62
9680	3-1/2"		55.20	.290		42	16.80		58.80	72
9682	4"		50	.320		64	18.55		82.55	99
9684	6" SDR 7.4		26.50	.604		88	35		123	151
9686	6" SDR 11		26.50	.604		71	35		106	132
9688	8" SDR 7.4	Q-2	16.30	1.472		88	88.50		176.50	234
9690	8" SDR 11		16.30	1.472		77.50	88.50		166	222
9692	10" SDR 7.4		14.90	1.611		132	97		229	295
9694	10" SDR 11		14.90	1.611		94	97		191	253
9800	Accessories and tools									
9802	Pipe clamps for suspension, not including rod or beam clamp									
9804	3/8"	1 Plum	74	.108	Ea.	2.25	6.95		9.20	13.25
9805	1/2"		70	.114		2.83	7.35		10.18	14.50
9806	3/4"		68	.118		2.69	7.60		10.29	14.65
9807	1"		66	.121		3.55	7.80		11.35	16
9808	1-1/4"		64	.125		3.64	8.05		11.69	16.45
9809	1-1/2"		62	.129		3.98	8.30		12.28	17.25
9810	2"	Q-1	110	.145		4.94	8.45		13.39	18.50
9811	2-1/2"		104	.154		6.35	8.90		15.25	21
9812	3"		98	.163		6.85	9.45		16.30	22
9813	3-1/2"		92	.174		7.50	10.10		17.60	24
9814	4"		86	.186		8.40	10.80		19.20	26
9815	6"		70	.229		10.20	13.25		23.45	32
9816	8"	Q-2	100	.240		36.50	14.45		50.95	63
9817	10"	"	94	.255		42	15.35		57.35	69.50
9820	Pipe cutter									
9822	For 3/8" thru 1-1/4"				Ea.	113			113	124
9824	For 1-1/2" thru 4"				"	296			296	325
9826	Note: Pipes may be cut with standard									
9827	iron saw with blades for plastic.									
9982	For plastic hangers see Line 22 05 29.10 8000									
9986	For copper/brass fittings see Line 22 11 13.25 7000									
9990	Minimum labor/equipment charge	1 Plum	4	2	Job		129		129	199

22 11 19 – Domestic Water Piping Specialties

22 11 19.10 Flexible Connectors

	22 11 19.10 Flexible Connectors	Crew	Daily Output	Labor-Hours	Unit	2020 Bare Costs Material	Labor	Equipment	Total	Total Incl O&P
0010	**FLEXIBLE CONNECTORS**, Corrugated, 5/8" OD, 3/4" ID									
0050	Gas, seamless brass, steel fittings									
0200	12" long	1 Plum	36	.222	Ea.	6.80	14.30		21.10	29.50
0220	18" long		36	.222		8.55	14.30		22.85	31.50
0240	24" long		34	.235		10	15.15		25.15	34.50
0260	30" long		34	.235		12.45	15.15		27.60	37
0280	36" long		32	.250		11.90	16.10		28	38
0320	48" long		30	.267		15.10	17.20		32.30	43
0340	60" long		30	.267		17.95	17.20		35.15	46.50
0360	72" long		30	.267		21	17.20		38.20	49.50
2000	Water, copper tubing, dielectric separators									
2100	12" long	1 Plum	36	.222	Ea.	8.20	14.30		22.50	31
2220	15" long		36	.222		9.15	14.30		23.45	32
2240	18" long		36	.222		9.10	14.30		23.40	32
2260	24" long		34	.235		12.20	15.15		27.35	37
9000	Minimum labor/equipment charge		4	2	Job		129		129	199

22 11 19.14 Flexible Metal Hose

	22 11 19.14 Flexible Metal Hose	Crew	Daily Output	Labor-Hours	Unit	Material	Labor	Equipment	Total	Total Incl O&P
0010	**FLEXIBLE METAL HOSE**, Connectors, standard lengths									
0100	Bronze braided, bronze ends									
0120	3/8" diameter x 12"	1 Stpi	26	.308	Ea.	28.50	20		48.50	62
0140	1/2" diameter x 12"		24	.333		29	22		51	66
0160	3/4" diameter x 12"		20	.400		39.50	26		65.50	84
0180	1" diameter x 18"		19	.421		51	27.50		78.50	98.50
0200	1-1/2" diameter x 18"		13	.615		57	40.50		97.50	125
0220	2" diameter x 18"		11	.727		93.50	47.50		141	177
1000	Carbon steel ends									
1020	1/4" diameter x 12"	1 Stpi	28	.286	Ea.	36	18.75		54.75	68.50
1040	3/8" diameter x 12"		26	.308		37	20		57	71.50
1060	1/2" diameter x 12"		24	.333		32	22		54	69.50
1080	1/2" diameter x 24"		24	.333		61.50	22		83.50	102
1120	3/4" diameter x 12"		20	.400		47	26		73	92
1140	3/4" diameter x 24"		20	.400		75	26		101	123
1160	3/4" diameter x 36"		20	.400		82	26		108	131
1180	1" diameter x 18"		19	.421		65.50	27.50		93	115
1200	1" diameter x 30"		19	.421		69.50	27.50		97	119
1220	1" diameter x 36"		19	.421		104	27.50		131.50	157
1240	1-1/4" diameter x 18"		15	.533		88.50	35		123.50	151
1260	1-1/4" diameter x 36"		15	.533		141	35		176	209
1280	1-1/2" diameter x 18"		13	.615		121	40.50		161.50	196
1300	1-1/2" diameter x 36"		13	.615		145	40.50		185.50	223
1320	2" diameter x 24"		11	.727		182	47.50		229.50	274
1340	2" diameter x 36"		11	.727		183	47.50		230.50	276
1360	2-1/2" diameter x 24"		9	.889		350	58.50		408.50	475
1380	2-1/2" diameter x 36"		9	.889		126	58.50		184.50	229
1400	3" diameter x 24"		7	1.143		465	75		540	625
1420	3" diameter x 36"		7	1.143		1,025	75		1,100	1,275
2000	Carbon steel braid, carbon steel solid ends									
2100	1/2" diameter x 12"	1 Stpi	24	.333	Ea.	50.50	22		72.50	90
2120	3/4" diameter x 12"		20	.400		78.50	26		104.50	127
2140	1" diameter x 12"		19	.421		112	27.50		139.50	166
2160	1-1/4" diameter x 12"		15	.533		55	35		90	115
2180	1-1/2" diameter x 12"		13	.615		60.50	40.50		101	129

22 11 19.14 Flexible Metal Hose		Crew	Daily Output	Labor-Hours	Unit	Material	2020 Bare Costs Labor	Equipment	Total	Total Incl O&P
3000	Stainless steel braid, welded on carbon steel ends									
3100	1/2" diameter x 12"	1 Stpi	24	.333	Ea.	74.50	22		96.50	116
3120	3/4" diameter x 12"		20	.400		90	26		116	140
3140	3/4" diameter x 24"		20	.400		103	26		129	154
3160	3/4" diameter x 36"		20	.400		119	26		145	172
3180	1" diameter x 12"		19	.421		108	27.50		135.50	162
3200	1" diameter x 24"		19	.421		115	27.50		142.50	169
3220	1" diameter x 36"		19	.421		147	27.50		174.50	205
3240	1-1/4" diameter x 12"		15	.533		151	35		186	220
3260	1-1/4" diameter x 24"		15	.533		178	35		213	250
3280	1-1/4" diameter x 36"		15	.533		200	35		235	274
3300	1-1/2" diameter x 12"		13	.615		62.50	40.50		103	132
3320	1-1/2" diameter x 24"		13	.615		186	40.50		226.50	268
3340	1-1/2" diameter x 36"	↓	13	.615	↓	226	40.50		266.50	310
3400	Metal stainless steel braid, over corrugated stainless steel, flanged ends									
3410	150 psi									
3420	1/2" diameter x 12"	1 Stpi	24	.333	Ea.	91	22		113	134
3430	1" diameter x 12"		20	.400		136	26		162	191
3440	1-1/2" diameter x 12"		15	.533		146	35		181	214
3450	2-1/2" diameter x 9"		12	.667		64.50	43.50		108	139
3460	3" diameter x 9"		9	.889		32	58.50		90.50	125
3470	4" diameter x 9"		7	1.143		41	75		116	161
3480	4" diameter x 30"		5	1.600		410	105		515	610
3490	4" diameter x 36"		4.80	1.667		430	109		539	640
3500	6" diameter x 11"		5	1.600		75	105		180	244
3510	6" diameter x 36"		3.80	2.105		525	138		663	790
3520	8" diameter x 12"		4	2		296	131		427	530
3530	10" diameter x 13"	↓	3	2.667		213	175		388	505
3540	12" diameter x 14"	Q-5	4	4	↓	335	236		571	735

22 11 19.18 Mixing Valve		Crew	Daily Output	Labor-Hours	Unit	Material	Labor	Equipment	Total	Total Incl O&P
0010	**MIXING VALVE**, Automatic, water tempering.									
0040	1/2" size	1 Stpi	19	.421	Ea.	625	27.50		652.50	735
0050	3/4" size		18	.444		660	29		689	770
0100	1" size		16	.500		930	33		963	1,075
0120	1-1/4" size		13	.615		1,250	40.50		1,290.50	1,450
0140	1-1/2" size		10	.800		1,525	52.50		1,577.50	1,750
0160	2" size		8	1		1,900	65.50		1,965.50	2,200
0170	2-1/2" size		6	1.333		1,900	87.50		1,987.50	2,225
0180	3" size		4	2		4,550	131		4,681	5,200
0190	4" size		3	2.667	↓	4,550	175		4,725	5,275
9000	Minimum labor/equipment charge	↓	5	1.600	Job		105		105	162

22 11 19.22 Pressure Reducing Valve		Crew	Daily Output	Labor-Hours	Unit	Material	Labor	Equipment	Total	Total Incl O&P
0010	**PRESSURE REDUCING VALVE**, Steam, pilot operated.									
0100	Threaded, iron body									
0200	1-1/2" size	1 Stpi	8	1	Ea.	2,125	65.50		2,190.50	2,425
0220	2" size	"	5	1.600	"	2,450	105		2,555	2,850
1000	Flanged, iron body, 125 lb. flanges									
1020	2" size	1 Stpi	8	1	Ea.	3,400	65.50		3,465.50	3,850
1040	2-1/2" size	"	4	2		2,525	131		2,656	2,975
1060	3" size	Q-5	4.50	3.556		3,225	210		3,435	3,875
1080	4" size	"	3	5.333		4,050	315		4,365	4,925
1500	For 250 lb. flanges, add				↓	5%				

22 11 19 – Domestic Water Piping Specialties

22 11 19.26 Pressure Regulators		Crew	Daily Output	Labor-Hours	Unit	Material	Labor	Equipment	Total	Total Incl O&P
0010	**PRESSURE REGULATORS**									
0200	Oil, light, hot water, ordinary steam, threaded									
0220	Bronze body, 1/4" size	1 Stpi	24	.333	Ea.	197	22		219	251
0230	3/8" size		24	.333		190	22		212	243
0240	1/2" size		24	.333		233	22		255	291
0250	3/4" size		20	.400		278	26		304	345
0260	1" size		19	.421		430	27.50		457.50	515
0270	1-1/4" size		15	.533		575	35		610	685
0280	1-1/2" size		13	.615		595	40.50		635.50	720
0290	2" size		11	.727		1,250	47.50		1,297.50	1,450
0320	Iron body, 1/4" size		24	.333		128	22		150	174
0330	3/8" size		24	.333		172	22		194	224
0340	1/2" size		24	.333		157	22		179	207
0350	3/4" size		20	.400		191	26		217	251
0360	1" size		19	.421		247	27.50		274.50	315
0370	1-1/4" size		15	.533		345	35		380	435
0380	1-1/2" size		13	.615		380	40.50		420.50	485
0390	2" size	↓	11	.727	↓	550	47.50		597.50	680
0500	Oil, heavy, viscous fluids, threaded									
0520	Bronze body, 3/8" size	1 Stpi	24	.333	Ea.	300	22		322	365
0530	1/2" size		24	.333		365	22		387	435
0540	3/4" size		20	.400		430	26		456	515
0550	1" size		19	.421		505	27.50		532.50	605
0560	1-1/4" size		15	.533		760	35		795	890
0570	1-1/2" size		13	.615		870	40.50		910.50	1,025
0600	Iron body, 3/8" size		24	.333		239	22		261	297
0620	1/2" size		24	.333		290	22		312	355
0630	3/4" size		20	.400		330	26		356	400
0640	1" size		19	.421		365	27.50		392.50	450
0650	1-1/4" size		15	.533		560	35		595	675
0660	1-1/2" size	↓	13	.615	↓	570	40.50		610.50	690
0800	Process steam, wet or super heated, monel trim, threaded									
0820	Bronze body, 1/4" size	1 Stpi	24	.333	Ea.	625	22		647	725
0830	3/8" size		24	.333		625	22		647	725
0840	1/2" size		24	.333		720	22		742	825
0850	3/4" size		20	.400		845	26		871	970
0860	1" size		19	.421		1,100	27.50		1,127.50	1,250
0870	1-1/4" size		15	.533		1,350	35		1,385	1,550
0880	1-1/2" size		13	.615		1,625	40.50		1,665.50	1,875
0920	Iron body, max 125 PSIG press out, 1/4" size		24	.333		820	22		842	935
0930	3/8" size		24	.333		845	22		867	965
0940	1/2" size		24	.333		1,050	22		1,072	1,175
0950	3/4" size		20	.400		1,275	26		1,301	1,450
0960	1" size		19	.421		1,625	27.50		1,652.50	1,825
0970	1-1/4" size		15	.533		1,950	35		1,985	2,200
0980	1-1/2" size	↓	13	.615	↓	2,350	40.50		2,390.50	2,650
3000	Steam, high capacity, bronze body, stainless steel trim									
3020	Threaded, 1/2" diameter	1 Stpi	24	.333	Ea.	3,000	22		3,022	3,325
3030	3/4" diameter		24	.333		3,050	22		3,072	3,400
3040	1" diameter		19	.421		3,425	27.50		3,452.50	3,800
3060	1-1/4" diameter		15	.533		3,475	35		3,510	3,875
3080	1-1/2" diameter	↓	13	.615	↓	4,325	40.50		4,365.50	4,825

(Labor, Equipment and Total columns fall under the heading "2020 Bare Costs".)

22 11 Facility Water Distribution

22 11 19 – Domestic Water Piping Specialties

22 11 19.26 Pressure Regulators		Crew	Daily Output	Labor-Hours	Unit	Material	2020 Bare Costs Labor	Equipment	Total	Total Incl O&P
3100	2" diameter	1 Stpi	11	.727	Ea.	5,275	47.50		5,322.50	5,900
3120	2-1/2" diameter	Q-5	12	1.333		6,575	78.50		6,653.50	7,350
3140	3" diameter	"	11	1.455	↓	7,450	86		7,536	8,325
3500	Flanged connection, iron body, 125 lb. W.S.P.									
3520	3" diameter	Q-5	11	1.455	Ea.	8,275	86		8,361	9,225
3540	4" diameter	"	5	3.200	"	10,400	189		10,589	11,800
9002	For water pressure regulators, see Section 22 05 23.20									

22 11 19.30 Pressure and Temperature Safety Plug		Crew	Daily Output	Labor-Hours	Unit	Material	Labor	Equipment	Total	Total Incl O&P
0010	**PRESSURE & TEMPERATURE SAFETY PLUG**									
1000	3/4" external thread, 3/8" diam. element									
1020	Carbon steel									
1050	7-1/2" insertion	1 Stpi	32	.250	Ea.	50.50	16.40		66.90	81
1120	304 stainless steel									
1150	7-1/2" insertion	1 Stpi	32	.250	Ea.	48	16.40		64.40	78
1220	316 stainless steel									
1250	7-1/2" insertion	1 Stpi	32	.250	Ea.	58.50	16.40		74.90	89.50

22 11 19.32 Pressure and Temperature Measurement Plug		Crew	Daily Output	Labor-Hours	Unit	Material	Labor	Equipment	Total	Total Incl O&P
0010	**PRESSURE & TEMPERATURE MEASUREMENT PLUG**									
0020	A permanent access port for insertion of									
0030	a pressure or temperature measuring probe									
0100	Plug, brass									
0110	1/4" MNPT, 1-1/2" long	1 Stpi	32	.250	Ea.	6.55	16.40		22.95	32.50
0120	3" long		31	.258		12.20	16.90		29.10	39.50
0140	1/2" MNPT, 1-1/2" long		30	.267		9	17.50		26.50	37
0150	3" long	↓	29	.276	↓	14.80	18.10		32.90	44.50
0200	Pressure gauge probe adapter									
0210	1/8" diameter, 1-1/2" probe				Ea.	21			21	23.50
0220	3" probe				"	35			35	38.50
0300	Temperature gauge, 5" stem									
0310	Analog				Ea.	10.90			10.90	12
0330	Digital				"	29.50			29.50	32.50
0400	Pressure gauge, compound									
0410	1/4" MNPT				Ea.	32.50			32.50	35.50
0500	Pressure and temperature test kit									
0510	Contains 2 thermometers and 2 pressure gauges									
0520	Kit				Ea.	355			355	390

22 11 19.34 Sleeves and Escutcheons		Crew	Daily Output	Labor-Hours	Unit	Material	Labor	Equipment	Total	Total Incl O&P
0010	**SLEEVES & ESCUTCHEONS**									
0100	Pipe sleeve									
0110	Steel, w/water stop, 12" long, with link seal									
0120	2" diam. for 1/2" carrier pipe	1 Plum	8.40	.952	Ea.	64.50	61.50		126	166
0130	2-1/2" diam. for 3/4" carrier pipe		8	1		73	64.50		137.50	180
0140	2-1/2" diam. for 1" carrier pipe		8	1		69	64.50		133.50	176
0150	3" diam. for 1-1/4" carrier pipe		7.20	1.111		89	71.50		160.50	209
0160	3-1/2" diam. for 1-1/2" carrier pipe		6.80	1.176		97	76		173	223
0170	4" diam. for 2" carrier pipe		6	1.333		98	86		184	241
0180	4" diam. for 2-1/2" carrier pipe		6	1.333		98	86		184	241
0190	5" diam. for 3" carrier pipe		5.40	1.481		115	95.50		210.50	275
0200	6" diam. for 4" carrier pipe	↓	4.80	1.667		138	107		245	315
0210	10" diam. for 6" carrier pipe	Q-1	8	2		271	116		387	475
0220	12" diam. for 8" carrier pipe		7.20	2.222		370	129		499	610
0230	14" diam. for 10" carrier pipe	↓	6.40	2.500	↓	385	145		530	645

	22 11 19.34 Sleeves and Escutcheons	Crew	Daily Output	Labor-Hours	Unit	2020 Bare Costs Material	Labor	Equipment	Total	Total Incl O&P
0240	16" diam. for 12" carrier pipe	Q-1	5.80	2.759	Ea.	425	160		585	715
0250	18" diam. for 14" carrier pipe		5.20	3.077		780	178		958	1,125
0260	24" diam. for 18" carrier pipe		4	4		1,125	232		1,357	1,600
0270	24" diam. for 20" carrier pipe		4	4		975	232		1,207	1,425
0280	30" diam. for 24" carrier pipe		3.20	5		1,400	290		1,690	1,975
0500	Wall sleeve									
0510	Ductile iron with rubber gasket seal									
0520	3"	1 Plum	8.40	.952	Ea.	213	61.50		274.50	330
0530	4"		7.20	1.111		227	71.50		298.50	360
0540	6"		6	1.333		276	86		362	440
0550	8"		4	2		1,225	129		1,354	1,550
0560	10"		3	2.667		1,500	172		1,672	1,925
0570	12"		2.40	3.333		1,750	215		1,965	2,250
5000	Escutcheon									
5100	Split ring, pipe									
5110	Chrome plated									
5120	1/2"	1 Plum	160	.050	Ea.	1.10	3.22		4.32	6.20
5130	3/4"		160	.050		1.29	3.22		4.51	6.40
5140	1"		135	.059		1.43	3.82		5.25	7.45
5150	1-1/2"		115	.070		1.86	4.48		6.34	9
5160	2"		100	.080		2.16	5.15		7.31	10.35
5170	4"		80	.100		5	6.45		11.45	15.45
5180	6"		68	.118		7.90	7.60		15.50	20.50
5400	Shallow flange type									
5410	Chrome plated steel									
5420	1/2" CTS	1 Plum	180	.044	Ea.	.33	2.86		3.19	4.79
5430	3/4" CTS		180	.044		.37	2.86		3.23	4.84
5440	1/2" IPS		180	.044		.40	2.86		3.26	4.87
5450	3/4" IPS		180	.044		.44	2.86		3.30	4.91
5460	1" IPS		175	.046		.48	2.95		3.43	5.10
5470	1-1/2" IPS		170	.047		.81	3.03		3.84	5.60
5480	2" IPS		160	.050		.99	3.22		4.21	6.05

	22 11 19.38 Water Supply Meters	Crew	Daily Output	Labor-Hours	Unit	Material	Labor	Equipment	Total	Total Incl O&P
0010	**WATER SUPPLY METERS**									
1000	Detector, serves dual systems such as fire and domestic or									
1020	process water, wide range cap., UL and FM approved									
1100	3" mainline x 2" by-pass, 400 GPM	Q-1	3.60	4.444	Ea.	8,025	258		8,283	9,225
1140	4" mainline x 2" by-pass, 700 GPM	"	2.50	6.400		8,025	370		8,395	9,400
1180	6" mainline x 3" by-pass, 1,600 GPM	Q-2	2.60	9.231		12,300	555		12,855	14,400
1220	8" mainline x 4" by-pass, 2,800 GPM		2.10	11.429		18,200	685		18,885	21,100
1260	10" mainline x 6" by-pass, 4,400 GPM		2	12		26,000	720		26,720	29,700
1300	10" x 12" mainlines x 6" by-pass, 5,400 GPM		1.70	14.118		35,300	850		36,150	40,100
2000	Domestic/commercial, bronze									
2020	Threaded									
2060	5/8" diameter, to 20 GPM	1 Plum	16	.500	Ea.	54	32		86	110
2080	3/4" diameter, to 30 GPM		14	.571		98.50	37		135.50	165
2100	1" diameter, to 50 GPM		12	.667		149	43		192	231
2300	Threaded/flanged									
2340	1-1/2" diameter, to 100 GPM	1 Plum	8	1	Ea.	365	64.50		429.50	500
2360	2" diameter, to 160 GPM	"	6	1.333	"	495	86		581	680
2600	Flanged, compound									
2640	3" diameter, 320 GPM	Q-1	3	5.333	Ea.	2,650	310		2,960	3,400

22 11 19 – Domestic Water Piping Specialties

22 11 19.38 Water Supply Meters		Crew	Daily Output	Labor-Hours	Unit	Material	Labor	Equipment	Total	Total Incl O&P
						2020 Bare Costs				
2660	4" diameter, to 500 GPM	Q-1	1.50	10.667	Ea.	4,250	620		4,870	5,625
2680	6" diameter, to 1,000 GPM		1	16		6,850	930		7,780	8,975
2700	8" diameter, to 1,800 GPM		.80	20		10,700	1,150		11,850	13,600
7000	Turbine									
7260	Flanged									
7300	2" diameter, to 160 GPM	1 Plum	7	1.143	Ea.	615	73.50		688.50	790
7320	3" diameter, to 450 GPM	Q-1	3.60	4.444		1,100	258		1,358	1,600
7340	4" diameter, to 650 GPM	"	2.50	6.400		1,825	370		2,195	2,575
7360	6" diameter, to 1,800 GPM	Q-2	2.60	9.231		3,000	555		3,555	4,150
7380	8" diameter, to 2,500 GPM		2.10	11.429		5,150	685		5,835	6,750
7400	10" diameter, to 5,500 GPM		1.70	14.118		6,925	850		7,775	8,950
9000	Minimum labor/equipment charge	1 Plum	3.25	2.462	Job		159		159	245

22 11 19.42 Backflow Preventers		Crew	Daily Output	Labor-Hours	Unit	Material	Labor	Equipment	Total	Total Incl O&P
0010	**BACKFLOW PREVENTERS**, Includes valves									
0020	and four test cocks, corrosion resistant, automatic operation									
1000	Double check principle									
1010	Threaded, with ball valves									
1020	3/4" pipe size	1 Plum	16	.500	Ea.	241	32		273	315
1030	1" pipe size		14	.571		253	37		290	335
1040	1-1/2" pipe size		10	.800		620	51.50		671.50	765
1050	2" pipe size		7	1.143		665	73.50		738.50	850
1080	Threaded, with gate valves									
1100	3/4" pipe size	1 Plum	16	.500	Ea.	1,150	32		1,182	1,300
1120	1" pipe size		14	.571		1,150	37		1,187	1,325
1140	1-1/2" pipe size		10	.800		1,225	51.50		1,276.50	1,425
1160	2" pipe size		7	1.143		1,825	73.50		1,898.50	2,125
1300	Flanged, valves are OS&Y									
1380	3" pipe size	Q-1	4.50	3.556	Ea.	2,850	206		3,056	3,475
1400	4" pipe size	"	3	5.333		4,675	310		4,985	5,600
1420	6" pipe size	Q-2	3	8		6,700	480		7,180	8,125
4000	Reduced pressure principle									
4100	Threaded, bronze, valves are ball									
4120	3/4" pipe size	1 Plum	16	.500	Ea.	530	32		562	635
4140	1" pipe size		14	.571		560	37		597	670
4150	1-1/4" pipe size		12	.667		1,125	43		1,168	1,300
4160	1-1/2" pipe size		10	.800		1,150	51.50		1,201.50	1,350
4180	2" pipe size		7	1.143		1,325	73.50		1,398.50	1,600
5000	Flanged, bronze, valves are OS&Y									
5060	2-1/2" pipe size	Q-1	5	3.200	Ea.	5,475	186		5,661	6,300
5080	3" pipe size		4.50	3.556		6,600	206		6,806	7,575
5100	4" pipe size		3	5.333		7,725	310		8,035	8,975
5120	6" pipe size	Q-2	3	8		12,000	480		12,480	13,900
5600	Flanged, iron, valves are OS&Y									
5660	2-1/2" pipe size	Q-1	5	3.200	Ea.	3,300	186		3,486	3,900
5680	3" pipe size		4.50	3.556		3,275	206		3,481	3,925
5700	4" pipe size		3	5.333		4,725	310		5,035	5,675
5720	6" pipe size	Q-2	3	8		5,825	480		6,305	7,175
5740	8" pipe size		2	12		10,300	720		11,020	12,400
5760	10" pipe size		1	24		19,800	1,450		21,250	24,000
9010	Minimum labor/equipment charge	1 Plum	2	4	Job		258		258	400

22 11 19 – Domestic Water Piping Specialties

22 11 19.50 Vacuum Breakers		Crew	Daily Output	Labor-Hours	Unit	Material	2020 Bare Costs Labor	Equipment	Total	Total Incl O&P
0010	**VACUUM BREAKERS**									
0013	See also backflow preventers Section 22 11 19.42									
1000	Anti-siphon continuous pressure type									
1010	Max. 150 psi - 210°F									
1020	Bronze body									
1030	1/2" size	1 Stpi	24	.333	Ea.	192	22		214	245
1040	3/4" size		20	.400		192	26		218	252
1050	1" size		19	.421		197	27.50		224.50	259
1060	1-1/4" size		15	.533		390	35		425	485
1070	1-1/2" size		13	.615		475	40.50		515.50	590
1080	2" size		11	.727		490	47.50		537.50	615
1200	Max. 125 psi with atmospheric vent									
1210	Brass, in-line construction									
1220	1/4" size	1 Stpi	24	.333	Ea.	143	22		165	191
1230	3/8" size	"	24	.333		143	22		165	191
1260	For polished chrome finish, add					13%				
2000	Anti-siphon, non-continuous pressure type									
2010	Hot or cold water 125 psi - 210°F									
2020	Bronze body									
2030	1/4" size	1 Stpi	24	.333	Ea.	88.50	22		110.50	131
2040	3/8" size		24	.333		88.50	22		110.50	131
2050	1/2" size		24	.333		99	22		121	143
2060	3/4" size		20	.400		119	26		145	171
2070	1" size		19	.421		183	27.50		210.50	245
2080	1-1/4" size		15	.533		320	35		355	410
2090	1-1/2" size		13	.615		350	40.50		390.50	450
2100	2" size		11	.727		585	47.50		632.50	720
2110	2-1/2" size		8	1		1,675	65.50		1,740.50	1,950
2120	3" size		6	1.333		2,225	87.50		2,312.50	2,575
2150	For polished chrome finish, add					50%				
3000	Air gap fitting									
3020	1/2" NPT size	1 Plum	19	.421	Ea.	61.50	27		88.50	110
3030	1" NPT size	"	15	.533		70.50	34.50		105	131
3040	2" NPT size	Q-1	21	.762		138	44		182	221
3050	3" NPT size		14	1.143		273	66.50		339.50	400
3060	4" NPT size		10	1.600		273	93		366	445

22 11 19.54 Water Hammer Arresters/Shock Absorbers

		Crew	Daily Output	Labor-Hours	Unit	Material	Labor	Equipment	Total	Total Incl O&P
0010	**WATER HAMMER ARRESTERS/SHOCK ABSORBERS**									
0490	Copper									
0500	3/4" male IPS for 1 to 11 fixtures	1 Plum	12	.667	Ea.	31	43		74	101
0600	1" male IPS for 12 to 32 fixtures		8	1		51.50	64.50		116	156
0700	1-1/4" male IPS for 33 to 60 fixtures		8	1		50.50	64.50		115	156
0800	1-1/2" male IPS for 61 to 113 fixtures		8	1		73	64.50		137.50	180
0900	2" male IPS for 114 to 154 fixtures		8	1		108	64.50		172.50	218
1000	2-1/2" male IPS for 155 to 330 fixtures		4	2		330	129		459	565
9000	Minimum labor/equipment charge		3.50	2.286	Job		147		147	228

22 11 19.64 Hydrants

		Crew	Daily Output	Labor-Hours	Unit	Material	Labor	Equipment	Total	Total Incl O&P
0010	**HYDRANTS**									
0050	Wall type, moderate climate, bronze, encased									
0200	3/4" IPS connection	1 Plum	16	.500	Ea.	1,025	32		1,057	1,175
0300	1" IPS connection		14	.571		1,125	37		1,162	1,300
0500	Anti-siphon type, 3/4" connection		16	.500		815	32		847	945

22 11 19.64 Hydrants		Crew	Daily Output	Labor-Hours	Unit	2020 Bare Costs Material	Labor	Equipment	Total	Total Incl O&P
1000	Non-freeze, bronze, exposed									
1100	3/4" IPS connection, 4" to 9" thick wall	1 Plum	14	.571	Ea.	470	37		507	570
1120	10" to 14" thick wall		12	.667		445	43		488	555
1140	15" to 19" thick wall		12	.667		550	43		593	670
1160	20" to 24" thick wall		10	.800		830	51.50		881.50	990
1200	For 1" IPS connection, add					15%	10%			
1240	For 3/4" adapter type vacuum breaker, add				Ea.	83			83	91.50
1280	For anti-siphon type, add				"	176			176	194
2000	Non-freeze bronze, encased, anti-siphon type									
2100	3/4" IPS connection, 5" to 9" thick wall	1 Plum	14	.571	Ea.	1,275	37		1,312	1,450
2120	10" to 14" thick wall		12	.667		1,625	43		1,668	1,875
2140	15" to 19" thick wall		12	.667		1,725	43		1,768	1,975
2160	20" to 24" thick wall		10	.800		1,800	51.50		1,851.50	2,050
2200	For 1" IPS connection, add					10%	10%			
3000	Ground box type, bronze frame, 3/4" IPS connection									
3080	Non-freeze, all bronze, polished face, set flush									
3100	2' depth of bury	1 Plum	8	1	Ea.	1,325	64.50		1,389.50	1,550
3120	3' depth of bury		8	1		1,425	64.50		1,489.50	1,675
3140	4' depth of bury		8	1		1,525	64.50		1,589.50	1,775
3160	5' depth of bury		7	1.143		1,625	73.50		1,698.50	1,900
3180	6' depth of bury		7	1.143		1,700	73.50		1,773.50	1,975
3200	7' depth of bury		6	1.333		1,675	86		1,761	1,975
3220	8' depth of bury		5	1.600		1,775	103		1,878	2,100
3240	9' depth of bury		4	2		1,875	129		2,004	2,275
3260	10' depth of bury		4	2		1,975	129		2,104	2,350
3400	For 1" IPS connection, add					15%	10%			
3450	For 1-1/4" IPS connection, add					325%	14%			
3500	For 1-1/2" IPS connection, add					370%	18%			
3550	For 2" IPS connection, add					445%	24%			
3600	For tapped drain port in box, add					107			107	118
4000	Non-freeze, CI body, bronze frame & scoriated cover									
4010	with hose storage									
4100	2' depth of bury	1 Plum	7	1.143	Ea.	2,200	73.50		2,273.50	2,550
4120	3' depth of bury		7	1.143		2,300	73.50		2,373.50	2,650
4140	4' depth of bury		7	1.143		2,375	73.50		2,448.50	2,725
4160	5' depth of bury		6.50	1.231		2,425	79.50		2,504.50	2,775
4180	6' depth of bury		6	1.333		2,450	86		2,536	2,825
4200	7' depth of bury		5.50	1.455		2,550	94		2,644	2,950
4220	8' depth of bury		5	1.600		2,625	103		2,728	3,025
4240	9' depth of bury		4.50	1.778		2,825	115		2,940	3,300
4260	10' depth of bury		4	2		2,825	129		2,954	3,300
4280	For 1" IPS connection, add					505			505	555
4300	For tapped drain port in box, add					113			113	124
5000	Moderate climate, all bronze, polished face									
5020	and scoriated cover, set flush									
5100	3/4" IPS connection	1 Plum	16	.500	Ea.	830	32		862	965
5120	1" IPS connection	"	14	.571		2,000	37		2,037	2,275
5200	For tapped drain port in box, add					113			113	124
6000	Ground post type, all non-freeze, all bronze, aluminum casing									
6010	guard, exposed head, 3/4" IPS connection									
6100	2' depth of bury	1 Plum	8	1	Ea.	1,275	64.50		1,339.50	1,500
6120	3' depth of bury		8	1		1,300	64.50		1,364.50	1,550
6140	4' depth of bury		8	1		1,475	64.50		1,539.50	1,725

22 11 Facility Water Distribution

22 11 19 – Domestic Water Piping Specialties

22 11 19.64 Hydrants		Crew	Daily Output	Labor-Hours	Unit	Material	Labor	Equipment	Total	Total Incl O&P
						2020 Bare Costs				
6160	5′ depth of bury	1 Plum	7	1.143	Ea.	1,450	73.50		1,523.50	1,725
6180	6′ depth of bury		7	1.143		1,775	73.50		1,848.50	2,075
6200	7′ depth of bury		6	1.333		1,675	86		1,761	1,950
6220	8′ depth of bury		5	1.600		1,775	103		1,878	2,100
6240	9′ depth of bury		4	2		1,875	129		2,004	2,275
6260	10′ depth of bury		4	2		2,000	129		2,129	2,375
6300	For 1″ IPS connection, add					40%	10%			
6350	For 1-1/4″ IPS connection, add					140%	14%			
6400	For 1-1/2″ IPS connection, add					225%	18%			
6450	For 2″ IPS connection, add					315%	24%			
9000	Minimum labor/equipment charge	1 Plum	3	2.667	Job		172		172	266

22 11 23 – Domestic Water Pumps

22 11 23.10 General Utility Pumps		Crew	Daily Output	Labor-Hours	Unit	Material	Labor	Equipment	Total	Total Incl O&P
0010	**GENERAL UTILITY PUMPS**									
2000	Single stage									
3000	Double suction,									
3140	50 HP, 5″ D x 6″ S	Q-2	.33	72.727	Ea.	9,900	4,375		14,275	17,700
3180	60 HP, 6″ D x 8″ S	Q-3	.30	107		17,500	6,550		24,050	29,400
3190	75 HP, to 2,500 GPM		.28	114		16,500	7,000		23,500	28,900
3220	100 HP, to 3,000 GPM		.26	123		20,300	7,550		27,850	34,000
3240	150 HP, to 4,000 GPM		.24	133		23,600	8,175		31,775	38,600

22 11 23.13 Domestic-Water Packaged Booster Pumps		Crew	Daily Output	Labor-Hours	Unit	Material	Labor	Equipment	Total	Total Incl O&P
0010	**DOMESTIC-WATER PACKAGED BOOSTER PUMPS**									
0200	Pump system, with diaphragm tank, control, press. switch									
0300	1 HP pump	Q-1	1.30	12.308	Ea.	7,500	715		8,215	9,350
0400	1-1/2 HP pump		1.25	12.800		7,575	740		8,315	9,500
0420	2 HP pump		1.20	13.333		7,300	775		8,075	9,225
0440	3 HP pump		1.10	14.545		7,900	845		8,745	9,975
0460	5 HP pump	Q-2	1.50	16		8,200	960		9,160	10,500
0480	7-1/2 HP pump		1.42	16.901		9,725	1,025		10,750	12,300
0500	10 HP pump		1.34	17.910		10,100	1,075		11,175	12,900
2000	Pump system, variable speed, base, controls, starter									
2010	Duplex, 100′ head									
2020	400 GPM, 7-1/2 HP, 4″ discharge	Q-2	.70	34.286	Ea.	44,200	2,050		46,250	52,000
2025	Triplex, 100′ head									
2030	1,000 GPM, 15 HP, 6″ discharge	Q-2	.50	48	Ea.	49,200	2,875		52,075	58,500
2040	1,700 GPM, 30 HP, 6″ discharge	″	.30	80	″	72,000	4,800		76,800	86,500

22 12 Facility Potable-Water Storage Tanks

22 12 21 – Facility Underground Potable-Water Storage Tanks

22 12 21.13 Fiberglass, Undrgrnd Pot.-Water Storage Tanks		Crew	Daily Output	Labor-Hours	Unit	Material	Labor	Equipment	Total	Total Incl O&P
0010	**FIBERGLASS, UNDERGROUND POTABLE-WATER STORAGE TANKS**									
0020	Excludes excavation, backfill & piping									
0030	Single wall									
2000	600 gallon capacity	B-21B	3.75	10.667	Ea.	4,125	490	126	4,741	5,475
2010	1,000 gallon capacity		3.50	11.429		5,275	525	135	5,935	6,775
2020	2,000 gallon capacity		3.25	12.308		7,500	565	145	8,210	9,300
2030	4,000 gallon capacity		3	13.333		10,100	610	157	10,867	12,200
2040	6,000 gallon capacity		2.65	15.094		11,400	695	178	12,273	13,900
2050	8,000 gallon capacity		2.30	17.391		13,900	800	205	14,905	16,800

22 12 Facility Potable-Water Storage Tanks

22 12 21 – Facility Underground Potable-Water Storage Tanks

22 12 21.13 Fiberglass, Undrgrnd Pot.-Water Storage Tanks		Crew	Daily Output	Labor-Hours	Unit	2020 Bare Costs Material	Labor	Equipment	Total	Total Incl O&P
2060	10,000 gallon capacity	B-21B	2	20	Ea.	15,400	920	236	16,556	18,700
2070	12,000 gallon capacity		1.50	26.667		21,600	1,225	315	23,140	26,000
2080	15,000 gallon capacity		1	40		24,500	1,825	470	26,795	30,400
2090	20,000 gallon capacity		.75	53.333		31,800	2,450	630	34,880	39,600
2100	25,000 gallon capacity		.50	80		47,400	3,675	940	52,015	59,000
2110	30,000 gallon capacity		.35	114		57,000	5,250	1,350	63,600	72,500
2120	40,000 gallon capacity	↓	.30	133	↓	80,000	6,125	1,575	87,700	99,500

22 12 23 – Facility Indoor Potable-Water Storage Tanks

22 12 23.13 Facility Steel, Indoor Pot.-Water Storage Tanks		Crew	Daily Output	Labor-Hours	Unit	Material	Labor	Equipment	Total	Total Incl O&P
0010	**FACILITY STEEL, INDOOR POT.-WATER STORAGE TANKS**									
2000	Galvanized steel, 15 gal., 14″ diam. x 26″ LOA	1 Plum	12	.667	Ea.	1,500	43		1,543	1,725
2060	30 gal., 14″ diam. x 49″ LOA		11	.727		1,775	47		1,822	2,025
2080	80 gal., 20″ diam. x 64″ LOA		9	.889		2,800	57.50		2,857.50	3,175
2100	135 gal., 24″ diam. x 75″ LOA		6	1.333		4,025	86		4,111	4,550
2120	240 gal., 30″ diam. x 86″ LOA		4	2		7,850	129		7,979	8,825
2140	300 gal., 36″ diam. x 76″ LOA	↓	3	2.667		11,200	172		11,372	12,600
2160	400 gal., 36″ diam. x 100″ LOA	Q-1	4	4		13,700	232		13,932	15,500
2180	500 gal., 36″ diam. x 126″ LOA	″	3	5.333		17,000	310		17,310	19,200
3000	Glass lined, P.E., 80 gal., 20″ diam. x 60″ LOA	1 Plum	9	.889		4,050	57.50		4,107.50	4,550
3060	140 gal., 24″ diam. x 75″ LOA		6	1.333		5,575	86		5,661	6,250
3080	200 gal., 30″ diam. x 71″ LOA		4	2		8,050	129		8,179	9,050
3100	350 gal., 36″ diam. x 86″ LOA	↓	3	2.667		9,250	172		9,422	10,500
3120	450 gal., 42″ diam. x 79″ LOA	Q-1	4	4		13,300	232		13,532	15,100
3140	600 gal., 48″ diam. x 81″ LOA		3	5.333		15,900	310		16,210	18,000
3160	750 gal., 48″ diam. x 105″ LOA		3	5.333		17,800	310		18,110	20,100
3180	900 gal., 54″ diam. x 95″ LOA		2.50	6.400		25,500	370		25,870	28,700
3200	1,500 gal., 54″ diam. x 153″ LOA		2	8		32,200	465		32,665	36,100
3220	1,800 gal., 54″ diam. x 181″ LOA		1.50	10.667		35,700	620		36,320	40,300
3240	2,000 gal., 60″ diam. x 165″ LOA	↓	1	16		41,000	930		41,930	46,500
3260	3,500 gal., 72″ diam. x 201″ LOA	Q-2	1.50	16	↓	51,000	960		51,960	57,500

22 13 Facility Sanitary Sewerage

22 13 16 – Sanitary Waste and Vent Piping

22 13 16.20 Pipe, Cast Iron		Crew	Daily Output	Labor-Hours	Unit	Material	Labor	Equipment	Total	Total Incl O&P
0010	**PIPE, CAST IRON**, Soil, on clevis hanger assemblies, 5′ OC R221113-50									
0020	Single hub, service wt., lead & oakum joints 10′ OC									
2120	2″ diameter	Q-1	63	.254	L.F.	17.10	14.75		31.85	42
2140	3″ diameter		60	.267		26.50	15.45		41.95	53
2160	4″ diameter	↓	55	.291		24.50	16.85		41.35	53
2180	5″ diameter	Q-2	76	.316		31	19		50	63.50
2200	6″ diameter	″	73	.329		42.50	19.80		62.30	77
2220	8″ diameter	Q-3	59	.542		64.50	33.50		98	123
2240	10″ diameter		54	.593		102	36.50		138.50	169
2260	12″ diameter		48	.667		145	41		186	223
2261	15″ diameter	↓	40	.800		173	49		222	266
2320	For service weight, double hub, add					10%				
2340	For extra heavy, single hub, add					48%	4%			
2360	For extra heavy, double hub, add				↓	71%	4%			
2400	Lead for caulking (1#/diam. in.)	Q-1	160	.100	Lb.	1.01	5.80		6.81	10.05
2420	Oakum for caulking (1/8#/diam. in.)	″	40	.400	″	4.37	23		27.37	41

22 13 16.20 Pipe, Cast Iron		Crew	Daily Output	Labor-Hours	Unit	Material (2020 Bare Costs)	Labor (2020 Bare Costs)	Equipment (2020 Bare Costs)	Total (2020 Bare Costs)	Total Incl O&P
2960	To delete hangers, subtract									
2970	2″ diam. to 4″ diam.					16%	19%			
2980	5″ diam. to 8″ diam.					14%	14%			
2990	10″ diam. to 15″ diam.					13%	19%			
3000	Single hub, service wt., push-on gasket joints 10′ OC									
3010	2″ diameter	Q-1	66	.242	L.F.	18.30	14.05		32.35	41.50
3020	3″ diameter		63	.254		28	14.75		42.75	54
3030	4″ diameter		57	.281		26.50	16.30		42.80	54
3040	5″ diameter	Q-2	79	.304		34	18.25		52.25	66
3050	6″ diameter	″	75	.320		45.50	19.25		64.75	80
3060	8″ diameter	Q-3	62	.516		71	31.50		102.50	127
3070	10″ diameter		56	.571		114	35		149	179
3080	12″ diameter		49	.653		160	40		200	238
3082	15″ diameter		40	.800		191	49		240	286
3100	For service weight, double hub, add					65%				
3110	For extra heavy, single hub, add					48%	4%			
3120	For extra heavy, double hub, add					29%	4%			
3130	To delete hangers, subtract									
3140	2″ diam. to 4″ diam.					12%	21%			
3150	5″ diam. to 8″ diam.					10%	16%			
3160	10″ diam. to 15″ diam.					9%	21%			
4000	No hub, couplings 10′ OC									
4100	1-1/2″ diameter	Q-1	71	.225	L.F.	16.85	13.05		29.90	38.50
4120	2″ diameter		67	.239		20	13.85		33.85	43.50
4140	3″ diameter		64	.250		27	14.50		41.50	52
4160	4″ diameter		58	.276		25	16		41	52
4180	5″ diameter	Q-2	83	.289		36.50	17.40		53.90	67.50
4200	6″ diameter	″	79	.304		43.50	18.25		61.75	76.50
4220	8″ diameter	Q-3	69	.464		71	28.50		99.50	122
4240	10″ diameter		61	.525		117	32		149	179
4244	12″ diameter		58	.552		135	34		169	202
4248	15″ diameter		52	.615		207	38		245	287
4280	To delete hangers, subtract									
4290	1-1/2″ diam. to 6″ diam.					22%	47%			
4300	8″ diam. to 10″ diam.					21%	44%			
4310	12″ diam. to 15″ diam.					19%	40%			
9000	Minimum labor/equipment charge	1 Plum	4	2	Job		129		129	199

22 13 16.30 Pipe Fittings, Cast Iron		Crew	Daily Output	Labor-Hours	Unit	Material	Labor	Equipment	Total	Total Incl O&P
0010	**PIPE FITTINGS, CAST IRON**, Soil									
0040	Hub and spigot, service weight, lead & oakum joints									
0080	1/4 bend, 2″	Q-1	16	1	Ea.	23.50	58		81.50	115
0120	3″		14	1.143		31	66.50		97.50	136
0140	4″		13	1.231		49	71.50		120.50	164
0160	5″	Q-2	18	1.333		68	80		148	199
0180	6″	″	17	1.412		85	85		170	225
0200	8″	Q-3	11	2.909		256	178		434	560
0220	10″		10	3.200		370	196		566	710
0224	12″		9	3.556		510	218		728	895
0266	Closet bend, 3″ diameter with flange 10″ x 16″	Q-1	14	1.143		128	66.50		194.50	243
0268	16″ x 16″		12	1.333		139	77.50		216.50	273
0270	Closet bend, 4″ diameter, 2-1/2″ x 4″ ring, 6″ x 16″		13	1.231		110	71.50		181.50	231
0280	8″ x 16″		13	1.231		97.50	71.50		169	217

22 13 16.30 Pipe Fittings, Cast Iron		Crew	Daily Output	Labor-Hours	Unit	2020 Bare Costs Material	Labor	Equipment	Total	Total Incl O&P
0290	10″ x 12″	Q-1	12	1.333	Ea.	129	77.50		206.50	262
0300	10″ x 18″		11	1.455		130	84.50		214.50	273
0310	12″ x 16″		11	1.455		110	84.50		194.50	251
0330	16″ x 16″		10	1.600		143	93		236	300
0340	1/8 bend, 2″		16	1		16.65	58		74.65	108
0350	3″		14	1.143		26	66.50		92.50	131
0360	4″		13	1.231		38	71.50		109.50	152
0380	5″	Q-2	18	1.333		54	80		134	184
0400	6″	″	17	1.412		64.50	85		149.50	202
0420	8″	Q-3	11	2.909		193	178		371	490
0440	10″		10	3.200		277	196		473	610
0460	12″		9	3.556		525	218		743	915
0500	Sanitary tee, 2″	Q-1	10	1.600		32.50	93		125.50	179
0540	3″		9	1.778		53	103		156	217
0620	4″		8	2		65.50	116		181.50	251
0700	5″	Q-2	12	2		129	120		249	330
0800	6″	″	11	2.182		146	131		277	365
0880	8″	Q-3	7	4.571		385	280		665	860
1000	Tee, 2″	Q-1	10	1.600		47	93		140	195
1060	3″		9	1.778		70.50	103		173.50	237
1120	4″		8	2		88.50	116		204.50	276
1200	5″	Q-2	12	2		187	120		307	390
1300	6″	″	11	2.182		186	131		317	410
1380	8″	Q-3	7	4.571		370	280		650	840
1400	Combination Y and 1/8 bend									
1420	2″	Q-1	10	1.600	Ea.	41	93		134	188
1460	3″		9	1.778		62	103		165	227
1520	4″		8	2		85.50	116		201.50	273
1540	5″	Q-2	12	2		162	120		282	365
1560	6″		11	2.182		205	131		336	430
1580	8″		7	3.429		505	206		711	880
1582	12″	Q-3	6	5.333		1,025	325		1,350	1,625
1600	Double Y, 2″	Q-1	8	2		72	116		188	258
1610	3″		7	2.286		90	133		223	305
1620	4″		6.50	2.462		118	143		261	350
1630	5″	Q-2	9	2.667		210	160		370	480
1640	6″	″	8	3		310	180		490	620
1650	8″	Q-3	5.50	5.818		745	355		1,100	1,375
1660	10″		5	6.400		1,675	395		2,070	2,450
1670	12″		4.50	7.111		1,925	435		2,360	2,800
1740	Reducer, 3″ x 2″	Q-1	15	1.067		23	62		85	121
1750	4″ x 2″		14.50	1.103		26	64		90	128
1760	4″ x 3″		14	1.143		29.50	66.50		96	135
1770	5″ x 2″		14	1.143		61.50	66.50		128	170
1780	5″ x 3″		13.50	1.185		66	68.50		134.50	179
1790	5″ x 4″		13	1.231		38.50	71.50		110	152
1800	6″ x 2″		13.50	1.185		59	68.50		127.50	171
1810	6″ x 3″		13	1.231		61	71.50		132.50	177
1830	6″ x 4″		12.50	1.280		60	74		134	181
1840	6″ x 5″		11	1.455		64.50	84.50		149	201
1880	8″ x 3″	Q-2	13.50	1.778		116	107		223	293
1900	8″ x 4″		13	1.846		100	111		211	282
1920	8″ x 5″		12	2		106	120		226	305

22 13 16.30 Pipe Fittings, Cast Iron		Crew	Daily Output	Labor-Hours	Unit	2020 Bare Costs Material	Labor	Equipment	Total	Total Incl O&P
1940	8" x 6"	Q-2	12	2	Ea.	100	120		220	296
1960	Increaser, 2" x 3"	Q-1	15	1.067		55	62		117	156
1980	2" x 4"		14	1.143		55	66.50		121.50	162
2000	2" x 5"		13	1.231		62.50	71.50		134	179
2020	3" x 4"		13	1.231		60	71.50		131.50	176
2040	3" x 5"		13	1.231		62.50	71.50		134	179
2060	3" x 6"		12	1.333		80.50	77.50		158	209
2070	4" x 5"		13	1.231		71	71.50		142.50	188
2080	4" x 6"		12	1.333		81.50	77.50		159	210
2090	4" x 8"	Q-2	13	1.846		170	111		281	360
2100	5" x 6"	Q-1	11	1.455		118	84.50		202.50	259
2110	5" x 8"	Q-2	12	2		183	120		303	385
2120	6" x 8"		12	2		195	120		315	400
2130	6" x 10"		8	3		350	180		530	665
2140	8" x 10"		6.50	3.692		360	222		582	745
2150	10" x 12"		5.50	4.364		640	262		902	1,100
2500	Y, 2"	Q-1	10	1.600		29.50	93		122.50	176
2510	3"		9	1.778		55	103		158	220
2520	4"		8	2		72	116		188	258
2530	5"	Q-2	12	2		128	120		248	325
2540	6"	"	11	2.182		170	131		301	390
2550	8"	Q-3	7	4.571		415	280		695	890
2560	10"		6	5.333		670	325		995	1,250
2570	12"		5	6.400		1,550	395		1,945	2,325
2580	15"		4	8		3,350	490		3,840	4,450
3000	For extra heavy, add					44%	4%			
3600	Hub and spigot, service weight gasket joint									
3605	Note: gaskets and joint labor have									
3606	been included with all listed fittings.									
3610	1/4 bend, 2"	Q-1	20	.800	Ea.	35	46.50		81.50	110
3620	3"		17	.941		46.50	54.50		101	136
3630	4"		15	1.067		68	62		130	171
3640	5"	Q-2	21	1.143		97.50	68.50		166	213
3650	6"	"	19	1.263		116	76		192	244
3660	8"	Q-3	12	2.667		325	164		489	610
3670	10"		11	2.909		485	178		663	805
3680	12"		10	3.200		655	196		851	1,025
3700	Closet bend, 3" diameter with ring 10" x 16"	Q-1	17	.941		143	54.50		197.50	242
3710	16" x 16"		15	1.067		155	62		217	266
3730	Closet bend, 4" diameter, 1" x 4" ring, 6" x 16"		15	1.067		129	62		191	238
3740	8" x 16"		15	1.067		117	62		179	224
3750	10" x 12"		14	1.143		148	66.50		214.50	265
3760	10" x 18"		13	1.231		149	71.50		220.50	274
3770	12" x 16"		13	1.231		129	71.50		200.50	252
3780	16" x 16"		12	1.333		162	77.50		239.50	299
3800	1/8 bend, 2"		20	.800		28.50	46.50		75	103
3810	3"		17	.941		41	54.50		95.50	130
3820	4"		15	1.067		57	62		119	159
3830	5"	Q-2	21	1.143		84	68.50		152.50	199
3840	6"	"	19	1.263		95.50	76		171.50	222
3850	8"	Q-3	12	2.667		261	164		425	540
3860	10"		11	2.909		390	178		568	705
3870	12"		10	3.200		670	196		866	1,050

22 13 16 – Sanitary Waste and Vent Piping

22 13 16.30	Pipe Fittings, Cast Iron	Crew	Daily Output	Labor-Hours	Unit	Material	Labor	Equipment	Total	Total Incl O&P
						2020 Bare Costs				
3900	Sanitary tee, 2"	Q-1	12	1.333	Ea.	56	77.50		133.50	182
3910	3"		10	1.600		83	93		176	235
3920	4"		9	1.778		103	103		206	273
3930	5"	Q-2	13	1.846		188	111		299	380
3940	6"	"	11	2.182		208	131		339	430
3950	8"	Q-3	8.50	3.765		520	231		751	930
3980	Tee, 2"	Q-1	12	1.333		70.50	77.50		148	198
3990	3"		10	1.600		101	93		194	254
4000	4"		9	1.778		126	103		229	298
4010	5"	Q-2	13	1.846		246	111		357	445
4020	6"	"	11	2.182		248	131		379	475
4030	8"	Q-3	8	4		505	245		750	935
4060	Combination Y and 1/8 bend									
4070	2"	Q-1	12	1.333	Ea.	64.50	77.50		142	191
4080	3"		10	1.600		92.50	93		185.50	245
4090	4"		9	1.778		124	103		227	295
4100	5"	Q-2	13	1.846		222	111		333	415
4110	6"	"	11	2.182		267	131		398	495
4120	8"	Q-3	8	4		640	245		885	1,075
4121	12"	"	7	4.571		1,325	280		1,605	1,875
4160	Double Y, 2"	Q-1	10	1.600		107	93		200	261
4170	3"		8	2		135	116		251	330
4180	4"		7	2.286		175	133		308	400
4190	5"	Q-2	10	2.400		300	144		444	555
4200	6"	"	9	2.667		405	160		565	695
4210	8"	Q-3	6	5.333		950	325		1,275	1,550
4220	10"		5	6.400		2,025	395		2,420	2,825
4230	12"		4.50	7.111		2,375	435		2,810	3,275
4260	Reducer, 3" x 2"	Q-1	17	.941		50	54.50		104.50	140
4270	4" x 2"		16.50	.970		56.50	56		112.50	150
4280	4" x 3"		16	1		64	58		122	160
4290	5" x 2"		16	1		103	58		161	204
4300	5" x 3"		15.50	1.032		111	60		171	215
4310	5" x 4"		15	1.067		87	62		149	192
4320	6" x 2"		15.50	1.032		102	60		162	205
4330	6" x 3"		15	1.067		107	62		169	214
4336	6" x 4"		14	1.143		110	66.50		176.50	223
4340	6" x 5"		13	1.231		125	71.50		196.50	248
4360	8" x 3"	Q-2	15	1.600		199	96		295	370
4370	8" x 4"		15	1.600		187	96		283	355
4380	8" x 5"		14	1.714		204	103		307	385
4390	8" x 6"		14	1.714		199	103		302	380
4430	Increaser, 2" x 3"	Q-1	17	.941		70	54.50		124.50	162
4440	2" x 4"		16	1		74	58		132	171
4450	2" x 5"		15	1.067		92.50	62		154.50	198
4460	3" x 4"		15	1.067		79	62		141	183
4470	3" x 5"		15	1.067		92.50	62		154.50	198
4480	3" x 6"		14	1.143		111	66.50		177.50	225
4490	4" x 5"		15	1.067		101	62		163	207
4500	4" x 6"		14	1.143		112	66.50		178.50	226
4510	4" x 8"	Q-2	15	1.600		237	96		333	410
4520	5" x 6"	Q-1	13	1.231		149	71.50		220.50	273
4530	5" x 8"	Q-2	14	1.714		250	103		353	435

22 13 16 – Sanitary Waste and Vent Piping

22 13 16.30 Pipe Fittings, Cast Iron		Crew	Daily Output	Labor-Hours	Unit	Material	2020 Bare Costs Labor	Equipment	Total	Total Incl O&P
4540	6" x 8"	Q-2	14	1.714	Ea.	263	103		366	450
4550	6" x 10"		10	2.400		465	144		609	735
4560	8" x 10"		8.50	2.824		475	170		645	790
4570	10" x 12"		7.50	3.200		785	192		977	1,175
4600	Y, 2"	Q-1	12	1.333		53.50	77.50		131	179
4610	3"		10	1.600		85.50	93		178.50	237
4620	4"		9	1.778		110	103		213	280
4630	5"	Q-2	13	1.846		188	111		299	380
4640	6"	"	11	2.182		232	131		363	460
4650	8"	Q-3	8	4		550	245		795	985
4660	10"		7	4.571		900	280		1,180	1,425
4670	12"		6	5.333		1,850	325		2,175	2,525
4672	15"		5	6.400		3,700	395		4,095	4,675
4900	For extra heavy, add					44%	4%			
4940	Gasket and making push-on joint									
4950	2"	Q-1	40	.400	Ea.	11.75	23		34.75	49
4960	3"		35	.457		15.15	26.50		41.65	57.50
4970	4"		32	.500		19.05	29		48.05	66
4980	5"	Q-2	43	.558		30	33.50		63.50	85
4990	6"	"	40	.600		31	36		67	90
5000	8"	Q-3	32	1		67.50	61.50		129	170
5010	10"		29	1.103		114	67.50		181.50	231
5020	12"		25	1.280		146	78.50		224.50	282
5022	15"		21	1.524		175	93.50		268.50	335
5030	Note: gaskets and joint labor have									
5040	been included with all listed fittings.									
5990	No hub									
6000	Cplg. & labor required at joints not incl. in fitting									
6010	price. Add 1 coupling per joint for installed price									
6020	1/4 bend, 1-1/2"				Ea.	11.65			11.65	12.80
6060	2"					12.60			12.60	13.85
6080	3"					17.65			17.65	19.40
6120	4"					26			26	28.50
6140	5"					65			65	71.50
6160	6"					63			63	69.50
6180	8"					178			178	195
6184	1/4 bend, long sweep, 1-1/2"					29.50			29.50	32.50
6186	2"					28			28	30.50
6188	3"					33.50			33.50	37
6189	4"					53			53	58.50
6190	5"					103			103	114
6191	6"					118			118	129
6192	8"					310			310	345
6193	10"					655			655	725
6200	1/8 bend, 1-1/2"					9.70			9.70	10.70
6210	2"					10.90			10.90	12
6212	3"					14.55			14.55	16
6214	4"					19.15			19.15	21
6216	5"					40.50			40.50	44.50
6218	6"					42.50			42.50	46.50
6220	8"					122			122	134
6222	10"					233			233	256
6380	Sanitary tee, tapped, 1-1/2"					23.50			23.50	26

22 13 16.30 Pipe Fittings, Cast Iron		Crew	Daily Output	Labor-Hours	Unit	Material	2020 Bare Costs Labor	Equipment	Total	Total Incl O&P
6382	2" x 1-1/2"				Ea.	21			21	23
6384	2"					22			22	24
6386	3" x 2"					33.50			33.50	36.50
6388	3"					56			56	62
6390	4" x 1-1/2"					30			30	32.50
6392	4" x 2"					34			34	37
6393	4"					34			34	37
6394	6" x 1-1/2"					76.50			76.50	84
6396	6" x 2"					77.50			77.50	85
6459	Sanitary tee, 1-1/2"					16.25			16.25	17.85
6460	2"					17.40			17.40	19.15
6470	3"					21.50			21.50	23.50
6472	4"					41			41	45
6474	5"					97			97	107
6476	6"					97			97	106
6478	8"					395			395	435
6730	Y, 1-1/2"					16.45			16.45	18.10
6740	2"					16.10			16.10	17.70
6750	3"					23.50			23.50	26
6760	4"					37.50			37.50	41.50
6762	5"					90.50			90.50	99.50
6764	6"					100			100	110
6768	8"					239			239	263
6769	10"					540			540	595
6770	12"					1,025			1,025	1,150
6771	15"					2,325			2,325	2,550
6791	Y, reducing, 3" x 2"					17.40			17.40	19.15
6792	4" x 2"					25			25	27.50
6793	5" x 2"					56			56	61.50
6794	6" x 2"					62.50			62.50	68.50
6795	6" x 4"					79.50			79.50	87.50
6796	8" x 4"					139			139	153
6797	8" x 6"					168			168	185
6798	10" x 6"					385			385	425
6799	10" x 8"					455			455	500
6800	Double Y, 2"					26			26	28.50
6920	3"					48			48	52.50
7000	4"					96			96	106
7100	6"					173			173	190
7120	8"					490			490	540
7200	Combination Y and 1/8 bend									
7220	1-1/2"				Ea.	17.80			17.80	19.60
7260	2"					18.40			18.40	20
7320	3"					29.50			29.50	32.50
7400	4"					56.50			56.50	62.50
7480	5"					116			116	127
7500	6"					153			153	169
7520	8"					360			360	395
7800	Reducer, 3" x 2"					9.05			9.05	9.95
7820	4" x 2"					13.70			13.70	15.05
7840	4" x 3"					13.70			13.70	15.05
7842	6" x 3"					37			37	41
7844	6" x 4"					37.50			37.50	41

22 13 16 – Sanitary Waste and Vent Piping

22 13 16.30 Pipe Fittings, Cast Iron		Crew	Daily Output	Labor-Hours	Unit	Material	2020 Bare Costs Labor	Equipment	Total	Total Incl O&P
7846	6" x 5"				Ea.	38			38	41.50
7848	8" x 2"					58.50			58.50	64.50
7850	8" x 3"					54.50			54.50	60
7852	8" x 4"					57			57	63
7854	8" x 5"					64.50			64.50	71
7856	8" x 6"					63.50			63.50	70
7858	10" x 4"					112			112	123
7860	10" x 6"					119			119	131
7862	10" x 8"					139			139	153
7864	12" x 4"					235			235	259
7866	12" x 6"					250			250	275
7868	12" x 8"					257			257	283
7870	12" x 10"					262			262	289
7872	15" x 4"					490			490	540
7874	15" x 6"					465			465	510
7876	15" x 8"					550			550	605
7878	15" x 10"					550			550	605
7880	15" x 12"				↓	555			555	610
8000	Coupling, standard (by CISPI Mfrs.)									
8020	1-1/2"	Q-1	48	.333	Ea.	16.70	19.35		36.05	48.50
8040	2"		44	.364		17.80	21		38.80	52
8080	3"		38	.421		20	24.50		44.50	60
8120	4"	↓	33	.485		23	28		51	69
8160	5"	Q-2	44	.545		58	33		91	114
8180	6"	"	40	.600		60	36		96	122
8200	8"	Q-3	33	.970		111	59.50		170.50	214
8220	10"	"	26	1.231	↓	149	75.50		224.50	281
8300	Coupling, cast iron clamp & neoprene gasket (by MG)									
8310	1-1/2"	Q-1	48	.333	Ea.	9.30	19.35		28.65	40.50
8320	2"		44	.364		12.90	21		33.90	46.50
8330	3"		38	.421		11.55	24.50		36.05	50.50
8340	4"	↓	33	.485		18.85	28		46.85	64
8350	5"	Q-2	44	.545		33	33		66	86.50
8360	6"	"	40	.600		36	36		72	95.50
8380	8"	Q-3	33	.970		110	59.50		169.50	213
8400	10"	"	26	1.231	↓	181	75.50		256.50	315
8600	Coupling, stainless steel, heavy duty									
8620	1-1/2"	Q-1	48	.333	Ea.	5.90	19.35		25.25	36.50
8630	2"		44	.364		6.40	21		27.40	39.50
8640	2" x 1-1/2"		44	.364		11.45	21		32.45	45
8650	3"		38	.421		6.70	24.50		31.20	45.50
8660	4"		33	.485		7.55	28		35.55	52
8670	4" x 3"	↓	33	.485		18.85	28		46.85	64
8680	5"	Q-2	44	.545		16.20	33		49.20	68.50
8690	6"	"	40	.600		18.15	36		54.15	76
8700	8"	Q-3	33	.970		30.50	59.50		90	126
8710	10"		26	1.231		39	75.50		114.50	160
8712	12"		22	1.455		69.50	89		158.50	215
8715	15"	↓	18	1.778	↓	111	109		220	291
9000	Minimum labor/equipment charge	1 Plum	4	2	Job		129		129	199

22 13 Facility Sanitary Sewerage

22 13 16 – Sanitary Waste and Vent Piping

22 13 16.40 Pipe Fittings, Cast Iron for Drainage		Crew	Daily Output	Labor-Hours	Unit	Material	2020 Bare Costs Labor	Equipment	Total	Total Incl O&P
0010	**PIPE FITTINGS, CAST IRON FOR DRAINAGE**, Special									
1000	Drip pan elbow (safety valve discharge elbow)									
1010	Cast iron, threaded inlet									
1014	2-1/2"	Q-1	8	2	Ea.	1,600	116		1,716	1,950
1015	3"		6.40	2.500		1,725	145		1,870	2,125
1017	4"		4.80	3.333		2,325	193		2,518	2,850
1018	Cast iron, flanged inlet									
1019	6"	Q-2	3.60	6.667	Ea.	2,525	400		2,925	3,400
1020	8"	"	2.60	9.231	"	2,675	555		3,230	3,800

22 13 16.50 Shower Drains

Line	Description	Crew	Daily Output	Labor-Hours	Unit	Material	Labor	Equipment	Total	Total Incl O&P
0010	**SHOWER DRAINS**									
2780	Shower, with strainer, uniform diam. trap, bronze top									
2800	2" and 3" pipe size	Q-1	8	2	Ea.	360	116		476	575
2820	4" pipe size	"	7	2.286		415	133		548	660
2840	For galvanized body, add					239			239	263
2860	With strainer, backwater valve, drum trap									
2880	1-1/2", 2" & 3" pipe size	Q-1	8	2	Ea.	405	116		521	625
2890	4" pipe size	"	7	2.286		565	133		698	825
2900	For galvanized body, add					189			189	208

22 13 16.60 Traps

Line	Description	Crew	Daily Output	Labor-Hours	Unit	Material	Labor	Equipment	Total	Total Incl O&P
0010	**TRAPS**									
0030	Cast iron, service weight									
0050	Running P trap, without vent									
1100	2"	Q-1	16	1	Ea.	167	58		225	274
1140	3"		14	1.143		167	66.50		233.50	286
1150	4"		13	1.231		167	71.50		238.50	294
1160	6"	Q-2	17	1.412		770	85		855	980
1180	Running trap, single hub, with vent									
2080	3" pipe size, 3" vent	Q-1	14	1.143	Ea.	133	66.50		199.50	248
2120	4" pipe size, 4" vent	"	13	1.231		187	71.50		258.50	315
2140	5" pipe size, 4" vent	Q-2	11	2.182		299	131		430	535
2160	6" pipe size, 4" vent		10	2.400		805	144		949	1,100
2180	6" pipe size, 6" vent		8	3		865	180		1,045	1,225
2200	8" pipe size, 4" vent	Q-3	10	3.200		3,500	196		3,696	4,150
2220	8" pipe size, 6" vent	"	8	4		2,725	245		2,970	3,375
2300	For double hub, vent, add					10%	20%			
2800	S trap,									
2850	4" pipe size	Q-1	13	1.231	Ea.	89	71.50		160.50	208
3000	P trap, B&S, 2" pipe size		16	1		41.50	58		99.50	135
3040	3" pipe size		14	1.143		61.50	66.50		128	170
3060	4" pipe size		13	1.231		88.50	71.50		160	208
3080	5" pipe size	Q-2	18	1.333		197	80		277	340
3100	6" pipe size	"	17	1.412		274	85		359	430
3120	8" pipe size	Q-3	11	2.909		825	178		1,003	1,175
3130	10" pipe size	"	10	3.200		1,625	196		1,821	2,100
3150	P trap, no hub, 1-1/2" pipe size	Q-1	17	.941		21.50	54.50		76	108
3160	2" pipe size		16	1		20	58		78	112
3170	3" pipe size		14	1.143		39.50	66.50		106	146
3180	4" pipe size		13	1.231		82	71.50		153.50	200
3190	6" pipe size	Q-2	17	1.412		199	85		284	350
3350	Deep seal trap, B&S									
3400	1-1/4" pipe size	Q-1	14	1.143	Ea.	69	66.50		135.50	178

22 13 16 – Sanitary Waste and Vent Piping

22 13 16.60 Traps		Crew	Daily Output	Labor-Hours	Unit	Material	Labor	Equipment	Total	Total Incl O&P
						2020 Bare Costs				
3410	1-1/2" pipe size	Q-1	14	1.143	Ea.	69	66.50		135.50	178
3420	2" pipe size		14	1.143		59.50	66.50		126	167
3440	3" pipe size		12	1.333		76	77.50		153.50	204
3460	4" pipe size		11	1.455		120	84.50		204.50	262
3500	For trap primer connection, add					170			170	187
3540	For trap with floor cleanout, add					70%	5%			
3580	For trap with adjustable cleanout, add	Q-1	10	1.600	Ea.	197	93		290	360
4700	Copper, drainage, drum trap									
4800	3" x 5" solid, 1-1/2" pipe size	1 Plum	16	.500	Ea.	156	32		188	222
4840	3" x 6" swivel, 1-1/2" pipe size	"	16	.500	"	315	32		347	395
5100	P trap, standard pattern									
5200	1-1/4" pipe size	1 Plum	18	.444	Ea.	102	28.50		130.50	158
5240	1-1/2" pipe size		17	.471		111	30.50		141.50	170
5260	2" pipe size		15	.533		184	34.50		218.50	255
5280	3" pipe size		11	.727		525	47		572	655
5340	With cleanout, swivel joint and slip joint									
5360	1-1/4" pipe size	1 Plum	18	.444	Ea.	124	28.50		152.50	181
5400	1-1/2" pipe size		17	.471		199	30.50		229.50	266
5420	2" pipe size		15	.533		176	34.50		210.50	247
5750	Chromed brass, tubular, P trap, without cleanout, 20 ga.									
5800	1-1/4" pipe size	1 Plum	18	.444	Ea.	15.45	28.50		43.95	61.50
5840	1-1/2" pipe size	"	17	.471	"	17.60	30.50		48.10	66.50
5900	With cleanout, 20 ga.									
5940	1-1/4" pipe size	1 Plum	18	.444	Ea.	22.50	28.50		51	69.50
6000	1-1/2" pipe size	"	17	.471	"	27	30.50		57.50	76.50
6350	S trap, without cleanout, 20 ga.									
6400	1-1/4" pipe size	1 Plum	18	.444	Ea.	70	28.50		98.50	122
6440	1-1/2" pipe size	"	17	.471	"	43	30.50		73.50	94
6550	With cleanout, 20 ga.									
6600	1-1/4" pipe size	1 Plum	18	.444	Ea.	60.50	28.50		89	111
6640	1-1/2" pipe size	"	17	.471		48	30.50		78.50	99.50
6660	Corrosion resistant, glass, P trap, 1-1/2" pipe size	Q-1	17	.941		82.50	54.50		137	176
6670	2" pipe size		16	1		108	58		166	209
6680	3" pipe size		14	1.143		222	66.50		288.50	345
6690	4" pipe size		13	1.231		330	71.50		401.50	475
6700	6" pipe size	Q-2	17	1.412		1,275	85		1,360	1,525
6710	ABS DWV P trap, solvent weld joint									
6720	1-1/2" pipe size	1 Plum	18	.444	Ea.	10.50	28.50		39	56
6722	2" pipe size		17	.471		14.05	30.50		44.55	62.50
6724	3" pipe size		15	.533		55.50	34.50		90	114
6726	4" pipe size		14	.571		114	37		151	183
6732	PVC DWV P trap, solvent weld joint									
6733	1-1/2" pipe size	1 Plum	18	.444	Ea.	8.95	28.50		37.45	54.50
6734	2" pipe size		17	.471		11.15	30.50		41.65	59.50
6735	3" pipe size		15	.533		38	34.50		72.50	94.50
6736	4" pipe size		14	.571		85	37		122	151
6760	PP DWV, dilution trap, 1-1/2" pipe size		16	.500		274	32		306	350
6770	P trap, 1-1/2" pipe size		17	.471		68	30.50		98.50	122
6780	2" pipe size		16	.500		101	32		133	161
6790	3" pipe size		14	.571		186	37		223	261
6800	4" pipe size		13	.615		310	39.50		349.50	405
6830	S trap, 1-1/2" pipe size		16	.500		55.50	32		87.50	111
6840	2" pipe size		15	.533		83	34.50		117.50	145

22 13 16.60 Traps		Crew	Daily Output	Labor-Hours	Unit	Material	2020 Bare Costs Labor	Equipment	Total	Total Incl O&P
6850	Universal trap, 1-1/2" pipe size	1 Plum	14	.571	Ea.	112	37		149	180
6860	PVC DWV hub x hub, basin trap, 1-1/4" pipe size		18	.444		54	28.50		82.50	104
6870	Sink P trap, 1-1/2" pipe size		18	.444		14.45	28.50		42.95	60.50
6880	Tubular S trap, 1-1/2" pipe size		17	.471		26.50	30.50		57	76
6890	PVC sch. 40 DWV, drum trap									
6900	1-1/2" pipe size	1 Plum	16	.500	Ea.	37	32		69	90.50
6910	P trap, 1-1/2" pipe size		18	.444		8.80	28.50		37.30	54
6920	2" pipe size		17	.471		11.85	30.50		42.35	60
6930	3" pipe size		15	.533		40.50	34.50		75	98
6940	4" pipe size		14	.571		91.50	37		128.50	158
6950	P trap w/clean out, 1-1/2" pipe size		18	.444		14.90	28.50		43.40	61
6960	2" pipe size		17	.471		24.50	30.50		55	74
6970	P trap adjustable, 1-1/2" pipe size		17	.471		11.60	30.50		42.10	60
6980	P trap adj. w/union & cleanout, 1-1/2" pipe size		16	.500		47.50	32		79.50	103
7000	Trap primer, flow through type, 1/2" diameter		24	.333		47	21.50		68.50	85
7100	With sediment strainer		22	.364		51	23.50		74.50	92
7450	Trap primer distribution unit									
7500	2 openings	1 Plum	18	.444	Ea.	29.50	28.50		58	77
7540	3 openings		17	.471		32.50	30.50		63	83
7560	4 openings		16	.500		37.50	32		69.50	91
7850	Trap primer manifold									
7900	2 outlet	1 Plum	18	.444	Ea.	62.50	28.50		91	114
7940	4 outlet		16	.500		100	32		132	160
7960	6 outlet		15	.533		139	34.50		173.50	206
7980	8 outlet		13	.615		178	39.50		217.50	257
9000	Minimum labor/equipment charge		3	2.667	Job		172		172	266

22 13 16.80 Vent Flashing and Caps		Crew	Daily Output	Labor-Hours	Unit	Material	Labor	Equipment	Total	Total Incl O&P
0010	**VENT FLASHING AND CAPS**									
0120	Vent caps									
0140	Cast iron									
0180	2-1/2" to 3-5/8" pipe	1 Plum	21	.381	Ea.	52	24.50		76.50	95.50
0190	4" to 4-1/8" pipe	"	19	.421	"	75	27		102	125
0900	Vent flashing									
1000	Aluminum with lead ring									
1020	1-1/4" pipe	1 Plum	20	.400	Ea.	5.95	26		31.95	46.50
1030	1-1/2" pipe		20	.400		5.75	26		31.75	46.50
1040	2" pipe		18	.444		5.65	28.50		34.15	50.50
1050	3" pipe		17	.471		6.25	30.50		36.75	54
1060	4" pipe		16	.500		7.55	32		39.55	58.50
1350	Copper with neoprene ring									
1400	1-1/4" pipe	1 Plum	20	.400	Ea.	74	26		100	122
1430	1-1/2" pipe		20	.400		74	26		100	122
1440	2" pipe		18	.444		74	28.50		102.50	126
1450	3" pipe		17	.471		89.50	30.50		120	146
1460	4" pipe		16	.500		89.50	32		121.50	149
2000	Galvanized with neoprene ring									
2020	1-1/4" pipe	1 Plum	20	.400	Ea.	16.40	26		42.40	58
2030	1-1/2" pipe		20	.400		17.45	26		43.45	59
2040	2" pipe		18	.444		16	28.50		44.50	62
2050	3" pipe		17	.471		19.30	30.50		49.80	68
2060	4" pipe		16	.500		17.65	32		49.65	69.50
2980	Neoprene, one piece									

22 13 16 – Sanitary Waste and Vent Piping

22 13 16.80 Vent Flashing and Caps		Crew	Daily Output	Labor-Hours	Unit	Material	Labor	Equipment	Total	Total Incl O&P
						2020 Bare Costs				
3000	1-1/4" pipe	1 Plum	24	.333	Ea.	3.29	21.50		24.79	36.50
3030	1-1/2" pipe		24	.333		3.35	21.50		24.85	36.50
3040	2" pipe		23	.348		4.79	22.50		27.29	40
3050	3" pipe		21	.381		7	24.50		31.50	45.50
3060	4" pipe		20	.400		10.45	26		36.45	51.50
4000	Lead, 4#, 8" skirt, vent through roof									
4100	2" pipe	1 Plum	18	.444	Ea.	43	28.50		71.50	92
4110	3" pipe		17	.471		48.50	30.50		79	101
4120	4" pipe		16	.500		56	32		88	112
4130	6" pipe		14	.571		84	37		121	150
9000	Minimum labor/equipment charge		4	2	Job		129		129	199

22 13 19 – Sanitary Waste Piping Specialties

22 13 19.13 Sanitary Drains		Crew	Daily Output	Labor-Hours	Unit	Material	Labor	Equipment	Total	Total Incl O&P
0010	**SANITARY DRAINS**									
0400	Deck, auto park, CI, 13" top									
0440	3", 4", 5", and 6" pipe size	Q-1	8	2	Ea.	1,875	116		1,991	2,225
0480	For galvanized body, add				"	1,125			1,125	1,250
0800	Promenade, heelproof grate, CI, 14" top									
0840	2", 3", and 4" pipe size	Q-1	10	1.600	Ea.	750	93		843	970
0860	5" and 6" pipe size		9	1.778		935	103		1,038	1,175
0880	8" pipe size		8	2		1,050	116		1,166	1,325
0940	For galvanized body, add					590			590	645
0960	With polished bronze top, 2"-3"-4" diam.					1,325			1,325	1,450
1200	Promenade, heelproof grate, CI, lateral, 14" top									
1240	2", 3" and 4" pipe size	Q-1	10	1.600	Ea.	985	93		1,078	1,225
1260	5" and 6" pipe size		9	1.778		1,100	103		1,203	1,375
1280	8" pipe size		8	2		1,275	116		1,391	1,575
1340	For galvanized body, add					645			645	710
1360	For polished bronze top, add					1,000			1,000	1,100
1500	Promenade, slotted grate, CI, 11" top									
1540	2", 3", 4", 5", and 6" pipe size	Q-1	12	1.333	Ea.	555	77.50		632.50	730
1600	For galvanized body, add					335			335	365
1640	With polished bronze top					920			920	1,000
2000	Floor, medium duty, CI, deep flange, 7" diam. top									
2040	2" and 3" pipe size	Q-1	12	1.333	Ea.	227	77.50		304.50	370
2080	For galvanized body, add					140			140	154
2120	With polished bronze top					375			375	410
2160	Heavy duty, CI, 12" diam. anti-tilt grate									
2180	2", 3", 4", 5" and 6" pipe size	Q-1	10	1.600	Ea.	665	93		758	875
2220	For galvanized body, add					425			425	470
2240	With polished bronze top					1,000			1,000	1,100
2300	Extra-heavy duty, CI, 15" anti-tilt grate									
2320	4", 5", 6", and 8" pipe size	Q-1	8	2	Ea.	1,625	116		1,741	1,950
2360	For galvanized body, add					700			700	770
2380	With polished bronze top					1,975			1,975	2,175
2400	Heavy duty, with sediment bucket, CI, 12" diam. loose grate									
2420	2", 3", 4", 5", and 6" pipe size	Q-1	9	1.778	Ea.	935	103		1,038	1,175
2440	For galvanized body, add					635			635	700
2460	With polished bronze top					1,150			1,150	1,250
2500	Heavy duty, cleanout & trap w/bucket, CI, 15" top									
2540	2", 3", and 4" pipe size	Q-1	6	2.667	Ea.	8,125	155		8,280	9,200
2560	For galvanized body, add					2,425			2,425	2,675

22 13 19 – Sanitary Waste Piping Specialties

22 13 19.13 Sanitary Drains		Crew	Daily Output	Labor-Hours	Unit	Material	Labor (2020 Bare Costs)	Equipment	Total	Total Incl O&P
2580	With polished bronze top				Ea.	9,700			9,700	10,700
2600	Medium duty, with perforated SS basket, CI, body,									
2610	18" top for refuse container washing area									
2620	2" thru 6" pipe size	Q-1	4	4	Ea.	4,525	232		4,757	5,325
2630	Acid resistant									
2638	PVC									
2640	2", 3" and 4" pipe size	Q-1	16	1	Ea.	385	58		443	515
2644	Cast iron, epoxy coated									
2646	2", 3" and 4" pipe size	Q-1	14	1.143	Ea.	1,375	66.50		1,441.50	1,600
2650	PVC or ABS thermoplastic									
2660	3" and 4" pipe size	Q-1	16	1	Ea.	415	58		473	545
2680	Extra heavy duty, oil intercepting, gas seal cone,									
2690	with cleanout, loose grate, CI, body 16" top									
2700	3" and 4" diameter outlet, 4" slab depth	Q-1	4	4	Ea.	9,225	232		9,457	10,500
2720	4" diameter outlet, 8" slab depth		3	5.333		8,550	310		8,860	9,875
2740	4" diam. outlet, 10"-12" slab depth, 16" top		2	8		11,600	465		12,065	13,400
2910	Prison cell, vandal-proof, 1-1/2", and 2" diam. pipe		12	1.333		575	77.50		652.50	755
2920	3" pipe size		10	1.600		570	93		663	770
2930	Trap drain, light duty, backwater valve CI top									
2950	8" diameter top, 2" pipe size	Q-1	12	1.333	Ea.	480	77.50		557.50	645
2960	10" diameter top, 3" pipe size		10	1.600		655	93		748	870
2970	12" diameter top, 4" pipe size		8	2		920	116		1,036	1,175

22 13 19.14 Floor Receptors

		Crew	Daily Output	Labor-Hours	Unit	Material	Labor	Equipment	Total	Total Incl O&P
0010	**FLOOR RECEPTORS**, For connection to 2", 3" & 4" diameter pipe									
0200	12-1/2" square top, 25 sq. in. open area	Q-1	10	1.600	Ea.	980	93		1,073	1,225
0300	For grate with 4" diameter x 3-3/4" high funnel, add					330			330	360
0400	For grate with 6" diameter x 6" high funnel, add					264			264	290
0500	For full hinged grate with open center, add					103			103	114
0600	For aluminum bucket, add					189			189	208
0700	For acid-resisting bucket, add					325			325	355
0900	For stainless steel mesh bucket liner, add					282			282	310
1000	For bronze antisplash dome strainer, add					135			135	148
1100	For partial solid cover, add					68.50			68.50	75.50
1200	For trap primer connection, add					112			112	123
2000	12-5/8" diameter top, 40 sq. in. open area	Q-1	10	1.600		960	93		1,053	1,200
2100	For options, add same prices as square top									
3000	8" x 4" rectangular top, 7.5 sq. in. open area	Q-1	14	1.143	Ea.	1,050	66.50		1,116.50	1,250
3100	For trap primer connections, add					110			110	121
4000	24" x 16" rectangular top, 70 sq. in. open area	Q-1	4	4		5,375	232		5,607	6,250
4100	For trap primer connection, add					226			226	248
9000	Minimum labor/equipment charge	Q-1	3	5.333	Job		310		310	480

22 13 19.15 Sink Waste Treatment

		Crew	Daily Output	Labor-Hours	Unit	Material	Labor	Equipment	Total	Total Incl O&P
0010	**SINK WASTE TREATMENT**, System for commercial kitchens									
0100	includes clock timer & fittings									
0200	System less chemical, wall mounted cabinet	1 Plum	16	.500	Ea.	505	32		537	605
2000	Chemical, 1 gallon, add					41			41	45
2100	6 gallons, add					185			185	203
2200	15 gallons, add					500			500	550
2300	30 gallons, add					940			940	1,025
2400	55 gallons, add					1,600			1,600	1,750

22 13 19 – Sanitary Waste Piping Specialties

22 13 19.39 Floor Drain Trap Seal

	22 13 19.39 Floor Drain Trap Seal	Crew	Daily Output	Labor-Hours	Unit	2020 Bare Costs Material	Labor	Equipment	Total	Total Incl O&P
0010	**FLOOR DRAIN TRAP SEAL**									
0100	Inline									
0110	2″	1 Plum	20	.400	Ea.	35	26		61	78.50
0120	3″		16	.500		43	32		75	97
0130	3.5″		14	.571		42.50	37		79.50	104
0140	4″		10	.800		53	51.50		104.50	138

22 13 23 – Sanitary Waste Interceptors

22 13 23.10 Interceptors

	22 13 23.10 Interceptors	Crew	Daily Output	Labor-Hours	Unit	Material	Labor	Equipment	Total	Total Incl O&P
0010	**INTERCEPTORS**									
0150	Grease, fabricated steel, 4 GPM, 8 lb. fat capacity	1 Plum	4	2	Ea.	1,575	129		1,704	1,925
0200	7 GPM, 14 lb. fat capacity		4	2		2,050	129		2,179	2,475
1000	10 GPM, 20 lb. fat capacity		4	2		2,575	129		2,704	3,025
1040	15 GPM, 30 lb. fat capacity		4	2		3,575	129		3,704	4,150
1060	20 GPM, 40 lb. fat capacity		3	2.667		4,650	172		4,822	5,400
1080	25 GPM, 50 lb. fat capacity	Q-1	3.50	4.571		4,925	265		5,190	5,825
1100	35 GPM, 70 lb. fat capacity		3	5.333		4,900	310		5,210	5,875
1120	50 GPM, 100 lb. fat capacity		2	8		8,075	465		8,540	9,600
1140	75 GPM, 150 lb. fat capacity		2	8		14,500	465		14,965	16,600
1160	100 GPM, 200 lb. fat capacity		2	8		20,100	465		20,565	22,900
1180	150 GPM, 300 lb. fat capacity		2	8		21,700	465		22,165	24,600
1200	200 GPM, 400 lb. fat capacity		1.50	10.667		31,200	620		31,820	35,400
1220	250 GPM, 500 lb. fat capacity		1.30	12.308		34,300	715		35,015	38,800
1240	300 GPM, 600 lb. fat capacity		1	16		40,400	930		41,330	45,900
1260	400 GPM, 800 lb. fat capacity	Q-2	1.20	20		49,000	1,200		50,200	56,000
1280	500 GPM, 1,000 lb. fat capacity	″	1	24		58,500	1,450		59,950	66,500
1580	For seepage pan, add					7%				
3000	Hair, cast iron, 1-1/4″ and 1-1/2″ pipe connection	1 Plum	8	1	Ea.	560	64.50		624.50	715
3100	For chrome-plated cast iron, add					375			375	410
3200	For polished bronze, add					1,025			1,025	1,125
3400	Lint interceptor, fabricated steel									
3410	Size based on 10 GPM per machine									
3420	30 GPM, 2″ pipe size	Q-1	3	5.333	Ea.	6,500	310		6,810	7,650
3430	70 GPM, 3″ pipe size		2.50	6.400		7,575	370		7,945	8,900
3440	100 GPM, 4″ pipe size		2	8		8,925	465		9,390	10,500
3450	200 GPM, 4″ pipe size		1.50	10.667		9,950	620		10,570	11,900
3460	300 GPM, 6″ pipe size		1	16		12,500	930		13,430	15,200
3470	400 GPM, 6″ pipe size	Q-2	1.20	20		14,300	1,200		15,500	17,600
3480	500 GPM, 6″ pipe size	″	1	24		14,800	1,450		16,250	18,500
4000	Oil, fabricated steel, 10 GPM, 2″ pipe size	1 Plum	4	2		3,300	129		3,429	3,850
4100	15 GPM, 2″ or 3″ pipe size		4	2		4,525	129		4,654	5,200
4120	20 GPM, 2″ or 3″ pipe size		3	2.667		6,150	172		6,322	7,025
4140	25 GPM, 2″ or 3″ pipe size	Q-1	3.50	4.571		5,950	265		6,215	6,950
4160	35 GPM, 2″, 3″, or 4″ pipe size		3	5.333		7,250	310		7,560	8,450
4180	50 GPM, 2″, 3″, or 4″ pipe size		2	8		9,750	465		10,215	11,400
4200	75 GPM, 3″ pipe size		2	8		18,600	465		19,065	21,200
4220	100 GPM, 3″ pipe size		2	8		18,300	465		18,765	20,800
4240	150 GPM, 4″ pipe size		2	8		22,700	465		23,165	25,600
4260	200 GPM, 4″ pipe size		1.50	10.667		32,000	620		32,620	36,200
4280	250 GPM, 5″ pipe size		1.30	12.308		37,100	715		37,815	41,900
4300	300 GPM, 5″ pipe size		1	16		42,000	930		42,930	47,600
4320	400 GPM, 6″ pipe size	Q-2	1.20	20		54,000	1,200		55,200	61,500
4340	500 GPM, 6″ pipe size	″	1	24		68,000	1,450		69,450	77,000

22 13 23 – Sanitary Waste Interceptors

22 13 23.10 Interceptors		Crew	Daily Output	Labor-Hours	Unit	Material	Labor	Equipment	Total	Total Incl O&P
						2020 Bare Costs				
5000	Sand interceptor, fabricated steel									
5020	20 GPM, 4" pipe size	Q-1	3	5.333	Ea.	8,975	310		9,285	10,400
5030	50 GPM, 4" pipe size		2.50	6.400		12,500	370		12,870	14,300
5040	150 GPM, 4" pipe size		2	8		37,700	465		38,165	42,200
5050	250 GPM, 6" pipe size		1.30	12.308		39,200	715		39,915	44,200
5060	500 GPM, 6" pipe size	Q-2	1	24		50,500	1,450		51,950	57,500
6000	Solids, precious metals recovery, CI, 1-1/4" to 2" pipe	1 Plum	4	2		755	129		884	1,025
6100	Dental lab., large, CI, 1-1/2" to 2" pipe		3	2.667		2,650	172		2,822	3,175
9000	Minimum labor/equipment charge		3	2.667	Job		172		172	266

22 13 26 – Sanitary Waste Separators

22 13 26.10 Separators		Crew	Daily Output	Labor-Hours	Unit	Material	Labor	Equipment	Total	Total Incl O&P
0010	**SEPARATORS**, Entrainment eliminator, steel body, 150 PSIG									
0100	1/4" size	1 Stpi	24	.333	Ea.	276	22		298	340
0120	1/2" size		24	.333		284	22		306	350
0140	3/4" size		20	.400		284	26		310	350
0160	1" size		19	.421		310	27.50		337.50	385
0180	1-1/4" size		15	.533		325	35		360	415
0200	1-1/2" size		13	.615		360	40.50		400.50	460
0220	2" size		11	.727		400	47.50		447.50	515
0240	2-1/2" size	Q-5	15	1.067		1,725	63		1,788	2,000
0260	3" size		13	1.231		1,900	72.50		1,972.50	2,200
0280	4" size		10	1.600		2,375	94.50		2,469.50	2,775
0300	5" size		6	2.667		2,800	157		2,957	3,325
0320	6" size		3	5.333		3,050	315		3,365	3,850
0340	8" size	Q-6	4.40	5.455		3,525	335		3,860	4,400
0360	10" size	"	4	6		5,125	365		5,490	6,225
1000	For 300 PSIG, add					15%				

22 13 29 – Sanitary Sewerage Pumps

22 13 29.13 Wet-Pit-Mounted, Vertical Sewerage Pumps		Crew	Daily Output	Labor-Hours	Unit	Material	Labor	Equipment	Total	Total Incl O&P
0010	**WET-PIT-MOUNTED, VERTICAL SEWERAGE PUMPS**									
0020	Controls incl. alarm/disconnect panel w/wire. Excavation not included									
0260	Simplex, 9 GPM at 60 PSIG, 91 gal. tank				Ea.	3,525			3,525	3,900
0300	Unit with manway, 26" ID, 18" high					4,150			4,150	4,575
0340	26" ID, 36" high					4,075			4,075	4,500
0380	43" ID, 4' high					4,250			4,250	4,675
0600	Simplex, 9 GPM at 60 PSIG, 150 gal. tank, indoor					4,275			4,275	4,700
0700	Unit with manway, 26" ID, 36" high					4,975			4,975	5,475
0740	26" ID, 4' high					5,025			5,025	5,525
2000	Duplex, 18 GPM at 60 PSIG, 150 gal. tank, indoor					8,175			8,175	9,000
2060	Unit with manway, 43" ID, 4' high					8,975			8,975	9,875
2400	For core only					2,050			2,050	2,250
3000	Indoor residential type installation									
3020	Simplex, 9 GPM at 60 PSIG, 91 gal. HDPE tank				Ea.	3,625			3,625	3,975

22 13 29.14 Sewage Ejector Pumps		Crew	Daily Output	Labor-Hours	Unit	Material	Labor	Equipment	Total	Total Incl O&P
0010	**SEWAGE EJECTOR PUMPS**, With operating and level controls									
0100	Simplex system incl. tank, cover, pump 15' head									
0500	37 gal. PE tank, 12 GPM, 1/2 HP, 2" discharge	Q-1	3.20	5	Ea.	505	290		795	1,000
0510	3" discharge		3.10	5.161		550	299		849	1,075
0530	87 GPM, .7 HP, 2" discharge		3.20	5		770	290		1,060	1,300
0540	3" discharge		3.10	5.161		835	299		1,134	1,375
0600	45 gal. coated stl. tank, 12 GPM, 1/2 HP, 2" discharge		3	5.333		900	310		1,210	1,475

22 13 29.14 Sewage Ejector Pumps		Crew	Daily Output	Labor-Hours	Unit	Material	2020 Bare Costs Labor	Equipment	Total	Total Incl O&P
0610	3" discharge	Q-1	2.90	5.517	Ea.	935	320		1,255	1,525
0630	87 GPM, .7 HP, 2" discharge		3	5.333		1,150	310		1,460	1,750
0640	3" discharge		2.90	5.517		1,225	320		1,545	1,850
0660	134 GPM, 1 HP, 2" discharge		2.80	5.714		1,250	330		1,580	1,875
0680	3" discharge		2.70	5.926		1,325	345		1,670	1,975
0700	70 gal. PE tank, 12 GPM, 1/2 HP, 2" discharge		2.60	6.154		970	355		1,325	1,625
0710	3" discharge		2.40	6.667		1,050	385		1,435	1,750
0730	87 GPM, .7 HP, 2" discharge		2.50	6.400		1,275	370		1,645	2,000
0740	3" discharge		2.30	6.957		1,325	405		1,730	2,100
0760	134 GPM, 1 HP, 2" discharge		2.20	7.273		1,375	420		1,795	2,150
0770	3" discharge		2	8		1,450	465		1,915	2,325
0800	75 gal. coated stl. tank, 12 GPM, 1/2 HP, 2" discharge		2.40	6.667		1,075	385		1,460	1,775
0810	3" discharge		2.20	7.273		1,125	420		1,545	1,875
0830	87 GPM, .7 HP, 2" discharge		2.30	6.957		1,350	405		1,755	2,125
0840	3" discharge		2.10	7.619		1,425	440		1,865	2,250
0860	134 GPM, 1 HP, 2" discharge		2	8		1,450	465		1,915	2,325
0880	3" discharge		1.80	8.889		1,525	515		2,040	2,475
1040	Duplex system incl. tank, covers, pumps									
1060	110 gal. fiberglass tank, 24 GPM, 1/2 HP, 2" discharge	Q-1	1.60	10	Ea.	1,950	580		2,530	3,050
1080	3" discharge		1.40	11.429		2,050	665		2,715	3,275
1100	174 GPM, .7 HP, 2" discharge		1.50	10.667		2,525	620		3,145	3,725
1120	3" discharge		1.30	12.308		2,625	715		3,340	3,975
1140	268 GPM, 1 HP, 2" discharge		1.20	13.333		2,725	775		3,500	4,200
1160	3" discharge		1	16		2,825	930		3,755	4,525
1260	135 gal. coated stl. tank, 24 GPM, 1/2 HP, 2" discharge	Q-2	1.70	14.118		2,000	850		2,850	3,525
2000	3" discharge		1.60	15		2,125	900		3,025	3,750
2640	174 GPM, .7 HP, 2" discharge		1.60	15		2,625	900		3,525	4,275
2660	3" discharge		1.50	16		2,775	960		3,735	4,550
2700	268 GPM, 1 HP, 2" discharge		1.30	18.462		2,850	1,100		3,950	4,850
3040	3" discharge		1.10	21.818		3,025	1,300		4,325	5,350
3060	275 gal. coated stl. tank, 24 GPM, 1/2 HP, 2" discharge		1.50	16		2,500	960		3,460	4,250
3080	3" discharge		1.40	17.143		2,550	1,025		3,575	4,400
3100	174 GPM, .7 HP, 2" discharge		1.40	17.143		3,225	1,025		4,250	5,150
3120	3" discharge		1.30	18.462		3,475	1,100		4,575	5,550
3140	268 GPM, 1 HP, 2" discharge		1.10	21.818		3,550	1,300		4,850	5,925
3160	3" discharge		.90	26.667		3,750	1,600		5,350	6,575
3260	Pump system accessories, add									
3300	Alarm horn and lights, 115 V mercury switch	Q-1	8	2	Ea.	102	116		218	291
3340	Switch, mag. contactor, alarm bell, light, 3 level control		5	3.200		495	186		681	830
3380	Alternator, mercury switch activated		4	4		895	232		1,127	1,350
9000	Minimum labor/equipment charge		2.50	6.400	Job		370		370	575

22 14 23 – Storm Drainage Piping Specialties

22 14 23.33 Backwater Valves		Crew	Daily Output	Labor-Hours	Unit	Material	2020 Bare Costs Labor	Equipment	Total	Total Incl O&P
0010	**BACKWATER VALVES**, CI Body									
6980	Bronze gate and automatic flapper valves									
7000	3″ and 4″ pipe size	Q-1	13	1.231	Ea.	2,525	71.50		2,596.50	2,875
7100	5″ and 6″ pipe size	″	13	1.231	″	3,850	71.50		3,921.50	4,350
7240	Bronze flapper valve, bolted cover									
7260	2″ pipe size	Q-1	16	1	Ea.	735	58		793	900
7280	3″ pipe size		14.50	1.103		1,150	64		1,214	1,375
7300	4″ pipe size		13	1.231		1,300	71.50		1,371.50	1,525
7320	5″ pipe size	Q-2	18	1.333		1,600	80		1,680	1,875
7340	6″ pipe size	″	17	1.412		2,275	85		2,360	2,625
7360	8″ pipe size	Q-3	10	3.200		3,225	196		3,421	3,825
7380	10″ pipe size	″	9	3.556		5,250	218		5,468	6,100
7500	For threaded cover, same cost									
7540	Revolving disk type, same cost as flapper type									

22 14 26 – Facility Storm Drains

22 14 26.13 Roof Drains		Crew	Daily Output	Labor-Hours	Unit	Material	Labor	Equipment	Total	Total Incl O&P
0010	**ROOF DRAINS**									
0140	Cornice, CI, 45° or 90° outlet									
0200	3″ and 4″ pipe size	Q-1	12	1.333	Ea.	385	77.50		462.50	545
0260	For galvanized body, add					111			111	122
0280	For polished bronze dome, add					114			114	126
3860	Roof, flat metal deck, CI body, 12″ CI dome									
3880	2″ pipe size	Q-1	15	1.067	Ea.	340	62		402	470
3890	3″ pipe size		14	1.143		410	66.50		476.50	555
3900	4″ pipe size		13	1.231		510	71.50		581.50	670
3910	5″ pipe size		12	1.333		660	77.50		737.50	850
3920	6″ pipe size		10	1.600		865	93		958	1,100
4280	Integral expansion joint, CI body, 12″ CI dome									
4300	2″ pipe size	Q-1	8	2	Ea.	765	116		881	1,025
4320	3″ pipe size		7	2.286		790	133		923	1,075
4340	4″ pipe size		6	2.667		845	155		1,000	1,175
4360	5″ pipe size		4	4		970	232		1,202	1,425
4380	6″ pipe size		3	5.333		1,050	310		1,360	1,650
4400	8″ pipe size		3	5.333		1,600	310		1,910	2,250
4440	For galvanized body, add					435			435	475
4620	Main, all aluminum, 12″ low profile dome									
4640	2″, 3″ and 4″ pipe size	Q-1	14	1.143	Ea.	590	66.50		656.50	745
4660	5″ and 6″ pipe size		13	1.231		780	71.50		851.50	965
4680	8″ pipe size		10	1.600		785	93		878	1,000
4690	Main, CI body, 12″ poly. dome, 2″, 3″, & 4″ pipe		8	2		435	116		551	660
4710	5″ and 6″ pipe size		6	2.667		625	155		780	930
4720	8″ pipe size		4	4		815	232		1,047	1,250
4730	For underdeck clamp, add		22	.727		300	42		342	395
4740	For vandalproof dome, add					75.50			75.50	83
4750	For galvanized body, add					615			615	680
4760	Main, ABS body and dome, 2″ pipe size	Q-1	14	1.143		126	66.50		192.50	240
4780	3″ pipe size		14	1.143		126	66.50		192.50	241
4800	4″ pipe size		14	1.143		126	66.50		192.50	240
4820	For underdeck clamp, add		24	.667		28	38.50		66.50	90.50
4900	Terrace planting area, with perforated overflow, CI									
4920	2″, 3″ and 4″ pipe size	Q-1	8	2	Ea.	765	116		881	1,025
9000	Minimum labor/equipment charge	1 Plum	4	2	Job		129		129	199

22 14 26 – Facility Storm Drains

22 14 26.16 Facility Area Drains		Crew	Daily Output	Labor-Hours	Unit	Material	Labor	Equipment	Total	Total Incl O&P
						2020 Bare Costs				
0010	**FACILITY AREA DRAINS**									
4980	Scupper floor, oblique strainer, CI									
5000	6" x 7" top, 2", 3" and 4" pipe size	Q-1	16	1	Ea.	405	58		463	535
5100	8" x 12" top, 5" and 6" pipe size	"	14	1.143		790	66.50		856.50	965
5160	For galvanized body, add					40%				
5200	For polished bronze strainer, add					85%				

22 14 26.19 Facility Trench Drains

Line	Description	Crew	Daily Output	Labor-Hours	Unit	Material	Labor	Equipment	Total	Total Incl O&P
0010	**FACILITY TRENCH DRAINS**									
5980	Trench, floor, heavy duty, modular, CI, 12" x 12" top									
6000	2", 3", 4", 5" & 6" pipe size	Q-1	8	2	Ea.	1,100	116		1,216	1,400
6100	For unit with polished bronze top		8	2		1,625	116		1,741	1,975
6200	For 12" extension section, CI top		8	2		1,100	116		1,216	1,400
6240	For 12" extension section, polished bronze top		8	2		1,825	116		1,941	2,175
6600	Trench, floor, for cement concrete encasement									
6610	Not including trenching or concrete									
6640	Polyester polymer concrete									
6650	4" internal width, with grate									
6660	Light duty steel grate	Q-1	120	.133	L.F.	68.50	7.75		76.25	87
6670	Medium duty steel grate		115	.139		129	8.05		137.05	155
6680	Heavy duty iron grate		110	.145		122	8.45		130.45	147
6700	12" internal width, with grate									
6770	Heavy duty galvanized grate	Q-1	80	.200	L.F.	171	11.60		182.60	206
6800	Fiberglass									
6810	8" internal width, with grate									
6820	Medium duty galvanized grate	Q-1	115	.139	L.F.	122	8.05		130.05	147
6830	Heavy duty iron grate	"	110	.145	"	201	8.45		209.45	234

22 14 29 – Sump Pumps

22 14 29.13 Wet-Pit-Mounted, Vertical Sump Pumps

Line	Description	Crew	Daily Output	Labor-Hours	Unit	Material	Labor	Equipment	Total	Total Incl O&P
0010	**WET-PIT-MOUNTED, VERTICAL SUMP PUMPS**									
0400	Molded PVC base, 21 GPM at 15' head, 1/3 HP	1 Plum	5	1.600	Ea.	141	103		244	315
0800	Iron base, 21 GPM at 15' head, 1/3 HP		5	1.600		155	103		258	330
1200	Solid brass, 21 GPM at 15' head, 1/3 HP		5	1.600		243	103		346	425
2000	Sump pump, single stage									
2010	25 GPM, 1 HP, 1-1/2" discharge	Q-1	1.80	8.889	Ea.	3,825	515		4,340	5,025
2020	75 GPM, 1-1/2 HP, 2" discharge		1.50	10.667		4,050	620		4,670	5,400
2030	100 GPM, 2 HP, 2-1/2" discharge		1.30	12.308		4,125	715		4,840	5,650
2040	150 GPM, 3 HP, 3" discharge		1.10	14.545		4,125	845		4,970	5,850
2050	200 GPM, 3 HP, 3" discharge		1	16		4,375	930		5,305	6,250
2060	300 GPM, 10 HP, 4" discharge	Q-2	1.20	20		4,725	1,200		5,925	7,050
2070	500 GPM, 15 HP, 5" discharge		1.10	21.818		5,375	1,300		6,675	7,925
2080	800 GPM, 20 HP, 6" discharge		1	24		6,350	1,450		7,800	9,200
2090	1,000 GPM, 30 HP, 6" discharge		.85	28.235		6,975	1,700		8,675	10,300
2100	1,600 GPM, 50 HP, 8" discharge		.72	33.333		10,900	2,000		12,900	15,100
2110	2,000 GPM, 60 HP, 8" discharge	Q-3	.85	37.647		11,100	2,300		13,400	15,800
2202	For general purpose float switch, copper coated float, add	Q-1	5	3.200		107	186		293	405

22 14 29.16 Submersible Sump Pumps

Line	Description	Crew	Daily Output	Labor-Hours	Unit	Material	Labor	Equipment	Total	Total Incl O&P
0010	**SUBMERSIBLE SUMP PUMPS**									
1000	Elevator sump pumps, automatic									
1010	Complete systems, pump, oil detector, controls and alarm									
1020	1-1/2" discharge, does not include the sump pit/tank									
1040	1/3 HP, 115 V	1 Plum	4.40	1.818	Ea.	1,925	117		2,042	2,275

22 14 Facility Storm Drainage

22 14 29 – Sump Pumps

22 14 29.16 Submersible Sump Pumps		Crew	Daily Output	Labor-Hours	Unit	2020 Bare Costs Material	Labor	Equipment	Total	Total Incl O&P
1050	1/2 HP, 115 V	1 Plum	4	2	Ea.	2,025	129		2,154	2,425
1060	1/2 HP, 230 V		4	2		2,000	129		2,129	2,400
1070	3/4 HP, 115 V		3.60	2.222		2,100	143		2,243	2,550
1080	3/4 HP, 230 V		3.60	2.222		2,075	143		2,218	2,500
1100	Sump pump only									
1110	1/3 HP, 115 V	1 Plum	6.40	1.250	Ea.	179	80.50		259.50	320
1120	1/2 HP, 115 V		5.80	1.379		254	89		343	415
1130	1/2 HP, 230 V		5.80	1.379		294	89		383	460
1140	3/4 HP, 115 V		5.40	1.481		340	95.50		435.50	525
1150	3/4 HP, 230 V		5.40	1.481		380	95.50		475.50	570
1200	Oil detector, control and alarm only									
1210	115 V	1 Plum	8	1	Ea.	1,775	64.50		1,839.50	2,050
1220	230 V	"	8	1	"	1,775	64.50		1,839.50	2,050
7000	Sump pump, automatic									
7100	Plastic, 1-1/4" discharge, 1/4 HP	1 Plum	6.40	1.250	Ea.	168	80.50		248.50	310
7140	1/3 HP		6	1.333		224	86		310	380
7160	1/2 HP		5.40	1.481		245	95.50		340.50	420
7180	1-1/2" discharge, 1/2 HP		5.20	1.538		325	99		424	510
7500	Cast iron, 1-1/4" discharge, 1/4 HP		6	1.333		225	86		311	380
7540	1/3 HP		6	1.333		264	86		350	425
7560	1/2 HP		5	1.600		320	103		423	510
9000	Minimum labor/equipment charge		4	2	Job		129		129	199

22 14 53 – Rainwater Storage Tanks

22 14 53.13 Fiberglass, Rainwater Storage Tank		Crew	Daily Output	Labor-Hours	Unit	Material	Labor	Equipment	Total	Total Incl O&P
0010	**FIBERGLASS, RAINWATER STORAGE TANK**									
2000	600 gallon	B-21B	3.75	10.667	Ea.	4,125	490	126	4,741	5,475
2010	1,000 gallon		3.50	11.429		5,275	525	135	5,935	6,775
2020	2,000 gallon		3.25	12.308		7,500	565	145	8,210	9,300
2030	4,000 gallon		3	13.333		10,100	610	157	10,867	12,200
2040	6,000 gallon		2.65	15.094		11,400	695	178	12,273	13,900
2050	8,000 gallon		2.30	17.391		13,900	800	205	14,905	16,800
2060	10,000 gallon		2	20		15,400	920	236	16,556	18,700
2070	12,000 gallon		1.50	26.667		21,600	1,225	315	23,140	26,000
2080	15,000 gallon		1	40		24,500	1,825	470	26,795	30,400
2090	20,000 gallon		.75	53.333		31,800	2,450	630	34,880	39,600
2100	25,000 gallon		.50	80		47,400	3,675	940	52,015	59,000
2110	30,000 gallon		.35	114		57,000	5,250	1,350	63,600	72,500
2120	40,000 gallon		.30	133		80,000	6,125	1,575	87,700	99,500

22 15 General Service Compressed-Air Systems

22 15 13 – General Service Compressed-Air Piping

22 15 13.10 Compressor Accessories		Crew	Daily Output	Labor-Hours	Unit	Material	Labor	Equipment	Total	Total Incl O&P
0010	**COMPRESSOR ACCESSORIES**									
1700	Refrigerated air dryers with ambient air filters									
1710	10 CFM	Q-5	8	2	Ea.	785	118		903	1,050
1715	15 CFM		7.30	2.192		1,075	129		1,204	1,375
1716	20 CFM		6.90	2.319		2,150	137		2,287	2,575
1720	25 CFM		6.60	2.424		1,575	143		1,718	1,950
1730	50 CFM		6.20	2.581		1,875	152		2,027	2,300
1740	75 CFM		5.80	2.759		1,925	163		2,088	2,350

22 15 13 – General Service Compressed-Air Piping

22 15 13.10 Compressor Accessories		Crew	Daily Output	Labor-Hours	Unit	Material	Labor	Equipment	Total	Total Incl O&P
						2020 Bare Costs				
1750	100 CFM	Q-5	5.60	2.857	Ea.	3,100	169		3,269	3,650
1751	125 CFM		5.40	2.963		3,300	175		3,475	3,900
1752	175 CFM		5	3.200		3,575	189		3,764	4,225
1753	200 CFM		4.80	3.333		6,000	197		6,197	6,900
1754	300 CFM		4.40	3.636		6,900	215		7,115	7,925
1755	400 CFM		4	4		8,825	236		9,061	10,100
1756	500 CFM		3.80	4.211		9,200	248		9,448	10,500
1757	600 CFM		3.40	4.706		11,700	278		11,978	13,300
1758	800 CFM		3	5.333		19,600	315		19,915	22,000
1759	1,000 CFM		2.60	6.154		20,800	365		21,165	23,400
3460	Air filter, regulator, lubricator combination									
3470	Flush mount									
3480	Adjustable range 0-140 psi									
3500	1/8" NPT, 34 SCFM	1 Stpi	17	.471	Ea.	107	31		138	165
3510	1/4" NPT, 61 SCFM		17	.471		192	31		223	259
3520	3/8" NPT, 85 SCFM		16	.500		196	33		229	267
3530	1/2" NPT, 150 SCFM		15	.533		210	35		245	285
3540	3/4" NPT, 171 SCFM		14	.571		260	37.50		297.50	345
3550	1" NPT, 150 SCFM		13	.615		415	40.50		455.50	520
4000	Couplers, air line, sleeve type									
4010	Female, connection size NPT									
4020	1/4"	1 Stpi	38	.211	Ea.	7.80	13.80		21.60	30
4030	3/8"		36	.222		12.60	14.55		27.15	36.50
4040	1/2"		35	.229		20.50	15		35.50	45.50
4050	3/4"		34	.235		22.50	15.40		37.90	49
4100	Male									
4110	1/4"	1 Stpi	38	.211	Ea.	7.95	13.80		21.75	30.50
4120	3/8"		36	.222		11.25	14.55		25.80	35
4130	1/2"		35	.229		19.90	15		34.90	45
4140	3/4"		34	.235		22.50	15.40		37.90	48.50
4150	Coupler, combined male and female halves									
4160	1/2"	1 Stpi	17	.471	Ea.	40.50	31		71.50	92
4170	3/4"	"	15	.533	"	45	35		80	104

22 15 19 – General Service Packaged Air Compressors and Receivers

22 15 19.10 Air Compressors		Crew	Daily Output	Labor-Hours	Unit	Material	Labor	Equipment	Total	Total Incl O&P
0010	**AIR COMPRESSORS**									
5250	Air, reciprocating air cooled, splash lubricated, tank mounted									
5300	Single stage, 1 phase, 140 psi									
5303	1/2 HP, 17 gal. tank	1 Stpi	3	2.667	Ea.	2,000	175		2,175	2,475
5305	3/4 HP, 30 gal. tank		2.60	3.077		1,700	202		1,902	2,175
5307	1 HP, 30 gal. tank		2.20	3.636		2,075	238		2,313	2,675
5309	2 HP, 30 gal. tank	Q-5	4	4		2,575	236		2,811	3,225
5310	3 HP, 30 gal. tank		3.60	4.444		2,700	262		2,962	3,375
5314	3 HP, 60 gal. tank		3.50	4.571		3,175	270		3,445	3,900
5320	5 HP, 60 gal. tank		3.20	5		3,375	295		3,670	4,175
5330	5 HP, 80 gal. tank		3	5.333		3,325	315		3,640	4,125
5340	7.5 HP, 80 gal. tank		2.60	6.154		4,550	365		4,915	5,550
5600	2 stage pkg., 3 phase									
5650	6 CFM at 125 psi, 1-1/2 HP, 60 gal. tank	Q-5	3	5.333	Ea.	2,575	315		2,890	3,325
5670	10.9 CFM at 125 psi, 3 HP, 80 gal. tank		1.50	10.667		3,300	630		3,930	4,625
5680	38.7 CFM at 125 psi, 10 HP, 120 gal. tank		.60	26.667		6,275	1,575		7,850	9,325
5690	105 CFM at 125 psi, 25 HP, 250 gal. tank	Q-6	.60	40		16,300	2,450		18,750	21,700

22 15 General Service Compressed-Air Systems

22 15 19 – General Service Packaged Air Compressors and Receivers

22 15 19.10 Air Compressors		Crew	Daily Output	Labor-Hours	Unit	2020 Bare Costs Material	Labor	Equipment	Total	Total Incl O&P
5800	With single stage pump									
5850	8.3 CFM at 125 psi, 2 HP, 80 gal. tank	Q-6	3.50	6.857	Ea.	3,175	420		3,595	4,150
5860	38.7 CFM at 125 psi, 10 HP, 120 gal. tank	"	.90	26.667	"	5,800	1,625		7,425	8,900
6000	Reciprocating, 2 stage, tank mtd, 3 ph, cap rated @ 175 PSIG									
6050	Pressure lubricated, hvy. duty, 9.7 CFM, 3 HP, 120 gal. tank	Q-5	1.30	12.308	Ea.	4,975	725		5,700	6,600
6054	5 CFM, 1-1/2 HP, 80 gal. tank		2.80	5.714		3,100	335		3,435	3,925
6056	6.4 CFM, 2 HP, 80 gal. tank		2	8		3,175	470		3,645	4,225
6058	8.1 CFM, 3 HP, 80 gal. tank		1.70	9.412		3,400	555		3,955	4,575
6059	14.8 CFM, 5 HP, 80 gal. tank		1	16		3,425	945		4,370	5,200
6060	16.5 CFM, 5 HP, 120 gal. tank		1	16		5,625	945		6,570	7,625
6063	13 CFM, 6 HP, 80 gal. tank		.90	17.778		5,225	1,050		6,275	7,375
6066	19.8 CFM, 7.5 HP, 80 gal. tank		.80	20		5,150	1,175		6,325	7,475
6070	25.8 CFM, 7-1/2 HP, 120 gal. tank		.80	20		5,400	1,175		6,575	7,750
6078	34.8 CFM, 10 HP, 80 gal. tank		.70	22.857		7,800	1,350		9,150	10,700
6080	34.8 CFM, 10 HP, 120 gal. tank		.60	26.667		8,325	1,575		9,900	11,600
6090	53.7 CFM, 15 HP, 120 gal. tank	Q-6	.80	30		8,750	1,825		10,575	12,500
6100	76.7 CFM, 20 HP, 120 gal. tank		.70	34.286		12,600	2,100		14,700	17,100
6104	76.7 CFM, 20 HP, 240 gal. tank		.68	35.294		12,200	2,150		14,350	16,800
6110	90.1 CFM, 25 HP, 120 gal. tank		.63	38.095		12,700	2,325		15,025	17,600
6120	101 CFM, 30 HP, 120 gal. tank		.57	42.105		11,300	2,575		13,875	16,400
6130	101 CFM, 30 HP, 250 gal. tank		.52	46.154		12,800	2,825		15,625	18,500
6200	Oil-less, 13.6 CFM, 5 HP, 120 gal. tank	Q-5	.88	18.182		13,700	1,075		14,775	16,700
6210	13.6 CFM, 5 HP, 250 gal. tank		.80	20		14,600	1,175		15,775	17,900
6220	18.2 CFM, 7.5 HP, 120 gal. tank		.73	21.918		13,600	1,300		14,900	17,000
6230	18.2 CFM, 7.5 HP, 250 gal. tank		.67	23.881		14,600	1,400		16,000	18,300
6250	30.5 CFM, 10 HP, 120 gal. tank		.57	28.070		16,100	1,650		17,750	20,300
6260	30.5 CFM, 10 HP, 250 gal. tank		.53	30.189		17,000	1,775		18,775	21,600
6270	41.3 CFM, 15 HP, 120 gal. tank	Q-6	.70	34.286		17,400	2,100		19,500	22,500
6280	41.3 CFM, 15 HP, 250 gal. tank	"	.67	35.821		18,500	2,200		20,700	23,700

22 31 Domestic Water Softeners

22 31 13 – Residential Domestic Water Softeners

22 31 13.10 Residential Water Softeners

Line	Description	Crew	Daily Output	Labor-Hours	Unit	Material	Labor	Equipment	Total	Total Incl O&P
0010	**RESIDENTIAL WATER SOFTENERS**									
7350	Water softener, automatic, to 30 grains per gallon	2 Plum	5	3.200	Ea.	395	206		601	755
7400	To 100 grains per gallon	"	4	4	"	940	258		1,198	1,425

22 31 16 – Commercial Domestic Water Softeners

22 31 16.10 Water Softeners

Line	Description	Crew	Daily Output	Labor-Hours	Unit	Material	Labor	Equipment	Total	Total Incl O&P
0010	**WATER SOFTENERS**									
5800	Softener systems, automatic, intermediate sizes									
5820	available, may be used in multiples.									
6000	Hardness capacity between regenerations and flow									
6100	150,000 grains, 37 GPM cont., 51 GPM peak	Q-1	1.20	13.333	Ea.	4,925	775		5,700	6,625
6200	300,000 grains, 81 GPM cont., 113 GPM peak		1	16		8,525	930		9,455	10,800
6300	750,000 grains, 160 GPM cont., 230 GPM peak		.80	20		12,600	1,150		13,750	15,700
6400	900,000 grains, 185 GPM cont., 270 GPM peak		.70	22.857		20,300	1,325		21,625	24,400

22 32 Domestic Water Filtration Equipment

22 32 19 – Domestic-Water Off-Floor Cartridge Filters

22 32 19.10 Water Filters		Crew	Daily Output	Labor-Hours	Unit	Material	2020 Bare Costs Labor	Equipment	Total	Total Incl O&P
0010	**WATER FILTERS**, Purification and treatment									
1000	Cartridge style, dirt and rust type	1 Plum	12	.667	Ea.	80.50	43		123.50	156
1200	Replacement cartridge		32	.250		16.05	16.10		32.15	42.50
1600	Taste and odor type		12	.667		115	43		158	193
1700	Replacement cartridge		32	.250		42	16.10		58.10	71
3000	Central unit, dirt/rust/odor/taste/scale		4	2		271	129		400	495
3100	Replacement cartridge, standard		20	.400		92.50	26		118.50	142
3600	Replacement cartridge, heavy duty	↓	20	.400	↓	84	26		110	132
8000	Commercial, fully automatic or push button automatic									
8200	Iron removal, 660 GPH, 1" pipe size	Q-1	1.50	10.667	Ea.	1,750	620		2,370	2,875
8240	1,500 GPH, 1-1/4" pipe size		1	16		2,950	930		3,880	4,675
8280	2,340 GPH, 1-1/2" pipe size		.80	20		3,250	1,150		4,400	5,375
8320	3,420 GPH, 2" pipe size		.60	26.667		5,975	1,550		7,525	8,975
8360	4,620 GPH, 2-1/2" pipe size		.50	32		9,500	1,850		11,350	13,400
8500	Neutralizer for acid water, 780 GPH, 1" pipe size		1.50	10.667		1,700	620		2,320	2,800
8540	1,140 GPH, 1-1/4" pipe size		1	16		1,900	930		2,830	3,525
8580	1,740 GPH, 1-1/2" pipe size		.80	20		2,775	1,150		3,925	4,850
8620	2,520 GPH, 2" pipe size		.60	26.667		3,600	1,550		5,150	6,350
8660	3,480 GPH, 2-1/2" pipe size		.50	32		5,975	1,850		7,825	9,450
8800	Sediment removal, 780 GPH, 1" pipe size		1.50	10.667		1,625	620		2,245	2,725
8840	1,140 GPH, 1-1/4" pipe size		1	16		1,925	930		2,855	3,550
8880	1,740 GPH, 1-1/2" pipe size		.80	20		2,550	1,150		3,700	4,625
8920	2,520 GPH, 2" pipe size		.60	26.667		3,700	1,550		5,250	6,475
8960	3,480 GPH, 2-1/2" pipe size		.50	32		5,825	1,850		7,675	9,275
9200	Taste and odor removal, 660 GPH, 1" pipe size		1.50	10.667		2,250	620		2,870	3,425
9240	1,500 GPH, 1-1/4" pipe size		1	16		3,875	930		4,805	5,675
9280	2,340 GPH, 1-1/2" pipe size		.80	20		4,425	1,150		5,575	6,650
9320	3,420 GPH, 2" pipe size		.60	26.667		6,775	1,550		8,325	9,850
9360	4,620 GPH, 2-1/2" pipe size	↓	.50	32	↓	9,050	1,850		10,900	12,800

22 33 Electric Domestic Water Heaters

22 33 13 – Instantaneous Electric Domestic Water Heaters

22 33 13.10 Hot Water Dispensers		Crew	Daily Output	Labor-Hours	Unit	Material	Labor	Equipment	Total	Total Incl O&P
0010	**HOT WATER DISPENSERS**									
0160	Commercial, 100 cup, 11.3 amp	1 Plum	14	.571	Ea.	460	37		497	560
3180	Household, 60 cup	"	14	.571	"	244	37		281	325

22 33 13.20 Instantaneous Elec. Point-Of-Use Water Heaters		Crew	Daily Output	Labor-Hours	Unit	Material	Labor	Equipment	Total	Total Incl O&P
0010	**INSTANTANEOUS ELECTRIC POINT-OF-USE WATER HEATERS**									
8965	Point of use, electric, glass lined									
8969	Energy saver									
8970	2.5 gal. single element G	1 Plum	2.80	2.857	Ea.	370	184		554	690
8971	4 gal. single element G		2.80	2.857		375	184		559	695
8974	6 gal. single element G		2.50	3.200		415	206		621	775
8975	10 gal. single element G		2.50	3.200		470	206		676	835
8976	15 gal. single element G		2.40	3.333		545	215		760	930
8977	20 gal. single element G		2.40	3.333		675	215		890	1,075
8978	30 gal. single element G		2.30	3.478		700	224		924	1,125
8979	40 gal. single element G	↓	2.20	3.636	↓	1,250	234		1,484	1,725
8988	Commercial (ASHRAE energy std. 90)									
8989	6 gallon G	1 Plum	2.50	3.200	Ea.	1,000	206		1,206	1,425
8990	10 gallon G	↓	2.50	3.200	↓	1,175	206		1,381	1,600

22 33 Electric Domestic Water Heaters

22 33 13 – Instantaneous Electric Domestic Water Heaters

22 33 13.20	Instantaneous Elec. Point-Of-Use Water Heaters	Crew	Daily Output	Labor-Hours	Unit	Material	2020 Bare Costs Labor	Equipment	Total	Total Incl O&P
8991	15 gallon G	1 Plum	2.40	3.333	Ea.	1,150	215		1,365	1,600
8992	20 gallon G		2.40	3.333		1,350	215		1,565	1,800
8993	30 gallon G		2.30	3.478		2,825	224		3,049	3,450
8995	Under the sink, copper, w/bracket									
8996	2.5 gallon G	1 Plum	4	2	Ea.	405	129		534	645
9000	Minimum labor/equipment charge	"	1.75	4.571	Job		295		295	455

22 33 30 – Residential, Electric Domestic Water Heaters

22 33 30.13	Residential, Small-Capacity Elec. Water Heaters	Crew	Daily Output	Labor-Hours	Unit	Material	Labor	Equipment	Total	Total Incl O&P
0010	RESIDENTIAL, SMALL-CAPACITY ELECTRIC DOMESTIC WATER HEATERS									
1000	Residential, electric, glass lined tank, 5 yr., 10 gal., single element	1 Plum	2.30	3.478	Ea.	470	224		694	860
1040	20 gallon, single element		2.20	3.636		675	234		909	1,100
1060	30 gallon, double element		2.20	3.636		1,075	234		1,309	1,525
1080	40 gallon, double element		2	4		1,250	258		1,508	1,775
1100	52 gallon, double element		2	4		1,425	258		1,683	1,975
1120	66 gallon, double element		1.80	4.444		1,925	286		2,211	2,575
1140	80 gallon, double element		1.60	5		2,175	320		2,495	2,900
1180	120 gallon, double element		1.40	5.714		3,000	370		3,370	3,875

22 33 33 – Light-Commercial Electric Domestic Water Heaters

22 33 33.10	Commercial Electric Water Heaters	Crew	Daily Output	Labor-Hours	Unit	Material	Labor	Equipment	Total	Total Incl O&P
0010	COMMERCIAL ELECTRIC WATER HEATERS									
4000	Commercial, 100° rise. NOTE: for each size tank, a range of									
4010	heaters between the ones shown is available									
4020	Electric									
4100	5 gal., 3 kW, 12 GPH, 208 volt	1 Plum	2	4	Ea.	4,600	258		4,858	5,450
4120	10 gal., 6 kW, 25 GPH, 208 volt		2	4		5,025	258		5,283	5,950
4130	30 gal., 24 kW, 98 GPH, 208 volt		1.92	4.167		8,100	269		8,369	9,325
4136	40 gal., 36 kW, 148 GPH, 208 volt		1.88	4.255		11,300	274		11,574	12,800
4140	50 gal., 9 kW, 37 GPH, 208 volt		1.80	4.444		6,900	286		7,186	8,050
4160	50 gal., 36 kW, 148 GPH, 208 volt		1.80	4.444		11,100	286		11,386	12,600
4180	80 gal., 12 kW, 49 GPH, 208 volt		1.50	5.333		8,500	345		8,845	9,875
4200	80 gal., 36 kW, 148 GPH, 208 volt		1.50	5.333		12,100	345		12,445	13,800
4220	100 gal., 36 kW, 148 GPH, 208 volt		1.20	6.667		13,200	430		13,630	15,200
4240	120 gal., 36 kW, 148 GPH, 208 volt		1.20	6.667		13,700	430		14,130	15,800
4260	150 gal., 15 kW, 61 GPH, 480 volt		1	8		34,500	515		35,015	38,700
4280	150 gal., 120 kW, 490 GPH, 480 volt		1	8		47,300	515		47,815	53,000
4300	200 gal., 15 kW, 61 GPH, 480 volt	Q-1	1.70	9.412		36,700	545		37,245	41,200
4320	200 gal., 120 kW, 490 GPH, 480 volt		1.70	9.412		48,400	545		48,945	54,000
4340	250 gal., 15 kW, 61 GPH, 480 volt		1.50	10.667		38,000	620		38,620	42,800
4360	250 gal., 150 kW, 615 GPH, 480 volt		1.50	10.667		48,100	620		48,720	54,000
4380	300 gal., 30 kW, 123 GPH, 480 volt		1.30	12.308		41,900	715		42,615	47,100
4400	300 gal., 180 kW, 738 GPH, 480 volt		1.30	12.308		67,500	715		68,215	75,500
4420	350 gal., 30 kW, 123 GPH, 480 volt		1.10	14.545		39,600	845		40,445	44,800
4440	350 gal., 180 kW, 738 GPH, 480 volt		1.10	14.545		56,000	845		56,845	63,000
4460	400 gal., 30 kW, 123 GPH, 480 volt		1	16		50,500	930		51,430	57,000
4480	400 gal., 210 kW, 860 GPH, 480 volt		1	16		80,500	930		81,430	90,000
4500	500 gal., 30 kW, 123 GPH, 480 volt		.80	20		57,500	1,150		58,650	65,000
4520	500 gal., 240 kW, 984 GPH, 480 volt		.80	20		97,500	1,150		98,650	109,500
4540	600 gal., 30 kW, 123 GPH, 480 volt	Q-2	1.20	20		66,500	1,200		67,700	75,000
4560	600 gal., 300 kW, 1,230 GPH, 480 volt		1.20	20		112,500	1,200		113,700	125,500
4580	700 gal., 30 kW, 123 GPH, 480 volt		1	24		63,500	1,450		64,950	71,500
4600	700 gal., 300 kW, 1,230 GPH, 480 volt		1	24		95,000	1,450		96,450	106,500
4620	800 gal., 60 kW, 245 GPH, 480 volt		.90	26.667		79,000	1,600		80,600	89,000

22 33 Electric Domestic Water Heaters

22 33 33 – Light-Commercial Electric Domestic Water Heaters

22 33 33.10 Commercial Electric Water Heaters		Crew	Daily Output	Labor-Hours	Unit	2020 Bare Costs Material	Labor	Equipment	Total	Total Incl O&P
4640	800 gal., 300 kW, 1,230 GPH, 480 volt	Q-2	.90	26.667	Ea.	97,000	1,600		98,600	109,500
4660	1,000 gal., 60 kW, 245 GPH, 480 volt		.70	34.286		75,000	2,050		77,050	85,500
4680	1,000 gal., 480 kW, 1,970 GPH, 480 volt		.70	34.286		124,500	2,050		126,550	140,000
4700	1,200 gal., 60 kW, 245 GPH, 480 volt		.60	40		81,500	2,400		83,900	93,500
4720	1,200 gal., 480 kW, 1,970 GPH, 480 volt		.60	40		128,000	2,400		130,400	144,000
4740	1,500 gal., 60 kW, 245 GPH, 480 volt		.50	48		107,500	2,875		110,375	122,500
4760	1,500 gal., 480 kW, 1,970 GPH, 480 volt		.50	48		154,500	2,875		157,375	174,500
5400	Modulating step control for under 90 kW, 2-5 steps	1 Elec	5.30	1.509		880	92.50		972.50	1,100
5440	For above 90 kW, 1 through 5 steps beyond standard, add		3.20	2.500		272	153		425	535
5460	For above 90 kW, 6 through 10 steps beyond standard, add		2.70	2.963		550	182		732	885
5480	For above 90 kW, 11 through 18 steps beyond standard, add		1.60	5		820	305		1,125	1,375

22 34 Fuel-Fired Domestic Water Heaters

22 34 13 – Instantaneous, Tankless, Gas Domestic Water Heaters

22 34 13.10 Instantaneous, Tankless, Gas Water Heaters		Crew	Daily Output	Labor-Hours	Unit	Material	Labor	Equipment	Total	Total Incl O&P
0010	**INSTANTANEOUS, TANKLESS, GAS WATER HEATERS**									
9410	Natural gas/propane, 3.2 GPM G	1 Plum	2	4	Ea.	575	258		833	1,025
9420	6.4 GPM G		1.90	4.211		780	271		1,051	1,275
9430	8.4 GPM G		1.80	4.444		880	286		1,166	1,400
9440	9.5 GPM G		1.60	5		1,050	320		1,370	1,650

22 34 30 – Residential Gas Domestic Water Heaters

22 34 30.13 Residential, Atmos, Gas Domestic Wtr Heaters		Crew	Daily Output	Labor-Hours	Unit	Material	Labor	Equipment	Total	Total Incl O&P
0010	**RESIDENTIAL, ATMOSPHERIC, GAS DOMESTIC WATER HEATERS**									
2000	Gas fired, foam lined tank, 10 yr., vent not incl.									
2040	30 gallon	1 Plum	2	4	Ea.	2,100	258		2,358	2,725
2060	40 gallon		1.90	4.211		1,825	271		2,096	2,425
2080	50 gallon		1.80	4.444		1,950	286		2,236	2,600
2090	60 gallon		1.70	4.706		1,900	305		2,205	2,575
2100	75 gallon		1.50	5.333		2,650	345		2,995	3,450
2120	100 gallon		1.30	6.154		3,150	395		3,545	4,100
2900	Water heater, safety-drain pan, 26" round		20	.400		21.50	26		47.50	63.50
3000	Tank leak safety, water & gas shut off see 22 05 23.20 8800									

22 34 36 – Commercial Gas Domestic Water Heaters

22 34 36.13 Commercial, Atmos., Gas Domestic Water Htrs.		Crew	Daily Output	Labor-Hours	Unit	Material	Labor	Equipment	Total	Total Incl O&P
0010	**COMMERCIAL, ATMOSPHERIC, GAS DOMESTIC WATER HEATERS**									
6000	Gas fired, flush jacket, std. controls, vent not incl.									
6040	75 MBH input, 73 GPH	1 Plum	1.40	5.714	Ea.	3,725	370		4,095	4,675
6060	98 MBH input, 95 GPH		1.40	5.714		8,800	370		9,170	10,200
6080	120 MBH input, 110 GPH		1.20	6.667		9,000	430		9,430	10,600
6100	120 MBH input, 115 GPH		1.10	7.273		8,725	470		9,195	10,300
6120	140 MBH input, 130 GPH		1	8		10,800	515		11,315	12,600
6140	155 MBH input, 150 GPH		.80	10		12,100	645		12,745	14,300
6160	180 MBH input, 170 GPH		.70	11.429		11,700	735		12,435	14,100
6180	200 MBH input, 192 GPH		.60	13.333		12,100	860		12,960	14,700
6200	250 MBH input, 245 GPH		.50	16		12,700	1,025		13,725	15,600
6220	260 MBH input, 250 GPH	Q-1	.80	20		13,100	1,150		14,250	16,200
6240	360 MBH input, 360 GPH		.80	20		15,400	1,150		16,550	18,700
6260	500 MBH input, 480 GPH		.70	22.857		21,600	1,325		22,925	25,900
6280	725 MBH input, 690 GPH		.60	26.667		26,900	1,550		28,450	32,000
6900	For low water cutoff, add	1 Plum	8	1		385	64.50		449.50	525

22 34 Fuel-Fired Domestic Water Heaters

22 34 36 – Commercial Gas Domestic Water Heaters

22 34 36.13 Commercial, Atmos., Gas Domestic Water Htrs.		Crew	Daily Output	Labor-Hours	Unit	Material	2020 Bare Costs Labor	Equipment	Total	Total Incl O&P
6960	For bronze body hot water circulator, add	1 Plum	4	2	Ea.	2,250	129		2,379	2,675

22 34 46 – Oil-Fired Domestic Water Heaters

22 34 46.10 Residential Oil-Fired Water Heaters

		Crew	Daily Output	Labor-Hours	Unit	Material	Labor	Equipment	Total	Total Incl O&P
0010	**RESIDENTIAL OIL-FIRED WATER HEATERS**									
3000	Oil fired, glass lined tank, 5 yr., vent not included, 30 gallon	1 Plum	2	4	Ea.	1,400	258		1,658	1,950
3040	50 gallon		1.80	4.444		1,375	286		1,661	1,975
3060	70 gallon		1.50	5.333		2,375	345		2,720	3,150

22 34 46.20 Commercial Oil-Fired Water Heaters

		Crew	Daily Output	Labor-Hours	Unit	Material	Labor	Equipment	Total	Total Incl O&P
0010	**COMMERCIAL OIL-FIRED WATER HEATERS**									
8000	Oil fired, glass lined, UL listed, std. controls, vent not incl.									
8060	140 gal., 140 MBH input, 134 GPH	Q-1	2.13	7.512	Ea.	25,200	435		25,635	28,400
8080	140 gal., 199 MBH input, 191 GPH		2	8		26,100	465		26,565	29,400
8100	140 gal., 255 MBH input, 247 GPH		1.60	10		26,800	580		27,380	30,400
8120	140 gal., 270 MBH input, 259 GPH		1.20	13.333		33,100	775		33,875	37,700
8140	140 gal., 400 MBH input, 384 GPH		1	16		34,000	930		34,930	38,800
8160	140 gal., 540 MBH input, 519 GPH		.96	16.667		35,600	965		36,565	40,600
8180	140 gal., 720 MBH input, 691 GPH		.92	17.391		36,200	1,000		37,200	41,400
8200	221 gal., 300 MBH input, 288 GPH		.88	18.182		47,900	1,050		48,950	54,000
8220	221 gal., 600 MBH input, 576 GPH		.86	18.605		61,500	1,075		62,575	69,000
8240	221 gal., 800 MBH input, 768 GPH		.82	19.512		54,000	1,125		55,125	61,000
8260	201 gal., 1,000 MBH input, 960 GPH	Q-2	1.26	19.048		55,000	1,150		56,150	62,500
8280	201 gal., 1,250 MBH input, 1,200 GPH		1.22	19.672		56,000	1,175		57,175	63,500
8300	201 gal., 1,500 MBH input, 1,441 GPH		1.16	20.690		60,500	1,250		61,750	68,500
8320	411 gal., 600 MBH input, 576 GPH		1.12	21.429		61,000	1,300		62,300	69,000
8340	411 gal., 800 MBH input, 768 GPH		1.08	22.222		63,500	1,325		64,825	71,500
8360	411 gal., 1,000 MBH input, 960 GPH		1.04	23.077		64,500	1,400		65,900	73,000
8380	411 gal., 1,250 MBH input, 1,200 GPH		.98	24.490		66,000	1,475		67,475	75,500
8400	397 gal., 1,500 MBH input, 1,441 GPH		.92	26.087		70,000	1,575		71,575	79,500
8420	397 gal., 1,750 MBH input, 1,681 GPH		.86	27.907		72,000	1,675		73,675	82,000
8430	397 gal., 2,000 MBH input, 1,921 GPH		.82	29.268		78,000	1,750		79,750	88,500
8440	375 gal., 2,250 MBH input, 2,161 GPH		.76	31.579		80,500	1,900		82,400	91,500
8450	375 gal., 2,500 MBH input, 2,401 GPH		.82	29.268		83,500	1,750		85,250	94,000
8900	For low water cutoff, add	1 Plum	8	1		405	64.50		469.50	550
8960	For bronze body hot water circulator, add	"	4	2		1,050	129		1,179	1,350

22 35 Domestic Water Heat Exchangers

22 35 30 – Water Heating by Steam

22 35 30.10 Water Heating Transfer Package

		Crew	Daily Output	Labor-Hours	Unit	Material	Labor	Equipment	Total	Total Incl O&P
0010	**WATER HEATING TRANSFER PACKAGE**, Complete controls,									
0020	expansion tank, converter, air separator									
1000	Hot water, 180°F enter, 200°F leaving, 15# steam									
1010	One pump system, 28 GPM	Q-6	.75	32	Ea.	21,800	1,950		23,750	27,000
1020	35 GPM		.70	34.286		24,700	2,100		26,800	30,500
1040	55 GPM		.65	36.923		28,000	2,250		30,250	34,300
1060	130 GPM		.55	43.636		35,500	2,675		38,175	43,200
1080	255 GPM		.40	60		46,600	3,675		50,275	56,500
1100	550 GPM		.30	80		67,000	4,900		71,900	81,500
1120	800 GPM		.25	96		74,000	5,875		79,875	90,500
1220	Two pump system, 28 GPM		.70	34.286		29,700	2,100		31,800	35,900
1240	35 GPM		.65	36.923		36,100	2,250		38,350	43,200

22 35 Domestic Water Heat Exchangers

22 35 30 – Water Heating by Steam

22 35 30.10 Water Heating Transfer Package		Crew	Daily Output	Labor-Hours	Unit	Material	Labor	Equipment	Total	Total Incl O&P
						2020 Bare Costs				
1260	55 GPM	Q-6	.60	40	Ea.	35,900	2,450		38,350	43,200
1280	130 GPM		.50	48		49,400	2,925		52,325	59,000
1300	255 GPM		.35	68.571		65,000	4,200		69,200	78,000
1320	550 GPM		.25	96		79,500	5,875		85,375	96,000
1340	800 GPM	↓	.20	120	↓	102,500	7,350		109,850	124,000

22 41 Residential Plumbing Fixtures

22 41 06 – Plumbing Fixtures General

22 41 06.10 Plumbing Fixture Notes

Line	Description	Crew	Daily Output	Labor-Hours	Unit	Material	Labor	Equipment	Total	Total Incl O&P
0010	**PLUMBING FIXTURE NOTES**, Incl. trim fittings unless otherwise noted									
0080	For rough-in, supply, waste, and vent, see add for each type									
0122	For electric water coolers, see Section 22 47 16.10									
0160	For color, unless otherwise noted, add				Ea.	20%				

22 41 13 – Residential Water Closets, Urinals, and Bidets

22 41 13.13 Water Closets

Line	Description	Crew	Daily Output	Labor-Hours	Unit	Material	Labor	Equipment	Total	Total Incl O&P
0010	**WATER CLOSETS** R224000-30									
0022	For seats, see Section 22 41 13.44									
0032	For automatic flush, see Line 22 42 39.10 0972									
0150	Tank type, vitreous china, incl. seat, supply pipe w/stop, 1.6 gpf or noted									
0200	Wall hung									
0400	Two piece, close coupled	Q-1	5.30	3.019	Ea.	420	175		595	730
0960	For rough-in, supply, waste, vent and carrier	"	2.73	5.861	"	1,275	340		1,615	1,925
0999	Floor mounted									
1020	One piece, low profile	Q-1	5.30	3.019	Ea.	900	175		1,075	1,250
1050	One piece		5.30	3.019		945	175		1,120	1,325
1100	Two piece, close coupled		5.30	3.019		219	175		394	510
1102	Economy		5.30	3.019		121	175		296	405
1110	Two piece, close coupled, dual flush		5.30	3.019		280	175		455	580
1140	Two piece, close coupled, 1.28 gpf, ADA G	↓	5.30	3.019	↓	320	175		495	620
1960	For color, add					30%				
1980	For rough-in, supply, waste and vent	Q-1	3.05	5.246	Ea.	340	305		645	845

22 41 13.19 Bidets

Line	Description	Crew	Daily Output	Labor-Hours	Unit	Material	Labor	Equipment	Total	Total Incl O&P
0010	**BIDETS**									
0180	Vitreous china, with trim on fixture	Q-1	5	3.200	Ea.	685	186		871	1,050
0200	With trim for wall mounting		5	3.200		805	186		991	1,175
9600	For rough-in, supply, waste and vent, add	↓	1.78	8.989	↓	430	520		950	1,275

22 41 13.44 Toilet Seats

Line	Description	Crew	Daily Output	Labor-Hours	Unit	Material	Labor	Equipment	Total	Total Incl O&P
0010	**TOILET SEATS**									
0100	Molded composition, white									
0150	Industrial, w/o cover, open front, regular bowl	1 Plum	24	.333	Ea.	22	21.50		43.50	57.50
0200	With self-sustaining hinge		24	.333		23.50	21.50		45	59
0220	With self-sustaining check hinge		24	.333		23	21.50		44.50	58
0240	Extra heavy, with check hinge	↓	24	.333	↓	22.50	21.50		44	58
0260	Elongated bowl, same price									
0300	Junior size, w/o cover, open front	1 Plum	24	.333	Ea.	43.50	21.50		65	81
0320	Regular primary bowl, open front		24	.333		43.50	21.50		65	81
0340	Regular baby bowl, open front, check hinge		24	.333		37.50	21.50		59	74.50
0380	Open back & front, w/o cover, reg. or elongated bowl	↓	24	.333	↓	21	21.50		42.50	56
0400	Residential									
0420	Regular bowl, w/cover, closed front	1 Plum	24	.333	Ea.	29.50	21.50		51	65

22 41 Residential Plumbing Fixtures

22 41 13 – Residential Water Closets, Urinals, and Bidets

22 41 13.44 Toilet Seats		Crew	Daily Output	Labor-Hours	Unit	Material	Labor	Equipment	Total	Total Incl O&P
0440	Open front	1 Plum	24	.333	Ea.	26.50	21.50		48	62
0460	Elongated bowl, add					25%				
0500	Self-raising hinge, w/o cover, open front									
0520	Regular bowl	1 Plum	24	.333	Ea.	103	21.50		124.50	146
0540	Elongated bowl	"	24	.333	"	25	21.50		46.50	60.50
0700	Molded wood, white, with cover									
0720	Closed front, regular bowl, square back	1 Plum	24	.333	Ea.	10.15	21.50		31.65	44
0740	Extended back		24	.333		14.10	21.50		35.60	48.50
0780	Elongated bowl, square back		24	.333		13.50	21.50		35	48
0800	Open front	↓	24	.333	↓	14.20	21.50		35.70	48.50
0850	Decorator styles									
0890	Vinyl top, patterned	1 Plum	24	.333	Ea.	21.50	21.50		43	57
0900	Vinyl padded, plain colors, regular bowl		24	.333		21.50	21.50		43	57
0930	Elongated bowl	↓	24	.333	↓	25.50	21.50		47	61
1000	Solid plastic, white									
1030	Industrial, w/o cover, open front, regular bowl	1 Plum	24	.333	Ea.	26.50	21.50		48	62.50
1080	Extra heavy, concealed check hinge		24	.333		21	21.50		42.50	56.50
1100	Self-sustaining hinge		24	.333		25	21.50		46.50	60.50
1150	Elongated bowl		24	.333		26	21.50		47.50	61.50
1170	Concealed check		24	.333		17.60	21.50		39.10	52.50
1190	Self-sustaining hinge, concealed check		24	.333		55.50	21.50		77	94
1220	Residential, with cover, closed front, regular bowl		24	.333		41.50	21.50		63	78.50
1240	Elongated bowl		24	.333		50	21.50		71.50	88
1260	Open front, regular bowl		24	.333		43	21.50		64.50	80
1280	Elongated bowl	↓	24	.333	↓	52	21.50		73.50	90.50

22 41 16 – Residential Lavatories and Sinks

22 41 16.13 Lavatories		Crew	Daily Output	Labor-Hours	Unit	Material	Labor	Equipment	Total	Total Incl O&P
0010	**LAVATORIES**, With trim, white unless noted otherwise R224000-30									
0500	Vanity top, porcelain enamel on cast iron									
0600	20" x 18"	Q-1	6.40	2.500	Ea.	305	145		450	565
0640	33" x 19" oval		6.40	2.500		520	145		665	800
0680	20" x 17" oval		6.40	2.500		117	145		262	355
0720	19" round		6.40	2.500		455	145		600	725
0760	20" x 12" triangular bowl	↓	6.40	2.500	↓	239	145		384	485
0860	For color, add					25%				
1000	Cultured marble, 19" x 17", single bowl	Q-1	6.40	2.500	Ea.	124	145		269	360
1040	25" x 19", single bowl		6.40	2.500		150	145		295	390
1080	31" x 19", single bowl		6.40	2.500		166	145		311	405
1120	25" x 22", single bowl		6.40	2.500		164	145		309	405
1160	37" x 22", single bowl		6.40	2.500		203	145		348	450
1200	49" x 22", single bowl	↓	6.40	2.500	↓	242	145		387	490
1580	For color, same price									
1900	Stainless steel, self-rimming, 25" x 22", single bowl, ledge	Q-1	6.40	2.500	Ea.	315	145		460	575
1960	17" x 22", single bowl		6.40	2.500		305	145		450	560
2040	18-3/4" round		6.40	2.500		820	145		965	1,125
2600	Steel, enameled, 20" x 17", single bowl		5.80	2.759		126	160		286	385
2660	19" round		5.80	2.759		162	160		322	425
2720	18" round	↓	5.80	2.759	↓	90	160		250	345
2860	For color, add					10%				
2900	Vitreous china, 20" x 16", single bowl	Q-1	5.40	2.963	Ea.	211	172		383	500
2960	20" x 17", single bowl		5.40	2.963		121	172		293	400
3020	19" round, single bowl	↓	5.40	2.963	↓	117	172		289	395

22 41 16 – Residential Lavatories and Sinks

22 41 16.13 Lavatories		Crew	Daily Output	Labor-Hours	Unit	2020 Bare Costs Material	Labor	Equipment	Total	Total Incl O&P
3080	19″ x 16″, single bowl	Q-1	5.40	2.963	Ea.	220	172		392	510
3140	17″ x 14″, single bowl		5.40	2.963		154	172		326	435
3200	22″ x 13″, single bowl		5.40	2.963		216	172		388	505
3560	For color, add					50%				
3580	Rough-in, supply, waste and vent for all above lavatories	Q-1	2.30	6.957	Ea.	278	405		683	930
4000	Wall hung									
4040	Porcelain enamel on cast iron, 16″ x 14″, single bowl	Q-1	8	2	Ea.	435	116		551	655
4060	18″ x 15″, single bowl		8	2		350	116		466	565
4120	19″ x 17″, single bowl		8	2		405	116		521	625
4180	20″ x 18″, single bowl		8	2		246	116		362	450
4240	22″ x 19″, single bowl		8	2		690	116		806	940
4580	For color, add					30%				
6000	Vitreous china, 18″ x 15″, single bowl with backsplash	Q-1	7	2.286	Ea.	170	133		303	390
6060	19″ x 17″, single bowl		7	2.286		122	133		255	340
6120	20″ x 18″, single bowl		7	2.286		247	133		380	475
6210	27″ x 20″, ADA compliant		7	2.286		930	133		1,063	1,225
6500	For color, add					30%				
6960	Rough-in, supply, waste and vent for above lavatories	Q-1	1.66	9.639	Ea.	475	560		1,035	1,375
7000	Pedestal type									
7600	Vitreous china, 27″ x 21″, white	Q-1	6.60	2.424	Ea.	685	141		826	970
7610	27″ x 21″, colored		6.60	2.424		875	141		1,016	1,175
7620	27″ x 21″, premium color		6.60	2.424		995	141		1,136	1,325
7660	26″ x 20″, white		6.60	2.424		665	141		806	945
7670	26″ x 20″, colored		6.60	2.424		855	141		996	1,150
7680	26″ x 20″, premium color		6.60	2.424		1,025	141		1,166	1,350
7700	24″ x 20″, white		6.60	2.424		465	141		606	725
7710	24″ x 20″, colored		6.60	2.424		580	141		721	850
7720	24″ x 20″, premium color		6.60	2.424		640	141		781	920
7760	21″ x 18″, white		6.60	2.424		258	141		399	500
7770	21″ x 18″, colored		6.60	2.424		289	141		430	530
7990	Rough-in, supply, waste and vent for pedestal lavatories		1.66	9.639		475	560		1,035	1,375
9000	Minimum labor/equipment charge	1 Plum	3	2.667	Job		172		172	266

22 41 16.16 Sinks		Crew	Daily Output	Labor-Hours	Unit	Material	Labor	Equipment	Total	Total Incl O&P
0010	**SINKS**, With faucets and drain R224000-30									
2000	Kitchen, counter top style, PE on CI, 24″ x 21″ single bowl	Q-1	5.60	2.857	Ea.	310	166		476	595
2100	31″ x 22″ single bowl		5.60	2.857		860	166		1,026	1,200
2200	32″ x 21″ double bowl		4.80	3.333		390	193		583	730
3000	Stainless steel, self rimming, 19″ x 18″ single bowl		5.60	2.857		625	166		791	940
3100	25″ x 22″ single bowl		5.60	2.857		690	166		856	1,025
3200	33″ x 22″ double bowl		4.80	3.333		1,000	193		1,193	1,400
3300	43″ x 22″ double bowl		4.80	3.333		1,150	193		1,343	1,575
3400	22″ x 43″ triple bowl		4.40	3.636		1,225	211		1,436	1,675
3500	Corner double bowl each 14″ x 16″		4.80	3.333		805	193		998	1,175
4000	Steel, enameled, with ledge, 24″ x 21″ single bowl		5.60	2.857		535	166		701	845
4100	32″ x 21″ double bowl		4.80	3.333		520	193		713	870
4960	For color sinks except stainless steel, add					10%				
4980	For rough-in, supply, waste and vent, counter top sinks	Q-1	2.14	7.477		320	435		755	1,025
5000	Kitchen, raised deck, PE on CI									
5100	32″ x 21″, dual level, double bowl	Q-1	2.60	6.154	Ea.	485	355		840	1,075
5200	42″ x 21″, double bowl & disposer well	″	2.20	7.273		1,000	420		1,420	1,750
5700	For color, add					20%				
5790	For rough-in, supply, waste & vent, sinks	Q-1	1.85	8.649		320	500		820	1,125

22 41 Residential Plumbing Fixtures

22 41 19 – Residential Bathtubs

22 41 19.10 Baths			Crew	Daily Output	Labor-Hours	Unit	Material	2020 Bare Costs Labor	Equipment	Total	Total Incl O&P
0010	**BATHS**	R224000-30									
0100	Tubs, recessed porcelain enamel on cast iron, with trim										
0180	48" x 42"		Q-1	4	4	Ea.	3,150	232		3,382	3,800
0220	72" x 36"		"	3	5.333	"	2,975	310		3,285	3,750
0300	Mat bottom										
0340	4'-6" long		Q-1	5	3.200	Ea.	1,500	186		1,686	1,925
0380	5' long			4.40	3.636		1,275	211		1,486	1,725
0420	5'-6" long			4	4		1,975	232		2,207	2,500
0480	Above floor drain, 5' long			4	4		825	232		1,057	1,275
0560	Corner 48" x 44"		↓	4.40	3.636	↓	2,925	211		3,136	3,525
0750	For color, add						30%				
2000	Enameled formed steel, 4'-6" long		Q-1	5.80	2.759	Ea.	515	160		675	810
2300	Above floor drain, 5' long		"	5.50	2.909	"	590	169		759	910
2350	For color, add						10%				
4000	Soaking, acrylic, w/pop-up drain 66" x 36" x 20" deep		Q-1	5.50	2.909	Ea.	1,725	169		1,894	2,150
4100	60" x 42" x 20" deep			5	3.200		1,325	186		1,511	1,725
4200	72" x 42" x 23" deep		↓	4.80	3.333	↓	2,200	193		2,393	2,725
4600	Module tub & showerwall surround, molded fiberglass										
4610	5' long x 34" wide x 76" high		Q-1	4	4	Ea.	790	232		1,022	1,225
4750	ADA compliant with 1-1/2" OD grab bar, antiskid bottom										
4760	60" x 32-3/4" x 72" high		Q-1	4	4	Ea.	595	232		827	1,025
4770	60" x 30" x 71" high with molded seat			3.50	4.571		755	265		1,020	1,250
9600	Rough-in, supply, waste and vent, for all above tubs, add			2.07	7.729	↓	445	450		895	1,175
9900	Minimum labor/equipment charge		↓	3	5.333	Job		310		310	480

22 41 23 – Residential Showers

22 41 23.20 Showers

			Crew	Daily Output	Labor-Hours	Unit	Material	Labor	Equipment	Total	Total Incl O&P
0010	**SHOWERS**	R224000-30									
1500	Stall, with drain only. Add for valve and door/curtain										
1510	Baked enamel, molded stone receptor, 30" square		Q-1	5.20	3.077	Ea.	1,300	178		1,478	1,700
1520	32" square			5	3.200		1,175	186		1,361	1,550
1530	36" square			4.80	3.333		2,950	193		3,143	3,525
1540	Terrazzo receptor, 32" square			5	3.200		1,375	186		1,561	1,800
1560	36" square			4.80	3.333		1,650	193		1,843	2,125
1580	36" corner angle		↓	4.80	3.333	↓	2,050	193		2,243	2,550
1600	For color, add						10%				
3000	Fiberglass, one piece, with 3 walls, 32" x 32" square		Q-1	5.50	2.909	Ea.	355	169		524	650
3100	36" x 36" square		"	5.50	2.909	"	465	169		634	770
3200	ADA compliant, 1-1/2" OD grab bars, nonskid floor										
3210	48" x 34-1/2" x 72" corner seat		Q-1	5	3.200	Ea.	665	186		851	1,025
3220	60" x 34-1/2" x 72" corner seat			4	4		765	232		997	1,200
3230	48" x 34-1/2" x 72" fold up seat			5	3.200		1,150	186		1,336	1,550
3250	64" x 65-3/4" x 81-1/2" fold. seat, ADA			3.80	4.211		1,475	244		1,719	2,000
4000	Polypropylene, stall only, w/molded-stone floor, 30" x 30"			2	8		720	465		1,185	1,500
4100	32" x 32"			2	8		735	465		1,200	1,525
4200	Rough-in, supply, waste and vent for above showers		↓	2.05	7.805	↓	410	455		865	1,150

22 41 23.40 Shower System Components

			Crew	Daily Output	Labor-Hours	Unit	Material	Labor	Equipment	Total	Total Incl O&P
0010	**SHOWER SYSTEM COMPONENTS**										
4500	Receptor only										
4510	For tile, 36" x 36"		1 Plum	4	2	Ea.	390	129		519	630
4520	Fiberglass receptor only, 32" x 32"			8	1		107	64.50		171.50	218
4530	34" x 34"		↓	7.80	1.026	↓	128	66		194	243

22 41 Residential Plumbing Fixtures

22 41 23 – Residential Showers

22 41 23.40 Shower System Components		Crew	Daily Output	Labor-Hours	Unit	Material	Labor	Equipment	Total	Total Incl O&P
						2020 Bare Costs				
4540	36" x 36"	1 Plum	7.60	1.053	Ea.	125	68		193	243
4600	Rectangular									
4620	32" x 48"	1 Plum	7.40	1.081	Ea.	154	69.50		223.50	277
4630	34" x 54"		7.20	1.111		176	71.50		247.50	305
4640	34" x 60"		7	1.143		187	73.50		260.50	320
5000	Built-in, head, arm, 2.5 GPM valve		4	2		104	129		233	315
5200	Head, arm, by-pass, integral stops, handles		3.60	2.222		296	143		439	545
5500	Head, water economizer, 1.6 GPM G		24	.333		53	21.50		74.50	91.50
5800	Mixing valve, built-in		6	1.333		144	86		230	292
5900	Exposed		6	1.333		710	86		796	920

22 41 36 – Residential Laundry Trays

22 41 36.10 Laundry Sinks

		Crew	Daily Output	Labor-Hours	Unit	Material	Labor	Equipment	Total	Total Incl O&P
0010	**LAUNDRY SINKS**, With trim									
0020	Porcelain enamel on cast iron, black iron frame									
0050	24" x 21", single compartment	Q-1	6	2.667	Ea.	615	155		770	915
0100	26" x 21", single compartment	"	6	2.667	"	630	155		785	935
2000	Molded stone, on wall hanger or legs									
2020	22" x 23", single compartment	Q-1	6	2.667	Ea.	176	155		331	435
2100	45" x 21", double compartment	"	5	3.200	"	360	186		546	680
3000	Plastic, on wall hanger or legs									
3020	18" x 23", single compartment	Q-1	6.50	2.462	Ea.	145	143		288	380
3100	20" x 24", single compartment		6.50	2.462		165	143		308	400
3200	36" x 23", double compartment		5.50	2.909		219	169		388	500
3300	40" x 24", double compartment		5.50	2.909		287	169		456	575
5000	Stainless steel, counter top, 22" x 17" single compartment		6	2.667		77.50	155		232.50	325
5200	33" x 22", double compartment		5	3.200		93	186		279	390
9600	Rough-in, supply, waste and vent, for all laundry sinks		2.14	7.477		320	435		755	1,025
9810	Minimum labor/equipment charge	1 Plum	3	2.667	Job		172		172	266

22 41 39 – Residential Faucets, Supplies and Trim

22 41 39.10 Faucets and Fittings

		Crew	Daily Output	Labor-Hours	Unit	Material	Labor	Equipment	Total	Total Incl O&P
0010	**FAUCETS AND FITTINGS**									
0150	Bath, faucets, diverter spout combination, sweat	1 Plum	8	1	Ea.	87	64.50		151.50	195
0200	For integral stops, IPS unions, add					111			111	123
0300	Three valve combinations, spout, head, arm, flange, sweat	1 Plum	6	1.333		98.50	86		184.50	242
0400	For integral stops, IPS unions, add				Pr.	63.50			63.50	70
0420	Bath, press-bal mix valve w/diverter, spout, shower head, arm/flange	1 Plum	8	1	Ea.	185	64.50		249.50	305
0500	Drain, central lift, 1-1/2" IPS male		20	.400		50.50	26		76.50	95.50
0600	Trip lever, 1-1/2" IPS male		20	.400		60.50	26		86.50	107
0700	Pop up, 1-1/2" IPS male		18	.444		69	28.50		97.50	121
0800	Chain and stopper, 1-1/2" IPS male		24	.333		33	21.50		54.50	69.50
0810	Bidet									
0812	Fitting, over the rim, swivel spray/pop-up drain	1 Plum	8	1	Ea.	279	64.50		343.50	405
1000	Kitchen sink faucets, top mount, cast spout		10	.800		84	51.50		135.50	172
1100	For spray, add		24	.333		17.25	21.50		38.75	52
1110	For basket strainer w/tail piece, add		24	.333		15.50	21.50		37	50
1200	Wall type, swing tube spout		10	.800		74.50	51.50		126	162
1240	For soap dish, add					3.67			3.67	4.04
1250	For basket strainer w/tail piece, add					45.50			45.50	50
1300	Single control lever handle									
1310	With pull out spray									
1320	Polished chrome	1 Plum	10	.800	Ea.	196	51.50		247.50	296
2000	Laundry faucets, shelf type, IPS or copper unions		12	.667		62	43		105	135

22 41 39 – Residential Faucets, Supplies and Trim

22 41 39.10 Faucets and Fittings		Crew	Daily Output	Labor-Hours	Unit	Material	Labor (2020 Bare Costs)	Equipment	Total	Total Incl O&P
2100	Lavatory faucet, centerset, without drain	1 Plum	10	.800	Ea.	67.50	51.50		119	154
2120	With pop-up drain		6.66	1.201		54.50	77.50		132	180
2130	For acrylic handles, add					5.25			5.25	5.80
2150	Concealed, 12" centers	1 Plum	10	.800		90	51.50		141.50	179
2160	With pop-up drain	"	6.66	1.201		108	77.50		185.50	239
2210	Porcelain cross handles and pop-up drain									
2220	Polished chrome	1 Plum	6.66	1.201	Ea.	222	77.50		299.50	365
2230	Polished brass	"	6.66	1.201	"	293	77.50		370.50	445
2260	Single lever handle and pop-up drain									
2280	Satin nickel	1 Plum	6.66	1.201	Ea.	280	77.50		357.50	430
2290	Polished chrome		6.66	1.201		200	77.50		277.50	340
2600	Shelfback, 4" to 6" centers, 17 ga. tailpiece		10	.800		81	51.50		132.50	169
2650	With pop-up drain		6.66	1.201		101	77.50		178.50	231
2700	Shampoo faucet with supply tube		24	.333		47.50	21.50		69	85.50
2800	Self-closing, center set		10	.800		151	51.50		202.50	246
2810	Automatic sensor and operator, with faucet head G		6.15	1.301		495	84		579	670
4000	Shower by-pass valve with union		18	.444		57.50	28.50		86	108
4100	Shower arm with flange and head		22	.364		21	23.50		44.50	59.50
4140	Shower, hand held, pin mount, massage action, chrome		22	.364		82	23.50		105.50	127
4142	Polished brass		22	.364		162	23.50		185.50	214
4144	Shower, hand held, wall mtd, adj. spray, 2 wall mounts, chrome		20	.400		116	26		142	168
4146	Polished brass		20	.400		237	26		263	300
4148	Shower, hand held head, bar mounted 24", adj. spray, chrome		20	.400		180	26		206	238
4150	Polished brass		20	.400		370	26		396	445
4200	Shower thermostatic mixing valve, concealed, with shower head trim kit		8	1		385	64.50		449.50	520
4220	Shower pressure balancing mixing valve									
4230	With shower head, arm, flange and diverter tub spout									
4240	Chrome	1 Plum	6.14	1.303	Ea.	415	84		499	585
4250	Satin nickel		6.14	1.303		560	84		644	745
4260	Polished graphite		6.14	1.303		545	84		629	730
5000	Sillcock, compact, brass, IPS or copper to hose		24	.333		12.05	21.50		33.55	46.50
6000	Stop and waste valves, bronze									
6100	Angle, solder end 1/2"	1 Plum	24	.333	Ea.	28	21.50		49.50	64
6110	3/4"		20	.400		36	26		62	79.50
6300	Straightway, solder end 3/8"		24	.333		20.50	21.50		42	55.50
6310	1/2"		24	.333		20.50	21.50		42	55.50
6320	3/4"		20	.400		21	26		47	63
6410	Straightway, threaded 1/2"		24	.333		22	21.50		43.50	57
6420	3/4"		20	.400		24.50	26		50.50	67
6430	1"		19	.421		27	27		54	71.50
7800	Water closet, wax gasket		96	.083		1.67	5.35		7.02	10.15
7820	Gasket toilet tank to bowl		32	.250		3.15	16.10		19.25	28.50
7830	Replacement diaphragm washer assy for ballcock valve		12	.667		3.15	43		46.15	70
7850	Dual flush valve		12	.667		133	43		176	214
8000	Water supply stops, polished chrome plate									
8200	Angle, 3/8"	1 Plum	24	.333	Ea.	9.30	21.50		30.80	43
8300	1/2"		22	.364		10.05	23.50		33.55	47
8400	Straight, 3/8"		26	.308		9.40	19.85		29.25	41
8500	1/2"		24	.333		9.75	21.50		31.25	43.50
8600	Water closet, angle, w/flex riser, 3/8"		24	.333		34	21.50		55.50	70.50
9000	Minimum labor/equipment charge		4	2	Job		129		129	199

22 41 Residential Plumbing Fixtures

22 41 39 – Residential Faucets, Supplies and Trim

22 41 39.70 Washer/Dryer Accessories		Crew	Daily Output	Labor-Hours	Unit	Material	2020 Bare Costs Labor	Equipment	Total	Total Incl O&P
0010	WASHER/DRYER ACCESSORIES									
1020	Valves ball type single lever									
1030	1/2" diam., IPS	1 Plum	21	.381	Ea.	65	24.50		89.50	110
1040	1/2" diam., solder	"	21	.381	"	65	24.50		89.50	110
1050	Recessed box, 16 ga., two hose valves and drain									
1060	1/2" size, 1-1/2" drain	1 Plum	18	.444	Ea.	152	28.50		180.50	212
1070	1/2" size, 2" drain	"	17	.471	"	135	30.50		165.50	196
1080	With grounding electric receptacle									
1090	1/2" size, 1-1/2" drain	1 Plum	18	.444	Ea.	167	28.50		195.50	228
1100	1/2" size, 2" drain	"	17	.471	"	179	30.50		209.50	244
1110	With grounding and dryer receptacle									
1120	1/2" size, 1-1/2" drain	1 Plum	18	.444	Ea.	205	28.50		233.50	270
1130	1/2" size, 2" drain	"	17	.471	"	207	30.50		237.50	275
1140	Recessed box, 16 ga., ball valves with single lever and drain									
1150	1/2" size, 1-1/2" drain	1 Plum	19	.421	Ea.	286	27		313	355
1160	1/2" size, 2" drain	"	18	.444	"	247	28.50		275.50	315
1170	With grounding electric receptacle									
1180	1/2" size, 1-1/2" drain	1 Plum	19	.421	Ea.	305	27		332	375
1190	1/2" size, 2" drain	"	18	.444	"	275	28.50		303.50	350
1200	With grounding and dryer receptacles									
1210	1/2" size, 1-1/2" drain	1 Plum	19	.421	Ea.	272	27		299	340
1220	1/2" size, 2" drain	"	18	.444	"	300	28.50		328.50	375
1300	Recessed box, 20 ga., two hose valves and drain (economy type)									
1310	1/2" size, 1-1/2" drain	1 Plum	19	.421	Ea.	118	27		145	172
1320	1/2" size, 2" drain		18	.444		110	28.50		138.50	166
1330	Box with drain only		24	.333		68.50	21.50		90	108
1340	1/2" size, 1-1/2" ABS/PVC drain		19	.421		126	27		153	180
1350	1/2" size, 2" ABS/PVC drain		18	.444		134	28.50		162.50	193
1352	Box with drain and 15 A receptacle		24	.333		72.50	21.50		94	113
1360	1/2" size, 2" drain ABS/PVC, 15 A receptacle	↓	24	.333	↓	134	21.50		155.50	180
1400	Wall mounted									
1410	1/2" size, 1-1/2" plastic drain	1 Plum	19	.421	Ea.	34.50	27		61.50	79.50
1420	1/2" size, 2" plastic drain	"	18	.444	"	20	28.50		48.50	66.50
1500	Dryer vent kit									
1510	8' flex duct, clamps and outside hood	1 Plum	20	.400	Ea.	13.95	26		39.95	55.50
1980	Rough-in, supply, waste, and vent for washer boxes		3.46	2.310		355	149		504	620
9605	Washing machine valve assembly, hot & cold water supply, recessed		8	1		82.50	64.50		147	191
9610	Washing machine valve assembly, hot & cold water supply, mounted	↓	8	1	↓	65	64.50		129.50	171

22 42 Commercial Plumbing Fixtures

22 42 13 – Commercial Water Closets, Urinals, and Bidets

22 42 13.13 Water Closets		Crew	Daily Output	Labor-Hours	Unit	Material	Labor	Equipment	Total	Total Incl O&P
0010	WATER CLOSETS									
3000	Bowl only, with flush valve, seat, 1.6 gpf unless noted									
3100	Wall hung	Q-1	5.80	2.759	Ea.	1,100	160		1,260	1,450
3200	For rough-in, supply, waste and vent, single WC		2.56	6.250		1,300	365		1,665	2,000
3300	Floor mounted		5.80	2.759		360	160		520	640
3350	With wall outlet		5.80	2.759		570	160		730	870
3360	With floor outlet, 1.28 gpf G		5.80	2.759		550	160		710	850
3362	With floor outlet, 1.28 gpf, ADA G	↓	5.80	2.759	↓	570	160		730	875

22 42 13 – Commercial Water Closets, Urinals, and Bidets

22 42 13.13 Water Closets		Crew	Daily Output	Labor-Hours	Unit	Material	Labor	Equipment	Total	Total Incl O&P
3370	For rough-in, supply, waste and vent, single WC	Q-1	2.84	5.634	Ea.	385	325		710	925
3390	Floor mounted children's size, 10-3/4" high									
3392	With automatic flush sensor, 1.6 gpf	Q-1	6.20	2.581	Ea.	660	150		810	955
3396	With automatic flush sensor, 1.28 gpf		6.20	2.581		610	150		760	900
3400	For rough-in, supply, waste and vent, single WC		2.84	5.634		385	325		710	925
3500	Gang side by side carrier system, rough-in, supply, waste & vent									
3510	For single hook-up	Q-1	1.97	8.122	Ea.	1,550	470		2,020	2,425
3520	For each additional hook-up, add	"	2.14	7.477	"	1,475	435		1,910	2,275
3550	Gang back to back carrier system, rough-in, supply, waste & vent									
3560	For pair hook-up	Q-1	1.76	9.091	Pr.	2,075	525		2,600	3,100
3570	For each additional pair hook-up, add		1.81	8.840	"	2,000	515		2,515	3,000
9000	Minimum labor/equipment charge		4	4	Job		232		232	360

22 42 13.16 Urinals		Crew	Daily Output	Labor-Hours	Unit	Material	Labor	Equipment	Total	Total Incl O&P
0010	**URINALS** R224000-30									
0102	For automatic flush see Line 22 42 39.10 0972									
3000	Wall hung, vitreous china, with self-closing valve									
3100	Siphon jet type	Q-1	3	5.333	Ea.	315	310		625	825
3120	Blowout type		3	5.333		480	310		790	1,000
3140	Water saving .5 gpf G		3	5.333		595	310		905	1,125
3300	Rough-in, supply, waste & vent		2.83	5.654		755	330		1,085	1,350
5000	Stall type, vitreous china, includes valve		2.50	6.400		830	370		1,200	1,475
6980	Rough-in, supply, waste and vent		1.99	8.040		510	465		975	1,275
8000	Waterless (no flush) urinal									
8010	Wall hung									
8014	Fiberglass reinforced polyester									
8020	Standard unit G	Q-1	21.30	.751	Ea.	450	43.50		493.50	565
8030	ADA compliant unit G	"	21.30	.751		420	43.50		463.50	530
8070	For solid color, add G					64			64	70
8080	For 2" brass flange (new const.), add G	Q-1	96	.167		20.50	9.65		30.15	37.50
8200	Vitreous china									
8220	ADA compliant unit, 14" G	Q-1	21.30	.751	Ea.	211	43.50		254.50	300
8250	ADA compliant unit, 15.5"	"	21.30	.751		272	43.50		315.50	365
8270	For solid color, add G					64			64	70
8290	Rough-in, supply, waste & vent G	Q-1	2.92	5.479		720	320		1,040	1,275
8400	Trap liquid									
8410	1 quart G				Ea.	19.20			19.20	21
8420	1 gallon G				"	63			63	69.50
9000	Minimum labor/equipment charge	Q-1	4	4	Job		232		232	360

22 42 16 – Commercial Lavatories and Sinks

22 42 16.13 Lavatories		Crew	Daily Output	Labor-Hours	Unit	Material	Labor	Equipment	Total	Total Incl O&P
0010	**LAVATORIES**, With trim, white unless noted otherwise									
0020	Commercial lavatories same as residential. See Section 22 41 16									

22 42 16.34 Laboratory Countertops and Sinks		Crew	Daily Output	Labor-Hours	Unit	Material	Labor	Equipment	Total	Total Incl O&P
0010	**LABORATORY COUNTERTOPS AND SINKS**									
0050	Laboratory sinks, corrosion resistant									
1000	Stainless steel sink, bench mounted, with									
1020	plug & waste fitting with 1-1/2" straight threads									
1030	Single bowl, 2 drainboards, backnut & strainer									
1050	18-1/2" x 15-1/2" x 12-1/2" sink, 54" x 24" OD	Q-1	3	5.333	Ea.	1,150	310		1,460	1,725
1100	Single bowl, single drainboard, backnut & strainer									
1130	18-1/2" x 15-1/2" x 12-1/2" sink, 47" x 24" OD	Q-1	3	5.333	Ea.	920	310		1,230	1,475
1146	Double bowl, single drainboard, backnut & strainer									

22 42 16 – Commercial Lavatories and Sinks

22 42 16.34 Laboratory Countertops and Sinks		Crew	Daily Output	Labor-Hours	Unit	Material	2020 Bare Costs Labor	Equipment	Total	Total Incl O&P
1150	18-1/2" x 15-1/2" x 12-1/2" sink, 70" x 24" OD	Q-1	3	5.333	Ea.	1,250	310		1,560	1,850
1280	Polypropylene									
1290	Flanged 1-1/4" wide, rectangular with strainer									
1300	plug & waste fitting, 1-1/2" straight threads									
1320	12" x 12" x 8" sink, 14-1/2" x 14-1/2" OD	Q-1	4	4	Ea.	267	232		499	655
1340	16" x 16" x 8" sink, 18-1/2" x 18-1/2" OD		4	4		375	232		607	775
1360	21" x 18" x 10" sink, 23-1/2" x 20-1/2" OD		4	4		395	232		627	795
1490	For rough-in, supply, waste & vent, add		2.02	7.921		229	460		689	960
1600	Polypropylene									
1620	Cup sink, oval, integral strainers									
1640	6" x 3" I.D., 7" x 4" OD	Q-1	6	2.667	Ea.	149	155		304	405
1660	9" x 3" I.D., 10" x 4-1/2" OD	"	6	2.667		176	155		331	435
1740	1-1/2" diam. x 11" long					44.50			44.50	48.50
1980	For rough-in, supply, waste & vent, add	Q-1	1.70	9.412		256	545		801	1,125

22 42 16.40 Service Sinks		Crew	Daily Output	Labor-Hours	Unit	Material	Labor	Equipment	Total	Total Incl O&P
0010	**SERVICE SINKS**									
6650	Service, floor, corner, PE on CI, 28" x 28"	Q-1	4.40	3.636	Ea.	1,125	211		1,336	1,575
6670	Service, floor, molded resin., 12" x 12", 10" deep, w/drn, strnr	"	4.40	3.636		1,400	211		1,611	1,875
6750	Vinyl coated rim guard, add					65.50			65.50	72
6755	Mop sink, molded stone, 22" x 18"	1 Plum	3.33	2.402		545	155		700	840
6760	Mop sink, molded stone, 24" x 36"		3.33	2.402		279	155		434	545
6770	Mop sink, molded stone, 24" x 36", w/rim 3 sides		3.33	2.402		264	155		419	530
6790	For rough-in, supply, waste & vent, floor service sinks	Q-1	1.64	9.756		1,075	565		1,640	2,050
7000	Service, wall, PE on CI, roll rim, 22" x 18"		4	4		930	232		1,162	1,375
7100	24" x 20"		4	4		950	232		1,182	1,400
7600	For stainless steel rim guard, two sides only, add					49			49	53.50
7800	For stainless steel rim guard, front only, add					57			57	62.50
8600	Vitreous china, 22" x 20"	Q-1	4	4		875	232		1,107	1,325
8960	For stainless steel rim guard, front or one side, add					65.50			65.50	72
8980	For rough-in, supply, waste & vent, wall service sinks	Q-1	1.30	12.308		1,425	715		2,140	2,675
9000	Minimum labor/equipment charge	"	4	4	Job		232		232	360

22 42 23 – Commercial Showers

22 42 23.30 Group Showers		Crew	Daily Output	Labor-Hours	Unit	Material	Labor	Equipment	Total	Total Incl O&P
0010	**GROUP SHOWERS**									
6000	Group, w/pressure balancing valve, rough-in and rigging not included									
6800	Column, 6 heads, no receptors, less partitions	Q-1	3	5.333	Ea.	9,675	310		9,985	11,100
6900	With stainless steel partitions		1	16		12,600	930		13,530	15,300
7600	5 heads, no receptors, less partitions		3	5.333		6,650	310		6,960	7,775
7620	4 heads (1 ADA compliant) no receptors, less partitions		3	5.333		6,050	310		6,360	7,125
7700	With stainless steel partitions		1	16		6,600	930		7,530	8,700
8000	Wall, 2 heads, no receptors, less partitions		4	4		2,825	232		3,057	3,450
8100	With stainless steel partitions		2	8		6,200	465		6,665	7,525
9000	Minimum labor/equipment charge		4	4	Job		232		232	360

22 42 33 – Wash Fountains

22 42 33.20 Commercial Wash Fountains		Crew	Daily Output	Labor-Hours	Unit	Material	Labor	Equipment	Total	Total Incl O&P
0010	**COMMERCIAL WASH FOUNTAINS**									
1900	Group, foot control									
2000	Precast terrazzo, circular, 36" diam., 5 or 6 persons	Q-2	3	8	Ea.	7,675	480		8,155	9,175
2100	54" diam. for 8 or 10 persons		2.50	9.600		10,600	575		11,175	12,500
2400	Semi-circular, 36" diam. for 3 persons		3	8		6,250	480		6,730	7,625
2500	54" diam. for 4 or 5 persons		2.50	9.600		9,700	575		10,275	11,600

22 42 Commercial Plumbing Fixtures

22 42 33 – Wash Fountains

22 42 33.20 Commercial Wash Fountains		Crew	Daily Output	Labor-Hours	Unit	2020 Bare Costs Material	Labor	Equipment	Total	Total Incl O&P
2700	Quarter circle (corner), 54" diam. for 3 persons	Q-2	3.50	6.857	Ea.	7,675	410		8,085	9,100
3000	Stainless steel, circular, 36" diameter		3.50	6.857		6,775	410		7,185	8,100
3100	54" diameter		2.80	8.571		8,575	515		9,090	10,200
3400	Semi-circular, 36" diameter		3.50	6.857		5,625	410		6,035	6,850
3500	54" diameter		2.80	8.571		7,150	515		7,665	8,675
5000	Thermoplastic, pre-assembled, circular, 36" diameter		6	4		4,700	241		4,941	5,550
5100	54" diameter		4	6		5,450	360		5,810	6,550
5400	Semi-circular, 36" diameter		6	4		4,750	241		4,991	5,600
5600	54" diameter	↓	4	6	↓	6,250	360		6,610	7,425
5610	Group, infrared control, barrier free									
5614	Precast terrazzo									
5620	Semi-circular 36" diam. for 3 persons	Q-2	3	8	Ea.	8,475	480		8,955	10,100
5630	46" diam. for 4 persons		2.80	8.571		9,100	515		9,615	10,800
5640	Circular, 54" diam. for 8 persons, button control	↓	2.50	9.600		11,200	575		11,775	13,200
5700	Rough-in, supply, waste and vent for above wash fountains	Q-1	1.82	8.791		505	510		1,015	1,350
6200	Duo for small washrooms, stainless steel		2	8		3,300	465		3,765	4,350
6400	Bowl with backsplash		2	8		2,400	465		2,865	3,350
6500	Rough-in, supply, waste & vent for duo fountains	↓	2.02	7.921	↓	268	460		728	1,000
9000	Minimum labor/equipment charge	Q-2	3	8	Job		480		480	745

22 42 39 – Commercial Faucets, Supplies, and Trim

22 42 39.10 Faucets and Fittings		Crew	Daily Output	Labor-Hours	Unit	Material	Labor	Equipment	Total	Total Incl O&P
0010	**FAUCETS AND FITTINGS**									
0840	Flush valves, with vacuum breaker									
0850	Water closet									
0860	Exposed, rear spud	1 Plum	8	1	Ea.	146	64.50		210.50	261
0870	Top spud		8	1		197	64.50		261.50	315
0880	Concealed, rear spud		8	1		213	64.50		277.50	335
0890	Top spud		8	1		173	64.50		237.50	290
0900	Wall hung		8	1		199	64.50		263.50	320
0910	Dual flush flushometer		12	.667		245	43		288	335
0912	Flushometer retrofit kit	↓	18	.444	↓	18.15	28.50		46.65	64.50
0920	Urinal									
0930	Exposed, stall	1 Plum	8	1	Ea.	197	64.50		261.50	315
0940	Wall (washout)		8	1		156	64.50		220.50	272
0950	Pedestal, top spud		8	1		137	64.50		201.50	250
0960	Concealed, stall		8	1		170	64.50		234.50	287
0970	Wall (washout)	↓	8	1	↓	183	64.50		247.50	300
0971	Automatic flush sensor and operator for									
0972	urinals or water closets, standard G	1 Plum	8	1	Ea.	485	64.50		549.50	635
0980	High efficiency water saving									
0984	Water closets, 1.28 gpf G	1 Plum	8	1	Ea.	425	64.50		489.50	570
0988	Urinals, .5 gpf G	"	8	1	"	425	64.50		489.50	570
2790	Faucets for lavatories									
2800	Self-closing, center set	1 Plum	10	.800	Ea.	151	51.50		202.50	246
2810	Automatic sensor and operator, with faucet head		6.15	1.301		495	84		579	670
3000	Service sink faucet, cast spout, pail hook, hose end	↓	14	.571	↓	76.50	37		113.50	141

22 42 39.30 Carriers and Supports		Crew	Daily Output	Labor-Hours	Unit	Material	Labor	Equipment	Total	Total Incl O&P
0010	**CARRIERS AND SUPPORTS**, For plumbing fixtures									
0500	Drinking fountain, wall mounted									
0600	Plate type with studs, top back plate	1 Plum	7	1.143	Ea.	61.50	73.50		135	182
0700	Top front and back plate		7	1.143		153	73.50		226.50	282
0800	Top & bottom, front & back plates, w/bearing jacks	↓	7	1.143	↓	181	73.50		254.50	315

22 42 39 – Commercial Faucets, Supplies, and Trim

22 42 39.30 Carriers and Supports		Crew	Daily Output	Labor-Hours	Unit	2020 Bare Costs Material	Labor	Equipment	Total	Total Incl O&P
3000	Lavatory, concealed arm									
3050	Floor mounted, single									
3100	High back fixture	1 Plum	6	1.333	Ea.	655	86		741	855
3200	Flat slab fixture		6	1.333		575	86		661	770
3220	ADA compliant		6	1.333		665	86		751	870
3250	Floor mounted, back to back									
3300	High back fixtures	1 Plum	5	1.600	Ea.	1,075	103		1,178	1,325
3400	Flat slab fixtures		5	1.600		1,325	103		1,428	1,600
3430	ADA compliant		5	1.600		810	103		913	1,050
3500	Wall mounted, in stud or masonry									
3600	High back fixture	1 Plum	6	1.333	Ea.	345	86		431	515
3700	Flat slab fixture	″	6	1.333	″	275	86		361	440
4000	Exposed arm type, floor mounted									
4100	Single high back or flat slab fixture	1 Plum	6	1.333	Ea.	950	86		1,036	1,175
4200	Back to back, high back or flat slab fixtures		5	1.600		1,700	103		1,803	2,000
4300	Wall mounted, high back or flat slab lavatory		6	1.333		750	86		836	960
4600	Sink, floor mounted									
4650	Exposed arm system									
4700	Single heavy fixture	1 Plum	5	1.600	Ea.	645	103		748	870
4750	Single heavy sink with slab		5	1.600		1,600	103		1,703	1,900
4800	Back to back, standard fixtures		5	1.600		845	103		948	1,100
4850	Back to back, heavy fixtures		5	1.600		1,300	103		1,403	1,575
4900	Back to back, heavy sink with slab		5	1.600		1,825	103		1,928	2,175
4950	Exposed offset arm system									
5000	Single heavy deep fixture	1 Plum	5	1.600	Ea.	1,200	103		1,303	1,475
5100	Plate type system									
5200	With bearing jacks, single fixture	1 Plum	5	1.600	Ea.	1,600	103		1,703	1,900
5300	With exposed arms, single heavy fixture		5	1.600		1,250	103		1,353	1,550
5400	Wall mounted, exposed arms, single heavy fixture		5	1.600		450	103		553	655
6000	Urinal, floor mounted, 2″ or 3″ coupling, blowout type		6	1.333		805	86		891	1,025
6100	With fixture or hanger bolts, blowout or washout		6	1.333		575	86		661	765
6200	With bearing plate		6	1.333		640	86		726	840
6300	Wall mounted, plate type system		6	1.333		430	86		516	605
6980	Water closet, siphon jet									
7000	Horizontal, adjustable, caulk									
7040	Single, 4″ pipe size	1 Plum	5.33	1.501	Ea.	1,125	96.50		1,221.50	1,375
7050	4″ pipe size, ADA compliant		5.33	1.501		825	96.50		921.50	1,050
7060	5″ pipe size		5.33	1.501		1,125	96.50		1,221.50	1,375
7100	Double, 4″ pipe size		5	1.600		1,625	103		1,728	1,925
7110	4″ pipe size, ADA compliant		5	1.600		1,625	103		1,728	1,950
7120	5″ pipe size		5	1.600		1,925	103		2,028	2,275
7160	Horizontal, adjustable, extended, caulk									
7180	Single, 4″ pipe size	1 Plum	5.33	1.501	Ea.	1,500	96.50		1,596.50	1,800
7200	5″ pipe size		5.33	1.501		1,875	96.50		1,971.50	2,225
7240	Double, 4″ pipe size		5	1.600		2,150	103		2,253	2,500
7260	5″ pipe size		5	1.600		2,850	103		2,953	3,275
7400	Vertical, adjustable, caulk or thread									
7440	Single, 4″ pipe size	1 Plum	5.33	1.501	Ea.	1,250	96.50		1,346.50	1,525
7460	5″ pipe size		5.33	1.501		1,575	96.50		1,671.50	1,875
7480	6″ pipe size		5	1.600		1,625	103		1,728	1,950
7520	Double, 4″ pipe size		5	1.600		2,300	103		2,403	2,675
7540	5″ pipe size		5	1.600		2,325	103		2,428	2,700
7560	6″ pipe size		4	2		2,575	129		2,704	3,025

22 42 Commercial Plumbing Fixtures

22 42 39 – Commercial Faucets, Supplies, and Trim

22 42 39.30 Carriers and Supports		Crew	Daily Output	Labor-Hours	Unit	Material	2020 Bare Costs Labor	Equipment	Total	Total Incl O&P
7600	Vertical, adjustable, extended, caulk									
7620	Single, 4" pipe size	1 Plum	5.33	1.501	Ea.	1,125	96.50		1,221.50	1,400
7640	5" pipe size		5.33	1.501		890	96.50		986.50	1,125
7680	6" pipe size		5	1.600		1,725	103		1,828	2,050
7720	Double, 4" pipe size		5	1.600		2,300	103		2,403	2,700
7740	5" pipe size		5	1.600		1,250	103		1,353	1,525
7760	6" pipe size	↓	4	2	↓	1,350	129		1,479	1,700
7780	Water closet, blow out									
7800	Vertical offset, caulk or thread									
7820	Single, 4" pipe size	1 Plum	5.33	1.501	Ea.	1,225	96.50		1,321.50	1,500
7840	Double, 4" pipe size	"	5	1.600	"	2,075	103		2,178	2,450
7880	Vertical offset, extended, caulk									
7900	Single, 4" pipe size	1 Plum	5.33	1.501	Ea.	1,525	96.50		1,621.50	1,825
7920	Double, 4" pipe size	"	5	1.600	"	2,400	103		2,503	2,775
7960	Vertical, for floor mounted back-outlet									
7980	Single, 4" thread, 2" vent	1 Plum	5.33	1.501	Ea.	790	96.50		886.50	1,025
8000	Double, 4" thread, 2" vent	"	6	1.333	"	2,325	86		2,411	2,675
8040	Vertical, for floor mounted back-outlet, extended									
8060	Single, 4" caulk, 2" vent	1 Plum	6	1.333	Ea.	790	86		876	1,000
8080	Double, 4" caulk, 2" vent	"	6	1.333	"	2,325	86		2,411	2,675
8200	Water closet, residential									
8220	Vertical centerline, floor mount									
8240	Single, 3" caulk, 2" or 3" vent	1 Plum	6	1.333	Ea.	875	86		961	1,100
8260	4" caulk, 2" or 4" vent		6	1.333		1,125	86		1,211	1,375
8280	3" copper sweat, 3" vent		6	1.333		785	86		871	1,000
8300	4" copper sweat, 4" vent	↓	6	1.333	↓	950	86		1,036	1,175
8400	Vertical offset, floor mount									
8420	Single, 3" or 4" caulk, vent	1 Plum	4	2	Ea.	1,100	129		1,229	1,400
8440	3" or 4" copper sweat, vent		5	1.600		1,100	103		1,203	1,350
8460	Double, 3" or 4" caulk, vent		4	2		1,875	129		2,004	2,250
8480	3" or 4" copper sweat, vent	↓	5	1.600	↓	1,875	103		1,978	2,200
9000	Water cooler (electric), floor mounted									
9100	Plate type with bearing plate, single	1 Plum	6	1.333	Ea.	480	86		566	665
9140	Plate type with bearing plate, back to back		4	2	"	635	129		764	895
9990	Minimum labor/equipment charge	↓	3.50	2.286	Job		147		147	228

22 43 Healthcare Plumbing Fixtures

22 43 13 – Healthcare Water Closets

22 43 13.40 Water Closets		Crew	Daily Output	Labor-Hours	Unit	Material	Labor	Equipment	Total	Total Incl O&P
0010	**WATER CLOSETS**									
1000	Bowl only, 1 piece, w/seat and flush valve, ADA compliant, 18" high									
1030	Floor mounted									
1150	With wall outlet ♿	Q-1	5.30	3.019	Ea.	320	175		495	620
1180	For rough-in, supply, waste and vent		2.84	5.634		385	325		710	925
1200	With floor outlet		5.30	3.019		370	175		545	675
1800	For rough-in, supply, waste and vent		3.05	5.246		340	305		645	845
3100	Wall hung ♿	↓	5.80	2.759	↓	1,100	160		1,260	1,450
3150	Hospital type, slotted rim for bed pan									
3156	Elongated bowl, top spud	Q-1	5.80	2.759	Ea.	650	160		810	960
3160	Elongated bowl, rear spud		5.80	2.759		395	160		555	675
3200	For rough-in, supply, waste and vent, single WC	↓	2.56	6.250	↓	1,300	365		1,665	2,000

22 43 Healthcare Plumbing Fixtures

22 43 13 – Healthcare Water Closets

22 43 13.40 Water Closets		Crew	Daily Output	Labor-Hours	Unit	Material	Labor	Equipment	Total	Total Incl O&P
3300	Floor mounted									
3320	Bariatric (1,200 lb. capacity), elongated bowl, ADA compliant	Q-1	4.60	3.478	Ea.	2,875	202		3,077	3,450
3360	Hospital type, slotted rim for bed pan									
3370	Elongated bowl, top spud	Q-1	5	3.200	Ea.	330	186		516	645
3380	Elongated bowl, rear spud		5	3.200		405	186		591	730
3500	For rough-in, supply, waste and vent		3.05	5.246		340	305		645	845

22 43 16 – Healthcare Sinks

22 43 16.10 Sinks

		Crew	Daily Output	Labor-Hours	Unit	Material	Labor	Equipment	Total	Total Incl O&P
0010	**SINKS**									
0020	Vitreous china									
6702	Hospital type, without trim (see Section 22 41 39.10)									
6710	20" x 18", contoured splash shield	Q-1	8	2	Ea.	96.50	116		212.50	285
6730	28" x 20", surgeon, side decks		8	2		560	116		676	800
6740	28" x 22", surgeon scrub-up, deep bowl		8	2		855	116		971	1,125
6750	20" x 27", patient, ADA compliant		7	2.286		490	133		623	745
6760	30" x 22", all purpose		7	2.286		780	133		913	1,050
6770	30" x 22", plaster work		7	2.286		710	133		843	985
6820	20" x 24" clinic service, liquid/solid waste		6	2.667		985	155		1,140	1,325

22 43 19 – Healthcare Bathtubs

22 43 19.10 Bathtubs

		Crew	Daily Output	Labor-Hours	Unit	Material	Labor	Equipment	Total	Total Incl O&P
0010	**BATHTUBS**									
5002	Hospital type, with trim, see Section 22 41 39.10									
5050	Bathing pool, porcelain enamel on cast iron, grab bars									
5060	pop-up drain, 72" x 36"	Q-1	3	5.333	Ea.	3,850	310		4,160	4,725
5100	Perineal (sitz), vitreous china		3	5.333		1,300	310		1,610	1,900
5120	For pedestal, vitreous china, add		8	2		261	116		377	465
5300	Whirlpool, porcelain enamel on cast iron, 72" x 36"		1	16		4,775	930		5,705	6,675

22 43 23 – Healthcare Showers

22 43 23.10 Showers

		Crew	Daily Output	Labor-Hours	Unit	Material	Labor	Equipment	Total	Total Incl O&P
0010	**SHOWERS**									
5950	Module, ADA compl, SS panel, fixed & hand held head, control									
5960	valves, grab bar, curtain & rod, folding seat	1 Plum	4	2	Ea.	1,625	129		1,754	1,975

22 43 39 – Healthcare Faucets

22 43 39.10 Faucets and Fittings

		Crew	Daily Output	Labor-Hours	Unit	Material	Labor	Equipment	Total	Total Incl O&P
0010	**FAUCETS AND FITTINGS**									
2850	Medical, bedpan cleanser, with pedal valve,	1 Plum	12	.667	Ea.	815	43		858	960
2860	With screwdriver stop valve		12	.667		420	43		463	530
2870	With self-closing spray valve		12	.667		260	43		303	355
2900	Faucet, gooseneck spout, wrist handles, grid drain		10	.800		202	51.50		253.50	300
2940	Mixing valve, knee action, screwdriver stops		4	2		450	129		579	695

22 45 Emergency Plumbing Fixtures

22 45 13 – Emergency Showers

22 45 13.10 Emergency Showers		Crew	Daily Output	Labor-Hours	Unit	Material	2020 Bare Costs Labor	Equipment	Total	Total Incl O&P
0010	**EMERGENCY SHOWERS**, Rough-in not included									
5000	Shower, single head, drench, ball valve, pull, freestanding	Q-1	4	4	Ea.	390	232		622	790
5200	Horizontal or vertical supply		4	4		615	232		847	1,025
6000	Multi-nozzle, eye/face wash combination		4	4		760	232		992	1,200
6400	Multi-nozzle, 12 spray, shower only		4	4		2,100	232		2,332	2,675
6600	For freeze-proof, add		6	2.667		505	155		660	795
8000	Walk-thru decontamination with eye-face wash		2	8		4,150	465		4,615	5,300
8200	For freeze proof, add		4	4		600	232		832	1,025
9000	Minimum labor/equipment charge		3	5.333	Job		310		310	480

22 45 16 – Eyewash Equipment

22 45 16.10 Eyewash Safety Equipment		Crew	Daily Output	Labor-Hours	Unit	Material	Labor	Equipment	Total	Total Incl O&P
0010	**EYEWASH SAFETY EQUIPMENT**, Rough-in not included									
1000	Eye wash fountain									
1400	Plastic bowl, pedestal mounted	Q-1	4	4	Ea.	335	232		567	725
1600	Unmounted		4	4		259	232		491	645
1800	Wall mounted		4	4		490	232		722	900
2000	Stainless steel, pedestal mounted		4	4		360	232		592	755
2200	Unmounted		4	4		298	232		530	690
2400	Wall mounted		4	4		325	232		557	715

22 45 19 – Self-Contained Eyewash Equipment

22 45 19.10 Self-Contained Eyewash Safety Equipment		Crew	Daily Output	Labor-Hours	Unit	Material	Labor	Equipment	Total	Total Incl O&P
0010	**SELF-CONTAINED EYEWASH SAFETY EQUIPMENT**									
3000	Eye wash, portable, self-contained				Ea.	1,950			1,950	2,150

22 45 26 – Eye/Face Wash Equipment

22 45 26.10 Eye/Face Wash Safety Equipment		Crew	Daily Output	Labor-Hours	Unit	Material	Labor	Equipment	Total	Total Incl O&P
0010	**EYE/FACE WASH SAFETY EQUIPMENT**, Rough-in not included									
4000	Eye and face wash, combination fountain									
4200	Stainless steel, pedestal mounted	Q-1	4	4	Ea.	1,150	232		1,382	1,600
4400	Unmounted		4	4		297	232		529	685
4600	Wall mounted		4	4		284	232		516	675

22 46 Security Plumbing Fixtures

22 46 13 – Security Water Closets and Urinals

22 46 13.10 Security Water Closets and Urinals		Crew	Daily Output	Labor-Hours	Unit	Material	Labor	Equipment	Total	Total Incl O&P
0010	**SECURITY WATER CLOSETS AND URINALS**, Stainless steel									
2000	Urinal, back supply and flush									
2200	Wall hung	Q-1	4	4	Ea.	3,250	232		3,482	3,925
2240	Stall		2.50	6.400		4,300	370		4,670	5,300
2300	For urinal rough-in, supply, waste and vent		1.49	10.738		395	625		1,020	1,400
3000	Water closet, integral seat, back supply and flush									
3300	Wall hung, wall outlet	Q-1	5.80	2.759	Ea.	1,700	160		1,860	2,125
3400	Floor mount, wall outlet		5.80	2.759		2,150	160		2,310	2,600
3440	Floor mount, floor outlet		5.80	2.759		2,500	160		2,660	3,000
3480	For recessed tissue holder, add					172			172	189
3500	For water closet rough-in, supply, waste and vent	Q-1	1.19	13.445		380	780		1,160	1,625
5000	Water closet and lavatory units, push button filler valves,									
5010	soap & paper holders, seat									
5300	Wall hung	Q-1	5	3.200	Ea.	3,775	186		3,961	4,425
5400	Floor mount		5	3.200		3,775	186		3,961	4,425

22 46 Security Plumbing Fixtures

22 46 13 – Security Water Closets and Urinals

22 46 13.10 Security Water Closets and Urinals		Crew	Daily Output	Labor-Hours	Unit	2020 Bare Costs Material	Labor	Equipment	Total	Total Incl O&P
6300	For unit rough-in, supply, waste and vent	Q-1	1	16	Ea.	445	930		1,375	1,900

22 46 16 – Security Lavatories and Sinks

22 46 16.13 Security Lavatories		Crew	Daily Output	Labor-Hours	Unit	Material	Labor	Equipment	Total	Total Incl O&P
0010	**SECURITY LAVATORIES**, Stainless steel									
1000	Lavatory, wall hung, push button filler valve									
1100	Rectangular bowl	Q-1	8	2	Ea.	1,350	116		1,466	1,650
1200	Oval bowl		8	2		1,325	116		1,441	1,625
1240	Oval bowl, corner mount		8	2		1,650	116		1,766	2,000
1300	For lavatory rough-in, supply, waste and vent	↓	1.50	10.667	↓	375	620		995	1,375

22 46 63 – Security Service Sink

22 46 63.10 Security Service Sink		Crew	Daily Output	Labor-Hours	Unit	Material	Labor	Equipment	Total	Total Incl O&P
0010	**SECURITY SERVICE SINK**, Stainless steel									
1700	Service sink, with soap dish									
1740	24" x 19" size	Q-1	3	5.333	Ea.	2,650	310		2,960	3,400
1790	For sink rough-in, supply, waste and vent	"	.89	17.978	"	760	1,050		1,810	2,425

22 46 73 – Security Shower

22 46 73.10 Security Shower		Crew	Daily Output	Labor-Hours	Unit	Material	Labor	Equipment	Total	Total Incl O&P
0010	**SECURITY SHOWER**, Stainless steel									
1800	Shower cabinet, unitized									
1840	36" x 36" x 88"	Q-1	2.20	7.273	Ea.	10,100	420		10,520	11,800
1900	Shower package for built-in									
1940	Hot & cold valves, recessed soap dish	Q-1	6	2.667	Ea.	440	155		595	725

22 47 Drinking Fountains and Water Coolers

22 47 13 – Drinking Fountains

22 47 13.10 Drinking Water Fountains		Crew	Daily Output	Labor-Hours	Unit	Material	Labor	Equipment	Total	Total Incl O&P
0010	**DRINKING WATER FOUNTAINS**, For connection to cold water supply R224000-30									
0802	For remote water chiller, see Section 22 47 23.10									
1000	Wall mounted, non-recessed									
1200	Aluminum,									
1280	Dual bubbler type	1 Plum	3.20	2.500	Ea.	2,625	161		2,786	3,125
1400	Bronze, with no back		4	2		1,125	129		1,254	1,425
1600	Cast iron, enameled, low back, single bubbler		4	2		1,225	129		1,354	1,550
1640	Dual bubbler type		3.20	2.500		1,725	161		1,886	2,150
1680	Triple bubbler type		3.20	2.500		2,150	161		2,311	2,625
1800	Cast aluminum, enameled, for correctional institutions		4	2		1,400	129		1,529	1,750
2000	Fiberglass, 12" back, single bubbler unit		4	2		2,000	129		2,129	2,425
2040	Dual bubbler		3.20	2.500		2,575	161		2,736	3,075
2080	Triple bubbler		3.20	2.500		2,825	161		2,986	3,350
2200	Polymarble, no back, single bubbler		4	2		935	129		1,064	1,225
2240	Dual bubbler		3.20	2.500		2,250	161		2,411	2,725
2280	Triple bubbler		3.20	2.500		2,175	161		2,336	2,650
2400	Precast stone, no back		4	2		1,050	129		1,179	1,350
2700	Stainless steel, single bubbler, no back		4	2		915	129		1,044	1,200
2740	With back		4	2		1,100	129		1,229	1,400
2780	Dual handle, ADA compliant		4	2		760	129		889	1,025
2820	Dual level, ADA compliant		3.20	2.500		1,600	161		1,761	2,000
2840	Vandal resistant type	↓	4	2	↓	680	129		809	945
3300	Vitreous china									

22 47 Drinking Fountains and Water Coolers

22 47 13 – Drinking Fountains

22 47 13.10 Drinking Water Fountains		Crew	Daily Output	Labor-Hours	Unit	Material	2020 Bare Costs Labor	Equipment	Total	Total Incl O&P
3340	7" back	1 Plum	4	2	Ea.	575	129		704	835
3940	For vandal-resistant bottom plate, add					75			75	82.50
3960	For freeze-proof valve system, add	1 Plum	2	4		770	258		1,028	1,250
3980	For rough-in, supply and waste, add	"	2.21	3.620		229	233		462	610
4000	Wall mounted, semi-recessed									
4200	Poly-marble, single bubbler	1 Plum	4	2	Ea.	960	129		1,089	1,250
4600	Stainless steel, satin finish, single bubbler		4	2		1,400	129		1,529	1,750
4900	Vitreous china, single bubbler		4	2		935	129		1,064	1,225
5980	For rough-in, supply and waste, add		1.83	4.372		229	282		511	685
6000	Wall mounted, fully recessed									
6400	Poly-marble, single bubbler	1 Plum	4	2	Ea.	1,775	129		1,904	2,150
6440	For water glass filler, add					95.50			95.50	105
6800	Stainless steel, single bubbler	1 Plum	4	2		1,750	129		1,879	2,125
6900	Fountain and cuspidor combination		2	4		3,150	258		3,408	3,875
7560	For freeze-proof valve system, add		2	4		985	258		1,243	1,475
7580	For rough-in, supply and waste, add		1.83	4.372		229	282		511	685
7600	Floor mounted, pedestal type									
7700	Aluminum, architectural style, CI base	1 Plum	2	4	Ea.	2,650	258		2,908	3,325
7780	ADA compliant unit ♿		2	4		1,675	258		1,933	2,250
8000	Bronze, architectural style		2	4		2,625	258		2,883	3,275
8040	Enameled steel cylindrical column style		2	4		2,325	258		2,583	2,975
8200	Precast stone/concrete, cylindrical column		1	8		1,625	515		2,140	2,575
8240	ADA compliant unit ♿		1	8		3,250	515		3,765	4,375
8400	Stainless steel, architectural style		2	4		2,150	258		2,408	2,775
8600	Enameled iron, heavy duty service, 2 bubblers		2	4		3,675	258		3,933	4,450
8660	4 bubblers		2	4		4,650	258		4,908	5,500
8880	For freeze-proof valve system, add		2	4		700	258		958	1,175
8900	For rough-in, supply and waste, add		1.83	4.372		229	282		511	685
9000	Minimum labor/equipment charge		2	4	Job		258		258	400
9100	Deck mounted									
9500	Stainless steel, circular receptor	1 Plum	4	2	Ea.	455	129		584	705
9540	14" x 9" receptor		4	2		380	129		509	615
9580	25" x 17" deep receptor, with water glass filler		3	2.667		300	172		472	595
9760	White enameled steel, 14" x 9" receptor		4	2		400	129		529	640
9860	White enameled cast iron, 24" x 16" receptor		3	2.667		525	172		697	845
9980	For rough-in, supply and waste, add		1.83	4.372		229	282		511	685

22 47 16 – Pressure Water Coolers

22 47 16.10 Electric Water Coolers		Crew	Daily Output	Labor-Hours	Unit	Material	Labor	Equipment	Total	Total Incl O&P
0010	ELECTRIC WATER COOLERS R224000-30									
0100	Wall mounted, non-recessed									
0140	4 GPH	Q-1	4	4	Ea.	700	232		932	1,125
0160	8 GPH, barrier free, sensor operated ♿		4	4		1,125	232		1,357	1,600
0180	8.2 GPH		4	4		1,025	232		1,257	1,475
0220	14.3 GPH		4	4		1,075	232		1,307	1,525
0600	8 GPH hot and cold water		4	4		1,050	232		1,282	1,500
0640	For stainless steel cabinet, add					93			93	102
1000	Dual height, 8.2 GPH	Q-1	3.80	4.211		2,025	244		2,269	2,600
1040	14.3 GPH	"	3.80	4.211		1,750	244		1,994	2,300
1240	For stainless steel cabinet, add					213			213	235
2600	ADA compliant, 8 GPH ♿	Q-1	4	4		1,100	232		1,332	1,550
3000	Simulated recessed, 8 GPH		4	4		795	232		1,027	1,225
3040	11.5 GPH		4	4		1,025	232		1,257	1,475

22 47 Drinking Fountains and Water Coolers

22 47 16 – Pressure Water Coolers

22 47 16.10 Electric Water Coolers		Crew	Daily Output	Labor-Hours	Unit	Material	2020 Bare Costs Labor	Equipment	Total	Total Incl O&P
3200	For glass filler, add				Ea.	102			102	112
3240	For stainless steel cabinet, add					86.50			86.50	95.50
3300	Semi-recessed, 8.1 GPH	Q-1	4	4		915	232		1,147	1,350
3320	12 GPH	"	4	4		1,025	232		1,257	1,475
3340	For glass filler, add					172			172	189
3360	For stainless steel cabinet, add					161			161	177
3400	Full recessed, stainless steel, 8 GPH	Q-1	3.50	4.571		2,325	265		2,590	2,975
3420	11.5 GPH	"	3.50	4.571		1,775	265		2,040	2,350
3460	For glass filler, add					197			197	216
3600	For mounting can only					218			218	240
4600	Floor mounted, flush-to-wall									
4640	4 GPH	1 Plum	3	2.667	Ea.	840	172		1,012	1,175
4680	8.2 GPH		3	2.667		935	172		1,107	1,300
4720	14.3 GPH		3	2.667		1,075	172		1,247	1,450
4960	14 GPH hot and cold water		3	2.667		1,200	172		1,372	1,600
4980	For stainless steel cabinet, add					141			141	155
5000	Dual height, 8.2 GPH	1 Plum	2	4		1,225	258		1,483	1,750
5040	14.3 GPH	"	2	4		1,275	258		1,533	1,800
5120	For stainless steel cabinet, add					209			209	229
5600	Explosion proof, 16 GPH	1 Plum	3	2.667		2,525	172		2,697	3,050
6000	Refrigerator compartment type, 4.5 GPH		3	2.667		1,550	172		1,722	2,000
6600	Bottle supply type, 1.0 GPH		4	2		415	129		544	655
6640	Hot and cold, 1.0 GPH		4	2		570	129		699	830
9000	Minimum labor/equipment charge		2	4	Job		258		258	400
9800	For supply, waste & vent, all coolers		2.21	3.620	Ea.	229	233		462	610

22 47 23 – Remote Water Coolers

22 47 23.10 Remote Water Coolers		Crew	Daily Output	Labor-Hours	Unit	Material	Labor	Equipment	Total	Total Incl O&P
0010	**REMOTE WATER COOLERS**, 80°F inlet									
0100	Air cooled, 50°F outlet, 115 V, 4.1 GPH	1 Plum	6	1.333	Ea.	580	86		666	775
0200	5.7 GPH		5.50	1.455		1,075	94		1,169	1,325
0300	8.0 GPH		5	1.600		825	103		928	1,075
0400	10.0 GPH		4.50	1.778		1,075	115		1,190	1,350
0500	13.4 GPH		4	2		1,800	129		1,929	2,175
0700	29 GPH	Q-1	5	3.200		1,925	186		2,111	2,400
1000	230 V, 32 GPH	"	5	3.200		2,000	186		2,186	2,475

22 51 Swimming Pool Plumbing Systems

22 51 19 – Swimming Pool Water Treatment Equipment

22 51 19.50 Swimming Pool Filtration Equipment		Crew	Daily Output	Labor-Hours	Unit	Material	Labor	Equipment	Total	Total Incl O&P
0010	**SWIMMING POOL FILTRATION EQUIPMENT**									
0900	Filter system, sand or diatomite type, incl. pump, 6,000 gal./hr.	2 Plum	1.80	8.889	Total	2,375	575		2,950	3,500
1020	Add for chlorination system, 800 S.F. pool		3	5.333	Ea.	235	345		580	790
1040	5,000 S.F. pool		3	5.333	"	1,950	345		2,295	2,650

22 52 Fountain Plumbing Systems

22 52 16 – Fountain Pumps

22 52 16.10 Fountain Water Pumps		Crew	Daily Output	Labor-Hours	Unit	Material	2020 Bare Costs Labor	Equipment	Total	Total Incl O&P
0010	**FOUNTAIN WATER PUMPS**									
0100	Pump w/controls									
0200	Single phase, 100′ cord, 1/2 HP pump	2 Skwk	4.40	3.636	Ea.	1,325	199		1,524	1,775
0300	3/4 HP pump		4.30	3.721		1,375	204		1,579	1,850
0400	1 HP pump		4.20	3.810		1,925	209		2,134	2,450
0500	1-1/2 HP pump		4.10	3.902		2,575	214		2,789	3,200
0600	2 HP pump		4	4		4,350	219		4,569	5,125
0700	Three phase, 200′ cord, 5 HP pump		3.90	4.103		6,250	225		6,475	7,225
0800	7-1/2 HP pump		3.80	4.211		12,600	231		12,831	14,200
0900	10 HP pump		3.70	4.324		15,200	237		15,437	17,200
1000	15 HP pump	↓	3.60	4.444	↓	21,500	244		21,744	24,000
2000	DESIGN NOTE: Use two horsepower per surface acre.									

22 52 33 – Fountain Ancillary

22 52 33.10 Fountain Miscellaneous		Crew	Daily Output	Labor-Hours	Unit	Material	Labor	Equipment	Total	Total Incl O&P
0010	**FOUNTAIN MISCELLANEOUS**									
1300	Lights w/mounting kits, 200 watt	2 Skwk	18	.889	Ea.	1,200	49		1,249	1,400
1400	300 watt		18	.889		1,350	49		1,399	1,575
1500	500 watt		18	.889		1,525	49		1,574	1,750
1600	Color blender	↓	12	1.333	↓	600	73		673	775

22 62 Vacuum Systems for Laboratory and Healthcare Facilities

22 62 19 – Vacuum Equipment for Laboratory and Healthcare Facilities

22 62 19.70 Healthcare Vacuum Equipment		Crew	Daily Output	Labor-Hours	Unit	Material	Labor	Equipment	Total	Total Incl O&P
0010	**HEALTHCARE VACUUM EQUIPMENT**									
0300	Dental oral									
0310	Duplex									
0330	165 SCFM with 77 gal. separator	Q-2	1.30	18.462	Ea.	52,500	1,100		53,600	59,000
1100	Vacuum system									
1110	Vacuum outlet alarm panel	1 Plum	3.20	2.500	Ea.	1,100	161		1,261	1,450
2000	Medical, with receiver									
2100	Rotary vane type, lubricated, with controls									
2110	Simplex									
2120	1.5 HP, 80 gal. tank	Q-1	8	2	Ea.	6,650	116		6,766	7,500
2130	2 HP, 80 gal. tank		7	2.286		6,900	133		7,033	7,775
2140	3 HP, 80 gal. tank		6.60	2.424		7,275	141		7,416	8,225
2150	5 HP, 80 gal. tank	↓	6	2.667	↓	7,925	155		8,080	8,975
2200	Duplex									
2210	1 HP, 80 gal. tank	Q-1	8.40	1.905	Ea.	10,600	110		10,710	11,900
2220	1.5 HP, 80 gal. tank		7.80	2.051		10,900	119		11,019	12,200
2230	2 HP, 80 gal. tank		6.80	2.353		11,800	136		11,936	13,200
2240	3 HP, 120 gal. tank		6	2.667		12,800	155		12,955	14,300
2250	5 HP, 120 gal. tank	↓	5.40	2.963		14,600	172		14,772	16,300
2260	7.5 HP, 200 gal. tank	Q-2	7	3.429		20,900	206		21,106	23,200
2270	10 HP, 200 gal. tank		6.40	3.750		24,100	226		24,326	26,900
2280	15 HP, 200 gal. tank		5.80	4.138		43,100	249		43,349	47,800
2290	20 HP, 200 gal. tank		5	4.800		49,100	289		49,389	54,500
2300	25 HP, 200 gal. tank	↓	4	6	↓	57,000	360		57,360	63,500
2400	Triplex									
2410	7.5 HP, 200 gal. tank	Q-2	6.40	3.750	Ea.	33,100	226		33,326	36,800
2420	10 HP, 200 gal. tank	↓	5.70	4.211	↓	38,500	253		38,753	42,800

22 62 19.70 Healthcare Vacuum Equipment		Crew	Daily Output	Labor-Hours	Unit	2020 Bare Costs Material	Labor	Equipment	Total	Total Incl O&P
2430	15 HP, 200 gal. tank	Q-2	4.90	4.898	Ea.	65,000	295		65,295	72,000
2440	20 HP, 200 gal. tank		4	6		74,500	360		74,860	82,000
2450	25 HP, 200 gal. tank	↓	3.70	6.486	↓	86,000	390		86,390	95,000
2500	Quadruplex									
2510	7.5 HP, 200 gal. tank	Q-2	5.30	4.528	Ea.	41,200	272		41,472	45,700
2520	10 HP, 200 gal. tank		4.70	5.106		47,500	305		47,805	53,000
2530	15 HP, 200 gal. tank		4	6		86,000	360		86,360	95,000
2540	20 HP, 200 gal. tank		3.60	6.667		98,000	400		98,400	108,000
2550	25 HP, 200 gal. tank	↓	3.20	7.500	↓	113,500	450		113,950	125,500
4000	Liquid ring type, water sealed, with controls									
4200	Duplex									
4210	1.5 HP, 120 gal. tank	Q-1	6	2.667	Ea.	21,000	155		21,155	23,300
4220	3 HP, 120 gal. tank	"	5.50	2.909		21,900	169		22,069	24,400
4230	4 HP, 120 gal. tank	Q-2	6.80	3.529		23,400	212		23,612	26,000
4240	5 HP, 120 gal. tank		6.50	3.692		24,700	222		24,922	27,500
4250	7.5 HP, 200 gal. tank		6.20	3.871		30,000	233		30,233	33,400
4260	10 HP, 200 gal. tank		5.80	4.138		37,500	249		37,749	41,600
4270	15 HP, 200 gal. tank		5.10	4.706		45,400	283		45,683	50,500
4280	20 HP, 200 gal. tank		4.60	5.217		58,500	315		58,815	65,000
4290	30 HP, 200 gal. tank	↓	4	6	↓	70,000	360		70,360	77,500

22 63 Gas Systems for Laboratory and Healthcare Facilities

22 63 13 – Gas Piping for Laboratory and Healthcare Facilities

22 63 13.70 Healthcare Gas Piping

		Crew	Daily Output	Labor-Hours	Unit	Material	Labor	Equipment	Total	Total Incl O&P
0010	**HEALTHCARE GAS PIPING**									
0030	Air compressor intake filter									
0034	Rooftop									
0036	Filter/silencer									
0040	1"	1 Stpi	10	.800	Ea.	254	52.50		306.50	360
0044	1-1/4"		9.60	.833		261	54.50		315.50	370
0048	1-1/2"		9.20	.870		280	57		337	400
0052	2"		8.80	.909		280	59.50		339.50	400
0056	2-1/2"		8.40	.952		375	62.50		437.50	505
0060	3"		8	1		450	65.50		515.50	595
0064	4"		7.60	1.053		465	69		534	620
0068	5"	↓	7.40	1.081	↓	655	71		726	830
0076	Inline									
0080	Filter/silencer									
0084	2"	1 Stpi	8.20	.976	Ea.	850	64		914	1,025
0088	2-1/2"		8	1		870	65.50		935.50	1,050
0090	3"		7.60	1.053		965	69		1,034	1,150
0092	4"		7.20	1.111		1,125	73		1,198	1,375
0094	5"		6.80	1.176		1,225	77		1,302	1,475
0096	6"	↓	6.40	1.250	↓	1,375	82		1,457	1,625
1000	Nitrogen or oxygen system									
1010	Cylinder manifold									
1020	5 cylinder	1 Plum	.80	10	Ea.	5,900	645		6,545	7,475
1026	10 cylinder	"	.40	20		7,100	1,300		8,400	9,800
1050	Nitrogen generator, 30 LPM	Q-5	.40	40	↓	28,900	2,350		31,250	35,500
1900	Vaporizers									
1910	LOX vaporizers									

22 63 13.70 Healthcare Gas Piping		Crew	Daily Output	Labor-Hours	Unit	2020 Bare Costs Material	Labor	Equipment	Total	Total Incl O&P
1920	Nominal capacity									
1930	1410 SCFM	Q-1	2	8	Ea.	1,900	465		2,365	2,825
1940	5650 SCFM		1.40	11.429		4,400	665		5,065	5,875
1950	12,703 SCFM	↓	.80	20	↓	6,750	1,150		7,900	9,225
1980	Removal of LOX vaporizers									
1982	Nominal capacity									
1986	1410 SCFM	Q-1	4	4	Ea.		232		232	360
1990	5650 SCFM		2.80	5.714			330		330	510
1994	12,703 SCFM	↓	1.60	10	↓		580		580	895
3000	Outlets and valves									
3010	Recessed, wall mounted									
3012	Single outlet	1 Plum	3.20	2.500	Ea.	70.50	161		231.50	325
3100	Ceiling outlet									
3190	Zone valve with box									
3192	Cleaned for oxygen service, not including gauges									
3194	1/2" valve size	1 Plum	4.60	1.739	Ea.	222	112		334	415
3196	3/4" valve size		4.30	1.860		245	120		365	455
3198	1" valve size		4	2		271	129		400	495
3202	1-1/4" valve size		3.80	2.105		305	136		441	545
3206	1-1/2" valve size		3.60	2.222		340	143		483	590
3210	2" valve size		3.20	2.500		415	161		576	705
3214	2-1/2" valve size		3.10	2.581		1,000	166		1,166	1,350
3218	3" valve size	↓	3	2.667	↓	1,400	172		1,572	1,825
3224	Gauges for zone valve box									
3226	0-100 psi (O2, air, N2, CO_2)	1 Plum	16	.500	Ea.	17.10	32		49.10	69
3228	Vacuum, WAGD (Waste Anesthesia Gas)		16	.500		17.10	32		49.10	69
3230	0-300 psi (Nitrogen)	↓	16	.500	↓	17.10	32		49.10	69
4000	Alarm panel, medical gases and vacuum									
4010	Alarm panel	1 Plum	3.20	2.500	Ea.	1,100	161		1,261	1,450
4030	Master alarm panel									
4034	Can also monitor area alarms									
4038	and communicate with PC-based alarm monitor.									
4040	10 signal	1 Elec	3.20	2.500	Ea.	1,025	153		1,178	1,350
4044	20 signal		3	2.667		1,175	164		1,339	1,550
4048	30 signal		2.80	2.857		1,575	175		1,750	2,000
4052	40 signal		2.60	3.077		1,775	189		1,964	2,250
4056	50 signal		2.40	3.333		1,900	205		2,105	2,425
4060	60 signal	↓	2.20	3.636	↓	2,225	223		2,448	2,800
4100	Area alarm panel									
4104	Does not include specific gas transducers.									
4108	3 module alarm panel									
4112	P-P-P	1 Elec	2.80	2.857	Ea.	1,375	175		1,550	1,775
4116	P-P-V		2.80	2.857		930	175		1,105	1,300
4120	D-D-D	↓	2.80	2.857	↓	1,025	175		1,200	1,400
4130	6 module alarm panel									
4132	4-P, 3-V	1 Elec	2.20	3.636	Ea.	1,450	223		1,673	1,950
4136	3-P, 2-V, B		2.20	3.636		1,075	223		1,298	1,550
4140	5-D, B		2.20	3.636		1,425	223		1,648	1,925
4144	6-D		2.20	3.636		1,900	223		2,123	2,450
4148	3-P, 3-V	↓	2.20	3.636	↓	1,775	223		1,998	2,300
4170	Note: P=pressure, V=vacuum, B=blank, D=dual display									
4180	Alarm transducers, gas specific									
4182	Oxygen	1 Elec	24	.333	Ea.	157	20.50		177.50	205

22 63 13 – Gas Piping for Laboratory and Healthcare Facilities

22 63 13.70 Healthcare Gas Piping		Crew	Daily Output	Labor-Hours	Unit	2020 Bare Costs Material	Labor	Equipment	Total	Total Incl O&P
4184	Vacuum	1 Elec	24	.333	Ea.	153	20.50		173.50	201
4186	Nitrous oxide		24	.333		153	20.50		173.50	201
4188	Medical air		24	.333		157	20.50		177.50	205
4190	Carbon dioxide		24	.333		157	20.50		177.50	205
4192	Nitrogen		24	.333		157	20.50		177.50	205
4194	WAGD		24	.333		167	20.50		187.50	216
4300	Ball valves cleaned for oxygen service									
4310	with copper extensions and gauge port									
4320	1/4" diam.	1 Plum	24	.333	Ea.	63.50	21.50		85	103
4330	1/2" diam.		22	.364		60	23.50		83.50	102
4334	3/4" diam.		20	.400		77.50	26		103.50	126
4338	1" diam.		19	.421		112	27		139	165
4342	1-1/4" diam.		15	.533		135	34.50		169.50	202
4346	1-1/2" diam.		13	.615		179	39.50		218.50	258
4350	2" diam.		11	.727		305	47		352	410
4354	2-1/2" diam.	Q-1	15	1.067		845	62		907	1,025
4358	3" diam.		13	1.231		1,250	71.50		1,321.50	1,475
4362	4" diam.		10	1.600		2,400	93		2,493	2,775
5000	Manifold									
5010	Automatic switchover type									
5020	Note: Both a control panel and header assembly are required									
5030	Control panel									
5040	Oxygen	Q-1	4	4	Ea.	4,825	232		5,057	5,675
5060	Header assembly									
5066	Oxygen									
5070	2 x 2	Q-1	6	2.667	Ea.	790	155		945	1,100
5074	3 x 3		5.50	2.909		930	169		1,099	1,275
5078	4 x 4		5	3.200		1,225	186		1,411	1,625
5082	5 x 5		4.50	3.556		1,350	206		1,556	1,825
5086	6 x 6		4	4		1,700	232		1,932	2,225
5090	7 x 7		3.50	4.571		1,850	265		2,115	2,425
7000	Medical air compressors									
7020	Oil-less with inlet filter, duplexed dryers and aftercoolers									
7030	Duplex systems, horizontal tank, 208/230/460/575 V, 3 Ph.									
7040	1 HP, 80 gal. tank, 3.8 ACFM @50PSIG, 2.7 ACFM @100PSIG	Q-5	4	4	Ea.	29,900	236		30,136	33,300
7050	1.5 HP, 80 gal. tank, 6.8 ACFM @50PSIG, 4.5 ACFM @100PSIG		3.80	4.211		35,700	248		35,948	39,600
7060	2 HP, 80 gal. tank, 8.2 ACFM @50PSIG, 6.3 ACFM @100PSIG		3.60	4.444		35,800	262		36,062	39,800
7070	3 HP, 120 gal. tank, 11.2 ACFM @50PSIG, 9.4 ACFM @100PSIG		3.20	5		27,700	295		27,995	31,000
7080	5 HP, 120 gal. tank, 17.4 ACFM @50PSIG, 15.6 ACFM @100PSIG		2.80	5.714		46,200	335		46,535	51,500
7090	7.5 HP, 250 gal. tank, 31.6 ACFM @50PSIG, 25.9 ACFM @100PSIG		2.40	6.667		61,500	395		61,895	68,000
7100	10 HP, 250 gal. tank, 43 ACFM @50PSIG, 35.2 ACFM @100PSIG		2	8		66,500	470		66,970	73,500
7110	15 HP, 250 gal. tank, 69 ACFM @50PSIG, 56.5 ACFM @100PSIG		1.80	8.889		80,000	525		80,525	89,000
7200	Aftercooler, air-cooled									
7210	Steel manifold, copper tube, aluminum fins									
7220	35 SCFM @ 100PSIG	Q-5	4	4	Ea.	1,025	236		1,261	1,500
7300	Air dryer system									
7310	Refrigerated type									
7320	Flow @ 125 psi									
7330	20 SCFM	Q-5	6	2.667	Ea.	1,525	157		1,682	1,925
7332	25 SCFM		5.80	2.759		1,600	163		1,763	2,000
7334	35 SCFM		5.40	2.963		1,950	175		2,125	2,425
7336	50 SCFM		5	3.200		2,425	189		2,614	2,975
7338	75 SCFM		4.60	3.478		3,025	205		3,230	3,650

22 63 Gas Systems for Laboratory and Healthcare Facilities

22 63 13 – Gas Piping for Laboratory and Healthcare Facilities

22 63 13.70 Healthcare Gas Piping		Crew	Daily Output	Labor-Hours	Unit	2020 Bare Costs Material	Labor	Equipment	Total	Total Incl O&P
7340	100 SCFM	Q-5	4.30	3.721	Ea.	3,625	220		3,845	4,350
7342	125 SCFM	↓	4	4	↓	4,675	236		4,911	5,525
7360	Desiccant type									
7362	Flow with energy saving controls									
7366	40 SCFM	Q-5	3.60	4.444	Ea.	6,200	262		6,462	7,225
7368	60 SCFM		3.40	4.706		6,750	278		7,028	7,875
7370	90 SCFM		3.20	5		8,025	295		8,320	9,275
7372	115 SCFM		3	5.333		8,625	315		8,940	9,950
7374	165 SCFM	↓	2.90	5.517		9,275	325		9,600	10,700
7378	260 SCFM	Q-6	3.60	6.667		10,900	410		11,310	12,600
7382	370 SCFM		3.40	7.059		12,700	430		13,130	14,700
7386	590 SCFM		3.20	7.500		15,800	460		16,260	18,000
7390	1130 SCFM	↓	3	8	↓	24,400	490		24,890	27,600
7460	Dew point monitor									
7466	LCD readout, high dew point alarm, probe included									
7470	Monitor	1 Stpi	2	4	Ea.	3,250	262		3,512	3,975

22 66 Chemical-Waste Systems for Lab. and Healthcare Facilities

22 66 53 – Laboratory Chemical-Waste and Vent Piping

22 66 53.30 Glass Pipe		Crew	Daily Output	Labor-Hours	Unit	Material	Labor	Equipment	Total	Total Incl O&P
0010	**GLASS PIPE**, Borosilicate, couplings & clevis hanger assemblies, 10′ OC									
0020	Drainage									
1100	1-1/2″ diameter	Q-1	52	.308	L.F.	15.30	17.85		33.15	44.50
1120	2″ diameter		44	.364		19.50	21		40.50	54
1140	3″ diameter		39	.410		26	24		50	65.50
1160	4″ diameter		30	.533		42.50	31		73.50	95
1180	6″ diameter	↓	26	.615	↓	75	35.50		110.50	138
1870	To delete coupling & hanger, subtract									
1880	1-1/2″ diam. to 2″ diam.					19%	22%			
1890	3″ diam. to 6″ diam.					20%	17%			
2000	Process supply (pressure), beaded joints									
2040	1/2″ diameter	1 Plum	36	.222	L.F.	9.55	14.30		23.85	32.50
2060	3/4″ diameter		31	.258		10.55	16.65		27.20	37
2080	1″ diameter	↓	27	.296		24.50	19.10		43.60	56.50
2100	1-1/2″ diameter	Q-1	47	.340		17	19.75		36.75	49
2120	2″ diameter		39	.410		21	24		45	60
2140	3″ diameter		34	.471		29.50	27.50		57	74.50
2160	4″ diameter		25	.640		42.50	37		79.50	105
2180	6″ diameter	↓	21	.762	↓	128	44		172	210
2860	To delete coupling & hanger, subtract									
2870	1/2″ diam. to 1″ diam.					25%	33%			
2880	1-1/2″ diam. to 3″ diam.					22%	21%			
2890	4″ diam. to 6″ diam.					23%	15%			
3800	Conical joint, transparent									
3980	6″ diameter	Q-1	21	.762	L.F.	168	44		212	254
4500	To delete couplings & hangers, subtract									
4530	6″ diam.					22%	26%			
9000	Minimum labor/equipment charge	1 Plum	4	2	Job		129		129	199

22 66 53.40 Pipe Fittings, Glass		Crew	Daily Output	Labor-Hours	Unit	2020 Bare Costs Material	Labor	Equipment	Total	Total Incl O&P
0010	**PIPE FITTINGS, GLASS**									
0020	Drainage, beaded ends									
0040	Coupling & labor required at joints not incl. in fitting									
0050	price. Add 1 per joint for installed price									
0070	90° bend or sweep, 1-1/2"				Ea.	36.50			36.50	40
0090	2"					46.50			46.50	51.50
0100	3"					76			76	83.50
0110	4"					121			121	134
0120	6" (sweep only)					360			360	395
0200	45° bend or sweep same as 90°									
0350	Tee, single sanitary, 1-1/2"				Ea.	59			59	65
0370	2"					59			59	65
0380	3"					86			86	94.50
0390	4"					157			157	173
0400	6"					420			420	465
0410	Tee, straight, 1-1/2"					73			73	80
0430	2"					73			73	80
0440	3"					106			106	116
0450	4"					139			139	153
0460	6"					455			455	500
0500	Coupling, stainless steel, TFE seal ring									
0520	1-1/2"	Q-1	32	.500	Ea.	27.50	29		56.50	75.50
0530	2"		30	.533		35	31		66	86.50
0540	3"		25	.640		33.50	37		70.50	94
0550	4"		23	.696		80.50	40.50		121	151
0560	6"		20	.800		181	46.50		227.50	271
0600	Coupling, stainless steel, bead to plain end									
0610	1-1/2"	Q-1	36	.444	Ea.	36.50	26		62.50	80.50
0620	2"		34	.471		44.50	27.50		72	90.50
0630	3"		29	.552		75.50	32		107.50	133
0640	4"		27	.593		112	34.50		146.50	176
0650	6"		24	.667		380	38.50		418.50	475
2350	Coupling, Viton liner, for temperatures to 400°F									
2370	1/2"	Q-1	40	.400	Ea.	71	23		94	114
2380	3/4"		37	.432		81	25		106	128
2390	1"		35	.457		135	26.50		161.50	189
2400	1-1/2"		32	.500		27.50	29		56.50	75.50
2410	2"		30	.533		34.50	31		65.50	86
2420	3"		25	.640		46	37		83	108
2430	4"		23	.696		79.50	40.50		120	150
2440	6"		20	.800		201	46.50		247.50	293
2550	For beaded joint armored fittings, add					200%				
2600	Conical ends. Flange set, gasket & labor not incl. in fitting									
2620	price. Add 1 per joint for installed price.									
2650	90° sweep elbow, 1"				Ea.	118			118	130
2670	1-1/2"					275			275	305
2680	2"					246			246	270
2690	3"					445			445	485
2700	4"					795			795	875
2710	6"					1,175			1,175	1,300
2750	Cross (straight), add					55%				
2850	Tee, add					20%				

22 66 53 – Laboratory Chemical-Waste and Vent Piping

22 66 53.40 Pipe Fittings, Glass		Crew	Daily Output	Labor-Hours	Unit	Material	Labor (2020 Bare Costs)	Equipment	Total	Total Incl O&P
9000	Minimum labor/equipment charge	1 Plum	4	2	Job		129		129	199

22 66 53.60 Corrosion Resistant Pipe

Line	Description	Crew	Daily Output	Labor-Hours	Unit	Material	Labor	Equipment	Total	Total Incl O&P
0010	**CORROSION RESISTANT PIPE**, No couplings or hangers									
0020	Iron alloy, drain, mechanical joint									
1000	1-1/2" diameter	Q-1	70	.229	L.F.	84.50	13.25		97.75	114
1100	2" diameter		66	.242		87	14.05		101.05	117
1120	3" diameter		60	.267		95	15.45		110.45	129
1140	4" diameter		52	.308		117	17.85		134.85	157
1980	Iron alloy, drain, B&S joint									
2000	2" diameter	Q-1	54	.296	L.F.	85.50	17.20		102.70	121
2100	3" diameter		52	.308		82.50	17.85		100.35	118
2120	4" diameter		48	.333		101	19.35		120.35	141
2140	6" diameter	Q-2	59	.407		152	24.50		176.50	205
2160	8" diameter	"	54	.444		315	26.50		341.50	385
2980	Plastic, epoxy, fiberglass filament wound, B&S joint									
3000	2" diameter	Q-1	62	.258	L.F.	12.35	14.95		27.30	36.50
3100	3" diameter		51	.314		14.40	18.20		32.60	44
3120	4" diameter		45	.356		20.50	20.50		41	54.50
3140	6" diameter		32	.500		29	29		58	76.50
3160	8" diameter	Q-2	38	.632		45	38		83	108
3180	10" diameter		32	.750		62	45		107	138
3200	12" diameter		28	.857		74.50	51.50		126	162
3980	Polyester, fiberglass filament wound, B&S joint									
4000	2" diameter	Q-1	62	.258	L.F.	13.40	14.95		28.35	38
4100	3" diameter		51	.314		17.55	18.20		35.75	47.50
4120	4" diameter		45	.356		25.50	20.50		46	60.50
4140	6" diameter		32	.500		37	29		66	86
4160	8" diameter	Q-2	38	.632		88	38		126	156
4180	10" diameter		32	.750		118	45		163	200
4200	12" diameter		28	.857		130	51.50		181.50	223
4980	Polypropylene, acid resistant, fire retardant, Schedule 40									
5000	1-1/2" diameter	Q-1	68	.235	L.F.	9.05	13.65		22.70	31
5100	2" diameter		62	.258		14.55	14.95		29.50	39
5120	3" diameter		51	.314		25	18.20		43.20	55.50
5140	4" diameter		45	.356		32	20.50		52.50	67
5160	6" diameter		32	.500		63.50	29		92.50	115
5980	Proxylene, fire retardant, Schedule 40									
6000	1-1/2" diameter	Q-1	68	.235	L.F.	14	13.65		27.65	36.50
6100	2" diameter		62	.258		19.20	14.95		34.15	44
6120	3" diameter		51	.314		34.50	18.20		52.70	66
6140	4" diameter		45	.356		49	20.50		69.50	86
6160	6" diameter		32	.500		83	29		112	137
6820	For Schedule 80, add					35%	2%			
9800	Minimum labor/equipment charge	1 Plum	4	2	Job		129		129	199

22 66 53.70 Pipe Fittings, Corrosion Resistant

Line	Description	Crew	Daily Output	Labor-Hours	Unit	Material	Labor	Equipment	Total	Total Incl O&P
0010	**PIPE FITTINGS, CORROSION RESISTANT**									
0030	Iron alloy									
0050	Mechanical joint									
0060	1/4 bend, 1-1/2"	Q-1	12	1.333	Ea.	125	77.50		202.50	257
0080	2"		10	1.600		204	93		297	370
0090	3"		9	1.778		245	103		348	430
0100	4"		8	2		282	116		398	490

22 66 53 – Laboratory Chemical-Waste and Vent Piping

22 66 53.70 Pipe Fittings, Corrosion Resistant		Crew	Daily Output	Labor-Hours	Unit	Material	Labor	Equipment	Total	Total Incl O&P
						2020 Bare Costs				
0110	1/8 bend, 1-1/2"	Q-1	12	1.333	Ea.	79.50	77.50		157	208
0130	2"		10	1.600		136	93		229	293
0140	3"		9	1.778		182	103		285	360
0150	4"		8	2		243	116		359	445
0160	Tee and Y, sanitary, straight									
0170	1-1/2"	Q-1	8	2	Ea.	136	116		252	330
0180	2"		7	2.286		182	133		315	405
0190	3"		6	2.667		282	155		437	550
0200	4"		5	3.200		520	186		706	855
0360	Coupling, 1-1/2"		14	1.143		75	66.50		141.50	185
0380	2"		12	1.333		85	77.50		162.50	214
0390	3"		11	1.455		89.50	84.50		174	229
0400	4"		10	1.600		101	93		194	254
0500	Bell & Spigot									
0510	1/4 and 1/16 bend, 2"	Q-1	16	1	Ea.	119	58		177	221
0520	3"		14	1.143		278	66.50		344.50	405
0530	4"		13	1.231		285	71.50		356.50	425
0540	6"	Q-2	17	1.412		565	85		650	750
0550	8"	"	12	2		2,325	120		2,445	2,725
0620	1/8 bend, 2"	Q-1	16	1		134	58		192	237
0640	3"		14	1.143		248	66.50		314.50	375
0650	4"		13	1.231		246	71.50		317.50	380
0660	6"	Q-2	17	1.412		470	85		555	650
0680	8"	"	12	2		1,850	120		1,970	2,200
0700	Tee, sanitary, 2"	Q-1	10	1.600		254	93		347	425
0710	3"		9	1.778		825	103		928	1,075
0720	4"		8	2		685	116		801	935
0730	6"	Q-2	11	2.182		855	131		986	1,150
0740	8"	"	8	3		2,350	180		2,530	2,850
1800	Y, sanitary, 2"	Q-1	10	1.600		268	93		361	440
1820	3"		9	1.778		480	103		583	685
1830	4"		8	2		430	116		546	655
1840	6"	Q-2	11	2.182		1,425	131		1,556	1,775
1850	8"	"	8	3		3,950	180		4,130	4,625
3000	Epoxy, filament wound									
3030	Quick-lock joint									
3040	90° elbow, 2"	Q-1	28	.571	Ea.	100	33		133	163
3060	3"		16	1		116	58		174	217
3070	4"		13	1.231		158	71.50		229.50	283
3080	6"		8	2		230	116		346	430
3090	8"	Q-2	9	2.667		425	160		585	715
3100	10"		7	3.429		535	206		741	905
3110	12"		6	4		765	241		1,006	1,200
3120	45° elbow, 2"	Q-1	28	.571		77.50	33		110.50	137
3130	3"		16	1		119	58		177	221
3140	4"		13	1.231		122	71.50		193.50	244
3150	6"		8	2		230	116		346	430
3160	8"	Q-2	9	2.667		425	160		585	715
3170	10"		7	3.429		535	206		741	905
3180	12"		6	4		765	241		1,006	1,200
3190	Tee, 2"	Q-1	19	.842		239	49		288	340
3200	3"		11	1.455		289	84.50		373.50	450
3210	4"		9	1.778		350	103		453	545

22 66 53 – Laboratory Chemical-Waste and Vent Piping

22 66 53.70 Pipe Fittings, Corrosion Resistant		Crew	Daily Output	Labor-Hours	Unit	2020 Bare Costs Material	Labor	Equipment	Total	Total Incl O&P
3220	6"	Q-1	5	3.200	Ea.	585	186		771	930
3230	8"	Q-2	6	4		665	241		906	1,100
3240	10"		5	4.800		925	289		1,214	1,475
3250	12"		4	6		1,425	360		1,785	2,125
4000	Polypropylene, acid resistant									
4020	Non-pressure, electrofusion joints									
4050	1/4 bend, 1-1/2"	1 Plum	16	.500	Ea.	16.55	32		48.55	68.50
4060	2"	Q-1	28	.571		33	33		66	87.50
4080	3"		17	.941		35.50	54.50		90	124
4090	4"		14	1.143		58	66.50		124.50	166
4110	6"		8	2		138	116		254	330
4150	1/4 bend, long sweep									
4170	1-1/2"	1 Plum	16	.500	Ea.	18.70	32		50.70	70.50
4180	2"	Q-1	28	.571		33	33		66	87.50
4200	3"		17	.941		41	54.50		95.50	130
4210	4"		14	1.143		59.50	66.50		126	168
4250	1/8 bend, 1-1/2"	1 Plum	16	.500		17.90	32		49.90	69.50
4260	2"	Q-1	28	.571		20	33		53	73.50
4280	3"		17	.941		37.50	54.50		92	126
4290	4"		14	1.143		42	66.50		108.50	148
4310	6"		8	2		116	116		232	305
4400	Tee, sanitary									
4420	1-1/2"	1 Plum	10	.800	Ea.	20.50	51.50		72	102
4430	2"	Q-1	17	.941		25	54.50		79.50	112
4450	3"		11	1.455		50	84.50		134.50	185
4460	4"		9	1.778		75	103		178	242
4480	6"		5	3.200		495	186		681	830
4490	Tee, sanitary reducing, 2" x 2" x 1-1/2"		17	.941		25	54.50		79.50	112
4492	3" x 3" x 2"		11	1.455		50	84.50		134.50	185
4494	4" x 4" x 3"		9	1.778		73	103		176	240
4496	6" x 6" x 4"		5	3.200		240	186		426	550
4650	Wye 45°, 1-1/2"	1 Plum	10	.800		23	51.50		74.50	105
4652	2"	Q-1	17	.941		31.50	54.50		86	119
4653	3"		11	1.455		53.50	84.50		138	189
4654	4"		9	1.778		77.50	103		180.50	244
4656	6"		5	3.200		200	186		386	505
4678	Combination Y & 1/8 bend									
4681	1-1/2"	1 Plum	10	.800	Ea.	28	51.50		79.50	110
4683	2"	Q-1	17	.941		35.50	54.50		90	124
4684	3"		11	1.455		60.50	84.50		145	197
4685	4"		9	1.778		83	103		186	251
9000	Minimum labor/equipment charge	1 Plum	4	2	Job		129		129	199

22 66 83 – Chemical-Waste Tanks

22 66 83.13 Chemical-Waste Dilution Tanks		Crew	Daily Output	Labor-Hours	Unit	Material	Labor	Equipment	Total	Total Incl O&P
0010	**CHEMICAL-WASTE DILUTION TANKS**									
7000	Tanks, covers included									
7800	Polypropylene									
7810	Continuous service to 200°F									
7830	2 gallon, 8" x 8" x 8"	Q-1	20	.800	Ea.	121	46.50		167.50	206
7850	7 gallon, 12" x 12" x 12"		20	.800		189	46.50		235.50	280
7870	16 gallon, 18" x 12" x 18"		17	.941		247	54.50		301.50	355
8010	33 gallon, 24" x 18" x 18"		12	1.333		320	77.50		397.50	470

22 66 83 – Chemical-Waste Tanks

22 66 83.13 Chemical-Waste Dilution Tanks		Crew	Daily Output	Labor-Hours	Unit	2020 Bare Costs Material	Labor	Equipment	Total	Total Incl O&P
8070	44 gallon, 24" x 18" x 24"	Q-1	10	1.600	Ea.	385	93		478	570
8080	89 gallon, 36" x 24" x 24"	↓	8	2	↓	540	116		656	775
8150	Polyethylene, heavy duty walls									
8160	Continuous service to 180°F									
8180	5 gallon, 12" x 6" x 18"	Q-1	20	.800	Ea.	53	46.50		99.50	130
8210	15 gallon, 14" I.D. x 27" deep		17	.941		78.50	54.50		133	171
8230	55 gallon, 22" I.D. x 36" deep		10	1.600		180	93		273	340
8250	100 gallon, 28" I.D. x 42" deep		8	2		390	116		506	610
8270	200 gallon, 36" I.D. x 48" deep		6	2.667		610	155		765	910
8290	360 gallon, 48" I.D. x 48" deep	↓	5	3.200	↓	750	186		936	1,100

Division Notes

	CREW	DAILY OUTPUT	LABOR-HOURS	UNIT	BARE COSTS MAT.	LABOR	EQUIP.	TOTAL	TOTAL INCL O&P

Division 23 Heating, Ventilating & Air Conditioning

Estimating Tips

The labor adjustment factors listed in Subdivision 22 01 02.20 also apply to Division 23.

23 10 00 Facility Fuel Systems

- The prices in this subdivision for above- and below-ground storage tanks do not include foundations or hold-down slabs, unless noted. The estimator should refer to Divisions 3 and 31 for foundation system pricing. In addition to the foundations, required tank accessories, such as tank gauges, leak detection devices, and additional manholes and piping, must be added to the tank prices.

23 50 00 Central Heating Equipment

- When estimating the cost of an HVAC system, check to see who is responsible for providing and installing the temperature control system. It is possible to overlook controls, assuming that they would be included in the electrical estimate.
- When looking up a boiler, be careful on specified capacity. Some manufacturers rate their products on output while others use input.
- Include HVAC insulation for pipe, boiler, and duct (wrap and liner).
- Be careful when looking up mechanical items to get the correct pressure rating and connection type (thread, weld, flange).

23 70 00 Central HVAC Equipment

- Combination heating and cooling units are sized by the air conditioning requirements. (See Reference No. R236000-20 for the preliminary sizing guide.)
- A ton of air conditioning is nominally 400 CFM.
- Rectangular duct is taken off by the linear foot for each size, but its cost is usually estimated by the pound. Remember that SMACNA standards now base duct on internal pressure.
- Prefabricated duct is estimated and purchased like pipe: straight sections and fittings.
- Note that cranes or other lifting equipment are not included on any lines in Division 23. For example, if a crane is required to lift a heavy piece of pipe into place high above a gym floor, or to put a rooftop unit on the roof of a four-story building, etc., it must be added. Due to the potential for extreme variation—from nothing additional required to a major crane or helicopter—we feel that including a nominal amount for "lifting contingency" would be useless and detract from the accuracy of the estimate. When using equipment rental cost data from RSMeans, do not forget to include the cost of the operator(s).

Reference Numbers

Reference numbers are shown at the beginning of some major classifications. These numbers refer to related items in the Reference Section. The reference information may be an estimating procedure, an alternate pricing method, or technical information.

Note: Not all subdivisions listed here necessarily appear. ■

Note: "Powered in part by CINX™, based on liscensed proprietary information of Harrison Publishing House, Inc."

Note: Trade Service, in part, has been used as a reference source for some of the material prices used in Division 23.

23 05 Common Work Results for HVAC

23 05 02 – HVAC General

23 05 02.10 Air Conditioning, General		Crew	Daily Output	Labor-Hours	Unit	Material	2020 Bare Costs Labor	Equipment	Total	Total Incl O&P
0010	**AIR CONDITIONING, GENERAL** Prices are for standard efficiencies (SEER 13)									
0020	for upgrade to SEER 14 add					10%				

23 05 05 – Selective Demolition for HVAC

23 05 05.10 HVAC Demolition

		Crew	Daily Output	Labor-Hours	Unit	Material	Labor	Equipment	Total	Total Incl O&P
0010	**HVAC DEMOLITION** R220105-10									
0100	Air conditioner, split unit, 3 ton	Q-5	2	8	Ea.		470		470	730
0150	Package unit, 3 ton R024119-10	Q-6	3	8			490		490	755
0190	Rooftop, self contained, up to 5 ton	1 Plum	1.20	6.667			430		430	665
0250	Air curtain	Q-9	20	.800	L.F.		45		45	70.50
0254	Air filters, up thru 16,000 CFM		20	.800	Ea.		45		45	70.50
0256	20,000 thru 60,000 CFM		16	1			56		56	88
0297	Boiler blowdown	Q-5	8	2			118		118	182
0298	Boilers									
0300	Electric, up thru 148 kW	Q-19	2	12	Ea.		715		715	1,100
0310	150 thru 518 kW	"	1	24			1,425		1,425	2,225
0320	550 thru 2,000 kW	Q-21	.40	80			4,900		4,900	7,575
0330	2,070 kW and up	"	.30	107			6,525		6,525	10,100
0340	Gas and/or oil, up thru 150 MBH	Q-7	2.20	14.545			910		910	1,400
0350	160 thru 2,000 MBH		.80	40			2,500		2,500	3,850
0360	2,100 thru 4,500 MBH		.50	64			4,000		4,000	6,175
0370	4,600 thru 7,000 MBH		.30	107			6,650		6,650	10,300
0380	7,100 thru 12,000 MBH		.16	200			12,500		12,500	19,300
0390	12,200 thru 25,000 MBH		.12	267			16,600		16,600	25,700
0400	Central station air handler unit, up thru 15 ton	Q-5	1.60	10			590		590	910
0410	17.5 thru 30 ton	"	.80	20			1,175		1,175	1,825
0430	Computer room unit									
0434	Air cooled split, up thru 10 ton	Q-5	.67	23.881	Ea.		1,400		1,400	2,175
0436	12 thru 23 ton		.53	30.189			1,775		1,775	2,750
0440	Chilled water, up thru 10 ton		1.30	12.308			725		725	1,125
0444	12 thru 23 ton		1	16			945		945	1,450
0450	Glycol system, up thru 10 ton		.53	30.189			1,775		1,775	2,750
0454	12 thru 23 ton		.40	40			2,350		2,350	3,650
0460	Water cooled, not including condenser, up thru 10 ton		.80	20			1,175		1,175	1,825
0464	12 thru 23 ton		.60	26.667			1,575		1,575	2,425
0600	Condenser, up thru 50 ton		1	16			945		945	1,450
0610	51 thru 100 ton	Q-6	.90	26.667			1,625		1,625	2,525
0620	101 thru 1,000 ton	"	.70	34.286			2,100		2,100	3,250
0660	Condensing unit, up thru 10 ton	Q-5	1.25	12.800			755		755	1,175
0670	11 thru 50 ton	"	.40	40			2,350		2,350	3,650
0680	60 thru 100 ton	Q-6	.30	80			4,900		4,900	7,575
0700	Cooling tower, up thru 400 ton		.80	30			1,825		1,825	2,850
0710	450 thru 600 ton		.53	45.283			2,775		2,775	4,275
0720	700 thru 1,300 ton		.40	60			3,675		3,675	5,675
0780	Dehumidifier, up thru 155 lb./hr.	Q-1	8	2			116		116	179
0790	240 lb./hr. and up	"	2	8			465		465	715
1560	Ductwork									
1570	Metal, steel, sst, fabricated	Q-9	1000	.016	Lb.		.90		.90	1.41
1580	Aluminum, fabricated		485	.033	"		1.85		1.85	2.90
1590	Spiral, prefabricated		400	.040	L.F.		2.24		2.24	3.51
1600	Fiberglass, prefabricated		400	.040			2.24		2.24	3.51
1610	Flex, prefabricated		500	.032			1.79		1.79	2.81
1620	Glass fiber reinforced plastic, prefabricated		280	.057			3.20		3.20	5

23 05 05.10 HVAC Demolition		Crew	Daily Output	Labor-Hours	Unit	2020 Bare Costs Material	Labor	Equipment	Total	Total Incl O&P
1630	Diffusers, registers or grills, up thru 20" max dimension	1 Shee	50	.160	Ea.		9.95		9.95	15.60
1640	21 thru 36" max dimension		36	.222			13.85		13.85	21.50
1650	Above 36" max dimension		30	.267			16.60		16.60	26
1700	Evaporator, up thru 12,000 BTUH	Q-5	5.30	3.019			178		178	275
1710	12,500 thru 30,000 BTUH	"	2.70	5.926			350		350	540
1720	31,000 BTUH and up	Q-6	1.50	16			980		980	1,525
1730	Evaporative cooler, up thru 5 HP	Q-9	2.70	5.926			330		330	520
1740	10 thru 30 HP	"	.67	23.881			1,350		1,350	2,100
1750	Exhaust systems									
1760	Exhaust components	1 Shee	8	1	System		62.50		62.50	97.50
1770	Weld fume hoods	"	20	.400	Ea.		25		25	39
1850	Minimum labor/equipment charge	1 Clab	3	2.667	Job		112		112	180
2120	Fans, up thru 1 HP or 2,000 CFM	Q-9	8	2	Ea.		112		112	176
2124	1-1/2 thru 10 HP or 20,000 CFM		5.30	3.019			169		169	265
2128	15 thru 30 HP or above 20,000 CFM		4	4			224		224	350
2150	Fan coil air conditioner, chilled water, up thru 7.5 ton	Q-5	14	1.143			67.50		67.50	104
2154	Direct expansion, up thru 10 ton		8	2			118		118	182
2158	11 thru 30 ton		2	8			470		470	730
2170	Flue shutter damper	Q-9	8	2			112		112	176
2200	Furnace, electric	Q-20	2	10			570		570	890
2300	Gas or oil, under 120 MBH	Q-9	4	4			224		224	350
2340	Over 120 MBH	"	3	5.333			299		299	470
2730	Heating and ventilating unit	Q-5	2.70	5.926			350		350	540
2740	Heater, electric, wall, baseboard and quartz	1 Elec	10	.800			49		49	75.50
2750	Heater, electric, unit, cabinet, fan and convector	"	8	1			61.50		61.50	94.50
2760	Heat exchanger, shell and tube type	Q-5	1.60	10			590		590	910
2770	Plate type	Q-6	.60	40			2,450		2,450	3,775
2810	Heat pump									
2820	Air source, split, 4 thru 10 ton	Q-5	.90	17.778	Ea.		1,050		1,050	1,625
2830	15 thru 25 ton	Q-6	.80	30			1,825		1,825	2,850
2850	Single package, up thru 12 ton	Q-5	1	16			945		945	1,450
2860	Water source, up thru 15 ton	"	.90	17.778			1,050		1,050	1,625
2870	20 thru 50 ton	Q-6	.80	30			1,825		1,825	2,850
2910	Heat recovery package, up thru 20,000 CFM	Q-5	2	8			470		470	730
2920	25,000 CFM and up		1.20	13.333			785		785	1,225
2930	Heat transfer package, up thru 130 GPM		.80	20			1,175		1,175	1,825
2934	255 thru 800 GPM		.42	38.095			2,250		2,250	3,475
2940	Humidifier		10.60	1.509			89		89	138
2961	Hydronic unit heaters, up thru 200 MBH		14	1.143			67.50		67.50	104
2962	Above 200 MBH		8	2			118		118	182
2964	Valance units		32	.500			29.50		29.50	45.50
2966	Radiant floor heating									
2967	System valves, controls, manifolds	Q-5	16	1	Ea.		59		59	91
2968	Per room distribution		8	2			118		118	182
2970	Hydronic heating, baseboard radiation		16	1			59		59	91
2976	Convectors and free standing radiators		18	.889			52.50		52.50	81
2980	Induced draft fan, up thru 1 HP	Q-9	4.60	3.478			195		195	305
2984	1-1/2 HP thru 7-1/2 HP	"	2.20	7.273			410		410	640
2988	Infrared unit	Q-5	16	1			59		59	91
2992	Louvers	1 Shee	46	.174	S.F.		10.85		10.85	16.95
3000	Mechanical equipment, light items. Unit is weight, not cooling.	Q-5	.90	17.778	Ton		1,050		1,050	1,625
3600	Heavy items		1.10	14.545	"		860		860	1,325
3720	Make-up air unit, up thru 6,000 CFM		3	5.333	Ea.		315		315	485

23 05 Common Work Results for HVAC

23 05 05 – Selective Demolition for HVAC

23 05 05.10 HVAC Demolition		Crew	Daily Output	Labor-Hours	Unit	2020 Bare Costs Material	Labor	Equipment	Total	Total Incl O&P
3730	6,500 thru 30,000 CFM	Q-5	1.60	10	Ea.		590		590	910
3740	35,000 thru 75,000 CFM	Q-6	1	24			1,475		1,475	2,275
3800	Mixing boxes, constant and VAV	Q-9	18	.889			50		50	78
4000	Packaged terminal air conditioner, up thru 18,000 BTUH	Q-5	8	2			118		118	182
4010	24,000 thru 48,000 BTUH	"	2.80	5.714			335		335	520
5000	Refrigerant compressor, reciprocating or scroll									
5010	Up thru 5 ton	1 Stpi	6	1.333	Ea.		87.50		87.50	135
5020	5.08 thru 10 ton	Q-5	6	2.667			157		157	243
5030	15 thru 50 ton	"	3	5.333			315		315	485
5040	60 thru 130 ton	Q-6	2.80	8.571			525		525	810
5090	Remove refrigerant from system	1 Stpi	40	.200	Lb.		13.10		13.10	20.50
5100	Rooftop air conditioner, up thru 10 ton	Q-5	1.40	11.429	Ea.		675		675	1,050
5110	12 thru 40 ton	Q-6	1	24			1,475		1,475	2,275
5120	50 thru 140 ton		.50	48			2,925		2,925	4,550
5130	150 thru 300 ton		.30	80			4,900		4,900	7,575
6000	Self contained single package air conditioner, up thru 10 ton	Q-5	1.60	10			590		590	910
6010	15 thru 60 ton	Q-6	1.20	20			1,225		1,225	1,900
6100	Space heaters, up thru 200 MBH	Q-5	10	1.600			94.50		94.50	146
6110	Over 200 MBH		5	3.200			189		189	292
6200	Split ductless, both sections		8	2			118		118	182
6300	Steam condensate meter	1 Stpi	11	.727			47.50		47.50	73.50
6600	Thru-the-wall air conditioner	L-2	8	2			93		93	150
7000	Vent chimney, prefabricated, up thru 12" diameter	Q-9	94	.170	V.L.F.		9.55		9.55	14.95
7010	14" thru 36" diameter		40	.400			22.50		22.50	35
7020	38" thru 48" diameter		32	.500			28		28	44
7030	54" thru 60" diameter	Q-10	14	1.714			99.50		99.50	156
7400	Ventilators, up thru 14" neck diameter	Q-9	58	.276	Ea.		15.45		15.45	24
7410	16" thru 50" neck diameter		40	.400			22.50		22.50	35
7450	Relief vent, up thru 24" x 96"		22	.727			41		41	64
7460	48" x 60" thru 96" x 144"		10	1.600			89.50		89.50	141
8000	Water chiller up thru 10 ton	Q-5	2.50	6.400			380		380	585
8010	15 thru 100 ton	Q-6	.48	50			3,050		3,050	4,725
8020	110 thru 500 ton	Q-7	.29	110			6,875		6,875	10,600
8030	600 thru 1000 ton		.23	139			8,675		8,675	13,400
8040	1100 ton and up		.20	160			9,975		9,975	15,400
8400	Window air conditioner	1 Carp	16	.500			26.50		26.50	42.50
8401	HVAC demo, water chiller, 1100 ton and up	"	12	.667	Ton		35.50		35.50	57
9000	Minimum labor/equipment charge	Q-6	3	8	Job		490		490	755

23 05 23 – General-Duty Valves for HVAC Piping

23 05 23.20 Valves, Bronze/Brass		Crew	Daily Output	Labor-Hours	Unit	Material	Labor	Equipment	Total	Total Incl O&P
0010	**VALVES, BRONZE/BRASS**									
0020	Brass									
1300	Ball combination valves, shut-off and union									
1310	Solder, with strainer, drain and PT ports									
1320	1/2"	1 Stpi	17	.471	Ea.	70	31		101	125
1330	3/4"		16	.500		76.50	33		109.50	135
1340	1"		14	.571		104	37.50		141.50	173
1350	1-1/4"		12	.667		133	43.50		176.50	214
1360	1-1/2"		10	.800		195	52.50		247.50	296
1370	2"		8	1		245	65.50		310.50	370
1410	Threaded, with strainer, drain and PT ports									
1420	1/2"	1 Stpi	20	.400	Ea.	70	26		96	118

23 05 23.20 Valves, Bronze/Brass		Crew	Daily Output	Labor-Hours	Unit	Material	2020 Bare Costs Labor	Equipment	Total	Total Incl O&P
1430	3/4"	1 Stpi	18	.444	Ea.	76.50	29		105.50	130
1440	1"		17	.471		104	31		135	163
1450	1-1/4"		12	.667		133	43.50		176.50	214
1460	1-1/2"		11	.727		195	47.50		242.50	289
1470	2"		9	.889		245	58.50		303.50	360

23 05 23.30 Valves, Iron Body		Crew	Daily Output	Labor-Hours	Unit	Material	Labor	Equipment	Total	Total Incl O&P
0010	**VALVES, IRON BODY** R220523-90									
0022	For grooved joint, see Section 22 11 13.48									
0560	Butterfly, lug type, pneumatic operator, 2" size	1 Stpi	14	.571	Ea.	680	37.50		717.50	805
0570	3"	Q-1	8	2		720	116		836	975
0580	4"	"	5	3.200		775	186		961	1,150
0590	6"	Q-2	5	4.800		1,050	289		1,339	1,600
0600	8"		4.50	5.333		1,225	320		1,545	1,850
0610	10"		4	6		1,575	360		1,935	2,275
0620	12"		3	8		1,875	480		2,355	2,800
0630	14"		2.30	10.435		2,575	630		3,205	3,800
0640	18"		1.50	16		4,400	960		5,360	6,325
0650	20"		1	24		5,625	1,450		7,075	8,425
0790	Butterfly, lug type, gear operated, 2" size, 200 lb. except noted	1 Plum	14	.571		370	37		407	460
0800	2-1/2"	Q-1	9	1.778		370	103		473	570
0810	3"		8	2		390	116		506	610
0820	4"		5	3.200		455	186		641	785
0830	5"	Q-2	5	4.800		690	289		979	1,200
0840	6"		5	4.800		625	289		914	1,125
0850	8"		4.50	5.333		815	320		1,135	1,400
0860	10"		4	6		1,375	360		1,735	2,050
0870	12"		3	8		1,550	480		2,030	2,475
0880	14", 150 lb.		2.30	10.435		2,500	630		3,130	3,725
0890	16", 150 lb.		1.75	13.714		3,825	825		4,650	5,475
0900	18", 150 lb.		1.50	16		4,625	960		5,585	6,600
0910	20", 150 lb.		1	24		6,200	1,450		7,650	9,050
0930	24", 150 lb.		.75	32		10,500	1,925		12,425	14,600
1020	Butterfly, wafer type, gear actuator, 200 lb.									
1030	2"	1 Plum	14	.571	Ea.	118	37		155	187
1040	2-1/2"	Q-1	9	1.778		125	103		228	296
1050	3"		8	2		124	116		240	315
1060	4"		5	3.200		139	186		325	440
1070	5"	Q-2	5	4.800		157	289		446	615
1080	6"		5	4.800		177	289		466	640
1090	8"		4.50	5.333		220	320		540	735
1100	10"		4	6		295	360		655	885
1110	12"		3	8		685	480		1,165	1,500
1200	Wafer type, lever actuator, 200 lb.									
1220	2"	1 Plum	14	.571	Ea.	203	37		240	280
1230	2-1/2"	Q-1	9	1.778		206	103		309	385
1240	3"		8	2		258	116		374	460
1250	4"		5	3.200		265	186		451	580
1260	5"	Q-2	5	4.800		395	289		684	880
1270	6"		5	4.800		445	289		734	935
1280	8"		4.50	5.333		700	320		1,020	1,275
1290	10"		4	6		740	360		1,100	1,375
1300	12"		3	8		1,250	480		1,730	2,125

23 05 23.30 Valves, Iron Body		Crew	Daily Output	Labor-Hours	Unit	Material	Labor	Equipment	Total	Total Incl O&P
						2020 Bare Costs				
1650	Gate, 125 lb., N.R.S.									
2150	Flanged									
2200	2"	1 Plum	5	1.600	Ea.	610	103		713	835
2240	2-1/2"	Q-1	5	3.200		630	186		816	975
2260	3"		4.50	3.556		710	206		916	1,100
2280	4"		3	5.333		1,000	310		1,310	1,575
2290	5"	Q-2	3.40	7.059		1,725	425		2,150	2,550
2300	6"		3	8		1,725	480		2,205	2,650
2320	8"		2.50	9.600		2,950	575		3,525	4,150
2340	10"		2.20	10.909		5,175	655		5,830	6,725
2360	12"		1.70	14.118		7,125	850		7,975	9,150
2370	14"		1.30	18.462		10,700	1,100		11,800	13,500
2380	16"		1	24		14,800	1,450		16,250	18,500
2420	For 250 lb. flanged, add					200%	10%			
3550	OS&Y, 125 lb., flanged									
3600	2"	1 Plum	5	1.600	Ea.	400	103		503	600
3640	2-1/2"	Q-1	5	3.200		430	186		616	755
3660	3"		4.50	3.556		460	206		666	830
3670	3-1/2"		3	5.333		770	310		1,080	1,325
3680	4"		3	5.333		685	310		995	1,225
3690	5"	Q-2	3.40	7.059		1,075	425		1,500	1,850
3700	6"		3	8		1,075	480		1,555	1,950
3720	8"		2.50	9.600		1,925	575		2,500	3,025
3740	10"		2.20	10.909		3,525	655		4,180	4,900
3760	12"		1.70	14.118		4,800	850		5,650	6,600
3770	14"		1.30	18.462		10,400	1,100		11,500	13,100
3780	16"		1	24		14,800	1,450		16,250	18,500
3790	18"		.80	30		24,200	1,800		26,000	29,400
3800	20"		.60	40		25,300	2,400		27,700	31,500
3830	24"		.50	48		38,000	2,875		40,875	46,300
3900	For 175 lb., flanged, add					200%	10%			
4350	Globe, OS&Y									
4540	Class 125, flanged									
4550	2"	1 Plum	5	1.600	Ea.	935	103		1,038	1,175
4560	2-1/2"	Q-1	5	3.200		940	186		1,126	1,300
4570	3"		4.50	3.556		1,150	206		1,356	1,575
4580	4"		3	5.333		1,625	310		1,935	2,275
4590	5"	Q-2	3.40	7.059		2,975	425		3,400	3,925
4600	6"		3	8		2,975	480		3,455	4,025
4610	8"		2.50	9.600		5,850	575		6,425	7,325
5040	Class 250, flanged									
5050	2"	1 Plum	4.50	1.778	Ea.	1,500	115		1,615	1,825
5060	2-1/2"	Q-1	4.50	3.556		1,950	206		2,156	2,475
5070	3"		4	4		2,025	232		2,257	2,575
5080	4"		2.70	5.926		2,950	345		3,295	3,775
5090	5"	Q-2	3	8		6,225	480		6,705	7,600
5100	6"		2.70	8.889		6,475	535		7,010	7,950
5110	8"		2.20	10.909		8,975	655		9,630	10,900
5120	10"		2	12		12,900	720		13,620	15,300
5130	12"		1.60	15		24,200	900		25,100	28,100
5450	Swing check, 125 lb., threaded									
5500	2"	1 Plum	11	.727	Ea.	575	47		622	705
5540	2-1/2"	Q-1	15	1.067		710	62		772	875

23 05 23.30 Valves, Iron Body

	23 05 23.30 Valves, Iron Body	Crew	Daily Output	Labor-Hours	Unit	Material	Labor (2020 Bare Costs)	Equipment	Total	Total Incl O&P
5550	3″	Q-1	13	1.231	Ea.	770	71.50		841.50	960
5560	4″		10	1.600		1,225	93		1,318	1,500
5950	Flanged									
6000	2″	1 Plum	5	1.600	Ea.	510	103		613	725
6040	2-1/2″	Q-1	5	3.200		365	186		551	685
6050	3″		4.50	3.556		490	206		696	860
6060	4″		3	5.333		785	310		1,095	1,350
6070	6″	Q-2	3	8		1,325	480		1,805	2,200
6080	8″		2.50	9.600		2,525	575		3,100	3,675
6090	10″		2.20	10.909		4,275	655		4,930	5,725
6100	12″		1.70	14.118		7,000	850		7,850	9,025
6160	For 250 lb. flanged, add					200%	20%			
6600	Silent check, bronze trim									
6610	Compact wafer type, for 125 or 150 lb. flanges									
6630	1-1/2″	1 Plum	11	.727	Ea.	162	47		209	251
6640	2″	"	9	.889		198	57.50		255.50	305
6650	2-1/2″	Q-1	9	1.778		215	103		318	395
6660	3″		8	2		236	116		352	440
6670	4″		5	3.200		290	186		476	605
6680	5″	Q-2	6	4		390	241		631	795
6690	6″		6	4		530	241		771	955
6700	8″		4.50	5.333		885	320		1,205	1,475
6710	10″		4	6		1,550	360		1,910	2,250
6720	12″		3	8		2,875	480		3,355	3,900
9000	Minimum labor/equipment charge	1 Plum	3	2.667	Job		172		172	266

23 05 23.70 Valves, Semi-Steel

	Description	Crew	Daily Output	Labor-Hours	Unit	Material	Labor	Equipment	Total	Total Incl O&P
0010	**VALVES, SEMI-STEEL**									
1020	Lubricated plug valve, threaded, 200 psi									
1030	1/2″	1 Plum	18	.444	Ea.	119	28.50		147.50	176
1040	3/4″		16	.500		125	32		157	188
1050	1″		14	.571		135	37		172	205
1060	1-1/4″		12	.667		159	43		202	242
1070	1-1/2″		11	.727		154	47		201	243
1080	2″		8	1		208	64.50		272.50	330
1090	2-1/2″	Q-1	5	3.200		299	186		485	615
1100	3″	"	4.50	3.556		425	206		631	790
6990	Flanged, 200 psi									
7000	2″	1 Plum	8	1	Ea.	277	64.50		341.50	405
7010	2-1/2″	Q-1	5	3.200		390	186		576	715
7020	3″		4.50	3.556		470	206		676	840
7030	4″		3	5.333		585	310		895	1,125
7036	5″		2.50	6.400		1,325	370		1,695	2,025
7040	6″	Q-2	3	8		1,675	480		2,155	2,575
7050	8″		2.50	9.600		2,225	575		2,800	3,350
7060	10″		2.20	10.909		3,650	655		4,305	5,050
7070	12″		1.70	14.118		5,625	850		6,475	7,525

23 05 23.80 Valves, Steel

	Description	Crew	Daily Output	Labor-Hours	Unit	Material	Labor	Equipment	Total	Total Incl O&P
0010	**VALVES, STEEL** R220523-90									
0800	Cast									
1350	Check valve, swing type, 150 lb., flanged									
1370	1″	1 Plum	10	.800	Ea.	360	51.50		411.50	480
1400	2″	"	8	1		765	64.50		829.50	940

23 05 23 – General-Duty Valves for HVAC Piping

23 05 23.80 Valves, Steel		Crew	Daily Output	Labor-Hours	Unit	Material	Labor	Equipment	Total	Total Incl O&P
						2020 Bare Costs				
1440	2-1/2"	Q-1	5	3.200	Ea.	885	186		1,071	1,250
1450	3"		4.50	3.556		900	206		1,106	1,300
1460	4"		3	5.333		1,100	310		1,410	1,700
1470	6"	Q-2	3	8		1,725	480		2,205	2,650
1480	8"		2.50	9.600		2,925	575		3,500	4,100
1490	10"		2.20	10.909		5,100	655		5,755	6,625
1500	12"		1.70	14.118		6,275	850		7,125	8,225
1510	14"		1.30	18.462		9,350	1,100		10,450	12,000
1520	16"		1	24		10,700	1,450		12,150	14,000
1540	For 300 lb., flanged, add					50%	15%			
1548	For 600 lb., flanged, add					110%	20%			
1571	300 lb., 2"	1 Plum	7.40	1.081		680	69.50		749.50	860
1572	2-1/2"	Q-1	4.20	3.810		1,025	221		1,246	1,475
1573	3"		4	4		1,025	232		1,257	1,475
1574	4"		2.80	5.714		1,400	330		1,730	2,050
1575	6"	Q-2	2.90	8.276		2,700	500		3,200	3,750
1576	8"		2.40	10		4,000	600		4,600	5,325
1577	10"		2.10	11.429		6,000	685		6,685	7,650
1578	12"		1.60	15		8,675	900		9,575	11,000
1579	14"		1.20	20		12,900	1,200		14,100	16,100
1581	16"		.90	26.667		16,400	1,600		18,000	20,600
1950	Gate valve, 150 lb., flanged									
2000	2"	1 Plum	8	1	Ea.	645	64.50		709.50	805
2040	2-1/2"	Q-1	5	3.200		1,075	186		1,261	1,450
2050	3"		4.50	3.556		910	206		1,116	1,325
2060	4"		3	5.333		1,150	310		1,460	1,725
2070	6"	Q-2	3	8		1,825	480		2,305	2,750
2080	8"		2.50	9.600		2,800	575		3,375	3,975
2090	10"		2.20	10.909		4,175	655		4,830	5,625
2100	12"		1.70	14.118		5,700	850		6,550	7,600
2110	14"		1.30	18.462		9,200	1,100		10,300	11,800
2120	16"		1	24		11,900	1,450		13,350	15,300
2130	18"		.80	30		15,600	1,800		17,400	19,900
2140	20"		.60	40		18,100	2,400		20,500	23,600
2650	300 lb., flanged									
2700	2"	1 Plum	7.40	1.081	Ea.	1,025	69.50		1,094.50	1,225
2740	2-1/2"	Q-1	4.20	3.810		1,425	221		1,646	1,900
2750	3"		4	4		1,200	232		1,432	1,675
2760	4"		2.80	5.714		1,700	330		2,030	2,350
2770	6"	Q-2	2.90	8.276		2,950	500		3,450	4,025
2780	8"		2.40	10		4,625	600		5,225	6,000
2790	10"		2.10	11.429		6,000	685		6,685	7,675
2800	12"		1.60	15		8,650	900		9,550	10,900
2810	14"		1.20	20		17,000	1,200		18,200	20,700
2820	16"		.90	26.667		21,400	1,600		23,000	26,000
2830	18"		.70	34.286		30,500	2,050		32,550	36,800
2840	20"		.50	48		34,200	2,875		37,075	42,200
3650	Globe valve, 150 lb., flanged									
3700	2"	1 Plum	8	1	Ea.	950	64.50		1,014.50	1,150
3740	2-1/2"	Q-1	5	3.200		1,200	186		1,386	1,600
3750	3"		4.50	3.556		1,200	206		1,406	1,650
3760	4"		3	5.333		1,750	310		2,060	2,400
3770	6"	Q-2	3	8		2,750	480		3,230	3,775

23 05 23 – General-Duty Valves for HVAC Piping

23 05 23.80 Valves, Steel		Crew	Daily Output	Labor-Hours	Unit	Material	2020 Bare Costs Labor	Equipment	Total	Total Incl O&P
3780	8"	Q-2	2.50	9.600	Ea.	5,175	575		5,750	6,575
3790	10"		2.20	10.909		9,475	655		10,130	11,400
3800	12"		1.70	14.118		10,700	850		11,550	13,100
4080	300 lb., flanged									
4100	2"	1 Plum	7.40	1.081	Ea.	1,275	69.50		1,344.50	1,500
4140	2-1/2"	Q-1	4.20	3.810		1,725	221		1,946	2,250
4150	3"		4	4		1,725	232		1,957	2,250
4160	4"		2.80	5.714		2,375	330		2,705	3,125
4170	6"	Q-2	2.90	8.276		4,300	500		4,800	5,525
4180	8"		2.40	10		6,175	600		6,775	7,725
4190	10"		2.10	11.429		14,000	685		14,685	16,500
4200	12"		1.60	15		16,900	900		17,800	19,900
4680	600 lb., flanged									
4700	2"	1 Plum	7	1.143	Ea.	1,775	73.50		1,848.50	2,075
4740	2-1/2"	Q-1	4	4		2,850	232		3,082	3,475
4750	3"		3.60	4.444		2,875	258		3,133	3,550
4760	4"		2.50	6.400		4,375	370		4,745	5,375
4770	6"	Q-2	2.60	9.231		9,250	555		9,805	11,100
4780	8"	"	2.10	11.429		12,900	685		13,585	15,300
5150	Forged									
5340	Ball valve, 1,500 psi, threaded, 1/4" size	1 Plum	24	.333	Ea.	75.50	21.50		97	116
5350	3/8"		24	.333		74	21.50		95.50	115
5360	1/2"		24	.333		103	21.50		124.50	147
5370	3/4"		20	.400		121	26		147	173
5380	1"		19	.421		154	27		181	211
5390	1-1/4"		15	.533		203	34.50		237.50	276
5400	1-1/2"		13	.615		250	39.50		289.50	335
5410	2"		11	.727		355	47		402	465
5460	Ball valve, 800 lb., socket weld, 1/4" size	Q-15	19	.842		75.50	49	5.60	130.10	165
5470	3/8"		19	.842		75.50	49	5.60	130.10	165
5480	1/2"		19	.842		94	49	5.60	148.60	185
5490	3/4"		19	.842		125	49	5.60	179.60	219
5500	1"		15	1.067		156	62	7.10	225.10	274
5510	1-1/4"		13	1.231		211	71.50	8.20	290.70	350
5520	1-1/2"		11	1.455		267	84.50	9.65	361.15	435
5530	2"		8.50	1.882		350	109	12.50	471.50	570
5650	Check valve, class 800, horizontal									
5651	Socket									
5652	1/4"	Q-15	19	.842	Ea.	86	49	5.60	140.60	176
5654	3/8"		19	.842		86	49	5.60	140.60	176
5656	1/2"		19	.842		86	49	5.60	140.60	176
5658	3/4"		19	.842		91.50	49	5.60	146.10	183
5660	1"		15	1.067		108	62	7.10	177.10	222
5662	1-1/4"		13	1.231		212	71.50	8.20	291.70	350
5664	1-1/2"		11	1.455		212	84.50	9.65	306.15	375
5666	2"		8.50	1.882		297	109	12.50	418.50	510
5698	Threaded									
5700	1/4"	1 Plum	24	.333	Ea.	86	21.50		107.50	128
5720	3/8"		24	.333		86	21.50		107.50	128
5730	1/2"		24	.333		86	21.50		107.50	128
5740	3/4"		20	.400		92	26		118	141
5750	1"		19	.421		108	27		135	161
5760	1-1/4"		15	.533		212	34.50		246.50	286

23 05 23.80 Valves, Steel		Crew	Daily Output	Labor-Hours	Unit	Material	Labor	Equipment	Total	Total Incl O&P
						2020 Bare Costs				
5770	1-1/2"	1 Plum	13	.615	Ea.	212	39.50		251.50	295
5780	2"	↓	11	.727		297	47		344	400
5840	For class 150, flanged, add					100%	15%			
5860	For class 300, flanged, add				↓	120%	20%			
6100	Gate, class 800, OS&Y, socket									
6102	3/8"	Q-15	19	.842	Ea.	56.50	49	5.60	111.10	144
6103	1/2"		19	.842		56.50	49	5.60	111.10	144
6104	3/4"		19	.842		62	49	5.60	116.60	150
6105	1"		15	1.067		75	62	7.10	144.10	186
6106	1-1/4"		13	1.231		142	71.50	8.20	221.70	275
6107	1-1/2"		11	1.455		142	84.50	9.65	236.15	297
6108	2"	↓	8.50	1.882	↓	188	109	12.50	309.50	390
6118	Threaded									
6120	3/8"	1 Plum	24	.333	Ea.	56.50	21.50		78	95
6130	1/2"		24	.333		56.50	21.50		78	95
6140	3/4"		20	.400		62	26		88	109
6150	1"		19	.421		75	27		102	125
6160	1-1/4"		15	.533		142	34.50		176.50	209
6170	1-1/2"		13	.615		142	39.50		181.50	218
6180	2"	↓	11	.727		188	47		235	280
6260	For OS&Y, flanged, add				↓	100%	20%			
6700	Globe, OS&Y, class 800, socket									
6710	1/4"	Q-15	19	.842	Ea.	87	49	5.60	141.60	177
6720	3/8"		19	.842		87	49	5.60	141.60	177
6730	1/2"		19	.842		87	49	5.60	141.60	177
6740	3/4"		19	.842		100	49	5.60	154.60	192
6750	1"		15	1.067		130	62	7.10	199.10	246
6760	1-1/4"		13	1.231		255	71.50	8.20	334.70	400
6770	1-1/2"		11	1.455		255	84.50	9.65	349.15	420
6780	2"	↓	8.50	1.882	↓	325	109	12.50	446.50	545

23 05 23.90 Valves, Stainless Steel		Crew	Daily Output	Labor-Hours	Unit	Material	Labor	Equipment	Total	Total Incl O&P
0010	**VALVES, STAINLESS STEEL** R220523-90									
1610	Ball, threaded 1/4"	1 Stpi	24	.333	Ea.	47.50	22		69.50	86
1620	3/8"		24	.333		53.50	22		75.50	92.50
1630	1/2"		22	.364		47.50	24		71.50	89
1640	3/4"		20	.400		88.50	26		114.50	138
1650	1"		19	.421		96	27.50		123.50	148
1660	1-1/4"		15	.533		193	35		228	267
1670	1-1/2"		13	.615		240	40.50		280.50	325
1680	2"	↓	11	.727	↓	291	47.50		338.50	395
1700	Check, 200 lb., threaded									
1710	1/4"	1 Plum	24	.333	Ea.	145	21.50		166.50	192
1720	1/2"		22	.364		145	23.50		168.50	195
1730	3/4"		20	.400		149	26		175	204
1750	1"		19	.421		194	27		221	256
1760	1-1/2"		13	.615		365	39.50		404.50	460
1770	2"	↓	11	.727	↓	620	47		667	755
1800	150 lb., flanged									
1810	2-1/2"	Q-1	5	3.200	Ea.	1,925	186		2,111	2,400
1820	3"		4.50	3.556		1,975	206		2,181	2,500
1830	4"	↓	3	5.333		2,950	310		3,260	3,700
1840	6"	Q-2	3	8	↓	5,200	480		5,680	6,475

23 05 23 – General-Duty Valves for HVAC Piping

23 05 23.90 Valves, Stainless Steel		Crew	Daily Output	Labor-Hours	Unit	Material	Labor	Equipment	Total	Total Incl O&P
1850	8"	Q-2	2.50	9.600	Ea.	10,600	575		11,175	12,600
2100	Gate, OS&Y, 150 lb., flanged									
2120	1/2"	1 Plum	18	.444	Ea.	515	28.50		543.50	610
2140	3/4"		16	.500		495	32		527	595
2150	1"		14	.571		625	37		662	745
2160	1-1/2"		11	.727		1,200	47		1,247	1,400
2170	2"		8	1		1,425	64.50		1,489.50	1,675
2180	2-1/2"	Q-1	5	3.200		1,800	186		1,986	2,275
2190	3"		4.50	3.556		1,950	206		2,156	2,475
2200	4"		3	5.333		2,675	310		2,985	3,425
2205	5"		2.80	5.714		5,125	330		5,455	6,150
2210	6"	Q-2	3	8		5,000	480		5,480	6,250
2220	8"		2.50	9.600		8,675	575		9,250	10,400
2230	10"		2.30	10.435		15,200	630		15,830	17,700
2240	12"		1.90	12.632		20,200	760		20,960	23,500
2260	For 300 lb., flanged, add					120%	15%			
2600	600 lb., flanged									
2620	1/2"	1 Plum	16	.500	Ea.	184	32		216	252
2640	3/4"		14	.571		199	37		236	276
2650	1"		12	.667		239	43		282	330
2660	1-1/2"		10	.800		380	51.50		431.50	500
2670	2"		7	1.143		525	73.50		598.50	695
2680	2-1/2"	Q-1	4	4		6,625	232		6,857	7,650
2690	3"	"	3.60	4.444		6,625	258		6,883	7,700
3100	Globe, OS&Y, 150 lb., flanged									
3120	1/2"	1 Plum	18	.444	Ea.	495	28.50		523.50	585
3140	3/4"		16	.500		535	32		567	640
3150	1"		14	.571		700	37		737	825
3160	1-1/2"		11	.727		1,100	47		1,147	1,275
3170	2"		8	1		1,425	64.50		1,489.50	1,675
3180	2-1/2"	Q-1	5	3.200		3,075	186		3,261	3,650
3190	3"		4.50	3.556		3,075	206		3,281	3,700
3200	4"		3	5.333		4,925	310		5,235	5,875
3210	6"	Q-2	3	8		8,275	480		8,755	9,850

(2020 Bare Costs: Material, Labor, Equipment, Total)

23 05 23.94 Hospital Type Valves		Crew	Daily Output	Labor-Hours	Unit	Material	Labor	Equipment	Total	Total Incl O&P
0010	**HOSPITAL TYPE VALVES**									
0300	Chiller valves									
0330	Manual operation balancing valve									
0340	2-1/2" line size	Q-1	9	1.778	Ea.	620	103		723	840
0350	3" line size		8	2		665	116		781	910
0360	4" line size		5	3.200		920	186		1,106	1,275
0370	5" line size	Q-2	5	4.800		1,150	289		1,439	1,725
0380	6" line size		5	4.800		1,425	289		1,714	2,025
0390	8" line size		4.50	5.333		3,100	320		3,420	3,925
0400	10" line size		4	6		5,075	360		5,435	6,125
0410	12" line size		3	8		6,725	480		7,205	8,150
0600	Automatic flow limiting valve									
0620	2-1/2" line size	Q-1	8	2	Ea.	620	116		736	865
0630	3" line size		7	2.286		1,000	133		1,133	1,300
0640	4" line size		6	2.667		1,425	155		1,580	1,825
0645	5" line size	Q-2	5	4.800		1,800	289		2,089	2,425
0650	6" line size		5	4.800		2,400	289		2,689	3,075

23 05 Common Work Results for HVAC

23 05 23 – General-Duty Valves for HVAC Piping

23 05 23.94 Hospital Type Valves		Crew	Daily Output	Labor-Hours	Unit	Material	2020 Bare Costs Labor	Equipment	Total	Total Incl O&P
0660	8″ line size	Q-2	4.50	5.333	Ea.	3,600	320		3,920	4,450
0670	10″ line size		4	6		5,625	360		5,985	6,750
0680	12″ line size		3	8		9,550	480		10,030	11,200
0140	0-250 L.F.	1 Stpi	1.33	6.015			395		395	610
0160	250-500 L.F.	″	.80	10			655		655	1,025
0180	500-1000 L.F.	Q-5	1.14	14.035			830		830	1,275
0200	1000-2000 L.F.		.80	20			1,175		1,175	1,825
0320	0-250 L.F.		1	16			945		945	1,450
0340	250-500 L.F.		.73	21.918			1,300		1,300	2,000
0360	500-1000 L.F.		.53	30.189			1,775		1,775	2,750
0380	1000-2000 L.F.		.38	42.105			2,475		2,475	3,850
2110	2″ diam.	1 Stpi	8	1		17.25	65.50		82.75	120
2120	3″ diam.		8	1		17.25	65.50		82.75	120
2130	4″ diam.		8	1		26	65.50		91.50	130
2140	6″ diam.		8	1		26	65.50		91.50	130
2150	8″ diam.		6.60	1.212		26	79.50		105.50	152
2160	10″ diam.		6	1.333		34.50	87.50		122	173

23 07 HVAC Insulation

23 07 13 – Duct Insulation

23 07 13.10 Duct Thermal Insulation		Crew	Daily Output	Labor-Hours	Unit	Material	Labor	Equipment	Total	Total Incl O&P
0010	**DUCT THERMAL INSULATION**									
0110	Insulation req'd. is based on the surface size/area to be covered									
3000	Ductwork									
3020	Blanket type, fiberglass, flexible									
3030	Fire rated for grease and hazardous exhaust ducts									
3060	1-1/2″ thick	Q-14	84	.190	S.F.	4.49	10.05		14.54	21
3090	Fire rated for plenums									
3100	1/2″ x 24″ x 25′	Q-14	1.94	8.247	Roll	177	435		612	885
3110	1/2″ x 24″ x 25′		98	.163	S.F.	3.54	8.60		12.14	17.55
3120	1/2″ x 48″ x 25′		1.04	15.385	Roll	350	810		1,160	1,650
3126	1/2″ x 48″ x 25′		104	.154	S.F.	3.49	8.10		11.59	16.70
3140	FSK vapor barrier wrap, .75 lb. density									
3160	1″ thick G	Q-14	350	.046	S.F.	.22	2.41		2.63	4.07
3170	1-1/2″ thick G		320	.050		.27	2.64		2.91	4.48
3180	2″ thick G		300	.053		.33	2.81		3.14	4.82
3190	3″ thick G		260	.062		.51	3.25		3.76	5.70
3200	4″ thick G		242	.066		.69	3.49		4.18	6.30
3210	Vinyl jacket, same as FSK									
3280	Unfaced, 1 lb. density									
3310	1″ thick G	Q-14	360	.044	S.F.	.28	2.35		2.63	4.03
3320	1-1/2″ thick G		330	.048		.38	2.56		2.94	4.48
3330	2″ thick G		310	.052		.50	2.72		3.22	4.87
3400	FSK facing, 1 lb. density									
3420	1-1/2″ thick G	Q-14	310	.052	S.F.	.38	2.72		3.10	4.74
3430	2″ thick G	″	300	.053	″	.51	2.81		3.32	5
3450	FSK facing, 1.5 lb. density									
3470	1-1/2″ thick G	Q-14	300	.053	S.F.	.54	2.81		3.35	5.05
3480	2″ thick G	″	290	.055	″	.66	2.91		3.57	5.35
3730	Sheet insulation									
3760	Polyethylene foam, closed cell, UV resistant									

23 07 HVAC Insulation

23 07 13 – Duct Insulation

23 07 13.10 Duct Thermal Insulation		Crew	Daily Output	Labor-Hours	Unit	Material	2020 Bare Costs Labor	Equipment	Total	Total Incl O&P
3770	Standard temperature (-90°F to +212°F)									
3771	1/4" thick G	Q-14	450	.036	S.F.	1.83	1.88		3.71	4.99
3772	3/8" thick G		440	.036		2.61	1.92		4.53	5.90
3773	1/2" thick G		420	.038		3.21	2.01		5.22	6.70
3774	3/4" thick G		400	.040		4.58	2.11		6.69	8.40
3775	1" thick G		380	.042		6.20	2.22		8.42	10.30
3776	1-1/2" thick G		360	.044		9.80	2.35		12.15	14.45
3777	2" thick G		340	.047		12.95	2.48		15.43	18.20
3778	2-1/2" thick G	↓	320	.050	↓	16.60	2.64		19.24	22.50
3779	Adhesive (see line 7878)									
3780	Foam, rubber									
3782	1" thick G	1 Stpi	50	.160	S.F.	3.05	10.50		13.55	19.55
3795	Finishes									
3800	Stainless steel woven mesh	Q-14	100	.160	S.F.	.99	8.45		9.44	14.50
3810	For .010" stainless steel, add		160	.100		3.26	5.30		8.56	11.95
3820	18 oz. fiberglass cloth, pasted on		170	.094		1.02	4.97		5.99	9
3900	8 oz. canvas, pasted on		180	.089		.30	4.69		4.99	7.80
3940	For .016" aluminum jacket, add	↓	200	.080	↓	1.03	4.22		5.25	7.85
7878	Contact cement, quart can				Ea.	13.20			13.20	14.55
9600	Minimum labor/equipment charge	1 Stpi	4	2	Job		131		131	203

23 07 16 – HVAC Equipment Insulation

23 07 16.10 HVAC Equipment Thermal Insulation		Crew	Daily Output	Labor-Hours	Unit	Material	Labor	Equipment	Total	Total Incl O&P
0010	**HVAC EQUIPMENT THERMAL INSULATION**									
0110	Insulation req'd. is based on the surface size/area to be covered									
1000	Boiler, 1-1/2" calcium silicate only G	Q-14	110	.145	S.F.	4.47	7.70		12.17	17.05
1020	Plus 2" fiberglass G	"	80	.200	"	6.05	10.55		16.60	23.50
2000	Breeching, 2" calcium silicate									
2020	Rectangular G	Q-14	42	.381	S.F.	8.65	20		28.65	41.50
2040	Round G	"	38.70	.413	"	9	22		31	44.50
2300	Calcium silicate block, +200°F to +1,200°F									
2310	On irregular surfaces, valves and fittings									
2340	1" thick G	Q-14	30	.533	S.F.	3.90	28		31.90	49
2360	1-1/2" thick G		25	.640		4.40	34		38.40	58.50
2380	2" thick G		22	.727		4.19	38.50		42.69	65.50
2400	3" thick G	↓	18	.889	↓	6.45	47		53.45	81.50
2410	On plane surfaces									
2420	1" thick G	Q-14	126	.127	S.F.	3.90	6.70		10.60	14.95
2430	1-1/2" thick G		120	.133		4.40	7.05		11.45	16
2440	2" thick G		100	.160		4.19	8.45		12.64	18
2450	3" thick G	↓	70	.229	↓	6.45	12.05		18.50	26.50
9610	Minimum labor/equipment charge	1 Stpi	4	2	Job		131		131	203

23 09 13 – Instrumentation and Control Devices for HVAC

23 09 13.60 Water Level Controls		Crew	Daily Output	Labor-Hours	Unit	Material (2020 Bare Costs)	Labor (2020 Bare Costs)	Equipment (2020 Bare Costs)	Total (2020 Bare Costs)	Total Incl O&P
0010	WATER LEVEL CONTROLS									
1000	Electric water feeder	1 Stpi	12	.667	Ea.	375	43.50		418.50	485
2000	Feeder cut-off combination									
2100	Steam system up to 5,000 sq. ft.	1 Stpi	12	.667	Ea.	710	43.50		753.50	850
2200	Steam system above 5,000 sq. ft.		12	.667		1,000	43.50		1,043.50	1,175
2300	Steam and hot water, high pressure		10	.800		1,200	52.50		1,252.50	1,375
3000	Low water cut-off for hot water boiler, 50 psi maximum									
3100	1″ top & bottom equalizing pipes, manual reset	1 Stpi	14	.571	Ea.	460	37.50		497.50	565
3200	1″ top & bottom equalizing pipes		14	.571		420	37.50		457.50	520
3300	2-1/2″ side connection for nipple-to-boiler		14	.571		415	37.50		452.50	515
4000	Low water cut-off for low pressure steam with quick hook-up ftgs.									
4100	For installation in gauge glass tappings	1 Stpi	16	.500	Ea.	340	33		373	425
4200	Built-in type, 2-1/2″ tap, 3-1/8″ insertion		16	.500		271	33		304	350
4300	Built-in type, 2-1/2″ tap, 1-3/4″ insertion		16	.500		298	33		331	375
4400	Side connection to 2-1/2″ tapping		16	.500		320	33		353	400
5000	Pump control, low water cut-off and alarm switch		14	.571		865	37.50		902.50	1,025

23 09 43 – Pneumatic Control System for HVAC

23 09 43.10 Pneumatic Control Systems		Crew	Daily Output	Labor-Hours	Unit	Material (2020 Bare Costs)	Labor (2020 Bare Costs)	Equipment (2020 Bare Costs)	Total (2020 Bare Costs)	Total Incl O&P
0010	PNEUMATIC CONTROL SYSTEMS									
0011	Including a nominal 50′ of tubing. Add control panelboard if req'd.									
0100	Heating and ventilating, split system									
0200	Mixed air control, economizer cycle, panel readout, tubing									
0220	Up to 10 tons G	Q-19	.68	35.294	Ea.	4,600	2,100		6,700	8,325
0240	For 10 to 20 tons G		.63	37.915		4,950	2,275		7,225	8,950
0260	For over 20 tons G		.58	41.096		5,375	2,450		7,825	9,700
0270	Enthalpy cycle, up to 10 tons		.50	48.387		5,100	2,900		8,000	10,100
0280	For 10 to 20 tons		.46	52.174		5,500	3,125		8,625	10,900
0290	For over 20 tons		.42	56.604		5,750	3,375		9,125	11,600
0300	Heating coil, hot water, 3 way valve,									
0320	Freezestat, limit control on discharge, readout	Q-5	.69	23.088	Ea.	3,425	1,350		4,775	5,850
0500	Cooling coil, chilled water, room									
0520	Thermostat, 3 way valve	Q-5	2	8	Ea.	1,475	470		1,945	2,350
0600	Cooling tower, fan cycle, damper control,									
0620	Control system including water readout in/out at panel	Q-19	.67	35.821	Ea.	6,050	2,150		8,200	9,975
1000	Unit ventilator, day/night operation,									
1100	freezestat, ASHRAE, cycle 2	Q-19	.91	26.374	Ea.	3,250	1,575		4,825	6,000
2000	Compensated hot water from boiler, valve control,									
2100	readout and reset at panel, up to 60 GPM	Q-19	.55	43.956	Ea.	6,300	2,625		8,925	11,000
2120	For 120 GPM		.51	47.059		6,475	2,825		9,300	11,500
2140	For 240 GPM		.49	49.180		7,050	2,950		10,000	12,300
3000	Boiler room combustion air, damper to 5 S.F., controls		1.37	17.582		3,025	1,050		4,075	4,950
3500	Fan coil, heating and cooling valves, 4 pipe control system		3	8		1,375	480		1,855	2,250
3600	Heat exchanger system controls		.86	27.907		2,850	1,675		4,525	5,700
3900	Multizone control (one per zone), includes thermostat, damper									
3910	motor and reset of discharge temperature	Q-5	.51	31.373	Ea.	2,925	1,850		4,775	6,075
4000	Pneumatic thermostat, including controlling room radiator valve	"	2.43	6.593		880	390		1,270	1,575
4040	Program energy saving optimizer G	Q-19	1.21	19.786		7,575	1,175		8,750	10,200
4060	Pump control system	"	3	8		1,350	480		1,830	2,225
4080	Reheat coil control system, not incl. coil	Q-5	2.43	6.593		1,150	390		1,540	1,850
4500	Air supply for pneumatic control system									
4600	Tank mounted duplex compressor, starter, alternator,									
4620	piping, dryer, PRV station and filter									

23 09 43 – Pneumatic Control System for HVAC

23 09 43.10 Pneumatic Control Systems		Crew	Daily Output	Labor-Hours	Unit	Material	2020 Bare Costs Labor	Equipment	Total	Total Incl O&P
4630	1/2 HP	Q-19	.68	35.139	Ea.	11,300	2,100		13,400	15,700
4640	3/4 HP		.64	37.383		11,400	2,225		13,625	16,000
4650	1 HP		.61	39.539		12,900	2,375		15,275	17,900
4660	1-1/2 HP		.58	41.739		13,300	2,500		15,800	18,500
4680	3 HP		.55	43.956		18,100	2,625		20,725	24,000
4690	5 HP		.42	57.143		32,800	3,425		36,225	41,400
4800	Main air supply, includes 3/8" copper main and labor	Q-5	1.82	8.791	C.L.F.	380	520		900	1,225
7000	Static pressure control for air handling unit, includes pressure									
7010	sensor, receiver controller, readout and damper motors	Q-19	.64	37.383	Ea.	9,000	2,225		11,225	13,400
8600	VAV boxes, incl. thermostat, damper motor, reheat coil & tubing	Q-5	1.46	10.989	"	1,350	650		2,000	2,500

23 09 53 – Pneumatic and Electric Control System for HVAC

23 09 53.10 Control Components		Crew	Daily Output	Labor-Hours	Unit	Material	Labor	Equipment	Total	Total Incl O&P
0010	**CONTROL COMPONENTS**									
0600	Carbon monoxide detector system									
0606	Panel	1 Stpi	4	2	Ea.	960	131		1,091	1,250
0610	Sensor		7.30	1.096		645	72		717	820
0680	Controller for VAV box, includes actuator		7.30	1.096		281	72		353	420
0700	Controller, receiver									
0730	Pneumatic, panel mount, single input	1 Plum	8	1	Ea.	460	64.50		524.50	605
0740	With conversion mounting bracket		8	1		460	64.50		524.50	605
0750	Dual input, with control point adjustment		7	1.143		635	73.50		708.50	815
0850	Electric, single snap switch	1 Elec	4	2		485	123		608	725
0860	Dual snap switches	"	3	2.667		655	164		819	970
1000	Enthalpy control, boiler water temperature control									
1010	governed by outdoor temperature, with timer	1 Elec	3	2.667	Ea.	355	164		519	640
2000	Gauges, pressure or vacuum									
2100	2" diameter dial	1 Stpi	32	.250	Ea.	10.55	16.40		26.95	37
2200	2-1/2" diameter dial		32	.250		12.80	16.40		29.20	39.50
2300	3-1/2" diameter dial		32	.250		18.15	16.40		34.55	45.50
2400	4-1/2" diameter dial		32	.250		20	16.40		36.40	47.50
2700	Flanged iron case, black ring									
2800	3-1/2" diameter dial	1 Stpi	32	.250	Ea.	101	16.40		117.40	138
2900	4-1/2" diameter dial		32	.250		105	16.40		121.40	141
3000	6" diameter dial		32	.250		167	16.40		183.40	210
3300	For compound pressure-vacuum, add					18%				
3350	Humidistat									
3390	Electric operated	1 Shee	8	1	Ea.	50	62.50		112.50	153
3400	Relays									
3430	Pneumatic/electric	1 Plum	16	.500	Ea.	310	32		342	390
3440	Pneumatic proportioning		8	1		280	64.50		344.50	410
3450	Pneumatic switching		12	.667		151	43		194	234
3460	Selector, 3 point		6	1.333		108	86		194	252
3470	Pneumatic time delay		8	1		345	64.50		409.50	480
3500	Sensor, air operated									
3520	Humidity	1 Plum	16	.500	Ea.	440	32		472	535
3540	Pressure		16	.500		66	32		98	123
3560	Temperature		12	.667		162	43		205	246
3600	Electric operated									
3620	Humidity	1 Elec	8	1	Ea.	284	61.50		345.50	405
3650	Pressure		8	1		325	61.50		386.50	455
3680	Temperature		10	.800		121	49		170	209
4000	Thermometers									

23 09 53 – Pneumatic and Electric Control System for HVAC

23 09 53.10 Control Components		Crew	Daily Output	Labor-Hours	Unit	Material	Labor	Equipment	Total	Total Incl O&P
4100	Dial type, 3-1/2" diameter, vapor type, union connection	1 Stpi	32	.250	Ea.	226	16.40		242.40	275
4120	Liquid type, union connection		32	.250		435	16.40		451.40	505
4500	Stem type, 6-1/2" case, 2" stem, 1/2" NPT		32	.250		60	16.40		76.40	91.50
4520	4" stem, 1/2" NPT		32	.250		73.50	16.40		89.90	107
4600	9" case, 3-1/2" stem, 3/4" NPT		28	.286		93.50	18.75		112.25	132
4620	6" stem, 3/4" NPT		28	.286		111	18.75		129.75	152
4640	8" stem, 3/4" NPT		28	.286		187	18.75		205.75	235
4660	12" stem, 1" NPT	↓	26	.308	↓	164	20		184	212
5000	Thermostats									
5030	Manual	1 Stpi	8	1	Ea.	42.50	65.50		108	148
5040	1 set back, electric, timed G		8	1		77.50	65.50		143	187
5050	2 set back, electric, timed G		8	1		257	65.50		322.50	385
5100	Locking cover	↓	20	.400		20	26		46	62.50
5200	24 hour, automatic, clock G	1 Shee	8	1		204	62.50		266.50	325
5220	Electric, low voltage, 2 wire	1 Elec	13	.615		58	38		96	122
5230	3 wire	"	10	.800	↓	58	49		107	139
5300	Transmitter, pneumatic									
5320	Temperature averaging element	Q-1	8	2	Ea.	151	116		267	345
5350	Pressure differential	1 Plum	7	1.143		1,275	73.50		1,348.50	1,525
5370	Humidity, duct		8	1		400	64.50		464.50	540
5380	Room		12	.667		405	43		448	510
5390	Temperature, with averaging element	↓	6	1.333	↓	175	86		261	325
6000	Valves, motorized zone									
6100	Sweat connections, 1/2" C x C	1 Stpi	20	.400	Ea.	136	26		162	190
6110	3/4" C x C		20	.400		138	26		164	193
6120	1" C x C		19	.421		185	27.50		212.50	246
6140	1/2" C x C, with end switch, 2 wire		20	.400		137	26		163	192
6150	3/4" C x C, with end switch, 2 wire		20	.400		118	26		144	171
6160	1" C x C, with end switch, 2 wire	↓	19	.421	↓	179	27.50		206.50	239
7090	Valves, motor controlled, including actuator									
7100	Electric motor actuated									
7200	Brass, two way, screwed									
7210	1/2" pipe size	L-6	36	.333	Ea.	385	21		406	460
7220	3/4" pipe size		30	.400		625	25.50		650.50	725
7230	1" pipe size		28	.429		620	27		647	720
7240	1-1/2" pipe size		19	.632		960	40		1,000	1,100
7250	2" pipe size	↓	16	.750	↓	1,250	47.50		1,297.50	1,450
7350	Brass, three way, screwed									
7360	1/2" pipe size	L-6	33	.364	Ea.	340	23		363	410
7370	3/4" pipe size		27	.444		395	28		423	480
7380	1" pipe size		25.50	.471		410	30		440	495
7384	1-1/4" pipe size		21	.571		450	36		486	550
7390	1-1/2" pipe size		17	.706		495	45		540	615
7400	2" pipe size	↓	14	.857	↓	780	54.50		834.50	940
7550	Iron body, two way, flanged									
7560	2-1/2" pipe size	L-6	4	3	Ea.	1,175	190		1,365	1,575
7570	3" pipe size		3	4		1,200	254		1,454	1,725
7580	4" pipe size	↓	2	6	↓	2,175	380		2,555	2,975
7850	Iron body, three way, flanged									
7860	2-1/2" pipe size	L-6	3	4	Ea.	1,175	254		1,429	1,700
7870	3" pipe size		2.50	4.800		1,325	305		1,630	1,950
7880	4" pipe size	↓	2	6	↓	1,700	380		2,080	2,475
8000	Pneumatic, air operated									

23 09 Instrumentation and Control for HVAC

23 09 53 – Pneumatic and Electric Control System for HVAC

23 09 53.10 Control Components		Crew	Daily Output	Labor-Hours	Unit	Material	Labor	Equipment	Total	Total Incl O&P
						2020 Bare Costs				
8050	Brass, two way, screwed									
8060	1/2" pipe size, class 250	1 Plum	24	.333	Ea.	174	21.50		195.50	224
8070	3/4" pipe size, class 250		20	.400		208	26		234	269
8080	1" pipe size, class 250		19	.421		243	27		270	310
8090	1-1/4" pipe size, class 125		15	.533		300	34.50		334.50	385
8100	1-1/2" pipe size, class 125		13	.615		385	39.50		424.50	485
8110	2" pipe size, class 125		11	.727		445	47		492	565
8180	Brass, three way, screwed									
8190	1/2" pipe size, class 250	1 Plum	22	.364	Ea.	184	23.50		207.50	238
8200	3/4" pipe size, class 250		18	.444		209	28.50		237.50	275
8210	1" pipe size, class 250		17	.471		244	30.50		274.50	315
8214	1-1/4" pipe size, class 250		14	.571		330	37		367	415
8220	1-1/2" pipe size, class 125		11	.727		390	47		437	500
8230	2" pipe size, class 125		9	.889		455	57.50		512.50	590
8450	Iron body, two way, flanged									
8560	Iron body, three way, flanged									
8570	2-1/2" pipe size, class 125	Q-1	4.50	3.556	Ea.	1,000	206		1,206	1,425
8580	3" pipe size, class 125		4	4		1,200	232		1,432	1,650
8590	4" pipe size, class 125		2.50	6.400		2,350	370		2,720	3,150
8600	6" pipe size, class 125	Q-2	3	8		3,550	480		4,030	4,675
9000	Minimum labor/equipment charge	1 Plum	4	2	Job		129		129	199

23 11 Facility Fuel Piping

23 11 13 – Facility Fuel-Oil Piping

23 11 13.10 Fuel Oil Specialties		Crew	Daily Output	Labor-Hours	Unit	Material	Labor	Equipment	Total	Total Incl O&P
0010	**FUEL OIL SPECIALTIES**									
0020	Foot valve, single poppet, metal to metal construction									
0040	Bevel seat, 1/2" diameter	1 Stpi	20	.400	Ea.	80.50	26		106.50	129
0060	3/4" diameter		18	.444		113	29		142	169
0080	1" diameter		16	.500		113	33		146	175
0100	1-1/4" diameter		15	.533		165	35		200	236
0120	1-1/2" diameter		13	.615		209	40.50		249.50	292
0140	2" diameter		11	.727		260	47.50		307.50	360
0400	Fuel fill box, flush type									
0408	Nonlocking, watertight									
0410	1-1/2" diameter	1 Stpi	12	.667	Ea.	23	43.50		66.50	92.50
0440	2" diameter		10	.800		18.80	52.50		71.30	102
0450	2-1/2" diameter		9	.889		66	58.50		124.50	163
0460	3" diameter		7	1.143		54	75		129	175
0470	4" diameter		5	1.600		77	105		182	247
0500	Locking inner cover									
0510	2" diameter	1 Stpi	8	1	Ea.	83	65.50		148.50	193
0520	2-1/2" diameter		7	1.143		111	75		186	238
0530	3" diameter		5	1.600		119	105		224	293
0540	4" diameter		4	2		154	131		285	370
0600	Fuel system components									
0620	Spill container	1 Stpi	4	2	Ea.	1,000	131		1,131	1,325
0640	Fill adapter, 4", straight drop		8	1		60	65.50		125.50	167
0680	Fill cap, 4"		30	.267		42.50	17.50		60	73.50
0700	Extractor fitting, 4" x 1-1/2"		8	1		545	65.50		610.50	700
0740	Vapor hose adapter, 4"		8	1		128	65.50		193.50	241

23 11 13 – Facility Fuel-Oil Piping

23 11 13.10 Fuel Oil Specialties		Crew	Daily Output	Labor-Hours	Unit	Material	2020 Bare Costs Labor	Equipment	Total	Total Incl O&P
0760	Wood gage stick, 10′				Ea.	13.25			13.25	14.60
1000	Oil filters, 3/8″ IPT, 20 gal. per hour	1 Stpi	20	.400		43	26		69	88
1020	32 gal. per hour		18	.444		58.50	29		87.50	110
1040	40 gal. per hour		16	.500		56.50	33		89.50	113
1060	50 gal. per hour		14	.571		61.50	37.50		99	126
2000	Remote tank gauging system, self contained									
2100	Single tank kit/8 sensors inputs w/printer	1 Stpi	2.50	3.200	Ea.	4,225	210		4,435	4,975
2120	Two tank kit/8 sensors inputs w/printer		2	4		5,750	262		6,012	6,725
3000	Valve, ball check, globe type, 3/8″ diameter		24	.333		13.05	22		35.05	48.50
3500	Fusible, 3/8″ diameter		24	.333		14.70	22		36.70	50
3600	1/2″ diameter		24	.333		40	22		62	78
3610	3/4″ diameter		20	.400		79	26		105	128
3620	1″ diameter		19	.421		229	27.50		256.50	295
4000	Nonfusible, 3/8″ diameter		24	.333		27	22		49	64
4500	Shutoff, gate type, lever handle, spring-fusible kit									
4520	1/4″ diameter	1 Stpi	14	.571	Ea.	44	37.50		81.50	106
4540	3/8″ diameter		12	.667		42.50	43.50		86	115
4560	1/2″ diameter		10	.800		60.50	52.50		113	148
4570	3/4″ diameter		8	1		113	65.50		178.50	225
4580	Lever handle, requires weight and fusible kit									
4600	1″ diameter	1 Stpi	9	.889	Ea.	400	58.50		458.50	530
4620	1-1/4″ diameter		8	1		545	65.50		610.50	700
4640	1-1/2″ diameter		7	1.143		455	75		530	615
4660	2″ diameter		6	1.333		905	87.50		992.50	1,125
4680	For fusible link, weight and braided wire, add					5%				
5000	Vent alarm, whistling signal					38			38	42
5500	Vent protector/breather, 1-1/4″ diameter	1 Stpi	32	.250		22.50	16.40		38.90	50
5520	1-1/2″ diameter		32	.250		26	16.40		42.40	54
5540	2″ diameter		32	.250		53.50	16.40		69.90	84.50
5560	3″ diameter		28	.286		75	18.75		93.75	112
5580	4″ diameter		24	.333		91.50	22		113.50	135
5600	Dust cap, breather, 2″		40	.200		12.05	13.10		25.15	34
8000	Fuel oil and tank heaters									
8020	Electric, capacity rated at 230 volts									
8040	Immersion element in steel manifold									
8060	96 GPH at 50°F rise	Q-5	6.40	2.500	Ea.	1,350	148		1,498	1,700
8070	128 GPH at 50°F rise		6.20	2.581		1,350	152		1,502	1,700
8080	160 GPH at 50°F rise		5.90	2.712		1,550	160		1,710	1,950
8090	192 GPH at 50°F rise		5.50	2.909		1,700	172		1,872	2,125
8100	240 GPH at 50°F rise		5.10	3.137		1,950	185		2,135	2,425
8110	288 GPH at 50°F rise		4.60	3.478		2,125	205		2,330	2,650
8120	384 GPH at 50°F rise		3.10	5.161		2,475	305		2,780	3,200
8130	480 GPH at 50°F rise		2.30	6.957		3,125	410		3,535	4,050
8140	576 GPH at 50°F rise		2.10	7.619		3,125	450		3,575	4,125
8300	Suction stub, immersion type									
8320	75″ long, 750 watts	1 Stpi	14	.571	Ea.	890	37.50		927.50	1,025
8330	99″ long, 2000 watts		12	.667		1,275	43.50		1,318.50	1,500
8340	123″ long, 3000 watts		10	.800		1,375	52.50		1,427.50	1,600
8660	Steam, cross flow, rated at 5 PSIG									
8680	42 GPH	Q-5	7	2.286	Ea.	2,050	135		2,185	2,450
8690	73 GPH		6.70	2.388		2,475	141		2,616	2,925
8700	112 GPH		6.20	2.581		2,475	152		2,627	2,950
8710	158 GPH		5.80	2.759		2,700	163		2,863	3,225

23 11 Facility Fuel Piping

23 11 13 – Facility Fuel-Oil Piping

23 11 13.10 Fuel Oil Specialties		Crew	Daily Output	Labor-Hours	Unit	Material	Labor	Equipment	Total	Total Incl O&P
						2020 Bare Costs				
8720	187 GPH	Q-5	4	4	Ea.	3,825	236		4,061	4,575
8730	270 GPH		3.60	4.444		4,000	262		4,262	4,800
8740	365 GPH	Q-6	4.90	4.898		4,850	300		5,150	5,800
8750	635 GPH		3.70	6.486		5,925	395		6,320	7,150
8760	845 GPH		2.50	9.600		8,950	585		9,535	10,800
8770	1420 GPH		1.60	15		14,600	920		15,520	17,400
8780	2100 GPH		1.10	21.818		20,000	1,325		21,325	24,100

23 11 23 – Facility Natural-Gas Piping

23 11 23.10 Gas Meters

		Crew	Daily Output	Labor-Hours	Unit	Material	Labor	Equipment	Total	Total Incl O&P
0010	**GAS METERS**									
4000	Residential									
4010	Gas meter, residential, 3/4" pipe size	1 Plum	14	.571	Ea.	305	37		342	390
4020	Gas meter, residential, 1" pipe size		12	.667		290	43		333	385
4030	Gas meter, residential, 1-1/4" pipe size		10	.800		254	51.50		305.50	360

23 11 23.20 Gas Piping, Flexible (Csst)

		Crew	Daily Output	Labor-Hours	Unit	Material	Labor	Equipment	Total	Total Incl O&P
0010	**GAS PIPING, FLEXIBLE (CSST)**									
0100	Tubing with lightning protection									
0110	3/8"	1 Stpi	65	.123	L.F.	3.22	8.05		11.27	16
0120	1/2"		62	.129		3.62	8.45		12.07	17.10
0130	3/4"		60	.133		4.71	8.75		13.46	18.70
0140	1"		55	.145		7.25	9.55		16.80	22.50
0150	1-1/4"		50	.160		8.40	10.50		18.90	25.50
0160	1-1/2"		45	.178		14.75	11.65		26.40	34.50
0170	2"		40	.200		21	13.10		34.10	44
0200	Tubing for underground/underslab burial									
0210	3/8"	1 Stpi	65	.123	L.F.	4.48	8.05		12.53	17.40
0220	1/2"		62	.129		5.10	8.45		13.55	18.75
0230	3/4"		60	.133		6.50	8.75		15.25	20.50
0240	1"		55	.145		8.75	9.55		18.30	24.50
0250	1-1/4"		50	.160		11.75	10.50		22.25	29
0260	1-1/2"		45	.178		22	11.65		33.65	42
0270	2"		40	.200		26	13.10		39.10	49.50
3000	Fittings									
3010	Straight									
3100	Tube to NPT									
3110	3/8"	1 Stpi	29	.276	Ea.	13.35	18.10		31.45	42.50
3120	1/2"		27	.296		14.35	19.40		33.75	46
3130	3/4"		25	.320		19.60	21		40.60	54
3140	1"		23	.348		30.50	23		53.50	69
3150	1-1/4"		20	.400		68	26		94	116
3160	1-1/2"		17	.471		138	31		169	200
3170	2"		15	.533		235	35		270	315
3200	Coupling									
3210	3/8"	1 Stpi	29	.276	Ea.	23	18.10		41.10	53.50
3220	1/2"		27	.296		26.50	19.40		45.90	59
3230	3/4"		25	.320		36.50	21		57.50	72.50
3240	1"		23	.348		56.50	23		79.50	97.50
3250	1-1/4"		20	.400		125	26		151	179
3260	1-1/2"		17	.471		262	31		293	335
3270	2"		15	.533		445	35		480	545
3300	Flange fitting									
3310	3/8"	1 Stpi	25	.320	Ea.	20.50	21		41.50	55

23 11 Facility Fuel Piping

23 11 23 – Facility Natural-Gas Piping

23 11 23.20 Gas Piping, Flexible (Csst)		Crew	Daily Output	Labor-Hours	Unit	Material	Labor	Equipment	Total	Total Incl O&P
						2020 Bare Costs				
3320	1/2"	1 Stpi	22	.364	Ea.	19.50	24		43.50	58.50
3330	3/4"		19	.421		25.50	27.50		53	70.50
3340	1"		16	.500		35	33		68	89
3350	1-1/4"	↓	12	.667	↓	81	43.50		124.50	157
3400	90° flange valve									
3410	3/8"	1 Stpi	25	.320	Ea.	39	21		60	75.50
3420	1/2"		22	.364		39.50	24		63.50	80.50
3430	3/4"	↓	19	.421	↓	48.50	27.50		76	96
4000	Tee									
4120	1/2"	1 Stpi	20.50	.390	Ea.	41.50	25.50		67	85.50
4130	3/4"		19	.421		55.50	27.50		83	104
4140	1"	↓	17.50	.457	↓	99.50	30		129.50	157
5000	Reducing									
5110	Tube to NPT									
5120	3/4" to 1/2" NPT	1 Stpi	26	.308	Ea.	21	20		41	54
5130	1" to 3/4" NPT	"	24	.333	"	32.50	22		54.50	70
5200	Reducing tee									
5210	1/2" x 3/8" x 3/8"	1 Stpi	21	.381	Ea.	55	25		80	99
5220	3/4" x 1/2" x 1/2"		20	.400		51.50	26		77.50	97.50
5230	1" x 3/4" x 1/2"		18	.444		85.50	29		114.50	139
5240	1-1/4" x 1-1/4" x 1"		15.60	.513		228	33.50		261.50	305
5250	1-1/2" x 1-1/2" x 1-1/4"		13.30	.602		350	39.50		389.50	445
5260	2" x 2" x 1-1/2"	↓	11.80	.678	↓	525	44.50		569.50	645
5300	Manifold with four ports and mounting bracket									
5302	Labor to mount manifold does not include making pipe									
5304	connections which are included in fitting labor.									
5310	3/4" x 1/2" x 1/2" (4)	1 Stpi	76	.105	Ea.	42	6.90		48.90	57
5330	1-1/4" x 1" x 3/4" (4)		72	.111		60.50	7.30		67.80	78
5350	2" x 1-1/2" x 1" (4)	↓	68	.118	↓	76	7.70		83.70	95.50
5600	Protective striker plate									
5610	Quarter plate, 3" x 2"	1 Stpi	88	.091	Ea.	.93	5.95		6.88	10.20
5620	Half plate, 3" x 7"		82	.098		2.30	6.40		8.70	12.45
5630	Full plate, 3" x 12"	↓	78	.103	↓	3.92	6.70		10.62	14.70

23 12 Facility Fuel Pumps

23 12 13 – Facility Fuel-Oil Pumps

23 12 13.10 Pump and Motor Sets		Crew	Daily Output	Labor-Hours	Unit	Material	Labor	Equipment	Total	Total Incl O&P
0010	**PUMP AND MOTOR SETS**									
1810	Light fuel and diesel oils									
1820	20 GPH, 1/3 HP	Q-5	6	2.667	Ea.	1,600	157		1,757	2,025
1830	27 GPH, 1/3 HP		6	2.667		1,600	157		1,757	2,025
1840	80 GPH, 1/3 HP		5	3.200		1,600	189		1,789	2,050
1850	145 GPH, 1/2 HP		4	4		1,700	236		1,936	2,250
1860	277 GPH, 1 HP		4	4		1,900	236		2,136	2,450
1870	700 GPH, 1-1/2 HP		3	5.333		4,400	315		4,715	5,325
1880	1000 GPH, 2 HP		3	5.333		5,000	315		5,315	5,975
1890	1800 GPH, 5 HP	↓	1.80	8.889	↓	8,100	525		8,625	9,700

23 13 13.09 Single-Wall Steel Fuel-Oil Tanks		Crew	Daily Output	Labor-Hours	Unit	2020 Bare Costs Material	Labor	Equipment	Total	Total Incl O&P
0010	**SINGLE-WALL STEEL FUEL-OIL TANKS**									
5000	Tanks, steel ugnd., sti-p3, not incl. hold-down bars									
5500	Excavation, pad, pumps and piping not included									
5510	Single wall, 500 gallon capacity, 7 ga. shell	Q-5	2.70	5.926	Ea.	1,875	350		2,225	2,625
5520	1,000 gallon capacity, 7 ga. shell	"	2.50	6.400		2,925	380		3,305	3,800
5530	2,000 gallon capacity, 1/4" thick shell	Q-7	4.60	6.957		3,725	435		4,160	4,775
5535	2,500 gallon capacity, 7 ga. shell	Q-5	3	5.333		5,675	315		5,990	6,700
5540	5,000 gallon capacity, 1/4" thick shell	Q-7	3.20	10		13,400	625		14,025	15,700
5560	10,000 gallon capacity, 1/4" thick shell		2	16		10,800	1,000		11,800	13,500
5580	15,000 gallon capacity, 5/16" thick shell		1.70	18.824		11,200	1,175		12,375	14,100
5600	20,000 gallon capacity, 5/16" thick shell		1.50	21.333		22,600	1,325		23,925	27,000
5610	25,000 gallon capacity, 3/8" thick shell		1.30	24.615		29,600	1,525		31,125	35,000
5620	30,000 gallon capacity, 3/8" thick shell		1.10	29.091		35,000	1,825		36,825	41,300
5630	40,000 gallon capacity, 3/8" thick shell		.90	35.556		44,500	2,225		46,725	52,500
5640	50,000 gallon capacity, 3/8" thick shell		.80	40		49,800	2,500		52,300	58,500

23 13 13.13 Dbl-Wall Steel, Undrgrnd Fuel-Oil, Stor. Tanks		Crew	Daily Output	Labor-Hours	Unit	Material	Labor	Equipment	Total	Total Incl O&P
0010	**DOUBLE-WALL STEEL, UNDERGROUND FUEL-OIL, STORAGE TANKS**									
6200	Steel, underground, 360°, double wall, UL listed,									
6210	with sti-P3 corrosion protection,									
6220	(dielectric coating, cathodic protection, electrical									
6230	isolation) 30 year warranty,									
6240	not incl. manholes or hold-downs.									
6250	500 gallon capacity	Q-5	2.40	6.667	Ea.	3,325	395		3,720	4,250
6260	1,000 gallon capacity	"	2.25	7.111		3,675	420		4,095	4,675
6270	2,000 gallon capacity	Q-7	4.16	7.692		4,850	480		5,330	6,075
6280	3,000 gallon capacity		3.90	8.205		6,950	510		7,460	8,450
6290	4,000 gallon capacity		3.64	8.791		8,200	550		8,750	9,875
6300	5,000 gallon capacity		2.91	10.997		8,800	685		9,485	10,800
6310	6,000 gallon capacity		2.42	13.223		11,600	825		12,425	14,100
6320	8,000 gallon capacity		2.08	15.385		11,200	960		12,160	13,800
6330	10,000 gallon capacity		1.82	17.582		13,900	1,100		15,000	17,000
6340	12,000 gallon capacity		1.70	18.824		16,600	1,175		17,775	20,100
6350	15,000 gallon capacity		1.33	24.060		20,700	1,500		22,200	25,100
6360	20,000 gallon capacity		1.33	24.060		34,200	1,500		35,700	40,000
6370	25,000 gallon capacity		1.16	27.586		38,100	1,725		39,825	44,600
6380	30,000 gallon capacity		1.03	31.068		37,800	1,950		39,750	44,600
6390	40,000 gallon capacity		.80	40		115,000	2,500		117,500	130,000
6395	50,000 gallon capacity		.73	43.836		139,500	2,725		142,225	157,500
6400	For hold-downs 500-2,000 gal., add		16	2	Set	226	125		351	440
6410	For hold-downs 3,000-6,000 gal., add		12	2.667		385	166		551	675
6420	For hold-downs 8,000-12,000 gal., add		11	2.909		455	182		637	780
6430	For hold-downs 15,000 gal., add		9	3.556		675	222		897	1,100
6440	For hold-downs 20,000 gal., add		8	4		780	250		1,030	1,250
6450	For hold-downs 20,000 gal. plus, add		6	5.333		1,000	335		1,335	1,625
6500	For manways, add				Ea.	1,050			1,050	1,175
6600	In place with hold-downs									
6652	550 gallon capacity	Q-5	1.84	8.696	Ea.	3,550	515		4,065	4,700

23 13 13.23 Glass-Fiber-Reinfcd-Plastic, Fuel-Oil, Storage		Crew	Daily Output	Labor-Hours	Unit	Material	2020 Bare Costs Labor	Equipment	Total	Total Incl O&P
0010	**GLASS-FIBER-REINFCD-PLASTIC, UNDERGRND. FUEL-OIL, STORAGE**									
0210	Fiberglass, underground, single wall, UL listed, not including									
0220	manway or hold-down strap									
0225	550 gallon capacity	Q-5	2.67	5.993	Ea.	4,075	355		4,430	5,050
0230	1,000 gallon capacity	"	2.46	6.504		5,275	385		5,660	6,400
0240	2,000 gallon capacity	Q-7	4.57	7.002		8,275	435		8,710	9,775
0245	3,000 gallon capacity		3.90	8.205		8,400	510		8,910	10,000
0250	4,000 gallon capacity		3.55	9.014		11,100	560		11,660	13,100
0255	5,000 gallon capacity		3.20	10		10,400	625		11,025	12,400
0260	6,000 gallon capacity		2.67	11.985		11,800	750		12,550	14,200
0270	8,000 gallon capacity		2.29	13.974		14,900	870		15,770	17,800
0280	10,000 gallon capacity		2	16		15,600	1,000		16,600	18,800
0282	12,000 gallon capacity		1.88	17.021		20,900	1,050		21,950	24,700
0284	15,000 gallon capacity		1.68	19.048		27,900	1,200		29,100	32,600
0290	20,000 gallon capacity		1.45	22.069		33,200	1,375		34,575	38,600
0300	25,000 gallon capacity		1.28	25		58,500	1,550		60,050	67,000
0320	30,000 gallon capacity		1.14	28.070		104,500	1,750		106,250	117,500
0340	40,000 gallon capacity		.89	35.955		114,000	2,250		116,250	129,000
0360	48,000 gallon capacity		.81	39.506		177,000	2,475		179,475	198,500
0500	For manway, fittings and hold-downs, add					20%	15%			
0600	For manways, add					2,525			2,525	2,775
1000	For helical heating coil, add	Q-5	2.50	6.400		5,450	380		5,830	6,575
1020	Fiberglass, underground, double wall, UL listed									
1030	includes manways, not incl. hold-down straps									
1040	600 gallon capacity	Q-5	2.42	6.612	Ea.	9,100	390		9,490	10,600
1050	1,000 gallon capacity	"	2.25	7.111		12,400	420		12,820	14,400
1060	2,500 gallon capacity	Q-7	4.16	7.692		17,300	480		17,780	19,700
1070	3,000 gallon capacity		3.90	8.205		19,400	510		19,910	22,100
1080	4,000 gallon capacity		3.64	8.791		19,600	550		20,150	22,400
1090	6,000 gallon capacity		2.42	13.223		26,000	825		26,825	29,900
1100	8,000 gallon capacity		2.08	15.385		33,000	960		33,960	37,800
1110	10,000 gallon capacity		1.82	17.582		38,800	1,100		39,900	44,300
1120	12,000 gallon capacity		1.70	18.824		48,300	1,175		49,475	55,000
1122	15,000 gallon capacity		1.52	21.053		67,000	1,325		68,325	76,000
1124	20,000 gallon capacity		1.33	24.060		79,500	1,500		81,000	90,000
1126	25,000 gallon capacity		1.16	27.586		92,000	1,725		93,725	103,500
1128	30,000 gallon capacity		1.03	31.068		110,500	1,950		112,450	124,500
1140	For hold-down straps, add					2%	10%			
1150	For hold-downs 500-4,000 gal., add	Q-7	16	2	Set	505	125		630	750
1160	For hold-downs 5,000-15,000 gal., add		8	4		1,000	250		1,250	1,475
1170	For hold-downs 20,000 gal., add		5.33	6.004		1,500	375		1,875	2,225
1180	For hold-downs 25,000 gal., add		4	8		2,025	500		2,525	3,000
1190	For hold-downs 30,000 gal., add		2.60	12.308		3,025	770		3,795	4,525
2210	Fiberglass, underground, single wall, UL listed, including									
2220	hold-down straps, no manways									
2225	550 gallon capacity	Q-5	2	8	Ea.	4,600	470		5,070	5,775
2230	1,000 gallon capacity	"	1.88	8.511		5,775	500		6,275	7,125
2240	2,000 gallon capacity	Q-7	3.55	9.014		8,775	560		9,335	10,500
2250	4,000 gallon capacity		2.90	11.034		11,600	690		12,290	13,900
2260	6,000 gallon capacity		2	16		12,800	1,000		13,800	15,700
2270	8,000 gallon capacity		1.78	17.978		15,900	1,125		17,025	19,200
2280	10,000 gallon capacity		1.60	20		16,600	1,250		17,850	20,200

23 13 Facility Fuel-Storage Tanks

23 13 13 – Facility Underground Fuel-Oil, Storage Tanks

23 13 13.23 Glass-Fiber-Reinfcd-Plastic, Fuel-Oil, Storage		Crew	Daily Output	Labor-Hours	Unit	2020 Bare Costs Material	Labor	Equipment	Total	Total Incl O&P
2282	12,000 gallon capacity	Q-7	1.52	21.053	Ea.	21,900	1,325		23,225	26,100
2284	15,000 gallon capacity		1.39	23.022		28,900	1,425		30,325	34,000
2290	20,000 gallon capacity		1.14	28.070		34,700	1,750		36,450	40,900
2300	25,000 gallon capacity		.96	33.333		60,500	2,075		62,575	69,500
2320	30,000 gallon capacity		.80	40		107,500	2,500		110,000	122,000
3020	Fiberglass, underground, double wall, UL listed									
3030	includes manways and hold-down straps									
3040	600 gallon capacity	Q-5	1.86	8.602	Ea.	9,600	510		10,110	11,400
3050	1,000 gallon capacity	"	1.70	9.412		12,900	555		13,455	15,100
3060	2,500 gallon capacity	Q-7	3.29	9.726		17,800	605		18,405	20,500
3070	3,000 gallon capacity		3.13	10.224		19,900	640		20,540	22,800
3080	4,000 gallon capacity		2.93	10.922		20,100	680		20,780	23,200
3090	6,000 gallon capacity		1.86	17.204		27,000	1,075		28,075	31,400
3100	8,000 gallon capacity		1.65	19.394		34,000	1,200		35,200	39,300
3110	10,000 gallon capacity		1.48	21.622		39,800	1,350		41,150	45,900
3120	12,000 gallon capacity		1.40	22.857		49,300	1,425		50,725	56,000
3122	15,000 gallon capacity		1.28	25		68,000	1,550		69,550	77,500
3124	20,000 gallon capacity		1.06	30.189		81,000	1,875		82,875	92,000
3126	25,000 gallon capacity		.90	35.556		94,000	2,225		96,225	107,000
3128	30,000 gallon capacity		.74	43.243		113,500	2,700		116,200	129,000

23 13 23 – Facility Aboveground Fuel-Oil, Storage Tanks

23 13 23.16 Horizontal, Stl, Abvgrd Fuel-Oil, Storage Tanks

		Crew	Daily Output	Labor-Hours	Unit	Material	Labor	Equipment	Total	Total Incl O&P
0010	**HORIZONTAL, STEEL, ABOVEGROUND FUEL-OIL, STORAGE TANKS**									
3000	Steel, storage, aboveground, including cradles, coating,									
3020	fittings, not including foundation, pumps or piping									
3040	Single wall, 275 gallon	Q-5	5	3.200	Ea.	510	189		699	850
3060	550 gallon	"	2.70	5.926		4,400	350		4,750	5,400
3080	1,000 gallon	Q-7	5	6.400		7,175	400		7,575	8,500
3100	1,500 gallon		4.75	6.737		10,100	420		10,520	11,800
3120	2,000 gallon		4.60	6.957		13,500	435		13,935	15,600
3140	5,000 gallon		3.20	10		21,400	625		22,025	24,600
3150	10,000 gallon		2	16		40,800	1,000		41,800	46,500
3160	15,000 gallon		1.70	18.824		57,000	1,175		58,175	65,000
3170	20,000 gallon		1.45	22.069		74,000	1,375		75,375	83,500
3180	25,000 gallon		1.30	24.615		84,500	1,525		86,025	95,500
3190	30,000 gallon		1.10	29.091		104,000	1,825		105,825	117,500
3320	Double wall, 500 gallon capacity	Q-5	2.40	6.667		2,000	395		2,395	2,800
3330	2,000 gallon capacity	Q-7	4.15	7.711		6,650	480		7,130	8,075
3340	4,000 gallon capacity		3.60	8.889		14,500	555		15,055	16,800
3350	6,000 gallon capacity		2.40	13.333		16,400	830		17,230	19,400
3360	8,000 gallon capacity		2	16		19,500	1,000		20,500	23,100
3370	10,000 gallon capacity		1.80	17.778		30,500	1,100		31,600	35,200
3380	15,000 gallon capacity		1.50	21.333		41,100	1,325		42,425	47,300
3390	20,000 gallon capacity		1.30	24.615		48,400	1,525		49,925	55,500
3400	25,000 gallon capacity		1.15	27.826		60,000	1,725		61,725	68,500
3410	30,000 gallon capacity		1	32		67,000	2,000		69,000	76,500

23 13 23.26 Horizontal, Conc., Abvgrd Fuel-Oil, Stor. Tanks

		Crew	Daily Output	Labor-Hours	Unit	Material	Labor	Equipment	Total	Total Incl O&P
0010	**HORIZONTAL, CONCRETE, ABOVEGROUND FUEL-OIL, STORAGE TANKS**									
0050	Concrete, storage, aboveground, including pad & pump									
0100	500 gallon	F-3	2	20	Ea.	10,500	1,075	236	11,811	13,500
0200	1,000 gallon	"	2	20		14,700	1,075	236	16,011	18,100
0300	2,000 gallon	F-4	2	24		18,900	1,300	490	20,690	23,400

23 13 Facility Fuel-Storage Tanks

23 13 23 – Facility Aboveground Fuel-Oil, Storage Tanks

23 13 23.26 Horizontal, Conc., Abvgrd Fuel-Oil, Stor. Tanks		Crew	Daily Output	Labor-Hours	Unit	Material	Labor	Equipment	Total	Total Incl O&P
						2020 Bare Costs				
0400	4,000 gallon	F-4	2	24	Ea.	24,100	1,300	490	25,890	29,100
0500	8,000 gallon		2	24		37,700	1,300	490	39,490	44,100
0600	12,000 gallon	↓	2	24	↓	50,500	1,300	490	52,290	58,000

23 21 Hydronic Piping and Pumps

23 21 20 – Hydronic HVAC Piping Specialties

		Crew	Daily Output	Labor-Hours	Unit	Material	Labor	Equipment	Total	Total Incl O&P
23 21 20.10 Air Control										
0010	**AIR CONTROL**									
0030	Air separator, with strainer									
0040	2" diameter	Q-5	6	2.667	Ea.	1,575	157		1,732	1,975
0080	2-1/2" diameter		5	3.200		1,750	189		1,939	2,225
0100	3" diameter		4	4		2,700	236		2,936	3,350
0120	4" diameter	↓	3	5.333		3,900	315		4,215	4,750
0130	5" diameter	Q-6	3.60	6.667		4,950	410		5,360	6,075
0140	6" diameter		3.40	7.059		5,950	430		6,380	7,225
0160	8" diameter		3	8		8,875	490		9,365	10,500
0180	10" diameter		2.20	10.909		13,700	665		14,365	16,100
0200	12" diameter	↓	1.70	14.118	↓	22,800	865		23,665	26,400
23 21 20.18 Automatic Air Vent										
0010	**AUTOMATIC AIR VENT**									
0020	Cast iron body, stainless steel internals, float type									
0060	1/2" NPT inlet, 300 psi	1 Stpi	12	.667	Ea.	180	43.50		223.50	266
0140	3/4" NPT inlet, 300 psi		12	.667		180	43.50		223.50	266
0180	1/2" NPT inlet, 250 psi		10	.800		375	52.50		427.50	495
0220	3/4" NPT inlet, 250 psi		10	.800		375	52.50		427.50	495
0260	1" NPT inlet, 250 psi	↓	10	.800		550	52.50		602.50	685
0340	1-1/2" NPT inlet, 250 psi	Q-5	12	1.333		1,175	78.50		1,253.50	1,400
0380	2" NPT inlet, 250 psi	"	12	1.333	↓	1,175	78.50		1,253.50	1,400
0600	Forged steel body, stainless steel internals, float type									
0640	1/2" NPT inlet, 750 psi	1 Stpi	12	.667	Ea.	1,250	43.50		1,293.50	1,450
0680	3/4" NPT inlet, 750 psi		12	.667		1,250	43.50		1,293.50	1,450
0760	3/4" NPT inlet, 1,000 psi	↓	10	.800		1,875	52.50		1,927.50	2,125
0800	1" NPT inlet, 1,000 psi	Q-5	12	1.333		1,875	78.50		1,953.50	2,175
0880	1-1/2" NPT inlet, 1,000 psi		10	1.600		5,250	94.50		5,344.50	5,925
0920	2" NPT inlet, 1,000 psi	↓	10	1.600	↓	5,250	94.50		5,344.50	5,925
1100	Formed steel body, noncorrosive									
1110	1/8" NPT inlet, 150 psi	1 Stpi	32	.250	Ea.	15.70	16.40		32.10	43
1120	1/4" NPT inlet, 150 psi		32	.250		53.50	16.40		69.90	84.50
1130	3/4" NPT inlet, 150 psi	↓	32	.250	↓	53.50	16.40		69.90	84.50
1300	Chrome plated brass, automatic/manual, for radiators									
1310	1/8" NPT inlet, nickel plated brass	1 Stpi	32	.250	Ea.	9.25	16.40		25.65	35.50
23 21 20.22 Circuit Sensor										
0010	**CIRCUIT SENSOR**, Flow meter									
0020	Metering stations									
0040	Wafer orifice insert type									
0060	2-1/2" pipe size	Q-5	12	1.333	Ea.	300	78.50		378.50	450
0100	3" pipe size		11	1.455		340	86		426	510
0140	4" pipe size		8	2		395	118		513	615
0180	5" pipe size		7.30	2.192		495	129		624	745
0220	6" pipe size	↓	6.40	2.500	↓	590	148		738	880

23 21 20.22 Circuit Sensor

	23 21 20.22 Circuit Sensor	Crew	Daily Output	Labor-Hours	Unit	2020 Bare Costs Material	Labor	Equipment	Total	Total Incl O&P
0260	8" pipe size	Q-6	5.30	4.528	Ea.	815	277		1,092	1,325
0280	10" pipe size		4.60	5.217		945	320		1,265	1,550
0360	12" pipe size		4.20	5.714		1,525	350		1,875	2,225

23 21 20.26 Circuit Setter

	23 21 20.26 Circuit Setter	Crew	Daily Output	Labor-Hours	Unit	Material	Labor	Equipment	Total	Total Incl O&P
0010	**CIRCUIT SETTER**, Balance valve									
0018	Threaded									
0020	3/4" pipe size	1 Stpi	20	.400	Ea.	90	26		116	140
0040	1" pipe size		18	.444		117	29		146	173
0060	1-1/2" pipe size		12	.667		202	43.50		245.50	290
0080	2" pipe size		10	.800		290	52.50		342.50	400
0100	2-1/2" pipe size	Q-5	15	1.067		655	63		718	820
0120	3" pipe size	"	10	1.600		920	94.50		1,014.50	1,175
0130	Cast iron body, flanged									
0136	3" pipe size, flanged	Q-5	4	4	Ea.	925	236		1,161	1,400
0140	4" pipe size	"	3	5.333		1,375	315		1,690	2,000
0200	For differential meter, accurate to 1%, add					915			915	1,000

23 21 20.34 Dielectric Unions

	23 21 20.34 Dielectric Unions	Crew	Daily Output	Labor-Hours	Unit	Material	Labor	Equipment	Total	Total Incl O&P
0010	**DIELECTRIC UNIONS**, Standard gaskets for water and air									
0020	250 psi maximum pressure									
0280	Female IPT to sweat, straight									
0300	1/2" pipe size	1 Plum	24	.333	Ea.	8.40	21.50		29.90	42.50
0340	3/4" pipe size		20	.400		9.40	26		35.40	50.50
0360	1" pipe size		19	.421		12.65	27		39.65	56
0380	1-1/4" pipe size		15	.533		17.35	34.50		51.85	72
0400	1-1/2" pipe size		13	.615		29.50	39.50		69	94
0420	2" pipe size		11	.727		40	47		87	117
0580	Female IPT to brass pipe thread, straight									
0600	1/2" pipe size	1 Plum	24	.333	Ea.	18.10	21.50		39.60	53
0640	3/4" pipe size		20	.400		19.85	26		45.85	62
0660	1" pipe size		19	.421		37	27		64	82.50
0680	1-1/4" pipe size		15	.533		37.50	34.50		72	94
0700	1-1/2" pipe size		13	.615		54.50	39.50		94	122
0720	2" pipe size		11	.727		110	47		157	194
0780	Female IPT to female IPT, straight									
0800	1/2" pipe size	1 Plum	24	.333	Ea.	15.90	21.50		37.40	50.50
0840	3/4" pipe size		20	.400		18.25	26		44.25	60
0860	1" pipe size		19	.421		24	27		51	68
0880	1-1/4" pipe size		15	.533		31.50	34.50		66	88
0900	1-1/2" pipe size		13	.615		63.50	39.50		103	131
0920	2" pipe size		11	.727		94	47		141	176
2000	175 psi maximum pressure									
2180	Female IPT to sweat									
2240	2" pipe size	1 Plum	9	.889	Ea.	188	57.50		245.50	295
2260	2-1/2" pipe size	Q-1	15	1.067		222	62		284	340
2280	3" pipe size		14	1.143		305	66.50		371.50	435
2300	4" pipe size		11	1.455		805	84.50		889.50	1,025
2480	Female IPT to brass pipe									
2500	1-1/2" pipe size	1 Plum	11	.727	Ea.	218	47		265	310
2540	2" pipe size	"	9	.889		229	57.50		286.50	340
2560	2-1/2" pipe size	Q-1	15	1.067		320	62		382	445
2580	3" pipe size		14	1.143		330	66.50		396.50	460
2600	4" pipe size		11	1.455		470	84.50		554.50	650

23 21 20.34 Dielectric Unions		Crew	Daily Output	Labor-Hours	Unit	2020 Bare Costs Material	Labor	Equipment	Total	Total Incl O&P
9000	Minimum labor/equipment charge	1 Plum	4	2	Job		129		129	199

23 21 20.38 Expansion Couplings		Crew	Daily Output	Labor-Hours	Unit	Material	Labor	Equipment	Total	Total Incl O&P
0010	**EXPANSION COUPLINGS**, Hydronic									
0100	Copper to copper, sweat									
1000	Baseboard riser fitting, 5" stub by coupling 12" long									
1020	1/2" diameter	1 Stpi	24	.333	Ea.	23.50	22		45.50	60
1040	3/4" diameter		20	.400		34.50	26		60.50	78.50
1060	1" diameter		19	.421		46.50	27.50		74	93.50
1080	1-1/4" diameter	↓	15	.533	↓	61	35		96	121
1180	9" stub by tubing 8" long									
1200	1/2" diameter	1 Stpi	24	.333	Ea.	21	22		43	57
1220	3/4" diameter		20	.400		28	26		54	71.50
1240	1" diameter		19	.421		39	27.50		66.50	85
1260	1-1/4" diameter	↓	15	.533	↓	58	35		93	118

23 21 20.42 Expansion Joints		Crew	Daily Output	Labor-Hours	Unit	Material	Labor	Equipment	Total	Total Incl O&P
0010	**EXPANSION JOINTS**									
0100	Bellows type, neoprene cover, flanged spool									
0140	6" face to face, 1-1/4" diameter	1 Stpi	11	.727	Ea.	264	47.50		311.50	365
0160	1-1/2" diameter	"	10.60	.755		264	49.50		313.50	365
0180	2" diameter	Q-5	13.30	1.203		267	71		338	405
0190	2-1/2" diameter		12.40	1.290		276	76		352	425
0200	3" diameter		11.40	1.404		310	83		393	470
0480	10" face to face, 2" diameter		13	1.231		385	72.50		457.50	535
0500	2-1/2" diameter		12	1.333		405	78.50		483.50	565
0520	3" diameter		11	1.455		410	86		496	590
0540	4" diameter		8	2		470	118		588	695
0560	5" diameter		7	2.286		560	135		695	825
0580	6" diameter		6	2.667		580	157		737	880
0600	8" diameter		5	3.200		690	189		879	1,050
0620	10" diameter		4.60	3.478		760	205		965	1,150
0640	12" diameter		4	4		945	236		1,181	1,425
0660	14" diameter	↓	3.80	4.211	↓	1,175	248		1,423	1,650

23 21 20.46 Expansion Tanks		Crew	Daily Output	Labor-Hours	Unit	Material	Labor	Equipment	Total	Total Incl O&P
0010	**EXPANSION TANKS**									
1400	Plastic, corrosion resistant, see Plumbing Costs									
1507	Underground fuel-oil storage tanks, see Section 23 13 13									
1512	Tank leak detection systems, see Section 28 33 33.50									
2000	Steel, liquid expansion, ASME, painted, 15 gallon capacity	Q-5	17	.941	Ea.	855	55.50		910.50	1,025
2020	24 gallon capacity		14	1.143		760	67.50		827.50	940
2040	30 gallon capacity		12	1.333		985	78.50		1,063.50	1,200
2060	40 gallon capacity		10	1.600		1,150	94.50		1,244.50	1,425
2080	60 gallon capacity		8	2		1,375	118		1,493	1,700
2100	80 gallon capacity		7	2.286		1,475	135		1,610	1,825
2120	100 gallon capacity		6	2.667		2,000	157		2,157	2,450
2130	120 gallon capacity		5	3.200		2,150	189		2,339	2,650
2140	135 gallon capacity		4.50	3.556		2,225	210		2,435	2,775
2150	175 gallon capacity		4	4		3,525	236		3,761	4,250
2160	220 gallon capacity		3.60	4.444		3,975	262		4,237	4,775
2170	240 gallon capacity		3.30	4.848		4,150	286		4,436	5,000
2180	305 gallon capacity		3	5.333		5,850	315		6,165	6,900
2190	400 gallon capacity	↓	2.80	5.714	↓	7,175	335		7,510	8,425
2360	Galvanized									

23 21 20.46 Expansion Tanks		Crew	Daily Output	Labor-Hours	Unit	Material	2020 Bare Costs Labor	Equipment	Total	Total Incl O&P
2370	15 gallon capacity	Q-5	17	.941	Ea.	1,275	55.50		1,330.50	1,475
2380	24 gallon capacity		14	1.143		1,550	67.50		1,617.50	1,825
2390	30 gallon capacity		12	1.333		1,600	78.50		1,678.50	1,900
2400	40 gallon capacity		10	1.600		1,875	94.50		1,969.50	2,200
2410	60 gallon capacity		8	2		2,175	118		2,293	2,575
2420	80 gallon capacity		7	2.286		2,725	135		2,860	3,200
2430	100 gallon capacity		6	2.667		3,550	157		3,707	4,150
2440	120 gallon capacity		5	3.200		3,375	189		3,564	4,025
2450	135 gallon capacity		4.50	3.556		4,000	210		4,210	4,725
2460	175 gallon capacity		4	4		4,900	236		5,136	5,775
2470	220 gallon capacity		3.60	4.444		6,725	262		6,987	7,800
2480	240 gallon capacity		3.30	4.848		7,050	286		7,336	8,200
2490	305 gallon capacity		3	5.333		8,425	315		8,740	9,750
2500	400 gallon capacity		2.80	5.714		12,600	335		12,935	14,300
3000	Steel ASME expansion, rubber diaphragm, 19 gal. cap. accept.		12	1.333		3,200	78.50		3,278.50	3,650
3020	31 gallon capacity		8	2		3,575	118		3,693	4,125
3040	61 gallon capacity		6	2.667		5,225	157		5,382	6,000
3060	79 gallon capacity		5	3.200		5,150	189		5,339	5,950
3080	119 gallon capacity		4	4		5,425	236		5,661	6,325
3100	158 gallon capacity		3.80	4.211		7,525	248		7,773	8,650
3120	211 gallon capacity		3.30	4.848		8,700	286		8,986	10,000
3140	317 gallon capacity		2.80	5.714		11,400	335		11,735	13,000
3160	422 gallon capacity		2.60	6.154		16,800	365		17,165	19,100
3180	528 gallon capacity		2.40	6.667		18,500	395		18,895	20,900
9000	Minimum labor/equipment charge		4	4	Job		236		236	365

23 21 20.58 Hydronic Heating Control Valves		Crew	Daily Output	Labor-Hours	Unit	Material	Labor	Equipment	Total	Total Incl O&P
0010	**HYDRONIC HEATING CONTROL VALVES**									
0050	Hot water, nonelectric, thermostatic									
0100	Radiator supply, 1/2" diameter	1 Stpi	24	.333	Ea.	78.50	22		100.50	120
0120	3/4" diameter		20	.400		74.50	26		100.50	123
0140	1" diameter		19	.421		88	27.50		115.50	140
0160	1-1/4" diameter		15	.533		108	35		143	173
0500	For low pressure steam, add					25%				
1000	Manual, radiator supply									
1010	1/2" pipe size, angle union	1 Stpi	24	.333	Ea.	71	22		93	113
1020	3/4" pipe size, angle union		20	.400		89.50	26		115.50	139
1030	1" pipe size, angle union		19	.421		114	27.50		141.50	168
1100	Radiator, balancing, straight, sweat connections									
1110	1/2" pipe size	1 Stpi	24	.333	Ea.	25	22		47	61.50
1120	3/4" pipe size		20	.400		34.50	26		60.50	78.50
1130	1" pipe size		19	.421		56.50	27.50		84	105
1140	Balance and stop valve 1/2" size		22	.364		68.50	24		92.50	113
1150	3/4" size		20	.400		74.50	26		100.50	123
1160	1" size		19	.421		86	27.50		113.50	137
1170	1-1/4" size		15	.533		110	35		145	175
1200	Steam, radiator, supply									
1210	1/2" pipe size, angle union	1 Stpi	24	.333	Ea.	67	22		89	108
1220	3/4" pipe size, angle union		20	.400		73.50	26		99.50	122
1230	1" pipe size, angle union		19	.421		84	27.50		111.50	135
1240	1-1/4" pipe size, angle union		15	.533		109	35		144	174
8000	System balancing and shut-off									
8020	Butterfly, quarter turn, calibrated, threaded or solder									

23 21 20 – Hydronic HVAC Piping Specialties

23 21 20.58 Hydronic Heating Control Valves		Crew	Daily Output	Labor-Hours	Unit	Material	Labor	Equipment	Total	Total Incl O&P
						2020 Bare Costs				
8040	Bronze, -30°F to +350°F, pressure to 175 psi									
8060	1/2" size	1 Stpi	22	.364	Ea.	21	24		45	60.50
8070	3/4" size		20	.400		29	26		55	72.50
8080	1" size		19	.421		36	27.50		63.50	82
8090	1-1/4" size		15	.533		69	35		104	130
8100	1-1/2" size		13	.615		73.50	40.50		114	144
8110	2" size		11	.727		93	47.50		140.50	177

23 21 20.70 Steam Traps		Crew	Daily Output	Labor-Hours	Unit	Material	Labor	Equipment	Total	Total Incl O&P
0010	**STEAM TRAPS**									
0030	Cast iron body, threaded									
0040	Inverted bucket									
0050	1/2" pipe size	1 Stpi	12	.667	Ea.	170	43.50		213.50	255
0070	3/4" pipe size		10	.800		280	52.50		332.50	390
0100	1" pipe size		9	.889		430	58.50		488.50	565
0120	1-1/4" pipe size		8	1		650	65.50		715.50	815
1000	Float & thermostatic, 15 psi									
1010	3/4" pipe size	1 Stpi	16	.500	Ea.	168	33		201	235
1020	1" pipe size		15	.533		193	35		228	266
1030	1-1/4" pipe size		13	.615		239	40.50		279.50	325
1040	1-1/2" pipe size		9	.889		360	58.50		418.50	485
1060	2" pipe size		6	1.333		1,000	87.50		1,087.50	1,225
1290	Brass body, threaded									
1300	Thermostatic, angle union, 25 psi									
1310	1/2" pipe size	1 Stpi	24	.333	Ea.	70	22		92	111
1320	3/4" pipe size		20	.400		107	26		133	159
1330	1" pipe size		19	.421		182	27.50		209.50	244
9000	Minimum labor/equipment charge		4	2	Job		131		131	203

23 21 20.74 Strainers, Basket Type		Crew	Daily Output	Labor-Hours	Unit	Material	Labor	Equipment	Total	Total Incl O&P
0010	**STRAINERS, BASKET TYPE**, Perforated stainless steel basket									
0100	Brass or monel available									
2000	Simplex style									
2300	Bronze body									
2320	Screwed, 3/8" pipe size	1 Stpi	22	.364	Ea.	191	24		215	247
2340	1/2" pipe size		20	.400		199	26		225	260
2360	3/4" pipe size		17	.471		305	31		336	385
2380	1" pipe size		15	.533		380	35		415	475
2400	1-1/4" pipe size		13	.615		415	40.50		455.50	525
2420	1-1/2" pipe size		12	.667		425	43.50		468.50	535
2440	2" pipe size		10	.800		625	52.50		677.50	765
2460	2-1/2" pipe size	Q-5	15	1.067		1,050	63		1,113	1,250
2480	3" pipe size	"	14	1.143		1,550	67.50		1,617.50	1,800
2600	Flanged, 2" pipe size	1 Stpi	6	1.333		860	87.50		947.50	1,075
2620	2-1/2" pipe size	Q-5	4.50	3.556		1,500	210		1,710	1,975
2640	3" pipe size		3.50	4.571		1,700	270		1,970	2,275
2660	4" pipe size		3	5.333		2,575	315		2,890	3,300
2680	5" pipe size	Q-6	3.40	7.059		4,100	430		4,530	5,200
2700	6" pipe size		3	8		5,000	490		5,490	6,250
2710	8" pipe size		2.50	9.600		8,150	585		8,735	9,875
3600	Iron body									
3700	Screwed, 3/8" pipe size	1 Stpi	22	.364	Ea.	165	24		189	218
3720	1/2" pipe size		20	.400		170	26		196	228
3740	3/4" pipe size		17	.471		217	31		248	286

23 21 20 – Hydronic HVAC Piping Specialties

23 21 20.74 Strainers, Basket Type		Crew	Daily Output	Labor-Hours	Unit	Material	Labor	Equipment	Total	Total Incl O&P
						2020 Bare Costs				
3760	1" pipe size	1 Stpi	15	.533	Ea.	221	35		256	297
3780	1-1/4" pipe size		13	.615		289	40.50		329.50	385
3800	1-1/2" pipe size		12	.667		320	43.50		363.50	420
3820	2" pipe size		10	.800		380	52.50		432.50	500
3840	2-1/2" pipe size	Q-5	15	1.067		510	63		573	660
3860	3" pipe size	"	14	1.143		615	67.50		682.50	785
4000	Flanged, 2" pipe size	1 Stpi	6	1.333		560	87.50		647.50	750
4020	2-1/2" pipe size	Q-5	4.50	3.556		755	210		965	1,150
4040	3" pipe size		3.50	4.571		790	270		1,060	1,275
4060	4" pipe size		3	5.333		1,225	315		1,540	1,825
4080	5" pipe size	Q-6	3.40	7.059		1,825	430		2,255	2,700
4100	6" pipe size		3	8		2,350	490		2,840	3,325
4120	8" pipe size		2.50	9.600		4,225	585		4,810	5,550
4140	10" pipe size		2.20	10.909		9,300	665		9,965	11,200
6000	Cast steel body									
6400	Screwed, 1" pipe size	1 Stpi	15	.533	Ea.	360	35		395	450
6410	1-1/4" pipe size		13	.615		540	40.50		580.50	655
6420	1-1/2" pipe size		12	.667		560	43.50		603.50	690
6440	2" pipe size		10	.800		740	52.50		792.50	895
6460	2-1/2" pipe size	Q-5	15	1.067		1,075	63		1,138	1,275
6480	3" pipe size	"	14	1.143		1,375	67.50		1,442.50	1,600
6560	Flanged, 2" pipe size	1 Stpi	6	1.333		1,450	87.50		1,537.50	1,725
6580	2-1/2" pipe size	Q-5	4.50	3.556		2,150	210		2,360	2,700
6600	3" pipe size		3.50	4.571		2,450	270		2,720	3,125
6620	4" pipe size		3	5.333		2,600	315		2,915	3,325
6640	6" pipe size	Q-6	3	8		4,750	490		5,240	5,975
6660	8" pipe size	"	2.50	9.600		9,625	585		10,210	11,500
7000	Stainless steel body									
7200	Screwed, 1" pipe size	1 Stpi	15	.533	Ea.	465	35		500	570
7210	1-1/4" pipe size		13	.615		725	40.50		765.50	865
7220	1-1/2" pipe size		12	.667		725	43.50		768.50	870
7240	2" pipe size		10	.800		1,075	52.50		1,127.50	1,275
7260	2-1/2" pipe size	Q-5	15	1.067		1,525	63		1,588	1,775
7280	3" pipe size	"	14	1.143		2,100	67.50		2,167.50	2,425
7400	Flanged, 2" pipe size	1 Stpi	6	1.333		1,650	87.50		1,737.50	1,950
7420	2-1/2" pipe size	Q-5	4.50	3.556		3,225	210		3,435	3,850
7440	3" pipe size		3.50	4.571		3,300	270		3,570	4,050
7460	4" pipe size		3	5.333		5,175	315		5,490	6,175
7480	6" pipe size	Q-6	3	8		10,900	490		11,390	12,800
7500	8" pipe size	"	2.50	9.600		15,300	585		15,885	17,800
8100	Duplex style									
8200	Bronze body									
8240	Screwed, 3/4" pipe size	1 Stpi	16	.500	Ea.	1,325	33		1,358	1,525
8260	1" pipe size		14	.571		1,325	37.50		1,362.50	1,525
8280	1-1/4" pipe size		12	.667		2,700	43.50		2,743.50	3,050
8300	1-1/2" pipe size		11	.727		2,700	47.50		2,747.50	3,050
8320	2" pipe size		9	.889		4,225	58.50		4,283.50	4,750
8340	2-1/2" pipe size	Q-5	14	1.143		5,450	67.50		5,517.50	6,100
8420	Flanged, 2" pipe size	1 Stpi	6	1.333		5,225	87.50		5,312.50	5,875
8440	2-1/2" pipe size	Q-5	4.50	3.556		7,225	210		7,435	8,275
8460	3" pipe size		3.50	4.571		7,875	270		8,145	9,075
8480	4" pipe size		3	5.333		11,500	315		11,815	13,200
8500	5" pipe size	Q-6	3.40	7.059		25,900	430		26,330	29,100

23 21 20.74 Strainers, Basket Type		Crew	Daily Output	Labor-Hours	Unit	Material	Labor	Equipment	Total	Total Incl O&P
						2020 Bare Costs				
8520	6" pipe size	Q-6	3	8	Ea.	25,100	490		25,590	28,500
8700	Iron body									
8740	Screwed, 3/4" pipe size	1 Stpi	16	.500	Ea.	1,675	33		1,708	1,900
8760	1" pipe size		14	.571		1,675	37.50		1,712.50	1,900
8780	1-1/4" pipe size		12	.667		1,850	43.50		1,893.50	2,100
8800	1-1/2" pipe size		11	.727		1,850	47.50		1,897.50	2,100
8820	2" pipe size	↓	9	.889		3,100	58.50		3,158.50	3,525
8840	2-1/2" pipe size	Q-5	14	1.143		3,425	67.50		3,492.50	3,875
9000	Flanged, 2" pipe size	1 Stpi	6	1.333		2,825	87.50		2,912.50	3,225
9020	2-1/2" pipe size	Q-5	4.50	3.556		3,550	210		3,760	4,225
9040	3" pipe size		3.50	4.571		3,875	270		4,145	4,675
9060	4" pipe size	↓	3	5.333		6,525	315		6,840	7,650
9080	5" pipe size	Q-6	3.40	7.059		14,300	430		14,730	16,400
9100	6" pipe size		3	8		14,300	490		14,790	16,500
9120	8" pipe size		2.50	9.600		25,100	585		25,685	28,500
9140	10" pipe size		2.20	10.909		40,700	665		41,365	45,800
9160	12" pipe size		1.70	14.118		44,800	865		45,665	50,500
9170	14" pipe size		1.40	17.143		52,500	1,050		53,550	59,500
9180	16" pipe size	↓	1	24	↓	64,500	1,475		65,975	73,500
9300	Cast steel body									
9340	Screwed, 1" pipe size	1 Stpi	14	.571	Ea.	2,125	37.50		2,162.50	2,400
9360	1-1/2" pipe size		11	.727		3,425	47.50		3,472.50	3,850
9380	2" pipe size		9	.889		4,550	58.50		4,608.50	5,100
9460	Flanged, 2" pipe size	↓	6	1.333		3,950	87.50		4,037.50	4,450
9480	2-1/2" pipe size	Q-5	4.50	3.556		5,775	210		5,985	6,700
9500	3" pipe size		3.50	4.571		6,300	270		6,570	7,350
9520	4" pipe size	↓	3	5.333		7,775	315		8,090	9,025
9540	6" pipe size	Q-6	3	8		14,600	490		15,090	16,800
9560	8" pipe size	"	2.50	9.600	↓	33,400	585		33,985	37,600
9700	Stainless steel body									
9740	Screwed, 1" pipe size	1 Stpi	14	.571	Ea.	3,250	37.50		3,287.50	3,625
9760	1-1/2" pipe size		11	.727		4,875	47.50		4,922.50	5,425
9780	2" pipe size		9	.889		7,175	58.50		7,233.50	8,000
9860	Flanged, 2" pipe size	↓	6	1.333		7,400	87.50		7,487.50	8,275
9880	2-1/2" pipe size	Q-5	4.50	3.556		12,400	210		12,610	14,000
9900	3" pipe size		3.50	4.571		13,500	270		13,770	15,300
9920	4" pipe size	↓	3	5.333		17,600	315		17,915	19,900
9940	6" pipe size	Q-6	3	8		26,900	490		27,390	30,400
9960	8" pipe size	"	2.50	9.600	↓	69,000	585		69,585	77,000

23 21 20.76 Strainers, Y Type, Bronze Body		Crew	Daily Output	Labor-Hours	Unit	Material	Labor	Equipment	Total	Total Incl O&P
0010	**STRAINERS, Y TYPE, BRONZE BODY**									
0050	Screwed, 125 lb., 1/4" pipe size	1 Stpi	24	.333	Ea.	44	22		66	82.50
0070	3/8" pipe size		24	.333		45.50	22		67.50	84.50
0100	1/2" pipe size		20	.400		45.50	26		71.50	90.50
0120	3/4" pipe size		19	.421		57	27.50		84.50	106
0140	1" pipe size		17	.471		66.50	31		97.50	121
0150	1-1/4" pipe size		15	.533		134	35		169	202
0160	1-1/2" pipe size		14	.571		144	37.50		181.50	216
0180	2" pipe size		13	.615		190	40.50		230.50	272
0182	3" pipe size	↓	12	.667		1,200	43.50		1,243.50	1,400
0200	300 lb., 2-1/2" pipe size	Q-5	17	.941		865	55.50		920.50	1,025
0220	3" pipe size	↓	16	1	↓	1,325	59		1,384	1,575

23 21 20 – Hydronic HVAC Piping Specialties

23 21 20.76 Strainers, Y Type, Bronze Body		Crew	Daily Output	Labor-Hours	Unit	Material	2020 Bare Costs Labor	Equipment	Total	Total Incl O&P
0240	4" pipe size	Q-5	15	1.067	Ea.	2,425	63		2,488	2,775
0500	For 300 lb. rating 1/4" thru 2", add					15%				
1000	Flanged, 150 lb., 1-1/2" pipe size	1 Stpi	11	.727	Ea.	545	47.50		592.50	675
1020	2" pipe size	"	8	1		630	65.50		695.50	790
1030	2-1/2" pipe size	Q-5	5	3.200		1,125	189		1,314	1,550
1040	3" pipe size		4.50	3.556		1,375	210		1,585	1,850
1060	4" pipe size		3	5.333		2,125	315		2,440	2,825
1080	5" pipe size	Q-6	3.40	7.059		3,025	430		3,455	4,000
1100	6" pipe size		3	8		4,100	490		4,590	5,275
1106	8" pipe size		2.60	9.231		4,450	565		5,015	5,775
1500	For 300 lb. rating, add					40%				
9000	Minimum labor/equipment charge	1 Stpi	3.75	2.133	Job		140		140	216

23 21 20.78 Strainers, Y Type, Iron Body		Crew	Daily Output	Labor-Hours	Unit	Material	Labor	Equipment	Total	Total Incl O&P
0010	**STRAINERS, Y TYPE, IRON BODY**									
0050	Screwed, 250 lb., 1/4" pipe size	1 Stpi	20	.400	Ea.	20	26		46	62.50
0070	3/8" pipe size		20	.400		20	26		46	62.50
0100	1/2" pipe size		20	.400		20	26		46	62.50
0120	3/4" pipe size		18	.444		23.50	29		52.50	70.50
0140	1" pipe size		16	.500		27.50	33		60.50	80.50
0150	1-1/4" pipe size		15	.533		43.50	35		78.50	102
0160	1-1/2" pipe size		12	.667		55.50	43.50		99	129
0180	2" pipe size		8	1		83	65.50		148.50	192
0200	2-1/2" pipe size	Q-5	12	1.333		288	78.50		366.50	435
0220	3" pipe size		11	1.455		310	86		396	480
0240	4" pipe size		5	3.200		530	189		719	870
0500	For galvanized body, add					50%				
1000	Flanged, 125 lb., 1-1/2" pipe size	1 Stpi	11	.727	Ea.	167	47.50		214.50	258
1020	2" pipe size	"	8	1		177	65.50		242.50	296
1030	2-1/2" pipe size	Q-5	5	3.200		161	189		350	470
1040	3" pipe size		4.50	3.556		188	210		398	530
1060	4" pipe size		3	5.333		375	315		690	900
1080	5" pipe size	Q-6	3.40	7.059		405	430		835	1,125
1100	6" pipe size		3	8		675	490		1,165	1,500
1120	8" pipe size		2.50	9.600		1,300	585		1,885	2,325
1140	10" pipe size		2	12		2,300	735		3,035	3,650
1160	12" pipe size		1.70	14.118		3,525	865		4,390	5,200
1170	14" pipe size		1.30	18.462		6,500	1,125		7,625	8,900
1180	16" pipe size		1	24		9,225	1,475		10,700	12,500
1500	For 250 lb. rating, add					20%				
2000	For galvanized body, add					50%				
2500	For steel body, add					40%				

23 21 20.84 Thermoflo Indicator		Crew	Daily Output	Labor-Hours	Unit	Material	Labor	Equipment	Total	Total Incl O&P
0010	**THERMOFLO INDICATOR**, For balancing									
1000	Sweat connections, 1-1/4" pipe size	1 Stpi	12	.667	Ea.	790	43.50		833.50	940
1020	1-1/2" pipe size		10	.800		805	52.50		857.50	965
1040	2" pipe size		8	1		845	65.50		910.50	1,025
1060	2-1/2" pipe size		7	1.143		1,275	75		1,350	1,550
2000	Flange connections, 3" pipe size	Q-5	5	3.200		1,550	189		1,739	2,000
2020	4" pipe size		4	4		1,850	236		2,086	2,400
2030	5" pipe size		3.50	4.571		2,325	270		2,595	2,975
2040	6" pipe size		3	5.333		2,475	315		2,790	3,200
2060	8" pipe size		2	8		3,025	470		3,495	4,050

23 21 20 – Hydronic HVAC Piping Specialties

23 21 20.88 Venturi Flow		Crew	Daily Output	Labor-Hours	Unit	Material	2020 Bare Costs Labor	Equipment	Total	Total Incl O&P
0010	**VENTURI FLOW**, Measuring device									
0050	1/2" diameter	1 Stpi	24	.333	Ea.	310	22		332	375
0100	3/4" diameter		20	.400		284	26		310	355
0120	1" diameter		19	.421		305	27.50		332.50	380
0140	1-1/4" diameter		15	.533		380	35		415	470
0160	1-1/2" diameter		13	.615		380	40.50		420.50	480
0180	2" diameter		11	.727		405	47.50		452.50	520
0200	2-1/2" diameter	Q-5	16	1		555	59		614	700
0220	3" diameter		14	1.143		570	67.50		637.50	735
0240	4" diameter		11	1.455		855	86		941	1,075
0260	5" diameter	Q-6	4	6		1,125	365		1,490	1,800
0280	6" diameter		3.50	6.857		1,250	420		1,670	2,025
0300	8" diameter		3	8		1,600	490		2,090	2,500
0320	10" diameter		2	12		3,775	735		4,510	5,275
0500	For meter, add					2,425			2,425	2,650

23 21 20.94 Weld End Ball Joints		Crew	Daily Output	Labor-Hours	Unit	Material	Labor	Equipment	Total	Total Incl O&P
0010	**WELD END BALL JOINTS**, Steel									
0050	2-1/2" diameter	Q-17	13	1.231	Ea.	840	72.50	8.20	920.70	1,050
0100	3" diameter		12	1.333		990	78.50	8.85	1,077.35	1,225
0120	4" diameter		11	1.455		1,450	86	9.65	1,545.65	1,750
0140	5" diameter	Q-18	14	1.714		2,000	105	7.60	2,112.60	2,375
0160	6" diameter		12	2		2,300	122	8.85	2,430.85	2,725
0180	8" diameter		9	2.667		3,350	163	11.80	3,524.80	3,975
0200	10" diameter		8	3		4,900	184	13.30	5,097.30	5,675
0220	12" diameter		6	4		6,850	245	17.70	7,112.70	7,925

23 21 23 – Hydronic Pumps

23 21 23.13 In-Line Centrifugal Hydronic Pumps		Crew	Daily Output	Labor-Hours	Unit	Material	Labor	Equipment	Total	Total Incl O&P
0010	**IN-LINE CENTRIFUGAL HYDRONIC PUMPS**									
0600	Bronze, sweat connections, 1/40 HP, in line									
0640	3/4" size	Q-1	16	1	Ea.	275	58		333	390
1000	Flange connection, 3/4" to 1-1/2" size									
1040	1/12 HP	Q-1	6	2.667	Ea.	710	155		865	1,025
1060	1/8 HP		6	2.667		1,250	155		1,405	1,625
1100	1/3 HP		6	2.667		1,400	155		1,555	1,775
1140	2" size, 1/6 HP		5	3.200		1,750	186		1,936	2,200
1180	2-1/2" size, 1/4 HP		5	3.200		2,200	186		2,386	2,700
1220	3" size, 1/4 HP		4	4		2,400	232		2,632	2,975
1260	1/3 HP		4	4		2,825	232		3,057	3,475
1300	1/2 HP		4	4		2,950	232		3,182	3,575
1340	3/4 HP		4	4		3,000	232		3,232	3,650
1380	1 HP		4	4		4,775	232		5,007	5,600
2000	Cast iron, flange connection									
2040	3/4" to 1-1/2" size, in line, 1/12 HP	Q-1	6	2.667	Ea.	470	155		625	760
2060	1/8 HP		6	2.667		780	155		935	1,100
2100	1/3 HP		6	2.667		870	155		1,025	1,200
2140	2" size, 1/6 HP		5	3.200		955	186		1,141	1,325
2180	2-1/2" size, 1/4 HP		5	3.200		1,150	186		1,336	1,525
2220	3" size, 1/4 HP		4	4		1,150	232		1,382	1,625
2260	1/3 HP		4	4		1,525	232		1,757	2,025
2300	1/2 HP		4	4		1,575	232		1,807	2,075
2340	3/4 HP		4	4		1,825	232		2,057	2,350

23 21 23 – Hydronic Pumps

23 21 23.13 In-Line Centrifugal Hydronic Pumps		Crew	Daily Output	Labor-Hours	Unit	Material	2020 Bare Costs Labor	Equipment	Total	Total Incl O&P
2380	1 HP	Q-1	4	4	Ea.	2,625	232		2,857	3,225
2600	For nonferrous impeller, add					3%				
3000	High head, bronze impeller									
3030	1-1/2" size, 1/2 HP	Q-1	5	3.200	Ea.	1,400	186		1,586	1,825
3040	1-1/2" size, 3/4 HP		5	3.200		1,625	186		1,811	2,075
3050	2" size, 1 HP		4	4		1,975	232		2,207	2,525
3090	2" size, 1-1/2 HP	↓	4	4	↓	2,400	232		2,632	2,975
4000	Close coupled, end suction, bronze impeller									
4040	1-1/2" size, 1-1/2 HP, to 40 GPM	Q-1	3	5.333	Ea.	2,500	310		2,810	3,225
4090	2" size, 2 HP, to 50 GPM		3	5.333		3,025	310		3,335	3,800
4100	2" size, 3 HP, to 90 GPM		2.30	6.957		3,075	405		3,480	4,000
4190	2-1/2" size, 3 HP, to 150 GPM		2	8		3,350	465		3,815	4,425
4300	3" size, 5 HP, to 225 GPM		1.80	8.889		3,900	515		4,415	5,100
4410	3" size, 10 HP, to 350 GPM		1.60	10		6,100	580		6,680	7,600
4420	4" size, 7-1/2 HP, to 350 GPM	↓	1.60	10		6,250	580		6,830	7,775
4520	4" size, 10 HP, to 600 GPM	Q-2	1.70	14.118		6,475	850		7,325	8,450
4530	5" size, 15 HP, to 1,000 GPM		1.70	14.118		6,500	850		7,350	8,475
4610	5" size, 20 HP, to 1,350 GPM		1.50	16		6,925	960		7,885	9,100
4620	5" size, 25 HP, to 1,550 GPM	↓	1.50	16	↓	8,325	960		9,285	10,700
5000	Base mounted, bronze impeller, coupling guard									
5040	1-1/2" size, 1-1/2 HP, to 40 GPM	Q-1	2.30	6.957	Ea.	7,300	405		7,705	8,650
5090	2" size, 2 HP, to 50 GPM		2.30	6.957		8,175	405		8,580	9,625
5100	2" size, 3 HP, to 90 GPM		2	8		9,000	465		9,465	10,600
5190	2-1/2" size, 3 HP, to 150 GPM		1.80	8.889		10,400	515		10,915	12,200
5300	3" size, 5 HP, to 225 GPM		1.60	10		11,300	580		11,880	13,300
5410	4" size, 5 HP, to 350 GPM		1.50	10.667		13,100	620		13,720	15,400
5420	4" size, 7-1/2 HP, to 350 GPM	↓	1.50	10.667		14,100	620		14,720	16,500
5520	5" size, 10 HP, to 600 GPM	Q-2	1.60	15		19,400	900		20,300	22,700
5530	5" size, 15 HP, to 1,000 GPM		1.60	15		20,900	900		21,800	24,400
5610	6" size, 20 HP, to 1,350 GPM		1.40	17.143		22,600	1,025		23,625	26,500
5620	6" size, 25 HP, to 1,550 GPM	↓	1.40	17.143	↓	26,600	1,025		27,625	30,900
9000	Minimum labor/equipment charge	Q-1	3.25	4.923	Job		286		286	440

23 21 29 – Automatic Condensate Pump Units

23 21 29.10 Condensate Removal Pump System		Crew	Daily Output	Labor-Hours	Unit	Material	Labor	Equipment	Total	Total Incl O&P
0010	**CONDENSATE REMOVAL PUMP SYSTEM**									
0020	Pump with 1 gal. ABS tank									
0100	115 V									
0120	1/50 HP, 200 GPH G	1 Stpi	12	.667	Ea.	188	43.50		231.50	274
0140	1/18 HP, 270 GPH G		10	.800		186	52.50		238.50	286
0160	1/5 HP, 450 GPH G	↓	8	1	↓	425	65.50		490.50	570
0200	230 V									
0240	1/18 HP, 270 GPH	1 Stpi	10	.800	Ea.	208	52.50		260.50	310
0260	1/5 HP, 450 GPH G	"	8	1	"	495	65.50		560.50	645

23 22 Steam and Condensate Piping and Pumps

23 22 13 – Steam and Condensate Heating Piping

23 22 13.23 Aboveground Steam and Condensate Piping		Crew	Daily Output	Labor-Hours	Unit	2020 Bare Costs Material	Labor	Equipment	Total	Total Incl O&P
0010	**ABOVEGROUND STEAM AND CONDENSATE HEATING PIPING**									
0020	Condensate meter									
0100	500 lb. per hour	1 Stpi	14	.571	Ea.	4,300	37.50		4,337.50	4,775
0140	1500 lb. per hour		7	1.143		4,825	75		4,900	5,450
0160	3000 lb. per hour		5	1.600		5,675	105		5,780	6,375
0200	12,000 lb. per hour	Q-5	3.50	4.571		6,475	270		6,745	7,525

23 22 23 – Steam Condensate Pumps

23 22 23.10 Condensate Return System

Code	Description	Crew	Daily Output	Labor-Hours	Unit	Material	Labor	Equipment	Total	Total Incl O&P
0010	**CONDENSATE RETURN SYSTEM**									
2000	Simplex									
2010	With pump, motor, CI receiver, float switch									
2020	3/4 HP, 15 GPM	Q-1	1.80	8.889	Ea.	7,175	515		7,690	8,675
2100	Duplex									
2110	With 2 pumps and motors, CI receiver, float switch, alternator									
2120	3/4 HP, 15 GPM, 15 gal. CI receiver	Q-1	1.40	11.429	Ea.	7,825	665		8,490	9,625
2130	1 HP, 25 GPM		1.20	13.333		9,400	775		10,175	11,500
2140	1-1/2 HP, 45 GPM		1	16		11,100	930		12,030	13,600
2150	1-1/2 HP, 60 GPM		1	16		12,200	930		13,130	14,900

23 23 Refrigerant Piping

23 23 13 – Refrigerant Piping Valves

23 23 13.10 Valves

Code	Description	Crew	Daily Output	Labor-Hours	Unit	Material	Labor	Equipment	Total	Total Incl O&P
0010	**VALVES**									
8100	Check valve, soldered									
8110	5/8"	1 Stpi	36	.222	Ea.	94	14.55		108.55	127
8114	7/8"		26	.308		114	20		134	157
8118	1-1/8"		18	.444		129	29		158	187
8122	1-3/8"		14	.571		200	37.50		237.50	278
8126	1-5/8"		13	.615		250	40.50		290.50	340
8130	2-1/8"		12	.667		285	43.50		328.50	385
8134	2-5/8"	Q-5	22	.727		405	43		448	515
8138	3-1/8"	"	20	.800		540	47		587	670
8500	Refrigeration valve, packless, soldered									
8510	1/2"	1 Stpi	38	.211	Ea.	47	13.80		60.80	73.50
8514	5/8"		36	.222		48.50	14.55		63.05	76
8518	7/8"		26	.308		183	20		203	232
8520	Packed, soldered									
8522	1-1/8"	1 Stpi	18	.444	Ea.	237	29		266	305
8526	1-3/8"		14	.571		385	37.50		422.50	485
8530	1-5/8"		13	.615		370	40.50		410.50	470
8534	2-1/8"		12	.667		835	43.50		878.50	990
8538	2-5/8"	Q-5	22	.727		1,250	43		1,293	1,450
8542	3-1/8"		20	.800		1,250	47		1,297	1,450
8546	4-1/8"		18	.889		2,075	52.50		2,127.50	2,350
8600	Solenoid valve, flange/solder									
8610	1/2"	1 Stpi	38	.211	Ea.	261	13.80		274.80	310
8614	5/8"		36	.222		355	14.55		369.55	420
8618	3/4"		30	.267		455	17.50		472.50	525
8622	7/8"		26	.308		535	20		555	620
8626	1-1/8"		18	.444		645	29		674	755

23 23 13 – Refrigerant Piping Valves

23 23 13.10 Valves		Crew	Daily Output	Labor-Hours	Unit	Material	2020 Bare Costs Labor	Equipment	Total	Total Incl O&P
8630	1-3/8"	1 Stpi	14	.571	Ea.	780	37.50		817.50	915
8634	1-5/8"		13	.615		1,100	40.50		1,140.50	1,275
8638	2-1/8"		12	.667		1,200	43.50		1,243.50	1,400
8800	Thermostatic valve, flange/solder									
8810	1/2-3 ton, 3/8" x 5/8"	1 Stpi	9	.889	Ea.	127	58.50		185.50	229
8814	4-5 ton, 1/2" x 7/8"		7	1.143		151	75		226	282
8818	6-8 ton, 5/8" x 7/8"		5	1.600		280	105		385	470
8822	7-12 ton, 7/8" x 1-1/8"		4	2		330	131		461	570
8826	15-20 ton, 7/8" x 1-3/8"		3.20	2.500		288	164		452	570

23 23 16 – Refrigerant Piping Specialties

23 23 16.10 Refrigerant Piping Component Specialties

		Crew	Daily Output	Labor-Hours	Unit	Material	Labor	Equipment	Total	Total Incl O&P
0010	REFRIGERANT PIPING COMPONENT SPECIALTIES									
0600	Accumulator									
0610	3/4"	1 Stpi	8.80	.909	Ea.	52.50	59.50		112	150
0614	7/8"		6.40	1.250		67.50	82		149.50	202
0618	1-1/8"		4.80	1.667		125	109		234	305
0622	1-3/8"		4	2		123	131		254	340
0626	1-5/8"		3.20	2.500		129	164		293	395
0630	2-1/8"		2.40	3.333		345	219		564	720
0700	Condensate drip/drain pan, electric, 10" x 9" x 2-3/4"		24	.333		106	22		128	151
1000	Filter dryer									
1010	Replaceable core type, solder									
1020	1/2"	1 Stpi	20	.400	Ea.	69	26		95	117
1030	5/8"		19	.421		68	27.50		95.50	118
1040	7/8"		18	.444		38	29		67	87
1050	1-1/8"		15	.533		101	35		136	165
1060	1-3/8"		14	.571		157	37.50		194.50	231
1070	1-5/8"		12	.667		157	43.50		200.50	241
1080	2-1/8"		10	.800		460	52.50		512.50	585
1090	2-5/8"		9	.889		685	58.50		743.50	840
1100	3-1/8"		8	1		845	65.50		910.50	1,025
1200	Sealed in-line, solder									
1210	1/4"	1 Stpi	22	.364	Ea.	30.50	24		54.50	70.50
1220	3/8"		21	.381		25.50	25		50.50	66.50
1230	1/2"		20	.400		30.50	26		56.50	74
1260	5/8"		19	.421		41.50	27.50		69	88
1270	7/8"		18	.444		45.50	29		74.50	95
1290	1-1/8"		15	.533		69	35		104	130
4000	P-trap, suction line, solder									
4010	5/8"	1 Stpi	19	.421	Ea.	49	27.50		76.50	96
4020	3/4"		19	.421		69	27.50		96.50	119
4030	7/8"		18	.444		48.50	29		77.50	98.50
4040	1-1/8"		15	.533		79	35		114	141
4050	1-3/8"		14	.571		132	37.50		169.50	203
4060	1-5/8"		12	.667		247	43.50		290.50	340
4070	2-1/8"		10	.800		330	52.50		382.50	445
5000	Sightglass									
5010	Moisture and liquid indicator, solder									
5020	1/4"	1 Stpi	22	.364	Ea.	18.25	24		42.25	57
5030	3/8"		21	.381		18.30	25		43.30	58.50
5040	1/2"		20	.400		23	26		49	66
5050	5/8"		19	.421		24	27.50		51.50	69

23 23 16 – Refrigerant Piping Specialties

23 23 16.10 Refrigerant Piping Component Specialties		Crew	Daily Output	Labor-Hours	Unit	Material	2020 Bare Costs Labor	Equipment	Total	Total Incl O&P
5060	7/8"	1 Stpi	18	.444	Ea.	35.50	29		64.50	84
5070	1-1/8"		15	.533		38.50	35		73.50	96
5080	1-3/8"		14	.571		76	37.50		113.50	142
5090	1-5/8"		12	.667		85.50	43.50		129	162
5100	2-1/8"		10	.800		89.50	52.50		142	180
7400	Vacuum pump set									
7410	Two stage, high vacuum continuous duty	1 Stpi	8	1	Ea.	4,000	65.50		4,065.50	4,475

23 23 16.16 Refrigerant Line Sets		Crew	Daily Output	Labor-Hours	Unit	Material	Labor	Equipment	Total	Total Incl O&P
0010	**REFRIGERANT LINE SETS**, Standard									
0100	Copper tube									
0110	1/2" insulation, both tubes									
0120	Combination 1/4" and 1/2" tubes									
0130	10′ set	Q-5	42	.381	Ea.	48.50	22.50		71	88.50
0135	15′ set		42	.381		70.50	22.50		93	113
0140	20′ set		40	.400		76	23.50		99.50	120
0150	30′ set		37	.432		107	25.50		132.50	158
0160	40′ set		35	.457		134	27		161	189
0170	50′ set		32	.500		166	29.50		195.50	229
0180	100′ set		22	.727		360	43		403	460
0300	Combination 1/4" and 3/4" tubes									
0310	10′ set	Q-5	40	.400	Ea.	54.50	23.50		78	96.50
0320	20′ set		38	.421		96.50	25		121.50	145
0330	30′ set		35	.457		144	27		171	200
0340	40′ set		33	.485		188	28.50		216.50	251
0350	50′ set		30	.533		237	31.50		268.50	310
0380	100′ set		20	.800		540	47		587	670
0500	Combination 3/8" & 3/4" tubes									
0510	10′ set	Q-5	28	.571	Ea.	62.50	33.50		96	121
0520	20′ set		36	.444		97.50	26		123.50	148
0530	30′ set		34	.471		132	28		160	189
0540	40′ set		31	.516		174	30.50		204.50	238
0550	50′ set		28	.571		204	33.50		237.50	277
0580	100′ set		18	.889		605	52.50		657.50	745
0700	Combination 3/8" & 1-1/8" tubes									
0710	10′ set	Q-5	36	.444	Ea.	109	26		135	161
0720	20′ set		33	.485		183	28.50		211.50	245
0730	30′ set		31	.516		246	30.50		276.50	315
0740	40′ set		28	.571		370	33.50		403.50	455
0750	50′ set		26	.615		355	36.50		391.50	450
0900	Combination 1/2" & 3/4" tubes									
0910	10′ set	Q-5	37	.432	Ea.	66	25.50		91.50	112
0920	20′ set		35	.457		117	27		144	171
0930	30′ set		33	.485		176	28.50		204.50	238
0940	40′ set		30	.533		232	31.50		263.50	305
0950	50′ set		27	.593		292	35		327	375
0980	100′ set		17	.941		660	55.50		715.50	810
2100	Combination 1/2" & 1-1/8" tubes									
2110	10′ set	Q-5	35	.457	Ea.	113	27		140	166
2120	20′ set		31	.516		193	30.50		223.50	259
2130	30′ set		29	.552		290	32.50		322.50	370
2140	40′ set		25	.640		390	38		428	490
2150	50′ set		14	1.143		475	67.50		542.50	630

23 23 Refrigerant Piping

23 23 16 – Refrigerant Piping Specialties

	23 23 16.16 Refrigerant Line Sets	Crew	Daily Output	Labor-Hours	Unit	2020 Bare Costs Material	Labor	Equipment	Total	Total Incl O&P
2300	For 1" thick insulation add					30%	15%			
3000	Refrigerant line sets, min-split, flared									
3100	Combination 1/4" & 3/8" tubes									
3120	15′ set	Q-5	41	.390	Ea.	69.50	23		92.50	112
3140	25′ set		38.50	.416		98.50	24.50		123	146
3160	35′ set		37	.432		124	25.50		149.50	177
3180	50′ set		30	.533		169	31.50		200.50	235
3200	Combination 1/4" & 1/2" tubes									
3220	15′ set	Q-5	41	.390	Ea.	74.50	23		97.50	118
3240	25′ set		38.50	.416		101	24.50		125.50	149
3260	35′ set		37	.432		130	25.50		155.50	183
3280	50′ set		30	.533		176	31.50		207.50	242

23 23 23 – Refrigerants

	23 23 23.10 Anti-Freeze	Crew	Daily Output	Labor-Hours	Unit	Material	Labor	Equipment	Total	Total Incl O&P
0010	**ANTI-FREEZE**, Inhibited									
0900	Ethylene glycol concentrated									
1000	55 gallon drums, small quantities				Gal.	12.90			12.90	14.20
1200	Large quantities					11.80			11.80	12.95
2000	Propylene glycol, for solar heat, small quantities					23			23	25.50
2100	Large quantities					9.90			9.90	10.90

	23 23 23.20 Refrigerant	Crew	Daily Output	Labor-Hours	Unit	Material	Labor	Equipment	Total	Total Incl O&P
0010	**REFRIGERANT**									
4420	R-22, 30 lb. disposable cylinder				Lb.	22			22	24.50
4428	R-134A, 30 lb. disposable cylinder					14.20			14.20	15.60
4434	R-407C, 30 lb. disposable cylinder					10.65			10.65	11.70
4440	R-408A, 25 lb. disposable cylinder					26			26	29
4450	R-410A, 25 lb. disposable cylinder					11.80			11.80	12.95
4470	R-507, 25 lb. disposable cylinder					7.05			7.05	7.75

23 31 HVAC Ducts and Casings

23 31 13 – Metal Ducts

	23 31 13.13 Rectangular Metal Ducts	Crew	Daily Output	Labor-Hours	Unit	Material	Labor	Equipment	Total	Total Incl O&P
0010	**RECTANGULAR METAL DUCTS** R233100-40									
0020	Fabricated rectangular, includes fittings, joints, supports,									
0021	allowance for flexible connections and field sketches.									
0030	Does not include "as-built dwgs." or insulation.									
0031	NOTE: Fabrication and installation are combined									
0040	as LABOR cost. Approx. 25% fittings assumed.									
0042	Fabrication/Inst. is to commercial quality standards									
0043	(SMACNA or equiv.) for structure, sealing, leak testing, etc.									
0050	Add to labor for elevated installation									
0051	of fabricated ductwork									
0052	10′ to 15′ high						6%			
0053	15′ to 20′ high						12%			
0054	20′ to 25′ high						15%			
0055	25′ to 30′ high						21%			
0056	30′ to 35′ high						24%			
0057	35′ to 40′ high						30%			
0058	Over 40′ high						33%			
0072	For duct insulation see Line 23 07 13.10 3000									

23 31 13.13 Rectangular Metal Ducts		Crew	Daily Output	Labor-Hours	Unit	Material	2020 Bare Costs Labor	Equipment	Total	Total Incl O&P
0100	Aluminum, alloy 3003-H14, under 100 lb.	Q-10	75	.320	Lb.	3.20	18.60		21.80	32.50
0110	100 to 500 lb.		80	.300		1.89	17.45		19.34	29.50
0120	500 to 1,000 lb.		95	.253		1.91	14.70		16.61	25
0140	1,000 to 2,000 lb.		120	.200		1.86	11.65		13.51	20.50
0150	2,000 to 5,000 lb.		130	.185		1.80	10.75		12.55	18.80
0160	Over 5,000 lb.		145	.166		1.86	9.65		11.51	17.15
0500	Galvanized steel, under 200 lb.		235	.102		.60	5.95		6.55	9.95
0520	200 to 500 lb.		245	.098		.57	5.70		6.27	9.55
0540	500 to 1,000 lb.		255	.094		.54	5.45		5.99	9.15
0560	1,000 to 2,000 lb.		265	.091		.55	5.25		5.80	8.85
0570	2,000 to 5,000 lb.		275	.087		.56	5.05		5.61	8.55
0580	Over 5,000 lb.		285	.084		.55	4.90		5.45	8.25
0600	For large quantities special prices available from supplier									
1000	Stainless steel, type 304, under 100 lb.	Q-10	165	.145	Lb.	4.78	8.45		13.23	18.50
1020	100 to 500 lb.		175	.137		3.88	7.95		11.83	16.75
1030	500 to 1,000 lb.		190	.126		2.77	7.35		10.12	14.55
1040	1,000 to 2,000 lb.		200	.120		2.46	7		9.46	13.65
1050	2,000 to 5,000 lb.		225	.107		2.17	6.20		8.37	12.10
1060	Over 5,000 lb.		235	.102		2.67	5.95		8.62	12.25
1080	Note: Minimum order cost exceeds per lb. cost for min. wt.									
1100	For medium pressure ductwork, add				Lb.		15%			
1200	For high pressure ductwork, add						40%			
1210	For welded ductwork, add						85%			
1220	For 30% fittings, add						11%			
1224	For 40% fittings, add						34%			
1228	For 50% fittings, add						56%			
1232	For 60% fittings, add						79%			
1236	For 70% fittings, add						101%			
1240	For 80% fittings, add						124%			
1244	For 90% fittings, add						147%			
1248	For 100% fittings, add						169%			
1252	Note: Fittings add includes time for detailing and installation.									

23 31 13.16 Round and Flat-Oval Spiral Ducts		Crew	Daily Output	Labor-Hours	Unit	Material	Labor	Equipment	Total	Total Incl O&P
0010	**ROUND AND FLAT-OVAL SPIRAL DUCTS**									
0020	Fabricated round and flat oval spiral, includes hangers,									
0021	supports and field sketches.									
1280	Add to labor for elevated installation									
1282	of prefabricated (purchased) ductwork									
1283	10′ to 15′ high						10%			
1284	15′ to 20′ high						20%			
1285	20′ to 25′ high						25%			
1286	25′ to 30′ high						35%			
1287	30′ to 35′ high						40%			
1288	35′ to 40′ high						50%			
1289	Over 40′ high						55%			
5400	Spiral preformed, steel, galv., straight lengths, max 10″ spwg.									
5410	4″ diameter, 26 ga.	Q-9	360	.044	L.F.	1.93	2.49		4.42	6
5416	5″ diameter, 26 ga.		320	.050		1.98	2.80		4.78	6.55
5420	6″ diameter, 26 ga.		280	.057		2.03	3.20		5.23	7.25
5425	7″ diameter, 26 ga.		240	.067		2.27	3.74		6.01	8.35
5430	8″ diameter, 26 ga.		200	.080		2.90	4.49		7.39	10.25
5440	10″ diameter, 26 ga.		160	.100		3.54	5.60		9.14	12.70

23 31 13.16 Round and Flat-Oval Spiral Ducts		Crew	Daily Output	Labor-Hours	Unit	Material	2020 Bare Costs Labor	Equipment	Total	Total Incl O&P
5450	12" diameter, 26 ga.	Q-9	120	.133	L.F.	4.24	7.50		11.74	16.35
5460	14" diameter, 26 ga.		80	.200		4.63	11.20		15.83	22.50
5480	16" diameter, 24 ga.		60	.267		6.40	14.95		21.35	30.50
5490	18" diameter, 24 ga.		50	.320		7.75	17.95		25.70	36.50
5500	20" diameter, 24 ga.	Q-10	65	.369		8.60	21.50		30.10	43
5510	22" diameter, 24 ga.		60	.400		9.45	23.50		32.95	47
5520	24" diameter, 24 ga.		55	.436		10.30	25.50		35.80	51.50
5540	30" diameter, 22 ga.		45	.533		13.90	31		44.90	64
5600	36" diameter, 22 ga.		40	.600		19.10	35		54.10	75.50
5800	Connector, 4" diameter	Q-9	100	.160	Ea.	3.31	8.95		12.26	17.70
5810	5" diameter		94	.170		3.55	9.55		13.10	18.85
5820	6" diameter		88	.182		3.72	10.20		13.92	20
5840	8" diameter		78	.205		4.47	11.50		15.97	23
5860	10" diameter		70	.229		4.48	12.80		17.28	25
5880	12" diameter		50	.320		5.60	17.95		23.55	34
5900	14" diameter		44	.364		5.45	20.50		25.95	38
5920	16" diameter		40	.400		6	22.50		28.50	41.50
5930	18" diameter		37	.432		6.55	24.50		31.05	45
5940	20" diameter		34	.471		8.15	26.50		34.65	50.50
5950	22" diameter		31	.516		8.95	29		37.95	55.50
5960	24" diameter		28	.571		9.50	32		41.50	60.50
5980	30" diameter		22	.727		15.25	41		56.25	81
6000	36" diameter		18	.889		17.45	50		67.45	97
6300	Elbow, 45°, 4" diameter		60	.267		6.40	14.95		21.35	30.50
6310	5" diameter		52	.308		6.75	17.25		24	34.50
6320	6" diameter		44	.364		6.90	20.50		27.40	39.50
6340	8" diameter		28	.571		8.15	32		40.15	59
6360	10" diameter		18	.889		8.45	50		58.45	87.50
6380	12" diameter		13	1.231		9.50	69		78.50	118
6400	14" diameter		11	1.455		13.15	81.50		94.65	142
6420	16" diameter		10	1.600		21	89.50		110.50	164
6430	18" diameter		9.60	1.667		52	93.50		145.50	203
6440	20" diameter	Q-10	14	1.714		58	99.50		157.50	220
6450	22" diameter		13	1.846		74	107		181	250
6460	24" diameter		12	2		77	116		193	267
6480	30" diameter		9	2.667		121	155		276	375
6500	36" diameter		7	3.429		149	199		348	475
6600	Elbow, 90°, 4" diameter	Q-9	60	.267		4.61	14.95		19.56	28.50
6610	5" diameter		52	.308		4.87	17.25		22.12	32.50
6620	6" diameter		44	.364		5	20.50		25.50	37.50
6625	7" diameter		36	.444		5.40	25		30.40	45
6630	8" diameter		28	.571		6.15	32		38.15	57
6640	10" diameter		18	.889		7.65	50		57.65	86.50
6650	12" diameter		13	1.231		9.55	69		78.55	119
6660	14" diameter		11	1.455		16.45	81.50		97.95	146
6670	16" diameter		10	1.600		17.05	89.50		106.55	160
6676	18" diameter		9.60	1.667		69.50	93.50		163	223
6680	20" diameter	Q-10	14	1.714		77	99.50		176.50	241
6684	22" diameter		13	1.846		72.50	107		179.50	248
6690	24" diameter		12	2		83	116		199	274
6700	30" diameter		9	2.667		168	155		323	430
6710	36" diameter		7	3.429		223	199		422	555
6800	Reducing coupling, 6" x 4"	Q-9	46	.348		17.40	19.50		36.90	49.50

23 31 13.16 Round and Flat-Oval Spiral Ducts		Crew	Daily Output	Labor-Hours	Unit	2020 Bare Costs Material	Labor	Equipment	Total	Total Incl O&P
6820	8″ x 6″	Q-9	40	.400	Ea.	20	22.50		42.50	57
6840	10″ x 8″		32	.500		22	28		50	68
6860	12″ x 10″		24	.667		22	37.50		59.50	83
6880	14″ x 12″		20	.800		21	45		66	93.50
6900	16″ x 14″		18	.889		26.50	50		76.50	107
6920	18″ x 16″		16	1		38.50	56		94.50	131
6940	20″ x 18″	Q-10	24	1		49.50	58		107.50	146
6950	22″ x 20″		23	1.043		49.50	60.50		110	149
6960	24″ x 22″		22	1.091		56	63.50		119.50	161
6980	30″ x 28″		18	1.333		111	77.50		188.50	243
7000	36″ x 34″		16	1.500		153	87		240	305
7100	Tee, 90°, 4″ diameter	Q-9	40	.400		13.70	22.50		36.20	50
7110	5″ diameter		34.60	.462		14.20	26		40.20	56
7120	6″ diameter		30	.533		14.85	30		44.85	63.50
7130	8″ diameter		19	.842		18.70	47		65.70	94.50
7140	10″ diameter		12	1.333		28	75		103	148
7150	12″ diameter		9	1.778		33.50	99.50		133	193
7160	14″ diameter		7	2.286		45	128		173	250
7170	16″ diameter		6.70	2.388		72.50	134		206.50	290
7176	18″ diameter		6	2.667		106	150		256	350
7180	20″ diameter	Q-10	8.70	2.759		126	160		286	390
7190	24″ diameter		7	3.429		188	199		387	515
7200	30″ diameter		6	4		252	233		485	640
7210	36″ diameter		5	4.800		320	279		599	790
7250	Saddle tap									
7254	4″ diameter on 6″ diameter	Q-9	36	.444	Ea.	20.50	25		45.50	61.50
7256	5″ diameter on 7″ diameter		35.80	.447		22.50	25		47.50	64
7258	6″ diameter on 8″ diameter		35.40	.452		22.50	25.50		48	64
7260	6″ diameter on 10″ diameter		35.20	.455		22.50	25.50		48	64.50
7262	6″ diameter on 12″ diameter		35	.457		22.50	25.50		48	64.50
7264	7″ diameter on 9″ diameter		35.20	.455		25.50	25.50		51	68
7266	8″ diameter on 10″ diameter		35	.457		25.50	25.50		51	68
7268	8″ diameter on 12″ diameter		34.80	.460		25.50	26		51.50	68.50
7270	8″ diameter on 14″ diameter		34.50	.464		26	26		52	69
7272	10″ diameter on 12″ diameter		34.20	.468		28.50	26		54.50	72
7274	10″ diameter on 14″ diameter		33.80	.473		28.50	26.50		55	72.50
7276	10″ diameter on 16″ diameter		33.40	.479		28.50	27		55.50	73
7278	10″ diameter on 18″ diameter		33	.485		31	27		58	77
7280	12″ diameter on 14″ diameter		32.60	.491		34.50	27.50		62	81
7282	12″ diameter on 18″ diameter		32.20	.497		36	28		64	83.50
7284	14″ diameter on 16″ diameter		31.60	.506		51	28.50		79.50	101
7286	14″ diameter on 20″ diameter		31	.516		51	29		80	102
7288	16″ diameter on 20″ diameter		30	.533		69	30		99	123
7290	18″ diameter on 20″ diameter		28	.571		85	32		117	144
7292	20″ diameter on 22″ diameter		26	.615		110	34.50		144.50	175
7294	22″ diameter on 24″ diameter		24	.667		122	37.50		159.50	193
7400	Steel, PVC coated both sides, straight lengths									
7410	4″ diameter, 26 ga.	Q-9	360	.044	L.F.	4.43	2.49		6.92	8.75
7416	5″ diameter, 26 ga.		240	.067		4.65	3.74		8.39	10.95
7420	6″ diameter, 26 ga.		280	.057		4.42	3.20		7.62	9.85
7440	8″ diameter, 26 ga.		200	.080		5.95	4.49		10.44	13.60
7460	10″ diameter, 26 ga.		160	.100		7.50	5.60		13.10	17.05
7480	12″ diameter, 26 ga.		120	.133		8.45	7.50		15.95	21

23 31 13.16 Round and Flat-Oval Spiral Ducts		Crew	Daily Output	Labor-Hours	Unit	2020 Bare Costs Material	Labor	Equipment	Total	Total Incl O&P
7500	14" diameter, 26 ga.	Q-9	80	.200	L.F.	10.05	11.20		21.25	28.50
7520	16" diameter, 24 ga.		60	.267		12.35	14.95		27.30	37
7540	18" diameter, 24 ga.		45	.356		13.85	19.95		33.80	46.50
7560	20" diameter, 24 ga.	Q-10	65	.369		15.45	21.50		36.95	50.50
7580	24" diameter, 24 ga.		55	.436		18.60	25.50		44.10	60.50
7600	30" diameter, 22 ga.		45	.533		28	31		59	79
7620	36" diameter, 22 ga.		40	.600		33.50	35		68.50	91.50
7890	Connector, 4" diameter	Q-9	100	.160	Ea.	2.71	8.95		11.66	17.05
7896	5" diameter		83	.193		3.02	10.80		13.82	20.50
7900	6" diameter		88	.182		3.43	10.20		13.63	19.70
7920	8" diameter		78	.205		4.06	11.50		15.56	22.50
7940	10" diameter		70	.229		4.71	12.80		17.51	25
7960	12" diameter		50	.320		5.55	17.95		23.50	34
7980	14" diameter		44	.364		5.90	20.50		26.40	38.50
8000	16" diameter		40	.400		6.95	22.50		29.45	42.50
8020	18" diameter		37	.432		7.75	24.50		32.25	46.50
8040	20" diameter		34	.471		9.35	26.50		35.85	52
8060	24" diameter		28	.571		11.55	32		43.55	62.50
8080	30" diameter		22	.727		15.40	41		56.40	81
8100	36" diameter		18	.889		18.80	50		68.80	98.50
8390	Elbow, 45°, 4" diameter		60	.267		26	14.95		40.95	52
8396	5" diameter		36	.444		25	25		50	66.50
8400	6" diameter		44	.364		25	20.50		45.50	59.50
8420	8" diameter		28	.571		23	32		55	75
8440	10" diameter		18	.889		23.50	50		73.50	104
8460	12" diameter		13	1.231		24	69		93	134
8480	14" diameter		11	1.455		24.50	81.50		106	155
8500	16" diameter		10	1.600		23.50	89.50		113	167
8520	18" diameter		9	1.778		26.50	99.50		126	185
8540	20" diameter	Q-10	13	1.846		83.50	107		190.50	260
8560	24" diameter		11	2.182		104	127		231	315
8580	30" diameter		9	2.667		172	155		327	435
8600	36" diameter		7	3.429		202	199		401	535
8800	Elbow, 90°, 4" diameter	Q-9	60	.267		27	14.95		41.95	53
8804	5" diameter		52	.308		26.50	17.25		43.75	56
8806	6" diameter		44	.364		26.50	20.50		47	61
8808	8" diameter		28	.571		25.50	32		57.50	78
8810	10" diameter		18	.889		25.50	50		75.50	107
8812	12" diameter		13	1.231		26	69		95	137
8814	14" diameter		11	1.455		29.50	81.50		111	161
8816	16" diameter		10	1.600		30.50	89.50		120	175
8818	18" diameter		9.60	1.667		37.50	93.50		131	188
8820	20" diameter	Q-10	14	1.714		47.50	99.50		147	208
8822	24" diameter		12	2		109	116		225	300
8824	30" diameter		9	2.667		213	155		368	475
8826	36" diameter		7	3.429		335	199		534	680
9000	Reducing coupling									
9040	6" x 4"	Q-9	46	.348	Ea.	19.40	19.50		38.90	52
9060	8" x 6"		40	.400		21	22.50		43.50	58
9080	10" x 8"		32	.500		22.50	28		50.50	69
9100	12" x 10"		24	.667		23.50	37.50		61	84
9120	14" x 12"		20	.800		28.50	45		73.50	102
9140	16" x 14"		18	.889		40	50		90	122

23 31 13 – Metal Ducts

23 31 13.16 Round and Flat-Oval Spiral Ducts		Crew	Daily Output	Labor-Hours	Unit	2020 Bare Costs Material	Labor	Equipment	Total	Total Incl O&P
9160	18" x 16"	Q-9	16	1	Ea.	50	56		106	143
9180	20" x 18"	Q-10	24	1		57	58		115	154
9200	24" x 22"		22	1.091		76.50	63.50		140	184
9220	30" x 28"		18	1.333		118	77.50		195.50	251
9240	36" x 34"		16	1.500		171	87		258	325
9800	Steel, stainless, straight lengths									
9810	3" diameter, 26 ga.	Q-9	400	.040	L.F.	15.25	2.24		17.49	20.50
9820	4" diameter, 26 ga.		360	.044		17.70	2.49		20.19	23.50
9830	5" diameter, 26 ga.		320	.050		19.20	2.80		22	25.50
9840	6" diameter, 26 ga.		280	.057		19.35	3.20		22.55	26.50
9850	7" diameter, 26 ga.		240	.067		35.50	3.74		39.24	45
9860	8" diameter, 26 ga.		200	.080		26.50	4.49		30.99	36
9870	9" diameter, 26 ga.		180	.089		38.50	4.99		43.49	50.50
9880	10" diameter, 26 ga.		160	.100		50.50	5.60		56.10	64.50
9890	12" diameter, 26 ga.		120	.133		72	7.50		79.50	90.50
9900	14" diameter, 26 ga.		80	.200		75	11.20		86.20	100
9990	Minimum labor/equipment charge	1 Shee	3	2.667	Job		166		166	260

23 31 13.17 Round Grease Duct		Crew	Daily Output	Labor-Hours	Unit	Material	Labor	Equipment	Total	Total Incl O&P
0010	**ROUND GREASE DUCT**									
1280	Add to labor for elevated installation									
1282	of prefabricated (purchased) ductwork									
1283	10' to 15' high						10%			
1284	15' to 20' high						20%			
1285	20' to 25' high						25%			
1286	25' to 30' high						35%			
1287	30' to 35' high						40%			
1288	35' to 40' high						50%			
1289	Over 40' high						55%			
4020	Zero clearance, 2 hr. fire rated, UL listed, 3" double wall									
4030	304 interior, aluminized exterior, all sizes are ID									
4032	For 316 interior and aluminized exterior, add					6%				
4034	For 304 interior and 304 exterior, add					21%				
4036	For 316 interior and 316 exterior, add					40%				
4040	Straight									
4052	6"	Q-9	34.20	.468	L.F.	77	26		103	126
4054	8"		29.64	.540		88.50	30.50		119	145
4056	10"		27.36	.585		99	33		132	161
4058	12"		25.10	.637		113	36		149	180
4060	14"		23.94	.668		128	37.50		165.50	200
4062	16"		22.80	.702		139	39.50		178.50	214
4064	18"		21.66	.739		163	41.50		204.50	245
4066	20"	Q-10	20.52	1.170		186	68		254	310
4068	22"		19.38	1.238		210	72		282	345
4070	24"		18.24	1.316		238	76.50		314.50	380
4072	26"		17.67	1.358		255	79		334	405
4074	28"		17.10	1.404		272	81.50		353.50	425
4076	30"		15.96	1.504		287	87.50		374.50	450
4078	32"		15.39	1.559		310	90.50		400.50	480
4080	36"		14.25	1.684		345	98		443	535
4100	Adjustable section, 30" long									
4104	6"	Q-9	17.10	.936	Ea.	234	52.50		286.50	340
4106	8"		14.82	1.080		287	60.50		347.50	410

23 31 13 – Metal Ducts

23 31 13.17 Round Grease Duct		Crew	Daily Output	Labor-Hours	Unit	Material	Labor	Equipment	Total	Total Incl O&P
						2020 Bare Costs				
4108	10"	Q-9	13.68	1.170	Ea.	325	65.50		390.50	460
4110	12"		12.54	1.276		360	71.50		431.50	505
4112	14"		11.97	1.337		405	75		480	560
4114	16"		11.40	1.404		465	78.50		543.50	635
4116	18"		10.83	1.477		515	83		598	700
4118	20"	Q-10	10.26	2.339		605	136		741	880
4120	22"		9.69	2.477		680	144		824	975
4122	24"		9.12	2.632		775	153		928	1,100
4124	26"		8.83	2.718		830	158		988	1,150
4126	28"		8.55	2.807		860	163		1,023	1,200
4128	30"		8.26	2.906		945	169		1,114	1,325
4130	32"		7.98	3.008		1,000	175		1,175	1,375
4132	36"		6.84	3.509		1,125	204		1,329	1,550
4140	Tee, 90°, grease									
4144	6"	Q-9	13.68	1.170	Ea.	415	65.50		480.50	560
4146	8"		12.54	1.276		455	71.50		526.50	610
4148	10"		11.97	1.337		505	75		580	670
4150	12"		11.40	1.404		585	78.50		663.50	770
4152	14"		10.26	1.559		665	87.50		752.50	870
4154	16"		9.12	1.754		735	98.50		833.50	965
4156	18"		7.98	2.005		865	112		977	1,125
4158	20"		9.69	1.651		1,000	92.50		1,092.50	1,250
4160	22"		8.26	1.937		1,125	109		1,234	1,400
4162	24"		6.84	2.339		1,225	131		1,356	1,550
4164	26"		6.55	2.443		1,350	137		1,487	1,725
4166	28"	Q-10	6.27	3.828		1,500	223		1,723	2,000
4168	30"		5.98	4.013		1,775	233		2,008	2,350
4170	32"		5.70	4.211		2,100	245		2,345	2,675
4172	36"		5.13	4.678		2,325	272		2,597	2,975
4180	Cleanout tee cap									
4184	6"	Q-9	21.09	.759	Ea.	84.50	42.50		127	160
4186	8"		19.38	.826		89	46.50		135.50	171
4188	10"		18.24	.877		96	49		145	182
4190	12"		17.10	.936		103	52.50		155.50	195
4192	14"		15.96	1.003		108	56		164	207
4194	16"		14.25	1.123		120	63		183	231
4196	18"		13.68	1.170		127	65.50		192.50	243
4198	20"	Q-10	15.39	1.559		141	90.50		231.50	297
4200	22"		12.54	1.914		152	111		263	340
4202	24"		11.97	2.005		166	117		283	365
4204	26"		11.40	2.105		181	122		303	390
4206	28"		10.83	2.216		197	129		326	420
4208	30"		10.26	2.339		214	136		350	450
4210	32"		9.69	2.477		234	144		378	485
4212	36"		8.55	2.807		274	163		437	555
4220	Elbow, 45°									
4224	6"	Q-9	17.10	.936	Ea.	288	52.50		340.50	395
4226	8"		14.82	1.080		325	60.50		385.50	450
4228	10"		13.68	1.170		370	65.50		435.50	510
4230	12"		12.54	1.276		420	71.50		491.50	570
4232	14"		11.97	1.337		470	75		545	635
4234	16"		11.40	1.404		535	78.50		613.50	710
4236	18"		10.83	1.477		605	83		688	795

23 31 13.17 Round Grease Duct		Crew	Daily Output	Labor-Hours	Unit	2020 Bare Costs Material	Labor	Equipment	Total	Total Incl O&P
4238	20″	Q-10	10.26	2.339	Ea.	685	136		821	965
4240	22″		9.69	2.477		765	144		909	1,075
4242	24″		9.12	2.632		875	153		1,028	1,200
4244	26″		8.83	2.718		925	158		1,083	1,275
4246	28″		8.55	2.807		1,100	163		1,263	1,475
4248	30″		8.26	2.906		1,100	169		1,269	1,500
4250	32″		7.98	3.008		1,325	175		1,500	1,725
4252	36″		6.84	3.509		1,775	204		1,979	2,275
4260	Elbow, 90°									
4264	6″	Q-9	17.10	.936	Ea.	565	52.50		617.50	700
4266	8″		14.82	1.080		635	60.50		695.50	795
4268	10″		13.68	1.170		775	65.50		840.50	955
4270	12″		12.54	1.276		875	71.50		946.50	1,075
4272	14″		11.97	1.337		940	75		1,015	1,150
4274	16″		11.40	1.404		1,050	78.50		1,128.50	1,300
4276	18″		10.83	1.477		1,275	83		1,358	1,525
4278	20″	Q-10	10.26	2.339		1,350	136		1,486	1,725
4280	22″		9.69	2.477		1,525	144		1,669	1,900
4282	24″		9.12	2.632		1,700	153		1,853	2,125
4284	26″		8.83	2.718		1,825	158		1,983	2,275
4286	28″		8.55	2.807		1,925	163		2,088	2,350
4288	30″		8.26	2.906		2,225	169		2,394	2,725
4290	32″		7.98	3.008		2,575	175		2,750	3,125
4292	36″		6.84	3.509		3,475	204		3,679	4,150
4300	Support strap									
4304	6″	Q-9	25.50	.627	Ea.	98	35		133	163
4306	8″		22	.727		114	41		155	190
4308	10″		20.50	.780		118	44		162	198
4310	12″		18.80	.851		126	47.50		173.50	214
4312	14″		17.90	.894		136	50		186	228
4314	16″		17	.941		137	53		190	233
4316	18″		16.20	.988		139	55.50		194.50	240
4318	20″	Q-10	15.40	1.558		154	90.50		244.50	310
4320	22″	″	14.50	1.655		170	96.50		266.50	340
4340	Plate support assembly									
4344	6″	Q-9	14.82	1.080	Ea.	141	60.50		201.50	250
4346	8″		12.54	1.276		166	71.50		237.50	295
4348	10″		11.40	1.404		178	78.50		256.50	320
4350	12″		10.26	1.559		190	87.50		277.50	345
4352	14″		9.69	1.651		223	92.50		315.50	390
4354	16″		9.12	1.754		234	98.50		332.50	410
4356	18″		8.55	1.871		248	105		353	435
4358	20″	Q-10	9.12	2.632		260	153		413	525
4360	22″		8.55	2.807		266	163		429	550
4362	24″		7.98	3.008		269	175		444	570
4364	26″		7.69	3.121		300	181		481	615
4366	28″		7.41	3.239		325	188		513	655
4368	30″		7.12	3.371		365	196		561	705
4370	32″		6.84	3.509		405	204		609	770
4372	36″		5.70	4.211		485	245		730	920
4380	Wall guide assembly									
4384	6″	Q-9	19.10	.838	Ea.	183	47		230	276
4386	8″		16.50	.970		209	54.50		263.50	315

23 31 13 – Metal Ducts

23 31 13.17 Round Grease Duct		Crew	Daily Output	Labor-Hours	Unit	Material	2020 Bare Costs Labor	Equipment	Total	Total Incl O&P
4388	10"	Q-9	15.30	1.046	Ea.	221	58.50		279.50	335
4390	12"		14.10	1.135		231	63.50		294.50	355
4392	14"		13.40	1.194		250	67		317	380
4394	16"		12.75	1.255		253	70.50		323.50	390
4396	18"		12.10	1.322		256	74		330	400
4398	20"	Q-10	11.50	2.087		286	121		407	505
4400	22"		10.90	2.202		310	128		438	540
4402	24"		10.20	2.353		345	137		482	595
4404	26"		9.90	2.424		380	141		521	640
4406	28"		9.60	2.500		395	145		540	665
4408	30"		9.30	2.581		450	150		600	730
4410	32"		9	2.667		480	155		635	770
4412	36"		7.60	3.158		525	184		709	865
4420	Hood transition, flanged or unflanged									
4424	6"	Q-9	15.60	1.026	Ea.	78.50	57.50		136	177
4426	8"		13.50	1.185		89.50	66.50		156	203
4428	10"		12.50	1.280		99.50	72		171.50	221
4430	12"		11.40	1.404		119	78.50		197.50	254
4432	14"		10.90	1.468		128	82.50		210.50	270
4434	16"		10.40	1.538		138	86.50		224.50	287
4436	18"		9.80	1.633		158	91.50		249.50	315
4438	20"	Q-10	9.40	2.553		167	148		315	415
4440	22"		8.80	2.727		181	159		340	445
4442	24"		8.30	2.892		203	168		371	485
4444	26"		8.10	2.963		208	172		380	500
4446	28"		7.80	3.077		221	179		400	525
4448	30"		7.50	3.200		240	186		426	555
4450	32"		7.30	3.288		251	191		442	575
4452	36"		6.20	3.871		279	225		504	660
4460	Fan adapter									
4464	6"	Q-9	15	1.067	Ea.	227	60		287	345
4466	8"		13	1.231		257	69		326	390
4468	10"		12	1.333		281	75		356	425
4470	12"		11	1.455		315	81.50		396.50	475
4472	14"		10.50	1.524		340	85.50		425.50	510
4474	16"		10	1.600		370	89.50		459.50	550
4476	18"		9.50	1.684		405	94.50		499.50	595
4478	20"	Q-10	9	2.667		430	155		585	715
4480	22"		8.50	2.824		460	164		624	760
4482	24"		8	3		485	174		659	810
4484	26"		7.70	3.117		515	181		696	850
4486	28"		7.50	3.200		540	186		726	885
4488	30"		7.20	3.333		570	194		764	935
4490	32"		7	3.429		600	199		799	970
4492	36"		6	4		665	233		898	1,100
4500	Tapered increaser/reducer									
4530	6" x 10" diam.	Q-9	13.40	1.194	Ea.	290	67		357	425
4540	6" x 20" diam.		10.10	1.584		293	89		382	465
4550	8" x 10" diam.		13.20	1.212		350	68		418	490
4560	8" x 20" diam.		9.90	1.616		350	90.50		440.50	525
4570	10" x 20" diam.		10.60	1.509		410	84.50		494.50	585
4580	10" x 28" diam.	Q-10	8.60	2.791		410	162		572	705
4590	12" x 20" diam.		8.80	2.727		470	159		629	765

23 31 13 – Metal Ducts

23 31 13.17 Round Grease Duct		Crew	Daily Output	Labor-Hours	Unit	Material	2020 Bare Costs Labor	Equipment	Total	Total Incl O&P
4600	14" x 28" diam.	Q-10	8.40	2.857	Ea.	535	166		701	850
4610	16" x 20" diam.		8.60	2.791		595	162		757	905
4620	16" x 30" diam.		8.30	2.892		590	168		758	915
4630	18" x 24" diam.		8.80	2.727		645	159		804	960
4640	18" x 30" diam.		8.10	2.963		650	172		822	985
4650	20" x 24" diam.		8.30	2.892		705	168		873	1,050
4660	20" x 32" diam.		8	3		705	174		879	1,050
4670	24" x 26" diam.		8	3		820	174		994	1,175
4680	24" x 36" diam.		7.30	3.288		830	191		1,021	1,200
4690	28" x 30" diam.		7.90	3.038		945	177		1,122	1,325
4700	28" x 36" diam.		7.10	3.380		945	197		1,142	1,350
4710	30" x 36" diam.		6.90	3.478		1,000	202		1,202	1,425
4730	Many intermediate standard sizes of tapered fittings are available									

23 31 13.19 Metal Duct Fittings		Crew	Daily Output	Labor-Hours	Unit	Material	Labor	Equipment	Total	Total Incl O&P
0010	**METAL DUCT FITTINGS**									
2000	Fabrics for flexible connections, with metal edge	1 Shee	100	.080	L.F.	3.89	4.98		8.87	12.10
2100	Without metal edge	"	160	.050	"	3.53	3.12		6.65	8.75

23 31 16 – Nonmetal Ducts

23 31 16.13 Fibrous-Glass Ducts		Crew	Daily Output	Labor-Hours	Unit	Material	Labor	Equipment	Total	Total Incl O&P
0010	**FIBROUS-GLASS DUCTS** R233100-40									
1280	Add to labor for elevated installation									
1282	of prefabricated (purchased) ductwork									
1283	10' to 15' high						10%			
1284	15' to 20' high						20%			
1285	20' to 25' high						25%			
1286	25' to 30' high						35%			
1287	30' to 35' high						40%			
1288	35' to 40' high						50%			
1289	Over 40' high						55%			
3490	Rigid fiberglass duct board, foil reinf. kraft facing									
3500	Rectangular, 1" thick, alum. faced, (FRK), std. weight	Q-10	350	.069	SF Surf	.89	3.99		4.88	7.25
9990	Minimum labor/equipment charge	1 Shee	3	2.667	Job		166		166	260

23 31 16.16 Thermoset Fiberglass-Reinforced Plastic Ducts		Crew	Daily Output	Labor-Hours	Unit	Material	Labor	Equipment	Total	Total Incl O&P
0010	**THERMOSET FIBERGLASS-REINFORCED PLASTIC DUCTS**									
1280	Add to labor for elevated installation									
1282	of prefabricated (purchased) ductwork									
1283	10' to 15' high						10%			
1284	15' to 20' high						20%			
1285	20' to 25' high						25%			
1286	25' to 30' high						35%			
1287	30' to 35' high						40%			
1288	35' to 40' high						50%			
1289	Over 40' high						55%			
3550	Rigid fiberglass reinforced plastic, FM approved									
3552	for acid fume and smoke exhaust system, nonflammable									
3554	Straight, 4" diameter	Q-9	200	.080	L.F.	10.35	4.49		14.84	18.45
3555	6" diameter		146	.110		13.60	6.15		19.75	24.50
3556	8" diameter		106	.151		16.45	8.45		24.90	31.50
3557	10" diameter		85	.188		19.75	10.55		30.30	38
3558	12" diameter		68	.235		23	13.20		36.20	46
3561	18" diameter		31	.516		42.50	29		71.50	92
3564	24" diameter	Q-10	37	.649		54.50	37.50		92	119

23 31 16 – Nonmetal Ducts

23 31 16.16 Thermoset Fiberglass-Reinforced Plastic Ducts		Crew	Daily Output	Labor-Hours	Unit	2020 Bare Costs Material	Labor	Equipment	Total	Total Incl O&P
3584	Note: joints are cemented with									
3586	fiberglass resin, included in material cost.									
3590	Elbow, 90°, 4" diameter	Q-9	20.30	.788	Ea.	50	44		94	124
3591	6" diameter		13.80	1.159		70	65		135	179
3592	8" diameter		10	1.600		77	89.50		166.50	226
3593	10" diameter		7.80	2.051		101	115		216	291
3594	12" diameter		6.50	2.462		128	138		266	355
3597	18" diameter		4.40	3.636		268	204		472	615
3600	24" diameter	Q-10	4.90	4.898		400	285		685	885
3626	Elbow, 45°, 4" diameter	Q-9	22.30	.717		48	40		88	116
3627	6" diameter		15.20	1.053		66.50	59		125.50	166
3628	8" diameter		11	1.455		70.50	81.50		152	206
3629	10" diameter		8.60	1.860		74.50	104		178.50	245
3630	12" diameter		7.20	2.222		82	125		207	285
3633	18" diameter		4.84	3.306		140	185		325	445
3636	24" diameter	Q-10	5.40	4.444		251	258		509	680
3660	Tee, 90°, 4" diameter	Q-9	14.06	1.138		27.50	64		91.50	130
3661	6" diameter		9.52	1.681		42	94.50		136.50	194
3662	8" diameter		6.98	2.292		57.50	129		186.50	264
3663	10" diameter		5.45	2.936		74.50	165		239.50	340
3664	12" diameter		4.60	3.478		93.50	195		288.50	410
3667	18" diameter		3.05	5.246		192	294		486	670
3670	24" diameter	Q-10	3.46	6.936		289	405		694	950
3690	For Y @ 45°, add					22%				

23 31 16.19 PVC Ducts		Crew	Daily Output	Labor-Hours	Unit	Material	Labor	Equipment	Total	Total Incl O&P
0010	**PVC DUCTS**									
4000	Rigid plastic, corrosive fume resistant PVC									
4020	Straight, 6" diameter	Q-9	220	.073	L.F.	14.15	4.08		18.23	22
4040	8" diameter		160	.100		20.50	5.60		26.10	31.50
4060	10" diameter		120	.133		26	7.50		33.50	40
4070	12" diameter		100	.160		30	8.95		38.95	47
4080	14" diameter		70	.229		37.50	12.80		50.30	61.50
4090	16" diameter		62	.258		43.50	14.45		57.95	70.50
4100	18" diameter		58	.276		60	15.45		75.45	90
4110	20" diameter	Q-10	75	.320		54.50	18.60		73.10	89
4130	24" diameter	"	55	.436		74.50	25.50		100	122
4250	Coupling, 6" diameter	Q-9	88	.182	Ea.	29	10.20		39.20	48
4270	8" diameter		78	.205		38	11.50		49.50	60
4290	10" diameter		70	.229		43.50	12.80		56.30	67.50
4300	12" diameter		55	.291		47	16.30		63.30	77.50
4310	14" diameter		44	.364		60.50	20.50		81	98.50
4320	16" diameter		40	.400		64	22.50		86.50	106
4330	18" diameter		38	.421		83	23.50		106.50	129
4340	20" diameter	Q-10	55	.436		98.50	25.50		124	148
4360	24" diameter	"	45	.533		125	31		156	186
4470	Elbow, 90°, 6" diameter	Q-9	44	.364		132	20.50		152.50	178
4490	8" diameter		28	.571		148	32		180	212
4510	10" diameter		18	.889		185	50		235	281
4520	12" diameter		15	1.067		179	60		239	291
4530	14" diameter		11	1.455		248	81.50		329.50	400
4540	16" diameter		10	1.600		315	89.50		404.50	485
4550	18" diameter	Q-10	15	1.600		380	93		473	565

23 31 HVAC Ducts and Casings

23 31 16 – Nonmetal Ducts

23 31 16.19 PVC Ducts		Crew	Daily Output	Labor-Hours	Unit	2020 Bare Costs Material	Labor	Equipment	Total	Total Incl O&P
4560	20″ diameter	Q-10	14	1.714	Ea.	640	99.50		739.50	855
4580	24″ diameter	↓	12	2	↓	785	116		901	1,050
4750	Elbow 45°, use 90° and deduct					25%				
9990	Minimum labor/equipment charge	1 Shee	3	2.667	Job		166		166	260

23 31 16.21 Polypropylene Ducts		Crew	Daily Output	Labor-Hours	Unit	Material	Labor	Equipment	Total	Total Incl O&P
0010	**POLYPROPYLENE DUCTS**									
0020	Rigid plastic, corrosive fume resistant, flame retardant, PP									
0110	Straight, 4″ diameter	Q-9	200	.080	L.F.	7.95	4.49		12.44	15.80
0120	6″ diameter		146	.110		12.15	6.15		18.30	23
0140	8″ diameter		106	.151		15.20	8.45		23.65	30
0160	10″ diameter		85	.188		22.50	10.55		33.05	41
0170	12″ diameter		68	.235		40.50	13.20		53.70	65
0180	14″ diameter		50	.320		45.50	17.95		63.45	78.50
0190	16″ diameter		40	.400		61	22.50		83.50	103
0200	18″ diameter	↓	31	.516		98	29		127	154
0210	20″ diameter	Q-10	44	.545		112	31.50		143.50	173
0220	22″ diameter		40	.600		126	35		161	193
0230	24″ diameter		37	.649		174	37.50		211.50	250
0250	28″ diameter		28	.857		234	50		284	335
0270	32″ diameter	↓	24	1	↓	264	58		322	380
0400	Coupling									
0410	4″ diameter	Q-9	23	.696	Ea.	22.50	39		61.50	85.50
0420	6″ diameter		16	1		27	56		83	118
0430	8″ diameter		11.40	1.404		31.50	78.50		110	158
0440	10″ diameter		9	1.778		42	99.50		141.50	203
0450	12″ diameter		8	2		49.50	112		161.50	231
0460	14″ diameter		6.60	2.424		82.50	136		218.50	305
0470	16″ diameter		5.70	2.807		88.50	157		245.50	345
0480	18″ diameter	↓	5.20	3.077		117	173		290	400
0490	20″ diameter	Q-10	6.80	3.529		128	205		333	460
0500	22″ diameter		6.20	3.871		345	225		570	735
0510	24″ diameter		5.80	4.138		435	241		676	850
0520	28″ diameter		5.50	4.364		390	254		644	830
0530	32″ diameter	↓	5.20	4.615	↓	425	268		693	890
0800	Elbow, 90°									
0810	4″ diameter	Q-9	22	.727	Ea.	34.50	41		75.50	102
0820	6″ diameter		15.40	1.039		48.50	58.50		107	145
0830	8″ diameter		11	1.455		59	81.50		140.50	193
0840	10″ diameter		8.60	1.860		85.50	104		189.50	257
0850	12″ diameter		7.70	2.078		220	117		337	425
0860	14″ diameter		6.40	2.500		254	140		394	500
0870	16″ diameter	↓	5.40	2.963		271	166		437	560
0880	18″ diameter	Q-10	7.50	3.200		790	186		976	1,150
0890	20″ diameter		6.60	3.636		790	211		1,001	1,200
0896	22″ diameter		6.10	3.934		1,075	229		1,304	1,525
0900	24″ diameter		5.60	4.286		1,275	249		1,524	1,800
0910	28″ diameter		5.40	4.444		2,175	258		2,433	2,800
0920	32″ diameter	↓	5.20	4.615	↓	2,850	268		3,118	3,575
1000	Elbow, 45°									
1020	4″ diameter	Q-9	22	.727	Ea.	26	41		67	92.50
1030	6″ diameter		15.40	1.039		41.50	58.50		100	137
1040	8″ diameter	↓	11	1.455	↓	47	81.50		128.50	180

23 31 HVAC Ducts and Casings

23 31 16 – Nonmetal Ducts

23 31 16.21 Polypropylene Ducts		Crew	Daily Output	Labor-Hours	Unit	2020 Bare Costs Material	Labor	Equipment	Total	Total Incl O&P
1050	10″ diameter	Q-9	8.60	1.860	Ea.	63	104		167	233
1060	12″ diameter		7.70	2.078		170	117		287	370
1070	14″ diameter		6.40	2.500		191	140		331	430
1080	16″ diameter		5.40	2.963		205	166		371	485
1090	18″ diameter	Q-10	7.50	3.200		475	186		661	810
1100	20″ diameter		6.60	3.636		580	211		791	965
1106	22″ diameter		6.10	3.934		815	229		1,044	1,250
1110	24″ diameter		5.60	4.286		920	249		1,169	1,400
1120	28″ diameter		5.40	4.444		1,450	258		1,708	2,000
1130	32″ diameter		5.20	4.615		1,800	268		2,068	2,425
1200	Tee									
1210	4″ diameter	Q-9	16.50	.970	Ea.	151	54.50		205.50	252
1220	6″ diameter		11.60	1.379		154	77.50		231.50	290
1230	8″ diameter		8.30	1.928		191	108		299	380
1240	10″ diameter		6.50	2.462		270	138		408	515
1250	12″ diameter		5.80	2.759		375	155		530	655
1260	14″ diameter		4.80	3.333		475	187		662	815
1270	16″ diameter		4.05	3.951		545	222		767	945
1280	18″ diameter	Q-10	5.60	4.286		685	249		934	1,150
1290	20″ diameter		4.90	4.898		750	285		1,035	1,275
1300	22″ diameter		4.40	5.455		925	315		1,240	1,525
1310	24″ diameter		4.20	5.714		1,675	330		2,005	2,350
1320	28″ diameter		4.10	5.854		2,250	340		2,590	3,000
1330	32″ diameter		4	6		2,775	350		3,125	3,600
23 31 16.22 Duct Board										
0010	**DUCT BOARD**									
3800	Rigid, resin bonded fibrous glass, FSK									
3810	Temperature, bacteria and fungi resistant									
3820	1″ thick	Q-14	150	.107	S.F.	1.68	5.65		7.33	10.80
3830	1-1/2″ thick		130	.123		2.22	6.50		8.72	12.75
3840	2″ thick		120	.133		3.82	7.05		10.87	15.35

23 33 Air Duct Accessories

23 33 13 – Dampers

23 33 13.13 Volume-Control Dampers		Crew	Daily Output	Labor-Hours	Unit	Material	Labor	Equipment	Total	Total Incl O&P
0010	**VOLUME-CONTROL DAMPERS**									
5990	Multi-blade dampers, opposed blade, 8″ x 6″	1 Shee	24	.333	Ea.	32	21		53	67.50
5994	8″ x 8″		22	.364		34	22.50		56.50	73
5996	10″ x 10″		21	.381		33.50	23.50		57	74
6000	12″ x 12″		21	.381		46	23.50		69.50	87.50
6020	12″ x 18″		18	.444		61.50	27.50		89	111
6030	14″ x 10″		20	.400		45	25		70	88.50
6031	14″ x 14″		17	.471		44.50	29.50		74	95
6033	16″ x 12″		17	.471		44.50	29.50		74	95
6035	16″ x 16″		16	.500		55.50	31		86.50	110
6037	18″ x 16″		15	.533		75	33		108	135
6038	18″ x 18″		15	.533		80	33		113	140
6040	18″ x 24″		12	.667		104	41.50		145.50	180
6060	18″ x 28″		10	.800		117	50		167	206
6070	20″ x 16″		14	.571		80	35.50		115.50	144
6072	20″ x 20″		13	.615		79	38.50		117.50	147

23 33 13 – Dampers

23 33 13.13 Volume-Control Dampers		Crew	Daily Output	Labor-Hours	Unit	Material	2020 Bare Costs Labor	Equipment	Total	Total Incl O&P
6074	22" x 18"	1 Shee	14	.571	Ea.	83	35.50		118.50	148
6076	24" x 16"		11	.727		94.50	45.50		140	175
6078	24" x 20"		8	1		114	62.50		176.50	224
6080	24" x 24"		8	1		121	62.50		183.50	231
6100	24" x 28"		6	1.333		141	83		224	285
6110	26" x 26"		6	1.333		143	83		226	287
6120	28" x 28"	Q-9	11	1.455		134	81.50		215.50	276
6130	30" x 18"		10	1.600		109	89.50		198.50	261
6132	30" x 24"		7	2.286		154	128		282	370
6133	30" x 30"		6.60	2.424		216	136		352	450
6135	32" x 32"		6.40	2.500		236	140		376	480
6151	36" x 12"		10	1.600		106	89.50		195.50	258
6152	36" x 16"		8	2		172	112		284	365
6158	36" x 36"		6	2.667		340	150		490	605
6160	44" x 28"		5.80	2.759		286	155		441	555
6180	48" x 36"		5.60	2.857		400	160		560	690
6200	56" x 36"		5.40	2.963		470	166		636	775
6220	60" x 36"		5.20	3.077		505	173		678	825
6240	60" x 44"		5	3.200		585	179		764	920
7500	Variable volume modulating motorized damper, incl. elect. mtr.									
7504	8" x 6"	1 Shee	15	.533	Ea.	177	33		210	246
7506	10" x 6"		14	.571		177	35.50		212.50	250
7510	10" x 10"		13	.615		184	38.50		222.50	263
7520	12" x 12"		12	.667		190	41.50		231.50	274
7522	12" x 16"		11	.727		191	45.50		236.50	281
7524	16" x 10"		12	.667		188	41.50		229.50	271
7526	16" x 14"		10	.800		197	50		247	294
7528	16" x 18"		9	.889		203	55.50		258.50	310
7540	18" x 12"		10	.800		197	50		247	295
7542	18" x 18"		8	1		206	62.50		268.50	325
7544	20" x 14"		8	1		213	62.50		275.50	335
7546	20" x 18"		7	1.143		220	71		291	355
7560	24" x 12"		8	1		215	62.50		277.50	335
7562	24" x 18"		7	1.143		232	71		303	365
7564	24" x 24"		6	1.333		242	83		325	395
7568	28" x 10"		7	1.143		215	71		286	350
7580	28" x 16"		6	1.333		215	83		298	365
7590	30" x 14"		5	1.600		221	99.50		320.50	400
7600	30" x 18"		4	2		296	125		421	520
7610	30" x 24"		3.80	2.105		395	131		526	640
7700	For thermostat, add		8	1		48.50	62.50		111	151
7800	For transformer 40 VA capacity, add		16	.500		29.50	31		60.50	81.50
8000	Multi-blade dampers, parallel blade									
8100	8" x 8"	1 Shee	24	.333	Ea.	119	21		140	164
8120	12" x 8"		22	.364		119	22.50		141.50	167
8140	16" x 10"		20	.400		152	25		177	207
8160	18" x 12"		18	.444		168	27.50		195.50	228
8180	22" x 12"		15	.533		176	33		209	246
8200	24" x 16"		11	.727		194	45.50		239.50	285
8220	28" x 16"		10	.800		218	50		268	315
8240	30" x 16"		8	1		225	62.50		287.50	345
8260	30" x 18"		7	1.143		270	71		341	410

23 33 13 – Dampers

23 33 13.16 Fire Dampers		Crew	Daily Output	Labor-Hours	Unit	Material	2020 Bare Costs Labor	Equipment	Total	Total Incl O&P
0010	**FIRE DAMPERS**									
3000	Fire damper, curtain type, 1-1/2 hr. rated, vertical, 6" x 6"	1 Shee	24	.333	Ea.	35	21		56	71
3020	8" x 6"		22	.364		35	22.50		57.50	74
3040	12" x 6"		22	.364		35	22.50		57.50	74
3060	20" x 6"		18	.444		39.50	27.50		67	87
3080	12" x 8"		22	.364		29.50	22.50		52	68
3100	24" x 8"		16	.500		44.50	31		75.50	98
3120	12" x 10"		21	.381		35	23.50		58.50	75.50
3140	24" x 10"		15	.533		62	33		95	120
3160	36" x 10"		12	.667		80.50	41.50		122	154
3180	16" x 12"		20	.400		53.50	25		78.50	97.50
3200	24" x 12"		13	.615		50	38.50		88.50	115
3220	48" x 12"		10	.800		78	50		128	164
3240	16" x 14"		18	.444		59	27.50		86.50	109
3260	24" x 14"		12	.667		71	41.50		112.50	143
3280	30" x 14"		11	.727		78.50	45.50		124	158
3300	18" x 16"		17	.471		67	29.50		96.50	120
3320	24" x 16"		11	.727		56.50	45.50		102	133
3340	36" x 16"		10	.800		70.50	50		120.50	156
3360	24" x 18"		10	.800		84	50		134	171
3380	48" x 18"		8	1		121	62.50		183.50	232
3400	24" x 20"		8	1		66	62.50		128.50	170
3420	36" x 20"		7	1.143		79	71		150	199
3440	24" x 22"		7	1.143		90	71		161	211
3460	30" x 22"		6	1.333		99	83		182	239
3480	26" x 24"		7	1.143		88.50	71		159.50	210
3500	48" x 24"	Q-9	12	1.333		106	75		181	234
3520	28" x 26"		13	1.231		99	69		168	217
3540	30" x 28"		12	1.333		110	75		185	238
3560	48" x 30"		11	1.455		160	81.50		241.50	305
3580	48" x 36"		10	1.600		166	89.50		255.50	325
3600	44" x 44"		10	1.600		175	89.50		264.50	335
3620	48" x 48"		8	2		196	112		308	390
3700	UL label included in above									
3800	For horizontal operation, add				Ea.	20%				
3900	For cap for blades out of air stream, add					20%				
4000	For 10" 22 ga., UL approved sleeve, add					35%				
4100	For 10" 22 ga., UL sleeve, 100% free area, add					65%				
4200	For oversize openings group dampers									
4502	Power open-spring close, sleeve and actuator motor mounted, 1-1/2 hour									
4510	8" x 8"	1 Shee	22	.364	Ea.	291	22.50		313.50	355
4520	16" x 8"		20	.400		360	25		385	435
4540	18" x 8"		18	.444		365	27.50		392.50	450
4560	20" x 8"		16	.500		365	31		396	450
4580	10" x 10"		21	.381		350	23.50		373.50	420
4600	24" x 10"		15	.533		415	33		448	505
4620	30" x 10"		12	.667		515	41.50		556.50	630
4640	12" x 12"		20	.400		400	25		425	480
4660	18" x 12"		18	.444		375	27.50		402.50	460
4680	24" x 12"		13	.615		425	38.50		463.50	530
4700	30" x 12"		11	.727		420	45.50		465.50	535
4720	14" x 14"		17	.471		430	29.50		459.50	515

23 33 Air Duct Accessories

23 33 13 – Dampers

23 33 13.16 Fire Dampers		Crew	Daily Output	Labor-Hours	Unit	Material	2020 Bare Costs Labor	Equipment	Total	Total Incl O&P
4740	16" x 14"	1 Shee	18	.444	Ea.	430	27.50		457.50	515
4760	20" x 14"		14	.571		410	35.50		445.50	505
4780	24" x 14"		12	.667		435	41.50		476.50	545
4800	30" x 14"		10	.800		430	50		480	550
4820	16" x 16"		16	.500		440	31		471	535
4840	20" x 16"		14	.571		430	35.50		465.50	530
4860	24" x 16"		11	.727		435	45.50		480.50	550
4880	30" x 16"		8	1		475	62.50		537.50	620
4900	18" x 18"		15	.533		465	33		498	560
5000	24" x 18"		10	.800		465	50		515	595
5020	36" x 18"		7	1.143		480	71		551	635
5040	20" x 20"		13	.615		460	38.50		498.50	565
5060	24" x 20"		8	1		465	62.50		527.50	610
5080	30" x 20"		8	1		485	62.50		547.50	630
5100	36" x 20"		7	1.143		565	71		636	735
5120	24" x 24"		8	1		505	62.50		567.50	655
5130	30" x 24"	↓	7.60	1.053		520	65.50		585.50	675
5140	36" x 24"	Q-9	12	1.333		530	75		605	700
5141	36" x 30"		10	1.600		625	89.50		714.50	825
5142	40" x 36"		8	2		1,125	112		1,237	1,425
5143	48" x 48"	↓	4	4		2,025	224		2,249	2,575
5150	Damper operator motor, 24 or 120 volt	1 Shee	16	.500	↓	495	31		526	595

23 33 13.28 Splitter Damper Assembly		Crew	Daily Output	Labor-Hours	Unit	Material	Labor	Equipment	Total	Total Incl O&P
0010	**SPLITTER DAMPER ASSEMBLY**									
7000	Self locking, 1' rod	1 Shee	24	.333	Ea.	29.50	21		50.50	65
7020	3' rod		22	.364		40.50	22.50		63	80
7040	4' rod		20	.400		45	25		70	88.50
7060	6' rod	↓	18	.444	↓	54.50	27.50		82	104

23 33 19 – Duct Silencers

23 33 19.10 Duct Silencers		Crew	Daily Output	Labor-Hours	Unit	Material	Labor	Equipment	Total	Total Incl O&P
0010	**DUCT SILENCERS**									
9000	Silencers, noise control for air flow, duct				MCFM	87			87	96

23 33 23 – Turning Vanes

23 33 23.13 Air Turning Vanes		Crew	Daily Output	Labor-Hours	Unit	Material	Labor	Equipment	Total	Total Incl O&P
0010	**AIR TURNING VANES**									
9400	Turning vane components									
9410	Turning vane rail	1 Shee	160	.050	L.F.	.84	3.12		3.96	5.80
9420	Double thick, factory fab. vane		300	.027		1.26	1.66		2.92	3.99
9428	12" high set		170	.047		1.92	2.93		4.85	6.70
9432	14" high set		160	.050		2.24	3.12		5.36	7.35
9434	16" high set		150	.053		2.56	3.32		5.88	8
9436	18" high set		144	.056		2.88	3.46		6.34	8.55
9438	20" high set		138	.058		3.20	3.61		6.81	9.15
9440	22" high set		130	.062		3.52	3.83		7.35	9.85
9442	24" high set		124	.065		3.84	4.02		7.86	10.50
9444	26" high set		116	.069		4.16	4.30		8.46	11.35
9446	30" high set		112	.071	↓	4.80	4.45		9.25	12.25
9900	Minimum labor/equipment charge	↓	4	2	Job		125		125	195

23 33 33 – Duct-Mounting Access Doors

23 33 33.13 Duct Access Doors		Crew	Daily Output	Labor-Hours	Unit	Material	2020 Bare Costs Labor	Equipment	Total	Total Incl O&P
0010	**DUCT ACCESS DOORS**									
1000	Duct access door, insulated, 6" x 6"	1 Shee	14	.571	Ea.	22.50	35.50		58	81
1020	10" x 10"		11	.727		23.50	45.50		69	96.50
1040	12" x 12"		10	.800		28	50		78	109
1050	12" x 18"		9	.889		48	55.50		103.50	140
1060	16" x 12"		9	.889		39	55.50		94.50	130
1070	18" x 18"		8	1		40	62.50		102.50	142
1074	24" x 18"		8	1		69	62.50		131.50	174
1080	24" x 24"		8	1		60	62.50		122.50	164

23 33 46 – Flexible Ducts

23 33 46.10 Flexible Air Ducts		Crew	Daily Output	Labor-Hours	Unit	Material	2020 Bare Costs Labor	Equipment	Total	Total Incl O&P
0010	**FLEXIBLE AIR DUCTS** R233100-40									
1280	Add to labor for elevated installation									
1282	of prefabricated (purchased) ductwork									
1283	10' to 15' high						10%			
1284	15' to 20' high						20%			
1285	20' to 25' high						25%			
1286	25' to 30' high						35%			
1287	30' to 35' high						40%			
1288	35' to 40' high						50%			
1289	Over 40' high						55%			
1300	Flexible, coated fiberglass fabric on corr. resist. metal helix									
1400	pressure to 12" (WG) UL-181									
1500	Noninsulated, 3" diameter	Q-9	400	.040	L.F.	1.36	2.24		3.60	5
1520	4" diameter		360	.044		1.45	2.49		3.94	5.50
1540	5" diameter		320	.050		1.57	2.80		4.37	6.10
1560	6" diameter		280	.057		1.85	3.20		5.05	7.05
1580	7" diameter		240	.067		2.05	3.74		5.79	8.10
1600	8" diameter		200	.080		2.32	4.49		6.81	9.60
1620	9" diameter		180	.089		2.51	4.99		7.50	10.55
1640	10" diameter		160	.100		2.97	5.60		8.57	12.05
1660	12" diameter		120	.133		3.50	7.50		11	15.55
1680	14" diameter		80	.200		4.27	11.20		15.47	22.50
1700	16" diameter		60	.267		6.15	14.95		21.10	30.50
1800	For plastic cable tie, add				Ea.	.61			.61	.67
1900	Insulated, 1" thick, PE jacket, 3" diameter G	Q-9	380	.042	L.F.	2.68	2.36		5.04	6.65
1910	4" diameter G		340	.047		3.26	2.64		5.90	7.70
1920	5" diameter G		300	.053		3.42	2.99		6.41	8.45
1940	6" diameter G		260	.062		3.61	3.45		7.06	9.35
1960	7" diameter G		220	.073		4.16	4.08		8.24	11
1980	8" diameter G		180	.089		4.04	4.99		9.03	12.25
2000	9" diameter G		160	.100		4.74	5.60		10.34	14
2020	10" diameter G		140	.114		5.15	6.40		11.55	15.75
2040	12" diameter G		100	.160		6.05	8.95		15	21
2060	14" diameter G		80	.200		7.40	11.20		18.60	25.50
2080	16" diameter G		60	.267		10.55	14.95		25.50	35
2100	18" diameter G		45	.356		11.90	19.95		31.85	44
2120	20" diameter G	Q-10	65	.369		10.90	21.50		32.40	45.50
2500	Insulated, heavy duty, coated fiberglass fabric									
2520	4" diameter G	Q-9	340	.047	L.F.	4.40	2.64		7.04	8.95
2540	5" diameter G		300	.053		5.25	2.99		8.24	10.45
2560	6" diameter G		260	.062		5.20	3.45		8.65	11.15

23 33 46.10 Flexible Air Ducts			Crew	Daily Output	Labor-Hours	Unit	Material	Labor	Equipment	Total	Total Incl O&P
							2020 Bare Costs				
2580	7″ diameter	G	Q-9	220	.073	L.F.	6.55	4.08		10.63	13.60
2600	8″ diameter	G		180	.089		6.50	4.99		11.49	14.95
2620	9″ diameter	G		160	.100		8.10	5.60		13.70	17.70
2640	10″ diameter	G		140	.114		8	6.40		14.40	18.85
2660	12″ diameter	G		100	.160		9.45	8.95		18.40	24.50
2680	14″ diameter	G		80	.200		11.40	11.20		22.60	30
2700	16″ diameter	G		60	.267		15.25	14.95		30.20	40.50
2720	18″ diameter	G	↓	45	.356	↓	17.15	19.95		37.10	50
5000	Flexible, aluminum, acoustical, pressure to 2″ (WG), NFPA-90A										
5010	Fiberglass insulation 1-1/2″ thick, 1/2 lb. density										
5020	Polyethylene jacket, UL approved										
5026	5″ diameter	G	Q-9	300	.053	L.F.	5.30	2.99		8.29	10.55
5030	6″ diameter	G		260	.062		6.40	3.45		9.85	12.40
5034	7″ diameter	G		220	.073		7.55	4.08		11.63	14.70
5038	8″ diameter	G		180	.089		7.80	4.99		12.79	16.35
5042	9″ diameter	G		160	.100		9.10	5.60		14.70	18.80
5046	10″ diameter	G		140	.114		10.65	6.40		17.05	22
5050	12″ diameter	G		100	.160		12.65	8.95		21.60	28
5054	14″ diameter	G		80	.200		14.60	11.20		25.80	33.50
5058	16″ diameter	G	↓	60	.267	↓	20.50	14.95		35.45	46
5072	Hospital grade, PE jacket, UL approved										
5076	5″ diameter	G	Q-9	300	.053	L.F.	5.75	2.99		8.74	11
5080	6″ diameter	G		260	.062		6.90	3.45		10.35	13
5084	7″ diameter	G		220	.073		8.15	4.08		12.23	15.35
5088	8″ diameter	G		180	.089		8.40	4.99		13.39	17.05
5092	9″ diameter	G		160	.100		9.75	5.60		15.35	19.55
5096	10″ diameter	G		140	.114		11.50	6.40		17.90	22.50
5100	12″ diameter	G		100	.160		13.65	8.95		22.60	29
5104	14″ diameter	G		80	.200		15.75	11.20		26.95	35
5108	16″ diameter	G	↓	60	.267	↓	22.50	14.95		37.45	48
5140	Flexible, aluminum, silencer, pressure to 12″ (WG), NFPA-90A										
5144	Fiberglass insulation 1-1/2″ thick, 1/2 lb. density										
5148	Aluminum outer shell, UL approved										
5152	5″ diameter	G	Q-9	290	.055	L.F.	14.80	3.09		17.89	21
5156	6″ diameter	G		250	.064		13.65	3.59		17.24	20.50
5160	7″ diameter	G		210	.076		20.50	4.27		24.77	29
5164	8″ diameter	G		170	.094		20.50	5.30		25.80	31
5168	9″ diameter	G		150	.107		21	6		27	33
5172	10″ diameter	G		130	.123		21	6.90		27.90	34
5176	12″ diameter	G		90	.178		25	9.95		34.95	43
5180	14″ diameter	G		70	.229		30.50	12.80		43.30	53.50
5184	16″ diameter	G	↓	50	.320	↓	39	17.95		56.95	70.50
5200	Hospital grade, aluminum shell, UL approved										
5204	5″ diameter	G	Q-9	280	.057	L.F.	15.40	3.20		18.60	22
5208	6″ diameter	G		240	.067		14.25	3.74		17.99	21.50
5212	7″ diameter	G		200	.080		21	4.49		25.49	30
5216	8″ diameter	G		160	.100		21	5.60		26.60	32
5220	9″ diameter	G		140	.114		22	6.40		28.40	34.50
5224	10″ diameter	G		130	.123		22	6.90		28.90	35
5228	12″ diameter	G		80	.200		26	11.20		37.20	46
5232	14″ diameter	G		60	.267		32	14.95		46.95	58.50
5236	16″ diameter	G	↓	40	.400	↓	40.50	22.50		63	79.50
9990	Minimum labor/equipment charge		1 Shee	3	2.667	Job		166		166	260

23 33 Air Duct Accessories

23 33 53 – Duct Liners

23 33 53.10 Duct Liner Board			Crew	Daily Output	Labor-Hours	Unit	Material	Labor	Equipment	Total	Total Incl O&P
							2020 Bare Costs				
0010	**DUCT LINER BOARD**										
3340	Board type fiberglass liner, FSK, 1-1/2 lb. density										
3344	1″ thick	G	Q-14	150	.107	S.F.	1.06	5.65		6.71	10.10
3345	1-1/2″ thick	G		130	.123		1.16	6.50		7.66	11.60
3346	2″ thick	G		120	.133		1.34	7.05		8.39	12.60
3348	3″ thick	G		110	.145		1.75	7.70		9.45	14.10
3350	4″ thick	G		100	.160		2.13	8.45		10.58	15.75
3356	3 lb. density, 1″ thick	G		150	.107		1.34	5.65		6.99	10.40
3358	1-1/2″ thick	G		130	.123		1.49	6.50		7.99	11.95
3360	2″ thick	G		120	.133		2.07	7.05		9.12	13.45
3362	2-1/2″ thick	G		110	.145		2.44	7.70		10.14	14.85
3364	3″ thick	G		100	.160		2.80	8.45		11.25	16.50
3366	4″ thick	G		90	.178		3.50	9.40		12.90	18.75
3370	6 lb. density, 1″ thick	G		140	.114		1.90	6.05		7.95	11.65
3374	1-1/2″ thick	G		120	.133		2.54	7.05		9.59	13.95
3378	2″ thick	G		100	.160		3.20	8.45		11.65	16.90
3490	Board type, fiberglass liner, 3 lb. density										
3680	No finish										
3700	1″ thick	G	Q-14	170	.094	S.F.	.73	4.97		5.70	8.70
3710	1-1/2″ thick	G		140	.114		1.10	6.05		7.15	10.75
3720	2″ thick	G		130	.123		1.49	6.50		7.99	11.95
3940	Board type, non-fibrous foam										
3950	Temperature, bacteria and fungi resistant										
3960	1″ thick	G	Q-14	150	.107	S.F.	2.76	5.65		8.41	12
3970	1-1/2″ thick	G		130	.123		4.89	6.50		11.39	15.70
3980	2″ thick	G		120	.133		4.13	7.05		11.18	15.70

23 34 HVAC Fans

23 34 13 – Axial HVAC Fans

23 34 13.10 Axial Flow HVAC Fans		Crew	Daily Output	Labor-Hours	Unit	Material	Labor	Equipment	Total	Total Incl O&P
0010	**AXIAL FLOW HVAC FANS**									
0020	Air conditioning and process air handling									
1500	Vaneaxial, low pressure, 2,000 CFM, 1/2 HP	Q-20	3.60	5.556	Ea.	2,400	315		2,715	3,125
1520	4,000 CFM, 1 HP		3.20	6.250		2,500	355		2,855	3,300
1540	8,000 CFM, 2 HP		2.80	7.143		3,550	410		3,960	4,525
1560	16,000 CFM, 5 HP		2.40	8.333		5,150	475		5,625	6,425

23 34 14 – Blower HVAC Fans

23 34 14.10 Blower Type HVAC Fans		Crew	Daily Output	Labor-Hours	Unit	Material	Labor	Equipment	Total	Total Incl O&P
0010	**BLOWER TYPE HVAC FANS**									
2000	Blowers, direct drive with motor, complete									
2020	1,045 CFM @ 0.5″ S.P., 1/5 HP	Q-20	18	1.111	Ea.	350	63.50		413.50	485
2040	1,385 CFM @ 0.5″ S.P., 1/4 HP		18	1.111		365	63.50		428.50	500
2060	1,640 CFM @ 0.5″ S.P., 1/3 HP		18	1.111		370	63.50		433.50	510
2080	1,760 CFM @ 0.5″ S.P., 1/2 HP		18	1.111		360	63.50		423.50	500
2090	4 speed									
2100	1,164 to 1,739 CFM @ 0.5″ S.P., 1/3 HP	Q-20	16	1.250	Ea.	335	71.50		406.50	480
2120	1,467 to 2,218 CFM @ 1.0″ S.P., 3/4 HP	″	14	1.429	″	385	81.50		466.50	545
2500	Ceiling fan, right angle, extra quiet, 0.10″ S.P.									
2520	95 CFM	Q-20	20	1	Ea.	305	57		362	425
2540	210 CFM		19	1.053		360	60		420	490

23 34 14 – Blower HVAC Fans

23 34 14.10 Blower Type HVAC Fans		Crew	Daily Output	Labor-Hours	Unit	2020 Bare Costs Material	Labor	Equipment	Total	Total Incl O&P
2560	385 CFM	Q-20	18	1.111	Ea.	455	63.50		518.50	605
2580	885 CFM		16	1.250		900	71.50		971.50	1,100
2600	1,650 CFM		13	1.538		1,250	88		1,338	1,500
2620	2,960 CFM		11	1.818		1,675	104		1,779	1,975
2640	For wall or roof cap, add	1 Shee	16	.500		305	31		336	385
2660	For straight thru fan, add					10%				
2680	For speed control switch, add	1 Elec	16	.500		166	30.50		196.50	230
7500	Utility set, steel construction, pedestal, 1/4" S.P.									
7520	Direct drive, 150 CFM, 1/8 HP	Q-20	6.40	3.125	Ea.	975	179		1,154	1,350
7540	485 CFM, 1/6 HP		5.80	3.448		1,100	197		1,297	1,500
7560	1,950 CFM, 1/2 HP		4.80	4.167		1,425	238		1,663	1,950
7580	2,410 CFM, 3/4 HP		4.40	4.545		2,650	260		2,910	3,325
7600	3,328 CFM, 1-1/2 HP		3	6.667		2,625	380		3,005	3,500
7680	V-belt drive, drive cover, 3 phase									
7700	800 CFM, 1/4 HP	Q-20	6	3.333	Ea.	1,050	190		1,240	1,475
7720	1,300 CFM, 1/3 HP		5	4		1,125	229		1,354	1,575
7740	2,000 CFM, 1 HP		4.60	4.348		1,325	248		1,573	1,850
7760	2,900 CFM, 3/4 HP		4.20	4.762		1,775	272		2,047	2,375
7780	3,600 CFM, 3/4 HP		4	5		2,200	286		2,486	2,850
7800	4,800 CFM, 1 HP		3.50	5.714		2,575	325		2,900	3,350
7820	6,700 CFM, 1-1/2 HP		3	6.667		3,175	380		3,555	4,100
7830	7,500 CFM, 2 HP		2.50	8		4,325	455		4,780	5,475
7840	11,000 CFM, 3 HP		2	10		5,775	570		6,345	7,250
7860	13,000 CFM, 3 HP		1.60	12.500		5,875	715		6,590	7,575
7880	15,000 CFM, 5 HP		1	20		6,075	1,150		7,225	8,450
7900	17,000 CFM, 7-1/2 HP		.80	25		6,500	1,425		7,925	9,375
7920	20,000 CFM, 7-1/2 HP		.80	25		7,750	1,425		9,175	10,800

23 34 16 – Centrifugal HVAC Fans

23 34 16.10 Centrifugal Type HVAC Fans		Crew	Daily Output	Labor-Hours	Unit	Material	Labor	Equipment	Total	Total Incl O&P
0010	**CENTRIFUGAL TYPE HVAC FANS**									
0200	In-line centrifugal, supply/exhaust booster									
0220	aluminum wheel/hub, disconnect switch, 1/4" S.P.									
0240	500 CFM, 10" diameter connection	Q-20	3	6.667	Ea.	1,625	380		2,005	2,375
0260	1,380 CFM, 12" diameter connection		2	10		1,750	570		2,320	2,825
0280	1,520 CFM, 16" diameter connection		2	10		1,700	570		2,270	2,775
0300	2,560 CFM, 18" diameter connection		1	20		1,850	1,150		3,000	3,825
0320	3,480 CFM, 20" diameter connection		.80	25		2,175	1,425		3,600	4,625
0326	5,080 CFM, 20" diameter connection		.75	26.667		2,375	1,525		3,900	4,975
3500	Centrifugal, airfoil, motor and drive, complete									
3520	1,000 CFM, 1/2 HP	Q-20	2.50	8	Ea.	2,100	455		2,555	3,025
3540	2,000 CFM, 1 HP		2	10		2,250	570		2,820	3,400
3560	4,000 CFM, 3 HP		1.80	11.111		3,000	635		3,635	4,300
3580	8,000 CFM, 7-1/2 HP		1.40	14.286		4,525	815		5,340	6,250
3600	12,000 CFM, 10 HP		1	20		5,875	1,150		7,025	8,225
4000	Single width, belt drive, not incl. motor, capacities									
4020	at 2,000 fpm, 2.5" S.P. for indicated motor									
4040	6,900 CFM, 5 HP	Q-9	2.40	6.667	Ea.	4,225	375		4,600	5,225
4060	10,340 CFM, 7-1/2 HP		2.20	7.273		5,900	410		6,310	7,125
4080	15,320 CFM, 10 HP		2	8		6,425	450		6,875	7,750
4100	22,780 CFM, 15 HP		1.80	8.889		10,300	500		10,800	12,200
4120	33,840 CFM, 20 HP		1.60	10		13,700	560		14,260	16,000
4140	41,400 CFM, 25 HP		1.40	11.429		16,200	640		16,840	18,800

23 34 16 – Centrifugal HVAC Fans

23 34 16.10 Centrifugal Type HVAC Fans		Crew	Daily Output	Labor-Hours	Unit	2020 Bare Costs Material	Labor	Equipment	Total	Total Incl O&P
4160	50,100 CFM, 30 HP	Q-9	.80	20	Ea.	22,400	1,125		23,525	26,400
4200	Double width wheel, 12,420 CFM, 7-1/2 HP		2.20	7.273		5,600	410		6,010	6,800
4220	18,620 CFM, 15 HP		2	8		8,675	450		9,125	10,200
4240	27,580 CFM, 20 HP		1.80	8.889		9,800	500		10,300	11,600
4260	40,980 CFM, 25 HP		1.50	10.667		15,300	600		15,900	17,800
4280	60,920 CFM, 40 HP		1	16		22,400	895		23,295	26,000
4300	74,520 CFM, 50 HP		.80	20		24,500	1,125		25,625	28,700
4320	90,160 CFM, 50 HP		.70	22.857		36,200	1,275		37,475	41,900
4340	110,300 CFM, 60 HP		.50	32		49,000	1,800		50,800	57,000
4360	134,960 CFM, 75 HP	↓	.40	40	↓	65,000	2,250		67,250	75,000
5000	Utility set, centrifugal, V belt drive, motor									
5020	1/4" S.P., 1,200 CFM, 1/4 HP	Q-20	6	3.333	Ea.	1,925	190		2,115	2,425
5040	1,520 CFM, 1/3 HP		5	4		2,575	229		2,804	3,175
5060	1,850 CFM, 1/2 HP		4	5		2,550	286		2,836	3,275
5080	2,180 CFM, 3/4 HP		3	6.667		2,950	380		3,330	3,850
5100	1/2" S.P., 3,600 CFM, 1 HP		2	10		3,100	570		3,670	4,300
5120	4,250 CFM, 1-1/2 HP		1.60	12.500		3,800	715		4,515	5,300
5140	4,800 CFM, 2 HP		1.40	14.286		4,625	815		5,440	6,375
5160	6,920 CFM, 5 HP		1.30	15.385		5,575	880		6,455	7,525
5180	7,700 CFM, 7-1/2 HP	↓	1.20	16.667	↓	6,775	950		7,725	8,925
5200	For explosion proof motor, add					15%				
5500	Fans, industrial exhauster, for air which may contain granular matl.									
5520	1,000 CFM, 1-1/2 HP	Q-20	2.50	8	Ea.	3,300	455		3,755	4,375
5540	2,000 CFM, 3 HP		2	10		4,000	570		4,570	5,300
5560	4,000 CFM, 7-1/2 HP		1.80	11.111		5,750	635		6,385	7,325
5580	8,000 CFM, 15 HP		1.40	14.286		7,425	815		8,240	9,450
5600	12,000 CFM, 30 HP	↓	1	20	↓	11,500	1,150		12,650	14,400
7000	Roof exhauster, centrifugal, aluminum housing, 12" galvanized									
7020	curb, bird screen, back draft damper, 1/4" S.P.									
7100	Direct drive, 320 CFM, 11" sq. damper	Q-20	7	2.857	Ea.	785	163		948	1,125
7120	600 CFM, 11" sq. damper		6	3.333		880	190		1,070	1,275
7140	815 CFM, 13" sq. damper		5	4		1,050	229		1,279	1,525
7160	1,450 CFM, 13" sq. damper		4.20	4.762		1,750	272		2,022	2,350
7180	2,050 CFM, 16" sq. damper		4	5		1,900	286		2,186	2,525
7200	V-belt drive, 1,650 CFM, 12" sq. damper		6	3.333		1,425	190		1,615	1,875
7220	2,750 CFM, 21" sq. damper		5	4		1,800	229		2,029	2,350
7230	3,500 CFM, 21" sq. damper		4.50	4.444		2,025	254		2,279	2,650
7240	4,910 CFM, 23" sq. damper		4	5		2,500	286		2,786	3,200
7260	8,525 CFM, 28" sq. damper		3	6.667		3,250	380		3,630	4,175
7280	13,760 CFM, 35" sq. damper		2	10		4,500	570		5,070	5,850
7300	20,558 CFM, 43" sq. damper	↓	1	20		8,950	1,150		10,100	11,600
7320	For 2 speed winding, add					15%				
7340	For explosion proof motor, add					745			745	820
7360	For belt driven, top discharge, add				↓	15%				
8500	Wall exhausters, centrifugal, auto damper, 1/8" S.P.									
8520	Direct drive, 610 CFM, 1/20 HP	Q-20	14	1.429	Ea.	425	81.50		506.50	595
8540	796 CFM, 1/12 HP		13	1.538		1,075	88		1,163	1,300
8560	822 CFM, 1/6 HP		12	1.667		1,075	95		1,170	1,325
8580	1,320 CFM, 1/4 HP		12	1.667		1,375	95		1,470	1,650
8600	1,756 CFM, 1/4 HP		11	1.818		1,275	104		1,379	1,550
8620	1,983 CFM, 1/4 HP		10	2		1,300	114		1,414	1,600
8640	2,900 CFM, 1/2 HP		9	2.222		1,375	127		1,502	1,725
8660	3,307 CFM, 3/4 HP	↓	8	2.500	↓	1,500	143		1,643	1,875

23 34 16 – Centrifugal HVAC Fans

23 34 16.10 Centrifugal Type HVAC Fans			Crew	Daily Output	Labor-Hours	Unit	Material	Labor	Equipment	Total	Total Incl O&P
							2020 Bare Costs				
9500	V-belt drive, 3 phase										
9520	2,800 CFM, 1/4 HP		Q-20	9	2.222	Ea.	1,925	127		2,052	2,325
9540	3,740 CFM, 1/2 HP			8	2.500		2,000	143		2,143	2,425
9560	4,400 CFM, 3/4 HP			7	2.857		2,025	163		2,188	2,475
9580	5,700 CFM, 1-1/2 HP			6	3.333		2,125	190		2,315	2,625
9900	Minimum labor/equipment charge		1 Elec	4	2	Job		123		123	189

23 34 23 – HVAC Power Ventilators

23 34 23.10 HVAC Power Circulators and Ventilators			Crew	Daily Output	Labor-Hours	Unit	Material	Labor	Equipment	Total	Total Incl O&P
0010	**HVAC POWER CIRCULATORS AND VENTILATORS**										
3000	Paddle blade air circulator, 3 speed switch										
3020	42", 5,000 CFM high, 3,000 CFM low	G	1 Elec	2.40	3.333	Ea.	105	205		310	430
3040	52", 6,500 CFM high, 4,000 CFM low	G	"	2.20	3.636	"	144	223		367	505
3100	For antique white motor, same cost										
3200	For brass plated motor, same cost										
3300	For light adaptor kit, add	G				Ea.	40			40	44
4000	High volume, low speed (HVLS), paddle blade air circulator										
4010	Variable speed, reversible, 1 HP motor, motor control panel,										
4020	motor drive cable, control cable with remote, and safety cable.										
4140	8' diameter		Q-5	1.50	10.667	Ea.	4,925	630		5,555	6,375
4150	10' diameter			1.45	11.034		4,975	650		5,625	6,450
4160	12' diameter			1.40	11.429		5,175	675		5,850	6,750
4170	14' diameter			1.35	11.852		5,275	700		5,975	6,875
4180	16' diameter			1.30	12.308		5,325	725		6,050	6,975
4190	18' diameter			1.25	12.800		5,350	755		6,105	7,075
4200	20' diameter			1.20	13.333		5,725	785		6,510	7,525
4210	24' diameter			1	16		5,850	945		6,795	7,875
6000	Propeller exhaust, wall shutter										
6020	Direct drive, one speed, 0.075" S.P.										
6100	653 CFM, 1/30 HP		Q-20	10	2	Ea.	236	114		350	440
6120	1,033 CFM, 1/20 HP			9	2.222		325	127		452	555
6140	1,323 CFM, 1/15 HP			8	2.500		390	143		533	655
6160	2,444 CFM, 1/4 HP			7	2.857		425	163		588	725
6300	V-belt drive, 3 phase										
6320	6,175 CFM, 3/4 HP		Q-20	5	4	Ea.	1,200	229		1,429	1,650
6340	7,500 CFM, 3/4 HP			5	4		1,275	229		1,504	1,750
6360	10,100 CFM, 1 HP			4.50	4.444		1,500	254		1,754	2,075
6380	14,300 CFM, 1-1/2 HP			4	5		1,775	286		2,061	2,400
6400	19,800 CFM, 2 HP			3	6.667		2,400	380		2,780	3,250
6420	26,250 CFM, 3 HP			2.60	7.692		2,425	440		2,865	3,325
6440	38,500 CFM, 5 HP			2.20	9.091		2,975	520		3,495	4,075
6460	46,000 CFM, 7-1/2 HP			2	10		3,200	570		3,770	4,425
6480	51,500 CFM, 10 HP			1.80	11.111		3,175	635		3,810	4,475
6490	V-belt drive, 115 V, residential, whole house										
6500	Ceiling-wall, 5,200 CFM, 1/4 HP, 30" x 30"		1 Shee	6	1.333	Ea.	605	83		688	800
6530	13,200 CFM, 1/3 HP, 48" x 48"			4	2		735	125		860	1,000
6540	15,445 CFM, 1/2 HP, 48" x 48"			4	2		765	125		890	1,025
6550	17,025 CFM, 1/2 HP, 54" x 54"			4	2		1,475	125		1,600	1,825
6560	For two speed motor, add						20%				
6570	Shutter, automatic, ceiling/wall										
6580	30" x 30"		1 Shee	8	1	Ea.	198	62.50		260.50	315
6590	36" x 36"			8	1		226	62.50		288.50	345
6600	42" x 42"			8	1		282	62.50		344.50	410

23 34 23.10 HVAC Power Circulators and Ventilators		Crew	Daily Output	Labor-Hours	Unit	2020 Bare Costs Material	Labor	Equipment	Total	Total Incl O&P
6610	48″ x 48″	1 Shee	7	1.143	Ea.	305	71		376	445
6620	54″ x 54″		6	1.333		405	83		488	575
6630	Timer, shut off, to 12 hour		20	.400		65	25		90	111
6650	Residential, bath exhaust, grille, back draft damper									
6660	50 CFM	Q-20	24	.833	Ea.	61	47.50		108.50	142
6670	110 CFM		22	.909		116	52		168	209
6680	Light combination, squirrel cage, 100 watt, 70 CFM		24	.833		123	47.50		170.50	210
6700	Light/heater combination, ceiling mounted									
6710	70 CFM, 1,450 watt	Q-20	24	.833	Ea.	139	47.50		186.50	228
6800	Heater combination, recessed, 70 CFM		24	.833		83.50	47.50		131	167
6820	With 2 infrared bulbs		23	.870		100	49.50		149.50	188
6900	Kitchen exhaust, grille, complete, 160 CFM		22	.909		104	52		156	196
6910	180 CFM		20	1		97	57		154	196
6920	270 CFM		18	1.111		214	63.50		277.50	335
6930	350 CFM		16	1.250		146	71.50		217.50	272
6940	Residential roof jacks and wall caps									
6944	Wall cap with back draft damper									
6946	3″ & 4″ diam. round duct	1 Shee	11	.727	Ea.	26.50	45.50		72	100
6948	6″ diam. round duct	″	11	.727	″	76	45.50		121.50	155
6958	Roof jack with bird screen and back draft damper									
6960	3″ & 4″ diam. round duct	1 Shee	11	.727	Ea.	18.85	45.50		64.35	92
6962	3-1/4″ x 10″ rectangular duct	″	10	.800	″	35	50		85	117
6980	Transition									
6982	3-1/4″ x 10″ to 6″ diam. round	1 Shee	20	.400	Ea.	36	25		61	78.50
8020	Attic, roof type									
8030	Aluminum dome, damper & curb									
8080	12″ diameter, 1,000 CFM (gravity)	1 Elec	10	.800	Ea.	640	49		689	780
8090	16″ diameter, 1,500 CFM (gravity)		9	.889		775	54.50		829.50	935
8100	20″ diameter, 2,500 CFM (gravity)		8	1		950	61.50		1,011.50	1,150
8110	26″ diameter, 4,000 CFM (gravity)		7	1.143		1,150	70		1,220	1,375
8120	32″ diameter, 6,500 CFM (gravity)		6	1.333		1,575	82		1,657	1,850
8130	38″ diameter, 8,000 CFM (gravity)		5	1.600		2,350	98		2,448	2,725
8140	50″ diameter, 13,000 CFM (gravity)		4	2		3,400	123		3,523	3,925
8160	Plastic, ABS dome									
8180	1,050 CFM	1 Elec	14	.571	Ea.	189	35		224	261
8200	1,600 CFM	″	12	.667	″	282	41		323	375
8240	Attic, wall type, with shutter, one speed									
8250	12″ diameter, 1,000 CFM	1 Elec	14	.571	Ea.	430	35		465	530
8260	14″ diameter, 1,500 CFM		12	.667		465	41		506	580
8270	16″ diameter, 2,000 CFM		9	.889		530	54.50		584.50	665
8290	Whole house, wall type, with shutter, one speed									
8300	30″ diameter, 4,800 CFM	1 Elec	7	1.143	Ea.	1,125	70		1,195	1,350
8310	36″ diameter, 7,000 CFM		6	1.333		1,225	82		1,307	1,475
8320	42″ diameter, 10,000 CFM		5	1.600		1,375	98		1,473	1,675
8330	48″ diameter, 16,000 CFM		4	2		1,700	123		1,823	2,075
8340	For two speed, add					103			103	113
8350	Whole house, lay-down type, with shutter, one speed									
8360	30″ diameter, 4,500 CFM	1 Elec	8	1	Ea.	1,200	61.50		1,261.50	1,425
8370	36″ diameter, 6,500 CFM		7	1.143		1,300	70		1,370	1,525
8380	42″ diameter, 9,000 CFM		6	1.333		1,425	82		1,507	1,675
8390	48″ diameter, 12,000 CFM		5	1.600		1,600	98		1,698	1,925
8440	For two speed, add					77.50			77.50	85
8450	For 12 hour timer switch, add	1 Elec	32	.250		77.50	15.35		92.85	109

23 34 33 – Air Curtains

	23 34 33.10 Air Barrier Curtains	Crew	Daily Output	Labor-Hours	Unit	2020 Bare Costs Material	Labor	Equipment	Total	Total Incl O&P
0010	**AIR BARRIER CURTAINS**, Incl. motor starters, transformers,									
0050	and door switches									
2450	Conveyor openings or service windows									
3000	Service window, 5′ high x 25″ wide	2 Shee	5	3.200	Ea.	345	199		544	690
3100	Environmental separation									
3110	Door heights up to 8′, low profile, super quiet									
3120	Unheated, variable speed									
3130	36″ wide	2 Shee	4	4	Ea.	765	249		1,014	1,225
3134	42″ wide		3.80	4.211		780	262		1,042	1,275
3138	48″ wide		3.60	4.444		810	277		1,087	1,325
3142	60″ wide		3.40	4.706		865	293		1,158	1,425
3146	72″ wide	Q-3	4.60	6.957		1,050	425		1,475	1,800
3150	96″ wide		4.40	7.273		1,625	445		2,070	2,475
3154	120″ wide		4.20	7.619		1,600	465		2,065	2,475
3158	144″ wide		4	8		2,000	490		2,490	2,950
3200	Door heights up to 10′									
3210	Unheated									
3230	36″ wide	2 Shee	3.80	4.211	Ea.	735	262		997	1,225
3234	42″ wide		3.60	4.444		795	277		1,072	1,300
3238	48″ wide		3.40	4.706		780	293		1,073	1,325
3242	60″ wide		3.20	5		1,100	310		1,410	1,725
3246	72″ wide	Q-3	4.40	7.273		1,325	445		1,770	2,175
3250	96″ wide		4.20	7.619		1,575	465		2,040	2,450
3254	120″ wide		4	8		2,250	490		2,740	3,225
3258	144″ wide		3.80	8.421		2,275	515		2,790	3,300
3300	Door heights up to 12′									
3310	Unheated									
3334	42″ wide	2 Shee	3.40	4.706	Ea.	1,125	293		1,418	1,675
3338	48″ wide		3.20	5		1,125	310		1,435	1,750
3342	60″ wide		3	5.333		1,150	330		1,480	1,775
3346	72″ wide	Q-3	4.20	7.619		1,925	465		2,390	2,825
3350	96″ wide		4	8		2,225	490		2,715	3,200
3354	120″ wide		3.80	8.421		2,525	515		3,040	3,575
3358	144″ wide		3.60	8.889		3,650	545		4,195	4,850
3400	Door heights up to 16′									
3410	Unheated									
3438	48″ wide	2 Shee	3	5.333	Ea.	1,475	330		1,805	2,150
3442	60″ wide	″	2.80	5.714		1,550	355		1,905	2,250
3446	72″ wide	Q-3	3.80	8.421		2,700	515		3,215	3,775
3450	96″ wide		3.60	8.889		3,175	545		3,720	4,325
3454	120″ wide		3.40	9.412		3,625	575		4,200	4,900
3458	144″ wide		3.20	10		3,625	615		4,240	4,950
3470	Heated, electric									
3474	48″ wide	2 Shee	2.90	5.517	Ea.	2,500	345		2,845	3,300
3478	60″ wide	″	2.70	5.926		2,550	370		2,920	3,375
3482	72″ wide	Q-3	3.70	8.649		4,275	530		4,805	5,525
3486	96″ wide		3.50	9.143		4,150	560		4,710	5,450
3490	120″ wide		3.30	9.697		4,725	595		5,320	6,125
3494	144″ wide		3.10	10.323		5,550	635		6,185	7,075

23 34 HVAC Fans

23 34 39 – Ceiling Fans

23 34 39.10 Powerfoil		Crew	Daily Output	Labor-Hours	Unit	Material	2020 Bare Costs Labor	Equipment	Total	Total Incl O&P
0010	POWERFOIL									
2301	Ceiling fan, Powerfoil, 8′ diameter	R-1A	2.40	6.667	Ea.	8,650	370		9,020	10,100
2302	10′		2.40	6.667		8,750	370		9,120	10,200
2303	12′		2.40	6.667		8,850	370		9,220	10,300
2304	14′		2.40	6.667		8,950	370		9,320	10,400
2305	16′		2	8		9,050	440		9,490	10,600
2306	18′		2	8		9,150	440		9,590	10,800
2307	20′		2	8		9,250	440		9,690	10,900
2308	24′	↓	2	8	↓	9,450	440		9,890	11,100
2400	Powerfoil with Hybrid Airfoil									
2401	12′	R-1A	2	8	Ea.	9,350	440		9,790	11,000
2402	14′		2	8		9,450	440		9,890	11,100
2403	16′		1.75	9.143		9,550	505		10,055	11,300
2404	18′		1.75	9.143		9,650	505		10,155	11,400
2405	20′		1.75	9.143		9,750	505		10,255	11,500
2406	24′	↓	1.75	9.143	↓	9,975	505		10,480	11,800

23 35 Special Exhaust Systems

23 35 16 – Engine Exhaust Systems

23 35 16.10 Engine Exhaust Removal Systems		Crew	Daily Output	Labor-Hours	Unit	Material	Labor	Equipment	Total	Total Incl O&P
0010	ENGINE EXHAUST REMOVAL SYSTEMS									
0500	Engine exhaust, garage, in-floor system									
0510	Single tube outlet assemblies									
0520	For transite pipe ducting, self-storing tube									
0530	3″ tubing adapter plate	1 Shee	16	.500	Ea.	262	31		293	335
0540	4″ tubing adapter plate		16	.500		266	31		297	340
0550	5″ tubing adapter plate	↓	16	.500	↓	264	31		295	340
0600	For vitrified tile ducting									
0610	3″ tubing adapter plate, self-storing tube	1 Shee	16	.500	Ea.	260	31		291	335
0620	4″ tubing adapter plate, self-storing tube		16	.500		265	31		296	340
0660	5″ tubing adapter plate, self-storing tube	↓	16	.500	↓	265	31		296	340
0800	Two tube outlet assemblies									
0810	For transite pipe ducting, self-storing tube									
0820	3″ tubing, dual exhaust adapter plate	1 Shee	16	.500	Ea.	267	31		298	340
0850	For vitrified tile ducting									
0860	3″ tubing, dual exhaust, self-storing tube	1 Shee	16	.500	Ea.	300	31		331	380
0870	3″ tubing, double outlet, non-storing tubes	″	16	.500	″	300	31		331	380
0900	Accessories for metal tubing (overhead systems also)									
0910	Adapters, for metal tubing end									
0920	3″ tail pipe type				Ea.	51			51	56.50
0930	4″ tail pipe type					56			56	62
0940	5″ tail pipe type					57			57	63
0990	5″ diesel stack type					283			283	310
1000	6″ diesel stack type				↓	335			335	370
1100	Bullnose (guide) required for in-floor assemblies									
1110	3″ tubing size				Ea.	30.50			30.50	33.50
1120	4″ tubing size					31.50			31.50	35
1130	5″ tubing size				↓	33.50			33.50	36.50
1150	Plain rings, for tubing end									
1160	3″ tubing size				Ea.	23			23	25

23 35 16.10 Engine Exhaust Removal Systems		Crew	Daily Output	Labor-Hours	Unit	2020 Bare Costs Material	Labor	Equipment	Total	Total Incl O&P
1170	4" tubing size				Ea.	38.50			38.50	42.50
1200	Tubing, galvanized, flexible (for overhead systems also)									
1210	3" ID				L.F.	10.80			10.80	11.85
1220	4" ID					13.20			13.20	14.55
1230	5" ID					15.55			15.55	17.15
1240	6" ID					18			18	19.80
1250	Stainless steel, flexible (for overhead system also)									
1260	3" ID				L.F.	24			24	26.50
1270	4" ID					33			33	36
1280	5" ID					37.50			37.50	41
1290	6" ID					43.50			43.50	47.50
1500	Engine exhaust, garage, overhead components, for neoprene tubing									
1510	Alternate metal tubing & accessories see above									
1550	Adapters, for neoprene tubing end									
1560	3" tail pipe, adjustable, neoprene				Ea.	58.50			58.50	64
1570	3" tail pipe, heavy wall neoprene					66			66	72.50
1580	4" tail pipe, heavy wall neoprene					103			103	113
1590	5" tail pipe, heavy wall neoprene					117			117	129
1650	Connectors, tubing									
1660	3" interior, aluminum				Ea.	23			23	25.50
1670	4" interior, aluminum					38.50			38.50	42.50
1710	5" interior, neoprene					59			59	64.50
1750	3" spiralock, neoprene					23			23	25
1760	4" spiralock, neoprene					38.50			38.50	42.50
1780	Y for 3" ID tubing, neoprene, dual exhaust					194			194	213
1790	Y for 4" ID tubing, aluminum, dual exhaust					191			191	210
1850	Elbows, aluminum, splice into tubing for strap									
1860	3" neoprene tubing size				Ea.	54			54	59.50
1870	4" neoprene tubing size					36			36	40
1900	Flange assemblies, connect tubing to overhead duct					69.50			69.50	76.50
2000	Hardware and accessories									
2020	Cable, galvanized, 1/8" diameter				L.F.	.47			.47	.52
2040	Cleat, tie down cable or rope				Ea.	5.90			5.90	6.50
2060	Pulley					8.15			8.15	9
2080	Pulley hook, universal					5.90			5.90	6.50
2100	Rope, nylon, 1/4" diameter				L.F.	.41			.41	.45
2120	Winch, 1" diameter				Ea.	119			119	131
2150	Lifting strap, mounts on neoprene									
2160	3" tubing size				Ea.	29.50			29.50	32.50
2170	4" tubing size					29.50			29.50	32.50
2180	5" tubing size					29.50			29.50	32.50
2190	6" tubing size					29.50			29.50	32.50
2200	Tubing, neoprene, 11' lengths									
2210	3" ID				L.F.	12.25			12.25	13.50
2220	4" ID					16.90			16.90	18.55
2230	5" ID					29			29	32
2500	Engine exhaust, thru-door outlet									
2510	3" tube size	1 Carp	16	.500	Ea.	64	26.50		90.50	113
2530	4" tube size	"	16	.500	"	69.50	26.50		96	119

23 35 Special Exhaust Systems

23 35 43 – Welding Fume Elimination Systems

23 35 43.10 Welding Fume Elimination System Components		Crew	Daily Output	Labor-Hours	Unit	2020 Bare Costs Material	Labor	Equipment	Total	Total Incl O&P
0010	**WELDING FUME ELIMINATION SYSTEM COMPONENTS**									
7500	Welding fume elimination accessories for garage exhaust systems									
7600	Cut off (blast gate)									
7610	3" tubing size, 3" x 6" opening	1 Shee	24	.333	Ea.	27.50	21		48.50	63
7620	4" tubing size, 4" x 8" opening		24	.333		28	21		49	63.50
7630	5" tubing size, 5" x 10" opening		24	.333		32	21		53	68
7640	6" tubing size		24	.333		33.50	21		54.50	69.50
7650	8" tubing size		24	.333		43.50	21		64.50	80
7700	Hoods, magnetic, with handle & screen									
7710	3" tubing size, 3" x 6" opening	1 Shee	24	.333	Ea.	94.50	21		115.50	137
7720	4" tubing size, 4" x 8" opening		24	.333		102	21		123	146
7730	5" tubing size, 5" x 10" opening		24	.333		102	21		123	146

23 36 Air Terminal Units

23 36 13 – Constant-Air-Volume Units

23 36 13.10 Constant Volume Mixing Boxes		Crew	Daily Output	Labor-Hours	Unit	Material	Labor	Equipment	Total	Total Incl O&P
0010	**CONSTANT VOLUME MIXING BOXES**									
5180	Mixing box, includes electric or pneumatic motor									
5192	Recommend use with silencer, see Line 23 33 19.10 0010									
5200	Constant volume, 150 to 270 CFM	Q-9	12	1.333	Ea.	1,200	75		1,275	1,450
5210	270 to 600 CFM		11	1.455		1,150	81.50		1,231.50	1,400
5230	550 to 1,000 CFM		9	1.778		1,150	99.50		1,249.50	1,425
5240	1,000 to 1,600 CFM		8	2		1,250	112		1,362	1,575
5250	1,300 to 1,900 CFM		6	2.667		1,300	150		1,450	1,650
5260	550 to 2,640 CFM		5.60	2.857		1,425	160		1,585	1,800
5270	650 to 3,120 CFM		5.20	3.077		1,425	173		1,598	1,850

23 36 16 – Variable-Air-Volume Units

23 36 16.10 Variable Volume Mixing Boxes		Crew	Daily Output	Labor-Hours	Unit	Material	Labor	Equipment	Total	Total Incl O&P
0010	**VARIABLE VOLUME MIXING BOXES**									
5180	Mixing box, includes electric or pneumatic motor									
5192	Recommend use with attenuator, see Line 23 33 19.10 0010									
5500	VAV cool only, pneumatic, pressure independent 300 to 600 CFM	Q-9	11	1.455	Ea.	815	81.50		896.50	1,025
5510	500 to 1,000 CFM		9	1.778		835	99.50		934.50	1,075
5520	800 to 1,600 CFM		9	1.778		865	99.50		964.50	1,100
5530	1,100 to 2,000 CFM		8	2		885	112		997	1,150
5540	1,500 to 3,000 CFM		7	2.286		940	128		1,068	1,225
5550	2,000 to 4,000 CFM		6	2.667		950	150		1,100	1,275
5600	VAV, HW coils, damper, actuator and thermostat									
5610	200 CFM	Q-9	11	1.455	Ea.	1,275	81.50		1,356.50	1,525
5620	400 CFM		10	1.600		1,275	89.50		1,364.50	1,550
5630	600 CFM		10	1.600		1,275	89.50		1,364.50	1,550
5640	800 CFM		8	2		1,325	112		1,437	1,625
5650	1,000 CFM		8	2		1,325	112		1,437	1,625
5660	1,250 CFM		6	2.667		1,450	150		1,600	1,800
5670	1,500 CFM		6	2.667		1,450	150		1,600	1,800
5680	2,000 CFM		4	4		1,675	224		1,899	2,175
5684	3,000 CFM		3.80	4.211		1,825	236		2,061	2,375
5700	VAV cool only, fan powered, damper, actuator, thermostat									
5710	200 CFM	Q-9	10	1.600	Ea.	1,725	89.50		1,814.50	2,050
5720	400 CFM		9	1.778		1,975	99.50		2,074.50	2,300

23 36 Air Terminal Units

23 36 16 – Variable-Air-Volume Units

23 36 16.10 Variable Volume Mixing Boxes		Crew	Daily Output	Labor-Hours	Unit	2020 Bare Costs Material	Labor	Equipment	Total	Total Incl O&P
5730	600 CFM	Q-9	9	1.778	Ea.	1,975	99.50		2,074.50	2,300
5740	800 CFM		7	2.286		2,100	128		2,228	2,525
5750	1,000 CFM		7	2.286		2,100	128		2,228	2,525
5760	1,250 CFM		5	3.200		2,275	179		2,454	2,775
5770	1,500 CFM		5	3.200		2,275	179		2,454	2,775
5780	2,000 CFM		4	4		2,500	224		2,724	3,100
5800	VAV fan powr'd, with HW coils, dampers, actuators, thermostat									
5810	200 CFM	Q-9	10	1.600	Ea.	2,275	89.50		2,364.50	2,650
5820	400 CFM		9	1.778		2,350	99.50		2,449.50	2,750
5830	600 CFM		9	1.778		2,350	99.50		2,449.50	2,750
5840	800 CFM		7	2.286		2,500	128		2,628	2,950
5850	1,000 CFM		7	2.286		2,500	128		2,628	2,950
5860	1,250 CFM		5	3.200		2,825	179		3,004	3,400
5870	1,500 CFM		5	3.200		2,825	179		3,004	3,400
5880	2,000 CFM		3.50	4.571		3,050	256		3,306	3,750

23 37 Air Outlets and Inlets

23 37 13 – Diffusers, Registers, and Grilles

23 37 13.10 Diffusers

		Crew	Daily Output	Labor-Hours	Unit	Material	Labor	Equipment	Total	Total Incl O&P
0010	**DIFFUSERS**, Aluminum, opposed blade damper unless noted									
0100	Ceiling, linear, also for sidewall									
0120	2" wide	1 Shee	32	.250	L.F.	21	15.60		36.60	47.50
0140	3" wide		30	.267		24	16.60		40.60	52.50
0160	4" wide		26	.308		24	19.15		43.15	56.50
0180	6" wide		24	.333		30	21		51	65.50
0200	8" wide		22	.364		40.50	22.50		63	80
0220	10" wide		20	.400		39.50	25		64.50	82.50
0240	12" wide		18	.444		50.50	27.50		78	99
0260	For floor or sill application, add					15%				
0500	Perforated, 24" x 24" lay-in panel size, 6" x 6"	1 Shee	16	.500	Ea.	166	31		197	231
0520	8" x 8"		15	.533		181	33		214	251
0530	9" x 9"		14	.571		177	35.50		212.50	250
0540	10" x 10"		14	.571		177	35.50		212.50	251
0560	12" x 12"		12	.667		184	41.50		225.50	268
0580	15" x 15"		11	.727		196	45.50		241.50	287
0590	16" x 16"		11	.727		207	45.50		252.50	299
0600	18" x 18"		10	.800		222	50		272	320
0610	20" x 20"		10	.800		239	50		289	340
0620	24" x 24"		9	.889		284	55.50		339.50	395
1000	Rectangular, 1 to 4 way blow, 6" x 6"		16	.500		43.50	31		74.50	97
1010	8" x 8"		15	.533		63	33		96	121
1014	9" x 9"		15	.533		53.50	33		86.50	111
1016	10" x 10"		15	.533		83.50	33		116.50	144
1020	12" x 6"		15	.533		78	33		111	138
1040	12" x 9"		14	.571		84.50	35.50		120	149
1060	12" x 12"		12	.667		76	41.50		117.50	149
1070	14" x 6"		13	.615		85	38.50		123.50	154
1074	14" x 14"		12	.667		133	41.50		174.50	212
1080	18" x 12"		11	.727		150	45.50		195.50	236
1120	15" x 15"		10	.800		94	50		144	182
1140	18" x 15"		9	.889		172	55.50		227.50	276

23 37 13 – Diffusers, Registers, and Grilles

23 37 13.10 Diffusers		Crew	Daily Output	Labor-Hours	Unit	Material	Labor	Equipment	Total	Total Incl O&P
						2020 Bare Costs				
1150	18" x 18"	1 Shee	9	.889	Ea.	129	55.50		184.50	228
1160	21" x 21"		8	1		232	62.50		294.50	355
1170	24" x 12"		10	.800		177	50		227	272
1180	24" x 24"		7	1.143		298	71		369	440
1500	Round, butterfly damper, steel, diffuser size, 6" diameter		18	.444		13.80	27.50		41.30	58.50
1520	8" diameter		16	.500		14.70	31		45.70	65
1540	10" diameter		14	.571		18.55	35.50		54.05	76.50
1560	12" diameter		12	.667		25	41.50		66.50	92.50
1580	14" diameter		10	.800		31.50	50		81.50	113
1600	18" diameter		9	.889		77.50	55.50		133	172
2000	T-bar mounting, 24" x 24" lay-in frame, 6" x 6"		16	.500		73.50	31		104.50	130
2020	8" x 8"		14	.571		73.50	35.50		109	137
2040	12" x 12"		12	.667		90	41.50		131.50	164
2060	16" x 16"		11	.727		125	45.50		170.50	208
2080	18" x 18"	↓	10	.800	↓	123	50		173	213
2500	Combination supply and return									
2520	21" x 21" supply, 15" x 15" return	Q-9	10	1.600	Ea.	175	89.50		264.50	335
2540	24" x 24" supply, 18" x 18" return		9.50	1.684		231	94.50		325.50	400
2560	27" x 27" supply, 18" x 18" return		9	1.778		255	99.50		354.50	435
2580	30" x 30" supply, 21" x 21" return		8.50	1.882		345	106		451	545
2600	33" x 33" supply, 21" x 21" return		8	2		375	112		487	585
2620	36" x 36" supply, 24" x 24" return	↓	7.50	2.133	↓	440	120		560	670
3000	Baseboard, white enameled steel									
3100	18" long	1 Shee	20	.400	Ea.	7.65	25		32.65	47.50
3120	24" long		18	.444		15.55	27.50		43.05	60.50
3140	48" long	↓	16	.500		28.50	31		59.50	80.50
3400	For matching return, deduct				↓	10%				
4000	Floor, steel, adjustable pattern									
4100	2" x 10"	1 Shee	34	.235	Ea.	10.70	14.65		25.35	35
4120	2" x 12"		32	.250		11.75	15.60		27.35	37.50
4140	2" x 14"		30	.267		13.30	16.60		29.90	40.50
4200	4" x 10"		28	.286		11.95	17.80		29.75	41
4220	4" x 12"		26	.308		13.20	19.15		32.35	44.50
4240	4" x 14"		25	.320		15.45	19.95		35.40	48
4260	6" x 10"		26	.308		18.65	19.15		37.80	50.50
4280	6" x 12"		24	.333		18.85	21		39.85	53
4300	6" x 14"	↓	22	.364	↓	20.50	22.50		43	58
5000	Sidewall, aluminum, 3 way dispersion									
5100	8" x 4"	1 Shee	24	.333	Ea.	6.60	21		27.60	40
5110	8" x 6"		24	.333		7.10	21		28.10	40.50
5120	10" x 4"		22	.364		6.75	22.50		29.25	43
5130	10" x 6"		22	.364		7.20	22.50		29.70	43.50
5140	10" x 8"		20	.400		10.20	25		35.20	50
5160	12" x 4"		18	.444		7.50	27.50		35	52
5170	12" x 6"		17	.471		8	29.50		37.50	55
5180	12" x 8"		16	.500		10.75	31		41.75	61
5200	14" x 4"		15	.533		8.35	33		41.35	61
5220	14" x 6"		14	.571		9.20	35.50		44.70	66
5240	14" x 8"		13	.615		12	38.50		50.50	73
5260	16" x 6"	↓	12.50	.640		13.15	40		53.15	77
6000	For steel diffusers instead of aluminum, deduct				↓	10%				
9000	Minimum labor/equipment charge	1 Shee	4	2	Job		125		125	195

23 37 13.30 Grilles		Crew	Daily Output	Labor-Hours	Unit	2020 Bare Costs Material	Labor	Equipment	Total	Total Incl O&P
0010	**GRILLES**									
0020	Aluminum, unless noted otherwise									
0100	Air supply, single deflection, adjustable									
0120	8″ x 4″	1 Shee	30	.267	Ea.	12.90	16.60		29.50	40
0140	8″ x 8″		28	.286		15.65	17.80		33.45	45
0160	10″ x 4″		24	.333		14.25	21		35.25	48
0180	10″ x 10″		23	.348		18	21.50		39.50	54
0200	12″ x 6″		23	.348		11.70	21.50		33.20	47
0220	12″ x 12″		22	.364		20.50	22.50		43	58
0230	14″ x 6″		23	.348		14.45	21.50		35.95	50
0240	14″ x 8″		23	.348		17.85	21.50		39.35	53.50
0260	14″ x 14″		22	.364		27	22.50		49.50	65.50
0270	18″ x 8″		23	.348		26	21.50		47.50	62.50
0280	18″ x 10″		23	.348		25.50	21.50		47	62
0300	18″ x 18″		21	.381		35.50	23.50		59	76
0320	20″ x 12″		22	.364		21	22.50		43.50	58.50
0340	20″ x 20″		21	.381		44.50	23.50		68	86
0350	24″ x 8″		20	.400		19.85	25		44.85	61
0360	24″ x 14″		18	.444		36	27.50		63.50	83.50
0380	24″ x 24″		15	.533		49.50	33		82.50	106
0400	30″ x 8″		20	.400		41	25		66	84
0420	30″ x 10″		19	.421		27	26		53	70.50
0440	30″ x 12″		18	.444		29.50	27.50		57	76
0460	30″ x 16″		17	.471		44	29.50		73.50	94.50
0480	30″ x 18″		17	.471		58.50	29.50		88	110
0500	30″ x 30″		14	.571		90	35.50		125.50	155
0520	36″ x 12″		17	.471		50	29.50		79.50	101
0540	36″ x 16″		16	.500		54	31		85	108
0560	36″ x 18″		15	.533		70.50	33		103.50	130
0570	36″ x 20″		15	.533		66.50	33		99.50	125
0580	36″ x 24″		14	.571		100	35.50		135.50	166
0600	36″ x 28″		13	.615		88.50	38.50		127	158
0620	36″ x 30″		12	.667		135	41.50		176.50	213
0640	36″ x 32″		12	.667		109	41.50		150.50	185
0660	36″ x 34″		11	.727		139	45.50		184.50	224
0680	36″ x 36″		11	.727		175	45.50		220.50	263
0700	For double deflecting, add					70%				
1000	Air return, steel, 6″ x 6″	1 Shee	26	.308		23.50	19.15		42.65	56
1020	10″ x 6″		24	.333		23.50	21		44.50	58.50
1040	14″ x 6″		23	.348		28	21.50		49.50	65
1060	10″ x 8″		23	.348		28	21.50		49.50	65
1080	16″ x 8″		22	.364		35	22.50		57.50	74
1100	12″ x 12″		22	.364		33.50	22.50		56	72
1120	24″ x 12″		18	.444		43.50	27.50		71	91.50
1140	30″ x 12″		16	.500		56.50	31		87.50	112
1160	14″ x 14″		22	.364		38	22.50		60.50	77.50
1180	16″ x 16″		22	.364		42	22.50		64.50	82
1200	18″ x 18″		21	.381		48.50	23.50		72	90.50
1220	24″ x 18″		16	.500		54.50	31		85.50	109
1240	36″ x 18″		15	.533		84.50	33		117.50	145
1260	24″ x 24″		15	.533		65.50	33		98.50	124
1280	36″ x 24″		14	.571		93	35.50		128.50	159

23 37 13 – Diffusers, Registers, and Grilles

23 37 13.30 Grilles		Crew	Daily Output	Labor-Hours	Unit	2020 Bare Costs Material	Labor	Equipment	Total	Total Incl O&P
1300	48" x 24"	1 Shee	12	.667	Ea.	156	41.50		197.50	237
1320	48" x 30"		11	.727		203	45.50		248.50	294
1340	36" x 36"		13	.615		123	38.50		161.50	195
1360	48" x 36"		11	.727		255	45.50		300.50	350
1380	48" x 48"		8	1		279	62.50		341.50	405
2000	Door grilles, 12" x 12"		22	.364		72	22.50		94.50	115
2020	18" x 12"		22	.364		83	22.50		105.50	127
2040	24" x 12"		18	.444		93.50	27.50		121	147
2060	18" x 18"		18	.444		115	27.50		142.50	171
2080	24" x 18"		16	.500		131	31		162	193
2100	24" x 24"		15	.533		185	33		218	255
3000	Filter grille with filter, 12" x 12"		24	.333		65	21		86	104
3020	18" x 12"		20	.400		83	25		108	131
3040	24" x 18"		18	.444		105	27.50		132.50	159
3060	24" x 24"		16	.500		126	31		157	188
3080	30" x 24"		14	.571		140	35.50		175.50	210
3100	30" x 30"		13	.615		165	38.50		203.50	242
3950	Eggcrate, framed, 6" x 6" opening		26	.308		16.95	19.15		36.10	48.50
3954	8" x 8" opening		24	.333		23.50	21		44.50	58.50
3960	10" x 10" opening		23	.348		25	21.50		46.50	61.50
3970	12" x 12" opening		22	.364		29.50	22.50		52	68
3980	14" x 14" opening		22	.364		36	22.50		58.50	75
3984	16" x 16" opening		21	.381		44	23.50		67.50	85
3990	18" x 18" opening		21	.381		50.50	23.50		74	93
4020	22" x 22" opening		17	.471		59.50	29.50		89	112
4040	24" x 24" opening		15	.533		76.50	33		109.50	136
4044	28" x 28" opening		14	.571		114	35.50		149.50	181
4048	36" x 36" opening		12	.667		130	41.50		171.50	208
4050	48" x 24" opening		12	.667		165	41.50		206.50	246
4060	Eggcrate, lay-in, T-bar system									
4070	48" x 24" sheet	1 Shee	40	.200	Ea.	77.50	12.45		89.95	105
5000	Transfer grille, vision proof, 8" x 4"		30	.267		15.50	16.60		32.10	43
5020	8" x 6"		28	.286		23	17.80		40.80	53.50
5040	8" x 8"		26	.308		25.50	19.15		44.65	58
5060	10" x 6"		24	.333		24	21		45	59
5080	10" x 10"		23	.348		24.50	21.50		46	61
5090	12" x 6"		23	.348		20	21.50		41.50	56
5100	12" x 10"		22	.364		26.50	22.50		49	65
5110	12" x 12"		22	.364		29	22.50		51.50	67.50
5120	14" x 10"		22	.364		36	22.50		58.50	75
5140	16" x 8"		21	.381		36	23.50		59.50	76.50
5160	18" x 12"		20	.400		38	25		63	80.50
5170	18" x 18"		21	.381		53.50	23.50		77	95.50
5180	20" x 12"		20	.400		41	25		66	84
5200	20" x 20"		19	.421		64.50	26		90.50	112
5220	24" x 12"		17	.471		44	29.50		73.50	94
5240	24" x 24"		16	.500		88.50	31		119.50	147
5260	30" x 6"		15	.533		41	33		74	97
5280	30" x 8"		15	.533		46	33		79	103
5300	30" x 12"		14	.571		59.50	35.50		95	122
5320	30" x 16"		13	.615		70.50	38.50		109	138
5340	30" x 20"		12	.667		89	41.50		130.50	163
5360	30" x 24"		11	.727		110	45.50		155.50	192

23 37 13 – Diffusers, Registers, and Grilles

23 37 13.30 Grilles		Crew	Daily Output	Labor-Hours	Unit	Material	Labor	Equipment	Total	Total Incl O&P
						2020 Bare Costs				
5380	30" x 30"	1 Shee	10	.800	Ea.	140	50		190	232
6000	For steel grilles instead of aluminum in above, deduct				Ea.	10%				
6200	Plastic, eggcrate, lay-in, T-bar system									
6210	48" x 24" sheet	1 Shee	50	.160	Ea.	22	9.95		31.95	39.50
6250	Steel door louver									
6270	With fire link, steel only									
6350	12" x 18"	1 Shee	22	.364	Ea.	225	22.50		247.50	283
6410	12" x 24"		21	.381		250	23.50		273.50	310
6470	18" x 18"		20	.400		250	25		275	315
6530	18" x 24"		19	.421		288	26		314	355
6660	24" x 24"		15	.533		291	33		324	370
9000	Minimum labor/equipment charge		4	2	Job		125		125	195

23 37 13.60 Registers		Crew	Daily Output	Labor-Hours	Unit	Material	Labor	Equipment	Total	Total Incl O&P
0010	**REGISTERS**									
0980	Air supply									
1000	Ceiling/wall, O.B. damper, anodized aluminum									
1010	One or two way deflection, adj. curved face bars									
1014	6" x 6"	1 Shee	24	.333	Ea.	28	21		49	63
1020	8" x 4"		26	.308		30.50	19.15		49.65	63.50
1040	8" x 8"		24	.333		35	21		56	71
1060	10" x 6"		20	.400		21.50	25		46.50	62.50
1080	10" x 10"		19	.421		28.50	26		54.50	72.50
1100	12" x 6"		19	.421		32.50	26		58.50	76.50
1120	12" x 12"		18	.444		37	27.50		64.50	84
1140	14" x 8"		17	.471		22	29.50		51.50	70.50
1160	14" x 14"		18	.444		39.50	27.50		67	87
1170	16" x 16"		17	.471		43.50	29.50		73	94
1180	18" x 8"		18	.444		49.50	27.50		77	98
1200	18" x 18"		17	.471		90.50	29.50		120	146
1220	20" x 4"		19	.421		20	26		46	63
1240	20" x 6"		18	.444		30.50	27.50		58	77
1260	20" x 8"		18	.444		50	27.50		77.50	98.50
1280	20" x 20"		17	.471		110	29.50		139.50	167
1290	22" x 22"		15	.533		126	33		159	191
1300	24" x 4"		17	.471		25	29.50		54.50	73.50
1320	24" x 6"		16	.500		56	31		87	111
1340	24" x 8"		13	.615		61.50	38.50		100	128
1350	24" x 18"		12	.667		113	41.50		154.50	189
1360	24" x 24"		11	.727		112	45.50		157.50	195
1380	30" x 4"		16	.500		31	31		62	83
1400	30" x 6"		15	.533		70.50	33		103.50	130
1420	30" x 8"		14	.571		80.50	35.50		116	145
1440	30" x 24"		12	.667		200	41.50		241.50	285
1460	30" x 30"		10	.800		256	50		306	360
1504	4 way deflection, adjustable curved face bars									
1510	6" x 6"	1 Shee	26	.308	Ea.	33.50	19.15		52.65	67
1514	8" x 8"		24	.333		42	21		63	78.50
1518	10" x 10"		19	.421		34.50	26		60.50	79
1522	12" x 6"		19	.421		39	26		65	84
1526	12" x 12"		18	.444		44.50	27.50		72	92.50
1530	14" x 14"		18	.444		47	27.50		74.50	95.50
1534	16" x 16"		17	.471		52.50	29.50		82	104

23 37 13.60 Registers		Crew	Daily Output	Labor-Hours	Unit	Material	Labor (2020 Bare Costs)	Equipment	Total	Total Incl O&P
1538	18" x 18"	1 Shee	17	.471	Ea.	108	29.50		137.50	165
1542	22" x 22"		16	.500		152	31		183	216
1980	One way deflection, adj. vert. or horiz. face bars									
1990	6" x 6"	1 Shee	26	.308	Ea.	24.50	19.15		43.65	57
2000	8" x 4"		26	.308		23	19.15		42.15	55.50
2020	8" x 8"		24	.333		27.50	21		48.50	62.50
2040	10" x 6"		20	.400		23.50	25		48.50	64.50
2060	10" x 10"		19	.421		27	26		53	71
2080	12" x 6"		19	.421		25	26		51	68.50
2100	12" x 12"		18	.444		33	27.50		60.50	79.50
2120	14" x 8"		17	.471		30.50	29.50		60	79.50
2140	14" x 12"		18	.444		53	27.50		80.50	102
2160	14" x 14"		18	.444		60	27.50		87.50	110
2180	16" x 6"		18	.444		30.50	27.50		58	77
2200	16" x 12"		18	.444		58.50	27.50		86	108
2220	16" x 16"		17	.471		62.50	29.50		92	115
2240	18" x 8"		18	.444		51.50	27.50		79	100
2260	18" x 12"		17	.471		42.50	29.50		72	93
2280	18" x 18"		16	.500		91	31		122	149
2300	20" x 10"		18	.444		61	27.50		88.50	111
2320	20" x 16"		16	.500		85	31		116	143
2340	20" x 20"		16	.500		111	31		142	171
2360	24" x 12"		15	.533		77	33		110	137
2380	24" x 16"		14	.571		103	35.50		138.50	169
2400	24" x 20"		12	.667		128	41.50		169.50	206
2420	24" x 24"		10	.800		121	50		171	211
2440	30" x 12"		13	.615		99.50	38.50		138	170
2460	30" x 16"		13	.615		90.50	38.50		129	160
2480	30" x 24"		11	.727		143	45.50		188.50	229
2500	30" x 30"		9	.889		259	55.50		314.50	370
2520	36" x 12"		11	.727		124	45.50		169.50	207
2540	36" x 24"		10	.800		250	50		300	355
2560	36" x 36"		8	1		420	62.50		482.50	560
2600	For 2 way deflect., adj. vert. or horiz. face bars, add					40%				
2700	Above registers in steel instead of aluminum, deduct					10%				
3000	Baseboard, hand adj. damper, enameled steel									
3012	8" x 6"	1 Shee	26	.308	Ea.	8.10	19.15		27.25	39
3020	10" x 6"		24	.333		8.80	21		29.80	42
3040	12" x 5"		23	.348		7.70	21.50		29.20	42.50
3060	12" x 6"		23	.348		11.45	21.50		32.95	46.50
3080	12" x 8"		22	.364		13.85	22.50		36.35	51
3100	14" x 6"		20	.400		10.40	25		35.40	50.50
4000	Floor, toe operated damper, enameled steel									
4020	4" x 8"	1 Shee	32	.250	Ea.	14.65	15.60		30.25	40.50
4040	4" x 12"		26	.308		17.20	19.15		36.35	49
4060	6" x 8"		28	.286		15.70	17.80		33.50	45.50
4080	6" x 14"		22	.364		19.90	22.50		42.40	57.50
4100	8" x 10"		22	.364		17.90	22.50		40.40	55
4120	8" x 16"		20	.400		15.80	25		40.80	56.50
4140	10" x 10"		20	.400		21.50	25		46.50	62.50
4160	10" x 16"		18	.444		26.50	27.50		54	72.50
4180	12" x 12"		18	.444		26.50	27.50		54	72.50
4200	12" x 24"		16	.500		30	31		61	82

23 37 13 – Diffusers, Registers, and Grilles

23 37 13.60 Registers		Crew	Daily Output	Labor-Hours	Unit	Material	2020 Bare Costs Labor	Equipment	Total	Total Incl O&P
4220	14" x 14"	1 Shee	16	.500	Ea.	56.50	31		87.50	112
4240	14" x 20"	↓	15	.533	↓	53	33		86	110
4300	Spiral pipe supply register									
4310	Steel, with air scoop									
4320	4" x 12", for 8" thru 13" diameter duct	1 Shee	25	.320	Ea.	84	19.95		103.95	124
4330	4" x 18", for 8" thru 13" diameter duct		18	.444		99.50	27.50		127	153
4340	6" x 12", for 14" thru 21" diameter duct		19	.421		91	26		117	141
4350	6" x 16", for 14" thru 21" diameter duct		18	.444		102	27.50		129.50	156
4360	6" x 20", for 14" thru 21" diameter duct		17	.471		109	29.50		138.50	166
4370	6" x 24", for 14" thru 21" diameter duct		16	.500		132	31		163	194
4380	8" x 16", for 22" thru 31" diameter duct		19	.421		104	26		130	155
4390	8" x 18", for 22" thru 31" diameter duct		18	.444		112	27.50		139.50	167
4400	8" x 24", for 22" thru 31" diameter duct	↓	15	.533	↓	140	33		173	206
4980	Air return									
5000	Ceiling or wall, fixed 45° face blades									
5010	Adjustable O.B. damper, anodized aluminum									
5020	4" x 8"	1 Shee	26	.308	Ea.	19.95	19.15		39.10	52
5040	6" x 8"		24	.333		26	21		47	61
5060	6" x 10"		19	.421		24.50	26		50.50	68
5080	6" x 16"		18	.444		30.50	27.50		58	77.50
5100	8" x 10"		19	.421		30.50	26		56.50	74.50
5120	8" x 12"		16	.500		33.50	31		64.50	86
5140	10" x 10"		18	.444		28	27.50		55.50	74
5160	10" x 16"		17	.471		52	29.50		81.50	103
5180	12" x 18"		18	.444		43	27.50		70.50	91
5200	12" x 30"		12	.667		68.50	41.50		110	141
5220	16" x 16"		17	.471		74	29.50		103.50	128
5240	18" x 18"		16	.500		62	31		93	118
5260	18" x 36"		10	.800		171	50		221	266
5280	24" x 24"		11	.727		113	45.50		158.50	195
5300	24" x 36"		8	1		179	62.50		241.50	295
5320	24" x 48"	↓	6	1.333	↓	230	83		313	385
6000	For steel construction instead of aluminum, deduct					10%				
9000	Minimum labor/equipment charge	1 Shee	4	2	Job		125		125	195

23 37 15 – Air Outlets and Inlets, HVAC Louvers

23 37 15.40 HVAC Louvers		Crew	Daily Output	Labor-Hours	Unit	Material	Labor	Equipment	Total	Total Incl O&P
0010	**HVAC LOUVERS**									
0100	Aluminum, extruded, with screen, mill finish									
1002	Brick vent, see also Section 04 05 23.19									
1100	Standard, 4" deep, 8" wide, 5" high	1 Shee	24	.333	Ea.	37	21		58	73.50
1200	Modular, 4" deep, 7-3/4" wide, 5" high		24	.333		38.50	21		59.50	75
1300	Speed brick, 4" deep, 11-5/8" wide, 3-7/8" high		24	.333		38.50	21		59.50	75
1400	Fuel oil brick, 4" deep, 8" wide, 5" high		24	.333	↓	66.50	21		87.50	106
2000	Cooling tower and mechanical equip., screens, light weight		40	.200	S.F.	16.65	12.45		29.10	38
2020	Standard weight		35	.229		44	14.25		58.25	71
2500	Dual combination, automatic, intake or exhaust		20	.400		60.50	25		85.50	106
2520	Manual operation		20	.400		45	25		70	88.50
2540	Electric or pneumatic operation		20	.400	↓	45	25		70	88.50
2560	Motor, for electric or pneumatic	↓	14	.571	Ea.	520	35.50		555.50	625
3000	Fixed blade, continuous line									
3100	Mullion type, stormproof	1 Shee	28	.286	S.F.	45	17.80		62.80	77.50
3200	Stormproof	↓	28	.286	↓	45	17.80		62.80	77.50

23 37 Air Outlets and Inlets

23 37 15 – Air Outlets and Inlets, HVAC Louvers

23 37 15.40 HVAC Louvers		Crew	Daily Output	Labor-Hours	Unit	2020 Bare Costs Material	Labor	Equipment	Total	Total Incl O&P
3300	Vertical line	1 Shee	28	.286	S.F.	53.50	17.80		71.30	86.50
3500	For damper to use with above, add					50%	30%			
3520	Motor, for damper, electric or pneumatic	1 Shee	14	.571	Ea.	520	35.50		555.50	625
4000	Operating, 45°, manual, electric or pneumatic		24	.333	S.F.	60	21		81	98.50
4100	Motor, for electric or pneumatic		14	.571	Ea.	520	35.50		555.50	625
4200	Penthouse, roof		56	.143	S.F.	26.50	8.90		35.40	43.50
4300	Walls		40	.200		62.50	12.45		74.95	88.50
5000	Thinline, under 4" thick, fixed blade		40	.200		26	12.45		38.45	48
5010	Finishes, applied by mfr. at additional cost, available in colors									
5020	Prime coat only, add				S.F.	3.57			3.57	3.93
5040	Baked enamel finish coating, add					6.60			6.60	7.25
5060	Anodized finish, add					7.15			7.15	7.85
5080	Duranodic finish, add					12.95			12.95	14.25
5100	Fluoropolymer finish coating, add					20.50			20.50	22.50
9000	Stainless steel, fixed blade, continuous line									
9010	Hospital grade									
9110	20 ga.	1 Shee	22	.364	S.F.	43	22.50		65.50	83
9980	For small orders (under 10 pieces), add				"	25%				

23 37 23 – HVAC Gravity Ventilators

23 37 23.10 HVAC Gravity Air Ventilators		Crew	Daily Output	Labor-Hours	Unit	Material	Labor	Equipment	Total	Total Incl O&P
0010	**HVAC GRAVITY AIR VENTILATORS**, Includes base									
1280	Rotary ventilators, wind driven, galvanized									
1300	4" neck diameter	Q-9	20	.800	Ea.	52	45		97	128
1320	5" neck diameter		18	.889		55	50		105	139
1340	6" neck diameter		16	1		52.50	56		108.50	146
1360	8" neck diameter		14	1.143		61	64		125	167
1380	10" neck diameter		12	1.333		67	75		142	191
1400	12" neck diameter		10	1.600		73	89.50		162.50	221
1420	14" neck diameter		10	1.600		124	89.50		213.50	277
1440	16" neck diameter		9	1.778		144	99.50		243.50	315
1460	18" neck diameter		9	1.778		174	99.50		273.50	350
1480	20" neck diameter		8	2		209	112		321	405
1500	24" neck diameter		8	2		320	112		432	530
1520	30" neck diameter		7	2.286		375	128		503	615
1540	36" neck diameter		6	2.667		605	150		755	900
1600	For aluminum, add					30%				
2000	Stationary, gravity, syphon, galvanized									
2100	3" neck diameter, 40 CFM	Q-9	24	.667	Ea.	30.50	37.50		68	92.50
2120	4" neck diameter, 50 CFM		20	.800		32.50	45		77.50	107
2140	5" neck diameter, 58 CFM		18	.889		35.50	50		85.50	117
2160	6" neck diameter, 66 CFM		16	1		37	56		93	129
2180	7" neck diameter, 86 CFM		15	1.067		50	60		110	149
2200	8" neck diameter, 110 CFM		14	1.143		51	64		115	156
2220	10" neck diameter, 140 CFM		12	1.333		68.50	75		143.50	193
2240	12" neck diameter, 160 CFM		10	1.600		85.50	89.50		175	235
2260	14" neck diameter, 250 CFM		10	1.600		136	89.50		225.50	291
2280	16" neck diameter, 380 CFM		9	1.778		175	99.50		274.50	350
2300	18" neck diameter, 500 CFM		9	1.778		197	99.50		296.50	370
2320	20" neck diameter, 625 CFM		8	2		235	112		347	435
2340	24" neck diameter, 900 CFM		8	2		290	112		402	495
2360	30" neck diameter, 1,375 CFM		7	2.286		405	128		533	645
2380	36" neck diameter, 2,000 CFM		6	2.667		405	150		555	680

23 37 23 – HVAC Gravity Ventilators

23 37 23.10 HVAC Gravity Air Ventilators		Crew	Daily Output	Labor-Hours	Unit	Material	Labor	Equipment	Total	Total Incl O&P
						2020 Bare Costs				
2400	42" neck diameter, 3,000 CFM	Q-9	4	4	Ea.	605	224		829	1,025
2500	For aluminum, add					30%				
2520	For stainless steel, add					60%				
3000	Rotating chimney cap, galvanized, 4" neck diameter	Q-9	20	.800		21	45		66	94
3020	5" neck diameter		18	.889		21	50		71	102
3040	6" neck diameter		16	1		21	56		77	112
3060	7" neck diameter		15	1.067		27.50	60		87.50	124
3080	8" neck diameter		14	1.143		27.50	64		91.50	130
3100	10" neck diameter		12	1.333		36	75		111	157
3600	Stationary chimney rain cap, galvanized, 3" neck diameter		24	.667		3.50	37.50		41	62.50
3620	4" neck diameter		20	.800		3.83	45		48.83	74.50
3640	6" neck diameter		16	1		4.55	56		60.55	93
3680	8" neck diameter		14	1.143		5.70	64		69.70	106
3700	10" neck diameter		12	1.333		6.90	75		81.90	125
3720	12" neck diameter		10	1.600		10.15	89.50		99.65	152
3740	14" neck diameter		10	1.600		16.10	89.50		105.60	159
3760	16" neck diameter		9	1.778		16.95	99.50		116.45	175
4200	Stationary mushroom, aluminum, 16" orifice diameter		10	1.600		720	89.50		809.50	930
4220	26" orifice diameter		6.15	2.602		1,050	146		1,196	1,400
4230	30" orifice diameter		5.71	2.802		1,550	157		1,707	1,950
4240	38" orifice diameter		5	3.200		2,225	179		2,404	2,725
4250	42" orifice diameter		4.70	3.404		2,950	191		3,141	3,550
4260	50" orifice diameter	↓	4.44	3.604	↓	3,500	202		3,702	4,175
5000	Relief vent									
5500	Rectangular, aluminum, galvanized curb									
5510	intake/exhaust, 0.033" SP									
5580	500 CFM, 12" x 12"	Q-9	8.60	1.860	Ea.	790	104		894	1,025
5600	600 CFM, 12" x 16"		8	2		880	112		992	1,150
5620	750 CFM, 12" x 20"		7.20	2.222		945	125		1,070	1,225
5640	1,000 CFM, 12" x 24"		6.60	2.424		985	136		1,121	1,300
5660	1,500 CFM, 12" x 36"		5.80	2.759		1,325	155		1,480	1,725
5680	3,000 CFM, 20" x 42"		4	4		1,725	224		1,949	2,250
5700	5,000 CFM, 20" x 72"		3	5.333		2,500	299		2,799	3,225
5720	6,000 CFM, 24" x 72"		2.30	6.957		2,700	390		3,090	3,575
5740	10,000 CFM, 48" x 60"		1.80	8.889		3,425	500		3,925	4,525
5760	12,000 CFM, 48" x 72"		1.60	10		4,025	560		4,585	5,300
5780	13,750 CFM, 60" x 66"		1.40	11.429		4,450	640		5,090	5,900
5800	15,000 CFM, 60" x 72"		1.30	12.308		4,775	690		5,465	6,325
5820	18,000 CFM, 72" x 72"	↓	1.20	13.333	↓	5,400	750		6,150	7,125
5880	Size is throat area, volume is at 500 fpm									
7000	Note: sizes based on exhaust. Intake, with 0.125" SP									
7100	loss, approximately twice listed capacity.									
9000	Minimum labor/equipment charge	1 Plum	2	4	Job		258		258	400

23 38 Ventilation Hoods

23 38 13 – Commercial-Kitchen Hoods

23 38 13.10 Hood and Ventilation Equipment		Crew	Daily Output	Labor-Hours	Unit	Material	2020 Bare Costs Labor	Equipment	Total	Total Incl O&P
0010	**HOOD AND VENTILATION EQUIPMENT**									
2970	Exhaust hood, sst, gutter on all sides, 4′ x 4′ x 2′	1 Carp	1.80	4.444	Ea.	4,075	236		4,311	4,850
2980	4′ x 4′ x 7′	"	1.60	5		6,750	266		7,016	7,850
7800	Vent hood, wall canopy with fire protection, 30″	L-3A	9	1.333		390	76.50		466.50	550
7810	Without fire protection, 36″		10	1.200		430	69		499	580
7820	Island canopy with fire protection, 30″		7	1.714		760	98.50		858.50	990
7830	Without fire protection, 36″		8	1.500		785	86.50		871.50	995
7840	Back shelf with fire protection, 30″		11	1.091		410	63		473	550
7850	Without fire protection, black, 36″		12	1		450	57.50		507.50	590
7852	Without fire protection, stainless steel, 36″		12	1		470	57.50		527.50	605
7860	Range hood & CO_2 system, 30″	1 Carp	2.50	3.200		4,325	170		4,495	5,025
7950	Hood fire protection system, electric stove	Q-1	3	5.333		2,025	310		2,335	2,700
7952	Hood fire protection system, gas stove	"	3	5.333		2,350	310		2,660	3,050

23 41 Particulate Air Filtration

23 41 13 – Panel Air Filters

23 41 13.10 Panel Type Air Filters

		Crew	Daily Output	Labor-Hours	Unit	Material	Labor	Equipment	Total	Total Incl O&P
0010	**PANEL TYPE AIR FILTERS**									
2950	Mechanical media filtration units									
3000	High efficiency type, with frame, non-supported G				MCFM	41			41	45
3100	Supported type G				"	52.50			52.50	58
5500	Throwaway glass or paper media type, 12″ x 36″ x 1″				Ea.	2.48			2.48	2.73

23 41 16 – Renewable-Media Air Filters

23 41 16.10 Disposable Media Air Filters

		Crew	Daily Output	Labor-Hours	Unit	Material	Labor	Equipment	Total	Total Incl O&P
0010	**DISPOSABLE MEDIA AIR FILTERS**									
5000	Renewable disposable roll				C.S.F.	5.25			5.25	5.75

23 41 19 – Washable Air Filters

23 41 19.10 Permanent Air Filters

		Crew	Daily Output	Labor-Hours	Unit	Material	Labor	Equipment	Total	Total Incl O&P
0010	**PERMANENT AIR FILTERS**									
4500	Permanent washable G				MCFM	25			25	27.50

23 41 23 – Extended Surface Filters

23 41 23.10 Expanded Surface Filters

		Crew	Daily Output	Labor-Hours	Unit	Material	Labor	Equipment	Total	Total Incl O&P
0010	**EXPANDED SURFACE FILTERS**									
4000	Medium efficiency, extended surface G				MCFM	6.65			6.65	7.35

23 42 Gas-Phase Air Filtration

23 42 13 – Activated-Carbon Air Filtration

23 42 13.10 Charcoal Type Air Filtration

		Crew	Daily Output	Labor-Hours	Unit	Material	Labor	Equipment	Total	Total Incl O&P
0010	**CHARCOAL TYPE AIR FILTRATION**									
0050	Activated charcoal type, full flow				MCFM	650			650	715
0060	Full flow, impregnated media 12″ deep					225			225	248
0070	HEPA filter & frame for field erection					410			410	450
0080	HEPA filter-diffuser, ceiling install.					340			340	375

23 43 Electronic Air Cleaners

23 43 13 – Washable Electronic Air Cleaners

23 43 13.10 Electronic Air Cleaners		Crew	Daily Output	Labor-Hours	Unit	Material	Labor	Equipment	Total	Total Incl O&P
						2020 Bare Costs				
0010	ELECTRONIC AIR CLEANERS									
2000	Electronic air cleaner, duct mounted									
2150	1,000 CFM	1 Shee	4	2	Ea.	425	125		550	665
2200	1,200 CFM		3.80	2.105		570	131		701	830
2250	1,400 CFM	↓	3.60	2.222	↓	600	138		738	875

23 51 Breechings, Chimneys, and Stacks

23 51 13 – Draft Control Devices

23 51 13.13 Draft-Induction Fans		Crew	Daily Output	Labor-Hours	Unit	Material	Labor	Equipment	Total	Total Incl O&P
0010	DRAFT-INDUCTION FANS									
1000	Breeching installation									
1800	Hot gas, 600°F, variable pitch pulley and motor									
1840	6" diam. inlet, 1/4 HP, 1 phase, 400 CFM	Q-9	6	2.667	Ea.	1,725	150		1,875	2,100
1860	8" diam. inlet, 1/4 HP, 1 phase, 1,120 CFM		4	4		2,150	224		2,374	2,700
1870	9" diam. inlet, 3/4 HP, 1 phase, 1,440 CFM		3.60	4.444		2,650	249		2,899	3,300
1880	10" diam. inlet, 3/4 HP, 1 phase, 2,000 CFM		3.30	4.848		2,625	272		2,897	3,325
1900	12" diam. inlet, 3/4 HP, 3 phase, 2,960 CFM		3	5.333		2,925	299		3,224	3,700
1910	14" diam. inlet, 1 HP, 3 phase, 4,160 CFM		2.60	6.154		2,900	345		3,245	3,725
1920	16" diam. inlet, 2 HP, 3 phase, 5,500 CFM		2.30	6.957		3,200	390		3,590	4,125
1950	20" diam. inlet, 3 HP, 3 phase, 9,760 CFM		1.50	10.667		4,675	600		5,275	6,075
1960	22" diam. inlet, 5 HP, 3 phase, 13,360 CFM		1	16		8,500	895		9,395	10,800
1980	24" diam. inlet, 7-1/2 HP, 3 phase, 17,760 CFM	↓	.80	20		9,625	1,125		10,750	12,400
2300	For multi-blade damper at fan inlet, add				↓	20%				
3600	Chimney-top installation									
3700	6" size	1 Shee	8	1	Ea.	1,425	62.50		1,487.50	1,650
3740	8" size		7	1.143		1,425	71		1,496	1,675
3750	10" size		6.50	1.231		2,075	76.50		2,151.50	2,400
3780	13" size	↓	6	1.333		1,925	83		2,008	2,225
3880	For speed control switch, add					138			138	152
3920	For thermal fan control, add				↓	138			138	152
5500	Flue blade style damper for draft control,									
5510	locking quadrant blade									
5550	8" size	Q-9	8	2	Ea.	470	112		582	695
5560	9" size		7.50	2.133		475	120		595	710
5570	10" size		7	2.286		490	128		618	735
5580	12" size		6.50	2.462		500	138		638	765
5590	14" size		6	2.667		490	150		640	775
5600	16" size		5.50	2.909		505	163		668	815
5610	18" size		5	3.200		530	179		709	865
5620	20" size		4.50	3.556		565	199		764	930
5630	22" size		4	4		570	224		794	975
5640	24" size		3.50	4.571		600	256		856	1,050
5650	27" size		3	5.333		630	299		929	1,175
5660	30" size		2.50	6.400		715	360		1,075	1,350
5670	32" size		2	8		750	450		1,200	1,525
5680	36" size	↓	1.50	10.667	↓	830	600		1,430	1,850

23 51 13.16 Vent Dampers		Crew	Daily Output	Labor-Hours	Unit	Material	Labor	Equipment	Total	Total Incl O&P
0010	VENT DAMPERS									
5000	Vent damper, bi-metal, gas, 3" diameter	Q-9	24	.667	Ea.	71	37.50		108.50	137
5010	4" diameter		24	.667		70	37.50		107.50	136
5020	5" diameter	↓	23	.696	↓	71	39		110	139

23 51 13 – Draft Control Devices

23 51 13.16 Vent Dampers		Crew	Daily Output	Labor-Hours	Unit	Material	2020 Bare Costs Labor	Equipment	Total	Total Incl O&P
5030	6″ diameter	Q-9	22	.727	Ea.	70	41		111	141
5040	7″ diameter		21	.762		72.50	42.50		115	147
5050	8″ diameter		20	.800		77.50	45		122.50	156
9000	Minimum labor/equipment charge	1 Shee	4	2	Job		125		125	195

23 51 13.19 Barometric Dampers		Crew	Daily Output	Labor-Hours	Unit	Material	Labor	Equipment	Total	Total Incl O&P
0010	**BAROMETRIC DAMPERS**									
1000	Barometric, gas fired system only, 6″ size for 5″ and 6″ pipes	1 Shee	20	.400	Ea.	105	25		130	155
1020	7″ size, for 6″ and 7″ pipes		19	.421		115	26		141	167
1040	8″ size, for 7″ and 8″ pipes		18	.444		148	27.50		175.50	207
1060	9″ size, for 8″ and 9″ pipes		16	.500		164	31		195	229
2000	All fuel, oil, oil/gas, coal									
2020	10″ for 9″ and 10″ pipes	1 Shee	15	.533	Ea.	255	33		288	335
2040	12″ for 11″ and 12″ pipes		15	.533		335	33		368	415
2060	14″ for 13″ and 14″ pipes		14	.571		435	35.50		470.50	530
2080	16″ for 15″ and 16″ pipes		13	.615		605	38.50		643.50	725
2100	18″ for 17″ and 18″ pipes		12	.667		805	41.50		846.50	950
2120	20″ for 19″ and 21″ pipes		10	.800		965	50		1,015	1,125
2140	24″ for 22″ and 25″ pipes	Q-9	12	1.333		1,175	75		1,250	1,425
2160	28″ for 26″ and 30″ pipes		10	1.600		1,475	89.50		1,564.50	1,750
2180	32″ for 31″ and 34″ pipes		8	2		1,875	112		1,987	2,225
3260	For thermal switch for above, add	1 Shee	24	.333		105	21		126	149

23 51 23 – Gas Vents

23 51 23.10 Gas Chimney Vents		Crew	Daily Output	Labor-Hours	Unit	Material	Labor	Equipment	Total	Total Incl O&P
0010	**GAS CHIMNEY VENTS**, Prefab metal, UL listed									
0020	Gas, double wall, galvanized steel									
0080	3″ diameter	Q-9	72	.222	V.L.F.	7.60	12.45		20.05	28
0100	4″ diameter		68	.235		9.35	13.20		22.55	31
0120	5″ diameter		64	.250		10.60	14		24.60	33.50
0140	6″ diameter		60	.267		12.65	14.95		27.60	37.50
0160	7″ diameter		56	.286		24.50	16		40.50	52
0180	8″ diameter		52	.308		26	17.25		43.25	55.50
0200	10″ diameter		48	.333		49.50	18.70		68.20	84
0220	12″ diameter		44	.364		58.50	20.50		79	96.50
0240	14″ diameter		42	.381		97.50	21.50		119	141
0260	16″ diameter		40	.400		140	22.50		162.50	189
0280	18″ diameter		38	.421		173	23.50		196.50	227
0300	20″ diameter	Q-10	36	.667		205	39		244	287
0320	22″ diameter		34	.706		261	41		302	350
0340	24″ diameter		32	.750		320	43.50		363.50	420
0600	For 4″, 5″ and 6″ oval, add					50%				
0650	Gas, double wall, galvanized steel, fittings									
0660	Elbow 45°, 3″ diameter	Q-9	36	.444	Ea.	13.45	25		38.45	54
0670	4″ diameter		34	.471		16.45	26.50		42.95	59.50
0680	5″ diameter		32	.500		19.05	28		47.05	65
0690	6″ diameter		30	.533		23.50	30		53.50	73
0700	7″ diameter		28	.571		38	32		70	92
0710	8″ diameter		26	.615		50	34.50		84.50	109
0720	10″ diameter		24	.667		106	37.50		143.50	176
0730	12″ diameter		22	.727		113	41		154	188
0740	14″ diameter		21	.762		178	42.50		220.50	263
0750	16″ diameter		20	.800		231	45		276	325
0760	18″ diameter		19	.842		305	47		352	410

23 51 Breechings, Chimneys, and Stacks

23 51 23 – Gas Vents

23 51 23.10 Gas Chimney Vents		Crew	Daily Output	Labor-Hours	Unit	2020 Bare Costs Material	Labor	Equipment	Total	Total Incl O&P
0770	20″ diameter	Q-10	18	1.333	Ea.	340	77.50		417.50	490
0780	22″ diameter		17	1.412		545	82		627	730
0790	24″ diameter		16	1.500		695	87		782	900
0950	Elbow 90°, adjustable, 3″ diameter	Q-9	36	.444		23	25		48	64.50
0960	4″ diameter		34	.471		27	26.50		53.50	71
0970	5″ diameter		32	.500		32.50	28		60.50	80
0980	6″ diameter		30	.533		39	30		69	90
0990	7″ diameter		28	.571		67	32		99	124
1010	8″ diameter		26	.615		68.50	34.50		103	129
1020	Wall thimble, 4 to 7″ adjustable, 3″ diameter		36	.444		14.40	25		39.40	55
1022	4″ diameter		34	.471		16.20	26.50		42.70	59.50
1024	5″ diameter		32	.500		19.40	28		47.40	65.50
1026	6″ diameter		30	.533		20	30		50	69
1028	7″ diameter		28	.571		41.50	32		73.50	96
1030	8″ diameter		26	.615		51	34.50		85.50	110
1040	Roof flashing, 3″ diameter		36	.444		7.85	25		32.85	47.50
1050	4″ diameter		34	.471		9.15	26.50		35.65	51.50
1060	5″ diameter		32	.500		27.50	28		55.50	74.50
1070	6″ diameter		30	.533		22	30		52	71
1080	7″ diameter		28	.571		29	32		61	81.50
1090	8″ diameter		26	.615		31	34.50		65.50	88
1100	10″ diameter		24	.667		41.50	37.50		79	104
1110	12″ diameter		22	.727		58	41		99	128
1120	14″ diameter		20	.800		132	45		177	216
1130	16″ diameter		18	.889		158	50		208	252
1140	18″ diameter		16	1		222	56		278	330
1150	20″ diameter	Q-10	18	1.333		288	77.50		365.50	435
1160	22″ diameter		14	1.714		360	99.50		459.50	550
1170	24″ diameter		12	2		415	116		531	635
1200	Tee, 3″ diameter	Q-9	27	.593		35.50	33		68.50	91
1210	4″ diameter		26	.615		38	34.50		72.50	95.50
1220	5″ diameter		25	.640		40	36		76	100
1230	6″ diameter		24	.667		45.50	37.50		83	109
1240	7″ diameter		23	.696		65	39		104	133
1250	8″ diameter		22	.727		72.50	41		113.50	144
1260	10″ diameter		21	.762		191	42.50		233.50	277
1270	12″ diameter		20	.800		195	45		240	286
1280	14″ diameter		18	.889		350	50		400	465
1290	16″ diameter		16	1		520	56		576	660
1300	18″ diameter		14	1.143		625	64		689	785
1310	20″ diameter	Q-10	17	1.412		855	82		937	1,075
1320	22″ diameter		13	1.846		1,100	107		1,207	1,375
1330	24″ diameter		12	2		1,275	116		1,391	1,575
1460	Tee cap, 3″ diameter	Q-9	45	.356		2.38	19.95		22.33	33.50
1470	4″ diameter		42	.381		2.57	21.50		24.07	36.50
1490	6″ diameter		37	.432		4.71	24.50		29.21	43
1510	8″ diameter		34	.471		8.65	26.50		35.15	51
1530	12″ diameter		30	.533		69	30		99	123
1550	16″ diameter		25	.640		74	36		110	138
1570	20″ diameter	Q-10	27	.889		97.50	51.50		149	188
1590	24″ diameter	″	21	1.143		200	66.50		266.50	325
1750	Top, 3″ diameter	Q-9	46	.348		17.75	19.50		37.25	50
1760	4″ diameter		44	.364		18.65	20.50		39.15	52.50

23 51 23 – Gas Vents

23 51 23.10 Gas Chimney Vents		Crew	Daily Output	Labor-Hours	Unit	Material	Labor	Equipment	Total	Total Incl O&P
						2020 Bare Costs				
1780	6″ diameter	Q-9	40	.400	Ea.	23.50	22.50		46	61
1800	8″ diameter		36	.444		63	25		88	109
1820	12″ diameter		32	.500		129	28		157	186
1840	16″ diameter		28	.571		260	32		292	335
1860	20″ diameter	Q-10	28	.857		555	50		605	690
1880	24″ diameter	"	20	1.200		1,050	70		1,120	1,250

23 51 26 – All-Fuel Vent Chimneys

23 51 26.10 All-Fuel Vent Chimneys, Press. Tight, Dbl. Wall		Crew	Daily Output	Labor-Hours	Unit	Material	Labor	Equipment	Total	Total Incl O&P
0010	**ALL-FUEL VENT CHIMNEYS, PRESSURE TIGHT, DOUBLE WALL**									
3200	All fuel, pressure tight, double wall, 1″ insulation, UL listed, 1,400°F.									
3210	304 stainless steel liner, aluminized steel outer jacket									
3220	6″ diameter	Q-9	60	.267	L.F.	60.50	14.95		75.45	90
3221	8″ diameter		52	.308		65	17.25		82.25	98.50
3222	10″ diameter		48	.333		72.50	18.70		91.20	110
3223	12″ diameter		44	.364		70.50	20.50		91	110
3224	14″ diameter		42	.381		90.50	21.50		112	133
3225	16″ diameter		40	.400		106	22.50		128.50	152
3226	18″ diameter		38	.421		114	23.50		137.50	163
3227	20″ diameter	Q-10	36	.667		136	39		175	211
3228	24″ diameter		32	.750		174	43.50		217.50	260
3229	28″ diameter		30	.800		200	46.50		246.50	293
3230	32″ diameter		27	.889		227	51.50		278.50	330
3231	36″ diameter		25	.960		251	56		307	365
3232	42″ diameter		22	1.091		285	63.50		348.50	415
3233	48″ diameter		19	1.263		335	73.50		408.50	485
3260	For 316 stainless steel liner, add					30%				
3280	All fuel, pressure tight, double wall fittings									
3284	304 stainless steel inner, aluminized steel jacket									
3288	Adjustable 20″/29″ section									
3292	6″ diameter	Q-9	30	.533	Ea.	204	30		234	271
3293	8″ diameter		26	.615		218	34.50		252.50	294
3294	10″ diameter		24	.667		245	37.50		282.50	330
3295	12″ diameter		22	.727		279	41		320	370
3296	14″ diameter		21	.762		295	42.50		337.50	390
3297	16″ diameter		20	.800		345	45		390	450
3298	18″ diameter		19	.842		395	47		442	510
3299	20″ diameter	Q-10	18	1.333		430	77.50		507.50	595
3300	24″ diameter		16	1.500		555	87		642	745
3301	28″ diameter		15	1.600		615	93		708	820
3302	32″ diameter		14	1.714		715	99.50		814.50	945
3303	36″ diameter		12	2		840	116		956	1,100
3304	42″ diameter		11	2.182		960	127		1,087	1,250
3305	48″ diameter		10	2.400		1,100	140		1,240	1,425
3350	Elbow 90° fixed									
3354	6″ diameter	Q-9	30	.533	Ea.	420	30		450	510
3355	8″ diameter		26	.615		475	34.50		509.50	575
3356	10″ diameter		24	.667		540	37.50		577.50	650
3357	12″ diameter		22	.727		610	41		651	735
3358	14″ diameter		21	.762		690	42.50		732.50	825
3359	16″ diameter		20	.800		780	45		825	930
3360	18″ diameter		19	.842		880	47		927	1,050
3361	20″ diameter	Q-10	18	1.333		995	77.50		1,072.50	1,225

23 51 26 – All-Fuel Vent Chimneys

23 51 26.10 All-Fuel Vent Chimneys, Press. Tight, Dbl. Wall		Crew	Daily Output	Labor-Hours	Unit	Material	Labor	Equipment	Total	Total Incl O&P
						2020 Bare Costs				
3362	24" diameter	Q-10	16	1.500	Ea.	1,275	87		1,362	1,525
3363	28" diameter		15	1.600		1,450	93		1,543	1,750
3364	32" diameter		14	1.714		1,975	99.50		2,074.50	2,300
3365	36" diameter		12	2		2,600	116		2,716	3,050
3366	42" diameter		11	2.182		3,125	127		3,252	3,650
3367	48" diameter		10	2.400		3,525	140		3,665	4,100
3380	For 316 stainless steel liner, add					30%				
3400	Elbow 45°									
3404	6" diameter	Q-9	30	.533	Ea.	212	30		242	280
3405	8" diameter		26	.615		238	34.50		272.50	315
3406	10" diameter		24	.667		271	37.50		308.50	360
3407	12" diameter		22	.727		310	41		351	405
3408	14" diameter		21	.762		350	42.50		392.50	450
3409	16" diameter		20	.800		395	45		440	500
3410	18" diameter		19	.842		445	47		492	565
3411	20" diameter	Q-10	18	1.333		515	77.50		592.50	690
3412	24" diameter		16	1.500		660	87		747	860
3413	28" diameter		15	1.600		740	93		833	955
3414	32" diameter		14	1.714		995	99.50		1,094.50	1,250
3415	36" diameter		12	2		1,325	116		1,441	1,650
3416	42" diameter		11	2.182		1,550	127		1,677	1,900
3417	48" diameter		10	2.400		1,750	140		1,890	2,150
3430	For 316 stainless steel liner, add					30%				
3450	Tee 90°									
3454	6" diameter	Q-9	24	.667	Ea.	252	37.50		289.50	335
3455	8" diameter		22	.727		275	41		316	365
3456	10" diameter		21	.762		310	42.50		352.50	405
3457	12" diameter		20	.800		360	45		405	465
3458	14" diameter		18	.889		405	50		455	530
3459	16" diameter		16	1		455	56		511	595
3460	18" diameter		14	1.143		535	64		599	690
3461	20" diameter	Q-10	17	1.412		610	82		692	800
3462	24" diameter		12	2		760	116		876	1,025
3463	28" diameter		11	2.182		920	127		1,047	1,225
3464	32" diameter		10	2.400		1,275	140		1,415	1,625
3465	36" diameter		9	2.667		1,350	155		1,505	1,750
3466	42" diameter		6	4		1,675	233		1,908	2,200
3467	48" diameter		5	4.800		1,925	279		2,204	2,525
3480	For tee cap, add					35%	20%			
3500	For 316 stainless steel liner, add					30%				
3520	Plate support, galvanized									
3524	6" diameter	Q-9	26	.615	Ea.	125	34.50		159.50	192
3525	8" diameter		22	.727		147	41		188	225
3526	10" diameter		20	.800		160	45		205	247
3527	12" diameter		18	.889		168	50		218	263
3528	14" diameter		17	.941		198	53		251	300
3529	16" diameter		16	1		209	56		265	320
3530	18" diameter		15	1.067		220	60		280	335
3531	20" diameter	Q-10	16	1.500		231	87		318	390
3532	24" diameter		14	1.714		240	99.50		339.50	420
3533	28" diameter		13	1.846		295	107		402	495
3534	32" diameter		12	2		360	116		476	575
3535	36" diameter		10	2.400		445	140		585	710

23 51 26 – All-Fuel Vent Chimneys

23 51 26.10 All-Fuel Vent Chimneys, Press. Tight, Dbl. Wall		Crew	Daily Output	Labor-Hours	Unit	Material	Labor	Equipment	Total	Total Incl O&P
						2020 Bare Costs				
3536	42" diameter	Q-10	9	2.667	Ea.	510	155		665	805
3537	48" diameter	↓	8	3	↓	585	174		759	920
3570	Bellows, lined									
3574	6" diameter	Q-9	30	.533	Ea.	1,350	30		1,380	1,525
3575	8" diameter		26	.615		1,400	34.50		1,434.50	1,600
3576	10" diameter		24	.667		1,425	37.50		1,462.50	1,625
3577	12" diameter		22	.727		1,475	41		1,516	1,700
3578	14" diameter		21	.762		1,525	42.50		1,567.50	1,750
3579	16" diameter		20	.800		1,575	45		1,620	1,825
3580	18" diameter	↓	19	.842		1,650	47		1,697	1,875
3581	20" diameter	Q-10	18	1.333		1,675	77.50		1,752.50	1,975
3582	24" diameter		16	1.500		1,975	87		2,062	2,300
3583	28" diameter		15	1.600		2,100	93		2,193	2,450
3584	32" diameter		14	1.714		2,350	99.50		2,449.50	2,725
3585	36" diameter		12	2		2,575	116		2,691	3,000
3586	42" diameter		11	2.182		2,850	127		2,977	3,325
3587	48" diameter	↓	10	2.400		3,350	140		3,490	3,900
3590	For all 316 stainless steel construction, add				↓	55%				
3600	Ventilated roof thimble, 304 stainless steel									
3620	6" diameter	Q-9	26	.615	Ea.	264	34.50		298.50	345
3624	8" diameter		22	.727		259	41		300	350
3625	10" diameter		20	.800		272	45		317	370
3626	12" diameter		18	.889		282	50		332	390
3627	14" diameter		17	.941		305	53		358	420
3628	16" diameter		16	1		325	56		381	450
3629	18" diameter	↓	15	1.067		350	60		410	480
3630	20" diameter	Q-10	16	1.500		370	87		457	545
3631	24" diameter		14	1.714		410	99.50		509.50	610
3632	28" diameter		13	1.846		435	107		542	645
3633	32" diameter		12	2		470	116		586	695
3634	36" diameter		10	2.400		505	140		645	775
3635	42" diameter		9	2.667		565	155		720	865
3636	48" diameter	↓	8	3	↓	620	174		794	960
3650	For 316 stainless steel, add					30%				
3670	Exit cone, 316 stainless steel only									
3674	6" diameter	Q-9	46	.348	Ea.	199	19.50		218.50	250
3675	8" diameter		42	.381		204	21.50		225.50	259
3676	10" diameter		40	.400		210	22.50		232.50	266
3677	12" diameter		38	.421		230	23.50		253.50	290
3678	14" diameter		37	.432		239	24.50		263.50	300
3679	16" diameter		36	.444		294	25		319	365
3680	18" diameter	↓	35	.457		320	25.50		345.50	395
3681	20" diameter	Q-10	28	.857		385	50		435	505
3682	24" diameter		26	.923		510	53.50		563.50	645
3683	28" diameter		25	.960		600	56		656	755
3684	32" diameter		24	1		745	58		803	910
3685	36" diameter		22	1.091		905	63.50		968.50	1,100
3686	42" diameter		21	1.143		1,050	66.50		1,116.50	1,250
3687	48" diameter	↓	20	1.200	↓	1,200	70		1,270	1,425
3720	Roof guide, 304 stainless steel									
3724	6" diameter	Q-9	25	.640	Ea.	93	36		129	158
3725	8" diameter		21	.762		108	42.50		150.50	186
3726	10" diameter	↓	19	.842	↓	119	47		166	204

23 51 26 – All-Fuel Vent Chimneys

23 51 26.10 All-Fuel Vent Chimneys, Press. Tight, Dbl. Wall		Crew	Daily Output	Labor-Hours	Unit	Material	2020 Bare Costs Labor	Equipment	Total	Total Incl O&P
3727	12" diameter	Q-9	17	.941	Ea.	123	53		176	218
3728	14" diameter		16	1		143	56		199	245
3729	16" diameter		15	1.067		151	60		211	260
3730	18" diameter		14	1.143		160	64		224	277
3731	20" diameter	Q-10	15	1.600		169	93		262	330
3732	24" diameter		13	1.846		176	107		283	360
3733	28" diameter		12	2		215	116		331	420
3734	32" diameter		11	2.182		264	127		391	490
3735	36" diameter		9	2.667		315	155		470	590
3736	42" diameter		8	3		370	174		544	680
3737	48" diameter		7	3.429		420	199		619	775
3750	For 316 stainless steel, add					30%				
3770	Rain cap with bird screen									
3774	6" diameter	Q-9	46	.348	Ea.	305	19.50		324.50	365
3775	8" diameter		42	.381		350	21.50		371.50	420
3776	10" diameter		40	.400		410	22.50		432.50	485
3777	12" diameter		38	.421		475	23.50		498.50	555
3778	14" diameter		37	.432		545	24.50		569.50	640
3779	16" diameter		36	.444		620	25		645	720
3780	18" diameter		35	.457		705	25.50		730.50	815
3781	20" diameter	Q-10	28	.857		800	50		850	960
3782	24" diameter		26	.923		965	53.50		1,018.50	1,125
3783	28" diameter		25	.960		1,150	56		1,206	1,350
3784	32" diameter		24	1		1,300	58		1,358	1,525
3785	36" diameter		22	1.091		1,525	63.50		1,588.50	1,800
3786	42" diameter		21	1.143		1,825	66.50		1,891.50	2,100
3787	48" diameter		20	1.200		2,075	70		2,145	2,375

23 51 26.30 All-Fuel Vent Chimneys, Double Wall, St. Stl.

		Crew	Daily Output	Labor-Hours	Unit	Material	Labor	Equipment	Total	Total Incl O&P
0010	**ALL-FUEL VENT CHIMNEYS, DOUBLE WALL, STAINLESS STEEL**									
7780	All fuel, pressure tight, double wall, 4" insulation, UL listed, 1,400°F.									
7790	304 stainless steel liner, aluminized steel outer jacket									
7800	6" diameter	Q-9	60	.267	V.L.F.	64.50	14.95		79.45	94.50
7804	8" diameter		52	.308		77	17.25		94.25	112
7806	10" diameter		48	.333		87.50	18.70		106.20	126
7808	12" diameter		44	.364		101	20.50		121.50	143
7810	14" diameter		42	.381		113	21.50		134.50	158
7880	For 316 stainless steel liner, add				L.F.	30%				
8000	All fuel, double wall, stainless steel fittings									
8010	Roof support, 6" diameter	Q-9	30	.533	Ea.	120	30		150	179
8030	8" diameter		26	.615		136	34.50		170.50	203
8040	10" diameter		24	.667		151	37.50		188.50	225
8050	12" diameter		22	.727		157	41		198	236
8060	14" diameter		21	.762		167	42.50		209.50	250
8100	Elbow 45°, 6" diameter		30	.533		252	30		282	325
8140	8" diameter		26	.615		286	34.50		320.50	370
8160	10" diameter		24	.667		330	37.50		367.50	425
8180	12" diameter		22	.727		375	41		416	475
8200	14" diameter		21	.762		410	42.50		452.50	515
8300	Insulated tee, 6" diameter		30	.533		296	30		326	370
8360	8" diameter		26	.615		330	34.50		364.50	420
8380	10" diameter		24	.667		350	37.50		387.50	445
8400	12" diameter		22	.727		420	41		461	525

23 51 Breechings, Chimneys, and Stacks

23 51 26 – All-Fuel Vent Chimneys

23 51 26.30 All-Fuel Vent Chimneys, Double Wall, St. Stl.		Crew	Daily Output	Labor-Hours	Unit	2020 Bare Costs Material	Labor	Equipment	Total	Total Incl O&P
8420	14″ diameter	Q-9	20	.800	Ea.	490	45		535	610
8500	Boot tee, 6″ diameter		28	.571		610	32		642	720
8520	8″ diameter		24	.667		670	37.50		707.50	795
8530	10″ diameter		22	.727		770	41		811	910
8540	12″ diameter		20	.800		905	45		950	1,075
8550	14″ diameter		18	.889		945	50		995	1,125
8600	Rain cap with bird screen, 6″ diameter		30	.533		297	30		327	370
8640	8″ diameter		26	.615		350	34.50		384.50	440
8660	10″ diameter		24	.667		410	37.50		447.50	510
8680	12″ diameter		22	.727		470	41		511	585
8700	14″ diameter		21	.762		535	42.50		577.50	655
8800	Flat roof flashing, 6″ diameter		30	.533		111	30		141	169
8840	8″ diameter		26	.615		121	34.50		155.50	187
8860	10″ diameter		24	.667		132	37.50		169.50	204
8880	12″ diameter		22	.727		145	41		186	224
8900	14″ diameter	↓	21	.762	↓	147	42.50		189.50	229

23 52 Heating Boilers

23 52 13 – Electric Boilers

23 52 13.10 Electric Boilers, ASME

		Crew	Daily Output	Labor-Hours	Unit	Material	Labor	Equipment	Total	Total Incl O&P
0010	**ELECTRIC BOILERS, ASME**, Standard controls and trim									
1000	Steam, 6 KW, 20.5 MBH	Q-19	1.20	20	Ea.	4,450	1,200		5,650	6,725
1040	9 KW, 30.7 MBH		1.20	20		4,400	1,200		5,600	6,700
1080	24 KW, 81.8 MBH		1.10	21.818		5,250	1,300		6,550	7,800
1160	60 KW, 205 MBH		1	24		7,250	1,425		8,675	10,200
1240	148 KW, 505 MBH		.65	36.923		10,500	2,200		12,700	14,900
1280	222 KW, 758 MBH		.55	43.636		24,000	2,600		26,600	30,400
1320	300 KW, 1,023 MBH		.40	60		26,600	3,575		30,175	34,800
1360	444 KW, 1,515 MBH	↓	.30	80		31,600	4,775		36,375	42,200
1400	592 KW, 2,020 MBH	Q-21	.34	94.118		36,000	5,775		41,775	48,500
1480	814 KW, 2,778 MBH		.25	128		40,100	7,825		47,925	56,000
1520	1,036 KW, 3,536 MBH		.20	160		48,600	9,800		58,400	68,500
1560	2,070 KW, 7,063 MBH		.18	178		69,500	10,900		80,400	93,500
1600	2,340 KW, 7,984 MBH	↓	.16	200		88,000	12,200		100,200	115,500
2000	Hot water, 7.5 KW, 25.6 MBH	Q-19	1.30	18.462		5,425	1,100		6,525	7,650
2040	30 KW, 102 MBH		1.20	20		5,850	1,200		7,050	8,275
2070	60 KW, 205 MBH		1.20	20		5,800	1,200		7,000	8,250
2100	90 KW, 307 MBH		1.10	21.818		6,125	1,300		7,425	8,775
2140	120 KW, 410 MBH		.90	26.667		6,875	1,600		8,475	10,000
2180	150 KW, 512 MBH		.65	36.923		7,925	2,200		10,125	12,100
2220	296 KW, 1,010 MBH		.55	43.636		15,300	2,600		17,900	20,900
2300	444 KW, 1,515 MBH	↓	.35	68.571		20,600	4,100		24,700	29,000
2340	518 KW, 1,768 MBH	Q-21	.44	72.727		22,900	4,450		27,350	32,000
2420	740 KW, 2,526 MBH		.39	82.051		27,100	5,025		32,125	37,600
2460	888 KW, 3,031 MBH		.37	86.486		31,400	5,300		36,700	42,800
2500	1,036 KW, 3,536 MBH		.34	94.118		34,700	5,775		40,475	47,100
2540	1,440 KW, 4,915 MBH		.32	100		42,400	6,125		48,525	56,000
2580	1,680 KW, 5,733 MBH		.30	107		51,500	6,525		58,025	66,500
2620	1,980 KW, 6,757 MBH		.28	114		58,000	7,000		65,000	74,500
2660	2,220 KW, 7,576 MBH		.26	123		66,500	7,525		74,025	85,000
2700	2,610 KW, 8,905 MBH	↓	.24	133	↓	74,000	8,175		82,175	94,000

23 52 Heating Boilers

23 52 13 – Electric Boilers

23 52 13.10 Electric Boilers, ASME			Crew	Daily Output	Labor-Hours	Unit	2020 Bare Costs Material	Labor	Equipment	Total	Total Incl O&P
2740	2,970 KW, 10,133 MBH		Q-21	.21	152	Ea.	79,500	9,325		88,825	101,500
2780	3,240 KW, 11,055 MBH			.18	178		86,000	10,900		96,900	112,000
2820	3,600 KW, 12,283 MBH			.16	200		96,500	12,200		108,700	125,000
9000	Minimum labor/equipment charge		Q-20	1	20	Job		1,150		1,150	1,775

23 52 16 – Condensing Boilers

23 52 16.24 Condensing Boilers			Crew	Daily Output	Labor-Hours	Unit	Material	Labor	Equipment	Total	Total Incl O&P
0010	**CONDENSING BOILERS**, Cast iron, high efficiency										
0020	Packaged with standard controls, circulator and trim										
0030	Intermittent (spark) pilot, natural or LP gas										
0040	Hot water, DOE MBH output (AFUE %)										
0100	42 MBH (84.0%)	G	Q-5	1.80	8.889	Ea.	1,625	525		2,150	2,600
0120	57 MBH (84.3%)	G		1.60	10		1,775	590		2,365	2,850
0140	85 MBH (84.0%)	G		1.40	11.429		1,975	675		2,650	3,225
0160	112 MBH (83.7%)	G		1.20	13.333		2,225	785		3,010	3,675
0180	140 MBH (83.3%)	G	Q-6	1.60	15		2,500	920		3,420	4,175
0200	167 MBH (83.0%)	G		1.40	17.143		2,825	1,050		3,875	4,750
0220	194 MBH (82.7%)	G		1.20	20		3,125	1,225		4,350	5,325

23 52 19 – Pulse Combustion Boilers

23 52 19.20 Pulse Type Combustion Boilers			Crew	Daily Output	Labor-Hours	Unit	Material	Labor	Equipment	Total	Total Incl O&P
0010	**PULSE TYPE COMBUSTION BOILERS**, High efficiency										
7990	Special feature gas fired boilers										
8000	Pulse combustion, standard controls/trim										
8010	Hot water, DOE MBH output (AFUE %)										
8030	71 MBH (95.2%)		Q-5	1.60	10	Ea.	3,975	590		4,565	5,250
8050	94 MBH (95.3%)	G		1.40	11.429		4,250	675		4,925	5,725
8080	139 MBH (95.6%)	G		1.20	13.333		4,950	785		5,735	6,675
8090	207 MBH (95.4%)			1.16	13.793		5,600	815		6,415	7,425
8120	270 MBH (96.4%)			1.12	14.286		7,750	845		8,595	9,825
8130	365 MBH (91.7%)			1.07	14.953		8,750	880		9,630	11,000

23 52 23 – Cast-Iron Boilers

23 52 23.20 Gas-Fired Boilers		Crew	Daily Output	Labor-Hours	Unit	Material	Labor	Equipment	Total	Total Incl O&P
0010	**GAS-FIRED BOILERS**, Natural or propane, standard controls, packaged									
1000	Cast iron, with insulated jacket									
2000	Steam, gross output, 81 MBH	Q-7	1.40	22.857	Ea.	2,750	1,425		4,175	5,225
2020	102 MBH		1.30	24.615		2,625	1,525		4,150	5,275
2040	122 MBH		1	32		3,250	2,000		5,250	6,675
2060	163 MBH		.90	35.556		3,400	2,225		5,625	7,175
2080	203 MBH		.90	35.556		3,950	2,225		6,175	7,775
2100	240 MBH		.85	37.647		4,025	2,350		6,375	8,050
2120	280 MBH		.80	40		4,525	2,500		7,025	8,825
2140	320 MBH		.70	45.714		5,200	2,850		8,050	10,100
2160	360 MBH		.63	51.200		5,475	3,200		8,675	11,000
2180	400 MBH		.56	56.838		5,825	3,550		9,375	11,900
2200	440 MBH		.51	62.500		6,250	3,900		10,150	12,900
2220	544 MBH		.45	71.588		9,000	4,475		13,475	16,800
2240	765 MBH		.43	74.419		11,300	4,650		15,950	19,600
2260	892 MBH		.38	84.211		13,300	5,250		18,550	22,800
2280	1,275 MBH		.34	94.118		16,700	5,875		22,575	27,500
2300	1,530 MBH		.32	100		19,000	6,250		25,250	30,600
2320	1,875 MBH		.30	107		25,200	6,650		31,850	38,100
2340	2,170 MBH		.26	122		27,000	7,625		34,625	41,500

23 52 23 – Cast-Iron Boilers

23 52 23.20 Gas-Fired Boilers		Crew	Daily Output	Labor-Hours	Unit	Material	2020 Bare Costs Labor	Equipment	Total	Total Incl O&P
2360	2,675 MBH	Q-7	.20	163	Ea.	28,800	10,200		39,000	47,500
2380	3,060 MBH		.19	172		30,000	10,700		40,700	49,600
2400	3,570 MBH		.18	182		35,800	11,300		47,100	57,000
2420	4,207 MBH		.16	205		39,200	12,800		52,000	63,000
2440	4,720 MBH		.15	208		63,500	13,000		76,500	90,000
2460	5,660 MBH		.15	221		71,500	13,800		85,300	100,500
2480	6,100 MBH		.13	246		86,500	15,400		101,900	119,000
2500	6,390 MBH		.12	267		90,500	16,600		107,100	125,000
2520	6,680 MBH		.11	291		92,500	18,200		110,700	130,000
2540	6,970 MBH		.10	320		97,000	20,000		117,000	137,500
3000	Hot water, gross output, 80 MBH		1.46	21.918		1,975	1,375		3,350	4,300
3020	100 MBH		1.35	23.704		2,575	1,475		4,050	5,100
3040	122 MBH		1.10	29.091		2,825	1,825		4,650	5,900
3060	163 MBH		1	32		3,275	2,000		5,275	6,700
3080	203 MBH		1	32		3,950	2,000		5,950	7,450
3100	240 MBH		.95	33.684		3,825	2,100		5,925	7,475
3120	280 MBH		.90	35.556		4,375	2,225		6,600	8,250
3140	320 MBH		.80	40		4,275	2,500		6,775	8,550
3160	360 MBH		.71	45.070		4,900	2,800		7,700	9,725
3180	400 MBH		.64	50		5,175	3,125		8,300	10,500
3200	440 MBH		.58	54.983		5,550	3,425		8,975	11,400
3220	544 MBH		.51	62.992		8,600	3,925		12,525	15,600
3240	765 MBH		.46	70.022		10,800	4,375		15,175	18,700
3260	1,088 MBH		.40	80		13,900	5,000		18,900	23,000
3280	1,275 MBH		.36	89.888		16,300	5,600		21,900	26,700
3300	1,530 MBH		.31	105		18,700	6,550		25,250	30,600
3320	2,000 MBH		.26	125		23,300	7,800		31,100	37,700
3340	2,312 MBH		.22	148		27,600	9,250		36,850	44,700
3360	2,856 MBH		.20	160		30,100	9,975		40,075	48,500
3380	3,264 MBH		.18	180		31,900	11,200		43,100	52,500
3400	3,996 MBH		.16	195		34,900	12,200		47,100	57,000
3420	4,488 MBH		.15	211		38,000	13,100		51,100	62,000
3440	4,720 MBH		.15	221		69,000	13,800		82,800	97,500
3460	5,520 MBH		.14	229		100,000	14,300		114,300	132,500
3480	6,100 MBH		.13	250		139,500	15,600		155,100	177,500
3500	6,390 MBH		.11	286		117,000	17,800		134,800	156,500
3520	6,680 MBH		.10	311		122,000	19,400		141,400	164,000
3540	6,970 MBH		.09	360		128,000	22,400		150,400	175,500
7000	For tankless water heater, add					10%				
7050	For additional zone valves up to 312 MBH, add					199			199	219
9900	Minimum labor/equipment charge	Q-6	1	24	Job		1,475		1,475	2,275

23 52 23.30 Gas/Oil Fired Boilers		Crew	Daily Output	Labor-Hours	Unit	Material	Labor	Equipment	Total	Total Incl O&P
0010	**GAS/OIL FIRED BOILERS**, Combination with burners and controls, packaged									
1000	Cast iron with insulated jacket									
2000	Steam, gross output, 720 MBH	Q-7	.43	74.074	Ea.	15,200	4,625		19,825	23,900
2020	810 MBH		.38	83.990		15,200	5,250		20,450	24,800
2040	1,084 MBH		.34	93.023		17,500	5,800		23,300	28,300
2060	1,360 MBH		.33	98.160		20,000	6,125		26,125	31,500
2080	1,600 MBH		.30	107		20,700	6,675		27,375	33,100
2100	2,040 MBH		.25	130		24,100	8,125		32,225	39,100
2120	2,450 MBH		.21	156		25,400	9,750		35,150	43,000
2140	2,700 MBH		.19	166		26,800	10,300		37,100	45,500

23 52 23.30 Gas/Oil Fired Boilers		Crew	Daily Output	Labor-Hours	Unit	2020 Bare Costs Material	Labor	Equipment	Total	Total Incl O&P
2160	3,000 MBH	Q-7	.18	176	Ea.	29,000	11,000		40,000	48,900
2180	3,270 MBH		.17	184		34,000	11,500		45,500	55,000
2200	3,770 MBH		.17	192		69,500	12,000		81,500	95,000
2220	4,070 MBH		.16	200		73,500	12,500		86,000	100,000
2240	4,650 MBH		.15	211		77,500	13,100		90,600	105,500
2260	5,230 MBH		.14	224		85,500	14,000		99,500	115,500
2280	5,520 MBH		.14	235		93,500	14,700		108,200	125,000
2300	5,810 MBH		.13	248		95,000	15,500		110,500	128,500
2320	6,100 MBH		.12	260		97,000	16,200		113,200	131,500
2340	6,390 MBH		.11	296		100,000	18,500		118,500	138,500
2360	6,680 MBH		.10	320		99,500	20,000		119,500	140,500
2380	6,970 MBH		.09	372		106,000	23,200		129,200	152,500
2900	Hot water, gross output									
2910	200 MBH	Q-6	.62	39.024	Ea.	11,400	2,400		13,800	16,200
2920	300 MBH		.49	49.080		11,400	3,000		14,400	17,200
2930	400 MBH		.41	57.971		13,300	3,550		16,850	20,200
2940	500 MBH		.36	67.039		14,300	4,100		18,400	22,200
3000	584 MBH	Q-7	.44	72.072		14,100	4,500		18,600	22,600
3020	876 MBH		.41	79.012		21,000	4,925		25,925	30,700
3040	1,168 MBH		.31	104		28,500	6,475		34,975	41,300
3060	1,460 MBH		.28	113		33,200	7,050		40,250	47,400
3080	2,044 MBH		.26	122		39,900	7,625		47,525	55,500
3100	2,628 MBH		.21	150		47,800	9,375		57,175	67,000
3120	3,210 MBH		.18	175		52,500	10,900		63,400	74,500
3140	3,796 MBH		.17	186		58,500	11,600		70,100	82,500
3160	4,088 MBH		.16	195		66,500	12,200		78,700	92,500
3180	4,672 MBH		.16	204		72,500	12,700		85,200	99,500
3200	5,256 MBH		.15	218		82,500	13,600		96,100	111,500
3220	6,000 MBH, 179 BHP		.13	256		116,500	16,000		132,500	153,000
3240	7,130 MBH, 213 BHP		.08	386		122,000	24,100		146,100	171,500
3260	9,800 MBH, 286 BHP		.06	533		140,500	33,300		173,800	206,000
3280	10,900 MBH, 325.6 BHP		.05	593		170,500	37,000		207,500	244,500
3290	12,200 MBH, 364.5 BHP		.05	667		186,000	41,600		227,600	269,500
3300	13,500 MBH, 403.3 BHP		.04	727		206,000	45,400		251,400	296,500
23 52 23.40 Oil-Fired Boilers										
0010	**OIL-FIRED BOILERS**, Standard controls, flame retention burner, packaged									
1000	Cast iron, with insulated flush jacket									
2000	Steam, gross output, 109 MBH	Q-7	1.20	26.667	Ea.	2,375	1,675		4,050	5,175
2020	144 MBH		1.10	29.091		2,650	1,825		4,475	5,725
2040	173 MBH		1	32		3,000	2,000		5,000	6,400
2060	207 MBH		.90	35.556		3,225	2,225		5,450	6,950
2080	236 MBH		.85	37.647		3,600	2,350		5,950	7,600
2100	300 MBH		.70	45.714		4,600	2,850		7,450	9,450
2120	480 MBH		.50	64		6,550	4,000		10,550	13,400
2140	665 MBH		.45	71.111		8,500	4,425		12,925	16,200
2160	794 MBH		.41	78.049		9,450	4,875		14,325	17,900
2180	1,084 MBH		.38	85.106		10,800	5,300		16,100	20,100
2200	1,360 MBH		.33	98.160		13,300	6,125		19,425	24,100
2220	1,600 MBH		.26	122		17,700	7,625		25,325	31,300
2240	2,175 MBH		.24	134		17,400	8,350		25,750	32,000
2260	2,480 MBH		.21	156		20,300	9,750		30,050	37,400
2280	3,000 MBH		.19	170		23,400	10,600		34,000	42,100

23 52 Heating Boilers

23 52 23 – Cast-Iron Boilers

23 52 23.40 Oil-Fired Boilers		Crew	Daily Output	Labor-Hours	Unit	Material	2020 Bare Costs Labor	Equipment	Total	Total Incl O&P
2300	3,550 MBH	Q-7	.17	187	Ea.	27,800	11,700		39,500	48,700
2320	3,820 MBH		.16	200		37,200	12,500		49,700	60,000
2340	4,360 MBH		.15	215		40,800	13,400		54,200	65,500
2360	4,940 MBH		.14	225		68,000	14,100		82,100	96,000
2380	5,520 MBH		.14	235		80,000	14,700		94,700	110,500
2400	6,100 MBH		.13	256		91,500	16,000		107,500	125,000
2420	6,390 MBH		.11	291		96,500	18,200		114,700	134,000
2440	6,680 MBH		.10	314		99,000	19,600		118,600	139,500
2460	6,970 MBH		.09	364		102,500	22,700		125,200	148,000
3000	Hot water, same price as steam									
4000	For tankless coil in smaller sizes, add				Ea.	15%				

23 52 23.60 Solid-Fuel Boilers		Crew	Daily Output	Labor-Hours	Unit	Material	Labor	Equipment	Total	Total Incl O&P
0010	**SOLID-FUEL BOILERS**									
3000	Stoker fired (coal/wood/biomass) cast iron with flush jacket and									
3400	insulation, steam or water, gross output, 1,280 MBH	Q-6	.36	66.667	Ea.	173,500	4,075		177,575	197,500
3420	1,460 MBH		.30	80		180,000	4,900		184,900	205,500
3440	1,640 MBH		.28	85.714		193,000	5,250		198,250	220,500
3460	1,820 MBH		.26	92.308		209,000	5,650		214,650	238,000
3480	2,000 MBH		.25	96		225,500	5,875		231,375	257,000
3500	2,360 MBH		.23	104		226,500	6,375		232,875	259,000
3540	2,725 MBH		.20	120		234,000	7,350		241,350	269,000
3800	2,950 MBH	Q-7	.16	200		247,500	12,500		260,000	292,000
3820	3,210 MBH		.15	213		261,000	13,300		274,300	308,000
3840	3,480 MBH		.14	229		263,000	14,300		277,300	311,500
3860	3,745 MBH		.14	229		261,000	14,300		275,300	309,000
3880	4,000 MBH		.13	246		263,000	15,400		278,400	313,500
3900	4,200 MBH		.13	246		265,500	15,400		280,900	316,000
3920	4,400 MBH		.12	267		267,500	16,600		284,100	320,000
3940	4,600 MBH		.12	267		271,000	16,600		287,600	323,500

23 52 26 – Steel Boilers

23 52 26.40 Oil-Fired Boilers		Crew	Daily Output	Labor-Hours	Unit	Material	Labor	Equipment	Total	Total Incl O&P
0010	**OIL-FIRED BOILERS**, Standard controls, flame retention burner									
5000	Steel, with insulated flush jacket									
7000	Hot water, gross output, 103 MBH	Q-6	1.60	15	Ea.	2,000	920		2,920	3,625
7020	122 MBH		1.45	16.506		2,150	1,000		3,150	3,925
7040	137 MBH		1.36	17.595		2,325	1,075		3,400	4,225
7060	168 MBH		1.30	18.405		2,275	1,125		3,400	4,275
7080	225 MBH		1.22	19.704		4,000	1,200		5,200	6,275
7100	315 MBH		.96	25.105		5,075	1,525		6,600	7,950
7120	420 MBH		.70	34.483		5,925	2,100		8,025	9,775
7140	525 MBH		.57	42.403		7,650	2,600		10,250	12,400
7180	735 MBH		.48	50.104		8,125	3,075		11,200	13,700
7220	1,050 MBH		.37	65.753		21,300	4,025		25,325	29,600
7280	2,310 MBH		.21	115		28,200	7,025		35,225	41,900
7320	3,150 MBH		.13	185		45,400	11,300		56,700	67,500
7340	For tankless coil in steam or hot water, add					7%				
9000	Minimum labor/equipment charge	Q-6	1.75	13.714	Job		840		840	1,300

23 52 26.70 Packaged Water Tube Boilers		Crew	Daily Output	Labor-Hours	Unit	Material	Labor	Equipment	Total	Total Incl O&P
0010	**PACKAGED WATER TUBE BOILERS**									
2000	Packaged water tube, #2 oil, steam or hot water, gross output									
2040	1,200 MBH	Q-7	.50	64	Ea.	28,300	4,000		32,300	37,400
2060	1,600 MBH		.40	80		40,100	5,000		45,100	52,000

23 52 Heating Boilers

23 52 26 – Steel Boilers

23 52 26.70 Packaged Water Tube Boilers		Crew	Daily Output	Labor-Hours	Unit	Material	Labor	Equipment	Total	Total Incl O&P
						2020 Bare Costs				
2080	2,400 MBH	Q-7	.30	107	Ea.	54,000	6,650		60,650	70,000
2100	3,200 MBH		.25	128		61,500	7,975		69,475	80,500
2120	4,800 MBH		.20	160		80,500	9,975		90,475	104,000
2200	Gas fired									
2204	200 MBH	Q-6	1.50	16	Ea.	8,625	980		9,605	11,000
2208	275 MBH		1.40	17.143		9,025	1,050		10,075	11,600
2212	360 MBH		1.10	21.818		9,300	1,325		10,625	12,300
2216	520 MBH		.65	36.923		8,375	2,250		10,625	12,700
2220	600 MBH		.60	40		10,700	2,450		13,150	15,600
2224	720 MBH		.55	43.636		10,300	2,675		12,975	15,400
2228	960 MBH		.48	50		18,700	3,050		21,750	25,300
2232	1,220 MBH	Q-7	.50	64		19,700	4,000		23,700	27,800
2236	1,440 MBH		.45	71.111		20,600	4,425		25,025	29,600
2240	1,680 MBH		.40	80		32,700	5,000		37,700	43,700
2244	1,920 MBH		.35	91.429		29,800	5,700		35,500	41,600
2248	2,160 MBH		.33	96.970		33,500	6,050		39,550	46,200
2252	2,400 MBH		.30	107		37,600	6,650		44,250	51,500

23 52 28 – Swimming Pool Boilers

23 52 28.10 Swimming Pool Heaters

		Crew	Daily Output	Labor-Hours	Unit	Material	Labor	Equipment	Total	Total Incl O&P
0010	**SWIMMING POOL HEATERS**, Not including wiring, external									
0020	piping, base or pad									
0160	Gas fired, input, 155 MBH	Q-6	1.50	16	Ea.	1,850	980		2,830	3,550
0200	199 MBH		1	24		1,975	1,475		3,450	4,450
0220	250 MBH		.70	34.286		2,175	2,100		4,275	5,625
0240	300 MBH		.60	40		2,275	2,450		4,725	6,275
0260	399 MBH		.50	48		2,550	2,925		5,475	7,350
0280	500 MBH		.40	60		8,500	3,675		12,175	15,000
0300	650 MBH		.35	68.571		9,050	4,200		13,250	16,400
0320	750 MBH		.33	72.727		9,875	4,450		14,325	17,800
0360	990 MBH		.22	109		13,300	6,675		19,975	24,900
0370	1,260 MBH		.21	114		17,700	7,000		24,700	30,300
0380	1,440 MBH		.19	126		19,000	7,725		26,725	32,800
0400	1,800 MBH		.14	171		21,000	10,500		31,500	39,300
0410	2,070 MBH		.13	185		24,800	11,300		36,100	44,800
2000	Electric, 12 KW, 4,800 gallon pool	Q-19	3	8		2,200	480		2,680	3,175
2020	15 KW, 7,200 gallon pool		2.80	8.571		2,225	510		2,735	3,250
2040	24 KW, 9,600 gallon pool		2.40	10		2,575	600		3,175	3,775
2060	30 KW, 12,000 gallon pool		2	12		2,650	715		3,365	4,025
2080	36 KW, 14,400 gallon pool		1.60	15		3,075	895		3,970	4,750
2100	57 KW, 24,000 gallon pool		1.20	20		3,875	1,200		5,075	6,100
9000	To select pool heater: 12 BTUH x S.F. pool area									
9010	X temperature differential = required output									
9050	For electric, KW = gallons x 2.5 divided by 1,000									
9100	For family home type pool, double the									
9110	Rated gallon capacity = 1/2°F rise per hour									

23 52 39 – Fire-Tube Boilers

23 52 39.13 Scotch Marine Boilers

		Crew	Daily Output	Labor-Hours	Unit	Material	Labor	Equipment	Total	Total Incl O&P
0010	**SCOTCH MARINE BOILERS**									
1000	Packaged fire tube, #2 oil, gross output									
1006	15 psi steam									
1020	3,348 MBH, 100 HP	Q-7	.21	152	Ea.	105,500	9,500		115,000	130,500
1040	6,696 MBH, 200 HP		.14	224		157,500	14,000		171,500	194,500

23 52 Heating Boilers

23 52 39 – Fire-Tube Boilers

23 52 39.13 Scotch Marine Boilers

23 52 39.13 Scotch Marine Boilers		Crew	Daily Output	Labor-Hours	Unit	2020 Bare Costs Material	Labor	Equipment	Total	Total Incl O&P
1060	10,044 MBH, 300 HP	Q-7	.13	252	Ea.	181,500	15,700		197,200	224,000
1080	16,740 MBH, 500 HP		.08	381		246,500	23,800		270,300	308,500
1100	23,435 MBH, 700 HP		.07	485		302,000	30,300		332,300	379,500
1102	26,780 MBH, 800 HP		.07	485		330,000	30,300		360,300	410,000
1104	30,125 MBH, 900 HP		.06	500		602,000	31,200		633,200	710,000
1106	33,475 MBH, 1,000 HP		.06	525		637,000	32,700		669,700	751,000
1107	36,825 MBH, 1,100 HP		.06	552		683,500	34,400		717,900	805,000
1108	40,170 MBH, 1,200 HP		.06	582		717,000	36,300		753,300	844,500
1110	46,865 MBH, 1,400 HP		.05	604		755,000	37,700		792,700	889,000
1112	53,560 MBH, 1,600 HP		.05	627		820,500	39,200		859,700	963,000
1114	60,250 MBH, 1,800 HP		.05	667		847,500	41,600		889,100	997,000
1116	66,950 MBH, 2,000 HP		.05	696		897,500	43,400		940,900	1,054,500
1118	73,650 MBH, 2,200 HP		.04	744		970,000	46,400		1,016,400	1,138,500
1120	To fire #6, add		.83	38.554		18,300	2,400		20,700	23,800
1140	To fire #6, and gas, add		.42	76.190		26,900	4,750		31,650	37,000
1142	6,696 MBH, 200 HP		.14	224		162,500	14,000		176,500	200,500
1143	10,044 MBH, 300 HP		.13	252		200,000	15,700		215,700	244,500
1144	16,740 MBH, 500 HP		.08	381		262,500	23,800		286,300	325,500
1145	23,435 MBH, 700 HP		.07	485		316,000	30,300		346,300	394,500
1146	30,125 MBH, 900 HP		.06	500		555,000	31,200		586,200	658,500
1160	For high pressure, add					29,900			29,900	32,800
1180	For duplex package feed system									
1200	To 3,348 MBH boiler, add	Q-7	.54	59.259	Ea.	14,500	3,700		18,200	21,600
1220	To 6,696 MBH boiler, add		.41	78.049		17,100	4,875		21,975	26,400
1240	To 10,044 MBH boiler, add		.38	84.211		19,700	5,250		24,950	29,700
1260	To 16,740 MBH boiler, add		.28	114		22,400	7,125		29,525	35,700
1280	To 23,435 MBH boiler, add		.25	128		26,500	7,975		34,475	41,500

23 52 84 – Boiler Blowdown

23 52 84.10 Boiler Blowdown Systems

23 52 84.10 Boiler Blowdown Systems		Crew	Daily Output	Labor-Hours	Unit	Material	Labor	Equipment	Total	Total Incl O&P
0010	**BOILER BLOWDOWN SYSTEMS**									
1010	Boiler blowdown, auto/manual to 2,000 MBH	Q-5	3.75	4.267	Ea.	6,975	252		7,227	8,075
1020	7,300 MBH	"	3	5.333	"	7,525	315		7,840	8,775

23 52 88 – Burners

23 52 88.10 Replacement Type Burners

23 52 88.10 Replacement Type Burners		Crew	Daily Output	Labor-Hours	Unit	Material	Labor	Equipment	Total	Total Incl O&P
0010	**REPLACEMENT TYPE BURNERS**									
0990	Residential, conversion, gas fired, LP or natural									
1000	Gun type, atmospheric input 50 to 225 MBH	Q-1	2.50	6.400	Ea.	1,025	370		1,395	1,700
1020	100 to 400 MBH		2	8		1,725	465		2,190	2,625
1040	300 to 1,000 MBH		1.70	9.412		5,125	545		5,670	6,500
2000	Commercial and industrial, gas/oil, input									
2050	400 MBH	Q-1	1.50	10.667	Ea.	4,400	620		5,020	5,800
2090	670 MBH		1.40	11.429		4,425	665		5,090	5,900
2140	1,155 MBH		1.30	12.308		4,925	715		5,640	6,525
2200	1,800 MBH		1.20	13.333		4,950	775		5,725	6,650
2260	3,000 MBH		1.10	14.545		5,125	845		5,970	6,950
2320	4,100 MBH		1	16		6,850	930		7,780	8,975
3000	Flame retention oil fired assembly, input									
3020	.50 to 2.25 GPH	Q-1	2.40	6.667	Ea.	360	385		745	995
3040	2.0 to 5.0 GPH		2	8		335	465		800	1,075
3060	3.0 to 7.0 GPH		1.80	8.889		625	515		1,140	1,475
3080	6.0 to 12.0 GPH		1.60	10		1,000	580		1,580	2,000
4600	Gas safety, shut off valve, 3/4" threaded	1 Stpi	20	.400		190	26		216	249

23 52 Heating Boilers

23 52 88 – Burners

23 52 88.10 Replacement Type Burners		Crew	Daily Output	Labor-Hours	Unit	Material	2020 Bare Costs Labor	Equipment	Total	Total Incl O&P
4610	1" threaded	1 Stpi	19	.421	Ea.	183	27.50		210.50	244
4620	1-1/4" threaded		15	.533		206	35		241	281
4630	1-1/2" threaded		13	.615		223	40.50		263.50	310
4640	2" threaded		11	.727		250	47.50		297.50	350
4650	2-1/2" threaded	Q-1	15	1.067		286	62		348	410
4660	3" threaded		13	1.231		395	71.50		466.50	545
4670	4" flanged		3	5.333		2,800	310		3,110	3,550
4680	6" flanged	Q-2	3	8		6,250	480		6,730	7,625

23 54 Furnaces

23 54 13 – Electric-Resistance Furnaces

23 54 13.10 Electric Furnaces		Crew	Daily Output	Labor-Hours	Unit	Material	Labor	Equipment	Total	Total Incl O&P
0010	**ELECTRIC FURNACES**, Hot air, blowers, std. controls									
0011	not including gas, oil or flue piping									
1000	Electric, UL listed									
1070	10.2 MBH	Q-20	4.40	4.545	Ea.	350	260		610	790
1080	17.1 MBH		4.60	4.348		430	248		678	860
1100	34.1 MBH		4.40	4.545		535	260		795	995
1120	51.6 MBH		4.20	4.762		590	272		862	1,075
1140	68.3 MBH		4	5		610	286		896	1,125
1160	85.3 MBH		3.80	5.263		650	300		950	1,175

23 54 16 – Fuel-Fired Furnaces

23 54 16.13 Gas-Fired Furnaces		Crew	Daily Output	Labor-Hours	Unit	Material	Labor	Equipment	Total	Total Incl O&P
0010	**GAS-FIRED FURNACES**									
3000	Gas, AGA certified, upflow, direct drive models									
3020	45 MBH input	Q-9	4	4	Ea.	700	224		924	1,125
3040	60 MBH input		3.80	4.211		730	236		966	1,175
3060	75 MBH input		3.60	4.444		835	249		1,084	1,300
3100	100 MBH input		3.20	5		825	280		1,105	1,350
3120	125 MBH input		3	5.333		855	299		1,154	1,400
3130	150 MBH input		2.80	5.714		1,750	320		2,070	2,425
3140	200 MBH input		2.60	6.154		3,525	345		3,870	4,425
3160	300 MBH input		2.30	6.957		4,125	390		4,515	5,125
3180	400 MBH input		2	8		5,500	450		5,950	6,750

23 54 16.14 Condensing Furnaces		Crew	Daily Output	Labor-Hours	Unit	Material	Labor	Equipment	Total	Total Incl O&P
0010	**CONDENSING FURNACES**, High efficiency									
0020	Oil fired, packaged, complete									
0030	Upflow									
0040	Output @ 95% A.F.U.E.									
0100	49 MBH @ 1,000 CFM	Q-9	3.70	4.324	Ea.	7,375	243		7,618	8,475
0110	73.5 MBH @ 2,000 CFM		3.60	4.444		7,825	249		8,074	9,025
0120	96 MBH @ 2,000 CFM		3.40	4.706		7,825	264		8,089	9,050
0130	115.6 MBH @ 2,000 CFM		3.40	4.706		7,825	264		8,089	9,050
0140	147 MBH @ 2,000 CFM		3.30	4.848		18,400	272		18,672	20,700
0150	192 MBH @ 4,000 CFM		2.60	6.154		19,000	345		19,345	21,400
0170	231.5 MBH @ 4,000 CFM		2.30	6.957		19,000	390		19,390	21,500
0260	For variable speed motor, add					785			785	865
0270	Note: Also available in horizontal, counterflow and lowboy configurations.									

23 54 16 – Fuel-Fired Furnaces

23 54 16.16 Oil-Fired Furnaces		Crew	Daily Output	Labor-Hours	Unit	Material	2020 Bare Costs Labor	Equipment	Total	Total Incl O&P
0010	OIL-FIRED FURNACES									
6000	Oil, UL listed, atomizing gun type burner									
6020	56 MBH output	Q-9	3.60	4.444	Ea.	3,275	249		3,524	4,000
6030	84 MBH output		3.50	4.571		3,400	256		3,656	4,150
6040	95 MBH output		3.40	4.706		3,400	264		3,664	4,150
6060	134 MBH output		3.20	5		3,325	280		3,605	4,125
6080	151 MBH output		3	5.333		3,700	299		3,999	4,550
6100	200 MBH input		2.60	6.154		4,175	345		4,520	5,150
6120	300 MBH input		2.30	6.957		4,375	390		4,765	5,400
6140	400 MBH input		2	8		5,000	450		5,450	6,200
9000	Minimum labor/equipment charge		2.75	5.818	Job		325		325	510

23 54 24 – Furnace Components for Cooling

23 54 24.10 Furnace Components and Combinations		Crew	Daily Output	Labor-Hours	Unit	Material	Labor	Equipment	Total	Total Incl O&P
0010	FURNACE COMPONENTS AND COMBINATIONS									
0080	Coils, A.C. evaporator, for gas or oil furnaces									
0090	Add-on, with holding charge									
0100	Upflow									
0120	1-1/2 ton cooling	Q-5	4	4	Ea.	244	236		480	635
0130	2 ton cooling		3.70	4.324		305	255		560	730
0140	3 ton cooling		3.30	4.848		430	286		716	910
0150	4 ton cooling		3	5.333		570	315		885	1,100
0160	5 ton cooling		2.70	5.926		615	350		965	1,225
0300	Downflow									
0330	2-1/2 ton cooling	Q-5	3	5.333	Ea.	365	315		680	885
0340	3-1/2 ton cooling		2.60	6.154		495	365		860	1,100
0350	5 ton cooling		2.20	7.273		510	430		940	1,225
0600	Horizontal									
0630	2 ton cooling	Q-5	3.90	4.103	Ea.	440	242		682	860
0640	3 ton cooling		3.50	4.571		450	270		720	910
0650	4 ton cooling		3.20	5		540	295		835	1,050
0660	5 ton cooling		2.90	5.517		540	325		865	1,100
2000	Cased evaporator coils for air handlers									
2100	1-1/2 ton cooling	Q-5	4.40	3.636	Ea.	310	215		525	675
2110	2 ton cooling		4.10	3.902		375	230		605	765
2120	2-1/2 ton cooling		3.90	4.103		400	242		642	815
2130	3 ton cooling		3.70	4.324		470	255		725	910
2140	3-1/2 ton cooling		3.50	4.571		600	270		870	1,075
2150	4 ton cooling		3.20	5		630	295		925	1,150
2160	5 ton cooling		2.90	5.517		625	325		950	1,200
3010	Air handler, modular									
3100	With cased evaporator cooling coil									
3120	1-1/2 ton cooling	Q-5	3.80	4.211	Ea.	1,050	248		1,298	1,525
3130	2 ton cooling		3.50	4.571		1,100	270		1,370	1,650
3140	2-1/2 ton cooling		3.30	4.848		1,175	286		1,461	1,750
3150	3 ton cooling		3.10	5.161		1,450	305		1,755	2,075
3160	3-1/2 ton cooling		2.90	5.517		1,700	325		2,025	2,375
3170	4 ton cooling		2.50	6.400		1,650	380		2,030	2,400
3180	5 ton cooling		2.10	7.619		2,125	450		2,575	3,025
3500	With no cooling coil									
3520	1-1/2 ton coil size	Q-5	12	1.333	Ea.	1,025	78.50		1,103.50	1,250
3530	2 ton coil size		10	1.600		1,100	94.50		1,194.50	1,350
3540	2-1/2 ton coil size		10	1.600		1,075	94.50		1,169.50	1,325

23 54 Furnaces

23 54 24 – Furnace Components for Cooling

23 54 24.10 Furnace Components and Combinations		Crew	Daily Output	Labor-Hours	Unit	Material	2020 Bare Costs Labor	Equipment	Total	Total Incl O&P
3554	3 ton coil size	Q-5	9	1.778	Ea.	1,075	105		1,180	1,325
3560	3-1/2 ton coil size		9	1.778		1,150	105		1,255	1,400
3570	4 ton coil size		8.50	1.882		1,275	111		1,386	1,575
3580	5 ton coil size		8	2		1,575	118		1,693	1,925
4000	With heater									
4120	5 kW, 17.1 MBH	Q-5	16	1	Ea.	485	59		544	625
4130	7.5 kW, 25.6 MBH		15.60	1.026		830	60.50		890.50	1,000
4140	10 kW, 34.2 MBH		15.20	1.053		1,225	62		1,287	1,450
4150	12.5 kW, 42.7 MBH		14.80	1.081		1,375	64		1,439	1,625
4160	15 kW, 51.2 MBH		14.40	1.111		1,825	65.50		1,890.50	2,100
4170	25 kW, 85.4 MBH		14	1.143		2,500	67.50		2,567.50	2,850
4180	30 kW, 102 MBH		13	1.231		2,900	72.50		2,972.50	3,275

23 55 Fuel-Fired Heaters

23 55 13 – Fuel-Fired Duct Heaters

23 55 13.16 Gas-Fired Duct Heaters

		Crew	Daily Output	Labor-Hours	Unit	Material	Labor	Equipment	Total	Total Incl O&P
0010	**GAS-FIRED DUCT HEATERS**, Includes burner, controls, stainless steel									
0020	heat exchanger. Gas fired, electric ignition									
0030	Indoor installation									
0080	100 MBH output	Q-5	5	3.200	Ea.	2,975	189		3,164	3,575
0100	120 MBH output		4	4		3,275	236		3,511	3,975
0130	200 MBH output		2.70	5.926		4,225	350		4,575	5,175
0140	240 MBH output		2.30	6.957		5,250	410		5,660	6,400
0160	280 MBH output		2	8		4,775	470		5,245	5,975
0180	320 MBH output		1.60	10		5,225	590		5,815	6,650
0300	For powered venter and adapter, add					535			535	585
0502	For required flue pipe, see Section 23 51 23.10									
1000	Outdoor installation, with power venter									
1020	75 MBH output	Q-5	4	4	Ea.	3,550	236		3,786	4,275
1040	94 MBH output		4	4		3,775	236		4,011	4,525
1060	120 MBH output		4	4		4,075	236		4,311	4,875
1080	157 MBH output		3.50	4.571		4,275	270		4,545	5,125
1100	187 MBH output		3	5.333		5,500	315		5,815	6,525
1120	225 MBH output		2.50	6.400		5,725	380		6,105	6,875
1140	300 MBH output		1.80	8.889		6,375	525		6,900	7,800
1160	375 MBH output		1.60	10		6,800	590		7,390	8,375
1180	450 MBH output		1.40	11.429		8,300	675		8,975	10,200
1200	600 MBH output		1	16		10,500	945		11,445	13,000
1300	Aluminized exchanger, subtract					15%				
1500	For two stage gas valve, add					870			870	955

23 55 23 – Gas-Fired Radiant Heaters

23 55 23.10 Infrared Type Heating Units

		Crew	Daily Output	Labor-Hours	Unit	Material	Labor	Equipment	Total	Total Incl O&P
0010	**INFRARED TYPE HEATING UNITS**									
0020	Gas fired, unvented, electric ignition, 100% shutoff.									
0030	Piping and wiring not included									
0100	Input, 30 MBH	Q-5	6	2.667	Ea.	1,050	157		1,207	1,425
0120	45 MBH		5	3.200		995	189		1,184	1,400
0140	50 MBH		4.50	3.556		1,000	210		1,210	1,425
0160	60 MBH		4	4		1,025	236		1,261	1,500
0180	75 MBH		3	5.333		1,025	315		1,340	1,600

23 55 Fuel-Fired Heaters

23 55 23 – Gas-Fired Radiant Heaters

23 55 23.10 Infrared Type Heating Units		Crew	Daily Output	Labor-Hours	Unit	Material	Labor	Equipment	Total	Total Incl O&P
						2020 Bare Costs				
0200	90 MBH	Q-5	2.50	6.400	Ea.	1,050	380		1,430	1,725
0220	105 MBH		2	8		1,100	470		1,570	1,950
0240	120 MBH		2	8		1,625	470		2,095	2,525
1000	Gas fired, vented, electric ignition, tubular									
1020	Piping and wiring not included, 20′ to 80′ lengths									
1030	Single stage, input, 60 MBH	Q-6	4.50	5.333	Ea.	1,550	325		1,875	2,200
1040	80 MBH		3.90	6.154		1,550	375		1,925	2,275
1050	100 MBH		3.40	7.059		1,550	430		1,980	2,375
1060	125 MBH		2.90	8.276		1,525	505		2,030	2,450
1070	150 MBH		2.70	8.889		1,525	545		2,070	2,525
1080	170 MBH		2.50	9.600		1,525	585		2,110	2,575
1090	200 MBH		2.20	10.909		1,725	665		2,390	2,925
1100	Note: Final pricing may vary due to									
1110	tube length and configuration package selected									
1130	Two stage, input, 60 MBH high, 45 MBH low	Q-6	4.50	5.333	Ea.	1,900	325		2,225	2,575
1140	80 MBH high, 60 MBH low		3.90	6.154		1,900	375		2,275	2,650
1150	100 MBH high, 65 MBH low		3.40	7.059		1,900	430		2,330	2,750
1160	125 MBH high, 95 MBH low		2.90	8.276		1,925	505		2,430	2,875
1170	150 MBH high, 100 MBH low		2.70	8.889		1,925	545		2,470	2,950
1180	170 MBH high, 125 MBH low		2.50	9.600		2,125	585		2,710	3,250
1190	200 MBH high, 150 MBH low		2.20	10.909		2,125	665		2,790	3,375
1220	Note: Final pricing may vary due to									
1230	tube length and configuration package selected									

23 55 33 – Fuel-Fired Unit Heaters

23 55 33.13 Oil-Fired Unit Heaters

Line	Description	Crew	Daily Output	Labor-Hours	Unit	Material	Labor	Equipment	Total	Total Incl O&P
0010	**OIL-FIRED UNIT HEATERS**, Cabinet, grilles, fan, ctrl., burner, no piping									
6000	Oil fired, suspension mounted, 94 MBH output	Q-5	4	4	Ea.	4,875	236		5,111	5,725
6040	140 MBH output		3	5.333		5,200	315		5,515	6,200
6060	184 MBH output		3	5.333		5,600	315		5,915	6,625

23 55 33.16 Gas-Fired Unit Heaters

Line	Description	Crew	Daily Output	Labor-Hours	Unit	Material	Labor	Equipment	Total	Total Incl O&P
0010	**GAS-FIRED UNIT HEATERS**, Cabinet, grilles, fan, ctrls., burner, no piping									
0022	thermostat, no piping. For flue see Section 23 51 23.10									
1000	Gas fired, floor mounted									
1100	60 MBH output	Q-5	10	1.600	Ea.	765	94.50		859.50	990
1120	80 MBH output		9	1.778		775	105		880	1,025
1140	100 MBH output		8	2		845	118		963	1,100
1160	120 MBH output		7	2.286		970	135		1,105	1,275
1180	180 MBH output		6	2.667		1,200	157		1,357	1,575
1500	Rooftop mounted, power vent, stainless steel exchanger									
1520	75 MBH input	Q-6	4	6	Ea.	5,925	365		6,290	7,075
1540	100 MBH input		3.60	6.667		6,175	410		6,585	7,425
1560	125 MBH input		3.30	7.273		6,450	445		6,895	7,800
1580	150 MBH input		3	8		6,950	490		7,440	8,375
1600	175 MBH input		2.60	9.231		7,125	565		7,690	8,700
1620	225 MBH input		2.30	10.435		8,100	640		8,740	9,900
1640	300 MBH input		1.90	12.632		9,350	775		10,125	11,500
1660	350 MBH input		1.40	17.143		10,000	1,050		11,050	12,600
1680	450 MBH input		1.20	20		10,700	1,225		11,925	13,700
1720	700 MBH input		.80	30		13,700	1,825		15,525	18,000
1760	1,200 MBH input		.30	80		13,800	4,900		18,700	22,800
1900	For aluminized steel exchanger, subtract					10%				
2000	Suspension mounted, propeller fan, 20 MBH output	Q-5	8.50	1.882		1,725	111		1,836	2,075

23 55 Fuel-Fired Heaters

23 55 33 – Fuel-Fired Unit Heaters

23 55 33.16 Gas-Fired Unit Heaters		Crew	Daily Output	Labor-Hours	Unit	Material	2020 Bare Costs Labor	Equipment	Total	Total Incl O&P
2020	40 MBH output	Q-5	7.50	2.133	Ea.	1,650	126		1,776	2,025
2040	60 MBH output		7	2.286		1,850	135		1,985	2,225
2060	80 MBH output		6	2.667		2,000	157		2,157	2,450
2080	100 MBH output		5.50	2.909		2,150	172		2,322	2,625
2100	130 MBH output		5	3.200		2,375	189		2,564	2,900
2120	140 MBH output		4.50	3.556		2,575	210		2,785	3,175
2140	160 MBH output		4	4		2,600	236		2,836	3,225
2160	180 MBH output		3.50	4.571		3,150	270		3,420	3,900
2180	200 MBH output		3	5.333		3,100	315		3,415	3,875
2200	240 MBH output		2.70	5.926		3,400	350		3,750	4,300
2220	280 MBH output		2.30	6.957		3,775	410		4,185	4,775
2240	320 MBH output		2	8		4,400	470		4,870	5,575
2500	For powered venter and adapter, add					400			400	440
3000	Suspension mounted, blower type, 40 MBH output	Q-5	6.80	2.353		810	139		949	1,100
3020	60 MBH output		6.60	2.424		935	143		1,078	1,250
3040	84 MBH output		5.80	2.759		875	163		1,038	1,225
3060	104 MBH output		5.20	3.077		945	182		1,127	1,325
3080	140 MBH output		4.30	3.721		1,075	220		1,295	1,525
3100	180 MBH output		3.30	4.848		1,500	286		1,786	2,100
3120	240 MBH output		2.50	6.400		1,675	380		2,055	2,400
3140	280 MBH output		2	8		1,475	470		1,945	2,350
4000	Suspension mounted, sealed combustion system,									
4020	Aluminized steel exchanger, powered vent									
4040	100 MBH output	Q-5	5	3.200	Ea.	3,125	189		3,314	3,725
4060	120 MBH output		4.70	3.404		3,725	201		3,926	4,400
4080	160 MBH output		3.70	4.324		4,350	255		4,605	5,175
4100	200 MBH output		2.90	5.517		5,250	325		5,575	6,275
4120	240 MBH output		2.50	6.400		5,700	380		6,080	6,850
4140	320 MBH output		1.70	9.412		6,825	555		7,380	8,375
5000	Wall furnace, 17.5 MBH output		6	2.667		710	157		867	1,025
5020	24 MBH output		5	3.200		805	189		994	1,175
5040	35 MBH output		4	4		1,150	236		1,386	1,625
9000	Minimum labor/equipment charge		3.50	4.571	Job		270		270	415

23 56 Solar Energy Heating Equipment

23 56 16 – Packaged Solar Heating Equipment

23 56 16.40 Solar Heating Systems		Crew	Daily Output	Labor-Hours	Unit	Material	Labor	Equipment	Total	Total Incl O&P
0010	**SOLAR HEATING SYSTEMS**									
0020	System/package prices, not including connecting									
0030	pipe, insulation, or special heating/plumbing fixtures									
0152	For solar ultraviolet pipe insulation see Section 22 07 19.10									
0500	Hot water, standard package, low temperature									
0540	1 collector, circulator, fittings, 65 gal. tank G	Q-1	.50	32	Ea.	4,400	1,850		6,250	7,700
0580	2 collectors, circulator, fittings, 120 gal. tank G		.40	40		5,750	2,325		8,075	9,900
0620	3 collectors, circulator, fittings, 120 gal. tank G		.34	47.059		7,850	2,725		10,575	12,900
0700	Medium temperature package									
0720	1 collector, circulator, fittings, 80 gal. tank G	Q-1	.50	32	Ea.	5,450	1,850		7,300	8,850
0740	2 collectors, circulator, fittings, 120 gal. tank G		.40	40		7,400	2,325		9,725	11,700
0780	3 collectors, circulator, fittings, 120 gal. tank G		.30	53.333		8,250	3,100		11,350	13,900
0980	For each additional 120 gal. tank, add G					1,775			1,775	1,975

23 56 19.50 Solar Heating Ancillary			Crew	Daily Output	Labor-Hours	Unit	2020 Bare Costs Material	Labor	Equipment	Total	Total Incl O&P
0010	**SOLAR HEATING ANCILLARY**										
2300	Circulators, air										
2310	Blowers										
2330	100-300 S.F. system, 1/10 HP	G	Q-9	16	1	Ea.	266	56		322	380
2340	300-500 S.F. system, 1/5 HP	G		15	1.067		355	60		415	485
2350	Two speed, 100-300 S.F., 1/10 HP	G		14	1.143		147	64		211	262
2400	Reversible fan, 20″ diameter, 2 speed	G		18	.889		116	50		166	206
2550	Booster fan 6″ diameter, 120 CFM	G		16	1		38	56		94	130
2570	6″ diameter, 225 CFM	G		16	1		46.50	56		102.50	139
2580	8″ diameter, 150 CFM	G		16	1		42.50	56		98.50	135
2590	8″ diameter, 310 CFM	G		14	1.143		65.50	64		129.50	172
2600	8″ diameter, 425 CFM	G		14	1.143		73.50	64		137.50	181
2650	Rheostat	G		32	.500		15.95	28		43.95	61.50
2660	Shutter/damper	G		12	1.333		58.50	75		133.50	182
2670	Shutter motor	G		16	1		146	56		202	249
2800	Circulators, liquid, 1/25 HP, 5.3 GPM	G	Q-1	14	1.143		205	66.50		271.50	325
2820	1/20 HP, 17 GPM	G		12	1.333		284	77.50		361.50	430
2850	1/20 HP, 17 GPM, stainless steel	G		12	1.333		262	77.50		339.50	410
2870	1/12 HP, 30 GPM	G		10	1.600		360	93		453	540
3000	Collector panels, air with aluminum absorber plate										
3010	Wall or roof mount										
3040	Flat black, plastic glazing										
3080	4′ x 8′	G	Q-9	6	2.667	Ea.	685	150		835	990
3100	4′ x 10′	G		5	3.200	"	840	179		1,019	1,200
3200	Flush roof mount, 10′ to 16′ x 22″ wide	G		96	.167	L.F.	165	9.35		174.35	196
3210	Manifold, by L.F. width of collectors	G		160	.100	"	161	5.60		166.60	187
3300	Collector panels, liquid with copper absorber plate										
3320	Black chrome, tempered glass glazing										
3330	Alum. frame, 4′ x 8′, 5/32″ single glazing	G	Q-1	9.50	1.684	Ea.	1,075	97.50		1,172.50	1,325
3390	Alum. frame, 4′ x 10′, 5/32″ single glazing	G		6	2.667		1,225	155		1,380	1,600
3450	Flat black, alum. frame, 3.5′ x 7.5′	G		9	1.778		815	103		918	1,050
3500	4′ x 8′	G		5.50	2.909		1,025	169		1,194	1,375
3520	4′ x 10′	G		10	1.600		1,225	93		1,318	1,500
3540	4′ x 12.5′	G		5	3.200		1,300	186		1,486	1,700
3550	Liquid with fin tube absorber plate										
3560	Alum. frame 4′ x 8′ tempered glass	G	Q-1	10	1.600	Ea.	605	93		698	810
3580	Liquid with vacuum tubes, 4′ x 6′-10″	G		9	1.778		915	103		1,018	1,150
3600	Liquid, full wetted, plastic, alum. frame, 4′ x 10′	G		5	3.200		335	186		521	655
3650	Collector panel mounting, flat roof or ground rack	G		7	2.286		253	133		386	485
3670	Roof clamps	G		70	.229	Set	3.27	13.25		16.52	24
3700	Roof strap, teflon	G	1 Plum	205	.039	L.F.	25	2.51		27.51	31.50
3900	Differential controller with two sensors										
3930	Thermostat, hard wired	G	1 Plum	8	1	Ea.	112	64.50		176.50	223
3950	Line cord and receptacle	G		12	.667		154	43		197	236
4050	Pool valve system	G		2.50	3.200		182	206		388	520
4070	With 12 VAC actuator	G		2	4		335	258		593	765
4080	Pool pump system, 2″ pipe size	G		6	1.333		205	86		291	360
4100	Five station with digital read-out	G		3	2.667		271	172		443	565
4150	Sensors										
4200	Brass plug, 1/2″ MPT	G	1 Plum	32	.250	Ea.	16.70	16.10		32.80	43.50
4210	Brass plug, reversed	G		32	.250		26.50	16.10		42.60	54
4220	Freeze prevention	G		32	.250		25	16.10		41.10	52.50

23 56 19.50 Solar Heating Ancillary			Crew	Daily Output	Labor-Hours	Unit	2020 Bare Costs Material	Labor	Equipment	Total	Total Incl O&P
4240	Screw attached	G	1 Plum	32	.250	Ea.	10.05	16.10		26.15	36
4250	Brass, immersion	G		32	.250		15.50	16.10		31.60	42
4300	Heat exchanger										
4315	includes coil, blower, circulator										
4316	and controller for DHW and space hot air										
4330	Fluid to air coil, up flow, 45 MBH	G	Q-1	4	4	Ea.	330	232		562	725
4380	70 MBH	G		3.50	4.571		375	265		640	820
4400	80 MBH	G		3	5.333		495	310		805	1,025
4580	Fluid to fluid package includes two circulating pumps										
4590	expansion tank, check valve, relief valve										
4600	controller, high temperature cutoff and sensors	G	Q-1	2.50	6.400	Ea.	810	370		1,180	1,475
4650	Heat transfer fluid										
4700	Propylene glycol, inhibited anti-freeze	G	1 Plum	28	.286	Gal.	23	18.40		41.40	54
4800	Solar storage tanks, knocked down										
4810	Air, galvanized steel clad, double wall, 4" fiberglass insulation										
5120	45 mil reinforced polypropylene lining,										
5140	4' high, 4' x 4' = 64 C.F./450 gallons	G	Q-9	2	8	Ea.	3,700	450		4,150	4,775
5150	4' x 8' = 128 C.F./900 gallons	G		1.50	10.667		5,875	600		6,475	7,400
5160	4' x 12' = 190 C.F./1,300 gallons	G		1.30	12.308		7,425	690		8,115	9,250
5170	8' x 8' = 250 C.F./1,700 gallons	G		1	16		7,700	895		8,595	9,875
5190	6'-3" high, 7' x 7' = 306 C.F./2,000 gallons	G	Q-10	1.20	20		13,900	1,175		15,075	17,100
5200	7' x 10'-6" = 459 C.F./3,000 gallons	G		.80	30		17,300	1,750		19,050	21,700
5210	7' x 14' = 613 C.F./4,000 gallons	G		.60	40		22,400	2,325		24,725	28,300
5220	10'-6" x 10'-6" = 689 C.F./4,500 gallons	G		.50	48		20,800	2,800		23,600	27,200
5230	10'-6" x 14' = 919 C.F./6,000 gallons	G		.40	60		24,200	3,500		27,700	32,200
5240	14' x 14' = 1,225 C.F./8,000 gallons	G	Q-11	.40	80		29,000	4,750		33,750	39,300
5250	14' x 17'-6" = 1,531 C.F./10,000 gallons	G		.30	107		31,400	6,325		37,725	44,500
5260	17'-6" x 17'-6" = 1,914 C.F./12,500 gallons	G		.25	128		35,500	7,600		43,100	51,000
5270	17'-6" x 21' = 2,297 C.F./15,000 gallons	G		.20	160		40,300	9,500		49,800	59,500
5280	21' x 21' = 2,756 C.F./18,000 gallons	G		.18	178		43,100	10,500		53,600	64,000
5290	30 mil reinforced Hypalon lining, add						.02%				
7000	Solar control valves and vents										
7050	Air purger, 1" pipe size	G	1 Plum	12	.667	Ea.	46.50	43		89.50	118
7070	Air eliminator, automatic 3/4" size	G		32	.250		31	16.10		47.10	59
7090	Air vent, automatic, 1/8" fitting	G		32	.250		17.90	16.10		34	44.50
7100	Manual, 1/8" NPT	G		32	.250		2.69	16.10		18.79	28
7120	Backflow preventer, 1/2" pipe size	G		16	.500		90.50	32		122.50	150
7130	3/4" pipe size	G		16	.500		89	32		121	148
7150	Balancing valve, 3/4" pipe size	G		20	.400		59.50	26		85.50	105
7180	Draindown valve, 1/2" copper tube	G		9	.889		219	57.50		276.50	330
7200	Flow control valve, 1/2" pipe size	G		22	.364		140	23.50		163.50	190
7220	Expansion tank, up to 5 gal.	G		32	.250		70.50	16.10		86.60	103
7250	Hydronic controller (aquastat)	G		8	1		217	64.50		281.50	340
7400	Pressure gauge, 2" dial	G		32	.250		25	16.10		41.10	52.50
7450	Relief valve, temp. and pressure 3/4" pipe size	G		30	.267		25.50	17.20		42.70	54.50
7500	Solenoid valve, normally closed										
7520	Brass, 3/4" NPT, 24V	G	1 Plum	9	.889	Ea.	137	57.50		194.50	239
7530	1" NPT, 24V	G		9	.889		224	57.50		281.50	335
7750	Vacuum relief valve, 3/4" pipe size	G		32	.250		31	16.10		47.10	59
7800	Thermometers										
7820	Digital temperature monitoring, 4 locations	G	1 Plum	2.50	3.200	Ea.	138	206		344	470
7900	Upright, 1/2" NPT	G		8	1		39.50	64.50		104	143
7970	Remote probe, 2" dial	G		8	1		34.50	64.50		99	138

23 56 Solar Energy Heating Equipment

23 56 19 – Solar Heating Components

23 56 19.50 Solar Heating Ancillary			Crew	Daily Output	Labor-Hours	Unit	Material	2020 Bare Costs Labor	Equipment	Total	Total Incl O&P
7990	Stem, 2" dial, 9" stem	G	1 Plum	16	.500	Ea.	22.50	32		54.50	75
8250	Water storage tank with heat exchanger and electric element										
8270	66 gal. with 2" x 2 lb. density insulation	G	1 Plum	1.60	5	Ea.	2,150	320		2,470	2,875
8300	80 gal. with 2" x 2 lb. density insulation	G		1.60	5		1,950	320		2,270	2,650
8380	120 gal. with 2" x 2 lb. density insulation	G		1.40	5.714		1,900	370		2,270	2,675
8400	120 gal. with 2" x 2 lb. density insul., 40 S.F. heat coil	G	↓	1.40	5.714	↓	2,825	370		3,195	3,700
8500	Water storage module, plastic										
8600	Tubular, 12" diameter, 4' high	G	1 Carp	48	.167	Ea.	99	8.85		107.85	123
8610	12" diameter, 8' high	G		40	.200		157	10.65		167.65	189
8620	18" diameter, 5' high	G		38	.211		176	11.20		187.20	211
8630	18" diameter, 10' high	G	↓	32	.250		263	13.30		276.30	310
8640	58" diameter, 5' high	G	2 Carp	32	.500		815	26.50		841.50	940
8650	Cap, 12" diameter	G					20			20	22
8660	18" diameter	G				↓	25.50			25.50	28
9000	Minimum labor/equipment charge		1 Plum	2	4	Job		258		258	400

23 57 Heat Exchangers for HVAC

23 57 16 – Steam-to-Water Heat Exchangers

23 57 16.10 Shell/Tube Type Steam-to-Water Heat Exch.		Crew	Daily Output	Labor-Hours	Unit	Material	Labor	Equipment	Total	Total Incl O&P
0010	**SHELL AND TUBE TYPE STEAM-TO-WATER HEAT EXCHANGERS**									
0016	Shell & tube type, 2 or 4 pass, 3/4" OD copper tubes,									
0020	C.I. heads, C.I. tube sheet, steel shell									
0100	Hot water 40°F to 180°F, by steam at 10 psi									
0120	8 GPM	Q-5	6	2.667	Ea.	2,750	157		2,907	3,275
0140	10 GPM		5	3.200		4,125	189		4,314	4,850
0160	40 GPM		4	4		6,400	236		6,636	7,400
0180	64 GPM		2	8		9,350	470		9,820	11,000
0200	96 GPM	↓	1	16		12,500	945		13,445	15,300
0220	120 GPM	Q-6	1.50	16		16,000	980		16,980	19,100
0240	168 GPM		1	24		20,100	1,475		21,575	24,500
0260	240 GPM		.80	30		31,500	1,825		33,325	37,500
0300	600 GPM	↓	.70	34.286	↓	69,500	2,100		71,600	80,000
0500	For bronze head and tube sheet, add					50%				

23 57 19 – Liquid-to-Liquid Heat Exchangers

23 57 19.13 Plate-Type, Liquid-to-Liquid Heat Exchangers		Crew	Daily Output	Labor-Hours	Unit	Material	Labor	Equipment	Total	Total Incl O&P
0010	**PLATE-TYPE, LIQUID-TO-LIQUID HEAT EXCHANGERS**									
3000	Plate type,									
3100	400 GPM	Q-6	.80	30	Ea.	45,700	1,825		47,525	53,500
3120	800 GPM	"	.50	48		79,000	2,925		81,925	91,000
3140	1,200 GPM	Q-7	.34	94.118		117,000	5,875		122,875	138,000
3160	1,800 GPM	"	.24	133	↓	155,500	8,325		163,825	184,000

23 57 19.16 Shell-Type, Liquid-to-Liquid Heat Exchangers		Crew	Daily Output	Labor-Hours	Unit	Material	Labor	Equipment	Total	Total Incl O&P
0010	**SHELL-TYPE, LIQUID-TO-LIQUID HEAT EXCHANGERS**									
1000	Hot water 40°F to 140°F, by water at 200°F									
1020	7 GPM	Q-5	6	2.667	Ea.	3,225	157		3,382	3,775
1040	16 GPM		5	3.200		4,550	189		4,739	5,300
1060	34 GPM		4	4		6,875	236		7,111	7,950
1080	55 GPM		3	5.333		9,750	315		10,065	11,200
1100	74 GPM		1.50	10.667		12,500	630		13,130	14,700
1120	86 GPM	↓	1.40	11.429	↓	16,700	675		17,375	19,500

23 57 Heat Exchangers for HVAC

23 57 19 – Liquid-to-Liquid Heat Exchangers

23 57 19.16 Shell-Type, Liquid-to-Liquid Heat Exchangers		Crew	Daily Output	Labor-Hours	Unit	Material	2020 Bare Costs Labor	Equipment	Total	Total Incl O&P
1140	112 GPM	Q-6	2	12	Ea.	20,800	735		21,535	24,000
1160	126 GPM		1.80	13.333		25,900	815		26,715	29,800
1180	152 GPM		1	24		33,100	1,475		34,575	38,700

23 61 Refrigerant Compressors

23 61 15 – Rotary Refrigerant Compressors

23 61 15.10 Rotary Compressors		Crew	Daily Output	Labor-Hours	Unit	Material	Labor	Equipment	Total	Total Incl O&P
0010	**ROTARY COMPRESSORS**									
0100	Refrigeration, hermetic, switches and protective devices									
0210	1.25 ton	1 Stpi	3	2.667	Ea.	251	175		426	545
0220	1.42 ton		3	2.667		254	175		429	550
0230	1.68 ton		2.80	2.857		276	187		463	595
0240	2.00 ton		2.60	3.077		284	202		486	625
0250	2.37 ton		2.50	3.200		320	210		530	680
0260	2.67 ton		2.40	3.333		335	219		554	710
0270	3.53 ton		2.30	3.478		370	228		598	760
0280	4.43 ton		2.20	3.636		470	238		708	885
0290	5.08 ton		2.10	3.810		520	250		770	955
0300	5.22 ton		2	4		730	262		992	1,200
0310	7.31 ton	Q-5	3	5.333		1,175	315		1,490	1,750
0320	9.95 ton		2.60	6.154		1,375	365		1,740	2,050
0330	11.6 ton		2.50	6.400		1,750	380		2,130	2,500
0340	15.25 ton		2.30	6.957		2,425	410		2,835	3,275
0350	17.7 ton		2.10	7.619		2,575	450		3,025	3,550

23 61 16 – Reciprocating Refrigerant Compressors

23 61 16.10 Reciprocating Compressors		Crew	Daily Output	Labor-Hours	Unit	Material	Labor	Equipment	Total	Total Incl O&P
0010	**RECIPROCATING COMPRESSORS**									
0990	Refrigeration, recip. hermetic, switches & protective devices									
1000	10 ton	Q-5	1	16	Ea.	15,200	945		16,145	18,200
1100	20 ton	Q-6	.72	33.333		23,700	2,050		25,750	29,300
1200	30 ton		.64	37.500		28,700	2,300		31,000	35,200
1300	40 ton		.44	54.545		33,000	3,325		36,325	41,500
1400	50 ton		.20	120		34,300	7,350		41,650	49,200
1500	75 ton	Q-7	.27	119		38,000	7,400		45,400	53,000
1600	130 ton	"	.21	152		43,400	9,500		52,900	62,500

23 61 19 – Scroll Refrigerant Compressors

23 61 19.10 Scroll Compressors		Crew	Daily Output	Labor-Hours	Unit	Material	Labor	Equipment	Total	Total Incl O&P
0010	**SCROLL COMPRESSORS**									
1800	Refrigeration, scroll type									
1810	1.5 ton	1 Stpi	2.70	2.963	Ea.	935	194		1,129	1,325
1820	2.5 ton		2.50	3.200		930	210		1,140	1,350
1830	2.75 ton		2.35	3.404		760	223		983	1,175
1840	3 ton		2.30	3.478		980	228		1,208	1,425
1850	3.5 ton		2.25	3.556		1,000	233		1,233	1,450
1860	4 ton		2.20	3.636		1,200	238		1,438	1,700
1870	5 ton		2.10	3.810		1,275	250		1,525	1,775
1880	6 ton	Q-5	3.70	4.324		1,450	255		1,705	2,000
1900	7.5 ton		3.20	5		2,625	295		2,920	3,350
1910	10 ton		3.10	5.161		2,800	305		3,105	3,550
1920	13 ton		2.90	5.517		2,850	325		3,175	3,650

23 61 Refrigerant Compressors

23 61 19 – Scroll Refrigerant Compressors

23 61 19.10 Scroll Compressors		Crew	Daily Output	Labor-Hours	Unit	Material	2020 Bare Costs Labor	Equipment	Total	Total Incl O&P
1960	25 ton	Q-5	2.20	7.273	Ea.	5,175	430		5,605	6,375

23 62 Packaged Compressor and Condenser Units

23 62 13 – Packaged Air-Cooled Refrigerant Compressor and Condenser Units

23 62 13.10 Packaged Air-Cooled Refrig. Condensing Units

		Crew	Daily Output	Labor-Hours	Unit	Material	Labor	Equipment	Total	Total Incl O&P
0010	PACKAGED AIR-COOLED REFRIGERANT CONDENSING UNITS									
0020	Condensing unit									
0030	Air cooled, compressor, standard controls									
0050	1.5 ton	Q-5	2.50	6.400	Ea.	1,025	380		1,405	1,700
0100	2 ton		2.10	7.619		1,050	450		1,500	1,850
0200	2.5 ton		1.70	9.412		1,225	555		1,780	2,200
0300	3 ton		1.30	12.308		1,250	725		1,975	2,500
0350	3.5 ton		1.10	14.545		1,525	860		2,385	3,000
0400	4 ton		.90	17.778		1,700	1,050		2,750	3,500
0500	5 ton		.60	26.667		2,025	1,575		3,600	4,650
0550	7.5 ton		.55	29.091		3,350	1,725		5,075	6,325
0560	8.5 ton		.53	30.189		3,475	1,775		5,250	6,575
0600	10 ton		.50	32		3,525	1,900		5,425	6,800
0620	12.5 ton		.48	33.333		4,475	1,975		6,450	7,975
0650	15 ton		.40	40		7,850	2,350		10,200	12,300
0700	20 ton	Q-6	.40	60		9,350	3,675		13,025	16,000
0720	25 ton		.35	68.571		14,300	4,200		18,500	22,300
0750	30 ton		.30	80		16,600	4,900		21,500	25,800
0800	40 ton		.20	120		20,100	7,350		27,450	33,600
0840	50 ton		.18	133		21,700	8,150		29,850	36,400
0860	60 ton		.16	150		23,000	9,175		32,175	39,500
0900	70 ton		.14	171		28,700	10,500		39,200	47,800
1000	80 ton		.12	200		31,900	12,200		44,100	54,000
1010	90 ton		.09	267		34,700	16,300		51,000	63,500
1100	100 ton		.09	267		40,300	16,300		56,600	69,500
1102	120 ton		.09	267		45,300	16,300		61,600	75,000

23 62 23 – Packaged Water-Cooled Refrigerant Compressor and Condenser Units

23 62 23.10 Pckgd. Water-Cooled Refrigerant Condns. Units

		Crew	Daily Output	Labor-Hours	Unit	Material	Labor	Equipment	Total	Total Incl O&P
0010	PACKAGED WATER-COOLED REFRIGERANT CONDENSING UNITS									
2000	Water cooled, compressor, heat exchanger, controls									
2100	5 ton	Q-5	.70	22.857	Ea.	16,600	1,350		17,950	20,400
2200	15 ton	"	.50	32		29,700	1,900		31,600	35,500
2300	20 ton	Q-6	.40	60		33,300	3,675		36,975	42,400
2400	40 ton	"	.20	120		35,200	7,350		42,550	50,000
9000	Minimum labor/equipment charge	Q-5	1.50	10.667	Job		630		630	975

23 63 Refrigerant Condensers

23 63 13 – Air-Cooled Refrigerant Condensers

23 63 13.10 Air-Cooled Refrig. Condensers		Crew	Daily Output	Labor-Hours	Unit	2020 Bare Costs Material	Labor	Equipment	Total	Total Incl O&P
0010	AIR-COOLED REFRIG. CONDENSERS									
0080	Air cooled, belt drive, propeller fan									
0220	45 ton	Q-6	.70	34.286	Ea.	10,500	2,100		12,600	14,800
0240	50 ton		.69	34.985		11,200	2,150		13,350	15,700
0260	54 ton		.64	37.795		12,200	2,300		14,500	17,000
0280	59 ton		.58	41.308		13,400	2,525		15,925	18,600
0300	65 ton		.53	45.541		15,900	2,775		18,675	21,800
0320	73 ton		.47	51.173		17,500	3,125		20,625	24,100
0340	81 ton		.42	56.738		18,900	3,475		22,375	26,200
0360	86 ton		.40	60.302		19,900	3,700		23,600	27,600
0380	88 ton		.39	61.697		21,400	3,775		25,175	29,300
0400	101 ton	Q-7	.45	70.640		24,500	4,400		28,900	33,700
0500	159 ton		.31	103		26,300	6,425		32,725	38,900
0600	228 ton		.22	148		54,500	9,250		63,750	74,500
0650	320 ton		.16	206		52,500	12,900		65,400	78,000
0660	475 ton		.11	296		113,500	18,500		132,000	153,500
1500	May be specified single or multi-circuit									
1550	Air cooled, direct drive, propeller fan									
1590	1 ton	Q-5	3.80	4.211	Ea.	1,825	248		2,073	2,375
1600	1-1/2 ton		3.60	4.444		2,225	262		2,487	2,850
1620	2 ton		3.20	5		2,400	295		2,695	3,075
1630	3 ton		2.40	6.667		2,675	395		3,070	3,550
1640	5 ton		2	8		5,100	470		5,570	6,325
1650	8 ton		1.80	8.889		6,150	525		6,675	7,575
1660	10 ton		1.40	11.429		7,475	675		8,150	9,275
1670	12 ton		1.30	12.308		8,450	725		9,175	10,400
1680	14 ton		1.20	13.333		9,650	785		10,435	11,800
1690	16 ton		1.10	14.545		10,800	860		11,660	13,200
1700	21 ton		1	16		12,200	945		13,145	14,900
1720	26 ton		.84	19.002		13,700	1,125		14,825	16,800
1740	30 ton		.70	22.792		17,100	1,350		18,450	20,900
1760	41 ton	Q-6	.77	31.008		18,800	1,900		20,700	23,600
1780	52 ton		.66	36.419		26,800	2,225		29,025	33,000
1800	63 ton		.55	44.037		29,600	2,700		32,300	36,800
1820	76 ton		.45	52.980		33,600	3,250		36,850	42,000
1840	86 ton		.40	60		41,500	3,675		45,175	51,500
1860	97 ton		.35	67.989		47,300	4,150		51,450	58,500
1880	105 ton	Q-7	.44	73.563		54,500	4,600		59,100	67,000
1890	118 ton		.39	82.687		58,500	5,150		63,650	72,500
1900	126 ton		.36	88.154		68,500	5,500		74,000	83,500
1910	136 ton		.34	95.238		76,500	5,950		82,450	93,000
1920	142 ton		.32	99.379		78,500	6,200		84,700	95,500

23 63 30 – Lubricants

23 63 30.10 Lubricant Oils		Crew	Daily Output	Labor-Hours	Unit	Material	Labor	Equipment	Total	Total Incl O&P
0010	LUBRICANT OILS									
8000	Oils									
8100	Lubricating									
8120	Oil, lubricating				Oz.	.32			.32	.35
8500	Refrigeration									
8520	Oil, refrigeration				Gal.	48			48	53
8525	Oil, refrigeration				Qt.	12			12	13.20

23 63 Refrigerant Condensers

23 63 33 – Evaporative Refrigerant Condensers

23 63 33.10 Evaporative Condensers		Crew	Daily Output	Labor-Hours	Unit	Material	2020 Bare Costs Labor	Equipment	Total	Total Incl O&P
0010	**EVAPORATIVE CONDENSERS**									
3400	Evaporative, copper coil, pump, fan motor									
3440	10 ton	Q-5	.54	29.630	Ea.	12,600	1,750		14,350	16,500
3460	15 ton		.50	32		12,700	1,900		14,600	16,900
3480	20 ton		.47	34.043		13,000	2,000		15,000	17,400
3500	25 ton		.45	35.556		10,200	2,100		12,300	14,500
3520	30 ton		.42	38.095		10,700	2,250		12,950	15,300
3540	40 ton	Q-6	.49	48.980		11,500	3,000		14,500	17,200
3560	50 ton		.39	61.538		13,900	3,775		17,675	21,100
3580	65 ton		.35	68.571		14,200	4,200		18,400	22,100
3600	80 ton		.33	72.727		15,700	4,450		20,150	24,200
3620	90 ton		.29	82.759		17,200	5,075		22,275	26,700
3640	100 ton	Q-7	.36	88.889		19,600	5,550		25,150	30,200
3660	110 ton		.33	96.970		25,600	6,050		31,650	37,500
3680	125 ton		.30	107		27,700	6,650		34,350	40,700
3700	135 ton		.28	114		29,300	7,125		36,425	43,300
3720	150 ton		.25	128		31,900	7,975		39,875	47,500
3740	165 ton		.23	139		33,000	8,675		41,675	49,700
3760	185 ton		.22	145		37,900	9,075		46,975	55,500
3860	For fan damper control, add	Q-5	2	8		890	470		1,360	1,700

23 64 Packaged Water Chillers

23 64 13 – Absorption Water Chillers

23 64 13.13 Direct-Fired Absorption Water Chillers

		Crew	Daily Output	Labor-Hours	Unit	Material	Labor	Equipment	Total	Total Incl O&P
0010	**DIRECT-FIRED ABSORPTION WATER CHILLERS**									
3000	Gas fired, air cooled									
3220	5 ton	Q-5	.60	26.667	Ea.	8,850	1,575		10,425	12,200
3270	10 ton	"	.40	40	"	23,200	2,350		25,550	29,200
4000	Water cooled, duplex									
4130	100 ton	Q-7	.13	246	Ea.	144,000	15,400		159,400	182,500
4140	200 ton		.11	283		188,000	17,700		205,700	234,500
4150	300 ton		.11	299		232,500	18,700		251,200	285,000
4160	400 ton		.10	317		302,500	19,800		322,300	363,000
4170	500 ton		.10	330		381,000	20,600		401,600	451,000
4180	600 ton		.09	340		450,000	21,200		471,200	528,000
4190	700 ton		.09	360		499,000	22,400		521,400	583,000
4200	800 ton		.08	381		602,000	23,800		625,800	699,500
4210	900 ton		.08	400		698,000	25,000		723,000	806,000
4220	1,000 ton		.08	421		775,500	26,300		801,800	893,500

23 64 13.16 Indirect-Fired Absorption Water Chillers

		Crew	Daily Output	Labor-Hours	Unit	Material	Labor	Equipment	Total	Total Incl O&P
0010	**INDIRECT-FIRED ABSORPTION WATER CHILLERS**									
0020	Steam or hot water, water cooled									
0050	100 ton	Q-7	.13	240	Ea.	132,500	15,000		147,500	169,000
0100	148 ton		.12	258		245,500	16,100		261,600	295,000
0200	200 ton		.12	276		273,000	17,200		290,200	327,000
0300	354 ton		.11	305		375,500	19,000		394,500	442,500
0400	420 ton		.10	323		414,500	20,200		434,700	487,000
0500	665 ton		.09	340		576,000	21,200		597,200	666,500
0600	750 ton		.09	364		670,000	22,700		692,700	772,000
0700	850 ton		.08	386		718,500	24,100		742,600	827,000

23 64 13 – Absorption Water Chillers

23 64 13.16 Indirect-Fired Absorption Water Chillers		Crew	Daily Output	Labor-Hours	Unit	Material	Labor	Equipment	Total	Total Incl O&P
						2020 Bare Costs				
0800	955 ton	Q-7	.08	410	Ea.	768,000	25,600		793,600	884,500
0900	1,125 ton		.08	421		887,000	26,300		913,300	1,016,000
1000	1,250 ton		.07	444		949,500	27,700		977,200	1,087,000
1100	1,465 ton		.07	464		1,130,500	28,900		1,159,400	1,288,000
1200	1,660 ton		.07	478		1,330,000	29,800		1,359,800	1,509,000
2000	For two stage unit, add					80%	25%			
9000	Minimum labor/equipment charge	Q-5	1	16	Job		945		945	1,450

23 64 16 – Centrifugal Water Chillers

23 64 16.10 Centrifugal Type Water Chillers		Crew	Daily Output	Labor-Hours	Unit	Material	Labor	Equipment	Total	Total Incl O&P
0010	**CENTRIFUGAL TYPE WATER CHILLERS**, With standard controls									
0020	Centrifugal liquid chiller, water cooled									
0030	not including water tower									
0100	2,000 ton (twin 1,000 ton units)	Q-7	.07	478	Ea.	808,500	29,800		838,300	935,500
0274	Centrifugal, packaged unit, water cooled, not incl. tower									
0278	200 ton	Q-7	.19	172	Ea.	131,000	10,700		141,700	161,000
0279	300 ton		.14	234		152,500	14,600		167,100	190,500
0280	400 ton		.11	283		150,500	17,700		168,200	193,000
0320	1,000 ton		.09	372		404,500	23,200		427,700	481,000
0340	1,300 ton		.08	410		438,500	25,600		464,100	522,000
0360	1,500 ton		.08	427		544,500	26,600		571,100	639,500

23 64 19 – Reciprocating Water Chillers

23 64 19.10 Reciprocating Type Water Chillers		Crew	Daily Output	Labor-Hours	Unit	Material	Labor	Equipment	Total	Total Incl O&P
0010	**RECIPROCATING TYPE WATER CHILLERS**, With standard controls									
0494	Water chillers, integral air cooled condenser									
0554	90 ton cooling	Q-7	.25	126	Ea.	60,000	7,850		67,850	78,000
0640	150 ton cooling		.23	138		99,500	8,600		108,100	123,000
0654	190 ton cooling		.22	144		122,000	9,000		131,000	148,000
0660	210 ton cooling		.22	148		123,500	9,250		132,750	150,500
0664	275 ton cooling		.21	156		154,500	9,750		164,250	185,000
0666	300 ton cooling		.20	160		167,500	9,975		177,475	200,000
0668	330 ton cooling		.20	164		197,500	10,200		207,700	233,500
0670	360 ton cooling		.19	168		218,500	10,500		229,000	257,000
0672	390 ton cooling		.19	173		229,500	10,800		240,300	269,000
0980	Water cooled, multiple compressor, semi-hermetic, tower not incl.									
1090	45 ton cooling	Q-7	.29	111	Ea.	27,200	6,925		34,125	40,600
1130	75 ton cooling		.23	139		45,100	8,675		53,775	63,000
1140	85 ton cooling		.21	152		52,000	9,475		61,475	71,500
1150	95 ton cooling		.19	165		51,000	10,300		61,300	72,000
1160	100 ton cooling		.18	180		49,800	11,200		61,000	72,500
1170	115 ton cooling		.17	190		53,500	11,900		65,400	77,500
1180	125 ton cooling		.16	196		54,000	12,300		66,300	78,500
1200	145 ton cooling		.16	203		61,500	12,600		74,100	87,000
1210	155 ton cooling		.15	211		64,500	13,100		77,600	91,500
1300	Water cooled, single compressor, semi-hermetic, tower not incl.									
1320	40 ton cooling	Q-7	.30	108	Ea.	17,200	6,750		23,950	29,300
1340	50 ton cooling		.28	114		21,400	7,100		28,500	34,500
1360	60 ton cooling		.25	126		23,800	7,850		31,650	38,300
4000	Packaged chiller, remote air cooled condensers not incl.									
4020	15 ton cooling	Q-7	.30	108	Ea.	20,500	6,750		27,250	32,900
4030	20 ton cooling		.28	116		21,200	7,225		28,425	34,600
4040	25 ton cooling		.25	126		26,700	7,850		34,550	41,500
4050	35 ton cooling		.24	134		32,700	8,350		41,050	48,900

23 64 Packaged Water Chillers

23 64 19 – Reciprocating Water Chillers

23 64 19.10 Reciprocating Type Water Chillers		Crew	Daily Output	Labor-Hours	Unit	Material	Labor	Equipment	Total	Total Incl O&P
						2020 Bare Costs				
4060	40 ton cooling	Q-7	.22	144	Ea.	29,000	9,000		38,000	45,800
4066	45 ton cooling		.21	150		26,300	9,325		35,625	43,300
4070	50 ton cooling		.21	154		31,300	9,600		40,900	49,200
4080	60 ton cooling		.20	164		32,900	10,200		43,100	52,000
4090	75 ton cooling		.18	174		62,000	10,900		72,900	85,000
4100	85 ton cooling		.17	184		63,500	11,500		75,000	87,000
4110	95 ton cooling		.17	194		54,000	12,100		66,100	77,500
4120	105 ton cooling		.16	204		69,500	12,700		82,200	95,500
4130	115 ton cooling		.15	213		61,000	13,300		74,300	87,500
4140	125 ton cooling		.14	224		64,000	14,000		78,000	91,500
4150	145 ton cooling		.14	234		73,500	14,600		88,100	103,500

23 64 23 – Scroll Water Chillers

23 64 23.10 Scroll Water Chillers		Crew	Daily Output	Labor-Hours	Unit	Material	Labor	Equipment	Total	Total Incl O&P
0010	**SCROLL WATER CHILLERS**, With standard controls									
0480	Packaged w/integral air cooled condenser									
0482	10 ton cooling	Q-7	.34	94.118	Ea.	15,200	5,875		21,075	25,800
0490	15 ton cooling		.37	86.486		18,300	5,400		23,700	28,600
0500	20 ton cooling		.34	94.118		23,300	5,875		29,175	34,800
0510	25 ton cooling		.34	94.118		27,800	5,875		33,675	39,700
0515	30 ton cooling		.31	102		29,800	6,350		36,150	42,600
0517	35 ton cooling		.31	105		32,600	6,550		39,150	46,000
0520	40 ton cooling		.30	108		34,800	6,750		41,550	48,700
0528	45 ton cooling		.29	110		38,300	6,850		45,150	52,500
0536	50 ton cooling		.28	113		40,700	7,050		47,750	55,500
0540	70 ton cooling		.27	119		55,000	7,400		62,400	72,000
0590	100 ton cooling		.25	128		72,500	7,975		80,475	92,500
0600	110 ton cooling		.24	133		74,500	8,325		82,825	95,000
0630	130 ton cooling		.24	133		88,500	8,325		96,825	110,500
0670	170 ton cooling		.23	139		111,000	8,675		119,675	135,500
0671	210 ton cooling		.22	145		123,500	9,075		132,575	150,000
0675	250 ton cooling		.21	152		143,500	9,500		153,000	172,000
0676	275 ton cooling		.21	152		154,500	9,500		164,000	184,500
0677	300 ton cooling		.20	160		167,500	9,975		177,475	200,000
0678	330 ton cooling		.20	160		197,500	9,975		207,475	233,000
0679	390 ton cooling		.19	168		229,500	10,500		240,000	269,000
0680	Scroll water cooled, single compressor, hermetic, tower not incl.									
0700	2 ton cooling	Q-5	.57	28.070	Ea.	3,475	1,650		5,125	6,375
0710	5 ton cooling		.57	28.070		4,150	1,650		5,800	7,125
0720	6 ton cooling		.42	38.005		5,250	2,250		7,500	9,250
0740	8 ton cooling		.31	52.117		5,725	3,075		8,800	11,100
0760	10 ton cooling	Q-6	.36	67.039		6,575	4,100		10,675	13,600
0780	15 ton cooling	"	.36	66.667		10,300	4,075		14,375	17,600
0800	20 ton cooling	Q-7	.38	83.990		11,800	5,250		17,050	21,000
0820	30 ton cooling	"	.33	96.096		13,200	6,000		19,200	23,800
0900	Scroll water cooled, multiple compressor, hermetic, tower not incl.									
0910	15 ton cooling	Q-6	.36	66.667	Ea.	20,500	4,075		24,575	28,800
0930	30 ton cooling	Q-7	.36	88.889		26,700	5,550		32,250	37,900
0935	35 ton cooling		.31	103		32,700	6,450		39,150	46,000
0940	40 ton cooling		.30	107		29,000	6,650		35,650	42,200
0950	50 ton cooling		.28	114		31,300	7,125		38,425	45,400
0960	60 ton cooling		.25	128		32,900	7,975		40,875	48,600
0962	75 ton cooling		.23	139		62,000	8,675		70,675	81,500

23 64 Packaged Water Chillers

23 64 23 – Scroll Water Chillers

23 64 23.10 Scroll Water Chillers		Crew	Daily Output	Labor-Hours	Unit	Material	Labor	Equipment	Total	Total Incl O&P
						2020 Bare Costs				
0964	85 ton cooling	Q-7	.23	139	Ea.	63,500	8,675		72,175	83,000
0970	100 ton cooling		.18	178		69,500	11,100		80,600	93,000
0980	115 ton cooling		.17	188		61,000	11,700		72,700	85,000
0990	125 ton cooling		.16	200		64,000	12,500		76,500	89,500
0995	145 ton cooling		.16	200		73,500	12,500		86,000	100,500

23 64 26 – Rotary-Screw Water Chillers

23 64 26.10 Rotary-Screw Type Water Chillers

		Crew	Daily Output	Labor-Hours	Unit	Material	Labor	Equipment	Total	Total Incl O&P
0010	**ROTARY-SCREW TYPE WATER CHILLERS**, With standard controls									
0110	Screw, liquid chiller, air cooled, insulated evaporator									
0120	130 ton	Q-7	.14	229	Ea.	98,500	14,300		112,800	130,500
0124	160 ton		.13	246		125,000	15,400		140,400	161,500
0128	180 ton		.13	250		140,500	15,600		156,100	178,500
0132	210 ton		.12	258		147,000	16,100		163,100	186,500
0136	270 ton		.12	267		177,000	16,600		193,600	220,000
0140	320 ton		.12	276		221,500	17,200		238,700	270,000
0200	Packaged unit, water cooled, not incl. tower									
0240	200 ton	Q-7	.13	252	Ea.	83,500	15,700		99,200	116,500
1450	Water cooled, tower not included									
1560	135 ton cooling, screw compressors	Q-7	.14	235	Ea.	59,000	14,700		73,700	87,500
1580	150 ton cooling, screw compressors		.13	241		68,500	15,000		83,500	98,500
1620	200 ton cooling, screw compressors		.13	250		94,500	15,600		110,100	128,000
1660	291 ton cooling, screw compressors		.12	260		98,500	16,200		114,700	133,500

23 64 33 – Modular Water Chillers

23 64 33.10 Direct Expansion Type Water Chillers

		Crew	Daily Output	Labor-Hours	Unit	Material	Labor	Equipment	Total	Total Incl O&P
0010	**DIRECT EXPANSION TYPE WATER CHILLERS**, With standard controls									
8000	Direct expansion, shell and tube type, for built up systems									
8020	1 ton	Q-5	2	8	Ea.	8,125	470		8,595	9,650
8030	5 ton		1.90	8.421		13,500	495		13,995	15,600
8040	10 ton		1.70	9.412		16,700	555		17,255	19,300
9000	Minimum labor/equipment charge	Q-6	1	24	Job		1,475		1,475	2,275

23 65 Cooling Towers

23 65 13 – Forced-Draft Cooling Towers

23 65 13.10 Forced-Draft Type Cooling Towers

		Crew	Daily Output	Labor-Hours	Unit	Material	Labor	Equipment	Total	Total Incl O&P
0010	**FORCED-DRAFT TYPE COOLING TOWERS**, Packaged units									
0070	Galvanized steel									
0080	Induced draft, crossflow									
0100	Vertical, belt drive, 61 tons	Q-6	90	.267	TonAC	203	16.30		219.30	248
0150	100 ton		100	.240		195	14.70		209.70	238
0200	115 ton		109	.220		170	13.45		183.45	208
0250	131 ton		120	.200		204	12.25		216.25	243
0260	162 ton		132	.182		165	11.10		176.10	198
1000	For higher capacities, use multiples									
1500	Induced air, double flow									
1900	Vertical, gear drive, 167 ton	Q-6	126	.190	TonAC	160	11.65		171.65	194
2000	297 ton		129	.186		98.50	11.40		109.90	127
2100	582 ton		132	.182		55	11.10		66.10	77.50
2150	849 ton		142	.169		74	10.35		84.35	97
2200	1,016 ton		150	.160		74.50	9.80		84.30	97
3000	For higher capacities, use multiples									

23 65 Cooling Towers

23 65 13 – Forced-Draft Cooling Towers

23 65 13.10 Forced-Draft Type Cooling Towers		Crew	Daily Output	Labor-Hours	Unit	Material	2020 Bare Costs Labor	Equipment	Total	Total Incl O&P
3500	For pumps and piping, add	Q-6	38	.632	TonAC	101	38.50		139.50	171
4000	For absorption systems, add				"	75%	75%			
4100	Cooling water chemical feeder	Q-5	3	5.333	Ea.	425	315		740	955
5000	Fiberglass tower on galvanized steel support structure									
5010	Draw thru									
5100	100 ton	Q-6	1.40	17.143	Ea.	14,000	1,050		15,050	17,000
5120	120 ton		1.20	20		16,700	1,225		17,925	20,300
5140	140 ton		1	24		18,100	1,475		19,575	22,200
5160	160 ton		.80	30		22,900	1,825		24,725	28,100
5180	180 ton		.65	36.923		26,400	2,250		28,650	32,600
5251	600 ton		.16	150		77,500	9,175		86,675	99,000
5252	1,000 ton		.10	250		129,000	15,300		144,300	165,000
5300	For stainless steel support structure, add					30%				
5360	For higher capacities, use multiples of each size									
6000	Stainless steel									
6010	Induced draft, crossflow, horizontal, belt drive									
6100	57 ton	Q-6	1.50	16	Ea.	29,900	980		30,880	34,400
6120	91 ton		.99	24.242		36,700	1,475		38,175	42,700
6140	111 ton		.43	55.814		49,600	3,425		53,025	60,000
6160	126 ton		.22	109		54,000	6,675		60,675	70,000
6170	Induced draft, crossflow, vertical, gear drive									
6172	167 ton	Q-6	.75	32	Ea.	61,500	1,950		63,450	70,500
6174	297 ton		.43	55.814		68,000	3,425		71,425	80,000
6176	582 ton		.23	104		107,000	6,375		113,375	127,500
6178	849 ton		.17	141		151,000	8,625		159,625	180,000
6180	1,016 ton		.15	160		185,500	9,800		195,300	219,500
9000	Minimum labor/equipment charge		1	24	Job		1,475		1,475	2,275

23 65 33 – Liquid Coolers

23 65 33.10 Liquid Cooler, Closed Circuit Type Cooling Towers		Crew	Daily Output	Labor-Hours	Unit	Material	Labor	Equipment	Total	Total Incl O&P
0010	**LIQUID COOLER, CLOSED CIRCUIT TYPE COOLING TOWERS**									
0070	Packaged options include: vibration switch, sump sweeper piping,									
0076	grooved coil connections, fan motor inverter capable,									
0080	304 st. stl. cold water basin, external platform w/ladder.									
0100	50 ton	Q-6	2	12	Ea.	36,500	735		37,235	41,200
0150	100 ton	"	1.13	21.239		73,000	1,300		74,300	82,500
0200	150 ton	Q-7	.98	32.653		109,500	2,050		111,550	123,500
0250	200 ton		.78	41.026		146,000	2,550		148,550	164,500
0300	250 ton		.63	50.794		182,500	3,175		185,675	205,500

23 72 Air-to-Air Energy Recovery Equipment

23 72 13 – Heat-Wheel Air-to-Air Energy-Recovery Equipment

23 72 13.10 Heat-Wheel Air-to-Air Energy Recovery Equip.			Crew	Daily Output	Labor-Hours	Unit	Material	Labor	Equipment	Total	Total Incl O&P
0010	**HEAT-WHEEL AIR-TO-AIR ENERGY RECOVERY EQUIPMENT**										
0100	Air to air										
4000	Enthalpy recovery wheel										
4010	1,000 max CFM	G	Q-9	1.20	13.333	Ea.	7,675	750		8,425	9,625
4020	2,000 max CFM	G		1	16		8,550	895		9,445	10,800
4030	4,000 max CFM	G		.80	20		9,875	1,125		11,000	12,700
4040	6,000 max CFM	G		.70	22.857		11,500	1,275		12,775	14,700
4050	8,000 max CFM	G	Q-10	1	24		13,400	1,400		14,800	17,000

23 72 Air-to-Air Energy Recovery Equipment

23 72 13 – Heat-Wheel Air-to-Air Energy-Recovery Equipment

23 72 13.10 Heat-Wheel Air-to-Air Energy Recovery Equip.		Crew	Daily Output	Labor-Hours	Unit	Material	Labor	Equipment	Total	Total Incl O&P
4060	10,000 max CFM G	Q-10	.90	26.667	Ea.	16,000	1,550		17,550	20,100
4070	20,000 max CFM G		.80	30		27,700	1,750		29,450	33,100
4080	25,000 max CFM G		.70	34.286		35,600	2,000		37,600	42,200
4090	30,000 max CFM G		.50	48		39,500	2,800		42,300	47,800
4100	40,000 max CFM G		.45	53.333		52,000	3,100		55,100	62,000
4110	50,000 max CFM G		.40	60		60,000	3,500		63,500	71,500
9000	Minimum labor/equipment charge	1 Shee	4	2	Job		125		125	195

23 73 Indoor Central-Station Air-Handling Units

23 73 13 – Modular Indoor Central-Station Air-Handling Units

23 73 13.20 Air-Handling Units, Packaged Indoor Type		Crew	Daily Output	Labor-Hours	Unit	Material	Labor	Equipment	Total	Total Incl O&P
0010	**AIR-HANDLING UNITS, PACKAGED INDOOR TYPE**									
0040	Cooling coils may be chilled water or DX									
0050	Heating coils may be hot water, steam or electric									
1090	Constant volume									
1200	2,000 CFM	Q-5	1.10	14.545	Ea.	10,900	860		11,760	13,300
1400	5,000 CFM	Q-6	.80	30		17,700	1,825		19,525	22,300
1550	10,000 CFM		.54	44.444		26,000	2,725		28,725	32,800
1670	15,000 CFM		.38	63.158		53,000	3,875		56,875	64,500
1700	20,000 CFM		.31	77.419		67,500	4,725		72,225	82,000
1710	30,000 CFM		.27	88.889		92,000	5,450		97,450	109,500
2300	Variable air volume									
2330	2,000 CFM	Q-5	1	16	Ea.	15,600	945		16,545	18,700
2340	5,000 CFM	Q-6	.70	34.286		21,600	2,100		23,700	27,000
2350	10,000 CFM		.50	48		41,900	2,925		44,825	50,500
2360	15,000 CFM		.40	60		61,000	3,675		64,675	73,000
2370	20,000 CFM		.29	82.759		78,500	5,075		83,575	94,000
2380	30,000 CFM		.20	120		94,500	7,350		101,850	115,500

23 73 39 – Indoor, Direct Gas-Fired Heating and Ventilating Units

23 73 39.10 Make-Up Air Unit		Crew	Daily Output	Labor-Hours	Unit	Material	Labor	Equipment	Total	Total Incl O&P
0010	**MAKE-UP AIR UNIT**									
0020	Indoor suspension, natural/LP gas, direct fired,									
0032	standard control. For flue see Section 23 51 23.10									
0040	70°F temperature rise, MBH is input									
0100	75 MBH input	Q-6	3.60	6.667	Ea.	5,225	410		5,635	6,350
0120	100 MBH input		3.40	7.059		5,400	430		5,830	6,625
0140	125 MBH input		3.20	7.500		5,750	460		6,210	7,025
0160	150 MBH input		3	8		6,325	490		6,815	7,700
0180	175 MBH input		2.80	8.571		6,475	525		7,000	7,925
0200	200 MBH input		2.60	9.231		6,825	565		7,390	8,400
0220	225 MBH input		2.40	10		7,075	610		7,685	8,750
0240	250 MBH input		2	12		8,175	735		8,910	10,100
0260	300 MBH input		1.90	12.632		8,375	775		9,150	10,400
0280	350 MBH input		1.75	13.714		10,200	840		11,040	12,500
0300	400 MBH input		1.60	15		16,500	920		17,420	19,600
0320	840 MBH input		1.20	20		26,700	1,225		27,925	31,300
0340	1,200 MBH input		.88	27.273		36,800	1,675		38,475	43,100
0600	For discharge louver assembly, add					5%				
0700	For filters, add					10%				
0800	For air shut-off damper section, add					30%				

23 74 Packaged Outdoor HVAC Equipment

23 74 13 – Packaged, Outdoor, Central-Station Air-Handling Units

23 74 13.10 Packaged, Outdoor Type, Central-Station AHU		Crew	Daily Output	Labor-Hours	Unit	Material	2020 Bare Costs Labor	Equipment	Total	Total Incl O&P
0010	**PACKAGED, OUTDOOR TYPE, CENTRAL-STATION AIR-HANDLING UNITS**									
2900	Cooling coils may be chilled water or DX									
2910	Heating coils may be hot water, steam or electric									
3000	Weathertight, ground or rooftop									
3010	Constant volume									
3100	2,000 CFM	Q-6	1	24	Ea.	23,900	1,475		25,375	28,600
3140	5,000 CFM		.79	30.380		29,100	1,850		30,950	35,000
3150	10,000 CFM		.52	46.154		63,500	2,825		66,325	74,500
3160	15,000 CFM		.41	58.537		92,000	3,575		95,575	106,500
3170	20,000 CFM		.30	80		127,000	4,900		131,900	147,500
3180	30,000 CFM		.24	100		145,500	6,125		151,625	170,000
3200	Variable air volume									
3210	2,000 CFM	Q-6	.96	25	Ea.	15,200	1,525		16,725	19,200
3240	5,000 CFM		.73	32.877		30,600	2,000		32,600	36,800
3250	10,000 CFM		.50	48		65,000	2,925		67,925	76,000
3260	15,000 CFM		.40	60		100,000	3,675		103,675	115,500
3270	20,000 CFM		.29	82.759		129,500	5,075		134,575	150,500
3280	30,000 CFM		.20	120		153,000	7,350		160,350	180,000

23 74 23 – Packaged, Outdoor, Heating-Only Makeup-Air Units

23 74 23.16 Packaged, Ind.-Fired, In/Outdoor

23 74 23.16 Packaged, Ind.-Fired, In/Outdoor		Crew	Daily Output	Labor-Hours	Unit	Material	Labor	Equipment	Total	Total Incl O&P
0010	**PACKAGED, IND.-FIRED, IN/OUTDOOR**, Heat-Only Makeup-Air Units									
1000	Rooftop unit, natural gas, gravity vent, S.S. exchanger									
1010	70°F temperature rise, MBH is input									
1020	250 MBH	Q-6	4	6	Ea.	16,000	365		16,365	18,300
1040	400 MBH		3.60	6.667		18,300	410		18,710	20,800
1060	550 MBH		3.30	7.273		21,200	445		21,645	24,100
1080	750 MBH		3	8		23,500	490		23,990	26,700
1100	1,000 MBH		2.60	9.231		26,300	565		26,865	29,800
1120	1,750 MBH		2.30	10.435		30,900	640		31,540	35,000
1140	2,500 MBH		1.90	12.632		45,700	775		46,475	51,500
1160	3,250 MBH		1.40	17.143		50,500	1,050		51,550	57,000
1180	4,000 MBH		1.20	20		68,000	1,225		69,225	77,000
1200	6,000 MBH		1	24		86,000	1,475		87,475	97,000
1600	For cleanable filters, add					5%				
1700	For electric modulating gas control, add					10%				
9000	Minimum labor/equipment charge	Q-6	2.75	8.727	Job		535		535	825

23 74 33 – Dedicated Outdoor-Air Units

23 74 33.10 Rooftop Air Conditioners

23 74 33.10 Rooftop Air Conditioners		Crew	Daily Output	Labor-Hours	Unit	Material	Labor	Equipment	Total	Total Incl O&P
0010	**ROOFTOP AIR CONDITIONERS**, Standard controls, curb, economizer									
1000	Single zone, electric cool, gas heat									
1100	3 ton cooling, 60 MBH heating	Q-5	.70	22.857	Ea.	2,900	1,350		4,250	5,250
1120	4 ton cooling, 95 MBH heating		.61	26.403		3,750	1,550		5,300	6,525
1140	5 ton cooling, 112 MBH heating		.56	28.521		4,475	1,675		6,150	7,525
1145	6 ton cooling, 140 MBH heating		.52	30.769		5,125	1,825		6,950	8,450
1150	7.5 ton cooling, 170 MBH heating		.50	32.258		5,725	1,900		7,625	9,250
1156	8.5 ton cooling, 170 MBH heating		.46	34.783		7,200	2,050		9,250	11,100
1160	10 ton cooling, 200 MBH heating	Q-6	.67	35.982		9,675	2,200		11,875	14,000
1170	12.5 ton cooling, 230 MBH heating		.63	37.975		10,600	2,325		12,925	15,300
1180	15 ton cooling, 270 MBH heating		.57	42.032		13,300	2,575		15,875	18,600
1190	17.5 ton cooling, 330 MBH heating		.52	45.889		14,000	2,800		16,800	19,800
1200	20 ton cooling, 360 MBH heating	Q-7	.67	47.976		23,400	3,000		26,400	30,300
1210	25 ton cooling, 450 MBH heating		.56	57.554		34,800	3,600		38,400	43,900

23 74 33 – Dedicated Outdoor-Air Units

23 74 33.10 Rooftop Air Conditioners		Crew	Daily Output	Labor-Hours	Unit	Material	2020 Bare Costs Labor	Equipment	Total	Total Incl O&P
1220	30 ton cooling, 540 MBH heating	Q-7	.47	68.376	Ea.	36,300	4,275		40,575	46,500
1240	40 ton cooling, 675 MBH heating	↓	.35	91.168	↓	44,400	5,700		50,100	57,500
2000	Multizone, electric cool, gas heat, economizer									
2100	15 ton cooling, 360 MBH heating	Q-7	.61	52.545	Ea.	66,500	3,275		69,775	78,000
2120	20 ton cooling, 360 MBH heating		.53	60.038		72,000	3,750		75,750	85,500
2140	25 ton cooling, 450 MBH heating		.45	71.910		87,500	4,475		91,975	103,000
2160	28 ton cooling, 450 MBH heating		.41	79.012		99,500	4,925		104,425	116,500
2180	30 ton cooling, 540 MBH heating		.37	85.562		110,500	5,350		115,850	130,000
2200	40 ton cooling, 540 MBH heating		.28	114		126,500	7,100		133,600	150,000
2210	50 ton cooling, 540 MBH heating		.23	142		161,000	8,875		169,875	190,500
2220	70 ton cooling, 1,500 MBH heating		.16	199		171,500	12,400		183,900	207,500
2240	80 ton cooling, 1,500 MBH heating		.14	229		195,500	14,300		209,800	237,500
2260	90 ton cooling, 1,500 MBH heating		.13	256		200,500	16,000		216,500	245,000
2280	105 ton cooling, 1,500 MBH heating	↓	.11	291		223,000	18,200		241,200	273,500
2400	For hot water heat coil, deduct					5%				
2500	For steam heat coil, deduct					2%				
2600	For electric heat, deduct				↓	3%	5%			
5000	Single zone, electric cool only									
5050	3 ton cooling	Q-5	.88	18.203	Ea.	2,750	1,075		3,825	4,675
5060	4 ton cooling		.76	21.108		3,575	1,250		4,825	5,850
5070	5 ton cooling		.70	22.792		3,650	1,350		5,000	6,100
5080	6 ton cooling		.65	24.502		4,350	1,450		5,800	7,000
5090	7.5 ton cooling		.62	25.806		6,425	1,525		7,950	9,425
5100	8.5 ton cooling	↓	.59	26.981		7,025	1,600		8,625	10,200
5110	10 ton cooling	Q-6	.83	28.812		8,400	1,775		10,175	12,000
5120	12.5 ton cooling		.80	30		11,500	1,825		13,325	15,500
5130	15 ton cooling		.71	33.613		15,300	2,050		17,350	20,100
5140	17.5 ton cooling	↓	.65	36.697		15,800	2,250		18,050	20,900
5150	20 ton cooling	Q-7	.83	38.415		22,400	2,400		24,800	28,300
5160	25 ton cooling		.70	45.977		22,700	2,875		25,575	29,400
5170	30 ton cooling		.59	54.701		24,500	3,425		27,925	32,300
5180	40 ton cooling		.44	73.059		33,000	4,550		37,550	43,400
5190	50 ton cooling		.38	84.211		38,500	5,250		43,750	50,500
5200	60 ton cooling	↓	.28	114		50,000	7,125		57,125	66,000
5400	For low heat, add					7%				
5410	For high heat, add				↓	10%				
6000	Single zone electric cooling with variable volume distribution									
6020	20 ton cooling	Q-7	.72	44.199	Ea.	23,600	2,750		26,350	30,300
6030	25 ton cooling		.61	52.893		26,000	3,300		29,300	33,700
6040	30 ton cooling		.51	62.868		30,100	3,925		34,025	39,200
6050	40 ton cooling		.38	83.990		34,700	5,250		39,950	46,300
6060	50 ton cooling		.31	105		37,100	6,550		43,650	51,000
6070	60 ton cooling	↓	.25	126		49,700	7,850		57,550	66,500
6200	For low heat, add					7%				
6210	For high heat, add				↓	10%				
7000	Multizone, cool/heat, variable volume distribution									
7100	50 ton cooling	Q-7	.20	164	Ea.	117,500	10,200		127,700	145,500
7110	70 ton cooling		.14	229		164,500	14,300		178,800	203,000
7120	90 ton cooling		.11	296		180,000	18,500		198,500	226,500
7130	105 ton cooling		.10	333		214,000	20,800		234,800	267,500
7140	120 ton cooling		.08	381		244,500	23,800		268,300	306,000
7150	140 ton cooling	↓	.07	444	↓	261,000	27,700		288,700	330,000
7300	Penthouse unit, cool/heat, variable volume distribution									

23 74 Packaged Outdoor HVAC Equipment

23 74 33 – Dedicated Outdoor-Air Units

23 74 33.10 Rooftop Air Conditioners		Crew	Daily Output	Labor-Hours	Unit	2020 Bare Costs Material	Labor	Equipment	Total	Total Incl O&P
7400	150 ton cooling	Q-7	.10	311	Ea.	410,000	19,400		429,400	481,000
7410	170 ton cooling		.10	320		450,500	20,000		470,500	526,500
7420	200 ton cooling		.10	327		530,000	20,400		550,400	614,000
7430	225 ton cooling		.09	340		557,500	21,200		578,700	646,000
7440	250 ton cooling		.09	360		595,000	22,400		617,400	689,000
7450	270 ton cooling		.09	364		643,000	22,700		665,700	742,000
7460	300 ton cooling		.09	372		714,500	23,200		737,700	821,500
9000	Minimum labor/equipment charge	Q-5	1.50	10.667	Job		630		630	975
9400	Sub assemblies for assembly systems									
9410	Ductwork, per ton, rooftop 1 zone units	Q-9	.96	16.667	Ton	290	935		1,225	1,800
9420	Ductwork, per ton, rooftop multizone units		.50	32		375	1,800		2,175	3,225
9430	Ductwork, VAV cooling/ton, rooftop multizone, 1 zone/ton		.32	50		455	2,800		3,255	4,900
9440	Ductwork per ton, packaged water & air cooled units		1.26	12.698		61.50	710		771.50	1,200
9450	Ductwork per ton, split system remote condensing units		1.32	12.121		58	680		738	1,150
9500	Ductwork package for residential units, 1 ton		.80	20	Ea.	410	1,125		1,535	2,200
9520	2 ton		.40	40		825	2,250		3,075	4,425
9530	3 ton		.27	59.259		1,225	3,325		4,550	6,550
9540	4 ton	Q-10	.30	80		1,650	4,650		6,300	9,100
9550	5 ton		.23	104		2,050	6,075		8,125	11,800
9560	6 ton		.20	120		2,475	6,975		9,450	13,600
9570	7 ton		.17	141		2,875	8,200		11,075	16,100
9580	8 ton		.15	160		3,300	9,300		12,600	18,200

23 76 Evaporative Air-Cooling Equipment

23 76 13 – Direct Evaporative Air Coolers

23 76 13.10 Evaporative Coolers

			Crew	Daily Output	Labor-Hours	Unit	Material	Labor	Equipment	Total	Total Incl O&P
0010	**EVAPORATIVE COOLERS** (Swamp Coolers) ducted, not incl. duct.										
0100	Side discharge style, capacities at 0.25" S.P.										
0120	1,785 CFM, 1/3 HP, 115 V	G	Q-9	5	3.200	Ea.	465	179		644	790
0140	2,740 CFM, 1/3 HP, 115 V	G		4.50	3.556		565	199		764	930
0160	3,235 CFM, 1/2 HP, 115 V	G		4	4		575	224		799	980
0180	3,615 CFM, 1/2 HP, 230 V	G		3.60	4.444		695	249		944	1,150
0200	4,215 CFM, 3/4 HP, 230 V	G		3.20	5		745	280		1,025	1,250
0220	5,255 CFM, 1 HP, 115/230 V	G		3	5.333		1,150	299		1,449	1,725
0240	6,090 CFM, 1 HP, 230/460 V	G		2.80	5.714		2,225	320		2,545	2,950
0260	8,300 CFM, 1-1/2 HP, 230/460 V	G		2.60	6.154		1,950	345		2,295	2,700
0280	8,360 CFM, 1-1/2 HP, 230/460 V	G		2.20	7.273		2,000	410		2,410	2,850
0300	9,725 CFM, 2 HP, 230/460 V	G		1.80	8.889		2,200	500		2,700	3,200
0320	11,715 CFM, 3 HP, 230/460 V	G		1.40	11.429		2,550	640		3,190	3,800
0340	14,410 CFM, 5 HP, 230/460 V	G		1	16		3,000	895		3,895	4,700
0400	For two-speed motor, add						5%				
0500	For down discharge style, add						10%				

23 76 26 – Fan Coil Evaporators for Coolers/Freezers

23 76 26.10 Evaporators

		Crew	Daily Output	Labor-Hours	Unit	Material	Labor	Equipment	Total	Total Incl O&P
0010	**EVAPORATORS**, DX coils, remote compressors not included.									
1000	Coolers, reach-in type, above freezing temperatures									
1300	Shallow depth, wall mount, 7 fins per inch, air defrost									
1310	600 BTUH, 8" fan, rust-proof core	Q-5	3.80	4.211	Ea.	710	248		958	1,175
1320	900 BTUH, 8" fan, rust-proof core		3.30	4.848		825	286		1,111	1,350
1330	1,200 BTUH, 8" fan, rust-proof core		3	5.333		1,050	315		1,365	1,625

23 76 26 – Fan Coil Evaporators for Coolers/Freezers

23 76 26.10 Evaporators		Crew	Daily Output	Labor-Hours	Unit	Material	Labor	Equipment	Total	Total Incl O&P
						2020 Bare Costs				
1340	1,800 BTUH, 8" fan, rust-proof core	Q-5	2.40	6.667	Ea.	1,050	395		1,445	1,750
1350	2,500 BTUH, 10" fan, rust-proof core		2.20	7.273		1,100	430		1,530	1,900
1360	3,500 BTUH, 10" fan		2	8		1,175	470		1,645	2,000
1370	4,500 BTUH, 10" fan		1.90	8.421		1,225	495		1,720	2,125
1390	For pan drain, add					63.50			63.50	70
1600	Undercounter refrigerators, ceiling or wall mount,									
1610	8 fins per inch, air defrost, rust-proof core									
1630	800 BTUH, one 6" fan	Q-5	5	3.200	Ea.	287	189		476	605
1640	1,300 BTUH, two 6" fans		4	4		370	236		606	770
1650	1,700 BTUH, two 6" fans		3.60	4.444		415	262		677	860
2000	Coolers, reach-in and walk-in types, above freezing temperatures									
2600	Two-way discharge, ceiling mount, 150 to 4,100 CFM,									
2610	above 34°F applications, air defrost									
2630	900 BTUH, 7 fins per inch, 8" fan	Q-5	4	4	Ea.	565	236		801	990
2660	2,500 BTUH, 7 fins per inch, 10" fan		2.60	6.154		860	365		1,225	1,500
2690	5,500 BTUH, 7 fins per inch, 12" fan		1.70	9.412		1,425	555		1,980	2,425
2720	8,500 BTUH, 8 fins per inch, 16" fan		1.20	13.333		1,700	785		2,485	3,100
2750	15,000 BTUH, 7 fins per inch, 18" fan		1.10	14.545		2,650	860		3,510	4,225
2770	24,000 BTUH, 7 fins per inch, two 16" fans		1	16		3,950	945		4,895	5,800
2790	30,000 BTUH, 7 fins per inch, two 18" fans		.90	17.778		4,900	1,050		5,950	7,025
2850	Two-way discharge, low profile, ceiling mount,									
2860	8 fins per inch, 200 to 570 CFM, air defrost									
2880	800 BTUH, one 6" fan	Q-5	4	4	Ea.	340	236		576	740
2890	1,300 BTUH, two 6" fans		3.50	4.571		410	270		680	865
2900	1,800 BTUH, two 6" fans		3.30	4.848		495	286		781	985
2910	2,700 BTUH, three 6" fans		2.70	5.926		595	350		945	1,200
3000	Coolers, walk-in type, above freezing temperatures									
3300	General use, ceiling mount, 108 to 2,080 CFM									
3320	600 BTUH, 7 fins per inch, 6" fan	Q-5	5.20	3.077	Ea.	345	182		527	660
3340	1,200 BTUH, 7 fins per inch, 8" fan		3.70	4.324		480	255		735	920
3360	1,800 BTUH, 7 fins per inch, 10" fan		2.70	5.926		670	350		1,020	1,275
3380	3,500 BTUH, 7 fins per inch, 12" fan		2.40	6.667		985	395		1,380	1,675
3400	5,500 BTUH, 8 fins per inch, 12" fan		2	8		1,100	470		1,570	1,925
3430	8,500 BTUH, 8 fins per inch, 16" fan		1.40	11.429		1,650	675		2,325	2,850
3460	15,000 BTUH, 7 fins per inch, two 16" fans		1.10	14.545		2,625	860		3,485	4,225
3640	Low velocity, high latent load, ceiling mount,									
3650	1,050 to 2,420 CFM, air defrost, 6 fins per inch									
3670	6,700 BTUH, two 10" fans	Q-5	1.30	12.308	Ea.	1,775	725		2,500	3,075
3680	10,000 BTUH, three 10" fans		1	16		2,250	945		3,195	3,925
3690	13,500 BTUH, three 10" fans		1	16		2,850	945		3,795	4,575
3700	18,000 BTUH, four 10" fans		1	16		3,550	945		4,495	5,350
3710	26,500 BTUH, four 10" fans		.80	20		4,500	1,175		5,675	6,775
3730	For electric defrost, add					17%				
5000	Freezers and coolers, reach-in type, above 34°F									
5030	to sub-freezing temperature range, low latent load,									
5050	air defrost, 7 fins per inch									
5070	1,200 BTUH, 8" fan, rustproof core	Q-5	4.40	3.636	Ea.	730	215		945	1,125
5080	1,500 BTUH, 8" fan, rustproof core		4.20	3.810		755	225		980	1,175
5090	1,800 BTUH, 8" fan, rustproof core		3.70	4.324		780	255		1,035	1,250
5100	2,500 BTUH, 10" fan, rustproof core		2.90	5.517		855	325		1,180	1,450
5110	3,500 BTUH, 10" fan, rustproof core		2.50	6.400		870	380		1,250	1,550
5120	4,500 BTUH, 12" fan, rustproof core		2.10	7.619		1,075	450		1,525	1,875
6000	Freezers and coolers, walk-in type									

23 76 26.10 Evaporators		Crew	Daily Output	Labor-Hours	Unit	2020 Bare Costs Material	Labor	Equipment	Total	Total Incl O&P
6050	1,960 to 18,000 CFM, medium profile									
6060	Standard motor, 6 fins per inch, to -30°F, all aluminum									
6080	10,500 BTUH, air defrost, one 18" fan	Q-5	1.40	11.429	Ea.	1,675	675		2,350	2,875
6090	12,500 BTUH, air defrost, one 18" fan		1.40	11.429		1,800	675		2,475	3,025
6100	16,400 BTUH, air defrost, one 18" fan		1.30	12.308		1,875	725		2,600	3,200
6110	20,900 BTUH, air defrost, one 18" fan		1	16		2,200	945		3,145	3,875
6120	27,000 BTUH, air defrost, two 18" fans		1	16		3,100	945		4,045	4,850
6130	32,900 BTUH, air defrost, two 20" fans	Q-6	1.30	18.462		3,450	1,125		4,575	5,525
6140	39,000 BTUH, air defrost, two 20" fans		1.25	19.200		3,975	1,175		5,150	6,200
6150	44,100 BTUH, air defrost, three 20" fans		1	24		4,600	1,475		6,075	7,350
6155	53,000 BTUH, air defrost, three 20" fans		1	24		4,850	1,475		6,325	7,625
6160	66,200 BTUH, air defrost, four 20" fans		1	24		5,100	1,475		6,575	7,875
6170	78,000 BTUH, air defrost, four 20" fans		.80	30		5,875	1,825		7,700	9,300
6180	88,200 BTUH, air defrost, four 24" fans		.60	40		6,325	2,450		8,775	10,800
6190	110,000 BTUH, air defrost, four 24" fans		.50	48		7,650	2,925		10,575	13,000
6330	Hot gas defrost, standard motor units, add					35%				
6370	Electric defrost, 230V, standard motor, add					16%				
6410	460V, standard or high capacity motor, add					25%				
6800	Eight fins per inch, increases BTUH 75%									
6810	Standard capacity, add				Ea.	5%				
6830	Hot gas & electric defrost not recommended									
7000	800 to 4,150 CFM, 12" fans									
7010	Four fins per inch									
7030	3,400 BTUH, 1 fan	Q-5	2.80	5.714	Ea.	560	335		895	1,125
7040	4,200 BTUH, 1 fan		2.40	6.667		675	395		1,070	1,350
7050	5,300 BTUH, 1 fan		1.90	8.421		785	495		1,280	1,625
7060	6,800 BTUH, 1 fan		1.70	9.412		900	555		1,455	1,850
7070	8,400 BTUH, 2 fans		1.60	10		805	590		1,395	1,800
7080	10,500 BTUH, 2 fans		1.40	11.429		1,025	675		1,700	2,175
7090	13,000 BTUH, 3 fans		1.30	12.308		1,650	725		2,375	2,950
7100	17,000 BTUH, 4 fans		1.20	13.333		2,000	785		2,785	3,425
7110	21,500 BTUH, 5 fans		1	16		2,300	945		3,245	3,975
7160	Six fins per inch increases BTUH 35%, add					5%				
7180	Eight fins per inch increases BTUH 53%,									
7190	not recommended for below 35°F, add				Ea.	8%				
7210	For 12°F temperature differential, add					20%				
7230	For adjustable thermostat control, add					131			131	144
7240	For hot gas defrost, except 8 fin models									
7250	on applications below 35°F, add				Ea.	58%				
7260	For electric defrost, except 8 fin models									
7270	on applications below 35°F, add				Ea.	36%				
8000	Freezers, pass-thru door uprights, walk-in storage									
8300	Low temperature, thin profile, electric defrost									
8310	138 to 1,800 CFM, 5 fins per inch									
8320	900 BTUH, one 6" fan	Q-5	3.30	4.848	Ea.	435	286		721	920
8340	1,500 BTUH, two 6" fans		2.20	7.273		655	430		1,085	1,375
8360	2,600 BTUH, three 6" fans		1.90	8.421		915	495		1,410	1,775
8370	3,300 BTUH, four 6" fans		1.70	9.412		1,025	555		1,580	1,975
8380	4,400 BTUH, five 6" fans		1.40	11.429		1,425	675		2,100	2,625
8400	7,000 BTUH, three 10" fans		1.30	12.308		1,800	725		2,525	3,100
8410	8,700 BTUH, four 10" fans		1.10	14.545		2,250	860		3,110	3,800
8450	For air defrost, deduct					271			271	298

23 81 Decentralized Unitary HVAC Equipment

23 81 13 – Packaged Terminal Air-Conditioners

23 81 13.10 Packaged Cabinet Type Air-Conditioners		Crew	Daily Output	Labor-Hours	Unit	2020 Bare Costs Material	Labor	Equipment	Total	Total Incl O&P
0010	**PACKAGED CABINET TYPE AIR-CONDITIONERS**, Cabinet, wall sleeve,									
0100	louver, electric heat, thermostat, manual changeover, 208 V									
0200	6,000 BTUH cooling, 8,800 BTU heat	Q-5	6	2.667	Ea.	725	157		882	1,050
0220	9,000 BTUH cooling, 13,900 BTU heat		5	3.200		975	189		1,164	1,375
0240	12,000 BTUH cooling, 13,900 BTU heat		4	4		1,550	236		1,786	2,075
0260	15,000 BTUH cooling, 13,900 BTU heat		3	5.333		1,525	315		1,840	2,175
0280	18,000 BTUH cooling, 10 KW heat		2.40	6.667		2,700	395		3,095	3,575
0300	24,000 BTUH cooling, 10 KW heat		1.90	8.421		2,725	495		3,220	3,775
0320	30,000 BTUH cooling, 10 KW heat		1.40	11.429		2,900	675		3,575	4,250
0340	36,000 BTUH cooling, 10 KW heat		1.25	12.800		3,075	755		3,830	4,550
0360	42,000 BTUH cooling, 10 KW heat		1	16		3,225	945		4,170	5,000
0380	48,000 BTUH cooling, 10 KW heat		.90	17.778		3,450	1,050		4,500	5,425
0500	For hot water coil, increase heat by 10%, add					5%	10%			
1000	For steam, increase heat output by 30%, add					8%	10%			

23 81 19 – Self-Contained Air-Conditioners

23 81 19.10 Window Unit Air Conditioners		Crew	Daily Output	Labor-Hours	Unit	Material	Labor	Equipment	Total	Total Incl O&P
0010	**WINDOW UNIT AIR CONDITIONERS**									
4000	Portable/window, 15 amp, 125 V grounded receptacle required									
4060	5,000 BTUH	1 Carp	8	1	Ea.	305	53		358	425
4340	6,000 BTUH		8	1		239	53		292	350
4480	8,000 BTUH		6	1.333		385	71		456	540
4500	10,000 BTUH		6	1.333		650	71		721	830
4520	12,000 BTUH	L-2	8	2		2,000	93		2,093	2,350
4600	Window/thru-the-wall, 15 amp, 230 V grounded receptacle required									
4780	18,000 BTUH	L-2	6	2.667	Ea.	715	124		839	985
4940	25,000 BTUH		4	4		1,075	186		1,261	1,500
4960	29,000 BTUH		4	4		1,050	186		1,236	1,450
9000	Minimum labor/equipment charge	1 Carp	2	4	Job		213		213	340

23 81 19.20 Self-Contained Single Package		Crew	Daily Output	Labor-Hours	Unit	Material	Labor	Equipment	Total	Total Incl O&P
0010	**SELF-CONTAINED SINGLE PACKAGE**									
0100	Air cooled, for free blow or duct, not incl. remote condenser									
0110	Constant volume									
0200	3 ton cooling	Q-5	1	16	Ea.	4,050	945		4,995	5,900
0210	4 ton cooling	"	.80	20		4,400	1,175		5,575	6,675
0220	5 ton cooling	Q-6	1.20	20		4,525	1,225		5,750	6,875
0230	7.5 ton cooling	"	.90	26.667		5,650	1,625		7,275	8,750
0240	10 ton cooling	Q-7	1	32		8,350	2,000		10,350	12,300
0250	15 ton cooling		.95	33.684		10,300	2,100		12,400	14,600
0260	20 ton cooling		.90	35.556		13,800	2,225		16,025	18,600
0270	25 ton cooling		.85	37.647		20,700	2,350		23,050	26,300
0280	30 ton cooling		.80	40		26,400	2,500		28,900	33,000
0300	40 ton cooling		.60	53.333		37,300	3,325		40,625	46,200
0320	50 ton cooling		.50	64		50,500	4,000		54,500	62,000
0340	60 ton cooling	Q-8	.40	80		61,000	5,000	266	66,266	75,000
0490	For duct mounting, no price change									
0500	For steam heating coils, add				Ea.	10%	10%			
1000	Water cooled for free blow or duct, not including tower									
1010	Constant volume									
1100	3 ton cooling	Q-6	1	24	Ea.	3,750	1,475		5,225	6,400
1120	5 ton cooling		1	24		4,875	1,475		6,350	7,650
1130	7.5 ton cooling		.80	30		6,625	1,825		8,450	10,200
1140	10 ton cooling	Q-7	.90	35.556		8,725	2,225		10,950	13,000

23 81 19 – Self-Contained Air-Conditioners

23 81 19.20 Self-Contained Single Package		Crew	Daily Output	Labor-Hours	Unit	2020 Bare Costs Material	Labor	Equipment	Total	Total Incl O&P
1150	15 ton cooling	Q-7	.85	37.647	Ea.	12,400	2,350		14,750	17,300
1160	20 ton cooling		.80	40		27,700	2,500		30,200	34,300
1170	25 ton cooling		.75	42.667		32,600	2,650		35,250	39,900
1180	30 ton cooling		.70	45.714		36,600	2,850		39,450	44,700
1200	40 ton cooling		.40	80		44,500	5,000		49,500	56,500
1220	50 ton cooling		.30	107		50,500	6,650		57,150	66,000
1240	60 ton cooling	Q-8	.30	107		58,000	6,650	355	65,005	74,500
1300	For hot water or steam heat coils, add					12%	10%			
9000	Minimum labor/equipment charge	Q-5	1	16	Job		945		945	1,450

23 81 23 – Computer-Room Air-Conditioners

23 81 23.10 Computer Room Units		Crew	Daily Output	Labor-Hours	Unit	Material	Labor	Equipment	Total	Total Incl O&P
0010	**COMPUTER ROOM UNITS**									
1000	Air cooled, includes remote condenser but not									
1020	interconnecting tubing or refrigerant R236000-30									
1080	3 ton	Q-5	.50	32	Ea.	27,600	1,900		29,500	33,300
1120	5 ton		.45	35.556		29,500	2,100		31,600	35,700
1160	6 ton		.30	53.333		55,000	3,150		58,150	65,500
1200	8 ton		.27	59.259		55,000	3,500		58,500	66,000
1240	10 ton		.25	64		57,500	3,775		61,275	69,500
1280	15 ton		.22	72.727		63,500	4,300		67,800	76,000
1290	18 ton		.20	80		71,500	4,725		76,225	86,000
1300	20 ton	Q-6	.26	92.308		76,000	5,650		81,650	92,000
1320	22 ton		.24	100		77,000	6,125		83,125	94,000
1360	30 ton		.21	114		94,500	7,000		101,500	115,000
2200	Chilled water, for connection to									
2220	existing chiller system of adequate capacity									
2260	5 ton	Q-5	.74	21.622	Ea.	20,000	1,275		21,275	24,000
2280	6 ton		.52	30.769		19,800	1,825		21,625	24,600
2300	8 ton		.50	32		20,600	1,900		22,500	25,600
2320	10 ton		.49	32.653		20,700	1,925		22,625	25,800
2330	12 ton		.49	32.990		21,800	1,950		23,750	26,900
2360	15 ton		.48	33.333		22,600	1,975		24,575	28,000
2400	20 ton		.46	34.783		24,100	2,050		26,150	29,700
2440	25 ton	Q-6	.63	38.095		25,600	2,325		27,925	31,800
2460	30 ton		.57	42.105		27,800	2,575		30,375	34,600
2480	35 ton		.52	46.154		28,100	2,825		30,925	35,300
2500	45 ton		.46	52.174		29,700	3,200		32,900	37,500
2520	50 ton		.42	57.143		34,000	3,500		37,500	42,900
2540	60 ton		.38	63.158		36,200	3,875		40,075	45,800
2560	77 ton		.35	68.571		57,000	4,200		61,200	69,500
2580	95 ton		.32	75		66,500	4,600		71,100	80,500
4000	Glycol system, complete except for interconnecting tubing									
4060	3 ton	Q-5	.40	40	Ea.	34,700	2,350		37,050	41,800
4100	5 ton		.38	42.105		37,900	2,475		40,375	45,600
4120	6 ton		.25	64		53,500	3,775		57,275	65,000
4140	8 ton		.23	69.565		61,000	4,100		65,100	73,500
4160	10 ton		.21	76.190		65,000	4,500		69,500	78,500
4180	12 ton		.19	84.211		65,500	4,975		70,475	79,500
4200	15 ton	Q-6	.26	92.308		80,500	5,650		86,150	97,000
4240	20 ton		.24	100		88,000	6,125		94,125	106,500
4280	22 ton		.23	104		92,500	6,375		98,875	112,000
4430	30 ton		.22	109		115,000	6,675		121,675	137,000

23 81 Decentralized Unitary HVAC Equipment

23 81 23 – Computer-Room Air-Conditioners

23 81 23.10 Computer Room Units		Crew	Daily Output	Labor-Hours	Unit	Material	Labor	Equipment	Total	Total Incl O&P
						2020 Bare Costs				
8000	Water cooled system, not including condenser,									
8020	water supply or cooling tower									
8060	3 ton	Q-5	.62	25.806	Ea.	28,100	1,525		29,625	33,300
8100	5 ton		.54	29.630		31,700	1,750		33,450	37,600
8120	6 ton		.35	45.714		40,600	2,700		43,300	48,900
8140	8 ton		.33	48.485		50,000	2,850		52,850	59,500
8160	10 ton		.31	51.613		52,000	3,050		55,050	61,500
8180	12 ton		.29	55.172		52,500	3,250		55,750	62,500
8200	15 ton		.27	59.259		61,000	3,500		64,500	72,500
8240	20 ton	Q-6	.38	63.158		65,500	3,875		69,375	78,000
8280	22 ton		.35	68.571		68,500	4,200		72,700	82,000
8300	30 ton		.30	80		85,500	4,900		90,400	101,500

23 81 26 – Split-System Air-Conditioners

23 81 26.10 Split Ductless Systems

		Crew	Daily Output	Labor-Hours	Unit	Material	Labor	Equipment	Total	Total Incl O&P
0010	**SPLIT DUCTLESS SYSTEMS**									
0100	Cooling only, single zone									
0110	Wall mount									
0120	3/4 ton cooling	Q-5	2	8	Ea.	1,125	470		1,595	1,950
0130	1 ton cooling		1.80	8.889		1,250	525		1,775	2,175
0140	1-1/2 ton cooling		1.60	10		1,975	590		2,565	3,075
0150	2 ton cooling		1.40	11.429		2,275	675		2,950	3,550
0160	3 ton cooling		1.20	13.333		2,275	785		3,060	3,725
1000	Ceiling mount									
1020	2 ton cooling	Q-5	1.40	11.429	Ea.	2,000	675		2,675	3,250
1030	3 ton cooling	"	1.20	13.333	"	2,575	785		3,360	4,075
3000	Multizone									
3010	Wall mount									
3020	2 @ 3/4 ton cooling	Q-5	1.80	8.889	Ea.	3,425	525		3,950	4,575
5000	Cooling/Heating									
5010	Wall mount									
5110	1 ton cooling	Q-5	1.70	9.412	Ea.	1,300	555		1,855	2,275
5120	1-1/2 ton cooling	"	1.50	10.667	"	2,000	630		2,630	3,175
7000	Accessories for all split ductless systems									
7010	Add for ambient frost control	Q-5	8	2	Ea.	129	118		247	325
7020	Add for tube/wiring kit (line sets)									
7030	15′ kit	Q-5	32	.500	Ea.	91	29.50		120.50	146
7036	25′ kit		28	.571		158	33.50		191.50	225
7040	35′ kit		24	.667		205	39.50		244.50	287
7050	50′ kit		20	.800		206	47		253	299

23 81 29 – Variable Refrigerant Flow HVAC Systems

23 81 29.10 Heat Pump, Gas Driven

		Crew	Daily Output	Labor-Hours	Unit	Material	Labor	Equipment	Total	Total Incl O&P
0010	**HEAT PUMP, GAS DRIVEN**, Variable refrigerant volume (VRV) type									
0020	Not including interconnecting tubing or multi-zone controls									
1000	For indoor fan VRV type AHU see 23 82 19.40									
1010	Multi-zone split									
1020	Outdoor unit									
1100	8 tons cooling, for up to 17 zones	Q-5	1.30	12.308	Ea.	30,200	725		30,925	34,300
1110	Isolation rails		2.60	6.154	Pair	1,075	365		1,440	1,725
1160	15 tons cooling, for up to 33 zones		1	16	Ea.	38,600	945		39,545	43,900
1170	Isolation rails		2.60	6.154	Pair	1,250	365		1,615	1,925
2000	Packaged unit									
2020	Outdoor unit									

23 81 29 – Variable Refrigerant Flow HVAC Systems

23 81 29.10 Heat Pump, Gas Driven		Crew	Daily Output	Labor-Hours	Unit	Material	2020 Bare Costs Labor	Equipment	Total	Total Incl O&P
2200	11 tons cooling	Q-5	1.30	12.308	Ea.	38,900	725		39,625	43,900
2210	Roof curb adapter	"	2.60	6.154		675	365		1,040	1,300
2220	Thermostat	1 Stpi	1.30	6.154		405	405		810	1,075

23 81 43 – Air-Source Unitary Heat Pumps

23 81 43.10 Air-Source Heat Pumps

		Crew	Daily Output	Labor-Hours	Unit	Material	Labor	Equipment	Total	Total Incl O&P
0010	**AIR-SOURCE HEAT PUMPS**, Not including interconnecting tubing									
1000	Air to air, split system, not including curbs, pads, fan coil and ductwork									
1012	Outside condensing unit only, for fan coil see Section 23 82 19.10									
1015	1.5 ton cooling, 7 MBH heat @ 0°F	Q-5	2.40	6.667	Ea.	1,625	395		2,020	2,400
1020	2 ton cooling, 8.5 MBH heat @ 0°F		2	8		1,775	470		2,245	2,675
1030	2.5 ton cooling, 10 MBH heat @ 0°F		1.60	10		1,950	590		2,540	3,050
1040	3 ton cooling, 13 MBH heat @ 0°F		1.20	13.333		2,150	785		2,935	3,575
1050	3.5 ton cooling, 18 MBH heat @ 0°F		1	16		2,275	945		3,220	3,950
1054	4 ton cooling, 24 MBH heat @ 0°F		.80	20		2,475	1,175		3,650	4,550
1060	5 ton cooling, 27 MBH heat @ 0°F		.50	32		2,700	1,900		4,600	5,900
1080	7.5 ton cooling, 33 MBH heat @ 0°F		.45	35.556		3,975	2,100		6,075	7,625
1100	10 ton cooling, 50 MBH heat @ 0°F	Q-6	.64	37.500		6,575	2,300		8,875	10,800
1120	15 ton cooling, 64 MBH heat @ 0°F		.50	48		10,400	2,925		13,325	16,000
1130	20 ton cooling, 85 MBH heat @ 0°F		.35	68.571		18,600	4,200		22,800	27,000
1140	25 ton cooling, 119 MBH heat @ 0°F		.25	96		21,800	5,875		27,675	33,100
1500	Single package, not including curbs, pads, or plenums									
1502	0.5 ton cooling, supplementary heat included	Q-5	8	2	Ea.	3,550	118		3,668	4,075
1504	0.75 ton cooling, supplementary heat included		6	2.667		4,150	157		4,307	4,825
1506	1 ton cooling, supplementary heat included		4	4		3,425	236		3,661	4,150
1510	1.5 ton cooling, 5 MBH heat @ 0°F		1.55	10.323		3,375	610		3,985	4,675
1520	2 ton cooling, 6.5 MBH heat @ 0°F		1.50	10.667		3,325	630		3,955	4,650
1540	2.5 ton cooling, 8 MBH heat @ 0°F		1.40	11.429		3,300	675		3,975	4,675
1560	3 ton cooling, 10 MBH heat @ 0°F		1.20	13.333		3,775	785		4,560	5,375
1570	3.5 ton cooling, 11 MBH heat @ 0°F		1	16		4,125	945		5,070	6,000
1580	4 ton cooling, 13 MBH heat @ 0°F		.96	16.667		4,425	985		5,410	6,400
1620	5 ton cooling, 27 MBH heat @ 0°F		.65	24.615		5,125	1,450		6,575	7,900
1640	7.5 ton cooling, 35 MBH heat @ 0°F		.40	40		7,900	2,350		10,250	12,300
1648	10 ton cooling, 45 MBH heat @ 0°F	Q-6	.40	60		10,600	3,675		14,275	17,400
1652	12 ton cooling, 50 MBH heat @ 0°F	"	.36	66.667		11,300	4,075		15,375	18,700
1696	Supplementary electric heat coil incl., unless noted otherwise									
6000	Air to water, single package, excluding storage tank and ductwork									
6010	Includes circulating water pump, air duct connections, digital temperature									
6020	controller with remote tank temp. probe and sensor for storage tank.									
6040	Water heating - air cooling capacity									
6110	35.5 MBH heat water, 2.3 ton cool air	Q-5	1.60	10	Ea.	17,300	590		17,890	20,000
6120	58 MBH heat water, 3.8 ton cool air		1.10	14.545		19,800	860		20,660	23,000
6130	76 MBH heat water, 4.9 ton cool air		.87	18.391		23,600	1,075		24,675	27,600
6140	98 MBH heat water, 6.5 ton cool air		.62	25.806		30,500	1,525		32,025	35,900
6150	113 MBH heat water, 7.4 ton cool air		.59	27.119		33,300	1,600		34,900	39,100
6160	142 MBH heat water, 9.2 ton cool air		.52	30.769		40,500	1,825		42,325	47,400
6170	171 MBH heat water, 11.1 ton cool air		.49	32.653		47,500	1,925		49,425	55,000

23 81 46 – Water-Source Unitary Heat Pumps

23 81 46.10 Water Source Heat Pumps

		Crew	Daily Output	Labor-Hours	Unit	Material	Labor	Equipment	Total	Total Incl O&P
0010	**WATER SOURCE HEAT PUMPS**, Not incl. connecting tubing or water source									
2000	Water source to air, single package									
2100	1 ton cooling, 13 MBH heat @ 75°F	Q-5	2	8	Ea.	2,125	470		2,595	3,075
2120	1.5 ton cooling, 17 MBH heat @ 75°F		1.80	8.889		2,075	525		2,600	3,075

23 81 Decentralized Unitary HVAC Equipment

23 81 46 – Water-Source Unitary Heat Pumps

23 81 46.10 Water Source Heat Pumps		Crew	Daily Output	Labor-Hours	Unit	2020 Bare Costs Material	Labor	Equipment	Total	Total Incl O&P
2140	2 ton cooling, 19 MBH heat @ 75°F	Q-5	1.70	9.412	Ea.	2,475	555		3,030	3,575
2160	2.5 ton cooling, 25 MBH heat @ 75°F		1.60	10		2,700	590		3,290	3,875
2180	3 ton cooling, 27 MBH heat @ 75°F		1.40	11.429		2,750	675		3,425	4,075
2190	3.5 ton cooling, 29 MBH heat @ 75°F		1.30	12.308		2,975	725		3,700	4,400
2200	4 ton cooling, 31 MBH heat @ 75°F		1.20	13.333		3,125	785		3,910	4,650
2220	5 ton cooling, 29 MBH heat @ 75°F		.90	17.778		3,400	1,050		4,450	5,350
2240	7.5 ton cooling, 35 MBH heat @ 75°F		.60	26.667		6,900	1,575		8,475	10,000
2250	8 ton cooling, 40 MBH heat @ 75°F		.58	27.586		7,000	1,625		8,625	10,200
2260	10 ton cooling, 50 MBH heat @ 75°F		.53	30.189		7,675	1,775		9,450	11,200
2280	15 ton cooling, 64 MBH heat @ 75°F	Q-6	.47	51.064		14,900	3,125		18,025	21,100
2300	20 ton cooling, 100 MBH heat @ 75°F		.41	58.537		16,200	3,575		19,775	23,400
2310	25 ton cooling, 100 MBH heat @ 75°F		.32	75		22,100	4,600		26,700	31,400
2320	30 ton cooling, 128 MBH heat @ 75°F		.24	102		24,200	6,250		30,450	36,300
2340	40 ton cooling, 200 MBH heat @ 75°F		.21	117		34,200	7,175		41,375	48,700
2360	50 ton cooling, 200 MBH heat @ 75°F		.15	160		38,600	9,800		48,400	57,500
3960	For supplementary heat coil, add					10%				
4000	For increase in capacity thru use									
4020	of solar collector, size boiler at 60%									
9000	Minimum labor/equipment charge	Q-5	1.75	9.143	Job		540		540	835

23 82 Convection Heating and Cooling Units

23 82 13 – Valance Heating and Cooling Units

23 82 13.16 Valance Units

Code	Description	Crew	Daily Output	Labor-Hours	Unit	Material	Labor	Equipment	Total	Total Incl O&P
0010	**VALANCE UNITS**									
6000	Valance units, complete with 1/2" cooling coil, enclosure									
6020	2 tube	Q-5	18	.889	L.F.	37	52.50		89.50	122
6040	3 tube		16	1		44.50	59		103.50	140
6060	4 tube		16	1		52	59		111	149
6080	5 tube		15	1.067		57.50	63		120.50	161
6100	6 tube		15	1.067		62.50	63		125.50	166
6120	8 tube		14	1.143		91	67.50		158.50	204
6200	For 3/4" cooling coil, add					10%				

23 82 16 – Air Coils

23 82 16.10 Flanged Coils

Code	Description	Crew	Daily Output	Labor-Hours	Unit	Material	Labor	Equipment	Total	Total Incl O&P
0010	**FLANGED COILS**									
0500	Chilled water cooling, 6 rows, 24" x 48"	Q-5	3.20	5	Ea.	4,650	295		4,945	5,550
1000	Direct expansion cooling, 6 rows, 24" x 48"		2.80	5.714		5,175	335		5,510	6,225
1500	Hot water heating, 1 row, 24" x 48"		4	4		1,850	236		2,086	2,425
2000	Steam heating, 1 row, 24" x 48"		3.06	5.229		2,625	310		2,935	3,350
9000	Minimum labor/equipment charge	1 Plum	1	8	Job		515		515	795

23 82 16.20 Duct Heaters

Code	Description	Crew	Daily Output	Labor-Hours	Unit	Material	Labor	Equipment	Total	Total Incl O&P
0010	**DUCT HEATERS**, Electric, 480 V, 3 Ph.									
0020	Finned tubular insert, 500°F									
0100	8" wide x 6" high, 4.0 kW	Q-20	16	1.250	Ea.	825	71.50		896.50	1,025
0120	12" high, 8.0 kW		15	1.333		1,375	76		1,451	1,625
0140	18" high, 12.0 kW		14	1.429		1,925	81.50		2,006.50	2,225
0160	24" high, 16.0 kW		13	1.538		2,475	88		2,563	2,850
0180	30" high, 20.0 kW		12	1.667		3,025	95		3,120	3,475
0300	12" wide x 6" high, 6.7 kW		15	1.333		875	76		951	1,075
0320	12" high, 13.3 kW		14	1.429		1,425	81.50		1,506.50	1,675

23 82 16 – Air Coils

23 82 16.20 Duct Heaters		Crew	Daily Output	Labor-Hours	Unit	Material	Labor	Equipment	Total	Total Incl O&P
						2020 Bare Costs				
0340	18" high, 20.0 kW	Q-20	13	1.538	Ea.	1,975	88		2,063	2,300
0360	24" high, 26.7 kW		12	1.667		2,550	95		2,645	2,950
0380	30" high, 33.3 kW		11	1.818		3,125	104		3,229	3,575
0500	18" wide x 6" high, 13.3 kW		14	1.429		965	81.50		1,046.50	1,200
0520	12" high, 26.7 kW		13	1.538		1,625	88		1,713	1,900
0540	18" high, 40.0 kW		12	1.667		2,150	95		2,245	2,525
0560	24" high, 53.3 kW		11	1.818		2,875	104		2,979	3,300
0580	30" high, 66.7 kW		10	2		3,575	114		3,689	4,125
0700	24" wide x 6" high, 17.8 kW		13	1.538		1,075	88		1,163	1,300
0720	12" high, 35.6 kW		12	1.667		1,750	95		1,845	2,075
0740	18" high, 53.3 kW		11	1.818		2,525	104		2,629	2,925
0760	24" high, 71.1 kW		10	2		3,175	114		3,289	3,675
0780	30" high, 88.9 kW		9	2.222		3,900	127		4,027	4,500
0900	30" wide x 6" high, 22.2 kW		12	1.667		1,125	95		1,220	1,400
0920	12" high, 44.4 kW		11	1.818		1,850	104		1,954	2,200
0940	18" high, 66.7 kW		10	2		2,675	114		2,789	3,125
0960	24" high, 88.9 kW		9	2.222		3,500	127		3,627	4,050
0980	30" high, 111.0 kW		8	2.500		4,125	143		4,268	4,775
1400	Note decreased kW available for									
1410	each duct size at same cost									
1420	See line 5000 for modifications and accessories									
2000	Finned tubular flange with insulated									
2020	terminal box, 500°F									
2100	12" wide x 36" high, 54 kW	Q-20	10	2	Ea.	4,075	114		4,189	4,650
2120	40" high, 60 kW		9	2.222		4,675	127		4,802	5,350
2200	24" wide x 36" high, 118.8 kW		9	2.222		5,175	127		5,302	5,875
2220	40" high, 132 kW		8	2.500		5,250	143		5,393	6,000
2400	36" wide x 8" high, 40 kW		11	1.818		2,100	104		2,204	2,475
2420	16" high, 80 kW		10	2		2,750	114		2,864	3,225
2440	24" high, 120 kW		9	2.222		3,625	127		3,752	4,175
2460	32" high, 160 kW		8	2.500		4,625	143		4,768	5,325
2480	36" high, 180 kW		7	2.857		5,700	163		5,863	6,525
2500	40" high, 200 kW		6	3.333		6,325	190		6,515	7,250
2600	40" wide x 8" high, 45 kW		11	1.818		2,150	104		2,254	2,525
2620	16" high, 90 kW		10	2		2,975	114		3,089	3,450
2640	24" high, 135 kW		9	2.222		3,775	127		3,902	4,350
2660	32" high, 180 kW		8	2.500		5,025	143		5,168	5,750
2680	36" high, 202.5 kW		7	2.857		5,800	163		5,963	6,625
2700	40" high, 225 kW		6	3.333		6,600	190		6,790	7,550
2800	48" wide x 8" high, 54.8 kW		10	2		2,275	114		2,389	2,675
2820	16" high, 109.8 kW		9	2.222		3,150	127		3,277	3,650
2840	24" high, 164.4 kW		8	2.500		4,000	143		4,143	4,625
2860	32" high, 219.2 kW		7	2.857		5,325	163		5,488	6,125
2880	36" high, 246.6 kW		6	3.333		6,125	190		6,315	7,050
2900	40" high, 274 kW		5	4		6,925	229		7,154	7,975
3000	56" wide x 8" high, 64 kW		9	2.222		2,575	127		2,702	3,050
3020	16" high, 128 kW		8	2.500		3,575	143		3,718	4,150
3040	24" high, 192 kW		7	2.857		4,325	163		4,488	5,025
3060	32" high, 256 kW		6	3.333		6,050	190		6,240	6,975
3080	36" high, 288 kW		5	4		6,950	229		7,179	8,000
3100	40" high, 320 kW		4	5		7,675	286		7,961	8,900
3200	64" wide x 8" high, 74 kW		8	2.500		2,650	143		2,793	3,150
3220	16" high, 148 kW		7	2.857		3,675	163		3,838	4,300

23 82 16 – Air Coils

23 82 16.20 Duct Heaters		Crew	Daily Output	Labor-Hours	Unit	Material	2020 Bare Costs Labor	Equipment	Total	Total Incl O&P
3240	24" high, 222 kW	Q-20	6	3.333	Ea.	4,675	190		4,865	5,450
3260	32" high, 296 kW		5	4		6,275	229		6,504	7,250
3280	36" high, 333 kW		4	5		7,475	286		7,761	8,675
3300	40" high, 370 kW		3	6.667		8,250	380		8,630	9,675
3800	Note decreased kW available for									
3820	each duct size at same cost									
5000	Duct heater modifications and accessories									
5120	T.C.O. limit auto or manual reset	Q-20	42	.476	Ea.	135	27		162	192
5140	Thermostat		28	.714		580	41		621	700
5160	Overheat thermocouple (removable)		7	2.857		820	163		983	1,150
5180	Fan interlock relay		18	1.111		197	63.50		260.50	315
5200	Air flow switch		20	1		171	57		228	277
5220	Split terminal box cover		100	.200		56.50	11.45		67.95	80
8000	To obtain BTU multiply kW by 3413									
9900	Minimum labor/equipment charge	1 Shee	4	2	Job		125		125	195

23 82 19 – Fan Coil Units

23 82 19.10 Fan Coil Air Conditioning		Crew	Daily Output	Labor-Hours	Unit	Material	Labor	Equipment	Total	Total Incl O&P
0010	**FAN COIL AIR CONDITIONING**									
0030	Fan coil AC, cabinet mounted, filters and controls									
0100	Chilled water, 1/2 ton cooling	Q-5	8	2	Ea.	575	118		693	815
0110	3/4 ton cooling		7	2.286		700	135		835	980
0120	1 ton cooling		6	2.667		820	157		977	1,150
0140	1-1/2 ton cooling		5.50	2.909		865	172		1,037	1,225
0150	2 ton cooling		5.25	3.048		1,175	180		1,355	1,550
0160	2-1/2 ton cooling		5	3.200		1,800	189		1,989	2,275
0180	3 ton cooling		4	4		1,975	236		2,211	2,550
0262	For hot water coil, add					40%	10%			
0320	1 ton cooling	Q-5	6	2.667	Ea.	1,750	157		1,907	2,175
0940	Direct expansion, for use w/air cooled condensing unit, 1-1/2 ton cooling		5	3.200		685	189		874	1,050
0950	2 ton cooling		4.80	3.333		670	197		867	1,050
0960	2-1/2 ton cooling		4.40	3.636		785	215		1,000	1,200
0970	3 ton cooling		3.80	4.211		960	248		1,208	1,425
0980	3-1/2 ton cooling		3.60	4.444		990	262		1,252	1,500
0990	4 ton cooling		3.40	4.706		995	278		1,273	1,525
1000	5 ton cooling		3	5.333		1,250	315		1,565	1,850
1020	7-1/2 ton cooling		3	5.333		1,775	315		2,090	2,425
1040	10 ton cooling	Q-6	2.60	9.231		2,250	565		2,815	3,375
1042	12-1/2 ton cooling		2.30	10.435		2,875	640		3,515	4,125
1050	15 ton cooling		1.60	15		2,950	920		3,870	4,675
1060	20 ton cooling		.70	34.286		4,325	2,100		6,425	8,000
1070	25 ton cooling		.65	36.923		5,225	2,250		7,475	9,250
1080	30 ton cooling		.60	40		5,675	2,450		8,125	10,000
1500	For hot water coil, add					40%	10%			
1512	For condensing unit add see Section 23 62									
3000	Chilled water, horizontal unit, housing, 2 pipe, fan control, no valves									
3100	1/2 ton cooling	Q-5	8	2	Ea.	1,425	118		1,543	1,750
3110	1 ton cooling		6	2.667		1,675	157		1,832	2,100
3120	1-1/2 ton cooling		5.50	2.909		1,875	172		2,047	2,350
3130	2 ton cooling		5.25	3.048		2,250	180		2,430	2,750
3140	3 ton cooling		4	4		2,775	236		3,011	3,425
3150	3-1/2 ton cooling		3.80	4.211		3,075	248		3,323	3,750
3160	4 ton cooling		3.80	4.211		3,075	248		3,323	3,750

23 82 19.10 Fan Coil Air Conditioning		Crew	Daily Output	Labor-Hours	Unit	Material	Labor	Equipment	Total	Total Incl O&P
3170	5 ton cooling	Q-5	3.40	4.706	Ea.	3,550	278		3,828	4,325
3180	6 ton cooling		3.40	4.706		3,550	278		3,828	4,325
3190	7 ton cooling		2.90	5.517		3,875	325		4,200	4,750
3200	8 ton cooling		2.90	5.517		3,875	325		4,200	4,750
3210	10 ton cooling	Q-6	2.80	8.571		4,050	525		4,575	5,250
3212	12-1/2 ton cooling		2.50	9.600		3,350	585		3,935	4,575
3214	15 ton cooling		2.30	10.435		3,975	640		4,615	5,350
3216	20 ton cooling		2.10	11.429		4,575	700		5,275	6,125
3218	25 ton cooling		1.80	13.333		6,775	815		7,590	8,700
3220	30 ton cooling		1.60	15		8,325	920		9,245	10,600
3240	For hot water coil, add					40%	10%			
9000	Minimum labor/equipment charge	Q-5	3.75	4.267	Job		252		252	390

23 82 19.20 Heating and Ventilating Units		Crew	Daily Output	Labor-Hours	Unit	Material	Labor	Equipment	Total	Total Incl O&P
0010	**HEATING AND VENTILATING UNITS**, Classroom units									
0020	Includes filter, heating/cooling coils, standard controls									
0080	750 CFM, 2 tons cooling	Q-6	2	12	Ea.	4,750	735		5,485	6,375
0100	1,000 CFM, 2-1/2 tons cooling		1.60	15		5,550	920		6,470	7,525
0120	1,250 CFM, 3 tons cooling		1.40	17.143		5,725	1,050		6,775	7,925
0140	1,500 CFM, 4 tons cooling		.80	30		6,125	1,825		7,950	9,600
0160	2,000 CFM, 5 tons cooling		.50	48		7,200	2,925		10,125	12,500
0500	For electric heat, add					35%				
1000	For no cooling, deduct					25%	10%			

23 82 19.40 Fan Coil Air Conditioning			Crew	Daily Output	Labor-Hours	Unit	Material	Labor	Equipment	Total	Total Incl O&P
0010	**FAN COIL AIR CONDITIONING**, Variable refrigerant volume (VRV) type										
0020	Not including interconnecting tubing or multi-zone controls										
0030	For VRV condensing unit see section 23 81 29.10										
0050	Indoor type, ducted										
0100	Vertical concealed										
0130	1 ton cooling	G	Q-5	2.97	5.387	Ea.	2,650	320		2,970	3,400
0140	1.5 ton cooling	G		2.77	5.776		2,675	340		3,015	3,475
0150	2 ton cooling	G		2.70	5.926		2,900	350		3,250	3,725
0160	2.5 ton cooling	G		2.60	6.154		3,050	365		3,415	3,900
0170	3 ton cooling	G		2.50	6.400		3,100	380		3,480	3,975
0180	3.5 ton cooling	G		2.30	6.957		3,250	410		3,660	4,200
0190	4 ton cooling	G		2.19	7.306		3,275	430		3,705	4,275
0200	4.5 ton cooling	G		2.08	7.692		3,600	455		4,055	4,650
0230	Outside air connection possible										
1100	Ceiling concealed										
1130	0.6 ton cooling	G	Q-5	2.60	6.154	Ea.	1,725	365		2,090	2,450
1140	0.75 ton cooling	G		2.60	6.154		1,775	365		2,140	2,525
1150	1 ton cooling	G		2.60	6.154		1,875	365		2,240	2,625
1152	Screening door	G	1 Stpi	5.20	1.538		104	101		205	271
1160	1.5 ton cooling	G	Q-5	2.60	6.154		1,950	365		2,315	2,700
1162	Screening door	G	1 Stpi	5.20	1.538		121	101		222	289
1170	2 ton cooling	G	Q-5	2.60	6.154		2,300	365		2,665	3,075
1172	Screening door	G	1 Stpi	5.20	1.538		104	101		205	271
1180	2.5 ton cooling	G	Q-5	2.60	6.154		2,650	365		3,015	3,475
1182	Screening door	G	1 Stpi	5.20	1.538		181	101		282	355
1190	3 ton cooling	G	Q-5	2.60	6.154		2,825	365		3,190	3,650
1192	Screening door	G	1 Stpi	5.20	1.538		181	101		282	355
1200	4 ton cooling	G	Q-5	2.30	6.957		2,675	410		3,085	3,575
1202	Screening door	G	1 Stpi	5.20	1.538		181	101		282	355

23 82 19 – Fan Coil Units

23 82 19.40 Fan Coil Air Conditioning			Crew	Daily Output	Labor-Hours	Unit	Material	2020 Bare Costs Labor	Equipment	Total	Total Incl O&P
1220	6 ton cooling	G	Q-5	2.08	7.692	Ea.	4,900	455		5,355	6,100
1230	8 ton cooling	G	"	1.89	8.466	↓	5,500	500		6,000	6,825
1260	Outside air connection possible										
4050	Indoor type, duct-free										
4100	Ceiling mounted cassette										
4130	0.75 ton cooling	G	Q-5	3.46	4.624	Ea.	2,050	273		2,323	2,675
4140	1 ton cooling	G		2.97	5.387		2,175	320		2,495	2,900
4150	1.5 ton cooling	G		2.77	5.776		2,250	340		2,590	2,975
4160	2 ton cooling	G		2.60	6.154		2,375	365		2,740	3,175
4170	2.5 ton cooling	G		2.50	6.400		2,450	380		2,830	3,275
4180	3 ton cooling	G		2.30	6.957		2,625	410		3,035	3,500
4190	4 ton cooling	G	↓	2.08	7.692		2,925	455		3,380	3,925
4198	For ceiling cassette decoration panel, add	G	1 Stpi	5.20	1.538	↓	310	101		411	500
4230	Outside air connection possible										
4500	Wall mounted										
4530	0.6 ton cooling	G	Q-5	5.20	3.077	Ea.	1,050	182		1,232	1,425
4540	0.75 ton cooling	G		5.20	3.077		1,100	182		1,282	1,475
4550	1 ton cooling	G		4.16	3.846		1,250	227		1,477	1,725
4560	1.5 ton cooling	G		3.77	4.244		1,375	250		1,625	1,900
4570	2 ton cooling	G	↓	3.46	4.624		1,500	273		1,773	2,075
4590	Condensate pump for the above air handlers	G	1 Stpi	6.46	1.238	↓	225	81		306	375
4700	Floor standing unit										
4730	1 ton cooling	G	Q-5	2.60	6.154	Ea.	1,775	365		2,140	2,500
4740	1.5 ton cooling	G		2.60	6.154		2,000	365		2,365	2,775
4750	2 ton cooling	G	↓	2.60	6.154	↓	2,150	365		2,515	2,925
4800	Floor standing unit, concealed										
4830	1 ton cooling	G	Q-5	2.60	6.154	Ea.	1,700	365		2,065	2,425
4840	1.5 ton cooling	G		2.60	6.154		1,925	365		2,290	2,675
4850	2 ton cooling	G	↓	2.60	6.154	↓	2,075	365		2,440	2,825
4880	Outside air connection possible										
8000	Accessories										
8100	Branch divergence pipe fitting										
8110	Capacity under 76 MBH		1 Stpi	2.50	3.200	Ea.	164	210		374	505
8120	76 MBH to 112 MBH			2.50	3.200		188	210		398	530
8130	112 MBH to 234 MBH		↓	2.50	3.200	↓	530	210		740	905
8200	Header pipe fitting										
8210	Max 4 branches										
8220	Capacity under 76 MBH		1 Stpi	1.70	4.706	Ea.	233	310		543	730
8260	Max 8 branches										
8270	76 MBH to 112 MBH		1 Stpi	1.70	4.706	Ea.	475	310		785	1,000
8280	112 MBH to 234 MBH		"	1.30	6.154	"	755	405		1,160	1,450

23 82 29 – Radiators

23 82 29.10 Hydronic Heating		Crew	Daily Output	Labor-Hours	Unit	Material	Labor	Equipment	Total	Total Incl O&P
0010	**HYDRONIC HEATING**, Terminal units, not incl. main supply pipe									
1000	Radiation									
1100	Panel, baseboard, C.I., including supports, no covers	Q-5	46	.348	L.F.	49	20.50		69.50	85.50
3000	Radiators, cast iron									
3100	Free standing or wall hung, 6 tube, 25" high	Q-5	96	.167	Section	61	9.85		70.85	82
3150	4 tube, 25" high		96	.167		47.50	9.85		57.35	67
3200	4 tube, 19" high	↓	96	.167	↓	42	9.85		51.85	61
3250	Adj. brackets, 2 per wall radiator up to 30 sections	1 Stpi	32	.250	Ea.	66	16.40		82.40	98
3500	Recessed, 20" high x 5" deep, without grille	Q-5	60	.267	Section	51.50	15.75		67.25	81.50

23 82 29 – Radiators

23 82 29.10 Hydronic Heating		Crew	Daily Output	Labor-Hours	Unit	Material	2020 Bare Costs Labor	Equipment	Total	Total Incl O&P
3525	Free standing or wall hung, 30" high	Q-5	60	.267	Section	46.50	15.75		62.25	75.50
3600	For inlet grille, add				↓	6.40			6.40	7.05
9000	Minimum labor/equipment charge	Q-5	3.20	5	Job		295		295	455
9500	To convert SFR to BTU rating: Hot water, 150 x SFR									
9510	Forced hot water, 180 x SFR; steam, 240 x SFR									

23 82 33 – Convectors

23 82 33.10 Convector Units		Crew	Daily Output	Labor-Hours	Unit	Material	Labor	Equipment	Total	Total Incl O&P
0010	**CONVECTOR UNITS**, Terminal units, not incl. main supply pipe									
2204	Convector, multifin, 2 pipe w/cabinet									
2210	17" H x 24" L	Q-5	10	1.600	Ea.	110	94.50		204.50	267
2214	17" H x 36" L		8.60	1.860		165	110		275	350
2218	17" H x 48" L		7.40	2.162		220	128		348	440
2222	21" H x 24" L		9	1.778		124	105		229	298
2226	21" H x 36" L		8.20	1.951		185	115		300	380
2228	21" H x 48" L	↓	6.80	2.353	↓	247	139		386	485
2240	For knob operated damper, add					140%				
2241	For metal trim strips, add	Q-5	64	.250	Ea.	14.25	14.75		29	38.50
2243	For snap-on inlet grille, add					10%	10%			
2245	For hinged access door, add	Q-5	64	.250	Ea.	40.50	14.75		55.25	67.50
2246	For air chamber, auto-venting, add	"	58	.276	"	8.75	16.30		25.05	34.50

23 82 36 – Finned-Tube Radiation Heaters

23 82 36.10 Finned Tube Radiation		Crew	Daily Output	Labor-Hours	Unit	Material	Labor	Equipment	Total	Total Incl O&P
0010	**FINNED TUBE RADIATION**, Terminal units, not incl. main supply pipe									
1150	Fin tube, wall hung, 14" slope top cover, with damper									
1200	1-1/4" copper tube, 4-1/4" alum. fin	Q-5	38	.421	L.F.	52.50	25		77.50	96.50
1250	1-1/4" steel tube, 4-1/4" steel fin		36	.444		43	26		69	88
1255	2" steel tube, 4-1/4" steel fin		32	.500		47	29.50		76.50	97
1310	Baseboard, pkgd, 1/2" copper tube, alum. fin, 7" high		60	.267		11.95	15.75		27.70	37.50
1320	3/4" copper tube, alum. fin, 7" high		58	.276		8.05	16.30		24.35	34
1340	1" copper tube, alum. fin, 8-7/8" high		56	.286		21	16.85		37.85	49
1360	1-1/4" copper tube, alum. fin, 8-7/8" high		54	.296		31	17.50		48.50	61.50
1380	1-1/4" IPS steel tube with steel fins	↓	52	.308	↓	32.50	18.15		50.65	63.50
1500	Note: fin tube may also require corners, caps, etc.									

23 82 39 – Unit Heaters

23 82 39.16 Propeller Unit Heaters		Crew	Daily Output	Labor-Hours	Unit	Material	Labor	Equipment	Total	Total Incl O&P
0010	**PROPELLER UNIT HEATERS**									
3950	Unit heaters, propeller, 115 V 2 psi steam, 60°F entering air									
4000	Horizontal, 12 MBH	Q-5	12	1.333	Ea.	415	78.50		493.50	580
4020	28.2 MBH		10	1.600		540	94.50		634.50	740
4040	36.5 MBH		8	2		625	118		743	865
4060	43.9 MBH		8	2		625	118		743	865
4080	56.5 MBH		7.50	2.133		690	126		816	955
4100	65.6 MBH		7	2.286		710	135		845	990
4120	87.6 MBH		6.50	2.462		750	145		895	1,050
4140	96.8 MBH		6	2.667		880	157		1,037	1,225
4160	133.3 MBH		5	3.200		995	189		1,184	1,400
4180	157.6 MBH		4	4		1,225	236		1,461	1,725
4200	197.7 MBH		3	5.333		1,400	315		1,715	2,000
4220	257.2 MBH		2.50	6.400		1,600	380		1,980	2,350
4240	286.9 MBH		2	8		1,825	470		2,295	2,725
4260	364 MBH	↓	1.80	8.889	↓	2,325	525		2,850	3,350

23 82 Convection Heating and Cooling Units

23 82 39 – Unit Heaters

23 82 39.16 Propeller Unit Heaters		Crew	Daily Output	Labor-Hours	Unit	Material	2020 Bare Costs Labor	Equipment	Total	Total Incl O&P
4270	404 MBH	Q-5	1.60	10	Ea.	2,375	590		2,965	3,525
4300	Vertical diffuser same price									
4310	Vertical flow, 40 MBH	Q-5	11	1.455	Ea.	640	86		726	840
4314	58.5 MBH		8	2		635	118		753	880
4318	92 MBH		7	2.286		790	135		925	1,075
4322	109.7 MBH		6	2.667		955	157		1,112	1,300
4326	131 MBH		4	4		955	236		1,191	1,425
4330	160 MBH		3	5.333		1,025	315		1,340	1,600
4334	194 MBH		2.20	7.273		1,175	430		1,605	1,975
4338	212 MBH		2.10	7.619		1,450	450		1,900	2,300
4342	247 MBH		1.96	8.163		1,675	480		2,155	2,600
4346	297 MBH		1.80	8.889		1,900	525		2,425	2,900
4350	333 MBH	Q-6	1.90	12.632		2,000	775		2,775	3,400
4354	420 MBH (460 V)		1.80	13.333		2,525	815		3,340	4,025
4358	500 MBH (460 V)		1.71	14.035		3,125	860		3,985	4,775
4362	570 MBH (460 V)		1.40	17.143		4,075	1,050		5,125	6,100
4366	620 MBH (460 V)		1.30	18.462		4,725	1,125		5,850	6,950
4370	960 MBH (460 V)		1.10	21.818		8,200	1,325		9,525	11,100

23 83 Radiant Heating Units

23 83 16 – Radiant-Heating Hydronic Piping

23 83 16.10 Radiant Floor Heating		Crew	Daily Output	Labor-Hours	Unit	Material	Labor	Equipment	Total	Total Incl O&P
0010	**RADIANT FLOOR HEATING**									
0100	Tubing, PEX (cross-linked polyethylene)									
0110	Oxygen barrier type for systems with ferrous materials									
0120	1/2"	Q-5	800	.020	L.F.	1.05	1.18		2.23	2.98
0130	3/4"		535	.030		1.55	1.76		3.31	4.44
0140	1"		400	.040		2.32	2.36		4.68	6.20
0200	Non barrier type for ferrous free systems									
0210	1/2"	Q-5	800	.020	L.F.	.57	1.18		1.75	2.45
0220	3/4"		535	.030		1	1.76		2.76	3.83
0230	1"		400	.040		1.79	2.36		4.15	5.60
1000	Manifolds									
1110	Brass									
1120	With supply and return valves, flow meter, thermometer,									
1122	auto air vent and drain/fill valve.									
1130	1", 2 circuit	Q-5	14	1.143	Ea.	355	67.50		422.50	500
1140	1", 3 circuit		13.50	1.185		405	70		475	555
1150	1", 4 circuit		13	1.231		435	72.50		507.50	590
1154	1", 5 circuit		12.50	1.280		525	75.50		600.50	690
1158	1", 6 circuit		12	1.333		590	78.50		668.50	765
1162	1", 7 circuit		11.50	1.391		640	82		722	825
1166	1", 8 circuit		11	1.455		685	86		771	890
1172	1", 9 circuit		10.50	1.524		760	90		850	975
1174	1", 10 circuit		10	1.600		795	94.50		889.50	1,025
1178	1", 11 circuit		9.50	1.684		820	99.50		919.50	1,050
1182	1", 12 circuit		9	1.778		930	105		1,035	1,175
1610	Copper manifold header (cut to size)									
1620	1" header, 12 circuit 1/2" sweat outlets	Q-5	3.33	4.805	Ea.	117	283		400	570
1630	1-1/4" header, 12 circuit 1/2" sweat outlets		3.20	5		136	295		431	605
1640	1-1/4" header, 12 circuit 3/4" sweat outlets		3	5.333		143	315		458	640

23 83 16 – Radiant-Heating Hydronic Piping

23 83 16.10 Radiant Floor Heating		Crew	Daily Output	Labor-Hours	Unit	Material	2020 Bare Costs Labor	Equipment	Total	Total Incl O&P
1650	1-1/2" header, 12 circuit 3/4" sweat outlets	Q-5	3.10	5.161	Ea.	173	305		478	660
1660	2" header, 12 circuit 3/4" sweat outlets	↓	2.90	5.517	↓	257	325		582	790
3000	Valves									
3110	Thermostatic zone valve actuator with end switch	Q-5	40	.400	Ea.	51.50	23.50		75	93
3114	Thermostatic zone valve actuator	"	36	.444	"	98	26		124	149
3120	Motorized straight zone valve with operator complete									
3130	3/4"	Q-5	35	.457	Ea.	160	27		187	218
3140	1"		32	.500		174	29.50		203.50	237
3150	1-1/4"	↓	29.60	.541	↓	219	32		251	291
3500	4 way mixing valve, manual, brass									
3530	1"	Q-5	13.30	1.203	Ea.	230	71		301	365
3540	1-1/4"		11.40	1.404		249	83		332	400
3550	1-1/2"		11	1.455		315	86		401	480
3560	2"		10.60	1.509		455	89		544	640
3800	Mixing valve motor, 4 way for valves, 1" and 1-1/4"		34	.471		390	28		418	470
3810	Mixing valve motor, 4 way for valves, 1-1/2" and 2"	↓	30	.533	↓	435	31.50		466.50	530
5000	Radiant floor heating, zone control panel									
5120	4 zone actuator valve control, expandable	Q-5	20	.800	Ea.	163	47		210	253
5130	6 zone actuator valve control, expandable		18	.889		271	52.50		323.50	380
6070	Thermal track, straight panel for long continuous runs, 5.333 S.F.		40	.400		36	23.50		59.50	76.50
6080	Thermal track, utility panel, for direction reverse at run end, 5.333 S.F.		40	.400		36	23.50		59.50	76.50
6090	Combination panel, for direction reverse plus straight run, 5.333 S.F.	↓	40	.400	↓	36	23.50		59.50	76.50
7000	PEX tubing fittings									
7100	Compression type									
7116	Coupling									
7120	1/2" x 1/2"	1 Stpi	27	.296	Ea.	6.95	19.40		26.35	37.50
7124	3/4" x 3/4"	"	23	.348	"	16.35	23		39.35	53.50
7130	Adapter									
7132	1/2" x female sweat 1/2"	1 Stpi	27	.296	Ea.	4.51	19.40		23.91	35
7134	1/2" x female sweat 3/4"		26	.308		5.05	20		25.05	36.50
7136	5/8" x female sweat 3/4"	↓	24	.333	↓	7.25	22		29.25	42
7140	Elbow									
7142	1/2" x female sweat 1/2"	1 Stpi	27	.296	Ea.	7.30	19.40		26.70	38
7144	1/2" x female sweat 3/4"		26	.308		8.55	20		28.55	40.50
7146	5/8" x female sweat 3/4"	↓	24	.333	↓	9.60	22		31.60	44.50
7200	Insert type									
7206	PEX x male NPT									
7210	1/2" x 1/2"	1 Stpi	29	.276	Ea.	2.94	18.10		21.04	31
7220	3/4" x 3/4"		27	.296		4.35	19.40		23.75	35
7230	1" x 1"	↓	26	.308	↓	7.30	20		27.30	39
7300	PEX coupling									
7310	1/2" x 1/2"	1 Stpi	30	.267	Ea.	.64	17.50		18.14	27.50
7320	3/4" x 3/4"		29	.276		.81	18.10		18.91	29
7330	1" x 1"	↓	28	.286	↓	1.40	18.75		20.15	30.50
7400	PEX stainless crimp ring									
7410	1/2" x 1/2"	1 Stpi	86	.093	Ea.	.56	6.10		6.66	10.05
7420	3/4" x 3/4"		84	.095		.77	6.25		7.02	10.50
7430	1" x 1"	↓	82	.098	↓	1.11	6.40		7.51	11.10

23 83 33 – Electric Radiant Heaters

23 83 33.10 Electric Heating		Crew	Daily Output	Labor-Hours	Unit	Material	2020 Bare Costs Labor	Equipment	Total	Total Incl O&P
0010	**ELECTRIC HEATING**, not incl. conduit or feed wiring									
1100	Rule of thumb: Baseboard units, including control	1 Elec	4.40	1.818	kW	116	112		228	300
1300	Baseboard heaters, 2′ long, 350 watt		8	1	Ea.	29	61.50		90.50	126
1400	3′ long, 750 watt		8	1		35	61.50		96.50	133
1600	4′ long, 1,000 watt		6.70	1.194		40.50	73.50		114	158
1800	5′ long, 935 watt		5.70	1.404		44	86		130	182
2000	6′ long, 1,500 watt		5	1.600		57	98		155	214
2200	7′ long, 1,310 watt		4.40	1.818		61	112		173	240
2400	8′ long, 2,000 watt		4	2		70	123		193	266
2600	9′ long, 1,680 watt		3.60	2.222		86.50	136		222.50	305
2800	10′ long, 1,875 watt	↓	3.30	2.424	↓	162	149		311	405
2950	Wall heaters with fan, 120 to 277 volt									
3160	Recessed, residential, 750 watt	1 Elec	6	1.333	Ea.	94.50	82		176.50	230
3170	1,000 watt		6	1.333		94.50	82		176.50	230
3180	1,250 watt		5	1.600		107	98		205	269
3190	1,500 watt		4	2		107	123		230	305
3210	2,000 watt		4	2		99	123		222	298
3230	2,500 watt		3.50	2.286		236	140		376	475
3240	3,000 watt		3	2.667		365	164		529	655
3250	4,000 watt		2.70	2.963		370	182		552	685
3260	Commercial, 750 watt		6	1.333		211	82		293	360
3270	1,000 watt		6	1.333		211	82		293	360
3280	1,250 watt		5	1.600		211	98		309	385
3290	1,500 watt		4	2		195	123		318	405
3300	2,000 watt		4	2		211	123		334	420
3310	2,500 watt		3.50	2.286		211	140		351	450
3320	3,000 watt		3	2.667		425	164		589	720
3330	4,000 watt		2.70	2.963		500	182		682	830
3600	Thermostats, integral		16	.500		29	30.50		59.50	79
3800	Line voltage, 1 pole		8	1		16.30	61.50		77.80	112
3810	2 pole	↓	8	1	↓	23.50	61.50		85	121
4000	Heat trace system, 400 degree									
4020	115 V, 2.5 watts/L.F.	1 Elec	530	.015	L.F.	11	.93		11.93	13.55
4030	5 watts/L.F.		530	.015		11	.93		11.93	13.55
4050	10 watts/L.F.		530	.015		9.20	.93		10.13	11.55
4060	208 V, 5 watts/L.F.		530	.015		11	.93		11.93	13.55
4080	480 V, 8 watts/L.F.	↓	530	.015	↓	11	.93		11.93	13.55
4200	Heater raceway									
4260	Heat transfer cement									
4280	1 gallon				Ea.	64.50			64.50	71
4300	5 gallon				"	252			252	277
4320	Cable tie									
4340	3/4″ pipe size	1 Elec	470	.017	Ea.	.02	1.04		1.06	1.63
4360	1″ pipe size		444	.018		.02	1.11		1.13	1.72
4380	1-1/4″ pipe size		400	.020		.02	1.23		1.25	1.91
4400	1-1/2″ pipe size		355	.023		.02	1.38		1.40	2.15
4420	2″ pipe size		320	.025		.05	1.53		1.58	2.42
4440	3″ pipe size		160	.050		.05	3.07		3.12	4.78
4460	4″ pipe size		100	.080		.06	4.91		4.97	7.60
4480	Thermostat NEMA 3R, 22 amp, 0-150 degree, 10′ cap.		8	1		193	61.50		254.50	310
4500	Thermostat NEMA 4X, 25 amp, 40 degree, 5-1/2′ cap.		7	1.143		248	70		318	380
4520	Thermostat NEMA 4X, 22 amp, 25-325 degree, 10′ cap.	↓	7	1.143	↓	660	70		730	835

23 83 33 – Electric Radiant Heaters

23 83 33.10 Electric Heating		Crew	Daily Output	Labor-Hours	Unit	Material	Labor	Equipment	Total	Total Incl O&P
						2020 Bare Costs				
4540	Thermostat NEMA 4X, 22 amp, 15-140 degree	1 Elec	6	1.333	Ea.	560	82		642	740
4580	Thermostat NEMA 4, 7, 9, 22 amp, 25-325 degree, 10′ cap.		3.60	2.222		810	136		946	1,100
4600	Thermostat NEMA 4, 7, 9, 22 amp, 15-140 degree		3	2.667		815	164		979	1,150
4720	Fiberglass application tape, 36 yard roll		11	.727		75	44.50		119.50	151
5000	Radiant heating ceiling panels, 2′ x 4′, 500 watt		16	.500		405	30.50		435.50	490
5050	750 watt		16	.500		430	30.50		460.50	520
5200	For recessed plaster frame, add		32	.250		138	15.35		153.35	176
5300	Infrared quartz heaters, 120 volts, 1,000 watt		6.70	1.194		350	73.50		423.50	500
5350	1,500 watt		5	1.600		335	98		433	520
5400	240 volts, 1,500 watt		5	1.600		405	98		503	595
5450	2,000 watt		4	2		335	123		458	560
5500	3,000 watt		3	2.667		330	164		494	615
5550	4,000 watt		2.60	3.077		380	189		569	710
5570	Modulating control	↓	.80	10	↓	141	615		756	1,100
5600	Unit heaters, heavy duty, with fan & mounting bracket									
5650	Single phase, 208-240-277 volt, 3 kW	1 Elec	6	1.333	Ea.	705	82		787	900
5750	5 kW		5.50	1.455		800	89		889	1,025
5800	7 kW		5	1.600		1,375	98		1,473	1,650
5850	10 kW		4	2		1,100	123		1,223	1,400
5950	15 kW		3.80	2.105		1,850	129		1,979	2,225
6000	480 volt, 3 kW		6	1.333		470	82		552	640
6020	4 kW		5.80	1.379		700	84.50		784.50	900
6040	5 kW		5.50	1.455		710	89		799	915
6060	7 kW		5	1.600		1,150	98		1,248	1,425
6080	10 kW		4	2		1,100	123		1,223	1,425
6100	13 kW		3.80	2.105		1,950	129		2,079	2,350
6120	15 kW		3.70	2.162		2,150	133		2,283	2,575
6140	20 kW		3.50	2.286		3,075	140		3,215	3,625
6300	3 phase, 208-240 volt, 5 kW		5.50	1.455		575	89		664	770
6320	7 kW		5	1.600		940	98		1,038	1,175
6340	10 kW		4	2		805	123		928	1,075
6360	15 kW		3.70	2.162		1,600	133		1,733	1,950
6380	20 kW		3.50	2.286		2,500	140		2,640	2,975
6400	25 kW		3.30	2.424		3,725	149		3,874	4,325
6500	480 volt, 5 kW		5.50	1.455		650	89		739	850
6520	7 kW		5	1.600		930	98		1,028	1,175
6540	10 kW		4	2		860	123		983	1,125
6560	13 kW		3.80	2.105		1,650	129		1,779	2,025
6580	15 kW		3.70	2.162		1,475	133		1,608	1,825
6600	20 kW		3.50	2.286		2,325	140		2,465	2,775
6620	25 kW		3.30	2.424		3,450	149		3,599	4,025
6630	30 kW		3	2.667		3,025	164		3,189	3,575
6640	40 kW		2	4		4,050	245		4,295	4,825
6650	50 kW	↓	1.60	5	↓	6,675	305		6,980	7,825
6800	Vertical discharge heaters, with fan									
6820	Single phase, 208-240-277 volt, 10 kW	1 Elec	4	2	Ea.	860	123		983	1,125
6840	15 kW		3.70	2.162		1,650	133		1,783	2,000
6900	3 phase, 208-240 volt, 10 kW		4	2		1,050	123		1,173	1,375
6920	15 kW		3.70	2.162		1,600	133		1,733	1,975
6940	20 kW		3.50	2.286		2,925	140		3,065	3,450
6960	25 kW		3.30	2.424		3,175	149		3,324	3,725
6980	30 kW		3	2.667		4,350	164		4,514	5,025
7000	40 kW	↓	2	4	↓	4,325	245		4,570	5,125

23 83 Radiant Heating Units

23 83 33 – Electric Radiant Heaters

23 83 33.10 Electric Heating		Crew	Daily Output	Labor-Hours	Unit	Material	Labor	Equipment	Total	Total Incl O&P
						2020 Bare Costs				
7020	50 kW	1 Elec	1.60	5	Ea.	6,950	305		7,255	8,125
7100	480 volt, 10 kW		4	2		1,200	123		1,323	1,525
7120	15 kW		3.70	2.162		1,950	133		2,083	2,350
7140	20 kW		3.50	2.286		2,650	140		2,790	3,150
7160	25 kW		3.30	2.424		3,450	149		3,599	4,025
7180	30 kW		3	2.667		3,800	164		3,964	4,425
7200	40 kW		2	4		3,925	245		4,170	4,700
7410	Sill height convector heaters, 5" high x 2' long, 500 watt		6.70	1.194		365	73.50		438.50	520
7420	3' long, 750 watt		6.50	1.231		435	75.50		510.50	590
7430	4' long, 1,000 watt		6.20	1.290		500	79		579	670
7440	5' long, 1,250 watt		5.50	1.455		570	89		659	760
7450	6' long, 1,500 watt		4.80	1.667		645	102		747	865
7460	8' long, 2,000 watt		3.60	2.222		875	136		1,011	1,175
7470	10' long, 2,500 watt		3	2.667		1,100	164		1,264	1,450
7900	Cabinet convector heaters, 240 volt, three phase,									
7920	3' long, 2,000 watt	1 Elec	5.30	1.509	Ea.	2,300	92.50		2,392.50	2,675
7940	3,000 watt		5.30	1.509		2,400	92.50		2,492.50	2,800
7960	4,000 watt		5.30	1.509		2,475	92.50		2,567.50	2,850
7980	6,000 watt		4.60	1.739		2,675	107		2,782	3,125
8000	8,000 watt		4.60	1.739		2,750	107		2,857	3,200
8020	4' long, 4,000 watt		4.60	1.739		2,500	107		2,607	2,925
8040	6,000 watt		4	2		2,600	123		2,723	3,050
8060	8,000 watt		4	2		2,675	123		2,798	3,125
8080	10,000 watt		4	2		2,700	123		2,823	3,175
8100	Available also in 208 or 277 volt									
8200	Cabinet unit heaters, 120 to 277 volt, 1 pole,									
8220	wall mounted, 2 kW	1 Elec	4.60	1.739	Ea.	2,075	107		2,182	2,475
8230	3 kW		4.60	1.739		2,075	107		2,182	2,450
8240	4 kW		4.40	1.818		2,150	112		2,262	2,550
8250	5 kW		4.40	1.818		2,275	112		2,387	2,675
8260	6 kW		4.20	1.905		2,250	117		2,367	2,650
8270	8 kW		4	2		2,300	123		2,423	2,725
8280	10 kW		3.80	2.105		2,350	129		2,479	2,775
8290	12 kW		3.50	2.286		2,575	140		2,715	3,075
8300	13.5 kW		2.90	2.759		3,650	169		3,819	4,275
8310	16 kW		2.70	2.963		3,025	182		3,207	3,625
8320	20 kW		2.30	3.478		3,825	213		4,038	4,525
8330	24 kW		1.90	4.211		4,000	258		4,258	4,825
8350	Recessed, 2 kW		4.40	1.818		2,275	112		2,387	2,675
8370	3 kW		4.40	1.818		2,325	112		2,437	2,725
8380	4 kW		4.20	1.905		2,200	117		2,317	2,600
8390	5 kW		4.20	1.905		2,300	117		2,417	2,700
8400	6 kW		4	2		2,350	123		2,473	2,775
8410	8 kW		3.80	2.105		2,525	129		2,654	3,000
8420	10 kW		3.50	2.286		3,150	140		3,290	3,700
8430	12 kW		2.90	2.759		3,175	169		3,344	3,750
8440	13.5 kW		2.70	2.963		3,050	182		3,232	3,625
8450	16 kW		2.30	3.478		3,250	213		3,463	3,900
8460	20 kW		1.90	4.211		3,775	258		4,033	4,575
8470	24 kW		1.60	5		3,725	305		4,030	4,575
8490	Ceiling mounted, 2 kW		3.20	2.500		2,100	153		2,253	2,525
8510	3 kW		3.20	2.500		2,200	153		2,353	2,650
8520	4 kW		3	2.667		2,250	164		2,414	2,725

23 83 33.10 Electric Heating		Crew	Daily Output	Labor-Hours	Unit	Material	2020 Bare Costs Labor	Equipment	Total	Total Incl O&P
8530	5 kW	1 Elec	3	2.667	Ea.	2,325	164		2,489	2,800
8540	6 kW		2.80	2.857		2,325	175		2,500	2,850
8550	8 kW		2.40	3.333		2,450	205		2,655	3,025
8560	10 kW		2.20	3.636		3,050	223		3,273	3,700
8570	12 kW		2	4		2,625	245		2,870	3,275
8580	13.5 kW		1.50	5.333		2,550	325		2,875	3,300
8590	16 kW		1.30	6.154		2,675	380		3,055	3,500
8600	20 kW		.90	8.889		3,700	545		4,245	4,925
8610	24 kW		.60	13.333		3,725	820		4,545	5,350
8630	208 to 480 V, 3 pole									
8650	Wall mounted, 2 kW	1 Elec	4.60	1.739	Ea.	2,025	107		2,132	2,425
8670	3 kW		4.60	1.739		2,375	107		2,482	2,800
8680	4 kW		4.40	1.818		2,425	112		2,537	2,850
8690	5 kW		4.40	1.818		2,500	112		2,612	2,925
8700	6 kW		4.20	1.905		2,525	117		2,642	2,950
8710	8 kW		4	2		2,625	123		2,748	3,075
8720	10 kW		3.80	2.105		2,775	129		2,904	3,250
8730	12 kW		3.50	2.286		2,825	140		2,965	3,325
8740	13.5 kW		2.90	2.759		2,825	169		2,994	3,350
8750	16 kW		2.70	2.963		2,925	182		3,107	3,500
8760	20 kW		2.30	3.478		4,475	213		4,688	5,250
8770	24 kW		1.90	4.211		4,200	258		4,458	5,000
8790	Recessed, 2 kW		4.40	1.818		2,250	112		2,362	2,650
8810	3 kW		4.40	1.818		2,300	112		2,412	2,725
8820	4 kW		4.20	1.905		2,400	117		2,517	2,825
8830	5 kW		4.20	1.905		2,425	117		2,542	2,850
8840	6 kW		4	2		2,350	123		2,473	2,775
8850	8 kW		3.80	2.105		2,425	129		2,554	2,875
8860	10 kW		3.50	2.286		2,450	140		2,590	2,925
8870	12 kW		2.90	2.759		2,750	169		2,919	3,250
8880	13.5 kW		2.70	2.963		2,375	182		2,557	2,900
8890	16 kW		2.30	3.478		5,175	213		5,388	6,025
8900	20 kW		1.90	4.211		4,300	258		4,558	5,125
8920	24 kW		1.60	5		4,350	305		4,655	5,275
8940	Ceiling mount, 2 kW		3.20	2.500		2,325	153		2,478	2,775
8950	3 kW		3.20	2.500		2,150	153		2,303	2,575
8960	4 kW		3	2.667		2,475	164		2,639	2,975
8970	5 kW		3	2.667		2,175	164		2,339	2,625
8980	6 kW		2.80	2.857		2,475	175		2,650	3,000
8990	8 kW		2.40	3.333		2,550	205		2,755	3,150
9000	10 kW		2.20	3.636		2,350	223		2,573	2,950
9020	13.5 kW		1.50	5.333		3,150	325		3,475	3,975
9030	16 kW		1.30	6.154		4,300	380		4,680	5,325
9040	20 kW		.90	8.889		4,300	545		4,845	5,575
9060	24 kW		.60	13.333		5,325	820		6,145	7,100
9230	13.5 kW, 40,956 BTU		2.20	3.636		2,075	223		2,298	2,625
9250	24 kW, 81,912 BTU		2	4		2,025	245		2,270	2,600
9990	Minimum labor/equipment charge		4	2	Job		123		123	189

23 84 Humidity Control Equipment

23 84 13 – Humidifiers

23 84 13.10 Humidifier Units		Crew	Daily Output	Labor-Hours	Unit	Material	Labor	Equipment	Total	Total Incl O&P
0010	**HUMIDIFIER UNITS**									
0520	Steam, room or duct, filter, regulators, auto. controls, 220 V									
0540	11 lb./hr. R236000-30	Q-5	6	2.667	Ea.	2,775	157		2,932	3,300
0560	22 lb./hr.		5	3.200		3,050	189		3,239	3,675
0580	33 lb./hr.		4	4		3,150	236		3,386	3,825
0600	50 lb./hr.		4	4		3,850	236		4,086	4,600
0620	100 lb./hr.		3	5.333		5,175	315		5,490	6,175
0640	150 lb./hr.		2.50	6.400		6,550	380		6,930	7,775
0660	200 lb./hr.	↓	2	8	↓	8,225	470		8,695	9,775
0700	With blower									
0720	11 lb./hr.	Q-5	5.50	2.909	Ea.	4,175	172		4,347	4,875
0740	22 lb./hr.		4.75	3.368		4,800	199		4,999	5,575
0760	33 lb./hr.		3.75	4.267		4,650	252		4,902	5,525
0780	50 lb./hr.		3.50	4.571		5,450	270		5,720	6,425
0800	100 lb./hr.		2.75	5.818		6,075	345		6,420	7,225
0820	150 lb./hr.		2	8		9,075	470		9,545	10,700
0840	200 lb./hr.	↓	1.50	10.667	↓	11,300	630		11,930	13,500
5000	Furnace type, wheel bypass									
5020	10 GPD	1 Stpi	4	2	Ea.	203	131		334	425
5040	14 GPD		3.80	2.105		211	138		349	445
5060	19 GPD	↓	3.60	2.222	↓	305	146		451	565
9000	Minimum labor/equipment charge	Q-5	3.50	4.571	Job		270		270	415

23 84 16 – Mechanical Dehumidification Units

23 84 16.10 Dehumidifier Units		Crew	Daily Output	Labor-Hours	Unit	Material	Labor	Equipment	Total	Total Incl O&P
0010	**DEHUMIDIFIER UNITS** R236000-30									
6000	Self contained with filters and standard controls									
6040	1.5 lb./hr., 50 CFM	1 Plum	8	1	Ea.	8,575	64.50		8,639.50	9,525
6060	3 lb./hr., 150 CFM	Q-1	12	1.333		9,200	77.50		9,277.50	10,200
6065	6 lb./hr., 150 CFM		9	1.778		15,000	103		15,103	16,700
6070	16 to 20 lb./hr., 600 CFM		5	3.200		28,300	186		28,486	31,500
6080	30 to 40 lb./hr., 1,125 CFM		4	4		44,100	232		44,332	48,900
6090	60 to 75 lb./hr., 2,250 CFM		3	5.333		57,500	310		57,810	64,000
6100	120 to 155 lb./hr., 4,500 CFM		2	8		102,000	465		102,465	112,500
6110	240 to 310 lb./hr., 9,000 CFM	↓	1.50	10.667		146,000	620		146,620	161,500
6120	400 to 515 lb./hr., 15,000 CFM	Q-2	1.60	15		183,000	900		183,900	202,500
6130	530 to 690 lb./hr., 20,000 CFM		1.40	17.143		198,000	1,025		199,025	219,500
6140	800 to 1030 lb./hr., 30,000 CFM		1.20	20		225,500	1,200		226,700	250,000
6150	1060 to 1375 lb./hr., 40,000 CFM	↓	1	24	↓	314,000	1,450		315,450	347,500

Estimating Tips

26 05 00 Common Work Results for Electrical

- Conduit should be taken off in three main categories—power distribution, branch power, and branch lighting—so the estimator can concentrate on systems and components, therefore making it easier to ensure all items have been accounted for.
- For cost modifications for elevated conduit installation, add the percentages to labor according to the height of installation and only to the quantities exceeding the different height levels, not to the total conduit quantities. Refer to subdivision 26 01 02.20 for labor adjustment factors.
- Remember that aluminum wiring of equal ampacity is larger in diameter than copper and may require larger conduit.
- If more than three wires at a time are being pulled, deduct percentages from the labor hours of that grouping of wires.
- When taking off grounding systems, identify separately the type and size of wire, and list each unique type of ground connection.
- The estimator should take the weights of materials into consideration when completing a takeoff. Topics to consider include: How will the materials be supported? What methods of support are available? How high will the support structure have to reach? Will the final support structure be able to withstand the total burden? Is the support material included or separate from the fixture, equipment, and material specified?
- Do not overlook the costs for equipment used in the installation. If scaffolding or highlifts are available in the field, contractors may use them in lieu of the proposed ladders and rolling staging.

26 20 00 Low-Voltage Electrical Transmission

- Supports and concrete pads may be shown on drawings for the larger equipment, or the support system may be only a piece of plywood for the back of a panelboard. In either case, they must be included in the costs.

26 40 00 Electrical and Cathodic Protection

- When taking off cathodic protection systems, identify the type and size of cable, and list each unique type of anode connection.

26 50 00 Lighting

- Fixtures should be taken off room by room using the fixture schedule, specifications, and the ceiling plan. For large concentrations of lighting fixtures in the same area, deduct the percentages from labor hours.

Reference Numbers

Reference numbers are shown at the beginning of some major classifications. These numbers refer to related items in the Reference Section. The reference information may be an estimating procedure, an alternate pricing method, or technical information.

Note: Not all subdivisions listed here necessarily appear. ■

Note: "Powered in part by CINX™, based on liscensed proprietary information of Harrison Publishing House, Inc."

Note: The following companies, in part, have been used as a reference source for some of the material prices used in Division 26:

Electriflex

Trade Service

26 01 02 – Labor Adjustment

	26 01 02.20 Labor Adjustment Factors	Crew	Daily Output	Labor-Hours	Unit	Material	2020 Bare Costs Labor	Equipment	Total	Total Incl O&P
0010	**LABOR ADJUSTMENT FACTORS** (For Div. 26, 27 and 28) R260519-90									
0100	Subtract from labor for Economy of Scale for Wire									
0110	4-5 wires						25%			
0120	6-10 wires						30%			
0130	11-15 wires						35%			
0140	over 15 wires						40%			
0150	Labor adjustment factors (For Div. 26, 27, 28 and 48)									
0200	Labor factors: The below are reasonable suggestions, but									
0210	each project must be evaluated for its own peculiarities, and									
0220	the adjustments be increased or decreased depending on the									
0230	severity of the special conditions.									
1000	Add to labor for elevated installation (above floor level)									
1010	10′ to 14.5′ high						10%			
1020	15′ to 19.5′ high						20%			
1030	20′ to 24.5′ high						25%			
1040	25′ to 29.5′ high						35%			
1050	30′ to 34.5′ high						40%			
1060	35′ to 39.5′ high						50%			
1070	40′ and higher						55%			
2000	Add to labor for crawl space									
2010	3′ high						40%			
2020	4′ high						30%			
3000	Add to labor for multi-story building									
3100	For new construction (No elevator available)									
3110	Add for floors 3 thru 10						5%			
3120	Add for floors 11 thru 15						10%			
3130	Add for floors 16 thru 20						15%			
3140	Add for floors 21 thru 30						20%			
3150	Add for floors 31 and up						30%			
3200	For existing structure (Elevator available)									
3210	Add for work on floor 3 and above						2%			
4000	Add to labor for working in existing occupied buildings									
4010	Hospital						35%			
4020	Office building						25%			
4030	School						20%			
4040	Factory or warehouse						15%			
4050	Multi-dwelling						15%			
5000	Add to labor, miscellaneous									
5010	Cramped shaft						35%			
5020	Congested area						15%			
5030	Excessive heat or cold						30%			
6000	Labor factors: the above are reasonable suggestions, but									
6100	each project should be evaluated for its own peculiarities									
6200	Other factors to be considered are:									
6210	Movement of material and equipment through finished areas						10%			
6220	Equipment room min security direct access w/authorization						15%			
6230	Attic space						25%			
6240	No service road						25%			
6250	Poor unloading/storage area, no hydraulic lifts or jacks						20%			
6260	Congested site area/heavy traffic						20%			
7000	Correctional facilities (no compounding division 1 adjustment factors)									
7010	Minimum security w/facilities escort						30%			
7020	Medium security w/facilities and correctional officer escort						40%			

26 01 Operation and Maintenance of Electrical Systems

26 01 02 – Labor Adjustment

26 01 02.20 Labor Adjustment Factors		Crew	Daily Output	Labor-Hours	Unit	Material	2020 Bare Costs Labor	Equipment	Total	Total Incl O&P
7030	Max security w/facilities & correctional officer escort (no inmate contact)						50%			

26 01 40 – Operation and Maintenance of Electrical Protection Systems

26 01 40.51 Electrical Systems Repair and Replacement

		Crew	Daily Output	Labor-Hours	Unit	Material	Labor	Equipment	Total	Total Incl O&P
0010	**ELECTRICAL SYSTEMS REPAIR AND REPLACEMENT**									
3000	Remove and replace (reinstall), switch cover	1 Elec	60	.133	Ea.		8.20		8.20	12.60
3020	Outlet cover	″	60	.133	″		8.20		8.20	12.60

26 01 50 – Operation and Maintenance of Lighting

26 01 50.81 Luminaire Replacement

		Crew	Daily Output	Labor-Hours	Unit	Material	Labor	Equipment	Total	Total Incl O&P
0010	**LUMINAIRE REPLACEMENT**									
0962	Remove and install new replacement lens, 2′ x 2′ troffer	1 Elec	16	.500	Ea.	18	30.50		48.50	67
0964	2′ x 4′ troffer		16	.500		14.85	30.50		45.35	63.50
3200	Remove and replace (reinstall), lighting fixture		4	2			123		123	189

26 05 Common Work Results for Electrical

26 05 05 – Selective Demolition for Electrical

26 05 05.10 Electrical Demolition

		Crew	Daily Output	Labor-Hours	Unit	Material	Labor	Equipment	Total	Total Incl O&P
0010	**ELECTRICAL DEMOLITION** R260105-30									
0020	Electrical demolition, conduit to 10′ high, incl. fittings & hangers									
0100	Rigid galvanized steel, 1/2″ to 1″ diameter R024119-10	1 Elec	242	.033	L.F.		2.03		2.03	3.12
0120	1-1/4″ to 2″	″	200	.040			2.45		2.45	3.78
0140	2-1/2″ to 3-1/2″	2 Elec	302	.053			3.25		3.25	5
0160	4″ to 6″	″	160	.100			6.15		6.15	9.45
0170	PVC #40, 1/2″ to 1″	1 Elec	410	.020			1.20		1.20	1.84
0172	1-1/4″ to 2″		350	.023			1.40		1.40	2.16
0174	2-1/2″		250	.032			1.96		1.96	3.02
0176	3″ to 3-1/2″	2 Elec	340	.047			2.89		2.89	4.44
0178	4″ to 6″	″	230	.070			4.27		4.27	6.55
0200	Electric metallic tubing (EMT), 1/2″ to 1″	1 Elec	394	.020			1.25		1.25	1.92
0220	1-1/4″ to 1-1/2″		326	.025			1.51		1.51	2.32
0240	2″ to 3″		236	.034			2.08		2.08	3.20
0260	3-1/2″ to 4″	2 Elec	310	.052			3.17		3.17	4.87
0270	Armored cable (BX) avg. 50′ runs									
0280	#14, 2 wire	1 Elec	690	.012	L.F.		.71		.71	1.09
0290	#14, 3 wire		571	.014			.86		.86	1.32
0300	#12, 2 wire		605	.013			.81		.81	1.25
0310	#12, 3 wire		514	.016			.95		.95	1.47
0320	#10, 2 wire		514	.016			.95		.95	1.47
0330	#10, 3 wire		425	.019			1.15		1.15	1.78
0340	#8, 3 wire		342	.023			1.44		1.44	2.21
0350	Non metallic sheathed cable (Romex)									
0360	#14, 2 wire	1 Elec	720	.011	L.F.		.68		.68	1.05
0370	#14, 3 wire		657	.012			.75		.75	1.15
0380	#12, 2 wire		629	.013			.78		.78	1.20
0390	#10, 3 wire		450	.018			1.09		1.09	1.68
0400	Wiremold raceway, including fittings & hangers									
0420	No. 3000	1 Elec	250	.032	L.F.		1.96		1.96	3.02
0440	No. 4000		217	.037			2.26		2.26	3.48
0460	No. 6000		166	.048			2.96		2.96	4.55
0462	Plugmold with receptacle		114	.070			4.31		4.31	6.65
0465	Telephone/power pole		12	.667	Ea.		41		41	63

26 05 05 – Selective Demolition for Electrical

26 05 05.10 Electrical Demolition		Crew	Daily Output	Labor-Hours	Unit	Material	2020 Bare Costs Labor	Equipment	Total	Total Incl O&P
0470	Non-metallic, straight section	1 Elec	480	.017	L.F.		1.02		1.02	1.57
0500	Channels, steel, including fittings & hangers									
0520	3/4" x 1-1/2"	1 Elec	308	.026	L.F.		1.59		1.59	2.45
0540	1-1/2" x 1-1/2"		269	.030			1.82		1.82	2.81
0560	1-1/2" x 1-7/8"		229	.035			2.14		2.14	3.30
0600	Copper bus duct, indoor, 3 phase									
0610	Including hangers & supports									
0620	225 amp	2 Elec	135	.119	L.F.		7.25		7.25	11.20
0640	400 amp		106	.151			9.25		9.25	14.25
0660	600 amp		86	.186			11.40		11.40	17.55
0680	1,000 amp		60	.267			16.35		16.35	25
0700	1,600 amp		40	.400			24.50		24.50	38
0720	3,000 amp		10	1.600			98		98	151
0800	Plug-in switches, 600 V 3 ph., incl. disconnecting									
0820	wire, conduit terminations, 30 amp	1 Elec	15.50	.516	Ea.		31.50		31.50	49
0840	60 amp		13.90	.576			35.50		35.50	54.50
0850	100 amp		10.40	.769			47		47	72.50
0860	200 amp		6.20	1.290			79		79	122
0880	400 amp	2 Elec	5.40	2.963			182		182	280
0900	600 amp		3.40	4.706			289		289	445
0920	800 amp		2.60	6.154			380		380	580
0940	1,200 amp		2	8			490		490	755
0960	1,600 amp		1.70	9.412			575		575	890
1010	Safety switches, 250 or 600 V, incl. disconnection									
1050	of wire & conduit terminations									
1100	30 amp	1 Elec	12.30	.650	Ea.		40		40	61.50
1120	60 amp		8.80	.909			56		56	86
1140	100 amp		7.30	1.096			67		67	104
1160	200 amp		5	1.600			98		98	151
1180	400 amp	2 Elec	6.80	2.353			144		144	222
1200	600 amp	"	4.60	3.478			213		213	330
1202	Explosion proof, 60 amp	1 Elec	8.80	.909			56		56	86
1204	100 amp		5	1.600			98		98	151
1206	200 amp		3.33	2.402			147		147	227
1210	Panel boards, incl. removal of all breakers,									
1220	conduit terminations & wire connections									
1230	3 wire, 120/240 V, 100A, to 20 circuits	1 Elec	2.60	3.077	Ea.		189		189	291
1240	200 amps, to 42 circuits	2 Elec	2.60	6.154			380		380	580
1241	225 amps, to 42 circuits		2.40	6.667			410		410	630
1250	400 amps, to 42 circuits		2.20	7.273			445		445	685
1260	4 wire, 120/208 V, 125A, to 20 circuits	1 Elec	2.40	3.333			205		205	315
1270	200 amps, to 42 circuits	2 Elec	2.40	6.667			410		410	630
1280	400 amps, to 42 circuits		1.92	8.333			510		510	785
1285	600 amps, to 42 circuits		1.60	10			615		615	945
1300	Transformer, dry type, 1 phase, incl. removal of									
1320	supports, wire & conduit terminations									
1340	1 kVA	1 Elec	7.70	1.039	Ea.		63.50		63.50	98
1360	5 kVA		4.70	1.702			104		104	161
1380	10 kVA		3.60	2.222			136		136	210
1400	37.5 kVA	2 Elec	3	5.333			325		325	505
1420	75 kVA	"	2.50	6.400			395		395	605
1440	3 phase to 600 V, primary									
1460	3 kVA	1 Elec	3.87	2.067	Ea.		127		127	195

26 05 05 – Selective Demolition for Electrical

26 05 05.10 Electrical Demolition		Crew	Daily Output	Labor-Hours	Unit	Material	2020 Bare Costs Labor	Equipment	Total	Total Incl O&P
1480	15 kVA	2 Elec	3.67	4.360	Ea.		267		267	410
1490	25 kVA		3.51	4.558			280		280	430
1500	30 kVA		3.42	4.678			287		287	440
1510	45 kVA		3.18	5.031			310		310	475
1520	75 kVA		2.69	5.948			365		365	560
1530	112.5 kVA	R-3	2.90	6.897			420	65	485	720
1540	150 kVA		2.70	7.407			455	70	525	775
1560	500 kVA		1.40	14.286			875	135	1,010	1,500
1570	750 kVA		1.10	18.182			1,100	172	1,272	1,925
1600	Pull boxes & cabinets, sheet metal, incl. removal									
1620	of supports and conduit terminations									
1640	6" x 6" x 4"	1 Elec	31.10	.257	Ea.		15.80		15.80	24.50
1660	12" x 12" x 4"		23.30	.343			21		21	32.50
1680	24" x 24" x 6"		12.30	.650			40		40	61.50
1700	36" x 36" x 8"		7.70	1.039			63.50		63.50	98
1720	Junction boxes, 4" sq. & oct.		80	.100			6.15		6.15	9.45
1740	Handy box		107	.075			4.59		4.59	7.05
1760	Switch box		107	.075			4.59		4.59	7.05
1780	Receptacle & switch plates		257	.031			1.91		1.91	2.94
1790	Receptacles & switches, 15 to 30 amp		135	.059			3.64		3.64	5.60
1800	Wire, THW-THWN-THHN, removed from									
1810	in place conduit, to 10' high									
1830	#14	1 Elec	65	.123	C.L.F.		7.55		7.55	11.60
1840	#12		55	.145			8.90		8.90	13.75
1850	#10		45.50	.176			10.80		10.80	16.60
1860	#8		40.40	.198			12.15		12.15	18.70
1870	#6		32.60	.245			15.05		15.05	23
1880	#4	2 Elec	53	.302			18.50		18.50	28.50
1890	#3		50	.320			19.65		19.65	30
1900	#2		44.60	.359			22		22	34
1910	1/0		33.20	.482			29.50		29.50	45.50
1920	2/0		29.20	.548			33.50		33.50	52
1930	3/0		25	.640			39.50		39.50	60.50
1940	4/0		22	.727			44.50		44.50	68.50
1950	250 kcmil		20	.800			49		49	75.50
1960	300 kcmil		19	.842			51.50		51.50	79.50
1970	350 kcmil		18	.889			54.50		54.50	84
1980	400 kcmil		17	.941			57.50		57.50	89
1990	500 kcmil		16.20	.988			60.50		60.50	93.50
2000	Interior fluorescent fixtures, incl. supports									
2010	& whips, to 10' high									
2100	Recessed drop-in 2' x 2', 2 lamp	2 Elec	35	.457	Ea.		28		28	43
2110	2' x 2', 4 lamp		30	.533			32.50		32.50	50.50
2120	2' x 4', 2 lamp		33	.485			30		30	46
2140	2' x 4', 4 lamp		30	.533			32.50		32.50	50.50
2160	4' x 4', 4 lamp		20	.800			49		49	75.50
2180	Surface mount, acrylic lens & hinged frame									
2200	1' x 4', 2 lamp	2 Elec	44	.364	Ea.		22.50		22.50	34.50
2210	6" x 4', 2 lamp		44	.364			22.50		22.50	34.50
2220	2' x 2', 2 lamp		44	.364			22.50		22.50	34.50
2260	2' x 4', 4 lamp		33	.485			30		30	46
2280	4' x 4', 4 lamp		23	.696			42.50		42.50	65.50
2281	4' x 4', 6 lamp		23	.696			42.50		42.50	65.50

26 05 Common Work Results for Electrical

26 05 05 – Selective Demolition for Electrical

26 05 05.10 Electrical Demolition		Crew	Daily Output	Labor-Hours	Unit	Material	2020 Bare Costs Labor	Equipment	Total	Total Incl O&P
2300	Strip fixtures, surface mount									
2320	4′ long, 1 lamp	2 Elec	53	.302	Ea.		18.50		18.50	28.50
2340	4′ long, 2 lamp		50	.320			19.65		19.65	30
2360	8′ long, 1 lamp		42	.381			23.50		23.50	36
2380	8′ long, 2 lamp		40	.400			24.50		24.50	38
2400	Pendant mount, industrial, incl. removal									
2410	of chain or rod hangers, to 10′ high									
2420	4′ long, 2 lamp	2 Elec	35	.457	Ea.		28		28	43
2421	4′ long, 4 lamp		35	.457			28		28	43
2440	8′ long, 2 lamp		27	.593			36.50		36.50	56
2460	Interior incandescent, surface, ceiling									
2470	or wall mount, to 10′ high									
2480	Metal cylinder type, 75 Watt	2 Elec	62	.258	Ea.		15.85		15.85	24.50
2500	150 Watt		62	.258			15.85		15.85	24.50
2502	300 Watt		53	.302			18.50		18.50	28.50
2520	Metal halide, high bay									
2540	400 Watt	2 Elec	15	1.067	Ea.		65.50		65.50	101
2560	1,000 Watt		12	1.333			82		82	126
2580	150 Watt, low bay		20	.800			49		49	75.50
2585	Globe type, ceiling mount		62	.258			15.85		15.85	24.50
2586	Can type, recessed mount		62	.258			15.85		15.85	24.50
2600	Exterior fixtures, incandescent, wall mount									
2620	100 Watt	2 Elec	50	.320	Ea.		19.65		19.65	30
2640	Quartz, 500 Watt		33	.485			30		30	46
2660	1,500 Watt		27	.593			36.50		36.50	56
2680	Wall pack, mercury vapor									
2700	175 Watt	2 Elec	25	.640	Ea.		39.50		39.50	60.50
2720	250 Watt	"	25	.640			39.50		39.50	60.50
7000	Weatherhead/mast, 2″	1 Elec	16	.500			30.50		30.50	47
7002	3″		10	.800			49		49	75.50
7004	3-1/2″		8.50	.941			57.50		57.50	89
7006	4″		8	1			61.50		61.50	94.50
7100	Service entry cable, #6, +#6 neutral		420	.019	L.F.		1.17		1.17	1.80
7102	#4, +#4 neutral		360	.022			1.36		1.36	2.10
7104	#2, +#4 neutral		345	.023			1.42		1.42	2.19
7106	#2, +#2 neutral		330	.024			1.49		1.49	2.29
9000	Minimum labor/equipment charge		4	2	Job		123		123	189
9900	Add to labor for higher elevated installation									
9905	10′ to 14.5′ high, add						10%			
9910	15′ to 20′ high, add						20%			
9920	20′ to 25′ high, add						25%			
9930	25′ to 30′ high, add						35%			
9940	30′ to 35′ high, add						40%			
9950	35′ to 40′ high, add						50%			
9960	Over 40′ high, add						55%			

26 05 05.15 Electrical Demolition, Grounding

Line	Description	Crew	Daily Output	Labor-Hours	Unit	Material	Labor	Equipment	Total	Total Incl O&P
0010	**ELECTRICAL DEMOLITION, GROUNDING** Addition									
0100	Ground clamp, bronze	1 Elec	64	.125	Ea.		7.65		7.65	11.80
0140	Water pipe ground clamp, bronze, heavy duty		24	.333			20.50		20.50	31.50
0150	Ground rod, 8′ to 10′		13	.615			38		38	58
0200	Ground wire, bare armored	2 Elec	900	.018	L.F.		1.09		1.09	1.68
0240	Bare copper or aluminum	"	2800	.006	"		.35		.35	.54

26 05 05.20 Electrical Demolition, Wiring Methods		Crew	Daily Output	Labor-Hours	Unit	Material	2020 Bare Costs Labor	Equipment	Total	Total Incl O&P
0010	**ELECTRICAL DEMOLITION, WIRING METHODS** Addition R024119-10									
0100	Armored cable, w/PVC jacket, in cable tray									
0110	#6	1 Elec	9.30	.860	C.L.F.		53		53	81.50
0120	#4	2 Elec	16.20	.988			60.50		60.50	93.50
0130	#2		13.80	1.159			71		71	110
0140	#1		12	1.333			82		82	126
0150	1/0		10.80	1.481			91		91	140
0160	2/0		10.20	1.569			96		96	148
0170	3/0		9.60	1.667			102		102	157
0180	4/0		9	1.778			109		109	168
0190	250 kcmil	3 Elec	10.80	2.222			136		136	210
0210	350 kcmil		9.90	2.424			149		149	229
0230	500 kcmil		9	2.667			164		164	252
0240	750 kcmil		8.10	2.963			182		182	280
1100	Control cable, 600 V or less									
1110	3 wires	1 Elec	24	.333	C.L.F.		20.50		20.50	31.50
1120	5 wires		20	.400			24.50		24.50	38
1130	7 wires		16.50	.485			30		30	46
1140	9 wires		15	.533			32.50		32.50	50.50
1150	12 wires	2 Elec	26	.615			38		38	58
1160	15 wires		22	.727			44.50		44.50	68.50
1170	19 wires		19	.842			51.50		51.50	79.50
1180	25 wires		16	1			61.50		61.50	94.50
1500	Mineral insulated (MI) cable, 600 V									
1510	#10	1 Elec	4.80	1.667	C.L.F.		102		102	157
1520	#8		4.50	1.778			109		109	168
1530	#6		4.20	1.905			117		117	180
1540	#4	2 Elec	7.20	2.222			136		136	210
1550	#2		6.60	2.424			149		149	229
1560	#1		6.30	2.540			156		156	240
1570	1/0		6	2.667			164		164	252
1580	2/0		5.70	2.807			172		172	265
1590	3/0		5.40	2.963			182		182	280
1600	4/0		4.80	3.333			205		205	315
1610	250 kcmil	3 Elec	7.20	3.333			205		205	315
1620	500 kcmil	"	5.90	4.068			250		250	385
2000	Shielded cable, XLP shielding, to 35 kV									
2010	#4	2 Elec	13.20	1.212	C.L.F.		74.50		74.50	114
2020	#1		12	1.333			82		82	126
2030	1/0		11.40	1.404			86		86	133
2040	2/0		10.80	1.481			91		91	140
2050	4/0		9.60	1.667			102		102	157
2060	250 kcmil	3 Elec	13.50	1.778			109		109	168
2070	350 kcmil		11.70	2.051			126		126	194
2080	500 kcmil		11.20	2.143			131		131	202
2090	750 kcmil		10.80	2.222			136		136	210
3000	Modular flexible wiring									
3010	Cable set	1 Elec	120	.067	Ea.		4.09		4.09	6.30
3020	Conversion module		48	.167			10.25		10.25	15.75
3030	Switching assembly		96	.083			5.10		5.10	7.85
3200	Undercarpet									
3210	Power or telephone, flat cable	1 Elec	1320	.006	L.F.		.37		.37	.57

26 05 05.20 Electrical Demolition, Wiring Methods		Crew	Daily Output	Labor-Hours	Unit	Material	2020 Bare Costs Labor	Equipment	Total	Total Incl O&P
3220	Transition block assemblies w/fitting	1 Elec	75	.107	Ea.		6.55		6.55	10.10
3230	Floor box with fitting		60	.133			8.20		8.20	12.60
3300	Data system, cable with connection		50	.160			9.80		9.80	15.10
4000	Cable tray, including fitting & support									
4010	Galvanized steel, 6" wide	2 Elec	310	.052	L.F.		3.17		3.17	4.87
4020	9" wide		295	.054			3.33		3.33	5.10
4030	12" wide		285	.056			3.44		3.44	5.30
4040	18" wide		270	.059			3.64		3.64	5.60
4050	24" wide		260	.062			3.78		3.78	5.80
4060	30" wide		240	.067			4.09		4.09	6.30
4070	36" wide		220	.073			4.46		4.46	6.85
4110	Aluminum, 6" wide		420	.038			2.34		2.34	3.60
4120	9" wide		415	.039			2.37		2.37	3.64
4130	12" wide		390	.041			2.52		2.52	3.88
4140	18" wide		370	.043			2.65		2.65	4.08
4150	24" wide		350	.046			2.80		2.80	4.32
4160	30" wide		325	.049			3.02		3.02	4.65
4170	36" wide		300	.053			3.27		3.27	5.05
4310	Cable channel, aluminum, 4" wide straight	1 Elec	240	.033			2.04		2.04	3.15
4320	Cable tray fittings, 6" to 9" wide	2 Elec	27	.593	Ea.		36.50		36.50	56
4330	12" to 18" wide	"	19	.842	"		51.50		51.50	79.50
5000	Conduit nipples, with locknuts and bushings									
5020	1/2"	1 Elec	108	.074	Ea.		4.54		4.54	7
5040	3/4"		96	.083			5.10		5.10	7.85
5060	1"		81	.099			6.05		6.05	9.35
5080	1-1/4"		69	.116			7.10		7.10	10.95
5100	1-1/2"		60	.133			8.20		8.20	12.60
5120	2"		54	.148			9.10		9.10	14
5140	2-1/2"		45	.178			10.90		10.90	16.80
5160	3"		36	.222			13.65		13.65	21
5180	3-1/2"		33	.242			14.85		14.85	23
5200	4"		27	.296			18.20		18.20	28
5220	5"		21	.381			23.50		23.50	36
5240	6"		18	.444			27.50		27.50	42
5500	Electric nonmetallic tubing (ENT), flexible, 1/2" to 1" diameter		690	.012	L.F.		.71		.71	1.09
5510	1-1/4" to 2" diameter		300	.027			1.64		1.64	2.52
5600	Flexible metallic tubing, steel, 3/8" to 3/4" diameter		600	.013			.82		.82	1.26
5610	1" to 1-1/4" diameter		260	.031			1.89		1.89	2.91
5620	1-1/2" to 2" diameter		140	.057			3.51		3.51	5.40
5630	2-1/2" diameter		100	.080			4.91		4.91	7.55
5640	3" to 3-1/2" diameter	2 Elec	150	.107			6.55		6.55	10.10
5650	4" diameter	"	100	.160			9.80		9.80	15.10
5700	Sealtite flexible conduit, 3/8" to 3/4" diameter	1 Elec	420	.019			1.17		1.17	1.80
5710	1" to 1-1/4" diameter		180	.044			2.73		2.73	4.20
5720	1-1/2" to 2" diameter		120	.067			4.09		4.09	6.30
5730	2-1/2" diameter		80	.100			6.15		6.15	9.45
5740	3" diameter	2 Elec	150	.107			6.55		6.55	10.10
5750	4" diameter	"	100	.160			9.80		9.80	15.10
5800	Wiring duct, plastic, 1-1/2" to 2-1/2" wide	1 Elec	360	.022			1.36		1.36	2.10
5810	3" wide		330	.024			1.49		1.49	2.29
5820	4" wide		300	.027			1.64		1.64	2.52
6000	Floor box and carpet flange		9	.889	Ea.		54.50		54.50	84
6300	Wireway, with fittings and supports, to 10' high									

26 05 05 – Selective Demolition for Electrical

26 05 05.20 Electrical Demolition, Wiring Methods		Crew	Daily Output	Labor-Hours	Unit	Material	2020 Bare Costs Labor	Equipment	Total	Total Incl O&P
6310	2-1/2" x 2-1/2"	1 Elec	180	.044	L.F.		2.73		2.73	4.20
6320	4" x 4"	"	160	.050			3.07		3.07	4.72
6330	6" x 6"	2 Elec	240	.067			4.09		4.09	6.30
6340	8" x 8"		160	.100			6.15		6.15	9.45
6350	10" x 10"		120	.133			8.20		8.20	12.60
6360	12" x 12"		80	.200			12.25		12.25	18.90
6400	Cable reel with receptacle, 120 V/208 V		13	1.231	Ea.		75.50		75.50	116
6500	Equipment connection, to 10 HP	1 Elec	23	.348			21.50		21.50	33
6510	To 15 to 30 HP		18	.444			27.50		27.50	42
6520	To 40 to 60 HP		15	.533			32.50		32.50	50.50
6530	To 75 to 100 HP		10	.800			49		49	75.50
7008	Receptacle, explosion proof, 120 V, to 30 A		16	.500			30.50		30.50	47
7020	Dimmer switch, 2,000 W or less		62	.129			7.90		7.90	12.20

26 05 05.25 Electrical Demolition, Electrical Power		Crew	Daily Output	Labor-Hours	Unit	Material	2020 Bare Costs Labor	Equipment	Total	Total Incl O&P
0010	**ELECTRICAL DEMOLITION, ELECTRICAL POWER** R024119-10									
0100	Meter centers and sockets									
0120	Meter socket, 4 terminal R260105-30	1 Elec	8.40	.952	Ea.		58.50		58.50	90
0140	Trans-socket, 13 terminal, 400 A	"	3.60	2.222			136		136	210
0160	800 A	2 Elec	4.40	3.636			223		223	345
0200	Meter center, 400 A		5.80	2.759			169		169	261
0210	600 A		4	4			245		245	380
0220	800 A		3.30	4.848			297		297	460
0230	1,200 A		2.80	5.714			350		350	540
0240	1,600 A		2.50	6.400			395		395	605
0300	Base meter devices, 3 meter		3.60	4.444			273		273	420
0310	4 meter		3.30	4.848			297		297	460
0320	5 meter		2.90	5.517			340		340	520
0330	6 meter		2.20	7.273			445		445	685
0340	7 meter		2	8			490		490	755
0350	8 meter		1.90	8.421			515		515	795
0400	Branch meter devices									
0410	Socket w/circuit breaker 200 A, 2 meter	2 Elec	3.30	4.848	Ea.		297		297	460
0420	3 meter		2.90	5.517			340		340	520
0430	4 meter		2.60	6.154			380		380	580
0450	Main circuit breaker, 400 A		5.80	2.759			169		169	261
0460	600 A		4	4			245		245	380
0470	800 A		3.30	4.848			297		297	460
0480	1,200 A		2.80	5.714			350		350	540
0490	1,600 A		2.50	6.400			395		395	605
0500	Main lug terminal box, 800 A		3.40	4.706			289		289	445
0510	1,200 A		2.60	6.154			380		380	580
1000	Motors, 230/460 V, 60 Hz, 3/4 HP	1 Elec	10.70	.748			46		46	70.50
1010	5 HP		9	.889			54.50		54.50	84
1020	10 HP		8	1			61.50		61.50	94.50
1030	15 HP		6.40	1.250			76.50		76.50	118
1040	20 HP	2 Elec	10.40	1.538			94.50		94.50	145
1050	50 HP		9.60	1.667			102		102	157
1060	75 HP		5.60	2.857			175		175	270
1070	100 HP	3 Elec	5.40	4.444			273		273	420
1080	150 HP		3.60	6.667			410		410	630
1090	200 HP		3	8			490		490	755
1200	Variable frequency drive, 460 V, for 5 HP motor size	1 Elec	3.20	2.500			153		153	236

26 05 05 – Selective Demolition for Electrical

26 05 05.25 Electrical Demolition, Electrical Power		Crew	Daily Output	Labor-Hours	Unit	Material	2020 Bare Costs Labor	Equipment	Total	Total Incl O&P
1210	10 HP motor size	1 Elec	2.70	2.963	Ea.		182		182	280
1220	20 HP motor size	2 Elec	3.60	4.444			273		273	420
1230	50 HP motor size	"	2.10	7.619			465		465	720
1240	75 HP motor size	R-3	2.20	9.091			555	86	641	955
1250	100 HP motor size		2	10			610	94.50	704.50	1,050
1260	150 HP motor size		2	10			610	94.50	704.50	1,050
1270	200 HP motor size		1.70	11.765			720	111	831	1,225
2000	Generator set w/accessories, 3 phase 4 wire, 277/480 V									
2010	7.5 kW	3 Elec	2	12	Ea.		735		735	1,125
2020	20 kW		1.70	14.118			865		865	1,325
2040	30 kW		1.33	18.045			1,100		1,100	1,700
2060	50 kW		1	24			1,475		1,475	2,275
2080	100 kW		.75	32			1,975		1,975	3,025
2100	150 kW		.63	38.095			2,325		2,325	3,600
2120	250 kW		.59	40.678			2,500		2,500	3,850
2140	400 kW		.50	48			2,950		2,950	4,525
2160	500 kW		.44	54.545			3,350		3,350	5,150
2180	750 kW	4 Elec	.56	57.143			3,500		3,500	5,400
2200	1,000 kW	"	.46	69.565			4,275		4,275	6,575
3000	Uninterruptible power supply system (UPS)									
3010	Single phase, 120 V, 1 kVA	1 Elec	3.20	2.500	Ea.		153		153	236
3020	2 kVA	2 Elec	3.60	4.444			273		273	420
3030	5 kVA	3 Elec	2.50	9.600			590		590	905
3040	10 kVA		2.30	10.435			640		640	985
3050	15 kVA		1.80	13.333			820		820	1,250
4000	Transformer, incl. support, wire & conduit termination									
4010	Buck-boost, single phase, 120/240 V, 0.1 kVA	1 Elec	25	.320	Ea.		19.65		19.65	30
4020	0.5 kVA		12.50	.640			39.50		39.50	60.50
4030	1 kVA		6.30	1.270			78		78	120
4040	5 kVA		3.80	2.105			129		129	199
4800	5 kV or 15 kV primary, 277/480 V second, 112.5 kVA	R-3	3.50	5.714			350	54	404	600
4801	5 kV or 15 kV primary, 277/480 V second, 112.5 kVA		400	.050	kVA		3.06	.47	3.53	5.25
4810	150 kVA		2.70	7.407	Ea.		455	70	525	775
4811	150 kVA		412	.049	kVA		2.97	.46	3.43	5.10
4820	225 kVA		2.30	8.696	Ea.		530	82	612	910
4821	225 kVA		512	.039	kVA		2.39	.37	2.76	4.10
4830	500 kVA		1.45	13.793	Ea.		845	130	975	1,450
4831	500 kVA		700	.029	kVA		1.75	.27	2.02	3
4840	750 kVA		1.33	15.038	Ea.		920	142	1,062	1,575
4841	750 kVA		1065	.019	kVA		1.15	.18	1.33	1.97
4850	1,000 kVA		1.25	16	Ea.		980	151	1,131	1,675
4851	1,000 kVA		1250	.016	kVA		.98	.15	1.13	1.68
4860	2,000 kVA		1.05	19.048	Ea.		1,175	180	1,355	2,000
4861	2,000 kVA		20000	.001	kVA		.06	.01	.07	.10
4870	3,000 kVA		.75	26.667	Ea.		1,625	252	1,877	2,800
4871	3,000 kVA		2200	.009	kVA		.56	.09	.65	.95
6010	Isolation panel, 3 kVA	1 Elec	2.30	3.478	Ea.		213		213	330
6020	5 kVA		2.20	3.636			223		223	345
6030	7.5 kVA		2.10	3.810			234		234	360
6040	10 kVA		1.80	4.444			273		273	420
6050	15 kVA		1.40	5.714			350		350	540
7000	Power filters & conditioners									
7100	Automatic voltage regulator	2 Elec	6	2.667	Ea.		164		164	252

26 05 05 – Selective Demolition for Electrical

26 05 05.25 Electrical Demolition, Electrical Power		Crew	Daily Output	Labor-Hours	Unit	Material	2020 Bare Costs Labor	Equipment	Total	Total Incl O&P
7200	Capacitor, 1 kVAR	1 Elec	8.60	.930	Ea.		57		57	88
7210	5 kVAR		5.80	1.379			84.50		84.50	130
7220	10 kVAR		4.80	1.667			102		102	157
7230	15 kVAR		4.20	1.905			117		117	180
7240	20 kVAR		3.50	2.286			140		140	216
7250	30 kVAR		3.40	2.353			144		144	222
7260	50 kVAR	↓	3.20	2.500	↓		153		153	236
7400	Computer isolator transformer									
7410	Single phase, 120/240 V, 0.5 kVAR	1 Elec	12.80	.625	Ea.		38.50		38.50	59
7420	1 kVAR		8.50	.941			57.50		57.50	89
7430	2.5 kVAR		6.40	1.250			76.50		76.50	118
7440	5 kVAR	↓	3.70	2.162	↓		133		133	204
7500	Computer regulator transformer									
7510	Single phase, 240 V, 0.5 kVAR	1 Elec	8.50	.941	Ea.		57.50		57.50	89
7520	1 kVAR		6.40	1.250			76.50		76.50	118
7530	2 kVAR		3.20	2.500			153		153	236
7540	Single phase, plug-in unit 120 V, 0.5 kVAR		26	.308			18.90		18.90	29
7550	1 kVAR	↓	17	.471	↓		29		29	44.50
7600	Power conditioner transformer									
7610	Single phase 115 V - 240 V, 3 kVA	2 Elec	5.10	3.137	Ea.		192		192	296
7620	5 kVA	"	3.70	4.324			265		265	410
7630	7.5 kVA	3 Elec	4.80	5			305		305	470
7640	10 kVA	"	4.30	5.581	↓		340		340	525
7700	Transient suppressor/voltage regulator									
7710	Single phase 115 or 220 V, 1 kVA	1 Elec	8.50	.941	Ea.		57.50		57.50	89
7720	2 kVA		7.30	1.096			67		67	104
7730	4 kVA	↓	6.80	1.176	↓		72		72	111
7800	Transient voltage suppressor transformer									
7810	Single phase, 115 or 220 V, 3.6 kVA	1 Elec	12.80	.625	Ea.		38.50		38.50	59
7820	7.2 kVA		11.50	.696			42.50		42.50	65.50
7830	14.4 kVA		10.20	.784			48		48	74
7840	Single phase, plug-in, 120 V, 1.8 kVA	↓	26	.308	↓		18.90		18.90	29
8000	Power measurement & control									
8010	Switchboard instruments, 3 phase 4 wire, indicating unit	1 Elec	25	.320	Ea.		19.65		19.65	30
8020	Recording unit		12	.667			41		41	63
8100	3 current transformers, 3 phase 4 wire, 5 to 800 A		6.40	1.250			76.50		76.50	118
8110	1,000 to 1,500 A		4.20	1.905			117		117	180
8120	2,000 to 4,000 A	↓	3.20	2.500	↓		153		153	236

26 05 05.30 Electrical Demolition, Transmission and Distribution		Crew	Daily Output	Labor-Hours	Unit	Material	Labor	Equipment	Total	Total Incl O&P
0010	ELECTRICAL DEMOLITION, TRANSMISSION & DISTRIBUTION									
0100	Load interrupter switch, 600 A, NEMA 1, 4.8 kV	R-3	1.33	15.038	Ea.		920	142	1,062	1,575
0120	13.8 kV R024119-10	"	1.27	15.748			965	149	1,114	1,650
0200	Lightning arrester, 4.8 kV	1 Elec	9	.889			54.50		54.50	84
0220	13.8 kV R260105-30		6.67	1.199			73.50		73.50	113
0300	Alarm or option items	↓	3.33	2.402	↓		147		147	227

26 05 05.35 Electrical Demolition, L.V. Distribution		Crew	Daily Output	Labor-Hours	Unit	Material	Labor	Equipment	Total	Total Incl O&P
0010	**ELECTRICAL DEMOLITION, L.V. DISTRIBUTION** Addition R024119-10									
0100	Circuit breakers in enclosure									
0120	Enclosed (NEMA 1), 600 V, 3 pole, 30 A R260105-30	1 Elec	12.30	.650	Ea.		40		40	61.50
0140	60 A		8.80	.909			56		56	86
0160	100 A		7.30	1.096			67		67	104
0180	225 A	↓	5	1.600	↓		98		98	151

26 05 05.35 Electrical Demolition, L.V. Distribution		Crew	Daily Output	Labor-Hours	Unit	Material	2020 Bare Costs Labor	Equipment	Total	Total Incl O&P
0200	400 A	2 Elec	6.80	2.353	Ea.		144		144	222
0220	600 A		4.60	3.478			213		213	330
0240	800 A		3.60	4.444			273		273	420
0260	1,200 A		3.10	5.161			315		315	485
0280	1,600 A		2.80	5.714			350		350	540
0300	2,000 A		2.50	6.400			395		395	605
0400	Enclosed (NEMA 7), 600 V, 3 pole, 50 A	1 Elec	8.80	.909			56		56	86
0410	100 A		5.80	1.379			84.50		84.50	130
0420	150 A		3.80	2.105			129		129	199
0430	250 A	2 Elec	6.20	2.581			158		158	244
0440	400 A	"	4.60	3.478			213		213	330
0460	Manual motor starter, NEMA 1	1 Elec	24	.333			20.50		20.50	31.50
0480	NEMA 4 or NEMA 7		15	.533			32.50		32.50	50.50
0500	Time switches, single pole single throw		15.40	.519			32		32	49
0520	Photo cell		30	.267			16.35		16.35	25
0540	Load management device, 4 loads		7.70	1.039			63.50		63.50	98
0550	8 loads		3.80	2.105			129		129	199
0560	Master light control panel	2 Elec	2	8			490		490	755
0600	Transfer switches, enclosed, 30 A	1 Elec	12	.667			41		41	63
0610	60 A		9.50	.842			51.50		51.50	79.50
0620	100 A		6.50	1.231			75.50		75.50	116
0630	150 A	2 Elec	12	1.333			82		82	126
0640	260 A		10	1.600			98		98	151
0660	400 A		8	2			123		123	189
0670	600 A		5	3.200			196		196	300
0680	800 A		4	4			245		245	380
0690	1,200 A		3.50	4.571			280		280	430
0700	1,600 A		3	5.333			325		325	505
0710	2,000 A		2.50	6.400			395		395	605
0800	Air terminal with base and connector	1 Elec	24	.333			20.50		20.50	31.50
1000	Enclosed controller, NEMA 1, 30 A		13.80	.580			35.50		35.50	55
1020	60 A		11.50	.696			42.50		42.50	65.50
1040	100 A		9.60	.833			51		51	78.50
1060	150 A		7.70	1.039			63.50		63.50	98
1080	200 A		5.40	1.481			91		91	140
1100	400 A	2 Elec	6.90	2.319			142		142	219
1120	600 A		4.60	3.478			213		213	330
1140	800 A		3.80	4.211			258		258	400
1160	1,200 A		3.10	5.161			315		315	485
1200	Control station, NEMA 1	1 Elec	30	.267			16.35		16.35	25
1220	NEMA 7		23	.348			21.50		21.50	33
1230	Control switches, push button		69	.116			7.10		7.10	10.95
1240	Indicating light unit		120	.067			4.09		4.09	6.30
1250	Relay		15	.533			32.50		32.50	50.50
2000	Motor control center components									
2020	Starter, NEMA 1, size 1	1 Elec	9	.889	Ea.		54.50		54.50	84
2040	Size 2	2 Elec	13.30	1.203			74		74	114
2060	Size 3		6.70	2.388			147		147	226
2080	Size 4		5.30	3.019			185		185	285
2100	Size 5		3.30	4.848			297		297	460
2120	NEMA 7, size 1	1 Elec	8.70	.920			56.50		56.50	87
2140	Size 2	2 Elec	12.70	1.260			77.50		77.50	119
2160	Size 3		6.30	2.540			156		156	240

26 05 05 – Selective Demolition for Electrical

26 05 05.35 Electrical Demolition, L.V. Distribution		Crew	Daily Output	Labor-Hours	Unit	Material	2020 Bare Costs Labor	Equipment	Total	Total Incl O&P
2180	Size 4	2 Elec	5	3.200	Ea.		196		196	300
2200	Size 5	↓	3.20	5			305		305	470
2300	Fuse, light contactor, NEMA 1, 30 A	1 Elec	9	.889			54.50		54.50	84
2310	60 A		6.70	1.194			73.50		73.50	113
2320	100 A		3.30	2.424			149		149	229
2330	200 A	↓	2.70	2.963			182		182	280
2350	Motor control center, incoming section	2 Elec	4	4			245		245	380
2352	Structure per section	"	4.20	3.810			234		234	360
2400	Starter & structure, 10 HP	1 Elec	9	.889			54.50		54.50	84
2410	25 HP	2 Elec	13.30	1.203			74		74	114
2420	50 HP		6.70	2.388			147		147	226
2430	75 HP		5.30	3.019			185		185	285
2440	100 HP		4.70	3.404			209		209	320
2450	200 HP		3.30	4.848			297		297	460
2460	400 HP	↓	2.70	5.926	↓		365		365	560
2500	Motor starter & control									
2510	Motor starter, NEMA 1, 5 HP	1 Elec	8.90	.899	Ea.		55		55	85
2520	10 HP	"	6.20	1.290			79		79	122
2530	25 HP	2 Elec	8.40	1.905			117		117	180
2540	50 HP		6.90	2.319			142		142	219
2550	100 HP		4.60	3.478			213		213	330
2560	200 HP		3.50	4.571			280		280	430
2570	400 HP	↓	3.10	5.161			315		315	485
2610	Motor starter, NEMA 7, 5 HP	1 Elec	6.20	1.290			79		79	122
2620	10 HP	"	4.20	1.905			117		117	180
2630	25 HP	2 Elec	6.90	2.319			142		142	219
2640	50 HP		4.60	3.478			213		213	330
2650	100 HP		3.50	4.571			280		280	430
2660	200 HP	↓	1.90	8.421			515		515	795
2710	Combination control unit, NEMA 1, 5 HP	1 Elec	6.90	1.159			71		71	110
2720	10 HP	"	5	1.600			98		98	151
2730	25 HP	2 Elec	7.70	2.078			127		127	196
2740	50 HP		5.10	3.137			192		192	296
2750	100 HP	↓	3.10	5.161			315		315	485
2810	NEMA 7, 5 HP	1 Elec	5	1.600			98		98	151
2820	10 HP	"	3.90	2.051			126		126	194
2830	25 HP	2 Elec	5.10	3.137			192		192	296
2840	50 HP		3.10	5.161			315		315	485
2850	100 HP		2.30	6.957			425		425	655
2860	200 HP	↓	1.50	10.667	↓		655		655	1,000
3000	Panelboard or load center circuit breaker									
3010	Bolt-on or plug in, 15 A to 50 A	1 Elec	20	.400	Ea.		24.50		24.50	38
3020	60 A to 70 A		16	.500			30.50		30.50	47
3030	Bolt-on, 80 A to 100 A		14	.571			35		35	54
3040	Up to 250 A		7.30	1.096			67		67	104
3050	Motor operated, 30 A		13	.615			38		38	58
3060	60 A		10	.800			49		49	75.50
3070	100 A	↓	8	1	↓		61.50		61.50	94.50
3200	Switchboard circuit breaker									
3210	15 A to 60 A	1 Elec	19	.421	Ea.		26		26	40
3220	70 A to 100 A		14	.571			35		35	54
3230	125 A to 400 A		10	.800			49		49	75.50
3240	450 A to 600 A	↓	5.30	1.509	↓		92.50		92.50	143

26 05 05.35 Electrical Demolition, L.V. Distribution		Crew	Daily Output	Labor-Hours	Unit	Material	2020 Bare Costs Labor	2020 Bare Costs Equipment	2020 Bare Costs Total	Total Incl O&P
3250	700 A to 800 A	1 Elec	4.30	1.860	Ea.		114		114	176
3260	1,000 A		3.30	2.424			149		149	229
3270	1,200 A		2.70	2.963			182		182	280
3500	Switchboard, incoming section, 400 A	2 Elec	3.70	4.324			265		265	410
3510	600 A		3.30	4.848			297		297	460
3520	800 A		2.90	5.517			340		340	520
3530	1,200 A		2.40	6.667			410		410	630
3540	1,600 A		2.20	7.273			445		445	685
3550	2,000 A		2.10	7.619			465		465	720
3560	3,000 A		1.90	8.421			515		515	795
3570	4,000 A		1.70	9.412			575		575	890
3610	Distribution section, 600 A		4	4			245		245	380
3620	800 A		3.60	4.444			273		273	420
3630	1,200 A		3.10	5.161			315		315	485
3640	1,600 A		2.90	5.517			340		340	520
3650	2,000 A		2.70	5.926			365		365	560
3710	Transition section, 600 A		3.80	4.211			258		258	400
3720	800 A		3.30	4.848			297		297	460
3730	1,200 A		2.70	5.926			365		365	560
3740	1,600 A		2.40	6.667			410		410	630
3750	2,000 A		2.20	7.273			445		445	685
3760	2,500 A		2.10	7.619			465		465	720
3770	3,000 A		1.90	8.421			515		515	795
4000	Bus duct, aluminum or copper, 30 A	1 Elec	200	.040	L.F.		2.45		2.45	3.78
4020	60 A		160	.050			3.07		3.07	4.72
4040	100 A		140	.057			3.51		3.51	5.40
5000	Feedrail, trolley busway, up to 60 A		160	.050			3.07		3.07	4.72
5020	100 A		120	.067			4.09		4.09	6.30
5040	Busway, 50 A		200	.040			2.45		2.45	3.78
6000	Fuse, 30 A		133	.060	Ea.		3.69		3.69	5.70
6010	60 A		133	.060			3.69		3.69	5.70
6020	100 A		105	.076			4.67		4.67	7.20
6030	200 A		95	.084			5.15		5.15	7.95
6040	400 A		80	.100			6.15		6.15	9.45
6050	600 A		53	.151			9.25		9.25	14.25
6060	601 A to 1,200 A		42	.190			11.70		11.70	18
6070	1,500 A to 1,600 A		34	.235			14.45		14.45	22
6080	1,800 A to 2,500 A		26	.308			18.90		18.90	29
6090	4,000 A		21	.381			23.50		23.50	36
6100	4,500 A to 5,000 A		18	.444			27.50		27.50	42
6120	6,000 A		15	.533			32.50		32.50	50.50
6200	Fuse plug or fustat		100	.080			4.91		4.91	7.55

26 05 05.50 Electrical Demolition, Lighting		Crew	Daily Output	Labor-Hours	Unit	Material	Labor	Equipment	Total	Total Incl O&P
0010	**ELECTRICAL DEMOLITION, LIGHTING** Addition R024119-10									
0100	Fixture hanger, flexible, 1/2" diameter	1 Elec	36	.222	Ea.		13.65		13.65	21
0120	3/4" diameter R260105-30	"	30	.267	"		16.35		16.35	25
3000	Light pole, anchor base, excl. concrete bases									
3010	Metal light pole, 10'	2 Elec	24	.667	Ea.		41		41	63
3020	16'	"	18	.889			54.50		54.50	84
3030	20'	R-3	8.70	2.299			141	21.50	162.50	241
3035	25'		8.60	2.326			142	22	164	244
3038	30'		6	3.333			204	31.50	235.50	350

26 05 05 – Selective Demolition for Electrical

26 05 05.50 Electrical Demolition, Lighting		Crew	Daily Output	Labor-Hours	Unit	Material	2020 Bare Costs Labor	Equipment	Total	Total Incl O&P
3039	35′	R-3	6	3.333	Ea.		204	31.50	235.50	350
3040	40′		6	3.333			204	31.50	235.50	350
3100	Wood light pole, 10′	2 Elec	36	.444			27.50		27.50	42
3120	20′	"	24	.667			41		41	63
3140	Bollard light, 42″	1 Elec	9	.889			54.50		54.50	84
3160	Walkway luminaire	"	8.10	.988			60.50		60.50	93.50
4000	Explosion proof									
4010	Metal halide, 175 W	1 Elec	8.70	.920	Ea.		56.50		56.50	87
4020	250 W	"	8.10	.988			60.50		60.50	93.50
4030	400 W	2 Elec	14.40	1.111			68		68	105
4050	High pressure sodium, 70 W	1 Elec	9	.889			54.50		54.50	84
4060	100 W		9	.889			54.50		54.50	84
4070	150 W		8.10	.988			60.50		60.50	93.50
4100	Incandescent		8.70	.920			56.50		56.50	87
4200	Fluorescent		8.10	.988			60.50		60.50	93.50
5000	Ballast, fluorescent fixture		24	.333			20.50		20.50	31.50
5040	High intensity discharge fixture		24	.333			20.50		20.50	31.50
5300	Exit and emergency lighting									
5310	Exit light	1 Elec	24	.333	Ea.		20.50		20.50	31.50
5320	Emergency battery pack lighting unit		12	.667			41		41	63
5330	Remote lamp only		80	.100			6.15		6.15	9.45
5340	Self-contained fluorescent lamp pack		30	.267			16.35		16.35	25
5500	Track lighting, 8′ section	2 Elec	40	.400			24.50		24.50	38
5510	Track lighting fixture	1 Elec	64	.125			7.65		7.65	11.80
5800	Energy saving devices									
5810	Occupancy sensor	1 Elec	21	.381	Ea.		23.50		23.50	36
5820	Automatic wall switch		72	.111			6.80		6.80	10.50
5830	Remote power pack		30	.267			16.35		16.35	25
5840	Photoelectric control		24	.333			20.50		20.50	31.50
5850	Fixture whip		100	.080			4.91		4.91	7.55
6000	Lamps									
6010	Fluorescent	1 Elec	200	.040	Ea.		2.45		2.45	3.78
6030	High intensity discharge lamp, up to 400 W		68	.118			7.20		7.20	11.10
6040	Up to 1,000 W		45	.178			10.90		10.90	16.80
6050	Quartz		90	.089			5.45		5.45	8.40
6070	Incandescent		360	.022			1.36		1.36	2.10
6080	Exterior, PAR		290	.028			1.69		1.69	2.61
6090	Guards for fluorescent lamp		24	.333			20.50		20.50	31.50

26 05 13 – Medium-Voltage Cables

26 05 13.10 Cable Terminations		Crew	Daily Output	Labor-Hours	Unit	Material	Labor	Equipment	Total	Total Incl O&P
0010	**CABLE TERMINATIONS**, 5 kV to 35 kV									
0100	Indoor, insulation diameter range 0.64″ to 1.08″									
0300	Padmount, 5 kV	1 Elec	8	1	Ea.	87.50	61.50		149	191
0400	10 kV		6.40	1.250		109	76.50		185.50	237
0500	15 kV		6	1.333		142	82		224	282
0600	25 kV		5.60	1.429		210	87.50		297.50	365
0700	Insulation diameter range 1.05″ to 1.8″									
0800	Padmount, 5 kV	1 Elec	8	1	Ea.	36.50	61.50		98	135
0900	10 kV		6	1.333		163	82		245	305
1000	15 kV		5.60	1.429		169	87.50		256.50	320
1100	25 kV		5.30	1.509		242	92.50		334.50	410
1200	Insulation diameter range 1.53″ to 2.32″									

26 05 13.10 Cable Terminations		Crew	Daily Output	Labor-Hours	Unit	2020 Bare Costs Material	Labor	Equipment	Total	Total Incl O&P
1300	Padmount, 5 kV	1 Elec	7.40	1.081	Ea.	152	66.50		218.50	269
1400	10 kV		5.60	1.429		175	87.50		262.50	330
1500	15 kV		5.30	1.509		194	92.50		286.50	355
1600	25 kV	↓	5	1.600	↓	315	98		413	500
1700	Outdoor systems, #4 stranded to 1/0 stranded									
1800	5 kV	1 Elec	7.40	1.081	Ea.	108	66.50		174.50	220
1900	15 kV		5.30	1.509		120	92.50		212.50	275
2000	25 kV		5	1.600		178	98		276	345
2100	35 kV		4.80	1.667		325	102		427	515
2200	#1 solid to 4/0 stranded, 5 kV		6.90	1.159		146	71		217	270
2300	15 kV		5	1.600		184	98		282	355
2400	25 kV		4.80	1.667		254	102		356	435
2500	35 kV		4.60	1.739		355	107		462	555
2600	2/0 solid to 350 kcmil stranded, 5 kV		6.40	1.250		193	76.50		269.50	330
2700	15 kV		4.80	1.667		210	102		312	390
2800	25 kV		4.60	1.739		294	107		401	490
2900	35 kV		4.40	1.818		405	112		517	615
3000	400 kcmil compact to 750 kcmil stranded, 5 kV		6	1.333		175	82		257	320
3100	15 kV		4.60	1.739		231	107		338	420
3200	25 kV		4.40	1.818		335	112		447	540
3300	35 kV		4.20	1.905		340	117		457	550
3400	1,000 kcmil, 5 kV		5.60	1.429		250	87.50		337.50	410
3500	15 kV		4.40	1.818		390	112		502	595
3600	25 kV		4.20	1.905		340	117		457	550
3700	35 kV	↓	4	2	↓	470	123		593	710

26 05 13.16 Medium-Voltage, Single Cable		Crew	Daily Output	Labor-Hours	Unit	Material	Labor	Equipment	Total	Total Incl O&P
0010	**MEDIUM-VOLTAGE, SINGLE CABLE** Splicing & terminations not included									
0040	Copper, XLP shielding, 5 kV, #6	2 Elec	4.40	3.636	C.L.F.	171	223		394	535
0050	#4		4.40	3.636		222	223		445	590
0100	#2		4	4		254	245		499	660
0200	#1		4	4		340	245		585	755
0400	1/0		3.80	4.211		335	258		593	765
0600	2/0		3.60	4.444		410	273		683	870
0800	4/0	↓	3.20	5		545	305		850	1,075
1000	250 kcmil	3 Elec	4.50	5.333		655	325		980	1,225
1200	350 kcmil		3.90	6.154		850	380		1,230	1,525
1400	500 kcmil	↓	3.60	6.667		1,100	410		1,510	1,825
1600	15 kV, ungrounded neutral, #1	2 Elec	4	4		365	245		610	785
1800	1/0		3.80	4.211		440	258		698	880
2000	2/0		3.60	4.444		500	273		773	970
2200	4/0	↓	3.20	5		660	305		965	1,200
2400	250 kcmil	3 Elec	4.50	5.333		735	325		1,060	1,300
2600	350 kcmil		3.90	6.154		925	380		1,305	1,600
2800	500 kcmil	↓	3.60	6.667		1,050	410		1,460	1,775
3000	25 kV, grounded neutral, 1/0	2 Elec	3.60	4.444		605	273		878	1,100
3200	2/0		3.40	4.706		670	289		959	1,175
3400	4/0	↓	3	5.333		840	325		1,165	1,425
3600	250 kcmil	3 Elec	4.20	5.714		1,050	350		1,400	1,700
3800	350 kcmil		3.60	6.667		1,225	410		1,635	1,975
3900	500 kcmil	↓	3.30	7.273		1,425	445		1,870	2,250
4000	35 kV, grounded neutral, 1/0	2 Elec	3.40	4.706		595	289		884	1,100
4200	2/0	↓	3.20	5	↓	755	305		1,060	1,300

26 05 13 – Medium-Voltage Cables

26 05 13.16 Medium-Voltage, Single Cable		Crew	Daily Output	Labor-Hours	Unit	Material	2020 Bare Costs Labor	Equipment	Total	Total Incl O&P
4400	4/0	2 Elec	2.80	5.714	C.L.F.	950	350		1,300	1,600
4600	250 kcmil	3 Elec	3.90	6.154		1,125	380		1,505	1,800
4800	350 kcmil		3.30	7.273		1,350	445		1,795	2,150
5000	500 kcmil		3	8		1,575	490		2,065	2,475
5050	Aluminum, XLP shielding, 5 kV, #2	2 Elec	5	3.200		242	196		438	565
5070	#1		4.40	3.636		251	223		474	620
5090	1/0		4	4		290	245		535	700
5100	2/0		3.80	4.211		325	258		583	760
5150	4/0		3.60	4.444		385	273		658	845
5200	250 kcmil	3 Elec	4.80	5		465	305		770	985
5220	350 kcmil		4.50	5.333		550	325		875	1,100
5240	500 kcmil		3.90	6.154		705	380		1,085	1,350
5260	750 kcmil		3.60	6.667		930	410		1,340	1,650
5300	15 kV aluminum, XLP, #1	2 Elec	4.40	3.636		310	223		533	685
5320	1/0		4	4		320	245		565	735
5340	2/0		3.80	4.211		345	258		603	780
5360	4/0		3.60	4.444		425	273		698	885
5380	250 kcmil	3 Elec	4.80	5		510	305		815	1,025
5400	350 kcmil		4.50	5.333		570	325		895	1,125
5420	500 kcmil		3.90	6.154		785	380		1,165	1,450
5440	750 kcmil		3.60	6.667		1,050	410		1,460	1,775

26 05 19 – Low-Voltage Electrical Power Conductors and Cables

26 05 19.13 Undercarpet Electrical Power Cables

		Crew	Daily Output	Labor-Hours	Unit	Material	Labor	Equipment	Total	Total Incl O&P
0010	**UNDERCARPET ELECTRICAL POWER CABLES** R260519-80									
0020	Power System									
0100	Cable flat, 3 conductor, #12, w/attached bottom shield	1 Elec	982	.008	L.F.	5.35	.50		5.85	6.65
0200	Shield, top, steel		1768	.005	"	4.83	.28		5.11	5.75
0250	Splice, 3 conductor		48	.167	Ea.	14.65	10.25		24.90	32
0300	Top shield		96	.083		1.35	5.10		6.45	9.35
0350	Tap		40	.200		18.80	12.25		31.05	39.50
0400	Insulating patch, splice, tap & end		48	.167		47	10.25		57.25	67.50
0450	Fold		230	.035			2.13		2.13	3.29
0500	Top shield, tap & fold		96	.083		1.35	5.10		6.45	9.35
0700	Transition, block assembly		77	.104		68.50	6.35		74.85	85.50
0750	Receptacle frame & base		32	.250		37.50	15.35		52.85	65
0800	Cover receptacle		120	.067		3.17	4.09		7.26	9.80
0850	Cover blank		160	.050		3.74	3.07		6.81	8.85
0860	Receptacle, direct connected, single		25	.320		80.50	19.65		100.15	119
0870	Dual		16	.500		132	30.50		162.50	192
0880	Combination high & low, tension		21	.381		97.50	23.50		121	143
0900	Box, floor with cover		20	.400		81.50	24.50		106	128
0920	Floor service w/barrier		4	2		231	123		354	445
1000	Wall, surface, with cover		20	.400		53.50	24.50		78	97
1100	Wall, flush, with cover		20	.400		37.50	24.50		62	79.50
1450	Cable flat, 5 conductor #12, w/attached bottom shield		800	.010	L.F.	8.80	.61		9.41	10.60
1550	Shield, top, steel		1768	.005	"	8.70	.28		8.98	10.05
1600	Splice, 5 conductor		48	.167	Ea.	23.50	10.25		33.75	42
1650	Top shield		96	.083		1.35	5.10		6.45	9.35
1700	Tap		48	.167		31.50	10.25		41.75	50.50
1750	Insulating patch, splice tap & end		83	.096		55	5.90		60.90	69.50
1800	Transition, block assembly		77	.104		49.50	6.35		55.85	64
1850	Box, wall, flush with cover		20	.400		50	24.50		74.50	93

	26 05 19.13 Undercarpet Electrical Power Cables	Crew	Daily Output	Labor-Hours	Unit	2020 Bare Costs Material	Labor	Equipment	Total	Total Incl O&P
1900	Cable flat, 4 conductor #12	1 Elec	933	.009	L.F.	6.15	.53		6.68	7.60
1950	3 conductor #10		982	.008		5.95	.50		6.45	7.30
1960	4 conductor #10		933	.009		7.80	.53		8.33	9.40
1970	5 conductor #10		884	.009		9.50	.56		10.06	11.30
2500	Telephone System									
2510	Transition fitting wall box, surface	1 Elec	24	.333	Ea.	60.50	20.50		81	98
2520	Flush		24	.333		60.50	20.50		81	98
2530	Flush, for PC board		24	.333		60.50	20.50		81	98
2540	Floor service box		4	2		238	123		361	450
2550	Cover, surface					22			22	24
2560	Flush					22			22	24
2570	Flush for PC board					22			22	24
2700	Floor fitting w/duplex jack & cover	1 Elec	21	.381		55	23.50		78.50	96.50
2720	Low profile		53	.151		18.15	9.25		27.40	34
2740	Miniature w/duplex jack		53	.151		34	9.25		43.25	52
2760	25 pair kit		21	.381		51	23.50		74.50	92
2780	Low profile		53	.151		18.55	9.25		27.80	35
2800	Call director kit for 5 cable		19	.421		95	26		121	144
2820	4 pair kit		19	.421		96.50	26		122.50	146
2840	3 pair kit		19	.421		120	26		146	172
2860	Comb. 25 pair & 3 conductor power		21	.381		95	23.50		118.50	140
2880	5 conductor power		21	.381		93	23.50		116.50	138
2900	PC board, 8 per 3 pair		161	.050		70.50	3.05		73.55	82
2920	6 per 4 pair		161	.050		70.50	3.05		73.55	82
2940	3 pair adapter		161	.050		64	3.05		67.05	75
2950	Plug		77	.104		3.21	6.35		9.56	13.35
2960	Couplers		321	.025		9.05	1.53		10.58	12.30
3000	Bottom shield for 25 pair cable		4420	.002	L.F.	.86	.11		.97	1.12
3020	4 pair		4420	.002		.41	.11		.52	.62
3040	Top shield for 25 pair cable		4420	.002		.86	.11		.97	1.12
3100	Cable assembly, double-end, 50′, 25 pair		11.80	.678	Ea.	248	41.50		289.50	335
3110	3 pair		23.60	.339		78.50	21		99.50	119
3120	4 pair		23.60	.339		95	21		116	136
3140	Bulk 3 pair		1473	.005	L.F.	1.34	.33		1.67	1.98
3160	4 pair		1473	.005	″	1.66	.33		1.99	2.34
3500	Data System									
3520	Cable 25 conductor w/connection 40′, 75 ohm	1 Elec	14.50	.552	Ea.	85.50	34		119.50	146
3530	Single lead		22	.364		234	22.50		256.50	292
3540	Dual lead		22	.364		248	22.50		270.50	310
3560	Shields same for 25 conductor as 25 pair telephone									
3570	Single & dual, none required									
3590	BNC coax connectors, plug	1 Elec	40	.200	Ea.	15.30	12.25		27.55	35.50
3600	TNC coax connectors, plug	″	40	.200	″	15.40	12.25		27.65	36
3700	Cable-bulk									
3710	Single lead	1 Elec	1473	.005	L.F.	3.70	.33		4.03	4.58
3720	Dual lead	″	1473	.005	″	5.35	.33		5.68	6.40
3730	Hand tool crimp				Ea.	520			520	570
3740	Hand tool notch				″	20.50			20.50	22.50
3750	Boxes & floor fitting same as telephone									
3790	Data cable notching, 90°	1 Elec	97	.082	Ea.		5.05		5.05	7.80
3800	180°		60	.133			8.20		8.20	12.60
8100	Drill floor		160	.050		4.42	3.07		7.49	9.60
8200	Marking floor		1600	.005	L.F.		.31		.31	.47

26 05 19.13 Undercarpet Electrical Power Cables		Crew	Daily Output	Labor-Hours	Unit	2020 Bare Costs Material	Labor	Equipment	Total	Total Incl O&P
8300	Tape, hold down	1 Elec	6400	.001	L.F.	.18	.08		.26	.32
8350	Tape primer, 500′ per can		96	.083	Ea.	49	5.10		54.10	62
8400	Tool, splicing				"	167			167	184

26 05 19.20 Armored Cable

		Crew	Daily Output	Labor-Hours	Unit	Material	Labor	Equipment	Total	Total Incl O&P
0010	**ARMORED CABLE**									
0050	600 volt, copper (BX), #14, 2 conductor, solid	1 Elec	2.40	3.333	C.L.F.	39	205		244	360
0100	3 conductor, solid		2.20	3.636		65.50	223		288.50	415
0120	4 conductor, solid		2	4		93	245		338	480
0150	#12, 2 conductor, solid		2.30	3.478		41	213		254	375
0200	3 conductor, solid		2	4		69	245		314	455
0220	4 conductor, solid		1.80	4.444		95	273		368	525
0250	#10, 2 conductor, solid		2	4		81.50	245		326.50	470
0300	3 conductor, solid		1.60	5		109	305		414	590
0320	4 conductor, solid		1.40	5.714		168	350		518	725
0340	#8, 2 conductor, stranded		1.50	5.333		257	325		582	790
0350	3 conductor, stranded		1.30	6.154		257	380		637	865
0370	4 conductor, stranded		1.10	7.273		365	445		810	1,100
0380	#6, 2 conductor, stranded		1.30	6.154		251	380		631	855
0390	#4, 3 conductor, stranded		1.40	5.714		475	350		825	1,050
0400	3 conductor with PVC jacket, in cable tray, #6		3.10	2.581		530	158		688	830
0450	#4	2 Elec	5.40	2.963		650	182		832	995
0500	#2		4.60	3.478		810	213		1,023	1,225
0550	#1		4	4		1,025	245		1,270	1,500
0600	1/0		3.60	4.444		1,100	273		1,373	1,625
0650	2/0		3.40	4.706		1,125	289		1,414	1,700
0700	3/0		3.20	5		1,750	305		2,055	2,400
0750	4/0		3	5.333		2,100	325		2,425	2,825
0800	250 kcmil	3 Elec	3.60	6.667		2,625	410		3,035	3,525
0850	350 kcmil		3.30	7.273		3,300	445		3,745	4,300
0900	500 kcmil		3	8		4,550	490		5,040	5,750
0910	4 conductor with PVC jacket, in cable tray, #6	1 Elec	2.70	2.963		1,375	182		1,557	1,800
0920	#4	2 Elec	4.60	3.478		1,500	213		1,713	1,975
0930	#2		4	4		1,200	245		1,445	1,700
0940	#1		3.60	4.444		1,525	273		1,798	2,100
0950	1/0		3.40	4.706		1,425	289		1,714	2,000
0960	2/0		3.20	5		1,750	305		2,055	2,400
0970	3/0		3	5.333		2,300	325		2,625	3,050
0980	4/0		2.40	6.667		2,100	410		2,510	2,925
0990	250 kcmil	3 Elec	3.30	7.273		3,200	445		3,645	4,200
1000	350 kcmil		3	8		4,300	490		4,790	5,475
1010	500 kcmil		2.70	8.889		5,975	545		6,520	7,400
1050	5 kV, copper, 3 conductor with PVC jacket,									
1060	non-shielded, in cable tray, #4	2 Elec	380	.042	L.F.	7.35	2.58		9.93	12.10
1100	#2		360	.044		9.55	2.73		12.28	14.70
1200	#1		300	.053		12.15	3.27		15.42	18.45
1400	1/0		290	.055		14.05	3.38		17.43	20.50
1600	2/0		260	.062		16.20	3.78		19.98	23.50
2000	4/0		240	.067		21.50	4.09		25.59	30.50
2100	250 kcmil	3 Elec	330	.073		29.50	4.46		33.96	39.50
2150	350 kcmil		315	.076		36.50	4.67		41.17	47
2200	500 kcmil		270	.089		54	5.45		59.45	67.50
2400	15 kV, copper, 3 conductor with PVC jacket galv., steel armored									

26 05 19.20 Armored Cable		Crew	Daily Output	Labor-Hours	Unit	Material	2020 Bare Costs Labor	Equipment	Total	Total Incl O&P
2500	grounded neutral, in cable tray, #2	2 Elec	300	.053	L.F.	16	3.27		19.27	22.50
2600	#1		280	.057		16.45	3.51		19.96	23.50
2800	1/0		260	.062		19.35	3.78		23.13	27.50
2900	2/0		220	.073		25	4.46		29.46	34.50
3000	4/0		190	.084		28.50	5.15		33.65	39
3100	250 kcmil	3 Elec	270	.089		31.50	5.45		36.95	43
3150	350 kcmil		240	.100		37	6.15		43.15	50.50
3200	500 kcmil		210	.114		50	7		57	66
3400	15 kV, copper, 3 conductor with PVC jacket,									
3450	ungrounded neutral, in cable tray, #2	2 Elec	260	.062	L.F.	16.65	3.78		20.43	24
3500	#1		230	.070		18.40	4.27		22.67	27
3600	1/0		200	.080		21	4.91		25.91	30.50
3700	2/0		190	.084		26	5.15		31.15	36.50
3800	4/0		160	.100		31	6.15		37.15	44
4000	250 kcmil	3 Elec	210	.114		36.50	7		43.50	51
4050	350 kcmil		195	.123		48	7.55		55.55	64
4100	500 kcmil		180	.133		58.50	8.20		66.70	76.50
4200	600 volt, aluminum, 3 conductor in cable tray with PVC jacket									
4300	#2	2 Elec	540	.030	L.F.	3.97	1.82		5.79	7.15
4400	#1		460	.035		4.70	2.13		6.83	8.45
4500	1/0		400	.040		5.85	2.45		8.30	10.25
4600	2/0		360	.044		5.95	2.73		8.68	10.70
4700	3/0		340	.047		6.95	2.89		9.84	12.10
4800	4/0		320	.050		8.40	3.07		11.47	13.95
4900	250 kcmil	3 Elec	450	.053		10.10	3.27		13.37	16.15
5000	350 kcmil		360	.067		12.05	4.09		16.14	19.55
5200	500 kcmil		330	.073		15	4.46		19.46	23.50
5300	750 kcmil		285	.084		19.30	5.15		24.45	29
5400	600 volt, aluminum, 4 conductor in cable tray with PVC jacket									
5410	#2	2 Elec	520	.031	L.F.	4.70	1.89		6.59	8.05
5430	#1		440	.036		5.75	2.23		7.98	9.75
5450	1/0		380	.042		6.70	2.58		9.28	11.40
5470	2/0		340	.047		6.85	2.89		9.74	12
5480	3/0		320	.050		8.05	3.07		11.12	13.60
5500	4/0		300	.053		9.45	3.27		12.72	15.45
5520	250 kcmil	3 Elec	420	.057		10.15	3.51		13.66	16.60
5540	350 kcmil		330	.073		13.15	4.46		17.61	21.50
5560	500 kcmil		300	.080		16.45	4.91		21.36	25.50
5580	750 kcmil		270	.089		22.50	5.45		27.95	33.50
5600	5 kV, aluminum, unshielded in cable tray, #2 with PVC jacket	2 Elec	380	.042		5.95	2.58		8.53	10.50
5700	#1 with PVC jacket		360	.044		6.65	2.73		9.38	11.50
5800	1/0 with PVC jacket		300	.053		6.80	3.27		10.07	12.55
6000	2/0 with PVC jacket		290	.055		6.95	3.38		10.33	12.85
6200	3/0 with PVC jacket		260	.062		8.35	3.78		12.13	15
6300	4/0 with PVC jacket		240	.067		9.80	4.09		13.89	17.10
6400	250 kcmil with PVC jacket	3 Elec	330	.073		10.75	4.46		15.21	18.70
6500	350 kcmil with PVC jacket		315	.076		12.60	4.67		17.27	21
6600	500 kcmil with PVC jacket		300	.080		14.90	4.91		19.81	24
6800	750 kcmil with PVC jacket		270	.089		18.40	5.45		23.85	28.50
6900	15 kV, aluminum, shielded-grounded, #2 with PVC jacket	2 Elec	320	.050		13.50	3.07		16.57	19.55
7000	#1 with PVC jacket		300	.053		13.90	3.27		17.17	20.50
7200	1/0 with PVC jacket		280	.057		15	3.51		18.51	22
7300	2/0 with PVC jacket		260	.062		15.20	3.78		18.98	22.50

26 05 19 – Low-Voltage Electrical Power Conductors and Cables

26 05 19.20 Armored Cable		Crew	Daily Output	Labor-Hours	Unit	Material	2020 Bare Costs Labor	Equipment	Total	Total Incl O&P
7400	3/0 with PVC jacket	2 Elec	240	.067	L.F.	17.10	4.09		21.19	25
7500	4/0 with PVC jacket		220	.073		17.55	4.46		22.01	26
7600	250 kcmil with PVC jacket	3 Elec	300	.080		19.25	4.91		24.16	28.50
7700	350 kcmil with PVC jacket		270	.089		22.50	5.45		27.95	33.50
7800	500 kcmil with PVC jacket		240	.100		27	6.15		33.15	39
8000	750 kcmil with PVC jacket		204	.118		32	7.20		39.20	46.50
8200	15 kV, aluminum, shielded-ungrounded, #1 with PVC jacket	2 Elec	250	.064		16.95	3.93		20.88	24.50
8300	1/0 with PVC jacket		230	.070		17.55	4.27		21.82	26
8400	2/0 with PVC jacket		210	.076		19.25	4.67		23.92	28
8500	3/0 with PVC jacket		200	.080		19.55	4.91		24.46	29
8600	4/0 with PVC jacket		190	.084		21	5.15		26.15	31.50
8700	250 kcmil with PVC jacket	3 Elec	270	.089		23	5.45		28.45	33.50
8800	350 kcmil with PVC jacket		240	.100		26	6.15		32.15	38
8900	500 kcmil with PVC jacket		210	.114		31.50	7		38.50	45.50
8950	750 kcmil with PVC jacket		174	.138		39	8.45		47.45	56
9010	600 volt, copper (MC) steel clad, #14, 2 wire	R-1A	8.16	1.961	C.L.F.	41.50	108		149.50	213
9020	3 wire		7.21	2.219		77	123		200	274
9030	4 wire		6.69	2.392		200	132		332	425
9040	#12, 2 wire		7.21	2.219		57	123		180	252
9050	3 wire		6.69	2.392		67.50	132		199.50	278
9060	4 wire		6.04	2.649		116	146		262	355
9070	#10, 2 wire		6.04	2.649		103	146		249	340
9080	3 wire		5.65	2.832		132	156		288	385
9090	4 wire		5.23	3.059		208	169		377	490
9100	#8, 2 wire, stranded		3.94	4.061		161	224		385	520
9110	3 wire, stranded		3.40	4.706		185	260		445	605
9120	4 wire, stranded		2.80	5.714		305	315		620	820
9130	#6, 2 wire, stranded		3.45	4.638		223	256		479	640
9200	600 volt, copper (MC) aluminum clad, #14, 2 wire		8.16	1.961		41	108		149	212
9210	3 wire		7.21	2.219		74.50	123		197.50	271
9220	4 wire		6.69	2.392		92.50	132		224.50	305
9230	#12, 2 wire		7.21	2.219		42	123		165	235
9240	3 wire		6.69	2.392		72	132		204	282
9250	4 wire		6.04	2.649		148	146		294	385
9260	#10, 2 wire		6.04	2.649		89	146		235	325
9270	3 wire		5.65	2.832		114	156		270	365
9280	4 wire		5.23	3.059		217	169		386	500
9600	Alum (MC) aluminum clad, #6, 3 conductor w/#6 grnd		2.75	5.818		292	320		612	815
9610	4 conductor w/#6 grnd		2.29	6.987		261	385		646	880
9620	#4, 3 conductor w/#6 grnd	R-1B	3.13	7.668		249	410		659	905
9630	4 conductor w/#6 grnd		2.67	8.989		269	480		749	1,025
9640	#2, 3 conductor w/#4 grnd		2.59	9.266		445	495		940	1,250
9650	4 conductor w/#4 grnd		2.21	10.860		500	580		1,080	1,450
9660	#1, 3 conductor w/#4 grnd		2.15	11.163		370	595		965	1,325
9670	4 conductor w/#4 grnd		1.85	12.973		680	690		1,370	1,800
9680	1/0, 3 conductor w/#4 grnd		2.16	11.111		420	590		1,010	1,375
9690	4 conductor w/#4 grnd		1.73	13.873		510	740		1,250	1,700
9700	2/0, 3 conductor w/#4 grnd		1.91	12.565		690	670		1,360	1,775
9710	4 conductor w/#4 grnd		1.53	15.686		490	835		1,325	1,825
9720	3/0, 3 conductor w/#4 grnd		1.75	13.714		885	730		1,615	2,100
9730	4 conductor w/#4 grnd		1.40	17.143		610	910		1,520	2,075
9740	4/0, 3 conductor w/#2 grnd		1.54	15.584		990	830		1,820	2,375
9750	4 conductor w/#2 grnd		1.23	19.512		430	1,050		1,480	2,075

26 05 19.20 Armored Cable		Crew	Daily Output	Labor-Hours	Unit	Material	2020 Bare Costs Labor	Equipment	Total	Total Incl O&P
9760	250 kcmil, 3 conductor w/#1 grnd	R-1C	1.90	16.842	C.L.F.	1,125	930	53.50	2,108.50	2,700
9770	4 conductor w/#1 grnd		1.52	21.053		1,525	1,175	67	2,767	3,550
9780	350 kcmil, 3 conductor w/1/0 grnd		1.76	18.182		1,500	1,000	58	2,558	3,275
9790	4 conductor w/1/0 grnd		1.41	22.695		1,150	1,250	72	2,472	3,275
9800	500 kcmil, 3 conductor w/#1 grnd		1.64	19.512		1,900	1,075	62	3,037	3,825
9810	4 conductor w/2/0 grnd		1.31	24.427		1,575	1,350	77.50	3,002.50	3,875
9840	750 kcmil, 3 conductor w/1/0 grnd		1.31	24.427		3,025	1,350	77.50	4,452.50	5,475
9850	4 conductor w/3/0 grnd		1.04	30.769		3,125	1,700	98	4,923	6,175
9900	Minimum labor/equipment charge		4	8	Job		440	25.50	465.50	710

26 05 19.25 Cable Connectors		Crew	Daily Output	Labor-Hours	Unit	Material	Labor	Equipment	Total	Total Incl O&P
0010	**CABLE CONNECTORS**									
0100	600 volt, nonmetallic, #14-2 wire	1 Elec	160	.050	Ea.	1.71	3.07		4.78	6.60
0200	#14-3 wire to #12-2 wire		133	.060		1.71	3.69		5.40	7.60
0300	#12-3 wire to #10-2 wire		114	.070		1.71	4.31		6.02	8.55
0400	#10-3 wire to #14-4 and #12-4 wire		100	.080		1.71	4.91		6.62	9.45
0500	#8-3 wire to #10-4 wire		80	.100		3.24	6.15		9.39	13
0600	#6-3 wire		40	.200		4.15	12.25		16.40	23.50
0800	SER, 3 #8 insulated + 1 #8 ground		32	.250		3.15	15.35		18.50	27
0900	3 #6 + 1 #6 ground		24	.333		3.94	20.50		24.44	36
1000	3 #4 + 1 #6 ground		22	.364		4.91	22.50		27.41	40
1100	3 #2 + 1 #4 ground		20	.400		10.55	24.50		35.05	49.50
1200	3 1/0 + 1 #2 ground		18	.444		21.50	27.50		49	65.50
1400	3 2/0 + 1 #1 ground		16	.500		29.50	30.50		60	79.50
1600	3 4/0 + 1 #2/0 ground		14	.571		29.50	35		64.50	86
1800	600 volt, armored, #14-2 wire		80	.100		1.26	6.15		7.41	10.85
2200	#14-4, #12-3 and #10-2 wire		40	.200		.79	12.25		13.04	19.75
2400	#12-4, #10-3 and #8-2 wire		32	.250		3.48	15.35		18.83	27.50
2600	#8-3 and #10-4 wire		26	.308		4.10	18.90		23	33.50
2650	#8-4 wire		22	.364		6.10	22.50		28.60	41
2652	Non-PVC jacket connector, #6-3 wire, #6-4 wire		22	.364		52.50	22.50		75	92.50
2660	1/0-3 wire		11	.727		73.50	44.50		118	150
2670	300 kcmil-3 wire		6	1.333		126	82		208	264
2680	700 kcmil-3 wire		4	2		187	123		310	395
2700	PVC jacket connector, #6-3 wire, #6-4 wire		16	.500		6	30.50		36.50	53.50
2800	#4-3 wire, #4-4 wire		16	.500		6	30.50		36.50	53.50
2900	#2-3 wire		12	.667		6	41		47	69.50
3000	#1-3 wire, #2-4 wire		12	.667		13.55	41		54.55	78
3200	1/0-3 wire		11	.727		13.55	44.50		58.05	83.50
3400	2/0-3 wire, 1/0-4 wire		10	.800		13.55	49		62.55	90.50
3500	3/0-3 wire, 2/0-4 wire		9	.889		17.65	54.50		72.15	103
3600	4/0-3 wire, 3/0-4 wire		7	1.143		17.65	70		87.65	127
3800	250 kcmil-3 wire, 4/0-4 wire		6	1.333		34.50	82		116.50	164
4000	350 kcmil-3 wire, 250 kcmil-4 wire		5	1.600		34.50	98		132.50	189
4100	350 kcmil-4 wire		4	2		198	123		321	405
4200	500 kcmil-3 wire		4	2		198	123		321	405
4250	500 kcmil-4 wire, 750 kcmil-3 wire		3.50	2.286		219	140		359	455
4300	750 kcmil-4 wire		3	2.667		219	164		383	495
4400	5 kV, armored, #4		8	1		67	61.50		128.50	168
4600	#2		8	1		67	61.50		128.50	168
4800	#1		8	1		87.50	61.50		149	191
5000	1/0		6.40	1.250		108	76.50		184.50	237
5200	2/0		5.30	1.509		108	92.50		200.50	262

26 05 19 – Low-Voltage Electrical Power Conductors and Cables

	26 05 19.25 Cable Connectors	Crew	Daily Output	Labor-Hours	Unit	Material	2020 Bare Costs Labor	Equipment	Total	Total Incl O&P
5500	4/0	1 Elec	4	2	Ea.	145	123		268	350
5600	250 kcmil		3.60	2.222		145	136		281	370
5650	350 kcmil		3.20	2.500		171	153		324	425
5700	500 kcmil		2.50	3.200		223	196		419	545
5720	750 kcmil		2.20	3.636		268	223		491	640
5750	1,000 kcmil		2	4		320	245		565	730
5800	15 kV, armored, #1		4	2		139	123		262	340
5900	1/0		4	2		174	123		297	380
6000	3/0		3.60	2.222		254	136		390	490
6100	4/0		3.40	2.353		281	144		425	530
6200	250 kcmil		3.20	2.500		350	153		503	620
6300	350 kcmil		2.70	2.963		315	182		497	625
6400	500 kcmil	↓	2	4	↓	450	245		695	875

	26 05 19.30 Cable Splicing	Crew	Daily Output	Labor-Hours	Unit	Material	Labor	Equipment	Total	Total Incl O&P
0010	**CABLE SPLICING** URD or similar, ideal conditions									
0100	#6 stranded to #1 stranded, 5 kV	1 Elec	4	2	Ea.	166	123		289	370
0120	15 kV		3.60	2.222		224	136		360	455
0140	25 kV		3.20	2.500		272	153		425	535
0200	#1 stranded to 4/0 stranded, 5 kV		3.60	2.222		171	136		307	400
0210	15 kV		3.20	2.500		245	153		398	505
0220	25 kV		2.80	2.857		272	175		447	570
0300	4/0 stranded to 500 kcmil stranded, 5 kV		3.30	2.424		171	149		320	415
0310	15 kV		2.90	2.759		405	169		574	705
0320	25 kV		2.50	3.200		465	196		661	810
0400	500 kcmil, 5 kV		3.20	2.500		228	153		381	485
0410	15 kV		2.80	2.857		405	175		580	715
0420	25 kV		2.30	3.478		465	213		678	840
0500	600 kcmil, 5 kV		2.90	2.759		228	169		397	510
0510	15 kV		2.40	3.333		430	205		635	785
0520	25 kV		2	4		495	245		740	925
0600	750 kcmil, 5 kV		2.60	3.077		228	189		417	540
0610	15 kV		2.20	3.636		430	223		653	815
0620	25 kV		1.90	4.211		495	258		753	945
0700	1,000 kcmil, 5 kV		2.30	3.478		228	213		441	580
0710	15 kV		1.90	4.211		465	258		723	915
0720	25 kV	↓	1.60	5	↓	475	305		780	995

	26 05 19.35 Cable Terminations	Crew	Daily Output	Labor-Hours	Unit	Material	Labor	Equipment	Total	Total Incl O&P
0010	**CABLE TERMINATIONS**									
0015	Wire connectors, screw type, #22 to #14	1 Elec	260	.031	Ea.	.06	1.89		1.95	2.98
0020	#18 to #12		240	.033		.07	2.04		2.11	3.23
0035	#16 to #10		230	.035		.26	2.13		2.39	3.58
0040	#14 to #8		210	.038		.33	2.34		2.67	3.96
0045	#12 to #6		180	.044		.47	2.73		3.20	4.72
0050	Terminal lugs, solderless, #16 to #10		50	.160		.37	9.80		10.17	15.50
0100	#8 to #4		30	.267		.74	16.35		17.09	26
0150	#2 to #1		22	.364		.79	22.50		23.29	35.50
0200	1/0 to 2/0		16	.500		1.58	30.50		32.08	48.50
0250	3/0		12	.667		4.66	41		45.66	68
0300	4/0		11	.727		4.19	44.50		48.69	73
0350	250 kcmil		9	.889		3.13	54.50		57.63	87.50
0400	350 kcmil		7	1.143		4.26	70		74.26	113
0450	500 kcmil	↓	6	1.333	↓	7.85	82		89.85	135

26 05 19.35 Cable Terminations		Crew	Daily Output	Labor-Hours	Unit	Material	2020 Bare Costs Labor	Equipment	Total	Total Incl O&P
0500	600 kcmil	1 Elec	5.80	1.379	Ea.	9.15	84.50		93.65	140
0550	750 kcmil		5.20	1.538		10.15	94.50		104.65	156
0600	Split bolt connectors, tapped, #6		16	.500		3.64	30.50		34.14	51
0650	#4		14	.571		3.41	35		38.41	58
0700	#2		12	.667		6.30	41		47.30	70
0750	#1		11	.727		7.05	44.50		51.55	76.50
0800	1/0		10	.800		7.05	49		56.05	83.50
0850	2/0		9	.889		12.30	54.50		66.80	97.50
0900	3/0		7.20	1.111		23	68		91	130
1000	4/0		6.40	1.250		28	76.50		104.50	149
1100	250 kcmil		5.70	1.404		28	86		114	164
1200	300 kcmil		5.30	1.509		33.50	92.50		126	180
1400	350 kcmil		4.60	1.739		33.50	107		140.50	201
1500	500 kcmil		4	2		77.50	123		200.50	275
1600	Crimp 1 hole lugs, copper or aluminum, 600 volt									
1620	#14	1 Elec	60	.133	Ea.	.59	8.20		8.79	13.25
1630	#12		50	.160		1	9.80		10.80	16.20
1640	#10		45	.178		1	10.90		11.90	17.90
1780	#8		36	.222		1.51	13.65		15.16	22.50
1800	#6		30	.267		1.75	16.35		18.10	27
2000	#4		27	.296		4.02	18.20		22.22	32.50
2200	#2		24	.333		2.78	20.50		23.28	34.50
2400	#1		20	.400		4.99	24.50		29.49	43.50
2500	1/0		17.50	.457		6.05	28		34.05	49.50
2600	2/0		15	.533		8.10	32.50		40.60	59.50
2800	3/0		12	.667		7.95	41		48.95	71.50
3000	4/0		11	.727		6.40	44.50		50.90	75.50
3200	250 kcmil		9	.889		7.20	54.50		61.70	92
3400	300 kcmil		8	1		11.10	61.50		72.60	107
3500	350 kcmil		7	1.143		11.70	70		81.70	121
3600	400 kcmil		6.50	1.231		15.40	75.50		90.90	133
3800	500 kcmil		6	1.333		20.50	82		102.50	149
4000	600 kcmil		5.80	1.379		27	84.50		111.50	160
4200	700 kcmil		5.50	1.455		32.50	89		121.50	173
4400	750 kcmil		5.20	1.538		32.50	94.50		127	181
4500	Crimp 2-way connectors, copper or alum., 600 volt									
4510	#14	1 Elec	60	.133	Ea.	2.53	8.20		10.73	15.40
4520	#12		50	.160		2.69	9.80		12.49	18.05
4530	#10		45	.178		2.84	10.90		13.74	19.90
4540	#8		27	.296		2.87	18.20		21.07	31
4600	#6		25	.320		4.07	19.65		23.72	34.50
4800	#4		23	.348		3.97	21.50		25.47	37.50
5000	#2		20	.400		5.65	24.50		30.15	44
5200	#1		16	.500		5.70	30.50		36.20	53.50
5400	1/0		13	.615		5.70	38		43.70	64.50
5420	2/0		12	.667		7.85	41		48.85	71.50
5440	3/0		11	.727		7.40	44.50		51.90	76.50
5460	4/0		10	.800		7.70	49		56.70	84
5480	250 kcmil		9	.889		16.80	54.50		71.30	102
5500	300 kcmil		8.50	.941		16.20	57.50		73.70	107
5520	350 kcmil		8	1		15.35	61.50		76.85	111
5540	400 kcmil		7.30	1.096		19.85	67		86.85	126
5560	500 kcmil		6.20	1.290		23.50	79		102.50	148

26 05 19 – Low-Voltage Electrical Power Conductors and Cables

26 05 19.35 Cable Terminations		Crew	Daily Output	Labor-Hours	Unit	Material	Labor	Equipment	Total	Total Incl O&P
						2020 Bare Costs				
5580	600 kcmil	1 Elec	5.50	1.455	Ea.	47.50	89		136.50	190
5600	700 kcmil		4.50	1.778		50	109		159	223
5620	750 kcmil		4	2		55.50	123		178.50	250
7000	Compression equipment adapter, aluminum wire, #6		30	.267		7.55	16.35		23.90	33.50
7020	#4		27	.296		7.65	18.20		25.85	36.50
7040	#2		24	.333		7.05	20.50		27.55	39.50
7060	#1		20	.400		8.10	24.50		32.60	47
7080	1/0		18	.444		13.45	27.50		40.95	57
7100	2/0		15	.533		14.10	32.50		46.60	66
7140	4/0		11	.727		16.40	44.50		60.90	86.50
7160	250 kcmil		9	.889		17.65	54.50		72.15	103
7180	300 kcmil		8	1		18.35	61.50		79.85	115
7200	350 kcmil		7	1.143		25.50	70		95.50	136
7220	400 kcmil		6.50	1.231		30	75.50		105.50	149
7240	500 kcmil		6	1.333		30.50	82		112.50	160
7260	600 kcmil		5.80	1.379		34.50	84.50		119	168
7280	750 kcmil	↓	5.20	1.538		42	94.50		136.50	191
8000	Compression tool, hand					1,550			1,550	1,700
8100	Hydraulic					1,925			1,925	2,100
8500	Hydraulic dies				↓	360			360	395

26 05 19.50 Mineral Insulated Cable		Crew	Daily Output	Labor-Hours	Unit	Material	Labor	Equipment	Total	Total Incl O&P
0010	**MINERAL INSULATED CABLE** 600 volt									
0100	1 conductor, #12	1 Elec	1.60	5	C.L.F.	355	305		660	860
0200	#10		1.60	5		535	305		840	1,050
0400	#8		1.50	5.333		675	325		1,000	1,250
0500	#6	↓	1.40	5.714		710	350		1,060	1,325
0600	#4	2 Elec	2.40	6.667		1,050	410		1,460	1,800
0800	#2		2.20	7.273		1,450	445		1,895	2,250
0900	#1		2.10	7.619		1,650	465		2,115	2,550
1000	1/0		2	8		1,925	490		2,415	2,875
1100	2/0		1.90	8.421		2,425	515		2,940	3,475
1200	3/0		1.80	8.889		2,950	545		3,495	4,075
1400	4/0	↓	1.60	10		3,400	615		4,015	4,675
1410	250 kcmil	3 Elec	2.40	10		3,825	615		4,440	5,150
1420	350 kcmil		1.95	12.308		4,400	755		5,155	5,975
1430	500 kcmil	↓	1.95	12.308		5,550	755		6,305	7,250
1500	2 conductor, #12	1 Elec	1.40	5.714		925	350		1,275	1,575
1600	#10		1.20	6.667		1,150	410		1,560	1,875
1800	#8		1.10	7.273		1,300	445		1,745	2,100
2000	#6	↓	1.05	7.619		1,800	465		2,265	2,700
2100	#4	2 Elec	2	8		2,550	490		3,040	3,550
2200	3 conductor, #12	1 Elec	1.20	6.667		1,075	410		1,485	1,825
2400	#10		1.10	7.273		1,225	445		1,670	2,025
2600	#8		1.05	7.619		1,650	465		2,115	2,525
2800	#6	↓	1	8		2,175	490		2,665	3,125
3000	#4	2 Elec	1.80	8.889		2,850	545		3,395	3,975
3100	4 conductor, #12	1 Elec	1.20	6.667		1,150	410		1,560	1,875
3200	#10		1.10	7.273		1,375	445		1,820	2,175
3400	#8		1	8		1,925	490		2,415	2,875
3600	#6		.90	8.889		2,600	545		3,145	3,725
3620	7 conductor, #12		1.10	7.273		1,575	445		2,020	2,400
3640	#10	↓	1	8	↓	2,150	490		2,640	3,125

26 05 19 – Low-Voltage Electrical Power Conductors and Cables

26 05 19.50 Mineral Insulated Cable

	26 05 19.50 Mineral Insulated Cable	Crew	Daily Output	Labor-Hours	Unit	Material	Labor	Equipment	Total	Total Incl O&P
						2020 Bare Costs				
3800	Terminations, 600 volt, 1 conductor, #12	1 Elec	8	1	Ea.	18.30	61.50		79.80	115
4000	#10		7.60	1.053		18.30	64.50		82.80	120
4100	#8		7.30	1.096		18.30	67		85.30	124
4200	#6		6.70	1.194		18.30	73.50		91.80	133
4400	#4		6.20	1.290		18.30	79		97.30	142
4600	#2		5.70	1.404		27.50	86		113.50	164
4800	#1		5.30	1.509		27.50	92.50		120	174
5000	1/0		5	1.600		27.50	98		125.50	182
5100	2/0		4.70	1.702		27.50	104		131.50	192
5200	3/0		4.30	1.860		27.50	114		141.50	207
5400	4/0		4	2		61	123		184	256
5410	250 kcmil		4	2		61	123		184	256
5420	350 kcmil		4	2		105	123		228	305
5430	500 kcmil		4	2		105	123		228	305
5500	2 conductor, #12		6.70	1.194		18.30	73.50		91.80	133
5600	#10		6.40	1.250		27.50	76.50		104	149
5800	#8		6.20	1.290		27.50	79		106.50	153
6000	#6		5.70	1.404		27.50	86		113.50	164
6200	#4		5.30	1.509		61	92.50		153.50	210
6400	3 conductor, #12		5.70	1.404		27.50	86		113.50	164
6500	#10		5.50	1.455		27.50	89		116.50	168
6600	#8		5.20	1.538		27.50	94.50		122	176
6800	#6		4.80	1.667		27.50	102		129.50	188
7200	#4		4.60	1.739		61	107		168	231
7400	4 conductor, #12		4.60	1.739		31	107		138	198
7500	#10		4.40	1.818		31	112		143	206
7600	#8		4.20	1.905		31	117		148	214
8400	#6		4	2		64	123		187	260
8500	7 conductor, #12		3.50	2.286		30.50	140		170.50	250
8600	#10	↓	3	2.667		66.50	164		230.50	325
8800	Crimping tool, plier type					77			77	84.50
9000	Stripping tool					291			291	320
9200	Hand vise				↓	86.50			86.50	95.50
9500	Minimum labor/equipment charge	1 Elec	4	2	Job		123		123	189

26 05 19.55 Non-Metallic Sheathed Cable

	26 05 19.55 Non-Metallic Sheathed Cable	Crew	Daily Output	Labor-Hours	Unit	Material	Labor	Equipment	Total	Total Incl O&P
0010	**NON-METALLIC SHEATHED CABLE** 600 volt									
0100	Copper with ground wire (Romex)									
0150	#14, 2 conductor	1 Elec	2.70	2.963	C.L.F.	18.40	182		200.40	300
0200	3 conductor		2.40	3.333		26.50	205		231.50	345
0220	4 conductor		2.20	3.636		42	223		265	390
0250	#12, 2 conductor		2.50	3.200		26.50	196		222.50	330
0300	3 conductor		2.20	3.636		41	223		264	390
0320	4 conductor		2	4		65	245		310	450
0350	#10, 2 conductor		2.20	3.636		45	223		268	395
0400	3 conductor		1.80	4.444		66.50	273		339.50	495
0420	4 conductor		1.60	5		105	305		410	585
0430	#8, 2 conductor		1.60	5		87.50	305		392.50	565
0450	3 conductor		1.50	5.333		117	325		442	635
0500	#6, 3 conductor	↓	1.40	5.714		179	350		529	735
0520	#4, 3 conductor	2 Elec	2.40	6.667		400	410		810	1,075
0540	#2, 3 conductor	"	2.20	7.273	↓	910	445		1,355	1,675
0550	SE type SER aluminum cable, 3 RHW and									

26 05 19.55 Non-Metallic Sheathed Cable		Crew	Daily Output	Labor-Hours	Unit	Material	Labor	Equipment	Total	Total Incl O&P
						2020 Bare Costs				
0600	1 bare neutral, 3 #8 & 1 #8	1 Elec	1.60	5	C.L.F.	63	305		368	540
0650	3 #6 & 1 #6	"	1.40	5.714		70.50	350		420.50	620
0700	3 #4 & 1 #6	2 Elec	2.40	6.667		78.50	410		488.50	715
0750	3 #2 & 1 #4		2.20	7.273		143	445		588	845
0800	3 #1/0 & 1 #2		2	8		167	490		657	940
0850	3 #2/0 & 1 #1		1.80	8.889		199	545		744	1,050
0900	3 #4/0 & 1 #2/0		1.60	10		292	615		907	1,275
1000	URD - triplex underground distribution cable, alum. 2 #4 + #4 neutral		2.80	5.714		84	350		434	635
1010	2 #2 + #4 neutral		2.65	6.038		106	370		476	685
1020	2 #2 + #2 neutral		2.55	6.275		105	385		490	710
1030	2 1/0 + #2 neutral		2.40	6.667		130	410		540	775
1040	2 1/0 + 1/0 neutral		2.30	6.957		147	425		572	815
1050	2 2/0 + #1 neutral		2.20	7.273		146	445		591	845
1060	2 2/0 + 2/0 neutral		2.10	7.619		154	465		619	890
1070	2 3/0 + 1/0 neutral		1.95	8.205		194	505		699	990
1080	2 3/0 + 3/0 neutral		1.95	8.205		198	505		703	995
1090	2 4/0 + 2/0 neutral		1.85	8.649		188	530		718	1,025
1100	2 4/0 + 4/0 neutral		1.85	8.649		226	530		756	1,075
1450	UF underground feeder cable, copper with ground, #14, 2 conductor	1 Elec	4	2		36	123		159	229
1500	#12, 2 conductor		3.50	2.286		33.50	140		173.50	253
1550	#10, 2 conductor		3	2.667		53.50	164		217.50	310
1600	#14, 3 conductor		3.50	2.286		36.50	140		176.50	257
1650	#12, 3 conductor		3	2.667		49.50	164		213.50	305
1700	#10, 3 conductor		2.50	3.200		80	196		276	390
1710	#8, 3 conductor		2	4		139	245		384	535
1720	#6, 3 conductor		1.80	4.444		213	273		486	655
2400	SEU service entrance cable, copper 2 conductors, #8 + #8 neutral		1.50	5.333		98.50	325		423.50	615
2600	#6 + #8 neutral		1.30	6.154		139	380		519	730
2800	#6 + #6 neutral		1.30	6.154		156	380		536	750
3000	#4 + #6 neutral	2 Elec	2.20	7.273		222	445		667	930
3200	#4 + #4 neutral		2.20	7.273		234	445		679	940
3400	#3 + #5 neutral		2.10	7.619		320	465		785	1,075
3600	#3 + #3 neutral		2.10	7.619		305	465		770	1,050
3800	#2 + #4 neutral		2	8		360	490		850	1,150
4000	#1 + #1 neutral		1.90	8.421		375	515		890	1,200
4200	1/0 + 1/0 neutral		1.80	8.889		850	545		1,395	1,775
4400	2/0 + 2/0 neutral		1.70	9.412		810	575		1,385	1,775
4600	3/0 + 3/0 neutral		1.60	10		905	615		1,520	1,950
4620	4/0 + 4/0 neutral		1.45	11.034		1,775	675		2,450	3,000
4800	Aluminum 2 conductors, #8 + #8 neutral	1 Elec	1.60	5		56	305		361	530
5000	#6 + #6 neutral	"	1.40	5.714		57	350		407	605
5100	#4 + #6 neutral	2 Elec	2.50	6.400		70	395		465	680
5200	#4 + #4 neutral		2.40	6.667		81	410		491	720
5300	#2 + #4 neutral		2.30	6.957		105	425		530	770
5400	#2 + #2 neutral		2.20	7.273		108	445		553	805
5450	1/0 + #2 neutral		2.10	7.619		160	465		625	895
5500	1/0 + 1/0 neutral		2	8		170	490		660	940
5550	2/0 + #1 neutral		1.90	8.421		180	515		695	990
5600	2/0 + 2/0 neutral		1.80	8.889		180	545		725	1,050
5800	3/0 + 1/0 neutral		1.70	9.412		242	575		817	1,150
6000	3/0 + 3/0 neutral		1.70	9.412		198	575		773	1,100
6200	4/0 + 2/0 neutral		1.60	10		199	615		814	1,175
6400	4/0 + 4/0 neutral		1.60	10		220	615		835	1,175

26 05 19 – Low-Voltage Electrical Power Conductors and Cables

26 05 19.55 Non-Metallic Sheathed Cable		Crew	Daily Output	Labor-Hours	Unit	Material	2020 Bare Costs Labor	Equipment	Total	Total Incl O&P
6500	Service entrance cap for copper SEU									
6600	100 amp	1 Elec	12	.667	Ea.	7.85	41		48.85	71.50
6700	150 amp		10	.800		11.75	49		60.75	88.50
6800	200 amp		8	1		17.85	61.50		79.35	114
9000	Minimum labor/equipment charge		4	2	Job		123		123	189

26 05 19.70 Portable Cord

		Crew	Daily Output	Labor-Hours	Unit	Material	Labor	Equipment	Total	Total Incl O&P
0010	**PORTABLE CORD** 600 volt									
0100	Type SO, #18, 2 conductor	1 Elec	980	.008	L.F.	.35	.50		.85	1.16
0110	3 conductor		980	.008		.34	.50		.84	1.14
0120	#16, 2 conductor		840	.010		.42	.58		1	1.36
0130	3 conductor		840	.010		4.32	.58		4.90	5.65
0140	4 conductor		840	.010		.53	.58		1.11	1.48
0240	#14, 2 conductor		840	.010		.53	.58		1.11	1.48
0250	3 conductor		840	.010		.56	.58		1.14	1.52
0260	4 conductor		840	.010		1.11	.58		1.69	2.12
0280	#12, 2 conductor		840	.010		.73	.58		1.31	1.70
0290	3 conductor		840	.010		.90	.58		1.48	1.89
0300	4 conductor		840	.010		1.09	.58		1.67	2.10
0320	#10, 2 conductor		765	.010		.99	.64		1.63	2.08
0330	3 conductor		765	.010		1.46	.64		2.10	2.60
0340	4 conductor		765	.010		1.59	.64		2.23	2.74
0360	#8, 2 conductor		555	.014		1.71	.88		2.59	3.24
0370	3 conductor		540	.015		2.06	.91		2.97	3.67
0380	4 conductor		525	.015		2.15	.94		3.09	3.81
0400	#6, 2 conductor		525	.015		2.60	.94		3.54	4.30
0410	3 conductor		490	.016		3.08	1		4.08	4.93
0420	4 conductor		415	.019		3.25	1.18		4.43	5.40
0440	#4, 2 conductor	2 Elec	830	.019		3.82	1.18		5	6
0450	3 conductor		700	.023		4	1.40		5.40	6.55
0460	4 conductor		660	.024		5.30	1.49		6.79	8.15
0480	#2, 2 conductor		450	.036		4.99	2.18		7.17	8.85
0490	3 conductor		350	.046		7.15	2.80		9.95	12.20
0500	4 conductor		280	.057		9.20	3.51		12.71	15.50
2000	See 26 27 26.20 for Wiring Devices Elements									

26 05 19.75 Modular Flexible Wiring System

		Crew	Daily Output	Labor-Hours	Unit	Material	Labor	Equipment	Total	Total Incl O&P
0010	**MODULAR FLEXIBLE WIRING SYSTEM**									
0020	Commercial system grid ceiling									
0100	Conversion Module	1 Elec	32	.250	Ea.	8.20	15.35		23.55	32.50
0120	Fixture cable, for fixture to fixture, 3 conductor, 15′ long		40	.200		33.50	12.25		45.75	56
0150	Extender cable, 3 conductor, 15′ long		48	.167		33	10.25		43.25	52
0200	Switch drop, 1 level, 9′ long		32	.250		35.50	15.35		50.85	62.50
0220	2 level, 9′ long		32	.250		34.50	15.35		49.85	61.50
0250	Power tee, 9′ long		32	.250		28.50	15.35		43.85	55
1020	Industrial system open ceiling									
1100	Converter, interface between hardwiring and modular wiring	1 Elec	32	.250	Ea.	19.90	15.35		35.25	45.50
1120	Fixture cable, for fixture to fixture, 3 conductor, 21′ long		32	.250		114	15.35		129.35	150
1125	3 conductor, 25′ long		24	.333		147	20.50		167.50	194
1130	3 conductor, 31′ long		19	.421		130	26		156	183
1150	Fixture cord drop, 10′ long		40	.200		35	12.25		47.25	58

26 05 19.90 Wire			Crew	Daily Output	Labor-Hours	Unit	2020 Bare Costs Material	Labor	Equipment	Total	Total Incl O&P
0010	**WIRE**, normal installation conditions in wireway, conduit, cable tray										
0020	600 volt, copper type THW, solid, #14	R260519-90	1 Elec	13	.615	C.L.F.	6.25	38		44.25	65
0030	#12			11	.727		9.85	44.50		54.35	79.50
0040	#10			10	.800		15.95	49		64.95	93
0050	Stranded, #14	R260519-92		13	.615		7.60	38		45.60	66.50
0100	#12			11	.727		13.75	44.50		58.25	83.50
0120	#10	R260533-22		10	.800		22	49		71	99.50
0140	#8			8	1		34.50	61.50		96	133
0160	#6			6.50	1.231		57.50	75.50		133	180
0180	#4		2 Elec	10.60	1.509		87.50	92.50		180	240
0200	#3			10	1.600		118	98		216	280
0220	#2			9	1.778		136	109		245	320
0240	#1			8	2		174	123		297	380
0260	1/0			6.60	2.424		226	149		375	475
0280	2/0			5.80	2.759		390	169		559	690
0300	3/0			5	3.200		278	196		474	605
0350	4/0			4.40	3.636		450	223		673	840
0400	250 kcmil		3 Elec	6	4		490	245		735	915
0420	300 kcmil			5.70	4.211		685	258		943	1,150
0450	350 kcmil			5.40	4.444		670	273		943	1,150
0480	400 kcmil			5.10	4.706		1,200	289		1,489	1,775
0490	500 kcmil			4.80	5		965	305		1,270	1,550
0500	600 kcmil			3.90	6.154		1,275	380		1,655	1,975
0510	750 kcmil			3.30	7.273		1,150	445		1,595	1,950
0520	1,000 kcmil			2.70	8.889		1,875	545		2,420	2,900
0540	600 volt, aluminum type THHN, stranded, #6		1 Elec	8	1		20.50	61.50		82	117
0560	#4		2 Elec	13	1.231		25	75.50		100.50	144
0580	#2			10.60	1.509		48	92.50		140.50	196
0600	#1			9	1.778		57	109		166	231
0620	1/0			8	2		80	123		203	278
0640	2/0			7.20	2.222		82	136		218	300
0680	3/0			6.60	2.424		103	149		252	340
0700	4/0			6.20	2.581		131	158		289	390
0720	250 kcmil		3 Elec	8.70	2.759		137	169		306	410
0740	300 kcmil			8.10	2.963		190	182		372	490
0760	350 kcmil			7.50	3.200		192	196		388	510
0780	400 kcmil			6.90	3.478		219	213		432	570
0800	500 kcmil			6	4		246	245		491	650
0850	600 kcmil			5.70	4.211		315	258		573	745
0880	700 kcmil			5.10	4.706		385	289		674	870
0900	750 kcmil			4.80	5		375	305		680	885
0910	1,000 kcmil			3.78	6.349		665	390		1,055	1,325
0920	600 volt, copper type THWN-THHN, solid, #14		1 Elec	13	.615		6.65	38		44.65	65.50
0940	#12			11	.727		10.05	44.50		54.55	79.50
0960	#10			10	.800		15.60	49		64.60	92.50
1000	Stranded, #14			13	.615		7.70	38		45.70	66.50
1200	#12			11	.727		10.60	44.50		55.10	80
1250	#10			10	.800		20.50	49		69.50	98
1300	#8			8	1		27	61.50		88.50	125
1350	#6			6.50	1.231		44	75.50		119.50	165
1400	#4		2 Elec	10.60	1.509		70	92.50		162.50	220
1450	#3			10	1.600		87.50	98		185.50	248

26 05 19.90 Wire		Crew	Daily Output	Labor-Hours	Unit	Material	Labor	Equipment	Total	Total Incl O&P
						2020 Bare Costs				
1500	#2	2 Elec	9	1.778	C.L.F.	109	109		218	287
1550	#1		8	2		138	123		261	340
1600	1/0		6.60	2.424		200	149		349	450
1650	2/0		5.80	2.759		208	169		377	490
1700	3/0		5	3.200		259	196		455	585
2000	4/0		4.40	3.636		330	223		553	705
2200	250 kcmil	3 Elec	6	4		380	245		625	800
2400	300 kcmil		5.70	4.211		515	258		773	965
2600	350 kcmil		5.40	4.444		535	273		808	1,000
2700	400 kcmil		5.10	4.706		605	289		894	1,100
2800	500 kcmil		4.80	5		775	305		1,080	1,325
2802	600 kcmil		3.90	6.154		780	380		1,160	1,425
2804	750 kcmil		3.30	7.273		1,525	445		1,970	2,350
2805	1,000 kcmil		1.94	12.371		2,475	760		3,235	3,900
2900	600 volt, copper type XHHW, solid, #14	1 Elec	13	.615		11.05	38		49.05	70
2920	#12		11	.727		17.50	44.50		62	88
2940	#10		10	.800		27	49		76	106
3000	Stranded, #14		13	.615		9.70	38		47.70	68.50
3020	#12		11	.727		13.70	44.50		58.20	83.50
3040	#10		10	.800		20.50	49		69.50	98
3060	#8		8	1		34	61.50		95.50	132
3080	#6		6.50	1.231		54.50	75.50		130	176
3100	#4	2 Elec	10.60	1.509		77.50	92.50		170	228
3110	#3		10	1.600		97	98		195	257
3120	#2		9	1.778		120	109		229	300
3140	#1		8	2		159	123		282	365
3160	1/0		6.60	2.424		197	149		346	445
3180	2/0		5.80	2.759		246	169		415	530
3200	3/0		5	3.200		305	196		501	635
3220	4/0		4.40	3.636		385	223		608	765
3240	250 kcmil	3 Elec	6	4		450	245		695	875
3260	300 kcmil		5.70	4.211		535	258		793	985
3280	350 kcmil		5.40	4.444		585	273		858	1,050
3300	400 kcmil		5.10	4.706		715	289		1,004	1,225
3320	500 kcmil		4.80	5		835	305		1,140	1,375
3340	600 kcmil		3.90	6.154		1,025	380		1,405	1,725
3360	750 kcmil		3.30	7.273		1,825	445		2,270	2,675
3380	1,000 kcmil		2.40	10		2,450	615		3,065	3,650
5020	600 volt, aluminum type XHHW, stranded, #6	1 Elec	8	1		26	61.50		87.50	123
5040	#4	2 Elec	13	1.231		32	75.50		107.50	152
5060	#2		10.60	1.509		40	92.50		132.50	187
5080	#1		9	1.778		64	109		173	239
5100	1/0		8	2		76	123		199	273
5120	2/0		7.20	2.222		93.50	136		229.50	315
5140	3/0		6.60	2.424		112	149		261	350
5160	4/0		6.20	2.581		101	158		259	355
5180	250 kcmil	3 Elec	8.70	2.759		157	169		326	435
5200	300 kcmil		8.10	2.963		209	182		391	510
5220	350 kcmil		7.50	3.200		221	196		417	545
5240	400 kcmil		6.90	3.478		247	213		460	600
5260	500 kcmil		6	4		271	245		516	680
5280	600 kcmil		5.70	4.211		350	258		608	785
5300	700 kcmil		5.40	4.444		395	273		668	855

26 05 19 – Low-Voltage Electrical Power Conductors and Cables

26 05 19.90 Wire		Crew	Daily Output	Labor-Hours	Unit	Material	Labor	Equipment	Total	Total Incl O&P
						2020 Bare Costs				
5320	750 kcmil	3 Elec	5.10	4.706	C.L.F.	375	289		664	860
5340	1,000 kcmil		3.60	6.667		595	410		1,005	1,275
5390	600 volt, copper type XLPE-USE (RHW), solid, #14	1 Elec	12	.667		11.85	41		52.85	76
5400	#12		11	.727		18.10	44.50		62.60	88.50
5420	#10		10	.800		23.50	49		72.50	101
5440	Stranded, #14		13	.615		15.05	38		53.05	74.50
5460	#12		11	.727		18.05	44.50		62.55	88.50
5480	#10		10	.800		28	49		77	107
5500	#8		8	1		34	61.50		95.50	132
5520	#6		6.50	1.231		56.50	75.50		132	178
5540	#4	2 Elec	10.60	1.509		86.50	92.50		179	238
5560	#2		9	1.778		140	109		249	320
5580	#1		8	2		188	123		311	395
5600	1/0		6.60	2.424		236	149		385	490
5620	2/0		5.80	2.759		310	169		479	600
5640	3/0		5	3.200		390	196		586	725
5660	4/0		4.40	3.636		420	223		643	805
5680	250 kcmil	3 Elec	6	4		510	245		755	940
5700	300 kcmil		5.70	4.211		585	258		843	1,050
5720	350 kcmil		5.40	4.444		655	273		928	1,150
5740	400 kcmil		5.10	4.706		765	289		1,054	1,300
5760	500 kcmil		4.80	5		830	305		1,135	1,375
5780	600 kcmil		3.90	6.154		1,250	380		1,630	1,950
5800	750 kcmil		3.30	7.273		1,950	445		2,395	2,800
5820	1,000 kcmil		2.70	8.889		2,600	545		3,145	3,725
5840	600 volt, aluminum type XLPE-USE (RHW), stranded, #6	1 Elec	8	1		32	61.50		93.50	130
5860	#4	2 Elec	13	1.231		37.50	75.50		113	158
5880	#2		10.60	1.509		53	92.50		145.50	201
5900	#1		9	1.778		75	109		184	251
5920	1/0		8	2		88	123		211	286
5940	2/0		7.20	2.222		103	136		239	325
5960	3/0		6.60	2.424		121	149		270	360
5980	4/0		6.20	2.581		135	158		293	395
6000	250 kcmil	3 Elec	8.70	2.759		183	169		352	460
6020	300 kcmil		8.10	2.963		238	182		420	540
6040	350 kcmil		7.50	3.200		241	196		437	565
6060	400 kcmil		6.90	3.478		293	213		506	650
6080	500 kcmil		6	4		325	245		570	740
6100	600 kcmil		5.70	4.211		420	258		678	860
6110	700 kcmil		5.40	4.444		455	273		728	920
6120	750 kcmil		5.10	4.706		490	289		779	980
9000	Minimum labor/equipment charge	1 Elec	4	2	Job		123		123	189

26 05 23 – Control-Voltage Electrical Power Cables

26 05 23.10 Control Cable		Crew	Daily Output	Labor-Hours	Unit	Material	Labor	Equipment	Total	Total Incl O&P
0010	**CONTROL CABLE**									
0020	600 volt, copper, #14 THWN wire with PVC jacket, 2 wires	1 Elec	9	.889	C.L.F.	44.50	54.50		99	133
0030	3 wires		8	1		52.50	61.50		114	152
0100	4 wires		7	1.143		54	70		124	168
0150	5 wires		6.50	1.231		74.50	75.50		150	198
0200	6 wires		6	1.333		99.50	82		181.50	236
0300	8 wires		5.30	1.509		140	92.50		232.50	296
0400	10 wires		4.80	1.667		166	102		268	340

26 05 23.10 Control Cable		Crew	Daily Output	Labor-Hours	Unit	2020 Bare Costs Material	Labor	Equipment	Total	Total Incl O&P
0500	12 wires	1 Elec	4.30	1.860	C.L.F.	213	114		327	410
0600	14 wires		3.80	2.105		226	129		355	450
0700	16 wires		3.50	2.286		258	140		398	500
0800	18 wires		3.30	2.424		299	149		448	560
0810	19 wires		3.10	2.581		279	158		437	550
0900	20 wires		3	2.667		295	164		459	575
1000	22 wires		2.80	2.857		355	175		530	660

26 05 23.20 Special Wires and Fittings

		Crew	Daily Output	Labor-Hours	Unit	Material	Labor	Equipment	Total	Total Incl O&P
0010	**SPECIAL WIRES & FITTINGS**									
0100	Fixture TFFN 600 volt 90°C stranded, #18	1 Elec	13	.615	C.L.F.	9.35	38		47.35	68.50
0150	#16		13	.615		12.35	38		50.35	71.50
0500	Thermostat, jacket non-plenum, twisted, #18-2 conductor		8	1		10.15	61.50		71.65	106
0550	#18-3 conductor		7	1.143		12.65	70		82.65	122
0600	#18-4 conductor		6.50	1.231		20	75.50		95.50	138
0650	#18-5 conductor		6	1.333		24.50	82		106.50	153
0700	#18-6 conductor		5.50	1.455		28	89		117	168
0750	#18-7 conductor		5	1.600		36	98		134	191
0800	#18-8 conductor		4.80	1.667		35.50	102		137.50	196
2460	Tray cable, type TC, copper, #16-2 conductor		9.40	.851		31	52		83	115
2464	#16-3 conductor		8.40	.952		51	58.50		109.50	147
2468	#16-4 conductor		7.30	1.096		54.50	67		121.50	164
2472	#16-5 conductor		6.70	1.194		42.50	73.50		116	160
2476	#16-7 conductor		5.50	1.455		58.50	89		147.50	201
2480	#16-9 conductor		5	1.600		73	98		171	232
2484	#16-12 conductor	2 Elec	8.80	1.818		127	112		239	310
2488	#16-15 conductor		7.20	2.222		127	136		263	350
2492	#16-19 conductor		6.60	2.424		156	149		305	400
2496	#16-25 conductor		6.40	2.500		221	153		374	480
2500	#14-2 conductor	1 Elec	9	.889		35.50	54.50		90	123
2520	#14-3 conductor		8	1		34.50	61.50		96	132
2540	#14-4 conductor		7	1.143		49	70		119	162
2560	#14-5 conductor		6.50	1.231		74	75.50		149.50	198
2564	#14-7 conductor		5.30	1.509		92	92.50		184.50	244
2568	#14-9 conductor	2 Elec	9.60	1.667		106	102		208	274
2572	#14-12 conductor		8.60	1.860		148	114		262	340
2576	#14-15 conductor		7.20	2.222		174	136		310	400
2578	#14-19 conductor		6.20	2.581		219	158		377	485
2582	#14-25 conductor		5.30	3.019		291	185		476	605
2590	#12-2 conductor	1 Elec	8.40	.952		52	58.50		110.50	147
2592	#12-3 conductor		7.60	1.053		56	64.50		120.50	162
2594	#12-4 conductor		6.60	1.212		105	74.50		179.50	229
2596	#12-5 conductor		6.20	1.290		114	79		193	248
2598	#12-7 conductor	2 Elec	10.40	1.538		120	94.50		214.50	277
2602	#12-9 conductor		8.80	1.818		217	112		329	410
2604	#12-12 conductor		7.80	2.051		220	126		346	435
2606	#12-15 conductor		6.80	2.353		261	144		405	510
2608	#12-19 conductor		6	2.667		320	164		484	600
2610	#12-25 conductor	3 Elec	7.70	3.117		435	191		626	775
2618	#10-2 conductor	1 Elec	8	1		84.50	61.50		146	188
2622	#10-3 conductor	″	7.30	1.096		84.50	67		151.50	197
2624	#10-4 conductor	2 Elec	12.80	1.250		160	76.50		236.50	294
2626	#10-5 conductor		11.80	1.356		136	83		219	278

26 05 23.20	Special Wires and Fittings	Crew	Daily Output	Labor-Hours	Unit	Material	Labor	Equipment	Total	Total Incl O&P
						2020 Bare Costs				
2628	#10-7 conductor	2 Elec	9.40	1.702	C.L.F.	196	104		300	375
2630	#10-9 conductor		8.40	1.905		248	117		365	455
2632	#10-12 conductor		7.20	2.222		275	136		411	515
2640	300 V, copper braided shield, PVC jacket									
2650	2 conductor #18 stranded	1 Elec	7	1.143	C.L.F.	30	70		100	141
2660	3 conductor #18 stranded		6	1.333		44.50	82		126.50	175
2670	4 conductor #18 stranded		6	1.333		161	82		243	305
3000	Strain relief grip for cable									
3050	Cord, top, #12-3	1 Elec	40	.200	Ea.	10.70	12.25		22.95	30.50
3060	#12-4		40	.200		10.70	12.25		22.95	30.50
3070	#12-5		39	.205		12.65	12.60		25.25	33.50
3100	#10-3		39	.205		12.65	12.60		25.25	33.50
3110	#10-4		38	.211		12.65	12.90		25.55	34
3120	#10-5		38	.211		15.30	12.90		28.20	37
3200	Bottom, #12-3		40	.200		25.50	12.25		37.75	47
3210	#12-4		40	.200		25.50	12.25		37.75	47
3220	#12-5		39	.205		26.50	12.60		39.10	49
3230	#10-3		39	.205		26.50	12.60		39.10	49
3300	#10-4		38	.211		26.50	12.90		39.40	49.50
3310	#10-5		38	.211		33	12.90		45.90	56.50
3400	Cable ties, standard, 4" length		190	.042		.15	2.58		2.73	4.15
3410	7" length		160	.050		.18	3.07		3.25	4.92
3420	14.5" length		90	.089		.41	5.45		5.86	8.85
3430	Heavy, 14.5" length		80	.100		.59	6.15		6.74	10.10
3500	Cable gland, nylon, 1/4" NPT thread		60	.133		2.11	8.20		10.31	14.90
3510	Cable gland, nylon, 3/8" NPT thread		60	.133		2.43	8.20		10.63	15.25
3520	Cable gland, nylon, 1/2" NPT thread		60	.133		3.04	8.20		11.24	15.95
3530	Cable gland, nylon, 3/4" NPT thread		60	.133		4.59	8.20		12.79	17.65
3540	Cable gland, nylon, 1" NPT thread		60	.133		7.25	8.20		15.45	20.50
3550	Cable gland, nylon, 1-1/4" NPT thread		60	.133		8.55	8.20		16.75	22
3970	U-cable guards									
3980	.75" x 5' U-cable guard,14 gauge	1 Elec	24	.333	Ea.	11.30	20.50		31.80	44
3990	.75" x 8' U-cable guard,14 gauge		24	.333		21	20.50		41.50	54.50
4000	1-1/8" x 5' U-cable guard		24	.333		16.40	20.50		36.90	49.50
4010	1-1/8" x 8' U-cable guard		24	.333		19.30	20.50		39.80	52.50
4020	2-3/16" x 5' U-cable guard		24	.333		33.50	20.50		54	68
4030	2-3/16" x 8' U-cable guard		24	.333		32.50	20.50		53	67.50
4040	3-3/16" x 5' U-cable guard		24	.333		37	20.50		57.50	72.50
4050	3-3/16" x 8' U-cable guard		24	.333		48	20.50		68.50	84.50
4060	4" x 8' U-cable guard		24	.333		42	20.50		62.50	77.50
4070	2" x 9' flanged-cable guard		20	.400		98.50	24.50		123	146
4080	3" x 9' flanged-cable guard		20	.400		125	24.50		149.50	176
4090	3.5" x 9' flanged-cable guard		20	.400		151	24.50		175.50	204
5000	2" x 5' flanged-cable guard, extension		20	.400		66.50	24.50		91	112
5010	3" x 5' flanged-cable guard, extension		20	.400		91	24.50		115.50	138
5020	.05" x 8' U-cable guard, black plastic		26	.308		5.55	18.90		24.45	35
5030	.75" x 8' U-cable guard, black plastic		26	.308		7.50	18.90		26.40	37.50
5040	1" x 8' U-cable guard, black plastic		26	.308		11.95	18.90		30.85	42
9000	Minimum labor/equipment charge		4	2	Job		123		123	189

	26 05 26.80 Grounding	Crew	Daily Output	Labor-Hours	Unit	Material	Labor	Equipment	Total	Total Incl O&P
						2020 Bare Costs				
0010	**GROUNDING**									
0030	Rod, copper clad, 8' long, 1/2" diameter	1 Elec	5.50	1.455	Ea.	22.50	89		111.50	162
0040	5/8" diameter		5.50	1.455		22.50	89		111.50	162
0050	3/4" diameter		5.30	1.509		36	92.50		128.50	183
0080	10' long, 1/2" diameter		4.80	1.667		24.50	102		126.50	184
0090	5/8" diameter		4.60	1.739		22.50	107		129.50	189
0100	3/4" diameter		4.40	1.818		53	112		165	231
0130	15' long, 3/4" diameter		4	2		55	123		178	250
0150	Coupling, bronze, 1/2" diameter					5.05			5.05	5.55
0160	5/8" diameter					5.45			5.45	6
0170	3/4" diameter					14.75			14.75	16.20
0190	Drive studs, 1/2" diameter					12.75			12.75	14.05
0210	5/8" diameter					20.50			20.50	22.50
0220	3/4" diameter					23			23	25
0230	Clamp, bronze, 1/2" diameter	1 Elec	32	.250		5.25	15.35		20.60	29.50
0240	5/8" diameter		32	.250		4.64	15.35		19.99	28.50
0250	3/4" diameter		32	.250		5.65	15.35		21	30
0260	Wire ground bare armored, #8-1 conductor		2	4	C.L.F.	74.50	245		319.50	460
0270	#6-1 conductor		1.80	4.444		84	273		357	515
0280	#4-1 conductor		1.60	5		146	305		451	630
0320	Bare copper wire, #14 solid		14	.571		7.10	35		42.10	62
0330	#12		13	.615		12.85	38		50.85	72
0340	#10		12	.667		17.70	41		58.70	82.50
0350	#8		11	.727		25	44.50		69.50	96
0360	#6		10	.800		39.50	49		88.50	119
0370	#4		8	1		98	61.50		159.50	203
0380	#2		5	1.600		127	98		225	290
0390	Bare copper wire, stranded, #8		11	.727		40	44.50		84.50	113
0400	#6		10	.800		42	49		91	122
0450	#4	2 Elec	16	1		85.50	61.50		147	189
0600	#2		10	1.600		89	98		187	249
0650	#1		9	1.778		130	109		239	310
0700	1/0		8	2		156	123		279	360
0750	2/0		7.20	2.222		253	136		389	490
0800	3/0		6.60	2.424		305	149		454	565
1000	4/0		5.70	2.807		420	172		592	730
1200	250 kcmil	3 Elec	7.20	3.333		440	205		645	795
1210	300 kcmil		6.60	3.636		560	223		783	965
1220	350 kcmil		6	4		550	245		795	985
1230	400 kcmil		5.70	4.211		605	258		863	1,075
1240	500 kcmil		5.10	4.706		785	289		1,074	1,300
1260	750 kcmil		3.60	6.667		1,300	410		1,710	2,050
1270	1,000 kcmil		3	8		1,575	490		2,065	2,475
1360	Bare aluminum, stranded, #6	1 Elec	9	.889		14.55	54.50		69.05	100
1370	#4	2 Elec	16	1		31.50	61.50		93	129
1380	#2		13	1.231		45.50	75.50		121	166
1390	#1		10.60	1.509		40	92.50		132.50	187
1400	1/0		9	1.778		50.50	109		159.50	224
1410	2/0		8	2		77.50	123		200.50	275
1420	3/0		7.20	2.222		86	136		222	305
1430	4/0		6.60	2.424		109	149		258	350
1440	250 kcmil	3 Elec	9.30	2.581		115	158		273	370

26 05 26.80 Grounding		Crew	Daily Output	Labor-Hours	Unit	Material	Labor	Equipment	Total	Total Incl O&P
						2020 Bare Costs				
1450	300 kcmil	3 Elec	8.70	2.759	C.L.F.	121	169		290	395
1460	400 kcmil		7.50	3.200		159	196		355	475
1470	500 kcmil		6.90	3.478		193	213		406	540
1480	600 kcmil		6	4		230	245		475	635
1490	700 kcmil		5.70	4.211		247	258		505	670
1500	750 kcmil		5.10	4.706		273	289		562	745
1510	1,000 kcmil		4.80	5		350	305		655	855
1800	Water pipe ground clamps, heavy duty									
2000	Bronze, 1/2" to 1" diameter	1 Elec	8	1	Ea.	29	61.50		90.50	126
2100	1-1/4" to 2" diameter		8	1		31	61.50		92.50	129
2200	2-1/2" to 3" diameter		6	1.333		60.50	82		142.50	193
2740	4/0 wire to building steel		7	1.143		9.45	70		79.45	118
2750	4/0 wire to motor frame		7	1.143		9.45	70		79.45	118
2760	4/0 wire to 4/0 wire		7	1.143		9.45	70		79.45	118
2770	4/0 wire to #4 wire		7	1.143		9.45	70		79.45	118
2780	4/0 wire to #8 wire		7	1.143		9.45	70		79.45	118
2790	Mold, reusable, for above					141			141	155
2800	Brazed connections, #6 wire	1 Elec	12	.667		16.15	41		57.15	81
3000	#2 wire		10	.800		21.50	49		70.50	99.50
3100	3/0 wire		8	1		32	61.50		93.50	130
3200	4/0 wire		7	1.143		36.50	70		106.50	149
3400	250 kcmil wire		5	1.600		43	98		141	198
3600	500 kcmil wire		4	2		52.50	123		175.50	247
3700	Insulated ground wire, copper #14		13	.615	C.L.F.	7.70	38		45.70	66.50
3710	#12		11	.727		10.60	44.50		55.10	80
3720	#10		10	.800		20.50	49		69.50	98
3730	#8		8	1		27	61.50		88.50	125
3740	#6		6.50	1.231		44	75.50		119.50	165
3750	#4	2 Elec	10.60	1.509		70	92.50		162.50	220
3770	#2		9	1.778		109	109		218	287
3780	#1		8	2		138	123		261	340
3790	1/0		6.60	2.424		200	149		349	450
3800	2/0		5.80	2.759		208	169		377	490
3810	3/0		5	3.200		259	196		455	585
3820	4/0		4.40	3.636		330	223		553	705
3830	250 kcmil	3 Elec	6	4		380	245		625	800
3840	300 kcmil		5.70	4.211		515	258		773	965
3850	350 kcmil		5.40	4.444		535	273		808	1,000
3860	400 kcmil		5.10	4.706		605	289		894	1,100
3870	500 kcmil		4.80	5		775	305		1,080	1,325
3880	600 kcmil		3.90	6.154		1,275	380		1,655	1,975
3890	750 kcmil		3.30	7.273		1,150	445		1,595	1,950
3900	1,000 kcmil		2.70	8.889		1,875	545		2,420	2,900
3960	Insulated ground wire, aluminum, #6	1 Elec	8	1		20.50	61.50		82	117
3970	#4	2 Elec	13	1.231		25	75.50		100.50	144
3980	#2		10.60	1.509		48	92.50		140.50	196
3990	#1		9	1.778		57	109		166	231
4000	1/0		8	2		80	123		203	278
4010	2/0		7.20	2.222		82	136		218	300
4020	3/0		6.60	2.424		103	149		252	340
4030	4/0		6.20	2.581		131	158		289	390
4040	250 kcmil	3 Elec	8.70	2.759		137	169		306	410
4050	300 kcmil		8.10	2.963		190	182		372	490

26 05 26 – Grounding and Bonding for Electrical Systems

26 05 26.80 Grounding		Crew	Daily Output	Labor-Hours	Unit	Material	Labor	Equipment	Total	Total Incl O&P
						2020 Bare Costs				
4060	350 kcmil	3 Elec	7.50	3.200	C.L.F.	192	196		388	510
4070	400 kcmil		6.90	3.478		219	213		432	570
4080	500 kcmil		6	4		246	245		491	650
4090	600 kcmil		5.70	4.211		315	258		573	745
4100	700 kcmil		5.10	4.706		385	289		674	870
4110	750 kcmil	↓	4.80	5	↓	375	305		680	885
5000	Copper electrolytic ground rod system									
5010	Includes augering hole, mixing bentonite clay,									
5020	Installing rod, and terminating ground wire									
5100	Straight vertical type, 2″ diam.									
5120	8.5′ long, clamp connection	1 Elec	2.67	2.996	Ea.	720	184		904	1,075
5130	With exothermic weld connection		1.95	4.103		730	252		982	1,175
5140	10′ long		2.35	3.404		710	209		919	1,100
5150	With exothermic weld connection		1.78	4.494		890	276		1,166	1,400
5160	12′ long		2.16	3.704		1,025	227		1,252	1,475
5170	With exothermic weld connection		1.67	4.790		1,125	294		1,419	1,675
5180	20′ long		1.74	4.598		1,400	282		1,682	1,950
5190	With exothermic weld connection	↓	1.40	5.714		1,375	350		1,725	2,075
5195	40′ long with exothermic weld connection	2 Elec	2	8	↓	3,050	490		3,540	4,125
5200	L-shaped, 2″ diam.									
5220	4′ vert. x 10′ horiz., clamp connection	1 Elec	5.33	1.501	Ea.	1,325	92		1,417	1,625
5230	With exothermic weld connection	″	3.08	2.597	″	1,325	159		1,484	1,700
5300	Protective box at grade level, with breather slots									
5320	Round 12″ long, fiberlyte	1 Elec	32	.250	Ea.	70	15.35		85.35	101
5330	Concrete	″	16	.500		112	30.50		142.50	170
5400	Bentonite clay, 50# bag, 1 per 10′ of rod					56			56	61.50
5500	Equipotential earthing bar	1 Elec	2	4	↓	138	245		383	530
9000	Minimum labor/equipment charge	″	4	2	Job		123		123	189

26 05 29 – Hangers and Supports for Electrical Systems

26 05 29.20 Hangers		Crew	Daily Output	Labor-Hours	Unit	Material	Labor	Equipment	Total	Total Incl O&P
0010	**HANGERS** R260533-21									
0015	See section 22 05 29.10 for additional items									
0030	Conduit supports									
0050	Strap w/2 holes, rigid steel conduit									
0100	1/2″ diameter	1 Elec	470	.017	Ea.	.09	1.04		1.13	1.71
0150	3/4″ diameter		440	.018		.10	1.12		1.22	1.83
0200	1″ diameter		400	.020		.14	1.23		1.37	2.04
0300	1-1/4″ diameter		355	.023		.23	1.38		1.61	2.38
0350	1-1/2″ diameter		320	.025		.30	1.53		1.83	2.69
0400	2″ diameter		266	.030		.40	1.85		2.25	3.28
0500	2-1/2″ diameter		160	.050		.62	3.07		3.69	5.40
0550	3″ diameter		133	.060		.83	3.69		4.52	6.60
0600	3-1/2″ diameter		100	.080		2.18	4.91		7.09	9.95
0650	4″ diameter		80	.100		.84	6.15		6.99	10.35
0700	EMT, 1/2″ diameter		470	.017		.16	1.04		1.20	1.79
0800	3/4″ diameter		440	.018		.15	1.12		1.27	1.89
0850	1″ diameter		400	.020		.29	1.23		1.52	2.21
0900	1-1/4″ diameter		355	.023		.59	1.38		1.97	2.78
0950	1-1/2″ diameter		320	.025		.79	1.53		2.32	3.23
1000	2″ diameter		266	.030		.80	1.85		2.65	3.72
1100	2-1/2″ diameter		160	.050		1.19	3.07		4.26	6.05
1150	3″ diameter	↓	133	.060	↓	1.48	3.69		5.17	7.35

26 05 29 – Hangers and Supports for Electrical Systems

26 05 29.20 Hangers		Crew	Daily Output	Labor-Hours	Unit	2020 Bare Costs Material	Labor	Equipment	Total	Total Incl O&P
1200	3-1/2" diameter	1 Elec	100	.080	Ea.	3.29	4.91		8.20	11.15
1250	4" diameter		80	.100		2.37	6.15		8.52	12.05
1400	Hanger, with bolt, 1/2" diameter		200	.040		.35	2.45		2.80	4.17
1450	3/4" diameter		190	.042		.37	2.58		2.95	4.39
1500	1" diameter		176	.045		.39	2.79		3.18	4.72
1550	1-1/4" diameter		160	.050		.49	3.07		3.56	5.25
1600	1-1/2" diameter		140	.057		.65	3.51		4.16	6.10
1650	2" diameter		130	.062		1.13	3.78		4.91	7.05
1700	2-1/2" diameter		100	.080		1.45	4.91		6.36	9.15
1750	3" diameter		64	.125		1.90	7.65		9.55	13.90
1800	3-1/2" diameter		50	.160		1.87	9.80		11.67	17.15
1850	4" diameter		40	.200		2.88	12.25		15.13	22
1900	Riser clamps, conduit, 1/2" diameter		40	.200		7.15	12.25		19.40	27
1950	3/4" diameter		36	.222		6.85	13.65		20.50	28.50
2000	1" diameter		30	.267		6.20	16.35		22.55	32
2100	1-1/4" diameter		27	.296		9.40	18.20		27.60	38.50
2150	1-1/2" diameter		27	.296		8.75	18.20		26.95	37.50
2200	2" diameter		20	.400		9	24.50		33.50	48
2250	2-1/2" diameter		20	.400		9.85	24.50		34.35	49
2300	3" diameter		18	.444		11	27.50		38.50	54
2350	3-1/2" diameter		18	.444		10.20	27.50		37.70	53
2400	4" diameter		14	.571		13.20	35		48.20	68.50
2500	Threaded rod, painted, 1/4" diameter		260	.031	L.F.	1.57	1.89		3.46	4.64
2600	3/8" diameter		200	.040		5.90	2.45		8.35	10.30
2700	1/2" diameter		140	.057		6.75	3.51		10.26	12.80
2800	5/8" diameter		100	.080		8.55	4.91		13.46	16.95
2900	3/4" diameter		60	.133		6.20	8.20		14.40	19.45
2940	Couplings painted, 1/4" diameter				C	310			310	340
2960	3/8" diameter					455			455	500
2970	1/2" diameter					385			385	425
2980	5/8" diameter					765			765	840
2990	3/4" diameter					960			960	1,050
3000	Nuts, galvanized, 1/4" diameter					10.40			10.40	11.45
3050	3/8" diameter					16.80			16.80	18.50
3100	1/2" diameter					46.50			46.50	51
3150	5/8" diameter					177			177	195
3200	3/4" diameter					125			125	138
3250	Washers, galvanized, 1/4" diameter					14.10			14.10	15.50
3300	3/8" diameter					16.10			16.10	17.70
3350	1/2" diameter					41			41	45
3400	5/8" diameter					81.50			81.50	89.50
3450	3/4" diameter					143			143	157
3500	Lock washers, galvanized, 1/4" diameter					9.15			9.15	10.05
3550	3/8" diameter					15.60			15.60	17.15
3600	1/2" diameter					23.50			23.50	25.50
3650	5/8" diameter					41			41	45.50
3700	3/4" diameter					80.50			80.50	88.50
3705	Metal channel slotted, 304 stainless steel, aluminum									
3710	304 stainless steel, 1' long, 12GA, 1-5/8" x 1-5/8"	1 Elec	50	.160	Ea.	18.05	9.80		27.85	35
3715	304 stainless steel, 1'6" long, 12GA, 1-5/8" x 1-5/8"		33.33	.240		24.50	14.75		39.25	49.50
3720	304 stainless steel, 2' long, 12GA, 1-5/8" x 1-5/8"		25	.320		38	19.65		57.65	71.50
3725	304 stainless steel, 3' long, 12GA, 1-5/8" x 1-5/8"		16.67	.480		48.50	29.50		78	99
3730	304 stainless steel, 4' long, 12GA, 1-5/8" x 1-5/8"		12.50	.640		60	39.50		99.50	127

26 05 29.20 Hangers		Crew	Daily Output	Labor-Hours	Unit	Material	Labor (2020 Bare Costs)	Equipment	Total	Total Incl O&P
3735	304 stainless steel, 5′ long, 12GA, 1-5/8″ x 1-5/8″	1 Elec	10	.800	Ea.	90.50	49		139.50	175
3740	304 stainless steel, 10′ long, 12GA, 1-5/8″ x 1-5/8″		5	1.600		137	98		235	300
3745	Aluminum, 1′ length, 12GA, 1-5/8″ x 1-5/8″		50	.160		5.45	9.80		15.25	21
3750	Aluminum, 1′6″ length, 12GA, 1-5/8″ x 1-5/8″		33.33	.240		8.10	14.70		22.80	31.50
3755	Aluminum, 2′ length, 12GA, 1-5/8″ x 1-5/8″		25	.320		10.25	19.65		29.90	41.50
3760	Aluminum, 3′ length, 12GA, 1-5/8″ x 1-5/8″		16.67	.480		15.95	29.50		45.45	63
3765	Aluminum, 4′ length, 12GA, 1-5/8″ x 1-5/8″		12.50	.640		21.50	39.50		61	84
3770	Aluminum, 5′ length, 12GA, 1-5/8″ x 1-5/8″		10	.800		25	49		74	103
3775	Aluminum, 10′ length, 12GA, 1-5/8″ x 1-5/8″		5	1.600		46	98		144	202
3800	Channels, steel, 3/4″ x 1-1/2″, 14 ga.		80	.100	L.F.	2.98	6.15		9.13	12.75
3900	1-1/2″ x 1-1/2″, 12 ga.		70	.114		3.68	7		10.68	14.85
4000	1-7/8″ x 1-1/2″		60	.133		21	8.20		29.20	36
4100	3″ x 1-1/2″		50	.160		19.85	9.80		29.65	37
4110	1-5/8″ x 13/16″, 16 ga.		80	.100		5.65	6.15		11.80	15.70
4114	Trapeze channel support, 12″ wide, steel, 12 ga.		8.80	.909	Ea.	26	56		82	115
4116	18″ wide, steel		8.30	.964	″	27.50	59		86.50	122
4120	1-5/8″ x 1-5/8″, 14 ga.		70	.114	L.F.	5.55	7		12.55	16.95
4130	1-5/8″ x 7/8″, 12 ga.		70	.114		5.75	7		12.75	17.10
4140	1-5/8″ x 1-3/8″, 12 ga.		60	.133		7.30	8.20		15.50	20.50
4150	1-5/8″ x 1-5/8″, 12 ga.		50	.160		6.90	9.80		16.70	22.50
4160	Flat plate fitting, 2 hole, 3-1/2″, material only				Ea.	2.88			2.88	3.17
4170	3 hole, 1-7/16″ x 4-1/8″					3.81			3.81	4.19
4200	Spring nuts, long, 1/4″	1 Elec	120	.067		.65	4.09		4.74	7
4250	3/8″		100	.080		.75	4.91		5.66	8.40
4300	1/2″		80	.100		.83	6.15		6.98	10.35
4350	Spring nuts, short, 1/4″		120	.067		1.36	4.09		5.45	7.80
4400	3/8″		100	.080		1.70	4.91		6.61	9.40
4450	1/2″		80	.100		1.81	6.15		7.96	11.45
4500	Closure strip		200	.040	L.F.	3.66	2.45		6.11	7.80
4550	End cap		60	.133	Ea.	1.21	8.20		9.41	13.95
4600	End connector 3/4″ conduit		40	.200		7.15	12.25		19.40	27
4650	Junction box, 1 channel		16	.500		37.50	30.50		68	88
4700	2 channel		14	.571		51.50	35		86.50	111
4750	3 channel		12	.667		65	41		106	135
4800	4 channel		10	.800		86	49		135	170
4850	Splice plate		40	.200		9.40	12.25		21.65	29.50
4900	Continuous concrete insert, 1-1/2″ deep, 1′ long		16	.500		19.90	30.50		50.40	69
4950	2′ long		14	.571		24	35		59	80.50
5000	3′ long		12	.667		24	41		65	89.50
5050	4′ long		10	.800		34.50	49		83.50	114
5100	6′ long		8	1		52	61.50		113.50	152
5150	3/4″ deep, 1′ long		16	.500		20.50	30.50		51	69.50
5200	2′ long		14	.571		20.50	35		55.50	76.50
5250	3′ long		12	.667		21.50	41		62.50	87
5300	4′ long		10	.800		35	49		84	114
5350	6′ long		8	1		52.50	61.50		114	152
5400	90° angle fitting 2-1/8″ x 2-1/8″		60	.133		3.78	8.20		11.98	16.75
5450	Supports, suspension rod type, small		60	.133		61.50	8.20		69.70	80
5500	Large		40	.200		65.50	12.25		77.75	91
5550	Beam clamp, small		60	.133		28.50	8.20		36.70	44
5600	Large		40	.200		25.50	12.25		37.75	47
5650	U-support, small		60	.133		5.25	8.20		13.45	18.40
5700	Large		40	.200		10.85	12.25		23.10	31

26 05 29 – Hangers and Supports for Electrical Systems

26 05 29.20 Hangers		Crew	Daily Output	Labor-Hours	Unit	Material	Labor	Equipment	Total	Total Incl O&P
						2020 Bare Costs				
5750	Concrete insert, cast, for up to 1/2" threaded rod	1 Elec	16	.500	Ea.	11.80	30.50		42.30	60
5800	Beam clamp, 1/4" clamp, for 1/4" threaded drop rod		32	.250		2.72	15.35		18.07	26.50
5900	3/8" clamp, for 3/8" threaded drop rod		32	.250		3.18	15.35		18.53	27
6000	Strap, rigid conduit, 1/2" diameter		540	.015		1.20	.91		2.11	2.72
6050	3/4" diameter		440	.018		1.25	1.12		2.37	3.10
6100	1" diameter		420	.019		1.49	1.17		2.66	3.44
6150	1-1/4" diameter		400	.020		1.75	1.23		2.98	3.82
6200	1-1/2" diameter		400	.020		2.02	1.23		3.25	4.11
6250	2" diameter		267	.030		2.31	1.84		4.15	5.35
6300	2-1/2" diameter		267	.030		2.22	1.84		4.06	5.25
6350	3" diameter		160	.050		2.86	3.07		5.93	7.85
6400	3-1/2" diameter		133	.060		3.16	3.69		6.85	9.20
6450	4" diameter		100	.080		3.70	4.91		8.61	11.60
6500	5" diameter		80	.100		9.10	6.15		15.25	19.45
6550	6" diameter		60	.133		31	8.20		39.20	46.50
6600	EMT, 1/2" diameter		540	.015		1.01	.91		1.92	2.51
6650	3/4" diameter		440	.018		1.06	1.12		2.18	2.89
6700	1" diameter		420	.019		1.11	1.17		2.28	3.02
6750	1-1/4" diameter		400	.020		1.41	1.23		2.64	3.44
6800	1-1/2" diameter		400	.020		1.80	1.23		3.03	3.87
6850	2" diameter		267	.030		1.72	1.84		3.56	4.72
6900	2-1/2" diameter		267	.030		1.85	1.84		3.69	4.87
6950	3" diameter		160	.050		2.34	3.07		5.41	7.30
6970	3-1/2" diameter		133	.060		3.12	3.69		6.81	9.15
6990	4" diameter		100	.080		3.49	4.91		8.40	11.40
7000	Clip, 1 hole for rigid conduit, 1/2" diameter		500	.016		.06	.98		1.04	1.58
7050	3/4" diameter		470	.017		.74	1.04		1.78	2.42
7100	1" diameter		440	.018		1.81	1.12		2.93	3.71
7150	1-1/4" diameter		400	.020		1.66	1.23		2.89	3.72
7200	1-1/2" diameter		355	.023		2.06	1.38		3.44	4.40
7250	2" diameter		320	.025		1.59	1.53		3.12	4.11
7300	2-1/2" diameter		266	.030		4.81	1.85		6.66	8.15
7350	3" diameter		160	.050		4.07	3.07		7.14	9.20
7400	3-1/2" diameter		133	.060		6	3.69		9.69	12.30
7450	4" diameter		100	.080		12.65	4.91		17.56	21.50
7500	5" diameter		80	.100		97	6.15		103.15	116
7550	6" diameter		60	.133		213	8.20		221.20	247
7610	1/4" hole threaded,1/4" flange, rod hanger		60	.133		1.36	8.20		9.56	14.10
7620	3/8" hole threaded,1/4" flange, rod hanger		60	.133		1.42	8.20		9.62	14.15
7630	Z purlin clip, 1/4" & 3/8" flange threaded rod hanger		60	.133		1.09	8.20		9.29	13.80
7640	1/2" rigid, 3/4" conduit, 1" flange		60	.133		1.43	8.20		9.63	14.15
7650	1" conduit, 1" flange		60	.133		1.62	8.20		9.82	14.40
7660	1-1/4" conduit, 1" flange		60	.133		1.91	8.20		10.11	14.70
7680	3/4" EMT, 1/2" rigid		60	.133		1.25	8.20		9.45	14
7690	Push-in type tandem conduit clip 3/4" - 1" EMT, 1/2" - 3/4" rigid		60	.133		1.33	8.20		9.53	14.05
7820	Conduit hangers, with bolt & 12" rod, 1/2" diameter		150	.053		6.25	3.27		9.52	11.95
7830	3/4" diameter		145	.055		6.25	3.38		9.63	12.10
7840	1" diameter		135	.059		6.30	3.64		9.94	12.50
7850	1-1/4" diameter		120	.067		6.40	4.09		10.49	13.35
7860	1-1/2" diameter		110	.073		6.55	4.46		11.01	14.05
7870	2" diameter		100	.080		7.85	4.91		12.76	16.20
7880	2-1/2" diameter		80	.100		8.20	6.15		14.35	18.45
7890	3" diameter	↓	60	.133	↓	8.65	8.20		16.85	22

26 05 29 – Hangers and Supports for Electrical Systems

26 05 29.20 Hangers		Crew	Daily Output	Labor-Hours	Unit	Material	Labor	Equipment	Total	Total Incl O&P
						2020 Bare Costs				
7900	3-1/2" diameter	1 Elec	45	.178	Ea.	8.60	10.90		19.50	26.50
7910	4" diameter		35	.229		9.60	14		23.60	32
7920	5" diameter		30	.267		11.40	16.35		27.75	37.50
7930	6" diameter		25	.320		12.55	19.65		32.20	44
7950	Jay clamp, 1/2" diameter		32	.250		7.50	15.35		22.85	32
7960	3/4" diameter		32	.250		9.85	15.35		25.20	34.50
7970	1" diameter		32	.250		11	15.35		26.35	35.50
7980	1-1/4" diameter		30	.267		10.90	16.35		27.25	37
7990	1-1/2" diameter		30	.267		12.20	16.35		28.55	38.50
8000	2" diameter		30	.267		18.15	16.35		34.50	45
8010	2-1/2" diameter		28	.286		22.50	17.55		40.05	51.50
8020	3" diameter		28	.286		22.50	17.55		40.05	51.50
8030	3-1/2" diameter		25	.320		63	19.65		82.65	99.50
8040	4" diameter		25	.320		63	19.65		82.65	99.50
8050	5" diameter		20	.400		113	24.50		137.50	162
8060	6" diameter		16	.500		215	30.50		245.50	284
8070	Channels, 3/4" x 1-1/2" w/12" rods for 1/2" to 1" conduit		30	.267		10.35	16.35		26.70	36.50
8080	1-1/2" x 1-1/2" w/12" rods for 1-1/4" to 2" conduit		28	.286		10.50	17.55		28.05	38.50
8090	1-1/2" x 1-1/2" w/12" rods for 2-1/2" to 4" conduit		26	.308		13.15	18.90		32.05	43.50
8100	1-1/2" x 1-7/8" w/12" rods for 5" to 6" conduit		24	.333		48.50	20.50		69	85
8110	Beam clamp, conduit, plastic coated steel, 1/2" diam.		30	.267		34.50	16.35		50.85	63
8120	3/4" diameter		30	.267		34	16.35		50.35	62.50
8130	1" diameter		30	.267		40	16.35		56.35	69
8140	1-1/4" diameter		28	.286		49.50	17.55		67.05	81.50
8150	1-1/2" diameter		28	.286		59	17.55		76.55	92
8160	2" diameter		28	.286		58.50	17.55		76.05	91.50
8170	2-1/2" diameter		26	.308		81	18.90		99.90	118
8180	3" diameter		26	.308		90	18.90		108.90	128
8190	3-1/2" diameter		23	.348		67.50	21.50		89	107
8200	4" diameter		23	.348		93	21.50		114.50	135
8210	5" diameter		18	.444		200	27.50		227.50	262
8220	Channels, plastic coated									
8250	3/4" x 1-1/2", w/12" rods for 1/2" to 1" conduit	1 Elec	28	.286	Ea.	35	17.55		52.55	65.50
8260	1-1/2" x 1-1/2", w/12" rods for 1-1/4" to 2" conduit		26	.308		43.50	18.90		62.40	77
8270	1-1/2" x 1-1/2", w/12" rods for 2-1/2" to 3-1/2" conduit		24	.333		47	20.50		67.50	83.50
8280	1-1/2" x 1-7/8", w/12" rods for 4" to 5" conduit		22	.364		78.50	22.50		101	121
8290	1-1/2" x 1-7/8", w/12" rods for 6" conduit		20	.400		86	24.50		110.50	133
8320	Conduit hangers, plastic coated steel, with bolt & 12" rod, 1/2" diam.		140	.057		22.50	3.51		26.01	30
8330	3/4" diameter		135	.059		23	3.64		26.64	31
8340	1" diameter		125	.064		23.50	3.93		27.43	32
8350	1-1/4" diameter		110	.073		25	4.46		29.46	34.50
8360	1-1/2" diameter		100	.080		28	4.91		32.91	38
8370	2" diameter		90	.089		30.50	5.45		35.95	42.50
8380	2-1/2" diameter		70	.114		36.50	7		43.50	51
8390	3" diameter		50	.160		44	9.80		53.80	63.50
8400	3-1/2" diameter		35	.229		40	14		54	65.50
8410	4" diameter		25	.320		65.50	19.65		85.15	102
8420	5" diameter		20	.400		71.50	24.50		96	117
9000	Parallel type, conduit beam clamp, 1/2"		32	.250		6.20	15.35		21.55	30.50
9010	3/4"		32	.250		7.15	15.35		22.50	31.50
9020	1"		32	.250		7.25	15.35		22.60	31.50
9030	1-1/4"		30	.267		10.30	16.35		26.65	36.50
9040	1-1/2"		30	.267		12.70	16.35		29.05	39

26 05 Common Work Results for Electrical

26 05 29 – Hangers and Supports for Electrical Systems

26 05 29.20 Hangers		Crew	Daily Output	Labor-Hours	Unit	Material	Labor	Equipment	Total	Total Incl O&P
						2020 Bare Costs				
9050	2"	1 Elec	30	.267	Ea.	15.15	16.35		31.50	41.50
9060	2-1/2"		28	.286		19.10	17.55		36.65	48
9070	3"		28	.286		23	17.55		40.55	52.50
9090	4"		25	.320		15.25	19.65		34.90	47
9110	Right angle, conduit beam clamp, 1/2"		32	.250		2.52	15.35		17.87	26.50
9120	3/4"		32	.250		2.71	15.35		18.06	26.50
9130	1"		32	.250		2.44	15.35		17.79	26
9140	1-1/4"		30	.267		3.73	16.35		20.08	29
9150	1-1/2"		30	.267		3.81	16.35		20.16	29
9160	2"		30	.267		5.55	16.35		21.90	31
9170	2-1/2"		28	.286		6.85	17.55		24.40	34.50
9180	3"		28	.286		7.45	17.55		25	35
9190	3-1/2"		25	.320		10.35	19.65		30	41.50
9200	4"		25	.320		11.55	19.65		31.20	42.50
9230	Adjustable, conduit hanger, 1/2"		32	.250		7.55	15.35		22.90	32
9240	3/4"		32	.250		5.60	15.35		20.95	29.50
9250	1"		32	.250		7.25	15.35		22.60	31.50
9260	1-1/4"		30	.267		7.95	16.35		24.30	33.50
9270	1-1/2"		30	.267		9.45	16.35		25.80	35.50
9280	2"		30	.267		3.93	16.35		20.28	29.50
9290	2-1/2"		28	.286		22.50	17.55		40.05	52
9300	3"		28	.286		26.50	17.55		44.05	56
9310	3-1/2"		25	.320		26.50	19.65		46.15	59
9320	4"		25	.320		31	19.65		50.65	64.50
9330	5"		20	.400		37.50	24.50		62	79.50
9340	6"		16	.500		47.50	30.50		78	99
9350	Combination conduit hanger, 3/8"		32	.250		14.75	15.35		30.10	39.50
9360	Adjustable flange, 3/8"	↓	32	.250	↓	17.25	15.35		32.60	42.50

26 05 29.30 Fittings and Channel Support		Crew	Daily Output	Labor-Hours	Unit	Material	Labor	Equipment	Total	Total Incl O&P
0010	**FITTINGS & CHANNEL SUPPORT**									
0020	Rooftop channel support									
0200	2-7/8" L x 1-5/8" W, 12 ga. pre galv., dbl. base	1 Elec	46	.174	Ea.	3.82	10.65		14.47	20.50
0210	2-7/8" L x 1-5/8" W, 12 ga. hot dip galv., dbl. base		46	.174		2.22	10.65		12.87	18.90
0220	2-7/16" L x 1-5/8" W, 12 ga. pre galv., sngl base		48	.167		1.83	10.25		12.08	17.75
0230	2-7/16" L x 1-5/8" W, 12 ga. hot dip galv., sngl base		48	.167		6.50	10.25		16.75	23
0240	13/16" L x 1-5/8" W, 14 ga. pre galv., sngl base		56	.143		1.56	8.75		10.31	15.20
0250	13/16" L x 1-5/8" W, 14 ga. hot dip galv., sngl base		56	.143		6.35	8.75		15.10	20.50
0260	1-5/8" L x 1-5/8" W, 12 ga. pre galv., sngl base		54	.148		1.69	9.10		10.79	15.85
0270	1-5/8" L x 1-5/8" W, 12 ga. hot dip galv., sngl base		54	.148		1.79	9.10		10.89	15.95
0280	1-5/8" x 28", 12 ga. pre galv., dbl. H block base		42	.190		3.87	11.70		15.57	22.50
0290	1-5/8" x 36", 12 ga. pre galv., dbl. H block base		40	.200		4.09	12.25		16.34	23.50
0300	1-5/8" x 42", 12 ga. pre galv., dbl. H block base		34	.235		4.59	14.45		19.04	27
0310	1-5/8" x 50", 12 ga. pre galv., dbl. H block base		38	.211		4.90	12.90		17.80	25.50
0320	1-5/8" x 60", 12 ga. pre galv., dbl. H block base		36	.222		5.40	13.65		19.05	27
0330	13/16" x 8" H, threaded rod pre galv., H block base		35	.229		2.59	14		16.59	24.50
0340	13/16" x 12" H, threaded rod pre galv., H block base		34.60	.231		3.24	14.20		17.44	25.50
0350	1-5/8" x 16" H, threaded rod dbl. H block base		34	.235		4.85	14.45		19.30	27.50
0360	1-5/8" x 12" H, pre galv., dbl. H block base		34.60	.231		9.25	14.20		23.45	32
0370	1-5/8" x 24"H, pre galv., dbl. H block base		34	.235		9.80	14.45		24.25	33
0380	3-1/4" x 24" H, pre galv., dbl. H block base		32	.250		22	15.35		37.35	48
0390	3-1/4" x 36" H, pre galv., dbl. H block base	↓	30	.267	↓	24.50	16.35		40.85	51.50

26 05 33.13 Conduit		Crew	Daily Output	Labor-Hours	Unit	2020 Bare Costs Material	Labor	Equipment	Total	Total Incl O&P
0010	**CONDUIT** To 10' high, includes 2 terminations, 2 elbows, R260533-20									
0020	11 beam clamps, and 11 couplings per 100 L.F.									
0300	Aluminum, 1/2" diameter	1 Elec	100	.080	L.F.	1.82	4.91		6.73	9.55
0500	3/4" diameter		90	.089		2.69	5.45		8.14	11.35
0700	1" diameter		80	.100		3.57	6.15		9.72	13.40
1000	1-1/4" diameter		70	.114		4.36	7		11.36	15.60
1030	1-1/2" diameter		65	.123		5.30	7.55		12.85	17.45
1050	2" diameter		60	.133		7.25	8.20		15.45	20.50
1070	2-1/2" diameter		50	.160		9.15	9.80		18.95	25
1100	3" diameter	2 Elec	90	.178		12.85	10.90		23.75	31
1130	3-1/2" diameter		80	.200		16.15	12.25		28.40	36.50
1140	4" diameter		70	.229		19.30	14		33.30	42.50
1150	5" diameter		50	.320		36	19.65		55.65	70
1160	6" diameter		40	.400		64	24.50		88.50	109
1161	Field bends, 45° to 90°, 1/2" diameter	1 Elec	53	.151	Ea.		9.25		9.25	14.25
1162	3/4" diameter		47	.170			10.45		10.45	16.10
1163	1" diameter		44	.182			11.15		11.15	17.15
1164	1-1/4" diameter		23	.348			21.50		21.50	33
1165	1-1/2" diameter		21	.381			23.50		23.50	36
1166	2" diameter		16	.500			30.50		30.50	47
1170	Elbows, 1/2" diameter		40	.200		4.88	12.25		17.13	24.50
1200	3/4" diameter		32	.250		6.35	15.35		21.70	30.50
1230	1" diameter		28	.286		9.05	17.55		26.60	37
1250	1-1/4" diameter		24	.333		18.95	20.50		39.45	52.50
1270	1-1/2" diameter		20	.400		25	24.50		49.50	65.50
1300	2" diameter		16	.500		28	30.50		58.50	77.50
1330	2-1/2" diameter		12	.667		53.50	41		94.50	122
1350	3" diameter		8	1		87	61.50		148.50	190
1370	3-1/2" diameter		6	1.333		128	82		210	267
1400	4" diameter		5	1.600		147	98		245	315
1410	5" diameter		4	2		445	123		568	680
1420	6" diameter		2.50	3.200		610	196		806	975
1430	Couplings, 1/2" diameter		320	.025		1.78	1.53		3.31	4.32
1450	3/4" diameter		192	.042		2.70	2.56		5.26	6.90
1470	1" diameter		160	.050		3.50	3.07		6.57	8.55
1500	1-1/4" diameter		120	.067		3.75	4.09		7.84	10.45
1530	1-1/2" diameter		96	.083		4.92	5.10		10.02	13.25
1550	2" diameter		80	.100		7.05	6.15		13.20	17.25
1570	2-1/2" diameter		69	.116		12.05	7.10		19.15	24.50
1600	3" diameter		64	.125		20.50	7.65		28.15	34.50
1630	3-1/2" diameter		56	.143		23	8.75		31.75	38.50
1650	4" diameter		53	.151		34	9.25		43.25	52
1670	5" diameter		48	.167		84.50	10.25		94.75	109
1690	6" diameter		46	.174		197	10.65		207.65	233
1691	See note on line 26 05 33.13 9995 R260533-30									
1750	Rigid galvanized steel, 1/2" diameter	1 Elec	90	.089	L.F.	2.77	5.45		8.22	11.45
1770	3/4" diameter		80	.100		5.15	6.15		11.30	15.10
1800	1" diameter		65	.123		7.70	7.55		15.25	20
1830	1-1/4" diameter		60	.133		5.35	8.20		13.55	18.45
1850	1-1/2" diameter		55	.145		9.35	8.90		18.25	24
1870	2" diameter		45	.178		9.80	10.90		20.70	27.50
1900	2-1/2" diameter		35	.229		14.25	14		28.25	37

26 05 33 – Raceway and Boxes for Electrical Systems

26 05 33.13 Conduit		Crew	Daily Output	Labor-Hours	Unit	Material	2020 Bare Costs Labor	Equipment	Total	Total Incl O&P
1930	3" diameter	2 Elec	50	.320	L.F.	15.85	19.65		35.50	47.50
1950	3-1/2" diameter		44	.364		20	22.50		42.50	57
1970	4" diameter		40	.400		24.50	24.50		49	65
1980	5" diameter		30	.533		39	32.50		71.50	93.50
1990	6" diameter		20	.800		60.50	49		109.50	142
1991	Field bends, 45° to 90°, 1/2" diameter	1 Elec	44	.182	Ea.		11.15		11.15	17.15
1992	3/4" diameter		40	.200			12.25		12.25	18.90
1993	1" diameter		36	.222			13.65		13.65	21
1994	1-1/4" diameter		19	.421			26		26	40
1995	1-1/2" diameter		18	.444			27.50		27.50	42
1996	2" diameter		13	.615			38		38	58
2000	Elbows, 1/2" diameter		32	.250		3.87	15.35		19.22	28
2030	3/4" diameter		28	.286		4.47	17.55		22.02	32
2050	1" diameter		24	.333		5.85	20.50		26.35	38
2070	1-1/4" diameter		18	.444		11.35	27.50		38.85	54.50
2100	1-1/2" diameter		16	.500		19	30.50		49.50	68
2130	2" diameter		12	.667		27.50	41		68.50	93.50
2150	2-1/2" diameter		8	1		30	61.50		91.50	128
2170	3" diameter		6	1.333		39.50	82		121.50	170
2200	3-1/2" diameter		4.20	1.905		65	117		182	252
2220	4" diameter		4	2		73.50	123		196.50	270
2230	5" diameter		3.50	2.286		224	140		364	465
2240	6" diameter		2	4		315	245		560	725
2250	Couplings, 1/2" diameter		267	.030		1.04	1.84		2.88	3.97
2270	3/4" diameter		160	.050		1.53	3.07		4.60	6.40
2300	1" diameter		133	.060		2.67	3.69		6.36	8.65
2330	1-1/4" diameter		100	.080		2.58	4.91		7.49	10.40
2350	1-1/2" diameter		80	.100		5.45	6.15		11.60	15.45
2370	2" diameter		67	.119		7.15	7.35		14.50	19.15
2400	2-1/2" diameter		57	.140		10.50	8.60		19.10	25
2430	3" diameter		53	.151		13.55	9.25		22.80	29
2450	3-1/2" diameter		47	.170		17.60	10.45		28.05	35.50
2470	4" diameter		44	.182		29	11.15		40.15	49
2480	5" diameter		40	.200		34	12.25		46.25	56.50
2490	6" diameter		38	.211		56	12.90		68.90	81.50
2491	See note on line 26 05 33.13 9995 R260533-30									
2500	Steel, intermediate conduit (IMC), 1/2" diameter	1 Elec	100	.080	L.F.	1.91	4.91		6.82	9.65
2530	3/4" diameter		90	.089		2.64	5.45		8.09	11.30
2550	1" diameter		70	.114		3.57	7		10.57	14.70
2570	1-1/4" diameter		65	.123		3.95	7.55		11.50	15.95
2600	1-1/2" diameter		60	.133		5.75	8.20		13.95	18.90
2630	2" diameter		50	.160		6.85	9.80		16.65	22.50
2650	2-1/2" diameter		40	.200		9.05	12.25		21.30	29
2670	3" diameter	2 Elec	60	.267		13.55	16.35		29.90	40
2700	3-1/2" diameter		54	.296		18.85	18.20		37.05	48.50
2730	4" diameter		50	.320		12.05	19.65		31.70	43.50
2731	Field bends, 45° to 90°, 1/2" diameter	1 Elec	44	.182	Ea.		11.15		11.15	17.15
2732	3/4" diameter		40	.200			12.25		12.25	18.90
2733	1" diameter		36	.222			13.65		13.65	21
2734	1-1/4" diameter		19	.421			26		26	40
2735	1-1/2" diameter		18	.444			27.50		27.50	42
2736	2" diameter		13	.615			38		38	58
2750	Elbows, 1/2" diameter		32	.250		5.45	15.35		20.80	29.50

26 05 33.13 Conduit		Crew	Daily Output	Labor-Hours	Unit	Material	2020 Bare Costs Labor	Equipment	Total	Total Incl O&P
2770	3/4" diameter	1 Elec	28	.286	Ea.	11.75	17.55		29.30	40
2800	1" diameter		24	.333		13.30	20.50		33.80	46
2830	1-1/4" diameter		18	.444		17.85	27.50		45.35	61.50
2850	1-1/2" diameter		16	.500		38.50	30.50		69	89.50
2870	2" diameter		12	.667		60.50	41		101.50	130
2900	2-1/2" diameter		8	1		40	61.50		101.50	139
2930	3" diameter		6	1.333		55.50	82		137.50	187
2950	3-1/2" diameter		4.20	1.905		197	117		314	395
2970	4" diameter		4	2		73.50	123		196.50	270
3000	Couplings, 1/2" diameter		293	.027		1.04	1.67		2.71	3.72
3030	3/4" diameter		176	.045		1.53	2.79		4.32	5.95
3050	1" diameter		147	.054		2.67	3.34		6.01	8.10
3070	1-1/4" diameter		110	.073		2.58	4.46		7.04	9.70
3100	1-1/2" diameter		88	.091		5.45	5.60		11.05	14.60
3130	2" diameter		73	.110		7.15	6.70		13.85	18.20
3150	2-1/2" diameter		63	.127		10.50	7.80		18.30	23.50
3170	3" diameter		59	.136		13.55	8.30		21.85	27.50
3200	3-1/2" diameter		52	.154		17.60	9.45		27.05	34
3230	4" diameter		49	.163		29	10		39	47.50
3231	See note on line 26 05 33.13 9995 R260533-30									
4100	Rigid steel, plastic coated, 40 mil thick									
4130	1/2" diameter	1 Elec	80	.100	L.F.	9.70	6.15		15.85	20
4150	3/4" diameter		70	.114		10	7		17	22
4170	1" diameter		55	.145		12.55	8.90		21.45	27.50
4200	1-1/4" diameter		50	.160		17.50	9.80		27.30	34.50
4230	1-1/2" diameter		45	.178		19.05	10.90		29.95	38
4250	2" diameter		35	.229		22.50	14		36.50	46.50
4270	2-1/2" diameter		25	.320		38	19.65		57.65	71.50
4300	3" diameter	2 Elec	44	.364		41.50	22.50		64	80.50
4330	3-1/2" diameter		40	.400		50	24.50		74.50	93
4350	4" diameter		36	.444		62.50	27.50		90	111
4370	5" diameter		30	.533		121	32.50		153.50	184
4400	Elbows, 1/2" diameter	1 Elec	28	.286	Ea.	19.90	17.55		37.45	49
4430	3/4" diameter		24	.333		18.05	20.50		38.55	51.50
4450	1" diameter		18	.444		21	27.50		48.50	65
4470	1-1/4" diameter		16	.500		31	30.50		61.50	81
4500	1-1/2" diameter		12	.667		32	41		73	98.50
4530	2" diameter		8	1		42	61.50		103.50	141
4550	2-1/2" diameter		6	1.333		85.50	82		167.50	221
4570	3" diameter		4.20	1.905		134	117		251	330
4600	3-1/2" diameter		4	2		211	123		334	420
4630	4" diameter		3.80	2.105		196	129		325	415
4650	5" diameter		3.50	2.286		495	140		635	760
4680	Couplings, 1/2" diameter		213	.038		5.80	2.30		8.10	9.95
4700	3/4" diameter		128	.063		5.65	3.83		9.48	12.15
4730	1" diameter		107	.075		6.85	4.59		11.44	14.60
4750	1-1/4" diameter		80	.100		10.10	6.15		16.25	20.50
4770	1-1/2" diameter		64	.125		10.10	7.65		17.75	23
4800	2" diameter		53	.151		14.35	9.25		23.60	30
4830	2-1/2" diameter		46	.174		41	10.65		51.65	61.50
4850	3" diameter		43	.186		39	11.40		50.40	60
4870	3-1/2" diameter		38	.211		58	12.90		70.90	83.50
4900	4" diameter		36	.222		57.50	13.65		71.15	84

26 05 33.13 Conduit		Crew	Daily Output	Labor-Hours	Unit	Material	Labor	Equipment	Total	Total Incl O&P
						2020 Bare Costs				
4950	5″ diameter	1 Elec	32	.250	Ea.	214	15.35		229.35	259
4951	See note on line 26 05 33.13 9995 R260533-30									
5000	Electric metallic tubing (EMT), 1/2″ diameter	1 Elec	170	.047	L.F.	.76	2.89		3.65	5.25
5020	3/4″ diameter		130	.062		1.07	3.78		4.85	7
5040	1″ diameter		115	.070		1.80	4.27		6.07	8.55
5060	1-1/4″ diameter		100	.080		2.94	4.91		7.85	10.80
5080	1-1/2″ diameter		90	.089		3.46	5.45		8.91	12.20
5100	2″ diameter		80	.100		4.37	6.15		10.52	14.25
5120	2-1/2″ diameter		60	.133		4.85	8.20		13.05	17.95
5140	3″ diameter	2 Elec	100	.160		6.15	9.80		15.95	22
5160	3-1/2″ diameter		90	.178		7.70	10.90		18.60	25.50
5180	4″ diameter		80	.200		13.55	12.25		25.80	34
5200	Field bends, 45° to 90°, 1/2″ diameter	1 Elec	89	.090	Ea.		5.50		5.50	8.50
5220	3/4″ diameter		80	.100			6.15		6.15	9.45
5240	1″ diameter		73	.110			6.70		6.70	10.35
5260	1-1/4″ diameter		38	.211			12.90		12.90	19.90
5280	1-1/2″ diameter		36	.222			13.65		13.65	21
5300	2″ diameter		26	.308			18.90		18.90	29
5320	Offsets, 1/2″ diameter		65	.123			7.55		7.55	11.60
5340	3/4″ diameter		62	.129			7.90		7.90	12.20
5360	1″ diameter		53	.151			9.25		9.25	14.25
5380	1-1/4″ diameter		30	.267			16.35		16.35	25
5400	1-1/2″ diameter		28	.286			17.55		17.55	27
5420	2″ diameter		20	.400			24.50		24.50	38
5700	Elbows, 1″ diameter		40	.200		3.26	12.25		15.51	22.50
5720	1-1/4″ diameter		32	.250		4.36	15.35		19.71	28.50
5740	1-1/2″ diameter		24	.333		5.10	20.50		25.60	37
5760	2″ diameter		20	.400		7.30	24.50		31.80	46
5780	2-1/2″ diameter		12	.667		17.90	41		58.90	82.50
5800	3″ diameter		9	.889		24.50	54.50		79	111
5820	3-1/2″ diameter		7	1.143		34	70		104	145
5840	4″ diameter		6	1.333		39.50	82		121.50	170
6200	Couplings, set screw, steel, 1/2″ diameter		235	.034		1.61	2.09		3.70	4.99
6220	3/4″ diameter		188	.043		2.28	2.61		4.89	6.55
6240	1″ diameter		157	.051		3.48	3.13		6.61	8.65
6260	1-1/4″ diameter		118	.068		7.45	4.16		11.61	14.60
6280	1-1/2″ diameter		94	.085		1.63	5.20		6.83	9.85
6300	2″ diameter		78	.103		14.50	6.30		20.80	25.50
6320	2-1/2″ diameter		67	.119		6.35	7.35		13.70	18.30
6340	3″ diameter		59	.136		6	8.30		14.30	19.40
6360	3-1/2″ diameter		47	.170		8.40	10.45		18.85	25.50
6380	4″ diameter		43	.186		58.50	11.40		69.90	82
6381	See note on line 26 05 33.13 9995 R260533-30									
6500	Box connectors, set screw, steel, 1/2″ diameter	1 Elec	120	.067	Ea.	.36	4.09		4.45	6.70
6520	3/4″ diameter		110	.073		.59	4.46		5.05	7.50
6540	1″ diameter		90	.089		1.11	5.45		6.56	9.60
6560	1-1/4″ diameter		70	.114		1.98	7		8.98	13
6580	1-1/2″ diameter		60	.133		3.30	8.20		11.50	16.25
6600	2″ diameter		50	.160		3.93	9.80		13.73	19.40
6620	2-1/2″ diameter		36	.222		7.90	13.65		21.55	29.50
6640	3″ diameter		27	.296		34.50	18.20		52.70	66
6680	3-1/2″ diameter		21	.381		10.70	23.50		34.20	48
6700	4″ diameter		16	.500		12.80	30.50		43.30	61

26 05 33.13 Conduit		Crew	Daily Output	Labor-Hours	Unit	Material	Labor	Equipment	Total	Total Incl O&P
						2020 Bare Costs				
6740	Insulated box connectors, set screw, steel, 1/2" diameter	1 Elec	120	.067	Ea.	.68	4.09		4.77	7.05
6760	3/4" diameter		110	.073		1.51	4.46		5.97	8.50
6780	1" diameter		90	.089		2.90	5.45		8.35	11.60
6800	1-1/4" diameter		70	.114		3.45	7		10.45	14.60
6820	1-1/2" diameter		60	.133		4.95	8.20		13.15	18.05
6840	2" diameter		50	.160		6.65	9.80		16.45	22.50
6860	2-1/2" diameter		36	.222		30	13.65		43.65	54
6880	3" diameter		27	.296		25.50	18.20		43.70	56
6900	3-1/2" diameter		21	.381		48	23.50		71.50	89
6920	4" diameter		16	.500		35.50	30.50		66	86.50
7000	EMT to conduit adapters, 1/2" diameter (compression)		70	.114		3.49	7		10.49	14.65
7020	3/4" diameter		60	.133		5.45	8.20		13.65	18.60
7040	1" diameter		50	.160		8.60	9.80		18.40	24.50
7060	1-1/4" diameter		40	.200		16.85	12.25		29.10	37.50
7080	1-1/2" diameter		30	.267		20	16.35		36.35	47
7100	2" diameter		25	.320		31.50	19.65		51.15	64.50
7200	EMT to Greenfield adapters, 1/2" to 3/8" diameter (compression)		90	.089		2.16	5.45		7.61	10.80
7220	1/2" diameter		90	.089		4.86	5.45		10.31	13.75
7240	3/4" diameter		80	.100		5.95	6.15		12.10	15.95
7260	1" diameter		70	.114		15.55	7		22.55	28
7270	1-1/4" diameter		60	.133		19.60	8.20		27.80	34
7280	1-1/2" diameter		50	.160		22	9.80		31.80	39.50
7290	2" diameter		40	.200		33	12.25		45.25	55.50
7400	EMT, LB, LR or LL fittings with covers, 1/2" diameter, set screw		24	.333		4.83	20.50		25.33	37
7420	3/4" diameter		20	.400		6.75	24.50		31.25	45.50
7440	1" diameter		16	.500		10.05	30.50		40.55	58
7450	1-1/4" diameter		13	.615		14.95	38		52.95	74.50
7460	1-1/2" diameter		11	.727		19.95	44.50		64.45	90.50
7470	2" diameter		9	.889		33	54.50		87.50	120
7600	EMT, "T" fittings with covers, 1/2" diameter, set screw		16	.500		5.60	30.50		36.10	53
7620	3/4" diameter		15	.533		7.25	32.50		39.75	58.50
7640	1" diameter		12	.667		9.15	41		50.15	73
7650	1-1/4" diameter		11	.727		13.60	44.50		58.10	83.50
7660	1-1/2" diameter		10	.800		17.45	49		66.45	94.50
7670	2" diameter		8	1		24	61.50		85.50	121
8000	EMT, expansion fittings, no jumper, 1/2" diameter		24	.333		85	20.50		105.50	125
8020	3/4" diameter		20	.400		125	24.50		149.50	175
8040	1" diameter		16	.500		134	30.50		164.50	194
8060	1-1/4" diameter		13	.615		152	38		190	226
8080	1-1/2" diameter		11	.727		202	44.50		246.50	291
8100	2" diameter		9	.889		335	54.50		389.50	455
8110	2-1/2" diameter		7	1.143		410	70		480	560
8120	3" diameter		6	1.333		555	82		637	735
8140	4" diameter		5	1.600		790	98		888	1,025
8200	Split adapter, 1/2" diameter		110	.073		3.27	4.46		7.73	10.45
8210	3/4" diameter		90	.089		3.14	5.45		8.59	11.85
8220	1" diameter		70	.114		4.88	7		11.88	16.15
8230	1-1/4" diameter		60	.133		8.30	8.20		16.50	21.50
8240	1-1/2" diameter		50	.160		11.60	9.80		21.40	28
8250	2" diameter		36	.222		32.50	13.65		46.15	56.50
8300	1 hole clips, 1/2" diameter		500	.016		.42	.98		1.40	1.97
8320	3/4" diameter		470	.017		.19	1.04		1.23	1.82
8340	1" diameter	↓	444	.018	↓	.28	1.11		1.39	2.01

26 05 33.13 Conduit		Crew	Daily Output	Labor-Hours	Unit	Material	Labor	Equipment	Total	Total Incl O&P
						2020 Bare Costs				
8360	1-1/4" diameter	1 Elec	400	.020	Ea.	.40	1.23		1.63	2.33
8380	1-1/2" diameter		355	.023		.99	1.38		2.37	3.22
8400	2" diameter		320	.025		1.46	1.53		2.99	3.97
8420	2-1/2" diameter		266	.030		3.43	1.85		5.28	6.60
8440	3" diameter		160	.050		3.85	3.07		6.92	8.95
8460	3-1/2" diameter		133	.060		7.45	3.69		11.14	13.90
8480	4" diameter		100	.080		7.35	4.91		12.26	15.65
8500	Clamp back spacers, 1/2" diameter		500	.016		.81	.98		1.79	2.40
8510	3/4" diameter		470	.017		1.01	1.04		2.05	2.72
8520	1" diameter		444	.018		1.89	1.11		3	3.78
8530	1-1/4" diameter		400	.020		3.94	1.23		5.17	6.20
8540	1-1/2" diameter		355	.023		7.65	1.38		9.03	10.55
8550	2" diameter		320	.025		5.45	1.53		6.98	8.35
8560	2-1/2" diameter		266	.030		10.25	1.85		12.10	14.10
8570	3" diameter		160	.050		27	3.07		30.07	34
8580	3-1/2" diameter		133	.060		42	3.69		45.69	52
8590	4" diameter		100	.080		102	4.91		106.91	120
8600	Offset connectors, 1/2" diameter		40	.200		2.78	12.25		15.03	22
8610	3/4" diameter		32	.250		3.64	15.35		18.99	27.50
8620	1" diameter		24	.333		5.05	20.50		25.55	37
8650	90° pulling elbows, female, 1/2" diameter, with gasket		24	.333		7.80	20.50		28.30	40
8660	3/4" diameter		20	.400		13.80	24.50		38.30	53
8700	Couplings, compression, 1/2" diameter, steel		78	.103		1.31	6.30		7.61	11.15
8710	3/4" diameter		67	.119		1.42	7.35		8.77	12.85
8720	1" diameter		59	.136		2.31	8.30		10.61	15.35
8730	1-1/4" diameter		47	.170		5.05	10.45		15.50	21.50
8740	1-1/2" diameter		38	.211		7.50	12.90		20.40	28
8750	2" diameter		31	.258		13.45	15.85		29.30	39.50
8760	2-1/2" diameter		24	.333		38	20.50		58.50	73
8770	3" diameter		21	.381		27.50	23.50		51	66.50
8780	3-1/2" diameter		19	.421		111	26		137	162
8790	4" diameter		16	.500		68	30.50		98.50	122
8791	See note on line 26 05 33.13 9995 R260533-30									
8800	Box connectors, compression, 1/2" diam., steel	1 Elec	120	.067	Ea.	2.45	4.09		6.54	9
8810	3/4" diameter		110	.073		3.36	4.46		7.82	10.55
8820	1" diameter		90	.089		4.75	5.45		10.20	13.65
8830	1-1/4" diameter		70	.114		12	7		19	24
8840	1-1/2" diameter		60	.133		14.65	8.20		22.85	29
8850	2" diameter		50	.160		20	9.80		29.80	37
8860	2-1/2" diameter		36	.222		48	13.65		61.65	73.50
8870	3" diameter		27	.296		64.50	18.20		82.70	99
8880	3-1/2" diameter		21	.381		101	23.50		124.50	147
8890	4" diameter		16	.500		102	30.50		132.50	159
8900	Box connectors, insulated compression, 1/2" diam., steel		120	.067		1.78	4.09		5.87	8.25
8910	3/4" diameter		110	.073		2.03	4.46		6.49	9.10
8920	1" diameter		90	.089		3.76	5.45		9.21	12.55
8930	1-1/4" diameter		70	.114		7.85	7		14.85	19.40
8940	1-1/2" diameter		60	.133		11.60	8.20		19.80	25.50
8950	2" diameter		50	.160		16.90	9.80		26.70	33.50
8960	2-1/2" diameter		36	.222		66	13.65		79.65	93.50
8970	3" diameter		27	.296		59	18.20		77.20	93
8980	3-1/2" diameter		21	.381		100	23.50		123.50	146
8990	4" diameter		16	.500		88	30.50		118.50	144

26 05 33.13 Conduit		Crew	Daily Output	Labor-Hours	Unit	Material	2020 Bare Costs Labor	Equipment	Total	Total Incl O&P
9100	PVC, schedule 40, 1/2" diameter	1 Elec	190	.042	L.F.	.87	2.58		3.45	4.93
9110	3/4" diameter		145	.055		1.03	3.38		4.41	6.35
9120	1" diameter		125	.064		1.54	3.93		5.47	7.75
9130	1-1/4" diameter		110	.073		1.99	4.46		6.45	9.05
9140	1-1/2" diameter		100	.080		2.39	4.91		7.30	10.20
9150	2" diameter		90	.089		3.36	5.45		8.81	12.10
9160	2-1/2" diameter		65	.123		4.60	7.55		12.15	16.65
9170	3" diameter	2 Elec	110	.145		4.84	8.90		13.74	19.10
9180	3-1/2" diameter		100	.160		6.40	9.80		16.20	22
9190	4" diameter		90	.178		5.75	10.90		16.65	23
9200	5" diameter		70	.229		8.35	14		22.35	30.50
9210	6" diameter		60	.267		12.50	16.35		28.85	39
9220	Elbows, 1/2" diameter	1 Elec	50	.160	Ea.	.58	9.80		10.38	15.75
9225	3/4" diameter		42	.190		.60	11.70		12.30	18.65
9230	1" diameter		35	.229		1.21	14		15.21	23
9235	1-1/4" diameter		28	.286		2.72	17.55		20.27	30
9240	1-1/2" diameter		20	.400		2.71	24.50		27.21	41
9245	2" diameter	R-1A	36.40	.440		3.54	24.50		28.04	41.50
9250	2-1/2" diameter		26.70	.599		7.75	33		40.75	59.50
9255	3" diameter		22.90	.699		12.25	38.50		50.75	73
9260	3-1/2" diameter		22.20	.721		18.65	40		58.65	82
9265	4" diameter		18.20	.879		23	48.50		71.50	99.50
9270	5" diameter		12.10	1.322		34.50	73		107.50	150
9275	6" diameter		11.10	1.441		35	79.50		114.50	162
9312	Couplings, 1/2" diameter	1 Elec	50	.160		.11	9.80		9.91	15.20
9314	3/4" diameter		42	.190		.31	11.70		12.01	18.35
9316	1" diameter		35	.229		.25	14		14.25	22
9318	1-1/4" diameter		28	.286		.35	17.55		17.90	27.50
9320	1-1/2" diameter		20	.400		.54	24.50		25.04	38.50
9322	2" diameter	R-1A	36.40	.440		1.26	24.50		25.76	39
9324	2-1/2" diameter		26.70	.599		1.80	33		34.80	53
9326	3" diameter		22.90	.699		1.48	38.50		39.98	61
9328	3-1/2" diameter		22.20	.721		2.05	40		42.05	64
9330	4" diameter		18.20	.879		2.65	48.50		51.15	77.50
9332	5" diameter		12.10	1.322		7	73		80	120
9334	6" diameter		11.10	1.441		10.20	79.50		89.70	134
9335	See note on line 26 05 33.13 9995 R260533-30									
9340	Field bends, 45° & 90°, 1/2" diameter	1 Elec	45	.178	Ea.		10.90		10.90	16.80
9350	3/4" diameter		40	.200			12.25		12.25	18.90
9360	1" diameter		35	.229			14		14	21.50
9370	1-1/4" diameter		32	.250			15.35		15.35	23.50
9380	1-1/2" diameter		27	.296			18.20		18.20	28
9390	2" diameter		20	.400			24.50		24.50	38
9400	2-1/2" diameter		16	.500			30.50		30.50	47
9410	3" diameter		13	.615			38		38	58
9420	3-1/2" diameter		12	.667			41		41	63
9430	4" diameter		10	.800			49		49	75.50
9440	5" diameter		9	.889			54.50		54.50	84
9450	6" diameter		8	1			61.50		61.50	94.50
9460	PVC adapters, 1/2" diameter		50	.160		.26	9.80		10.06	15.40
9470	3/4" diameter		42	.190		.37	11.70		12.07	18.40
9480	1" diameter		38	.211		.39	12.90		13.29	20.50
9490	1-1/4" diameter		35	.229		.76	14		14.76	22.50

26 05 33 – Raceway and Boxes for Electrical Systems

26 05 33.13 Conduit		Crew	Daily Output	Labor-Hours	Unit	Material	2020 Bare Costs Labor	Equipment	Total	Total Incl O&P
9500	1-1/2" diameter	1 Elec	32	.250	Ea.	.71	15.35		16.06	24.50
9510	2" diameter		27	.296		1.39	18.20		19.59	29.50
9520	2-1/2" diameter		23	.348		2.01	21.50		23.51	35
9530	3" diameter		18	.444		2.85	27.50		30.35	45
9540	3-1/2" diameter		13	.615		5.20	38		43.20	63.50
9550	4" diameter		11	.727		4.80	44.50		49.30	74
9560	5" diameter		8	1		9.70	61.50		71.20	105
9570	6" diameter		6	1.333		18.90	82		100.90	147
9580	PVC-LB, LR or LL fittings & covers									
9590	1/2" diameter	1 Elec	20	.400	Ea.	2.35	24.50		26.85	40.50
9600	3/4" diameter		16	.500		4.78	30.50		35.28	52.50
9610	1" diameter		12	.667		4.63	41		45.63	68
9620	1-1/4" diameter		9	.889		8.65	54.50		63.15	93.50
9630	1-1/2" diameter		7	1.143		10.25	70		80.25	119
9640	2" diameter		6	1.333		14.85	82		96.85	142
9650	2-1/2" diameter		6	1.333		80.50	82		162.50	215
9660	3" diameter		5	1.600		42.50	98		140.50	198
9670	3-1/2" diameter		4	2		56.50	123		179.50	252
9680	4" diameter		3	2.667		62	164		226	320
9690	PVC-tee fitting & cover									
9700	1/2"	1 Elec	14	.571	Ea.	5.30	35		40.30	60
9710	3/4"		13	.615		5.70	38		43.70	64.50
9720	1"		10	.800		5.35	49		54.35	81.50
9730	1-1/4"		9	.889		10.95	54.50		65.45	96
9740	1-1/2"		8	1		13.40	61.50		74.90	109
9750	2"		7	1.143		18.10	70		88.10	128
9760	PVC-reducers, 3/4" x 1/2" diameter					1.01			1.01	1.11
9770	1" x 1/2" diameter					3.19			3.19	3.51
9780	1" x 3/4" diameter					2.80			2.80	3.08
9790	1-1/4" x 3/4" diameter					2.40			2.40	2.64
9800	1-1/4" x 1" diameter					4.32			4.32	4.75
9810	1-1/2" x 1-1/4" diameter					3.44			3.44	3.78
9820	2" x 1-1/4" diameter					4.51			4.51	4.96
9830	2-1/2" x 2" diameter					18.30			18.30	20
9840	3" x 2" diameter					21			21	23
9850	4" x 3" diameter					23			23	25.50
9860	Cement, quart					15.55			15.55	17.10
9870	Gallon					71			71	78
9880	Heat bender, to 6" diameter					1,500			1,500	1,650
9900	Add to labor for higher elevated installation									
9905	10' to 14.5' high, add						10%			
9910	15' to 20' high, add						10%			
9920	20' to 25' high, add						20%			
9930	25' to 30' high, add						25%			
9940	30' to 35' high, add						30%			
9950	35' to 40' high, add						35%			
9960	Over 40' high, add						40%			
9990	Minimum labor/equipment charge	1 Elec	4	2	Job		123		123	189
9995	Do not include labor when adding couplings to a fitting installation R260533-30									

26 05 33.14 Conduit

	26 05 33.14 Conduit	Crew	Daily Output	Labor-Hours	Unit	Material	2020 Bare Costs Labor	Equipment	Total	Total Incl O&P
0010	**CONDUIT** To 10′ high, includes 11 couplings per 100′									
0200	Electric metallic tubing, 1/2″ diameter	1 Elec	435	.018	L.F.	.59	1.13		1.72	2.39
0220	3/4″ diameter		253	.032		.87	1.94		2.81	3.95
0240	1″ diameter		207	.039		1.50	2.37		3.87	5.30
0260	1-1/4″ diameter		173	.046		2.61	2.84		5.45	7.25
0280	1-1/2″ diameter		153	.052		3.03	3.21		6.24	8.25
0300	2″ diameter		130	.062		3.85	3.78		7.63	10.05
0320	2-1/2″ diameter		92	.087		3.95	5.35		9.30	12.55
0340	3″ diameter	2 Elec	148	.108		4.57	6.65		11.22	15.25
0360	3-1/2″ diameter		134	.119		6.40	7.35		13.75	18.35
0380	4″ diameter		114	.140		12.05	8.60		20.65	26.50
0500	Steel rigid galvanized, 1/2″ diameter	1 Elec	146	.055		1.85	3.36		5.21	7.20
0520	3/4″ diameter		125	.064		3.95	3.93		7.88	10.40
0540	1″ diameter		93	.086		6.35	5.30		11.65	15.10
0560	1-1/4″ diameter		88	.091		3.87	5.60		9.47	12.85
0580	1-1/2″ diameter		80	.100		7.55	6.15		13.70	17.75
0600	2″ diameter		65	.123		7.20	7.55		14.75	19.50
0620	2-1/2″ diameter		48	.167		12.20	10.25		22.45	29
0640	3″ diameter	2 Elec	64	.250		13.45	15.35		28.80	38.50
0660	3-1/2″ diameter		60	.267		17.45	16.35		33.80	44
0680	4″ diameter		52	.308		21	18.90		39.90	52.50
0700	5″ diameter		50	.320		31.50	19.65		51.15	64.50
0720	6″ diameter		48	.333		48.50	20.50		69	84.50
1000	Steel intermediate conduit (IMC), 1/2″ diameter	1 Elec	155	.052		1.14	3.17		4.31	6.15
1010	3/4″ diameter		130	.062		1.39	3.78		5.17	7.35
1020	1″ diameter		100	.080		2.15	4.91		7.06	9.90
1030	1-1/4″ diameter		93	.086		2.44	5.30		7.74	10.80
1040	1-1/2″ diameter		85	.094		3.69	5.75		9.44	12.95
1050	2″ diameter		70	.114		4.21	7		11.21	15.45
1060	2-1/2″ diameter		53	.151		7.55	9.25		16.80	22.50
1070	3″ diameter	2 Elec	80	.200		11.60	12.25		23.85	31.50
1080	3-1/2″ diameter		70	.229		13.90	14		27.90	37
1090	4″ diameter		60	.267		9.30	16.35		25.65	35.50

26 05 33.15 Conduit Nipples

	26 05 33.15 Conduit Nipples	Crew	Daily Output	Labor-Hours	Unit	Material	Labor	Equipment	Total	Total Incl O&P
0010	**CONDUIT NIPPLES** With locknuts and bushings									
0100	Aluminum, 1/2″ diameter, close	1 Elec	36	.222	Ea.	6.75	13.65		20.40	28.50
0120	1-1/2″ long		36	.222		6.35	13.65		20	28
0140	2″ long		36	.222		7.85	13.65		21.50	29.50
0160	2-1/2″ long		36	.222		9.95	13.65		23.60	32
0180	3″ long		36	.222		8.55	13.65		22.20	30.50
0200	3-1/2″ long		36	.222		6.85	13.65		20.50	28.50
0220	4″ long		36	.222		7.20	13.65		20.85	29
0240	5″ long		36	.222		9.40	13.65		23.05	31.50
0260	6″ long		36	.222		9.90	13.65		23.55	32
0280	8″ long		36	.222		9.75	13.65		23.40	32
0300	10″ long		36	.222		10.70	13.65		24.35	33
0320	12″ long		36	.222		17.75	13.65		31.40	40.50
0340	3/4″ diameter, close		32	.250		9.25	15.35		24.60	33.50
0360	1-1/2″ long		32	.250		9.25	15.35		24.60	33.50
0380	2″ long		32	.250		9.25	15.35		24.60	33.50
0400	2-1/2″ long		32	.250		11	15.35		26.35	35.50
0420	3″ long		32	.250		11.05	15.35		26.40	35.50

26 05 33.15 Conduit Nipples		Crew	Daily Output	Labor-Hours	Unit	Material	2020 Bare Costs Labor	Equipment	Total	Total Incl O&P
0440	3-1/2" long	1 Elec	32	.250	Ea.	15.10	15.35		30.45	40
0460	4" long		32	.250		11.40	15.35		26.75	36
0480	5" long		32	.250		12.60	15.35		27.95	37.50
0500	6" long		32	.250		12.45	15.35		27.80	37
0520	8" long		32	.250		12.65	15.35		28	37.50
0540	10" long		32	.250		14.35	15.35		29.70	39.50
0560	12" long		32	.250		19.70	15.35		35.05	45
0580	1" diameter, close		27	.296		13.20	18.20		31.40	42.50
0600	2" long		27	.296		19.55	18.20		37.75	49.50
0620	2-1/2" long		27	.296		13.95	18.20		32.15	43.50
0640	3" long		27	.296		16.55	18.20		34.75	46.50
0660	3-1/2" long		27	.296		24	18.20		42.20	54.50
0680	4" long		27	.296		18.70	18.20		36.90	48.50
0700	5" long		27	.296		18.25	18.20		36.45	48
0720	6" long		27	.296		18	18.20		36.20	48
0740	8" long		27	.296		27	18.20		45.20	57.50
0760	10" long		27	.296		48	18.20		66.20	81
0780	12" long		27	.296		45	18.20		63.20	77.50
0800	1-1/4" diameter, close		23	.348		20.50	21.50		42	55.50
0820	2" long		23	.348		26.50	21.50		48	62
0840	2-1/2" long		23	.348		18.85	21.50		40.35	54
0860	3" long		23	.348		23	21.50		44.50	58.50
0880	3-1/2" long		23	.348		23.50	21.50		45	58.50
0900	4" long		23	.348		20.50	21.50		42	56
0920	5" long		23	.348		26.50	21.50		48	62
0940	6" long		23	.348		30	21.50		51.50	66
0960	8" long		23	.348		46	21.50		67.50	84
0980	10" long		23	.348		28.50	21.50		50	64.50
1000	12" long		23	.348		38	21.50		59.50	74.50
1020	1-1/2" diameter, close		20	.400		27	24.50		51.50	67.50
1040	2" long		20	.400		28.50	24.50		53	69
1060	2-1/2" long		20	.400		28.50	24.50		53	69.50
1080	3" long		20	.400		28.50	24.50		53	69.50
1100	3-1/2" long		20	.400		28	24.50		52.50	69
1120	4" long		20	.400		28	24.50		52.50	68.50
1140	5" long		20	.400		34	24.50		58.50	75.50
1160	6" long		20	.400		35.50	24.50		60	77
1180	8" long		20	.400		35	24.50		59.50	76.50
1200	10" long		20	.400		40.50	24.50		65	82.50
1220	12" long		20	.400		52	24.50		76.50	95
1240	2" diameter, close		18	.444		38.50	27.50		66	84
1260	2-1/2" long		18	.444		37	27.50		64.50	82.50
1280	3" long		18	.444		39.50	27.50		67	85.50
1300	3-1/2" long		18	.444		39.50	27.50		67	85.50
1320	4" long		18	.444		41.50	27.50		69	87.50
1340	5" long		18	.444		46	27.50		73.50	92.50
1360	6" long		18	.444		42	27.50		69.50	88.50
1380	8" long		18	.444		60.50	27.50		88	109
1400	10" long		18	.444		60.50	27.50		88	109
1420	12" long		18	.444		72.50	27.50		100	122
1440	2-1/2" diameter, close		15	.533		84.50	32.50		117	144
1460	3" long		15	.533		78.50	32.50		111	137
1480	3-1/2" long		15	.533		80.50	32.50		113	139

26 05 33.15 Conduit Nipples		Crew	Daily Output	Labor-Hours	Unit	Material	2020 Bare Costs Labor	Equipment	Total	Total Incl O&P
1500	4″ long	1 Elec	15	.533	Ea.	89.50	32.50		122	149
1520	5″ long		15	.533		114	32.50		146.50	176
1540	6″ long		15	.533		85.50	32.50		118	145
1560	8″ long		15	.533		89.50	32.50		122	149
1580	10″ long		15	.533		112	32.50		144.50	174
1600	12″ long		15	.533		101	32.50		133.50	162
1620	3″ diameter, close		12	.667		90.50	41		131.50	163
1640	3″ long		12	.667		95	41		136	168
1660	3-1/2″ long		12	.667		115	41		156	189
1680	4″ long		12	.667		119	41		160	193
1700	5″ long		12	.667		90	41		131	162
1720	6″ long		12	.667		99	41		140	172
1740	8″ long		12	.667		124	41		165	199
1760	10″ long		12	.667		149	41		190	227
1780	12″ long		12	.667		143	41		184	220
1800	3-1/2″ diameter, close		11	.727		199	44.50		243.50	288
1820	4″ long		11	.727		226	44.50		270.50	320
1840	5″ long		11	.727		206	44.50		250.50	295
1860	6″ long		11	.727		216	44.50		260.50	305
1880	8″ long		11	.727		201	44.50		245.50	290
1900	10″ long		11	.727		227	44.50		271.50	320
1920	12″ long		11	.727		288	44.50		332.50	385
1940	4″ diameter, close		9	.889		224	54.50		278.50	330
1960	4″ long		9	.889		240	54.50		294.50	350
1980	5″ long		9	.889		241	54.50		295.50	350
2000	6″ long		9	.889		284	54.50		338.50	395
2020	8″ long		9	.889		277	54.50		331.50	390
2040	10″ long		9	.889		289	54.50		343.50	405
2060	12″ long		9	.889		262	54.50		316.50	370
2080	5″ diameter, close		7	1.143		445	70		515	600
2100	5″ long		7	1.143		525	70		595	685
2120	6″ long		7	1.143		525	70		595	690
2140	8″ long		7	1.143		565	70		635	730
2160	10″ long		7	1.143		470	70		540	630
2180	12″ long		7	1.143		490	70		560	650
2200	6″ diameter, close		6	1.333		680	82		762	875
2220	5″ long		6	1.333		730	82		812	925
2240	6″ long		6	1.333		665	82		747	860
2260	8″ long		6	1.333		790	82		872	995
2280	10″ long		6	1.333		840	82		922	1,050
2300	12″ long		6	1.333		865	82		947	1,075
2320	Rigid galvanized steel, 1/2″ diameter, close		32	.250		2.42	15.35		17.77	26
2340	1-1/2″ long		32	.250		2.59	15.35		17.94	26.50
2360	2″ long		32	.250		2.67	15.35		18.02	26.50
2380	2-1/2″ long		32	.250		2.79	15.35		18.14	26.50
2400	3″ long		32	.250		2.91	15.35		18.26	26.50
2420	3-1/2″ long		32	.250		3.02	15.35		18.37	27
2440	4″ long		32	.250		3.14	15.35		18.49	27
2460	5″ long		32	.250		3.31	15.35		18.66	27
2480	6″ long		32	.250		3.66	15.35		19.01	27.50
2500	8″ long		32	.250		5.15	15.35		20.50	29
2520	10″ long		32	.250		6	15.35		21.35	30
2540	12″ long		32	.250		6.30	15.35		21.65	30.50

26 05 33 – Raceway and Boxes for Electrical Systems

26 05 33.15 Conduit Nipples		Crew	Daily Output	Labor-Hours	Unit	Material	2020 Bare Costs Labor	Equipment	Total	Total Incl O&P
2560	3/4" diameter, close	1 Elec	27	.296	Ea.	3.51	18.20		21.71	32
2580	2" long		27	.296		3.71	18.20		21.91	32
2600	2-1/2" long		27	.296		3.84	18.20		22.04	32
2620	3" long		27	.296		3.97	18.20		22.17	32.50
2640	3-1/2" long		27	.296		4.03	18.20		22.23	32.50
2660	4" long		27	.296		4.24	18.20		22.44	32.50
2680	5" long		27	.296		4.49	18.20		22.69	33
2700	6" long		27	.296		4.82	18.20		23.02	33.50
2720	8" long		27	.296		6.35	18.20		24.55	35
2740	10" long		27	.296		7.65	18.20		25.85	36.50
2760	12" long		27	.296		7.75	18.20		25.95	36.50
2780	1" diameter, close		23	.348		5.45	21.50		26.95	39
2800	2" long		23	.348		5.55	21.50		27.05	39
2820	2-1/2" long		23	.348		5.70	21.50		27.20	39.50
2840	3" long		23	.348		5.90	21.50		27.40	39.50
2860	3-1/2" long		23	.348		6.25	21.50		27.75	40
2880	4" long		23	.348		6.40	21.50		27.90	40
2900	5" long		23	.348		6.70	21.50		28.20	40.50
2920	6" long		23	.348		7	21.50		28.50	40.50
2940	8" long		23	.348		8.80	21.50		30.30	42.50
2960	10" long		23	.348		10.25	21.50		31.75	44.50
2980	12" long		23	.348		11.10	21.50		32.60	45
3000	1-1/4" diameter, close		20	.400		7.50	24.50		32	46.50
3020	2" long		20	.400		7.65	24.50		32.15	46.50
3040	3" long		20	.400		8.05	24.50		32.55	47
3060	3-1/2" long		20	.400		8.35	24.50		32.85	47
3080	4" long		20	.400		8.60	24.50		33.10	47.50
3100	5" long		20	.400		9.10	24.50		33.60	48
3120	6" long		20	.400		9.55	24.50		34.05	48.50
3140	8" long		20	.400		12.10	24.50		36.60	51.50
3160	10" long		20	.400		13.85	24.50		38.35	53.50
3180	12" long		20	.400		15.10	24.50		39.60	54.50
3200	1-1/2" diameter, close		18	.444		10.85	27.50		38.35	54
3220	2" long		18	.444		11	27.50		38.50	54
3240	2-1/2" long		18	.444		11.30	27.50		38.80	54.50
3260	3" long		18	.444		11.50	27.50		39	54.50
3280	3-1/2" long		18	.444		12	27.50		39.50	55
3300	4" long		18	.444		12.30	27.50		39.80	55.50
3320	5" long		18	.444		12.85	27.50		40.35	56
3340	6" long		18	.444		13.85	27.50		41.35	57
3360	8" long		18	.444		16.65	27.50		44.15	60.50
3380	10" long		18	.444		18.35	27.50		45.85	62
3400	12" long		18	.444		19.15	27.50		46.65	63
3420	2" diameter, close		16	.500		14	30.50		44.50	62.50
3440	2-1/2" long		16	.500		14.45	30.50		44.95	63
3460	3" long		16	.500		15.10	30.50		45.60	63.50
3480	3-1/2" long		16	.500		15.65	30.50		46.15	64
3500	4" long		16	.500		16.10	30.50		46.60	64.50
3520	5" long		16	.500		17	30.50		47.50	65.50
3540	6" long		16	.500		17.75	30.50		48.25	66.50
3560	8" long		16	.500		21	30.50		51.50	70
3580	10" long		16	.500		22.50	30.50		53	72
3600	12" long		16	.500		24.50	30.50		55	74

26 05 33.15 Conduit Nipples		Crew	Daily Output	Labor-Hours	Unit	Material	2020 Bare Costs Labor	Equipment	Total	Total Incl O&P
3620	2-1/2" diameter, close	1 Elec	13	.615	Ea.	48	38		86	111
3640	3" long		13	.615		51.50	38		89.50	115
3660	3-1/2" long		13	.615		53	38		91	117
3680	4" long		13	.615		53.50	38		91.50	117
3700	5" long		13	.615		56	38		94	120
3720	6" long		13	.615		57	38		95	121
3740	8" long		13	.615		61.50	38		99.50	126
3760	10" long		13	.615		64.50	38		102.50	129
3780	12" long		13	.615		68.50	38		106.50	133
3800	3" diameter, close		12	.667		48.50	41		89.50	117
3820	3" long		12	.667		52.50	41		93.50	121
3900	3-1/2" long		12	.667		50	41		91	118
3920	4" long		12	.667		51.50	41		92.50	120
3940	5" long		12	.667		53.50	41		94.50	122
3960	6" long		12	.667		55.50	41		96.50	124
3980	8" long		12	.667		60.50	41		101.50	130
4000	10" long		12	.667		64.50	41		105.50	134
4020	12" long		12	.667		70.50	41		111.50	141
4040	3-1/2" diameter, close		10	.800		84	49		133	168
4060	4" long		10	.800		86.50	49		135.50	171
4080	5" long		10	.800		89	49		138	173
4100	6" long		10	.800		91.50	49		140.50	177
4120	8" long		10	.800		96	49		145	182
4140	10" long		10	.800		101	49		150	187
4160	12" long		10	.800		107	49		156	193
4180	4" diameter, close		8	1		134	61.50		195.50	242
4200	4" long		8	1		135	61.50		196.50	244
4220	5" long		8	1		137	61.50		198.50	246
4240	6" long		8	1		139	61.50		200.50	248
4260	8" long		8	1		142	61.50		203.50	252
4280	10" long		8	1		146	61.50		207.50	256
4300	12" long		8	1		151	61.50		212.50	261
4320	5" diameter, close		6	1.333		192	82		274	335
4340	5" long		6	1.333		203	82		285	350
4360	6" long		6	1.333		207	82		289	355
4380	8" long		6	1.333		209	82		291	355
4400	10" long		6	1.333		224	82		306	370
4420	12" long		6	1.333		238	82		320	390
4440	6" diameter, close		5	1.600		335	98		433	520
4460	5" long		5	1.600		355	98		453	540
4480	6" long		5	1.600		360	98		458	545
4500	8" long		5	1.600		370	98		468	555
4520	10" long		5	1.600		385	98		483	570
4540	12" long		5	1.600		395	98		493	585
4560	Plastic coated, 40 mil thick, 1/2" diameter, 2" long		32	.250		23	15.35		38.35	48.50
4580	2-1/2" long		32	.250		20	15.35		35.35	45.50
4600	3" long		32	.250		20.50	15.35		35.85	46
4680	3-1/2" long		32	.250		24	15.35		39.35	50
4700	4" long		32	.250		22	15.35		37.35	47.50
4720	5" long		32	.250		24.50	15.35		39.85	50.50
4740	6" long		32	.250		20.50	15.35		35.85	46
4760	8" long		32	.250		28.50	15.35		43.85	55
4780	10" long		32	.250		31.50	15.35		46.85	58.50

26 05 33.15 Conduit Nipples		Crew	Daily Output	Labor-Hours	Unit	2020 Bare Costs Material	Labor	Equipment	Total	Total Incl O&P
4800	12" long	1 Elec	32	.250	Ea.	35.50	15.35		50.85	62.50
4820	3/4" diameter, 2" long		26	.308		24.50	18.90		43.40	56
4840	2-1/2" long		26	.308		23.50	18.90		42.40	54.50
4860	3" long		26	.308		26	18.90		44.90	57.50
4880	3-1/2" long		26	.308		25	18.90		43.90	56.50
4900	4" long		26	.308		27	18.90		45.90	58.50
4920	5" long		26	.308		26	18.90		44.90	58
4940	6" long		26	.308		32.50	18.90		51.40	65
4960	8" long		26	.308		29	18.90		47.90	60.50
4980	10" long		26	.308		26.50	18.90		45.40	58
5000	12" long		26	.308		25	18.90		43.90	57
5020	1" diameter, 2" long		22	.364		26.50	22.50		49	63.50
5040	2-1/2" long		22	.364		25.50	22.50		48	62.50
5060	3" long		22	.364		29.50	22.50		52	67
5080	3-1/2" long		22	.364		29.50	22.50		52	67
5100	4" long		22	.364		25	22.50		47.50	61.50
5120	5" long		22	.364		23.50	22.50		46	60.50
5140	6" long		22	.364		27.50	22.50		50	65
5160	8" long		22	.364		32	22.50		54.50	69.50
5180	10" long		22	.364		34	22.50		56.50	72
5200	12" long		22	.364		31	22.50		53.50	68.50
5220	1-1/4" diameter, 2" long		18	.444		30	27.50		57.50	75
5240	2-1/2" long		18	.444		33	27.50		60.50	78
5260	3" long		18	.444		33.50	27.50		61	78.50
5280	3-1/2" long		18	.444		36	27.50		63.50	82
5300	4" long		18	.444		40.50	27.50		68	87
5320	5" long		18	.444		37	27.50		64.50	82.50
5340	6" long		18	.444		36	27.50		63.50	82
5360	8" long		18	.444		38.50	27.50		66	84.50
5380	10" long		18	.444		43	27.50		70.50	89.50
5400	12" long		18	.444		42.50	27.50		70	89
5420	1-1/2" diameter, 2" long		16	.500		37	30.50		67.50	87.50
5440	2-1/2" long		16	.500		37	30.50		67.50	87.50
5460	3" long		16	.500		35	30.50		65.50	85.50
5480	3-1/2" long		16	.500		39	30.50		69.50	90
5500	4" long		16	.500		33	30.50		63.50	83.50
5520	5" long		16	.500		33	30.50		63.50	83.50
5540	6" long		16	.500		39.50	30.50		70	90.50
5560	8" long		16	.500		46.50	30.50		77	98
5580	10" long		16	.500		62	30.50		92.50	115
5600	12" long		16	.500		56	30.50		86.50	109
5620	2" diameter, 2-1/2" long		14	.571		38	35		73	96
5640	3" long		14	.571		41	35		76	99
5660	3-1/2" long		14	.571		47	35		82	106
5680	4" long		14	.571		47	35		82	106
5700	5" long		14	.571		51	35		86	110
5720	6" long		14	.571		48.50	35		83.50	108
5740	8" long		14	.571		58.50	35		93.50	119
5760	10" long		14	.571		76.50	35		111.50	138
5780	12" long		14	.571		75	35		110	137
5800	2-1/2" diameter, 3-1/2" long		12	.667		99.50	41		140.50	172
5820	4" long		12	.667		102	41		143	175
5840	5" long	↓	12	.667	↓	121	41		162	196

26 05 33.15 Conduit Nipples		Crew	Daily Output	Labor-Hours	Unit	Material	Labor	Equipment	Total	Total Incl O&P
						2020 Bare Costs				
5860	6" long	1 Elec	12	.667	Ea.	101	41		142	174
5880	8" long		12	.667		133	41		174	210
5900	10" long		12	.667		145	41		186	223
5920	12" long		12	.667		147	41		188	225
5940	3" diameter, 3-1/2" long		11	.727		112	44.50		156.50	193
5960	4" long		11	.727		101	44.50		145.50	180
5980	5" long		11	.727		117	44.50		161.50	198
6000	6" long		11	.727		128	44.50		172.50	210
6020	8" long		11	.727		159	44.50		203.50	244
6040	10" long		11	.727		177	44.50		221.50	264
6060	12" long		11	.727		174	44.50		218.50	261
6080	3-1/2" diameter, 4" long		9	.889		159	54.50		213.50	259
6100	5" long		9	.889		161	54.50		215.50	261
6120	6" long		9	.889		149	54.50		203.50	248
6140	8" long		9	.889		195	54.50		249.50	299
6160	10" long		9	.889		228	54.50		282.50	335
6180	12" long		9	.889		260	54.50		314.50	370
6200	4" diameter, 4" long		7.50	1.067		208	65.50		273.50	330
6220	5" long		7.50	1.067		223	65.50		288.50	345
6240	6" long		7.50	1.067		241	65.50		306.50	365
6260	8" long		7.50	1.067		258	65.50		323.50	385
6280	10" long		7.50	1.067		315	65.50		380.50	445
6300	12" long		7.50	1.067		350	65.50		415.50	485
6320	5" diameter, 5" long		5.50	1.455		212	89		301	370
6340	6" long		5.50	1.455		261	89		350	425
6360	8" long		5.50	1.455		270	89		359	435
6380	10" long		5.50	1.455		284	89		373	445
6400	12" long		5.50	1.455		315	89		404	480
6420	6" diameter, 5" long		4.50	1.778		415	109		524	625
6440	6" long		4.50	1.778		425	109		534	640
6460	8" long		4.50	1.778		445	109		554	660
6480	10" long		4.50	1.778		465	109		574	680
6500	12" long		4.50	1.778		490	109		599	705

26 05 33.16 Boxes for Electrical Systems		Crew	Daily Output	Labor-Hours	Unit	Material	Labor	Equipment	Total	Total Incl O&P
0010	**BOXES FOR ELECTRICAL SYSTEMS**									
0020	Pressed steel, octagon, 4"	1 Elec	20	.400	Ea.	2.96	24.50		27.46	41.50
0040	For Romex or BX		20	.400		5.70	24.50		30.20	44.50
0050	For Romex or BX, with bracket		20	.400		8.55	24.50		33.05	47.50
0060	Covers, blank		64	.125		.87	7.65		8.52	12.75
0100	Extension rings		40	.200		4.59	12.25		16.84	24
0150	Square, 4"		20	.400		6.85	24.50		31.35	45.50
0160	For Romex or BX		20	.400		12	24.50		36.50	51
0170	For Romex or BX, with bracket		20	.400		9.10	24.50		33.60	48
0200	Extension rings		40	.200		4.85	12.25		17.10	24.50
0220	2-1/8" deep, 1" KO		20	.400		3.37	24.50		27.87	41.50
0250	Covers, blank		64	.125		.77	7.65		8.42	12.65
0260	Raised device		64	.125		1.57	7.65		9.22	13.55
0300	Plaster rings		64	.125		1.66	7.65		9.31	13.65
0350	Square, 4-11/16"		20	.400		4.38	24.50		28.88	43
0370	2-1/8" deep, 3/4" to 1-1/4" KO		20	.400		5.90	24.50		30.40	44.50
0400	Extension rings		40	.200		10.60	12.25		22.85	30.50
0450	Covers, blank		53	.151		1.13	9.25		10.38	15.50

26 05 33 – Raceway and Boxes for Electrical Systems

26 05 33.16 Boxes for Electrical Systems		Crew	Daily Output	Labor-Hours	Unit	Material	Labor	Equipment	Total	Total Incl O&P
						2020 Bare Costs				
0460	Raised device	1 Elec	53	.151	Ea.	7.20	9.25		16.45	22
0500	Plaster rings		53	.151		6.10	9.25		15.35	21
0550	Handy box		27	.296		2.52	18.20		20.72	31
0560	Covers, device		64	.125		1.01	7.65		8.66	12.90
0600	Extension rings		54	.148		3.60	9.10		12.70	17.95
0652	Switchbox		24	.333		5.60	20.50		26.10	37.50
0660	Romex or BX		27	.296		7.80	18.20		26	36.50
0670	with bracket		27	.296		7.20	18.20		25.40	36
0680	Partition, metal		27	.296		3.36	18.20		21.56	31.50
0700	Masonry, 1 gang, 2-1/2" deep		27	.296		8.35	18.20		26.55	37
0710	3-1/2" deep		27	.296		8.50	18.20		26.70	37.50
0750	2 gang, 2-1/2" deep		20	.400		17.70	24.50		42.20	57.50
0760	3-1/2" deep		20	.400		11.75	24.50		36.25	51
0800	3 gang, 2-1/2" deep		13	.615		18.60	38		56.60	78.50
0850	4 gang, 2-1/2" deep		10	.800		25	49		74	104
0860	5 gang, 2-1/2" deep		9	.889		37	54.50		91.50	125
0870	6 gang, 2-1/2" deep		8	1		61.50	61.50		123	162
0880	Masonry thru-the-wall, 1 gang, 4" block		16	.500		35.50	30.50		66	86
0890	6" block		16	.500		46.50	30.50		77	98
0900	8" block		16	.500		62	30.50		92.50	115
0920	2 gang, 6" block		16	.500		69.50	30.50		100	124
0940	Bar hanger with 3/8" stud, for wood and masonry boxes		53	.151		5.30	9.25		14.55	20
0950	Concrete, set flush, 4" deep		20	.400		8.20	24.50		32.70	47
1000	Plate with 3/8" stud		80	.100		8.35	6.15		14.50	18.60
1100	Concrete, floor, 1 gang		5.30	1.509		112	92.50		204.50	266
1150	2 gang		4	2		167	123		290	375
1200	3 gang		2.70	2.963		269	182		451	575
1250	For duplex receptacle, pedestal mounted, add		24	.333		117	20.50		137.50	160
1270	Flush mounted, add		27	.296		34.50	18.20		52.70	66
1300	For telephone, pedestal mounted, add		30	.267		113	16.35		129.35	149
1350	Carpet flange, 1 gang		53	.151		57.50	9.25		66.75	77.50
1400	Cast, 1 gang, FS (2" deep), 1/2" hub		12	.667		25.50	41		66.50	91
1410	3/4" hub		12	.667		29	41		70	95
1420	FD (2-11/16" deep), 1/2" hub		12	.667		18.70	41		59.70	83.50
1430	3/4" hub		12	.667		22	41		63	87.50
1450	2 gang, FS, 1/2" hub		10	.800		50	49		99	131
1460	3/4" hub		10	.800		45	49		94	125
1470	FD, 1/2" hub		10	.800		61	49		110	143
1480	3/4" hub		10	.800		55	49		104	136
1500	3 gang, FS, 3/4" hub		9	.889		81	54.50		135.50	173
1510	Switch cover, 1 gang, FS		64	.125		6.40	7.65		14.05	18.85
1520	2 gang		53	.151		10.10	9.25		19.35	25.50
1530	Duplex receptacle cover, 1 gang, FS		64	.125		6.95	7.65		14.60	19.45
1540	2 gang, FS		53	.151		10.60	9.25		19.85	26
1542	Weatherproof blank cover, 1 gang		64	.125		1.13	7.65		8.78	13.05
1544	2 gang		53	.151		2.28	9.25		11.53	16.75
1550	Weatherproof switch cover, 1 gang		64	.125		5.85	7.65		13.50	18.25
1554	2 gang		53	.151		9.75	9.25		19	25
1600	Weatherproof receptacle cover, 1 gang		64	.125		7.15	7.65		14.80	19.70
1604	2 gang		53	.151		8.95	9.25		18.20	24
1620	Weatherproof receptacle cover, tamper resistant, 1 gang		58	.138		13.60	8.45		22.05	28
1624	2 gang		48	.167		26.50	10.25		36.75	45
1750	FSC, 1 gang, 1/2" hub	↓	11	.727	↓	27.50	44.50		72	98.50

26 05 33.16 Boxes for Electrical Systems		Crew	Daily Output	Labor-Hours	Unit	Material	Labor	Equipment	Total	Total Incl O&P
						2020 Bare Costs				
1760	3/4" hub	1 Elec	11	.727	Ea.	33.50	44.50		78	106
1770	2 gang, 1/2" hub		9	.889		51	54.50		105.50	140
1780	3/4" hub		9	.889		55	54.50		109.50	145
1790	FDC, 1 gang, 1/2" hub		11	.727		29.50	44.50		74	101
1800	3/4" hub		11	.727		38	44.50		82.50	110
1810	2 gang, 1/2" hub		9	.889		66.50	54.50		121	157
1820	3/4" hub		9	.889		66	54.50		120.50	157
1850	Weatherproof in-use cover, 1 gang		64	.125		24	7.65		31.65	38.50
1870	2 gang		53	.151		31	9.25		40.25	48.50
2000	Poke-thru fitting, fire rated, for 3-3/4" floor		6.80	1.176		169	72		241	297
2040	For 7" floor		6.80	1.176		167	72		239	295
2100	Pedestal, 15 amp, duplex receptacle & blank plate		5.25	1.524		152	93.50		245.50	310
2120	Duplex receptacle and telephone plate		5.25	1.524		152	93.50		245.50	310
2140	Pedestal, 20 amp, duplex recept. & phone plate		5	1.600		153	98		251	320
2160	Telephone plate, both sides		5.25	1.524		143	93.50		236.50	300
2200	Abandonment plate		32	.250		44.50	15.35		59.85	72.50
9000	Minimum labor/equipment charge		4	2	Job		123		123	189

26 05 33.17 Outlet Boxes, Plastic		Crew	Daily Output	Labor-Hours	Unit	Material	Labor	Equipment	Total	Total Incl O&P
0010	**OUTLET BOXES, PLASTIC**									
0050	4" diameter, round with 2 mounting nails	1 Elec	25	.320	Ea.	2.89	19.65		22.54	33
0100	Bar hanger mounted		25	.320		5.05	19.65		24.70	35.50
0200	4", square with 2 mounting nails		25	.320		5.55	19.65		25.20	36
0300	Plaster ring		64	.125		2.10	7.65		9.75	14.10
0400	Switch box with 2 mounting nails, 1 gang		30	.267		3.21	16.35		19.56	28.50
0500	2 gang		25	.320		3.95	19.65		23.60	34.50
0600	3 gang		20	.400		5.50	24.50		30	44
0700	Old work box		30	.267		4.59	16.35		20.94	30
1400	PVC, FSS, 1 gang, 1/2" hub		14	.571		13.10	35		48.10	68.50
1410	3/4" hub		14	.571		11.40	35		46.40	66.50
1420	FD, 1 gang for variable terminations		14	.571		7.95	35		42.95	62.50
1450	FS, 2 gang for variable terminations		12	.667		13.45	41		54.45	78
1480	Weatherproof blank cover, FS, 1 gang		64	.125		3.85	7.65		11.50	16.05
1500	2 gang		53	.151		5.10	9.25		14.35	19.90
1510	Weatherproof switch cover, FS, 1 gang		64	.125		8.20	7.65		15.85	21
1520	2 gang		53	.151		16.60	9.25		25.85	32.50
1530	Weatherproof duplex receptacle cover, FS, 1 gang		64	.125		12.85	7.65		20.50	26
1540	2 gang		53	.151		13.60	9.25		22.85	29
1750	FSC, 1 gang, 1/2" hub		13	.615		8.05	38		46.05	67
1760	3/4" hub		13	.615		8.45	38		46.45	67.50
1770	FSC, 2 gang, 1/2" hub		11	.727		15.05	44.50		59.55	85
1780	3/4" hub		11	.727		11.65	44.50		56.15	81.50
1790	FDC, 1 gang, 1/2" hub		13	.615		10.45	38		48.45	69.50
1800	3/4" hub		13	.615		10.60	38		48.60	69.50
1810	Weatherproof, T box w/3 holes		14	.571		10.95	35		45.95	66
1820	4" diameter round w/5 holes		14	.571		10.25	35		45.25	65.50
1850	In-use cover, 1 gang		64	.125		9.05	7.65		16.70	22
1870	2 gang		53	.151		12.90	9.25		22.15	28.50
9000	Minimum labor/equipment charge		4	2	Job		123		123	189

26 05 33.18 Pull Boxes		Crew	Daily Output	Labor-Hours	Unit	Material	Labor	Equipment	Total	Total Incl O&P
0010	**PULL BOXES**									
0100	Steel, pull box, NEMA 1, type SC, 6" W x 6" H x 4" D	1 Elec	8	1	Ea.	9.85	61.50		71.35	105
0180	8" W x 6" H x 4" D		8	1		23	61.50		84.50	120

26 05 33.18 Pull Boxes		Crew	Daily Output	Labor-Hours	Unit	Material	Labor	Equipment	Total	Total Incl O&P
						2020 Bare Costs				
0200	8" W x 8" H x 4" D	1 Elec	8	1	Ea.	12.90	61.50		74.40	109
0210	10" W x 10" H x 4" D		7	1.143		25	70		95	136
0220	12" W x 12" H x 4" D		6.50	1.231		23.50	75.50		99	142
0230	15" W x 15" H x 4" D		5.20	1.538		43.50	94.50		138	193
0240	18" W x 18" H x 4" D		4.40	1.818		39.50	112		151.50	216
0250	6" W x 6" H x 6" D		8	1		14.70	61.50		76.20	111
0260	8" W x 8" H x 6" D		7.50	1.067		15.20	65.50		80.70	118
0270	10" W x 10" H x 6" D		5.50	1.455		20.50	89		109.50	160
0300	10" W x 12" H x 6" D		5.30	1.509		30.50	92.50		123	177
0310	12" W x 12" H x 6" D		5.20	1.538		25	94.50		119.50	173
0320	15" W x 15" H x 6" D		4.60	1.739		39	107		146	207
0330	18" W x 18" H x 6" D		4.20	1.905		47	117		164	232
0340	24" W x 24" H x 6" D		3.20	2.500		93.50	153		246.50	340
0350	12" W x 12" H x 8" D		5	1.600		30.50	98		128.50	185
0360	15" W x 15" H x 8" D		4.50	1.778		45.50	109		154.50	218
0370	18" W x 18" H x 8" D		4	2		60.50	123		183.50	256
0380	24" W x 18" H x 6" D		3.70	2.162		116	133		249	330
0400	16" W x 20" H x 8" D		4	2		82.50	123		205.50	280
0500	20" W x 24" H x 8" D		3.20	2.500		110	153		263	355
0510	24" W x 24" H x 8" D		3	2.667		105	164		269	365
0600	24" W x 36" H x 8" D		2.70	2.963		153	182		335	450
0610	30" W x 30" H x 8" D		2.70	2.963		202	182		384	500
0620	36" W x 36" H x 8" D		2	4		216	245		461	620
0630	24" W x 24" H x 10" D		2.50	3.200		186	196		382	505
0650	Pull box, hinged, NEMA 1, 6" W x 6" H x 4" D		8	1		16.95	61.50		78.45	113
0660	8" W x 8" H x 4" D		8	1		23.50	61.50		85	121
0670	10" W x 10" H x 4" D		7	1.143		37.50	70		107.50	150
0680	12" W x 12" H x 4" D		6	1.333		39.50	82		121.50	170
0690	15" W x 15" H x 4" D		5.20	1.538		44	94.50		138.50	194
0700	18" W x 18" H x 4" D		4.40	1.818		50.50	112		162.50	228
0710	6" W x 6" H x 6" D		8	1		19.65	61.50		81.15	116
0720	8" W x 8" H x 6" D		7.50	1.067		26.50	65.50		92	131
0730	10" W x 10" H x 6" D		5.50	1.455		35.50	89		124.50	176
0740	12" W x 12" H x 6" D		5.20	1.538		44.50	94.50		139	194
0800	12" W x 16" H x 6" D		4.70	1.702		49.50	104		153.50	216
0820	18" W x 18" H x 6" D		4.20	1.905		77.50	117		194.50	265
1000	20" W x 20" H x 6" D		3.60	2.222		107	136		243	330
1010	24" W x 24" H x 6" D		3.20	2.500		163	153		316	415
1020	12" W x 12" H x 8" D		5	1.600		80.50	98		178.50	240
1030	15" W x 15" H x 8" D		4.50	1.778		92.50	109		201.50	270
1040	18" W x 18" H x 8" D		4	2		130	123		253	330
1200	20" W x 20" H x 8" D		3.20	2.500		163	153		316	415
1210	24" W x 24" H x 8" D		3	2.667		183	164		347	455
1220	30" W x 30" H x 8" D		2.70	2.963		265	182		447	570
1400	24" W x 36" H x 8" D		2.70	2.963		257	182		439	565
1600	24" W x 42" H x 8" D		2	4		380	245		625	795
1610	36" W x 36" H x 8" D		2	4		335	245		580	750
2100	Pull box, NEMA 3R, type SC, raintight & weatherproof									
2150	6" L x 6" W x 6" D	1 Elec	10	.800	Ea.	17.25	49		66.25	94.50
2200	8" L x 6" W x 6" D		8	1		25	61.50		86.50	122
2250	10" L x 6" W x 6" D		7	1.143		32	70		102	143
2300	12" L x 12" W x 6" D		5	1.600		67.50	98		165.50	225
2350	16" L x 16" W x 6" D		4.50	1.778		92	109		201	269

26 05 33.18 Pull Boxes		Crew	Daily Output	Labor-Hours	Unit	Material	Labor	Equipment	Total	Total Incl O&P
						2020 Bare Costs				
2400	20" L x 20" W x 6" D	1 Elec	4	2	Ea.	86	123		209	284
2450	24" L x 18" W x 8" D		3	2.667		153	164		317	420
2500	24" L x 24" W x 10" D		2.50	3.200		288	196		484	615
2550	30" L x 24" W x 12" D		2	4		400	245		645	815
2600	36" L x 36" W x 12" D	↓	1.50	5.333	↓	695	325		1,020	1,275
2800	Cast iron, pull boxes for surface mounting									
3000	NEMA 4, watertight & dust tight									
3050	6" L x 6" W x 6" D	1 Elec	4	2	Ea.	276	123		399	495
3100	8" L x 6" W x 6" D		3.20	2.500		420	153		573	695
3150	10" L x 6" W x 6" D		2.50	3.200		440	196		636	785
3200	12" L x 12" W x 6" D		2.30	3.478		745	213		958	1,150
3250	16" L x 16" W x 6" D		1.30	6.154		840	380		1,220	1,500
3300	20" L x 20" W x 6" D		.80	10		965	615		1,580	2,000
3350	24" L x 18" W x 8" D		.70	11.429		2,600	700		3,300	3,925
3400	24" L x 24" W x 10" D		.50	16		4,725	980		5,705	6,700
3450	30" L x 24" W x 12" D		.40	20		6,725	1,225		7,950	9,300
3500	36" L x 36" W x 12" D		.20	40		8,275	2,450		10,725	12,900
3510	NEMA 4 clamp cover, 6" L x 6" W x 4" D		4	2		273	123		396	490
3520	8" L x 6" W x 4" D	↓	4	2	↓	330	123		453	555
4000	NEMA 7, explosion proof									
4050	6" L x 6" W x 6" D	1 Elec	2	4	Ea.	750	245		995	1,200
4100	8" L x 6" W x 6" D		1.80	4.444		1,050	273		1,323	1,575
4150	10" L x 6" W x 6" D		1.60	5		1,325	305		1,630	1,925
4200	12" L x 12" W x 6" D		1	8		2,400	490		2,890	3,400
4250	16" L x 14" W x 6" D		.60	13.333		3,500	820		4,320	5,100
4300	18" L x 18" W x 8" D		.50	16		6,475	980		7,455	8,625
4350	24" L x 18" W x 8" D		.40	20		7,675	1,225		8,900	10,300
4400	24" L x 24" W x 10" D		.30	26.667		11,200	1,625		12,825	14,800
4450	30" L x 24" W x 12" D		.20	40		14,400	2,450		16,850	19,600
5000	NEMA 9, dust tight 6" L x 6" W x 6" D		3.20	2.500		415	153		568	690
5050	8" L x 6" W x 6" D		2.70	2.963		490	182		672	820
5100	10" L x 6" W x 6" D		2	4		655	245		900	1,100
5150	12" L x 12" W x 6" D		1.60	5		1,250	305		1,555	1,850
5200	16" L x 16" W x 6" D		1	8		2,075	490		2,565	3,025
5250	18" L x 18" W x 8" D		.70	11.429		3,250	700		3,950	4,650
5300	24" L x 18" W x 8" D		.60	13.333		4,600	820		5,420	6,300
5350	24" L x 24" W x 10" D		.40	20		6,175	1,225		7,400	8,700
5400	30" L x 24" W x 12" D	↓	.30	26.667	↓	9,325	1,625		10,950	12,700
6000	J.I.C. wiring boxes, NEMA 12, dust tight & drip tight									
6050	6" L x 8" W x 4" D	1 Elec	10	.800	Ea.	77.50	49		126.50	161
6100	8" L x 10" W x 4" D		8	1		129	61.50		190.50	237
6150	12" L x 14" W x 6" D		5.30	1.509		97.50	92.50		190	250
6200	14" L x 16" W x 6" D		4.70	1.702		127	104		231	300
6250	16" L x 20" W x 6" D		4.40	1.818		330	112		442	535
6300	24" L x 30" W x 6" D		3.20	2.500		420	153		573	695
6350	24" L x 30" W x 8" D		2.90	2.759		490	169		659	795
6400	24" L x 36" W x 8" D		2.70	2.963		400	182		582	720
6450	24" L x 42" W x 8" D		2.30	3.478		540	213		753	925
6500	24" L x 48" W x 8" D	↓	2	4	↓	595	245		840	1,025

26 05 33.23 Wireway		Crew	Daily Output	Labor-Hours	Unit	2020 Bare Costs Material	Labor	Equipment	Total	Total Incl O&P
0010	**WIREWAY** to 10′ high									
0020	For higher elevations, see Section 26 05 36.40									
0100	NEMA 1, screw cover w/fittings and supports, 2-1/2″ x 2-1/2″	1 Elec	45	.178	L.F.	12.45	10.90		23.35	30.50
0200	4″ x 4″	″	40	.200		13.40	12.25		25.65	33.50
0400	6″ x 6″	2 Elec	60	.267		23.50	16.35		39.85	51
0600	8″ x 8″		40	.400		28	24.50		52.50	69
0620	10″ x 10″		30	.533		55.50	32.50		88	112
0640	12″ x 12″		20	.800		64	49		113	146
0800	Elbows, 90°, 2-1/2″	1 Elec	24	.333	Ea.	32.50	20.50		53	67
1000	4″		20	.400		38	24.50		62.50	79.50
1200	6″		18	.444		42	27.50		69.50	88
1400	8″		16	.500		66.50	30.50		97	121
1420	10″		12	.667		88	41		129	160
1440	12″		10	.800		124	49		173	212
1500	Elbows, 45°, 2-1/2″		24	.333		33.50	20.50		54	68.50
1510	4″		20	.400		39.50	24.50		64	81.50
1520	6″		18	.444		45.50	27.50		73	92
1530	8″		16	.500		66.50	30.50		97	121
1540	10″		12	.667		88.50	41		129.50	161
1550	12″		10	.800		169	49		218	261
1600	″T″ box, 2-1/2″		18	.444		37	27.50		64.50	83
1800	4″		16	.500		49	30.50		79.50	101
2000	6″		14	.571		54.50	35		89.50	114
2200	8″		12	.667		93	41		134	165
2220	10″		10	.800		114	49		163	202
2240	12″		8	1		163	61.50		224.50	274
2300	Cross, 2-1/2″		16	.500		41.50	30.50		72	92.50
2310	4″		14	.571		50.50	35		85.50	110
2320	6″		12	.667		62.50	41		103.50	132
2400	Panel adapter, 2-1/2″		24	.333		6.95	20.50		27.45	39
2600	4″		20	.400		8.85	24.50		33.35	47.50
2800	6″		18	.444		12.65	27.50		40.15	56
3000	8″		16	.500		18.25	30.50		48.75	67
3020	10″		14	.571		28	35		63	85
3040	12″		12	.667		41	41		82	109
3200	Reducer, 4″ to 2-1/2″		24	.333		12.50	20.50		33	45.50
3400	6″ to 4″		20	.400		28	24.50		52.50	68.50
3600	8″ to 6″		18	.444		34	27.50		61.50	79.50
3620	10″ to 8″		16	.500		44	30.50		74.50	95
3640	12″ to 10″		14	.571		57	35		92	117
3780	End cap, 2-1/2″		24	.333		5.20	20.50		25.70	37.50
3800	4″		20	.400		6.05	24.50		30.55	44.50
4000	6″		18	.444		7.15	27.50		34.65	50
4200	8″		16	.500		9.15	30.50		39.65	57
4220	10″		14	.571		16.10	35		51.10	71.50
4240	12″		12	.667		21	41		62	86.50
4300	U-connector, 2-1/2″		200	.040		4.46	2.45		6.91	8.70
4320	4″		200	.040		5.35	2.45		7.80	9.70
4340	6″		180	.044		6.30	2.73		9.03	11.10
4360	8″		170	.047		10.95	2.89		13.84	16.50
4380	10″		150	.053		15.65	3.27		18.92	22.50
4400	12″		130	.062		20	3.78		23.78	28

26 05 33 – Raceway and Boxes for Electrical Systems

26 05 33.23 Wireway		Crew	Daily Output	Labor-Hours	Unit	Material	Labor	Equipment	Total	Total Incl O&P
						2020 Bare Costs				
4420	Hanger, 2-1/2"	1 Elec	100	.080	Ea.	12.25	4.91		17.16	21
4430	4"		100	.080		12.30	4.91		17.21	21
4440	6"		80	.100		14.55	6.15		20.70	25.50
4450	8"		65	.123		24.50	7.55		32.05	38.50
4460	10"		50	.160		44.50	9.80		54.30	64
4470	12"		40	.200		65	12.25		77.25	90.50
4475	NEMA 3R, screw cover w/fittings and supports, 4" x 4"		36	.222	L.F.	16.65	13.65		30.30	39.50
4480	6" x 6"	2 Elec	55	.291		18.50	17.85		36.35	48
4485	8" x 8"		36	.444		32.50	27.50		60	78
4490	12" x 12"		18	.889		60.50	54.50		115	151
4500	Hinged cover, with fittings and supports, 2-1/2" x 2-1/2"	1 Elec	60	.133		20.50	8.20		28.70	35
4520	4" x 4"	"	45	.178		13.85	10.90		24.75	32
4540	6" x 6"	2 Elec	80	.200		21	12.25		33.25	42
4560	8" x 8"		60	.267		37	16.35		53.35	65.50
4580	10" x 10"		50	.320		48	19.65		67.65	82.50
4600	12" x 12"		24	.667		85	41		126	157
4700	Elbows 90°, 2-1/2" x 2-1/2"	1 Elec	32	.250	Ea.	46.50	15.35		61.85	75
4720	4"		27	.296		76.50	18.20		94.70	112
4730	6"		23	.348		59	21.50		80.50	98
4740	8"		18	.444		71.50	27.50		99	121
4750	10"		14	.571		111	35		146	176
4760	12"		12	.667		180	41		221	261
4800	Tee box, hinged cover, 2-1/2" x 2-1/2"		23	.348		61.50	21.50		83	101
4810	4"		20	.400		88	24.50		112.50	135
4820	6"		18	.444		68	27.50		95.50	117
4830	8"		16	.500		162	30.50		192.50	225
4840	10"		12	.667		165	41		206	244
4860	12"		10	.800		213	49		262	310
4880	Cross box, hinged cover, 2-1/2" x 2-1/2"		18	.444		62	27.50		89.50	110
4900	4"		16	.500		88	30.50		118.50	144
4920	6"		13	.615		107	38		145	176
4940	8"		11	.727		153	44.50		197.50	238
4960	10"		10	.800		243	49		292	345
4980	12"		9	.889		235	54.50		289.50	345
5000	NEMA 12, hinged cover, 2-1/2" x 2-1/2"		40	.200	L.F.	24.50	12.25		36.75	46
5020	4" x 4"		35	.229		28	14		42	52
5040	6" x 6"	2 Elec	60	.267		41.50	16.35		57.85	70.50
5060	8" x 8"	"	50	.320		55.50	19.65		75.15	91
5120	Elbows 90°, flanged, 2-1/2" x 2-1/2"	1 Elec	23	.348	Ea.	100	21.50		121.50	143
5140	4"		20	.400		126	24.50		150.50	176
5160	6"		18	.444		150	27.50		177.50	207
5180	8"		15	.533		139	32.50		171.50	204
5240	Tee box, flanged, 2-1/2" x 2-1/2"		18	.444		130	27.50		157.50	185
5260	4"		16	.500		170	30.50		200.50	234
5280	6"		15	.533		137	32.50		169.50	202
5300	8"		13	.615		193	38		231	270
5360	Cross box, flanged, 2-1/2" x 2-1/2"		15	.533		171	32.50		203.50	239
5380	4"		13	.615		141	38		179	213
5400	6"		12	.667		182	41		223	263
5420	8"		10	.800		227	49		276	325
5480	Flange gasket, 2-1/2"		160	.050		2.57	3.07		5.64	7.55
5500	4"		80	.100		3.52	6.15		9.67	13.30
5520	6"		53	.151		5.70	9.25		14.95	20.50

26 05 33.23 Wireway		Crew	Daily Output	Labor-Hours	Unit	Material	2020 Bare Costs Labor	Equipment	Total	Total Incl O&P
5530	8″	1 Elec	40	.200	Ea.	6.25	12.25		18.50	26

26 05 33.25 Conduit Fittings for Rigid Galvanized Steel

		Crew	Daily Output	Labor-Hours	Unit	Material	Labor	Equipment	Total	Total Incl O&P
0010	**CONDUIT FITTINGS FOR RIGID GALVANIZED STEEL**									
0050	Standard, locknuts, 1/2″ diameter				Ea.	.20			.20	.22
0100	3/4″ diameter					.40			.40	.44
0300	1″ diameter					.62			.62	.68
0500	1-1/4″ diameter					.79			.79	.87
0700	1-1/2″ diameter					1.32			1.32	1.45
1000	2″ diameter					1.85			1.85	2.04
1030	2-1/2″ diameter					4.56			4.56	5
1050	3″ diameter					6.45			6.45	7.10
1070	3-1/2″ diameter					11.15			11.15	12.25
1100	4″ diameter					15.10			15.10	16.65
1110	5″ diameter					30.50			30.50	33.50
1120	6″ diameter					54			54	59
1130	Bushings, plastic, 1/2″ diameter	1 Elec	40	.200		.09	12.25		12.34	19
1150	3/4″ diameter		32	.250		.10	15.35		15.45	23.50
1170	1″ diameter		28	.286		.20	17.55		17.75	27
1200	1-1/4″ diameter		24	.333		.25	20.50		20.75	32
1230	1-1/2″ diameter		18	.444		.35	27.50		27.85	42.50
1250	2″ diameter		15	.533		.62	32.50		33.12	51
1270	2-1/2″ diameter		13	.615		1.15	38		39.15	59.50
1300	3″ diameter		12	.667		1.34	41		42.34	64.50
1330	3-1/2″ diameter		11	.727		1.60	44.50		46.10	70.50
1350	4″ diameter		9	.889		1.77	54.50		56.27	86
1360	5″ diameter		7	1.143		12.40	70		82.40	122
1370	6″ diameter		5	1.600		45.50	98		143.50	201
1390	Steel, 1/2″ diameter		40	.200		.61	12.25		12.86	19.55
1400	3/4″ diameter		32	.250		.83	15.35		16.18	24.50
1430	1″ diameter		28	.286		1.30	17.55		18.85	28.50
1450	Steel insulated, 1-1/4″ diameter		24	.333		4.89	20.50		25.39	37
1470	1-1/2″ diameter		18	.444		5.95	27.50		33.45	48.50
1500	2″ diameter		15	.533		7.10	32.50		39.60	58.50
1530	2-1/2″ diameter		13	.615		16.80	38		54.80	76.50
1550	3″ diameter		12	.667		19.40	41		60.40	84.50
1570	3-1/2″ diameter		11	.727		25.50	44.50		70	96.50
1600	4″ diameter		9	.889		31.50	54.50		86	119
1610	5″ diameter		7	1.143		66	70		136	181
1620	6″ diameter		5	1.600		148	98		246	315
1630	Sealing locknuts, 1/2″ diameter		40	.200		1.73	12.25		13.98	21
1650	3/4″ diameter		32	.250		1.58	15.35		16.93	25
1670	1″ diameter		28	.286		4.58	17.55		22.13	32
1700	1-1/4″ diameter		24	.333		5	20.50		25.50	37
1730	1-1/2″ diameter		18	.444		5.70	27.50		33.20	48.50
1750	2″ diameter		15	.533		7.25	32.50		39.75	58.50
1760	Grounding bushing, insulated, 1/2″ diameter		32	.250		4.92	15.35		20.27	29
1770	3/4″ diameter		28	.286		6.80	17.55		24.35	34.50
1780	1″ diameter		20	.400		10.05	24.50		34.55	49
1800	1-1/4″ diameter		18	.444		11.75	27.50		39.25	55
1830	1-1/2″ diameter		16	.500		13.70	30.50		44.20	62
1850	2″ diameter		13	.615		12.50	38		50.50	72
1870	2-1/2″ diameter		12	.667		29.50	41		70.50	95.50

26 05 33.25 Conduit Fittings for Rigid Galvanized Steel		Crew	Daily Output	Labor-Hours	Unit	Material	2020 Bare Costs Labor	Equipment	Total	Total Incl O&P
1900	3" diameter	1 Elec	11	.727	Ea.	31	44.50		75.50	103
1930	3-1/2" diameter		9	.889		25	54.50		79.50	112
1950	4" diameter		8	1		58.50	61.50		120	159
1960	5" diameter		6	1.333		76.50	82		158.50	210
1970	6" diameter		4	2		110	123		233	310
1990	Coupling, with set screw, 1/2" diameter		50	.160		4.31	9.80		14.11	19.85
2000	3/4" diameter		40	.200		5.40	12.25		17.65	25
2030	1" diameter		35	.229		8.60	14		22.60	31
2050	1-1/4" diameter		28	.286		11.65	17.55		29.20	40
2070	1-1/2" diameter		23	.348		13.50	21.50		35	48
2090	2" diameter		20	.400		31	24.50		55.50	72
2100	2-1/2" diameter		18	.444		70	27.50		97.50	119
2110	3" diameter		15	.533		88.50	32.50		121	148
2120	3-1/2" diameter		12	.667		152	41		193	230
2130	4" diameter		10	.800		175	49		224	269
2140	5" diameter		9	.889		244	54.50		298.50	350
2150	6" diameter		8	1		415	61.50		476.50	555
2160	Box connector with set screw, plain, 1/2" diameter		70	.114		2.22	7		9.22	13.25
2170	3/4" diameter		60	.133		3.94	8.20		12.14	16.95
2180	1" diameter		50	.160		7	9.80		16.80	23
2190	Insulated, 1-1/4" diameter		40	.200		10.10	12.25		22.35	30
2200	1-1/2" diameter		30	.267		14.20	16.35		30.55	40.50
2210	2" diameter		20	.400		29.50	24.50		54	70.50
2220	2-1/2" diameter		18	.444		87	27.50		114.50	138
2230	3" diameter		15	.533		140	32.50		172.50	205
2240	3-1/2" diameter		12	.667		173	41		214	253
2250	4" diameter		10	.800		180	49		229	274
2260	5" diameter		9	.889		395	54.50		449.50	520
2270	6" diameter		8	1		405	61.50		466.50	540
2280	LB, LR or LL fittings & covers, 1/2" diameter		16	.500		7.35	30.50		37.85	55
2290	3/4" diameter		13	.615		8.80	38		46.80	67.50
2300	1" diameter		11	.727		13.25	44.50		57.75	83
2330	1-1/4" diameter		8	1		26	61.50		87.50	123
2350	1-1/2" diameter		6	1.333		25	82		107	154
2370	2" diameter		5	1.600		61	98		159	218
2380	2-1/2" diameter		4	2		113	123		236	315
2390	3" diameter		3.50	2.286		149	140		289	380
2400	3-1/2" diameter		3	2.667		183	164		347	455
2410	4" diameter		2.50	3.200		222	196		418	545
2420	T fittings, with cover, 1/2" diameter		12	.667		10.20	41		51.20	74
2430	3/4" diameter		11	.727		11.10	44.50		55.60	80.50
2440	1" diameter		9	.889		17.05	54.50		71.55	103
2450	1-1/4" diameter		6	1.333		22	82		104	150
2470	1-1/2" diameter		5	1.600		38.50	98		136.50	194
2500	2" diameter		4	2		40	123		163	233
2510	2-1/2" diameter		3.50	2.286		92.50	140		232.50	320
2520	3" diameter		3	2.667		122	164		286	385
2530	3-1/2" diameter		2.50	3.200		225	196		421	545
2540	4" diameter		2	4		242	245		487	645
2550	Nipples chase, plain, 1/2" diameter		40	.200		.38	12.25		12.63	19.30
2560	3/4" diameter		32	.250		1.49	15.35		16.84	25
2570	1" diameter		28	.286		2.14	17.55		19.69	29.50
2600	Insulated, 1-1/4" diameter		24	.333		7.90	20.50		28.40	40

26 05 33.25 Conduit Fittings for Rigid Galvanized Steel		Crew	Daily Output	Labor-Hours	Unit	Material	Labor	Equipment	Total	Total Incl O&P
						2020 Bare Costs				
2630	1-1/2" diameter	1 Elec	18	.444	Ea.	12.35	27.50		39.85	55.50
2650	2" diameter		15	.533		13.85	32.50		46.35	66
2660	2-1/2" diameter		12	.667		41	41		82	108
2670	3" diameter		10	.800		39	49		88	119
2680	3-1/2" diameter		9	.889		62.50	54.50		117	153
2690	4" diameter		8	1		86	61.50		147.50	189
2700	5" diameter		7	1.143		281	70		351	420
2710	6" diameter		6	1.333		440	82		522	610
2720	Nipples offset, plain, 1/2" diameter		40	.200		5.75	12.25		18	25.50
2730	3/4" diameter		32	.250		5.75	15.35		21.10	30
2740	1" diameter		24	.333		8.70	20.50		29.20	41
2750	Insulated, 1-1/4" diameter		20	.400		7.25	24.50		31.75	46
2760	1-1/2" diameter		18	.444		8.80	27.50		36.30	51.50
2770	2" diameter		16	.500		12.50	30.50		43	61
2780	3" diameter		14	.571		82.50	35		117.50	145
2850	Coupling, expansion, 1/2" diameter		12	.667		49.50	41		90.50	118
2880	3/4" diameter		10	.800		58	49		107	140
2900	1" diameter		8	1		76.50	61.50		138	179
2920	1-1/4" diameter		6.40	1.250		96.50	76.50		173	224
2940	1-1/2" diameter		5.30	1.509		132	92.50		224.50	288
2960	2" diameter		4.60	1.739		192	107		299	375
2980	2-1/2" diameter		3.60	2.222		320	136		456	565
3000	3" diameter		3	2.667		375	164		539	660
3020	3-1/2" diameter		2.80	2.857		560	175		735	885
3040	4" diameter		2.40	3.333		670	205		875	1,050
3060	5" diameter		2	4		1,050	245		1,295	1,525
3080	6" diameter		1.80	4.444		1,800	273		2,073	2,400
3100	Expansion deflection, 1/2" diameter		12	.667		264	41		305	355
3120	3/4" diameter		12	.667		297	41		338	390
3140	1" diameter		10	.800		345	49		394	455
3160	1-1/4" diameter		6.40	1.250		550	76.50		626.50	725
3180	1-1/2" diameter		5.30	1.509		665	92.50		757.50	875
3200	2" diameter		4.60	1.739		785	107		892	1,025
3220	2-1/2" diameter		3.60	2.222		875	136		1,011	1,175
3240	3" diameter		3	2.667		1,325	164		1,489	1,725
3260	3-1/2" diameter		2.80	2.857		1,325	175		1,500	1,725
3280	4" diameter		2.40	3.333		1,500	205		1,705	1,975
3300	5" diameter		2	4		2,850	245		3,095	3,525
3320	6" diameter		1.80	4.444		4,100	273		4,373	4,925
3340	Ericson, 1/2" diameter		16	.500		3.78	30.50		34.28	51
3360	3/4" diameter		14	.571		4.60	35		39.60	59
3380	1" diameter		11	.727		7.65	44.50		52.15	77
3400	1-1/4" diameter		8	1		14.75	61.50		76.25	111
3420	1-1/2" diameter		7	1.143		18.95	70		88.95	129
3440	2" diameter		5	1.600		35.50	98		133.50	191
3460	2-1/2" diameter		4	2		64.50	123		187.50	260
3480	3" diameter		3.50	2.286		105	140		245	330
3500	3-1/2" diameter		3	2.667		138	164		302	405
3520	4" diameter		2.70	2.963		161	182		343	455
3540	5" diameter		2.50	3.200		330	196		526	665
3560	6" diameter		2.30	3.478		420	213		633	795
3580	Split, 1/2" diameter		32	.250		4.20	15.35		19.55	28
3600	3/4" diameter		27	.296		5.15	18.20		23.35	33.50

26 05 33.25 Conduit Fittings for Rigid Galvanized Steel		Crew	Daily Output	Labor-Hours	Unit	Material	Labor	Equipment	Total	Total Incl O&P
						2020 Bare Costs				
3620	1" diameter	1 Elec	20	.400	Ea.	10.85	24.50		35.35	50
3640	1-1/4" diameter		16	.500		14.10	30.50		44.60	62.50
3660	1-1/2" diameter		14	.571		14.90	35		49.90	70.50
3680	2" diameter		12	.667		30.50	41		71.50	96.50
3700	2-1/2" diameter		10	.800		79	49		128	162
3720	3" diameter		9	.889		121	54.50		175.50	217
3740	3-1/2" diameter		8	1		189	61.50		250.50	300
3760	4" diameter		7	1.143		184	70		254	310
3780	5" diameter		6	1.333		370	82		452	535
3800	6" diameter		5	1.600		495	98		593	695
4600	Reducing bushings, 3/4" to 1/2" diameter		54	.148		2.45	9.10		11.55	16.70
4620	1" to 3/4" diameter		46	.174		2.84	10.65		13.49	19.55
4640	1-1/4" to 1" diameter		40	.200		4.77	12.25		17.02	24
4660	1-1/2" to 1-1/4" diameter		36	.222		5.90	13.65		19.55	27.50
4680	2" to 1-1/2" diameter		32	.250		13	15.35		28.35	38
4740	2-1/2" to 2" diameter		30	.267		16.15	16.35		32.50	43
4760	3" to 2-1/2" diameter		28	.286		19.55	17.55		37.10	48.50
4800	Through-wall seal, 1/2" diameter		8	1		66	61.50		127.50	167
4820	3/4" diameter		7.50	1.067		281	65.50		346.50	410
4840	1" diameter		6.50	1.231		274	75.50		349.50	415
4860	1-1/4" diameter		5.50	1.455		289	89		378	455
4880	1-1/2" diameter		5	1.600		410	98		508	605
4900	2" diameter		4.20	1.905		291	117		408	500
4920	2-1/2" diameter		3.50	2.286		525	140		665	790
4940	3" diameter		3	2.667		560	164		724	865
4960	3-1/2" diameter		2.50	3.200		1,075	196		1,271	1,475
4980	4" diameter		2	4		840	245		1,085	1,300
5000	5" diameter		1.50	5.333		825	325		1,150	1,425
5020	6" diameter		1	8		825	490		1,315	1,675
5100	Cable supports, 2 or more wires									
5120	1-1/2" diameter	1 Elec	8	1	Ea.	164	61.50		225.50	275
5140	2" diameter		6	1.333		205	82		287	350
5160	2-1/2" diameter		4	2		220	123		343	430
5180	3" diameter		3.50	2.286		284	140		424	525
5200	3-1/2" diameter		2.60	3.077		410	189		599	740
5220	4" diameter		2	4		465	245		710	890
5240	5" diameter		1.50	5.333		640	325		965	1,200
5260	6" diameter		1	8		1,025	490		1,515	1,875
5280	Service entrance cap, 1/2" diameter		16	.500		5.40	30.50		35.90	53
5300	3/4" diameter		13	.615		5.90	38		43.90	64.50
5320	1" diameter		10	.800		4.79	49		53.79	81
5340	1-1/4" diameter		8	1		4.85	61.50		66.35	100
5360	1-1/2" diameter		6.50	1.231		9.70	75.50		85.20	127
5380	2" diameter		5.50	1.455		21.50	89		110.50	161
5400	2-1/2" diameter		4	2		82	123		205	280
5420	3" diameter		3.40	2.353		89	144		233	320
5440	3-1/2" diameter		3	2.667		141	164		305	405
5460	4" diameter		2.70	2.963		186	182		368	485
5750	90° pull elbows steel, female, 1/2" diameter		16	.500		7.90	30.50		38.40	55.50
5760	3/4" diameter		13	.615		9.95	38		47.95	69
5780	1" diameter		11	.727		14.80	44.50		59.30	85
5800	1-1/4" diameter		8	1		24	61.50		85.50	121
5820	1-1/2" diameter		6	1.333		33	82		115	163

26 05 33 – Raceway and Boxes for Electrical Systems

26 05 33.25 Conduit Fittings for Rigid Galvanized Steel		Crew	Daily Output	Labor-Hours	Unit	Material	Labor	Equipment	Total	Total Incl O&P
						2020 Bare Costs				
5840	2" diameter	1 Elec	5	1.600	Ea.	53.50	98		151.50	210
6000	Explosion proof, flexible coupling									
6010	1/2" diameter, 4" long	1 Elec	12	.667	Ea.	134	41		175	210
6020	6" long		12	.667		122	41		163	197
6050	12" long		12	.667		165	41		206	245
6070	18" long		12	.667		213	41		254	297
6090	24" long		12	.667		287	41		328	380
6110	30" long		12	.667		272	41		313	360
6130	36" long		12	.667		305	41		346	400
6140	3/4" diameter, 4" long		10	.800		141	49		190	232
6150	6" long		10	.800		150	49		199	241
6180	12" long		10	.800		195	49		244	290
6200	18" long		10	.800		254	49		303	355
6220	24" long		10	.800		310	49		359	415
6240	30" long		10	.800		310	49		359	415
6260	36" long		10	.800		385	49		434	495
6270	1" diameter, 6" long		8	1		285	61.50		346.50	410
6300	12" long		8	1		365	61.50		426.50	495
6320	18" long		8	1		410	61.50		471.50	545
6340	24" long		8	1		615	61.50		676.50	770
6360	30" long		8	1		750	61.50		811.50	920
6380	36" long		8	1		670	61.50		731.50	835
6390	1-1/4" diameter, 12" long		6.40	1.250		490	76.50		566.50	660
6410	18" long		6.40	1.250		605	76.50		681.50	785
6430	24" long		6.40	1.250		795	76.50		871.50	995
6450	30" long		6.40	1.250		815	76.50		891.50	1,025
6470	36" long		6.40	1.250		1,350	76.50		1,426.50	1,625
6480	1-1/2" diameter, 12" long		5.30	1.509		650	92.50		742.50	860
6500	18" long		5.30	1.509		780	92.50		872.50	1,000
6520	24" long		5.30	1.509		955	92.50		1,047.50	1,200
6540	30" long		5.30	1.509		1,475	92.50		1,567.50	1,775
6560	36" long		5.30	1.509		1,200	92.50		1,292.50	1,475
6570	2" diameter, 12" long		4.60	1.739		880	107		987	1,125
6590	18" long		4.60	1.739		1,125	107		1,232	1,400
6610	24" long		4.60	1.739		1,200	107		1,307	1,500
6630	30" long		4.60	1.739		1,425	107		1,532	1,750
6650	36" long		4.60	1.739		1,625	107		1,732	1,950
7000	Close up plug, 1/2" diameter, explosion proof		40	.200		2.48	12.25		14.73	21.50
7010	3/4" diameter		32	.250		2.92	15.35		18.27	26.50
7020	1" diameter		28	.286		3.44	17.55		20.99	31
7030	1-1/4" diameter		24	.333		3.83	20.50		24.33	35.50
7040	1-1/2" diameter		18	.444		5.05	27.50		32.55	47.50
7050	2" diameter		15	.533		8.65	32.50		41.15	60
7060	2-1/2" diameter		13	.615		13.20	38		51.20	72.50
7070	3" diameter		12	.667		20	41		61	85
7080	3-1/2" diameter		11	.727		23.50	44.50		68	94
7090	4" diameter		9	.889		40	54.50		94.50	128
7091	Elbow female, 45°, 1/2"		16	.500		15.75	30.50		46.25	64.50
7092	3/4"		13	.615		20.50	38		58.50	80.50
7093	1"		11	.727		25	44.50		69.50	96
7094	1-1/4"		8	1		31.50	61.50		93	129
7095	1-1/2"		6	1.333		25.50	82		107.50	155
7096	2"	↓	5	1.600	↓	39	98		137	194

26 05 33.25 Conduit Fittings for Rigid Galvanized Steel		Crew	Daily Output	Labor-Hours	Unit	2020 Bare Costs Material	Labor	Equipment	Total	Total Incl O&P
7097	2-1/2"	1 Elec	4.50	1.778	Ea.	102	109		211	280
7098	3"		4.20	1.905		117	117		234	310
7099	3-1/2"		4	2		177	123		300	385
7100	4"		3.80	2.105		217	129		346	440
7101	90°, 1/2"		16	.500		15.40	30.50		45.90	64
7102	3/4"		13	.615		18.25	38		56.25	78
7103	1"		11	.727		22.50	44.50		67	93.50
7104	1-1/4"		8	1		34	61.50		95.50	132
7105	1-1/2"		6	1.333		51	82		133	182
7106	2"		5	1.600		96	98		194	256
7107	2-1/2"		4.50	1.778		170	109		279	355
7110	Elbows 90°, long male & female, 1/2" diameter, explosion proof		16	.500		22.50	30.50		53	72
7120	3/4" diameter		13	.615		25	38		63	85.50
7130	1" diameter		11	.727		33.50	44.50		78	106
7140	1-1/4" diameter		8	1		52	61.50		113.50	152
7150	1-1/2" diameter		6	1.333		43.50	82		125.50	174
7160	2" diameter		5	1.600		59	98		157	216
7170	Capped elbow, 1/2" diameter, explosion proof		11	.727		22.50	44.50		67	93.50
7180	3/4" diameter		8	1		30.50	61.50		92	128
7190	1" diameter		6	1.333		33	82		115	162
7200	1-1/4" diameter		5	1.600		71.50	98		169.50	230
7210	Pulling elbow, 1/2" diameter, explosion proof		11	.727		140	44.50		184.50	223
7220	3/4" diameter		8	1		192	61.50		253.50	305
7230	1" diameter		6	1.333		244	82		326	395
7240	1-1/4" diameter		5	1.600		315	98		413	500
7250	1-1/2" diameter		5	1.600		390	98		488	575
7260	2" diameter		4	2		455	123		578	690
7270	2-1/2" diameter		3.50	2.286		765	140		905	1,050
7280	3" diameter		3	2.667		1,200	164		1,364	1,575
7290	3-1/2" diameter		2.50	3.200		1,625	196		1,821	2,075
7300	4" diameter		2.20	3.636		1,600	223		1,823	2,125
7310	LB conduit body, 1/2" diameter		11	.727		59.50	44.50		104	134
7320	3/4" diameter		8	1		79	61.50		140.50	181
7330	T conduit body, 1/2" diameter		9	.889		60	54.50		114.50	150
7340	3/4" diameter		6	1.333		75	82		157	209
7350	Explosion proof, round box w/cover, 3 threaded hubs, 1/2" diameter		8	1		58.50	61.50		120	159
7351	3/4" diameter		8	1		78.50	61.50		140	181
7352	1" diameter		7.50	1.067		84	65.50		149.50	194
7353	1-1/4" diameter		7	1.143		88	70		158	205
7354	1-1/2" diameter		7	1.143		237	70		307	370
7355	2" diameter		6	1.333		236	82		318	385
7356	Round box w/cover & mtng flange, 3 threaded hubs, 1/2" diameter		8	1		61.50	61.50		123	162
7357	3/4" diameter		8	1		55.50	61.50		117	156
7358	4 threaded hubs, 1" diameter		7	1.143		66.50	70		136.50	182
7400	Unions, 1/2" diameter		20	.400		12.10	24.50		36.60	51.50
7410	3/4" - 1/2" diameter		16	.500		19.30	30.50		49.80	68.50
7420	3/4" diameter		16	.500		20.50	30.50		51	69.50
7430	1" diameter		14	.571		37.50	35		72.50	95
7440	1-1/4" diameter		12	.667		56.50	41		97.50	125
7450	1-1/2" diameter		10	.800		70.50	49		119.50	154
7460	2" diameter		8.50	.941		93.50	57.50		151	192
7480	2-1/2" diameter		8	1		141	61.50		202.50	250
7490	3" diameter		7	1.143		207	70		277	335

26 05 33.25 Conduit Fittings for Rigid Galvanized Steel		Crew	Daily Output	Labor-Hours	Unit	Material	Labor	Equipment	Total	Total Incl O&P
						2020 Bare Costs				
7500	3-1/2" diameter	1 Elec	6	1.333	Ea.	400	82		482	565
7510	4" diameter		5	1.600		370	98		468	555
7680	Reducer, 3/4" to 1/2"		54	.148		5.40	9.10		14.50	19.95
7690	1" to 1/2"		46	.174		5.15	10.65		15.80	22
7700	1" to 3/4"		46	.174		6.50	10.65		17.15	23.50
7710	1-1/4" to 3/4"		40	.200		8.40	12.25		20.65	28
7720	1-1/4" to 1"		40	.200		10	12.25		22.25	30
7730	1-1/2" to 1"		36	.222		12.80	13.65		26.45	35
7740	1-1/2" to 1-1/4"		36	.222		12.90	13.65		26.55	35
7750	2" to 3/4"		32	.250		18.15	15.35		33.50	43.50
7760	2" to 1-1/4"		32	.250		20.50	15.35		35.85	46
7770	2" to 1-1/2"		32	.250		26	15.35		41.35	52.50
7780	2-1/2" to 1-1/2"		30	.267		22.50	16.35		38.85	49.50
7790	3" to 2"		30	.267		21.50	16.35		37.85	48.50
7800	3-1/2" to 2-1/2"		28	.286		67.50	17.55		85.05	102
7810	4" to 3"		28	.286		70	17.55		87.55	104
7820	Sealing fitting, vertical/horizontal, 1/2" diameter		14.50	.552		19.35	34		53.35	73.50
7830	3/4" diameter		13.30	.602		22.50	37		59.50	82
7840	1" diameter		11.40	.702		28	43		71	97.50
7850	1-1/4" diameter		10	.800		28	49		77	107
7860	1-1/2" diameter		8.80	.909		45	56		101	136
7870	2" diameter		8	1		55.50	61.50		117	156
7880	2-1/2" diameter		6.70	1.194		82	73.50		155.50	204
7890	3" diameter		5.70	1.404		112	86		198	256
7900	3-1/2" diameter		4.70	1.702		293	104		397	480
7910	4" diameter		4	2		490	123		613	725
7920	Sealing hubs, 1" by 1-1/2"		12	.667		28.50	41		69.50	94.50
7930	1-1/4" by 2"		10	.800		54	49		103	135
7940	1-1/2" by 2"		9	.889		74.50	54.50		129	166
7950	2" by 2-1/2"		8	1		99.50	61.50		161	204
7960	3" by 4"		7	1.143		178	70		248	305
7970	4" by 5"		6	1.333		380	82		462	540
7980	Drain, 1/2"		32	.250		110	15.35		125.35	145
7990	Breather, 1/2"		32	.250		118	15.35		133.35	154
7991	Explosion proof sealant compound, hub, fittings, 60 min set time, 1lb. pail		48	.167		22	10.25		32.25	40
7992	5 lb. pail		48	.167		62.50	10.25		72.75	85
7993	2 oz. tube		50	.160		67.50	9.80		77.30	89.50
7994	6 oz. tube		50	.160		124	9.80		133.80	152
7995	2.0 oz. cartridge		50	.160		44.50	9.80		54.30	64
7996	6.0 oz. cartridge		50	.160		87	9.80		96.80	111
7997	2.0 oz. box		50	.160		29	9.80		38.80	47
7998	8.0 oz. box		50	.160		89	9.80		98.80	113
7999	1.0 lb. box		50	.160		147	9.80		156.80	176
8000	Plastic coated 40 mil thick									
8010	LB, LR or LL conduit body w/cover, 1/2" diameter	1 Elec	13	.615	Ea.	54.50	38		92.50	118
8020	3/4" diameter		11	.727		70	44.50		114.50	146
8030	1" diameter		8	1		83	61.50		144.50	186
8040	1-1/4" diameter		6	1.333		123	82		205	261
8050	1-1/2" diameter		5	1.600		129	98		227	293
8060	2" diameter		4.50	1.778		219	109		328	410
8070	2-1/2" diameter		4	2		430	123		553	665
8080	3" diameter		3.50	2.286		430	140		570	685
8090	3-1/2" diameter		3	2.667		775	164		939	1,100

26 05 33.25 Conduit Fittings for Rigid Galvanized Steel		Crew	Daily Output	Labor-Hours	Unit	Material	Labor	Equipment	Total	Total Incl O&P
						2020 Bare Costs				
8100	4" diameter	1 Elec	2.50	3.200	Ea.	890	196		1,086	1,275
8150	T conduit body with cover, 1/2" diameter		11	.727		64	44.50		108.50	139
8160	3/4" diameter		9	.889		82	54.50		136.50	174
8170	1" diameter		6	1.333		96	82		178	231
8180	1-1/4" diameter		5	1.600		144	98		242	310
8190	1-1/2" diameter		4.50	1.778		163	109		272	345
8200	2" diameter		4	2		209	123		332	420
8210	2-1/2" diameter		3.50	2.286		405	140		545	660
8220	3" diameter		3	2.667		550	164		714	855
8230	3-1/2" diameter		2.50	3.200		950	196		1,146	1,350
8240	4" diameter		2	4		925	245		1,170	1,400
8300	FS conduit body, 1 gang, 3/4" diameter		11	.727		69.50	44.50		114	145
8310	1" diameter		10	.800		59	49		108	141
8350	2 gang, 3/4" diameter		9	.889		106	54.50		160.50	201
8360	1" diameter		8	1		154	61.50		215.50	264
8400	Duplex receptacle cover		64	.125		52	7.65		59.65	69
8410	Switch cover		64	.125		62	7.65		69.65	80
8420	Switch, vaportight cover		53	.151		189	9.25		198.25	222
8430	Blank, cover		64	.125		41	7.65		48.65	57
8520	FSC conduit body, 1 gang, 3/4" diameter		10	.800		68	49		117	151
8530	1" diameter		9	.889		70	54.50		124.50	162
8550	2 gang, 3/4" diameter		8	1		118	61.50		179.50	225
8560	1" diameter		7	1.143		163	70		233	288
8590	Conduit hubs, 1/2" diameter		18	.444		40.50	27.50		68	86.50
8600	3/4" diameter		16	.500		46	30.50		76.50	97.50
8610	1" diameter		14	.571		59	35		94	119
8620	1-1/4" diameter		12	.667		71	41		112	141
8630	1-1/2" diameter		10	.800		79	49		128	163
8640	2" diameter		8.80	.909		111	56		167	208
8650	2-1/2" diameter		8.50	.941		152	57.50		209.50	257
8660	3" diameter		8	1		238	61.50		299.50	355
8670	3-1/2" diameter		7.50	1.067		330	65.50		395.50	465
8680	4" diameter		7	1.143		325	70		395	470
8690	5" diameter	▼	6	1.333	▼	395	82		477	560
8700	Plastic coated 40 mil thick									
8710	Pipe strap, stamped 1 hole, 1/2" diameter	1 Elec	470	.017	Ea.	14.50	1.04		15.54	17.55
8720	3/4" diameter		440	.018		12.90	1.12		14.02	15.85
8730	1" diameter		400	.020		13.85	1.23		15.08	17.10
8740	1-1/4" diameter		355	.023		22.50	1.38		23.88	26.50
8750	1-1/2" diameter		320	.025		23	1.53		24.53	27.50
8760	2" diameter		266	.030		30.50	1.85		32.35	36.50
8770	2-1/2" diameter		200	.040		32	2.45		34.45	39
8780	3" diameter		133	.060		43.50	3.69		47.19	53.50
8790	3-1/2" diameter		110	.073		100	4.46		104.46	117
8800	4" diameter		90	.089		104	5.45		109.45	122
8810	5" diameter		70	.114		142	7		149	167
8840	Clamp back spacers, 3/4" diameter		440	.018		19.85	1.12		20.97	23.50
8850	1" diameter		400	.020		27	1.23		28.23	32
8860	1-1/4" diameter		355	.023		26	1.38		27.38	31
8870	1-1/2" diameter		320	.025		39	1.53		40.53	45.50
8880	2" diameter		266	.030		59.50	1.85		61.35	68.50
8900	3" diameter		133	.060		88	3.69		91.69	103
8920	4" diameter	▼	90	.089	▼	71.50	5.45		76.95	87.50

26 05 33.25 Conduit Fittings for Rigid Galvanized Steel

	26 05 33.25 Conduit Fittings for Rigid Galvanized Steel	Crew	Daily Output	Labor-Hours	Unit	Material	2020 Bare Costs Labor	Equipment	Total	Total Incl O&P
8950	Touch-up plastic coating, spray, 12 oz.				Ea.	66.50			66.50	73.50
8960	Sealing fittings, 1/2" diameter	1 Elec	11	.727		66.50	44.50		111	142
8970	3/4" diameter		9	.889		100	54.50		154.50	194
8980	1" diameter		7.50	1.067		75.50	65.50		141	185
8990	1-1/4" diameter		6.50	1.231		114	75.50		189.50	242
9000	1-1/2" diameter		5.50	1.455		140	89		229	291
9010	2" diameter		4.80	1.667		160	102		262	335
9020	2-1/2" diameter		4	2		290	123		413	510
9030	3" diameter		3.50	2.286		291	140		431	535
9040	3-1/2" diameter		3	2.667		715	164		879	1,025
9050	4" diameter		2.50	3.200		1,100	196		1,296	1,525
9060	5" diameter		1.70	4.706		2,025	289		2,314	2,675
9070	Unions, 1/2" diameter		18	.444		53.50	27.50		81	101
9080	3/4" diameter		15	.533		68.50	32.50		101	126
9090	1" diameter		13	.615		71	38		109	136
9100	1-1/4" diameter		11	.727		123	44.50		167.50	204
9110	1-1/2" diameter		9.50	.842		185	51.50		236.50	284
9120	2" diameter		8	1		219	61.50		280.50	335
9130	2-1/2" diameter		7.50	1.067		288	65.50		353.50	415
9140	3" diameter		6.80	1.176		320	72		392	465
9150	3-1/2" diameter		5.80	1.379		485	84.50		569.50	665
9160	4" diameter		4.80	1.667		590	102		692	805
9170	5" diameter		4	2		930	123		1,053	1,225

26 05 33.30 Electrical Nonmetallic Tubing (ENT)

	26 05 33.30 Electrical Nonmetallic Tubing (ENT)	Crew	Daily Output	Labor-Hours	Unit	Material	Labor	Equipment	Total	Total Incl O&P
0010	**ELECTRICAL NONMETALLIC TUBING (ENT)**									
0050	Flexible, 1/2" diameter	1 Elec	270	.030	L.F.	.55	1.82		2.37	3.41
0100	3/4" diameter		230	.035		.70	2.13		2.83	4.06
0200	1" diameter		145	.055		1.47	3.38		4.85	6.80
0210	1-1/4" diameter		125	.064		1.42	3.93		5.35	7.60
0220	1-1/2" diameter		100	.080		2.46	4.91		7.37	10.25
0230	2" diameter		75	.107		2.95	6.55		9.50	13.35
0300	Connectors, to outlet box, 1/2" diameter		230	.035	Ea.	1.02	2.13		3.15	4.41
0310	3/4" diameter		210	.038		1.92	2.34		4.26	5.70
0320	1" diameter		200	.040		2.40	2.45		4.85	6.40
0400	Couplings, to conduit, 1/2" diameter		145	.055		1	3.38		4.38	6.30
0410	3/4" diameter		130	.062		1.16	3.78		4.94	7.10
0420	1" diameter		125	.064		2.19	3.93		6.12	8.45

26 05 33.35 Flexible Metallic Conduit

	26 05 33.35 Flexible Metallic Conduit	Crew	Daily Output	Labor-Hours	Unit	Material	Labor	Equipment	Total	Total Incl O&P
0010	**FLEXIBLE METALLIC CONDUIT**									
0050	Steel, 3/8" diameter	1 Elec	200	.040	L.F.	.38	2.45		2.83	4.20
0100	1/2" diameter		200	.040		.45	2.45		2.90	4.28
0200	3/4" diameter		160	.050		.61	3.07		3.68	5.40
0250	1" diameter		100	.080		1.20	4.91		6.11	8.85
0300	1-1/4" diameter		70	.114		1.39	7		8.39	12.35
0350	1-1/2" diameter		50	.160		2.55	9.80		12.35	17.90
0370	2" diameter		40	.200		3.23	12.25		15.48	22.50
0380	2-1/2" diameter		30	.267		3.50	16.35		19.85	29
0390	3" diameter	2 Elec	50	.320		6	19.65		25.65	36.50
0400	3-1/2" diameter		40	.400		6.75	24.50		31.25	45.50
0410	4" diameter		30	.533		7.75	32.50		40.25	59
0420	Connectors, plain, 3/8" diameter	1 Elec	100	.080	Ea.	1.86	4.91		6.77	9.60
0430	1/2" diameter		80	.100		2.02	6.15		8.17	11.65

26 05 33.35 Flexible Metallic Conduit		Crew	Daily Output	Labor-Hours	Unit	2020 Bare Costs Material	Labor	Equipment	Total	Total Incl O&P
0440	3/4" diameter	1 Elec	70	.114	Ea.	2.12	7		9.12	13.15
0450	1" diameter		50	.160		4.40	9.80		14.20	19.95
0452	1-1/4" diameter		45	.178		5.70	10.90		16.60	23
0454	1-1/2" diameter		40	.200		8.35	12.25		20.60	28
0456	2" diameter		28	.286		12.10	17.55		29.65	40.50
0458	2-1/2" diameter		25	.320		20	19.65		39.65	52
0460	3" diameter		20	.400		28.50	24.50		53	69
0462	3-1/2" diameter		16	.500		91.50	30.50		122	148
0464	4" diameter		13	.615		101	38		139	169
0490	Insulated, 1" diameter		40	.200		8.20	12.25		20.45	28
0500	1-1/4" diameter		40	.200		13.20	12.25		25.45	33.50
0550	1-1/2" diameter		32	.250		24.50	15.35		39.85	50
0600	2" diameter		23	.348		26	21.50		47.50	61.50
0610	2-1/2" diameter		20	.400		57.50	24.50		82	101
0620	3" diameter		17	.471		117	29		146	173
0630	3-1/2" diameter		13	.615		282	38		320	370
0640	4" diameter		10	.800		300	49		349	405
0650	Connectors 90°, plain, 3/8" diameter		80	.100		2.93	6.15		9.08	12.65
0660	1/2" diameter		60	.133		5.35	8.20		13.55	18.50
0700	3/4" diameter		50	.160		7.40	9.80		17.20	23.50
0750	1" diameter		40	.200		12.05	12.25		24.30	32
0790	Insulated, 1" diameter		40	.200		17.35	12.25		29.60	38
0800	1-1/4" diameter		30	.267		23.50	16.35		39.85	51
0850	1-1/2" diameter		23	.348		46	21.50		67.50	84
0900	2" diameter		18	.444		67	27.50		94.50	116
0910	2-1/2" diameter		16	.500		162	30.50		192.50	225
0920	3" diameter		14	.571		211	35		246	287
0930	3-1/2" diameter		11	.727		550	44.50		594.50	675
0940	4" diameter		8	1		685	61.50		746.50	850
0960	Couplings, to flexible conduit, 1/2" diameter		50	.160		.92	9.80		10.72	16.10
0970	3/4" diameter		40	.200		1.53	12.25		13.78	20.50
0980	1" diameter		35	.229		2.49	14		16.49	24
0990	1-1/4" diameter		28	.286		6.05	17.55		23.60	33.50
1000	1-1/2" diameter		23	.348		7.15	21.50		28.65	41
1010	2" diameter		20	.400		14.25	24.50		38.75	53.50
1020	2-1/2" diameter		18	.444		37.50	27.50		65	83.50
1030	3" diameter		15	.533		95	32.50		127.50	156
1032	Aluminum, 3/8" diameter		210	.038	L.F.	.43	2.34		2.77	4.07
1034	1/2" diameter		210	.038		.58	2.34		2.92	4.24
1036	3/4" diameter		165	.048		.83	2.97		3.80	5.50
1038	1" diameter		105	.076		1.49	4.67		6.16	8.85
1040	1-1/4" diameter		75	.107		1.73	6.55		8.28	12
1042	1-1/2" diameter		53	.151		3.03	9.25		12.28	17.60
1044	2" diameter		42	.190		4.37	11.70		16.07	23
1046	2-1/2" diameter		32	.250		4.22	15.35		19.57	28
1048	3" diameter	2 Elec	53	.302		7.35	18.50		25.85	36.50
1050	3-1/2" diameter		42	.381		6	23.50		29.50	42.50
1052	4" diameter		32	.500		8.85	30.50		39.35	56.50
1070	Sealtite, 3/8" diameter	1 Elec	140	.057		.64	3.51		4.15	6.10
1080	1/2" diameter		140	.057		1.08	3.51		4.59	6.60
1090	3/4" diameter		100	.080		1.52	4.91		6.43	9.20
1100	1" diameter		70	.114		2.23	7		9.23	13.25
1200	1-1/4" diameter		50	.160		2.98	9.80		12.78	18.40

26 05 33.35 Flexible Metallic Conduit

	26 05 33.35 Flexible Metallic Conduit	Crew	Daily Output	Labor-Hours	Unit	2020 Bare Costs Material	Labor	Equipment	Total	Total Incl O&P
1300	1-1/2" diameter	1 Elec	40	.200	L.F.	3.32	12.25		15.57	22.50
1400	2" diameter		30	.267		4.30	16.35		20.65	29.50
1410	2-1/2" diameter		27	.296		7.65	18.20		25.85	36.50
1420	3" diameter	2 Elec	50	.320		9.75	19.65		29.40	41
1440	4" diameter	"	30	.533		20	32.50		52.50	72.50
1490	Connectors, plain, 3/8" diameter	1 Elec	70	.114	Ea.	2.30	7		9.30	13.35
1500	1/2" diameter		70	.114		3.02	7		10.02	14.10
1700	3/4" diameter		50	.160		4.33	9.80		14.13	19.85
1900	1" diameter		40	.200		6.90	12.25		19.15	26.50
1910	Insulated, 1" diameter		40	.200		9.50	12.25		21.75	29.50
2000	1-1/4" diameter		32	.250		11.30	15.35		26.65	36
2100	1-1/2" diameter		27	.296		21	18.20		39.20	51
2200	2" diameter		20	.400		48	24.50		72.50	91
2210	2-1/2" diameter		15	.533		103	32.50		135.50	165
2220	3" diameter		12	.667		147	41		188	225
2240	4" diameter		8	1		194	61.50		255.50	310
2290	Connectors, 90°, 3/8" diameter		70	.114		4.95	7		11.95	16.25
2300	1/2" diameter		70	.114		6.65	7		13.65	18.10
2400	3/4" diameter		50	.160		7.20	9.80		17	23
2600	1" diameter		40	.200		14.10	12.25		26.35	34.50
2790	Insulated, 1" diameter		40	.200		11.90	12.25		24.15	32
2800	1-1/4" diameter		32	.250		13.85	15.35		29.20	38.50
3000	1-1/2" diameter		27	.296		24.50	18.20		42.70	55
3100	2" diameter		20	.400		25.50	24.50		50	66
3110	2-1/2" diameter		14	.571		123	35		158	189
3120	3" diameter		11	.727		345	44.50		389.50	450
3140	4" diameter		7	1.143		163	70		233	287
4300	Coupling sealtite to rigid, 1/2" diameter		20	.400		4.15	24.50		28.65	42.50
4500	3/4" diameter		18	.444		5.85	27.50		33.35	48.50
4800	1" diameter		14	.571		6.60	35		41.60	61.50
4900	1-1/4" diameter		12	.667		16.25	41		57.25	81
5000	1-1/2" diameter		11	.727		20	44.50		64.50	90.50
5100	2" diameter		10	.800		31.50	49		80.50	110
5110	2-1/2" diameter		9.50	.842		147	51.50		198.50	242
5120	3" diameter		9	.889		150	54.50		204.50	249
5130	3-1/2" diameter		9	.889		174	54.50		228.50	275
5140	4" diameter		8.50	.941		198	57.50		255.50	305

26 05 33.95 Cutting and Drilling

	26 05 33.95 Cutting and Drilling	Crew	Daily Output	Labor-Hours	Unit	Material	Labor	Equipment	Total	Total Incl O&P
0010	**CUTTING AND DRILLING**									
0100	Hole drilling to 10' high, concrete wall									
0110	8" thick, 1/2" pipe size	R-31	12	.667	Ea.	.27	41	5.20	46.47	69
0120	3/4" pipe size		12	.667		.27	41	5.20	46.47	69
0130	1" pipe size		9.50	.842		.35	51.50	6.60	58.45	87
0140	1-1/4" pipe size		9.50	.842		.35	51.50	6.60	58.45	87
0150	1-1/2" pipe size		9.50	.842		.35	51.50	6.60	58.45	87
0160	2" pipe size		4.40	1.818		.53	112	14.20	126.73	188
0170	2-1/2" pipe size		4.40	1.818		.53	112	14.20	126.73	188
0180	3" pipe size		4.40	1.818		.53	112	14.20	126.73	188
0190	3-1/2" pipe size		3.30	2.424		.60	149	18.95	168.55	251
0200	4" pipe size		3.30	2.424		.60	149	18.95	168.55	251
0500	12" thick, 1/2" pipe size		9.40	.851		.40	52	6.65	59.05	88
0520	3/4" pipe size		9.40	.851		.40	52	6.65	59.05	88

26 05 33 – Raceway and Boxes for Electrical Systems

26 05 33.95 Cutting and Drilling		Crew	Daily Output	Labor-Hours	Unit	Material	Labor	Equipment	Total	Total Incl O&P
						2020 Bare Costs				
0540	1" pipe size	R-31	7.30	1.096	Ea.	.52	67	8.55	76.07	114
0560	1-1/4" pipe size		7.30	1.096		.52	67	8.55	76.07	114
0570	1-1/2" pipe size		7.30	1.096		.52	67	8.55	76.07	114
0580	2" pipe size		3.60	2.222		.80	136	17.35	154.15	230
0590	2-1/2" pipe size		3.60	2.222		.80	136	17.35	154.15	230
0600	3" pipe size		3.60	2.222		.80	136	17.35	154.15	230
0610	3-1/2" pipe size		2.80	2.857		.90	175	22.50	198.40	295
0630	4" pipe size		2.50	3.200		.90	196	25	221.90	330
0650	16" thick, 1/2" pipe size		7.60	1.053		.54	64.50	8.20	73.24	109
0670	3/4" pipe size		7	1.143		.54	70	8.95	79.49	118
0690	1" pipe size		6	1.333		.69	82	10.40	93.09	138
0710	1-1/4" pipe size		5.50	1.455		.69	89	11.35	101.04	150
0730	1-1/2" pipe size		5.50	1.455		.69	89	11.35	101.04	150
0750	2" pipe size		3	2.667		1.06	164	21	186.06	276
0770	2-1/2" pipe size		2.70	2.963		1.06	182	23	206.06	305
0790	3" pipe size		2.50	3.200		1.06	196	25	222.06	330
0810	3-1/2" pipe size		2.30	3.478		1.20	213	27	241.20	360
0830	4" pipe size		2	4		1.20	245	31	277.20	415
0850	20" thick, 1/2" pipe size		6.40	1.250		.67	76.50	9.75	86.92	129
0870	3/4" pipe size		6	1.333		.67	82	10.40	93.07	138
0890	1" pipe size		5	1.600		.86	98	12.50	111.36	166
0910	1-1/4" pipe size		4.80	1.667		.86	102	13	115.86	172
0930	1-1/2" pipe size		4.60	1.739		.86	107	13.60	121.46	180
0950	2" pipe size		2.70	2.963		1.33	182	23	206.33	305
0970	2-1/2" pipe size		2.40	3.333		1.33	205	26	232.33	345
0990	3" pipe size		2.20	3.636		1.33	223	28.50	252.83	375
1010	3-1/2" pipe size		2	4		1.50	245	31	277.50	415
1030	4" pipe size		1.70	4.706		1.50	289	37	327.50	485
1050	24" thick, 1/2" pipe size		5.50	1.455		.81	89	11.35	101.16	150
1070	3/4" pipe size		5.10	1.569		.81	96	12.25	109.06	162
1090	1" pipe size		4.30	1.860		1.04	114	14.55	129.59	193
1110	1-1/4" pipe size		4	2		1.04	123	15.60	139.64	207
1130	1-1/2" pipe size		4	2		1.04	123	15.60	139.64	207
1150	2" pipe size		2.40	3.333		1.59	205	26	232.59	345
1170	2-1/2" pipe size		2.20	3.636		1.59	223	28.50	253.09	380
1190	3" pipe size		2	4		1.59	245	31	277.59	415
1210	3-1/2" pipe size		1.80	4.444		1.80	273	34.50	309.30	460
1230	4" pipe size		1.50	5.333		1.80	325	41.50	368.30	555
1500	Brick wall, 8" thick, 1/2" pipe size		18	.444		.27	27.50	3.47	31.24	46
1520	3/4" pipe size		18	.444		.27	27.50	3.47	31.24	46
1540	1" pipe size		13.30	.602		.35	37	4.70	42.05	62.50
1560	1-1/4" pipe size		13.30	.602		.35	37	4.70	42.05	62.50
1580	1-1/2" pipe size		13.30	.602		.35	37	4.70	42.05	62.50
1600	2" pipe size		5.70	1.404		.53	86	10.95	97.48	146
1620	2-1/2" pipe size		5.70	1.404		.53	86	10.95	97.48	146
1640	3" pipe size		5.70	1.404		.53	86	10.95	97.48	146
1660	3-1/2" pipe size		4.40	1.818		.60	112	14.20	126.80	188
1680	4" pipe size		4	2		.60	123	15.60	139.20	207
1700	12" thick, 1/2" pipe size		14.50	.552		.40	34	4.31	38.71	57
1720	3/4" pipe size		14.50	.552		.40	34	4.31	38.71	57
1740	1" pipe size		11	.727		.52	44.50	5.70	50.72	75.50
1760	1-1/4" pipe size		11	.727		.52	44.50	5.70	50.72	75.50
1780	1-1/2" pipe size		11	.727		.52	44.50	5.70	50.72	75.50

26 05 33.95 Cutting and Drilling		Crew	Daily Output	Labor-Hours	Unit	Material	Labor	Equipment	Total	Total Incl O&P
						2020 Bare Costs				
1800	2″ pipe size	R-31	5	1.600	Ea.	.80	98	12.50	111.30	166
1820	2-1/2″ pipe size		5	1.600		.80	98	12.50	111.30	166
1840	3″ pipe size		5	1.600		.80	98	12.50	111.30	166
1860	3-1/2″ pipe size		3.80	2.105		.90	129	16.45	146.35	218
1880	4″ pipe size		3.30	2.424		.90	149	18.95	168.85	251
1900	16″ thick, 1/2″ pipe size		12.30	.650		.54	40	5.10	45.64	67.50
1920	3/4″ pipe size		12.30	.650		.54	40	5.10	45.64	67.50
1940	1″ pipe size		9.30	.860		.69	53	6.70	60.39	89.50
1960	1-1/4″ pipe size		9.30	.860		.69	53	6.70	60.39	89.50
1980	1-1/2″ pipe size		9.30	.860		.69	53	6.70	60.39	89.50
2000	2″ pipe size		4.40	1.818		1.06	112	14.20	127.26	189
2010	2-1/2″ pipe size		4.40	1.818		1.06	112	14.20	127.26	189
2030	3″ pipe size		4.40	1.818		1.06	112	14.20	127.26	189
2050	3-1/2″ pipe size		3.30	2.424		1.20	149	18.95	169.15	251
2070	4″ pipe size		3	2.667		1.20	164	21	186.20	276
2090	20″ thick, 1/2″ pipe size		10.70	.748		.67	46	5.85	52.52	77.50
2110	3/4″ pipe size		10.70	.748		.67	46	5.85	52.52	77.50
2130	1″ pipe size		8	1		.86	61.50	7.80	70.16	104
2150	1-1/4″ pipe size		8	1		.86	61.50	7.80	70.16	104
2170	1-1/2″ pipe size		8	1		.86	61.50	7.80	70.16	104
2190	2″ pipe size		4	2		1.33	123	15.60	139.93	208
2210	2-1/2″ pipe size		4	2		1.33	123	15.60	139.93	208
2230	3″ pipe size		4	2		1.33	123	15.60	139.93	208
2250	3-1/2″ pipe size		3	2.667		1.50	164	21	186.50	277
2270	4″ pipe size		2.70	2.963		1.50	182	23	206.50	305
2290	24″ thick, 1/2″ pipe size		9.40	.851		.81	52	6.65	59.46	88.50
2310	3/4″ pipe size		9.40	.851		.81	52	6.65	59.46	88.50
2330	1″ pipe size		7.10	1.127		1.04	69	8.80	78.84	117
2350	1-1/4″ pipe size		7.10	1.127		1.04	69	8.80	78.84	117
2370	1-1/2″ pipe size		7.10	1.127		1.04	69	8.80	78.84	117
2390	2″ pipe size		3.60	2.222		1.59	136	17.35	154.94	231
2410	2-1/2″ pipe size		3.60	2.222		1.59	136	17.35	154.94	231
2430	3″ pipe size		3.60	2.222		1.59	136	17.35	154.94	231
2450	3-1/2″ pipe size		2.80	2.857		1.80	175	22.50	199.30	296
2470	4″ pipe size	▼	2.50	3.200	▼	1.80	196	25	222.80	330
3000	Knockouts to 8′ high, metal boxes & enclosures									
3020	With hole saw, 1/2″ pipe size	1 Elec	53	.151	Ea.		9.25		9.25	14.25
3040	3/4″ pipe size		47	.170			10.45		10.45	16.10
3050	1″ pipe size		40	.200			12.25		12.25	18.90
3060	1-1/4″ pipe size		36	.222			13.65		13.65	21
3070	1-1/2″ pipe size		32	.250			15.35		15.35	23.50
3080	2″ pipe size		27	.296			18.20		18.20	28
3090	2-1/2″ pipe size		20	.400			24.50		24.50	38
4010	3″ pipe size		16	.500			30.50		30.50	47
4030	3-1/2″ pipe size		13	.615			38		38	58
4050	4″ pipe size	▼	11	.727	▼		44.50		44.50	68.50

26 05 36.10 Cable Tray Ladder Type		Crew	Daily Output	Labor-Hours	Unit	Material	Labor	Equipment	Total	Total Incl O&P
						2020 Bare Costs				
0010	**CABLE TRAY LADDER TYPE** w/ftngs. & supports, 4" dp., to 15' elev.									
0100	For higher elevations, see Section 26 05 36.40									
0160	Galvanized steel tray									
0170	4" rung spacing, 6" wide	2 Elec	98	.163	L.F.	14.90	10		24.90	32
0180	9" wide		92	.174		16.30	10.65		26.95	34.50
0200	12" wide		86	.186		17.95	11.40		29.35	37.50
0400	18" wide		82	.195		21	11.95		32.95	41.50
0600	24" wide		78	.205		24	12.60		36.60	46
0650	30" wide		68	.235		30.50	14.45		44.95	55.50
0800	6" rung spacing, 6" wide		100	.160		13.60	9.80		23.40	30
0850	9" wide		94	.170		14.90	10.45		25.35	32.50
0860	12" wide		88	.182		15.65	11.15		26.80	34.50
0870	18" wide		84	.190		17.90	11.70		29.60	37.50
0880	24" wide		80	.200		19.95	12.25		32.20	41
0890	30" wide		70	.229		23.50	14		37.50	47.50
0910	9" rung spacing, 6" wide		102	.157		12.65	9.60		22.25	29
0920	9" wide		98	.163		13	10		23	29.50
0930	12" wide		94	.170		14.90	10.45		25.35	32.50
0940	18" wide		90	.178		16.85	10.90		27.75	35.50
0950	24" wide		86	.186		19.15	11.40		30.55	38.50
0960	30" wide		80	.200		21	12.25		33.25	42
0980	12" rung spacing, 6" wide		106	.151		12.45	9.25		21.70	28
0990	9" wide		104	.154		12.75	9.45		22.20	28.50
1000	12" wide		100	.160		13.50	9.80		23.30	30
1010	18" wide		96	.167		14.50	10.25		24.75	31.50
1020	24" wide		94	.170		15.55	10.45		26	33
1030	30" wide		88	.182		17.65	11.15		28.80	36.50
1041	18" rung spacing, 6" wide		108	.148		12.40	9.10		21.50	27.50
1042	9" wide		106	.151		12.55	9.25		21.80	28
1043	12" wide		102	.157		13.35	9.60		22.95	29.50
1044	18" wide		98	.163		14.05	10		24.05	31
1045	24" wide		96	.167		14.50	10.25		24.75	31.50
1046	30" wide		90	.178		15.25	10.90		26.15	33.50
1050	Elbows horiz. 9" rung spacing, 90°, 12" radius, 6" wide		9.60	1.667	Ea.	57.50	102		159.50	221
1060	9" wide		8.40	1.905		61	117		178	247
1070	12" wide		7.60	2.105		66	129		195	272
1080	18" wide		6.20	2.581		87	158		245	340
1090	24" wide		5.40	2.963		97.50	182		279.50	385
1100	30" wide		4.80	3.333		129	205		334	455
1120	90°, 24" radius, 6" wide		9.20	1.739		119	107		226	294
1130	9" wide		8	2		122	123		245	325
1140	12" wide		7.20	2.222		125	136		261	350
1150	18" wide		5.80	2.759		136	169		305	410
1160	24" wide		5	3.200		146	196		342	460
1170	30" wide		4.40	3.636		159	223		382	520
1190	90°, 36" radius, 6" wide		8.80	1.818		131	112		243	315
1200	9" wide		7.60	2.105		138	129		267	350
1210	12" wide		6.80	2.353		145	144		289	380
1220	18" wide		5.40	2.963		166	182		348	460
1230	24" wide		4.60	3.478		181	213		394	530
1240	30" wide		4	4		218	245		463	620
1260	45°, 12" radius, 6" wide		13.20	1.212		44.50	74.50		119	163

26 05 36.10 Cable Tray Ladder Type		Crew	Daily Output	Labor-Hours	Unit	Material	Labor	Equipment	Total	Total Incl O&P
						2020 Bare Costs				
1270	9" wide	2 Elec	11	1.455	Ea.	46	89		135	188
1280	12" wide		9.60	1.667		48	102		150	210
1290	18" wide		7.60	2.105		50.50	129		179.50	255
1300	24" wide		6.20	2.581		65.50	158		223.50	315
1310	30" wide		5.40	2.963		83.50	182		265.50	370
1330	45°, 24" radius, 6" wide		12.80	1.250		56	76.50		132.50	180
1340	9" wide		10.60	1.509		60	92.50		152.50	209
1350	12" wide		9.20	1.739		67	107		174	238
1360	18" wide		7.20	2.222		69	136		205	286
1370	24" wide		5.80	2.759		83.50	169		252.50	355
1380	30" wide		5	3.200		99.50	196		295.50	410
1400	45°, 36" radius, 6" wide		12.40	1.290		80.50	79		159.50	211
1410	9" wide		10.20	1.569		85.50	96		181.50	242
1420	12" wide		8.80	1.818		87	112		199	268
1430	18" wide		6.80	2.353		94	144		238	325
1440	24" wide		5.40	2.963		99.50	182		281.50	390
1450	30" wide	▼	4.60	3.478	▼	129	213		342	470
1470	Elbows horizontal, 4" rung spacing, use 9" rung x 1.50									
1480	6" rung spacing, use 9" rung x 1.20									
1490	12" rung spacing, use 9" rung x 0.93									
1500	Elbows vertical, 9" rung spacing, 90°, 12" radius, 6" wide	2 Elec	9.60	1.667	Ea.	79	102		181	244
1510	9" wide		8.40	1.905		80.50	117		197.50	269
1520	12" wide		7.60	2.105		83.50	129		212.50	291
1530	18" wide		6.20	2.581		87	158		245	340
1540	24" wide		5.40	2.963		90.50	182		272.50	380
1550	30" wide		4.80	3.333		97.50	205		302.50	420
1570	24" radius, 6" wide		9.20	1.739		112	107		219	287
1580	9" wide		8	2		115	123		238	315
1590	12" wide		7.20	2.222		119	136		255	340
1600	18" wide		5.80	2.759		125	169		294	400
1610	24" wide		5	3.200		132	196		328	445
1620	30" wide		4.40	3.636		138	223		361	495
1640	36" radius, 6" wide		8.80	1.818		150	112		262	335
1650	9" wide		7.60	2.105		155	129		284	370
1660	12" wide		6.80	2.353		157	144		301	395
1670	18" wide		5.40	2.963		171	182		353	470
1680	24" wide		4.60	3.478		173	213		386	520
1690	30" wide	▼	4	4	▼	195	245		440	595
1710	Elbows vertical, 4" rung spacing, use 9" rung x 1.25									
1720	6" rung spacing, use 9" rung x 1.15									
1730	12" rung spacing, use 9" rung x 0.90									
1740	Tee horizontal, 9" rung spacing, 12" radius, 6" wide	2 Elec	5	3.200	Ea.	122	196		318	435
1750	9" wide		4.60	3.478		131	213		344	475
1760	12" wide		4.40	3.636		138	223		361	495
1770	18" wide		4	4		157	245		402	555
1780	24" wide		3.60	4.444		185	273		458	625
1790	30" wide		3.40	4.706		221	289		510	690
1810	24" radius, 6" wide		4.60	3.478		190	213		403	540
1820	9" wide		4.20	3.810		199	234		433	580
1830	12" wide		4	4		206	245		451	605
1840	18" wide		3.60	4.444		223	273		496	665
1850	24" wide		3.20	5		242	305		547	735
1860	30" wide	▼	3	5.333	▼	263	325		588	795

26 05 36.10 Cable Tray Ladder Type		Crew	Daily Output	Labor-Hours	Unit	Material	Labor	Equipment	Total	Total Incl O&P
						2020 Bare Costs				
1880	36″ radius, 6″ wide	2 Elec	4.20	3.810	Ea.	284	234		518	670
1890	9″ wide		3.80	4.211		296	258		554	725
1900	12″ wide		3.60	4.444		315	273		588	765
1910	18″ wide		3.20	5		340	305		645	840
1920	24″ wide		2.80	5.714		380	350		730	960
1930	30″ wide		2.60	6.154		415	380		795	1,050
1980	Tee vertical, 9″ rung spacing, 12″ radius, 6″ wide		5.40	2.963		207	182		389	510
1990	9″ wide		5.20	3.077		209	189		398	520
2000	12″ wide		5	3.200		211	196		407	530
2010	18″ wide		4.60	3.478		213	213		426	565
2020	24″ wide		4.40	3.636		218	223		441	585
2030	30″ wide		4	4		232	245		477	635
2050	24″ radius, 6″ wide		5	3.200		375	196		571	710
2060	9″ wide		4.80	3.333		380	205		585	730
2070	12″ wide		4.60	3.478		385	213		598	750
2080	18″ wide		4.20	3.810		395	234		629	795
2090	24″ wide		4	4		405	245		650	825
2100	30″ wide		3.60	4.444		415	273		688	880
2120	36″ radius, 6″ wide		4.60	3.478		695	213		908	1,100
2130	9″ wide		4.40	3.636		710	223		933	1,125
2140	12″ wide		4.20	3.810		715	234		949	1,150
2150	18″ wide		3.80	4.211		725	258		983	1,200
2160	24″ wide		3.60	4.444		735	273		1,008	1,225
2170	30″ wide		3.20	5		760	305		1,065	1,300
2190	Tee, 4″ rung spacing, use 9″ rung x 1.30									
2200	6″ rung spacing, use 9″ rung x 1.20									
2210	12″ rung spacing, use 9″ rung x 0.90									
2220	Cross horizontal, 9″ rung spacing, 12″ radius, 6″ wide	2 Elec	4	4	Ea.	150	245		395	545
2230	9″ wide		3.80	4.211		158	258		416	575
2240	12″ wide		3.60	4.444		173	273		446	610
2250	18″ wide		3.40	4.706		188	289		477	650
2260	24″ wide		3	5.333		209	325		534	735
2270	30″ wide		2.80	5.714		253	350		603	820
2290	24″ radius, 6″ wide		3.60	4.444		293	273		566	740
2300	9″ wide		3.40	4.706		300	289		589	775
2310	12″ wide		3.20	5		310	305		615	810
2320	18″ wide		3	5.333		330	325		655	870
2330	24″ wide		2.60	6.154		350	380		730	965
2340	30″ wide		2.40	6.667		375	410		785	1,050
2360	36″ radius, 6″ wide		3.20	5		410	305		715	920
2370	9″ wide		3	5.333		425	325		750	975
2380	12″ wide		2.80	5.714		445	350		795	1,025
2390	18″ wide		2.60	6.154		490	380		870	1,125
2400	24″ wide		2.20	7.273		545	445		990	1,275
2410	30″ wide		2	8		590	490		1,080	1,400
2430	Cross horizontal, 4″ rung spacing, use 9″ rung x 1.30									
2440	6″ rung spacing, use 9″ rung x 1.20									
2450	12″ rung spacing, use 9″ rung x 0.90									
2460	Reducer, 9″ to 6″ wide tray	2 Elec	13	1.231	Ea.	83.50	75.50		159	208
2470	12″ to 9″ wide tray		12	1.333		85.50	82		167.50	220
2480	18″ to 12″ wide tray		10.40	1.538		85.50	94.50		180	239
2490	24″ to 18″ wide tray		9	1.778		87	109		196	264
2500	30″ to 24″ wide tray		8	2		89	123		212	287

26 05 36 – Cable Trays for Electrical Systems

26 05 36.10 Cable Tray Ladder Type		Crew	Daily Output	Labor-Hours	Unit	Material	Labor	Equipment	Total	Total Incl O&P
						2020 Bare Costs				
2510	36" to 30" wide tray	2 Elec	7	2.286	Ea.	99.50	140		239.50	325
2511	Reducer, 18" to 6" wide tray		10.40	1.538		89	94.50		183.50	243
2512	24" to 12" wide tray		9	1.778		90.50	109		199.50	268
2513	30" to 18" wide tray		8	2		93	123		216	291
2514	30" to 12" wide tray		8	2		98.50	123		221.50	297
2515	36" to 24" wide tray		7	2.286		101	140		241	325
2516	36" to 18" wide tray		7	2.286		101	140		241	325
2517	36" to 12" wide tray		7	2.286		101	140		241	325
2520	Dropout or end plate, 6" wide		32	.500		9.35	30.50		39.85	57.50
2530	9" wide		28	.571		11	35		46	66
2540	12" wide		26	.615		12.05	38		50.05	71.50
2550	18" wide		22	.727		14.10	44.50		58.60	84
2560	24" wide		20	.800		17.10	49		66.10	94.50
2570	30" wide		18	.889		18.70	54.50		73.20	105
2590	Tray connector		48	.333		16.55	20.50		37.05	49.50
3200	Aluminum tray, 4" deep, 6" rung spacing, 6" wide		134	.119	L.F.	15.75	7.35		23.10	28.50
3210	9" wide		128	.125		16.75	7.65		24.40	30
3220	12" wide		124	.129		17.65	7.90		25.55	31.50
3230	18" wide		114	.140		19.70	8.60		28.30	35
3240	24" wide		106	.151		23	9.25		32.25	39.50
3250	30" wide		100	.160		25	9.80		34.80	43
3270	9" rung spacing, 6" wide		140	.114		12.80	7		19.80	25
3280	9" wide		134	.119		13.50	7.35		20.85	26
3290	12" wide		130	.123		13.70	7.55		21.25	26.50
3300	18" wide		122	.131		16	8.05		24.05	30
3310	24" wide		116	.138		18.60	8.45		27.05	33.50
3320	30" wide		108	.148		20.50	9.10		29.60	36.50
3340	12" rung spacing, 6" wide		146	.110		13.65	6.70		20.35	25.50
3350	9" wide		140	.114		13.80	7		20.80	26
3360	12" wide		134	.119		14.15	7.35		21.50	27
3370	18" wide		128	.125		15.10	7.65		22.75	28.50
3380	24" wide		124	.129		16.75	7.90		24.65	30.50
3390	30" wide		114	.140		17.95	8.60		26.55	33
3401	18" rung spacing, 6" wide		150	.107		13.55	6.55		20.10	25
3402	9" wide tray		144	.111		13.70	6.80		20.50	25.50
3403	12" wide tray		140	.114		14.15	7		21.15	26.50
3404	18" wide tray		134	.119		15	7.35		22.35	28
3405	24" wide tray		130	.123		16.15	7.55		23.70	29.50
3406	30" wide tray		120	.133		16.75	8.20		24.95	31
3410	Elbows horiz., 9" rung spacing, 90°, 12" radius, 6" wide		9.60	1.667	Ea.	60.50	102		162.50	224
3420	9" wide		8.40	1.905		62.50	117		179.50	249
3430	12" wide		7.60	2.105		74	129		203	281
3440	18" wide		6.20	2.581		83.50	158		241.50	335
3450	24" wide		5.40	2.963		103	182		285	395
3460	30" wide		4.80	3.333		109	205		314	435
3480	24" radius, 6" wide		9.20	1.739		108	107		215	283
3490	9" wide		8	2		112	123		235	315
3500	12" wide		7.20	2.222		118	136		254	340
3510	18" wide		5.80	2.759		128	169		297	400
3520	24" wide		5	3.200		141	196		337	455
3530	30" wide		4.40	3.636		153	223		376	515
3550	90°, 36" radius, 6" wide		8.80	1.818		119	112		231	300
3560	9" wide		7.60	2.105		122	129		251	335

26 05 36 – Cable Trays for Electrical Systems

26 05 36.10 Cable Tray Ladder Type		Crew	Daily Output	Labor-Hours	Unit	Material	2020 Bare Costs Labor	Equipment	Total	Total Incl O&P
3570	12″ wide	2 Elec	6.80	2.353	Ea.	136	144		280	370
3580	18″ wide		5.40	2.963		153	182		335	450
3590	24″ wide		4.60	3.478		172	213		385	520
3600	30″ wide		4	4		190	245		435	590
3620	45°, 12″ radius, 6″ wide		13.20	1.212		40	74.50		114.50	158
3630	9″ wide		11	1.455		41.50	89		130.50	183
3640	12″ wide		9.60	1.667		43	102		145	205
3650	18″ wide		7.60	2.105		51	129		180	255
3660	24″ wide		6.20	2.581		63	158		221	315
3670	30″ wide		5.40	2.963		72.50	182		254.50	360
3690	45°, 24″ radius, 6″ wide		12.80	1.250		53.50	76.50		130	177
3700	9″ wide		10.60	1.509		60.50	92.50		153	210
3710	12″ wide		9.20	1.739		62	107		169	233
3720	18″ wide		7.20	2.222		66.50	136		202.50	284
3730	24″ wide		5.80	2.759		77	169		246	345
3740	30″ wide		5	3.200		92	196		288	400
3760	45°, 36″ radius, 6″ wide		12.40	1.290		72.50	79		151.50	202
3770	9″ wide		10.20	1.569		74	96		170	230
3780	12″ wide		8.80	1.818		75.50	112		187.50	255
3790	18″ wide		6.80	2.353		83	144		227	315
3800	24″ wide		5.40	2.963		102	182		284	395
3810	30″ wide	↓	4.60	3.478	↓	108	213		321	450
3830	Elbows horizontal, 4″ rung spacing, use 9″ rung x 1.50									
3840	6″ rung spacing, use 9″ rung x 1.20									
3850	12″ rung spacing, use 9″ rung x 0.93									
3860	Elbows vertical, 9″ rung spacing, 90°, 12″ radius, 6″ wide	2 Elec	9.60	1.667	Ea.	84.50	102		186.50	250
3870	9″ wide		8.40	1.905		89	117		206	278
3880	12″ wide		7.60	2.105		90.50	129		219.50	299
3890	18″ wide		6.20	2.581		93.50	158		251.50	345
3900	24″ wide		5.40	2.963		99.50	182		281.50	390
3910	30″ wide		4.80	3.333		104	205		309	430
3930	24″ radius, 6″ wide		9.20	1.739		92	107		199	265
3940	9″ wide		8	2		95	123		218	293
3950	12″ wide		7.20	2.222		98	136		234	320
3960	18″ wide		5.80	2.759		104	169		273	375
3970	24″ wide		5	3.200		110	196		306	420
3980	30″ wide		4.40	3.636		116	223		339	470
4000	36″ radius, 6″ wide		8.80	1.818		142	112		254	330
4010	9″ wide		7.60	2.105		145	129		274	360
4020	12″ wide		6.80	2.353		154	144		298	390
4030	18″ wide		5.40	2.963		160	182		342	455
4040	24″ wide		4.60	3.478		166	213		379	515
4050	30″ wide	↓	4	4	↓	179	245		424	575
4070	Elbows vertical, 4″ rung spacing, use 9″ rung x 1.25									
4080	6″ rung spacing, use 9″ rung x 1.15									
4090	12″ rung spacing, use 9″ rung x 0.90									
4100	Tee horizontal, 9″ rung spacing, 12″ radius, 6″ wide	2 Elec	5	3.200	Ea.	94	196		290	405
4110	9″ wide		4.60	3.478		101	213		314	440
4120	12″ wide		4.40	3.636		96	223		319	450
4130	18″ wide		4.20	3.810		122	234		356	495
4140	24″ wide		4	4		133	245		378	525
4150	30″ wide		3.60	4.444		156	273		429	590
4170	24″ radius, 6″ wide	↓	4.60	3.478	↓	160	213		373	505

26 05 36 – Cable Trays for Electrical Systems

26 05 36.10 Cable Tray Ladder Type		Crew	Daily Output	Labor-Hours	Unit	Material	2020 Bare Costs Labor	Equipment	Total	Total Incl O&P
4180	9" wide	2 Elec	4.20	3.810	Ea.	169	234		403	545
4190	12" wide		4	4		173	245		418	570
4200	18" wide		3.80	4.211		190	258		448	610
4210	24" wide		3.60	4.444		207	273		480	650
4220	30" wide		3.20	5		225	305		530	720
4240	36" radius, 6" wide		4.20	3.810		227	234		461	610
4250	9" wide		3.80	4.211		233	258		491	655
4260	12" wide		3.60	4.444		255	273		528	700
4270	18" wide		3.40	4.706		268	289		557	740
4280	24" wide		3.20	5		296	305		601	795
4290	30" wide		2.80	5.714		375	350		725	950
4310	Tee vertical, 9" rung spacing, 12" radius, 6" wide		5.40	2.963		210	182		392	510
4320	9" wide		5.20	3.077		218	189		407	530
4330	12" wide		5	3.200		218	196		414	540
4340	18" wide		4.60	3.478		224	213		437	575
4350	24" wide		4.40	3.636		242	223		465	610
4360	30" wide		4.20	3.810		246	234		480	630
4380	24" radius, 6" wide		5	3.200		355	196		551	695
4390	9" wide		4.80	3.333		360	205		565	715
4400	12" wide		4.60	3.478		370	213		583	735
4410	18" wide		4.20	3.810		375	234		609	775
4420	24" wide		4	4		385	245		630	805
4430	30" wide		3.80	4.211		395	258		653	835
4450	36" radius, 6" wide		4.60	3.478		785	213		998	1,200
4460	9" wide		4.40	3.636		800	223		1,023	1,225
4470	12" wide		4.20	3.810		835	234		1,069	1,275
4480	18" wide		3.80	4.211		840	258		1,098	1,325
4490	24" wide		3.60	4.444		860	273		1,133	1,375
4500	30" wide	↓	3.40	4.706	↓	870	289		1,159	1,400
4520	Tees, 4" rung spacing, use 9" rung x 1.30									
4530	6" rung spacing, use 9" rung x 1.20									
4540	12" rung spacing, use 9" rung x 0.90									
4550	Cross horizontal, 9" rung spacing, 12" radius, 6" wide	2 Elec	4.40	3.636	Ea.	130	223		353	490
4560	9" wide		4.20	3.810		145	234		379	520
4570	12" wide		4	4		139	245		384	535
4580	18" wide		3.60	4.444		178	273		451	615
4590	24" wide		3.40	4.706		184	289		473	645
4600	30" wide		3	5.333		216	325		541	745
4620	24" radius, 6" wide		4	4		250	245		495	655
4630	9" wide		3.80	4.211		259	258		517	685
4640	12" wide		3.60	4.444		267	273		540	715
4650	18" wide		3.20	5		280	305		585	780
4660	24" wide		3	5.333		305	325		630	840
4670	30" wide		2.60	6.154		325	380		705	940
4690	36" radius, 6" wide		3.60	4.444		296	273		569	745
4700	9" wide		3.40	4.706		310	289		599	785
4710	12" wide		3.20	5		325	305		630	830
4720	18" wide		2.80	5.714		350	350		700	925
4730	24" wide		2.60	6.154		430	380		810	1,050
4740	30" wide	↓	2.20	7.273	↓	475	445		920	1,200
4760	Cross horizontal, 4" rung spacing, use 9" rung x 1.30									
4770	6" rung spacing, use 9" rung x 1.20									
4780	12" rung spacing, use 9" rung x 0.90									

26 05 36 – Cable Trays for Electrical Systems

26 05 36.10 Cable Tray Ladder Type		Crew	Daily Output	Labor-Hours	Unit	Material	2020 Bare Costs Labor	Equipment	Total	Total Incl O&P
4790	Reducer, 9" to 6" wide tray	2 Elec	16	1	Ea.	69.50	61.50		131	171
4800	12" to 9" wide tray		14	1.143		72.50	70		142.50	188
4810	18" to 12" wide tray		12.40	1.290		72.50	79		151.50	202
4820	24" to 18" wide tray		10.60	1.509		74	92.50		166.50	225
4830	30" to 24" wide tray		9.20	1.739		75.50	107		182.50	247
4840	36" to 30" wide tray		8	2		80	123		203	277
4841	Reducer, 18" to 6" wide tray		12.40	1.290		72.50	79		151.50	202
4842	24" to 12" wide tray		10.60	1.509		74	92.50		166.50	225
4843	30" to 18" wide tray		9.20	1.739		75.50	107		182.50	247
4844	30" to 12" wide tray		9.20	1.739		75.50	107		182.50	247
4845	36" to 24" wide tray		8	2		80	123		203	277
4846	36" to 18" wide tray		8	2		80	123		203	277
4847	36" to 12" wide tray		8	2		80	123		203	277
4850	Dropout or end plate, 6" wide		32	.500		9.10	30.50		39.60	57
4860	9" wide tray		28	.571		9.85	35		44.85	65
4870	12" wide tray		26	.615		10.60	38		48.60	69.50
4880	18" wide tray		22	.727		14.30	44.50		58.80	84
4890	24" wide tray		20	.800		16.75	49		65.75	94
4900	30" wide tray		18	.889		19.85	54.50		74.35	106
4920	Tray connector		48	.333		13.30	20.50		33.80	46
9200	Splice plate	1 Elec	48	.167	Pr.	13.95	10.25		24.20	31
9210	Expansion joint		48	.167		16.75	10.25		27	34
9220	Horizontal hinged		48	.167		13.95	10.25		24.20	31
9230	Vertical hinged		48	.167		16.75	10.25		27	34
9240	Ladder hanger, vertical		28	.286	Ea.	4.18	17.55		21.73	31.50
9250	Ladder to channel connector		24	.333		48.50	20.50		69	85
9260	Ladder to box connector, 30" wide		19	.421		48.50	26		74.50	93.50
9270	24" wide		20	.400		48.50	24.50		73	91.50
9280	18" wide		21	.381		47	23.50		70.50	87.50
9290	12" wide		22	.364		44	22.50		66.50	83
9300	9" wide		23	.348		39.50	21.50		61	76
9310	6" wide		24	.333		31	20.50		51.50	66
9320	Ladder floor flange		24	.333		31.50	20.50		52	66
9330	Cable roller for tray, 30" wide		10	.800		275	49		324	380
9340	24" wide		11	.727		234	44.50		278.50	325
9350	18" wide		12	.667		223	41		264	310
9360	12" wide		13	.615		175	38		213	251
9370	9" wide		14	.571		155	35		190	225
9380	6" wide		15	.533		126	32.50		158.50	190
9390	Pulley, single wheel		12	.667		288	41		329	380
9400	Triple wheel		10	.800		575	49		624	705
9440	Nylon cable tie, 14" long		80	.100		.57	6.15		6.72	10.10
9450	Ladder, hold down clamp		60	.133		8.45	8.20		16.65	22
9460	Cable clamp		60	.133		9	8.20		17.20	22.50
9470	Wall bracket, 30" wide tray		19	.421		51	26		77	96
9480	24" wide tray		20	.400		39.50	24.50		64	81.50
9490	18" wide tray		21	.381		35.50	23.50		59	75
9500	12" wide tray		22	.364		33	22.50		55.50	71
9510	9" wide tray		23	.348		29	21.50		50.50	65
9520	6" wide tray		24	.333		28.50	20.50		49	62.50

26 05 36 – Cable Trays for Electrical Systems

26 05 36.20 Cable Tray Solid Bottom		Crew	Daily Output	Labor-Hours	Unit	Material	Labor	Equipment	Total	Total Incl O&P
						2020 Bare Costs				
0010	**CABLE TRAY SOLID BOTTOM** w/ftngs. & supports, 3" deep, to 15' high									
0200	For higher elevations, see Section 26 05 36.40									
0220	Galvanized steel, tray, 6" wide	2 Elec	120	.133	L.F.	11.80	8.20		20	25.50
0240	12" wide		100	.160		15.15	9.80		24.95	32
0260	18" wide		70	.229		18.40	14		32.40	42
0280	24" wide		60	.267		21.50	16.35		37.85	48.50
0300	30" wide		50	.320		25.50	19.65		45.15	58.50
0340	Elbow horizontal, 90°, 12" radius, 6" wide		9.60	1.667	Ea.	89.50	102		191.50	256
0360	12" wide		6.80	2.353		100	144		244	330
0370	18" wide		5.40	2.963		115	182		297	405
0380	24" wide		4.40	3.636		140	223		363	500
0390	30" wide		3.80	4.211		167	258		425	585
0420	24" radius, 6" wide		9.20	1.739		128	107		235	305
0440	12" wide		6.40	2.500		148	153		301	400
0450	18" wide		5	3.200		172	196		368	490
0460	24" wide		4	4		200	245		445	600
0470	30" wide		3.40	4.706		234	289		523	705
0500	36" radius, 6" wide		8.80	1.818		186	112		298	375
0520	12" wide		6	2.667		210	164		374	485
0530	18" wide		4.60	3.478		257	213		470	615
0540	24" wide		3.60	4.444		278	273		551	725
0550	30" wide		3	5.333		335	325		660	875
0580	Elbow vertical, 90°, 12" radius, 6" wide		9.60	1.667		109	102		211	276
0600	12" wide		6.80	2.353		119	144		263	350
0610	18" wide		5.40	2.963		134	182		316	430
0620	24" wide		4.40	3.636		136	223		359	495
0630	30" wide		3.80	4.211		145	258		403	560
0670	24" radius, 6" wide		9.20	1.739		153	107		260	335
0690	12" wide		6.40	2.500		169	153		322	420
0700	18" wide		5	3.200		183	196		379	500
0710	24" wide		4	4		198	245		443	600
0720	30" wide		3.40	4.706		210	289		499	675
0750	36" radius, 6" wide		8.80	1.818		210	112		322	405
0770	12" wide		6.60	2.424		234	149		383	485
0780	18" wide		4.60	3.478		257	213		470	615
0790	24" wide		3.60	4.444		281	273		554	730
0800	30" wide		3	5.333		305	325		630	845
0840	Tee horizontal, 12" radius, 6" wide		5	3.200		126	196		322	440
0860	12" wide		4	4		136	245		381	530
0870	18" wide		3.40	4.706		162	289		451	625
0880	24" wide		2.80	5.714		181	350		531	740
0890	30" wide		2.60	6.154		207	380		587	810
0940	24" radius, 6" wide		4.60	3.478		195	213		408	545
0960	12" wide		3.60	4.444		228	273		501	670
0970	18" wide		3	5.333		253	325		578	785
0980	24" wide		2.40	6.667		345	410		755	1,000
0990	30" wide		2.20	7.273		370	445		815	1,100
1020	36" radius, 6" wide		4.20	3.810		305	234		539	700
1040	12" wide		3.20	5		345	305		650	845
1050	18" wide		2.60	6.154		370	380		750	990
1060	24" wide		2.20	7.273		480	445		925	1,200
1070	30" wide		2	8		520	490		1,010	1,325

26 05 36 – Cable Trays for Electrical Systems

26 05 36.20 Cable Tray Solid Bottom		Crew	Daily Output	Labor-Hours	Unit	Material	Labor	Equipment	Total	Total Incl O&P
						2020 Bare Costs				
1100	Tee vertical, 12″ radius, 6″ wide	2 Elec	5	3.200	Ea.	207	196		403	530
1120	12″ wide		4	4		216	245		461	615
1130	18″ wide		3.60	4.444		219	273		492	660
1140	24″ wide		3.40	4.706		240	289		529	710
1150	30″ wide		3	5.333		257	325		582	790
1180	24″ radius, 6″ wide		4.60	3.478		305	213		518	665
1200	12″ wide		3.60	4.444		320	273		593	770
1210	18″ wide		3.20	5		335	305		640	840
1220	24″ wide		3	5.333		350	325		675	890
1230	30″ wide		2.60	6.154		385	380		765	1,000
1260	36″ radius, 6″ wide		4.20	3.810		475	234		709	885
1280	12″ wide		3.20	5		485	305		790	1,000
1290	18″ wide		2.80	5.714		520	350		870	1,125
1300	24″ wide		2.60	6.154		540	380		920	1,175
1310	30″ wide		2.20	7.273		630	445		1,075	1,375
1340	Cross horizontal, 12″ radius, 6″ wide		4	4		150	245		395	545
1360	12″ wide		3.40	4.706		164	289		453	625
1370	18″ wide		2.80	5.714		191	350		541	750
1380	24″ wide		2.40	6.667		216	410		626	865
1390	30″ wide		2	8		240	490		730	1,025
1420	24″ radius, 6″ wide		3.60	4.444		274	273		547	720
1440	12″ wide		3	5.333		305	325		630	845
1450	18″ wide		2.40	6.667		345	410		755	1,000
1460	24″ wide		2	8		440	490		930	1,250
1470	30″ wide		1.80	8.889		475	545		1,020	1,375
1500	36″ radius, 6″ wide		3.20	5		450	305		755	965
1520	12″ wide		2.60	6.154		495	380		875	1,125
1530	18″ wide		2	8		540	490		1,030	1,350
1540	24″ wide		1.80	8.889		635	545		1,180	1,550
1550	30″ wide		1.60	10		690	615		1,305	1,700
1580	Drop out or end plate, 6″ wide		32	.500		18.30	30.50		48.80	67
1600	12″ wide		26	.615		23	38		61	83
1610	18″ wide		22	.727		24.50	44.50		69	95.50
1620	24″ wide		20	.800		29	49		78	107
1630	30″ wide		18	.889		32	54.50		86.50	120
1660	Reducer, 12″ to 6″ wide		12	1.333		88	82		170	223
1680	18″ to 12″ wide		10.60	1.509		89.50	92.50		182	242
1700	18″ to 6″ wide		10.60	1.509		89.50	92.50		182	242
1720	24″ to 18″ wide		9.20	1.739		93	107		200	266
1740	24″ to 12″ wide		9.20	1.739		93	107		200	266
1760	30″ to 24″ wide		8	2		100	123		223	299
1780	30″ to 18″ wide		8	2		100	123		223	299
1800	30″ to 12″ wide		8	2		100	123		223	299
1820	36″ to 30″ wide		7.20	2.222		103	136		239	325
1840	36″ to 24″ wide		7.20	2.222		103	136		239	325
1860	36″ to 18″ wide		7.20	2.222		103	136		239	325
1880	36″ to 12″ wide		7.20	2.222		103	136		239	325
2000	Aluminum tray, 6″ wide		150	.107	L.F.	10.95	6.55		17.50	22
2020	12″ wide		130	.123		14.40	7.55		21.95	27.50
2030	18″ wide		100	.160		18.05	9.80		27.85	35
2040	24″ wide		90	.178		23	10.90		33.90	42
2050	30″ wide		70	.229		27	14		41	51
2080	Elbow horizontal, 90°, 12″ radius, 6″ wide		9.60	1.667	Ea.	88.50	102		190.50	254

26 05 36.20 Cable Tray Solid Bottom		Crew	Daily Output	Labor-Hours	Unit	Material	Labor	Equipment	Total	Total Incl O&P
						2020 Bare Costs				
2100	12″ wide	2 Elec	7.60	2.105	Ea.	99.50	129		228.50	310
2110	18″ wide		6.80	2.353		122	144		266	355
2120	24″ wide		5.80	2.759		139	169		308	415
2130	30″ wide		5	3.200		170	196		366	485
2160	24″ radius, 6″ wide		9.20	1.739		126	107		233	305
2180	12″ wide		7.20	2.222		148	136		284	375
2190	18″ wide		6.40	2.500		170	153		323	425
2200	24″ wide		5.40	2.963		209	182		391	510
2210	30″ wide		4.60	3.478		240	213		453	595
2240	36″ radius, 6″ wide		8.80	1.818		186	112		298	375
2260	12″ wide		6.80	2.353		223	144		367	465
2270	18″ wide		6	2.667		265	164		429	545
2280	24″ wide		5	3.200		284	196		480	610
2290	30″ wide		4.20	3.810		320	234		554	715
2320	Elbow vertical, 90°, 12″ radius, 6″ wide		9.60	1.667		105	102		207	272
2340	12″ wide		7.60	2.105		109	129		238	320
2350	18″ wide		6.80	2.353		123	144		267	360
2360	24″ wide		5.80	2.759		130	169		299	405
2370	30″ wide		5	3.200		138	196		334	450
2400	24″ radius, 6″ wide		9.20	1.739		148	107		255	325
2420	12″ wide		7.20	2.222		162	136		298	390
2430	18″ wide		6.40	2.500		174	153		327	425
2440	24″ wide		5.40	2.963		188	182		370	485
2450	30″ wide		4.60	3.478		200	213		413	550
2480	36″ radius, 6″ wide		8.80	1.818		196	112		308	390
2500	12″ wide		6.80	2.353		220	144		364	465
2510	18″ wide		6	2.667		240	164		404	515
2520	24″ wide		5	3.200		249	196		445	575
2530	30″ wide		4.20	3.810		276	234		510	665
2560	Tee horizontal, 12″ radius, 6″ wide		5	3.200		137	196		333	450
2580	12″ wide		4.40	3.636		166	223		389	525
2590	18″ wide		4	4		186	245		431	585
2600	24″ wide		3.60	4.444		208	273		481	650
2610	30″ wide		3	5.333		240	325		565	770
2640	24″ radius, 6″ wide		4.60	3.478		225	213		438	580
2660	12″ wide		4	4		253	245		498	660
2670	18″ wide		3.60	4.444		286	273		559	735
2680	24″ wide		3	5.333		350	325		675	890
2690	30″ wide		2.40	6.667		400	410		810	1,075
2720	36″ radius, 6″ wide		4.20	3.810		365	234		599	760
2740	12″ wide		3.60	4.444		400	273		673	860
2750	18″ wide		3.20	5		460	305		765	975
2760	24″ wide		2.60	6.154		540	380		920	1,175
2770	30″ wide		2	8		590	490		1,080	1,400
2800	Tee vertical, 12″ radius, 6″ wide		5	3.200		195	196		391	515
2820	12″ wide		4.40	3.636		199	223		422	565
2830	18″ wide		4.20	3.810		207	234		441	585
2840	24″ wide		4	4		220	245		465	620
2850	30″ wide		3.60	4.444		240	273		513	685
2880	24″ radius, 6″ wide		4.60	3.478		286	213		499	645
2900	12″ wide		4	4		305	245		550	715
2910	18″ wide		3.80	4.211		325	258		583	755
2920	24″ wide		3.60	4.444		350	273		623	805

26 05 36.20 Cable Tray Solid Bottom

	26 05 36.20 Cable Tray Solid Bottom	Crew	Daily Output	Labor-Hours	Unit	2020 Bare Costs Material	Labor	Equipment	Total	Total Incl O&P
2930	30″ wide	2 Elec	3.20	5	Ea.	385	305		690	895
2960	36″ radius, 6″ wide		4.20	3.810		450	234		684	855
2980	12″ wide		3.40	4.706		470	289		759	960
2990	18″ wide		3.40	4.706		495	289		784	990
3000	24″ wide		3.20	5		510	305		815	1,025
3010	30″ wide		2.80	5.714		550	350		900	1,150
3040	Cross horizontal, 12″ radius, 6″ wide		4.40	3.636		174	223		397	535
3060	12″ wide		4	4		196	245		441	595
3070	18″ wide		3.40	4.706		223	289		512	690
3080	24″ wide		2.80	5.714		253	350		603	820
3090	30″ wide		2.60	6.154		286	380		666	895
3120	24″ radius, 6″ wide		4	4		305	245		550	715
3140	12″ wide		3.60	4.444		350	273		623	805
3150	18″ wide		3	5.333		380	325		705	925
3160	24″ wide		2.40	6.667		460	410		870	1,125
3170	30″ wide		2.20	7.273		510	445		955	1,250
3200	36″ radius, 6″ wide		3.60	4.444		530	273		803	1,000
3220	12″ wide		3.20	5		560	305		865	1,075
3230	18″ wide		2.60	6.154		620	380		1,000	1,250
3240	24″ wide		2	8		675	490		1,165	1,500
3250	30″ wide		1.80	8.889		730	545		1,275	1,650
3280	Dropout, or end plate, 6″ wide		32	.500		17.45	30.50		47.95	66
3300	12″ wide		26	.615		19.75	38		57.75	79.50
3310	18″ wide		22	.727		23.50	44.50		68	94.50
3320	24″ wide		20	.800		26	49		75	104
3330	30″ wide		18	.889		30.50	54.50		85	118
3380	Reducer, 12″ to 6″ wide		14	1.143		85.50	70		155.50	202
3400	18″ to 12″ wide		12	1.333		88.50	82		170.50	223
3420	18″ to 6″ wide		12	1.333		88.50	82		170.50	223
3440	24″ to 18″ wide		10.60	1.509		93.50	92.50		186	246
3460	24″ to 12″ wide		10.60	1.509		93.50	92.50		186	246
3480	30″ to 24″ wide		9.20	1.739		102	107		209	276
3500	30″ to 18″ wide		9.20	1.739		102	107		209	276
3520	30″ to 12″ wide		9.20	1.739		102	107		209	276
3540	36″ to 30″ wide		8	2		107	123		230	305
3560	36″ to 24″ wide		8	2		109	123		232	310
3580	36″ to 18″ wide		8	2		109	123		232	310
3600	36″ to 12″ wide		8	2		111	123		234	310

26 05 36.30 Cable Tray Trough

	26 05 36.30 Cable Tray Trough	Crew	Daily Output	Labor-Hours	Unit	Material	Labor	Equipment	Total	Total Incl O&P
0010	**CABLE TRAY TROUGH** vented, w/ftngs. & supports, 6″ deep, to 10′ high									
0020	For higher elevations, see Section 26 05 36.40									
0200	Galvanized steel, tray, 6″ wide	2 Elec	90	.178	L.F.	15.55	10.90		26.45	34
0240	12″ wide		80	.200		14.20	12.25		26.45	34.50
0260	18″ wide		70	.229		21.50	14		35.50	45
0280	24″ wide		60	.267		48.50	16.35		64.85	78.50
0300	30″ wide		50	.320		47.50	19.65		67.15	82.50
0340	Elbow horizontal, 90°, 12″ radius, 6″ wide		7.60	2.105	Ea.	124	129		253	335
0360	12″ wide		5.60	2.857		147	175		322	430
0370	18″ wide		4.40	3.636		170	223		393	530
0380	24″ wide		3.60	4.444		210	273		483	650
0390	30″ wide		3.20	5		237	305		542	730
0420	24″ radius, 6″ wide		7.20	2.222		183	136		319	410

26 05 36 – Cable Trays for Electrical Systems

26 05 36.30 Cable Tray Trough		Crew	Daily Output	Labor-Hours	Unit	Material	Labor	Equipment	Total	Total Incl O&P
						2020 Bare Costs				
0440	12″ wide	2 Elec	5.20	3.077	Ea.	212	189		401	525
0450	18″ wide		4	4		255	245		500	660
0460	24″ wide		3.20	5		288	305		593	785
0470	30″ wide		2.80	5.714		335	350		685	905
0500	36″ radius, 6″ wide		6.80	2.353		239	144		383	485
0520	12″ wide		4.80	3.333		272	205		477	615
0530	18″ wide		3.60	4.444		325	273		598	780
0540	24″ wide		2.80	5.714		335	350		685	910
0550	30″ wide		2.40	6.667		375	410		785	1,050
0580	Elbow vertical, 90°, 12″ radius, 6″ wide		7.60	2.105		136	129		265	350
0600	12″ wide		5.60	2.857		151	175		326	435
0610	18″ wide		4.40	3.636		152	223		375	515
0620	24″ wide		3.60	4.444		175	273		448	615
0630	30″ wide		3.20	5		179	305		484	665
0660	24″ radius, 6″ wide		7.20	2.222		194	136		330	425
0680	12″ wide		5.20	3.077		205	189		394	515
0690	18″ wide		4	4		228	245		473	630
0700	24″ wide		3.20	5		235	305		540	730
0710	30″ wide		2.80	5.714		264	350		614	830
0740	36″ radius, 6″ wide		6.80	2.353		262	144		406	510
0760	12″ wide		4.80	3.333		279	205		484	620
0770	18″ wide		3.60	4.444		310	273		583	760
0780	24″ wide		2.80	5.714		330	350		680	905
0790	30″ wide		2.40	6.667		340	410		750	1,000
0820	Tee horizontal, 12″ radius, 6″ wide		4	4		158	245		403	555
0840	12″ wide		3.20	5		177	305		482	665
0850	18″ wide		2.80	5.714		200	350		550	760
0860	24″ wide		2.40	6.667		232	410		642	885
0870	30″ wide		2.20	7.273		264	445		709	975
0900	24″ radius, 6″ wide		3.60	4.444		262	273		535	710
0920	12″ wide		2.80	5.714		286	350		636	855
0930	18″ wide		2.40	6.667		320	410		730	980
0940	24″ wide		2	8		415	490		905	1,200
0950	30″ wide		1.80	8.889		450	545		995	1,325
0980	36″ radius, 6″ wide		3.20	5		375	305		680	885
1000	12″ wide		2.40	6.667		430	410		840	1,100
1010	18″ wide		2	8		470	490		960	1,275
1020	24″ wide		1.60	10		565	615		1,180	1,575
1030	30″ wide		1.40	11.429		600	700		1,300	1,725
1060	Tee vertical, 12″ radius, 6″ wide		4	4		249	245		494	655
1080	12″ wide		3.20	5		253	305		558	750
1090	18″ wide		3	5.333		260	325		585	790
1100	24″ wide		2.80	5.714		274	350		624	840
1110	30″ wide		2.60	6.154		290	380		670	900
1140	24″ radius, 6″ wide		3.60	4.444		335	273		608	790
1160	12″ wide		2.80	5.714		350	350		700	925
1170	18″ wide		2.60	6.154		370	380		750	985
1180	24″ wide		2.40	6.667		395	410		805	1,075
1190	30″ wide		2.20	7.273		415	445		860	1,150
1220	36″ radius, 6″ wide		3.20	5		545	305		850	1,075
1240	12″ wide		2.40	6.667		560	410		970	1,250
1250	18″ wide		2.20	7.273		580	445		1,025	1,325
1260	24″ wide		2	8		610	490		1,100	1,425

26 05 36.30 Cable Tray Trough		Crew	Daily Output	Labor-Hours	Unit	Material	Labor	Equipment	Total	Total Incl O&P
						2020 Bare Costs				
1270	30" wide	2 Elec	1.80	8.889	Ea.	710	545		1,255	1,625
1300	Cross horizontal, 12" radius, 6" wide		3.20	5		196	305		501	685
1320	12" wide		2.80	5.714		198	350		548	760
1330	18" wide		2.40	6.667		222	410		632	875
1340	24" wide		2	8		233	490		723	1,000
1350	30" wide		1.80	8.889		262	545		807	1,125
1380	24" radius, 6" wide		2.80	5.714		310	350		660	880
1400	12" wide		2.40	6.667		325	410		735	985
1410	18" wide		2	8		355	490		845	1,150
1420	24" wide		1.60	10		450	615		1,065	1,450
1430	30" wide		1.40	11.429		500	700		1,200	1,625
1460	36" radius, 6" wide		2.40	6.667		530	410		940	1,200
1480	12" wide		2	8		545	490		1,035	1,350
1490	18" wide		1.60	10		560	615		1,175	1,550
1500	24" wide		1.20	13.333		670	820		1,490	1,975
1510	30" wide		1	16		740	980		1,720	2,300
1540	Dropout or end plate, 6" wide		26	.615		23.50	38		61.50	84
1560	12" wide		22	.727		29	44.50		73.50	100
1580	18" wide		20	.800		31.50	49		80.50	111
1600	24" wide		18	.889		36.50	54.50		91	124
1620	30" wide		16	1		39	61.50		100.50	138
1660	Reducer, 12" to 6" wide		9.40	1.702		95.50	104		199.50	266
1680	18" to 12" wide		8.40	1.905		99	117		216	289
1700	18" to 6" wide		8.40	1.905		99	117		216	289
1720	24" to 18" wide		7.20	2.222		106	136		242	325
1740	24" to 12" wide		7.20	2.222		106	136		242	325
1760	30" to 24" wide		6.40	2.500		110	153		263	355
1780	30" to 18" wide		6.40	2.500		110	153		263	355
1800	30" to 12" wide		6.40	2.500		110	153		263	355
1820	36" to 30" wide		5.80	2.759		117	169		286	390
1840	36" to 24" wide		5.80	2.759		117	169		286	390
1860	36" to 18" wide		5.80	2.759		119	169		288	390
1880	36" to 12" wide		5.80	2.759		119	169		288	390
2000	Aluminum, tray, vented, 6" wide		120	.133	L.F.	16.10	8.20		24.30	30.50
2010	9" wide		110	.145		18.25	8.90		27.15	34
2020	12" wide		100	.160		23.50	9.80		33.30	41
2030	18" wide		90	.178		24.50	10.90		35.40	44
2040	24" wide		80	.200		29	12.25		41.25	50.50
2050	30" wide		70	.229		38.50	14		52.50	63.50
2080	Elbow horiz., 90°, 12" radius, 6" wide		7.60	2.105	Ea.	102	129		231	310
2090	9" wide		7	2.286		110	140		250	335
2100	12" wide		6.20	2.581		117	158		275	375
2110	18" wide		5.60	2.857		137	175		312	420
2120	24" wide		4.60	3.478		158	213		371	505
2130	30" wide		4	4		195	245		440	595
2160	24" radius, 6" wide		7.20	2.222		147	136		283	370
2180	12" wide		5.80	2.759		173	169		342	450
2190	18" wide		5.20	3.077		198	189		387	510
2200	24" wide		4.20	3.810		226	234		460	610
2210	30" wide		3.60	4.444		259	273		532	705
2240	36" radius, 6" wide		6.80	2.353		212	144		356	455
2260	12" wide		5.40	2.963		241	182		423	545
2270	18" wide		4.80	3.333		275	205		480	615

26 05 36 – Cable Trays for Electrical Systems

26 05 36.30 Cable Tray Trough		Crew	Daily Output	Labor-Hours	Unit	Material	Labor	Equipment	Total	Total Incl O&P
						2020 Bare Costs				
2280	24" wide	2 Elec	3.80	4.211	Ea.	300	258		558	730
2290	30" wide		3.40	4.706		355	289		644	835
2320	Elbow vertical, 90°, 12" radius, 6" wide		7.60	2.105		123	129		252	335
2330	9" wide		7	2.286		132	140		272	360
2340	12" wide		6.20	2.581		133	158		291	390
2350	18" wide		5.60	2.857		137	175		312	420
2360	24" wide		4.60	3.478		151	213		364	495
2370	30" wide		4	4		158	245		403	555
2400	24" radius, 6" wide		7.20	2.222		167	136		303	395
2420	12" wide		5.80	2.759		181	169		350	460
2430	18" wide		5.20	3.077		197	189		386	505
2440	24" wide		4.20	3.810		199	234		433	580
2450	30" wide		3.60	4.444		211	273		484	650
2480	36" radius, 6" wide		6.80	2.353		211	144		355	455
2500	12" wide		5.40	2.963		225	182		407	530
2510	18" wide		4.80	3.333		253	205		458	595
2520	24" wide		3.80	4.211		273	258		531	700
2530	30" wide		3.40	4.706		300	289		589	775
2560	Tee horizontal, 12" radius, 6" wide		4	4		154	245		399	550
2570	9" wide		3.80	4.211		158	258		416	575
2580	12" wide		3.60	4.444		173	273		446	610
2590	18" wide		3.20	5		199	305		504	690
2600	24" wide		2.80	5.714		229	350		579	790
2610	30" wide		2.40	6.667		246	410		656	900
2640	24" radius, 6" wide		3.60	4.444		235	273		508	680
2660	12" wide		3.20	5		267	305		572	765
2670	18" wide		2.80	5.714		294	350		644	865
2680	24" wide		2.40	6.667		370	410		780	1,025
2690	30" wide		2	8		395	490		885	1,200
2720	36" radius, 6" wide		3.20	5		380	305		685	885
2740	12" wide		2.80	5.714		435	350		785	1,025
2750	18" wide		2.40	6.667		490	410		900	1,175
2760	24" wide		2	8		575	490		1,065	1,375
2770	30" wide		1.60	10		610	615		1,225	1,625
2800	Tee vertical, 12" radius, 6" wide		4	4		217	245		462	620
2810	9" wide		3.80	4.211		218	258		476	640
2820	12" wide		3.60	4.444		234	273		507	680
2830	18" wide		3.40	4.706		238	289		527	705
2840	24" wide		3.20	5		241	305		546	735
2850	30" wide		3	5.333		253	325		578	785
2880	24" radius, 6" wide		3.60	4.444		305	273		578	755
2900	12" wide		3.20	5		330	305		635	835
2910	18" wide		3	5.333		355	325		680	895
2920	24" wide		2.80	5.714		375	350		725	955
2930	30" wide		2.60	6.154		405	380		785	1,025
2960	36" radius, 6" wide		3.20	5		465	305		770	980
2980	12" wide		2.80	5.714		490	350		840	1,075
2990	18" wide		2.60	6.154		510	380		890	1,150
3000	24" wide		2.40	6.667		545	410		955	1,225
3010	30" wide		2.20	7.273		590	445		1,035	1,325
3040	Cross horizontal, 12" radius, 6" wide		3.60	4.444		192	273		465	630
3050	9" wide		3.40	4.706		207	289		496	670
3060	12" wide		3.20	5		212	305		517	705

26 05 36.30 Cable Tray Trough		Crew	Daily Output	Labor-Hours	Unit	Material	Labor	Equipment	Total	Total Incl O&P
						2020 Bare Costs				
3070	18" wide	2 Elec	2.80	5.714	Ea.	222	350		572	785
3080	24" wide		2.40	6.667		248	410		658	900
3090	30" wide		2.20	7.273		315	445		760	1,025
3120	24" radius, 6" wide		3.20	5		345	305		650	850
3140	12" wide		2.80	5.714		380	350		730	955
3150	18" wide		2.40	6.667		405	410		815	1,075
3160	24" wide		2	8		470	490		960	1,275
3170	30" wide		1.80	8.889		510	545		1,055	1,400
3200	36" radius, 6" wide		2.80	5.714		600	350		950	1,200
3220	12" wide		2.40	6.667		630	410		1,040	1,325
3230	18" wide		2	8		685	490		1,175	1,500
3240	24" wide		1.60	10		795	615		1,410	1,825
3250	30" wide		1.40	11.429		860	700		1,560	2,025
3280	Dropout, or end plate, 6" wide		26	.615		19.65	38		57.65	79.50
3300	12" wide		22	.727		23.50	44.50		68	94
3310	18" wide		20	.800		28	49		77	107
3320	24" wide		18	.889		34	54.50		88.50	121
3330	30" wide		16	1		36	61.50		97.50	134
3370	Reducer, 9" to 6" wide		12	1.333		96.50	82		178.50	232
3380	12" to 6" wide		11.40	1.404		101	86		187	244
3390	12" to 9" wide		11.40	1.404		101	86		187	244
3400	18" to 12" wide		9.60	1.667		108	102		210	276
3420	18" to 6" wide		9.60	1.667		108	102		210	276
3430	18" to 9" wide		9.60	1.667		108	102		210	276
3440	24" to 18" wide		8.40	1.905		116	117		233	305
3460	24" to 12" wide		8.40	1.905		116	117		233	305
3470	24" to 9" wide		8.40	1.905		117	117		234	310
3475	24" to 6" wide		8.40	1.905		117	117		234	310
3480	30" to 24" wide		7.20	2.222		119	136		255	340
3500	30" to 18" wide		7.20	2.222		119	136		255	340
3520	30" to 12" wide		7.20	2.222		121	136		257	345
3540	36" to 30" wide		6.40	2.500		123	153		276	370
3560	36" to 24" wide		6.40	2.500		123	153		276	370
3580	36" to 18" wide		6.40	2.500		123	153		276	370
3600	36" to 12" wide		6.40	2.500		123	153		276	370
3610	Elbow horizontal, 60°, 12" radius, 6" wide		7.80	2.051		84	126		210	286
3620	9" wide		7.20	2.222		91.50	136		227.50	310
3630	12" wide		6.40	2.500		101	153		254	345
3640	18" wide		5.80	2.759		114	169		283	385
3650	24" wide		4.80	3.333		133	205		338	460
3680	Elbow horizontal, 45°, 12" radius, 6" wide		8	2		75.50	123		198.50	272
3690	9" wide		7.40	2.162		76.50	133		209.50	289
3700	12" wide		6.60	2.424		81	149		230	320
3710	18" wide		6	2.667		91	164		255	350
3720	24" wide		5	3.200		106	196		302	415
3750	Elbow horizontal, 30°, 12" radius, 6" wide		8.20	1.951		63	120		183	254
3760	9" wide		7.60	2.105		66	129		195	272
3770	12" wide		6.80	2.353		72	144		216	300
3780	18" wide		6.20	2.581		76.50	158		234.50	330
3790	24" wide		5.20	3.077		84	189		273	385
3820	Elbow vertical, 60° in/outside, 12" radius, 6" wide		7.80	2.051		104	126		230	310
3830	9" wide		7.20	2.222		106	136		242	325
3840	12" wide		6.40	2.500		107	153		260	355

26 05 36.30 Cable Tray Trough		Crew	Daily Output	Labor-Hours	Unit	2020 Bare Costs Material	Labor	Equipment	Total	Total Incl O&P
3850	18" wide	2 Elec	5.80	2.759	Ea.	111	169		280	385
3860	24" wide		4.80	3.333		116	205		321	440
3890	Elbow vertical, 45° in/outside, 12" radius, 6" wide		8	2		88.50	123		211.50	287
3900	9" wide		7.40	2.162		85.50	133		218.50	298
3910	12" wide		6.60	2.424		91.50	149		240.50	330
3920	18" wide		6	2.667		94.50	164		258.50	355
3930	24" wide		5	3.200		104	196		300	415
3960	Elbow vertical, 30° in/outside, 12" radius, 6" wide		8.20	1.951		72	120		192	264
3970	9" wide		7.60	2.105		76.50	129		205.50	284
3980	12" wide		6.80	2.353		78	144		222	310
3990	18" wide		6.20	2.581		81	158		239	335
4000	24" wide		5.20	3.077		82.50	189		271.50	380
4250	Reducer, left or right hand, 24" to 18" wide		8.40	1.905		110	117		227	300
4260	24" to 12" wide		8.40	1.905		110	117		227	300
4270	24" to 9" wide		8.40	1.905		110	117		227	300
4280	24" to 6" wide		8.40	1.905		111	117		228	300
4290	18" to 12" wide		9.60	1.667		103	102		205	270
4300	18" to 9" wide		9.60	1.667		103	102		205	270
4310	18" to 6" wide		9.60	1.667		103	102		205	270
4320	12" to 9" wide		11.40	1.404		98.50	86		184.50	241
4330	12" to 6" wide		11.40	1.404		98.50	86		184.50	241
4340	9" to 6" wide		12	1.333		95.50	82		177.50	231
4350	Splice plate	1 Elec	48	.167		9.10	10.25		19.35	26
4360	Splice plate, expansion joint		48	.167		9.35	10.25		19.60	26
4370	Splice plate, hinged, horizontal		48	.167		7.50	10.25		17.75	24
4380	Vertical		48	.167		10.70	10.25		20.95	27.50
4390	Trough, hanger, vertical		28	.286		30	17.55		47.55	60
4400	Box connector, 24" wide		20	.400		36	24.50		60.50	77.50
4410	18" wide		21	.381		34	23.50		57.50	73
4420	12" wide		22	.364		32.50	22.50		55	70
4430	9" wide		23	.348		31	21.50		52.50	67
4440	6" wide		24	.333		29.50	20.50		50	64
4450	Floor flange		24	.333		31	20.50		51.50	65.50
4460	Hold down clamp		60	.133		3.53	8.20		11.73	16.50
4520	Wall bracket, 24" wide tray		20	.400		35	24.50		59.50	76.50
4530	18" wide tray		21	.381		34	23.50		57.50	73.50
4540	12" wide tray		22	.364		18.25	22.50		40.75	54.50
4550	9" wide tray		23	.348		16.30	21.50		37.80	51
4560	6" wide tray		24	.333		14.85	20.50		35.35	48
5000	Cable channel aluminum, vented, 1-1/4" deep, 4" wide, straight		80	.100	L.F.	8	6.15		14.15	18.25
5010	Elbow horizontal, 36" radius, 90°		5	1.600	Ea.	145	98		243	310
5020	60°		5.50	1.455		111	89		200	259
5030	45°		6	1.333		90.50	82		172.50	226
5040	30°		6.50	1.231		77.50	75.50		153	201
5050	Adjustable		6	1.333		73.50	82		155.50	207
5060	Elbow vertical, 36" radius, 90°		5	1.600		156	98		254	320
5070	60°		5.50	1.455		120	89		209	269
5080	45°		6	1.333		98.50	82		180.50	234
5090	30°		6.50	1.231		87.50	75.50		163	212
5100	Adjustable		6	1.333		73.50	82		155.50	207
5110	Splice plate, hinged, horizontal		48	.167		10.20	10.25		20.45	27
5120	Splice plate, hinged, vertical		48	.167		14.45	10.25		24.70	31.50
5130	Hanger, vertical		28	.286		15.95	17.55		33.50	44.50

26 05 Common Work Results for Electrical

26 05 36 – Cable Trays for Electrical Systems

26 05 36.30 Cable Tray Trough		Crew	Daily Output	Labor-Hours	Unit	2020 Bare Costs Material	Labor	Equipment	Total	Total Incl O&P
5140	Single	1 Elec	28	.286	Ea.	24.50	17.55		42.05	53.50
5150	Double		20	.400		25	24.50		49.50	65.50
5160	Channel to box connector		24	.333		33	20.50		53.50	68
5170	Hold down clip		80	.100		4.60	6.15		10.75	14.50
5180	Wall bracket, single		28	.286		11.80	17.55		29.35	40
5190	Double		20	.400		15.15	24.50		39.65	54.50
5200	Cable roller		16	.500		186	30.50		216.50	251
5210	Splice plate	↓	48	.167	↓	5.70	10.25		15.95	22

26 05 36.40 Cable Tray, Covers and Dividers		Crew	Daily Output	Labor-Hours	Unit	Material	Labor	Equipment	Total	Total Incl O&P
0010	**CABLE TRAY, COVERS AND DIVIDERS** To 10′ high									
0011	For higher elevations, see lines 9900 – 9960									
0100	Covers, ventilated galv. steel, straight, 6″ wide tray size	2 Elec	520	.031	L.F.	8.20	1.89		10.09	11.90
0200	9″ wide tray size		460	.035		8.70	2.13		10.83	12.90
0300	12″ wide tray size		400	.040		14.35	2.45		16.80	19.55
0400	18″ wide tray size		300	.053		34	3.27		37.27	42.50
0500	24″ wide tray size		220	.073		58.50	4.46		62.96	71.50
0600	30″ wide tray size		180	.089	↓	26.50	5.45		31.95	37.50
1000	Elbow horizontal, 90°, 12″ radius, 6″ wide tray size		150	.107	Ea.	43.50	6.55		50.05	58
1020	9″ wide tray size		128	.125		48	7.65		55.65	65
1040	12″ wide tray size		108	.148		51	9.10		60.10	70
1060	18″ wide tray size		84	.190		70.50	11.70		82.20	95.50
1080	24″ wide tray size		66	.242		85	14.85		99.85	117
1100	30″ wide tray size		60	.267		108	16.35		124.35	144
1160	24″ radius, 6″ wide tray size		136	.118		72.50	7.20		79.70	90.50
1180	9″ wide tray size		116	.138		74	8.45		82.45	94.50
1200	12″ wide tray size		96	.167		82.50	10.25		92.75	106
1220	18″ wide tray size		76	.211		102	12.90		114.90	132
1240	24″ wide tray size		60	.267		124	16.35		140.35	161
1260	30″ wide tray size		52	.308		163	18.90		181.90	208
1320	36″ radius, 6″ wide tray size		120	.133		108	8.20		116.20	132
1340	9″ wide tray size		104	.154		119	9.45		128.45	146
1360	12″ wide tray size		84	.190		127	11.70		138.70	158
1380	18″ wide tray size		72	.222		162	13.65		175.65	199
1400	24″ wide tray size		52	.308		193	18.90		211.90	241
1420	30″ wide tray size		46	.348		229	21.50		250.50	285
1480	Elbow horizontal, 45°, 12″ radius, 6″ wide tray size		150	.107		32.50	6.55		39.05	46
1500	9″ wide tray size		128	.125		39	7.65		46.65	55
1520	12″ wide tray size		108	.148		42.50	9.10		51.60	61
1540	18″ wide tray size		88	.182		52	11.15		63.15	74
1560	24″ wide tray size		76	.211		60.50	12.90		73.40	86.50
1580	30″ wide tray size		66	.242		71.50	14.85		86.35	102
1640	24″ radius, 6″ wide tray size		136	.118		45.50	7.20		52.70	61
1660	9″ wide tray size		116	.138		51	8.45		59.45	69
1680	12″ wide tray size		96	.167		57.50	10.25		67.75	79
1700	18″ wide tray size		80	.200		66.50	12.25		78.75	92
1720	24″ wide tray size		70	.229		76	14		90	105
1740	30″ wide tray size		60	.267		96.50	16.35		112.85	131
1800	36″ radius, 6″ wide tray size		120	.133		67	8.20		75.20	86
1820	9″ wide tray size		104	.154		75.50	9.45		84.95	97.50
1840	12″ wide tray size		84	.190		83	11.70		94.70	110
1860	18″ wide tray size		76	.211		95.50	12.90		108.40	125
1880	24″ wide tray size	↓	62	.258	↓	116	15.85		131.85	153

26 05 36.40 Cable Tray, Covers and Dividers		Crew	Daily Output	Labor-Hours	Unit	Material	2020 Bare Costs Labor	Equipment	Total	Total Incl O&P
1900	30" wide tray size	2 Elec	52	.308	Ea.	126	18.90		144.90	168
1960	Elbow vertical, 90°, 12" radius, 6" wide tray size		150	.107		38.50	6.55		45.05	52.50
1980	9" wide tray size		128	.125		39	7.65		46.65	55
2000	12" wide tray size		108	.148		42	9.10		51.10	60.50
2020	18" wide tray size		88	.182		48	11.15		59.15	70
2040	24" wide tray size		68	.235		49	14.45		63.45	76
2060	30" wide tray size		60	.267		54	16.35		70.35	84
2120	24" radius, 6" wide tray size		136	.118		47	7.20		54.20	63
2140	9" wide tray size		116	.138		52	8.45		60.45	70
2160	12" wide tray size		96	.167		54	10.25		64.25	75
2180	18" wide tray size		80	.200		71.50	12.25		83.75	97.50
2200	24" wide tray size		62	.258		80	15.85		95.85	113
2220	30" wide tray size		52	.308		88	18.90		106.90	126
2280	36" radius, 6" wide tray size		120	.133		53	8.20		61.20	71
2300	9" wide tray size		104	.154		68	9.45		77.45	89.50
2320	12" wide tray size		84	.190		75.50	11.70		87.20	101
2340	18" wide tray size		76	.211		95.50	12.90		108.40	125
2350	24" wide tray size		54	.296		106	18.20		124.20	144
2360	30" wide tray size		46	.348		127	21.50		148.50	172
2400	Tee horizontal, 12" radius, 6" wide tray size		92	.174		69.50	10.65		80.15	93
2410	9" wide tray size		80	.200		71.50	12.25		83.75	97.50
2420	12" wide tray size		68	.235		81	14.45		95.45	111
2430	18" wide tray size		60	.267		98	16.35		114.35	133
2440	24" wide tray size		52	.308		123	18.90		141.90	165
2460	30" wide tray size		36	.444		145	27.50		172.50	201
2500	24" radius, 6" wide tray size		88	.182		112	11.15		123.15	140
2510	9" wide tray size		76	.211		119	12.90		131.90	150
2520	12" wide tray size		64	.250		122	15.35		137.35	158
2530	18" wide tray size		56	.286		157	17.55		174.55	199
2540	24" wide tray size		48	.333		240	20.50		260.50	296
2560	30" wide tray size		32	.500		274	30.50		304.50	345
2600	36" radius, 6" wide tray size		84	.190		191	11.70		202.70	228
2610	9" wide tray size		72	.222		194	13.65		207.65	235
2620	12" wide tray size		60	.267		216	16.35		232.35	263
2630	18" wide tray size		52	.308		253	18.90		271.90	305
2640	24" wide tray size		44	.364		335	22.50		357.50	400
2660	30" wide tray size		28	.571		360	35		395	455
2700	Cross horizontal, 12" radius, 6" wide tray size		68	.235		104	14.45		118.45	136
2710	9" wide tray size		64	.250		110	15.35		125.35	145
2720	12" wide tray size		60	.267		122	16.35		138.35	160
2730	18" wide tray size		52	.308		145	18.90		163.90	188
2740	24" wide tray size		36	.444		171	27.50		198.50	230
2760	30" wide tray size		30	.533		194	32.50		226.50	265
2800	24" radius, 6" wide tray size		64	.250		187	15.35		202.35	230
2810	9" wide tray size		60	.267		203	16.35		219.35	248
2820	12" wide tray size		56	.286		219	17.55		236.55	268
2830	18" wide tray size		48	.333		262	20.50		282.50	320
2840	24" wide tray size		32	.500		320	30.50		350.50	395
2860	30" wide tray size		26	.615		350	38		388	445
2900	36" radius, 6" wide tray size		60	.267		320	16.35		336.35	375
2910	9" wide tray size		56	.286		320	17.55		337.55	375
2920	12" wide tray size		52	.308		345	18.90		363.90	410
2930	18" wide tray size	↓	44	.364	↓	390	22.50		412.50	465

26 05 36.40 Cable Tray, Covers and Dividers		Crew	Daily Output	Labor-Hours	Unit	Material	Labor	Equipment	Total	Total Incl O&P
						2020 Bare Costs				
2940	24" wide tray size	2 Elec	28	.571	Ea.	490	35		525	595
2960	30" wide tray size		22	.727		525	44.50		569.50	650
3000	Reducer, 9" to 6" wide tray size		128	.125		44.50	7.65		52.15	61
3010	12" to 6" wide tray size		108	.148		46.50	9.10		55.60	65
3020	12" to 9" wide tray size		108	.148		46.50	9.10		55.60	65
3030	18" to 12" wide tray size		88	.182		49.50	11.15		60.65	71.50
3050	18" to 6" wide tray size		88	.182		49.50	11.15		60.65	71.50
3060	24" to 18" wide tray size		80	.200		69.50	12.25		81.75	95
3070	24" to 12" wide tray size		80	.200		63.50	12.25		75.75	89
3090	30" to 24" wide tray size		70	.229		73.50	14		87.50	102
3100	30" to 18" wide tray size		70	.229		73.50	14		87.50	102
3110	30" to 12" wide tray size		70	.229		63.50	14		77.50	91.50
3140	36" to 30" wide tray size		64	.250		80	15.35		95.35	112
3150	36" to 24" wide tray size		64	.250		80	15.35		95.35	112
3160	36" to 18" wide tray size		64	.250		80	15.35		95.35	112
3170	36" to 12" wide tray size		64	.250		80	15.35		95.35	112
3250	Covers, aluminum, straight, 6" wide tray size		520	.031	L.F.	4.79	1.89		6.68	8.15
3270	9" wide tray size		460	.035		5.80	2.13		7.93	9.65
3290	12" wide tray size		400	.040		6.90	2.45		9.35	11.40
3310	18" wide tray size		320	.050		9.15	3.07		12.22	14.75
3330	24" wide tray size		260	.062		11.40	3.78		15.18	18.35
3350	30" wide tray size		200	.080		12.60	4.91		17.51	21.50
3400	Elbow horizontal, 90°, 12" radius, 6" wide tray size		150	.107	Ea.	36	6.55		42.55	49.50
3410	9" wide tray size		128	.125		37.50	7.65		45.15	53
3420	12" wide tray size		108	.148		40.50	9.10		49.60	58.50
3430	18" wide tray size		88	.182		52.50	11.15		63.65	75
3440	24" wide tray size		70	.229		66.50	14		80.50	94.50
3460	30" wide tray size		64	.250		79.50	15.35		94.85	111
3500	24" radius, 6" wide tray size		136	.118		46.50	7.20		53.70	62
3510	9" wide tray size		116	.138		57.50	8.45		65.95	76
3520	12" wide tray size		96	.167		63	10.25		73.25	85.50
3530	18" wide tray size		80	.200		75	12.25		87.25	101
3540	24" wide tray size		64	.250		92.50	15.35		107.85	126
3560	30" wide tray size		56	.286		113	17.55		130.55	152
3600	36" radius, 6" wide tray size		120	.133		81	8.20		89.20	102
3610	9" wide tray size		104	.154		89	9.45		98.45	112
3620	12" wide tray size		84	.190		101	11.70		112.70	129
3630	18" wide tray size		76	.211		120	12.90		132.90	152
3640	24" wide tray size		56	.286		146	17.55		163.55	188
3660	30" wide tray size		50	.320		168	19.65		187.65	214
3700	Elbow horizontal, 45°, 12" radius, 6" wide tray size		150	.107		26.50	6.55		33.05	39
3710	9" wide tray size		128	.125		27	7.65		34.65	41.50
3720	12" wide tray size		108	.148		30.50	9.10		39.60	47.50
3730	18" wide tray size		88	.182		35	11.15		46.15	55
3740	24" wide tray size		80	.200		38.50	12.25		50.75	61.50
3760	30" wide tray size		70	.229		48.50	14		62.50	75
3800	24" radius, 6" wide tray size		136	.118		29	7.20		36.20	43
3810	9" wide tray size		116	.138		37	8.45		45.45	53.50
3820	12" wide tray size		96	.167		38.50	10.25		48.75	58
3830	18" wide tray size		80	.200		46.50	12.25		58.75	70
3840	24" wide tray size		72	.222		57.50	13.65		71.15	84
3860	30" wide tray size		64	.250		66	15.35		81.35	96.50
3900	36" radius, 6" wide tray size		120	.133		51	8.20		59.20	69

26 05 36 – Cable Trays for Electrical Systems

26 05 36.40 Cable Tray, Covers and Dividers		Crew	Daily Output	Labor-Hours	Unit	Material	2020 Bare Costs Labor	Equipment	Total	Total Incl O&P
3910	9" wide tray size	2 Elec	104	.154	Ea.	55.50	9.45		64.95	75.50
3920	12" wide tray size		84	.190		58.50	11.70		70.20	82.50
3930	18" wide tray size		76	.211		70.50	12.90		83.40	97.50
3940	24" wide tray size		64	.250		85.50	15.35		100.85	118
3960	30" wide tray size		56	.286		97.50	17.55		115.05	134
3970	36" wide tray size		50	.320		111	19.65		130.65	152
4000	Elbow vertical, 90°, 12" radius, 6" wide tray size		150	.107		30.50	6.55		37.05	43.50
4010	9" wide tray size		128	.125		30.50	7.65		38.15	45.50
4020	12" wide tray size		108	.148		33.50	9.10		42.60	50.50
4030	18" wide tray size		88	.182		38.50	11.15		49.65	59.50
4040	24" wide tray size		70	.229		40	14		54	65.50
4060	30" wide tray size		64	.250		41	15.35		56.35	68.50
4070	36" wide tray size		54	.296		50	18.20		68.20	83
4100	24" radius, 6" wide tray size		136	.118		35	7.20		42.20	49
4110	9" wide tray size		116	.138		38	8.45		46.45	54.50
4120	12" wide tray size		96	.167		41	10.25		51.25	61
4130	18" wide tray size		80	.200		49.50	12.25		61.75	73.50
4140	24" wide tray size		64	.250		54	15.35		69.35	83
4160	30" wide tray size		56	.286		66	17.55		83.55	100
4170	36" wide tray size		48	.333		72	20.50		92.50	111
4200	36" radius, 6" wide tray size		120	.133		39	8.20		47.20	55.50
4210	9" wide tray size		104	.154		48	9.45		57.45	67.50
4220	12" wide tray size		84	.190		55.50	11.70		67.20	79
4230	18" wide tray size		76	.211		67.50	12.90		80.40	94.50
4240	24" wide tray size		56	.286		79.50	17.55		97.05	115
4260	30" wide tray size		50	.320		101	19.65		120.65	141
4270	36" wide tray size		44	.364		108	22.50		130.50	154
4300	Tee horizontal, 12" radius, 6" wide tray size		108	.148		49.50	9.10		58.60	68.50
4310	9" wide tray size		88	.182		52.50	11.15		63.65	75
4320	12" wide tray size		80	.200		58.50	12.25		70.75	83.50
4330	18" wide tray size		68	.235		69.50	14.45		83.95	98.50
4340	24" wide tray size		56	.286		88	17.55		105.55	124
4360	30" wide tray size		44	.364		103	22.50		125.50	149
4370	36" wide tray size		36	.444		125	27.50		152.50	179
4400	24" radius, 6" wide tray size		96	.167		80.50	10.25		90.75	104
4410	9" wide tray size		80	.200		85.50	12.25		97.75	113
4420	12" wide tray size		72	.222		95.50	13.65		109.15	126
4430	18" wide tray size		60	.267		112	16.35		128.35	148
4440	24" wide tray size		48	.333		178	20.50		198.50	228
4460	30" wide tray size		40	.400		197	24.50		221.50	254
4470	36" wide tray size		32	.500		219	30.50		249.50	288
4500	36" radius, 6" wide tray size		88	.182		143	11.15		154.15	174
4510	9" wide tray size		72	.222		146	13.65		159.65	182
4520	12" wide tray size		64	.250		160	15.35		175.35	200
4530	18" wide tray size		56	.286		182	17.55		199.55	227
4540	24" wide tray size		44	.364		231	22.50		253.50	289
4560	30" wide tray size		36	.444		260	27.50		287.50	330
4570	36" wide tray size		28	.571		293	35		328	380
4600	Cross horizontal, 12" radius, 6" wide tray size		80	.200		75.50	12.25		87.75	102
4610	9" wide tray size		72	.222		80	13.65		93.65	109
4620	12" wide tray size		64	.250		89	15.35		104.35	122
4630	18" wide tray size		56	.286		108	17.55		125.55	146
4640	24" wide tray size	↓	48	.333	↓	125	20.50		145.50	169

26 05 36 – Cable Trays for Electrical Systems

26 05 36.40 Cable Tray, Covers and Dividers		Crew	Daily Output	Labor-Hours	Unit	Material	2020 Bare Costs Labor	Equipment	Total	Total Incl O&P
4660	30" wide tray size	2 Elec	40	.400	Ea.	148	24.50		172.50	201
4670	36" wide tray size		32	.500		171	30.50		201.50	235
4700	24" radius, 6" wide tray size		72	.222		141	13.65		154.65	177
4710	9" wide tray size		64	.250		151	15.35		166.35	191
4720	12" wide tray size		56	.286		163	17.55		180.55	206
4730	18" wide tray size		48	.333		190	20.50		210.50	241
4740	24" wide tray size		40	.400		226	24.50		250.50	287
4760	30" wide tray size		32	.500		265	30.50		295.50	340
4770	36" wide tray size		24	.667		287	41		328	380
4800	36" radius, 6" wide tray size		64	.250		231	15.35		246.35	278
4810	9" wide tray size		56	.286		241	17.55		258.55	292
4820	12" wide tray size		50	.320		267	19.65		286.65	325
4830	18" wide tray size		44	.364		293	22.50		315.50	360
4840	24" wide tray size		36	.444		360	27.50		387.50	440
4860	30" wide tray size		28	.571		415	35		450	515
4870	36" wide tray size		22	.727		455	44.50		499.50	570
4900	Reducer, 9" to 6" wide tray size		128	.125		38	7.65		45.65	54
4910	12" to 6" wide tray size		108	.148		39.50	9.10		48.60	57.50
4920	12" to 9" wide tray size		108	.148		39.50	9.10		48.60	57.50
4930	18" to 12" wide tray size		88	.182		43.50	11.15		54.65	65
4950	18" to 6" wide tray size		88	.182		43.50	11.15		54.65	65
4960	24" to 18" wide tray size		80	.200		56	12.25		68.25	80.50
4970	24" to 12" wide tray size		80	.200		48	12.25		60.25	72
4990	30" to 24" wide tray size		70	.229		58.50	14		72.50	85.50
5000	30" to 18" wide tray size		70	.229		58.50	14		72.50	85.50
5010	30" to 12" wide tray size		70	.229		58.50	14		72.50	85.50
5040	36" to 30" wide tray size		64	.250		64.50	15.35		79.85	94.50
5050	36" to 24" wide tray size		64	.250		64.50	15.35		79.85	94.50
5060	36" to 18" wide tray size		64	.250		64.50	15.35		79.85	94.50
5070	36" to 12" wide tray size		64	.250		64.50	15.35		79.85	94.50
5710	Tray cover hold down clamp	1 Elec	60	.133		10.35	8.20		18.55	24
8000	Divider strip, straight, galvanized, 3" deep		200	.040	L.F.	5.55	2.45		8	9.90
8020	4" deep		180	.044		6.75	2.73		9.48	11.65
8040	6" deep		160	.050		8.85	3.07		11.92	14.45
8060	Aluminum, straight, 3" deep		210	.038		5.55	2.34		7.89	9.70
8080	4" deep		190	.042		6.85	2.58		9.43	11.55
8100	6" deep		170	.047		8.75	2.89		11.64	14.10
8110	Divider strip, vertical fitting, 3" deep									
8120	12" radius, galvanized, 30°	1 Elec	28	.286	Ea.	26.50	17.55		44.05	56.50
8140	45°		27	.296		33.50	18.20		51.70	65
8160	60°		26	.308		35.50	18.90		54.40	68
8180	90°		25	.320		44.50	19.65		64.15	79
8200	Aluminum, 30°		29	.276		17.80	16.90		34.70	45.50
8220	45°		28	.286		21	17.55		38.55	50
8240	60°		27	.296		24.50	18.20		42.70	55
8260	90°		26	.308		31	18.90		49.90	63
8280	24" radius, galvanized, 30°		25	.320		40	19.65		59.65	74
8300	45°		24	.333		42.50	20.50		63	78.50
8320	60°		23	.348		54	21.50		75.50	92
8340	90°		22	.364		73	22.50		95.50	115
8360	Aluminum, 30°		26	.308		28	18.90		46.90	60
8380	45°		25	.320		32.50	19.65		52.15	65.50
8400	60°		24	.333		40.50	20.50		61	76

26 05 36.40 Cable Tray, Covers and Dividers		Crew	Daily Output	Labor-Hours	Unit	Material	2020 Bare Costs Labor	Equipment	Total	Total Incl O&P
8420	90°	1 Elec	23	.348	Ea.	55	21.50		76.50	93.50
8440	36" radius, galvanized, 30°		22	.364		51	22.50		73.50	90.50
8460	45°		21	.381		59.50	23.50		83	101
8480	60°		20	.400		69.50	24.50		94	115
8500	90°		19	.421		96.50	26		122.50	146
8520	Aluminum, 30°		23	.348		42.50	21.50		64	80
8540	45°		22	.364		55	22.50		77.50	95
8560	60°		21	.381		71.50	23.50		95	115
8570	90°		20	.400		93.50	24.50		118	141
8590	Divider strip, vertical fitting, 4" deep									
8600	12" radius, galvanized, 30°	1 Elec	27	.296	Ea.	32	18.20		50.20	63
8610	45°		26	.308		37	18.90		55.90	69.50
8620	60°		25	.320		41.50	19.65		61.15	76
8630	90°		24	.333		51	20.50		71.50	87.50
8640	Aluminum, 30°		28	.286		24	17.55		41.55	53.50
8650	45°		27	.296		27.50	18.20		45.70	58.50
8660	60°		26	.308		31.50	18.90		50.40	64
8670	90°		25	.320		37.50	19.65		57.15	71
8680	24" radius, galvanized, 30°		24	.333		51	20.50		71.50	87.50
8690	45°		23	.348		64	21.50		85.50	104
8700	60°		22	.364		72.50	22.50		95	114
8710	90°		21	.381		96.50	23.50		120	142
8720	Aluminum, 30°		25	.320		37.50	19.65		57.15	71
8730	45°		24	.333		45.50	20.50		66	81.50
8740	60°		23	.348		54	21.50		75.50	92
8750	90°		22	.364		73.50	22.50		96	116
8760	36" radius, galvanized, 30°		23	.348		60.50	21.50		82	99.50
8770	45°		22	.364		69	22.50		91.50	111
8780	60°		21	.381		87	23.50		110.50	132
8790	90°		20	.400		116	24.50		140.50	165
8800	Aluminum, 30°		24	.333		58.50	20.50		79	96
8810	45°		23	.348		75	21.50		96.50	116
8820	60°		22	.364		91.50	22.50		114	136
8830	90°		21	.381		116	23.50		139.50	163
8840	Divider strip, vertical fitting, 6" deep									
8850	12" radius, galvanized, 30°	1 Elec	24	.333	Ea.	36.50	20.50		57	72
8860	45°		23	.348		41	21.50		62.50	78
8870	60°		22	.364		47	22.50		69.50	86.50
8880	90°		21	.381		59	23.50		82.50	101
8890	Aluminum, 30°		25	.320		27.50	19.65		47.15	60
8900	45°		24	.333		32.50	20.50		53	67
8910	60°		23	.348		34	21.50		55.50	70.50
8920	90°		22	.364		40	22.50		62.50	78.50
8930	24" radius, galvanized, 30°		23	.348		51	21.50		72.50	89
8940	45°		22	.364		64	22.50		86.50	105
8950	60°		21	.381		73	23.50		96.50	117
8960	90°		20	.400		96.50	24.50		121	144
8970	Aluminum, 30°		24	.333		38.50	20.50		59	73.50
8980	45°		23	.348		52	21.50		73.50	90.50
8990	60°		22	.364		57.50	22.50		80	97.50
9000	90°		21	.381		78	23.50		101.50	122
9010	36" radius, galvanized, 30°		22	.364		59.50	22.50		82	99.50
9020	45°		21	.381		73	23.50		96.50	117

26 05 Common Work Results for Electrical

26 05 36 – Cable Trays for Electrical Systems

26 05 36.40 Cable Tray, Covers and Dividers		Crew	Daily Output	Labor-Hours	Unit	2020 Bare Costs Material	Labor	Equipment	Total	Total Incl O&P
9030	60°	1 Elec	20	.400	Ea.	96.50	24.50		121	144
9040	90°		19	.421		127	26		153	179
9050	Aluminum, 30°		23	.348		59.50	21.50		81	98.50
9060	45°		22	.364		79.50	22.50		102	122
9070	60°		21	.381		94	23.50		117.50	139
9080	90°		20	.400		113	24.50		137.50	162
9120	Divider strip, horizontal fitting, galvanized, 3″ deep		33	.242		36.50	14.85		51.35	63
9130	4″ deep		30	.267		40.50	16.35		56.85	70
9140	6″ deep		27	.296		52	18.20		70.20	85
9150	Aluminum, 3″ deep		35	.229		27.50	14		41.50	51.50
9160	4″ deep		32	.250		30	15.35		45.35	56.50
9170	6″ deep		29	.276		40	16.90		56.90	70
9300	Divider strip protector		300	.027	L.F.	2.96	1.64		4.60	5.80
9310	Fastener, ladder tray				Ea.	.52			.52	.57
9320	Trough or solid bottom tray				"	.39			.39	.43
9900	Add to labor for higher elevated installation									
9905	10′ to 14.5′ high, add						10%			
9910	15′ to 20′ high, add						20%			
9920	20′ to 25′ high, add						25%			
9930	25′ to 30′ high, add						35%			
9940	30′ to 35′ high, add						40%			
9950	35′ to 40′ high, add						50%			
9960	Over 40′ high, add						55%			

26 05 39 – Underfloor Raceways for Electrical Systems

26 05 39.30 Conduit In Concrete Slab

Line	Description	Crew	Daily Output	Labor-Hours	Unit	Material	Labor	Equipment	Total	Total Incl O&P
0010	**CONDUIT IN CONCRETE SLAB** Including terminations,									
0020	fittings and supports									
3230	PVC, schedule 40, 1/2″ diameter	1 Elec	270	.030	L.F.	.55	1.82		2.37	3.40
3250	3/4″ diameter		230	.035		.59	2.13		2.72	3.94
3270	1″ diameter		200	.040		.78	2.45		3.23	4.63
3300	1-1/4″ diameter		170	.047		1.05	2.89		3.94	5.60
3330	1-1/2″ diameter		140	.057		1.27	3.51		4.78	6.80
3350	2″ diameter		120	.067		1.59	4.09		5.68	8.05
3370	2-1/2″ diameter		90	.089		2.62	5.45		8.07	11.30
3400	3″ diameter	2 Elec	160	.100		3.47	6.15		9.62	13.25
3430	3-1/2″ diameter		120	.133		4.50	8.20		12.70	17.55
3440	4″ diameter		100	.160		4.76	9.80		14.56	20.50
3450	5″ diameter		80	.200		7.45	12.25		19.70	27
3460	6″ diameter		60	.267		9.65	16.35		26	35.50
3530	Sweeps, 1″ diameter, 30″ radius	1 Elec	32	.250	Ea.	9.55	15.35		24.90	34
3550	1-1/4″ diameter		24	.333		8.55	20.50		29.05	41
3570	1-1/2″ diameter		21	.381		9.35	23.50		32.85	46.50
3600	2″ diameter		18	.444		11.25	27.50		38.75	54.50
3630	2-1/2″ diameter		14	.571		80	35		115	142
3650	3″ diameter		10	.800		68	49		117	151
3670	3-1/2″ diameter		8	1		52	61.50		113.50	152
3700	4″ diameter		7	1.143		25	70		95	136
3710	5″ diameter		6	1.333		55	82		137	187
3730	Couplings, 1/2″ diameter					.12			.12	.13
3750	3/4″ diameter					.15			.15	.17
3770	1″ diameter					.23			.23	.25
3800	1-1/4″ diameter					.38			.38	.42

26 05 39 – Underfloor Raceways for Electrical Systems

26 05 39.30 Conduit In Concrete Slab		Crew	Daily Output	Labor-Hours	Unit	2020 Bare Costs Material	Labor	Equipment	Total	Total Incl O&P
3830	1-1/2" diameter				Ea.	.43			.43	.47
3850	2" diameter					.58			.58	.64
3870	2-1/2" diameter					1.11			1.11	1.22
3900	3" diameter					1.76			1.76	1.94
3930	3-1/2" diameter					2.45			2.45	2.70
3950	4" diameter					2.74			2.74	3.01
3960	5" diameter					5.65			5.65	6.20
3970	6" diameter					8.85			8.85	9.75
4030	End bells, 1" diameter, PVC	1 Elec	60	.133		2.68	8.20		10.88	15.55
4050	1-1/4" diameter		53	.151		4.54	9.25		13.79	19.25
4100	1-1/2" diameter		48	.167		3.10	10.25		13.35	19.15
4150	2" diameter		34	.235		4.36	14.45		18.81	27
4170	2-1/2" diameter		27	.296		4.05	18.20		22.25	32.50
4200	3" diameter		20	.400		6.30	24.50		30.80	45
4250	3-1/2" diameter		16	.500		6.90	30.50		37.40	54.50
4300	4" diameter		14	.571		7.75	35		42.75	62.50
4310	5" diameter		12	.667		12.05	41		53.05	76.50
4320	6" diameter		9	.889		12.35	54.50		66.85	97.50
4350	Rigid galvanized steel, 1/2" diameter		200	.040	L.F.	2.34	2.45		4.79	6.35
4400	3/4" diameter		170	.047		4.45	2.89		7.34	9.35
4450	1" diameter		130	.062		6.90	3.78		10.68	13.40
4500	1-1/4" diameter		110	.073		4.53	4.46		8.99	11.85
4600	1-1/2" diameter		100	.080		8.40	4.91		13.31	16.80
4800	2" diameter		90	.089		8.20	5.45		13.65	17.45
9000	Minimum labor/equipment charge		4	2	Job		123		123	189

26 05 39.40 Conduit In Trench		Crew	Daily Output	Labor-Hours	Unit	Material	Labor	Equipment	Total	Total Incl O&P
0010	**CONDUIT IN TRENCH** Includes terminations and fittings									
0020	Does not include excavation or backfill, see Section 31 23 16									
0200	Rigid galvanized steel, 2" diameter	1 Elec	150	.053	L.F.	7.80	3.27		11.07	13.65
0400	2-1/2" diameter	"	100	.080		13	4.91		17.91	22
0600	3" diameter	2 Elec	160	.100		14.55	6.15		20.70	25.50
0800	3-1/2" diameter		140	.114		19.20	7		26.20	32
1000	4" diameter		100	.160		23.50	9.80		33.30	40.50
1200	5" diameter		80	.200		37.50	12.25		49.75	60
1400	6" diameter		60	.267		57.50	16.35		73.85	88.50
9000	Minimum labor/equipment charge	1 Elec	4	2	Job		123		123	189

26 05 43 – Underground Ducts and Raceways for Electrical Systems

26 05 43.10 Trench Duct		Crew	Daily Output	Labor-Hours	Unit	Material	Labor	Equipment	Total	Total Incl O&P
0010	**TRENCH DUCT** Steel with cover									
0020	Standard adjustable, depths to 4"									
0100	Straight, single compartment, 9" wide	2 Elec	40	.400	L.F.	289	24.50		313.50	360
0200	12" wide		32	.500		365	30.50		395.50	445
0400	18" wide		26	.615		395	38		433	495
0600	24" wide		22	.727		415	44.50		459.50	525
0700	27" wide		21	.762		215	46.50		261.50	310
0800	30" wide		20	.800		228	49		277	325
1000	36" wide		16	1		261	61.50		322.50	380
1020	Two compartment, 9" wide		38	.421		132	26		158	185
1030	12" wide		30	.533		157	32.50		189.50	224
1040	18" wide		24	.667		199	41		240	282
1050	24" wide		20	.800		247	49		296	345
1060	30" wide		18	.889		315	54.50		369.50	435

26 05 43.10 Trench Duct		Crew	Daily Output	Labor-Hours	Unit	Material	Labor	Equipment	Total	Total Incl O&P
						2020 Bare Costs				
1070	36" wide	2 Elec	14	1.143	L.F.	350	70		420	495
1090	Three compartment, 9" wide		36	.444		152	27.50		179.50	209
1100	12" wide		28	.571		175	35		210	247
1110	18" wide		22	.727		213	44.50		257.50	305
1120	24" wide		18	.889		267	54.50		321.50	375
1130	30" wide		16	1		330	61.50		391.50	460
1140	36" wide		12	1.333		380	82		462	540
1200	Horizontal elbow, 9" wide		5.40	2.963	Ea.	455	182		637	780
1400	12" wide		4.60	3.478		440	213		653	815
1600	18" wide		4	4		625	245		870	1,075
1800	24" wide		3.20	5		895	305		1,200	1,450
1900	27" wide		3	5.333		1,075	325		1,400	1,700
2000	30" wide		2.60	6.154		1,175	380		1,555	1,875
2200	36" wide		2.40	6.667		1,575	410		1,985	2,350
2220	Two compartment, 9" wide		3.80	4.211		735	258		993	1,200
2230	12" wide		3	5.333		810	325		1,135	1,400
2240	18" wide		2.40	6.667		975	410		1,385	1,700
2250	24" wide		2	8		1,175	490		1,665	2,025
2260	30" wide		1.80	8.889		1,550	545		2,095	2,575
2270	36" wide		1.60	10		1,900	615		2,515	3,025
2290	Three compartment, 9" wide		3.60	4.444		725	273		998	1,225
2300	12" wide		2.80	5.714		810	350		1,160	1,425
2310	18" wide		2.20	7.273		985	445		1,430	1,750
2320	24" wide		1.80	8.889		1,250	545		1,795	2,225
2330	30" wide		1.60	10		1,625	615		2,240	2,725
2350	36" wide		1.40	11.429		1,975	700		2,675	3,250
2400	Vertical elbow, 9" wide		5.40	2.963		179	182		361	475
2600	12" wide		4.60	3.478		172	213		385	520
2800	18" wide		4	4		197	245		442	595
3000	24" wide		3.20	5		245	305		550	740
3100	27" wide		3	5.333		256	325		581	785
3200	30" wide		2.60	6.154		271	380		651	880
3400	36" wide		2.40	6.667		297	410		707	955
3600	Cross, 9" wide		4	4		745	245		990	1,200
3800	12" wide		3.20	5		785	305		1,090	1,325
4000	18" wide		2.60	6.154		940	380		1,320	1,600
4200	24" wide		2.20	7.273		1,175	445		1,620	1,950
4300	27" wide		2.20	7.273		1,325	445		1,770	2,125
4400	30" wide		2	8		1,450	490		1,940	2,350
4600	36" wide		1.80	8.889		1,825	545		2,370	2,850
4620	Two compartment, 9" wide		3.80	4.211		730	258		988	1,200
4630	12" wide		3	5.333		765	325		1,090	1,350
4640	18" wide		2.40	6.667		930	410		1,340	1,650
4650	24" wide		2	8		1,175	490		1,665	2,050
4660	30" wide		1.80	8.889		1,550	545		2,095	2,575
4670	36" wide		1.60	10		1,875	615		2,490	3,025
4690	Three compartment, 9" wide		3.60	4.444		740	273		1,013	1,225
4700	12" wide		2.80	5.714		850	350		1,200	1,475
4710	18" wide		2.20	7.273		1,000	445		1,445	1,775
4720	24" wide		1.80	8.889		1,250	545		1,795	2,225
4730	30" wide		1.60	10		1,625	615		2,240	2,750
4740	36" wide		1.40	11.429		2,000	700		2,700	3,275
4800	End closure, 9" wide		14.40	1.111		44	68		112	153

26 05 43.10 Trench Duct		Crew	Daily Output	Labor-Hours	Unit	Material	Labor	Equipment	Total	Total Incl O&P
						2020 Bare Costs				
5000	12" wide	2 Elec	12	1.333	Ea.	50	82		132	181
5200	18" wide		10	1.600		77	98		175	236
5400	24" wide		8	2		101	123		224	300
5500	27" wide		7	2.286		117	140		257	345
5600	30" wide		6.60	2.424		156	149		305	400
5800	36" wide		5.80	2.759		151	169		320	425
6000	Tees, 9" wide		4	4		425	245		670	850
6200	12" wide		3.60	4.444		495	273		768	965
6400	18" wide		3.20	5		635	305		940	1,175
6600	24" wide		3	5.333		910	325		1,235	1,500
6700	27" wide		2.80	5.714		1,000	350		1,350	1,650
6800	30" wide		2.60	6.154		1,175	380		1,555	1,875
7000	36" wide		2	8		1,550	490		2,040	2,450
7020	Two compartment, 9" wide		3.80	4.211		495	258		753	945
7030	12" wide		3.40	4.706		530	289		819	1,025
7040	18" wide		3	5.333		710	325		1,035	1,275
7050	24" wide		2.80	5.714		960	350		1,310	1,600
7060	30" wide		2.40	6.667		1,300	410		1,710	2,050
7070	36" wide		1.90	8.421		1,625	515		2,140	2,600
7090	Three compartment, 9" wide		3.60	4.444		565	273		838	1,050
7100	12" wide		3.20	5		595	305		900	1,125
7110	18" wide		2.80	5.714		750	350		1,100	1,375
7120	24" wide		2.60	6.154		1,025	380		1,405	1,725
7130	30" wide		2.20	7.273		1,350	445		1,795	2,175
7140	36" wide		1.80	8.889		1,725	545		2,270	2,750
7200	Riser, and cabinet connector, 9" wide		5.40	2.963		186	182		368	485
7400	12" wide		4.60	3.478		217	213		430	570
7600	18" wide		4	4		228	245		473	630
7800	24" wide		3.20	5		325	305		630	825
7900	27" wide		3	5.333		300	325		625	835
8000	30" wide		2.60	6.154		370	380		750	990
8200	36" wide		2	8		430	490		920	1,225
8400	Insert assembly, cell to conduit adapter, 1-1/4"	1 Elec	16	.500		74	30.50		104.50	129
8500	Adjustable partition	"	320	.025	L.F.	26	1.53		27.53	31.50
8600	Depth of duct over 4", per 1", add					12.10			12.10	13.30
8700	Support post	1 Elec	240	.033		27.50	2.04		29.54	33.50
8800	Cover double tile trim, 2 sides					43			43	47
8900	4 sides					120			120	132
9160	Trench duct 3-1/2" x 4-1/2", add					11.50			11.50	12.65
9170	Trench duct 4" x 5", add					11.50			11.50	12.65
9200	For carpet trim, add					39			39	43
9210	For double carpet trim, add					119			119	131

26 05 43.20 Underfloor Duct		Crew	Daily Output	Labor-Hours	Unit	Material	Labor	Equipment	Total	Total Incl O&P
0010	**UNDERFLOOR DUCT**									
0100	Duct, 1-3/8" x 3-1/8" blank, standard	2 Elec	160	.100	L.F.	12.85	6.15		19	23.50
0200	1-3/8" x 7-1/4" blank, super duct		120	.133		30	8.20		38.20	45.50
0400	7/8" or 1-3/8" insert type, 24" OC, 1-3/8" x 3-1/8", std.		140	.114		20	7		27	33
0600	1-3/8" x 7-1/4", super duct		100	.160		35	9.80		44.80	53.50
0800	Junction box, single duct, 1 level, 3-1/8"	1 Elec	4	2	Ea.	445	123		568	680
0820	3-1/8" x 7-1/4"		4	2		575	123		698	825
0840	2 level, 3-1/8" upper & lower		3.20	2.500		520	153		673	805
0860	3-1/8" upper, 7-1/4" lower		2.70	2.963		510	182		692	840

26 05 43 – Underground Ducts and Raceways for Electrical Systems

26 05 43.20 Underfloor Duct		Crew	Daily Output	Labor-Hours	Unit	Material	2020 Bare Costs Labor	Equipment	Total	Total Incl O&P
0880	Carpet pan for above	1 Elec	80	.100	Ea.	370	6.15		376.15	420
0900	Terrazzo pan for above		67	.119		880	7.35		887.35	980
1000	Junction box, single duct, 1 level, 7-1/4"		2.70	2.963		520	182		702	850
1020	2 level, 7-1/4" upper & lower		2.70	2.963		595	182		777	935
1040	2 duct, two 3-1/8" upper & lower		3.20	2.500		805	153		958	1,125
1200	1 level, 2 duct, 3-1/8"		3.20	2.500		590	153		743	885
1220	Carpet pan for above boxes		80	.100		365	6.15		371.15	410
1240	Terrazzo pan for above boxes		67	.119		850	7.35		857.35	945
1260	Junction box, 1 level, two 3-1/8" x one 3-1/8" + one 7-1/4"		2.30	3.478		995	213		1,208	1,425
1280	2 level, two 3-1/8" upper, one 3-1/8" + one 7-1/4" lower		2	4		1,100	245		1,345	1,575
1300	Carpet pan for above boxes		80	.100		365	6.15		371.15	410
1320	Terrazzo pan for above boxes		67	.119		850	7.35		857.35	945
1400	Junction box, 1 level, 2 duct, 7-1/4"		2.30	3.478		1,475	213		1,688	1,950
1420	Two 3-1/8" + one 7-1/4"		2	4		1,500	245		1,745	2,025
1440	Carpet pan for above		80	.100		365	6.15		371.15	410
1460	Terrazzo pan for above		67	.119		850	7.35		857.35	945
1580	Junction box, 1 level, one 3-1/8" + one 7-1/4" x same		2.30	3.478		1,000	213		1,213	1,425
1600	Triple duct, 3-1/8"		2.30	3.478		1,000	213		1,213	1,425
1700	Junction box, 1 level, one 3-1/8" + two 7-1/4"		2	4		1,675	245		1,920	2,225
1720	Carpet pan for above		80	.100		365	6.15		371.15	410
1740	Terrazzo pan for above		67	.119		850	7.35		857.35	945
1800	Insert to conduit adapter, 3/4" & 1"		32	.250		36.50	15.35		51.85	64
2000	Support, single cell		27	.296		54.50	18.20		72.70	88
2200	Super duct		16	.500		55	30.50		85.50	108
2400	Double cell		16	.500		55	30.50		85.50	108
2600	Triple cell		11	.727		64	44.50		108.50	139
2800	Vertical elbow, standard duct		10	.800		98.50	49		147.50	184
3000	Super duct		8	1		98.50	61.50		160	203
3200	Cabinet connector, standard duct		32	.250		78.50	15.35		93.85	110
3400	Super duct		27	.296		76.50	18.20		94.70	113
3600	Conduit adapter, 1" to 1-1/4"		32	.250		73.50	15.35		88.85	105
3800	2" to 1-1/4"		27	.296		88.50	18.20		106.70	125
4000	Outlet, low tension (tele, computer, etc.)		8	1		104	61.50		165.50	209
4200	High tension, receptacle (120 volt)		8	1		106	61.50		167.50	212
4300	End closure, standard duct		160	.050		4.44	3.07		7.51	9.60
4310	Super duct		160	.050		8.40	3.07		11.47	13.95
4350	Elbow, horiz., standard duct		26	.308		252	18.90		270.90	305
4360	Super duct		26	.308		240	18.90		258.90	293
4380	Elbow, offset, standard duct		26	.308		98.50	18.90		117.40	137
4390	Super duct		26	.308		100	18.90		118.90	139
4400	Marker screw assembly for inserts		50	.160		19.40	9.80		29.20	36.50
4410	Y take off, standard duct		26	.308		153	18.90		171.90	198
4420	Super duct		26	.308		153	18.90		171.90	198
4430	Box opening plug, standard duct		160	.050		18.70	3.07		21.77	25
4440	Super duct		160	.050		18.70	3.07		21.77	25
4450	Sleeve coupling, standard duct		160	.050		49.50	3.07		52.57	59
4460	Super duct		160	.050		49.50	3.07		52.57	59
4470	Conduit adapter, standard duct, 3/4"		32	.250		80	15.35		95.35	111
4480	1" or 1-1/4"		32	.250		76	15.35		91.35	107
4500	1-1/2"		32	.250		74	15.35		89.35	105

26 05 83 – Wiring Connections

26 05 83.10 Motor Connections		Crew	Daily Output	Labor-Hours	Unit	Material	2020 Bare Costs Labor	Equipment	Total	Total Incl O&P
0010	**MOTOR CONNECTIONS**									
0020	Flexible conduit and fittings, 115 volt, 1 phase, up to 1 HP motor	1 Elec	8	1	Ea.	5.90	61.50		67.40	101
0050	2 HP motor		6.50	1.231		10.40	75.50		85.90	127
0100	3 HP motor		5.50	1.455		9.20	89		98.20	147
0110	230 volt, 3 phase, 3 HP motor		6.78	1.180		7.15	72.50		79.65	119
0112	5 HP motor		5.47	1.463		6.05	89.50		95.55	145
0114	7-1/2 HP motor		4.61	1.735		9.35	106		115.35	174
0120	10 HP motor		4.20	1.905		17.40	117		134.40	199
0150	15 HP motor		3.30	2.424		17.40	149		166.40	248
0200	25 HP motor		2.70	2.963		29.50	182		211.50	315
0400	50 HP motor		2.20	3.636		54.50	223		277.50	405
0600	100 HP motor		1.50	5.333		121	325		446	640
1500	460 volt, 5 HP motor, 3 phase		8	1		6.35	61.50		67.85	101
1520	10 HP motor		8	1		6.35	61.50		67.85	101
1530	25 HP motor		6	1.333		11.25	82		93.25	138
1540	30 HP motor		6	1.333		11.25	82		93.25	138
1550	40 HP motor		5	1.600		16.70	98		114.70	169
1560	50 HP motor		5	1.600		22.50	98		120.50	176
1570	60 HP motor		3.80	2.105		25	129		154	227
1580	75 HP motor		3.50	2.286		31.50	140		171.50	251
1590	100 HP motor		2.50	3.200		53.50	196		249.50	360
1600	125 HP motor		2	4		58.50	245		303.50	445
1610	150 HP motor		1.80	4.444		60	273		333	485
1620	200 HP motor		1.50	5.333		99.50	325		424.50	615
2005	460 volt, 5 HP motor, 3 phase, w/sealtite		8	1		9.30	61.50		70.80	105
2010	10 HP motor		8	1		9.30	61.50		70.80	105
2015	25 HP motor		6	1.333		17	82		99	145
2020	30 HP motor		6	1.333		17	82		99	145
2025	40 HP motor		5	1.600		30	98		128	185
2030	50 HP motor		5	1.600		34.50	98		132.50	189
2035	60 HP motor		3.80	2.105		38.50	129		167.50	241
2040	75 HP motor		3.50	2.286		45	140		185	266
2045	100 HP motor		2.50	3.200		80	196		276	390
2055	150 HP motor		1.80	4.444		86.50	273		359.50	515
2060	200 HP motor		1.50	5.333		273	325		598	805
9000	Minimum labor/equipment charge		4	2	Job		123		123	189

26 05 90 – Residential Applications

26 05 90.10 Residential Wiring		Crew	Daily Output	Labor-Hours	Unit	Material	Labor	Equipment	Total	Total Incl O&P
0010	**RESIDENTIAL WIRING**									
0020	20' avg. runs and #14/2 wiring incl. unless otherwise noted									
1000	Service & panel, includes 24' SE-AL cable, service eye, meter,									
1010	Socket, panel board, main bkr., ground rod, 15 or 20 amp									
1020	1-pole circuit breakers, and misc. hardware									
1100	100 amp, with 10 branch breakers	1 Elec	1.19	6.723	Ea.	335	410		745	1,000
1110	With PVC conduit and wire		.92	8.696		370	535		905	1,225
1120	With RGS conduit and wire		.73	10.959		570	670		1,240	1,650
1150	150 amp, with 14 branch breakers		1.03	7.767		805	475		1,280	1,625
1170	With PVC conduit and wire		.82	9.756		875	600		1,475	1,875
1180	With RGS conduit and wire		.67	11.940		1,200	735		1,935	2,450
1200	200 amp, with 18 branch breakers	2 Elec	1.80	8.889		1,025	545		1,570	1,975
1220	With PVC conduit and wire		1.46	10.959		1,100	670		1,770	2,225
1230	With RGS conduit and wire		1.24	12.903		1,500	790		2,290	2,875

26 05 90.10 Residential Wiring		Crew	Daily Output	Labor-Hours	Unit	Material	2020 Bare Costs Labor	Equipment	Total	Total Incl O&P
1800	Lightning surge suppressor	1 Elec	32	.250	Ea.	85	15.35		100.35	117
2000	Switch devices									
2100	Single pole, 15 amp, ivory, with a 1-gang box, cover plate,									
2110	Type NM (Romex) cable	1 Elec	17.10	.468	Ea.	15.55	28.50		44.05	61
2120	Type MC cable		14.30	.559		26	34.50		60.50	82
2130	EMT & wire		5.71	1.401		36	86		122	172
2150	3-way, #14/3, type NM cable		14.55	.550		10.10	33.50		43.60	63
2170	Type MC cable		12.31	.650		24.50	40		64.50	88.50
2180	EMT & wire		5	1.600		30.50	98		128.50	185
2200	4-way, #14/3, type NM cable		14.55	.550		18.20	33.50		51.70	72
2220	Type MC cable		12.31	.650		32.50	40		72.50	97.50
2230	EMT & wire		5	1.600		38.50	98		136.50	194
2250	S.P., 20 amp, #12/2, type NM cable		13.33	.600		12.30	37		49.30	70
2270	Type MC cable		11.43	.700		22	43		65	90
2280	EMT & wire		4.85	1.649		35	101		136	195
2290	S.P. rotary dimmer, 600 W, no wiring		17	.471		32	29		61	79.50
2300	S.P. rotary dimmer, 600 W, type NM cable		14.55	.550		35.50	33.50		69	91
2320	Type MC cable		12.31	.650		46	40		86	113
2330	EMT & wire		5	1.600		57.50	98		155.50	214
2350	3-way rotary dimmer, type NM cable		13.33	.600		24	37		61	83
2370	Type MC cable		11.43	.700		34.50	43		77.50	104
2380	EMT & wire	↓	4.85	1.649	↓	46	101		147	207
2400	Interval timer wall switch, 20 amp, 1-30 min., #12/2									
2410	Type NM cable	1 Elec	14.55	.550	Ea.	59	33.50		92.50	117
2420	Type MC cable		12.31	.650		64.50	40		104.50	133
2430	EMT & wire	↓	5	1.600	↓	81.50	98		179.50	241
2500	Decorator style									
2510	S.P., 15 amp, type NM cable	1 Elec	17.10	.468	Ea.	21	28.50		49.50	67
2520	Type MC cable		14.30	.559		32	34.50		66.50	88
2530	EMT & wire		5.71	1.401		41.50	86		127.50	178
2550	3-way, #14/3, type NM cable		14.55	.550		15.65	33.50		49.15	69.50
2570	Type MC cable		12.31	.650		30	40		70	94.50
2580	EMT & wire		5	1.600		36	98		134	191
2600	4-way, #14/3, type NM cable		14.55	.550		24	33.50		57.50	78
2620	Type MC cable		12.31	.650		38	40		78	104
2630	EMT & wire		5	1.600		44	98		142	200
2650	S.P., 20 amp, #12/2, type NM cable		13.33	.600		17.85	37		54.85	76
2670	Type MC cable		11.43	.700		27.50	43		70.50	96
2680	EMT & wire		4.85	1.649		40.50	101		141.50	201
2700	S.P., slide dimmer, type NM cable		17.10	.468		37.50	28.50		66	85.50
2720	Type MC cable		14.30	.559		48	34.50		82.50	106
2730	EMT & wire		5.71	1.401		59.50	86		145.50	198
2750	S.P., touch dimmer, type NM cable		17.10	.468		53	28.50		81.50	103
2770	Type MC cable		14.30	.559		64	34.50		98.50	123
2780	EMT & wire		5.71	1.401		75	86		161	215
2800	3-way touch dimmer, type NM cable		13.33	.600		49.50	37		86.50	111
2820	Type MC cable		11.43	.700		60.50	43		103.50	133
2830	EMT & wire	↓	4.85	1.649	↓	71.50	101		172.50	235
3000	Combination devices									
3100	S.P. switch/15 amp recpt., ivory, 1-gang box, plate									
3110	Type NM cable	1 Elec	11.43	.700	Ea.	22	43		65	90
3120	Type MC cable		10	.800		32.50	49		81.50	112
3130	EMT & wire	↓	4.40	1.818	↓	44	112		156	220

26 05 90.10 Residential Wiring		Crew	Daily Output	Labor-Hours	Unit	Material	Labor	Equipment	Total	Total Incl O&P
						2020 Bare Costs				
3150	S.P. switch/pilot light, type NM cable	1 Elec	11.43	.700	Ea.	23.50	43		66.50	91.50
3170	Type MC cable		10	.800		34	49		83	113
3180	EMT & wire		4.43	1.806		45	111		156	221
3190	2-S.P. switches, 2-#14/2, no wiring		14	.571		13	35		48	68.50
3200	2-S.P. switches, 2-#14/2, type NM cables		10	.800		24	49		73	102
3220	Type MC cable		8.89	.900		38.50	55		93.50	128
3230	EMT & wire		4.10	1.951		46	120		166	235
3250	3-way switch/15 amp recpt., #14/3, type NM cable		10	.800		30	49		79	109
3270	Type MC cable		8.89	.900		44	55		99	134
3280	EMT & wire		4.10	1.951		50	120		170	239
3300	2-3 way switches, 2-#14/3, type NM cables		8.89	.900		38	55		93	127
3320	Type MC cable		8	1		60	61.50		121.50	161
3330	EMT & wire		4	2		56.50	123		179.50	252
3350	S.P. switch/20 amp recpt., #12/2, type NM cable		10	.800		40	49		89	120
3370	Type MC cable		8.89	.900		45.50	55		100.50	135
3380	EMT & wire	↓	4.10	1.951	↓	62.50	120		182.50	253
3400	Decorator style									
3410	S.P. switch/15 amp recpt., type NM cable	1 Elec	11.43	.700	Ea.	27.50	43		70.50	96
3420	Type MC cable		10	.800		38	49		87	118
3430	EMT & wire		4.40	1.818		49.50	112		161.50	227
3450	S.P. switch/pilot light, type NM cable		11.43	.700		29	43		72	97.50
3470	Type MC cable		10	.800		39.50	49		88.50	119
3480	EMT & wire		4.40	1.818		50.50	112		162.50	228
3500	2-S.P. switches, 2-#14/2, type NM cables		10	.800		29.50	49		78.50	108
3520	Type MC cable		8.89	.900		44.50	55		99.50	134
3530	EMT & wire		4.10	1.951		51.50	120		171.50	241
3550	3-way/15 amp recpt., #14/3, type NM cable		10	.800		35.50	49		84.50	115
3570	Type MC cable		8.89	.900		49.50	55		104.50	140
3580	EMT & wire		4.10	1.951		55.50	120		175.50	245
3650	2-3 way switches, 2-#14/3, type NM cables		8.89	.900		43.50	55		98.50	133
3670	Type MC cable		8	1		65.50	61.50		127	167
3680	EMT & wire		4	2		62	123		185	258
3700	S.P. switch/20 amp recpt., #12/2, type NM cable		10	.800		45.50	49		94.50	126
3720	Type MC cable		8.89	.900		51	55		106	141
3730	EMT & wire	↓	4.10	1.951	↓	68	120		188	259
4000	Receptacle devices									
4010	Duplex outlet, 15 amp recpt., ivory, 1-gang box, plate									
4015	Type NM cable	1 Elec	14.55	.550	Ea.	8.90	33.50		42.40	62
4020	Type MC cable		12.31	.650		19.55	40		59.55	83
4030	EMT & wire		5.33	1.501		29.50	92		121.50	175
4050	With #12/2, type NM cable		12.31	.650		10.50	40		50.50	73
4070	Type MC cable		10.67	.750		19.95	46		65.95	93
4080	EMT & wire		4.71	1.699		33	104		137	196
4100	20 amp recpt., #12/2, type NM cable		12.31	.650		19.50	40		59.50	83
4120	Type MC cable		10.67	.750		29	46		75	103
4130	EMT & wire	↓	4.71	1.699	↓	42	104		146	206
4140	For GFI see Section 26 05 90.10 line 4300 below									
4150	Decorator style, 15 amp recpt., type NM cable	1 Elec	14.55	.550	Ea.	14.45	33.50		47.95	68
4170	Type MC cable		12.31	.650		25	40		65	89
4180	EMT & wire		5.33	1.501		35	92		127	181
4200	With #12/2, type NM cable		12.31	.650		16.05	40		56.05	79
4220	Type MC cable		10.67	.750		25.50	46		71.50	99
4230	EMT & wire	↓	4.71	1.699	↓	38.50	104		142.50	203

26 05 90.10 Residential Wiring		Crew	Daily Output	Labor-Hours	Unit	2020 Bare Costs Material	Labor	Equipment	Total	Total Incl O&P
4250	20 amp recpt., #12/2, type NM cable	1 Elec	12.31	.650	Ea.	25	40		65	89
4270	Type MC cable		10.67	.750		34.50	46		80.50	109
4280	EMT & wire		4.71	1.699		47.50	104		151.50	213
4300	GFI, 15 amp recpt., type NM cable		12.31	.650		21	40		61	84.50
4320	Type MC cable		10.67	.750		31.50	46		77.50	106
4330	EMT & wire		4.71	1.699		41.50	104		145.50	206
4350	GFI with #12/2, type NM cable		10.67	.750		22.50	46		68.50	95.50
4370	Type MC cable		9.20	.870		32	53.50		85.50	117
4380	EMT & wire		4.21	1.900		45	117		162	229
4400	20 amp recpt., #12/2, type NM cable		10.67	.750		54	46		100	131
4420	Type MC cable		9.20	.870		63.50	53.50		117	152
4430	EMT & wire		4.21	1.900		76.50	117		193.50	263
4500	Weather-proof cover for above receptacles, add		32	.250		2.02	15.35		17.37	25.50
4550	Air conditioner outlet, 20 amp-240 volt recpt.									
4560	30′ of #12/2, 2 pole circuit breaker									
4570	Type NM cable	1 Elec	10	.800	Ea.	61	49		110	143
4580	Type MC cable		9	.889		72	54.50		126.50	163
4590	EMT & wire		4	2		84	123		207	281
4600	Decorator style, type NM cable		10	.800		66	49		115	148
4620	Type MC cable		9	.889		77	54.50		131.50	169
4630	EMT & wire		4	2		88.50	123		211.50	287
4650	Dryer outlet, 30 amp-240 volt recpt., 20′ of #10/3									
4660	2 pole circuit breaker									
4670	Type NM cable	1 Elec	6.41	1.248	Ea.	54.50	76.50		131	178
4680	Type MC cable		5.71	1.401		62.50	86		148.50	201
4690	EMT & wire		3.48	2.299		73	141		214	298
4700	Range outlet, 50 amp-240 volt recpt., 30′ of #8/3									
4710	Type NM cable	1 Elec	4.21	1.900	Ea.	82.50	117		199.50	270
4720	Type MC cable		4	2		133	123		256	335
4730	EMT & wire		2.96	2.703		105	166		271	370
4750	Central vacuum outlet, type NM cable		6.40	1.250		58	76.50		134.50	182
4770	Type MC cable		5.71	1.401		70.50	86		156.50	210
4780	EMT & wire		3.48	2.299		88	141		229	315
4800	30 amp-110 volt locking recpt., #10/2 circ. bkr.									
4810	Type NM cable	1 Elec	6.20	1.290	Ea.	67	79		146	196
4820	Type MC cable		5.40	1.481		83	91		174	232
4830	EMT & wire		3.20	2.500		98	153		251	345
4900	Low voltage outlets									
4910	Telephone recpt., 20′ of 4/C phone wire	1 Elec	26	.308	Ea.	8.80	18.90		27.70	38.50
4920	TV recpt., 20′ of RG59U coax wire, F type connector	"	16	.500	"	17.60	30.50		48.10	66.50
4950	Door bell chime, transformer, 2 buttons, 60′ of bellwire									
4970	Economy model	1 Elec	11.50	.696	Ea.	57.50	42.50		100	129
4980	Custom model		11.50	.696		111	42.50		153.50	188
4990	Luxury model, 3 buttons		9.50	.842		188	51.50		239.50	287
6000	Lighting outlets									
6050	Wire only (for fixture), type NM cable	1 Elec	32	.250	Ea.	5.95	15.35		21.30	30
6070	Type MC cable		24	.333		11.30	20.50		31.80	44
6080	EMT & wire		10	.800		20.50	49		69.50	98
6100	Box (4″), and wire (for fixture), type NM cable		25	.320		15.05	19.65		34.70	46.50
6120	Type MC cable		20	.400		20.50	24.50		45	60.50
6130	EMT & wire		11	.727		29.50	44.50		74	101
6200	Fixtures (use with line 6050 or 6100 above)									
6210	Canopy style, economy grade	1 Elec	40	.200	Ea.	22	12.25		34.25	43.50

26 05 90 – Residential Applications

26 05 90.10 Residential Wiring		Crew	Daily Output	Labor-Hours	Unit	Material	2020 Bare Costs Labor	Equipment	Total	Total Incl O&P
6220	Custom grade	1 Elec	40	.200	Ea.	53	12.25		65.25	77
6250	Dining room chandelier, economy grade		19	.421		82	26		108	130
6260	Custom grade		19	.421		320	26		346	395
6270	Luxury grade		15	.533		1,300	32.50		1,332.50	1,475
6310	Kitchen fixture (fluorescent), economy grade		30	.267		72.50	16.35		88.85	105
6320	Custom grade		25	.320		147	19.65		166.65	192
6350	Outdoor, wall mounted, economy grade		30	.267		30.50	16.35		46.85	58.50
6360	Custom grade		30	.267		119	16.35		135.35	155
6370	Luxury grade		25	.320		247	19.65		266.65	300
6410	Outdoor PAR floodlights, 1 lamp, 150 watt		20	.400		28	24.50		52.50	68.50
6420	2 lamp, 150 watt each		20	.400		45	24.50		69.50	87.50
6425	Motion sensing, 2 lamp, 150 watt each		20	.400		109	24.50		133.50	158
6430	For infrared security sensor, add		32	.250		95	15.35		110.35	128
6450	Outdoor, quartz-halogen, 300 watt flood		20	.400		40	24.50		64.50	82.50
6600	Recessed downlight, round, pre-wired, 50 or 75 watt trim		30	.267		68	16.35		84.35	100
6610	With shower light trim		30	.267		93.50	16.35		109.85	128
6620	With wall washer trim		28	.286		94	17.55		111.55	130
6630	With eye-ball trim		28	.286		85	17.55		102.55	121
6700	Porcelain lamp holder		40	.200		2.76	12.25		15.01	22
6710	With pull switch		40	.200		10.70	12.25		22.95	30.50
6750	Fluorescent strip, 2-20 watt tube, wrap around diffuser, 24"		24	.333		45.50	20.50		66	81.50
6760	1-34 watt tube, 48"		24	.333		122	20.50		142.50	166
6770	2-34 watt tubes, 48"		20	.400		160	24.50		184.50	214
6800	Bathroom heat lamp, 1-250 watt		28	.286		33.50	17.55		51.05	64
6810	2-250 watt lamps	↓	28	.286	↓	64.50	17.55		82.05	97.50
6820	For timer switch, see Section 26 05 90.10 line 2400									
6900	Outdoor post lamp, incl. post, fixture, 35' of #14/2									
6910	Type NM cable	1 Elec	3.50	2.286	Ea.	325	140		465	570
6920	Photo-eye, add		27	.296		27	18.20		45.20	58
6950	Clock dial time switch, 24 hr., w/enclosure, type NM cable		11.43	.700		72.50	43		115.50	146
6970	Type MC cable		11	.727		83	44.50		127.50	160
6980	EMT & wire	↓	4.85	1.649	↓	93	101		194	258
7000	Alarm systems									
7050	Smoke detectors, box, #14/3, type NM cable	1 Elec	14.55	.550	Ea.	33.50	33.50		67	89
7070	Type MC cable		12.31	.650		44	40		84	110
7080	EMT & wire	↓	5	1.600		50	98		148	206
7090	For relay output to security system, add				↓	10.25			10.25	11.25
8000	Residential equipment									
8050	Disposal hook-up, incl. switch, outlet box, 3' of flex									
8060	20 amp-1 pole circ. bkr., and 25' of #12/2									
8070	Type NM cable	1 Elec	10	.800	Ea.	29	49		78	108
8080	Type MC cable		8	1		39	61.50		100.50	138
8090	EMT & wire	↓	5	1.600	↓	55	98		153	211
8100	Trash compactor or dishwasher hook-up, incl. outlet box,									
8110	3' of flex, 15 amp-1 pole circ. bkr., and 25' of #14/2									
8120	Type NM cable	1 Elec	10	.800	Ea.	16.05	49		65.05	93
8130	Type MC cable		8	1		28	61.50		89.50	126
8140	EMT & wire	↓	5	1.600	↓	41.50	98		139.50	197
8150	Hot water sink dispenser hook-up, use line 8100									
8200	Vent/exhaust fan hook-up, type NM cable	1 Elec	32	.250	Ea.	5.95	15.35		21.30	30
8220	Type MC cable		24	.333		11.30	20.50		31.80	44
8230	EMT & wire	↓	10	.800	↓	20.50	49		69.50	98
8250	Bathroom vent fan, 50 CFM (use with above hook-up)									

26 05 90.10 Residential Wiring		Crew	Daily Output	Labor-Hours	Unit	2020 Bare Costs Material	Labor	Equipment	Total	Total Incl O&P
8260	Economy model	1 Elec	15	.533	Ea.	18.80	32.50		51.30	71
8270	Low noise model		15	.533		47.50	32.50		80	103
8280	Custom model		12	.667		117	41		158	192
8300	Bathroom or kitchen vent fan, 110 CFM									
8310	Economy model	1 Elec	15	.533	Ea.	66.50	32.50		99	124
8320	Low noise model	"	15	.533	"	96.50	32.50		129	157
8350	Paddle fan, variable speed (w/o lights)									
8360	Economy model (AC motor)	1 Elec	10	.800	Ea.	136	49		185	225
8362	With light kit		10	.800		176	49		225	269
8370	Custom model (AC motor)		10	.800		345	49		394	455
8372	With light kit		10	.800		385	49		434	500
8380	Luxury model (DC motor)		8	1		315	61.50		376.50	440
8382	With light kit		8	1		355	61.50		416.50	485
8390	Remote speed switch for above, add		12	.667		40.50	41		81.50	108
8500	Whole house exhaust fan, ceiling mount, 36", variable speed									
8510	Remote switch, incl. shutters, 20 amp-1 pole circ. bkr.									
8520	30' of #12/2, type NM cable	1 Elec	4	2	Ea.	1,375	123		1,498	1,725
8530	Type MC cable		3.50	2.286		1,400	140		1,540	1,775
8540	EMT & wire		3	2.667		1,425	164		1,589	1,800
8600	Whirlpool tub hook-up, incl. timer switch, outlet box									
8610	3' of flex, 20 amp-1 pole GFI circ. bkr.									
8620	30' of #12/2, type NM cable	1 Elec	5	1.600	Ea.	129	98		227	293
8630	Type MC cable		4.20	1.905		136	117		253	330
8640	EMT & wire		3.40	2.353		149	144		293	385
8650	Hot water heater hook-up, incl. 1-2 pole circ. bkr., box;									
8660	3' of flex, 20' of #10/2, type NM cable	1 Elec	5	1.600	Ea.	29	98		127	183
8670	Type MC cable		4.20	1.905		42	117		159	226
8680	EMT & wire		3.40	2.353		48	144		192	275
9000	Heating/air conditioning									
9050	Furnace/boiler hook-up, incl. firestat, local on-off switch									
9060	Emergency switch, and 40' of type NM cable	1 Elec	4	2	Ea.	57	123		180	252
9070	Type MC cable		3.50	2.286		72	140		212	295
9080	EMT & wire		1.50	5.333		93.50	325		418.50	610
9100	Air conditioner hook-up, incl. local 60 amp disc. switch									
9110	3' sealtite, 40 amp, 2 pole circuit breaker									
9130	40' of #8/2, type NM cable	1 Elec	3.50	2.286	Ea.	144	140		284	375
9140	Type MC cable		3	2.667		212	164		376	485
9150	EMT & wire		1.30	6.154		185	380		565	785
9200	Heat pump hook-up, 1-40 & 1-100 amp 2 pole circ. bkr.									
9210	Local disconnect switch, 3' sealtite									
9220	40' of #8/2 & 30' of #3/2									
9230	Type NM cable	1 Elec	1.30	6.154	Ea.	520	380		900	1,150
9240	Type MC cable		1.08	7.407		550	455		1,005	1,300
9250	EMT & wire		.94	8.511		535	520		1,055	1,400
9500	Thermostat hook-up, using low voltage wire									
9520	Heating only, 25' of #18-3	1 Elec	24	.333	Ea.	6.75	20.50		27.25	39
9530	Heating/cooling, 25' of #18-4	"	20	.400	"	8.60	24.50		33.10	47.50

26 09 Instrumentation and Control for Electrical Systems

26 09 13 – Electrical Power Monitoring

26 09 13.10 Switchboard Instruments			Crew	Daily Output	Labor-Hours	Unit	2020 Bare Costs Material	Labor	Equipment	Total	Total Incl O&P
0010	**SWITCHBOARD INSTRUMENTS** 3 phase, 4 wire										
0100	AC indicating, ammeter & switch		1 Elec	8	1	Ea.	2,925	61.50		2,986.50	3,325
0200	Voltmeter & switch			8	1		3,300	61.50		3,361.50	3,725
0300	Wattmeter			8	1		4,150	61.50		4,211.50	4,650
0400	AC recording, ammeter			4	2		7,375	123		7,498	8,325
0500	Voltmeter			4	2		7,375	123		7,498	8,325
0600	Ground fault protection, zero sequence			2.70	2.963		6,525	182		6,707	7,450
0700	Ground return path			2.70	2.963		6,525	182		6,707	7,450
0800	3 current transformers, 5 to 800 amp			2	4		3,025	245		3,270	3,725
0900	1,000 to 1,500 amp			1.30	6.154		4,375	380		4,755	5,375
1200	2,000 to 4,000 amp			1	8		5,150	490		5,640	6,425
1300	Fused potential transformer, maximum 600 volt			8	1		1,150	61.50		1,211.50	1,350

26 09 13.20 Voltage Monitor Systems			Crew	Daily Output	Labor-Hours	Unit	Material	Labor	Equipment	Total	Total Incl O&P
0010	**VOLTAGE MONITOR SYSTEMS** (test equipment)										
0100	AC voltage monitor system, 120/240 V, one-channel					Ea.	2,875			2,875	3,175
0110	Modem adapter						360			360	395
0120	Add-on detector only						1,500			1,500	1,675
0150	AC voltage remote monitor sys., 3 channel, 120, 230, or 480 V						5,225			5,225	5,750
0160	With internal modem						5,525			5,525	6,075
0170	Combination temperature and humidity probe						810			810	890
0180	Add-on detector only						3,800			3,800	4,175
0190	With internal modem						4,125			4,125	4,550

26 09 13.30 Smart Metering			Crew	Daily Output	Labor-Hours	Unit	Material	Labor	Equipment	Total	Total Incl O&P
0010	**SMART METERING,** In panel										
0100	Single phase, 120/208 volt, 100 amp	G	1 Elec	8.78	.911	Ea.	380	56		436	505
0120	200 amp	G		8.78	.911		385	56		441	510
0200	277 volt, 100 amp	G		8.78	.911		410	56		466	535
0220	200 amp	G		8.78	.911		420	56		476	545
1100	Three phase, 120/208 volt, 100 amp	G		4.69	1.706		675	105		780	900
1120	200 amp	G		4.69	1.706		945	105		1,050	1,200
1130	400 amp	G		4.69	1.706		875	105		980	1,125
1140	800 amp	G		4.69	1.706		775	105		880	1,000
1150	1,600 amp	G		4.69	1.706		790	105		895	1,025
1200	277/480 volt, 100 amp	G		4.69	1.706		1,000	105		1,105	1,250
1220	200 amp	G		4.69	1.706		865	105		970	1,100
1230	400 amp	G		4.69	1.706		880	105		985	1,125
1240	800 amp	G		4.69	1.706		810	105		915	1,050
1250	1,600 amp	G		4.69	1.706		845	105		950	1,075
2000	Data recorder, 8 meters	G		10.97	.729		1,600	44.50		1,644.50	1,825
2100	16 meters	G		8.53	.938		3,625	57.50		3,682.50	4,100
3000	Software package, per meter, basic	G					261			261	287
3100	Premium	G					690			690	755

26 09 23 – Lighting Control Devices

26 09 23.10 Energy Saving Lighting Devices			Crew	Daily Output	Labor-Hours	Unit	Material	Labor	Equipment	Total	Total Incl O&P
0010	**ENERGY SAVING LIGHTING DEVICES**										
0100	Occupancy sensors, passive infrared ceiling mounted	G	1 Elec	7	1.143	Ea.	75	70		145	191
0110	Ultrasonic ceiling mounted	G		7	1.143		105	70		175	224
0120	Dual technology ceiling mounted	G		6.50	1.231		135	75.50		210.50	265
0150	Automatic wall switches	G		24	.333		68.50	20.50		89	107
0160	Daylighting sensor, manual control, ceiling mounted	G		7	1.143		169	70		239	294
0170	Remote and dimming control with remote controller	G		6.50	1.231		202	75.50		277.50	340
0200	Passive infrared ceiling mounted			6.50	1.231		35	75.50		110.50	155

26 09 Instrumentation and Control for Electrical Systems

26 09 23 – Lighting Control Devices

26 09 23.10 Energy Saving Lighting Devices			Crew	Daily Output	Labor-Hours	Unit	2020 Bare Costs Material	Labor	Equipment	Total	Total Incl O&P
0400	Remote power pack	G	1 Elec	10	.800	Ea.	35	49		84	114
0450	Photoelectric control, S.P.S.T. 120 V	G		8	1		24.50	61.50		86	122
0500	S.P.S.T. 208 V/277 V	G		8	1		28.50	61.50		90	126
0550	D.P.S.T. 120 V	G		6	1.333		207	82		289	355
0600	D.P.S.T. 208 V/277 V	G		6	1.333		234	82		316	385
0650	S.P.D.T. 208 V/277 V	G		6	1.333		219	82		301	365
0660	Daylight level sensor, wall mounted, on/off or dimming	G		8	1		168	61.50		229.50	280

26 09 26 – Lighting Control Panelboards

26 09 26.10 Lighting Control Relay Panel			Crew	Daily Output	Labor-Hours	Unit	Material	Labor	Equipment	Total	Total Incl O&P
0010	**LIGHTING CONTROL RELAY PANEL** with timeclock										
0100	4 Relay	G	1 Elec	2.50	3.200	Ea.	1,025	196		1,221	1,425
0110	8 Relay	G		2.30	3.478		1,400	213		1,613	1,850
0120	16 Relay	G		1.80	4.444		1,625	273		1,898	2,200
0130	24 Relay	G		1.50	5.333		2,450	325		2,775	3,175
0140	48 Relay	G		1	8		1,650	490		2,140	2,575
0200	Room Controller, switching only										
0210	1 Relay	G	1 Elec	3	2.667	Ea.	1,450	164		1,614	1,825
0220	2 Relay	G		3	2.667		1,375	164		1,539	1,750
0230	3 Relay	G		3	2.667		1,650	164		1,814	2,075
0240	Dimming										
0250	1 Relay	G	1 Elec	3	2.667	Ea.	1,275	164		1,439	1,675
0260	2 Relay	G		3	2.667		1,350	164		1,514	1,750
0270	3 Relay	G		3	2.667		1,650	164		1,814	2,075

26 09 36 – Modular Dimming Controls

26 09 36.13 Manual Modular Dimming Controls			Crew	Daily Output	Labor-Hours	Unit	Material	Labor	Equipment	Total	Total Incl O&P
0010	**MANUAL MODULAR DIMMING CONTROLS**										
2000	Lighting control module	G	1 Elec	2	4	Ea.	315	245		560	725

26 12 Medium-Voltage Transformers

26 12 19 – Pad-Mounted, Liquid-Filled, Medium-Voltage Transformers

26 12 19.10 Transformer, Oil-Filled		Crew	Daily Output	Labor-Hours	Unit	Material	Labor	Equipment	Total	Total Incl O&P
0010	**TRANSFORMER, OIL-FILLED** primary delta or Y,									
0050	Pad mounted 5 kV or 15 kV, with taps, 277/480 V secondary, 3 phase									
0100	150 kVA	R-3	.65	30.769	Ea.	9,875	1,875	291	12,041	14,100
0110	225 kVA		.55	36.364		16,900	2,225	345	19,470	22,400
0200	300 kVA		.45	44.444		14,800	2,725	420	17,945	20,900
0300	500 kVA		.40	50		21,900	3,050	475	25,425	29,300
0400	750 kVA		.38	52.632		26,600	3,225	495	30,320	34,700
0500	1,000 kVA		.26	76.923		31,500	4,700	725	36,925	42,700
0600	1,500 kVA		.23	86.957		37,400	5,325	820	43,545	50,000
0700	2,000 kVA		.20	100		47,200	6,100	945	54,245	62,500
0710	2,500 kVA		.19	105		57,000	6,425	995	64,420	74,000
0720	3,000 kVA		.17	118		69,000	7,200	1,100	77,300	88,500
0800	3,750 kVA		.16	125		91,000	7,650	1,175	99,825	113,500

26 12 19.20 Transformer, Liquid-Filled		Crew	Daily Output	Labor-Hours	Unit	Material	Labor	Equipment	Total	Total Incl O&P
0010	**TRANSFORMER, LIQUID-FILLED** Pad mounted									
0020	5 kV or 15 kV primary, 277/480 volt secondary, 3 phase									
0050	225 kVA	R-3	.55	36.364	Ea.	14,000	2,225	345	16,570	19,300
0100	300 kVA		.45	44.444		16,700	2,725	420	19,845	23,100
0200	500 kVA		.40	50		21,100	3,050	475	24,625	28,400

26 12 Medium-Voltage Transformers

26 12 19 – Pad-Mounted, Liquid-Filled, Medium-Voltage Transformers

26 12 19.20 Transformer, Liquid-Filled		Crew	Daily Output	Labor-Hours	Unit	Material	2020 Bare Costs Labor	Equipment	Total	Total Incl O&P
0250	750 kVA	R-3	.38	52.632	Ea.	27,200	3,225	495	30,920	35,400
0300	1,000 kVA		.26	76.923		31,600	4,700	725	37,025	42,900
0350	1,500 kVA		.23	86.957		36,900	5,325	820	43,045	49,700
0400	2,000 kVA		.20	100		45,700	6,100	945	52,745	60,500
0450	2,500 kVA		.19	105		55,500	6,425	995	62,920	72,000

26 13 Medium-Voltage Switchgear

26 13 16 – Medium-Voltage Fusible Interrupter Switchgear

26 13 16.10 Switchgear		Crew	Daily Output	Labor-Hours	Unit	Material	Labor	Equipment	Total	Total Incl O&P
0010	**SWITCHGEAR**, Incorporate switch with cable connections, transformer,									
0100	& low voltage section									
0200	Load interrupter switch, 600 amp, 2 position									
0300	NEMA 1, 4.8 kV, 300 kVA & below w/CLF fuses	R-3	.40	50	Ea.	20,200	3,050	475	23,725	27,400
0400	400 kVA & above w/CLF fuses		.38	52.632		23,700	3,225	495	27,420	31,600
0500	Non fusible		.41	48.780		18,100	2,975	460	21,535	25,000
0600	13.8 kV, 300 kVA & below w/CLF fuses		.38	52.632		29,800	3,225	495	33,520	38,200
0700	400 kVA & above w/CLF fuses		.36	55.556		29,800	3,400	525	33,725	38,500
0800	Non fusible		.40	50		22,100	3,050	475	25,625	29,500
0900	Cable lugs for 2 feeders 4.8 kV or 13.8 kV	1 Elec	8	1		710	61.50		771.50	875
1000	Pothead, one 3 conductor or three 1 conductor		4	2		3,400	123		3,523	3,950
1100	Two 3 conductor or six 1 conductor		2	4		6,725	245		6,970	7,775
1200	Key interlocks		8	1		785	61.50		846.50	960
1300	Lightning arresters, distribution class (no charge)									
1400	Intermediate class or line type 4.8 kV	1 Elec	2.70	2.963	Ea.	3,825	182		4,007	4,475
1500	13.8 kV		2	4		5,075	245		5,320	5,950
1600	Station class, 4.8 kV		2.70	2.963		6,550	182		6,732	7,475
1700	13.8 kV		2	4		11,300	245		11,545	12,800
1800	Transformers, 4,800 volts to 480/277 volts, 75 kVA	R-3	.68	29.412		19,900	1,800	278	21,978	25,000
1900	112.5 kVA		.65	30.769		24,300	1,875	291	26,466	30,000
2000	150 kVA		.57	35.088		27,700	2,150	330	30,180	34,200
2100	225 kVA		.48	41.667		31,800	2,550	395	34,745	39,400
2200	300 kVA		.41	48.780		35,600	2,975	460	39,035	44,200
2300	500 kVA		.36	55.556		46,800	3,400	525	50,725	57,500
2400	750 kVA		.29	68.966		53,000	4,225	650	57,875	65,500
2500	13,800 volts to 480/277 volts, 75 kVA		.61	32.787		28,100	2,000	310	30,410	34,300
2600	112.5 kVA		.55	36.364		37,300	2,225	345	39,870	44,800
2700	150 kVA		.49	40.816		37,600	2,500	385	40,485	45,700
2800	225 kVA		.41	48.780		43,400	2,975	460	46,835	53,000
2900	300 kVA		.37	54.054		44,300	3,300	510	48,110	54,500
3000	500 kVA		.31	64.516		49,100	3,950	610	53,660	61,000
3100	750 kVA		.26	76.923		54,000	4,700	725	59,425	67,500
3200	Forced air cooling & temperature alarm	1 Elec	1	8		4,350	490		4,840	5,525
3300	Low voltage components									
3400	Maximum panel height 49-1/2", single or twin row									
3500	Breaker heights, type FA or FH, 6"									
3600	type KA or KH, 8"									
3700	type LA, 11"									
3800	type MA, 14"									
3900	Breakers, 2 pole, 15 to 60 amp, type FA	1 Elec	5.60	1.429	Ea.	750	87.50		837.50	955
4000	70 to 100 amp, type FA		4.20	1.905		960	117		1,077	1,225
4100	15 to 60 amp, type FH		5.60	1.429		1,125	87.50		1,212.50	1,375

26 13 Medium-Voltage Switchgear

26 13 16 – Medium-Voltage Fusible Interrupter Switchgear

26 13 16.10 Switchgear		Crew	Daily Output	Labor-Hours	Unit	2020 Bare Costs Material	Labor	Equipment	Total	Total Incl O&P
4200	70 to 100 amp, type FH	1 Elec	4.20	1.905	Ea.	1,475	117		1,592	1,800
4300	125 to 225 amp, type KA		3.40	2.353		1,400	144		1,544	1,775
4400	125 to 225 amp, type KH		3.40	2.353		2,350	144		2,494	2,825
4500	125 to 400 amp, type LA		2.50	3.200		2,775	196		2,971	3,350
4600	125 to 600 amp, type MA		1.80	4.444		5,750	273		6,023	6,750
4700	700 & 800 amp, type MA		1.50	5.333		7,225	325		7,550	8,450
4800	3 pole, 15 to 60 amp, type FA		5.30	1.509		900	92.50		992.50	1,125
4900	70 to 100 amp, type FA		4	2		1,125	123		1,248	1,450
5000	15 to 60 amp, type FH		5.30	1.509		1,525	92.50		1,617.50	1,825
5100	70 to 100 amp, type FH		4	2		1,775	123		1,898	2,150
5200	125 to 225 amp, type KA		3.20	2.500		2,150	153		2,303	2,600
5300	125 to 225 amp, type KH		3.20	2.500		3,900	153		4,053	4,500
5400	125 to 400 amp, type LA		2.30	3.478		3,575	213		3,788	4,275
5500	125 to 600 amp, type MA		1.60	5		5,150	305		5,455	6,150
5600	700 & 800 amp, type MA		1.30	6.154		6,400	380		6,780	7,625

26 22 Low-Voltage Transformers

26 22 13 – Low-Voltage Distribution Transformers

26 22 13.10 Transformer, Dry-Type		Crew	Daily Output	Labor-Hours	Unit	Material	Labor	Equipment	Total	Total Incl O&P
0010	**TRANSFORMER, DRY-TYPE**									
0050	Single phase, 240/480 volt primary, 120/240 volt secondary									
0100	1 kVA	1 Elec	2	4	Ea.	390	245		635	805
0300	2 kVA		1.60	5		700	305		1,005	1,250
0500	3 kVA		1.40	5.714		845	350		1,195	1,475
0700	5 kVA		1.20	6.667		1,300	410		1,710	2,050
0900	7.5 kVA	2 Elec	2.20	7.273		1,250	445		1,695	2,050
1100	10 kVA		1.60	10		1,625	615		2,240	2,750
1300	15 kVA		1.20	13.333		2,075	820		2,895	3,525
1500	25 kVA		1	16		2,475	980		3,455	4,225
1700	37.5 kVA		.80	20		3,325	1,225		4,550	5,575
1900	50 kVA		.70	22.857		3,125	1,400		4,525	5,575
2100	75 kVA		.65	24.615		5,225	1,500		6,725	8,075
2110	100 kVA	R-3	.90	22.222		6,825	1,350	210	8,385	9,825
2120	167 kVA	"	.80	25		11,300	1,525	236	13,061	15,000
2190	480 V primary, 120/240 V secondary, nonvent., 15 kVA	2 Elec	1.20	13.333		1,625	820		2,445	3,025
2200	25 kVA		.90	17.778		2,950	1,100		4,050	4,925
2210	37 kVA		.75	21.333		3,375	1,300		4,675	5,750
2220	50 kVA		.65	24.615		3,975	1,500		5,475	6,700
2230	75 kVA		.60	26.667		5,275	1,625		6,900	8,350
2240	100 kVA		.50	32		6,875	1,975		8,850	10,600
2250	Low operating temperature (80°C), 25 kVA		1	16		5,150	980		6,130	7,175
2260	37 kVA		.80	20		5,550	1,225		6,775	8,000
2270	50 kVA		.70	22.857		7,250	1,400		8,650	10,100
2280	75 kVA		.65	24.615		11,500	1,500		13,000	14,900
2290	100 kVA		.55	29.091		12,000	1,775		13,775	16,000
2300	3 phase, 480 volt primary, 120/208 volt secondary									
2310	Ventilated, 3 kVA	1 Elec	1	8	Ea.	1,225	490		1,715	2,100
2700	6 kVA		.80	10		1,225	615		1,840	2,300
2900	9 kVA		.70	11.429		1,450	700		2,150	2,650
3100	15 kVA	2 Elec	1.10	14.545		1,900	890		2,790	3,450
3300	30 kVA		.90	17.778		1,800	1,100		2,900	3,650

26 22 Low-Voltage Transformers

26 22 13 – Low-Voltage Distribution Transformers

26 22 13.10 Transformer, Dry-Type		Crew	Daily Output	Labor-Hours	Unit	Material (2020 Bare Costs)	Labor (2020 Bare Costs)	Equipment (2020 Bare Costs)	Total (2020 Bare Costs)	Total Incl O&P
3500	45 kVA	2 Elec	.80	20	Ea.	1,850	1,225		3,075	3,950
3700	75 kVA		.70	22.857		2,500	1,400		3,900	4,900
3900	112.5 kVA	R-3	.90	22.222		4,075	1,350	210	5,635	6,800
4100	150 kVA		.85	23.529		5,000	1,450	222	6,672	7,975
4300	225 kVA		.65	30.769		8,975	1,875	291	11,141	13,100
4500	300 kVA		.55	36.364		9,325	2,225	345	11,895	14,100
4700	500 kVA		.45	44.444		18,200	2,725	420	21,345	24,700
4800	750 kVA		.35	57.143		22,100	3,500	540	26,140	30,300
4820	1,000 kVA		.32	62.500		26,000	3,825	590	30,415	35,300
4850	K-4 rated, 15 kVA	2 Elec	1.10	14.545		3,550	890		4,440	5,275
4855	30 kVA		.90	17.778		5,825	1,100		6,925	8,075
4860	45 kVA		.80	20		7,450	1,225		8,675	10,100
4865	75 kVA		.70	22.857		10,200	1,400		11,600	13,400
4870	112.5 kVA	R-3	.90	22.222		12,300	1,350	210	13,860	15,800
4875	150 kVA		.85	23.529		16,100	1,450	222	17,772	20,200
4880	225 kVA		.65	30.769		22,400	1,875	291	24,566	27,800
4885	300 kVA		.55	36.364		30,700	2,225	345	33,270	37,600
4890	500 kVA		.45	44.444		42,900	2,725	420	46,045	52,000
4900	K-13 rated, 15 kVA	2 Elec	1.10	14.545		4,025	890		4,915	5,800
4905	30 kVA		.90	17.778		5,775	1,100		6,875	8,050
4910	45 kVA		.80	20		6,975	1,225		8,200	9,550
4915	75 kVA		.70	22.857		10,600	1,400		12,000	13,900
4920	112.5 kVA	R-3	.90	22.222		13,900	1,350	210	15,460	17,600
4925	150 kVA		.85	23.529		28,800	1,450	222	30,472	34,100
4930	225 kVA		.65	30.769		25,000	1,875	291	27,166	30,700
4935	300 kVA		.55	36.364		32,900	2,225	345	35,470	40,000
4940	500 kVA		.45	44.444		56,000	2,725	420	59,145	66,000
5020	480 volt primary, 120/208 volt secondary									
5030	Nonventilated, 15 kVA	2 Elec	1.10	14.545	Ea.	3,450	890		4,340	5,150
5040	30 kVA		.80	20		5,000	1,225		6,225	7,400
5050	45 kVA		.70	22.857		5,900	1,400		7,300	8,625
5060	75 kVA		.65	24.615		11,300	1,500		12,800	14,800
5070	112.5 kVA	R-3	.85	23.529		20,600	1,450	222	22,272	25,200
5081	150 kVA		.85	23.529		21,500	1,450	222	23,172	26,100
5090	225 kVA		.60	33.333		26,500	2,025	315	28,840	32,600
5100	300 kVA		.50	40		23,800	2,450	380	26,630	30,400
5200	Low operating temperature (80°C), 30 kVA	2 Elec	.90	17.778		6,475	1,100		7,575	8,800
5210	45 kVA		.80	20		5,625	1,225		6,850	8,100
5220	75 kVA		.70	22.857		8,450	1,400		9,850	11,500
5230	112.5 kVA	R-3	.90	22.222		11,200	1,350	210	12,760	14,600
5240	150 kVA		.85	23.529		14,300	1,450	222	15,972	18,300
5250	225 kVA		.65	30.769		19,500	1,875	291	21,666	24,600
5260	300 kVA		.55	36.364		32,400	2,225	345	34,970	39,400
5270	500 kVA		.45	44.444		35,200	2,725	420	38,345	43,400
5380	3 phase, 5 kV primary, 277/480 volt secondary									
5400	High voltage, 112.5 kVA	R-3	.85	23.529	Ea.	18,200	1,450	222	19,872	22,500
5410	150 kVA		.65	30.769		21,000	1,875	291	23,166	26,300
5420	225 kVA		.55	36.364		24,700	2,225	345	27,270	31,000
5430	300 kVA		.45	44.444		30,500	2,725	420	33,645	38,300
5440	500 kVA		.35	57.143		39,400	3,500	540	43,440	49,300
5450	750 kVA		.32	62.500		61,000	3,825	590	65,415	73,500
5460	1,000 kVA		.30	66.667		71,500	4,075	630	76,205	85,500
5470	1,500 kVA		.27	74.074		83,000	4,525	700	88,225	99,500

26 22 13.10 Transformer, Dry-Type		Crew	Daily Output	Labor-Hours	Unit	2020 Bare Costs Material	Labor	Equipment	Total	Total Incl O&P
5480	2,000 kVA	R-3	.25	80	Ea.	97,500	4,900	755	103,155	115,500
5490	2,500 kVA		.20	100		110,000	6,100	945	117,045	131,500
5500	3,000 kVA		.18	111		143,500	6,800	1,050	151,350	169,500
5590	15 kV primary, 277/480 volt secondary									
5600	High voltage, 112.5 kVA	R-3	.85	23.529	Ea.	28,400	1,450	222	30,072	33,700
5610	150 kVA		.65	30.769		33,200	1,875	291	35,366	39,700
5620	225 kVA		.55	36.364		36,600	2,225	345	39,170	44,100
5630	300 kVA		.45	44.444		43,100	2,725	420	46,245	52,000
5640	500 kVA		.35	57.143		54,000	3,500	540	58,040	65,500
5650	750 kVA		.32	62.500		71,000	3,825	590	75,415	84,500
5660	1,000 kVA		.30	66.667		80,500	4,075	630	85,205	95,500
5670	1,500 kVA		.27	74.074		93,000	4,525	700	98,225	110,000
5680	2,000 kVA		.25	80		103,000	4,900	755	108,655	121,500
5690	2,500 kVA		.20	100		119,500	6,100	945	126,545	141,500
5700	3,000 kVA		.18	111		142,000	6,800	1,050	149,850	168,000
6000	2400 volt primary, 480 volt secondary, 300 kVA		.45	44.444		30,500	2,725	420	33,645	38,300
6010	500 kVA		.35	57.143		39,400	3,500	540	43,440	49,300
6020	750 kVA		.32	62.500		56,000	3,825	590	60,415	68,500
9000	Minimum labor/equipment charge	1 Elec	1	8	Job		490		490	755

26 22 13.20 Isolating Panels		Crew	Daily Output	Labor-Hours	Unit	Material	Labor	Equipment	Total	Total Incl O&P
0010	**ISOLATING PANELS** used with isolating transformers									
0020	For hospital applications									
0100	Critical care area, 8 circuit, 3 kVA	1 Elec	.58	13.793	Ea.	7,075	845		7,920	9,075
0200	5 kVA		.54	14.815		7,325	910		8,235	9,450
0400	7.5 kVA		.52	15.385		4,900	945		5,845	6,850
0600	10 kVA		.44	18.182		8,075	1,125		9,200	10,600
0800	Operating room power & lighting, 8 circuit, 3 kVA		.58	13.793		5,300	845		6,145	7,125
1000	5 kVA		.54	14.815		5,775	910		6,685	7,750
1200	7.5 kVA		.52	15.385		4,050	945		4,995	5,900
1400	10 kVA		.44	18.182		4,950	1,125		6,075	7,175
1600	X-ray systems, 15 kVA, 90 amp		.44	18.182		15,800	1,125		16,925	19,100
1800	25 kVA, 125 amp		.36	22.222		16,600	1,375		17,975	20,400

26 22 13.30 Isolating Transformer		Crew	Daily Output	Labor-Hours	Unit	Material	Labor	Equipment	Total	Total Incl O&P
0010	**ISOLATING TRANSFORMER**									
0100	Single phase, 120/240 volt primary, 120/240 volt secondary									
0200	0.50 kVA	1 Elec	4	2	Ea.	410	123		533	640
0400	1 kVA		2	4		945	245		1,190	1,425
0600	2 kVA		1.60	5		1,525	305		1,830	2,150
0800	3 kVA		1.40	5.714		860	350		1,210	1,475
1000	5 kVA		1.20	6.667		1,125	410		1,535	1,875
1200	7.5 kVA		1.10	7.273		1,425	445		1,870	2,250
1400	10 kVA		.80	10		1,825	615		2,440	2,975
1600	15 kVA		.60	13.333		2,350	820		3,170	3,825
1800	25 kVA		.50	16		3,375	980		4,355	5,200
1810	37.5 kVA	2 Elec	.80	20		5,575	1,225		6,800	8,025
1820	75 kVA	"	.65	24.615		8,375	1,500		9,875	11,600
1830	3 phase, 120/240 V primary, 120/240 V secondary, 112.5 kVA	R-3	.90	22.222		11,200	1,350	210	12,760	14,600
1840	150 kVA		.85	23.529		14,200	1,450	222	15,872	18,200
1850	225 kVA		.65	30.769		19,800	1,875	291	21,966	25,000
1860	300 kVA		.55	36.364		26,400	2,225	345	28,970	32,800
1870	500 kVA		.45	44.444		43,900	2,725	420	47,045	53,000
1880	750 kVA		.35	57.143		44,400	3,500	540	48,440	55,000

26 22 13 – Low-Voltage Distribution Transformers

26 22 13.90 Transformer Handling		Crew	Daily Output	Labor-Hours	Unit	2020 Bare Costs Material	Labor	Equipment	Total	Total Incl O&P
0010	**TRANSFORMER HANDLING** Add to normal labor cost in restricted areas									
5000	Transformers									
5150	15 kVA, approximately 200 pounds	2 Elec	2.70	5.926	Ea.		365		365	560
5160	25 kVA, approximately 300 pounds		2.50	6.400			395		395	605
5170	37.5 kVA, approximately 400 pounds		2.30	6.957			425		425	655
5180	50 kVA, approximately 500 pounds		2	8			490		490	755
5190	75 kVA, approximately 600 pounds		1.80	8.889			545		545	840
5200	100 kVA, approximately 700 pounds		1.60	10			615		615	945
5210	112.5 kVA, approximately 800 pounds	3 Elec	2.20	10.909			670		670	1,025
5220	125 kVA, approximately 900 pounds		2	12			735		735	1,125
5230	150 kVA, approximately 1,000 pounds		1.80	13.333			820		820	1,250
5240	167 kVA, approximately 1,200 pounds		1.60	15			920		920	1,425
5250	200 kVA, approximately 1,400 pounds		1.40	17.143			1,050		1,050	1,625
5260	225 kVA, approximately 1,600 pounds		1.30	18.462			1,125		1,125	1,750
5270	250 kVA, approximately 1,800 pounds		1.10	21.818			1,350		1,350	2,050
5280	300 kVA, approximately 2,000 pounds		1	24			1,475		1,475	2,275
5290	500 kVA, approximately 3,000 pounds		.75	32			1,975		1,975	3,025
5300	600 kVA, approximately 3,500 pounds		.67	35.821			2,200		2,200	3,375
5310	750 kVA, approximately 4,000 pounds		.60	40			2,450		2,450	3,775
5320	1,000 kVA, approximately 5,000 pounds		.50	48			2,950		2,950	4,525

26 22 16 – Low-Voltage Buck-Boost Transformers

26 22 16.10 Buck-Boost Transformer		Crew	Daily Output	Labor-Hours	Unit	Material	Labor	Equipment	Total	Total Incl O&P
0010	**BUCK-BOOST TRANSFORMER**									
0100	Single phase, 120/240 V primary, 12/24 V secondary									
0200	0.10 kVA	1 Elec	8	1	Ea.	100	61.50		161.50	205
0400	0.25 kVA		5.70	1.404		170	86		256	320
0600	0.50 kVA		4	2		192	123		315	400
0800	0.75 kVA		3.10	2.581		305	158		463	585
1000	1.0 kVA		2	4		258	245		503	665
1200	1.5 kVA		1.80	4.444		350	273		623	805
1400	2.0 kVA		1.60	5		580	305		885	1,100
1600	3.0 kVA		1.40	5.714		560	350		910	1,150
1800	5.0 kVA		1.20	6.667		1,100	410		1,510	1,850
2000	3 phase, 240 V primary, 208/120 V secondary, 15 kVA	2 Elec	2.40	6.667		2,150	410		2,560	3,000
2200	30 kVA		1.60	10		2,800	615		3,415	4,025
2400	45 kVA		1.40	11.429		3,375	700		4,075	4,775
2600	75 kVA		1.20	13.333		4,075	820		4,895	5,725
2800	112.5 kVA	R-3	1.40	14.286		5,075	875	135	6,085	7,075
3000	150 kVA		1.10	18.182		6,725	1,100	172	7,997	9,325
3200	225 kVA		1	20		8,775	1,225	189	10,189	11,800
3400	300 kVA		.90	22.222		11,900	1,350	210	13,460	15,400

26 24 13.10 Incoming Switchboards		Crew	Daily Output	Labor-Hours	Unit	2020 Bare Costs Material	Labor	Equipment	Total	Total Incl O&P
0010	**INCOMING SWITCHBOARDS** main service section									
0100	Aluminum bus bars, not including CT's or PT's									
0200	No main disconnect, includes CT compartment									
0300	120/208 volt, 4 wire, 600 amp	2 Elec	1	16	Ea.	4,625	980		5,605	6,575
0400	800 amp		.88	18.182		4,625	1,125		5,750	6,800
0500	1,000 amp		.80	20		5,550	1,225		6,775	8,000
0600	1,200 amp		.72	22.222		5,550	1,375		6,925	8,200
0700	1,600 amp		.66	24.242		5,550	1,475		7,025	8,400
0800	2,000 amp		.62	25.806		6,000	1,575		7,575	9,025
1000	3,000 amp		.56	28.571		7,900	1,750		9,650	11,400
1200	277/480 volt, 4 wire, 600 amp		1	16		4,625	980		5,605	6,575
1300	800 amp		.88	18.182		4,625	1,125		5,750	6,800
1400	1,000 amp		.80	20		5,875	1,225		7,100	8,350
1500	1,200 amp		.72	22.222		5,875	1,375		7,250	8,550
1600	1,600 amp		.66	24.242		5,875	1,475		7,350	8,750
1700	2,000 amp		.62	25.806		6,000	1,575		7,575	9,025
1800	3,000 amp		.56	28.571		7,900	1,750		9,650	11,400
1900	4,000 amp		.52	30.769		9,775	1,900		11,675	13,700
2000	Fused switch & CT compartment									
2100	120/208 volt, 4 wire, 400 amp	2 Elec	1.12	14.286	Ea.	2,700	875		3,575	4,325
2200	600 amp		.94	17.021		3,200	1,050		4,250	5,125
2300	800 amp		.84	19.048		12,500	1,175		13,675	15,600
2400	1,200 amp		.68	23.529		16,300	1,450		17,750	20,100
2500	277/480 volt, 4 wire, 400 amp		1.14	14.035		3,800	860		4,660	5,500
2600	600 amp		.94	17.021		5,950	1,050		7,000	8,150
2700	800 amp		.84	19.048		12,500	1,175		13,675	15,600
2800	1,200 amp		.68	23.529		16,300	1,450		17,750	20,100
2900	Pressure switch & CT compartment									
3000	120/208 volt, 4 wire, 800 amp	2 Elec	.80	20	Ea.	11,200	1,225		12,425	14,300
3100	1,200 amp		.66	24.242		21,800	1,475		23,275	26,200
3200	1,600 amp		.62	25.806		23,100	1,575		24,675	27,800
3300	2,000 amp		.56	28.571		24,600	1,750		26,350	29,700
3310	2,500 amp		.50	32		30,300	1,975		32,275	36,300
3320	3,000 amp		.44	36.364		40,800	2,225		43,025	48,200
3330	4,000 amp		.40	40		52,500	2,450		54,950	61,500
3340	120/208 volt, 4 wire, 800 amp, with ground fault		.80	20		18,600	1,225		19,825	22,300
3350	1,200 amp, with ground fault		.66	24.242		24,000	1,475		25,475	28,700
3360	1,600 amp, with ground fault		.62	25.806		26,100	1,575		27,675	31,100
3370	2,000 amp, with ground fault		.56	28.571		28,100	1,750		29,850	33,700
3400	277/480 volt, 4 wire, 800 amp, with ground fault		.80	20		18,600	1,225		19,825	22,300
3600	1,200 amp, with ground fault		.66	24.242		24,000	1,475		25,475	28,700
4000	1,600 amp, with ground fault		.62	25.806		26,100	1,575		27,675	31,100
4200	2,000 amp, with ground fault		.56	28.571		28,100	1,750		29,850	33,700
4400	Circuit breaker, molded case & CT compartment									
4600	3 pole, 4 wire, 600 amp	2 Elec	.94	17.021	Ea.	9,675	1,050		10,725	12,200
4800	800 amp		.84	19.048		11,600	1,175		12,775	14,500
5000	1,200 amp		.68	23.529		15,800	1,450		17,250	19,600
5100	Copper bus bars, not incl. CT's or PT's, add, minimum					15%				

26 24 13.20 In Plant Distribution Switchboards		Crew	Daily Output	Labor-Hours	Unit	Material	Labor	Equipment	Total	Total Incl O&P
0010	**IN PLANT DISTRIBUTION SWITCHBOARDS**									
0100	Main lugs only, to 600 volt, 3 pole, 3 wire, 200 amp	2 Elec	1.20	13.333	Ea.	1,250	820		2,070	2,625
0110	400 amp		1.20	13.333		1,250	820		2,070	2,625

26 24 13.20 In Plant Distribution Switchboards		Crew	Daily Output	Labor-Hours	Unit	Material	2020 Bare Costs Labor	Equipment	Total	Total Incl O&P
0120	600 amp	2 Elec	1.20	13.333	Ea.	1,300	820		2,120	2,675
0130	800 amp		1.08	14.815		1,400	910		2,310	2,950
0140	1,200 amp		.92	17.391		1,700	1,075		2,775	3,525
0150	1,600 amp		.86	18.605		2,225	1,150		3,375	4,200
0160	2,000 amp		.82	19.512		2,475	1,200		3,675	4,575
0250	To 480 volt, 3 pole, 4 wire, 200 amp		1.20	13.333		1,100	820		1,920	2,450
0260	400 amp		1.20	13.333		1,250	820		2,070	2,625
0270	600 amp		1.20	13.333		1,375	820		2,195	2,750
0280	800 amp		1.08	14.815		1,500	910		2,410	3,050
0290	1,200 amp		.92	17.391		1,825	1,075		2,900	3,675
0300	1,600 amp		.86	18.605		2,100	1,150		3,250	4,050
0310	2,000 amp		.82	19.512		2,450	1,200		3,650	4,550
0400	Main circuit breaker, to 600 volt, 3 pole, 3 wire, 200 amp		1.20	13.333		3,425	820		4,245	5,025
0410	400 amp		1.14	14.035		3,450	860		4,310	5,125
0420	600 amp		1.10	14.545		4,350	890		5,240	6,150
0430	800 amp		1.04	15.385		7,275	945		8,220	9,450
0440	1,200 amp		.88	18.182		9,525	1,125		10,650	12,200
0450	1,600 amp		.84	19.048		15,300	1,175		16,475	18,600
0460	2,000 amp		.80	20		16,300	1,225		17,525	19,800
0550	277/480 volt, 3 pole, 4 wire, 200 amp		1.20	13.333		3,625	820		4,445	5,225
0560	400 amp		1.14	14.035		3,625	860		4,485	5,300
0570	600 amp		1.10	14.545		4,525	890		5,415	6,350
0580	800 amp		1.04	15.385		7,650	945		8,595	9,875
0590	1,200 amp		.88	18.182		9,875	1,125		11,000	12,600
0600	1,600 amp		.84	19.048		15,300	1,175		16,475	18,700
0610	2,000 amp		.80	20		16,400	1,225		17,625	19,900
0700	Main fusible switch w/fuse, 208/240 volt, 3 pole, 3 wire, 200 amp		1.20	13.333		3,775	820		4,595	5,425
0710	400 amp		1.14	14.035		3,750	860		4,610	5,450
0720	600 amp		1.10	14.545		4,600	890		5,490	6,450
0730	800 amp		1.04	15.385		9,425	945		10,370	11,900
0740	1,200 amp		.88	18.182		11,000	1,125		12,125	13,800
0800	120/208, 120/240 volt, 3 pole, 4 wire, 200 amp		1.20	13.333		3,375	820		4,195	4,950
0810	400 amp		1.14	14.035		3,450	860		4,310	5,125
0820	600 amp		1.10	14.545		4,400	890		5,290	6,200
0830	800 amp		1.04	15.385		6,975	945		7,920	9,125
0840	1,200 amp		.88	18.182		8,125	1,125		9,250	10,700
0900	480 or 600 volt, 3 pole, 3 wire, 200 amp		1.20	13.333		3,650	820		4,470	5,250
0910	400 amp		1.14	14.035		3,575	860		4,435	5,275
0920	600 amp		1.10	14.545		4,375	890		5,265	6,200
0930	800 amp		1.04	15.385		6,750	945		7,695	8,875
0940	1,200 amp		.88	18.182		7,825	1,125		8,950	10,400
1000	277 or 480 volt, 3 pole, 4 wire, 200 amp		1.20	13.333		3,775	820		4,595	5,425
1010	400 amp		1.14	14.035		3,725	860		4,585	5,425
1020	600 amp		1.10	14.545		4,575	890		5,465	6,400
1030	800 amp		1.04	15.385		7,000	945		7,945	9,150
1040	1,200 amp		.88	18.182		8,125	1,125		9,250	10,700
1120	1,600 amp		.76	21.053		14,800	1,300		16,100	18,300
1130	2,000 amp		.68	23.529		19,500	1,450		20,950	23,700
1150	Pressure switch, bolted, 3 pole, 208/240 volt, 3 wire, 800 amp		.96	16.667		11,000	1,025		12,025	13,700
1160	1,200 amp		.80	20		14,000	1,225		15,225	17,300
1170	1,600 amp		.76	21.053		16,000	1,300		17,300	19,600
1180	2,000 amp		.68	23.529		18,200	1,450		19,650	22,300
1200	120/208 or 120/240 volt, 3 pole, 4 wire, 800 amp		.96	16.667		8,675	1,025		9,700	11,100

26 24 13.20 In Plant Distribution Switchboards		Crew	Daily Output	Labor-Hours	Unit	Material	Labor	Equipment	Total	Total Incl O&P
						2020 Bare Costs				
1210	1,200 amp	2 Elec	.80	20	Ea.	10,100	1,225		11,325	13,100
1220	1,600 amp		.76	21.053		16,000	1,300		17,300	19,600
1230	2,000 amp		.68	23.529		18,200	1,450		19,650	22,300
1300	480 or 600 volt, 3 wire, 800 amp		.96	16.667		11,000	1,025		12,025	13,700
1310	1,200 amp		.80	20		15,400	1,225		16,625	18,800
1320	1,600 amp		.76	21.053		17,300	1,300		18,600	21,000
1330	2,000 amp		.68	23.529		19,500	1,450		20,950	23,700
1400	277/480 volt, 4 wire, 800 amp		.96	16.667		11,000	1,025		12,025	13,700
1410	1,200 amp		.80	20		15,400	1,225		16,625	18,800
1420	1,600 amp		.76	21.053		17,300	1,300		18,600	21,000
1430	2,000 amp		.68	23.529		19,500	1,450		20,950	23,700
1500	Main ground fault protector, 1,200-2,000 amp		5.40	2.963		3,225	182		3,407	3,825
1600	Busway connection, 200 amp		5.40	2.963		535	182		717	870
1610	400 amp		4.60	3.478		535	213		748	920
1620	600 amp		4	4		535	245		780	970
1630	800 amp		3.20	5		535	305		840	1,050
1640	1,200 amp		2.60	6.154		535	380		915	1,175
1650	1,600 amp		2.40	6.667		1,100	410		1,510	1,850
1660	2,000 amp		2	8		1,100	490		1,590	1,975
1700	Shunt trip for remote operation, 200 amp		8	2		605	123		728	860
1710	400 amp		8	2		970	123		1,093	1,275
1720	600 amp		8	2		1,150	123		1,273	1,450
1730	800 amp		8	2		1,475	123		1,598	1,825
1740	1,200-2,000 amp		8	2		3,300	123		3,423	3,825
1800	Motor operated main breaker, 200 amp		8	2		3,625	123		3,748	4,175
1810	400 amp		8	2		3,650	123		3,773	4,225
1820	600 amp		8	2		3,375	123		3,498	3,925
1830	800 amp		8	2		3,400	123		3,523	3,950
1840	1,200-2,000 amp		8	2		3,175	123		3,298	3,675
1900	Current/potential transformer metering compartment, 200-800 amp		5.40	2.963		2,150	182		2,332	2,650
1940	1,200 amp		5.40	2.963		6,000	182		6,182	6,875
1950	1,600-2,000 amp		5.40	2.963		9,875	182		10,057	11,200
2000	With watt meter, 200-800 amp		4	4		9,275	245		9,520	10,600
2040	1,200 amp		4	4		10,800	245		11,045	12,200
2050	1,600-2,000 amp		4	4		11,500	245		11,745	13,000
2100	Split bus, 60-200 amp	1 Elec	5.30	1.509		194	92.50		286.50	355
2130	400 amp	2 Elec	4.60	3.478		320	213		533	680
2140	600 amp		3.60	4.444		390	273		663	845
2150	800 amp		2.60	6.154		495	380		875	1,125
2170	1,200 amp		2	8		565	490		1,055	1,375
2250	Contactor control, 60 amp	1 Elec	2	4		1,275	245		1,520	1,775
2260	100 amp		1.50	5.333		1,350	325		1,675	1,975
2270	200 amp		1	8		2,025	490		2,515	3,000
2280	400 amp	2 Elec	1	16		6,550	980		7,530	8,700
2290	600 amp		.84	19.048		7,325	1,175		8,500	9,850
2300	800 amp		.72	22.222		8,200	1,375		9,575	11,100
2500	Modifier, two distribution sections, add		.80	20		2,850	1,225		4,075	5,025
2520	Three distribution sections, add		.40	40		5,475	2,450		7,925	9,800
2560	Auxiliary pull section, 20", add		2	8		1,025	490		1,515	1,875
2580	24", add		1.80	8.889		1,025	545		1,570	1,975
2600	30", add		1.60	10		1,025	615		1,640	2,075
2620	36", add		1.40	11.429		1,225	700		1,925	2,425
2640	Dog house, 12", add		2.40	6.667		212	410		622	865

	26 24 13.20 In Plant Distribution Switchboards	Crew	Daily Output	Labor-Hours	Unit	2020 Bare Costs Material	Labor	Equipment	Total	Total Incl O&P
2660	18", add	2 Elec	2	8	Ea.	420	490		910	1,225
3000	Transition section between switchboard and transformer									
3050	or motor control center, 4 wire alum. bus, 600 amp	2 Elec	1.14	14.035	Ea.	2,175	860		3,035	3,725
3100	800 amp		1	16		2,450	980		3,430	4,175
3150	1,000 amp		.88	18.182		2,750	1,125		3,875	4,750
3200	1,200 amp		.80	20		3,025	1,225		4,250	5,225
3250	1,600 amp		.72	22.222		3,575	1,375		4,950	6,025
3300	2,000 amp		.66	24.242		4,100	1,475		5,575	6,800
3350	2,500 amp		.62	25.806		4,775	1,575		6,350	7,675
3400	3,000 amp		.56	28.571		5,500	1,750		7,250	8,750
4000	Weatherproof construction, per vertical section		1.76	9.091		2,525	560		3,085	3,625

	26 24 13.30 Distribution Switchboards Section	Crew	Daily Output	Labor-Hours	Unit	Material	Labor	Equipment	Total	Total Incl O&P
0010	**DISTRIBUTION SWITCHBOARDS SECTION**									
0100	Aluminum bus bars, not including breakers									
0160	Subfeed lug-rated at 60 amp	2 Elec	1.30	12.308	Ea.	1,200	755		1,955	2,475
0170	100 amp		1.26	12.698		1,525	780		2,305	2,875
0180	200 amp		1.20	13.333		1,450	820		2,270	2,850
0190	400 amp		1.10	14.545		1,425	890		2,315	2,950
0195	120/208 or 277/480 volt, 4 wire, 400 amp		1.10	14.545		1,425	890		2,315	2,950
0200	600 amp		1	16		1,700	980		2,680	3,375
0300	800 amp		.88	18.182		2,175	1,125		3,300	4,125
0400	1,000 amp		.80	20		2,500	1,225		3,725	4,650
0500	1,200 amp		.72	22.222		2,950	1,375		4,325	5,350
0600	1,600 amp		.66	24.242		4,400	1,475		5,875	7,150
0700	2,000 amp		.62	25.806		5,500	1,575		7,075	8,475
0800	2,500 amp		.60	26.667		5,950	1,625		7,575	9,075
0900	3,000 amp		.56	28.571		7,850	1,750		9,600	11,400
0950	4,000 amp		.52	30.769		7,875	1,900		9,775	11,600

	26 24 13.40 Switchboards Feeder Section	Crew	Daily Output	Labor-Hours	Unit	Material	Labor	Equipment	Total	Total Incl O&P
0010	**SWITCHBOARDS FEEDER SECTION** group mounted devices									
0030	Circuit breakers									
0160	FA frame, 15 to 60 amp, 240 volt, 1 pole	1 Elec	8	1	Ea.	130	61.50		191.50	238
0170	2 pole		7	1.143		380	70		450	530
0180	3 pole		5.30	1.509		580	92.50		672.50	785
0210	480 volt, 1 pole		8	1		181	61.50		242.50	294
0220	2 pole		7	1.143		455	70		525	615
0230	3 pole		5.30	1.509		585	92.50		677.50	790
0260	600 volt, 2 pole		7	1.143		885	70		955	1,075
0270	3 pole		5.30	1.509		535	92.50		627.50	730
0280	FA frame, 70 to 100 amp, 240 volt, 1 pole		7	1.143		229	70		299	360
0310	2 pole		5	1.600		615	98		713	825
0320	3 pole		4	2		595	123		718	845
0330	480 volt, 1 pole		7	1.143		345	70		415	490
0360	2 pole		5	1.600		585	98		683	795
0370	3 pole		4	2		690	123		813	950
0380	600 volt, 2 pole		5	1.600		665	98		763	880
0410	3 pole		4	2		830	123		953	1,100
0420	KA frame, 70 to 225 amp		3.20	2.500		1,525	153		1,678	1,900
0430	LA frame, 125 to 400 amp		2.30	3.478		3,900	213		4,113	4,625
0460	MA frame, 450 to 600 amp		1.60	5		6,200	305		6,505	7,300
0470	700 to 800 amp		1.30	6.154		8,075	380		8,455	9,450
0480	MAL frame, 1,000 amp		1	8		8,375	490		8,865	9,975

26 24 Switchboards and Panelboards

26 24 13 – Switchboards

26 24 13.40 Switchboards Feeder Section		Crew	Daily Output	Labor-Hours	Unit	2020 Bare Costs Material	Labor	Equipment	Total	Total Incl O&P
0490	PA frame, 1,200 amp	1 Elec	.80	10	Ea.	17,000	615		17,615	19,600
0500	Branch circuit, fusible switch, 600 volt, double 30/30 amp		4	2		855	123		978	1,125
0550	60/60 amp		3.20	2.500		875	153		1,028	1,200
0600	100/100 amp		2.70	2.963		1,100	182		1,282	1,500
0650	Single, 30 amp		5.30	1.509		790	92.50		882.50	1,025
0700	60 amp		4.70	1.702		815	104		919	1,050
0750	100 amp		4	2		1,300	123		1,423	1,650
0800	200 amp		2.70	2.963		1,350	182		1,532	1,750
0850	400 amp		2.30	3.478		2,475	213		2,688	3,050
0900	600 amp		1.80	4.444		3,025	273		3,298	3,750
0950	800 amp		1.30	6.154		5,075	380		5,455	6,150
1000	1,200 amp		.80	10		5,800	615		6,415	7,350
1080	Branch circuit, circuit breakers, high interrupting capacity									
1100	60 amp, 240, 480 or 600 volt, 1 pole	1 Elec	8	1	Ea.	535	61.50		596.50	685
1120	2 pole		7	1.143		1,325	70		1,395	1,550
1140	3 pole		5.30	1.509		660	92.50		752.50	870
1150	100 amp, 240, 480 or 600 volt, 1 pole		7	1.143		595	70		665	765
1160	2 pole		5	1.600		1,475	98		1,573	1,775
1180	3 pole		4	2		705	123		828	970
1200	225 amp, 240, 480 or 600 volt, 2 pole		3.50	2.286		2,275	140		2,415	2,725
1220	3 pole		3.20	2.500		2,200	153		2,353	2,650
1240	400 amp, 240, 480 or 600 volt, 2 pole		2.50	3.200		3,250	196		3,446	3,875
1260	3 pole		2.30	3.478		2,975	213		3,188	3,600
1280	600 amp, 240, 480 or 600 volt, 2 pole		1.80	4.444		5,275	273		5,548	6,225
1300	3 pole		1.60	5		3,625	305		3,930	4,475
1320	800 amp, 240, 480 or 600 volt, 2 pole		1.50	5.333		4,650	325		4,975	5,625
1340	3 pole		1.30	6.154		5,450	380		5,830	6,550
1360	1,000 amp, 240, 480 or 600 volt, 2 pole		1.10	7.273		7,325	445		7,770	8,750
1380	3 pole		1	8		6,050	490		6,540	7,400
1400	1,200 amp, 240, 480 or 600 volt, 2 pole		.90	8.889		7,425	545		7,970	9,025
1420	3 pole		.80	10		7,675	615		8,290	9,375
1700	Fusible switch, 240 V, 60 amp, 2 pole		3.20	2.500		415	153		568	690
1720	3 pole		3	2.667		570	164		734	875
1740	100 amp, 2 pole		2.70	2.963		420	182		602	745
1760	3 pole		2.50	3.200		640	196		836	1,000
1780	200 amp, 2 pole		2	4		880	245		1,125	1,350
1800	3 pole		1.90	4.211		1,025	258		1,283	1,525
1820	400 amp, 2 pole		1.50	5.333		1,650	325		1,975	2,325
1840	3 pole		1.30	6.154		2,075	380		2,455	2,850
1860	600 amp, 2 pole		1	8		2,350	490		2,840	3,325
1880	3 pole		.90	8.889		2,875	545		3,420	4,025
1900	240-600 V, 800 amp, 2 pole		.70	11.429		5,225	700		5,925	6,825
1920	3 pole		.60	13.333		6,400	820		7,220	8,300
2000	600 V, 60 amp, 2 pole		3.20	2.500		630	153		783	925
2040	100 amp, 2 pole		2.70	2.963		645	182		827	990
2080	200 amp, 2 pole		2	4		1,125	245		1,370	1,600
2120	400 amp, 2 pole		1.50	5.333		2,225	325		2,550	2,950
2160	600 amp, 2 pole		1	8		2,700	490		3,190	3,725
2500	Branch circuit, circuit breakers, 60 amp, 600 volt, 3 pole		5.30	1.509		465	92.50		557.50	655
2520	240, 480 or 600 volt, 1 pole		8	1		97	61.50		158.50	202
2540	240 volt, 2 pole		7	1.143		173	70		243	298
2560	480 or 600 volt, 2 pole		7	1.143		355	70		425	500
2580	240 volt, 3 pole		5.30	1.509		410	92.50		502.50	595

26 24 13 – Switchboards

26 24 13.40 Switchboards Feeder Section		Crew	Daily Output	Labor-Hours	Unit	Material	Labor	Equipment	Total	Total Incl O&P
						2020 Bare Costs				
2600	480 volt, 3 pole	1 Elec	5.30	1.509	Ea.	470	92.50		562.50	665
2620	100 amp, 600 volt, 2 pole		5	1.600		450	98		548	645
2640	3 pole		4	2		570	123		693	815
2660	480 volt, 2 pole		5	1.600		390	98		488	580
2680	240 volt, 2 pole		5	1.600		222	98		320	395
2700	3 pole		4	2		355	123		478	585
2720	480 volt, 3 pole		4	2		525	123		648	765
2740	225 amp, 240, 480 or 600 volt, 2 pole		3.50	2.286		650	140		790	930
2760	3 pole		3.20	2.500		680	153		833	985
2780	400 amp, 240, 480 or 600 volt, 2 pole		2.50	3.200		1,375	196		1,571	1,800
2800	3 pole		2.30	3.478		1,575	213		1,788	2,075
2820	600 amp, 240 or 480 volt, 2 pole		1.80	4.444		2,225	273		2,498	2,875
2840	3 pole		1.60	5		2,725	305		3,030	3,475
2860	800 amp, 240, 480 or 600 volt, 2 pole		1.50	5.333		3,350	325		3,675	4,175
2880	3 pole		1.30	6.154		3,900	380		4,280	4,875
2900	1,000 amp, 240, 480 or 600 volt, 2 pole		1.10	7.273		4,100	445		4,545	5,200
2920	480 or 600 volt, 3 pole		1	8		4,700	490		5,190	5,925
2940	1,200 amp, 240, 480 or 600 volt, 2 pole		.90	8.889		5,775	545		6,320	7,200
2960	3 pole		.80	10		6,300	615		6,915	7,900
2980	600 volt, 3 pole	↓	.80	10	↓	6,500	615		7,115	8,100

26 24 16 – Panelboards

26 24 16.10 Load Centers		Crew	Daily Output	Labor-Hours	Unit	Material	Labor	Equipment	Total	Total Incl O&P
0010	**LOAD CENTERS** (residential type) R262416-50									
0100	3 wire, 120/240 V, 1 phase, including 1 pole plug-in breakers									
0200	100 amp main lugs, indoor, 8 circuits	1 Elec	1.40	5.714	Ea.	101	350		451	650
0300	12 circuits		1.20	6.667		127	410		537	770
0400	Rainproof, 8 circuits		1.40	5.714		130	350		480	685
0500	12 circuits	↓	1.20	6.667		156	410		566	800
0600	200 amp main lugs, indoor, 16 circuits	R-1A	1.80	8.889		218	490		708	995
0700	20 circuits		1.50	10.667		208	590		798	1,125
0800	24 circuits		1.30	12.308		266	680		946	1,350
0900	30 circuits		1.20	13.333		285	735		1,020	1,450
1000	40 circuits		.80	20		435	1,100		1,535	2,175
1200	Rainproof, 16 circuits		1.80	8.889		250	490		740	1,025
1300	20 circuits		1.50	10.667		325	590		915	1,275
1400	24 circuits		1.30	12.308		340	680		1,020	1,425
1500	30 circuits		1.20	13.333		380	735		1,115	1,550
1600	40 circuits		.80	20		490	1,100		1,590	2,250
1800	400 amp main lugs, indoor, 42 circuits		.72	22.222		1,200	1,225		2,425	3,225
1900	Rainproof, 42 circuits	↓	.72	22.222	↓	1,625	1,225		2,850	3,675
2200	Plug in breakers, 20 amp, 1 pole, 4 wire, 120/208 volts									
2210	125 amp main lugs, indoor, 12 circuits	1 Elec	1.20	6.667	Ea.	194	410		604	845
2300	18 circuits		.80	10		261	615		876	1,225
2400	Rainproof, 12 circuits		1.20	6.667		215	410		625	865
2500	18 circuits	↓	.80	10		315	615		930	1,300
2600	200 amp main lugs, indoor, 24 circuits	R-1A	1.30	12.308		380	680		1,060	1,475
2700	30 circuits		1.20	13.333		370	735		1,105	1,525
2800	36 circuits		1	16		380	885		1,265	1,775
2900	42 circuits		.80	20		540	1,100		1,640	2,300
3000	Rainproof, 24 circuits		1.30	12.308		287	680		967	1,375
3100	30 circuits		1.20	13.333		305	735		1,040	1,450
3200	36 circuits	↓	1	16	↓	465	885		1,350	1,875

26 24 16.10 Load Centers		Crew	Daily Output	Labor-Hours	Unit	2020 Bare Costs Material	Labor	Equipment	Total	Total Incl O&P
3300	42 circuits	R-1A	.80	20	Ea.	615	1,100		1,715	2,375
3500	400 amp main lugs, indoor, 42 circuits		.72	22.222		1,250	1,225		2,475	3,275
3600	Rainproof, 42 circuits		.72	22.222		1,575	1,225		2,800	3,625
3700	Plug-in breakers, 20 amp, 1 pole, 3 wire, 120/240 volts									
3800	100 amp main breaker, indoor, 12 circuits	1 Elec	1.20	6.667	Ea.	159	410		569	805
3900	18 circuits	"	.80	10		197	615		812	1,150
4000	200 amp main breaker, indoor, 20 circuits	R-1A	1.50	10.667		279	590		869	1,200
4200	24 circuits		1.30	12.308		305	680		985	1,375
4300	30 circuits		1.20	13.333		330	735		1,065	1,475
4400	40 circuits		.90	17.778		390	980		1,370	1,925
4500	Rainproof, 20 circuits		1.50	10.667		335	590		925	1,275
4600	24 circuits		1.30	12.308		390	680		1,070	1,475
4700	30 circuits		1.20	13.333		440	735		1,175	1,600
4800	40 circuits		.90	17.778		560	980		1,540	2,125
5000	400 amp main breaker, indoor, 42 circuits		.72	22.222		2,975	1,225		4,200	5,175
5100	Rainproof, 42 circuits		.72	22.222		3,150	1,225		4,375	5,350
5300	Plug in breakers, 20 amp, 1 pole, 4 wire, 120/208 volts									
5400	200 amp main breaker, indoor, 30 circuits	R-1A	1.20	13.333	Ea.	420	735		1,155	1,575
5500	42 circuits		.80	20		490	1,100		1,590	2,250
5600	Rainproof, 30 circuits		1.20	13.333		895	735		1,630	2,100
5700	42 circuits		.80	20		575	1,100		1,675	2,325

26 24 16.20 Panelboard and Load Center Circuit Breakers		Crew	Daily Output	Labor-Hours	Unit	Material	Labor	Equipment	Total	Total Incl O&P
0010	**PANELBOARD AND LOAD CENTER CIRCUIT BREAKERS**									
0050	Bolt-on, 10,000 amp I.C., 120 volt, 1 pole									
0100	15-50 amp	1 Elec	10	.800	Ea.	18.60	49		67.60	96
0200	60 amp		8	1		20.50	61.50		82	117
0300	70 amp		8	1		28	61.50		89.50	125
0350	240 volt, 2 pole									
0400	15-50 amp	1 Elec	8	1	Ea.	56.50	61.50		118	157
0500	60 amp		7.50	1.067		37.50	65.50		103	142
0600	80-100 amp		5	1.600		111	98		209	273
0700	3 pole, 15-60 amp		6.20	1.290		135	79		214	271
0800	70 amp		5	1.600		172	98		270	340
0900	80-100 amp		3.60	2.222		205	136		341	435
1000	22,000 amp I.C., 240 volt, 2 pole, 70-225 amp		2.70	2.963		555	182		737	890
1100	3 pole, 70-225 amp		2.30	3.478		345	213		558	705
1200	14,000 amp I.C., 277 volts, 1 pole, 15-30 amp		8	1		37	61.50		98.50	136
1300	22,000 amp I.C., 480 volts, 2 pole, 70-225 amp		2.70	2.963		525	182		707	855
1400	3 pole, 70-225 amp		2.30	3.478		1,050	213		1,263	1,475
2000	Plug-in panel or load center, 120/240 volt, to 60 amp, 1 pole		12	.667		6.40	41		47.40	70
2004	Circuit breaker, 120/240 volt, 20 A, 1 pole with NM cable		6.50	1.231		11.65	75.50		87.15	129
2006	30 A, 1 pole with NM cable		6.50	1.231		11.65	75.50		87.15	129
2010	2 pole		9	.889		25.50	54.50		80	112
2014	50 A, 2 pole with NM cable		5.50	1.455		30.50	89		119.50	171
2020	3 pole		7.50	1.067		97.50	65.50		163	208
2030	100 amp, 2 pole		6	1.333		105	82		187	242
2040	3 pole		4.50	1.778		118	109		227	297
2050	150-200 amp, 2 pole		3	2.667		256	164		420	535
2060	Plug-in tandem, 120/240 V, 2-15 A, 1 pole		11	.727		30	44.50		74.50	102
2070	1-15 A & 1-20 A		11	.727		18.35	44.50		62.85	88.50
2080	2-20 A		11	.727		18.40	44.50		62.90	88.50
2082	Arc fault circuit interrupter, 120/240 V, 1-15 A & 1-20 A, 1 pole		11	.727		57	44.50		101.50	131

26 24 16.20 Panelboard and Load Center Circuit Breakers		Crew	Daily Output	Labor-Hours	Unit	Material	2020 Bare Costs Labor	Equipment	Total	Total Incl O&P
2100	High interrupting capacity, 120/240 volt, plug-in, 30 amp, 1 pole	1 Elec	12	.667	Ea.	24	41		65	89.50
2110	60 amp, 2 pole		9	.889		26	54.50		80.50	113
2120	3 pole		7.50	1.067		257	65.50		322.50	385
2130	100 amp, 2 pole		6	1.333		138	82		220	278
2140	3 pole		4.50	1.778		430	109		539	640
2150	125 amp, 2 pole		3	2.667		870	164		1,034	1,200
2200	Bolt-on, 30 amp, 1 pole		10	.800		61.50	49		110.50	143
2210	60 amp, 2 pole		7.50	1.067		103	65.50		168.50	215
2220	3 pole		6.20	1.290		271	79		350	420
2230	100 amp, 2 pole		5	1.600		293	98		391	470
2240	3 pole		3.60	2.222		470	136		606	730
2300	Ground fault, 240 volt, 30 amp, 1 pole		7	1.143		92.50	70		162.50	210
2310	2 pole		6	1.333		170	82		252	315
2350	Key operated, 240 volt, 1 pole, 30 amp		7	1.143		135	70		205	257
2360	Switched neutral, 240 volt, 30 amp, 2 pole		6	1.333		40	82		122	170
2370	3 pole		5.50	1.455		65.50	89		154.50	209
2400	Shunt trip, for 240 volt breaker, 60 amp, 1 pole		4	2		81.50	123		204.50	279
2410	2 pole		3.50	2.286		81.50	140		221.50	305
2420	3 pole		3	2.667		81.50	164		245.50	340
2430	100 amp, 2 pole		3	2.667		81.50	164		245.50	340
2440	3 pole		2.50	3.200		81.50	196		277.50	390
2450	150 amp, 2 pole		2	4		235	245		480	640
2500	Auxiliary switch, for 240 volt breaker, 60 amp, 1 pole		4	2		90	123		213	288
2510	2 pole		3.50	2.286		90	140		230	315
2520	3 pole		3	2.667		90	164		254	350
2530	100 amp, 2 pole		3	2.667		90	164		254	350
2540	3 pole		2.50	3.200		90	196		286	400
2550	150 amp, 2 pole		2	4		123	245		368	515
2600	Panel or load center, 277/480 volt, plug-in, 30 amp, 1 pole		12	.667		71	41		112	142
2610	60 amp, 2 pole		9	.889		218	54.50		272.50	325
2620	3 pole		7.50	1.067		320	65.50		385.50	450
2650	Bolt-on, 60 amp, 2 pole		7.50	1.067		218	65.50		283.50	340
2660	3 pole		6.20	1.290		320	79		399	470
2700	I-line, 277/480 volt, 30 amp, 1 pole		8	1		68.50	61.50		130	170
2710	60 amp, 2 pole		7.50	1.067		241	65.50		306.50	365
2720	3 pole		6.20	1.290		375	79		454	530
2730	100 amp, 1 pole		7.50	1.067		161	65.50		226.50	278
2740	2 pole		5	1.600		375	98		473	560
2750	3 pole		3.50	2.286		440	140		580	700
2800	High interrupting capacity, 277/480 volt, plug-in, 30 amp, 1 pole		12	.667		420	41		461	525
2810	60 amp, 2 pole		9	.889		650	54.50		704.50	800
2820	3 pole		7	1.143		755	70		825	940
2830	Bolt-on, 30 amp, 1 pole		8	1		420	61.50		481.50	555
2840	60 amp, 2 pole		7.50	1.067		570	65.50		635.50	725
2850	3 pole		6.20	1.290		635	79		714	820
2900	I-line, 30 amp, 1 pole		8	1		420	61.50		481.50	555
2910	60 amp, 2 pole		7.50	1.067		585	65.50		650.50	745
2920	3 pole		6.20	1.290		665	79		744	850
2930	100 amp, 1 pole		7.50	1.067		680	65.50		745.50	845
2940	2 pole		5	1.600		475	98		573	675
2950	3 pole		3.60	2.222		735	136		871	1,025
2960	Shunt trip, 277/480 volt breaker, remote oper., 30 amp, 1 pole		4	2		440	123		563	675
2970	60 amp, 2 pole		3.50	2.286		645	140		785	925

26 24 16.20 Panelboard and Load Center Circuit Breakers		Crew	Daily Output	Labor-Hours	Unit	Material	2020 Bare Costs Labor	Equipment	Total	Total Incl O&P
2980	3 pole	1 Elec	3	2.667	Ea.	730	164		894	1,050
2990	100 amp, 1 pole		3.50	2.286		525	140		665	795
3000	2 pole		3	2.667		730	164		894	1,050
3010	3 pole		2.50	3.200		670	196		866	1,050
3050	Under voltage trip, 277/480 volt breaker, 30 amp, 1 pole		4	2		440	123		563	675
3060	60 amp, 2 pole		3.50	2.286		645	140		785	925
3070	3 pole		3	2.667		730	164		894	1,050
3080	100 amp, 1 pole		3.50	2.286		525	140		665	795
3090	2 pole		3	2.667		730	164		894	1,050
3100	3 pole		2.50	3.200		670	196		866	1,050
3150	Motor operated, 277/480 volt breaker, 30 amp, 1 pole		4	2		785	123		908	1,050
3160	60 amp, 2 pole		3.50	2.286		820	140		960	1,125
3170	3 pole		3	2.667		1,100	164		1,264	1,450
3180	100 amp, 1 pole		3.50	2.286		875	140		1,015	1,175
3190	2 pole		3	2.667		1,100	164		1,264	1,450
3200	3 pole		2.50	3.200		950	196		1,146	1,350
3250	Panelboard spacers, per pole		40	.200		3.79	12.25		16.04	23
5110	NEMA 1 enclosure only, 600V, 3 p, 14k AIC, 100A		2.20	3.636	↓	160	223		383	520
9000	Minimum labor/equipment charge	↓	3	2.667	Job		164		164	252

26 24 16.30 Panelboards Commercial Applications		Crew	Daily Output	Labor-Hours	Unit	Material	Labor	Equipment	Total	Total Incl O&P
0010	**PANELBOARDS COMMERCIAL APPLICATIONS** R262416-50									
0050	NQOD, w/20 amp 1 pole bolt-on circuit breakers									
0100	3 wire, 120/240 volts, 100 amp main lugs									
0150	10 circuits	1 Elec	1	8	Ea.	935	490		1,425	1,775
0200	14 circuits		.88	9.091		1,050	560		1,610	2,000
0250	18 circuits		.75	10.667		1,150	655		1,805	2,250
0300	20 circuits	↓	.65	12.308		1,275	755		2,030	2,575
0350	225 amp main lugs, 24 circuits	2 Elec	1.20	13.333		1,450	820		2,270	2,850
0400	30 circuits		.90	17.778		1,675	1,100		2,775	3,525
0450	36 circuits		.80	20		1,925	1,225		3,150	4,000
0500	38 circuits		.72	22.222		2,075	1,375		3,450	4,400
0550	42 circuits	↓	.66	24.242		2,150	1,475		3,625	4,650
0600	4 wire, 120/208 volts, 100 amp main lugs, 12 circuits	1 Elec	1	8		1,025	490		1,515	1,875
0650	16 circuits		.75	10.667		1,175	655		1,830	2,275
0700	20 circuits		.65	12.308		1,350	755		2,105	2,625
0750	24 circuits		.60	13.333		1,275	820		2,095	2,650
0800	30 circuits	↓	.53	15.094		1,675	925		2,600	3,275
0850	225 amp main lugs, 32 circuits	2 Elec	.90	17.778		1,875	1,100		2,975	3,750
0900	34 circuits		.84	19.048		1,925	1,175		3,100	3,900
0950	36 circuits		.80	20		1,975	1,225		3,200	4,050
1000	42 circuits		.68	23.529		2,200	1,450		3,650	4,650
1010	400 amp main lugs, 42 circs		.68	23.529		2,200	1,450		3,650	4,650
1040	225 amp main lugs, NEMA 7, 12 circuits		1	16		5,925	980		6,905	8,025
1100	24 circuits	↓	.40	40	↓	7,175	2,450		9,625	11,700
1200	NEHB, w/20 amp, 1 pole bolt-on circuit breakers									
1250	4 wire, 277/480 volts, 100 amp main lugs, 12 circuits	1 Elec	.88	9.091	Ea.	1,475	560		2,035	2,475
1300	20 circuits	"	.60	13.333		2,200	820		3,020	3,675
1350	225 amp main lugs, 24 circuits	2 Elec	.90	17.778		2,500	1,100		3,600	4,425
1400	30 circuits		.80	20		2,975	1,225		4,200	5,175
1448	32 circuits		4.90	3.265		3,475	200		3,675	4,100
1450	36 circuits		.72	22.222		3,475	1,375		4,850	5,900
1500	42 circuits	↓	.60	26.667	↓	3,950	1,625		5,575	6,875

26 24 16.30 Panelboards Commercial Applications		Crew	Daily Output	Labor-Hours	Unit	Material	2020 Bare Costs Labor	Equipment	Total	Total Incl O&P
1510	225 amp main lugs, NEMA 7, 12 circuits	2 Elec	.90	17.778	Ea.	6,175	1,100		7,275	8,450
1590	24 circuits	↓	.30	53.333	↓	8,275	3,275		11,550	14,100
1600	NQOD panel, w/20 amp, 1 pole, circuit breakers									
1650	3 wire, 120/240 volt with main circuit breaker									
1700	100 amp main, 12 circuits	1 Elec	.80	10	Ea.	1,275	615		1,890	2,350
1750	20 circuits	"	.60	13.333		1,625	820		2,445	3,025
1800	225 amp main, 30 circuits	2 Elec	.68	23.529		3,050	1,450		4,500	5,575
1801	225 amp main, 32 circuits		5	3.200		3,050	196		3,246	3,650
1850	42 circuits		.52	30.769		3,525	1,900		5,425	6,775
1900	400 amp main, 30 circuits		.54	29.630		4,225	1,825		6,050	7,450
1950	42 circuits	↓	.50	32	↓	4,725	1,975		6,700	8,225
2000	4 wire, 120/208 volts with main circuit breaker									
2050	100 amp main, 24 circuits	1 Elec	.47	17.021	Ea.	1,825	1,050		2,875	3,625
2100	30 circuits	"	.40	20		2,125	1,225		3,350	4,250
2200	225 amp main, 32 circuits	2 Elec	.72	22.222		3,550	1,375		4,925	6,025
2250	42 circuits		.56	28.571		4,125	1,750		5,875	7,225
2300	400 amp main, 42 circuits		.48	33.333		5,225	2,050		7,275	8,900
2350	600 amp main, 42 circuits	↓	.40	40	↓	7,725	2,450		10,175	12,300
2400	NEHB, with 20 amp, 1 pole circuit breaker									
2450	4 wire, 277/480 volts with main circuit breaker									
2500	100 amp main, 24 circuits	1 Elec	.42	19.048	Ea.	2,875	1,175		4,050	4,950
2550	30 circuits	"	.38	21.053		3,350	1,300		4,650	5,700
2600	225 amp main, 30 circuits	2 Elec	.72	22.222		4,200	1,375		5,575	6,725
2650	42 circuits		.56	28.571		5,175	1,750		6,925	8,400
2700	400 amp main, 42 circuits		.46	34.783		6,225	2,125		8,350	10,100
2750	600 amp main, 42 circuits	↓	.38	42.105	↓	8,525	2,575		11,100	13,400
2900	Note: the following line items don't include branch circuit breakers									
2910	For branch circuit breakers information, see Section 26 24 16.20									
3010	Main lug, no main breaker, 240 volt, 1 pole, 3 wire, 100 amp	1 Elec	2.30	3.478	Ea.	605	213		818	1,000
3020	225 amp	2 Elec	2.40	6.667		660	410		1,070	1,350
3030	400 amp	"	1.80	8.889		1,125	545		1,670	2,100
3060	3 pole, 3 wire, 100 amp	1 Elec	2.30	3.478		655	213		868	1,050
3070	225 amp	2 Elec	2.40	6.667		700	410		1,110	1,400
3080	400 amp		1.80	8.889		1,225	545		1,770	2,200
3090	600 amp	↓	1.60	10		1,425	615		2,040	2,525
3110	3 pole, 4 wire, 100 amp	1 Elec	2.30	3.478		720	213		933	1,125
3120	225 amp	2 Elec	2.40	6.667		825	410		1,235	1,525
3130	400 amp		1.80	8.889		1,225	545		1,770	2,200
3140	600 amp	↓	1.60	10		1,425	615		2,040	2,525
3160	480 volt, 3 pole, 3 wire, 100 amp	1 Elec	2.30	3.478		785	213		998	1,200
3170	225 amp	2 Elec	2.40	6.667		1,100	410		1,510	1,850
3180	400 amp		1.80	8.889		1,550	545		2,095	2,550
3190	600 amp	↓	1.60	10		1,725	615		2,340	2,850
3210	277/480 volt, 3 pole, 4 wire, 100 amp	1 Elec	2.30	3.478		765	213		978	1,175
3220	225 amp	2 Elec	2.40	6.667		930	410		1,340	1,650
3230	400 amp		1.80	8.889		1,500	545		2,045	2,500
3240	600 amp	↓	1.60	10		1,675	615		2,290	2,800
3260	Main circuit breaker, 240 volt, 1 pole, 3 wire, 100 amp	1 Elec	2	4		740	245		985	1,200
3270	225 amp	2 Elec	2	8		1,775	490		2,265	2,700
3280	400 amp	"	1.60	10		2,775	615		3,390	4,000
3310	3 pole, 3 wire, 100 amp	1 Elec	2	4		890	245		1,135	1,350
3320	225 amp	2 Elec	2	8		2,025	490		2,515	2,975
3330	400 amp	↓	1.60	10	↓	3,175	615		3,790	4,425

26 24 Switchboards and Panelboards

26 24 16 – Panelboards

26 24 16.30 Panelboards Commercial Applications		Crew	Daily Output	Labor-Hours	Unit	2020 Bare Costs Material	Labor	Equipment	Total	Total Incl O&P
3360	120/208 volt, 3 pole, 4 wire, 100 amp	2 Elec	4	4	Ea.	850	245		1,095	1,325
3370	225 amp		2	8		2,025	490		2,515	2,975
3380	400 amp		1.60	10		3,175	615		3,790	4,425
3410	480 volt, 3 pole, 3 wire, 100 amp	1 Elec	2	4		1,375	245		1,620	1,900
3420	225 amp	2 Elec	2	8		2,375	490		2,865	3,375
3430	400 amp		1.60	10		3,550	615		4,165	4,850
3460	277/480 volt, 3 pole, 4 wire, 100 amp		4	4		1,350	245		1,595	1,875
3470	225 amp		2	8		2,325	490		2,815	3,300
3480	400 amp		1.60	10		3,575	615		4,190	4,875
3510	Main circuit breaker, HIC, 240 volt, 1 pole, 3 wire, 100 amp	1 Elec	2	4		1,325	245		1,570	1,825
3520	225 amp	2 Elec	2	8		3,450	490		3,940	4,550
3530	400 amp	"	1.60	10		4,625	615		5,240	6,025
3560	3 pole, 3 wire, 100 amp	1 Elec	2	4		1,475	245		1,720	2,000
3570	225 amp	2 Elec	2	8		3,900	490		4,390	5,025
3580	400 amp	"	1.60	10		5,150	615		5,765	6,625
3610	120/208 volt, 3 pole, 4 wire, 100 amp	1 Elec	2	4		1,475	245		1,720	2,000
3620	225 amp	2 Elec	2	8		3,900	490		4,390	5,025
3630	400 amp	"	1.60	10		5,150	615		5,765	6,625
3660	480 volt, 3 pole, 3 wire, 100 amp	1 Elec	2	4		2,200	245		2,445	2,775
3670	225 amp	2 Elec	2	8		4,325	490		4,815	5,500
3680	400 amp	"	1.60	10		5,500	615		6,115	7,000
3710	277/480 volt, 3 pole, 4 wire, 100 amp	1 Elec	2	4		2,075	245		2,320	2,675
3720	225 amp	2 Elec	2	8		4,125	490		4,615	5,275
3730	400 amp	"	1.60	10		5,450	615		6,065	6,950
3760	Main circuit breaker, shunt trip, 100 amp	1 Elec	1.20	6.667		1,100	410		1,510	1,825
3770	225 amp	2 Elec	1.60	10		2,500	615		3,115	3,700
3780	400 amp	"	1.40	11.429		3,625	700		4,325	5,075
9000	Minimum labor/equipment charge	1 Elec	1	8	Job		490		490	755

26 24 19 – Motor-Control Centers

26 24 19.20 Motor Control Center Components		Crew	Daily Output	Labor-Hours	Unit	2020 Bare Costs Material	Labor	Equipment	Total	Total Incl O&P
0010	**MOTOR CONTROL CENTER COMPONENTS**									
0100	Starter, size 1, FVNR, NEMA 1, type A, fusible	1 Elec	2.70	2.963	Ea.	1,650	182		1,832	2,100
0120	Circuit breaker		2.70	2.963		1,800	182		1,982	2,250
0140	Type B, fusible		2.70	2.963		1,825	182		2,007	2,275
0160	Circuit breaker		2.70	2.963		1,975	182		2,157	2,450
0180	NEMA 12, type A, fusible		2.60	3.077		1,700	189		1,889	2,150
0200	Circuit breaker		2.60	3.077		1,825	189		2,014	2,325
0220	Type B, fusible		2.60	3.077		1,850	189		2,039	2,350
0240	Circuit breaker		2.60	3.077		2,025	189		2,214	2,525
0300	Starter, size 1, FVR, NEMA 1, type A, fusible		2	4		2,375	245		2,620	3,000
0320	Circuit breaker		2	4		2,375	245		2,620	3,000
0340	Type B, fusible		2	4		2,625	245		2,870	3,250
0360	Circuit breaker		2	4		2,625	245		2,870	3,250
0380	NEMA 12, type A, fusible		1.90	4.211		2,400	258		2,658	3,050
0400	Circuit breaker		1.90	4.211		2,400	258		2,658	3,050
0420	Type B, fusible		1.90	4.211		2,650	258		2,908	3,325
0440	Circuit breaker		1.90	4.211		2,650	258		2,908	3,325
0490	Starter size 1, 2 speed, separate winding									
0500	NEMA 1, type A, fusible	1 Elec	2.60	3.077	Ea.	3,125	189		3,314	3,750
0520	Circuit breaker		2.60	3.077		3,125	189		3,314	3,750
0540	Type B, fusible		2.60	3.077		3,450	189		3,639	4,100
0560	Circuit breaker		2.60	3.077		3,450	189		3,639	4,100

26 24 19.20 Motor Control Center Components		Crew	Daily Output	Labor-Hours	Unit	Material	Labor	Equipment	Total	Total Incl O&P
						2020 Bare Costs				
0580	NEMA 12, type A, fusible	1 Elec	2.50	3.200	Ea.	3,200	196		3,396	3,825
0600	Circuit breaker		2.50	3.200		3,200	196		3,396	3,825
0620	Type B, fusible		2.50	3.200		3,500	196		3,696	4,150
0640	Circuit breaker		2.50	3.200		3,500	196		3,696	4,150
0650	Starter size 1, 2 speed, consequent pole									
0660	NEMA 1, type A, fusible	1 Elec	2.60	3.077	Ea.	3,125	189		3,314	3,750
0680	Circuit breaker		2.60	3.077		3,125	189		3,314	3,750
0700	Type B, fusible		2.60	3.077		3,450	189		3,639	4,100
0720	Circuit breaker		2.60	3.077		3,450	189		3,639	4,100
0740	NEMA 12, type A, fusible		2.50	3.200		3,200	196		3,396	3,825
0760	Circuit breaker		2.50	3.200		3,200	196		3,396	3,825
0780	Type B, fusible		2.50	3.200		3,475	196		3,671	4,125
0800	Circuit breaker		2.50	3.200		3,500	196		3,696	4,150
0810	Starter size 1, 2 speed, space only									
0820	NEMA 1, type A, fusible	1 Elec	16	.500	Ea.	705	30.50		735.50	820
0840	Circuit breaker		16	.500		705	30.50		735.50	820
0860	Type B, fusible		16	.500		705	30.50		735.50	820
0880	Circuit breaker		16	.500		705	30.50		735.50	820
0900	NEMA 12, type A, fusible		15	.533		735	32.50		767.50	855
0920	Circuit breaker		15	.533		735	32.50		767.50	855
0940	Type B, fusible		15	.533		735	32.50		767.50	855
0960	Circuit breaker		15	.533		735	32.50		767.50	855
1100	Starter size 2, FVNR, NEMA 1, type A, fusible	2 Elec	4	4		1,875	245		2,120	2,450
1120	Circuit breaker		4	4		2,050	245		2,295	2,650
1140	Type B, fusible		4	4		2,075	245		2,320	2,650
1160	Circuit breaker		4	4		2,250	245		2,495	2,850
1180	NEMA 12, type A, fusible		3.80	4.211		1,900	258		2,158	2,500
1200	Circuit breaker		3.80	4.211		2,075	258		2,333	2,700
1220	Type B, fusible		3.80	4.211		2,075	258		2,333	2,700
1240	Circuit breaker		3.80	4.211		2,275	258		2,533	2,900
1300	FVR, NEMA 1, type A, fusible		3.20	5		3,175	305		3,480	3,975
1320	Circuit breaker		3.20	5		3,175	305		3,480	3,975
1340	Type B, fusible		3.20	5		3,500	305		3,805	4,325
1360	Circuit breaker		3.20	5		3,500	305		3,805	4,325
1380	NEMA type 12, type A, fusible		3	5.333		3,225	325		3,550	4,050
1400	Circuit breaker		3	5.333		3,225	325		3,550	4,050
1420	Type B, fusible		3	5.333		3,550	325		3,875	4,400
1440	Circuit breaker		3	5.333		3,550	325		3,875	4,400
1490	Starter size 2, 2 speed, separate winding									
1500	NEMA 1, type A, fusible	2 Elec	3.80	4.211	Ea.	3,550	258		3,808	4,300
1520	Circuit breaker		3.80	4.211		3,550	258		3,808	4,300
1540	Type B, fusible		3.80	4.211		3,900	258		4,158	4,700
1560	Circuit breaker		3.80	4.211		3,900	258		4,158	4,700
1570	NEMA 12, type A, fusible		3.60	4.444		3,600	273		3,873	4,400
1580	Circuit breaker		3.60	4.444		3,600	273		3,873	4,400
1600	Type B, fusible		3.60	4.444		3,950	273		4,223	4,775
1620	Circuit breaker		3.60	4.444		3,950	273		4,223	4,775
1630	Starter size 2, 2 speed, consequent pole									
1640	NEMA 1, type A, fusible	2 Elec	3.80	4.211	Ea.	4,025	258		4,283	4,825
1660	Circuit breaker		3.80	4.211		4,025	258		4,283	4,825
1680	Type B, fusible		3.80	4.211		4,250	258		4,508	5,075
1700	Circuit breaker		3.80	4.211		4,250	258		4,508	5,075
1720	NEMA 12, type A, fusible		3.80	4.211		4,050	258		4,308	4,850

26 24 19.20 Motor Control Center Components		Crew	Daily Output	Labor-Hours	Unit	2020 Bare Costs Material	Labor	Equipment	Total	Total Incl O&P
1740	Circuit breaker	2 Elec	3.60	4.444	Ea.	4,075	273		4,348	4,925
1760	Type B, fusible		3.60	4.444		4,325	273		4,598	5,175
1780	Circuit breaker		3.60	4.444		4,325	273		4,598	5,175
1830	Starter size 2, autotransformer									
1840	NEMA 1, type A, fusible	2 Elec	3.40	4.706	Ea.	6,275	289		6,564	7,350
1860	Circuit breaker		3.40	4.706		6,425	289		6,714	7,525
1880	Type B, fusible		3.40	4.706		6,850	289		7,139	8,000
1900	Circuit breaker		3.40	4.706		6,850	289		7,139	8,000
1920	NEMA 12, type A, fusible		3.20	5		6,400	305		6,705	7,500
1940	Circuit breaker		3.20	5		6,400	305		6,705	7,500
1960	Type B, fusible		3.20	5		6,975	305		7,280	8,150
1980	Circuit breaker		3.20	5		6,975	305		7,280	8,150
2030	Starter size 2, space only									
2040	NEMA 1, type A, fusible	1 Elec	16	.500	Ea.	705	30.50		735.50	820
2060	Circuit breaker		16	.500		705	30.50		735.50	820
2080	Type B, fusible		16	.500		705	30.50		735.50	820
2100	Circuit breaker		16	.500		705	30.50		735.50	820
2120	NEMA 12, type A, fusible		15	.533		735	32.50		767.50	855
2140	Circuit breaker		15	.533		735	32.50		767.50	855
2160	Type B, fusible		15	.533		735	32.50		767.50	855
2180	Circuit breaker		15	.533		735	32.50		767.50	855
2300	Starter size 3, FVNR, NEMA 1, type A, fusible	2 Elec	2	8		3,575	490		4,065	4,675
2320	Circuit breaker		2	8		3,175	490		3,665	4,250
2340	Type B, fusible		2	8		3,925	490		4,415	5,075
2360	Circuit breaker		2	8		3,500	490		3,990	4,600
2380	NEMA 12, type A, fusible		1.90	8.421		3,650	515		4,165	4,800
2400	Circuit breaker		1.90	8.421		3,225	515		3,740	4,350
2420	Type B, fusible		1.90	8.421		4,075	515		4,590	5,275
2440	Circuit breaker		1.90	8.421		3,550	515		4,065	4,725
2500	Starter size 3, FVR, NEMA 1, type A, fusible		1.60	10		4,875	615		5,490	6,300
2520	Circuit breaker		1.60	10		4,650	615		5,265	6,075
2540	Type B, fusible		1.60	10		5,300	615		5,915	6,775
2560	Circuit breaker		1.60	10		5,100	615		5,715	6,550
2580	NEMA 12, type A, fusible		1.50	10.667		4,975	655		5,630	6,475
2600	Circuit breaker		1.50	10.667		4,750	655		5,405	6,225
2620	Type B, fusible		1.50	10.667		5,400	655		6,055	6,950
2640	Circuit breaker		1.50	10.667		5,600	655		6,255	7,175
2690	Starter size 3, 2 speed, separate winding									
2700	NEMA 1, type A, fusible	2 Elec	2	8	Ea.	5,600	490		6,090	6,925
2720	Circuit breaker		2	8		4,950	490		5,440	6,200
2740	Type B, fusible		2	8		6,125	490		6,615	7,500
2760	Circuit breaker		2	8		5,425	490		5,915	6,700
2780	NEMA 12, type A, fusible		1.90	8.421		5,725	515		6,240	7,100
2800	Circuit breaker		1.90	8.421		5,050	515		5,565	6,350
2820	Type B, fusible		1.90	8.421		6,250	515		6,765	7,675
2840	Circuit breaker		1.90	8.421		5,525	515		6,040	6,875
2850	Starter size 3, 2 speed, consequent pole									
2860	NEMA 1, type A, fusible	2 Elec	2	8	Ea.	6,250	490		6,740	7,625
2880	Circuit breaker		2	8		5,600	490		6,090	6,900
2900	Type B, fusible		2	8		6,775	490		7,265	8,225
2920	Circuit breaker		2	8		6,075	490		6,565	7,425
2940	NEMA 12, type A, fusible		1.90	8.421		6,375	515		6,890	7,800
2960	Circuit breaker		1.90	8.421		5,700	515		6,215	7,075

26 24 19 – Motor-Control Centers

26 24 19.20 Motor Control Center Components		Crew	Daily Output	Labor-Hours	Unit	Material	Labor	Equipment	Total	Total Incl O&P
						2020 Bare Costs				
2980	Type B, fusible	2 Elec	1.90	8.421	Ea.	6,900	515		7,415	8,400
3000	Circuit breaker		1.90	8.421		6,175	515		6,690	7,575
3100	Starter size 3, autotransformer, NEMA 1, type A, fusible		1.60	10		8,250	615		8,865	10,000
3120	Circuit breaker		1.60	10		7,775	615		8,390	9,500
3140	Type B, fusible		1.60	10		8,500	615		9,115	10,300
3160	Circuit breaker		1.60	10		8,500	615		9,115	10,300
3180	NEMA 12, type A, fusible		1.50	10.667		7,925	655		8,580	9,725
3200	Circuit breaker		1.50	10.667		7,925	655		8,580	9,725
3220	Type B, fusible		1.50	10.667		8,650	655		9,305	10,500
3240	Circuit breaker		1.50	10.667		8,650	655		9,305	10,500
3260	Starter size 3, space only, NEMA 1, type A, fusible	1 Elec	15	.533		1,200	32.50		1,232.50	1,375
3280	Circuit breaker		15	.533		940	32.50		972.50	1,075
3300	Type B, fusible		15	.533		1,200	32.50		1,232.50	1,375
3320	Circuit breaker		15	.533		940	32.50		972.50	1,075
3340	NEMA 12, type A, fusible		14	.571		1,275	35		1,310	1,450
3360	Circuit breaker		14	.571		995	35		1,030	1,150
3380	Type B, fusible		14	.571		1,275	35		1,310	1,450
3400	Circuit breaker		14	.571		995	35		1,030	1,150
3500	Starter size 4, FVNR, NEMA 1, type A, fusible	2 Elec	1.60	10		4,650	615		5,265	6,075
3520	Circuit breaker		1.60	10		4,225	615		4,840	5,600
3540	Type B, fusible		1.60	10		5,100	615		5,715	6,550
3560	Circuit breaker		1.60	10		4,650	615		5,265	6,075
3580	NEMA 12, type A, fusible		1.50	10.667		4,775	655		5,430	6,250
3600	Circuit breaker		1.50	10.667		4,300	655		4,955	5,725
3620	Type B, fusible		1.50	10.667		5,200	655		5,855	6,725
3640	Circuit breaker		1.50	10.667		4,725	655		5,380	6,200
3700	Starter size 4, FVR, NEMA 1, type A, fusible		1.20	13.333		6,375	820		7,195	8,250
3720	Circuit breaker		1.20	13.333		5,750	820		6,570	7,550
3740	Type B, fusible		1.20	13.333		6,950	820		7,770	8,900
3760	Circuit breaker		1.20	13.333		6,325	820		7,145	8,200
3780	NEMA 12, type A, fusible		1.16	13.793		6,500	845		7,345	8,450
3800	Circuit breaker		1.16	13.793		5,825	845		6,670	7,700
3820	Type B, fusible		1.16	13.793		5,875	845		6,720	7,775
3840	Circuit breaker		1.16	13.793		6,400	845		7,245	8,350
3890	Starter size 4, 2 speed, separate windings									
3900	NEMA 1, type A, fusible	2 Elec	1.60	10	Ea.	8,025	615		8,640	9,775
3920	Circuit breaker		1.60	10		6,025	615		6,640	7,600
3940	Type B, fusible		1.60	10		8,825	615		9,440	10,600
3960	Circuit breaker		1.60	10		6,600	615		7,215	8,225
3980	NEMA 12, type A, fusible		1.50	10.667		8,175	655		8,830	10,000
4000	Circuit breaker		1.50	10.667		6,125	655		6,780	7,750
4020	Type B, fusible		1.50	10.667		8,975	655		9,630	10,900
4040	Circuit breaker		1.50	10.667		6,725	655		7,380	8,375
4050	Starter size 4, 2 speed, consequent pole									
4060	NEMA 1, type A, fusible	2 Elec	1.60	10	Ea.	9,375	615		9,990	11,200
4080	Circuit breaker		1.60	10		6,975	615		7,590	8,625
4100	Type B, fusible		1.60	10		10,300	615		10,915	12,300
4120	Circuit breaker		1.60	10		7,650	615		8,265	9,375
4140	NEMA 12, type A, fusible		1.50	10.667		9,525	655		10,180	11,500
4160	Circuit breaker		1.50	10.667		7,075	655		7,730	8,800
4180	Type B, fusible		1.50	10.667		10,500	655		11,155	12,500
4200	Circuit breaker		1.50	10.667		7,750	655		8,405	9,525
4300	Starter size 4, autotransformer, NEMA 1, type A, fusible		1.30	12.308		9,250	755		10,005	11,400

26 24 19.20 Motor Control Center Components		Crew	Daily Output	Labor-Hours	Unit	2020 Bare Costs Material	Labor	Equipment	Total	Total Incl O&P
4320	Circuit breaker	2 Elec	1.30	12.308	Ea.	9,325	755		10,080	11,400
4340	Type B, fusible		1.30	12.308		10,200	755		10,955	12,400
4360	Circuit breaker		1.30	12.308		10,200	755		10,955	12,400
4380	NEMA 12, type A, fusible		1.24	12.903		9,400	790		10,190	11,500
4400	Circuit breaker		1.24	12.903		9,450	790		10,240	11,600
4420	Type B, fusible		1.24	12.903		10,300	790		11,090	12,600
4440	Circuit breaker		1.24	12.903		10,300	790		11,090	12,500
4500	Starter size 4, space only, NEMA 1, type A, fusible	1 Elec	14	.571		1,675	35		1,710	1,900
4520	Circuit breaker		14	.571		1,200	35		1,235	1,375
4540	Type B, fusible		14	.571		1,675	35		1,710	1,900
4560	Circuit breaker		14	.571		1,200	35		1,235	1,375
4580	NEMA 12, type A, fusible		13	.615		1,775	38		1,813	2,000
4600	Circuit breaker		13	.615		1,275	38		1,313	1,450
4620	Type B, fusible		13	.615		1,775	38		1,813	2,000
4640	Circuit breaker		13	.615		1,275	38		1,313	1,450
4800	Starter size 5, FVNR, NEMA 1, type A, fusible	2 Elec	1	16		9,600	980		10,580	12,100
4820	Circuit breaker		1	16		6,825	980		7,805	9,000
4840	Type B, fusible		1	16		10,500	980		11,480	13,100
4860	Circuit breaker		1	16		7,500	980		8,480	9,750
4880	NEMA 12, type A, fusible		.96	16.667		9,750	1,025		10,775	12,300
4900	Circuit breaker		.96	16.667		6,950	1,025		7,975	9,225
4920	Type B, fusible		.96	16.667		10,700	1,025		11,725	13,300
4940	Circuit breaker		.96	16.667		7,625	1,025		8,650	9,975
5000	Starter size 5, FVR, NEMA 1, type A, fusible		.80	20		14,200	1,225		15,425	17,500
5020	Circuit breaker		.80	20		11,100	1,225		12,325	14,100
5040	Type B, fusible		.80	20		15,500	1,225		16,725	19,000
5060	Circuit breaker		.80	20		12,200	1,225		13,425	15,300
5080	NEMA 12, type A, fusible		.76	21.053		14,400	1,300		15,700	17,900
5100	Circuit breaker		.76	21.053		11,300	1,300		12,600	14,400
5120	Type B, fusible		.76	21.053		15,800	1,300		17,100	19,400
5140	Circuit breaker		.76	21.053		12,400	1,300		13,700	15,600
5190	Starter size 5, 2 speed, separate windings									
5200	NEMA 1, type A, fusible	2 Elec	1	16	Ea.	19,500	980		20,480	23,000
5220	Circuit breaker		1	16		14,600	980		15,580	17,500
5240	Type B, fusible		1	16		21,400	980		22,380	25,100
5260	Circuit breaker		1	16		16,000	980		16,980	19,100
5280	NEMA 12, type A, fusible		.96	16.667		19,800	1,025		20,825	23,400
5300	Circuit breaker		.96	16.667		14,700	1,025		15,725	17,800
5320	Type B, fusible		.96	16.667		21,700	1,025		22,725	25,500
5340	Circuit breaker		.96	16.667		16,200	1,025		17,225	19,400
5400	Starter size 5, autotransformer, NEMA 1, type A, fusible		.70	22.857		16,000	1,400		17,400	19,800
5420	Circuit breaker		.70	22.857		13,300	1,400		14,700	16,800
5440	Type B, fusible		.70	22.857		17,500	1,400		18,900	21,500
5460	Circuit breaker		.70	22.857		14,600	1,400		16,000	18,200
5480	NEMA 12, type A, fusible		.68	23.529		16,200	1,450		17,650	20,000
5500	Circuit breaker		.68	23.529		13,400	1,450		14,850	16,900
5520	Type B, fusible		.68	23.529		17,700	1,450		19,150	21,700
5540	Circuit breakers		.68	23.529		14,700	1,450		16,150	18,400
5600	Starter size 5, space only, NEMA 1, type A, fusible	1 Elec	12	.667		2,150	41		2,191	2,425
5620	Circuit breaker		12	.667		2,150	41		2,191	2,425
5640	Type B, fusible		12	.667		2,150	41		2,191	2,425
5660	Circuit breaker		12	.667		1,400	41		1,441	1,625
5680	NEMA 12, type A, fusible		11	.727		2,275	44.50		2,319.50	2,575

26 24 19.20 Motor Control Center Components		Crew	Daily Output	Labor-Hours	Unit	Material	Labor	Equipment	Total	Total Incl O&P
						2020 Bare Costs				
5700	Circuit breaker	1 Elec	11	.727	Ea.	1,500	44.50		1,544.50	1,725
5720	Type B, fusible		11	.727		2,275	44.50		2,319.50	2,575
5740	Circuit breaker		11	.727		1,500	44.50		1,544.50	1,725
5800	Fuse, light contactor NEMA 1, type A, 30 amp		2.70	2.963		1,500	182		1,682	1,925
5820	60 amp		2	4		1,700	245		1,945	2,250
5840	100 amp		1	8		3,250	490		3,740	4,325
5860	200 amp		.80	10		7,525	615		8,140	9,250
5880	Type B, 30 amp		2.70	2.963		1,650	182		1,832	2,075
5900	60 amp		2	4		1,825	245		2,070	2,400
5920	100 amp		1	8		3,575	490		4,065	4,675
5940	200 amp	2 Elec	1.60	10		8,250	615		8,865	10,000
5960	NEMA 12, type A, 30 amp	1 Elec	2.60	3.077		1,525	189		1,714	1,975
5980	60 amp		1.90	4.211		1,725	258		1,983	2,300
6000	100 amp		.95	8.421		3,300	515		3,815	4,450
6020	200 amp	2 Elec	1.50	10.667		7,650	655		8,305	9,425
6040	Type B, 30 amp	1 Elec	2.60	3.077		1,675	189		1,864	2,125
6060	60 amp		1.90	4.211		1,850	258		2,108	2,450
6080	100 amp		.95	8.421		3,625	515		4,140	4,800
6100	200 amp	2 Elec	1.50	10.667		8,375	655		9,030	10,200
6200	Circuit breaker, light contactor NEMA 1, type A, 30 amp	1 Elec	2.70	2.963		1,650	182		1,832	2,075
6220	60 amp		2	4		1,875	245		2,120	2,425
6240	100 amp		1	8		2,875	490		3,365	3,925
6260	200 amp	2 Elec	1.60	10		6,250	615		6,865	7,825
6280	Type B, 30 amp	1 Elec	2.70	2.963		1,800	182		1,982	2,250
6300	60 amp		2	4		2,075	245		2,320	2,650
6320	100 amp		1	8		3,150	490		3,640	4,200
6340	200 amp	2 Elec	1.60	10		6,825	615		7,440	8,450
6360	NEMA 12, type A, 30 amp	1 Elec	2.60	3.077		1,675	189		1,864	2,125
6380	60 amp		1.90	4.211		1,900	258		2,158	2,475
6400	100 amp		.95	8.421		2,925	515		3,440	4,025
6420	200 amp	2 Elec	1.50	10.667		6,275	655		6,930	7,900
6440	Type B, 30 amp	1 Elec	2.60	3.077		1,950	189		2,139	2,450
6460	60 amp		1.90	4.211		2,100	258		2,358	2,700
6480	100 amp		.95	8.421		3,200	515		3,715	4,325
6500	200 amp	2 Elec	1.50	10.667		6,900	655		7,555	8,575
6600	Fusible switch, NEMA 1, type A, 30 amp	1 Elec	5.30	1.509		1,100	92.50		1,192.50	1,350
6620	60 amp		5	1.600		1,175	98		1,273	1,450
6640	100 amp		4	2		1,300	123		1,423	1,625
6660	200 amp		3.20	2.500		2,250	153		2,403	2,675
6680	400 amp	2 Elec	4.60	3.478		5,275	213		5,488	6,125
6700	600 amp		3.20	5		5,700	305		6,005	6,750
6720	800 amp		2.60	6.154		15,600	380		15,980	17,800
6740	NEMA 12, type A, 30 amp	1 Elec	5.20	1.538		1,125	94.50		1,219.50	1,400
6760	60 amp		4.90	1.633		1,200	100		1,300	1,475
6780	100 amp		3.90	2.051		1,325	126		1,451	1,650
6800	200 amp		3.10	2.581		2,300	158		2,458	2,775
6820	400 amp	2 Elec	4.40	3.636		5,375	223		5,598	6,275
6840	600 amp		3	5.333		5,875	325		6,200	6,950
6860	800 amp		2.40	6.667		15,800	410		16,210	18,000
6900	Circuit breaker, NEMA 1, type A, 30 amp	1 Elec	5.30	1.509		1,000	92.50		1,092.50	1,250
6920	60 amp		5	1.600		1,000	98		1,098	1,250
6940	100 amp		4	2		1,000	123		1,123	1,300
6960	225 amp		3.20	2.500		1,800	153		1,953	2,200

26 24 19.20 Motor Control Center Components		Crew	Daily Output	Labor-Hours	Unit	Material	Labor	Equipment	Total	Total Incl O&P
						2020 Bare Costs				
6980	400 amp	2 Elec	4.60	3.478	Ea.	3,425	213		3,638	4,100
7000	600 amp		3.20	5		4,000	305		4,305	4,875
7020	800 amp		2.60	6.154		8,375	380		8,755	9,775
7040	NEMA 12, type A, 30 amp	1 Elec	5.20	1.538		1,050	94.50		1,144.50	1,300
7060	60 amp		4.90	1.633		1,050	100		1,150	1,300
7080	100 amp		3.90	2.051		1,050	126		1,176	1,350
7100	225 amp		3.10	2.581		1,825	158		1,983	2,275
7120	400 amp	2 Elec	4.40	3.636		3,475	223		3,698	4,175
7140	600 amp		3	5.333		4,075	325		4,400	4,975
7160	800 amp		2.40	6.667		8,525	410		8,935	10,000
7300	Incoming line, main lug only, 600 amp, alum., NEMA 1		1.60	10		1,275	615		1,890	2,350
7320	NEMA 12		1.50	10.667		1,300	655		1,955	2,425
7340	Copper, NEMA 1		1.60	10		1,350	615		1,965	2,425
7360	800 amp, alum., NEMA 1		1.50	10.667		3,200	655		3,855	4,525
7380	NEMA 12		1.40	11.429		3,250	700		3,950	4,650
7400	Copper, NEMA 1		1.50	10.667		3,325	655		3,980	4,675
7420	1,200 amp, copper, NEMA 1		1.40	11.429		3,450	700		4,150	4,875
7440	Incoming line, fusible switch, 400 amp, alum., NEMA 1		1.20	13.333		4,400	820		5,220	6,100
7460	NEMA 12		1.10	14.545		4,500	890		5,390	6,325
7480	Copper, NEMA 1		1.20	13.333		4,475	820		5,295	6,175
7500	600 amp, alum., NEMA 1		1.10	14.545		5,375	890		6,265	7,300
7520	NEMA 12		1	16		5,475	980		6,455	7,525
7540	Copper, NEMA 1		1.10	14.545		5,450	890		6,340	7,375
7560	Incoming line, circuit breaker, 225 amp, alum., NEMA 1		1.20	13.333		2,300	820		3,120	3,775
7580	NEMA 12		1.10	14.545		2,350	890		3,240	3,950
7600	Copper, NEMA 1		1.20	13.333		2,375	820		3,195	3,850
7620	400 amp, alum., NEMA 1		1.20	13.333		3,425	820		4,245	5,025
7640	NEMA 12		1.10	14.545		3,475	890		4,365	5,200
7660	Copper, NEMA 1		1.20	13.333		3,500	820		4,320	5,100
7680	600 amp, alum., NEMA 1		1.10	14.545		4,000	890		4,890	5,775
7700	NEMA 12		1	16		4,075	980		5,055	5,975
7720	Copper, NEMA 1		1.10	14.545		4,075	890		4,965	5,850
7740	800 amp, copper, NEMA 1		.90	17.778		8,375	1,100		9,475	10,900
7760	Incoming line, for copper bus, add					126			126	139
7780	For 65,000 amp bus bracing, add					188			188	206
7800	For NEMA 3R enclosure, add					5,900			5,900	6,475
7820	For NEMA 12 enclosure, add					162			162	179
7840	For 1/4" x 1" ground bus, add	1 Elec	16	.500		105	30.50		135.50	162
7860	For 1/4" x 2" ground bus, add	"	12	.667		105	41		146	178
7900	Main rating basic section, alum., NEMA 1, 800 amp	2 Elec	1.40	11.429		345	700		1,045	1,450
7920	1,200 amp	"	1.20	13.333		675	820		1,495	2,000
7940	For copper bus, add					505			505	555
7960	For 65,000 amp bus bracing, add					340			340	375
7980	For NEMA 3R enclosure, add					5,900			5,900	6,475
8000	For NEMA 12, enclosure, add					162			162	179
8020	For 1/4" x 1" ground bus, add	1 Elec	16	.500		105	30.50		135.50	162
8040	For 1/4" x 2" ground bus, add		12	.667		105	41		146	178
8060	Unit devices, pilot light, standard		16	.500		90	30.50		120.50	147
8080	Pilot light, push to test		16	.500		126	30.50		156.50	186
8100	Pilot light, standard, and push button		12	.667		217	41		258	300
8120	Pilot light, push to test, and push button		12	.667		253	41		294	340
8140	Pilot light, standard, and select switch		12	.667		217	41		258	300
8160	Pilot light, push to test, and select switch		12	.667		253	41		294	340

26 24 19.30 Motor Control Center		Crew	Daily Output	Labor-Hours	Unit	Material	Labor	Equipment	Total	Total Incl O&P
						2020 Bare Costs				
0010	**MOTOR CONTROL CENTER** Consists of starters & structures									
0050	Starters, class 1, type B, comb. MCP, FVNR, with									
0100	control transformer, 10 HP, size 1, 12" high	1 Elec	2.70	2.963	Ea.	1,800	182		1,982	2,250
0200	25 HP, size 2, 18" high	2 Elec	4	4		2,050	245		2,295	2,650
0300	50 HP, size 3, 24" high		2	8		3,175	490		3,665	4,250
0350	75 HP, size 4, 24" high		1.60	10		4,225	615		4,840	5,600
0400	100 HP, size 4, 30" high		1.40	11.429		5,575	700		6,275	7,200
0500	200 HP, size 5, 48" high		1	16		8,325	980		9,305	10,700
0600	400 HP, size 6, 72" high		.80	20		17,100	1,225		18,325	20,700
0800	Structures, 600 amp, 22,000 rms, takes any									
0900	combination of starters up to 72" high	2 Elec	1.60	10	Ea.	1,925	615		2,540	3,050
1000	Back to back, 72" front & 66" back	"	1.20	13.333		2,625	820		3,445	4,125
1100	For copper bus, add per structure					278			278	305
1200	For NEMA 12, add per structure					160			160	176
1300	For 42,000 rms, add per structure					220			220	242
1400	For 100,000 rms, size 1 & 2, add					715			715	785
1500	Size 3, add					1,150			1,150	1,250
1600	Size 4, add					925			925	1,025
1700	For pilot lights, add per starter	1 Elec	16	.500		125	30.50		155.50	184
1800	For push button, add per starter		16	.500		125	30.50		155.50	184
1900	For auxiliary contacts, add per starter		16	.500		221	30.50		251.50	290

26 24 19.40 Motor Starters and Controls		Crew	Daily Output	Labor-Hours	Unit	Material	Labor	Equipment	Total	Total Incl O&P
0010	**MOTOR STARTERS AND CONTROLS**									
0050	Magnetic, FVNR, with enclosure and heaters, 480 volt									
0080	2 HP, size 00	1 Elec	3.50	2.286	Ea.	196	140		336	430
0100	5 HP, size 0		2.30	3.478		345	213		558	710
0200	10 HP, size 1		1.60	5		262	305		567	760
0300	25 HP, size 2	2 Elec	2.20	7.273		490	445		935	1,225
0400	50 HP, size 3		1.80	8.889		800	545		1,345	1,725
0500	100 HP, size 4		1.20	13.333		1,775	820		2,595	3,200
0600	200 HP, size 5		.90	17.778		4,150	1,100		5,250	6,250
0610	400 HP, size 6		.80	20		18,300	1,225		19,525	22,000
0620	NEMA 7, 5 HP, size 0	1 Elec	1.60	5		1,375	305		1,680	1,975
0630	10 HP, size 1	"	1.10	7.273		1,425	445		1,870	2,250
0640	25 HP, size 2	2 Elec	1.80	8.889		2,300	545		2,845	3,375
0650	50 HP, size 3		1.20	13.333		3,450	820		4,270	5,050
0660	100 HP, size 4		.90	17.778		5,575	1,100		6,675	7,800
0670	200 HP, size 5		.50	32		13,300	1,975		15,275	17,600
0700	Combination, with motor circuit protectors, 5 HP, size 0	1 Elec	1.80	4.444		985	273		1,258	1,500
0800	10 HP, size 1	"	1.30	6.154		1,025	380		1,405	1,700
0900	25 HP, size 2	2 Elec	2	8		1,425	490		1,915	2,325
1000	50 HP, size 3		1.32	12.121		2,075	745		2,820	3,425
1200	100 HP, size 4		.80	20		4,450	1,225		5,675	6,800
1220	NEMA 7, 5 HP, size 0	1 Elec	1.30	6.154		3,550	380		3,930	4,500
1230	10 HP, size 1	"	1	8		3,650	490		4,140	4,750
1240	25 HP, size 2	2 Elec	1.32	12.121		4,875	745		5,620	6,500
1250	50 HP, size 3		.80	20		8,050	1,225		9,275	10,800
1260	100 HP, size 4		.60	26.667		12,500	1,625		14,125	16,300
1270	200 HP, size 5		.40	40		27,200	2,450		29,650	33,800
1400	Combination, with fused switch, 5 HP, size 0	1 Elec	1.80	4.444		590	273		863	1,075
1600	10 HP, size 1	"	1.30	6.154		630	380		1,010	1,275
1800	25 HP, size 2	2 Elec	2	8		1,025	490		1,515	1,875

26 24 19.40 Motor Starters and Controls		Crew	Daily Output	Labor-Hours	Unit	Material	Labor	Equipment	Total	Total Incl O&P
						2020 Bare Costs				
2000	50 HP, size 3	2 Elec	1.32	12.121	Ea.	1,725	745		2,470	3,050
2200	100 HP, size 4		.80	20		3,025	1,225		4,250	5,225
2610	NEMA 4, with start-stop push button, size 1	1 Elec	1.30	6.154		1,575	380		1,955	2,325
2620	Size 2	2 Elec	2	8		2,125	490		2,615	3,075
2630	Size 3		1.32	12.121		3,350	745		4,095	4,850
2640	Size 4		.80	20		5,150	1,225		6,375	7,575
2650	NEMA 4, FVNR, including control transformer									
2660	Size 1	2 Elec	2.60	6.154	Ea.	1,525	380		1,905	2,250
2670	Size 2		2	8		2,175	490		2,665	3,150
2680	Size 3		1.32	12.121		3,475	745		4,220	4,975
2690	Size 4		.80	20		5,350	1,225		6,575	7,800
2710	Magnetic, FVR, control circuit transformer, NEMA 1, size 1	1 Elec	1.30	6.154		900	380		1,280	1,575
2720	Size 2	2 Elec	2	8		1,450	490		1,940	2,325
2730	Size 3		1.32	12.121		2,175	745		2,920	3,550
2740	Size 4		.80	20		4,750	1,225		5,975	7,125
2760	NEMA 4, size 1	1 Elec	1.10	7.273		1,300	445		1,745	2,100
2770	Size 2	2 Elec	1.60	10		2,075	615		2,690	3,250
2780	Size 3		1.20	13.333		3,125	820		3,945	4,700
2790	Size 4		.70	22.857		6,425	1,400		7,825	9,225
2820	NEMA 12, size 1	1 Elec	1.10	7.273		1,050	445		1,495	1,850
2830	Size 2	2 Elec	1.60	10		1,675	615		2,290	2,800
2840	Size 3		1.20	13.333		2,650	820		3,470	4,150
2850	Size 4		.70	22.857		5,425	1,400		6,825	8,125
2870	Combination FVR, fused, w/control XFMR & PB, NEMA 1, size 1	1 Elec	1	8		1,550	490		2,040	2,450
2880	Size 2	2 Elec	1.50	10.667		2,200	655		2,855	3,425
2890	Size 3		1.10	14.545		3,275	890		4,165	5,000
2900	Size 4		.70	22.857		6,650	1,400		8,050	9,475
2910	NEMA 4, size 1	1 Elec	.90	8.889		2,300	545		2,845	3,375
2920	Size 2	2 Elec	1.40	11.429		3,350	700		4,050	4,775
2930	Size 3		1	16		5,300	980		6,280	7,325
2940	Size 4		.60	26.667		8,775	1,625		10,400	12,200
2950	NEMA 12, size 1	1 Elec	1	8		1,750	490		2,240	2,675
2960	Size 2	2 Elec	1.40	11.429		2,475	700		3,175	3,800
2970	Size 3		1	16		3,650	980		4,630	5,500
2980	Size 4		.60	26.667		7,225	1,625		8,850	10,500
3010	Manual, single phase, w/pilot, 1 pole, 120 V, NEMA 1	1 Elec	6.40	1.250		91	76.50		167.50	218
3020	NEMA 4		4	2		365	123		488	590
3030	2 pole, 120/240 V, NEMA 1		6.40	1.250		123	76.50		199.50	253
3040	NEMA 4		4	2		330	123		453	550
3041	3 phase, 3 pole, 600 V, NEMA 1		5.50	1.455		246	89		335	410
3042	NEMA 4		3.50	2.286		420	140		560	680
3043	NEMA 12		3.50	2.286		281	140		421	525
3070	Auxiliary contact, normally open					96			96	106
3500	Magnetic FVNR with NEMA 12, enclosure & heaters, 480 volt									
3600	5 HP, size 0	1 Elec	2.20	3.636	Ea.	237	223		460	605
3700	10 HP, size 1	"	1.50	5.333		355	325		680	895
3800	25 HP, size 2	2 Elec	2	8		665	490		1,155	1,500
3900	50 HP, size 3		1.60	10		1,025	615		1,640	2,075
4000	100 HP, size 4		1	16		2,450	980		3,430	4,200
4100	200 HP, size 5		.80	20		5,900	1,225		7,125	8,375
4200	Combination, with motor circuit protectors, 5 HP, size 0	1 Elec	1.70	4.706		785	289		1,074	1,300
4300	10 HP, size 1	"	1.20	6.667		815	410		1,225	1,525
4400	25 HP, size 2	2 Elec	1.80	8.889		1,225	545		1,770	2,200

26 24 19.40 Motor Starters and Controls		Crew	Daily Output	Labor-Hours	Unit	Material	Labor	Equipment	Total	Total Incl O&P
						2020 Bare Costs				
4500	50 HP, size 3	2 Elec	1.20	13.333	Ea.	2,000	820		2,820	3,425
4600	100 HP, size 4		.74	21.622		4,475	1,325		5,800	6,975
4700	Combination, with fused switch, 5 HP, size 0	1 Elec	1.70	4.706		745	289		1,034	1,250
4800	10 HP, size 1	″	1.20	6.667		775	410		1,185	1,475
4900	25 HP, size 2	2 Elec	1.80	8.889		1,175	545		1,720	2,150
5000	50 HP, size 3		1.20	13.333		1,900	820		2,720	3,325
5100	100 HP, size 4		.74	21.622		3,825	1,325		5,150	6,275
5200	Factory installed controls, adders to size 0 thru 5									
5300	Start-stop push button	1 Elec	32	.250	Ea.	49.50	15.35		64.85	78
5400	Hand-off-auto-selector switch		32	.250		49.50	15.35		64.85	78
5500	Pilot light		32	.250		93	15.35		108.35	126
5600	Start-stop-pilot		32	.250		143	15.35		158.35	181
5700	Auxiliary contact, NO or NC		32	.250		68	15.35		83.35	98.50
5800	NO-NC		32	.250		136	15.35		151.35	174
5810	Magnetic FVR, NEMA 7, w/heaters, size 1		.66	12.121		3,100	745		3,845	4,550
5830	Size 2	2 Elec	1.10	14.545		5,100	890		5,990	6,975
5840	Size 3		.70	22.857		8,200	1,400		9,600	11,200
5850	Size 4		.60	26.667		9,275	1,625		10,900	12,700
5860	Combination w/circuit breakers, heaters, control XFMR PB, size 1	1 Elec	.60	13.333		1,650	820		2,470	3,050
5870	Size 2	2 Elec	.80	20		2,100	1,225		3,325	4,225
5880	Size 3		.50	32		2,825	1,975		4,800	6,125
5890	Size 4		.40	40		5,575	2,450		8,025	9,900
5900	Manual, 240 volt, 0.75 HP motor	1 Elec	4	2		52.50	123		175.50	247
5910	2 HP motor		4	2		143	123		266	345
6000	Magnetic, 240 volt, 1 or 2 pole, 0.75 HP motor		4	2		198	123		321	405
6020	2 HP motor		4	2		218	123		341	430
6040	5 HP motor		3	2.667		315	164		479	595
6060	10 HP motor		2.30	3.478		775	213		988	1,175
6100	3 pole, 0.75 HP motor		3	2.667		196	164		360	470
6120	5 HP motor		2.30	3.478		267	213		480	625
6140	10 HP motor		1.60	5		500	305		805	1,025
6160	15 HP motor		1.60	5		500	305		805	1,025
6180	20 HP motor		1.10	7.273		820	445		1,265	1,575
6200	25 HP motor	2 Elec	2.20	7.273		820	445		1,265	1,575
6210	30 HP motor		1.80	8.889		820	545		1,365	1,750
6220	40 HP motor		1.80	8.889		1,825	545		2,370	2,850
6230	50 HP motor		1.80	8.889		1,825	545		2,370	2,850
6240	60 HP motor		1.20	13.333		4,250	820		5,070	5,925
6250	75 HP motor		1.20	13.333		4,250	820		5,070	5,925
6260	100 HP motor		1.20	13.333		4,250	820		5,070	5,925
6270	125 HP motor		.90	17.778		11,900	1,100		13,000	14,800
6280	150 HP motor		.90	17.778		11,900	1,100		13,000	14,800
6290	200 HP motor		.90	17.778		11,900	1,100		13,000	14,800
6400	Starter & nonfused disconnect, 240 volt, 1-2 pole,0 .75 HP motor	1 Elec	2	4		245	245		490	650
6410	2 HP motor		2	4		266	245		511	675
6420	5 HP motor		1.80	4.444		360	273		633	815
6430	10 HP motor		1.40	5.714		840	350		1,190	1,475
6440	3 pole, 0.75 HP motor		1.60	5		244	305		549	740
6450	5 HP motor		1.40	5.714		315	350		665	885
6460	10 HP motor		1.10	7.273		565	445		1,010	1,300
6470	15 HP motor		1	8		565	490		1,055	1,375
6480	20 HP motor	2 Elec	1.50	10.667		965	655		1,620	2,075
6490	25 HP motor		1.50	10.667		965	655		1,620	2,075

26 24 19.40 Motor Starters and Controls		Crew	Daily Output	Labor-Hours	Unit	2020 Bare Costs Material	Labor	Equipment	Total	Total Incl O&P
6500	30 HP motor	2 Elec	1.30	12.308	Ea.	965	755		1,720	2,225
6510	40 HP motor		1.24	12.903		2,100	790		2,890	3,525
6520	50 HP motor		1.12	14.286		2,100	875		2,975	3,650
6530	60 HP motor		.90	17.778		4,525	1,100		5,625	6,650
6540	75 HP motor		.76	21.053		4,525	1,300		5,825	6,975
6550	100 HP motor		.70	22.857		4,525	1,400		5,925	7,125
6560	125 HP motor		.60	26.667		12,700	1,625		14,325	16,400
6570	150 HP motor		.52	30.769		12,700	1,900		14,600	16,800
6580	200 HP motor		.50	32		12,700	1,975		14,675	16,900
6600	Starter & fused disconnect, 240 volt, 1-2 pole, 0.75 HP motor	1 Elec	2	4		257	245		502	665
6610	2 HP motor		2	4		278	245		523	685
6620	5 HP motor		1.80	4.444		370	273		643	830
6630	10 HP motor		1.40	5.714		875	350		1,225	1,500
6640	3 pole, 0.75 HP motor		1.60	5		256	305		561	750
6650	5 HP motor		1.40	5.714		325	350		675	900
6660	10 HP motor		1.10	7.273		605	445		1,050	1,350
6690	15 HP motor		1	8		605	490		1,095	1,425
6700	20 HP motor	2 Elec	1.60	10		995	615		1,610	2,050
6710	25 HP motor		1.60	10		995	615		1,610	2,050
6720	30 HP motor		1.40	11.429		995	700		1,695	2,175
6730	40 HP motor		1.20	13.333		2,200	820		3,020	3,650
6740	50 HP motor		1.20	13.333		2,200	820		3,020	3,650
6750	60 HP motor		.90	17.778		4,625	1,100		5,725	6,750
6760	75 HP motor		.90	17.778		4,625	1,100		5,725	6,750
6770	100 HP motor		.70	22.857		4,625	1,400		6,025	7,225
6780	125 HP motor		.54	29.630		12,900	1,825		14,725	17,000
6790	Combination starter & nonfusible disconnect									
6800	240 volt, 1-2 pole, 0.75 HP motor	1 Elec	2	4	Ea.	600	245		845	1,050
6810	2 HP motor		2	4		625	245		870	1,075
6820	5 HP motor		1.50	5.333		645	325		970	1,225
6830	10 HP motor		1.20	6.667		975	410		1,385	1,700
6840	3 pole, 0.75 HP motor		1.80	4.444		595	273		868	1,075
6850	5 HP motor		1.30	6.154		625	380		1,005	1,275
6860	10 HP motor		1	8		975	490		1,465	1,825
6870	15 HP motor		1	8		975	490		1,465	1,825
6880	20 HP motor	2 Elec	1.32	12.121		1,600	745		2,345	2,925
6890	25 HP motor		1.32	12.121		1,600	745		2,345	2,925
6900	30 HP motor		1.32	12.121		1,600	745		2,345	2,925
6910	40 HP motor		.80	20		3,075	1,225		4,300	5,275
6920	50 HP motor		.80	20		3,075	1,225		4,300	5,275
6930	60 HP motor		.70	22.857		6,850	1,400		8,250	9,700
6940	75 HP motor		.70	22.857		6,850	1,400		8,250	9,700
6950	100 HP motor		.70	22.857		6,850	1,400		8,250	9,700
6960	125 HP motor		.60	26.667		18,000	1,625		19,625	22,300
6970	150 HP motor		.60	26.667		18,000	1,625		19,625	22,300
6980	200 HP motor		.60	26.667		18,000	1,625		19,625	22,300
6990	Combination starter and fused disconnect									
7000	240 volt, 1-2 pole, 0.75 HP motor	1 Elec	2	4	Ea.	655	245		900	1,100
7010	2 HP motor		2	4		655	245		900	1,100
7020	5 HP motor		1.50	5.333		685	325		1,010	1,250
7030	10 HP motor		1.20	6.667		1,075	410		1,485	1,800
7040	3 pole, 0.75 HP motor		1.80	4.444		655	273		928	1,150
7050	5 HP motor		1.30	6.154		685	380		1,065	1,325

26 24 19 – Motor-Control Centers

26 24 19.40 Motor Starters and Controls		Crew	Daily Output	Labor-Hours	Unit	Material	2020 Bare Costs Labor	Equipment	Total	Total Incl O&P
7060	10 HP motor	1 Elec	1	8	Ea.	1,075	490		1,565	1,925
7070	15 HP motor		1	8		1,075	490		1,565	1,925
7080	20 HP motor	2 Elec	1.32	12.121		1,775	745		2,520	3,100
7090	25 HP motor		1.32	12.121		1,775	745		2,520	3,100
7100	30 HP motor		1.32	12.121		1,775	745		2,520	3,100
7110	40 HP motor		.80	20		3,375	1,225		4,600	5,625
7120	50 HP motor		.80	20		3,375	1,225		4,600	5,625
7130	60 HP motor		.80	20		7,550	1,225		8,775	10,200
7140	75 HP motor		.70	22.857		7,550	1,400		8,950	10,500
7150	100 HP motor		.70	22.857		7,550	1,400		8,950	10,500
7160	125 HP motor		.70	22.857		19,900	1,400		21,300	24,100
7170	150 HP motor		.60	26.667		19,900	1,625		21,525	24,400
7180	200 HP motor		.60	26.667		19,900	1,625		21,525	24,400
7190	Combination starter & circuit breaker disconnect									
7200	240 volt, 1-2 pole, 0.75 HP motor	1 Elec	2	4	Ea.	620	245		865	1,050
7210	2 HP motor		2	4		620	245		865	1,050
7220	5 HP motor		1.50	5.333		650	325		975	1,225
7230	10 HP motor		1.20	6.667		995	410		1,405	1,725
7240	3 pole, 0.75 HP motor		1.80	4.444		640	273		913	1,125
7250	5 HP motor		1.30	6.154		670	380		1,050	1,325
7260	10 HP motor		1	8		1,000	490		1,490	1,875
7270	15 HP motor		1	8		1,000	490		1,490	1,875
7280	20 HP motor	2 Elec	1.32	12.121		1,725	745		2,470	3,050
7290	25 HP motor		1.32	12.121		1,725	745		2,470	3,050
7300	30 HP motor		1.32	12.121		1,725	745		2,470	3,050
7310	40 HP motor		.80	20		3,750	1,225		4,975	6,025
7320	50 HP motor		.80	20		3,750	1,225		4,975	6,025
7330	60 HP motor		.80	20		8,650	1,225		9,875	11,400
7340	75 HP motor		.70	22.857		8,650	1,400		10,050	11,700
7350	100 HP motor		.70	22.857		8,650	1,400		10,050	11,700
7360	125 HP motor		.70	22.857		18,800	1,400		20,200	22,800
7370	150 HP motor		.60	26.667		18,800	1,625		20,425	23,100
7380	200 HP motor		.60	26.667		18,800	1,625		20,425	23,100
7400	Magnetic FVNR with enclosure & heaters, 2 pole,									
7410	230 volt, 1 HP, size 00	1 Elec	4	2	Ea.	179	123		302	385
7420	2 HP, size 0		4	2		199	123		322	410
7430	3 HP, size 1		3	2.667		229	164		393	505
7440	5 HP, size 1P		3	2.667		293	164		457	570
7450	115 volt, 1/3 HP, size 00		4	2		179	123		302	385
7460	1 HP, size 0		4	2		199	123		322	410
7470	2 HP, size 1		3	2.667		229	164		393	505
7480	3 HP, size 1P		3	2.667		284	164		448	565
7500	3 pole, 480 volt, 600 HP, size 7	2 Elec	.70	22.857		15,800	1,400		17,200	19,600
7590	Magnetic FVNR with heater, NEMA 1									
7600	600 volt, 3 pole, 5 HP motor	1 Elec	2.30	3.478	Ea.	237	213		450	590
7610	10 HP motor	"	1.60	5		267	305		572	765
7620	25 HP motor	2 Elec	2.20	7.273		500	445		945	1,225
7630	30 HP motor		1.80	8.889		820	545		1,365	1,750
7640	40 HP motor		1.80	8.889		820	545		1,365	1,750
7650	50 HP motor		1.80	8.889		820	545		1,365	1,750
7660	60 HP motor		1.20	13.333		1,825	820		2,645	3,250
7670	75 HP motor		1.20	13.333		2,200	820		3,020	3,675
7680	100 HP motor		1.20	13.333		2,550	820		3,370	4,050

26 24 19.40 Motor Starters and Controls		Crew	Daily Output	Labor-Hours	Unit	Material	2020 Bare Costs Labor	Equipment	Total	Total Incl O&P
7690	125 HP motor	2 Elec	.90	17.778	Ea.	4,150	1,100		5,250	6,250
7700	150 HP motor		.90	17.778		4,350	1,100		5,450	6,450
7710	200 HP motor		.90	17.778		4,675	1,100		5,775	6,825
7750	Starter & nonfused disconnect, 600 volt, 3 pole, 5 HP motor	1 Elec	1.40	5.714		325	350		675	895
7760	10 HP motor	"	1.10	7.273		355	445		800	1,075
7770	25 HP motor	2 Elec	1.50	10.667		585	655		1,240	1,650
7780	30 HP motor		1.30	12.308		970	755		1,725	2,225
7790	40 HP motor		1.30	12.308		970	755		1,725	2,225
7800	50 HP motor		1.30	12.308		970	755		1,725	2,225
7810	60 HP motor		.92	17.391		2,050	1,075		3,125	3,925
7820	75 HP motor		.92	17.391		2,450	1,075		3,525	4,350
7830	100 HP motor		.84	19.048		2,800	1,175		3,975	4,875
7840	125 HP motor		.70	22.857		4,525	1,400		5,925	7,125
7850	150 HP motor		.70	22.857		4,700	1,400		6,100	7,325
7860	200 HP motor		.60	26.667		5,050	1,625		6,675	8,075
7870	Starter & fused disconnect, 600 volt, 3 pole, 5 HP motor	1 Elec	1.40	5.714		405	350		755	985
7880	10 HP motor	"	1.10	7.273		435	445		880	1,150
7890	25 HP motor	2 Elec	1.50	10.667		665	655		1,320	1,725
7900	30 HP motor		1.30	12.308		1,025	755		1,780	2,275
7910	40 HP motor		1.30	12.308		1,025	755		1,780	2,275
7920	50 HP motor		1.30	12.308		1,025	755		1,780	2,275
7930	60 HP motor		.92	17.391		2,175	1,075		3,250	4,050
7940	75 HP motor		.92	17.391		2,575	1,075		3,650	4,475
7950	100 HP motor		.84	19.048		2,925	1,175		4,100	5,000
7960	125 HP motor		.70	22.857		4,675	1,400		6,075	7,300
7970	150 HP motor		.70	22.857		4,850	1,400		6,250	7,500
7980	200 HP motor		.60	26.667		5,200	1,625		6,825	8,250
7990	Combination starter and nonfusible disconnect									
8000	600 volt, 3 pole, 5 HP motor	1 Elec	1.80	4.444	Ea.	620	273		893	1,100
8010	10 HP motor	"	1.30	6.154		830	380		1,210	1,500
8020	25 HP motor	2 Elec	2	8		1,000	490		1,490	1,850
8030	30 HP motor		1.32	12.121		1,675	745		2,420	2,975
8040	40 HP motor		1.32	12.121		1,675	745		2,420	2,975
8050	50 HP motor		1.32	12.121		2,000	745		2,745	3,350
8060	60 HP motor		.80	20		3,175	1,225		4,400	5,400
8070	75 HP motor		.80	20		3,175	1,225		4,400	5,400
8080	100 HP motor		.80	20		3,975	1,225		5,200	6,250
8090	125 HP motor		.70	22.857		7,325	1,400		8,725	10,200
8100	150 HP motor		.70	22.857		8,525	1,400		9,925	11,500
8110	200 HP motor		.70	22.857		7,675	1,400		9,075	10,600
8140	Combination starter and fused disconnect									
8150	600 volt, 3 pole, 5 HP motor	1 Elec	1.80	4.444	Ea.	655	273		928	1,150
8160	10 HP motor	"	1.30	6.154		715	380		1,095	1,375
8170	25 HP motor	2 Elec	2	8		1,150	490		1,640	2,025
8180	30 HP motor		1.32	12.121		1,800	745		2,545	3,150
8190	40 HP motor		1.32	12.121		1,950	745		2,695	3,300
8200	50 HP motor		1.32	12.121		1,950	745		2,695	3,300
8210	60 HP motor		.80	20		3,400	1,225		4,625	5,650
8220	75 HP motor		.80	20		3,400	1,225		4,625	5,650
8230	100 HP motor		.80	20		3,400	1,225		4,625	5,650
8240	125 HP motor		.70	22.857		7,550	1,400		8,950	10,500
8250	150 HP motor		.70	22.857		7,550	1,400		8,950	10,500
8260	200 HP motor		.70	22.857		7,550	1,400		8,950	10,500

26 24 19.40 Motor Starters and Controls		Crew	Daily Output	Labor-Hours	Unit	Material	2020 Bare Costs Labor	Equipment	Total	Total Incl O&P
8290	Combination starter & circuit breaker disconnect									
8300	600 volt, 3 pole, 5 HP motor	1 Elec	1.80	4.444	Ea.	930	273		1,203	1,450
8310	10 HP motor	"	1.30	6.154		910	380		1,290	1,575
8320	25 HP motor	2 Elec	2	8		1,350	490		1,840	2,250
8330	30 HP motor		1.32	12.121		1,725	745		2,470	3,050
8340	40 HP motor		1.32	12.121		1,725	745		2,470	3,050
8350	50 HP motor		1.32	12.121		1,775	745		2,520	3,125
8360	60 HP motor		.80	20		3,750	1,225		4,975	6,025
8370	75 HP motor		.80	20		3,750	1,225		4,975	6,025
8380	100 HP motor		.80	20		3,875	1,225		5,100	6,175
8390	125 HP motor		.70	22.857		8,650	1,400		10,050	11,700
8400	150 HP motor		.70	22.857		8,300	1,400		9,700	11,300
8410	200 HP motor		.70	22.857		8,975	1,400		10,375	12,000
8430	Starter & circuit breaker disconnect									
8440	600 volt, 3 pole, 5 HP motor	1 Elec	1.40	5.714	Ea.	765	350		1,115	1,375
8450	10 HP motor	"	1.10	7.273		795	445		1,240	1,550
8460	25 HP motor	2 Elec	1.50	10.667		1,025	655		1,680	2,125
8470	30 HP motor		1.30	12.308		1,450	755		2,205	2,750
8480	40 HP motor		1.30	12.308		1,450	755		2,205	2,750
8490	50 HP motor		1.30	12.308		1,450	755		2,205	2,750
8500	60 HP motor		.92	17.391		3,525	1,075		4,600	5,525
8510	75 HP motor		.92	17.391		3,900	1,075		4,975	5,950
8520	100 HP motor		.84	19.048		4,250	1,175		5,425	6,475
8530	125 HP motor		.70	22.857		5,850	1,400		7,250	8,600
8540	150 HP motor		.70	22.857		6,050	1,400		7,450	8,800
8550	200 HP motor		.60	26.667		7,575	1,625		9,200	10,900
8900	240 volt, 1-2 pole, 0.75 HP motor	1 Elec	2	4		725	245		970	1,175
8910	2 HP motor		2	4		745	245		990	1,200
8920	5 HP motor		1.80	4.444		840	273		1,113	1,350
8930	10 HP motor		1.40	5.714		1,425	350		1,775	2,100
8950	3 pole, 0.75 HP motor		1.60	5		725	305		1,030	1,275
8970	5 HP motor		1.40	5.714		795	350		1,145	1,425
8980	10 HP motor		1.10	7.273		1,150	445		1,595	1,925
8990	15 HP motor		1	8		1,150	490		1,640	2,000
9100	20 HP motor	2 Elec	1.50	10.667		1,550	655		2,205	2,700
9110	25 HP motor		1.50	10.667		1,550	655		2,205	2,700
9120	30 HP motor		1.30	12.308		1,550	755		2,305	2,850
9130	40 HP motor		1.24	12.903		3,525	790		4,315	5,100
9140	50 HP motor		1.12	14.286		3,525	875		4,400	5,225
9150	60 HP motor		.90	17.778		7,150	1,100		8,250	9,550
9160	75 HP motor		.76	21.053		7,150	1,300		8,450	9,875
9170	100 HP motor		.70	22.857		7,150	1,400		8,550	10,000
9180	125 HP motor		.60	26.667		16,100	1,625		17,725	20,200
9190	150 HP motor		.52	30.769		16,100	1,900		18,000	20,600
9200	200 HP motor		.50	32		16,100	1,975		18,075	20,700

26 25 13.10 Aluminum Bus Duct		Crew	Daily Output	Labor-Hours	Unit	Material	2020 Bare Costs Labor	Equipment	Total	Total Incl O&P
0010	**ALUMINUM BUS DUCT** 10 ft. long R262513-10									
0050	Indoor 3 pole 4 wire, plug-in, straight section, 225 amp	2 Elec	44	.364	L.F.	172	22.50		194.50	224
0100	400 amp		36	.444		216	27.50		243.50	279
0150	600 amp		32	.500		298	30.50		328.50	375
0200	800 amp		26	.615		345	38		383	440
0250	1,000 amp		24	.667		435	41		476	545
0270	1,200 amp		23	.696		550	42.50		592.50	670
0300	1,350 amp		22	.727		305	44.50		349.50	405
0310	1,600 amp		18	.889		330	54.50		384.50	445
0320	2,000 amp		16	1		425	61.50		486.50	565
0330	2,500 amp		14	1.143		460	70		530	615
0340	3,000 amp		12	1.333		595	82		677	780
0350	Feeder, 600 amp		34	.471		113	29		142	170
0400	800 amp		28	.571		430	35		465	525
0450	1,000 amp		26	.615		535	38		573	645
0455	1,200 amp		25	.640		650	39.50		689.50	775
0500	1,350 amp		24	.667		236	41		277	325
0550	1,600 amp		20	.800		900	49		949	1,075
0600	2,000 amp		18	.889		1,100	54.50		1,154.50	1,275
0620	2,500 amp		14	1.143		1,350	70		1,420	1,575
0630	3,000 amp		12	1.333		1,575	82		1,657	1,850
0640	4,000 amp		10	1.600		680	98		778	900
0650	Elbow, 225 amp		4.40	3.636	Ea.	835	223		1,058	1,250
0700	400 amp		3.80	4.211		840	258		1,098	1,325
0750	600 amp		3.40	4.706		845	289		1,134	1,375
0800	800 amp		3	5.333		875	325		1,200	1,475
0850	1,000 amp		2.80	5.714		1,125	350		1,475	1,800
0870	1,200 amp		2.70	5.926		1,400	365		1,765	2,100
0900	1,350 amp		2.60	6.154		1,350	380		1,730	2,050
0950	1,600 amp		2.40	6.667		1,425	410		1,835	2,200
1000	2,000 amp		2	8		1,575	490		2,065	2,475
1020	2,500 amp		1.80	8.889		1,850	545		2,395	2,875
1030	3,000 amp		1.60	10		2,125	615		2,740	3,300
1040	4,000 amp		1.40	11.429		3,450	700		4,150	4,875
1100	Cable tap box end, 225 amp		3.60	4.444		1,775	273		2,048	2,375
1150	400 amp		3.20	5		2,175	305		2,480	2,875
1200	600 amp		2.60	6.154		1,475	380		1,855	2,200
1250	800 amp		2.20	7.273		1,825	445		2,270	2,675
1300	1,000 amp		2	8		3,500	490		3,990	4,600
1320	1,200 amp		2	8		1,875	490		2,365	2,825
1350	1,350 amp		1.60	10		1,550	615		2,165	2,650
1400	1,600 amp		1.40	11.429		1,825	700		2,525	3,075
1450	2,000 amp		1.20	13.333		2,075	820		2,895	3,525
1460	2,500 amp		1	16		2,475	980		3,455	4,200
1470	3,000 amp		.80	20		2,725	1,225		3,950	4,900
1480	4,000 amp		.60	26.667		3,225	1,625		4,850	6,075
1500	Switchboard stub, 225 amp		5.80	2.759		1,850	169		2,019	2,275
1550	400 amp		5.40	2.963		1,875	182		2,057	2,350
1600	600 amp		4.60	3.478		2,100	213		2,313	2,650
1650	800 amp		4	4		1,900	245		2,145	2,475
1700	1,000 amp		3.20	5		1,700	305		2,005	2,350
1720	1,200 amp		3.10	5.161		1,750	315		2,065	2,400
1750	1,350 amp		3	5.333	L.F.	1,725	325		2,050	2,400

26 25 Low-Voltage Enclosed Bus Assemblies

26 25 13 – Low-Voltage Busways

26 25 13.10 Aluminum Bus Duct		Crew	Daily Output	Labor-Hours	Unit	2020 Bare Costs Material	Labor	Equipment	Total	Total Incl O&P
1800	1,600 amp	2 Elec	2.60	6.154	Ea.	1,825	380		2,205	2,575
1850	2,000 amp		2.40	6.667		1,950	410		2,360	2,775
1860	2,500 amp		2.20	7.273		2,175	445		2,620	3,050
1870	3,000 amp		2	8		2,275	490		2,765	3,250
1880	4,000 amp		1.80	8.889		2,525	545		3,070	3,625
1890	Tee fittings, 225 amp		3.20	5		895	305		1,200	1,450
1900	400 amp		2.80	5.714		895	350		1,245	1,525
1950	600 amp		2.60	6.154		895	380		1,275	1,575
2000	800 amp		2.40	6.667		950	410		1,360	1,675
2050	1,000 amp		2.20	7.273		1,000	445		1,445	1,775
2070	1,200 amp		2.10	7.619		1,175	465		1,640	2,025
2100	1,350 amp		2	8		1,625	490		2,115	2,550
2150	1,600 amp		1.60	10		1,975	615		2,590	3,100
2200	2,000 amp		1.20	13.333		2,175	820		2,995	3,650
2220	2,500 amp		1	16		2,600	980		3,580	4,350
2230	3,000 amp		.80	20		2,975	1,225		4,200	5,175
2240	4,000 amp		.60	26.667		4,925	1,625		6,550	7,950
2300	Wall flange, 600 amp		20	.800		246	49		295	345
2310	800 amp		16	1		246	61.50		307.50	365
2320	1,000 amp		13	1.231		246	75.50		321.50	385
2325	1,200 amp		12	1.333		246	82		328	395
2330	1,350 amp		10.80	1.481		246	91		337	410
2340	1,600 amp		9	1.778		246	109		355	440
2350	2,000 amp		8	2		246	123		369	460
2360	2,500 amp		6.60	2.424		246	149		395	500
2370	3,000 amp		5.40	2.963		246	182		428	550
2380	4,000 amp		4	4		390	245		635	810
2390	5,000 amp		3	5.333		390	325		715	935
2400	Vapor barrier		8	2		460	123		583	695
2420	Roof flange kit		4	4		885	245		1,130	1,350
2600	Expansion fitting, 225 amp		10	1.600		1,400	98		1,498	1,675
2610	400 amp		8	2		1,400	123		1,523	1,725
2620	600 amp		6	2.667		1,400	164		1,564	1,775
2630	800 amp		4.60	3.478		1,675	213		1,888	2,175
2640	1,000 amp		4	4		1,800	245		2,045	2,350
2650	1,350 amp		3.60	4.444		2,075	273		2,348	2,725
2660	1,600 amp		3.20	5		2,500	305		2,805	3,225
2670	2,000 amp		2.80	5.714		2,775	350		3,125	3,600
2680	2,500 amp		2.40	6.667		3,350	410		3,760	4,325
2690	3,000 amp		2	8		3,875	490		4,365	5,000
2700	4,000 amp		1.60	10		5,100	615		5,715	6,550
2800	Reducer nonfused, 400 amp		8	2		820	123		943	1,100
2810	600 amp		6	2.667		820	164		984	1,150
2820	800 amp		4.60	3.478		985	213		1,198	1,400
2830	1,000 amp		4	4		1,150	245		1,395	1,650
2840	1,350 amp		3.60	4.444		1,500	273		1,773	2,075
2850	1,600 amp		3.20	5		2,050	305		2,355	2,725
2860	2,000 amp		2.80	5.714		2,350	350		2,700	3,125
2870	2,500 amp		2.40	6.667		2,950	410		3,360	3,875
2880	3,000 amp		2	8		3,400	490		3,890	4,500
2890	4,000 amp		1.60	10		4,550	615		5,165	5,950
2950	Reducer fuse included, 225 amp		4.40	3.636		2,400	223		2,623	3,000
2960	400 amp		4.20	3.810		2,450	234		2,684	3,050

26 25 13.10 Aluminum Bus Duct		Crew	Daily Output	Labor-Hours	Unit	2020 Bare Costs Material	Labor	Equipment	Total	Total Incl O&P
2970	600 amp	2 Elec	3.60	4.444	Ea.	2,900	273		3,173	3,600
2980	800 amp		3.20	5		4,600	305		4,905	5,525
2990	1,000 amp		3	5.333		5,250	325		5,575	6,275
3000	1,200 amp		2.80	5.714		5,250	350		5,600	6,325
3010	1,600 amp		2.20	7.273		12,000	445		12,445	13,900
3020	2,000 amp		1.80	8.889		13,300	545		13,845	15,400
3100	Reducer circuit breaker, 225 amp		4.40	3.636		2,375	223		2,598	2,950
3110	400 amp		4.20	3.810		2,875	234		3,109	3,525
3120	600 amp		3.60	4.444		4,100	273		4,373	4,925
3130	800 amp		3.20	5		4,800	305		5,105	5,750
3140	1,000 amp		3	5.333		5,450	325		5,775	6,500
3150	1,200 amp		2.80	5.714		6,550	350		6,900	7,750
3160	1,600 amp		2.20	7.273		9,675	445		10,120	11,400
3170	2,000 amp		1.80	8.889		10,600	545		11,145	12,500
3250	Reducer circuit breaker, 75,000 AIC, 225 amp		4.40	3.636		3,700	223		3,923	4,425
3260	400 amp		4.20	3.810		3,700	234		3,934	4,425
3270	600 amp		3.60	4.444		4,950	273		5,223	5,875
3280	800 amp		3.20	5		5,450	305		5,755	6,450
3290	1,000 amp		3	5.333		8,750	325		9,075	10,100
3300	1,200 amp		2.80	5.714		8,750	350		9,100	10,200
3310	1,600 amp		2.20	7.273		9,675	445		10,120	11,400
3320	2,000 amp		1.80	8.889		10,600	545		11,145	12,500
3400	Reducer circuit breaker CLF 225 amp		4.40	3.636		3,800	223		4,023	4,550
3410	400 amp		4.20	3.810		4,525	234		4,759	5,325
3420	600 amp		3.60	4.444		6,775	273		7,048	7,875
3430	800 amp		3.20	5		7,075	305		7,380	8,250
3440	1,000 amp		3	5.333		7,375	325		7,700	8,600
3450	1,200 amp		2.80	5.714		9,625	350		9,975	11,100
3460	1,600 amp		2.20	7.273		9,675	445		10,120	11,400
3470	2,000 amp		1.80	8.889		10,600	545		11,145	12,500
3550	Ground bus added to bus duct, 225 amp		320	.050	L.F.	33.50	3.07		36.57	41.50
3560	400 amp		320	.050		33.50	3.07		36.57	41.50
3570	600 amp		280	.057		33.50	3.51		37.01	42.50
3580	800 amp		240	.067		33.50	4.09		37.59	43.50
3590	1,000 amp		200	.080		33.50	4.91		38.41	44.50
3600	1,350 amp		180	.089		33.50	5.45		38.95	45.50
3610	1,600 amp		160	.100		33.50	6.15		39.65	46.50
3620	2,000 amp		160	.100		33.50	6.15		39.65	46.50
3630	2,500 amp		140	.114		33.50	7		40.50	48
3640	3,000 amp		120	.133		33.50	8.20		41.70	49.50
3650	4,000 amp		100	.160		33.50	9.80		43.30	52
3810	High short circuit, 400 amp		36	.444		208	27.50		235.50	271
3820	600 amp		32	.500		360	30.50		390.50	440
3830	800 amp		26	.615		415	38		453	515
3840	1,000 amp		24	.667		520	41		561	640
3850	1,350 amp		22	.727		189	44.50		233.50	277
3860	1,600 amp		18	.889		217	54.50		271.50	325
3870	2,000 amp		16	1		255	61.50		316.50	375
3880	2,500 amp		14	1.143		425	70		495	575
3890	3,000 amp		12	1.333		480	82		562	655
3920	Cross, 225 amp		5.60	2.857	Ea.	1,350	175		1,525	1,750
3930	400 amp		4.60	3.478		1,350	213		1,563	1,800
3940	600 amp		4	4		1,350	245		1,595	1,850

26 25 13.10 Aluminum Bus Duct		Crew	Daily Output	Labor-Hours	Unit	Material	Labor	Equipment	Total	Total Incl O&P
						2020 Bare Costs				
3950	800 amp	2 Elec	3.40	4.706	Ea.	1,400	289		1,689	2,000
3960	1,000 amp		3	5.333		1,475	325		1,800	2,125
3970	1,350 amp		2.80	5.714		2,400	350		2,750	3,200
3980	1,600 amp		2.20	7.273		2,850	445		3,295	3,825
3990	2,000 amp		1.80	8.889		3,125	545		3,670	4,300
4000	2,500 amp		1.60	10		3,700	615		4,315	5,025
4010	3,000 amp		1.20	13.333		4,275	820		5,095	5,950
4020	4,000 amp		1	16		6,525	980		7,505	8,675
4040	Cable tap box center, 225 amp		3.60	4.444		2,975	273		3,248	3,700
4050	400 amp		3.20	5		3,375	305		3,680	4,200
4060	600 amp		2.60	6.154		4,475	380		4,855	5,475
4070	800 amp		2.20	7.273		4,925	445		5,370	6,075
4080	1,000 amp		2	8		6,200	490		6,690	7,575
4090	1,350 amp		1.60	10		1,450	615		2,065	2,550
4100	1,600 amp		1.40	11.429		1,625	700		2,325	2,875
4110	2,000 amp		1.20	13.333		1,875	820		2,695	3,300
4120	2,500 amp		1	16		2,275	980		3,255	4,000
4130	3,000 amp		.80	20		2,500	1,225		3,725	4,650
4140	4,000 amp		.60	26.667		3,350	1,625		4,975	6,200
4500	Weatherproof 3 pole 4 wire, feeder, 600 amp		30	.533	L.F.	136	32.50		168.50	200
4520	800 amp		24	.667		159	41		200	238
4540	1,000 amp		22	.727		181	44.50		225.50	268
4550	1,200 amp		21	.762		251	46.50		297.50	350
4560	1,350 amp		20	.800		283	49		332	385
4580	1,600 amp		17	.941		330	57.50		387.50	450
4600	2,000 amp		16	1		385	61.50		446.50	520
4620	2,500 amp		12	1.333		500	82		582	675
4640	3,000 amp		10	1.600		565	98		663	775
4660	4,000 amp		8	2		815	123		938	1,075
5000	Indoor 3 pole, 3 wire, feeder, 600 amp		40	.400		104	24.50		128.50	152
5010	800 amp		32	.500		355	30.50		385.50	435
5020	1,000 amp		30	.533		395	32.50		427.50	485
5025	1,200 amp		29	.552		460	34		494	560
5030	1,350 amp		28	.571		179	35		214	251
5040	1,600 amp		24	.667		755	41		796	895
5050	2,000 amp		20	.800		900	49		949	1,075
5060	2,500 amp		16	1		1,100	61.50		1,161.50	1,300
5070	3,000 amp		14	1.143		1,250	70		1,320	1,475
5080	4,000 amp		12	1.333		490	82		572	665
5200	Plug-in type, 225 amp		50	.320		139	19.65		158.65	183
5210	400 amp		42	.381		170	23.50		193.50	223
5220	600 amp		36	.444		211	27.50		238.50	274
5230	800 amp		30	.533		295	32.50		327.50	375
5240	1,000 amp		28	.571		335	35		370	420
5245	1,200 amp		27	.593		236	36.50		272.50	315
5250	1,350 amp		26	.615		189	38		227	266
5260	1,600 amp		20	.800		217	49		266	315
5270	2,000 amp		18	.889		255	54.50		309.50	365
5280	2,500 amp		16	1		350	61.50		411.50	480
5290	3,000 amp		14	1.143		415	70		485	565
5300	4,000 amp		12	1.333		500	82		582	675
5330	High short circuit, 400 amp		42	.381		204	23.50		227.50	260
5340	600 amp		36	.444		252	27.50		279.50	320

26 25 Low-Voltage Enclosed Bus Assemblies

26 25 13 – Low-Voltage Busways

26 25 13.10 Aluminum Bus Duct		Crew	Daily Output	Labor-Hours	Unit	Material	Labor	Equipment	Total	Total Incl O&P
						2020 Bare Costs				
5350	800 amp	2 Elec	30	.533	L.F.	355	32.50		387.50	445
5360	1,000 amp		28	.571		405	35		440	500
5370	1,350 amp		26	.615		189	38		227	266
5380	1,600 amp		20	.800		217	49		266	315
5390	2,000 amp		18	.889		255	54.50		309.50	365
5400	2,500 amp		16	1		400	61.50		461.50	535
5410	3,000 amp		14	1.143		415	70		485	565
5440	Elbow, 225 amp		5	3.200	Ea.	655	196		851	1,025
5450	400 amp		4.40	3.636		655	223		878	1,075
5460	600 amp		4	4		655	245		900	1,100
5470	800 amp		3.40	4.706		700	289		989	1,225
5480	1,000 amp		3.20	5		715	305		1,020	1,250
5485	1,200 amp		3.10	5.161		760	315		1,075	1,325
5490	1,350 amp		3	5.333		730	325		1,055	1,300
5500	1,600 amp		2.80	5.714		1,125	350		1,475	1,775
5510	2,000 amp		2.40	6.667		1,225	410		1,635	1,975
5520	2,500 amp		2	8		1,500	490		1,990	2,400
5530	3,000 amp		1.80	8.889		1,800	545		2,345	2,825
5540	4,000 amp		1.60	10		2,400	615		3,015	3,575
5560	Tee fittings, 225 amp		3.60	4.444		800	273		1,073	1,300
5570	400 amp		3.20	5		800	305		1,105	1,350
5580	600 amp		3	5.333		800	325		1,125	1,400
5590	800 amp		2.80	5.714		860	350		1,210	1,475
5600	1,000 amp		2.60	6.154		890	380		1,270	1,550
5605	1,200 amp		2.50	6.400		945	395		1,340	1,650
5610	1,350 amp		2.40	6.667		1,325	410		1,735	2,075
5620	1,600 amp		1.80	8.889		1,550	545		2,095	2,550
5630	2,000 amp		1.40	11.429		1,725	700		2,425	2,975
5640	2,500 amp		1.20	13.333		2,150	820		2,970	3,600
5650	3,000 amp		1	16		2,525	980		3,505	4,275
5660	4,000 amp		.70	22.857		3,700	1,400		5,100	6,200
5680	Cross, 225 amp		6.40	2.500		1,300	153		1,453	1,650
5690	400 amp		5.40	2.963		1,300	182		1,482	1,700
5700	600 amp		4.60	3.478		1,300	213		1,513	1,750
5710	800 amp		4	4		1,375	245		1,620	1,875
5720	1,000 amp		3.60	4.444		1,400	273		1,673	1,975
5730	1,350 amp		3.20	5		2,100	305		2,405	2,800
5740	1,600 amp		2.60	6.154		2,450	380		2,830	3,275
5750	2,000 amp		2.20	7.273		2,675	445		3,120	3,625
5760	2,500 amp		1.80	8.889		3,250	545		3,795	4,425
5770	3,000 amp		1.40	11.429		3,875	700		4,575	5,325
5780	4,000 amp		1.20	13.333		5,375	820		6,195	7,175
5800	Expansion fitting, 225 amp		11.60	1.379		1,100	84.50		1,184.50	1,350
5810	400 amp		9.20	1.739		1,100	107		1,207	1,400
5820	600 amp		7	2.286		1,100	140		1,240	1,450
5830	800 amp		5.20	3.077		1,325	189		1,514	1,750
5840	1,000 amp		4.60	3.478		1,450	213		1,663	1,900
5850	1,350 amp		4.20	3.810		1,475	234		1,709	1,975
5860	1,600 amp		3.60	4.444		1,825	273		2,098	2,450
5870	2,000 amp		3.20	5		2,175	305		2,480	2,875
5880	2,500 amp		2.80	5.714		2,450	350		2,800	3,250
5890	3,000 amp		2.40	6.667		2,875	410		3,285	3,775
5900	4,000 amp		1.80	8.889		3,925	545		4,470	5,175

26 25 13 – Low-Voltage Busways

26 25 13.10 Aluminum Bus Duct		Crew	Daily Output	Labor-Hours	Unit	Material	Labor	Equipment	Total	Total Incl O&P
5940	Reducer, nonfused, 400 amp	2 Elec	9.20	1.739	Ea.	705	107		812	940
5950	600 amp		7	2.286		705	140		845	990
5960	800 amp		5.20	3.077		740	189		929	1,100
5970	1,000 amp		4.60	3.478		910	213		1,123	1,325
5980	1,350 amp		4.20	3.810		1,350	234		1,584	1,825
5990	1,600 amp		3.60	4.444		1,525	273		1,798	2,125
6000	2,000 amp		3.20	5		1,800	305		2,105	2,450
6010	2,500 amp		2.80	5.714		2,300	350		2,650	3,075
6020	3,000 amp		2.20	7.273		2,675	445		3,120	3,625
6030	4,000 amp		1.80	8.889		3,675	545		4,220	4,900
6050	Reducer, fuse included, 225 amp		5	3.200		1,850	196		2,046	2,325
6060	400 amp		4.80	3.333		2,450	205		2,655	3,025
6070	600 amp		4.20	3.810		3,125	234		3,359	3,800
6080	800 amp		3.60	4.444		4,850	273		5,123	5,775
6090	1,000 amp		3.40	4.706		5,300	289		5,589	6,275
6100	1,350 amp		3.20	5		9,675	305		9,980	11,200
6110	1,600 amp		2.60	6.154		11,500	380		11,880	13,200
6120	2,000 amp		2	8		13,300	490		13,790	15,400
6160	Reducer, circuit breaker, 225 amp		5	3.200		2,275	196		2,471	2,800
6170	400 amp		4.80	3.333		2,775	205		2,980	3,375
6180	600 amp		4.20	3.810		4,000	234		4,234	4,750
6190	800 amp		3.60	4.444		4,650	273		4,923	5,550
6200	1,000 amp		3.40	4.706		5,300	289		5,589	6,275
6210	1,350 amp		3.20	5		6,400	305		6,705	7,525
6220	1,600 amp		2.60	6.154		9,550	380		9,930	11,100
6230	2,000 amp		2	8		10,400	490		10,890	12,300
6270	Cable tap box center, 225 amp		4.20	3.810		1,025	234		1,259	1,475
6280	400 amp		3.60	4.444		1,025	273		1,298	1,550
6290	600 amp		3	5.333		1,025	325		1,350	1,625
6300	800 amp		2.60	6.154		1,125	380		1,505	1,800
6310	1,000 amp		2.40	6.667		1,200	410		1,610	1,925
6320	1,350 amp		1.80	8.889		1,250	545		1,795	2,250
6330	1,600 amp		1.60	10		1,425	615		2,040	2,500
6340	2,000 amp		1.40	11.429		1,625	700		2,325	2,850
6350	2,500 amp		1.20	13.333		2,025	820		2,845	3,475
6360	3,000 amp		1	16		2,300	980		3,280	4,025
6370	4,000 amp		.70	22.857		2,725	1,400		4,125	5,125
6390	Cable tap box end, 225 amp		4.20	3.810		625	234		859	1,050
6400	400 amp		3.60	4.444		625	273		898	1,100
6410	600 amp		3	5.333		625	325		950	1,200
6420	800 amp		2.60	6.154		685	380		1,065	1,325
6430	1,000 amp		2.40	6.667		740	410		1,150	1,450
6435	1,200 amp		2.10	7.619		805	465		1,270	1,600
6440	1,350 amp		1.80	8.889		785	545		1,330	1,700
6450	1,600 amp		1.60	10		890	615		1,505	1,925
6460	2,000 amp		1.40	11.429		1,000	700		1,700	2,175
6470	2,500 amp		1.20	13.333		1,175	820		1,995	2,525
6480	3,000 amp		1	16		1,375	980		2,355	3,025
6490	4,000 amp		.70	22.857		1,650	1,400		3,050	3,975
7000	Weatherproof 3 pole 3 wire, feeder, 600 amp		34	.471	L.F.	134	29		163	192
7020	800 amp		28	.571		148	35		183	216
7040	1,000 amp		26	.615		159	38		197	233
7050	1,200 amp		25	.640		185	39.50		224.50	265

26 25 Low-Voltage Enclosed Bus Assemblies

26 25 13 – Low-Voltage Busways

26 25 13.10 Aluminum Bus Duct		Crew	Daily Output	Labor-Hours	Unit	Material	2020 Bare Costs Labor	Equipment	Total	Total Incl O&P
7060	1,350 amp	2 Elec	24	.667	L.F.	215	41		256	300
7080	1,600 amp		20	.800		249	49		298	350
7100	2,000 amp		18	.889		294	54.50		348.50	410
7120	2,500 amp		14	1.143		410	70		480	560
7140	3,000 amp		12	1.333		485	82		567	660
7160	4,000 amp		10	1.600		590	98		688	800

26 25 13.20 Bus Duct		Crew	Daily Output	Labor-Hours	Unit	Material	Labor	Equipment	Total	Total Incl O&P
0010	**BUS DUCT** 100 amp and less, aluminum or copper, plug-in									
0080	Bus duct, 3 pole 3 wire, 100 amp	1 Elec	42	.190	L.F.	78	11.70		89.70	104
0110	Elbow		4	2	Ea.	132	123		255	335
0120	Tee		2	4		187	245		432	585
0130	Wall flange		8	1		27.50	61.50		89	125
0140	Ground kit		16	.500		61.50	30.50		92	115
0180	3 pole 4 wire, 100 amp		40	.200	L.F.	84.50	12.25		96.75	111
0200	Cable tap box		3.10	2.581	Ea.	258	158		416	530
0300	End closure		16	.500		34.50	30.50		65	85
0400	Elbow		4	2		216	123		339	425
0500	Tee		2	4		315	245		560	725
0600	Hangers		10	.800		24.50	49		73.50	103
0700	Circuit breakers, 15 to 50 amp, 1 pole		8	1		670	61.50		731.50	835
0800	15 to 60 amp, 2 pole		6.70	1.194		360	73.50		433.50	510
0900	3 pole		5.30	1.509		450	92.50		542.50	645
1000	60 to 100 amp, 1 pole		6.70	1.194		705	73.50		778.50	890
1100	70 to 100 amp, 2 pole		5.30	1.509		1,300	92.50		1,392.50	1,575
1200	3 pole		4.50	1.778		1,750	109		1,859	2,100
1220	Switch, nonfused, 3 pole, 4 wire		8	1		219	61.50		280.50	335
1240	Fused, 3 fuses, 4 wire, 30 amp		8	1		360	61.50		421.50	495
1260	60 amp		5.30	1.509		425	92.50		517.50	610
1280	100 amp		4.50	1.778		570	109		679	795
1300	Plug, fusible, 3 pole 250 volt, 30 amp		5.30	1.509		535	92.50		627.50	730
1310	60 amp		5.30	1.509		600	92.50		692.50	805
1320	100 amp		4.50	1.778		815	109		924	1,075
1330	3 pole 480 volt, 30 amp		5.30	1.509		540	92.50		632.50	740
1340	60 amp		5.30	1.509		585	92.50		677.50	785
1350	100 amp		4.50	1.778		845	109		954	1,100
1360	Circuit breaker, 3 pole 250 volt, 60 amp		5.30	1.509		960	92.50		1,052.50	1,200
1370	3 pole 480 volt, 100 amp		4.50	1.778		960	109		1,069	1,225
2000	Bus duct, 2 wire, 250 volt, 30 amp		60	.133	L.F.	8.35	8.20		16.55	22
2100	60 amp		50	.160		8.35	9.80		18.15	24.50
2200	300 volt, 30 amp		60	.133		8.35	8.20		16.55	22
2300	60 amp		50	.160		8.35	9.80		18.15	24.50
2400	3 wire, 250 volt, 30 amp		60	.133		10.80	8.20		19	24.50
2500	60 amp		50	.160		10.55	9.80		20.35	26.50
2600	480/277 volt, 30 amp		60	.133		10.80	8.20		19	24.50
2700	60 amp		50	.160		10.80	9.80		20.60	27
2750	End feed, 300 volt 2 wire max. 30 amp		6	1.333	Ea.	75.50	82		157.50	209
2800	60 amp		5.50	1.455		75.50	89		164.50	220
2850	30 amp miniature		6	1.333		75.50	82		157.50	209
2900	3 wire, 30 amp		6	1.333		95	82		177	230
2950	60 amp		5.50	1.455		95	89		184	241
3000	30 amp miniature		6	1.333		95	82		177	230
3050	Center feed, 300 volt 2 wire, 30 amp		6	1.333	L.F.	104	82		186	241

26 25 13.20 Bus Duct		Crew	Daily Output	Labor-Hours	Unit	Material	2020 Bare Costs Labor	Equipment	Total	Total Incl O&P
3100	60 amp	1 Elec	5.50	1.455	Ea.	104	89		193	252
3150	3 wire, 30 amp		6	1.333		119	82		201	256
3200	60 amp		5.50	1.455		119	89		208	267
3220	Elbow, 30 amp		6	1.333		40.50	82		122.50	171
3240	60 amp		5.50	1.455		40.50	89		129.50	182
3260	End cap		40	.200		9.90	12.25		22.15	30
3280	Strength beam, 10'		15	.533		24.50	32.50		57	77.50
3300	Hanger		24	.333		5.25	20.50		25.75	37.50
3320	Tap box, nonfusible		6.30	1.270		84.50	78		162.50	213
3340	Fusible switch 30 amp, 1 fuse		6	1.333		465	82		547	635
3360	2 fuse		6	1.333		485	82		567	660
3380	3 fuse		6	1.333		530	82		612	710
3400	Circuit breaker handle on cover, 1 pole		6	1.333		62	82		144	195
3420	2 pole		6	1.333		580	82		662	760
3440	3 pole		6	1.333		710	82		792	910
3460	Circuit breaker external operhandle, 1 pole		6	1.333		74	82		156	208
3480	2 pole		6	1.333		750	82		832	950
3500	3 pole		6	1.333		815	82		897	1,025
3520	Terminal plug only		16	.500		99.50	30.50		130	157
3540	Terminal with receptacle		16	.500		109	30.50		139.50	167
3560	Fixture plug		16	.500		73.50	30.50		104	128
4000	Copper bus duct, lighting, 2 wire 300 volt, 20 amp		70	.114	L.F.	7.55	7		14.55	19.10
4020	35 amp		60	.133		7.55	8.20		15.75	21
4040	50 amp		55	.145		7.55	8.90		16.45	22
4060	60 amp		50	.160		7.55	9.80		17.35	23.50
4080	3 wire 300 volt, 20 amp		70	.114		7.15	7		14.15	18.65
4100	35 amp		60	.133		6.95	8.20		15.15	20
4120	50 amp		55	.145		7.15	8.90		16.05	21.50
4140	60 amp		50	.160		7.15	9.80		16.95	23
4160	Feeder in box, end, 1 circuit		6	1.333	Ea.	101	82		183	237
4180	2 circuit		5.50	1.455		106	89		195	253
4200	Center, 1 circuit		6	1.333		140	82		222	280
4220	2 circuit		5.50	1.455		145	89		234	296
4240	End cap		40	.200		17.05	12.25		29.30	37.50
4260	Hanger, surface mount		24	.333		10.20	20.50		30.70	42.50
4280	Coupling		40	.200		13	12.25		25.25	33

26 25 13.30 Copper Bus Duct		Crew	Daily Output	Labor-Hours	Unit	Material	Labor	Equipment	Total	Total Incl O&P
0010	**COPPER BUS DUCT**									
0100	Weatherproof 3 pole 4 wire, feeder duct, 600 amp	2 Elec	24	.667	L.F.	315	41		356	410
0110	800 amp		18	.889		410	54.50		464.50	535
0120	1,000 amp		17	.941		520	57.50		577.50	660
0125	1,200 amp		16.50	.970		555	59.50		614.50	705
0130	1,350 amp		16	1		580	61.50		641.50	730
0140	1,600 amp		12	1.333		640	82		722	830
0150	2,000 amp		10	1.600		740	98		838	965
0160	2,500 amp		7	2.286		855	140		995	1,150
0170	3,000 amp		5	3.200		1,125	196		1,321	1,525
0180	4,000 amp		3.60	4.444		1,750	273		2,023	2,350
0200	Indoor 3 pole 4 wire, plug-in, bus duct high short circuit, 400 amp		32	.500		450	30.50		480.50	540
0210	600 amp		26	.615		515	38		553	630
0220	800 amp		20	.800		785	49		834	940
0230	1,000 amp		18	.889		905	54.50		959.50	1,075

26 25 13.30 Copper Bus Duct		Crew	Daily Output	Labor-Hours	Unit	Material	2020 Bare Costs Labor	Equipment	Total	Total Incl O&P
0240	1,350 amp	2 Elec	16	1	L.F.	525	61.50		586.50	670
0250	1,600 amp		12	1.333		590	82		672	775
0260	2,000 amp		10	1.600		750	98		848	975
0270	2,500 amp		8	2		925	123		1,048	1,225
0280	3,000 amp		6	2.667		945	164		1,109	1,300
0310	Cross, 225 amp		3	5.333	Ea.	3,350	325		3,675	4,200
0320	400 amp		2.80	5.714		3,350	350		3,700	4,250
0330	600 amp		2.60	6.154		3,350	380		3,730	4,275
0340	800 amp		2.20	7.273		3,650	445		4,095	4,675
0350	1,000 amp		2	8		4,075	490		4,565	5,225
0360	1,350 amp		1.80	8.889		4,525	545		5,070	5,825
0370	1,600 amp		1.70	9.412		5,025	575		5,600	6,425
0380	2,000 amp		1.60	10		8,250	615		8,865	10,000
0390	2,500 amp		1.40	11.429		10,000	700		10,700	12,100
0400	3,000 amp		1.20	13.333		9,750	820		10,570	12,000
0410	4,000 amp		1	16		12,700	980		13,680	15,500
0430	Expansion fitting, 225 amp		5.40	2.963		2,150	182		2,332	2,650
0440	400 amp		4.60	3.478		2,425	213		2,638	3,000
0450	600 amp		4	4		3,025	245		3,270	3,700
0460	800 amp		3.40	4.706		3,550	289		3,839	4,350
0470	1,000 amp		3	5.333		4,075	325		4,400	5,000
0480	1,350 amp		2.80	5.714		4,125	350		4,475	5,075
0490	1,600 amp		2.60	6.154		5,750	380		6,130	6,900
0500	2,000 amp		2.20	7.273		6,600	445		7,045	7,950
0510	2,500 amp		1.80	8.889		8,050	545		8,595	9,700
0520	3,000 amp		1.60	10		8,050	615		8,665	9,800
0530	4,000 amp		1.20	13.333		10,400	820		11,220	12,700
0550	Reducer nonfused, 225 amp		5.40	2.963		1,925	182		2,107	2,375
0560	400 amp		4.60	3.478		1,925	213		2,138	2,425
0570	600 amp		4	4		1,925	245		2,170	2,475
0580	800 amp		3.40	4.706		2,325	289		2,614	3,000
0590	1,000 amp		3	5.333		2,800	325		3,125	3,575
0600	1,350 amp		2.80	5.714		4,325	350		4,675	5,300
0610	1,600 amp		2.60	6.154		4,975	380		5,355	6,050
0620	2,000 amp		2.20	7.273		5,975	445		6,420	7,250
0630	2,500 amp		1.80	8.889		7,550	545		8,095	9,150
0640	3,000 amp		1.60	10		8,225	615		8,840	10,000
0650	4,000 amp		1.20	13.333		10,700	820		11,520	13,100
0670	Reducer fuse included, 225 amp		4.40	3.636		4,100	223		4,323	4,875
0680	400 amp		4.20	3.810		5,150	234		5,384	6,000
0690	600 amp		3.60	4.444		6,350	273		6,623	7,400
0700	800 amp		3.20	5		8,950	305		9,255	10,300
0710	1,000 amp		3	5.333		11,200	325		11,525	12,800
0720	1,350 amp		2.80	5.714		17,800	350		18,150	20,100
0730	1,600 amp		2.20	7.273		25,400	445		25,845	28,700
0740	2,000 amp		1.80	8.889		28,600	545		29,145	32,300
0790	Reducer, circuit breaker, 225 amp		4.40	3.636		5,200	223		5,423	6,075
0800	400 amp		4.20	3.810		6,075	234		6,309	7,025
0810	600 amp		3.60	4.444		8,625	273		8,898	9,900
0820	800 amp		3.20	5		10,100	305		10,405	11,600
0830	1,000 amp		3	5.333		11,500	325		11,825	13,100
0840	1,350 amp		2.80	5.714		14,100	350		14,450	16,000
0850	1,600 amp		2.20	7.273		20,800	445		21,245	23,600

26 25 13 – Low-Voltage Busways

26 25 13.30 Copper Bus Duct		Crew	Daily Output	Labor-Hours	Unit	Material	Labor	Equipment	Total	Total Incl O&P
						2020 Bare Costs				
0860	2,000 amp	2 Elec	1.80	8.889	Ea.	22,800	545		23,345	25,900
0910	Cable tap box, center, 225 amp		3.20	5		2,075	305		2,380	2,750
0920	400 amp		2.60	6.154		2,075	380		2,455	2,850
0930	600 amp		2.20	7.273		2,075	445		2,520	2,950
0940	800 amp		2	8		2,300	490		2,790	3,275
0950	1,000 amp		1.60	10		2,475	615		3,090	3,675
0960	1,350 amp		1.40	11.429		3,025	700		3,725	4,400
0970	1,600 amp		1.20	13.333		3,375	820		4,195	4,950
0980	2,000 amp		1	16		4,075	980		5,055	5,975
1040	2,500 amp		.80	20		4,800	1,225		6,025	7,200
1060	3,000 amp		.60	26.667		4,700	1,625		6,325	7,700
1080	4,000 amp		.40	40		5,925	2,450		8,375	10,300
1800	Weatherproof 3 pole 3 wire, feeder duct, 600 amp		28	.571	L.F.	298	35		333	385
1820	800 amp		22	.727		360	44.50		404.50	470
1840	1,000 amp		20	.800		405	49		454	520
1850	1,200 amp		19	.842		420	51.50		471.50	540
1860	1,350 amp		18	.889		585	54.50		639.50	730
1880	1,600 amp		14	1.143		670	70		740	845
1900	2,000 amp		12	1.333		855	82		937	1,075
1920	2,500 amp		8	2		1,075	123		1,198	1,375
1940	3,000 amp		6	2.667		1,100	164		1,264	1,450
1960	4,000 amp		4	4		1,450	245		1,695	1,975
2000	Indoor 3 pole 3 wire, feeder duct, 600 amp		32	.500		248	30.50		278.50	320
2010	800 amp		26	.615		620	38		658	745
2020	1,000 amp		24	.667		335	41		376	435
2025	1,200 amp		22	.727		445	44.50		489.50	560
2030	1,350 amp		20	.800		490	49		539	610
2040	1,600 amp		16	1		1,075	61.50		1,136.50	1,275
2050	2,000 amp		14	1.143		1,375	70		1,445	1,625
2060	2,500 amp		10	1.600		890	98		988	1,125
2070	3,000 amp		8	2		885	123		1,008	1,175
2080	4,000 amp		6	2.667		1,200	164		1,364	1,575
2090	5,000 amp		5	3.200		1,475	196		1,671	1,925
2200	Indoor 3 pole 3 wire, bus duct plug-in, 225 amp		46	.348		172	21.50		193.50	222
2210	400 amp		36	.444		263	27.50		290.50	330
2220	600 amp		30	.533		330	32.50		362.50	415
2230	800 amp		24	.667		500	41		541	615
2240	1,000 amp		20	.800		540	49		589	670
2250	1,350 amp		18	.889		850	54.50		904.50	1,025
2260	1,600 amp		14	1.143		890	70		960	1,100
2270	2,000 amp		12	1.333		750	82		832	950
2280	2,500 amp		10	1.600		925	98		1,023	1,175
2290	3,000 amp		8	2		945	123		1,068	1,250
2330	High short circuit, 400 amp		36	.444		325	27.50		352.50	395
2340	600 amp		30	.533		405	32.50		437.50	495
2350	800 amp		24	.667		585	41		626	710
2360	1,000 amp		20	.800		635	49		684	775
2370	1,350 amp		18	.889		950	54.50		1,004.50	1,125
2380	1,600 amp		14	1.143		1,025	70		1,095	1,225
2390	2,000 amp		12	1.333		750	82		832	950
2400	2,500 amp		10	1.600		925	98		1,023	1,175
2410	3,000 amp		8	2		945	123		1,068	1,250
2440	Elbows, 225 amp		4.60	3.478	Ea.	1,600	213		1,813	2,100

26 25 13.30 Copper Bus Duct		Crew	Daily Output	Labor-Hours	Unit	Material	2020 Bare Costs Labor	Equipment	Total	Total Incl O&P
2450	400 amp	2 Elec	4.20	3.810	Ea.	1,450	234		1,684	1,950
2460	600 amp		3.60	4.444		1,450	273		1,723	2,025
2470	800 amp		3.20	5		1,575	305		1,880	2,200
2480	1,000 amp		3	5.333		1,650	325		1,975	2,300
2485	1,200 amp		2.90	5.517		1,725	340		2,065	2,425
2490	1,350 amp		2.80	5.714		1,850	350		2,200	2,600
2500	1,600 amp		2.60	6.154		2,000	380		2,380	2,775
2510	2,000 amp		2	8		2,450	490		2,940	3,450
2520	2,500 amp		1.80	8.889		3,725	545		4,270	4,950
2530	3,000 amp		1.60	10		3,650	615		4,265	4,950
2540	4,000 amp		1.40	11.429		4,650	700		5,350	6,200
2560	Tee fittings, 225 amp		2.80	5.714		1,675	350		2,025	2,400
2570	400 amp		2.40	6.667		1,675	410		2,085	2,475
2580	600 amp		2	8		1,675	490		2,165	2,600
2590	800 amp		1.80	8.889		1,825	545		2,370	2,850
2600	1,000 amp		1.60	10		2,025	615		2,640	3,200
2605	1,200 amp		1.50	10.667		2,175	655		2,830	3,400
2610	1,350 amp		1.40	11.429		2,350	700		3,050	3,650
2620	1,600 amp		1.20	13.333		2,725	820		3,545	4,250
2630	2,000 amp		1	16		4,550	980		5,530	6,500
2640	2,500 amp		.70	22.857		5,325	1,400		6,725	8,000
2650	3,000 amp		.60	26.667		5,225	1,625		6,850	8,275
2660	4,000 amp		.50	32		6,750	1,975		8,725	10,500
2680	Cross, 225 amp		3.60	4.444		2,550	273		2,823	3,225
2690	400 amp		3.20	5		2,550	305		2,855	3,275
2700	600 amp		3	5.333		2,550	325		2,875	3,300
2710	800 amp		2.60	6.154		3,050	380		3,430	3,925
2720	1,000 amp		2.40	6.667		3,200	410		3,610	4,125
2730	1,350 amp		2.20	7.273		3,775	445		4,220	4,825
2740	1,600 amp		2	8		4,050	490		4,540	5,200
2750	2,000 amp		1.80	8.889		6,400	545		6,945	7,900
2760	2,500 amp		1.60	10		7,450	615		8,065	9,150
2770	3,000 amp		1.40	11.429		7,275	700		7,975	9,075
2780	4,000 amp		1	16		9,300	980		10,280	11,700
2800	Expansion fitting, 225 amp		6.40	2.500		2,275	153		2,428	2,725
2810	400 amp		5.40	2.963		5,025	182		5,207	5,800
2820	600 amp		4.60	3.478		2,275	213		2,488	2,825
2830	800 amp		4	4		2,700	245		2,945	3,350
2840	1,000 amp		3.60	4.444		2,975	273		3,248	3,675
2850	1,350 amp		3.20	5		3,100	305		3,405	3,900
2860	1,600 amp		3	5.333		3,400	325		3,725	4,250
2870	2,000 amp		2.60	6.154		4,350	380		4,730	5,350
2880	2,500 amp		2.20	7.273		6,075	445		6,520	7,350
2890	3,000 amp		1.80	8.889		6,050	545		6,595	7,500
2900	4,000 amp		1.40	11.429		7,725	700		8,425	9,575
2920	Reducer nonfused, 225 amp		6.40	2.500		1,575	153		1,728	1,950
2930	400 amp		5.40	2.963		1,575	182		1,757	2,000
2940	600 amp		4.60	3.478		1,575	213		1,788	2,050
2950	800 amp		4	4		1,850	245		2,095	2,400
2960	1,000 amp		3.60	4.444		2,075	273		2,348	2,700
2970	1,350 amp		3.20	5		2,800	305		3,105	3,550
2980	1,600 amp		3	5.333		3,175	325		3,500	3,975
2990	2,000 amp	↓	2.60	6.154	↓	3,800	380		4,180	4,750

26 25 13.30 Copper Bus Duct		Crew	Daily Output	Labor-Hours	Unit	2020 Bare Costs Material	Labor	Equipment	Total	Total Incl O&P
3000	2,500 amp	2 Elec	2.20	7.273	Ea.	5,575	445		6,020	6,800
3010	3,000 amp		1.80	8.889		6,025	545		6,570	7,475
3020	4,000 amp		1.40	11.429		7,825	700		8,525	9,675
3040	Reducer fuse included, 225 amp		5	3.200		3,725	196		3,921	4,400
3050	400 amp		4.80	3.333		4,975	205		5,180	5,775
3060	600 amp		4.20	3.810		5,925	234		6,159	6,850
3070	800 amp		3.60	4.444		8,425	273		8,698	9,700
3080	1,000 amp		3.40	4.706		9,975	289		10,264	11,400
3090	1,350 amp		3.20	5		10,400	305		10,705	11,900
3100	1,600 amp		2.60	6.154		24,200	380		24,580	27,200
3110	2,000 amp		2	8		27,400	490		27,890	31,000
3160	Reducer circuit breaker, 225 amp		5	3.200		4,775	196		4,971	5,550
3170	400 amp		4.80	3.333		5,850	205		6,055	6,750
3180	600 amp		4.20	3.810		8,375	234		8,609	9,575
3190	800 amp		3.60	4.444		9,800	273		10,073	11,200
3200	1,000 amp		3.40	4.706		11,100	289		11,389	12,700
3210	1,350 amp		3.20	5		13,800	305		14,105	15,600
3220	1,600 amp		2.60	6.154		20,500	380		20,880	23,200
3230	2,000 amp		2	8		22,400	490		22,890	25,500
3280	3 pole, 3 wire, cable tap box center, 225 amp		3.60	4.444		2,350	273		2,623	3,025
3290	400 amp		3	5.333		2,350	325		2,675	3,100
3300	600 amp		2.60	6.154		2,350	380		2,730	3,175
3310	800 amp		2.40	6.667		2,625	410		3,035	3,525
3320	1,000 amp		1.80	8.889		2,850	545		3,395	4,000
3330	1,350 amp		1.60	10		3,475	615		4,090	4,775
3340	1,600 amp		1.40	11.429		3,875	700		4,575	5,350
3350	2,000 amp		1.20	13.333		4,800	820		5,620	6,550
3360	2,500 amp		1	16		5,725	980		6,705	7,800
3370	3,000 amp		.70	22.857		5,650	1,400		7,050	8,350
3380	4,000 amp		.50	32		7,150	1,975		9,125	10,900
3400	Cable tap box end, 225 amp		3.60	4.444		1,275	273		1,548	1,825
3410	400 amp		3	5.333		1,200	325		1,525	1,825
3420	600 amp		2.60	6.154		1,425	380		1,805	2,150
3430	800 amp		2.40	6.667		1,425	410		1,835	2,200
3440	1,000 amp		1.80	8.889		1,850	545		2,395	2,875
3445	1,200 amp		1.70	9.412		1,750	575		2,325	2,825
3450	1,350 amp		1.60	10		1,875	615		2,490	3,000
3460	1,600 amp		1.40	11.429		2,150	700		2,850	3,425
3470	2,000 amp		1.20	13.333		2,500	820		3,320	4,000
3480	2,500 amp		1	16		2,975	980		3,955	4,775
3490	3,000 amp		.70	22.857		2,925	1,400		4,325	5,350
3500	4,000 amp		.50	32		3,700	1,975		5,675	7,100
4600	Plug-in, fusible switch w/3 fuses, 3 pole, 250 volt, 30 amp	1 Elec	4	2		465	123		588	705
4610	60 amp		3.60	2.222		610	136		746	880
4620	100 amp		2.70	2.963		875	182		1,057	1,250
4630	200 amp	2 Elec	3.20	5		1,475	305		1,780	2,075
4640	400 amp		1.40	11.429		3,825	700		4,525	5,275
4650	600 amp		.90	17.778		5,300	1,100		6,400	7,500
4700	4 pole, 120/208 volt, 30 amp	1 Elec	3.90	2.051		640	126		766	895
4710	60 amp		3.50	2.286		690	140		830	970
4720	100 amp		2.60	3.077		960	189		1,149	1,350
4730	200 amp	2 Elec	3	5.333		1,600	325		1,925	2,275
4740	400 amp		1.30	12.308		3,775	755		4,530	5,300

26 25 13.30 Copper Bus Duct		Crew	Daily Output	Labor-Hours	Unit	Material	2020 Bare Costs Labor	Equipment	Total	Total Incl O&P
4750	600 amp	2 Elec	.80	20	Ea.	5,300	1,225		6,525	7,725
4800	3 pole, 480 volt, 30 amp	1 Elec	4	2		480	123		603	715
4810	60 amp		3.60	2.222		505	136		641	765
4820	100 amp		2.70	2.963		855	182		1,037	1,225
4830	200 amp	2 Elec	3.20	5		1,475	305		1,780	2,100
4840	400 amp		1.40	11.429		3,425	700		4,125	4,825
4850	600 amp		.90	17.778		4,850	1,100		5,950	7,025
4860	800 amp		.66	24.242		20,100	1,475		21,575	24,400
4870	1,000 amp		.60	26.667		20,600	1,625		22,225	25,100
4880	1,200 amp		.50	32		22,400	1,975		24,375	27,700
4890	1,600 amp		.44	36.364		23,600	2,225		25,825	29,400
4900	4 pole, 277/480 volt, 30 amp	1 Elec	3.90	2.051		690	126		816	955
4910	60 amp		3.50	2.286		735	140		875	1,025
4920	100 amp		2.60	3.077		1,075	189		1,264	1,475
4930	200 amp	2 Elec	3	5.333		2,150	325		2,475	2,850
4940	400 amp		1.30	12.308		4,050	755		4,805	5,600
4950	600 amp		.80	20		5,525	1,225		6,750	7,975
5050	800 amp		.60	26.667		18,200	1,625		19,825	22,500
5060	1,000 amp		.56	28.571		20,900	1,750		22,650	25,700
5070	1,200 amp		.48	33.333		21,100	2,050		23,150	26,400
5080	1,600 amp		.42	38.095		25,100	2,325		27,425	31,200
5150	Fusible with starter, 3 pole 250 volt, 30 amp	1 Elec	3.50	2.286		3,600	140		3,740	4,175
5160	60 amp		3.20	2.500		3,800	153		3,953	4,400
5170	100 amp		2.50	3.200		4,300	196		4,496	5,025
5180	200 amp	2 Elec	2.80	5.714		7,100	350		7,450	8,350
5200	3 pole 480 volt, 30 amp	1 Elec	3.50	2.286		3,600	140		3,740	4,175
5210	60 amp		3.20	2.500		3,800	153		3,953	4,400
5220	100 amp		2.50	3.200		4,300	196		4,496	5,025
5230	200 amp	2 Elec	2.80	5.714		7,100	350		7,450	8,350
5300	Fusible with contactor, 3 pole 250 volt, 30 amp	1 Elec	3.50	2.286		3,500	140		3,640	4,075
5310	60 amp		3.20	2.500		4,450	153		4,603	5,125
5320	100 amp		2.50	3.200		6,225	196		6,421	7,150
5330	200 amp	2 Elec	2.80	5.714		7,125	350		7,475	8,400
5400	3 pole 480 volt, 30 amp	1 Elec	3.50	2.286		3,775	140		3,915	4,375
5410	60 amp		3.20	2.500		5,325	153		5,478	6,075
5420	100 amp		2.50	3.200		7,300	196		7,496	8,325
5430	200 amp	2 Elec	2.80	5.714		7,475	350		7,825	8,775
5450	Fusible with capacitor, 3 pole 250 volt, 30 amp	1 Elec	3	2.667		9,100	164		9,264	10,300
5460	60 amp		2	4		10,600	245		10,845	12,000
5500	3 pole 480 volt, 30 amp		3	2.667		7,625	164		7,789	8,625
5510	60 amp		2	4		9,525	245		9,770	10,900
5600	Circuit breaker, 3 pole, 250 volt, 60 amp		4.50	1.778		655	109		764	890
5610	100 amp		3.20	2.500		805	153		958	1,125
5650	4 pole, 120/208 volt, 60 amp		4.40	1.818		740	112		852	985
5660	100 amp		3.10	2.581		880	158		1,038	1,200
5700	3 pole, 4 wire 277/480 volt, 60 amp		4.30	1.860		1,000	114		1,114	1,275
5710	100 amp		3	2.667		1,100	164		1,264	1,475
5720	225 amp	2 Elec	3.20	5		2,450	305		2,755	3,175
5730	400 amp		1.20	13.333		5,100	820		5,920	6,875
5740	600 amp		.96	16.667		6,875	1,025		7,900	9,125
5750	700 amp		.60	26.667		8,700	1,625		10,325	12,100
5760	800 amp		.60	26.667		8,700	1,625		10,325	12,100
5770	900 amp		.54	29.630		11,500	1,825		13,325	15,400

26 25 13 – Low-Voltage Busways

26 25 13.30 Copper Bus Duct		Crew	Daily Output	Labor-Hours	Unit	Material	Labor	Equipment	Total	Total Incl O&P
						2020 Bare Costs				
5780	1,000 amp	2 Elec	.54	29.630	Ea.	11,500	1,825		13,325	15,400
5790	1,200 amp		.42	38.095		13,800	2,325		16,125	18,800
5810	Circuit breaker w/HIC fuses, 3 pole 480 volt, 60 amp	1 Elec	4.40	1.818		1,250	112		1,362	1,550
5820	100 amp	"	3.10	2.581		1,375	158		1,533	1,775
5830	225 amp	2 Elec	3.40	4.706		4,400	289		4,689	5,275
5840	400 amp		1.40	11.429		6,975	700		7,675	8,750
5850	600 amp		1	16		7,075	980		8,055	9,275
5860	700 amp		.64	25		9,375	1,525		10,900	12,700
5870	800 amp		.64	25		9,375	1,525		10,900	12,700
5880	900 amp		.56	28.571		20,300	1,750		22,050	25,000
5890	1,000 amp		.56	28.571		20,300	1,750		22,050	25,000
5950	3 pole 4 wire 277/480 volt, 60 amp	1 Elec	4.30	1.860		1,250	114		1,364	1,550
5960	100 amp	"	3	2.667		1,375	164		1,539	1,775
5970	225 amp	2 Elec	3	5.333		4,400	325		4,725	5,325
5980	400 amp		1.10	14.545		6,975	890		7,865	9,050
5990	600 amp		.94	17.021		7,075	1,050		8,125	9,375
6000	700 amp		.58	27.586		9,375	1,700		11,075	12,900
6010	800 amp		.58	27.586		9,375	1,700		11,075	12,900
6020	900 amp		.52	30.769		20,300	1,900		22,200	25,200
6030	1,000 amp		.52	30.769		20,300	1,900		22,200	25,200
6040	1,200 amp		.40	40		20,300	2,450		22,750	26,100
6100	Circuit breaker with starter, 3 pole 250 volt, 60 amp	1 Elec	3.20	2.500		1,875	153		2,028	2,300
6110	100 amp	"	2.50	3.200		2,900	196		3,096	3,500
6120	225 amp	2 Elec	3	5.333		4,125	325		4,450	5,025
6130	3 pole 480 volt, 60 amp	1 Elec	3.20	2.500		2,100	153		2,253	2,525
6140	100 amp	"	2.50	3.200		2,900	196		3,096	3,500
6150	225 amp	2 Elec	3	5.333		3,500	325		3,825	4,350
6200	Circuit breaker with contactor, 3 pole 250 volt, 60 amp	1 Elec	3.20	2.500		1,975	153		2,128	2,400
6210	100 amp	"	2.50	3.200		2,700	196		2,896	3,275
6220	225 amp	2 Elec	3	5.333		3,825	325		4,150	4,700
6250	3 pole 480 volt, 60 amp	1 Elec	3.20	2.500		1,975	153		2,128	2,400
6260	100 amp	"	2.50	3.200		2,700	196		2,896	3,275
6270	225 amp	2 Elec	3	5.333		3,600	325		3,925	4,450
6300	Circuit breaker with capacitor, 3 pole 250 volt, 60 amp	1 Elec	2	4		10,700	245		10,945	12,200
6310	3 pole 480 volt, 60 amp		2	4		11,700	245		11,945	13,300
6400	Add control transformer with pilot light to above starter		16	.500		590	30.50		620.50	695
6410	Switch, fusible, mechanically held contactor optional		16	.500		1,575	30.50		1,605.50	1,775
6430	Circuit breaker, mechanically held contactor optional		16	.500		1,575	30.50		1,605.50	1,775
6450	Ground neutralizer, 3 pole		16	.500		72	30.50		102.50	127

26 25 13.40 Copper Bus Duct		Crew	Daily Output	Labor-Hours	Unit	Material	Labor	Equipment	Total	Total Incl O&P
0010	**COPPER BUS DUCT** 10' long									
0050	Indoor 3 pole 4 wire, plug-in, straight section, 225 amp	2 Elec	40	.400	L.F.	222	24.50		246.50	283
1000	400 amp		32	.500		385	30.50		415.50	470
1500	600 amp		26	.615		455	38		493	560
2400	800 amp		20	.800		620	49		669	755
2450	1,000 amp		18	.889		305	54.50		359.50	420
2470	1,200 amp		17	.941		395	57.50		452.50	525
2500	1,350 amp		16	1		415	61.50		476.50	555
2510	1,600 amp		12	1.333		470	82		552	645
2520	2,000 amp		10	1.600		595	98		693	805
2530	2,500 amp		8	2		735	123		858	1,000
2540	3,000 amp		6	2.667		855	164		1,019	1,200

26 25 Low-Voltage Enclosed Bus Assemblies

26 25 13 – Low-Voltage Busways

26 25 13.40 Copper Bus Duct		Crew	Daily Output	Labor-Hours	Unit	2020 Bare Costs Material	Labor	Equipment	Total	Total Incl O&P
2550	Feeder, 600 amp	2 Elec	28	.571	L.F.	204	35		239	278
2600	800 amp		22	.727		755	44.50		799.50	900
2700	1,000 amp		20	.800		905	49		954	1,075
2750	1,200 amp		19	.842		1,100	51.50		1,151.50	1,275
2800	1,350 amp		18	.889		390	54.50		444.50	510
2900	1,600 amp		14	1.143		1,450	70		1,520	1,700
3000	2,000 amp		12	1.333		1,800	82		1,882	2,100
3010	2,500 amp		8	2		705	123		828	970
3020	3,000 amp		6	2.667		830	164		994	1,150
3030	4,000 amp		4	4		1,100	245		1,345	1,575
3040	5,000 amp		2	8		1,325	490		1,815	2,225
3100	Elbows, 225 amp		4	4	Ea.	1,575	245		1,820	2,125
3200	400 amp		3.60	4.444		1,575	273		1,848	2,175
3300	600 amp		3.20	5		1,575	305		1,880	2,225
3400	800 amp		2.80	5.714		1,725	350		2,075	2,450
3500	1,000 amp		2.60	6.154		1,925	380		2,305	2,675
3550	1,200 amp		2.50	6.400		1,875	395		2,270	2,675
3600	1,350 amp		2.40	6.667		1,900	410		2,310	2,700
3700	1,600 amp		2.20	7.273		2,075	445		2,520	2,950
3800	2,000 amp		1.80	8.889		2,550	545		3,095	3,650
3810	2,500 amp		1.60	10		4,025	615		4,640	5,375
3820	3,000 amp		1.40	11.429		4,500	700		5,200	6,025
3830	4,000 amp		1.20	13.333		5,825	820		6,645	7,650
3840	5,000 amp		1	16		9,400	980		10,380	11,800
4000	End box, 225 amp		34	.471		171	29		200	233
4100	400 amp		32	.500		189	30.50		219.50	255
4200	600 amp		28	.571		189	35		224	262
4300	800 amp		26	.615		189	38		227	266
4400	1,000 amp		24	.667		181	41		222	262
4410	1,200 amp		23	.696		189	42.50		231.50	274
4500	1,350 amp		22	.727		180	44.50		224.50	267
4600	1,600 amp		20	.800		180	49		229	274
4700	2,000 amp		18	.889		221	54.50		275.50	325
4710	2,500 amp		16	1		221	61.50		282.50	340
4720	3,000 amp		14	1.143		213	70		283	340
4730	4,000 amp		12	1.333		258	82		340	410
4740	5,000 amp		10	1.600		258	98		356	435
4800	Cable tap box end, 225 amp		3.20	5		1,225	305		1,530	1,825
5000	400 amp		2.60	6.154		1,225	380		1,605	1,900
5100	600 amp		2.20	7.273		1,625	445		2,070	2,450
5200	800 amp		2	8		1,775	490		2,265	2,700
5300	1,000 amp		1.60	10		1,725	615		2,340	2,850
5350	1,200 amp		1.50	10.667		2,025	655		2,680	3,225
5400	1,350 amp		1.40	11.429		2,200	700		2,900	3,500
5500	1,600 amp		1.20	13.333		2,475	820		3,295	3,975
5600	2,000 amp		1	16		2,775	980		3,755	4,550
5610	2,500 amp		.80	20		3,175	1,225		4,400	5,400
5620	3,000 amp		.60	26.667		3,700	1,625		5,325	6,600
5630	4,000 amp		.40	40		4,275	2,450		6,725	8,475
5640	5,000 amp		.20	80		5,200	4,900		10,100	13,300
5700	Switchboard stub, 225 amp		5.40	2.963		1,225	182		1,407	1,625
5800	400 amp		4.60	3.478		1,300	213		1,513	1,750
5900	600 amp		4	4		1,375	245		1,620	1,900

26 25 13.40	Copper Bus Duct	Crew	Daily Output	Labor-Hours	Unit	Material	2020 Bare Costs Labor	Equipment	Total	Total Incl O&P
6000	800 amp	2 Elec	3.20	5	Ea.	1,675	305		1,980	2,300
6100	1,000 amp		3	5.333		1,950	325		2,275	2,625
6150	1,200 amp		2.80	5.714		2,275	350		2,625	3,050
6200	1,350 amp		2.60	6.154		2,400	380		2,780	3,225
6300	1,600 amp		2.40	6.667		2,700	410		3,110	3,600
6400	2,000 amp		2	8		3,275	490		3,765	4,350
6410	2,500 amp		1.80	8.889		4,000	545		4,545	5,250
6420	3,000 amp		1.60	10		4,500	615		5,115	5,900
6430	4,000 amp		1.40	11.429		5,850	700		6,550	7,525
6440	5,000 amp		1.20	13.333		7,200	820		8,020	9,150
6490	Tee fittings, 225 amp		2.40	6.667		2,275	410		2,685	3,125
6500	400 amp		2	8		2,275	490		2,765	3,250
6600	600 amp		1.80	8.889		2,275	545		2,820	3,350
6700	800 amp		1.60	10		2,350	615		2,965	3,525
6750	1,000 amp		1.40	11.429		2,575	700		3,275	3,925
6770	1,200 amp		1.30	12.308		2,900	755		3,655	4,350
6800	1,350 amp		1.20	13.333		3,075	820		3,895	4,625
7000	1,600 amp		1	16		3,475	980		4,455	5,325
7100	2,000 amp		.80	20		4,125	1,225		5,350	6,450
7110	2,500 amp		.60	26.667		5,075	1,625		6,700	8,100
7120	3,000 amp		.50	32		5,700	1,975		7,675	9,300
7130	4,000 amp		.40	40		7,300	2,450		9,750	11,800
7140	5,000 amp		.20	80		8,625	4,900		13,525	17,100
7200	Plug-in fusible switches w/3 fuses, 600 volt, 3 pole, 30 amp	1 Elec	4	2		885	123		1,008	1,150
7300	60 amp		3.60	2.222		990	136		1,126	1,300
7400	100 amp		2.70	2.963		1,425	182		1,607	1,825
7500	200 amp	2 Elec	3.20	5		2,500	305		2,805	3,225
7600	400 amp		1.40	11.429		7,075	700		7,775	8,850
7700	600 amp		.90	17.778		8,400	1,100		9,500	10,900
7800	800 amp		.66	24.242		12,800	1,475		14,275	16,400
7900	1,200 amp		.50	32		24,100	1,975		26,075	29,500
7910	1,600 amp		.44	36.364		22,600	2,225		24,825	28,300
8000	Plug-in circuit breakers, molded case, 15 to 50 amp	1 Elec	4.40	1.818		835	112		947	1,100
8100	70 to 100 amp	"	3.10	2.581		930	158		1,088	1,275
8200	150 to 225 amp	2 Elec	3.40	4.706		2,525	289		2,814	3,225
8300	250 to 400 amp		1.40	11.429		4,425	700		5,125	5,925
8400	500 to 600 amp		1	16		5,950	980		6,930	8,050
8500	700 to 800 amp		.64	25		7,350	1,525		8,875	10,400
8600	900 to 1,000 amp		.56	28.571		10,500	1,750		12,250	14,300
8700	1,200 amp		.44	36.364		12,600	2,225		14,825	17,300
8720	1,400 amp		.40	40		17,700	2,450		20,150	23,200
8730	1,600 amp		.40	40		19,400	2,450		21,850	25,100
8750	Circuit breakers, with current limiting fuse, 15 to 50 amp	1 Elec	4.40	1.818		1,675	112		1,787	2,025
8760	70 to 100 amp	"	3.10	2.581		1,975	158		2,133	2,425
8770	150 to 225 amp	2 Elec	3.40	4.706		4,250	289		4,539	5,125
8780	250 to 400 amp		1.40	11.429		6,575	700		7,275	8,325
8790	500 to 600 amp		1	16		7,575	980		8,555	9,850
8800	700 to 800 amp		.64	25		12,500	1,525		14,025	16,100
8810	900 to 1,000 amp		.56	28.571		14,200	1,750		15,950	18,300
8850	Combination starter FVNR, fusible switch, NEMA size 0, 30 amp	1 Elec	2	4		2,250	245		2,495	2,850
8860	NEMA size 1, 60 amp		1.80	4.444		2,375	273		2,648	3,050
8870	NEMA size 2, 100 amp		1.30	6.154		3,000	380		3,380	3,875
8880	NEMA size 3, 200 amp	2 Elec	2	8		4,800	490		5,290	6,025

26 25 13.40 Copper Bus Duct		Crew	Daily Output	Labor-Hours	Unit	2020 Bare Costs Material	Labor	Equipment	Total	Total Incl O&P
8900	Circuit breaker, NEMA size 0, 30 amp	1 Elec	2	4	Ea.	2,325	245		2,570	2,925
8910	NEMA size 1, 60 amp		1.80	4.444		2,400	273		2,673	3,075
8920	NEMA size 2, 100 amp		1.30	6.154		3,450	380		3,830	4,375
8930	NEMA size 3, 200 amp	2 Elec	2	8		4,375	490		4,865	5,575
8950	Combination contactor, fusible switch, NEMA size 0, 30 amp	1 Elec	2	4		1,325	245		1,570	1,825
8960	NEMA size 1, 60 amp		1.80	4.444		1,350	273		1,623	1,925
8970	NEMA size 2, 100 amp		1.30	6.154		2,025	380		2,405	2,800
8980	NEMA size 3, 200 amp	2 Elec	2	8		2,325	490		2,815	3,300
9000	Circuit breaker, NEMA size 0, 30 amp	1 Elec	2	4		1,500	245		1,745	2,025
9010	NEMA size 1, 60 amp		1.80	4.444		1,550	273		1,823	2,125
9020	NEMA size 2, 100 amp		1.30	6.154		2,400	380		2,780	3,200
9030	NEMA size 3, 200 amp	2 Elec	2	8		2,900	490		3,390	3,950
9050	Control transformer for above, NEMA size 0, 30 amp	1 Elec	8	1		256	61.50		317.50	375
9060	NEMA size 1, 60 amp		8	1		256	61.50		317.50	375
9070	NEMA size 2, 100 amp		7	1.143		355	70		425	500
9080	NEMA size 3, 200 amp	2 Elec	14	1.143		495	70		565	655
9100	Comb. fusible switch & lighting control, electrically held, 30 amp	1 Elec	2	4		1,050	245		1,295	1,525
9110	60 amp		1.80	4.444		1,500	273		1,773	2,075
9120	100 amp		1.30	6.154		1,925	380		2,305	2,700
9130	200 amp	2 Elec	2	8		4,750	490		5,240	6,000
9150	Mechanically held, 30 amp	1 Elec	2	4		1,300	245		1,545	1,825
9160	60 amp		1.80	4.444		1,950	273		2,223	2,575
9170	100 amp		1.30	6.154		2,500	380		2,880	3,325
9180	200 amp	2 Elec	2	8		5,100	490		5,590	6,375
9200	Ground bus added to bus duct, 225 amp		320	.050	L.F.	44.50	3.07		47.57	53.50
9210	400 amp		240	.067		44.50	4.09		48.59	55.50
9220	600 amp		240	.067		44.50	4.09		48.59	55.50
9230	800 amp		160	.100		52	6.15		58.15	67
9240	1,000 amp		160	.100		59.50	6.15		65.65	75
9250	1,350 amp		140	.114		85	7		92	104
9260	1,600 amp		120	.133		92	8.20		100.20	114
9270	2,000 amp		110	.145		120	8.90		128.90	146
9280	2,500 amp		100	.160		149	9.80		158.80	178
9290	3,000 amp		90	.178		170	10.90		180.90	204
9300	4,000 amp		80	.200		225	12.25		237.25	266
9310	5,000 amp		70	.229		270	14		284	320
9320	High short circuit bracing, add					18.60			18.60	20.50

26 25 13.60 Copper or Aluminum Bus Duct Fittings		Crew	Daily Output	Labor-Hours	Unit	Material	Labor	Equipment	Total	Total Incl O&P
0010	**COPPER OR ALUMINUM BUS DUCT FITTINGS**									
0100	Flange, wall, with vapor barrier, 225 amp	2 Elec	6.20	2.581	Ea.	845	158		1,003	1,175
0110	400 amp		6	2.667		795	164		959	1,125
0120	600 amp		5.80	2.759		845	169		1,014	1,200
0130	800 amp		5.40	2.963		845	182		1,027	1,200
0140	1,000 amp		5	3.200		845	196		1,041	1,225
0145	1,200 amp		4.80	3.333		845	205		1,050	1,250
0150	1,350 amp		4.60	3.478		845	213		1,058	1,250
0160	1,600 amp		4.20	3.810		845	234		1,079	1,300
0170	2,000 amp		4	4		845	245		1,090	1,300
0180	2,500 amp		3.60	4.444		845	273		1,118	1,350
0190	3,000 amp		3.20	5		845	305		1,150	1,400
0200	4,000 amp		2.60	6.154		820	380		1,200	1,475
0300	Roof, 225 amp		6.20	2.581		1,050	158		1,208	1,400

26 25 13.60 Copper or Aluminum Bus Duct Fittings		Crew	Daily Output	Labor-Hours	Unit	Material	2020 Bare Costs Labor	Equipment	Total	Total Incl O&P
0310	400 amp	2 Elec	6	2.667	Ea.	1,050	164		1,214	1,400
0320	600 amp		5.80	2.759		1,050	169		1,219	1,400
0330	800 amp		5.40	2.963		1,050	182		1,232	1,425
0340	1,000 amp		5	3.200		1,050	196		1,246	1,450
0345	1,200 amp		4.80	3.333		1,050	205		1,255	1,475
0350	1,350 amp		4.60	3.478		1,050	213		1,263	1,475
0360	1,600 amp		4.20	3.810		1,050	234		1,284	1,500
0370	2,000 amp		4	4		1,050	245		1,295	1,525
0380	2,500 amp		3.60	4.444		1,050	273		1,323	1,575
0390	3,000 amp		3.20	5		1,050	305		1,355	1,625
0400	4,000 amp		2.60	6.154		1,050	380		1,430	1,725
0420	Support, floor mounted, 225 amp		20	.800		171	49		220	264
0430	400 amp		20	.800		171	49		220	264
0440	600 amp		18	.889		171	54.50		225.50	272
0450	800 amp		16	1		171	61.50		232.50	283
0460	1,000 amp		13	1.231		171	75.50		246.50	305
0465	1,200 amp		11.80	1.356		171	83		254	315
0470	1,350 amp		10.60	1.509		171	92.50		263.50	330
0480	1,600 amp		9.20	1.739		171	107		278	350
0490	2,000 amp		8	2		171	123		294	375
0500	2,500 amp		6.40	2.500		171	153		324	425
0510	3,000 amp		5.40	2.963		171	182		353	470
0520	4,000 amp		4	4		171	245		416	570
0540	Weather stop, 225 amp		12	1.333		530	82		612	705
0550	400 amp		10	1.600		530	98		628	730
0560	600 amp		9	1.778		530	109		639	750
0570	800 amp		8	2		530	123		653	770
0580	1,000 amp		6.40	2.500		530	153		683	815
0585	1,200 amp		5.90	2.712		530	166		696	835
0590	1,350 amp		5.40	2.963		530	182		712	860
0600	1,600 amp		4.60	3.478		530	213		743	910
0610	2,000 amp		4	4		530	245		775	960
0620	2,500 amp		3.20	5		530	305		835	1,050
0630	3,000 amp		2.60	6.154		530	380		910	1,150
0640	4,000 amp		2	8		530	490		1,020	1,325
0660	End closure, 225 amp		34	.471		198	29		227	263
0670	400 amp		32	.500		198	30.50		228.50	265
0680	600 amp		28	.571		198	35		233	272
0690	800 amp		26	.615		172	38		210	247
0700	1,000 amp		24	.667		198	41		239	281
0705	1,200 amp		23	.696		170	42.50		212.50	253
0710	1,350 amp		22	.727		170	44.50		214.50	256
0720	1,600 amp		20	.800		198	49		247	294
0730	2,000 amp		18	.889		277	54.50		331.50	390
0740	2,500 amp		16	1		253	61.50		314.50	375
0750	3,000 amp		14	1.143		253	70		323	385
0760	4,000 amp		12	1.333		255	82		337	405
0780	Switchboard stub, 3 pole 3 wire, 225 amp		6	2.667		1,025	164		1,189	1,375
0790	400 amp		5.20	3.077		1,025	189		1,214	1,425
0800	600 amp		4.60	3.478		1,025	213		1,238	1,450
0810	800 amp		3.60	4.444		1,250	273		1,523	1,800
0820	1,000 amp		3.40	4.706		1,450	289		1,739	2,050
0825	1,200 amp		3.20	5		1,450	305		1,755	2,075

26 25 13 – Low-Voltage Busways

26 25 13.60 Copper or Aluminum Bus Duct Fittings		Crew	Daily Output	Labor-Hours	Unit	Material	2020 Bare Costs Labor	Equipment	Total	Total Incl O&P
0830	1,350 amp	2 Elec	3	5.333	Ea.	1,725	325		2,050	2,400
0840	1,600 amp		2.80	5.714		2,050	350		2,400	2,825
0850	2,000 amp		2.40	6.667		2,400	410		2,810	3,250
0860	2,500 amp		2	8		2,925	490		3,415	3,975
0870	3,000 amp		1.80	8.889		3,325	545		3,870	4,500
0880	4,000 amp		1.60	10		4,275	615		4,890	5,675
0890	5,000 amp		1.40	11.429		5,300	700		6,000	6,925
0900	3 pole 4 wire, 225 amp		5.40	2.963		1,275	182		1,457	1,675
0910	400 amp		4.60	3.478		1,275	213		1,488	1,725
0920	600 amp		4	4		1,275	245		1,520	1,775
0930	800 amp		3.20	5		1,525	305		1,830	2,175
0940	1,000 amp		3	5.333		1,800	325		2,125	2,475
0950	1,350 amp		2.60	6.154		2,300	380		2,680	3,100
0960	1,600 amp		2.40	6.667		2,625	410		3,035	3,500
0970	2,000 amp		2	8		3,150	490		3,640	4,200
0980	2,500 amp		1.80	8.889		3,875	545		4,420	5,100
0990	3,000 amp		1.60	10		4,475	615		5,090	5,875
1000	4,000 amp		1.40	11.429		5,875	700		6,575	7,525
1050	Service head, weatherproof, 3 pole 3 wire, 225 amp		3	5.333		1,725	325		2,050	2,400
1060	400 amp		2.80	5.714		1,725	350		2,075	2,450
1070	600 amp		2.60	6.154		1,725	380		2,105	2,475
1080	800 amp		2.40	6.667		1,950	410		2,360	2,775
1090	1,000 amp		2	8		2,100	490		2,590	3,075
1100	1,350 amp		1.80	8.889		2,725	545		3,270	3,850
1110	1,600 amp		1.60	10		3,050	615		3,665	4,325
1120	2,000 amp		1.40	11.429		3,750	700		4,450	5,200
1130	2,500 amp		1.20	13.333		4,475	820		5,295	6,175
1140	3,000 amp		.90	17.778		5,175	1,100		6,275	7,375
1150	4,000 amp		.70	22.857		6,600	1,400		8,000	9,400
1200	3 pole 4 wire, 225 amp		2.60	6.154		1,950	380		2,330	2,725
1210	400 amp		2.40	6.667		1,950	410		2,360	2,775
1220	600 amp		2.20	7.273		1,950	445		2,395	2,825
1230	800 amp		2	8		2,250	490		2,740	3,225
1240	1,000 amp		1.70	9.412		2,600	575		3,175	3,750
1250	1,350 amp		1.50	10.667		3,100	655		3,755	4,400
1260	1,600 amp		1.40	11.429		3,400	700		4,100	4,800
1270	2,000 amp		1.20	13.333		4,500	820		5,320	6,200
1280	2,500 amp		1	16		5,550	980		6,530	7,600
1290	3,000 amp		.80	20		6,550	1,225		7,775	9,100
1300	4,000 amp		.60	26.667		8,425	1,625		10,050	11,800
1350	Flanged end, 3 pole 3 wire, 225 amp		6	2.667		945	164		1,109	1,300
1360	400 amp		5.20	3.077		945	189		1,134	1,350
1370	600 amp		4.60	3.478		945	213		1,158	1,375
1380	800 amp		3.60	4.444		1,050	273		1,323	1,600
1390	1,000 amp		3.40	4.706		1,200	289		1,489	1,750
1395	1,200 amp		3.20	5		1,325	305		1,630	1,925
1400	1,350 amp		3	5.333		1,400	325		1,725	2,050
1410	1,600 amp		2.80	5.714		1,600	350		1,950	2,325
1420	2,000 amp		2.40	6.667		1,900	410		2,310	2,725
1430	2,500 amp		2	8		2,200	490		2,690	3,175
1440	3,000 amp		1.80	8.889		2,525	545		3,070	3,625
1450	4,000 amp		1.60	10		3,125	615		3,740	4,375
1500	3 pole 4 wire, 225 amp	↓	5.40	2.963	↓	1,100	182		1,282	1,475

26 25 13.60 Copper or Aluminum Bus Duct Fittings		Crew	Daily Output	Labor-Hours	Unit	Material	Labor	Equipment	Total	Total Incl O&P
						2020 Bare Costs				
1510	400 amp	2 Elec	4.60	3.478	Ea.	1,100	213		1,313	1,525
1520	600 amp		4	4		1,100	245		1,345	1,575
1530	800 amp		3.20	5		1,300	305		1,605	1,900
1540	1,000 amp		3	5.333		1,425	325		1,750	2,075
1545	1,200 amp		2.80	5.714		1,625	350		1,975	2,350
1550	1,350 amp		2.60	6.154		1,750	380		2,130	2,500
1560	1,600 amp		2.40	6.667		2,000	410		2,410	2,825
1570	2,000 amp		2	8		2,375	490		2,865	3,350
1580	2,500 amp		1.80	8.889		2,800	545		3,345	3,950
1590	3,000 amp		1.60	10		3,225	615		3,840	4,500
1600	4,000 amp		1.40	11.429		4,175	700		4,875	5,675
1650	Hanger, standard, 225 amp		64	.250		26.50	15.35		41.85	52.50
1660	400 amp		48	.333		26.50	20.50		47	60.50
1670	600 amp		40	.400		26.50	24.50		51	67
1680	800 amp		32	.500		26.50	30.50		57	76
1690	1,000 amp		24	.667		26.50	41		67.50	92
1695	1,200 amp		22	.727		26.50	44.50		71	97.50
1700	1,350 amp		20	.800		26.50	49		75.50	105
1710	1,600 amp		20	.800		26.50	49		75.50	105
1720	2,000 amp		18	.889		26.50	54.50		81	113
1730	2,500 amp		16	1		26.50	61.50		88	124
1740	3,000 amp		16	1		26.50	61.50		88	124
1750	4,000 amp		16	1		26.50	61.50		88	124
1800	Spring type, 225 amp		16	1		89	61.50		150.50	193
1810	400 amp		14	1.143		89	70		159	206
1820	600 amp		14	1.143		89	70		159	206
1830	800 amp		14	1.143		89	70		159	206
1840	1,000 amp		14	1.143		89	70		159	206
1845	1,200 amp		14	1.143		89	70		159	206
1850	1,350 amp		14	1.143		89	70		159	206
1860	1,600 amp		12	1.333		89	82		171	224
1870	2,000 amp		12	1.333		89	82		171	224
1880	2,500 amp		12	1.333		89	82		171	224
1890	3,000 amp		10	1.600		89	98		187	249
1900	4,000 amp	↓	10	1.600	↓	89	98		187	249

26 25 13.70 Feedrail		Crew	Daily Output	Labor-Hours	Unit	Material	Labor	Equipment	Total	Total Incl O&P
0010	**FEEDRAIL**, 12′ mounting									
0050	Trolley busway, 3 pole									
0100	300 volt 60 amp, plain, 10′ lengths	1 Elec	50	.160	L.F.	22	9.80		31.80	39
0300	Door track		50	.160		41.50	9.80		51.30	61
0500	Curved track	↓	30	.267	↓	16.35	16.35		32.70	43
0700	Coupling				Ea.	12.75			12.75	14.05
0900	Center feed	1 Elec	5.30	1.509		47	92.50		139.50	195
1100	End feed		5.30	1.509		52	92.50		144.50	200
1300	Hanger set		24	.333	↓	2.89	20.50		23.39	34.50
3000	600 volt 100 amp, plain, 10′ lengths		35	.229	L.F.	59	14		73	86.50
3300	Door track	↓	35	.229	"	93	14		107	125
3700	Coupling				Ea.	53			53	58.50
4000	End cap	1 Elec	40	.200		67	12.25		79.25	92.50
4200	End feed		4	2		239	123		362	450
4500	Trolley, 600 volt, 20 amp		5.30	1.509		263	92.50		355.50	430
4700	30 amp	↓	5.30	1.509	↓	263	92.50		355.50	430

26 25 Low-Voltage Enclosed Bus Assemblies

26 25 13 – Low-Voltage Busways

26 25 13.70 Feedrail		Crew	Daily Output	Labor-Hours	Unit	2020 Bare Costs Material	Labor	Equipment	Total	Total Incl O&P
4900	Duplex, 40 amp	1 Elec	4	2	Ea.	920	123		1,043	1,225
5000	60 amp		4	2		895	123		1,018	1,175
5300	Fusible, 20 amp		4	2		540	123		663	785
5500	30 amp		4	2		540	123		663	785
5900	300 volt, 20 amp		5.30	1.509		249	92.50		341.50	415
6000	30 amp		5.30	1.509		325	92.50		417.50	505
6300	Fusible, 20 amp		4.70	1.702		350	104		454	545
6500	30 amp		4.70	1.702		490	104		594	700
7300	Busway, 250 volt 50 amp, 2 wire		70	.114	L.F.	20	7		27	33.50
7330	Coupling				Ea.	44			44	48.50
7340	Center feed	1 Elec	6	1.333		510	82		592	685
7350	End feed		6	1.333		124	82		206	262
7360	End cap		40	.200		29	12.25		41.25	51
7370	Hanger set		24	.333		3.12	20.50		23.62	35
7400	125/250 volt 50 amp, 3 wire		60	.133	L.F.	20.50	8.20		28.70	35
7430	Coupling		6	1.333	Ea.	61.50	82		143.50	194
7440	Center feed		6	1.333		505	82		587	680
7450	End feed		6	1.333		111	82		193	248
7460	End cap		40	.200		29.50	12.25		41.75	51.50
7470	Hanger set		24	.333		4.22	20.50		24.72	36
7480	Trolley, 250 volt, 2 pole, 20 amp		6	1.333		38.50	82		120.50	169
7490	30 amp		6	1.333		38.50	82		120.50	169
7500	125/250 volt, 3 pole, 20 amp		6	1.333		37.50	82		119.50	167
7510	30 amp		6	1.333		37.50	82		119.50	167
8000	Cleaning tools, 300 volt, dust remover					107			107	118
8100	Bus bar cleaner					201			201	221
8300	600 volt, dust remover, 60 amp					310			310	340
8400	100 amp					655			655	720
8600	Bus bar cleaner, 60 amp					805			805	885
8700	100 amp					805			805	885

26 27 Low-Voltage Distribution Equipment

26 27 13 – Electricity Metering

26 27 13.10 Meter Centers and Sockets		Crew	Daily Output	Labor-Hours	Unit	Material	Labor	Equipment	Total	Total Incl O&P
0010	**METER CENTERS AND SOCKETS**									
0100	Sockets, single position, 4 terminal, 100 amp	1 Elec	3.20	2.500	Ea.	52.50	153		205.50	294
0200	150 amp		2.30	3.478		62.50	213		275.50	400
0300	200 amp		1.90	4.211		110	258		368	520
0400	Transformer rated, 20 amp		3.20	2.500		179	153		332	435
0500	Double position, 4 terminal, 100 amp		2.80	2.857		240	175		415	535
0600	150 amp		2.10	3.810		290	234		524	680
0700	200 amp		1.70	4.706		550	289		839	1,050
0800	Trans-socket, 13 terminal, 3 CT mounts, 400 amp		1	8		1,225	490		1,715	2,100
0900	800 amp	2 Elec	1.20	13.333		1,525	820		2,345	2,925
1100	Meter centers and sockets, three phase, single pos, 7 terminal, 100 amp	1 Elec	2.80	2.857		136	175		311	420
1200	200 amp		2.10	3.810		239	234		473	625
1400	400 amp		1.70	4.706		825	289		1,114	1,350
2000	Meter center, main fusible switch, 1P 3W 120/240 V									
2030	400 amp	2 Elec	1.60	10	Ea.	855	615		1,470	1,875
2040	600 amp		1.10	14.545		1,300	890		2,190	2,825
2050	800 amp		.90	17.778		5,000	1,100		6,100	7,175

26 27 Low-Voltage Distribution Equipment

26 27 13 – Electricity Metering

26 27 13.10 Meter Centers and Sockets		Crew	Daily Output	Labor-Hours	Unit	2020 Bare Costs Material	Labor	Equipment	Total	Total Incl O&P
2060	Rainproof 1P 3W 120/240 V, 400 A	2 Elec	1.60	10	Ea.	1,950	615		2,565	3,100
2070	600 amp		1.10	14.545		3,400	890		4,290	5,100
2080	800 amp		.90	17.778		5,325	1,100		6,425	7,525
2100	3P 4W 120/208 V, 400 amp		1.60	10		865	615		1,480	1,900
2110	600 amp		1.10	14.545		1,475	890		2,365	2,975
2120	800 amp		.90	17.778		2,300	1,100		3,400	4,200
2130	Rainproof 3P 4W 120/208 V, 400 amp		1.60	10		2,225	615		2,840	3,400
2140	600 amp		1.10	14.545		3,575	890		4,465	5,325
2150	800 amp		.90	17.778		7,750	1,100		8,850	10,200
2170	Main circuit breaker, 1P 3W 120/240 V									
2180	400 amp	2 Elec	1.60	10	Ea.	1,475	615		2,090	2,575
2190	600 amp		1.10	14.545		1,800	890		2,690	3,350
2200	800 amp		.90	17.778		2,775	1,100		3,875	4,750
2210	1,000 amp		.80	20		3,200	1,225		4,425	5,425
2220	1,200 amp		.76	21.053		4,300	1,300		5,600	6,725
2230	1,600 amp		.68	23.529		18,800	1,450		20,250	22,900
2240	Rainproof 1P 3W 120/240 V, 400 amp		1.60	10		3,000	615		3,615	4,250
2250	600 amp		1.10	14.545		4,700	890		5,590	6,550
2260	800 amp		.90	17.778		5,475	1,100		6,575	7,700
2270	1,000 amp		.80	20		7,550	1,225		8,775	10,200
2280	1,200 amp		.76	21.053		10,200	1,300		11,500	13,200
2300	3P 4W 120/208 V, 400 amp		1.60	10		3,350	615		3,965	4,650
2310	600 amp		1.10	14.545		5,600	890		6,490	7,525
2320	800 amp		.90	17.778		6,650	1,100		7,750	9,000
2330	1,000 amp		.80	20		8,725	1,225		9,950	11,500
2340	1,200 amp		.76	21.053		11,200	1,300		12,500	14,300
2350	1,600 amp		.68	23.529		22,900	1,450		24,350	27,400
2360	Rainproof 3P 4W 120/208 V, 400 amp		1.60	10		3,975	615		4,590	5,325
2370	600 amp		1.10	14.545		5,600	890		6,490	7,525
2380	800 amp		.90	17.778		6,650	1,100		7,750	9,000
2390	1,000 amp		.76	21.053		8,725	1,300		10,025	11,600
2400	1,200 amp		.68	23.529		11,200	1,450		12,650	14,500
2420	Main lugs terminal box, 1P 3W 120/240 V									
2430	800 amp	2 Elec	.94	17.021	Ea.	565	1,050		1,615	2,225
2440	1,200 amp		.72	22.222		595	1,375		1,970	2,750
2450	Rainproof 1P 3W 120/240 V, 225 amp		2.40	6.667		445	410		855	1,125
2460	800 amp		.94	17.021		655	1,050		1,705	2,325
2470	1,200 amp		.72	22.222		1,325	1,375		2,700	3,550
2500	3P 4W 120/208 V, 800 amp		.94	17.021		725	1,050		1,775	2,400
2510	1,200 amp		.72	22.222		1,475	1,375		2,850	3,725
2520	Rainproof 3P 4W 120/208 V, 225 amp		2.40	6.667		445	410		855	1,125
2530	800 amp		.94	17.021		725	1,050		1,775	2,400
2540	1,200 amp		.72	22.222		1,475	1,375		2,850	3,725
2590	Basic meter device									
2600	1P 3W 120/240 V 4 jaw 125A sockets, 3 meter	2 Elec	1	16	Ea.	355	980		1,335	1,900
2610	4 meter		.90	17.778		535	1,100		1,635	2,275
2620	5 meter		.80	20		630	1,225		1,855	2,600
2630	6 meter		.60	26.667		570	1,625		2,195	3,150
2640	7 meter		.56	28.571		1,650	1,750		3,400	4,525
2650	8 meter		.52	30.769		1,800	1,900		3,700	4,875
2660	10 meter		.48	33.333		2,250	2,050		4,300	5,625
2680	Rainproof 1P 3W 120/240 V 4 jaw 125A sockets									
2690	3 meter	2 Elec	1	16	Ea.	750	980		1,730	2,325

	26 27 13.10 Meter Centers and Sockets	Crew	Daily Output	Labor-Hours	Unit	Material	2020 Bare Costs Labor	Equipment	Total	Total Incl O&P
2700	4 meter	2 Elec	.90	17.778	Ea.	900	1,100		2,000	2,675
2710	6 meter		.60	26.667		1,300	1,625		2,925	3,950
2720	7 meter		.56	28.571		1,650	1,750		3,400	4,525
2730	8 meter		.52	30.769		1,800	1,900		3,700	4,875
2750	1P 3W 120/240 V 4 jaw sockets									
2760	with 125A circuit breaker, 3 meter	2 Elec	1	16	Ea.	1,400	980		2,380	3,050
2770	4 meter		.90	17.778		1,775	1,100		2,875	3,625
2780	5 meter		.80	20		2,225	1,225		3,450	4,325
2790	6 meter		.60	26.667		2,600	1,625		4,225	5,375
2800	7 meter		.56	28.571		3,175	1,750		4,925	6,200
2810	8 meter		.52	30.769		3,550	1,900		5,450	6,800
2820	10 meter		.48	33.333		4,425	2,050		6,475	8,025
2830	Rainproof 1P 3W 120/240 V 4 jaw sockets									
2840	with 125A circuit breaker, 3 meter	2 Elec	1	16	Ea.	1,400	980		2,380	3,050
2850	4 meter		.90	17.778		1,775	1,100		2,875	3,625
2870	6 meter		.60	26.667		2,600	1,625		4,225	5,375
2880	7 meter		.56	28.571		3,175	1,750		4,925	6,200
2890	8 meter		.52	30.769		3,550	1,900		5,450	6,800
2920	1P 3W on 3P 4W 120/208 V system 5 jaw									
2930	125A sockets, 3 meter	2 Elec	1	16	Ea.	750	980		1,730	2,325
2940	4 meter		.90	17.778		900	1,100		2,000	2,675
2950	5 meter		.80	20		1,125	1,225		2,350	3,125
2960	6 meter		.60	26.667		1,300	1,625		2,925	3,950
2970	7 meter		.56	28.571		1,650	1,750		3,400	4,525
2980	8 meter		.52	30.769		1,800	1,900		3,700	4,875
2990	10 meter		.48	33.333		2,250	2,050		4,300	5,625
3000	Rainproof 1P 3W on 3P 4W 120/208 V system									
3020	5 jaw 125A sockets, 3 meter	2 Elec	1	16	Ea.	750	980		1,730	2,325
3030	4 meter		.90	17.778		900	1,100		2,000	2,675
3050	6 meter		.60	26.667		1,300	1,625		2,925	3,950
3060	7 meter		.56	28.571		1,650	1,750		3,400	4,525
3070	8 meter		.52	30.769		1,800	1,900		3,700	4,875
3090	1P 3W on 3P 4W 120/208 V system 5 jaw sockets									
3100	With 125A circuit breaker, 3 meter	2 Elec	1	16	Ea.	1,400	980		2,380	3,050
3110	4 meter		.90	17.778		1,775	1,100		2,875	3,625
3120	5 meter		.80	20		2,225	1,225		3,450	4,325
3130	6 meter		.60	26.667		2,600	1,625		4,225	5,375
3140	7 meter		.56	28.571		3,175	1,750		4,925	6,200
3150	8 meter		.52	30.769		3,550	1,900		5,450	6,800
3160	10 meter		.48	33.333		4,425	2,050		6,475	8,025
3170	Rainproof 1P 3W on 3P 4W 120/208 V system									
3180	5 jaw sockets w/125A circuit breaker, 3 meter	2 Elec	1	16	Ea.	1,400	980		2,380	3,050
3190	4 meter		.90	17.778		1,775	1,100		2,875	3,625
3210	6 meter		.60	26.667		2,600	1,625		4,225	5,375
3220	7 meter		.56	28.571		3,175	1,750		4,925	6,200
3230	8 meter		.52	30.769		3,550	1,900		5,450	6,800
3250	1P 3W 120/240 V 4 jaw sockets									
3260	with 200A circuit breaker, 3 meter	2 Elec	1	16	Ea.	2,100	980		3,080	3,825
3270	4 meter		.90	17.778		2,850	1,100		3,950	4,800
3290	6 meter		.60	26.667		4,200	1,625		5,825	7,150
3300	7 meter		.56	28.571		4,950	1,750		6,700	8,150
3310	8 meter		.56	28.571		5,700	1,750		7,450	8,950
3330	Rainproof 1P 3W 120/240 V 4 jaw sockets									

26 27 13.10	Meter Centers and Sockets	Crew	Daily Output	Labor-Hours	Unit	Material	2020 Bare Costs Labor	Equipment	Total	Total Incl O&P
3350	with 200A circuit breaker, 3 meter	2 Elec	1	16	Ea.	2,100	980		3,080	3,825
3360	4 meter		.90	17.778		2,850	1,100		3,950	4,800
3380	6 meter		.60	26.667		4,200	1,625		5,825	7,150
3390	7 meter		.56	28.571		4,950	1,750		6,700	8,150
3400	8 meter		.52	30.769		5,700	1,900		7,600	9,150
3420	1P 3W on 3P 4W 120/208 V 5 jaw sockets									
3430	with 200A circuit breaker, 3 meter	2 Elec	1	16	Ea.	2,100	980		3,080	3,825
3440	4 meter		.90	17.778		2,850	1,100		3,950	4,800
3460	6 meter		.60	26.667		4,225	1,625		5,850	7,175
3470	7 meter		.56	28.571		4,950	1,750		6,700	8,150
3480	8 meter		.52	30.769		5,700	1,900		7,600	9,150
3500	Rainproof 1P 3W on 3P 4W 120/208 V 5 jaw socket									
3510	with 200A circuit breaker, 3 meter	2 Elec	1	16	Ea.	2,100	980		3,080	3,825
3520	4 meter		.90	17.778		2,850	1,100		3,950	4,800
3540	6 meter		.60	26.667		4,200	1,625		5,825	7,150
3550	7 meter		.56	28.571		4,950	1,750		6,700	8,150
3560	8 meter		.52	30.769		5,700	1,900		7,600	9,150
3600	Automatic circuit closing, add					81.50			81.50	89.50
3610	Manual circuit closing, add					93			93	102
3650	Branch meter device									
3660	3P 4W 208/120 or 240/120 V 7 jaw sockets									
3670	with 200A circuit breaker, 2 meter	2 Elec	.90	17.778	Ea.	3,525	1,100		4,625	5,550
3680	3 meter		.80	20		5,275	1,225		6,500	7,700
3690	4 meter		.70	22.857		7,025	1,400		8,425	9,875
3700	Main circuit breaker 42,000 rms, 400 amp		1.60	10		2,350	615		2,965	3,550
3710	600 amp		1.10	14.545		2,675	890		3,565	4,300
3720	800 amp		.90	17.778		3,700	1,100		4,800	5,750
3730	Rainproof main circ. breaker 42,000 rms, 400 amp		1.60	10		3,075	615		3,690	4,325
3740	600 amp		1.10	14.545		4,950	890		5,840	6,825
3750	800 amp		.90	17.778		6,650	1,100		7,750	9,000
3760	Main circuit breaker 65,000 rms, 400 amp		1.60	10		4,175	615		4,790	5,550
3770	600 amp		1.10	14.545		5,875	890		6,765	7,825
3780	800 amp		.90	17.778		6,650	1,100		7,750	9,000
3790	1,000 amp		.80	20		8,725	1,225		9,950	11,500
3800	1,200 amp		.76	21.053		11,200	1,300		12,500	14,300
3810	1,600 amp		.68	23.529		22,900	1,450		24,350	27,400
3820	Rainproof main circ. breaker 65,000 rms, 400 amp		1.60	10		4,175	615		4,790	5,550
3830	600 amp		1.10	14.545		5,875	890		6,765	7,825
3840	800 amp		.90	17.778		6,650	1,100		7,750	8,975
3850	1,000 amp		.80	20		8,725	1,225		9,950	11,500
3860	1,200 amp		.76	21.053		11,200	1,300		12,500	14,300
3880	Main circuit breaker 100,000 rms, 400 amp		1.60	10		4,175	615		4,790	5,550
3890	600 amp		1.10	14.545		5,875	890		6,765	7,825
3900	800 amp		.90	17.778		6,900	1,100		8,000	9,250
3910	Rainproof main circ. breaker 100,000 rms, 400 amp		1.60	10		4,175	615		4,790	5,550
3920	600 amp		1.10	14.545		5,875	890		6,765	7,825
3930	800 amp		.90	17.778		6,900	1,100		8,000	9,250
3940	Main lugs terminal box, 800 amp		.94	17.021		725	1,050		1,775	2,400
3950	1,600 amp		.72	22.222		2,825	1,375		4,200	5,200
3960	Rainproof, 800 amp		.94	17.021		725	1,050		1,775	2,400
3970	1,600 amp		.72	22.222		2,825	1,375		4,200	5,225
9000	Minimum labor/equipment charge	1 Elec	3	2.667	Job		164		164	252

26 27 16.10 Cabinets		Crew	Daily Output	Labor-Hours	Unit	Material	Labor	Equipment	Total	Total Incl O&P
						2020 Bare Costs				
0010	**CABINETS**									
7000	Cabinets, current transformer									
7050	Single door, 24" H x 24" W x 10" D	1 Elec	1.60	5	Ea.	172	305		477	660
7100	30" H x 24" W x 10" D		1.30	6.154		165	380		545	760
7150	36" H x 24" W x 10" D		1.10	7.273		410	445		855	1,125
7200	30" H x 30" W x 10" D		1	8		226	490		716	1,000
7250	36" H x 30" W x 10" D		.90	8.889		299	545		844	1,175
7300	36" H x 36" W x 10" D		.80	10		310	615		925	1,275
7500	Double door, 48" H x 36" W x 10" D		.60	13.333		575	820		1,395	1,875
7550	24" H x 24" W x 12" D		1	8		190	490		680	965
8000	NEMA 12, double door, floor mounted									
8020	54" H x 42" W x 8" D	2 Elec	6	2.667	Ea.	1,000	164		1,164	1,350
8040	60" H x 48" W x 8" D		5.40	2.963		1,375	182		1,557	1,775
8060	60" H x 48" W x 10" D		5.40	2.963		1,375	182		1,557	1,775
8080	60" H x 60" W x 10" D		5	3.200		1,575	196		1,771	2,025
8100	72" H x 60" W x 10" D		4	4		1,800	245		2,045	2,350
8120	72" H x 72" W x 10" D		3.40	4.706		1,625	289		1,914	2,225
8140	60" H x 48" W x 12" D		3.40	4.706		1,400	289		1,689	2,000
8160	60" H x 60" W x 12" D		3.20	5		1,575	305		1,880	2,200
8180	72" H x 60" W x 12" D		3	5.333		2,750	325		3,075	3,525
8200	72" H x 72" W x 12" D		3	5.333		3,050	325		3,375	3,850
8220	60" H x 48" W x 16" D		3.20	5		1,500	305		1,805	2,125
8240	72" H x 72" W x 16" D		2.60	6.154		2,125	380		2,505	2,900
8260	60" H x 48" W x 20" D		3	5.333		1,625	325		1,950	2,300
8280	72" H x 72" W x 20" D		2.20	7.273		2,300	445		2,745	3,225
8300	60" H x 48" W x 24" D		2.60	6.154		1,700	380		2,080	2,450
8320	72" H x 72" W x 24" D		2	8		2,425	490		2,915	3,425
8340	Pushbutton enclosure, oiltight									
8360	3-1/2" H x 3-1/4" W x 2-3/4" D, for 1 P.B.	1 Elec	12	.667	Ea.	43	41		84	111
8380	5-3/4" H x 3-1/4" W x 2-3/4" D, for 2 P.B.		11	.727		45.50	44.50		90	119
8400	8" H x 3-1/4" W x 2-3/4" D, for 3 P.B.		10.50	.762		53.50	46.50		100	131
8420	10-1/4" H x 3-1/4" W x 2-3/4" D, for 4 P.B.		10.50	.762		57.50	46.50		104	135
8460	12-1/2" H x 3-1/4" W x 3" D, for 5 P.B.		9	.889		106	54.50		160.50	201
8480	9-1/2" H x 6-1/4" W x 3" D, for 6 P.B.		8.50	.941		73	57.50		130.50	170
8500	9-1/2" H x 8-1/2" W x 3" D, for 9 P.B.		8	1		90	61.50		151.50	194
8510	11-3/4" H x 8-1/2" W x 3" D, for 12 P.B.		7	1.143		94.50	70		164.50	212
8520	11-3/4" H x 10-3/4" W x 3" D, for 16 P.B.		6.50	1.231		99.50	75.50		175	226
8540	14" H x 10-3/4" W x 3" D, for 20 P.B.		5	1.600		162	98		260	330
8560	14" H x 13" W x 3" D, for 25 P.B.		4.50	1.778		177	109		286	360
8580	Sloping front pushbutton enclosures									
8600	3-1/2" H x 7-3/4" W x 4-7/8" D, for 3 P.B.	1 Elec	10	.800	Ea.	71.50	49		120.50	155
8620	7-1/4" H x 8-1/2" W x 6-3/4" D, for 6 P.B.		8	1		105	61.50		166.50	211
8640	9-1/2" H x 8-1/2" W x 7-7/8" D, for 9 P.B.		7	1.143		140	70		210	262
8660	11-1/4" H x 8-1/2" W x 9" D, for 12 P.B.		5	1.600		140	98		238	305
8680	11-3/4" H x 10" W x 9" D, for 16 P.B.		5	1.600		159	98		257	325
8700	11-3/4" H x 13" W x 9" D, for 20 P.B.		5	1.600		182	98		280	350
8720	14" H x 13" W x 10-1/8" D, for 25 P.B.		4.50	1.778		203	109		312	390
8740	Pedestals, not including P.B. enclosure or base									
8760	Straight column 4" x 4"	1 Elec	4.50	1.778	Ea.	191	109		300	380
8780	6" x 6"		4	2		297	123		420	515
8800	Angled column 4" x 4"		4.50	1.778		220	109		329	410
8820	6" x 6"		4	2		330	123		453	550

26 27 16 – Electrical Cabinets and Enclosures

26 27 16.10 Cabinets		Crew	Daily Output	Labor-Hours	Unit	2020 Bare Costs Material	Labor	Equipment	Total	Total Incl O&P
8840	Pedestal, base 18" x 18"	1 Elec	10	.800	Ea.	113	49		162	200
8860	24" x 24"	↓	9	.889	↓	259	54.50		313.50	370
8900	Electronic rack enclosures									
8920	72" H x 19" W x 24" D	1 Elec	1.50	5.333	Ea.	2,150	325		2,475	2,850
8940	72" H x 23" W x 24" D		1.50	5.333		2,325	325		2,650	3,050
8960	72" H x 19" W x 30" D		1.30	6.154		2,350	380		2,730	3,175
8980	72" H x 19" W x 36" D		1.20	6.667		2,750	410		3,160	3,650
9000	72" H x 23" W x 36" D	↓	1.20	6.667	↓	3,025	410		3,435	3,975
9020	NEMA 12 & 4 enclosure panels									
9040	12" x 24"	1 Elec	20	.400	Ea.	44	24.50		68.50	86.50
9060	16" x 12"		20	.400		31.50	24.50		56	72.50
9080	20" x 16"		20	.400		45.50	24.50		70	88
9100	20" x 20"		19	.421		55	26		81	101
9120	24" x 20"		18	.444		71.50	27.50		99	121
9140	24" x 24"		17	.471		82	29		111	135
9160	30" x 20"		16	.500		85	30.50		115.50	141
9180	30" x 24"		16	.500		100	30.50		130.50	157
9200	36" x 24"		15	.533		120	32.50		152.50	183
9220	36" x 30"		15	.533		158	32.50		190.50	225
9240	42" x 24"		15	.533		142	32.50		174.50	208
9260	42" x 30"		14	.571		180	35		215	252
9280	42" x 36"		14	.571		213	35		248	288
9300	48" x 24"		14	.571		105	35		140	169
9320	48" x 30"		14	.571		137	35		172	204
9340	48" x 36"		13	.615		239	38		277	320
9360	60" x 36"	↓	12	.667	↓	297	41		338	390
9400	Wiring trough steel JIC, clamp cover									
9490	4" x 4", 12" long	1 Elec	12	.667	Ea.	68.50	41		109.50	138
9510	24" long		10	.800		87.50	49		136.50	172
9530	36" long		8	1		113	61.50		174.50	219
9540	48" long		7	1.143		125	70		195	246
9550	60" long		6	1.333		153	82		235	294
9560	6" x 6", 12" long		11	.727		95.50	44.50		140	174
9580	24" long		9	.889		124	54.50		178.50	220
9600	36" long		7	1.143		147	70		217	270
9610	48" long		6	1.333		187	82		269	330
9620	60" long		5	1.600	↓	220	98		318	395
9990	Minimum labor/equipment charge	↓	2	4	Job		245		245	380

26 27 16.20 Cabinets and Enclosures		Crew	Daily Output	Labor-Hours	Unit	Material	Labor	Equipment	Total	Total Incl O&P
0010	**CABINETS AND ENCLOSURES** Nonmetallic									
0080	Enclosures fiberglass NEMA 4X									
0100	Wall mount, quick release latch door, 20" H x 16" W x 6" D	1 Elec	4.80	1.667	Ea.	730	102		832	955
0110	20" H x 20" W x 6" D		4.50	1.778		845	109		954	1,100
0120	24" H x 20" W x 6" D		4.20	1.905		900	117		1,017	1,175
0130	20" H x 16" W x 8" D		4.50	1.778		795	109		904	1,050
0140	20" H x 20" W x 8" D		4.20	1.905		885	117		1,002	1,150
0150	24" H x 24" W x 8" D		3.80	2.105		1,000	129		1,129	1,300
0160	30" H x 24" W x 8" D		3.20	2.500		1,150	153		1,303	1,475
0170	36" H x 30" W x 8" D		3	2.667		1,600	164		1,764	2,000
0180	20" H x 16" W x 10" D		3.50	2.286		955	140		1,095	1,275
0190	20" H x 20" W x 10" D		3.20	2.500		1,025	153		1,178	1,350
0200	24" H x 20" W x 10" D	↓	3	2.667	↓	1,100	164		1,264	1,450

26 27 16 – Electrical Cabinets and Enclosures

26 27 16.20 Cabinets and Enclosures		Crew	Daily Output	Labor-Hours	Unit	2020 Bare Costs Material	Labor	Equipment	Total	Total Incl O&P
0210	30″ H x 24″ W x 10″ D	1 Elec	2.80	2.857	Ea.	1,275	175		1,450	1,675
0220	20″ H x 16″ W x 12″ D		3	2.667		1,050	164		1,214	1,400
0230	20″ H x 20″ W x 12″ D		2.80	2.857		1,100	175		1,275	1,475
0240	24″ H x 24″ W x 12″ D		2.60	3.077		1,175	189		1,364	1,600
0250	30″ H x 24″ W x 12″ D		2.40	3.333		1,350	205		1,555	1,800
0260	36″ H x 30″ W x 12″ D		2.20	3.636		1,825	223		2,048	2,375
0270	36″ H x 36″ W x 12″ D		2.10	3.810		2,025	234		2,259	2,575
0280	48″ H x 36″ W x 12″ D		2	4		2,375	245		2,620	2,975
0290	60″ H x 36″ W x 12″ D		1.80	4.444		1,725	273		1,998	2,325
0300	30″ H x 24″ W x 16″ D		1.40	5.714		1,575	350		1,925	2,275
0310	48″ H x 36″ W x 16″ D		1.20	6.667		2,625	410		3,035	3,500
0320	60″ H x 36″ W x 16″ D		1	8		2,950	490		3,440	4,000
0480	Freestanding, one door, 72″ H x 25″ W x 25″ D		.80	10		3,350	615		3,965	4,650
0490	Two doors with two panels, 72″ H x 49″ W x 24″ D		.50	16		7,775	980		8,755	10,100
0500	Floor stand kits, for NEMA 4 & 12, 20″ W or more, 6″ H x 8″ D		24	.333		116	20.50		136.50	159
0510	6″ H x 10″ D		24	.333		127	20.50		147.50	171
0520	6″ H x 12″ D		24	.333		144	20.50		164.50	190
0530	6″ H x 18″ D		24	.333		174	20.50		194.50	224
0540	12″ H x 8″ D		22	.364		148	22.50		170.50	197
0550	12″ H x 10″ D		22	.364		268	22.50		290.50	330
0560	12″ H x 12″ D		22	.364		277	22.50		299.50	340
0570	12″ H x 16″ D		22	.364		193	22.50		215.50	247
0580	12″ H x 18″ D		22	.364		209	22.50		231.50	265
0590	12″ H x 20″ D		22	.364		251	22.50		273.50	310
0600	18″ H x 8″ D		20	.400		173	24.50		197.50	228
0610	18″ H x 10″ D		20	.400		225	24.50		249.50	285
0620	18″ H x 12″ D		20	.400		239	24.50		263.50	300
0630	18″ H x 16″ D		20	.400		265	24.50		289.50	330
0640	24″ H x 8″ D		16	.500		215	30.50		245.50	283
0650	24″ H x 10″ D		16	.500		265	30.50		295.50	340
0660	24″ H x 12″ D		16	.500		282	30.50		312.50	355
0670	24″ H x 16″ D		16	.500		305	30.50		335.50	380
0680	Small, screw cover, 5-1/2″ H x 4″ W x 4-15/16″ D		12	.667		117	41		158	192
0690	7-1/2″ H x 4″ W x 4-15/16″ D		12	.667		76.50	41		117.50	147
0700	7-1/2″ H x 6″ W x 5-3/16″ D		10	.800		141	49		190	231
0710	9-1/2″ H x 6″ W x 5-11/16″ D		10	.800		148	49		197	238
0720	11-1/2″ H x 8″ W x 6-11/16″ D		8	1		217	61.50		278.50	335
0730	13-1/2″ H x 10″ W x 7-3/16″ D		7	1.143		264	70		334	400
0740	15-1/2″ H x 12″ W x 8-3/16″ D		6	1.333		340	82		422	495
0750	17-1/2″ H x 14″ W x 8-11/16″ D		5	1.600		400	98		498	590
0760	Screw cover with window, 6″ H x 4″ W x 5″ D		12	.667		149	41		190	227
0770	8″ H x 4″ W x 5″ D		11	.727		196	44.50		240.50	285
0780	8″ H x 6″ W x 5″ D		11	.727		287	44.50		331.50	385
0790	10″ H x 6″ W x 6″ D		10	.800		298	49		347	405
0800	12″ H x 8″ W x 7″ D		8	1		385	61.50		446.50	515
0810	14″ H x 10″ W x 7″ D		7	1.143		495	70		565	655
0820	16″ H x 12″ W x 8″ D		6	1.333		585	82		667	765
0830	18″ H x 14″ W x 9″ D		5	1.600		655	98		753	870
0840	Quick-release latch cover, 5-1/2″ H x 4″ W x 5″ D		12	.667		153	41		194	231
0850	7-1/2″ H x 4″ W x 5″ D		12	.667		156	41		197	235
0860	7-1/2″ H x 6″ W x 5-1/4″ D		10	.800		182	49		231	276
0870	9-1/2″ H x 6″ W x 5-3/4″ D		10	.800		112	49		161	199
0880	11-1/2″ H x 8″ W x 6-3/4″ D		8	1		177	61.50		238.50	289

26 27 16 – Electrical Cabinets and Enclosures

26 27 16.20 Cabinets and Enclosures		Crew	Daily Output	Labor-Hours	Unit	Material	Labor	Equipment	Total	Total Incl O&P
						2020 Bare Costs				
0890	13-1/2" H x 10" W x 7-1/4" D	1 Elec	7	1.143	Ea.	201	70		271	330
0900	15-1/2" H x 12" W x 8-1/4" D		6	1.333		450	82		532	620
0910	17-1/2" H x 14" W x 8-3/4" D		5	1.600		555	98		653	765
0920	Pushbutton, 1 hole 5-1/2" H x 4" W x 4-15/16" D		12	.667		435	41		476	545
0930	2 hole 7-1/2" H x 4" W x 4-15/16" D		11	.727		605	44.50		649.50	735
0940	4 hole 7-1/2" H x 6" W x 5-3/16" D		10.50	.762		425	46.50		471.50	540
0950	6 hole 9-1/2" H x 6" W x 5-11/16" D		9	.889		515	54.50		569.50	655
0960	8 hole 11-1/2" H x 8" W x 6-11/16" D		8.50	.941		660	57.50		717.50	815
0970	12 hole 13-1/2" H x 10" W x 7-3/16" D		8	1		865	61.50		926.50	1,050
0980	20 hole 15-1/2" H x 12" W x 8-3/16" D		5	1.600		815	98		913	1,050
0990	30 hole 17-1/2" H x 14" W x 8-11/16" D	↓	4.50	1.778	↓	965	109		1,074	1,225
1450	Enclosures polyester NEMA 4X									
1460	Small, screw cover,									
1500	3-15/16" H x 3-15/16" W x 3-1/16" D	1 Elec	12	.667	Ea.	64.50	41		105.50	134
1510	5-3/16" H x 3-5/16" W x 3-1/16" D		12	.667		63.50	41		104.50	133
1520	5-7/8" H x 3-7/8" W x 4-3/16" D		12	.667		67	41		108	137
1530	5-7/8" H x 5-7/8" W x 4-3/16" D		12	.667		75	41		116	146
1540	7-5/8" H x 3-5/16" W x 3-1/16" D		12	.667		69.50	41		110.50	140
1550	10-3/16" H x 3-5/16" W x 3-1/16" D		10	.800		86.50	49		135.50	171
1560	Clear cover, 3-15/16" H x 3-15/16" W x 2-7/8" D		12	.667		73	41		114	143
1570	5-3/16" H x 3-5/16" W x 2-7/8" D		12	.667		77.50	41		118.50	149
1580	5-7/8" H x 3-7/8" W x 4" D		12	.667		96.50	41		137.50	169
1590	5-7/8" H x 5-7/8" W x 4" D		12	.667		119	41		160	194
1600	7-5/8" H x 3-5/16" W x 2-7/8" D		12	.667		87	41		128	159
1610	10-3/16" H x 3-5/16" W x 2-7/8" D		10	.800		109	49		158	196
1620	Pushbutton, 1 hole, 5-5/16" H x 3-5/16" W x 3-1/16" D		12	.667		59.50	41		100.50	128
1630	2 hole, 7-5/8" H x 3-5/16" W x 3-1/8" D		11	.727		63.50	44.50		108	139
1640	3 hole, 10-3/16" H x 3-5/16" W x 3-1/16" D		10.50	.762		75	46.50		121.50	155
8000	Wireway fiberglass, straight sect. screwcover, 12" L, 4" W x 4" D		40	.200		220	12.25		232.25	261
8010	6" W x 6" D		30	.267		305	16.35		321.35	360
8020	24" L, 4" W x 4" D		20	.400		279	24.50		303.50	345
8030	6" W x 6" D		15	.533		430	32.50		462.50	525
8040	36" L, 4" W x 4" D		13.30	.602		340	37		377	430
8050	6" W x 6" D		10	.800		540	49		589	670
8060	48" L, 4" W x 4" D		10	.800		415	49		464	530
8070	6" W x 6" D		7.50	1.067		675	65.50		740.50	840
8080	60" L, 4" W x 4" D		8	1		505	61.50		566.50	650
8090	6" W x 6" D		6	1.333		790	82		872	990
8100	Elbow, 90°, 4" W x 4" D		20	.400		220	24.50		244.50	280
8110	6" W x 6" D		18	.444		440	27.50		467.50	525
8120	Elbow, 45°, 4" W x 4" D		20	.400		218	24.50		242.50	278
8130	6" W x 6" D		18	.444		445	27.50		472.50	530
8140	Tee, 4" W x 4" D		16	.500		280	30.50		310.50	350
8150	6" W x 6" D		14	.571		520	35		555	630
8160	Cross, 4" W x 4" D		14	.571		305	35		340	390
8170	6" W x 6" D		12	.667		745	41		786	885
8180	Cut-off fitting, w/flange & adhesive, 4" W x 4" D		18	.444		173	27.50		200.50	233
8190	6" W x 6" D		16	.500		355	30.50		385.50	435
8200	Flexible ftng., hvy. neoprene coated nylon, 4" W x 4" D		20	.400		325	24.50		349.50	395
8210	6" W x 6" D		18	.444		470	27.50		497.50	555
8220	Closure plate, fiberglass, 4" W x 4" D		20	.400		61.50	24.50		86	106
8230	6" W x 6" D		18	.444		68.50	27.50		96	118
8240	Box connector, stainless steel type 304, 4" W x 4" D	↓	20	.400	↓	106	24.50		130.50	155

26 27 Low-Voltage Distribution Equipment

26 27 16 – Electrical Cabinets and Enclosures

26 27 16.20 Cabinets and Enclosures		Crew	Daily Output	Labor-Hours	Unit	Material	2020 Bare Costs Labor	Equipment	Total	Total Incl O&P
8250	6" W x 6" D	1 Elec	18	.444	Ea.	121	27.50		148.50	175
8260	Hanger, 4" W x 4" D		100	.080		26.50	4.91		31.41	36.50
8270	6" W x 6" D		80	.100		33.50	6.15		39.65	46.50
8280	Straight tube section fiberglass, 4" W x 4" D, 12" long		40	.200		198	12.25		210.25	237
8290	24" long		20	.400		240	24.50		264.50	300
8300	36" long		13.30	.602		220	37		257	299
8310	48" long		10	.800		295	49		344	400
8320	60" long		8	1		345	61.50		406.50	475
8330	120" long		4	2		545	123		668	790

26 27 19 – Multi-Outlet Assemblies

26 27 19.10 Wiring Duct		Crew	Daily Output	Labor-Hours	Unit	Material	Labor	Equipment	Total	Total Incl O&P
0010	**WIRING DUCT** Plastic									
1250	PVC, snap-in slots, adhesive backed									
1270	1-1/2" W x 2" H	2 Elec	120	.133	L.F.	5.05	8.20		13.25	18.15
1280	1-1/2" W x 3" H		120	.133		6.60	8.20		14.80	19.85
1290	1-1/2" W x 4" H		120	.133		7.05	8.20		15.25	20.50
1300	2" W x 1" H		120	.133		4.53	8.20		12.73	17.60
1310	2" W x 1-1/2" H		120	.133		4.79	8.20		12.99	17.85
1320	2" W x 2" H		120	.133		4.81	8.20		13.01	17.90
1340	2" W x 3" H		120	.133		6.60	8.20		14.80	19.85
1350	2" W x 4" H		120	.133		7.95	8.20		16.15	21.50
1360	2-1/2" W x 3" H		120	.133		7.15	8.20		15.35	20.50
1370	3" W x 1" H		110	.145		5.30	8.90		14.20	19.60
1390	3" W x 2" H		110	.145		5.55	8.90		14.45	19.85
1400	3" W x 3" H		110	.145		6.70	8.90		15.60	21
1410	3" W x 4" H		110	.145		8.40	8.90		17.30	23
1420	3" W x 5" H		110	.145		11	8.90		19.90	26
1430	4" W x 1-1/2" H		100	.160		7.75	9.80		17.55	23.50
1440	4" W x 2" H		100	.160		6.50	9.80		16.30	22.50
1450	4" W x 3" H		100	.160		7.35	9.80		17.15	23
1460	4" W x 4" H		100	.160		8.85	9.80		18.65	25
1470	4" W x 5" H		100	.160		12.50	9.80		22.30	29
1550	Cover, 1-1/2" W		200	.080		.93	4.91		5.84	8.55
1560	2" W		200	.080		1.02	4.91		5.93	8.65
1570	2-1/2" W		200	.080		1.43	4.91		6.34	9.10
1580	3" W		200	.080		1.55	4.91		6.46	9.25
1590	4" W		200	.080		1.87	4.91		6.78	9.60

26 27 23 – Indoor Service Poles

26 27 23.40 Surface Raceway		Crew	Daily Output	Labor-Hours	Unit	Material	Labor	Equipment	Total	Total Incl O&P
0010	**SURFACE RACEWAY**									
0090	Metal, straight section									
0100	No. 500	1 Elec	100	.080	L.F.	1.30	4.91		6.21	9
0110	No. 700		100	.080		1.41	4.91		6.32	9.10
0400	No. 1500, small pancake		90	.089		2.44	5.45		7.89	11.10
0600	No. 2000, base & cover, blank		90	.089		2.60	5.45		8.05	11.25
0610	Receptacle, 6" OC		40	.200		21	12.25		33.25	42.50
0620	12" OC		44	.182		13.85	11.15		25	32.50
0630	18" OC		46	.174		8.15	10.65		18.80	25.50
0650	30" OC		50	.160		4.86	9.80		14.66	20.50
0660	60" OC		50	.160		4.50	9.80		14.30	20
0670	No. 2400, base & cover, blank		80	.100		2.28	6.15		8.43	11.95
0680	Receptacle, 6" OC		42	.190		43	11.70		54.70	65.50

26 27 23.40 Surface Raceway		Crew	Daily Output	Labor-Hours	Unit	Material	2020 Bare Costs Labor	Equipment	Total	Total Incl O&P
0690	12" OC	1 Elec	53	.151	L.F.	29.50	9.25		38.75	47
0700	18" OC		55	.145		26	8.90		34.90	43
0710	24" OC		57	.140		10	8.60		18.60	24.50
0720	30" OC		59	.136		6.70	8.30		15	20
0730	60" OC		61	.131		5.40	8.05		13.45	18.35
0800	No. 3000, base & cover, blank		75	.107		4.85	6.55		11.40	15.45
0810	Receptacle, 6" OC		45	.178		44	10.90		54.90	65
0820	12" OC		62	.129		25	7.90		32.90	39
0830	18" OC		64	.125		20	7.65		27.65	34
0840	24" OC		66	.121		15	7.45		22.45	28
0850	30" OC		68	.118		13.50	7.20		20.70	26
0860	60" OC		70	.114		9.85	7		16.85	21.50
1000	No. 4000, base & cover, blank		65	.123		7.90	7.55		15.45	20.50
1010	Receptacle, 6" OC		41	.195		51	11.95		62.95	74.50
1020	12" OC		52	.154		36.50	9.45		45.95	54.50
1030	18" OC		54	.148		30	9.10		39.10	47
1040	24" OC		56	.143		24.50	8.75		33.25	40.50
1050	30" OC		58	.138		22.50	8.45		30.95	38
1060	60" OC		60	.133		17	8.20		25.20	31.50
1200	No. 6000, base & cover, blank		50	.160		13.85	9.80		23.65	30.50
1210	Receptacle, 6" OC		30	.267		62.50	16.35		78.85	93.50
1220	12" OC		37	.216		47.50	13.25		60.75	73
1230	18" OC		39	.205		40	12.60		52.60	64
1240	24" OC		41	.195		33	11.95		44.95	55
1250	30" OC		43	.186		31.50	11.40		42.90	52
1260	60" OC		45	.178		24.50	10.90		35.40	43.50
2400	Fittings, elbows, No. 500		40	.200	Ea.	2.34	12.25		14.59	21.50
2800	Elbow cover, No. 2000		40	.200		4.23	12.25		16.48	23.50
2880	Tee, No. 500		42	.190		4.19	11.70		15.89	22.50
2900	No. 2000		27	.296		13.65	18.20		31.85	43
3000	Switch box, No. 500		16	.500		14.85	30.50		45.35	63.50
3400	Telephone outlet, No. 1500		16	.500		17.55	30.50		48.05	66.50
3600	Junction box, No. 1500		16	.500		11.55	30.50		42.05	59.50
3800	Plugmold wired sections, No. 2000									
4000	1 circuit, 6 outlets, 3' long	1 Elec	8	1	Ea.	45	61.50		106.50	144
4100	2 circuits, 8 outlets, 6' long		5.30	1.509		65.50	92.50		158	215
4110	Tele-power pole, alum, w/2 recept, 10'		4	2		250	123		373	465
4120	12'		3.85	2.078		296	127		423	520
4130	15'		3.70	2.162		375	133		508	615
4140	Steel, w/2 recept, 10'		4	2		154	123		277	360
4150	One phone fitting, 10'		4	2		168	123		291	375
4160	Alum, 4 outlets, 10'		3.70	2.162		360	133		493	600
4300	Overhead distribution systems, 125 volt									
4800	No. 2000, entrance end fitting	1 Elec	20	.400	Ea.	6.80	24.50		31.30	45.50
5000	Blank end fitting		40	.200		2.28	12.25		14.53	21.50
5200	Supporting clip		40	.200		1.35	12.25		13.60	20.50
5800	No. 3000, entrance end fitting		20	.400		10.75	24.50		35.25	50
6000	Blank end fitting		40	.200		2.98	12.25		15.23	22
6020	Internal elbow		20	.400		14.10	24.50		38.60	53.50
6030	External elbow		20	.400		19.95	24.50		44.45	60
6040	Device bracket		53	.151		4.79	9.25		14.04	19.50
6400	Hanger clamp		32	.250		7.30	15.35		22.65	31.50
7000	No. 4000 base		90	.089	L.F.	5.15	5.45		10.60	14.05

26 27 23.40 Surface Raceway		Crew	Daily Output	Labor-Hours	Unit	Material	Labor	Equipment	Total	Total Incl O&P
						2020 Bare Costs				
7200	Divider	1 Elec	100	.080	L.F.	.97	4.91		5.88	8.60
7400	Entrance end fitting		16	.500	Ea.	25.50	30.50		56	75
7600	Blank end fitting		40	.200		7.05	12.25		19.30	26.50
7610	Recpt. & tele. cover		53	.151		12.30	9.25		21.55	28
7620	External elbow		16	.500		40.50	30.50		71	91.50
7630	Coupling		53	.151		6.15	9.25		15.40	21
7640	Divider clip & coupling		80	.100		1.22	6.15		7.37	10.80
7650	Panel connector		16	.500		25.50	30.50		56	75
7800	Take off connector		16	.500		85	30.50		115.50	141
8000	No. 6000, take off connector		16	.500		101	30.50		131.50	158
8100	Take off fitting		16	.500		74.50	30.50		105	129
8200	Hanger clamp		32	.250		18.95	15.35		34.30	44.50
8230	Coupling					10.50			10.50	11.55
8240	One gang device plate	1 Elec	53	.151		9.65	9.25		18.90	25
8250	Two gang device plate		40	.200		11.85	12.25		24.10	32
8260	Blank end fitting		40	.200		10.55	12.25		22.80	30.50
8270	Combination elbow		14	.571		42	35		77	100
8300	Panel connector		16	.500		18.40	30.50		48.90	67
8500	Chan-L-Wire system installed in 1-5/8" x 1-5/8" strut. Strut									
8600	not incl., 30 amp, 4 wire, 3 phase	1 Elec	200	.040	L.F.	5.10	2.45		7.55	9.40
8700	Junction box		8	1	Ea.	34	61.50		95.50	132
8800	Insulating end cap		40	.200		9.10	12.25		21.35	29
8900	Strut splice plate		40	.200		12.05	12.25		24.30	32
9000	Tap		40	.200		24.50	12.25		36.75	46
9100	Fixture hanger		60	.133		10.75	8.20		18.95	24.50
9200	Pulling tool					92			92	101
9300	Non-metallic, straight section									
9310	7/16" x 7/8", base & cover, blank	1 Elec	160	.050	L.F.	2	3.07		5.07	6.90
9320	Base & cover w/adhesive		160	.050		1.71	3.07		4.78	6.60
9340	7/16" x 1-5/16", base & cover, blank		145	.055		2.20	3.38		5.58	7.60
9350	Base & cover w/adhesive		145	.055		2.26	3.38		5.64	7.70
9370	11/16" x 2-1/4", base & cover, blank		130	.062		3.02	3.78		6.80	9.10
9380	Base & cover w/adhesive		130	.062		3.24	3.78		7.02	9.35
9385	1-11/16" x 5-1/4", two compartment base & cover w/screws		80	.100		10.40	6.15		16.55	21
9400	Fittings, elbows, 7/16" x 7/8"		50	.160	Ea.	1.97	9.80		11.77	17.25
9410	7/16" x 1-5/16"		45	.178		2.04	10.90		12.94	19.05
9420	11/16" x 2-1/4"		40	.200		2.21	12.25		14.46	21.50
9425	1-11/16" x 5-1/4"		28	.286		12.45	17.55		30	40.50
9430	Tees, 7/16" x 7/8"		35	.229		2.53	14		16.53	24.50
9440	7/16" x 1-5/16"		32	.250		2.59	15.35		17.94	26.50
9450	11/16" x 2-1/4"		30	.267		2.69	16.35		19.04	28
9455	1-11/16" x 5-1/4"		24	.333		19.95	20.50		40.45	53.50
9460	Cover clip, 7/16" x 7/8"		80	.100		.51	6.15		6.66	10
9470	7/16" x 1-5/16"		72	.111		.46	6.80		7.26	11
9480	11/16" x 2-1/4"		64	.125		.84	7.65		8.49	12.70
9484	1-11/16" x 5-1/4"		42	.190		2.94	11.70		14.64	21
9486	Wire clip, 1-11/16" x 5-1/4"		68	.118		.58	7.20		7.78	11.75
9490	Blank end, 7/16" x 7/8"		50	.160		.77	9.80		10.57	15.95
9500	7/16" x 1-5/16"		45	.178		.86	10.90		11.76	17.75
9510	11/16" x 2-1/4"		40	.200		1.28	12.25		13.53	20.50
9515	1-11/16" x 5-1/4"		38	.211		6.70	12.90		19.60	27.50
9520	Round fixture box, 5.5" diam. x 1"		25	.320		11.55	19.65		31.20	42.50
9530	Device box, 1 gang		30	.267		5.25	16.35		21.60	31

26 27 23 – Indoor Service Poles

26 27 23.40 Surface Raceway		Crew	Daily Output	Labor-Hours	Unit	Material	2020 Bare Costs Labor	Equipment	Total	Total Incl O&P
9540	2 gang	1 Elec	25	.320	Ea.	7.65	19.65		27.30	38.50
9990	Minimum labor/equipment charge	↓	5	1.600	Job		98		98	151

26 27 26 – Wiring Devices

26 27 26.10 Low Voltage Switching

		Crew	Daily Output	Labor-Hours	Unit	Material	Labor	Equipment	Total	Total Incl O&P
0010	**LOW VOLTAGE SWITCHING**									
3600	Relays, 120 V or 277 V standard	1 Elec	12	.667	Ea.	44	41		85	112
3800	Flush switch, standard		40	.200		13.45	12.25		25.70	33.50
4000	Interchangeable		40	.200		17.55	12.25		29.80	38
4100	Surface switch, standard		40	.200		8.35	12.25		20.60	28
4200	Transformer 115 V to 25 V		12	.667		128	41		169	204
4400	Master control, 12 circuit, manual		4	2		139	123		262	340
4500	25 circuit, motorized		4	2		161	123		284	365
4600	Rectifier, silicon		12	.667		50.50	41		91.50	119
4800	Switchplates, 1 gang, 1, 2 or 3 switch, plastic		80	.100		5.25	6.15		11.40	15.25
5000	Stainless steel		80	.100		11.35	6.15		17.50	22
5400	2 gang, 3 switch, stainless steel		53	.151		23.50	9.25		32.75	40.50
5500	4 switch, plastic		53	.151		10.35	9.25		19.60	25.50
5600	2 gang, 4 switch, stainless steel		53	.151		22	9.25		31.25	39
5700	6 switch, stainless steel		53	.151		44	9.25		53.25	63
5800	3 gang, 9 switch, stainless steel		32	.250		67.50	15.35		82.85	97.50
5900	Receptacle, triple, 1 return, 1 feed		26	.308		47	18.90		65.90	80.50
6000	2 feed		20	.400		47	24.50		71.50	89.50
6100	Relay gang boxes, flush or surface, 6 gang		5.30	1.509		92	92.50		184.50	244
6200	12 gang		4.70	1.702		109	104		213	281
6400	18 gang		4	2		120	123		243	320
6500	Frame, to hold up to 6 relays		12	.667	↓	89.50	41		130.50	162
7200	Control wire, 2 conductor		6.30	1.270	C.L.F.	35.50	78		113.50	159
7400	3 conductor		5	1.600		34.50	98		132.50	189
7600	19 conductor		2.50	3.200		320	196		516	650
7800	26 conductor		2	4		435	245		680	860
8000	Weatherproof, 3 conductor	↓	5	1.600	↓	72	98		170	230

26 27 26.20 Wiring Devices Elements

		Crew	Daily Output	Labor-Hours	Unit	Material	Labor	Equipment	Total	Total Incl O&P
0010	**WIRING DEVICES ELEMENTS** R262726-90									
0200	Toggle switch, quiet type, single pole, 15 amp	1 Elec	40	.200	Ea.	.52	12.25		12.77	19.45
0500	20 amp		27	.296		3.46	18.20		21.66	32
0510	30 amp		23	.348		23.50	21.50		45	58.50
0530	Lock handle, 20 amp		27	.296		30	18.20		48.20	60.50
0540	Security key, 20 amp		26	.308		85.50	18.90		104.40	123
0550	Rocker, 15 amp		40	.200		2.82	12.25		15.07	22
0560	20 amp		27	.296		10.85	18.20		29.05	40
0600	3 way, 15 amp		23	.348		1.83	21.50		23.33	35
0800	20 amp		18	.444		3.54	27.50		31.04	46
0810	30 amp		9	.889		24.50	54.50		79	111
0830	Lock handle, 20 amp		18	.444		28.50	27.50		56	73.50
0840	Security key, 20 amp		17	.471		124	29		153	182
0850	Rocker, 15 amp		23	.348		8.25	21.50		29.75	42
0860	20 amp		18	.444		12.80	27.50		40.30	56
0900	4 way, 15 amp		15	.533		10.60	32.50		43.10	62
1000	20 amp		11	.727		48	44.50		92.50	121
1020	Lock handle, 20 amp		11	.727		39.50	44.50		84	112
1030	Rocker, 15 amp		15	.533		10	32.50		42.50	61.50
1040	20 amp	↓	11	.727	↓	19.15	44.50		63.65	89.50

26 27 26.20 Wiring Devices Elements			Crew	Daily Output	Labor-Hours	Unit	Material	Labor (2020 Bare Costs)	Equipment	Total	Total Incl O&P
1100	Toggle switch, quiet type, double pole, 15 amp		1 Elec	15	.533	Ea.	13.95	32.50		46.45	66
1200	20 amp			11	.727		19.25	44.50		63.75	89.50
1210	30 amp			9	.889		34	54.50		88.50	122
1230	Lock handle, 20 amp			11	.727		35	44.50		79.50	107
1250	Security key, 20 amp			10	.800		91	49		140	176
1420	Toggle switch quiet type, 1 pole, 2 throw center off, 15 amp			23	.348		44.50	21.50		66	82
1440	20 amp			18	.444		70.50	27.50		98	120
1460	2 pole, 2 throw center off, lock handle, 20 amp			11	.727		85	44.50		129.50	162
1480	1 pole, momentary contact, 15 amp			23	.348		20	21.50		41.50	55
1500	20 amp			18	.444		28.50	27.50		56	73.50
1520	Momentary contact, lock handle, 20 amp			18	.444		32	27.50		59.50	77
1650	Dimmer switch, 120 volt, incandescent, 600 watt, 1 pole	G		16	.500		22.50	30.50		53	72
1700	600 watt, 3 way	G		12	.667		11.10	41		52.10	75
1750	1,000 watt, 1 pole	G		16	.500		44	30.50		74.50	95.50
1800	1,000 watt, 3 way	G		12	.667		73.50	41		114.50	144
2000	1,500 watt, 1 pole	G		11	.727		102	44.50		146.50	181
2100	2,000 watt, 1 pole	G		8	1		147	61.50		208.50	257
2110	Fluorescent, 600 watt	G		15	.533		111	32.50		143.50	173
2120	1,000 watt	G		15	.533		155	32.50		187.50	221
2130	1,500 watt	G		10	.800		286	49		335	390
2160	Explosion proof, toggle switch, wall, single pole 20 amp			5.30	1.509		269	92.50		361.50	440
2180	Receptacle, single outlet, 20 amp			5.30	1.509		615	92.50		707.50	825
2190	30 amp			4	2		985	123		1,108	1,275
2290	60 amp			2.50	3.200		810	196		1,006	1,200
2360	Plug, 20 amp			16	.500		197	30.50		227.50	264
2370	30 amp			12	.667		375	41		416	480
2380	60 amp			8	1		535	61.50		596.50	685
2410	Furnace, thermal cutoff switch with plate			26	.308		17.05	18.90		35.95	48
2460	Receptacle, duplex, 120 volt, grounded, 15 amp			40	.200		1.62	12.25		13.87	20.50
2470	20 amp			27	.296		10.65	18.20		28.85	40
2480	Ground fault interrupting, 15 amp			27	.296		13.55	18.20		31.75	43
2482	20 amp			27	.296		45	18.20		63.20	77.50
2490	Dryer, 30 amp			15	.533		4.55	32.50		37.05	55.50
2500	Range, 50 amp			11	.727		11.20	44.50		55.70	81
2540	Isolated ground receptacle, duplex, 20 amp			27	.296		27	18.20		45.20	58
2542	Quad, 20 amp			20	.400		33.50	24.50		58	75
2550	Simplex, 20 amp			27	.296		29.50	18.20		47.70	60.50
2560	Simplex, 30 amp			15	.533		38.50	32.50		71	93
2570	Cable reel w/receptacle 50' w/3#12, 120 V, 20 A		2 Elec	2.67	5.999		1,100	370		1,470	1,800
2600	Wall plates, stainless steel, 1 gang		1 Elec	80	.100		2.57	6.15		8.72	12.30
2800	2 gang			53	.151		4.38	9.25		13.63	19.05
3000	3 gang			32	.250		11.50	15.35		26.85	36
3100	4 gang			27	.296		11.05	18.20		29.25	40
3110	Brown plastic, 1 gang			80	.100		.38	6.15		6.53	9.85
3120	2 gang			53	.151		.75	9.25		10	15.10
3130	3 gang			32	.250		1.11	15.35		16.46	24.50
3140	4 gang			27	.296		2.91	18.20		21.11	31
3150	Brushed brass, 1 gang			80	.100		5.95	6.15		12.10	15.95
3160	Anodized aluminum, 1 gang			80	.100		3.03	6.15		9.18	12.80
3170	Switch cover, weatherproof, 1 gang			60	.133		5.25	8.20		13.45	18.40
3180	Vandal proof lock, 1 gang			60	.133		15.25	8.20		23.45	29.50
3200	Lampholder, keyless			26	.308		19.35	18.90		38.25	50.50
3400	Pullchain with receptacle			22	.364		23	22.50		45.50	59.50

26 27 26.20 Wiring Devices Elements		Crew	Daily Output	Labor-Hours	Unit	Material	Labor	Equipment	Total	Total Incl O&P
						2020 Bare Costs				
3500	Pilot light, neon with jewel	1 Elec	27	.296	Ea.	10.25	18.20		28.45	39.50
3600	Receptacle, 20 amp, 250 volt, NEMA 6		27	.296		24	18.20		42.20	54.50
3620	277 volt NEMA 7		27	.296		23	18.20		41.20	53.50
3640	125/250 volt NEMA 10		27	.296		23	18.20		41.20	53.50
3680	125/250 volt NEMA 14		25	.320		30	19.65		49.65	63
3700	3 pole, 250 volt NEMA 15		25	.320		31.50	19.65		51.15	65
3720	120/208 volt NEMA 18		25	.320		34	19.65		53.65	67.50
3740	30 amp, 125 volt NEMA 5		15	.533		19.55	32.50		52.05	72
3760	250 volt NEMA 6		15	.533		26.50	32.50		59	79.50
3780	277 volt NEMA 7		15	.533		31	32.50		63.50	84.50
3820	125/250 volt NEMA 14		14	.571		66	35		101	127
3840	3 pole, 250 volt NEMA 15		14	.571		60.50	35		95.50	121
3880	50 amp, 125 volt NEMA 5		11	.727		32	44.50		76.50	104
3900	250 volt NEMA 6		11	.727		30.50	44.50		75	102
3920	277 volt NEMA 7		11	.727		36	44.50		80.50	108
3960	125/250 volt NEMA 14		10	.800		94.50	49		143.50	180
3980	3 pole, 250 volt NEMA 15		10	.800		87	49		136	171
4020	60 amp, 125/250 volt, NEMA 14		8	1		94	61.50		155.50	198
4040	3 pole, 250 volt NEMA 15		8	1		110	61.50		171.50	216
4060	120/208 volt NEMA 18		8	1		83.50	61.50		145	187
4100	Receptacle locking, 20 amp, 125 volt NEMA L5		27	.296		25	18.20		43.20	55.50
4120	250 volt NEMA L6		27	.296		23	18.20		41.20	53.50
4140	277 volt NEMA L7		27	.296		26.50	18.20		44.70	57.50
4150	3 pole, 250 volt, NEMA L11		27	.296		22	18.20		40.20	52
4160	20 amp, 480 volt NEMA L8		27	.296		30	18.20		48.20	61
4180	600 volt NEMA L9		27	.296		38	18.20		56.20	69.50
4200	125/250 volt NEMA L10		27	.296		29	18.20		47.20	60
4230	125/250 volt NEMA L14		25	.320		31	19.65		50.65	64
4280	250 volt NEMA L15		25	.320		34.50	19.65		54.15	67.50
4300	480 volt NEMA L16		25	.320		28	19.65		47.65	61
4320	3 phase, 120/208 volt NEMA L18		25	.320		36	19.65		55.65	69.50
4340	277/480 volt NEMA L19		25	.320		37.50	19.65		57.15	71.50
4360	347/600 volt NEMA L20		25	.320		36.50	19.65		56.15	70.50
4380	120/208 volt NEMA L21		23	.348		38	21.50		59.50	75
4400	277/480 volt NEMA L22		23	.348		38.50	21.50		60	75.50
4420	347/600 volt NEMA L23		23	.348		47.50	21.50		69	85
4440	30 amp, 125 volt NEMA L5		15	.533		32.50	32.50		65	86.50
4460	250 volt NEMA L6		15	.533		35	32.50		67.50	89
4480	277 volt NEMA L7		15	.533		38	32.50		70.50	92.50
4500	480 volt NEMA L8		15	.533		36	32.50		68.50	90
4520	600 volt NEMA L9		15	.533		40.50	32.50		73	95
4540	125/250 volt NEMA L10		15	.533		45	32.50		77.50	100
4560	3 phase, 250 volt NEMA L11		15	.533		32	32.50		64.50	85.50
4620	125/250 volt NEMA L14		14	.571		50	35		85	109
4640	250 volt NEMA L15		14	.571		51.50	35		86.50	111
4660	480 volt NEMA L16		14	.571		54	35		89	114
4680	600 volt NEMA L17		14	.571		56	35		91	116
4700	120/208 volt NEMA L18		14	.571		56	35		91	116
4720	277/480 volt NEMA L19		14	.571		58.50	35		93.50	118
4740	347/600 volt NEMA L20		14	.571		61	35		96	121
4760	120/208 volt NEMA L21		13	.615		54	38		92	118
4780	277/480 volt NEMA L22		13	.615		54.50	38		92.50	118
4800	347/600 volt NEMA L23	↓	13	.615	↓	63	38		101	127

26 27 26.20 Wiring Devices Elements		Crew	Daily Output	Labor-Hours	Unit	2020 Bare Costs Material	Labor	Equipment	Total	Total Incl O&P
4840	Receptacle, corrosion resistant, 15 or 20 amp, 125 volt NEMA L5	1 Elec	27	.296	Ea.	35.50	18.20		53.70	67
4860	250 volt NEMA L6		27	.296		22.50	18.20		40.70	53
4900	Receptacle, cover plate, phenolic plastic, NEMA 5 & 6		80	.100		.62	6.15		6.77	10.15
4910	NEMA 7-23		80	.100		.67	6.15		6.82	10.20
4920	Stainless steel, NEMA 5 & 6		80	.100		2.72	6.15		8.87	12.45
4930	NEMA 7-23		80	.100		3.05	6.15		9.20	12.80
4940	Brushed brass NEMA 5 & 6		80	.100		5.45	6.15		11.60	15.45
4950	NEMA 7-23		80	.100		5.95	6.15		12.10	16
4960	Anodized aluminum, NEMA 5 & 6		80	.100		3.57	6.15		9.72	13.40
4970	NEMA 7-23		80	.100		5.90	6.15		12.05	15.95
4980	Weatherproof NEMA 7-23		60	.133		42.50	8.20		50.70	59.50
5000	Duplex receptacle, combo 15A/125V, 3 wire w/2-5V 0.7A, port USB, AL		27	.296		19	18.20		37.20	49
5002	15A/125V, 3 wire w/2-5V 0.7A, port USB, BK		27	.296		18.30	18.20		36.50	48
5004	15A/125V, 3 wire w/2-5V 0.7A, port USB, LA		27	.296		18.15	18.20		36.35	48
5006	15A/125V, 3 wire w/2-5V 0.7A, port USB, IV		27	.296		30	18.20		48.20	61
5008	15A/125V, 3 wire w/2-5V 0.7A, port USB, WH		27	.296		22.50	18.20		40.70	52.50
5010	Duplex receptacle, combo 15A/125V, 3 wire w/2-5V 2.1A, port USB, AL		27	.296		58.50	18.20		76.70	92
5012	15A/125V, 3 wire w/2-5V 2.1A, port USB, BK		27	.296		57	18.20		75.20	90.50
5014	15A/125V, 3 wire w/2-5V 2.1A, port USB, LA		27	.296		50	18.20		68.20	83
5016	15A/125V, 3 wire w/2-5V 2.1A, port USB, IV		27	.296		50.50	18.20		68.70	83.50
5018	15A/125V, 3 wire w/2-5V 2.1A, port USB, WH		27	.296		31	18.20		49.20	62
5020	Duplex receptacle, combo 20A/125V, 3 wire w/2-5V 2.1A, port USB, AL		27	.296		51.50	18.20		69.70	84.50
5022	20A/125V, 3 wire w/2-5V 2.1A, port USB, BK		27	.296		64.50	18.20		82.70	99
5024	20A/125V, 3 wire w/2-5V 2.1A, port USB, LA		27	.296		35.50	18.20		53.70	67
5026	20A/125V, 3 wire w/2-5V 2.1A, port USB, IV		27	.296		35	18.20		53.20	66.50
5028	20A/125V, 3 wire w/2-5V 2.1A, port USB, WH		27	.296		35	18.20		53.20	66.50
5100	Plug, 20 amp, 250 volt NEMA 6		30	.267		19.85	16.35		36.20	47
5110	277 volt NEMA 7		30	.267		22	16.35		38.35	49
5120	3 pole, 120/250 volt NEMA 10		26	.308		24	18.90		42.90	55.50
5130	125/250 volt NEMA 14		26	.308		50	18.90		68.90	84
5140	250 volt NEMA 15		26	.308		50.50	18.90		69.40	84.50
5150	120/208 volt NEMA 8		26	.308		58	18.90		76.90	92.50
5160	30 amp, 125 volt NEMA 5		13	.615		51	38		89	114
5170	250 volt NEMA 6		13	.615		50	38		88	113
5180	277 volt NEMA 7		13	.615		46.50	38		84.50	109
5190	125/250 volt NEMA 14		13	.615		57	38		95	121
5200	3 pole, 250 volt NEMA 15		12	.667		61	41		102	130
5210	50 amp, 125 volt NEMA 5		9	.889		77	54.50		131.50	169
5220	250 volt NEMA 6		9	.889		80.50	54.50		135	173
5230	277 volt NEMA 7		9	.889		86	54.50		140.50	179
5240	125/250 volt NEMA 14		9	.889		74.50	54.50		129	166
5250	3 pole, 250 volt NEMA 15		8	1		73.50	61.50		135	176
5260	60 amp, 125/250 volt NEMA 14		7	1.143		80.50	70		150.50	197
5270	3 pole, 250 volt NEMA 15		7	1.143		85.50	70		155.50	202
5280	120/208 volt NEMA 18		7	1.143		103	70		173	222
5300	Plug angle, 20 amp, 250 volt NEMA 6		30	.267		30	16.35		46.35	58
5310	30 amp, 125 volt NEMA 5		13	.615		55.50	38		93.50	120
5320	250 volt NEMA 6		13	.615		58	38		96	122
5330	277 volt NEMA 7		13	.615		68.50	38		106.50	133
5340	125/250 volt NEMA 14		13	.615		63.50	38		101.50	128
5350	3 pole, 250 volt NEMA 15		12	.667		68.50	41		109.50	139
5360	50 amp, 125 volt NEMA 5		9	.889		64.50	54.50		119	155
5370	250 volt NEMA 6		9	.889		60.50	54.50		115	151

26 27 26.20 Wiring Devices Elements		Crew	Daily Output	Labor-Hours	Unit	Material	Labor	Equipment	Total	Total Incl O&P
5380	277 volt NEMA 7	1 Elec	9	.889	Ea.	76.50	54.50		131	168
5390	125/250 volt NEMA 14		9	.889		77	54.50		131.50	169
5400	3 pole, 250 volt NEMA 15		8	1		81.50	61.50		143	185
5410	60 amp, 125/250 volt NEMA 14		7	1.143		91.50	70		161.50	209
5420	3 pole, 250 volt NEMA 15		7	1.143		93.50	70		163.50	211
5430	120/208 volt NEMA 18		7	1.143		96.50	70		166.50	214
5500	Plug, locking, 20 amp, 125 volt NEMA L5		30	.267		18.15	16.35		34.50	45
5510	250 volt NEMA L6		30	.267		18.15	16.35		34.50	45
5520	277 volt NEMA L7		30	.267		17.90	16.35		34.25	44.50
5530	480 volt NEMA L8		30	.267		19.95	16.35		36.30	47
5540	600 volt NEMA L9		30	.267		22	16.35		38.35	49
5550	3 pole, 125/250 volt NEMA L10		26	.308		23	18.90		41.90	54.50
5560	250 volt NEMA L11		26	.308		23	18.90		41.90	54.50
5570	480 volt NEMA L12		26	.308		27	18.90		45.90	58.50
5580	125/250 volt NEMA L14		26	.308		27.50	18.90		46.40	59
5590	250 volt NEMA L15		26	.308		34	18.90		52.90	66.50
5600	480 volt NEMA L16		26	.308		30.50	18.90		49.40	62.50
5610	4 pole, 120/208 volt NEMA L18		24	.333		33	20.50		53.50	68
5620	277/480 volt NEMA L19		24	.333		39.50	20.50		60	75
5630	347/600 volt NEMA L20		24	.333		34.50	20.50		55	69.50
5640	120/208 volt NEMA L21		24	.333		35	20.50		55.50	70
5650	277/480 volt NEMA L22		24	.333		37	20.50		57.50	72.50
5660	347/600 volt NEMA L23		24	.333		41.50	20.50		62	77
5670	30 amp, 125 volt NEMA L5		13	.615		28	38		66	88.50
5680	250 volt NEMA L6		13	.615		28.50	38		66.50	89.50
5690	277 volt NEMA L7		13	.615		28.50	38		66.50	89.50
5700	480 volt NEMA L8		13	.615		29.50	38		67.50	90.50
5710	600 volt NEMA L9		13	.615		30.50	38		68.50	91.50
5720	3 pole, 125/250 volt NEMA L10		11	.727		25.50	44.50		70	96.50
5730	250 volt NEMA L11		11	.727		25	44.50		69.50	96
5760	125/250 volt NEMA L14		11	.727		37.50	44.50		82	110
5770	250 volt NEMA L15		11	.727		37.50	44.50		82	110
5780	480 volt NEMA L16		11	.727		38.50	44.50		83	111
5790	600 volt NEMA L17		11	.727		38.50	44.50		83	111
5800	4 pole, 120/208 volt NEMA L18		10	.800		43	49		92	123
5810	120/208 volt NEMA L19		10	.800		43.50	49		92.50	124
5820	347/600 volt NEMA L20		10	.800		45.50	49		94.50	126
5830	120/208 volt NEMA L21		10	.800		40.50	49		89.50	120
5840	277/480 volt NEMA L22		10	.800		46	49		95	126
5850	347/600 volt NEMA L23		10	.800		47	49		96	127
6000	Connector, 20 amp, 250 volt NEMA 6		30	.267		32	16.35		48.35	60
6010	277 volt NEMA 7		30	.267		32	16.35		48.35	60
6020	3 pole, 120/250 volt NEMA 10		26	.308		36	18.90		54.90	68.50
6030	125/250 volt NEMA 14		26	.308		36	18.90		54.90	68.50
6040	250 volt NEMA 15		26	.308		36.50	18.90		55.40	69
6050	120/208 volt NEMA 18		26	.308		39.50	18.90		58.40	72.50
6060	30 amp, 125 volt NEMA 5		13	.615		62.50	38		100.50	127
6070	250 volt NEMA 6		13	.615		62.50	38		100.50	127
6080	277 volt NEMA 7		13	.615		62.50	38		100.50	127
6110	50 amp, 125 volt NEMA 5		9	.889		86	54.50		140.50	179
6120	250 volt NEMA 6		9	.889		86	54.50		140.50	179
6130	277 volt NEMA 7		9	.889		86	54.50		140.50	179
6200	Connector, locking, 20 amp, 125 volt NEMA L5		30	.267		28	16.35		44.35	56

(Material, Labor, Equipment and Total are 2020 Bare Costs.)

26 27 26.20 Wiring Devices Elements		Crew	Daily Output	Labor-Hours	Unit	2020 Bare Costs Material	Labor	Equipment	Total	Total Incl O&P
6210	250 volt NEMA L6	1 Elec	30	.267	Ea.	28	16.35		44.35	56
6220	277 volt NEMA L7		30	.267		27	16.35		43.35	55
6230	480 volt NEMA L8		30	.267		32	16.35		48.35	60
6240	600 volt NEMA L9		30	.267		37.50	16.35		53.85	66
6250	3 pole, 125/250 volt NEMA L10		26	.308		37	18.90		55.90	69.50
6260	250 volt NEMA L11		26	.308		37	18.90		55.90	69.50
6280	125/250 volt NEMA L14		26	.308		38	18.90		56.90	71
6290	250 volt NEMA L15		26	.308		38.50	18.90		57.40	71
6300	480 volt NEMA L16		26	.308		41	18.90		59.90	74
6310	4 pole, 120/208 volt NEMA L18		24	.333		47.50	20.50		68	84
6320	277/480 volt NEMA L19		24	.333		49	20.50		69.50	85.50
6330	347/600 volt NEMA L20		24	.333		49.50	20.50		70	86
6340	120/208 volt NEMA L21		24	.333		56.50	20.50		77	93.50
6350	277/480 volt NEMA L22		24	.333		63.50	20.50		84	101
6360	347/600 volt NEMA L23		24	.333		71	20.50		91.50	110
6370	30 amp, 125 volt NEMA L5		13	.615		55	38		93	119
6380	250 volt NEMA L6		13	.615		55.50	38		93.50	119
6390	277 volt NEMA L7		13	.615		58.50	38		96.50	123
6400	480 volt NEMA L8		13	.615		58	38		96	122
6410	600 volt NEMA L9		13	.615		65	38		103	130
6420	3 pole, 125/250 volt NEMA L10		11	.727		71	44.50		115.50	147
6430	250 volt NEMA L11		11	.727		71	44.50		115.50	147
6460	125/250 volt NEMA L14		11	.727		77	44.50		121.50	153
6470	250 volt NEMA L15		11	.727		76.50	44.50		121	153
6480	480 volt NEMA L16		11	.727		82	44.50		126.50	159
6490	600 volt NEMA L17		11	.727		81.50	44.50		126	159
6500	4 pole, 120/208 volt NEMA L18		10	.800		88.50	49		137.50	173
6510	120/208 volt NEMA L19		10	.800		88.50	49		137.50	173
6520	347/600 volt NEMA L20		10	.800		90	49		139	175
6530	120/208 volt NEMA L21		10	.800		75	49		124	158
6540	277/480 volt NEMA L22		10	.800		80.50	49		129.50	164
6550	347/600 volt NEMA L23		10	.800		86.50	49		135.50	171
7000	Receptacle computer, 250 volt, 15 amp, 3 pole 4 wire		8	1		93	61.50		154.50	197
7010	20 amp, 2 pole 3 wire		8	1		111	61.50		172.50	217
7020	30 amp, 2 pole 3 wire		6.50	1.231		178	75.50		253.50	310
7030	30 amp, 3 pole 4 wire		6.50	1.231		193	75.50		268.50	330
7040	60 amp, 3 pole 4 wire		4.50	1.778		320	109		429	520
7050	100 amp, 3 pole 4 wire		3	2.667		395	164		559	685
7100	Connector computer, 250 volt, 15 amp, 3 pole 4 wire		27	.296		153	18.20		171.20	196
7110	20 amp, 2 pole 3 wire		27	.296		152	18.20		170.20	195
7120	30 amp, 2 pole 3 wire		15	.533		210	32.50		242.50	282
7130	30 amp, 3 pole 4 wire		15	.533		237	32.50		269.50	310
7140	60 amp, 3 pole 4 wire		8	1		400	61.50		461.50	535
7150	100 amp, 3 pole 4 wire		4	2		540	123		663	785
7200	Plug, computer, 250 volt, 15 amp, 3 pole 4 wire		27	.296		136	18.20		154.20	178
7210	20 amp, 2 pole 3 wire		27	.296		139	18.20		157.20	181
7220	30 amp, 2 pole 3 wire		15	.533		218	32.50		250.50	291
7230	30 amp, 3 pole 4 wire		15	.533		216	32.50		248.50	289
7240	60 amp, 3 pole 4 wire		8	1		325	61.50		386.50	450
7250	100 amp, 3 pole 4 wire		4	2		395	123		518	625
7300	Connector adapter to flexible conduit, 1/2"		60	.133		3.54	8.20		11.74	16.50
7310	3/4"		50	.160		4.52	9.80		14.32	20
7320	1-1/4"		30	.267		13.85	16.35		30.20	40.50

26 27 Low-Voltage Distribution Equipment

26 27 26 – Wiring Devices

26 27 26.20 Wiring Devices Elements		Crew	Daily Output	Labor-Hours	Unit	Material	2020 Bare Costs Labor	Equipment	Total	Total Incl O&P
7330	1-1/2"	1 Elec	23	.348	Ea.	18.40	21.50		39.90	53
9000	Minimum labor/equipment charge	↓	4	2	Job		123		123	189

26 27 33 – Power Distribution Units

26 27 33.20 Power Distribution Unit		Crew	Daily Output	Labor-Hours	Unit	Material	Labor	Equipment	Total	Total Incl O&P
0010	**POWER DISTRIBUTION UNIT**									
0050	3PH, 60 kVA, PDU									
0100	208V-208V/120V	2 Elec	2	8	Ea.	14,300	490		14,790	16,500
0110	480V-208V/120V		2	8		14,100	490		14,590	16,300
0120	600V-208V/120V	↓	2	8	↓	14,700	490		15,190	16,900
0125	3PH, 80 kVA, PDU									
0130	208V-208V/120V	2 Elec	2	8	Ea.	27,700	490		28,190	31,300
0140	480V-208V/120V		2	8		28,100	490		28,590	31,800
0150	600V-208V/120V	↓	2	8	↓	27,900	490		28,390	31,500

26 27 73 – Door Chimes

26 27 73.10 Doorbell System		Crew	Daily Output	Labor-Hours	Unit	Material	Labor	Equipment	Total	Total Incl O&P
0010	**DOORBELL SYSTEM**, incl. transformer, button & signal									
0100	6" bell	1 Elec	4	2	Ea.	158	123		281	365
0200	Buzzer		4	2		127	123		250	330
1000	Door chimes, 2 notes		16	.500		32	30.50		62.50	82
1020	with ambient light		12	.667		116	41		157	191
1100	Tube type, 3 tube system		12	.667		233	41		274	320
1180	4 tube system		10	.800		495	49		544	620
1900	For transformer & button, add		5	1.600		16.15	98		114.15	169
3000	For push button only		24	.333		.92	20.50		21.42	32.50
3200	Bell transformer		16	.500	↓	24	30.50		54.50	73.50
9000	Minimum labor/equipment charge	↓	4	2	Job		123		123	189

26 28 Low-Voltage Circuit Protective Devices

26 28 13 – Fuses

26 28 13.10 Fuse Elements		Crew	Daily Output	Labor-Hours	Unit	Material	Labor	Equipment	Total	Total Incl O&P
0010	**FUSE ELEMENTS**									
0020	Cartridge, nonrenewable									
0050	250 volt, 30 amp	1 Elec	50	.160	Ea.	2.29	9.80		12.09	17.60
0100	60 amp		50	.160		4.17	9.80		13.97	19.70
0150	100 amp		40	.200		16.35	12.25		28.60	37
0200	200 amp		36	.222		37	13.65		50.65	62
0250	400 amp		30	.267		113	16.35		129.35	149
0300	600 amp		24	.333		197	20.50		217.50	248
0400	600 volt, 30 amp		40	.200		10.30	12.25		22.55	30.50
0450	60 amp		40	.200		15.35	12.25		27.60	36
0500	100 amp		36	.222		32.50	13.65		46.15	57
0550	200 amp		30	.267		61	16.35		77.35	92
0600	400 amp		24	.333		122	20.50		142.50	166
0650	600 amp		20	.400		193	24.50		217.50	250
0800	Dual element, time delay, 250 volt, 30 amp		50	.160		9.90	9.80		19.70	26
0840	50 amp		50	.160		11.35	9.80		21.15	27.50
0850	60 amp		50	.160		11.60	9.80		21.40	28
0900	100 amp		40	.200		43.50	12.25		55.75	67
0950	200 amp		36	.222		94	13.65		107.65	124
1000	400 amp	↓	30	.267	↓	115	16.35		131.35	151

26 28 Low-Voltage Circuit Protective Devices

26 28 13 – Fuses

26 28 13.10 Fuse Elements		Crew	Daily Output	Labor-Hours	Unit	Material	2020 Bare Costs Labor	Equipment	Total	Total Incl O&P
1050	600 amp	1 Elec	24	.333	Ea.	189	20.50		209.50	240
1300	600 volt, 15 to 30 amp		40	.200		21.50	12.25		33.75	43
1350	35 to 60 amp		40	.200		31.50	12.25		43.75	53.50
1400	70 to 100 amp		36	.222		66.50	13.65		80.15	94
1450	110 to 200 amp		30	.267		131	16.35		147.35	169
1500	225 to 400 amp		24	.333		283	20.50		303.50	340
1550	600 amp		20	.400		360	24.50		384.50	440
1800	Class RK1, high capacity, 250 volt, 30 amp		50	.160		8.80	9.80		18.60	25
1850	60 amp		50	.160		16.85	9.80		26.65	33.50
1900	100 amp		40	.200		26	12.25		38.25	47.50
1950	200 amp		36	.222		66.50	13.65		80.15	94.50
2000	400 amp		30	.267		151	16.35		167.35	191
2050	600 amp		24	.333		284	20.50		304.50	340
2200	600 volt, 30 amp		40	.200		12.65	12.25		24.90	33
2250	60 amp		40	.200		23	12.25		35.25	44.50
2300	100 amp		36	.222		79.50	13.65		93.15	109
2350	200 amp		30	.267		112	16.35		128.35	149
2400	400 amp		24	.333		196	20.50		216.50	247
2450	600 amp		20	.400		405	24.50		429.50	485
2700	Class J, current limiting, 250 or 600 volt, 30 amp		40	.200		19.80	12.25		32.05	41
2750	60 amp		40	.200		31.50	12.25		43.75	53.50
2800	100 amp		36	.222		64	13.65		77.65	91.50
2850	200 amp		30	.267		97.50	16.35		113.85	132
2900	400 amp		24	.333		296	20.50		316.50	355
2950	600 amp		20	.400		310	24.50		334.50	380
3100	Class L, current limiting, 250 or 600 volt, 601 to 1,200 amp		16	.500		740	30.50		770.50	860
3150	1,500 to 1,600 amp		13	.615		715	38		753	850
3200	1,800 to 2,000 amp		10	.800		1,500	49		1,549	1,725
3250	2,500 amp		10	.800		1,225	49		1,274	1,425
3300	3,000 amp		8	1		1,950	61.50		2,011.50	2,250
3350	3,500 to 4,000 amp		8	1		1,900	61.50		1,961.50	2,175
3400	4,500 to 5,000 amp		6.70	1.194		1,850	73.50		1,923.50	2,150
3450	6,000 amp		5.70	1.404		5,475	86		5,561	6,125
3600	Plug, 120 volt, 1 to 10 amp		50	.160		4.47	9.80		14.27	20
3650	15 to 30 amp		50	.160		4.17	9.80		13.97	19.70
3700	Dual element 0.3 to 14 amp		50	.160		6.60	9.80		16.40	22.50
3750	15 to 30 amp		50	.160		6.55	9.80		16.35	22.50
3800	Fustat, 120 volt, 15 to 30 amp		50	.160		4.84	9.80		14.64	20.50
3850	0.3 to 14 amp		50	.160		5.70	9.80		15.50	21.50
3900	Adapters 0.3 to 10 amp		50	.160		6.25	9.80		16.05	22
3950	15 to 30 amp		50	.160		8.85	9.80		18.65	25
4000	F-frame current limiting fuse, 14 to 2 AWG, 3 ampere, 3P, aluminum terminal		40	.200		1,925	12.25		1,937.25	2,150
4010	7 ampere, 3P, aluminum terminal		40	.200		1,925	12.25		1,937.25	2,125
4020	15 AMP, 3P, aluminum terminal		40	.200		1,975	12.25		1,987.25	2,200
4030	30 AMP, 3P, aluminum terminal		40	.200		1,925	12.25		1,937.25	2,150
4040	50 AMP, 3P, aluminum terminal		40	.200		1,625	12.25		1,637.25	1,800
4050	F-frame current limiting fuse, 1 to 4/0 AWG, 100 AMP, 3P, aluminum terminal		40	.200		1,525	12.25		1,537.25	1,700
4060	150 AMP, 3P, aluminum terminal	▼	40	.200	▼	2,250	12.25		2,262.25	2,475

26 28 16.10 Circuit Breakers		Crew	Daily Output	Labor-Hours	Unit	Material	2020 Bare Costs Labor	Equipment	Total	Total Incl O&P
0010	**CIRCUIT BREAKERS** (in enclosure)									
0100	Enclosed (NEMA 1), 600 volt, 3 pole, 30 amp	1 Elec	3.20	2.500	Ea.	530	153		683	815
0200	60 amp		2.80	2.857		645	175		820	975
0400	100 amp		2.30	3.478		735	213		948	1,150
0500	200 amp		1.50	5.333		1,875	325		2,200	2,575
0600	225 amp		1.50	5.333		1,700	325		2,025	2,375
0700	400 amp	2 Elec	1.60	10		2,900	615		3,515	4,150
0800	600 amp		1.20	13.333		4,200	820		5,020	5,875
1000	800 amp		.94	17.021		5,475	1,050		6,525	7,625
1200	1,000 amp		.84	19.048		6,925	1,175		8,100	9,400
1220	1,200 amp		.80	20		8,850	1,225		10,075	11,600
1240	1,600 amp		.72	22.222		16,100	1,375		17,475	19,800
1260	2,000 amp		.64	25		17,400	1,525		18,925	21,600
1400	1,200 amp with ground fault		.80	20		15,500	1,225		16,725	19,000
1600	1,600 amp with ground fault		.72	22.222		18,300	1,375		19,675	22,200
1800	2,000 amp with ground fault		.64	25		19,600	1,525		21,125	23,900
2000	Disconnect, 240 volt 3 pole, 5 HP motor	1 Elec	3.20	2.500		450	153		603	730
2020	10 HP motor		3.20	2.500		450	153		603	730
2040	15 HP motor		2.80	2.857		450	175		625	765
2060	20 HP motor		2.30	3.478		545	213		758	930
2080	25 HP motor		2.30	3.478		545	213		758	930
2100	30 HP motor		2.30	3.478		545	213		758	930
2120	40 HP motor		2	4		935	245		1,180	1,400
2140	50 HP motor		1.50	5.333		935	325		1,260	1,525
2160	60 HP motor		1.50	5.333		2,150	325		2,475	2,875
2180	75 HP motor	2 Elec	2	8		2,150	490		2,640	3,125
2200	100 HP motor		1.60	10		2,150	615		2,765	3,325
2220	125 HP motor		1.60	10		2,150	615		2,765	3,325
2240	150 HP motor		1.20	13.333		4,200	820		5,020	5,875
2260	200 HP motor		1.20	13.333		5,475	820		6,295	7,275
2300	Enclosed (NEMA 7), explosion proof, 600 volt 3 pole, 50 amp	1 Elec	2.30	3.478		1,925	213		2,138	2,450
2350	100 amp		1.50	5.333		2,000	325		2,325	2,700
2400	150 amp		1	8		4,825	490		5,315	6,075
2450	250 amp	2 Elec	1.60	10		6,050	615		6,665	7,600
2500	400 amp	"	1.20	13.333		6,700	820		7,520	8,625
9000	Minimum labor/equipment charge	1 Elec	4	2	Job		123		123	189

26 28 16.20 Safety Switches		Crew	Daily Output	Labor-Hours	Unit	Material	Labor	Equipment	Total	Total Incl O&P
0010	**SAFETY SWITCHES**									
0100	General duty 240 volt, 3 pole NEMA 1, fusible, 30 amp	1 Elec	3.20	2.500	Ea.	59.50	153		212.50	300
0200	60 amp		2.30	3.478		102	213		315	440
0300	100 amp		1.90	4.211		175	258		433	590
0400	200 amp		1.30	6.154		375	380		755	990
0500	400 amp	2 Elec	1.80	8.889		985	545		1,530	1,925
0600	600 amp	"	1.20	13.333		1,925	820		2,745	3,350
0610	Nonfusible, 30 amp	1 Elec	3.20	2.500		48	153		201	289
0650	60 amp		2.30	3.478		63.50	213		276.50	400
0700	100 amp		1.90	4.211		149	258		407	565
0750	200 amp		1.30	6.154		274	380		654	880
0800	400 amp	2 Elec	1.80	8.889		745	545		1,290	1,650
0850	600 amp	"	1.20	13.333		1,550	820		2,370	2,950
1100	Heavy duty, 600 volt, 3 pole NEMA 1 nonfused									
1110	30 amp	1 Elec	3.20	2.500	Ea.	86	153		239	330

26 28 16.20 Safety Switches		Crew	Daily Output	Labor-Hours	Unit	2020 Bare Costs Material	Labor	Equipment	Total	Total Incl O&P
1500	60 amp	1 Elec	2.30	3.478	Ea.	154	213		367	500
1700	100 amp		1.90	4.211		243	258		501	665
1900	200 amp		1.30	6.154		365	380		745	985
2100	400 amp	2 Elec	1.80	8.889		890	545		1,435	1,825
2300	600 amp		1.20	13.333		1,675	820		2,495	3,075
2500	800 amp		.94	17.021		3,000	1,050		4,050	4,900
2700	1,200 amp		.80	20		3,575	1,225		4,800	5,825
2900	Heavy duty, 240 volt, 3 pole NEMA 1 fusible									
2910	30 amp	1 Elec	3.20	2.500	Ea.	102	153		255	350
3000	60 amp		2.30	3.478		170	213		383	520
3300	100 amp		1.90	4.211		263	258		521	690
3500	200 amp		1.30	6.154		455	380		835	1,075
3700	400 amp	2 Elec	1.80	8.889		1,225	545		1,770	2,200
3900	600 amp		1.20	13.333		1,900	820		2,720	3,325
4100	800 amp		.94	17.021		4,700	1,050		5,750	6,775
4300	1,200 amp		.80	20		6,325	1,225		7,550	8,875
4340	2 pole fusible, 30 amp	1 Elec	3.50	2.286		77.50	140		217.50	300
4350	600 volt, 3 pole, fusible, 30 amp		3.20	2.500		166	153		319	420
4380	60 amp		2.30	3.478		202	213		415	555
4400	100 amp		1.90	4.211		365	258		623	805
4420	200 amp		1.30	6.154		520	380		900	1,150
4440	400 amp	2 Elec	1.80	8.889		1,400	545		1,945	2,375
4450	600 amp		1.20	13.333		2,475	820		3,295	3,975
4460	800 amp		.94	17.021		4,700	1,050		5,750	6,775
4480	1,200 amp		.80	20		6,325	1,225		7,550	8,875
4500	240 volt 3 pole NEMA 3R (no hubs), fusible									
4510	30 amp	1 Elec	3.10	2.581	Ea.	178	158		336	440
4700	60 amp		2.20	3.636		283	223		506	655
4900	100 amp		1.80	4.444		410	273		683	870
5100	200 amp		1.20	6.667		555	410		965	1,250
5300	400 amp	2 Elec	1.60	10		1,275	615		1,890	2,350
5500	600 amp	″	1	16		2,575	980		3,555	4,325
5510	Heavy duty, 600 volt, 3 pole 3 ph. NEMA 3R fusible, 30 amp	1 Elec	3.10	2.581		279	158		437	550
5520	60 amp		2.20	3.636		340	223		563	720
5530	100 amp		1.80	4.444		500	273		773	970
5540	200 amp		1.20	6.667		695	410		1,105	1,400
5550	400 amp	2 Elec	1.60	10		1,700	615		2,315	2,825
5700	600 volt, 3 pole NEMA 3R nonfused									
5710	30 amp	1 Elec	3.10	2.581	Ea.	152	158		310	410
5900	60 amp		2.20	3.636		266	223		489	635
6100	100 amp		1.80	4.444		375	273		648	835
6300	200 amp		1.20	6.667		450	410		860	1,125
6500	400 amp	2 Elec	1.60	10		1,175	615		1,790	2,250
6700	600 amp	″	1	16		2,450	980		3,430	4,175
6900	600 volt, 6 pole NEMA 3R nonfused, 30 amp	1 Elec	2.70	2.963		1,025	182		1,207	1,400
7100	60 amp		2	4		1,950	245		2,195	2,525
7300	100 amp		1.50	5.333		1,875	325		2,200	2,575
7500	200 amp		1.20	6.667		3,500	410		3,910	4,450
7600	600 volt, 3 pole NEMA 7 explosion proof nonfused									
7610	30 amp	1 Elec	2.20	3.636	Ea.	1,425	223		1,648	1,900
7620	60 amp		1.80	4.444		1,700	273		1,973	2,300
7630	100 amp		1.20	6.667		2,850	410		3,260	3,750
7640	200 amp		.80	10		6,125	615		6,740	7,700

26 28 16.20 Safety Switches		Crew	Daily Output	Labor-Hours	Unit	Material	Labor	Equipment	Total	Total Incl O&P
7710	600 volt 6 pole, NEMA 3R fusible, 30 amp	1 Elec	2.70	2.963	Ea.	1,525	182		1,707	1,950
7900	60 amp		2	4		2,150	245		2,395	2,725
8100	100 amp		1.50	5.333		1,725	325		2,050	2,375
8110	240 volt 3 pole, NEMA 12 fusible, 30 amp		3.10	2.581		257	158		415	525
8120	60 amp		2.20	3.636		455	223		678	845
8130	100 amp		1.80	4.444		550	273		823	1,025
8140	200 amp		1.20	6.667		615	410		1,025	1,300
8150	400 amp	2 Elec	1.60	10		1,675	615		2,290	2,800
8160	600 amp	"	1	16		2,775	980		3,755	4,550
8180	600 volt 3 pole, NEMA 12 fusible, 30 amp	1 Elec	3.10	2.581		370	158		528	655
8190	60 amp		2.20	3.636		465	223		688	855
8200	100 amp		1.80	4.444		695	273		968	1,175
8210	200 amp		1.20	6.667		955	410		1,365	1,675
8220	400 amp	2 Elec	1.60	10		2,400	615		3,015	3,575
8230	600 amp	"	1	16		3,975	980		4,955	5,875
8240	600 volt 3 pole, NEMA 12 nonfused, 30 amp	1 Elec	3.10	2.581		232	158		390	500
8250	60 amp		2.20	3.636		360	223		583	740
8260	100 amp		1.80	4.444		520	273		793	990
8270	200 amp		1.20	6.667		565	410		975	1,250
8280	400 amp	2 Elec	1.60	10		1,425	615		2,040	2,525
8290	600 amp	"	1	16		2,900	980		3,880	4,675
8310	600 volt, 3 pole NEMA 4 fusible, 30 amp	1 Elec	3	2.667		775	164		939	1,100
8320	60 amp		2.20	3.636		920	223		1,143	1,375
8330	100 amp		1.80	4.444		1,900	273		2,173	2,500
8340	200 amp		1.20	6.667		2,500	410		2,910	3,375
8350	400 amp	2 Elec	1.60	10		4,975	615		5,590	6,425
8360	600 volt 3 pole NEMA 4 nonfused, 30 amp	1 Elec	3	2.667		690	164		854	1,000
8370	60 amp		2.20	3.636		820	223		1,043	1,250
8380	100 amp		1.80	4.444		2,825	273		3,098	3,525
8390	200 amp		1.20	6.667		2,325	410		2,735	3,175
8400	400 amp	2 Elec	1.60	10		4,250	615		4,865	5,625
8490	Motor starters, manual, single phase, NEMA 1	1 Elec	6.40	1.250		99	76.50		175.50	227
8500	NEMA 4		4	2		259	123		382	475
8700	NEMA 7		4	2		310	123		433	535
8900	NEMA 1 with pilot		6.40	1.250		123	76.50		199.50	254
8920	3 pole, NEMA 1, 230/460 volt, 5 HP, size 0		3.50	2.286		202	140		342	440
8940	10 HP, size 1		2	4		239	245		484	645
9010	Disc. switch, 600 volt 3 pole fusible, 30 amp, to 10 HP motor		3.20	2.500		350	153		503	620
9050	60 amp, to 30 HP motor		2.30	3.478		800	213		1,013	1,225
9070	100 amp, to 60 HP motor		1.90	4.211		800	258		1,058	1,275
9100	200 amp, to 125 HP motor		1.30	6.154		1,200	380		1,580	1,900
9110	400 amp, to 200 HP motor	2 Elec	1.80	8.889		3,050	545		3,595	4,200
9990	Minimum labor/equipment charge	1 Elec	3	2.667	Job		164		164	252

26 28 16.40 Time Switches		Crew	Daily Output	Labor-Hours	Unit	Material	Labor	Equipment	Total	Total Incl O&P
0010	TIME SWITCHES									
0100	Single pole, single throw, 24 hour dial	1 Elec	4	2	Ea.	131	123		254	335
0200	24 hour dial with reserve power		3.60	2.222		675	136		811	950
0300	Astronomic dial		3.60	2.222		256	136		392	490
0400	Astronomic dial with reserve power		3.30	2.424		940	149		1,089	1,250
0500	7 day calendar dial		3.30	2.424		218	149		367	470
0600	7 day calendar dial with reserve power		3.20	2.500		212	153		365	470
0700	Photo cell 2,000 watt		8	1		28.50	61.50		90	126

26 28 Low-Voltage Circuit Protective Devices

26 28 16 – Enclosed Switches and Circuit Breakers

26 28 16.40 Time Switches		Crew	Daily Output	Labor-Hours	Unit	2020 Bare Costs Material	Labor	Equipment	Total	Total Incl O&P
1080	Load management device, 4 loads	1 Elec	2	4	Ea.	1,150	245		1,395	1,625
1100	8 loads		1	8		2,400	490		2,890	3,375
9000	Minimum labor/equipment charge		3.50	2.286	Job		140		140	216

26 29 Low-Voltage Controllers

26 29 13 – Enclosed Controllers

26 29 13.10 Contactors, AC		Crew	Daily Output	Labor-Hours	Unit	2020 Bare Costs Material	Labor	Equipment	Total	Total Incl O&P
0010	**CONTACTORS, AC** Enclosed (NEMA 1)									
0050	Lighting, 600 volt 3 pole, electrically held									
0100	20 amp	1 Elec	4	2	Ea.	375	123		498	600
0200	30 amp		3.60	2.222		300	136		436	540
0300	60 amp		3	2.667		625	164		789	940
0400	100 amp		2.50	3.200		1,000	196		1,196	1,425
0500	200 amp		1.40	5.714		2,525	350		2,875	3,325
0600	300 amp	2 Elec	1.60	10		4,525	615		5,140	5,925
0800	600 volt 3 pole, mechanically held, 30 amp	1 Elec	3.60	2.222		590	136		726	860
0900	60 amp		3	2.667		1,175	164		1,339	1,550
1000	75 amp		2.80	2.857		1,500	175		1,675	1,925
1100	100 amp		2.50	3.200		1,700	196		1,896	2,175
1200	150 amp		2	4		3,275	245		3,520	3,975
1300	200 amp		1.40	5.714		5,650	350		6,000	6,775
1500	Magnetic with auxiliary contact, size 00, 9 amp		4	2		189	123		312	395
1600	Size 0, 18 amp		4	2		225	123		348	435
1700	Size 1, 27 amp		3.60	2.222		256	136		392	490
1800	Size 2, 45 amp		3	2.667		475	164		639	775
1900	Size 3, 90 amp		2.50	3.200		770	196		966	1,150
2000	Size 4, 135 amp		2.30	3.478		1,750	213		1,963	2,250
2100	Size 5, 270 amp	2 Elec	1.80	8.889		3,675	545		4,220	4,900
2200	Size 6, 540 amp		1.20	13.333		10,800	820		11,620	13,200
2300	Size 7, 810 amp		1	16		14,500	980		15,480	17,500
2310	Size 8, 1,215 amp		.80	20		22,600	1,225		23,825	26,700
2500	Magnetic, 240 volt, 1-2 pole, 0.75 HP motor	1 Elec	4	2		152	123		275	355
2520	2 HP motor		3.60	2.222		201	136		337	430
2540	5 HP motor		2.50	3.200		415	196		611	755
2560	10 HP motor		1.40	5.714		680	350		1,030	1,300
2600	240 volt or less, 3 pole,0.75 HP motor		4	2		152	123		275	355
2620	5 HP motor		3.60	2.222		189	136		325	415
2640	10 HP motor		3.60	2.222		219	136		355	450
2660	15 HP motor		2.50	3.200		440	196		636	780
2700	25 HP motor		2.50	3.200		440	196		636	780
2720	30 HP motor	2 Elec	2.80	5.714		730	350		1,080	1,350
2740	40 HP motor		2.80	5.714		730	350		1,080	1,350
2760	50 HP motor		1.60	10		730	615		1,345	1,750
2800	75 HP motor		1.60	10		1,725	615		2,340	2,825
2820	100 HP motor		1	16		1,725	980		2,705	3,375
2860	150 HP motor		1	16		3,650	980		4,630	5,525
2880	200 HP motor		1	16		3,650	980		4,630	5,525
3000	600 volt, 3 pole, 5 HP motor	1 Elec	4	2		189	123		312	395
3020	10 HP motor		3.60	2.222		219	136		355	450
3040	25 HP motor		3	2.667		440	164		604	730
3100	50 HP motor		2.50	3.200		730	196		926	1,100

26 29 13.10 Contactors, AC		Crew	Daily Output	Labor-Hours	Unit	Material	2020 Bare Costs Labor	Equipment	Total	Total Incl O&P
3160	100 HP motor	2 Elec	2.80	5.714	Ea.	1,725	350		2,075	2,425
3220	200 HP motor	"	1.60	10		3,650	615		4,265	4,975

26 29 13.20 Control Stations

		Crew	Daily Output	Labor-Hours	Unit	Material	Labor	Equipment	Total	Total Incl O&P
0010	**CONTROL STATIONS**									
0050	NEMA 1, heavy duty, stop/start	1 Elec	8	1	Ea.	153	61.50		214.50	263
0100	Stop/start, pilot light		6.20	1.290		208	79		287	350
0200	Hand/off/automatic		6.20	1.290		113	79		192	246
0400	Stop/start/reverse		5.30	1.509		206	92.50		298.50	370
0500	NEMA 7, heavy duty, stop/start		6	1.333		555	82		637	735
0600	Stop/start, pilot light		4	2		680	123		803	935
0700	NEMA 7 or 9, 1 element		6	1.333		450	82		532	625
0800	2 element		6	1.333		585	82		667	765
0900	3 element		4	2		1,250	123		1,373	1,575
0910	Selector switch, 2 position		6	1.333		455	82		537	625
0920	3 position		4	2		455	123		578	690
0930	Oiltight, 1 element		8	1		112	61.50		173.50	219
0940	2 element		6.20	1.290		156	79		235	293
0950	3 element		5.30	1.509		142	92.50		234.50	300
0960	Selector switch, 2 position		6.20	1.290		113	79		192	247
0970	3 position		5.30	1.509		118	92.50		210.50	273

26 29 13.30 Control Switches

		Crew	Daily Output	Labor-Hours	Unit	Material	Labor	Equipment	Total	Total Incl O&P
0010	**CONTROL SWITCHES** Field installed									
6000	Push button 600 V 10A, momentary contact									
6150	Standard operator with colored button	1 Elec	34	.235	Ea.	18.90	14.45		33.35	43
6160	With single block 1NO 1NC		18	.444		42	27.50		69.50	88.50
6170	With double block 2NO 2NC		15	.533		61	32.50		93.50	118
6180	Std operator w/mushroom button 1-9/16" diam.		34	.235		39	14.45		53.45	64.50
6190	Std operator w/mushroom button 2-1/4" diam.									
6200	With single block 1NO 1NC	1 Elec	18	.444	Ea.	57	27.50		84.50	105
6210	With double block 2NO 2NC		15	.533		77	32.50		109.50	135
6500	Maintained contact, selector operator		34	.235		57	14.45		71.45	85
6510	With single block 1NO 1NC		18	.444		77	27.50		104.50	127
6520	With double block 2NO 2NC		15	.533		96.50	32.50		129	157
6560	Spring-return selector operator		34	.235		57	14.45		71.45	85
6570	With single block 1NO 1NC		18	.444		77	27.50		104.50	127
6580	With double block 2NO 2NC		15	.533		96.50	32.50		129	157
6620	Transformer operator w/illuminated									
6630	button 6 V #12 lamp	1 Elec	32	.250	Ea.	96.50	15.35		111.85	130
6640	With single block 1NO 1NC w/guard		16	.500		116	30.50		146.50	175
6650	With double block 2NO 2NC w/guard		13	.615		136	38		174	207
6690	Combination operator		34	.235		57	14.45		71.45	85
6700	With single block 1NO 1NC		18	.444		77	27.50		104.50	127
6710	With double block 2NO 2NC		15	.533		96.50	32.50		129	157
9000	Indicating light unit, full voltage									
9010	110-125 V front mount	1 Elec	32	.250	Ea.	77	15.35		92.35	108
9020	130 V resistor type		32	.250		63.50	15.35		78.85	93
9030	6 V transformer type		32	.250		70	15.35		85.35	101

26 29 13.40 Relays

		Crew	Daily Output	Labor-Hours	Unit	Material	Labor	Equipment	Total	Total Incl O&P
0010	**RELAYS** Enclosed (NEMA 1)									
0050	600 volt AC, 1 pole, 12 amp	1 Elec	5.30	1.509	Ea.	89	92.50		181.50	241
0100	2 pole, 12 amp		5	1.600		89	98		187	249
0200	4 pole, 10 amp		4.50	1.778		119	109		228	299

26 29 13 – Enclosed Controllers

26 29 13.40 Relays		Crew	Daily Output	Labor-Hours	Unit	Material	Labor (2020 Bare Costs)	Equipment	Total	Total Incl O&P
0500	250 volt DC, 1 pole, 15 amp	1 Elec	5.30	1.509	Ea.	117	92.50		209.50	271
0600	2 pole, 10 amp		5	1.600		111	98		209	273
0700	4 pole, 4 amp		4.50	1.778		147	109		256	330

26 29 23 – Variable-Frequency Motor Controllers

26 29 23.10 Variable Frequency Drives/Adj. Frequency Drives			Crew	Daily Output	Labor-Hours	Unit	Material	Labor (2020 Bare Costs)	Equipment	Total	Total Incl O&P
0010	VARIABLE FREQUENCY DRIVES/ADJ. FREQUENCY DRIVES										
0100	Enclosed (NEMA 1), 460 volt, for 3 HP motor size	G	1 Elec	.80	10	Ea.	1,825	615		2,440	2,950
0110	5 HP motor size	G		.80	10		2,050	615		2,665	3,200
0120	7.5 HP motor size	G		.67	11.940		2,450	735		3,185	3,800
0130	10 HP motor size	G		.67	11.940		2,775	735		3,510	4,175
0140	15 HP motor size	G	2 Elec	.89	17.978		3,450	1,100		4,550	5,500
0150	20 HP motor size	G		.89	17.978		4,350	1,100		5,450	6,475
0160	25 HP motor size	G		.67	23.881		5,125	1,475		6,600	7,900
0170	30 HP motor size	G		.67	23.881		6,250	1,475		7,725	9,125
0180	40 HP motor size	G		.67	23.881		7,225	1,475		8,700	10,200
0190	50 HP motor size	G		.53	30.189		9,650	1,850		11,500	13,500
0200	60 HP motor size	G	R-3	.56	35.714		11,700	2,175	340	14,215	16,600
0210	75 HP motor size	G		.56	35.714		13,700	2,175	340	16,215	18,800
0220	100 HP motor size	G		.50	40		16,800	2,450	380	19,630	22,600
0230	125 HP motor size	G		.50	40		18,200	2,450	380	21,030	24,200
0240	150 HP motor size	G		.50	40		23,600	2,450	380	26,430	30,200
0250	200 HP motor size	G		.42	47.619		30,600	2,900	450	33,950	38,700
1100	Custom-engineered, 460 volt, for 3 HP motor size	G	1 Elec	.56	14.286		3,425	875		4,300	5,125
1110	5 HP motor size	G		.56	14.286		3,450	875		4,325	5,150
1120	7.5 HP motor size	G		.47	17.021		3,100	1,050		4,150	5,000
1130	10 HP motor size	G		.47	17.021		3,250	1,050		4,300	5,200
1140	15 HP motor size	G	2 Elec	.62	25.806		4,700	1,575		6,275	7,600
1150	20 HP motor size	G		.62	25.806		4,350	1,575		5,925	7,200
1160	25 HP motor size	G		.47	34.043		5,375	2,100		7,475	9,150
1170	30 HP motor size	G		.47	34.043		6,625	2,100		8,725	10,500
1180	40 HP motor size	G		.47	34.043		7,925	2,100		10,025	12,000
1190	50 HP motor size	G		.37	43.243		10,500	2,650		13,150	15,600
1200	60 HP motor size	G	R-3	.39	51.282		15,800	3,125	485	19,410	22,800
1210	75 HP motor size	G		.39	51.282		14,600	3,125	485	18,210	21,500
1220	100 HP motor size	G		.35	57.143		18,500	3,500	540	22,540	26,300
1230	125 HP motor size	G		.35	57.143		19,800	3,500	540	23,840	27,800
1240	150 HP motor size	G		.35	57.143		22,500	3,500	540	26,540	30,800
1250	200 HP motor size	G		.29	68.966		25,700	4,225	650	30,575	35,400
2000	For complex & special design systems to meet specific										
2010	requirements, obtain quote from vendor.										

26 31 Photovoltaic Collectors

26 31 13 – Photovoltaics

26 31 13.50 Solar Energy - Photovoltaics			Crew	Daily Output	Labor-Hours	Unit	2020 Bare Costs Material	Labor	Equipment	Total	Total Incl O&P
0010	**SOLAR ENERGY - PHOTOVOLTAICS**										
0220	Alt. energy source, photovoltaic module, 6 watt, 15 V	G	1 Elec	8	1	Ea.	54	61.50		115.50	154
0230	10 watt, 16.3 V	G		8	1		119	61.50		180.50	226
0240	20 watt, 14.5 V	G		8	1		176	61.50		237.50	289
0250	36 watt, 17 V	G		8	1		175	61.50		236.50	288
0260	55 watt, 17 V	G		8	1		238	61.50		299.50	355
0270	75 watt, 17 V	G		8	1		400	61.50		461.50	535
0280	130 watt, 33 V	G		8	1		605	61.50		666.50	765
0290	140 watt, 33 V	G		8	1		495	61.50		556.50	640
0300	150 watt, 33 V	G		8	1		450	61.50		511.50	590
0310	DC to AC inverter for, 12 V, 2,000 watt	G		4	2		1,425	123		1,548	1,750
0320	12 V, 2,500 watt	G		4	2		1,100	123		1,223	1,425
0330	24 V, 2,500 watt	G		4	2		1,675	123		1,798	2,050
0340	12 V, 3,000 watt	G		3	2.667		1,325	164		1,489	1,700
0350	24 V, 3,000 watt	G		3	2.667		2,400	164		2,564	2,900
0360	24 V, 4,000 watt	G		2	4		3,675	245		3,920	4,425
0370	48 V, 4,000 watt	G		2	4		3,050	245		3,295	3,750
0380	48 V, 5,500 watt	G		2	4		3,000	245		3,245	3,675
0390	PV components, combiner box, 10 lug, NEMA 3R enclosure	G		4	2		270	123		393	485
0400	Fuse, 15 A for combiner box	G		40	.200		21.50	12.25		33.75	43
0410	Battery charger controller w/temperature sensor	G		4	2		450	123		573	685
0420	Digital readout panel, displays hours, volts, amps, etc.	G		4	2		219	123		342	430
0430	Deep cycle solar battery, 6 V, 180 Ah (C/20)	G		8	1		335	61.50		396.50	465
0440	Battery interconn, 15" AWG #2/0, sealed w/copper ring lugs	G		16	.500		16.60	30.50		47.10	65.50
0442	Battery interconn, 24" AWG #2/0, sealed w/copper ring lugs	G		16	.500		26.50	30.50		57	76
0444	Battery interconn, 60" AWG #2/0, sealed w/copper ring lugs	G		16	.500		53	30.50		83.50	105
0446	Batt temp computer probe, RJ11 jack, 15' cord	G		16	.500		22	30.50		52.50	71
0450	System disconnect, DC 175 amp circuit breaker	G		8	1		223	61.50		284.50	340
0460	Conduit box for inverter	G		8	1		63.50	61.50		125	164
0470	Low voltage disconnect	G		8	1		63.50	61.50		125	165
0480	Vented battery enclosure, wood	G	1 Carp	2	4		285	213		498	655
0490	PV rack system, roof, non-penetrating ballast, 1 panel	G	R-1A	30.50	.525		1,000	29		1,029	1,175
0500	Penetrating surface mount, on steel framing, 1 panel			5.13	3.122		58	172		230	330
0510	On wood framing, 1 panel			11	1.455		57	80.50		137.50	187
0520	With standoff, 1 panel			11	1.455		66	80.50		146.50	197
0530	Ground, ballast, fixed, 3 panel			20.50	.780		1,100	43		1,143	1,275
0540	4 panel			20.50	.780		1,500	43		1,543	1,725
0550	5 panel			20.50	.780		1,975	43		2,018	2,250
0560	6 panel			20.50	.780		2,275	43		2,318	2,575
0570	Adjustable, 3 panel			20.50	.780		1,175	43		1,218	1,350
0580	4 panel			20.50	.780		1,600	43		1,643	1,825
0590	5 panel			20.50	.780		2,100	43		2,143	2,400
0600	6 panel			20.50	.780		2,425	43		2,468	2,750
0605	Top of Pole, see 26 56 13.10 3800+, 5400+ & 33 71 16.33 7400+ for poles										
0610	Passive tracking, 1 panel		R-1A	20.50	.780	Ea.	685	43		728	815
0620	2 panel			20.50	.780		1,350	43		1,393	1,575
0630	3 panel			20.50	.780		1,950	43		1,993	2,225
0640	4 panel			20.50	.780		2,100	43		2,143	2,375
0650	6 panel			20.50	.780		2,300	43		2,343	2,600
0660	8 panel			20.50	.780		3,425	43		3,468	3,825
1020	Photovoltaic module, Thin film		1 Elec	8	1		176	61.50		237.50	289
1040	Photovoltaic module, Cadium telluride			8	1		238	61.50		299.50	355
1060	Photovoltaic module, Polycrystalline			8	1		176	61.50		237.50	289

26 31 Photovoltaic Collectors

26 31 13 – Photovoltaics

26 31 13.50 Solar Energy - Photovoltaics		Crew	Daily Output	Labor-Hours	Unit	2020 Bare Costs Material	Labor	Equipment	Total	Total Incl O&P
1080	Photovoltaic module, Monocrystalline	1 Elec	8	1	Ea.	605	61.50		666.50	765

26 32 Packaged Generator Assemblies

26 32 13 – Engine Generators

26 32 13.13 Diesel-Engine-Driven Generator Sets

		Crew	Daily Output	Labor-Hours	Unit	Material	Labor	Equipment	Total	Total Incl O&P
0010	**DIESEL-ENGINE-DRIVEN GENERATOR SETS**									
2000	Diesel engine, including battery, charger,									
2010	muffler, & day tank, 30 kW	R-3	.55	36.364	Ea.	11,200	2,225	345	13,770	16,100
2100	50 kW		.42	47.619		21,800	2,900	450	25,150	29,000
2110	60 kW		.39	51.282		22,000	3,125	485	25,610	29,600
2200	75 kW		.35	57.143		25,300	3,500	540	29,340	33,800
2300	100 kW		.31	64.516		27,600	3,950	610	32,160	37,200
2400	125 kW		.29	68.966		29,000	4,225	650	33,875	39,100
2500	150 kW		.26	76.923		45,800	4,700	725	51,225	58,500
2501	Generator set, dsl eng in alum encl, incl btry, chgr, muf & day tank,150 kW		.26	76.923		41,500	4,700	725	46,925	53,500
2600	175 kW		.25	80		48,900	4,900	755	54,555	62,000
2700	200 kW		.24	83.333		49,900	5,100	790	55,790	63,500
2800	250 kW		.23	86.957		52,500	5,325	820	58,645	67,000
2850	275 kW		.22	90.909		58,000	5,550	860	64,410	73,500
2900	300 kW		.22	90.909		58,000	5,550	860	64,410	73,000
3000	350 kW		.20	100		67,000	6,100	945	74,045	84,500
3100	400 kW		.19	105		78,000	6,425	995	85,420	97,000
3200	500 kW		.18	111		106,500	6,800	1,050	114,350	128,500
3220	600 kW		.17	118		128,500	7,200	1,100	136,800	154,000
3230	650 kW	R-13	.38	111		170,500	6,550	620	177,670	198,500
3240	750 kW		.38	111		160,000	6,550	620	167,170	187,000
3250	800 kW		.36	117		156,500	6,925	655	164,080	183,500
3260	900 kW		.31	135		195,500	8,025	765	204,290	228,000
3270	1,000 kW		.27	156		205,000	9,225	875	215,100	240,500

26 32 13.16 Gas-Engine-Driven Generator Sets

		Crew	Daily Output	Labor-Hours	Unit	Material	Labor	Equipment	Total	Total Incl O&P
0010	**GAS-ENGINE-DRIVEN GENERATOR SETS**									
0020	Gas or gasoline operated, includes battery,									
0050	charger, & muffler									
0200	7.5 kW	R-3	.83	24.096	Ea.	8,750	1,475	228	10,453	12,200
0300	11.5 kW		.71	28.169		12,400	1,725	266	14,391	16,500
0400	20 kW		.63	31.746		14,600	1,950	300	16,850	19,400
0500	35 kW		.55	36.364		17,400	2,225	345	19,970	22,900
0520	60 kW		.50	40		22,900	2,450	380	25,730	29,400
0600	80 kW	R-13	.40	105		28,500	6,225	590	35,315	41,700
0700	100 kW		.33	127		31,200	7,550	715	39,465	46,900
0800	125 kW		.28	150		64,000	8,900	845	73,745	85,000
0900	185 kW		.25	168		84,500	9,950	945	95,395	109,500

26 33 Battery Equipment

26 33 43 – Battery Chargers

26 33 43.55 Electric Vehicle Charging		Crew	Daily Output	Labor-Hours	Unit	Material	2020 Bare Costs Labor	Equipment	Total	Total Incl O&P
0010	**ELECTRIC VEHICLE CHARGING**									
0020	Level 2, wall mounted									
2200	Heavy duty G	R-1A	15.36	1.042	Ea.	1,975	57.50		2,032.50	2,275
2210	with RFID G		12.29	1.302		3,875	72		3,947	4,375
2300	Free standing, single connector G		10.24	1.563		2,000	86.50		2,086.50	2,325
2310	with RFID G		8.78	1.822		4,600	101		4,701	5,200
2320	Double connector G		7.68	2.083		3,700	115		3,815	4,225
2330	with RFID G		6.83	2.343		7,825	129		7,954	8,800

26 33 53 – Static Uninterruptible Power Supply

26 33 53.10 Uninterruptible Power Supply/Conditioner Trans.		Crew	Daily Output	Labor-Hours	Unit	Material	Labor	Equipment	Total	Total Incl O&P
0010	**UNINTERRUPTIBLE POWER SUPPLY/CONDITIONER TRANSFORMERS**									
0100	Volt. regulating, isolating transf., w/invert. & 10 min. battery pack									
0110	Single-phase, 120 V, 0.35 kVA	1 Elec	2.29	3.493	Ea.	1,150	214		1,364	1,575
0120	0.5 kVA		2	4		1,200	245		1,445	1,700
0130	For additional 55 min. battery, add to 0.35 kVA		2.29	3.493		675	214		889	1,075
0140	Add to 0.5 kVA		1.14	7.018		705	430		1,135	1,450
0150	Single-phase, 120 V, 0.75 kVA		.80	10		1,525	615		2,140	2,625
0160	1.0 kVA		.80	10		2,175	615		2,790	3,325
0170	1.5 kVA	2 Elec	1.14	14.035		3,725	860		4,585	5,425
0180	2 kVA	"	.89	17.978		4,050	1,100		5,150	6,150
0190	3 kVA	R-3	.63	31.746		5,000	1,950	300	7,250	8,825
0200	5 kVA		.42	47.619		7,100	2,900	450	10,450	12,800
0210	7.5 kVA		.33	60.606		9,075	3,700	575	13,350	16,300
0220	10 kVA		.28	71.429		12,200	4,375	675	17,250	21,000
0230	15 kVA		.22	90.909		15,700	5,550	860	22,110	26,700
0240	3 phase, 120/208 V input 120/208 V output, 20 kVA, incl 17 min. battery		.21	95.238		29,000	5,825	900	35,725	42,000
0242	30 kVA, incl 11 min. battery		.20	100		32,500	6,100	945	39,545	46,300
0250	40 kVA, incl 15 min. battery		.20	100		43,400	6,100	945	50,445	58,000
0260	480 V input 277/480 V output, 60 kVA, incl 6 min. battery		.19	105		48,800	6,425	995	56,220	64,500
0262	80 kVA, incl 4 min. battery		.19	105		58,000	6,425	995	65,420	75,000
0400	For additional 34 min./15 min. battery, add to 40 kVA		.71	28.011		15,100	1,700	265	17,065	19,600
0600	For complex & special design systems to meet specific									
0610	requirements, obtain quote from vendor									

26 35 Power Filters and Conditioners

26 35 13 – Capacitors

26 35 13.10 Capacitors Indoor		Crew	Daily Output	Labor-Hours	Unit	Material	Labor	Equipment	Total	Total Incl O&P
0010	**CAPACITORS INDOOR**									
0020	240 volts, single & 3 phase, 0.5 kVAR	1 Elec	2.70	2.963	Ea.	600	182		782	940
0100	1.0 kVAR		2.70	2.963		725	182		907	1,075
0150	2.5 kVAR		2	4		815	245		1,060	1,275
0200	5.0 kVAR		1.80	4.444		795	273		1,068	1,300
0250	7.5 kVAR		1.60	5		1,625	305		1,930	2,275
0300	10 kVAR		1.50	5.333		1,750	325		2,075	2,425
0350	15 kVAR		1.30	6.154		2,075	380		2,455	2,850
0400	20 kVAR		1.10	7.273		2,850	445		3,295	3,800
0450	25 kVAR		1	8		3,350	490		3,840	4,425
1000	480 volts, single & 3 phase, 1 kVAR		2.70	2.963		545	182		727	880
1050	2 kVAR		2.70	2.963		630	182		812	975
1100	5 kVAR		2	4		795	245		1,040	1,250

26 35 13 – Capacitors

26 35 13.10 Capacitors Indoor		Crew	Daily Output	Labor-Hours	Unit	Material	2020 Bare Costs Labor	Equipment	Total	Total Incl O&P
1150	7.5 kVAR	1 Elec	2	4	Ea.	855	245		1,100	1,325
1200	10 kVAR		2	4		1,200	245		1,445	1,700
1250	15 kVAR		2	4		1,525	245		1,770	2,050
1300	20 kVAR		1.60	5		1,600	305		1,905	2,225
1350	30 kVAR		1.50	5.333		1,575	325		1,900	2,250
1400	40 kVAR		1.20	6.667		2,275	410		2,685	3,150
1450	50 kVAR		1.10	7.273		2,875	445		3,320	3,850
2000	600 volts, single & 3 phase, 1 kVAR		2.70	2.963		565	182		747	900
2050	2 kVAR		2.70	2.963		650	182		832	995
2100	5 kVAR		2	4		795	245		1,040	1,250
2150	7.5 kVAR		2	4		855	245		1,100	1,325
2200	10 kVAR		2	4		1,225	245		1,470	1,700
2250	15 kVAR		1.60	5		1,550	305		1,855	2,175
2300	20 kVAR		1.60	5		1,675	305		1,980	2,300
2350	25 kVAR		1.50	5.333		1,825	325		2,150	2,500
2400	35 kVAR		1.40	5.714		2,225	350		2,575	3,000
2450	50 kVAR		1.30	6.154		2,850	380		3,230	3,725

26 35 26 – Harmonic Filters

26 35 26.10 Computer Isolation Transformer

Line	Description	Crew	Daily Output	Labor-Hours	Unit	Material	Labor	Equipment	Total	Total Incl O&P
0010	**COMPUTER ISOLATION TRANSFORMER**									
0100	Computer grade									
0110	Single-phase, 120/240 V, 0.5 kVA	1 Elec	4	2	Ea.	455	123		578	690
0120	1.0 kVA		2.67	2.996		645	184		829	995
0130	2.5 kVA		2	4		985	245		1,230	1,450
0140	5 kVA		1.14	7.018		1,125	430		1,555	1,900

26 35 26.20 Computer Regulator Transformer

Line	Description	Crew	Daily Output	Labor-Hours	Unit	Material	Labor	Equipment	Total	Total Incl O&P
0010	**COMPUTER REGULATOR TRANSFORMER**									
0100	Ferro-resonant, constant voltage, variable transformer									
0110	Single-phase, 240 V, 0.5 kVA	1 Elec	2.67	2.996	Ea.	600	184		784	945
0120	1.0 kVA		2	4		825	245		1,070	1,300
0130	2.0 kVA		1	8		1,425	490		1,915	2,300
0210	Plug-in unit 120 V, 0.14 kVA		8	1		350	61.50		411.50	475
0220	0.25 kVA		8	1		405	61.50		466.50	540
0230	0.5 kVA		8	1		600	61.50		661.50	755
0240	1.0 kVA		5.33	1.501		825	92		917	1,050
0250	2.0 kVA		4	2		1,425	123		1,548	1,750

26 35 26.30 Power Conditioner Transformer

Line	Description	Crew	Daily Output	Labor-Hours	Unit	Material	Labor	Equipment	Total	Total Incl O&P
0010	**POWER CONDITIONER TRANSFORMER**									
0100	Electronic solid state, buck-boost, transformer, w/tap switch									
0110	Single-phase, 115 V, 3.0 kVA, + or - 3% accuracy	2 Elec	1.60	10	Ea.	3,250	615		3,865	4,525
0120	208, 220, 230, or 240 V, 5.0 kVA, + or - 1.5% accuracy	3 Elec	1.60	15		4,200	920		5,120	6,050
0130	5.0 kVA, + or - 6% accuracy	2 Elec	1.14	14.035		3,775	860		4,635	5,475
0140	7.5 kVA, + or - 1.5% accuracy	3 Elec	1.50	16		5,350	980		6,330	7,375
0150	7.5 kVA, + or - 6% accuracy		1.60	15		4,450	920		5,370	6,325
0160	10.0 kVA, + or - 1.5% accuracy		1.33	18.045		7,125	1,100		8,225	9,525
0170	10.0 kVA, + or - 6% accuracy		1.41	17.021		6,050	1,050		7,100	8,250

26 35 26.40 Transient Voltage Suppressor Transformer

Line	Description	Crew	Daily Output	Labor-Hours	Unit	Material	Labor	Equipment	Total	Total Incl O&P
0010	**TRANSIENT VOLTAGE SUPPRESSOR TRANSFORMER**									
0110	Single-phase, 120 V, 1.8 kVA	1 Elec	4	2	Ea.	1,625	123		1,748	2,000
0120	3.6 kVA		4	2		2,625	123		2,748	3,075
0130	7.2 kVA		3.20	2.500		3,150	153		3,303	3,700

26 35 Power Filters and Conditioners

26 35 26 – Harmonic Filters

26 35 26.40 Transient Voltage Suppressor Transformer		Crew	Daily Output	Labor-Hours	Unit	Material	2020 Bare Costs Labor	Equipment	Total	Total Incl O&P
0150	240 V, 3.6 kVA	1 Elec	4	2	Ea.	3,275	123		3,398	3,825
0160	7.2 kVA		4	2		3,825	123		3,948	4,400
0170	14.4 kVA		3.20	2.500		5,350	153		5,503	6,100
0210	Plug-in unit, 120 V, 1.8 kVA		8	1		1,125	61.50		1,186.50	1,350

26 35 53 – Voltage Regulators

26 35 53.10 Automatic Voltage Regulators

		Crew	Daily Output	Labor-Hours	Unit	Material	Labor	Equipment	Total	Total Incl O&P
0010	**AUTOMATIC VOLTAGE REGULATORS**									
0100	Computer grade, solid state, variable transf. volt. regulator									
0110	Single-phase, 120 V, 8.6 kVA	2 Elec	1.33	12.030	Ea.	6,725	740		7,465	8,525
0120	17.3 kVA		1.14	14.035		7,950	860		8,810	10,100
0130	208/240 V, 7.5/8.6 kVA		1.33	12.030		6,725	740		7,465	8,525
0140	13.5/15.6 kVA		1.33	12.030		7,950	740		8,690	9,875
0150	27.0/31.2 kVA		1.14	14.035		10,000	860		10,860	12,300
0210	Two-phase, single control, 208/240 V, 15.0/17.3 kVA		1.14	14.035		7,950	860		8,810	10,100
0220	Individual phase control, 15.0/17.3 kVA		1.14	14.035		7,950	860		8,810	10,100
0230	30.0/34.6 kVA	3 Elec	1.33	18.045		10,000	1,100		11,100	12,700
0310	Three-phase single control, 208/240 V, 26/30 kVA	2 Elec	1	16		7,950	980		8,930	10,300
0320	380/480 V, 24/30 kVA	"	1	16		7,950	980		8,930	10,300
0330	43/54 kVA	3 Elec	1.33	18.045		14,400	1,100		15,500	17,600
0340	Individual phase control, 208 V, 26 kVA	"	1.33	18.045		7,950	1,100		9,050	10,500
0350	52 kVA	R-3	.91	21.978		10,000	1,350	208	11,558	13,300
0360	340/480 V, 24/30 kVA	2 Elec	1	16		7,950	980		8,930	10,300
0370	43/54 kVA	"	1	16		10,000	980		10,980	12,500
0380	48/60 kVA	3 Elec	1.33	18.045		14,500	1,100		15,600	17,700
0390	86/108 kVA	R-3	.91	21.978		16,100	1,350	208	17,658	20,100
0500	Standard grade, solid state, variable transformer volt. regulator									
0510	Single-phase, 115 V, 2.3 kVA	1 Elec	2.29	3.493	Ea.	2,125	214		2,339	2,675
0520	4.2 kVA		2	4		3,400	245		3,645	4,125
0530	6.6 kVA		1.14	7.018		4,175	430		4,605	5,275
0540	13.0 kVA		1.14	7.018		7,225	430		7,655	8,625
0550	16.6 kVA	2 Elec	1.23	13.008		8,525	800		9,325	10,600
0610	230 V, 8.3 kVA		1.33	12.030		7,225	740		7,965	9,075
0620	21.4 kVA		1.23	13.008		8,525	800		9,325	10,600
0630	29.9 kVA		1.23	13.008		8,525	800		9,325	10,600
0710	460 V, 9.2 kVA		1.33	12.030		7,225	740		7,965	9,075
0720	20.7 kVA		1.23	13.008		8,525	800		9,325	10,600
0810	Three-phase, 230 V, 13.1 kVA	3 Elec	1.41	17.021		7,225	1,050		8,275	9,550
0820	19.1 kVA		1.41	17.021		8,525	1,050		9,575	11,000
0830	25.1 kVA		1.23	19.512		8,525	1,200		9,725	11,200
0840	57.8 kVA	R-3	.95	21.053		15,500	1,275	199	16,974	19,200
0850	74.9 kVA	"	.91	21.978		15,500	1,350	208	17,058	19,300
0910	460 V, 14.3 kVA	3 Elec	1.41	17.021		7,225	1,050		8,275	9,550
0920	19.1 kVA		1.41	17.021		8,525	1,050		9,575	11,000
0930	27.9 kVA		1.23	19.512		8,525	1,200		9,725	11,200
0940	59.8 kVA	R-3	1	20		15,500	1,225	189	16,914	19,100
0950	79.7 kVA		.95	21.053		17,300	1,275	199	18,774	21,300
0960	118 kVA		.95	21.053		18,200	1,275	199	19,674	22,300
1000	Laboratory grade, precision, electronic voltage regulator									
1110	Single-phase, 115 V, 0.5 kVA	1 Elec	2.29	3.493	Ea.	1,650	214		1,864	2,150
1120	1.0 kVA		2	4		1,750	245		1,995	2,300
1130	3.0 kVA		.80	10		2,450	615		3,065	3,650
1140	6.0 kVA	2 Elec	1.46	10.959		4,550	670		5,220	6,025

26 35 Power Filters and Conditioners

26 35 53 – Voltage Regulators

26 35 53.10 Automatic Voltage Regulators		Crew	Daily Output	Labor-Hours	Unit	2020 Bare Costs Material	Labor	Equipment	Total	Total Incl O&P
1150	10.0 kVA	3 Elec	1	24	Ea.	5,925	1,475		7,400	8,775
1160	15.0 kVA	″	1.50	16		6,750	980		7,730	8,925
1210	230 V, 3.0 kVA	1 Elec	.80	10		2,775	615		3,390	4,000
1220	6.0 kVA	2 Elec	1.46	10.959		4,650	670		5,320	6,125
1230	10.0 kVA	3 Elec	1.71	14.035		6,175	860		7,035	8,125
1240	15.0 kVA	″	1.60	15		6,975	920		7,895	9,100

26 35 53.30 Transient Suppressor/Voltage Regulator		Crew	Daily Output	Labor-Hours	Unit	Material	Labor	Equipment	Total	Total Incl O&P
0010	**TRANSIENT SUPPRESSOR/VOLTAGE REGULATOR** (without isolation)									
0110	Single-phase, 115 V, 1.0 kVA	1 Elec	2.67	2.996	Ea.	1,100	184		1,284	1,500
0120	2.0 kVA		2.29	3.493		1,525	214		1,739	2,000
0130	4.0 kVA		2.13	3.756		1,875	230		2,105	2,400
0140	220 V, 1.0 kVA		2.67	2.996		1,100	184		1,284	1,500
0150	2.0 kVA		2.29	3.493		1,550	214		1,764	2,050
0160	4.0 kVA		2.13	3.756		1,975	230		2,205	2,525
0210	Plug-in unit, 120 V, 1.0 kVA		8	1		1,050	61.50		1,111.50	1,275
0220	2.0 kVA		8	1		1,500	61.50		1,561.50	1,750

26 36 Transfer Switches

26 36 13 – Manual Transfer Switches

26 36 13.10 Non-Automatic Transfer Switches		Crew	Daily Output	Labor-Hours	Unit	Material	Labor	Equipment	Total	Total Incl O&P
0010	**NON-AUTOMATIC TRANSFER SWITCHES** enclosed									
0100	Manual operated, 480 volt 3 pole, 30 amp	1 Elec	2.30	3.478	Ea.	1,100	213		1,313	1,550
0150	60 amp		1.90	4.211		2,025	258		2,283	2,625
0200	100 amp		1.30	6.154		3,525	380		3,905	4,475
0250	200 amp	2 Elec	2	8		4,450	490		4,940	5,650
0300	400 amp		1.60	10		8,350	615		8,965	10,100
0350	600 amp		1	16		5,475	980		6,455	7,525
1000	250 volt 3 pole, 30 amp	1 Elec	2.30	3.478		1,050	213		1,263	1,475
1100	60 amp		1.90	4.211		1,700	258		1,958	2,250
1150	100 amp		1.30	6.154		2,875	380		3,255	3,750
1200	200 amp	2 Elec	2	8		4,150	490		4,640	5,300
1300	600 amp	″	1	16		10,500	980		11,480	13,000
1500	Electrically operated, 480 volt 3 pole, 60 amp	1 Elec	1.90	4.211		1,650	258		1,908	2,225
1600	100 amp	″	1.30	6.154		1,650	380		2,030	2,400
1650	200 amp	2 Elec	2	8		2,725	490		3,215	3,725
1700	400 amp		1.60	10		3,800	615		4,415	5,125
1750	600 amp		1	16		5,475	980		6,455	7,525
2000	250 volt 3 pole, 30 amp	1 Elec	2.30	3.478		1,800	213		2,013	2,300
2050	60 amp	″	1.90	4.211		2,150	258		2,408	2,750
2150	200 amp	2 Elec	2	8		3,525	490		4,015	4,625
2200	400 amp		1.60	10		4,925	615		5,540	6,350
2250	600 amp		1	16		7,075	980		8,055	9,300
2500	NEMA 3R, 480 volt 3 pole, 60 amp	1 Elec	1.80	4.444		2,550	273		2,823	3,225
2550	100 amp	″	1.20	6.667		3,875	410		4,285	4,875
2600	200 amp	2 Elec	1.80	8.889		5,300	545		5,845	6,675
2650	400 amp	″	1.40	11.429		4,150	700		4,850	5,650
2800	NEMA 3R, 250 volt 3 pole solid state, 100 amp	1 Elec	1.20	6.667		3,975	410		4,385	5,000
2850	150 amp	2 Elec	1.80	8.889		5,275	545		5,820	6,675
2900	250 volt 2 pole solid state, 100 amp	1 Elec	1.30	6.154		3,875	380		4,255	4,825
2950	150 amp	2 Elec	2	8		5,150	490		5,640	6,400

26 36 Transfer Switches

26 36 23 – Automatic Transfer Switches

26 36 23.10 Automatic Transfer Switch Devices		Crew	Daily Output	Labor-Hours	Unit	Material	2020 Bare Costs Labor	Equipment	Total	Total Incl O&P
0010	AUTOMATIC TRANSFER SWITCH DEVICES									
0015	Switches, enclosed 120/240 volt, 2 pole, 30 amp	1 Elec	2.40	3.333	Ea.	2,025	205		2,230	2,550
0020	70 amp		2	4		2,025	245		2,270	2,600
0030	100 amp		1.35	5.926		2,025	365		2,390	2,775
0040	225 amp	2 Elec	2.10	7.619		2,975	465		3,440	4,000
0050	400 amp		1.70	9.412		4,525	575		5,100	5,900
0060	600 amp		1.06	15.094		9,575	925		10,500	11,900
0070	800 amp		.84	19.048		11,300	1,175		12,475	14,200
0100	Switches, enclosed 480 volt, 3 pole, 30 amp	1 Elec	2.30	3.478		3,425	213		3,638	4,100
0200	60 amp		1.90	4.211		3,425	258		3,683	4,175
0300	100 amp		1.30	6.154		3,425	380		3,805	4,350
0400	150 amp	2 Elec	2.40	6.667		4,200	410		4,610	5,250
0500	225 amp		2	8		5,400	490		5,890	6,700
0600	260 amp		2	8		6,250	490		6,740	7,625
0700	400 amp		1.60	10		7,850	615		8,465	9,575
0800	600 amp		1	16		11,300	980		12,280	13,900
0900	800 amp		.80	20		13,300	1,225		14,525	16,500
1000	1,000 amp		.76	21.053		18,500	1,300		19,800	22,400
1100	1,200 amp		.70	22.857		25,400	1,400		26,800	30,200
1200	1,600 amp		.60	26.667		28,900	1,625		30,525	34,300
1300	2,000 amp		.50	32		32,400	1,975		34,375	38,600
1600	Accessories, time delay on engine starting					284			284	310
1700	Adjustable time delay on retransfer					284			284	310
1800	Shunt trips for customer connections					505			505	555
1900	Maintenance select switch					115			115	127
2000	Auxiliary contact when normal fails					103			103	114
2100	Pilot light-emergency					115			115	127
2200	Pilot light-normal					115			115	127
2300	Auxiliary contact-closed on normal					133			133	147
2400	Auxiliary contact-closed on emergency					133			133	147
2500	Emergency source sensing, frequency relay					585			585	645

26 41 Facility Lightning Protection

26 41 13 – Lightning Protection for Structures

26 41 13.13 Lightning Protection for Buildings		Crew	Daily Output	Labor-Hours	Unit	Material	Labor	Equipment	Total	Total Incl O&P
0010	LIGHTNING PROTECTION FOR BUILDINGS									
0200	Air terminals & base, copper									
0400	3/8" diameter x 10" (to 75' high)	1 Elec	8	1	Ea.	26	61.50		87.50	123
0500	1/2" diameter x 12" (over 75' high)		8	1		30	61.50		91.50	128
0520	1/2" diameter x 24"		7.30	1.096		40	67		107	148
0540	1/2" diameter x 60"		6.70	1.194		73.50	73.50		147	194
1000	Aluminum, 1/2" diameter x 12" (to 75' high)		8	1		17.15	61.50		78.65	113
1020	1/2" diameter x 24"		7.30	1.096		18.95	67		85.95	125
1040	1/2" diameter x 60"		6.70	1.194		25	73.50		98.50	141
1100	5/8" diameter x 12" (over 75' high)		8	1		18.25	61.50		79.75	115
2000	Cable, copper, 220 lb. per thousand ft. (to 75' high)		320	.025	L.F.	3.18	1.53		4.71	5.85
2100	375 lb. per thousand ft. (over 75' high)		230	.035		5.75	2.13		7.88	9.65
2500	Aluminum, 101 lb. per thousand ft. (to 75' high)		280	.029		.91	1.75		2.66	3.70
2600	199 lb. per thousand ft. (over 75' high)		240	.033		1.33	2.04		3.37	4.61
3000	Arrester, 175 volt AC to ground		8	1	Ea.	163	61.50		224.50	274

26 41 13 – Lightning Protection for Structures

26 41 13.13 Lightning Protection for Buildings		Crew	Daily Output	Labor-Hours	Unit	Material	2020 Bare Costs Labor	Equipment	Total	Total Incl O&P
3100	650 volt AC to ground	1 Elec	6.70	1.194	Ea.	152	73.50		225.50	281
4000	Air terminals, copper									
4010	3/8" x 10"	1 Elec	50	.160	Ea.	7.55	9.80		17.35	23.50
4020	1/2" x 12"		38	.211		11.60	12.90		24.50	32.50
4030	5/8" x 12"		33	.242		18.65	14.85		33.50	43.50
4040	1/2" x 24"		30	.267		21.50	16.35		37.85	49
4050	1/2" x 60"		19	.421		55	26		81	101
4060	Air terminals, aluminum									
4070	3/8" x 10"	1 Elec	50	.160	Ea.	3.32	9.80		13.12	18.75
4080	1/2" x 12"		38	.211		3.55	12.90		16.45	24
4090	5/8" x 12"		33	.242		4.62	14.85		19.47	28
4100	1/2" x 24"		30	.267		5.30	16.35		21.65	31
4110	1/2" x 60"		19	.421		11.55	26		37.55	53
4200	Air terminal bases, copper									
4210	Adhesive bases, 1/2"	1 Elec	15	.533	Ea.	18.45	32.50		50.95	71
4215	Adhesive base, lightning air terminal					18.45			18.45	20.50
4220	Bolted bases	1 Elec	9	.889		18.35	54.50		72.85	104
4230	Hinged bases		7	1.143		31.50	70		101.50	143
4240	Side mounted bases		9	.889		19.05	54.50		73.55	105
4250	Offset point support		9	.889		31	54.50		85.50	118
4260	Concealed base assembly		5	1.600		16.30	98		114.30	169
4270	Tee connector base		13	.615		48	38		86	111
4280	Tripod base, 36" for 60" air terminal		7	1.143		40	70		110	152
4290	Intermediate base, 1/2"		15	.533		30.50	32.50		63	84
4300	Air terminal bases, aluminum									
4310	Adhesive bases, 1/2"	1 Elec	15	.533	Ea.	9.15	32.50		41.65	60.50
4320	Bolted bases		9	.889		13.60	54.50		68.10	99
4330	Hinged bases		7	1.143		22.50	70		92.50	133
4340	Side mounted bases		9	.889		12.15	54.50		66.65	97.50
4350	Offset point support		9	.889		15	54.50		69.50	101
4360	Concealed base assembly		5	1.600		8.70	98		106.70	161
4370	Tee connector base		13	.615		15	38		53	74.50
4380	Tripod base, 36" for 60" air terminal		7	1.143		31.50	70		101.50	143
4390	Intermediate base, 1/2"		15	.533		14.20	32.50		46.70	66
4400	Connector cable, copper									
4410	Through wall connector	1 Elec	7	1.143	Ea.	83.50	70		153.50	200
4420	Through roof connector		5	1.600		122	98		220	285
4430	Beam connector		7	1.143		31	70		101	143
4440	Double bolt connector		38	.211		10.45	12.90		23.35	31.50
4450	Bar, copper		40	.200		13.25	12.25		25.50	33.50
4500	Connector cable, aluminum									
4510	Through wall connector	1 Elec	7	1.143	Ea.	33	70		103	145
4520	Through roof connector		5	1.600		32.50	98		130.50	187
4530	Beam connector		7	1.143		19.90	70		89.90	130
4540	Double bolt connector		38	.211		5.55	12.90		18.45	26
4600	Bonding plates, copper									
4610	I-beam, 8" square	1 Elec	7	1.143	Ea.	22	70		92	132
4620	Purlin, 8" square		7	1.143		25.50	70		95.50	136
4630	Large heavy duty, 16" square		5	1.600		32	98		130	186
4640	Aluminum, I-beam, 8" square		7	1.143		17	70		87	127
4650	Purlin, 8" square		7	1.143		17.10	70		87.10	127
4660	Large heavy duty, 16" square		5	1.600		18.95	98		116.95	172
4700	Cable support, copper, loop fastener		38	.211		1.04	12.90		13.94	21

26 41 Facility Lightning Protection

26 41 13 – Lightning Protection for Structures

26 41 13.13 Lightning Protection for Buildings		Crew	Daily Output	Labor-Hours	Unit	Material	Labor (2020 Bare Costs)	Equipment	Total	Total Incl O&P
4710	Adhesive cable holder	1 Elec	38	.211	Ea.	1.37	12.90		14.27	21.50
4720	Aluminum, loop fastener		38	.211		.78	12.90		13.68	21
4730	Adhesive cable holder		38	.211		.67	12.90		13.57	20.50
4740	Bonding strap, copper, 3/4" x 9-1/2"		15	.533		11.20	32.50		43.70	63
4750	Aluminum, 3/4" x 9-1/2"		15	.533		4.98	32.50		37.48	56
4760	Swivel adapter, copper, 1/2"		73	.110		15.45	6.70		22.15	27.50
4770	Aluminum, 1/2"		73	.110		8.30	6.70		15	19.45

26 51 Interior Lighting

26 51 13 – Interior Lighting Fixtures, Lamps, and Ballasts

26 51 13.10 Fixture Hangers

Line	Description	Crew	Daily Output	Labor-Hours	Unit	Material	Labor	Equipment	Total	Total Incl O&P
0010	**FIXTURE HANGERS**									
0220	Box hub cover	1 Elec	32	.250	Ea.	4.13	15.35		19.48	28
0240	Canopy		12	.667		9.05	41		50.05	73
0260	Connecting block		40	.200		2.83	12.25		15.08	22
0280	Cushion hanger		16	.500		23.50	30.50		54	73
0300	Box hanger, with mounting strap		8	1		9.60	61.50		71.10	105
0320	Connecting block		40	.200		2.63	12.25		14.88	22
0340	Flexible, 1/2" diameter, 4" long		12	.667		15.75	41		56.75	80.50
0360	6" long		12	.667		17.10	41		58.10	82
0380	8" long		12	.667		18.95	41		59.95	84
0400	10" long		12	.667		20	41		61	85
0420	12" long		12	.667		21	41		62	86.50
0440	15" long		12	.667		22.50	41		63.50	87.50
0460	18" long		12	.667		26	41		67	91.50
0480	3/4" diameter, 4" long		10	.800		20	49		69	97.50
0500	6" long		10	.800		22	49		71	100
0520	8" long		10	.800		22.50	49		71.50	100
0540	10" long		10	.800		25	49		74	103
0560	12" long		10	.800		27	49		76	105
0580	15" long		10	.800		30	49		79	109
0600	18" long		10	.800		32.50	49		81.50	112

26 51 13.40 Interior HID Fixtures

Line	Description	Crew	Daily Output	Labor-Hours	Unit	Material	Labor	Equipment	Total	Total Incl O&P
0010	**INTERIOR HID FIXTURES** Incl. lamps and mounting hardware									
0700	High pressure sodium, recessed, round, 70 watt	1 Elec	3.50	2.286	Ea.	585	140		725	860
0720	100 watt		3.50	2.286		705	140		845	990
0740	150 watt		3.20	2.500		725	153		878	1,025
0760	Square, 70 watt		3.60	2.222		625	136		761	900
0780	100 watt		3.60	2.222		705	136		841	985
0820	250 watt		3	2.667		890	164		1,054	1,225
0840	1,000 watt	2 Elec	4.80	3.333		1,775	205		1,980	2,275
0860	Surface, round, 70 watt	1 Elec	3	2.667		960	164		1,124	1,300
0880	100 watt		3	2.667		985	164		1,149	1,325
0900	150 watt		2.70	2.963		955	182		1,137	1,325
0920	Square, 70 watt		3	2.667		830	164		994	1,175
0940	100 watt		3	2.667		865	164		1,029	1,200
0980	250 watt		2.50	3.200		695	196		891	1,075
1040	Pendent, round, 70 watt		3	2.667		980	164		1,144	1,325
1060	100 watt		3	2.667		860	164		1,024	1,200
1080	150 watt		2.70	2.963		960	182		1,142	1,325
1100	Square, 70 watt		3	2.667		970	164		1,134	1,325

26 51 13 – Interior Lighting Fixtures, Lamps, and Ballasts

26 51 13.40 Interior HID Fixtures		Crew	Daily Output	Labor-Hours	Unit	Material	Labor	Equipment	Total	Total Incl O&P
						2020 Bare Costs				
1120	100 watt	1 Elec	3	2.667	Ea.	985	164		1,149	1,325
1140	150 watt		2.70	2.963		1,000	182		1,182	1,400
1160	250 watt		2.50	3.200		1,375	196		1,571	1,800
1180	400 watt		2.40	3.333		1,450	205		1,655	1,900
1220	Wall, round, 70 watt		3	2.667		860	164		1,024	1,200
1240	100 watt		3	2.667		855	164		1,019	1,200
1260	150 watt		2.70	2.963		850	182		1,032	1,225
1300	Square, 70 watt		3	2.667		880	164		1,044	1,225
1320	100 watt		3	2.667		965	164		1,129	1,325
1340	150 watt		2.70	2.963		950	182		1,132	1,325
1360	250 watt		2.50	3.200		960	196		1,156	1,350
1380	400 watt	2 Elec	4.80	3.333		1,175	205		1,380	1,625
1400	1,000 watt	"	3.60	4.444		1,775	273		2,048	2,375
1500	Metal halide, recessed, round, 175 watt	1 Elec	3.40	2.353		445	144		589	710
1520	250 watt	"	3.20	2.500		545	153		698	835
1540	400 watt	2 Elec	5.80	2.759		815	169		984	1,150
1580	Square, 175 watt	1 Elec	3.40	2.353		460	144		604	725
1640	Surface, round, 175 watt		2.90	2.759		600	169		769	920
1660	250 watt		2.70	2.963		1,225	182		1,407	1,625
1680	400 watt	2 Elec	4.80	3.333		1,225	205		1,430	1,675
1720	Square, 175 watt	1 Elec	2.90	2.759		655	169		824	980
1800	Pendent, round, 175 watt		2.90	2.759		875	169		1,044	1,225
1820	250 watt		2.70	2.963		1,325	182		1,507	1,750
1840	400 watt	2 Elec	4.80	3.333		1,225	205		1,430	1,675
1880	Square, 175 watt	1 Elec	2.90	2.759		500	169		669	810
1900	250 watt	"	2.70	2.963		740	182		922	1,100
1920	400 watt	2 Elec	4.80	3.333		1,250	205		1,455	1,700
1980	Wall, round, 175 watt	1 Elec	2.90	2.759		975	169		1,144	1,325
2000	250 watt	"	2.70	2.963		1,175	182		1,357	1,550
2020	400 watt	2 Elec	4.80	3.333		1,025	205		1,230	1,475
2060	Square, 175 watt	1 Elec	2.90	2.759		515	169		684	825
2080	250 watt	"	2.70	2.963		725	182		907	1,075
2100	400 watt	2 Elec	4.80	3.333		1,025	205		1,230	1,450
2800	High pressure sodium, vaporproof, recessed, 70 watt	1 Elec	3.50	2.286		765	140		905	1,050
2820	100 watt		3.50	2.286		780	140		920	1,075
2840	150 watt		3.20	2.500		800	153		953	1,125
2900	Surface, 70 watt		3	2.667		860	164		1,024	1,200
2920	100 watt		3	2.667		900	164		1,064	1,250
2940	150 watt		2.70	2.963		935	182		1,117	1,300
3000	Pendent, 70 watt		3	2.667		850	164		1,014	1,175
3020	100 watt		3	2.667		875	164		1,039	1,225
3040	150 watt		2.70	2.963		930	182		1,112	1,300
3100	Wall, 70 watt		3	2.667		915	164		1,079	1,250
3120	100 watt		3	2.667		955	164		1,119	1,300
3140	150 watt		2.70	2.963		995	182		1,177	1,375
3200	Metal halide, vaporproof, recessed, 175 watt		3.40	2.353		790	144		934	1,100
3220	250 watt		3.20	2.500		720	153		873	1,025
3240	400 watt	2 Elec	5.80	2.759		900	169		1,069	1,250
3260	1,000 watt	"	4.80	3.333		1,650	205		1,855	2,125
3280	Surface, 175 watt	1 Elec	2.90	2.759		1,150	169		1,319	1,500
3300	250 watt	"	2.70	2.963		1,025	182		1,207	1,400
3320	400 watt	2 Elec	4.80	3.333		1,275	205		1,480	1,725
3340	1,000 watt	"	3.60	4.444		1,850	273		2,123	2,475

26 51 13.40 Interior HID Fixtures		Crew	Daily Output	Labor-Hours	Unit	2020 Bare Costs Material	Labor	Equipment	Total	Total Incl O&P
3360	Pendent, 175 watt	1 Elec	2.90	2.759	Ea.	1,250	169		1,419	1,625
3380	250 watt	″	2.70	2.963		1,050	182		1,232	1,425
3400	400 watt	2 Elec	4.80	3.333		1,250	205		1,455	1,700
3420	1,000 watt	″	3.60	4.444		2,025	273		2,298	2,650
3440	Wall, 175 watt	1 Elec	2.90	2.759		1,325	169		1,494	1,700
3460	250 watt	″	2.70	2.963		1,100	182		1,282	1,500
3480	400 watt	2 Elec	4.80	3.333		1,325	205		1,530	1,775
3500	1,000 watt	″	3.60	4.444		2,100	273		2,373	2,750

26 51 13.50 Interior Lighting Fixtures		Crew	Daily Output	Labor-Hours	Unit	Material	Labor	Equipment	Total	Total Incl O&P
0010	**INTERIOR LIGHTING FIXTURES** Including lamps, mounting R265113-40									
0030	hardware and connections									
0100	Fluorescent, C.W. lamps, troffer, recess mounted in grid, RS									
0130	Grid ceiling mount									
0200	Acrylic lens, 1′ W x 4′ L, two 40 watt	1 Elec	5.70	1.404	Ea.	53.50	86		139.50	192
0210	1′ W x 4′ L, three 40 watt		5.40	1.481		60.50	91		151.50	207
0300	2′ W x 2′ L, two U40 watt		5.70	1.404		57.50	86		143.50	196
0400	2′ W x 4′ L, two 40 watt		5.30	1.509		56	92.50		148.50	205
0500	2′ W x 4′ L, three 40 watt		5	1.600		62	98		160	219
0600	2′ W x 4′ L, four 40 watt		4.70	1.702		64.50	104		168.50	232
0700	4′ W x 4′ L, four 40 watt	2 Elec	6.40	2.500		294	153		447	560
0800	4′ W x 4′ L, six 40 watt		6.20	2.581		310	158		468	585
0900	4′ W x 4′ L, eight 40 watt		5.80	2.759		350	169		519	645
0910	Acrylic lens, 1′ W x 4′ L, two 32 watt T8 G	1 Elec	5.70	1.404		76	86		162	217
0930	2′ W x 2′ L, two U32 watt T8 G		5.70	1.404		115	86		201	260
0940	2′ W x 4′ L, two 32 watt T8 G		5.30	1.509		87	92.50		179.50	239
0950	2′ W x 4′ L, three 32 watt T8 G		5	1.600		80.50	98		178.50	240
0960	2′ W x 4′ L, four 32 watt T8 G		4.70	1.702		75	104		179	244
1000	Surface mounted, RS									
1030	Acrylic lens with hinged & latched door frame									
1100	1′ W x 4′ L, two 40 watt	1 Elec	7	1.143	Ea.	68.50	70		138.50	184
1110	1′ W x 4′ L, three 40 watt		6.70	1.194		71	73.50		144.50	191
1200	2′ W x 2′ L, two U40 watt		7	1.143		73.50	70		143.50	189
1300	2′ W x 4′ L, two 40 watt		6.20	1.290		83.50	79		162.50	214
1400	2′ W x 4′ L, three 40 watt		5.70	1.404		85	86		171	227
1500	2′ W x 4′ L, four 40 watt		5.30	1.509		87	92.50		179.50	239
1501	2′ W x 4′ L, six 40 watt T8		5.20	1.538		87	94.50		181.50	241
1600	4′ W x 4′ L, four 40 watt	2 Elec	7.20	2.222		430	136		566	685
1700	4′ W x 4′ L, six 40 watt		6.60	2.424		465	149		614	745
1800	4′ W x 4′ L, eight 40 watt		6.20	2.581		485	158		643	780
1900	2′ W x 8′ L, four 40 watt		6.40	2.500		171	153		324	425
2000	2′ W x 8′ L, eight 40 watt		6.20	2.581		184	158		342	445
2010	Acrylic wrap around lens									
2020	6″ W x 4′ L, one 40 watt	1 Elec	8	1	Ea.	69.50	61.50		131	171
2030	6″ W x 8′ L, two 40 watt	2 Elec	8	2		76.50	123		199.50	273
2040	11″ W x 4′ L, two 40 watt	1 Elec	7	1.143		46.50	70		116.50	160
2050	11″ W x 8′ L, four 40 watt	2 Elec	6.60	2.424		77	149		226	315
2060	16″ W x 4′ L, four 40 watt	1 Elec	5.30	1.509		77	92.50		169.50	228
2070	16″ W x 8′ L, eight 40 watt	2 Elec	6.40	2.500		168	153		321	420
2080	2′ W x 2′ L, two U40 watt	1 Elec	7	1.143		94	70		164	212
2100	Strip fixture									
2130	Surface mounted									
2200	4′ long, one 40 watt, RS	1 Elec	8.50	.941	Ea.	31.50	57.50		89	124

26 51 13.50 Interior Lighting Fixtures			Crew	Daily Output	Labor-Hours	Unit	2020 Bare Costs Material	Labor	Equipment	Total	Total Incl O&P
2300	4′ long, two 40 watt, RS		1 Elec	8	1	Ea.	49	61.50		110.50	149
2310	4′ long, two 32 watt T8, RS	G		8	1		75	61.50		136.50	177
2400	4′ long, one 40 watt, SL			8	1		54	61.50		115.50	154
2500	4′ long, two 40 watt, SL			7	1.143		73	70		143	189
2580	8′ long, one 60 watt T8, SL	G	2 Elec	13.40	1.194		113	73.50		186.50	237
2590	8′ long, two 60 watt T8, SL	G		12.40	1.290		99.50	79		178.50	231
2600	8′ long, one 75 watt, SL			13.40	1.194		57.50	73.50		131	176
2700	8′ long, two 75 watt, SL			12.40	1.290		71	79		150	200
2800	4′ long, two 60 watt, HO		1 Elec	6.70	1.194		112	73.50		185.50	236
2810	4′ long, two 54 watt, T5HO	G	"	6.70	1.194		176	73.50		249.50	305
2900	8′ long, two 110 watt, HO		2 Elec	10.60	1.509		110	92.50		202.50	264
2910	4′ long, two 115 watt, VHO		1 Elec	6.50	1.231		145	75.50		220.50	275
2920	8′ long, two 215 watt, VHO		2 Elec	10.40	1.538		157	94.50		251.50	315
2950	High bay pendent mounted, 16″ W x 4′ L, four 54 watt, T5HO	G		8.90	1.798		298	110		408	500
2952	2′ W x 4′ L, six 54 watt, T5HO	G		8.50	1.882		305	115		420	515
2954	2′ W x 4′ L, six 32 watt, T8	G		8.50	1.882		178	115		293	375
3000	Strip, pendent mounted, industrial, white porcelain enamel										
3100	4′ long, two 40 watt, RS		1 Elec	5.70	1.404	Ea.	53	86		139	191
3110	4′ long, two 32 watt T8, RS	G		5.70	1.404		83	86		169	225
3200	4′ long, two 60 watt, HO			5	1.600		82.50	98		180.50	242
3290	8′ long, two 60 watt T8, SL	G	2 Elec	8.80	1.818		121	112		233	305
3300	8′ long, two 75 watt, SL			8.80	1.818		101	112		213	283
3400	8′ long, two 110 watt, HO			8	2		130	123		253	330
3410	Acrylic finish, 4′ long, two 40 watt, RS		1 Elec	5.70	1.404		87	86		173	229
3420	4′ long, two 60 watt, HO			5	1.600		161	98		259	330
3430	4′ long, two 115 watt, VHO			4.80	1.667		210	102		312	390
3440	8′ long, two 75 watt, SL		2 Elec	8.80	1.818		172	112		284	360
3450	8′ long, two 110 watt, HO			8	2		203	123		326	410
3460	8′ long, two 215 watt, VHO			7.60	2.105		287	129		416	515
3470	Troffer, air handling, 2′ W x 4′ L with four 32 watt T8	G	1 Elec	4	2		119	123		242	320
3480	2′ W x 2′ L with two U32 watt T8	G		5.50	1.455		110	89		199	258
3490	Air connector insulated, 5″ diameter			20	.400		69.50	24.50		94	115
3500	6″ diameter			20	.400		68.50	24.50		93	114
3502	Troffer, direct/indirect, 2′ W x 4′ L with two 32 W T8	G		5.30	1.509		286	92.50		378.50	460
3510	Troffer parabolic lay-in, 1′ W x 4′ L with one 32 W T8	G		5.70	1.404		119	86		205	264
3520	1′ W x 4′ L with two 32 W T8	G		5.30	1.509		141	92.50		233.50	298
3525	2′ W x 2′ L with two U32 W T8	G		5.70	1.404		124	86		210	269
3530	2′ W x 4′ L with three 32 W T8	G		5	1.600		132	98		230	296
3531	Intr fxtr, fluor, troffer prismatic lay-in, 2′ W x 4′l w/three 32 W T8			5	1.600		152	98		250	320
3535	Downlight, recess mounted	G		8	1		148	61.50		209.50	258
3540	Wall wash reflector, recess mounted	G		8	1		113	61.50		174.50	220
3550	Direct/indirect, 4′ long, stl., pendent mtd.	G		5	1.600		162	98		260	330
3560	4′ long, alum., pendent mtd.	G		5	1.600		335	98		433	515
3565	Prefabricated cove, 4′ long, stl. continuous row	G		5	1.600		210	98		308	380
3570	4′ long, alum. continuous row	G		5	1.600		345	98		443	530
3580	Wet location, recess mounted, 2′ W x 4′ L with two 32 watt T8	G		5.30	1.509		238	92.50		330.50	405
3590	Pendent mounted, 2′ W x 4′ L with two 32 watt T8	G		5.70	1.404		385	86		471	555
4000	Induction lamp, integral ballast, ceiling mounted										
4110	High bay, aluminum reflector, 160 watt		1 Elec	3.20	2.500	Ea.	920	153		1,073	1,225
4120	320 watt		"	3	2.667		1,650	164		1,814	2,075
4130	480 watt		2 Elec	5.80	2.759		2,500	169		2,669	3,000
4150	Low bay, aluminum reflector, 250 watt		1 Elec	3.20	2.500		935	153		1,088	1,250
4170	Garage, aluminum reflector, 80 watt			3.60	2.222		840	136		976	1,125

26 51 13 – Interior Lighting Fixtures, Lamps, and Ballasts

26 51 13.50 Interior Lighting Fixtures		Crew	Daily Output	Labor-Hours	Unit	Material (2020 Bare Costs)	Labor (2020 Bare Costs)	Equipment (2020 Bare Costs)	Total (2020 Bare Costs)	Total Incl O&P
4180	Vandalproof, aluminum reflector, 100 watt	1 Elec	3.20	2.500	Ea.	615	153		768	915
4220	Metal halide, integral ballast, ceiling, recess mounted									
4230	prismatic glass lens, floating door									
4240	2′ W x 2′ L, 250 watt	1 Elec	3.20	2.500	Ea.	300	153		453	565
4250	2′ W x 2′ L, 400 watt	2 Elec	5.80	2.759		370	169		539	665
4260	Surface mounted, 2′ W x 2′ L, 250 watt	1 Elec	2.70	2.963		340	182		522	655
4270	400 watt	2 Elec	4.80	3.333		405	205		610	760
4280	High bay, aluminum reflector,									
4290	Single unit, 400 watt	2 Elec	4.60	3.478	Ea.	425	213		638	795
4300	Single unit, 1,000 watt		4	4		610	245		855	1,050
4310	Twin unit, 400 watt		3.20	5		820	305		1,125	1,375
4320	Low bay, aluminum reflector, 250W DX lamp	1 Elec	3.20	2.500		360	153		513	635
4330	400 watt lamp	2 Elec	5	3.200		545	196		741	900
4340	High pressure sodium integral ballast ceiling, recess mounted									
4350	prismatic glass lens, floating door									
4360	2′ W x 2′ L, 150 watt lamp	1 Elec	3.20	2.500	Ea.	400	153		553	675
4370	2′ W x 2′ L, 400 watt lamp	2 Elec	5.80	2.759		475	169		644	785
4380	Surface mounted, 2′ W x 2′ L, 150 watt lamp	1 Elec	2.70	2.963		485	182		667	815
4390	400 watt lamp	2 Elec	4.80	3.333		545	205		750	915
4400	High bay, aluminum reflector,									
4410	Single unit, 400 watt lamp	2 Elec	4.60	3.478	Ea.	390	213		603	760
4430	Single unit, 1,000 watt lamp	″	4	4		545	245		790	980
4440	Low bay, aluminum reflector, 150 watt lamp	1 Elec	3.20	2.500		340	153		493	605
4445	High bay H.I.D. quartz restrike	″	16	.500		169	30.50		199.50	232
4450	Incandescent, high hat can, round alzak reflector, prewired									
4470	100 watt	1 Elec	8	1	Ea.	68	61.50		129.50	169
4480	150 watt		8	1		104	61.50		165.50	209
4500	300 watt		6.70	1.194		241	73.50		314.50	380
4520	Round with reflector and baffles, 150 watt		8	1		54	61.50		115.50	154
4540	Round with concentric louver, 150 watt PAR		8	1		78.50	61.50		140	181
4600	Square glass lens with metal trim, prewired									
4630	100 watt	1 Elec	6.70	1.194	Ea.	55.50	73.50		129	174
4680	150 watt		6.70	1.194		97.50	73.50		171	220
4700	200 watt		6.70	1.194		97.50	73.50		171	220
4800	300 watt		5.70	1.404		146	86		232	293
4810	500 watt		5	1.600		286	98		384	465
4900	Ceiling/wall, surface mounted, metal cylinder, 75 watt		10	.800		51.50	49		100.50	132
4920	150 watt		10	.800		89	49		138	174
4930	300 watt		8	1		173	61.50		234.50	285
5000	500 watt		6.70	1.194		365	73.50		438.50	515
5010	Square, 100 watt		8	1		124	61.50		185.50	231
5020	150 watt		8	1		124	61.50		185.50	232
5030	300 watt		7	1.143		340	70		410	485
5040	500 watt		6	1.333		360	82		442	525
5200	Ceiling, surface mounted, opal glass drum									
5300	8″, one 60 watt lamp	1 Elec	10	.800	Ea.	67	49		116	150
5400	10″, two 60 watt lamps		8	1		74.50	61.50		136	177
5500	12″, four 60 watt lamps		6.70	1.194		104	73.50		177.50	227
5510	Pendent, round, 100 watt		8	1		124	61.50		185.50	231
5520	150 watt		8	1		124	61.50		185.50	231
5530	300 watt		6.70	1.194		171	73.50		244.50	300
5540	500 watt		5.50	1.455		330	89		419	500
5550	Square, 100 watt		6.70	1.194		152	73.50		225.50	280

26 51 13.50 Interior Lighting Fixtures		Crew	Daily Output	Labor-Hours	Unit	2020 Bare Costs Material	Labor	Equipment	Total	Total Incl O&P
5560	150 watt	1 Elec	6.70	1.194	Ea.	159	73.50		232.50	288
5570	300 watt		5.70	1.404		237	86		323	395
5580	500 watt		5	1.600		320	98		418	505
5600	Wall, round, 100 watt		8	1		66	61.50		127.50	167
5620	300 watt		8	1		133	61.50		194.50	242
5630	500 watt		6.70	1.194		390	73.50		463.50	545
5640	Square, 100 watt		8	1		123	61.50		184.50	231
5650	150 watt		8	1		108	61.50		169.50	214
5660	300 watt		7	1.143		167	70		237	292
5670	500 watt		6	1.333		278	82		360	430
6010	Vapor tight, incandescent, ceiling mounted, 200 watt		6.20	1.290		78.50	79		157.50	209
6020	Recessed, 200 watt		6.70	1.194		127	73.50		200.50	253
6030	Pendent, 200 watt		6.70	1.194		78.50	73.50		152	200
6040	Wall, 200 watt		8	1		80.50	61.50		142	184
6100	Fluorescent, surface mounted, 2 lamps, 4′ L, RS, 40 watt		3.20	2.500		119	153		272	365
6110	Industrial, 2 lamps, 4′ L in tandem, 430 MA		2.20	3.636		212	223		435	580
6130	2 lamps, 4′ L, 800 MA		1.90	4.211		181	258		439	600
6160	Pendent, indust, 2 lamps, 4′ L in tandem, 430 MA		1.90	4.211		247	258		505	670
6170	2 lamps, 4′ L, 430 MA		2.30	3.478		165	213		378	510
6180	2 lamps, 4′ L, 800 MA		1.70	4.706		208	289		497	675
6850	Vandalproof, surface mounted, fluorescent, two 32 watt T8 G		3.20	2.500		278	153		431	540
6860	Incandescent, one 150 watt		8	1		94.50	61.50		156	199
6900	Mirror light, fluorescent, RS, acrylic enclosure, two 40 watt		8	1		115	61.50		176.50	222
6910	One 40 watt		8	1		99.50	61.50		161	204
6920	One 20 watt		12	.667		84	41		125	156
7000	Low bay, aluminum reflector, 70 watt, high pressure sodium		4	2		277	123		400	495
7010	250 watt		3.20	2.500		370	153		523	640
7020	400 watt	2 Elec	5	3.200		430	196		626	775
7500	Ballast replacement, by weight of ballast, to 15′ high									
7520	Indoor fluorescent, less than 2 lb.	1 Elec	10	.800	Ea.	25	49		74	103
7540	Two 40W, watt reducer, 2 to 5 lb.		9.40	.851		71.50	52		123.50	159
7560	Two F96 slimline, over 5 lb.		8	1		111	61.50		172.50	217
7580	Vaportite ballast, less than 2 lb.		9.40	.851		25	52		77	108
7600	2 lb. to 5 lb.		8.90	.899		71.50	55		126.50	164
7620	Over 5 lb.		7.60	1.053		111	64.50		175.50	222
7630	Electronic ballast for two tubes		8	1		41.50	61.50		103	140
7640	Dimmable ballast one-lamp G		8	1		105	61.50		166.50	211
7650	Dimmable ballast two-lamp G		7.60	1.053		108	64.50		172.50	219
7660	Dimmable ballast three-lamp		6.90	1.159		123	71		194	245
7690	Emergency ballast (factory installed in fixture)					152			152	167
7990	Decorator									
8000	Pendent RLM in colors, shallow dome, 12″ diam., 100 watt	1 Elec	8	1	Ea.	80	61.50		141.50	183
8010	Regular dome, 12″ diam., 100 watt		8	1		82.50	61.50		144	186
8020	16″ diam., 200 watt		7	1.143		84	70		154	201
8030	18″ diam., 300 watt		6	1.333		93.50	82		175.50	229
8100	Picture framing light		16	.500		98.50	30.50		129	155
8150	Miniature low voltage, recessed, pinhole		8	1		134	61.50		195.50	243
8160	Star		8	1		133	61.50		194.50	241
8170	Adjustable cone		8	1		169	61.50		230.50	281
8180	Eyeball		8	1		109	61.50		170.50	214
8190	Cone		8	1		129	61.50		190.50	237
8200	Coilex baffle		8	1		140	61.50		201.50	249
8210	Surface mounted, adjustable cylinder		8	1		140	61.50		201.50	249

26 51 13 – Interior Lighting Fixtures, Lamps, and Ballasts

26 51 13.50 Interior Lighting Fixtures		Crew	Daily Output	Labor-Hours	Unit	2020 Bare Costs Material	Labor	Equipment	Total	Total Incl O&P
8250	Chandeliers, incandescent									
8260	24" diam. x 42" high, 6 light candle	1 Elec	6	1.333	Ea.	440	82		522	605
8270	24" diam. x 42" high, 6 light candle w/glass shade		6	1.333		430	82		512	600
8280	17" diam. x 12" high, 8 light w/glass panels		8	1		282	61.50		343.50	405
8300	27" diam. x 29" high, 10 light bohemian lead crystal		4	2		460	123		583	695
8310	21" diam. x 9" high, 6 light sculptured ice crystal		8	1		420	61.50		481.50	555
8500	Accent lights, on floor or edge, 0.5 W low volt incandescent									
8520	incl. transformer & fastenings, based on 100' lengths									
8550	Lights in clear tubing, 12" OC	1 Elec	230	.035	L.F.	8.50	2.13		10.63	12.65
8560	6" OC		160	.050		11.10	3.07		14.17	16.90
8570	4" OC		130	.062		16.95	3.78		20.73	24.50
8580	3" OC		125	.064		18.85	3.93		22.78	26.50
8590	2" OC		100	.080		27.50	4.91		32.41	37.50
8600	Carpet, lights both sides 6" OC, in alum. extrusion		270	.030		25.50	1.82		27.32	31
8610	In bronze extrusion		270	.030		29	1.82		30.82	35
8620	Carpet-bare floor, lights 18" OC, in alum. extrusion		270	.030		20.50	1.82		22.32	25.50
8630	In bronze extrusion		270	.030		24	1.82		25.82	29.50
8640	Carpet edge-wall, lights 6" OC in alum. extrusion		270	.030		25.50	1.82		27.32	31
8650	In bronze extrusion		270	.030		29	1.82		30.82	35
8660	Bare floor, lights 18" OC, in alum. extrusion		300	.027		20.50	1.64		22.14	25
8670	In bronze extrusion		300	.027		24	1.64		25.64	29
8680	Bare floor conduit, alum. extrusion		300	.027		6.80	1.64		8.44	9.95
8690	In bronze extrusion		300	.027		13.55	1.64		15.19	17.45
8700	Step edge to 36", lights 6" OC, in alum. extrusion		100	.080	Ea.	68.50	4.91		73.41	82.50
8710	In bronze extrusion		100	.080		71	4.91		75.91	86
8720	Step edge to 54", lights 6" OC, in alum. extrusion		100	.080		102	4.91		106.91	121
8730	In bronze extrusion		100	.080		108	4.91		112.91	127
8740	Step edge to 72", lights 6" OC, in alum. extrusion		100	.080		136	4.91		140.91	158
8750	In bronze extrusion		100	.080		149	4.91		153.91	172
8760	Connector, male		32	.250		2.53	15.35		17.88	26.50
8770	Female with pigtail		32	.250		5.30	15.35		20.65	29.50
8780	Clamps		400	.020		.51	1.23		1.74	2.45
8790	Transformers, 50 watt		8	1		100	61.50		161.50	205
8800	250 watt		4	2		320	123		443	540
8810	1,000 watt		2.70	2.963		350	182		532	665
9000	Minimum labor/equipment charge		3	2.667	Job		164		164	252

26 51 13.55 Interior LED Fixtures

26 51 13.55 Interior LED Fixtures		Crew	Daily Output	Labor-Hours	Unit	Material	Labor	Equipment	Total	Total Incl O&P
0010	**INTERIOR LED FIXTURES** Incl. lamps and mounting hardware									
0100	Downlight, recess mounted, 7.5" diameter, 25 watt G	1 Elec	8	1	Ea.	340	61.50		401.50	470
0120	10" diameter, 36 watt G		8	1		360	61.50		421.50	490
0160	cylinder, 10 watts G		8	1		104	61.50		165.50	209
0180	20 watts G		8	1		135	61.50		196.50	244
0900	Interior LED fixts, troffer, recess mounted, 2' x 2', 3,500K		8.50	.941		129	57.50		186.50	230
0910	2' x 2', 4,000K		8.50	.941		130	57.50		187.50	232
1000	Troffer, recess mounted, 2' x 4', 3,200 lumens G		5.30	1.509		138	92.50		230.50	295
1010	4,800 lumens G		5	1.600		150	98		248	315
1020	6,400 lumens G		4.70	1.702		189	104		293	370
1100	Troffer retrofit lamp, 38 watt G		21	.381		63.50	23.50		87	106
1110	60 watt G		20	.400		156	24.50		180.50	209
1120	100 watt G		18	.444		140	27.50		167.50	196
1200	Troffer, volumetric recess mounted, 2' x 2' G		5.70	1.404		345	86		431	515
2000	Strip, surface mounted, one light bar 4' long, 3,500 K G		8.50	.941		305	57.50		362.50	430

26 51 13.55 Interior LED Fixtures			Crew	Daily Output	Labor-Hours	Unit	2020 Bare Costs Material	Labor	Equipment	Total	Total Incl O&P
2010	5,000 K	G	1 Elec	8	1	Ea.	265	61.50		326.50	385
2020	Two light bar 4′ long, 5,000 K	G		7	1.143		415	70		485	570
3000	Linear, suspended mounted, one light bar 4′ long, 37 watt	G		6.70	1.194		171	73.50		244.50	300
3010	One light bar 8′ long, 74 watt	G	2 Elec	12.20	1.311		310	80.50		390.50	465
3020	Two light bar 4′ long, 74 watt	G	1 Elec	5.70	1.404		335	86		421	500
3030	Two light bar 8′ long, 148 watt	G	2 Elec	8.80	1.818		360	112		472	565
4000	High bay, surface mounted, round, 150 watts	G		5.41	2.959		465	182		647	790
4010	2 bars, 164 watts	G		5.41	2.959		385	182		567	700
4020	3 bars, 246 watts	G		5.01	3.197		515	196		711	865
4030	4 bars, 328 watts	G		4.60	3.478		735	213		948	1,150
4040	5 bars, 410 watts	G	3 Elec	4.20	5.716		830	350		1,180	1,450
4050	6 bars, 492 watts	G		3.80	6.324		930	390		1,320	1,625
4060	7 bars, 574 watts	G		3.39	7.075		1,000	435		1,435	1,775
4070	8 bars, 656 watts	G		2.99	8.029		1,050	495		1,545	1,900
5000	Track, lighthead, 6 watt	G	1 Elec	32	.250		55.50	15.35		70.85	84.50
5010	9 watt	G	″	32	.250		62.50	15.35		77.85	92
6000	Garage, surface mounted, 103 watts	G	2 Elec	6.50	2.462		985	151		1,136	1,300
6100	pendent mounted, 80 watts	G		6.50	2.462		690	151		841	990
6200	95 watts	G		6.50	2.462		815	151		966	1,125
6300	125 watts	G		6.50	2.462		850	151		1,001	1,175
7000	Downlight, recess mtd., low profile, 4″ diam., 9W, 3,000K		1 Elec	8.50	.941		24.50	57.50		82	116
7010	4,000K			8.50	.941		23	57.50		80.50	115
7020	5,000K			8.50	.941		24	57.50		81.50	115
26 51 13.70 Residential Fixtures											
0010	**RESIDENTIAL FIXTURES**										
0400	Fluorescent, interior, surface, circline, 32 watt & 40 watt		1 Elec	20	.400	Ea.	164	24.50		188.50	218
0500	2′ x 2′, two U-tube 32 watt T8			8	1		118	61.50		179.50	225
0700	Shallow under cabinet, two 20 watt			16	.500		70.50	30.50		101	125
0900	Wall mounted, 4′ L, two 32 watt T8, with baffle			10	.800		165	49		214	258
2000	Incandescent, exterior lantern, wall mounted, 60 watt			16	.500		61.50	30.50		92	115
2100	Post light, 150 W, with 7′ post			4	2		291	123		414	510
2500	Lamp holder, weatherproof with 150 W PAR			16	.500		35.50	30.50		66	86
2550	With reflector and guard			12	.667		66	41		107	136
2600	Interior pendent, globe with shade, 150 W			20	.400		190	24.50		214.50	247
9000	Minimum labor/equipment charge			4	2	Job		123		123	189
26 51 13.90 Ballast, Replacement HID											
0010	**BALLAST, REPLACEMENT HID**										
7510	Multi-tap 120/208/240/277 V										
7550	High pressure sodium, 70 watt		1 Elec	10	.800	Ea.	183	49		232	277
7560	100 watt			9.40	.851		130	52		182	224
7570	150 watt			9	.889		205	54.50		259.50	310
7580	250 watt			8.50	.941		305	57.50		362.50	425
7590	400 watt			7	1.143		269	70		339	405
7600	1,000 watt			6	1.333		287	82		369	440
7610	Metal halide, 175 watt			8	1		97	61.50		158.50	202
7620	250 watt			8	1		128	61.50		189.50	236
7630	400 watt			7	1.143		158	70		228	282
7640	1,000 watt			6	1.333		258	82		340	410
7650	1,500 watt			5	1.600		245	98		343	420

26 52 Safety Lighting

26 52 13 – Emergency and Exit Lighting

26 52 13.10 Emergency Lighting and Battery Units		Crew	Daily Output	Labor-Hours	Unit	2020 Bare Costs Material	Labor	Equipment	Total	Total Incl O&P
0010	EMERGENCY LIGHTING AND BATTERY UNITS									
0300	Emergency light units, battery operated									
0350	Twin sealed beam light, 25 W, 6 V each									
0500	Lead battery operated	1 Elec	4	2	Ea.	140	123		263	345
0700	Nickel cadmium battery operated		4	2		330	123		453	555
0780	Additional remote mount, sealed beam, 25 W 6 V		26.70	.300		29	18.40		47.40	60.50
0781	Additional remote mount, sealed beam, 25 W 6 V		26.70	.300		29	18.40		47.40	60.50
0790	Twin sealed beam light, 25 W 6 V each		26.70	.300		62.50	18.40		80.90	97
0900	Self-contained fluorescent lamp pack		10	.800		181	49		230	275
9000	Minimum labor/equipment charge		4	2	Job		123		123	189

26 52 13.16 Exit Signs

		Crew	Daily Output	Labor-Hours	Unit	Material	Labor	Equipment	Total	Total Incl O&P
0010	EXIT SIGNS									
0080	Exit light ceiling or wall mount, incandescent, single face	1 Elec	8	1	Ea.	73	61.50		134.50	175
0100	Double face		6.70	1.194		50.50	73.50		124	169
0120	Explosion proof		3.80	2.105		585	129		714	840
0150	Fluorescent, single face		8	1		92.50	61.50		154	197
0160	Double face		6.70	1.194		76.50	73.50		150	198
0200	LED standard, single face G		8	1		49.50	61.50		111	149
0220	Double face G		6.70	1.194		52	73.50		125.50	170
0230	LED vandal-resistant, single face G		7.27	1.100		218	67.50		285.50	345
0240	LED w/battery unit, single face G		4.40	1.818		192	112		304	385
0260	Double face G		4	2		223	123		346	435
0262	LED w/battery unit, vandal-resistant, single face G		4.40	1.818		258	112		370	455
0270	Combination emergency light units and exit sign		4	2		179	123		302	385
0290	LED retrofit kits G		60	.133		52.50	8.20		60.70	70
1500	Exit sign, 12 V, 1 face, remote end mounted (Type 602A1)	R-19	18	1.111		73	68.50		141.50	186
1780	With emergency battery, explosion proof	"	7.70	2.597		4,675	160		4,835	5,375
9000	Minimum labor/equipment charge	1 Elec	4	2	Job		123		123	189

26 54 Classified Location Lighting

26 54 13 – Incandescent Classified Location Lighting

26 54 13.20 Explosion Proof

		Crew	Daily Output	Labor-Hours	Unit	Material	Labor	Equipment	Total	Total Incl O&P
0010	**EXPLOSION PROOF**, incl. lamps, mounting hardware and connections									
6310	Metal halide with ballast, ceiling, surface mounted, 175 watt	1 Elec	2.90	2.759	Ea.	1,400	169		1,569	1,800
6320	250 watt	"	2.70	2.963		1,675	182		1,857	2,125
6330	400 watt	2 Elec	4.80	3.333		1,800	205		2,005	2,300
6340	Ceiling, pendent mounted, 175 watt	1 Elec	2.60	3.077		1,325	189		1,514	1,750
6350	250 watt	"	2.40	3.333		1,600	205		1,805	2,100
6360	400 watt	2 Elec	4.20	3.810		1,725	234		1,959	2,250
6370	Wall, surface mounted, 175 watt	1 Elec	2.90	2.759		1,500	169		1,669	1,900
6380	250 watt	"	2.70	2.963		1,775	182		1,957	2,225
6390	400 watt	2 Elec	4.80	3.333		1,900	205		2,105	2,425
6400	High pressure sodium, ceiling surface mounted, 70 watt	1 Elec	3	2.667		2,275	164		2,439	2,775
6410	100 watt		3	2.667		2,375	164		2,539	2,850
6420	150 watt		2.70	2.963		2,500	182		2,682	3,025
6430	Pendent mounted, 70 watt		2.70	2.963		2,175	182		2,357	2,675
6440	100 watt		2.70	2.963		2,275	182		2,457	2,775
6450	150 watt		2.40	3.333		2,050	205		2,255	2,575
6460	Wall mounted, 70 watt		3	2.667		2,450	164		2,614	2,925
6470	100 watt		3	2.667		2,575	164		2,739	3,075
6480	150 watt		2.70	2.963		2,600	182		2,782	3,125

26 54 Classified Location Lighting

26 54 13 – Incandescent Classified Location Lighting

26 54 13.20 Explosion Proof		Crew	Daily Output	Labor-Hours	Unit	Material	2020 Bare Costs Labor	Equipment	Total	Total Incl O&P
6510	Incandescent, ceiling mounted, 200 watt	1 Elec	4	2	Ea.	1,650	123		1,773	2,000
6520	Pendent mounted, 200 watt		3.50	2.286		1,400	140		1,540	1,775
6530	Wall mounted, 200 watt		4	2		1,625	123		1,748	1,975
6600	Fluorescent, RS, 4' long, ceiling mounted, two 40 watt		2.70	2.963		4,975	182		5,157	5,750
6610	Three 40 watt		2.20	3.636		7,200	223		7,423	8,275
6620	Four 40 watt		1.90	4.211		9,250	258		9,508	10,600
6630	Pendent mounted, two 40 watt		2.30	3.478		5,750	213		5,963	6,650
6640	Three 40 watt		1.90	4.211		8,175	258		8,433	9,400
6650	Four 40 watt	↓	1.70	4.706	↓	10,800	289		11,089	12,200

26 55 Special Purpose Lighting

26 55 33 – Hazard Warning Lighting

26 55 33.10 Warning Beacons		Crew	Daily Output	Labor-Hours	Unit	Material	Labor	Equipment	Total	Total Incl O&P
0010	**WARNING BEACONS**									
0015	Surface mount with colored or clear lens									
0100	Rotating beacon									
0110	120V, 40 watt halogen	1 Elec	3.50	2.286	Ea.	107	140		247	335
0120	24V, 20 watt halogen	"	3.50	2.286	"	254	140		394	495
0200	Steady beacon									
0210	120V, 40 watt halogen	1 Elec	3.50	2.286	Ea.	109	140		249	335
0220	24V, 20 watt		3.50	2.286		113	140		253	340
0230	12V DC, incandescent	↓	3.50	2.286	↓	114	140		254	340
0300	Flashing beacon									
0310	120V, 40 watt halogen	1 Elec	3.50	2.286	Ea.	106	140		246	335
0320	24V, 20 watt halogen		3.50	2.286		107	140		247	335
0410	12V DC with two 6V lantern batteries	↓	7	1.143	↓	108	70		178	227

26 55 59 – Display Lighting

26 55 59.10 Track Lighting		Crew	Daily Output	Labor-Hours	Unit	Material	Labor	Equipment	Total	Total Incl O&P
0010	**TRACK LIGHTING**									
0080	Track, 1 circuit, 4' section	1 Elec	6.70	1.194	Ea.	43	73.50		116.50	161
0100	8' section	2 Elec	10.60	1.509		76	92.50		168.50	227
0200	12' section	"	8.80	1.818		112	112		224	295
0300	3 circuits, 4' section	1 Elec	6.70	1.194		117	73.50		190.50	241
0400	8' section	2 Elec	10.60	1.509		130	92.50		222.50	286
0500	12' section	"	8.80	1.818		168	112		280	355
1000	Feed kit, surface mounting	1 Elec	16	.500		15.85	30.50		46.35	64.50
1100	End cover		24	.333		8.30	20.50		28.80	40.50
1200	Feed kit, stem mounting, 1 circuit		16	.500		52.50	30.50		83	105
1300	3 circuit		16	.500		52.50	30.50		83	105
2000	Electrical joiner, for continuous runs, 1 circuit		32	.250		37.50	15.35		52.85	64.50
2100	3 circuit		32	.250		76.50	15.35		91.85	108
2200	Fixtures, spotlight, 75 W PAR halogen		16	.500		48	30.50		78.50	99.50
2210	50 W MR16 halogen		16	.500		175	30.50		205.50	239
3000	Wall washer, 250 W tungsten halogen		16	.500		134	30.50		164.50	194
3100	Low voltage, 25/50 W, 1 circuit		16	.500		135	30.50		165.50	195
3120	3 circuit		16	.500	↓	193	30.50		223.50	260
9000	Minimum labor/equipment charge	↓	3	2.667	Job		164		164	252

26 55 Special Purpose Lighting

26 55 61 – Theatrical Lighting

26 55 61.10 Lights		Crew	Daily Output	Labor-Hours	Unit	Material	2020 Bare Costs Labor	Equipment	Total	Total Incl O&P
0010	**LIGHTS**									
2000	Lights, border, quartz, reflector, vented,									
2100	colored or white	1 Elec	20	.400	L.F.	187	24.50		211.50	244
2500	Spotlight, follow spot, with transformer, 2,100 watt	"	4	2	Ea.	3,575	123		3,698	4,125
2600	For no transformer, deduct					990			990	1,100
3000	Stationary spot, fresnel quartz, 6" lens	1 Elec	4	2		233	123		356	445
3100	8" lens		4	2		251	123		374	465
3500	Ellipsoidal quartz, 1,000 watt, 6" lens		4	2		375	123		498	600
3600	12" lens		4	2		665	123		788	925
4000	Strobe light, 1 to 15 flashes per second, quartz		3	2.667		875	164		1,039	1,225
4500	Color wheel, portable, five hole, motorized		4	2		226	123		349	435

26 55 63 – Detention Lighting

26 55 63.10 Detention Lighting Fixtures

Line	Description	Crew	Daily Output	Labor-Hours	Unit	Material	Labor	Equipment	Total	Total Incl O&P
0010	**DETENTION LIGHTING FIXTURES**									
1000	Surface mounted, cold rolled steel, 14 ga., 1' x 4'	1 Elec	5.40	1.481	Ea.	835	91		926	1,050
1010	2' x 2'		5.40	1.481		845	91		936	1,075
1020	2' x 4'		5	1.600		975	98		1,073	1,225
1100	12 ga., 1' x 4'		5.40	1.481		1,050	91		1,141	1,300
1110	2' x 2'		5.40	1.481		845	91		936	1,075
1120	2' x 4'		5	1.600		1,050	98		1,148	1,300

26 56 Exterior Lighting

26 56 13 – Lighting Poles and Standards

26 56 13.10 Lighting Poles

Line	Description	Crew	Daily Output	Labor-Hours	Unit	Material	Labor	Equipment	Total	Total Incl O&P
0010	**LIGHTING POLES**									
0100	Exterior, light poles, concrete, 30' above 5' below, 13.5" Base, 5.5" Tip	2 Elec	4.60	3.478	Ea.	1,575	213		1,788	2,050
0110	39' above 6' below, 15.5" Base, 5.25" Tip		4.60	3.478		1,850	213		2,063	2,375
0120	43' above 7' below, 17.25" Base, 6.5" Tip		4.60	3.478		1,875	213		2,088	2,400
0130	43' above 7' below, 19.5" Base, 8.25" Tip		4.60	3.478		1,900	213		2,113	2,425
2800	Light poles, anchor base									
2820	not including concrete bases									
2840	Aluminum pole, 8' high	1 Elec	4	2	Ea.	810	123		933	1,075
2850	10' high		4	2		855	123		978	1,125
2860	12' high		3.80	2.105		890	129		1,019	1,175
2870	14' high		3.40	2.353		920	144		1,064	1,225
2880	16' high		3	2.667		1,025	164		1,189	1,375
3000	20' high	R-3	2.90	6.897		1,100	420	65	1,585	1,925
3200	30' high		2.60	7.692		2,075	470	72.50	2,617.50	3,075
3400	35' high		2.30	8.696		2,275	530	82	2,887	3,400
3600	40' high		2	10		2,750	610	94.50	3,454.50	4,075
3800	Bracket arms, 1 arm	1 Elec	8	1		141	61.50		202.50	250
4000	2 arms		8	1		280	61.50		341.50	405
4200	3 arms		5.30	1.509		425	92.50		517.50	610
4400	4 arms		4.80	1.667		565	102		667	780
4500	Steel pole, galvanized, 8' high		3.80	2.105		675	129		804	940
4510	10' high		3.70	2.162		700	133		833	980
4520	12' high		3.40	2.353		755	144		899	1,050
4530	14' high		3.10	2.581		805	158		963	1,125
4540	16' high		2.90	2.759		855	169		1,024	1,200
4550	18' high		2.70	2.963		900	182		1,082	1,275

26 56 Exterior Lighting

26 56 13 – Lighting Poles and Standards

26 56 13.10 Lighting Poles		Crew	Daily Output	Labor-Hours	Unit	Material	Labor	Equipment	Total	Total Incl O&P
						2020 Bare Costs				
4600	20′ high	R-3	2.60	7.692	Ea.	1,275	470	72.50	1,817.50	2,200
4800	30′ high		2.30	8.696		1,275	530	82	1,887	2,300
5000	35′ high		2.20	9.091		1,550	555	86	2,191	2,675
5200	40′ high		1.70	11.765		1,725	720	111	2,556	3,100
5400	Bracket arms, 1 arm	1 Elec	8	1		214	61.50		275.50	330
5600	2 arms		8	1		310	61.50		371.50	435
5800	3 arms		5.30	1.509		215	92.50		307.50	380
6000	4 arms		5.30	1.509		350	92.50		442.50	530
6100	Fiberglass pole, 1 or 2 fixtures, 20′ high	R-3	4	5		860	305	47.50	1,212.50	1,475
6200	30′ high		3.60	5.556		980	340	52.50	1,372.50	1,650
6300	35′ high		3.20	6.250		1,675	380	59	2,114	2,500
6400	40′ high		2.80	7.143		1,850	435	67.50	2,352.50	2,775
6420	Wood pole, 4-1/2″ x 5-1/8″, 8′ high	1 Elec	6	1.333		385	82		467	550
6430	10′ high		6	1.333		450	82		532	625
6440	12′ high		5.70	1.404		570	86		656	760
6450	15′ high		5	1.600		665	98		763	880
6460	20′ high		4	2		810	123		933	1,075
6461	Light poles,anchor base,w/o conc base, pwdr ct stl, 16′ H	2 Elec	3.10	5.161		855	315		1,170	1,425
6462	20′ high	R-3	2.90	6.897		1,275	420	65	1,760	2,125
6463	30′ high		2.30	8.696		1,275	530	82	1,887	2,300
6464	35′ high		2.40	8.333		1,550	510	79	2,139	2,600
6465	25′ high		2.70	7.407		1,275	455	70	1,800	2,175
6470	Light pole conc base, max 6′ buried, 2′ exposed, 18″ diam., average cost	C-6	6	8		190	350	8.95	548.95	780
7300	Transformer bases, not including concrete bases									
7320	Maximum pole size, steel, 40′ high	1 Elec	2	4	Ea.	1,600	245		1,845	2,150
7340	Cast aluminum, 30′ high		3	2.667		850	164		1,014	1,175
7350	40′ high		2.50	3.200		1,300	196		1,496	1,725
8000	Line cover protective devices									
8015	Refer to 26 01 02.20 for labor adjustment factors as they apply									
8100	MVLC 1″ W x 1-1/2″ H, 14/5′	1 Elec	60	.133	Ea.	4.70	8.20		12.90	17.75
8110	8′		58	.138		5.25	8.45		13.70	18.80
8120	MVLC 1″ W x 1-1/2″ H, 18/5′		60	.133		5.95	8.20		14.15	19.15
8130	8′		58	.138		6.65	8.45		15.10	20.50
8140	MVLC 1″ W x 1-1/2″ H, 38/5′		60	.133		10.60	8.20		18.80	24.50
8150	8′		58	.138		12.90	8.45		21.35	27.50

26 56 19 – LED Exterior Lighting

26 56 19.60 Parking LED Lighting		Crew	Daily Output	Labor-Hours	Unit	Material	Labor	Equipment	Total	Total Incl O&P
0010	**PARKING LED LIGHTING**									
0100	Round pole mounting, 88 lamp watts G	1 Elec	2	4	Ea.	1,200	245		1,445	1,700

26 56 21 – HID Exterior Lighting

26 56 21.20 Roadway Luminaire		Crew	Daily Output	Labor-Hours	Unit	Material	Labor	Equipment	Total	Total Incl O&P
0010	**ROADWAY LUMINAIRE**									
2650	Roadway area luminaire, low pressure sodium, 135 watt	1 Elec	2	4	Ea.	835	245		1,080	1,300
2700	180 watt	″	2	4		995	245		1,240	1,475
2750	Metal halide, 400 watt	2 Elec	4.40	3.636		690	223		913	1,100
2760	1,000 watt		4	4		775	245		1,020	1,225
2780	High pressure sodium, 400 watt		4.40	3.636		825	223		1,048	1,250
2790	1,000 watt		4	4		940	245		1,185	1,400

26 56 Exterior Lighting

26 56 23 – Area Lighting

26 56 23.10 Exterior Fixtures

	26 56 23.10 Exterior Fixtures	Crew	Daily Output	Labor-Hours	Unit	2020 Bare Costs Material	Labor	Equipment	Total	Total Incl O&P
0010	**EXTERIOR FIXTURES** With lamps									
0200	Wall mounted, incandescent, 100 watt	1 Elec	8	1	Ea.	51.50	61.50		113	151
0400	Quartz, 500 watt		5.30	1.509		62	92.50		154.50	212
0420	1,500 watt		4.20	1.905		108	117		225	298
1100	Wall pack, low pressure sodium, 35 watt		4	2		205	123		328	415
1150	55 watt		4	2		241	123		364	455
1160	High pressure sodium, 70 watt		4	2		197	123		320	405
1170	150 watt		4	2		218	123		341	430
1175	High pressure sodium, 250 watt		4	2		218	123		341	430
1180	Metal halide, 175 watt		4	2		221	123		344	435
1190	250 watt		4	2		251	123		374	465
1195	400 watt		4	2		410	123		533	645
1250	Induction lamp, 40 watt		4	2		495	123		618	735
1260	80 watt		4	2		590	123		713	835
1278	LED, poly lens, 26 watt		4	2		310	123		433	530
1280	110 watt		4	2		830	123		953	1,100
1500	LED, glass lens, 13 watt		4	2		310	123		433	530

26 56 23.55 Exterior LED Fixtures

	26 56 23.55 Exterior LED Fixtures	Crew	Daily Output	Labor-Hours	Unit	Material	Labor	Equipment	Total	Total Incl O&P
0010	**EXTERIOR LED FIXTURES**									
0100	Wall mounted, indoor/outdoor, 12 watt G	1 Elec	10	.800	Ea.	199	49		248	295
0110	32 watt G		10	.800		455	49		504	575
0120	66 watt G		10	.800		480	49		529	600
0200	outdoor, 110 watt G		10	.800		1,150	49		1,199	1,350
0210	220 watt G		10	.800		1,775	49		1,824	2,025
0300	modular, type IV, 120 V, 50 lamp watts G		9	.889		1,325	54.50		1,379.50	1,550
0310	101 lamp watts G		9	.889		1,325	54.50		1,379.50	1,525
0320	126 lamp watts G		9	.889		1,650	54.50		1,704.50	1,875
0330	202 lamp watts G		9	.889		1,900	54.50		1,954.50	2,175
0340	240 V, 50 lamp watts G		8	1		1,225	61.50		1,286.50	1,450
0350	101 lamp watts G		8	1		1,375	61.50		1,436.50	1,600
0360	126 lamp watts G		8	1		1,275	61.50		1,336.50	1,500
0370	202 lamp watts G		8	1		1,925	61.50		1,986.50	2,200
0400	wall pack, glass, 13 lamp watts G		4	2		450	123		573	685
0410	poly w/photocell, 26 lamp watts G		4	2		330	123		453	555
0420	50 lamp watts G		4	2		705	123		828	965
0430	replacement, 40 watts G		4	2		385	123		508	615
0440	60 watts G		4	2		435	123		558	670

26 56 26 – Landscape Lighting

26 56 26.20 Landscape Fixtures

	26 56 26.20 Landscape Fixtures	Crew	Daily Output	Labor-Hours	Unit	Material	Labor	Equipment	Total	Total Incl O&P
0010	**LANDSCAPE FIXTURES**									
7380	Landscape recessed uplight, incl. housing, ballast, transformer									
7390	& reflector, not incl. conduit, wire, trench									
7420	Incandescent, 250 watt	1 Elec	5	1.600	Ea.	665	98		763	880
7440	Quartz, 250 watt		5	1.600		630	98		728	845
7460	500 watt		4	2		650	123		773	905

26 56 26.50 Landscape LED Fixtures

	26 56 26.50 Landscape LED Fixtures	Crew	Daily Output	Labor-Hours	Unit	Material	Labor	Equipment	Total	Total Incl O&P
0010	**LANDSCAPE LED FIXTURES**									
0100	12 volt alum bullet hooded-BLK	1 Elec	5	1.600	Ea.	97.50	98		195.50	258
0200	12 volt alum bullet hooded-BRZ		5	1.600		97.50	98		195.50	258
0300	12 volt alum bullet hooded-GRN		5	1.600		97.50	98		195.50	258
1000	12 volt alum large bullet hooded-BLK		5	1.600		72	98		170	230

26 56 Exterior Lighting

26 56 26 – Landscape Lighting

26 56 26.50 Landscape LED Fixtures		Crew	Daily Output	Labor-Hours	Unit	2020 Bare Costs Material	Labor	Equipment	Total	Total Incl O&P
1100	12 volt alum large bullet hooded-BRZ	1 Elec	5	1.600	Ea.	72	98		170	230
1200	12 volt alum large bullet hooded-GRN		5	1.600		72	98		170	230
2000	12 volt large bullet landscape light fixture		5	1.600		72	98		170	230
2100	12 volt alum light large bullet		5	1.600		72	98		170	230
2200	12 volt alum bullet light		5	1.600		72	98		170	230

26 56 33 – Walkway Lighting

26 56 33.10 Walkway Luminaire

		Crew	Daily Output	Labor-Hours	Unit	Material	Labor	Equipment	Total	Total Incl O&P
0010	**WALKWAY LUMINAIRE**									
6500	Bollard light, lamp & ballast, 42" high with polycarbonate lens									
6800	Metal halide, 175 watt	1 Elec	3	2.667	Ea.	985	164		1,149	1,325
6900	High pressure sodium, 70 watt		3	2.667		950	164		1,114	1,300
7000	100 watt		3	2.667		950	164		1,114	1,300
7100	150 watt		3	2.667		925	164		1,089	1,275
7200	Incandescent, 150 watt		3	2.667		675	164		839	990
7810	Walkway luminaire, square 16", metal halide 250 watt		2.70	2.963		770	182		952	1,125
7820	High pressure sodium, 70 watt		3	2.667		880	164		1,044	1,225
7830	100 watt		3	2.667		895	164		1,059	1,225
7840	150 watt		3	2.667		895	164		1,059	1,225
7850	200 watt		3	2.667		900	164		1,064	1,250
7910	Round 19", metal halide, 250 watt		2.70	2.963		1,250	182		1,432	1,650
7920	High pressure sodium, 70 watt		3	2.667		1,375	164		1,539	1,750
7930	100 watt		3	2.667		1,375	164		1,539	1,750
7940	150 watt		3	2.667		1,375	164		1,539	1,750
7950	250 watt		2.70	2.963		1,300	182		1,482	1,700
8000	Sphere 14" opal, incandescent, 200 watt		4	2		355	123		478	580
8020	Sphere 18" opal, incandescent, 300 watt		3.50	2.286		430	140		570	685
8040	Sphere 16" clear, high pressure sodium, 70 watt		3	2.667		740	164		904	1,075
8050	100 watt		3	2.667		790	164		954	1,125
8100	Cube 16" opal, incandescent, 300 watt		3.50	2.286		470	140		610	730
8120	High pressure sodium, 70 watt		3	2.667		685	164		849	1,000
8130	100 watt		3	2.667		705	164		869	1,025
8230	Lantern, high pressure sodium, 70 watt		3	2.667		615	164		779	925
8240	100 watt		3	2.667		660	164		824	980
8250	150 watt		3	2.667		620	164		784	930
8260	250 watt		2.70	2.963		865	182		1,047	1,225
8270	Incandescent, 300 watt		3.50	2.286		455	140		595	720
8330	Reflector 22" w/globe, high pressure sodium, 70 watt		3	2.667		555	164		719	860
8340	100 watt		3	2.667		565	164		729	870
8350	150 watt		3	2.667		570	164		734	880
8360	250 watt		2.70	2.963		730	182		912	1,075
9000	Minimum labor/equipment charge		3.75	2.133	Job		131		131	201

26 56 36 – Flood Lighting

26 56 36.20 Floodlights

		Crew	Daily Output	Labor-Hours	Unit	Material	Labor	Equipment	Total	Total Incl O&P
0010	**FLOODLIGHTS** with ballast and lamp,									
1290	floor mtd, mount with swivel bracket									
1300	Induction lamp, 40 watt	1 Elec	3	2.667	Ea.	565	164		729	870
1310	80 watt		3	2.667		750	164		914	1,075
1320	150 watt		3	2.667		1,350	164		1,514	1,750
1400	Pole mounted, pole not included									
1950	Metal halide, 175 watt	1 Elec	2.70	2.963	Ea.	213	182		395	515
2000	400 watt	2 Elec	4.40	3.636		219	223		442	585
2200	1,000 watt		4	4		900	245		1,145	1,375

26 56 Exterior Lighting

26 56 36 – Flood Lighting

26 56 36.20 Floodlights			Crew	Daily Output	Labor-Hours	Unit	Material	Labor	Equipment	Total	Total Incl O&P
							2020 Bare Costs				
2210	1,500 watt		2 Elec	3.70	4.324	Ea.	425	265		690	875
2250	Low pressure sodium, 55 watt		1 Elec	2.70	2.963		460	182		642	785
2270	90 watt			2	4		620	245		865	1,050
2290	180 watt			2	4		680	245		925	1,125
2340	High pressure sodium, 70 watt			2.70	2.963		254	182		436	560
2360	100 watt			2.70	2.963		261	182		443	565
2380	150 watt			2.70	2.963		293	182		475	605
2400	400 watt		2 Elec	4.40	3.636		320	223		543	700
2600	1,000 watt		"	4	4		635	245		880	1,075
9005	Solar powered floodlight, w/motion det, incl. batt pack for cloudy days	G	1 Elec	8	1		71	61.50		132.50	173
9020	Replacement battery pack	G	"	8	1		20	61.50		81.50	117

26 56 36.55 LED Floodlights			Crew	Daily Output	Labor-Hours	Unit	Material	Labor	Equipment	Total	Total Incl O&P
0010	**LED FLOODLIGHTS** with ballast and lamp,										
0020	Pole mounted, pole not included										
0100	11 watt	G	1 Elec	4	2	Ea.	440	123		563	675
0110	46 watt	G		4	2		1,500	123		1,623	1,850
0120	90 watt	G		4	2		2,175	123		2,298	2,600
0130	288 watt	G		4	2		2,000	123		2,123	2,400

26 61 Lighting Systems and Accessories

26 61 13 – Lighting Accessories

26 61 13.30 Fixture Whips		Crew	Daily Output	Labor-Hours	Unit	Material	Labor	Equipment	Total	Total Incl O&P
0010	**FIXTURE WHIPS**									
0080	3/8" Greenfield, 2 connectors, 6' long									
0100	TFFN wire, three #18	1 Elec	32	.250	Ea.	7.65	15.35		23	32
0150	Four #18		28	.286		8.25	17.55		25.80	36
0200	Three #16		32	.250		7.90	15.35		23.25	32
0250	Four #16		28	.286		8.60	17.55		26.15	36.50
0300	THHN wire, three #14		32	.250		5.45	15.35		20.80	29.50
0350	Four #14		28	.286		7.60	17.55		25.15	35.50
0360	Three #12		32	.250		10.50	15.35		25.85	35

26 61 23 – Lamps Applications

26 61 23.10 Lamps			Crew	Daily Output	Labor-Hours	Unit	Material	Labor	Equipment	Total	Total Incl O&P
0010	**LAMPS**										
0080	Fluorescent, rapid start, cool white, 2' long, 20 watt		1 Elec	1	8	C	330	490		820	1,125
0100	4' long, 40 watt			.90	8.889		276	545		821	1,150
0120	3' long, 30 watt			.90	8.889		405	545		950	1,275
0125	3' long, 25 watt energy saver	G		.90	8.889		1,275	545		1,820	2,275
0150	U-40 watt			.80	10		1,125	615		1,740	2,200
0155	U-34 watt energy saver	G		.80	10		11.75	615		626.75	960
0170	4' long, 34 watt energy saver	G		.90	8.889		790	545		1,335	1,700
0176	2' long, T8, 17 watt energy saver	G		1	8		350	490		840	1,150
0178	3' long, T8, 25 watt energy saver	G		.90	8.889		370	545		915	1,250
0180	4' long, T8, 32 watt energy saver	G		.90	8.889		215	545		760	1,075
0200	Slimline, 4' long, 40 watt			.90	8.889		1,050	545		1,595	2,000
0210	4' long, 30 watt energy saver	G		.90	8.889		1,050	545		1,595	2,000
0300	8' long, 75 watt			.80	10		1,000	615		1,615	2,075
0350	8' long, 60 watt energy saver	G		.80	10		395	615		1,010	1,375
0400	High output, 4' long, 60 watt			.90	8.889		610	545		1,155	1,500
0410	8' long, 95 watt energy saver	G		.80	10		605	615		1,220	1,600

26 61 23.10 Lamps		Crew	Daily Output	Labor-Hours	Unit	Material	2020 Bare Costs Labor	Equipment	Total	Total Incl O&P
0500	8' long, 110 watt	1 Elec	.80	10	C	605	615		1,220	1,600
0512	2' long, T5, 14 watt energy saver G		1	8		178	490		668	950
0514	3' long, T5, 21 watt energy saver G		.90	8.889		173	545		718	1,025
0516	4' long, T5, 28 watt energy saver G		.90	8.889		173	545		718	1,025
0517	4' long, T5, 54 watt energy saver G		.90	8.889		510	545		1,055	1,400
0520	Very high output, 4' long, 110 watt		.90	8.889		1,600	545		2,145	2,625
0525	8' long, 195 watt energy saver G		.70	11.429		1,375	700		2,075	2,600
0550	8' long, 215 watt		.70	11.429		1,325	700		2,025	2,525
0554	Full spectrum, 4' long, 60 watt		.90	8.889		660	545		1,205	1,575
0556	6' long, 85 watt		.90	8.889		740	545		1,285	1,650
0558	8' long, 110 watt		.80	10		1,950	615		2,565	3,100
0560	Twin tube compact lamp G		.90	8.889		435	545		980	1,325
0570	Double twin tube compact lamp G		.80	10		950	615		1,565	2,000
0600	Mercury vapor, mogul base, deluxe white, 100 watt		.30	26.667		5,150	1,625		6,775	8,200
0650	175 watt		.30	26.667		2,725	1,625		4,350	5,525
0700	250 watt		.30	26.667		5,000	1,625		6,625	8,025
0800	400 watt		.30	26.667		4,575	1,625		6,200	7,575
0900	1,000 watt		.20	40		12,700	2,450		15,150	17,800
1000	Metal halide, mogul base, 175 watt		.30	26.667		1,050	1,625		2,675	3,675
1100	250 watt		.30	26.667		1,650	1,625		3,275	4,325
1200	400 watt		.30	26.667		2,000	1,625		3,625	4,725
1300	1,000 watt		.20	40		3,925	2,450		6,375	8,075
1320	1,000 watt, 125,000 initial lumens		.20	40		16,100	2,450		18,550	21,500
1330	1,500 watt		.20	40		3,575	2,450		6,025	7,700
1350	High pressure sodium, 70 watt		.30	26.667		1,700	1,625		3,325	4,400
1360	100 watt		.30	26.667		1,750	1,625		3,375	4,450
1370	150 watt		.30	26.667		1,625	1,625		3,250	4,325
1380	250 watt		.30	26.667		1,950	1,625		3,575	4,675
1400	400 watt		.30	26.667		1,650	1,625		3,275	4,325
1450	1,000 watt		.20	40		4,500	2,450		6,950	8,725
1500	Low pressure sodium, 35 watt		.30	26.667		15,900	1,625		17,525	19,900
1550	55 watt		.30	26.667		18,100	1,625		19,725	22,400
1600	90 watt		.30	26.667		6,375	1,625		8,000	9,525
1650	135 watt		.20	40		26,300	2,450		28,750	32,700
1700	180 watt		.20	40		39,400	2,450		41,850	47,100
1750	Quartz line, clear, 500 watt		1.10	7.273		745	445		1,190	1,500
1760	1,500 watt		.20	40		2,125	2,450		4,575	6,100
1762	Spot, MR 16, 50 watt		1.30	6.154		1,050	380		1,430	1,725
1770	Tungsten halogen, T4, 400 watt		1.10	7.273		3,850	445		4,295	4,900
1775	T3, 1,200 watt		.30	26.667		4,900	1,625		6,525	7,925
1778	PAR 30, 50 watt		1.30	6.154		1,075	380		1,455	1,750
1780	PAR 38, 90 watt		1.30	6.154		9,275	380		9,655	10,800
1800	Incandescent, interior, A21, 100 watt		1.60	5		2,700	305		3,005	3,450
1900	A21, 150 watt		1.60	5		16,000	305		16,305	18,100
2000	A23, 200 watt		1.60	5		320	305		625	825
2200	PS 35, 300 watt		1.60	5		965	305		1,270	1,525
2210	PS 35, 500 watt		1.60	5		1,500	305		1,805	2,125
2230	PS 52, 1,000 watt		1.30	6.154		2,425	380		2,805	3,250
2240	PS 52, 1,500 watt		1.30	6.154		6,575	380		6,955	7,825
2300	R30, 75 watt		1.30	6.154		695	380		1,075	1,350
2400	R40, 100 watt		1.30	6.154		695	380		1,075	1,350
2500	Exterior, PAR 38, 75 watt		1.30	6.154		1,825	380		2,205	2,575
2600	PAR 38, 150 watt	↓	1.30	6.154	↓	2,050	380		2,430	2,825

26 61 23 – Lamps Applications

26 61 23.10 Lamps		Crew	Daily Output	Labor-Hours	Unit	Material	2020 Bare Costs Labor	Equipment	Total	Total Incl O&P
2700	PAR 46, 200 watt	1 Elec	1.10	7.273	C	3,725	445		4,170	4,775
2800	PAR 56, 300 watt		1.10	7.273		2,550	445		2,995	3,475
3000	Guards, fluorescent lamp, 4' long		100	.080	Ea.	14	4.91		18.91	23
3200	8' long		90	.089	"	28	5.45		33.45	39.50
9000	Minimum labor/equipment charge		4	2	Job		123		123	189

26 61 23.55 LED Lamps		Crew	Daily Output	Labor-Hours	Unit	Material	Labor	Equipment	Total	Total Incl O&P
0010	**LED LAMPS**									
0100	LED lamp, interior, shape A60, equal to 60 W G	1 Elec	160	.050	Ea.	18.60	3.07		21.67	25
0110	7 W LED decorative c, ca, f, g shape		160	.050		33	3.07		36.07	41
0120	12V mini lamp LED		160	.050		5.30	3.07		8.37	10.50
0200	Globe frosted A60, equal to 60 W G		160	.050		11.10	3.07		14.17	16.90
0205	LED lamp, interior, globe		160	.050		15.55	3.07		18.62	22
0210	2.2 W LED LMP		160	.050		63.50	3.07		66.57	74.50
0220	2.2 W LED replacement decorative lamp		160	.050		8.75	3.07		11.82	14.35
0230	3.5 W LED replacement decorative lamp		160	.050		10.20	3.07		13.27	15.90
0240	4.5 W 120V LED replacement decorative lamp		160	.050		12.25	3.07		15.32	18.15
0250	4.5 W 120V LED, 2700k replacement decorative lamp		160	.050		10.80	3.07		13.87	16.55
0260	4.9 W 120V LED, 2700k replacement decorative lamp candelabra base		160	.050		9.65	3.07		12.72	15.35
0270	4.9 W 120V LED, 3000k replacement decorative lamp candelabra base		160	.050		9.05	3.07		12.12	14.65
0280	5 W LED PAR 20 parabolic reflector lamp FL		140	.057		58	3.51		61.51	69
0300	Globe earth, equal to 100 W G		140	.057		27	3.51		30.51	35
0305	7 W LED reflector lamp 3000k DIM		140	.057		56	3.51		59.51	67
0310	10 W omni LED warm white light bulb E26 medium base 120 volt card		140	.057		16.75	3.51		20.26	24
0315	7 W LED reflector lamp WFL 2700k DIM		140	.057		56	3.51		59.51	67
0320	9 W omni LED warm white light bulb E26 medium base 120 volt card		140	.057		8.95	3.51		12.46	15.25
0500	8 W LED, A19 lamp, equal to 40 W		140	.057		5.70	3.51		9.21	11.65
0505	9 W LED, A19 lamp, 5000K dimmable, equal to 60 W		140	.057		7.10	3.51		10.61	13.20
0510	10.5 W LED, A19 lamp, dimmable, equal to 60 W		140	.057		8.05	3.51		11.56	14.30
0515	10 W LED, A19 lamp, frosted, dimmable, equal to 60 W		140	.057		12.95	3.51		16.46	19.60
0520	10 W LED, A19 lamp, omni-directional, dimmable, equal to 60 W		140	.057		7.55	3.51		11.06	13.70
0525	12 W LED, A19 lamp, dimmable, equal to 75 W		140	.057		10.95	3.51		14.46	17.45
1100	MR16, 3 W, replacement of halogen lamp 25 W G		130	.062		18.65	3.78		22.43	26.50
1200	6 W replacement of halogen lamp 45 W G		130	.062		19.60	3.78		23.38	27.50
2100	10 W, PAR20, equal to 60 W G		130	.062		26	3.78		29.78	34.50
2200	15 W, PAR30, equal to 100 W G		130	.062		49	3.78		52.78	60
2210	50 W LED lamp		130	.062		550	3.78		553.78	610
2220	11 Watt reflector dimmable warm white LED light bulb with medium base		130	.062		40	3.78		43.78	50
2221	12 W A-Line LED lamp DIM		130	.062		75.50	3.78		79.28	89
2225	13 Watt reflector LED warm white e26 with medium base 120 volt box		130	.062		37.50	3.78		41.28	47.50
2226	13 W br30 LED lamp		130	.062		65.50	3.78		69.28	78
2227	15 W 120V br30 inc LED lamp, 2700k		130	.062		61	3.78		64.78	73
2228	15 W 120V br30 inc LED lamp, 4000k		130	.062		58	3.78		61.78	69.50
2230	3 Watt dimmable warm white decorative LED lamp with medium base		160	.050		11.80	3.07		14.87	17.70
2240	.43 W night light LED daylight bulb E12 candelabra base 120 volt 2 pack		160	.050		4.55	3.07		7.62	9.70
2250	15 W omni-directional LED warm white e26 medium base 120 volt box		160	.050		38.50	3.07		41.57	47
2251	11 W omni-directional LED warm white e26 medium base 120 volt box		160	.050		27.50	3.07		30.57	35
2252	10 W omni-directional LED warm white e26 medium base 120 volt box		160	.050		27	3.07		30.07	34
2253	7 W omni A19 LED warm white e26 medium base 120 volt box		160	.050		20	3.07		23.07	27
2255	8 PAR 20 parabolic reflector 2700 LED lamp		160	.050		18.15	3.07		21.22	24.50
2256	8 PAR 20 parabolic reflector 3000 LED lamp		160	.050		18	3.07		21.07	24.50
2260	12 W PAR 38 120V LED 15 degree directional lamp		160	.050		195	3.07		198.07	220
2270	12 W PAR 38 120V LED 25 degree directional lamp		160	.050		30	3.07		33.07	37.50

26 61 23.55 LED Lamps		Crew	Daily Output	Labor-Hours	Unit	Material	Labor	Equipment	Total	Total Incl O&P
						2020 Bare Costs				
2280	12 W PAR 38 120V LED 40 degree directional lamp	1 Elec	160	.050	Ea.	195	3.07		198.07	220
2285	16 W PAR 38 120V 2700k LED parabolic reflector lamp		160	.050		29.50	3.07		32.57	37
2290	16 W PAR 38 120V 3000k LED parabolic reflector lamp		160	.050		22	3.07		25.07	29
3000	15 W PAR 30 LED daylight E26 medium base 120V box		160	.050		58	3.07		61.07	68
3100	17 W LED 3000k PAR 38 100 W replacement		160	.050		45	3.07		48.07	54
3110	17 W LED PAR 38 100 W replacement		160	.050		34	3.07		37.07	41.50
3120	17 W LED PAR 38 100 W replacement parabolic reflector		160	.050		35.50	3.07		38.57	43.50
3130	24 W LED T8 PW straight fluorescent lamp		8.89	.900		278	55		333	390
3135	Linear fluorescent LED lamp 120V		8.89	.900		128	55		183	226
3200	30 W LED 2700K recessed 8055E PAR 38 high power		160	.050		128	3.07		131.07	146
3210	30 W LED 4200K 120 degree 8055E PAR 38 high power		160	.050		129	3.07		132.07	147
3220	30 W LED 5700K 8055E PAR 38 high power		160	.050		130	3.07		133.07	148
3230	50 W LED 2700K 8045M PAR 38 277V high power		160	.050		330	3.07		333.07	365
3240	50 W LED 4200K 8045M PAR 38 277V high power retro fit		160	.050		174	3.07		177.07	196
3250	50 W LED 5700K 8045M PAR 38 277V high power retro fit		160	.050		330	3.07		333.07	365
8000	10 W LED PAR 30/fl 10 pk		160	.050		92.50	3.07		95.57	107
8010	Gen 3 PAR 30 15 W short neck power LED 120 VAC E26 80 +cri 300k dimm		160	.050		12.45	3.07		15.52	18.40
8020	12 PAR 30 2700K parabolic reflector LED lamp		160	.050		40.50	3.07		43.57	49
8030	12 PAR 30 3000K parabolic reflector LED lamp		160	.050		31	3.07		34.07	38.50
8040	3500K LED advantage T8 9 W 800LM 2ft linear 2 BD frosted		69	.116		10.10	7.10		17.20	22
8050	3500K LED litespan T8 9 W 900LM 2ft linear frosted		69	.116		12.70	7.10		19.80	25
8060	4000K LED advantage T8 9 W 800LM 2ft linear 2BD frosted		69	.116		12.10	7.10		19.20	24.50
8070	4000K LED litespan T8 9 W 900LM 2ft linear frosted		69	.116		13.30	7.10		20.40	25.50
8080	5000K LED advantage T8 9 W 800LM 2ft linear 2BD frosted		69	.116		10.85	7.10		17.95	23
8090	5000K LED litespan T8 9 W 900LM 2ft linear frosted		69	.116		10.60	7.10		17.70	22.50
8100	3500K LED advantage T8 18 W 1600LM 4ft linear 2 BD frosted		69	.116		10	7.10		17.10	22
8105	18 W LED 4ft T8 4000K frost 1600L linear lamp		69	.116		17.80	7.10		24.90	30.50
8108	18 W LED 48 inch T8 4100K 1890LM linear lamp		69	.116		55	7.10		62.10	71.50
8110	3500K LED advantage T8 18 W 1800LM 4ft linear 2 BD frosted		69	.116		28.50	7.10		35.60	42.50
8120	4000K LED advantage T8 18 W 1600LM 4ft linear 2 BD frosted		69	.116		19.45	7.10		26.55	32.50
8130	4000K LED litespan T8 18 W 1800LM 4ft linear frosted		69	.116		20.50	7.10		27.60	33.50
8140	5000K LED advantage T8 18 W 1600LM 4ft linear 2BD frosted		69	.116		12.70	7.10		19.80	25
8150	5000K LED litespan T8 18 W 1800LM 4ft linear 2BD frosted		69	.116		20.50	7.10		27.60	33.50
8200	16.5 T8 3000 IF-6U U-shape fluorescent LED lamp		65	.123		20.50	7.55		28.05	34.50
8210	16.5 T8 3500 IF-6U U-shape fluorescent LED lamp		65	.123		21	7.55		28.55	35
8220	16.5 T8 4000 IF-6U U-shape fluorescent LED lamp		65	.123		22	7.55		29.55	35.50
8230	16.5 T8 5000 IF-6U U-shape fluorescent LED lamp		65	.123		20.50	7.55		28.05	34
8240	18 W 6 inch T8 4100K U-shape frosted fluorescent LED lamp		60	.133		11.55	8.20		19.75	25.50
8250	18 W 6 inch T8 5000K U-shape frosted fluorescent LED lamp		60	.133		11.55	8.20		19.75	25.50
8260	Circular 12 W linear fluorescent LED 2700K MOD lamp		62.40	.128		291	7.85		298.85	330
8270	Circular 12 W linear fluorescent LED 3000K MOD lamp		62.40	.128		291	7.85		298.85	330
8280	Circular 12 W linear fluorescent LED 3500K MOD lamp		62.40	.128		263	7.85		270.85	300
8290	Circular 18 W linear fluorescent LED 3000K MOD lamp		62.40	.128		360	7.85		367.85	405
8300	Circular 18 W linear fluorescent LED 3500K MOD lamp		62.40	.128		395	7.85		402.85	445
8310	Circular 18 W linear fluorescent LED 5000K MOD lamp		62.40	.128		360	7.85		367.85	405
8320	Circular 18 W linear fluorescent LED 4100K MOD lamp		62.40	.128		380	7.85		387.85	430

26 71 13.10 Handling		Crew	Daily Output	Labor-Hours	Unit	Material	Labor	Equipment	Total	Total Incl O&P
						2020 Bare Costs				
0010	**HANDLING** Add to normal labor cost for restricted areas									
5000	Motors									
5100	1/2 HP, 23 pounds	1 Elec	4	2	Ea.		123		123	189
5110	3/4 HP, 28 pounds		4	2			123		123	189
5120	1 HP, 33 pounds		4	2			123		123	189
5130	1-1/2 HP, 44 pounds		3.20	2.500			153		153	236
5140	2 HP, 56 pounds		3	2.667			164		164	252
5150	3 HP, 71 pounds		2.30	3.478			213		213	330
5160	5 HP, 82 pounds		1.90	4.211			258		258	400
5170	7-1/2 HP, 124 pounds		1.50	5.333			325		325	505
5180	10 HP, 144 pounds		1.20	6.667			410		410	630
5190	15 HP, 185 pounds		1	8			490		490	755
5200	20 HP, 214 pounds	2 Elec	1.50	10.667			655		655	1,000
5210	25 HP, 266 pounds		1.40	11.429			700		700	1,075
5220	30 HP, 310 pounds		1.20	13.333			820		820	1,250
5230	40 HP, 400 pounds		1	16			980		980	1,500
5240	50 HP, 450 pounds		.90	17.778			1,100		1,100	1,675
5250	75 HP, 680 pounds		.80	20			1,225		1,225	1,900
5260	100 HP, 870 pounds	3 Elec	1	24			1,475		1,475	2,275
5270	125 HP, 940 pounds		.80	30			1,850		1,850	2,825
5280	150 HP, 1,200 pounds		.70	34.286			2,100		2,100	3,250
5290	175 HP, 1,300 pounds		.60	40			2,450		2,450	3,775
5300	200 HP, 1,400 pounds		.50	48			2,950		2,950	4,525

26 71 13.20 Motors		Crew	Daily Output	Labor-Hours	Unit	Material	Labor	Equipment	Total	Total Incl O&P
0010	**MOTORS** 230/460 V, 60 HZ									
0050	Dripproof, premium efficiency, 1.15 service factor									
0060	1,800 RPM, 1/4 HP	1 Elec	5.33	1.501	Ea.	280	92		372	450
0070	1/3 HP		5.33	1.501		247	92		339	415
0080	1/2 HP		5.33	1.501		207	92		299	370
0090	3/4 HP		5.33	1.501		320	92		412	490
0100	1 HP		4.50	1.778		345	109		454	550
0150	2 HP		4.50	1.778		370	109		479	580
0200	3 HP		4.50	1.778		805	109		914	1,050
0250	5 HP		4.50	1.778		595	109		704	825
0300	7.5 HP		4.20	1.905		985	117		1,102	1,250
0350	10 HP		4	2		1,175	123		1,298	1,500
0400	15 HP		3.20	2.500		1,600	153		1,753	2,000
0450	20 HP	2 Elec	5.20	3.077		2,125	189		2,314	2,625
0500	25 HP		5	3.200		2,575	196		2,771	3,125
0550	30 HP		4.80	3.333		2,725	205		2,930	3,300
0600	40 HP		4	4		3,625	245		3,870	4,375
0650	50 HP		3.20	5		3,800	305		4,105	4,675
0700	60 HP		2.80	5.714		4,725	350		5,075	5,750
0750	75 HP		2.40	6.667		5,825	410		6,235	7,025
0800	100 HP	3 Elec	2.70	8.889		5,950	545		6,495	7,400
0850	125 HP		2.10	11.429		7,175	700		7,875	8,950
0900	150 HP		1.80	13.333		9,800	820		10,620	12,100
0950	200 HP		1.50	16		10,800	980		11,780	13,400
1000	1,200 RPM, 1 HP	1 Elec	4.50	1.778		480	109		589	700
1050	2 HP		4.50	1.778		585	109		694	815
1100	3 HP		4.50	1.778		735	109		844	980
1150	5 HP		4.50	1.778		955	109		1,064	1,225

26 71 13 – Motors Applications

26 71 13.20 Motors		Crew	Daily Output	Labor-Hours	Unit	2020 Bare Costs Material	Labor	Equipment	Total	Total Incl O&P
1200	3,600 RPM, 2 HP	1 Elec	4.50	1.778	Ea.	515	109		624	735
1250	3 HP		4.50	1.778		575	109		684	805
1300	5 HP		4.50	1.778		600	109		709	830
1350	Totally enclosed, premium efficiency 1.15 service factor									
1360	1,800 RPM, 1/4 HP	1 Elec	5.33	1.501	Ea.	395	92		487	575
1370	1/3 HP		5.33	1.501		277	92		369	445
1380	1/2 HP		5.33	1.501		385	92		477	565
1390	3/4 HP		5.33	1.501		430	92		522	610
1400	1 HP		4.50	1.778		685	109		794	925
1450	2 HP		4.50	1.778		670	109		779	905
1500	3 HP		4.50	1.778		695	109		804	935
1550	5 HP		4.50	1.778		810	109		919	1,050
1600	7.5 HP		4.20	1.905		1,025	117		1,142	1,300
1650	10 HP		4	2		1,400	123		1,523	1,750
1700	15 HP		3.20	2.500		2,500	153		2,653	2,975
1750	20 HP	2 Elec	5.20	3.077		2,450	189		2,639	3,000
1800	25 HP		5	3.200		3,875	196		4,071	4,550
1850	30 HP		4.80	3.333		3,575	205		3,780	4,250
1900	40 HP		4	4		4,050	245		4,295	4,825
1950	50 HP		3.20	5		4,775	305		5,080	5,725
2000	60 HP		2.80	5.714		6,150	350		6,500	7,325
2050	75 HP		2.40	6.667		7,850	410		8,260	9,250
2100	100 HP	3 Elec	2.70	8.889		8,775	545		9,320	10,500
2150	125 HP		2.10	11.429		12,300	700		13,000	14,700
2200	150 HP		1.80	13.333		12,900	820		13,720	15,400
2250	200 HP		1.50	16		16,500	980		17,480	19,600
2300	1,200 RPM, 1 HP	1 Elec	4.50	1.778		475	109		584	695
2350	2 HP		4.50	1.778		600	109		709	830
2400	3 HP		4.50	1.778		780	109		889	1,025
2450	5 HP		4.50	1.778		970	109		1,079	1,250
2500	3,600 RPM, 2 HP		4.50	1.778		420	109		529	630
2550	3 HP		4.50	1.778		555	109		664	780
2600	5 HP		4.50	1.778		700	109		809	940

26 71 13.40 Motors Explosion Proof

Line	Description	Crew	Daily Output	Labor-Hours	Unit	Material	Labor	Equipment	Total	Total Incl O&P
0010	**MOTORS EXPLOSION PROOF**, 208-230/460 V, 60 HZ									
0020	1,800 RPM, 1/4 HP	1 Elec	5	1.600	Ea.	565	98		663	775
0030	1/3 HP		5	1.600		450	98		548	645
0040	1/2 HP		5	1.600		610	98		708	820
0050	3/4 HP		5.33	1.501		1,075	92		1,167	1,325
0060	1 HP		4.20	1.905		660	117		777	910
0070	2 HP		4.20	1.905		625	117		742	865
0080	3 HP		4.20	1.905		880	117		997	1,150
0090	5 HP		4.20	1.905		875	117		992	1,150
0100	7.5 HP		4	2		1,075	123		1,198	1,400
0110	10 HP		3.70	2.162		1,275	133		1,408	1,600
0120	15 HP		3.20	2.500		1,625	153		1,778	2,000
0130	20 HP	2 Elec	5	3.200		1,975	196		2,171	2,475
0140	25 HP		4.80	3.333		2,350	205		2,555	2,900
0150	30 HP		4.40	3.636		2,650	223		2,873	3,275
0160	40 HP		3.80	4.211		4,325	258		4,583	5,175
0170	50 HP		3.20	5		4,425	305		4,730	5,350
0180	60 HP		2.60	6.154		7,325	380		7,705	8,625

26 71 13.40 Motors Explosion Proof		Crew	Daily Output	Labor-Hours	Unit	Material	Labor	Equipment	Total	Total Incl O&P
						2020 Bare Costs				
0190	75 HP	2 Elec	2	8	Ea.	9,000	490		9,490	10,700
0200	100 HP	3 Elec	2.50	9.600		11,800	590		12,390	13,800
0210	125 HP		2.20	10.909		14,900	670		15,570	17,400
0220	150 HP		1.60	15		17,700	920		18,620	20,900
0230	200 HP		1.20	20		21,700	1,225		22,925	25,800
1000	1,200 RPM, 1 HP	1 Elec	4.20	1.905		500	117		617	730
1010	2 HP		4.20	1.905		575	117		692	815
1020	3 HP		4.20	1.905		785	117		902	1,050
1030	5 HP		4.20	1.905		1,075	117		1,192	1,350
2000	3,600 RPM, 1/4 HP		5	1.600		425	98		523	620
2010	1/3 HP		5	1.600		445	98		543	640
2020	1/2 HP		5	1.600		565	98		663	770
2030	3/4 HP		5	1.600		600	98		698	810
2040	1 HP		4.20	1.905		650	117		767	895
2050	2 HP		4.20	1.905		815	117		932	1,075
2060	3 HP		4.20	1.905		890	117		1,007	1,150
2070	5 HP		4.20	1.905		1,200	117		1,317	1,475
2080	7.5 HP		4	2		1,350	123		1,473	1,675
2090	10 HP		3.70	2.162		1,550	133		1,683	1,900
2100	15 HP		3.20	2.500		2,125	153		2,278	2,550
2110	20 HP	2 Elec	5	3.200		2,600	196		2,796	3,175
2120	25 HP		4.80	3.333		3,225	205		3,430	3,875
2130	30 HP		4.40	3.636		3,775	223		3,998	4,500
2140	40 HP		3.80	4.211		4,775	258		5,033	5,650
2150	50 HP		3.20	5		4,400	305		4,705	5,300
2160	60 HP		2.60	6.154		7,400	380		7,780	8,700
2170	75 HP		2	8		9,025	490		9,515	10,700
2180	100 HP	3 Elec	2.50	9.600		12,600	590		13,190	14,700
2190	125 HP		2.20	10.909		15,500	670		16,170	18,000
2200	150 HP		1.60	15		18,900	920		19,820	22,200
2210	200 HP		1.20	20		24,100	1,225		25,325	28,400
2220	250 HP		1.20	20		19,500	1,225		20,725	23,300

Division Notes

		CREW	DAILY OUTPUT	LABOR-HOURS	UNIT	BARE COSTS MAT.	LABOR	EQUIP.	TOTAL	TOTAL INCL O&P

Estimating Tips

27 20 00 Data Communications
27 30 00 Voice Communications
27 40 00 Audio-Video Communications

- When estimating material costs for special systems, it is always prudent to obtain manufacturers' quotations for equipment prices and special installation requirements that may affect the total cost.
- For cost modifications for elevated tray installation, add the percentages to labor according to the height of the installation and only to the quantities exceeding the different height levels, not to the total tray quantities. Refer to subdivision 26 01 02.20 for labor adjustment factors.
- Do not overlook the costs for equipment used in the installation. If scissor lifts and boom lifts are available in the field, contractors may use them in lieu of the proposed ladders and rolling staging.

Reference Numbers

Reference numbers are shown at the beginning of some major classifications. These numbers refer to related items in the Reference Section. The reference information may be an estimating procedure, an alternate pricing method, or technical information.

Note: Not all subdivisions listed here necessarily appear. ■

Note: Trade Service, in part, has been used as a reference source for some of the material prices used in Division 27.

27 01 Operation and Maintenance of Communications Systems

27 01 30 – Operation and Maintenance of Voice Communications

27 01 30.51 Operation and Maintenance of Voice Equipment		Crew	Daily Output	Labor-Hours	Unit	2020 Bare Costs Material	Labor	Equipment	Total	Total Incl O&P
0010	**OPERATION AND MAINTENANCE OF VOICE EQUIPMENT**									
3400	Remove and replace (reinstall), speaker	1 Elec	6	1.333	Ea.		82		82	126

27 05 Common Work Results for Communications

27 05 05 – Selective Demolition for Communications

27 05 05.20 Electrical Demolition, Communications

27 05 05.20 Electrical Demolition, Communications		Crew	Daily Output	Labor-Hours	Unit	Material	Labor	Equipment	Total	Total Incl O&P
0010	**ELECTRICAL DEMOLITION, COMMUNICATIONS** R024119-10									
0100	Fiber optics									
0120	Cable R260105-30	1 Elec	2400	.003	L.F.		.20		.20	.31
0160	Multi-channel rack enclosure		6	1.333	Ea.		82		82	126
0180	Patch panel		18	.444	"		27.50		27.50	42
0200	Communication cables & fittings									
0220	Voice/data outlet	1 Elec	140	.057	Ea.		3.51		3.51	5.40
0240	Telephone cable		2800	.003	L.F.		.18		.18	.27
0260	Phone jack		135	.059	Ea.		3.64		3.64	5.60
0280	Data jack		135	.059	"		3.64		3.64	5.60
0300	High performance cable, 2 pair		3000	.003	L.F.		.16		.16	.25
0320	4 pair		2100	.004			.23		.23	.36
0340	25 pair		900	.009			.55		.55	.84
0400	Terminal cabinet		5	1.600	Ea.		98		98	151
1000	Nurse call system									
1020	Station	1 Elec	24	.333	Ea.		20.50		20.50	31.50
1040	Standard call button		24	.333			20.50		20.50	31.50
1060	Corridor dome light or zone indicator		24	.333			20.50		20.50	31.50
1080	Master control station	2 Elec	2	8			490		490	755

27 05 05.30 Electrical Demolition, Sound and Video

27 05 05.30 Electrical Demolition, Sound and Video		Crew	Daily Output	Labor-Hours	Unit	Material	Labor	Equipment	Total	Total Incl O&P
0010	**ELECTRICAL DEMOLITION, SOUND & VIDEO** R024119-10									
0100	Cables									
0120	TV antenna lead-in cable R260105-30	1 Elec	2100	.004	L.F.		.23		.23	.36
0140	Sound cable		2400	.003			.20		.20	.31
0160	Microphone cable		2400	.003			.20		.20	.31
0180	Coaxial cable		2400	.003			.20		.20	.31
0200	Doorbell system, not including wires, cables, and conduit		16	.500	Ea.		30.50		30.50	47
0220	Door chime or devices		36	.222	"		13.65		13.65	21
0300	Public address system, not including wires, cables, and conduit									
0320	Conventional office	1 Elec	16	.500	Speaker		30.50		30.50	47
0340	Conventional industrial	"	8	1	"		61.50		61.50	94.50
0352	PA cabinet/panel	2 Elec	5	3.200	Ea.		196		196	300
0400	Sound system, not including wires, cables, and conduit									
0410	Components	1 Elec	24	.333	Ea.		20.50		20.50	31.50
0412	Speaker		24	.333			20.50		20.50	31.50
0416	Volume control		24	.333			20.50		20.50	31.50
0420	Intercom, master station		6	1.333			82		82	126
0440	Remote station		24	.333			20.50		20.50	31.50
0460	Intercom outlets		24	.333			20.50		20.50	31.50
0480	Handset		12	.667			41		41	63
0500	Emergency call system, not including wires, cables, and conduit									
0520	Annunciator	1 Elec	4	2	Ea.		123		123	189
0540	Devices	"	16	.500			30.50		30.50	47
0600	Master door, buzzer type unit	2 Elec	1.60	10			615		615	945
0800	TV system, not including wires, cables, and conduit									

27 05 Common Work Results for Communications

27 05 05 – Selective Demolition for Communications

27 05 05.30 Electrical Demolition, Sound and Video		Crew	Daily Output	Labor-Hours	Unit	2020 Bare Costs Material	Labor	Equipment	Total	Total Incl O&P
0820	Master TV antenna system, per outlet	1 Elec	39	.205	Outlet		12.60		12.60	19.35
0840	School application, per outlet		16	.500	"		30.50		30.50	47
0860	Amplifier		12	.667	Ea.		41		41	63
0880	Antenna		6	1.333	"		82		82	126
0900	One camera & one monitor	2 Elec	7.80	2.051	Total		126		126	194
0920	One camera	1 Elec	8	1	Ea.		61.50		61.50	94.50

27 05 29 – Hangers and Supports for Communications Systems

27 05 29.10 Cable Support

Line	Description	Crew	Daily Output	Labor-Hours	Unit	Material	Labor	Equipment	Total	Total Incl O&P
0010	**CABLE SUPPORT**									
0110	J-hook, single tier, single sided, 1" diam.	1 Elec	68.57	.117	Ea.	2.19	7.15		9.34	13.40
0120	1-1/2" diam.		67.80	.118		2.85	7.25		10.10	14.30
0130	2" diam.		67.30	.119		3.31	7.30		10.61	14.90
0140	4" diam.		66.80	.120		6.55	7.35		13.90	18.50
0330	Double tier, single sided, 2" diam.		67	.119		11.85	7.35		19.20	24.50
0340	4" diam.		66.20	.121		16.80	7.40		24.20	30
0430	Double sided, 2" diam.		65.80	.122		18.25	7.45		25.70	31.50
0440	4" diam.		65.40	.122		29.50	7.50		37	44
0530	Triple tier, single sided, 2" diam.		66.40	.120		15.45	7.40		22.85	28.50
0540	4" diam.		65.85	.121		23.50	7.45		30.95	37
0630	Double sided, 2" diam.		65.10	.123		21	7.55		28.55	34.50
0640	4" diam.		64.70	.124		35	7.60		42.60	50

27 11 Communications Equipment Room Fittings

27 11 16 – Communications Cabinets, Racks, Frames and Enclosures

27 11 16.10 Public Phone

Line	Description	Crew	Daily Output	Labor-Hours	Unit	Material	Labor	Equipment	Total	Total Incl O&P
0010	**PUBLIC PHONE**									
7600	Telephone with wood backboard									
7620	Single door, 12" H x 12" W x 4" D	1 Elec	5.30	1.509	Ea.	95	92.50		187.50	248
7650	18" H x 12" W x 4" D		4.70	1.702		116	104		220	289
7700	24" H x 12" W x 4" D		4.20	1.905		176	117		293	375
7720	18" H x 18" W x 4" D		4.20	1.905		154	117		271	350
7750	24" H x 18" W x 4" D		4	2		218	123		341	430
7780	36" H x 36" W x 4" D		3.60	2.222		211	136		347	440
7800	24" H x 24" W x 6" D		3.60	2.222		294	136		430	535
7820	30" H x 24" W x 6" D		3.20	2.500		310	153		463	575
7850	30" H x 30" W x 6" D		2.70	2.963		470	182		652	795
7880	36" H x 30" W x 6" D		2.50	3.200		500	196		696	850
7900	48" H x 36" W x 6" D		2.20	3.636		800	223		1,023	1,225
7920	Double door, 48" H x 36" W x 6" D		2	4		1,100	245		1,345	1,575

27 11 16.20 Rack Mount Cabinet

Line	Description	Crew	Daily Output	Labor-Hours	Unit	Material	Labor	Equipment	Total	Total Incl O&P
0010	**RACK MOUNT CABINET**									
0100	80" H x 24" W x 40" D									
0110	No sides	1 Elec	9.60	.833	Ea.	1,375	51		1,426	1,575
0120	One side		8	1		1,450	61.50		1,511.50	1,700
0130	Two sides		7.86	1.018		1,550	62.50		1,612.50	1,825
0200	80" H x 24" W x 42" D									
0210	No sides	1 Elec	9.60	.833	Ea.	1,400	51		1,451	1,600
0220	One side		8	1		1,425	61.50		1,486.50	1,675
0230	Two sides		7.86	1.018		1,450	62.50		1,512.50	1,700
0300	80" H x 24" W x 48" D									

27 11 Communications Equipment Room Fittings

27 11 16 – Communications Cabinets, Racks, Frames and Enclosures

27 11 16.20 Rack Mount Cabinet		Crew	Daily Output	Labor-Hours	Unit	Material	Labor	Equipment	Total	Total Incl O&P
						2020 Bare Costs				
0310	No sides	1 Elec	9.60	.833	Ea.	1,575	51		1,626	1,825
0320	One side		8	1		1,575	61.50		1,636.50	1,825
0330	Two sides		7.86	1.018		1,650	62.50		1,712.50	1,925
0400	80″ H x 30″ W x 40″ D									
0410	No sides	1 Elec	9.60	.833	Ea.	1,500	51		1,551	1,700
0420	One side		8	1		1,550	61.50		1,611.50	1,800
0430	Two sides		7.86	1.018		1,625	62.50		1,687.50	1,875
0500	80″ H x 30″ W x 42″ D									
0510	No sides	1 Elec	9.60	.833	Ea.	1,525	51		1,576	1,750
0520	One side		8	1		1,550	61.50		1,611.50	1,800
0530	Two sides		7.86	1.018		1,575	62.50		1,637.50	1,825
0600	80″ H x 30″ W x 48″ D									
0610	No sides	1 Elec	9.60	.833	Ea.	1,675	51		1,726	1,900
0620	One side		8	1		1,750	61.50		1,811.50	2,025
0630	Two sides		7.86	1.018		1,850	62.50		1,912.50	2,150
0700	80″ H x 32″ W x 32″ D									
0710	No sides	1 Elec	9.60	.833	Ea.	1,475	51		1,526	1,675
0720	One side		8	1		1,500	61.50		1,561.50	1,750
0730	Two sides		7.86	1.018		1,525	62.50		1,587.50	1,775
0800	80″ H x 32″ W x 40″ D									
0810	No sides	1 Elec	9.60	.833	Ea.	1,625	51		1,676	1,850
0820	One side		8	1		1,675	61.50		1,736.50	1,950
0830	Two sides		7.86	1.018		1,750	62.50		1,812.50	2,025
0900	80″ H x 32″ W x 42″ D									
0910	No sides	1 Elec	9.60	.833	Ea.	1,625	51		1,676	1,850
0920	One side		8	1		1,650	61.50		1,711.50	1,900
0930	Two sides		7.86	1.018		1,675	62.50		1,737.50	1,950
1000	80″ H x 32″ W x 48″ D									
1010	No sides	1 Elec	9.60	.833	Ea.	1,800	51		1,851	2,050
1020	One side		8	1		1,900	61.50		1,961.50	2,175
1030	Two sides		7.86	1.018		1,350	62.50		1,412.50	1,575

27 11 19 – Communications Termination Blocks and Patch Panels

27 11 19.10 Termination Blocks and Patch Panels		Crew	Daily Output	Labor-Hours	Unit	Material	Labor	Equipment	Total	Total Incl O&P
0010	**TERMINATION BLOCKS AND PATCH PANELS**									
2960	Patch panel, RJ-45/110 type, 24 ports	2 Elec	6	2.667	Ea.	183	164		347	455
3000	48 ports	3 Elec	6	4		320	245		565	730
3040	96 ports	"	4	6		515	370		885	1,125
3100	Punch down termination per port	1 Elec	107	.075			4.59		4.59	7.05

27 13 Communications Backbone Cabling

27 13 23 – Communications Optical Fiber Backbone Cabling

27 13 23.13 Communications Optical Fiber		Crew	Daily Output	Labor-Hours	Unit	Material	Labor	Equipment	Total	Total Incl O&P
0010	**COMMUNICATIONS OPTICAL FIBER**									
0040	Specialized tools & techniques cause installation costs to vary.									
0070	Fiber optic, cable, bulk simplex, single mode R271323-40	1 Elec	8	1	C.L.F.	23	61.50		84.50	120
0080	Multi mode		8	1		30	61.50		91.50	128
0090	4 strand, single mode		7.34	1.090		38.50	67		105.50	146
0095	Multi mode		7.34	1.090		50.50	67		117.50	159
0100	12 strand, single mode		6.67	1.199		86	73.50		159.50	208
0105	Multi mode		6.67	1.199		99	73.50		172.50	222

27 13 Communications Backbone Cabling

27 13 23 – Communications Optical Fiber Backbone Cabling

27 13 23.13 Communications Optical Fiber		Crew	Daily Output	Labor-Hours	Unit	2020 Bare Costs Material	Labor	Equipment	Total	Total Incl O&P
0150	Jumper				Ea.	34.50			34.50	38
0200	Pigtail					36			36	40
0300	Connector	1 Elec	24	.333		26.50	20.50		47	60.50
0350	Finger splice		32	.250		37.50	15.35		52.85	65
0400	Transceiver (low cost bi-directional)		8	1		455	61.50		516.50	600
0450	Rack housing, 4 rack spaces, 12 panels (144 fibers)		2	4		615	245		860	1,050
0500	Patch panel, 12 ports		6	1.333		295	82		377	450
0600	Cable connector panel, 6 fiber		6	1.333		295	82		377	450
1000	Cable, 62.5 microns, direct burial, 4 fiber	R-15	1200	.040	L.F.	.93	2.40	.23	3.56	4.99
1020	Indoor, 2 fiber	R-19	1000	.020		.44	1.23		1.67	2.37
1040	Outdoor, aerial/duct	"	1670	.012		.65	.74		1.39	1.85
1060	50 microns, direct burial, 8 fiber	R-22	4000	.009		1.30	.52		1.82	2.24
1080	12 fiber		4000	.009		2.43	.52		2.95	3.48
1100	Indoor, 12 fiber		759	.049		2.01	2.76		4.77	6.45
1140	Cable splice	R-19	40	.500	Ea.	17.50	30.50		48	67
1160	125 micron cable, transmission		16	1.250		15.45	77		92.45	135
1180	Receiver, 1.2 mile range		20	1		282	61.50		343.50	405
1200	1.9 mile range		20	1		249	61.50		310.50	370
1220	6.2 mile range		5	4		325	246		571	740
1240	Transmitter, 1.2 mile range		20	1		315	61.50		376.50	440
1260	1.9 mile range		20	1		284	61.50		345.50	405
1280	6.2 mile range		5	4		435	246		681	855
1300	Modem, 1.2 mile range		5	4		183	246		429	580
1320	6.2 mile range		5	4		325	246		571	735
1340	1.9 mile range, 12 channel		5	4		2,075	246		2,321	2,650
1360	Repeater, 1.2 mile range		10	2		370	123		493	595
1380	1.9 mile range		10	2		495	123		618	730
1400	6.2 mile range		5	4		910	246		1,156	1,375
1420	1.2 mile range, digital		5	4		455	246		701	880

27 15 Communications Horizontal Cabling

27 15 01 – Communications Horizontal Cabling Applications

27 15 01.19 Fire Alarm Communications Conductors & Cables

Line	Description	Crew	Daily Output	Labor-Hours	Unit	Material	Labor	Equipment	Total	Total Incl O&P
0010	**FIRE ALARM COMMUNICATIONS CONDUCTORS AND CABLES**									
1500	Fire alarm FEP teflon 150 V to 200°C									
1550	#22, 1 pair	1 Elec	10	.800	C.L.F.	94	49		143	179
1600	2 pair		8	1		96	61.50		157.50	200
1650	4 pair		7	1.143		350	70		420	495
1700	6 pair		6	1.333		495	82		577	665
1750	8 pair		5.50	1.455		840	89		929	1,050
1800	10 pair		5	1.600		775	98		873	1,000
1850	#18, 1 pair	2 Elec	16	1		67.50	61.50		129	169
1900	2 pair		13	1.231		224	75.50		299.50	365
1950	4 pair		9.60	1.667		206	102		308	385
2000	6 pair		8	2		440	123		563	675
2050	8 pair		7	2.286		320	140		460	570
2100	10 pair		6	2.667		480	164		644	780

27 15 10 – Special Communications Cabling

27 15 10.23 Sound and Video Cables and Fittings		Crew	Daily Output	Labor-Hours	Unit	Material	Labor	Equipment	Total	Total Incl O&P
						2020 Bare Costs				
0010	**SOUND AND VIDEO CABLES & FITTINGS**									
0900	TV antenna lead-in, 300 ohm, #20-2 conductor	1 Elec	7	1.143	C.L.F.	22	70		92	132
0950	Coaxial, feeder outlet		7	1.143		20.50	70		90.50	131
1000	Coaxial, main riser		6	1.333		14.85	82		96.85	142
1100	Sound, shielded with drain, #22-2 conductor		8	1		8.50	61.50		70	104
1150	#22-3 conductor		7.50	1.067		12.45	65.50		77.95	115
1200	#22-4 conductor		6.50	1.231		17.25	75.50		92.75	135
1250	Nonshielded, #22-2 conductor		10	.800		11.85	49		60.85	88.50
1300	#22-3 conductor		9	.889		21.50	54.50		76	108
1350	#22-4 conductor		8	1		24	61.50		85.50	121
1400	Microphone cable		8	1		27.50	61.50		89	125

27 15 13 – Communications Copper Horizontal Cabling

27 15 13.13 Communication Cables and Fittings		Crew	Daily Output	Labor-Hours	Unit	Material	Labor	Equipment	Total	Total Incl O&P
0010	**COMMUNICATION CABLES AND FITTINGS**									
2200	Telephone twisted, PVC insulation, #22-2 conductor	1 Elec	10	.800	C.L.F.	10.40	49		59.40	87
2250	#22-3 conductor		9	.889		12.60	54.50		67.10	98
2300	#22-4 conductor		8	1		14	61.50		75.50	110
2350	#18-2 conductor		9	.889		20.50	54.50		75	107
2370	Telephone jack, eight pins		32	.250	Ea.	3.49	15.35		18.84	27.50
5000	High performance unshielded twisted pair (UTP)									
5100	Cable, category 3, #24, 2 pair solid, PVC jacket R271513-75	1 Elec	10	.800	C.L.F.	9.15	49		58.15	85.50
5200	4 pair solid, PVC jacket		7	1.143		13.35	70		83.35	123
5300	25 pair solid, PVC jacket		3	2.667		75.50	164		239.50	335
5400	2 pair solid, plenum		10	.800		11.20	49		60.20	88
5500	4 pair solid, plenum		7	1.143		12.40	70		82.40	122
5600	25 pair solid, plenum		3	2.667		111	164		275	375
5700	4 pair stranded, PVC jacket		7	1.143		13.50	70		83.50	123
7000	Category 5, #24, 4 pair solid, PVC jacket		7	1.143		9.60	70		79.60	119
7100	4 pair solid, plenum		7	1.143		14.60	70		84.60	124
7200	4 pair stranded, PVC jacket		7	1.143		25	70		95	136
7210	Category 5e, #24, 4 pair solid, PVC jacket		7	1.143		15.60	70		85.60	125
7212	4 pair solid, plenum		7	1.143		21.50	70		91.50	132
7214	4 pair stranded, PVC jacket		7	1.143		22	70		92	132
7240	Category 6, #24, 4 pair solid, PVC jacket		7	1.143		16.65	70		86.65	126
7242	4 pair solid, plenum		7	1.143		14	70		84	123
7244	4 pair stranded, PVC jacket		7	1.143		15.45	70		85.45	125
7300	Connector, RJ45, category 5		80	.100	Ea.	.76	6.15		6.91	10.30
7302	Shielded RJ45, category 5		72	.111		1.73	6.80		8.53	12.40
7310	Jack, UTP RJ45, category 3		72	.111		.44	6.80		7.24	11
7312	Category 5		65	.123		6.25	7.55		13.80	18.50
7314	Category 5e		65	.123		4.98	7.55		12.53	17.10
7316	Category 6		65	.123		3.44	7.55		10.99	15.40
7322	Jack, shielded RJ45, category 5		60	.133		6.60	8.20		14.80	19.85
7324	Category 5e		60	.133		6.60	8.20		14.80	19.85
7326	Category 6		60	.133		6.90	8.20		15.10	20
7400	Voice/data expansion module, category 5e		8	1		47.50	61.50		109	147
7401	Modular jack, cat 6 keystone, RJ45, office white		16.50	.485		6	30		36	52.50
7402	RJ45, green		16.50	.485		5.90	30		35.90	52.50
7404	RJ45, grey		16.50	.485		6.70	30		36.70	53.50
7408	RJ45, orange		16.50	.485		6.15	30		36.15	53
7420	5M, RJ45, 568B, cord set		16.50	.485		35.50	30		65.50	85
7422	3M, RJ45, 568B, cord set		16.50	.485		32	30		62	81.50

27 15 13 – Communications Copper Horizontal Cabling

27 15 13.13 Communication Cables and Fittings		Crew	Daily Output	Labor-Hours	Unit	2020 Bare Costs Material	Labor	Equipment	Total	Total Incl O&P
7424	1M, RJ45, 568B, cord set	1 Elec	16.50	.485	Ea.	28	30		58	77
7450	M12 TO RJ45, 2M ultra-loch D-mode, cord set		15.50	.516		40	31.50		71.50	93
8000	Multipair unshielded non-plenum cable, 150 V PVC jacket									
8002	#22, 2 pair	1 Elec	8.40	.952	C.L.F.	15.25	58.50		73.75	107
8003	3 pair		8	1		25	61.50		86.50	122
8004	4 pair		7.30	1.096		57	67		124	167
8006	6 pair		6.20	1.290		34.50	79		113.50	160
8008	8 pair		5.70	1.404		79.50	86		165.50	221
8010	10 pair		5.30	1.509		62.50	92.50		155	212
8012	12 pair		4.20	1.905		158	117		275	355
8015	15 pair		3.80	2.105		203	129		332	425
8020	20 pair	2 Elec	6.60	2.424		235	149		384	490
8025	25 pair		6	2.667		290	164		454	570
8030	30 pair		5.60	2.857		475	175		650	790
8040	40 pair		5.20	3.077		420	189		609	750
8050	50 pair		4.90	3.265		595	200		795	965
8100	Multipair unshielded non-plenum cable, 300 V PVC jacket									
8102	#20, 2 pair	1 Elec	7.30	1.096	C.L.F.	23.50	67		90.50	130
8103	3 pair		6.70	1.194		31.50	73.50		105	148
8104	4 pair		6	1.333		35.50	82		117.50	165
8106	6 pair		5.30	1.509		76.50	92.50		169	227
8108	8 pair		4.40	1.818		87.50	112		199.50	269
8110	10 pair	2 Elec	7.30	2.192		96.50	134		230.50	315
8112	12 pair		6.20	2.581		120	158		278	375
8115	15 pair		5.90	2.712		151	166		317	420
8201	#18, 1 pair	1 Elec	8	1		73	61.50		134.50	175
8202	2 pair		6.50	1.231		77.50	75.50		153	202
8203	3 pair		5.60	1.429		136	87.50		223.50	284
8204	4 pair		5	1.600		177	98		275	345
8206	6 pair		4.40	1.818		249	112		361	445
8208	8 pair		4	2		247	123		370	460
8215	15 pair	2 Elec	5	3.200		640	196		836	1,000
8300	Multipair shielded non-plenum cable, 300 V PVC jacket									
8303	#22, 3 pair	1 Elec	6.70	1.194	C.L.F.	59	73.50		132.50	178
8306	6 pair		5.70	1.404		114	86		200	258
8309	9 pair		5	1.600		140	98		238	305
8312	12 pair	2 Elec	8	2		440	123		563	675
8315	15 pair		7.30	2.192		535	134		669	790
8317	17 pair		6.80	2.353		635	144		779	920
8319	19 pair		6.40	2.500		640	153		793	935
8327	27 pair		5.90	2.712		1,125	166		1,291	1,475
8402	#20, 2 pair	1 Elec	6.70	1.194		106	73.50		179.50	230
8403	3 pair		6.20	1.290		128	79		207	263
8406	6 pair		4.40	1.818		325	112		437	525
8409	9 pair		3.60	2.222		277	136		413	515
8412	12 pair	2 Elec	5.90	2.712		530	166		696	840
8415	15 pair	"	5.70	2.807		485	172		657	800
8502	#18, 2 pair	1 Elec	6.20	1.290		120	79		199	254
8503	3 pair		5.30	1.509		188	92.50		280.50	350
8504	4 pair		4.70	1.702		221	104		325	405
8506	6 pair		4	2		350	123		473	575
8509	9 pair		3.40	2.353		530	144		674	805
8515	15 pair	2 Elec	4.80	3.333		890	205		1,095	1,300

27 15 Communications Horizontal Cabling

27 15 33 – Communications Coaxial Horizontal Cabling

27 15 33.10 Coaxial Cable and Fittings		Crew	Daily Output	Labor-Hours	Unit	Material	2020 Bare Costs Labor	Equipment	Total	Total Incl O&P
0010	**COAXIAL CABLE & FITTINGS**									
3500	Coaxial connectors, 50 ohm impedance quick disconnect									
3540	BNC plug, for RG A/U #58 cable	1 Elec	42	.190	Ea.	2.79	11.70		14.49	21
3550	RG A/U #59 cable		42	.190		2.70	11.70		14.40	21
3560	RG A/U #62 cable		42	.190		2.73	11.70		14.43	21
3600	BNC jack, for RG A/U #58 cable		42	.190		3.12	11.70		14.82	21.50
3610	RG A/U #59 cable		42	.190		3.56	11.70		15.26	22
3620	RG A/U #62 cable		42	.190		3.46	11.70		15.16	22
3660	BNC panel jack, for RG A/U #58 cable		40	.200		7.95	12.25		20.20	27.50
3670	RG A/U #59 cable		40	.200		6.55	12.25		18.80	26
3680	RG A/U #62 cable		40	.200		6.55	12.25		18.80	26
3720	BNC bulkhead jack, for RG A/U #58 cable		40	.200		6.45	12.25		18.70	26
3730	RG A/U #59 cable		40	.200		6.45	12.25		18.70	26
3740	RG A/U #62 cable		40	.200		6.45	12.25		18.70	26
3850	Coaxial cable, RG A/U #58, 50 ohm		8	1	C.L.F.	49	61.50		110.50	149
3860	RG A/U #59, 75 ohm		8	1		39.50	61.50		101	138
3870	RG A/U #62, 93 ohm		8	1		47.50	61.50		109	147
3875	RG 6/U, 75 ohm		8	1		35	61.50		96.50	133
3950	Fire rated, RG A/U #58, 50 ohm		8	1		94.50	61.50		156	199
3960	RG A/U #59, 75 ohm		8	1		124	61.50		185.50	231
3970	RG A/U #62, 93 ohm		8	1		109	61.50		170.50	215

27 15 43 – Communications Faceplates and Connectors

27 15 43.13 Communication Outlets

		Crew	Daily Output	Labor-Hours	Unit	Material	Labor	Equipment	Total	Total Incl O&P
0010	**COMMUNICATION OUTLETS**									
0100	Voice/data devices not included									
0120	Voice/data outlets, single opening	1 Elec	48	.167	Ea.	7.60	10.25		17.85	24
0140	Two jack openings		48	.167		2.45	10.25		12.70	18.45
0160	One jack & one 3/4" round opening		48	.167		7.40	10.25		17.65	24
0180	One jack & one twinaxial opening		48	.167		7.55	10.25		17.80	24
0200	One jack & one connector cabling opening		48	.167		7.40	10.25		17.65	24
0220	Two 3/8" coaxial openings		48	.167		7.40	10.25		17.65	24
0300	Data outlets, single opening		48	.167		7.40	10.25		17.65	24
0320	One 25-pin subminiature opening		48	.167		7.40	10.25		17.65	24
1000	Voice/data wall plate, plastic, 1 gang, 1-port		72	.111		2.62	6.80		9.42	13.40
1020	2-port		72	.111		2.78	6.80		9.58	13.55
1040	3-port		72	.111		2.33	6.80		9.13	13.05
1060	4-port		72	.111		2.63	6.80		9.43	13.40
1080	6-port		72	.111		2.25	6.80		9.05	13
1100	2 gang, 6-port		48	.167		7.30	10.25		17.55	24
1120	Voice/data wall plate, stainless steel, 1 gang, 1-port		72	.111		8.60	6.80		15.40	19.95
1140	2-port		72	.111		8.20	6.80		15	19.50
1160	3-port		72	.111		7.95	6.80		14.75	19.25
1180	4-port		72	.111		8.25	6.80		15.05	19.60
1200	2 gang, 6-port		48	.167		12.65	10.25		22.90	29.50

27 21 Data Communications Network Equipment

27 21 29 – Data Communications Switches and Hubs

27 21 29.10 Switching and Routing Equipment		Crew	Daily Output	Labor-Hours	Unit	Material	2020 Bare Costs Labor	Equipment	Total	Total Incl O&P
0010	SWITCHING AND ROUTING EQUIPMENT									
1100	Network hub, dual speed, 24 ports, includes cabinet	3 Elec	.66	36.364	Ea.	960	2,225		3,185	4,475
1300	Network switch, 50/60 HZ, 8 port, multi-platform, analog KVM	1 Elec	6.85	1.168		1,175	71.50		1,246.50	1,375
1310	16 port, multi-platform, analog KVM		6	1.333		1,400	82		1,482	1,650
1320	Network switch, 0x2x16,CAT5, analog KVM		6	1.333		1,225	82		1,307	1,450
1330	2x1x16, digital KVM, w/VM		16	.500		3,850	30.50		3,880.50	4,275
1340	2x1x32, digital KVM, w/VM		16	.500		4,575	30.50		4,605.50	5,075
1350	8x1x32, digital KVM, w/VM		16	.500		5,125	30.50		5,155.50	5,675
1500	KVM,1 RMU 16 port, no keyboard		16	.500		1,550	30.50		1,580.50	1,750
1510	KVM,1 RMU, 17" LCD, 16 port, US keyboard		15.15	.528		2,975	32.50		3,007.50	3,325
1520	KVM,1 RMU, 17" LCD, 16 port, UK keyboard		15.15	.528		2,975	32.50		3,007.50	3,325
2000	10/100/1000 Mbps, 24 ports		14	.571		620	35		655	735
2040	10/100/1000 Mbps, 48 ports		12	.667		1,400	41		1,441	1,625
2050	10/100/1000 Mbps, 5 port, industrial ethernet type		32	.250		330	15.35		345.35	390
2060	10/100/1000 Mbps, 6 port, industrial ethernet type		32	.250		1,300	15.35		1,315.35	1,450
2070	10/100/1000 Mbps, 10 port, X307-3		28	.286		3,075	17.55		3,092.55	3,425
2080	10/100/1000 Mbps, 10 port, X308-2		28	.286		2,850	17.55		2,867.55	3,175
3000	10/100/1000 Mbps, 20 port, front ports		20	.400		4,425	24.50		4,449.50	4,925
3010	10/100/1000 Mbps, 20 port, rear ports		20	.400		4,425	24.50		4,449.50	4,925
3040	KVM, 10/100/1000/10000 Mbps, 28 port, rear ports		16	.500		9,150	30.50		9,180.50	10,100
3050	KVM, 10/100/1000/10000 Mbps, 28 port, front ports		16	.500		10,700	30.50		10,730.50	11,800
3070	KVM, 10/100/1000/10000 Mbps, 52 port, front ports	2 Elec	12	1.333		11,300	82		11,382	12,500
3080	KVM, 10/100/1000/10000 Mbps, 52 port, rear ports	"	12	1.333		11,300	82		11,382	12,500

27 32 Voice Communications Terminal Equipment

27 32 36 – TTY Equipment

27 32 36.10 TTY Telephone Equipment

		Crew	Daily Output	Labor-Hours	Unit	Material	Labor	Equipment	Total	Total Incl O&P
0010	TTY TELEPHONE EQUIPMENT									
1620	Telephone, TTY, compact, pocket type				Ea.	355			355	390
1630	Advanced, desk type	2 Elec	20	.800		495	49		544	620
1640	Full-featured public, wall type	"	4	4		705	245		950	1,150

27 41 Audio-Video Systems

27 41 33 – Master Antenna Television Systems

27 41 33.10 TV Systems

		Crew	Daily Output	Labor-Hours	Unit	Material	Labor	Equipment	Total	Total Incl O&P
0010	TV SYSTEMS, not including rough-in wires, cables & conduits									
0100	Master TV antenna system									
0200	VHF reception & distribution, 12 outlets	1 Elec	6	1.333	Outlet	126	82		208	265
0400	30 outlets		10	.800		254	49		303	355
0600	100 outlets		13	.615		295	38		333	385
0800	VHF & UHF reception & distribution, 12 outlets		6	1.333		247	82		329	400
1000	30 outlets		10	.800		169	49		218	262
1200	100 outlets		13	.615		168	38		206	243
1400	School and deluxe systems, 12 outlets		2.40	3.333		325	205		530	675
1600	30 outlets		4	2		286	123		409	505
1800	80 outlets		5.30	1.509		275	92.50		367.50	445
1900	Amplifier		4	2	Ea.	820	123		943	1,100

27 42 Electronic Digital Systems

27 42 13 – Point of Sale Systems

27 42 13.10 Bar Code Scanner		Crew	Daily Output	Labor-Hours	Unit	Material	2020 Bare Costs Labor	Equipment	Total	Total Incl O&P
0010	**BAR CODE SCANNER**									
0100	CR3600, w/B4 battery, handle	1 Elec	6	1.333	Outlet	930	82		1,012	1,150
0110	Palm		12	.667		820	41		861	965
0120	Batch, B4 batt, handle, chgstn, 3′ USB		9	.889		950	54.50		1,004.50	1,125
0130	Palm, chgstn, 3′ USB		12	.667		860	41		901	1,000
0140	CR2500, w/H2 handle, no cable		10	.800		625	49		674	765
0150	Core unit, palm, no cable	↓	16	.500	↓	580	30.50		610.50	685

27 51 Distributed Audio-Video Communications Systems

27 51 16 – Public Address Systems

27 51 16.10 Public Address System		Crew	Daily Output	Labor-Hours	Unit	Material	Labor	Equipment	Total	Total Incl O&P
0010	**PUBLIC ADDRESS SYSTEM**									
0100	Conventional, office	1 Elec	5.33	1.501	Speaker	171	92		263	330
0200	Industrial		2.70	2.963	"	330	182		512	640
9000	Minimum labor/equipment charge	↓	3.50	2.286	Job		140		140	216

27 51 19 – Sound Masking Systems

27 51 19.10 Sound System		Crew	Daily Output	Labor-Hours	Unit	Material	Labor	Equipment	Total	Total Incl O&P
0010	**SOUND SYSTEM**, not including rough-in wires, cables & conduits									
0100	Components, projector outlet	1 Elec	8	1	Ea.	61.50	61.50		123	163
0200	Microphone		4	2		115	123		238	315
0400	Speakers, ceiling or wall		8	1		157	61.50		218.50	268
0600	Trumpets		4	2		291	123		414	510
0800	Privacy switch		8	1		117	61.50		178.50	223
1000	Monitor panel		4	2		520	123		643	760
1200	Antenna, AM/FM		4	2		155	123		278	360
1400	Volume control		8	1		60	61.50		121.50	161
1600	Amplifier, 250 W		1	8		1,400	490		1,890	2,300
1800	Cabinets	↓	1	8		1,125	490		1,615	2,000
2000	Intercom, 30 station capacity, master station	2 Elec	2	8		2,400	490		2,890	3,400
2020	10 station capacity	"	4	4		1,475	245		1,720	2,000
2200	Remote station	1 Elec	8	1		217	61.50		278.50	335
2400	Intercom outlets		8	1		128	61.50		189.50	235
2600	Handset		4	2		425	123		548	655
2800	Emergency call system, 12 zones, annunciator		1.30	6.154		1,275	380		1,655	1,975
3000	Bell		5.30	1.509		132	92.50		224.50	288
3200	Light or relay		8	1		66	61.50		127.50	167
3400	Transformer		4	2		291	123		414	510
3600	House telephone, talking station		1.60	5		625	305		930	1,150
3800	Press to talk, release to listen	↓	5.30	1.509		145	92.50		237.50	300
4000	System-on button					87			87	95.50
4200	Door release	1 Elec	4	2		155	123		278	360
4400	Combination speaker and microphone		8	1		265	61.50		326.50	385
4600	Termination box		3.20	2.500		83.50	153		236.50	330
4800	Amplifier or power supply		5.30	1.509	↓	955	92.50		1,047.50	1,200
5000	Vestibule door unit		16	.500	Name	161	30.50		191.50	224
5200	Strip cabinet		27	.296	Ea.	330	18.20		348.20	395
5400	Directory	↓	16	.500		156	30.50		186.50	218
6000	Master door, button buzzer type, 100 unit	2 Elec	.54	29.630		1,600	1,825		3,425	4,550
6020	200 unit		.30	53.333		3,000	3,275		6,275	8,325
6040	300 unit	↓	.20	80	↓	4,575	4,900		9,475	12,600

27 51 Distributed Audio-Video Communications Systems

27 51 19 – Sound Masking Systems

27 51 19.10 Sound System		Crew	Daily Output	Labor-Hours	Unit	2020 Bare Costs Material	Labor	Equipment	Total	Total Incl O&P
6060	Transformer	1 Elec	8	1	Ea.	40.50	61.50		102	139
6080	Door opener		5.30	1.509		58.50	92.50		151	207
6100	Buzzer with door release and plate		4	2		58.50	123		181.50	253
6200	Intercom type, 100 unit	2 Elec	.54	29.630		1,975	1,825		3,800	4,975
6220	200 unit		.30	53.333		3,875	3,275		7,150	9,275
6240	300 unit		.20	80		5,825	4,900		10,725	14,000
6260	Amplifier	1 Elec	2	4		291	245		536	700
6280	Speaker with door release		4	2		87	123		210	285
9000	Minimum labor/equipment charge		3.50	2.286	Job		140		140	216

27 52 Healthcare Communications and Monitoring Systems

27 52 23 – Nurse Call/Code Blue Systems

27 52 23.10 Nurse Call Systems		Crew	Daily Output	Labor-Hours	Unit	Material	Labor	Equipment	Total	Total Incl O&P
0010	**NURSE CALL SYSTEMS**									
0100	Single bedside call station	1 Elec	8	1	Ea.	182	61.50		243.50	295
0200	Ceiling speaker station		8	1		90	61.50		151.50	194
0400	Emergency call station		8	1		92.50	61.50		154	197
0600	Pillow speaker		8	1		223	61.50		284.50	340
0800	Double bedside call station		4	2		169	123		292	375
1000	Duty station		4	2		137	123		260	340
1200	Standard call button		8	1		125	61.50		186.50	232
1400	Lights, corridor, dome or zone indicator		8	1		64.50	61.50		126	166
1600	Master control station for 20 stations	2 Elec	.65	24.615	Total	3,725	1,500		5,225	6,425

27 53 Distributed Systems

27 53 13 – Clock Systems

27 53 13.50 Clock Equipments		Crew	Daily Output	Labor-Hours	Unit	Material	Labor	Equipment	Total	Total Incl O&P
0010	**CLOCK EQUIPMENTS**, not including wires & conduits									
0100	Time system components, master controller	1 Elec	.33	24.242	Ea.	1,575	1,475		3,050	4,025
0200	Program bell		8	1		114	61.50		175.50	220
0400	Combination clock & speaker		3.20	2.500		200	153		353	455
0600	Frequency generator		2	4		2,725	245		2,970	3,350
0800	Job time automatic stamp recorder		4	2		560	123		683	805
1600	Master time clock system, clocks & bells, 20 room	4 Elec	.20	160		6,025	9,825		15,850	21,700
1800	50 room	"	.08	400		12,700	24,500		37,200	52,000
1900	Time clock	1 Elec	3.20	2.500		375	153		528	650
2000	100 cards in & out, 1 color					10.45			10.45	11.50
2200	2 colors					9.90			9.90	10.90
2800	Metal rack for 25 cards	1 Elec	7	1.143		50	70		120	163
4000	Wireless time systems component, master controller	"	2	4		630	245		875	1,075
4010	For transceiver and antenna, see Section 28 47 12.10									
4100	Wireless analog clock w/battery operated	1 Elec	8	1	Ea.	110	61.50		171.50	216
4200	Wireless digital clock w/battery operated	"	8	1	"	425	61.50		486.50	560

Division Notes

	CREW	DAILY OUTPUT	LABOR-HOURS	UNIT	BARE COSTS MAT.	LABOR	EQUIP.	TOTAL	TOTAL INCL O&P

Estimating Tips

- When estimating material costs for electronic safety and security systems, it is always prudent to obtain manufacturers' quotations for equipment prices and special installation requirements that may affect the total cost.
- Fire alarm systems consist of control panels, annunciator panels, batteries with rack, charger, and fire alarm actuating and indicating devices. Some fire alarm systems include speakers, telephone lines, door closer controls, and other components. Be careful not to overlook the costs related to installation for these items. Also be aware of costs for integrated automation instrumentation and terminal devices, control equipment, control wiring, and programming. Insurance underwriters may have specific requirements for the type of materials to be installed or design requirements based on the hazard to be protected. Local jurisdictions may have requirements not covered by code. It is advisable to be aware of any special conditions.
- Security equipment includes items such as CCTV, access control, and other detection and identification systems to perform alert and alarm functions. Be sure to consider the costs related to installation for this security equipment, such as for integrated automation instrumentation and terminal devices, control equipment, control wiring, and programming.

Reference Numbers

Reference numbers are shown at the beginning of some major classifications. These numbers refer to related items in the Reference Section. The reference information may be an estimating procedure, an alternate pricing method, or technical information.

Note: Trade Service, in part, has been used as a reference source for some of the material prices used in Division 28.

28 01 Operation and Maint. of Electronic Safety and Security

28 01 80 – Operation and Maintenance of Fire Detection and Alarm

28 01 80.51 Maintenance & Admin. of Fire Detection & Alarm		Crew	Daily Output	Labor-Hours	Unit	Material	2020 Bare Costs Labor	Equipment	Total	Total Incl O&P
0010	**MAINTENANCE AND ADMINISTRATION OF FIRE DETECTION AND ALARM**									
3300	Remove and replace (reinstall), fire alarm device	1 Elec	5.33	1.501	Ea.		92		92	142

28 05 Common Work Results for Electronic Safety and Security

28 05 05 – Selective Demolition for Electronic Safety and Security

28 05 05.10 Safety and Security Demolition

		Crew	Daily Output	Labor-Hours	Unit	Material	Labor	Equipment	Total	Total Incl O&P
0010	**SAFETY AND SECURITY DEMOLITION**									
1050	Finger print/card reader	1 Elec	8	1	Ea.		61.50		61.50	94.50
1051	Electrical dml, CA 3000 card reader sys w/25 users		.31	25.600			1,575		1,575	2,425
1060	Video camera		8	1			61.50		61.50	94.50
1070	Motion detector, multi-channel		7	1.143			70		70	108
1090	Infrared detector		7	1.143			70		70	108
1130	Video monitor, 19"		8	1			61.50		61.50	94.50
1210	Fire alarm horn and strobe light		16	.500			30.50		30.50	47
1215	Remote fire alarm indicator light		23	.348			21.50		21.50	33
1220	Flame detector		23	.348			21.50		21.50	33
1230	Duct detector		10	.800			49		49	75.50
1240	Smoke detector		23	.348			21.50		21.50	33
1241	Fire alarm duct smoke remote test station		21.60	.370			22.50		22.50	35
1250	Fire alarm pull station		23	.348			21.50		21.50	33
1260	Fire alarm control panel, 4 to 8 zone	2 Elec	4	4			245		245	380
1270	12 to 16 zone	"	2.67	5.993			370		370	565
1280	Fire alarm annunciation panel, 4 to 8 zone	1 Elec	6	1.333			82		82	126
1290	12 to 16 zone	2 Elec	6	2.667			164		164	252
1300	Electrical dml, tamper switch	1 Elec	38	.211			12.90		12.90	19.90
1310	Electrical dml, flow switch	"	38	.211			12.90		12.90	19.90

28 05 19 – Storage Appliances for Electronic Safety and Security

28 05 19.11 Digital Video Recorder (DVR)

		Crew	Daily Output	Labor-Hours	Unit	Material	Labor	Equipment	Total	Total Incl O&P
0010	**DIGITAL VIDEO RECORDERS**									
0100	Pentaplex hybrid, internet protocol, and hard drive									
0200	4 channel	1 Elec	1.33	6.015	Ea.	1,800	370		2,170	2,550
0300	8 channel		1	8		3,925	490		4,415	5,075
0400	16 channel		1	8		3,400	490		3,890	4,500

28 15 Access Control Hardware Devices

28 15 11 – Integrated Credential Readers and Field Entry Management

28 15 11.11 Standard Card Readers

		Crew	Daily Output	Labor-Hours	Unit	Material	Labor	Equipment	Total	Total Incl O&P
0010	**STANDARD CARD READERS**									
0015	Card key access									
0020	Computerized system, processor, proximity reader and cards									
0030	Does not include door hardware, lockset or wiring									
0040	Card key system for 1 door				Ea.	1,525			1,525	1,675
0060	Card key system for 2 doors					2,350			2,350	2,600
0080	Card key system for 4 doors					2,500			2,500	2,750
0100	Processor for card key access system					1,175			1,175	1,300
0160	Magnetic lock for electric access, 600 pound holding force					185			185	203
0170	Magnetic lock for electric access, 1200 pound holding force					185			185	203
0200	Proximity card reader					188			188	206
0210	Entrance card systems									

28 15 Access Control Hardware Devices

28 15 11 – Integrated Credential Readers and Field Entry Management

28 15 11.11 Standard Card Readers		Crew	Daily Output	Labor-Hours	Unit	2020 Bare Costs Material	Labor	Equipment	Total	Total Incl O&P
0300	Entrance card, barium ferrite				Ea.	5			5	5.50
0320	Credential					8.25			8.25	9.05
0340	Proximity					3.80			3.80	4.18
0360	Weigand					9.70			9.70	10.70
0700	Entrance card reader, barium ferrite	R-19	4	5		350	305		655	860
0720	Credential		4	5		220	305		525	715
0740	Proximity		4	5		300	305		605	805
0760	Weigand		4	5		535	305		840	1,050
0850	Scanner, eye retina		4	5		2,675	305		2,980	3,425
0900	Gate opener, cantilever	R-18	3	8.667		2,250	465		2,715	3,200
0950	Switch, tamper	R-19	6	3.333		39.50	205		244.50	360
1100	Accessories, electric door strike/bolt		5	4		83	246		329	470
1120	Electromagnetic lock		5	4		161	246		407	555
1140	Keypad for card reader		4	5		130	305		435	620

28 15 11.13 Keypads		Crew	Daily Output	Labor-Hours	Unit	Material	Labor	Equipment	Total	Total Incl O&P
0010	**KEYPADS**									
0340	Digital keypad, int./ext., basic, excl. striker/power/wiring	1 Elec	3	2.667	Ea.	112	164		276	375
0350	Lockset, mechanical push-button type, complete, incl. hardware	1 Carp	4	2	"	420	106		526	635

28 15 11.15 Biometric Identity Devices		Crew	Daily Output	Labor-Hours	Unit	Material	Labor	Equipment	Total	Total Incl O&P
0010	**BIOMETRIC IDENTITY DEVICES**									
0200	Fingerprint scanner unit, excl. striker/power supply	1 Elec	4	2	Ea.	1,575	123		1,698	1,925
0210	Fingerprint scanner unit, for computer keyboard access		8	1		1,350	61.50		1,411.50	1,575
0220	Hand geometry scanner, mem of 512 users, excl. striker/power		3	2.667		2,450	164		2,614	2,950
0230	Memory upgrade for, adds 9,700 user profiles		8	1		410	61.50		471.50	545
0240	Adds 32,500 user profiles		8	1		695	61.50		756.50	860
0250	Prison type, memory of 256 users, excl. striker/power		3	2.667		2,775	164		2,939	3,300
0260	Memory upgrade for, adds 3,300 user profiles		8	1		385	61.50		446.50	520
0270	Adds 9,700 user profiles		8	1		565	61.50		626.50	720
0280	Adds 27,900 user profiles		8	1		755	61.50		816.50	925
0290	All weather, mem of 512 users, excl. striker/power		3	2.667		4,375	164		4,539	5,050
0300	Facial & fingerprint scanner, combination unit, excl. striker/power		3	2.667		930	164		1,094	1,275
0310	Access for, for initial setup, excl. striker/power		3	2.667		1,350	164		1,514	1,725

28 15 11.19 Security Access Control Accessories		Crew	Daily Output	Labor-Hours	Unit	Material	Labor	Equipment	Total	Total Incl O&P
0010	**SECURITY ACCESS CONTROL ACCESSORIES**									
0360	Scanner/reader access, power supply/transf, 110V to 12/24V	1 Elec	4	2	Ea.	232	123		355	445
0370	Elec/mag strikers, 12/24V		4	2		94	123		217	293
0380	1 hour battery backup power supply		4	2		54.50	123		177.50	249
0390	Deadbolt, digital, batt-operated, indoor/outdoor, complete, incl. hardware	1 Carp	4	2		385	106		491	595

28 18 Security Access Detection Equipment

28 18 11 – Security Access Metal Detectors

28 18 11.13 Security Access Metal Detectors		Crew	Daily Output	Labor-Hours	Unit	Material	Labor	Equipment	Total	Total Incl O&P
0010	**SECURITY ACCESS METAL DETECTORS**									
0240	Metal detector, hand-held, wand type, unit only				Ea.	129			129	138
0250	Metal detector, walk through portal type, single zone	1 Elec	2	4		3,750	245		3,995	4,500
0260	Multi-zone	"	2	4		4,875	245		5,120	5,725

28 18 Security Access Detection Equipment

28 18 13 – Security Access X-Ray Equipment

28 18 13.16 Security Access X-Ray Equipment		Crew	Daily Output	Labor-Hours	Unit	Material	2020 Bare Costs Labor	Equipment	Total	Total Incl O&P
0010	SECURITY ACCESS X-RAY EQUIPMENT									
0290	X-ray machine, desk top, for mail/small packages/letters	1 Elec	4	2	Ea.	3,425	123		3,548	3,975
0300	Conveyor type, incl. monitor		2	4		15,700	245		15,945	17,700
0310	Includes additional features		2	4		27,500	245		27,745	30,700
0320	X-ray machine, large unit, for airports, incl. monitor	2 Elec	1	16		39,400	980		40,380	44,800
0330	Full console	″	.50	32		67,500	1,975		69,475	77,000

28 18 15 – Security Access Explosive Detection Equipment

28 18 15.23 Security Access Explosive Detection Equipment		Crew	Daily Output	Labor-Hours	Unit	Material	Labor	Equipment	Total	Total Incl O&P
0010	SECURITY ACCESS EXPLOSIVE DETECTION EQUIPMENT									
0270	Explosives detector, walk through portal type	1 Elec	2	4	Ea.	48,800	245		49,045	54,000
0280	Hand-held, battery operated				″				25,500	28,100

28 23 Video Management System

28 23 13 – Video Management System Interfaces

28 23 13.10 Closed Circuit Television System		Crew	Daily Output	Labor-Hours	Unit	Material	Labor	Equipment	Total	Total Incl O&P
0010	CLOSED CIRCUIT TELEVISION SYSTEM									
2000	Surveillance, one station (camera & monitor)	2 Elec	2.60	6.154	Total	665	380		1,045	1,325
2200	For additional camera stations, add	1 Elec	2.70	2.963	Ea.	320	182		502	630
2400	Industrial quality, one station (camera & monitor)	2 Elec	2.60	6.154	Total	1,650	380		2,030	2,400
2600	For additional camera stations, add	1 Elec	2.70	2.963	Ea.	690	182		872	1,050
2610	For low light, add		2.70	2.963		525	182		707	855
2620	For very low light, add		2.70	2.963		3,450	182		3,632	4,075
2800	For weatherproof camera station, add		1.30	6.154		540	380		920	1,175
3000	For pan and tilt, add		1.30	6.154		1,575	380		1,955	2,300
3200	For zoom lens - remote control, add		2	4		1,800	245		2,045	2,350
3400	Extended zoom lens		2	4		5,025	245		5,270	5,900
3410	For automatic iris for low light, add		2	4		1,025	245		1,270	1,500
3600	Educational TV studio, basic 3 camera system, black & white,									
3800	electrical & electronic equip. only	4 Elec	.80	40	Total	8,275	2,450		10,725	12,900
4000	Full console		.28	114		25,300	7,000		32,300	38,700
4100	As above, but color system		.28	114		50,500	7,000		57,500	66,500
4120	Full console		.12	267		190,000	16,400		206,400	234,000
4200	For film chain, black & white, add	1 Elec	1	8	Ea.	12,500	490		12,990	14,500
4250	Color, add		.25	32		8,925	1,975		10,900	12,900
4400	For video recorders, add		1	8		2,450	490		2,940	3,425
4600	Premium	4 Elec	.40	80		16,000	4,900		20,900	25,200

28 23 23 – Video Surveillance Systems Infrastructure

28 23 23.50 Video Surveillance Equipments		Crew	Daily Output	Labor-Hours	Unit	Material	Labor	Equipment	Total	Total Incl O&P
0010	VIDEO SURVEILLANCE EQUIPMENTS									
0200	Video cameras, wireless, hidden in exit signs, clocks, etc., incl. receiver	1 Elec	3	2.667	Ea.	171	164		335	440
0210	Accessories for video recorder, single camera		3	2.667		195	164		359	465
0220	For multiple cameras		3	2.667		1,675	164		1,839	2,100
0230	Video cameras, wireless, for under vehicle searching, complete		2	4		15,200	245		15,445	17,100
0234	Master monitor station, 3 doors x 5 color monitor with tilt feature		2	4		985	245		1,230	1,450
0400	Internet protocol network camera, day/night, color & power supply		2.60	3.077		1,250	189		1,439	1,675
0500	Monitor, color flat screen, liquid crystal display (LCD), 15″		2.70	2.963		535	182		717	870
0520	17″		2.70	2.963		470	182		652	800
0540	19″		2.70	2.963		590	182		772	925

28 31 Intrusion Detection

28 31 16 – Intrusion Detection Systems Infrastructure

28 31 16.50 Intrusion Detection		Crew	Daily Output	Labor-Hours	Unit	Material	Labor	Equipment	Total	Total Incl O&P
						2020 Bare Costs				
0010	**INTRUSION DETECTION**, not including wires & conduits									
0100	Burglar alarm, battery operated, mechanical trigger	1 Elec	4	2	Ea.	289	123		412	510
0200	Electrical trigger		4	2		345	123		468	570
0400	For outside key control, add		8	1		89.50	61.50		151	193
0800	Card reader, flush type, standard		2.70	2.963		725	182		907	1,075
1000	Multi-code		2.70	2.963		1,250	182		1,432	1,650
1010	Card reader, proximity type		2.70	2.963		300	182		482	610
1200	Door switches, hinge switch		5.30	1.509		64.50	92.50		157	214
1400	Magnetic switch		5.30	1.509		105	92.50		197.50	259
1600	Exit control locks, horn alarm		4	2		218	123		341	430
1800	Flashing light alarm		4	2		247	123		370	460
2000	Indicating panels, 1 channel		2.70	2.963		244	182		426	550
2200	10 channel	2 Elec	3.20	5		1,175	305		1,480	1,775
2400	20 channel		2	8		2,525	490		3,015	3,525
2600	40 channel		1.14	14.035		4,650	860		5,510	6,450
2800	Ultrasonic motion detector, 12 V	1 Elec	2.30	3.478		200	213		413	550
3000	Infrared photoelectric detector		4	2		133	123		256	335
3200	Passive infrared detector		4	2		237	123		360	450
3400	Glass break alarm switch		8	1		85.50	61.50		147	189
3420	Switchmats, 30" x 5'		5.30	1.509		116	92.50		208.50	271
3440	30" x 25'		4	2		212	123		335	420
3460	Police connect panel		4	2		283	123		406	500
3480	Telephone dialer		5.30	1.509		400	92.50		492.50	585
3500	Alarm bell		4	2		106	123		229	305
3520	Siren		4	2		152	123		275	355
3540	Microwave detector, 10' to 200'		2	4		585	245		830	1,025
3560	10' to 350'		2	4		1,775	245		2,020	2,325

28 33 Security Monitoring and Control

28 33 11 – Electronic Structural Monitoring Systems

28 33 11.10 Load Cell Units		Crew	Daily Output	Labor-Hours	Unit	Material	Labor	Equipment	Total	Total Incl O&P
0010	**LOAD CELL UNITS**									
0100	Load cell, stainless steel, 0.5/1 ton rated, mounting unit	2 Elec	1.33	12.030	Ea.	665	740		1,405	1,850
0110	2/3 to 5/5 ton rated, mounting unit		1.20	13.333		705	820		1,525	2,025
0120	0.5/1/2/3.5/5 ton, self aligning bottom base		1.66	9.639		195	590		785	1,125
0130	Combo mounting unit, 10/25 ton rated		.86	18.605		1,825	1,150		2,975	3,750
0140	40/60 ton rated		.66	24.242		2,975	1,475		4,450	5,575
0150	60kg/3m, rated		1.84	8.696		1,550	535		2,085	2,525
0160	130kg/3m, rated		1.55	10.323		1,475	635		2,110	2,575
0170	280 kg, rated		1.22	13.115		1,475	805		2,280	2,850

28 41 Radiation Detection and Alarm

28 41 15 – Radiation Detection Sensors

28 41 15.50 Radiation Detection Systems		Crew	Daily Output	Labor-Hours	Unit	Material	2020 Bare Costs Labor	Equipment	Total	Total Incl O&P
0010	**RADIATION DETECTION SYSTEMS**									
9410	Minimum labor/equipment charge	1 Elec	4	2	Job		123		123	189

28 42 Gas Detection and Alarm

28 42 15 – Gas Detection Sensors

28 42 15.50 Tank Leak Detection Systems

Line	Description	Crew	Daily Output	Labor-Hours	Unit	Material	Labor	Equipment	Total	Total Incl O&P
0010	**TANK LEAK DETECTION SYSTEMS** Liquid and vapor									
0100	For hydrocarbons and hazardous liquids/vapors									
0120	Controller, data acquisition, incl. printer, modem, RS232 port									
0140	24 channel, for use with all probes				Ea.	3,125			3,125	3,450
0160	9 channel, for external monitoring				"	1,225			1,225	1,350
0200	Probes									
0210	Well monitoring									
0220	Liquid phase detection				Ea.	725			725	800
0230	Hydrocarbon vapor, fixed position					870			870	960
0240	Hydrocarbon vapor, float mounted					615			615	675
0250	Both liquid and vapor hydrocarbon					645			645	710
0300	Secondary containment, liquid phase									
0310	Pipe trench/manway sump				Ea.	765			765	840
0320	Double wall pipe and manual sump					640			640	705
0330	Double wall fiberglass annular space					430			430	470
0340	Double wall steel tank annular space					282			282	310
0500	Accessories									
0510	Modem, non-dedicated phone line				Ea.	281			281	310
0600	Monitoring, internal									
0610	Automatic tank gauge, incl. overfill				Ea.	1,325			1,325	1,450
0620	Product line				"	1,375			1,375	1,525
0700	Monitoring, special									
0710	Cathodic protection				Ea.	695			695	765
0720	Annular space chemical monitor				"	945			945	1,050

28 46 Fire Detection and Alarm

28 46 11 – Fire Sensors and Detectors

28 46 11.27 Other Sensors

Line	Description	Crew	Daily Output	Labor-Hours	Unit	Material	Labor	Equipment	Total	Total Incl O&P
0010	**OTHER SENSORS**									
5200	Smoke detector, ceiling type	1 Elec	6.20	1.290	Ea.	120	79		199	255
5240	Smoke detector, addressable type		6	1.333		224	82		306	370
5400	Duct type		3.20	2.500		295	153		448	560
5420	Duct addressable type		3.20	2.500		520	153		673	810

28 46 11.50 Fire and Heat Detectors

Line	Description	Crew	Daily Output	Labor-Hours	Unit	Material	Labor	Equipment	Total	Total Incl O&P
0010	**FIRE & HEAT DETECTORS**									
5000	Detector, rate of rise	1 Elec	8	1	Ea.	52	61.50		113.50	152
5010	Heat addressable type		7.25	1.103		225	67.50		292.50	350
5100	Fixed temp fire alarm		7	1.143		42	70		112	155
5400	104/85 DB, indoor/outdoor, ceiling/wall, red		10.25	.780		73	48		121	154
5410	Ceiling/wall, white		10.25	.780		71.50	48		119.50	153
5420	102/98 DB, lumin 110cd, indoor/wall, red		10.25	.780		113	48		161	199
5430	Lumin15/75cd, indoor/wall, red		10.25	.780		110	48		158	195
5440	HI/LO DB, 24 V, fire marking, red		10.25	.780		81	48		129	163

28 46 Fire Detection and Alarm

28 46 20 – Fire Alarm

28 46 20.50 Alarm Panels and Devices		Crew	Daily Output	Labor-Hours	Unit	Material	Labor	Equipment	Total	Total Incl O&P
						2020 Bare Costs				
0010	**ALARM PANELS AND DEVICES**, not including wires & conduits									
2200	Intercom remote station	1 Elec	8	1	Ea.	89.50	61.50		151	193
2600	Sound system, intercom handset	"	8	1		485	61.50		546.50	630
3600	4 zone	2 Elec	2	8		410	490		900	1,200
3800	8 zone		1	16		875	980		1,855	2,475
3810	5 zone		1.50	10.667		700	655		1,355	1,775
3900	10 zone		1.25	12.800		1,025	785		1,810	2,350
4000	12 zone		.67	23.988		2,350	1,475		3,825	4,850
4020	Alarm device, tamper, flow	1 Elec	8	1		230	61.50		291.50	350
4050	Actuating device	"	8	1		340	61.50		401.50	465
4160	Alarm control panel, addressable w/o voice, up to 200 points	2 Elec	1.14	13.998		4,575	860		5,435	6,350
4170	addressable w/voice, up to 400 points	"	.73	22.008		9,700	1,350		11,050	12,800
4175	Addressable interface device	1 Elec	7.25	1.103		149	67.50		216.50	268
4200	Battery and rack		4	2		460	123		583	700
4400	Automatic charger		8	1		525	61.50		586.50	675
4600	Signal bell		8	1		66.50	61.50		128	168
4610	Fire alarm signal bell 10" red 20-24 V P		8	1		154	61.50		215.50	264
4800	Trouble buzzer or manual station		8	1		88.50	61.50		150	192
5425	Duct smoke and heat detector 2 wire		8	1		125	61.50		186.50	233
5430	Fire alarm duct detector controller		3	2.667		254	164		418	530
5435	Fire alarm duct detector sensor kit		8	1		75	61.50		136.50	177
5440	Remote test station for smoke detector duct type		5.30	1.509		54.50	92.50		147	203
5460	Remote fire alarm indicator light		5.30	1.509		25	92.50		117.50	171
5600	Strobe and horn		5.30	1.509		137	92.50		229.50	294
5610	Strobe and horn (ADA type)		5.30	1.509		165	92.50		257.50	325
5620	Visual alarm (ADA type)		6.70	1.194		119	73.50		192.50	244
5800	electric bell		6.70	1.194		54	73.50		127.50	173
6000	Door holder, electro-magnetic		4	2		92	123		215	290
6200	Combination holder and closer		3.20	2.500		146	153		299	395
6600	Drill switch		8	1		400	61.50		461.50	535
6800	Master box		2.70	2.963		6,675	182		6,857	7,600
7000	Break glass station		8	1		58	61.50		119.50	158
7010	Break glass station, addressable		7.25	1.103		171	67.50		238.50	292
7800	Remote annunciator, 8 zone lamp		1.80	4.444		212	273		485	655
8000	12 zone lamp	2 Elec	2.60	6.154		415	380		795	1,025
8200	16 zone lamp	"	2.20	7.273		370	445		815	1,100

28 47 Mass Notification

28 47 12 – Notification Systems

28 47 12.10 Mass Notification System

Code	Description	Crew	Daily Output	Labor-Hours	Unit	Material	Labor	Equipment	Total	Total Incl O&P
0010	**MASS NOTIFICATION SYSTEM**									
0100	Wireless command center, 10,000 devices	2 Elec	1.33	12.030	Ea.	2,325	740		3,065	3,675
0200	Option, email notification					2,100			2,100	2,325
0210	Remote device supervision & monitor					2,650			2,650	2,900
0300	Antenna VHF or UHF, for medium range	1 Elec	4	2		93	123		216	292
0310	For high-power transmitter		2	4		1,125	245		1,370	1,625
0400	Transmitter, 25 watt		4	2		1,750	123		1,873	2,125
0410	40 watt		2.66	3.008		2,025	185		2,210	2,500
0420	100 watt		1.33	6.015		6,600	370		6,970	7,825
0500	Wireless receiver/control module for speaker		8	1		286	61.50		347.50	410

28 47 Mass Notification

28 47 12 – Notification Systems

28 47 12.10 Mass Notification System		Crew	Daily Output	Labor-Hours	Unit	2020 Bare Costs Material	Labor	Equipment	Total	Total Incl O&P
0600	Desktop paging controller, stand alone	1 Elec	4	2	Ea.	850	123		973	1,125

28 52 Detention Security Systems

28 52 11 – Detention Monitoring and Control Systems

28 52 11.10 Detention Control Systems		Crew	Daily Output	Labor-Hours	Unit	Material	Labor	Equipment	Total	Total Incl O&P
0010	**DETENTION CONTROL SYSTEMS**									
1000	Desk top control systems for 10 doors, 10 intercoms, and 10 lights	2 Elec	.30	53.333	Ea.	63,000	3,275		66,275	74,000
1020	Push button control panel systems for 10 doors, 10 intercoms, and 10 lights	"	.32	50	"	56,500	3,075		59,575	66,500

Division 31

Earthwork

Estimating Tips

31 05 00 Common Work Results for Earthwork

- Estimating the actual cost of performing earthwork requires careful consideration of the variables involved. This includes items such as type of soil, whether water will be encountered, dewatering, whether banks need bracing, disposal of excavated earth, and length of haul to fill or spoil sites, etc. If the project has large quantities of cut or fill, consider raising or lowering the site to reduce costs, while paying close attention to the effect on site drainage and utilities.
- If the project has large quantities of fill, creating a borrow pit on the site can significantly lower the costs.
- It is very important to consider what time of year the project is scheduled for completion. Bad weather can create large cost overruns from dewatering, site repair, and lost productivity from cold weather.

Reference Numbers

Reference numbers are shown at the beginning of some major classifications. These numbers refer to related items in the Reference Section. The reference information may be an estimating procedure, an alternate pricing method, or technical information.

Note: Not all subdivisions listed here necessarily appear. ■

31 05 Common Work Results for Earthwork

31 05 23 – Cement and Concrete for Earthwork

31 05 23.30 Plant Mixed Bituminous Concrete		Crew	Daily Output	Labor-Hours	Unit	Material	2020 Bare Costs Labor	Equipment	Total	Total Incl O&P
0010	**PLANT MIXED BITUMINOUS CONCRETE**									
0020	Asphaltic concrete plant mix (145 lb./C.F.)				Ton	68			68	74.50
0040	Asphaltic concrete less than 300 tons add trucking costs									
0200	All weather patching mix, hot				Ton	66			66	73
0250	Cold patch					69.50			69.50	76.50
0300	Berm mix					68			68	75
0400	Base mix					68			68	74.50
0500	Binder mix					68			68	74.50
0600	Sand or sheet mix					65			65	71.50

31 05 23.40 Recycled Plant Mixed Bituminous Concrete		Crew	Daily Output	Labor-Hours	Unit	Material	Labor	Equipment	Total	Total Incl O&P
0010	**RECYCLED PLANT MIXED BITUMINOUS CONCRETE**									
0200	Reclaimed pavement in stockpile G				Ton	26			26	29
0400	Recycled pavement, at plant, ratio old:new, 70:30 G					38			38	42
0600	Ratio old:new, 30:70 G					54			54	59.50

31 06 Schedules for Earthwork

31 06 60 – Schedules for Special Foundations and Load Bearing Elements

31 06 60.14 Piling Special Costs		Crew	Daily Output	Labor-Hours	Unit	Material	Labor	Equipment	Total	Total Incl O&P
0010	**PILING SPECIAL COSTS**									
0011	Piling special costs, pile caps, see Section 03 30 53.40									
0500	Cutoffs, concrete piles, plain	1 Pile	5.50	1.455	Ea.		79		79	126
0600	With steel thin shell, add		38	.211			11.40		11.40	18.30
0700	Steel pile or "H" piles		19	.421			23		23	36.50
0800	Wood piles		38	.211			11.40		11.40	18.30
0900	Pre-augering up to 30′ deep, average soil, 24″ diameter	B-43	180	.267	L.F.		12.45	4.28	16.73	24.50
0920	36″ diameter		115	.417			19.50	6.70	26.20	38.50
0960	48″ diameter		70	.686			32	11	43	63
0980	60″ diameter		50	.960			45	15.40	60.40	88
1000	Testing, any type piles, test load is twice the design load									
1050	50 ton design load, 100 ton test				Ea.				14,000	15,500
1100	100 ton design load, 200 ton test								20,000	22,000
1150	150 ton design load, 300 ton test								26,000	28,500
1200	200 ton design load, 400 ton test								28,000	31,000
1250	400 ton design load, 800 ton test								32,000	35,000
1500	Wet conditions, soft damp ground									
1600	Requiring mats for crane, add								40%	40%
1700	Barge mounted driving rig, add								30%	30%

31 06 60.15 Mobilization		Crew	Daily Output	Labor-Hours	Unit	Material	Labor	Equipment	Total	Total Incl O&P
0010	**MOBILIZATION**									
0020	Set up & remove, air compressor, 600 CFM	A-5	3.30	5.455	Ea.		233	14.85	247.85	390
0100	1,200 CFM	"	2.20	8.182			350	22.50	372.50	585
0200	Crane, with pile leads and pile hammer, 75 ton	B-19	.60	107			5,900	3,350	9,250	13,000
0300	150 ton	"	.36	178			9,825	5,575	15,400	21,800
0500	Drill rig, for caissons, to 36″, minimum	B-43	2	24			1,125	385	1,510	2,200
0520	Maximum		.50	96			4,475	1,550	6,025	8,825
0600	Up to 84″		1	48			2,250	770	3,020	4,400
0800	Auxiliary boiler, for steam small	A-5	1.66	10.843			465	29.50	494.50	775
0900	Large	"	.83	21.687			925	59	984	1,550
1100	Rule of thumb: complete pile driving set up, small	B-19	.45	142			7,850	4,475	12,325	17,400
1200	Large	"	.27	237			13,100	7,450	20,550	29,000

31 06 Schedules for Earthwork

31 06 60 – Schedules for Special Foundations and Load Bearing Elements

31 06 60.15 Mobilization		Crew	Daily Output	Labor-Hours	Unit	Material	2020 Bare Costs Labor	Equipment	Total	Total Incl O&P
1500	Mobilization, barge, by tug boat	B-83	25	.640	Mile		31.50	28.50	60	81.50
1600	Standby time for shore pile driving crew	B-19	8	8	Hr.		440	251	691	975
1700	Standby time for barge driving rig	B-76	8	9	"		495	460	955	1,300

31 11 Clearing and Grubbing

31 11 10 – Clearing and Grubbing Land

31 11 10.10 Clear and Grub Site

		Crew	Daily Output	Labor-Hours	Unit	Material	Labor	Equipment	Total	Total Incl O&P
0010	**CLEAR AND GRUB SITE**									
0020	Cut & chip light trees to 6" diam.	B-7	1	48	Acre		2,150	1,650	3,800	5,225
0150	Grub stumps and remove	B-30	2	12			620	915	1,535	1,975
0200	Cut & chip medium trees to 12" diam.	B-7	.70	68.571			3,075	2,350	5,425	7,475
0250	Grub stumps and remove	B-30	1	24			1,225	1,825	3,050	3,975
0300	Cut & chip heavy trees to 24" diam.	B-7	.30	160			7,175	5,475	12,650	17,400
0350	Grub stumps and remove	B-30	.50	48			2,475	3,675	6,150	7,950
0400	If burning is allowed, deduct cut & chip								40%	40%
3000	Chipping stumps, to 18" deep, 12" diam.	B-86	20	.400	Ea.		22.50	9.70	32.20	46
3040	18" diameter		16	.500			28.50	12.15	40.65	58
3080	24" diameter		14	.571			32.50	13.90	46.40	66
3100	30" diameter		12	.667			38	16.20	54.20	77
3120	36" diameter		10	.800			45.50	19.45	64.95	92.50
3160	48" diameter		8	1			57	24.50	81.50	116
5000	Tree thinning, feller buncher, conifer									
5080	Up to 8" diameter	B-93	240	.033	Ea.		1.89	2.45	4.34	5.65
5120	12" diameter		160	.050			2.84	3.68	6.52	8.50
5240	Hardwood, up to 4" diameter		240	.033			1.89	2.45	4.34	5.65
5280	8" diameter		180	.044			2.52	3.27	5.79	7.55
5320	12" diameter		120	.067			3.78	4.91	8.69	11.30
7000	Tree removal, congested area, aerial lift truck									
7040	8" diameter	B-85	7	5.714	Ea.		265	115	380	545
7080	12" diameter		6	6.667			310	134	444	640
7120	18" diameter		5	8			370	161	531	765
7160	24" diameter		4	10			465	201	666	960
7240	36" diameter		3	13.333			620	268	888	1,275
7280	48" diameter		2	20			930	400	1,330	1,925

31 13 Selective Tree and Shrub Removal and Trimming

31 13 13 – Selective Tree and Shrub Removal

31 13 13.20 Selective Tree Removal

		Crew	Daily Output	Labor-Hours	Unit	Material	Labor	Equipment	Total	Total Incl O&P
0010	**SELECTIVE TREE REMOVAL**									
0011	With tractor, large tract, firm									
0020	level terrain, no boulders, less than 12" diam. trees									
0300	300 HP dozer, up to 400 trees/acre, up to 25% hardwoods	B-10M	.75	16	Acre		830	2,350	3,180	3,875
0340	25% to 50% hardwoods		.60	20			1,025	2,950	3,975	4,850
0370	75% to 100% hardwoods		.45	26.667			1,375	3,925	5,300	6,475
0400	500 trees/acre, up to 25% hardwoods		.60	20			1,025	2,950	3,975	4,850
0440	25% to 50% hardwoods		.48	25			1,300	3,675	4,975	6,100
0470	75% to 100% hardwoods		.36	33.333			1,725	4,900	6,625	8,125
0500	More than 600 trees/acre, up to 25% hardwoods		.52	23.077			1,200	3,400	4,600	5,600
0540	25% to 50% hardwoods		.42	28.571			1,475	4,200	5,675	6,950

31 13 Selective Tree and Shrub Removal and Trimming

31 13 13 – Selective Tree and Shrub Removal

31 13 13.20 Selective Tree Removal		Crew	Daily Output	Labor-Hours	Unit	Material	2020 Bare Costs Labor	Equipment	Total	Total Incl O&P
0570	75% to 100% hardwoods	B-10M	.31	38.710	Acre		2,000	5,700	7,700	9,400
0900	Large tract clearing per tree									
1500	300 HP dozer, to 12" diameter, softwood	B-10M	320	.038	Ea.		1.95	5.50	7.45	9.10
1550	Hardwood		100	.120			6.20	17.65	23.85	29
1600	12" to 24" diameter, softwood		200	.060			3.11	8.80	11.91	14.60
1650	Hardwood		80	.150			7.80	22	29.80	37
1700	24" to 36" diameter, softwood		100	.120			6.20	17.65	23.85	29
1750	Hardwood		50	.240			12.45	35.50	47.95	58.50
1800	36" to 48" diameter, softwood		70	.171			8.90	25	33.90	41.50
1850	Hardwood	↓	35	.343	↓		17.80	50.50	68.30	83.50
2000	Stump removal on site by hydraulic backhoe, 1-1/2 C.Y.									
2040	4" to 6" diameter	B-17	60	.533	Ea.		25	10.70	35.70	51.50
2050	8" to 12" diameter	B-30	33	.727			37.50	55.50	93	121
2100	14" to 24" diameter		25	.960			49.50	73.50	123	159
2150	26" to 36" diameter	↓	16	1.500	↓		77.50	115	192.50	249
3000	Remove selective trees, on site using chain saws and chipper,									
3050	not incl. stumps, up to 6" diameter	B-7	18	2.667	Ea.		120	91.50	211.50	291
3100	8" to 12" diameter		12	4			180	137	317	435
3150	14" to 24" diameter		10	4.800			215	164	379	525
3200	26" to 36" diameter	↓	8	6			269	206	475	655
3300	Machine load, 2 mile haul to dump, 12" diam. tree	A-3B	8	2	↓		106	159	265	340

31 14 Earth Stripping and Stockpiling

31 14 13 – Soil Stripping and Stockpiling

31 14 13.23 Topsoil Stripping and Stockpiling		Crew	Daily Output	Labor-Hours	Unit	Material	Labor	Equipment	Total	Total Incl O&P
0010	**TOPSOIL STRIPPING AND STOCKPILING**									
0020	200 HP dozer, ideal conditions	B-10B	2300	.005	C.Y.		.27	.65	.92	1.15
0100	Adverse conditions	"	1150	.010			.54	1.31	1.85	2.29
0200	300 HP dozer, ideal conditions	B-10M	3000	.004			.21	.59	.80	.98
0300	Adverse conditions	"	1650	.007			.38	1.07	1.45	1.77
0400	400 HP dozer, ideal conditions	B-10X	3900	.003			.16	.71	.87	1.03
0500	Adverse conditions	"	2000	.006			.31	1.39	1.70	2.02
0600	Clay, dry and soft, 200 HP dozer, ideal conditions	B-10B	1600	.008			.39	.94	1.33	1.64
0700	Adverse conditions	"	800	.015			.78	1.88	2.66	3.30
1000	Medium hard, 300 HP dozer, ideal conditions	B-10M	2000	.006			.31	.88	1.19	1.46
1100	Adverse conditions	"	1100	.011			.57	1.60	2.17	2.65
1200	Very hard, 400 HP dozer, ideal conditions	B-10X	2600	.005			.24	1.07	1.31	1.56
1300	Adverse conditions	"	1340	.009	↓		.46	2.07	2.53	3.01
1500	Loam or topsoil, remove/stockpile on site									
1510	By hand, 6" deep, 50' haul, less than 100 S.Y.	B-1	100	.240	S.Y.		10.25		10.25	16.45
1520	By skid steer, 6" deep, 100' haul, 101-500 S.Y.	B-62	500	.048			2.20	.36	2.56	3.88
1530	100' haul, 501-900 S.Y.	"	900	.027			1.22	.20	1.42	2.16
1540	200' haul, 901-1,100 S.Y.	B-63	1000	.040			1.77	.18	1.95	3.02
1550	By dozer, 200' haul, 1,101-4,000 S.Y.	B-10B	4000	.003	↓		.16	.38	.54	.66

31 22 Grading

31 22 13 – Rough Grading

31 22 13.20 Rough Grading Sites		Crew	Daily Output	Labor-Hours	Unit	Material	2020 Bare Costs Labor	Equipment	Total	Total Incl O&P
0010	**ROUGH GRADING SITES**									
0100	Rough grade sites 400 S.F. or less, hand labor	B-1	2	12	Ea.		515		515	820
0120	410-1,000 S.F.	"	1	24			1,025		1,025	1,650
0130	1,100-3,000 S.F., skid steer & labor	B-62	1.50	16			730	118	848	1,275
0140	3,100-5,000 S.F.	"	1	24			1,100	178	1,278	1,950
0150	5,100-8,000 S.F.	B-63	1	40			1,775	178	1,953	3,025
0160	8,100-10,000 S.F.	"	.75	53.333			2,350	237	2,587	4,025
0170	8,100-10,000 S.F., dozer	B-10L	1	12			620	400	1,020	1,425
0200	Rough grade open sites 10,000-20,000 S.F., grader	B-11L	1.80	8.889			440	590	1,030	1,350
0210	20,100-25,000 S.F.		1.40	11.429			565	760	1,325	1,725
0220	25,100-30,000 S.F.		1.20	13.333			660	885	1,545	2,025
0230	30,100-35,000 S.F.		1	16			790	1,050	1,840	2,425
0240	35,100-40,000 S.F.		.90	17.778			880	1,175	2,055	2,700
0250	40,100-45,000 S.F.		.80	20			990	1,325	2,315	3,025
0260	45,100-50,000 S.F.		.72	22.222			1,100	1,475	2,575	3,350
0270	50,100-75,000 S.F.		.50	32			1,575	2,125	3,700	4,825
0280	75,100-100,000 S.F.		.36	44.444			2,200	2,950	5,150	6,725

31 22 16 – Fine Grading

31 22 16.10 Finish Grading

		Crew	Daily Output	Labor-Hours	Unit	Material	Labor	Equipment	Total	Total Incl O&P
0010	**FINISH GRADING**									
0012	Finish grading area to be paved with grader, small area	B-11L	400	.040	S.Y.		1.98	2.66	4.64	6.05
0100	Large area		2000	.008			.40	.53	.93	1.21
0200	Grade subgrade for base course, roadways		3500	.005			.23	.30	.53	.69
1020	For large parking lots	B-32C	5000	.010			.48	.56	1.04	1.38
1050	For small irregular areas	"	2000	.024			1.19	1.40	2.59	3.43
1100	Fine grade for slab on grade, machine	B-11L	1040	.015			.76	1.02	1.78	2.32
1150	Hand grading	B-18	700	.034			1.47	.24	1.71	2.61
1200	Fine grade granular base for sidewalks and bikeways	B-62	1200	.020			.91	.15	1.06	1.61
2550	Hand grade select gravel	2 Clab	60	.267	C.S.F.		11.25		11.25	18
3000	Hand grade select gravel, including compaction, 4" deep	B-18	555	.043	S.Y.		1.85	.30	2.15	3.29
3100	6" deep		400	.060			2.57	.41	2.98	4.56
3120	8" deep		300	.080			3.42	.55	3.97	6.10
3300	Finishing grading slopes, gentle	B-11L	8900	.002			.09	.12	.21	.27
3310	Steep slopes		7100	.002			.11	.15	.26	.34
3500	Finish grading lagoon bottoms		4	4	M.S.F.		198	266	464	605
9000	Minimum labor/equipment charge, hand grading	1 Clab	2	4	Job		168		168	270
9100	Minimum labor/equipment charge, machine grading	B-11L	2	8	"		395	530	925	1,200

31 23 Excavation and Fill

31 23 16 – Excavation

31 23 16.13 Excavating, Trench

		Crew	Daily Output	Labor-Hours	Unit	Material	Labor	Equipment	Total	Total Incl O&P
0010	**EXCAVATING, TRENCH**									
0011	Or continuous footing									
0020	Common earth with no sheeting or dewatering included									
0050	1' to 4' deep, 3/8 C.Y. excavator	B-11C	150	.107	B.C.Y.		5.25	1.43	6.68	9.90
0060	1/2 C.Y. excavator	B-11M	200	.080			3.95	1.16	5.11	7.55
0090	4' to 6' deep, 1/2 C.Y. excavator	"	200	.080			3.95	1.16	5.11	7.55
0100	5/8 C.Y. excavator	B-12Q	250	.064			3.24	2.39	5.63	7.75
0300	1/2 C.Y. excavator, truck mounted	B-12J	200	.080			4.05	4.23	8.28	11.05
0500	6' to 10' deep, 3/4 C.Y. excavator	B-12F	225	.071			3.60	3.09	6.69	9.10

31 23 16 – Excavation

31 23 16.13 Excavating, Trench		Crew	Daily Output	Labor-Hours	Unit	Material	2020 Bare Costs Labor	Equipment	Total	Total Incl O&P
0600	1 C.Y. excavator, truck mounted	B-12K	400	.040	B.C.Y.		2.03	2.43	4.46	5.90
0900	10′ to 14′ deep, 3/4 C.Y. excavator	B-12F	200	.080			4.05	3.47	7.52	10.20
1000	1-1/2 C.Y. excavator	B-12B	540	.030			1.50	1.27	2.77	3.77
1300	14′ to 20′ deep, 1 C.Y. excavator	B-12A	320	.050			2.53	2.49	5.02	6.75
1340	20′ to 24′ deep, 1 C.Y. excavator	″	288	.056			2.81	2.76	5.57	7.50
1352	4′ to 6′ deep, 1/2 C.Y. excavator w/trench box	B-13H	188	.085			4.31	5.15	9.46	12.45
1354	5/8 C.Y. excavator	″	235	.068			3.45	4.10	7.55	9.95
1362	6′ to 10′ deep, 3/4 C.Y. excavator w/trench box	B-13G	212	.075			3.82	3.83	7.65	10.25
1374	10′ to 14′ deep, 3/4 C.Y. excavator w/trench box	″	188	.085			4.31	4.32	8.63	11.55
1376	1-1/2 C.Y. excavator	B-13E	508	.032			1.60	1.59	3.19	4.26
1381	14′ to 20′ deep, 1 C.Y. excavator w/trench box	B-13D	301	.053			2.69	3.03	5.72	7.60
1386	20′ to 24′ deep, 1 C.Y. excavator w/trench box	″	271	.059			2.99	3.37	6.36	8.45
1391	Shoring by S.F./day trench wall protected loose mat., 4′ W	B-6	3200	.008	SF Wall	.51	.34	.07	.92	1.17
1392	Rent shoring per week per S.F. wall protected, loose mat., 4′ W					1.40			1.40	1.54
1395	Hydraulic shoring, S.F. trench wall protected stable mat., 4′ W	2 Clab	2700	.006		.20	.25		.45	.62
1397	semi-stable material, 4′ W	″	2400	.007		.28	.28		.56	.76
1398	Rent hydraulic shoring per day/S.F. wall, stable mat., 4′ W					.32			.32	.36
1399	semi-stable material					.41			.41	.45
1400	By hand with pick and shovel 2′ to 6′ deep, light soil	1 Clab	8	1	B.C.Y.		42		42	67.50
1500	Heavy soil	″	4	2	″		84		84	135
1700	For tamping backfilled trenches, air tamp, add	A-1G	100	.080	E.C.Y.		3.37	.54	3.91	6
1900	Vibrating plate, add	B-18	180	.133	″		5.70	.92	6.62	10.15
2100	Trim sides and bottom for concrete pours, common earth		1500	.016	S.F.		.68	.11	.79	1.22
2300	Hardpan		600	.040	″		1.71	.28	1.99	3.04
2400	Pier and spread footing excavation, add to above				B.C.Y.				30%	30%
3000	Backfill trench, F.E. loader, wheel mtd., 1 C.Y. bucket									
3020	Minimal haul	B-10R	400	.030	L.C.Y.		1.56	.76	2.32	3.28
3040	100′ haul		200	.060			3.11	1.51	4.62	6.55
3060	200′ haul		100	.120			6.20	3.02	9.22	13.10
5020	Loam & sandy clay with no sheeting or dewatering included									
5050	1′ to 4′ deep, 3/8 C.Y. tractor loader/backhoe	B-11C	162	.099	B.C.Y.		4.88	1.32	6.20	9.15
5060	1/2 C.Y. excavator	B-11M	216	.074			3.66	1.08	4.74	7
5080	4′ to 6′ deep, 1/2 C.Y. excavator	″	216	.074			3.66	1.08	4.74	7
5090	5/8 C.Y. excavator	B-12Q	276	.058			2.94	2.17	5.11	7
5130	1/2 C.Y. excavator, truck mounted	B-12J	216	.074			3.75	3.92	7.67	10.25
5140	6′ to 10′ deep, 3/4 C.Y. excavator	B-12F	243	.066			3.33	2.86	6.19	8.40
5160	1 C.Y. excavator, truck mounted	B-12K	432	.037			1.88	2.25	4.13	5.45
5190	10′ to 14′ deep, 3/4 C.Y. excavator	B-12F	216	.074			3.75	3.21	6.96	9.50
5210	1-1/2 C.Y. excavator	B-12B	583	.027			1.39	1.18	2.57	3.50
5250	14′ to 20′ deep, 1 C.Y. excavator	B-12A	346	.046			2.34	2.30	4.64	6.25
5300	20′ to 24′ deep, 1 C.Y. excavator	″	311	.051			2.61	2.56	5.17	6.95
5352	4′ to 6′ deep, 1/2 C.Y. excavator w/trench box	B-13H	205	.078			3.95	4.70	8.65	11.40
5354	5/8 C.Y. excavator	″	257	.062			3.15	3.75	6.90	9.10
5362	6′ to 10′ deep, 3/4 C.Y. excavator w/trench box	B-13G	231	.069			3.51	3.51	7.02	9.40
5370	10′ to 14′ deep, 3/4 C.Y. excavator w/trench box	″	205	.078			3.95	3.96	7.91	10.60
5374	1-1/2 C.Y. excavator	B-13E	554	.029			1.46	1.45	2.91	3.91
5382	14′ to 20′ deep, 1 C.Y. excavator w/trench box	B-13D	329	.049			2.46	2.78	5.24	6.95
5392	20′ to 24′ deep, 1 C.Y. excavator w/trench box	″	295	.054			2.75	3.10	5.85	7.75
6020	Sand & gravel with no sheeting or dewatering included									
6050	1′ to 4′ deep, 3/8 C.Y. excavator	B-11C	165	.097	B.C.Y.		4.79	1.30	6.09	9.05
6060	1/2 C.Y. excavator	B-11M	220	.073			3.60	1.06	4.66	6.85
6080	4′ to 6′ deep, 1/2 C.Y. excavator	″	220	.073			3.60	1.06	4.66	6.85
6090	5/8 C.Y. excavator	B-12Q	275	.058			2.95	2.18	5.13	7.05

31 23 16.13 Excavating, Trench

	31 23 16.13 Excavating, Trench	Crew	Daily Output	Labor-Hours	Unit	Material	2020 Bare Costs Labor	Equipment	Total	Total Incl O&P
6130	1/2 C.Y. excavator, truck mounted	B-12J	220	.073	B.C.Y.		3.68	3.85	7.53	10.05
6140	6′ to 10′ deep, 3/4 C.Y. excavator	B-12F	248	.065			3.27	2.80	6.07	8.25
6160	1 C.Y. excavator, truck mounted	B-12K	440	.036			1.84	2.21	4.05	5.35
6190	10′ to 14′ deep, 3/4 C.Y. excavator	B-12F	220	.073			3.68	3.16	6.84	9.25
6210	1-1/2 C.Y. excavator	B-12B	594	.027			1.36	1.16	2.52	3.43
6250	14′ to 20′ deep, 1 C.Y. excavator	B-12A	352	.045			2.30	2.26	4.56	6.15
6300	20′ to 24′ deep, 1 C.Y. excavator	″	317	.050			2.56	2.51	5.07	6.80
6352	4′ to 6′ deep, 1/2 C.Y. excavator w/trench box	B-13H	209	.077			3.88	4.61	8.49	11.25
6354	5/8 C.Y. excavator	″	261	.061			3.10	3.69	6.79	8.95
6362	6′ to 10′ deep, 3/4 C.Y. excavator w/trench box	B-13G	236	.068			3.43	3.44	6.87	9.25
6370	10′ to 14′ deep, 3/4 C.Y. excavator w/trench box	″	209	.077			3.88	3.89	7.77	10.40
6374	1-1/2 C.Y. excavator	B-13E	564	.028			1.44	1.43	2.87	3.84
6382	14′ to 20′ deep, 1 C.Y. excavator w/trench box	B-13D	334	.048			2.43	2.73	5.16	6.85
6392	20′ to 24′ deep, 1 C.Y. excavator w/trench box	″	301	.053	↓		2.69	3.03	5.72	7.60
7020	Dense hard clay with no sheeting or dewatering included									
7050	1′ to 4′ deep, 3/8 C.Y. excavator	B-11C	132	.121	B.C.Y.		6	1.62	7.62	11.25
7060	1/2 C.Y. excavator	B-11M	176	.091			4.49	1.32	5.81	8.55
7080	4′ to 6′ deep, 1/2 C.Y. excavator	″	176	.091			4.49	1.32	5.81	8.55
7090	5/8 C.Y. excavator	B-12Q	220	.073			3.68	2.72	6.40	8.80
7130	1/2 C.Y. excavator, truck mounted	B-12J	176	.091			4.60	4.81	9.41	12.60
7140	6′ to 10′ deep, 3/4 C.Y. excavator	B-12F	198	.081			4.09	3.51	7.60	10.30
7160	1 C.Y. excavator, truck mounted	B-12K	352	.045			2.30	2.77	5.07	6.70
7190	10′ to 14′ deep, 3/4 C.Y. excavator	B-12F	176	.091			4.60	3.94	8.54	11.65
7210	1-1/2 C.Y. excavator	B-12B	475	.034			1.71	1.45	3.16	4.29
7250	14′ to 20′ deep, 1 C.Y. excavator	B-12A	282	.057			2.87	2.82	5.69	7.65
7300	20′ to 24′ deep, 1 C.Y. excavator	″	254	.063	↓		3.19	3.13	6.32	8.50
9000	Minimum labor/equipment charge	1 Clab	4	2	Job		84		84	135

31 23 16.14 Excavating, Utility Trench

	31 23 16.14 Excavating, Utility Trench	Crew	Daily Output	Labor-Hours	Unit	Material	Labor	Equipment	Total	Total Incl O&P
0010	**EXCAVATING, UTILITY TRENCH**									
0011	Common earth									
0050	Trenching with chain trencher, 12 HP, operator walking									
0100	4″ wide trench, 12″ deep	B-53	800	.010	L.F.		.53	.20	.73	1.05
0150	18″ deep		750	.011			.57	.21	.78	1.12
0200	24″ deep		700	.011			.61	.22	.83	1.20
0300	6″ wide trench, 12″ deep		650	.012			.65	.24	.89	1.29
0350	18″ deep		600	.013			.71	.26	.97	1.40
0400	24″ deep		550	.015			.77	.29	1.06	1.52
0450	36″ deep		450	.018			.94	.35	1.29	1.85
0600	8″ wide trench, 12″ deep		475	.017			.89	.33	1.22	1.76
0650	18″ deep		400	.020			1.06	.39	1.45	2.09
0700	24″ deep		350	.023			1.21	.45	1.66	2.39
0750	36″ deep		300	.027	↓		1.41	.52	1.93	2.79
0900	Minimum labor/equipment charge	↓	2	4	Job		212	78.50	290.50	415
1000	Backfill by hand including compaction, add									
1050	4″ wide trench, 12″ deep	A-1G	800	.010	L.F.		.42	.07	.49	.74
1100	18″ deep		530	.015			.64	.10	.74	1.13
1150	24″ deep		400	.020			.84	.14	.98	1.50
1300	6″ wide trench, 12″ deep		540	.015			.62	.10	.72	1.11
1350	18″ deep		405	.020			.83	.13	.96	1.48
1400	24″ deep		270	.030			1.25	.20	1.45	2.22
1450	36″ deep		180	.044			1.87	.30	2.17	3.33
1600	8″ wide trench, 12″ deep	↓	400	.020	↓		.84	.14	.98	1.50

31 23 16.14 Excavating, Utility Trench		Crew	Daily Output	Labor-Hours	Unit	Material	2020 Bare Costs Labor	Equipment	Total	Total Incl O&P
1650	18" deep	A-1G	265	.030	L.F.		1.27	.20	1.47	2.26
1700	24" deep		200	.040			1.68	.27	1.95	3
1750	36" deep		135	.059			2.49	.40	2.89	4.44
2000	Chain trencher, 40 HP operator riding									
2050	6" wide trench and backfill, 12" deep	B-54	1200	.007	L.F.		.35	.34	.69	.92
2100	18" deep		1000	.008			.42	.40	.82	1.10
2150	24" deep		975	.008			.44	.41	.85	1.13
2200	36" deep		900	.009			.47	.45	.92	1.23
2250	48" deep		750	.011			.57	.54	1.11	1.48
2300	60" deep		650	.012			.65	.62	1.27	1.70
2400	8" wide trench and backfill, 12" deep		1000	.008			.42	.40	.82	1.10
2450	18" deep		950	.008			.45	.42	.87	1.17
2500	24" deep		900	.009			.47	.45	.92	1.23
2550	36" deep		800	.010			.53	.50	1.03	1.38
2600	48" deep		650	.012			.65	.62	1.27	1.70
2700	12" wide trench and backfill, 12" deep		975	.008			.44	.41	.85	1.13
2750	18" deep		860	.009			.49	.47	.96	1.28
2800	24" deep		800	.010			.53	.50	1.03	1.38
2850	36" deep		725	.011			.58	.55	1.13	1.52
3000	16" wide trench and backfill, 12" deep		835	.010			.51	.48	.99	1.32
3050	18" deep		750	.011			.57	.54	1.11	1.48
3100	24" deep		700	.011			.61	.57	1.18	1.58
3200	Compaction with vibratory plate, add								35%	35%
5100	Hand excavate and trim for pipe bells after trench excavation									
5200	8" pipe	1 Clab	155	.052	L.F.		2.17		2.17	3.48
5300	18" pipe	"	130	.062	"		2.59		2.59	4.15
9000	Minimum labor/equipment charge	A-1G	4	2	Job		84	13.50	97.50	150

31 23 16.16 Structural Excavation for Minor Structures

		Crew	Daily Output	Labor-Hours	Unit	Material	Labor	Equipment	Total	Total Incl O&P
0010	**STRUCTURAL EXCAVATION FOR MINOR STRUCTURES**									
0015	Hand, pits to 6' deep, sandy soil	1 Clab	8	1	B.C.Y.		42		42	67.50
0100	Heavy soil or clay		4	2			84		84	135
0300	Pits 6' to 12' deep, sandy soil		5	1.600			67.50		67.50	108
0500	Heavy soil or clay		3	2.667			112		112	180
0700	Pits 12' to 18' deep, sandy soil		4	2			84		84	135
0900	Heavy soil or clay		2	4			168		168	270
1100	Hand loading trucks from stock pile, sandy soil		12	.667			28		28	45
1300	Heavy soil or clay		8	1			42		42	67.50
1500	For wet or muck hand excavation, add to above								50%	50%
1550	Excavation rock by hand/air tool	B-9	3.40	11.765	B.C.Y.		500	105	605	915
9000	Minimum labor/equipment charge	1 Clab	4	2	Job		84		84	135

31 23 16.26 Rock Removal

		Crew	Daily Output	Labor-Hours	Unit	Material	Labor	Equipment	Total	Total Incl O&P
0010	**ROCK REMOVAL**									
0015	Drilling only rock, 2" hole for rock bolts	B-47	316	.076	L.F.		3.52	4.92	8.44	11
0800	2-1/2" hole for pre-splitting		250	.096			4.45	6.20	10.65	13.90
4600	Quarry operations, 2-1/2" to 3-1/2" diameter		240	.100			4.64	6.50	11.14	14.50

31 23 16.30 Drilling and Blasting Rock

		Crew	Daily Output	Labor-Hours	Unit	Material	Labor	Equipment	Total	Total Incl O&P
0010	**DRILLING AND BLASTING ROCK**									
0020	Rock, open face, under 1,500 C.Y.	B-47	225	.107	B.C.Y.	4.54	4.95	6.90	16.39	20.50
0100	Over 1,500 C.Y.		300	.080		4.54	3.71	5.20	13.45	16.60
0200	Areas where blasting mats are required, under 1,500 C.Y.		175	.137		4.54	6.35	8.90	19.79	25
0250	Over 1,500 C.Y.		250	.096		4.54	4.45	6.20	15.19	18.90
0300	Bulk drilling and blasting, can vary greatly, average								9.65	12.20

31 23 16.30 Drilling and Blasting Rock		Crew	Daily Output	Labor-Hours	Unit	2020 Bare Costs Material	Labor	Equipment	Total	Total Incl O&P
0500	Pits, average				B.C.Y.				25.50	31.50
1300	Deep hole method, up to 1,500 C.Y.	B-47	50	.480		4.54	22.50	31	58.04	74.50
1400	Over 1,500 C.Y.		66	.364		4.54	16.85	23.50	44.89	58
1900	Restricted areas, up to 1,500 C.Y.		13	1.846		4.54	85.50	120	210.04	273
2000	Over 1,500 C.Y.		20	1.200		4.54	55.50	78	138.04	179
2200	Trenches, up to 1,500 C.Y.		22	1.091		13.15	50.50	70.50	134.15	173
2300	Over 1,500 C.Y.		26	.923		13.15	43	60	116.15	149
2500	Pier holes, up to 1,500 C.Y.		22	1.091		4.54	50.50	70.50	125.54	163
2600	Over 1,500 C.Y.		31	.774		4.54	36	50	90.54	117
2800	Boulders under 1/2 C.Y., loaded on truck, no hauling	B-100	80	.150			7.80	12	19.80	25.50
2900	Boulders, drilled, blasted	B-47	100	.240		4.54	11.15	15.55	31.24	40
3100	Jackhammer operators with foreman compressor, air tools	B-9	1	40	Day		1,700	355	2,055	3,125
3300	Track drill, compressor, operator and foreman	B-47	1	24	"		1,125	1,550	2,675	3,475
3500	Blasting caps				Ea.	6.65			6.65	7.30
3700	Explosives					.54			.54	.59
3800	Blasting mats, for purchase, no mobilization, 10′ x 15′ x 12″					1,275			1,275	1,400
3900	Blasting mats, rent, for first day					211			211	232
4000	Per added day					65.50			65.50	72
4200	Preblast survey for 6 room house, individual lot, minimum	A-6	2.40	6.667			355	12.85	367.85	580
4300	Maximum	"	1.35	11.852			625	23	648	1,025
4500	City block within zone of influence, minimum	A-8	25200	.001	S.F.		.07		.07	.11
4600	Maximum	"	15100	.002	"		.12		.12	.19
5000	Excavate and load boulders, less than 0.5 C.Y.	B-10T	80	.150	B.C.Y.		7.80	8.80	16.60	22
5020	0.5 C.Y. to 1 C.Y.	B-10U	100	.120			6.20	9.55	15.75	20.50
5200	Excavate and load blasted rock, 3 C.Y. power shovel	B-12T	1530	.010			.53	1.38	1.91	2.36
5400	Haul boulders, 25 ton off-highway dump, 1 mile round trip	B-34E	330	.024			1.19	4.18	5.37	6.50
5420	2 mile round trip		275	.029			1.42	5	6.42	7.80
5440	3 mile round trip		225	.036			1.74	6.15	7.89	9.55
5460	4 mile round trip		200	.040			1.96	6.90	8.86	10.75
5600	Bury boulders on site, less than 0.5 C.Y., 300 HP dozer									
5620	150′ haul	B-10M	310	.039	B.C.Y.		2.01	5.70	7.71	9.40
5640	300′ haul		210	.057			2.96	8.40	11.36	13.90
5800	0.5 to 1 C.Y., 300 HP dozer, 150′ haul		300	.040			2.07	5.90	7.97	9.70
5820	300′ haul		200	.060			3.11	8.80	11.91	14.60

31 23 16.32 Ripping

		Crew	Daily Output	Labor-Hours	Unit	Material	Labor	Equipment	Total	Total Incl O&P
0010	**RIPPING**									
0020	Ripping, trap rock, soft, 300 HP dozer, ideal conditions	B-11S	700	.017	B.C.Y.		.89	2.65	3.54	4.31
1500	Adverse conditions		660	.018			.94	2.81	3.75	4.57
1600	Medium hard, 300 HP dozer, ideal conditions		600	.020			1.04	3.09	4.13	5.05
1700	Adverse conditions		540	.022			1.15	3.43	4.58	5.60
2000	Very hard, 410 HP dozer, ideal conditions	B-11T	350	.034			1.78	8.35	10.13	11.95
2100	Adverse conditions	"	310	.039			2.01	9.40	11.41	13.50
2200	Shale, soft, 300 HP dozer, ideal conditions	B-11S	1500	.008			.42	1.24	1.66	2.01
2300	Adverse conditions		1350	.009			.46	1.37	1.83	2.24
2400	Medium hard, 300 HP dozer, ideal conditions		1200	.010			.52	1.55	2.07	2.52
2500	Adverse conditions		1080	.011			.58	1.72	2.30	2.80
2600	Very hard, 410 HP dozer, ideal conditions	B-11T	800	.015			.78	3.64	4.42	5.25
2700	Adverse conditions	"	720	.017			.86	4.05	4.91	5.80
3000	Dozing ripped material, 200 HP, 100′ haul	B-10B	700	.017			.89	2.15	3.04	3.76
3050	300′ haul	"	250	.048			2.49	6	8.49	10.50
3200	300 HP, 100′ haul	B-10M	1150	.010			.54	1.53	2.07	2.54
3250	300′ haul	"	400	.030			1.56	4.41	5.97	7.30

31 23 16.32 Ripping		Crew	Daily Output	Labor-Hours	Unit	Material	2020 Bare Costs Labor	Equipment	Total	Total Incl O&P
3400	410 HP, 100′ haul	B-10X	1680	.007	B.C.Y.		.37	1.65	2.02	2.40
3450	300′ haul	"	600	.020	↓		1.04	4.63	5.67	6.75

31 23 16.42 Excavating, Bulk Bank Measure		Crew	Daily Output	Labor-Hours	Unit	Material	Labor	Equipment	Total	Total Incl O&P
0010	**EXCAVATING, BULK BANK MEASURE**									
0011	Common earth piled									
0020	For loading onto trucks, add								15%	15%
0050	For mobilization and demobilization, see Section 01 54 36.50									
0100	For hauling, see Section 31 23 23.20									
0200	Excavator, hydraulic, crawler mtd., 1 C.Y. cap. = 100 C.Y./hr.	B-12A	800	.020	B.C.Y.		1.01	.99	2	2.69
0250	1-1/2 C.Y. cap. = 125 C.Y./hr.	B-12B	1000	.016			.81	.69	1.50	2.04
0260	2 C.Y. cap. = 165 C.Y./hr.	B-12C	1320	.012			.61	.72	1.33	1.76
0300	3 C.Y. cap. = 260 C.Y./hr.	B-12D	2080	.008			.39	1.04	1.43	1.76
0305	3-1/2 C.Y. cap. = 300 C.Y./hr.	"	2400	.007			.34	.90	1.24	1.52
0310	Wheel mounted, 1/2 C.Y. cap. = 40 C.Y./hr.	B-12E	320	.050			2.53	1.41	3.94	5.55
0360	3/4 C.Y. cap. = 60 C.Y./hr.	B-12F	480	.033			1.69	1.45	3.14	4.26
0500	Clamshell, 1/2 C.Y. cap. = 20 C.Y./hr.	B-12G	160	.100			5.05	5.35	10.40	13.90
0550	1 C.Y. cap. = 35 C.Y./hr.	B-12H	280	.057			2.89	4.28	7.17	9.25
0950	Dragline, 1/2 C.Y. cap. = 30 C.Y./hr.	B-12I	240	.067			3.38	4.26	7.64	10.05
1000	3/4 C.Y. cap. = 35 C.Y./hr.	"	280	.057			2.89	3.65	6.54	8.60
1050	1-1/2 C.Y. cap. = 65 C.Y./hr.	B-12P	520	.031			1.56	2.44	4	5.15
1100	3 C.Y. cap. = 112 C.Y./hr.	B-12V	900	.018			.90	2.30	3.20	3.95
1200	Front end loader, track mtd., 1-1/2 C.Y. cap. = 70 C.Y./hr.	B-10N	560	.021			1.11	1.01	2.12	2.86
1250	2-1/2 C.Y. cap. = 95 C.Y./hr.	B-10O	760	.016			.82	1.26	2.08	2.68
1300	3 C.Y. cap. = 130 C.Y./hr.	B-10P	1040	.012			.60	1.12	1.72	2.18
1350	5 C.Y. cap. = 160 C.Y./hr.	B-10Q	1280	.009			.49	1.16	1.65	2.05
1500	Wheel mounted, 3/4 C.Y. cap. = 45 C.Y./hr.	B-10R	360	.033			1.73	.84	2.57	3.64
1550	1-1/2 C.Y. cap. = 80 C.Y./hr.	B-10S	640	.019			.97	.66	1.63	2.26
1601	3 C.Y. cap. = 140 C.Y./hr.	B-10T	1120	.011			.56	.63	1.19	1.56
1650	5 C.Y. cap. = 185 C.Y./hr.	B-10U	1480	.008			.42	.65	1.07	1.37
1800	Hydraulic excavator, truck mtd. 1/2 C.Y. = 30 C.Y./hr.	B-12J	240	.067			3.38	3.53	6.91	9.25
1850	48″ bucket, 1 C.Y. = 45 C.Y./hr.	B-12K	360	.044			2.25	2.70	4.95	6.55
3700	Shovel, 1/2 C.Y. cap. = 55 C.Y./hr.	B-12L	440	.036			1.84	1.95	3.79	5.05
3750	3/4 C.Y. cap. = 85 C.Y./hr.	B-12M	680	.024			1.19	1.52	2.71	3.55
3800	1 C.Y. cap. = 120 C.Y./hr.	B-12N	960	.017			.84	1.26	2.10	2.71
3850	1-1/2 C.Y. cap. = 160 C.Y./hr.	B-12O	1280	.013			.63	1.01	1.64	2.11
3900	3 C.Y. cap. = 250 C.Y./hr.	B-12T	2000	.008			.41	1.06	1.47	1.80
4000	For soft soil or sand, deduct								15%	15%
4100	For heavy soil or stiff clay, add								60%	60%
4200	For wet excavation with clamshell or dragline, add								100%	100%
4250	All other equipment, add								50%	50%
4400	Clamshell in sheeting or cofferdam, minimum	B-12H	160	.100			5.05	7.50	12.55	16.25
4450	Maximum	"	60	.267	↓		13.50	19.95	33.45	43.50
5000	Excavating, bulk bank measure, sandy clay & loam piled									
5020	For loading onto trucks, add								15%	15%
5100	Excavator, hydraulic, crawler mtd., 1 C.Y. cap. = 120 C.Y./hr.	B-12A	960	.017	B.C.Y.		.84	.83	1.67	2.24
5150	1-1/2 C.Y. cap. = 150 C.Y./hr.	B-12B	1200	.013			.68	.57	1.25	1.70
5300	2 C.Y. cap. = 195 C.Y./hr.	B-12C	1560	.010			.52	.61	1.13	1.49
5400	3 C.Y. cap. = 300 C.Y./hr.	B-12D	2400	.007			.34	.90	1.24	1.52
5500	3-1/2 C.Y. cap. = 350 C.Y./hr.	"	2800	.006			.29	.77	1.06	1.31
5610	Wheel mounted, 1/2 C.Y. cap. = 44 C.Y./hr.	B-12E	352	.045			2.30	1.28	3.58	5.05
5660	3/4 C.Y. cap. = 66 C.Y./hr.	B-12F	528	.030	↓		1.53	1.31	2.84	3.88
8000	For hauling excavated material, see Section 31 23 23.20									

31 23 16.42 Excavating, Bulk Bank Measure		Crew	Daily Output	Labor-Hours	Unit	Material	2020 Bare Costs Labor	Equipment	Total	Total Incl O&P
9000	Minimum labor/equipment charge	B-10L	2	6	Job		310	201	511	710

31 23 16.46 Excavating, Bulk, Dozer

Line	Description	Crew	Daily Output	Labor-Hours	Unit	Material	Labor	Equipment	Total	Total Incl O&P
0010	**EXCAVATING, BULK, DOZER**									
0011	Open site									
2000	80 HP, 50′ haul, sand & gravel	B-10L	460	.026	B.C.Y.		1.35	.87	2.22	3.09
2010	Sandy clay & loam		440	.027			1.41	.91	2.32	3.23
2020	Common earth		400	.030			1.56	1	2.56	3.55
2040	Clay		250	.048			2.49	1.61	4.10	5.70
2200	150′ haul, sand & gravel		230	.052			2.71	1.75	4.46	6.20
2210	Sandy clay & loam		220	.055			2.83	1.82	4.65	6.45
2220	Common earth		200	.060			3.11	2.01	5.12	7.10
2240	Clay		125	.096			4.98	3.21	8.19	11.40
2400	300′ haul, sand & gravel		120	.100			5.20	3.35	8.55	11.85
2410	Sandy clay & loam		115	.104			5.40	3.49	8.89	12.35
2420	Common earth		100	.120			6.20	4.01	10.21	14.20
2440	Clay		65	.185			9.60	6.20	15.80	22
3000	105 HP, 50′ haul, sand & gravel	B-10W	700	.017			.89	.91	1.80	2.40
3010	Sandy clay & loam		680	.018			.92	.93	1.85	2.47
3020	Common earth		610	.020			1.02	1.04	2.06	2.75
3040	Clay		385	.031			1.62	1.65	3.27	4.36
3200	150′ haul, sand & gravel		310	.039			2.01	2.04	4.05	5.40
3210	Sandy clay & loam		300	.040			2.07	2.11	4.18	5.60
3220	Common earth		270	.044			2.31	2.35	4.66	6.20
3240	Clay		170	.071			3.66	3.73	7.39	9.85
3300	300′ haul, sand & gravel		140	.086			4.45	4.53	8.98	12
3310	Sandy clay & loam		135	.089			4.61	4.70	9.31	12.40
3320	Common earth		120	.100			5.20	5.30	10.50	13.95
3340	Clay		100	.120			6.20	6.35	12.55	16.75
4000	200 HP, 50′ haul, sand & gravel	B-10B	1400	.009			.44	1.07	1.51	1.88
4010	Sandy clay & loam		1360	.009			.46	1.11	1.57	1.94
4020	Common earth		1230	.010			.51	1.22	1.73	2.15
4040	Clay		770	.016			.81	1.95	2.76	3.42
4200	150′ haul, sand & gravel		595	.020			1.05	2.53	3.58	4.43
4210	Sandy clay & loam		580	.021			1.07	2.59	3.66	4.54
4220	Common earth		516	.023			1.21	2.92	4.13	5.10
4240	Clay		325	.037			1.92	4.63	6.55	8.10
4400	300′ haul, sand & gravel		310	.039			2.01	4.85	6.86	8.50
4410	Sandy clay & loam		300	.040			2.07	5	7.07	8.75
4420	Common earth		270	.044			2.31	5.55	7.86	9.80
4440	Clay		170	.071			3.66	8.85	12.51	15.50
5000	300 HP, 50′ haul, sand & gravel	B-10M	1900	.006			.33	.93	1.26	1.54
5010	Sandy clay & loam		1850	.006			.34	.95	1.29	1.58
5020	Common earth		1650	.007			.38	1.07	1.45	1.77
5040	Clay		1025	.012			.61	1.72	2.33	2.85
5200	150′ haul, sand & gravel		920	.013			.68	1.92	2.60	3.18
5210	Sandy clay & loam		895	.013			.70	1.97	2.67	3.27
5220	Common earth		800	.015			.78	2.21	2.99	3.66
5240	Clay		500	.024			1.24	3.53	4.77	5.85
5400	300′ haul, sand & gravel		470	.026			1.32	3.75	5.07	6.20
5410	Sandy clay & loam		455	.026			1.37	3.88	5.25	6.40
5420	Common earth		410	.029			1.52	4.30	5.82	7.10
5440	Clay		250	.048			2.49	7.05	9.54	11.65

	31 23 16.50 Excavation, Bulk, Scrapers	Crew	Daily Output	Labor-Hours	Unit	Material	2020 Bare Costs Labor	Equipment	Total	Total Incl O&P
0010	**EXCAVATION, BULK, SCRAPERS**									
0100	Elev. scraper 11 C.Y., sand & gravel 1,500' haul, 1/4 dozer	B-33F	690	.020	B.C.Y.		1.07	2.28	3.35	4.19
0150	3,000' haul		610	.023			1.21	2.58	3.79	4.74
0200	5,000' haul		505	.028			1.46	3.12	4.58	5.70
0300	Common earth, 1,500' haul		600	.023			1.23	2.62	3.85	4.82
0350	3,000' haul		530	.026			1.39	2.97	4.36	5.45
0400	5,000' haul		440	.032			1.67	3.58	5.25	6.55
0410	Sandy clay & loam, 1,500' haul		648	.022			1.14	2.43	3.57	4.46
0420	3,000' haul		572	.024			1.29	2.75	4.04	5.05
0430	5,000' haul		475	.029			1.55	3.31	4.86	6.10
0500	Clay, 1,500' haul		375	.037			1.96	4.20	6.16	7.70
0550	3,000' haul		330	.042			2.23	4.77	7	8.75
0600	5,000' haul		275	.051			2.68	5.70	8.38	10.50
1000	Self propelled scraper, 14 C.Y., 1/4 push dozer									
1050	Sand and gravel, 1,500' haul	B-33D	920	.015	B.C.Y.		.80	3.08	3.88	4.65
1100	3,000' haul		805	.017			.91	3.52	4.43	5.30
1200	5,000' haul		645	.022			1.14	4.40	5.54	6.65
1300	Common earth, 1,500' haul		800	.018			.92	3.55	4.47	5.35
1350	3,000' haul		700	.020			1.05	4.05	5.10	6.10
1400	5,000' haul		560	.025			1.31	5.05	6.36	7.60
1420	Sandy clay & loam, 1,500' haul		864	.016			.85	3.28	4.13	4.95
1430	3,000' haul		786	.018			.94	3.61	4.55	5.45
1440	5,000' haul		605	.023			1.22	4.69	5.91	7.05
1500	Clay, 1,500' haul		500	.028			1.47	5.65	7.12	8.55
1550	3,000' haul		440	.032			1.67	6.45	8.12	9.75
1600	5,000' haul		350	.040			2.10	8.10	10.20	12.20
2000	21 C.Y., 1/4 push dozer, sand & gravel, 1,500' haul	B-33E	1180	.012			.62	2.60	3.22	3.84
2100	3,000' haul		910	.015			.81	3.37	4.18	4.98
2200	5,000' haul		750	.019			.98	4.09	5.07	6.05
2300	Common earth, 1,500' haul		1030	.014			.71	2.98	3.69	4.40
2350	3,000' haul		790	.018			.93	3.88	4.81	5.75
2400	5,000' haul		650	.022			1.13	4.72	5.85	7
2420	Sandy clay & loam, 1,500' haul		1112	.013			.66	2.76	3.42	4.08
2430	3,000' haul		854	.016			.86	3.59	4.45	5.30
2440	5,000' haul		702	.020			1.05	4.37	5.42	6.45
2500	Clay, 1,500' haul		645	.022			1.14	4.76	5.90	7.05
2550	3,000' haul		495	.028			1.49	6.20	7.69	9.15
2600	5,000' haul		405	.035			1.82	7.60	9.42	11.20
2700	Towed, 10 C.Y., 1/4 push dozer, sand & gravel, 1,500' haul	B-33B	560	.025			1.31	4.22	5.53	6.70
2720	3,000' haul		450	.031			1.64	5.25	6.89	8.35
2730	5,000' haul		365	.038			2.02	6.50	8.52	10.30
2750	Common earth, 1,500' haul		420	.033			1.75	5.65	7.40	8.95
2770	3,000' haul		400	.035			1.84	5.90	7.74	9.40
2780	5,000' haul		310	.045			2.37	7.65	10.02	12.15
2785	Sandy clay & loam, 1,500' haul		454	.031			1.62	5.20	6.82	8.30
2790	3,000' haul		432	.032			1.70	5.45	7.15	8.70
2795	5,000' haul		340	.041			2.16	6.95	9.11	11.05
2800	Clay, 1,500' haul		315	.044			2.34	7.50	9.84	11.95
2820	3,000' haul		300	.047			2.45	7.90	10.35	12.50
2840	5,000' haul		225	.062			3.27	10.50	13.77	16.70
2900	15 C.Y., 1/4 push dozer, sand & gravel, 1,500' haul	B-33C	800	.018			.92	2.98	3.90	4.73
2920	3,000' haul		640	.022			1.15	3.72	4.87	5.90

31 23 Excavation and Fill

31 23 16 – Excavation

31 23 16.50 Excavation, Bulk, Scrapers		Crew	Daily Output	Labor-Hours	Unit	2020 Bare Costs Material	Labor	Equipment	Total	Total Incl O&P
2940	5,000′ haul	B-33C	520	.027	B.C.Y.		1.41	4.58	5.99	7.30
2960	Common earth, 1,500′ haul		600	.023			1.23	3.97	5.20	6.30
2980	3,000′ haul		560	.025			1.31	4.25	5.56	6.75
3000	5,000′ haul		440	.032			1.67	5.40	7.07	8.60
3005	Sandy clay & loam, 1,500′ haul		648	.022			1.14	3.67	4.81	5.85
3010	3,000′ haul		605	.023			1.22	3.94	5.16	6.25
3015	5,000′ haul		475	.029			1.55	5	6.55	7.95
3020	Clay, 1,500′ haul		450	.031			1.64	5.30	6.94	8.35
3040	3,000′ haul		420	.033			1.75	5.65	7.40	9
3060	5,000′ haul		320	.044			2.30	7.45	9.75	11.80

31 23 19 – Dewatering

31 23 19.20 Dewatering Systems

Line	Description	Crew	Daily Output	Labor-Hours	Unit	Material	Labor	Equipment	Total	Total Incl O&P
0010	**DEWATERING SYSTEMS**									
0020	Excavate drainage trench, 2′ wide, 2′ deep	B-11C	90	.178	C.Y.		8.80	2.38	11.18	16.50
0100	2′ wide, 3′ deep, with backhoe loader	″	135	.119			5.85	1.58	7.43	11
0200	Excavate sump pits by hand, light soil	1 Clab	7.10	1.127			47.50		47.50	76
0300	Heavy soil	″	3.50	2.286			96		96	154
0500	Pumping 8 hrs., attended 2 hrs./day, incl. 20 L.F.									
0550	of suction hose & 100 L.F. discharge hose									
0600	2″ diaphragm pump used for 8 hrs.	B-10H	4	3	Day		156	24.50	180.50	272
0650	4″ diaphragm pump used for 8 hrs.	B-10I	4	3			156	37	193	286
0800	8 hrs. attended, 2″ diaphragm pump	B-10H	1	12			620	98.50	718.50	1,100
0900	3″ centrifugal pump	B-10J	1	12			620	92	712	1,075
1000	4″ diaphragm pump	B-10I	1	12			620	148	768	1,150
1100	6″ centrifugal pump	B-10K	1	12			620	294	914	1,300
1300	CMP, incl. excavation 3′ deep, 12″ diameter	B-6	115	.209	L.F.	12.40	9.55	1.86	23.81	31
1400	18″ diameter		100	.240	″	18.80	11	2.14	31.94	40.50
1600	Sump hole construction, incl. excavation and gravel, pit		1250	.019	C.F.	1.10	.88	.17	2.15	2.79
1700	With 12″ gravel collar, 12″ pipe, corrugated, 16 ga.		70	.343	L.F.	23	15.70	3.05	41.75	54
1800	15″ pipe, corrugated, 16 ga.		55	.436		30	19.95	3.89	53.84	69
1900	18″ pipe, corrugated, 16 ga.		50	.480		35	22	4.28	61.28	78
2000	24″ pipe, corrugated, 14 ga.		40	.600		42	27.50	5.35	74.85	95.50
2200	Wood lining, up to 4′ x 4′, add		300	.080	SFCA	16.40	3.66	.71	20.77	24.50
9950	See Section 31 23 19.40 for wellpoints									
9960	See Section 31 23 19.30 for deep well systems									

31 23 19.30 Wells

Line	Description	Crew	Daily Output	Labor-Hours	Unit	Material	Labor	Equipment	Total	Total Incl O&P
0010	**WELLS**									
0011	For dewatering 10′ to 20′ deep, 2′ diameter									
0020	with steel casing, minimum	B-6	165	.145	V.L.F.	37.50	6.65	1.30	45.45	53.50
0050	Average		98	.245		44.50	11.20	2.18	57.88	69
0100	Maximum		49	.490		48	22.50	4.36	74.86	93

31 23 19.40 Wellpoints

Line	Description	Crew	Daily Output	Labor-Hours	Unit	Material	Labor	Equipment	Total	Total Incl O&P
0010	**WELLPOINTS**									
0011	For equipment rental, see 01 54 33 in Reference Section									
0100	Installation and removal of single stage system									
0110	Labor only, 0.75 labor-hours per L.F.	1 Clab	10.70	.748	LF Hdr		31.50		31.50	50.50
0200	2.0 labor-hours per L.F.	″	4	2	″		84		84	135
0400	Pump operation, 4 @ 6 hr. shifts									
0410	Per 24 hr. day	4 Eqlt	1.27	25.197	Day		1,325		1,325	2,100
0500	Per 168 hr. week, 160 hr. straight, 8 hr. double time		.18	178	Week		9,425		9,425	14,700
0550	Per 4.3 week month		.04	800	Month		42,400		42,400	66,500
0600	Complete installation, operation, equipment rental, fuel &									

31 23 Excavation and Fill

31 23 19 – Dewatering

31 23 19.40 Wellpoints		Crew	Daily Output	Labor-Hours	Unit	Material	Labor	Equipment	Total	Total Incl O&P
						2020 Bare Costs				
0610	removal of system with 2" wellpoints 5' OC									
0700	100' long header, 6" diameter, first month	4 Eqlt	3.23	9.907	LF Hdr	160	525		685	995
0800	Thereafter, per month		4.13	7.748		128	410		538	785
1000	200' long header, 8" diameter, first month		6	5.333		154	283		437	610
1100	Thereafter, per month		8.39	3.814		72	202		274	395
1300	500' long header, 8" diameter, first month		10.63	3.010		56	160		216	310
1400	Thereafter, per month		20.91	1.530		40	81		121	171
1600	1,000' long header, 10" diameter, first month		11.62	2.754		48	146		194	281
1700	Thereafter, per month		41.81	.765		24	40.50		64.50	90
1900	Note: above figures include pumping 168 hrs. per week,									
1910	the pump operator, and one stand-by pump.									

31 23 23 – Fill

31 23 23.13 Backfill

		Crew	Daily Output	Labor-Hours	Unit	Material	Labor	Equipment	Total	Total Incl O&P
0010	**BACKFILL**									
0015	By hand, no compaction, light soil	1 Clab	14	.571	L.C.Y.		24		24	38.50
0100	Heavy soil		11	.727	"		30.50		30.50	49
0300	Compaction in 6" layers, hand tamp, add to above		20.60	.388	E.C.Y.		16.35		16.35	26
0400	Roller compaction operator walking, add	B-10A	100	.120			6.20	1.66	7.86	11.65
0500	Air tamp, add	B-9D	190	.211			8.95	1.67	10.62	16.20
0600	Vibrating plate, add	A-1D	60	.133			5.60	.53	6.13	9.60
0800	Compaction in 12" layers, hand tamp, add to above	1 Clab	34	.235			9.90		9.90	15.85
0900	Roller compaction operator walking, add	B-10A	150	.080			4.15	1.11	5.26	7.75
1000	Air tamp, add	B-9	285	.140			5.95	1.25	7.20	10.95
1100	Vibrating plate, add	A-1E	90	.089			3.74	1.84	5.58	8

31 23 23.14 Backfill, Structural

		Crew	Daily Output	Labor-Hours	Unit	Material	Labor	Equipment	Total	Total Incl O&P
0010	**BACKFILL, STRUCTURAL**									
0011	Dozer or F.E. loader									
0020	From existing stockpile, no compaction									
1000	55 HP wheeled loader, 50' haul, common earth	B-11C	200	.080	L.C.Y.		3.95	1.07	5.02	7.45
2000	80 HP, 50' haul, sand & gravel	B-10L	1100	.011			.57	.36	.93	1.29
2010	Sandy clay & loam		1070	.011			.58	.38	.96	1.33
2020	Common earth		975	.012			.64	.41	1.05	1.46
2040	Clay		850	.014			.73	.47	1.20	1.67
2200	150' haul, sand & gravel		550	.022			1.13	.73	1.86	2.58
2210	Sandy clay & loam		535	.022			1.16	.75	1.91	2.66
2220	Common earth		490	.024			1.27	.82	2.09	2.90
2240	Clay		425	.028			1.46	.94	2.40	3.35
2400	300' haul, sand & gravel		370	.032			1.68	1.08	2.76	3.84
2410	Sandy clay & loam		360	.033			1.73	1.11	2.84	3.95
2420	Common earth		330	.036			1.89	1.22	3.11	4.31
2440	Clay		290	.041			2.15	1.38	3.53	4.90
3000	105 HP, 50' haul, sand & gravel	B-10W	1350	.009			.46	.47	.93	1.25
3010	Sandy clay & loam		1325	.009			.47	.48	.95	1.27
3020	Common earth		1225	.010			.51	.52	1.03	1.37
3040	Clay		1100	.011			.57	.58	1.15	1.52
3200	150' haul, sand & gravel		670	.018			.93	.95	1.88	2.50
3210	Sandy clay & loam		655	.018			.95	.97	1.92	2.56
3220	Common earth		610	.020			1.02	1.04	2.06	2.75
3240	Clay		550	.022			1.13	1.15	2.28	3.05
3300	300' haul, sand & gravel		465	.026			1.34	1.36	2.70	3.61
3310	Sandy clay & loam		455	.026			1.37	1.39	2.76	3.68
3320	Common earth		415	.029			1.50	1.53	3.03	4.04

31 23 23.14 Backfill, Structural		Crew	Daily Output	Labor-Hours	Unit	Material	2020 Bare Costs Labor	Equipment	Total	Total Incl O&P
3340	Clay	B-10W	370	.032	L.C.Y.		1.68	1.71	3.39	4.53
4000	200 HP, 50' haul, sand & gravel	B-10B	2500	.005			.25	.60	.85	1.05
4010	Sandy clay & loam		2435	.005			.26	.62	.88	1.08
4020	Common earth		2200	.005			.28	.68	.96	1.20
4040	Clay		1950	.006			.32	.77	1.09	1.35
4200	150' haul, sand & gravel		1225	.010			.51	1.23	1.74	2.15
4210	Sandy clay & loam		1200	.010			.52	1.25	1.77	2.20
4220	Common earth		1100	.011			.57	1.37	1.94	2.39
4240	Clay		975	.012			.64	1.54	2.18	2.71
4400	300' haul, sand & gravel		805	.015			.77	1.87	2.64	3.28
4410	Sandy clay & loam		790	.015			.79	1.90	2.69	3.33
4420	Common earth		735	.016			.85	2.05	2.90	3.58
4440	Clay		660	.018			.94	2.28	3.22	3.99
5000	300 HP, 50' haul, sand & gravel	B-10M	3170	.004			.20	.56	.76	.92
5010	Sandy clay & loam		3110	.004			.20	.57	.77	.94
5020	Common earth		2900	.004			.21	.61	.82	1.01
5040	Clay		2700	.004			.23	.65	.88	1.08
5200	150' haul, sand & gravel		2200	.005			.28	.80	1.08	1.33
5210	Sandy clay & loam		2150	.006			.29	.82	1.11	1.36
5220	Common earth		1950	.006			.32	.90	1.22	1.49
5240	Clay		1700	.007			.37	1.04	1.41	1.72
5400	300' haul, sand & gravel		1500	.008			.42	1.18	1.60	1.94
5410	Sandy clay & loam		1470	.008			.42	1.20	1.62	1.99
5420	Common earth		1350	.009			.46	1.31	1.77	2.17
5440	Clay		1225	.010			.51	1.44	1.95	2.38

31 23 23.15 Borrow, Loading and/or Spreading		Crew	Daily Output	Labor-Hours	Unit	Material	Labor	Equipment	Total	Total Incl O&P
0010	**BORROW, LOADING AND/OR SPREADING**									
0020	Material only, bank run gravel				Ton	13.30			13.30	14.60
0100	Crushed stone, 1-1/2" to 3/4" size					16.25			16.25	17.90
0150	3/8" size					23			23	25.50
0200	Dead or bank run sand					18.25			18.25	20
0500	Haul 2 mi. spread, 200 HP dozer, bank run gravel	B-15	1100	.025			1.28	2.41	3.69	4.68
0540	Crushed stone, 1-1/2" to 3/4" size		1150	.024			1.22	2.30	3.52	4.47
0560	3/8" size		1150	.024			1.22	2.30	3.52	4.47
0600	Bank run or dead sand		1000	.028			1.41	2.65	4.06	5.15
1000	Hand spread, bank run gravel	A-5	33	.545			23.50	1.48	24.98	39
1040	Crushed stone, 1-1/2" to 3/4" size		38	.474			20	1.29	21.29	34
1060	3/8" size		40	.450			19.20	1.22	20.42	32.50
1100	Bank run or dead sand		33	.545			23.50	1.48	24.98	39
1800	Delivery charge, minimum 20 tons, 1 hr. round trip, add	B-34B	130	.062			3.01	4.41	7.42	9.65
1820	1-1/2 hr. round trip, add		93	.086			4.21	6.15	10.36	13.55
1840	2 hr. round trip, add		65	.123			6	8.80	14.80	19.35
7000	Topsoil or loam from stockpile, shovel, 1 C.Y. bucket	B-12N	840	.019	B.C.Y.	26.50	.96	1.44	28.90	32
7010	1-1/2 C.Y. bucket	B-12O	1135	.014		26.50	.71	1.14	28.35	31.50
7020	3 C.Y. bucket	B-12T	1800	.009		26.50	.45	1.18	28.13	31
7030	Front end loader, wheel mounted									
7050	3/4 C.Y. bucket	B-10R	550	.022	B.C.Y.	26.50	1.13	.55	28.18	31.50
7060	1-1/2 C.Y. bucket	B-10S	970	.012		26.50	.64	.44	27.58	30.50
7070	3 C.Y. bucket	B-10T	1575	.008		26.50	.40	.45	27.35	30
7080	5 C.Y. bucket	B-10U	2600	.005		26.50	.24	.37	27.11	30

31 23 23.16 Fill By Borrow and Utility Bedding		Crew	Daily Output	Labor-Hours	Unit	Material	Labor	Equipment	Total	Total Incl O&P
						2020 Bare Costs				
0010	**FILL BY BORROW AND UTILITY BEDDING**									
0015	Fill by borrow, load, 1 mile haul, spread with dozer									
0020	for embankments	B-15	1200	.023	L.C.Y.	12.60	1.17	2.21	15.98	18.15
0035	Select fill for shoulders & embankments	"	1200	.023	"	24.50	1.17	2.21	27.88	31.50
0040	Fill,for hauling over 1 mile,add to above per C.Y., see Section 31 23 23.20				Mile				1.41	1.73
0049	Utility bedding, for pipe & conduit, not incl. compaction									
0050	Crushed or screened bank run gravel	B-6	150	.160	L.C.Y.	21	7.30	1.43	29.73	36
0100	Crushed stone 3/4" to 1/2"		150	.160		27.50	7.30	1.43	36.23	43.50
0200	Sand, dead or bank		150	.160		18.25	7.30	1.43	26.98	33
0500	Compacting bedding in trench	A-1D	90	.089	E.C.Y.		3.74	.35	4.09	6.40
0600	If material source exceeds 2 miles, add for extra mileage.									
0610	See Section 31 23 23.20 for hauling mileage add.									

31 23 23.17 General Fill		Crew	Daily Output	Labor-Hours	Unit	Material	Labor	Equipment	Total	Total Incl O&P
0010	**GENERAL FILL**									
0011	Spread dumped material, no compaction									
0020	By dozer	B-10B	1000	.012	L.C.Y.		.62	1.50	2.12	2.63
0100	By hand	1 Clab	12	.667	"		28		28	45
0150	Spread fill, from stockpile with 2-1/2 C.Y. F.E. loader									
0170	130 HP, 300' haul	B-10P	600	.020	L.C.Y.		1.04	1.95	2.99	3.77
0190	With dozer 300 HP, 300' haul	B-10M	600	.020	"		1.04	2.94	3.98	4.86
0500	Gravel fill, compacted, under floor slabs, 4" deep	B-37	10000	.005	S.F.	.49	.21	.03	.73	.91
0600	6" deep		8600	.006		.74	.25	.03	1.02	1.23
0700	9" deep		7200	.007		1.23	.30	.04	1.57	1.86
0800	12" deep		6000	.008		1.72	.35	.04	2.11	2.50
1000	Alternate pricing method, 4" deep		120	.400	E.C.Y.	37	17.70	2.13	56.83	71
1100	6" deep		160	.300		37	13.30	1.60	51.90	63.50
1200	9" deep		200	.240		37	10.60	1.28	48.88	59
1300	12" deep		220	.218		37	9.65	1.16	47.81	57
1500	For fill under exterior paving, see Section 32 11 23.23									
1600	For flowable fill, see Section 03 31 13.35									
9000	Minimum labor/equipment charge	1 Clab	4	2	Job		84		84	135

31 23 23.20 Hauling		Crew	Daily Output	Labor-Hours	Unit	Material	Labor	Equipment	Total	Total Incl O&P
0010	**HAULING**									
0011	Excavated or borrow, loose cubic yards									
0012	no loading equipment, including hauling, waiting, loading/dumping									
0013	time per cycle (wait, load, travel, unload or dump & return)									
0014	8 C.Y. truck, 15 MPH avg., cycle 0.5 miles, 10 min. wait/ld./uld.	B-34A	320	.025	L.C.Y.		1.22	1.34	2.56	3.43
0016	cycle 1 mile		272	.029			1.44	1.57	3.01	4.03
0018	cycle 2 miles		208	.038			1.88	2.06	3.94	5.25
0020	cycle 4 miles		144	.056			2.72	2.97	5.69	7.60
0022	cycle 6 miles		112	.071			3.50	3.82	7.32	9.80
0024	cycle 8 miles		88	.091			4.45	4.86	9.31	12.45
0026	20 MPH avg., cycle 0.5 mile		336	.024			1.17	1.27	2.44	3.26
0028	cycle 1 mile		296	.027			1.32	1.45	2.77	3.71
0030	cycle 2 miles		240	.033			1.63	1.78	3.41	4.57
0032	cycle 4 miles		176	.045			2.22	2.43	4.65	6.25
0034	cycle 6 miles		136	.059			2.88	3.15	6.03	8.05
0036	cycle 8 miles		112	.071			3.50	3.82	7.32	9.80
0044	25 MPH avg., cycle 4 miles		192	.042			2.04	2.23	4.27	5.70
0046	cycle 6 miles		160	.050			2.45	2.68	5.13	6.85
0048	cycle 8 miles		128	.063			3.06	3.34	6.40	8.55
0050	30 MPH avg., cycle 4 miles		216	.037			1.81	1.98	3.79	5.10

31 23 23.20 Hauling		Crew	Daily Output	Labor-Hours	Unit	Material	2020 Bare Costs Labor	Equipment	Total	Total Incl O&P
0052	cycle 6 miles	B-34A	176	.045	L.C.Y.		2.22	2.43	4.65	6.25
0054	cycle 8 miles		144	.056			2.72	2.97	5.69	7.60
0114	15 MPH avg., cycle 0.5 mile, 15 min. wait/ld./uld.		224	.036			1.75	1.91	3.66	4.89
0116	cycle 1 mile		200	.040			1.96	2.14	4.10	5.50
0118	cycle 2 miles		168	.048			2.33	2.55	4.88	6.55
0120	cycle 4 miles		120	.067			3.26	3.57	6.83	9.10
0122	cycle 6 miles		96	.083			4.08	4.46	8.54	11.40
0124	cycle 8 miles		80	.100			4.90	5.35	10.25	13.75
0126	20 MPH avg., cycle 0.5 mile		232	.034			1.69	1.85	3.54	4.73
0128	cycle 1 mile		208	.038			1.88	2.06	3.94	5.25
0130	cycle 2 miles		184	.043			2.13	2.33	4.46	5.95
0132	cycle 4 miles		144	.056			2.72	2.97	5.69	7.60
0134	cycle 6 miles		112	.071			3.50	3.82	7.32	9.80
0136	cycle 8 miles		96	.083			4.08	4.46	8.54	11.40
0144	25 MPH avg., cycle 4 miles		152	.053			2.58	2.82	5.40	7.20
0146	cycle 6 miles		128	.063			3.06	3.34	6.40	8.55
0148	cycle 8 miles		112	.071			3.50	3.82	7.32	9.80
0150	30 MPH avg., cycle 4 miles		168	.048			2.33	2.55	4.88	6.55
0152	cycle 6 miles		144	.056			2.72	2.97	5.69	7.60
0154	cycle 8 miles		120	.067			3.26	3.57	6.83	9.10
0214	15 MPH avg., cycle 0.5 mile, 20 min. wait/ld./uld.		176	.045			2.22	2.43	4.65	6.25
0216	cycle 1 mile		160	.050			2.45	2.68	5.13	6.85
0218	cycle 2 miles		136	.059			2.88	3.15	6.03	8.05
0220	cycle 4 miles		104	.077			3.77	4.12	7.89	10.55
0222	cycle 6 miles		88	.091			4.45	4.86	9.31	12.45
0224	cycle 8 miles		72	.111			5.45	5.95	11.40	15.25
0226	20 MPH avg., cycle 0.5 mile		176	.045			2.22	2.43	4.65	6.25
0228	cycle 1 mile		168	.048			2.33	2.55	4.88	6.55
0230	cycle 2 miles		144	.056			2.72	2.97	5.69	7.60
0232	cycle 4 miles		120	.067			3.26	3.57	6.83	9.10
0234	cycle 6 miles		96	.083			4.08	4.46	8.54	11.40
0236	cycle 8 miles		88	.091			4.45	4.86	9.31	12.45
0244	25 MPH avg., cycle 4 miles		128	.063			3.06	3.34	6.40	8.55
0246	cycle 6 miles		112	.071			3.50	3.82	7.32	9.80
0248	cycle 8 miles		96	.083			4.08	4.46	8.54	11.40
0250	30 MPH avg., cycle 4 miles		136	.059			2.88	3.15	6.03	8.05
0252	cycle 6 miles		120	.067			3.26	3.57	6.83	9.10
0254	cycle 8 miles		104	.077			3.77	4.12	7.89	10.55
0314	15 MPH avg., cycle 0.5 mile, 25 min. wait/ld./uld.		144	.056			2.72	2.97	5.69	7.60
0316	cycle 1 mile		128	.063			3.06	3.34	6.40	8.55
0318	cycle 2 miles		112	.071			3.50	3.82	7.32	9.80
0320	cycle 4 miles		96	.083			4.08	4.46	8.54	11.40
0322	cycle 6 miles		80	.100			4.90	5.35	10.25	13.75
0324	cycle 8 miles		64	.125			6.10	6.70	12.80	17.15
0326	20 MPH avg., cycle 0.5 mile		144	.056			2.72	2.97	5.69	7.60
0328	cycle 1 mile		136	.059			2.88	3.15	6.03	8.05
0330	cycle 2 miles		120	.067			3.26	3.57	6.83	9.10
0332	cycle 4 miles		104	.077			3.77	4.12	7.89	10.55
0334	cycle 6 miles		88	.091			4.45	4.86	9.31	12.45
0336	cycle 8 miles		80	.100			4.90	5.35	10.25	13.75
0344	25 MPH avg., cycle 4 miles		112	.071			3.50	3.82	7.32	9.80
0346	cycle 6 miles		96	.083			4.08	4.46	8.54	11.40
0348	cycle 8 miles		88	.091			4.45	4.86	9.31	12.45

31 23 23.20 Hauling		Crew	Daily Output	Labor-Hours	Unit	Material	2020 Bare Costs Labor	Equipment	Total	Total Incl O&P
0350	30 MPH avg., cycle 4 miles	B-34A	112	.071	L.C.Y.		3.50	3.82	7.32	9.80
0352	cycle 6 miles		104	.077			3.77	4.12	7.89	10.55
0354	cycle 8 miles		96	.083			4.08	4.46	8.54	11.40
0414	15 MPH avg., cycle 0.5 mile, 30 min. wait/ld./uld.		120	.067			3.26	3.57	6.83	9.10
0416	cycle 1 mile		112	.071			3.50	3.82	7.32	9.80
0418	cycle 2 miles		96	.083			4.08	4.46	8.54	11.40
0420	cycle 4 miles		80	.100			4.90	5.35	10.25	13.75
0422	cycle 6 miles		72	.111			5.45	5.95	11.40	15.25
0424	cycle 8 miles		64	.125			6.10	6.70	12.80	17.15
0426	20 MPH avg., cycle 0.5 mile		120	.067			3.26	3.57	6.83	9.10
0428	cycle 1 mile		112	.071			3.50	3.82	7.32	9.80
0430	cycle 2 miles		104	.077			3.77	4.12	7.89	10.55
0432	cycle 4 miles		88	.091			4.45	4.86	9.31	12.45
0434	cycle 6 miles		80	.100			4.90	5.35	10.25	13.75
0436	cycle 8 miles		72	.111			5.45	5.95	11.40	15.25
0444	25 MPH avg., cycle 4 miles		96	.083			4.08	4.46	8.54	11.40
0446	cycle 6 miles		88	.091			4.45	4.86	9.31	12.45
0448	cycle 8 miles		80	.100			4.90	5.35	10.25	13.75
0450	30 MPH avg., cycle 4 miles		96	.083			4.08	4.46	8.54	11.40
0452	cycle 6 miles		88	.091			4.45	4.86	9.31	12.45
0454	cycle 8 miles		80	.100			4.90	5.35	10.25	13.75
0514	15 MPH avg., cycle 0.5 mile, 35 min. wait/ld./uld.		104	.077			3.77	4.12	7.89	10.55
0516	cycle 1 mile		96	.083			4.08	4.46	8.54	11.40
0518	cycle 2 miles		88	.091			4.45	4.86	9.31	12.45
0520	cycle 4 miles		72	.111			5.45	5.95	11.40	15.25
0522	cycle 6 miles		64	.125			6.10	6.70	12.80	17.15
0524	cycle 8 miles		56	.143			7	7.65	14.65	19.60
0526	20 MPH avg., cycle 0.5 mile		104	.077			3.77	4.12	7.89	10.55
0528	cycle 1 mile		96	.083			4.08	4.46	8.54	11.40
0530	cycle 2 miles		96	.083			4.08	4.46	8.54	11.40
0532	cycle 4 miles		80	.100			4.90	5.35	10.25	13.75
0534	cycle 6 miles		72	.111			5.45	5.95	11.40	15.25
0536	cycle 8 miles		64	.125			6.10	6.70	12.80	17.15
0544	25 MPH avg., cycle 4 miles		88	.091			4.45	4.86	9.31	12.45
0546	cycle 6 miles		80	.100			4.90	5.35	10.25	13.75
0548	cycle 8 miles		72	.111			5.45	5.95	11.40	15.25
0550	30 MPH avg., cycle 4 miles		88	.091			4.45	4.86	9.31	12.45
0552	cycle 6 miles		80	.100			4.90	5.35	10.25	13.75
0554	cycle 8 miles		72	.111			5.45	5.95	11.40	15.25
1014	12 C.Y. truck, cycle 0.5 mile, 15 MPH avg., 15 min. wait/ld./uld.	B-34B	336	.024			1.17	1.71	2.88	3.74
1016	cycle 1 mile		300	.027			1.31	1.91	3.22	4.19
1018	cycle 2 miles		252	.032			1.55	2.27	3.82	4.98
1020	cycle 4 miles		180	.044			2.18	3.18	5.36	7
1022	cycle 6 miles		144	.056			2.72	3.98	6.70	8.75
1024	cycle 8 miles		120	.067			3.26	4.77	8.03	10.45
1025	cycle 10 miles		96	.083			4.08	5.95	10.03	13.05
1026	20 MPH avg., cycle 0.5 mile		348	.023			1.13	1.65	2.78	3.61
1028	cycle 1 mile		312	.026			1.26	1.84	3.10	4.03
1030	cycle 2 miles		276	.029			1.42	2.08	3.50	4.55
1032	cycle 4 miles		216	.037			1.81	2.65	4.46	5.80
1034	cycle 6 miles		168	.048			2.33	3.41	5.74	7.50
1036	cycle 8 miles		144	.056			2.72	3.98	6.70	8.75
1038	cycle 10 miles		120	.067			3.26	4.77	8.03	10.45

31 23 23.20 Hauling		Crew	Daily Output	Labor-Hours	Unit	Material	2020 Bare Costs Labor	Equipment	Total	Total Incl O&P
1040	25 MPH avg., cycle 4 miles	B-34B	228	.035	L.C.Y.		1.72	2.51	4.23	5.50
1042	cycle 6 miles		192	.042			2.04	2.98	5.02	6.55
1044	cycle 8 miles		168	.048			2.33	3.41	5.74	7.50
1046	cycle 10 miles		144	.056			2.72	3.98	6.70	8.75
1050	30 MPH avg., cycle 4 miles		252	.032			1.55	2.27	3.82	4.98
1052	cycle 6 miles		216	.037			1.81	2.65	4.46	5.80
1054	cycle 8 miles		180	.044			2.18	3.18	5.36	7
1056	cycle 10 miles		156	.051			2.51	3.67	6.18	8.05
1060	35 MPH avg., cycle 4 miles		264	.030			1.48	2.17	3.65	4.76
1062	cycle 6 miles		228	.035			1.72	2.51	4.23	5.50
1064	cycle 8 miles		204	.039			1.92	2.81	4.73	6.15
1066	cycle 10 miles		180	.044			2.18	3.18	5.36	7
1068	cycle 20 miles		120	.067			3.26	4.77	8.03	10.45
1069	cycle 30 miles		84	.095			4.66	6.80	11.46	14.95
1070	cycle 40 miles		72	.111			5.45	7.95	13.40	17.45
1072	40 MPH avg., cycle 6 miles		240	.033			1.63	2.39	4.02	5.25
1074	cycle 8 miles		216	.037			1.81	2.65	4.46	5.80
1076	cycle 10 miles		192	.042			2.04	2.98	5.02	6.55
1078	cycle 20 miles		120	.067			3.26	4.77	8.03	10.45
1080	cycle 30 miles		96	.083			4.08	5.95	10.03	13.05
1082	cycle 40 miles		72	.111			5.45	7.95	13.40	17.45
1084	cycle 50 miles		60	.133			6.55	9.55	16.10	21
1094	45 MPH avg., cycle 8 miles		216	.037			1.81	2.65	4.46	5.80
1096	cycle 10 miles		204	.039			1.92	2.81	4.73	6.15
1098	cycle 20 miles		132	.061			2.97	4.34	7.31	9.50
1100	cycle 30 miles		108	.074			3.63	5.30	8.93	11.65
1102	cycle 40 miles		84	.095			4.66	6.80	11.46	14.95
1104	cycle 50 miles		72	.111			5.45	7.95	13.40	17.45
1106	50 MPH avg., cycle 10 miles		216	.037			1.81	2.65	4.46	5.80
1108	cycle 20 miles		144	.056			2.72	3.98	6.70	8.75
1110	cycle 30 miles		108	.074			3.63	5.30	8.93	11.65
1112	cycle 40 miles		84	.095			4.66	6.80	11.46	14.95
1114	cycle 50 miles		72	.111			5.45	7.95	13.40	17.45
1214	15 MPH avg., cycle 0.5 mile, 20 min. wait/ld./uld.		264	.030			1.48	2.17	3.65	4.76
1216	cycle 1 mile		240	.033			1.63	2.39	4.02	5.25
1218	cycle 2 miles		204	.039			1.92	2.81	4.73	6.15
1220	cycle 4 miles		156	.051			2.51	3.67	6.18	8.05
1222	cycle 6 miles		132	.061			2.97	4.34	7.31	9.50
1224	cycle 8 miles		108	.074			3.63	5.30	8.93	11.65
1225	cycle 10 miles		96	.083			4.08	5.95	10.03	13.05
1226	20 MPH avg., cycle 0.5 mile		264	.030			1.48	2.17	3.65	4.76
1228	cycle 1 mile		252	.032			1.55	2.27	3.82	4.98
1230	cycle 2 miles		216	.037			1.81	2.65	4.46	5.80
1232	cycle 4 miles		180	.044			2.18	3.18	5.36	7
1234	cycle 6 miles		144	.056			2.72	3.98	6.70	8.75
1236	cycle 8 miles		132	.061			2.97	4.34	7.31	9.50
1238	cycle 10 miles		108	.074			3.63	5.30	8.93	11.65
1240	25 MPH avg., cycle 4 miles		192	.042			2.04	2.98	5.02	6.55
1242	cycle 6 miles		168	.048			2.33	3.41	5.74	7.50
1244	cycle 8 miles		144	.056			2.72	3.98	6.70	8.75
1246	cycle 10 miles		132	.061			2.97	4.34	7.31	9.50
1250	30 MPH avg., cycle 4 miles		204	.039			1.92	2.81	4.73	6.15
1252	cycle 6 miles		180	.044			2.18	3.18	5.36	7

31 23 23.20 Hauling		Crew	Daily Output	Labor-Hours	Unit	Material	2020 Bare Costs Labor	Equipment	Total	Total Incl O&P
1254	cycle 8 miles	B-34B	156	.051	L.C.Y.		2.51	3.67	6.18	8.05
1256	cycle 10 miles		144	.056			2.72	3.98	6.70	8.75
1260	35 MPH avg., cycle 4 miles		216	.037			1.81	2.65	4.46	5.80
1262	cycle 6 miles		192	.042			2.04	2.98	5.02	6.55
1264	cycle 8 miles		168	.048			2.33	3.41	5.74	7.50
1266	cycle 10 miles		156	.051			2.51	3.67	6.18	8.05
1268	cycle 20 miles		108	.074			3.63	5.30	8.93	11.65
1269	cycle 30 miles		72	.111			5.45	7.95	13.40	17.45
1270	cycle 40 miles		60	.133			6.55	9.55	16.10	21
1272	40 MPH avg., cycle 6 miles		192	.042			2.04	2.98	5.02	6.55
1274	cycle 8 miles		180	.044			2.18	3.18	5.36	7
1276	cycle 10 miles		156	.051			2.51	3.67	6.18	8.05
1278	cycle 20 miles		108	.074			3.63	5.30	8.93	11.65
1280	cycle 30 miles		84	.095			4.66	6.80	11.46	14.95
1282	cycle 40 miles		72	.111			5.45	7.95	13.40	17.45
1284	cycle 50 miles		60	.133			6.55	9.55	16.10	21
1294	45 MPH avg., cycle 8 miles		180	.044			2.18	3.18	5.36	7
1296	cycle 10 miles		168	.048			2.33	3.41	5.74	7.50
1298	cycle 20 miles		120	.067			3.26	4.77	8.03	10.45
1300	cycle 30 miles		96	.083			4.08	5.95	10.03	13.05
1302	cycle 40 miles		72	.111			5.45	7.95	13.40	17.45
1304	cycle 50 miles		60	.133			6.55	9.55	16.10	21
1306	50 MPH avg., cycle 10 miles		180	.044			2.18	3.18	5.36	7
1308	cycle 20 miles		132	.061			2.97	4.34	7.31	9.50
1310	cycle 30 miles		96	.083			4.08	5.95	10.03	13.05
1312	cycle 40 miles		84	.095			4.66	6.80	11.46	14.95
1314	cycle 50 miles		72	.111			5.45	7.95	13.40	17.45
1414	15 MPH avg., cycle 0.5 mile, 25 min. wait/ld./uld.		204	.039			1.92	2.81	4.73	6.15
1416	cycle 1 mile		192	.042			2.04	2.98	5.02	6.55
1418	cycle 2 miles		168	.048			2.33	3.41	5.74	7.50
1420	cycle 4 miles		132	.061			2.97	4.34	7.31	9.50
1422	cycle 6 miles		120	.067			3.26	4.77	8.03	10.45
1424	cycle 8 miles		96	.083			4.08	5.95	10.03	13.05
1425	cycle 10 miles		84	.095			4.66	6.80	11.46	14.95
1426	20 MPH avg., cycle 0.5 mile		216	.037			1.81	2.65	4.46	5.80
1428	cycle 1 mile		204	.039			1.92	2.81	4.73	6.15
1430	cycle 2 miles		180	.044			2.18	3.18	5.36	7
1432	cycle 4 miles		156	.051			2.51	3.67	6.18	8.05
1434	cycle 6 miles		132	.061			2.97	4.34	7.31	9.50
1436	cycle 8 miles		120	.067			3.26	4.77	8.03	10.45
1438	cycle 10 miles		96	.083			4.08	5.95	10.03	13.05
1440	25 MPH avg., cycle 4 miles		168	.048			2.33	3.41	5.74	7.50
1442	cycle 6 miles		144	.056			2.72	3.98	6.70	8.75
1444	cycle 8 miles		132	.061			2.97	4.34	7.31	9.50
1446	cycle 10 miles		108	.074			3.63	5.30	8.93	11.65
1450	30 MPH avg., cycle 4 miles		168	.048			2.33	3.41	5.74	7.50
1452	cycle 6 miles		156	.051			2.51	3.67	6.18	8.05
1454	cycle 8 miles		132	.061			2.97	4.34	7.31	9.50
1456	cycle 10 miles		120	.067			3.26	4.77	8.03	10.45
1460	35 MPH avg., cycle 4 miles		180	.044			2.18	3.18	5.36	7
1462	cycle 6 miles		156	.051			2.51	3.67	6.18	8.05
1464	cycle 8 miles		144	.056			2.72	3.98	6.70	8.75
1466	cycle 10 miles		132	.061			2.97	4.34	7.31	9.50

31 23 23 – Fill

31 23 23.20 Hauling		Crew	Daily Output	Labor-Hours	Unit	Material	2020 Bare Costs Labor	Equipment	Total	Total Incl O&P
1468	cycle 20 miles	B-34B	96	.083	L.C.Y.		4.08	5.95	10.03	13.05
1469	cycle 30 miles		72	.111			5.45	7.95	13.40	17.45
1470	cycle 40 miles		60	.133			6.55	9.55	16.10	21
1472	40 MPH avg., cycle 6 miles		168	.048			2.33	3.41	5.74	7.50
1474	cycle 8 miles		156	.051			2.51	3.67	6.18	8.05
1476	cycle 10 miles		144	.056			2.72	3.98	6.70	8.75
1478	cycle 20 miles		96	.083			4.08	5.95	10.03	13.05
1480	cycle 30 miles		84	.095			4.66	6.80	11.46	14.95
1482	cycle 40 miles		60	.133			6.55	9.55	16.10	21
1484	cycle 50 miles		60	.133			6.55	9.55	16.10	21
1494	45 MPH avg., cycle 8 miles		156	.051			2.51	3.67	6.18	8.05
1496	cycle 10 miles		144	.056			2.72	3.98	6.70	8.75
1498	cycle 20 miles		108	.074			3.63	5.30	8.93	11.65
1500	cycle 30 miles		84	.095			4.66	6.80	11.46	14.95
1502	cycle 40 miles		72	.111			5.45	7.95	13.40	17.45
1504	cycle 50 miles		60	.133			6.55	9.55	16.10	21
1506	50 MPH avg., cycle 10 miles		156	.051			2.51	3.67	6.18	8.05
1508	cycle 20 miles		120	.067			3.26	4.77	8.03	10.45
1510	cycle 30 miles		96	.083			4.08	5.95	10.03	13.05
1512	cycle 40 miles		72	.111			5.45	7.95	13.40	17.45
1514	cycle 50 miles		60	.133			6.55	9.55	16.10	21
1614	15 MPH avg., cycle 0.5 mile, 30 min. wait/ld./uld.		180	.044			2.18	3.18	5.36	7
1616	cycle 1 mile		168	.048			2.33	3.41	5.74	7.50
1618	cycle 2 miles		144	.056			2.72	3.98	6.70	8.75
1620	cycle 4 miles		120	.067			3.26	4.77	8.03	10.45
1622	cycle 6 miles		108	.074			3.63	5.30	8.93	11.65
1624	cycle 8 miles		84	.095			4.66	6.80	11.46	14.95
1625	cycle 10 miles		84	.095			4.66	6.80	11.46	14.95
1626	20 MPH avg., cycle 0.5 mile		180	.044			2.18	3.18	5.36	7
1628	cycle 1 mile		168	.048			2.33	3.41	5.74	7.50
1630	cycle 2 miles		156	.051			2.51	3.67	6.18	8.05
1632	cycle 4 miles		132	.061			2.97	4.34	7.31	9.50
1634	cycle 6 miles		120	.067			3.26	4.77	8.03	10.45
1636	cycle 8 miles		108	.074			3.63	5.30	8.93	11.65
1638	cycle 10 miles		96	.083			4.08	5.95	10.03	13.05
1640	25 MPH avg., cycle 4 miles		144	.056			2.72	3.98	6.70	8.75
1642	cycle 6 miles		132	.061			2.97	4.34	7.31	9.50
1644	cycle 8 miles		108	.074			3.63	5.30	8.93	11.65
1646	cycle 10 miles		108	.074			3.63	5.30	8.93	11.65
1650	30 MPH avg., cycle 4 miles		144	.056			2.72	3.98	6.70	8.75
1652	cycle 6 miles		132	.061			2.97	4.34	7.31	9.50
1654	cycle 8 miles		120	.067			3.26	4.77	8.03	10.45
1656	cycle 10 miles		108	.074			3.63	5.30	8.93	11.65
1660	35 MPH avg., cycle 4 miles		156	.051			2.51	3.67	6.18	8.05
1662	cycle 6 miles		144	.056			2.72	3.98	6.70	8.75
1664	cycle 8 miles		132	.061			2.97	4.34	7.31	9.50
1666	cycle 10 miles		120	.067			3.26	4.77	8.03	10.45
1668	cycle 20 miles		84	.095			4.66	6.80	11.46	14.95
1669	cycle 30 miles		72	.111			5.45	7.95	13.40	17.45
1670	cycle 40 miles		60	.133			6.55	9.55	16.10	21
1672	40 MPH avg., cycle 6 miles		144	.056			2.72	3.98	6.70	8.75
1674	cycle 8 miles		132	.061			2.97	4.34	7.31	9.50
1676	cycle 10 miles		120	.067			3.26	4.77	8.03	10.45

31 23 23.20 Hauling		Crew	Daily Output	Labor-Hours	Unit	Material	2020 Bare Costs Labor	Equipment	Total	Total Incl O&P
1678	cycle 20 miles	B-34B	96	.083	L.C.Y.		4.08	5.95	10.03	13.05
1680	cycle 30 miles		72	.111			5.45	7.95	13.40	17.45
1682	cycle 40 miles		60	.133			6.55	9.55	16.10	21
1684	cycle 50 miles		48	.167			8.15	11.95	20.10	26
1694	45 MPH avg., cycle 8 miles		144	.056			2.72	3.98	6.70	8.75
1696	cycle 10 miles		132	.061			2.97	4.34	7.31	9.50
1698	cycle 20 miles		96	.083			4.08	5.95	10.03	13.05
1700	cycle 30 miles		84	.095			4.66	6.80	11.46	14.95
1702	cycle 40 miles		60	.133			6.55	9.55	16.10	21
1704	cycle 50 miles		60	.133			6.55	9.55	16.10	21
1706	50 MPH avg., cycle 10 miles		132	.061			2.97	4.34	7.31	9.50
1708	cycle 20 miles		108	.074			3.63	5.30	8.93	11.65
1710	cycle 30 miles		84	.095			4.66	6.80	11.46	14.95
1712	cycle 40 miles		72	.111			5.45	7.95	13.40	17.45
1714	cycle 50 miles		60	.133			6.55	9.55	16.10	21
2000	Hauling, 8 C.Y. truck, small project cost per hour	B-34A	8	1	Hr.		49	53.50	102.50	138
2100	12 C.Y. truck	B-34B	8	1			49	71.50	120.50	158
2150	16.5 C.Y. truck	B-34C	8	1			49	78.50	127.50	165
2175	18 C.Y. 8 wheel truck	B-34I	8	1			49	89	138	177
2200	20 C.Y. truck	B-34D	8	1			49	80.50	129.50	167
2300	Grading at dump, or embankment if required, by dozer	B-10B	1000	.012	L.C.Y.		.62	1.50	2.12	2.63
2310	Spotter at fill or cut, if required	1 Clab	8	1	Hr.		42		42	67.50
9014	18 C.Y. truck, 8 wheels,15 min. wait/ld./uld.,15 MPH, cycle 0.5 mi.	B-34I	504	.016	L.C.Y.		.78	1.42	2.20	2.80
9016	cycle 1 mile		450	.018			.87	1.59	2.46	3.14
9018	cycle 2 miles		378	.021			1.04	1.89	2.93	3.74
9020	cycle 4 miles		270	.030			1.45	2.64	4.09	5.25
9022	cycle 6 miles		216	.037			1.81	3.31	5.12	6.55
9024	cycle 8 miles		180	.044			2.18	3.97	6.15	7.85
9025	cycle 10 miles		144	.056			2.72	4.96	7.68	9.80
9026	20 MPH avg., cycle 0.5 mile		522	.015			.75	1.37	2.12	2.70
9028	cycle 1 mile		468	.017			.84	1.52	2.36	3.02
9030	cycle 2 miles		414	.019			.95	1.72	2.67	3.41
9032	cycle 4 miles		324	.025			1.21	2.20	3.41	4.35
9034	cycle 6 miles		252	.032			1.55	2.83	4.38	5.60
9036	cycle 8 miles		216	.037			1.81	3.31	5.12	6.55
9038	cycle 10 miles		180	.044			2.18	3.97	6.15	7.85
9040	25 MPH avg., cycle 4 miles		342	.023			1.14	2.09	3.23	4.13
9042	cycle 6 miles		288	.028			1.36	2.48	3.84	4.90
9044	cycle 8 miles		252	.032			1.55	2.83	4.38	5.60
9046	cycle 10 miles		216	.037			1.81	3.31	5.12	6.55
9050	30 MPH avg., cycle 4 miles		378	.021			1.04	1.89	2.93	3.74
9052	cycle 6 miles		324	.025			1.21	2.20	3.41	4.35
9054	cycle 8 miles		270	.030			1.45	2.64	4.09	5.25
9056	cycle 10 miles		234	.034			1.67	3.05	4.72	6.05
9060	35 MPH avg., cycle 4 miles		396	.020			.99	1.80	2.79	3.56
9062	cycle 6 miles		342	.023			1.14	2.09	3.23	4.13
9064	cycle 8 miles		288	.028			1.36	2.48	3.84	4.90
9066	cycle 10 miles		270	.030			1.45	2.64	4.09	5.25
9068	cycle 20 miles		162	.049			2.42	4.41	6.83	8.70
9070	cycle 30 miles		126	.063			3.11	5.65	8.76	11.20
9072	cycle 40 miles		90	.089			4.35	7.95	12.30	15.65
9074	40 MPH avg., cycle 6 miles		360	.022			1.09	1.98	3.07	3.92
9076	cycle 8 miles		324	.025			1.21	2.20	3.41	4.35

31 23 23.20 Hauling		Crew	Daily Output	Labor-Hours	Unit	Material	2020 Bare Costs Labor	Equipment	Total	Total Incl O&P
9078	cycle 10 miles	B-34I	288	.028	L.C.Y.		1.36	2.48	3.84	4.90
9080	cycle 20 miles		180	.044			2.18	3.97	6.15	7.85
9082	cycle 30 miles		144	.056			2.72	4.96	7.68	9.80
9084	cycle 40 miles		108	.074			3.63	6.60	10.23	13.05
9086	cycle 50 miles		90	.089			4.35	7.95	12.30	15.65
9094	45 MPH avg., cycle 8 miles		324	.025			1.21	2.20	3.41	4.35
9096	cycle 10 miles		306	.026			1.28	2.33	3.61	4.62
9098	cycle 20 miles		198	.040			1.98	3.60	5.58	7.15
9100	cycle 30 miles		144	.056			2.72	4.96	7.68	9.80
9102	cycle 40 miles		126	.063			3.11	5.65	8.76	11.20
9104	cycle 50 miles		108	.074			3.63	6.60	10.23	13.05
9106	50 MPH avg., cycle 10 miles		324	.025			1.21	2.20	3.41	4.35
9108	cycle 20 miles		216	.037			1.81	3.31	5.12	6.55
9110	cycle 30 miles		162	.049			2.42	4.41	6.83	8.70
9112	cycle 40 miles		126	.063			3.11	5.65	8.76	11.20
9114	cycle 50 miles		108	.074			3.63	6.60	10.23	13.05
9214	20 min. wait/ld./uld.,15 MPH, cycle 0.5 mi.		396	.020			.99	1.80	2.79	3.56
9216	cycle 1 mile		360	.022			1.09	1.98	3.07	3.92
9218	cycle 2 miles		306	.026			1.28	2.33	3.61	4.62
9220	cycle 4 miles		234	.034			1.67	3.05	4.72	6.05
9222	cycle 6 miles		198	.040			1.98	3.60	5.58	7.15
9224	cycle 8 miles		162	.049			2.42	4.41	6.83	8.70
9225	cycle 10 miles		144	.056			2.72	4.96	7.68	9.80
9226	20 MPH avg., cycle 0.5 mile		396	.020			.99	1.80	2.79	3.56
9228	cycle 1 mile		378	.021			1.04	1.89	2.93	3.74
9230	cycle 2 miles		324	.025			1.21	2.20	3.41	4.35
9232	cycle 4 miles		270	.030			1.45	2.64	4.09	5.25
9234	cycle 6 miles		216	.037			1.81	3.31	5.12	6.55
9236	cycle 8 miles		198	.040			1.98	3.60	5.58	7.15
9238	cycle 10 miles		162	.049			2.42	4.41	6.83	8.70
9240	25 MPH avg., cycle 4 miles		288	.028			1.36	2.48	3.84	4.90
9242	cycle 6 miles		252	.032			1.55	2.83	4.38	5.60
9244	cycle 8 miles		216	.037			1.81	3.31	5.12	6.55
9246	cycle 10 miles		198	.040			1.98	3.60	5.58	7.15
9250	30 MPH avg., cycle 4 miles		306	.026			1.28	2.33	3.61	4.62
9252	cycle 6 miles		270	.030			1.45	2.64	4.09	5.25
9254	cycle 8 miles		234	.034			1.67	3.05	4.72	6.05
9256	cycle 10 miles		216	.037			1.81	3.31	5.12	6.55
9260	35 MPH avg., cycle 4 miles		324	.025			1.21	2.20	3.41	4.35
9262	cycle 6 miles		288	.028			1.36	2.48	3.84	4.90
9264	cycle 8 miles		252	.032			1.55	2.83	4.38	5.60
9266	cycle 10 miles		234	.034			1.67	3.05	4.72	6.05
9268	cycle 20 miles		162	.049			2.42	4.41	6.83	8.70
9270	cycle 30 miles		108	.074			3.63	6.60	10.23	13.05
9272	cycle 40 miles		90	.089			4.35	7.95	12.30	15.65
9274	40 MPH avg., cycle 6 miles		288	.028			1.36	2.48	3.84	4.90
9276	cycle 8 miles		270	.030			1.45	2.64	4.09	5.25
9278	cycle 10 miles		234	.034			1.67	3.05	4.72	6.05
9280	cycle 20 miles		162	.049			2.42	4.41	6.83	8.70
9282	cycle 30 miles		126	.063			3.11	5.65	8.76	11.20
9284	cycle 40 miles		108	.074			3.63	6.60	10.23	13.05
9286	cycle 50 miles		90	.089			4.35	7.95	12.30	15.65
9294	45 MPH avg., cycle 8 miles		270	.030			1.45	2.64	4.09	5.25

31 23 23.20 Hauling		Crew	Daily Output	Labor-Hours	Unit	Material	2020 Bare Costs Labor	Equipment	Total	Total Incl O&P
9296	cycle 10 miles	B-34I	252	.032	L.C.Y.		1.55	2.83	4.38	5.60
9298	cycle 20 miles		180	.044			2.18	3.97	6.15	7.85
9300	cycle 30 miles		144	.056			2.72	4.96	7.68	9.80
9302	cycle 40 miles		108	.074			3.63	6.60	10.23	13.05
9304	cycle 50 miles		90	.089			4.35	7.95	12.30	15.65
9306	50 MPH avg., cycle 10 miles		270	.030			1.45	2.64	4.09	5.25
9308	cycle 20 miles		198	.040			1.98	3.60	5.58	7.15
9310	cycle 30 miles		144	.056			2.72	4.96	7.68	9.80
9312	cycle 40 miles		126	.063			3.11	5.65	8.76	11.20
9314	cycle 50 miles		108	.074			3.63	6.60	10.23	13.05
9414	25 min. wait/ld./uld.,15 MPH, cycle 0.5 mi.		306	.026			1.28	2.33	3.61	4.62
9416	cycle 1 mile		288	.028			1.36	2.48	3.84	4.90
9418	cycle 2 miles		252	.032			1.55	2.83	4.38	5.60
9420	cycle 4 miles		198	.040			1.98	3.60	5.58	7.15
9422	cycle 6 miles		180	.044			2.18	3.97	6.15	7.85
9424	cycle 8 miles		144	.056			2.72	4.96	7.68	9.80
9425	cycle 10 miles		126	.063			3.11	5.65	8.76	11.20
9426	20 MPH avg., cycle 0.5 mile		324	.025			1.21	2.20	3.41	4.35
9428	cycle 1 mile		306	.026			1.28	2.33	3.61	4.62
9430	cycle 2 miles		270	.030			1.45	2.64	4.09	5.25
9432	cycle 4 miles		234	.034			1.67	3.05	4.72	6.05
9434	cycle 6 miles		198	.040			1.98	3.60	5.58	7.15
9436	cycle 8 miles		180	.044			2.18	3.97	6.15	7.85
9438	cycle 10 miles		144	.056			2.72	4.96	7.68	9.80
9440	25 MPH avg., cycle 4 miles		252	.032			1.55	2.83	4.38	5.60
9442	cycle 6 miles		216	.037			1.81	3.31	5.12	6.55
9444	cycle 8 miles		198	.040			1.98	3.60	5.58	7.15
9446	cycle 10 miles		180	.044			2.18	3.97	6.15	7.85
9450	30 MPH avg., cycle 4 miles		252	.032			1.55	2.83	4.38	5.60
9452	cycle 6 miles		234	.034			1.67	3.05	4.72	6.05
9454	cycle 8 miles		198	.040			1.98	3.60	5.58	7.15
9456	cycle 10 miles		180	.044			2.18	3.97	6.15	7.85
9460	35 MPH avg., cycle 4 miles		270	.030			1.45	2.64	4.09	5.25
9462	cycle 6 miles		234	.034			1.67	3.05	4.72	6.05
9464	cycle 8 miles		216	.037			1.81	3.31	5.12	6.55
9466	cycle 10 miles		198	.040			1.98	3.60	5.58	7.15
9468	cycle 20 miles		144	.056			2.72	4.96	7.68	9.80
9470	cycle 30 miles		108	.074			3.63	6.60	10.23	13.05
9472	cycle 40 miles		90	.089			4.35	7.95	12.30	15.65
9474	40 MPH avg., cycle 6 miles		252	.032			1.55	2.83	4.38	5.60
9476	cycle 8 miles		234	.034			1.67	3.05	4.72	6.05
9478	cycle 10 miles		216	.037			1.81	3.31	5.12	6.55
9480	cycle 20 miles		144	.056			2.72	4.96	7.68	9.80
9482	cycle 30 miles		126	.063			3.11	5.65	8.76	11.20
9484	cycle 40 miles		90	.089			4.35	7.95	12.30	15.65
9486	cycle 50 miles		90	.089			4.35	7.95	12.30	15.65
9494	45 MPH avg., cycle 8 miles		234	.034			1.67	3.05	4.72	6.05
9496	cycle 10 miles		216	.037			1.81	3.31	5.12	6.55
9498	cycle 20 miles		162	.049			2.42	4.41	6.83	8.70
9500	cycle 30 miles		126	.063			3.11	5.65	8.76	11.20
9502	cycle 40 miles		108	.074			3.63	6.60	10.23	13.05
9504	cycle 50 miles		90	.089			4.35	7.95	12.30	15.65
9506	50 MPH avg., cycle 10 miles		234	.034			1.67	3.05	4.72	6.05

31 23 23.20 Hauling		Crew	Daily Output	Labor-Hours	Unit	Material	2020 Bare Costs Labor	Equipment	Total	Total Incl O&P
9508	cycle 20 miles	B-34I	180	.044	L.C.Y.		2.18	3.97	6.15	7.85
9510	cycle 30 miles		144	.056			2.72	4.96	7.68	9.80
9512	cycle 40 miles		108	.074			3.63	6.60	10.23	13.05
9514	cycle 50 miles		90	.089			4.35	7.95	12.30	15.65
9614	30 min. wait/ld./uld.,15 MPH, cycle 0.5 mi.		270	.030			1.45	2.64	4.09	5.25
9616	cycle 1 mile		252	.032			1.55	2.83	4.38	5.60
9618	cycle 2 miles		216	.037			1.81	3.31	5.12	6.55
9620	cycle 4 miles		180	.044			2.18	3.97	6.15	7.85
9622	cycle 6 miles		162	.049			2.42	4.41	6.83	8.70
9624	cycle 8 miles		126	.063			3.11	5.65	8.76	11.20
9625	cycle 10 miles		126	.063			3.11	5.65	8.76	11.20
9626	20 MPH avg., cycle 0.5 mile		270	.030			1.45	2.64	4.09	5.25
9628	cycle 1 mile		252	.032			1.55	2.83	4.38	5.60
9630	cycle 2 miles		234	.034			1.67	3.05	4.72	6.05
9632	cycle 4 miles		198	.040			1.98	3.60	5.58	7.15
9634	cycle 6 miles		180	.044			2.18	3.97	6.15	7.85
9636	cycle 8 miles		162	.049			2.42	4.41	6.83	8.70
9638	cycle 10 miles		144	.056			2.72	4.96	7.68	9.80
9640	25 MPH avg., cycle 4 miles		216	.037			1.81	3.31	5.12	6.55
9642	cycle 6 miles		198	.040			1.98	3.60	5.58	7.15
9644	cycle 8 miles		180	.044			2.18	3.97	6.15	7.85
9646	cycle 10 miles		162	.049			2.42	4.41	6.83	8.70
9650	30 MPH avg., cycle 4 miles		216	.037			1.81	3.31	5.12	6.55
9652	cycle 6 miles		198	.040			1.98	3.60	5.58	7.15
9654	cycle 8 miles		180	.044			2.18	3.97	6.15	7.85
9656	cycle 10 miles		162	.049			2.42	4.41	6.83	8.70
9660	35 MPH avg., cycle 4 miles		234	.034			1.67	3.05	4.72	6.05
9662	cycle 6 miles		216	.037			1.81	3.31	5.12	6.55
9664	cycle 8 miles		198	.040			1.98	3.60	5.58	7.15
9666	cycle 10 miles		180	.044			2.18	3.97	6.15	7.85
9668	cycle 20 miles		126	.063			3.11	5.65	8.76	11.20
9670	cycle 30 miles		108	.074			3.63	6.60	10.23	13.05
9672	cycle 40 miles		90	.089			4.35	7.95	12.30	15.65
9674	40 MPH avg., cycle 6 miles		216	.037			1.81	3.31	5.12	6.55
9676	cycle 8 miles		198	.040			1.98	3.60	5.58	7.15
9678	cycle 10 miles		180	.044			2.18	3.97	6.15	7.85
9680	cycle 20 miles		144	.056			2.72	4.96	7.68	9.80
9682	cycle 30 miles		108	.074			3.63	6.60	10.23	13.05
9684	cycle 40 miles		90	.089			4.35	7.95	12.30	15.65
9686	cycle 50 miles		72	.111			5.45	9.90	15.35	19.60
9694	45 MPH avg., cycle 8 miles		216	.037			1.81	3.31	5.12	6.55
9696	cycle 10 miles		198	.040			1.98	3.60	5.58	7.15
9698	cycle 20 miles		144	.056			2.72	4.96	7.68	9.80
9700	cycle 30 miles		126	.063			3.11	5.65	8.76	11.20
9702	cycle 40 miles		108	.074			3.63	6.60	10.23	13.05
9704	cycle 50 miles		90	.089			4.35	7.95	12.30	15.65
9706	50 MPH avg., cycle 10 miles		198	.040			1.98	3.60	5.58	7.15
9708	cycle 20 miles		162	.049			2.42	4.41	6.83	8.70
9710	cycle 30 miles		126	.063			3.11	5.65	8.76	11.20
9712	cycle 40 miles		108	.074			3.63	6.60	10.23	13.05
9714	cycle 50 miles	▼	90	.089	▼		4.35	7.95	12.30	15.65

31 23 23.23 Compaction

	31 23 23.23 Compaction	Crew	Daily Output	Labor-Hours	Unit	Material	2020 Bare Costs Labor	Equipment	Total	Total Incl O&P
0010	**COMPACTION**									
5000	Riding, vibrating roller, 6" lifts, 2 passes	B-10Y	3000	.004	E.C.Y.		.21	.19	.40	.54
5020	3 passes		2300	.005			.27	.25	.52	.71
5040	4 passes		1900	.006			.33	.30	.63	.86
5050	8" lifts, 2 passes		4100	.003			.15	.14	.29	.40
5060	12" lifts, 2 passes		5200	.002			.12	.11	.23	.31
5080	3 passes		3500	.003			.18	.17	.35	.46
5100	4 passes		2600	.005			.24	.22	.46	.63
5600	Sheepsfoot or wobbly wheel roller, 6" lifts, 2 passes	B-10G	2400	.005			.26	.56	.82	1.03
5620	3 passes		1735	.007			.36	.78	1.14	1.43
5640	4 passes		1300	.009			.48	1.04	1.52	1.89
5680	12" lifts, 2 passes		5200	.002			.12	.26	.38	.48
5700	3 passes		3500	.003			.18	.39	.57	.70
5720	4 passes		2600	.005			.24	.52	.76	.95
7000	Walk behind, vibrating plate 18" wide, 6" lifts, 2 passes	A-1D	200	.040			1.68	.16	1.84	2.87
7020	3 passes		185	.043			1.82	.17	1.99	3.11
7040	4 passes		140	.057			2.41	.23	2.64	4.10
7200	12" lifts, 2 passes, 21" wide	A-1E	560	.014			.60	.30	.90	1.29
7220	3 passes		375	.021			.90	.44	1.34	1.93
7240	4 passes		280	.029			1.20	.59	1.79	2.58
7500	Vibrating roller 24" wide, 6" lifts, 2 passes	B-10A	420	.029			1.48	.40	1.88	2.77
7520	3 passes		280	.043			2.22	.59	2.81	4.15
7540	4 passes		210	.057			2.96	.79	3.75	5.55
7600	12" lifts, 2 passes		840	.014			.74	.20	.94	1.39
7620	3 passes		560	.021			1.11	.30	1.41	2.08
7640	4 passes		420	.029			1.48	.40	1.88	2.77
8000	Rammer tamper, 6" to 11", 4" lifts, 2 passes	A-1F	130	.062			2.59	.36	2.95	4.54
8050	3 passes		97	.082			3.47	.48	3.95	6.10
8100	4 passes		65	.123			5.20	.72	5.92	9.10
8200	8" lifts, 2 passes		260	.031			1.30	.18	1.48	2.28
8250	3 passes		195	.041			1.73	.24	1.97	3.03
8300	4 passes		130	.062			2.59	.36	2.95	4.54
8400	13" to 18", 4" lifts, 2 passes	A-1G	390	.021			.86	.14	1	1.53
8450	3 passes		290	.028			1.16	.19	1.35	2.07
8500	4 passes		195	.041			1.73	.28	2.01	3.07
8600	8" lifts, 2 passes		780	.010			.43	.07	.50	.77
8650	3 passes		585	.014			.58	.09	.67	1.02
8700	4 passes		390	.021			.86	.14	1	1.53
9000	Water, 3,000 gal. truck, 3 mile haul	B-45	1888	.008		1.22	.45	.44	2.11	2.53
9010	6 mile haul		1444	.011		1.22	.59	.57	2.38	2.90
9020	12 mile haul		1000	.016		1.22	.85	.82	2.89	3.59
9030	6,000 gal. wagon, 3 mile haul	B-59	2000	.004		1.22	.20	.23	1.65	1.90
9040	6 mile haul	"	1600	.005		1.22	.24	.29	1.75	2.05
9900	Minimum labor/equipment charge	1 Clab	4	2	Job		84		84	135

31 23 23.24 Compaction, Structural

	31 23 23.24 Compaction, Structural	Crew	Daily Output	Labor-Hours	Unit	Material	Labor	Equipment	Total	Total Incl O&P
0010	**COMPACTION, STRUCTURAL**									
0020	Steel wheel tandem roller, 5 tons	B-10E	8	1.500	Hr.		78	32	110	158

31 25 Erosion and Sedimentation Controls

31 25 14 – Stabilization Measures for Erosion and Sedimentation Control

31 25 14.16 Rolled Erosion Control Mats and Blankets		Crew	Daily Output	Labor-Hours	Unit	2020 Bare Costs Material	Labor	Equipment	Total	Total Incl O&P
0010	**ROLLED EROSION CONTROL MATS AND BLANKETS**									
0020	Jute mesh, 100 S.Y. per roll, 4' wide, stapled G	B-80A	2400	.010	S.Y.	1.03	.42	.34	1.79	2.18
0060	Polyethylene 3 dimensional geomatrix, 50 mil thick G		700	.034		3.28	1.44	1.17	5.89	7.20
0062	120 mil thick G		515	.047		6.20	1.96	1.59	9.75	11.75
0070	Paper biodegradable mesh G	B-1	2500	.010		.18	.41		.59	.86
0100	Plastic netting, stapled, 2" x 1" mesh, 20 mil G	"	2500	.010		.27	.41		.68	.96
0120	Revegetation mat, webbed G	2 Clab	1000	.016		2.96	.67		3.63	4.34
0200	Polypropylene mesh, stapled, 6.5 oz./S.Y. G	B-1	2500	.010		1.70	.41		2.11	2.53
0300	Tobacco netting, or jute mesh #2, stapled G	"	2500	.010		.31	.41		.72	1
0600	Straw in polymeric netting, biodegradable log	A-2	1000	.024	L.F.	6.45	1.05	.20	7.70	9
0705	Sediment Log, Filter Sock, 9"		1000	.024		3.50	1.05	.20	4.75	5.75
0710	Sediment Log, Filter Sock, 12"		1000	.024		4.95	1.05	.20	6.20	7.35
1000	Silt fence, install and maintain, remove G	B-62	1300	.018		.45	.84	.14	1.43	1.99

31 31 Soil Treatment

31 31 16 – Termite Control

31 31 16.13 Chemical Termite Control		Crew	Daily Output	Labor-Hours	Unit	Material	Labor	Equipment	Total	Total Incl O&P
0010	**CHEMICAL TERMITE CONTROL**									
0020	Slab and walls, residential	1 Skwk	1200	.007	SF Flr.	.33	.37		.70	.94
0030	SS mesh, no chemicals, avg 1,400 S.F. home, min G		1000	.008		.33	.44		.77	1.06
0100	Commercial, minimum		2496	.003		.35	.18		.53	.67
0200	Maximum		1645	.005		.53	.27		.80	1
0390	Minimum labor/equipment charge		4	2	Job		110		110	175
0400	Insecticides for termite control, minimum		14.20	.563	Gal.	70.50	31		101.50	127
0500	Maximum		11	.727	"	120	40		160	196
3000	Soil poisoning (sterilization)	1 Clab	4496	.002	S.F.	.36	.07		.43	.52
3100	Herbicide application from truck	B-59	19000	.001	S.Y.		.02	.02	.04	.06

31 32 Soil Stabilization

31 32 13 – Soil Mixing Stabilization

31 32 13.13 Asphalt Soil Stabilization		Crew	Daily Output	Labor-Hours	Unit	Material	Labor	Equipment	Total	Total Incl O&P
0010	**ASPHALT SOIL STABILIZATION**									
0011	Including scarifying and compaction									
0020	Asphalt, 1-1/2" deep, 1/2 gal./S.Y.	B-75	4000	.014	S.Y.	1.11	.72	1.35	3.18	3.85
0040	3/4 gal./S.Y.		4000	.014		1.67	.72	1.35	3.74	4.46
0100	3" deep, 1 gal./S.Y.		3500	.016		2.22	.83	1.55	4.60	5.45
0140	1-1/2 gal./S.Y.		3500	.016		3.33	.83	1.55	5.71	6.65
0200	6" deep, 2 gal./S.Y.		3000	.019		4.44	.97	1.81	7.22	8.40
0240	3 gal./S.Y.		3000	.019		6.65	.97	1.81	9.43	10.85
0300	8" deep, 2-2/3 gal./S.Y.		2800	.020		5.95	1.03	1.94	8.92	10.25
0340	4 gal./S.Y.		2800	.020		8.90	1.03	1.94	11.87	13.50
0500	12" deep, 4 gal./S.Y.		5000	.011		8.90	.58	1.08	10.56	11.85
0540	6 gal./S.Y.		2600	.022		13.30	1.11	2.08	16.49	18.70

31 32 13.16 Cement Soil Stabilization		Crew	Daily Output	Labor-Hours	Unit	Material	Labor	Equipment	Total	Total Incl O&P
0010	**CEMENT SOIL STABILIZATION**									
0011	Including scarifying and compaction									
1020	Cement, 4% mix, by volume, 6" deep	B-74	1100	.058	S.Y.	1.88	2.99	5.40	10.27	12.75
1030	8" deep		1050	.061		2.45	3.13	5.65	11.23	13.90
1060	12" deep		960	.067		3.68	3.43	6.20	13.31	16.30

31 32 Soil Stabilization

31 32 13 – Soil Mixing Stabilization

31 32 13.16 Cement Soil Stabilization		Crew	Daily Output	Labor-Hours	Unit	Material	Labor	Equipment	Total	Total Incl O&P
						2020 Bare Costs				
1100	6% mix, 6″ deep	B-74	1100	.058	S.Y.	2.70	2.99	5.40	11.09	13.65
1120	8″ deep		1050	.061		3.51	3.13	5.65	12.29	15.05
1160	12″ deep		960	.067		5.30	3.43	6.20	14.93	18.10
1200	9% mix, 6″ deep		1100	.058		4.09	2.99	5.40	12.48	15.15
1220	8″ deep		1050	.061		5.30	3.13	5.65	14.08	17.05
1260	12″ deep		960	.067		8	3.43	6.20	17.63	21
1300	12% mix, 6″ deep		1100	.058		5.30	2.99	5.40	13.69	16.55
1320	8″ deep		1050	.061		7.10	3.13	5.65	15.88	19
1360	12″ deep		960	.067		10.60	3.43	6.20	20.23	24

31 32 13.19 Lime Soil Stabilization

		Crew	Daily Output	Labor-Hours	Unit	Material	Labor	Equipment	Total	Total Incl O&P
0010	**LIME SOIL STABILIZATION**									
0011	Including scarifying and compaction									
2020	Hydrated lime, for base, 2% mix by weight, 6″ deep	B-74	1800	.036	S.Y.	.76	1.83	3.31	5.90	7.35
2030	8″ deep		1700	.038		1.01	1.93	3.50	6.44	8.05
2060	12″ deep		1550	.041		1.51	2.12	3.84	7.47	9.25
2100	4% mix, 6″ deep		1800	.036		1.51	1.83	3.31	6.65	8.20
2120	8″ deep		1700	.038		2.03	1.93	3.50	7.46	9.15
2160	12″ deep		1550	.041		3.02	2.12	3.84	8.98	10.90
2200	6% mix, 6″ deep		1800	.036		2.27	1.83	3.31	7.41	9
2220	8″ deep		1700	.038		3.04	1.93	3.50	8.47	10.25
2260	12″ deep		1550	.041		4.54	2.12	3.84	10.50	12.55

31 32 13.30 Calcium Chloride

		Crew	Daily Output	Labor-Hours	Unit	Material	Labor	Equipment	Total	Total Incl O&P
0010	**CALCIUM CHLORIDE**									
0020	Calcium chloride, delivered, 100 lb. bags, truckload lots				Ton	635			635	700
0030	Solution, 4 lb. flake per gallon, tank truck delivery				Gal.	1.67			1.67	1.84

31 32 19 – Geosynthetic Soil Stabilization and Layer Separation

31 32 19.16 Geotextile Soil Stabilization

		Crew	Daily Output	Labor-Hours	Unit	Material	Labor	Equipment	Total	Total Incl O&P
0010	**GEOTEXTILE SOIL STABILIZATION**									
1500	Geotextile fabric, woven, 200 lb. tensile strength	2 Clab	2500	.006	S.Y.	.96	.27		1.23	1.49
1510	Heavy duty, 600 lb. tensile strength		2400	.007		1.80	.28		2.08	2.43
1550	Non-woven, 120 lb. tensile strength		2500	.006		.86	.27		1.13	1.38

31 33 Rock Stabilization

31 33 13 – Rock Bolting and Grouting

31 33 13.10 Rock Bolting

		Crew	Daily Output	Labor-Hours	Unit	Material	Labor	Equipment	Total	Total Incl O&P
0010	**ROCK BOLTING**									
2020	Hollow core, prestressable anchor, 1″ diameter, 5′ long	2 Skwk	32	.500	Ea.	190	27.50		217.50	254
2025	10′ long		24	.667		320	36.50		356.50	410
2060	2″ diameter, 5′ long		32	.500		705	27.50		732.50	820
2065	10′ long		24	.667		1,250	36.50		1,286.50	1,425
2100	Super high-tensile, 3/4″ diameter, 5′ long		32	.500		51.50	27.50		79	100
2105	10′ long		24	.667		127	36.50		163.50	198
2160	2″ diameter, 5′ long		32	.500		400	27.50		427.50	485
2165	10′ long		24	.667		705	36.50		741.50	835
4400	Drill hole for rock bolt, 1-3/4″ diam., 5′ long (for 3/4″ bolt)	B-56	17	.941			45	89.50	134.50	169
4405	10′ long		9	1.778			84.50	169	253.50	320
4420	2″ diameter, 5′ long (for 1″ bolt)		13	1.231			58.50	117	175.50	221
4425	10′ long		7	2.286			109	217	326	410
4460	3-1/2″ diameter, 5′ long (for 2″ bolt)		10	1.600			76	152	228	287
4465	10′ long		5	3.200			152	305	457	575

31 36 Gabions

31 36 13 – Gabion Boxes

31 36 13.10 Gabion Box Systems		Crew	Daily Output	Labor-Hours	Unit	2020 Bare Costs Material	Labor	Equipment	Total	Total Incl O&P
0010	**GABION BOX SYSTEMS**									
0400	Gabions, galvanized steel mesh mats or boxes, stone filled, 6″ deep	B-13	200	.280	S.Y.	18.30	12.90	2.90	34.10	43.50
0500	9″ deep		163	.344		25.50	15.80	3.56	44.86	57.50
0600	12″ deep		153	.366		37	16.85	3.80	57.65	72
0700	18″ deep		102	.549		47.50	25.50	5.70	78.70	98.50
0800	36″ deep		60	.933		80	43	9.70	132.70	167

31 37 Riprap

31 37 13 – Machined Riprap

31 37 13.10 Riprap and Rock Lining		Crew	Daily Output	Labor-Hours	Unit	Material	Labor	Equipment	Total	Total Incl O&P
0010	**RIPRAP AND ROCK LINING**									
0011	Random, broken stone									
0100	Machine placed for slope protection	B-12G	62	.258	L.C.Y.	32	13.05	13.85	58.90	71
0110	3/8 to 1/4 C.Y. pieces, grouted	B-13	80	.700	S.Y.	64.50	32	7.25	103.75	130
0200	18″ minimum thickness, not grouted	″	53	1.057	″	19.85	48.50	10.95	79.30	112
0300	Dumped, 50 lb. average	B-11A	800	.020	Ton	28.50	.99	1.88	31.37	35
0350	100 lb. average		700	.023		28.50	1.13	2.15	31.78	35.50
0370	300 lb. average		600	.027		28.50	1.32	2.51	32.33	36.50

31 41 Shoring

31 41 13 – Timber Shoring

31 41 13.10 Building Shoring		Crew	Daily Output	Labor-Hours	Unit	Material	Labor	Equipment	Total	Total Incl O&P
0010	**BUILDING SHORING**									
0020	Shoring, existing building, with timber, no salvage allowance	B-51	2.20	21.818	M.B.F.	860	945	89	1,894	2,575
1000	On cribbing with 35 ton screw jacks, per box and jack		3.60	13.333	Jack	65	575	54.50	694.50	1,050
1090	Minimum labor/equipment charge		2	24	Ea.		1,050	98	1,148	1,775

31 41 16 – Sheet Piling

31 41 16.10 Sheet Piling Systems		Crew	Daily Output	Labor-Hours	Unit	Material	Labor	Equipment	Total	Total Incl O&P
0010	**SHEET PILING SYSTEMS**									
0020	Sheet piling, 50,000 psi steel, not incl. wales, 22 psf, left in place	B-40	10.81	5.920	Ton	1,875	325	320	2,520	2,925
0100	Drive, extract & salvage		6	10.667		525	590	580	1,695	2,150
0300	20′ deep excavation, 27 psf, left in place		12.95	4.942		1,875	273	269	2,417	2,775
0400	Drive, extract & salvage		6.55	9.771		525	540	530	1,595	2,025
0600	25′ deep excavation, 38 psf, left in place		19	3.368		1,875	186	183	2,244	2,550
0700	Drive, extract & salvage		10.50	6.095		525	335	330	1,190	1,475
0900	40′ deep excavation, 38 psf, left in place		21.20	3.019		1,875	167	164	2,206	2,500
1000	Drive, extract & salvage		12.25	5.224		525	289	284	1,098	1,350
1200	15′ deep excavation, 22 psf, left in place		983	.065	S.F.	21.50	3.60	3.54	28.64	33.50
1300	Drive, extract & salvage		545	.117		5.90	6.50	6.40	18.80	24
1500	20′ deep excavation, 27 psf, left in place		960	.067		27.50	3.68	3.62	34.80	40
1600	Drive, extract & salvage		485	.132		7.65	7.30	7.15	22.10	28
1800	25′ deep excavation, 38 psf, left in place		1000	.064		40	3.54	3.48	47.02	53.50
1900	Drive, extract & salvage		553	.116		10.50	6.40	6.30	23.20	28.50
2100	Rent steel sheet piling and wales, first month				Ton	335			335	370
2200	Per added month					33.50			33.50	37
2300	Rental piling left in place, add to rental					1,200			1,200	1,325
2500	Wales, connections & struts, 2/3 salvage					535			535	585
2700	High strength piling, 60,000 psi, add					187			187	206
2800	65,000 psi, add					280			280	310

31 41 Shoring

31 41 16 – Sheet Piling

31 41 16.10 Sheet Piling Systems		Crew	Daily Output	Labor-Hours	Unit	2020 Bare Costs Material	Labor	Equipment	Total	Total Incl O&P
3000	Tie rod, not upset, 1-1/2" to 4" diameter with turnbuckle				Ton	2,275			2,275	2,500
3100	No turnbuckle					1,650			1,650	1,800
3300	Upset, 1-3/4" to 4" diameter with turnbuckle					2,400			2,400	2,650
3400	No turnbuckle					2,125			2,125	2,325
3600	Lightweight, 18" to 28" wide, 7 ga., 9.22 psf, and									
3610	9 ga., 8.6 psf, minimum				Lb.	.73			.73	.80
3700	Average					.92			.92	1.01
3750	Maximum					1.02			1.02	1.12
3900	Wood, solid sheeting, incl. wales, braces and spacers,									
3910	drive, extract & salvage, 8' deep excavation	B-31	330	.121	S.F.	2.06	5.40	.78	8.24	11.80
4000	10' deep, 50 S.F./hr. in & 150 S.F./hr. out		300	.133		2.12	5.95	.86	8.93	12.80
4100	12' deep, 45 S.F./hr. in & 135 S.F./hr. out		270	.148		2.18	6.60	.95	9.73	14.05
4200	14' deep, 42 S.F./hr. in & 126 S.F./hr. out		250	.160		2.24	7.15	1.03	10.42	15.05
4300	16' deep, 40 S.F./hr. in & 120 S.F./hr. out		240	.167		2.31	7.45	1.07	10.83	15.65
4400	18' deep, 38 S.F./hr. in & 114 S.F./hr. out		230	.174		2.39	7.80	1.12	11.31	16.30
4500	20' deep, 35 S.F./hr. in & 105 S.F./hr. out		210	.190		2.47	8.50	1.22	12.19	17.70
4520	Left in place, 8' deep, 55 S.F./hr.		440	.091		3.70	4.06	.58	8.34	11.20
4540	10' deep, 50 S.F./hr.		400	.100		3.90	4.47	.64	9.01	12.15
4560	12' deep, 45 S.F./hr.		360	.111		4.11	4.97	.71	9.79	13.25
4565	14' deep, 42 S.F./hr.		335	.119		4.35	5.35	.77	10.47	14.20
4570	16' deep, 40 S.F./hr.		320	.125		4.63	5.60	.80	11.03	14.95
4580	18' deep, 38 S.F./hr.		305	.131		4.94	5.85	.84	11.63	15.80
4590	20' deep, 35 S.F./hr.		280	.143		5.30	6.40	.92	12.62	17.05
4700	Alternate pricing, left in place, 8' deep		1.76	22.727	M.B.F.	830	1,025	146	2,001	2,700
4800	Drive, extract and salvage, 8' deep		1.32	30.303	"	740	1,350	195	2,285	3,200
4990	Minimum labor/equipment charge		2	20	Job		895	129	1,024	1,575
5000	For treated lumber add cost of treatment to lumber									

31 43 Concrete Raising

31 43 13 – Pressure Grouting

31 43 13.13 Concrete Pressure Grouting		Crew	Daily Output	Labor-Hours	Unit	Material	Labor	Equipment	Total	Total Incl O&P
0010	**CONCRETE PRESSURE GROUTING**									
0020	Grouting, pressure, cement & sand, 1:1 mix, minimum	B-61	124	.323	Bag	19.30	14.40	2.49	36.19	47
0100	Maximum		51	.784	"	19.30	35	6.05	60.35	84
0200	Cement and sand, 1:1 mix, minimum		250	.160	C.F.	38.50	7.15	1.23	46.88	55.50
0300	Maximum		100	.400		58	17.85	3.08	78.93	96
0400	Epoxy cement grout, minimum		137	.292		785	13.05	2.25	800.30	890
0500	Maximum		57	.702		785	31.50	5.40	821.90	920
0700	Alternate pricing method: (Add for materials)									
0710	5 person crew and equipment	B-61	1	40	Day		1,775	310	2,085	3,200

31 45 Vibroflotation and Densification

31 45 13 – Vibroflotation

31 45 13.10 Vibroflotation Densification		Crew	Daily Output	Labor-Hours	Unit	Material	2020 Bare Costs Labor	Equipment	Total	Total Incl O&P
0010	VIBROFLOTATION DENSIFICATION									
0900	Vibroflotation compacted sand cylinder, minimum	B-60	750	.075	V.L.F.		3.67	3	6.67	9.10
0950	Maximum		325	.172			8.45	6.90	15.35	21
1100	Vibro replacement compacted stone cylinder, minimum		500	.112			5.50	4.49	9.99	13.65
1150	Maximum		250	.224			11	9	20	27.50
1300	Mobilization and demobilization, minimum		.47	119	Total		5,850	4,775	10,625	14,500
1400	Maximum		.14	400	"		19,700	16,000	35,700	48,600

31 46 Needle Beams

31 46 13 – Cantilever Needle Beams

31 46 13.10 Needle Beams		Crew	Daily Output	Labor-Hours	Unit	Material	Labor	Equipment	Total	Total Incl O&P
0010	NEEDLE BEAMS									
0011	Incl. wood shoring 10′ x 10′ opening									
0400	Block, concrete, 8″ thick	B-9	7.10	5.634	Ea.	59	239	50.50	348.50	505
0420	12″ thick		6.70	5.970		68	254	53.50	375.50	540
0800	Brick, 4″ thick with 8″ backup block		5.70	7.018		68	298	62.50	428.50	625
1000	Brick, solid, 8″ thick		6.20	6.452		59	274	57.50	390.50	570
1040	12″ thick		4.90	8.163		68	345	73	486	710
1080	16″ thick		4.50	8.889		87	380	79.50	546.50	790
2000	Add for additional floors of shoring	B-1	6	4		59	171		230	340
9000	Minimum labor/equipment charge	"	2	12	Job		515		515	820

31 48 Underpinning

31 48 13 – Underpinning Piers

31 48 13.10 Underpinning Foundations		Crew	Daily Output	Labor-Hours	Unit	Material	Labor	Equipment	Total	Total Incl O&P
0010	UNDERPINNING FOUNDATIONS									
0011	Including excavation,									
0020	forming, reinforcing, concrete and equipment									
0100	5′ to 16′ below grade, 100 to 500 C.Y.	B-52	2.30	24.348	C.Y.	340	1,175	254	1,769	2,550
0200	Over 500 C.Y.		2.50	22.400		305	1,100	234	1,639	2,350
0400	16′ to 25′ below grade, 100 to 500 C.Y.		2	28		370	1,375	292	2,037	2,900
0500	Over 500 C.Y.		2.10	26.667		350	1,300	278	1,928	2,775
0700	26′ to 40′ below grade, 100 to 500 C.Y.		1.60	35		405	1,700	365	2,470	3,575
0800	Over 500 C.Y.		1.80	31.111		370	1,525	325	2,220	3,200
0900	For under 50 C.Y., add					10%	40%			
1000	For 50 C.Y. to 100 C.Y., add					5%	20%			

31 52 Cofferdams

31 52 16 – Timber Cofferdams

31 52 16.10 Cofferdams		Crew	Daily Output	Labor-Hours	Unit	Material	Labor	Equipment	Total	Total Incl O&P
0010	COFFERDAMS									
0011	Incl. mobilization and temporary sheeting									
0080	Soldier beams & lagging H-piles with 3″ wood sheeting									
0090	horizontal between piles, including removal of wales & braces									
0100	No hydrostatic head, 15′ deep, 1 line of braces, minimum	B-50	545	.206	S.F.	8.60	10.75	4.72	24.07	32
0200	Maximum		495	.226		9.55	11.85	5.20	26.60	35
0400	15′ to 22′ deep with 2 lines of braces, 10″ H, minimum		360	.311		10.15	16.30	7.15	33.60	45

31 52 Cofferdams

31 52 16 – Timber Cofferdams

31 52 16.10 Cofferdams		Crew	Daily Output	Labor-Hours	Unit	Material	2020 Bare Costs Labor	Equipment	Total	Total Incl O&P
0500	Maximum	B-50	330	.339	S.F.	11.50	17.75	7.80	37.05	49.50
0700	23' to 35' deep with 3 lines of braces, 12" H, minimum		325	.345		13.25	18.05	7.90	39.20	52.50
0800	Maximum		295	.380		14.35	19.90	8.70	42.95	57
1000	36' to 45' deep with 4 lines of braces, 14" H, minimum		290	.386		14.85	20	8.85	43.70	58
1100	Maximum		265	.423		15.65	22	9.70	47.35	63.50
1300	No hydrostatic head, left in place, 15' deep, 1 line of braces, min.		635	.176		11.50	9.25	4.05	24.80	32
1400	Maximum		575	.195		12.30	10.20	4.47	26.97	34.50
1600	15' to 22' deep with 2 lines of braces, minimum		455	.246		17.20	12.90	5.65	35.75	45.50
1700	Maximum		415	.270		19.15	14.15	6.20	39.50	50.50
1900	23' to 35' deep with 3 lines of braces, minimum		420	.267		20.50	13.95	6.10	40.55	52
2000	Maximum		380	.295		22.50	15.45	6.75	44.70	57
2200	36' to 45' deep with 4 lines of braces, minimum		385	.291		24.50	15.25	6.70	46.45	59
2300	Maximum		350	.320		28.50	16.75	7.35	52.60	66
2350	Lagging only, 3" thick wood between piles 8' OC, minimum	B-46	400	.120		1.91	5.80	.10	7.81	11.55
2370	Maximum		250	.192		2.87	9.30	.17	12.34	18.25
2400	Open sheeting no bracing, for trenches to 10' deep, min.		1736	.028		.86	1.34	.02	2.22	3.13
2450	Maximum		1510	.032		.96	1.54	.03	2.53	3.55
2500	Tie-back method, add to open sheeting, add, minimum								20%	20%
2550	Maximum								60%	60%
2700	Tie-backs only, based on tie-backs total length, minimum	B-46	86.80	.553	L.F.	17.25	27	.48	44.73	62.50
2750	Maximum		38.50	1.247	"	30.50	60.50	1.07	92.07	132
3500	Tie-backs only, typical average, 25' long		2	24	Ea.	760	1,175	20.50	1,955.50	2,725
3600	35' long		1.58	30.380	"	1,025	1,475	26	2,526	3,500
4500	Trench box, 7' deep, 16' x 8', see 01 54 33 in Reference Section				Day				191	210
4600	20' x 10', see 01 54 33 in Reference Section				"				240	265
5200	Wood sheeting, in trench, jacks at 4' OC, 8' deep	B-1	800	.030	S.F.	.66	1.28		1.94	2.79
5250	12' deep		700	.034		.78	1.47		2.25	3.21
5300	15' deep		600	.040		1.08	1.71		2.79	3.92
6000	See also Section 31 41 16.10									

31 56 Slurry Walls

31 56 23 – Lean Concrete Slurry Walls

31 56 23.20 Slurry Trench		Crew	Daily Output	Labor-Hours	Unit	Material	Labor	Equipment	Total	Total Incl O&P
0010	**SLURRY TRENCH**									
0011	Excavated slurry trench in wet soils									
0020	backfilled with 3,000 psi concrete, no reinforcing steel									
0050	Minimum	C-7	333	.216	C.F.	9.55	9.90	3.24	22.69	30
0100	Maximum		200	.360	"	16	16.45	5.40	37.85	49.50
0200	Alternate pricing method, minimum		150	.480	S.F.	19.10	22	7.20	48.30	64
0300	Maximum		120	.600		28.50	27.50	9	65	85
0500	Reinforced slurry trench, minimum	B-48	177	.316		14.30	15.05	5.90	35.25	46.50
0600	Maximum	"	69	.812		47.50	38.50	15.20	101.20	130
0800	Haul for disposal, 2 mile haul, excavated material, add	B-34B	99	.081	C.Y.		3.96	5.80	9.76	12.65
0900	Haul bentonite castings for disposal, add	"	40	.200	"		9.80	14.30	24.10	31.50

31 62 Driven Piles

31 62 13 – Concrete Piles

	31 62 13.23 Prestressed Concrete Piles	Crew	Daily Output	Labor-Hours	Unit	2020 Bare Costs Material	Labor	Equipment	Total	Total Incl O&P
0010	**PRESTRESSED CONCRETE PILES**, 200 piles									
0020	Unless specified otherwise, not incl. pile caps or mobilization									
2200	Precast, prestressed, 50′ long, cylinder, 12″ diam., 2-3/8″ wall	B-19	720	.089	V.L.F.	29.50	4.91	2.79	37.20	43
2300	14″ diameter, 2-1/2″ wall		680	.094		38	5.20	2.96	46.16	53.50
2500	16″ diameter, 3″ wall		640	.100		50.50	5.50	3.14	59.14	68
2600	18″ diameter, 3-1/2″ wall	B-19A	600	.107		63.50	5.90	4.91	74.31	85
2800	20″ diameter, 4″ wall		560	.114		55.50	6.30	5.25	67.05	77
2900	24″ diameter, 5″ wall		520	.123		75.50	6.80	5.65	87.95	100
2920	36″ diameter, 5-1/2″ wall	B-19	400	.160		99.50	8.85	5	113.35	129
2940	54″ diameter, 6″ wall		340	.188		236	10.40	5.90	252.30	283
2950	60″ diameter, 6″ wall		280	.229		251	12.65	7.20	270.85	305
2960	66″ diameter, 6-1/2″ wall		220	.291		340	16.05	9.15	365.20	405
3100	Precast, prestressed, 40′ long, 10″ thick, square		700	.091		14.10	5.05	2.87	22.02	26.50
3200	12″ thick, square		680	.094		26	5.20	2.96	34.16	40
3400	14″ thick, square		600	.107		29	5.90	3.35	38.25	45
3500	Octagonal		640	.100		28	5.50	3.14	36.64	43.50
3700	16″ thick, square		560	.114		30	6.30	3.59	39.89	47
3800	Octagonal		600	.107		33.50	5.90	3.35	42.75	50
4000	18″ thick, square	B-19A	520	.123		52	6.80	5.65	64.45	74
4100	Octagonal	B-19	560	.114		45	6.30	3.59	54.89	63.50
4300	20″ thick, square	B-19A	480	.133		47.50	7.35	6.15	61	71
4400	Octagonal	B-19	520	.123		50.50	6.80	3.86	61.16	70.50
4600	24″ thick, square	B-19A	440	.145		57.50	8.05	6.70	72.25	83
4700	Octagonal	B-19	480	.133		59.50	7.35	4.19	71.04	82
4730	Precast, prestressed, 60′ long, 10″ thick, square		700	.091		14.85	5.05	2.87	22.77	27.50
4740	12″ thick, square (60′ long)		680	.094		27	5.20	2.96	35.16	41
4750	Mobilization for 10,000 L.F. pile job, add		3300	.019			1.07	.61	1.68	2.37
4800	25,000 L.F. pile job, add		8500	.008			.42	.24	.66	.92

31 62 16 – Steel Piles

	31 62 16.13 Steel Piles	Crew	Daily Output	Labor-Hours	Unit	2020 Bare Costs Material	Labor	Equipment	Total	Total Incl O&P
0010	**STEEL PILES**									
0100	Step tapered, round, concrete filled									
0110	8″ tip, 12″ butt, 60 ton capacity, 30′ depth	B-19	760	.084	V.L.F.	18.85	4.65	2.64	26.14	31
0120	60′ depth with extension		740	.086		36.50	4.78	2.72	44	51
0130	80′ depth with extensions		700	.091		57	5.05	2.87	64.92	73.50
0250	"H" Sections, 50′ long, HP8 x 36		640	.100		18.35	5.50	3.14	26.99	32.50
0400	HP10 x 42		610	.105		20.50	5.80	3.29	29.59	35.50
0500	HP10 x 57		610	.105		28	5.80	3.29	37.09	44
0700	HP12 x 53		590	.108		27.50	6	3.41	36.91	43.50
0800	HP12 x 74	B-19A	590	.108		35	6	4.99	45.99	53.50
1000	HP14 x 73		540	.119		37	6.55	5.45	49	57
1100	HP14 x 89		540	.119		47.50	6.55	5.45	59.50	68.50
1300	HP14 x 102		510	.125		54	6.95	5.80	66.75	77
1400	HP14 x 117		510	.125		62.50	6.95	5.80	75.25	86
1600	Splice on standard points, not in leads, 8″ or 10″	1 Sswl	5	1.600	Ea.	111	92		203	273
1700	12″ or 14″		4	2		155	115		270	360
1900	Heavy duty points, not in leads, 10″ wide		4	2		186	115		301	395
2100	14″ wide		3.50	2.286		228	131		359	465

31 62 19 – Timber Piles

31 62 19.10 Wood Piles		Crew	Daily Output	Labor-Hours	Unit	Material	2020 Bare Costs Labor	Equipment	Total	Total Incl O&P
0010	**WOOD PILES**									
0011	Friction or end bearing, not including									
0050	mobilization or demobilization									
0100	ACZA Treated piles, 1.0 lb/C.F., up to 30′ long, 12″ butts, 8″ points	B-19	625	.102	V.L.F.	13.70	5.65	3.22	22.57	27.50
0200	30′ to 39′ long, 12″ butts, 8″ points		700	.091		14.45	5.05	2.87	22.37	27
0300	40′ to 49′ long, 12″ butts, 7″ points		720	.089		15.20	4.91	2.79	22.90	27.50
0400	50′ to 59′ long, 13″ butts, 7″ points		800	.080		21.50	4.42	2.51	28.43	33.50
0500	60′ to 69′ long, 13″ butts, 7″ points		840	.076		24.50	4.21	2.39	31.10	36.50
0600	70′ to 80′ long, 13″ butts, 6″ points	↓	840	.076	↓	25	4.21	2.39	31.60	36.50
0800	ACZA Treated piles, 1.5 lb./C.F.									
0810	friction or end bearing, ASTM class B									
1000	Up to 30′ long, 12″ butts, 8″ points	B-19	625	.102	V.L.F.	18.20	5.65	3.22	27.07	32.50
1100	30′ to 39′ long, 12″ butts, 8″ points		700	.091		18.90	5.05	2.87	26.82	32
1200	40′ to 49′ long, 12″ butts, 7″ points		720	.089		20.50	4.91	2.79	28.20	33.50
1300	50′ to 59′ long, 13″ butts, 7″ points	↓	800	.080		23.50	4.42	2.51	30.43	36
1400	60′ to 69′ long, 13″ butts, 6″ points	B-19A	840	.076		25.50	4.21	3.51	33.22	38.50
1500	70′ to 80′ long, 13″ butts, 6″ points	″	840	.076	↓	28	4.21	3.51	35.72	41
1600	ACZA Treated piles, 2.5 lb./C.F.									
1610	8″ butts, 10′ long	B-19	400	.160	V.L.F.	7.35	8.85	5	21.20	27.50
1620	11′ to 16′ long		500	.128		7.35	7.05	4.02	18.42	23.50
1630	17′ to 20′ long		575	.111		7.35	6.15	3.49	16.99	21.50
1640	10″ butts, 10′ to 16′ long		500	.128		22	7.05	4.02	33.07	39.50
1650	17′ to 20′ long		575	.111		22	6.15	3.49	31.64	37.50
1660	21′ to 40′ long		700	.091		22	5.05	2.87	29.92	35
1670	12″ butts, 10′ to 20′ long		575	.111		23	6.15	3.49	32.64	39
1680	21′ to 35′ long		650	.098		23	5.45	3.09	31.54	37.50
1690	36′ to 40′ long		700	.091		23.50	5.05	2.87	31.42	37
1695	14″ butts, to 40′ long	↓	700	.091	↓	16.90	5.05	2.87	24.82	30
1700	Boot for pile tip, minimum	1 Pile	27	.296	Ea.	45.50	16.05		61.55	76
1800	Maximum		21	.381		136	20.50		156.50	182
2000	Point for pile tip, minimum		20	.400		45.50	21.50		67	85
2100	Maximum	↓	15	.533		163	29		192	226
2300	Splice for piles over 50′ long, minimum	B-46	35	1.371		59	66.50	1.18	126.68	173
2400	Maximum		20	2.400	↓	75.50	116	2.06	193.56	272
2600	Concrete encasement with wire mesh and tube	↓	331	.145	V.L.F.	75.50	7.05	.12	82.67	94.50
2700	Mobilization for 10,000 L.F. pile job, add	B-19	3300	.019			1.07	.61	1.68	2.37
2800	25,000 L.F. pile job, add	″	8500	.008	↓		.42	.24	.66	.92

31 62 23 – Composite Piles

31 62 23.13 Concrete-Filled Steel Piles		Crew	Daily Output	Labor-Hours	Unit	Material	Labor	Equipment	Total	Total Incl O&P
0010	**CONCRETE-FILLED STEEL PILES** no mobilization or demobilization									
2600	Pipe piles, 50′ L, 8″ diam., 29 lb./L.F., no concrete	B-19	500	.128	V.L.F.	21.50	7.05	4.02	32.57	39
2700	Concrete filled		460	.139		25.50	7.70	4.37	37.57	45
2900	10″ diameter, 34 lb./L.F., no concrete		500	.128		24	7.05	4.02	35.07	42
3000	Concrete filled		450	.142		30	7.85	4.47	42.32	50.50
3200	12″ diameter, 44 lb./L.F., no concrete		475	.135		31.50	7.45	4.23	43.18	51
3300	Concrete filled		415	.154		35.50	8.50	4.84	48.84	58
3500	14″ diameter, 46 lb./L.F., no concrete		430	.149		33.50	8.20	4.67	46.37	55
3600	Concrete filled		355	.180		41.50	9.95	5.65	57.10	67.50
3800	16″ diameter, 52 lb./L.F., no concrete		385	.166		38	9.20	5.20	52.40	62
3900	Concrete filled		335	.191		52	10.55	6	68.55	81
4100	18″ diameter, 59 lb./L.F., no concrete		355	.180		47	9.95	5.65	62.60	73.50
4200	Concrete filled	↓	310	.206	↓	61.50	11.40	6.50	79.40	93.50

31 62 Driven Piles

31 62 23 – Composite Piles

31 62 23.13 Concrete-Filled Steel Piles		Crew	Daily Output	Labor-Hours	Unit	2020 Bare Costs Material	Labor	Equipment	Total	Total Incl O&P
4400	Splices for pipe piles, stl., not in leads, 8″ diameter	1 Sswl	5	1.600	Ea.	81	92		173	240
4410	10″ diameter		4.75	1.684		88.50	97		185.50	256
4430	12″ diameter		4.50	1.778		106	102		208	283
4500	14″ diameter		4.25	1.882		146	108		254	340
4600	16″ diameter		4	2		165	115		280	370
4650	18″ diameter		3.75	2.133		274	123		397	500
4710	Steel pipe pile backing rings, w/spacer, 8″ diameter		12	.667		9.10	38.50		47.60	73
4720	10″ diameter		12	.667		11.90	38.50		50.40	76
4730	12″ diameter		10	.800		14.35	46		60.35	91.50
4740	14″ diameter		9	.889		18.35	51		69.35	104
4750	16″ diameter		8	1		21.50	57.50		79	118
4760	18″ diameter		6	1.333		24	76.50		100.50	152
4800	Points, standard, 8″ diameter		4.61	1.735		93	99.50		192.50	265
4840	10″ diameter		4.45	1.798		123	103		226	305
4880	12″ diameter		4.25	1.882		164	108		272	355
4900	14″ diameter		4.05	1.975		209	113		322	415
5000	16″ diameter		3.37	2.374		296	136		432	550
5050	18″ diameter		3.50	2.286		385	131		516	635
5200	Points, heavy duty, 10″ diameter		2.90	2.759		242	158		400	525
5240	12″ diameter		2.95	2.712		305	156		461	590
5260	14″ diameter		2.95	2.712		305	156		461	590
5280	16″ diameter		2.95	2.712		350	156		506	640
5290	18″ diameter		2.80	2.857		425	164		589	740
5500	For reinforcing steel, add		1150	.007	Lb.	.79	.40		1.19	1.53
5700	For thick wall sections, add				″	.68			.68	.75
6020	Steel pipe pile end plates, 8″ diameter	1 Sswl	14	.571	Ea.	31	33		64	88
6050	10″ diameter		14	.571		40.50	33		73.50	98.50
6100	12″ diameter		12	.667		52	38.50		90.50	120
6150	14″ diameter		10	.800		60.50	46		106.50	142
6200	16″ diameter		9	.889		76	51		127	167
6250	18″ diameter		8	1		100	57.50		157.50	204
6300	Steel pipe pile shoes, 8″ diameter		12	.667		65.50	38.50		104	135
6350	10″ diameter		12	.667		82	38.50		120.50	153
6400	12″ diameter		10	.800		97	46		143	183
6450	14″ diameter		9	.889		143	51		194	241
6500	16″ diameter		8	1		149	57.50		206.50	258
6550	18″ diameter		6	1.333		217	76.50		293.50	365

31 63 Bored Piles

31 63 26 – Drilled Caissons

31 63 26.13 Fixed End Caisson Piles		Crew	Daily Output	Labor-Hours	Unit	Material	Labor	Equipment	Total	Total Incl O&P
0010	**FIXED END CAISSON PILES**									
0015	Including excavation, concrete, 50 lb. reinforcing									
0020	per C.Y., not incl. mobilization, boulder removal, disposal									
0100	Open style, machine drilled, to 50′ deep, in stable ground, no									
0110	casings or ground water, 18″ diam., 0.065 C.Y./L.F.	B-43	200	.240	V.L.F.	10.20	11.20	3.85	25.25	33
0200	24″ diameter, 0.116 C.Y./L.F.		190	.253		18.20	11.80	4.06	34.06	43
0300	30″ diameter, 0.182 C.Y./L.F.		150	.320		28.50	14.95	5.15	48.60	60.50
0400	36″ diameter, 0.262 C.Y./L.F.		125	.384		41	17.95	6.15	65.10	80.50
0500	48″ diameter, 0.465 C.Y./L.F.		100	.480		73	22.50	7.70	103.20	125
0600	60″ diameter, 0.727 C.Y./L.F.		90	.533		114	25	8.55	147.55	175

31 63 Bored Piles

31 63 26 – Drilled Caissons

	31 63 26.13 Fixed End Caisson Piles	Crew	Daily Output	Labor-Hours	Unit	Material	2020 Bare Costs Labor	Equipment	Total	Total Incl O&P
0700	72" diameter, 1.05 C.Y./L.F.	B-43	80	.600	V.L.F.	165	28	9.65	202.65	236
0800	84" diameter, 1.43 C.Y./L.F.		75	.640		225	30	10.30	265.30	305
1000	For bell excavation and concrete, add									
1020	4' bell diameter, 24" shaft, 0.444 C.Y.	B-43	20	2.400	Ea.	57	112	38.50	207.50	284
1040	6' bell diameter, 30" shaft, 1.57 C.Y.		5.70	8.421		202	395	135	732	995
1060	8' bell diameter, 36" shaft, 3.72 C.Y.		2.40	20		480	935	320	1,735	2,350
1080	9' bell diameter, 48" shaft, 4.48 C.Y.		2	24		575	1,125	385	2,085	2,825
1100	10' bell diameter, 60" shaft, 5.24 C.Y.		1.70	28.235		675	1,325	455	2,455	3,350
1120	12' bell diameter, 72" shaft, 8.74 C.Y.		1	48		1,125	2,250	770	4,145	5,650
1140	14' bell diameter, 84" shaft, 13.6 C.Y.		.70	68.571		1,750	3,200	1,100	6,050	8,200
1200	Open style, machine drilled, to 50' deep, in wet ground, pulled									
1300	casing and pumping, 18" diameter, 0.065 C.Y./L.F.	B-48	160	.350	V.L.F.	10.20	16.65	6.55	33.40	45
1400	24" diameter, 0.116 C.Y./L.F.		125	.448		18.20	21.50	8.40	48.10	63
1500	30" diameter, 0.182 C.Y./L.F.		85	.659		28.50	31.50	12.30	72.30	94.50
1600	36" diameter, 0.262 C.Y./L.F.		60	.933		41	44.50	17.45	102.95	135
1700	48" diameter, 0.465 C.Y./L.F.	B-49	55	1.600		73	80	29.50	182.50	240
1800	60" diameter, 0.727 C.Y./L.F.		35	2.514		114	126	46.50	286.50	375
1900	72" diameter, 1.05 C.Y./L.F.		30	2.933		165	147	54.50	366.50	475
2000	84" diameter, 1.43 C.Y./L.F.		25	3.520		225	176	65	466	600
2100	For bell excavation and concrete, add									
2120	4' bell diameter, 24" shaft, 0.444 C.Y.	B-48	19.80	2.828	Ea.	57	135	53	245	335
2140	6' bell diameter, 30" shaft, 1.57 C.Y.		5.70	9.825		202	470	184	856	1,175
2160	8' bell diameter, 36" shaft, 3.72 C.Y.		2.40	23.333		480	1,100	435	2,015	2,750
2180	9' bell diameter, 48" shaft, 4.48 C.Y.	B-49	3.30	26.667		575	1,325	495	2,395	3,300
2200	10' bell diameter, 60" shaft, 5.24 C.Y.		2.80	31.429		675	1,575	580	2,830	3,875
2220	12' bell diameter, 72" shaft, 8.74 C.Y.		1.60	55		1,125	2,750	1,025	4,900	6,750
2240	14' bell diameter, 84" shaft, 13.6 C.Y.		1	88		1,750	4,400	1,625	7,775	10,700
2300	Open style, machine drilled, to 50' deep, in soft rocks and									
2400	medium hard shales, 18" diameter, 0.065 C.Y./L.F.	B-49	50	1.760	V.L.F.	10.20	88	32.50	130.70	187
2500	24" diameter, 0.116 C.Y./L.F.		30	2.933		18.20	147	54.50	219.70	315
2600	30" diameter, 0.182 C.Y./L.F.		20	4.400		28.50	221	81.50	331	470
2700	36" diameter, 0.262 C.Y./L.F.		15	5.867		41	294	109	444	630
2800	48" diameter, 0.465 C.Y./L.F.		10	8.800		73	440	163	676	960
2900	60" diameter, 0.727 C.Y./L.F.		7	12.571		114	630	233	977	1,375
3000	72" diameter, 1.05 C.Y./L.F.		6	14.667		165	735	271	1,171	1,650
3100	84" diameter, 1.43 C.Y./L.F.		5	17.600		225	880	325	1,430	2,000
3200	For bell excavation and concrete, add									
3220	4' bell diameter, 24" shaft, 0.444 C.Y.	B-49	10.90	8.073	Ea.	57	405	149	611	865
3240	6' bell diameter, 30" shaft, 1.57 C.Y.		3.10	28.387		202	1,425	525	2,152	3,050
3260	8' bell diameter, 36" shaft, 3.72 C.Y.		1.30	67.692		480	3,400	1,250	5,130	7,275
3280	9' bell diameter, 48" shaft, 4.48 C.Y.		1.10	80		575	4,000	1,475	6,050	8,600
3300	10' bell diameter, 60" shaft, 5.24 C.Y.		.90	97.778		675	4,900	1,800	7,375	10,500
3320	12' bell diameter, 72" shaft, 8.74 C.Y.		.60	147		1,125	7,350	2,725	11,200	15,800
3340	14' bell diameter, 84" shaft, 13.6 C.Y.		.40	220		1,750	11,000	4,075	16,825	23,900
3600	For rock excavation, sockets, add, minimum		120	.733	C.F.		37	13.55	50.55	73
3650	Average		95	.926			46.50	17.15	63.65	92.50
3700	Maximum		48	1.833			92	34	126	184
3900	For 50' to 100' deep, add				V.L.F.				7%	7%
4000	For 100' to 150' deep, add								25%	25%
4100	For 150' to 200' deep, add								30%	30%
4200	For casings left in place, add				Lb.	1.38			1.38	1.52
4300	For other than 50 lb. reinf. per C.Y., add or deduct				"	1.29			1.29	1.42
4400	For steel I-beam cores, add	B-49	8.30	10.602	Ton	2,300	530	196	3,026	3,575

31 63 Bored Piles

31 63 26 – Drilled Caissons

31 63 26.13 Fixed End Caisson Piles		Crew	Daily Output	Labor-Hours	Unit	Material	2020 Bare Costs Labor	Equipment	Total	Total Incl O&P
4500	Load and haul excess excavation, 2 miles	B-34B	178	.045	L.C.Y.		2.20	3.22	5.42	7.05

31 63 26.16 Concrete Caissons for Marine Construction

		Crew	Daily Output	Labor-Hours	Unit	Material	Labor	Equipment	Total	Total Incl O&P
0010	CONCRETE CAISSONS FOR MARINE CONSTRUCTION									
0100	Caissons, incl. mobilization and demobilization, up to 50 miles									
0200	Uncased shafts, 30 to 80 tons cap., 17" diam., 10' depth	B-44	88	.727	V.L.F.	26	39.50	23	88.50	117
0300	25' depth		165	.388		18.40	21	12.30	51.70	67
0400	80 to 150 ton capacity, 22" diameter, 10' depth		80	.800		32	43.50	25.50	101	133
0500	20' depth		130	.492		26	26.50	15.65	68.15	88
0700	Cased shafts, 10 to 30 ton capacity, 10-5/8" diam., 20' depth		175	.366		18.40	19.80	11.60	49.80	64.50
0800	30' depth		240	.267		17.15	14.45	8.45	40.05	51
0850	30 to 60 ton capacity, 12" diameter, 20' depth		160	.400		26	21.50	12.70	60.20	77
0900	40' depth		230	.278		19.80	15.10	8.85	43.75	55.50
1000	80 to 100 ton capacity, 16" diameter, 20' depth		160	.400		37	21.50	12.70	71.20	89
1100	40' depth		230	.278		34.50	15.10	8.85	58.45	71.50
1200	110 to 140 ton capacity, 17-5/8" diameter, 20' depth		160	.400		39.50	21.50	12.70	73.70	92
1300	40' depth		230	.278		37	15.10	8.85	60.95	74
1400	140 to 175 ton capacity, 19" diameter, 20' depth		130	.492		43	26.50	15.65	85.15	107
1500	40' depth	↓	210	.305	↓	39.50	16.50	9.70	65.70	80.50
1700	Over 30' long, L.F. cost tends to be lower									
1900	Maximum depth is about 90'									

31 63 29 – Drilled Concrete Piers and Shafts

31 63 29.13 Uncased Drilled Concrete Piers

		Crew	Daily Output	Labor-Hours	Unit	Material	Labor	Equipment	Total	Total Incl O&P
0010	UNCASED DRILLED CONCRETE PIERS									
0020	Unless specified otherwise, not incl. pile caps or mobilization									
0100	Cast in place, thin wall shell pile, straight sided,									
0110	not incl. reinforcing, 8" diam., 16 ga., 5.8 lb./L.F.	B-19	700	.091	V.L.F.	9.50	5.05	2.87	17.42	21.50
0200	10" diameter, 16 ga. corrugated, 7.3 lb./L.F.		650	.098		12.45	5.45	3.09	20.99	26
0300	12" diameter, 16 ga. corrugated, 8.7 lb./L.F.		600	.107		16.20	5.90	3.35	25.45	31
0400	14" diameter, 16 ga. corrugated, 10.0 lb./L.F.		550	.116		19.05	6.45	3.65	29.15	35
0500	16" diameter, 16 ga. corrugated, 11.6 lb./L.F.	↓	500	.128	↓	23.50	7.05	4.02	34.57	41
0800	Cast in place friction pile, 50' long, fluted,									
0810	tapered steel, 4,000 psi concrete, no reinforcing									
0900	12" diameter, 7 ga.	B-19	600	.107	V.L.F.	29.50	5.90	3.35	38.75	45.50
1000	14" diameter, 7 ga.		560	.114		32	6.30	3.59	41.89	49.50
1100	16" diameter, 7 ga.		520	.123		38	6.80	3.86	48.66	56.50
1200	18" diameter, 7 ga.	↓	480	.133	↓	44.50	7.35	4.19	56.04	65
1300	End bearing, fluted, constant diameter,									
1320	4,000 psi concrete, no reinforcing									
1340	12" diameter, 7 ga.	B-19	600	.107	V.L.F.	31	5.90	3.35	40.25	47
1360	14" diameter, 7 ga.		560	.114		39	6.30	3.59	48.89	56.50
1380	16" diameter, 7 ga.		520	.123		45	6.80	3.86	55.66	64.50
1400	18" diameter, 7 ga.	↓	480	.133	↓	49.50	7.35	4.19	61.04	71

31 63 29.20 Cast In Place Piles, Adds

		Crew	Daily Output	Labor-Hours	Unit	Material	Labor	Equipment	Total	Total Incl O&P
0010	CAST IN PLACE PILES, ADDS									
1500	For reinforcing steel, add				Lb.	1.13			1.13	1.24
1700	For ball or pedestal end, add	B-19	11	5.818	C.Y.	150	320	183	653	875
1900	For lengths above 60', concrete, add	"	11	5.818	"	157	320	183	660	885
2000	For steel thin shell, pipe only				Lb.	1.67			1.67	1.84

31 63 33 – Drilled Micropiles

	31 63 33.10 Drilled Micropiles Metal Pipe	Crew	Daily Output	Labor-Hours	Unit	Material	2020 Bare Costs Labor	Equipment	Total	Total Incl O&P
0010	**DRILLED MICROPILES METAL PIPE**									
0011	No mobilization or demobilization									
5000	Pressure grouted pin pile, 5" diam., cased, up to 50 ton,									
5040	End bearing, less than 20'	B-48	160	.350	V.L.F.	41	16.65	6.55	64.20	79
5080	More than 40'		240	.233		39	11.10	4.36	54.46	65.50
5120	Friction, loose sand and gravel		240	.233		41	11.10	4.36	56.46	68
5160	Dense sand and gravel		240	.233		41	11.10	4.36	56.46	68
5200	Uncased, up to 10 ton capacity, 20'	▼	200	.280	▼	22	13.35	5.25	40.60	51.50

Estimating Tips

32 01 00 Operations and Maintenance of Exterior Improvements

- Recycling of asphalt pavement is becoming very popular and is an alternative to removal and replacement. It can be a good value engineering proposal if removed pavement can be recycled, either at the project site or at another site that is reasonably close to the project site. Sections on repair of flexible and rigid pavement are included.

32 10 00 Bases, Ballasts, and Paving

- When estimating paving, keep in mind the project schedule. Also note that prices for asphalt and concrete are generally higher in the cold seasons. Lines for pavement markings, including tactile warning systems and fence lines, are included.

32 90 00 Planting

- The timing of planting and guarantee specifications often dictate the costs for establishing tree and shrub growth and a stand of grass or ground cover. Establish the work performance schedule to coincide with the local planting season. Maintenance and growth guarantees can add 20–100% to the total landscaping cost and can be contractually cumbersome. The cost to replace trees and shrubs can be as high as 5% of the total cost, depending on the planting zone, soil conditions, and time of year.

Reference Numbers

Reference numbers are shown at the beginning of some major classifications. These numbers refer to related items in the Reference Section. The reference information may be an estimating procedure, an alternate pricing method, or technical information.

Note: Not all subdivisions listed here necessarily appear. ■

32 01 Operation and Maintenance of Exterior Improvements

32 01 11 – Paving Cleaning

32 01 11.51 Rubber and Paint Removal From Paving		Crew	Daily Output	Labor-Hours	Unit	2020 Bare Costs Material	Labor	Equipment	Total	Total Incl O&P
0010	RUBBER AND PAINT REMOVAL FROM PAVING									
0015	Does not include traffic control costs									
0020	See other items in Section 32 17 23.13									
0100	Remove permanent painted traffic lines and markings	B-78A	500	.016	C.L.F.		.85	1.83	2.68	3.35
0200	Temporary traffic line tape	2 Clab	1500	.011	L.F.		.45		.45	.72
0300	Thermoplastic traffic lines and markings	B-79A	500	.024	C.L.F.		1.27	2.78	4.05	5.05
0400	Painted pavement markings	B-78B	500	.036	S.F.		1.56	.81	2.37	3.38
1000	Prepare and clean, 25,000 S.Y.	B-91A	7.36	2.174	Mile		103	112	215	288
2000	Minimum labor/equipment charge	1 Clab	2	4	Job		168		168	270

32 01 13 – Flexible Paving Surface Treatment

32 01 13.61 Slurry Seal (Latex Modified)

Line	Description	Crew	Daily Output	Labor-Hours	Unit	Material	Labor	Equipment	Total	Total Incl O&P
0010	SLURRY SEAL (LATEX MODIFIED)									
0011	Chip seal, slurry seal, and microsurfacing, see section 32 12 36									

32 01 13.62 Asphalt Surface Treatment

Line	Description	Crew	Daily Output	Labor-Hours	Unit	Material	Labor	Equipment	Total	Total Incl O&P
0010	ASPHALT SURFACE TREATMENT									
3000	Pavement overlay, polypropylene									
3040	6 oz./S.Y., ideal conditions	B-63	10000	.004	S.Y.	1.09	.18	.02	1.29	1.50
3190	Prime coat, bituminous, 0.28 gallon/S.Y.	B-45	2400	.007	C.S.F.	8.20	.35	.34	8.89	10
3200	Tack coat, emulsion, 0.05 gal./S.Y., 1,000 S.Y.		2500	.006	S.Y.	.35	.34	.33	1.02	1.28
3240	10,000 S.Y.		10000	.002		.28	.08	.08	.44	.53
3275	10,000 S.Y.		10000	.002		.53	.08	.08	.69	.80
3280	0.15 gal./S.Y., 1,000 S.Y.		2500	.006		.96	.34	.33	1.63	1.95
3320	10,000 S.Y.		10000	.002		.77	.08	.08	.93	1.07
3780	Rubberized asphalt (latex) seal		5000	.003		1.39	.17	.16	1.72	1.98
5400	Thermoplastic coal-tar, Type I, small or irregular area	B-90	2400	.027		4.32	1.25	.92	6.49	7.75
5450	Roadway or large area		8000	.008		4.32	.37	.27	4.96	5.65
5500	Type II, small or irregular area		2400	.027		5.50	1.25	.92	7.67	9.05
6000	Gravel surfacing on asphalt, screened and rolled	B-11L	160	.100	C.Y.	23	4.94	6.65	34.59	40.50
7000	For subbase treatment, see Section 31 32 13									

32 01 13.64 Sand Seal

Line	Description	Crew	Daily Output	Labor-Hours	Unit	Material	Labor	Equipment	Total	Total Incl O&P
0010	SAND SEAL									
2080	Sand sealing, sharp sand, asphalt emulsion, small area	B-91	10000	.006	S.Y.	1.48	.32	.23	2.03	2.39
2120	Roadway or large area	"	18000	.004	"	1.27	.18	.13	1.58	1.82
3000	Sealing random cracks, min. 1/2" wide, to 1-1/2", 1,000 L.F.	B-77	2800	.014	L.F.	1.65	.62	.37	2.64	3.22
3040	10,000 L.F.		4000	.010	"	1.14	.44	.26	1.84	2.24
3080	Alternate method, 1,000 L.F.		200	.200	Gal.	45	8.70	5.15	58.85	69
3120	10,000 L.F.		325	.123	"	36	5.35	3.16	44.51	52
3200	Multi-cracks (flooding), 1 coat, small area	B-92	460	.070	S.Y.	2.55	2.96	2.57	8.08	10.40
3320	Large area		1425	.022	"	14.35	.96	.83	16.14	18.25
3360	Alternate method, small area		115	.278	Gal.	15.55	11.85	10.30	37.70	47.50
3400	Large area		715	.045	"	14.60	1.91	1.65	18.16	21
3500	Sealing, roads, resealing joints in concrete	B-77	525	.076	L.F.	.39	3.32	1.96	5.67	7.90

32 01 13.66 Fog Seal

Line	Description	Crew	Daily Output	Labor-Hours	Unit	Material	Labor	Equipment	Total	Total Incl O&P
0010	FOG SEAL									
0012	Sealcoating, 2 coat coal tar pitch emulsion over 10,000 S.Y.	B-45	5000	.003	S.Y.	.84	.17	.16	1.17	1.37
0030	1,000 to 10,000 S.Y.	"	3000	.005		.84	.28	.27	1.39	1.67
0100	Under 1,000 S.Y.	B-1	1050	.023		.84	.98		1.82	2.49
0300	Petroleum resistant, over 10,000 S.Y.	B-45	5000	.003		1.40	.17	.16	1.73	1.99
0320	1,000 to 10,000 S.Y.	"	3000	.005		1.40	.28	.27	1.95	2.29
0400	Under 1,000 S.Y.	B-1	1050	.023		1.40	.98		2.38	3.11
0600	Non-skid pavement renewal, over 10,000 S.Y.	B-45	5000	.003		1.37	.17	.16	1.70	1.96

32 01 13 – Flexible Paving Surface Treatment

32 01 13.66 Fog Seal		Crew	Daily Output	Labor-Hours	Unit	Material	Labor	Equipment	Total	Total Incl O&P
						2020 Bare Costs				
0620	1,000 to 10,000 S.Y.	B-45	3000	.005	S.Y.	1.37	.28	.27	1.92	2.26
0700	Under 1,000 S.Y.	B-1	1050	.023		1.37	.98		2.35	3.08
0800	Prepare and clean surface for above	A-2	8545	.003			.12	.02	.14	.23
1000	Hand seal asphalt curbing	B-1	4420	.005	L.F.	.59	.23		.82	1.02
1900	Asphalt surface treatment, single course, small area									
1901	0.30 gal./S.Y. asphalt material, 20 lb./S.Y. aggregate	B-91	5000	.013	S.Y.	1.41	.65	.45	2.51	3.07
1910	Roadway or large area		10000	.006		1.30	.32	.23	1.85	2.19
1950	Asphalt surface treatment, dbl. course for small area		3000	.021		2.97	1.08	.75	4.80	5.80
1960	Roadway or large area		6000	.011		2.67	.54	.38	3.59	4.20
1980	Asphalt surface treatment, single course, for shoulders		7500	.009		1.46	.43	.30	2.19	2.62

32 01 16 – Flexible Paving Rehabilitation

32 01 16.71 Cold Milling Asphalt Paving

Line	Description	Crew	Daily Output	Labor-Hours	Unit	Material	Labor	Equipment	Total	Total Incl O&P
0010	**COLD MILLING ASPHALT PAVING**									
5200	Cold planing & cleaning, 1″ to 3″ asphalt pavmt., over 25,000 S.Y.	B-71	6000	.009	S.Y.		.45	.76	1.21	1.56
5280	5,000 S.Y. to 10,000 S.Y.		4000	.014	″		.68	1.15	1.83	2.34
5285	5,000 S.Y. to 10,000 S.Y.		36000	.002	S.F.		.08	.13	.21	.26
5300	Asphalt pavement removal from conc. base, no haul									
5320	Rip, load & sweep 1″ to 3″	B-70	8000	.007	S.Y.		.34	.29	.63	.86
5330	3″ to 6″ deep	″	5000	.011			.55	.46	1.01	1.36
5340	Profile grooving, asphalt pavement load & sweep, 1″ deep	B-71	12500	.004			.22	.37	.59	.75
5350	3″ deep		9000	.006			.30	.51	.81	1.04
5360	6″ deep		5000	.011			.55	.92	1.47	1.87

32 01 16.73 In Place Cold Reused Asphalt Paving

Line	Description	Crew	Daily Output	Labor-Hours	Unit	Material	Labor	Equipment	Total	Total Incl O&P
0010	**IN PLACE COLD REUSED ASPHALT PAVING**									
5000	Reclamation, pulverizing and blending with existing base									
5040	Aggregate base, 4″ thick pavement, over 15,000 S.Y.	B-73	2400	.027	S.Y.		1.37	1.87	3.24	4.23
5080	5,000 S.Y. to 15,000 S.Y.		2200	.029			1.50	2.04	3.54	4.60
5120	8″ thick pavement, over 15,000 S.Y.		2200	.029			1.50	2.04	3.54	4.60
5160	5,000 S.Y. to 15,000 S.Y.		2000	.032			1.65	2.24	3.89	5.05

32 01 16.74 In Place Hot Reused Asphalt Paving

Line	Description	Crew	Daily Output	Labor-Hours	Unit	Material	Labor	Equipment	Total	Total Incl O&P
0010	**IN PLACE HOT REUSED ASPHALT PAVING**									
5500	Recycle asphalt pavement at site									
5520	Remove, rejuvenate and spread 4″ deep G	B-72	2500	.026	S.Y.	4.58	1.27	3.29	9.14	10.70
5521	6″ deep G	″	2000	.032	″	6.70	1.59	4.11	12.40	14.45

32 01 17 – Flexible Paving Repair

32 01 17.10 Repair of Asphalt Pavement Holes

Line	Description	Crew	Daily Output	Labor-Hours	Unit	Material	Labor	Equipment	Total	Total Incl O&P
0010	**REPAIR OF ASPHALT PAVEMENT HOLES** (cold patch)									
0100	Flexible pavement repair holes, roadway, light traffic, 1 C.F. size each	B-37A	24	1	Ea.	9.50	44	14.70	68.20	96.50
0150	Group of two, 1 C.F. size each		16	1.500	Set	19	65.50	22	106.50	150
0200	Group of three, 1 C.F. size each		12	2	″	28.50	87.50	29.50	145.50	204
0300	Medium traffic, 1 C.F. size each	B-37B	24	1.333	Ea.	9.50	58	14.70	82.20	119
0350	Group of two, 1 C.F. size each		16	2	Set	19	87	22	128	184
0400	Group of three, 1 C.F. size each		12	2.667	″	28.50	116	29.50	174	249
0500	Highway/heavy traffic, 1 C.F. size each	B-37C	24	1.333	Ea.	9.50	59.50	23	92	131
0550	Group of two, 1 C.F. size each		16	2	Set	19	89.50	34.50	143	202
0600	Group of three, 1 C.F. size each		12	2.667	″	28.50	119	45.50	193	273
0700	Add police officer and car for traffic control				Hr.	54.50			54.50	60
1000	Flexible pavement repair holes, parking lot, bag material, 1 C.F. size each	B-37D	18	.889	Ea.	59	39.50	6.15	104.65	135
1010	Economy bag material, 1 C.F. size each		18	.889	″	37	39.50	6.15	82.65	111
1100	Group of two, 1 C.F. size each		12	1.333	Set	118	59.50	9.25	186.75	235
1110	Group of two, economy bag, 1 C.F. size each		12	1.333		73.50	59.50	9.25	142.25	187

32 01 17.10 Repair of Asphalt Pavement Holes

	32 01 17.10 Repair of Asphalt Pavement Holes	Crew	Daily Output	Labor-Hours	Unit	2020 Bare Costs Material	Labor	Equipment	Total	Total Incl O&P
1200	Group of three, 1 C.F. size each	B-37D	10	1.600	Set	176	71.50	11.10	258.60	320
1210	Economy bag, group of three, 1 C.F. size each		10	1.600		110	71.50	11.10	192.60	247
1300	Flexible pavement repair holes, parking lot, 1 C.F. single hole	A-3A	4	2	Ea.	59	106	44	209	279
1310	Economy material, 1 C.F. single hole		4	2	"	37	106	44	187	255
1400	Flexible pavement repair holes, parking lot, 1 C.F. four holes		2	4	Set	235	212	88	535	685
1410	Economy material, 1 C.F. four holes		2	4	"	147	212	88	447	590
1500	Flexible pavement repair holes, large parking lot, bulk matl, 1 C.F. size	B-37A	32	.750	Ea.	9.50	33	11	53.50	75

32 01 17.20 Repair of Asphalt Pavement Patches

	32 01 17.20 Repair of Asphalt Pavement Patches	Crew	Daily Output	Labor-Hours	Unit	Material	Labor	Equipment	Total	Total Incl O&P
0010	**REPAIR OF ASPHALT PAVEMENT PATCHES**									
0100	Flexible pavement patches, roadway, light traffic, sawcut 10-25 S.F.	B-89	12	1.333	Ea.		67	86	153	201
0150	Sawcut 26-60 S.F.		10	1.600			80	103	183	241
0200	Sawcut 61-100 S.F.		8	2			100	129	229	300
0210	Sawcut groups of small size patches		24	.667			33.50	43	76.50	101
0220	Large size patches		16	1			50	64.50	114.50	151
0300	Flexible pavement patches, roadway, light traffic, digout 10-25 S.F.	B-6	16	1.500			68.50	13.35	81.85	124
0350	Digout 26-60 S.F.		12	2			91.50	17.80	109.30	165
0400	Digout 61-100 S.F.		8	3			137	26.50	163.50	248
0450	Add 8 C.Y. truck, small project debris haulaway	B-34A	8	1	Hr.		49	53.50	102.50	138
0460	Add 12 C.Y. truck, small project debris haulaway	B-34B	8	1			49	71.50	120.50	158
0480	Add flagger for non-intersection medium traffic	1 Clab	8	1			42		42	67.50
0490	Add flasher truck for intersection medium traffic or heavy traffic	A-2B	8	1			47.50	24.50	72	103
0500	Flexible pavement patches, roadway, repave, cold, 15 S.F., 4" D	B-37	8	6	Ea.	45	266	32	343	510
0510	6" depth		8	6		68	266	32	366	535
0520	Repave, cold, 20 S.F., 4" depth		8	6		60	266	32	358	525
0530	6" depth		8	6		90.50	266	32	388.50	560
0540	Repave, cold, 25 S.F., 4" depth		8	6		74.50	266	32	372.50	540
0550	6" depth		8	6		113	266	32	411	585
0600	Repave, cold, 30 S.F., 4" depth		8	6		89.50	266	32	387.50	560
0610	6" depth		8	6		136	266	32	434	610
0640	Repave, cold, 40 S.F., 4" depth		8	6		119	266	32	417	590
0650	6" depth		8	6		181	266	32	479	660
0680	Repave, cold, 50 S.F., 4" depth		8	6		149	266	32	447	625
0690	6" depth		8	6		226	266	32	524	710
0720	Repave, cold, 60 S.F., 4" depth		8	6		179	266	32	477	655
0730	6" depth		8	6		271	266	32	569	760
0800	Repave, cold, 70 S.F., 4" depth		7	6.857		209	305	36.50	550.50	755
0810	6" depth		7	6.857		315	305	36.50	656.50	875
0820	Repave, cold, 75 S.F., 4" depth		7	6.857		222	305	36.50	563.50	770
0830	6" depth		7	6.857		340	305	36.50	681.50	900
0900	Repave, cold, 80 S.F., 4" depth		6	8		239	355	42.50	636.50	875
0910	6" depth		6	8		360	355	42.50	757.50	1,000
0940	Repave, cold, 90 S.F., 4" depth		6	8		268	355	42.50	665.50	905
0950	6" depth		6	8		405	355	42.50	802.50	1,050
0980	Repave, cold, 100 S.F., 4" depth		6	8		298	355	42.50	695.50	940
0990	6" depth		6	8		450	355	42.50	847.50	1,100
1000	Add flasher truck for paving operations in medium or heavy traffic	A-2B	8	1	Hr.		47.50	24.50	72	103
1100	Prime coat for repair 15-40 S.F., 25% overspray	B-37A	48	.500	Ea.	8.75	22	7.35	38.10	53
1150	41-60 S.F.		40	.600		17.50	26.50	8.80	52.80	71
1175	61-80 S.F.		32	.750		23.50	33	11	67.50	90.50
1200	81-100 S.F.		32	.750		29	33	11	73	96.50
1210	Groups of patches w/25% overspray		3600	.007	S.F.	.29	.29	.10	.68	.90
1300	Flexible pavement repair patches, street repave, hot, 60 S.F., 4" D	B-37	8	6	Ea.	97	266	32	395	565

32 01 17 – Flexible Paving Repair

32 01 17.20 Repair of Asphalt Pavement Patches		Crew	Daily Output	Labor-Hours	Unit	2020 Bare Costs Material	Labor	Equipment	Total	Total Incl O&P
1310	6" depth	B-37	8	6	Ea.	147	266	32	445	620
1320	Repave, hot, 70 S.F., 4" depth		7	6.857		113	305	36.50	454.50	650
1330	6" depth		7	6.857		172	305	36.50	513.50	715
1340	Repave, hot, 75 S.F., 4" depth		7	6.857		121	305	36.50	462.50	660
1350	6" depth		7	6.857		184	305	36.50	525.50	725
1360	Repave, hot, 80 S.F., 4" depth		6	8		130	355	42.50	527.50	755
1370	6" depth		6	8		196	355	42.50	593.50	830
1380	Repave, hot, 90 S.F., 4" depth		6	8		146	355	42.50	543.50	770
1390	6" depth		6	8		221	355	42.50	618.50	855
1400	Repave, hot, 100 S.F., 4" depth		6	8		162	355	42.50	559.50	790
1410	6" depth		6	8		245	355	42.50	642.50	880
1420	Pave hot groups of patches, 4" depth		900	.053	S.F.	1.62	2.36	.28	4.26	5.85
1430	6" depth		900	.053	"	2.46	2.36	.28	5.10	6.75
1500	Add 8 C.Y. truck for hot asphalt paving operations	B-34A	1	8	Day		390	430	820	1,100
1550	Add flasher truck for hot paving operations in med/heavy traffic	A-2B	8	1	Hr.		47.50	24.50	72	103
2000	Add police officer and car for traffic control				"	54.50			54.50	60

32 01 17.61 Sealing Cracks In Asphalt Paving		Crew	Daily Output	Labor-Hours	Unit	Material	Labor	Equipment	Total	Total Incl O&P
0010	**SEALING CRACKS IN ASPHALT PAVING**									
0100	Sealing cracks in asphalt paving, 1/8" wide x 1/2" depth, slow set	B-37A	2000	.012	L.F.	.17	.53	.18	.88	1.22
0110	1/4" wide x 1/2" depth		2000	.012		.20	.53	.18	.91	1.25
0130	3/8" wide x 1/2" depth		2000	.012		.22	.53	.18	.93	1.28
0140	1/2" wide x 1/2" depth		1800	.013		.25	.58	.20	1.03	1.44
0150	3/4" wide x 1/2" depth		1600	.015		.31	.66	.22	1.19	1.63
0160	1" wide x 1/2" depth		1600	.015		.37	.66	.22	1.25	1.69
0165	1/8" wide x 1" depth, rapid set	B-37F	2000	.016		.18	.69	.26	1.13	1.60
0170	1/4" wide x 1" depth		2000	.016		.36	.69	.26	1.31	1.79
0175	3/8" wide x 1" depth		1800	.018		.53	.77	.29	1.59	2.15
0180	1/2" wide x 1" depth		1800	.018		.71	.77	.29	1.77	2.34
0185	5/8" wide x 1" depth		1800	.018		.89	.77	.29	1.95	2.54
0190	3/4" wide x 1" depth		1600	.020		1.07	.87	.33	2.27	2.93
0195	1" wide x 1" depth		1600	.020		1.42	.87	.33	2.62	3.32
0200	1/8" wide x 2" depth, rapid set		2000	.016		.36	.69	.26	1.31	1.79
0210	1/4" wide x 2" depth		2000	.016		.71	.69	.26	1.66	2.18
0220	3/8" wide x 2" depth		1800	.018		1.07	.77	.29	2.13	2.74
0230	1/2" wide x 2" depth		1800	.018		1.42	.77	.29	2.48	3.13
0240	5/8" wide x 2" depth		1800	.018		1.78	.77	.29	2.84	3.52
0250	3/4" wide x 2" depth		1600	.020		2.14	.87	.33	3.34	4.10
0260	1" wide x 2" depth		1600	.020		2.85	.87	.33	4.05	4.89
0300	Add flagger for non-intersection medium traffic	1 Clab	8	1	Hr.		42		42	67.50
0400	Add flasher truck for intersection medium traffic or heavy traffic	A-2B	8	1	"		47.50	24.50	72	103

32 01 19 – Rigid Paving Surface Treatment

32 01 19.61 Sealing of Joints In Rigid Paving		Crew	Daily Output	Labor-Hours	Unit	Material	Labor	Equipment	Total	Total Incl O&P
0010	**SEALING OF JOINTS IN RIGID PAVING**									
2000	Waterproofing, membrane, tar and fabric, small area	B-63	233	.172	S.Y.	15.40	7.60	.76	23.76	30
2500	Large area		1435	.028		13.60	1.23	.12	14.95	17.05
3000	Preformed rubberized asphalt, small area		100	.400		19.90	17.70	1.78	39.38	52
3500	Large area		367	.109		18.10	4.83	.48	23.41	28
9000	Minimum labor/equipment charge	1 Clab	2	4	Job		168		168	270

32 01 29.61 Partial Depth Patching of Rigid Pavement

	32 01 29.61 Partial Depth Patching of Rigid Pavement	Crew	Daily Output	Labor-Hours	Unit	Material	Labor	Equipment	Total	Total Incl O&P
						2020 Bare Costs				
0010	**PARTIAL DEPTH PATCHING OF RIGID PAVEMENT**									
0100	Rigid pavement repair, roadway, light traffic, 25% pitting 15 S.F.	B-37F	16	2	Ea.	24	87	32.50	143.50	201
0110	Pitting 20 S.F.		16	2		31.50	87	32.50	151	210
0120	Pitting 25 S.F.		16	2		39.50	87	32.50	159	219
0130	Pitting 30 S.F.		16	2		47.50	87	32.50	167	228
0140	Pitting 40 S.F.		16	2		63.50	87	32.50	183	245
0150	Pitting 50 S.F.		12	2.667		79	116	43.50	238.50	320
0160	Pitting 60 S.F.		12	2.667		95	116	43.50	254.50	340
0170	Pitting 70 S.F.		12	2.667		111	116	43.50	270.50	355
0180	Pitting 75 S.F.		12	2.667		119	116	43.50	278.50	365
0190	Pitting 80 S.F.		8	4		127	174	65	366	490
0200	Pitting 90 S.F.		8	4		143	174	65	382	505
0210	Pitting 100 S.F.		8	4		158	174	65	397	525
0300	Roadway, light traffic, 50% pitting 15 S.F.		12	2.667		47.50	116	43.50	207	286
0310	Pitting 20 S.F.		12	2.667		63.50	116	43.50	223	305
0320	Pitting 25 S.F.		12	2.667		79	116	43.50	238.50	320
0330	Pitting 30 S.F.		12	2.667		95	116	43.50	254.50	340
0340	Pitting 40 S.F.		12	2.667		127	116	43.50	286.50	370
0350	Pitting 50 S.F.		8	4		158	174	65	397	525
0360	Pitting 60 S.F.		8	4		190	174	65	429	560
0370	Pitting 70 S.F.		8	4		222	174	65	461	595
0380	Pitting 75 S.F.		8	4		238	174	65	477	610
0390	Pitting 80 S.F.		6	5.333		253	231	87	571	745
0400	Pitting 90 S.F.		6	5.333		285	231	87	603	780
0410	Pitting 100 S.F.		6	5.333		315	231	87	633	815
1000	Rigid pavement repair, light traffic, surface patch, 2″ deep, 15 S.F.		8	4		95	174	65	334	455
1010	Surface patch, 2″ deep, 20 S.F.		8	4		127	174	65	366	490
1020	Surface patch, 2″ deep, 25 S.F.		8	4		158	174	65	397	525
1030	Surface patch, 2″ deep, 30 S.F.		8	4		190	174	65	429	560
1040	Surface patch, 2″ deep, 40 S.F.		8	4		253	174	65	492	630
1050	Surface patch, 2″ deep, 50 S.F.		6	5.333		315	231	87	633	815
1060	Surface patch, 2″ deep, 60 S.F.		6	5.333		380	231	87	698	885
1070	Surface patch, 2″ deep, 70 S.F.		6	5.333		445	231	87	763	955
1080	Surface patch, 2″ deep, 75 S.F.		6	5.333		475	231	87	793	990
1090	Surface patch, 2″ deep, 80 S.F.		5	6.400		505	278	104	887	1,125
1100	Surface patch, 2″ deep, 90 S.F.		5	6.400		570	278	104	952	1,175
1200	Surface patch, 2″ deep, 100 S.F.		5	6.400		635	278	104	1,017	1,250
2000	Add flagger for non-intersection medium traffic	1 Clab	8	1	Hr.		42		42	67.50
2100	Add flasher truck for intersection medium traffic or heavy traffic	A-2B	8	1			47.50	24.50	72	103
2200	Add police officer and car for traffic control					54.50			54.50	60

32 01 29.70 Full Depth Patching of Rigid Pavement

	32 01 29.70 Full Depth Patching of Rigid Pavement	Crew	Daily Output	Labor-Hours	Unit	Material	Labor	Equipment	Total	Total Incl O&P
0010	**FULL DEPTH PATCHING OF RIGID PAVEMENT**									
0015	Pavement preparation includes sawcut, remove pavement and replace									
0020	6″ of subbase and one layer of reinforcement									
0030	Trucking and haulaway of debris excluded									
0100	Rigid pavement replace, light traffic, prep, 15 S.F., 6″ depth	B-37E	16	3.500	Ea.	23	165	58	246	350
0110	Replacement preparation 20 S.F., 6″ depth		16	3.500		30.50	165	58	253.50	360
0115	25 S.F., 6″ depth		16	3.500		38	165	58	261	370
0120	30 S.F., 6″ depth		14	4		46	189	66	301	425
0125	35 S.F., 6″ depth		14	4		53.50	189	66	308.50	430
0130	40 S.F., 6″ depth		14	4		61	189	66	316	440
0135	45 S.F., 6″ depth		14	4		69	189	66	324	450

	32 01 29.70 Full Depth Patching of Rigid Pavement	Crew	Daily Output	Labor-Hours	Unit	2020 Bare Costs Material	Labor	Equipment	Total	Total Incl O&P
0140	50 S.F., 6" depth	B-37E	12	4.667	Ea.	76.50	220	77	373.50	520
0145	55 S.F., 6" depth		12	4.667		84	220	77	381	530
0150	60 S.F., 6" depth		10	5.600		92	264	92.50	448.50	625
0155	65 S.F., 6" depth		10	5.600		99.50	264	92.50	456	630
0160	70 S.F., 6" depth		10	5.600		107	264	92.50	463.50	640
0165	75 S.F., 6" depth		10	5.600		115	264	92.50	471.50	650
0170	80 S.F., 6" depth		8	7		122	330	116	568	785
0175	85 S.F., 6" depth		8	7		130	330	116	576	795
0180	90 S.F., 6" depth		8	7		138	330	116	584	805
0185	95 S.F., 6" depth		8	7		145	330	116	591	810
0190	100 S.F., 6" depth		8	7		153	330	116	599	820
0200	Pavement preparation for 8" rigid paving same as 6"									
0290	Pavement preparation for 9" rigid paving same as 10"									
0300	Rigid pavement replace, light traffic, prep, 15 S.F., 10" depth	B-37E	16	3.500	Ea.	38	165	58	261	370
0302	Pavement preparation 10" includes sawcut, remove pavement and									
0304	replace 6" of subbase and two layers of reinforcement									
0306	Trucking and haulaway of debris excluded									
0310	Replacement preparation 20 S.F., 10" depth	B-37E	16	3.500	Ea.	51	165	58	274	385
0315	25 S.F., 10" depth		16	3.500		63.50	165	58	286.50	395
0320	30 S.F., 10" depth		14	4		76.50	189	66	331.50	455
0325	35 S.F., 10" depth		14	4		58.50	189	66	313.50	435
0330	40 S.F., 10" depth		14	4		102	189	66	357	485
0335	45 S.F., 10" depth		12	4.667		115	220	77	412	560
0340	50 S.F., 10" depth		12	4.667		127	220	77	424	575
0345	55 S.F., 10" depth		12	4.667		140	220	77	437	590
0350	60 S.F., 10" depth		10	5.600		153	264	92.50	509.50	690
0355	65 S.F., 10" depth		10	5.600		166	264	92.50	522.50	705
0360	70 S.F., 10" depth		10	5.600		178	264	92.50	534.50	720
0365	75 S.F., 10" depth		8	7		191	330	116	637	860
0370	80 S.F., 10" depth		8	7		204	330	116	650	875
0375	85 S.F., 10" depth		8	7		216	330	116	662	890
0380	90 S.F., 10" depth		8	7		229	330	116	675	905
0385	95 S.F., 10" depth		8	7		242	330	116	688	920
0390	100 S.F., 10" depth		8	7		255	330	116	701	930
0500	Add 8 C.Y. truck, small project debris haulaway	B-34A	8	1	Hr.		49	53.50	102.50	138
0550	Add 12 C.Y. truck, small project debris haulaway	B-34B	8	1			49	71.50	120.50	158
0600	Add flagger for non-intersection medium traffic	1 Clab	8	1			42		42	67.50
0650	Add flasher truck for intersection medium traffic or heavy traffic	A-2B	8	1			47.50	24.50	72	103
0700	Add police officer and car for traffic control					54.50			54.50	60
0990	Concrete will be replaced using quick set mix with aggregate									
1000	Rigid pavement replace, light traffic, repour, 15 S.F., 6" depth	B-37F	18	1.778	Ea.	174	77	29	280	350
1010	20 S.F., 6" depth		18	1.778		232	77	29	338	410
1015	25 S.F., 6" depth		18	1.778		290	77	29	396	475
1020	30 S.F., 6" depth		16	2		350	87	32.50	469.50	560
1025	35 S.F., 6" depth		16	2		405	87	32.50	524.50	620
1030	40 S.F., 6" depth		16	2		465	87	32.50	584.50	685
1035	45 S.F., 6" depth		16	2		525	87	32.50	644.50	750
1040	50 S.F., 6" depth		16	2		580	87	32.50	699.50	815
1045	55 S.F., 6" depth		12	2.667		640	116	43.50	799.50	940
1050	60 S.F., 6" depth		12	2.667		695	116	43.50	854.50	1,000
1055	65 S.F., 6" depth		12	2.667		755	116	43.50	914.50	1,075
1060	70 S.F., 6" depth		12	2.667		815	116	43.50	974.50	1,125
1065	75 S.F., 6" depth		10	3.200		870	139	52	1,061	1,250

32 01 29 – Rigid Paving Repair

32 01 29.70 Full Depth Patching of Rigid Pavement		Crew	Daily Output	Labor-Hours	Unit	Material	Labor	Equipment	Total	Total Incl O&P
						2020 Bare Costs				
1070	80 S.F., 6" depth	B-37F	10	3.200	Ea.	930	139	52	1,121	1,300
1075	85 S.F., 6" depth		8	4		990	174	65	1,229	1,425
1080	90 S.F., 6" depth		8	4		1,050	174	65	1,289	1,500
1085	95 S.F., 6" depth		8	4		1,100	174	65	1,339	1,575
1090	100 S.F., 6" depth		8	4		1,150	174	65	1,389	1,625
1099	Concrete will be replaced using 4,500 psi concrete ready mix									
1100	Rigid pavement replace, light traffic, repour, 15 S.F., 6" depth	A-2	12	2	Ea.	345	87.50	16.35	448.85	540
1110	20-50 S.F., 6" depth		12	2		345	87.50	16.35	448.85	540
1120	55-65 S.F., 6" depth		10	2.400		380	105	19.60	504.60	610
1130	70-80 S.F., 6" depth		10	2.400		415	105	19.60	539.60	645
1140	85-100 S.F., 6" depth		8	3		485	131	24.50	640.50	770
1200	Repour 15-40 S.F., 8" depth		12	2		345	87.50	16.35	448.85	540
1210	45-50 S.F., 8" depth		12	2		380	87.50	16.35	483.85	580
1220	55-60 S.F., 8" depth		10	2.400		415	105	19.60	539.60	645
1230	65-80 S.F., 8" depth		10	2.400		485	105	19.60	609.60	725
1240	85-90 S.F., 8" depth		8	3		525	131	24.50	680.50	810
1250	95-100 S.F., 8" depth		8	3		555	131	24.50	710.50	845
1300	Repour 15-30 S.F., 10" depth		12	2		345	87.50	16.35	448.85	540
1310	35-40 S.F., 10" depth		12	2		380	87.50	16.35	483.85	580
1320	45 S.F., 10" depth		12	2		415	87.50	16.35	518.85	615
1330	50-65 S.F., 10" depth		10	2.400		485	105	19.60	609.60	725
1340	70 S.F., 10" depth		10	2.400		525	105	19.60	649.60	765
1350	75-80 S.F., 10" depth		10	2.400		555	105	19.60	679.60	800
1360	85-95 S.F., 10" depth		8	3		625	131	24.50	780.50	925
1370	100 S.F., 10" depth		8	3		665	131	24.50	820.50	965

32 01 30 – Operation and Maintenance of Site Improvements

32 01 30.10 Site Maintenance

Line	Description	Crew	Daily Output	Labor-Hours	Unit	Material	Labor	Equipment	Total	Total Incl O&P
0010	**SITE MAINTENANCE** R019313-10									
0800	Flower bed maintenance									
0810	Cultivate bed, no mulch	1 Clab	14	.571	M.S.F.		24		24	38.50
0830	Fall clean-up of flower bed, including pick-up mulch for re-use		1	8			335		335	540
0840	Fertilize flower bed, dry granular 3 lb./M.S.F.		85	.094		1.23	3.96		5.19	7.70
1000	Mulch see Section 32 91 13.16									
1100	Plant bed preparation see Section 32 91 13.16									
1110	Plant from flats	1 Clab	800	.010	S.F.		.42		.42	.67
1120	Planting, ground cover see Section 32 93 13.20									
1130	Police, hand pickup	1 Clab	30	.267	M.S.F.		11.25		11.25	18
1140	Vacuum (outside)		48	.167			7		7	11.25
1200	Spring prepare		2	4			168		168	270
1300	Weed mulched bed		20	.400			16.85		16.85	27
1310	Unmulched bed		8	1			42		42	67.50
1550	General site work maintenance									
1560	Clearing brush with brush saw & rake	1 Clab	565	.014	S.Y.		.60		.60	.96
1570	By hand	"	280	.029			1.20		1.20	1.93
1580	With dozer, ball and chain, light clearing	B-11A	3675	.004			.22	.41	.63	.79
1590	Medium clearing	"	3110	.005			.25	.48	.73	.93
1600	Ground cover planting see Section 32 93 13.20									
1610	Grounds, policing, incl. picking up trash & other debris	1 Clab	300	.027	M.S.F.		1.12		1.12	1.80
3000	Lawn maintenance									
3010	Aerate lawn, 18" cultivating width, walk behind	A-1K	95	.084	M.S.F.		3.55	1.07	4.62	6.85
3040	48" cultivating width	B-66	750	.011			.57	.32	.89	1.24
3060	72" cultivating width	"	1100	.007			.39	.22	.61	.84

32 01 30 – Operation and Maintenance of Site Improvements

32 01 30.10 Site Maintenance		Crew	Daily Output	Labor-Hours	Unit	Material (2020 Bare Costs)	Labor (2020 Bare Costs)	Equipment (2020 Bare Costs)	Total (2020 Bare Costs)	Total Incl O&P
3100	Edge lawn, by hand at walks	1 Clab	16	.500	C.L.F.		21		21	33.50
3150	At planting beds		7	1.143			48		48	77
3200	Using gas powered edger at walks		88	.091			3.83		3.83	6.15
3250	At planting beds		24	.333			14.05		14.05	22.50
3260	Vacuum, 30″ gas, outdoors with hose		96	.083	M.L.F.		3.51		3.51	5.60
3350	Edging material for lawns & planting beds see Section 32 94 13.20									
3400	Weed lawn, by hand	1 Clab	3	2.667	M.S.F.		112		112	180
3800	Lawn bed preparation see Section 32 91 13.23									
4500	Rake leaves or lawn, by hand	1 Clab	7.50	1.067	M.S.F.		45		45	72
4510	Power rake	″	45	.178	″		7.50		7.50	12
4700	Seeding lawn, see Section 32 92 19.14									
4750	Sodding, see Section 32 92 23.10									
5900	Road & walk maintenance									
5910	Asphaltic concrete paving, cold patch, 2″ thick	B-37	350	.137	S.Y.	13.60	6.05	.73	20.38	25.50
5913	3″ thick	″	260	.185	″	23.50	8.15	.98	32.63	40
5915	De-icing roads and walks									
5920	Calcium chloride in truckload lots see Section 31 32 13.30									
6000	Ice melting comp., 90% calc. chlor., effec. to -30°F									
6010	50-80 lb. poly bags, med. applic. 19 lb./M.S.F., by hand	1 Clab	60	.133	M.S.F.	197	5.60		202.60	225
6050	With hand operated rotary spreader		110	.073		197	3.06		200.06	221
6100	Rock salt, med. applic. on road & walkway, by hand		60	.133		108	5.60		113.60	128
6110	With hand operated rotary spreader		110	.073		108	3.06		111.06	124
6130	Hosing, sidewalks & other paved areas		30	.267			11.25		11.25	18
6150	Painting lines on pavement see Section 32 17 23.13									
6250	Seal coating bituminous surfaces see Section 32 01 13.61									
6260	Sidewalk, brick pavers, steam cleaning	A-1H	950	.008	S.F.	.11	.35	.08	.54	.78
6400	Sweep walk by hand	1 Clab	15	.533	M.S.F.		22.50		22.50	36
6410	Power vacuum		100	.080			3.37		3.37	5.40
6420	Drives & parking areas with power vacuum		120	.067			2.81		2.81	4.50
6600	Shrub maintenance									
6640	Shrub bed fertilize dry granular 3 lb./M.S.F.	1 Clab	85	.094	M.S.F.	1.23	3.96		5.19	7.70
6800	Weed, by handhoe		8	1			42		42	67.50
6810	Spray out		32	.250			10.55		10.55	16.85
6820	Spray after mulch		48	.167			7		7	11.25
7100	Tree maintenance									
7140	Clear and grub trees, see Section 31 11 10.10									
7160	Cutting and piling trees, see Section 31 13 13.20									
7200	Fertilize, tablets, slow release, 30 gram/tree	1 Clab	100	.080	Ea.	2.30	3.37		5.67	7.95
7280	Guying, including stakes, guy wire & wrap, see Section 32 94 50.10									
7300	Planting, trees, Deciduous, in prep. beds, see Section 32 93 43.20									
7400	Removal, trees see Section 32 96 43.20									
7420	Pest control, spray	1 Clab	24	.333	Ea.	28.50	14.05		42.55	54
7430	Systemic	″	48	.167	″	29	7		36	43

32 01 30.20 Snow Removal		Crew	Daily Output	Labor-Hours	Unit	Material	Labor	Equipment	Total	Total Incl O&P
0010	**SNOW REMOVAL**									
0020	Plowing, 12 ton truck, 2″-4″ deep	B-34A	250	.032	M.S.F.		1.57	1.71	3.28	4.38
0040	4″-10″ deep		200	.040			1.96	2.14	4.10	5.50
0060	10″-15″ deep		150	.053			2.61	2.85	5.46	7.30
0080	Pickup truck, 2″-4″ deep	A-3A	175	.046			2.42	1	3.42	4.90
0100	4″-10″ deep		130	.062			3.26	1.35	4.61	6.60
0120	10″-15″ deep		75	.107			5.65	2.34	7.99	11.45
0140	Load and haul snow, 1 mile round trip	A-3B	230	.070	C.Y.		3.68	5.55	9.23	11.90

32 01 Operation and Maintenance of Exterior Improvements

32 01 30 – Operation and Maintenance of Site Improvements

32 01 30.20 Snow Removal		Crew	Daily Output	Labor-Hours	Unit	Material	2020 Bare Costs Labor	Equipment	Total	Total Incl O&P
0160	2 mile round trip	A-3B	175	.091	C.Y.		4.83	7.30	12.13	15.65
0180	3 mile round trip		150	.107			5.65	8.50	14.15	18.25
0200	4 mile round trip		130	.123			6.50	9.80	16.30	21
0220	5 mile round trip		120	.133			7.05	10.65	17.70	23
0240	Clearing with wheeled skid steer loader, 1 C.Y.	A-3C	240	.033			1.77	1.67	3.44	4.59
0260	Spread sand and salt mix	B-34A	375	.021	M.S.F.	7.70	1.04	1.14	9.88	11.40
0280	Sidewalks and drives, by hand	1 Clab	1200	.007	C.F.		.28		.28	.45
0300	Power, 24" blower	A-1M	7000	.001	"		.05	.01	.06	.09
0320	2"-4" deep, single driveway (10' x 50')		16	.500	Ea.		21	4.22	25.22	38
0340	Double driveway (20' x 50')		16	.500			21	4.22	25.22	38
0360	4"-10" deep, single driveway		16	.500			21	4.22	25.22	38
0380	Double driveway		16	.500			21	4.22	25.22	38
0400	10"-15" deep, single driveway		12	.667			28	5.65	33.65	51
0420	Double driveway		12	.667			28	5.65	33.65	51
0440	For heavy wet snow, add								20%	20%
9000	Minimum labor and equipment charge	A-3A	2	4	Job		212	88	300	425

32 01 90 – Operation and Maintenance of Planting

32 01 90.13 Fertilizing		Crew	Daily Output	Labor-Hours	Unit	Material	Labor	Equipment	Total	Total Incl O&P
0010	**FERTILIZING**									
0100	Dry granular, 4 lb./M.S.F., hand spread	1 Clab	24	.333	M.S.F.	2.32	14.05		16.37	25
0110	Push rotary		140	.057	"	2.32	2.41		4.73	6.40
0112	Push rotary, per 1076 feet squared		130	.062	Ea.	2.32	2.59		4.91	6.70
0120	Tractor towed spreader, 8'	B-66	500	.016	M.S.F.	2.32	.85	.48	3.65	4.41
0130	12' spread		800	.010		2.32	.53	.30	3.15	3.71
0140	Truck whirlwind spreader		1200	.007		2.32	.35	.20	2.87	3.32
0180	Water soluble, hydro spread, 1.5 lb./M.S.F.	B-64	600	.027		2.44	1.19	.70	4.33	5.35
0190	Add for weed control					.30			.30	.33

32 01 90.19 Mowing		Crew	Daily Output	Labor-Hours	Unit	Material	Labor	Equipment	Total	Total Incl O&P
0010	**MOWING**									
1650	Mowing brush, tractor with rotary mower									
1660	Light density	B-84	22	.364	M.S.F.		20.50	16.65	37.15	51
1670	Medium density		13	.615			35	28	63	85.50
1680	Heavy density		9	.889			50.50	41	91.50	124
2000	Mowing, brush/grass, tractor, rotary mower, highway/airport median		13	.615			35	28	63	85.50
2010	Traffic safety flashing truck for highway/airport median mowing	A-2B	1	8	Day		380	196	576	820
4050	Lawn mowing, power mower, 18" - 22"	1 Clab	65	.123	M.S.F.		5.20		5.20	8.30
4100	22" - 30"		110	.073			3.06		3.06	4.91
4150	30" - 32"		140	.057			2.41		2.41	3.85
4160	Riding mower, 36" - 44"	B-66	300	.027			1.41	.80	2.21	3.10
4170	48" - 58"	"	480	.017			.88	.50	1.38	1.93
4175	Mowing with tractor & attachments									
4180	3 gang reel, 7'	B-66	930	.009	M.S.F.		.46	.26	.72	1
4190	5 gang reel, 12'		1200	.007			.35	.20	.55	.77
4200	Cutter or sickle-bar, 5', rough terrain		210	.038			2.02	1.15	3.17	4.42
4210	Cutter or sickle-bar, 5', smooth terrain		340	.024			1.25	.71	1.96	2.73
4220	Drainage channel, 5' sickle bar		5	1.600	Mile		85	48.50	133.50	186
4250	Lawn mower, rotary type, sharpen (all sizes)	1 Clab	10	.800	Ea.		33.50		33.50	54
4260	Repair or replace part		7	1.143	"		48		48	77
5000	Edge trimming with weed whacker		5760	.001	L.F.		.06		.06	.09

32 01 90.23 Pruning

	32 01 90.23 Pruning	Crew	Daily Output	Labor-Hours	Unit	Material	2020 Bare Costs Labor	Equipment	Total	Total Incl O&P
0010	**PRUNING**									
0020	1-1/2" caliper	1 Clab	84	.095	Ea.		4.01		4.01	6.40
0030	2" caliper		70	.114			4.81		4.81	7.70
0040	2-1/2" caliper		50	.160			6.75		6.75	10.80
0050	3" caliper		30	.267			11.25		11.25	18
0060	4" caliper, by hand	2 Clab	21	.762			32		32	51.50
0070	Aerial lift equipment	B-85	38	1.053			49	21	70	101
0100	6" caliper, by hand	2 Clab	12	1.333			56		56	90
0110	Aerial lift equipment	B-85	20	2			93	40	133	192
0200	9" caliper, by hand	2 Clab	7.50	2.133			90		90	144
0210	Aerial lift equipment	B-85	12.50	3.200			148	64	212	305
0300	12" caliper, by hand	2 Clab	6.50	2.462			104		104	166
0310	Aerial lift equipment	B-85	10.80	3.704			172	74.50	246.50	355
0400	18" caliper by hand	2 Clab	5.60	2.857			120		120	193
0410	Aerial lift equipment	B-85	9.30	4.301			200	86.50	286.50	415
0500	24" caliper, by hand	2 Clab	4.60	3.478			146		146	235
0510	Aerial lift equipment	B-85	7.70	5.195			241	104	345	500
0600	30" caliper, by hand	2 Clab	3.70	4.324			182		182	292
0610	Aerial lift equipment	B-85	6.20	6.452			299	129	428	615
0700	36" caliper, by hand	2 Clab	2.70	5.926			249		249	400
0710	Aerial lift equipment	B-85	4.50	8.889			410	178	588	850
0800	48" caliper, by hand	2 Clab	1.70	9.412			395		395	635
0810	Aerial lift equipment	B-85	2.80	14.286			665	287	952	1,375

32 01 90.24 Shrub Pruning

	32 01 90.24 Shrub Pruning	Crew	Daily Output	Labor-Hours	Unit	Material	Labor	Equipment	Total	Total Incl O&P
0010	**SHRUB PRUNING**									
6700	Prune, shrub bed	1 Clab	7	1.143	M.S.F.		48		48	77
6710	Shrub under 3' height		190	.042	Ea.		1.77		1.77	2.84
6720	4' height		90	.089			3.74		3.74	6
6730	Over 6'		50	.160			6.75		6.75	10.80
7350	Prune trees from ground		20	.400			16.85		16.85	27
7360	High work		8	1			42		42	67.50

32 01 90.26 Watering

	32 01 90.26 Watering	Crew	Daily Output	Labor-Hours	Unit	Material	Labor	Equipment	Total	Total Incl O&P
0010	**WATERING**									
4900	Water lawn or planting bed with hose, 1" of water	1 Clab	16	.500	M.S.F.		21		21	33.50
4910	50' soaker hoses, in place		82	.098			4.11		4.11	6.60
4920	60' soaker hoses, in place		89	.090			3.78		3.78	6.05
7500	Water trees or shrubs, under 1" caliper		32	.250	Ea.		10.55		10.55	16.85
7550	1" - 3" caliper		17	.471			19.80		19.80	31.50
7600	3" - 4" caliper		12	.667			28		28	45
7650	Over 4" caliper		10	.800			33.50		33.50	54
9000	For sprinkler irrigation systems, see Section 32 84 23.10									

32 01 90.29 Topsoil Preservation

	32 01 90.29 Topsoil Preservation	Crew	Daily Output	Labor-Hours	Unit	Material	Labor	Equipment	Total	Total Incl O&P
0010	**TOPSOIL PRESERVATION**									
0100	Weed planting bed	1 Clab	800	.010	S.Y.		.42		.42	.67

32 06 10.10 Sidewalks, Driveways and Patios

	32 06 10.10 Sidewalks, Driveways and Patios	Crew	Daily Output	Labor-Hours	Unit	2020 Bare Costs Material	Labor	Equipment	Total	Total Incl O&P
0010	**SIDEWALKS, DRIVEWAYS AND PATIOS** No base									
0020	Asphaltic concrete, 2" thick	B-37	720	.067	S.Y.	7.20	2.95	.36	10.51	13
0100	2-1/2" thick	"	660	.073	"	9.10	3.22	.39	12.71	15.65
0110	Bedding for brick or stone, mortar, 1" thick	D-1	300	.053	S.F.	.82	2.50		3.32	4.94
0120	2" thick	"	200	.080		2.05	3.75		5.80	8.30
0130	Sand, 2" thick	B-18	8000	.003		.34	.13	.02	.49	.61
0140	4" thick	"	4000	.006		.68	.26	.04	.98	1.21
0300	Concrete, 3,000 psi, CIP, 6 x 6 - W1.4 x W1.4 mesh,									
0310	broomed finish, no base, 4" thick	B-24	600	.040	S.F.	2.31	1.94		4.25	5.60
0350	5" thick		545	.044		2.86	2.13		4.99	6.55
0400	6" thick		510	.047		3.34	2.28		5.62	7.30
0450	For bank run gravel base, 4" thick, add	B-18	2500	.010		.43	.41	.07	.91	1.21
0520	8" thick, add	"	1600	.015		.87	.64	.10	1.61	2.10
0550	Exposed aggregate finish, add to above, minimum	B-24	1875	.013		.13	.62		.75	1.13
0600	Maximum	"	455	.053		.43	2.55		2.98	4.53
0850	Splash block, precast concrete	1 Clab	150	.053	Ea.	11.95	2.25		14.20	16.70
0950	Concrete tree grate, 5' square	B-6	25	.960		440	44	8.55	492.55	565
0955	Tree well & cover, concrete, 3' square		25	.960		245	44	8.55	297.55	350
0960	Cast iron tree grate with frame, 2 piece, round, 5' diameter		25	.960		1,075	44	8.55	1,127.55	1,250
0980	Square, 5' side		25	.960		1,075	44	8.55	1,127.55	1,250
1000	Crushed stone, 1" thick, white marble	2 Clab	1700	.009	S.F.	.49	.40		.89	1.16
1050	Bluestone		1700	.009		.18	.40		.58	.83
1070	Granite chips		1700	.009		.24	.40		.64	.89
1660	Limestone pavers, 3" thick	D-1	72	.222		10.75	10.40		21.15	28.50
1670	4" thick		70	.229		14.25	10.70		24.95	33
1680	5" thick		68	.235		17.85	11.05		28.90	37.50
1700	Redwood, prefabricated, 4' x 4' sections	2 Carp	316	.051		4.89	2.69		7.58	9.65
1750	Redwood planks, 1" thick, on sleepers	"	240	.067		4.89	3.54		8.43	11.05
1830	1-1/2" thick	B-28	167	.144		6.05	7.10		13.15	18.05
1840	2" thick		167	.144		7.75	7.10		14.85	19.95
1850	3" thick		150	.160		11.45	7.90		19.35	25.50
1860	4" thick		150	.160		15.05	7.90		22.95	29.50
1870	5" thick		150	.160		19.10	7.90		27	33.50
2100	River or beach stone, stock	B-1	18	1.333	Ton	33	57		90	128
2150	Quarried	"	18	1.333	"	58	57		115	156
2160	Load, dump, and spread stone with skid steer, 100' haul	B-62	24	1	C.Y.		45.50	7.40	52.90	80.50
2165	200' haul		18	1.333			61	9.85	70.85	108
2168	300' haul		12	2			91.50	14.80	106.30	161
2170	Shale paver, 2-1/4" thick	D-1	200	.080	S.F.	3.78	3.75		7.53	10.20
2200	Coarse washed sand bed, 1" thick	B-62	1350	.018	S.Y.	1.03	.81	.13	1.97	2.57
2250	Stone dust, 4" thick	"	900	.027	"	5.25	1.22	.20	6.67	7.95
2300	Tile thinset pavers, 3/8" thick	D-1	300	.053	S.F.	3.74	2.50		6.24	8.15
2350	3/4" thick	"	280	.057	"	5.95	2.68		8.63	10.90
2400	Wood rounds, cypress	B-1	175	.137	Ea.	10.20	5.85		16.05	20.50
9000	Minimum labor/equipment charge	D-1	2	8	Job		375		375	605

32 06 10.20 Steps

	32 06 10.20 Steps	Crew	Daily Output	Labor-Hours	Unit	Material	Labor	Equipment	Total	Total Incl O&P
0010	**STEPS**									
0011	Incl. excav., borrow & concrete base as required									
0100	Brick steps	B-24	35	.686	LF Riser	17.95	33		50.95	73
0200	Railroad ties	2 Clab	25	.640		3.75	27		30.75	47
0300	Bluestone treads, 12" x 2" or 12" x 1-1/2"	B-24	30	.800		44.50	38.50		83	111
0490	Minimum labor/equipment charge	D-1	2	8	Job		375		375	605

32 06 Schedules for Exterior Improvements

32 06 10 – Schedules for Bases, Ballasts, and Paving

32 06 10.20 Steps		Crew	Daily Output	Labor-Hours	Unit	Material	2020 Bare Costs Labor	Equipment	Total	Total Incl O&P
0600	Precast concrete, see Section 03 41 23.50									
4000	Edging, redwood, 2" x 4"	2 Carp	330	.048	L.F.	3.06	2.58		5.64	7.50
4025	Steel edge strips, incl. stakes, 1/4" x 5"	B-1	390	.062		4.58	2.63		7.21	9.25
4050	Edging, landscape timber or railroad ties, 6" x 8"	2 Carp	170	.094	↓	2.48	5		7.48	10.75

32 11 Base Courses

32 11 23 – Aggregate Base Courses

32 11 23.23 Base Course Drainage Layers

		Crew	Daily Output	Labor-Hours	Unit	Material	Labor	Equipment	Total	Total Incl O&P
0010	**BASE COURSE DRAINAGE LAYERS**									
0011	For Soil Stabilization, see Section 31 32									
0012	For roadways and large areas									
0050	Crushed 3/4" stone base, compacted, 3" deep	B-36C	5200	.008	S.Y.	2.70	.41	.82	3.93	4.50
0100	6" deep		5000	.008		5.40	.42	.85	6.67	7.55
0200	9" deep		4600	.009		8.10	.46	.92	9.48	10.65
0300	12" deep	↓	4200	.010		10.80	.50	1.01	12.31	13.80
0301	Crushed 1-1/2" stone base, compacted to 4" deep	B-36B	6000	.011		4.87	.54	.82	6.23	7.10
0302	6" deep		5400	.012		7.30	.60	.91	8.81	10
0303	8" deep		4500	.014		9.75	.72	1.09	11.56	13.05
0304	12" deep	↓	3800	.017	↓	14.60	.85	1.29	16.74	18.85
0310	Minimum labor/equipment charge	B-36	1	40	Job		1,925	1,800	3,725	5,050
0350	Bank run gravel, spread and compacted									
0370	6" deep	B-32	6000	.005	S.Y.	3.62	.28	.47	4.37	4.93
0390	9" deep		4900	.007		5.45	.35	.57	6.37	7.15
0400	12" deep	↓	4200	.008	↓	7.25	.40	.67	8.32	9.30
1500	Alternate method to figure base course									
1510	Crushed stone, 3/4", compacted, 3" deep	B-36C	435	.092	E.C.Y.	27.50	4.84	9.75	42.09	49
1511	6" deep	B-36B	835	.077		27.50	3.87	5.85	37.22	43
1512	9" deep		1150	.056		27.50	2.81	4.26	34.57	39.50
1513	12" deep		1400	.046		27.50	2.31	3.50	33.31	38
1520	Crushed stone, 1-1/2", compacted, 4" deep		665	.096		27.50	4.86	7.35	39.71	46.50
1521	6" deep		900	.071		27.50	3.59	5.45	36.54	42
1522	8" deep		1000	.064		27.50	3.23	4.90	35.63	41
1523	12" deep	↓	1265	.051		27.50	2.56	3.88	33.94	39
1530	Gravel, bank run, compacted, 6" deep	B-36C	835	.048		18.60	2.52	5.10	26.22	30
1531	9" deep		1150	.035		18.60	1.83	3.69	24.12	27.50
1532	12" deep	↓	1400	.029	↓	18.60	1.50	3.03	23.13	26
2010	Crushed stone, 3/4" maximum size, 3" deep	B-36	540	.074	Ton	18	3.58	3.35	24.93	29
2011	6" deep		1625	.025		18	1.19	1.11	20.30	23
2012	9" deep		1785	.022		18	1.08	1.01	20.09	22.50
2013	12" deep		1950	.021		18	.99	.93	19.92	22.50
2020	Crushed stone, 1-1/2" maximum size, 4" deep		720	.056		18	2.69	2.51	23.20	27
2021	6" deep		815	.049		18	2.37	2.22	22.59	26
2022	8" deep		835	.048		18	2.32	2.16	22.48	26
2023	12" deep	↓	975	.041		18	1.98	1.85	21.83	25
2030	Bank run gravel, 6" deep	B-32A	875	.027		12.50	1.42	1.97	15.89	18.15
2031	9" deep		970	.025		12.50	1.28	1.78	15.56	17.75
2032	12" deep	↓	1060	.023	↓	12.50	1.17	1.63	15.30	17.40
6000	Stabilization fabric, polypropylene, 6 oz./S.Y.	B-6	10000	.002	S.Y.	.91	.11	.02	1.04	1.19
6900	For small and irregular areas, add						50%	50%		
6990	Minimum labor/equipment charge	1 Clab	3	2.667	Job		112		112	180
7000	Prepare and roll sub-base, small areas to 2,500 S.Y.	B-32A	1500	.016	S.Y.		.83	1.15	1.98	2.58

32 11 Base Courses

32 11 23 – Aggregate Base Courses

32 11 23.23 Base Course Drainage Layers		Crew	Daily Output	Labor-Hours	Unit	Material	2020 Bare Costs Labor	Equipment	Total	Total Incl O&P
8000	Large areas over 2,500 S.Y.	B-32A	3500	.007	S.Y.		.36	.49	.85	1.10
8050	For roadways	B-32	4000	.008			.42	.70	1.12	1.44
9000	Minimum labor/equipment charge	1 Clab	4	2	Job		84		84	135

32 11 26 – Asphaltic Base Courses

32 11 26.13 Plant Mix Asphaltic Base Courses		Crew	Daily Output	Labor-Hours	Unit	Material	Labor	Equipment	Total	Total Incl O&P
0010	**PLANT MIX ASPHALTIC BASE COURSES**									
0011	For roadways and large paved areas									
0500	Bituminous concrete, 4" thick	B-25	4545	.019	S.Y.	15.30	.90	.59	16.79	18.85
0550	6" thick		3700	.024		22.50	1.10	.73	24.33	27
0560	8" thick		3000	.029		30	1.36	.90	32.26	36
0570	10" thick		2545	.035		37	1.60	1.06	39.66	44
2000	Alternate method to figure base course									
2005	Bituminous concrete, 4" thick	B-25	1000	.088	Ton	68	4.07	2.70	74.77	84
2006	6" thick		1220	.072		68	3.34	2.21	73.55	82
2007	8" thick		1320	.067		68	3.09	2.05	73.14	81.50
2008	10" thick		1400	.063		68	2.91	1.93	72.84	81
8900	For small and irregular areas, add						50%	50%		

32 11 26.19 Bituminous-Stabilized Base Courses		Crew	Daily Output	Labor-Hours	Unit	Material	Labor	Equipment	Total	Total Incl O&P
0010	**BITUMINOUS-STABILIZED BASE COURSES**									
0020	For roadways and large paved areas									
0700	Liquid application to gravel base, asphalt emulsion	B-45	6000	.003	Gal.	4.84	.14	.14	5.12	5.65
0800	Prime and seal, cut back asphalt		6000	.003	"	5.70	.14	.14	5.98	6.65
1000	Macadam penetration crushed stone, 2 gal./S.Y., 4" thick		6000	.003	S.Y.	9.70	.14	.14	9.98	11
1100	6" thick, 3 gal./S.Y.		4000	.004		14.50	.21	.21	14.92	16.50
1200	8" thick, 4 gal./S.Y.		3000	.005		19.35	.28	.27	19.90	22.50
8900	For small and irregular areas, add						50%	50%		

32 12 Flexible Paving

32 12 16 – Asphalt Paving

32 12 16.13 Plant-Mix Asphalt Paving		Crew	Daily Output	Labor-Hours	Unit	Material	Labor	Equipment	Total	Total Incl O&P
0010	**PLANT-MIX ASPHALT PAVING**									
0020	For highways and large paved areas, excludes hauling									
0025	See Section 31 23 23.20 for hauling costs									
0080	Binder course, 1-1/2" thick	B-25	7725	.011	S.Y.	5.55	.53	.35	6.43	7.30
0120	2" thick		6345	.014		7.40	.64	.43	8.47	9.60
0130	2-1/2" thick		5620	.016		9.25	.72	.48	10.45	11.85
0160	3" thick		4905	.018		11.05	.83	.55	12.43	14.15
0170	3-1/2" thick		4520	.019		12.90	.90	.60	14.40	16.30
0200	4" thick		4140	.021		14.75	.98	.65	16.38	18.55
0300	Wearing course, 1" thick	B-25B	10575	.009		3.66	.43	.28	4.37	5
0340	1-1/2" thick		7725	.012		6.15	.59	.38	7.12	8.10
0380	2" thick		6345	.015		8.25	.71	.46	9.42	10.75
0420	2-1/2" thick		5480	.018		10.20	.83	.54	11.57	13.10
0460	3" thick		4900	.020		12.15	.92	.60	13.67	15.50
0470	3-1/2" thick		4520	.021		14.25	1	.65	15.90	18
0480	4" thick		4140	.023		16.30	1.09	.71	18.10	20.50
0500	Open graded friction course	B-25C	5000	.010		2.55	.45	.47	3.47	4.04
0800	Alternate method of figuring paving costs									
0810	Binder course, 1-1/2" thick	B-25	630	.140	Ton	68	6.45	4.29	78.74	89.50
0811	2" thick		690	.128		68	5.90	3.91	77.81	88

32 12 16 – Asphalt Paving

	32 12 16.13 Plant-Mix Asphalt Paving	Crew	Daily Output	Labor-Hours	Unit	Material	Labor (2020 Bare Costs)	Equipment	Total	Total Incl O&P
0812	3" thick	B-25	800	.110	Ton	68	5.10	3.37	76.47	86.50
0813	4" thick		900	.098		68	4.53	3	75.53	85
0850	Wearing course, 1" thick	B-25B	575	.167		73.50	7.85	5.10	86.45	99
0851	1-1/2" thick		630	.152		73.50	7.20	4.66	85.36	97.50
0852	2" thick		690	.139		73.50	6.55	4.26	84.31	96
0853	2-1/2" thick		765	.125		73.50	5.90	3.84	83.24	94.50
0854	3" thick		800	.120		73.50	5.65	3.67	82.82	94
1000	Pavement replacement over trench, 2" thick	B-37	90	.533	S.Y.	7.60	23.50	2.84	33.94	49
1050	4" thick		70	.686		15.10	30.50	3.65	49.25	69
1080	6" thick		55	.873		24	38.50	4.65	67.15	93
3000	Prime coat, emulsion, 0.30 gal./S.Y., 1000 S.Y.	B-45	2500	.006		2.10	.34	.33	2.77	3.20
3100	Tack coat, emulsion, 0.10 gal./S.Y., 1000 S.Y.	"	2500	.006		.70	.34	.33	1.37	1.66

32 12 16.14 Asphaltic Concrete Paving

	32 12 16.14 Asphaltic Concrete Paving	Crew	Daily Output	Labor-Hours	Unit	Material	Labor	Equipment	Total	Total Incl O&P
0011	**ASPHALTIC CONCRETE PAVING**, parking lots & driveways									
0015	No asphalt hauling included									
0018	Use 6.05 C.Y. per inch per M.S.F. for hauling									
0020	6" stone base, 2" binder course, 1" topping	B-25C	9000	.005	S.F.	1.92	.25	.26	2.43	2.80
0025	2" binder course, 2" topping		9000	.005		2.36	.25	.26	2.87	3.29
0030	3" binder course, 2" topping		9000	.005		2.78	.25	.26	3.29	3.75
0035	4" binder course, 2" topping		9000	.005		3.19	.25	.26	3.70	4.19
0040	1-1/2" binder course, 1" topping		9000	.005		1.72	.25	.26	2.23	2.58
0042	3" binder course, 1" topping		9000	.005		2.33	.25	.26	2.84	3.26
0045	3" binder course, 3" topping		9000	.005		3.23	.25	.26	3.74	4.24
0050	4" binder course, 3" topping		9000	.005		3.63	.25	.26	4.14	4.69
0055	4" binder course, 4" topping		9000	.005		4.08	.25	.26	4.59	5.15
0300	Binder course, 1-1/2" thick		35000	.001		.62	.06	.07	.75	.85
0400	2" thick		25000	.002		.80	.09	.09	.98	1.12
0500	3" thick		15000	.003		1.23	.15	.16	1.54	1.77
0600	4" thick		10800	.004		1.62	.21	.22	2.05	2.35
0800	Sand finish course, 3/4" thick		41000	.001		.31	.06	.06	.43	.49
0900	1" thick		34000	.001		.38	.07	.07	.52	.61
1000	Fill pot holes, hot mix, 2" thick	B-16	4200	.008		.82	.34	.14	1.30	1.59
1100	4" thick		3500	.009		1.20	.41	.16	1.77	2.15
1120	6" thick		3100	.010		1.61	.46	.18	2.25	2.70
1140	Cold patch, 2" thick	B-51	3000	.016		.87	.69	.07	1.63	2.14
1160	4" thick		2700	.018		1.66	.77	.07	2.50	3.14
1180	6" thick		1900	.025		2.58	1.09	.10	3.77	4.70
3000	Prime coat, emulsion, 0.30 gal./S.Y., 1000 S.Y.	B-45	2500	.006	S.Y.	2.10	.34	.33	2.77	3.20
3100	Tack coat, emulsion, 0.10 gal./S.Y., 1000 S.Y.	"	2500	.006	"	.70	.34	.33	1.37	1.66

32 12 36 – Seal Coats

32 12 36.13 Chip Seal

	32 12 36.13 Chip Seal	Crew	Daily Output	Labor-Hours	Unit	Material	Labor	Equipment	Total	Total Incl O&P
0010	**CHIP SEAL**									
0011	Excludes crack repair and flush coat									
1000	Fine - PMCRS-2h (20lbs/sy, 1/4" (No.10), .30gal/sy app. rate)									
1010	Small, irregular areas	B-91	5000	.013	S.Y.	1.10	.65	.45	2.20	2.73
1020	Parking Lot	B-91D	15000	.007		1.10	.34	.22	1.66	2
1030	Roadway	"	30000	.003		1.10	.17	.11	1.38	1.60
1090	For Each .5% Latex Additive, Add					.23			.23	.25
1100	Medium Fine - PMCRS-2h (25lbs/sy, 5/16" (No.8), .35gal/sy app. rate)									
1110	Small, irregular areas	B-91	4000	.016	S.Y.	1.29	.81	.56	2.66	3.31
1120	Parking Lot	B-91D	12000	.009		1.29	.42	.28	1.99	2.39
1130	Roadway	"	24000	.004		1.29	.21	.14	1.64	1.90

32 12 36 – Seal Coats

32 12 36.13 Chip Seal		Crew	Daily Output	Labor-Hours	Unit	2020 Bare Costs Material	Labor	Equipment	Total	Total Incl O&P
1190	For Each .5% Latex Additive, Add				S.Y.	.28			.28	.31
1200	Medium - PMCRS-2h (30lbs/sy, 3/8" (No.6), .40gal/sy app. rate)									
1210	Small, irregular areas	B-91	3330	.019	S.Y.	1.41	.97	.68	3.06	3.83
1220	Parking Lot	B-91D	10000	.010		1.41	.51	.34	2.26	2.73
1230	Roadway	″	20000	.005		1.41	.25	.17	1.83	2.13
1290	For Each .5% Latex Additive, Add					.34			.34	.37
1300	Course - PMCRS-2h (30lbs/sy, 1/2" (No.4), .40gal/sy app. rate)									
1310	Small, irregular areas	B-91	2500	.026	S.Y.	1.39	1.29	.90	3.58	4.55
1320	Parking Lot	B-91D	7500	.014		1.39	.68	.45	2.52	3.09
1330	Roadway	″	15000	.007		1.39	.34	.22	1.95	2.31
1390	For Each .5% Latex Additive, Add					.34			.34	.37
1400	Double - PMCRS-2h (Course Base w/ Fine Top)									
1410	Small, irregular areas	B-91	2000	.032	S.Y.	2.36	1.62	1.13	5.11	6.40
1420	Parking Lot	B-91D	6000	.017		2.36	.85	.56	3.77	4.56
1430	Roadway	″	12000	.009		2.36	.42	.28	3.06	3.58
1490	For Each .5% Latex Additive, Add					.56			.56	.62

32 12 36.14 Flush Coat

		Crew	Daily Output	Labor-Hours	Unit	Material	Labor	Equipment	Total	Total Incl O&P
0010	**FLUSH COAT**									
0011	Fog Seal w/ Sand Cover.18 gal/sy, 6lbs/sy									
1010	Small, irregular areas	B-91	2000	.032	S.Y.	.52	1.62	1.13	3.27	4.38
1020	Parking lot		6000	.011		.52	.54	.38	1.44	1.84
1030	Roadway		12000	.005		.52	.27	.19	.98	1.22

32 12 36.33 Slurry Seal

		Crew	Daily Output	Labor-Hours	Unit	Material	Labor	Equipment	Total	Total Incl O&P
0010	**SLURRY SEAL**									
0011	Includes sweeping and cleaning of area									
1000	Type I-PMCQS-1h-EAS (12lbs/sy, 1/8", 20% asphalt emulsion)									
1010	Small, irregular areas	B-90	8000	.008	S.Y.	.98	.37	.27	1.62	1.97
1020	Parking Lot		25000	.003		.98	.12	.09	1.19	1.36
1030	Roadway		50000	.001		.98	.06	.04	1.08	1.22
1090	For Each .5% Latex Additive, Add					.15			.15	.16
1100	Type II-PMCQS-1h-EAS (15lbs/sy, 1/4", 18% asphalt emulsion)									
1110	Small, irregular areas	B-90	6000	.011	S.Y.	1.10	.50	.37	1.97	2.40
1120	Parking lot		20000	.003		1.10	.15	.11	1.36	1.57
1130	Roadway		40000	.002		1.10	.07	.05	1.22	1.39
1200	Type III-PMCQS-1h-EAS (25lbs/sy, 3/8", 15% asphalt emulsion)									
1210	Small, irregular areas	B-90	4000	.016	S.Y.	1.51	.75	.55	2.81	3.45
1220	Parking lot		12000	.005		1.51	.25	.18	1.94	2.26
1230	Roadway		24000	.003		1.51	.12	.09	1.72	1.96
1290	For Each .5% Latex Additive, Add					.29			.29	.32

32 12 36.36 Microsurfacing

		Crew	Daily Output	Labor-Hours	Unit	Material	Labor	Equipment	Total	Total Incl O&P
0010	**MICROSURFACING**									
1100	Type II-MSE (20lbs/sy, 1/4", 18% microsurfacing emulsion)									
1110	Small, irregular areas	B-90	5000	.013	S.Y.	1.65	.60	.44	2.69	3.24
1120	Parking lot		15000	.004		1.65	.20	.15	2	2.29
1130	Roadway		30000	.002		1.65	.10	.07	1.82	2.05
1200	Type IIIa-MSE (32lbs/sy, 3/8", 15% microsurfacing emulsion)									
1210	Small, irregular areas	B-90	3000	.021	S.Y.	2.29	1	.73	4.02	4.92
1220	Parking lot		9000	.007		2.29	.33	.24	2.86	3.32
1230	Roadway		18000	.004		2.29	.17	.12	2.58	2.92

32 13 13 – Concrete Paving

32 13 13.25 Concrete Pavement, Highways		Crew	Daily Output	Labor-Hours	Unit	Material	Labor	Equipment	Total	Total Incl O&P
						2020 Bare Costs				
0010	**CONCRETE PAVEMENT, HIGHWAYS**									
0015	Including joints, finishing and curing									
0020	Fixed form, 12′ pass, unreinforced, 6″ thick	B-26	3000	.029	S.Y.	26	1.38	1.18	28.56	32
0030	7″ thick		2850	.031		31.50	1.45	1.24	34.19	38
0100	8″ thick		2750	.032		35.50	1.50	1.29	38.29	43
0110	8″ thick, small area		1375	.064		35.50	3.01	2.57	41.08	46.50
0200	9″ thick		2500	.035		40	1.65	1.41	43.06	48.50
0300	10″ thick		2100	.042		44	1.97	1.68	47.65	53.50
0310	10″ thick, small area		1050	.084		44	3.94	3.37	51.31	58.50
0400	12″ thick		1800	.049		50.50	2.30	1.96	54.76	62
0410	Conc. pavement, w/jt., fnsh.& curing, fix form, 24′ pass, unreinforced, 6″T		6000	.015		24.50	.69	.59	25.78	28.50
0420	7″ thick		5700	.015		29.50	.73	.62	30.85	34.50
0430	8″ thick		5500	.016		33.50	.75	.64	34.89	39
0440	9″ thick		5000	.018		38.50	.83	.71	40.04	44
0450	10″ thick		4200	.021		42.50	.98	.84	44.32	49
0460	12″ thick		3600	.024		49	1.15	.98	51.13	56.50
0470	15″ thick		3000	.029		64	1.38	1.18	66.56	73.50
0500	Fixed form 12′ pass, 15″ thick	↓	1500	.059	↓	64.50	2.76	2.36	69.62	78
0510	For small irregular areas, add				%	10%	100%	100%		
0520	Welded wire fabric, sheets for rigid paving 2.33 lb./S.Y.	2 Rodm	389	.041	S.Y.	1.46	2.32		3.78	5.25
0530	Reinforcing steel for rigid paving 12 lb./S.Y.		666	.024		7.15	1.35		8.50	10
0540	Reinforcing steel for rigid paving 18 lb./S.Y.	↓	444	.036	↓	10.70	2.03		12.73	14.95
0610	For under 10′ pass, add				%	10%	100%	100%		
0620	Slip form, 12′ pass, unreinforced, 6″ thick	B-26A	5600	.016	S.Y.	25	.74	.65	26.39	29.50
0622	7″ thick		5600	.016		30.50	.74	.65	31.89	35.50
0624	8″ thick		5300	.017		34.50	.78	.69	35.97	39.50
0626	9″ thick		4820	.018		39	.86	.76	40.62	45
0628	10″ thick		4050	.022		43	1.02	.90	44.92	50
0630	12″ thick		3470	.025		49.50	1.19	1.05	51.74	57.50
0632	15″ thick		2890	.030		62.50	1.43	1.26	65.19	72.50
0640	Slip form, 24′ pass, unreinforced, 6″ thick		11200	.008		24.50	.37	.33	25.20	28
0642	7″ thick		11200	.008		29	.37	.33	29.70	33
0644	8″ thick		10600	.008		33	.39	.34	33.73	37
0646	9″ thick		9640	.009		38	.43	.38	38.81	42.50
0648	10″ thick		8100	.011		41.50	.51	.45	42.46	47.50
0650	12″ thick		6940	.013		48.50	.60	.53	49.63	54.50
0652	15″ thick	↓	5780	.015		60.50	.71	.63	61.84	69
0700	Finishing, broom finish small areas	2 Cefi	120	.133	↓		6.65		6.65	10.45
0710	Transverse joint support dowels	C-1	350	.091	Ea.	4.55	4.61		9.16	12.40
0720	Transverse contraction joints, saw cut & grind	A-1B	120	.067	L.F.		2.81	.93	3.74	5.50
0730	Transverse expansion joints, incl. premolded bit. jt. filler	C-1	150	.213		2.06	10.75		12.81	19.45
0740	Transverse construction joint using bulkhead	″	73	.438	↓	2.99	22		24.99	39
0750	Longitudinal joint tie bars, grouted	B-23	70	.571	Ea.	4.64	24.50	22.50	51.64	69
1000	Curing, with sprayed membrane by hand	2 Clab	1500	.011	S.Y.	1.12	.45		1.57	1.95

32 14 Unit Paving

32 14 13 – Precast Concrete Unit Paving

32 14 13.13 Interlocking Precast Concrete Unit Paving

	32 14 13.13 Interlocking Precast Concrete Unit Paving	Crew	Daily Output	Labor-Hours	Unit	2020 Bare Costs Material	Labor	Equipment	Total	Total Incl O&P
0010	**INTERLOCKING PRECAST CONCRETE UNIT PAVING**									
0020	"V" blocks for retaining soil	D-1	205	.078	S.F.	11.15	3.66		14.81	18.20

32 14 13.16 Precast Concrete Unit Paving Slabs

	32 14 13.16 Precast Concrete Unit Paving Slabs	Crew	Daily Output	Labor-Hours	Unit	Material	Labor	Equipment	Total	Total Incl O&P
0010	**PRECAST CONCRETE UNIT PAVING SLABS**									
0710	Precast concrete patio blocks, 2-3/8" thick, colors, 8" x 16"	D-1	265	.060	S.F.	10.65	2.83		13.48	16.35
0715	12" x 12"		300	.053		13.85	2.50		16.35	19.30
0720	16" x 16"		335	.048		15.25	2.24		17.49	20.50
0730	24" x 24"		510	.031		19	1.47		20.47	23.50
0740	Green, 8" x 16"		265	.060		13.75	2.83		16.58	19.75
0750	Exposed local aggregate, natural	2 Bric	250	.064		9.20	3.33		12.53	15.50
0800	Colors		250	.064		9.40	3.33		12.73	15.75
0850	Exposed granite or limestone aggregate		250	.064		9.10	3.33		12.43	15.40
0900	Exposed white tumblestone aggregate		250	.064		10.65	3.33		13.98	17.15

32 14 13.18 Precast Concrete Plantable Pavers

	32 14 13.18 Precast Concrete Plantable Pavers	Crew	Daily Output	Labor-Hours	Unit	Material	Labor	Equipment	Total	Total Incl O&P
0010	**PRECAST CONCRETE PLANTABLE PAVERS** (50% grass)									
0015	Subgrade preparation and grass planting not included									
0100	Precast concrete plantable pavers with topsoil, 24" x 16"	B-63	800	.050	S.F.	4.52	2.21	.22	6.95	8.75
0200	Less than 600 S.F. or irregular area	"	500	.080	"	4.52	3.54	.36	8.42	11
0300	3/4" crushed stone base for plantable pavers, 6" depth	B-62	1000	.024	S.Y.	4.20	1.10	.18	5.48	6.55
0400	8" depth		900	.027		5.60	1.22	.20	7.02	8.30
0500	10" depth		800	.030		7	1.37	.22	8.59	10.10
0600	12" depth		700	.034		8.40	1.57	.25	10.22	12
0700	Hydro seeding plantable pavers	B-81A	20	.800	M.S.F.	12.10	35.50	47	94.60	122
0800	Apply fertilizer and seed to plantable pavers	1 Clab	8	1	"	46	42		88	119

32 14 16 – Brick Unit Paving

32 14 16.10 Brick Paving

	32 14 16.10 Brick Paving	Crew	Daily Output	Labor-Hours	Unit	Material	Labor	Equipment	Total	Total Incl O&P
0010	**BRICK PAVING**									
0012	4" x 8" x 1-1/2", without joints (4.5 bricks/S.F.)	D-1	110	.145	S.F.	2.66	6.80		9.46	13.90
0100	Grouted, 3/8" joint (3.9 bricks/S.F.)		90	.178		2.17	8.35		10.52	15.85
0200	4" x 8" x 2-1/4", without joints (4.5 bricks/S.F.)		110	.145		2.51	6.80		9.31	13.75
0300	Grouted, 3/8" joint (3.9 bricks/S.F.)		90	.178		2.17	8.35		10.52	15.85
0455	Pervious brick paving, 4" x 8" x 3-1/4", without joints (4.5 bricks/S.F.)		110	.145		3.70	6.80		10.50	15.05
0500	Bedding, asphalt, 3/4" thick	B-25	5130	.017		.68	.79	.53	2	2.59
0540	Course washed sand bed, 1" thick	B-18	5000	.005		.38	.21	.03	.62	.79
0580	Mortar, 1" thick	D-1	300	.053		.68	2.50		3.18	4.79
0620	2" thick		200	.080		1.37	3.75		5.12	7.55
2000	Brick pavers, laid on edge, 7.2/S.F.		70	.229		4.61	10.70		15.31	22.50
2500	For 4" thick concrete bed and joints, add		595	.027		1.31	1.26		2.57	3.49
2800	For steam cleaning, add	A-1H	950	.008		.12	.35	.08	.55	.79
9000	Minimum labor/equipment charge	1 Bric	2	4	Job		208		208	335

32 14 23 – Asphalt Unit Paving

32 14 23.10 Asphalt Blocks

	32 14 23.10 Asphalt Blocks	Crew	Daily Output	Labor-Hours	Unit	Material	Labor	Equipment	Total	Total Incl O&P
0010	**ASPHALT BLOCKS**									
0020	Rectangular, 6" x 12" x 1-1/4", w/bed & neopr. adhesive	D-1	135	.119	S.F.	10.20	5.55		15.75	20
0100	3" thick		130	.123		14.30	5.75		20.05	25
0300	Hexagonal tile, 8" wide, 1-1/4" thick		135	.119		10.20	5.55		15.75	20
0400	2" thick		130	.123		14.30	5.75		20.05	25
0500	Square, 8" x 8", 1-1/4" thick		135	.119		10.20	5.55		15.75	20
0600	2" thick		130	.123		14.30	5.75		20.05	25
0900	For exposed aggregate (ground finish), add					.58			.58	.64
0910	For colors, add					.58			.58	.64

32 14 Unit Paving

32 14 23 – Asphalt Unit Paving

32 14 23.10 Asphalt Blocks		Crew	Daily Output	Labor-Hours	Unit	Material	2020 Bare Costs Labor	Equipment	Total	Total Incl O&P
9000	Minimum labor/equipment charge	1 Bric	2	4	Job		208		208	335

32 14 40 – Stone Paving

32 14 40.10 Stone Pavers

		Crew	Daily Output	Labor-Hours	Unit	Material	Labor	Equipment	Total	Total Incl O&P
0010	STONE PAVERS									
1100	Flagging, bluestone, irregular, 1" thick,	D-1	81	.198	S.F.	10.35	9.25		19.60	26.50
1110	1-1/2" thick		90	.178		12.20	8.35		20.55	27
1120	Pavers, 1/2" thick		110	.145		17.25	6.80		24.05	30
1130	3/4" thick		95	.168		22	7.90		29.90	37
1140	1" thick		81	.198		23.50	9.25		32.75	41
1150	Snapped random rectangular, 1" thick		92	.174		15.70	8.15		23.85	30.50
1200	1-1/2" thick		85	.188		18.85	8.85		27.70	35
1250	2" thick		83	.193		22	9.05		31.05	38.50
1300	Slate, natural cleft, irregular, 3/4" thick		92	.174		9.75	8.15		17.90	24
1310	1" thick		85	.188		11.35	8.85		20.20	27
1350	Random rectangular, gauged, 1/2" thick		105	.152		21	7.15		28.15	34.50
1400	Random rectangular, butt joint, gauged, 1/4" thick		150	.107		22.50	5		27.50	33
1450	For sand rubbed finish, add					9.65			9.65	10.65
1500	For interior setting, add								25%	25%
1550	Granite blocks, 3-1/2" x 3-1/2" x 3-1/2"	D-1	92	.174	S.F.	21.50	8.15		29.65	36.50
1560	4" x 4" x 4"		95	.168		22.50	7.90		30.40	38
1600	4" to 12" long, 3" to 5" wide, 3" to 5" thick		98	.163		17.95	7.65		25.60	32
1650	6" to 15" long, 3" to 6" wide, 3" to 5" thick		105	.152		9.60	7.15		16.75	22

32 16 Curbs, Gutters, Sidewalks, and Driveways

32 16 13 – Curbs and Gutters

32 16 13.13 Cast-in-Place Concrete Curbs and Gutters

		Crew	Daily Output	Labor-Hours	Unit	Material	Labor	Equipment	Total	Total Incl O&P
0010	CAST-IN-PLACE CONCRETE CURBS AND GUTTERS									
0290	Forms only, no concrete									
0300	Concrete, wood forms, 6" x 18", straight	C-2	500	.096	L.F.	3.16	4.96		8.12	11.40
0400	6" x 18", radius	"	200	.240	"	3.30	12.40		15.70	23.50
0402	Forms and concrete complete									
0404	Concrete, wood forms, 6" x 18", straight & concrete	C-2A	500	.096	L.F.	6.75	4.91		11.66	15.30
0406	6" x 18", radius	"	200	.240		6.90	12.25		19.15	27
0415	Machine formed, 6" x 18", straight	B-69A	2000	.024		5.15	1.11	.61	6.87	8.10
0416	6" x 18", radius	"	900	.053		5.20	2.46	1.35	9.01	11.15
0421	Curb and gutter, straight									
0422	with 6" high curb and 6" thick gutter, wood forms									
0430	24" wide, 0.055 C.Y./L.F.	C-2A	375	.128	L.F.	17.35	6.55		23.90	29.50
0435	30" wide, 0.066 C.Y./L.F.	"	340	.141	"	19.20	7.20		26.40	32.50

32 16 13.23 Precast Concrete Curbs and Gutters

		Crew	Daily Output	Labor-Hours	Unit	Material	Labor	Equipment	Total	Total Incl O&P
0010	PRECAST CONCRETE CURBS AND GUTTERS									
0550	Precast, 6" x 18", straight	B-29	700	.080	L.F.	9.15	3.68	1.21	14.04	17.30
0600	6" x 18", radius	"	325	.172	"	12.20	7.95	2.61	22.76	29

32 16 13.33 Asphalt Curbs

		Crew	Daily Output	Labor-Hours	Unit	Material	Labor	Equipment	Total	Total Incl O&P
0010	ASPHALT CURBS									
0012	Curbs, asphaltic, machine formed, 8" wide, 6" high, 40 L.F./ton	B-27	1000	.032	L.F.	1.64	1.36	.27	3.27	4.28
0100	8" wide, 8" high, 30 L.F./ton		900	.036		2.19	1.51	.30	4	5.15
0150	Asphaltic berm, 12" W, 3" to 6" H, 35 L.F./ton, before pavement		700	.046		.04	1.95	.39	2.38	3.60
0200	12" W, 1-1/2" to 4" H, 60 L.F./ton, laid with pavement	B-2	1050	.038		.02	1.62		1.64	2.62

32 16 Curbs, Gutters, Sidewalks, and Driveways

32 16 13 – Curbs and Gutters

	32 16 13.43 Stone Curbs	Crew	Daily Output	Labor-Hours	Unit	2020 Bare Costs Material	Labor	Equipment	Total	Total Incl O&P
0010	**STONE CURBS**									
1000	Granite, split face, straight, 5" x 16"	D-13	275	.175	L.F.	16.50	8.80	1.56	26.86	34
1300	Radius curbing, 6" x 18", over 10' radius	B-29	260	.215	"	26.50	9.90	3.26	39.66	48.50
1400	Corners, 2' radius	"	80	.700	Ea.	89	32	10.60	131.60	161
1600	Edging, 4-1/2" x 12", straight	D-13	300	.160	L.F.	8.25	8.05	1.43	17.73	23.50
1800	Curb inlets (guttermouth) straight	B-29	41	1.366	Ea.	198	63	20.50	281.50	340
2000	Indian granite (Belgian block)									
2100	Jumbo, 10-1/2" x 7-1/2" x 4", grey	D-1	150	.107	L.F.	9.65	5		14.65	18.70
2150	Pink		150	.107		9.55	5		14.55	18.60
2200	Regular, 9" x 4-1/2" x 4-1/2", grey		160	.100		4.64	4.69		9.33	12.70
2250	Pink		160	.100		6.75	4.69		11.44	15.05
2300	Cubes, 4" x 4" x 4", grey		175	.091		3.34	4.29		7.63	10.60
2350	Pink		175	.091		3.99	4.29		8.28	11.35
2400	6" x 6" x 6", pink		155	.103		12.60	4.84		17.44	21.50
2500	Alternate pricing method for Indian granite									
2550	Jumbo, 10-1/2" x 7-1/2" x 4" (30 lb.), grey				Ton	550			550	605
2600	Pink					555			555	610
2650	Regular, 9" x 4-1/2" x 4-1/2" (20 lb.), grey					325			325	360
2700	Pink					475			475	520
2750	Cubes, 4" x 4" x 4" (5 lb.), grey					405			405	445
2800	Pink					515			515	565
2850	6" x 6" x 6" (25 lb.), pink					490			490	540
2900	For pallets, add					22			22	24

32 17 Paving Specialties

32 17 13 – Parking Bumpers

	32 17 13.13 Metal Parking Bumpers	Crew	Daily Output	Labor-Hours	Unit	Material	Labor	Equipment	Total	Total Incl O&P
0010	**METAL PARKING BUMPERS**									
0015	Bumper rails for garages, 12 ga. rail, 6" wide, with steel									
0020	posts 12'-6" OC, minimum	E-4	190	.168	L.F.	22	9.80	.77	32.57	41
0030	Average		165	.194		27.50	11.30	.89	39.69	49.50
0100	Maximum		140	.229		33	13.30	1.05	47.35	59
0300	12" channel rail, minimum		160	.200		27.50	11.65	.92	40.07	50
0400	Maximum		120	.267		41	15.50	1.22	57.72	72.50
1300	Pipe bollards, conc. filled/paint, 8' L x 4' D hole, 6" diam.	B-6	20	1.200	Ea.	635	55	10.70	700.70	800
1400	8" diam.		15	1.600		740	73	14.25	827.25	945
1500	12" diam.		12	2		1,025	91.50	17.80	1,134.30	1,300
2030	Folding with individual padlocks	B-2	50	.800		205	34		239	280
8000	Parking lot control, see Section 11 12 13.10									
8900	Security bollards, SS, lighted, hyd., incl. controls, group of 3	L-7	.06	509	Ea.	43,000	26,000		69,000	89,000
8910	Group of 5	"	.04	683	"	75,000	34,900		109,900	138,000
9000	Minimum labor/equipment charge	E-4	2	16	Job		930	73.50	1,003.50	1,600
	32 17 13.16 Plastic Parking Bumpers									
0010	**PLASTIC PARKING BUMPERS**									
1200	Thermoplastic, 6" x 10" x 6'-0"	B-2	120	.333	Ea.	57	14.15		71.15	85
	32 17 13.19 Precast Concrete Parking Bumpers									
0010	**PRECAST CONCRETE PARKING BUMPERS**									
1000	Wheel stops, precast concrete incl. dowels, 6" x 10" x 6'-0"	B-2	120	.333	Ea.	63.50	14.15		77.65	92
1100	8" x 13" x 6'-0"	"	120	.333	"	74	14.15		88.15	104

32 17 13 – Parking Bumpers

32 17 13.26 Wood Parking Bumpers		Crew	Daily Output	Labor-Hours	Unit	2020 Bare Costs Material	Labor	Equipment	Total	Total Incl O&P
0010	**WOOD PARKING BUMPERS**									
0020	Parking barriers, timber w/saddles, treated type									
0100	4″ x 4″ for cars	B-2	520	.077	L.F.	3.16	3.27		6.43	8.75
0200	6″ x 6″ for trucks		520	.077	″	6.55	3.27		9.82	12.45
0600	Flexible fixed stanchion, 2′ high, 3″ diameter		100	.400	Ea.	44.50	17		61.50	76

32 17 23 – Pavement Markings

32 17 23.13 Painted Pavement Markings		Crew	Daily Output	Labor-Hours	Unit	Material	Labor	Equipment	Total	Total Incl O&P
0010	**PAINTED PAVEMENT MARKINGS**									
0020	Acrylic waterborne, white or yellow, 4″ wide, less than 3,000 L.F.	B-78	20000	.002	L.F.	.11	.10	.05	.26	.35
0200	6″ wide, less than 3,000 L.F.		11000	.004		.16	.19	.10	.45	.59
0500	8″ wide, less than 3,000 L.F.		10000	.005		.21	.21	.11	.53	.69
0600	12″ wide, less than 3,000 L.F.		4000	.012		.32	.52	.26	1.10	1.47
0620	Arrows or gore lines		2300	.021	S.F.	.22	.90	.46	1.58	2.20
0640	Temporary paint, white or yellow, less than 3,000 L.F.		15000	.003	L.F.	.12	.14	.07	.33	.43
0660	Removal	1 Clab	300	.027			1.12		1.12	1.80
0680	Temporary tape	2 Clab	1500	.011		.53	.45		.98	1.30
0710	Thermoplastic, white or yellow, 4″ wide, less than 6,000 L.F.	B-79	15000	.003		.42	.12	.12	.66	.78
0730	6″ wide, less than 6,000 L.F.		14000	.003		.63	.12	.12	.87	1.03
0740	8″ wide, less than 6,000 L.F.		12000	.003		.84	.15	.14	1.13	1.31
0750	12″ wide, less than 6,000 L.F.		6000	.007		1.24	.29	.29	1.82	2.15
0760	Arrows		660	.061	S.F.	.69	2.64	2.63	5.96	7.90
0770	Gore lines		2500	.016		.69	.70	.69	2.08	2.64
0780	Letters		660	.061		.70	2.64	2.63	5.97	7.90
0782	Thermoplastic material, small users				Ton	2,050			2,050	2,250
0784	Glass beads, highway use, add				Lb.	.68			.68	.75
0786	Thermoplastic material, highway departments				Ton	1,750			1,750	1,925
1000	Airport painted markings									
1050	Traffic safety flashing truck for airport painting	A-2B	1	8	Day		380	196	576	820
1100	Painting, white or yellow, taxiway markings	B-78	4000	.012	S.F.	.38	.52	.26	1.16	1.54
1110	with 12 lb. beads per 100 S.F.		4000	.012		.46	.52	.26	1.24	1.63
1200	Runway markings		3500	.014		.38	.59	.30	1.27	1.70
1210	with 12 lb. beads per 100 S.F.		3500	.014		.46	.59	.30	1.35	1.79
1300	Pavement location or direction signs		2500	.019		.38	.83	.42	1.63	2.22
1310	with 12 lb. beads per 100 S.F.		2500	.019		.46	.83	.42	1.71	2.31
1350	Mobilization airport pavement painting		4	12	Ea.		520	265	785	1,125
1400	Paint markings or pavement signs removal daytime	B-78B	400	.045	S.F.		1.95	1.01	2.96	4.22
1500	Removal nighttime		335	.054	″		2.33	1.20	3.53	5.05
1600	Mobilization pavement paint removal		4	4.500	Ea.		195	101	296	420

32 17 23.14 Pavement Parking Markings		Crew	Daily Output	Labor-Hours	Unit	Material	Labor	Equipment	Total	Total Incl O&P
0010	**PAVEMENT PARKING MARKINGS**									
0790	Layout of pavement marking	A-2	25000	.001	L.F.		.04	.01	.05	.08
0800	Lines on pvmt., parking stall, paint, white, 4″ wide	B-78B	400	.045	Stall	5.05	1.95	1.01	8.01	9.75
0825	Parking stall, small quantities	2 Pord	80	.200		10.10	8.90		19	25.50
0830	Lines on pvmt., parking stall, thermoplastic, white, 4″ wide	B-79	300	.133		20	5.80	5.80	31.60	37.50
1000	Street letters and numbers	B-78B	1600	.011	S.F.	.82	.49	.25	1.56	1.96
1100	Pavement marking letter, 6″	2 Pord	400	.040	Ea.	12.05	1.78		13.83	16.05
1110	12″ letter		272	.059		15.15	2.61		17.76	21
1120	24″ letter		160	.100		35.50	4.44		39.94	46.50
1130	36″ letter		84	.190		43	8.45		51.45	61
1140	42″ letter		84	.190		52	8.45		60.45	71
1150	72″ letter		40	.400		46	17.75		63.75	79

32 17 Paving Specialties

32 17 23 – Pavement Markings

32 17 23.14 Pavement Parking Markings		Crew	Daily Output	Labor-Hours	Unit	Material	2020 Bare Costs Labor	Equipment	Total	Total Incl O&P
1200	Handicap symbol	2 Pord	40	.400	Ea.	34	17.75		51.75	65
1210	Handicap parking sign 12" x 18" and post	A-2	12	2		135	87.50	16.35	238.85	305
1300	Pavement marking, thermoplastic tape including layout, 4" width	B-79B	320	.025	L.F.	2.51	1.05	.54	4.10	5.05
1310	12" width		192	.042	"	9.15	1.75	.89	11.79	13.85
1320	Letters including layout, 4"		240	.033	Ea.	16.30	1.40	.71	18.41	21
1330	6"		160	.050		18.35	2.11	1.07	21.53	24.50
1340	12"		120	.067		21	2.81	1.43	25.24	29
1350	48"		64	.125		79	5.25	2.68	86.93	98.50
1360	96"		32	.250		100	10.55	5.35	115.90	133
1380	4 letter words, 8' tall		8	1		370	42	21.50	433.50	495

32 17 26 – Tactile Warning Surfacing

32 17 26.10 Tactile Warning Surfacing		Crew	Daily Output	Labor-Hours	Unit	Material	Labor	Equipment	Total	Total Incl O&P
0010	**TACTILE WARNING SURFACING**									
0100	Tactile warning tiles S.F.	2 Clab	400	.040	S.F.	18.35	1.68		20.03	22.50

32 18 Athletic and Recreational Surfacing

32 18 13 – Synthetic Grass Surfacing

32 18 13.10 Artificial Grass Surfacing		Crew	Daily Output	Labor-Hours	Unit	Material	Labor	Equipment	Total	Total Incl O&P
0010	**ARTIFICIAL GRASS SURFACING**									
0015	Not including asphalt base or drainage,									
0020	but including cushion pad, over 50,000 S.F.									
0200	1/2" pile and 5/16" cushion pad, standard	C-17	3200	.025	S.F.	6.50	1.38		7.88	9.35
0300	Deluxe		2560	.031		6.30	1.73		8.03	9.70
0500	1/2" pile and 5/8" cushion pad, standard		2844	.028		6.20	1.55		7.75	9.30
0600	Deluxe		2327	.034		6.30	1.90		8.20	9.95
0800	For asphaltic concrete base, 2-1/2" thick,									
0900	with 6" crushed stone sub-base, add	B-25	12000	.007	S.F.	1.73	.34	.22	2.29	2.69

32 18 16 – Synthetic Resilient Surfacing

32 18 16.13 Playground Protective Surfacing		Crew	Daily Output	Labor-Hours	Unit	Material	Labor	Equipment	Total	Total Incl O&P
0010	**PLAYGROUND PROTECTIVE SURFACING**									
0100	Resilient rubber surface, 4" thick, black	2 Skwk	300	.053	S.F.	14.95	2.93		17.88	21
0150	2" thick topping, colors	"	2800	.006		7.55	.31		7.86	8.80
0200	Wood chip mulch, 6" deep	1 Clab	300	.027		.91	1.12		2.03	2.80

32 18 16.16 Epoxy Acrylic Coatings		Crew	Daily Output	Labor-Hours	Unit	Material	Labor	Equipment	Total	Total Incl O&P
0010	**EPOXY ACRYLIC COATINGS**									
0100	Surface Prep, Sand blasting to SSPC-SP6, 2.0#/S.F. sand	E-11A	1200	.027	S.F.	.34	1.25	.51	2.10	3
0150	Layout and line striping by others									
0200	Painting epoxy, Foundation/Prime coat, large area (100 S.F. per Gal)	E-11B	2400	.010	S.F.	.37	.45	.17	.99	1.33
0210	Epoxy intermediate coat		2800	.009		.37	.38	.14	.89	1.20
0220	Epoxy top coat		2800	.009		.37	.38	.14	.89	1.20
0300	Painting epoxy, Foundation/Prime coat, small areas (100 S.F. per Gal)		1400	.017		.37	.76	.28	1.41	2
0310	Epoxy intermediate coat		1500	.016		.37	.71	.27	1.35	1.89
0320	Epoxy top coat		1500	.016		.37	.71	.27	1.35	1.89
0330	Adhesive Promoter		1400	.017		.27	.76	.28	1.31	1.90
0340	Sealer Concentrate		2400	.010		.15	.45	.17	.77	1.09
0400	Install debris tarp for construction work under 600 S.F.	2 Clab	2100	.008		.76	.32		1.08	1.35
0410	600 S.F. or more	"	2400	.007		.61	.28		.89	1.12

32 18 Athletic and Recreational Surfacing

32 18 23 – Athletic Surfacing

32 18 23.33 Running Track Surfacing		Crew	Daily Output	Labor-Hours	Unit	Material	2020 Bare Costs Labor	Equipment	Total	Total Incl O&P
0010	**RUNNING TRACK SURFACING**									
0020	Running track, asphalt concrete pavement, 2-1/2"	B-37	300	.160	S.Y.	15.95	7.10	.85	23.90	30
0102	Surface, latex rubber system, 1/2" thick, black	B-20	115	.209		53.50	9.80		63.30	74
0152	Colors		115	.209		65.50	9.80		75.30	87.50
0302	Urethane rubber system, 1/2" thick, black		110	.218		41	10.25		51.25	62
0402	Color coating		110	.218		51	10.25		61.25	72.50

32 18 23.53 Tennis Court Surfacing		Crew	Daily Output	Labor-Hours	Unit	Material	Labor	Equipment	Total	Total Incl O&P
0010	**TENNIS COURT SURFACING**									
0020	Tennis court, asphalt, incl. base, 2-1/2" thick, one court	B-37	450	.107	S.Y.	33.50	4.72	.57	38.79	44.50
0200	Two courts		675	.071		19.35	3.15	.38	22.88	27
0300	Clay courts		360	.133		46	5.90	.71	52.61	60.50
0400	Pulverized natural greenstone with 4" base, fast dry		250	.192		39	8.50	1.02	48.52	57
0800	Rubber-acrylic base resilient pavement		600	.080		67	3.54	.43	70.97	80
1000	Colored sealer, acrylic emulsion, 3 coats	2 Clab	800	.020		7.05	.84		7.89	9.10
1100	3 coats, 2 colors	"	900	.018		9.80	.75		10.55	12
1200	For preparing old courts, add	1 Clab	825	.010			.41		.41	.65
1400	Posts for nets, 3-1/2" diameter with eye bolts	B-1	3.40	7.059	Pr.	282	300		582	795
1500	With pulley & reel		3.40	7.059	"	900	300		1,200	1,475
1700	Net, 42' long, nylon thread with binder		50	.480	Ea.	275	20.50		295.50	340
1800	All metal		6.50	3.692	"	560	158		718	870
2000	Paint markings on asphalt, 2 coats	1 Pord	1.78	4.494	Court	252	200		452	590
2200	Complete court with fence, etc., asphaltic conc., minimum	B-37	.20	240		34,500	10,600	1,275	46,375	56,000
2300	Maximum		.16	300		82,500	13,300	1,600	97,400	113,500
2800	Clay courts, minimum		.20	240		38,000	10,600	1,275	49,875	60,000
2900	Maximum		.16	300		69,500	13,300	1,600	84,400	99,500

32 31 Fences and Gates

32 31 11 – Gate Operators

32 31 11.10 Gate Operators		Crew	Daily Output	Labor-Hours	Unit	Material	Labor	Equipment	Total	Total Incl O&P
0010	**GATE OPERATORS**									
7810	Motor operators for gates (no elec wiring), 3' wide swing	2 Skwk	.50	32	Ea.	1,150	1,750		2,900	4,075
7815	Up to 20' wide swing		.50	32		1,450	1,750		3,200	4,400
7820	Up to 45' sliding		.50	32		3,300	1,750		5,050	6,450
7825	Overhead gate, 6' to 18' wide, sliding/cantilever		45	.356	L.F.	340	19.50		359.50	405
7830	Gate operators, digital receiver		7	2.286	Ea.	87	125		212	296
7835	Two button transmitter		24	.667		28	36.50		64.50	89.50
7840	3 button station		14	1.143		43	62.50		105.50	148
7845	Master slave system		4	4		189	219		408	560

32 31 13 – Chain Link Fences and Gates

32 31 13.10 Chain Link Gates and Fences		Crew	Daily Output	Labor-Hours	Unit	Material	Labor	Equipment	Total	Total Incl O&P
0010	**CHAIN LINK GATES AND FENCES**									
4750	Gate, transom for 10' fence, galv. steel, single, 3' x 7'	B-80A	52	.462	Ea.	400	19.45	15.80	435.25	490
4752	4' x 7'		10	2.400		430	101	82	613	720
4754	3' x 10'		8	3		365	126	103	594	715
4756	4' x 10'		10	2.400		390	101	82	573	680
4758	Double transom, 10' x 7'	B-80B	10	3.200		690	143	23.50	856.50	1,025
4760	12' x 7'		6	5.333		730	239	39.50	1,008.50	1,225
4762	14' x 7'		5	6.400		795	287	47	1,129	1,375
4764	10' x 10'		4	8		990	360	59	1,409	1,725
4766	12' x 10'		7	4.571		1,100	205	33.50	1,338.50	1,550

32 31 13 – Chain Link Fences and Gates

32 31 13.10 Chain Link Gates and Fences		Crew	Daily Output	Labor-Hours	Unit	2020 Bare Costs Material	Labor	Equipment	Total	Total Incl O&P
4768	14′ x 10′	B-80B	7	4.571	Ea.	1,175	205	33.50	1,413.50	1,625
4780	Vinyl-clad, single transom, 3′ x 7′		10	3.200		525	143	23.50	691.50	835
4782	4′ x 7′		10	3.200		570	143	23.50	736.50	885
4784	3′ x 10′		8	4		575	179	29.50	783.50	955
4786	4′ x 10′		8	4		650	179	29.50	858.50	1,025
4788	Double transom, 10′ x 7′		10	3.200		895	143	23.50	1,061.50	1,250
4790	12′ x 7′		6	5.333		920	239	39.50	1,198.50	1,450
4792	14′ x 7′		5	6.400		1,025	287	47	1,359	1,625
4794	10′ x 10′		4	8		995	360	59	1,414	1,725
4798	12′ x 12′		7	4.571		1,425	205	33.50	1,663.50	1,925
4799	14′ x 14′		7	4.571		1,900	205	33.50	2,138.50	2,425

32 31 13.20 Fence, Chain Link Industrial		Crew	Daily Output	Labor-Hours	Unit	Material	Labor	Equipment	Total	Total Incl O&P
0010	**FENCE, CHAIN LINK INDUSTRIAL**									
0011	Schedule 40, including concrete									
0020	3 strands barb wire, 2″ post @ 10′ OC, set in concrete, 6′ H									
0200	9 ga. wire, galv. steel, in concrete	B-80C	240	.100	L.F.	19.95	4.38	1.04	25.37	30
0248	Fence, add for vinyl coated fabric				S.F.	.71			.71	.78
0300	Aluminized steel	B-80C	240	.100	L.F.	22.50	4.38	1.04	27.92	33
0301	Fence, wrought iron		240	.100		43	4.38	1.04	48.42	55
0303	Fence, commercial 4′ high		240	.100		32.50	4.38	1.04	37.92	43.50
0304	Fence, commercial 6′ high		240	.100		58	4.38	1.04	63.42	72
0500	6 ga. wire, galv. steel		240	.100		23.50	4.38	1.04	28.92	34
0600	Aluminized steel		240	.100		31	4.38	1.04	36.42	42
0800	6 ga. wire, 6′ high but omit barbed wire, galv. steel		250	.096		21.50	4.21	1	26.71	31.50
0900	Aluminized steel, in concrete		250	.096		28.50	4.21	1	33.71	39
0920	8′ H, 6 ga. wire, 2-1/2″ line post, galv. steel, in concrete		180	.133		33.50	5.85	1.39	40.74	48
0940	Aluminized steel, in concrete		180	.133		41	5.85	1.39	48.24	56
1400	Gate for 6′ high fence, 1-5/8″ frame, 3′ wide, galv. steel		10	2.400	Ea.	209	105	25	339	425
1500	Aluminized steel, in concrete		10	2.400	″	218	105	25	348	435
2000	5′-0″ high fence, 9 ga., no barbed wire, 2″ line post, in concrete									
2010	10′ OC, 1-5/8″ top rail, in concrete									
2100	Galvanized steel, in concrete	B-80C	300	.080	L.F.	22.50	3.51	.83	26.84	31.50
2200	Aluminized steel, in concrete		300	.080	″	21	3.51	.83	25.34	29.50
2400	Gate, 4′ wide, 5′ high, 2″ frame, galv. steel, in concrete		10	2.400	Ea.	229	105	25	359	450
2500	Aluminized steel, in concrete		10	2.400	″	191	105	25	321	405
3100	Overhead slide gate, chain link, 6′ high, to 18′ wide, in concrete		38	.632	L.F.	95	27.50	6.60	129.10	157
3110	Cantilever type, in concrete	B-80	48	.667		168	31	21.50	220.50	258
3120	8′ high, in concrete		24	1.333		176	62	42.50	280.50	340
3130	10′ high, in concrete		18	1.778		218	83	56.50	357.50	435
5000	Double swing gates, incl. posts & hardware, in concrete									
5010	5′ high, 12′ opening, in concrete	B-80C	3.40	7.059	Opng.	535	310	73.50	918.50	1,175
5020	20′ opening, in concrete		2.80	8.571		595	375	89.50	1,059.50	1,350
5060	6′ high, 12′ opening, in concrete		3.20	7.500		465	330	78	873	1,125
5070	20′ opening, in concrete		2.60	9.231		710	405	96	1,211	1,525
5080	8′ high, 12′ opening, in concrete	B-80	2.13	15.002		545	700	480	1,725	2,225
5090	20′ opening, in concrete		1.45	22.069		845	1,025	705	2,575	3,325
5100	10′ high, 12′ opening, in concrete		1.31	24.427		830	1,150	780	2,760	3,575
5110	20′ opening, in concrete		1.03	31.068		880	1,450	990	3,320	4,375
5120	12′ high, 12′ opening, in concrete		1.05	30.476		1,300	1,425	970	3,695	4,750
5130	20′ opening, in concrete		.85	37.647		1,325	1,750	1,200	4,275	5,575
5190	For aluminized steel, add					20%				
7055	Braces, galv. steel	B-80A	960	.025	L.F.	2.72	1.05	.85	4.62	5.60

32 31 Fences and Gates

32 31 13 – Chain Link Fences and Gates

32 31 13.20 Fence, Chain Link Industrial		Crew	Daily Output	Labor-Hours	Unit	2020 Bare Costs Material	Labor	Equipment	Total	Total Incl O&P
7056	Aluminized steel	B-80A	960	.025	L.F.	3.26	1.05	.85	5.16	6.20
7071	Privacy slats, vertical, vinyl	1 Clab	500	.016	S.F.	1.76	.67		2.43	3.02
7072	Redwood		450	.018		1.51	.75		2.26	2.86
7073	Diagonal, aluminum		300	.027		4.86	1.12		5.98	7.15
9000	Minimum labor/equipment charge	B-80	2	16	Job		745	510	1,255	1,725

32 31 13.25 Fence, Chain Link Residential		Crew	Daily Output	Labor-Hours	Unit	Material	Labor	Equipment	Total	Total Incl O&P
0010	**FENCE, CHAIN LINK RESIDENTIAL**									
0011	Schedule 20, 11 ga. wire, 1-5/8" post									
0020	10' OC, 1-3/8" top rail, 2" corner post, galv. stl. 3' high	B-80C	500	.048	L.F.	4.95	2.10	.50	7.55	9.35
0050	4' high		400	.060		5.60	2.63	.63	8.86	11.05
0100	6' high		200	.120		7.25	5.25	1.25	13.75	17.80
0150	Add for gate 3' wide, 1-3/8" frame, 3' high		12	2	Ea.	84.50	87.50	21	193	256
0170	4' high		10	2.400		91.50	105	25	221.50	297
0190	6' high		10	2.400		104	105	25	234	310
0200	Add for gate 4' wide, 1-3/8" frame, 3' high		9	2.667		96	117	28	241	325
0220	4' high		9	2.667		103	117	28	248	330
0240	6' high		8	3		118	131	31.50	280.50	375
0350	Aluminized steel, 11 ga. wire, 3' high		500	.048	L.F.	8.60	2.10	.50	11.20	13.40
0380	4' high		400	.060		7.85	2.63	.63	11.11	13.55
0400	6' high		200	.120		11.15	5.25	1.25	17.65	22
0450	Add for gate 3' wide, 1-3/8" frame, 3' high		12	2	Ea.	104	87.50	21	212.50	277
0470	4' high		10	2.400		97	105	25	227	305
0490	6' high		10	2.400		137	105	25	267	345
0500	Add for gate 4' wide, 1-3/8" frame, 3' high		10	2.400		108	105	25	238	315
0520	4' high		9	2.667		119	117	28	264	350
0540	6' high		8	3		133	131	31.50	295.50	390
0620	Vinyl covered, 9 ga. wire, 3' high		500	.048	L.F.	7.70	2.10	.50	10.30	12.35
0640	4' high		400	.060		8.20	2.63	.63	11.46	13.90
0660	6' high		200	.120		9.35	5.25	1.25	15.85	20
0720	Add for gate 3' wide, 1-3/8" frame, 3' high		12	2	Ea.	88	87.50	21	196.50	260
0740	4' high		10	2.400		95.50	105	25	225.50	300
0760	6' high		10	2.400		132	105	25	262	340
0780	Add for gate 4' wide, 1-3/8" frame, 3' high		10	2.400		109	105	25	239	315
0800	4' high		9	2.667		100	117	28	245	330
0820	6' high		8	3		138	131	31.50	300.50	395
7076	Fence, for small jobs 100 L.F. fence or less w/or w/o gate, add				L.F.	20%				
9000	Minimum labor/equipment charge	B-1	2	12	Job		515		515	820

32 31 13.26 Tennis Court Fences and Gates		Crew	Daily Output	Labor-Hours	Unit	Material	Labor	Equipment	Total	Total Incl O&P
0010	**TENNIS COURT FENCES AND GATES**									
0860	Tennis courts, 11 ga. wire, 2-1/2" post set									
0870	in concrete, 10' OC, 1-5/8" top rail									
0900	10' high	B-80	190	.168	L.F.	22	7.85	5.35	35.20	42.50
0920	12' high		170	.188	"	23	8.75	6	37.75	45.50
1000	Add for gate 4' wide, 1-5/8" frame 7' high		10	3.200	Ea.	231	149	102	482	605
1040	Aluminized steel, 11 ga. wire 10' high		190	.168	L.F.	20.50	7.85	5.35	33.70	41
1100	12' high		170	.188	"	25.50	8.75	6	40.25	48.50
1140	Add for gate 4' wide, 1-5/8" frame, 7' high		10	3.200	Ea.	238	149	102	489	610
1250	Vinyl covered, 9 ga. wire, 10' high		190	.168	L.F.	21.50	7.85	5.35	34.70	42
1300	12' high		170	.188	"	25	8.75	6	39.75	48
1310	Fence, CL, tennis court, transom gate, single, galv., 4' x 7'	B-80A	8.72	2.752	Ea.	350	116	94	560	680
1400	Add for gate 4' wide, 1-5/8" frame, 7' high	B-80	10	3.200	"	315	149	102	566	695

32 31 Fences and Gates

32 31 13 – Chain Link Fences and Gates

32 31 13.30 Fence, Chain Link, Gates and Posts		Crew	Daily Output	Labor-Hours	Unit	2020 Bare Costs Material	Labor	Equipment	Total	Total Incl O&P
0010	**FENCE, CHAIN LINK, GATES & POSTS**									
0011	(1/3 post length in ground)									
0013	For Concrete, See Section 03 31									
6580	Line posts, galvanized, 2-1/2" OD, set in conc., 4'	B-80	80	.400	Ea.	49.50	18.65	12.75	80.90	98
6585	5'		76	.421		55	19.65	13.45	88.10	106
6590	6'		74	.432		62.50	20	13.80	96.30	116
6595	7'		72	.444		68.50	20.50	14.20	103.20	124
6600	8'		69	.464		80.50	21.50	14.80	116.80	139
6635	Vinyl coated, 2-1/2" OD, set in conc., 4'		79	.405		58.50	18.90	12.90	90.30	109
6640	5'		77	.416		62.50	19.35	13.25	95.10	115
6645	6'		74	.432		66	20	13.80	99.80	120
6650	7'		72	.444		78	20.50	14.20	112.70	135
6655	8'		69	.464		99.50	21.50	14.80	135.80	161
6660	End gate post, steel, 3" OD, set in conc., 4'		68	.471		55	22	15	92	112
6665	5'		65	.492		54.50	23	15.70	93.20	114
6670	6'		63	.508		56.50	23.50	16.20	96.20	117
6675	7'		61	.525		64.50	24.50	16.75	105.75	128
6680	8'		59	.542		77.50	25.50	17.30	120.30	144
6685	Vinyl, 4'		68	.471		35.50	22	15	72.50	90.50
6690	5'		65	.492		46	23	15.70	84.70	104
6695	6'		63	.508		54	23.50	16.20	93.70	115
6700	7'		61	.525		55.50	24.50	16.75	96.75	118
6705	8'		59	.542		65.50	25.50	17.30	108.30	131
6710	Corner post, galv. steel, 4" OD, set in conc., 4'		65	.492		70.50	23	15.70	109.20	131
6715	6'		63	.508		109	23.50	16.20	148.70	175
6720	7'		61	.525		131	24.50	16.75	172.25	202
6725	8'		65	.492		202	23	15.70	240.70	276
6730	Vinyl, 5'		65	.492		50	23	15.70	88.70	109
6735	6'		63	.508		91	23.50	16.20	130.70	155
6740	7'		61	.525		73.50	24.50	16.75	114.75	138
6745	8'		59	.542		96.50	25.50	17.30	139.30	165
7031	For corner, end & pull post bracing, add					20%	15%			
7770	Gates, sliding w/overhead support, 4' high	B-80B	35	.914	L.F.	142	41	6.75	189.75	229
7775	5' high		32	1		160	45	7.40	212.40	256
7780	6' high		28	1.143		150	51	8.45	209.45	256
7785	7' high		25	1.280		185	57.50	9.45	251.95	305
7790	8' high		23	1.391		193	62.50	10.25	265.75	325
7795	Cantilever, manual, exp. roller (pr), 40' wide x 8' high	B-22	1	30	Ea.	5,950	1,475	284	7,709	9,200
7800	30' wide x 8' high		1	30		4,575	1,475	284	6,334	7,675
7805	24' wide x 8' high		1	30		3,075	1,475	284	4,834	6,025
7900	Auger fence post hole, 3' deep, medium soil, by hand	1 Clab	30	.267			11.25		11.25	18
7925	By machine	B-80	175	.183			8.50	5.85	14.35	19.95
7950	Rock, with jackhammer	B-9	32	1.250			53	11.15	64.15	97.50
7975	With rock drill	B-47C	65	.246			11.70	25.50	37.20	47

32 31 13.33 Chain Link Backstops

		Crew	Daily Output	Labor-Hours	Unit	Material	Labor	Equipment	Total	Total Incl O&P
0010	**CHAIN LINK BACKSTOPS**									
0015	Backstops, baseball, prefabricated, 30' wide, 12' high & 1 overhang	B-1	1	24	Ea.	2,775	1,025		3,800	4,725
0100	40' wide, 12' high & 2 overhangs	"	.75	32		7,075	1,375		8,450	9,975
0110	Regulation, galvanized	B-13	2.40	23.333		17,300	1,075	242	18,617	21,000
0120	Vinyl coated	"	2.40	23.333		18,600	1,075	242	19,917	22,500
0180	Softball, prefabricated, no overhang	B-1	1.60	15		3,100	640		3,740	4,425
0200	Softball, regulation, galv., 3" posts @ 8', 6 & 9 ga. mesh, 14' H	B-13	8.80	6.364		3,050	293	66	3,409	3,900

32 31 13 – Chain Link Fences and Gates

32 31 13.33 Chain Link Backstops		Crew	Daily Output	Labor-Hours	Unit	Material	2020 Bare Costs Labor	Equipment	Total	Total Incl O&P
0205	18′ high	B-13	.75	74.667	Ea.	8,100	3,450	775	12,325	15,200
0210	20′ high		.72	77.778		8,825	3,575	805	13,205	16,300
0215	22′ high		.70	80		8,550	3,675	830	13,055	16,200
0220	24′ high		.60	92.838		8,475	4,275	965	13,715	17,200
0250	Vinyl coated, 14′ high		1.20	46.667		10,600	2,150	485	13,235	15,700
0255	18′ high		.75	74.667		10,400	3,450	775	14,625	17,700
0260	20′ high		.72	77.778		15,600	3,575	805	19,980	23,800
0265	22′ high		.70	80		16,400	3,675	830	20,905	24,800
0270	24′ high		.60	93.333		18,400	4,300	970	23,670	28,100
0300	Basketball, steel, single goal		3.04	18.421		1,775	850	191	2,816	3,500
0400	Double goal		1.92	29.167		2,550	1,350	300	4,200	5,250
0600	Tennis, wire mesh with pair of ends	B-1	2.48	9.677	Set	2,775	415		3,190	3,725
0700	Enclosed court	"	1.30	18.462	Ea.	9,200	790		9,990	11,400

32 31 13.40 Fence, Fabric and Accessories		Crew	Daily Output	Labor-Hours	Unit	Material	Labor	Equipment	Total	Total Incl O&P
0010	**FENCE, FABRIC & ACCESSORIES**									
1000	Fabric, 9 ga., galv., 1.2 oz. coat, 2″ chain link, 4′	B-80A	304	.079	L.F.	3.56	3.32	2.70	9.58	12.25
1150	5′		285	.084		4.30	3.55	2.88	10.73	13.60
1200	6′		266	.090		9.20	3.80	3.08	16.08	19.65
1250	7′		247	.097		10.05	4.09	3.32	17.46	21.50
1300	8′		228	.105		12.80	4.43	3.60	20.83	25
1400	9 ga., fused, 4′		304	.079		4.25	3.32	2.70	10.27	13
1450	5′		285	.084		4.81	3.55	2.88	11.24	14.15
1500	6′		266	.090		4.80	3.80	3.08	11.68	14.80
1550	7′		247	.097		6.05	4.09	3.32	13.46	16.90
1600	8′		228	.105		11	4.43	3.60	19.03	23
1650	Barbed wire, galv., cost per strand		2280	.011		.15	.44	.36	.95	1.28
1700	Vinyl coated		2280	.011		.17	.44	.36	.97	1.30
1750	Extension arms, 3 strands		143	.168	Ea.	4.18	7.05	5.75	16.98	22
1800	6 strands, 2-3/8″		119	.202		11.15	8.50	6.90	26.55	33.50
1850	Eye tops, 2-3/8″		143	.168		1.73	7.05	5.75	14.53	19.50
1900	Top rail, incl. tie wires, 1-5/8″, galv.		912	.026	L.F.	4.86	1.11	.90	6.87	8.10
1950	Vinyl coated		912	.026		6.35	1.11	.90	8.36	9.70
2100	Rail, middle/bottom, w/tie wire, 1-5/8″, galv.		912	.026		5.15	1.11	.90	7.16	8.45
2150	Vinyl coated		912	.026		5.45	1.11	.90	7.46	8.75
2200	Reinforcing wire, coiled spring, 7 ga. galv.		2279	.011		.11	.44	.36	.91	1.23
2250	9 ga., vinyl coated		2282	.011		.58	.44	.36	1.38	1.75
2300	Steel T-post, galvanized with clips, 5′, common earth, flat		200	.120	Ea.	10.05	5.05	4.10	19.20	23.50
2310	Clay		176	.136		10.05	5.75	4.66	20.46	25.50
2320	Soil & rock		144	.167		10.05	7	5.70	22.75	28.50
2330	5-1/2′, common earth, flat		200	.120		10.70	5.05	4.10	19.85	24.50
2340	Clay		176	.136		10.70	5.75	4.66	21.11	26
2350	Soil & rock		144	.167		10.70	7	5.70	23.40	29.50
2360	6′, common earth, flat		200	.120		10.75	5.05	4.10	19.90	24.50
2370	Clay		176	.136		10.75	5.75	4.66	21.16	26
2375	Soil & rock		144	.167		10.75	7	5.70	23.45	29.50
2600	Steel T-post, galvanized with clips, 5′, common earth, hills		180	.133		10.05	5.60	4.56	20.21	25
2610	Clay		160	.150		10.05	6.30	5.15	21.50	27
2620	Soil & rock		130	.185		10.05	7.75	6.30	24.10	30.50
2630	5-1/2′, common earth, hills		180	.133		10.70	5.60	4.56	20.86	26
2640	Clay		160	.150		10.70	6.30	5.15	22.15	27.50
2650	Soil & rock		130	.185		10.70	7.75	6.30	24.75	31
2660	6′, common earth, hills		180	.133		10.75	5.60	4.56	20.91	26

32 31 13 – Chain Link Fences and Gates

32 31 13.40 Fence, Fabric and Accessories		Crew	Daily Output	Labor-Hours	Unit	Material	Labor	Equipment	Total	Total Incl O&P
						2020 Bare Costs				
2670	Clay	B-80A	160	.150	Ea.	10.75	6.30	5.15	22.20	27.50
2675	Soil & rock	↓	130	.185	↓	10.75	7.75	6.30	24.80	31.50

32 31 13.53 High-Security Chain Link Fences, Gates and Sys.

		Crew	Daily Output	Labor-Hours	Unit	Material	Labor	Equipment	Total	Total Incl O&P
0010	**HIGH-SECURITY CHAIN LINK FENCES, GATES AND SYSTEMS**									
0100	Fence, chain link, security, 7′ H, standard FE-7, incl. excavation & posts	B-80C	480	.050	L.F.	46.50	2.19	.52	49.21	55
0200	Fence, barbed wire, security, 7′ H, with 3 wire barbed wire arm	"	400	.060	"	8.20	2.63	.63	11.46	13.90
0300	Complete systems, including material and installation									
0310	Taunt wire fence detection system				M.L.F.				25,100	27,600
0410	Microwave fence detection system								41,300	45,400
0510	Passive magnetic fence detection system								19,500	21,400
0610	Infrared fence detection system								12,900	14,400
0710	Strain relief fence detection system								25,100	27,600
0810	Electro-shock fence detection system								35,900	39,500
0910	Photo-electric fence detection system				↓				16,300	18,000

32 31 13.64 Chain Link Terminal Post

		Crew	Daily Output	Labor-Hours	Unit	Material	Labor	Equipment	Total	Total Incl O&P
0010	**CHAIN LINK TERMINAL POST**									
0110	16 ga., steel, 2-1/2" x 6′ x 0.065 wall, incl. post cap, excavation	B-80C	80	.300	Ea.	13.45	13.15	3.13	29.73	39
0120	2-1/2" x 7′-6" x 0.065 wall		80	.300		17.70	13.15	3.13	33.98	44
0130	2-1/2" x 8′-6" x 0.095 wall	↓	80	.300		26	13.15	3.13	42.28	53
0210	16 ga., steel, 2-1/2" x 6′ x 0.065 wall, incl. floor flange	B-80A	80	.300		32.50	12.65	10.25	55.40	67.50
0220	2-1/2" x 8′ x 0.065 wall		80	.300		37	12.65	10.25	59.90	72
0230	4" x 10′ x 0.160 wall		80	.300		118	12.65	10.25	140.90	161
0240	4" x 12′ x 0.160 wall	↓	80	.300		130	12.65	10.25	152.90	174
0310	16 ga., steel, 4" x 11′ x 0.226 wall, incl. post cap, excavation	B-80C	80	.300		192	13.15	3.13	208.28	235
0320	4" x 13′-6" x 0.226 wall		80	.300		223	13.15	3.13	239.28	270
0330	4" x 21′ x 0.226 wall	↓	80	.300	↓	305	13.15	3.13	321.28	360

32 31 13.65 Chain Link Line Post

		Crew	Daily Output	Labor-Hours	Unit	Material	Labor	Equipment	Total	Total Incl O&P
0010	**CHAIN LINK LINE POST**									
0110	16 ga., steel, 1-5/8" x 6′ x 0.065 wall, incl. post cap and excavation	B-80C	80	.300	Ea.	9.50	13.15	3.13	25.78	35
0120	1-5/8" x 7′-6" x 0.065 wall		80	.300		12.40	13.15	3.13	28.68	38
0130	2" x 8′-6" x 0.095 wall	↓	80	.300		19.75	13.15	3.13	36.03	46
0210	16 ga., steel, 2" x 6′ x 0.065 wall, incl. post top and floor flange	B-80A	80	.300		28.50	12.65	10.25	51.40	63
0220	2" x 8′ x 0.065 wall		80	.300		32	12.65	10.25	54.90	66.50
0230	3" x 10′ x 0.160 wall		80	.300		82.50	12.65	10.25	105.40	122
0240	3" x 12′ x 0.160 wall	↓	80	.300		94	12.65	10.25	116.90	134
0410	16 ga., steel, 3" x 11′ x 0.203 wall, incl. post cap and excavation	B-80C	80	.300		122	13.15	3.13	138.28	159
0420	3" x 13′-6" x 0.203 wall		80	.300		143	13.15	3.13	159.28	181
0430	3" x 21′ x 0.203 wall	↓	80	.300	↓	186	13.15	3.13	202.28	228

32 31 13.66 Chain Link Top Rail

		Crew	Daily Output	Labor-Hours	Unit	Material	Labor	Equipment	Total	Total Incl O&P
0010	**CHAIN LINK TOP RAIL**									
0110	Fence, rail tubing, 16 ga., 1-3/8" x 21′ x 0.065 wall incl. hardware	B-80A	266	.090	L.F.	1.18	3.80	3.08	8.06	10.80
0120	1-5/8" x 21′ x 0.065 wall, swedge		266	.090		1.49	3.80	3.08	8.37	11.15
0130	1-5/8" x 21′ x 0.111 wall		266	.090		2.25	3.80	3.08	9.13	11.95
0140	2" x 18′ x 0.145 wall	↓	266	.090	↓	4.88	3.80	3.08	11.76	14.85

32 31 13.68 Chain Link Fabric

		Crew	Daily Output	Labor-Hours	Unit	Material	Labor	Equipment	Total	Total Incl O&P
0010	**CHAIN LINK FABRIC**									
0110	Fence, fabric, steel, galv., 11-1/2 ga., 2-1/4" mesh, 4′ H, incl. hardware	B-80A	266	.090	L.F.	2.60	3.80	3.08	9.48	12.35
0120	5′ H		266	.090		2.95	3.80	3.08	9.83	12.75
0130	6′ H		266	.090		3.24	3.80	3.08	10.12	13.05
0210	Fence, fabric, steel, galv., 11 ga., 2" mesh, 6′ H, incl. hardware		266	.090		3.95	3.80	3.08	10.83	13.85
0220	8′ H	↓	266	.090	↓	5.20	3.80	3.08	12.08	15.20

32 31 Fences and Gates

32 31 13 – Chain Link Fences and Gates

32 31 13.68 Chain Link Fabric		Crew	Daily Output	Labor-Hours	Unit	2020 Bare Costs Material	Labor	Equipment	Total	Total Incl O&P
0230	10′ H	B-80A	266	.090	L.F.	9.75	3.80	3.08	16.63	20
0240	12′ H		266	.090		11.55	3.80	3.08	18.43	22
0310	Fence, fabric, steel, galv., 9 ga., 2″ mesh, 8′ H, incl. hardware		266	.090		5.25	3.80	3.08	12.13	15.25
0320	10′ H		266	.090		6.10	3.80	3.08	12.98	16.20
0330	12′ H		266	.090		7.15	3.80	3.08	14.03	17.35
0340	14′ H		266	.090		7.25	3.80	3.08	14.13	17.45
0410	Fence, fabric, steel, galv., 9 ga., 1″ mesh, 8′ H, incl. hardware		266	.090		11.40	3.80	3.08	18.28	22
0420	10′ H		266	.090		18.35	3.80	3.08	25.23	29.50
0430	12′ H		266	.090		22	3.80	3.08	28.88	33.50
0440	14′ H		266	.090		25.50	3.80	3.08	32.38	37.50

32 31 13.70 Chain Link Barbed Wire

		Crew	Daily Output	Labor-Hours	Unit	Material	Labor	Equipment	Total	Total Incl O&P
0010	**CHAIN LINK BARBED WIRE**									
0110	Fence, barbed wire, 12-1/2 ga., 1320′ roll incl. barb arm and brace band	B-80A	266	.090	L.F.	.63	3.80	3.08	7.51	10.20

32 31 13.80 Residential Chain Link Gate

		Crew	Daily Output	Labor-Hours	Unit	Material	Labor	Equipment	Total	Total Incl O&P
0010	**RESIDENTIAL CHAIN LINK GATE**									
0110	Residential 4′ gate, single incl. hardware and concrete	B-80C	10	2.400	Ea.	128	105	25	258	335
0120	5′		10	2.400		138	105	25	268	350
0130	6′		10	2.400		149	105	25	279	360
0510	Residential 4′ gate, double incl. hardware and concrete		10	2.400		228	105	25	358	445
0520	5′		10	2.400		244	105	25	374	465
0530	6′		10	2.400		284	105	25	414	505

32 31 13.82 Internal Chain Link Gate

		Crew	Daily Output	Labor-Hours	Unit	Material	Labor	Equipment	Total	Total Incl O&P
0010	**INTERNAL CHAIN LINK GATE**									
0110	Internal 6′ gate, single incl. post flange, hardware and concrete	B-80C	10	2.400	Ea.	279	105	25	409	500
0120	8′		10	2.400		320	105	25	450	545
0130	10′		10	2.400		435	105	25	565	675
0510	Internal 6′ gate, double incl. post flange, hardware and concrete		10	2.400		510	105	25	640	755
0520	8′		10	2.400		585	105	25	715	840
0530	10′		10	2.400		740	105	25	870	1,000

32 31 13.84 Industrial Chain Link Gate

		Crew	Daily Output	Labor-Hours	Unit	Material	Labor	Equipment	Total	Total Incl O&P
0010	**INDUSTRIAL CHAIN LINK GATE**									
0110	Industrial 8′ gate, single incl. hardware and concrete	B-80C	10	2.400	Ea.	475	105	25	605	720
0120	10′		10	2.400		540	105	25	670	790
0510	Industrial 8′ gate, double incl. hardware and concrete		10	2.400		740	105	25	870	1,000
0520	10′		10	2.400		845	105	25	975	1,125

32 31 13.88 Chain Link Transom

		Crew	Daily Output	Labor-Hours	Unit	Material	Labor	Equipment	Total	Total Incl O&P
0010	**CHAIN LINK TRANSOM**									
0110	Add for, single transom, 3′ wide, incl. components & hardware	B-80C	10	2.400	Ea.	118	105	25	248	325
0120	Add for, double transom, 6′ wide, incl. components & hardware	″	10	2.400	″	127	105	25	257	335

32 31 19 – Decorative Metal Fences and Gates

32 31 19.10 Decorative Fence

		Crew	Daily Output	Labor-Hours	Unit	Material	Labor	Equipment	Total	Total Incl O&P
0010	**DECORATIVE FENCE**									
5300	Tubular picket, steel, 6′ sections, 1-9/16″ posts, 4′ high	B-80C	300	.080	L.F.	37	3.51	.83	41.34	47.50
5400	2″ posts, 5′ high		240	.100		44	4.38	1.04	49.42	56
5600	2″ posts, 6′ high		200	.120		53	5.25	1.25	59.50	68.50
5700	Staggered picket, 1-9/16″ posts, 4′ high		300	.080		39.50	3.51	.83	43.84	50
5800	2″ posts, 5′ high		240	.100		45	4.38	1.04	50.42	57.50
5900	2″ posts, 6′ high		200	.120		53	5.25	1.25	59.50	68.50
6200	Gates, 4′ high, 3′ wide	B-1	10	2.400	Ea.	335	103		438	535
6300	5′ high, 3′ wide		10	2.400		245	103		348	435
6400	6′ high, 3′ wide		10	2.400		288	103		391	480

32 31 Fences and Gates

32 31 19 – Decorative Metal Fences and Gates

32 31 19.10 Decorative Fence		Crew	Daily Output	Labor-Hours	Unit	2020 Bare Costs Material	Labor	Equipment	Total	Total Incl O&P
6500	4′ wide	B-1	10	2.400	Ea.	345	103		448	545

32 31 23 – Plastic Fences and Gates

32 31 23.10 Fence, Vinyl

		Crew	Daily Output	Labor-Hours	Unit	Material	Labor	Equipment	Total	Total Incl O&P
0010	**FENCE, VINYL**									
0011	White, steel reinforced, stainless steel fasteners									
0020	Picket, 4″ x 4″ posts @ 6′-0″ OC, 3′ high	B-1	140	.171	L.F.	24.50	7.35		31.85	39
0030	4′ high		130	.185		26.50	7.90		34.40	42
0040	5′ high		120	.200		34	8.55		42.55	51
0100	Board (semi-privacy), 5″ x 5″ posts @ 7′-6″ OC, 5′ high		130	.185		27.50	7.90		35.40	42.50
0120	6′ high		125	.192		30	8.20		38.20	46
0200	Basket weave, 5″ x 5″ posts @ 7′-6″ OC, 5′ high		160	.150		25.50	6.40		31.90	39
0220	6′ high		150	.160		29	6.85		35.85	43
0300	Privacy, 5″ x 5″ posts @ 7′-6″ OC, 5′ high		130	.185		26.50	7.90		34.40	41.50
0320	6′ high		150	.160		30	6.85		36.85	44
0350	Gate, 5′ high		9	2.667	Ea.	330	114		444	545
0360	6′ high		9	2.667		375	114		489	595
0400	For posts set in concrete, add		25	.960		9.30	41		50.30	76
0500	Post and rail fence, 2 rail		150	.160	L.F.	6.25	6.85		13.10	17.85
0510	3 rail		150	.160		8.10	6.85		14.95	19.85
0515	4 rail		150	.160		11.05	6.85		17.90	23

32 31 26 – Wire Fences and Gates

32 31 26.10 Fences, Misc. Metal

		Crew	Daily Output	Labor-Hours	Unit	Material	Labor	Equipment	Total	Total Incl O&P
0010	**FENCES, MISC. METAL**									
0012	Chicken wire, posts @ 4′, 1″ mesh, 4′ high	B-80C	410	.059	L.F.	3.53	2.57	.61	6.71	8.65
0100	2″ mesh, 6′ high		350	.069		4.01	3	.71	7.72	10
0200	Galv. steel, 12 ga., 2″ x 4″ mesh, posts 5′ OC, 3′ high		300	.080		2.80	3.51	.83	7.14	9.60
0300	5′ high		300	.080		3.37	3.51	.83	7.71	10.25
0400	14 ga., 1″ x 2″ mesh, 3′ high		300	.080		3.48	3.51	.83	7.82	10.35
0500	5′ high		300	.080		4.50	3.51	.83	8.84	11.45
1000	Kennel fencing, 1-1/2″ mesh, 6′ long, 3′-6″ wide, 6′-2″ high	2 Clab	4	4	Ea.	495	168		663	815
1050	12′ long		4	4		700	168		868	1,050
1200	Top covers, 1-1/2″ mesh, 6′ long		15	1.067		141	45		186	227
1250	12′ long		12	1.333		210	56		266	320
1300	For kennel doors, see Section 08 31 13.40									
4492	Security fence, prison grade, barbed wire, set in concrete, 10′ high	B-80	22	1.455	L.F.	64	68	46.50	178.50	230
4494	Security fence, prison grade, razor wire, set in concrete, 10′ high		18	1.778		65.50	83	56.50	205	267
4500	Security fence, prison grade, set in concrete, 12′ high		25	1.280		60	59.50	41	160.50	206
4600	16′ high		20	1.600		79.50	74.50	51	205	263
4990	Security fence, prison grade, set in concrete, 10′ high		25	1.280		60	59.50	41	160.50	206

32 31 26.20 Wire Fencing, General

		Crew	Daily Output	Labor-Hours	Unit	Material	Labor	Equipment	Total	Total Incl O&P
0010	**WIRE FENCING, GENERAL**									
0015	Barbed wire, galvanized, domestic steel, hi-tensile 15-1/2 ga.				M.L.F.	149			149	164
0020	Standard, 12-3/4 ga.					164			164	180
0210	Barbless wire, 2-strand galvanized, 12-1/2 ga.					164			164	180
0500	Helical razor ribbon, stainless steel, 18″ diam. x 18″ spacing				C.L.F.	195			195	215
0600	Hardware cloth galv., 1/4″ mesh, 23 ga., 2′ wide				C.S.F.	43.50			43.50	48
0700	3′ wide					27.50			27.50	30.50
0900	1/2″ mesh, 19 ga., 2′ wide					36.50			36.50	40
1000	4′ wide					50			50	55
1200	Chain link fabric, steel, 2″ mesh, 6 ga., galvanized					61.50			61.50	67.50
1300	9 ga., galvanized					60			60	66

32 31 Fences and Gates

32 31 26 – Wire Fences and Gates

32 31 26.20 Wire Fencing, General		Crew	Daily Output	Labor-Hours	Unit	2020 Bare Costs Material	Labor	Equipment	Total	Total Incl O&P
1350	Vinyl coated				C.S.F.	32			32	35
1360	Aluminized					228			228	251
1400	2-1/4″ mesh, 11-1/2 ga., galvanized					48			48	53
1600	1-3/4″ mesh (tennis courts), 11-1/2 ga. (core), vinyl coated					56			56	61.50
1700	9 ga., galvanized					91			91	100
2100	Welded wire fabric, galvanized, 1″ x 2″, 14 ga.	2 Carp	1600	.010	S.F.	.67	.53		1.20	1.59
2200	2″ x 4″, 12-1/2 ga.				C.S.F.	39			39	43

32 31 29 – Wood Fences and Gates

32 31 29.10 Fence, Wood

		Crew	Daily Output	Labor-Hours	Unit	Material	Labor	Equipment	Total	Total Incl O&P
0010	**FENCE, WOOD**									
0011	Basket weave, 3/8″ x 4″ boards, 2″ x 4″									
0020	stringers on spreaders, 4″ x 4″ posts									
0050	No. 1 cedar, 6′ high	B-80C	160	.150	L.F.	26	6.55	1.56	34.11	40.50
0070	Treated pine, 6′ high	″	150	.160	″	38	7	1.67	46.67	54.50
0200	Board fence, 1″ x 4″ boards, 2″ x 4″ rails, 4″ x 4″ post									
0220	Preservative treated, 2 rail, 3′ high	B-80C	145	.166	L.F.	11.80	7.25	1.72	20.77	26.50
0240	4′ high		135	.178		13.45	7.80	1.85	23.10	29
0260	3 rail, 5′ high		130	.185		12.70	8.10	1.92	22.72	29
0300	6′ high		125	.192		16.30	8.40	2	26.70	33.50
0320	No. 2 grade western cedar, 2 rail, 3′ high		145	.166		12.75	7.25	1.72	21.72	27.50
0340	4′ high		135	.178		12.40	7.80	1.85	22.05	28
0360	3 rail, 5′ high		130	.185		14.65	8.10	1.92	24.67	31
0400	6′ high		125	.192		15.50	8.40	2	25.90	32.50
0420	No. 1 grade cedar, 2 rail, 3′ high		145	.166		13.90	7.25	1.72	22.87	29
0440	4′ high		135	.178		15.20	7.80	1.85	24.85	31
0460	3 rail, 5′ high		130	.185		18.25	8.10	1.92	28.27	35
0500	6′ high		125	.192		21.50	8.40	2	31.90	39.50
0540	Shadow box, 1″ x 6″ board, 2″ x 4″ rail, 4″ x 4″ post									
0560	Pine, pressure treated, 3 rail, 6′ high	B-80C	150	.160	L.F.	28	7	1.67	36.67	44
0600	Gate, 3′-6″ wide		8	3	Ea.	136	131	31.50	298.50	395
0620	No. 1 cedar, 3 rail, 4′ high		130	.185	L.F.	19.15	8.10	1.92	29.17	36
0640	6′ high		125	.192		26.50	8.40	2	36.90	44.50
0860	Open rail fence, split rails, 2 rail, 3′ high, no. 1 cedar		160	.150		9.25	6.55	1.56	17.36	22.50
0870	No. 2 cedar		160	.150		8.65	6.55	1.56	16.76	21.50
0880	3 rail, 4′ high, no. 1 cedar		150	.160		13.30	7	1.67	21.97	27.50
0890	No. 2 cedar		150	.160		9	7	1.67	17.67	23
0920	Rustic rails, 2 rail, 3′ high, no. 1 cedar		160	.150		13.85	6.55	1.56	21.96	27.50
0930	No. 2 cedar		160	.150		12.70	6.55	1.56	20.81	26
0940	3 rail, 4′ high		150	.160		13.85	7	1.67	22.52	28
0950	No. 2 cedar		150	.160		8	7	1.67	16.67	22
0960	Picket fence, gothic, pressure treated pine									
1000	2 rail, 3′ high	B-80C	140	.171	L.F.	8.90	7.50	1.79	18.19	24
1020	3 rail, 4′ high		130	.185	″	10.95	8.10	1.92	20.97	27
1040	Gate, 3′-6″ wide		9	2.667	Ea.	77.50	117	28	222.50	305
1060	No. 2 cedar, 2 rail, 3′ high		140	.171	L.F.	10.30	7.50	1.79	19.59	25.50
1100	3 rail, 4′ high		130	.185	″	10.40	8.10	1.92	20.42	26.50
1120	Gate, 3′-6″ wide		9	2.667	Ea.	82.50	117	28	227.50	310
1140	No. 1 cedar, 2 rail, 3′ high		140	.171	L.F.	14.45	7.50	1.79	23.74	30
1160	3 rail, 4′ high		130	.185	″	20.50	8.10	1.92	30.52	37.50
1170	Gate, 3′-6″ wide		9	2.667	Ea.	272	117	28	417	515
1200	Rustic picket, molded pine, 2 rail, 3′ high		140	.171	L.F.	9.65	7.50	1.79	18.94	24.50
1220	No. 1 cedar, 2 rail, 3′ high		140	.171		11.60	7.50	1.79	20.89	27

32 31 Fences and Gates

32 31 29 – Wood Fences and Gates

32 31 29.10 Fence, Wood		Crew	Daily Output	Labor-Hours	Unit	Material	2020 Bare Costs Labor	Equipment	Total	Total Incl O&P
1240	Stockade fence, no. 1 cedar, 3-1/4" rails, 6' high	B-80C	160	.150	L.F.	14.05	6.55	1.56	22.16	27.50
1260	8' high		155	.155		20.50	6.80	1.61	28.91	35
1300	No. 2 cedar, treated wood rails, 6' high		160	.150		14.75	6.55	1.56	22.86	28.50
1320	Gate, 3'-6" wide		8	3	Ea.	95.50	131	31.50	258	350
1360	Treated pine, treated rails, 6' high		160	.150	L.F.	14.85	6.55	1.56	22.96	28.50
1400	8' high		150	.160	"	20.50	7	1.67	29.17	35.50

32 31 29.20 Fence, Wood Rail		Crew	Daily Output	Labor-Hours	Unit	Material	Labor	Equipment	Total	Total Incl O&P
0010	**FENCE, WOOD RAIL**									
0012	Picket, No. 2 cedar, Gothic, 2 rail, 3' high	B-1	160	.150	L.F.	8.60	6.40		15	19.75
0050	Gate, 3'-6" wide	B-80C	9	2.667	Ea.	75	117	28	220	300
0400	3 rail, 4' high		150	.160	L.F.	9.65	7	1.67	18.32	23.50
0500	Gate, 3'-6" wide		9	2.667	Ea.	90	117	28	235	315
5000	Fence rail, redwood, 2" x 4", merch. grade, 8'	B-1	2400	.010	L.F.	2.76	.43		3.19	3.73
5050	Select grade, 8'		2400	.010	"	6.10	.43		6.53	7.45
6000	Fence post, select redwood, earth packed & treated, 4" x 4" x 6'		96	.250	Ea.	14.70	10.70		25.40	33.50
6010	4" x 4" x 8'		96	.250		22	10.70		32.70	41
6020	Set in concrete, 4" x 4" x 6'		50	.480		23	20.50		43.50	58.50
6030	4" x 4" x 8'		50	.480		24.50	20.50		45	60
6040	Wood post, 4' high, set in concrete, incl. concrete		50	.480		15.50	20.50		36	50
6050	Earth packed		96	.250		17.40	10.70		28.10	36.50
6060	6' high, set in concrete, incl. concrete		50	.480		20	20.50		40.50	55
6070	Earth packed		96	.250		13.20	10.70		23.90	31.50
9000	Minimum labor/equipment charge	1 Clab	2	4	Job		168		168	270

32 32 Retaining Walls

32 32 13 – Cast-in-Place Concrete Retaining Walls

32 32 13.10 Retaining Walls, Cast Concrete		Crew	Daily Output	Labor-Hours	Unit	Material	Labor	Equipment	Total	Total Incl O&P
0010	**RETAINING WALLS, CAST CONCRETE**									
1800	Concrete gravity wall with vertical face including excavation & backfill									
1850	No reinforcing									
1900	6' high, level embankment	C-17C	36	2.306	L.F.	82.50	128	15.45	225.95	310
2000	33° slope embankment		32	2.594		109	144	17.40	270.40	370
2200	8' high, no surcharge		27	3.074		117	170	20.50	307.50	425
2300	33° slope embankment		24	3.458		142	192	23	357	485
2500	10' high, level embankment		19	4.368		167	242	29.50	438.50	600
2600	33° slope embankment		18	4.611		232	255	31	518	695
2800	Reinforced concrete cantilever, incl. excavation, backfill & reinf.									
2900	6' high, 33° slope embankment	C-17C	35	2.371	L.F.	86	131	15.90	232.90	320
3000	8' high, 33° slope embankment		29	2.862		99	159	19.20	277.20	380
3100	10' high, 33° slope embankment		20	4.150		129	230	28	387	540
3200	20' high, 500 lb./L.F. surcharge		7.50	11.067		385	615	74.50	1,074.50	1,475
3500	Concrete cribbing, incl. excavation and backfill									
3700	12' high, open face	B-13	210	.267	S.F.	43	12.30	2.77	58.07	69.50
3900	Closed face	"	210	.267	"	40.50	12.30	2.77	55.57	67
4100	Concrete filled slurry trench, see Section 31 56 23.20									

32 32 23 – Segmental Retaining Walls

32 32 23.13 Segmental Conc. Unit Masonry Retaining Walls		Crew	Daily Output	Labor-Hours	Unit	2020 Bare Costs Material	Labor	Equipment	Total	Total Incl O&P
0010	**SEGMENTAL CONC. UNIT MASONRY RETAINING WALLS**									
7100	Segmental retaining wall system, incl. pins and void fill									
7120	base and backfill not included									
7140	Large unit, 8" high x 18" wide x 20" deep, 3 plane split	B-62	300	.080	S.F.	14.70	3.66	.59	18.95	22.50
7150	Straight split		300	.080		14.60	3.66	.59	18.85	22.50
7160	Medium, lt. wt., 8" high x 18" wide x 12" deep, 3 plane split		400	.060		7.50	2.74	.44	10.68	13.10
7170	Straight split		400	.060		10.60	2.74	.44	13.78	16.55
7180	Small unit, 4" x 18" x 10" deep, 3 plane split		400	.060		17.25	2.74	.44	20.43	24
7190	Straight split		400	.060		12.45	2.74	.44	15.63	18.55
7200	Cap unit, 3 plane split		300	.080		14.70	3.66	.59	18.95	22.50
7210	Cap unit, straight split	↓	300	.080		14.70	3.66	.59	18.95	22.50
7250	Geo-grid soil reinforcement 4' x 50'	2 Clab	22500	.001		.80	.03		.83	.93
7255	Geo-grid soil reinforcement 6' x 150'	"	22500	.001	↓	.62	.03		.65	.73

32 32 26 – Metal Crib Retaining Walls

32 32 26.10 Metal Bin Retaining Walls		Crew	Daily Output	Labor-Hours	Unit	Material	Labor	Equipment	Total	Total Incl O&P
0010	**METAL BIN RETAINING WALLS**									
0011	Aluminized steel bin, excavation									
0020	and backfill not included, 10' wide									
0100	4' high, 5.5' deep	B-13	650	.086	S.F.	29.50	3.97	.89	34.36	39.50
0200	8' high, 5.5' deep		615	.091		33.50	4.19	.94	38.63	44.50
0300	10' high, 7.7' deep		580	.097		37.50	4.45	1	42.95	49
0400	12' high, 7.7' deep		530	.106		40.50	4.86	1.10	46.46	53.50
0500	16' high, 7.7' deep		515	.109		42.50	5	1.13	48.63	55.50
0600	16' high, 9.9' deep		500	.112		45.50	5.15	1.16	51.81	59.50
0700	20' high, 9.9' deep		470	.119		51	5.50	1.24	57.74	66
0800	20' high, 12.1' deep		460	.122		45	5.60	1.26	51.86	60
0900	24' high, 12.1' deep		455	.123		47.50	5.65	1.28	54.43	63
1000	24' high, 14.3' deep		450	.124		57.50	5.75	1.29	64.54	74
1100	28' high, 14.3' deep	↓	440	.127	↓	60	5.85	1.32	67.17	77
1300	For plain galvanized bin type walls, deduct					10%				

32 32 29 – Timber Retaining Walls

32 32 29.10 Landscape Timber Retaining Walls		Crew	Daily Output	Labor-Hours	Unit	Material	Labor	Equipment	Total	Total Incl O&P
0010	**LANDSCAPE TIMBER RETAINING WALLS**									
0100	Treated timbers, 6" x 6"	1 Clab	265	.030	L.F.	2.76	1.27		4.03	5.10
0110	6" x 8'	"	200	.040	"	6.65	1.68		8.33	10
0120	Drilling holes in timbers for fastening, 1/2"	1 Carp	450	.018	Inch		.95		.95	1.51
0130	5/8"	"	450	.018	"		.95		.95	1.51
0140	Reinforcing rods for fastening, 1/2"	1 Clab	312	.026	L.F.	.41	1.08		1.49	2.19
0150	5/8"	"	312	.026	"	.65	1.08		1.73	2.44
0160	Reinforcing fabric	2 Clab	2500	.006	S.Y.	2.22	.27		2.49	2.87
0170	Gravel backfill		28	.571	C.Y.	16.90	24		40.90	57
0180	Perforated pipe, 4" diameter with silt sock	↓	1200	.013	L.F.	1.05	.56		1.61	2.06
0190	Galvanized 60d common nails	1 Clab	625	.013	Ea.	.21	.54		.75	1.09
0200	20d common nails	"	3800	.002	"	.05	.09		.14	.19

32 32 Retaining Walls

32 32 36 – Gabion Retaining Walls

32 32 36.10 Stone Gabion Retaining Walls		Crew	Daily Output	Labor-Hours	Unit	2020 Bare Costs Material	Labor	Equipment	Total	Total Incl O&P
0010	STONE GABION RETAINING WALLS									
4300	Stone filled gabions, not incl. excavation,									
4310	Stone, delivered, 3' wide									
4340	Galvanized, 6' long, 1' high	B-13	113	.496	Ea.	84	23	5.15	112.15	135
4400	1'-6" high		50	1.120		105	51.50	11.60	168.10	210
4490	3'-0" high		13	4.308		173	198	44.50	415.50	555
4590	9' long, 1' high		50	1.120		195	51.50	11.60	258.10	310
4650	1'-6" high		22	2.545		146	117	26.50	289.50	375
4690	3'-0" high		6	9.333		370	430	97	897	1,200
4890	12' long, 1' high		28	2		165	92	20.50	277.50	350
4950	1'-6" high		13	4.308		195	198	44.50	437.50	580
4990	3'-0" high		3	18.667		435	860	194	1,489	2,075
5200	PVC coated, 6' long, 1' high		113	.496		95	23	5.15	123.15	147
5250	1'-6" high		50	1.120		116	51.50	11.60	179.10	222
5300	3' high		13	4.308		173	198	44.50	415.50	555
5500	9' long, 1' high		50	1.120		124	51.50	11.60	187.10	232
5550	1'-6" high		22	2.545		153	117	26.50	296.50	385
5600	3' high		6	9.333		273	430	97	800	1,100
5800	12' long, 1' high		28	2		194	92	20.50	306.50	380
5850	1'-6" high		13	4.308		232	198	44.50	474.50	620
5900	3' high		3	18.667		330	860	194	1,384	1,950
6000	Galvanized, 6' long, 1' high		75	.747	C.Y.	77.50	34.50	7.75	119.75	149
6010	1'-6" high		50	1.120		105	51.50	11.60	168.10	210
6020	3'-0" high		25	2.240		86.50	103	23	212.50	285
6030	9' long, 1' high		50	1.120		195	51.50	11.60	258.10	310
6040	1'-6" high		33.30	1.682		97.50	77.50	17.45	192.45	249
6050	3'-0" high		16.70	3.353		123	154	35	312	420
6060	12' long, 1' high		37.50	1.493		123	69	15.50	207.50	262
6070	1'-6" high		25	2.240		97.50	103	23	223.50	297
6080	3'-0" high		12.50	4.480		109	206	46.50	361.50	500
6100	PVC coated, 6' long, 1' high		75	.747		143	34.50	7.75	185.25	220
6110	1'-6" high		50	1.120		116	51.50	11.60	179.10	222
6120	3' high		25	2.240		86.50	103	23	212.50	285
6130	9' long, 1' high		50	1.120		124	51.50	11.60	187.10	232
6140	1'-6" high		33.30	1.682		102	77.50	17.45	196.95	254
6150	3' high		16.67	3.359		91	155	35	281	385
6160	12' long, 1' high		37.50	1.493		145	69	15.50	229.50	286
6170	1'-6" high		25	2.240		104	103	23	230	305
6180	3' high		12.50	4.480		83	206	46.50	335.50	470

32 32 53 – Stone Retaining Walls

32 32 53.10 Retaining Walls, Stone		Crew	Daily Output	Labor-Hours	Unit	Material	Labor	Equipment	Total	Total Incl O&P
0010	RETAINING WALLS, STONE									
0015	Including excavation, concrete footing and									
0020	stone 3' below grade. Price is exposed face area.									
0200	Decorative random stone, to 6' high, 1'-6" thick, dry set	D-1	35	.457	S.F.	77	21.50		98.50	120
0300	Mortar set		40	.400		78.50	18.75		97.25	117
0500	Cut stone, to 6' high, 1'-6" thick, dry set		35	.457		80	21.50		101.50	123
0600	Mortar set		40	.400		80.50	18.75		99.25	119
0800	Random stone, 6' to 10' high, 2' thick, dry set		45	.356		80	16.70		96.70	115
0900	Mortar set		50	.320		83	15		98	116
1100	Cut stone, 6' to 10' high, 2' thick, dry set		45	.356		80.50	16.70		97.20	116
1200	Mortar set		50	.320		83.50	15		98.50	116

32 32 Retaining Walls

32 32 53 – Stone Retaining Walls

32 32 53.10 Retaining Walls, Stone		Crew	Daily Output	Labor-Hours	Unit	Material	2020 Bare Costs Labor	Equipment	Total	Total Incl O&P
5100	Setting stone, dry	D-1	100	.160	C.F.		7.50		7.50	12.10
5600	With mortar		120	.133	"		6.25		6.25	10.10
9000	Minimum labor/equipment charge		2	8	Job		375		375	605

32 33 Site Furnishings

32 33 33 – Site Manufactured Planters

32 33 33.10 Planters		Crew	Daily Output	Labor-Hours	Unit	Material	Labor	Equipment	Total	Total Incl O&P
0010	**PLANTERS**									
0012	Concrete, sandblasted, precast, 48" diameter, 24" high	2 Clab	15	1.067	Ea.	640	45		685	775
0100	Fluted, precast, 7' diameter, 36" high		10	1.600		1,625	67.50		1,692.50	1,875
0300	Fiberglass, circular, 36" diameter, 24" high		15	1.067		745	45		790	890
0320	36" diameter, 27" high		12	1.333		710	56		766	875
0330	33" high		15	1.067		765	45		810	910
0335	24" diameter, 36" high		15	1.067		420	45		465	535
0340	60" diameter, 39" high		8	2		1,375	84		1,459	1,625
0400	60" diameter, 24" high		10	1.600		1,175	67.50		1,242.50	1,375
0600	Square, 24" side, 36" high		15	1.067		680	45		725	815
0610	24" side, 27" high		12	1.333		740	56		796	905
0620	24" side, 16" high		20	.800		355	33.50		388.50	445
0700	48" side, 36" high		15	1.067		1,125	45		1,170	1,300
0900	Planter/bench, 72" square, 36" high		5	3.200		2,950	135		3,085	3,475
1000	96" square, 27" high		5	3.200		3,525	135		3,660	4,100
1200	Wood, square, 48" side, 24" high		15	1.067		1,625	45		1,670	1,850
1300	Circular, 48" diameter, 30" high		10	1.600		1,125	67.50		1,192.50	1,325
1500	72" diameter, 30" high		10	1.600		2,000	67.50		2,067.50	2,300
1600	Planter/bench, 72"		5	3.200		3,850	135		3,985	4,450
9000	Minimum labor/equipment charge	1 Clab	2	4	Job		168		168	270

32 33 43 – Site Seating and Tables

32 33 43.13 Site Seating		Crew	Daily Output	Labor-Hours	Unit	Material	Labor	Equipment	Total	Total Incl O&P
0010	**SITE SEATING**									
0012	Seating, benches, park, precast conc., w/backs, wood rails, 4' long	2 Clab	5	3.200	Ea.	610	135		745	885
0100	8' long		4	4		1,125	168		1,293	1,525
0300	Fiberglass, without back, one piece, 4' long		10	1.600		1,725	67.50		1,792.50	2,000
0400	8' long		7	2.286		2,300	96		2,396	2,675
0500	Steel barstock pedestals w/backs, 2" x 3" wood rails, 4' long		10	1.600		1,375	67.50		1,442.50	1,625
0510	8' long		7	2.286		1,725	96		1,821	2,050
0515	Powder coated steel, 4" x 4" plastic slats, 6' long		8	2		540	84		624	730
0520	3" x 8" wood plank, 4' long		10	1.600		1,525	67.50		1,592.50	1,775
0530	8' long		7	2.286		1,550	96		1,646	1,875
0540	Backless, 4" x 4" wood plank, 4' square		10	1.600		1,275	67.50		1,342.50	1,500
0550	8' long		7	2.286		1,175	96		1,271	1,450
0560	Powder coated steel, with back and 2 anti-vagrant dividers, 6' long		8	2		1,200	84		1,284	1,450
0600	Aluminum pedestals, with backs, aluminum slats, 8' long		8	2		480	84		564	665
0610	15' long		5	3.200		1,025	135		1,160	1,350
0620	Portable, aluminum slats, 8' long		8	2		470	84		554	655
0630	15' long		5	3.200		505	135		640	775
0800	Cast iron pedestals, back & arms, wood slats, 4' long		8	2		470	84		554	655
0820	8' long		5	3.200		1,050	135		1,185	1,375
0840	Backless, wood slats, 4' long		8	2		715	84		799	920
0860	8' long		5	3.200		1,200	135		1,335	1,550

32 33 Site Furnishings

32 33 43 – Site Seating and Tables

32 33 43.13 Site Seating		Crew	Daily Output	Labor-Hours	Unit	Material	2020 Bare Costs Labor	Equipment	Total	Total Incl O&P
1700	Steel frame, fir seat, 10′ long	2 Clab	10	1.600	Ea.	410	67.50		477.50	560
2000	Benches, park, with back, galv. stl. frame, 4″ x 4″ plastic slats, 6′ long		7	2.286		545	96		641	755
9000	Minimum labor/equipment charge		2	8	Job		335		335	540

32 34 Fabricated Bridges

32 34 13 – Fabricated Pedestrian Bridges

32 34 13.10 Bridges, Pedestrian

		Crew	Daily Output	Labor-Hours	Unit	Material	Labor	Equipment	Total	Total Incl O&P
0010	**BRIDGES, PEDESTRIAN**									
0011	Spans over streams, roadways, etc.									
0020	including erection, not including foundations									
0050	Precast concrete, complete in place, 8′ wide, 60′ span	E-2	215	.260	S.F.	161	14.90	7.90	183.80	210
0100	100′ span		185	.303		177	17.30	9.15	203.45	232
0150	120′ span		160	.350		192	20	10.60	222.60	255
0200	150′ span		145	.386		200	22	11.70	233.70	268
0300	Steel, trussed or arch spans, compl. in place, 8′ wide, 40′ span		320	.175		131	10	5.30	146.30	166
0400	50′ span		395	.142		118	8.10	4.29	130.39	147
0500	60′ span		465	.120		118	6.90	3.65	128.55	144
0600	80′ span		570	.098		140	5.60	2.98	148.58	166
0700	100′ span		465	.120		197	6.90	3.65	207.55	232
0800	120′ span		365	.153		249	8.75	4.65	262.40	293
0900	150′ span		310	.181		265	10.30	5.45	280.75	315
1000	160′ span		255	.220		265	12.55	6.65	284.20	320
1100	10′ wide, 80′ span		640	.088		131	5	2.65	138.65	155
1200	120′ span		415	.135		170	7.70	4.09	181.79	204
1300	150′ span		445	.126		191	7.20	3.81	202.01	226
1400	200′ span		205	.273		204	15.60	8.25	227.85	259
1600	Wood, laminated type, complete in place, 80′ span	C-12	203	.236		98	12.45	2.32	112.77	130
1700	130′ span	″	153	.314		102	16.50	3.08	121.58	142

32 84 Planting Irrigation

32 84 23 – Underground Sprinklers

32 84 23.10 Sprinkler Irrigation System

		Crew	Daily Output	Labor-Hours	Unit	Material	Labor	Equipment	Total	Total Incl O&P
0010	**SPRINKLER IRRIGATION SYSTEM**									
0011	For lawns									
0800	Residential system, custom, 1″ supply	B-20	2000	.012	S.F.	.27	.56		.83	1.20
0900	1-1/2″ supply	″	1800	.013	″	.42	.63		1.05	1.46
1020	Pop up spray head w/risers, hi-pop, full circle pattern, 4″	2 Skwk	76	.211	Ea.	4.85	11.55		16.40	24
1030	1/2 circle pattern, 4″		76	.211		7	11.55		18.55	26
1040	6″, full circle pattern		76	.211		12.40	11.55		23.95	32
1050	1/2 circle pattern, 6″		76	.211		12.95	11.55		24.50	32.50
1060	12″, full circle pattern		76	.211		15.75	11.55		27.30	36
1070	1/2 circle pattern, 12″		76	.211		15.95	11.55		27.50	36
1080	Pop up bubbler head w/risers, hi-pop bubbler head, 4″		76	.211		5.60	11.55		17.15	24.50
1110	Impact full/part circle sprinklers, 28′-54′ @ 25-60 psi		37	.432		21.50	23.50		45	61.50
1120	Spaced 37′-49′ @ 25-50 psi		37	.432		22	23.50		45.50	62
1130	Spaced 43′-61′ @ 30-60 psi		37	.432		67.50	23.50		91	112
1140	Spaced 54′-78′ @ 40-80 psi		37	.432		124	23.50		147.50	174
1145	Impact rotor pop-up full/part commercial circle sprinklers									
1150	Spaced 42′-65′ @ 35-80 psi	2 Skwk	25	.640	Ea.	17.70	35		52.70	75.50

32 84 23 – Underground Sprinklers

32 84 23.10 Sprinkler Irrigation System		Crew	Daily Output	Labor-Hours	Unit	Material	2020 Bare Costs Labor	Equipment	Total	Total Incl O&P
1160	Spaced 48'-76' @ 45-85 psi	2 Skwk	25	.640	Ea.	19.45	35		54.45	77.50
1165	Impact rotor pop-up part. circle comm., 53'-75', 55-100 psi, w/accessories									
1180	Sprinkler, premium, pop-up rotator, 50'-100'	2 Skwk	25	.640	Ea.	96.50	35		131.50	162
1250	Plastic case, 2 nozzle, metal cover		25	.640		105	35		140	172
1260	Rubber cover		25	.640		104	35		139	170
1270	Iron case, 2 nozzle, metal cover		22	.727		149	40		189	228
1280	Rubber cover		22	.727		148	40		188	227
1282	Impact rotor pop-up full circle commercial, 39'-99', 30-100 psi									
1284	Plastic case, metal cover	2 Skwk	25	.640	Ea.	74	35		109	138
1286	Rubber cover		25	.640		108	35		143	175
1288	Iron case, metal cover		22	.727		143	40		183	221
1290	Rubber cover		22	.727		168	40		208	249
1292	Plastic case, 2 nozzle, metal cover		22	.727		107	40		147	181
1294	Rubber cover		22	.727		109	40		149	184
1296	Iron case, 2 nozzle, metal cover		20	.800		142	44		186	227
1298	Rubber cover		20	.800		148	44		192	233
1305	Electric remote control valve, plastic, 3/4"		18	.889		20.50	49		69.50	100
1310	1"		18	.889		25	49		74	105
1320	1-1/2"		18	.889		99	49		148	187
1330	2"		18	.889		126	49		175	216
1335	Quick coupling valves, brass, locking cover									
1340	Inlet coupling valve, 3/4"	2 Skwk	18.75	.853	Ea.	24.50	47		71.50	102
1350	1"		18.75	.853		33.50	47		80.50	112
1360	Controller valve boxes, 6" round boxes		18.75	.853		7.55	47		54.55	83
1370	10" round boxes		14.25	1.123		17.95	61.50		79.45	118
1380	12" square box		9.75	1.641		19.10	90		109.10	164
1388	Electromech. control, 14 day 3-60 min., auto start to 23/day									
1390	4 station	2 Skwk	1.04	15.385	Ea.	82	845		927	1,450
1400	7 station		.64	25		156	1,375		1,531	2,350
1410	12 station		.40	40		184	2,200		2,384	3,700
1420	Dual programs, 18 station		.24	66.667		217	3,650		3,867	6,075
1430	23 station		.16	100		267	5,475		5,742	9,050
1435	Backflow preventer, bronze, 0-175 psi, w/valves, test cocks									
1440	3/4"	2 Skwk	6	2.667	Ea.	93.50	146		239.50	335
1450	1"		6	2.667		106	146		252	350
1460	1-1/2"		6	2.667		269	146		415	530
1470	2"		6	2.667		395	146		541	670
1475	Pressure vacuum breaker, brass, 15-150 psi									
1480	3/4"	2 Skwk	6	2.667	Ea.	24	146		170	260
1490	1"		6	2.667		69	146		215	310
1500	1-1/2"		6	2.667		85	146		231	325
1510	2"		6	2.667		157	146		303	405

32 91 Planting Preparation

32 91 13 – Soil Preparation

32 91 13.16 Mulching		Crew	Daily Output	Labor-Hours	Unit	Material	Labor (2020 Bare Costs)	Equipment	Total	Total Incl O&P
0010	**MULCHING**									
0100	Aged barks, 3″ deep, hand spread	1 Clab	100	.080	S.Y.	4.12	3.37		7.49	9.95
0150	Skid steer loader	B-63	13.50	2.963	M.S.F.	460	131	13.15	604.15	730
0160	Skid steer loader	″	1500	.027	S.Y.	4.12	1.18	.12	5.42	6.55
0200	Hay, 1″ deep, hand spread	1 Clab	475	.017	″	.51	.71		1.22	1.70
0250	Power mulcher, small	B-64	180	.089	M.S.F.	56.50	3.97	2.32	62.79	71.50
0350	Large	B-65	530	.030	″	56.50	1.35	.99	58.84	66
0400	Humus peat, 1″ deep, hand spread	1 Clab	700	.011	S.Y.	3.27	.48		3.75	4.37
0450	Push spreader	″	2500	.003	″	3.27	.13		3.40	3.82
0550	Tractor spreader	B-66	700	.011	M.S.F.	365	.61	.35	365.96	400
0600	Oat straw, 1″ deep, hand spread	1 Clab	475	.017	S.Y.	.58	.71		1.29	1.78
0650	Power mulcher, small	B-64	180	.089	M.S.F.	64.50	3.97	2.32	70.79	80
0700	Large	B-65	530	.030	″	64.50	1.35	.99	66.84	74.50
0750	Add for asphaltic emulsion	B-45	1770	.009	Gal.	5.90	.48	.47	6.85	7.75
0800	Peat moss, 1″ deep, hand spread	1 Clab	900	.009	S.Y.	5.15	.37		5.52	6.25
0850	Push spreader	″	2500	.003	″	5.15	.13		5.28	5.85
0950	Tractor spreader	B-66	700	.011	M.S.F.	570	.61	.35	570.96	625
1000	Polyethylene film, 6 mil	2 Clab	2000	.008	S.Y.	.51	.34		.85	1.10
1100	Redwood nuggets, 3″ deep, hand spread	1 Clab	150	.053	″	3.33	2.25		5.58	7.25
1150	Skid steer loader	B-63	13.50	2.963	M.S.F.	370	131	13.15	514.15	630
1200	Stone mulch, hand spread, ceramic chips, economy	1 Clab	125	.064	S.Y.	7.25	2.69		9.94	12.30
1250	Deluxe	″	95	.084	″	11.40	3.55		14.95	18.25
1300	Granite chips	B-1	10	2.400	C.Y.	80	103		183	252
1400	Marble chips		10	2.400		245	103		348	435
1600	Pea gravel		28	.857		115	36.50		151.50	185
1700	Quartz		10	2.400		197	103		300	380
1800	Tar paper, 15 lb. felt	1 Clab	800	.010	S.Y.	.47	.42		.89	1.19
1900	Wood chips, 2″ deep, hand spread	″	220	.036	″	1.68	1.53		3.21	4.30
1950	Skid steer loader	B-63	20.30	1.970	M.S.F.	187	87.50	8.75	283.25	355

32 91 13.23 Structural Soil Mixing		Crew	Daily Output	Labor-Hours	Unit	Material	Labor (2020 Bare Costs)	Equipment	Total	Total Incl O&P
0010	**STRUCTURAL SOIL MIXING**									
0100	Rake topsoil, site material, harley rock rake, ideal	B-6	33	.727	M.S.F.		33.50	6.50	40	60
0200	Adverse	″	7	3.429			157	30.50	187.50	283
0300	Screened loam, york rake and finish, ideal	B-62	24	1			45.50	7.40	52.90	80.50
0400	Adverse	″	20	1.200			55	8.90	63.90	97
1000	Remove topsoil & stock pile on site, 75 HP dozer, 6″ deep, 50′ haul	B-10L	30	.400			21	13.40	34.40	47
1050	300′ haul		6.10	1.967			102	66	168	234
1100	12″ deep, 50′ haul		15.50	.774			40	26	66	91.50
1150	300′ haul		3.10	3.871			201	129	330	455
1200	200 HP dozer, 6″ deep, 50′ haul	B-10B	125	.096			4.98	12.05	17.03	21
1250	300′ haul		30.70	.391			20.50	49	69.50	86
1300	12″ deep, 50′ haul		62	.194			10.05	24.50	34.55	42.50
1350	300′ haul		15.40	.779			40.50	97.50	138	171
1400	Alternate method, 75 HP dozer, 50′ haul	B-10L	860	.014	C.Y.		.72	.47	1.19	1.65
1450	300′ haul	″	114	.105			5.45	3.52	8.97	12.45
1500	200 HP dozer, 50′ haul	B-10B	2660	.005			.23	.57	.80	.99
1600	300′ haul	″	570	.021			1.09	2.64	3.73	4.62
1800	Rolling topsoil, hand push roller	1 Clab	3200	.003	S.F.		.11		.11	.17
1850	Tractor drawn roller	B-66	10666	.001	″		.04	.02	.06	.08
2000	Root raking and loading, residential, no boulders	B-6	53.30	.450	M.S.F.		20.50	4.01	24.51	37
2100	With boulders		32	.750			34.50	6.70	41.20	62
2200	Municipal, no boulders		200	.120			5.50	1.07	6.57	9.90

32 91 13.23 Structural Soil Mixing		Crew	Daily Output	Labor-Hours	Unit	Material	2020 Bare Costs Labor	Equipment	Total	Total Incl O&P
2300	With boulders	B-6	120	.200	M.S.F.		9.15	1.78	10.93	16.45
2400	Large commercial, no boulders	B-10B	400	.030			1.56	3.76	5.32	6.60
2500	With boulders	"	240	.050			2.59	6.25	8.84	11
3000	Scarify subsoil, residential, skid steer loader w/scarifiers, 50 HP	B-66	32	.250			13.25	7.55	20.80	29
3050	Municipal, skid steer loader w/scarifiers, 50 HP	"	120	.067			3.53	2.01	5.54	7.75
3100	Large commercial, 75 HP, dozer w/scarifier	B-10L	240	.050			2.59	1.67	4.26	5.90
3500	Screen topsoil from stockpile, vibrating screen, wet material (organic)	B-10P	200	.060	C.Y.		3.11	5.85	8.96	11.35
3550	Dry material	"	300	.040			2.07	3.90	5.97	7.55
3600	Mixing with conditioners, manure and peat	B-10R	550	.022			1.13	.55	1.68	2.38
3650	Mobilization add for 2 days or less operation	B-34K	3	2.667	Job		131	286	417	525
3800	Spread conditioned topsoil, 6" deep, by hand	B-1	360	.067	S.Y.	5.90	2.85		8.75	11.05
3850	300 HP dozer	B-10M	27	.444	M.S.F.	635	23	65.50	723.50	810
4000	Spread soil conditioners, alum. sulfate, 1#/S.Y., hand push spreader	1 Clab	17500	.001	S.Y.	23	.02		23.02	25.50
4050	Tractor spreader	B-66	700	.011	M.S.F.	2,575	.61	.35	2,575.96	2,825
4100	Fertilizer, 0.2#/S.Y., push spreader	1 Clab	17500	.001	S.Y.	.08	.02		.10	.12
4150	Tractor spreader	B-66	700	.011	M.S.F.	9.10	.61	.35	10.06	11.35
4200	Ground limestone, 1#/S.Y., push spreader	1 Clab	17500	.001	S.Y.	.14	.02		.16	.18
4250	Tractor spreader	B-66	700	.011	M.S.F.	15.55	.61	.35	16.51	18.45
4400	Manure, 18#/S.Y., push spreader	1 Clab	2500	.003	S.Y.	8.10	.13		8.23	9.10
4450	Tractor spreader	B-66	280	.029	M.S.F.	900	1.51	.86	902.37	995
4500	Perlite, 1" deep, push spreader	1 Clab	17500	.001	S.Y.	11.15	.02		11.17	12.30
4550	Tractor spreader	B-66	700	.011	M.S.F.	1,225	.61	.35	1,225.96	1,350
4600	Vermiculite, push spreader	1 Clab	17500	.001	S.Y.	8.45	.02		8.47	9.35
4650	Tractor spreader	B-66	700	.011	M.S.F.	940	.61	.35	940.96	1,025
5000	Spread topsoil, skid steer loader and hand dress	B-62	270	.089	C.Y.	26.50	4.06	.66	31.22	36
5100	Articulated loader and hand dress	B-100	320	.038		26.50	1.95	3	31.45	35.50
5200	Articulated loader and 75 HP dozer	B-10M	500	.024		26.50	1.24	3.53	31.27	35
5300	Road grader and hand dress	B-11L	1000	.016		26.50	.79	1.06	28.35	31.50
6000	Tilling topsoil, 20 HP tractor, disk harrow, 2" deep	B-66	450	.018	M.S.F.		.94	.54	1.48	2.06
6050	4" deep		360	.022			1.18	.67	1.85	2.58
6100	6" deep		270	.030			1.57	.89	2.46	3.44
6150	26" rototiller, 2" deep	A-1J	1250	.006	S.Y.		.27	.04	.31	.48
6200	4" deep		1000	.008			.34	.05	.39	.60
6250	6" deep		750	.011			.45	.07	.52	.80
7000	Lawn maintenance, see Section 32 01 30.10									

32 91 13.26 Planting Beds		Crew	Daily Output	Labor-Hours	Unit	Material	Labor	Equipment	Total	Total Incl O&P
0010	**PLANTING BEDS**									
0100	Backfill planting pit, by hand, on site topsoil	2 Clab	18	.889	C.Y.		37.50		37.50	60
0200	Prepared planting mix, by hand	"	24	.667			28		28	45
0300	Skid steer loader, on site topsoil	B-62	340	.071			3.23	.52	3.75	5.70
0400	Prepared planting mix	"	410	.059			2.68	.43	3.11	4.73
1000	Excavate planting pit, by hand, sandy soil	2 Clab	16	1			42		42	67.50
1100	Heavy soil or clay	"	8	2			84		84	135
1200	1/2 C.Y. backhoe, sandy soil	B-11C	150	.107			5.25	1.43	6.68	9.90
1300	Heavy soil or clay	"	115	.139			6.90	1.86	8.76	12.90
2000	Mix planting soil, incl. loam, manure, peat, by hand	2 Clab	60	.267		46.50	11.25		57.75	69
2100	Skid steer loader	B-62	150	.160		46.50	7.30	1.18	54.98	64
3000	Pile sod, skid steer loader	"	2800	.009	S.Y.		.39	.06	.45	.69
3100	By hand	2 Clab	400	.040			1.68		1.68	2.70
4000	Remove sod, F.E. loader	B-10S	2000	.006			.31	.21	.52	.72
4100	Sod cutter	B-12K	3200	.005			.25	.30	.55	.73
4200	By hand	2 Clab	240	.067			2.81		2.81	4.50

32 92 19.14 Seeding, Athletic Fields		Crew	Daily Output	Labor-Hours	Unit	Material	Labor	Equipment	Total	Total Incl O&P
						2020 Bare Costs				
0010	SEEDING, ATHLETIC FIELDS R329219-50									
0020	Seeding, athletic fields, athletic field mix, 8#/M.S.F. push spreader	1 Clab	8	1	M.S.F.	11	42		53	79.50
0100	Tractor spreader	B-66	52	.154		11	8.15	4.64	23.79	30
0200	Hydro or air seeding, with mulch and fertilizer	B-81	80	.300		12.10	14.80	7.30	34.20	45
0400	Birdsfoot trefoil, 0.45#/M.S.F., push spreader	1 Clab	8	1		4.84	42		46.84	73
0500	Tractor spreader	B-66	52	.154		4.84	8.15	4.64	17.63	23
0600	Hydro or air seeding, with mulch and fertilizer	B-81	80	.300		9.30	14.80	7.30	31.40	42
0800	Bluegrass, 4#/M.S.F., common, push spreader	1 Clab	8	1		15.40	42		57.40	84.50
0900	Tractor spreader	B-66	52	.154		15.40	8.15	4.64	28.19	35
1000	Hydro or air seeding, with mulch and fertilizer	B-81	80	.300		25.50	14.80	7.30	47.60	59.50
1100	Baron, push spreader	1 Clab	8	1		15.50	42		57.50	84.50
1200	Tractor spreader	B-66	52	.154		15.50	8.15	4.64	28.29	35
1300	Hydro or air seeding, with mulch and fertilizer	B-81	80	.300		21.50	14.80	7.30	43.60	55
1500	Clover, 0.67#/M.S.F., white, push spreader	1 Clab	8	1		3.14	42		45.14	71
1600	Tractor spreader	B-66	52	.154		3.14	8.15	4.64	15.93	21.50
1700	Hydro or air seeding, with mulch and fertilizer	B-81	80	.300		17.25	14.80	7.30	39.35	50.50
1800	Ladino, push spreader	1 Clab	8	1		3.62	42		45.62	71.50
1900	Tractor spreader	B-66	52	.154		3.62	8.15	4.64	16.41	22
2000	Hydro or air seeding, with mulch and fertilizer	B-81	80	.300		15.95	14.80	7.30	38.05	49
2200	Fescue 5.5#/M.S.F., tall, push spreader	1 Clab	8	1		9.75	42		51.75	78
2300	Tractor spreader	B-66	52	.154		9.75	8.15	4.64	22.54	28.50
2400	Hydro or air seeding, with mulch and fertilizer	B-81	80	.300		32	14.80	7.30	54.10	67
2500	Chewing, push spreader	1 Clab	8	1		8.95	42		50.95	77.50
2600	Tractor spreader	B-66	52	.154		8.95	8.15	4.64	21.74	27.50
2700	Hydro or air seeding, with mulch and fertilizer	B-81	80	.300		29.50	14.80	7.30	51.60	64
2900	Crown vetch, 4#/M.S.F., push spreader	1 Clab	8	1		128	42		170	209
3000	Tractor spreader	B-66	52	.154		128	8.15	4.64	140.79	159
3100	Hydro or air seeding, with mulch and fertilizer	B-81	80	.300		176	14.80	7.30	198.10	226
3300	Rye, 10#/M.S.F., annual, push spreader	1 Clab	8	1		17.25	42		59.25	86.50
3400	Tractor spreader	B-66	52	.154		17.25	8.15	4.64	30.04	37
3500	Hydro or air seeding, with mulch and fertilizer	B-81	80	.300		38	14.80	7.30	60.10	73.50
3600	Fine textured, push spreader	1 Clab	8	1		16.30	42		58.30	85.50
3700	Tractor spreader	B-66	52	.154		16.30	8.15	4.64	29.09	36
3800	Hydro or air seeding, with mulch and fertilizer	B-81	80	.300		36	14.80	7.30	58.10	71
4000	Shade mix, 6#/M.S.F., push spreader	1 Clab	8	1		10.65	42		52.65	79
4100	Tractor spreader	B-66	52	.154		10.65	8.15	4.64	23.44	29.50
4200	Hydro or air seeding, with mulch and fertilizer	B-81	80	.300		23.50	14.80	7.30	45.60	57.50
4400	Slope mix, 6#/M.S.F., push spreader	1 Clab	8	1		11.95	42		53.95	80.50
4500	Tractor spreader	B-66	52	.154		11.95	8.15	4.64	24.74	31
4600	Hydro or air seeding, with mulch and fertilizer	B-81	80	.300		30	14.80	7.30	52.10	64.50
4800	Turf mix, 4#/M.S.F., push spreader	1 Clab	8	1		13.55	42		55.55	82.50
4900	Tractor spreader	B-66	52	.154		13.55	8.15	4.64	26.34	33
5000	Hydro or air seeding, with mulch and fertilizer	B-81	80	.300		34	14.80	7.30	56.10	69
5200	Utility mix, 7#/M.S.F., push spreader	1 Clab	8	1		12.45	42		54.45	81
5300	Tractor spreader	B-66	52	.154		12.45	8.15	4.64	25.24	31.50
5400	Hydro or air seeding, with mulch and fertilizer	B-81	80	.300		46.50	14.80	7.30	68.60	83
5600	Wildflower, 0.10#/M.S.F., push spreader	1 Clab	8	1		1.80	42		43.80	69.50
5700	Tractor spreader	B-66	52	.154		1.80	8.15	4.64	14.59	19.85
5800	Hydro or air seeding, with mulch and fertilizer	B-81	80	.300	↓	9.90	14.80	7.30	32	42.50
7025	Fertilizer, mechanical spread	1 Clab	1.75	4.571	Acre	5.60	192		197.60	315
7060	Limestone, mechanical spread		1.74	4.598	"	89.50	194		283.50	410
9000	Minimum labor/equipment charge	↓	4	2	Job		84		84	135

32 92 Turf and Grasses

32 92 23 – Sodding

32 92 23.10 Sodding Systems		Crew	Daily Output	Labor-Hours	Unit	Material	Labor	Equipment	Total	Total Incl O&P
						2020 Bare Costs				
0010	**SODDING SYSTEMS**									
0020	Sodding, 1" deep, bluegrass sod, on level ground, over 8 M.S.F.	B-63	22	1.818	M.S.F.	223	80.50	8.05	311.55	385
0200	4 M.S.F.		17	2.353		286	104	10.45	400.45	495
0300	1,000 S.F.		13.50	2.963		335	131	13.15	479.15	595
0500	Sloped ground, over 8 M.S.F.		6	6.667		223	295	29.50	547.50	750
0600	4 M.S.F.		5	8		286	355	35.50	676.50	920
0700	1,000 S.F.		4	10		335	445	44.50	824.50	1,125
1000	Bent grass sod, on level ground, over 6 M.S.F.		20	2		305	88.50	8.90	402.40	485
1100	3 M.S.F.		18	2.222		320	98.50	9.85	428.35	525
1200	Sodding 1,000 S.F. or less		14	2.857		345	127	12.70	484.70	595
1500	Sloped ground, over 6 M.S.F.		15	2.667		305	118	11.85	434.85	535
1600	3 M.S.F.		13.50	2.963		320	131	13.15	464.15	580
1700	1,000 S.F.	▼	12	3.333	▼	345	148	14.80	507.80	630

32 93 Plants

32 93 10 – General Planting Costs

32 93 10.12 Travel

		Crew	Daily Output	Labor-Hours	Unit	Material	Labor	Equipment	Total	Total Incl O&P
0010	**TRAVEL** add to all nursery items									
0015	10 to 20 miles one way, add				All				5%	5%
0100	30 to 50 miles one way, add				"				10%	10%

32 93 13 – Ground Covers

32 93 13.20 Ground Cover and Vines

		Crew	Daily Output	Labor-Hours	Unit	Material	Labor	Equipment	Total	Total Incl O&P
0010	**GROUND COVER AND VINES** Planting only, no preparation									
0100	Ajuga, 1 year, bare root	B-1	9	2.667	C	370	114		484	590
0150	Potted, 2 year		6	4		815	171		986	1,175
0200	Berberis, potted, 2 year		6	4		1,675	171		1,846	2,125
0250	Cotoneaster, 15" to 18", shady areas, B&B		.60	40		3,975	1,700		5,675	7,125
0300	Boston ivy, on bank, 1 year, bare root		6	4		335	171		506	645
0350	Potted, 2 year		6	4		610	171		781	945
0400	English ivy, 1 year, bare root		9	2.667		102	114		216	295
0450	Potted, 2 year		6	4		360	171		531	670
0500	Halls honeysuckle, 1 year, bare root		5	4.800		380	205		585	750
0550	Potted, 2 year		4	6		900	257		1,157	1,400
0600	Memorial rose, 9" to 12", 1 gallon cont		3	8		132	340		472	695
0650	Potted, 2 gallon cont		2	12		222	515		737	1,075
0700	Pachysandra, 1 year, bare root		10	2.400		110	103		213	284
0750	Potted, 2 year		6	4		208	171		379	505
0800	Vinca minor, 1 year, bare root		10	2.400		157	103		260	335
0850	Potted, 2 year		6	4		263	171		434	565
0900	Woodbine, on bank, 1/2 year, bare root		6	4		223	171		394	520
0950	Potted, 2 year	▼	4	6	▼	585	257		842	1,050
2000	Alternate method of figuring									
2100	Ajuga, field division, 4,000/M.S.F.	B-1	.23	104	M.S.F.	5,450	4,475		9,925	13,200
2300	Boston ivy, 1 year, 60/M.S.F.		10	2.400		228	103		331	415
2400	English ivy, 1 year, 500/M.S.F.		1.80	13.333		600	570		1,170	1,575
2500	Halls honeysuckle, 1 year, 333/M.S.F.		1.50	16		505	685		1,190	1,650
2600	Memorial rose, 9" to 12", 1 gal., 333/M.S.F.		.90	26.667		1,075	1,150		2,225	3,000
2700	Pachysandra, 1 year, 4,000/M.S.F.		.25	96		1,775	4,100		5,875	8,525
2800	Vinca minor, rooted cutting, 2,000/M.S.F.		1	24		1,100	1,025		2,125	2,875
2900	Woodbine, 1 year, 60/M.S.F.	▼	10	2.400	▼	40	103		143	208

32 93 33.10 Shrubs and Trees		Crew	Daily Output	Labor-Hours	Unit	2020 Bare Costs Material	Labor	Equipment	Total	Total Incl O&P
0010	**SHRUBS AND TREES**									
0011	Evergreen, in prepared beds, B&B									
0100	Arborvitae pyramidal, 4′-5′	B-17	30	1.067	Ea.	104	49.50	21.50	175	218
0150	Globe, 12″-15″	B-1	96	.250		24	10.70		34.70	43
0200	Balsam, fraser, 6′-7′	B-17	30	1.067		143	49.50	21.50	214	260
0300	Cedar, blue, 8′-10′		18	1.778		241	82.50	35.50	359	435
0350	Japanese, 4′-5′		55	.582		132	27	11.65	170.65	201
0400	Cypress, hinoki, 15″-18″	B-1	80	.300		87.50	12.85		100.35	117
0500	Hemlock, Canadian, 2-1/2′-3′		36	.667		33	28.50		61.50	82
0550	Holly, Savannah, 8′-10′ H		9.68	2.479		294	106		400	495
0600	Juniper, andorra, 18″-24″		80	.300		58.50	12.85		71.35	85
0620	Wiltoni, 15″-18″		80	.300		28.50	12.85		41.35	52
0640	Skyrocket, 4-1/2′-5′	B-17	55	.582		111	27	11.65	149.65	178
0660	Blue pfitzer, 2′-2-1/2′	B-1	44	.545		41	23.50		64.50	83
0680	Ketleerie, 2-1/2′-3′		50	.480		58.50	20.50		79	97.50
0700	Pine, black, 2-1/2′-3′		50	.480		68	20.50		88.50	108
0720	Mugo, 18″-24″		60	.400		64.50	17.10		81.60	98.50
0740	White, 4′-5′	B-17	75	.427		55.50	19.85	8.55	83.90	102
0800	Spruce, blue, 18″-24″	B-1	60	.400		71.50	17.10		88.60	107
0820	Dwarf alberta, 18″-24″	"	60	.400		62	17.10		79.10	96
0840	Norway, 4′-5′	B-17	75	.427		93.50	19.85	8.55	121.90	144
0900	Yew, denisforma, 12″-15″	B-1	60	.400		38.50	17.10		55.60	70
1000	Capitata, 18″-24″		30	.800		34.50	34		68.50	93
1100	Hicksi, 2′-2-1/2′		30	.800		102	34		136	167

32 93 33.20 Shrubs		Crew	Daily Output	Labor-Hours	Unit	Material	Labor	Equipment	Total	Total Incl O&P
0010	**SHRUBS**									
0011	Broadleaf Evergreen, planted in prepared beds									
0100	Andromeda, 15″-18″, cont	B-1	96	.250	Ea.	33.50	10.70		44.20	54
0200	Azalea, 15″-18″, cont		96	.250		34	10.70		44.70	54.50
0300	Barberry, 9″-12″, cont		130	.185		18.60	7.90		26.50	33
0400	Boxwood, 15″-18″, B&B		96	.250		48	10.70		58.70	69.50
0500	Euonymus, emerald gaiety, 12″-15″, cont		115	.209		28	8.95		36.95	45.50
0600	Holly, 15″-18″, B&B		96	.250		41.50	10.70		52.20	62.50
0700	Leucothoe, 15″-18″, cont		96	.250		23	10.70		33.70	42
0800	Mahonia, 18″-24″, cont		80	.300		32	12.85		44.85	56
0900	Mount laurel, 18″-24″, B&B		80	.300		76.50	12.85		89.35	105
1000	Paxistema, 9″-12″ H		130	.185		23	7.90		30.90	38
1100	Rhododendron, 18″-24″, cont		48	.500		39.50	21.50		61	78
1200	Rosemary, 1 gal. cont		600	.040		19.80	1.71		21.51	24.50
2000	Deciduous, planted in prepared beds, amelanchier, 2′-3′, B&B		57	.421		132	18		150	174
2100	Azalea, 15″-18″, B&B		96	.250		30.50	10.70		41.20	50.50
2200	Barberry, 2′-3′, B&B		57	.421		24.50	18		42.50	56
2300	Bayberry, 2′-3′, B&B		57	.421		35	18		53	67.50
2400	Boston ivy, 2 year, cont		600	.040		21.50	1.71		23.21	26
2500	Corylus, 3′-4′, B&B	B-17	75	.427		27	19.85	8.55	55.40	71
2600	Cotoneaster, 15″-18″, B&B	B-1	80	.300		26.50	12.85		39.35	50
2700	Deutzia, 12″-15″, B&B	"	96	.250		14	10.70		24.70	32.50
2800	Dogwood, 3′-4′, B&B	B-17	40	.800		36	37	16.05	89.05	116
2900	Euonymus, alatus compacta, 15″-18″, cont	B-1	80	.300		28	12.85		40.85	51.50
3000	Flowering almond, 2′-3′, cont	"	36	.667		23	28.50		51.50	71
3100	Flowering currant, 3′-4′, cont	B-17	75	.427		29.50	19.85	8.55	57.90	73.50
3200	Forsythia, 2′-3′, cont	B-1	60	.400		17.65	17.10		34.75	47

32 93 33 – Shrubs

32 93 33.20 Shrubs		Crew	Daily Output	Labor-Hours	Unit	Material	2020 Bare Costs Labor	Equipment	Total	Total Incl O&P
3300	Hibiscus, 3'-4', B&B	B-17	75	.427	Ea.	45	19.85	8.55	73.40	90.50
3400	Honeysuckle, 3'-4', B&B	B-1	60	.400		29	17.10		46.10	59.50
3500	Hydrangea, 2'-3', B&B	"	57	.421		33	18		51	65
3600	Lilac, 3'-4', B&B	B-17	40	.800		22	37	16.05	75.05	101
3700	Mockorange, 3'-4', B&B	B-1	36	.667		28.50	28.50		57	77
3800	Osier willow, 2'-3', B&B		57	.421		38.50	18		56.50	71.50
3900	Privet, bare root, 18"-24"		80	.300		17.50	12.85		30.35	40
4000	Pyracantha, 2'-3', cont		80	.300		42.50	12.85		55.35	67
4100	Quince, 2'-3', B&B		57	.421		30.50	18		48.50	62.50
4200	Russian olive, 3'-4', B&B	B-17	75	.427		26	19.85	8.55	54.40	70
4300	Snowberry, 2'-3', B&B	B-1	57	.421		20.50	18		38.50	51.50
4400	Spirea, 3'-4', B&B	"	70	.343		21.50	14.65		36.15	47
4500	Viburnum, 3'-4', B&B	B-17	40	.800		27	37	16.05	80.05	107
4600	Weigela, 3'-4', B&B	B-1	70	.343		17.05	14.65		31.70	42.50

32 93 43 – Trees

32 93 43.10 Planting		Crew	Daily Output	Labor-Hours	Unit	Material	Labor	Equipment	Total	Total Incl O&P
0010	**PLANTING**									
0011	Trees, shrubs and ground cover									
0100	Light soil									
0110	Bare root seedlings, 3" to 5" height	1 Clab	960	.008	Ea.		.35		.35	.56
0120	6" to 10"		520	.015			.65		.65	1.04
0130	11" to 16"		370	.022			.91		.91	1.46
0140	17" to 24"		210	.038			1.60		1.60	2.57
0200	Potted, 2-1/4" diameter		840	.010			.40		.40	.64
0210	3" diameter		700	.011			.48		.48	.77
0220	4" diameter		620	.013			.54		.54	.87
0300	Container, 1 gallon	2 Clab	84	.190			8		8	12.85
0310	2 gallon		52	.308			12.95		12.95	21
0320	3 gallon		40	.400			16.85		16.85	27
0330	5 gallon		29	.552			23		23	37

32 93 43.20 Trees		Crew	Daily Output	Labor-Hours	Unit	Material	Labor	Equipment	Total	Total Incl O&P
0010	**TREES**									
0011	Deciduous, in prep. beds, balled & burlapped (B&B)									
0100	Ash, 2" caliper G	B-17	8	4	Ea.	213	186	80	479	620
0200	Beech, 5'-6' G		50	.640		240	30	12.85	282.85	325
0300	Birch, 6'-8', 3 stems G		20	1.600		175	74.50	32	281.50	345
0400	Cherry, 6'-8', 1" caliper G		24	1.333		82	62	27	171	218
0500	Crabapple, 6'-8' G		20	1.600		160	74.50	32	266.50	330
0600	Dogwood, 4'-5' G		40	.800		140	37	16.05	193.05	231
0700	Eastern redbud, 4'-5' G		40	.800		151	37	16.05	204.05	243
0800	Elm, 8'-10' G		20	1.600		350	74.50	32	456.50	540
0900	Ginkgo, 6'-7' G		24	1.333		161	62	27	250	305
1000	Hawthorn, 8'-10', 1" caliper G		20	1.600		157	74.50	32	263.50	325
1100	Honeylocust, 10'-12', 1-1/2" caliper G		10	3.200		227	149	64	440	555
1200	Laburnum, 6'-8', 1" caliper G		24	1.333		64.50	62	27	153.50	199
1300	Larch, 8' G		32	1		135	46.50	20	201.50	245
1400	Linden, 8'-10', 1" caliper G		20	1.600		169	74.50	32	275.50	340
1500	Magnolia, 4'-5' G		20	1.600		118	74.50	32	224.50	284
1600	Maple, red, 8'-10', 1-1/2" caliper G		10	3.200		200	149	64	413	530
1700	Mountain ash, 8'-10', 1" caliper G		16	2		196	93	40	329	410
1800	Oak, 2-1/2"-3" caliper G		6	5.333		355	248	107	710	905
1900	Pagoda, 6'-8' G		20	1.600		181	74.50	32	287.50	355

32 93 Plants

32 93 43 – Trees

32 93 43.20 Trees			Crew	Daily Output	Labor-Hours	Unit	Material	Labor (2020 Bare Costs)	Equipment	Total	Total Incl O&P
2000	Pear, 6′-8′, 1″ caliper	G	B-17	20	1.600	Ea.	127	74.50	32	233.50	293
2100	Planetree, 9′-11′, 1-1/4″ caliper	G		10	3.200		278	149	64	491	615
2200	Plum, 6′-8′, 1″ caliper	G		20	1.600		82	74.50	32	188.50	244
2300	Poplar, 9′-11′, 1-1/4″ caliper	G		10	3.200		102	149	64	315	420
2400	Shadbush, 4′-5′	G		60	.533		67	25	10.70	102.70	125
2500	Sumac, 2′-3′	G		75	.427		45.50	19.85	8.55	73.90	91
2600	Tupelo, 5′-6′	G		40	.800		95.50	37	16.05	148.55	182
2700	Tulip, 5′-6′	G		40	.800		48.50	37	16.05	101.55	130
2800	Willow, 6′-8′, 1″ caliper	G		20	1.600		103	74.50	32	209.50	267
9000	Minimum labor/equipment charge		1 Clab	4	2	Job		84		84	135

32 94 Planting Accessories

32 94 13 – Landscape Edging

32 94 13.20 Edging

Line	Description	Crew	Daily Output	Labor-Hours	Unit	Material	Labor	Equipment	Total	Total Incl O&P
0010	**EDGING**									
0050	Aluminum alloy, including stakes, 1/8″ x 4″, mill finish	B-1	390	.062	L.F.	2.15	2.63		4.78	6.60
0051	Black paint		390	.062		2.49	2.63		5.12	6.95
0052	Black anodized		390	.062		2.88	2.63		5.51	7.40
0060	3/16″ x 4″, mill finish		380	.063		3.10	2.70		5.80	7.75
0061	Black paint		380	.063		3.57	2.70		6.27	8.25
0062	Black anodized		380	.063		4.17	2.70		6.87	8.90
0070	1/8″ x 5-1/2″ mill finish		370	.065		3.12	2.77		5.89	7.85
0071	Black paint		370	.065		3.70	2.77		6.47	8.50
0072	Black anodized		370	.065		4.24	2.77		7.01	9.10
0080	3/16″ x 5-1/2″ mill finish		360	.067		4.17	2.85		7.02	9.15
0081	Black paint		360	.067		4.71	2.85		7.56	9.75
0082	Black anodized		360	.067		5.45	2.85		8.30	10.55
0100	Brick, set horizontally, 1-1/2 bricks per L.F.	D-1	370	.043		1.70	2.03		3.73	5.15
0150	Set vertically, 3 bricks per L.F.	″	135	.119		3.80	5.55		9.35	13.20
0200	Corrugated aluminum, roll, 4″ wide	1 Carp	650	.012		2.25	.65		2.90	3.53
0250	6″ wide	″	550	.015		2.81	.77		3.58	4.33
0300	Concrete, cast in place, see Section 03 30 53.40									
0350	Granite, 5″ x 16″, straight	B-29	300	.187	L.F.	16.50	8.60	2.82	27.92	35
0400	Polyethylene grass barrier, 5″ x 1/8″	D-1	400	.040		1.47	1.88		3.35	4.65
0500	Precast scallops, green, 2″ x 8″ x 16″		400	.040		2.18	1.88		4.06	5.45
0550	2″ x 8″ x 16″ other than green		400	.040		1.74	1.88		3.62	4.95
0600	Railroad ties, 6″ x 8″	2 Carp	170	.094		2.48	5		7.48	10.75
0650	7″ x 9″		136	.118		2.75	6.25		9	13.05
0750	Redwood 2″ x 4″		330	.048		2.31	2.58		4.89	6.65
0800	Steel edge strips, incl. stakes, 1/4″ x 5″	B-1	390	.062		4.58	2.63		7.21	9.25
0850	3/16″ x 4″	″	390	.062		3.62	2.63		6.25	8.20
0900	Hardwood, pressure treated, 4″ x 6″	2 Carp	250	.064		2.37	3.40		5.77	8.05
0940	6″ x 6″		200	.080		3.39	4.25		7.64	10.55
0980	6″ x 8″		170	.094		3.86	5		8.86	12.25
1000	Pine, pressure treated, 1″ x 4″		500	.032		.68	1.70		2.38	3.47
1040	2″ x 4″		330	.048		1.14	2.58		3.72	5.40
1080	4″ x 6″		250	.064		3.81	3.40		7.21	9.65
1100	6″ x 6″		200	.080		5.70	4.25		9.95	13.10
1140	6″ x 8″		170	.094		7.50	5		12.50	16.25
9000	Minimum labor/equipment charge	1 Carp	4	2	Job		106		106	170

32 94 Planting Accessories

32 94 50 – Tree Guying

32 94 50.10 Tree Guying Systems		Crew	Daily Output	Labor-Hours	Unit	Material	Labor	Equipment	Total	Total Incl O&P
						2020 Bare Costs				
0010	**TREE GUYING SYSTEMS**									
0015	Tree guying including stakes, guy wire and wrap									
0100	Less than 3″ caliper, 2 stakes	2 Clab	35	.457	Ea.	14.15	19.25		33.40	46.50
0200	3″ to 4″ caliper, 3 stakes	″	21	.762	″	20	32		52	73.50
1000	Including arrowhead anchor, cable, turnbuckles and wrap									
1100	Less than 3″ caliper, 3″ anchors	2 Clab	20	.800	Ea.	23.50	33.50		57	79.50
1200	3″ to 6″ caliper, 4″ anchors		15	1.067		34	45		79	109
1300	6″ caliper, 6″ anchors		12	1.333		27.50	56		83.50	120
1400	8″ caliper, 8″ anchors		9	1.778		115	75		190	246

32 96 Transplanting

32 96 23 – Plant and Bulb Transplanting

32 96 23.23 Planting		Crew	Daily Output	Labor-Hours	Unit	Material	Labor	Equipment	Total	Total Incl O&P
0010	**PLANTING**									
0012	Moving shrubs on site, 12″ ball	B-62	28	.857	Ea.		39	6.35	45.35	69.50
0100	24″ ball	″	22	1.091	″		50	8.05	58.05	88

32 96 23.43 Moving Trees		Crew	Daily Output	Labor-Hours	Unit	Material	Labor	Equipment	Total	Total Incl O&P
0010	**MOVING TREES**, On site									
0300	Moving trees on site, 36″ ball	B-6	3.75	6.400	Ea.		293	57	350	530
0400	60″ ball	″	1	24	″		1,100	214	1,314	1,975

32 96 43 – Tree Transplanting

32 96 43.20 Tree Removal		Crew	Daily Output	Labor-Hours	Unit	Material	Labor	Equipment	Total	Total Incl O&P
0010	**TREE REMOVAL**									
0100	Dig & lace, shrubs, broadleaf evergreen, 18″-24″ high	B-1	55	.436	Ea.		18.65		18.65	30
0200	2′-3′	″	35	.686			29.50		29.50	47
0300	3′-4′	B-6	30	.800			36.50	7.15	43.65	66
0400	4′-5′	″	20	1.200			55	10.70	65.70	99
1000	Deciduous, 12″-15″	B-1	110	.218			9.35		9.35	14.95
1100	18″-24″		65	.369			15.80		15.80	25.50
1200	2′-3′		55	.436			18.65		18.65	30
1300	3′-4′	B-6	50	.480			22	4.28	26.28	39.50
2000	Evergreen, 18″-24″	B-1	55	.436			18.65		18.65	30
2100	2′-0″ to 2′-6″		50	.480			20.50		20.50	33
2200	2′-6″ to 3′-0″		35	.686			29.50		29.50	47
2300	3′-0″ to 3′-6″		20	1.200			51.50		51.50	82
3000	Trees, deciduous, small, 2′-3′		55	.436			18.65		18.65	30
3100	3′-4′	B-6	50	.480			22	4.28	26.28	39.50
3200	4′-5′		35	.686			31.50	6.10	37.60	56.50
3300	5′-6′		30	.800			36.50	7.15	43.65	66
4000	Shade, 5′-6′		50	.480			22	4.28	26.28	39.50
4100	6′-8′		35	.686			31.50	6.10	37.60	56.50
4200	8′-10′		25	.960			44	8.55	52.55	79
4300	2″ caliper		12	2			91.50	17.80	109.30	165
5000	Evergreen, 4′-5′		35	.686			31.50	6.10	37.60	56.50
5100	5′-6′		25	.960			44	8.55	52.55	79
5200	6′-7′		19	1.263			58	11.25	69.25	104
5300	7′-8′		15	1.600			73	14.25	87.25	132
5400	8′-10′		11	2.182			100	19.45	119.45	180

Division Notes

		CREW	DAILY OUTPUT	LABOR-HOURS	UNIT	BARE COSTS MAT.	LABOR	EQUIP.	TOTAL	TOTAL INCL O&P

Division 33

Utilities

Estimating Tips

33 10 00 Water Utilities
33 30 00 Sanitary Sewerage Utilities
33 40 00 Storm Drainage Utilities

- Never assume that the water, sewer, and drainage lines will go in at the early stages of the project. Consider the site access needs before dividing the site in half with open trenches, loose pipe, and machinery obstructions. Always inspect the site to establish that the site drawings are complete. Check off all existing utilities on your drawings as you locate them. Be especially careful with underground utilities because appurtenances are sometimes buried during regrading or repaving operations. If you find any discrepancies, mark up the site plan for further research. Differing site conditions can be very costly if discovered later in the project.
- See also Section 33 01 00 for restoration of pipe where removal/ replacement may be undesirable. Use of new types of piping materials can reduce the overall project cost. Owners/design engineers should consider the installing contractor as a valuable source of current information on utility products and local conditions that could lead to significant cost savings.

Reference Numbers

Reference numbers are shown at the beginning of some major classifications. These numbers refer to related items in the Reference Section. The reference information may be an estimating procedure, an alternate pricing method, or technical information.

Note: Not all subdivisions listed here necessarily appear. ■

Note: "Powered in part by CINX™, based on liscensed proprietary information of Harrison Publishing House, Inc."

Note: Trade Service, in part, has been used as a reference source for some of the material prices used in Division 33

33 01 10 – Operation and Maintenance of Water Utilities

33 01 10.10 Corrosion Resistance		Crew	Daily Output	Labor-Hours	Unit	Material	Labor	Equipment	Total	Total Incl O&P
						2020 Bare Costs				
0010	**CORROSION RESISTANCE**									
0012	Wrap & coat, add to pipe, 4" diameter				L.F.	2.38			2.38	2.62
0020	5" diameter					3.04			3.04	3.34
0040	6" diameter					3.67			3.67	4.04
0060	8" diameter					5			5	5.50
0080	10" diameter					6.30			6.30	6.95
0100	12" diameter					7.30			7.30	8
0120	14" diameter					8.35			8.35	9.20
0140	16" diameter					10.55			10.55	11.60
0160	18" diameter					11			11	12.10
0180	20" diameter					12.25			12.25	13.50
0200	24" diameter					14.75			14.75	16.20
0220	Small diameter pipe, 1" diameter, add					1.01			1.01	1.11
0240	2" diameter					1.28			1.28	1.41
0260	2-1/2" diameter					1.61			1.61	1.77
0280	3" diameter					1.93			1.93	2.12
0300	Fittings, field covered, add				S.F.	9.40			9.40	10.35
0500	Coating, bituminous, per diameter inch, 1 coat, add				L.F.	.21			.21	.23
0540	3 coat					.57			.57	.63
0560	Coal tar epoxy, per diameter inch, 1 coat, add					.21			.21	.23
0600	3 coat					.66			.66	.73
1000	Polyethylene H.D. extruded, 0.025" thk., 1/2" diameter, add					.10			.10	.11
1020	3/4" diameter					.15			.15	.17
1040	1" diameter					.19			.19	.21
1060	1-1/4" diameter					.25			.25	.28
1080	1-1/2" diameter					.28			.28	.31
1100	0.030" thk., 2" diameter					.39			.39	.43
1120	2-1/2" diameter					.50			.50	.55
1140	0.035" thk., 3" diameter					.56			.56	.62
1160	3-1/2" diameter					.65			.65	.72
1180	4" diameter					.81			.81	.89
1200	0.040" thk., 5" diameter					.93			.93	1.02
1220	6" diameter					1.14			1.14	1.25
1240	8" diameter					1.46			1.46	1.61
1260	10" diameter					1.79			1.79	1.97
1280	12" diameter					2.15			2.15	2.37
1300	0.060" thk., 14" diameter					2.76			2.76	3.04
1320	16" diameter					3.16			3.16	3.48
1340	18" diameter					3.30			3.30	3.63
1360	20" diameter					3.67			3.67	4.04
1380	Fittings, field wrapped, add				S.F.	4.30			4.30	4.73

33 01 10.20 Pipe Repair		Crew	Daily Output	Labor-Hours	Unit	Material	Labor	Equipment	Total	Total Incl O&P
0010	**PIPE REPAIR**									
0020	Not including excavation or backfill									
0100	Clamp, stainless steel, lightweight, for steel pipe									
0110	3" long, 1/2" diameter pipe	1 Plum	34	.235	Ea.	11.15	15.15		26.30	36
0120	3/4" diameter pipe		32	.250		14.10	16.10		30.20	40.50
0130	1" diameter pipe		30	.267		12.50	17.20		29.70	40.50
0140	1-1/4" diameter pipe		28	.286		13.15	18.40		31.55	43
0150	1-1/2" diameter pipe		26	.308		16.85	19.85		36.70	49
0160	2" diameter pipe		24	.333		18.15	21.50		39.65	53
0170	2-1/2" diameter pipe		23	.348		17.40	22.50		39.90	53.50

33 01 10.20 Pipe Repair		Crew	Daily Output	Labor-Hours	Unit	2020 Bare Costs Material	Labor	Equipment	Total	Total Incl O&P
0180	3" diameter pipe	1 Plum	22	.364	Ea.	24.50	23.50		48	63
0190	3-1/2" diameter pipe		21	.381		25	24.50		49.50	65.50
0200	4" diameter pipe	B-20	44	.545		23	25.50		48.50	66
0210	5" diameter pipe		42	.571		26.50	27		53.50	72
0220	6" diameter pipe		38	.632		29.50	29.50		59	80
0230	8" diameter pipe		30	.800		35.50	37.50		73	99.50
0240	10" diameter pipe		28	.857		120	40.50		160.50	197
0250	12" diameter pipe		24	1		131	47		178	220
0260	14" diameter pipe		22	1.091		166	51.50		217.50	265
0270	16" diameter pipe		20	1.200		179	56.50		235.50	287
0280	18" diameter pipe		18	1.333		206	62.50		268.50	325
0290	20" diameter pipe		16	1.500		234	70.50		304.50	370
0300	24" diameter pipe		14	1.714		238	80.50		318.50	390
0360	For 6" long, add					100%	40%			
0370	For 9" long, add					200%	100%			
0380	For 12" long, add					300%	150%			
0390	For 18" long, add					500%	200%			
0400	Pipe freezing for live repairs of systems 3/8" to 6"									
0410	Note: Pipe freezing can also be used to install a valve into a live system									
0420	Pipe freezing each side 3/8"	2 Skwk	8	2	Ea.	550	110		660	780
0425	Pipe freezing each side 3/8", second location same kit		8	2		32	110		142	211
0430	Pipe freezing each side 3/4"		8	2		515	110		625	745
0435	Pipe freezing each side 3/4", second location same kit		8	2		32	110		142	211
0440	Pipe freezing each side 1-1/2"		6	2.667		515	146		661	805
0445	Pipe freezing each side 1-1/2" , second location same kit		6	2.667		32	146		178	269
0450	Pipe freezing each side 2"		6	2.667		945	146		1,091	1,275
0455	Pipe freezing each side 2", second location same kit		6	2.667		32	146		178	269
0460	Pipe freezing each side 2-1/2" to 3"		6	2.667		955	146		1,101	1,275
0465	Pipe freezing each side 2-1/2" to 3", second location same kit		6	2.667		64	146		210	305
0470	Pipe freezing each side 4"		4	4		1,575	219		1,794	2,075
0475	Pipe freezing each side 4", second location same kit		4	4		73	219		292	430
0480	Pipe freezing each side 5" to 6"		4	4		4,400	219		4,619	5,200
0485	Pipe freezing each side 5" to 6", second location same kit		4	4		218	219		437	590
0490	Pipe freezing extra 20 lb. CO_2 cylinders (3/8" to 2" - 1 ea, 3" - 2 ea)					231			231	254
0500	Pipe freezing extra 50 lb. CO_2 cylinders (4" - 2 ea, 5"-6" - 6 ea)					530			530	585
1000	Clamp, stainless steel, with threaded service tap									
1040	Full seal for iron, steel, PVC pipe									
1100	6" long, 2" diameter pipe	1 Plum	17	.471	Ea.	115	30.50		145.50	174
1110	2-1/2" diameter pipe		16	.500		114	32		146	176
1120	3" diameter pipe		15.60	.513		112	33		145	174
1130	3-1/2" diameter pipe		15	.533		135	34.50		169.50	201
1140	4" diameter pipe	B-20	32	.750		125	35.50		160.50	194
1150	6" diameter pipe		28	.857		164	40.50		204.50	245
1160	8" diameter pipe		21	1.143		176	53.50		229.50	280
1170	10" diameter pipe		20	1.200		232	56.50		288.50	345
1180	12" diameter pipe		17	1.412		266	66.50		332.50	400
1200	8" long, 2" diameter pipe	1 Plum	11.72	.683		117	44		161	197
1210	2-1/2" diameter pipe		11	.727		127	47		174	213
1220	3" diameter pipe		10.75	.744		125	48		173	212
1230	3-1/2" diameter pipe		10.34	.774		132	50		182	222
1240	4" diameter pipe	B-20	22	1.091		158	51.50		209.50	256
1250	6" diameter pipe		19.31	1.243		162	58.50		220.50	273
1260	8" diameter pipe		14.48	1.657		211	78		289	355

33 01 10.20 Pipe Repair		Crew	Daily Output	Labor-Hours	Unit	2020 Bare Costs Material	Labor	Equipment	Total	Total Incl O&P
1270	10″ diameter pipe	B-20	13.80	1.739	Ea.	243	82		325	400
1280	12″ diameter pipe		11.72	2.048		305	96.50		401.50	495
1300	12″ long, 2″ diameter pipe	1 Plum	9.44	.847		248	54.50		302.50	360
1310	2-1/2″ diameter pipe		8.89	.900		204	58		262	315
1320	3″ diameter pipe		8.67	.923		191	59.50		250.50	300
1330	3-1/2″ diameter pipe		8.33	.960		228	62		290	345
1340	4″ diameter pipe	B-20	17.78	1.350		237	63.50		300.50	360
1350	6″ diameter pipe		15.56	1.542		243	72.50		315.50	385
1360	8″ diameter pipe		11.67	2.057		288	96.50		384.50	470
1370	10″ diameter pipe		11.11	2.160		470	102		572	680
1380	12″ diameter pipe		9.44	2.542		475	120		595	715
1400	20″ long, 2″ diameter pipe	1 Plum	8.10	.988		231	63.50		294.50	355
1410	2-1/2″ diameter pipe		7.62	1.050		272	67.50		339.50	405
1420	3″ diameter pipe		7.43	1.077		345	69.50		414.50	485
1430	3-1/2″ diameter pipe		7.14	1.120		310	72		382	450
1440	4″ diameter pipe	B-20	15.24	1.575		340	74		414	495
1450	6″ diameter pipe		13.33	1.800		405	84.50		489.50	580
1460	8″ diameter pipe		10	2.400		495	113		608	725
1470	10″ diameter pipe		9.52	2.521		555	119		674	805
1480	12″ diameter pipe		8.10	2.963		660	139		799	950
1600	Clamp, stainless steel, single section									
1640	Full seal for iron, steel, PVC pipe									
1700	6″ long, 2″ diameter pipe	1 Plum	17	.471	Ea.	91	30.50		121.50	147
1710	2-1/2″ diameter pipe		16	.500		94	32		126	153
1720	3″ diameter pipe		15.60	.513		86.50	33		119.50	146
1730	3-1/2″ diameter pipe		15	.533		113	34.50		147.50	177
1740	4″ diameter pipe	B-20	32	.750		123	35.50		158.50	192
1750	6″ diameter pipe		27	.889		152	42		194	235
1760	8″ diameter pipe		21	1.143		161	53.50		214.50	263
1770	10″ diameter pipe		20	1.200		227	56.50		283.50	340
1780	12″ diameter pipe		17	1.412		218	66.50		284.50	345
1800	8″ long, 2″ diameter pipe	1 Plum	11.72	.683		126	44		170	207
1805	2-1/2″ diameter pipe		11.03	.725		129	46.50		175.50	215
1810	3″ diameter pipe		10.76	.743		136	48		184	224
1815	3-1/2″ diameter pipe		10.34	.774		124	50		174	214
1820	4″ diameter pipe	B-20	22.07	1.087		147	51		198	244
1825	6″ diameter pipe		19.31	1.243		187	58.50		245.50	300
1830	8″ diameter pipe		14.48	1.657		223	78		301	370
1835	10″ diameter pipe		13.79	1.740		300	82		382	465
1840	12″ diameter pipe		11.72	2.048		335	96.50		431.50	525
1850	12″ long, 2″ diameter pipe	1 Plum	9.44	.847		190	54.50		244.50	293
1855	2-1/2″ diameter pipe		8.89	.900		196	58		254	305
1860	3″ diameter pipe		8.67	.923		215	59.50		274.50	330
1865	3-1/2″ diameter pipe		8.33	.960		197	62		259	315
1870	4″ diameter pipe	B-20	17.78	1.350		224	63.50		287.50	350
1875	6″ diameter pipe		15.56	1.542		305	72.50		377.50	450
1880	8″ diameter pipe		11.67	2.057		330	96.50		426.50	520
1885	10″ diameter pipe		11.11	2.160		450	102		552	650
1890	12″ diameter pipe		9.44	2.542		535	120		655	780
1900	20″ long, 2″ diameter pipe	1 Plum	8.10	.988		197	63.50		260.50	315
1905	2-1/2″ diameter pipe		7.62	1.050		218	67.50		285.50	345
1910	3″ diameter pipe		7.43	1.077		250	69.50		319.50	380
1915	3-1/2″ diameter pipe		7.14	1.120		320	72		392	465

33 01 10 – Operation and Maintenance of Water Utilities

33 01 10.20 Pipe Repair		Crew	Daily Output	Labor-Hours	Unit	Material	2020 Bare Costs Labor	Equipment	Total	Total Incl O&P
1920	4″ diameter pipe	B-20	15.24	1.575	Ea.	425	74		499	585
1925	6″ diameter pipe		13.33	1.800		465	84.50		549.50	645
1930	8″ diameter pipe		10	2.400		555	113		668	790
1935	10″ diameter pipe		9.52	2.521		700	119		819	960
1940	12″ diameter pipe	↓	8.10	2.963	↓	830	139		969	1,125
2000	Clamp, stainless steel, two section									
2040	Full seal, for iron, steel, PVC pipe									
2100	8″ long, 4″ diameter pipe	B-20	24	1	Ea.	230	47		277	325
2110	6″ diameter pipe		20	1.200		264	56.50		320.50	380
2120	8″ diameter pipe		13	1.846		305	87		392	480
2130	10″ diameter pipe		12	2		293	94		387	470
2140	12″ diameter pipe		10	2.400		395	113		508	615
2200	10″ long, 4″ diameter pipe		16	1.500		300	70.50		370.50	445
2210	6″ diameter pipe		13	1.846		340	87		427	515
2220	8″ diameter pipe		9	2.667		380	125		505	615
2230	10″ diameter pipe		8	3		490	141		631	765
2240	12″ diameter pipe	↓	7	3.429	↓	565	161		726	885
2242	Clamp, stainless steel, three section									
2244	Full seal, for iron, steel, PVC pipe									
2250	10″ long, 14″ diameter pipe	B-20	6.40	3.750	Ea.	655	176		831	1,000
2260	16″ diameter pipe		6	4		695	188		883	1,075
2270	18″ diameter pipe		5	4.800		795	226		1,021	1,225
2280	20″ diameter pipe		4.60	5.217		905	245		1,150	1,375
2290	24″ diameter pipe	↓	4	6		1,250	282		1,532	1,825
2320	For 12″ long, add to 10″					15%	25%			
2330	For 20″ long, add to 10″				↓	70%	55%			
8000	For internal cleaning and inspection, see Section 33 01 30.11									
8100	For pipe testing, see Section 23 05 93.50									

33 01 30 – Operation and Maintenance of Sewer Utilities

33 01 30.11 Television Inspection of Sewers		Crew	Daily Output	Labor-Hours	Unit	Material	Labor	Equipment	Total	Total Incl O&P
0010	TELEVISION INSPECTION OF SEWERS									
0100	Pipe internal cleaning & inspection, cleaning, pressure pipe systems									
0120	Pig method, lengths 1000′ to 10,000′									
0140	4″ diameter thru 24″ diameter, minimum				L.F.				3.60	4.14
0160	Maximum				″				18	21
6000	Sewage/sanitary systems									
6100	Power rodder with header & cutters									
6110	Mobilization charge, minimum				Total				695	800
6120	Mobilization charge, maximum				″				9,125	10,600
6140	Cleaning 4″-12″ diameter				L.F.				6	6.60
6190	14″-24″ diameter								8	8.80
6240	30″ diameter								9.60	10.50
6250	36″ diameter								12	13.20
6260	48″ diameter								16	17.60
6270	60″ diameter								24	26.50
6280	72″ diameter				↓				48	53
9000	Inspection, television camera with video									
9060	up to 500 linear feet				Total				715	820

33 01 30 – Operation and Maintenance of Sewer Utilities

33 01 30.23 Pipe Bursting		Crew	Daily Output	Labor-Hours	Unit	Material	2020 Bare Costs Labor	Equipment	Total	Total Incl O&P
0010	**PIPE BURSTING**									
0011	300′ runs, replace with HDPE pipe									
0020	Not including excavation, backfill, shoring, or dewatering									
0100	6″ to 15″ diameter, minimum				L.F.				200	220
0200	Maximum								550	605
0300	18″ to 36″ diameter, minimum								650	715
0400	Maximum				▼				1,050	1,150
0500	Mobilize and demobilize, minimum				Job				3,000	3,300
0600	Maximum				″				32,100	35,400

33 01 30.72 Cured-In-Place Pipe Lining		Crew	Daily Output	Labor-Hours	Unit	Material	Labor	Equipment	Total	Total Incl O&P
0010	**CURED-IN-PLACE PIPE LINING**									
0011	Not incl. bypass or cleaning									
0020	Less than 10,000 L.F., urban, 6″ to 10″	C-17E	130	.615	L.F.	9.40	34	.28	43.68	64.50
0050	10″ to 12″		125	.640		11.60	35.50	.29	47.39	69.50
0070	12″ to 16″		115	.696		11.85	38.50	.31	50.66	74.50
0100	16″ to 20″		95	.842		13.95	46.50	.38	60.83	89.50
0200	24″ to 36″		90	.889		15.30	49	.40	64.70	95.50
0300	48″ to 72″		80	1		24	55.50	.45	79.95	115
0500	Rural, 6″ to 10″		180	.444		9.40	24.50	.20	34.10	49.50
0550	10″ to 12″		175	.457		11.60	25.50	.21	37.31	53
0570	12″ to 16″		160	.500		12.15	27.50	.23	39.88	57.50
0600	16″ to 20″		135	.593		14.95	32.50	.27	47.72	68.50
0700	24″ to 36″		125	.640		15.35	35.50	.29	51.14	73.50
0800	48″ to 72″		100	.800		24	44	.36	68.36	97.50
1000	Greater than 10,000 L.F., urban, 6″ to 10″		160	.500		9.40	27.50	.23	37.13	54.50
1050	10″ to 12″		155	.516		11.55	28.50	.23	40.28	58.50
1070	12″ to 16″		140	.571		11.85	31.50	.26	43.61	64
1100	16″ to 20″		120	.667		14.95	37	.30	52.25	75
1200	24″ to 36″		115	.696		15.35	38.50	.31	54.16	78
1300	48″ to 72″		95	.842		24	46.50	.38	70.88	101
1500	Rural, 6″ to 10″		215	.372		9.40	20.50	.17	30.07	43.50
1550	10″ to 12″		210	.381		11.60	21	.17	32.77	46.50
1570	12″ to 16″		185	.432		11.85	24	.19	36.04	51.50
1600	16″ to 20″		150	.533		14.95	29.50	.24	44.69	63.50
1700	24″ to 36″		140	.571		15.35	31.50	.26	47.11	67.50
1800	48″ to 72″	▼	120	.667	▼	24.50	37	.30	61.80	86
2000	Cured in place pipe, non-pressure, flexible felt resin									
2100	6″ diameter				L.F.				25.50	28
2200	8″ diameter								26.50	29
2201	8″ diameter									
2300	10″ diameter								29	32
2400	12″ diameter								34.50	38
2500	15″ diameter								59	65
2600	18″ diameter								82	90
2700	21″ diameter								105	115
2800	24″ diameter								177	195
2900	30″ diameter								195	215
3000	36″ diameter								205	225
3100	48″ diameter				▼				218	240

33 01 Operation and Maintenance of Utilities

33 01 30 – Operation and Maintenance of Sewer Utilities

33 01 30.74 Sliplining, Excludes Cleaning		Crew	Daily Output	Labor-Hours	Unit	2020 Bare Costs Material	Labor	Equipment	Total	Total Incl O&P
0010	**SLIPLINING, excludes cleaning** and video inspection									
0020	Pipe relined with one pipe size smaller than original (4" for 6")									
0100	6" diameter, original size	B-6B	600	.080	L.F.	5.60	3.42	1.67	10.69	13.55
0150	8" diameter, original size		600	.080		9.60	3.42	1.67	14.69	17.90
0200	10" diameter, original size		600	.080		11.25	3.42	1.67	16.34	19.75
0250	12" diameter, original size		400	.120		16	5.15	2.50	23.65	28.50
0300	14" diameter, original size		400	.120		15.60	5.15	2.50	23.25	28
0350	16" diameter, original size	B-6C	300	.160		18.30	6.85	7.75	32.90	39.50
0400	18" diameter, original size	"	300	.160		28	6.85	7.75	42.60	50.50
1000	Pipe HDPE lining, make service line taps	B-6	4	6	Ea.	87	274	53.50	414.50	590

33 05 Common Work Results for Utilities

33 05 07 – Trenchless Installation of Utility Piping

33 05 07.23 Utility Boring and Jacking		Crew	Daily Output	Labor-Hours	Unit	Material	Labor	Equipment	Total	Total Incl O&P
0010	**UTILITY BORING AND JACKING**									
0011	Casing only, 100' minimum,									
0020	not incl. jacking pits or dewatering									
0100	Roadwork, 1/2" thick wall, 24" diameter casing	B-42	20	3.200	L.F.	122	152	52	326	435
0200	36" diameter		16	4		215	190	65	470	615
0300	48" diameter		15	4.267		300	203	69.50	572.50	730
0500	Railroad work, 24" diameter		15	4.267		122	203	69.50	394.50	535
0600	36" diameter		14	4.571		215	217	74.50	506.50	665
0700	48" diameter		12	5.333		300	253	87	640	830
0900	For ledge, add								20%	20%
1000	Small diameter boring, 3", sandy soil	B-82	900	.018		22	.85	.20	23.05	26
1040	Rocky soil	"	500	.032		22	1.52	.36	23.88	27.50

33 05 07.36 Microtunneling		Crew	Daily Output	Labor-Hours	Unit	Material	Labor	Equipment	Total	Total Incl O&P
0010	**MICROTUNNELING**									
0011	Not including excavation, backfill, shoring,									
0020	or dewatering, average 50'/day, slurry method									
0100	24" to 48" outside diameter, minimum				L.F.				965	965
0110	Adverse conditions, add				"				500	500
1000	Rent microtunneling machine, average monthly lease				Month				97,500	107,000
1010	Operating technician				Day				630	705
1100	Mobilization and demobilization, minimum				Job				41,200	45,900
1110	Maximum				"				445,500	490,500

33 05 61 – Concrete Manholes

33 05 61.10 Storm Drainage Manholes, Frames and Covers		Crew	Daily Output	Labor-Hours	Unit	Material	Labor	Equipment	Total	Total Incl O&P
0010	**STORM DRAINAGE MANHOLES, FRAMES & COVERS**									
0020	Excludes footing, excavation, backfill (See line items for frame & cover)									
0050	Brick, 4' inside diameter, 4' deep	D-1	1	16	Ea.	615	750		1,365	1,875
0100	6' deep		.70	22.857		870	1,075		1,945	2,675
0150	8' deep		.50	32		1,125	1,500		2,625	3,650
0200	For depths over 8', add		4	4	V.L.F.	95.50	188		283.50	410
0400	Concrete blocks (radial), 4' ID, 4' deep		1.50	10.667	Ea.	430	500		930	1,275
0500	6' deep		1	16		565	750		1,315	1,825
0600	8' deep		.70	22.857		705	1,075		1,780	2,500
0700	For depths over 8', add		5.50	2.909	V.L.F.	71.50	136		207.50	299
0800	Concrete, cast in place, 4' x 4', 8" thick, 4' deep	C-14H	2	24	Ea.	580	1,250	13.45	1,843.45	2,625
0900	6' deep		1.50	32		830	1,650	17.90	2,497.90	3,550

33 05 61 – Concrete Manholes

33 05 61.10 Storm Drainage Manholes, Frames and Covers		Crew	Daily Output	Labor-Hours	Unit	2020 Bare Costs Material	Labor	Equipment	Total	Total Incl O&P
1000	8′ deep	C-14H	1	48	Ea.	1,200	2,475	27	3,702	5,300
1100	For depths over 8′, add		8	6	V.L.F.	133	310	3.36	446.36	645
1110	Precast, 4′ ID, 4′ deep	B-22	4.10	7.317	Ea.	880	360	69	1,309	1,625
1120	6′ deep		3	10		1,025	495	94.50	1,614.50	2,025
1130	8′ deep		2	15		1,200	740	142	2,082	2,650
1140	For depths over 8′, add		16	1.875	V.L.F.	134	92.50	17.70	244.20	315
1150	5′ ID, 4′ deep	B-6	3	8	Ea.	1,725	365	71.50	2,161.50	2,550
1160	6′ deep		2	12		2,075	550	107	2,732	3,275
1170	8′ deep		1.50	16		2,625	730	143	3,498	4,175
1180	For depths over 8′, add		12	2	V.L.F.	325	91.50	17.80	434.30	525
1190	6′ ID, 4′ deep		2	12	Ea.	2,600	550	107	3,257	3,850
1200	6′ deep		1.50	16		2,975	730	143	3,848	4,575
1210	8′ deep		1	24		3,675	1,100	214	4,989	6,025
1220	For depths over 8′, add		8	3	V.L.F.	410	137	26.50	573.50	700
1250	Slab tops, precast, 8″ thick									
1300	4′ diameter manhole	B-6	8	3	Ea.	282	137	26.50	445.50	560
1400	5′ diameter manhole		7.50	3.200		440	146	28.50	614.50	750
1500	6′ diameter manhole		7	3.429		705	157	30.50	892.50	1,075
3800	Steps, heavyweight cast iron, 7″ x 9″	1 Bric	40	.200		18.05	10.40		28.45	36.50
3900	8″ x 9″		40	.200		21.50	10.40		31.90	41
3928	12″ x 10-1/2″		40	.200		29.50	10.40		39.90	49.50
4000	Standard sizes, galvanized steel		40	.200		26.50	10.40		36.90	46
4100	Aluminum		40	.200		30	10.40		40.40	50
4150	Polyethylene		40	.200		32.50	10.40		42.90	53
4210	Rubber boot 6″ diam. or smaller	1 Clab	32	.250		104	10.55		114.55	131
4215	8″ diam.		24	.333		118	14.05		132.05	153
4220	10″ diam.		19	.421		135	17.75		152.75	178
4225	12″ diam.		16	.500		172	21		193	223
4230	16″ diam.		15	.533		226	22.50		248.50	284
4235	18″ diam.		15	.533		262	22.50		284.50	325
4240	24″ diam.		14	.571		286	24		310	355
4245	30″ diam.		12	.667		375	28		403	460

33 05 63 – Concrete Vaults and Chambers

33 05 63.13 Precast Concrete Utility Structures

		Crew	Daily Output	Labor-Hours	Unit	Material	Labor	Equipment	Total	Total Incl O&P
0010	**PRECAST CONCRETE UTILITY STRUCTURES**, 6″ thick									
0040	4′ x 6′ x 6′ high, ID	B-13	2	28	Ea.	1,500	1,300	290	3,090	4,025
0050	5′ x 10′ x 6′ high, ID		2	28		1,850	1,300	290	3,440	4,425
0100	6′ x 10′ x 6′ high, ID		2	28		1,925	1,300	290	3,515	4,500
0150	5′ x 12′ x 6′ high, ID		2	28		2,050	1,300	290	3,640	4,625
0200	6′ x 12′ x 6′ high, ID		1.80	31.111		2,275	1,425	325	4,025	5,150
0250	6′ x 13′ x 6′ high, ID		1.50	37.333		3,000	1,725	385	5,110	6,450
0300	8′ x 14′ x 7′ high, ID		1	56		3,250	2,575	580	6,405	8,325
0350	Hand hole, precast concrete, 1-1/2″ thick									
0400	1′-0″ x 2′-0″ x 1′-9″, ID, light duty	B-1	4	6	Ea.	540	257		797	1,000
0450	4′-6″ x 3′-2″ x 2′-0″, OD, heavy duty	B-6	3	8		1,625	365	71.50	2,061.50	2,425
0460	Meter pit, 4′ x 4′, 4′ deep		2	12		1,600	550	107	2,257	2,750
0470	6′ deep		1.60	15		2,275	685	134	3,094	3,750
0480	8′ deep		1.40	17.143		3,000	785	153	3,938	4,725
0490	10′ deep		1.20	20		3,800	915	178	4,893	5,850
0500	15′ deep		1	24		5,575	1,100	214	6,889	8,100
0510	6′ x 6′, 4′ deep		1.40	17.143		2,725	785	153	3,663	4,425
0520	6′ deep		1.20	20		4,100	915	178	5,193	6,150

33 05 Common Work Results for Utilities

33 05 63 – Concrete Vaults and Chambers

33 05 63.13 Precast Concrete Utility Structures		Crew	Daily Output	Labor-Hours	Unit	2020 Bare Costs Material	Labor	Equipment	Total	Total Incl O&P
0530	8′ deep	B-6	1	24	Ea.	5,450	1,100	214	6,764	7,975
0540	10′ deep		.80	30		6,825	1,375	267	8,467	9,975
0550	15′ deep		.60	40		10,400	1,825	355	12,580	14,700

33 05 97 – Identification and Signage for Utilities

33 05 97.05 Utility Connection

Line	Description	Crew	Daily Output	Labor-Hours	Unit	Material	Labor	Equipment	Total	Total Incl O&P
0010	**UTILITY CONNECTION**									
0020	Water, sanitary, stormwater, gas, single connection	B-14	1	48	Ea.	3,050	2,125	214	5,389	6,950
0030	Telecommunication	"	3	16	"	400	710	71	1,181	1,650

33 05 97.10 Utility Accessories

Line	Description	Crew	Daily Output	Labor-Hours	Unit	Material	Labor	Equipment	Total	Total Incl O&P
0010	**UTILITY ACCESSORIES**									
0400	Underground tape, detectable, reinforced, alum. foil core, 2″	1 Clab	150	.053	C.L.F.	9	2.25		11.25	13.50
0500	6″		140	.057	"	38	2.41		40.41	45.50
9000	Minimum labor/equipment charge		4	2	Job		84		84	135

33 11 Groundwater Sources

33 11 13 – Potable Water Supply Wells

33 11 13.10 Wells and Accessories

Line	Description	Crew	Daily Output	Labor-Hours	Unit	Material	Labor	Equipment	Total	Total Incl O&P
0010	**WELLS & ACCESSORIES**									
0011	Domestic									
0100	Drilled, 4″ to 6″ diameter	B-23	120	.333	L.F.		14.15	13.25	27.40	37
0200	8″ diameter	"	95.20	.420	"		17.85	16.70	34.55	47
0400	Gravel pack well, 40′ deep, incl. gravel & casing, complete									
0500	24″ diameter casing x 18″ diameter screen	B-23	.13	308	Total	38,500	13,100	12,200	63,800	77,000
0600	36″ diameter casing x 18″ diameter screen		.12	333	"	40,800	14,200	13,300	68,300	82,000
0800	Observation wells, 1-1/4″ riser pipe		163	.245	V.L.F.	19.90	10.45	9.75	40.10	49.50
0900	For flush Buffalo roadway box, add	1 Skwk	16.60	.482	Ea.	53	26.50		79.50	100
1200	Test well, 2-1/2″ diameter, up to 50′ deep (15 to 50 GPM)	B-23	1.51	26.490	"	830	1,125	1,050	3,005	3,875
1300	Over 50′ deep, add	"	121.80	.328	L.F.	22	13.95	13.05	49	61.50
1500	Pumps, installed in wells to 100′ deep, 4″ submersible									
1510	1/2 HP	Q-1	3.22	4.969	Ea.	660	288		948	1,175
1520	3/4 HP		2.66	6.015		795	350		1,145	1,425
1600	1 HP		2.29	6.987		975	405		1,380	1,700
1700	1-1/2 HP	Q-22	1.60	10		2,150	580	294	3,024	3,600
1800	2 HP		1.33	12.030		1,800	700	355	2,855	3,475
1900	3 HP		1.14	14.035		2,425	815	415	3,655	4,375
2000	5 HP		1.14	14.035		2,850	815	415	4,080	4,825
2050	Remove and install motor only, 4 HP		1.14	14.035		1,450	815	415	2,680	3,300
3000	Pump, 6″ submersible, 25′ to 150′ deep, 25 HP, 103 to 400 GPM		.89	17.978		9,325	1,050	530	10,905	12,500
3100	25′ to 500′ deep, 30 HP, 104 to 400 GPM		.73	21.918		11,000	1,275	645	12,920	14,800
8000	Steel well casing	B-23A	3020	.008	Lb.	1.29	.38	.29	1.96	2.34
8110	Well screen assembly, stainless steel, 2″ diameter		273	.088	L.F.	81.50	4.19	3.23	88.92	100
8120	3″ diameter		253	.095		143	4.52	3.49	151.01	168
8130	4″ diameter		200	.120		178	5.70	4.41	188.11	210
8140	5″ diameter		168	.143		184	6.80	5.25	196.05	219
8150	6″ diameter		126	.190		204	9.10	7	220.10	247
8160	8″ diameter		98.50	.244		274	11.60	8.95	294.55	330
8170	10″ diameter		73	.329		340	15.65	12.10	367.75	415
8180	12″ diameter		62.50	.384		395	18.30	14.10	427.40	480
8190	14″ diameter		54.30	.442		445	21	16.25	482.25	540
8200	16″ diameter		48.30	.497		490	23.50	18.25	531.75	600

33 11 Groundwater Sources

33 11 13 – Potable Water Supply Wells

33 11 13.10 Wells and Accessories		Crew	Daily Output	Labor-Hours	Unit	2020 Bare Costs Material	Labor	Equipment	Total	Total Incl O&P
8210	18″ diameter	B-23A	39.20	.612	L.F.	620	29	22.50	671.50	755
8220	20″ diameter		31.20	.769		700	36.50	28.50	765	860
8230	24″ diameter		23.80	1.008		855	48	37	940	1,050
8240	26″ diameter		21	1.143		960	54.50	42	1,056.50	1,175
8244	Well casing or drop pipe, PVC, 1/2″ diameter		550	.044		1.31	2.08	1.60	4.99	6.50
8245	3/4″ diameter		550	.044		1.32	2.08	1.60	5	6.50
8246	1″ diameter		550	.044		1.36	2.08	1.60	5.04	6.55
8247	1-1/4″ diameter		520	.046		1.71	2.20	1.70	5.61	7.25
8248	1-1/2″ diameter		490	.049		1.81	2.33	1.80	5.94	7.65
8249	1-3/4″ diameter		380	.063		1.85	3.01	2.32	7.18	9.35
8250	2″ diameter		280	.086		2.12	4.08	3.15	9.35	12.30
8252	3″ diameter		260	.092		4.16	4.40	3.39	11.95	15.30
8254	4″ diameter		205	.117		4.46	5.60	4.30	14.36	18.50
8255	5″ diameter		170	.141		4.48	6.75	5.20	16.43	21.50
8256	6″ diameter		130	.185		8.60	8.80	6.80	24.20	31
8258	8″ diameter		100	.240		13	11.45	8.80	33.25	42
8260	10″ diameter		73	.329		19.90	15.65	12.10	47.65	60.50
8262	12″ diameter		62	.387		24	18.45	14.25	56.70	71
8300	Slotted PVC, 1-1/4″ diameter		521	.046		2.89	2.20	1.69	6.78	8.50
8310	1-1/2″ diameter		488	.049		3.27	2.34	1.81	7.42	9.30
8320	2″ diameter		273	.088		4.09	4.19	3.23	11.51	14.70
8330	3″ diameter		253	.095		6.75	4.52	3.49	14.76	18.40
8340	4″ diameter		200	.120		8.35	5.70	4.41	18.46	23
8350	5″ diameter		168	.143		15.05	6.80	5.25	27.10	33
8360	6″ diameter		126	.190		16.30	9.10	7	32.40	40
8370	8″ diameter		98.50	.244		24	11.60	8.95	44.55	54.50
8400	Artificial gravel pack, 2″ screen, 6″ casing	B-23B	174	.138		4.68	6.55	6.40	17.63	22.50
8405	8″ casing		111	.216		6.30	10.30	10.05	26.65	34.50
8410	10″ casing		74.50	.322		8.80	15.35	14.95	39.10	50.50
8415	12″ casing		60	.400		14.05	19.05	18.60	51.70	66.50
8420	14″ casing		50.20	.478		14.65	23	22	59.65	76.50
8425	16″ casing		40.70	.590		18.55	28	27.50	74.05	95
8430	18″ casing		36	.667		22.50	32	31	85.50	109
8435	20″ casing		29.50	.814		26	39	38	103	132
8440	24″ casing		25.70	.934		30	44.50	43.50	118	151
8445	26″ casing		24.60	.976		33	46.50	45.50	125	161
8450	30″ casing		20	1.200		38	57	55.50	150.50	194
8455	36″ casing		16.40	1.463		40.50	69.50	68	178	231
8500	Develop well		8	3	Hr.	300	143	139	582	710
8550	Pump test well		8	3		96.50	143	139	378.50	485
8560	Standby well	B-23A	8	3		105	143	110	358	465
8570	Standby, drill rig		8	3			143	110	253	350
8580	Surface seal well, concrete filled		1	24	Ea.	1,025	1,150	880	3,055	3,950
8590	Well test pump, install & remove	B-23	1	40			1,700	1,600	3,300	4,475
8600	Well sterilization, chlorine	2 Clab	1	16		141	675		816	1,225
8610	Well water pressure switch	1 Clab	12	.667		98	28		126	153
8630	Well water pressure switch with manual reset	″	12	.667		130	28		158	188
9000	Minimum labor/equipment charge	B-21	1.80	15.556	Job		760	105	865	1,325
9950	See Section 31 23 19.40 for wellpoints									
9960	See Section 31 23 19.30 for drainage wells									

33 11 Groundwater Sources

33 11 13 – Potable Water Supply Wells

33 11 13.20 Water Supply Wells, Pumps		Crew	Daily Output	Labor-Hours	Unit	Material	Labor	Equipment	Total	Total Incl O&P
						2020 Bare Costs				
0010	WATER SUPPLY WELLS, PUMPS									
0011	With pressure control									
1000	Deep well, jet, 42 gal. galvanized tank									
1040	3/4 HP	1 Plum	.80	10	Ea.	1,125	645		1,770	2,250
3000	Shallow well, jet, 30 gal. galvanized tank									
3040	1/2 HP	1 Plum	2	4	Ea.	915	258		1,173	1,400

33 14 Water Utility Transmission and Distribution

33 14 13 – Public Water Utility Distribution Piping

33 14 13.10 Water Supply, Concrete Pipe		Crew	Daily Output	Labor-Hours	Unit	Material	Labor	Equipment	Total	Total Incl O&P
0010	WATER SUPPLY, CONCRETE PIPE									
0020	Not including excavation or backfill									
3000	Prestressed Concrete Pipe (PCCP), 150 psi, 12″ diameter	B-13	192	.292	L.F.	83	13.45	3.02	99.47	116
3010	24″ diameter	″	128	.438		83	20	4.54	107.54	128
3040	36″ diameter	B-13B	96	.583		129	27	10.20	166.20	196
3050	48″ diameter		64	.875		184	40.50	15.30	239.80	284
3070	72″ diameter		60	.933		360	43	16.35	419.35	480
3080	84″ diameter		40	1.400		465	64.50	24.50	554	640
3090	96″ diameter	B-13C	40	1.400		680	64.50	57	801.50	915
3100	108″ diameter		32	1.750		965	80.50	71.50	1,117	1,275
3102	120″ diameter		16	3.500		1,800	161	143	2,104	2,425
3104	144″ diameter		16	3.500		1,700	161	143	2,004	2,300
3110	Prestressed Concrete Pipe (PCCP), 150 psi, elbow, 90°, 12″ diameter	B-13	24	2.333	Ea.	955	107	24	1,086	1,250
3140	24″ diameter	″	6	9.333		1,950	430	97	2,477	2,925
3150	36″ diameter	B-13B	4	14		3,650	645	245	4,540	5,300
3160	48″ diameter		3	18.667		7,200	860	325	8,385	9,650
3180	72″ diameter		1.60	35		20,500	1,600	615	22,715	25,800
3190	84″ diameter		1.30	43.077		31,700	1,975	755	34,430	38,900
3200	96″ diameter		1	56		38,600	2,575	980	42,155	47,700
3210	108″ diameter	B-13C	.66	84.848		50,000	3,900	3,475	57,375	65,000
3220	120″ diameter		.40	140		56,000	6,450	5,725	68,175	78,000
3225	144″ diameter		.30	184		79,500	8,475	7,525	95,500	109,500
3230	Prestressed Concrete Pipe (PCCP), 150 psi, elbow, 45°, 12″ diameter	B-13	24	2.333		595	107	24	726	855
3250	24″ diameter	″	6	9.333		1,250	430	97	1,777	2,175
3260	36″ diameter	B-13B	4	14		2,175	645	245	3,065	3,700
3270	48″ diameter		3	18.667		4,150	860	325	5,335	6,300
3290	72″ diameter		1.60	35		12,700	1,600	615	14,915	17,200
3300	84″ diameter		1.30	42.945		19,500	1,975	750	22,225	25,400
3310	96″ diameter		1	56		24,100	2,575	980	27,655	31,700
3320	108″ diameter	B-13C	.66	84.337		30,400	3,875	3,450	37,725	43,400
3330	120″ diameter		.40	140		34,500	6,450	5,725	46,675	54,500
3340	144″ diameter		.30	184		49,200	8,475	7,525	65,200	76,000

33 14 13.15 Water Supply, Ductile Iron Pipe		Crew	Daily Output	Labor-Hours	Unit	Material	Labor	Equipment	Total	Total Incl O&P
0010	WATER SUPPLY, DUCTILE IRON PIPE R331113-80									
0011	Cement lined									
0020	Not including excavation or backfill									
2000	Pipe, class 50 water piping, 18′ lengths									
2020	Mechanical joint, 4″ diameter	B-21A	200	.200	L.F.	42.50	10.45	2.14	55.09	65.50
2040	6″ diameter		160	.250		51	13.05	2.68	66.73	79.50
2060	8″ diameter		133.33	.300		53.50	15.70	3.21	72.41	87

33 14 13.15 Water Supply, Ductile Iron Pipe		Crew	Daily Output	Labor-Hours	Unit	Material	2020 Bare Costs Labor	Equipment	Total	Total Incl O&P
2080	10″ diameter	B-21A	114.29	.350	L.F.	71	18.30	3.75	93.05	111
2100	12″ diameter		105.26	.380		90.50	19.85	4.07	114.42	135
2120	14″ diameter		100	.400		105	21	4.28	130.28	154
2140	16″ diameter		72.73	.550		112	29	5.90	146.90	175
2160	18″ diameter		68.97	.580		129	30.50	6.20	165.70	196
2170	20″ diameter		57.14	.700		132	36.50	7.50	176	211
2180	24″ diameter		47.06	.850		168	44.50	9.10	221.60	265
3000	Push-on joint, 4″ diameter		400	.100		23	5.25	1.07	29.32	35
3020	6″ diameter		333.33	.120		23.50	6.25	1.29	31.04	37.50
3040	8″ diameter		200	.200		32.50	10.45	2.14	45.09	55
3060	10″ diameter		181.82	.220		50	11.50	2.36	63.86	75.50
3080	12″ diameter		160	.250		53	13.05	2.68	68.73	81.50
3100	14″ diameter		133.33	.300		53	15.70	3.21	71.91	86
3120	16″ diameter		114.29	.350		55	18.30	3.75	77.05	93
3140	18″ diameter		100	.400		61	21	4.28	86.28	105
3160	20″ diameter		88.89	.450		63.50	23.50	4.82	91.82	112
3180	24″ diameter		76.92	.520		86.50	27	5.55	119.05	144
8000	Piping, fittings, mechanical joint, AWWA C110									
8006	90° bend, 4″ diameter	B-20A	16	2	Ea.	176	101		277	350
8020	6″ diameter		12.80	2.500		256	126		382	480
8040	8″ diameter		10.67	2.999		500	152		652	790
8060	10″ diameter	B-21A	11.43	3.500		730	183	37.50	950.50	1,125
8080	12″ diameter		10.53	3.799		960	199	40.50	1,199.50	1,400
8100	14″ diameter		10	4		1,375	209	43	1,627	1,900
8120	16″ diameter		7.27	5.502		1,750	288	59	2,097	2,450
8140	18″ diameter		6.90	5.797		2,400	305	62	2,767	3,200
8160	20″ diameter		5.71	7.005		2,875	365	75	3,315	3,800
8180	24″ diameter		4.70	8.511		4,975	445	91	5,511	6,275
8200	Wye or tee, 4″ diameter	B-20A	10.67	2.999		390	152		542	665
8220	6″ diameter		8.53	3.751		585	190		775	940
8240	8″ diameter		7.11	4.501		890	228		1,118	1,325
8260	10″ diameter	B-21A	7.62	5.249		1,175	274	56	1,505	1,800
8280	12″ diameter		7.02	5.698		1,975	298	61	2,334	2,700
8300	14″ diameter		6.67	5.997		2,425	315	64	2,804	3,225
8320	16″ diameter		4.85	8.247		2,750	430	88.50	3,268.50	3,800
8340	18″ diameter		4.60	8.696		4,500	455	93	5,048	5,775
8360	20″ diameter		3.81	10.499		6,525	550	112	7,187	8,150
8380	24″ diameter		3.14	12.739		8,800	665	136	9,601	10,900
8398	45° bend, 4″ diameter	B-20A	16	2		206	101		307	385
8400	6″ diameter		12.80	2.500		295	126		421	525
8405	8″ diameter		10.67	2.999		430	152		582	715
8410	12″ diameter	B-21A	10.53	3.799		875	199	40.50	1,114.50	1,325
8420	16″ diameter		7.27	5.502		1,725	288	59	2,072	2,425
8430	20″ diameter		5.71	7.005		2,450	365	75	2,890	3,350
8440	24″ diameter		4.70	8.511		3,450	445	91	3,986	4,600
8450	Decreaser, 6″ x 4″ diameter	B-20A	14.22	2.250		262	114		376	465
8460	8″ x 6″ diameter	″	11.64	2.749		350	139		489	605
8470	10″ x 6″ diameter	B-21A	13.33	3.001		425	157	32	614	750
8480	12″ x 6″ diameter		12.70	3.150		620	165	33.50	818.50	975
8490	16″ x 6″ diameter		10	4		1,075	209	43	1,327	1,550
8500	20″ x 6″ diameter		8.42	4.751		1,800	248	51	2,099	2,450
8552	For water utility valves see Section 33 14 19									
8700	Joint restraint, ductile iron mechanical joints									

33 14 13 – Public Water Utility Distribution Piping

33 14 13.15 Water Supply, Ductile Iron Pipe		Crew	Daily Output	Labor-Hours	Unit	Material	2020 Bare Costs Labor	Equipment	Total	Total Incl O&P
8710	4" diameter	B-20A	32	1	Ea.	32.50	50.50		83	115
8720	6" diameter		25.60	1.250		41	63		104	144
8730	8" diameter		21.33	1.500		59	76		135	184
8740	10" diameter		18.28	1.751		102	88.50		190.50	252
8750	12" diameter		16.84	1.900		117	96		213	280
8760	14" diameter		16	2		177	101		278	355
8770	16" diameter		11.64	2.749		190	139		329	425
8780	18" diameter		11.03	2.901		267	147		414	525
8785	20" diameter		9.14	3.501		330	177		507	645
8790	24" diameter		7.53	4.250		450	215		665	830
9600	Steel sleeve with tap, 4" diameter	B-20	3	8		505	375		880	1,150
9620	6" diameter		2	12		555	565		1,120	1,525
9630	8" diameter		2	12		720	565		1,285	1,700

33 14 13.20 Water Supply, Polyethylene Pipe, C901		Crew	Daily Output	Labor-Hours	Unit	Material	Labor	Equipment	Total	Total Incl O&P
0010	**WATER SUPPLY, POLYETHYLENE PIPE, C901** R331113-80									
0020	Not including excavation or backfill									
1000	Piping, 160 psi, 3/4" diameter	Q-1A	525	.019	L.F.	.40	1.24		1.64	2.35
1120	1" diameter		485	.021		.58	1.34		1.92	2.71
1140	1-1/2" diameter		450	.022		.96	1.44		2.40	3.29
1160	2" diameter		365	.027		1.44	1.78		3.22	4.33

33 14 13.25 Water Supply, Polyvinyl Chloride Pipe		Crew	Daily Output	Labor-Hours	Unit	Material	Labor	Equipment	Total	Total Incl O&P
0010	**WATER SUPPLY, POLYVINYL CHLORIDE PIPE** R331113-80									
0020	Not including excavation or backfill, unless specified									
2100	PVC pipe, Class 150, 1-1/2" diameter	Q-1A	750	.013	L.F.	.61	.86		1.47	2.01
2120	2" diameter		686	.015		.88	.95		1.83	2.43
2140	2-1/2" diameter		500	.020		1.18	1.30		2.48	3.31
2160	3" diameter	B-20	430	.056		1.59	2.62		4.21	5.95
3010	AWWA C905, PR 100, DR 25									
3030	14" diameter	B-20A	213	.150	L.F.	14.20	7.60		21.80	27.50
3040	16" diameter		200	.160		17.30	8.10		25.40	31.50
3050	18" diameter		160	.200		27.50	10.10		37.60	46.50
3060	20" diameter		133	.241		30.50	12.15		42.65	52.50
3070	24" diameter		107	.299		45	15.10		60.10	73
3080	30" diameter		80	.400		74.50	20		94.50	114
3090	36" diameter		80	.400		115	20		135	159
3100	42" diameter		60	.533		153	27		180	211
3200	48" diameter		60	.533		200	27		227	263
4520	Pressure pipe Class 150, SDR 18, AWWA C900, 4" diameter		380	.084		3.03	4.26		7.29	10.05
4530	6" diameter		316	.101		5	5.10		10.10	13.55
4540	8" diameter		264	.121		8.60	6.15		14.75	19.05
4550	10" diameter		220	.145		12.65	7.35		20	25.50
4560	12" diameter		186	.172		17.75	8.70		26.45	33
8000	Fittings with rubber gasket									
8003	Class 150, DR 18									
8006	90° bend , 4" diameter	B-20	100	.240	Ea.	42.50	11.30		53.80	64.50
8020	6" diameter		90	.267		74	12.55		86.55	102
8040	8" diameter		80	.300		139	14.10		153.10	176
8060	10" diameter		50	.480		365	22.50		387.50	435
8080	12" diameter		30	.800		445	37.50		482.50	550
8100	Tee, 4" diameter		90	.267		63	12.55		75.55	89
8120	6" diameter		80	.300		139	14.10		153.10	176
8140	8" diameter		70	.343		199	16.10		215.10	244

33 14 13.25 Water Supply, Polyvinyl Chloride Pipe		Crew	Daily Output	Labor-Hours	Unit	Material	2020 Bare Costs Labor	Equipment	Total	Total Incl O&P
8160	10″ diameter	B-20	40	.600	Ea.	660	28		688	770
8180	12″ diameter		20	1.200		905	56.50		961.50	1,075
8200	45° bend, 4″ diameter		100	.240		41.50	11.30		52.80	63.50
8220	6″ diameter		90	.267		77	12.55		89.55	105
8240	8″ diameter		50	.480		137	22.50		159.50	186
8260	10″ diameter		50	.480		305	22.50		327.50	370
8280	12″ diameter		30	.800		400	37.50		437.50	500
8300	Reducing tee 6″ x 4″		100	.240		126	11.30		137.30	157
8320	8″ x 6″		90	.267		202	12.55		214.55	242
8330	10″ x 6″		90	.267		340	12.55		352.55	395
8340	10″ x 8″		90	.267		365	12.55		377.55	420
8350	12″ x 6″		90	.267		420	12.55		432.55	480
8360	12″ x 8″		90	.267		440	12.55		452.55	505
8400	Tapped service tee (threaded type) 6″ x 6″ x 3/4″		100	.240		96	11.30		107.30	124
8430	6″ x 6″ x 1″		90	.267		96	12.55		108.55	126
8440	6″ x 6″ x 1-1/2″		90	.267		90.50	12.55		103.05	120
8450	6″ x 6″ x 2″		90	.267		96	12.55		108.55	126
8460	8″ x 8″ x 3/4″		90	.267		126	12.55		138.55	159
8470	8″ x 8″ x 1″		90	.267		126	12.55		138.55	159
8480	8″ x 8″ x 1-1/2″		90	.267		126	12.55		138.55	159
8490	8″ x 8″ x 2″		90	.267		126	12.55		138.55	159
8500	Repair coupling 4″		100	.240		23.50	11.30		34.80	44
8520	6″ diameter		90	.267		36.50	12.55		49.05	60.50
8540	8″ diameter		50	.480		90	22.50		112.50	135
8560	10″ diameter		50	.480		175	22.50		197.50	229
8580	12″ diameter		50	.480		265	22.50		287.50	325
8600	Plug end 4″		100	.240		21	11.30		32.30	41
8620	6″ diameter		90	.267		43	12.55		55.55	67
8640	8″ diameter		50	.480		63.50	22.50		86	106
8660	10″ diameter		50	.480		97.50	22.50		120	143
8680	12″ diameter	▼	50	.480	▼	120	22.50		142.50	168
8700	PVC pipe, joint restraint									
8710	4″ diameter	B-20A	32	1	Ea.	46.50	50.50		97	131
8720	6″ diameter		25.60	1.250		58	63		121	163
8730	8″ diameter		21.33	1.500		84.50	76		160.50	212
8740	10″ diameter		18.28	1.751		139	88.50		227.50	292
8750	12″ diameter		16.84	1.900		146	96		242	310
8760	14″ diameter		16	2		201	101		302	380
8770	16″ diameter		11.64	2.749		270	139		409	515
8780	18″ diameter		11.03	2.901		335	147		482	595
8785	20″ diameter		9.14	3.501		445	177		622	770
8790	24″ diameter	▼	7.53	4.250	▼	515	215		730	900

33 14 13.35 Water Supply, HDPE		Crew	Daily Output	Labor-Hours	Unit	Material	Labor	Equipment	Total	Total Incl O&P
0010	**WATER SUPPLY, HDPE** R331113-80									
0011	Butt fusion joints, SDR 21 40′ lengths not including excavation or backfill									
0100	4″ diameter	B-22A	400	.100	L.F.	2.59	4.85	2.01	9.45	12.75
0200	6″ diameter		380	.105		5.60	5.10	2.12	12.82	16.65
0300	8″ diameter		320	.125		9.60	6.05	2.52	18.17	23
0400	10″ diameter		300	.133		11.25	6.45	2.68	20.38	25.50
0500	12″ diameter	▼	260	.154		16	7.45	3.10	26.55	33
0600	14″ diameter	B-22B	220	.182		15.60	8.80	6.60	31	38.50
0700	16″ diameter	▼	180	.222	▼	18.30	10.75	8.10	37.15	46

33 14 13 – Public Water Utility Distribution Piping

33 14 13.35 Water Supply, HDPE		Crew	Daily Output	Labor-Hours	Unit	2020 Bare Costs Material	Labor	Equipment	Total	Total Incl O&P
0800	18″ diameter	B-22B	140	.286	L.F.	28	13.85	10.40	52.25	64.50
0850	20″ diameter		130	.308		34	14.90	11.20	60.10	73.50
0900	24″ diameter		100	.400		55.50	19.40	14.55	89.45	108
1000	Fittings									
1100	Elbows, 90 degrees									
1200	4″ diameter	B-22A	32	1.250	Ea.	17.25	60.50	25	102.75	143
1300	6″ diameter		28	1.429		47.50	69	28.50	145	194
1400	8″ diameter		24	1.667		113	81	33.50	227.50	290
1500	10″ diameter		18	2.222		254	108	44.50	406.50	500
1600	12″ diameter		12	3.333		297	162	67	526	655
1700	14″ diameter	B-22B	9	4.444		630	215	162	1,007	1,225
1800	16″ diameter		6	6.667		935	325	242	1,502	1,800
1900	18″ diameter		4	10		1,050	485	365	1,900	2,325
2000	24″ diameter		3	13.333		1,825	645	485	2,955	3,550
2100	Tees									
2200	4″ diameter	B-22A	30	1.333	Ea.	22.50	64.50	27	114	158
2300	6″ diameter		26	1.538		74	74.50	31	179.50	234
2400	8″ diameter		22	1.818		130	88	36.50	254.50	325
2500	10″ diameter		15	2.667		172	129	53.50	354.50	455
2600	12″ diameter		10	4		400	194	80.50	674.50	840
2700	14″ diameter	B-22B	8	5		475	242	182	899	1,100
2800	16″ diameter		6	6.667		555	325	242	1,122	1,400
2900	18″ diameter		4	10		670	485	365	1,520	1,900
3000	24″ diameter		2	20		895	970	725	2,590	3,325
4100	Caps									
4110	4″ diameter	B-22A	34	1.176	Ea.	15.40	57	23.50	95.90	133
4120	6″ diameter		30	1.333		33.50	64.50	27	125	170
4130	8″ diameter		26	1.538		57	74.50	31	162.50	216
4150	10″ diameter		20	2		139	97	40	276	350
4160	12″ diameter		14	2.857		168	138	57.50	363.50	470

33 14 13.45 Water Supply, Copper Pipe		Crew	Daily Output	Labor-Hours	Unit	Material	Labor	Equipment	Total	Total Incl O&P
0010	**WATER SUPPLY, COPPER PIPE**									
0020	Not including excavation or backfill									
2000	Tubing, type K, 20′ joints, 3/4″ diameter	Q-1	400	.040	L.F.	7.05	2.32		9.37	11.35
2200	1″ diameter		320	.050		9.70	2.90		12.60	15.15
3000	1-1/2″ diameter		265	.060		14.95	3.50		18.45	22
3020	2″ diameter		230	.070		23	4.04		27.04	32
3040	2-1/2″ diameter		146	.110		35	6.35		41.35	48.50
3060	3″ diameter		134	.119		48.50	6.95		55.45	63.50
4012	4″ diameter		95	.168		80	9.75		89.75	103
4016	6″ diameter	Q-2	80	.300		134	18.05		152.05	176
4018	8″ diameter	"	80	.300		325	18.05		343.05	385
5000	Tubing, type L									
5108	2″ diameter	Q-1	230	.070	L.F.	14.70	4.04		18.74	22.50
6010	3″ diameter		134	.119		36.50	6.95		43.45	51
6012	4″ diameter		95	.168		48	9.75		57.75	68
6016	6″ diameter	Q-2	80	.300		112	18.05		130.05	151
7165	Fittings, brass, corporation stops, no lead, 3/4″ diameter	1 Plum	19	.421	Ea.	76.50	27		103.50	126
7166	1″ diameter		16	.500		98	32		130	158
7167	1-1/2″ diameter		13	.615		208	39.50		247.50	291
7168	2″ diameter		11	.727		330	47		377	435
7170	Curb stops, no lead, 3/4″ diameter		19	.421		96	27		123	147

33 14 Water Utility Transmission and Distribution

33 14 13 – Public Water Utility Distribution Piping

33 14 13.45 Water Supply, Copper Pipe		Crew	Daily Output	Labor-Hours	Unit	Material	2020 Bare Costs Labor	Equipment	Total	Total Incl O&P
7171	1" diameter	1 Plum	16	.500	Ea.	142	32		174	206
7172	1-1/2" diameter		13	.615		269	39.50		308.50	360
7173	2" diameter		11	.727		325	47		372	430
7180	Curb box, cast iron, 1/2" to 1" curb stops		12	.667		62	43		105	135
7200	1-1/4" to 2" curb stops		8	1		116	64.50		180.50	228
7220	Saddles, 3/4" & 1" diameter, add					76.50			76.50	84
7240	1-1/2" to 2" diameter, add					92			92	101

33 14 13.90 Water Supply, Thrust Blocks		Crew	Daily Output	Labor-Hours	Unit	Material	Labor	Equipment	Total	Total Incl O&P
0010	WATER SUPPLY, THRUST BLOCKS									
0110	Thrust block for 90 degree elbow, 4" diameter	C-30	41	.195	Ea.	21.50	8.20	3.42	33.12	41
0115	6" diameter		23	.348		37.50	14.65	6.10	58.25	71
0120	8" diameter		14	.571		58	24	10	92	113
0125	10" diameter		9	.889		83.50	37.50	15.60	136.60	169
0130	12" diameter		7	1.143		114	48	20	182	224
0135	14" diameter		5	1.600		154	67.50	28	249.50	310
0140	16" diameter		4	2		194	84	35	313	390
0145	18" diameter		3	2.667		240	112	47	399	495
0150	20" diameter		2.50	3.200		290	135	56	481	600
0155	24" diameter		2	4		415	168	70	653	805
0210	Thrust block for tee or deadend, 4" diameter		65	.123		14.75	5.20	2.16	22.11	27
0215	6" diameter		35	.229		26.50	9.60	4.01	40.11	49
0220	8" diameter		21	.381		41	16.05	6.70	63.75	78.50
0225	10" diameter		14	.571		60	24	10	94	115
0230	12" diameter		10	.800		82	33.50	14.05	129.55	159
0235	14" diameter		7	1.143		110	48	20	178	220
0240	16" diameter		5	1.600		139	67.50	28	234.50	292
0245	18" diameter		4	2		172	84	35	291	365
0250	20" diameter		3.50	2.286		208	96	40	344	425
0255	24" diameter		2.50	3.200		295	135	56	486	605

33 14 17 – Site Water Utility Service Laterals

33 14 17.15 Tapping, Crosses and Sleeves		Crew	Daily Output	Labor-Hours	Unit	Material	Labor	Equipment	Total	Total Incl O&P
0010	TAPPING, CROSSES AND SLEEVES									
4000	Drill and tap pressurized main (labor only)									
4100	6" main, 1" to 2" service	Q-1	3	5.333	Ea.		310		310	480
4150	8" main, 1" to 2" service	"	2.75	5.818	"		335		335	520
4500	Tap and insert gate valve									
4600	8" main, 4" branch	B-21	3.20	8.750	Ea.		425	59	484	745
4650	6" branch		2.70	10.370			505	70	575	880
4700	10" main, 4" branch		2.70	10.370			505	70	575	880
4750	6" branch		2.35	11.915			580	80.50	660.50	1,025
4800	12" main, 6" branch		2.35	11.915			580	80.50	660.50	1,025
4850	8" branch		2.35	11.915			580	80.50	660.50	1,025
7000	Tapping crosses, sleeves, valves; with rubber gaskets									
7020	Crosses, 4" x 4"	B-21	37	.757	Ea.	1,450	37	5.10	1,492.10	1,650
7060	8" x 6"		21	1.333		2,050	65	9	2,124	2,375
7080	8" x 8"		21	1.333		2,175	65	9	2,249	2,525
7100	10" x 6"		21	1.333		4,150	65	9	4,224	4,700
7160	12" x 12"		18	1.556		5,300	76	10.50	5,386.50	5,950
7180	14" x 6"		16	1.750		10,200	85.50	11.80	10,297.30	11,300
7240	16" x 10"		14	2		11,100	97.50	13.50	11,211	12,400
7280	18" x 6"		10	2.800		16,400	137	18.90	16,555.90	18,200
7320	18" x 18"		10	2.800		17,100	137	18.90	17,255.90	19,000

33 14 17 – Site Water Utility Service Laterals

33 14 17.15 Tapping, Crosses and Sleeves		Crew	Daily Output	Labor-Hours	Unit	Material	2020 Bare Costs Labor	Equipment	Total	Total Incl O&P
7340	20″ x 6″	B-21	8	3.500	Ea.	13,200	171	23.50	13,394.50	14,900
7360	20″ x 12″		8	3.500		14,200	171	23.50	14,394.50	15,900
7420	24″ x 12″		6	4.667		17,400	228	31.50	17,659.50	19,600
7440	24″ x 18″		6	4.667		27,100	228	31.50	27,359.50	30,200
7600	Cut-in sleeves with rubber gaskets, 4″		18	1.556		560	76	10.50	646.50	750
7640	8″		10	2.800		895	137	18.90	1,050.90	1,225
7680	12″		9	3.111		1,800	152	21	1,973	2,250
7800	Cut-in valves with rubber gaskets, 4″		18	1.556		560	76	10.50	646.50	750
7840	8″		10	2.800		1,000	137	18.90	1,155.90	1,350
7880	12″		9	3.111		2,350	152	21	2,523	2,850
7900	Tapping valve 4″, MJ, ductile iron		18	1.556		455	76	10.50	541.50	635
7920	6″, MJ, ductile iron		12	2.333		620	114	15.75	749.75	880
8000	Sleeves with rubber gaskets, 4″ x 4″		37	.757		1,050	37	5.10	1,092.10	1,225
8030	8″ x 4″		21	1.333		1,025	65	9	1,099	1,250
8040	8″ x 6″		21	1.333		910	65	9	984	1,125
8060	8″ x 8″		21	1.333		1,350	65	9	1,424	1,600
8070	10″ x 4″		21	1.333		830	65	9	904	1,025
8080	10″ x 6″		21	1.333		1,025	65	9	1,099	1,250
8090	10″ x 8″		21	1.333		1,525	65	9	1,599	1,825
8140	12″ x 12″		18	1.556		2,725	76	10.50	2,811.50	3,125
8160	14″ x 6″		16	1.750		1,600	85.50	11.80	1,697.30	1,900
8220	16″ x 10″		14	2		3,750	97.50	13.50	3,861	4,300
8260	18″ x 6″		10	2.800		1,150	137	18.90	1,305.90	1,525
8300	18″ x 18″		10	2.800		8,650	137	18.90	8,805.90	9,775
8320	20″ x 6″		8	3.500		2,425	171	23.50	2,619.50	2,975
8340	20″ x 12″		8	3.500		4,750	171	23.50	4,944.50	5,525
8400	24″ x 12″		6	4.667		6,350	228	31.50	6,609.50	7,375
8420	24″ x 18″		6	4.667		11,300	228	31.50	11,559.50	12,800
8800	Hydrant valve box, 6′ long	B-20	20	1.200		230	56.50		286.50	345
8820	8′ long		18	1.333		305	62.50		367.50	435
8830	Valve box w/lid 4′ deep		14	1.714		96.50	80.50		177	235
8840	Valve box and large base w/lid		14	1.714		325	80.50		405.50	490

33 14 19 – Valves and Hydrants for Water Utility Service

33 14 19.10 Valves		Crew	Daily Output	Labor-Hours	Unit	Material	Labor	Equipment	Total	Total Incl O&P
0010	**VALVES**, water distribution									
0011	See Sections 22 05 23.20 and 22 05 23.60									
3000	Butterfly valves with boxes, cast iron, mechanical joint									
3100	4″ diameter	B-6	6	4	Ea.	550	183	35.50	768.50	940
3180	8″ diameter		6	4		910	183	35.50	1,128.50	1,325
3340	12″ diameter		6	4		1,650	183	35.50	1,868.50	2,150
3400	14″ diameter		4	6		2,825	274	53.50	3,152.50	3,600
3460	18″ diameter		4	6		4,950	274	53.50	5,277.50	5,950
3480	20″ diameter		4	6		6,525	274	53.50	6,852.50	7,675
3500	24″ diameter		4	6		10,800	274	53.50	11,127.50	12,400
3510	30″ diameter		4	6		11,600	274	53.50	11,927.50	13,300
3520	36″ diameter		4	6		14,400	274	53.50	14,727.50	16,300
3530	42″ diameter		4	6		18,900	274	53.50	19,227.50	21,300
3540	48″ diameter		4	6		24,300	274	53.50	24,627.50	27,200
3600	With lever operator									
3610	4″ diameter	B-6	6	4	Ea.	455	183	35.50	673.50	830
3616	8″ diameter		6	4		815	183	35.50	1,033.50	1,225
3620	12″ diameter		6	4		1,550	183	35.50	1,768.50	2,050

33 14 19.10 Valves		Crew	Daily Output	Labor-Hours	Unit	2020 Bare Costs Material	Labor	Equipment	Total	Total Incl O&P
3624	16" diameter	B-6	4	6	Ea.	3,825	274	53.50	4,152.50	4,700
3630	24" diameter		4	6		10,500	274	53.50	10,827.50	12,100
3700	Check valves, flanged									
3710	4" diameter	B-6	6	4	Ea.	785	183	35.50	1,003.50	1,200
3714	6" diameter		6	4		1,325	183	35.50	1,543.50	1,775
3716	8" diameter		6	4		2,525	183	35.50	2,743.50	3,100
3720	12" diameter		6	4		7,000	183	35.50	7,218.50	8,025
3726	18" diameter		4	6		25,900	274	53.50	26,227.50	29,000
3730	24" diameter		4	6		66,000	274	53.50	66,327.50	73,000
3800	Gate valves, C.I., 125 psi, mechanical joint, w/boxes									
3810	4" diameter	B-6	6	4	Ea.	780	183	35.50	998.50	1,200
3814	6" diameter		6	4		1,175	183	35.50	1,393.50	1,625
3816	8" diameter		6	4		2,025	183	35.50	2,243.50	2,550
3820	12" diameter		6	4		4,900	183	35.50	5,118.50	5,700
3824	16" diameter		4	6		15,100	274	53.50	15,427.50	17,100
3828	20" diameter		4	6		25,600	274	53.50	25,927.50	28,700
3830	24" diameter		4	6		38,300	274	53.50	38,627.50	42,700

33 14 19.20 Valves

33 14 19.20 Valves		Crew	Daily Output	Labor-Hours	Unit	Material	Labor	Equipment	Total	Total Incl O&P
0010	**VALVES**									
0011	Special trim or use									
0150	Altitude valve, single acting, modulating, 2-1/2" diameter	B-6	6	4	Ea.	3,500	183	35.50	3,718.50	4,175
0155	3" diameter		6	4		3,575	183	35.50	3,793.50	4,250
0160	4" diameter		6	4		4,200	183	35.50	4,418.50	4,950
0165	6" diameter		6	4		5,150	183	35.50	5,368.50	5,975
0170	8" diameter		6	4		7,525	183	35.50	7,743.50	8,600
0175	10" diameter		6	4		8,225	183	35.50	8,443.50	9,375
0180	12" diameter		6	4		12,600	183	35.50	12,818.50	14,100
0250	Altitude valve, single acting, non-modulating, 2-1/2" diameter		6	4		3,525	183	35.50	3,743.50	4,200
0255	3" diameter		6	4		3,625	183	35.50	3,843.50	4,325
0260	4" diameter		6	4		4,325	183	35.50	4,543.50	5,075
0265	6" diameter		6	4		5,400	183	35.50	5,618.50	6,275
0270	8" diameter		6	4		7,550	183	35.50	7,768.50	8,650
0275	10" diameter		6	4		8,350	183	35.50	8,568.50	9,525
0280	12" diameter		6	4		12,900	183	35.50	13,118.50	14,500
0350	Altitude valve, double acting, non-modulating, 2-1/2" diameter		6	4		4,275	183	35.50	4,493.50	5,025
0355	3" diameter		6	4		4,450	183	35.50	4,668.50	5,225
0360	4" diameter		6	4		5,250	183	35.50	5,468.50	6,075
0365	6" diameter		6	4		5,900	183	35.50	6,118.50	6,825
0370	8" diameter		6	4		7,500	183	35.50	7,718.50	8,575
0375	10" diameter		6	4		10,200	183	35.50	10,418.50	11,500
0380	12" diameter		6	4		15,500	183	35.50	15,718.50	17,300
1000	Air release valve for water, 1/2" inlet	1 Plum	16	.500		83.50	32		115.50	142
1005	3/4" inlet		16	.500		83.50	32		115.50	142
1010	1" inlet		14	.571		83.50	37		120.50	149
1020	2" inlet		9	.889		240	57.50		297.50	355
1100	Air release & vacuum valve for water, 1/2" inlet		16	.500		345	32		377	430
1105	3/4" inlet		16	.500		365	32		397	450
1110	1" inlet		14	.571		485	37		522	585
1120	2" inlet		9	.889		820	57.50		877.50	995
1130	3" inlet	Q-1	8	2		2,650	116		2,766	3,100
1140	4" inlet		5	3.200		2,500	186		2,686	3,025
1150	6" inlet		5	3.200		4,625	186		4,811	5,350

33 14 19.20 Valves		Crew	Daily Output	Labor-Hours	Unit	2020 Bare Costs Material	Labor	Equipment	Total	Total Incl O&P
1160	8" inlet	Q-1	5	3.200	Ea.	11,400	186		11,586	12,900
1170	10" inlet		5	3.200		12,800	186		12,986	14,400
9000	Valves, gate valve, N.R.S. PIV with post, 4" diameter	B-6	6	4		2,000	183	35.50	2,218.50	2,525
9040	8" diameter		6	4		3,250	183	35.50	3,468.50	3,875
9080	12" diameter		6	4		6,125	183	35.50	6,343.50	7,050
9120	OS&Y, 4" diameter		6	4		920	183	35.50	1,138.50	1,325
9160	8" diameter		6	4		1,650	183	35.50	1,868.50	2,150
9200	12" diameter		6	4		4,025	183	35.50	4,243.50	4,750
9220	14" diameter		4	6		6,150	274	53.50	6,477.50	7,275
9400	Check valves, rubber disc, 2-1/2" diameter		6	4		475	183	35.50	693.50	855
9440	4" diameter		6	4		785	183	35.50	1,003.50	1,200
9500	8" diameter		6	4		2,525	183	35.50	2,743.50	3,100
9540	12" diameter		6	4		7,000	183	35.50	7,218.50	8,025
9700	Detector check valves, reducing, 4" diameter		6	4		1,125	183	35.50	1,343.50	1,575
9740	8" diameter		6	4		2,875	183	35.50	3,093.50	3,500
9800	Galvanized, 4" diameter		6	4		2,025	183	35.50	2,243.50	2,550
9840	8" diameter		6	4		4,025	183	35.50	4,243.50	4,750

33 14 19.30 Fire Hydrants		Crew	Daily Output	Labor-Hours	Unit	Material	Labor	Equipment	Total	Total Incl O&P
0010	**FIRE HYDRANTS**									
0020	Mechanical joints unless otherwise noted									
1000	Fire hydrants, two way; excavation and backfill not incl.									
1100	4-1/2" valve size, depth 2'-0"	B-21	10	2.800	Ea.	2,900	137	18.90	3,055.90	3,425
1200	4'-6"		9	3.111		3,000	152	21	3,173	3,575
1260	6'-0"		7	4		3,400	195	27	3,622	4,100
1340	8'-0"		6	4.667		2,775	228	31.50	3,034.50	3,450
1420	10'-0"		5	5.600		4,000	273	38	4,311	4,900
2000	5-1/4" valve size, depth 2'-0"		10	2.800		2,375	137	18.90	2,530.90	2,875
2080	4'-0"		9	3.111		2,700	152	21	2,873	3,250
2160	6'-0"		7	4		2,900	195	27	3,122	3,525
2240	8'-0"		6	4.667		3,325	228	31.50	3,584.50	4,075
2320	10'-0"		5	5.600		3,700	273	38	4,011	4,550
2350	For threeway valves, add					7%				
2400	Lower barrel extensions with stems, 1'-0"	B-20	14	1.714		360	80.50		440.50	525
2440	2'-0"		13	1.846		440	87		527	625
2480	3'-0"		12	2		915	94		1,009	1,150
2520	4'-0"		10	2.400		835	113		948	1,100
5000	Indicator post									
5020	Adjustable, valve size 4" to 14", 4' bury	B-21	10	2.800	Ea.	1,475	137	18.90	1,630.90	1,875
5060	8' bury		7	4		1,325	195	27	1,547	1,800
5080	10' bury		6	4.667		1,775	228	31.50	2,034.50	2,350
5100	12' bury		5	5.600		1,900	273	38	2,211	2,575
5120	14' bury		4	7		1,825	340	47.50	2,212.50	2,600
5500	Non-adjustable, valve size 4" to 14", 3' bury		10	2.800		925	137	18.90	1,080.90	1,275
5520	3'-6" bury		10	2.800		925	137	18.90	1,080.90	1,275
5540	4' bury		9	3.111		925	152	21	1,098	1,300

33 16 Water Utility Storage Tanks

33 16 11 – Elevated Composite Water Storage Tanks

33 16 11.50 Elevated Water Storage Tanks		Crew	Daily Output	Labor-Hours	Unit	Material	2020 Bare Costs Labor	Equipment	Total	Total Incl O&P
0010	ELEVATED WATER STORAGE TANKS									
0011	Not incl. pipe, pumps or foundation									
3000	Elevated water tanks, 100′ to bottom capacity line, incl. painting									
3010	50,000 gallons				Ea.				185,000	204,000
3300	100,000 gallons								280,000	307,500
3400	250,000 gallons								751,500	826,500
3600	500,000 gallons								1,336,000	1,470,000
3700	750,000 gallons								1,622,000	1,783,500
3900	1,000,000 gallons								2,322,000	2,556,000

33 16 23 – Ground-Level Steel Water Storage Tanks

33 16 23.13 Steel Water Storage Tanks		Crew	Daily Output	Labor-Hours	Unit	Material	Labor	Equipment	Total	Total Incl O&P
0010	STEEL WATER STORAGE TANKS									
0910	Steel, ground level, ht./diam. less than 1, not incl. fdn., 100,000 gallons				Ea.				202,000	244,500
1000	250,000 gallons								295,500	324,000
1200	500,000 gallons								417,000	458,500
1250	750,000 gallons								538,000	591,500
1300	1,000,000 gallons								558,000	725,500
1500	2,000,000 gallons								1,043,000	1,148,000
1600	4,000,000 gallons								2,121,000	2,333,000
1800	6,000,000 gallons								3,095,000	3,405,000
1850	8,000,000 gallons								4,068,000	4,475,000
1910	10,000,000 gallons								5,050,000	5,554,500
2100	Steel standpipes, ht./diam. more than 1,100′ to overflow, no fdn.									
2200	500,000 gallons				Ea.				546,500	600,500
2400	750,000 gallons								722,500	794,500
2500	1,000,000 gallons								1,060,500	1,167,000
2700	1,500,000 gallons								1,749,000	1,923,000
2800	2,000,000 gallons								2,327,000	2,559,000

33 16 36 – Ground-Level Reinforced Concrete Water Storage Tanks

33 16 36.16 Prestressed Conc. Water Storage Tanks		Crew	Daily Output	Labor-Hours	Unit	Material	Labor	Equipment	Total	Total Incl O&P
0010	PRESTRESSED CONC. WATER STORAGE TANKS									
0020	Not including fdn., pipe or pumps, 250,000 gallons				Ea.				299,000	329,500
0100	500,000 gallons								487,000	536,000
0300	1,000,000 gallons								707,000	807,500
0400	2,000,000 gallons								1,072,000	1,179,000
0600	4,000,000 gallons								1,706,000	1,877,000
0700	6,000,000 gallons								2,266,000	2,493,000
0750	8,000,000 gallons								2,924,000	3,216,000
0800	10,000,000 gallons								3,533,000	3,886,000

33 16 56 – Ground-Level Plastic Water Storage Cisterns

33 16 56.23 Plastic-Coated Fabric Pillow Water Tanks		Crew	Daily Output	Labor-Hours	Unit	Material	Labor	Equipment	Total	Total Incl O&P
0010	PLASTIC-COATED FABRIC PILLOW WATER TANKS									
7000	Water tanks, vinyl coated fabric pillow tanks, freestanding, 5,000 gallons	4 Clab	4	8	Ea.	3,800	335		4,135	4,725
7100	Supporting embankment not included, 25,000 gallons	6 Clab	2	24		13,500	1,000		14,500	16,400
7200	50,000 gallons	8 Clab	1.50	42.667		20,500	1,800		22,300	25,500
7300	100,000 gallons	9 Clab	.90	80		37,700	3,375		41,075	46,800
7400	150,000 gallons		.50	144		44,400	6,050		50,450	58,500
7500	200,000 gallons		.40	180		84,500	7,575		92,075	105,000
7600	250,000 gallons		.30	240		89,000	10,100		99,100	114,000

33 31 Sanitary Sewerage Piping

33 31 11 – Public Sanitary Sewerage Gravity Piping

33 31 11.13 Sewage Collection, Vent Cast Iron Pipe

	33 31 11.13 Sewage Collection, Vent Cast Iron Pipe	Crew	Daily Output	Labor-Hours	Unit	2020 Bare Costs Material	Labor	Equipment	Total	Total Incl O&P
0010	**SEWAGE COLLECTION, VENT CAST IRON PIPE**									
0020	Not including excavation or backfill									
2022	Sewage vent cast iron, B&S, 4″ diameter	Q-1	66	.242	L.F.	23	14.05		37.05	46.50
2024	5″ diameter	Q-2	88	.273		30	16.40		46.40	58.50
2026	6″ diameter	″	84	.286		39	17.20		56.20	69
2028	8″ diameter	Q-3	70	.457		62.50	28		90.50	113
2030	10″ diameter		66	.485		104	30		134	160
2032	12″ diameter		57	.561		149	34.50		183.50	218
2034	15″ diameter		49	.653		170	40		210	249
8001	Fittings, bends and elbows									
8110	4″ diameter	Q-1	13	1.231	Ea.	68	71.50		139.50	185
8112	5″ diameter	Q-2	18	1.333		97.50	80		177.50	231
8114	6″ diameter	″	17	1.412		116	85		201	258
8116	8″ diameter	Q-3	11	2.909		325	178		503	630
8118	10″ diameter		10	3.200		485	196		681	835
8120	12″ diameter		9	3.556		655	218		873	1,050
8122	15″ diameter		7	4.571		1,925	280		2,205	2,550
8500	Wyes and tees									
8510	4″ diameter	Q-1	8	2	Ea.	110	116		226	300
8512	5″ diameter	Q-2	12	2		188	120		308	395
8514	6″ diameter	″	11	2.182		232	131		363	460
8516	8″ diameter	Q-3	7	4.571		550	280		830	1,050
8518	10″ diameter		6	5.333		900	325		1,225	1,500
8520	12″ diameter		4	8		1,850	490		2,340	2,775
8522	15″ diameter		3	10.667		3,700	655		4,355	5,075

33 31 11.15 Sewage Collection, Concrete Pipe

0010	**SEWAGE COLLECTION, CONCRETE PIPE**									
0020	See Section 33 42 11.60 for sewage/drainage collection, concrete pipe									

33 31 11.20 Sewage Collection, Plastic Pipe

0010	**SEWAGE COLLECTION, PLASTIC PIPE**									
0020	Not including excavation & backfill									
1100	Piping, DWV Sch 40 ABS, 4″ diameter	B-20	375	.064	L.F.	4.14	3.01		7.15	9.35
1110	6″ diameter		350	.069	″	6.45	3.22		9.67	12.25
1120	Fitting, 1/4 bend, 4″		19	1.263	Ea.	38.50	59.50		98	138
1130	6″		15	1.600		27.50	75		102.50	151
1140	Tee, 4″		12	2		38.50	94		132.50	193
3000	Piping, HDPE Corrugated Type S with watertight gaskets, 4″ diameter		425	.056	L.F.	1.08	2.66		3.74	5.45
3020	6″ diameter		400	.060		3.13	2.82		5.95	7.95
3040	8″ diameter		380	.063		4.12	2.97		7.09	9.30
3060	10″ diameter		370	.065		5.50	3.05		8.55	10.95
3080	12″ diameter		340	.071		7.90	3.32		11.22	14
3100	15″ diameter		300	.080		11.65	3.76		15.41	18.85
3120	18″ diameter	B-21	275	.102		14.35	4.96	.69	20	24.50
3130	21″ diameter		250	.112		23	5.45	.76	29.21	34.50
3140	24″ diameter		250	.112		23	5.45	.76	29.21	34.50
3160	30″ diameter		200	.140		29	6.85	.95	36.80	44
3180	36″ diameter		180	.156		42	7.60	1.05	50.65	60
3200	42″ diameter		175	.160		49.50	7.80	1.08	58.38	67.50
3220	48″ diameter		170	.165		69.50	8.05	1.11	78.66	90.50
3240	54″ diameter		160	.175		90.50	8.55	1.18	100.23	114
3260	60″ diameter		150	.187		167	9.10	1.26	177.36	199
3300	Watertight elbows 12″ diameter	B-20	11	2.182	Ea.	82.50	103		185.50	255

33 31 Sanitary Sewerage Piping

33 31 11 – Public Sanitary Sewerage Gravity Piping

33 31 11.20 Sewage Collection, Plastic Pipe		Crew	Daily Output	Labor-Hours	Unit	2020 Bare Costs Material	Labor	Equipment	Total	Total Incl O&P
3320	15″ diameter	B-20	9	2.667	Ea.	125	125		250	340
3340	18″ diameter	B-21	9	3.111		206	152	21	379	490
3350	21″ diameter		9	3.111		297	152	21	470	590
3360	24″ diameter		9	3.111		435	152	21	608	740
3380	30″ diameter		8	3.500		690	171	23.50	884.50	1,050
3400	36″ diameter		8	3.500		890	171	23.50	1,084.50	1,275
3420	42″ diameter		6	4.667		1,125	228	31.50	1,384.50	1,625
3440	48″ diameter		6	4.667		2,000	228	31.50	2,259.50	2,600
3460	Watertight tee 12″ diameter	B-20	7	3.429		109	161		270	380
3480	15″ diameter	″	6	4		296	188		484	625
3500	18″ diameter	B-21	6	4.667		430	228	31.50	689.50	870
3520	24″ diameter		5	5.600		795	273	38	1,106	1,350
3540	30″ diameter		5	5.600		1,350	273	38	1,661	1,975
3560	36″ diameter		4	7		1,300	340	47.50	1,687.50	2,050
3580	42″ diameter		4	7		1,500	340	47.50	1,887.50	2,250
3600	48″ diameter		4	7		1,600	340	47.50	1,987.50	2,350

33 31 11.25 Sewage Collection, Polyvinyl Chloride Pipe

		Crew	Daily Output	Labor-Hours	Unit	Material	Labor	Equipment	Total	Total Incl O&P
0010	**SEWAGE COLLECTION, POLYVINYL CHLORIDE PIPE**									
0020	Not including excavation or backfill									
2000	20′ lengths, SDR 35, B&S, 4″ diameter	B-20	375	.064	L.F.	1.67	3.01		4.68	6.65
2040	6″ diameter		350	.069		4.05	3.22		7.27	9.60
2080	13′ lengths, SDR 35, B&S, 8″ diameter		335	.072		6.70	3.37		10.07	12.75
2120	10″ diameter	B-21	330	.085		11.30	4.14	.57	16.01	19.70
2160	12″ diameter		320	.088		15.90	4.27	.59	20.76	25
2170	14″ diameter		280	.100		19.45	4.88	.68	25.01	30
2200	15″ diameter		240	.117		22	5.70	.79	28.49	34
2250	16″ diameter		220	.127		29	6.20	.86	36.06	42.50
3040	Fittings, bends or elbows, 4″ diameter	2 Skwk	24	.667	Ea.	11.30	36.50		47.80	71
3080	6″ diameter		24	.667		36.50	36.50		73	98.50
3120	Tees, 4″ diameter		16	1		18.85	55		73.85	108
3160	6″ diameter		16	1		59.50	55		114.50	153
3200	Wyes, 4″ diameter		16	1		19.05	55		74.05	109
3240	6″ diameter		16	1		52	55		107	145
4000	Piping, DWV PVC, no exc./bkfill., 10′ L, Sch 40, 4″ diameter	B-20	375	.064	L.F.	3.83	3.01		6.84	9.05
4010	6″ diameter		350	.069		8.25	3.22		11.47	14.20
4020	8″ diameter		335	.072		13.40	3.37		16.77	20

33 34 Onsite Wastewater Disposal

33 34 13 – Septic Tanks

33 34 13.13 Concrete Septic Tanks

		Crew	Daily Output	Labor-Hours	Unit	Material	Labor	Equipment	Total	Total Incl O&P
0010	**CONCRETE SEPTIC TANKS**									
0011	Not including excavation or piping									
0015	Septic tanks, precast, 1,000 gallon	B-21	8	3.500	Ea.	1,050	171	23.50	1,244.50	1,475
0020	1,250 gallon		8	3.500		1,425	171	23.50	1,619.50	1,875
0060	1,500 gallon		7	4		1,625	195	27	1,847	2,150
0100	2,000 gallon		5	5.600		2,500	273	38	2,811	3,225
0140	2,500 gallon		5	5.600		2,375	273	38	2,686	3,075
0180	4,000 gallon		4	7		5,850	340	47.50	6,237.50	7,050
0220	5,000 gallon, 4 piece	B-13	3	18.667		9,025	860	194	10,079	11,500
0300	15,000 gallon, 4 piece	B-13B	1.70	32.941		23,700	1,525	575	25,800	29,100
0400	25,000 gallon, 4 piece		1.10	50.909		47,900	2,350	890	51,140	57,000

33 34 Onsite Wastewater Disposal

33 34 13 – Septic Tanks

33 34 13.13 Concrete Septic Tanks		Crew	Daily Output	Labor-Hours	Unit	Material	Labor	Equipment	Total	Total Incl O&P
						2020 Bare Costs				
0500	40,000 gallon, 4 piece	B-13B	.80	70	Ea.	53,500	3,225	1,225	57,950	65,000
0520	50,000 gallon, 5 piece	B-13C	.60	93.333		61,500	4,300	3,800	69,600	78,500
0640	75,000 gallon, cast in place	C-14C	.25	448		74,500	22,600	108	97,208	118,000
0660	100,000 gallon	"	.15	747		92,500	37,600	179	130,279	161,500
1150	Leaching field chambers, 13′ x 3′-7″ x 1′-4″, standard	B-13	16	3.500		500	161	36.50	697.50	845
1200	Heavy duty, 8′ x 4′ x 1′-6″		14	4		310	184	41.50	535.50	680
1300	13′ x 3′-9″ x 1′-6″		12	4.667		1,350	215	48.50	1,613.50	1,875
1350	20′ x 4′ x 1′-6″		5	11.200		1,225	515	116	1,856	2,300
1600	Leaching pit, 6′-6″ diameter, 6′ deep	B-21	5	5.600		1,075	273	38	1,386	1,650
1620	8′ deep		4	7		1,175	340	47.50	1,562.50	1,900
1700	8′ diameter, H-20 load, 6′ deep		4	7		1,475	340	47.50	1,862.50	2,200
1720	8′ deep		3	9.333		2,700	455	63	3,218	3,775
2000	Velocity reducing pit, precast conc., 6′ diameter, 3′ deep		4.70	5.957		1,825	290	40	2,155	2,525

33 34 13.33 Polyethylene Septic Tanks		Crew	Daily Output	Labor-Hours	Unit	Material	Labor	Equipment	Total	Total Incl O&P
0010	**POLYETHYLENE SEPTIC TANKS**									
0015	High density polyethylene, 1,000 gallon	B-21	8	3.500	Ea.	1,475	171	23.50	1,669.50	1,925
0020	1,250 gallon		8	3.500		1,325	171	23.50	1,519.50	1,750
0025	1,500 gallon		7	4		1,375	195	27	1,597	1,875

33 34 16 – Septic Tank Effluent Filters

33 34 16.13 Septic Tank Gravity Effluent Filters		Crew	Daily Output	Labor-Hours	Unit	Material	Labor	Equipment	Total	Total Incl O&P
0010	**SEPTIC TANK GRAVITY EFFLUENT FILTERS**									
3000	Effluent filter, 4″ diameter	1 Skwk	8	1	Ea.	44	55		99	136
3020	6″ diameter		7	1.143		42.50	62.50		105	147
3030	8″ diameter		7	1.143		252	62.50		314.50	375
3040	8″ diameter, very fine		7	1.143		500	62.50		562.50	650
3050	10″ diameter, very fine		6	1.333		237	73		310	380
3060	10″ diameter		6	1.333		276	73		349	420
3080	12″ diameter		6	1.333		655	73		728	835
3090	15″ diameter		5	1.600		1,075	88		1,163	1,350

33 34 51 – Drainage Field Systems

33 34 51.10 Drainage Field Excavation and Fill		Crew	Daily Output	Labor-Hours	Unit	Material	Labor	Equipment	Total	Total Incl O&P
0010	**DRAINAGE FIELD EXCAVATION AND FILL**									
2200	Septic tank & drainage field excavation with 3/4 C.Y. backhoe	B-12F	145	.110	C.Y.		5.60	4.79	10.39	14.10
2400	4′ trench for disposal field, 3/4 C.Y. backhoe	"	335	.048	L.F.		2.42	2.07	4.49	6.10
2600	Gravel fill, run of bank	B-6	150	.160	C.Y.	16.90	7.30	1.43	25.63	32
2800	Crushed stone, 3/4″	"	150	.160	"	42	7.30	1.43	50.73	59

33 34 51.13 Utility Septic Tank Tile Drainage Field		Crew	Daily Output	Labor-Hours	Unit	Material	Labor	Equipment	Total	Total Incl O&P
0010	**UTILITY SEPTIC TANK TILE DRAINAGE FIELD**									
0015	Distribution box, concrete, 5 outlets	2 Clab	20	.800	Ea.	93.50	33.50		127	157
0020	7 outlets		16	1		103	42		145	182
0025	9 outlets		8	2		545	84		629	735
0115	Distribution boxes, HDPE, 5 outlets		20	.800		76.50	33.50		110	138
0117	6 outlets		15	1.067		81	45		126	161
0118	7 outlets		15	1.067		75	45		120	155
0120	8 outlets		10	1.600		79	67.50		146.50	195
0240	Distribution boxes, outlet flow leveler	1 Clab	50	.160		2.53	6.75		9.28	13.60
0300	Precast concrete, galley, 4′ x 4′ x 4′	B-21	16	1.750		246	85.50	11.80	343.30	420
0350	HDPE infiltration chamber 12″ H x 15″ W	2 Clab	300	.053	L.F.	7	2.25		9.25	11.30
0351	12″ H x 15″ W end cap	1 Clab	32	.250	Ea.	18.90	10.55		29.45	38
0355	chamber 12″ H x 22″ W	2 Clab	300	.053	L.F.	6.65	2.25		8.90	10.95
0356	12″ H x 22″ W end cap	1 Clab	32	.250	Ea.	16.60	10.55		27.15	35

33 34 Onsite Wastewater Disposal

33 34 51 – Drainage Field Systems

33 34 51.13 Utility Septic Tank Tile Drainage Field		Crew	Daily Output	Labor-Hours	Unit	2020 Bare Costs Material	Labor	Equipment	Total	Total Incl O&P
0360	chamber 13″ H x 34″ W	2 Clab	300	.053	L.F.	14.30	2.25		16.55	19.35
0361	13″ H x 34″ W end cap	1 Clab	32	.250	Ea.	50.50	10.55		61.05	73
0365	chamber 16″ H x 34″ W	2 Clab	300	.053	L.F.	19.25	2.25		21.50	24.50
0366	16″ H x 34″ W end cap	1 Clab	32	.250	Ea.	16.55	10.55		27.10	35
0370	chamber 8″ H x 16″ W	2 Clab	300	.053	L.F.	9.90	2.25		12.15	14.50
0371	8″ H x 16″ W end cap	1 Clab	32	.250	Ea.	11.10	10.55		21.65	29

33 41 Subdrainage

33 41 16 – Subdrainage Piping

33 41 16.10 Piping, Subdrainage, Vitrified Clay

Line	Description	Crew	Daily Output	Labor-Hours	Unit	Material	Labor	Equipment	Total	Total Incl O&P
0010	**PIPING, SUBDRAINAGE, VITRIFIED CLAY**									
0020	Not including excavation and backfill									
1000	Foundation drain, 6″ diameter	B-14	315	.152	L.F.	6.25	6.75	.68	13.68	18.40
1010	8″ diameter		290	.166		9.30	7.30	.74	17.34	22.50
1020	12″ diameter		275	.175		20	7.70	.78	28.48	35
3000	Perforated, 5′ lengths, C700, 4″ diameter		400	.120		4.70	5.30	.53	10.53	14.20
3020	6″ diameter		315	.152		6.65	6.75	.68	14.08	18.85
3040	8″ diameter		290	.166		7.25	7.30	.74	15.29	20.50
3060	12″ diameter	↓	275	.175	↓	20.50	7.70	.78	28.98	35.50

33 41 16.25 Piping, Subdrainage, Corrugated Metal

Line	Description	Crew	Daily Output	Labor-Hours	Unit	Material	Labor	Equipment	Total	Total Incl O&P
0010	**PIPING, SUBDRAINAGE, CORRUGATED METAL**									
0021	Not including excavation and backfill									
2010	Aluminum, perforated									
2020	6″ diameter, 18 ga.	B-20	380	.063	L.F.	6.40	2.97		9.37	11.80
2200	8″ diameter, 16 ga.	″	370	.065		8.80	3.05		11.85	14.60
2220	10″ diameter, 16 ga.	B-21	360	.078		11	3.79	.53	15.32	18.75
2240	12″ diameter, 16 ga.		285	.098		12.30	4.79	.66	17.75	22
2260	18″ diameter, 16 ga.	↓	205	.137	↓	18.50	6.65	.92	26.07	32
3000	Uncoated galvanized, perforated									
3020	6″ diameter, 18 ga.	B-20	380	.063	L.F.	6.65	2.97		9.62	12.05
3200	8″ diameter, 16 ga.	″	370	.065		8.10	3.05		11.15	13.80
3220	10″ diameter, 16 ga.	B-21	360	.078		8.60	3.79	.53	12.92	16.10
3240	12″ diameter, 16 ga.		285	.098		9.55	4.79	.66	15	18.95
3260	18″ diameter, 16 ga.	↓	205	.137	↓	14.65	6.65	.92	22.22	27.50
4000	Steel, perforated, asphalt coated									
4020	6″ diameter, 18 ga.	B-20	380	.063	L.F.	6.35	2.97		9.32	11.75
4030	8″ diameter, 18 ga.	″	370	.065		9	3.05		12.05	14.80
4040	10″ diameter, 16 ga.	B-21	360	.078		10.05	3.79	.53	14.37	17.70
4050	12″ diameter, 16 ga.		285	.098		11.30	4.79	.66	16.75	21
4060	18″ diameter, 16 ga.	↓	205	.137	↓	17.80	6.65	.92	25.37	31

33 41 16.30 Piping, Subdrainage, Plastic

Line	Description	Crew	Daily Output	Labor-Hours	Unit	Material	Labor	Equipment	Total	Total Incl O&P
0010	**PIPING, SUBDRAINAGE, PLASTIC**									
0020	Not including excavation and backfill									
2100	Perforated PVC, 4″ diameter	B-14	314	.153	L.F.	1.67	6.75	.68	9.10	13.40
2110	6″ diameter		300	.160		4.05	7.10	.71	11.86	16.55
2120	8″ diameter		290	.166		6.10	7.30	.74	14.14	19.20
2130	10″ diameter		280	.171		10.45	7.60	.76	18.81	24.50
2140	12″ diameter	↓	270	.178	↓	14.75	7.85	.79	23.39	29.50

33 41 Subdrainage

33 41 23 – Drainage Layers

	33 41 23.19 Geosynthetic Drainage Layers	Crew	Daily Output	Labor-Hours	Unit	2020 Bare Costs Material	Labor	Equipment	Total	Total Incl O&P
0010	**GEOSYNTHETIC DRAINAGE LAYERS**									
0100	Fabric, laid in trench, polypropylene, ideal conditions	2 Clab	2400	.007	S.Y.	1.57	.28		1.85	2.18
0110	Adverse conditions		1600	.010	"	1.57	.42		1.99	2.40
0170	Fabric ply bonded to 3 dimen. nylon mat, 0.4" thick, ideal conditions		2000	.008	S.F.	.27	.34		.61	.83
0180	Adverse conditions		1200	.013	"	.33	.56		.89	1.27
0185	Soil drainage mat on vertical wall, 0.44" thick		265	.060	S.Y.	1.80	2.54		4.34	6.05
0188	0.25" thick		300	.053	"	.96	2.25		3.21	4.66
0190	0.8" thick, ideal conditions		2400	.007	S.F.	.28	.28		.56	.76
0200	Adverse conditions		1600	.010	"	.48	.42		.90	1.20
0300	Drainage material, 3/4" gravel fill in trench	B-6	260	.092	C.Y.	20.50	4.22	.82	25.54	30
0400	Pea stone	"	260	.092	"	20.50	4.22	.82	25.54	30

33 42 Stormwater Conveyance

33 42 11 – Stormwater Gravity Piping

33 42 11.40 Piping, Storm Drainage, Corrugated Metal

	33 42 11.40 Piping, Storm Drainage, Corrugated Metal	Crew	Daily Output	Labor-Hours	Unit	Material	Labor	Equipment	Total	Total Incl O&P
0010	**PIPING, STORM DRAINAGE, CORRUGATED METAL**									
0020	Not including excavation or backfill									
2000	Corrugated metal pipe, galvanized									
2020	Bituminous coated with paved invert, 20' lengths									
2040	8" diameter, 16 ga.	B-14	330	.145	L.F.	8.35	6.45	.65	15.45	20
2060	10" diameter, 16 ga.		260	.185		8.65	8.15	.82	17.62	23.50
2080	12" diameter, 16 ga.		210	.229		12.40	10.10	1.02	23.52	31
2100	15" diameter, 16 ga.		200	.240		14.60	10.60	1.07	26.27	34
2120	18" diameter, 16 ga.		190	.253		18.80	11.20	1.12	31.12	39.50
2140	24" diameter, 14 ga.		160	.300		24.50	13.30	1.34	39.14	49.50
2160	30" diameter, 14 ga.	B-13	120	.467		29	21.50	4.84	55.34	71.50
2180	36" diameter, 12 ga.		120	.467		35	21.50	4.84	61.34	78
2200	48" diameter, 12 ga.		100	.560		46.50	26	5.80	78.30	98.50
2220	60" diameter, 10 ga.	B-13B	75	.747		80	34.50	13.05	127.55	156
2240	72" diameter, 8 ga.	"	45	1.244		84.50	57.50	22	164	208
2250	End sections, 8" diameter, 16 ga.	B-14	20	2.400	Ea.	42	106	10.70	158.70	227
2255	10" diameter, 16 ga.		20	2.400		52	106	10.70	168.70	238
2260	12" diameter, 16 ga.		18	2.667		107	118	11.85	236.85	320
2265	15" diameter, 16 ga.		18	2.667		206	118	11.85	335.85	430
2270	18" diameter, 16 ga.		16	3		240	133	13.35	386.35	490
2275	24" diameter, 16 ga.	B-13	16	3.500		315	161	36.50	512.50	645
2280	30" diameter, 16 ga.		14	4		520	184	41.50	745.50	910
2285	36" diameter, 14 ga.		14	4		665	184	41.50	890.50	1,075
2290	48" diameter, 14 ga.		10	5.600		1,850	258	58	2,166	2,525
2292	60" diameter, 14 ga.		6	9.333		1,875	430	97	2,402	2,875
2294	72" diameter, 14 ga.	B-13B	5	11.200		2,925	515	196	3,636	4,250
2300	Bends or elbows, 8" diameter	B-14	28	1.714		110	76	7.65	193.65	250
2320	10" diameter		25	1.920		137	85	8.55	230.55	294
2340	12" diameter, 16 ga.		23	2.087		162	92.50	9.30	263.80	335
2342	18" diameter, 16 ga.		20	2.400		231	106	10.70	347.70	435
2344	24" diameter, 14 ga.		16	3		360	133	13.35	506.35	620
2346	30" diameter, 14 ga.		15	3.200		430	142	14.25	586.25	715
2348	36" diameter, 14 ga.	B-13	15	3.733		590	172	38.50	800.50	965
2350	48" diameter, 12 ga.	"	12	4.667		760	215	48.50	1,023.50	1,225
2352	60" diameter, 10 ga.	B-13B	10	5.600		1,025	258	98	1,381	1,650

	33 42 11.40 Piping, Storm Drainage, Corrugated Metal	Crew	Daily Output	Labor-Hours	Unit	Material	Labor	Equipment	Total	Total Incl O&P
						2020 Bare Costs				
2354	72" diameter, 10 ga.	B-13B	6	9.333	Ea.	1,275	430	163	1,868	2,275
2360	Wyes or tees, 8" diameter	B-14	25	1.920		156	85	8.55	249.55	315
2380	10" diameter		21	2.286		194	101	10.15	305.15	385
2400	12" diameter, 16 ga.		19	2.526		222	112	11.25	345.25	435
2410	18" diameter, 16 ga.		16	3		305	133	13.35	451.35	565
2412	24" diameter, 14 ga.		16	3		525	133	13.35	671.35	800
2414	30" diameter, 14 ga.	B-13	12	4.667		665	215	48.50	928.50	1,125
2416	36" diameter, 14 ga.		11	5.091		815	234	53	1,102	1,325
2418	48" diameter, 12 ga.		10	5.600		1,100	258	58	1,416	1,675
2420	60" diameter, 10 ga.	B-13B	8	7		1,600	320	123	2,043	2,425
2422	72" diameter, 10 ga.	"	5	11.200		1,925	515	196	2,636	3,150
2500	Galvanized, uncoated, 20' lengths									
2520	8" diameter, 16 ga.	B-14	355	.135	L.F.	7.85	6	.60	14.45	18.80
2540	10" diameter, 16 ga.		280	.171		8.40	7.60	.76	16.76	22
2560	12" diameter, 16 ga.		220	.218		9.75	9.65	.97	20.37	27
2580	15" diameter, 16 ga.		220	.218		11.05	9.65	.97	21.67	28.50
2600	18" diameter, 16 ga.		205	.234		13.50	10.35	1.04	24.89	32.50
2620	24" diameter, 14 ga.		175	.274		23	12.15	1.22	36.37	46
2640	30" diameter, 14 ga.	B-13	130	.431		32	19.85	4.47	56.32	72
2660	36" diameter, 12 ga.		130	.431		39	19.85	4.47	63.32	79
2680	48" diameter, 12 ga.		110	.509		54	23.50	5.30	82.80	102
2690	60" diameter, 10 ga.	B-13B	78	.718		84	33	12.55	129.55	158
2695	72" diameter, 10 ga.	"	60	.933		109	43	16.35	168.35	206
2711	Bends or elbows, 12" diameter, 16 ga.	B-14	30	1.600	Ea.	130	71	7.10	208.10	264
2712	15" diameter, 16 ga.		25.04	1.917		167	85	8.55	260.55	330
2714	18" diameter, 16 ga.		20	2.400		187	106	10.70	303.70	385
2716	24" diameter, 14 ga.		16	3		281	133	13.35	427.35	535
2718	30" diameter, 14 ga.		15	3.200		350	142	14.25	506.25	625
2720	36" diameter, 14 ga.	B-13	15	3.733		485	172	38.50	695.50	850
2722	48" diameter, 12 ga.		12	4.667		655	215	48.50	918.50	1,125
2724	60" diameter, 10 ga.		10	5.600		1,025	258	58	1,341	1,600
2726	72" diameter, 10 ga.		6	9.333		1,325	430	97	1,852	2,250
2728	Wyes or tees, 12" diameter, 16 ga.	B-14	22.48	2.135		168	94.50	9.50	272	345
2730	18" diameter, 16 ga.		15	3.200		261	142	14.25	417.25	530
2732	24" diameter, 14 ga.		15	3.200		435	142	14.25	591.25	720
2734	30" diameter, 14 ga.		14	3.429		565	152	15.25	732.25	880
2736	36" diameter, 14 ga.	B-13	14	4		715	184	41.50	940.50	1,125
2738	48" diameter, 12 ga.		12	4.667		1,050	215	48.50	1,313.50	1,550
2740	60" diameter, 10 ga.		10	5.600		1,475	258	58	1,791	2,100
2742	72" diameter, 10 ga.		6	9.333		1,800	430	97	2,327	2,775
2780	End sections, 8" diameter	B-14	35	1.371		78.50	60.50	6.10	145.10	190
2785	10" diameter		35	1.371		87	60.50	6.10	153.60	199
2790	12" diameter		35	1.371		100	60.50	6.10	166.60	214
2800	18" diameter		30	1.600		117	71	7.10	195.10	250
2810	24" diameter	B-13	25	2.240		197	103	23	323	405
2820	30" diameter		25	2.240		350	103	23	476	575
2825	36" diameter		20	2.800		470	129	29	628	750
2830	48" diameter		10	5.600		940	258	58	1,256	1,500
2835	60" diameter	B-13B	5	11.200		1,725	515	196	2,436	2,925
2840	72" diameter	"	4	14		2,175	645	245	3,065	3,700
2850	Couplings, 12" diameter					16.50			16.50	18.15
2855	18" diameter					21.50			21.50	24
2860	24" diameter					28			28	30.50

33 42 11.40 Piping, Storm Drainage, Corrugated Metal		Crew	Daily Output	Labor-Hours	Unit	2020 Bare Costs Material	Labor	Equipment	Total	Total Incl O&P
2865	30″ diameter				Ea.	29			29	32
2870	36″ diameter					38.50			38.50	42
2875	48″ diameter					56.50			56.50	62
2880	60″ diameter					62.50			62.50	69
2885	72″ diameter				↓	71			71	78

33 42 11.50 Piping, Drainage & Sewage, Corrug. HDPE Type S		Crew	Daily Output	Labor-Hours	Unit	Material	Labor	Equipment	Total	Total Incl O&P
0010	**PIPING, DRAINAGE & SEWAGE, CORRUGATED HDPE TYPE S**									
0020	Not including excavation & backfill, bell & spigot									
1000	With gaskets, 4″ diameter	B-20	425	.056	L.F.	.94	2.66		3.60	5.25
1010	6″ diameter		400	.060		2.72	2.82		5.54	7.50
1020	8″ diameter		380	.063		3.58	2.97		6.55	8.70
1030	10″ diameter		370	.065		4.77	3.05		7.82	10.15
1040	12″ diameter		340	.071		6.85	3.32		10.17	12.85
1050	15″ diameter	↓	300	.080		10.15	3.76		13.91	17.15
1060	18″ diameter	B-21	275	.102		12.45	4.96	.69	18.10	22.50
1070	24″ diameter		250	.112		19.90	5.45	.76	26.11	31.50
1080	30″ diameter		200	.140		25.50	6.85	.95	33.30	40
1090	36″ diameter		180	.156		36.50	7.60	1.05	45.15	54
1100	42″ diameter		175	.160		43	7.80	1.08	51.88	60.50
1110	48″ diameter		170	.165		60.50	8.05	1.11	69.66	80.50
1120	54″ diameter		160	.175		78.50	8.55	1.18	88.23	101
1130	60″ diameter	↓	150	.187	↓	145	9.10	1.26	155.36	175
1135	Add 15% to material pipe cost for water tight connection bell & spigot									
1140	HDPE type S, elbows 12″ diameter	B-20	11	2.182	Ea.	72	103		175	243
1150	15″ diameter	″	9	2.667		109	125		234	320
1160	18″ diameter	B-21	9	3.111		179	152	21	352	460
1170	24″ diameter		9	3.111		375	152	21	548	680
1180	30″ diameter		8	3.500		600	171	23.50	794.50	960
1190	36″ diameter		8	3.500		775	171	23.50	969.50	1,150
1200	42″ diameter		6	4.667		975	228	31.50	1,234.50	1,475
1220	48″ diameter	↓	6	4.667		1,750	228	31.50	2,009.50	2,325
1240	HDPE type S, Tee 12″ diameter	B-20	7	3.429		95	161		256	365
1260	15″ diameter	″	6	4		257	188		445	585
1280	18″ diameter	B-21	6	4.667		375	228	31.50	634.50	810
1300	24″ diameter		5	5.600		690	273	38	1,001	1,225
1320	30″ diameter		5	5.600		1,175	273	38	1,486	1,775
1340	36″ diameter		4	7		1,150	340	47.50	1,537.50	1,850
1360	42″ diameter		4	7		1,300	340	47.50	1,687.50	2,025
1380	48″ diameter	↓	4	7	↓	1,375	340	47.50	1,762.50	2,125
1400	Add to basic installation cost for each split coupling joint									
1402	HDPE type S, split coupling, 12″ diameter	B-20	17	1.412	Ea.	10.65	66.50		77.15	118
1420	15″ diameter		15	1.600		19.90	75		94.90	142
1440	18″ diameter		13	1.846		27.50	87		114.50	170
1460	24″ diameter		12	2		40.50	94		134.50	195
1480	30″ diameter		10	2.400		87.50	113		200.50	277
1500	36″ diameter		9	2.667		141	125		266	355
1520	42″ diameter		8	3		168	141		309	410
1540	48″ diameter	↓	8	3	↓	198	141		339	445

33 42 Stormwater Conveyance

33 42 11 – Stormwater Gravity Piping

33 42 11.60 Sewage/Drainage Collection, Concrete Pipe		Crew	Daily Output	Labor-Hours	Unit	2020 Bare Costs Material	Labor	Equipment	Total	Total Incl O&P
0010	**SEWAGE/DRAINAGE COLLECTION, CONCRETE PIPE**									
0020	Not including excavation or backfill									
0100	Box culvert, precast, base price, 8′ long, 6′ x 3′	B-69	140	.343	L.F.	265	16	10.60	291.60	330
0150	6′ x 7′		125	.384		320	17.95	11.90	349.85	390
0200	8′ x 3′		113	.425		350	19.85	13.15	383	430
0250	8′ x 8′		100	.480		415	22.50	14.85	452.35	505
0300	10′ x 3′		110	.436		405	20.50	13.50	439	490
0350	10′ x 8′		80	.600		605	28	18.60	651.60	730
0400	12′ x 3′		100	.480		825	22.50	14.85	862.35	960
0450	12′ x 8′		67	.716		940	33.50	22	995.50	1,100
0500	Set up charge at plant, add to base price				Job	6,225			6,225	6,825
0510	Inserts and keyway, add				Ea.	640			640	705
0520	Sloped or skewed end, add				″	895			895	985
1000	Non-reinforced pipe, extra strength, B&S or T&G joints									
1010	6″ diameter	B-14	265.04	.181	L.F.	7.80	8	.81	16.61	22
1020	8″ diameter		224	.214		8.55	9.50	.95	19	25.50
1030	10″ diameter		216	.222		9.50	9.85	.99	20.34	27
1040	12″ diameter		200	.240		10.05	10.60	1.07	21.72	29
1050	15″ diameter		180	.267		14.40	11.80	1.19	27.39	36
1060	18″ diameter		144	.333		18.05	14.75	1.48	34.28	45
1070	21″ diameter		112	.429		18.95	18.95	1.91	39.81	53
1080	24″ diameter		100	.480		22	21	2.14	45.14	60.50
1560	Reinforced culvert, class 2, no gaskets									
1590	27″ diameter	B-21	88	.318	L.F.	48	15.50	2.15	65.65	80
1592	30″ diameter	B-13	80	.700		49.50	32	7.25	88.75	114
1594	36″ diameter	″	72	.778		79.50	36	8.05	123.55	153
2000	Reinforced culvert, class 3, no gaskets									
2010	12″ diameter	B-14	150	.320	L.F.	16.15	14.15	1.42	31.72	42
2020	15″ diameter		150	.320		22	14.15	1.42	37.57	48
2030	18″ diameter		132	.364		24.50	16.10	1.62	42.22	54.50
2035	21″ diameter		120	.400		28.50	17.70	1.78	47.98	61.50
2040	24″ diameter		100	.480		28.50	21	2.14	51.64	68
2045	27″ diameter	B-13	92	.609		47.50	28	6.30	81.80	103
2050	30″ diameter		88	.636		52	29.50	6.60	88.10	111
2060	36″ diameter		72	.778		83	36	8.05	127.05	157
2070	42″ diameter	B-13B	68	.824		99.50	38	14.40	151.90	185
2080	48″ diameter		64	.875		100	40.50	15.30	155.80	191
2085	54″ diameter		56	1		147	46	17.50	210.50	253
2090	60″ diameter		48	1.167		181	53.50	20.50	255	305
2100	72″ diameter		40	1.400		273	64.50	24.50	362	430
2120	84″ diameter		32	1.750		405	80.50	30.50	516	605
2140	96″ diameter		24	2.333		485	107	41	633	750
2200	With gaskets, class 3, 12″ diameter	B-21	168	.167		17.80	8.15	1.13	27.08	33.50
2220	15″ diameter		160	.175		24	8.55	1.18	33.73	41.50
2230	18″ diameter		152	.184		27	9	1.24	37.24	45.50
2235	21″ diameter		152	.184		31.50	9	1.24	41.74	50
2240	24″ diameter		136	.206		35.50	10.05	1.39	46.94	56.50
2260	30″ diameter	B-13	88	.636		60	29.50	6.60	96.10	120
2270	36″ diameter	″	72	.778		92.50	36	8.05	136.55	168
2290	48″ diameter	B-13B	64	.875		114	40.50	15.30	169.80	206
2310	72″ diameter	″	40	1.400		294	64.50	24.50	383	455
2330	Flared ends, 12″ diameter	B-21	31	.903	Ea.	305	44	6.10	355.10	410

33 42 11 – Stormwater Gravity Piping

33 42 11.60 Sewage/Drainage Collection, Concrete Pipe		Crew	Daily Output	Labor-Hours	Unit	Material	2020 Bare Costs Labor	Equipment	Total	Total Incl O&P
2340	15″ diameter	B-21	25	1.120	Ea.	350	54.50	7.55	412.05	480
2400	18″ diameter		20	1.400		400	68.50	9.45	477.95	560
2420	24″ diameter		14	2		455	97.50	13.50	566	675
2440	36″ diameter	B-13	10	5.600		990	258	58	1,306	1,575
2500	Class 4									
2510	12″ diameter	B-21	168	.167	L.F.	18.80	8.15	1.13	28.08	34.50
2512	15″ diameter		160	.175		21	8.55	1.18	30.73	38
2514	18″ diameter		152	.184		21.50	9	1.24	31.74	39
2516	21″ diameter		144	.194		25.50	9.50	1.31	36.31	44.50
2518	24″ diameter		136	.206		30	10.05	1.39	41.44	50.50
2520	27″ diameter		120	.233		32	11.40	1.58	44.98	55
2522	30″ diameter	B-13	88	.636		40	29.50	6.60	76.10	98
2524	36″ diameter		72	.778		57.50	36	8.05	101.55	129
2528	42″ diameter		68	.824		70	38	8.55	116.55	147
2600	Class 5									
2610	12″ diameter	B-21	168	.167	L.F.	17.65	8.15	1.13	26.93	33.50
2612	15″ diameter		160	.175		22.50	8.55	1.18	32.23	40
2614	18″ diameter		152	.184		26.50	9	1.24	36.74	45
2616	21″ diameter		144	.194		35.50	9.50	1.31	46.31	56
2618	24″ diameter		136	.206		32	10.05	1.39	43.44	53
2620	27″ diameter		120	.233		54	11.40	1.58	66.98	79.50
2622	30″ diameter	B-13	88	.636		52.50	29.50	6.60	88.60	112
2624	36″ diameter	″	72	.778		90	36	8.05	134.05	165
2800	Add for rubber joints 12″ to 36″ diameter					12%				
3080	Radius pipe, add to pipe prices, 12″ to 60″ diameter					50%				
3090	Over 60″ diameter, add					20%				
3500	Reinforced elliptical, 8′ lengths, C507 class 3									
3520	14″ x 23″ inside, round equivalent 18″ diameter	B-21	82	.341	L.F.	40.50	16.65	2.30	59.45	73.50
3530	24″ x 38″ inside, round equivalent 30″ diameter	B-13	58	.966		69.50	44.50	10	124	158
3540	29″ x 45″ inside, round equivalent 36″ diameter		52	1.077		106	49.50	11.15	166.65	208
3550	38″ x 60″ inside, round equivalent 48″ diameter		38	1.474		171	68	15.30	254.30	315
3560	48″ x 76″ inside, round equivalent 60″ diameter		26	2.154		215	99	22.50	336.50	420
3570	58″ x 91″ inside, round equivalent 72″ diameter		22	2.545		330	117	26.50	473.50	575
3780	Concrete slotted pipe, class 4 mortar joint									
3800	12″ diameter	B-21	168	.167	L.F.	33	8.15	1.13	42.28	50
3840	18″ diameter	″	152	.184	″	37	9	1.24	47.24	56
3900	Concrete slotted pipe, Class 4 O-ring joint									
3940	12″ diameter	B-21	168	.167	L.F.	30.50	8.15	1.13	39.78	47.50
3960	18″ diameter	″	152	.184	″	32	9	1.24	42.24	50.50
6200	Gasket, conc. pipe joint, 12″				Ea.	4.21			4.21	4.63
6220	24″					6.90			6.90	7.60
6240	36″					9.80			9.80	10.75
6260	48″					13.50			13.50	14.85
6265	54″					15.15			15.15	16.70
6270	60″					16.35			16.35	17.95
6280	72″					20.50			20.50	22.50

33 42 13 – Stormwater Culverts

33 42 13.15 Oval Arch Culverts		Crew	Daily Output	Labor-Hours	Unit	Material	Labor	Equipment	Total	Total Incl O&P
0010	**OVAL ARCH CULVERTS**									
3000	Corrugated galvanized or aluminum, coated & paved									
3020	17″ x 13″, 16 ga., 15″ equivalent	B-14	200	.240	L.F.	14.80	10.60	1.07	26.47	34.50
3040	21″ x 15″, 16 ga., 18″ equivalent		150	.320		19.25	14.15	1.42	34.82	45

33 42 Stormwater Conveyance

33 42 13 – Stormwater Culverts

33 42 13.15 Oval Arch Culverts		Crew	Daily Output	Labor-Hours	Unit	Material	2020 Bare Costs Labor	Equipment	Total	Total Incl O&P
3060	28" x 20", 14 ga., 24" equivalent	B-14	125	.384	L.F.	23	17	1.71	41.71	54
3080	35" x 24", 14 ga., 30" equivalent		100	.480		30.50	21	2.14	53.64	70
3100	42" x 29", 12 ga., 36" equivalent	B-13	100	.560		36	26	5.80	67.80	87
3120	49" x 33", 12 ga., 42" equivalent		90	.622		43.50	28.50	6.45	78.45	101
3140	57" x 38", 12 ga., 48" equivalent		75	.747		56	34.50	7.75	98.25	125
3160	Steel, plain oval arch culverts, plain									
3180	17" x 13", 16 ga., 15" equivalent	B-14	225	.213	L.F.	13.25	9.45	.95	23.65	30.50
3200	21" x 15", 16 ga., 18" equivalent		175	.274		17.25	12.15	1.22	30.62	39.50
3220	28" x 20", 14 ga., 24" equivalent		150	.320		27	14.15	1.42	42.57	53.50
3240	35" x 24", 14 ga., 30" equivalent	B-13	108	.519		32	24	5.40	61.40	79
3260	42" x 29", 12 ga., 36" equivalent		108	.519		40	24	5.40	69.40	88
3280	49" x 33", 12 ga., 42" equivalent		92	.609		59.50	28	6.30	93.80	116
3300	57" x 38", 12 ga., 48" equivalent		75	.747		69	34.50	7.75	111.25	139
3320	End sections, 17" x 13"		22	2.545	Ea.	139	117	26.50	282.50	370
3340	42" x 29"		17	3.294	"	425	152	34	611	750
3360	Multi-plate arch, steel	B-20	1690	.014	Lb.	1.39	.67		2.06	2.60

33 42 33 – Stormwater Curbside Drains and Inlets

33 42 33.13 Catch Basins		Crew	Daily Output	Labor-Hours	Unit	Material	Labor	Equipment	Total	Total Incl O&P
0010	**CATCH BASINS**									
0011	Not including footing & excavation									
1600	Frames & grates, C.I., 24" square, 500 lb.	B-6	7.80	3.077	Ea.	375	141	27.50	543.50	665
1700	26" D shape, 600 lb.		7	3.429		720	157	30.50	907.50	1,075
1800	Light traffic, 18" diameter, 100 lb.		10	2.400		215	110	21.50	346.50	435
1900	24" diameter, 300 lb.		8.70	2.759		199	126	24.50	349.50	445
2000	36" diameter, 900 lb.		5.80	4.138		765	189	37	991	1,175
2100	Heavy traffic, 24" diameter, 400 lb.		7.80	3.077		295	141	27.50	463.50	580
2200	36" diameter, 1,150 lb.		3	8		935	365	71.50	1,371.50	1,675
2300	Mass. State standard, 26" diameter, 475 lb.		7	3.429		750	157	30.50	937.50	1,100
2400	30" diameter, 620 lb.		7	3.429		340	157	30.50	527.50	660
2500	Watertight, 24" diameter, 350 lb.		7.80	3.077		415	141	27.50	583.50	710
2600	26" diameter, 500 lb.		7	3.429		450	157	30.50	637.50	780
2700	32" diameter, 575 lb.		6	4		960	183	35.50	1,178.50	1,375
2800	3 piece cover & frame, 10" deep,									
2900	1,200 lb., for heavy equipment	B-6	3	8	Ea.	1,200	365	71.50	1,636.50	1,975
3000	Raised for paving 1-1/4" to 2" high									
3100	4 piece expansion ring									
3200	20" to 26" diameter	1 Clab	3	2.667	Ea.	213	112		325	415
3300	30" to 36" diameter	"	3	2.667	"	295	112		407	505
3320	Frames and covers, existing, raised for paving, 2", including									
3340	row of brick, concrete collar, up to 12" wide frame	B-6	18	1.333	Ea.	52	61	11.90	124.90	167
3360	20" to 26" wide frame		11	2.182		78.50	100	19.45	197.95	266
3380	30" to 36" wide frame		9	2.667		97	122	24	243	325
3400	Inverts, single channel brick	D-1	3	5.333		112	250		362	530
3500	Concrete		5	3.200		124	150		274	380
3600	Triple channel, brick		2	8		182	375		557	805
3700	Concrete		3	5.333		156	250		406	575

33 42 33.50 Stormwater Management		Crew	Daily Output	Labor-Hours	Unit	Material	Labor	Equipment	Total	Total Incl O&P
0010	**STORMWATER MANAGEMENT**									
0030	Add per S.F. of impervious surface	B-37	6000	.008	S.F.	2.63	.35	.04	3.02	3.51

33 52 Hydrocarbon Transmission and Distribution

33 52 13 – Liquid Hydrocarbon Piping

33 52 13.16 Gasoline Piping		Crew	Daily Output	Labor-Hours	Unit	Material	Labor	Equipment	Total	Total Incl O&P
						2020 Bare Costs				
0010	**GASOLINE PIPING**									
0020	Primary containment pipe, fiberglass-reinforced									
0030	Plastic pipe 15′ & 30′ lengths									
0040	2″ diameter	Q-6	425	.056	L.F.	7	3.45		10.45	13.05
0050	3″ diameter		400	.060		10.65	3.67		14.32	17.40
0060	4″ diameter		375	.064		13.85	3.92		17.77	21.50
0100	Fittings									
0110	Elbows, 90° & 45°, bell ends, 2″	Q-6	24	1	Ea.	40.50	61		101.50	140
0120	3″ diameter		22	1.091		56	66.50		122.50	165
0130	4″ diameter		20	1.200		70	73.50		143.50	192
0200	Tees, bell ends, 2″		21	1.143		57	70		127	171
0210	3″ diameter		18	1.333		68	81.50		149.50	201
0220	4″ diameter		15	1.600		81.50	98		179.50	241
0230	Flanges bell ends, 2″		24	1		31.50	61		92.50	129
0240	3″ diameter		22	1.091		38.50	66.50		105	145
0250	4″ diameter		20	1.200		43.50	73.50		117	162
0260	Sleeve couplings, 2″		21	1.143		12.05	70		82.05	121
0270	3″ diameter		18	1.333		18.05	81.50		99.55	146
0280	4″ diameter		15	1.600		22.50	98		120.50	176
0290	Threaded adapters, 2″		21	1.143		19.25	70		89.25	129
0300	3″ diameter		18	1.333		32	81.50		113.50	162
0310	4″ diameter		15	1.600		35.50	98		133.50	190
0320	Reducers, 2″		27	.889		27	54.50		81.50	114
0330	3″ diameter		22	1.091		27	66.50		93.50	133
0340	4″ diameter		20	1.200		36	73.50		109.50	154
1010	Gas station product line for secondary containment (double wall)									
1100	Fiberglass reinforced plastic pipe 25′ lengths									
1120	Pipe, plain end, 3″ diameter	Q-6	375	.064	L.F.	29.50	3.92		33.42	38.50
1130	4″ diameter		350	.069		33.50	4.20		37.70	43.50
1140	5″ diameter		325	.074		35.50	4.52		40.02	46.50
1150	6″ diameter		300	.080		39.50	4.89		44.39	51
1200	Fittings									
1230	Elbows, 90° & 45°, 3″ diameter	Q-6	18	1.333	Ea.	131	81.50		212.50	270
1240	4″ diameter		16	1.500		163	92		255	320
1250	5″ diameter		14	1.714		170	105		275	350
1260	6″ diameter		12	2		211	122		333	420
1270	Tees, 3″ diameter		15	1.600		157	98		255	325
1280	4″ diameter		12	2		184	122		306	390
1290	5″ diameter		9	2.667		299	163		462	580
1300	6″ diameter		6	4		340	245		585	755
1310	Couplings, 3″ diameter		18	1.333		54.50	81.50		136	186
1320	4″ diameter		16	1.500		117	92		209	271
1330	5″ diameter		14	1.714		218	105		323	400
1340	6″ diameter		12	2		320	122		442	540
1350	Cross-over nipples, 3″ diameter		18	1.333		10.30	81.50		91.80	137
1360	4″ diameter		16	1.500		12.55	92		104.55	156
1370	5″ diameter		14	1.714		16.30	105		121.30	180
1380	6″ diameter		12	2		17.55	122		139.55	208
1400	Telescoping, reducers, concentric 4″ x 3″		18	1.333		44.50	81.50		126	175
1410	5″ x 4″		17	1.412		94.50	86.50		181	238
1420	6″ x 5″		16	1.500		237	92		329	405

33 52 16.13 Steel Natural Gas Piping		Crew	Daily Output	Labor-Hours	Unit	2020 Bare Costs Material	Labor	Equipment	Total	Total Incl O&P
0010	**STEEL NATURAL GAS PIPING**									
0020	Not including excavation or backfill, tar coated and wrapped									
4000	Pipe schedule 40, plain end									
4040	1″ diameter	Q-4	300	.107	L.F.	5.10	6.55	.36	12.01	16.10
4080	2″ diameter		280	.114		8	7	.38	15.38	20
4120	3″ diameter		260	.123		13.20	7.55	.41	21.16	26.50
4160	4″ diameter	B-35	255	.188		17.15	9.90	3.14	30.19	38
4200	5″ diameter		220	.218		25	11.45	3.64	40.09	49.50
4240	6″ diameter		180	.267		30.50	14	4.45	48.95	60.50
4280	8″ diameter		140	.343		48.50	18	5.70	72.20	88
4320	10″ diameter		100	.480		114	25	8	147	175
4360	12″ diameter		80	.600		127	31.50	10	168.50	201
4400	14″ diameter		75	.640		136	33.50	10.70	180.20	214
4440	16″ diameter		70	.686		148	36	11.45	195.45	233
4480	18″ diameter		65	.738		191	39	12.30	242.30	285
4520	20″ diameter		60	.800		297	42	13.35	352.35	405
4560	24″ diameter		50	.960		340	50.50	16	406.50	470
6000	Schedule 80, plain end									
6002	4″ diameter	B-35	144	.333	L.F.	40	17.50	5.55	63.05	77.50
6006	6″ diameter		126	.381		85.50	20	6.35	111.85	133
6008	8″ diameter		108	.444		114	23.50	7.40	144.90	170
6012	12″ diameter		72	.667		228	35	11.10	274.10	315
8008	Elbow, weld joint, standard weight									
8020	4″ diameter	Q-16	6.80	3.529	Ea.	113	212	15.65	340.65	470
8026	8″ diameter		3.40	7.059		475	425	31.50	931.50	1,200
8030	12″ diameter		2.30	10.435		1,000	630	46	1,676	2,125
8034	16″ diameter		1.50	16		2,100	960	71	3,131	3,900
8038	20″ diameter		1.20	20		3,625	1,200	88.50	4,913.50	5,925
8040	24″ diameter		1.02	23.529		5,450	1,425	104	6,979	8,325
8100	Extra heavy									
8102	4″ diameter	Q-16	5.30	4.528	Ea.	226	272	20	518	690
8108	8″ diameter		2.60	9.231		710	555	41	1,306	1,700
8112	12″ diameter		1.80	13.333		1,325	800	59	2,184	2,800
8116	16″ diameter		1.20	20		2,800	1,200	88.50	4,088.50	5,050
8120	20″ diameter		.94	25.532		4,825	1,525	113	6,463	7,825
8122	24″ diameter		.80	30		7,275	1,800	133	9,208	10,900
8200	Malleable, standard weight									
8202	4″ diameter	B-20	12	2	Ea.	745	94		839	970
8208	8″ diameter		6	4		2,025	188		2,213	2,550
8212	12″ diameter		4	6		2,175	282		2,457	2,825
8300	Extra heavy									
8302	4″ diameter	B-20	12	2	Ea.	1,500	94		1,594	1,800
8308	8″ diameter	B-21	6	4.667		2,550	228	31.50	2,809.50	3,200
8312	12″ diameter	″	4	7		3,450	340	47.50	3,837.50	4,400
8500	Tee weld, standard weight									
8510	4″ diameter	Q-16	4.50	5.333	Ea.	186	320	23.50	529.50	725
8514	6″ diameter		3	8		320	480	35.50	835.50	1,125
8516	8″ diameter		2.30	10.435		555	630	46	1,231	1,625
8520	12″ diameter		1.50	16		1,525	960	71	2,556	3,250
8524	16″ diameter		1	24		3,000	1,450	106	4,556	5,650
8528	20″ diameter		.80	30		7,500	1,800	133	9,433	11,200
8530	24″ diameter		.70	34.286		9,675	2,050	152	11,877	14,000

33 52 Hydrocarbon Transmission and Distribution

33 52 16 – Gas Hydrocarbon Piping

33 52 16.13 Steel Natural Gas Piping		Crew	Daily Output	Labor-Hours	Unit	Material	Labor	Equipment	Total	Total Incl O&P
						2020 Bare Costs				
8810	Malleable, standard weight									
8812	4″ diameter	B-20	8	3	Ea.	1,275	141		1,416	1,625
8818	8″ diameter	B-21	4	7		2,175	340	47.50	2,562.50	3,000
8822	12″ diameter	″	2.70	10.370		4,000	505	70	4,575	5,275
8900	Extra heavy									
8902	4″ diameter	B-20	8	3	Ea.	1,900	141		2,041	2,325
8908	8″ diameter	B-21	4	7		2,725	340	47.50	3,112.50	3,600
8912	12″ diameter	″	2.70	10.370		4,000	505	70	4,575	5,275

33 52 16.20 Piping, Gas Service and Distribution, P.E.		Crew	Daily Output	Labor-Hours	Unit	Material	Labor	Equipment	Total	Total Incl O&P
0010	PIPING, GAS SERVICE AND DISTRIBUTION, POLYETHYLENE									
0020	Not including excavation or backfill									
1000	60 psi coils, compression coupling @ 100′, 1/2″ diameter, SDR 11	B-20A	608	.053	L.F.	.49	2.66		3.15	4.72
1010	1″ diameter, SDR 11		544	.059		1.14	2.97		4.11	5.90
1040	1-1/4″ diameter, SDR 11		544	.059		1.64	2.97		4.61	6.45
1100	2″ diameter, SDR 11		488	.066		2.48	3.31		5.79	7.90
1160	3″ diameter, SDR 11		408	.078		6.20	3.96		10.16	13
1500	60 psi 40′ joints with coupling, 3″ diameter, SDR 11	B-21A	408	.098		7.35	5.15	1.05	13.55	17.30
1540	4″ diameter, SDR 11		352	.114		13.35	5.95	1.22	20.52	25.50
1600	6″ diameter, SDR 11		328	.122		35.50	6.40	1.31	43.21	50.50
1640	8″ diameter, SDR 11		272	.147		54	7.70	1.58	63.28	73.50
9000	Minimum labor/equipment charge	B-20	2	12	Job		565		565	900

33 59 Hydrocarbon Utility Metering

33 59 33 – Natural-Gas Metering

33 59 33.10 Piping, Valves and Meters, Gas Distribution		Crew	Daily Output	Labor-Hours	Unit	Material	Labor	Equipment	Total	Total Incl O&P
0010	PIPING, VALVES & METERS, GAS DISTRIBUTION									
0020	Not including excavation or backfill									
0100	Gas stops, with or without checks									
0140	1-1/4″ size	1 Plum	12	.667	Ea.	70.50	43		113.50	144
0180	1-1/2″ size		10	.800		91.50	51.50		143	181
0200	2″ size		8	1		143	64.50		207.50	258
0600	Pressure regulator valves, iron and bronze									
0640	1-1/2″ diameter	1 Plum	13	.615	Ea.	174	39.50		213.50	254
0680	2″ diameter	″	11	.727		281	47		328	385
0700	3″ diameter	Q-1	13	1.231		650	71.50		721.50	825
0740	4″ diameter	″	8	2		2,200	116		2,316	2,600
2000	Lubricated semi-steel plug valve									
2040	3/4″ diameter	1 Plum	16	.500	Ea.	125	32		157	188
2080	1″ diameter		14	.571		135	37		172	205
2100	1-1/4″ diameter		12	.667		159	43		202	242
2140	1-1/2″ diameter		11	.727		154	47		201	243
2180	2″ diameter		8	1		208	64.50		272.50	330
2300	2-1/2″ diameter	Q-1	5	3.200		299	186		485	615
2340	3″ diameter	″	4.50	3.556		425	206		631	790

33 61 13.20 Pipe Conduit, Prefabricated/Preinsulated		Crew	Daily Output	Labor-Hours	Unit	Material	Labor	Equipment	Total	Total Incl O&P
						2020 Bare Costs				
0010	**PIPE CONDUIT, PREFABRICATED/PREINSULATED**									
0020	Does not include trenching, fittings or crane.									
0300	For cathodic protection, add 12 to 14%									
0310	of total built-up price (casing plus service pipe)									
0580	Polyurethane insulated system, 250°F max. temp.									
0620	Black steel service pipe, standard wt., 1/2" insulation									
0660	3/4" diam. pipe size	Q-17	54	.296	L.F.	88.50	17.50	1.97	107.97	127
0670	1" diam. pipe size		50	.320		97.50	18.90	2.13	118.53	138
0680	1-1/4" diam. pipe size		47	.340		109	20	2.26	131.26	153
0690	1-1/2" diam. pipe size		45	.356		118	21	2.36	141.36	165
0700	2" diam. pipe size		42	.381		122	22.50	2.53	147.03	173
0710	2-1/2" diam. pipe size		34	.471		124	28	3.13	155.13	183
0720	3" diam. pipe size		28	.571		144	33.50	3.80	181.30	214
0730	4" diam. pipe size		22	.727		182	43	4.84	229.84	272
0740	5" diam. pipe size		18	.889		232	52.50	5.90	290.40	345
0750	6" diam. pipe size	Q-18	23	1.043		269	64	4.62	337.62	400
0760	8" diam. pipe size		19	1.263		395	77.50	5.60	478.10	560
0770	10" diam. pipe size		16	1.500		500	92	6.65	598.65	700
0780	12" diam. pipe size		13	1.846		625	113	8.20	746.20	870
0790	14" diam. pipe size		11	2.182		695	133	9.65	837.65	980
0800	16" diam. pipe size		10	2.400		800	147	10.65	957.65	1,125
0810	18" diam. pipe size		8	3		920	184	13.30	1,117.30	1,300
0820	20" diam. pipe size		7	3.429		1,025	210	15.20	1,250.20	1,475
0830	24" diam. pipe size		6	4		1,250	245	17.70	1,512.70	1,775
0900	For 1" thick insulation, add					10%				
0940	For 1-1/2" thick insulation, add					13%				
0980	For 2" thick insulation, add					20%				
1500	Gland seal for system, 3/4" diam. pipe size	Q-17	32	.500	Ea.	1,025	29.50	3.33	1,057.83	1,175
1510	1" diam. pipe size		32	.500		1,025	29.50	3.33	1,057.83	1,175
1540	1-1/4" diam. pipe size		30	.533		1,100	31.50	3.55	1,135.05	1,250
1550	1-1/2" diam. pipe size		30	.533		1,100	31.50	3.55	1,135.05	1,250
1560	2" diam. pipe size		28	.571		1,300	33.50	3.80	1,337.30	1,475
1570	2-1/2" diam. pipe size		26	.615		1,400	36.50	4.09	1,440.59	1,600
1580	3" diam. pipe size		24	.667		1,500	39.50	4.43	1,543.93	1,725
1590	4" diam. pipe size		22	.727		1,775	43	4.84	1,822.84	2,025
1600	5" diam. pipe size		19	.842		2,200	49.50	5.60	2,255.10	2,475
1610	6" diam. pipe size	Q-18	26	.923		2,325	56.50	4.09	2,385.59	2,650
1620	8" diam. pipe size		25	.960		2,700	58.50	4.25	2,762.75	3,075
1630	10" diam. pipe size		23	1.043		3,175	64	4.62	3,243.62	3,600
1640	12" diam. pipe size		21	1.143		3,500	70	5.05	3,575.05	3,975
1650	14" diam. pipe size		19	1.263		3,925	77.50	5.60	4,008.10	4,450
1660	16" diam. pipe size		18	1.333		4,650	81.50	5.90	4,737.40	5,225
1670	18" diam. pipe size		16	1.500		4,925	92	6.65	5,023.65	5,575
1680	20" diam. pipe size		14	1.714		5,600	105	7.60	5,712.60	6,325
1690	24" diam. pipe size		12	2		6,200	122	8.85	6,330.85	7,025
2000	Elbow, 45° for system									
2020	3/4" diam. pipe size	Q-17	14	1.143	Ea.	675	67.50	7.60	750.10	850
2040	1" diam. pipe size		13	1.231		690	72.50	8.20	770.70	880
2050	1-1/4" diam. pipe size		11	1.455		785	86	9.65	880.65	1,000
2060	1-1/2" diam. pipe size		9	1.778		790	105	11.80	906.80	1,050
2070	2" diam. pipe size		6	2.667		845	157	17.75	1,019.75	1,200
2080	2-1/2" diam. pipe size		4	4		930	236	26.50	1,192.50	1,425
2090	3" diam. pipe size		3.50	4.571		1,075	270	30.50	1,375.50	1,625

33 61 Hydronic Energy Distribution

33 61 13 – Underground Hydronic Energy Distribution

33 61 13.20 Pipe Conduit, Prefabricated/Preinsulated		Crew	Daily Output	Labor-Hours	Unit	2020 Bare Costs Material	Labor	Equipment	Total	Total Incl O&P
2100	4" diam. pipe size	Q-17	3	5.333	Ea.	1,250	315	35.50	1,600.50	1,900
2110	5" diam. pipe size		2.80	5.714		1,600	335	38	1,973	2,325
2120	6" diam. pipe size	Q-18	4	6		1,825	365	26.50	2,216.50	2,600
2130	8" diam. pipe size		3	8		2,650	490	35.50	3,175.50	3,700
2140	10" diam. pipe size		2.40	10		3,250	610	44.50	3,904.50	4,575
2150	12" diam. pipe size		2	12		4,275	735	53	5,063	5,875
2160	14" diam. pipe size		1.80	13.333		5,325	815	59	6,199	7,175
2170	16" diam. pipe size		1.60	15		6,325	920	66.50	7,311.50	8,450
2180	18" diam. pipe size		1.30	18.462		7,925	1,125	82	9,132	10,600
2190	20" diam. pipe size		1	24		10,000	1,475	106	11,581	13,400
2200	24" diam. pipe size		.70	34.286		12,600	2,100	152	14,852	17,200
2260	For elbow, 90°, add					25%				
2300	For tee, straight, add					85%	30%			
2340	For tee, reducing, add					170%	30%			
2380	For weldolet, straight, add					50%				

33 63 Steam Energy Distribution

33 63 13 – Underground Steam and Condensate Distribution Piping

33 63 13.10 Calcium Silicate Insulated System		Crew	Daily Output	Labor-Hours	Unit	Material	Labor	Equipment	Total	Total Incl O&P
0010	**CALCIUM SILICATE INSULATED SYSTEM**									
0011	High temp. (1200 degrees F)									
2840	Steel casing with protective exterior coating									
2850	6-5/8" diameter	Q-18	52	.462	L.F.	130	28	2.04	160.04	189
2860	8-5/8" diameter		50	.480		141	29.50	2.13	172.63	203
2870	10-3/4" diameter		47	.511		165	31	2.26	198.26	233
2880	12-3/4" diameter		44	.545		180	33.50	2.42	215.92	252
2890	14" diameter		41	.585		203	36	2.59	241.59	282
2900	16" diameter		39	.615		218	37.50	2.73	258.23	300
2910	18" diameter		36	.667		240	41	2.95	283.95	330
2920	20" diameter		34	.706		271	43	3.13	317.13	370
2930	22" diameter		32	.750		375	46	3.32	424.32	485
2940	24" diameter		29	.828		430	50.50	3.67	484.17	555
2950	26" diameter		26	.923		490	56.50	4.09	550.59	630
2960	28" diameter		23	1.043		605	64	4.62	673.62	770
2970	30" diameter		21	1.143		650	70	5.05	725.05	830
2980	32" diameter		19	1.263		730	77.50	5.60	813.10	925
2990	34" diameter		18	1.333		735	81.50	5.90	822.40	945
3000	36" diameter		16	1.500		790	92	6.65	888.65	1,025
3040	For multi-pipe casings, add					10%				
3060	For oversize casings, add					2%				
3400	Steel casing gland seal, single pipe									
3420	6-5/8" diameter	Q-18	25	.960	Ea.	1,975	58.50	4.25	2,037.75	2,275
3440	8-5/8" diameter		23	1.043		2,325	64	4.62	2,393.62	2,650
3450	10-3/4" diameter		21	1.143		2,575	70	5.05	2,650.05	2,975
3460	12-3/4" diameter		19	1.263		3,050	77.50	5.60	3,133.10	3,475
3470	14" diameter		17	1.412		3,375	86.50	6.25	3,467.75	3,875
3480	16" diameter		16	1.500		3,950	92	6.65	4,048.65	4,500
3490	18" diameter		15	1.600		4,400	98	7.10	4,505.10	4,975
3500	20" diameter		13	1.846		4,850	113	8.20	4,971.20	5,525
3510	22" diameter		12	2		5,450	122	8.85	5,580.85	6,200
3520	24" diameter		11	2.182		6,125	133	9.65	6,267.65	6,975

33 63 13 – Underground Steam and Condensate Distribution Piping

33 63 13.10 Calcium Silicate Insulated System		Crew	Daily Output	Labor-Hours	Unit	2020 Bare Costs Material	Labor	Equipment	Total	Total Incl O&P
3530	26" diameter	Q-18	10	2.400	Ea.	7,000	147	10.65	7,157.65	7,950
3540	28" diameter		9.50	2.526		7,900	155	11.20	8,066.20	8,950
3550	30" diameter		9	2.667		8,050	163	11.80	8,224.80	9,125
3560	32" diameter		8.50	2.824		9,050	173	12.50	9,235.50	10,200
3570	34" diameter		8	3		9,850	184	13.30	10,047.30	11,100
3580	36" diameter		7	3.429		10,500	210	15.20	10,725.20	11,800
3620	For multi-pipe casings, add					5%				
4000	Steel casing anchors, single pipe									
4020	6-5/8" diameter	Q-18	8	3	Ea.	1,775	184	13.30	1,972.30	2,250
4040	8-5/8" diameter		7.50	3.200		1,850	196	14.20	2,060.20	2,350
4050	10-3/4" diameter		7	3.429		2,400	210	15.20	2,625.20	3,000
4060	12-3/4" diameter		6.50	3.692		2,575	226	16.35	2,817.35	3,225
4070	14" diameter		6	4		3,050	245	17.70	3,312.70	3,750
4080	16" diameter		5.50	4.364		3,525	267	19.35	3,811.35	4,300
4090	18" diameter		5	4.800		3,975	294	21.50	4,290.50	4,850
4100	20" diameter		4.50	5.333		4,400	325	23.50	4,748.50	5,350
4110	22" diameter		4	6		4,900	365	26.50	5,291.50	5,975
4120	24" diameter		3.50	6.857		5,325	420	30.50	5,775.50	6,525
4130	26" diameter		3	8		6,025	490	35.50	6,550.50	7,425
4140	28" diameter		2.50	9.600		6,625	585	42.50	7,252.50	8,250
4150	30" diameter		2	12		7,050	735	53	7,838	8,925
4160	32" diameter		1.50	16		8,400	980	71	9,451	10,900
4170	34" diameter		1	24		9,450	1,475	106	11,031	12,800
4180	36" diameter		1	24		10,300	1,475	106	11,881	13,700
4220	For multi-pipe, add					5%	20%			
4800	Steel casing elbow									
4820	6-5/8" diameter	Q-18	15	1.600	Ea.	2,525	98	7.10	2,630.10	2,925
4830	8-5/8" diameter		15	1.600		2,700	98	7.10	2,805.10	3,100
4850	10-3/4" diameter		14	1.714		3,250	105	7.60	3,362.60	3,750
4860	12-3/4" diameter		13	1.846		3,850	113	8.20	3,971.20	4,400
4870	14" diameter		12	2		4,125	122	8.85	4,255.85	4,750
4880	16" diameter		11	2.182		4,500	133	9.65	4,642.65	5,175
4890	18" diameter		10	2.400		5,150	147	10.65	5,307.65	5,900
4900	20" diameter		9	2.667		5,500	163	11.80	5,674.80	6,325
4910	22" diameter		8	3		5,900	184	13.30	6,097.30	6,800
4920	24" diameter		7	3.429		6,500	210	15.20	6,725.20	7,500
4930	26" diameter		6	4		7,125	245	17.70	7,387.70	8,250
4940	28" diameter		5	4.800		7,700	294	21.50	8,015.50	8,925
4950	30" diameter		4	6		7,725	365	26.50	8,116.50	9,100
4960	32" diameter		3	8		8,775	490	35.50	9,300.50	10,400
4970	34" diameter		2	12		9,650	735	53	10,438	11,800
4980	36" diameter		2	12		10,300	735	53	11,088	12,500
5500	Black steel service pipe, std. wt., 1" thick insulation									
5510	3/4" diameter pipe size	Q-17	54	.296	L.F.	79.50	17.50	1.97	98.97	116
5540	1" diameter pipe size		50	.320		81.50	18.90	2.13	102.53	121
5550	1-1/4" diameter pipe size		47	.340		104	20	2.26	126.26	147
5560	1-1/2" diameter pipe size		45	.356		109	21	2.36	132.36	154
5570	2" diameter pipe size		42	.381		118	22.50	2.53	143.03	168
5580	2-1/2" diameter pipe size		34	.471		124	28	3.13	155.13	182
5590	3" diameter pipe size		28	.571		108	33.50	3.80	145.30	175
5600	4" diameter pipe size		22	.727		121	43	4.84	168.84	205
5610	5" diameter pipe size		18	.889		136	52.50	5.90	194.40	238
5620	6" diameter pipe size	Q-18	23	1.043		147	64	4.62	215.62	265

33 63 13 – Underground Steam and Condensate Distribution Piping

33 63 13.10 Calcium Silicate Insulated System		Crew	Daily Output	Labor-Hours	Unit	Material	2020 Bare Costs Labor	Equipment	Total	Total Incl O&P
6000	Black steel service pipe, std. wt., 1-1/2" thick insul.									
6010	3/4" diameter pipe size	Q-17	54	.296	L.F.	94	17.50	1.97	113.47	132
6040	1" diameter pipe size		50	.320		99	18.90	2.13	120.03	140
6050	1-1/4" diameter pipe size		47	.340		107	20	2.26	129.26	150
6060	1-1/2" diameter pipe size		45	.356		106	21	2.36	129.36	151
6070	2" diameter pipe size		42	.381		113	22.50	2.53	138.03	163
6080	2-1/2" diameter pipe size		34	.471		104	28	3.13	135.13	160
6090	3" diameter pipe size		28	.571		115	33.50	3.80	152.30	183
6100	4" diameter pipe size		22	.727		132	43	4.84	179.84	217
6110	5" diameter pipe size		18	.889		157	52.50	5.90	215.40	261
6120	6" diameter pipe size	Q-18	23	1.043		164	64	4.62	232.62	284
6130	8" diameter pipe size		19	1.263		208	77.50	5.60	291.10	355
6140	10" diameter pipe size		16	1.500		278	92	6.65	376.65	455
6150	12" diameter pipe size		13	1.846		370	113	8.20	491.20	590
6190	For 2" thick insulation, add					15%				
6220	For 2-1/2" thick insulation, add					25%				
6260	For 3" thick insulation, add					30%				
6800	Black steel service pipe, ex. hvy. wt., 1" thick insul.									
6820	3/4" diameter pipe size	Q-17	50	.320	L.F.	95	18.90	2.13	116.03	136
6840	1" diameter pipe size		47	.340		101	20	2.26	123.26	144
6850	1-1/4" diameter pipe size		44	.364		112	21.50	2.42	135.92	159
6860	1-1/2" diameter pipe size		42	.381		114	22.50	2.53	139.03	163
6870	2" diameter pipe size		40	.400		125	23.50	2.66	151.16	176
6880	2-1/2" diameter pipe size		31	.516		102	30.50	3.43	135.93	164
6890	3" diameter pipe size		27	.593		118	35	3.94	156.94	188
6900	4" diameter pipe size		21	.762		129	45	5.05	179.05	217
6910	5" diameter pipe size		17	.941		156	55.50	6.25	217.75	265
6920	6" diameter pipe size	Q-18	22	1.091		166	66.50	4.83	237.33	291
7400	Black steel service pipe, ex. hvy. wt., 1-1/2" thick insul.									
7420	3/4" diameter pipe size	Q-17	50	.320	L.F.	70	18.90	2.13	91.03	108
7440	1" diameter pipe size		47	.340		88.50	20	2.26	110.76	130
7450	1-1/4" diameter pipe size		44	.364		97.50	21.50	2.42	121.42	143
7460	1-1/2" diameter pipe size		42	.381		96	22.50	2.53	121.03	144
7470	2" diameter pipe size		40	.400		93	23.50	2.66	119.16	141
7480	2-1/2" diameter pipe size		31	.516		95.50	30.50	3.43	129.43	156
7490	3" diameter pipe size		27	.593		105	35	3.94	143.94	174
7500	4" diameter pipe size		21	.762		134	45	5.05	184.05	222
7510	5" diameter pipe size		17	.941		185	55.50	6.25	246.75	296
7520	6" diameter pipe size	Q-18	22	1.091		197	66.50	4.83	268.33	325
7530	8" diameter pipe size		18	1.333		270	81.50	5.90	357.40	430
7540	10" diameter pipe size		15	1.600		300	98	7.10	405.10	495
7550	12" diameter pipe size		13	1.846		360	113	8.20	481.20	585
7590	For 2" thick insulation, add					13%				
7640	For 2-1/2" thick insulation, add					18%				
7680	For 3" thick insulation, add					24%				

33 71 16 – Electrical Utility Poles

33 71 16.33 Wood Electrical Utility Poles		Crew	Daily Output	Labor-Hours	Unit	Material	Labor	Equipment	Total	Total Incl O&P
						2020 Bare Costs				
0010	**WOOD ELECTRICAL UTILITY POLES**									
0011	Excludes excavation, backfill and cast-in-place concrete									
6200	Wood, class 3 Douglas Fir, penta-treated, 20′	R-3	3.10	6.452	Ea.	256	395	61	712	960
6600	30′		2.60	7.692		320	470	72.50	862.50	1,150
7000	40′		2.30	8.696		630	530	82	1,242	1,600
7200	45′		1.70	11.765		900	720	111	1,731	2,200
7400	Cross arms with hardware & insulators									
7600	4′ long	1 Elec	2.50	3.200	Ea.	156	196		352	470
7800	5′ long		2.40	3.333		168	205		373	500
8000	6′ long		2.20	3.636		190	223		413	555

33 71 19 – Electrical Underground Ducts and Manholes

33 71 19.15 Underground Ducts and Manholes		Crew	Daily Output	Labor-Hours	Unit	Material	Labor	Equipment	Total	Total Incl O&P
0010	**UNDERGROUND DUCTS AND MANHOLES**									
0011	Not incl. excavation, backfill, or concrete in slab and duct bank									
1000	Direct burial									
1010	PVC, schedule 40, w/coupling, 1/2″ diameter	1 Elec	340	.024	L.F.	.34	1.44		1.78	2.59
1020	3/4″ diameter		290	.028		.41	1.69		2.10	3.06
1030	1″ diameter		260	.031		.67	1.89		2.56	3.65
1040	1-1/2″ diameter		210	.038		.99	2.34		3.33	4.69
1050	2″ diameter		180	.044		1.29	2.73		4.02	5.60
1060	3″ diameter	2 Elec	240	.067		2.38	4.09		6.47	8.90
1070	4″ diameter		160	.100		3.16	6.15		9.31	12.95
1080	5″ diameter		120	.133		4.69	8.20		12.89	17.75
1090	6″ diameter		90	.178		6.20	10.90		17.10	23.50
1110	Elbows, 1/2″ diameter	1 Elec	48	.167	Ea.	.58	10.25		10.83	16.40
1120	3/4″ diameter		38	.211		.60	12.90		13.50	20.50
1130	1″ diameter		32	.250		1.21	15.35		16.56	25
1140	1-1/2″ diameter		21	.381		2.71	23.50		26.21	39
1150	2″ diameter		16	.500		3.54	30.50		34.04	51
1160	3″ diameter		12	.667		12.25	41		53.25	76.50
1170	4″ diameter		9	.889		23	54.50		77.50	109
1180	5″ diameter		8	1		34.50	61.50		96	133
1190	6″ diameter		5	1.600		35	98		133	190
1210	Adapters, 1/2″ diameter		52	.154		.26	9.45		9.71	14.85
1220	3/4″ diameter		43	.186		.37	11.40		11.77	17.95
1230	1″ diameter		39	.205		.39	12.60		12.99	19.80
1240	1-1/2″ diameter		35	.229		.71	14		14.71	22.50
1250	2″ diameter		26	.308		1.39	18.90		20.29	30.50
1260	3″ diameter		20	.400		2.85	24.50		27.35	41
1270	4″ diameter		14	.571		4.80	35		39.80	59.50
1280	5″ diameter		12	.667		9.70	41		50.70	73.50
1290	6″ diameter		9	.889		18.90	54.50		73.40	105
1340	Bell end & cap, 1-1/2″ diameter		35	.229		4.69	14		18.69	26.50
1350	Bell end & plug, 2″ diameter		26	.308		5.90	18.90		24.80	35.50
1360	3″ diameter		20	.400		8.25	24.50		32.75	47
1370	4″ diameter		14	.571		9.90	35		44.90	65
1380	5″ diameter		12	.667		15.95	41		56.95	80.50
1390	6″ diameter		9	.889		17.05	54.50		71.55	103
1450	Base spacer, 2″ diameter		56	.143		1.37	8.75		10.12	15
1460	3″ diameter		46	.174		1.78	10.65		12.43	18.40
1470	4″ diameter		41	.195		1.78	11.95		13.73	20.50
1480	5″ diameter		37	.216		1.99	13.25		15.24	22.50

33 71 19.15 Underground Ducts and Manholes		Crew	Daily Output	Labor-Hours	Unit	2020 Bare Costs Material	Labor	Equipment	Total	Total Incl O&P
1490	6" diameter	1 Elec	34	.235	Ea.	2.48	14.45		16.93	24.50
1550	Intermediate spacer, 2" diameter		60	.133		1.25	8.20		9.45	14
1560	3" diameter		46	.174		1.76	10.65		12.41	18.40
1570	4" diameter		41	.195		1.59	11.95		13.54	20
1580	5" diameter		37	.216		1.94	13.25		15.19	22.50
1590	6" diameter		34	.235		2.58	14.45		17.03	25
4010	PVC, schedule 80, w/coupling, 1/2" diameter		215	.037	L.F.	1.03	2.28		3.31	4.64
4020	3/4" diameter		180	.044		1.41	2.73		4.14	5.75
4030	1" diameter		145	.055		1.91	3.38		5.29	7.30
4040	1-1/2" diameter		120	.067		3.15	4.09		7.24	9.75
4050	2" diameter		100	.080		4.28	4.91		9.19	12.25
4060	3" diameter	2 Elec	130	.123		8.95	7.55		16.50	21.50
4070	4" diameter		90	.178		12.85	10.90		23.75	31
4080	5" diameter		70	.229		18.05	14		32.05	41.50
4090	6" diameter		50	.320		25.50	19.65		45.15	58
4110	Elbows, 1/2" diameter	1 Elec	29	.276	Ea.	2.47	16.90		19.37	28.50
4120	3/4" diameter		23	.348		5.70	21.50		27.20	39.50
4130	1" diameter		20	.400		6.80	24.50		31.30	45.50
4140	1-1/2" diameter		16	.500		11.40	30.50		41.90	59.50
4150	2" diameter		12	.667		15.25	41		56.25	80
4160	3" diameter		9	.889		38	54.50		92.50	126
4170	4" diameter		7	1.143		67	70		137	182
4180	5" diameter		6	1.333		168	82		250	310
4190	6" diameter		4	2		72.50	123		195.50	269
4210	Adapter, 1/2" diameter		39	.205		.26	12.60		12.86	19.65
4220	3/4" diameter		33	.242		.37	14.85		15.22	23.50
4230	1" diameter		29	.276		.39	16.90		17.29	26.50
4240	1-1/2" diameter		26	.308		.71	18.90		19.61	30
4250	2" diameter		23	.348		1.39	21.50		22.89	34.50
4260	3" diameter		18	.444		2.85	27.50		30.35	45
4270	4" diameter		13	.615		4.80	38		42.80	63.50
4280	5" diameter		11	.727		9.70	44.50		54.20	79
4290	6" diameter		8	1		18.90	61.50		80.40	116
4310	Bell end & cap, 1-1/2" diameter		26	.308		4.69	18.90		23.59	34
4320	Bell end & plug, 2" diameter		23	.348		5.90	21.50		27.40	39.50
4330	3" diameter		18	.444		8.25	27.50		35.75	51
4340	4" diameter		13	.615		9.90	38		47.90	69
4350	5" diameter		11	.727		15.95	44.50		60.45	86
4360	6" diameter		8	1		17.05	61.50		78.55	113
4370	Base spacer, 2" diameter		42	.190		1.37	11.70		13.07	19.50
4380	3" diameter		33	.242		1.78	14.85		16.63	25
4390	4" diameter		29	.276		1.78	16.90		18.68	28
4400	5" diameter		26	.308		1.99	18.90		20.89	31
4410	6" diameter		25	.320		2.48	19.65		22.13	32.50
4420	Intermediate spacer, 2" diameter		45	.178		1.25	10.90		12.15	18.20
4430	3" diameter		34	.235		1.76	14.45		16.21	24
4440	4" diameter		31	.258		1.59	15.85		17.44	26.50
4450	5" diameter		28	.286		1.94	17.55		19.49	29
4460	6" diameter		25	.320		2.58	19.65		22.23	33

33 71 Electrical Utility Transmission and Distribution

33 71 19 – Electrical Underground Ducts and Manholes

	33 71 19.17 Electric and Telephone Underground	Crew	Daily Output	Labor-Hours	Unit	Material	2020 Bare Costs Labor	Equipment	Total	Total Incl O&P
0010	**ELECTRIC AND TELEPHONE UNDERGROUND**									
0011	Not including excavation R337119-30									
0200	backfill and cast in place concrete									
0400	Hand holes, precast concrete, with concrete cover									
0600	2′ x 2′ x 3′ deep	R-3	2.40	8.333	Ea.	545	510	79	1,134	1,475
0800	3′ x 3′ x 3′ deep		1.90	10.526		705	645	99.50	1,449.50	1,875
1000	4′ x 4′ x 4′ deep		1.40	14.286		1,600	875	135	2,610	3,250
1200	Manholes, precast with iron racks & pulling irons, C.I. frame									
1400	and cover, 4′ x 6′ x 7′ deep	B-13	2	28	Ea.	3,000	1,300	290	4,590	5,675
1600	6′ x 8′ x 7′ deep		1.90	29.474		3,375	1,350	305	5,030	6,175
1800	6′ x 10′ x 7′ deep		1.80	31.111		3,775	1,425	325	5,525	6,775
4200	Underground duct, banks ready for concrete fill, min. of 7.5″									
4400	between conduits, center to center									
4580	PVC, type EB, 1 @ 2″ diameter	2 Elec	480	.033	L.F.	1.12	2.04		3.16	4.38
4600	2 @ 2″ diameter		240	.067		2.23	4.09		6.32	8.75
4800	4 @ 2″ diameter		120	.133		4.47	8.20		12.67	17.50
4900	1 @ 3″ diameter		400	.040		1.62	2.45		4.07	5.55
5000	2 @ 3″ diameter		200	.080		3.24	4.91		8.15	11.10
5200	4 @ 3″ diameter		100	.160		6.45	9.80		16.25	22
5300	1 @ 4″ diameter		320	.050		1.86	3.07		4.93	6.75
5400	2 @ 4″ diameter		160	.100		3.72	6.15		9.87	13.55
5600	4 @ 4″ diameter		80	.200		7.45	12.25		19.70	27
5800	6 @ 4″ diameter		54	.296		11.15	18.20		29.35	40.50
5810	1 @ 5″ diameter		260	.062		2.09	3.78		5.87	8.10
5820	2 @ 5″ diameter		130	.123		4.18	7.55		11.73	16.20
5840	4 @ 5″ diameter		70	.229		8.35	14		22.35	30.50
5860	6 @ 5″ diameter		50	.320		12.55	19.65		32.20	44
5870	1 @ 6″ diameter		200	.080		3.59	4.91		8.50	11.50
5880	2 @ 6″ diameter		100	.160		7.20	9.80		17	23
5900	4 @ 6″ diameter		50	.320		14.35	19.65		34	46
5920	6 @ 6″ diameter		30	.533		21.50	32.50		54	74
6200	Rigid galvanized steel, 2 @ 2″ diameter		180	.089		15.75	5.45		21.20	25.50
6400	4 @ 2″ diameter		90	.178		31.50	10.90		42.40	51.50
6800	2 @ 3″ diameter		100	.160		29	9.80		38.80	46.50
7000	4 @ 3″ diameter		50	.320		57.50	19.65		77.15	93.50
7200	2 @ 4″ diameter		70	.229		45.50	14		59.50	71.50
7400	4 @ 4″ diameter		34	.471		91.50	29		120.50	145
7600	6 @ 4″ diameter		22	.727		137	44.50		181.50	220
7620	2 @ 5″ diameter		60	.267		72	16.35		88.35	105
7640	4 @ 5″ diameter		30	.533		144	32.50		176.50	210
7660	6 @ 5″ diameter		18	.889		216	54.50		270.50	320
7680	2 @ 6″ diameter		40	.400		110	24.50		134.50	159
7700	4 @ 6″ diameter		20	.800		219	49		268	315
7720	6 @ 6″ diameter		14	1.143		330	70		400	470
7800	For cast-in-place concrete, add									
7810	Under 1 C.Y.	C-6	16	3	C.Y.	206	131	3.36	340.36	440
7820	1 C.Y. to 5 C.Y.		19.20	2.500		187	109	2.80	298.80	385
7830	Over 5 C.Y.		24	2		155	87.50	2.24	244.74	310
7850	For reinforcing rods, add									
7860	#4 to #7	2 Rodm	1.10	14.545	Ton	1,125	820		1,945	2,550
7870	#8 to #14	″	1.50	10.667	″	1,125	600		1,725	2,200
8000	Fittings, PVC type EB, elbow, 2″ diameter	1 Elec	16	.500	Ea.	15.85	30.50		46.35	64.50

33 71 Electrical Utility Transmission and Distribution

33 71 19 – Electrical Underground Ducts and Manholes

33 71 19.17 Electric and Telephone Underground		Crew	Daily Output	Labor-Hours	Unit	Material	Labor (2020 Bare Costs)	Equipment	Total	Total Incl O&P
8200	3″ diameter	1 Elec	14	.571	Ea.	18.35	35		53.35	74
8400	4″ diameter		12	.667		40	41		81	107
8420	5″ diameter		10	.800		123	49		172	211
8440	6″ diameter		9	.889		191	54.50		245.50	294
8500	Coupling, 2″ diameter					.85			.85	.94
8600	3″ diameter					4.14			4.14	4.55
8700	4″ diameter					4.48			4.48	4.93
8720	5″ diameter					8.55			8.55	9.40
8740	6″ diameter					24			24	26
8800	Adapter, 2″ diameter	1 Elec	26	.308		1.01	18.90		19.91	30
9000	3″ diameter		20	.400		3.07	24.50		27.57	41.50
9200	4″ diameter		16	.500		3.36	30.50		33.86	50.50
9220	5″ diameter		13	.615		8.55	38		46.55	67.50
9240	6″ diameter		10	.800		12.50	49		61.50	89.50
9400	End bell, 2″ diameter		16	.500		2.30	30.50		32.80	49.50
9600	3″ diameter		14	.571		4.08	35		39.08	58.50
9800	4″ diameter		12	.667		4.63	41		45.63	68
9810	5″ diameter		10	.800		8.10	49		57.10	84.50
9820	6″ diameter		8	1		10.15	61.50		71.65	106
9830	5° angle coupling, 2″ diameter		26	.308		7.80	18.90		26.70	37.50
9840	3″ diameter		20	.400		26	24.50		50.50	66.50
9850	4″ diameter		16	.500		19.15	30.50		49.65	68
9860	5″ diameter		13	.615		16.70	38		54.70	76.50
9870	6″ diameter		10	.800		26	49		75	104
9880	Expansion joint, 2″ diameter		16	.500		33.50	30.50		64	84
9890	3″ diameter		18	.444		70	27.50		97.50	119
9900	4″ diameter		12	.667		102	41		143	176
9910	5″ diameter		10	.800		203	49		252	299
9920	6″ diameter		8	1		176	61.50		237.50	289
9930	Heat bender, 2″ diameter					490			490	540
9940	6″ diameter					1,500			1,500	1,650
9950	Cement, quart					15.55			15.55	17.10
9960	Nylon polyethylene pull rope, 1/4″	2 Elec	2000	.008	L.F.	.18	.49		.67	.96
9990	Minimum labor/equipment charge	1 Elec	3.50	2.286	Job		140		140	216

33 81 Communications Structures

33 81 13 – Communications Transmission Towers

33 81 13.10 Radio Towers		Crew	Daily Output	Labor-Hours	Unit	Material	Labor (2020 Bare Costs)	Equipment	Total	Total Incl O&P
0010	**RADIO TOWERS**									
0020	Guyed, 50′ H, 40 lb. section, 70 MPH basic wind speed	2 Sswk	1	16	Ea.	2,875	920		3,795	4,675
0100	Wind load 90 MPH basic wind speed	"	1	16		3,325	920		4,245	5,150
0300	190′ high, 40 lb. section, wind load 70 MPH basic wind speed	K-2	.33	72.727		9,575	4,000	2,475	16,050	19,700
0400	200′ high, 70 lb. section, wind load 90 MPH basic wind speed		.33	72.727		15,100	4,000	2,475	21,575	25,800
0600	300′ high, 70 lb. section, wind load 70 MPH basic wind speed		.20	120		24,900	6,575	4,100	35,575	42,600
0700	270′ high, 90 lb. section, wind load 90 MPH basic wind speed		.20	120		28,100	6,575	4,100	38,775	46,100
0800	400′ high, 100 lb. section, wind load 70 MPH basic wind speed		.14	171		37,600	9,400	5,850	52,850	63,000
0900	Self-supporting, 60′ high, wind load 70 MPH basic wind speed		.80	30		4,750	1,650	1,025	7,425	9,025
0910	60′ high, wind load 90 MPH basic wind speed		.45	53.333		5,575	2,925	1,825	10,325	12,900
1000	120′ high, wind load 70 MPH basic wind speed		.40	60		10,100	3,300	2,050	15,450	18,700
1200	190′ high, wind load 90 MPH basic wind speed		.20	120		29,000	6,575	4,100	39,675	47,000
2000	For states west of Rocky Mountains, add for shipping					10%				

Division Notes

	CREW	DAILY OUTPUT	LABOR-HOURS	UNIT	BARE COSTS MAT.	BARE COSTS LABOR	BARE COSTS EQUIP.	BARE COSTS TOTAL	TOTAL INCL O&P

Estimating Tips

34 11 00 Rail Tracks

This subdivision includes items that may involve either repair of existing or construction of new railroad tracks. Additional preparation work, such as the roadbed earthwork, would be found in Division 31. Additional new construction siding and turnouts are found in Subdivision 34 72. Maintenance of railroads is found under 34 01 23 Operation and Maintenance of Railways.

34 40 00 Traffic Signals

This subdivision includes traffic signal systems. Other traffic control devices such as traffic signs are found in Subdivision 10 14 53 Traffic Signage.

34 70 00 Vehicle Barriers

This subdivision includes security vehicle barriers, guide and guard rails, crash barriers, and delineators. The actual maintenance and construction of concrete and asphalt pavement are found in Division 32.

Reference Numbers

Reference numbers are shown at the beginning of some major classifications. These numbers refer to related items in the Reference Section. The reference information may be an estimating procedure, an alternate pricing method, or technical information.

Note: Not all subdivisions listed here necessarily appear. ■

34 01 Operation and Maintenance of Transportation

34 01 23 – Operation and Maintenance of Railways

34 01 23.51 Maintenance of Railroads		Crew	Daily Output	Labor-Hours	Unit	2020 Bare Costs Material	Labor	Equipment	Total	Total Incl O&P
0010	**MAINTENANCE OF RAILROADS**									
0400	Resurface and realign existing track	B-14	200	.240	L.F.		10.60	1.07	11.67	18.15
0600	For crushed stone ballast, add	"	500	.096	"	14.65	4.25	.43	19.33	23.50

34 11 Rail Tracks

34 11 13 – Track Rails

34 11 13.23 Heavy Rail Track

		Crew	Daily Output	Labor-Hours	Unit	Material	Labor	Equipment	Total	Total Incl O&P
0010	**HEAVY RAIL TRACK** R347216-10									
1000	Rail, 100 lb. prime grade				L.F.	33			33	36
1500	Relay rail				"	16.40			16.40	18.05

34 11 33 – Track Cross Ties

34 11 33.13 Concrete Track Cross Ties

		Crew	Daily Output	Labor-Hours	Unit	Material	Labor	Equipment	Total	Total Incl O&P
0010	**CONCRETE TRACK CROSS TIES**									
1400	Ties, concrete, 8′-6″ long, 30″ OC	B-14	80	.600	Ea.	153	26.50	2.67	182.17	213

34 11 33.16 Timber Track Cross Ties

		Crew	Daily Output	Labor-Hours	Unit	Material	Labor	Equipment	Total	Total Incl O&P
0010	**TIMBER TRACK CROSS TIES**									
1600	Wood, pressure treated, 6″ x 8″ x 8′-6″, C.L. lots	B-14	90	.533	Ea.	56	23.50	2.37	81.87	102
1700	L.C.L. lots		90	.533		59	23.50	2.37	84.87	105
1900	Heavy duty, 7″ x 9″ x 8′-6″, C.L. lots		70	.686		61.50	30.50	3.05	95.05	120
2000	L.C.L. lots	↓	70	.686	↓	61.50	30.50	3.05	95.05	120

34 11 33.17 Timber Switch Ties

		Crew	Daily Output	Labor-Hours	Unit	Material	Labor	Equipment	Total	Total Incl O&P
0010	**TIMBER SWITCH TIES**									
1200	Switch timber, for a #8 switch, pressure treated	B-14	3.70	12.973	M.B.F.	3,375	575	57.50	4,007.50	4,675
1300	Complete set of timbers, 3.7 MBF for #8 switch	"	1	48	Total	13,000	2,125	214	15,339	17,900

34 11 93 – Track Appurtenances and Accessories

34 11 93.50 Track Accessories

		Crew	Daily Output	Labor-Hours	Unit	Material	Labor	Equipment	Total	Total Incl O&P
0010	**TRACK ACCESSORIES**									
0020	Car bumpers, test	B-14	2	24	Ea.	3,650	1,050	107	4,807	5,850
0100	Heavy duty		2	24		6,925	1,050	107	8,082	9,425
0200	Derails hand throw (sliding)		10	4.800		1,525	212	21.50	1,758.50	2,050
0300	Hand throw with standard timbers, open stand & target		8	6	↓	1,650	266	26.50	1,942.50	2,250
2400	Wheel stops, fixed		18	2.667	Pr.	920	118	11.85	1,049.85	1,225
2450	Hinged	↓	14	3.429	"	1,250	152	15.25	1,417.25	1,625

34 11 93.60 Track Material

		Crew	Daily Output	Labor-Hours	Unit	Material	Labor	Equipment	Total	Total Incl O&P
0010	**TRACK MATERIAL**									
0020	Track bolts				Ea.	3.64			3.64	4
0100	Joint bars				Pr.	90			90	98.50
0200	Spikes				Ea.	2.37			2.37	2.61
0300	Tie plates				"	15			15	16.50

34 41 Roadway Signaling and Control Equipment

34 41 13 – Traffic Signals

34 41 13.10 Traffic Signals Systems		Crew	Daily Output	Labor-Hours	Unit	Material	2020 Bare Costs Labor	Equipment	Total	Total Incl O&P
0010	**TRAFFIC SIGNALS SYSTEMS**									
0020	Component costs									
0600	Crew employs crane/directional driller as required									
1000	Vertical mast with foundation									
1010	Mast sized for single arm to 40′; no lighting or power function	R-11	.50	112	Signal	10,700	6,525	1,275	18,500	23,300
1100	Horizontal arm									
1110	Per linear foot of arm	R-11	50	1.120	Signal	230	65.50	12.75	308.25	370
1200	Traffic signal									
1210	Includes signal, bracket, sensor, and wiring	R-11	2.50	22.400	Signal	1,150	1,300	255	2,705	3,575
1300	Pedestrian signals and callers									
1310	Includes four signals with brackets and two call buttons	R-11	2.50	22.400	Signal	3,775	1,300	255	5,330	6,450
1400	Controller, design, and underground conduit									
1410	Includes miscellaneous signage and adjacent surface work	R-11	.25	224	Signal	25,700	13,100	2,550	41,350	51,500

34 43 Airfield Signaling and Control Equipment

34 43 13 – Airfield Signals and Lighting

34 43 13.16 Airfield Runway and Taxiway Inset Lighting

Line	Description	Crew	Daily Output	Labor-Hours	Unit	Material	Labor	Equipment	Total	Total Incl O&P
0010	**AIRFIELD RUNWAY AND TAXIWAY INSET LIGHTING**									
0100	Runway centerline, bidir., semi-flush, 200 W, w/shallow insert base	R-22	12.40	3.006	Ea.	2,350	169		2,519	2,850
0110	for mounting in base housing		18.60	2.004		1,125	113		1,238	1,425
0120	Flush, 200 W, w/shallow insert base		12.40	3.006		2,125	169		2,294	2,600
0130	for mounting in base housing		18.64	2		1,075	112		1,187	1,375
0140	45 W, for mounting in base housing		18.64	2		1,075	112		1,187	1,375
0150	Touchdown zone light, unidirectional, 200 W, w/shallow insert base		12.40	3.006		1,925	169		2,094	2,375
0160	115 W		12.40	3.006		1,825	169		1,994	2,275
0170	Bidirectional, 62 W		12.40	3.006		1,950	169		2,119	2,400
0180	Unidirectional, 200 W, for mounting in base housing		18.60	2.004		840	113		953	1,100
0190	115 W		18.60	2.004		905	113		1,018	1,175
0200	Bidirectional, 62 W		18.60	2.004		1,000	113		1,113	1,275
0210	Runway edge & threshold light, bidir., 200 W, for base housing		9.36	3.983		1,150	224		1,374	1,600
0220	300 W		9.36	3.983		1,600	224		1,824	2,100
0230	499 W		7.44	5.011		2,650	281		2,931	3,350
0240	Threshold & approach light, unidir., 200 W, for base housing		9.36	3.983		650	224		874	1,050
0250	499 W		7.44	5.011		805	281		1,086	1,325
0260	Runway edge, bidirectional, 2-115 W, for base housing		12.40	3.006		1,725	169		1,894	2,150
0270	2-185 W		12.40	3.006		1,900	169		2,069	2,350
0280	Runway threshold & end, bidir., 2-115 W, for base housing		12.40	3.006		1,900	169		2,069	2,325
0370	45 W, flush, for mounting in base housing		18.64	2		925	112		1,037	1,200
0380	115 W		18.64	2		950	112		1,062	1,225

34 43 23 – Weather Observation Equipment

34 43 23.16 Airfield Wind Cones

Line	Description	Crew	Daily Output	Labor-Hours	Unit	Material	Labor	Equipment	Total	Total Incl O&P
0010	**AIRFIELD WIND CONES**									
1200	Wind cone, 12′ lighted assembly, rigid, w/obstruction light	R-21	1.36	24.118	Ea.	11,900	1,475	84	13,459	15,400
1210	Without obstruction light		1.52	21.579		9,950	1,325	75.50	11,350.50	13,000
1220	Unlighted assembly, w/obstruction light		1.68	19.524		10,800	1,200	68	12,068	13,800
1230	Without obstruction light		1.84	17.826		9,950	1,100	62	11,112	12,600
1240	Wind cone slip fitter, 2-1/2″ pipe		21.84	1.502		132	92	5.25	229.25	293
1250	Wind cone sock, 12′ x 3′, cotton		6.56	5		720	305	17.45	1,042.45	1,300
1260	Nylon		6.56	5		690	305	17.45	1,012.45	1,250

34 71 13.17 Security Vehicle Barriers		Crew	Daily Output	Labor-Hours	Unit	Material	Labor	Equipment	Total	Total Incl O&P
						2020 Bare Costs				
0010	SECURITY VEHICLE BARRIERS									
0020	Security planters excludes filling material									
0100	Concrete security planter, exposed aggregate 36" diam. x 30" high	B-11M	8	2	Ea.	555	99	29	683	800
0200	48" diam. x 36" high		8	2		790	99	29	918	1,050
0300	53" diam. x 18" high		8	2		720	99	29	848	980
0400	72" diam. x 18" high with seats		8	2		1,975	99	29	2,103	2,375
0450	84" diam. x 36" high		8	2		2,075	99	29	2,203	2,500
0500	36" x 36" x 24" high square		8	2		460	99	29	588	695
0600	36" x 36" x 30" high square		8	2		580	99	29	708	825
0700	48" L x 24" W x 30" H rectangle		8	2		485	99	29	613	720
0800	72" L x 24" W x 30" H rectangle		8	2		580	99	29	708	825
0900	96" L x 24" W x 30" H rectangle		8	2		745	99	29	873	1,000
0950	Decorative geometric concrete barrier, 96" L x 24" W x 36" H		8	2		820	99	29	948	1,100
1000	Concrete security planter, filling with washed sand or gravel <1 C.Y.		8	2		34	99	29	162	225
1050	2 C.Y. or less		6	2.667		67.50	132	39	238.50	325
1200	Jersey concrete barrier, 10' L x 2' by 0.5' W x 30" H	B-21B	16	2.500		395	115	29.50	539.50	650
1300	10 or more same site		24	1.667		395	76.50	19.65	491.15	580
1400	10' L x 2' by 0.5' W x 32" H		16	2.500		820	115	29.50	964.50	1,125
1500	10 or more same site		24	1.667		820	76.50	19.65	916.15	1,050
1600	20' L x 2' by 0.5' W x 30" H		12	3.333		790	153	39.50	982.50	1,150
1700	10 or more same site		18	2.222		790	102	26	918	1,050
1800	20' L x 2' by 0.5' W x 32" H		12	3.333		695	153	39.50	887.50	1,050
1900	10 or more same site		18	2.222		695	102	26	823	955
2000	GFRC decorative security barrier per 10' section including concrete		4	10		3,625	460	118	4,203	4,850
2100	Per 12' section including concrete		4	10		4,350	460	118	4,928	5,650
2210	GFRC decorative security barrier will stop 30 MPH, 4,000 lb. vehicle									
2300	High security barrier base prep per 12' section on bare ground	B-11C	4	4	Ea.	25	198	53.50	276.50	400
2310	GFRC barrier base prep does not include haulaway of excavated matl.									
2400	GFRC decorative high security barrier per 12' section w/concrete	B-21B	3	13.333	Ea.	5,850	610	157	6,617	7,575
2410	GFRC decorative high security barrier will stop 50 MPH, 15,000 lb. vehicle									
2500	GFRC decorative impaler security barrier per 10' section w/prep	B-6	4	6	Ea.	2,275	274	53.50	2,602.50	3,000
2600	Per 12' section w/prep	"	4	6	"	2,750	274	53.50	3,077.50	3,525
2610	Impaler barrier should stop 50 MPH, 15,000 lb. vehicle w/some penetr.									
2700	Pipe bollards, steel, concrete filled/painted, 8' L x 4' D hole, 8" diam.	B-6	10	2.400	Ea.	865	110	21.50	996.50	1,150
2710	Schedule 80 concrete bollards will stop 4,000 lb. vehicle @ 30 MPH									
2800	GFRC decorative jersey barrier cover per 10' section excludes soil	B-6	8	3	Ea.	1,000	137	26.50	1,163.50	1,350
2900	Per 12' section excludes soil		8	3		1,225	137	26.50	1,388.50	1,575
3000	GFRC decorative 8" diameter bollard cover		12	2		390	91.50	17.80	499.30	595
3100	GFRC decorative barrier face 10' section excludes earth backing		10	2.400		980	110	21.50	1,111.50	1,275
3200	Drop arm crash barrier, 15,000 lb. vehicle @ 50 MPH									
3205	Includes all material, labor for complete installation									
3210	12' width				Ea.				64,500	71,000
3310	24' width								86,000	95,000
3410	Wedge crash barrier, 10' width								98,000	108,000
3510	12.5' width								108,000	119,000
3520	15' width								118,500	130,000
3610	Sliding crash barrier, 12' width								178,500	196,000
3710	Sliding roller crash barrier, 20' width								198,000	217,500
3810	Sliding cantilever crash barrier, 20' width								198,000	217,500
3890	Note: Raised bollard crash barriers should be used w/tire shredders									
3910	Raised bollard crash barrier, 10' width				Ea.				38,800	42,900
4010	12' width								44,900	49,000
4110	Raised bollard crash barrier, 10' width, solar powered								46,900	52,000

34 71 Roadway Construction

34 71 13 – Vehicle Barriers

34 71 13.17 Security Vehicle Barriers		Crew	Daily Output	Labor-Hours	Unit	Material	2020 Bare Costs Labor	Equipment	Total	Total Incl O&P
4210	12′ width				Ea.				53,000	58,000
4310	In ground tire shredder, 16′ width								43,900	47,900

34 71 13.26 Vehicle Guide Rails

		Crew	Daily Output	Labor-Hours	Unit	Material	Labor	Equipment	Total	Total Incl O&P
0010	**VEHICLE GUIDE RAILS**									
0012	Corrugated stl., galv. stl. posts, 6′-3″ OC	B-80	850	.038	L.F.	27.50	1.75	1.20	30.45	34
0200	End sections, galvanized, flared		50	.640	Ea.	99	30	20.50	149.50	179
0300	Wrap around end		50	.640	″	141	30	20.50	191.50	226
0400	Timber guide rail, 4″ x 8″ with 6″ x 8″ wood posts, treated		960	.033	L.F.	13.45	1.55	1.06	16.06	18.40
0600	Cable guide rail, 3 at 3/4″ cables, steel posts, single face		900	.036		11.75	1.66	1.13	14.54	16.85
0700	Wood posts		950	.034		13.30	1.57	1.07	15.94	18.30
0900	Guide rail, steel box beam, 6″ x 6″		120	.267		37.50	12.45	8.50	58.45	70
1100	Median barrier, steel box beam, 6″ x 8″		215	.149		52	6.95	4.75	63.70	73.50
1400	Resilient guide fence and light shield, 6′ high	B-2	130	.308		22	13.10		35.10	45.50
1500	Concrete posts, individual, 6′-5″, triangular	B-80	110	.291	Ea.	71.50	13.55	9.30	94.35	111
1550	Square	″	110	.291	″	77	13.55	9.30	99.85	117
2000	Median, precast concrete, 3′-6″ high, 2′ wide, single face	B-29	380	.147	L.F.	60.50	6.80	2.23	69.53	80
2200	Double face	″	340	.165	″	69	7.60	2.49	79.09	91
2400	Speed bumps, thermoplastic, 10-1/2″ x 2-1/4″ x 48″ long	B-2	120	.333	Ea.	112	14.15		126.15	146
3030	Impact barrier, MUTCD, barrel type	B-16	30	1.067	″	405	47.50	19.10	471.60	540

34 71 19 – Vehicle Delineators

34 71 19.13 Fixed Vehicle Delineators

		Crew	Daily Output	Labor-Hours	Unit	Material	Labor	Equipment	Total	Total Incl O&P
0010	**FIXED VEHICLE DELINEATORS**									
0020	Crash barriers									
0100	Traffic channelizing pavement markers, layout only	A-7	2000	.012	Ea.		.69	.02	.71	1.10
0110	13″ x 7-1/2″ x 2-1/2″ high, non-plowable, install	2 Clab	96	.167		29.50	7		36.50	44
0200	8″ x 8″ x 3-1/4″ high, non-plowable, install		96	.167		29	7		36	43.50
0230	4″ x 4″ x 3/4″ high, non-plowable, install		120	.133		2.68	5.60		8.28	11.95
0240	9-1/4″ x 5-7/8″ x 1/4″ high, plowable, concrete pavmt.	A-2A	70	.343		4.78	15	4.40	24.18	34
0250	9-1/4″ x 5-7/8″ x 1/4″ high, plowable, asphalt pavmt.	″	120	.200		3.93	8.75	2.56	15.24	21
0300	Barrier and curb delineators, reflectorized, 4″ x 6″	2 Clab	150	.107		3.66	4.49		8.15	11.25
0310	3″ x 5″	″	150	.107		5	4.49		9.49	12.70
0500	Rumble strip, polycarbonate									
0510	24″ x 3-1/2″ x 1/2″ high	2 Clab	50	.320	Ea.	10.25	13.45		23.70	33

34 72 Railway Construction

34 72 16 – Railway Siding

34 72 16.50 Railroad Sidings

		Crew	Daily Output	Labor-Hours	Unit	Material	Labor	Equipment	Total	Total Incl O&P
0010	**RAILROAD SIDINGS** R347216-10									
0800	Siding, yard spur, level grade									
0820	100 lb. new rail	B-14	57	.842	L.F.	132	37.50	3.75	173.25	209
1020	100 lb. new rail	″	22	2.182	″	193	96.50	9.70	299.20	375

34 72 16.60 Railroad Turnouts

		Crew	Daily Output	Labor-Hours	Unit	Material	Labor	Equipment	Total	Total Incl O&P
0010	**RAILROAD TURNOUTS**									
2200	Turnout, #8 complete, w/rails, plates, bars, frog, switch point,									
2250	timbers, and ballast to 6″ below bottom of ties									
2280	90 lb. rails	B-13	.25	224	Ea.	40,900	10,300	2,325	53,525	64,000
2290	90 lb. relay rails		.25	224		27,500	10,300	2,325	40,125	49,300
2300	100 lb. rails		.25	224		45,600	10,300	2,325	58,225	69,000
2310	100 lb. relay rails		.25	224		30,700	10,300	2,325	43,325	52,500
2320	110 lb. rails		.25	224		50,500	10,300	2,325	63,125	74,500

34 72 Railway Construction

34 72 16 – Railway Siding

34 72 16.60 Railroad Turnouts		Crew	Daily Output	Labor-Hours	Unit	2020 Bare Costs Material	Labor	Equipment	Total	Total Incl O&P
2330	110 lb. relay rails	B-13	.25	224	Ea.	33,800	10,300	2,325	46,425	56,000
2340	115 lb. rails		.25	224		55,000	10,300	2,325	67,625	79,500
2350	115 lb. relay rails		.25	224		35,800	10,300	2,325	48,425	58,500
2360	132 lb. rails		.25	224		63,000	10,300	2,325	75,625	88,000
2370	132 lb. relay rails	▼	.25	224	▼	41,700	10,300	2,325	54,325	65,000

34 77 Transportation Equipment

34 77 13 – Passenger Boarding Bridges

34 77 13.16 Movable Aircraft Passenger Boarding Bridges		Crew	Daily Output	Labor-Hours	Unit	Material	Labor	Equipment	Total	Total Incl O&P
0010	**MOVABLE AIRCRAFT PASSENGER BOARDING BRIDGES**									
0100	Aircraft movable passenger bridge	L-5B	.05	1440	Ea.	921,500	86,000	29,700	1,037,200	1,182,000

Estimating Tips

35 01 50 Operation and Maintenance of Marine Construction

Includes unit price lines for pile cleaning and pile wrapping for protection.

35 20 16 Hydraulic Gates

This subdivision includes various types of gates that are commonly used in waterway and canal construction. Various earthwork and structural support items are found in Division 31, and concrete work is found in Division 3.

35 20 23 Dredging

This subdivision includes barge and shore dredging systems for rivers, canals, and channels.

35 31 00 Shoreline Protection

This subdivision includes breakwaters, bulkheads, and revetments for ocean and river inlets. Additional earthwork may be required from Division 31, as well as concrete work from Division 3.

35 41 00 Levees

Contains information on levee construction, including the estimated cost of clay cone material.

35 49 00 Waterway Structures

This subdivision includes breakwaters and bulkheads for canals.

35 51 00 Floating Construction

This section includes floating piers, docks, and dock accessories. Fixed Pier Timber Construction is found in Section 06 13 33. Driven piles are found in Division 31, as are sheet piling, cofferdams, and riprap.

Reference Numbers

Reference numbers are shown at the beginning of some major classifications. These numbers refer to related items in the Reference Section. The reference information may be an estimating procedure, an alternate pricing method, or technical information.

Note: Not all subdivisions listed here necessarily appear. ■

35 22 Hydraulic Gates

35 22 26 – Sluice Gates

35 22 26.16 Hydraulic Sluice Gates		Crew	Daily Output	Labor-Hours	Unit	2020 Bare Costs Material	Labor	Equipment	Total	Total Incl O&P
0010	**HYDRAULIC SLUICE GATES**									
0100	Heavy duty, self contained w/crank oper. gate, 18" x 18"	L-5A	1.70	18.824	Ea.	7,625	1,100	675	9,400	10,900
0110	24" x 24"		1.20	26.667		11,200	1,550	955	13,705	15,900
0120	30" x 30"		1	32		12,400	1,875	1,150	15,425	17,900
0130	36" x 36"		.90	35.556		15,300	2,075	1,275	18,650	21,600
0140	42" x 42"		.80	40		17,500	2,350	1,425	21,275	24,600
0150	48" x 48"		.50	64		18,100	3,750	2,300	24,150	28,500
0160	54" x 54"		.40	80		29,500	4,675	2,850	37,025	43,100
0170	60" x 60"		.30	107		26,900	6,250	3,825	36,975	43,900
0180	66" x 66"		.30	107		31,600	6,250	3,825	41,675	49,100
0190	72" x 72"		.20	160		35,100	9,375	5,725	50,200	60,000
0200	78" x 78"		.20	160		47,600	9,375	5,725	62,700	74,000
0210	84" x 84"		.10	320		48,900	18,700	11,400	79,000	96,500
0220	90" x 90"	E-20	.30	213		58,000	12,200	6,750	76,950	91,000
0230	96" x 96"		.30	213		62,000	12,200	6,750	80,950	95,000
0240	108" x 108"		.20	320		70,500	18,300	10,100	98,900	118,500
0250	120" x 120"		.10	640		83,000	36,600	20,300	139,900	173,500
0260	132" x 132"		.10	640		97,500	36,600	20,300	154,400	189,500

35 22 63 – Through-Levee Access Gates

35 22 63.16 Canal Gates

		Crew	Daily Output	Labor-Hours	Unit	Material	Labor	Equipment	Total	Total Incl O&P
0010	**CANAL GATES**									
0011	Cast iron body, fabricated frame									
0100	12" diameter	L-5A	4.60	6.957	Ea.	1,200	405	249	1,854	2,250
0110	18" diameter		4	8		1,425	470	286	2,181	2,650
0120	24" diameter		3.50	9.143		2,075	535	325	2,935	3,500
0130	30" diameter		2.80	11.429		2,700	670	410	3,780	4,475
0140	36" diameter		2.30	13.913		5,075	815	495	6,385	7,450
0150	42" diameter		1.70	18.824		9,375	1,100	675	11,150	12,800
0160	48" diameter		1.20	26.667		10,500	1,550	955	13,005	15,200
0170	54" diameter		.90	35.556		12,000	2,075	1,275	15,350	18,100
0180	60" diameter		.50	64		12,600	3,750	2,300	18,650	22,500
0190	66" diameter		.50	64		17,100	3,750	2,300	23,150	27,400
0200	72" diameter		.40	80		21,300	4,675	2,850	28,825	34,100

35 22 66 – Flap Gates, Hydraulic

35 22 66.16 Flap Gates

		Crew	Daily Output	Labor-Hours	Unit	Material	Labor	Equipment	Total	Total Incl O&P
0010	**FLAP GATES**									
0100	Aluminum, 18" diameter	L-5A	5	6.400	Ea.	2,400	375	229	3,004	3,500
0110	24" diameter		4	8		2,800	470	286	3,556	4,150
0120	30" diameter		3.50	9.143		3,900	535	325	4,760	5,500
0130	36" diameter		2.80	11.429		4,775	670	410	5,855	6,775
0140	42" diameter		2.30	13.913		4,625	815	495	5,935	6,975
0150	48" diameter		1.70	18.824		5,425	1,100	675	7,200	8,500
0160	54" diameter		1.20	26.667		6,000	1,550	955	8,505	10,200
0170	60" diameter		.80	40		6,425	2,350	1,425	10,200	12,500
0180	66" diameter		.50	64		6,875	3,750	2,300	12,925	16,200
0190	72" diameter		.40	80		8,000	4,675	2,850	15,525	19,500

35 22 69 – Knife Gates, Hydraulic

35 22 69.16 Knife Gates

		Crew	Daily Output	Labor-Hours	Unit	Material	Labor	Equipment	Total	Total Incl O&P
0010	**KNIFE GATES**									
0100	Incl. handwheel operator for hub, 6" diameter	Q-23	7.70	3.117	Ea.	1,050	195	193	1,438	1,675
0110	8" diameter		7.20	3.333		1,575	209	207	1,991	2,275

35 22 Hydraulic Gates

35 22 69 – Knife Gates, Hydraulic

35 22 69.16 Knife Gates		Crew	Daily Output	Labor-Hours	Unit	Material	2020 Bare Costs Labor	Equipment	Total	Total Incl O&P
0120	10" diameter	Q-23	4.80	5	Ea.	2,400	315	310	3,025	3,450
0130	12" diameter		3.60	6.667		3,725	415	415	4,555	5,200
0140	14" diameter		3.40	7.059		4,425	440	440	5,305	6,050
0150	16" diameter		3.20	7.500		5,450	470	465	6,385	7,225
0160	18" diameter		3	8		10,300	500	495	11,295	12,600
0170	20" diameter		2.70	8.889		9,625	555	550	10,730	12,100
0180	24" diameter		2.40	10		15,000	625	620	16,245	18,200
0190	30" diameter		1.80	13.333		29,900	835	825	31,560	35,000
0200	36" diameter		1.20	20		45,100	1,250	1,250	47,600	53,000

35 22 73 – Slide Gates, Hydraulic

35 22 73.16 Slide Gates		Crew	Daily Output	Labor-Hours	Unit	Material	Labor	Equipment	Total	Total Incl O&P
0010	**SLIDE GATES**									
0100	Steel, self contained incl. anchor bolts and grout, 12" x 12"	L-5A	4.60	6.957	Ea.	4,700	405	249	5,354	6,100
0110	18" x 18"		4	8		5,250	470	286	6,006	6,850
0120	24" x 24"		3.50	9.143		5,575	535	325	6,435	7,350
0130	30" x 30"		2.80	11.429		6,550	670	410	7,630	8,750
0140	36" x 36"		2.30	13.913		7,000	815	495	8,310	9,575
0150	42" x 42"		1.70	18.824		7,350	1,100	675	9,125	10,600
0160	48" x 48"		1.20	26.667		7,650	1,550	955	10,155	12,000
0170	54" x 54"		.90	35.556		7,700	2,075	1,275	11,050	13,300
0180	60" x 60"		.55	58.182		9,025	3,400	2,075	14,500	17,800
0190	72" x 72"		.36	88.889		11,200	5,200	3,175	19,575	24,200

35 24 Dredging

35 24 13 – Suction Dredging

35 24 13.13 Cutter Suction Dredging		Crew	Daily Output	Labor-Hours	Unit	Material	Labor	Equipment	Total	Total Incl O&P
0010	**CUTTER SUCTION DREDGING**									
0015	Add, Marine Equipment Rental, See Section 01 54 33.80									
1000	Hydraulic method, pumped 1,000' to shore dump, minimum	B-57	460	.104	B.C.Y.		5.05	3.94	8.99	12.35
1100	Maximum		310	.155			7.50	5.85	13.35	18.35
1400	Into scows dumped 20 miles, minimum		425	.113			5.50	4.27	9.77	13.35
1500	Maximum		243	.198			9.60	7.45	17.05	23.50
1600	For inland rivers and canals in South, deduct								30%	30%

35 24 23 – Clamshell Dredging

35 24 23.13 Mechanical Dredging		Crew	Daily Output	Labor-Hours	Unit	Material	Labor	Equipment	Total	Total Incl O&P
0010	**MECHANICAL DREDGING**									
0015	Add, Marine Equipment Rental, See Section 01 54 33.80									
0020	Dredging mobilization and demobilization, add to below, minimum	B-8	.53	121	Total		5,900	5,475	11,375	15,400
0100	Maximum	"	.10	640	"		31,200	29,000	60,200	81,500
0300	Barge mounted clamshell excavation into scows									
0310	Dumped 20 miles at sea, minimum	B-57	310	.155	B.C.Y.		7.50	5.85	13.35	18.35
0400	Maximum	"	213	.225	"		10.95	8.50	19.45	26.50
0500	Barge mounted dragline or clamshell, hopper dumped,									
0510	pumped 1,000' to shore dump, minimum	B-57	340	.141	B.C.Y.		6.85	5.35	12.20	16.70
0525	All pumping uses 2,000 gallons of water per cubic yard									
0600	Maximum	B-57	243	.198	B.C.Y.		9.60	7.45	17.05	23.50

35 31 Shoreline Protection

35 31 16 – Seawalls

35 31 16.13 Concrete Seawalls		Crew	Daily Output	Labor-Hours	Unit	2020 Bare Costs Material	Labor	Equipment	Total	Total Incl O&P
0010	**CONCRETE SEAWALLS**									
0011	Reinforced concrete									
0015	include footing and tie-backs									
0020	Up to 6′ high, minimum	C-17C	28	2.964	L.F.	64.50	164	19.90	248.40	355
0060	Maximum		24.25	3.423		103	190	23	316	440
0100	12′ high, minimum		20	4.150		167	230	28	425	580
0160	Maximum		18.50	4.486		193	249	30	472	640

35 31 16.19 Steel Sheet Piling Seawalls

		Crew	Daily Output	Labor-Hours	Unit	Material	Labor	Equipment	Total	Total Incl O&P
0010	**STEEL SHEET PILING SEAWALLS**									
0200	Steel sheeting, with 4′ x 4′ x 8″ concrete deadmen, @ 10′ OC									
0210	12′ high, shore driven	B-40	27	2.370	L.F.	125	131	129	385	485
0260	Barge driven	B-76	15	4.800	″	187	265	246	698	895

35 31 19 – Revetments

35 31 19.18 Revetments, Concrete

		Crew	Daily Output	Labor-Hours	Unit	Material	Labor	Equipment	Total	Total Incl O&P
0010	**REVETMENTS, CONCRETE**									
0100	Concrete revetment matt 8′ x 20′ x 4-1/2″ excluding site preparation									
0110	Includes all labor, material and equip. for complete installation	B-57	5	9.600	Ea.	2,650	465	365	3,480	4,050

35 51 Floating Construction

35 51 13 – Floating Piers

35 51 13.23 Floating Wood Piers

		Crew	Daily Output	Labor-Hours	Unit	Material	Labor	Equipment	Total	Total Incl O&P
0010	**FLOATING WOOD PIERS**									
0020	Polyethylene encased polystyrene, no pilings included	F-3	330	.121	S.F.	30.50	6.60	1.43	38.53	45.50
0200	Pile supported, shore constructed, minimum		130	.308		29.50	16.75	3.62	49.87	63
0250	Maximum		120	.333		32.50	18.10	3.93	54.53	69
0400	Floating, small boat, prefab, no shore facilities, minimum		250	.160		27.50	8.70	1.88	38.08	46.50
0500	Maximum		150	.267		59.50	14.50	3.14	77.14	92
0700	Per slip, minimum (180 S.F. each)		1.59	25.157	Ea.	5,850	1,375	296	7,521	8,925
0800	Maximum		1.40	28.571	″	9,100	1,550	335	10,985	12,800

Estimating Tips

Products such as conveyors, material handling cranes and hoists, and other items specified in this division require trained installers. The general contractor may not have any choice as to who will perform the installation or when it will be performed. Long lead times are often required for these products, making early decisions in purchasing and scheduling necessary. The installation of this type of equipment may require the embedment of mounting hardware during the construction of floors, structural walls, or interior walls/partitions. Electrical connections will require coordination with the electrical contractor.

Reference Numbers

Reference numbers are shown at the beginning of some major classifications. These numbers refer to related items in the Reference Section. The reference information may be an estimating procedure, an alternate pricing method, or technical information.

41 21 Conveyors

41 21 23 – Piece Material Conveyors

41 21 23.16 Container Piece Material Conveyors		Crew	Daily Output	Labor-Hours	Unit	Material	Labor	Equipment	Total	Total Incl O&P
						2020 Bare Costs				
0010	CONTAINER PIECE MATERIAL CONVEYORS									
0020	Gravity fed, 2″ rollers, 3″ OC									
0050	10′ sections with 2 supports, 600 lb. capacity, 18″ wide				Ea.	530			530	580
0100	24″ wide					595			595	655
0150	1,400 lb. capacity, 18″ wide					705			705	775
0200	24″ wide					595			595	655
0350	Horizontal belt, center drive and takeup, 60 fpm									
0400	16″ belt, 26.5′ length	2 Mill	.50	32	Ea.	3,675	1,800		5,475	6,800
0450	24″ belt, 41.5′ length		.40	40		5,525	2,225		7,750	9,550
0500	61.5′ length		.30	53.333		7,700	2,975		10,675	13,100
0600	Inclined belt, 10′ rise with horizontal loader and									
0620	End idler assembly, 27.5′ length, 18″ belt	2 Mill	.30	53.333	Ea.	7,925	2,975		10,900	13,300
0700	24″ belt	″	.15	107	″	9,975	5,975		15,950	20,200
3600	Monorail, overhead, manual, channel type									
3700	125 lb./L.F.	1 Mill	26	.308	L.F.	20	17.20		37.20	48.50
3900	500 lb./L.F.	″	21	.381	″	19.60	21.50		41.10	54.50
4000	Trolleys for above, 2 wheel, 125 lb. capacity				Ea.	92			92	101
4200	4 wheel, 250 lb. capacity					405			405	445
4300	8 wheel, 500 lb. capacity					805			805	885

41 22 Cranes and Hoists

41 22 13 – Cranes

41 22 13.10 Crane Rail

		Crew	Daily Output	Labor-Hours	Unit	Material	Labor	Equipment	Total	Total Incl O&P
0010	CRANE RAIL									
0020	Box beam bridge, no equipment included	E-4	3400	.009	Lb.	1.44	.55	.04	2.03	2.53
0200	Running track only, 104 lb. per yard		5600	.006	″	.72	.33	.03	1.08	1.36
0210	Running track only, 104 lb. per yard, 20′ piece		160	.200	L.F.	25	11.65	.92	37.57	47.50

41 22 13.13 Bridge Cranes

		Crew	Daily Output	Labor-Hours	Unit	Material	Labor	Equipment	Total	Total Incl O&P
0010	BRIDGE CRANES									
0100	1 girder, 20′ span, 3 ton	M-3	1	34	Ea.	28,000	2,200	146	30,346	34,300
0125	5 ton		1	34		30,900	2,200	146	33,246	37,400
0150	7.5 ton		1	34		36,400	2,200	146	38,746	43,600
0175	10 ton		.80	42.500		48,900	2,725	182	51,807	58,000
0176	12.5 ton		.80	42.500		58,500	2,725	182	61,407	69,000
0200	15 ton		.80	42.500		63,000	2,725	182	65,907	74,000
0225	30′ span, 3 ton		1	34		29,200	2,200	146	31,546	35,600
0250	5 ton		1	34		32,200	2,200	146	34,546	39,000
0275	7.5 ton		1	34		38,300	2,200	146	40,646	45,700
0300	10 ton		.80	42.500		50,500	2,725	182	53,407	60,000
0325	15 ton		.80	42.500		66,000	2,725	182	68,907	77,000
0350	2 girder, 40′ span, 3 ton	M-4	.50	72		47,500	4,600	375	52,475	59,500
0375	5 ton		.50	72		49,800	4,600	375	54,775	62,500
0400	7.5 ton		.50	72		54,500	4,600	375	59,475	67,500
0425	10 ton		.40	90		52,000	5,750	470	58,220	67,000
0450	15 ton		.40	90		88,000	5,750	470	94,220	106,500
0475	25 ton		.30	120		103,000	7,650	630	111,280	125,500
0500	50′ span, 3 ton		.50	72		51,500	4,600	375	56,475	64,500
0525	5 ton		.50	72		56,500	4,600	375	61,475	70,000
0550	7.5 ton		.50	72		61,000	4,600	375	65,975	74,500
0575	10 ton		.40	90		70,500	5,750	470	76,720	87,000
0600	15 ton		.40	90		92,000	5,750	470	98,220	111,000

41 22 Cranes and Hoists

41 22 13 – Cranes

41 22 13.13 Bridge Cranes		Crew	Daily Output	Labor-Hours	Unit	Material	Labor	Equipment	Total	Total Incl O&P
						2020 Bare Costs				
0625	25 ton	M-4	.30	120	Ea.	108,000	7,650	630	116,280	131,000

41 22 13.19 Jib Cranes

		Crew	Daily Output	Labor-Hours	Unit	Material	Labor	Equipment	Total	Total Incl O&P
0010	**JIB CRANES**									
0020	Jib crane, wall cantilever, 500 lb. capacity, 8′ span	2 Mill	1	16	Ea.	1,325	895		2,220	2,825
0040	12′ span		1	16		1,450	895		2,345	2,975
0060	16′ span		1	16		1,700	895		2,595	3,225
0080	20′ span		1	16		2,025	895		2,920	3,600
0100	1,000 lb. capacity, 8′ span		1	16		1,425	895		2,320	2,925
0120	12′ span		1	16		1,550	895		2,445	3,075
0130	16′ span		1	16		2,150	895		3,045	3,725
0150	20′ span		1	16		2,400	895		3,295	4,000

41 22 23 – Hoists

41 22 23.10 Material Handling

		Crew	Daily Output	Labor-Hours	Unit	Material	Labor	Equipment	Total	Total Incl O&P
0010	**MATERIAL HANDLING**, cranes, hoists and lifts									
1500	Cranes, portable hydraulic, floor type, 2,000 lb. capacity				Ea.	3,700			3,700	4,075
1600	4,000 lb. capacity					5,075			5,075	5,575
1800	Movable gantry type, 12′ to 15′ range, 2,000 lb. capacity					4,775			4,775	5,250
1900	6,000 lb. capacity					7,450			7,450	8,200
2100	Hoists, electric overhead, chain, hook hung, 15′ lift, 1 ton cap.					2,825			2,825	3,100
2200	3 ton capacity					3,800			3,800	4,200
2500	5 ton capacity					8,350			8,350	9,200
2600	For hand-pushed trolley, add					15%				
2700	For geared trolley, add					30%				
2800	For motor trolley, add					75%				
3000	For lifts over 15′, 1 ton, add				L.F.	25.50			25.50	28
3100	5 ton, add				"	78.50			78.50	86
3110	Air powered hoist, 500 lb. capacity	2 Mill	2.67	5.993	Ea.	4,025	335		4,360	4,950
3120	1,000 lb. capacity		2.46	6.504		4,025	365		4,390	5,000
3130	2,000 lb. capacity		2.29	6.987		6,975	390		7,365	8,275
3140	4,000 lb. capacity		2	8		10,300	445		10,745	12,000
3160	6,000 lb. capacity		1.88	8.511		12,400	475		12,875	14,300
3180	8,000 lb. capacity		1.68	9.524		10,800	530		11,330	12,700
3190	10,000 lb. capacity		1.60	10		13,100	560		13,660	15,300
3300	Lifts, scissor type, portable, electric, 36″ high, 2,000 lb.					3,875			3,875	4,250
3400	48″ high, 4,000 lb.					4,700			4,700	5,175

Division Notes

		CREW	DAILY OUTPUT	LABOR-HOURS	UNIT	BARE COSTS MAT.	LABOR	EQUIP.	TOTAL	TOTAL INCL O&P

Estimating Tips

This section involves equipment and construction costs for air, noise, and odor pollution control systems. These systems may be interrelated and care must be taken that the complete systems are estimated. For example, air pollution equipment may include dust and air-entrained particles that have to be collected. The vacuum systems could be noisy, requiring silencers to reduce noise pollution, and the collected solids have to be disposed of to prevent solid pollution.

Reference Numbers

Reference numbers are shown at the beginning of some major classifications. These numbers refer to related items in the Reference Section. The reference information may be an estimating procedure, an alternate pricing method, or technical information.

44 11 16 – Fugitive Dust Barrier Systems

44 11 16.10 Dust Collection Systems		Crew	Daily Output	Labor-Hours	Unit	2020 Bare Costs Material	Labor	Equipment	Total	Total Incl O&P
0010	**DUST COLLECTION SYSTEMS** Commercial/Industrial									
0120	Central vacuum units									
0130	Includes stand, filters and motorized shaker									
0200	500 CFM, 10" inlet, 2 HP	Q-20	2.40	8.333	Ea.	4,850	475		5,325	6,075
0220	1,000 CFM, 10" inlet, 3 HP		2.20	9.091		5,100	520		5,620	6,400
0240	1,500 CFM, 10" inlet, 5 HP		2	10		5,775	570		6,345	7,275
0260	3,000 CFM, 13" inlet, 10 HP		1.50	13.333		16,100	760		16,860	18,900
0280	5,000 CFM, 16" inlet, 2 @ 10 HP		1	20		17,100	1,150		18,250	20,600
1000	Vacuum tubing, galvanized									
1100	2-1/8" OD, 16 ga.	Q-9	440	.036	L.F.	3.60	2.04		5.64	7.15
1110	2-1/2" OD, 16 ga.		420	.038		3.98	2.14		6.12	7.75
1120	3" OD, 16 ga.		400	.040		4.80	2.24		7.04	8.80
1130	3-1/2" OD, 16 ga.		380	.042		6.80	2.36		9.16	11.20
1140	4" OD, 16 ga.		360	.044		7.40	2.49		9.89	12
1150	5" OD, 14 ga.		320	.050		13.30	2.80		16.10	19
1160	6" OD, 14 ga.		280	.057		14.85	3.20		18.05	21.50
1170	8" OD, 14 ga.		200	.080		21.50	4.49		25.99	30.50
1180	10" OD, 12 ga.		160	.100		48.50	5.60		54.10	62.50
1190	12" OD, 12 ga.		120	.133		51.50	7.50		59	68
1200	14" OD, 12 ga.		80	.200		65	11.20		76.20	89
1940	Hose, flexible wire reinforced rubber									
1956	3" diam.	Q-9	400	.040	L.F.	6.90	2.24		9.14	11.05
1960	4" diam.		360	.044		8.10	2.49		10.59	12.80
1970	5" diam.		320	.050		10.05	2.80		12.85	15.45
1980	6" diam.		280	.057		10.70	3.20		13.90	16.75
2000	90° elbow, slip fit									
2110	2-1/8" diam.	Q-9	70	.229	Ea.	13.05	12.80		25.85	34.50
2120	2-1/2" diam.		65	.246		18.25	13.80		32.05	41.50
2130	3" diam.		60	.267		24.50	14.95		39.45	50.50
2140	3-1/2" diam.		55	.291		30.50	16.30		46.80	59
2150	4" diam.		50	.320		38	17.95		55.95	70
2160	5" diam.		45	.356		72	19.95		91.95	110
2170	6" diam.		40	.400		99	22.50		121.50	144
2180	8" diam.		30	.533		189	30		219	254
2400	45° elbow, slip fit									
2410	2-1/8" diam.	Q-9	70	.229	Ea.	11.45	12.80		24.25	32.50
2420	2-1/2" diam.		65	.246		16.85	13.80		30.65	40
2430	3" diam.		60	.267		20.50	14.95		35.45	46
2440	3-1/2" diam.		55	.291		25.50	16.30		41.80	53.50
2450	4" diam.		50	.320		33	17.95		50.95	64.50
2460	5" diam.		45	.356		56	19.95		75.95	92.50
2470	6" diam.		40	.400		75.50	22.50		98	118
2480	8" diam.		35	.457		148	25.50		173.50	203
2800	90° TY, slip fit thru 6" diam.									
2810	2-1/8" diam.	Q-9	42	.381	Ea.	21	21.50		42.50	57
2820	2-1/2" diam.		39	.410		27.50	23		50.50	66
2830	3" diam.		36	.444		36.50	25		61.50	79.50
2840	3-1/2" diam.		33	.485		46.50	27		73.50	94
2850	4" diam.		30	.533		45.50	30		75.50	97
2860	5" diam.		27	.593		74.50	33		107.50	134
2870	6" diam.		24	.667		111	37.50		148.50	181
2880	8" diam., butt end					180			180	198
2890	10" diam., butt end					565			565	625

44 11 16.10 Dust Collection Systems		Crew	Daily Output	Labor-Hours	Unit	2020 Bare Costs Material	Labor	Equipment	Total	Total Incl O&P
2900	12″ diam., butt end				Ea.	565			565	625
2910	14″ diam., butt end					1,025			1,025	1,125
2920	6″ x 4″ diam., butt end					127			127	140
2930	8″ x 4″ diam., butt end					241			241	265
2940	10″ x 4″ diam., butt end					315			315	350
2950	12″ x 4″ diam., butt end					360			360	395
3100	90° elbow, butt end, segmented									
3110	8″ diam., butt end, segmented				Ea.	158			158	174
3120	10″ diam., butt end, segmented					470			470	515
3130	12″ diam., butt end, segmented					595			595	655
3140	14″ diam., butt end, segmented					745			745	820
3200	45° elbow, butt end, segmented									
3210	8″ diam., butt end, segmented				Ea.	148			148	163
3220	10″ diam., butt end, segmented					340			340	375
3230	12″ diam., butt end, segmented					345			345	380
3240	14″ diam., butt end, segmented					640			640	705
3400	All butt end fittings require one coupling per joint.									
3410	Labor for fitting included with couplings.									
3460	Compression coupling, galvanized, neoprene gasket									
3470	2-1/8″ diam.	Q-9	44	.364	Ea.	13.55	20.50		34.05	47
3480	2-1/2″ diam.		44	.364		13.55	20.50		34.05	47
3490	3″ diam.		38	.421		19.45	23.50		42.95	58.50
3500	3-1/2″ diam.		35	.457		22	25.50		47.50	64
3510	4″ diam.		33	.485		24	27		51	68.50
3520	5″ diam.		29	.552		27	31		58	78.50
3530	6″ diam.		26	.615		33	34.50		67.50	90.50
3540	8″ diam.		22	.727		57.50	41		98.50	127
3550	10″ diam.		20	.800		93.50	45		138.50	174
3560	12″ diam.		18	.889		115	50		165	204
3570	14″ diam.		16	1		150	56		206	253
3800	Air gate valves, galvanized									
3810	2-1/8″ diam.	Q-9	30	.533	Ea.	122	30		152	181
3820	2-1/2″ diam.		28	.571		133	32		165	196
3830	3″ diam.		26	.615		141	34.50		175.50	210
3840	4″ diam.		23	.696		164	39		203	241
3850	6″ diam.		18	.889		228	50		278	330

Division Notes

		CREW	DAILY OUTPUT	LABOR-HOURS	UNIT	BARE COSTS MAT.	LABOR	EQUIP.	TOTAL	TOTAL INCL O&P

Estimating Tips

This division contains information about water and wastewater equipment and systems, which was formerly located in Division 44. The main areas of focus are total wastewater treatment plants and components of wastewater treatment plants. Also included in this section are oil/water separators for wastewater treatment.

Reference Numbers

Reference numbers are shown at the beginning of some major classifications. These numbers refer to related items in the Reference Section. The reference information may be an estimating procedure, an alternate pricing method, or technical information.

46 07 Packaged Water and Wastewater Treatment Equipment

46 07 53 – Packaged Wastewater Treatment Equipment

46 07 53.10 Biological Pkg. Wastewater Treatment Plants		Crew	Daily Output	Labor-Hours	Unit	2020 Bare Costs Material	Labor	Equipment	Total	Total Incl O&P
0010	**BIOLOGICAL PACKAGED WASTEWATER TREATMENT PLANTS**									
0011	Not including fencing or external piping									
0020	Steel packaged, blown air aeration plants									
0100	1,000 GPD				Gal.				55	60.50
0200	5,000 GPD								22	24
0300	15,000 GPD								22	24
0400	30,000 GPD								15.40	16.95
0500	50,000 GPD								11	12.10
0600	100,000 GPD								9.90	10.90
0700	200,000 GPD								8.80	9.70
0800	500,000 GPD				↓				7.70	8.45
1000	Concrete, extended aeration, primary and secondary treatment									
1010	10,000 GPD				Gal.				22	24
1100	30,000 GPD								15.40	16.95
1200	50,000 GPD								11	12.10
1400	100,000 GPD								9.90	10.90
1500	500,000 GPD				↓				7.70	8.45
1700	Municipal wastewater treatment facility									
1720	1.0 MGD				Gal.				11	12.10
1740	1.5 MGD								10.60	11.65
1760	2.0 MGD								10	11
1780	3.0 MGD								7.80	8.60
1800	5.0 MGD				↓				5.80	6.70
2000	Holding tank system, not incl. excavation or backfill									
2010	Recirculating chemical water closet	2 Plum	4	4	Ea.	550	258		808	1,000
2100	For voltage converter, add	"	16	1		320	64.50		384.50	450
2200	For high level alarm, add	1 Plum	7.80	1.026	↓	123	66		189	238

46 07 53.20 Wastewater Treatment System

		Crew	Daily Output	Labor-Hours	Unit	Material	Labor	Equipment	Total	Total Incl O&P
0010	**WASTEWATER TREATMENT SYSTEM**									
0020	Fiberglass, 1,000 gallon	B-21	1.29	21.705	Ea.	4,450	1,050	147	5,647	6,725
0100	1,500 gallon	"	1.03	27.184	"	9,700	1,325	184	11,209	13,000

46 23 Grit Removal And Handling Equipment

46 23 23 – Vortex Grit Removal Equipment

46 23 23.10 Rainwater Filters

		Crew	Daily Output	Labor-Hours	Unit	Material	Labor	Equipment	Total	Total Incl O&P
0010	**RAINWATER FILTERS**									
0100	42 gal./min	B-21	3.50	8	Ea.	40,600	390	54	41,044	45,400
0200	65 gal./min		3.50	8		40,700	390	54	41,144	45,400
0300	208 gal./min	↓	3.50	8	↓	40,700	390	54	41,144	45,500

46 25 13 – Coalescing Oil-Water Separators

46 25 13.20 Oil/Water Separators		Crew	Daily Output	Labor-Hours	Unit	2020 Bare Costs Material	Labor	Equipment	Total	Total Incl O&P
0010	**OIL/WATER SEPARATORS**									
0020	Underground, tank only									
0030	Excludes excavation, backfill & piping									
0100	200 GPM	B-21	3.50	8	Ea.	40,400	390	54	40,844	45,100
0110	400 GPM		3.25	8.615		85,500	420	58	85,978	94,500
0120	600 GPM		2.75	10.182		90,000	495	68.50	90,563.50	100,500
0130	800 GPM		2.50	11.200		108,000	545	75.50	108,620.50	119,500
0140	1,000 GPM		2	14		122,000	685	94.50	122,779.50	135,500
0150	1,200 GPM		1.50	18.667		134,000	910	126	135,036	148,500
0160	1,500 GPM	↓	1	28	↓	154,000	1,375	189	155,564	171,500

Division Notes

	CREW	DAILY OUTPUT	LABOR-HOURS	UNIT	BARE COSTS MAT.	LABOR	EQUIP.	TOTAL	TOTAL INCL O&P

Division 48 Electrical Power Generation

Estimating Tips

- When estimating costs for the installation of electrical power generation equipment, factors to review include access to the job site, access and setting up at the installation site, required connections, uncrating pads, anchors, leveling, final assembly of the components, and temporary protection from physical damage, such as environmental exposure.
- Be aware of the costs of equipment supports, concrete pads, and vibration isolators. Cross-reference them against other trades' specifications. Also, review site and structural drawings for items that must be included in the estimates.
- It is important to include items that are not documented in the plans and specifications but must be priced. These items include, but are not limited to, testing, dust protection, roof penetration, core drilling concrete floors and walls, patching, cleanup, and final adjustments. Add a contingency or allowance for utility company fees for power hookups, if needed.
- The project size and scope of electrical power generation equipment will have a significant impact on cost. The intent of RSMeans cost data is to provide a benchmark cost so that owners, engineers, and electrical contractors will have a comfortable number with which to start a project. Additionally, there are many websites available to use for research and to obtain a vendor's quote to finalize costs.

Reference Numbers

Reference numbers are shown at the beginning of some major classifications. These numbers refer to related items in the Reference Section. The reference information may be an estimating procedure, an alternate pricing method, or technical information.

Note: Trade Service, in part, has been used as a reference source for some of the material prices used in Division 48.

48 15 Wind Energy Electrical Power Generation Equipment

48 15 13 – Wind Turbines

48 15 13.50 Wind Turbines and Components		Crew	Daily Output	Labor-Hours	Unit	2020 Bare Costs Material	Labor	Equipment	Total	Total Incl O&P
0010	**WIND TURBINES & COMPONENTS**									
0500	Complete system, grid connected									
0501	Wind turbine marine rated, 160 W, 12 V, DC	2 Elec	2.20	7.273	Ea.	1,175	445		1,620	1,975
0510	160 W, 24 V, DC		2.20	7.273		1,200	445		1,645	2,000
0520	160 W, 48 V, DC		2.20	7.273		1,175	445		1,620	1,975
1000	20 kW, 31′ diam., incl. labor & material G				System				49,900	49,900
1001	Wind turbine, 300 W, 12 V, DC	2 Elec	2.20	7.273	Ea.	900	445		1,345	1,675
1011	300 W, 24 V, DC		2.20	7.273		915	445		1,360	1,675
1020	300 W, 48 V, DC		2.20	7.273		910	445		1,355	1,675
1030	Wind turbine, 400 W, 12 V, DC		2.20	7.273		910	445		1,355	1,675
1040	400 W, 24 V, DC		2.20	7.273		910	445		1,355	1,675
1050	400 W, 48 V, DC		2.20	7.273		915	445		1,360	1,675
2000	2.4 kW, 12′ diam., incl. labor & material G				System				18,000	18,000
2900	Component system									
3200	1,000 W, 9′ diam. G	1 Elec	2.05	3.902	Ea.	1,800	239		2,039	2,350
3300	Wind turbine/charge control, 2,900-7,900 W, 5.5″ H x 11′ W, 180 lb.	2 Elec	2	8		16,600	490		17,090	19,100
3320	Wind turbine/charge control, 4,000-10,800 W, 5.5″ H x 13′ W, 240 lb.	"	2	8		27,500	490		27,990	31,100
3400	Mounting hardware									
3401	Wind turbine tower, 2 stage steel, 8′ H x 6′ W	2 Elec	2	8	Ea.	2,075	490		2,565	3,050
3500	30′ guyed tower kit G	2 Clab	5.12	3.125		415	132		547	665
3505	3′ galvanized helical earth screw G	1 Clab	8	1		55.50	42		97.50	129
3510	Attic mount kit G	1 Rofc	2.56	3.125		224	144		368	500
3520	Roof mount kit G	1 Clab	3.41	2.346		240	99		339	425
4000	Anemometer, standard AC or 9V, 2-digit LCD, wood case, +/-2% of Input	2 Elec	2	8		177	490		667	950
8900	Equipment									
9100	DC to AC inverter for, 48 V, 4,000 W G	1 Elec	2	4	Ea.	2,275	245		2,520	2,875

Assemblies Section

Table of Contents

Table of Contents

RSMeans data: Assemblies—How They Work

Assemblies estimating provides a fast and reasonably accurate way to develop construction costs. An assembly is the grouping of individual work items—with appropriate quantities—to provide a cost for a major construction component in a convenient unit of measure.

An assemblies estimate is often used during early stages of design development to compare the cost impact of various design alternatives on total building cost.

Assemblies estimates are also used as an efficient tool to verify construction estimates.

Assemblies estimates do not require a completed design or detailed drawings. Instead, they are based on the general size of the structure and other known parameters of the project. The degree of accuracy of an assemblies estimate is generally within +/- 15%.

Most assemblies consist of three major elements: a graphic, the system components, and the cost data itself. The **Graphic** is a visual representation showing the typical appearance of the assembly

❶ Unique 12-character Identifier

Our assemblies are identified by a **unique 12-character identifier**. The assemblies are numbered using UNIFORMAT II, ASTM Standard E1557. The first 5 characters represent this system to Level 3. The last 7 characters represent further breakdown in order to arrange items in understandable groups of similar tasks. Line numbers are consistent across all of our publications, so a line number in any assemblies data set will always refer to the same item.

❷ Reference Box

Information is available in the Reference Section to assist the estimator with estimating procedures, alternate pricing methods, and additional technical information.

The **Reference Box** indicates the exact location of this information in the Reference Section. The "R" stands for "reference" and the remaining characters are the line numbers.

❸ Narrative Descriptions

Our assemblies descriptions appear in two formats: narrative and table. **Narrative descriptions** are shown in a hierarchical structure to make them readable. In order to read a complete description, read up through the indents to the top of the section. Include everything that is above and to the left that is not contradicted by information below.

Narrative Format

D40 Fire Protection

D4020 Standpipes

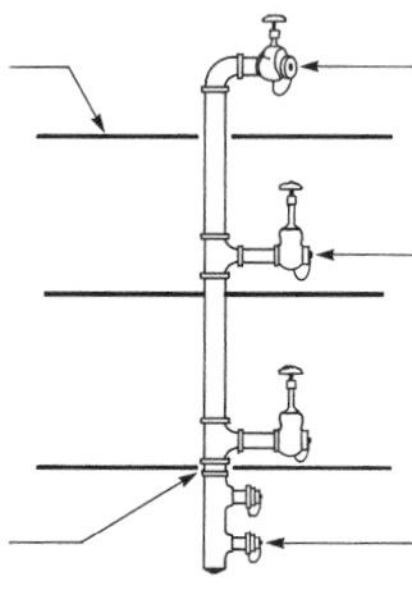

System Components ❶	QUANTITY	UNIT	COST PER FLOOR MAT.	INST.	TOTAL
SYSTEM D4020 310 0560					
WET STANDPIPE RISER, CLASS I, STEEL, BLACK, SCH. 40 PIPE, 10′ HEIGHT					
4″ DIAMETER PIPE, ONE FLOOR					
Pipe, steel, black, schedule 40, threaded, 4″ diam.	20.000	L.F.	510	800	1,310
Pipe, Tee, malleable iron, black, 150 lb. threaded, 4″ pipe size	2.000	Ea.	780	720	1,500
Pipe, 90° elbow, malleable iron, black, 150 lb threaded, 4″ pipe size	1.000	Ea.	216	239	455
Pipe, nipple, steel, black, schedule 40, 2-1/2″ pipe size x 3″ long	2.000	Ea.	28.20	179	207.20
Fire valve, gate, 300 lb., brass w/handwheel, 2-1/2″ pipe size	1.000	Ea.	239	112	351
Fire valve, pressure reducing rgh brs, 2-1/2″ pipe size	1.000	Ea.	980	224	1,204
Valve, swing check, w/ball drip, CI w/brs. ftngs., 4″ pipe size	1.000	Ea.	460	470	930
Standpipe conn wall dble. flush brs. w/plugs & chains 2-1/2″x2-1/2″x4″	1.000	Ea.	850	282	1,132
Valve, swing check, bronze, 125 lb, regrinding disc, 2-1/2″ pipe size	1.000	Ea.	1,025	95.50	1,120.50
Roof manifold, fire, w/valves & caps, horiz/vert brs 2-1/2″x2-1/2″x4″	1.000	Ea.	250	294	544
Fire, hydrolator, vent & drain, 2-1/2″ pipe size	1.000	Ea.	126	65.50	191.50
Valve, gate, iron body 125 lb., OS&Y, threaded, 4″ pipe size	1.000	Ea.	755	480	1,235
Tamper switch (valve supervisory switch)	1.000	Ea.	291	49	340
TOTAL			6,510.20	4,010	10,520.20

D4020 310	Wet Standpipe Risers, Class I	COST PER FLOOR MAT.	INST.	TOTAL
0550	Wet standpipe risers, Class I, steel, black, sch. 40, 10′ height			
0560	4″ diameter pipe, one floor	6,500	4,000	10,500
0580	❸ Additional floors	1,525	1,225	2,750
0600	6″ diameter pipe, one floor	10,200	6,975	17,175
0620	Additional floors ❷ R211226-20	2,700	1,950	4,650
0640	8″ diameter pipe, one floor	15,200	8,400	23,600
0660	Additional floors	3,900	2,350	6,250
0680				

D4020 310	Wet Standpipe Risers, Class II	COST PER FLOOR MAT.	INST.	TOTAL
1030	Wet standpipe risers, Class II, steel, black sch. 40, 10′ height			
1040	2″ diameter pipe, one floor	2,750	1,475	4,225
1060	Additional floors	880	550	1,430
1080	2-1/2″ diameter pipe, one floor	3,600	2,075	5,675
1100	Additional floors	975	640	1,615
1120				

For supplemental customizable square foot estimating forms, visit: **RSMeans.com/2020books**

in question. It is frequently accompanied by additional explanatory technical information describing the class of items. The **System Components** is a listing of the individual tasks that make up the assembly, including the quantity and unit of measure for each item, along with the cost of material and installation. The **Assemblies data** below lists prices for other similar systems with dimensional and/or size variations.

All of our assemblies costs represent the cost for the installing contractor. An allowance for profit has been added to all material, labor, and equipment rental costs. A markup for labor burdens, including workers' compensation, fixed overhead, and business overhead, is included with installation costs.

The information in RSMeans cost data represents a "national average" cost. This data should be modified to the project location using the **City Cost Indexes** or **Location Factors** tables found in the Reference Section.

Table Format

B10 Superstructure

B1010 Floor Construction

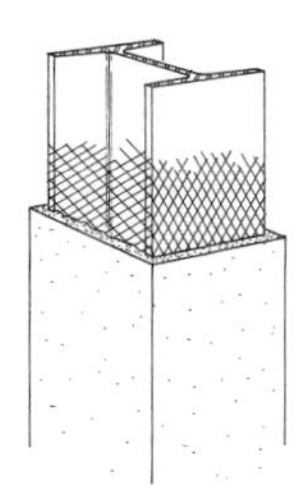

Listed below are costs per V.L.F. for fireproofing by material, column size, thickness and fire rating. Weights listed are for the fireproofing material only.

System Components ❺	QUANTITY	UNIT	COST PER V.L.F. ❹ MAT.	INST.	TOTAL
SYSTEM B1010 720 3000					
CONCRETE FIREPROOFING, 8″ STEEL COLUMN, 1″ THICK, 1 HR. FIRE RATING					
Forms in place, columns, plywood, 4 uses	3.330	SFCA	2.76	34.47	37.23
Welded wire fabric, 2 x 2 #14 galv. 21 lb./C.S.F., column wrap	2.700	S.F.	1.44	5.91	7.35
Concrete ready mix, regular weight, 3000 psi	.621	C.F.	3.27		3.27
Place and vibrate concrete, 12″ sq./round columns, pumped	.621	C.F.		1.96	1.96
TOTAL			7.47	42.34	49.81

B1010 720 — Steel Column Fireproofing

	ENCASEMENT SYSTEM	COLUMN SIZE (IN.)	THICKNESS (IN.)	FIRE RATING (HRS.)	WEIGHT (P.L.F.)	COST PER V.L.F. MAT.	INST.	TOTAL
3000	Concrete	8	1	1	110	7.45	42	49.45
3050	❻		1-1/2	2	133	8.65	46.50	55.15
3100			2	3	145	9.75	51	60.75
3150		10	1	1	145	9.85	52.50	62.35
3200			1-1/2	2	168	10.95	56.50	67.45
3250			2	3	196	12.25	60.50	72.75
3600	Gypsum board	8	1	3	14	5.35	33.50	38.85
3650	1/2″ fire rated	10	1	3	17	5.75	35.50	41.25
3700	2 layers	14	1	3	22	5.95	37	42.95
3750	Gypsum board	8	1-1/2	3	23	7.15	42.50	49.65
3800	1/2″ fire rated	10	1-1/2	3	27	8.05	47	55.05
3850	3 layers	14	1-1/2	3	35	8.90	51.50	60.40
3900	Sprayed fiber	8	1-1/2	2	6.3	4.41	8.60	13.01
3950	Direct application		2	3	8.3	6.10	11.85	17.95
4000			2-1/2	4	10.4	7.85	15.35	23.20
4050		10	1-1/2	2	7.9	5.30	10.40	15.70
4100			2	3	10.5	7.30	14.25	21.55
4150			2-1/2	4	13.1	9.40	18.30	27.70
4500	Perlite plaster	8	1	2	18	7.90	39	46.90
4550	On metal lath		1-3/8	3	23	8.65	43	51.65
4600			1-3/4	4	35	10.25	51.50	61.75
4650	Perlite plaster	10	1	2	21	9.10	45	54.10
4700			1-3/8	3	27	10.30	51.50	61.80
4750			1-3/4	4	41	11.60	58.50	70.10

❹ Unit of Measure

All RSMeans data: Assemblies include a typical **Unit of Measure** used for estimating that item. For instance, for continuous footings or foundation walls the unit is linear feet (L.F.). For spread footings the unit is each (Ea.). The estimator needs to take special care that the unit in the data matches the unit in the takeoff. Abbreviations and unit conversions can be found in the Reference Section.

❺ System Components

System components are listed separately to detail what is included in the development of the total system price.

❻ Table Descriptions

Table descriptions work similar to Narrative Descriptions, except that if there is a blank in the column at a particular line number, read up to the description above in the same column.

RSMeans data: Assemblies—How They Work (Continued)

Sample Estimate

This sample demonstrates the elements of an estimate, including a tally of the RSMeans data lines. Published assemblies costs include all markups for labor burden and profit for the installing contractor. This estimate adds a summary of the markups applied by a general contractor on the installing contractor's work. These figures represent the total cost to the owner. The location factor with RSMeans data is applied at the bottom of the estimate to adjust the cost of the work to a specific location.

Project Name:	Interior Fit-out, ABC Office			
Location:	Anywhere, USA	Date: 1/1/2020		R+R
Assembly Number	Description	Qty.	Unit	Subtotal
❶ C1010 124 1200	Wood partition, 2 x 4 @ 16" OC w/5/8" FR gypsum board	560	S.F.	$3,130.40
C1020 114 1800	Metal door & frame, flush hollow core, 3'-0" x 7'-0"	2	Ea.	$2,680.00
C3010 230 0080	Painting, brushwork, primer & 2 coats	1,120	S.F.	$1,545.60
C3020 410 0140	Carpet, tufted, nylon, roll goods, 12' wide, 26 oz	240	S.F.	$943.20
C3030 210 6000	Acoustic ceilings, 24" x 48" tile, tee grid suspension	200	S.F.	$1,352.00
D5020 125 0560	Receptacles incl plate, box, conduit, wire, 20 A duplex	8	Ea.	$2,500.00
D5020 125 0720	Light switch incl plate, box, conduit, wire, 20 A single pole	2	Ea.	$609.00
D5020 210 0560	Fluorescent fixtures, recess mounted, 20 per 1000 SF	200	S.F.	$2,390.00
	Assembly Subtotal			**$15,150.20**
	Sales Tax @ ❷	**5 %**		**$ 378.76**
	General Requirements @ ❸	**7 %**		**$ 1,060.51**
	Subtotal A			**$16,589.47**
	GC Overhead @ ❹	**5 %**		**$ 829.47**
	Subtotal B			**$17,418.94**
	GC Profit @ ❺	**5 %**		**$ 870.95**
	Subtotal C			**$18,289.89**
	Adjusted by Location Factor ❻	**115.5**		**$ 21,124.82**
	Architects Fee @ ❼	**8 %**		**$ 1,689.99**
	Contingency @ ❽	**15 %**		**$ 3,168.72**
	Project Total Cost			**$ 25,983.53**

This estimate is based on an interactive spreadsheet. You are free to download it and adjust it to your methodology.
A copy of this spreadsheet is available at **RSMeans.com/2020books.**

1 Work Performed

The body of the estimate shows the RSMeans data selected, including line numbers, a brief description of each item, its takeoff quantity and unit, and the total installed cost, including the installing contractor's overhead and profit.

2 Sales Tax

If the work is subject to state or local sales taxes, the amount must be added to the estimate. In a conceptual estimate it can be assumed that one half of the total represents material costs. Therefore, apply the sales tax rate to 50% of the assembly subtotal.

3 General Requirements

This item covers project-wide needs provided by the general contractor. These items vary by project but may include temporary facilities and utilities, security, testing, project cleanup, etc. In assemblies estimates a percentage is used—typically between 5% and 15% of project cost.

4 General Contractor Overhead

This entry represents the general contractor's markup on all work to cover project administration costs.

5 General Contractor Profit

This entry represents the GC's profit on all work performed. The value included here can vary widely by project and is influenced by the GC's perception of the project's financial risk and market conditions.

6 Location Factor

RSMeans published data are based on national average costs. If necessary, adjust the total cost of the project using a location factor from the "Location Factor" table or the "City Cost Indexes" table found in the Reference Section. Use location factors if the work is general, covering the work of multiple trades. If the work is by a single trade (e.g., masonry) use the more specific data found in the City Cost Indexes.

To adjust costs by location factors, multiply the base cost by the factor and divide by 100.

7 Architect's Fee

If appropriate, add the design cost to the project estimate. These fees vary based on project complexity and size. Typical design and engineering fees can be found in the Reference Section.

8 Contingency

A factor for contingency may be added to any estimate to represent the cost of unknowns that may occur between the time that the estimate is performed and the time the project is constructed. The amount of the allowance will depend on the stage of design at which the estimate is done, as well as the contractor's assessment of the risk involved.

A1010 Standard Foundations

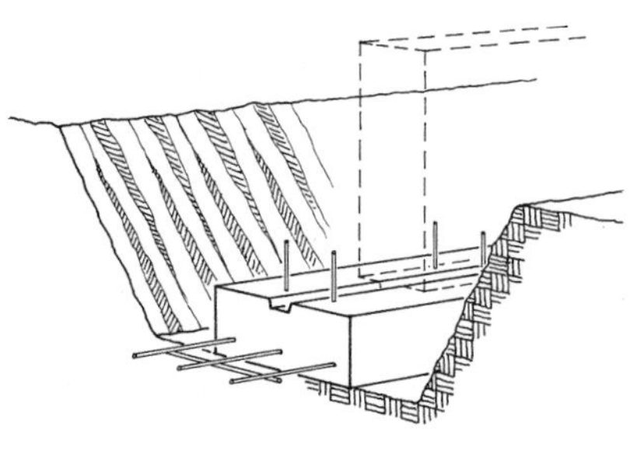

The Strip Footing System includes: excavation; hand trim; all forms needed for footing placement; forms for 2″ x 6″ keyway (four uses); dowels; and 3,000 p.s.i. concrete.

The footing size required varies for different soils. Soil bearing capacities are listed for 3 KSF and 6 KSF. Depths of the system range from 8″ and deeper. Widths range from 16″ and wider. Smaller strip footings may not require reinforcement.

Please see the reference section for further design and cost information.

System Components	QUANTITY	UNIT	COST PER L.F.		
			MAT.	INST.	TOTAL
SYSTEM A1010 110 2500					
STRIP FOOTING, LOAD 5.1 KLF, SOIL CAP. 3 KSF, 24″ WIDE X 12″ DEEP, REINF.					
Trench excavation	.148	C.Y.		1.47	1.47
Hand trim	2.000	S.F.		2.44	2.44
Compacted backfill	.074	C.Y.		.33	.33
Formwork, 4 uses	2.000	S.F.	4.92	10.70	15.62
Keyway form, 4 uses	1.000	L.F.	.37	1.36	1.73
Reinforcing, fy = 60000 psi	3.000	Lb.	1.95	2.04	3.99
Dowels	2.000	Ea.	1.82	5.92	7.74
Concrete, f'c = 3000 psi	.074	C.Y.	10.51		10.51
Place concrete, direct chute	.074	C.Y.		2.11	2.11
Screed finish	2.000	S.F.		.90	.90
TOTAL			19.57	27.27	46.84

A1010 110	Strip Footings	COST PER L.F.		
		MAT.	INST.	TOTAL
2100	Strip footing, load 2.6 KLF, soil capacity 3 KSF, 16″ wide x 8″ deep, plain	8.35	12.85	21.20
2300	Load 3.9 KLF, soil capacity 3 KSF, 24″ wide x 8″ deep, plain	10.75	14.55	25.30
2500	Load 5.1 KLF, soil capacity 3 KSF, 24″ wide x 12″ deep, reinf.	19.55	27	46.55
2700	Load 11.1 KLF, soil capacity 6 KSF, 24″ wide x 12″ deep, reinf.	19.55	27	46.55
2900	Load 6.8 KLF, soil capacity 3 KSF, 32″ wide x 12″ deep, reinf.	24	30	54
3100	Load 14.8 KLF, soil capacity 6 KSF, 32″ wide x 12″ deep, reinf.	24	30	54
3300	Load 9.3 KLF, soil capacity 3 KSF, 40″ wide x 12″ deep, reinf.	28	32.50	60.50
3500	Load 18.4 KLF, soil capacity 6 KSF, 40″ wide x 12″ deep, reinf.	28	32.50	60.50
3700	Load 10.1 KLF, soil capacity 3 KSF, 48″ wide x 12″ deep, reinf.	31.50	34.50	66
3900	Load 22.1 KLF, soil capacity 6 KSF, 48″ wide x 12″ deep, reinf.	33.50	36.50	70
4100	Load 11.8 KLF, soil capacity 3 KSF, 56″ wide x 12″ deep, reinf.	36.50	38.50	75
4300	Load 25.8 KLF, soil capacity 6 KSF, 56″ wide x 12″ deep, reinf.	39.50	41	80.50
4500	Load 10 KLF, soil capacity 3 KSF, 48″ wide x 16″ deep, reinf.	40	40.50	80.50
4700	Load 22 KLF, soil capacity 6 KSF, 48″ wide, 16″ deep, reinf.	41	41.50	82.50
4900	Load 11.6 KLF, soil capacity 3 KSF, 56″ wide x 16″ deep, reinf.	45.50	57.50	103
5100	Load 25.6 KLF, soil capacity 6 KSF, 56″ wide x 16″ deep, reinf.	47.50	60	107.50
5300	Load 13.3 KLF, soil capacity 3 KSF, 64″ wide x 16″ deep, reinf.	52	48	100
5500	Load 29.3 KLF, soil capacity 6 KSF, 64″ wide x 16″ deep, reinf.	55	51.50	106.50
5700	Load 15 KLF, soil capacity 3 KSF, 72″ wide x 20″ deep, reinf.	69.50	57.50	127
5900	Load 33 KLF, soil capacity 6 KSF, 72″ wide x 20″ deep, reinf.	73	61.50	134.50
6100	Load 18.3 KLF, soil capacity 3 KSF, 88″ wide x 24″ deep, reinf.	98	73	171
6300	Load 40.3 KLF, soil capacity 6 KSF, 88″ wide x 24″ deep, reinf.	106	81	187
6500	Load 20 KLF, soil capacity 3 KSF, 96″ wide x 24″ deep, reinf.	107	78	185
6700	Load 44 KLF, soil capacity 6 KSF, 96″ wide x 24″ deep, reinf.	112	84	196

A1010 Standard Foundations

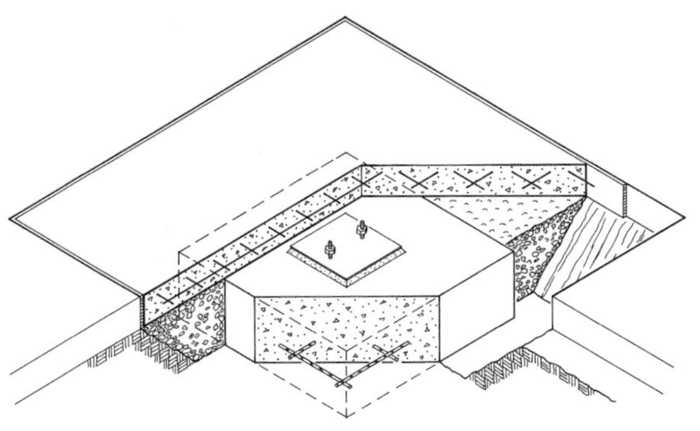

Column footing includes: Remove existing slab; excavation; forms for concrete; reinforcing steel; expansion joint; anchor bolts; grouting column base plate; and concrete placed and finished.

The footing size required varies for different soils.

Design assumptions:

Conc. slab f'c = 3.5 ksi
Conc. foundation f'c = 3.0 ksi
WWF 6 x 6 #10/#10
Reinforcement, fy = 60 ksi

System Components	QUANTITY	UNIT	COST EACH MAT.	COST EACH INST.	COST EACH TOTAL
SYSTEM A1010 260 0200					
CONC. COLUMN FOOTING IN EXISTING BUILDING, LOAD 10K, SOIL CAPACITY 3 KSF, 3′ SQ X 12″ DEEP, 3,000 PSI					
Saw cut slab	16.000	S.F.	14.40	235.20	249.60
Break up slab, 6″	16.000	S.F.		284.64	284.64
Remove concrete; hand	.755	C.Y.		33.98	33.98
Excavate; hand	.592	C.Y.		79.92	79.92
Forms, 1 use	12.000	SFCA	39	300	339
Reinforcing steel, #4 bar	15.000	Lb.	9.30	10.20	19.50
Concrete, 3,000 psi, for footing, incl. premium delv. chg.	.333	C.Y.	66.17		66.17
Place conc. footing	.333	C.Y.		28.93	28.93
Backfill by hand	.311	C.Y.		11.97	11.97
Compaction; hand tamp	.311	C.Y.		8.09	8.09
Welded wire fabric	.160	C.S.F.	2.86	6.48	9.34
Premolded, bituminous fiber, 1/2″ x 6″	16.000	L.F.	7.68	29.12	36.80
Concrete, 3500 psi, for slab, incl. premium delv. chg.	.166	C.Y.	31.78		31.78
Place conc. slab, 6″	.166	C.Y.		4.84	4.84
Finish slab; steel trowel	16.000	S.F.		17.76	17.76
Anchor bolts, 3/4″ x 12″	2.000	Ea.	11.50	10.90	22.40
Grout level plate	1.000	S.F.	8.35	17.95	26.30
TOTAL			191.04	1,079.98	1,271.02
COST per C.Y. of FOOTING			573.69	3,243.18	3,816.87

A1020 Special Foundations

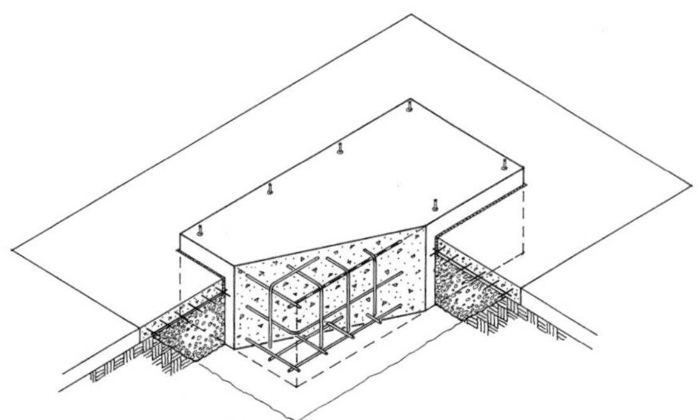

Equipment foundation includes: Remove existing slab; excavation; forms for concrete; reinforcing steel; expansion joint; anchor bolts; and concrete placed and screeded.

The foundation size required varies for different soils.

Design assumptions:

Conc. slab f'c = 3.5 ksi
Conc. foundation f'c = 3.0 ksi
WWF 6 x 6 #10/#10
Reinforcement, fy = 60 ksi

System Components	QUANTITY	UNIT	COST EACH		
			MAT.	INST.	TOTAL
SYSTEM A1020 810 0170					
CONC. EQUIPMENT FOUNDATION, 4′ X 8′ X 30″ DEEP, 3,000 PSI.					
Saw cut slab	45.000	S.F.	25.20	411.60	436.80
Break up slab; 6″	45.000	S.F.		800.55	800.55
Remove concrete; hand	3.133	C.Y.		140.99	140.99
Excavate; hand	2.500	C.Y.		337.50	337.50
Forms, 1 use	60.000	SFCA	195	1,500	1,695
Reinforcing steel, #6 bar	318.000	Lb.	197.16	216.24	413.40
Concrete, 3000 psi, for foundation, incl. premium delv. chg.	3.000	C.Y.	589.02		589.02
Anchor bolts, 3/4″ x 12″	8.000	Ea.	46	43.60	89.60
Place concrete foundation	3.000	C.Y.		118.17	118.17
Backfill by hand	.866	C.Y.		33.34	33.34
Compaction in 6″ layers, hand tamp	.866	C.Y.		22.52	22.52
Premolded, bituminous fiber, 1/2″x6″	24.000	L.F.	11.52	43.68	55.20
Welded wire fabric; 6x6-#10/10	.130	C.S.F.	2.32	5.27	7.59
Concrete, 3500 psi, for slab, incl. premium delv. chg.	.241	C.Y.	46.17		46.17
Place conc. slab	.241	C.Y.		7.03	7.03
Finish slab; steel trowel	13.000	S.F.		14.43	14.43
Finish equip. foundation; float finish	32.000	S.F.		14.40	14.40
TOTAL			1,112.39	3,709.32	4,821.71
COST per C.Y. of FOUNDATION			370.80	1,236.44	1,607.24

B1010 Floor Construction

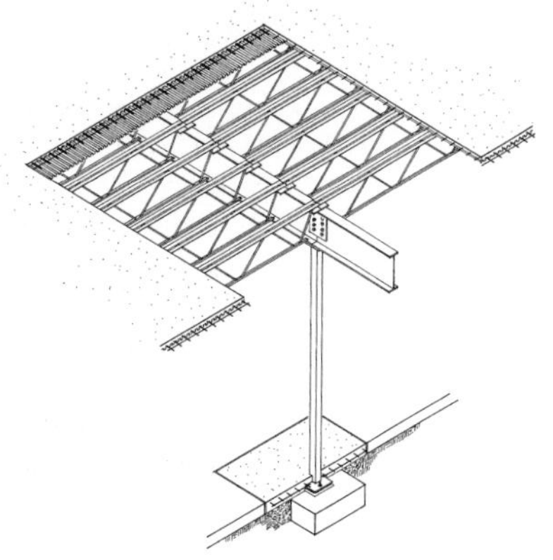

Mezzanine addition to existing building includes: Column footings; steel columns; structural steel; open web steel joists; uncoated 28 ga. steel slab forms; 2-1/2″ concrete slab reinforced with welded wire fabric; steel trowel finish.

Design assumptions:

Structural steel is A36, high strength bolted. Slab form is 28 gauge, galvanized.
WWF 6 x 6 #10/#10
Conc. slab f'c = 3 ksi

System Components	QUANTITY	UNIT	COST EACH		
			MAT.	INST.	TOTAL
SYSTEM B1010 310 0142					
MEZZANINE ADDITION TO EXISTING BUILDING; 100 PSF SUPERIMPOSED LOAD, SLAB FORM DECK, MTL. JOISTS, 2.5″ CONC. SLAB, 3,000 PSI, 3,000 S.F.					
Saw cut existing slab, 6″ thick	192.000	L.F.	172.80	2,822.40	2,995.20
Cut out existing slab, 4′ x 4′ openings	192.000	S.F.		3,415.68	3,415.68
Remove concrete debris	9.060	C.Y.		407.70	407.70
Excavate by hand	7.104	C.Y.		959.04	959.04
Backfill by hand	3.732	C.Y.		143.68	143.68
Compaction of backfill	3.732	C.Y.		97.03	97.03
Forms for footing, 3′ x 3′ x 1′	144.000	SFCA	468	3,600	4,068
Reinforcing bar for footing, 3#5 E.W. x 30″ long	180.000	Lb.	111.60	122.40	234
Anchor bolts, 12″, 2 per column	24.000	Ea.	138	130.80	268.80
Concrete, 3000 psi, footing	5.592	C.Y.	794.06		794.06
Place concrete in footing, direct chute	5.592	C.Y.		347.09	347.09
Premolded, bituminous fiber, 1/2″ x 6″	192.000	L.F.	92.16	349.44	441.60
Reinforcing mesh for slab cutouts	1.920	C.S.F.	34.27	77.76	112.03
Concrete, 3500 psi, slab cutouts	2.784	C.Y.	381.41		381.41
Place concrete slab cutouts, direct chute	2.784	C.Y.		58.07	58.07
Machine trowel slab cutouts	192.000	S.F.		213.12	213.12
Grout for leveling plates	12.000	S.F.	100.20	215.40	315.60
Column, 4″ x 4″ x 1/4″ x12′	12.000	Ea.	3,132	1,458	4,590
Structural steel W 21x50	80.000	L.F.	10,440	1,069.20	11,509.20
Structural steel W 16 x 31	80.000	L.F.	6,480	938.40	7,418.40
Open web joists, H or K series	6.800	Ton	21,675	8,180.40	29,855.40
Metal decking, slab form, steel, 28 gauge, 9/16″ deep, type UFS, uncoated	30.000	C.S.F.	6,270	2,400	8,670
Concrete 3,000 psi, 2 1/2″ slab, incl. premium delv. chg.	23.500	C.Y.	5,005.50		5,005.50
Welded wire fabric 6x6 #10/#10 (w1.4/w1.4)	30.000	C.S.F.	535.50	1,215	1,750.50
Place concrete	23.500	C.Y.		1,291.91	1,291.91
Monolithic steel trowel finish	30.000	C.S.F.		3,330	3,330
Curing with sprayed membrane curing compound	30.000	C.S.F.	411	340.50	751.50
TOTAL			56,241.50	33,183.02	89,424.52
COST PER S.F.			18.75	11.06	29.81

B1010 Floor Construction

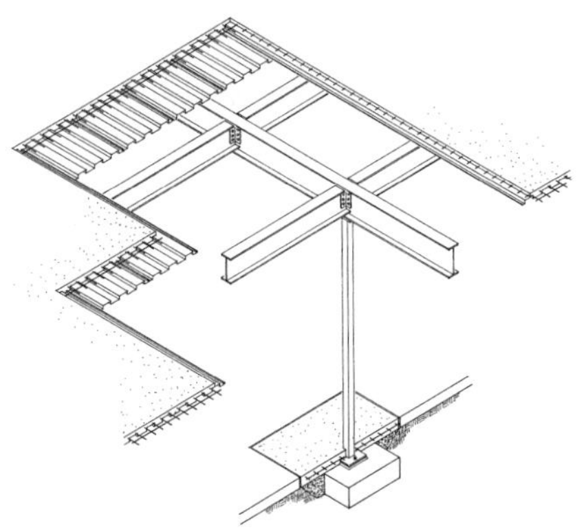

Mezzanine addition to existing building includes: Column footings; steel columns; structural steel; galvanized composite steel deck; 4″ concrete slab reinforced with welded wire fabric; steel trowel finish.

Design assumptions:

Structural steel is A36, high strength bolted.
Composite deck is 22 gauge, galvanized.
WWF 6 x 6 #10/#10
Conc. slab f'c = 3 ksi

System Components	QUANTITY	UNIT	COST EACH		
			MAT.	INST.	TOTAL
SYSTEM B1010 320 0152					
MEZZANINE ADDITION TO EXISTING BUILDING; 100 PSF SUPERIMPOSED LOAD					
COMPOSITE DECK, BEAMS, 4″ CONC. SLAB, 3,000 PSI, 3,000 S.F.					
Saw cut existing slab, 6″ thick	192.000	L.F.	172.80	2,822.40	2,995.20
Remove existing slab, 4′ x 4′ openings	192.000	S.F.		3,415.68	3,415.68
Remove concrete debris	9.060	C.Y.		407.70	407.70
Excavate by hand	7.104	C.Y.		959.04	959.04
Backfill by hand, no compaction, light soil	3.732	C.Y.		143.68	143.68
Compaction of backfill	3.732	C.Y.		97.03	97.03
Forms in place, footing, 3′ x 3′ x 1′	144.000	SFCA	468	3,600	4,068
Footings, #4 to #7 bars, 3#5 E.W. x 30″ long	180.000	Lb.	111.60	122.40	234
Anchor bolts, 12″ long, 2 per column	24.000	Ea.	138	130.80	268.80
Concrete, 3000 psi, footing	5.592	C.Y.	794.06		794.06
Place concrete in footings, direct chute	5.592	C.Y.		347.09	347.09
Premolded, bituminous fiber, 1/2″ x 6″	192.000	L.F.	92.16	349.44	441.60
Structural steel W 12x19	225.000	L.F.	8,662.50	1,804.50	10,467
Reinforcing mesh for slab cutouts	1.920	C.S.F.	34.27	77.76	112.03
Concrete, 3500 psi, for slab cutouts	2.784	C.Y.	381.41		381.41
Structural steel W 14 x 30	375.000	L.F.	29,250	4,398.76	33,648.76
Place concrete, slab cutouts, direct chute	2.784	C.Y.		58.07	58.07
Structural steel W 16 x 31	80.000	L.F.	6,480	938.40	7,418.40
Machine trowel slab cutouts	192.000	S.F.		213.12	213.12
Grout for leveling plates	12.000	S.F.	100.20	215.40	315.60
Column, 4″ x 4″ x 1/4″ x 12′	12.000	Ea.	3,132	1,458	4,590
Structural steel W 18 x 50	80.000	L.F.	10,440	1,250.40	11,690.40
Metal deck, galvanized, 2″ deep, 22 gauge.	30.000	C.S.F.	7,860	2,490	10,350
Concrete, 3,000 p.s.i., 4″ slab, incl. premium delv. chg.	37.000	C.Y.	7,881		7,881
Welded wire fabric 6x6 #10/#10 (w1.4/w1.4)	30.000	C.S.F.	535.50	1,215	1,750.50
Place concrete	37.000	C.Y.		2,034.08	2,034.08
Monolithic steel trowel finish	30.000	C.S.F.		3,330	3,330
Curing with sprayed membrane curing compound	30.000	C.S.F.	411	340.50	751.50
TOTAL			76,944.50	32,219.25	109,163.75
COST PER S.F.			25.65	10.74	36.39

B1010 Floor Construction

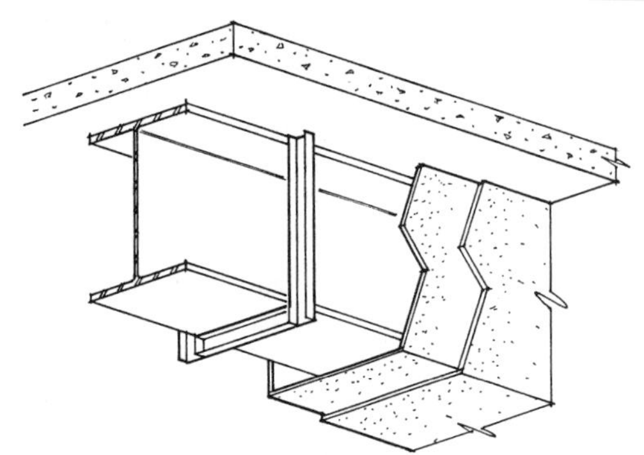

The table below lists fireproofing costs for steel beams by type, beam size, thickness and fire rating. Weights listed are for the fireproofing material only.

System Components	QUANTITY	UNIT	COST PER L.F. MAT.	COST PER L.F. INST.	COST PER L.F. TOTAL
SYSTEM B1010 710 1300					
FIREPROOFING, 5/8" F.R. GYP. BOARD, 12"X 4" BEAM, 2" THICK, 2 HR. F.R.					
1-1/4" x 1-1/4", galvanized	.020	C.L.F.	.38	3.90	4.28
L bead for drywall, galvanized	.020	C.L.F.	.49	4.54	5.03
Furring, beams & columns, 3/4" galv. channels, 24" OC	2.330	S.F.	.63	8.20	8.83
Drywall on beam, no finish, 2 layers at 5/8" thick	3.000	S.F.	2.85	13.62	16.47
Drywall, taping and finishing joints, add	3.000	S.F.	.18	2.04	2.22
TOTAL			4.53	32.30	36.83

B1010 710 Steel Beam Fireproofing

	ENCASEMENT SYSTEM	BEAM SIZE (IN.)	THICKNESS (IN.)	FIRE RATING (HRS.)	WEIGHT (P.L.F.)	COST PER L.F. MAT.	COST PER L.F. INST.	COST PER L.F. TOTAL
0400	Concrete	12x4	1	1	77	7.25	32	39.25
0450	3000 PSI		1-1/2	2	93	8.50	35	43.50
0500			2	3	121	9.55	38.50	48.05
0550		14x5	1	1	100	9.55	39.50	49.05
0600			1-1/2	2	122	11.05	44	55.05
0650			2	3	142	12.55	48.50	61.05
0700		16x7	1	1	147	11.20	42.50	53.70
0750			1-1/2	2	169	12.10	44	56.10
0800			2	3	195	13.70	48	61.70
0850		18x7-1/2	1	1	172	13	49	62
0900			1-1/2	2	196	14.65	53.50	68.15
0950			2	3	225	16.45	58	74.45
1000		24x9	1	1	264	17.60	59.50	77.10
1050			1-1/2	2	295	19.15	62.50	81.65
1100			2	3	328	20.50	66	86.50
1150		30x10-1/2	1	1	366	23	75	98
1200			1-1/2	2	404	25	79.50	104.50
1250			2	3	449	26.50	84	110.50
1300	5/8" fire rated	12x4	2	2	15	4.53	32.50	37.03
1350	Gypsum board		2-5/8	3	24	6.30	40	46.30
1400		14x5	2	2	17	4.86	33.50	38.36
1450			2-5/8	3	27	6.95	42.50	49.45
1500		16x7	2	2	20	5.45	37.50	42.95
1550			2-5/8	3	31	6.90	36	42.90
1600		18x7-1/2	2	2	22	5.90	40.50	46.40
1650			2-5/8	3	34	8.35	51	59.35

B1010 Floor Construction

B1010 710 Steel Beam Fireproofing

	ENCASEMENT SYSTEM	BEAM SIZE (IN.)	THICKNESS (IN.)	FIRE RATING (HRS.)	WEIGHT (P.L.F.)	COST PER L.F.		
						MAT.	INST.	TOTAL
1700	5/8" fire rated	24x9	2	2	27	7.30	50	57.30
1750	Gypsum board		2-5/8	3	42	10.30	62.50	72.80
1800		30x10-1/2	2	2	33	8.60	58.50	67.10
1850			2-5/8	3	51	12.15	73.50	85.65
1900	Gypsum	12x4	1-1/8	3	18	5.55	25	30.55
1950	Plaster on	14x5	1-1/8	3	21	6.30	28.50	34.80
2000	Metal lath	16x7	1-1/8	3	25	7.35	32	39.35
2050		18x7-1/2	1-1/8	3	32	8.35	36	44.35
2100		24x9	1-1/8	3	35	10.30	44	54.30
2150		30x10-1/2	1-1/8	3	44	12.40	52	64.40
2200	Perlite plaster	12x4	1-1/8	2	16	5.10	28	33.10
2250	On metal lath		1-1/4	3	20	5.45	29	34.45
2300			1-1/2	4	22	5.80	30.50	36.30
2350		14x5	1-1/8	2	18	6.65	34	40.65
2400			1-1/4	3	23	7	35.50	42.50
2450			1-1/2	4	25	7.75	38	45.75
2500		16x7	1-1/8	2	21	6.20	34	40.20
2550			1-1/4	3	26	6.55	35	41.55
2600			1-1/2	4	29	6.95	36.50	43.45
2650		18x7-1/2	1-1/8	2	26	7.70	39.50	47.20
2700			1-1/4	3	33	8.05	41	49.05
2750			1-1/2	4	36	8.80	43.50	52.30
2800		24x9	1-1/8	2	30	8.05	44	52.05
2850			1-1/4	3	38	8.40	45.50	53.90
2900			1-1/2	4	41	8.75	47	55.75
2950		30x10-1/2	1-1/8	2	37	10.45	55.50	65.95
3000			1-1/4	3	46	10.80	57	67.80
3050			1-1/2	4	51	11.55	59.50	71.05
3100	Sprayed Fiber	12x4	5/8	1	12	1.41	2.75	4.16
3150	(non asbestos)		1-1/8	2	21	2.54	4.96	7.50
3200			1-1/4	3	22	2.82	5.50	8.32
3250		14x5	5/8	1	14	1.41	2.75	4.16
3300			1-1/8	2	26	2.54	4.96	7.50
3350			1-1/4	3	29	2.82	5.50	8.32
3400		16x7	5/8	1	18	1.78	3.48	5.26
3450			1-1/8	2	32	3.21	6.30	9.51
3500			1-1/4	3	36	3.57	6.95	10.52
3550		18x7-1/2	5/8	1	20	1.97	3.85	5.82
3600			1-1/8	2	35	3.55	6.90	10.45
3650			1-1/4	3	39	3.91	7.65	11.56
3700		24x9	5/8	1	25	2.49	4.86	7.35
3750			1-1/8	2	45	4.50	8.80	13.30
3800			1-1/4	3	50	5	9.80	14.80
3850		30x10-1/2	5/8	1	31	3.01	5.90	8.91
3900			1-1/8	2	55	5.45	10.60	16.05
3950			1-1/4	3	61	6.05	11.80	17.85
4000	On Decking	Flat	1		6	.65	.80	1.45
4050	Per S.F.	Corrugated	1		7	.97	1.53	2.50
4100		Fluted	1		7	.97	1.53	2.50

B1010 Floor Construction

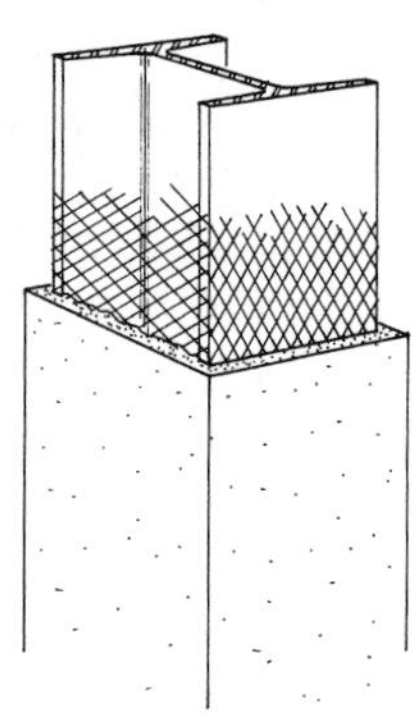

Listed below are costs per V.L.F. for fireproofing by material, column size, thickness and fire rating. Weights listed are for the fireproofing material only.

System Components	QUANTITY	UNIT	COST PER V.L.F. MAT.	COST PER V.L.F. INST.	COST PER V.L.F. TOTAL
SYSTEM B1010 720 3000					
CONCRETE FIREPROOFING, 8″ STEEL COLUMN, 1″ THICK, 1 HR. FIRE RATING					
Forms in place, columns, plywood, 4 uses	3.330	SFCA	2.76	34.47	37.23
Welded wire fabric, 2 x 2 #14 galv. 21 lb./C.S.F., column wrap	2.700	S.F.	1.44	5.91	7.35
Concrete ready mix, regular weight, 3000 psi	.621	C.F.	3.27		3.27
Place and vibrate concrete, 12″ sq./round columns, pumped	.621	C.F.		1.96	1.96
TOTAL			7.47	42.34	49.81

B1010 720 Steel Column Fireproofing

	ENCASEMENT SYSTEM	COLUMN SIZE (IN.)	THICKNESS (IN.)	FIRE RATING (HRS.)	WEIGHT (P.L.F.)	COST PER V.L.F. MAT.	COST PER V.L.F. INST.	COST PER V.L.F. TOTAL
3000	Concrete	8	1	1	110	7.45	42	49.45
3050			1-1/2	2	133	8.65	46.50	55.15
3100			2	3	145	9.75	51	60.75
3150		10	1	1	145	9.85	52.50	62.35
3200			1-1/2	2	168	10.95	56.50	67.45
3250			2	3	196	12.25	60.50	72.75
3600	Gypsum board	8	1	3	14	5.35	33.50	38.85
3650	1/2″ fire rated	10	1	3	17	5.75	35.50	41.25
3700	2 layers	14	1	3	22	5.95	37	42.95
3750	Gypsum board	8	1-1/2	3	23	7.15	42.50	49.65
3800	1/2″ fire rated	10	1-1/2	3	27	8.05	47	55.05
3850	3 layers	14	1-1/2	3	35	8.90	51.50	60.40
3900	Sprayed fiber	8	1-1/2	2	6.3	4.41	8.60	13.01
3950	Direct application		2	3	8.3	6.10	11.85	17.95
4000			2-1/2	4	10.4	7.85	15.35	23.20
4050		10	1-1/2	2	7.9	5.30	10.40	15.70
4100			2	3	10.5	7.30	14.25	21.55
4150			2-1/2	4	13.1	9.40	18.30	27.70
4500	Perlite plaster	8	1	2	18	7.90	39	46.90
4550	On metal lath		1-3/8	3	23	8.65	43	51.65
4600			1-3/4	4	35	10.25	51.50	61.75
4650	Perlite plaster	10	1	2	21	9.10	45	54.10
4700			1-3/8	3	27	10.30	51.50	61.80
4750			1-3/4	4	41	11.60	58.50	70.10

B1010 Floor Construction

B1010 720 Steel Column Fireproofing

	ENCASEMENT SYSTEM	COLUMN SIZE (IN.)	THICKNESS (IN.)	FIRE RATING (HRS.)	WEIGHT (P.L.F.)	COST PER V.L.F.		
						MAT.	INST.	TOTAL
5100	5/8" gypsum plaster	8	1	1-1/2	20	6.40	32.50	38.90
5150	On 3/8" gypsum lath	10	1	1-1/2	24	7.60	38.50	46.10
5200		14	1	1-1/2	33	9.30	46.50	55.80
5250	1"perlite plaster	8	1-3/8	2	23	7.95	34	41.95
5300	On 3/8" gypsum lath	10	1-3/8	2	28	9.15	38.50	47.65
5350		14	1-3/8	2	37	11.60	48.50	60.10
5400	1-3/8" perlite plaster	8	1-3/4	3	27	7.60	38	45.60
5450	On 3/8" gypsum lath	10	1-3/4	3	33	8.70	43	51.70
5500		14	1-3/4	3	43	10.90	54	64.90
5550	Concrete masonry	8	4-3/4	4	126	13.20	40	53.20
5600	Units 4" thick	10	4-3/4	4	166	16.55	50	66.55
5650	75% solid	14	4-3/4	4	262	19.80	60	79.80

B2010 Exterior Walls

This page illustrates masonry cleaning and restoration systems, including staging, cleaning and repointing.

System Components	QUANTITY	UNIT	COST PER S.F. MAT.	COST PER S.F. INST.	COST PER S.F. TOTAL
SYSTEM B2010 143 1300					
Repoint existing building, brick, common bond, high pressure cleaning, Water only, soft old mortar.					
Scaffolding,steel tubular,bldg ext wall face, labor only, 1 to 5 stories	.010	C.S.F.		2.55	2.55
Cleaning masonry, high pressure wash, water only, avg soil, bio staining	1.100	S.F.		1.89	1.89
Repoint, brick common bond	1.000	S.F.	.63	7	7.63
TOTAL			.63	11.44	12.07

B2010 143	Masonry Restoration - Cleaning	COST PER S.F. MAT.	COST PER S.F. INST.	COST PER S.F. TOTAL
1000	Repoint existing building, high pressure cleaning, water only,			
1100	soft old mortar			
1200	For alternate masonry surfaces:			
1300	Brick, common bond	.63	11.45	12.08
1400	Flemish bond	.66	11.95	12.61
1500	English bond	.66	12.65	13.31
1700	Stone, 2′ x 2′ blocks	.83	8.65	9.48
1800	2′ x 4′ blocks	.62	7.60	8.22
2000	Add to above prices for alternate cleaning systems:			
2100	Chemical brush and wash	.05	.76	.81
2200	High pressure chemical and water	.09	.65	.74
2300	Sandblasting, wet system	.43	1.82	2.25
2400	Dry system	.40	1.06	1.46
2500	Steam cleaning		.50	.50

B2020 Exterior Windows

B2020 108 — Vinyl Clad Windows

	MATERIAL	TYPE	GLAZING	SIZE	DETAIL	COST PER UNIT		
						MAT.	INST.	TOTAL
1000	Vinyl clad	casement	insul. glass	1'-4" x 4'-0"		455	157	612
1010				2'-0" x 3'-0"		375	157	532
1020				2'-0" x 4'-0"		430	165	595
1030				2'-0" x 5'-0"		465	174	639
1040				2'-0" x 6'-0"		515	174	689
1050				2'-4" x 4'-0"		465	174	639
1060				2'-6" x 5'-0"		590	169	759
1070				3'-0" x 5'-0"		820	174	994
1080				4'-0" x 3'-0"		910	174	1,084
1090				4'-0" x 4'-0"		775	174	949
1100				4'-8" x 4'-0"		845	174	1,019
1110				4'-8" x 5'-0"		980	203	1,183
1120				4'-8" x 6'-0"		1,075	203	1,278
1130				6'-0" x 4'-0"		980	203	1,183
1140				6'-0" x 5'-0"		1,200	203	1,403
3000		double-hung	insul. glass	2'-0" x 4'-0"		520	151	671
3050				2'-0" x 5'-0"		535	151	686
3100				2'-4" x 4'-0"		545	157	702
3150				2'-4" x 4'-8"		505	157	662
3200				2'-4" x 6'-0"		535	157	692
3250				2'-6" x 4'-0"		485	157	642
3300				2'-8" x 4'-0"		610	203	813
3350				2'-8" x 5'-0"		585	174	759
3375				2'-8" x 6'-0"		570	174	744
3400				3'-0" x 3'-6"		455	157	612
3450				3'-0" x 4'-0"		510	165	675
3500				3'-0" x 4'-8"		570	165	735
3550				3'-0" x 5'-0"		580	174	754
3600				3'-0" x 6'-0"		510	174	684
3700				4'-0" x 5'-0"		700	187	887
3800				4'-0" x 6'-0"		855	187	1,042

B3010 Roof Coverings

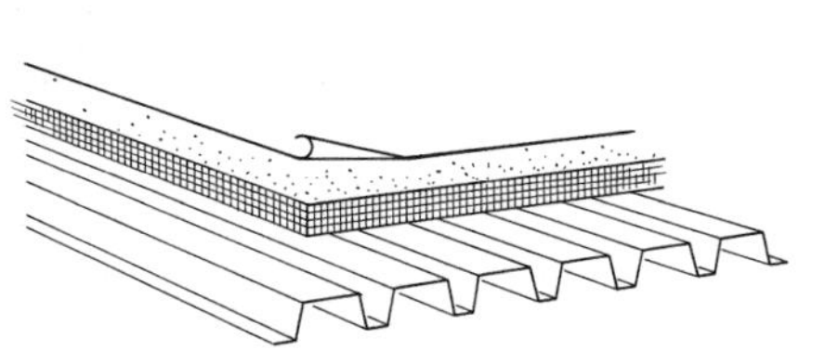

Fully Adhered

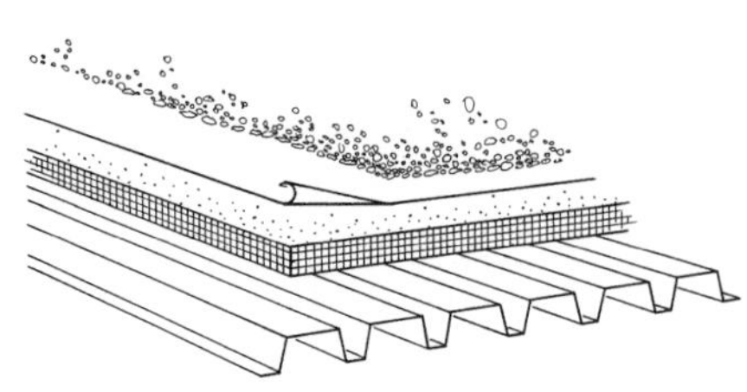

Ballasted

The systems listed below reflect only the cost for the single ply membrane.

B3010 120	Single Ply Membrane	COST PER S.F.		
		MAT.	INST.	TOTAL
1000	CSPE (Chlorosulfonated polyethylene), 45 mils, plate attached	2.74	.90	3.64
1100	Loosely laid with stone ballast	2.74	.90	3.64
2000	EPDM (Ethylene propylene diene monomer), 45 mils, fully adhered	1.32	1.21	2.53
2100	Loosely laid with stone ballast	1.01	.62	1.63
2200	Mechanically fastened with batten strips	.94	.90	1.84
3300	60 mils, fully adhered	1.53	1.21	2.74
3400	Loosely laid with stone ballast	1.24	.62	1.86
3500	Mechanically fastened with batten strips	1.15	.90	2.05
4000	Modified bit., SBS modified, granule surface cap sheet, mopped, 150 mils	1.50	2.43	3.93
4100	Smooth surface cap sheet, mopped, 145 mils	.91	2.31	3.22
4500	APP modified, granule surface cap sheet, torched, 180 mils	1.06	1.58	2.64
4600	Smooth surface cap sheet, torched, 170 mils	.84	1.50	2.34
5000	PIB (Polyisobutylene), 100 mils, fully adhered with contact cement	2.92	1.21	4.13
5100	Loosely laid with stone ballast	2.36	.62	2.98
5200	Partially adhered with adhesive	2.79	.90	3.69
5300	Hot asphalt attachment	2.66	.90	3.56
6000	Reinforced PVC, 48 mils, loose laid and ballasted with stone	1.40	.62	2.02
6100	Partially adhered with mechanical fasteners	1.32	.90	2.22
6200	Fully adhered with adhesive	1.82	1.21	3.03
6300	Reinforced PVC, 60 mils, loose laid and ballasted with stone	1.42	.62	2.04
6400	Partially adhered with mechanical fasteners	1.34	.90	2.24
6500	Fully adhered with adhesive	1.84	1.21	3.05

C1010 Partitions

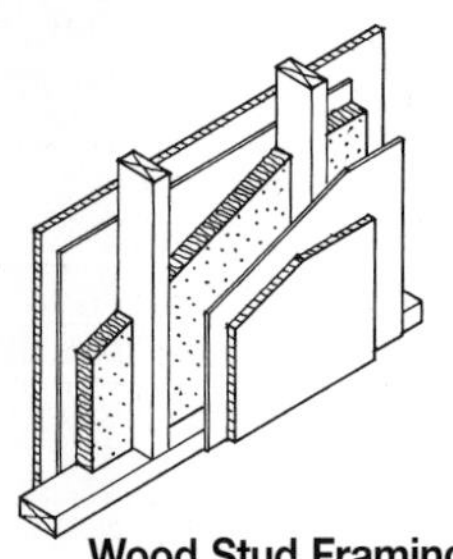
Wood Stud Framing

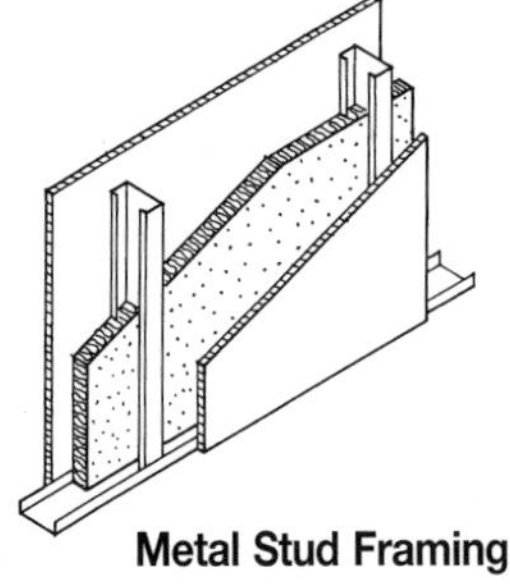
Metal Stud Framing

The Drywall Partitions/Stud Framing Systems are defined by type of drywall and number of layers, type and spacing of stud framing, and treatment on the opposite face. Components include taping and finishing.

Cost differences between regular and fire resistant drywall are negligible, and terminology is interchangeable. In some cases fiberglass insulation is included for additional sound deadening.

System Components	QUANTITY	UNIT	COST PER S.F. MAT.	INST.	TOTAL
SYSTEM C1010 124 1250					
DRYWALL PARTITION,5/8″ F.R.1 SIDE,5/8″ REG.1 SIDE,2″X4″STUDS,16″ O.C.					
Gypsum plasterboard, nailed/screwed to studs, 5/8″ fire resistant	1.000	S.F.	.42	.68	1.10
Gypsum plasterboard, nailed/screwed to studs, 5/8″ regular	1.000	S.F.	.36	.68	1.04
Taping and finishing joints	2.000	S.F.	.12	1.36	1.48
Framing, 2 x 4 studs @ 16″ O.C., 10′ high	1.000	S.F.	.55	1.36	1.91
TOTAL			1.45	4.08	5.53

C1010 124 Drywall Partitions/Wood Stud Framing

	FACE LAYER	BASE LAYER	FRAMING	OPPOSITE FACE	INSULATION	COST PER S.F. MAT.	INST.	TOTAL
1200	5/8″ FR drywall	none	2 x 4, @ 16″ O.C.	same	0	1.51	4.08	5.59
1250				5/8″ reg. drywall	0	1.45	4.08	5.53
1300				nothing	0	1.03	2.72	3.75
1400		1/4″ SD gypsum	2 x 4 @ 16″ O.C.	same	1-1/2″ fiberglass	2.90	6.30	9.20
1425					Sound attenuation	3.08	6.35	9.43
1450				5/8″ FR drywall	1-1/2″ fiberglass	2.44	5.50	7.94
1475					Sound attenuation	2.62	5.60	8.22
1500				nothing	1-1/2″ fiberglass	1.96	4.16	6.12
1600		resil. channels	2 x 4 @ 16″, O.C.	same	1-1/2″ fiberglass	2.34	7.95	10.29
1650				5/8″ FR drywall	1-1/2″ fiberglass	2.16	6.35	8.51
1700				nothing	1-1/2″ fiberglass	1.68	5	6.68
1800		5/8″ FR drywall	2 x 4 @ 24″ O.C.	same	0	2.22	5.15	7.37
1850				5/8″ FR drywall	0	1.80	4.49	6.29
1900				nothing	0	1.32	3.13	4.45
1950		5/8″ FR drywall	2 x 4, 16″ O.C.	same	0	2.35	5.45	7.80
1955				5/8″ FR drywall	0	1.93	4.76	6.69
2000				nothing	0	1.45	3.40	4.85
2010		5/8″ FR drywall	staggered, 6″ plate	same	0	2.90	6.85	9.75
2015				5/8″ FR drywall	0	2.48	6.15	8.63
2020				nothing	0	2	4.79	6.79
2200		5/8″ FR drywall	2 rows-2 x 4	same	2″ fiberglass	3.43	7.50	10.93
2250			16″O.C.	5/8″ FR drywall	2″ fiberglass	3.01	6.80	9.81
2300				nothing	2″ fiberglass	2.53	5.45	7.98
2400	5/8″ WR drywall	none	2 x 4, @ 16″ O.C.	same	0	1.63	4.08	5.71
2450				5/8″ FR drywall	0	1.57	4.08	5.65
2500				nothing	0	1.09	2.72	3.81
2600		5/8″ FR drywall	2 x 4, @ 24″ O.C.	same	0	2.34	5.15	7.49
2650				5/8″ FR drywall	0	1.86	4.49	6.35
2700				nothing	0	1.38	3.13	4.51

C1010 Partitions

C1010 128	Drywall Components	COST PER S.F.		
		MAT.	INST.	TOTAL
0140	Metal studs, 24" O.C. including track, load bearing, 18 gage, 2-1/2"	.65	1.28	1.93
0160	3-5/8"	.77	1.30	2.07
0180	4"	.68	1.33	2.01
0200	6"	1.03	1.35	2.38
0220	16 gage, 2-1/2"	.80	1.45	2.25
0240	3-5/8"	.94	1.48	2.42
0260	4"	.99	1.52	2.51
0280	6"	1.24	1.55	2.79
0300	Non load bearing, 25 gage, 1-5/8"	.21	.90	1.11
0340	3-5/8"	.33	.92	1.25
0360	4"	.37	.92	1.29
0380	6"	.44	.94	1.38
0400	20 gage, 2-1/2"	.36	1.14	1.50
0420	3-5/8"	.42	1.15	1.57
0440	4"	.51	1.15	1.66
0460	6"	.62	1.17	1.79
0542	Wood studs including blocking, shoe and double top plate, 2"x4", 12"O.C.	.68	1.71	2.39
0562	16" O.C.	.55	1.36	1.91
0582	24" O.C.	.42	1.09	1.51
0602	2"x6", 12" O.C.	.98	1.95	2.93
0622	16" O.C.	.79	1.52	2.31
0641	24" O.C.	.61	1.19	1.80
0642	Furring one side only, steel channels, 3/4", 12" O.C.	.46	2.77	3.23
0644	16" O.C.	.41	2.46	2.87
0646	24" O.C.	.27	1.86	2.13
0647	1-1/2" , 12" O.C.	.62	3.11	3.73
0648	16" O.C.	.55	2.72	3.27
0649	24" O.C.	.37	2.14	2.51
0656	Wood strips, 1" x 3", on wood, 12" O.C.	.48	1.24	1.72
0657	16" O.C.	.36	.93	1.29
0658	On masonry, 12" O.C.	.52	1.38	1.90
0659	16" O.C.	.39	1.04	1.43
0660	On concrete, 12" O.C.	.52	2.62	3.14
0661	16" O.C.	.39	1.97	2.36
0662	Gypsum board, one face only, exterior sheathing, 1/2"	.58	1.21	1.79
0680	Interior, fire resistant, 1/2"	.42	.68	1.10
0700	5/8"	.42	.68	1.10
0720	Sound deadening board 1/4"	.46	.76	1.22
0740	Standard drywall 3/8"	.39	.68	1.07
0760	1/2"	.36	.68	1.04
0780	5/8"	.36	.68	1.04
0800	Tongue & groove coreboard 1"	.98	2.84	3.82
0820	Water resistant, 1/2"	.45	.68	1.13
0840	5/8"	.48	.68	1.16
0860	Add for the following:, foil backing	.23		.23
0880	Fiberglass insulation, 3-1/2"	.52	.50	1.02
0900	6"	.69	.50	1.19
0920	Rigid insulation 1"	.63	.68	1.31
0940	Resilient furring @ 16" O.C.	.24	2.14	2.38
0960	Taping and finishing	.06	.68	.74
0980	Texture spray	.04	.82	.86
1000	Thin coat plaster	.11	.86	.97
1050	Sound wall framing, 2x6 plates, 2x4 staggered studs, 12" O.C.	.80	1.66	2.46

C1020 Interior Doors

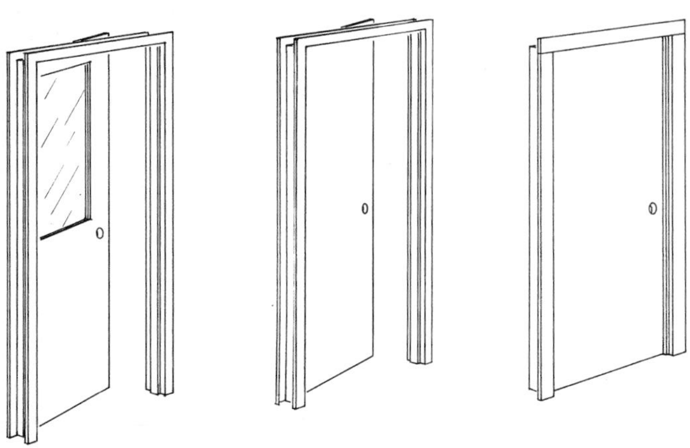

Steel Door, Half Glass
Steel Frame

Steel Door, Flush
Steel Frame

Wood Door, Flush
Wood Frame

The Metal Door/Metal Frame Systems are defined as follows: door type, design and size, frame type and depth. Included in the components for each system is painting the door and frame. No hardware has been included in the systems.

System Components	QUANTITY	UNIT	COST EACH		
			MAT.	INST.	TOTAL
SYSTEM C1020 114 1200					
STEEL DOOR, HOLLOW, 20 GA., HALF GLASS, 3'-0"X7'-0", D.W.FRAME, 4-7/8" DP					
Steel door, flush, hollow core, 1-3/4" thk, half gl, 20 Ga., 3'-0" x 7'-0"	1.000	Ea.	660	75.50	735.50
Steel frame, 16 ga., up to 4-7/8" deep, 3'-0" x 7'-0" single	1.000	Ea.	222	85	307
Hinges full mortise, avg. freq., steel base, USP, 4-1/2" x 4-1/2"	1.500	Pr.	78.75		78.75
Float glass, 3/16" thick, clear, tempered	5.000	S.F.	41.25	50.25	91.50
Paint door and frame each side, primer	1.000	Ea.	9.26	113	122.26
Paint door and frame each side, 2 coats	1.000	Ea.	10.90	188	198.90
TOTAL			1,022.16	511.75	1,533.91

C1020 114 Metal Door/Metal Frame

	TYPE	DESIGN	SIZE	FRAME	DEPTH	COST EACH		
						MAT.	INST.	TOTAL
1000	Flush-hollow	20 ga. full panel	3'-0" x 7'-0"	drywall K.D.	4-7/8"	885	465	1,350
1020				butt welded	8-3/4"	960	525	1,485
1160			6'-0" x 7'-0"	drywall K.D.	4-7/8"	1,550	740	2,290
1180				butt welded	8-3/4"	1,725	800	2,525
1200		20 ga. half glass	3'-0" x 7'-0"	drywall K.D.	4-7/8"	1,025	510	1,535
1220				butt welded	8-3/4"	1,100	570	1,670
1360			6'-0" x 7'-0"	drywall K.D.	4-7/8"	1,825	830	2,655
1380				butt welded	8-3/4"	2,000	890	2,890
1800		18 ga. full panel	3'-0" x 7'-0"	drywall K.D.	4-7/8"	875	465	1,340
1820				butt welded	8-3/4"	950	525	1,475
1960			6'-0" x 7'-0"	drywall K.D.	4-7/8"	1,525	740	2,265
1980				butt welded	8-3/4"	1,700	800	2,500
2000		18 ga. half glass	3'-0" x 7'-0"	drywall K.D.	4-7/8"	1,075	515	1,590
2020				butt welded	8-3/4"	1,150	575	1,725
2160			6'-0" x 7'-0"	drywall K.D.	4-7/8"	1,950	840	2,790
2180				butt welded	8-3/4"	2,125	900	3,025

C1020 Interior Doors

C1020 114 Metal Door/Metal Frame

	TYPE	DESIGN	SIZE	FRAME	DEPTH	COST EACH		
						MAT.	INST.	TOTAL
5000	Flush-hollow	16 ga. full panel	3'-0" x 7'-0"	drywall K.D.	4-7/8"	925	460	1,385
5020				butt welded	8-3/4"	1,000	520	1,520
5160			6'-0" x 7'-0"	drywall K.D.	4-7/8"	1,625	730	2,355
5180				butt welded	8-3/4"	1,800	790	2,590

C1020 Interior Doors

C1020 310	Hardware	COST EACH		
		MAT.	INST.	TOTAL
0060	**HINGES**			
0080				
0100	Full mortise, low frequency, steel base, 4-1/2" x 4-1/2", USP	13.75		13.75
0120	5" x 5" USP	30		30
0140	6" x 6" USP	53.50		53.50
0160	Average frequency, steel base, 4-1/2" x 4-1/2" USP	26.50		26.50
0180	5" x 5" USP	37.50		37.50
0200	6" x 6" USP	79		79
0220	High frequency, steel base, 4-1/2" x 4-1/2", USP	37		37
0240	5" x 5" USP	30		30
0260	6" x 6" USP	78.50		78.50
0280				
0300	**LOCKSETS**			
0320				
0340	Heavy duty cylindrical, passage door			
0360	Non-keyed, passage	89	57	146
0380	Privacy	85.50	57	142.50
0400	Keyed, single cylinder function	179	68	247
0420	Hotel	231	85	316
0440				
0460	For re-core cylinder, add	68.50		68.50
0480				
0500	**CLOSERS**			
0520				
0540	Rack & pinion			
0560	Adjustable backcheck, 3 way mount, all sizes, regular arm	232	114	346
0580	Hold open arm	232	114	346
0600	Fusible link	225	105	330
0620	Non-sized, regular arm	232	114	346
0640	4 way mount, non-sized, regular arm	232	114	346
0660	Hold open arm	109	114	223
0680				
0700	**PUSH, PULL**			
0720				
0740	Push plate, aluminum	17.05	57	74.05
0760	Bronze	29.50	57	86.50
0780	Pull handle, push bar, aluminum	137	62	199
0800	Bronze	186	68	254
0810				
0820	**PANIC DEVICES**			
0840				
0860	Narrow stile, rim mounted, bar, exit only	1,575	114	1,689
0880	Outside key & pull	900	136	1,036
0900	Bar and vertical rod, exit only	1,450	136	1,586
0920	Outside key & pull	1,450	170	1,620
0940	Bar and concealed rod, exit only	1,725	227	1,952
0960	Mortise, bar, exit only	1,525	170	1,695
0980	Touch bar, exit only	565	170	735
1000				
1020	**WEATHERSTRIPPING**			
1040				
1060	Interlocking, 3' x 7', zinc	64.50	227	291.50
1080	Spring type, 3' x 7', bronze	57	227	284

C3010 Wall Finishes

C3010 230	Paint & Covering	COST PER S.F.		
		MAT.	INST.	TOTAL
0060	Painting, interior on plaster and drywall, brushwork, primer & 1 coat	.15	.87	1.02
0080	Primer & 2 coats	.23	1.15	1.38
0100	Primer & 3 coats	.32	1.41	1.73
0120	Walls & ceilings, roller work, primer & 1 coat	.15	.58	.73
0140	Primer & 2 coats	.23	.74	.97
0160	Woodwork incl. puttying, brushwork, primer & 1 coat	.15	1.25	1.40
0180	Primer & 2 coats	.23	1.66	1.89
0200	Primer & 3 coats	.32	2.26	2.58
0260	Cabinets and casework, enamel, primer & 1 coat	.16	1.41	1.57
0280	Primer & 2 coats	.25	1.74	1.99
0300	Masonry or concrete, latex, brushwork, primer & 1 coat	.31	1.18	1.49
0320	Primer & 2 coats	.41	1.68	2.09
0340	Addition for block filler	.20	1.45	1.65
0380	Fireproof paints, intumescent, 1/8" thick 3/4 hour	2.39	1.15	3.54
0420	7/16" thick 2 hour	6.85	4.03	10.88
0440	1-1/16" thick 3 hour	11.20	8.05	19.25
0500	Gratings, primer & 1 coat	.39	1.76	2.15
0600	Pipes over 12" diameter	.88	5.65	6.53
0720	Heavy framing 50-100 S.F./Ton	.10	.99	1.09
0740	Spraywork, light framing 300-500 S.F./Ton	.11	.44	.55
0800	Varnish, interior wood trim, no sanding sealer & 1 coat	.08	1.41	1.49
0820	Hardwood floor, no sanding 2 coats	.17	.30	.47
0840	Wall coatings, acrylic glazed coatings, minimum	.42	1.07	1.49
0860	Maximum	.86	1.85	2.71
0880	Epoxy coatings, solvent based	.52	1.07	1.59
0900	Water based	.42	3.32	3.74
1100	High build epoxy 50 mil, solvent based	.84	1.45	2.29
1120	Water based	1.50	5.95	7.45
1140	Laminated epoxy with fiberglass solvent based	.95	1.91	2.86
1160	Water based	1.78	3.89	5.67
1180	Sprayed perlite or vermiculite 1/16" thick, solvent based	.31	.19	.50
1200	Water based	.96	.88	1.84
1260	Wall coatings, vinyl plastic, solvent based	.45	.77	1.22
1280	Water based	1.10	2.35	3.45
1300	Urethane on smooth surface, 2 coats, solvent based	.39	.50	.89
1320	Water based	.68	.85	1.53
1340	3 coats, solvent based	.48	.67	1.15
1360	Water based	1.08	1.20	2.28
1380	Ceramic-like glazed coating, cementitious, solvent based	.54	1.28	1.82
1400	Water based	1.09	1.63	2.72
1420	Resin base, solvent based	.37	.88	1.25
1440	Water based	.70	1.71	2.41
1460	Wall coverings, aluminum foil	1.16	2.06	3.22
1500	Vinyl backing	6.15	2.36	8.51
1940	Ceramic tile, thin set, 4-1/4" x 4-1/4"	3.41	5.85	9.26
1960	12" x 12"	5.50	6.90	12.40

C3020 Floor Finishes

C3020 410	Tile & Covering	COST PER S.F.		
		MAT.	INST.	TOTAL
0060	Carpet tile, nylon, fusion bonded, 18" x 18" or 24" x 24", 24 oz.	3.89	.86	4.75
0080	35 oz.	4.44	.86	5.30
0100	42 oz.	5.65	.86	6.51
0140	Carpet, tufted, nylon, roll goods, 12' wide, 26 oz.	3.01	.92	3.93
0160	36 oz.	4.56	.92	5.48
0180	Woven, wool, 36 oz.	12.45	.98	13.43
0200	42 oz.	13.65	.98	14.63
0220	Padding, add to above, 2.7 density	.73	.46	1.19
0240	13.0 density	.98	.46	1.44
0260	Composition flooring, acrylic, 1/4" thick	1.99	6.55	8.54
0280	3/8" thick	2.63	7.60	10.23
0300	Epoxy, 3/8" thick	3.54	5.05	8.59
0320	1/2" thick	4.97	6.95	11.92
0340	Epoxy terrazzo, granite chips	7.25	7.10	14.35
0360	Recycled porcelain	11.45	9.45	20.90
0380	Mastic, hot laid, 1-1/2" thick, minimum	5.20	4.95	10.15
0400	Maximum	6.70	6.55	13.25
0420	Neoprene 1/4" thick, minimum	5.15	6.25	11.40
0440	Maximum	7.05	7.95	15
0460	Polyacrylate with ground granite 1/4", granite chips	4.62	4.64	9.26
0480	Recycled porcelain	7.90	7.10	15
0500	Polyester with colored quartz chips 1/16", minimum	3.99	3.21	7.20
0520	Maximum	6.20	5.05	11.25
0540	Polyurethane with vinyl chips, clear	8.55	3.21	11.76
0560	Pigmented	12.60	3.97	16.57
0600	Concrete topping, granolithic concrete, 1/2" thick	.45	5.45	5.90
0620	1" thick	.91	5.60	6.51
0640	2" thick	1.81	6.45	8.26
0660	Heavy duty 3/4" thick, minimum	.61	8.45	9.06
0680	Maximum	1.05	10.10	11.15
0700	For colors, add to above, minimum	.49	1.95	2.44
0720	Maximum	.82	2.15	2.97
0740	Exposed aggregate finish, minimum	.24	1	1.24
0760	Maximum	.40	1.35	1.75
0780	Abrasives, .25 P.S.F. add to above, minimum	.65	.74	1.39
0800	Maximum	.93	.74	1.67
0820	Dust on coloring, add, minimum	.49	.48	.97
0840	Maximum	.82	1	1.82
0860	Floor coloring using 0.6 psf powdered color, 1/2" integral, minimum	5.10	5.45	10.55
0880	Maximum	5.40	5.45	10.85
1020	Integral topping and finish, 1:1:2 mix, 3/16" thick	.15	3.22	3.37
1040	1/2" thick	.41	3.39	3.80
1060	3/4" thick	.61	3.79	4.40
1080	1" thick	.82	4.29	5.11
1100	Terrazzo, minimum	4.22	18.90	23.12
1120	Maximum	7.85	23.50	31.35

C3030 Ceiling Finishes

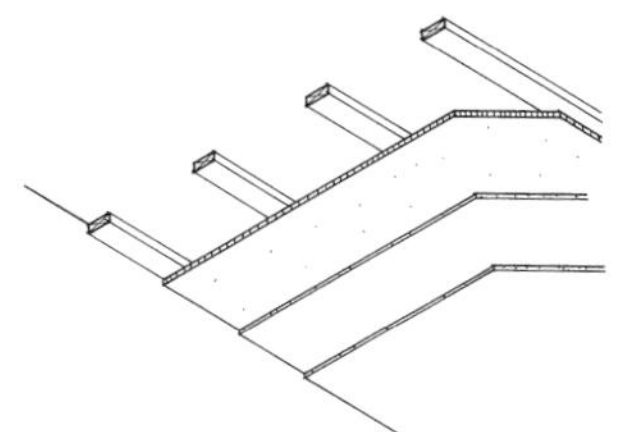

2 Coats of Plaster on Gypsum Lath on Wood Furring

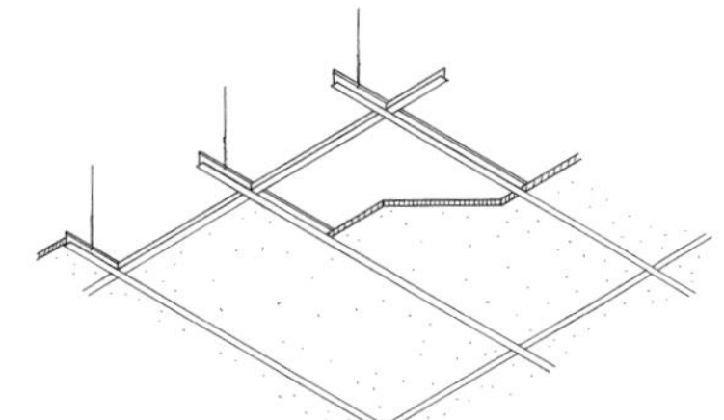

Fiberglass Board on Exposed Suspended Grid System

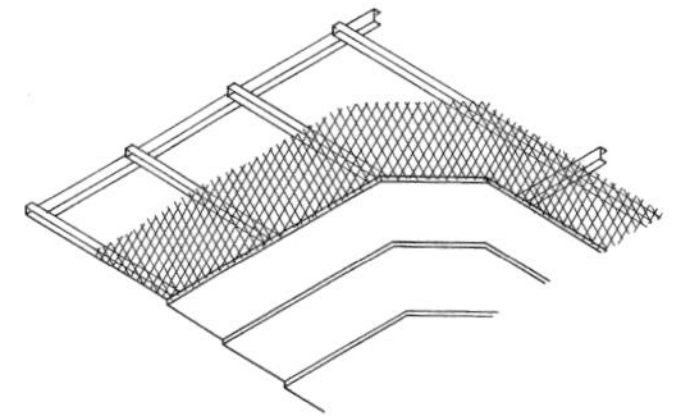

Plaster and Metal Lath on Metal Furring

System Components	QUANTITY	UNIT	COST PER S.F. MAT.	COST PER S.F. INST.	COST PER S.F. TOTAL
SYSTEM C3030 105 2400					
GYPSUM PLASTER, 2 COATS, 3/8"GYP. LATH, WOOD FURRING, FRAMING					
Gypsum plaster 2 coats no lath, on ceilings	.110	S.Y.	.48	3.71	4.19
Gypsum lath plain/perforated nailed, 3/8" thick	.110	S.Y.	.43	.85	1.28
Add for ceiling installation	.110	S.Y.		.34	.34
Furring, 1" x 3" wood strips on wood joists	.750	L.F.	.36	1.46	1.82
Paint, primer and one coat	1.000	S.F.	.15	.58	.73
TOTAL			1.42	6.94	8.36

C3030 105 Plaster Ceilings

	TYPE	LATH	FURRING	SUPPORT		COST PER S.F. MAT.	COST PER S.F. INST.	COST PER S.F. TOTAL
2400	2 coat gypsum	3/8" gypsum	1"x3" wood, 16" O.C.	wood		1.42	6.95	8.37
2500	Painted			masonry		1.45	7.05	8.50
2600				concrete		1.45	7.90	9.35
2700	3 coat gypsum	3.4# metal	1"x3" wood, 16" O.C.	wood		1.71	7.45	9.16
2800	Painted			masonry		1.74	7.55	9.29
2900				concrete		1.74	8.40	10.14
3000	2 coat perlite	3/8" gypsum	1"x3" wood, 16" O.C.	wood		1.67	7.20	8.87
3100	Painted			masonry		1.70	7.35	9.05
3200				concrete		1.70	8.15	9.85
3300	3 coat perlite	3.4# metal	1"x3" wood, 16" O.C.	wood		1.76	7.40	9.16
3400	Painted			masonry		1.79	7.50	9.29
3500				concrete		1.79	8.35	10.14
3600	2 coat gypsum	3/8" gypsum	3/4" CRC, 12" O.C.	1-1/2" CRC, 48"O.C.		1.52	8.60	10.12
3700	Painted		3/4" CRC, 16" O.C.	1-1/2" CRC, 48"O.C.		1.47	7.70	9.17
3800			3/4" CRC, 24" O.C.	1-1/2" CRC, 48"O.C.		1.33	7	8.33
3900	2 coat perlite	3/8" gypsum	3/4" CRC, 12" O.C.	1-1/2" CRC, 48"O.C		1.77	9.20	10.97
4000	Painted		3/4" CRC, 16" O.C.	1-1/2" CRC, 48"O.C.		1.72	8.35	10.07
4100			3/4" CRC, 24" O.C.	1-1/2" CRC, 48"O.C.		1.58	7.65	9.23
4200	3 coat gypsum	3.4# metal	3/4" CRC, 12" O.C.	1-1/2" CRC, 36" O.C.		2.01	11.10	13.11
4300	Painted		3/4" CRC, 16" O.C.	1-1/2" CRC, 36" O.C.		1.96	10.25	12.21
4400			3/4" CRC, 24" O.C.	1-1/2" CRC, 36" O.C.		1.82	9.55	11.37
4500	3 coat perlite	3.4# metal	3/4" CRC, 12" O.C.	1-1/2" CRC,36" O.C.		2.15	12.15	14.30
4600	Painted		3/4" CRC, 16" O.C.	1-1/2" CRC, 36" O.C.		2.10	11.30	13.40
4700			3/4" CRC, 24" O.C.	1-1/2" CRC, 36" O.C.		1.96	10.60	12.56

D1010 Elevators and Lifts

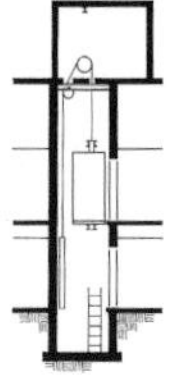

Geared traction elevators are the intermediate group both in cost and in operating areas. These are in buildings of four to fifteen floors and speed ranges from 200′ to 500′ per minute.

System Components	QUANTITY	UNIT	COST EACH		
			MAT.	INST.	TOTAL
SYSTEM D1010 140 1600					
PASSENGER, 2500 LB., 5 FLOORS, 200 FPM					
Passenger elevator, geared, 2000 lb. capacity, 4 stop, 200 FPM	1.000	Ea.	115,000	42,000	157,000
Over 40′ travel height, passenger elevator electric, add	10.000	V.L.F.	8,150	2,900	11,050
Passenger elevator, electric, 2,500 lb. cap., add	1.000	Ea.	4,925		4,925
Over 4 stops, passenger elevator, electric, add	1.000	Stop	3,550	7,775	11,325
Hall lantern	5.000	Ea.	3,000	1,315	4,315
Maintenance agreement for passenger elevator, 9 months	1.000	Ea.	4,825		4,825
Position indicator at lobby	1.000	Ea.	104	65.50	169.50
TOTAL			139,554	54,055.50	193,609.50

D1010 140	Traction Geared Elevators	COST EACH		
		MAT.	INST.	TOTAL
1300	Passenger, 2000 Lb., 5 floors, 200 FPM	134,500	54,000	188,500
1500	15 floors, 350 FPM	267,500	163,500	431,000
1600	2500 Lb., 5 floors, 200 FPM	139,500	54,000	193,500
1800	15 floors, 350 FPM	272,000	163,500	435,500
2200	3500 Lb., 5 floors, 200 FPM	141,500	54,000	195,500
2400	15 floors, 350 FPM	274,000	163,500	437,500
2500	4000 Lb., 5 floors, 200 FPM	142,500	54,000	196,500
2700	15 floors, 350 FPM	275,500	163,500	439,000
2800	4500 Lb., 5 floors, 200 FPM	145,500	54,000	199,500
3000	15 floors, 350 FPM	278,500	163,500	442,000
3100	5000 Lb., 5 floors, 200 FPM	149,000	54,000	203,000
3300	15 floors, 350 FPM	281,500	163,500	445,000
4000	Hospital, 3500 Lb., 5 floors, 200 FPM	146,500	54,000	200,500
4200	15 floors, 350 FPM	329,500	163,500	493,000
4300	4000 Lb., 5 floors, 200 FPM	146,500	54,000	200,500
4500	15 floors, 350 FPM	329,500	163,500	493,000
4600	4500 Lb., 5 floors, 200 FPM	153,000	54,000	207,000
4800	15 floors, 350 FPM	336,500	163,500	500,000
4900	5000 Lb., 5 floors, 200 FPM	155,000	54,000	209,000
5100	15 floors, 350 FPM	338,500	163,500	502,000
6000	Freight, 4000 Lb., 5 floors, 50 FPM class 'B'	162,500	56,000	218,500
6200	15 floors, 200 FPM class 'B'	350,500	194,500	545,000
6300	Freight, 8000 Lb., 5 floors, 50 FPM class 'B'	188,500	56,000	244,500
6500	15 floors, 200 FPM class 'B'	377,000	194,500	571,500
7000	10,000 Lb., 5 floors, 50 FPM class 'B'	219,500	56,000	275,500
7200	15 floors, 200 FPM class 'B'	615,000	194,500	809,500
8000	20,000 Lb., 5 floors, 50 FPM class 'B'	244,000	56,000	300,000
8200	15 floors, 200 FPM class 'B'	639,500	194,500	834,000

D1020 Escalators and Moving Walks

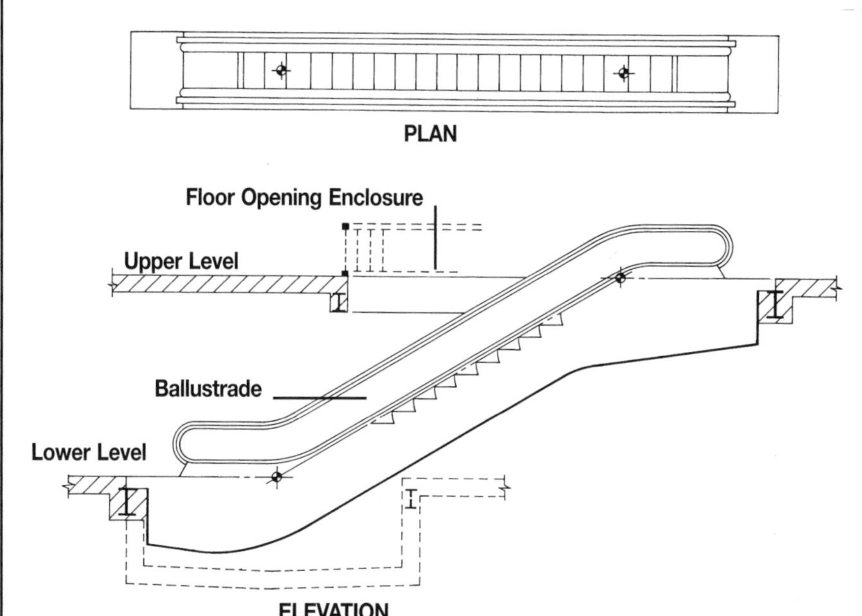

Moving stairs can be used for buildings where 600 or more people are to be carried to the second floor or beyond. Freight cannot be carried on escalators and at least one elevator must be available for this function.

Carrying capacity is 5000 to 8000 people per hour. Power requirement is 2 KW to 3 KW per hour and incline angle is 30°.

D1020 110 Moving Stairs

	TYPE	HEIGHT	WIDTH	BALUSTRADE MATERIAL		COST EACH		
						MAT.	INST.	TOTAL
0100	Escalator	10ft	32″	glass		99,500	58,000	157,500
0150				metal		108,000	58,000	166,000
0200			48″	glass		107,000	58,000	165,000
0250				metal		118,000	58,000	176,000
0300		15ft	32″	glass		105,000	67,500	172,500
0350				metal		115,500	67,500	183,000
0400			48″	glass		111,500	67,500	179,000
0450				metal		120,000	67,500	187,500
0500		20ft	32″	glass		111,500	82,500	194,000
0550				metal		122,500	82,500	205,000
0600			48″	glass		121,000	82,500	203,500
0650				metal		132,500	82,500	215,000
0700		25ft	32″	glass		123,500	101,500	225,000
0750				metal		135,000	101,500	236,500
0800			48″	glass		143,000	101,500	244,500
0850				metal		154,000	101,500	255,500

D1020 210 Moving Walks

	TYPE	STORY HEIGHT	DEGREE SLOPE	WIDTH	COST RANGE	COST PER L.F.		
						MAT.	INST.	TOTAL
1500	Ramp	10′-23′	12	3′-4″	minimum	2,825	1,150	3,975
1600					maximum	3,500	1,400	4,900
2000	Walk	0′	0	2′-0″	minimum	1,025	625	1,650
2500					maximum	1,425	910	2,335
3000				3′-4″	minimum	2,375	910	3,285
3500					maximum	2,725	1,075	3,800

D2010 Plumbing Fixtures

Systems are complete with trim seat and rough-in (supply, waste and vent) for connection to supply branches and waste mains.

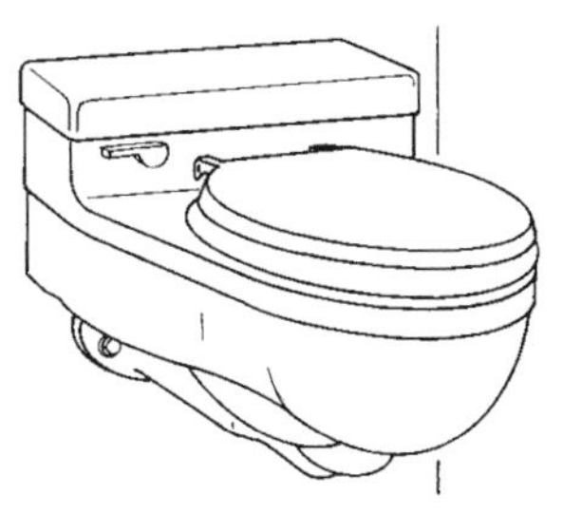

One Piece Wall Hung

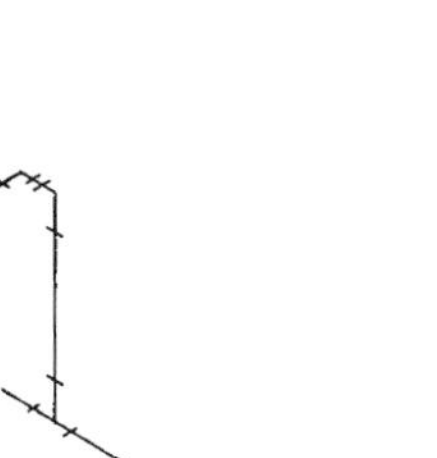

Supply

Waste/Vent

Floor Mount

System Components	QUANTITY	UNIT	COST EACH MAT.	COST EACH INST.	COST EACH TOTAL
SYSTEM D2010 110 1880					
WATER CLOSET, VITREOUS CHINA					
TANK TYPE, WALL HUNG, TWO PIECE					
Water closet, tank type vit china wall hung 2 pc. w/seat supply & stop	1.000	Ea.	460	271	731
Pipe Steel galvanized, schedule 40, threaded, 2″ diam.	4.000	L.F.	34.80	90	124.80
Pipe, CI soil, no hub, cplg 10′ OC, hanger 5′ OC, 4″ diam.	2.000	L.F.	55	49	104
Pipe, coupling, standard coupling, CI soil, no hub, 4″ diam.	2.000	Ea.	51	87	138
Copper tubing type L solder joint, hangar 10′ O.C., 1/2″ diam.	6.000	L.F.	24.30	59.10	83.40
Wrought copper 90° elbow for solder joints 1/2″ diam.	2.000	Ea.	2.78	80	82.78
Wrought copper Tee for solder joints 1/2″ diam.	1.000	Ea.	2.68	61.50	64.18
Supports/carrier, water closet, siphon jet, horiz, single, 4″ waste	1.000	Ea.	1,225	150	1,375
TOTAL			1,855.56	847.60	2,703.16

D2010 110	Water Closet Systems		COST EACH MAT.	COST EACH INST.	COST EACH TOTAL
1800	Water closet, vitreous china				
1840	Tank type, wall hung				
1880	Close coupled two piece	R221113 -40	1,850	850	2,700
1920	Floor mount, one piece		1,475	900	2,375
1960	One piece low profile	R224000 -30	1,425	900	2,325
2000	Two piece close coupled		675	900	1,575
2040	Bowl only with flush valve				
2080	Wall hung		2,775	960	3,735
2120	Floor mount		875	915	1,790
2160	Floor mount, ADA compliant with 18″ high bowl		885	940	1,825

D2010 Plumbing Fixtures

Systems are complete with trim, flush valve and rough-in (supply, waste and vent) for connection to supply branches and waste mains.

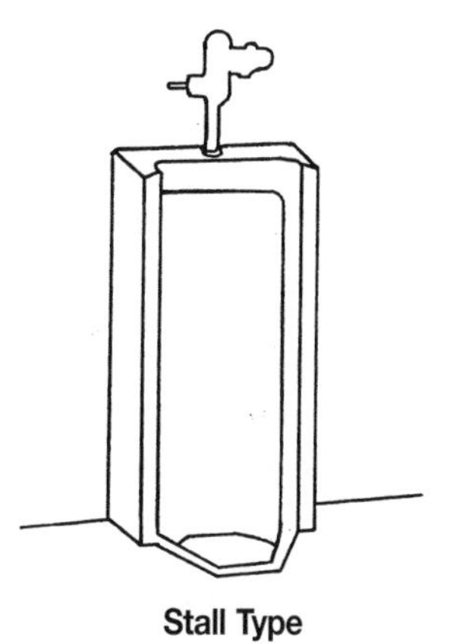

Stall Type

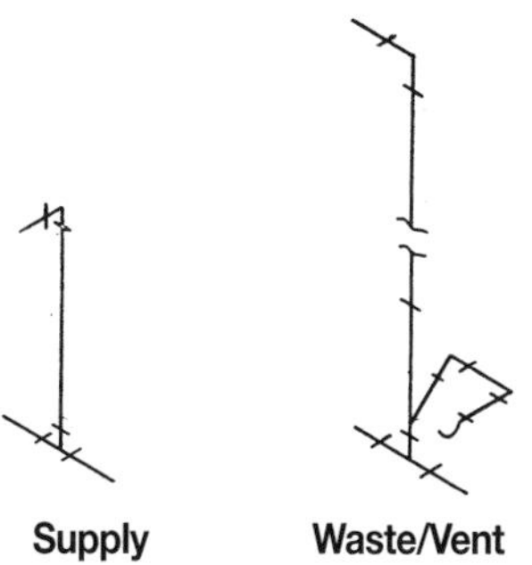

Supply Waste/Vent

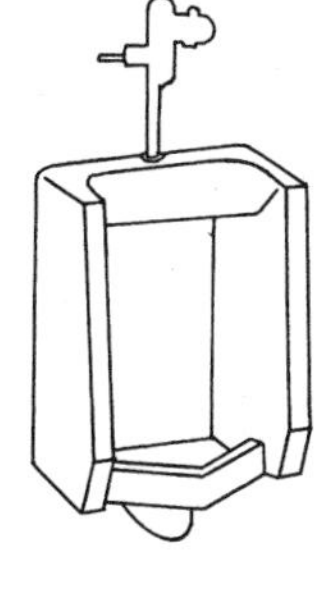

Wall Hung

System Components	QUANTITY	UNIT	COST EACH MAT.	COST EACH INST.	COST EACH TOTAL
SYSTEM D2010 210 2000					
URINAL, VITREOUS CHINA, WALL HUNG					
Urinal, wall hung, vitreous china, incl. hanger	1.000	Ea.	345	480	825
Pipe, steel, galvanized, schedule 40, threaded, 1-1/2" diam.	5.000	L.F.	52	89.75	141.75
Copper tubing type DWV, solder joint, hangers 10′ OC, 2" diam.	3.000	L.F.	60	54.30	114.30
Combination Y & 1/8 bend for CI soil pipe, no hub, 3" diam.	1.000	Ea.	23.50		23.50
Pipe, CI, no hub, cplg. 10′ OC, hanger 5′ OC, 3" diam.	4.000	L.F.	118	90	208
Pipe coupling standard, CI soil, no hub, 3" diam.	2.000	Ea.	44	76	120
Copper tubing type L, solder joint, hanger 10′ OC 3/4" diam.	5.000	L.F.	26	52.50	78.50
Wrought copper 90° elbow for solder joints 3/4" diam.	1.000	Ea.	2.95	42	44.95
Wrought copper Tee for solder joints, 3/4" diam.	1.000	Ea.	6.65	66.50	73.15
TOTAL			678.10	951.05	1,629.15

D2010 210	Urinal Systems		COST EACH MAT.	COST EACH INST.	COST EACH TOTAL
2000	Urinal, vitreous china, wall hung	R224000 -30	680	950	1,630
2040	Stall type		1,475	1,125	2,600

D2010 Plumbing Fixtures

Systems are complete with trim and rough-in (supply, waste and vent) to connect to supply branches and waste mains.

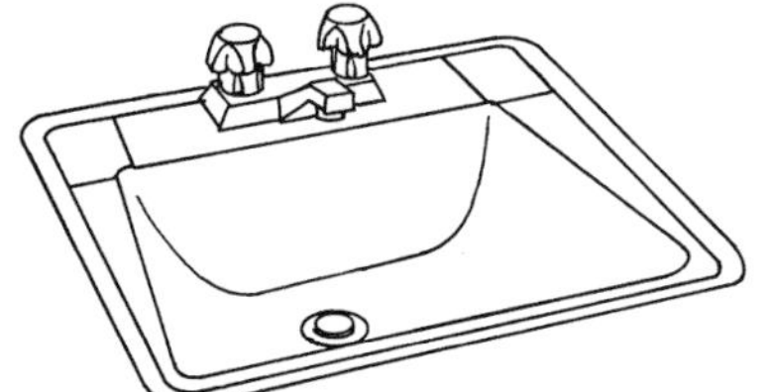

Vanity Top

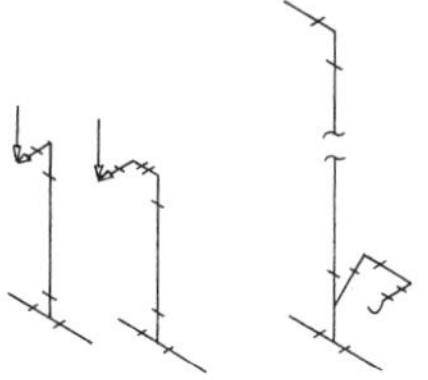

Supply Waste/Vent

Wall Hung

System Components	QUANTITY	UNIT	COST EACH MAT.	INST.	TOTAL
SYSTEM D2010 310 1560					
LAVATORY W/TRIM, VANITY TOP, P.E. ON C.I., 20″ X 18″					
Lavatory w/trim, PE on CI, white, vanity top, 20″ x 18″ oval	1.000	Ea.	340	224	564
Pipe, steel, galvanized, schedule 40, threaded, 1-1/4″ diam.	4.000	L.F.	38.80	64.40	103.20
Copper tubing type DWV, solder joint, hanger 10′ OC 1-1/4″ diam.	4.000	L.F.	54.80	53.20	108
Wrought copper DWV, Tee, sanitary, 1-1/4″ diam.	1.000	Ea.	31.50	88.50	120
P trap w/cleanout, 20 ga., 1-1/4″ diam.	1.000	Ea.	113	44.50	157.50
Copper tubing type L, solder joint, hanger 10′ OC 1/2″ diam.	10.000	L.F.	40.50	98.50	139
Wrought copper 90° elbow for solder joints 1/2″ diam.	2.000	Ea.	2.78	80	82.78
Wrought copper Tee for solder joints, 1/2″ diam.	2.000	Ea.	5.36	123	128.36
Stop, chrome, angle supply, 1/2″ diam.	2.000	Ea.	22.20	72	94.20
TOTAL			648.94	848.10	1,497.04

D2010 310	Lavatory Systems		COST EACH MAT.	INST.	TOTAL
1560	Lavatory w/trim, vanity top, PE on CI, 20″ x 18″, Vanity top by others.	R221113 -40	650	850	1,500
1600	19″ x 16″ oval		440	850	1,290
1640	18″ round		810	850	1,660
1680	Cultured marble, 19″ x 17″		445	850	1,295
1720	25″ x 19″		475	850	1,325
1760	Stainless, self-rimming, 25″ x 22″		660	850	1,510
1800	17″ x 22″	R224000 -30	645	850	1,495
1840	Steel enameled, 20″ x 17″		450	870	1,320
1880	19″ round		485	870	1,355
1920	Vitreous china, 20″ x 16″		540	890	1,430
1960	19″ x 16″		550	890	1,440
2000	22″ x 13″		545	890	1,435
2040	Wall hung, PE on CI, 18″ x 15″		910	935	1,845
2080	19″ x 17″		970	935	1,905
2120	20″ x 18″		795	935	1,730
2160	Vitreous china, 18″ x 15″		715	960	1,675
2200	19″ x 17″		660	960	1,620
2240	24″ x 20″		800	960	1,760
2300	20″ x 27″, handicap		1,600	1,025	2,625

D2010 Plumbing Fixtures

Systems are complete with trim and rough-in (supply, waste and vent) to connect to supply branches and waste mains.

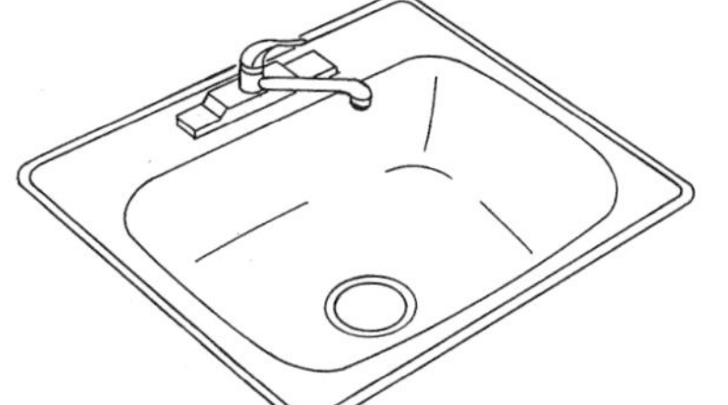

Countertop Single Bowl

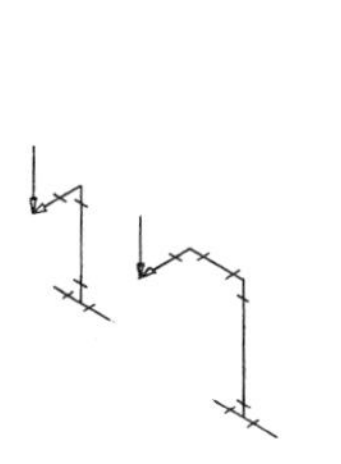

Supply

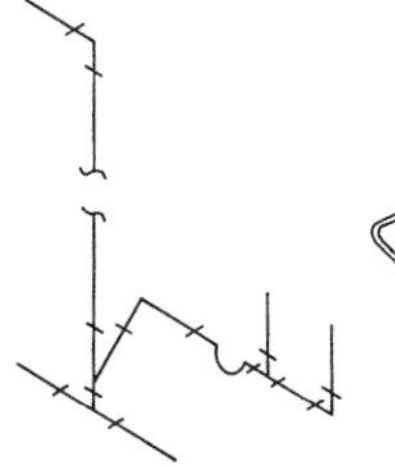

Waste/Vent

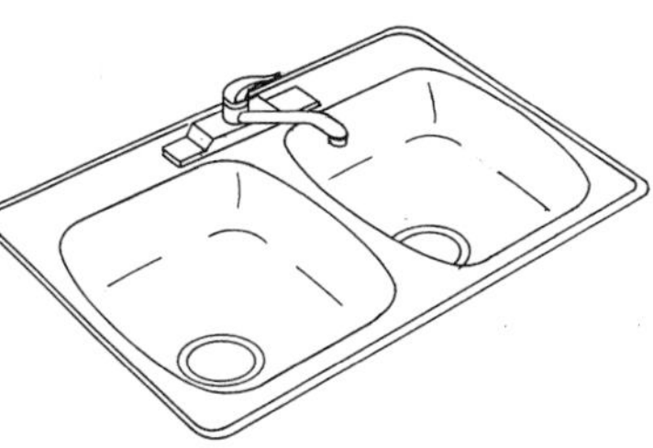

Countertop Double Bowl

System Components	QUANTITY	UNIT	COST EACH		
			MAT.	INST.	TOTAL
SYSTEM D2010 410 1720					
KITCHEN SINK W/TRIM, COUNTERTOP, P.E. ON C.I., 24" X 21", SINGLE BOWL					
Kitchen sink, counter top style, PE on CI, 24" x 21" single bowl	1.000	Ea.	340	256	596
Pipe, steel, galvanized, schedule 40, threaded, 1-1/4" diam.	4.000	L.F.	38.80	64.40	103.20
Copper tubing, type DWV, solder, hangers 10' OC 1-1/2" diam.	6.000	L.F.	83.10	88.50	171.60
Wrought copper, DWV, Tee, sanitary, 1-1/2" diam.	1.000	Ea.	39	99.50	138.50
P trap, standard, copper, 1-1/2" diam.	1.000	Ea.	123	47	170
Copper tubing, type L, solder joints, hangers 10' OC 1/2" diam.	10.000	L.F.	40.50	98.50	139
Wrought copper 90° elbow for solder joints 1/2" diam.	2.000	Ea.	2.78	80	82.78
Wrought copper Tee for solder joints, 1/2" diam.	2.000	Ea.	5.36	123	128.36
Stop, angle supply, chrome, 1/2" CTS	2.000	Ea.	22.20	72	94.20
TOTAL			694.74	928.90	1,623.64

D2010 410	Kitchen Sink Systems		COST EACH		
			MAT.	INST.	TOTAL
1720	Kitchen sink w/trim, countertop, PE on CI, 24"x21", single bowl	R221113 -40	695	930	1,625
1760	30" x 21" single bowl		1,300	930	2,230
1800	32" x 21" double bowl		810	1,000	1,810
1880	Stainless steel, 19" x 18" single bowl		1,050	930	1,980
1920	25" x 22" single bowl		1,125	930	2,055
1960	33" x 22" double bowl		1,475	1,000	2,475
2000	43" x 22" double bowl		1,675	1,025	2,700
2040	44" x 22" triple bowl		1,750	1,050	2,800
2080	44" x 24" corner double bowl		1,275	1,025	2,300
2120	Steel, enameled, 24" x 21" single bowl		945	930	1,875
2160	32" x 21" double bowl		950	1,000	1,950
2240	Raised deck, PE on CI, 32" x 21", dual level, double bowl		925	1,275	2,200
2280	42" x 21" dual level, triple bowl		1,500	1,375	2,875

D2010 Plumbing Fixtures

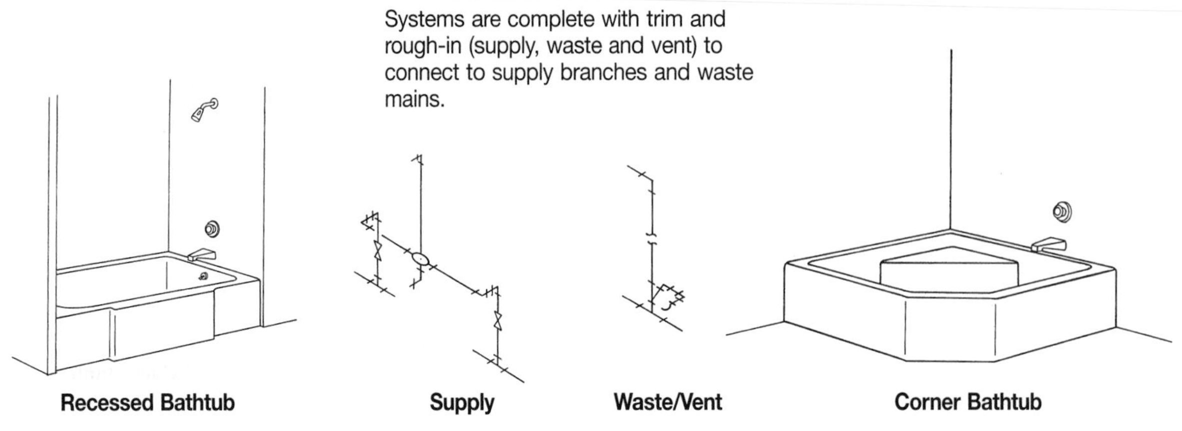

System Components	QUANTITY	UNIT	COST EACH MAT.	COST EACH INST.	COST EACH TOTAL
SYSTEM D2010 510 2000					
BATHTUB, RECESSED, PORCELAIN ENAMEL ON CAST IRON,, 48″ x 42″					
Bath tub, porcelain enamel on cast iron, w/fittings, 48″ x 42″	1.000	Ea.	3,450	360	3,810
Pipe, steel, galvanized, schedule 40, threaded, 1-1/4″ diam.	4.000	L.F.	38.80	64.40	103.20
Pipe, CI no hub soil w/couplings 10′ OC, hangers 5′ OC, 4″ diam.	3.000	L.F.	82.50	73.50	156
Combination Y and 1/8 bend for C.I. soil pipe, no hub, 4″ pipe size	1.000	Ea.	62.50		62.50
Drum trap, 3″ x 5″, copper, 1-1/2″ diam.	1.000	Ea.	172	50	222
Copper tubing type L, solder joints, hangers 10′ OC 1/2″ diam.	10.000	L.F.	40.50	98.50	139
Wrought copper 90° elbow, solder joints, 1/2″ diam.	2.000	Ea.	2.78	80	82.78
Wrought copper Tee, solder joints, 1/2″ diam.	2.000	Ea.	5.36	123	128.36
Stop, angle supply, 1/2″ diameter	2.000	Ea.	22.20	72	94.20
Copper tubing type DWV, solder joints, hanger 10′ OC 1-1/2″ diam.	3.000	L.F.	41.55	44.25	85.80
Pipe coupling, standard, C.I. soil no hub, 4″ pipe size	2.000	Ea.	51	87	138
TOTAL			3,969.19	1,052.65	5,021.84

D2010 510	Bathtub Systems		COST EACH MAT.	COST EACH INST.	COST EACH TOTAL
2000	Bathtub, recessed, P.E. on CI., 48″ x 42″	R221113 -40	3,975	1,050	5,025
2040	72″ x 36″		3,800	1,175	4,975
2080	Mat bottom, 5′ long	R224000 -30	1,925	1,025	2,950
2120	5′-6″ long		2,675	1,050	3,725
2160	Corner, 48″ x 42″		3,725	1,025	4,750
2200	Formed steel, enameled, 4′-6″ long		1,075	940	2,015

D2010 Plumbing Fixtures

Systems are complete with trim, flush valve and rough-in (supply, waste and vent) for connection to supply branches and waste mains.

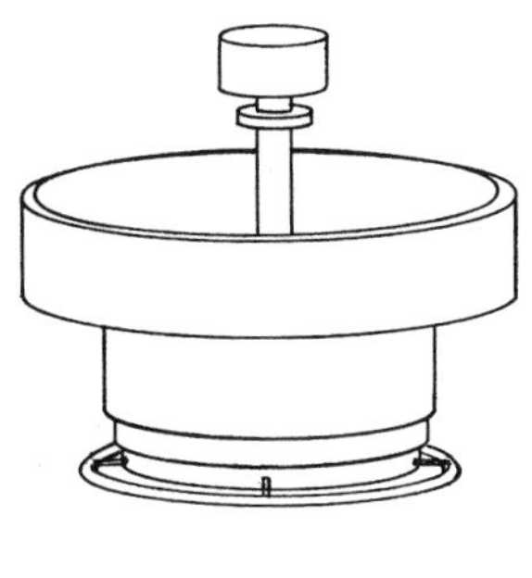

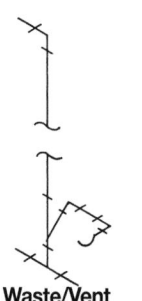

Circular Fountain

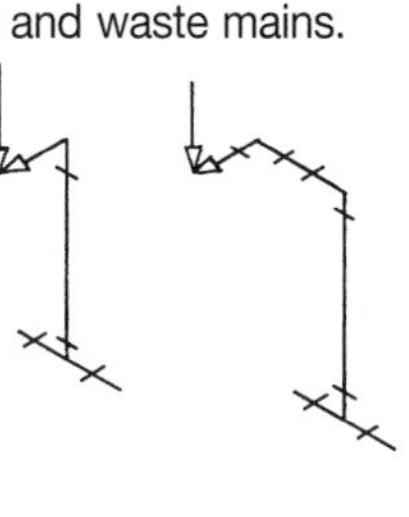

Supply

Waste/Vent

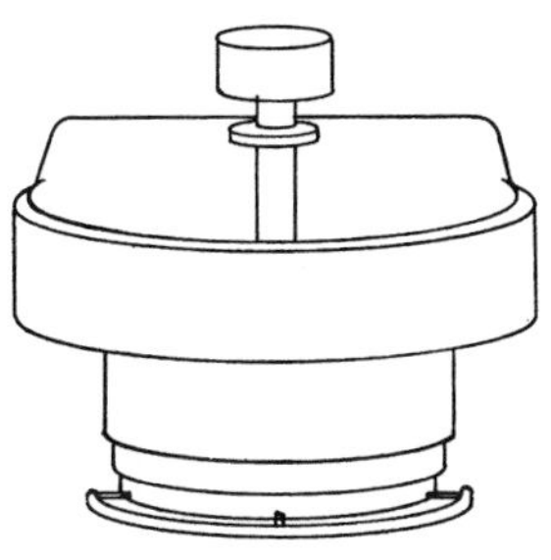

Semi-Circular Fountain

System Components	QUANTITY	UNIT	COST EACH MAT.	COST EACH INST.	COST EACH TOTAL
SYSTEM D2010 610 1760					
GROUP WASH FOUNTAIN, PRECAST TERRAZZO					
CIRCULAR, 36″ DIAMETER					
Wash fountain, group, precast terrazzo, foot control 36″ diam.	1.000	Ea.	8,425	745	9,170
Copper tubing type DWV, solder joint, hanger 10′ OC, 2″ diam.	10.000	L.F.	200	181	381
P trap, standard, copper, 2″ diam.	1.000	Ea.	202	53	255
Wrought copper, Tee, sanitary, 2″ diam.	1.000	Ea.	60.50	114	174.50
Copper tubing type L, solder joint, hanger 10′ OC 1/2″ diam.	20.000	L.F.	81	197	278
Wrought copper 90° elbow for solder joints 1/2″ diam.	3.000	Ea.	4.17	120	124.17
Wrought copper Tee for solder joints, 1/2″ diam.	2.000	Ea.	5.36	123	128.36
TOTAL			8,978.03	1,533	10,511.03

D2010 610	Group Wash Fountain Systems	COST EACH MAT.	COST EACH INST.	COST EACH TOTAL
1740	Group wash fountain, precast terrazzo			
1760	Circular, 36″ diameter (R221113-40)	8,975	1,525	10,500
1800	54″ diameter (R224000-30)	12,200	1,675	13,875
1840	Semi-circular, 36″ diameter	7,425	1,525	8,950
1880	54″ diameter	11,300	1,675	12,975
1960	Stainless steel, circular, 36″ diameter	8,000	1,425	9,425
2000	54″ diameter	9,975	1,575	11,550
2040	Semi-circular, 36″ diameter	6,750	1,425	8,175
2080	54″ diameter	8,425	1,575	10,000
2160	Thermoplastic, circular, 36″ diameter	5,725	1,150	6,875
2200	54″ diameter	6,550	1,350	7,900
2240	Semi-circular, 36″ diameter	5,775	1,150	6,925
2280	54″ diameter	7,425	1,350	8,775

D2010 Plumbing Fixtures

Three Wall

Systems are complete with trim and rough-in (supply, waste and vent) for connection to supply branches and waste mains.

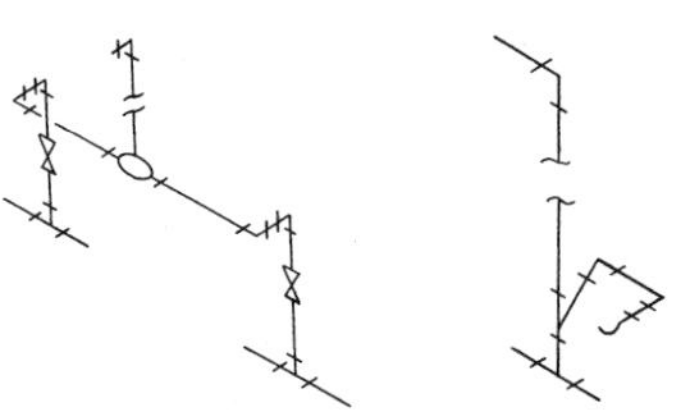
Supply

Waste/Vent

Corner Angle

System Components	QUANTITY	UNIT	COST EACH MAT.	COST EACH INST.	COST EACH TOTAL
SYSTEM D2010 710 1560					
SHOWER, STALL, BAKED ENAMEL, MOLDED STONE RECEPTOR, 30″ SQUARE					
Shower stall, enameled steel, molded stone receptor, 30″ square	1.000	Ea.	1,425	276	1,701
Copper tubing type DWV, solder joints, hangers 10′ OC, 2″ diam.	6.000	L.F.	83.10	88.50	171.60
Wrought copper DWV, Tee, sanitary, 2″ diam.	1.000	Ea.	39	99.50	138.50
Trap, standard, copper, 2″ diam.	1.000	Ea.	123	47	170
Copper tubing type L, solder joint, hanger 10′ OC 1/2″ diam.	16.000	L.F.	64.80	157.60	222.40
Wrought copper 90° elbow for solder joints 1/2″ diam.	3.000	Ea.	4.17	120	124.17
Wrought copper Tee for solder joints, 1/2″ diam.	2.000	Ea.	5.36	123	128.36
Stop and waste, straightway, bronze, solder joint 1/2″ diam.	2.000	Ea.	45	66	111
TOTAL			1,789.43	977.60	2,767.03

D2010 710	Shower Systems		COST EACH MAT.	COST EACH INST.	COST EACH TOTAL
1560	Shower, stall, baked enamel, molded stone receptor, 30″ square	R221113 -40	1,800	980	2,780
1600	32″ square		1,650	990	2,640
1640	Terrazzo receptor, 32″ square	R224000 -30	1,900	990	2,890
1680	36″ square		2,200	1,000	3,200
1720	36″ corner angle		2,425	410	2,835
1800	Fiberglass one piece, three walls, 32″ square		755	965	1,720
1840	36″ square		875	965	1,840
1880	Polypropylene, molded stone receptor, 30″ square		1,150	1,425	2,575
1920	32″ square		1,175	1,425	2,600
1960	Built-in head, arm, bypass, stops and handles		138	370	508

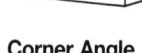

D2010 Plumbing Fixtures

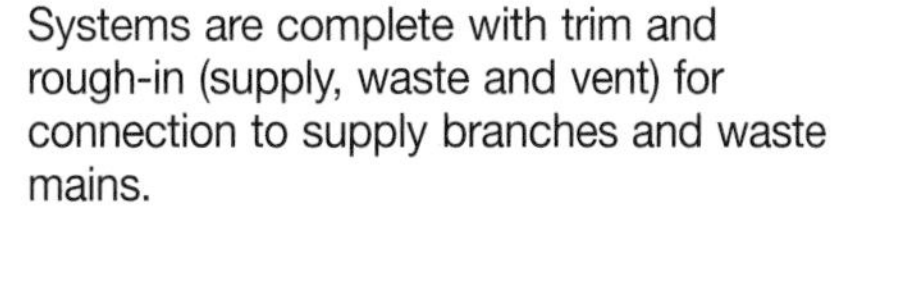

Systems are complete with trim and rough-in (supply, waste and vent) for connection to supply branches and waste mains.

Wall Hung

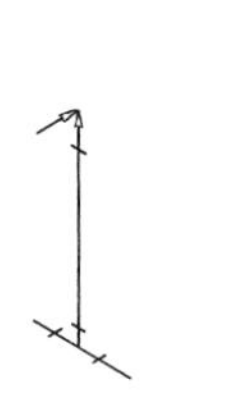

Supply

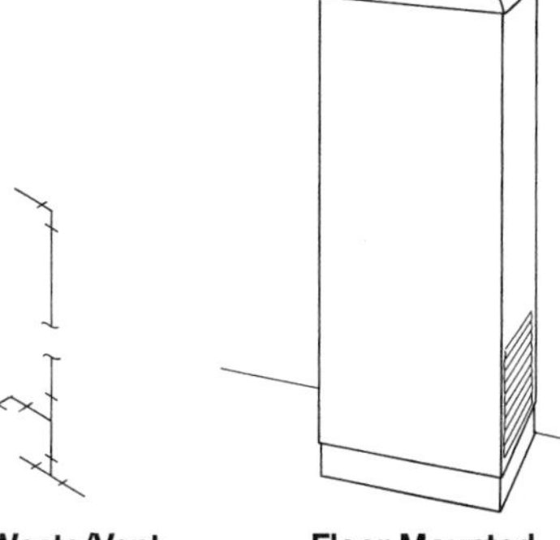

Waste/Vent

Floor Mounted

System Components	QUANTITY	UNIT	COST EACH MAT.	COST EACH INST.	COST EACH TOTAL
SYSTEM D2010 820 1840					
WATER COOLER, ELECTRIC, SELF CONTAINED, WALL HUNG, 8.2 G.P.H.					
Water cooler, wall mounted, 8.2 GPH	1.000	Ea.	1,125	360	1,485
Copper tubing type DWV, solder joint, hanger 10′ OC 1-1/4″ diam.	4.000	L.F.	54.80	53.20	108
Wrought copper DWV, Tee, sanitary 1-1/4″ diam.	1.000	Ea.	31.50	88.50	120
P trap, copper drainage, 1-1/4″ diam.	1.000	Ea.	113	44.50	157.50
Copper tubing type L, solder joint, hanger 10′ OC 3/8″ diam.	5.000	L.F.	18.35	47.50	65.85
Wrought copper 90° elbow for solder joints 3/8″ diam.	1.000	Ea.	4.57	36	40.57
Wrought copper Tee for solder joints, 3/8″ diam.	1.000	Ea.	7.55	57	64.55
Stop and waste, straightway, bronze, solder, 3/8″ diam.	1.000	Ea.	22.50	33	55.50
TOTAL			1,377.27	719.70	2,096.97

D2010 820	Water Cooler Systems		COST EACH MAT.	COST EACH INST.	COST EACH TOTAL
1840	Water cooler, electric, wall hung, 8.2 G.P.H.	R221113 -40	1,375	720	2,095
1880	Dual height, 14.3 G.P.H.		2,175	740	2,915
1920	Wheelchair type, 7.5 G.P.H.	R224000 -30	1,450	720	2,170
1960	Semi recessed, 8.1 G.P.H.		1,250	720	1,970
2000	Full recessed, 8 G.P.H.		2,825	770	3,595
2040	Floor mounted, 14.3 G.P.H.		1,425	625	2,050
2080	Dual height, 14.3 G.P.H.		1,650	760	2,410
2120	Refrigerated compartment type, 1.5 G.P.H.		1,975	625	2,600

D2010 Plumbing Fixtures

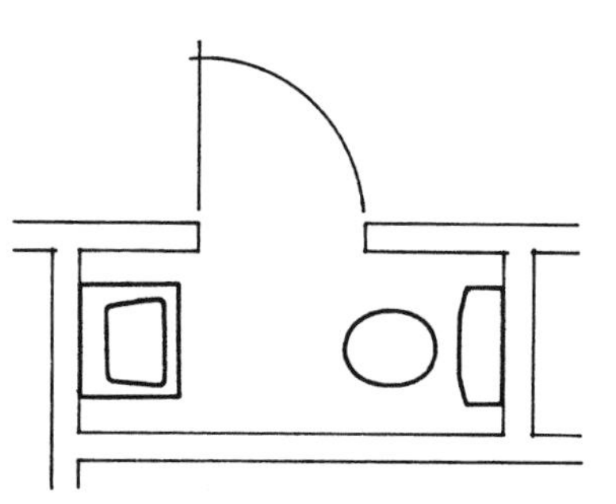

Two Fixture Bathroom Systems consisting of a lavatory, water closet, and rough-in service piping.

- Prices for plumbing and fixtures only.

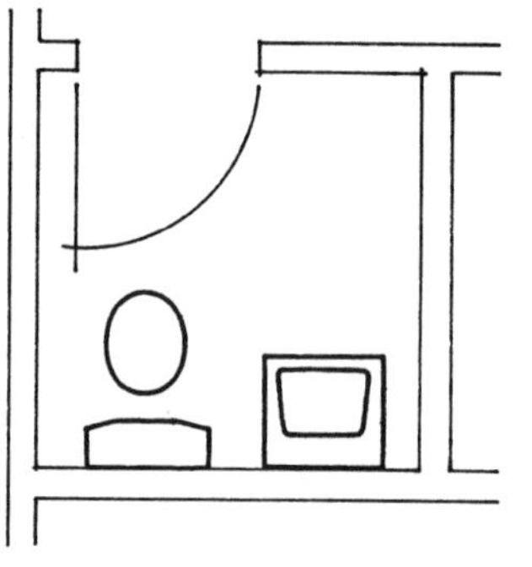

*Common wall is with an adjacent bathroom.

System Components	QUANTITY	UNIT	COST EACH MAT.	COST EACH INST.	COST EACH TOTAL
SYSTEM D2010 920 1180					
BATHROOM, LAVATORY & WATER CLOSET, 2 WALL PLUMBING, STAND ALONE					
Water closet, two piece, close coupled	1.000	Ea.	241	271	512
Water closet, rough-in waste & vent	1.000	Set	375	470	845
Lavatory w/ftngs., wall hung, white, PE on CI, 20″ x 18″	1.000	Ea.	271	179	450
Lavatory, rough-in waste & vent	1.000	Set	520	865	1,385
Copper tubing type L, solder joint, hanger 10′ OC 1/2″ diam.	10.000	L.F.	40.50	98.50	139
Pipe, steel, galvanized, schedule 40, threaded, 2″ diam.	12.000	L.F.	104.40	270	374.40
Pipe, CI soil, no hub, coupling 10′ OC, hanger 5′ OC, 4″ diam.	7.000	L.F.	189	182	371
TOTAL			1,740.90	2,335.50	4,076.40

D2010 920	Two Fixture Bathroom, Two Wall Plumbing		COST EACH MAT.	COST EACH INST.	COST EACH TOTAL
1180	Bathroom, lavatory & water closet, 2 wall plumbing, stand alone	R221113 -40	1,750	2,325	4,075
1200	Share common plumbing wall*		1,625	2,000	3,625

D2010 922	Two Fixture Bathroom, One Wall Plumbing		COST EACH MAT.	COST EACH INST.	COST EACH TOTAL
2220	Bathroom, lavatory & water closet, one wall plumbing, stand alone	R221113 -40	1,650	2,100	3,750
2240	Share common plumbing wall*		1,400	1,775	3,175

D3030 Cooling Generating Systems

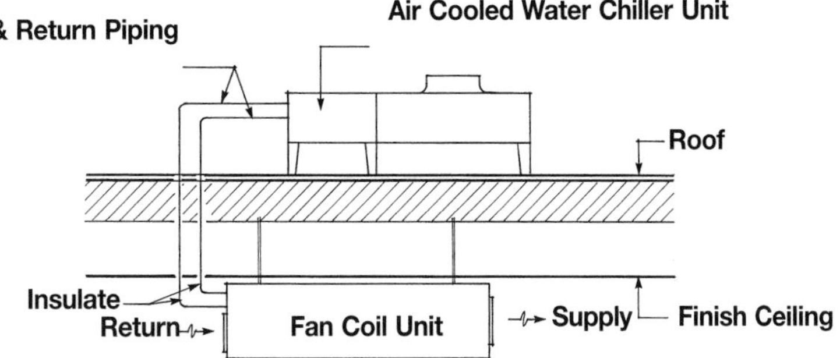

Design Assumptions: The chilled water, air cooled systems priced, utilize reciprocating hermetic compressors and propeller-type condenser fans. Piping with pumps and expansion tanks is included based on a two pipe system. No ducting is included and the fan-coil units are cooling only. Water treatment and balancing are not included. Chilled water piping is insulated. Area distribution is through the use of multiple fan coil units. Fewer but larger fan coil units with duct distribution would be approximately the same S.F. cost.

System Components	QUANTITY	UNIT	COST EACH		
			MAT.	INST.	TOTAL
SYSTEM D3030 110 1200					
PACKAGED CHILLER, AIR COOLED, WITH FAN COIL UNIT					
APARTMENT CORRIDORS, 3,000 S.F., 5.50 TON					
Fan coil air conditioning unit, cabinet mounted & filters chilled water	1.000	Ea.	3,986.78	669.05	4,655.83
Water chiller, air conditioning unit, air cooled	1.000	Ea.	7,067.50	2,495.63	9,563.13
Chilled water unit coil connections	1.000	Ea.	1,325	1,825	3,150
Chilled water distribution piping	440.000	L.F.	11,000	24,640	35,640
TOTAL			23,379.28	29,629.68	53,008.96
COST PER S.F.			7.79	9.88	17.67

D3030 110	Chilled Water, Air Cooled Condenser Systems	COST PER S.F.		
		MAT.	INST.	TOTAL
1180	Packaged chiller, air cooled, with fan coil unit			
1200	Apartment corridors, 3,000 S.F., 5.50 ton	7.80	9.87	17.67
1360	40,000 S.F., 73.33 ton	5.35	4.30	9.65
1440	Banks and libraries, 3,000 S.F., 12.50 ton	11.45	11.45	22.90
1560	20,000 S.F., 83.33 ton	9	5.80	14.80
1680	Bars and taverns, 3,000 S.F., 33.25 ton	19.55	13.40	32.95
1760	10,000 S.F., 110.83 ton	13.15	3.06	16.21
1920	Bowling alleys, 3,000 S.F., 17.00 ton	14.30	12.85	27.15
1960	6,000 S.F., 34.00 ton	10.95	8.95	19.90
2000	10,000 S.F., 56.66 ton	9.40	6.40	15.80
2040	20,000 S.F., 113.33 ton	9.95	5.85	15.80
2160	Department stores, 3,000 S.F., 8.75 ton	10.75	10.95	21.70
2320	40,000 S.F., 116.66 ton	6.55	4.42	10.97
2400	Drug stores, 3,000 S.F., 20.00 ton	16.20	13.25	29.45
2520	20,000 S.F., 133.33 ton	11.80	6.15	17.95
2640	Factories, 2,000 S.F., 10.00 ton	10.15	11.05	21.20
2800	40,000 S.F., 133.33 ton	7.05	4.53	11.58
2880	Food supermarkets, 3,000 S.F., 8.50 ton	10.60	10.95	21.55
3040	40,000 S.F., 113.33 ton	6.25	4.44	10.69
3120	Medical centers, 3,000 S.F., 7.00 ton	9.60	10.65	20.25
3280	40,000 S.F., 93.33 ton	5.85	4.36	10.21
3360	Offices, 3,000 S.F., 9.50 ton	9.65	10.85	20.50
3520	40,000 S.F., 126.66 ton	7	4.58	11.58
3600	Restaurants, 3,000 S.F., 15.00 ton	12.80	11.90	24.70
3720	20,000 S.F., 100.00 ton	10.25	6.15	16.40
3840	Schools and colleges, 3,000 S.F., 11.50 ton	10.95	11.30	22.25
3960	20,000 S.F., 76.66 ton	8.85	5.85	14.70

D3030 Cooling Generating Systems

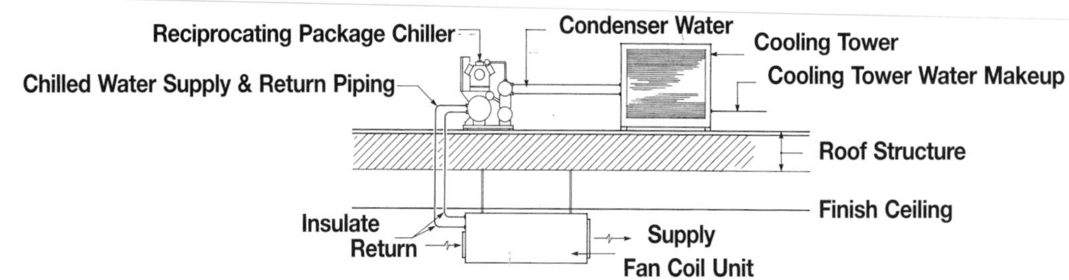

General: Water cooled chillers are available in the same sizes as air cooled units. They are also available in larger capacities.

Design Assumptions: The chilled water systems with water cooled condenser include reciprocating hermetic compressors, water cooling tower, pumps, piping and expansion tanks and are based on a two pipe system. Chilled water piping is insulated. No ducts are included and fan-coil units are cooling only. Area distribution is through use of multiple fan coil units. Fewer but larger fan coil units with duct distribution would be approximately the same S.F. cost. Water treatment and balancing are not included.

System Components	QUANTITY	UNIT	COST EACH		
			MAT.	INST.	TOTAL
SYSTEM D3030 115 1320					
PACKAGED CHILLER, WATER COOLED, WITH FAN COIL UNIT					
APARTMENT CORRIDORS, 4,000 S.F., 7.33 TON					
Fan coil air conditioner unit, cabinet mounted & filters, chilled water	2.000	Ea.	5,313.53	891.70	6,205.23
Water chiller, water cooled, 1 compressor, hermetic scroll,	1.000	Ea.	5,770.80	4,351	10,121.80
Cooling tower, draw thru single flow, belt drive	1.000	Ea.	1,634.59	183.25	1,817.84
Cooling tower pumps & piping	1.000	System	813.63	439.80	1,253.43
Chilled water unit coil connections	2.000	Ea.	2,650	3,650	6,300
Chilled water distribution piping	520.000	L.F.	13,000	29,120	42,120
TOTAL			29,182.55	38,635.75	67,818.30
COST PER S.F.			7.30	9.66	16.96

D3030 115	Chilled Water, Cooling Tower Systems	COST PER S.F.		
		MAT.	INST.	TOTAL
1300	Packaged chiller, water cooled, with fan coil unit			
1320	Apartment corridors, 4,000 S.F., 7.33 ton	7.30	9.65	16.95
1600	Banks and libraries, 4,000 S.F., 16.66 ton	13	10.45	23.45
1800	60,000 S.F., 250.00 ton	9.50	8.45	17.95
1880	Bars and taverns, 4,000 S.F., 44.33 ton	20.50	13	33.50
2000	20,000 S.F., 221.66 ton	21	10.90	31.90
2160	Bowling alleys, 4,000 S.F., 22.66 ton	14.95	11.75	26.70
2320	40,000 S.F., 226.66 ton	12.05	8	20.05
2440	Department stores, 4,000 S.F., 11.66 ton	8.10	10.40	18.50
2640	60,000 S.F., 175.00 ton	8.40	7.80	16.20
2720	Drug stores, 4,000 S.F., 26.66 ton	16.75	11.75	28.50
2880	40,000 S.F., 266.67 ton	12.55	9.05	21.60
3000	Factories, 4,000 S.F., 13.33 ton	11.25	10	21.25
3200	60,000 S.F., 200.00 ton	8.65	8.15	16.80
3280	Food supermarkets, 4,000 S.F., 11.33 ton	8	10.35	18.35
3480	60,000 S.F., 170.00 ton	8.35	7.80	16.15
3560	Medical centers, 4.000 S.F., 9.33 ton	7	9.45	16.45
3760	60,000 S.F., 140.00 ton	7.20	7.90	15.10
3840	Offices, 4,000 S.F., 12.66 ton	10.90	9.90	20.80
4040	60,000 S.F., 190.00 ton	8.35	8.05	16.40
4120	Restaurants, 4,000 S.F., 20.00 ton	13.45	10.90	24.35
4320	60,000 S.F., 300.00 ton	10.55	8.75	19.30
4400	Schools and colleges, 4,000 S.F., 15.33 ton	12.35	10.30	22.65
4600	60,000 S.F., 230.00 ton	8.80	8.15	16.95

D3050 Terminal & Package Units

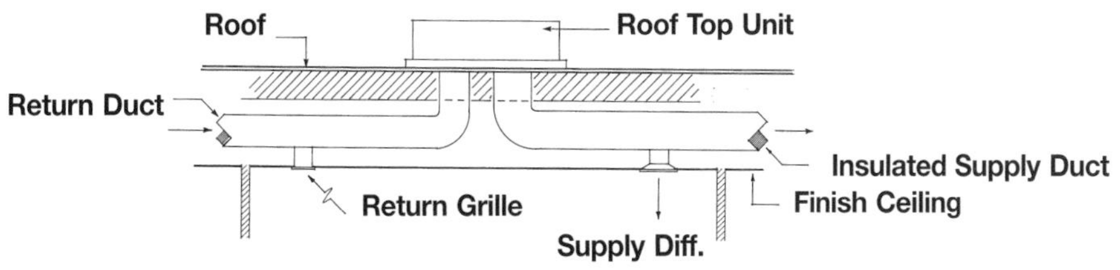

System Description: Rooftop single zone units are electric cooling and gas heat. Duct systems are low velocity, galvanized steel supply and return. Price variations between sizes are due to several factors. Jumps in the cost of the rooftop unit occur when the manufacturer shifts from the largest capacity unit on a small frame to the smallest capacity on the next larger frame, or changes from one compressor to two. As the unit capacity increases for larger areas the duct distribution grows in proportion. For most applications there is a tradeoff point where it is less expensive and more efficient to utilize smaller units with short simple distribution systems. Larger units also require larger initial supply and return ducts which can create a space problem. Supplemental heat may be desired in colder locations. The table below is based on one unit supplying the area listed. The 10,000 S.F. unit for bars and taverns is not listed because a nominal 110 ton unit would be required and this is above the normal single zone rooftop capacity.

System Components	QUANTITY	UNIT	COST EACH MAT.	COST EACH INST.	COST EACH TOTAL
SYSTEM D3050 150 1280					
ROOFTOP, SINGLE ZONE, AIR CONDITIONER					
APARTMENT CORRIDORS, 500 S.F., .92 TON					
Rooftop air conditioner, 1 zone, electric cool, standard controls, curb	1.000	Ea.	1,460.50	954.50	2,415
Ductwork package for rooftop single zone units	1.000	System	294.40	1,357	1,651.40
TOTAL			1,754.90	2,311.50	4,066.40
COST PER S.F.			3.51	4.62	8.13

D3050 150	Rooftop Single Zone Unit Systems	COST PER S.F. MAT.	COST PER S.F. INST.	COST PER S.F. TOTAL
1260	Rooftop, single zone, air conditioner			
1280	Apartment corridors, 500 S.F., .92 ton	3.50	4.60	8.10
1480	10,000 S.F., 18.33 ton	2.95	3.12	6.07
1560	Banks or libraries, 500 S.F., 2.08 ton	7.95	10.45	18.40
1760	10,000 S.F., 41.67 ton	6.40	7.05	13.45
1840	Bars and taverns, 500 S.F. 5.54 ton	12.70	13.95	26.65
2000	5,000 S.F., 55.42 ton	14.05	10.60	24.65
2080	Bowling alleys, 500 S.F., 2.83 ton	7.65	11.75	19.40
2280	10,000 S.F., 56.67 ton	8.10	9.60	17.70
2360	Department stores, 500 S.F., 1.46 ton	5.55	7.35	12.90
2560	10,000 S.F., 29.17 ton	4.81	4.94	9.75
2640	Drug stores, 500 S.F., 3.33 ton	9	13.80	22.80
2840	10,000 S.F., 66.67 ton	9.55	11.30	20.85
2920	Factories, 500 S.F., 1.67 ton	6.35	8.35	14.70
3120	10,000 S.F., 33.33 ton	5.50	5.65	11.15
3200	Food supermarkets, 500 S.F., 1.42 ton	5.40	7.15	12.55
3400	10,000 S.F., 28.33 ton	4.68	4.80	9.48
3480	Medical centers, 500 S.F., 1.17 ton	4.45	5.85	10.30
3680	10,000 S.F., 23.33 ton	3.75	3.98	7.73
3760	Offices, 500 S.F., 1.58 ton	6.05	7.95	14
3960	10,000 S.F., 31.67 ton	5.20	5.35	10.55
4000	Restaurants, 500 S.F., 2.50 ton	9.55	12.55	22.10
4200	10,000 S.F., 50.00 ton	7.15	8.50	15.65
4240	Schools and colleges, 500 S.F., 1.92 ton	7.30	9.65	16.95
4440	10,000 S.F., 38.33 ton	5.90	6.50	12.40

D3050 Terminal & Package Units

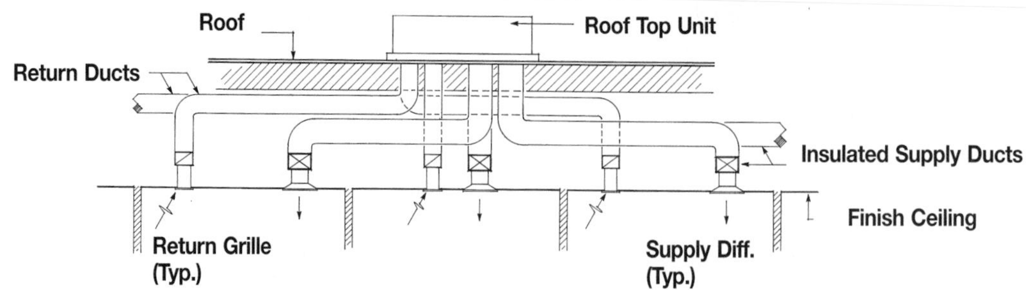

System Description: Rooftop units are multizone with up to 12 zones, and include electric cooling, gas heat, thermostats, filters, supply and return fans complete. Duct systems are low velocity, galvanized steel supply and return with insulated supplies.

Multizone units cost more per ton of cooling than single zone. However, they offer flexibility where load conditions are varied due to heat generating areas or exposure to radiational heating. For example, perimeter offices on the "sunny side" may require cooling at the same time "shady side" or central offices may require heating. It is possible to accomplish similar results using duct heaters in branches of the single zone unit. However, heater location could be a problem and total system operating energy efficiency could be lower.

System Components	QUANTITY	UNIT	COST EACH		
			MAT.	INST.	TOTAL
SYSTEM D3050 155 1280					
ROOFTOP, MULTIZONE, AIR CONDITIONER					
APARTMENT CORRIDORS, 3,000 S.F., 5.50 TON					
Rooftop multizone unit, standard controls, curb	1.000	Ea.	32,120	2,233	34,353
Ductwork package for rooftop multizone units	1.000	System	2,282.50	15,400	17,682.50
TOTAL			34,402.50	17,633	52,035.50
COST PER S.F.			11.47	5.88	17.35
Note A: Small single zone unit recommended.					

D3050 155	Rooftop Multizone Unit Systems	COST PER S.F.		
		MAT.	INST.	TOTAL
1240	Rooftop, multizone, air conditioner			
1260	Apartment corridors, 1,500 S.F., 2.75 ton. See Note A.			
1280	3,000 S.F., 5.50 ton	11.47	5.87	17.34
1440	25,000 S.F., 45.80 ton	7.65	5.70	13.35
1520	Banks or libraries, 1,500 S.F., 6.25 ton	26	13.35	39.35
1640	15,000 S.F., 62.50 ton	12.95	12.80	25.75
1720	25,000 S.F., 104.00 ton	11.45	12.75	24.20
1800	Bars and taverns, 1,500 S.F., 16.62 ton	56	19.25	75.25
1840	3,000 S.F., 33.24 ton	47	18.55	65.55
1880	10,000 S.F., 110.83 ton	28	18.50	46.50
2080	Bowling alleys, 1,500 S.F., 8.50 ton	35.50	18.15	53.65
2160	10,000 S.F., 56.70 ton	23.50	17.55	41.05
2240	20,000 S.F., 113.00 ton	15.55	17.35	32.90
2640	Drug stores, 1,500 S.F., 10.00 ton	41.50	21.50	63
2680	3,000 S.F., 20.00 ton	29.50	20.50	50
2760	15,000 S.F., 100.00 ton	18.35	20.50	38.85
3760	Offices, 1,500 S.F., 4.75 ton, See Note A.			
3880	15,000 S.F., 47.50 ton	13.20	9.80	23
3960	25,000 S.F., 79.16 ton	9.85	9.75	19.60
4000	Restaurants, 1,500 S.F., 7.50 ton	31.50	16.05	47.55
4080	10,000 S.F., 50.00 ton	21	15.50	36.50
4160	20,000 S.F., 100.00 ton	13.75	15.35	29.10
4240	Schools and colleges, 1,500 S.F., 5.75 ton	24	12.30	36.30
4360	15,000 S.F., 57.50 ton	11.90	11.80	23.70
4440	25,000 S.F., 95.83 ton	11	11.80	22.80

D3050 Terminal & Package Units

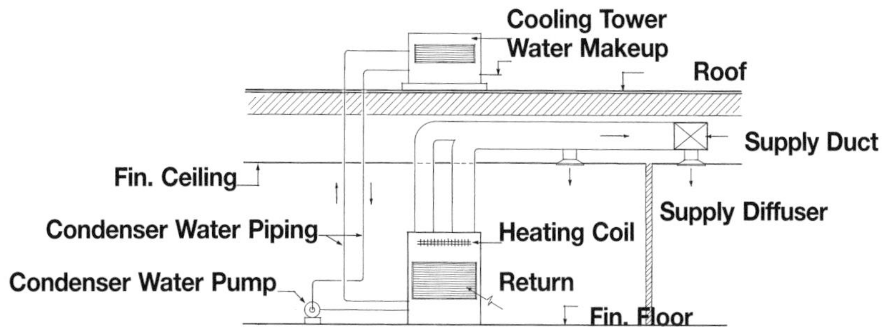

System Description: Self-contained, single package water cooled units include cooling tower, pump, piping allowance. Systems for 1000 S.F. and up include duct and diffusers to provide for even distribution of air. Smaller units distribute air through a supply air plenum, which is integral with the unit.

Returns are not ducted and supplies are not insulated.

Hot water or steam heating coils are included but piping to boiler and the boiler itself is not included.

Where local codes or conditions permit single pass cooling for the smaller units, deduct 10%.

System Components	QUANTITY	UNIT	COST EACH MAT.	COST EACH INST.	COST EACH TOTAL
SYSTEM D3050 160 1300					
SELF-CONTAINED, WATER COOLED UNIT					
APARTMENT CORRIDORS, 500 S.F., .92 TON					
Self-contained, water cooled, single package air conditioner unit	1.000	Ea.	1,386	764.40	2,150.40
Ductwork package for water or air cooled packaged units	1.000	System	62.10	1,035	1,097.10
Cooling tower, draw thru single flow, belt drive	1.000	Ea.	205.16	23	228.16
Cooling tower pumps & piping	1.000	System	102.12	55.20	157.32
TOTAL			1,755.38	1,877.60	3,632.98
COST PER S.F.			3.51	3.76	7.27

D3050 160	Self-contained, Water Cooled Unit Systems	COST PER S.F. MAT.	COST PER S.F. INST.	COST PER S.F. TOTAL
1280	Self-contained, water cooled unit	3.50	3.76	7.26
1300	Apartment corridors, 500 S.F., .92 ton	3.50	3.75	7.25
1440	10,000 S.F., 18.33 ton	3.80	2.61	6.41
1520	Banks or libraries, 500 S.F., 2.08 ton	7.70	3.82	11.52
1680	10,000 S.F., 41.66 ton	7.30	5.90	13.20
1760	Bars and taverns, 500 S.F., 5.54 ton	16.70	6.45	23.15
1920	10,000 S.F., 110.00 ton	16.55	9.25	25.80
2000	Bowling alleys, 500 S.F., 2.83 ton	10.45	5.20	15.65
2160	10,000 S.F., 56.66 ton	8.90	7.95	16.85
2200	Department stores, 500 S.F., 1.46 ton	5.40	2.67	8.07
2360	10,000 S.F., 29.17 ton	5.50	3.99	9.49
2440	Drug stores, 500 S.F., 3.33 ton	12.30	6.15	18.45
2600	10,000 S.F., 66.66 ton	10.50	9.40	19.90
2680	Factories, 500 S.F., 1.66 ton	6.15	3.06	9.21
2840	10,000 S.F., 33.33 ton	6.25	4.57	10.82
2920	Food supermarkets, 500 S.F., 1.42 ton	5.20	2.60	7.80
3080	10,000 S.F., 28.33 ton	5.30	3.89	9.19
3160	Medical centers, 500 S.F., 1.17 ton	4.30	2.14	6.44
3320	10,000 S.F., 23.33 ton	4.84	3.31	8.15
3400	Offices, 500 S.F., 1.58 ton	5.85	2.91	8.76
3560	10,000 S.F., 31.67 ton	5.95	4.34	10.29
3640	Restaurants, 500 S.F., 2.50 ton	9.25	4.60	13.85
3800	10,000 S.F., 50.00 ton	6.45	6.55	13
3880	Schools and colleges, 500 S.F., 1.92 ton	7.05	3.52	10.57
4040	10,000 S.F., 38.33 ton	6.70	5.45	12.15

D3050 Terminal & Package Units

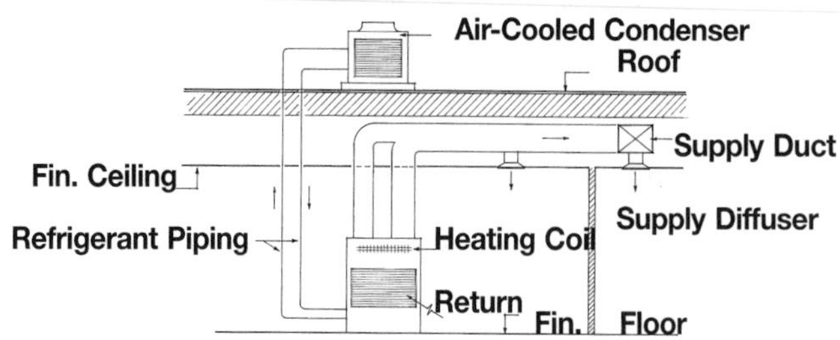

System Description: Self-contained air cooled units with remote air cooled condenser and interconnecting tubing. Systems for 1000 S.F. and up include duct and diffusers. Smaller units distribute air directly.

Returns are not ducted and supplies are not insulated.

Potential savings may be realized by using a single zone rooftop system or through-the-wall unit, especially in the smaller capacities, if the application permits.

Hot water or steam heating coils are included but piping to boiler and the boiler itself is not included.

Condenserless models are available for 15% less where remote refrigerant source is available.

System Components	QUANTITY	UNIT	COST EACH MAT.	COST EACH INST.	COST EACH TOTAL
SYSTEM D3050 165 1320					
SELF-CONTAINED, AIR COOLED UNIT					
APARTMENT CORRIDORS, 500 S.F., .92 TON					
Air cooled, package unit	1.000	Ea.	1,513	493	2,006
Ductwork package for water or air cooled packaged units	1.000	System	62.10	1,035	1,097.10
Refrigerant piping	1.000	System	381.80	671.60	1,053.40
Air cooled condenser, direct drive, propeller fan	1.000	Ea.	914.50	189.10	1,103.60
TOTAL			2,871.40	2,388.70	5,260.10
COST PER S.F.			5.74	4.78	10.52

D3050 165	Self-contained, Air Cooled Unit Systems	COST PER S.F. MAT.	COST PER S.F. INST.	COST PER S.F. TOTAL
1300	Self-contained, air cooled unit			
1320	Apartment corridors, 500 S.F., .92 ton	5.75	4.80	10.55
1480	10,000 S.F., 18.33 ton	3.59	3.88	7.47
1560	Banks or libraries, 500 S.F., 2.08 ton	12.50	6.05	18.55
1720	10,000 S.F., 41.66 ton	8.80	8.60	17.40
1800	Bars and taverns, 500 S.F., 5.54 ton	28	13.25	41.25
1960	10,000 S.F., 110.00 ton	24.50	16.55	41.05
2040	Bowling alleys, 500 S.F., 2.83 ton	17.15	8.30	25.45
2200	10,000 S.F., 56.66 ton	12.60	11.75	24.35
2240	Department stores, 500 S.F., 1.46 ton	8.85	4.27	13.12
2400	10,000 S.F., 29.17 ton	6.35	6	12.35
2480	Drug stores, 500 S.F., 3.33 ton	20	9.75	29.75
2640	10,000 S.F., 66.66 ton	15	13.80	28.80
2720	Factories, 500 S.F., 1.66 ton	10.25	4.92	15.17
2880	10,000 S.F., 33.33 ton	7.25	6.90	14.15
3200	Medical centers, 500 S.F., 1.17 ton	7.10	3.43	10.53
3360	10,000 S.F., 23.33 ton	4.58	4.92	9.50
3440	Offices, 500 S.F., 1.58 ton	9.60	4.64	14.24
3600	10,000 S.F., 31.66 ton	6.90	6.55	13.45
3680	Restaurants, 500 S.F., 2.50 ton	15.10	7.30	22.40
3840	10,000 S.F., 50.00 ton	11.40	10.30	21.70
3920	Schools and colleges, 500 S.F., 1.92 ton	11.60	5.60	17.20
4080	10,000 S.F., 38.33 ton	8.15	7.95	16.10

D3050 Terminal & Package Units

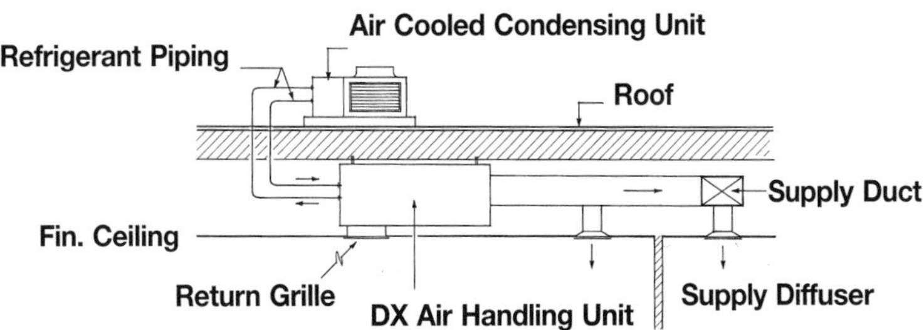

General: Split systems offer several important advantages which should be evaluated when a selection is to be made. They provide a greater degree of flexibility in component selection which permits an accurate match-up of the proper equipment size and type with the particular needs of the building. This allows for maximum use of modern energy saving concepts in heating and cooling. Outdoor installation of the air cooled condensing unit allows space savings in the building and also isolates the equipment operating sounds from building occupants.

Design Assumptions: The systems below are comprised of a direct expansion air handling unit and air cooled condensing unit with interconnecting copper tubing. Ducts and diffusers are also included for distribution of air. Systems are priced for cooling only. Heat can be added as desired either by putting hot water/steam coils into the air unit or into the duct supplying the particular area of need. Gas fired duct furnaces are also available. Refrigerant liquid line is insulated.

System Components	QUANTITY	UNIT	COST EACH MAT.	COST EACH INST.	COST EACH TOTAL
SYSTEM D3050 170 1280					
SPLIT SYSTEM, AIR COOLED CONDENSING UNIT					
APARTMENT CORRIDORS, 1,000 S.F., 1.80 TON					
Fan coil AC unit, cabinet mntd & filters direct expansion air cool	1.000	Ea.	503.25	177.51	680.76
Ductwork package, for split system, remote condensing unit	1.000	System	58.10	983.63	1,041.73
Refrigeration piping	1.000	System	404.43	1,116.30	1,520.73
Condensing unit, air cooled, incls compressor & standard controls	1.000	Ea.	1,372.50	713.70	2,086.20
TOTAL			2,338.28	2,991.14	5,329.42
COST PER S.F.			2.34	2.99	5.33
*Cooling requirements would lead to choosing a water cooled unit.					

D3050 170	Split Systems With Air Cooled Condensing Units	COST PER S.F. MAT.	COST PER S.F. INST.	COST PER S.F. TOTAL
1260	Split system, air cooled condensing unit			
1280	Apartment corridors, 1,000 S.F., 1.83 ton	2.35	3	5.35
1440	20,000 S.F., 36.66 ton	2.25	4.20	6.45
1520	Banks and libraries, 1,000 S.F., 4.17 ton	4.15	6.85	11
1680	20,000 S.F., 83.32 ton	6.20	9.90	16.10
1760	Bars and taverns, 1,000 S.F., 11.08 ton	8.65	13.55	22.20
1880	10,000 S.F., 110.84 ton	13.95	15.75	29.70
2000	Bowling alleys, 1,000 S.F., 5.66 ton	5.70	12.85	18.55
2160	20,000 S.F., 113.32 ton	9.55	14	23.55
2320	Department stores, 1,000 S.F., 2.92 ton	2.87	4.72	7.59
2480	20,000 S.F., 58.33 ton	3.56	6.70	10.26
2560	Drug stores, 1,000 S.F., 6.66 ton	6.70	15.10	21.80
2720	20,000 S.F., 133.32 ton*			
2800	Factories, 1,000 S.F., 3.33 ton	3.30	5.40	8.70
2960	20,000 S.F., 66.66 ton	4.28	7.95	12.23
3040	Food supermarkets, 1,000 S.F., 2.83 ton	2.80	4.58	7.38
3200	20,000 S.F., 56.66 ton	3.45	6.50	9.95
3280	Medical centers, 1,000 S.F., 2.33 ton	2.49	3.70	6.19
3440	20,000 S.F., 46.66 ton	2.85	5.35	8.20
3520	Offices, 1,000 S.F., 3.17 ton	3.12	5.15	8.27
3680	20,000 S.F., 63.32 ton	4.07	7.55	11.62
3760	Restaurants, 1,000 S.F., 5.00 ton	5.05	11.35	16.40
3920	20,000 S.F., 100.00 ton	8.40	12.35	20.75
4000	Schools and colleges, 1,000 S.F., 3.83 ton	3.82	6.35	10.17
4160	20,000 S.F., 76.66 ton	4.94	9.15	14.09

D3050 Terminal & Package Units

Computer rooms impose special requirements on air conditioning systems. A prime requirement is reliability, due to the potential monetary loss that could be incurred by a system failure. A second basic requirement is the tolerance of control with which temperature and humidity are regulated, and dust eliminated. As the air conditioning system reliability is so vital, the additional cost of reserve capacity and redundant components is often justified.

System Descriptions: Computer areas may be environmentally controlled by one of three methods as follows:

1. **Self-contained Units**
   These are units built to higher standards of performance and reliability. They usually contain alarms and controls to indicate component operation failure, filter change, etc. It should be remembered that these units in the room will occupy space that is relatively expensive to build and that all alterations and service of the equipment will also have to be accomplished within the computer area.
2. **Decentralized Air Handling Units**
   In operation these are similar to the self-contained units except that their cooling capability comes from remotely located refrigeration equipment as refrigerant or chilled water. As no compressors or refrigerating equipment are required in the air units, they are smaller and require less service than self-contained units. An added plus for this type of system occurs if some of the computer components themselves also require chilled water for cooling.
3. **Central System Supply**
   Cooling is obtained from a central source which, since it is not located within the computer room, may have excess capacity and permit greater flexibility without interfering with the computer components. System performance criteria must still be met.

Note: The costs shown below do not include an allowance for ductwork or piping.

D3050 190	Computer Room Cooling Units	COST EACH		
		MAT.	INST.	TOTAL
0560	Computer room unit, air cooled, includes remote condenser			
0580	3 ton	30,400	2,925	33,325
0600	5 ton	32,400	3,250	35,650
0620	8 ton	60,500	5,400	65,900
0640	10 ton	63,500	5,850	69,350
0660	15 ton	69,500	6,625	76,125
0680	20 ton	84,500	9,450	93,950
0700	23 ton	104,000	10,800	114,800
0800	Chilled water, for connection to existing chiller system			
0820	5 ton	22,000	1,975	23,975
0840	8 ton	22,700	2,925	25,625
0860	10 ton	22,800	2,975	25,775
0880	15 ton	24,900	3,050	27,950
0900	20 ton	26,500	3,175	29,675
0920	23 ton	28,200	3,600	31,800
1000	Glycol system, complete except for interconnecting tubing			
1020	3 ton	38,100	3,650	41,750
1040	5 ton	41,700	3,850	45,550
1060	8 ton	67,000	6,350	73,350
1080	10 ton	71,500	6,950	78,450
1100	15 ton	88,500	8,725	97,225
1120	20 ton	97,000	9,450	106,450
1140	23 ton	102,000	9,875	111,875
1240	Water cooled, not including condenser water supply or cooling tower			
1260	3 ton	30,900	2,350	33,250
1280	5 ton	34,900	2,700	37,600
1300	8 ton	55,000	4,425	59,425
1320	15 ton	67,000	5,400	72,400
1340	20 ton	72,000	5,975	77,975
1360	23 ton	75,500	6,475	81,975

D4010 Sprinklers

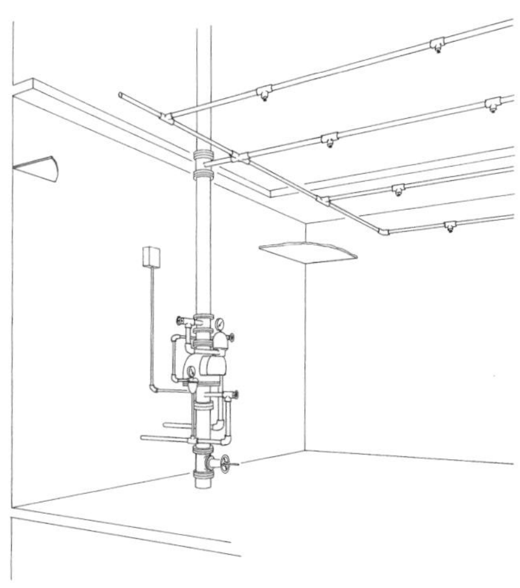

Dry Pipe System: A system employing automatic sprinklers attached to a piping system containing air under pressure, the release of which from the opening of sprinklers permits the water pressure to open a valve known as a "dry pipe valve". The water then flows into the piping system and out the opened sprinklers.

All areas are assumed to be open.

System Components	QUANTITY	UNIT	COST EACH MAT.	COST EACH INST.	COST EACH TOTAL
SYSTEM D4010 310 0580					
DRY PIPE SPRINKLER, STEEL, BLACK, SCH. 40 PIPE					
LIGHT HAZARD, ONE FLOOR, 2000 S.F.					
Valve, gate, iron body 125 lb., OS&Y, flanged, 4" pipe size	1.000	Ea.	755	480	1,235
4" pipe size	1.000	Ea.	5,125	480	5,605
Tamper switch (valve supervisory switch)	3.000	Ea.	873	147	1,020
Valve, swing check, bronze, 125 lb, regrinding disc, 2-1/2" pipe size	1.000	Ea.	1,025	95.50	1,120.50
Valve, angle, bronze, 150 lb., rising stem, threaded, 2" pipe size	1.000	Ea.	930	72.50	1,002.50
*Alarm valve, 2-1/2" pipe size	1.000	Ea.	1,900	470	2,370
Alarm, water motor, complete with gong	1.000	Ea.	530	196	726
Fire alarm horn, electric	1.000	Ea.	59.50	113	172.50
Valve swing check w/balldrip CI with brass trim, 4" pipe size	1.000	Ea.	460	470	930
Pipe, steel, black, schedule 40, 4" diam.	8.000	L.F.	212	337.36	549.36
Dry pipe valve, trim & gauges, 4" pipe size	1.000	Ea.	3,625	1,400	5,025
Pipe, steel, black, schedule 40, threaded, cplg & hngr 10'OC 2-1/2" diam.	15.000	L.F.	188.25	427.50	615.75
Pipe, steel, black, schedule 40, threaded, cplg & hngr 10'OC 2" diam.	9.375	L.F.	71.25	210.94	282.19
Pipe, steel, black, schedule 40, threaded, cplg & hngr 10'OC 1-1/4" diam.	28.125	L.F.	253.13	452.81	705.94
Pipe, steel, black, schedule 40, threaded, cplg & hngr 10'OC 1" diam.	84.000	L.F.	718.20	1,264.20	1,982.40
Pipe Tee, malleable iron black, 150 lb. threaded, 4" pipe size	2.000	Ea.	780	720	1,500
Pipe Tee, malleable iron black, 150 lb. threaded, 2-1/2" pipe size	2.000	Ea.	218	318	536
Pipe Tee, malleable iron black, 150 lb. threaded, 2" pipe size	1.000	Ea.	50.50	130	180.50
Pipe Tee, malleable iron black, 150 lb. threaded, 1-1/4" pipe size	4.000	Ea.	96	408	504
Pipe Tee, malleable iron black, 150 lb. threaded, 1" pipe size	3.000	Ea.	44.25	298.50	342.75
Pipe 90° elbow malleable iron black, 150 lb. threaded, 1" pipe size	5.000	Ea.	46	307.50	353.50
Sprinkler head dry K5.6, 1" NPT, 3" to 6" length	12.000	Ea.	1,920	672	2,592
Air compressor, 200 Gal sprinkler system capacity, 1/3 HP	1.000	Ea.	1,325	605	1,930
*Standpipe connection, wall, flush, brs. w/plug & chain 2-1/2"x2-1/2"	1.000	Ea.	205	282	487
Valve gate bronze, 300 psi, NRS, class 150, threaded, 1" pipe size	1.000	Ea.	148	42	190
TOTAL			21,558.08	10,399.81	31,957.89
COST PER S.F.			8.08	3.90	11.98
*Not included in systems under 2000 S.F.					

D4010 310	Dry Pipe Sprinkler Systems		COST PER S.F. MAT.	COST PER S.F. INST.	COST PER S.F. TOTAL
0520	Dry pipe sprinkler systems, steel, black, sch. 40 pipe				
0530	Light hazard, one floor, 500 S.F.		14.80	7.35	22.15
0560	1000 S.F.	R211313 -20	8.95	4.34	13.29
0580	2000 S.F.		8.10	3.90	12
0600	5000 S.F.		4.25	2.92	7.17
0620	10,000 S.F.		2.76	2.42	5.18

D4010 Sprinklers

D4010 310	Dry Pipe Sprinkler Systems	COST PER S.F.		
		MAT.	INST.	TOTAL
0640	50,000 S.F.	1.88	2.10	3.98
0660	Each additional floor, 500 S.F.	2.77	3.50	6.27
0680	1000 S.F.	2.38	2.86	5.24
0700	2000 S.F.	2.26	2.66	4.92
0720	5000 S.F.	1.85	2.26	4.11
0740	10,000 S.F.	1.59	2.07	3.66
0760	50,000 S.F.	1.45	1.86	3.31
1000	Ordinary hazard, one floor, 500 S.F.	15.05	7.45	22.50
1020	1000 S.F.	9.10	4.39	13.49
1040	2000 S.F.	8.65	4.63	13.28
1060	5000 S.F.	4.77	3.12	7.89
1080	10,000 S.F.	3.47	3.16	6.63
1100	50,000 S.F.	3.09	2.92	6.01
1140	Each additional floor, 500 S.F.	3.01	3.59	6.60
1160	1000 S.F.	2.60	3.22	5.82
1180	2000 S.F.	2.65	2.94	5.59
1200	5000 S.F.	2.44	2.51	4.95
1220	10,000 S.F.	2.18	2.44	4.62
1240	50,000 S.F.	2.13	2.10	4.23
1500	Extra hazard, one floor, 500 S.F.	18.45	9.25	27.70
1520	1000 S.F.	15.10	6.75	21.85
1540	2000 S.F.	9.45	5.85	15.30
1560	5000 S.F.	5.35	4.32	9.67
1580	10,000 S.F.	5.40	4.13	9.53
1600	50,000 S.F.	5.35	3.93	9.28
1660	Each additional floor, 500 S.F.	4.14	4.43	8.57
1680	1000 S.F.	3.64	4.17	7.81
1700	2000 S.F.	3.50	4.17	7.67
1720	5000 S.F.	2.97	3.65	6.62
1740	10,000 S.F.	3.64	3.34	6.98
1760	50,000 S.F.	3.67	3.20	6.87
2020	Grooved steel, black, sch. 40 pipe, light hazard, one floor, 2000 S.F.	8.35	3.83	12.18
2060	10,000 S.F.	2.81	2.10	4.91
2100	Each additional floor, 2000 S.F.	2.34	2.14	4.48
2150	10,000 S.F.	1.64	1.75	3.39
2200	Ordinary hazard, one floor, 2000 S.F.	8.65	4.07	12.72
2250	10,000 S.F.	3.41	2.70	6.11
2300	Each additional floor, 2000 S.F.	2.63	2.38	5.01
2350	10,000 S.F.	2.47	2.38	4.85
2400	Extra hazard, one floor, 2000 S.F.	9.50	5	14.50
2450	10,000 S.F.	4.82	3.46	8.28
2500	Each additional floor, 2000 S.F.	3.67	3.44	7.11
2550	10,000 S.F.	3.28	2.94	6.22
3050	Grooved steel, black, sch. 10 pipe, light hazard, one floor, 2000 S.F.	8.25	3.81	12.06
3100	10,000 S.F.	2.77	2.07	4.84
3150	Each additional floor, 2000 S.F.	2.26	2.12	4.38
3200	10,000 S.F.	1.60	1.72	3.32
3250	Ordinary hazard, one floor, 2000 S.F.	8.55	4.04	12.59
3300	10,000 S.F.	3.26	2.64	5.90
3350	Each additional floor, 2000 S.F.	2.56	2.35	4.91
3400	10,000 S.F.	2.17	2.31	4.48
3450	Extra hazard, one floor, 2000 S.F.	9.45	5	14.45
3500	10,000 S.F.	4.61	3.40	8.01
3550	Each additional floor, 2000 S.F.	3.61	3.42	7.03
3600	10,000 S.F.	3.18	2.90	6.08
4050	Copper tubing, type L, light hazard, one floor, 2000 S.F.	8.55	3.84	12.39
4100	10,000 S.F.	3.20	2.15	5.35
4150	Each additional floor, 2000 S.F.	2.56	2.19	4.75

D4010 Sprinklers

D4010 310	Dry Pipe Sprinkler Systems	COST PER S.F.		
		MAT.	INST.	TOTAL
4200	10,000 S.F.	2.03	1.81	3.84
4250	Ordinary hazard, one floor, 2000 S.F.	9	4.25	13.25
4300	10,000 S.F.	3.98	2.49	6.47
4350	Each additional floor, 2000 S.F.	3.47	2.47	5.94
4400	10,000 S.F.	2.76	2.10	4.86
4450	Extra hazard, one floor, 2000 S.F.	10	5.15	15.15
4500	10,000 S.F.	6.75	3.71	10.46
4550	Each additional floor, 2000 S.F.	4.33	3.51	7.84
4600	10,000 S.F.	4.53	3.17	7.70
5050	Copper tubing, type L, T-drill system, light hazard, one floor			
5060	2000 S.F.	8.60	3.62	12.22
5100	10,000 S.F.	3.08	1.80	4.88
5150	Each additional floor, 2000 S.F.	2.62	1.97	4.59
5200	10,000 S.F.	1.91	1.46	3.37
5250	Ordinary hazard, one floor, 2000 S.F.	8.80	3.70	12.50
5300	10,000 S.F.	3.75	2.27	6.02
5350	Each additional floor, 2000 S.F.	2.79	2.01	4.80
5400	10,000 S.F.	2.48	1.83	4.31
5450	Extra hazard, one floor, 2000 S.F.	9.45	4.31	13.76
5500	10,000 S.F.	5.45	2.82	8.27
5550	Each additional floor, 2000 S.F.	3.61	2.73	6.34
5600	10,000 S.F.	3.42	2.27	5.69

D4010 Sprinklers

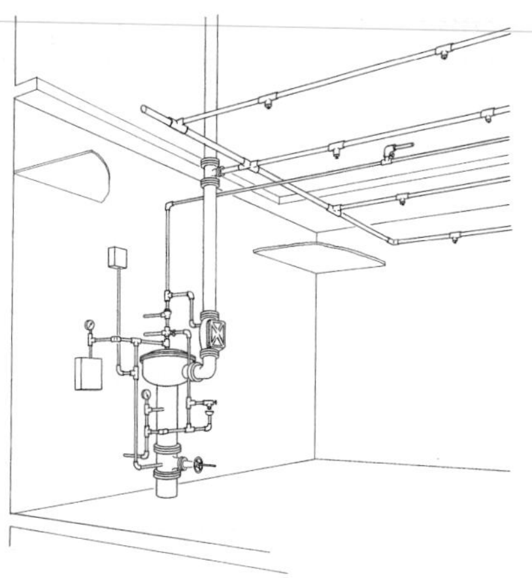

Pre-Action System: A system employing automatic sprinklers attached to a piping system containing air that may or may not be under pressure, with a supplemental heat responsive system of generally more sensitive characteristics than the automatic sprinklers themselves, installed in the same areas as the sprinklers. Actuation of the heat responsive system, as from a fire, opens a valve which permits water to flow into the sprinkler piping system and to be discharged from those sprinklers which were opened by heat from the fire.

All areas are assumed to be open.

System Components	QUANTITY	UNIT	COST EACH MAT.	COST EACH INST.	COST EACH TOTAL
SYSTEM D4010 350 0580					
PREACTION SPRINKLER SYSTEM, STEEL, BLACK, SCH. 40 PIPE					
LIGHT HAZARD, 1 FLOOR, 2000 S.F.					
Valve, gate, iron body 125 lb., OS&Y, flanged, 4″ pipe size	1.000	Ea.	755	480	1,235
4″ pipe size	1.000	Ea.	5,125	480	5,605
Tamper switch (valve supervisory switch)	3.000	Ea.	873	147	1,020
*Valve, swing check w/ball drip CI with brass trim 4″ pipe size	1.000	Ea.	460	470	930
Valve, swing check, bronze, 125 lb, regrinding disc, 2-1/2″ pipe size	1.000	Ea.	1,025	95.50	1,120.50
Valve, angle, bronze, 150 lb., rising stem, threaded, 2″ pipe size	1.000	Ea.	930	72.50	1,002.50
*Alarm valve, 2-1/2″ pipe size	1.000	Ea.	1,900	470	2,370
Alarm, water motor, complete with gong	1.000	Ea.	530	196	726
Fire alarm horn, electric	1.000	Ea.	59.50	113	172.50
Thermostatic release for release line	2.000	Ea.	1,760	78	1,838
Pipe, steel, black, schedule 40, 4″ diam.	8.000	L.F.	212	337.36	549.36
Dry pipe valve, trim & gauges, 4″ pipe size	1.000	Ea.	3,625	1,400	5,025
Pipe, steel, black, schedule 40, threaded, cplg. & hngr. 10′OC 2-1/2″ diam.	15.000	L.F.	188.25	427.50	615.75
Pipe steel black, schedule 40, threaded, cplg. & hngr. 10′OC 2″ diam.	9.375	L.F.	71.25	210.94	282.19
Pipe, steel, black, schedule 40, threaded, cplg. & hngr. 10′OC 1-1/4″ diam.	28.125	L.F.	253.13	452.81	705.94
Pipe, steel, black, schedule 40, threaded, cplg. & hngr. 10′OC 1″ diam.	84.000	L.F.	718.20	1,264.20	1,982.40
Pipe, Tee, malleable iron, black, 150 lb. threaded, 4″ diam.	2.000	Ea.	780	720	1,500
Pipe, Tee, malleable iron, black, 150 lb. threaded, 2-1/2″ pipe size	2.000	Ea.	218	318	536
Pipe, Tee, malleable iron, black, 150 lb. threaded, 2″ pipe size	1.000	Ea.	50.50	130	180.50
Pipe, Tee, malleable iron, black, 150 lb. threaded, 1-1/4″ pipe size	4.000	Ea.	96	408	504
Pipe, Tee, malleable iron, black, 150 lb. threaded, 1″ pipe size	3.000	Ea.	44.25	298.50	342.75
Pipe, 90° elbow, malleable iron, blk., 150 lb. threaded, 1″ pipe size	5.000	Ea.	46	307.50	353.50
Sprinkler head, std. spray, brass 135°-286°F 1/2″ NPT, 3/8″ orifice	12.000	Ea.	221.40	588	809.40
Air compressor auto complete 200 Gal sprinkler sys. cap., 1/3 HP	1.000	Ea.	1,325	605	1,930
*Standpipe conn.,wall, flush, brass w/plug & chain 2-1/2″ x 2-1/2″	1.000	Ea.	205	282	487
Valve, gate, bronze, 300 psi, NRS, class 150, threaded, 1″ pipe size	1.000	Ea.	148	42	190
TOTAL			21,619.48	10,393.81	32,013.29
COST PER S.F.			8.11	3.90	12.01

*Not included in systems under 2000 S.F.

D4010 350	Preaction Sprinkler Systems	COST PER S.F. MAT.	COST PER S.F. INST.	COST PER S.F. TOTAL
0520	Preaction sprinkler systems, steel, black, sch. 40 pipe R211313-20			
0530	Light hazard, one floor, 500 S.F.	15	5.95	20.95
0560	1000 S.F.	7.95	4.40	12.35
0580	2000 S.F.	8.10	3.90	12

D4010 Sprinklers

D4010 350	Preaction Sprinkler Systems	COST PER S.F.		
		MAT.	INST.	TOTAL
0600	5000 S.F.	4.10	2.91	7.01
0620	10,000 S.F.	2.63	2.41	5.04
0640	50,000 S.F.	1.74	2.09	3.83
0660	Each additional floor, 500 S.F.	4.16	3.13	7.29
0680	1000 S.F.	2.40	2.86	5.26
0700	2000 S.F.	2.28	2.66	4.94
0720	5000 S.F.	2.33	2.30	4.63
0740	10,000 S.F.	1.98	2.09	4.07
0760	50,000 S.F.	1.43	1.92	3.35
1000	Ordinary hazard, one floor, 500 S.F.	15.15	6.45	21.60
1020	1000 S.F.	8.90	4.37	13.27
1040	2000 S.F.	8.80	4.62	13.42
1060	5000 S.F.	4.45	3.09	7.54
1080	10,000 S.F.	3.08	3.13	6.21
1100	50,000 S.F.	2.71	2.90	5.61
1140	Each additional floor, 500 S.F.	3.48	3.60	7.08
1160	1000 S.F.	2.31	2.89	5.20
1180	2000 S.F.	2.13	2.90	5.03
1200	5000 S.F.	2.26	2.69	4.95
1220	10,000 S.F.	1.92	2.79	4.71
1240	50,000 S.F.	1.93	2.49	4.42
1500	Extra hazard, one floor, 500 S.F.	20	8.30	28.30
1520	1000 S.F.	14.30	6.20	20.50
1540	2000 S.F.	8.85	5.80	14.65
1560	5000 S.F.	4.97	4.65	9.62
1580	10,000 S.F.	4.82	4.55	9.37
1600	50,000 S.F.	4.69	4.36	9.05
1660	Each additional floor, 500 S.F.	4.19	4.43	8.62
1680	1000 S.F.	3.25	4.18	7.43
1700	2000 S.F.	3	4.16	7.16
1720	5000 S.F.	2.44	3.67	6.11
1740	10,000 S.F.	2.88	3.37	6.25
1760	50,000 S.F.	2.75	3.15	5.90
2020	Grooved steel, black, sch. 40 pipe, light hazard, one floor, 2000 S.F.	8.35	3.83	12.18
2060	10,000 S.F.	2.68	2.09	4.77
2100	Each additional floor of 2000 S.F.	2.36	2.14	4.50
2150	10,000 S.F.	1.51	1.74	3.25
2200	Ordinary hazard, one floor, 2000 S.F.	8.45	4.05	12.50
2250	10,000 S.F.	3.02	2.67	5.69
2300	Each additional floor, 2000 S.F.	2.44	2.36	4.80
2350	10,000 S.F.	1.86	2.33	4.19
2400	Extra hazard, one floor, 2000 S.F.	8.90	4.99	13.89
2450	10,000 S.F.	4.06	3.41	7.47
2500	Each additional floor, 2000 S.F.	3.06	3.41	6.47
2550	10,000 S.F.	2.46	2.89	5.35
3050	Grooved steel, black, sch. 10 pipe light hazard, one floor, 2000 S.F.	8.30	3.81	12.11
3100	10,000 S.F.	2.64	2.06	4.70
3150	Each additional floor, 2000 S.F.	2.28	2.12	4.40
3200	10,000 S.F.	1.47	1.71	3.18
3250	Ordinary hazard, one floor, 2000 S.F.	8.30	3.79	12.09
3300	10,000 S.F.	2.61	2.61	5.22
3350	Each additional floor, 2000 S.F.	2.37	2.33	4.70
3400	10,000 S.F.	1.78	2.28	4.06
3450	Extra hazard, one floor, 2000 S.F.	8.85	4.97	13.82
3500	10,000 S.F.	3.79	3.35	7.14
3550	Each additional floor, 2000 S.F.	3	3.39	6.39
3600	10,000 S.F.	2.36	2.85	5.21
4050	Copper tubing, type L, light hazard, one floor, 2000 S.F.	8.60	3.84	12.44

D4010 Sprinklers

D4010 350	Preaction Sprinkler Systems	COST PER S.F.		
		MAT.	INST.	TOTAL
4100	10,000 S.F.	3.07	2.14	5.21
4150	Each additional floor, 2000 S.F.	2.61	2.19	4.80
4200	10,000 S.F.	1.57	1.79	3.36
4250	Ordinary hazard, one floor, 2000 S.F.	8.90	4.18	13.08
4300	10,000 S.F.	3.59	2.46	6.05
4350	Each additional floor, 2000 S.F.	2.66	2.18	4.84
4400	10,000 S.F.	2.15	1.91	4.06
4450	Extra hazard, one floor, 2000 S.F.	9.55	5.05	14.60
4500	10,000 S.F.	5.85	3.66	9.51
4550	Each additional floor, 2000 S.F.	3.72	3.48	7.20
4600	10,000 S.F.	3.71	3.12	6.83
5050	Copper tubing, type L, T-drill system, light hazard, one floor			
5060	2000 S.F.	8.65	3.62	12.27
5100	10,000 S.F.	2.95	1.79	4.74
5150	Each additional floor, 2000 S.F.	2.64	1.97	4.61
5200	10,000 S.F.	1.78	1.45	3.23
5250	Ordinary hazard, one floor, 2000 S.F.	8.60	3.68	12.28
5300	10,000 S.F.	3.36	2.24	5.60
5350	Each additional floor, 2000 S.F.	2.60	2.01	4.61
5400	10,000 S.F.	2.20	1.90	4.10
5450	Extra hazard, one floor, 2000 S.F.	8.85	4.28	13.13
5500	10,000 S.F.	4.56	2.77	7.33
5550	Each additional floor, 2000 S.F.	3	2.70	5.70
5600	10,000 S.F.	2.60	2.22	4.82

D4010 Sprinklers

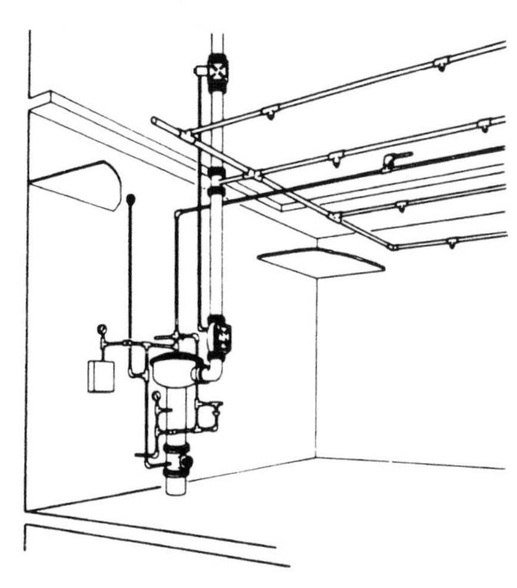

Deluge System: A system employing open sprinklers attached to a piping system connected to a water supply through a valve which is opened by the operation of a heat responsive system installed in the same areas as the sprinklers. When this valve opens, water flows into the piping system and discharges from all sprinklers attached thereto.

All areas are assumed to be open.

System Components	QUANTITY	UNIT	COST EACH		
			MAT.	INST.	TOTAL
SYSTEM D4010 370 0580					
DELUGE SPRINKLER SYSTEM, STEEL BLACK SCH. 40 PIPE					
LIGHT HAZARD, 1 FLOOR, 2000 S.F.					
Valve, gate, iron body 125 lb., OS&Y, flanged, 4″ pipe size	1.000	Ea.	755	480	1,235
4″ pipe size	1.000	Ea.	5,125	480	5,605
Tamper switch (valve supervisory switch)	3.000	Ea.	873	147	1,020
Valve, swing check w/ball drip, CI w/brass ftngs., 4″ pipe size	1.000	Ea.	460	470	930
Valve, swing check, bronze, 125 lb, regrinding disc, 2-1/2″ pipe size	1.000	Ea.	1,025	95.50	1,120.50
Valve, angle, bronze, 150 lb., rising stem, threaded, 2″ pipe size	1.000	Ea.	930	72.50	1,002.50
*Alarm valve, 2-1/2″ pipe size	1.000	Ea.	1,900	470	2,370
Alarm, water motor, complete with gong	1.000	Ea.	530	196	726
Fire alarm horn, electric	1.000	Ea.	59.50	113	172.50
Thermostatic release for release line	2.000	Ea.	1,760	78	1,838
Pipe, steel, black, schedule 40, 4″ diam.	8.000	L.F.	212	337.36	549.36
Deluge valve trim, pressure relief, emergency release, gauge, 4″ pipe size	1.000	Ea.	6,200	1,400	7,600
Deluge system, monitoring panel w/deluge valve & trim	1.000	Ea.	8,225	43.50	8,268.50
Pipe, steel, black, schedule 40, threaded, cplg & hngr 10′ OC 2-1/2″ diam.	15.000	L.F.	188.25	427.50	615.75
Pipe, steel, black, schedule 40, threaded, cplg & hngr 10′ OC 2″ diam.	9.375	L.F.	71.25	210.94	282.19
Pipe, steel, black, schedule 40, threaded, cplg & hngr 10′ OC 1-1/4″ diam.	28.125	L.F.	253.13	452.81	705.94
Pipe, steel, black, schedule 40, threaded, cplg & hngr 10′ OC 1″ diam.	84.000	L.F.	718.20	1,264.20	1,982.40
Pipe, Tee, malleable iron, black, 150 lb. threaded, 4″ pipe size	2.000	Ea.	780	720	1,500
Pipe, Tee, malleable iron, black, 150 lb. threaded, 2-1/2″ pipe size	2.000	Ea.	218	318	536
Pipe, Tee, malleable iron, black, 150 lb. threaded, 2″ pipe size	1.000	Ea.	50.50	130	180.50
Pipe, Tee, malleable iron, black, 150 lb. threaded, 1-1/4″ pipe size	4.000	Ea.	96	408	504
Pipe, Tee, malleable iron, black, 150 lb. threaded, 1″ pipe size	3.000	Ea.	44.25	298.50	342.75
Pipe, 90° elbow, malleable iron, black, 150 lb. threaded 1″ pipe size	5.000	Ea.	46	307.50	353.50
Sprinkler head, std spray, brass 135°-286°F 1/2″ NPT, 3/8″ orifice	12.000	Ea.	221.40	588	809.40
Air compressor, auto, complete, 200 Gal sprinkler sys. cap., 1/3 HP	1.000	Ea.	1,325	605	1,930
*Standpipe connection, wall, flush w/plug & chain 2-1/2″ x 2-1/2″	1.000	Ea.	205	282	487
Valve, gate, bronze, 300 psi, NRS, class 150, threaded, 1″ pipe size	1.000	Ea.	148	42	190
TOTAL			32,419.48	10,437.31	42,856.79
COST PER S.F.			12.16	3.91	16.07
*Not included in systems under 2000 S.F.					

D4010 370	Deluge Sprinkler Systems		COST PER S.F.		
			MAT.	INST.	TOTAL
0520	Deluge sprinkler systems, steel, black, sch. 40 pipe	R211313 -20			
0530	Light hazard, one floor, 500 S.F.		29.50	6.05	35.55

D4010 Sprinklers

D4010 370	Deluge Sprinkler Systems	COST PER S.F. MAT.	COST PER S.F. INST.	COST PER S.F. TOTAL
0560	1000 S.F.	16.20	4.27	20.47
0580	2000 S.F.	12.15	3.90	16.05
0600	5000 S.F.	5.70	2.92	8.62
0620	10,000 S.F.	3.45	2.41	5.86
0640	50,000 S.F.	1.90	2.09	3.99
0660	Each additional floor, 500 S.F.	3.34	3.13	6.47
0680	1000 S.F.	2.40	2.86	5.26
0700	2000 S.F.	2.28	2.66	4.94
0720	5000 S.F.	1.70	2.25	3.95
0740	10,000 S.F.	1.46	2.06	3.52
0760	50,000 S.F.	1.40	1.92	3.32
1000	Ordinary hazard, one floor, 500 S.F.	30.50	6.85	37.35
1020	1000 S.F.	16.15	4.40	20.55
1040	2000 S.F.	12.85	4.64	17.49
1060	5000 S.F.	6.05	3.10	9.15
1080	10,000 S.F.	3.90	3.13	7.03
1100	50,000 S.F.	2.92	2.94	5.86
1140	Each additional floor, 500 S.F.	3.48	3.60	7.08
1160	1000 S.F.	2.31	2.89	5.20
1180	2000 S.F.	2.13	2.90	5.03
1200	5000 S.F.	2.12	2.48	4.60
1220	10,000 S.F.	1.95	2.46	4.41
1240	50,000 S.F.	1.81	2.28	4.09
1500	Extra hazard, one floor, 500 S.F.	34.50	8.40	42.90
1520	1000 S.F.	22	6.45	28.45
1540	2000 S.F.	12.90	5.80	18.70
1560	5000 S.F.	6.35	4.29	10.64
1580	10,000 S.F.	5.45	4.16	9.61
1600	50,000 S.F.	5.15	4	9.15
1660	Each additional floor, 500 S.F.	4.19	4.43	8.62
1680	1000 S.F.	3.25	4.18	7.43
1700	2000 S.F.	3	4.16	7.16
1720	5000 S.F.	2.44	3.67	6.11
1740	10,000 S.F.	2.95	3.49	6.44
1760	50,000 S.F.	2.92	3.37	6.29
2000	Grooved steel, black, sch. 40 pipe, light hazard, one floor			
2020	2000 S.F.	12.40	3.85	16.25
2060	10,000 S.F.	3.53	2.11	5.64
2100	Each additional floor, 2,000 S.F.	2.36	2.14	4.50
2150	10,000 S.F.	1.51	1.74	3.25
2200	Ordinary hazard, one floor, 2000 S.F.	8.45	4.05	12.50
2250	10,000 S.F.	3.84	2.67	6.51
2300	Each additional floor, 2000 S.F.	2.44	2.36	4.80
2350	10,000 S.F.	1.86	2.33	4.19
2400	Extra hazard, one floor, 2000 S.F.	12.95	5	17.95
2450	10,000 S.F.	4.90	3.42	8.32
2500	Each additional floor, 2000 S.F.	3.06	3.41	6.47
2550	10,000 S.F.	2.46	2.89	5.35
3000	Grooved steel, black, sch. 10 pipe, light hazard, one floor			
3050	2000 S.F.	11.70	3.67	15.37
3100	10,000 S.F.	3.46	2.06	5.52
3150	Each additional floor, 2000 S.F.	2.28	2.12	4.40
3200	10,000 S.F.	1.47	1.71	3.18
3250	Ordinary hazard, one floor, 2000 S.F.	12.40	4.04	16.44
3300	10,000 S.F.	3.43	2.61	6.04
3350	Each additional floor, 2000 S.F.	2.37	2.33	4.70
3400	10,000 S.F.	1.78	2.28	4.06
3450	Extra hazard, one floor, 2000 S.F.	12.90	4.99	17.89

D4010 Sprinklers

D4010 370	Deluge Sprinkler Systems	COST PER S.F.		
		MAT.	INST.	TOTAL
3500	10,000 S.F.	4.61	3.35	7.96
3550	Each additional floor, 2000 S.F.	3	3.39	6.39
3600	10,000 S.F.	2.36	2.85	5.21
4000	Copper tubing, type L, light hazard, one floor			
4050	2000 S.F.	12.60	3.86	16.46
4100	10,000 S.F.	3.76	2.14	5.90
4150	Each additional floor, 2000 S.F.	2.58	2.19	4.77
4200	10,000 S.F.	1.57	1.79	3.36
4250	Ordinary hazard, one floor, 2000 S.F.	12.85	4.25	17.10
4300	10,000 S.F.	4.13	2.50	6.63
4350	Each additional floor, 2000 S.F.	2.56	2.23	4.79
4400	10,000 S.F.	2.03	1.94	3.97
4450	Extra hazard, one floor, 2000 S.F.	13.40	5.10	18.50
4500	10,000 S.F.	6.20	3.70	9.90
4550	Each additional floor, 2000 S.F.	3.53	3.52	7.05
4600	10,000 S.F.	3.51	3.14	6.65
5000	Copper tubing, type L, T-drill system, light hazard, one floor			
5050	2000 S.F.	12.65	3.64	16.29
5100	10,000 S.F.	3.77	1.79	5.56
5150	Each additional floor, 2000 S.F.	2.60	1.99	4.59
5200	10,000 S.F.	1.78	1.45	3.23
5250	Ordinary hazard, one floor, 2000 S.F.	10.40	3.46	13.86
5300	10,000 S.F.	4.18	2.24	6.42
5350	Each additional floor, 2000 S.F.	2.64	1.98	4.62
5400	10,000 S.F.	2.20	1.90	4.10
5450	Extra hazard, one floor, 2000 S.F.	10.65	4.06	14.71
5500	10,000 S.F.	5.40	2.77	8.17
5550	Each additional floor, 2000 S.F.	3	2.70	5.70
5600	10,000 S.F.	2.60	2.22	4.82

D4010 Sprinklers

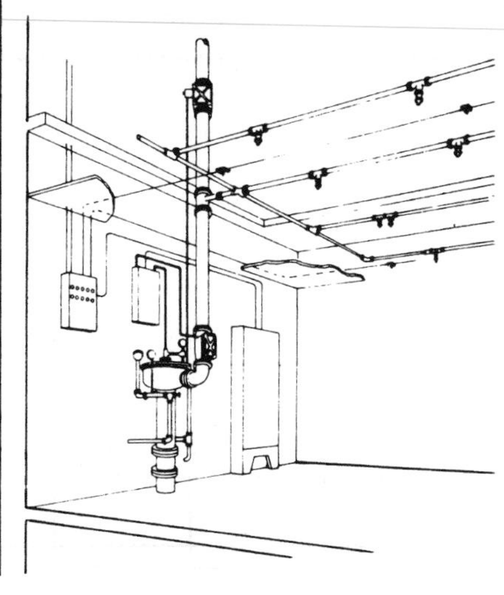

On-off multicycle sprinkler system is a fixed fire protection system utilizing water as its extinguishing agent. It is a time delayed, recycling, preaction type which automatically shuts the water off when heat is reduced below the detector operating temperature and turns the water back on when that temperature is exceeded.

The system senses a fire condition through a closed circuit electrical detector system which controls water flow to the fire automatically. Batteries supply up to 90 hour emergency power supply for system operation. The piping system is dry (until water is required) and is monitored with pressurized air. Should any leak in the system piping occur, an alarm will sound, but water will not enter the system until heat is sensed by a Firecycle detector.

All areas are assumed to be open.

System Components	QUANTITY	UNIT	COST EACH		
			MAT.	INST.	TOTAL
SYSTEM D4010 390 0580					
ON-OFF MULTICYCLE SPRINKLER SYSTEM, STEEL, BLACK, SCH. 40 PIPE					
LIGHT HAZARD, ONE FLOOR, 2000 S.F.					
Valve, gate, iron body 125 lb., OS&Y, flanged, 4″ pipe size	1.000	Ea.	755	480	1,235
4″ pipe size	1.000	Ea.	5,125	480	5,605
Tamper switch (valve supervisory switch)	3.000	Ea.	873	147	1,020
Valve, angle, bronze, 150 lb., rising stem, threaded, 2″ pipe size	1.000	Ea.	930	72.50	1,002.50
Valve, swing check, bronze, 125 lb, regrinding disc, 2-1/2″ pipe size	1.000	Ea.	1,025	95.50	1,120.50
*Alarm valve, 2-1/2″ pipe size	1.000	Ea.	1,900	470	2,370
Alarm, water motor, complete with gong	1.000	Ea.	530	196	726
Pipe, steel, black, schedule 40, 4″ diam.	8.000	L.F.	212	337.36	549.36
Fire alarm, horn, electric	1.000	Ea.	59.50	113	172.50
Pipe, steel, black, schedule 40, threaded, cplg & hngr 10′ OC 2-1/2″ diam.	15.000	L.F.	188.25	427.50	615.75
Pipe, steel, black, schedule 40, threaded, cplg & hngr 10′ OC 2″ diam.	9.375	L.F.	71.25	210.94	282.19
Pipe, steel, black, schedule 40, threaded, cplg & hngr 10′ OC 1-1/4″ diam.	28.125	L.F.	253.13	452.81	705.94
Pipe, steel, black, schedule 40, threaded, cplg & hngr 10′ OC 1″ diam.	84.000	L.F.	718.20	1,264.20	1,982.40
Pipe, Tee, malleable iron, black, 150 lb. threaded, 4″ pipe size	2.000	Ea.	780	720	1,500
Pipe, Tee, malleable iron, black, 150 lb. threaded, 2-1/2″ pipe size	2.000	Ea.	218	318	536
Pipe, Tee, malleable iron, black, 150 lb. threaded, 2″ pipe size	1.000	Ea.	50.50	130	180.50
Pipe, Tee, malleable iron, black, 150 lb. threaded, 1-1/4″ pipe size	4.000	Ea.	96	408	504
Pipe, Tee, malleable iron, black, 150 lb. threaded, 1″ pipe size	3.000	Ea.	44.25	298.50	342.75
Pipe, 90° elbow, malleable iron, black, 150 lb. threaded, 1″ pipe size	5.000	Ea.	46	307.50	353.50
Sprinkler head std spray, brass 135°-286°F 1/2″ NPT, 3/8″ orifice	12.000	Ea.	221.40	588	809.40
Firecycle controls, incls panel, battery, solenoid valves, press switches	1.000	Ea.	24,600	2,975	27,575
Detector, firecycle system	2.000	Ea.	1,760	98	1,858
Firecycle pkg, swing check & flow control valves w/trim 4″ pipe size	1.000	Ea.	7,400	1,400	8,800
Air compressor, auto, complete, 200 Gal sprinkler sys. cap., 1/3 HP	1.000	Ea.	1,325	605	1,930
*Standpipe connection, wall, flush, brass w/plug & chain 2-1/2″x2-1/2″	1.000	Ea.	205	282	487
Valve, gate, bronze 300 psi, NRS, class 150, threaded, 1″ diam.	1.000	Ea.	148	42	190
TOTAL			49,534.48	12,918.81	62,453.29
COST PER S.F.			18.58	4.84	23.42

*Not included in systems under 2000 S.F.

D4010 390	On-off Multicycle Sprinkler Systems		COST PER S.F.		
			MAT.	INST.	TOTAL
0520	On-off multicycle sprinkler systems, steel, black, sch. 40 pipe	R211313 -20			
0530	Light hazard, one floor, 500 S.F.		57.50	11.40	68.90

D4010 Sprinklers

D4010 390	On-off Multicycle Sprinkler Systems	COST PER S.F.		
		MAT.	INST.	TOTAL
0560	1000 S.F.	30	7.15	37.15
0580	2000 S.F.	18.56	4.84	23.40
0600	5000 S.F.	8.30	3.30	11.60
0620	10,000 S.F.	4.81	2.60	7.41
0640	50,000 S.F.	2.22	2.13	4.35
0660	Each additional floor of 500 S.F.	3.34	3.14	6.48
0680	1000 S.F.	2.40	2.87	5.27
0700	2000 S.F.	1.95	2.65	4.60
0720	5000 S.F.	1.70	2.26	3.96
0740	10,000 S.F.	1.53	2.07	3.60
0760	50,000 S.F.	1.46	1.92	3.38
1000	Ordinary hazard, one floor, 500 S.F.	54.50	11.55	66.05
1020	1000 S.F.	30	7.10	37.10
1040	2000 S.F.	18.95	5.55	24.50
1060	5000 S.F.	8.65	3.48	12.13
1080	10,000 S.F.	5.25	3.32	8.57
1100	50,000 S.F.	3.45	3.30	6.75
1140	Each additional floor, 500 S.F.	3.48	3.61	7.09
1160	1000 S.F.	2.31	2.90	5.21
1180	2000 S.F.	2.31	2.68	4.99
1200	5000 S.F.	2.12	2.49	4.61
1220	10,000 S.F.	1.86	2.42	4.28
1240	50,000 S.F.	1.80	2.17	3.97
1500	Extra hazard, one floor, 500 S.F.	62.50	13.70	76.20
1520	1000 S.F.	35	8.90	43.90
1540	2000 S.F.	19.30	6.75	26.05
1560	5000 S.F.	8.90	4.67	13.57
1580	10,000 S.F.	6.90	4.72	11.62
1600	50,000 S.F.	5.65	5.15	10.80
1660	Each additional floor, 500 S.F.	4.19	4.44	8.63
1680	1000 S.F.	3.25	4.19	7.44
1700	2000 S.F.	3	4.17	7.17
1720	5000 S.F.	2.44	3.68	6.12
1740	10,000 S.F.	2.95	3.38	6.33
1760	50,000 S.F.	2.91	3.26	6.17
2020	Grooved steel, black, sch. 40 pipe, light hazard, one floor			
2030	2000 S.F.	16.60	4.54	21.14
2060	10,000 S.F.	5.20	3.14	8.34
2100	Each additional floor, 2000 S.F.	2.36	2.15	4.51
2150	10,000 S.F.	1.58	1.75	3.33
2200	Ordinary hazard, one floor, 2000 S.F.	18.90	5	23.90
2250	10,000 S.F.	5.60	3.01	8.61
2300	Each additional floor, 2000 S.F.	2.44	2.37	4.81
2350	10,000 S.F.	1.93	2.34	4.27
2400	Extra hazard, one floor, 2000 S.F.	17.10	5.70	22.80
2450	10,000 S.F.	6.15	3.58	9.73
2500	Each additional floor, 2000 S.F.	3.06	3.42	6.48
2550	10,000 S.F.	2.53	2.90	5.43
3050	Grooved steel, black, sch. 10 pipe, light hazard, one floor,			
3060	2000 S.F.	18.75	4.76	23.51
3100	10,000 S.F.	4.82	2.25	7.07
3150	Each additional floor, 2000 S.F.	2.28	2.13	4.41
3200	10,000 S.F.	1.54	1.72	3.26
3250	Ordinary hazard, one floor, 2000 S.F.	18.85	4.97	23.82
3300	10,000 S.F.	5.10	2.81	7.91
3350	Each additional floor, 2000 S.F.	2.37	2.34	4.71
3400	10,000 S.F.	1.85	2.29	4.14
3450	Extra hazard, one floor, 2000 S.F.	19.30	5.90	25.20

D4010 Sprinklers

D4010 390	On-off Multicycle Sprinkler Systems	COST PER S.F.		
		MAT.	INST.	TOTAL
3500	10,000 S.F.	5.95	3.52	9.47
3550	Each additional floor, 2000 S.F.	3	3.40	6.40
3600	10,000 S.F.	2.43	2.86	5.29
4060	Copper tubing, type L, light hazard, one floor, 2000 S.F.	19.05	4.79	23.84
4100	10,000 S.F.	5.25	2.33	7.58
4150	Each additional floor, 2000 S.F.	2.58	2.20	4.78
4200	10,000 S.F.	1.97	1.81	3.78
4250	Ordinary hazard, one floor, 2000 S.F.	19.40	5.15	24.55
4300	10,000 S.F.	5.75	2.65	8.40
4350	Each additional floor, 2000 S.F.	2.66	2.19	4.85
4400	10,000 S.F.	2.19	1.89	4.08
4450	Extra hazard, one floor, 2000 S.F.	20	6	26
4500	10,000 S.F.	8.10	3.88	11.98
4550	Each additional floor, 2000 S.F.	3.72	3.49	7.21
4600	10,000 S.F.	3.78	3.13	6.91
5060	Copper tubing, type L, T-drill system, light hazard, one floor 2000 S.F.	19.10	4.57	23.67
5100	10,000 S.F.	5.15	1.97	7.12
5150	Each additional floor, 2000 S.F.	2.83	2.06	4.89
5200	10,000 S.F.	1.85	1.46	3.31
5250	Ordinary hazard, one floor, 2000 S.F.	19.20	4.58	23.78
5300	10,000 S.F.	5.65	2.39	8.04
5350	Each additional floor, 2000 S.F.	2.72	1.95	4.67
5400	10,000 S.F.	2.40	1.87	4.27
5450	Extra hazard, one floor, 2000 S.F.	19.45	5.20	24.65
5500	10,000 S.F.	7.10	2.90	10
5550	Each additional floor, 2000 S.F.	3.15	2.67	5.82
5600	10,000 S.F.	2.86	2.20	5.06

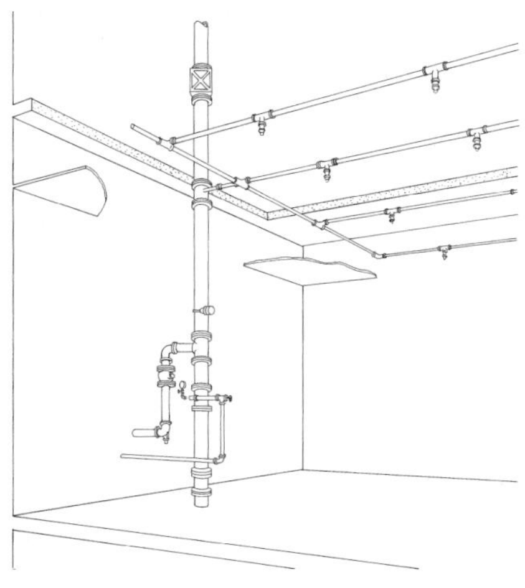

Wet Pipe System. A system employing automatic sprinklers attached to a piping system containing water and connected to a water supply so that water discharges immediately from sprinklers opened by heat from a fire.

All areas are assumed to be open.

System Components	QUANTITY	UNIT	COST EACH MAT.	COST EACH INST.	COST EACH TOTAL
SYSTEM D4010 410 0580					
WET PIPE SPRINKLER, STEEL, BLACK, SCH. 40 PIPE					
LIGHT HAZARD, ONE FLOOR, 2000 S.F.					
Valve, gate, iron body, 125 lb., OS&Y, flanged, 4″ diam.	1.000	Ea.	755	480	1,235
4″ pipe size	1.000	Ea.	5,125	480	5,605
Tamper switch (valve supervisory switch)	3.000	Ea.	873	147	1,020
Valve, swing check, bronze, 125 lb, regrinding disc, 2-1/2″ pipe size	1.000	Ea.	1,025	95.50	1,120.50
Valve, angle, bronze, 150 lb., rising stem, threaded, 2″ diam.	1.000	Ea.	930	72.50	1,002.50
*Alarm valve, 2-1/2″ pipe size	1.000	Ea.	1,900	470	2,370
Alarm, water motor, complete with gong	1.000	Ea.	530	196	726
Valve, swing check, w/balldrip CI with brass trim 4″ pipe size	1.000	Ea.	460	470	930
Pipe, steel, black, schedule 40, 4″ diam.	8.000	L.F.	212	337.36	549.36
Fire alarm horn, electric	1.000	Ea.	59.50	113	172.50
Pipe, steel, black, schedule 40, threaded, cplg & hngr 10′ OC, 2-1/2″ diam.	15.000	L.F.	188.25	427.50	615.75
Pipe, steel, black, schedule 40, threaded, cplg & hngr 10′ OC, 2″ diam.	9.375	L.F.	71.25	210.94	282.19
Pipe, steel, black, schedule 40, threaded, cplg & hngr 10′ OC, 1-1/4″ diam.	28.125	L.F.	253.13	452.81	705.94
Pipe, steel, black, schedule 40, threaded cplg & hngr 10′ OC, 1″ diam.	84.000	L.F.	718.20	1,264.20	1,982.40
Pipe Tee, malleable iron black, 150 lb. threaded, 4″ pipe size	2.000	Ea.	780	720	1,500
Pipe Tee, malleable iron black, 150 lb. threaded, 2-1/2″ pipe size	2.000	Ea.	218	318	536
Pipe Tee, malleable iron black, 150 lb. threaded, 2″ pipe size	1.000	Ea.	50.50	130	180.50
Pipe Tee, malleable iron black, 150 lb. threaded, 1-1/4″ pipe size	4.000	Ea.	96	408	504
Pipe Tee, malleable iron black, 150 lb. threaded, 1″ pipe size	3.000	Ea.	44.25	298.50	342.75
Pipe 90° elbow, malleable iron black, 150 lb. threaded, 1″ pipe size	5.000	Ea.	46	307.50	353.50
Sprinkler head, standard spray, brass 135°-286°F 1/2″ NPT, 3/8″ orifice	12.000	Ea.	221.40	588	809.40
Valve, gate, bronze, NRS, class 150, threaded, 1″ pipe size	1.000	Ea.	148	42	190
*Standpipe connection, wall, single, flush w/plug & chain 2-1/2″x2-1/2″	1.000	Ea.	205	282	487
TOTAL			14,909.48	8,310.81	23,220.29
COST PER S.F.			5.59	3.12	8.71

*Not included in systems under 2000 S.F.

D4010 410	Wet Pipe Sprinkler Systems		COST PER S.F. MAT.	COST PER S.F. INST.	COST PER S.F. TOTAL
0520	Wet pipe sprinkler systems, steel, black, sch. 40 pipe				
0530	Light hazard, one floor, 500 S.F.		6.60	3.93	10.53
0560	1000 S.F.	R211313 -20	4.83	3.42	8.25
0580	2000 S.F.		5.59	3.11	8.70
0600	5000 S.F.	R211313 -40	2.83	2.59	5.42
0620	10,000 S.F.		1.80	2.23	4.03

D4010 Sprinklers

D4010 410	Wet Pipe Sprinkler Systems	COST PER S.F. MAT.	INST.	TOTAL
0640	50,000 S.F.	1.25	2.04	3.29
0660	Each additional floor, 500 S.F.	2.05	3.14	5.19
0680	1000 S.F.	1.77	2.90	4.67
0700	2000 S.F.	1.62	2.63	4.25
0720	5000 S.F.	1.17	2.23	3.40
0740	10,000 S.F.	1	2.04	3.04
0760	50,000 S.F.	.83	1.61	2.44
1000	Ordinary hazard, one floor, 500 S.F.	6.70	4.19	10.89
1020	1000 S.F.	4.71	3.36	8.07
1040	2000 S.F.	5.95	3.83	9.78
1060	5000 S.F.	3.18	2.77	5.95
1080	10,000 S.F.	2.25	2.95	5.20
1100	50,000 S.F.	2.17	2.82	4.99
1140	Each additional floor, 500 S.F.	2.16	3.54	5.70
1160	1000 S.F.	1.65	2.86	4.51
1180	2000 S.F.	1.80	2.89	4.69
1200	5000 S.F.	1.75	2.72	4.47
1220	10,000 S.F.	1.46	2.77	4.23
1240	50,000 S.F.	1.51	2.47	3.98
1500	Extra hazard, one floor, 500 S.F.	9	5.40	14.40
1520	1000 S.F.	8.70	4.85	13.55
1540	2000 S.F.	6.35	5.15	11.50
1560	5000 S.F.	3.87	4.59	8.46
1580	10,000 S.F.	3.92	4.42	8.34
1600	50,000 S.F.	3.95	4.21	8.16
1660	Each additional floor, 500 S.F.	2.87	4.37	7.24
1680	1000 S.F.	2.59	4.15	6.74
1700	2000 S.F.	2.34	4.13	6.47
1720	5000 S.F.	1.91	3.65	5.56
1740	10,000 S.F.	2.42	3.35	5.77
1760	50,000 S.F.	2.39	3.19	5.58
2020	Grooved steel, black sch. 40 pipe, light hazard, one floor, 2000 S.F.	5.85	3.05	8.90
2060	10,000 S.F.	1.85	1.91	3.76
2100	Each additional floor, 2000 S.F.	1.70	2.11	3.81
2150	10,000 S.F.	1.05	1.72	2.77
2200	Ordinary hazard, one floor, 2000 S.F.	5.95	3.27	9.22
2250	10,000 S.F.	2.19	2.49	4.68
2300	Each additional floor, 2000 S.F.	1.78	2.33	4.11
2350	10,000 S.F.	1.40	2.31	3.71
2400	Extra hazard, one floor, 2000 S.F.	4.13	3.97	8.10
2450	10,000 S.F.	2.57	3.16	5.73
2500	Each additional floor, 2000 S.F.	2.40	3.38	5.78
2550	10,000 S.F.	2	2.87	4.87
3050	Grooved steel, black sch. 10 pipe, light hazard, one floor, 2000 S.F.	3.52	2.79	6.31
3100	10,000 S.F.	1.81	1.88	3.69
3150	Each additional floor, 2000 S.F.	1.62	2.09	3.71
3200	10,000 S.F.	1.01	1.69	2.70
3250	Ordinary hazard, one floor, 2000 S.F.	5.85	3.24	9.09
3300	10,000 S.F.	2.04	2.43	4.47
3350	Each additional floor, 2000 S.F.	1.71	2.30	4.01
3400	10,000 S.F.	1.32	2.26	3.58
3450	Extra hazard, one floor, 2000 S.F.	6.30	4.19	10.49
3500	10,000 S.F.	2.96	3.17	6.13
3550	Each additional floor, 2000 S.F.	2.34	3.36	5.70
3600	10,000 S.F.	1.90	2.83	4.73
4050	Copper tubing, type L, light hazard, one floor, 2000 S.F.	6.05	3.06	9.11
4100	10,000 S.F.	2.24	1.96	4.20
4150	Each additional floor, 2000 S.F.	1.92	2.16	4.08

D4010 Sprinklers

D4010 410	Wet Pipe Sprinkler Systems	COST PER S.F.		
		MAT.	INST.	TOTAL
4200	10,000 S.F.	1.44	1.78	3.22
4250	Ordinary hazard, one floor, 2000 S.F.	6.25	3.45	9.70
4300	10,000 S.F.	2.61	2.32	4.93
4350	Each additional floor, 2000 S.F.	2.17	2.38	4.55
4400	10,000 S.F.	1.76	2.09	3.85
4450	Extra hazard, one floor, 2000 S.F.	6.85	4.32	11.17
4500	10,000 S.F.	4.64	3.51	8.15
4550	Each additional floor, 2000 S.F.	2.87	3.49	6.36
4600	10,000 S.F.	3.05	3.12	6.17
5050	Copper tubing, type L, T-drill system, light hazard, one floor			
5060	2000 S.F.	6.10	2.84	8.94
5100	10,000 S.F.	2.12	1.61	3.73
5150	Each additional floor, 2000 S.F.	1.98	1.94	3.92
5200	10,000 S.F.	1.32	1.43	2.75
5250	Ordinary hazard, one floor, 2000 S.F.	3.84	2.66	6.50
5300	10,000 S.F.	2.53	2.06	4.59
5350	Each additional floor, 2000 S.F.	1.94	1.96	3.90
5400	10,000 S.F.	1.74	1.88	3.62
5450	Extra hazard, one floor, 2000 S.F.	6.30	3.50	9.80
5500	10,000 S.F.	3.73	2.59	6.32
5550	Each additional floor, 2000 S.F.	5.10	2.98	8.08
5600	10,000 S.F.	2.14	2.20	4.34

D4020 Standpipes

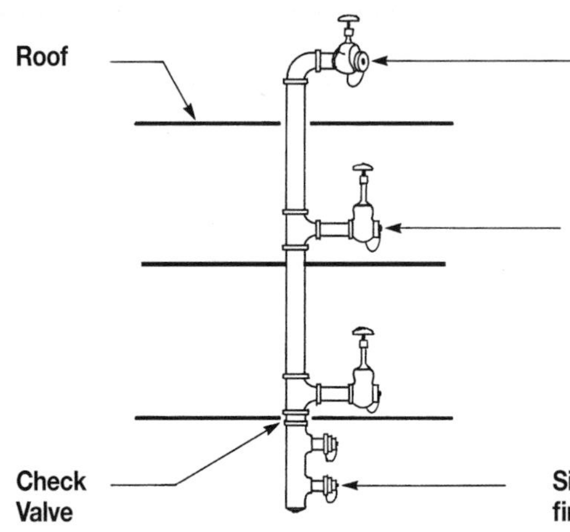

System Components	QUANTITY	UNIT	COST PER FLOOR MAT.	INST.	TOTAL
SYSTEM D4020 310 0560					
WET STANDPIPE RISER, CLASS I, STEEL, BLACK, SCH. 40 PIPE, 10′ HEIGHT					
4″ DIAMETER PIPE, ONE FLOOR					
Pipe, steel, black, schedule 40, threaded, 4″ diam.	20.000	L.F.	510	800	1,310
Pipe, Tee, malleable iron, black, 150 lb. threaded, 4″ pipe size	2.000	Ea.	780	720	1,500
Pipe, 90° elbow, malleable iron, black, 150 lb threaded, 4″ pipe size	1.000	Ea.	216	239	455
Pipe, nipple, steel, black, schedule 40, 2-1/2″ pipe size x 3″ long	2.000	Ea.	28.20	179	207.20
Fire valve, gate, 300 lb., brass w/handwheel, 2-1/2″ pipe size	1.000	Ea.	239	112	351
Fire valve, pressure reducing rgh brs, 2-1/2″ pipe size	1.000	Ea.	980	224	1,204
Valve, swing check, w/ball drip, CI w/brs. ftngs., 4″ pipe size	1.000	Ea.	460	470	930
Standpipe conn wall dble. flush brs. w/plugs & chains 2-1/2″x2-1/2″x4″	1.000	Ea.	850	282	1,132
Valve, swing check, bronze, 125 lb, regrinding disc, 2-1/2″ pipe size	1.000	Ea.	1,025	95.50	1,120.50
Roof manifold, fire, w/valves & caps, horiz/vert brs 2-1/2″x2-1/2″x4″	1.000	Ea.	250	294	544
Fire, hydrolator, vent & drain, 2-1/2″ pipe size	1.000	Ea.	126	65.50	191.50
Valve, gate, iron body 125 lb., OS&Y, threaded, 4″ pipe size	1.000	Ea.	755	480	1,235
Tamper switch (valve supervisory switch)	1.000	Ea.	291	49	340
TOTAL			6,510.20	4,010	10,520.20

D4020 310	Wet Standpipe Risers, Class I	COST PER FLOOR MAT.	INST.	TOTAL
0550	Wet standpipe risers, Class I, steel, black, sch. 40, 10′ height			
0560	4″ diameter pipe, one floor	6,500	4,000	10,500
0580	Additional floors	1,525	1,225	2,750
0600	6″ diameter pipe, one floor	10,200	6,975	17,175
0620	Additional floors R211226 -20	2,700	1,950	4,650
0640	8″ diameter pipe, one floor	15,200	8,400	23,600
0660	Additional floors	3,900	2,350	6,250
0680				

D4020 310	Wet Standpipe Risers, Class II	COST PER FLOOR MAT.	INST.	TOTAL
1030	Wet standpipe risers, Class II, steel, black sch. 40, 10′ height			
1040	2″ diameter pipe, one floor	2,750	1,475	4,225
1060	Additional floors	880	550	1,430
1080	2-1/2″ diameter pipe, one floor	3,600	2,075	5,675
1100	Additional floors	975	640	1,615
1120				

D4020 Standpipes

D4020 310	Wet Standpipe Risers, Class III	COST PER FLOOR		
		MAT.	INST.	TOTAL
1530	Wet standpipe risers, Class III, steel, black, sch. 40, 10′ height			
1540	4″ diameter pipe, one floor	6,675	4,000	10,675
1560	Additional floors	1,350	1,025	2,375
1580	6″ diameter pipe, one floor	10,400	6,975	17,375
1600	Additional floors	2,775	1,950	4,725
1620	8″ diameter pipe, one floor	15,300	8,400	23,700
1640	Additional floors	3,975	2,350	6,325

D4020 Standpipes

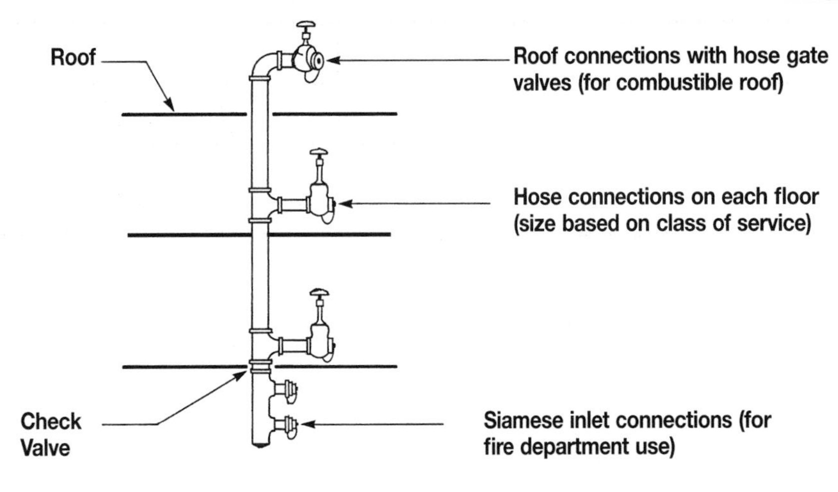

System Components	QUANTITY	UNIT	COST PER FLOOR		
			MAT.	INST.	TOTAL
SYSTEM D4020 330 0540					
DRY STANDPIPE RISER, CLASS I, PIPE, STEEL, BLACK, SCH 40, 10′ HEIGHT					
4″ DIAMETER PIPE, ONE FLOOR					
Pipe, steel, black, schedule 40, threaded, 4″ diam.	20.000	L.F.	510	800	1,310
Pipe, Tee, malleable iron, black, 150 lb. threaded, 4″ pipe size	2.000	Ea.	780	720	1,500
Pipe, 90° elbow, malleable iron, black, 150 lb threaded, 4″ pipe size	1.000	Ea.	216	239	455
Pipe, nipple, steel, black, schedule 40, 2-1/2″ pipe size x 3″ long	2.000	Ea.	28.20	179	207.20
Fire valve gate NRS 300 lb., brass w/handwheel, 2-1/2″ pipe size	1.000	Ea.	239	112	351
Tamper switch (valve supervisory switch)	1.000	Ea.	291	49	340
Fire valve, pressure reducing rgh brs, 2-1/2″ pipe size	1.000	Ea.	490	112	602
Standpipe conn wall dble. flush brs. w/plugs & chains 2-1/2″x2-1/2″x4″	1.000	Ea.	850	282	1,132
Valve swing check w/ball drip CI w/brs. ftngs., 4″pipe size	1.000	Ea.	460	470	930
Roof manifold, fire, w/valves & caps, horiz/vert brs 2-1/2″x2-1/2″x4″	1.000	Ea.	250	294	544
TOTAL			4,114.20	3,257	7,371.20

D4020 330	Dry Standpipe Risers, Class I	COST PER FLOOR		
		MAT.	INST.	TOTAL
0530	Dry standpipe riser, Class I, steel, black, sch. 40, 10′ height			
0540	4″ diameter pipe, one floor	4,125	3,250	7,375
0560	Additional floors R211226 -20	1,400	1,175	2,575
0580	6″ diameter pipe, one floor	7,775	5,525	13,300
0600	Additional floors	2,575	1,900	4,475
0620	8″ diameter pipe, one floor	11,700	6,700	18,400
0640	Additional floors	3,775	2,300	6,075
0660				

D4020 330	Dry Standpipe Risers, Class II	COST PER FLOOR		
		MAT.	INST.	TOTAL
1030	Dry standpipe risers, Class II, steel, black, sch. 40, 10′ height			
1040	2″ diameter pipe, one floor	2,200	1,475	3,675
1060	Additional floors	755	485	1,240
1080	2-1/2″ diameter pipe, one floor	3,000	1,725	4,725
1100	Additional floors	850	575	1,425
1120				

D4020 Standpipes

D4020 330	Dry Standpipe Risers, Class III	COST PER FLOOR		
		MAT.	INST.	TOTAL
1530	Dry standpipe risers, Class III, steel, black, sch. 40, 10′ height			
1540	4″ diameter pipe, one floor	3,875	3,150	7,025
1560	Additional floors	1,250	1,050	2,300
1580	6″ diameter pipe, one floor	7,850	5,525	13,375
1600	Additional floors	2,650	1,900	4,550
1620	8″ diameter pipe, one floor	11,800	6,700	18,500
1640	Additional floors	3,850	2,300	6,150

D4020 Standpipes

D4020 410	Fire Hose Equipment	COST EACH		
		MAT.	INST.	TOTAL
0100	Adapters, reducing, 1 piece, FxM, hexagon, cast brass, 2-1/2" x 1-1/2"	71		71
0200	Pin lug, 1-1/2" x 1"	50		50
0250	3" x 2-1/2"	156		156
0300	For polished chrome, add 75% mat.			
0400	Cabinets, D.S. glass in door, recessed, steel box, not equipped			
0500	Single extinguisher, steel door & frame	180	176	356
0550	Stainless steel door & frame	249	176	425
0600	Valve, 2-1/2" angle, steel door & frame	227	118	345
0650	Aluminum door & frame	276	118	394
0700	Stainless steel door & frame	370	118	488
0750	Hose rack assy, 2-1/2" x 1-1/2" valve & 100' hose, steel door & frame	425	235	660
0800	Aluminum door & frame	600	235	835
0850	Stainless steel door & frame	835	235	1,070
0900	Hose rack assy & extinguisher,2-1/2"x1-1/2" valve & hose,steel door & frame	330	282	612
0950	Aluminum	800	282	1,082
1000	Stainless steel	695	282	977
1550	Compressor, air, dry pipe system, automatic, 200 gal., 3/4 H.P.	1,325	605	1,930
1600	520 gal., 1 H.P.	1,750	605	2,355
1650	Alarm, electric pressure switch (circuit closer)	123	30	153
2500	Couplings, hose, rocker lug, cast brass, 1-1/2"	69.50		69.50
2550	2-1/2"	57		57
3000	Escutcheon plate, for angle valves, polished brass, 1-1/2"	18.25		18.25
3050	2-1/2"	26.50		26.50
3500	Fire pump, electric, w/controller, fittings, relief valve			
3550	4" pump, 30 HP, 500 GPM	16,900	4,400	21,300
3600	5" pump, 40 H.P., 1000 G.P.M.	22,300	4,975	27,275
3650	5" pump, 100 H.P., 1000 G.P.M.	27,500	5,525	33,025
3700	For jockey pump system, add	3,350	705	4,055
3800	Fire pump, diesel, w/controller, 6" pump, 140 HP, 1500 GPM, 100 psi	64,500	7,400	71,900
3900	Fire pump, diesel, w/controller, 10" pump, 300 HP, 3500 GPM, 100 psi	106,500	13,300	119,800
4100	Fire pump, electric, w/controller, 6" pump, 139 HP, 1500 GPM, 1770 RPM	36,300	7,400	43,700
4200	Fire pump, electric, w/controller, 10" pump, 300 HP, 3500 GPM, 1770 RPM	81,500	12,400	93,900
5000	Hose, per linear foot, synthetic jacket, lined,			
5100	300 lb. test, 1-1/2" diameter	3.64	.54	4.18
5150	2-1/2" diameter	6.50	.64	7.14
5200	500 lb. test, 1-1/2" diameter	2.85	.54	3.39
5250	2-1/2" diameter	6.65	.64	7.29
5500	Nozzle, plain stream, polished brass, 1-1/2" x 10"	63.50		63.50
5550	2-1/2" x 15" x 13/16" or 1-1/2"	119		119
5600	Heavy duty combination adjustable fog and straight stream w/handle 1-1/2"	485		485
5650	2-1/2" direct connection	520		520
6000	Rack, for 1-1/2" diameter hose 100 ft. long, steel	97.50	70.50	168
6050	Brass	168	70.50	238.50
6500	Reel, steel, for 50 ft. long 1-1/2" diameter hose	174	101	275
6550	For 75 ft. long 2-1/2" diameter hose	325	101	426
7050	Siamese, w/plugs & chains, polished brass, sidewalk, 4" x 2-1/2" x 2-1/2"	825	565	1,390
7100	6" x 2-1/2" x 2-1/2"	910	705	1,615
7200	Wall type, flush, 4" x 2-1/2" x 2-1/2"	850	282	1,132
7250	6" x 2-1/2" x 2-1/2"	940	305	1,245
7300	Projecting, 4" x 2-1/2" x 2-1/2"	670	282	952
7350	6" x 2-1/2" x 2-1/2"	1,125	305	1,430
7400	For chrome plate, add 15% mat.			
8000	Valves, angle, wheel handle, 300 Lb., rough brass, 1-1/2"	116	65.50	181.50
8050	2-1/2"	217	112	329
8100	Combination pressure restricting, 1-1/2"	121	65.50	186.50
8150	2-1/2"	222	112	334
8200	Pressure restricting, adjustable, satin brass, 1-1/2"	355	65.50	420.50
8250	2-1/2"	490	112	602

D4020 Standpipes

D4020 410	Fire Hose Equipment	COST EACH		
		MAT.	INST.	TOTAL
8300	Hydrolator, vent and drain, rough brass, 1-1/2"	126	65.50	191.50
8350	2-1/2"	126	65.50	191.50
8400	Cabinet assy, incls. adapter, rack, hose, and nozzle	1,025	425	1,450

D4090 Other Fire Protection Systems

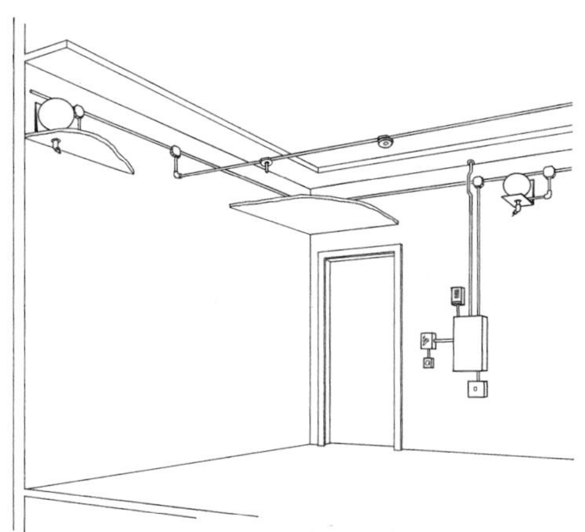

General: Automatic fire protection (suppression) systems other than water sprinklers may be desired for special environments, high risk areas, isolated locations or unusual hazards. Some typical applications would include:

Paint dip tanks
Securities vaults
Electronic data processing
Tape and data storage
Transformer rooms
Spray booths
Petroleum storage
High rack storage

Piping and wiring costs are dependent on the individual application and must be added to the component costs shown below.

All areas are assumed to be open.

D4090 910	Fire Suppression Unit Components	COST EACH		
		MAT.	INST.	TOTAL
0020	Detectors with brackets			
0040	Fixed temperature heat detector	46.50	108	154.50
0060	Rate of temperature rise detector	57	94.50	151.50
0080	Ion detector (smoke) detector	133	122	255
0200	Extinguisher agent			
0240	200 lb FM200, container	7,700	360	8,060
0280	75 lb carbon dioxide cylinder	1,625	239	1,864
0320	Dispersion nozzle			
0340	FM200 1-1/2" dispersion nozzle	260	57	317
0380	Carbon dioxide 3" x 5" dispersion nozzle	182	44.50	226.50
0420	Control station			
0440	Single zone control station with batteries	1,150	755	1,905
0470	Multizone (4) control station with batteries	3,375	1,500	4,875
0490				
0500	Electric mechanical release	1,150	390	1,540
0520				
0550	Manual pull station	102	133	235
0570				
0640	Battery standby power 10" x 10" x 17"	510	189	699
0700				
0740	Bell signalling device	73	94.50	167.50

D4090 920	FM200 Systems	COST PER C.F.		
		MAT.	INST.	TOTAL
0820	Average FM200 system, minimum			2.09
0840	Maximum			4.16

D5010 Electrical Service/Distribution

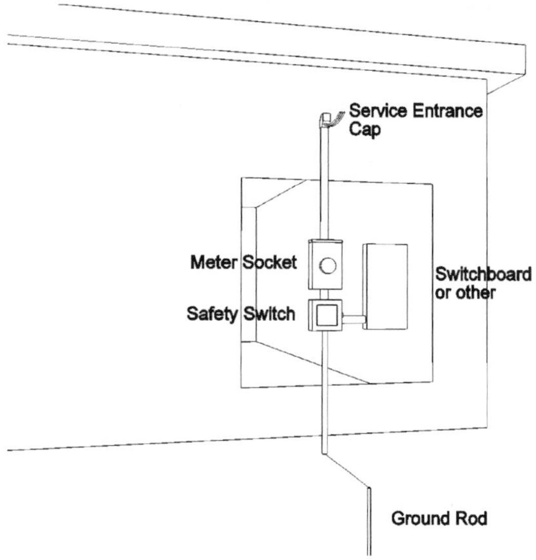

System Components	QUANTITY	UNIT	COST EACH		
			MAT.	INST.	TOTAL
SYSTEM D5010 120 0220					
SERVICE INSTALLATION, INCLUDES BREAKERS, METERING, 20′ CONDUIT & WIRE					
3 PHASE, 4 WIRE, 60 A					
Wire, 600 volt, copper type XHHW, stranded #6	.600	C.L.F.	36	69.60	105.60
Rigid galvanized steel conduit, 3/4″ including fittings	20.000	L.F.	113	189	302
Service entrance cap, 3/4″ diameter	1.000	Ea.	6.50	58	64.50
Conduit LB fitting with cover, 3/4″ diameter	1.000	Ea.	9.70	58	67.70
Meter socket, three phase, 100 A	1.000	Ea.	150	270	420
Safety switches, heavy duty, 240 volt, 3 pole NEMA 1 fusible, 60 amp	1.000	Ea.	188	330	518
Grounding, wire ground bare armored, #8-1 conductor	.200	C.L.F.	16.40	76	92.40
Grounding, clamp, bronze, 3/4″ diameter	1.000	Ea.	6.25	23.50	29.75
Grounding, rod, copper clad, 8′ long, 3/4″ diameter	1.000	Ea.	39.50	143	182.50
Wireway w/fittings, 2-1/2″ x 2-1/2″	1.000	L.F.	13.70	16.80	30.50
TOTAL			579.05	1,233.90	1,812.95

D5010 120	Overhead Electric Service, 3 Phase - 4 Wire	COST EACH		
		MAT.	INST.	TOTAL
0200	Service installation, includes breakers, metering, 20′ conduit & wire			
0220	3 phase, 4 wire, 120/208 volts, 60 A	580	1,225	1,805
0240	100 A	770	1,400	2,170
0245	100 A w/circuit breaker	1,575	1,725	3,300
0280	200 A	1,300	1,925	3,225
0285	200 A w/circuit breaker	3,375	2,425	5,800
0320	400 A	2,600	3,825	6,425
0325	400 A w/circuit breaker	5,800	4,775	10,575
0360	600 A	4,375	5,650	10,025
0365	600 A, w/switchboard	9,450	7,150	16,600
0400	800 A	6,750	6,775	13,525
0405	800 A, w/switchboard	11,800	8,500	20,300
0440	1000 A	8,200	8,350	16,550
0445	1000 A, w/switchboard	14,300	10,200	24,500
0480	1200 A	10,900	9,525	20,425
0485	1200 A, w/groundfault switchboard	37,300	11,800	49,100
0520	1600 A	13,700	12,500	26,200
0525	1600 A, w/groundfault switchboard	42,400	14,900	57,300

D5010 Electrical Service/Distribution

D5010 120	Overhead Electric Service, 3 Phase - 4 Wire	COST EACH		
		MAT.	INST.	TOTAL
0560	2000 A	18,200	15,100	33,300
0565	2000 A, w/groundfault switchboard	50,500	18,600	69,100
0610	1 phase, 3 wire, 120/240 volts, 100 A (no safety switch)	181	680	861
0615	100 A w/load center	355	1,300	1,655
0620	200 A	405	940	1,345
0625	200 A w/load center	740	2,000	2,740

D5010 Electrical Service/Distribution

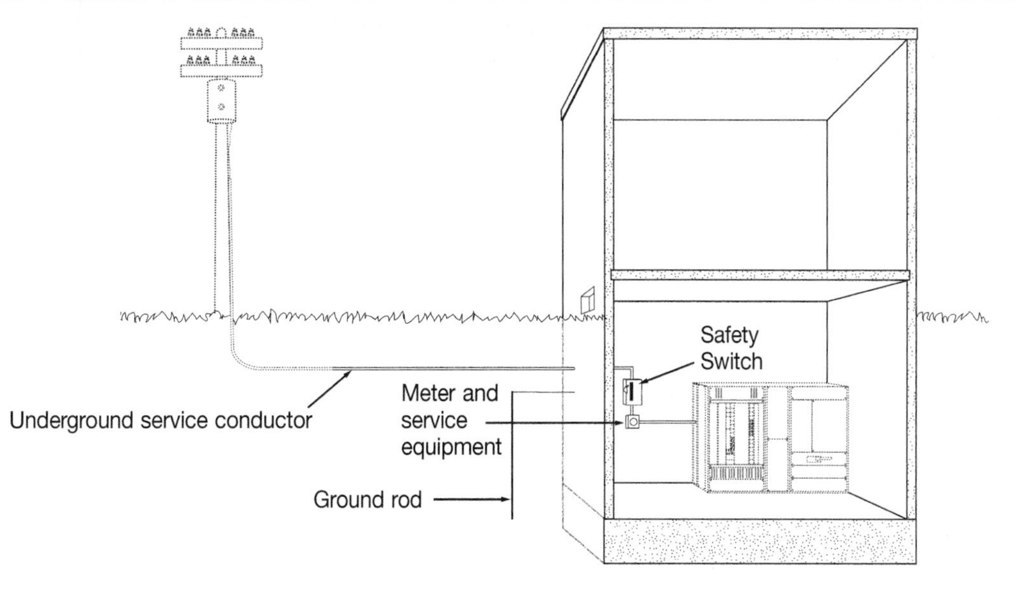

System Components	QUANTITY	UNIT	COST EACH MAT.	COST EACH INST.	COST EACH TOTAL
SYSTEM D5010 130 1000					
2000 AMP UNDERGROUND ELECTRIC SERVICE WITH SAFETY SWITCH					
INCLUDING EXCAVATION, BACKFILL AND COMPACTION					
Excavate Trench	44.440	B.C.Y.		336.42	336.42
4 inch conduit bank	108.000	L.F.	1,323	3,024	4,347
4 inch fitting	8.000	Ea.	352	504	856
4 inch bells	4.000	Ea.	20.40	252	272.40
Concrete material	16.580	C.Y.	2,354.36		2,354.36
Concrete placement	16.580	C.Y.		472.36	472.36
Backfill trench	33.710	L.C.Y.		110.57	110.57
Compact fill material in trench	25.930	E.C.Y.		143.65	143.65
Dispose of excess fill material on-site	24.070	L.C.Y.		315.80	315.80
500 kcmil power cable	18.000	C.L.F.	19,350	8,460	27,810
Wire, 600 volt, type THW, copper, stranded, 1/0	6.000	C.L.F.	1,488	1,374	2,862
Saw cutting, concrete walls, plain, per inch of depth	64.000	L.F.	2.56	745.60	748.16
Meter centers and sockets, single pos, 4 terminal, 400 amp	5.000	Ea.	4,550	2,225	6,775
Safety switch, 400 amp	5.000	Ea.	6,750	4,200	10,950
Ground rod clamp	6.000	Ea.	37.50	141	178.50
Wireway	1.000	L.F.	13.70	16.80	30.50
600 volt stranded copper wire, 1/0	1.200	C.L.F.	297.60	274.80	572.40
Flexible metallic conduit	120.000	L.F.	80.40	566.40	646.80
Ground rod	6.000	Ea.	363	1,134	1,497
TOTAL			36,982.52	24,296.40	61,278.92

D5010 130	Underground Electric Service	COST EACH MAT.	COST EACH INST.	COST EACH TOTAL
0950	Underground electric service including excavation, backfill and compaction			
1000	3 phase, 4 wire, 277/480 volts, 2000 A	37,000	24,300	61,300
1050	2000 A w/groundfault switchboard	68,000	27,000	95,000
1100	1600 A	29,200	19,600	48,800
1150	1600 A w/groundfault switchboard	58,000	22,000	80,000
1200	1200 A	22,500	15,600	38,100
1250	1200 A w/groundfault switchboard	48,900	17,900	66,800
1400	800 A	15,900	13,000	28,900

D5010 Electrical Service/Distribution

D5010 130	Underground Electric Service	COST EACH		
		MAT.	INST.	TOTAL
1450	800 A w/switchboard	21,000	14,700	35,700
1500	600 A	9,525	11,900	21,425
1550	600 A w/switchboard	14,600	13,400	28,000
1600	1 phase, 3 wire, 120/240 volts, 200 A	3,275	4,225	7,500
1650	200 A w/load center	3,600	5,275	8,875
1700	100 A	2,525	3,475	6,000
1750	100 A w/load center	2,875	4,525	7,400

D5010 Electrical Service/Distribution

Panelboard
Conduit
Safety switch

System Components	QUANTITY	UNIT	COST EACH MAT.	COST EACH INST.	COST EACH TOTAL
SYSTEM D5010 250 1020					
PANELBOARD INSTALLATION, INCLUDES PANELBOARD, CONDUCTOR, & CONDUIT					
Conduit, galvanized steel, 1-1/4"dia.	37.000	L.F.	216.45	466.20	682.65
Panelboards, NQOD, 4 wire, 120/208 volts, 100 amp main, 24 circuits	1.000	Ea.	2,025	1,600	3,625
Wire, 600 volt, type THW, copper, stranded, #3	1.480	C.L.F.	190.92	223.48	414.40
TOTAL			2,432.37	2,289.68	4,722.05

D5010 250	Panelboard	COST EACH MAT.	COST EACH INST.	COST EACH TOTAL
0900	Panelboards, NQOD, 4 wire, 120/208 volts w/conductor & conduit			
1000	100 A, 0 stories, 0′ horizontal	2,025	1,600	3,625
1020	1 stories, 25′ horizontal	2,425	2,300	4,725
1040	5 stories, 50′ horizontal	3,225	3,650	6,875
1060	10 stories, 75′ horizontal	4,175	5,225	9,400
1080	225A, 0 stories, 0′ horizontal	3,925	2,100	6,025
2000	1 stories, 25′ horizontal	5,100	3,350	8,450
2020	5 stories, 50′ horizontal	7,450	5,800	13,250
2040	10 stories, 75′ horizontal	10,200	8,650	18,850
2060	400A, 0 stories, 0′ horizontal	5,750	3,150	8,900
2080	1 stories, 25′ horizontal	7,300	5,175	12,475
3000	5 stories, 50′ horizontal	10,400	9,150	19,550
3020	10 stories, 75′ horizontal	17,900	17,800	35,700
3040	600 A, 0 stories, 0′ horizontal	8,500	3,775	12,275
3060	1 stories, 25′ horizontal	11,500	7,050	18,550
3080	5 stories, 50′ horizontal	17,500	13,500	31,000
4000	10 stories, 75′ horizontal	33,700	26,600	60,300
4010	Panelboards, NEHB, 4 wire, 277/480 volts w/conductor, conduit, & safety switch			
4020	100 A, 0 stories, 0′ horizontal, includes safety switch	3,450	2,200	5,650
4040	1 stories, 25′ horizontal	3,850	2,900	6,750
4060	5 stories, 50′ horizontal	4,650	4,250	8,900
4080	10 stories, 75′ horizontal	5,575	5,825	11,400
5000	225 A, 0 stories, 0′ horizontal	5,125	2,675	7,800
5020	1 stories, 25′ horizontal	6,300	3,925	10,225
5040	5 stories, 50′ horizontal	8,650	6,375	15,025
5060	10 stories, 75′ horizontal	11,400	9,225	20,625

D5010 Electrical Service/Distribution

D5010 250	Panelboard	COST EACH		
		MAT.	INST.	TOTAL
5080	400 A, 0 stories, 0' horizontal	8,200	4,125	12,325
6000	1 stories, 25' horizontal	9,750	6,125	15,875
6020	5 stories, 50' horizontal	12,800	10,100	22,900
6040	10 stories, 75' horizontal	20,300	18,800	39,100
6060	600 A, 0 stories, 0' horizontal	11,500	5,225	16,725
6080	1 stories, 25' horizontal	14,500	8,500	23,000
7000	5 stories, 50' horizontal	20,500	15,000	35,500
7020	10 stories, 75' horizontal	36,600	28,100	64,700

D5020 Lighting and Branch Wiring

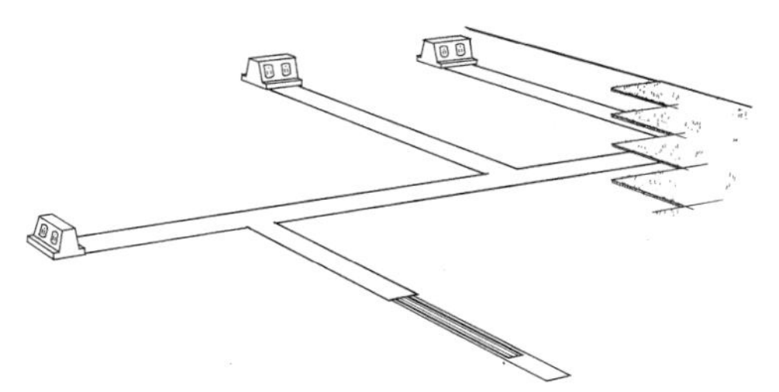

Underfloor Receptacle System

Description: Table D5020 115 includes installed costs of raceways and copper wire from panel to and including receptacle.

National Electrical Code prohibits use of undercarpet system in residential, school or hospital buildings. Can only be used with carpet squares.

Low density = (1) Outlet per 259 S.F. of floor area.

High density = (1) Outlet per 127 S.F. of floor area.

System Components	QUANTITY	UNIT	COST PER S.F. MAT.	COST PER S.F. INST.	COST PER S.F. TOTAL
SYSTEM D5020 115 0200					
RECEPTACLE SYSTEMS, UNDERFLOOR DUCT, 5′ ON CENTER, LOW DENSITY					
Underfloor duct 3-1/8″ x 7/8″ w/insert 24″ on center	.190	L.F.	4.18	2.05	6.23
Vertical elbow for underfloor duct, 3-1/8″, included					
Underfloor duct conduit adapter, 2″ x 1-1/4″, included					
Underfloor duct junction box, single duct, 3-1/8″	.003	Ea.	1.47	.57	2.04
Underfloor junction box carpet pan	.003	Ea.	1.23	.03	1.26
Underfloor duct outlet, high tension receptacle	.004	Ea.	.47	.38	.85
Wire 600V type THWN-THHN copper solid #12	.010	C.L.F.	.11	.69	.80
TOTAL			7.46	3.72	11.18

D5020 115	Receptacles, Floor	COST PER S.F. MAT.	COST PER S.F. INST.	COST PER S.F. TOTAL
0200	Receptacle systems, underfloor duct, 5′ on center, low density	7.45	3.72	11.17
0240	High density	8.05	4.78	12.83
0280	7′ on center, low density	5.90	3.20	9.10
0320	High density	6.50	4.26	10.76
0400	Poke thru fittings, low density	1.48	1.82	3.30
0440	High density	2.97	3.61	6.58
0520	Telepoles, using Romex, low density	1.61	1.12	2.73
0560	High density	3.22	2.23	5.45
0600	Using EMT, low density	1.69	1.49	3.18
0640	High density	3.40	2.94	6.34
0720	Conduit system with floor boxes, low density	1.38	1.28	2.66
0760	High density	2.75	2.53	5.28
0840	Undercarpet power system, 3 conductor with 5 conductor feeder, low density	1.50	.49	1.99
0880	High density	2.98	.94	3.92

D5020 Lighting and Branch Wiring

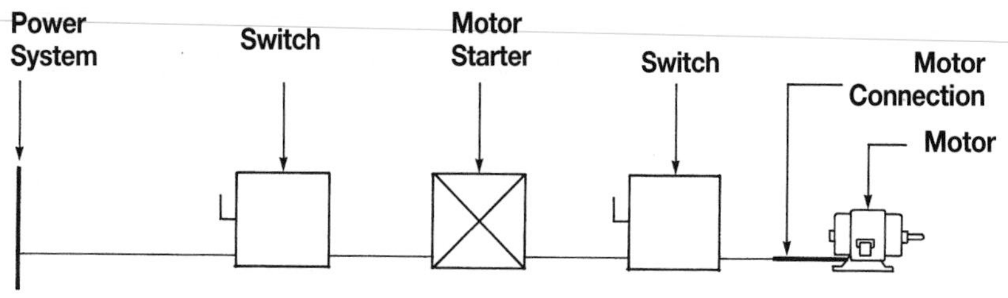

System D5020 145 installed cost of motor wiring using 50′ of rigid conduit and copper wire. **Cost and setting of motor not included.**

System Components	QUANTITY	UNIT	COST EACH MAT.	COST EACH INST.	COST EACH TOTAL
SYSTEM D5020 145 0200					
MOTOR INST., SINGLE PHASE, 115V, TO AND INCLUDING 1/3 HP MOTOR SIZE					
Wire 600V type THWN-THHN, copper solid #12	1.250	C.L.F.	13.81	85.63	99.44
Steel intermediate conduit, (IMC) 1/2″ diam.	50.000	L.F.	105	377.50	482.50
Magnetic FVNR, 115V, 1/3 HP, size 00 starter	1.000	Ea.	197	189	386
Safety switch, fused, heavy duty, 240V 2P 30 amp	1.000	Ea.	85	216	301
Safety switch, non fused, heavy duty, 600V, 3 phase, 30 A	1.000	Ea.	95	236	331
Flexible metallic conduit, Greenfield 1/2″ diam.	1.500	L.F.	.75	5.67	6.42
Connectors for flexible metallic conduit Greenfield 1/2″ diam.	1.000	Ea.	2.22	9.45	11.67
Coupling for Greenfield to conduit 1/2″ diam. flexible metallic conduit	1.000	Ea.	1.01	15.10	16.11
Fuse cartridge nonrenewable, 250V 30 amp	1.000	Ea.	2.52	15.10	17.62
TOTAL			502.31	1,149.45	1,651.76

D5020 145	Motor Installation	COST EACH MAT.	COST EACH INST.	COST EACH TOTAL
0200	Motor installation, single phase, 115V, 1/3 HP motor size	500	1,150	1,650
0240	1 HP motor size	525	1,150	1,675
0280	2 HP motor size	575	1,225	1,800
0320	3 HP motor size	650	1,250	1,900
0360	230V, 1 HP motor size	505	1,175	1,680
0400	2 HP motor size	545	1,175	1,720
0440	3 HP motor size	610	1,250	1,860
0520	Three phase, 200V, 1-1/2 HP motor size	585	1,275	1,860
0560	3 HP motor size	750	1,400	2,150
0600	5 HP motor size	690	1,550	2,240
0640	7-1/2 HP motor size	715	1,575	2,290
0680	10 HP motor size	1,100	1,975	3,075
0720	15 HP motor size	1,525	2,200	3,725
0760	20 HP motor size	1,975	2,550	4,525
0800	25 HP motor size	2,025	2,550	4,575
0840	30 HP motor size	3,100	3,000	6,100
0880	40 HP motor size	3,800	3,550	7,350
0920	50 HP motor size	6,725	4,125	10,850
0960	60 HP motor size	6,975	4,375	11,350
1000	75 HP motor size	8,775	5,000	13,775
1040	100 HP motor size	25,700	5,900	31,600
1080	125 HP motor size	25,900	6,475	32,375
1120	150 HP motor size	29,200	7,625	36,825
1160	200 HP motor size	29,900	8,975	38,875
1240	230V, 1-1/2 HP motor size	560	1,275	1,835
1280	3 HP motor size	725	1,375	2,100
1320	5 HP motor size	670	1,525	2,195
1360	7-1/2 HP motor size	670	1,525	2,195
1400	10 HP motor size	1,025	1,875	2,900
1440	15 HP motor size	1,175	2,050	3,225
1480	20 HP motor size	1,875	2,475	4,350
1520	25 HP motor size	1,975	2,550	4,525

D5020 Lighting and Branch Wiring

D5020 145	Motor Installation	COST EACH		
		MAT.	INST.	TOTAL
1560	30 HP motor size	1,975	2,550	4,525
1600	40 HP motor size	3,750	3,450	7,200
1640	50 HP motor size	4,300	3,650	7,950
1680	60 HP motor size	6,750	4,150	10,900
1720	75 HP motor size	8,025	4,700	12,725
1760	100 HP motor size	9,225	5,250	14,475
1800	125 HP motor size	26,300	6,100	32,400
1840	150 HP motor size	26,900	6,925	33,825
1880	200 HP motor size	28,300	7,675	35,975
1960	460V, 2 HP motor size	695	1,275	1,970
2000	5 HP motor size	860	1,400	2,260
2040	10 HP motor size	765	1,525	2,290
2080	15 HP motor size	1,025	1,750	2,775
2120	20 HP motor size	1,100	1,875	2,975
2160	25 HP motor size	1,175	1,975	3,150
2200	30 HP motor size	1,500	2,125	3,625
2240	40 HP motor size	1,925	2,300	4,225
2280	50 HP motor size	2,150	2,550	4,700
2320	60 HP motor size	3,225	2,975	6,200
2360	75 HP motor size	3,800	3,275	7,075
2400	100 HP motor size	4,500	3,675	8,175
2440	125 HP motor size	6,950	4,150	11,100
2480	150 HP motor size	8,750	4,675	13,425
2520	200 HP motor size	9,950	5,275	15,225
2600	575V, 2 HP motor size	695	1,275	1,970
2640	5 HP motor size	860	1,400	2,260
2680	10 HP motor size	765	1,525	2,290
2720	20 HP motor size	1,025	1,750	2,775
2760	25 HP motor size	1,100	1,875	2,975
2800	30 HP motor size	1,500	2,125	3,625
2840	50 HP motor size	1,625	2,225	3,850
2880	60 HP motor size	3,225	2,950	6,175
2920	75 HP motor size	3,225	2,975	6,200
2960	100 HP motor size	3,800	3,275	7,075
3000	125 HP motor size	7,125	4,100	11,225
3040	150 HP motor size	6,950	4,150	11,100
3080	200 HP motor size	8,750	4,725	13,475

D5020 Lighting and Branch Wiring

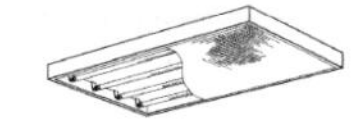

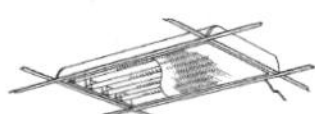
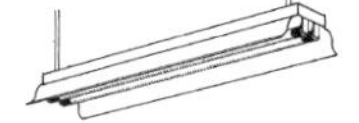

Design Assumptions:

1. A 100 footcandle average maintained level of illumination.
2. Ceiling heights range from 9′ to 11′.
3. Average reflectance values are assumed for ceilings, walls and floors.
4. Cool white (CW) fluorescent lamps with 3150 lumens for 40 watt lamps and 6300 lumens for 8′ slimline lamps.
5. Four 40 watt lamps per 4′ fixture and two 8′ lamps per 8′ fixture.
6. Average fixture efficiency values and spacing to mounting height ratios.
7. Installation labor is average U.S. rate as of January 1.

System Components	QUANTITY	UNIT	COST PER S.F. MAT.	INST.	TOTAL
SYSTEM D5020 208 0520					
FLUORESCENT FIXTURES MOUNTED 9′-11″ ABOVE FLOOR, 100 FC					
TYPE A, 8 FIXTURES PER 400 S.F.					
Conduit, steel intermediate, 1/2″ diam.	.185	L.F.	.39	1.40	1.79
Wire, 600V, type THWN-THHN, copper, solid, #12	.004	C.L.F.	.04	.25	.29
Fluorescent strip fixture 8′ long, surface mounted, two 75W SL	.020	Ea.	1.56	2.44	4
Steel outlet box 4″ concrete	.020	Ea.	.18	.76	.94
Steel outlet box plate with stud, 4″ concrete	.020	Ea.	.18	.19	.37
Fixture hangers, flexible, 1/2″ diameter, 4″ long	.040	Ea.	.69	2.52	3.21
Fixture whip, THHN wire, three #12, 3/8″ Greenfield	.040	Ea.	.46	.94	1.40
TOTAL			3.50	8.50	12

D5020 208	Fluorescent Fixtures (by Type)	COST PER S.F. MAT.	INST.	TOTAL
0520	Fluorescent fixtures, type A, 8 fixtures per 400 S.F.	3.50	8.50	12
0560	11 fixtures per 600 S.F.	3.19	7.75	10.94
0600	17 fixtures per 1000 S.F.	2.98	7.20	10.18
0640	23 fixtures per 1600 S.F.	2.49	6.05	8.54
0680	28 fixtures per 2000 S.F.	2.46	5.95	8.41
0720	41 fixtures per 3000 S.F.	2.42	5.80	8.22
0800	53 fixtures per 4000 S.F.	2.32	5.65	7.97
0840	64 fixtures per 5000 S.F.	2.28	5.50	7.78
0880	Type B, 11 fixtures per 400 S.F.	5.30	12.10	17.40
0920	15 fixtures per 600 S.F.	4.78	10.95	15.73
0960	24 fixtures per 1000 S.F.	4.58	10.55	15.13
1000	35 fixtures per 1600 S.F.	4.19	9.60	13.79
1040	42 fixtures per 2000 S.F.	4.01	9.15	13.16
1080	61 fixtures per 3000 S.F.	3.84	8.85	12.69
1160	80 fixtures per 4000 S.F.	2.77	5.60	8.37
1200	98 fixtures per 5000 S.F.	3.78	8.60	12.38
1240	Type C, 11 fixtures per 400 S.F.	4.61	12.60	17.21
1280	14 fixtures per 600 S.F.	3.85	10.55	14.40
1320	23 fixtures per 1000 S.F.	3.81	10.45	14.26
1360	34 fixtures per 1600 S.F.	3.53	9.65	13.18
1400	43 fixtures per 2000 S.F.	3.62	9.90	13.52
1440	63 fixtures per 3000 S.F.	3.49	9.55	13.04
1520	81 fixtures per 4000 S.F.	3.35	9.20	12.55
1560	101 fixtures per 5000 S.F.	3.32	9.10	12.42
1600	Type D, 8 fixtures per 400 S.F.	4.16	9.50	13.66
1640	12 fixtures per 600 S.F.	4.16	9.50	13.66
1680	19 fixtures per 1000 S.F.	3.96	9	12.96
1720	27 fixtures per 1600 S.F.	3.54	8.05	11.59
1760	34 fixtures per 2000 S.F.	3.54	8.05	11.59
1800	48 fixtures per 3000 S.F.	3.34	7.60	10.94
1880	64 fixtures per 4000 S.F.	3.34	7.60	10.94
1920	79 fixtures per 5000 S.F.	3.34	7.60	10.94

D5020 Lighting and Branch Wiring

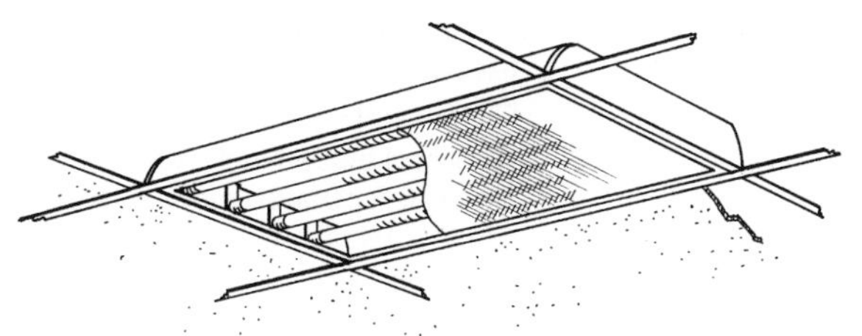

Type C. Recessed, mounted on grid ceiling suspension system, 2′ x 4′, four 40 watt lamps, acrylic prismatic diffusers.

5.3 watts per S.F. for 100 footcandles.

3 watts per S.F. for 57 footcandles.

System Components	QUANTITY	UNIT	COST PER S.F. MAT.	INST.	TOTAL
SYSTEM D5020 210 0200					
FLUORESCENT FIXTURES RECESS MOUNTED IN CEILING					
1 WATT PER S.F., 20 FC, 5 FIXTURES PER 1000 S.F.					
Steel intermediate conduit, (IMC) 1/2″ diam.	.128	L.F.	.27	.97	1.24
Wire, 600 volt, type THW, copper, solid, #12	.003	C.L.F.	.03	.21	.24
Fluorescent fixture, recessed, 2′x 4′, four 40W, w/lens, for grid ceiling	.005	Ea.	.36	.81	1.17
Steel outlet box 4″ square	.005	Ea.	.05	.19	.24
Fixture whip, Greenfield w/#12 THHN wire	.005	Ea.	.05	.05	.10
TOTAL			.76	2.23	2.99

D5020 210	Fluorescent Fixtures (by Wattage)	COST PER S.F. MAT.	INST.	TOTAL
0190	Fluorescent fixtures recess mounted in ceiling			
0195	T12, standard 40 watt lamps			
0200	1 watt per S.F., 20 FC, 5 fixtures @40 watts per 1000 S.F.	.76	2.23	2.99
0240	2 watt per S.F., 40 FC, 10 fixtures @40 watt per 1000 S.F.	1.48	4.35	5.83
0280	3 watt per S.F., 60 FC, 15 fixtures @40 watt per 1000 S.F	2.25	6.60	8.85
0320	4 watt per S.F., 80 FC, 20 fixtures @40 watt per 1000 S.F.	2.97	8.75	11.72
0400	5 watt per S.F., 100 FC, 25 fixtures @40 watt per 1000 S.F.	3.72	10.95	14.67
0450	T8, energy saver 32 watt lamps			
0500	0.8 watt per S.F., 20 FC, 5 fixtures @32 watt per 1000 S.F.	.81	2.23	3.04
0520	1.6 watt per S.F., 40 FC, 10 fixtures @32 watt per 1000 S.F.	1.61	4.35	5.96
0540	2.4 watt per S.F., 60 FC, 15 fixtures @ 32 watt per 1000 S.F	2.42	6.60	9.02
0560	3.2 watt per S.F., 80 FC, 20 fixtures @32 watt per 1000 S.F.	3.20	8.75	11.95
0580	4 watt per S.F., 100 FC, 25 fixtures @32 watt per 1000 S.F.	4	10.95	14.95

D5030 Communications and Security

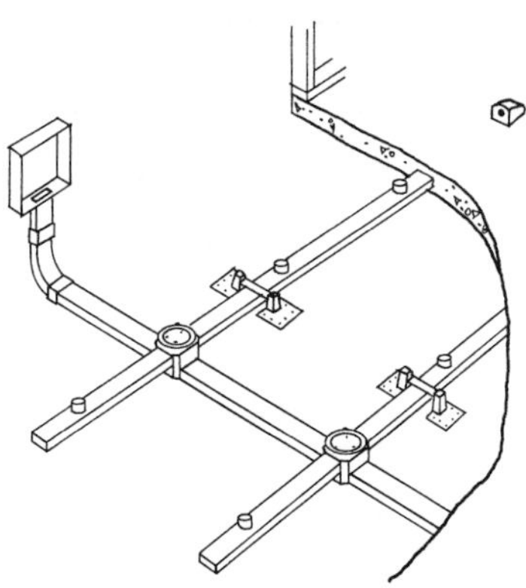

Description: System below includes telephone fitting installed. Does not include cable.

When poke thru fittings and telepoles are used for power, they can also be used for telephones at a negligible additional cost.

System Components	QUANTITY	UNIT	COST PER S.F. MAT.	COST PER S.F. INST.	COST PER S.F. TOTAL
SYSTEM D5030 310 0200					
TELEPHONE SYSTEMS, UNDERFLOOR DUCT, 5′ ON CENTER, LOW DENSITY					
Underfloor duct 7-1/4″ w/insert 2′ O.C. 1-3/8″ x 7-1/4″ super duct	.190	L.F.	7.32	2.87	10.19
Vertical elbow for underfloor superduct, 7-1/4″, included					
Underfloor duct conduit adapter, 2″ x 1-1/4″, included					
Underfloor duct junction box, single duct, 7-1/4″ x 3 1/8″	.003	Ea.	1.91	.57	2.48
Underfloor junction box carpet pan	.003	Ea.	1.23	.03	1.26
Underfloor duct outlet, low tension	.004	Ea.	.46	.38	.84
TOTAL			10.92	3.85	14.77

D5030 310	Telephone Systems	COST PER S.F. MAT.	COST PER S.F. INST.	COST PER S.F. TOTAL
0200	Telephone systems, underfloor duct, 5′ on center, low density	10.90	3.85	14.75
0240	5′ on center, high density	11.35	4.23	15.58
0280	7′ on center, low density	8.70	3.20	11.90
0320	7′ on center, high density	9.15	3.58	12.73
0400	Poke thru fittings, low density	1.41	1.36	2.77
0440	High density	2.83	2.72	5.55
0520	Telepoles, low density	1.58	.82	2.40
0560	High density	3.16	1.63	4.79
0640	Conduit system with floor boxes, low density	1.69	1.34	3.03
0680	High density	3.35	2.64	5.99
1020	Telephone wiring for offices & laboratories, 8 jacks/M.S.F.	.36	2.06	2.42

D5030 Communications and Security

D5030 810	Security & Detection Systems	COST EACH		
		MAT.	INST.	TOTAL
0200	Security system, head end equipment	30,800	5,025	35,825
0240	Security system, door/window contact biased, box, conduit & cable	181	570	751
0280	Security system, door/window contact balanced, box, conduit & cable	118	570	688
0440	Security system, proximity card reader, box, conduit & cable	445	760	1,205
1600	Card control entrance system, to 6 zone, including hardware for 100 doors	43,800	35,000	78,800

D5030 Communications and Security

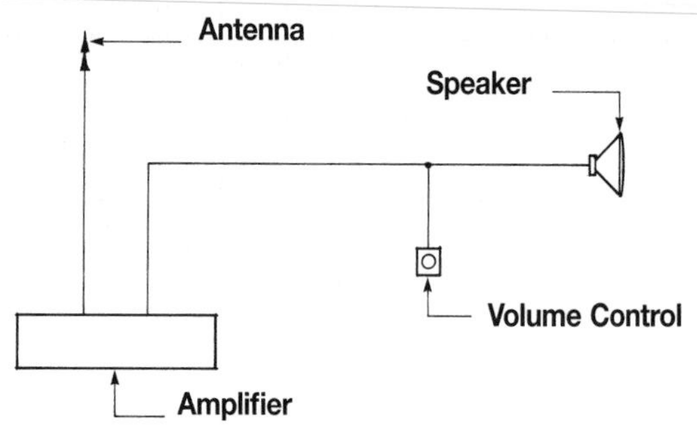

Sound System Includes AM–FM antenna, outlets, rigid conduit, and copper wire.
Fire Detection System Includes pull stations, signals, smoke and heat detectors, rigid conduit, and copper wire.
Intercom System Includes master and remote stations, rigid conduit, and copper wire.
Master Clock System Includes clocks, bells, rigid conduit, and copper wire.
Master TV Antenna Includes antenna, VHF–UHF reception and distribution, rigid conduit, and copper wire.

System Components	QUANTITY	UNIT	COST EACH MAT.	COST EACH INST.	COST EACH TOTAL
SYSTEM D5030 910 0220					
SOUND SYSTEM, INCLUDES OUTLETS, BOXES, CONDUIT & WIRE					
Steel intermediate conduit, (IMC) 1/2″ diam	1200.000	L.F.	2,520	9,060	11,580
Wire sound shielded w/drain, #22-2 conductor	15.500	C.L.F.	144.93	1,464.75	1,609.68
Sound system speakers ceiling or wall	12.000	Ea.	2,076	1,134	3,210
Sound system volume control	12.000	Ea.	792	1,134	1,926
Sound system amplifier, 250 W	1.000	Ea.	1,550	755	2,305
Sound system antenna, AM FM	1.000	Ea.	171	189	360
Sound system monitor panel	1.000	Ea.	570	189	759
Sound system cabinet	1.000	Ea.	1,250	755	2,005
Steel outlet box 4″ square	12.000	Ea.	90.60	456	546.60
Steel outlet box 4″ plaster rings	12.000	Ea.	21.96	141.60	163.56
TOTAL			9,186.49	15,278.35	24,464.84

D5030 910	Communication & Alarm Systems	COST EACH MAT.	COST EACH INST.	COST EACH TOTAL
0200	Communication & alarm systems, includes outlets, boxes, conduit & wire			
0210	Sound system, 6 outlets	6,450	9,550	16,000
0220	12 outlets	9,175	15,300	24,475
0240	30 outlets	16,100	28,900	45,000
0280	100 outlets	47,900	96,500	144,400
0320	Fire detection systems, non-addressable, 12 detectors	3,475	7,925	11,400
0360	25 detectors	6,150	13,400	19,550
0400	50 detectors	12,100	26,600	38,700
0440	100 detectors	22,600	48,300	70,900
0450	Addressable type, 12 detectors	5,125	7,975	13,100
0452	25 detectors	9,275	13,500	22,775
0454	50 detectors	17,800	26,500	44,300
0456	100 detectors	34,500	48,700	83,200
0458	Fire alarm control panel, 8 zone, excluding wire and conduit	965	1,500	2,465
0459	12 zone	2,575	2,275	4,850
0460	Fire alarm command center, addressable without voice, excl. wire & conduit	5,025	1,325	6,350
0462	Addressable with voice	10,700	2,075	12,775
0480	Intercom systems, 6 stations	4,575	6,525	11,100
0520	12 stations	8,550	13,000	21,550
0560	25 stations	14,600	25,000	39,600
0600	50 stations	28,400	47,000	75,400
0640	100 stations	56,000	92,000	148,000
0680	Master clock systems, 6 rooms	4,975	10,700	15,675
0720	12 rooms	8,025	18,200	26,225
0760	20 rooms	11,200	25,700	36,900
0800	30 rooms	18,600	47,400	66,000

D5030 Communications and Security

D5030 910	Communication & Alarm Systems	COST EACH		
		MAT.	INST.	TOTAL
0840	50 rooms	30,400	80,000	110,400
0880	100 rooms	59,000	158,500	217,500
0920	Master TV antenna systems, 6 outlets	2,525	6,750	9,275
0960	12 outlets	4,675	12,600	17,275
1000	30 outlets	15,500	29,100	44,600
1040	100 outlets	56,000	95,000	151,000

D5090 Other Electrical Systems

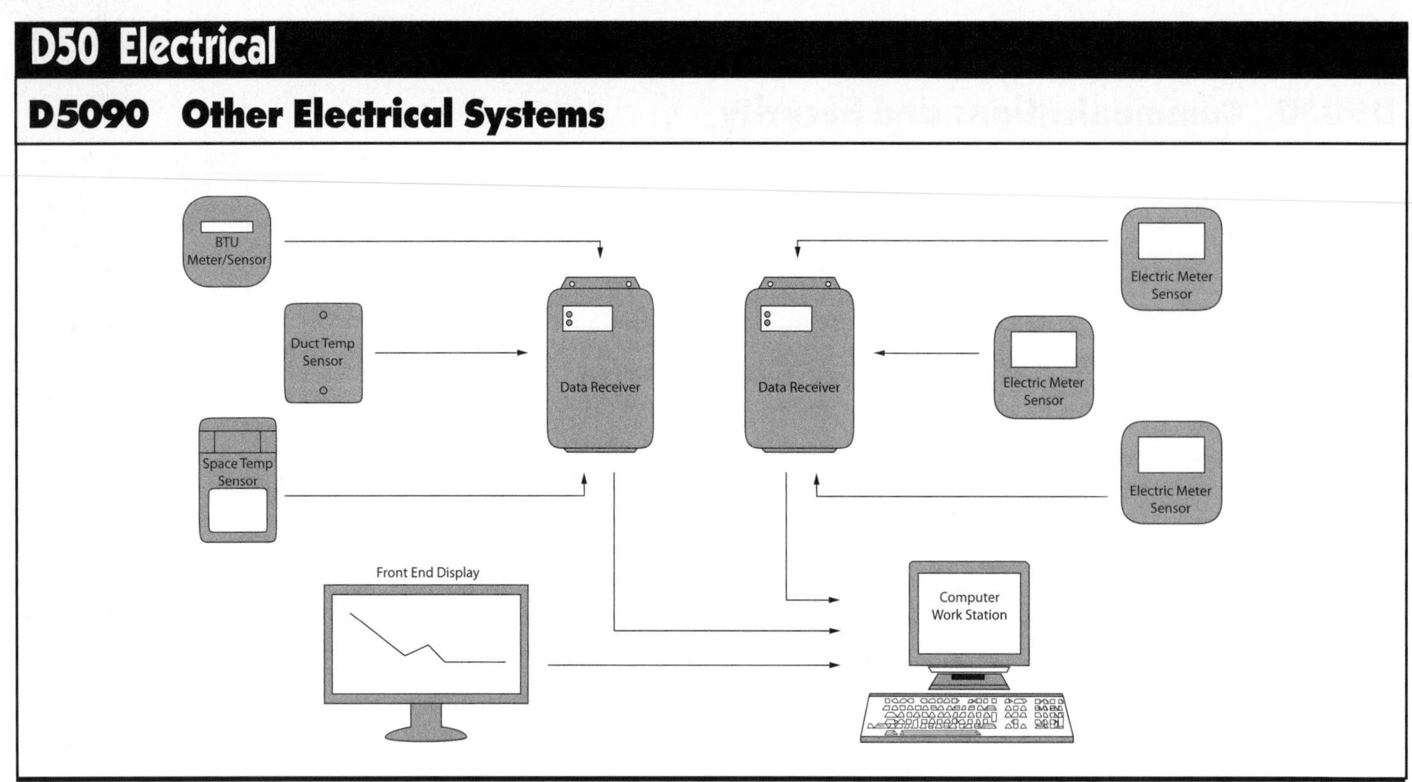

System Components	QUANTITY	UNIT	COST EACH MAT.	COST EACH INST.	COST EACH TOTAL
SYSTEM D5090 480 1100					
ELECTRICAL, SINGLE PHASE, 1 METER					
#18 twisted shielded pair in 1/2" EMT conduit	.050	C.L.F.	7.80	12.55	20.35
Wire, 600 volt, type THW, copper, solid, #12	.150	C.L.F.	1.63	10.28	11.91
Conduit (EMT) , to 10' H, incl 2 termn,2 elb&11 bm clp per 100', 3/4"	5.000	L.F.	5.90	29	34.90
Outlet boxes, pressed steel, handy box	1.000	Ea.	2.77	28	30.77
Outlet boxes, pressed steel, handy box, covers, device	1.000	Ea.	1.11	11.80	12.91
Wiring devices, receptacle, duplex, 120 volt, ground, 20 amp	1.000	Ea.	11.75	28	39.75
Single phase, 277 volt, 200 amp	1.000	Ea.	460	86	546
Data recorder, 8 meters	1.000	Ea.	1,750	69	1,819
Software package, per meter, premium	1.000	Ea.	755		755
TOTAL			2,995.96	274.63	3,270.59

D5090 480	Energy Monitoring Systems	COST EACH MAT.	COST EACH INST.	COST EACH TOTAL
1000	Electrical			
1100	Single phase, 1 meter	3,000	275	3,275
1110	4 meters	7,375	2,075	9,450
1120	8 meters	14,800	4,275	19,075
1200	Three phase, 1 meter	3,400	350	3,750
1210	5 meters	11,300	3,250	14,550
1220	10 meters	23,000	6,650	29,650
1230	25 meters	54,000	13,200	67,200
2000	Mechanical			
2100	BTU, 1 meter	3,700	985	4,685
2110	w/1 duct sensor	3,900	1,225	5,125
2120	& 1 space sensor	4,425	1,400	5,825
2130	& 5 space sensors	6,475	2,075	8,550
2140	& 10 space sensors	11,300	2,925	14,225
2200	BTU, 3 meters	7,600	2,825	10,425
2210	w/3 duct sensors	8,225	3,550	11,775
2220	& 3 space sensors	9,775	4,050	13,825
2230	& 15 space sensors	16,000	6,050	22,050
2240	& 30 space sensors	25,900	8,600	34,500
9000	Front end display	610	148	758
9100	Computer workstation	1,500	1,975	3,475

E2020 Moveable Furnishings

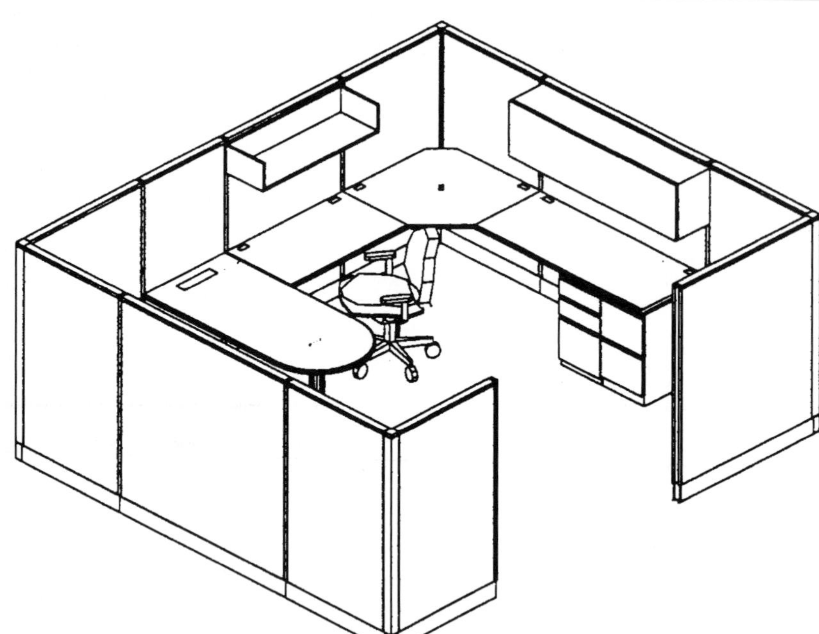

The work station shown here represents one of many types and sizes on the market. Costs can vary considerably due to the various parameters involved. For instance, wood systems can cost up to 40% more than metal or fabric systems. Generally, the cost per work station will decrease for "quantity" purchases.

Note: 4′ raceway included in panels (not pre-wired)

System Components	QUANTITY	UNIT	COST EACH MAT.	COST EACH INST.	COST EACH TOTAL
SYSTEM E2020 410 2600					
WORK STATION - MANAGER					
Acoustical panel, aluminum frame, fabric covered, 30″ wide, 64″ high	1.000	Ea.	580		580
36″ wide, 64″ high	6.000	Ea.	3,750		3,750
42″ wide, 64″ high	2.000	Ea.	1,450		1,450
48″ wide, 64″ high	1.000	Ea.	765		765
60″ wide, 64″ high	2.000	Ea.	1,690		1,690
Straight connector kit, 64″ high	7.000	Ea.	353.50		353.50
Ell connector kit, 64″ high, 90″ deep	4.000	Ea.	288		288
Panel end cover, 64″ high	2.000	Ea.	101		101
Worksurface, 24″ deep, 42″ wide	1.000	Ea.	296		296
Worksurface, 24″ deep, 60″ wide	1.000	Ea.	395		395
Corner worksurface, 24″ deep, 36″ wide	1.000	Ea.	580		580
Peninsula worksurface, 36″ wide, 66″ long	1.000	Ea.	755		755
Peninsula support column, 29-1/2″ high	1.000	Ea.	176		176
Cantilever bracket, 20″ deep, 24″ deep	3.000	Ea.	207		207
Pedestal spacer, 22″ deep, 15″ wide	2.000	Ea.	164		164
Pedestal, floorstanding, 2 box, 1 file	1.000	Ea.	575		575
Pedestal, floorstanding, 2 file, 22″ deep	1.000	Ea.	535		535
Overhead storage cabinet with door, 60″ wide	1.000	Ea.	720		720
Open bookshelf, 42″ wide	1.000	Ea.	213		213
Task light, recessed, 42″-48″ wide	1.000	Ea.	242		242
Task light, recessed, 60″ wide	1.000	Ea.	262		262
Electrical power harness, 36″ wide	2.000	Ea.	328		328
Electrical power harness, 42″ wide	1.000	Ea.	345		345
Electrical power harness, 60″ wide	1.000	Ea.	173		173
Base power in-feed cable	1.000	Ea.	183		183
Duplex receptacle, circuit 1	2.000	Ea.	50		50
Duplex receptacle, circuit 2	2.000	Ea.	50		50
Installation	1.000	Ea.		680	680
TOTAL			15,226.50	680	15,906.50

E2020 Moveable Furnishings

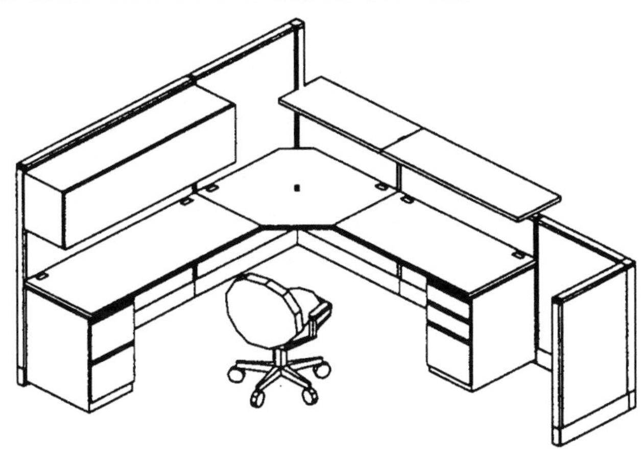

The work station shown here represents one of many types and sizes on the market. Costs can vary considerably due to the various parameters involved. For instance, wood systems can cost up to 40% more than metal or fabric systems. Generally, the cost per work station will decrease for "quantity" purchases.

System Components	QUANTITY	UNIT	COST EACH MAT.	COST EACH INST.	COST EACH TOTAL
SYSTEM E2020 420 2300					
WORK STATION - PROFESSIONAL/EXECUTIVE SECRETARY					
Acoustical panel, aluminum frame, fabric covered, 24" wide x 43" high	1.000	Ea.	440		440
30" wide x 43" high	1.000	Ea.	490		490
36" wide x 43" high	1.000	Ea.	545		545
48" wide x 43" high	1.000	Ea.	645		645
36" wide x 64" high	1.000	Ea.	625		625
60" wide x 64" high	1.000	Ea.	845		845
Straight connector kit, 43" high	2.000	Ea.	101		101
Straight connector kit, 64" high	1.000	Ea.	50.50		50.50
Ell connector kit, 43" high, 90" deep	1.000	Ea.	122		122
Ell connector kit, 64" high	1.000	Ea.	72		72
Panel end cover, 43" high	1.000	Ea.	50.50		50.50
Panel end cover, 64" high	1.000	Ea.	50.50		50.50
Finish end cover, variable height, 2-way	1.000	Ea.	75.50		75.50
Corner worksurface, 24" deep, 36" wide	1.000	Ea.	580		580
Worksurface, 24" deep, 48" wide	1.000	Ea.	315		315
Worksurface, 24" deep, 60" wide	1.000	Ea.	395		395
Cantilever bracket, 20" deep, 24" deep	2.000	Ea.	138		138
Countertop brackets, 1 pair	2.000	Ea.	57		57
Countertop, 15" deep, 36" wide	1.000	Ea.	232		232
Countertop, 15" deep, 48" wide	1.000	Ea.	258		258
Pedestal spacer, 22" deep, 15" wide	2.000	Ea.	164		164
Pedestal, floorstanding, 2 box, 1 file, 22" deep	1.000	Ea.	575		575
Pedestal, floorstanding, 2 file, 22" deep	1.000	Ea.	535		535
Overhead storage cabinet with door, 60" wide	1.000	Ea.	720		720
Task light, recessed, 60" wide	1.000	Ea.	262		262
Electrical power harness, 36" wide	2.000	Ea.	328		328
Electrical power harness, 60" wide	1.000	Ea.	173		173
Base power in-feed cable	1.000	Ea.	183		183
Duplex receptacle, circuit 1	2.000	Ea.	50		50
Duplex receptacle, circuit 2	1.000	Ea.	25		25
Installation	1.000	Ea.		415	415
TOTAL			9,102	415	9,517

E2020 Moveable Furnishings

The work station shown here represents one of many types and sizes on the market. Costs can vary considerably due to the various parameters involved. For instance, wood systems can cost up to 40% more than metal or fabric systems. Generally, the cost per work station will decrease for "quantity" purchases.

System Components	QUANTITY	UNIT	COST EACH MAT.	COST EACH INST.	COST EACH TOTAL
SYSTEM E2020 430 1500					
WORK STATION - SECRETARY/CLERK					
Acoustical panel, aluminum frame, fabric covered, 30″ wide x 64″ high	2.000	Ea.	1,160		1,160
36″ wide x 64″ high	2.000	Ea.	1,250		1,250
42″ wide x 64″ high	1.000	Ea.	725		725
Straight connector kit, 64″ high	2.000	Ea.	101		101
Ell connector kit, 64″ high, 90″ deep	2.000	Ea.	144		144
Panel end cover, 64″ high	2.000	Ea.	101		101
Worksurface, 24″ deep, 42″ wide	1.000	Ea.	296		296
Worksurface, 72″ wide, 30″ deep	1.000	Ea.	505		505
Worksurface bracket kit, pair	2.000	Ea.	57		57
Flat bracket, 20″ deep, 24″ deep	1.000	Ea.	35.50		35.50
Pedestal spacer 22″ deep, 15″ wide	1.000	Ea.	82		82
Pedestal spacer, 28″ deep, 15″ wide	1.000	Ea.	99		99
Pedestal, 2 box, 1 file, 28″ deep	1.000	Ea.	600		600
Pedestal, 2 file, 22″ deep	1.000	Ea.	535		535
Overhead storage cabinet with door, 42″ wide	1.000	Ea.	465		465
Task light, recessed, 42″-48″ wide	1.000	Ea.	242		242
Installation	1.000	Ea.		272	272
TOTAL			6,397.50	272	6,669.50

E2020 Moveable Furnishings

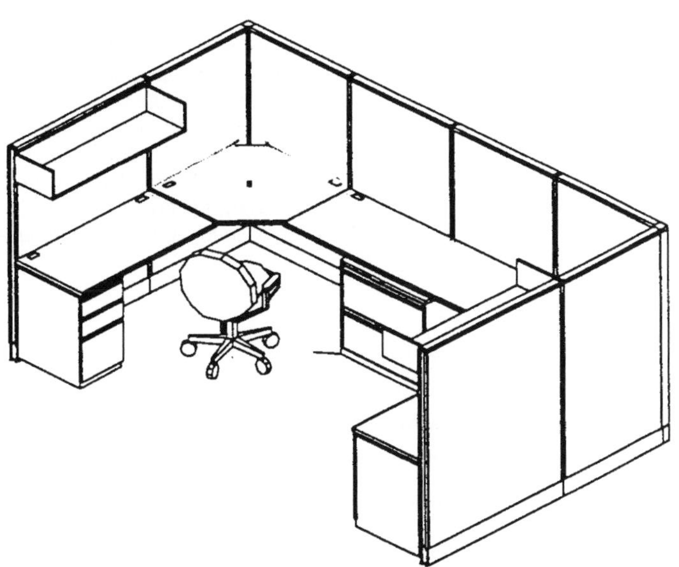

The work station shown here represents one of many types and sizes on the market. Costs can vary considerably due to the various parameters involved. For instance, wood systems can cost up to 40% more than metal or fabric systems. Generally, the cost per work station will decrease for "quantity" purchases.

System Components	QUANTITY	UNIT	COST EACH MAT.	COST EACH INST.	COST EACH TOTAL
SYSTEM E2020 440 1600					
WORK STATION - CLERICAL, TWO PERSON, UNIT					
Acoustical panel, aluminum frame, fabric covered, 36" wide, 64" high	6.000	Ea.	3,750		3,750
48" wide x 64" high	2.000	Ea.	1,530		1,530
Lateral file, 2 drawer, 30" wide	1.000	Ea.	760		760
Ell connector kit, 64" high, 90" deep	2.000	Ea.	144		144
Panel end cover, 64" high	2.000	Ea.	101		101
Corner worksurface, 36" wide x 24" deep	2.000	Ea.	1,160		1,160
Worksurface, 48" wide x 24" deep	2.000	Ea.	630		630
Worksurface, 72" wide x 24" deep	1.000	Ea.	445		445
Cantilever bracket, 20" deep, 24" deep	4.000	Ea.	276		276
Open bookshelf, 48" wide	2.000	Ea.	440		440
Pedestal spacer, 22" deep, 15" wide	2.000	Ea.	164		164
Pedestal, floorstanding, 2 box, 1 file, 22" deep	2.000	Ea.	1,150		1,150
Straight connector kit, 64" high	5.000	Ea.	252.50		252.50
Task light, recessed, 42"-48" wide	2.000	Ea.	484		484
Electrical power harness, 36" wide	4.000	Ea.	656		656
Base power in-feed cable	1.000	Ea.	183		183
Duplex receptacle, circuit 1	2.000	Ea.	50		50
Duplex receptacle, circuit 2	2.000	Ea.	50		50
Installation	1.000	Ea.		545	545
TOTAL			12,225.50	545	12,770.50

E2020 Moveable Furnishings

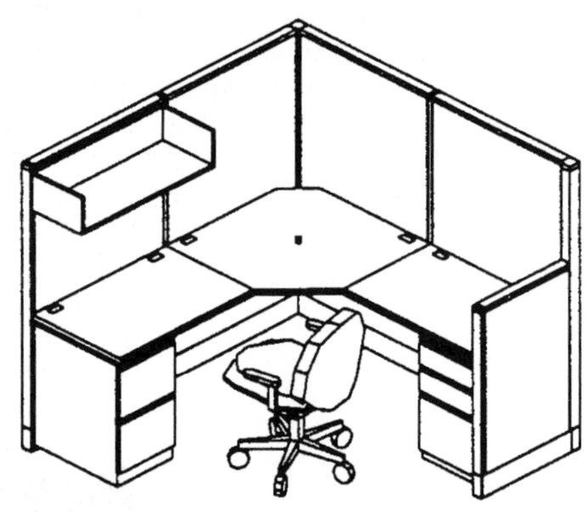

The work station shown here represents one of many types and sizes on the market. Costs can vary considerably due to the various parameters involved. For instance, wood systems can cost up to 40% more than metal or fabric systems. Generally, the cost per work station will decrease for "quantity" purchases.

System Components	QUANTITY	UNIT	COST EACH MAT.	COST EACH INST.	COST EACH TOTAL
SYSTEM E2020 450 1600					
WORK STATION - SECRETARY/WORD PROCESSING					
Acoustical panel, aluminum frame, fabric covered, 43" high, 24" wide	1.000	Ea.	440		440
Acoustical panel, 64" high, 36" wide	4.000	Ea.	2,500		2,500
Straight connector kit, 64" high	2.000	Ea.	101		101
Ell connector kit, 64" high, 90" deep	2.000	Ea.	144		144
Panel end cover, 43" high	1.000	Ea.	50.50		50.50
Panel end cover, 64" high	1.000	Ea.	50.50		50.50
Variable height finish end cover, 2-way	1.000	Ea.	75.50		75.50
Corner worksurface, 24" deep, 36" wide	1.000	Ea.	580		580
Worksurface, 24" deep, 36" wide	2.000	Ea.	480		480
Cantilever bracket, 20" deep, 24" deep	2.000	Ea.	138		138
Pedestal spacer, 22" deep, 15" wide	2.000	Ea.	164		164
Pedestal, floorstanding, 2 box, 1 file, 22" deep	1.000	Ea.	575		575
Pedestal, floorstanding, 2 file, 22" deep	1.000	Ea.	535		535
Open bookshelf, 36" wide	1.000	Ea.	203		203
Task light, recessed, 30"-36" wide	1.000	Ea.	223		223
Electrical power harness, 36" wide	3.000	Ea.	492		492
Base power in-feed cable	1.000	Ea.	183		183
Duplex receptacle, circuit 1	2.000	Ea.	50		50
Duplex receptacle, circuit 2	1.000	Ea.	25		25
Installation	1.000	Ea.		340	340
TOTAL			7,009.50	340	7,349.50

G1030 Site Earthwork

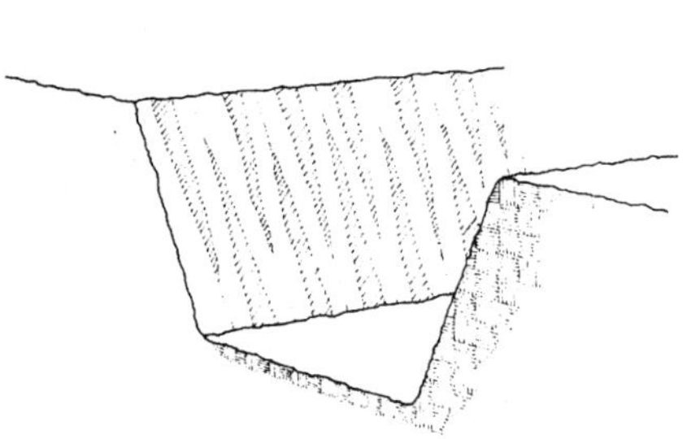

Trenching Systems are shown on a cost per linear foot basis. The systems include: excavation; backfill and removal of spoil; and compaction for various depths and trench bottom widths. The backfill has been reduced to accommodate a pipe of suitable diameter and bedding.

The slope for trench sides varies from none to 1:1.

The Expanded System Listing shows Trenching Systems that range from 2′ to 12′ in width. Depths range from 2′ to 25′.

System Components	QUANTITY	UNIT	COST PER L.F. EQUIP.	COST PER L.F. LABOR	COST PER L.F. TOTAL
SYSTEM G1030 805 1310					
TRENCHING, COMMON EARTH, NO SLOPE, 2′ WIDE, 2′ DP, 3/8 C.Y. BUCKET					
Excavation, trench, hyd. backhoe, track mtd., 3/8 C.Y. bucket	.148	B.C.Y.	.23	1.24	1.47
Backfill and load spoil, from stockpile	.153	L.C.Y.	.13	.37	.50
Compaction by vibrating plate, 6″ lifts, 4 passes	.118	E.C.Y.	.03	.45	.48
Remove excess spoil, 8 C.Y. dump truck, 2 mile roundtrip	.040	L.C.Y.	.16	.21	.37
TOTAL			.55	2.27	2.82

G1030 805	Trenching Common Earth	COST PER L.F. EQUIP.	COST PER L.F. LABOR	COST PER L.F. TOTAL
1310	Trenching, common earth, no slope, 2′ wide, 2′ deep, 3/8 C.Y. bucket	.55	2.27	2.82
1320	3′ deep, 3/8 C.Y. bucket	.77	3.41	4.18
1330	4′ deep, 3/8 C.Y. bucket	.98	4.55	5.53
1340	6′ deep, 3/8 C.Y. bucket	1.28	5.90	7.18
1350	8′ deep, 1/2 C.Y. bucket	1.65	7.80	9.45
1360	10′ deep, 1 C.Y. bucket	3.57	9.30	12.87
1400	4′ wide, 2′ deep, 3/8 C.Y. bucket	1.29	4.52	5.81
1410	3′ deep, 3/8 C.Y. bucket	1.72	6.80	8.52
1420	4′ deep, 1/2 C.Y. bucket	1.92	7.60	9.52
1430	6′ deep, 1/2 C.Y. bucket	3.18	12.15	15.33
1440	8′ deep, 1/2 C.Y. bucket	6.65	15.75	22.40
1450	10′ deep, 1 C.Y. bucket	8.10	19.55	27.65
1460	12′ deep, 1 C.Y. bucket	10.45	25	35.45
1470	15′ deep, 1-1/2 C.Y. bucket	7.65	22	29.65
1480	18′ deep, 2-1/2 C.Y. bucket	12.65	31	43.65
1520	6′ wide, 6′ deep, 5/8 C.Y. bucket w/trench box	9	18.20	27.20
1530	8′ deep, 3/4 C.Y. bucket	11.65	24	35.65
1540	10′ deep, 1 C.Y. bucket	11.35	24.50	35.85
1550	12′ deep, 1-1/2 C.Y. bucket	10.30	26.50	36.80
1560	16′ deep, 2-1/2 C.Y. bucket	16	33	49
1570	20′ deep, 3-1/2 C.Y. bucket	20.50	39.50	60
1580	24′ deep, 3-1/2 C.Y. bucket	24	47.50	71.50
1640	8′ wide, 12′ deep, 1-1/2 C.Y. bucket w/trench box	14.35	33.50	47.85
1650	15′ deep, 1-1/2 C.Y. bucket	18.60	44	62.60
1660	18′ deep, 2-1/2 C.Y. bucket	23	44.50	67.50
1680	24′ deep, 3-1/2 C.Y. bucket	32.50	61.50	94
1730	10′ wide, 20′ deep, 3-1/2 C.Y. bucket w/trench box	26.50	58.50	85
1740	24′ deep, 3-1/2 C.Y. bucket	39	70	109
1780	12′ wide, 20′ deep, 3-1/2 C.Y. bucket w/trench box	41.50	74	115.50
1790	25′ deep, bucket	51.50	94.50	146
1800	1/2 to 1 slope, 2′ wide, 2′ deep, 3/8 C.Y. bucket	.77	3.41	4.18
1810	3′ deep, 3/8 C.Y. bucket	1.25	6	7.25
1820	4′ deep, 3/8 C.Y. bucket	1.84	9.10	10.94
1840	6′ deep, 3/8 C.Y. bucket	3.02	14.75	17.77

G1030 Site Earthwork

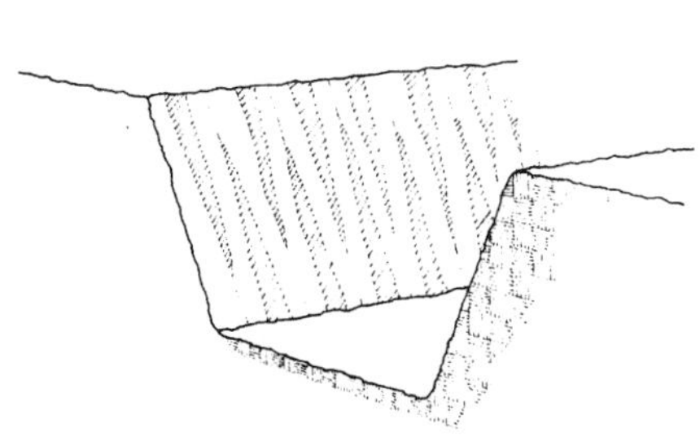

Trenching Systems are shown on a cost per linear foot basis. The systems include: excavation; backfill and removal of spoil; and compaction for various depths and trench bottom widths. The backfill has been reduced to accommodate a pipe of suitable diameter and bedding.

The slope for trench sides varies from none to 1:1.

The Expanded System Listing shows Trenching Systems that range from 2′ to 12′ in width. Depths range from 2′ to 25′.

System Components	QUANTITY	UNIT	COST PER L.F. EQUIP.	COST PER L.F. LABOR	COST PER L.F. TOTAL
SYSTEM G1030 806 1310					
TRENCHING, LOAM & SANDY CLAY, NO SLOPE, 2′ WIDE, 2′ DP, 3/8 C.Y. BUCKET					
Excavation, trench, hyd. backhoe, track mtd., 3/8 C.Y. bucket	.148	B.C.Y.	.21	1.14	1.35
Backfill and load spoil, from stockpile	.165	L.C.Y.	.14	.41	.55
Compaction by vibrating plate 18″ wide, 6″ lifts, 4 passes	.118	E.C.Y.	.03	.45	.48
Remove excess spoil, 8 C.Y. dump truck, 2 mile roundtrip	.042	L.C.Y.	.16	.22	.38
TOTAL			.54	2.22	2.76

G1030 806	Trenching Loam & Sandy Clay	COST PER L.F. EQUIP.	COST PER L.F. LABOR	COST PER L.F. TOTAL
1310	Trenching, loam & sandy clay, no slope, 2′ wide, 2′ deep, 3/8 C.Y. bucket	.54	2.22	2.76
1320	3′ deep, 3/8 C.Y. bucket	.81	3.64	4.45
1330	4′ deep, 3/8 C.Y. bucket	.97	4.43	5.40
1340	6′ deep, 3/8 C.Y. bucket	1.80	5.30	7.10
1350	8′ deep, 1/2 C.Y. bucket	2.35	7	9.35
1360	10′ deep, 1 C.Y. bucket	2.63	7.50	10.15
1400	4′ wide, 2′ deep, 3/8 C.Y. bucket	1.33	4.43	5.75
1410	3′ deep, 3/8 C.Y. bucket	1.74	6.65	8.40
1420	4′ deep, 1/2 C.Y. bucket	1.94	7.45	9.40
1430	6′ deep, 1/2 C.Y. bucket	4.26	10.95	15.20
1440	8′ deep, 1/2 C.Y. bucket	6.50	15.50	22
1450	10′ deep, 1 C.Y. bucket	6.25	15.85	22
1460	12′ deep, 1 C.Y. bucket	7.85	19.75	27.50
1470	15′ deep, 1-1/2 C.Y. bucket	7.95	23	31
1480	18′ deep, 2-1/2 C.Y. bucket	10.75	25	36
1520	6′ wide, 6′ deep, 5/8 C.Y. bucket w/trench box	8.75	17.90	26.50
1530	8′ deep, 3/4 C.Y. bucket	11.20	23.50	34.50
1540	10′ deep, 1 C.Y. bucket	10.40	23.50	34
1550	12′ deep, 1-1/2 C.Y. bucket	10.15	26.50	36.50
1560	16′ deep, 2-1/2 C.Y. bucket	15.65	33.50	49
1570	20′ deep, 3-1/2 C.Y. bucket	18.75	39.50	58.50
1580	24′ deep, 3-1/2 C.Y. bucket	23.50	48	71.50
1640	8′ wide, 12′ deep, 1-1/4 C.Y. bucket w/trench box	14.30	33.50	48
1650	15′ deep, 1-1/2 C.Y. bucket	17.50	43	60.50
1660	18′ deep, 2-1/2 C.Y. bucket	23.50	48	71.50
1680	24′ deep, 3-1/2 C.Y. bucket	32	62.50	94.50
1730	10′ wide, 20′ deep, 3-1/2 C.Y. bucket w/trench box	32	62.50	94.50
1740	24′ deep, 3-1/2 C.Y. bucket	40	77	117
1780	12′ wide, 20′ deep, 3-1/2 C.Y. bucket w/trench box	39	74.50	114
1790	25′ deep, 3-1/2 C.Y. bucket	50.50	96	147
1800	1/2:1 slope, 2′ wide, 2′ deep, 3/8 C.Y. bucket	.75	3.33	4.08
1810	3′ deep, 3/8 C.Y. bucket	1.23	5.85	7.10
1820	4′ deep, 3/8 C.Y. bucket	1.81	8.90	10.70
1840	6′ deep, 3/8 C.Y. bucket	4.33	13.25	17.60

G1030 Site Earthwork

G1030 806	Trenching Loam & Sandy Clay	COST PER L.F.		
		EQUIP.	LABOR	TOTAL
1860	8' deep, 1/2 C.Y. bucket	6.85	21	28
1880	10' deep, 1 C.Y. bucket	9	26.50	35.50
2300	4' wide, 2' deep, 3/8 C.Y. bucket	1.54	5.55	7.10
2310	3' deep, 3/8 C.Y. bucket	2.23	9.15	11.40
2320	4' deep, 1/2 C.Y. bucket	2.71	11.35	14.05
2340	6' deep, 1/2 C.Y. bucket	7.20	19.45	26.50
2360	8' deep, 1/2 C.Y. bucket	12.60	31.50	44
2380	10' deep, 1 C.Y. bucket	13.75	36.50	50.50
2400	12' deep, 1 C.Y. bucket	23	60	83
2430	15' deep, 1-1/2 C.Y. bucket	22.50	67	89.50
2460	18' deep, 2-1/2 C.Y. bucket	42.50	102	145
2840	6' wide, 6' deep, 5/8 C.Y. bucket w/trench box	12.65	27	39.50
2860	8' deep, 3/4 C.Y. bucket	18.20	39.50	57.50
2880	10' deep, 1 C.Y. bucket	18.70	44.50	63
2900	12' deep, 1-1/2 C.Y. bucket	19.80	54.50	74.50
2940	16' deep, 2-1/2 C.Y. bucket	36	79.50	116
2980	20' deep, 3-1/2 C.Y. bucket	49.50	108	158
3020	24' deep, 3-1/2 C.Y. bucket	70	147	217
3100	8' wide, 12' deep, 1-1/2 C.Y. bucket w/trench box	24	61.50	85.50
3120	15' deep, 1-1/2 C.Y. bucket	32.50	86.50	119
3140	18' deep, 2-1/2 C.Y. bucket	49	107	156
3180	24' deep, 3-1/2 C.Y. bucket	78.50	161	240
3270	10' wide, 20' deep, 3-1/2 C.Y. bucket w/trench box	63	131	194
3280	24' deep, 3-1/2 C.Y. bucket	86.50	176	263
3320	12' wide, 20' deep, 3-1/2 C.Y. bucket w/trench box	69.50	142	212
3380	25' deep, 3-1/2 C.Y. bucket w/trench box	95	191	286
3500	1:1 slope, 2' wide, 2' deep, 3/8 C.Y. bucket	.97	4.44	5.40
3520	3' deep, 3/8 C.Y. bucket	1.71	8.30	10
3540	4' deep, 3/8 C.Y. bucket	2.66	13.35	16
3560	6' deep, 1/2 C.Y. bucket	4.33	13.25	17.60
3580	8' deep, 1/2 C.Y. bucket	11.35	35.50	47
3600	10' deep, 1 C.Y. bucket	15.35	45.50	61
3800	4' wide, 2' deep, 3/8 C.Y. bucket	1.74	6.65	8.40
3820	3' deep, 1/2 C.Y. bucket	2.71	11.65	14.35
3840	4' deep, 1/2 C.Y. bucket	3.48	15.20	18.70
3860	6' deep, 1/2 C.Y. bucket	10.15	28	38
3880	8' deep, 1/2 C.Y. bucket	18.75	47	66
3900	10' deep, 1 C.Y. bucket	21.50	57	78.50
3920	12' deep, 1 C.Y. bucket	30.50	80.50	111
3940	15' deep, 1-1/2 C.Y. bucket	37	111	148
3960	18' deep, 2-1/2 C.Y. bucket	58.50	141	200
4030	6' wide, 6' deep, 5/8 C.Y. bucket w/trench box	16.60	36	52.50
4040	8' deep, 3/4 C.Y. bucket	25	55.50	80.50
4050	10' deep, 1 C.Y. bucket	27	65.50	92.50
4060	12' deep, 1-1/2 C.Y. bucket	29.50	82	112
4070	16' deep, 2-1/2 C.Y. bucket	56	126	182
4080	20' deep, 3-1/2 C.Y. bucket	80.50	176	257
4090	24' deep, 3-1/2 C.Y. bucket	117	246	365
4500	8' wide, 12' deep, 1-1/4 C.Y. bucket w/trench box	33.50	89	123
4550	15' deep, 1-1/2 C.Y. bucket	48	130	178
4600	18' deep, 2-1/2 C.Y. bucket	75	165	240
4650	24' deep, 3-1/2 C.Y. bucket	125	260	385
4800	10' wide, 20' deep, 3-1/2 C.Y. bucket w/trench box	93.50	199	293
4850	24' deep, 3-1/2 C.Y. bucket	133	275	410
4950	12' wide, 20' deep, 3-1/2 C.Y. bucket w/trench box	100	210	310
4980	25' deep, 3-1/2 C.Y. bucket	151	310	460

G1030 Site Earthwork

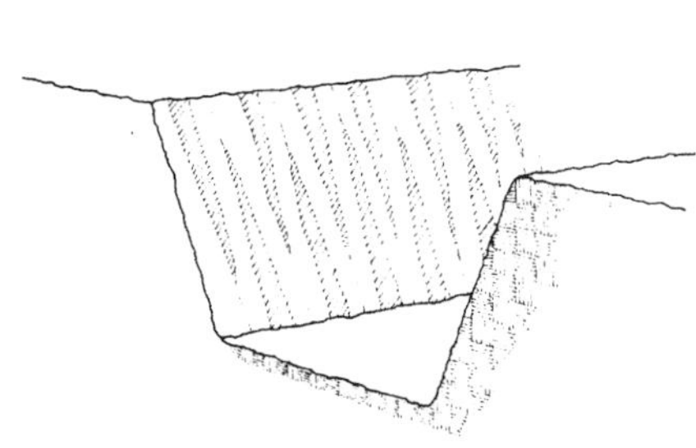

Trenching Systems are shown on a cost per linear foot basis. The systems include: excavation; backfill and removal of spoil; and compaction for various depths and trench bottom widths. The backfill has been reduced to accommodate a pipe of suitable diameter and bedding.

The slope for trench sides varies from none to 1:1.

The Expanded System Listing shows Trenching Systems that range from 2′ to 12′ in width. Depths range from 2′ to 25′.

System Components	QUANTITY	UNIT	COST PER L.F. EQUIP.	LABOR	TOTAL
SYSTEM G1030 807 1310					
TRENCHING, SAND & GRAVEL, NO SLOPE, 2′ WIDE, 2′ DEEP, 3/8 C.Y. BUCKET					
Excavation, trench, hyd. backhoe, track mtd., 3/8 C.Y. bucket	.148	B.C.Y.	.21	1.12	1.33
Backfill and load spoil, from stockpile	.140	L.C.Y.	.12	.34	.46
Compaction by vibrating plate 18″ wide, 6″ lifts, 4 passes	.118	E.C.Y.	.03	.45	.48
Remove excess spoil, 8 C.Y. dump truck, 2 mile roundtrip	.035	L.C.Y.	.14	.18	.32
TOTAL			.50	2.09	2.59

G1030 807	Trenching Sand & Gravel	COST PER L.F. EQUIP.	LABOR	TOTAL
1310	Trenching, sand & gravel, no slope, 2′ wide, 2′ deep, 3/8 C.Y. bucket	.50	2.09	2.59
1320	3′ deep, 3/8 C.Y. bucket	.75	3.51	4.26
1330	4′ deep, 3/8 C.Y. bucket	.89	4.22	5.10
1340	6′ deep, 3/8 C.Y. bucket	1.71	5.05	6.75
1350	8′ deep, 1/2 C.Y. bucket	2.23	6.70	8.95
1360	10′ deep, 1 C.Y. bucket	2.45	7.05	9.50
1400	4′ wide, 2′ deep, 3/8 C.Y. bucket	1.19	4.17	5.35
1410	3′ deep, 3/8 C.Y. bucket	1.57	6.30	7.85
1420	4′ deep, 1/2 C.Y. bucket	1.76	7.10	8.85
1430	6′ deep, 1/2 C.Y. bucket	4.03	10.45	14.50
1440	8′ deep, 1/2 C.Y. bucket	6.15	14.70	21
1450	10′ deep, 1 C.Y. bucket	5.85	14.95	21
1460	12′ deep, 1 C.Y. bucket	7.40	18.65	26
1470	15′ deep, 1-1/2 C.Y. bucket	7.45	21.50	29
1480	18′ deep, 2-1/2 C.Y. bucket	10.15	23.50	33.50
1520	6′ wide, 6′ deep, 5/8 C.Y. bucket w/trench box	8.25	16.95	25
1530	8′ deep, 3/4 C.Y. bucket	10.65	22.50	33
1540	10′ deep, 1 C.Y. bucket	9.80	22.50	32.50
1550	12′ deep, 1-1/2 C.Y. bucket	9.50	25	34.50
1560	16′ deep, 2 C.Y. bucket	14.90	31.50	46.50
1570	20′ deep, 3-1/2 C.Y. bucket	17.75	37	55
1580	24′ deep, 3-1/2 C.Y. bucket	22.50	45	67.50
1640	8′ wide, 12′ deep, 1-1/2 C.Y. bucket w/trench box	13.30	32	45.50
1650	15′ deep, 1-1/2 C.Y. bucket	16.25	40.50	57
1660	18′ deep, 2-1/2 C.Y. bucket	22.50	45	67.50
1680	24′ deep, 3-1/2 C.Y. bucket	30	58.50	88.50
1730	10′ wide, 20′ deep, 3-1/2 C.Y. bucket w/trench box	30	58.50	88.50
1740	24′ deep, 3-1/2 C.Y. bucket	37.50	72.50	110
1780	12′ wide, 20′ deep, 3-1/2 C.Y. bucket w/trench box	36.50	69.50	106
1790	25′ deep, 3-1/2 C.Y. bucket	47	89.50	137
1800	1/2:1 slope, 2′ wide, 2′ deep, 3/8 C.Y. bucket	.70	3.17	3.87
1810	3′ deep, 3/8 C.Y. bucket	1.14	5.55	6.70
1820	4′ deep, 3/8 C.Y. bucket	1.68	8.50	10.20
1840	6′ deep, 3/8 C.Y. bucket	4.12	12.65	16.75

G1030 Site Earthwork

G1030 807	Trenching Sand & Gravel	COST PER L.F.		
		EQUIP.	LABOR	TOTAL
1860	8' deep, 1/2 C.Y. bucket	6.50	20	26.50
1880	10' deep, 1 C.Y. bucket	8.40	25	33.50
2300	4' wide, 2' deep, 3/8 C.Y. bucket	1.38	5.25	6.65
2310	3' deep, 3/8 C.Y. bucket	2.01	8.70	10.70
2320	4' deep, 1/2 C.Y. bucket	2.46	10.75	13.20
2340	6' deep, 1/2 C.Y. bucket	6.85	18.60	25.50
2360	8' deep, 1/2 C.Y. bucket	11.95	29.50	41.50
2380	10' deep, 1 C.Y. bucket	12.95	34.50	47.50
2400	12' deep, 1 C.Y. bucket	22	57	79
2430	15' deep, 1-1/2 C.Y. bucket	21	63.50	84.50
2460	18' deep, 2-1/2 C.Y. bucket	40.50	96.50	137
2840	6' wide, 6' deep, 5/8 C.Y. bucket w/trench box	12	25.50	37.50
2860	8' deep, 3/4 C.Y. bucket	17.30	37.50	55
2880	10' deep, 1 C.Y. bucket	17.70	42	59.50
2900	12' deep, 1-1/2 C.Y. bucket	18.60	51	69.50
2940	16' deep, 2 C.Y. bucket	34	75	109
2980	20' deep, 3-1/2 C.Y. bucket	47	101	148
3020	24' deep, 3-1/2 C.Y. bucket	66.50	138	205
3100	8' wide, 12' deep, 1-1/4 C.Y. bucket w/trench box	22.50	58	80.50
3120	15' deep, 1-1/2 C.Y. bucket	30.50	81.50	112
3140	18' deep, 2-1/2 C.Y. bucket	46.50	100	147
3180	24' deep, 3-1/2 C.Y. bucket	74	152	226
3270	10' wide, 20' deep, 3-1/2 C.Y. bucket w/trench box	59	123	182
3280	24' deep, 3-1/2 C.Y. bucket	82	165	247
3370	12' wide, 20' deep, 3-1/2 C.Y. bucket w/trench box	66	136	202
3380	25' deep, 3-1/2 C.Y. bucket	95	191	286
3500	1:1 slope, 2' wide, 2' deep, 3/8 C.Y. bucket	1.81	5.35	7.15
3520	3' deep, 3/8 C.Y. bucket	1.57	7.95	9.50
3540	4' deep, 3/8 C.Y. bucket	2.46	12.75	15.20
3560	6' deep, 3/8 C.Y. bucket	4.13	12.65	16.80
3580	8' deep, 1/2 C.Y. bucket	10.80	33.50	44.50
3600	10' deep, 1 C.Y. bucket	14.35	43	57.50
3800	4' wide, 2' deep, 3/8 C.Y. bucket	1.58	6.30	7.90
3820	3' deep, 3/8 C.Y. bucket	2.47	11.10	13.55
3840	4' deep, 1/2 C.Y. bucket	3.17	14.45	17.60
3860	6' deep, 1/2 C.Y. bucket	9.70	27	36.50
3880	8' deep, 1/2 C.Y. bucket	17.80	45	63
3900	10' deep, 1 C.Y. bucket	20	54	74
3920	12' deep, 1 C.Y. bucket	29	76	105
3940	15' deep, 1-1/2 C.Y. bucket	35	105	140
3960	18' deep, 2-1/2 C.Y. bucket	55.50	133	189
4030	6' wide, 6' deep, 5/8 C.Y. bucket w/trench box	15.75	34.50	50.50
4040	8' deep, 3/4 C.Y. bucket	24	53	77
4050	10' deep, 1 C.Y. bucket	25.50	61.50	87
4060	12' deep, 1-1/2 C.Y. bucket	27.50	77.50	105
4070	16' deep, 2 C.Y. bucket	53.50	119	173
4080	20' deep, 3-1/2 C.Y. bucket	76	165	241
4090	24' deep, 3-1/2 C.Y. bucket	110	232	340
4500	8' wide, 12' deep, 1-1/2 C.Y. bucket w/trench box	31.50	84.50	116
4550	15' deep, 1-1/2 C.Y. bucket	44.50	122	167
4600	18' deep, 2-1/2 C.Y. bucket	71	155	226
4650	24' deep, 3-1/2 C.Y. bucket	118	245	365
4800	10' wide, 20' deep, 3-1/2 C.Y. bucket w/trench box	88.50	186	275
4850	24' deep, 3-1/2 C.Y. bucket	126	259	385
4950	12' wide, 20' deep, 3-1/2 C.Y. bucket w/trench box	94.50	197	292
4980	25' deep, 3-1/2 C.Y. bucket	143	292	435

G2020 Parking Lots

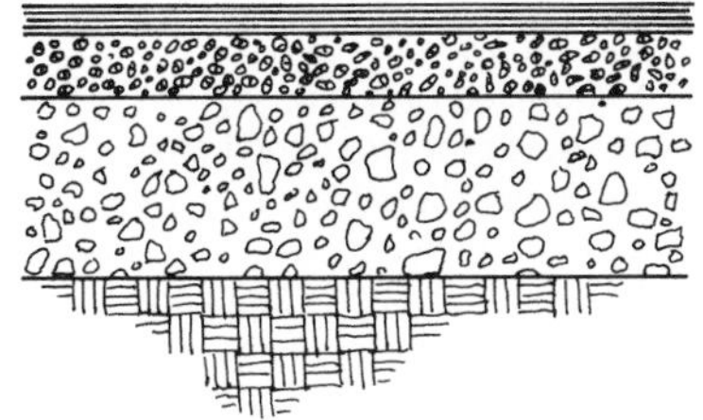

The Parking Lot System includes: compacted bank-run gravel; fine grading with a grader and roller; and bituminous concrete wearing course. All Parking Lot systems are on a cost per car basis. There are three basic types of systems: 90° angle, 60° angle, and 45° angle. The gravel base is compacted to 98%. Final stall design and lay-out of the parking lot with precast bumpers, sealcoating and white paint is also included.

The Expanded System Listing shows the three basic parking lot types with various depths of both gravel base and wearing course. The gravel base depths range from 6″ to 10″. The bituminous paving wearing course varies from a depth of 3″ to 6″.

System Components	QUANTITY	UNIT	COST PER CAR MAT.	COST PER CAR INST.	COST PER CAR TOTAL
SYSTEM G2020 220 1500					
PARKING LOT, 90° ANGLE PARKING, 3″ BITUMINOUS PAVING, 6″ GRAVEL BASE					
Surveying crew for layout, 4 man crew	.020	Day		57.18	57.18
Borrow, bank run gravel, haul 2 mi., spread w/dozer, no compaction	7.223	C.Y.	148.07	61.97	210.04
Grading, fine grade 3 passes with motor grader	43.333	S.Y.		133.10	133.10
Compact w/vibrating plate, 8″ lifts, granular mat'l. to 98%	7.223	C.Y.		55.33	55.33
Bituminous paving, 3″ thick	43.333	S.Y.	578.50	92.30	670.80
Seal coating, petroleum resistant under 1,000 S.Y.	43.333	S.Y.	66.73	68.03	134.76
Lines on pvmt., parking stall, paint, white, 4″ wide	1.000	Ea.	5.55	4.22	9.77
Precast concrete parking bar, 6″ x 10″ x 6′-0″	1.000	Ea.	69.50	22.50	92
TOTAL			868.35	494.63	1,362.98

G2020 220	Parking Lots	COST PER CAR MAT.	COST PER CAR INST.	COST PER CAR TOTAL
1500	Parking lot, 90° angle parking, 3″ bituminous paving, 6″ gravel base	870	495	1,365
1520	8″ gravel base	915	535	1,450
1540	10″ gravel base	965	575	1,540
1560	4″ bituminous paving, 6″ gravel base	1,025	495	1,520
1580	8″ gravel base	1,075	530	1,605
1600	10″ gravel base	1,125	570	1,695
1620	6″ bituminous paving, 6″ gravel base	1,350	515	1,865
1640	8″ gravel base	1,400	550	1,950
1661	10″ gravel base	1,450	590	2,040
1800	60° angle parking, 3″ bituminous paving, 6″ gravel base	870	495	1,365
1820	8″ gravel base	915	535	1,450
1840	10″ gravel base	965	575	1,540
1860	4″ bituminous paving, 6″ gravel base	1,025	495	1,520
1880	8″ gravel base	1,075	530	1,605
1900	10″ gravel base	1,125	570	1,695
1920	6″ bituminous paving, 6″ gravel base	1,350	515	1,865
1940	8″ gravel base	1,400	550	1,950
1961	10″ gravel base	1,450	590	2,040
2200	45° angle parking, 3″ bituminous paving, 6″ gravel base	890	510	1,400
2220	8″ gravel base	940	545	1,485
2240	10″ gravel base	990	590	1,580
2260	4″ bituminous paving, 6″ gravel base	1,050	505	1,555
2280	8″ gravel base	1,100	545	1,645
2300	10″ gravel base	1,150	585	1,735
2320	6″ bituminous paving, 6″ gravel base	1,375	530	1,905
2340	8″ gravel base	1,425	565	1,990
2361	10″ gravel base	1,475	605	2,080

G2030 Pedestrian Paving

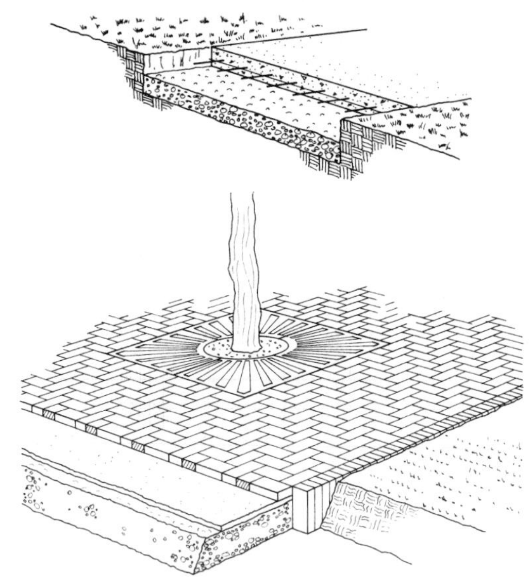

The Bituminous and Concrete Sidewalk Systems include excavation, hand grading and compacted gravel base. Pavements are shown for two conditions of each of the following variables; pavement thickness, gravel base thickness and pavement width. Costs are given on a linear foot basis.

The Plaza Systems listed include several brick and tile paving surfaces on two different bases: gravel and slab on grade. The type of bedding for the pavers depends on the base being used, and alternate bedding may be desirable. Also included in the paving costs are edging and precast grating costs and where concrete bases are involved, expansion joints. Costs are given on a square foot basis.

G2030 110	Bituminous Sidewalks	COST PER L.F.		
		MAT.	INST.	TOTAL
1580	Bituminous sidewalk, 1" thick paving, 4" gravel base, 3' width	2.08	5.80	7.88
1600	4' width	2.75	6.25	9
1640	6" gravel base, 3' width	2.47	6.05	8.52
1660	4' width	3.27	6.60	9.87
2120	2" thick paving, 4" gravel base, 3' width	3.39	6.90	10.29
2140	4' width	4.51	7.60	12.11
2180	6" gravel base, 3' width	3.78	7.15	10.93
2200	4' width	5.05	8	13.05

G2030 120	Concrete Sidewalks	COST PER L.F.		
		MAT.	INST.	TOTAL
1580	Concrete sidewalk, 4" thick, 4" gravel base, 3' wide	8.40	14.55	22.95
1600	4' wide	11.15	17.80	28.95
1640	6" gravel base, 3' wide	8.75	14.80	23.55
1660	4' wide	11.70	18.15	29.85
2120	6" thick concrete, 4" gravel base, 3' wide	11.75	16.50	28.25
2140	4' wide	15.70	20.50	36.20
2180	6" gravel base, 3' wide	12.15	16.75	28.90
2200	4' wide	16.20	20.50	36.70

G2030 150	Brick & Tile Plazas	COST PER S.F.		
		MAT.	INST.	TOTAL
2050	Brick pavers, 4" x 8" x 1-3/4", gravel base, stone dust bedding	7	7.05	14.05
2100	Slab on grade, asphalt bedding	9.65	10.10	19.75
3550	Concrete paving stone, 4" x 8" x 2-1/2", gravel base, sand bedding	5.25	4.99	10.24
3600	Slab on grade, asphalt bedding	7.35	7.40	14.75
4050	Concrete patio blocks, 8" x 16" x 2", gravel base, sand bedding	14.50	6.50	21
4100	Slab on grade, asphalt bedding	16.80	9.40	26.20
6050	Granite pavers, 3-1/2" x 3-1/2" x 3-1/2", gravel base, sand bedding	26.50	15.50	42
6100	Slab on grade, mortar bedding	30	24.50	54.50
7050	Limestone, 3" thick, gravel base, sand bedding	14.95	19.15	34.10
7100	Slab on grade, mortar bedding	17.25	24	41.25
8050	Slate flagging, 3/4" thick, gravel base, sand bedding	13.80	15.50	29.30
8100	Slab on grade, mastic bedding	29	22.50	51.50

G4020 Site Lighting

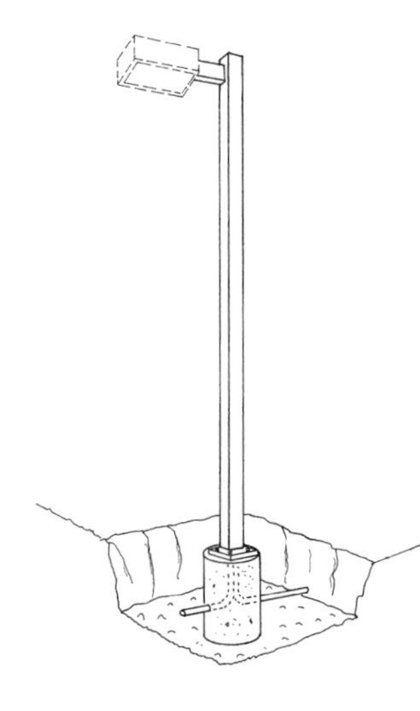

Table G4020 210 Procedure for Calculating Floodlights Required for Various Footcandles
Poles should not be spaced more than 4 times the fixture mounting height for good light distribution.

Estimating Chart
Select Lamp type.

Determine total square feet.

Chart will show quantity of fixtures to provide 1 footcandle initial, at intersection of lines. Multiply fixture quantity by desired footcandle level.

Chart based on use of wide beam luminaires in an area whose dimensions are large compared to mounting height and is approximate only.

To maintain 1 footcandle over a large area use these watts per square foot:

Incandescent	0.15
Metal Halide	0.032
Mercury Vapor	0.05
High Pressure Sodium	0.024

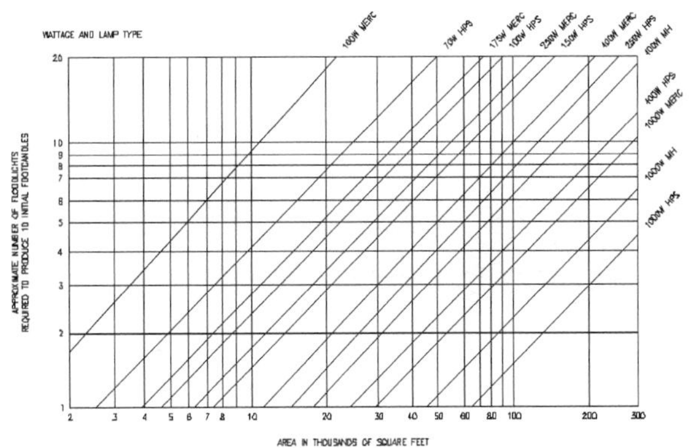

System Components	QUANTITY	UNIT	COST EACH MAT.	COST EACH INST.	COST EACH TOTAL
SYSTEM G4020 210 0200					
LIGHT POLES, ALUMINUM, 20′ HIGH, 1 ARM BRACKET					
Aluminum light pole, 20′, no concrete base	1.000	Ea.	1,200	721.50	1,921.50
Bracket arm for aluminum light pole	1.000	Ea.	155	94.50	249.50
Excavation by hand, pits to 6′ deep, heavy soil or clay	2.368	C.Y.		319.68	319.68
Footing, concrete incl forms, reinforcing, spread, under 1 C.Y.	.465	C.Y.	104.16	149.28	253.44
Backfill by hand	1.903	C.Y.		93.25	93.25
Compaction vibrating plate	1.903	C.Y.		15.26	15.26
TOTAL			1,459.16	1,393.47	2,852.63

G4020 210	Light Pole (Installed)	COST EACH MAT.	COST EACH INST.	COST EACH TOTAL
0200	Light pole, aluminum, 20′ high, 1 arm bracket	1,450	1,400	2,850
0240	2 arm brackets	1,625	1,400	3,025
0280	3 arm brackets	1,775	1,450	3,225
0320	4 arm brackets	1,925	1,450	3,375
0360	30′ high, 1 arm bracket	2,550	1,750	4,300
0400	2 arm brackets	2,725	1,750	4,475
0440	3 arm brackets	2,875	1,800	4,675
0480	4 arm brackets	3,025	1,825	4,850
0680	40′ high, 1 arm bracket	3,325	2,350	5,675
0720	2 arm brackets	3,500	2,350	5,850
0760	3 arm brackets	3,650	2,400	6,050
0800	4 arm brackets	3,800	2,425	6,225
0840	Steel, 20′ high, 1 arm bracket	1,750	1,475	3,225
0880	2 arm brackets	1,850	1,475	3,325
0920	3 arm brackets	1,750	1,525	3,275
0960	4 arm brackets	1,900	1,525	3,425
1000	30′ high, 1 arm bracket	1,775	1,875	3,650
1040	2 arm brackets	1,875	1,875	3,750
1080	3 arm brackets	1,775	1,925	3,700
1120	4 arm brackets	1,925	1,925	3,850
1320	40′ high, 1 arm bracket	2,275	2,525	4,800
1360	2 arm brackets	2,375	2,525	4,900
1400	3 arm brackets	2,275	2,575	4,850
1440	4 arm brackets	2,425	2,575	5,000

Reference Section

All the reference information is in one section, making it easy to find what you need to know . . . and easy to use the data set on a daily basis. This section is visually identified by a vertical black bar on the page edges.

In this Reference Section, we've included Equipment Rental Costs, a listing of rental and operating costs; Crew Listings, a full listing of all crews and equipment, and their costs; Historical Cost Indexes for cost comparisons over time; City Cost Indexes and Location Factors for adjusting costs to the region you are in; Reference Tables, where you will find explanations, estimating information and procedures, or technical data; Change Orders, information on pricing changes to contract documents; and an explanation of all the Abbreviations in the data set.

Table of Contents

Estimating Tips

- This section contains the average costs to rent and operate hundreds of pieces of construction equipment. This is useful information when one is estimating the time and material requirements of any particular operation in order to establish a unit or total cost. Bare equipment costs shown on a unit cost line include, not only rental, but also operating costs for equipment under normal use.

Rental Costs

- Equipment rental rates are obtained from the following industry sources throughout North America: contractors, suppliers, dealers, manufacturers, and distributors.
- Rental rates vary throughout the country, with larger cities generally having lower rates. Lease plans for new equipment are available for periods in excess of six months, with a percentage of payments applying toward purchase.
- Monthly rental rates vary from 2% to 5% of the purchase price of the equipment depending on the anticipated life of the equipment and its wearing parts.
- Weekly rental rates are about 1/3 of the monthly rates, and daily rental rates are about 1/3 of the weekly rate.
- Rental rates can also be treated as reimbursement costs for contractor-owned equipment. Owned equipment costs include depreciation, loan payments, interest, taxes, insurance, storage, and major repairs.

Operating Costs

- The operating costs include parts and labor for routine servicing, such as the repair and replacement of pumps, filters, and worn lines. Normal operating expendables, such as fuel, lubricants, tires, and electricity (where applicable), are also included.
- Extraordinary operating expendables with highly variable wear patterns, such as diamond bits and blades, are excluded. These costs can be found as material costs in the Unit Price section.
- The hourly operating costs listed do not include the operator's wages.

Equipment Cost/Day

- Any power equipment required by a crew is shown in the Crew Listings with a daily cost.
- This daily cost of equipment needed by a crew includes both the rental cost and the operating cost and is based on dividing the weekly rental rate by 5 (the number of working days in the week), then adding the hourly operating cost multiplied by 8 (the number of hours in a day). This "Equipment Cost/Day" is shown in the far right column of the Equipment Rental section.
- If equipment is needed for only one or two days, it is best to develop your own cost by including components for daily rent and hourly operating costs. This is important when the listed Crew for a task does not contain the equipment needed, such as a crane for lifting mechanical heating/cooling equipment up onto a roof.
- If the quantity of work is less than the crew's Daily Output shown for a Unit Price line item that includes a bare unit equipment cost, the recommendation is to estimate one day's rental cost and operating cost for equipment shown in the Crew Listing for that line item.
- Please note, in some cases the equipment description in the crew is followed by a time period in parenthesis. For example: (daily) or (monthly). In these cases the equipment cost/day is calculated by adding the rental cost per time period to the hourly operating cost multiplied by 8.

Mobilization, Demobilization Costs

- The cost to move construction equipment from an equipment yard or rental company to the job site and back again is not included in equipment rental costs listed in the Reference Section. It is also not included in the bare equipment cost of any Unit Price line item or in any equipment costs shown in the Crew Listings.
- Mobilization (to the site) and demobilization (from the site) costs can be found in the Unit Price section.
- If a piece of equipment is already at the job site, it is not appropriate to utilize mobilization or demobilization costs again in an estimate. ■

01 54 33 | Equipment Rental

			UNIT	HOURLY OPER. COST	RENT PER DAY	RENT PER WEEK	RENT PER MONTH	EQUIPMENT COST/DAY	
10	0010	**CONCRETE EQUIPMENT RENTAL** without operators R015433-10							10
	0200	Bucket, concrete lightweight, 1/2 C.Y.	Ea.	.87	38.50	115	345	30	
	0300	1 C.Y.		.98	63.50	190	570	45.80	
	0400	1-1/2 C.Y.		1.23	60	180	540	45.85	
	0500	2 C.Y.		1.34	73.50	220	660	54.70	
	0580	8 C.Y.		6.47	93.50	280	840	107.75	
	0600	Cart, concrete, self-propelled, operator walking, 10 C.F.		2.85	175	525	1,575	127.85	
	0700	Operator riding, 18 C.F.		4.81	192	575	1,725	153.45	
	0800	Conveyer for concrete, portable, gas, 16″ wide, 26′ long		10.61	160	480	1,450	180.85	
	0900	46′ long		10.98	175	525	1,575	192.85	
	1000	56′ long		11.15	192	575	1,725	204.15	
	1100	Core drill, electric, 2-1/2 H.P., 1″ to 8″ bit diameter		1.56	83.50	250	750	62.50	
	1150	11 H.P., 8″ to 18″ cores		5.38	119	356.32	1,075	114.30	
	1200	Finisher, concrete floor, gas, riding trowel, 96″ wide		9.64	153	459.60	1,375	169	
	1300	Gas, walk-behind, 3 blade, 36″ trowel		2.03	96	287.50	865	73.75	
	1400	4 blade, 48″ trowel		3.06	104	312.50	940	87	
	1500	Float, hand-operated (Bull float), 48″ wide		.08	12.35	37	111	8.05	
	1570	Curb builder, 14 H.P., gas, single screw		14.00	253	760	2,275	263.95	
	1590	Double screw		15.00	253	760	2,275	272	
	1600	Floor grinder, concrete and terrazzo, electric, 22″ path		3.03	134	401.75	1,200	104.60	
	1700	Edger, concrete, electric, 7″ path		1.18	57.50	172.50	520	43.95	
	1750	Vacuum pick-up system for floor grinders, wet/dry		1.61	102	305.71	915	74.05	
	1800	Mixer, powered, mortar and concrete, gas, 6 C.F., 18 H.P.		7.39	97	291	875	117.35	
	1900	10 C.F., 25 H.P.		8.97	114	342.50	1,025	140.30	
	2000	16 C.F.		9.33	144	432.50	1,300	161.15	
	2100	Concrete, stationary, tilt drum, 2 C.Y.		7.21	80	240	720	105.70	
	2120	Pump, concrete, truck mounted, 4″ line, 80′ boom		29.79	287	860	2,575	410.35	
	2140	5″ line, 110′ boom		37.34	287	860	2,575	470.75	
	2160	Mud jack, 50 C.F. per hr.		6.43	228	685	2,050	188.45	
	2180	225 C.F. per hr.		8.52	293	880	2,650	244.15	
	2190	Shotcrete pump rig, 12 C.Y./hr.		13.93	223	670	2,000	245.40	
	2200	35 C.Y./hr.		15.75	287	860	2,575	298	
	2600	Saw, concrete, manual, gas, 18 H.P.		5.52	112	337	1,000	111.55	
	2650	Self-propelled, gas, 30 H.P.		7.87	81	242.71	730	111.50	
	2675	V-groove crack chaser, manual, gas, 6 H.P.		1.64	100	300	900	73.10	
	2700	Vibrators, concrete, electric, 60 cycle, 2 H.P.		.47	73	218.50	655	47.45	
	2800	3 H.P.		.56	73.50	221	665	48.70	
	2900	Gas engine, 5 H.P.		1.54	16.85	50.61	152	22.45	
	3000	8 H.P.		2.08	17.05	51.12	153	26.85	
	3050	Vibrating screed, gas engine, 8 H.P.		2.80	88	263.50	790	75.10	
	3120	Concrete transit mixer, 6 x 4, 250 H.P., 8 C.Y., rear discharge		50.57	70	210	630	446.55	
	3200	Front discharge		58.71	135	405	1,225	550.65	
	3300	6 x 6, 285 H.P., 12 C.Y., rear discharge		57.97	150	450	1,350	553.80	
	3400	Front discharge		60.41	170	510	1,525	585.25	
20	0010	**EARTHWORK EQUIPMENT RENTAL** without operators R015433-10							20
	0040	Aggregate spreader, push type, 8′ to 12′ wide	Ea.	2.59	75	225	675	65.75	
	0045	Tailgate type, 8′ wide		2.54	63.50	190	570	58.30	
	0055	Earth auger, truck mounted, for fence & sign posts, utility poles		13.81	150	450	1,350	200.50	
	0060	For borings and monitoring wells		42.52	83.50	250	750	390.20	
	0070	Portable, trailer mounted		2.29	100	300	900	78.35	
	0075	Truck mounted, for caissons, water wells		85.14	150	450	1,350	771.10	
	0080	Horizontal boring machine, 12″ to 36″ diameter, 45 H.P.		22.70	104	312	935	244	
	0090	12″ to 48″ diameter, 65 H.P.		31.16	108	325	975	314.25	
	0095	Auger, for fence posts, gas engine, hand held		.45	84	251.50	755	53.85	
	0100	Excavator, diesel hydraulic, crawler mounted, 1/2 C.Y. cap.		21.66	465	1,394.28	4,175	452.10	
	0120	5/8 C.Y. capacity		28.95	610	1,833.22	5,500	598.25	
	0140	3/4 C.Y. capacity		32.56	725	2,168.88	6,500	694.25	
	0150	1 C.Y. capacity		41.09	780	2,333	7,000	795.30	

01 54 33 | Equipment Rental

		01 54 33 Equipment Rental	UNIT	HOURLY OPER. COST	RENT PER DAY	RENT PER WEEK	RENT PER MONTH	EQUIPMENT COST/DAY	
20	0200	1-1/2 C.Y. capacity	Ea.	48.44	500	1,500	4,500	687.55	20
	0300	2 C.Y. capacity		56.41	835	2,500	7,500	951.30	
	0320	2-1/2 C.Y. capacity		82.39	1,350	4,027.92	12,100	1,465	
	0325	3-1/2 C.Y. capacity		119.76	2,000	6,000	18,000	2,158	
	0330	4-1/2 C.Y. capacity		151.17	3,675	11,000	33,000	3,409	
	0335	6 C.Y. capacity		191.81	3,225	9,650	29,000	3,464	
	0340	7 C.Y. capacity		174.67	3,400	10,200	30,600	3,437	
	0342	Excavator attachments, bucket thumbs		3.39	258	774.60	2,325	182.05	
	0345	Grapples		3.13	222	666.16	2,000	158.30	
	0346	Hydraulic hammer for boom mounting, 4000 ft lb.		13.44	890	2,670	8,000	641.50	
	0347	5000 ft lb.		15.90	950	2,850	8,550	697.25	
	0348	8000 ft lb.		23.47	1,200	3,600	10,800	907.75	
	0349	12,000 ft lb.		25.64	1,100	3,333	10,000	871.75	
	0350	Gradall type, truck mounted, 3 ton @ 15′ radius, 5/8 C.Y.		43.31	835	2,500	7,500	846.50	
	0370	1 C.Y. capacity		59.22	835	2,500	7,500	973.80	
	0400	Backhoe-loader, 40 to 45 H.P., 5/8 C.Y. capacity		11.86	244	732.50	2,200	241.40	
	0450	45 H.P. to 60 H.P., 3/4 C.Y. capacity		17.97	117	350	1,050	213.75	
	0460	80 H.P., 1-1/4 C.Y. capacity		20.30	117	350	1,050	232.40	
	0470	112 H.P., 1-1/2 C.Y. capacity		32.89	610	1,833.22	5,500	629.75	
	0482	Backhoe-loader attachment, compactor, 20,000 lb.		6.42	155	464.76	1,400	144.30	
	0485	Hydraulic hammer, 750 ft lb.		3.67	107	320.17	960	93.40	
	0486	Hydraulic hammer, 1200 ft lb.		6.53	205	614.52	1,850	175.15	
	0500	Brush chipper, gas engine, 6″ cutter head, 35 H.P.		9.14	247	740	2,225	221.10	
	0550	Diesel engine, 12″ cutter head, 130 H.P.		23.60	340	1,020	3,050	392.80	
	0600	15″ cutter head, 165 H.P.		26.51	415	1,239.36	3,725	459.90	
	0750	Bucket, clamshell, general purpose, 3/8 C.Y.		1.40	91.50	275	825	66.15	
	0800	1/2 C.Y.		1.51	91.50	275	825	67.10	
	0850	3/4 C.Y.		1.64	91.50	275	825	68.10	
	0900	1 C.Y.		1.70	91.50	275	825	68.55	
	0950	1-1/2 C.Y.		2.78	91.50	275	825	77.25	
	1000	2 C.Y.		2.91	91.50	275	825	78.30	
	1010	Bucket, dragline, medium duty, 1/2 C.Y.		.82	91.50	275	825	61.55	
	1020	3/4 C.Y.		.78	91.50	275	825	61.25	
	1030	1 C.Y.		.80	91.50	275	825	61.35	
	1040	1-1/2 C.Y.		1.26	91.50	275	825	65.05	
	1050	2 C.Y.		1.29	91.50	275	825	65.30	
	1070	3 C.Y.		2.07	91.50	275	825	71.60	
	1200	Compactor, manually guided 2-drum vibratory smooth roller, 7.5 H.P.		7.20	181	542.50	1,625	166.10	
	1250	Rammer/tamper, gas, 8″		2.20	48	144.59	435	46.50	
	1260	15″		2.62	55	165.25	495	54	
	1300	Vibratory plate, gas, 18″ plate, 3000 lb. blow		2.12	24.50	72.81	218	31.55	
	1350	21″ plate, 5000 lb. blow		2.61	241	722.50	2,175	165.35	
	1370	Curb builder/extruder, 14 H.P., gas, single screw		13.99	253	760	2,275	263.95	
	1390	Double screw		14.99	253	760	2,275	271.95	
	1500	Disc harrow attachment, for tractor		.47	82.50	246.84	740	53.15	
	1810	Feller buncher, shearing & accumulating trees, 100 H.P.		39.08	460	1,380	4,150	588.60	
	1860	Grader, self-propelled, 25,000 lb.		33.25	1,100	3,333	10,000	932.60	
	1910	30,000 lb.		32.76	1,325	4,000	12,000	1,062	
	1920	40,000 lb.		51.73	1,550	4,667	14,000	1,347	
	1930	55,000 lb.		66.73	1,775	5,333	16,000	1,600	
	1950	Hammer, pavement breaker, self-propelled, diesel, 1000 to 1250 lb.		28.31	600	1,800	5,400	586.50	
	2000	1300 to 1500 lb.		42.67	1,000	3,020.94	9,075	945.55	
	2050	Pile driving hammer, steam or air, 4150 ft lb. @ 225 bpm		12.11	500	1,500	4,500	396.90	
	2100	8750 ft lb. @ 145 bpm		14.30	700	2,100	6,300	534.45	
	2150	15,000 ft lb. @ 60 bpm		14.63	835	2,500	7,500	617.05	
	2200	24,450 ft lb. @ 111 bpm		15.64	965	2,900	8,700	705.15	
	2250	Leads, 60′ high for pile driving hammers up to 20,000 ft lb.		3.66	300	900	2,700	209.25	
	2300	90′ high for hammers over 20,000 ft lb.	↓	5.43	540	1,620	4,850	367.45	

01 54 33 | Equipment Rental

		01 54 33 Equipment Rental	UNIT	HOURLY OPER. COST	RENT PER DAY	RENT PER WEEK	RENT PER MONTH	EQUIPMENT COST/DAY	
20	2350	Diesel type hammer, 22,400 ft lb.	Ea.	17.76	490	1,471.74	4,425	436.45	20
	2400	41,300 ft lb.		25.61	620	1,859.04	5,575	576.65	
	2450	141,000 ft lb.		41.20	980	2,943.48	8,825	918.35	
	2500	Vib. elec. hammer/extractor, 200 kW diesel generator, 34 H.P.		41.25	715	2,143.06	6,425	758.60	
	2550	80 H.P.		72.81	1,025	3,098.40	9,300	1,202	
	2600	150 H.P.		134.74	2,000	5,964.42	17,900	2,271	
	2700	Hydro Excavator w/EXT boom 12 C.Y., 1200 gallons		37.70	1,600	4,800	14,400	1,262	
	2800	Log chipper, up to 22" diameter, 600 H.P.		46.13	305	915	2,750	552	
	2850	Logger, for skidding & stacking logs, 150 H.P.		43.40	930	2,785	8,350	904.20	
	2860	Mulcher, diesel powered, trailer mounted		17.99	305	915	2,750	326.90	
	2900	Rake, spring tooth, with tractor		14.67	370	1,110.26	3,325	339.45	
	3000	Roller, vibratory, tandem, smooth drum, 20 H.P.		7.78	320	967	2,900	255.65	
	3050	35 H.P.		10.10	260	779.76	2,350	236.75	
	3100	Towed type vibratory compactor, smooth drum, 50 H.P.		25.20	520	1,566	4,700	514.75	
	3150	Sheepsfoot, 50 H.P.		25.56	385	1,161.90	3,475	436.90	
	3170	Landfill compactor, 220 H.P.		69.80	1,650	4,985	15,000	1,555	
	3200	Pneumatic tire roller, 80 H.P.		12.88	405	1,213.54	3,650	345.75	
	3250	120 H.P.		19.33	665	1,988.14	5,975	552.25	
	3300	Sheepsfoot vibratory roller, 240 H.P.		62.02	1,425	4,260.30	12,800	1,348	
	3320	340 H.P.		83.57	2,175	6,500	19,500	1,969	
	3350	Smooth drum vibratory roller, 75 H.P.		23.27	655	1,962.32	5,875	578.60	
	3400	125 H.P.		27.53	740	2,220.52	6,650	664.35	
	3410	Rotary mower, brush, 60", with tractor		18.73	360	1,084.44	3,250	366.75	
	3420	Rototiller, walk-behind, gas, 5 H.P.		2.13	60	180	540	53.05	
	3422	8 H.P.		2.80	132	395	1,175	101.40	
	3440	Scrapers, towed type, 7 C.Y. capacity		6.42	127	382.14	1,150	127.80	
	3450	10 C.Y. capacity		7.18	170	511.24	1,525	159.70	
	3500	15 C.Y. capacity		7.38	196	588.70	1,775	176.75	
	3525	Self-propelled, single engine, 14 C.Y. capacity		132.89	2,225	6,660	20,000	2,395	
	3550	Dual engine, 21 C.Y. capacity		140.95	2,500	7,500	22,500	2,628	
	3600	31 C.Y. capacity		187.28	3,625	10,844.40	32,500	3,667	
	3640	44 C.Y. capacity		231.98	4,650	13,942.80	41,800	4,644	
	3650	Elevating type, single engine, 11 C.Y. capacity		61.68	1,075	3,200	9,600	1,133	
	3700	22 C.Y. capacity		114.28	1,625	4,850	14,600	1,884	
	3710	Screening plant, 110 H.P. w/5' x 10' screen		21.07	645	1,933	5,800	555.15	
	3720	5' x 16' screen		26.60	1,325	4,000	12,000	1,013	
	3850	Shovel, crawler-mounted, front-loading, 7 C.Y. capacity		218.00	3,925	11,773.92	35,300	4,099	
	3855	12 C.Y. capacity		335.89	5,450	16,318.24	49,000	5,951	
	3860	Shovel/backhoe bucket, 1/2 C.Y.		2.68	73	218.95	655	65.25	
	3870	3/4 C.Y.		2.66	82	245.81	735	70.40	
	3880	1 C.Y.		2.75	91	272.66	820	76.50	
	3890	1-1/2 C.Y.		2.94	107	320.17	960	87.60	
	3910	3 C.Y.		3.43	145	433.78	1,300	114.15	
	3950	Stump chipper, 18" deep, 30 H.P.		6.88	232	697	2,100	194.45	
	4110	Dozer, crawler, torque converter, diesel 80 H.P.		25.18	335	1,000	3,000	401.40	
	4150	105 H.P.		34.23	600	1,800	5,400	633.85	
	4200	140 H.P.		41.15	720	2,166	6,500	762.40	
	4260	200 H.P.		62.97	1,675	5,000	15,000	1,504	
	4310	300 H.P.		80.49	1,875	5,600	16,800	1,764	
	4360	410 H.P.		106.42	3,200	9,630	28,900	2,777	
	4370	500 H.P.		132.98	3,900	11,670	35,000	3,398	
	4380	700 H.P.		229.47	5,475	16,421.52	49,300	5,120	
	4400	Loader, crawler, torque conv., diesel, 1-1/2 C.Y., 80 H.P.		29.45	550	1,651	4,950	565.80	
	4450	1-1/2 to 1-3/4 C.Y., 95 H.P.		30.18	695	2,091.42	6,275	659.75	
	4510	1-3/4 to 2-1/4 C.Y., 130 H.P.		47.61	965	2,900	8,700	960.90	
	4530	2-1/2 to 3-1/4 C.Y., 190 H.P.		57.61	1,175	3,540	10,600	1,169	
	4560	3-1/2 to 5 C.Y., 275 H.P.		71.19	1,525	4,595.96	13,800	1,489	
	4610	Front end loader, 4WD, articulated frame, diesel, 1 to 1-1/4 C.Y., 70 H.P.	↓	16.58	282	846.90	2,550	302	

01 54 33 | Equipment Rental

		01 54 33 Equipment Rental		UNIT	HOURLY OPER. COST	RENT PER DAY	RENT PER WEEK	RENT PER MONTH	EQUIPMENT COST/DAY	
20	4620	1-1/2 to 1-3/4 C.Y., 95 H.P.		Ea.	19.94	440	1,320	3,950	423.50	20
	4650	1-3/4 to 2 C.Y., 130 H.P.			21.00	395	1,187.72	3,575	405.55	
	4710	2-1/2 to 3-1/2 C.Y., 145 H.P.			29.44	780	2,333	7,000	702.10	
	4730	3 to 4-1/2 C.Y., 185 H.P.			31.99	890	2,667	8,000	789.35	
	4760	5-1/4 to 5-3/4 C.Y., 270 H.P.			53.03	890	2,666	8,000	957.50	
	4810	7 to 9 C.Y., 475 H.P.			90.90	2,550	7,667	23,000	2,261	
	4870	9 to 11 C.Y., 620 H.P.			131.52	2,700	8,107.48	24,300	2,674	
	4880	Skid-steer loader, wheeled, 10 C.F., 30 H.P. gas			9.54	169	506.07	1,525	177.55	
	4890	1 C.Y., 78 H.P., diesel			18.38	420	1,265.18	3,800	400.10	
	4892	Skid-steer attachment, auger			.74	145	433.50	1,300	92.65	
	4893	Backhoe			.74	122	366.64	1,100	79.25	
	4894	Broom			.70	140	420.25	1,250	89.70	
	4895	Forks			.15	32	96	288	20.45	
	4896	Grapple			.72	88.50	265.25	795	58.80	
	4897	Concrete hammer			1.05	183	550	1,650	118.40	
	4898	Tree spade			.60	103	309.84	930	66.75	
	4899	Trencher			.65	102	305	915	66.20	
	4900	Trencher, chain, boom type, gas, operator walking, 12 H.P.			4.16	206	618.25	1,850	156.95	
	4910	Operator riding, 40 H.P.			16.64	450	1,343.75	4,025	401.85	
	5000	Wheel type, diesel, 4′ deep, 12″ wide			68.50	965	2,891.84	8,675	1,126	
	5100	6′ deep, 20″ wide			87.32	1,050	3,127.50	9,375	1,324	
	5150	Chain type, diesel, 5′ deep, 8″ wide			16.25	360	1,084.44	3,250	346.90	
	5200	Diesel, 8′ deep, 16″ wide			89.39	1,925	5,783.68	17,400	1,872	
	5202	Rock trencher, wheel type, 6″ wide x 18″ deep			46.98	90	270	810	429.85	
	5206	Chain type, 18″ wide x 7′ deep			104.38	283	850	2,550	1,005	
	5210	Tree spade, self-propelled			13.65	230	690	2,075	247.20	
	5250	Truck, dump, 2-axle, 12 ton, 8 C.Y. payload, 220 H.P.			23.88	395	1,185	3,550	428.05	
	5300	Three axle dump, 16 ton, 12 C.Y. payload, 400 H.P.			44.50	360	1,084.44	3,250	572.90	
	5310	Four axle dump, 25 ton, 18 C.Y. payload, 450 H.P.			49.85	525	1,575.02	4,725	713.80	
	5350	Dump trailer only, rear dump, 16-1/2 C.Y.			5.73	151	454.43	1,375	136.70	
	5400	20 C.Y.			6.18	170	511.24	1,525	151.70	
	5450	Flatbed, single axle, 1-1/2 ton rating			19.00	73.50	221.02	665	196.15	
	5500	3 ton rating			23.05	1,050	3,180	9,550	820.40	
	5550	Off highway rear dump, 25 ton capacity			62.67	1,475	4,389.40	13,200	1,379	
	5600	35 ton capacity			66.90	665	2,000	6,000	935.20	
	5610	50 ton capacity			83.87	1,825	5,499.66	16,500	1,771	
	5620	65 ton capacity			89.56	2,000	5,990.24	18,000	1,915	
	5630	100 ton capacity			121.24	2,950	8,830.44	26,500	2,736	
	6000	Vibratory plow, 25 H.P., walking		↓	6.77	300	900	2,700	234.20	
40	0010	**GENERAL EQUIPMENT RENTAL** without operators	R015433 -10							40
	0020	Aerial lift, scissor type, to 20′ high, 1200 lb. capacity, electric		Ea.	3.48	129	385.75	1,150	105	
	0030	To 30′ high, 1200 lb. capacity			3.77	203	607.67	1,825	151.70	
	0040	Over 30′ high, 1500 lb. capacity			5.13	243	727.50	2,175	186.60	
	0070	Articulating boom, to 45′ high, 500 lb. capacity, diesel	R015433 -15		9.92	250	750	2,250	229.35	
	0075	To 60′ high, 500 lb. capacity			13.66	300	900	2,700	289.25	
	0080	To 80′ high, 500 lb. capacity			16.05	900	2,702.25	8,100	668.85	
	0085	To 125′ high, 500 lb. capacity			18.34	1,525	4,603.50	13,800	1,067	
	0100	Telescoping boom to 40′ high, 500 lb. capacity, diesel			11.24	315	945	2,825	278.90	
	0105	To 45′ high, 500 lb. capacity			12.51	320	965	2,900	293.05	
	0110	To 60′ high, 500 lb. capacity			16.36	300	900	2,700	310.90	
	0115	To 80′ high, 500 lb. capacity			21.27	355	1,067	3,200	383.55	
	0120	To 100′ high, 500 lb. capacity			28.71	865	2,587.75	7,775	747.25	
	0125	To 120′ high, 500 lb. capacity			29.16	1,450	4,348.50	13,000	1,103	
	0195	Air compressor, portable, 6.5 CFM, electric			.90	44.50	133	400	33.80	
	0196	Gasoline			.65	56	167.50	505	38.70	
	0200	Towed type, gas engine, 60 CFM			9.43	129	387.50	1,175	152.95	
	0300	160 CFM		↓	10.47	198	595	1,775	202.80	

01 54 | Construction Aids

	01 54 33 | Equipment Rental		UNIT	HOURLY OPER. COST	RENT PER DAY	RENT PER WEEK	RENT PER MONTH	EQUIPMENT COST/DAY	
40	0400	Diesel engine, rotary screw, 250 CFM	Ea.	12.08	175	524	1,575	201.50	40
	0500	365 CFM		16.00	310	937	2,800	315.40	
	0550	450 CFM		19.95	277	832.25	2,500	326.05	
	0600	600 CFM		34.10	248	743.62	2,225	421.50	
	0700	750 CFM		34.62	435	1,306.50	3,925	538.25	
	0930	Air tools, breaker, pavement, 60 lb.		.57	81.50	245	735	53.50	
	0940	80 lb.		.56	81	242.50	730	53	
	0950	Drills, hand (jackhammer), 65 lb.		.67	68	203.50	610	46.05	
	0960	Track or wagon, swing boom, 4" drifter		54.66	1,025	3,104	9,300	1,058	
	0970	5" drifter		63.30	1,025	3,104	9,300	1,127	
	0975	Track mounted quarry drill, 6" diameter drill		101.91	1,900	5,665	17,000	1,948	
	0980	Dust control per drill		1.04	25	75.50	227	23.40	
	0990	Hammer, chipping, 12 lb.		.60	46	138	415	32.40	
	1000	Hose, air with couplings, 50' long, 3/4" diameter		.07	12	36	108	7.75	
	1100	1" diameter		.08	12.35	37	111	8	
	1200	1-1/2" diameter		.22	37.50	112.50	340	24.25	
	1300	2" diameter		.24	45	135	405	28.90	
	1400	2-1/2" diameter		.36	57.50	172.50	520	37.35	
	1410	3" diameter		.42	58.50	175	525	38.35	
	1450	Drill, steel, 7/8" x 2'		.08	12.90	38.73	116	8.40	
	1460	7/8" x 6'		.12	19.60	58.87	177	12.70	
	1520	Moil points		.03	7	21	63	4.40	
	1525	Pneumatic nailer w/accessories		.48	39.50	118	355	27.40	
	1530	Sheeting driver for 60 lb. breaker		.04	7.75	23.24	69.50	5	
	1540	For 90 lb. breaker		.13	10.50	31.50	94.50	7.35	
	1550	Spade, 25 lb.		.50	7.40	22.21	66.50	8.45	
	1560	Tamper, single, 35 lb.		.59	48.50	145.75	435	33.85	
	1570	Triple, 140 lb.		.89	61.50	184.87	555	44.05	
	1580	Wrenches, impact, air powered, up to 3/4" bolt		.43	49.50	148.25	445	33.05	
	1590	Up to 1-1/4" bolt		.58	79.50	238.50	715	52.30	
	1600	Barricades, barrels, reflectorized, 1 to 99 barrels		.03	4	12	36	2.65	
	1610	100 to 200 barrels		.02	4.41	13.22	39.50	2.85	
	1620	Barrels with flashers, 1 to 99 barrels		.03	6.40	19.16	57.50	4.10	
	1630	100 to 200 barrels		.03	5.10	15.34	46	3.30	
	1640	Barrels with steady burn type C lights		.05	8.45	25.30	76	5.45	
	1650	Illuminated board, trailer mounted, with generator		3.28	139	418.28	1,250	109.85	
	1670	Portable barricade, stock, with flashers, 1 to 6 units		.03	6.35	19.11	57.50	4.10	
	1680	25 to 50 units		.03	5.95	17.82	53.50	3.85	
	1685	Butt fusion machine, wheeled, 1.5 HP electric, 2" - 8" diameter pipe		2.63	225	675	2,025	156.05	
	1690	Tracked, 20 HP diesel, 4"-12" diameter pipe		11.23	560	1,685	5,050	426.85	
	1695	83 HP diesel, 8" - 24" diameter pipe		51.32	1,100	3,325	9,975	1,076	
	1700	Carts, brick, gas engine, 1000 lb. capacity		2.94	65	195	585	62.55	
	1800	1500 lb., 7-1/2' lift		2.92	69.50	208	625	64.95	
	1822	Dehumidifier, medium, 6 lb./hr., 150 CFM		1.19	76.50	229.28	690	55.35	
	1824	Large, 18 lb./hr., 600 CFM		2.19	585	1,750	5,250	367.55	
	1830	Distributor, asphalt, trailer mounted, 2000 gal., 38 H.P. diesel		10.99	355	1,058.62	3,175	299.65	
	1840	3000 gal., 38 H.P. diesel		12.87	380	1,136.08	3,400	330.15	
	1850	Drill, rotary hammer, electric		1.11	71.50	214	640	51.70	
	1860	Carbide bit, 1-1/2" diameter, add to electric rotary hammer		.03	41.50	125	375	25.25	
	1865	Rotary, crawler, 250 H.P.		135.77	2,300	6,868.12	20,600	2,460	
	1870	Emulsion sprayer, 65 gal., 5 H.P. gas engine		2.77	107	320.17	960	86.15	
	1880	200 gal., 5 H.P. engine		7.22	179	537.06	1,600	165.20	
	1900	Floor auto-scrubbing machine, walk-behind, 28" path		5.62	222	667	2,000	178.40	
	1930	Floodlight, mercury vapor, or quartz, on tripod, 1000 watt		.46	36.50	110	330	25.65	
	1940	2000 watt		.59	28	84.69	254	21.65	
	1950	Floodlights, trailer mounted with generator, 1 - 300 watt light		3.54	78.50	235.48	705	75.45	
	1960	2 - 1000 watt lights		4.49	87.50	262.33	785	88.35	
	2000	4 - 300 watt lights		4.24	100	299.51	900	93.85	

01 54 33 | Equipment Rental

			UNIT	HOURLY OPER. COST	RENT PER DAY	RENT PER WEEK	RENT PER MONTH	EQUIPMENT COST/DAY	
40	2005	Foam spray rig, incl. box trailer, compressor, generator, proportioner	Ea.	25.46	535	1,600.84	4,800	523.85	40
	2015	Forklift, pneumatic tire, rough terr, straight mast, 5000 lb, 12′ lift, gas		18.59	219	655.83	1,975	279.90	
	2025	8000 lb, 12′ lift		22.68	360	1,084.44	3,250	398.35	
	2030	5000 lb, 12′ lift, diesel		15.41	244	733.29	2,200	269.90	
	2035	8000 lb, 12′ lift, diesel		16.70	277	831.40	2,500	299.90	
	2045	All terrain, telescoping boom, diesel, 5000 lb, 10′ reach, 19′ lift		17.20	233	700	2,100	277.60	
	2055	6600 lb, 29′ reach, 42′ lift		21.04	233	700	2,100	308.30	
	2065	10,000 lb, 31′ reach, 45′ lift		23.03	315	950	2,850	374.20	
	2070	Cushion tire, smooth floor, gas, 5000 lb capacity		8.23	247	741.50	2,225	214.10	
	2075	8000 lb capacity		11.33	275	826.25	2,475	255.90	
	2085	Diesel, 5000 lb capacity		7.73	210	629.25	1,900	187.65	
	2090	12,000 lb capacity		12.01	400	1,194.50	3,575	335	
	2095	20,000 lb capacity		17.20	660	1,980	5,950	533.60	
	2100	Generator, electric, gas engine, 1.5 kW to 3 kW		2.57	46.50	140	420	48.55	
	2200	5 kW		3.21	91	272.50	820	80.15	
	2300	10 kW		5.91	108	322.50	970	111.80	
	2400	25 kW		7.38	405	1,210	3,625	301.10	
	2500	Diesel engine, 20 kW		9.18	229	687.50	2,075	210.95	
	2600	50 kW		15.90	370	1,110	3,325	349.20	
	2700	100 kW		28.51	445	1,340.50	4,025	496.20	
	2800	250 kW		54.19	750	2,249.33	6,750	883.40	
	2850	Hammer, hydraulic, for mounting on boom, to 500 ft lb.		2.89	93.50	279.89	840	79.10	
	2860	1000 ft lb.		4.59	139	418.28	1,250	120.40	
	2900	Heaters, space, oil or electric, 50 MBH		1.46	46.50	140	420	39.65	
	3000	100 MBH		2.71	46.50	140	420	49.70	
	3100	300 MBH		7.90	135	405	1,225	144.20	
	3150	500 MBH		13.12	200	600	1,800	224.95	
	3200	Hose, water, suction with coupling, 20′ long, 2″ diameter		.02	5.65	17	51	3.55	
	3210	3″ diameter		.03	14.15	42.50	128	8.75	
	3220	4″ diameter		.03	28	84	252	17.05	
	3230	6″ diameter		.11	40.50	121.50	365	25.20	
	3240	8″ diameter		.27	53.50	160	480	34.15	
	3250	Discharge hose with coupling, 50′ long, 2″ diameter		.01	6.50	19.50	58.50	4	
	3260	3″ diameter		.01	7.35	22	66	4.50	
	3270	4″ diameter		.02	21	62.50	188	12.65	
	3280	6″ diameter		.06	29	87	261	17.90	
	3290	8″ diameter		.24	37.50	112.50	340	24.40	
	3295	Insulation blower		.83	117	350	1,050	76.65	
	3300	Ladders, extension type, 16′ to 36′ long		.18	41.50	125	375	26.45	
	3400	40′ to 60′ long		.64	120	360.50	1,075	77.20	
	3405	Lance for cutting concrete		2.20	65	195	585	56.60	
	3407	Lawn mower, rotary, 22″, 5 H.P.		1.05	38.50	115	345	31.40	
	3408	48″ self-propelled		2.89	138	415	1,250	106.10	
	3410	Level, electronic, automatic, with tripod and leveling rod		1.05	37.50	112	335	30.80	
	3430	Laser type, for pipe and sewer line and grade		2.17	117	350	1,050	87.35	
	3440	Rotating beam for interior control		.90	64	192.50	580	45.70	
	3460	Builder's optical transit, with tripod and rod		.10	37.50	112	335	23.20	
	3500	Light towers, towable, with diesel generator, 2000 watt		4.25	101	303.64	910	94.75	
	3600	4000 watt		4.50	165	495	1,475	135	
	3700	Mixer, powered, plaster and mortar, 6 C.F., 7 H.P.		2.05	83.50	250	750	66.40	
	3800	10 C.F., 9 H.P.		2.24	124	372.50	1,125	92.40	
	3850	Nailer, pneumatic		.48	33.50	100.18	300	23.85	
	3900	Paint sprayers complete, 8 CFM		.85	61.50	184.87	555	43.75	
	4000	17 CFM		1.60	110	330.50	990	78.85	
	4020	Pavers, bituminous, rubber tires, 8′ wide, 50 H.P., diesel		31.93	570	1,704.12	5,100	596.25	
	4030	10′ wide, 150 H.P.		95.62	1,950	5,835.32	17,500	1,932	
	4050	Crawler, 8′ wide, 100 H.P., diesel		87.59	2,050	6,170.98	18,500	1,935	
	4060	10′ wide, 150 H.P.	↓	103.97	2,350	7,048.86	21,100	2,241	

01 54 33 | Equipment Rental

			UNIT	HOURLY OPER. COST	RENT PER DAY	RENT PER WEEK	RENT PER MONTH	EQUIPMENT COST/DAY	
40	4070	Concrete paver, 12′ to 24′ wide, 250 H.P.	Ea.	87.62	1,675	5,060.72	15,200	1,713	40
	4080	Placer-spreader-trimmer, 24′ wide, 300 H.P.		117.51	2,550	7,668.54	23,000	2,474	
	4100	Pump, centrifugal gas pump, 1-1/2″ diam., 65 GPM		3.92	54	162.15	485	63.80	
	4200	2″ diameter, 130 GPM		4.98	44	132.50	400	66.35	
	4300	3″ diameter, 250 GPM		5.12	56	167.50	505	74.45	
	4400	6″ diameter, 1500 GPM		22.24	91.50	275	825	232.90	
	4500	Submersible electric pump, 1-1/4″ diameter, 55 GPM		.40	36	107.50	325	24.70	
	4600	1-1/2″ diameter, 83 GPM		.44	43.50	130	390	29.55	
	4700	2″ diameter, 120 GPM		1.64	61.50	185	555	50.15	
	4800	3″ diameter, 300 GPM		3.03	109	327.50	985	89.75	
	4900	4″ diameter, 560 GPM		14.75	61.50	185	555	155	
	5000	6″ diameter, 1590 GPM		22.08	65	195	585	215.60	
	5100	Diaphragm pump, gas, single, 1-1/2″ diameter		1.13	38.50	115	345	32	
	5200	2″ diameter		3.98	91.50	275	825	86.80	
	5300	3″ diameter		4.05	95	285	855	89.40	
	5400	Double, 4″ diameter		6.03	95	285	855	105.25	
	5450	Pressure washer 5 GPM, 3000 psi		3.87	110	330	990	96.95	
	5460	7 GPM, 3000 psi		4.94	85	255	765	90.50	
	5470	High pressure water jet 10 ksi		39.55	720	2,160	6,475	748.40	
	5480	40 ksi		27.88	980	2,940	8,825	811.05	
	5500	Trash pump, self-priming, gas, 2″ diameter		3.82	108	325	975	95.55	
	5600	Diesel, 4″ diameter		6.68	162	485	1,450	150.40	
	5650	Diesel, 6″ diameter		16.85	162	485	1,450	231.80	
	5655	Grout Pump		18.70	281	841.73	2,525	317.90	
	5700	Salamanders, L.P. gas fired, 100,000 BTU		2.88	57.50	172.50	520	57.55	
	5705	50,000 BTU		1.66	23.50	71	213	27.50	
	5720	Sandblaster, portable, open top, 3 C.F. capacity		.60	132	395	1,175	83.80	
	5730	6 C.F. capacity		1.00	132	395	1,175	87	
	5740	Accessories for above		.14	24	71.26	214	15.35	
	5750	Sander, floor		.77	73.50	220	660	50.15	
	5760	Edger		.52	35	105	315	25.15	
	5800	Saw, chain, gas engine, 18″ long		1.75	63.50	190	570	52	
	5900	Hydraulic powered, 36″ long		.78	58.50	175	525	41.25	
	5950	60″ long		.78	65	195	585	45.25	
	6000	Masonry, table mounted, 14″ diameter, 5 H.P.		1.32	76.50	230	690	56.55	
	6050	Portable cut-off, 8 H.P.		1.81	77.50	232.50	700	61	
	6100	Circular, hand held, electric, 7-1/4″ diameter		.23	13.85	41.50	125	10.10	
	6200	12″ diameter		.24	41	122.50	370	26.40	
	6250	Wall saw, w/hydraulic power, 10 H.P.		3.29	98.50	296	890	85.50	
	6275	Shot blaster, walk-behind, 20″ wide		4.73	281	841.73	2,525	206.20	
	6280	Sidewalk broom, walk-behind		2.24	82.50	247.87	745	67.50	
	6300	Steam cleaner, 100 gallons per hour		3.34	82.50	247.87	745	76.30	
	6310	200 gallons per hour		4.33	100	299.51	900	94.55	
	6340	Tar Kettle/Pot, 400 gallons		16.48	127	380	1,150	207.85	
	6350	Torch, cutting, acetylene-oxygen, 150′ hose, excludes gases		.45	15.30	45.96	138	12.80	
	6360	Hourly operating cost includes tips and gas		20.92	7	20.98	63	171.55	
	6410	Toilet, portable chemical		.13	23	69.20	208	14.90	
	6420	Recycle flush type		.16	28.50	85.72	257	18.45	
	6430	Toilet, fresh water flush, garden hose,		.19	34	102.25	305	22	
	6440	Hoisted, non-flush, for high rise		.16	28	83.66	251	18	
	6465	Tractor, farm with attachment		17.37	390	1,172.50	3,525	373.40	
	6480	Trailers, platform, flush deck, 2 axle, 3 ton capacity		1.69	94.50	284	850	70.30	
	6500	25 ton capacity		6.23	143	428.61	1,275	135.55	
	6600	40 ton capacity		8.04	203	609.35	1,825	186.20	
	6700	3 axle, 50 ton capacity		8.72	225	676.48	2,025	205.05	
	6800	75 ton capacity		11.08	300	898.54	2,700	268.30	
	6810	Trailer mounted cable reel for high voltage line work		5.89	28.50	85	255	64.10	
	6820	Trailer mounted cable tensioning rig	↓	11.67	28.50	85	255	110.40	

01 54 33 | Equipment Rental

		01 54 33 | Equipment Rental	UNIT	HOURLY OPER. COST	RENT PER DAY	RENT PER WEEK	RENT PER MONTH	EQUIPMENT COST/DAY
40	6830	Cable pulling rig	Ea.	73.77	28.50	85	255	607.15
	6850	Portable cable/wire puller, 8000 lb max pulling capacity		3.70	120	360	1,075	101.60
	6900	Water tank trailer, engine driven discharge, 5000 gallons		7.16	158	475.09	1,425	152.25
	6925	10,000 gallons		9.75	215	645.50	1,925	207.10
	6950	Water truck, off highway, 6000 gallons		71.75	835	2,504.54	7,525	1,075
	7010	Tram car for high voltage line work, powered, 2 conductor		6.88	28.50	85	255	72.05
	7020	Transit (builder's level) with tripod		.10	17.55	52.67	158	11.30
	7030	Trench box, 3000 lb., 6′ x 8′		.56	96.50	290.22	870	62.50
	7040	7200 lb., 6′ x 20′		.72	187	560	1,675	117.75
	7050	8000 lb., 8′ x 16′		1.08	186	557.71	1,675	120.15
	7060	9500 lb., 8′ x 20′		1.20	232	697.14	2,100	149.05
	7065	11,000 lb., 8′ x 24′		1.26	219	655.83	1,975	141.25
	7070	12,000 lb., 10′ x 20′		1.49	263	790.09	2,375	169.95
	7100	Truck, pickup, 3/4 ton, 2 wheel drive		9.24	61.50	184.87	555	110.85
	7200	4 wheel drive		9.48	167	500	1,500	175.85
	7250	Crew carrier, 9 passenger		12.66	108	325	975	166.25
	7290	Flat bed truck, 20,000 lb. GVW		15.26	133	397.63	1,200	201.60
	7300	Tractor, 4 x 2, 220 H.P.		22.25	215	645.50	1,925	307.10
	7410	330 H.P.		32.33	294	883.04	2,650	435.25
	7500	6 x 4, 380 H.P.		36.09	340	1,022.47	3,075	493.25
	7600	450 H.P.		44.23	415	1,239.36	3,725	601.75
	7610	Tractor, with A frame, boom and winch, 225 H.P.		24.74	293	877.88	2,625	373.50
	7620	Vacuum truck, hazardous material, 2500 gallons		12.79	310	929.52	2,800	288.25
	7625	5,000 gallons		13.02	440	1,316.82	3,950	367.55
	7650	Vacuum, HEPA, 16 gallon, wet/dry		.85	122	365	1,100	79.80
	7655	55 gallon, wet/dry		.78	25.50	76.50	230	21.50
	7660	Water tank, portable		.73	160	480.25	1,450	101.90
	7690	Sewer/catch basin vacuum, 14 C.Y., 1500 gallons		17.31	665	1,988.14	5,975	536.15
	7700	Welder, electric, 200 amp		3.81	33.50	100	300	50.50
	7800	300 amp		5.55	103	310	930	106.40
	7900	Gas engine, 200 amp		8.95	58.50	175	525	106.55
	8000	300 amp		10.13	110	330	990	147
	8100	Wheelbarrow, any size		.06	11.15	33.50	101	7.20
	8200	Wrecking ball, 4000 lb.	↓	2.50	60	180	540	56
50	0010	**HIGHWAY EQUIPMENT RENTAL** without operators (R015433-10)						
	0050	Asphalt batch plant, portable drum mixer, 100 ton/hr.	Ea.	88.41	1,550	4,621.78	13,900	1,632
	0060	200 ton/hr.		101.99	1,650	4,931.62	14,800	1,802
	0070	300 ton/hr.		119.86	1,925	5,783.68	17,400	2,116
	0100	Backhoe attachment, long stick, up to 185 H.P., 10.5′ long		.37	25.50	76.43	229	18.25
	0140	Up to 250 H.P., 12′ long		.41	28.50	85.72	257	20.45
	0180	Over 250 H.P., 15′ long		.56	39	116.71	350	27.85
	0200	Special dipper arm, up to 100 H.P., 32′ long		1.16	79.50	238.58	715	56.95
	0240	Over 100 H.P., 33′ long		1.44	100	299.51	900	71.45
	0280	Catch basin/sewer cleaning truck, 3 ton, 9 C.Y., 1000 gal.		35.39	420	1,265.18	3,800	536.15
	0300	Concrete batch plant, portable, electric, 200 C.Y./hr.		24.18	560	1,678.30	5,025	529.15
	0520	Grader/dozer attachment, ripper/scarifier, rear mounted, up to 135 H.P.		3.15	63.50	190.04	570	63.20
	0540	Up to 180 H.P.		4.13	95.50	287.12	860	90.50
	0580	Up to 250 H.P.		5.85	153	459.60	1,375	138.75
	0700	Pvmt. removal bucket, for hyd. excavator, up to 90 H.P.		2.16	58	174.54	525	52.20
	0740	Up to 200 H.P.		2.31	74.50	223.08	670	63.05
	0780	Over 200 H.P.		2.52	91	273.69	820	74.90
	0900	Aggregate spreader, self-propelled, 187 H.P.		50.60	740	2,220.52	6,650	848.90
	1000	Chemical spreader, 3 C.Y.		3.17	96.50	290	870	83.35
	1900	Hammermill, traveling, 250 H.P.		67.35	515	1,550	4,650	848.80
	2000	Horizontal borer, 3″ diameter, 13 H.P. gas driven		5.42	232	695	2,075	182.35
	2150	Horizontal directional drill, 20,000 lb. thrust, 78 H.P. diesel		27.58	530	1,590	4,775	538.65
	2160	30,000 lb. thrust, 115 H.P.		33.90	615	1,850	5,550	641.20
	2170	50,000 lb. thrust, 170 H.P.	↓	48.60	710	2,135	6,400	815.80

01 54 33 | Equipment Rental

		01 54 33 Equipment Rental	UNIT	HOURLY OPER. COST	RENT PER DAY	RENT PER WEEK	RENT PER MONTH	EQUIPMENT COST/DAY	
50	2190	Mud trailer for HDD, 1500 gallons, 175 H.P., gas	Ea.	25.50	175	525	1,575	309	50
	2200	Hydromulcher, diesel, 3000 gallon, for truck mounting		17.43	227	680	2,050	275.45	
	2300	Gas, 600 gallon		7.49	95	285	855	116.95	
	2400	Joint & crack cleaner, walk behind, 25 H.P.		3.16	45.50	136	410	52.45	
	2500	Filler, trailer mounted, 400 gallons, 20 H.P.		8.35	147	440	1,325	154.75	
	3000	Paint striper, self-propelled, 40 gallon, 22 H.P.		6.76	122	365	1,100	127.10	
	3100	120 gallon, 120 H.P.		19.23	380	1,140	3,425	381.80	
	3200	Post drivers, 6" I-Beam frame, for truck mounting		12.41	320	960	2,875	291.30	
	3400	Road sweeper, self-propelled, 8' wide, 90 H.P.		35.91	715	2,143.06	6,425	715.85	
	3450	Road sweeper, vacuum assisted, 4 C.Y., 220 gallons		58.28	670	2,013.96	6,050	869	
	4000	Road mixer, self-propelled, 130 H.P.		46.23	825	2,478.72	7,425	865.60	
	4100	310 H.P.		75.01	2,150	6,480.82	19,400	1,896	
	4220	Cold mix paver, incl. pug mill and bitumen tank, 165 H.P.		94.97	2,325	6,945.58	20,800	2,149	
	4240	Pavement brush, towed		3.43	100	299.51	900	87.30	
	4250	Paver, asphalt, wheel or crawler, 130 H.P., diesel		94.23	2,275	6,816.48	20,400	2,117	
	4300	Paver, road widener, gas, 1' to 6', 67 H.P.		46.66	975	2,917.66	8,750	956.80	
	4400	Diesel, 2' to 14', 88 H.P.		56.38	1,150	3,459.88	10,400	1,143	
	4600	Slipform pavers, curb and gutter, 2 track, 75 H.P.		57.83	1,250	3,769.72	11,300	1,217	
	4700	4 track, 165 H.P.		35.69	845	2,530.36	7,600	791.55	
	4800	Median barrier, 215 H.P.		58.43	1,350	4,027.92	12,100	1,273	
	4901	Trailer, low bed, 75 ton capacity		10.71	282	846.90	2,550	255.05	
	5000	Road planer, walk behind, 10" cutting width, 10 H.P.		2.45	243	730	2,200	165.65	
	5100	Self-propelled, 12" cutting width, 64 H.P.		8.26	190	570	1,700	180.05	
	5120	Traffic line remover, metal ball blaster, truck mounted, 115 H.P.		46.56	905	2,720	8,150	916.50	
	5140	Grinder, truck mounted, 115 H.P.		50.89	905	2,720	8,150	951.15	
	5160	Walk-behind, 11 H.P.		3.56	142	425	1,275	113.45	
	5200	Pavement profiler, 4' to 6' wide, 450 H.P.		216.58	1,275	3,800	11,400	2,493	
	5300	8' to 10' wide, 750 H.P.		331.58	1,325	3,975	11,900	3,448	
	5400	Roadway plate, steel, 1" x 8' x 20'		.09	61	182.50	550	37.20	
	5600	Stabilizer, self-propelled, 150 H.P.		41.14	1,025	3,100	9,300	949.10	
	5700	310 H.P.		76.18	1,300	3,900	11,700	1,389	
	5800	Striper, truck mounted, 120 gallon paint, 460 H.P.		48.74	340	1,015	3,050	592.95	
	5900	Thermal paint heating kettle, 115 gallons		7.71	61.50	185	555	98.65	
	6000	Tar kettle, 330 gallon, trailer mounted		12.27	96.50	290	870	156.20	
	7000	Tunnel locomotive, diesel, 8 to 12 ton		29.76	620	1,859.04	5,575	609.85	
	7005	Electric, 10 ton		29.25	705	2,117.24	6,350	657.40	
	7010	Muck cars, 1/2 C.Y. capacity		2.30	26.50	80.04	240	34.40	
	7020	1 C.Y. capacity		2.51	35	104.31	315	40.95	
	7030	2 C.Y. capacity		2.66	39	116.71	350	44.60	
	7040	Side dump, 2 C.Y. capacity		2.87	48	144.59	435	51.90	
	7050	3 C.Y. capacity		3.85	53	159.05	475	62.65	
	7060	5 C.Y. capacity		5.62	68.50	205.53	615	86.10	
	7100	Ventilating blower for tunnel, 7-1/2 H.P.		2.14	52.50	158.02	475	48.70	
	7110	10 H.P.		2.42	55	165.25	495	52.40	
	7120	20 H.P.		3.54	71.50	214.82	645	71.30	
	7140	40 H.P.		6.14	94.50	284.02	850	105.90	
	7160	60 H.P.		8.69	102	304.68	915	130.45	
	7175	75 H.P.		10.37	158	475.09	1,425	177.95	
	7180	200 H.P.		20.78	310	934.68	2,800	353.20	
	7800	Windrow loader, elevating		53.94	1,650	4,975	14,900	1,427	
60	0010	**LIFTING AND HOISTING EQUIPMENT RENTAL** without operators R015433-10							60
	0150	Crane, flatbed mounted, 3 ton capacity	Ea.	14.41	201	604.19	1,825	236.10	
	0200	Crane, climbing, 106' jib, 6000 lb. capacity, 410 fpm R312316-45		39.72	2,600	7,800	23,400	1,878	
	0300	101' jib, 10,250 lb. capacity, 270 fpm		46.43	2,275	6,800	20,400	1,731	
	0500	Tower, static, 130' high, 106' jib, 6200 lb. capacity at 400 fpm		45.16	2,250	6,715	20,100	1,704	
	0520	Mini crawler spider crane, up to 24" wide, 1990 lb. lifting capacity		12.50	550	1,652.48	4,950	430.50	
	0525	Up to 30" wide, 6450 lb. lifting capacity		14.52	655	1,962.32	5,875	508.65	
	0530	Up to 52" wide, 6680 lb. lifting capacity		23.10	800	2,401.26	7,200	665.05	

01 54 33 | Equipment Rental

		Unit	Hourly Oper. Cost	Rent Per Day	Rent Per Week	Rent Per Month	Equipment Cost/Day
60 0535	Up to 55″ wide, 8920 lb. lifting capacity	Ea.	25.79	885	2,659.46	7,975	738.25 60
0540	Up to 66″ wide, 13,350 lb. lifting capacity		34.92	1,375	4,131.20	12,400	1,106
0600	Crawler mounted, lattice boom, 1/2 C.Y., 15 tons at 12′ radius		36.96	830	2,483	7,450	792.30
0700	3/4 C.Y., 20 tons at 12′ radius		50.42	930	2,790	8,375	961.35
0800	1 C.Y., 25 tons at 12′ radius		67.42	985	2,950	8,850	1,129
0900	1-1/2 C.Y., 40 tons at 12′ radius		66.31	1,125	3,375	10,100	1,206
1000	2 C.Y., 50 tons at 12′ radius		88.77	1,325	4,000	12,000	1,510
1100	3 C.Y., 75 tons at 12′ radius		75.26	2,325	7,000	21,000	2,002
1200	100 ton capacity, 60′ boom		85.91	2,675	8,000	24,000	2,287
1300	165 ton capacity, 60′ boom		106.10	3,000	9,000	27,000	2,649
1400	200 ton capacity, 70′ boom		138.21	3,825	11,500	34,500	3,406
1500	350 ton capacity, 80′ boom		182.20	4,175	12,500	37,500	3,958
1600	Truck mounted, lattice boom, 6 x 4, 20 tons at 10′ radius		39.76	1,950	5,850	17,600	1,488
1700	25 tons at 10′ radius		42.73	2,325	7,000	21,000	1,742
1800	8 x 4, 30 tons at 10′ radius		45.54	2,500	7,500	22,500	1,864
1900	40 tons at 12′ radius		48.55	2,725	8,200	24,600	2,028
2000	60 tons at 15′ radius		53.69	1,650	4,950	14,900	1,419
2050	82 tons at 15′ radius		59.43	1,775	5,350	16,100	1,545
2100	90 tons at 15′ radius		66.39	1,950	5,825	17,500	1,696
2200	115 tons at 15′ radius		74.90	2,175	6,525	19,600	1,904
2300	150 tons at 18′ radius		81.09	2,700	8,100	24,300	2,269
2350	165 tons at 18′ radius		87.03	2,425	7,275	21,800	2,151
2400	Truck mounted, hydraulic, 12 ton capacity		29.50	390	1,175	3,525	471
2500	25 ton capacity		36.36	485	1,450	4,350	580.85
2550	33 ton capacity		50.67	900	2,700	8,100	945.35
2560	40 ton capacity		49.47	900	2,700	8,100	935.80
2600	55 ton capacity		53.78	915	2,750	8,250	980.20
2700	80 ton capacity		75.71	1,475	4,400	13,200	1,486
2720	100 ton capacity		74.96	1,550	4,675	14,000	1,535
2740	120 ton capacity		102.81	1,825	5,500	16,500	1,922
2760	150 ton capacity		109.92	2,050	6,125	18,400	2,104
2800	Self-propelled, 4 x 4, with telescoping boom, 5 ton		15.14	430	1,285	3,850	378.10
2900	12-1/2 ton capacity		21.42	430	1,285	3,850	428.30
3000	15 ton capacity		34.42	450	1,350	4,050	545.35
3050	20 ton capacity		24.02	650	1,950	5,850	582.20
3100	25 ton capacity		36.69	1,425	4,250	12,800	1,144
3150	40 ton capacity		44.90	660	1,975	5,925	754.20
3200	Derricks, guy, 20 ton capacity, 60′ boom, 75′ mast		22.74	1,425	4,250	12,800	1,032
3300	100′ boom, 115′ mast		36.04	2,000	6,000	18,000	1,488
3400	Stiffleg, 20 ton capacity, 70′ boom, 37′ mast		25.41	615	1,850	5,550	573.25
3500	100′ boom, 47′ mast		39.32	665	2,000	6,000	714.55
3550	Helicopter, small, lift to 1250 lb. maximum, w/pilot		99.14	2,150	6,435	19,300	2,080
3600	Hoists, chain type, overhead, manual, 3/4 ton		.14	10.25	30.70	92	7.30
3900	10 ton		.79	6.20	18.59	56	10
4000	Hoist and tower, 5000 lb. cap., portable electric, 40′ high		5.12	142	426	1,275	126.20
4100	For each added 10′ section, add		.12	31.50	95	285	19.95
4200	Hoist and single tubular tower, 5000 lb. electric, 100′ high		6.96	105	315	945	118.65
4300	For each added 6′-6″ section, add		.21	38.50	115	345	24.65
4400	Hoist and double tubular tower, 5000 lb., 100′ high		7.57	105	315	945	123.60
4500	For each added 6′-6″ section, add		.23	41.50	125	375	26.80
4550	Hoist and tower, mast type, 6000 lb., 100′ high		8.24	94.50	284	850	122.70
4570	For each added 10′ section, add		.13	31.50	95	285	20.05
4600	Hoist and tower, personnel, electric, 2000 lb., 100′ @ 125 fpm		17.50	25	75	225	155
4700	3000 lb., 100′ @ 200 fpm		20.02	25	75	225	175.15
4800	3000 lb., 150′ @ 300 fpm		22.22	25	75	225	192.75
4900	4000 lb., 100′ @ 300 fpm		22.98	25	75	225	198.85
5000	6000 lb., 100′ @ 275 fpm		24.70	25	75	225	212.60
5100	For added heights up to 500′, add	L.F.	.01	3.33	10	30	2.10

01 54 33 | Equipment Rental

		01 54 33 Equipment Rental	UNIT	HOURLY OPER. COST	RENT PER DAY	RENT PER WEEK	RENT PER MONTH	EQUIPMENT COST/DAY	
60	5200	Jacks, hydraulic, 20 ton	Ea.	.05	19.65	59	177	12.20	60
	5500	100 ton		.40	26	78.50	236	18.90	
	6100	Jacks, hydraulic, climbing w/50′ jackrods, control console, 30 ton cap.		2.17	31	93	279	35.90	
	6150	For each added 10′ jackrod section, add		.05	5	15	45	3.40	
	6300	50 ton capacity		3.48	33.50	100	300	47.85	
	6350	For each added 10′ jackrod section, add		.06	5	15	45	3.50	
	6500	125 ton capacity		9.10	51.50	155	465	103.85	
	6550	For each added 10′ jackrod section, add		.61	5	15	45	7.90	
	6600	Cable jack, 10 ton capacity with 200′ cable		1.82	35.50	107	320	35.95	
	6650	For each added 50′ of cable, add	▼	.22	15	45	135	10.75	
70	0010	**WELLPOINT EQUIPMENT RENTAL** without operators R015433-10							70
	0020	Based on 2 months rental							
	0100	Combination jetting & wellpoint pump, 60 H.P. diesel	Ea.	15.67	298	895	2,675	304.35	
	0200	High pressure gas jet pump, 200 H.P., 300 psi	″	33.83	275	825	2,475	435.65	
	0300	Discharge pipe, 8″ diameter	L.F.	.01	1.40	4.20	12.60	.90	
	0350	12″ diameter		.01	2.07	6.20	18.60	1.35	
	0400	Header pipe, flows up to 150 GPM, 4″ diameter		.01	.73	2.20	6.60	.50	
	0500	400 GPM, 6″ diameter		.01	1.07	3.20	9.60	.70	
	0600	800 GPM, 8″ diameter		.01	1.40	4.20	12.60	.95	
	0700	1500 GPM, 10″ diameter		.01	1.73	5.20	15.60	1.15	
	0800	2500 GPM, 12″ diameter		.03	2.07	6.20	18.60	1.45	
	0900	4500 GPM, 16″ diameter		.03	2.40	7.20	21.50	1.70	
	0950	For quick coupling aluminum and plastic pipe, add	▼	.03	9.35	28	84	5.85	
	1100	Wellpoint, 25′ long, with fittings & riser pipe, 1-1/2″ or 2″ diameter	Ea.	.07	132	395	1,175	79.55	
	1200	Wellpoint pump, diesel powered, 4″ suction, 20 H.P.		7.00	150	450	1,350	146	
	1300	6″ suction, 30 H.P.		9.39	167	500	1,500	175.15	
	1400	8″ suction, 40 H.P.		12.73	250	750	2,250	251.80	
	1500	10″ suction, 75 H.P.		18.77	265	795	2,375	309.20	
	1600	12″ suction, 100 H.P.		27.24	298	895	2,675	396.90	
	1700	12″ suction, 175 H.P.	▼	38.98	315	950	2,850	501.80	
80	0010	**MARINE EQUIPMENT RENTAL** without operators R015433-10							80
	0200	Barge, 400 Ton, 30′ wide x 90′ long	Ea.	17.63	1,200	3,588.98	10,800	858.85	
	0240	800 Ton, 45′ wide x 90′ long		22.14	1,475	4,415.22	13,200	1,060	
	2000	Tugboat, diesel, 100 H.P.		29.57	238	712.63	2,150	379.10	
	2040	250 H.P.		57.41	430	1,291	3,875	717.50	
	2080	380 H.P.		124.99	1,300	3,873	11,600	1,774	
	3000	Small work boat, gas, 16-foot, 50 H.P.		11.35	48	143.56	430	119.50	
	4000	Large, diesel, 48-foot, 200 H.P.	▼	74.68	1,375	4,105.38	12,300	1,418	

Crew No.	Bare Costs		Incl. Subs O&P		Cost Per Labor-Hour	
Crew A-1	**Hr.**	**Daily**	**Hr.**	**Daily**	**Bare Costs**	**Incl. O&P**
1 Building Laborer	$42.10	$336.80	$67.45	$539.60	$42.10	$67.45
1 Concrete Saw, Gas Manual		111.55		122.71	13.94	15.34
8 L.H., Daily Totals		$448.35		$662.30	$56.04	$82.79

Crew A-1A	Hr.	Daily	Hr.	Daily	Bare Costs	Incl. O&P
1 Skilled Worker	$54.85	$438.80	$87.40	$699.20	$54.85	$87.40
1 Shot Blaster, 20"		206.20		226.82	25.77	28.35
8 L.H., Daily Totals		$645.00		$926.02	$80.63	$115.75

Crew A-1B	Hr.	Daily	Hr.	Daily	Bare Costs	Incl. O&P
1 Building Laborer	$42.10	$336.80	$67.45	$539.60	$42.10	$67.45
1 Concrete Saw		111.50		122.65	13.94	15.33
8 L.H., Daily Totals		$448.30		$662.25	$56.04	$82.78

Crew A-1C	Hr.	Daily	Hr.	Daily	Bare Costs	Incl. O&P
1 Building Laborer	$42.10	$336.80	$67.45	$539.60	$42.10	$67.45
1 Chain Saw, Gas, 18"		52.00		57.20	6.50	7.15
8 L.H., Daily Totals		$388.80		$596.80	$48.60	$74.60

Crew A-1D	Hr.	Daily	Hr.	Daily	Bare Costs	Incl. O&P
1 Building Laborer	$42.10	$336.80	$67.45	$539.60	$42.10	$67.45
1 Vibrating Plate, Gas, 18"		31.55		34.70	3.94	4.34
8 L.H., Daily Totals		$368.35		$574.30	$46.04	$71.79

Crew A-1E	Hr.	Daily	Hr.	Daily	Bare Costs	Incl. O&P
1 Building Laborer	$42.10	$336.80	$67.45	$539.60	$42.10	$67.45
1 Vibrating Plate, Gas, 21"		165.35		181.88	20.67	22.74
8 L.H., Daily Totals		$502.15		$721.49	$62.77	$90.19

Crew A-1F	Hr.	Daily	Hr.	Daily	Bare Costs	Incl. O&P
1 Building Laborer	$42.10	$336.80	$67.45	$539.60	$42.10	$67.45
1 Rammer/Tamper, Gas, 8"		46.50		51.15	5.81	6.39
8 L.H., Daily Totals		$383.30		$590.75	$47.91	$73.84

Crew A-1G	Hr.	Daily	Hr.	Daily	Bare Costs	Incl. O&P
1 Building Laborer	$42.10	$336.80	$67.45	$539.60	$42.10	$67.45
1 Rammer/Tamper, Gas, 15"		54.00		59.40	6.75	7.42
8 L.H., Daily Totals		$390.80		$599.00	$48.85	$74.88

Crew A-1H	Hr.	Daily	Hr.	Daily	Bare Costs	Incl. O&P
1 Building Laborer	$42.10	$336.80	$67.45	$539.60	$42.10	$67.45
1 Exterior Steam Cleaner		76.30		83.93	9.54	10.49
8 L.H., Daily Totals		$413.10		$623.53	$51.64	$77.94

Crew A-1J	Hr.	Daily	Hr.	Daily	Bare Costs	Incl. O&P
1 Building Laborer	$42.10	$336.80	$67.45	$539.60	$42.10	$67.45
1 Cultivator, Walk-Behind, 5 H.P.		53.05		58.35	6.63	7.29
8 L.H., Daily Totals		$389.85		$597.96	$48.73	$74.74

Crew A-1K	Hr.	Daily	Hr.	Daily	Bare Costs	Incl. O&P
1 Building Laborer	$42.10	$336.80	$67.45	$539.60	$42.10	$67.45
1 Cultivator, Walk-Behind, 8 H.P.		101.40		111.54	12.68	13.94
8 L.H., Daily Totals		$438.20		$651.14	$54.77	$81.39

Crew A-1M	Hr.	Daily	Hr.	Daily	Bare Costs	Incl. O&P
1 Building Laborer	$42.10	$336.80	$67.45	$539.60	$42.10	$67.45
1 Snow Blower, Walk-Behind		67.50		74.25	8.44	9.28
8 L.H., Daily Totals		$404.30		$613.85	$50.54	$76.73

Crew No.	Bare Costs		Incl. Subs O&P		Cost Per Labor-Hour	
Crew A-2	**Hr.**	**Daily**	**Hr.**	**Daily**	**Bare Costs**	**Incl. O&P**
2 Laborers	$42.10	$673.60	$67.45	$1079.20	$43.82	$70.15
1 Truck Driver (light)	47.25	378.00	75.55	604.40		
1 Flatbed Truck, Gas, 1.5 Ton		196.15		215.76	8.17	8.99
24 L.H., Daily Totals		$1247.75		$1899.37	$51.99	$79.14

Crew A-2A	Hr.	Daily	Hr.	Daily	Bare Costs	Incl. O&P
2 Laborers	$42.10	$673.60	$67.45	$1079.20	$43.82	$70.15
1 Truck Driver (light)	47.25	378.00	75.55	604.40		
1 Flatbed Truck, Gas, 1.5 Ton		196.15		215.76		
1 Concrete Saw		111.50		122.65	12.82	14.10
24 L.H., Daily Totals		$1359.25		$2022.02	$56.64	$84.25

Crew A-2B	Hr.	Daily	Hr.	Daily	Bare Costs	Incl. O&P
1 Truck Driver (light)	$47.25	$378.00	$75.55	$604.40	$47.25	$75.55
1 Flatbed Truck, Gas, 1.5 Ton		196.15		215.76	24.52	26.97
8 L.H., Daily Totals		$574.15		$820.16	$71.77	$102.52

Crew A-3A	Hr.	Daily	Hr.	Daily	Bare Costs	Incl. O&P
1 Equip. Oper. (light)	$53.00	$424.00	$82.95	$663.60	$53.00	$82.95
1 Pickup Truck, 4x4, 3/4 Ton		175.85		193.44	21.98	24.18
8 L.H., Daily Totals		$599.85		$857.03	$74.98	$107.13

Crew A-3B	Hr.	Daily	Hr.	Daily	Bare Costs	Incl. O&P
1 Equip. Oper. (medium)	$56.75	$454.00	$88.80	$710.40	$52.85	$83.53
1 Truck Driver (heavy)	48.95	391.60	78.25	626.00		
1 Dump Truck, 12 C.Y., 400 H.P.		572.90		630.19		
1 F.E. Loader, W.M., 2.5 C.Y.		702.10		772.31	79.69	87.66
16 L.H., Daily Totals		$2120.60		$2738.90	$132.54	$171.18

Crew A-3C	Hr.	Daily	Hr.	Daily	Bare Costs	Incl. O&P
1 Equip. Oper. (light)	$53.00	$424.00	$82.95	$663.60	$53.00	$82.95
1 Loader, Skid Steer, 78 H.P.		400.10		440.11	50.01	55.01
8 L.H., Daily Totals		$824.10		$1103.71	$103.01	$137.96

Crew A-3D	Hr.	Daily	Hr.	Daily	Bare Costs	Incl. O&P
1 Truck Driver (light)	$47.25	$378.00	$75.55	$604.40	$47.25	$75.55
1 Pickup Truck, 4x4, 3/4 Ton		175.85		193.44		
1 Flatbed Trailer, 25 Ton		135.55		149.10	38.92	42.82
8 L.H., Daily Totals		$689.40		$946.94	$86.17	$118.37

Crew A-3E	Hr.	Daily	Hr.	Daily	Bare Costs	Incl. O&P
1 Equip. Oper. (crane)	$59.20	$473.60	$92.65	$741.20	$54.08	$85.45
1 Truck Driver (heavy)	48.95	391.60	78.25	626.00		
1 Pickup Truck, 4x4, 3/4 Ton		175.85		193.44	10.99	12.09
16 L.H., Daily Totals		$1041.05		$1560.64	$65.07	$97.54

Crew A-3F	Hr.	Daily	Hr.	Daily	Bare Costs	Incl. O&P
1 Equip. Oper. (crane)	$59.20	$473.60	$92.65	$741.20	$54.08	$85.45
1 Truck Driver (heavy)	48.95	391.60	78.25	626.00		
1 Pickup Truck, 4x4, 3/4 Ton		175.85		193.44		
1 Truck Tractor, 6x4, 380 H.P.		493.25		542.58		
1 Lowbed Trailer, 75 Ton		255.05		280.56	57.76	63.54
16 L.H., Daily Totals		$1789.35		$2383.76	$111.83	$148.99

Crew A-3G	Bare Costs Hr.	Bare Costs Daily	Incl. Subs O&P Hr.	Incl. Subs O&P Daily	Cost Per Labor-Hour Bare Costs	Cost Per Labor-Hour Incl. O&P
1 Equip. Oper. (crane)	$59.20	$473.60	$92.65	$741.20	$54.08	$85.45
1 Truck Driver (heavy)	48.95	391.60	78.25	626.00		
1 Pickup Truck, 4x4, 3/4 Ton		175.85		193.44		
1 Truck Tractor, 6x4, 450 H.P.		601.75		661.92		
1 Lowbed Trailer, 75 Ton		255.05		280.56	64.54	70.99
16 L.H., Daily Totals		$1897.85		$2503.11	$118.62	$156.44

Crew A-3H	Hr.	Daily	Hr.	Daily	Bare Costs	Incl. O&P
1 Equip. Oper. (crane)	$59.20	$473.60	$92.65	$741.20	$59.20	$92.65
1 Hyd. Crane, 12 Ton (Daily)		724.85		797.34	90.61	99.67
8 L.H., Daily Totals		$1198.45		$1538.54	$149.81	$192.32

Crew A-3I	Hr.	Daily	Hr.	Daily	Bare Costs	Incl. O&P
1 Equip. Oper. (crane)	$59.20	$473.60	$92.65	$741.20	$59.20	$92.65
1 Hyd. Crane, 25 Ton (Daily)		801.40		881.54	100.18	110.19
8 L.H., Daily Totals		$1275.00		$1622.74	$159.38	$202.84

Crew A-3J	Hr.	Daily	Hr.	Daily	Bare Costs	Incl. O&P
1 Equip. Oper. (crane)	$59.20	$473.60	$92.65	$741.20	$59.20	$92.65
1 Hyd. Crane, 40 Ton (Daily)		1272.00		1399.20	159.00	174.90
8 L.H., Daily Totals		$1745.60		$2140.40	$218.20	$267.55

Crew A-3K	Hr.	Daily	Hr.	Daily	Bare Costs	Incl. O&P
1 Equip. Oper. (crane)	$59.20	$473.60	$92.65	$741.20	$54.88	$85.88
1 Equip. Oper. (oiler)	50.55	404.40	79.10	632.80		
1 Hyd. Crane, 55 Ton (Daily)		1362.00		1498.20		
1 P/U Truck, 3/4 Ton (Daily)		142.15		156.37	94.01	103.41
16 L.H., Daily Totals		$2382.15		$3028.57	$148.88	$189.29

Crew A-3L	Hr.	Daily	Hr.	Daily	Bare Costs	Incl. O&P
1 Equip. Oper. (crane)	$59.20	$473.60	$92.65	$741.20	$54.88	$85.88
1 Equip. Oper. (oiler)	50.55	404.40	79.10	632.80		
1 Hyd. Crane, 80 Ton (Daily)		2101.00		2311.10		
1 P/U Truck, 3/4 Ton (Daily)		142.15		156.37	140.20	154.22
16 L.H., Daily Totals		$3121.15		$3841.47	$195.07	$240.09

Crew A-3M	Hr.	Daily	Hr.	Daily	Bare Costs	Incl. O&P
1 Equip. Oper. (crane)	$59.20	$473.60	$92.65	$741.20	$54.88	$85.88
1 Equip. Oper. (oiler)	50.55	404.40	79.10	632.80		
1 Hyd. Crane, 100 Ton (Daily)		2227.00		2449.70		
1 P/U Truck, 3/4 Ton (Daily)		142.15		156.37	148.07	162.88
16 L.H., Daily Totals		$3247.15		$3980.07	$202.95	$248.75

Crew A-3N	Hr.	Daily	Hr.	Daily	Bare Costs	Incl. O&P
1 Equip. Oper. (crane)	$59.20	$473.60	$92.65	$741.20	$59.20	$92.65
1 Tower Crane (monthly)		1693.00		1862.30	211.63	232.79
8 L.H., Daily Totals		$2166.60		$2603.50	$270.82	$325.44

Crew A-3P	Hr.	Daily	Hr.	Daily	Bare Costs	Incl. O&P
1 Equip. Oper. (light)	$53.00	$424.00	$82.95	$663.60	$53.00	$82.95
1 A.T. Forklift, 31' reach, 45' lift		374.20		411.62	46.77	51.45
8 L.H., Daily Totals		$798.20		$1075.22	$99.78	$134.40

Crew A-3Q	Hr.	Daily	Hr.	Daily	Bare Costs	Incl. O&P
1 Equip. Oper. (light)	$53.00	$424.00	$82.95	$663.60	$53.00	$82.95
1 Pickup Truck, 4x4, 3/4 Ton		175.85		193.44		
1 Flatbed Trailer, 3 Ton		70.30		77.33	30.77	33.85
8 L.H., Daily Totals		$670.15		$934.37	$83.77	$116.80

Crew A-3R	Hr.	Daily	Hr.	Daily	Bare Costs	Incl. O&P
1 Equip. Oper. (light)	$53.00	$424.00	$82.95	$663.60	$53.00	$82.95
1 Forklift, Smooth Floor, 8,000 Lb.		255.90		281.49	31.99	35.19
8 L.H., Daily Totals		$679.90		$945.09	$84.99	$118.14

Crew A-4	Hr.	Daily	Hr.	Daily	Bare Costs	Incl. O&P
2 Carpenters	$53.15	$850.40	$85.15	$1362.40	$50.23	$80.27
1 Painter, Ordinary	44.40	355.20	70.50	564.00		
24 L.H., Daily Totals		$1205.60		$1926.40	$50.23	$80.27

Crew A-5	Hr.	Daily	Hr.	Daily	Bare Costs	Incl. O&P
2 Laborers	$42.10	$673.60	$67.45	$1079.20	$42.67	$68.35
.25 Truck Driver (light)	47.25	94.50	75.55	151.10		
.25 Flatbed Truck, Gas, 1.5 Ton		49.04		53.94	2.72	3.00
18 L.H., Daily Totals		$817.14		$1284.24	$45.40	$71.35

Crew A-6	Hr.	Daily	Hr.	Daily	Bare Costs	Incl. O&P
1 Instrument Man	$54.85	$438.80	$87.40	$699.20	$52.92	$84.45
1 Rodman/Chainman	51.00	408.00	81.50	652.00		
1 Level, Electronic		30.80		33.88	1.93	2.12
16 L.H., Daily Totals		$877.60		$1385.08	$54.85	$86.57

Crew A-7	Hr.	Daily	Hr.	Daily	Bare Costs	Incl. O&P
1 Chief of Party	$66.05	$528.40	$102.30	$818.40	$57.30	$90.40
1 Instrument Man	54.85	438.80	87.40	699.20		
1 Rodman/Chainman	51.00	408.00	81.50	652.00		
1 Level, Electronic		30.80		33.88	1.28	1.41
24 L.H., Daily Totals		$1406.00		$2203.48	$58.58	$91.81

Crew A-8	Hr.	Daily	Hr.	Daily	Bare Costs	Incl. O&P
1 Chief of Party	$66.05	$528.40	$102.30	$818.40	$55.73	$88.17
1 Instrument Man	54.85	438.80	87.40	699.20		
2 Rodmen/Chainmen	51.00	816.00	81.50	1304.00		
1 Level, Electronic		30.80		33.88	.96	1.06
32 L.H., Daily Totals		$1814.00		$2855.48	$56.69	$89.23

Crew A-9	Hr.	Daily	Hr.	Daily	Bare Costs	Incl. O&P
1 Asbestos Foreman	$59.15	$473.20	$93.75	$750.00	$58.71	$93.09
7 Asbestos Workers	58.65	3284.40	93.00	5208.00		
64 L.H., Daily Totals		$3757.60		$5958.00	$58.71	$93.09

Crew A-10A	Hr.	Daily	Hr.	Daily	Bare Costs	Incl. O&P
1 Asbestos Foreman	$59.15	$473.20	$93.75	$750.00	$58.82	$93.25
2 Asbestos Workers	58.65	938.40	93.00	1488.00		
24 L.H., Daily Totals		$1411.60		$2238.00	$58.82	$93.25

Crew A-10B	Hr.	Daily	Hr.	Daily	Bare Costs	Incl. O&P
1 Asbestos Foreman	$59.15	$473.20	$93.75	$750.00	$58.77	$93.19
3 Asbestos Workers	58.65	1407.60	93.00	2232.00		
32 L.H., Daily Totals		$1880.80		$2982.00	$58.77	$93.19

Crew A-10C	Hr.	Daily	Hr.	Daily	Bare Costs	Incl. O&P
3 Asbestos Workers	$58.65	$1407.60	$93.00	$2232.00	$58.65	$93.00
1 Flatbed Truck, Gas, 1.5 Ton		196.15		215.76	8.17	8.99
24 L.H., Daily Totals		$1603.75		$2447.76	$66.82	$101.99

Crew No.	Bare Costs		Incl. Subs O&P		Cost Per Labor-Hour	

Crew A-10D	Hr.	Daily	Hr.	Daily	Bare Costs	Incl. O&P
2 Asbestos Workers	$58.65	$938.40	$93.00	$1488.00	$56.76	$89.44
1 Equip. Oper. (crane)	59.20	473.60	92.65	741.20		
1 Equip. Oper. (oiler)	50.55	404.40	79.10	632.80		
1 Hydraulic Crane, 33 Ton		945.35		1039.89	29.54	32.50
32 L.H., Daily Totals		$2761.75		$3901.89	$86.30	$121.93

Crew A-11	Hr.	Daily	Hr.	Daily	Bare Costs	Incl. O&P
1 Asbestos Foreman	$59.15	$473.20	$93.75	$750.00	$58.71	$93.09
7 Asbestos Workers	58.65	3284.40	93.00	5208.00		
2 Chip. Hammers, 12 Lb., Elec.		64.80		71.28	1.01	1.11
64 L.H., Daily Totals		$3822.40		$6029.28	$59.73	$94.21

Crew A-12	Hr.	Daily	Hr.	Daily	Bare Costs	Incl. O&P
1 Asbestos Foreman	$59.15	$473.20	$93.75	$750.00	$58.71	$93.09
7 Asbestos Workers	58.65	3284.40	93.00	5208.00		
1 Trk-Mtd Vac, 14 CY, 1500 Gal.		536.15		589.76		
1 Flatbed Truck, 20,000 GVW		201.60		221.76	11.53	12.68
64 L.H., Daily Totals		$4495.35		$6769.52	$70.24	$105.77

Crew A-13	Hr.	Daily	Hr.	Daily	Bare Costs	Incl. O&P
1 Equip. Oper. (light)	$53.00	$424.00	$82.95	$663.60	$53.00	$82.95
1 Trk-Mtd Vac, 14 CY, 1500 Gal.		536.15		589.76		
1 Flatbed Truck, 20,000 GVW		201.60		221.76	92.22	101.44
8 L.H., Daily Totals		$1161.75		$1475.13	$145.22	$184.39

Crew B-1	Hr.	Daily	Hr.	Daily	Bare Costs	Incl. O&P
1 Labor Foreman (outside)	$44.10	$352.80	$70.65	$565.20	$42.77	$68.52
2 Laborers	42.10	673.60	67.45	1079.20		
24 L.H., Daily Totals		$1026.40		$1644.40	$42.77	$68.52

Crew B-1A	Hr.	Daily	Hr.	Daily	Bare Costs	Incl. O&P
1 Labor Foreman (outside)	$44.10	$352.80	$70.65	$565.20	$42.77	$68.52
2 Laborers	42.10	673.60	67.45	1079.20		
2 Cutting Torches		25.60		28.16		
2 Sets of Gases		343.10		377.41	15.36	16.90
24 L.H., Daily Totals		$1395.10		$2049.97	$58.13	$85.42

Crew B-1B	Hr.	Daily	Hr.	Daily	Bare Costs	Incl. O&P
1 Labor Foreman (outside)	$44.10	$352.80	$70.65	$565.20	$46.88	$74.55
2 Laborers	42.10	673.60	67.45	1079.20		
1 Equip. Oper. (crane)	59.20	473.60	92.65	741.20		
2 Cutting Torches		25.60		28.16		
2 Sets of Gases		343.10		377.41		
1 Hyd. Crane, 12 Ton		471.00		518.10	26.24	28.86
32 L.H., Daily Totals		$2339.70		$3309.27	$73.12	$103.41

Crew B-1C	Hr.	Daily	Hr.	Daily	Bare Costs	Incl. O&P
1 Labor Foreman (outside)	$44.10	$352.80	$70.65	$565.20	$42.77	$68.52
2 Laborers	42.10	673.60	67.45	1079.20		
1 Telescoping Boom Lift, to 60'		310.90		341.99	12.95	14.25
24 L.H., Daily Totals		$1337.30		$1986.39	$55.72	$82.77

Crew B-1D	Hr.	Daily	Hr.	Daily	Bare Costs	Incl. O&P
2 Laborers	$42.10	$673.60	$67.45	$1079.20	$42.10	$67.45
1 Small Work Boat, Gas, 50 H.P.		119.50		131.45		
1 Pressure Washer, 7 GPM		90.50		99.55	13.13	14.44
16 L.H., Daily Totals		$883.60		$1310.20	$55.23	$81.89

Crew B-1E	Hr.	Daily	Hr.	Daily	Bare Costs	Incl. O&P
1 Labor Foreman (outside)	$44.10	$352.80	$70.65	$565.20	$42.60	$68.25
3 Laborers	42.10	1010.40	67.45	1618.80		
1 Work Boat, Diesel, 200 H.P.		1418.00		1559.80		
2 Pressure Washers, 7 GPM		181.00		199.10	49.97	54.97
32 L.H., Daily Totals		$2962.20		$3942.90	$92.57	$123.22

Crew B-1F	Hr.	Daily	Hr.	Daily	Bare Costs	Incl. O&P
2 Skilled Workers	$54.85	$877.60	$87.40	$1398.40	$50.60	$80.75
1 Laborer	42.10	336.80	67.45	539.60		
1 Small Work Boat, Gas, 50 H.P.		119.50		131.45		
1 Pressure Washer, 7 GPM		90.50		99.55	8.75	9.63
24 L.H., Daily Totals		$1424.40		$2169.00	$59.35	$90.38

Crew B-1G	Hr.	Daily	Hr.	Daily	Bare Costs	Incl. O&P
2 Laborers	$42.10	$673.60	$67.45	$1079.20	$42.10	$67.45
1 Small Work Boat, Gas, 50 H.P.		119.50		131.45	7.47	8.22
16 L.H., Daily Totals		$793.10		$1210.65	$49.57	$75.67

Crew B-1H	Hr.	Daily	Hr.	Daily	Bare Costs	Incl. O&P
2 Skilled Workers	$54.85	$877.60	$87.40	$1398.40	$50.60	$80.75
1 Laborer	42.10	336.80	67.45	539.60		
1 Small Work Boat, Gas, 50 H.P.		119.50		131.45	4.98	5.48
24 L.H., Daily Totals		$1333.90		$2069.45	$55.58	$86.23

Crew B-1J	Hr.	Daily	Hr.	Daily	Bare Costs	Incl. O&P
1 Labor Foreman (inside)	$42.60	$340.80	$68.25	$546.00	$42.35	$67.85
1 Laborer	42.10	336.80	67.45	539.60		
16 L.H., Daily Totals		$677.60		$1085.60	$42.35	$67.85

Crew B-1K	Hr.	Daily	Hr.	Daily	Bare Costs	Incl. O&P
1 Carpenter Foreman (inside)	$53.65	$429.20	$85.95	$687.60	$53.40	$85.55
1 Carpenter	53.15	425.20	85.15	681.20		
16 L.H., Daily Totals		$854.40		$1368.80	$53.40	$85.55

Crew B-2	Hr.	Daily	Hr.	Daily	Bare Costs	Incl. O&P
1 Labor Foreman (outside)	$44.10	$352.80	$70.65	$565.20	$42.50	$68.09
4 Laborers	42.10	1347.20	67.45	2158.40		
40 L.H., Daily Totals		$1700.00		$2723.60	$42.50	$68.09

Crew B-2A	Hr.	Daily	Hr.	Daily	Bare Costs	Incl. O&P
1 Labor Foreman (outside)	$44.10	$352.80	$70.65	$565.20	$42.77	$68.52
2 Laborers	42.10	673.60	67.45	1079.20		
1 Telescoping Boom Lift, to 60'		310.90		341.99	12.95	14.25
24 L.H., Daily Totals		$1337.30		$1986.39	$55.72	$82.77

Crew B-3	Hr.	Daily	Hr.	Daily	Bare Costs	Incl. O&P
1 Labor Foreman (outside)	$44.10	$352.80	$70.65	$565.20	$47.16	$75.14
2 Laborers	42.10	673.60	67.45	1079.20		
1 Equip. Oper. (medium)	56.75	454.00	88.80	710.40		
2 Truck Drivers (heavy)	48.95	783.20	78.25	1252.00		
1 Crawler Loader, 3 C.Y.		1169.00		1285.90		
2 Dump Trucks, 12 C.Y., 400 H.P.		1145.80		1260.38	48.23	53.05
48 L.H., Daily Totals		$4578.40		$6153.08	$95.38	$128.19

Crew B-3A	Hr.	Daily	Hr.	Daily	Bare Costs	Incl. O&P
4 Laborers	$42.10	$1347.20	$67.45	$2158.40	$45.03	$71.72
1 Equip. Oper. (medium)	56.75	454.00	88.80	710.40		
1 Hyd. Excavator, 1.5 C.Y.		687.55		756.30	17.19	18.91
40 L.H., Daily Totals		$2488.75		$3625.11	$62.22	$90.63

Crew No.	Bare Costs		Incl. Subs O&P		Cost Per Labor-Hour	
Crew B-3B	**Hr.**	**Daily**	**Hr.**	**Daily**	**Bare Costs**	**Incl. O&P**
2 Laborers	$42.10	$673.60	$67.45	$1079.20	$47.48	$75.49
1 Equip. Oper. (medium)	56.75	454.00	88.80	710.40		
1 Truck Driver (heavy)	48.95	391.60	78.25	626.00		
1 Backhoe Loader, 80 H.P.		232.40		255.64		
1 Dump Truck, 12 C.Y., 400 H.P.		572.90		630.19	25.17	27.68
32 L.H., Daily Totals		$2324.50		$3301.43	$72.64	$103.17
Crew B-3C	**Hr.**	**Daily**	**Hr.**	**Daily**	**Bare Costs**	**Incl. O&P**
3 Laborers	$42.10	$1010.40	$67.45	$1618.80	$45.76	$72.79
1 Equip. Oper. (medium)	56.75	454.00	88.80	710.40		
1 Crawler Loader, 4 C.Y.		1489.00		1637.90	46.53	51.18
32 L.H., Daily Totals		$2953.40		$3967.10	$92.29	$123.97
Crew B-4	**Hr.**	**Daily**	**Hr.**	**Daily**	**Bare Costs**	**Incl. O&P**
1 Labor Foreman (outside)	$44.10	$352.80	$70.65	$565.20	$43.58	$69.78
4 Laborers	42.10	1347.20	67.45	2158.40		
1 Truck Driver (heavy)	48.95	391.60	78.25	626.00		
1 Truck Tractor, 220 H.P.		307.10		337.81		
1 Flatbed Trailer, 40 Ton		186.20		204.82	10.28	11.30
48 L.H., Daily Totals		$2584.90		$3892.23	$53.85	$81.09
Crew B-5	**Hr.**	**Daily**	**Hr.**	**Daily**	**Bare Costs**	**Incl. O&P**
1 Labor Foreman (outside)	$44.10	$352.80	$70.65	$565.20	$46.57	$74.01
4 Laborers	42.10	1347.20	67.45	2158.40		
2 Equip. Oper. (medium)	56.75	908.00	88.80	1420.80		
1 Air Compressor, 250 cfm		201.50		221.65		
2 Breakers, Pavement, 60 lb.		107.00		117.70		
2 -50' Air Hoses, 1.5"		48.50		53.35		
1 Crawler Loader, 3 C.Y.		1169.00		1285.90	27.25	29.98
56 L.H., Daily Totals		$4134.00		$5823.00	$73.82	$103.98
Crew B-5D	**Hr.**	**Daily**	**Hr.**	**Daily**	**Bare Costs**	**Incl. O&P**
1 Labor Foreman (outside)	$44.10	$352.80	$70.65	$565.20	$46.87	$74.54
4 Laborers	42.10	1347.20	67.45	2158.40		
2 Equip. Oper. (medium)	56.75	908.00	88.80	1420.80		
1 Truck Driver (heavy)	48.95	391.60	78.25	626.00		
1 Air Compressor, 250 cfm		201.50		221.65		
2 Breakers, Pavement, 60 lb.		107.00		117.70		
2 -50' Air Hoses, 1.5"		48.50		53.35		
1 Crawler Loader, 3 C.Y.		1169.00		1285.90		
1 Dump Truck, 12 C.Y., 400 H.P.		572.90		630.19	32.80	36.07
64 L.H., Daily Totals		$5098.50		$7079.19	$79.66	$110.61
Crew B-5E	**Hr.**	**Daily**	**Hr.**	**Daily**	**Bare Costs**	**Incl. O&P**
1 Labor Foreman (outside)	$44.10	$352.80	$70.65	$565.20	$46.87	$74.54
4 Laborers	42.10	1347.20	67.45	2158.40		
2 Equip. Oper. (medium)	56.75	908.00	88.80	1420.80		
1 Truck Driver (heavy)	48.95	391.60	78.25	626.00		
1 Water Tank Trailer, 5000 Gal.		152.25		167.47		
1 High Pressure Water Jet, 40 KSI		811.05		892.16		
2 -50' Air Hoses, 1.5"		48.50		53.35		
1 Crawler Loader, 3 C.Y.		1169.00		1285.90		
1 Dump Truck, 12 C.Y., 400 H.P.		572.90		630.19	43.03	47.33
64 L.H., Daily Totals		$5753.21		$7799.38	$89.90	$121.87
Crew B-6	**Hr.**	**Daily**	**Hr.**	**Daily**	**Bare Costs**	**Incl. O&P**
2 Laborers	$42.10	$673.60	$67.45	$1079.20	$45.73	$72.62
1 Equip. Oper. (light)	53.00	424.00	82.95	663.60		
1 Backhoe Loader, 48 H.P.		213.75		235.13	8.91	9.80
24 L.H., Daily Totals		$1311.35		$1977.93	$54.64	$82.41

Crew No.	Bare Costs		Incl. Subs O&P		Cost Per Labor-Hour	
Crew B-6A	**Hr.**	**Daily**	**Hr.**	**Daily**	**Bare Costs**	**Incl. O&P**
.5 Labor Foreman (outside)	$44.10	$176.40	$70.65	$282.60	$48.36	$76.63
1 Laborer	42.10	336.80	67.45	539.60		
1 Equip. Oper. (medium)	56.75	454.00	88.80	710.40		
1 Vacuum Truck, 5000 Gal.		367.55		404.31	18.38	20.22
20 L.H., Daily Totals		$1334.75		$1936.91	$66.74	$96.85
Crew B-6B	**Hr.**	**Daily**	**Hr.**	**Daily**	**Bare Costs**	**Incl. O&P**
2 Labor Foremen (outside)	$44.10	$705.60	$70.65	$1130.40	$42.77	$68.52
4 Laborers	42.10	1347.20	67.45	2158.40		
1 S.P. Crane, 4x4, 5 Ton		378.10		415.91		
1 Flatbed Truck, Gas, 1.5 Ton		196.15		215.76		
1 Butt Fusion Mach., 4"-12" diam.		426.85		469.54	20.86	22.94
48 L.H., Daily Totals		$3053.90		$4390.01	$63.62	$91.46
Crew B-6C	**Hr.**	**Daily**	**Hr.**	**Daily**	**Bare Costs**	**Incl. O&P**
2 Labor Foremen (outside)	$44.10	$705.60	$70.65	$1130.40	$42.77	$68.52
4 Laborers	42.10	1347.20	67.45	2158.40		
1 S.P. Crane, 4x4, 12 Ton		428.30		471.13		
1 Flatbed Truck, Gas, 3 Ton		820.40		902.44		
1 Butt Fusion Mach., 8"-24" diam.		1076.00		1183.60	48.43	53.27
48 L.H., Daily Totals		$4377.50		$5845.97	$91.20	$121.79
Crew B-6D	**Hr.**	**Daily**	**Hr.**	**Daily**	**Bare Costs**	**Incl. O&P**
0.5 Labor Foreman (outside)	$44.10	$176.40	$70.65	$282.60	$48.36	$76.63
1 Laborer	42.10	336.80	67.45	539.60		
1 Equip. Oper. (medium)	56.75	454.00	88.80	710.40		
1 Hydro Excavator, 12 C.Y.		1262.00		1388.20	63.10	69.41
20 L.H., Daily Totals		$2228.80		$2920.36	$111.46	$146.04
Crew B-7	**Hr.**	**Daily**	**Hr.**	**Daily**	**Bare Costs**	**Incl. O&P**
1 Labor Foreman (outside)	$44.10	$352.80	$70.65	$565.20	$44.88	$71.54
4 Laborers	42.10	1347.20	67.45	2158.40		
1 Equip. Oper. (medium)	56.75	454.00	88.80	710.40		
1 Brush Chipper, 12", 130 H.P.		392.80		432.08		
1 Crawler Loader, 3 C.Y.		1169.00		1285.90		
2 Chain Saws, Gas, 36" Long		82.50		90.75	34.26	37.68
48 L.H., Daily Totals		$3798.30		$5242.73	$79.13	$109.22
Crew B-7A	**Hr.**	**Daily**	**Hr.**	**Daily**	**Bare Costs**	**Incl. O&P**
2 Laborers	$42.10	$673.60	$67.45	$1079.20	$45.73	$72.62
1 Equip. Oper. (light)	53.00	424.00	82.95	663.60		
1 Rake w/Tractor		339.45		373.39		
2 Chain Saws, Gas, 18"		104.00		114.40	18.48	20.32
24 L.H., Daily Totals		$1541.05		$2230.59	$64.21	$92.94
Crew B-7B	**Hr.**	**Daily**	**Hr.**	**Daily**	**Bare Costs**	**Incl. O&P**
1 Labor Foreman (outside)	$44.10	$352.80	$70.65	$565.20	$45.46	$72.50
4 Laborers	42.10	1347.20	67.45	2158.40		
1 Equip. Oper. (medium)	56.75	454.00	88.80	710.40		
1 Truck Driver (heavy)	48.95	391.60	78.25	626.00		
1 Brush Chipper, 12", 130 H.P.		392.80		432.08		
1 Crawler Loader, 3 C.Y.		1169.00		1285.90		
2 Chain Saws, Gas, 36" Long		82.50		90.75		
1 Dump Truck, 8 C.Y., 220 H.P.		428.05		470.86	37.01	40.71
56 L.H., Daily Totals		$4617.95		$6339.59	$82.46	$113.21

Crew B-7C	Hr.	Daily	Hr.	Daily	Bare Costs	Incl. O&P
1 Labor Foreman (outside)	$44.10	$352.80	$70.65	$565.20	$45.46	$72.50
4 Laborers	42.10	1347.20	67.45	2158.40		
1 Equip. Oper. (medium)	56.75	454.00	88.80	710.40		
1 Truck Driver (heavy)	48.95	391.60	78.25	626.00		
1 Brush Chipper, 12", 130 H.P.		392.80		432.08		
1 Crawler Loader, 3 C.Y.		1169.00		1285.90		
2 Chain Saws, Gas, 36" Long		82.50		90.75		
1 Dump Truck, 12 C.Y., 400 H.P.		572.90		630.19	39.59	43.55
56 L.H., Daily Totals		$4762.80		$6498.92	$85.05	$116.05

Crew B-8	Hr.	Daily	Hr.	Daily	Bare Costs	Incl. O&P
1 Labor Foreman (outside)	$44.10	$352.80	$70.65	$565.20	$48.78	$77.34
2 Laborers	42.10	673.60	67.45	1079.20		
2 Equip. Oper. (medium)	56.75	908.00	88.80	1420.80		
1 Equip. Oper. (oiler)	50.55	404.40	79.10	632.80		
2 Truck Drivers (heavy)	48.95	783.20	78.25	1252.00		
1 Hyd. Crane, 25 Ton		580.85		638.93		
1 Crawler Loader, 3 C.Y.		1169.00		1285.90		
2 Dump Trucks, 12 C.Y., 400 H.P.		1145.80		1260.38	45.24	49.77
64 L.H., Daily Totals		$6017.65		$8135.22	$94.03	$127.11

Crew B-9	Hr.	Daily	Hr.	Daily	Bare Costs	Incl. O&P
1 Labor Foreman (outside)	$44.10	$352.80	$70.65	$565.20	$42.50	$68.09
4 Laborers	42.10	1347.20	67.45	2158.40		
1 Air Compressor, 250 cfm		201.50		221.65		
2 Breakers, Pavement, 60 lb.		107.00		117.70		
2 -50' Air Hoses, 1.5"		48.50		53.35	8.93	9.82
40 L.H., Daily Totals		$2057.00		$3116.30	$51.42	$77.91

Crew B-9A	Hr.	Daily	Hr.	Daily	Bare Costs	Incl. O&P
2 Laborers	$42.10	$673.60	$67.45	$1079.20	$44.38	$71.05
1 Truck Driver (heavy)	48.95	391.60	78.25	626.00		
1 Water Tank Trailer, 5000 Gal.		152.25		167.47		
1 Truck Tractor, 220 H.P.		307.10		337.81		
2 -50' Discharge Hoses, 3"		9.00		9.90	19.51	21.47
24 L.H., Daily Totals		$1533.55		$2220.39	$63.90	$92.52

Crew B-9B	Hr.	Daily	Hr.	Daily	Bare Costs	Incl. O&P
2 Laborers	$42.10	$673.60	$67.45	$1079.20	$44.38	$71.05
1 Truck Driver (heavy)	48.95	391.60	78.25	626.00		
2 -50' Discharge Hoses, 3"		9.00		9.90		
1 Water Tank Trailer, 5000 Gal.		152.25		167.47		
1 Truck Tractor, 220 H.P.		307.10		337.81		
1 Pressure Washer		96.95		106.65	23.55	25.91
24 L.H., Daily Totals		$1630.50		$2327.03	$67.94	$96.96

Crew B-9D	Hr.	Daily	Hr.	Daily	Bare Costs	Incl. O&P
1 Labor Foreman (outside)	$44.10	$352.80	$70.65	$565.20	$42.50	$68.09
4 Common Laborers	42.10	1347.20	67.45	2158.40		
1 Air Compressor, 250 cfm		201.50		221.65		
2 -50' Air Hoses, 1.5"		48.50		53.35		
2 Air Powered Tampers		67.70		74.47	7.94	8.74
40 L.H., Daily Totals		$2017.70		$3073.07	$50.44	$76.83

Crew B-9E	Hr.	Daily	Hr.	Daily	Bare Costs	Incl. O&P
1 Cement Finisher	$49.95	$399.60	$78.45	$627.60	$46.02	$72.95
1 Laborer	42.10	336.80	67.45	539.60		
1 Chip. Hammers, 12 Lb., Elec.		32.40		35.64	2.02	2.23
16 L.H., Daily Totals		$768.80		$1202.84	$48.05	$75.18

Crew B-10	Hr.	Daily	Hr.	Daily	Bare Costs	Incl. O&P
1 Equip. Oper. (medium)	$56.75	$454.00	$88.80	$710.40	$51.87	$81.68
.5 Laborer	42.10	168.40	67.45	269.80		
12 L.H., Daily Totals		$622.40		$980.20	$51.87	$81.68

Crew B-10A	Hr.	Daily	Hr.	Daily	Bare Costs	Incl. O&P
1 Equip. Oper. (medium)	$56.75	$454.00	$88.80	$710.40	$51.87	$81.68
.5 Laborer	42.10	168.40	67.45	269.80		
1 Roller, 2-Drum, W.B., 7.5 H.P.		166.10		182.71	13.84	15.23
12 L.H., Daily Totals		$788.50		$1162.91	$65.71	$96.91

Crew B-10B	Hr.	Daily	Hr.	Daily	Bare Costs	Incl. O&P
1 Equip. Oper. (medium)	$56.75	$454.00	$88.80	$710.40	$51.87	$81.68
.5 Laborer	42.10	168.40	67.45	269.80		
1 Dozer, 200 H.P.		1504.00		1654.40	125.33	137.87
12 L.H., Daily Totals		$2126.40		$2634.60	$177.20	$219.55

Crew B-10C	Hr.	Daily	Hr.	Daily	Bare Costs	Incl. O&P
1 Equip. Oper. (medium)	$56.75	$454.00	$88.80	$710.40	$51.87	$81.68
.5 Laborer	42.10	168.40	67.45	269.80		
1 Dozer, 200 H.P.		1504.00		1654.40		
1 Vibratory Roller, Towed, 23 Ton		514.75		566.23	168.23	185.05
12 L.H., Daily Totals		$2641.15		$3200.82	$220.10	$266.74

Crew B-10D	Hr.	Daily	Hr.	Daily	Bare Costs	Incl. O&P
1 Equip. Oper. (medium)	$56.75	$454.00	$88.80	$710.40	$51.87	$81.68
.5 Laborer	42.10	168.40	67.45	269.80		
1 Dozer, 200 H.P.		1504.00		1654.40		
1 Sheepsft. Roller, Towed		436.90		480.59	161.74	177.92
12 L.H., Daily Totals		$2563.30		$3115.19	$213.61	$259.60

Crew B-10E	Hr.	Daily	Hr.	Daily	Bare Costs	Incl. O&P
1 Equip. Oper. (medium)	$56.75	$454.00	$88.80	$710.40	$51.87	$81.68
.5 Laborer	42.10	168.40	67.45	269.80		
1 Tandem Roller, 5 Ton		255.65		281.21	21.30	23.43
12 L.H., Daily Totals		$878.05		$1261.42	$73.17	$105.12

Crew B-10F	Hr.	Daily	Hr.	Daily	Bare Costs	Incl. O&P
1 Equip. Oper. (medium)	$56.75	$454.00	$88.80	$710.40	$51.87	$81.68
.5 Laborer	42.10	168.40	67.45	269.80		
1 Tandem Roller, 10 Ton		236.75		260.43	19.73	21.70
12 L.H., Daily Totals		$859.15		$1240.63	$71.60	$103.39

Crew B-10G	Hr.	Daily	Hr.	Daily	Bare Costs	Incl. O&P
1 Equip. Oper. (medium)	$56.75	$454.00	$88.80	$710.40	$51.87	$81.68
.5 Laborer	42.10	168.40	67.45	269.80		
1 Sheepsfoot Roller, 240 H.P.		1348.00		1482.80	112.33	123.57
12 L.H., Daily Totals		$1970.40		$2463.00	$164.20	$205.25

Crew B-10H	Hr.	Daily	Hr.	Daily	Bare Costs	Incl. O&P
1 Equip. Oper. (medium)	$56.75	$454.00	$88.80	$710.40	$51.87	$81.68
.5 Laborer	42.10	168.40	67.45	269.80		
1 Diaphragm Water Pump, 2"		86.80		95.48		
1 -20' Suction Hose, 2"		3.55		3.90		
2 -50' Discharge Hoses, 2"		8.00		8.80	8.20	9.02
12 L.H., Daily Totals		$720.75		$1088.39	$60.06	$90.70

Crew No.	Bare Costs		Incl. Subs O&P		Cost Per Labor-Hour	
Crew B-10I	**Hr.**	**Daily**	**Hr.**	**Daily**	**Bare Costs**	**Incl. O&P**
1 Equip. Oper. (medium)	$56.75	$454.00	$88.80	$710.40	$51.87	$81.68
.5 Laborer	42.10	168.40	67.45	269.80		
1 Diaphragm Water Pump, 4"		105.25		115.78		
1 -20' Suction Hose, 4"		17.05		18.75		
2 -50' Discharge Hoses, 4"		25.30		27.83	12.30	13.53
12 L.H., Daily Totals		$770.00		$1142.56	$64.17	$95.21

Crew B-10J	Hr.	Daily	Hr.	Daily	Bare Costs	Incl. O&P
1 Equip. Oper. (medium)	$56.75	$454.00	$88.80	$710.40	$51.87	$81.68
.5 Laborer	42.10	168.40	67.45	269.80		
1 Centrifugal Water Pump, 3"		74.45		81.89		
1 -20' Suction Hose, 3"		8.75		9.63		
2 -50' Discharge Hoses, 3"		9.00		9.90	7.68	8.45
12 L.H., Daily Totals		$714.60		$1081.62	$59.55	$90.14

Crew B-10K	Hr.	Daily	Hr.	Daily	Bare Costs	Incl. O&P
1 Equip. Oper. (medium)	$56.75	$454.00	$88.80	$710.40	$51.87	$81.68
.5 Laborer	42.10	168.40	67.45	269.80		
1 Centr. Water Pump, 6"		232.90		256.19		
1 -20' Suction Hose, 6"		25.20		27.72		
2 -50' Discharge Hoses, 6"		35.80		39.38	24.49	26.94
12 L.H., Daily Totals		$916.30		$1303.49	$76.36	$108.62

Crew B-10L	Hr.	Daily	Hr.	Daily	Bare Costs	Incl. O&P
1 Equip. Oper. (medium)	$56.75	$454.00	$88.80	$710.40	$51.87	$81.68
.5 Laborer	42.10	168.40	67.45	269.80		
1 Dozer, 80 H.P.		401.40		441.54	33.45	36.80
12 L.H., Daily Totals		$1023.80		$1421.74	$85.32	$118.48

Crew B-10M	Hr.	Daily	Hr.	Daily	Bare Costs	Incl. O&P
1 Equip. Oper. (medium)	$56.75	$454.00	$88.80	$710.40	$51.87	$81.68
.5 Laborer	42.10	168.40	67.45	269.80		
1 Dozer, 300 H.P.		1764.00		1940.40	147.00	161.70
12 L.H., Daily Totals		$2386.40		$2920.60	$198.87	$243.38

Crew B-10N	Hr.	Daily	Hr.	Daily	Bare Costs	Incl. O&P
1 Equip. Oper. (medium)	$56.75	$454.00	$88.80	$710.40	$51.87	$81.68
.5 Laborer	42.10	168.40	67.45	269.80		
1 F.E. Loader, T.M., 1.5 C.Y.		565.80		622.38	47.15	51.87
12 L.H., Daily Totals		$1188.20		$1602.58	$99.02	$133.55

Crew B-10O	Hr.	Daily	Hr.	Daily	Bare Costs	Incl. O&P
1 Equip. Oper. (medium)	$56.75	$454.00	$88.80	$710.40	$51.87	$81.68
.5 Laborer	42.10	168.40	67.45	269.80		
1 F.E. Loader, T.M., 2.25 C.Y.		960.90		1056.99	80.08	88.08
12 L.H., Daily Totals		$1583.30		$2037.19	$131.94	$169.77

Crew B-10P	Hr.	Daily	Hr.	Daily	Bare Costs	Incl. O&P
1 Equip. Oper. (medium)	$56.75	$454.00	$88.80	$710.40	$51.87	$81.68
.5 Laborer	42.10	168.40	67.45	269.80		
1 Crawler Loader, 3 C.Y.		1169.00		1285.90	97.42	107.16
12 L.H., Daily Totals		$1791.40		$2266.10	$149.28	$188.84

Crew B-10Q	Hr.	Daily	Hr.	Daily	Bare Costs	Incl. O&P
1 Equip. Oper. (medium)	$56.75	$454.00	$88.80	$710.40	$51.87	$81.68
.5 Laborer	42.10	168.40	67.45	269.80		
1 Crawler Loader, 4 C.Y.		1489.00		1637.90	124.08	136.49
12 L.H., Daily Totals		$2111.40		$2618.10	$175.95	$218.18

Crew B-10R	Hr.	Daily	Hr.	Daily	Bare Costs	Incl. O&P
1 Equip. Oper. (medium)	$56.75	$454.00	$88.80	$710.40	$51.87	$81.68
.5 Laborer	42.10	168.40	67.45	269.80		
1 F.E. Loader, W.M., 1 C.Y.		302.00		332.20	25.17	27.68
12 L.H., Daily Totals		$924.40		$1312.40	$77.03	$109.37

Crew B-10S	Hr.	Daily	Hr.	Daily	Bare Costs	Incl. O&P
1 Equip. Oper. (medium)	$56.75	$454.00	$88.80	$710.40	$51.87	$81.68
.5 Laborer	42.10	168.40	67.45	269.80		
1 F.E. Loader, W.M., 1.5 C.Y.		423.50		465.85	35.29	38.82
12 L.H., Daily Totals		$1045.90		$1446.05	$87.16	$120.50

Crew B-10T	Hr.	Daily	Hr.	Daily	Bare Costs	Incl. O&P
1 Equip. Oper. (medium)	$56.75	$454.00	$88.80	$710.40	$51.87	$81.68
.5 Laborer	42.10	168.40	67.45	269.80		
1 F.E. Loader, W.M., 2.5 C.Y.		702.10		772.31	58.51	64.36
12 L.H., Daily Totals		$1324.50		$1752.51	$110.38	$146.04

Crew B-10U	Hr.	Daily	Hr.	Daily	Bare Costs	Incl. O&P
1 Equip. Oper. (medium)	$56.75	$454.00	$88.80	$710.40	$51.87	$81.68
.5 Laborer	42.10	168.40	67.45	269.80		
1 F.E. Loader, W.M., 5.5 C.Y.		957.50		1053.25	79.79	87.77
12 L.H., Daily Totals		$1579.90		$2033.45	$131.66	$169.45

Crew B-10V	Hr.	Daily	Hr.	Daily	Bare Costs	Incl. O&P
1 Equip. Oper. (medium)	$56.75	$454.00	$88.80	$710.40	$51.87	$81.68
.5 Laborer	42.10	168.40	67.45	269.80		
1 Dozer, 700 H.P.		5120.00		5632.00	426.67	469.33
12 L.H., Daily Totals		$5742.40		$6612.20	$478.53	$551.02

Crew B-10W	Hr.	Daily	Hr.	Daily	Bare Costs	Incl. O&P
1 Equip. Oper. (medium)	$56.75	$454.00	$88.80	$710.40	$51.87	$81.68
.5 Laborer	42.10	168.40	67.45	269.80		
1 Dozer, 105 H.P.		633.85		697.24	52.82	58.10
12 L.H., Daily Totals		$1256.25		$1677.43	$104.69	$139.79

Crew B-10X	Hr.	Daily	Hr.	Daily	Bare Costs	Incl. O&P
1 Equip. Oper. (medium)	$56.75	$454.00	$88.80	$710.40	$51.87	$81.68
.5 Laborer	42.10	168.40	67.45	269.80		
1 Dozer, 410 H.P.		2777.00		3054.70	231.42	254.56
12 L.H., Daily Totals		$3399.40		$4034.90	$283.28	$336.24

Crew B-10Y	Hr.	Daily	Hr.	Daily	Bare Costs	Incl. O&P
1 Equip. Oper. (medium)	$56.75	$454.00	$88.80	$710.40	$51.87	$81.68
.5 Laborer	42.10	168.40	67.45	269.80		
1 Vibr. Roller, Towed, 12 Ton		578.60		636.46	48.22	53.04
12 L.H., Daily Totals		$1201.00		$1616.66	$100.08	$134.72

Crew B-11A	Hr.	Daily	Hr.	Daily	Bare Costs	Incl. O&P
1 Equipment Oper. (med.)	$56.75	$454.00	$88.80	$710.40	$49.42	$78.13
1 Laborer	42.10	336.80	67.45	539.60		
1 Dozer, 200 H.P.		1504.00		1654.40	94.00	103.40
16 L.H., Daily Totals		$2294.80		$2904.40	$143.43	$181.53

Crew B-11B	Hr.	Daily	Hr.	Daily	Bare Costs	Incl. O&P
1 Equipment Oper. (light)	$53.00	$424.00	$82.95	$663.60	$47.55	$75.20
1 Laborer	42.10	336.80	67.45	539.60		
1 Air Powered Tamper		33.85		37.23		
1 Air Compressor, 365 cfm		315.40		346.94		
2 -50' Air Hoses, 1.5"		48.50		53.35	24.86	27.35
16 L.H., Daily Totals		$1158.55		$1640.72	$72.41	$102.55

Crew B-11C	Hr.	Daily	Hr.	Daily	Bare Costs	Incl. O&P
1 Equipment Oper. (med.)	$56.75	$454.00	$88.80	$710.40	$49.42	$78.13
1 Laborer	42.10	336.80	67.45	539.60		
1 Backhoe Loader, 48 H.P.		213.75		235.13	13.36	14.70
16 L.H., Daily Totals		$1004.55		$1485.13	$62.78	$92.82

Crew B-11K	Hr.	Daily	Hr.	Daily	Bare Costs	Incl. O&P
1 Equipment Oper. (med.)	$56.75	$454.00	$88.80	$710.40	$49.42	$78.13
1 Laborer	42.10	336.80	67.45	539.60		
1 Trencher, Chain Type, 8' D		1872.00		2059.20	117.00	128.70
16 L.H., Daily Totals		$2662.80		$3309.20	$166.43	$206.82

Crew B-11L	Hr.	Daily	Hr.	Daily	Bare Costs	Incl. O&P
1 Equipment Oper. (med.)	$56.75	$454.00	$88.80	$710.40	$49.42	$78.13
1 Laborer	42.10	336.80	67.45	539.60		
1 Grader, 30,000 Lbs.		1062.00		1168.20	66.38	73.01
16 L.H., Daily Totals		$1852.80		$2418.20	$115.80	$151.14

Crew B-11M	Hr.	Daily	Hr.	Daily	Bare Costs	Incl. O&P
1 Equipment Oper. (med.)	$56.75	$454.00	$88.80	$710.40	$49.42	$78.13
1 Laborer	42.10	336.80	67.45	539.60		
1 Backhoe Loader, 80 H.P.		232.40		255.64	14.53	15.98
16 L.H., Daily Totals		$1023.20		$1505.64	$63.95	$94.10

Crew B-11S	Hr.	Daily	Hr.	Daily	Bare Costs	Incl. O&P
1 Equipment Operator (med.)	$56.75	$454.00	$88.80	$710.40	$51.87	$81.68
.5 Laborer	42.10	168.40	67.45	269.80		
1 Dozer, 300 H.P.		1764.00		1940.40		
1 Ripper, Beam & 1 Shank		90.50		99.55	154.54	170.00
12 L.H., Daily Totals		$2476.90		$3020.15	$206.41	$251.68

Crew B-11T	Hr.	Daily	Hr.	Daily	Bare Costs	Incl. O&P
1 Equipment Operator (med.)	$56.75	$454.00	$88.80	$710.40	$51.87	$81.68
.5 Laborer	42.10	168.40	67.45	269.80		
1 Dozer, 410 H.P.		2777.00		3054.70		
1 Ripper, Beam & 2 Shanks		138.75		152.63	242.98	267.28
12 L.H., Daily Totals		$3538.15		$4187.52	$294.85	$348.96

Crew B-11W	Hr.	Daily	Hr.	Daily	Bare Costs	Incl. O&P
1 Equipment Operator (med.)	$56.75	$454.00	$88.80	$710.40	$49.03	$78.23
1 Common Laborer	42.10	336.80	67.45	539.60		
10 Truck Drivers (heavy)	48.95	3916.00	78.25	6260.00		
1 Dozer, 200 H.P.		1504.00		1654.40		
1 Vibratory Roller, Towed, 23 Ton		514.75		566.23		
10 Dump Trucks, 8 C.Y., 220 H.P.		4280.50		4708.55	65.62	72.18
96 L.H., Daily Totals		$11006.05		$14439.17	$114.65	$150.41

Crew B-11Y	Hr.	Daily	Hr.	Daily	Bare Costs	Incl. O&P
1 Labor Foreman (outside)	$44.10	$352.80	$70.65	$565.20	$47.21	$74.92
5 Common Laborers	42.10	1684.00	67.45	2698.00		
3 Equipment Operators (med.)	56.75	1362.00	88.80	2131.20		
1 Dozer, 80 H.P.		401.40		441.54		
2 Rollers, 2-Drum, W.B., 7.5 H.P.		332.20		365.42		
4 Vibrating Plates, Gas, 21"		661.40		727.54	19.38	21.31
72 L.H., Daily Totals		$4793.80		$6928.90	$66.58	$96.23

Crew B-12A	Hr.	Daily	Hr.	Daily	Bare Costs	Incl. O&P
1 Equip. Oper. (crane)	$59.20	$473.60	$92.65	$741.20	$50.65	$80.05
1 Laborer	42.10	336.80	67.45	539.60		
1 Hyd. Excavator, 1 C.Y.		795.30		874.83	49.71	54.68
16 L.H., Daily Totals		$1605.70		$2155.63	$100.36	$134.73

Crew B-12B	Hr.	Daily	Hr.	Daily	Bare Costs	Incl. O&P
1 Equip. Oper. (crane)	$59.20	$473.60	$92.65	$741.20	$50.65	$80.05
1 Laborer	42.10	336.80	67.45	539.60		
1 Hyd. Excavator, 1.5 C.Y.		687.55		756.30	42.97	47.27
16 L.H., Daily Totals		$1497.95		$2037.11	$93.62	$127.32

Crew B-12C	Hr.	Daily	Hr.	Daily	Bare Costs	Incl. O&P
1 Equip. Oper. (crane)	$59.20	$473.60	$92.65	$741.20	$50.65	$80.05
1 Laborer	42.10	336.80	67.45	539.60		
1 Hyd. Excavator, 2 C.Y.		951.30		1046.43	59.46	65.40
16 L.H., Daily Totals		$1761.70		$2327.23	$110.11	$145.45

Crew B-12D	Hr.	Daily	Hr.	Daily	Bare Costs	Incl. O&P
1 Equip. Oper. (crane)	$59.20	$473.60	$92.65	$741.20	$50.65	$80.05
1 Laborer	42.10	336.80	67.45	539.60		
1 Hyd. Excavator, 3.5 C.Y.		2158.00		2373.80	134.88	148.36
16 L.H., Daily Totals		$2968.40		$3654.60	$185.53	$228.41

Crew B-12E	Hr.	Daily	Hr.	Daily	Bare Costs	Incl. O&P
1 Equip. Oper. (crane)	$59.20	$473.60	$92.65	$741.20	$50.65	$80.05
1 Laborer	42.10	336.80	67.45	539.60		
1 Hyd. Excavator, .5 C.Y.		452.10		497.31	28.26	31.08
16 L.H., Daily Totals		$1262.50		$1778.11	$78.91	$111.13

Crew B-12F	Hr.	Daily	Hr.	Daily	Bare Costs	Incl. O&P
1 Equip. Oper. (crane)	$59.20	$473.60	$92.65	$741.20	$50.65	$80.05
1 Laborer	42.10	336.80	67.45	539.60		
1 Hyd. Excavator, .75 C.Y.		694.25		763.67	43.39	47.73
16 L.H., Daily Totals		$1504.65		$2044.47	$94.04	$127.78

Crew B-12G	Hr.	Daily	Hr.	Daily	Bare Costs	Incl. O&P
1 Equip. Oper. (crane)	$59.20	$473.60	$92.65	$741.20	$50.65	$80.05
1 Laborer	42.10	336.80	67.45	539.60		
1 Crawler Crane, 15 Ton		792.30		871.53		
1 Clamshell Bucket, .5 C.Y.		67.10		73.81	53.71	59.08
16 L.H., Daily Totals		$1669.80		$2226.14	$104.36	$139.13

Crew B-12H	Hr.	Daily	Hr.	Daily	Bare Costs	Incl. O&P
1 Equip. Oper. (crane)	$59.20	$473.60	$92.65	$741.20	$50.65	$80.05
1 Laborer	42.10	336.80	67.45	539.60		
1 Crawler Crane, 25 Ton		1129.00		1241.90		
1 Clamshell Bucket, 1 C.Y.		68.55		75.41	74.85	82.33
16 L.H., Daily Totals		$2007.95		$2598.11	$125.50	$162.38

Crew B-12I	Hr.	Daily	Hr.	Daily	Bare Costs	Incl. O&P
1 Equip. Oper. (crane)	$59.20	$473.60	$92.65	$741.20	$50.65	$80.05
1 Laborer	42.10	336.80	67.45	539.60		
1 Crawler Crane, 20 Ton		961.35		1057.48		
1 Dragline Bucket, .75 C.Y.		61.25		67.38	63.91	70.30
16 L.H., Daily Totals		$1833.00		$2405.66	$114.56	$150.35

Crew B-12J	Hr.	Daily	Hr.	Daily	Bare Costs	Incl. O&P
1 Equip. Oper. (crane)	$59.20	$473.60	$92.65	$741.20	$50.65	$80.05
1 Laborer	42.10	336.80	67.45	539.60		
1 Gradall, 5/8 C.Y.		846.50		931.15	52.91	58.20
16 L.H., Daily Totals		$1656.90		$2211.95	$103.56	$138.25

Crew No.	Bare Costs		Incl. Subs O&P		Cost Per Labor-Hour	
Crew B-12K	**Hr.**	**Daily**	**Hr.**	**Daily**	**Bare Costs**	**Incl. O&P**
1 Equip. Oper. (crane)	$59.20	$473.60	$92.65	$741.20	$50.65	$80.05
1 Laborer	42.10	336.80	67.45	539.60		
1 Gradall, 3 Ton, 1 C.Y.		973.80		1071.18	60.86	66.95
16 L.H., Daily Totals		$1784.20		$2351.98	$111.51	$147.00

Crew No.	Bare Costs		Incl. Subs O&P		Cost Per Labor-Hour	
Crew B-12L	**Hr.**	**Daily**	**Hr.**	**Daily**	**Bare Costs**	**Incl. O&P**
1 Equip. Oper. (crane)	$59.20	$473.60	$92.65	$741.20	$50.65	$80.05
1 Laborer	42.10	336.80	67.45	539.60		
1 Crawler Crane, 15 Ton		792.30		871.53		
1 F.E. Attachment, .5 C.Y.		65.25		71.78	53.60	58.96
16 L.H., Daily Totals		$1667.95		$2224.11	$104.25	$139.01

Crew No.	Bare Costs		Incl. Subs O&P		Cost Per Labor-Hour	
Crew B-12M	**Hr.**	**Daily**	**Hr.**	**Daily**	**Bare Costs**	**Incl. O&P**
1 Equip. Oper. (crane)	$59.20	$473.60	$92.65	$741.20	$50.65	$80.05
1 Laborer	42.10	336.80	67.45	539.60		
1 Crawler Crane, 20 Ton		961.35		1057.48		
1 F.E. Attachment, .75 C.Y.		70.40		77.44	64.48	70.93
16 L.H., Daily Totals		$1842.15		$2415.72	$115.13	$150.98

Crew No.	Bare Costs		Incl. Subs O&P		Cost Per Labor-Hour	
Crew B-12N	**Hr.**	**Daily**	**Hr.**	**Daily**	**Bare Costs**	**Incl. O&P**
1 Equip. Oper. (crane)	$59.20	$473.60	$92.65	$741.20	$50.65	$80.05
1 Laborer	42.10	336.80	67.45	539.60		
1 Crawler Crane, 25 Ton		1129.00		1241.90		
1 F.E. Attachment, 1 C.Y.		76.50		84.15	75.34	82.88
16 L.H., Daily Totals		$2015.90		$2606.85	$125.99	$162.93

Crew No.	Bare Costs		Incl. Subs O&P		Cost Per Labor-Hour	
Crew B-12O	**Hr.**	**Daily**	**Hr.**	**Daily**	**Bare Costs**	**Incl. O&P**
1 Equip. Oper. (crane)	$59.20	$473.60	$92.65	$741.20	$50.65	$80.05
1 Laborer	42.10	336.80	67.45	539.60		
1 Crawler Crane, 40 Ton		1206.00		1326.60		
1 F.E. Attachment, 1.5 C.Y.		87.60		96.36	80.85	88.94
16 L.H., Daily Totals		$2104.00		$2703.76	$131.50	$168.99

Crew No.	Bare Costs		Incl. Subs O&P		Cost Per Labor-Hour	
Crew B-12P	**Hr.**	**Daily**	**Hr.**	**Daily**	**Bare Costs**	**Incl. O&P**
1 Equip. Oper. (crane)	$59.20	$473.60	$92.65	$741.20	$50.65	$80.05
1 Laborer	42.10	336.80	67.45	539.60		
1 Crawler Crane, 40 Ton		1206.00		1326.60		
1 Dragline Bucket, 1.5 C.Y.		65.05		71.56	79.44	87.38
16 L.H., Daily Totals		$2081.45		$2678.95	$130.09	$167.43

Crew No.	Bare Costs		Incl. Subs O&P		Cost Per Labor-Hour	
Crew B-12Q	**Hr.**	**Daily**	**Hr.**	**Daily**	**Bare Costs**	**Incl. O&P**
1 Equip. Oper. (crane)	$59.20	$473.60	$92.65	$741.20	$50.65	$80.05
1 Laborer	42.10	336.80	67.45	539.60		
1 Hyd. Excavator, 5/8 C.Y.		598.25		658.08	37.39	41.13
16 L.H., Daily Totals		$1408.65		$1938.88	$88.04	$121.18

Crew No.	Bare Costs		Incl. Subs O&P		Cost Per Labor-Hour	
Crew B-12S	**Hr.**	**Daily**	**Hr.**	**Daily**	**Bare Costs**	**Incl. O&P**
1 Equip. Oper. (crane)	$59.20	$473.60	$92.65	$741.20	$50.65	$80.05
1 Laborer	42.10	336.80	67.45	539.60		
1 Hyd. Excavator, 2.5 C.Y.		1465.00		1611.50	91.56	100.72
16 L.H., Daily Totals		$2275.40		$2892.30	$142.21	$180.77

Crew No.	Bare Costs		Incl. Subs O&P		Cost Per Labor-Hour	
Crew B-12T	**Hr.**	**Daily**	**Hr.**	**Daily**	**Bare Costs**	**Incl. O&P**
1 Equip. Oper. (crane)	$59.20	$473.60	$92.65	$741.20	$50.65	$80.05
1 Laborer	42.10	336.80	67.45	539.60		
1 Crawler Crane, 75 Ton		2002.00		2202.20		
1 F.E. Attachment, 3 C.Y.		114.15		125.57	132.26	145.49
16 L.H., Daily Totals		$2926.55		$3608.57	$182.91	$225.54

Crew No.	Bare Costs		Incl. Subs O&P		Cost Per Labor-Hour	
Crew B-12V	**Hr.**	**Daily**	**Hr.**	**Daily**	**Bare Costs**	**Incl. O&P**
1 Equip. Oper. (crane)	$59.20	$473.60	$92.65	$741.20	$50.65	$80.05
1 Laborer	42.10	336.80	67.45	539.60		
1 Crawler Crane, 75 Ton		2002.00		2202.20		
1 Dragline Bucket, 3 C.Y.		71.60		78.76	129.60	142.56
16 L.H., Daily Totals		$2884.00		$3561.76	$180.25	$222.61

Crew No.	Bare Costs		Incl. Subs O&P		Cost Per Labor-Hour	
Crew B-12Y	**Hr.**	**Daily**	**Hr.**	**Daily**	**Bare Costs**	**Incl. O&P**
1 Equip. Oper. (crane)	$59.20	$473.60	$92.65	$741.20	$47.80	$75.85
2 Laborers	42.10	673.60	67.45	1079.20		
1 Hyd. Excavator, 3.5 C.Y.		2158.00		2373.80	89.92	98.91
24 L.H., Daily Totals		$3305.20		$4194.20	$137.72	$174.76

Crew No.	Bare Costs		Incl. Subs O&P		Cost Per Labor-Hour	
Crew B-12Z	**Hr.**	**Daily**	**Hr.**	**Daily**	**Bare Costs**	**Incl. O&P**
1 Equip. Oper. (crane)	$59.20	$473.60	$92.65	$741.20	$47.80	$75.85
2 Laborers	42.10	673.60	67.45	1079.20		
1 Hyd. Excavator, 2.5 C.Y.		1465.00		1611.50	61.04	67.15
24 L.H., Daily Totals		$2612.20		$3431.90	$108.84	$143.00

Crew No.	Bare Costs		Incl. Subs O&P		Cost Per Labor-Hour	
Crew B-13	**Hr.**	**Daily**	**Hr.**	**Daily**	**Bare Costs**	**Incl. O&P**
1 Labor Foreman (outside)	$44.10	$352.80	$70.65	$565.20	$46.04	$73.17
4 Laborers	42.10	1347.20	67.45	2158.40		
1 Equip. Oper. (crane)	59.20	473.60	92.65	741.20		
1 Equip. Oper. (oiler)	50.55	404.40	79.10	632.80		
1 Hyd. Crane, 25 Ton		580.85		638.93	10.37	11.41
56 L.H., Daily Totals		$3158.85		$4736.53	$56.41	$84.58

Crew No.	Bare Costs		Incl. Subs O&P		Cost Per Labor-Hour	
Crew B-13A	**Hr.**	**Daily**	**Hr.**	**Daily**	**Bare Costs**	**Incl. O&P**
1 Labor Foreman (outside)	$44.10	$352.80	$70.65	$565.20	$48.53	$77.09
2 Laborers	42.10	673.60	67.45	1079.20		
2 Equipment Operators (med.)	56.75	908.00	88.80	1420.80		
2 Truck Drivers (heavy)	48.95	783.20	78.25	1252.00		
1 Crawler Crane, 75 Ton		2002.00		2202.20		
1 Crawler Loader, 4 C.Y.		1489.00		1637.90		
2 Dump Trucks, 8 C.Y., 220 H.P.		856.10		941.71	77.63	85.39
56 L.H., Daily Totals		$7064.70		$9099.01	$126.16	$162.48

Crew No.	Bare Costs		Incl. Subs O&P		Cost Per Labor-Hour	
Crew B-13B	**Hr.**	**Daily**	**Hr.**	**Daily**	**Bare Costs**	**Incl. O&P**
1 Labor Foreman (outside)	$44.10	$352.80	$70.65	$565.20	$46.04	$73.17
4 Laborers	42.10	1347.20	67.45	2158.40		
1 Equip. Oper. (crane)	59.20	473.60	92.65	741.20		
1 Equip. Oper. (oiler)	50.55	404.40	79.10	632.80		
1 Hyd. Crane, 55 Ton		980.20		1078.22	17.50	19.25
56 L.H., Daily Totals		$3558.20		$5175.82	$63.54	$92.43

Crew No.	Bare Costs		Incl. Subs O&P		Cost Per Labor-Hour	
Crew B-13C	**Hr.**	**Daily**	**Hr.**	**Daily**	**Bare Costs**	**Incl. O&P**
1 Labor Foreman (outside)	$44.10	$352.80	$70.65	$565.20	$46.04	$73.17
4 Laborers	42.10	1347.20	67.45	2158.40		
1 Equip. Oper. (crane)	59.20	473.60	92.65	741.20		
1 Equip. Oper. (oiler)	50.55	404.40	79.10	632.80		
1 Crawler Crane, 100 Ton		2287.00		2515.70	40.84	44.92
56 L.H., Daily Totals		$4865.00		$6613.30	$86.88	$118.09

Crew No.	Bare Costs		Incl. Subs O&P		Cost Per Labor-Hour	
Crew B-13D	**Hr.**	**Daily**	**Hr.**	**Daily**	**Bare Costs**	**Incl. O&P**
1 Laborer	$42.10	$336.80	$67.45	$539.60	$50.65	$80.05
1 Equip. Oper. (crane)	59.20	473.60	92.65	741.20		
1 Hyd. Excavator, 1 C.Y.		795.30		874.83		
1 Trench Box		117.75		129.53	57.07	62.77
16 L.H., Daily Totals		$1723.45		$2285.16	$107.72	$142.82

Crew No.	Bare Costs		Incl. Subs O&P		Cost Per Labor-Hour	

Crew B-13E	Hr.	Daily	Hr.	Daily	Bare Costs	Incl. O&P
1 Laborer	$42.10	$336.80	$67.45	$539.60	$50.65	$80.05
1 Equip. Oper. (crane)	59.20	473.60	92.65	741.20		
1 Hyd. Excavator, 1.5 C.Y.		687.55		756.30		
1 Trench Box		117.75		129.53	50.33	55.36
16 L.H., Daily Totals		$1615.70		$2166.63	$100.98	$135.41

Crew B-13F	Hr.	Daily	Hr.	Daily	Bare Costs	Incl. O&P
1 Laborer	$42.10	$336.80	$67.45	$539.60	$50.65	$80.05
1 Equip. Oper. (crane)	59.20	473.60	92.65	741.20		
1 Hyd. Excavator, 3.5 C.Y.		2158.00		2373.80		
1 Trench Box		117.75		129.53	142.23	156.46
16 L.H., Daily Totals		$3086.15		$3784.13	$192.88	$236.51

Crew B-13G	Hr.	Daily	Hr.	Daily	Bare Costs	Incl. O&P
1 Laborer	$42.10	$336.80	$67.45	$539.60	$50.65	$80.05
1 Equip. Oper. (crane)	59.20	473.60	92.65	741.20		
1 Hyd. Excavator, .75 C.Y.		694.25		763.67		
1 Trench Box		117.75		129.53	50.75	55.83
16 L.H., Daily Totals		$1622.40		$2174.00	$101.40	$135.88

Crew B-13H	Hr.	Daily	Hr.	Daily	Bare Costs	Incl. O&P
1 Laborer	$42.10	$336.80	$67.45	$539.60	$50.65	$80.05
1 Equip. Oper. (crane)	59.20	473.60	92.65	741.20		
1 Gradall, 5/8 C.Y.		846.50		931.15		
1 Trench Box		117.75		129.53	60.27	66.29
16 L.H., Daily Totals		$1774.65		$2341.47	$110.92	$146.34

Crew B-13I	Hr.	Daily	Hr.	Daily	Bare Costs	Incl. O&P
1 Laborer	$42.10	$336.80	$67.45	$539.60	$50.65	$80.05
1 Equip. Oper. (crane)	59.20	473.60	92.65	741.20		
1 Gradall, 3 Ton, 1 C.Y.		973.80		1071.18		
1 Trench Box		117.75		129.53	68.22	75.04
16 L.H., Daily Totals		$1901.95		$2481.51	$118.87	$155.09

Crew B-13J	Hr.	Daily	Hr.	Daily	Bare Costs	Incl. O&P
1 Laborer	$42.10	$336.80	$67.45	$539.60	$50.65	$80.05
1 Equip. Oper. (crane)	59.20	473.60	92.65	741.20		
1 Hyd. Excavator, 2.5 C.Y.		1465.00		1611.50		
1 Trench Box		117.75		129.53	98.92	108.81
16 L.H., Daily Totals		$2393.15		$3021.82	$149.57	$188.86

Crew B-13K	Hr.	Daily	Hr.	Daily	Bare Costs	Incl. O&P
2 Equip. Opers. (crane)	$59.20	$947.20	$92.65	$1482.40	$59.20	$92.65
1 Hyd. Excavator, .75 C.Y.		694.25		763.67		
1 Hyd. Hammer, 4000 ft-lb		641.50		705.65		
1 Hyd. Excavator, .75 C.Y.		694.25		763.67	126.88	139.56
16 L.H., Daily Totals		$2977.20		$3715.40	$186.07	$232.21

Crew B-13L	Hr.	Daily	Hr.	Daily	Bare Costs	Incl. O&P
2 Equip. Opers. (crane)	$59.20	$947.20	$92.65	$1482.40	$59.20	$92.65
1 Hyd. Excavator, 1.5 C.Y.		687.55		756.30		
1 Hyd. Hammer, 5000 ft-lb		697.25		766.98		
1 Hyd. Excavator, .75 C.Y.		694.25		763.67	129.94	142.93
16 L.H., Daily Totals		$3026.25		$3769.36	$189.14	$235.58

Crew No.	Bare Costs		Incl. Subs O&P		Cost Per Labor-Hour	

Crew B-13M	Hr.	Daily	Hr.	Daily	Bare Costs	Incl. O&P
2 Equip. Opers. (crane)	$59.20	$947.20	$92.65	$1482.40	$59.20	$92.65
1 Hyd. Excavator, 2.5 C.Y.		1465.00		1611.50		
1 Hyd. Hammer, 8000 ft-lb		907.75		998.52		
1 Hyd. Excavator, 1.5 C.Y.		687.55		756.30	191.27	210.40
16 L.H., Daily Totals		$4007.50		$4848.73	$250.47	$303.05

Crew B-13N	Hr.	Daily	Hr.	Daily	Bare Costs	Incl. O&P
2 Equip. Opers. (crane)	$59.20	$947.20	$92.65	$1482.40	$59.20	$92.65
1 Hyd. Excavator, 3.5 C.Y.		2158.00		2373.80		
1 Hyd. Hammer, 12,000 ft-lb		871.75		958.92		
1 Hyd. Excavator, 1.5 C.Y.		687.55		756.30	232.33	255.56
16 L.H., Daily Totals		$4664.50		$5571.43	$291.53	$348.21

Crew B-14	Hr.	Daily	Hr.	Daily	Bare Costs	Incl. O&P
1 Labor Foreman (outside)	$44.10	$352.80	$70.65	$565.20	$44.25	$70.57
4 Laborers	42.10	1347.20	67.45	2158.40		
1 Equip. Oper. (light)	53.00	424.00	82.95	663.60		
1 Backhoe Loader, 48 H.P.		213.75		235.13	4.45	4.90
48 L.H., Daily Totals		$2337.75		$3622.32	$48.70	$75.47

Crew B-14A	Hr.	Daily	Hr.	Daily	Bare Costs	Incl. O&P
1 Equip. Oper. (crane)	$59.20	$473.60	$92.65	$741.20	$53.50	$84.25
.5 Laborer	42.10	168.40	67.45	269.80		
1 Hyd. Excavator, 4.5 C.Y.		3409.00		3749.90	284.08	312.49
12 L.H., Daily Totals		$4051.00		$4760.90	$337.58	$396.74

Crew B-14B	Hr.	Daily	Hr.	Daily	Bare Costs	Incl. O&P
1 Equip. Oper. (crane)	$59.20	$473.60	$92.65	$741.20	$53.50	$84.25
.5 Laborer	42.10	168.40	67.45	269.80		
1 Hyd. Excavator, 6 C.Y.		3464.00		3810.40	288.67	317.53
12 L.H., Daily Totals		$4106.00		$4821.40	$342.17	$401.78

Crew B-14C	Hr.	Daily	Hr.	Daily	Bare Costs	Incl. O&P
1 Equip. Oper. (crane)	$59.20	$473.60	$92.65	$741.20	$53.50	$84.25
.5 Laborer	42.10	168.40	67.45	269.80		
1 Hyd. Excavator, 7 C.Y.		3437.00		3780.70	286.42	315.06
12 L.H., Daily Totals		$4079.00		$4791.70	$339.92	$399.31

Crew B-14F	Hr.	Daily	Hr.	Daily	Bare Costs	Incl. O&P
1 Equip. Oper. (crane)	$59.20	$473.60	$92.65	$741.20	$53.50	$84.25
.5 Laborer	42.10	168.40	67.45	269.80		
1 Hyd. Shovel, 7 C.Y.		4099.00		4508.90	341.58	375.74
12 L.H., Daily Totals		$4741.00		$5519.90	$395.08	$459.99

Crew B-14G	Hr.	Daily	Hr.	Daily	Bare Costs	Incl. O&P
1 Equip. Oper. (crane)	$59.20	$473.60	$92.65	$741.20	$53.50	$84.25
.5 Laborer	42.10	168.40	67.45	269.80		
1 Hyd. Shovel, 12 C.Y.		5951.00		6546.10	495.92	545.51
12 L.H., Daily Totals		$6593.00		$7557.10	$549.42	$629.76

Crew B-14J	Hr.	Daily	Hr.	Daily	Bare Costs	Incl. O&P
1 Equip. Oper. (medium)	$56.75	$454.00	$88.80	$710.40	$51.87	$81.68
.5 Laborer	42.10	168.40	67.45	269.80		
1 F.E. Loader, 8 C.Y.		2261.00		2487.10	188.42	207.26
12 L.H., Daily Totals		$2883.40		$3467.30	$240.28	$288.94

Crew No.	Bare Costs		Incl. Subs O&P		Cost Per Labor-Hour	
Crew B-14K	**Hr.**	**Daily**	**Hr.**	**Daily**	**Bare Costs**	**Incl. O&P**
1 Equip. Oper. (medium)	$56.75	$454.00	$88.80	$710.40	$51.87	$81.68
.5 Laborer	42.10	168.40	67.45	269.80		
1 F.E. Loader, 10 C.Y.		2674.00		2941.40	222.83	245.12
12 L.H., Daily Totals		$3296.40		$3921.60	$274.70	$326.80

Crew No.	Bare Costs		Incl. Subs O&P		Cost Per Labor-Hour	
Crew B-15	**Hr.**	**Daily**	**Hr.**	**Daily**	**Bare Costs**	**Incl. O&P**
1 Equipment Oper. (med.)	$56.75	$454.00	$88.80	$710.40	$50.20	$79.72
.5 Laborer	42.10	168.40	67.45	269.80		
2 Truck Drivers (heavy)	48.95	783.20	78.25	1252.00		
2 Dump Trucks, 12 C.Y., 400 H.P.		1145.80		1260.38		
1 Dozer, 200 H.P.		1504.00		1654.40	94.64	104.10
28 L.H., Daily Totals		$4055.40		$5146.98	$144.84	$183.82

Crew No.	Bare Costs		Incl. Subs O&P		Cost Per Labor-Hour	
Crew B-16	**Hr.**	**Daily**	**Hr.**	**Daily**	**Bare Costs**	**Incl. O&P**
1 Labor Foreman (outside)	$44.10	$352.80	$70.65	$565.20	$44.31	$70.95
2 Laborers	42.10	673.60	67.45	1079.20		
1 Truck Driver (heavy)	48.95	391.60	78.25	626.00		
1 Dump Truck, 12 C.Y., 400 H.P.		572.90		630.19	17.90	19.69
32 L.H., Daily Totals		$1990.90		$2900.59	$62.22	$90.64

Crew No.	Bare Costs		Incl. Subs O&P		Cost Per Labor-Hour	
Crew B-17	**Hr.**	**Daily**	**Hr.**	**Daily**	**Bare Costs**	**Incl. O&P**
2 Laborers	$42.10	$673.60	$67.45	$1079.20	$46.54	$74.03
1 Equip. Oper. (light)	53.00	424.00	82.95	663.60		
1 Truck Driver (heavy)	48.95	391.60	78.25	626.00		
1 Backhoe Loader, 48 H.P.		213.75		235.13		
1 Dump Truck, 8 C.Y., 220 H.P.		428.05		470.86	20.06	22.06
32 L.H., Daily Totals		$2131.00		$3074.78	$66.59	$96.09

Crew No.	Bare Costs		Incl. Subs O&P		Cost Per Labor-Hour	
Crew B-17A	**Hr.**	**Daily**	**Hr.**	**Daily**	**Bare Costs**	**Incl. O&P**
2 Labor Foremen (outside)	$44.10	$705.60	$70.65	$1130.40	$45.25	$72.39
6 Laborers	42.10	2020.80	67.45	3237.60		
1 Skilled Worker Foreman (out)	56.85	454.80	90.55	724.40		
1 Skilled Worker	54.85	438.80	87.40	699.20		
80 L.H., Daily Totals		$3620.00		$5791.60	$45.25	$72.39

Crew No.	Bare Costs		Incl. Subs O&P		Cost Per Labor-Hour	
Crew B-17B	**Hr.**	**Daily**	**Hr.**	**Daily**	**Bare Costs**	**Incl. O&P**
2 Laborers	$42.10	$673.60	$67.45	$1079.20	$46.54	$74.03
1 Equip. Oper. (light)	53.00	424.00	82.95	663.60		
1 Truck Driver (heavy)	48.95	391.60	78.25	626.00		
1 Backhoe Loader, 48 H.P.		213.75		235.13		
1 Dump Truck, 12 C.Y., 400 H.P.		572.90		630.19	24.58	27.04
32 L.H., Daily Totals		$2275.85		$3234.11	$71.12	$101.07

Crew No.	Bare Costs		Incl. Subs O&P		Cost Per Labor-Hour	
Crew B-18	**Hr.**	**Daily**	**Hr.**	**Daily**	**Bare Costs**	**Incl. O&P**
1 Labor Foreman (outside)	$44.10	$352.80	$70.65	$565.20	$42.77	$68.52
2 Laborers	42.10	673.60	67.45	1079.20		
1 Vibrating Plate, Gas, 21"		165.35		181.88	6.89	7.58
24 L.H., Daily Totals		$1191.75		$1826.29	$49.66	$76.10

Crew No.	Bare Costs		Incl. Subs O&P		Cost Per Labor-Hour	
Crew B-19	**Hr.**	**Daily**	**Hr.**	**Daily**	**Bare Costs**	**Incl. O&P**
1 Pile Driver Foreman (outside)	$56.20	$449.60	$90.15	$721.20	$55.24	$87.77
4 Pile Drivers	54.20	1734.40	86.90	2780.80		
2 Equip. Oper. (crane)	59.20	947.20	92.65	1482.40		
1 Equip. Oper. (oiler)	50.55	404.40	79.10	632.80		
1 Crawler Crane, 40 Ton		1206.00		1326.60		
1 Lead, 90' High		367.45		404.19		
1 Hammer, Diesel, 22k ft-lb		436.45		480.10	31.40	34.55
64 L.H., Daily Totals		$5545.50		$7828.09	$86.65	$122.31

Crew No.	Bare Costs		Incl. Subs O&P		Cost Per Labor-Hour	
Crew B-19A	**Hr.**	**Daily**	**Hr.**	**Daily**	**Bare Costs**	**Incl. O&P**
1 Pile Driver Foreman (outside)	$56.20	$449.60	$90.15	$721.20	$55.24	$87.77
4 Pile Drivers	54.20	1734.40	86.90	2780.80		
2 Equip. Oper. (crane)	59.20	947.20	92.65	1482.40		
1 Equip. Oper. (oiler)	50.55	404.40	79.10	632.80		
1 Crawler Crane, 75 Ton		2002.00		2202.20		
1 Lead, 90' High		367.45		404.19		
1 Hammer, Diesel, 41k ft-lb		576.65		634.32	46.03	50.64
64 L.H., Daily Totals		$6481.70		$8857.91	$101.28	$138.40

Crew No.	Bare Costs		Incl. Subs O&P		Cost Per Labor-Hour	
Crew B-19B	**Hr.**	**Daily**	**Hr.**	**Daily**	**Bare Costs**	**Incl. O&P**
1 Pile Driver Foreman (outside)	$56.20	$449.60	$90.15	$721.20	$55.24	$87.77
4 Pile Drivers	54.20	1734.40	86.90	2780.80		
2 Equip. Oper. (crane)	59.20	947.20	92.65	1482.40		
1 Equip. Oper. (oiler)	50.55	404.40	79.10	632.80		
1 Crawler Crane, 40 Ton		1206.00		1326.60		
1 Lead, 90' High		367.45		404.19		
1 Hammer, Diesel, 22k ft-lb		436.45		480.10		
1 Barge, 400 Ton		858.85		944.74	44.82	49.31
64 L.H., Daily Totals		$6404.35		$8772.83	$100.07	$137.08

Crew No.	Bare Costs		Incl. Subs O&P		Cost Per Labor-Hour	
Crew B-19C	**Hr.**	**Daily**	**Hr.**	**Daily**	**Bare Costs**	**Incl. O&P**
1 Pile Driver Foreman (outside)	$56.20	$449.60	$90.15	$721.20	$55.24	$87.77
4 Pile Drivers	54.20	1734.40	86.90	2780.80		
2 Equip. Oper. (crane)	59.20	947.20	92.65	1482.40		
1 Equip. Oper. (oiler)	50.55	404.40	79.10	632.80		
1 Crawler Crane, 75 Ton		2002.00		2202.20		
1 Lead, 90' High		367.45		404.19		
1 Hammer, Diesel, 41k ft-lb		576.65		634.32		
1 Barge, 400 Ton		858.85		944.74	59.45	65.40
64 L.H., Daily Totals		$7340.55		$9802.65	$114.70	$153.17

Crew No.	Bare Costs		Incl. Subs O&P		Cost Per Labor-Hour	
Crew B-20	**Hr.**	**Daily**	**Hr.**	**Daily**	**Bare Costs**	**Incl. O&P**
1 Labor Foreman (outside)	$44.10	$352.80	$70.65	$565.20	$47.02	$75.17
1 Skilled Worker	54.85	438.80	87.40	699.20		
1 Laborer	42.10	336.80	67.45	539.60		
24 L.H., Daily Totals		$1128.40		$1804.00	$47.02	$75.17

Crew No.	Bare Costs		Incl. Subs O&P		Cost Per Labor-Hour	
Crew B-20A	**Hr.**	**Daily**	**Hr.**	**Daily**	**Bare Costs**	**Incl. O&P**
1 Labor Foreman (outside)	$44.10	$352.80	$70.65	$565.20	$50.55	$79.36
1 Laborer	42.10	336.80	67.45	539.60		
1 Plumber	64.45	515.60	99.65	797.20		
1 Plumber Apprentice	51.55	412.40	79.70	637.60		
32 L.H., Daily Totals		$1617.60		$2539.60	$50.55	$79.36

Crew No.	Bare Costs		Incl. Subs O&P		Cost Per Labor-Hour	
Crew B-21	**Hr.**	**Daily**	**Hr.**	**Daily**	**Bare Costs**	**Incl. O&P**
1 Labor Foreman (outside)	$44.10	$352.80	$70.65	$565.20	$48.76	$77.66
1 Skilled Worker	54.85	438.80	87.40	699.20		
1 Laborer	42.10	336.80	67.45	539.60		
.5 Equip. Oper. (crane)	59.20	236.80	92.65	370.60		
.5 S.P. Crane, 4x4, 5 Ton		189.05		207.96	6.75	7.43
28 L.H., Daily Totals		$1554.25		$2382.55	$55.51	$85.09

Crew No.	Bare Costs		Incl. Subs O&P		Cost Per Labor-Hour	
Crew B-21A	**Hr.**	**Daily**	**Hr.**	**Daily**	**Bare Costs**	**Incl. O&P**
1 Labor Foreman (outside)	$44.10	$352.80	$70.65	$565.20	$52.28	$82.02
1 Laborer	42.10	336.80	67.45	539.60		
1 Plumber	64.45	515.60	99.65	797.20		
1 Plumber Apprentice	51.55	412.40	79.70	637.60		
1 Equip. Oper. (crane)	59.20	473.60	92.65	741.20		
1 S.P. Crane, 4x4, 12 Ton		428.30		471.13	10.71	11.78
40 L.H., Daily Totals		$2519.50		$3751.93	$62.99	$93.80

Crew No.	Bare Costs		Incl. Subs O&P		Cost Per Labor-Hour	
Crew B-21B	**Hr.**	**Daily**	**Hr.**	**Daily**	**Bare Costs**	**Incl. O&P**
1 Labor Foreman (outside)	$44.10	$352.80	$70.65	$565.20	$45.92	$73.13
3 Laborers	42.10	1010.40	67.45	1618.80		
1 Equip. Oper. (crane)	59.20	473.60	92.65	741.20		
1 Hyd. Crane, 12 Ton		471.00		518.10	11.78	12.95
40 L.H., Daily Totals		$2307.80		$3443.30	$57.70	$86.08
Crew B-21C	**Hr.**	**Daily**	**Hr.**	**Daily**	**Bare Costs**	**Incl. O&P**
1 Labor Foreman (outside)	$44.10	$352.80	$70.65	$565.20	$46.04	$73.17
4 Laborers	42.10	1347.20	67.45	2158.40		
1 Equip. Oper. (crane)	59.20	473.60	92.65	741.20		
1 Equip. Oper. (oiler)	50.55	404.40	79.10	632.80		
2 Cutting Torches		25.60		28.16		
2 Sets of Gases		343.10		377.41		
1 Lattice Boom Crane, 90 Ton		1696.00		1865.60	36.87	40.56
56 L.H., Daily Totals		$4642.70		$6368.77	$82.91	$113.73
Crew B-22	**Hr.**	**Daily**	**Hr.**	**Daily**	**Bare Costs**	**Incl. O&P**
1 Labor Foreman (outside)	$44.10	$352.80	$70.65	$565.20	$49.45	$78.66
1 Skilled Worker	54.85	438.80	87.40	699.20		
1 Laborer	42.10	336.80	67.45	539.60		
.75 Equip. Oper. (crane)	59.20	355.20	92.65	555.90		
.75 S.P. Crane, 4x4, 5 Ton		283.57		311.93	9.45	10.40
30 L.H., Daily Totals		$1767.18		$2671.83	$58.91	$89.06
Crew B-22A	**Hr.**	**Daily**	**Hr.**	**Daily**	**Bare Costs**	**Incl. O&P**
1 Labor Foreman (outside)	$44.10	$352.80	$70.65	$565.20	$48.47	$77.12
1 Skilled Worker	54.85	438.80	87.40	699.20		
2 Laborers	42.10	673.60	67.45	1079.20		
1 Equipment Operator, Crane	59.20	473.60	92.65	741.20		
1 S.P. Crane, 4x4, 5 Ton		378.10		415.91		
1 Butt Fusion Mach., 4"-12" diam.		426.85		469.54	20.12	22.14
40 L.H., Daily Totals		$2743.75		$3970.24	$68.59	$99.26
Crew B-22B	**Hr.**	**Daily**	**Hr.**	**Daily**	**Bare Costs**	**Incl. O&P**
1 Labor Foreman (outside)	$44.10	$352.80	$70.65	$565.20	$48.47	$77.12
1 Skilled Worker	54.85	438.80	87.40	699.20		
2 Laborers	42.10	673.60	67.45	1079.20		
1 Equip. Oper. (crane)	59.20	473.60	92.65	741.20		
1 S.P. Crane, 4x4, 5 Ton		378.10		415.91		
1 Butt Fusion Mach., 8"-24" diam.		1076.00		1183.60	36.35	39.99
40 L.H., Daily Totals		$3392.90		$4684.31	$84.82	$117.11
Crew B-22C	**Hr.**	**Daily**	**Hr.**	**Daily**	**Bare Costs**	**Incl. O&P**
1 Skilled Worker	$54.85	$438.80	$87.40	$699.20	$48.48	$77.42
1 Laborer	42.10	336.80	67.45	539.60		
1 Butt Fusion Mach., 2"-8" diam.		156.05		171.66	9.75	10.73
16 L.H., Daily Totals		$931.65		$1410.45	$58.23	$88.15
Crew B-23	**Hr.**	**Daily**	**Hr.**	**Daily**	**Bare Costs**	**Incl. O&P**
1 Labor Foreman (outside)	$44.10	$352.80	$70.65	$565.20	$42.50	$68.09
4 Laborers	42.10	1347.20	67.45	2158.40		
1 Drill Rig, Truck-Mounted		771.10		848.21		
1 Flatbed Truck, Gas, 3 Ton		820.40		902.44	39.79	43.77
40 L.H., Daily Totals		$3291.50		$4474.25	$82.29	$111.86

Crew No.	Bare Costs		Incl. Subs O&P		Cost Per Labor-Hour	
Crew B-23A	**Hr.**	**Daily**	**Hr.**	**Daily**	**Bare Costs**	**Incl. O&P**
1 Labor Foreman (outside)	$44.10	$352.80	$70.65	$565.20	$47.65	$75.63
1 Laborer	42.10	336.80	67.45	539.60		
1 Equip. Oper. (medium)	56.75	454.00	88.80	710.40		
1 Drill Rig, Truck-Mounted		771.10		848.21		
1 Pickup Truck, 3/4 Ton		110.85		121.94	36.75	40.42
24 L.H., Daily Totals		$2025.55		$2785.34	$84.40	$116.06
Crew B-23B	**Hr.**	**Daily**	**Hr.**	**Daily**	**Bare Costs**	**Incl. O&P**
1 Labor Foreman (outside)	$44.10	$352.80	$70.65	$565.20	$47.65	$75.63
1 Laborer	42.10	336.80	67.45	539.60		
1 Equip. Oper. (medium)	56.75	454.00	88.80	710.40		
1 Drill Rig, Truck-Mounted		771.10		848.21		
1 Pickup Truck, 3/4 Ton		110.85		121.94		
1 Centr. Water Pump, 6"		232.90		256.19	46.45	51.10
24 L.H., Daily Totals		$2258.45		$3041.53	$94.10	$126.73
Crew B-24	**Hr.**	**Daily**	**Hr.**	**Daily**	**Bare Costs**	**Incl. O&P**
1 Cement Finisher	$49.95	$399.60	$78.45	$627.60	$48.40	$77.02
1 Laborer	42.10	336.80	67.45	539.60		
1 Carpenter	53.15	425.20	85.15	681.20		
24 L.H., Daily Totals		$1161.60		$1848.40	$48.40	$77.02
Crew B-25	**Hr.**	**Daily**	**Hr.**	**Daily**	**Bare Costs**	**Incl. O&P**
1 Labor Foreman (outside)	$44.10	$352.80	$70.65	$565.20	$46.28	$73.56
7 Laborers	42.10	2357.60	67.45	3777.20		
3 Equip. Oper. (medium)	56.75	1362.00	88.80	2131.20		
1 Asphalt Paver, 130 H.P.		2117.00		2328.70		
1 Tandem Roller, 10 Ton		236.75		260.43		
1 Roller, Pneum. Whl., 12 Ton		345.75		380.32	30.68	33.74
88 L.H., Daily Totals		$6771.90		$9443.05	$76.95	$107.31
Crew B-25B	**Hr.**	**Daily**	**Hr.**	**Daily**	**Bare Costs**	**Incl. O&P**
1 Labor Foreman (outside)	$44.10	$352.80	$70.65	$565.20	$47.15	$74.83
7 Laborers	42.10	2357.60	67.45	3777.20		
4 Equip. Oper. (medium)	56.75	1816.00	88.80	2841.60		
1 Asphalt Paver, 130 H.P.		2117.00		2328.70		
2 Tandem Rollers, 10 Ton		473.50		520.85		
1 Roller, Pneum. Whl., 12 Ton		345.75		380.32	30.59	33.64
96 L.H., Daily Totals		$7462.65		$10413.88	$77.74	$108.48
Crew B-25C	**Hr.**	**Daily**	**Hr.**	**Daily**	**Bare Costs**	**Incl. O&P**
1 Labor Foreman (outside)	$44.10	$352.80	$70.65	$565.20	$47.32	$75.10
3 Laborers	42.10	1010.40	67.45	1618.80		
2 Equip. Oper. (medium)	56.75	908.00	88.80	1420.80		
1 Asphalt Paver, 130 H.P.		2117.00		2328.70		
1 Tandem Roller, 10 Ton		236.75		260.43	49.04	53.94
48 L.H., Daily Totals		$4624.95		$6193.93	$96.35	$129.04
Crew B-25D	**Hr.**	**Daily**	**Hr.**	**Daily**	**Bare Costs**	**Incl. O&P**
1 Labor Foreman (outside)	$44.10	$352.80	$70.65	$565.20	$47.54	$75.44
3 Laborers	42.10	1010.40	67.45	1618.80		
2.125 Equip. Oper. (medium)	56.75	964.75	88.80	1509.60		
.125 Truck Driver (heavy)	48.95	48.95	78.25	78.25		
.125 Truck Tractor, 6x4, 380 H.P.		61.66		67.82		
.125 Dist. Tanker, 3000 Gallon		41.27		45.40		
1 Asphalt Paver, 130 H.P.		2117.00		2328.70		
1 Tandem Roller, 10 Ton		236.75		260.43	49.13	54.05
50 L.H., Daily Totals		$4833.58		$6474.19	$96.67	$129.48

Crews - Renovation

Crew No.	Bare Costs		Incl. Subs O&P		Cost Per Labor-Hour	
Crew B-25E	**Hr.**	**Daily**	**Hr.**	**Daily**	**Bare Costs**	**Incl. O&P**
1 Labor Foreman (outside)	$44.10	$352.80	$70.65	$565.20	$47.74	$75.75
3 Laborers	42.10	1010.40	67.45	1618.80		
2.250 Equip. Oper. (medium)	56.75	1021.50	88.80	1598.40		
.25 Truck Driver (heavy)	48.95	97.90	78.25	156.50		
.25 Truck Tractor, 6x4, 380 H.P.		123.31		135.64		
.25 Dist. Tanker, 3000 Gallon		82.54		90.79		
1 Asphalt Paver, 130 H.P.		2117.00		2328.70		
1 Tandem Roller, 10 Ton		236.75		260.43	49.22	54.15
52 L.H., Daily Totals		$5042.20		$6754.46	$96.97	$129.89

Crew B-26	Hr.	Daily	Hr.	Daily	Bare Costs	Incl. O&P
1 Labor Foreman (outside)	$44.10	$352.80	$70.65	$565.20	$46.96	$74.57
6 Laborers	42.10	2020.80	67.45	3237.60		
2 Equip. Oper. (medium)	56.75	908.00	88.80	1420.80		
1 Rodman (reinf.)	56.40	451.20	88.85	710.80		
1 Cement Finisher	49.95	399.60	78.45	627.60		
1 Grader, 30,000 Lbs.		1062.00		1168.20		
1 Paving Mach. & Equip.		2474.00		2721.40	40.18	44.20
88 L.H., Daily Totals		$7668.40		$10451.60	$87.14	$118.77

Crew B-26A	Hr.	Daily	Hr.	Daily	Bare Costs	Incl. O&P
1 Labor Foreman (outside)	$44.10	$352.80	$70.65	$565.20	$46.96	$74.57
6 Laborers	42.10	2020.80	67.45	3237.60		
2 Equip. Oper. (medium)	56.75	908.00	88.80	1420.80		
1 Rodman (reinf.)	56.40	451.20	88.85	710.80		
1 Cement Finisher	49.95	399.60	78.45	627.60		
1 Grader, 30,000 Lbs.		1062.00		1168.20		
1 Paving Mach. & Equip.		2474.00		2721.40		
1 Concrete Saw		111.50		122.65	41.45	45.59
88 L.H., Daily Totals		$7779.90		$10574.25	$88.41	$120.16

Crew B-26B	Hr.	Daily	Hr.	Daily	Bare Costs	Incl. O&P
1 Labor Foreman (outside)	$44.10	$352.80	$70.65	$565.20	$47.77	$75.75
6 Laborers	42.10	2020.80	67.45	3237.60		
3 Equip. Oper. (medium)	56.75	1362.00	88.80	2131.20		
1 Rodman (reinf.)	56.40	451.20	88.85	710.80		
1 Cement Finisher	49.95	399.60	78.45	627.60		
1 Grader, 30,000 Lbs.		1062.00		1168.20		
1 Paving Mach. & Equip.		2474.00		2721.40		
1 Concrete Pump, 110' Boom		470.75		517.83	41.74	45.91
96 L.H., Daily Totals		$8593.15		$11679.83	$89.51	$121.66

Crew B-26C	Hr.	Daily	Hr.	Daily	Bare Costs	Incl. O&P
1 Labor Foreman (outside)	$44.10	$352.80	$70.65	$565.20	$45.98	$73.14
6 Laborers	42.10	2020.80	67.45	3237.60		
1 Equip. Oper. (medium)	56.75	454.00	88.80	710.40		
1 Rodman (reinf.)	56.40	451.20	88.85	710.80		
1 Cement Finisher	49.95	399.60	78.45	627.60		
1 Paving Mach. & Equip.		2474.00		2721.40		
1 Concrete Saw		111.50		122.65	32.32	35.55
80 L.H., Daily Totals		$6263.90		$8695.65	$78.30	$108.70

Crew B-27	Hr.	Daily	Hr.	Daily	Bare Costs	Incl. O&P
1 Labor Foreman (outside)	$44.10	$352.80	$70.65	$565.20	$42.60	$68.25
3 Laborers	42.10	1010.40	67.45	1618.80		
1 Berm Machine		271.95		299.14	8.50	9.35
32 L.H., Daily Totals		$1635.15		$2483.15	$51.10	$77.60

Crew B-28	Hr.	Daily	Hr.	Daily	Bare Costs	Incl. O&P
2 Carpenters	$53.15	$850.40	$85.15	$1362.40	$49.47	$79.25
1 Laborer	42.10	336.80	67.45	539.60		
24 L.H., Daily Totals		$1187.20		$1902.00	$49.47	$79.25

Crew B-29	Hr.	Daily	Hr.	Daily	Bare Costs	Incl. O&P
1 Labor Foreman (outside)	$44.10	$352.80	$70.65	$565.20	$46.04	$73.17
4 Laborers	42.10	1347.20	67.45	2158.40		
1 Equip. Oper. (crane)	59.20	473.60	92.65	741.20		
1 Equip. Oper. (oiler)	50.55	404.40	79.10	632.80		
1 Gradall, 5/8 C.Y.		846.50		931.15	15.12	16.63
56 L.H., Daily Totals		$3424.50		$5028.75	$61.15	$89.80

Crew B-30	Hr.	Daily	Hr.	Daily	Bare Costs	Incl. O&P
1 Equip. Oper. (medium)	$56.75	$454.00	$88.80	$710.40	$51.55	$81.77
2 Truck Drivers (heavy)	48.95	783.20	78.25	1252.00		
1 Hyd. Excavator, 1.5 C.Y.		687.55		756.30		
2 Dump Trucks, 12 C.Y., 400 H.P.		1145.80		1260.38	76.39	84.03
24 L.H., Daily Totals		$3070.55		$3979.09	$127.94	$165.80

Crew B-31	Hr.	Daily	Hr.	Daily	Bare Costs	Incl. O&P
1 Labor Foreman (outside)	$44.10	$352.80	$70.65	$565.20	$44.71	$71.63
3 Laborers	42.10	1010.40	67.45	1618.80		
1 Carpenter	53.15	425.20	85.15	681.20		
1 Air Compressor, 250 cfm		201.50		221.65		
1 Sheeting Driver		7.35		8.09		
2 -50' Air Hoses, 1.5"		48.50		53.35	6.43	7.08
40 L.H., Daily Totals		$2045.75		$3148.28	$51.14	$78.71

Crew B-32	Hr.	Daily	Hr.	Daily	Bare Costs	Incl. O&P
1 Laborer	$42.10	$336.80	$67.45	$539.60	$53.09	$83.46
3 Equip. Oper. (medium)	56.75	1362.00	88.80	2131.20		
1 Grader, 30,000 Lbs.		1062.00		1168.20		
1 Tandem Roller, 10 Ton		236.75		260.43		
1 Dozer, 200 H.P.		1504.00		1654.40	87.59	96.34
32 L.H., Daily Totals		$4501.55		$5753.82	$140.67	$179.81

Crew B-32A	Hr.	Daily	Hr.	Daily	Bare Costs	Incl. O&P
1 Laborer	$42.10	$336.80	$67.45	$539.60	$51.87	$81.68
2 Equip. Oper. (medium)	56.75	908.00	88.80	1420.80		
1 Grader, 30,000 Lbs.		1062.00		1168.20		
1 Roller, Vibratory, 25 Ton		664.35		730.78	71.93	79.12
24 L.H., Daily Totals		$2971.15		$3859.39	$123.80	$160.81

Crew B-32B	Hr.	Daily	Hr.	Daily	Bare Costs	Incl. O&P
1 Laborer	$42.10	$336.80	$67.45	$539.60	$51.87	$81.68
2 Equip. Oper. (medium)	56.75	908.00	88.80	1420.80		
1 Dozer, 200 H.P.		1504.00		1654.40		
1 Roller, Vibratory, 25 Ton		664.35		730.78	90.35	99.38
24 L.H., Daily Totals		$3413.15		$4345.59	$142.21	$181.07

Crew B-32C	Hr.	Daily	Hr.	Daily	Bare Costs	Incl. O&P
1 Labor Foreman (outside)	$44.10	$352.80	$70.65	$565.20	$49.76	$78.66
2 Laborers	42.10	673.60	67.45	1079.20		
3 Equip. Oper. (medium)	56.75	1362.00	88.80	2131.20		
1 Grader, 30,000 Lbs.		1062.00		1168.20		
1 Tandem Roller, 10 Ton		236.75		260.43		
1 Dozer, 200 H.P.		1504.00		1654.40	58.39	64.23
48 L.H., Daily Totals		$5191.15		$6858.63	$108.15	$142.89

Crew B-33A	Hr.	Daily	Hr.	Daily	Bare Costs	Incl. O&P
1 Equip. Oper. (medium)	$56.75	$454.00	$88.80	$710.40	$52.56	$82.70
.5 Laborer	42.10	168.40	67.45	269.80		
.25 Equip. Oper. (medium)	56.75	113.50	88.80	177.60		
1 Scraper, Towed, 7 C.Y.		127.80		140.58		
1.25 Dozers, 300 H.P.		2205.00		2425.50	166.63	183.29
14 L.H., Daily Totals		$3068.70		$3723.88	$219.19	$265.99

Crew B-33B	Hr.	Daily	Hr.	Daily	Bare Costs	Incl. O&P
1 Equip. Oper. (medium)	$56.75	$454.00	$88.80	$710.40	$52.56	$82.70
.5 Laborer	42.10	168.40	67.45	269.80		
.25 Equip. Oper. (medium)	56.75	113.50	88.80	177.60		
1 Scraper, Towed, 10 C.Y.		159.70		175.67		
1.25 Dozers, 300 H.P.		2205.00		2425.50	168.91	185.80
14 L.H., Daily Totals		$3100.60		$3758.97	$221.47	$268.50

Crew B-33C	Hr.	Daily	Hr.	Daily	Bare Costs	Incl. O&P
1 Equip. Oper. (medium)	$56.75	$454.00	$88.80	$710.40	$52.56	$82.70
.5 Laborer	42.10	168.40	67.45	269.80		
.25 Equip. Oper. (medium)	56.75	113.50	88.80	177.60		
1 Scraper, Towed, 15 C.Y.		176.75		194.43		
1.25 Dozers, 300 H.P.		2205.00		2425.50	170.13	187.14
14 L.H., Daily Totals		$3117.65		$3777.72	$222.69	$269.84

Crew B-33D	Hr.	Daily	Hr.	Daily	Bare Costs	Incl. O&P
1 Equip. Oper. (medium)	$56.75	$454.00	$88.80	$710.40	$52.56	$82.70
.5 Laborer	42.10	168.40	67.45	269.80		
.25 Equip. Oper. (medium)	56.75	113.50	88.80	177.60		
1 S.P. Scraper, 14 C.Y.		2395.00		2634.50		
.25 Dozer, 300 H.P.		441.00		485.10	202.57	222.83
14 L.H., Daily Totals		$3571.90		$4277.40	$255.14	$305.53

Crew B-33E	Hr.	Daily	Hr.	Daily	Bare Costs	Incl. O&P
1 Equip. Oper. (medium)	$56.75	$454.00	$88.80	$710.40	$52.56	$82.70
.5 Laborer	42.10	168.40	67.45	269.80		
.25 Equip. Oper. (medium)	56.75	113.50	88.80	177.60		
1 S.P. Scraper, 21 C.Y.		2628.00		2890.80		
.25 Dozer, 300 H.P.		441.00		485.10	219.21	241.14
14 L.H., Daily Totals		$3804.90		$4533.70	$271.78	$323.84

Crew B-33F	Hr.	Daily	Hr.	Daily	Bare Costs	Incl. O&P
1 Equip. Oper. (medium)	$56.75	$454.00	$88.80	$710.40	$52.56	$82.70
.5 Laborer	42.10	168.40	67.45	269.80		
.25 Equip. Oper. (medium)	56.75	113.50	88.80	177.60		
1 Elev. Scraper, 11 C.Y.		1133.00		1246.30		
.25 Dozer, 300 H.P.		441.00		485.10	112.43	123.67
14 L.H., Daily Totals		$2309.90		$2889.20	$164.99	$206.37

Crew B-33G	Hr.	Daily	Hr.	Daily	Bare Costs	Incl. O&P
1 Equip. Oper. (medium)	$56.75	$454.00	$88.80	$710.40	$52.56	$82.70
.5 Laborer	42.10	168.40	67.45	269.80		
.25 Equip. Oper. (medium)	56.75	113.50	88.80	177.60		
1 Elev. Scraper, 22 C.Y.		1884.00		2072.40		
.25 Dozer, 300 H.P.		441.00		485.10	166.07	182.68
14 L.H., Daily Totals		$3060.90		$3715.30	$218.64	$265.38

Crew B-33K	Hr.	Daily	Hr.	Daily	Bare Costs	Incl. O&P
1 Equipment Operator (med.)	$56.75	$454.00	$88.80	$710.40	$52.56	$82.70
.25 Equipment Operator (med.)	56.75	113.50	88.80	177.60		
.5 Laborer	42.10	168.40	67.45	269.80		
1 S.P. Scraper, 31 C.Y.		3667.00		4033.70		
.25 Dozer, 410 H.P.		694.25		763.67	311.52	342.67
14 L.H., Daily Totals		$5097.15		$5955.18	$364.08	$425.37

Crew B-34A	Hr.	Daily	Hr.	Daily	Bare Costs	Incl. O&P
1 Truck Driver (heavy)	$48.95	$391.60	$78.25	$626.00	$48.95	$78.25
1 Dump Truck, 8 C.Y., 220 H.P.		428.05		470.86	53.51	58.86
8 L.H., Daily Totals		$819.65		$1096.86	$102.46	$137.11

Crew B-34B	Hr.	Daily	Hr.	Daily	Bare Costs	Incl. O&P
1 Truck Driver (heavy)	$48.95	$391.60	$78.25	$626.00	$48.95	$78.25
1 Dump Truck, 12 C.Y., 400 H.P.		572.90		630.19	71.61	78.77
8 L.H., Daily Totals		$964.50		$1256.19	$120.56	$157.02

Crew B-34C	Hr.	Daily	Hr.	Daily	Bare Costs	Incl. O&P
1 Truck Driver (heavy)	$48.95	$391.60	$78.25	$626.00	$48.95	$78.25
1 Truck Tractor, 6x4, 380 H.P.		493.25		542.58		
1 Dump Trailer, 16.5 C.Y.		136.70		150.37	78.74	86.62
8 L.H., Daily Totals		$1021.55		$1318.94	$127.69	$164.87

Crew B-34D	Hr.	Daily	Hr.	Daily	Bare Costs	Incl. O&P
1 Truck Driver (heavy)	$48.95	$391.60	$78.25	$626.00	$48.95	$78.25
1 Truck Tractor, 6x4, 380 H.P.		493.25		542.58		
1 Dump Trailer, 20 C.Y.		151.70		166.87	80.62	88.68
8 L.H., Daily Totals		$1036.55		$1335.44	$129.57	$166.93

Crew B-34E	Hr.	Daily	Hr.	Daily	Bare Costs	Incl. O&P
1 Truck Driver (heavy)	$48.95	$391.60	$78.25	$626.00	$48.95	$78.25
1 Dump Truck, Off Hwy., 25 Ton		1379.00		1516.90	172.38	189.61
8 L.H., Daily Totals		$1770.60		$2142.90	$221.32	$267.86

Crew B-34F	Hr.	Daily	Hr.	Daily	Bare Costs	Incl. O&P
1 Truck Driver (heavy)	$48.95	$391.60	$78.25	$626.00	$48.95	$78.25
1 Dump Truck, Off Hwy., 35 Ton		935.20		1028.72	116.90	128.59
8 L.H., Daily Totals		$1326.80		$1654.72	$165.85	$206.84

Crew B-34G	Hr.	Daily	Hr.	Daily	Bare Costs	Incl. O&P
1 Truck Driver (heavy)	$48.95	$391.60	$78.25	$626.00	$48.95	$78.25
1 Dump Truck, Off Hwy., 50 Ton		1771.00		1948.10	221.38	243.51
8 L.H., Daily Totals		$2162.60		$2574.10	$270.32	$321.76

Crew B-34H	Hr.	Daily	Hr.	Daily	Bare Costs	Incl. O&P
1 Truck Driver (heavy)	$48.95	$391.60	$78.25	$626.00	$48.95	$78.25
1 Dump Truck, Off Hwy., 65 Ton		1915.00		2106.50	239.38	263.31
8 L.H., Daily Totals		$2306.60		$2732.50	$288.32	$341.56

Crew B-34I	Hr.	Daily	Hr.	Daily	Bare Costs	Incl. O&P
1 Truck Driver (heavy)	$48.95	$391.60	$78.25	$626.00	$48.95	$78.25
1 Dump Truck, 18 C.Y., 450 H.P.		713.80		785.18	89.22	98.15
8 L.H., Daily Totals		$1105.40		$1411.18	$138.18	$176.40

Crew B-34J	Hr.	Daily	Hr.	Daily	Bare Costs	Incl. O&P
1 Truck Driver (heavy)	$48.95	$391.60	$78.25	$626.00	$48.95	$78.25
1 Dump Truck, Off Hwy., 100 Ton		2736.00		3009.60	342.00	376.20
8 L.H., Daily Totals		$3127.60		$3635.60	$390.95	$454.45

Crew No.	Bare Costs		Incl. Subs O&P		Cost Per Labor-Hour	
Crew B-34K	**Hr.**	**Daily**	**Hr.**	**Daily**	**Bare Costs**	**Incl. O&P**
1 Truck Driver (heavy)	$48.95	$391.60	$78.25	$626.00	$48.95	$78.25
1 Truck Tractor, 6x4, 450 H.P.		601.75		661.92		
1 Lowbed Trailer, 75 Ton		255.05		280.56	107.10	117.81
8 L.H., Daily Totals		$1248.40		$1568.48	$156.05	$196.06

Crew B-34L	Hr.	Daily	Hr.	Daily	Bare Costs	Incl. O&P
1 Equip. Oper. (light)	$53.00	$424.00	$82.95	$663.60	$53.00	$82.95
1 Flatbed Truck, Gas, 1.5 Ton		196.15		215.76	24.52	26.97
8 L.H., Daily Totals		$620.15		$879.37	$77.52	$109.92

Crew B-34M	Hr.	Daily	Hr.	Daily	Bare Costs	Incl. O&P
1 Equip. Oper. (light)	$53.00	$424.00	$82.95	$663.60	$53.00	$82.95
1 Flatbed Truck, Gas, 3 Ton		820.40		902.44	102.55	112.81
8 L.H., Daily Totals		$1244.40		$1566.04	$155.55	$195.76

Crew B-34N	Hr.	Daily	Hr.	Daily	Bare Costs	Incl. O&P
1 Truck Driver (heavy)	$48.95	$391.60	$78.25	$626.00	$52.85	$83.53
1 Equip. Oper. (medium)	56.75	454.00	88.80	710.40		
1 Truck Tractor, 6x4, 380 H.P.		493.25		542.58		
1 Flatbed Trailer, 40 Ton		186.20		204.82	42.47	46.71
16 L.H., Daily Totals		$1525.05		$2083.80	$95.32	$130.24

Crew B-34P	Hr.	Daily	Hr.	Daily	Bare Costs	Incl. O&P
1 Pipe Fitter	$65.55	$524.40	$101.35	$810.80	$56.52	$88.57
1 Truck Driver (light)	47.25	378.00	75.55	604.40		
1 Equip. Oper. (medium)	56.75	454.00	88.80	710.40		
1 Flatbed Truck, Gas, 3 Ton		820.40		902.44		
1 Backhoe Loader, 48 H.P.		213.75		235.13	43.09	47.40
24 L.H., Daily Totals		$2390.55		$3263.17	$99.61	$135.97

Crew B-34Q	Hr.	Daily	Hr.	Daily	Bare Costs	Incl. O&P
1 Pipe Fitter	$65.55	$524.40	$101.35	$810.80	$57.33	$89.85
1 Truck Driver (light)	47.25	378.00	75.55	604.40		
1 Equip. Oper. (crane)	59.20	473.60	92.65	741.20		
1 Flatbed Trailer, 25 Ton		135.55		149.10		
1 Dump Truck, 8 C.Y., 220 H.P.		428.05		470.86		
1 Hyd. Crane, 25 Ton		580.85		638.93	47.69	52.45
24 L.H., Daily Totals		$2520.45		$3415.30	$105.02	$142.30

Crew B-34R	Hr.	Daily	Hr.	Daily	Bare Costs	Incl. O&P
1 Pipe Fitter	$65.55	$524.40	$101.35	$810.80	$57.33	$89.85
1 Truck Driver (light)	47.25	378.00	75.55	604.40		
1 Equip. Oper. (crane)	59.20	473.60	92.65	741.20		
1 Flatbed Trailer, 25 Ton		135.55		149.10		
1 Dump Truck, 8 C.Y., 220 H.P.		428.05		470.86		
1 Hyd. Crane, 25 Ton		580.85		638.93		
1 Hyd. Excavator, 1 C.Y.		795.30		874.83	80.82	88.91
24 L.H., Daily Totals		$3315.75		$4290.13	$138.16	$178.76

Crew B-34S	Hr.	Daily	Hr.	Daily	Bare Costs	Incl. O&P
2 Pipe Fitters	$65.55	$1048.80	$101.35	$1621.60	$59.81	$93.40
1 Truck Driver (heavy)	48.95	391.60	78.25	626.00		
1 Equip. Oper. (crane)	59.20	473.60	92.65	741.20		
1 Flatbed Trailer, 40 Ton		186.20		204.82		
1 Truck Tractor, 6x4, 380 H.P.		493.25		542.58		
1 Hyd. Crane, 80 Ton		1486.00		1634.60		
1 Hyd. Excavator, 2 C.Y.		951.30		1046.43	97.40	107.14
32 L.H., Daily Totals		$5030.75		$6417.23	$157.21	$200.54

Crew No.	Bare Costs		Incl. Subs O&P		Cost Per Labor-Hour	
Crew B-34T	**Hr.**	**Daily**	**Hr.**	**Daily**	**Bare Costs**	**Incl. O&P**
2 Pipe Fitters	$65.55	$1048.80	$101.35	$1621.60	$59.81	$93.40
1 Truck Driver (heavy)	48.95	391.60	78.25	626.00		
1 Equip. Oper. (crane)	59.20	473.60	92.65	741.20		
1 Flatbed Trailer, 40 Ton		186.20		204.82		
1 Truck Tractor, 6x4, 380 H.P.		493.25		542.58		
1 Hyd. Crane, 80 Ton		1486.00		1634.60	67.67	74.44
32 L.H., Daily Totals		$4079.45		$5370.80	$127.48	$167.84

Crew B-34U	Hr.	Daily	Hr.	Daily	Bare Costs	Incl. O&P
1 Truck Driver (heavy)	$48.95	$391.60	$78.25	$626.00	$50.98	$80.60
1 Equip. Oper. (light)	53.00	424.00	82.95	663.60		
1 Truck Tractor, 220 H.P.		307.10		337.81		
1 Flatbed Trailer, 25 Ton		135.55		149.10	27.67	30.43
16 L.H., Daily Totals		$1258.25		$1776.52	$78.64	$111.03

Crew B-34V	Hr.	Daily	Hr.	Daily	Bare Costs	Incl. O&P
1 Truck Driver (heavy)	$48.95	$391.60	$78.25	$626.00	$53.72	$84.62
1 Equip. Oper. (crane)	59.20	473.60	92.65	741.20		
1 Equip. Oper. (light)	53.00	424.00	82.95	663.60		
1 Truck Tractor, 6x4, 450 H.P.		601.75		661.92		
1 Equipment Trailer, 50 Ton		205.05		225.56		
1 Pickup Truck, 4x4, 3/4 Ton		175.85		193.44	40.94	45.04
24 L.H., Daily Totals		$2271.85		$3111.72	$94.66	$129.65

Crew B-34W	Hr.	Daily	Hr.	Daily	Bare Costs	Incl. O&P
5 Truck Drivers (heavy)	$48.95	$1958.00	$78.25	$3130.00	$51.60	$81.84
2 Equip. Opers. (crane)	59.20	947.20	92.65	1482.40		
1 Equip. Oper. (mechanic)	59.15	473.20	92.55	740.40		
1 Laborer	42.10	336.80	67.45	539.60		
4 Truck Tractors, 6x4, 380 H.P.		1973.00		2170.30		
2 Equipment Trailers, 50 Ton		410.10		451.11		
2 Flatbed Trailers, 40 Ton		372.40		409.64		
1 Pickup Truck, 4x4, 3/4 Ton		175.85		193.44		
1 S.P. Crane, 4x4, 20 Ton		582.20		640.42	48.80	53.68
72 L.H., Daily Totals		$7228.75		$9757.31	$100.40	$135.52

Crew B-35	Hr.	Daily	Hr.	Daily	Bare Costs	Incl. O&P
1 Labor Foreman (outside)	$44.10	$352.80	$70.65	$565.20	$52.54	$82.82
1 Skilled Worker	54.85	438.80	87.40	699.20		
1 Welder (plumber)	64.45	515.60	99.65	797.20		
1 Laborer	42.10	336.80	67.45	539.60		
1 Equip. Oper. (crane)	59.20	473.60	92.65	741.20		
1 Equip. Oper. (oiler)	50.55	404.40	79.10	632.80		
1 Welder, Electric, 300 amp		106.40		117.04		
1 Hyd. Excavator, .75 C.Y.		694.25		763.67	16.68	18.35
48 L.H., Daily Totals		$3322.65		$4855.92	$69.22	$101.16

Crew B-35A	Hr.	Daily	Hr.	Daily	Bare Costs	Incl. O&P
1 Labor Foreman (outside)	$44.10	$352.80	$70.65	$565.20	$51.05	$80.62
2 Laborers	42.10	673.60	67.45	1079.20		
1 Skilled Worker	54.85	438.80	87.40	699.20		
1 Welder (plumber)	64.45	515.60	99.65	797.20		
1 Equip. Oper. (crane)	59.20	473.60	92.65	741.20		
1 Equip. Oper. (oiler)	50.55	404.40	79.10	632.80		
1 Welder, Gas Engine, 300 amp		147.00		161.70		
1 Crawler Crane, 75 Ton		2002.00		2202.20	38.38	42.21
56 L.H., Daily Totals		$5007.80		$6878.70	$89.42	$122.83

Crew B-36	Bare Costs Hr.	Bare Costs Daily	Incl. Subs O&P Hr.	Incl. Subs O&P Daily	Cost Per Labor-Hour Bare Costs	Cost Per Labor-Hour Incl. O&P
1 Labor Foreman (outside)	$44.10	$352.80	$70.65	$565.20	$48.36	$76.63
2 Laborers	42.10	673.60	67.45	1079.20		
2 Equip. Oper. (medium)	56.75	908.00	88.80	1420.80		
1 Dozer, 200 H.P.		1504.00		1654.40		
1 Aggregate Spreader		65.75		72.33		
1 Tandem Roller, 10 Ton		236.75		260.43	45.16	49.68
40 L.H., Daily Totals		$3740.90		$5052.35	$93.52	$126.31

Crew B-36A	Hr.	Daily	Hr.	Daily	Bare Costs	Incl. O&P
1 Labor Foreman (outside)	$44.10	$352.80	$70.65	$565.20	$50.76	$80.11
2 Laborers	42.10	673.60	67.45	1079.20		
4 Equip. Oper. (medium)	56.75	1816.00	88.80	2841.60		
1 Dozer, 200 H.P.		1504.00		1654.40		
1 Aggregate Spreader		65.75		72.33		
1 Tandem Roller, 10 Ton		236.75		260.43		
1 Roller, Pneum. Whl., 12 Ton		345.75		380.32	38.43	42.28
56 L.H., Daily Totals		$4994.65		$6853.48	$89.19	$122.38

Crew B-36B	Hr.	Daily	Hr.	Daily	Bare Costs	Incl. O&P
1 Labor Foreman (outside)	$44.10	$352.80	$70.65	$565.20	$50.53	$79.88
2 Laborers	42.10	673.60	67.45	1079.20		
4 Equip. Oper. (medium)	56.75	1816.00	88.80	2841.60		
1 Truck Driver (heavy)	48.95	391.60	78.25	626.00		
1 Grader, 30,000 Lbs.		1062.00		1168.20		
1 F.E. Loader, Crl, 1.5 C.Y.		659.75		725.73		
1 Dozer, 300 H.P.		1764.00		1940.40		
1 Roller, Vibratory, 25 Ton		664.35		730.78		
1 Truck Tractor, 6x4, 450 H.P.		601.75		661.92		
1 Water Tank Trailer, 5000 Gal.		152.25		167.47	76.63	84.29
64 L.H., Daily Totals		$8138.10		$10506.51	$127.16	$164.16

Crew B-36C	Hr.	Daily	Hr.	Daily	Bare Costs	Incl. O&P
1 Labor Foreman (outside)	$44.10	$352.80	$70.65	$565.20	$52.66	$83.06
3 Equip. Oper. (medium)	56.75	1362.00	88.80	2131.20		
1 Truck Driver (heavy)	48.95	391.60	78.25	626.00		
1 Grader, 30,000 Lbs.		1062.00		1168.20		
1 Dozer, 300 H.P.		1764.00		1940.40		
1 Roller, Vibratory, 25 Ton		664.35		730.78		
1 Truck Tractor, 6x4, 450 H.P.		601.75		661.92		
1 Water Tank Trailer, 5000 Gal.		152.25		167.47	106.11	116.72
40 L.H., Daily Totals		$6350.75		$7991.19	$158.77	$199.78

Crew B-37	Hr.	Daily	Hr.	Daily	Bare Costs	Incl. O&P
1 Labor Foreman (outside)	$44.10	$352.80	$70.65	$565.20	$44.25	$70.57
4 Laborers	42.10	1347.20	67.45	2158.40		
1 Equip. Oper. (light)	53.00	424.00	82.95	663.60		
1 Tandem Roller, 5 Ton		255.65		281.21	5.33	5.86
48 L.H., Daily Totals		$2379.65		$3668.42	$49.58	$76.43

Crew B-37A	Hr.	Daily	Hr.	Daily	Bare Costs	Incl. O&P
2 Laborers	$42.10	$673.60	$67.45	$1079.20	$43.82	$70.15
1 Truck Driver (light)	47.25	378.00	75.55	604.40		
1 Flatbed Truck, Gas, 1.5 Ton		196.15		215.76		
1 Tar Kettle, T.M.		156.20		171.82	14.68	16.15
24 L.H., Daily Totals		$1403.95		$2071.18	$58.50	$86.30

Crew B-37B	Hr.	Daily	Hr.	Daily	Bare Costs	Incl. O&P
3 Laborers	$42.10	$1010.40	$67.45	$1618.80	$43.39	$69.47
1 Truck Driver (light)	47.25	378.00	75.55	604.40		
1 Flatbed Truck, Gas, 1.5 Ton		196.15		215.76		
1 Tar Kettle, T.M.		156.20		171.82	11.01	12.11
32 L.H., Daily Totals		$1740.75		$2610.78	$54.40	$81.59

Crew B-37C	Hr.	Daily	Hr.	Daily	Bare Costs	Incl. O&P
2 Laborers	$42.10	$673.60	$67.45	$1079.20	$44.67	$71.50
2 Truck Drivers (light)	47.25	756.00	75.55	1208.80		
2 Flatbed Trucks, Gas, 1.5 Ton		392.30		431.53		
1 Tar Kettle, T.M.		156.20		171.82	17.14	18.85
32 L.H., Daily Totals		$1978.10		$2891.35	$61.82	$90.35

Crew B-37D	Hr.	Daily	Hr.	Daily	Bare Costs	Incl. O&P
1 Laborer	$42.10	$336.80	$67.45	$539.60	$44.67	$71.50
1 Truck Driver (light)	47.25	378.00	75.55	604.40		
1 Pickup Truck, 3/4 Ton		110.85		121.94	6.93	7.62
16 L.H., Daily Totals		$825.65		$1265.93	$51.60	$79.12

Crew B-37E	Hr.	Daily	Hr.	Daily	Bare Costs	Incl. O&P
3 Laborers	$42.10	$1010.40	$67.45	$1618.80	$47.22	$75.03
1 Equip. Oper. (light)	53.00	424.00	82.95	663.60		
1 Equip. Oper. (medium)	56.75	454.00	88.80	710.40		
2 Truck Drivers (light)	47.25	756.00	75.55	1208.80		
4 Barrels w/ Flasher		16.40		18.04		
1 Concrete Saw		111.50		122.65		
1 Rotary Hammer Drill		51.70		56.87		
1 Hammer Drill Bit		25.25		27.77		
1 Loader, Skid Steer, 30 H.P.		177.55		195.31		
1 Conc. Hammer Attach.		118.40		130.24		
1 Vibrating Plate, Gas, 18"		31.55		34.70		
2 Flatbed Trucks, Gas, 1.5 Ton		392.30		431.53	16.51	18.16
56 L.H., Daily Totals		$3569.05		$5218.72	$63.73	$93.19

Crew B-37F	Hr.	Daily	Hr.	Daily	Bare Costs	Incl. O&P
3 Laborers	$42.10	$1010.40	$67.45	$1618.80	$43.39	$69.47
1 Truck Driver (light)	47.25	378.00	75.55	604.40		
4 Barrels w/ Flasher		16.40		18.04		
1 Concrete Mixer, 10 C.F.		140.30		154.33		
1 Air Compressor, 60 cfm		152.95		168.25		
1 -50' Air Hose, 3/4"		7.75		8.53		
1 Spade (Chipper)		8.45		9.29		
1 Flatbed Truck, Gas, 1.5 Ton		196.15		215.76	16.31	17.94
32 L.H., Daily Totals		$1910.40		$2797.40	$59.70	$87.42

Crew B-37G	Hr.	Daily	Hr.	Daily	Bare Costs	Incl. O&P
1 Labor Foreman (outside)	$44.10	$352.80	$70.65	$565.20	$44.25	$70.57
4 Laborers	42.10	1347.20	67.45	2158.40		
1 Equip. Oper. (light)	53.00	424.00	82.95	663.60		
1 Berm Machine		271.95		299.14		
1 Tandem Roller, 5 Ton		255.65		281.21	10.99	12.09
48 L.H., Daily Totals		$2651.60		$3967.56	$55.24	$82.66

Crew B-37H	Hr.	Daily	Hr.	Daily	Bare Costs	Incl. O&P
1 Labor Foreman (outside)	$44.10	$352.80	$70.65	$565.20	$44.25	$70.57
4 Laborers	42.10	1347.20	67.45	2158.40		
1 Equip. Oper. (light)	53.00	424.00	82.95	663.60		
1 Tandem Roller, 5 Ton		255.65		281.21		
1 Flatbed Truck, Gas, 1.5 Ton		196.15		215.76		
1 Tar Kettle, T.M.		156.20		171.82	12.67	13.93
48 L.H., Daily Totals		$2732.00		$4056.00	$56.92	$84.50

Crew B-37I	Bare Costs Hr.	Bare Costs Daily	Incl. Subs O&P Hr.	Incl. Subs O&P Daily	Cost Per Labor-Hour Bare Costs	Cost Per Labor-Hour Incl. O&P
3 Laborers	$42.10	$1010.40	$67.45	$1618.80	$47.22	$75.03
1 Equip. Oper. (light)	53.00	424.00	82.95	663.60		
1 Equip. Oper. (medium)	56.75	454.00	88.80	710.40		
2 Truck Drivers (light)	47.25	756.00	75.55	1208.80		
4 Barrels w/ Flasher		16.40		18.04		
1 Concrete Saw		111.50		122.65		
1 Rotary Hammer Drill		51.70		56.87		
1 Hammer Drill Bit		25.25		27.77		
1 Air Compressor, 60 cfm		152.95		168.25		
1 -50' Air Hose, 3/4"		7.75		8.53		
1 Spade (Chipper)		8.45		9.29		
1 Loader, Skid Steer, 30 H.P.		177.55		195.31		
1 Conc. Hammer Attach.		118.40		130.24		
1 Concrete Mixer, 10 C.F.		140.30		154.33		
1 Vibrating Plate, Gas, 18"		31.55		34.70		
2 Flatbed Trucks, Gas, 1.5 Ton		392.30		431.53	22.04	24.24
56 L.H., Daily Totals		$3878.50		$5559.11	$69.26	$99.27

Crew B-37J	Hr.	Daily	Hr.	Daily	Bare Costs	Incl. O&P
1 Labor Foreman (outside)	$44.10	$352.80	$70.65	$565.20	$44.25	$70.57
4 Laborers	42.10	1347.20	67.45	2158.40		
1 Equip. Oper. (light)	53.00	424.00	82.95	663.60		
1 Air Compressor, 60 cfm		152.95		168.25		
1 -50' Air Hose, 3/4"		7.75		8.53		
2 Concrete Mixers, 10 C.F.		280.60		308.66		
2 Flatbed Trucks, Gas, 1.5 Ton		392.30		431.53		
1 Shot Blaster, 20"		206.20		226.82	21.66	23.83
48 L.H., Daily Totals		$3163.80		$4530.98	$65.91	$94.40

Crew B-37K	Hr.	Daily	Hr.	Daily	Bare Costs	Incl. O&P
1 Labor Foreman (outside)	$44.10	$352.80	$70.65	$565.20	$44.25	$70.57
4 Laborers	42.10	1347.20	67.45	2158.40		
1 Equip. Oper. (light)	53.00	424.00	82.95	663.60		
1 Air Compressor, 60 cfm		152.95		168.25		
1 -50' Air Hose, 3/4"		7.75		8.53		
2 Flatbed Trucks, Gas, 1.5 Ton		392.30		431.53		
1 Shot Blaster, 20"		206.20		226.82	15.82	17.40
48 L.H., Daily Totals		$2883.20		$4222.32	$60.07	$87.97

Crew B-38	Hr.	Daily	Hr.	Daily	Bare Costs	Incl. O&P
1 Labor Foreman (outside)	$44.10	$352.80	$70.65	$565.20	$47.61	$75.46
2 Laborers	42.10	673.60	67.45	1079.20		
1 Equip. Oper. (light)	53.00	424.00	82.95	663.60		
1 Equip. Oper. (medium)	56.75	454.00	88.80	710.40		
1 Backhoe Loader, 48 H.P.		213.75		235.13		
1 Hyd. Hammer (1200 lb.)		175.15		192.66		
1 F.E. Loader, W.M., 4 C.Y.		789.35		868.28		
1 Pvmt. Rem. Bucket		63.05		69.36	31.03	34.14
40 L.H., Daily Totals		$3145.70		$4383.83	$78.64	$109.60

Crew B-39	Hr.	Daily	Hr.	Daily	Bare Costs	Incl. O&P
1 Labor Foreman (outside)	$44.10	$352.80	$70.65	$565.20	$44.25	$70.57
4 Laborers	42.10	1347.20	67.45	2158.40		
1 Equip. Oper. (light)	53.00	424.00	82.95	663.60		
1 Air Compressor, 250 cfm		201.50		221.65		
2 Breakers, Pavement, 60 lb.		107.00		117.70		
2 -50' Air Hoses, 1.5"		48.50		53.35	7.44	8.18
48 L.H., Daily Totals		$2481.00		$3779.90	$51.69	$78.75

Crew B-40	Hr.	Daily	Hr.	Daily	Bare Costs	Incl. O&P
1 Pile Driver Foreman (outside)	$56.20	$449.60	$90.15	$721.20	$55.24	$87.77
4 Pile Drivers	54.20	1734.40	86.90	2780.80		
2 Equip. Oper. (crane)	59.20	947.20	92.65	1482.40		
1 Equip. Oper. (oiler)	50.55	404.40	79.10	632.80		
1 Crawler Crane, 40 Ton		1206.00		1326.60		
1 Vibratory Hammer & Gen.		2271.00		2498.10	54.33	59.76
64 L.H., Daily Totals		$7012.60		$9441.90	$109.57	$147.53

Crew B-40B	Hr.	Daily	Hr.	Daily	Bare Costs	Incl. O&P
1 Labor Foreman (outside)	$44.10	$352.80	$70.65	$565.20	$46.69	$74.13
3 Laborers	42.10	1010.40	67.45	1618.80		
1 Equip. Oper. (crane)	59.20	473.60	92.65	741.20		
1 Equip. Oper. (oiler)	50.55	404.40	79.10	632.80		
1 Lattice Boom Crane, 40 Ton		2028.00		2230.80	42.25	46.48
48 L.H., Daily Totals		$4269.20		$5788.80	$88.94	$120.60

Crew B-41	Hr.	Daily	Hr.	Daily	Bare Costs	Incl. O&P
1 Labor Foreman (outside)	$44.10	$352.80	$70.65	$565.20	$43.63	$69.71
4 Laborers	42.10	1347.20	67.45	2158.40		
.25 Equip. Oper. (crane)	59.20	118.40	92.65	185.30		
.25 Equip. Oper. (oiler)	50.55	101.10	79.10	158.20		
.25 Crawler Crane, 40 Ton		301.50		331.65	6.85	7.54
44 L.H., Daily Totals		$2221.00		$3398.75	$50.48	$77.24

Crew B-42	Hr.	Daily	Hr.	Daily	Bare Costs	Incl. O&P
1 Labor Foreman (outside)	$44.10	$352.80	$70.65	$565.20	$47.46	$75.79
4 Laborers	42.10	1347.20	67.45	2158.40		
1 Equip. Oper. (crane)	59.20	473.60	92.65	741.20		
1 Equip. Oper. (oiler)	50.55	404.40	79.10	632.80		
1 Welder	57.45	459.60	94.15	753.20		
1 Hyd. Crane, 25 Ton		580.85		638.93		
1 Welder, Gas Engine, 300 amp		147.00		161.70		
1 Horz. Boring Csg. Mch.		314.25		345.68	16.28	17.91
64 L.H., Daily Totals		$4079.70		$5997.11	$63.75	$93.70

Crew B-43	Hr.	Daily	Hr.	Daily	Bare Costs	Incl. O&P
1 Labor Foreman (outside)	$44.10	$352.80	$70.65	$565.20	$46.69	$74.13
3 Laborers	42.10	1010.40	67.45	1618.80		
1 Equip. Oper. (crane)	59.20	473.60	92.65	741.20		
1 Equip. Oper. (oiler)	50.55	404.40	79.10	632.80		
1 Drill Rig, Truck-Mounted		771.10		848.21	16.06	17.67
48 L.H., Daily Totals		$3012.30		$4406.21	$62.76	$91.80

Crew B-44	Hr.	Daily	Hr.	Daily	Bare Costs	Incl. O&P
1 Pile Driver Foreman (outside)	$56.20	$449.60	$90.15	$721.20	$54.19	$86.31
4 Pile Drivers	54.20	1734.40	86.90	2780.80		
2 Equip. Oper. (crane)	59.20	947.20	92.65	1482.40		
1 Laborer	42.10	336.80	67.45	539.60		
1 Crawler Crane, 40 Ton		1206.00		1326.60		
1 Lead, 60' High		209.25		230.18		
1 Hammer, Diesel, 15K ft.-lbs.		617.05		678.76	31.75	34.93
64 L.H., Daily Totals		$5500.30		$7759.53	$85.94	$121.24

Crew B-45	Hr.	Daily	Hr.	Daily	Bare Costs	Incl. O&P
1 Equip. Oper. (medium)	$56.75	$454.00	$88.80	$710.40	$52.85	$83.53
1 Truck Driver (heavy)	48.95	391.60	78.25	626.00		
1 Dist. Tanker, 3000 Gallon		330.15		363.17		
1 Truck Tractor, 6x4, 380 H.P.		493.25		542.58	51.46	56.61
16 L.H., Daily Totals		$1669.00		$2242.14	$104.31	$140.13

Crew No.	Bare Costs		Incl. Subs O&P		Cost Per Labor-Hour	
Crew B-46	**Hr.**	**Daily**	**Hr.**	**Daily**	**Bare Costs**	**Incl. O&P**
1 Pile Driver Foreman (outside)	$56.20	$449.60	$90.15	$721.20	$48.48	$77.72
2 Pile Drivers	54.20	867.20	86.90	1390.40		
3 Laborers	42.10	1010.40	67.45	1618.80		
1 Chain Saw, Gas, 36" Long		41.25		45.38	.86	.95
48 L.H., Daily Totals		$2368.45		$3775.78	$49.34	$78.66
Crew B-47	**Hr.**	**Daily**	**Hr.**	**Daily**	**Bare Costs**	**Incl. O&P**
1 Blast Foreman (outside)	$44.10	$352.80	$70.65	$565.20	$46.40	$73.68
1 Driller	42.10	336.80	67.45	539.60		
1 Equip. Oper. (light)	53.00	424.00	82.95	663.60		
1 Air Track Drill, 4"		1058.00		1163.80		
1 Air Compressor, 600 cfm		421.50		463.65		
2 -50' Air Hoses, 3"		76.70		84.37	64.84	71.33
24 L.H., Daily Totals		$2669.80		$3480.22	$111.24	$145.01
Crew B-47A	**Hr.**	**Daily**	**Hr.**	**Daily**	**Bare Costs**	**Incl. O&P**
1 Drilling Foreman (outside)	$44.10	$352.80	$70.65	$565.20	$51.28	$80.80
1 Equip. Oper. (heavy)	59.20	473.60	92.65	741.20		
1 Equip. Oper. (oiler)	50.55	404.40	79.10	632.80		
1 Air Track Drill, 5"		1127.00		1239.70	46.96	51.65
24 L.H., Daily Totals		$2357.80		$3178.90	$98.24	$132.45
Crew B-47C	**Hr.**	**Daily**	**Hr.**	**Daily**	**Bare Costs**	**Incl. O&P**
1 Laborer	$42.10	$336.80	$67.45	$539.60	$47.55	$75.20
1 Equip. Oper. (light)	53.00	424.00	82.95	663.60		
1 Air Compressor, 750 cfm		538.25		592.08		
2 -50' Air Hoses, 3"		76.70		84.37		
1 Air Track Drill, 4"		1058.00		1163.80	104.56	115.02
16 L.H., Daily Totals		$2433.75		$3043.45	$152.11	$190.22
Crew B-47E	**Hr.**	**Daily**	**Hr.**	**Daily**	**Bare Costs**	**Incl. O&P**
1 Labor Foreman (outside)	$44.10	$352.80	$70.65	$565.20	$42.60	$68.25
3 Laborers	42.10	1010.40	67.45	1618.80		
1 Flatbed Truck, Gas, 3 Ton		820.40		902.44	25.64	28.20
32 L.H., Daily Totals		$2183.60		$3086.44	$68.24	$96.45
Crew B-47G	**Hr.**	**Daily**	**Hr.**	**Daily**	**Bare Costs**	**Incl. O&P**
1 Labor Foreman (outside)	$44.10	$352.80	$70.65	$565.20	$45.33	$72.13
2 Laborers	42.10	673.60	67.45	1079.20		
1 Equip. Oper. (light)	53.00	424.00	82.95	663.60		
1 Air Track Drill, 4"		1058.00		1163.80		
1 Air Compressor, 600 cfm		421.50		463.65		
2 -50' Air Hoses, 3"		76.70		84.37		
1 Gunite Pump Rig		317.90		349.69	58.57	64.42
32 L.H., Daily Totals		$3324.50		$4369.51	$103.89	$136.55
Crew B-47H	**Hr.**	**Daily**	**Hr.**	**Daily**	**Bare Costs**	**Incl. O&P**
1 Skilled Worker Foreman (out)	$56.85	$454.80	$90.55	$724.40	$55.35	$88.19
3 Skilled Workers	54.85	1316.40	87.40	2097.60		
1 Flatbed Truck, Gas, 3 Ton		820.40		902.44	25.64	28.20
32 L.H., Daily Totals		$2591.60		$3724.44	$80.99	$116.39

Crew No.	Bare Costs		Incl. Subs O&P		Cost Per Labor-Hour	
Crew B-48	**Hr.**	**Daily**	**Hr.**	**Daily**	**Bare Costs**	**Incl. O&P**
1 Labor Foreman (outside)	$44.10	$352.80	$70.65	$565.20	$47.59	$75.39
3 Laborers	42.10	1010.40	67.45	1618.80		
1 Equip. Oper. (crane)	59.20	473.60	92.65	741.20		
1 Equip. Oper. (oiler)	50.55	404.40	79.10	632.80		
1 Equip. Oper. (light)	53.00	424.00	82.95	663.60		
1 Centr. Water Pump, 6"		232.90		256.19		
1 -20' Suction Hose, 6"		25.20		27.72		
1 -50' Discharge Hose, 6"		17.90		19.69		
1 Drill Rig, Truck-Mounted		771.10		848.21	18.70	20.57
56 L.H., Daily Totals		$3712.30		$5373.41	$66.29	$95.95
Crew B-49	**Hr.**	**Daily**	**Hr.**	**Daily**	**Bare Costs**	**Incl. O&P**
1 Labor Foreman (outside)	$44.10	$352.80	$70.65	$565.20	$50.12	$79.39
3 Laborers	42.10	1010.40	67.45	1618.80		
2 Equip. Oper. (crane)	59.20	947.20	92.65	1482.40		
2 Equip. Oper. (oilers)	50.55	808.80	79.10	1265.60		
1 Equip. Oper. (light)	53.00	424.00	82.95	663.60		
2 Pile Drivers	54.20	867.20	86.90	1390.40		
1 Hyd. Crane, 25 Ton		580.85		638.93		
1 Centr. Water Pump, 6"		232.90		256.19		
1 -20' Suction Hose, 6"		25.20		27.72		
1 -50' Discharge Hose, 6"		17.90		19.69		
1 Drill Rig, Truck-Mounted		771.10		848.21	18.50	20.35
88 L.H., Daily Totals		$6038.35		$8776.75	$68.62	$99.74
Crew B-50	**Hr.**	**Daily**	**Hr.**	**Daily**	**Bare Costs**	**Incl. O&P**
2 Pile Driver Foremen (outside)	$56.20	$899.20	$90.15	$1442.40	$52.35	$83.46
6 Pile Drivers	54.20	2601.60	86.90	4171.20		
2 Equip. Oper. (crane)	59.20	947.20	92.65	1482.40		
1 Equip. Oper. (oiler)	50.55	404.40	79.10	632.80		
3 Laborers	42.10	1010.40	67.45	1618.80		
1 Crawler Crane, 40 Ton		1206.00		1326.60		
1 Lead, 60' High		209.25		230.18		
1 Hammer, Diesel, 15K ft.-lbs.		617.05		678.76		
1 Air Compressor, 600 cfm		421.50		463.65		
2 -50' Air Hoses, 3"		76.70		84.37		
1 Chain Saw, Gas, 36" Long		41.25		45.38	22.96	25.26
112 L.H., Daily Totals		$8434.55		$12176.53	$75.31	$108.72
Crew B-51	**Hr.**	**Daily**	**Hr.**	**Daily**	**Bare Costs**	**Incl. O&P**
1 Labor Foreman (outside)	$44.10	$352.80	$70.65	$565.20	$43.29	$69.33
4 Laborers	42.10	1347.20	67.45	2158.40		
1 Truck Driver (light)	47.25	378.00	75.55	604.40		
1 Flatbed Truck, Gas, 1.5 Ton		196.15		215.76	4.09	4.50
48 L.H., Daily Totals		$2274.15		$3543.76	$47.38	$73.83
Crew B-52	**Hr.**	**Daily**	**Hr.**	**Daily**	**Bare Costs**	**Incl. O&P**
1 Carpenter Foreman (outside)	$55.15	$441.20	$88.40	$707.20	$48.73	$77.60
1 Carpenter	53.15	425.20	85.15	681.20		
3 Laborers	42.10	1010.40	67.45	1618.80		
1 Cement Finisher	49.95	399.60	78.45	627.60		
.5 Rodman (reinf.)	56.40	225.60	88.85	355.40		
.5 Equip. Oper. (medium)	56.75	227.00	88.80	355.20		
.5 Crawler Loader, 3 C.Y.		584.50		642.95	10.44	11.48
56 L.H., Daily Totals		$3313.50		$4988.35	$59.17	$89.08
Crew B-53	**Hr.**	**Daily**	**Hr.**	**Daily**	**Bare Costs**	**Incl. O&P**
1 Equip. Oper. (light)	$53.00	$424.00	$82.95	$663.60	$53.00	$82.95
1 Trencher, Chain, 12 H.P.		156.95		172.65	19.62	21.58
8 L.H., Daily Totals		$580.95		$836.25	$72.62	$104.53

Crew No.	Bare Costs		Incl. Subs O&P		Cost Per Labor-Hour	
Crew B-54	**Hr.**	**Daily**	**Hr.**	**Daily**	**Bare Costs**	**Incl. O&P**
1 Equip. Oper. (light)	$53.00	$424.00	$82.95	$663.60	$53.00	$82.95
1 Trencher, Chain, 40 H.P.		401.85		442.04	50.23	55.25
8 L.H., Daily Totals		$825.85		$1105.64	$103.23	$138.20
Crew B-54A	**Hr.**	**Daily**	**Hr.**	**Daily**	**Bare Costs**	**Incl. O&P**
.17 Labor Foreman (outside)	$44.10	$59.98	$70.65	$96.08	$54.91	$86.16
1 Equipment Operator (med.)	56.75	454.00	88.80	710.40		
1 Wheel Trencher, 67 H.P.		1126.00		1238.60	120.30	132.33
9.36 L.H., Daily Totals		$1639.98		$2045.08	$175.21	$218.49
Crew B-54B	**Hr.**	**Daily**	**Hr.**	**Daily**	**Bare Costs**	**Incl. O&P**
.25 Labor Foreman (outside)	$44.10	$88.20	$70.65	$141.30	$54.22	$85.17
1 Equipment Operator (med.)	56.75	454.00	88.80	710.40		
1 Wheel Trencher, 150 H.P.		1324.00		1456.40	132.40	145.64
10 L.H., Daily Totals		$1866.20		$2308.10	$186.62	$230.81
Crew B-54D	**Hr.**	**Daily**	**Hr.**	**Daily**	**Bare Costs**	**Incl. O&P**
1 Laborer	$42.10	$336.80	$67.45	$539.60	$49.42	$78.13
1 Equipment Operator (med.)	56.75	454.00	88.80	710.40		
1 Rock Trencher, 6" Width		429.85		472.83	26.87	29.55
16 L.H., Daily Totals		$1220.65		$1722.84	$76.29	$107.68
Crew B-54E	**Hr.**	**Daily**	**Hr.**	**Daily**	**Bare Costs**	**Incl. O&P**
1 Laborer	$42.10	$336.80	$67.45	$539.60	$49.42	$78.13
1 Equipment Operator (med.)	56.75	454.00	88.80	710.40		
1 Rock Trencher, 18" Width		1005.00		1105.50	62.81	69.09
16 L.H., Daily Totals		$1795.80		$2355.50	$112.24	$147.22
Crew B-55	**Hr.**	**Daily**	**Hr.**	**Daily**	**Bare Costs**	**Incl. O&P**
2 Laborers	$42.10	$673.60	$67.45	$1079.20	$43.82	$70.15
1 Truck Driver (light)	47.25	378.00	75.55	604.40		
1 Truck-Mounted Earth Auger		390.20		429.22		
1 Flatbed Truck, Gas, 3 Ton		820.40		902.44	50.44	55.49
24 L.H., Daily Totals		$2262.20		$3015.26	$94.26	$125.64
Crew B-56	**Hr.**	**Daily**	**Hr.**	**Daily**	**Bare Costs**	**Incl. O&P**
1 Laborer	$42.10	$336.80	$67.45	$539.60	$47.55	$75.20
1 Equip. Oper. (light)	53.00	424.00	82.95	663.60		
1 Air Track Drill, 4"		1058.00		1163.80		
1 Air Compressor, 600 cfm		421.50		463.65		
1 -50' Air Hose, 3"		38.35		42.19	94.87	104.35
16 L.H., Daily Totals		$2278.65		$2872.84	$142.42	$179.55
Crew B-57	**Hr.**	**Daily**	**Hr.**	**Daily**	**Bare Costs**	**Incl. O&P**
1 Labor Foreman (outside)	$44.10	$352.80	$70.65	$565.20	$48.51	$76.71
2 Laborers	42.10	673.60	67.45	1079.20		
1 Equip. Oper. (crane)	59.20	473.60	92.65	741.20		
1 Equip. Oper. (light)	53.00	424.00	82.95	663.60		
1 Equip. Oper. (oiler)	50.55	404.40	79.10	632.80		
1 Crawler Crane, 25 Ton		1129.00		1241.90		
1 Clamshell Bucket, 1 C.Y.		68.55		75.41		
1 Centr. Water Pump, 6"		232.90		256.19		
1 -20' Suction Hose, 6"		25.20		27.72		
20 -50' Discharge Hoses, 6"		358.00		393.80	37.78	41.56
48 L.H., Daily Totals		$4142.05		$5677.02	$86.29	$118.27

Crew No.	Bare Costs		Incl. Subs O&P		Cost Per Labor-Hour	
Crew B-58	**Hr.**	**Daily**	**Hr.**	**Daily**	**Bare Costs**	**Incl. O&P**
2 Laborers	$42.10	$673.60	$67.45	$1079.20	$45.73	$72.62
1 Equip. Oper. (light)	53.00	424.00	82.95	663.60		
1 Backhoe Loader, 48 H.P.		213.75		235.13		
1 Small Helicopter, w/ Pilot		2080.00		2288.00	95.57	105.13
24 L.H., Daily Totals		$3391.35		$4265.93	$141.31	$177.75
Crew B-59	**Hr.**	**Daily**	**Hr.**	**Daily**	**Bare Costs**	**Incl. O&P**
1 Truck Driver (heavy)	$48.95	$391.60	$78.25	$626.00	$48.95	$78.25
1 Truck Tractor, 220 H.P.		307.10		337.81		
1 Water Tank Trailer, 5000 Gal.		152.25		167.47	57.42	63.16
8 L.H., Daily Totals		$850.95		$1131.29	$106.37	$141.41
Crew B-59A	**Hr.**	**Daily**	**Hr.**	**Daily**	**Bare Costs**	**Incl. O&P**
2 Laborers	$42.10	$673.60	$67.45	$1079.20	$44.38	$71.05
1 Truck Driver (heavy)	48.95	391.60	78.25	626.00		
1 Water Tank Trailer, 5000 Gal.		152.25		167.47		
1 Truck Tractor, 220 H.P.		307.10		337.81	19.14	21.05
24 L.H., Daily Totals		$1524.55		$2210.49	$63.52	$92.10
Crew B-60	**Hr.**	**Daily**	**Hr.**	**Daily**	**Bare Costs**	**Incl. O&P**
1 Labor Foreman (outside)	$44.10	$352.80	$70.65	$565.20	$49.15	$77.60
2 Laborers	42.10	673.60	67.45	1079.20		
1 Equip. Oper. (crane)	59.20	473.60	92.65	741.20		
2 Equip. Oper. (light)	53.00	848.00	82.95	1327.20		
1 Equip. Oper. (oiler)	50.55	404.40	79.10	632.80		
1 Crawler Crane, 40 Ton		1206.00		1326.60		
1 Lead, 60' High		209.25		230.18		
1 Hammer, Diesel, 15K ft.-lbs.		617.05		678.76		
1 Backhoe Loader, 48 H.P.		213.75		235.13	40.11	44.12
56 L.H., Daily Totals		$4998.45		$6816.26	$89.26	$121.72
Crew B-61	**Hr.**	**Daily**	**Hr.**	**Daily**	**Bare Costs**	**Incl. O&P**
1 Labor Foreman (outside)	$44.10	$352.80	$70.65	$565.20	$44.68	$71.19
3 Laborers	42.10	1010.40	67.45	1618.80		
1 Equip. Oper. (light)	53.00	424.00	82.95	663.60		
1 Cement Mixer, 2 C.Y.		105.70		116.27		
1 Air Compressor, 160 cfm		202.80		223.08	7.71	8.48
40 L.H., Daily Totals		$2095.70		$3186.95	$52.39	$79.67
Crew B-62	**Hr.**	**Daily**	**Hr.**	**Daily**	**Bare Costs**	**Incl. O&P**
2 Laborers	$42.10	$673.60	$67.45	$1079.20	$45.73	$72.62
1 Equip. Oper. (light)	53.00	424.00	82.95	663.60		
1 Loader, Skid Steer, 30 H.P.		177.55		195.31	7.40	8.14
24 L.H., Daily Totals		$1275.15		$1938.11	$53.13	$80.75
Crew B-62A	**Hr.**	**Daily**	**Hr.**	**Daily**	**Bare Costs**	**Incl. O&P**
2 Laborers	$42.10	$673.60	$67.45	$1079.20	$45.73	$72.62
1 Equip. Oper. (light)	53.00	424.00	82.95	663.60		
1 Loader, Skid Steer, 30 H.P.		177.55		195.31		
1 Trencher Attachment		66.20		72.82	10.16	11.17
24 L.H., Daily Totals		$1341.35		$2010.93	$55.89	$83.79
Crew B-63	**Hr.**	**Daily**	**Hr.**	**Daily**	**Bare Costs**	**Incl. O&P**
4 Laborers	$42.10	$1347.20	$67.45	$2158.40	$44.28	$70.55
1 Equip. Oper. (light)	53.00	424.00	82.95	663.60		
1 Loader, Skid Steer, 30 H.P.		177.55		195.31	4.44	4.88
40 L.H., Daily Totals		$1948.75		$3017.30	$48.72	$75.43

Crews - Renovation

Crew No.	Bare Costs		Incl. Subs O&P		Cost Per Labor-Hour	
Crew B-63B	**Hr.**	**Daily**	**Hr.**	**Daily**	**Bare Costs**	**Incl. O&P**
1 Labor Foreman (inside)	$42.60	$340.80	$68.25	$546.00	$44.95	$71.53
2 Laborers	42.10	673.60	67.45	1079.20		
1 Equip. Oper. (light)	53.00	424.00	82.95	663.60		
1 Loader, Skid Steer, 78 H.P.		400.10		440.11	12.50	13.75
32 L.H., Daily Totals		$1838.50		$2728.91	$57.45	$85.28

Crew B-64	Hr.	Daily	Hr.	Daily	Bare Costs	Incl. O&P
1 Laborer	$42.10	$336.80	$67.45	$539.60	$44.67	$71.50
1 Truck Driver (light)	47.25	378.00	75.55	604.40		
1 Power Mulcher (small)		221.10		243.21		
1 Flatbed Truck, Gas, 1.5 Ton		196.15		215.76	26.08	28.69
16 L.H., Daily Totals		$1132.05		$1602.97	$70.75	$100.19

Crew B-65	Hr.	Daily	Hr.	Daily	Bare Costs	Incl. O&P
1 Laborer	$42.10	$336.80	$67.45	$539.60	$44.67	$71.50
1 Truck Driver (light)	47.25	378.00	75.55	604.40		
1 Power Mulcher (Large)		326.90		359.59		
1 Flatbed Truck, Gas, 1.5 Ton		196.15		215.76	32.69	35.96
16 L.H., Daily Totals		$1237.85		$1719.36	$77.37	$107.46

Crew B-66	Hr.	Daily	Hr.	Daily	Bare Costs	Incl. O&P
1 Equip. Oper. (light)	$53.00	$424.00	$82.95	$663.60	$53.00	$82.95
1 Loader-Backhoe, 40 H.P.		241.40		265.54	30.18	33.19
8 L.H., Daily Totals		$665.40		$929.14	$83.17	$116.14

Crew B-67	Hr.	Daily	Hr.	Daily	Bare Costs	Incl. O&P
1 Millwright	$55.90	$447.20	$86.60	$692.80	$54.45	$84.78
1 Equip. Oper. (light)	53.00	424.00	82.95	663.60		
1 R.T. Forklift, 5,000 Lb., diesel		269.90		296.89	16.87	18.56
16 L.H., Daily Totals		$1141.10		$1653.29	$71.32	$103.33

Crew B-67B	Hr.	Daily	Hr.	Daily	Bare Costs	Incl. O&P
1 Millwright Foreman (inside)	$56.40	$451.20	$87.35	$698.80	$56.15	$86.97
1 Millwright	55.90	447.20	86.60	692.80		
16 L.H., Daily Totals		$898.40		$1391.60	$56.15	$86.97

Crew B-68	Hr.	Daily	Hr.	Daily	Bare Costs	Incl. O&P
2 Millwrights	$55.90	$894.40	$86.60	$1385.60	$54.93	$85.38
1 Equip. Oper. (light)	53.00	424.00	82.95	663.60		
1 R.T. Forklift, 5,000 Lb., diesel		269.90		296.89	11.25	12.37
24 L.H., Daily Totals		$1588.30		$2346.09	$66.18	$97.75

Crew B-68A	Hr.	Daily	Hr.	Daily	Bare Costs	Incl. O&P
1 Millwright Foreman (inside)	$56.40	$451.20	$87.35	$698.80	$56.07	$86.85
2 Millwrights	55.90	894.40	86.60	1385.60		
1 Forklift, Smooth Floor, 8,000 Lb.		255.90		281.49	10.66	11.73
24 L.H., Daily Totals		$1601.50		$2365.89	$66.73	$98.58

Crew B-68B	Hr.	Daily	Hr.	Daily	Bare Costs	Incl. O&P
1 Millwright Foreman (inside)	$56.40	$451.20	$87.35	$698.80	$59.97	$92.68
2 Millwrights	55.90	894.40	86.60	1385.60		
2 Electricians	61.35	981.60	94.45	1511.20		
2 Plumbers	64.45	1031.20	99.65	1594.40		
1 R.T. Forklift, 5,000 Lb., gas		279.90		307.89	5.00	5.50
56 L.H., Daily Totals		$3638.30		$5497.89	$64.97	$98.18

Crew No.	Bare Costs		Incl. Subs O&P		Cost Per Labor-Hour	
Crew B-68C	**Hr.**	**Daily**	**Hr.**	**Daily**	**Bare Costs**	**Incl. O&P**
1 Millwright Foreman (inside)	$56.40	$451.20	$87.35	$698.80	$59.52	$92.01
1 Millwright	55.90	447.20	86.60	692.80		
1 Electrician	61.35	490.80	94.45	755.60		
1 Plumber	64.45	515.60	99.65	797.20		
1 R.T. Forklift, 5,000 Lb., gas		279.90		307.89	8.75	9.62
32 L.H., Daily Totals		$2184.70		$3252.29	$68.27	$101.63

Crew B-68D	Hr.	Daily	Hr.	Daily	Bare Costs	Incl. O&P
1 Labor Foreman (inside)	$42.60	$340.80	$68.25	$546.00	$45.90	$72.88
1 Laborer	42.10	336.80	67.45	539.60		
1 Equip. Oper. (light)	53.00	424.00	82.95	663.60		
1 R.T. Forklift, 5,000 Lb., gas		279.90		307.89	11.66	12.83
24 L.H., Daily Totals		$1381.50		$2057.09	$57.56	$85.71

Crew B-68E	Hr.	Daily	Hr.	Daily	Bare Costs	Incl. O&P
1 Struc. Steel Foreman (inside)	$58.15	$465.20	$95.30	$762.40	$57.71	$94.56
3 Struc. Steel Workers	57.65	1383.60	94.45	2266.80		
1 Welder	57.45	459.60	94.15	753.20		
1 Forklift, Smooth Floor, 8,000 Lb.		255.90		281.49	6.40	7.04
40 L.H., Daily Totals		$2564.30		$4063.89	$64.11	$101.60

Crew B-68F	Hr.	Daily	Hr.	Daily	Bare Costs	Incl. O&P
1 Skilled Worker Foreman (out)	$56.85	$454.80	$90.55	$724.40	$55.52	$88.45
2 Skilled Workers	54.85	877.60	87.40	1398.40		
1 R.T. Forklift, 5,000 Lb., gas		279.90		307.89	11.66	12.83
24 L.H., Daily Totals		$1612.30		$2430.69	$67.18	$101.28

Crew B-68G	Hr.	Daily	Hr.	Daily	Bare Costs	Incl. O&P
2 Structural Steel Workers	$57.65	$922.40	$94.45	$1511.20	$57.65	$94.45
1 R.T. Forklift, 5,000 Lb., gas		279.90		307.89	17.49	19.24
16 L.H., Daily Totals		$1202.30		$1819.09	$75.14	$113.69

Crew B-69	Hr.	Daily	Hr.	Daily	Bare Costs	Incl. O&P
1 Labor Foreman (outside)	$44.10	$352.80	$70.65	$565.20	$46.69	$74.13
3 Laborers	42.10	1010.40	67.45	1618.80		
1 Equip. Oper. (crane)	59.20	473.60	92.65	741.20		
1 Equip. Oper. (oiler)	50.55	404.40	79.10	632.80		
1 Hyd. Crane, 80 Ton		1486.00		1634.60	30.96	34.05
48 L.H., Daily Totals		$3727.20		$5192.60	$77.65	$108.18

Crew B-69A	Hr.	Daily	Hr.	Daily	Bare Costs	Incl. O&P
1 Labor Foreman (outside)	$44.10	$352.80	$70.65	$565.20	$46.18	$73.38
3 Laborers	42.10	1010.40	67.45	1618.80		
1 Equip. Oper. (medium)	56.75	454.00	88.80	710.40		
1 Concrete Finisher	49.95	399.60	78.45	627.60		
1 Curb/Gutter Paver, 2-Track		1217.00		1338.70	25.35	27.89
48 L.H., Daily Totals		$3433.80		$4860.70	$71.54	$101.26

Crew B-69B	Hr.	Daily	Hr.	Daily	Bare Costs	Incl. O&P
1 Labor Foreman (outside)	$44.10	$352.80	$70.65	$565.20	$46.18	$73.38
3 Laborers	42.10	1010.40	67.45	1618.80		
1 Equip. Oper. (medium)	56.75	454.00	88.80	710.40		
1 Cement Finisher	49.95	399.60	78.45	627.60		
1 Curb/Gutter Paver, 4-Track		791.55		870.71	16.49	18.14
48 L.H., Daily Totals		$3008.35		$4392.70	$62.67	$91.51

Crew B-70	Bare Costs Hr.	Bare Costs Daily	Incl. Subs O&P Hr.	Incl. Subs O&P Daily	Cost Per Labor-Hour Bare Costs	Cost Per Labor-Hour Incl. O&P
1 Labor Foreman (outside)	$44.10	$352.80	$70.65	$565.20	$48.66	$77.06
3 Laborers	42.10	1010.40	67.45	1618.80		
3 Equip. Oper. (medium)	56.75	1362.00	88.80	2131.20		
1 Grader, 30,000 Lbs.		1062.00		1168.20		
1 Ripper, Beam & 1 Shank		90.50		99.55		
1 Road Sweeper, S.P., 8' wide		715.85		787.43		
1 F.E. Loader, W.M., 1.5 C.Y.		423.50		465.85	40.93	45.02
56 L.H., Daily Totals		$5017.05		$6836.23	$89.59	$122.08

Crew B-71	Hr.	Daily	Hr.	Daily	Bare Costs	Incl. O&P
1 Labor Foreman (outside)	$44.10	$352.80	$70.65	$565.20	$48.66	$77.06
3 Laborers	42.10	1010.40	67.45	1618.80		
3 Equip. Oper. (medium)	56.75	1362.00	88.80	2131.20		
1 Pvmt. Profiler, 750 H.P.		3448.00		3792.80		
1 Road Sweeper, S.P., 8' wide		715.85		787.43		
1 F.E. Loader, W.M., 1.5 C.Y.		423.50		465.85	81.92	90.11
56 L.H., Daily Totals		$7312.55		$9361.28	$130.58	$167.17

Crew B-72	Hr.	Daily	Hr.	Daily	Bare Costs	Incl. O&P
1 Labor Foreman (outside)	$44.10	$352.80	$70.65	$565.20	$49.67	$78.53
3 Laborers	42.10	1010.40	67.45	1618.80		
4 Equip. Oper. (medium)	56.75	1816.00	88.80	2841.60		
1 Pvmt. Profiler, 750 H.P.		3448.00		3792.80		
1 Hammermill, 250 H.P.		848.80		933.68		
1 Windrow Loader		1427.00		1569.70		
1 Mix Paver, 165 H.P.		2149.00		2363.90		
1 Roller, Pneum. Whl., 12 Ton		345.75		380.32	128.41	141.26
64 L.H., Daily Totals		$11397.75		$14066.00	$178.09	$219.78

Crew B-73	Hr.	Daily	Hr.	Daily	Bare Costs	Incl. O&P
1 Labor Foreman (outside)	$44.10	$352.80	$70.65	$565.20	$51.51	$81.19
2 Laborers	42.10	673.60	67.45	1079.20		
5 Equip. Oper. (medium)	56.75	2270.00	88.80	3552.00		
1 Road Mixer, 310 H.P.		1896.00		2085.60		
1 Tandem Roller, 10 Ton		236.75		260.43		
1 Hammermill, 250 H.P.		848.80		933.68		
1 Grader, 30,000 Lbs.		1062.00		1168.20		
.5 F.E. Loader, W.M., 1.5 C.Y.		211.75		232.93		
.5 Truck Tractor, 220 H.P.		153.55		168.91		
.5 Water Tank Trailer, 5000 Gal.		76.13		83.74	70.08	77.09
64 L.H., Daily Totals		$7781.38		$10129.87	$121.58	$158.28

Crew B-74	Hr.	Daily	Hr.	Daily	Bare Costs	Incl. O&P
1 Labor Foreman (outside)	$44.10	$352.80	$70.65	$565.20	$51.39	$81.22
1 Laborer	42.10	336.80	67.45	539.60		
4 Equip. Oper. (medium)	56.75	1816.00	88.80	2841.60		
2 Truck Drivers (heavy)	48.95	783.20	78.25	1252.00		
1 Grader, 30,000 Lbs.		1062.00		1168.20		
1 Ripper, Beam & 1 Shank		90.50		99.55		
2 Stabilizers, 310 H.P.		2778.00		3055.80		
1 Flatbed Truck, Gas, 3 Ton		820.40		902.44		
1 Chem. Spreader, Towed		83.35		91.69		
1 Roller, Vibratory, 25 Ton		664.35		730.78		
1 Water Tank Trailer, 5000 Gal.		152.25		167.47		
1 Truck Tractor, 220 H.P.		307.10		337.81	93.09	102.40
64 L.H., Daily Totals		$9246.75		$11752.15	$144.48	$183.63

Crew B-75	Hr.	Daily	Hr.	Daily	Bare Costs	Incl. O&P
1 Labor Foreman (outside)	$44.10	$352.80	$70.65	$565.20	$51.74	$81.65
1 Laborer	42.10	336.80	67.45	539.60		
4 Equip. Oper. (medium)	56.75	1816.00	88.80	2841.60		
1 Truck Driver (heavy)	48.95	391.60	78.25	626.00		
1 Grader, 30,000 Lbs.		1062.00		1168.20		
1 Ripper, Beam & 1 Shank		90.50		99.55		
2 Stabilizers, 310 H.P.		2778.00		3055.80		
1 Dist. Tanker, 3000 Gallon		330.15		363.17		
1 Truck Tractor, 6x4, 380 H.P.		493.25		542.58		
1 Roller, Vibratory, 25 Ton		664.35		730.78	96.75	106.43
56 L.H., Daily Totals		$8315.45		$10532.48	$148.49	$188.08

Crew B-76	Hr.	Daily	Hr.	Daily	Bare Costs	Incl. O&P
1 Dock Builder Foreman (outside)	$56.20	$449.60	$90.15	$721.20	$55.13	$87.67
5 Dock Builders	54.20	2168.00	86.90	3476.00		
2 Equip. Oper. (crane)	59.20	947.20	92.65	1482.40		
1 Equip. Oper. (oiler)	50.55	404.40	79.10	632.80		
1 Crawler Crane, 50 Ton		1510.00		1661.00		
1 Barge, 400 Ton		858.85		944.74		
1 Hammer, Diesel, 15K ft.-lbs.		617.05		678.76		
1 Lead, 60' High		209.25		230.18		
1 Air Compressor, 600 cfm		421.50		463.65		
2 -50' Air Hoses, 3"		76.70		84.37	51.30	56.43
72 L.H., Daily Totals		$7662.55		$10375.08	$106.42	$144.10

Crew B-76A	Hr.	Daily	Hr.	Daily	Bare Costs	Incl. O&P
1 Labor Foreman (outside)	$44.10	$352.80	$70.65	$565.20	$45.54	$72.46
5 Laborers	42.10	1684.00	67.45	2698.00		
1 Equip. Oper. (crane)	59.20	473.60	92.65	741.20		
1 Equip. Oper. (oiler)	50.55	404.40	79.10	632.80		
1 Crawler Crane, 50 Ton		1510.00		1661.00		
1 Barge, 400 Ton		858.85		944.74	37.01	40.71
64 L.H., Daily Totals		$5283.65		$7242.94	$82.56	$113.17

Crew B-77	Hr.	Daily	Hr.	Daily	Bare Costs	Incl. O&P
1 Labor Foreman (outside)	$44.10	$352.80	$70.65	$565.20	$43.53	$69.71
3 Laborers	42.10	1010.40	67.45	1618.80		
1 Truck Driver (light)	47.25	378.00	75.55	604.40		
1 Crack Cleaner, 25 H.P.		52.45		57.70		
1 Crack Filler, Trailer Mtd.		154.75		170.22		
1 Flatbed Truck, Gas, 3 Ton		820.40		902.44	25.69	28.26
40 L.H., Daily Totals		$2768.80		$3918.76	$69.22	$97.97

Crew B-78	Hr.	Daily	Hr.	Daily	Bare Costs	Incl. O&P
1 Labor Foreman (outside)	$44.10	$352.80	$70.65	$565.20	$43.29	$69.33
4 Laborers	42.10	1347.20	67.45	2158.40		
1 Truck Driver (light)	47.25	378.00	75.55	604.40		
1 Paint Striper, S.P., 40 Gallon		127.10		139.81		
1 Flatbed Truck, Gas, 3 Ton		820.40		902.44		
1 Pickup Truck, 3/4 Ton		110.85		121.94	22.05	24.25
48 L.H., Daily Totals		$3136.35		$4492.19	$65.34	$93.59

Crew B-78A	Hr.	Daily	Hr.	Daily	Bare Costs	Incl. O&P
1 Equip. Oper. (light)	$53.00	$424.00	$82.95	$663.60	$53.00	$82.95
1 Line Rem. (Metal Balls) 115 H.P.		916.50		1008.15	114.56	126.02
8 L.H., Daily Totals		$1340.50		$1671.75	$167.56	$208.97

Crew B-78B	Bare Costs Hr.	Bare Costs Daily	Incl. Subs O&P Hr.	Incl. Subs O&P Daily	Cost Per Labor-Hour Bare Costs	Cost Per Labor-Hour Incl. O&P
2 Laborers	$42.10	$673.60	$67.45	$1079.20	$43.31	$69.17
.25 Equip. Oper. (light)	53.00	106.00	82.95	165.90		
1 Pickup Truck, 3/4 Ton		110.85		121.94		
1 Line Rem.,11 H.P.,Walk Behind		113.45		124.80		
.25 Road Sweeper, S.P., 8' wide		178.96		196.86	22.40	24.64
18 L.H., Daily Totals		$1182.86		$1688.69	$65.71	$93.82

Crew B-78C	Hr.	Daily	Hr.	Daily	Bare Costs	Incl. O&P
1 Labor Foreman (outside)	$44.10	$352.80	$70.65	$565.20	$43.29	$69.33
4 Laborers	42.10	1347.20	67.45	2158.40		
1 Truck Driver (light)	47.25	378.00	75.55	604.40		
1 Paint Striper, T.M., 120 Gal.		592.95		652.25		
1 Flatbed Truck, Gas, 3 Ton		820.40		902.44		
1 Pickup Truck, 3/4 Ton		110.85		121.94	31.75	34.93
48 L.H., Daily Totals		$3602.20		$5004.62	$75.05	$104.26

Crew B-78D	Hr.	Daily	Hr.	Daily	Bare Costs	Incl. O&P
2 Labor Foremen (outside)	$44.10	$705.60	$70.65	$1130.40	$43.02	$68.90
7 Laborers	42.10	2357.60	67.45	3777.20		
1 Truck Driver (light)	47.25	378.00	75.55	604.40		
1 Paint Striper, T.M., 120 Gal.		592.95		652.25		
1 Flatbed Truck, Gas, 3 Ton		820.40		902.44		
3 Pickup Trucks, 3/4 Ton		332.55		365.81		
1 Air Compressor, 60 cfm		152.95		168.25		
1 -50' Air Hose, 3/4"		7.75		8.53		
1 Breaker, Pavement, 60 lb.		53.50		58.85	24.50	26.95
80 L.H., Daily Totals		$5401.30		$7668.11	$67.52	$95.85

Crew B-78E	Hr.	Daily	Hr.	Daily	Bare Costs	Incl. O&P
2 Labor Foremen (outside)	$44.10	$705.60	$70.65	$1130.40	$42.86	$68.66
9 Laborers	42.10	3031.20	67.45	4856.40		
1 Truck Driver (light)	47.25	378.00	75.55	604.40		
1 Paint Striper, T.M., 120 Gal.		592.95		652.25		
1 Flatbed Truck, Gas, 3 Ton		820.40		902.44		
4 Pickup Trucks, 3/4 Ton		443.40		487.74		
2 Air Compressors, 60 cfm		305.90		336.49		
2 -50' Air Hoses, 3/4"		15.50		17.05		
2 Breakers, Pavement, 60 lb.		107.00		117.70	23.80	26.18
96 L.H., Daily Totals		$6399.95		$9104.86	$66.67	$94.84

Crew B-78F	Hr.	Daily	Hr.	Daily	Bare Costs	Incl. O&P
2 Labor Foremen (outside)	$44.10	$705.60	$70.65	$1130.40	$42.75	$68.49
11 Laborers	42.10	3704.80	67.45	5935.60		
1 Truck Driver (light)	47.25	378.00	75.55	604.40		
1 Paint Striper, T.M., 120 Gal.		592.95		652.25		
1 Flatbed Truck, Gas, 3 Ton		820.40		902.44		
7 Pickup Trucks, 3/4 Ton		775.95		853.54		
3 Air Compressors, 60 cfm		458.85		504.74		
3 -50' Air Hoses, 3/4"		23.25		25.57		
3 Breakers, Pavement, 60 lb.		160.50		176.55	25.28	27.81
112 L.H., Daily Totals		$7620.30		$10785.49	$68.04	$96.30

Crew B-79	Hr.	Daily	Hr.	Daily	Bare Costs	Incl. O&P
1 Labor Foreman (outside)	$44.10	$352.80	$70.65	$565.20	$43.53	$69.71
3 Laborers	42.10	1010.40	67.45	1618.80		
1 Truck Driver (light)	47.25	378.00	75.55	604.40		
1 Paint Striper, T.M., 120 Gal.		592.95		652.25		
1 Heating Kettle, 115 Gallon		98.65		108.52		
1 Flatbed Truck, Gas, 3 Ton		820.40		902.44		
2 Pickup Trucks, 3/4 Ton		221.70		243.87	43.34	47.68
40 L.H., Daily Totals		$3474.90		$4695.47	$86.87	$117.39

Crew B-79A	Hr.	Daily	Hr.	Daily	Bare Costs	Incl. O&P
1.5 Equip. Oper. (light)	$53.00	$636.00	$82.95	$995.40	$53.00	$82.95
.5 Line Remov. (Grinder) 115 H.P.		475.57		523.13		
1 Line Rem. (Metal Balls) 115 H.P.		916.50		1008.15	116.01	127.61
12 L.H., Daily Totals		$2028.08		$2526.68	$169.01	$210.56

Crew B-79B	Hr.	Daily	Hr.	Daily	Bare Costs	Incl. O&P
1 Laborer	$42.10	$336.80	$67.45	$539.60	$42.10	$67.45
1 Set of Gases		171.55		188.71	21.44	23.59
8 L.H., Daily Totals		$508.35		$728.30	$63.54	$91.04

Crew B-79C	Hr.	Daily	Hr.	Daily	Bare Costs	Incl. O&P
1 Labor Foreman (outside)	$44.10	$352.80	$70.65	$565.20	$43.12	$69.06
5 Laborers	42.10	1684.00	67.45	2698.00		
1 Truck Driver (light)	47.25	378.00	75.55	604.40		
1 Paint Striper, T.M., 120 Gal.		592.95		652.25		
1 Heating Kettle, 115 Gallon		98.65		108.52		
1 Flatbed Truck, Gas, 3 Ton		820.40		902.44		
3 Pickup Trucks, 3/4 Ton		332.55		365.81		
1 Air Compressor, 60 cfm		152.95		168.25		
1 -50' Air Hose, 3/4"		7.75		8.53		
1 Breaker, Pavement, 60 lb.		53.50		58.85	36.76	40.44
56 L.H., Daily Totals		$4473.55		$6132.23	$79.88	$109.50

Crew B-79D	Hr.	Daily	Hr.	Daily	Bare Costs	Incl. O&P
2 Labor Foremen (outside)	$44.10	$705.60	$70.65	$1130.40	$43.24	$69.26
5 Laborers	42.10	1684.00	67.45	2698.00		
1 Truck Driver (light)	47.25	378.00	75.55	604.40		
1 Paint Striper, T.M., 120 Gal.		592.95		652.25		
1 Heating Kettle, 115 Gallon		98.65		108.52		
1 Flatbed Truck, Gas, 3 Ton		820.40		902.44		
4 Pickup Trucks, 3/4 Ton		443.40		487.74		
1 Air Compressor, 60 cfm		152.95		168.25		
1 -50' Air Hose, 3/4"		7.75		8.53		
1 Breaker, Pavement, 60 lb.		53.50		58.85	33.90	37.29
64 L.H., Daily Totals		$4937.20		$6819.36	$77.14	$106.55

Crew B-79E	Hr.	Daily	Hr.	Daily	Bare Costs	Incl. O&P
2 Labor Foremen (outside)	$44.10	$705.60	$70.65	$1130.40	$43.02	$68.90
7 Laborers	42.10	2357.60	67.45	3777.20		
1 Truck Driver (light)	47.25	378.00	75.55	604.40		
1 Paint Striper, T.M., 120 Gal.		592.95		652.25		
1 Heating Kettle, 115 Gallon		98.65		108.52		
1 Flatbed Truck, Gas, 3 Ton		820.40		902.44		
5 Pickup Trucks, 3/4 Ton		554.25		609.67		
2 Air Compressors, 60 cfm		305.90		336.49		
2 -50' Air Hoses, 3/4"		15.50		17.05		
2 Breakers, Pavement, 60 lb.		107.00		117.70	31.18	34.30
80 L.H., Daily Totals		$5935.85		$8256.11	$74.20	$103.20

Crew B-80	Hr.	Daily	Hr.	Daily	Bare Costs	Incl. O&P
1 Labor Foreman (outside)	$44.10	$352.80	$70.65	$565.20	$46.61	$74.15
1 Laborer	42.10	336.80	67.45	539.60		
1 Truck Driver (light)	47.25	378.00	75.55	604.40		
1 Equip. Oper. (light)	53.00	424.00	82.95	663.60		
1 Flatbed Truck, Gas, 3 Ton		820.40		902.44		
1 Earth Auger, Truck-Mtd.		200.50		220.55	31.90	35.09
32 L.H., Daily Totals		$2512.50		$3495.79	$78.52	$109.24

Crew No.	Bare Costs		Incl. Subs O&P		Cost Per Labor-Hour	

Crew B-80A	Hr.	Daily	Hr.	Daily	Bare Costs	Incl. O&P
3 Laborers	$42.10	$1010.40	$67.45	$1618.80	$42.10	$67.45
1 Flatbed Truck, Gas, 3 Ton		820.40		902.44	34.18	37.60
24 L.H., Daily Totals		$1830.80		$2521.24	$76.28	$105.05

Crew B-80B	Hr.	Daily	Hr.	Daily	Bare Costs	Incl. O&P
3 Laborers	$42.10	$1010.40	$67.45	$1618.80	$44.83	$71.33
1 Equip. Oper. (light)	53.00	424.00	82.95	663.60		
1 Crane, Flatbed Mounted, 3 Ton		236.10		259.71	7.38	8.12
32 L.H., Daily Totals		$1670.50		$2542.11	$52.20	$79.44

Crew B-80C	Hr.	Daily	Hr.	Daily	Bare Costs	Incl. O&P
2 Laborers	$42.10	$673.60	$67.45	$1079.20	$43.82	$70.15
1 Truck Driver (light)	47.25	378.00	75.55	604.40		
1 Flatbed Truck, Gas, 1.5 Ton		196.15		215.76		
1 Manual Fence Post Auger, Gas		53.85		59.23	10.42	11.46
24 L.H., Daily Totals		$1301.60		$1958.60	$54.23	$81.61

Crew B-81	Hr.	Daily	Hr.	Daily	Bare Costs	Incl. O&P
1 Laborer	$42.10	$336.80	$67.45	$539.60	$49.27	$78.17
1 Equip. Oper. (medium)	56.75	454.00	88.80	710.40		
1 Truck Driver (heavy)	48.95	391.60	78.25	626.00		
1 Hydromulcher, T.M., 3000 Gal.		275.45		303.00		
1 Truck Tractor, 220 H.P.		307.10		337.81	24.27	26.70
24 L.H., Daily Totals		$1764.95		$2516.80	$73.54	$104.87

Crew B-81A	Hr.	Daily	Hr.	Daily	Bare Costs	Incl. O&P
1 Laborer	$42.10	$336.80	$67.45	$539.60	$44.67	$71.50
1 Truck Driver (light)	47.25	378.00	75.55	604.40		
1 Hydromulcher, T.M., 600 Gal.		116.95		128.65		
1 Flatbed Truck, Gas, 3 Ton		820.40		902.44	58.58	64.44
16 L.H., Daily Totals		$1652.15		$2175.09	$103.26	$135.94

Crew B-82	Hr.	Daily	Hr.	Daily	Bare Costs	Incl. O&P
1 Laborer	$42.10	$336.80	$67.45	$539.60	$47.55	$75.20
1 Equip. Oper. (light)	53.00	424.00	82.95	663.60		
1 Horiz. Borer, 6 H.P.		182.35		200.59	11.40	12.54
16 L.H., Daily Totals		$943.15		$1403.79	$58.95	$87.74

Crew B-82A	Hr.	Daily	Hr.	Daily	Bare Costs	Incl. O&P
2 Laborers	$42.10	$673.60	$67.45	$1079.20	$47.55	$75.20
2 Equip. Opers. (light)	53.00	848.00	82.95	1327.20		
2 Dump Truck, 8 C.Y., 220 H.P.		856.10		941.71		
1 Flatbed Trailer, 25 Ton		135.55		149.10		
1 Horiz. Dir. Drill, 20k lb. Thrust		538.65		592.51		
1 Mud Trailer for HDD, 1500 Gal.		309.00		339.90		
1 Pickup Truck, 4x4, 3/4 Ton		175.85		193.44		
1 Flatbed Trailer, 3 Ton		70.30		77.33		
1 Loader, Skid Steer, 78 H.P.		400.10		440.11	77.67	85.44
32 L.H., Daily Totals		$4007.15		$5140.51	$125.22	$160.64

Crew B-82B	Hr.	Daily	Hr.	Daily	Bare Costs	Incl. O&P
2 Laborers	$42.10	$673.60	$67.45	$1079.20	$47.55	$75.20
2 Equip. Opers. (light)	53.00	848.00	82.95	1327.20		
2 Dump Truck, 8 C.Y., 220 H.P.		856.10		941.71		
1 Flatbed Trailer, 25 Ton		135.55		149.10		
1 Horiz. Dir. Drill, 30k lb. Thrust		641.20		705.32		
1 Mud Trailer for HDD, 1500 Gal.		309.00		339.90		
1 Pickup Truck, 4x4, 3/4 Ton		175.85		193.44		
1 Flatbed Trailer, 3 Ton		70.30		77.33		
1 Loader, Skid Steer, 78 H.P.		400.10		440.11	80.88	88.97
32 L.H., Daily Totals		$4109.70		$5253.31	$128.43	$164.17

Crew B-82C	Hr.	Daily	Hr.	Daily	Bare Costs	Incl. O&P
2 Laborers	$42.10	$673.60	$67.45	$1079.20	$47.55	$75.20
2 Equip. Opers. (light)	53.00	848.00	82.95	1327.20		
2 Dump Truck, 8 C.Y., 220 H.P.		856.10		941.71		
1 Flatbed Trailer, 25 Ton		135.55		149.10		
1 Horiz. Dir. Drill, 50k lb. Thrust		815.80		897.38		
1 Mud Trailer for HDD, 1500 Gal.		309.00		339.90		
1 Pickup Truck, 4x4, 3/4 Ton		175.85		193.44		
1 Flatbed Trailer, 3 Ton		70.30		77.33		
1 Loader, Skid Steer, 78 H.P.		400.10		440.11	86.33	94.97
32 L.H., Daily Totals		$4284.30		$5445.37	$133.88	$170.17

Crew B-82D	Hr.	Daily	Hr.	Daily	Bare Costs	Incl. O&P
1 Equip. Oper. (light)	$53.00	$424.00	$82.95	$663.60	$53.00	$82.95
1 Mud Trailer for HDD, 1500 Gal.		309.00		339.90	38.63	42.49
8 L.H., Daily Totals		$733.00		$1003.50	$91.63	$125.44

Crew B-83	Hr.	Daily	Hr.	Daily	Bare Costs	Incl. O&P
1 Tugboat Captain	$56.75	$454.00	$88.80	$710.40	$49.42	$78.13
1 Tugboat Hand	42.10	336.80	67.45	539.60		
1 Tugboat, 250 H.P.		717.50		789.25	44.84	49.33
16 L.H., Daily Totals		$1508.30		$2039.25	$94.27	$127.45

Crew B-84	Hr.	Daily	Hr.	Daily	Bare Costs	Incl. O&P
1 Equip. Oper. (medium)	$56.75	$454.00	$88.80	$710.40	$56.75	$88.80
1 Rotary Mower/Tractor		366.75		403.43	45.84	50.43
8 L.H., Daily Totals		$820.75		$1113.83	$102.59	$139.23

Crew B-85	Hr.	Daily	Hr.	Daily	Bare Costs	Incl. O&P
3 Laborers	$42.10	$1010.40	$67.45	$1618.80	$46.40	$73.88
1 Equip. Oper. (medium)	56.75	454.00	88.80	710.40		
1 Truck Driver (heavy)	48.95	391.60	78.25	626.00		
1 Telescoping Boom Lift, to 80'		383.55		421.90		
1 Brush Chipper, 12", 130 H.P.		392.80		432.08		
1 Pruning Saw, Rotary		26.40		29.04	20.07	22.08
40 L.H., Daily Totals		$2658.75		$3838.22	$66.47	$95.96

Crew B-86	Hr.	Daily	Hr.	Daily	Bare Costs	Incl. O&P
1 Equip. Oper. (medium)	$56.75	$454.00	$88.80	$710.40	$56.75	$88.80
1 Stump Chipper, S.P.		194.45		213.90	24.31	26.74
8 L.H., Daily Totals		$648.45		$924.29	$81.06	$115.54

Crew B-86A	Hr.	Daily	Hr.	Daily	Bare Costs	Incl. O&P
1 Equip. Oper. (medium)	$56.75	$454.00	$88.80	$710.40	$56.75	$88.80
1 Grader, 30,000 Lbs.		1062.00		1168.20	132.75	146.03
8 L.H., Daily Totals		$1516.00		$1878.60	$189.50	$234.82

Crew B-86B	Hr.	Daily	Hr.	Daily	Bare Costs	Incl. O&P
1 Equip. Oper. (medium)	$56.75	$454.00	$88.80	$710.40	$56.75	$88.80
1 Dozer, 200 H.P.		1504.00		1654.40	188.00	206.80
8 L.H., Daily Totals		$1958.00		$2364.80	$244.75	$295.60

Crew B-87	Hr.	Daily	Hr.	Daily	Bare Costs	Incl. O&P
1 Laborer	$42.10	$336.80	$67.45	$539.60	$53.82	$84.53
4 Equip. Oper. (medium)	56.75	1816.00	88.80	2841.60		
2 Feller Bunchers, 100 H.P.		1177.20		1294.92		
1 Log Chipper, 22" Tree		552.00		607.20		
1 Dozer, 105 H.P.		633.85		697.24		
1 Chain Saw, Gas, 36" Long		41.25		45.38	60.11	66.12
40 L.H., Daily Totals		$4557.10		$6025.93	$113.93	$150.65

Crew B-88	Hr.	Daily	Hr.	Daily	Bare Costs	Incl. O&P
1 Laborer	$42.10	$336.80	$67.45	$539.60	$54.66	$85.75
6 Equip. Oper. (medium)	56.75	2724.00	88.80	4262.40		
2 Feller Bunchers, 100 H.P.		1177.20		1294.92		
1 Log Chipper, 22" Tree		552.00		607.20		
2 Log Skidders, 50 H.P.		1808.40		1989.24		
1 Dozer, 105 H.P.		633.85		697.24		
1 Chain Saw, Gas, 36" Long		41.25		45.38	75.23	82.75
56 L.H., Daily Totals		$7273.50		$9435.97	$129.88	$168.50

Crew B-89	Hr.	Daily	Hr.	Daily	Bare Costs	Incl. O&P
1 Equip. Oper. (light)	$53.00	$424.00	$82.95	$663.60	$50.13	$79.25
1 Truck Driver (light)	47.25	378.00	75.55	604.40		
1 Flatbed Truck, Gas, 3 Ton		820.40		902.44		
1 Concrete Saw		111.50		122.65		
1 Water Tank, 65 Gal.		101.90		112.09	64.61	71.07
16 L.H., Daily Totals		$1835.80		$2405.18	$114.74	$150.32

Crew B-89A	Hr.	Daily	Hr.	Daily	Bare Costs	Incl. O&P
1 Skilled Worker	$54.85	$438.80	$87.40	$699.20	$48.48	$77.42
1 Laborer	42.10	336.80	67.45	539.60		
1 Core Drill (Large)		114.30		125.73	7.14	7.86
16 L.H., Daily Totals		$889.90		$1364.53	$55.62	$85.28

Crew B-89B	Hr.	Daily	Hr.	Daily	Bare Costs	Incl. O&P
1 Equip. Oper. (light)	$53.00	$424.00	$82.95	$663.60	$50.13	$79.25
1 Truck Driver (light)	47.25	378.00	75.55	604.40		
1 Wall Saw, Hydraulic, 10 H.P.		85.50		94.05		
1 Generator, Diesel, 100 kW		496.20		545.82		
1 Water Tank, 65 Gal.		101.90		112.09		
1 Flatbed Truck, Gas, 3 Ton		820.40		902.44	94.00	103.40
16 L.H., Daily Totals		$2306.00		$2922.40	$144.13	$182.65

Crew B-89C	Hr.	Daily	Hr.	Daily	Bare Costs	Incl. O&P
1 Cement Finisher	$49.95	$399.60	$78.45	$627.60	$49.95	$78.45
1 Masonry cut-off saw, gas		61.00		67.10	7.63	8.39
8 L.H., Daily Totals		$460.60		$694.70	$57.58	$86.84

Crew B-90	Hr.	Daily	Hr.	Daily	Bare Costs	Incl. O&P
1 Labor Foreman (outside)	$44.10	$352.80	$70.65	$565.20	$46.79	$74.42
3 Laborers	42.10	1010.40	67.45	1618.80		
2 Equip. Oper. (light)	53.00	848.00	82.95	1327.20		
2 Truck Drivers (heavy)	48.95	783.20	78.25	1252.00		
1 Road Mixer, 310 H.P.		1896.00		2085.60		
1 Dist. Truck, 2000 Gal.		299.65		329.62	34.31	37.74
64 L.H., Daily Totals		$5190.05		$7178.42	$81.09	$112.16

Crew B-90A	Hr.	Daily	Hr.	Daily	Bare Costs	Incl. O&P
1 Labor Foreman (outside)	$44.10	$352.80	$70.65	$565.20	$50.76	$80.11
2 Laborers	42.10	673.60	67.45	1079.20		
4 Equip. Oper. (medium)	56.75	1816.00	88.80	2841.60		
2 Graders, 30,000 Lbs.		2124.00		2336.40		
1 Tandem Roller, 10 Ton		236.75		260.43		
1 Roller, Pneum. Whl., 12 Ton		345.75		380.32	48.33	53.16
56 L.H., Daily Totals		$5548.90		$7463.15	$99.09	$133.27

Crew B-90B	Hr.	Daily	Hr.	Daily	Bare Costs	Incl. O&P
1 Labor Foreman (outside)	$44.10	$352.80	$70.65	$565.20	$49.76	$78.66
2 Laborers	42.10	673.60	67.45	1079.20		
3 Equip. Oper. (medium)	56.75	1362.00	88.80	2131.20		
1 Roller, Pneum. Whl., 12 Ton		345.75		380.32		
1 Road Mixer, 310 H.P.		1896.00		2085.60	46.70	51.37
48 L.H., Daily Totals		$4630.15		$6241.52	$96.46	$130.03

Crew B-90C	Hr.	Daily	Hr.	Daily	Bare Costs	Incl. O&P
1 Labor Foreman (outside)	$44.10	$352.80	$70.65	$565.20	$48.15	$76.51
4 Laborers	42.10	1347.20	67.45	2158.40		
3 Equip. Oper. (medium)	56.75	1362.00	88.80	2131.20		
3 Truck Drivers (heavy)	48.95	1174.80	78.25	1878.00		
3 Road Mixers, 310 H.P.		5688.00		6256.80	64.64	71.10
88 L.H., Daily Totals		$9924.80		$12989.60	$112.78	$147.61

Crew B-90D	Hr.	Daily	Hr.	Daily	Bare Costs	Incl. O&P
1 Labor Foreman (outside)	$44.10	$352.80	$70.65	$565.20	$47.22	$75.12
6 Laborers	42.10	2020.80	67.45	3237.60		
3 Equip. Oper. (medium)	56.75	1362.00	88.80	2131.20		
3 Truck Drivers (heavy)	48.95	1174.80	78.25	1878.00		
3 Road Mixers, 310 H.P.		5688.00		6256.80	54.69	60.16
104 L.H., Daily Totals		$10598.40		$14068.80	$101.91	$135.28

Crew B-90E	Hr.	Daily	Hr.	Daily	Bare Costs	Incl. O&P
1 Labor Foreman (outside)	$44.10	$352.80	$70.65	$565.20	$47.97	$76.12
4 Laborers	42.10	1347.20	67.45	2158.40		
3 Equip. Oper. (medium)	56.75	1362.00	88.80	2131.20		
1 Truck Driver (heavy)	48.95	391.60	78.25	626.00		
1 Road Mixer, 310 H.P.		1896.00		2085.60	26.33	28.97
72 L.H., Daily Totals		$5349.60		$7566.40	$74.30	$105.09

Crew B-91	Hr.	Daily	Hr.	Daily	Bare Costs	Incl. O&P
1 Labor Foreman (outside)	$44.10	$352.80	$70.65	$565.20	$50.53	$79.88
2 Laborers	42.10	673.60	67.45	1079.20		
4 Equip. Oper. (medium)	56.75	1816.00	88.80	2841.60		
1 Truck Driver (heavy)	48.95	391.60	78.25	626.00		
1 Dist. Tanker, 3000 Gallon		330.15		363.17		
1 Truck Tractor, 6x4, 380 H.P.		493.25		542.58		
1 Aggreg. Spreader, S.P.		848.90		933.79		
1 Roller, Pneum. Whl., 12 Ton		345.75		380.32		
1 Tandem Roller, 10 Ton		236.75		260.43	35.23	38.75
64 L.H., Daily Totals		$5488.80		$7592.28	$85.76	$118.63

Crew B-91B	Hr.	Daily	Hr.	Daily	Bare Costs	Incl. O&P
1 Laborer	$42.10	$336.80	$67.45	$539.60	$49.42	$78.13
1 Equipment Oper. (med.)	56.75	454.00	88.80	710.40		
1 Road Sweeper, Vac. Assist.		869.00		955.90	54.31	59.74
16 L.H., Daily Totals		$1659.80		$2205.90	$103.74	$137.87

Crew B-91C	Hr.	Daily	Hr.	Daily	Bare Costs	Incl. O&P
1 Laborer	$42.10	$336.80	$67.45	$539.60	$44.67	$71.50
1 Truck Driver (light)	47.25	378.00	75.55	604.40		
1 Catch Basin Cleaning Truck		536.15		589.76	33.51	36.86
16 L.H., Daily Totals		$1250.95		$1733.77	$78.18	$108.36

Crew No.	Bare Costs		Incl. Subs O&P		Cost Per Labor-Hour	
Crew B-91D	**Hr.**	**Daily**	**Hr.**	**Daily**	**Bare Costs**	**Incl. O&P**
1 Labor Foreman (outside)	$44.10	$352.80	$70.65	$565.20	$48.94	$77.57
5 Laborers	42.10	1684.00	67.45	2698.00		
5 Equip. Oper. (medium)	56.75	2270.00	88.80	3552.00		
2 Truck Drivers (heavy)	48.95	783.20	78.25	1252.00		
1 Aggreg. Spreader, S.P.		848.90		933.79		
2 Truck Tractors, 6x4, 380 H.P.		986.50		1085.15		
2 Dist. Tankers, 3000 Gallon		660.30		726.33		
2 Pavement Brushes, Towed		174.60		192.06		
2 Rollers Pneum. Whl., 12 Ton		691.50		760.65	32.33	35.56
104 L.H., Daily Totals		$8451.80		$11765.18	$81.27	$113.13

Crew B-92	**Hr.**	**Daily**	**Hr.**	**Daily**	**Bare Costs**	**Incl. O&P**
1 Labor Foreman (outside)	$44.10	$352.80	$70.65	$565.20	$42.60	$68.25
3 Laborers	42.10	1010.40	67.45	1618.80		
1 Crack Cleaner, 25 H.P.		52.45		57.70		
1 Air Compressor, 60 cfm		152.95		168.25		
1 Tar Kettle, T.M.		156.20		171.82		
1 Flatbed Truck, Gas, 3 Ton		820.40		902.44	36.94	40.63
32 L.H., Daily Totals		$2545.20		$3484.20	$79.54	$108.88

Crew B-93	**Hr.**	**Daily**	**Hr.**	**Daily**	**Bare Costs**	**Incl. O&P**
1 Equip. Oper. (medium)	$56.75	$454.00	$88.80	$710.40	$56.75	$88.80
1 Feller Buncher, 100 H.P.		588.60		647.46	73.58	80.93
8 L.H., Daily Totals		$1042.60		$1357.86	$130.32	$169.73

Crew B-94A	**Hr.**	**Daily**	**Hr.**	**Daily**	**Bare Costs**	**Incl. O&P**
1 Laborer	$42.10	$336.80	$67.45	$539.60	$42.10	$67.45
1 Diaphragm Water Pump, 2"		86.80		95.48		
1 -20' Suction Hose, 2"		3.55		3.90		
2 -50' Discharge Hoses, 2"		8.00		8.80	12.29	13.52
8 L.H., Daily Totals		$435.15		$647.78	$54.39	$80.97

Crew B-94B	**Hr.**	**Daily**	**Hr.**	**Daily**	**Bare Costs**	**Incl. O&P**
1 Laborer	$42.10	$336.80	$67.45	$539.60	$42.10	$67.45
1 Diaphragm Water Pump, 4"		105.25		115.78		
1 -20' Suction Hose, 4"		17.05		18.75		
2 -50' Discharge Hoses, 4"		25.30		27.83	18.45	20.30
8 L.H., Daily Totals		$484.40		$701.96	$60.55	$87.75

Crew B-94C	**Hr.**	**Daily**	**Hr.**	**Daily**	**Bare Costs**	**Incl. O&P**
1 Laborer	$42.10	$336.80	$67.45	$539.60	$42.10	$67.45
1 Centrifugal Water Pump, 3"		74.45		81.89		
1 -20' Suction Hose, 3"		8.75		9.63		
2 -50' Discharge Hoses, 3"		9.00		9.90	11.53	12.68
8 L.H., Daily Totals		$429.00		$641.02	$53.63	$80.13

Crew B-94D	**Hr.**	**Daily**	**Hr.**	**Daily**	**Bare Costs**	**Incl. O&P**
1 Laborer	$42.10	$336.80	$67.45	$539.60	$42.10	$67.45
1 Centr. Water Pump, 6"		232.90		256.19		
1 -20' Suction Hose, 6"		25.20		27.72		
2 -50' Discharge Hoses, 6"		35.80		39.38	36.74	40.41
8 L.H., Daily Totals		$630.70		$862.89	$78.84	$107.86

Crew C-1	**Hr.**	**Daily**	**Hr.**	**Daily**	**Bare Costs**	**Incl. O&P**
3 Carpenters	$53.15	$1275.60	$85.15	$2043.60	$50.39	$80.72
1 Laborer	42.10	336.80	67.45	539.60		
32 L.H., Daily Totals		$1612.40		$2583.20	$50.39	$80.72

Crew C-2	**Hr.**	**Daily**	**Hr.**	**Daily**	**Bare Costs**	**Incl. O&P**
1 Carpenter Foreman (outside)	$55.15	$441.20	$88.40	$707.20	$51.64	$82.74
4 Carpenters	53.15	1700.80	85.15	2724.80		
1 Laborer	42.10	336.80	67.45	539.60		
48 L.H., Daily Totals		$2478.80		$3971.60	$51.64	$82.74

Crew C-2A	**Hr.**	**Daily**	**Hr.**	**Daily**	**Bare Costs**	**Incl. O&P**
1 Carpenter Foreman (outside)	$55.15	$441.20	$88.40	$707.20	$51.11	$81.63
3 Carpenters	53.15	1275.60	85.15	2043.60		
1 Cement Finisher	49.95	399.60	78.45	627.60		
1 Laborer	42.10	336.80	67.45	539.60		
48 L.H., Daily Totals		$2453.20		$3918.00	$51.11	$81.63

Crew C-3	**Hr.**	**Daily**	**Hr.**	**Daily**	**Bare Costs**	**Incl. O&P**
1 Rodman Foreman (outside)	$58.40	$467.20	$92.00	$736.00	$52.65	$83.16
4 Rodmen (reinf.)	56.40	1804.80	88.85	2843.20		
1 Equip. Oper. (light)	53.00	424.00	82.95	663.60		
2 Laborers	42.10	673.60	67.45	1079.20		
3 Stressing Equipment		56.70		62.37		
.5 Grouting Equipment		122.08		134.28	2.79	3.07
64 L.H., Daily Totals		$3548.38		$5518.65	$55.44	$86.23

Crew C-4	**Hr.**	**Daily**	**Hr.**	**Daily**	**Bare Costs**	**Incl. O&P**
1 Rodman Foreman (outside)	$58.40	$467.20	$92.00	$736.00	$56.90	$89.64
3 Rodmen (reinf.)	56.40	1353.60	88.85	2132.40		
3 Stressing Equipment		56.70		62.37	1.77	1.95
32 L.H., Daily Totals		$1877.50		$2930.77	$58.67	$91.59

Crew C-4A	**Hr.**	**Daily**	**Hr.**	**Daily**	**Bare Costs**	**Incl. O&P**
2 Rodmen (reinf.)	$56.40	$902.40	$88.85	$1421.60	$56.40	$88.85
4 Stressing Equipment		75.60		83.16	4.72	5.20
16 L.H., Daily Totals		$978.00		$1504.76	$61.13	$94.05

Crew C-5	**Hr.**	**Daily**	**Hr.**	**Daily**	**Bare Costs**	**Incl. O&P**
1 Rodman Foreman (outside)	$58.40	$467.20	$92.00	$736.00	$56.25	$88.45
4 Rodmen (reinf.)	56.40	1804.80	88.85	2843.20		
1 Equip. Oper. (crane)	59.20	473.60	92.65	741.20		
1 Equip. Oper. (oiler)	50.55	404.40	79.10	632.80		
1 Hyd. Crane, 25 Ton		580.85		638.93	10.37	11.41
56 L.H., Daily Totals		$3730.85		$5592.14	$66.62	$99.86

Crew C-6	**Hr.**	**Daily**	**Hr.**	**Daily**	**Bare Costs**	**Incl. O&P**
1 Labor Foreman (outside)	$44.10	$352.80	$70.65	$565.20	$43.74	$69.82
4 Laborers	42.10	1347.20	67.45	2158.40		
1 Cement Finisher	49.95	399.60	78.45	627.60		
2 Gas Engine Vibrators		53.70		59.07	1.12	1.23
48 L.H., Daily Totals		$2153.30		$3410.27	$44.86	$71.05

Crew C-6A	**Hr.**	**Daily**	**Hr.**	**Daily**	**Bare Costs**	**Incl. O&P**
2 Cement Finishers	$49.95	$799.20	$78.45	$1255.20	$49.95	$78.45
1 Concrete Vibrator, Elec, 2 HP		47.45		52.20	2.97	3.26
16 L.H., Daily Totals		$846.65		$1307.40	$52.92	$81.71

Crew No.	Bare Costs		Incl. Subs O&P		Cost Per Labor-Hour	
Crew C-7	**Hr.**	**Daily**	**Hr.**	**Daily**	**Bare Costs**	**Incl. O&P**
1 Labor Foreman (outside)	$44.10	$352.80	$70.65	$565.20	$45.76	$72.69
5 Laborers	42.10	1684.00	67.45	2698.00		
1 Cement Finisher	49.95	399.60	78.45	627.60		
1 Equip. Oper. (medium)	56.75	454.00	88.80	710.40		
1 Equip. Oper. (oiler)	50.55	404.40	79.10	632.80		
2 Gas Engine Vibrators		53.70		59.07		
1 Concrete Bucket, 1 C.Y.		45.80		50.38		
1 Hyd. Crane, 55 Ton		980.20		1078.22	15.00	16.50
72 L.H., Daily Totals		$4374.50		$6421.67	$60.76	$89.19
Crew C-8	**Hr.**	**Daily**	**Hr.**	**Daily**	**Bare Costs**	**Incl. O&P**
1 Labor Foreman (outside)	$44.10	$352.80	$70.65	$565.20	$46.72	$74.10
3 Laborers	42.10	1010.40	67.45	1618.80		
2 Cement Finishers	49.95	799.20	78.45	1255.20		
1 Equip. Oper. (medium)	56.75	454.00	88.80	710.40		
1 Concrete Pump (Small)		410.35		451.38	7.33	8.06
56 L.H., Daily Totals		$3026.75		$4600.98	$54.05	$82.16
Crew C-8A	**Hr.**	**Daily**	**Hr.**	**Daily**	**Bare Costs**	**Incl. O&P**
1 Labor Foreman (outside)	$44.10	$352.80	$70.65	$565.20	$45.05	$71.65
3 Laborers	42.10	1010.40	67.45	1618.80		
2 Cement Finishers	49.95	799.20	78.45	1255.20		
48 L.H., Daily Totals		$2162.40		$3439.20	$45.05	$71.65
Crew C-8B	**Hr.**	**Daily**	**Hr.**	**Daily**	**Bare Costs**	**Incl. O&P**
1 Labor Foreman (outside)	$44.10	$352.80	$70.65	$565.20	$45.43	$72.36
3 Laborers	42.10	1010.40	67.45	1618.80		
1 Equip. Oper. (medium)	56.75	454.00	88.80	710.40		
1 Vibrating Power Screed		75.10		82.61		
1 Roller, Vibratory, 25 Ton		664.35		730.78		
1 Dozer, 200 H.P.		1504.00		1654.40	56.09	61.69
40 L.H., Daily Totals		$4060.65		$5362.19	$101.52	$134.05
Crew C-8C	**Hr.**	**Daily**	**Hr.**	**Daily**	**Bare Costs**	**Incl. O&P**
1 Labor Foreman (outside)	$44.10	$352.80	$70.65	$565.20	$46.18	$73.38
3 Laborers	42.10	1010.40	67.45	1618.80		
1 Cement Finisher	49.95	399.60	78.45	627.60		
1 Equip. Oper. (medium)	56.75	454.00	88.80	710.40		
1 Shotcrete Rig, 12 C.Y./hr		245.40		269.94		
1 Air Compressor, 160 cfm		202.80		223.08		
4 -50' Air Hoses, 1"		32.00		35.20		
4 -50' Air Hoses, 2"		115.60		127.16	12.41	13.65
48 L.H., Daily Totals		$2812.60		$4177.38	$58.60	$87.03
Crew C-8D	**Hr.**	**Daily**	**Hr.**	**Daily**	**Bare Costs**	**Incl. O&P**
1 Labor Foreman (outside)	$44.10	$352.80	$70.65	$565.20	$47.29	$74.88
1 Laborer	42.10	336.80	67.45	539.60		
1 Cement Finisher	49.95	399.60	78.45	627.60		
1 Equipment Oper. (light)	53.00	424.00	82.95	663.60		
1 Air Compressor, 250 cfm		201.50		221.65		
2 -50' Air Hoses, 1"		16.00		17.60	6.80	7.48
32 L.H., Daily Totals		$1730.70		$2635.25	$54.08	$82.35

Crew No.	Bare Costs		Incl. Subs O&P		Cost Per Labor-Hour	
Crew C-8E	**Hr.**	**Daily**	**Hr.**	**Daily**	**Bare Costs**	**Incl. O&P**
1 Labor Foreman (outside)	$44.10	$352.80	$70.65	$565.20	$45.56	$72.40
3 Laborers	42.10	1010.40	67.45	1618.80		
1 Cement Finisher	49.95	399.60	78.45	627.60		
1 Equipment Oper. (light)	53.00	424.00	82.95	663.60		
1 Shotcrete Rig, 35 C.Y./hr		298.00		327.80		
1 Air Compressor, 250 cfm		201.50		221.65		
4 -50' Air Hoses, 1"		32.00		35.20		
4 -50' Air Hoses, 2"		115.60		127.16	13.48	14.83
48 L.H., Daily Totals		$2833.90		$4187.01	$59.04	$87.23
Crew C-9	**Hr.**	**Daily**	**Hr.**	**Daily**	**Bare Costs**	**Incl. O&P**
1 Cement Finisher	$49.95	$399.60	$78.45	$627.60	$46.79	$74.08
2 Laborers	42.10	673.60	67.45	1079.20		
1 Equipment Oper. (light)	53.00	424.00	82.95	663.60		
1 Grout Pump, 50 C.F./hr.		188.45		207.29		
1 Air Compressor, 160 cfm		202.80		223.08		
2 -50' Air Hoses, 1"		16.00		17.60		
2 -50' Air Hoses, 2"		57.80		63.58	14.53	15.99
32 L.H., Daily Totals		$1962.25		$2881.95	$61.32	$90.06
Crew C-10	**Hr.**	**Daily**	**Hr.**	**Daily**	**Bare Costs**	**Incl. O&P**
1 Laborer	$42.10	$336.80	$67.45	$539.60	$47.33	$74.78
2 Cement Finishers	49.95	799.20	78.45	1255.20		
24 L.H., Daily Totals		$1136.00		$1794.80	$47.33	$74.78
Crew C-10B	**Hr.**	**Daily**	**Hr.**	**Daily**	**Bare Costs**	**Incl. O&P**
3 Laborers	$42.10	$1010.40	$67.45	$1618.80	$45.24	$71.85
2 Cement Finishers	49.95	799.20	78.45	1255.20		
1 Concrete Mixer, 10 C.F.		140.30		154.33		
2 Trowels, 48" Walk-Behind		174.00		191.40	7.86	8.64
40 L.H., Daily Totals		$2123.90		$3219.73	$53.10	$80.49
Crew C-10C	**Hr.**	**Daily**	**Hr.**	**Daily**	**Bare Costs**	**Incl. O&P**
1 Laborer	$42.10	$336.80	$67.45	$539.60	$47.33	$74.78
2 Cement Finishers	49.95	799.20	78.45	1255.20		
1 Trowel, 48" Walk-Behind		87.00		95.70	3.63	3.99
24 L.H., Daily Totals		$1223.00		$1890.50	$50.96	$78.77
Crew C-10D	**Hr.**	**Daily**	**Hr.**	**Daily**	**Bare Costs**	**Incl. O&P**
1 Laborer	$42.10	$336.80	$67.45	$539.60	$47.33	$74.78
2 Cement Finishers	49.95	799.20	78.45	1255.20		
1 Vibrating Power Screed		75.10		82.61		
1 Trowel, 48" Walk-Behind		87.00		95.70	6.75	7.43
24 L.H., Daily Totals		$1298.10		$1973.11	$54.09	$82.21
Crew C-10E	**Hr.**	**Daily**	**Hr.**	**Daily**	**Bare Costs**	**Incl. O&P**
1 Laborer	$42.10	$336.80	$67.45	$539.60	$47.33	$74.78
2 Cement Finishers	49.95	799.20	78.45	1255.20		
1 Vibrating Power Screed		75.10		82.61		
1 Cement Trowel, 96" Ride-On		169.00		185.90	10.17	11.19
24 L.H., Daily Totals		$1380.10		$2063.31	$57.50	$85.97
Crew C-10F	**Hr.**	**Daily**	**Hr.**	**Daily**	**Bare Costs**	**Incl. O&P**
1 Laborer	$42.10	$336.80	$67.45	$539.60	$47.33	$74.78
2 Cement Finishers	49.95	799.20	78.45	1255.20		
1 Telescoping Boom Lift, to 60'		310.90		341.99	12.95	14.25
24 L.H., Daily Totals		$1446.90		$2136.79	$60.29	$89.03

Crew C-11	Hr.	Daily	Hr.	Daily	Bare Costs	Incl. O&P
1 Struc. Steel Foreman (outside)	$59.65	$477.20	$97.75	$782.00	$57.26	$92.91
6 Struc. Steel Workers	57.65	2767.20	94.45	4533.60		
1 Equip. Oper. (crane)	59.20	473.60	92.65	741.20		
1 Equip. Oper. (oiler)	50.55	404.40	79.10	632.80		
1 Lattice Boom Crane, 150 Ton		2269.00		2495.90	31.51	34.67
72 L.H., Daily Totals		$6391.40		$9185.50	$88.77	$127.58

Crew C-12	Hr.	Daily	Hr.	Daily	Bare Costs	Incl. O&P
1 Carpenter Foreman (outside)	$55.15	$441.20	$88.40	$707.20	$52.65	$83.99
3 Carpenters	53.15	1275.60	85.15	2043.60		
1 Laborer	42.10	336.80	67.45	539.60		
1 Equip. Oper. (crane)	59.20	473.60	92.65	741.20		
1 Hyd. Crane, 12 Ton		471.00		518.10	9.81	10.79
48 L.H., Daily Totals		$2998.20		$4549.70	$62.46	$94.79

Crew C-13	Hr.	Daily	Hr.	Daily	Bare Costs	Incl. O&P
1 Struc. Steel Worker	$57.65	$461.20	$94.45	$755.60	$56.08	$91.25
1 Welder	57.45	459.60	94.15	753.20		
1 Carpenter	53.15	425.20	85.15	681.20		
1 Welder, Gas Engine, 300 amp		147.00		161.70	6.13	6.74
24 L.H., Daily Totals		$1493.00		$2351.70	$62.21	$97.99

Crew C-14	Hr.	Daily	Hr.	Daily	Bare Costs	Incl. O&P
1 Carpenter Foreman (outside)	$55.15	$441.20	$88.40	$707.20	$51.36	$81.56
5 Carpenters	53.15	2126.00	85.15	3406.00		
4 Laborers	42.10	1347.20	67.45	2158.40		
4 Rodmen (reinf.)	56.40	1804.80	88.85	2843.20		
2 Cement Finishers	49.95	799.20	78.45	1255.20		
1 Equip. Oper. (crane)	59.20	473.60	92.65	741.20		
1 Equip. Oper. (oiler)	50.55	404.40	79.10	632.80		
1 Hyd. Crane, 80 Ton		1486.00		1634.60	10.32	11.35
144 L.H., Daily Totals		$8882.40		$13378.60	$61.68	$92.91

Crew C-14A	Hr.	Daily	Hr.	Daily	Bare Costs	Incl. O&P
1 Carpenter Foreman (outside)	$55.15	$441.20	$88.40	$707.20	$52.88	$84.33
16 Carpenters	53.15	6803.20	85.15	10899.20		
4 Rodmen (reinf.)	56.40	1804.80	88.85	2843.20		
2 Laborers	42.10	673.60	67.45	1079.20		
1 Cement Finisher	49.95	399.60	78.45	627.60		
1 Equip. Oper. (medium)	56.75	454.00	88.80	710.40		
1 Gas Engine Vibrator		26.85		29.54		
1 Concrete Pump (Small)		410.35		451.38	2.19	2.40
200 L.H., Daily Totals		$11013.60		$17347.72	$55.07	$86.74

Crew C-14B	Hr.	Daily	Hr.	Daily	Bare Costs	Incl. O&P
1 Carpenter Foreman (outside)	$55.15	$441.20	$88.40	$707.20	$52.77	$84.11
16 Carpenters	53.15	6803.20	85.15	10899.20		
4 Rodmen (reinf.)	56.40	1804.80	88.85	2843.20		
2 Laborers	42.10	673.60	67.45	1079.20		
2 Cement Finishers	49.95	799.20	78.45	1255.20		
1 Equip. Oper. (medium)	56.75	454.00	88.80	710.40		
1 Gas Engine Vibrator		26.85		29.54		
1 Concrete Pump (Small)		410.35		451.38	2.10	2.31
208 L.H., Daily Totals		$11413.20		$17975.32	$54.87	$86.42

Crew C-14C	Hr.	Daily	Hr.	Daily	Bare Costs	Incl. O&P
1 Carpenter Foreman (outside)	$55.15	$441.20	$88.40	$707.20	$50.37	$80.38
6 Carpenters	53.15	2551.20	85.15	4087.20		
2 Rodmen (reinf.)	56.40	902.40	88.85	1421.60		
4 Laborers	42.10	1347.20	67.45	2158.40		
1 Cement Finisher	49.95	399.60	78.45	627.60		
1 Gas Engine Vibrator		26.85		29.54	.24	.26
112 L.H., Daily Totals		$5668.45		$9031.53	$50.61	$80.64

Crew C-14D	Hr.	Daily	Hr.	Daily	Bare Costs	Incl. O&P
1 Carpenter Foreman (outside)	$55.15	$441.20	$88.40	$707.20	$52.62	$84.04
18 Carpenters	53.15	7653.60	85.15	12261.60		
2 Rodmen (reinf.)	56.40	902.40	88.85	1421.60		
2 Laborers	42.10	673.60	67.45	1079.20		
1 Cement Finisher	49.95	399.60	78.45	627.60		
1 Equip. Oper. (medium)	56.75	454.00	88.80	710.40		
1 Gas Engine Vibrator		26.85		29.54		
1 Concrete Pump (Small)		410.35		451.38	2.19	2.40
200 L.H., Daily Totals		$10961.60		$17288.52	$54.81	$86.44

Crew C-14E	Hr.	Daily	Hr.	Daily	Bare Costs	Incl. O&P
1 Carpenter Foreman (outside)	$55.15	$441.20	$88.40	$707.20	$51.21	$81.35
2 Carpenters	53.15	850.40	85.15	1362.40		
4 Rodmen (reinf.)	56.40	1804.80	88.85	2843.20		
3 Laborers	42.10	1010.40	67.45	1618.80		
1 Cement Finisher	49.95	399.60	78.45	627.60		
1 Gas Engine Vibrator		26.85		29.54	.31	.34
88 L.H., Daily Totals		$4533.25		$7188.73	$51.51	$81.69

Crew C-14F	Hr.	Daily	Hr.	Daily	Bare Costs	Incl. O&P
1 Labor Foreman (outside)	$44.10	$352.80	$70.65	$565.20	$47.56	$75.14
2 Laborers	42.10	673.60	67.45	1079.20		
6 Cement Finishers	49.95	2397.60	78.45	3765.60		
1 Gas Engine Vibrator		26.85		29.54	.37	.41
72 L.H., Daily Totals		$3450.85		$5439.53	$47.93	$75.55

Crew C-14G	Hr.	Daily	Hr.	Daily	Bare Costs	Incl. O&P
1 Labor Foreman (outside)	$44.10	$352.80	$70.65	$565.20	$46.87	$74.19
2 Laborers	42.10	673.60	67.45	1079.20		
4 Cement Finishers	49.95	1598.40	78.45	2510.40		
1 Gas Engine Vibrator		26.85		29.54	.48	.53
56 L.H., Daily Totals		$2651.65		$4184.34	$47.35	$74.72

Crew C-14H	Hr.	Daily	Hr.	Daily	Bare Costs	Incl. O&P
1 Carpenter Foreman (outside)	$55.15	$441.20	$88.40	$707.20	$51.65	$82.24
2 Carpenters	53.15	850.40	85.15	1362.40		
1 Rodman (reinf.)	56.40	451.20	88.85	710.80		
1 Laborer	42.10	336.80	67.45	539.60		
1 Cement Finisher	49.95	399.60	78.45	627.60		
1 Gas Engine Vibrator		26.85		29.54	.56	.62
48 L.H., Daily Totals		$2506.05		$3977.14	$52.21	$82.86

Crew C-14L	Hr.	Daily	Hr.	Daily	Bare Costs	Incl. O&P
1 Carpenter Foreman (outside)	$55.15	$441.20	$88.40	$707.20	$49.37	$78.96
6 Carpenters	53.15	2551.20	85.15	4087.20		
4 Laborers	42.10	1347.20	67.45	2158.40		
1 Cement Finisher	49.95	399.60	78.45	627.60		
1 Gas Engine Vibrator		26.85		29.54	.28	.31
96 L.H., Daily Totals		$4766.05		$7609.94	$49.65	$79.27

Crew No.	Bare Costs		Incl. Subs O&P		Cost Per Labor-Hour	
Crew C-14M	**Hr.**	**Daily**	**Hr.**	**Daily**	**Bare Costs**	**Incl. O&P**
1 Carpenter Foreman (outside)	$55.15	$441.20	$88.40	$707.20	$51.09	$81.21
2 Carpenters	53.15	850.40	85.15	1362.40		
1 Rodman (reinf.)	56.40	451.20	88.85	710.80		
2 Laborers	42.10	673.60	67.45	1079.20		
1 Cement Finisher	49.95	399.60	78.45	627.60		
1 Equip. Oper. (medium)	56.75	454.00	88.80	710.40		
1 Gas Engine Vibrator		26.85		29.54		
1 Concrete Pump (Small)		410.35		451.38	6.83	7.51
64 L.H., Daily Totals		$3707.20		$5678.52	$57.93	$88.73
Crew C-15	**Hr.**	**Daily**	**Hr.**	**Daily**	**Bare Costs**	**Incl. O&P**
1 Carpenter Foreman (outside)	$55.15	$441.20	$88.40	$707.20	$49.34	$78.53
2 Carpenters	53.15	850.40	85.15	1362.40		
3 Laborers	42.10	1010.40	67.45	1618.80		
2 Cement Finishers	49.95	799.20	78.45	1255.20		
1 Rodman (reinf.)	56.40	451.20	88.85	710.80		
72 L.H., Daily Totals		$3552.40		$5654.40	$49.34	$78.53
Crew C-16	**Hr.**	**Daily**	**Hr.**	**Daily**	**Bare Costs**	**Incl. O&P**
1 Labor Foreman (outside)	$44.10	$352.80	$70.65	$565.20	$46.72	$74.10
3 Laborers	42.10	1010.40	67.45	1618.80		
2 Cement Finishers	49.95	799.20	78.45	1255.20		
1 Equip. Oper. (medium)	56.75	454.00	88.80	710.40		
1 Gunite Pump Rig		317.90		349.69		
2 -50' Air Hoses, 3/4"		15.50		17.05		
2 -50' Air Hoses, 2"		57.80		63.58	6.99	7.68
56 L.H., Daily Totals		$3007.60		$4579.92	$53.71	$81.78
Crew C-16A	**Hr.**	**Daily**	**Hr.**	**Daily**	**Bare Costs**	**Incl. O&P**
1 Laborer	$42.10	$336.80	$67.45	$539.60	$49.69	$78.29
2 Cement Finishers	49.95	799.20	78.45	1255.20		
1 Equip. Oper. (medium)	56.75	454.00	88.80	710.40		
1 Gunite Pump Rig		317.90		349.69		
2 -50' Air Hoses, 3/4"		15.50		17.05		
2 -50' Air Hoses, 2"		57.80		63.58		
1 Telescoping Boom Lift, to 60'		310.90		341.99	21.94	24.13
32 L.H., Daily Totals		$2292.10		$3277.51	$71.63	$102.42
Crew C-17	**Hr.**	**Daily**	**Hr.**	**Daily**	**Bare Costs**	**Incl. O&P**
2 Skilled Worker Foremen (out)	$56.85	$909.60	$90.55	$1448.80	$55.25	$88.03
8 Skilled Workers	54.85	3510.40	87.40	5593.60		
80 L.H., Daily Totals		$4420.00		$7042.40	$55.25	$88.03
Crew C-17A	**Hr.**	**Daily**	**Hr.**	**Daily**	**Bare Costs**	**Incl. O&P**
2 Skilled Worker Foremen (out)	$56.85	$909.60	$90.55	$1448.80	$55.30	$88.09
8 Skilled Workers	54.85	3510.40	87.40	5593.60		
.125 Equip. Oper. (crane)	59.20	59.20	92.65	92.65		
.125 Hyd. Crane, 80 Ton		185.75		204.32	2.29	2.52
81 L.H., Daily Totals		$4664.95		$7339.38	$57.59	$90.61
Crew C-17B	**Hr.**	**Daily**	**Hr.**	**Daily**	**Bare Costs**	**Incl. O&P**
2 Skilled Worker Foremen (out)	$56.85	$909.60	$90.55	$1448.80	$55.35	$88.14
8 Skilled Workers	54.85	3510.40	87.40	5593.60		
.25 Equip. Oper. (crane)	59.20	118.40	92.65	185.30		
.25 Hyd. Crane, 80 Ton		371.50		408.65		
.25 Trowel, 48" Walk-Behind		21.75		23.93	4.80	5.28
82 L.H., Daily Totals		$4931.65		$7660.27	$60.14	$93.42

Crew No.	Bare Costs		Incl. Subs O&P		Cost Per Labor-Hour	
Crew C-17C	**Hr.**	**Daily**	**Hr.**	**Daily**	**Bare Costs**	**Incl. O&P**
2 Skilled Worker Foremen (out)	$56.85	$909.60	$90.55	$1448.80	$55.39	$88.20
8 Skilled Workers	54.85	3510.40	87.40	5593.60		
.375 Equip. Oper. (crane)	59.20	177.60	92.65	277.95		
.375 Hyd. Crane, 80 Ton		557.25		612.98	6.71	7.39
83 L.H., Daily Totals		$5154.85		$7933.32	$62.11	$95.58
Crew C-17D	**Hr.**	**Daily**	**Hr.**	**Daily**	**Bare Costs**	**Incl. O&P**
2 Skilled Worker Foremen (out)	$56.85	$909.60	$90.55	$1448.80	$55.44	$88.25
8 Skilled Workers	54.85	3510.40	87.40	5593.60		
.5 Equip. Oper. (crane)	59.20	236.80	92.65	370.60		
.5 Hyd. Crane, 80 Ton		743.00		817.30	8.85	9.73
84 L.H., Daily Totals		$5399.80		$8230.30	$64.28	$97.98
Crew C-17E	**Hr.**	**Daily**	**Hr.**	**Daily**	**Bare Costs**	**Incl. O&P**
2 Skilled Worker Foremen (out)	$56.85	$909.60	$90.55	$1448.80	$55.25	$88.03
8 Skilled Workers	54.85	3510.40	87.40	5593.60		
1 Hyd. Jack with Rods		35.90		39.49	.45	.49
80 L.H., Daily Totals		$4455.90		$7081.89	$55.70	$88.52
Crew C-18	**Hr.**	**Daily**	**Hr.**	**Daily**	**Bare Costs**	**Incl. O&P**
.125 Labor Foreman (outside)	$44.10	$44.10	$70.65	$70.65	$42.32	$67.81
1 Laborer	42.10	336.80	67.45	539.60		
1 Concrete Cart, 10 C.F.		127.85		140.63	14.21	15.63
9 L.H., Daily Totals		$508.75		$750.88	$56.53	$83.43
Crew C-19	**Hr.**	**Daily**	**Hr.**	**Daily**	**Bare Costs**	**Incl. O&P**
.125 Labor Foreman (outside)	$44.10	$44.10	$70.65	$70.65	$42.32	$67.81
1 Laborer	42.10	336.80	67.45	539.60		
1 Concrete Cart, 18 C.F.		153.45		168.79	17.05	18.75
9 L.H., Daily Totals		$534.35		$779.04	$59.37	$86.56
Crew C-20	**Hr.**	**Daily**	**Hr.**	**Daily**	**Bare Costs**	**Incl. O&P**
1 Labor Foreman (outside)	$44.10	$352.80	$70.65	$565.20	$45.16	$71.89
5 Laborers	42.10	1684.00	67.45	2698.00		
1 Cement Finisher	49.95	399.60	78.45	627.60		
1 Equip. Oper. (medium)	56.75	454.00	88.80	710.40		
2 Gas Engine Vibrators		53.70		59.07		
1 Concrete Pump (Small)		410.35		451.38	7.25	7.98
64 L.H., Daily Totals		$3354.45		$5111.65	$52.41	$79.87
Crew C-21	**Hr.**	**Daily**	**Hr.**	**Daily**	**Bare Costs**	**Incl. O&P**
1 Labor Foreman (outside)	$44.10	$352.80	$70.65	$565.20	$45.16	$71.89
5 Laborers	42.10	1684.00	67.45	2698.00		
1 Cement Finisher	49.95	399.60	78.45	627.60		
1 Equip. Oper. (medium)	56.75	454.00	88.80	710.40		
2 Gas Engine Vibrators		53.70		59.07		
1 Concrete Conveyer		204.15		224.57	4.03	4.43
64 L.H., Daily Totals		$3148.25		$4884.84	$49.19	$76.33
Crew C-22	**Hr.**	**Daily**	**Hr.**	**Daily**	**Bare Costs**	**Incl. O&P**
1 Rodman Foreman (outside)	$58.40	$467.20	$92.00	$736.00	$56.71	$89.31
4 Rodmen (reinf.)	56.40	1804.80	88.85	2843.20		
.125 Equip. Oper. (crane)	59.20	59.20	92.65	92.65		
.125 Equip. Oper. (oiler)	50.55	50.55	79.10	79.10		
.125 Hyd. Crane, 25 Ton		72.61		79.87	1.73	1.90
42 L.H., Daily Totals		$2454.36		$3830.82	$58.44	$91.21

Crew No.	Bare Costs		Incl. Subs O&P		Cost Per Labor-Hour	
Crew C-23	**Hr.**	**Daily**	**Hr.**	**Daily**	**Bare Costs**	**Incl. O&P**
2 Skilled Worker Foremen (out)	$56.85	$909.60	$90.55	$1448.80	$55.26	$87.72
6 Skilled Workers	54.85	2632.80	87.40	4195.20		
1 Equip. Oper. (crane)	59.20	473.60	92.65	741.20		
1 Equip. Oper. (oiler)	50.55	404.40	79.10	632.80		
1 Lattice Boom Crane, 90 Ton		1696.00		1865.60	21.20	23.32
80 L.H., Daily Totals		$6116.40		$8883.60	$76.45	$111.05

Crew C-24	Hr.	Daily	Hr.	Daily	Bare Costs	Incl. O&P
2 Skilled Worker Foremen (out)	$56.85	$909.60	$90.55	$1448.80	$55.26	$87.72
6 Skilled Workers	54.85	2632.80	87.40	4195.20		
1 Equip. Oper. (crane)	59.20	473.60	92.65	741.20		
1 Equip. Oper. (oiler)	50.55	404.40	79.10	632.80		
1 Lattice Boom Crane, 150 Ton		2269.00		2495.90	28.36	31.20
80 L.H., Daily Totals		$6689.40		$9513.90	$83.62	$118.92

Crew C-25	Hr.	Daily	Hr.	Daily	Bare Costs	Incl. O&P
2 Rodmen (reinf.)	$56.40	$902.40	$88.85	$1421.60	$45.50	$74.70
2 Rodmen Helpers	34.60	553.60	60.55	968.80		
32 L.H., Daily Totals		$1456.00		$2390.40	$45.50	$74.70

Crew C-27	Hr.	Daily	Hr.	Daily	Bare Costs	Incl. O&P
2 Cement Finishers	$49.95	$799.20	$78.45	$1255.20	$49.95	$78.45
1 Concrete Saw		111.50		122.65	6.97	7.67
16 L.H., Daily Totals		$910.70		$1377.85	$56.92	$86.12

Crew C-28	Hr.	Daily	Hr.	Daily	Bare Costs	Incl. O&P
1 Cement Finisher	$49.95	$399.60	$78.45	$627.60	$49.95	$78.45
1 Portable Air Compressor, Gas		38.70		42.57	4.84	5.32
8 L.H., Daily Totals		$438.30		$670.17	$54.79	$83.77

Crew C-29	Hr.	Daily	Hr.	Daily	Bare Costs	Incl. O&P
1 Laborer	$42.10	$336.80	$67.45	$539.60	$42.10	$67.45
1 Pressure Washer		96.95		106.65	12.12	13.33
8 L.H., Daily Totals		$433.75		$646.25	$54.22	$80.78

Crew C-30	Hr.	Daily	Hr.	Daily	Bare Costs	Incl. O&P
1 Laborer	$42.10	$336.80	$67.45	$539.60	$42.10	$67.45
1 Concrete Mixer, 10 C.F.		140.30		154.33	17.54	19.29
8 L.H., Daily Totals		$477.10		$693.93	$59.64	$86.74

Crew C-31	Hr.	Daily	Hr.	Daily	Bare Costs	Incl. O&P
1 Cement Finisher	$49.95	$399.60	$78.45	$627.60	$49.95	$78.45
1 Grout Pump		317.90		349.69	39.74	43.71
8 L.H., Daily Totals		$717.50		$977.29	$89.69	$122.16

Crew C-32	Hr.	Daily	Hr.	Daily	Bare Costs	Incl. O&P
1 Cement Finisher	$49.95	$399.60	$78.45	$627.60	$46.02	$72.95
1 Laborer	42.10	336.80	67.45	539.60		
1 Crack Chaser Saw, Gas, 6 H.P.		73.10		80.41		
1 Vacuum Pick-Up System		74.05		81.45	9.20	10.12
16 L.H., Daily Totals		$883.55		$1329.07	$55.22	$83.07

Crew D-1	Hr.	Daily	Hr.	Daily	Bare Costs	Incl. O&P
1 Bricklayer	$52.05	$416.40	$84.10	$672.80	$46.90	$75.78
1 Bricklayer Helper	41.75	334.00	67.45	539.60		
16 L.H., Daily Totals		$750.40		$1212.40	$46.90	$75.78

Crew D-2	Hr.	Daily	Hr.	Daily	Bare Costs	Incl. O&P
3 Bricklayers	$52.05	$1249.20	$84.10	$2018.40	$48.40	$78.14
2 Bricklayer Helpers	41.75	668.00	67.45	1079.20		
.5 Carpenter	53.15	212.60	85.15	340.60		
44 L.H., Daily Totals		$2129.80		$3438.20	$48.40	$78.14

Crew D-3	Hr.	Daily	Hr.	Daily	Bare Costs	Incl. O&P
3 Bricklayers	$52.05	$1249.20	$84.10	$2018.40	$48.18	$77.81
2 Bricklayer Helpers	41.75	668.00	67.45	1079.20		
.25 Carpenter	53.15	106.30	85.15	170.30		
42 L.H., Daily Totals		$2023.50		$3267.90	$48.18	$77.81

Crew D-4	Hr.	Daily	Hr.	Daily	Bare Costs	Incl. O&P
1 Bricklayer	$52.05	$416.40	$84.10	$672.80	$47.14	$75.49
2 Bricklayer Helpers	41.75	668.00	67.45	1079.20		
1 Equip. Oper. (light)	53.00	424.00	82.95	663.60		
1 Grout Pump, 50 C.F./hr.		188.45		207.29	5.89	6.48
32 L.H., Daily Totals		$1696.85		$2622.90	$53.03	$81.97

Crew D-5	Hr.	Daily	Hr.	Daily	Bare Costs	Incl. O&P
1 Bricklayer	52.05	416.40	84.10	672.80	52.05	84.10
8 L.H., Daily Totals		$416.40		$672.80	$52.05	$84.10

Crew D-6	Hr.	Daily	Hr.	Daily	Bare Costs	Incl. O&P
3 Bricklayers	$52.05	$1249.20	$84.10	$2018.40	$47.15	$76.15
3 Bricklayer Helpers	41.75	1002.00	67.45	1618.80		
.25 Carpenter	53.15	106.30	85.15	170.30		
50 L.H., Daily Totals		$2357.50		$3807.50	$47.15	$76.15

Crew D-7	Hr.	Daily	Hr.	Daily	Bare Costs	Incl. O&P
1 Tile Layer	$49.50	$396.00	$77.55	$620.40	$44.15	$69.17
1 Tile Layer Helper	38.80	310.40	60.80	486.40		
16 L.H., Daily Totals		$706.40		$1106.80	$44.15	$69.17

Crew D-8	Hr.	Daily	Hr.	Daily	Bare Costs	Incl. O&P
3 Bricklayers	$52.05	$1249.20	$84.10	$2018.40	$47.93	$77.44
2 Bricklayer Helpers	41.75	668.00	67.45	1079.20		
40 L.H., Daily Totals		$1917.20		$3097.60	$47.93	$77.44

Crew D-9	Hr.	Daily	Hr.	Daily	Bare Costs	Incl. O&P
3 Bricklayers	$52.05	$1249.20	$84.10	$2018.40	$46.90	$75.78
3 Bricklayer Helpers	41.75	1002.00	67.45	1618.80		
48 L.H., Daily Totals		$2251.20		$3637.20	$46.90	$75.78

Crew D-10	Hr.	Daily	Hr.	Daily	Bare Costs	Incl. O&P
1 Bricklayer Foreman (outside)	$54.05	$432.40	$87.35	$698.80	$51.76	$82.89
1 Bricklayer	52.05	416.40	84.10	672.80		
1 Bricklayer Helper	41.75	334.00	67.45	539.60		
1 Equip. Oper. (crane)	59.20	473.60	92.65	741.20		
1 S.P. Crane, 4x4, 12 Ton		428.30		471.13	13.38	14.72
32 L.H., Daily Totals		$2084.70		$3123.53	$65.15	$97.61

Crew D-11	Hr.	Daily	Hr.	Daily	Bare Costs	Incl. O&P
1 Bricklayer Foreman (outside)	$54.05	$432.40	$87.35	$698.80	$49.28	$79.63
1 Bricklayer	52.05	416.40	84.10	672.80		
1 Bricklayer Helper	41.75	334.00	67.45	539.60		
24 L.H., Daily Totals		$1182.80		$1911.20	$49.28	$79.63

Crew No.	Bare Costs		Incl. Subs O&P		Cost Per Labor-Hour	
Crew D-12	**Hr.**	**Daily**	**Hr.**	**Daily**	**Bare Costs**	**Incl. O&P**
1 Bricklayer Foreman (outside)	$54.05	$432.40	$87.35	$698.80	$47.40	$76.59
1 Bricklayer	52.05	416.40	84.10	672.80		
2 Bricklayer Helpers	41.75	668.00	67.45	1079.20		
32 L.H., Daily Totals		$1516.80		$2450.80	$47.40	$76.59

Crew D-13	Hr.	Daily	Hr.	Daily	Bare Costs	Incl. O&P
1 Bricklayer Foreman (outside)	$54.05	$432.40	$87.35	$698.80	$50.33	$80.69
1 Bricklayer	52.05	416.40	84.10	672.80		
2 Bricklayer Helpers	41.75	668.00	67.45	1079.20		
1 Carpenter	53.15	425.20	85.15	681.20		
1 Equip. Oper. (crane)	59.20	473.60	92.65	741.20		
1 S.P. Crane, 4x4, 12 Ton		428.30		471.13	8.92	9.82
48 L.H., Daily Totals		$2843.90		$4344.33	$59.25	$90.51

Crew D-14	Hr.	Daily	Hr.	Daily	Bare Costs	Incl. O&P
3 Bricklayers	$52.05	$1249.20	$84.10	$2018.40	$49.48	$79.94
1 Bricklayer Helper	41.75	334.00	67.45	539.60		
32 L.H., Daily Totals		$1583.20		$2558.00	$49.48	$79.94

Crew E-1	Hr.	Daily	Hr.	Daily	Bare Costs	Incl. O&P
1 Welder Foreman (outside)	$59.45	$475.60	$97.40	$779.20	$56.63	$91.50
1 Welder	57.45	459.60	94.15	753.20		
1 Equip. Oper. (light)	53.00	424.00	82.95	663.60		
1 Welder, Gas Engine, 300 amp		147.00		161.70	6.13	6.74
24 L.H., Daily Totals		$1506.20		$2357.70	$62.76	$98.24

Crew E-2	Hr.	Daily	Hr.	Daily	Bare Costs	Incl. O&P
1 Struc. Steel Foreman (outside)	$59.65	$477.20	$97.75	$782.00	$57.14	$92.47
4 Struc. Steel Workers	57.65	1844.80	94.45	3022.40		
1 Equip. Oper. (crane)	59.20	473.60	92.65	741.20		
1 Equip. Oper. (oiler)	50.55	404.40	79.10	632.80		
1 Lattice Boom Crane, 90 Ton		1696.00		1865.60	30.29	33.31
56 L.H., Daily Totals		$4896.00		$7044.00	$87.43	$125.79

Crew E-3	Hr.	Daily	Hr.	Daily	Bare Costs	Incl. O&P
1 Struc. Steel Foreman (outside)	$59.65	$477.20	$97.75	$782.00	$58.25	$95.45
1 Struc. Steel Worker	57.65	461.20	94.45	755.60		
1 Welder	57.45	459.60	94.15	753.20		
1 Welder, Gas Engine, 300 amp		147.00		161.70	6.13	6.74
24 L.H., Daily Totals		$1545.00		$2452.50	$64.38	$102.19

Crew E-3A	Hr.	Daily	Hr.	Daily	Bare Costs	Incl. O&P
1 Struc. Steel Foreman (outside)	$59.65	$477.20	$97.75	$782.00	$58.25	$95.45
1 Struc. Steel Worker	57.65	461.20	94.45	755.60		
1 Welder	57.45	459.60	94.15	753.20		
1 Welder, Gas Engine, 300 amp		147.00		161.70		
1 Telescoping Boom Lift, to 40'		278.90		306.79	17.75	19.52
24 L.H., Daily Totals		$1823.90		$2759.29	$76.00	$114.97

Crew E-4	Hr.	Daily	Hr.	Daily	Bare Costs	Incl. O&P
1 Struc. Steel Foreman (outside)	$59.65	$477.20	$97.75	$782.00	$58.15	$95.28
3 Struc. Steel Workers	57.65	1383.60	94.45	2266.80		
1 Welder, Gas Engine, 300 amp		147.00		161.70	4.59	5.05
32 L.H., Daily Totals		$2007.80		$3210.50	$62.74	$100.33

Crew E-5	Hr.	Daily	Hr.	Daily	Bare Costs	Incl. O&P
2 Struc. Steel Foremen (outside)	$59.65	$954.40	$97.75	$1564.00	$57.48	$93.36
5 Struc. Steel Workers	57.65	2306.00	94.45	3778.00		
1 Equip. Oper. (crane)	59.20	473.60	92.65	741.20		
1 Welder	57.45	459.60	94.15	753.20		
1 Equip. Oper. (oiler)	50.55	404.40	79.10	632.80		
1 Lattice Boom Crane, 90 Ton		1696.00		1865.60		
1 Welder, Gas Engine, 300 amp		147.00		161.70	23.04	25.34
80 L.H., Daily Totals		$6441.00		$9496.50	$80.51	$118.71

Crew E-6	Hr.	Daily	Hr.	Daily	Bare Costs	Incl. O&P
3 Struc. Steel Foremen (outside)	$59.65	$1431.60	$97.75	$2346.00	$57.38	$93.26
9 Struc. Steel Workers	57.65	4150.80	94.45	6800.40		
1 Equip. Oper. (crane)	59.20	473.60	92.65	741.20		
1 Welder	57.45	459.60	94.15	753.20		
1 Equip. Oper. (oiler)	50.55	404.40	79.10	632.80		
1 Equip. Oper. (light)	53.00	424.00	82.95	663.60		
1 Lattice Boom Crane, 90 Ton		1696.00		1865.60		
1 Welder, Gas Engine, 300 amp		147.00		161.70		
1 Air Compressor, 160 cfm		202.80		223.08		
2 Impact Wrenches		104.60		115.06	16.80	18.48
128 L.H., Daily Totals		$9494.40		$14302.64	$74.17	$111.74

Crew E-7	Hr.	Daily	Hr.	Daily	Bare Costs	Incl. O&P
1 Struc. Steel Foreman (outside)	$59.65	$477.20	$97.75	$782.00	$57.44	$93.30
4 Struc. Steel Workers	57.65	1844.80	94.45	3022.40		
1 Equip. Oper. (crane)	59.20	473.60	92.65	741.20		
1 Equip. Oper. (oiler)	50.55	404.40	79.10	632.80		
1 Welder Foreman (outside)	59.45	475.60	97.40	779.20		
2 Welders	57.45	919.20	94.15	1506.40		
1 Lattice Boom Crane, 90 Ton		1696.00		1865.60		
2 Welder, Gas Engine, 300 amp		294.00		323.40	24.88	27.36
80 L.H., Daily Totals		$6584.80		$9653.00	$82.31	$120.66

Crew E-8	Hr.	Daily	Hr.	Daily	Bare Costs	Incl. O&P
1 Struc. Steel Foreman (outside)	$59.65	$477.20	$97.75	$782.00	$57.10	$92.63
4 Struc. Steel Workers	57.65	1844.80	94.45	3022.40		
1 Welder Foreman (outside)	59.45	475.60	97.40	779.20		
4 Welders	57.45	1838.40	94.15	3012.80		
1 Equip. Oper. (crane)	59.20	473.60	92.65	741.20		
1 Equip. Oper. (oiler)	50.55	404.40	79.10	632.80		
1 Equip. Oper. (light)	53.00	424.00	82.95	663.60		
1 Lattice Boom Crane, 90 Ton		1696.00		1865.60		
4 Welder, Gas Engine, 300 amp		588.00		646.80	21.96	24.16
104 L.H., Daily Totals		$8222.00		$12146.40	$79.06	$116.79

Crew E-9	Hr.	Daily	Hr.	Daily	Bare Costs	Incl. O&P
2 Struc. Steel Foremen (outside)	$59.65	$954.40	$97.75	$1564.00	$57.31	$93.16
5 Struc. Steel Workers	57.65	2306.00	94.45	3778.00		
1 Welder Foreman (outside)	59.45	475.60	97.40	779.20		
5 Welders	57.45	2298.00	94.15	3766.00		
1 Equip. Oper. (crane)	59.20	473.60	92.65	741.20		
1 Equip. Oper. (oiler)	50.55	404.40	79.10	632.80		
1 Equip. Oper. (light)	53.00	424.00	82.95	663.60		
1 Lattice Boom Crane, 90 Ton		1696.00		1865.60		
5 Welder, Gas Engine, 300 amp		735.00		808.50	18.99	20.89
128 L.H., Daily Totals		$9767.00		$14598.90	$76.30	$114.05

Crew No.	Bare Costs		Incl. Subs O&P		Cost Per Labor-Hour	
Crew E-10	**Hr.**	**Daily**	**Hr.**	**Daily**	**Bare Costs**	**Incl. O&P**
1 Welder Foreman (outside)	$59.45	$475.60	$97.40	$779.20	$58.45	$95.78
1 Welder	57.45	459.60	94.15	753.20		
1 Welder, Gas Engine, 300 amp		147.00		161.70		
1 Flatbed Truck, Gas, 3 Ton		820.40		902.44	60.46	66.51
16 L.H., Daily Totals		$1902.60		$2596.54	$118.91	$162.28
Crew E-11	**Hr.**	**Daily**	**Hr.**	**Daily**	**Bare Costs**	**Incl. O&P**
2 Painters, Struc. Steel	$45.85	$733.60	$79.00	$1264.00	$46.70	$77.10
1 Building Laborer	42.10	336.80	67.45	539.60		
1 Equip. Oper. (light)	53.00	424.00	82.95	663.60		
1 Air Compressor, 250 cfm		201.50		221.65		
1 Sandblaster, Portable, 3 C.F.		83.80		92.18		
1 Set Sand Blasting Accessories		15.35		16.89	9.40	10.33
32 L.H., Daily Totals		$1795.05		$2797.92	$56.10	$87.43
Crew E-11A	**Hr.**	**Daily**	**Hr.**	**Daily**	**Bare Costs**	**Incl. O&P**
2 Painters, Struc. Steel	$45.85	$733.60	$79.00	$1264.00	$46.70	$77.10
1 Building Laborer	42.10	336.80	67.45	539.60		
1 Equip. Oper. (light)	53.00	424.00	82.95	663.60		
1 Air Compressor, 250 cfm		201.50		221.65		
1 Sandblaster, Portable, 3 C.F.		83.80		92.18		
1 Set Sand Blasting Accessories		15.35		16.89		
1 Telescoping Boom Lift, to 60'		310.90		341.99	19.11	21.02
32 L.H., Daily Totals		$2105.95		$3139.91	$65.81	$98.12
Crew E-11B	**Hr.**	**Daily**	**Hr.**	**Daily**	**Bare Costs**	**Incl. O&P**
2 Painters, Struc. Steel	$45.85	$733.60	$79.00	$1264.00	$44.60	$75.15
1 Building Laborer	42.10	336.80	67.45	539.60		
2 Paint Sprayer, 8 C.F.M.		87.50		96.25		
1 Telescoping Boom Lift, to 60'		310.90		341.99	16.60	18.26
24 L.H., Daily Totals		$1468.80		$2241.84	$61.20	$93.41
Crew E-12	**Hr.**	**Daily**	**Hr.**	**Daily**	**Bare Costs**	**Incl. O&P**
1 Welder Foreman (outside)	$59.45	$475.60	$97.40	$779.20	$56.23	$90.17
1 Equip. Oper. (light)	53.00	424.00	82.95	663.60		
1 Welder, Gas Engine, 300 amp		147.00		161.70	9.19	10.11
16 L.H., Daily Totals		$1046.60		$1604.50	$65.41	$100.28
Crew E-13	**Hr.**	**Daily**	**Hr.**	**Daily**	**Bare Costs**	**Incl. O&P**
1 Welder Foreman (outside)	$59.45	$475.60	$97.40	$779.20	$57.30	$92.58
.5 Equip. Oper. (light)	53.00	212.00	82.95	331.80		
1 Welder, Gas Engine, 300 amp		147.00		161.70	12.25	13.48
12 L.H., Daily Totals		$834.60		$1272.70	$69.55	$106.06
Crew E-14	**Hr.**	**Daily**	**Hr.**	**Daily**	**Bare Costs**	**Incl. O&P**
1 Welder Foreman (outside)	$59.45	$475.60	$97.40	$779.20	$59.45	$97.40
1 Welder, Gas Engine, 300 amp		147.00		161.70	18.38	20.21
8 L.H., Daily Totals		$622.60		$940.90	$77.83	$117.61
Crew E-16	**Hr.**	**Daily**	**Hr.**	**Daily**	**Bare Costs**	**Incl. O&P**
1 Welder Foreman (outside)	$59.45	$475.60	$97.40	$779.20	$58.45	$95.78
1 Welder	57.45	459.60	94.15	753.20		
1 Welder, Gas Engine, 300 amp		147.00		161.70	9.19	10.11
16 L.H., Daily Totals		$1082.20		$1694.10	$67.64	$105.88
Crew E-17	**Hr.**	**Daily**	**Hr.**	**Daily**	**Bare Costs**	**Incl. O&P**
1 Struc. Steel Foreman (outside)	$59.65	$477.20	$97.75	$782.00	$58.65	$96.10
1 Structural Steel Worker	57.65	461.20	94.45	755.60		
16 L.H., Daily Totals		$938.40		$1537.60	$58.65	$96.10

Crew No.	Bare Costs		Incl. Subs O&P		Cost Per Labor-Hour	
Crew E-18	**Hr.**	**Daily**	**Hr.**	**Daily**	**Bare Costs**	**Incl. O&P**
1 Struc. Steel Foreman (outside)	$59.65	$477.20	$97.75	$782.00	$57.87	$93.98
3 Structural Steel Workers	57.65	1383.60	94.45	2266.80		
1 Equipment Operator (med.)	56.75	454.00	88.80	710.40		
1 Lattice Boom Crane, 20 Ton		1488.00		1636.80	37.20	40.92
40 L.H., Daily Totals		$3802.80		$5396.00	$95.07	$134.90
Crew E-19	**Hr.**	**Daily**	**Hr.**	**Daily**	**Bare Costs**	**Incl. O&P**
1 Struc. Steel Foreman (outside)	$59.65	$477.20	$97.75	$782.00	$56.77	$91.72
1 Structural Steel Worker	57.65	461.20	94.45	755.60		
1 Equip. Oper. (light)	53.00	424.00	82.95	663.60		
1 Lattice Boom Crane, 20 Ton		1488.00		1636.80	62.00	68.20
24 L.H., Daily Totals		$2850.40		$3838.00	$118.77	$159.92
Crew E-20	**Hr.**	**Daily**	**Hr.**	**Daily**	**Bare Costs**	**Incl. O&P**
1 Struc. Steel Foreman (outside)	$59.65	$477.20	$97.75	$782.00	$57.21	$92.72
5 Structural Steel Workers	57.65	2306.00	94.45	3778.00		
1 Equip. Oper. (crane)	59.20	473.60	92.65	741.20		
1 Equip. Oper. (oiler)	50.55	404.40	79.10	632.80		
1 Lattice Boom Crane, 40 Ton		2028.00		2230.80	31.69	34.86
64 L.H., Daily Totals		$5689.20		$8164.80	$88.89	$127.58
Crew E-22	**Hr.**	**Daily**	**Hr.**	**Daily**	**Bare Costs**	**Incl. O&P**
1 Skilled Worker Foreman (out)	$56.85	$454.80	$90.55	$724.40	$55.52	$88.45
2 Skilled Workers	54.85	877.60	87.40	1398.40		
24 L.H., Daily Totals		$1332.40		$2122.80	$55.52	$88.45
Crew E-24	**Hr.**	**Daily**	**Hr.**	**Daily**	**Bare Costs**	**Incl. O&P**
3 Structural Steel Workers	$57.65	$1383.60	$94.45	$2266.80	$57.42	$93.04
1 Equipment Operator (med.)	56.75	454.00	88.80	710.40		
1 Hyd. Crane, 25 Ton		580.85		638.93	18.15	19.97
32 L.H., Daily Totals		$2418.45		$3616.14	$75.58	$113.00
Crew E-25	**Hr.**	**Daily**	**Hr.**	**Daily**	**Bare Costs**	**Incl. O&P**
1 Welder Foreman (outside)	$59.45	$475.60	$97.40	$779.20	$59.45	$97.40
1 Cutting Torch		12.80		14.08	1.60	1.76
8 L.H., Daily Totals		$488.40		$793.28	$61.05	$99.16
Crew E-26	**Hr.**	**Daily**	**Hr.**	**Daily**	**Bare Costs**	**Incl. O&P**
1 Struc. Steel Foreman (outside)	$59.65	$477.20	$97.75	$782.00	$58.91	$95.68
1 Struc. Steel Worker	57.65	461.20	94.45	755.60		
1 Welder	57.45	459.60	94.15	753.20		
.25 Electrician	61.35	122.70	94.45	188.90		
.25 Plumber	64.45	128.90	99.65	199.30		
1 Welder, Gas Engine, 300 amp		147.00		161.70	5.25	5.78
28 L.H., Daily Totals		$1796.60		$2840.70	$64.16	$101.45
Crew E-27	**Hr.**	**Daily**	**Hr.**	**Daily**	**Bare Costs**	**Incl. O&P**
1 Struc. Steel Foreman (outside)	$59.65	$477.20	$97.75	$782.00	$57.21	$92.72
5 Struc. Steel Workers	57.65	2306.00	94.45	3778.00		
1 Equip. Oper. (crane)	59.20	473.60	92.65	741.20		
1 Equip. Oper. (oiler)	50.55	404.40	79.10	632.80		
1 Hyd. Crane, 12 Ton		471.00		518.10		
1 Hyd. Crane, 80 Ton		1486.00		1634.60	30.58	33.64
64 L.H., Daily Totals		$5618.20		$8086.70	$87.78	$126.35
Crew F-3	**Hr.**	**Daily**	**Hr.**	**Daily**	**Bare Costs**	**Incl. O&P**
4 Carpenters	$53.15	$1700.80	$85.15	$2724.80	$54.36	$86.65
1 Equip. Oper. (crane)	59.20	473.60	92.65	741.20		
1 Hyd. Crane, 12 Ton		471.00		518.10	11.78	12.95
40 L.H., Daily Totals		$2645.40		$3984.10	$66.14	$99.60

Crews - Renovation

Crew No.	Bare Costs		Incl. Subs O&P		Cost Per Labor-Hour	
Crew F-4	**Hr.**	**Daily**	**Hr.**	**Daily**	**Bare Costs**	**Incl. O&P**
4 Carpenters	$53.15	$1700.80	$85.15	$2724.80	$53.73	$85.39
1 Equip. Oper. (crane)	59.20	473.60	92.65	741.20		
1 Equip. Oper. (oiler)	50.55	404.40	79.10	632.80		
1 Hyd. Crane, 55 Ton		980.20		1078.22	20.42	22.46
48 L.H., Daily Totals		$3559.00		$5177.02	$74.15	$107.85
Crew F-5	**Hr.**	**Daily**	**Hr.**	**Daily**	**Bare Costs**	**Incl. O&P**
1 Carpenter Foreman (outside)	$55.15	$441.20	$88.40	$707.20	$53.65	$85.96
3 Carpenters	53.15	1275.60	85.15	2043.60		
32 L.H., Daily Totals		$1716.80		$2750.80	$53.65	$85.96
Crew F-6	**Hr.**	**Daily**	**Hr.**	**Daily**	**Bare Costs**	**Incl. O&P**
2 Carpenters	$53.15	$850.40	$85.15	$1362.40	$49.94	$79.57
2 Building Laborers	42.10	673.60	67.45	1079.20		
1 Equip. Oper. (crane)	59.20	473.60	92.65	741.20		
1 Hyd. Crane, 12 Ton		471.00		518.10	11.78	12.95
40 L.H., Daily Totals		$2468.60		$3700.90	$61.72	$92.52
Crew F-7	**Hr.**	**Daily**	**Hr.**	**Daily**	**Bare Costs**	**Incl. O&P**
2 Carpenters	$53.15	$850.40	$85.15	$1362.40	$47.63	$76.30
2 Building Laborers	42.10	673.60	67.45	1079.20		
32 L.H., Daily Totals		$1524.00		$2441.60	$47.63	$76.30
Crew G-1	**Hr.**	**Daily**	**Hr.**	**Daily**	**Bare Costs**	**Incl. O&P**
1 Roofer Foreman (outside)	$48.20	$385.60	$84.35	$674.80	$43.17	$75.55
4 Roofers Composition	46.20	1478.40	80.85	2587.20		
2 Roofer Helpers	34.60	553.60	60.55	968.80		
1 Application Equipment		192.85		212.13		
1 Tar Kettle/Pot		207.85		228.63		
1 Crew Truck		166.25		182.88	10.12	11.14
56 L.H., Daily Totals		$2984.55		$4854.44	$53.30	$86.69
Crew G-2	**Hr.**	**Daily**	**Hr.**	**Daily**	**Bare Costs**	**Incl. O&P**
1 Plasterer	$48.60	$388.80	$77.40	$619.20	$44.37	$70.78
1 Plasterer Helper	42.40	339.20	67.50	540.00		
1 Building Laborer	42.10	336.80	67.45	539.60		
1 Grout Pump, 50 C.F./hr.		188.45		207.29	7.85	8.64
24 L.H., Daily Totals		$1253.25		$1906.10	$52.22	$79.42
Crew G-2A	**Hr.**	**Daily**	**Hr.**	**Daily**	**Bare Costs**	**Incl. O&P**
1 Roofer Composition	$46.20	$369.60	$80.85	$646.80	$40.97	$69.62
1 Roofer Helper	34.60	276.80	60.55	484.40		
1 Building Laborer	42.10	336.80	67.45	539.60		
1 Foam Spray Rig, Trailer-Mtd.		523.85		576.24		
1 Pickup Truck, 3/4 Ton		110.85		121.94	26.45	29.09
24 L.H., Daily Totals		$1617.90		$2368.97	$67.41	$98.71
Crew G-3	**Hr.**	**Daily**	**Hr.**	**Daily**	**Bare Costs**	**Incl. O&P**
2 Sheet Metal Workers	$62.30	$996.80	$97.60	$1561.60	$52.20	$82.53
2 Building Laborers	42.10	673.60	67.45	1079.20		
32 L.H., Daily Totals		$1670.40		$2640.80	$52.20	$82.53
Crew G-4	**Hr.**	**Daily**	**Hr.**	**Daily**	**Bare Costs**	**Incl. O&P**
1 Labor Foreman (outside)	$44.10	$352.80	$70.65	$565.20	$42.77	$68.52
2 Building Laborers	42.10	673.60	67.45	1079.20		
1 Flatbed Truck, Gas, 1.5 Ton		196.15		215.76		
1 Air Compressor, 160 cfm		202.80		223.08	16.62	18.29
24 L.H., Daily Totals		$1425.35		$2083.24	$59.39	$86.80

Crew No.	Bare Costs		Incl. Subs O&P		Cost Per Labor-Hour	
Crew G-5	**Hr.**	**Daily**	**Hr.**	**Daily**	**Bare Costs**	**Incl. O&P**
1 Roofer Foreman (outside)	$48.20	$385.60	$84.35	$674.80	$41.96	$73.43
2 Roofers Composition	46.20	739.20	80.85	1293.60		
2 Roofer Helpers	34.60	553.60	60.55	968.80		
1 Application Equipment		192.85		212.13	4.82	5.30
40 L.H., Daily Totals		$1871.25		$3149.34	$46.78	$78.73
Crew G-6A	**Hr.**	**Daily**	**Hr.**	**Daily**	**Bare Costs**	**Incl. O&P**
2 Roofers Composition	$46.20	$739.20	$80.85	$1293.60	$46.20	$80.85
1 Small Compressor, Electric		33.80		37.18		
2 Pneumatic Nailers		54.80		60.28	5.54	6.09
16 L.H., Daily Totals		$827.80		$1391.06	$51.74	$86.94
Crew G-7	**Hr.**	**Daily**	**Hr.**	**Daily**	**Bare Costs**	**Incl. O&P**
1 Carpenter	$53.15	$425.20	$85.15	$681.20	$53.15	$85.15
1 Small Compressor, Electric		33.80		37.18		
1 Pneumatic Nailer		27.40		30.14	7.65	8.41
8 L.H., Daily Totals		$486.40		$748.52	$60.80	$93.56
Crew H-1	**Hr.**	**Daily**	**Hr.**	**Daily**	**Bare Costs**	**Incl. O&P**
2 Glaziers	$51.00	$816.00	$81.50	$1304.00	$54.33	$87.97
2 Struc. Steel Workers	57.65	922.40	94.45	1511.20		
32 L.H., Daily Totals		$1738.40		$2815.20	$54.33	$87.97
Crew H-2	**Hr.**	**Daily**	**Hr.**	**Daily**	**Bare Costs**	**Incl. O&P**
2 Glaziers	$51.00	$816.00	$81.50	$1304.00	$48.03	$76.82
1 Building Laborer	42.10	336.80	67.45	539.60		
24 L.H., Daily Totals		$1152.80		$1843.60	$48.03	$76.82
Crew H-3	**Hr.**	**Daily**	**Hr.**	**Daily**	**Bare Costs**	**Incl. O&P**
1 Glazier	$51.00	$408.00	$81.50	$652.00	$45.48	$73.22
1 Helper	39.95	319.60	64.95	519.60		
16 L.H., Daily Totals		$727.60		$1171.60	$45.48	$73.22
Crew H-4	**Hr.**	**Daily**	**Hr.**	**Daily**	**Bare Costs**	**Incl. O&P**
1 Carpenter	$53.15	$425.20	$85.15	$681.20	$49.51	$78.93
1 Carpenter Helper	39.95	319.60	64.95	519.60		
.5 Electrician	61.35	245.40	94.45	377.80		
20 L.H., Daily Totals		$990.20		$1578.60	$49.51	$78.93
Crew J-1	**Hr.**	**Daily**	**Hr.**	**Daily**	**Bare Costs**	**Incl. O&P**
3 Plasterers	$48.60	$1166.40	$77.40	$1857.60	$46.12	$73.44
2 Plasterer Helpers	42.40	678.40	67.50	1080.00		
1 Mixing Machine, 6 C.F.		117.35		129.09	2.93	3.23
40 L.H., Daily Totals		$1962.15		$3066.68	$49.05	$76.67
Crew J-2	**Hr.**	**Daily**	**Hr.**	**Daily**	**Bare Costs**	**Incl. O&P**
3 Plasterers	$48.60	$1166.40	$77.40	$1857.60	$47.11	$74.78
2 Plasterer Helpers	42.40	678.40	67.50	1080.00		
1 Lather	52.05	416.40	81.50	652.00		
1 Mixing Machine, 6 C.F.		117.35		129.09	2.44	2.69
48 L.H., Daily Totals		$2378.55		$3718.68	$49.55	$77.47
Crew J-3	**Hr.**	**Daily**	**Hr.**	**Daily**	**Bare Costs**	**Incl. O&P**
1 Terrazzo Worker	$49.40	$395.20	$77.40	$619.20	$44.95	$70.42
1 Terrazzo Helper	40.50	324.00	63.45	507.60		
1 Floor Grinder, 22" Path		104.60		115.06		
1 Terrazzo Mixer		161.15		177.26	16.61	18.27
16 L.H., Daily Totals		$984.95		$1419.13	$61.56	$88.70

Crews - Renovation

Crew No.	Bare Costs		Incl. Subs O&P		Cost Per Labor-Hour	
Crew J-4	**Hr.**	**Daily**	**Hr.**	**Daily**	**Bare Costs**	**Incl. O&P**
2 Cement Finishers	$49.95	$799.20	$78.45	$1255.20	$47.33	$74.78
1 Laborer	42.10	336.80	67.45	539.60		
1 Floor Grinder, 22" Path		104.60		115.06		
1 Floor Edger, 7" Path		43.95		48.34		
1 Vacuum Pick-Up System		74.05		81.45	9.28	10.20
24 L.H., Daily Totals		$1358.60		$2039.66	$56.61	$84.99

Crew J-4A	**Hr.**	**Daily**	**Hr.**	**Daily**	**Bare Costs**	**Incl. O&P**
2 Cement Finishers	$49.95	$799.20	$78.45	$1255.20	$46.02	$72.95
2 Laborers	42.10	673.60	67.45	1079.20		
1 Floor Grinder, 22" Path		104.60		115.06		
1 Floor Edger, 7" Path		43.95		48.34		
1 Vacuum Pick-Up System		74.05		81.45		
1 Floor Auto Scrubber		178.40		196.24	12.53	13.78
32 L.H., Daily Totals		$1873.80		$2775.50	$58.56	$86.73

Crew J-4B	**Hr.**	**Daily**	**Hr.**	**Daily**	**Bare Costs**	**Incl. O&P**
1 Laborer	$42.10	$336.80	$67.45	$539.60	$42.10	$67.45
1 Floor Auto Scrubber		178.40		196.24	22.30	24.53
8 L.H., Daily Totals		$515.20		$735.84	$64.40	$91.98

Crew J-6	**Hr.**	**Daily**	**Hr.**	**Daily**	**Bare Costs**	**Incl. O&P**
2 Painters	$44.40	$710.40	$70.50	$1128.00	$45.98	$72.85
1 Building Laborer	42.10	336.80	67.45	539.60		
1 Equip. Oper. (light)	53.00	424.00	82.95	663.60		
1 Air Compressor, 250 cfm		201.50		221.65		
1 Sandblaster, Portable, 3 C.F.		83.80		92.18		
1 Set Sand Blasting Accessories		15.35		16.89	9.40	10.33
32 L.H., Daily Totals		$1771.85		$2661.92	$55.37	$83.18

Crew J-7	**Hr.**	**Daily**	**Hr.**	**Daily**	**Bare Costs**	**Incl. O&P**
2 Painters	$44.40	$710.40	$70.50	$1128.00	$44.40	$70.50
1 Floor Belt Sander		50.15		55.16		
1 Floor Sanding Edger		25.15		27.66	4.71	5.18
16 L.H., Daily Totals		$785.70		$1210.83	$49.11	$75.68

Crew K-1	**Hr.**	**Daily**	**Hr.**	**Daily**	**Bare Costs**	**Incl. O&P**
1 Carpenter	$53.15	$425.20	$85.15	$681.20	$50.20	$80.35
1 Truck Driver (light)	47.25	378.00	75.55	604.40		
1 Flatbed Truck, Gas, 3 Ton		820.40		902.44	51.27	56.40
16 L.H., Daily Totals		$1623.60		$2188.04	$101.47	$136.75

Crew K-2	**Hr.**	**Daily**	**Hr.**	**Daily**	**Bare Costs**	**Incl. O&P**
1 Struc. Steel Foreman (outside)	$59.65	$477.20	$97.75	$782.00	$54.85	$89.25
1 Struc. Steel Worker	57.65	461.20	94.45	755.60		
1 Truck Driver (light)	47.25	378.00	75.55	604.40		
1 Flatbed Truck, Gas, 3 Ton		820.40		902.44	34.18	37.60
24 L.H., Daily Totals		$2136.80		$3044.44	$89.03	$126.85

Crew L-1	**Hr.**	**Daily**	**Hr.**	**Daily**	**Bare Costs**	**Incl. O&P**
1 Electrician	$61.35	$490.80	$94.45	$755.60	$62.90	$97.05
1 Plumber	64.45	515.60	99.65	797.20		
16 L.H., Daily Totals		$1006.40		$1552.80	$62.90	$97.05

Crew L-2	**Hr.**	**Daily**	**Hr.**	**Daily**	**Bare Costs**	**Incl. O&P**
1 Carpenter	$53.15	$425.20	$85.15	$681.20	$46.55	$75.05
1 Carpenter Helper	39.95	319.60	64.95	519.60		
16 L.H., Daily Totals		$744.80		$1200.80	$46.55	$75.05

Crew No.	Bare Costs		Incl. Subs O&P		Cost Per Labor-Hour	
Crew L-3	**Hr.**	**Daily**	**Hr.**	**Daily**	**Bare Costs**	**Incl. O&P**
1 Carpenter	$53.15	$425.20	$85.15	$681.20	$57.49	$90.59
.5 Electrician	61.35	245.40	94.45	377.80		
.5 Sheet Metal Worker	62.30	249.20	97.60	390.40		
16 L.H., Daily Totals		$919.80		$1449.40	$57.49	$90.59

Crew L-3A	**Hr.**	**Daily**	**Hr.**	**Daily**	**Bare Costs**	**Incl. O&P**
1 Carpenter Foreman (outside)	$55.15	$441.20	$88.40	$707.20	$57.53	$91.47
.5 Sheet Metal Worker	62.30	249.20	97.60	390.40		
12 L.H., Daily Totals		$690.40		$1097.60	$57.53	$91.47

Crew L-4	**Hr.**	**Daily**	**Hr.**	**Daily**	**Bare Costs**	**Incl. O&P**
2 Skilled Workers	$54.85	$877.60	$87.40	$1398.40	$49.88	$79.92
1 Helper	39.95	319.60	64.95	519.60		
24 L.H., Daily Totals		$1197.20		$1918.00	$49.88	$79.92

Crew L-5	**Hr.**	**Daily**	**Hr.**	**Daily**	**Bare Costs**	**Incl. O&P**
1 Struc. Steel Foreman (outside)	$59.65	$477.20	$97.75	$782.00	$58.16	$94.66
5 Struc. Steel Workers	57.65	2306.00	94.45	3778.00		
1 Equip. Oper. (crane)	59.20	473.60	92.65	741.20		
1 Hyd. Crane, 25 Ton		580.85		638.93	10.37	11.41
56 L.H., Daily Totals		$3837.65		$5940.14	$68.53	$106.07

Crew L-5A	**Hr.**	**Daily**	**Hr.**	**Daily**	**Bare Costs**	**Incl. O&P**
1 Struc. Steel Foreman (outside)	$59.65	$477.20	$97.75	$782.00	$58.54	$94.83
2 Structural Steel Workers	57.65	922.40	94.45	1511.20		
1 Equip. Oper. (crane)	59.20	473.60	92.65	741.20		
1 S.P. Crane, 4x4, 25 Ton		1144.00		1258.40	35.75	39.33
32 L.H., Daily Totals		$3017.20		$4292.80	$94.29	$134.15

Crew L-5B	**Hr.**	**Daily**	**Hr.**	**Daily**	**Bare Costs**	**Incl. O&P**
1 Struc. Steel Foreman (outside)	$59.65	$477.20	$97.75	$782.00	$59.83	$94.44
2 Structural Steel Workers	57.65	922.40	94.45	1511.20		
2 Electricians	61.35	981.60	94.45	1511.20		
2 Steamfitters/Pipefitters	65.55	1048.80	101.35	1621.60		
1 Equip. Oper. (crane)	59.20	473.60	92.65	741.20		
1 Equip. Oper. (oiler)	50.55	404.40	79.10	632.80		
1 Hyd. Crane, 80 Ton		1486.00		1634.60	20.64	22.70
72 L.H., Daily Totals		$5794.00		$8434.60	$80.47	$117.15

Crew L-6	**Hr.**	**Daily**	**Hr.**	**Daily**	**Bare Costs**	**Incl. O&P**
1 Plumber	$64.45	$515.60	$99.65	$797.20	$63.42	$97.92
.5 Electrician	61.35	245.40	94.45	377.80		
12 L.H., Daily Totals		$761.00		$1175.00	$63.42	$97.92

Crew L-7	**Hr.**	**Daily**	**Hr.**	**Daily**	**Bare Costs**	**Incl. O&P**
2 Carpenters	$53.15	$850.40	$85.15	$1362.40	$51.16	$81.42
1 Building Laborer	42.10	336.80	67.45	539.60		
.5 Electrician	61.35	245.40	94.45	377.80		
28 L.H., Daily Totals		$1432.60		$2279.80	$51.16	$81.42

Crew L-8	**Hr.**	**Daily**	**Hr.**	**Daily**	**Bare Costs**	**Incl. O&P**
2 Carpenters	$53.15	$850.40	$85.15	$1362.40	$55.41	$88.05
.5 Plumber	64.45	257.80	99.65	398.60		
20 L.H., Daily Totals		$1108.20		$1761.00	$55.41	$88.05

Crew L-9	Bare Costs Hr.	Bare Costs Daily	Incl. Subs O&P Hr.	Incl. Subs O&P Daily	Cost Per Labor-Hour Bare Costs	Cost Per Labor-Hour Incl. O&P
1 Labor Foreman (inside)	$42.60	$340.80	$68.25	$546.00	$47.81	$76.63
2 Building Laborers	42.10	673.60	67.45	1079.20		
1 Struc. Steel Worker	57.65	461.20	94.45	755.60		
.5 Electrician	61.35	245.40	94.45	377.80		
36 L.H., Daily Totals		$1721.00		$2758.60	$47.81	$76.63

Crew L-10	Hr.	Daily	Hr.	Daily	Bare Costs	Incl. O&P
1 Struc. Steel Foreman (outside)	$59.65	$477.20	$97.75	$782.00	$58.83	$94.95
1 Structural Steel Worker	57.65	461.20	94.45	755.60		
1 Equip. Oper. (crane)	59.20	473.60	92.65	741.20		
1 Hyd. Crane, 12 Ton		471.00		518.10	19.63	21.59
24 L.H., Daily Totals		$1883.00		$2796.90	$78.46	$116.54

Crew L-11	Hr.	Daily	Hr.	Daily	Bare Costs	Incl. O&P
2 Wreckers	$42.10	$673.60	$69.05	$1104.80	$49.10	$78.42
1 Equip. Oper. (crane)	59.20	473.60	92.65	741.20		
1 Equip. Oper. (light)	53.00	424.00	82.95	663.60		
1 Hyd. Excavator, 2.5 C.Y.		1465.00		1611.50		
1 Loader, Skid Steer, 78 H.P.		400.10		440.11	58.28	64.11
32 L.H., Daily Totals		$3436.30		$4561.21	$107.38	$142.54

Crew M-1	Hr.	Daily	Hr.	Daily	Bare Costs	Incl. O&P
3 Elevator Constructors	$85.55	$2053.20	$131.25	$3150.00	$81.28	$124.69
1 Elevator Apprentice	68.45	547.60	105.00	840.00		
5 Hand Tools		50.00		55.00	1.56	1.72
32 L.H., Daily Totals		$2650.80		$4045.00	$82.84	$126.41

Crew M-3	Hr.	Daily	Hr.	Daily	Bare Costs	Incl. O&P
1 Electrician Foreman (outside)	$63.35	$506.80	$97.50	$780.00	$64.39	$99.62
1 Common Laborer	42.10	336.80	67.45	539.60		
.25 Equipment Operator (med.)	56.75	113.50	88.80	177.60		
1 Elevator Constructor	85.55	684.40	131.25	1050.00		
1 Elevator Apprentice	68.45	547.60	105.00	840.00		
.25 S.P. Crane, 4x4, 20 Ton		145.55		160.10	4.28	4.71
34 L.H., Daily Totals		$2334.65		$3547.30	$68.67	$104.33

Crew M-4	Hr.	Daily	Hr.	Daily	Bare Costs	Incl. O&P
1 Electrician Foreman (outside)	$63.35	$506.80	$97.50	$780.00	$63.75	$98.70
1 Common Laborer	42.10	336.80	67.45	539.60		
.25 Equipment Operator, Crane	59.20	118.40	92.65	185.30		
.25 Equip. Oper. (oiler)	50.55	101.10	79.10	158.20		
1 Elevator Constructor	85.55	684.40	131.25	1050.00		
1 Elevator Apprentice	68.45	547.60	105.00	840.00		
.25 S.P. Crane, 4x4, 40 Ton		188.55		207.41	5.24	5.76
36 L.H., Daily Totals		$2483.65		$3760.51	$68.99	$104.46

Crew Q-1	Hr.	Daily	Hr.	Daily	Bare Costs	Incl. O&P
1 Plumber	$64.45	$515.60	$99.65	$797.20	$58.00	$89.67
1 Plumber Apprentice	51.55	412.40	79.70	637.60		
16 L.H., Daily Totals		$928.00		$1434.80	$58.00	$89.67

Crew Q-1A	Hr.	Daily	Hr.	Daily	Bare Costs	Incl. O&P
.25 Plumber Foreman (outside)	$66.45	$132.90	$102.75	$205.50	$64.85	$100.27
1 Plumber	64.45	515.60	99.65	797.20		
10 L.H., Daily Totals		$648.50		$1002.70	$64.85	$100.27

Crew Q-1C	Hr.	Daily	Hr.	Daily	Bare Costs	Incl. O&P
1 Plumber	$64.45	$515.60	$99.65	$797.20	$57.58	$89.38
1 Plumber Apprentice	51.55	412.40	79.70	637.60		
1 Equip. Oper. (medium)	56.75	454.00	88.80	710.40		
1 Trencher, Chain Type, 8' D		1872.00		2059.20	78.00	85.80
24 L.H., Daily Totals		$3254.00		$4204.40	$135.58	$175.18

Crew Q-2	Hr.	Daily	Hr.	Daily	Bare Costs	Incl. O&P
2 Plumbers	$64.45	$1031.20	$99.65	$1594.40	$60.15	$93.00
1 Plumber Apprentice	51.55	412.40	79.70	637.60		
24 L.H., Daily Totals		$1443.60		$2232.00	$60.15	$93.00

Crew Q-3	Hr.	Daily	Hr.	Daily	Bare Costs	Incl. O&P
1 Plumber Foreman (inside)	$64.95	$519.60	$100.45	$803.60	$61.35	$94.86
2 Plumbers	64.45	1031.20	99.65	1594.40		
1 Plumber Apprentice	51.55	412.40	79.70	637.60		
32 L.H., Daily Totals		$1963.20		$3035.60	$61.35	$94.86

Crew Q-4	Hr.	Daily	Hr.	Daily	Bare Costs	Incl. O&P
1 Plumber Foreman (inside)	$64.95	$519.60	$100.45	$803.60	$61.35	$94.86
1 Plumber	64.45	515.60	99.65	797.20		
1 Welder (plumber)	64.45	515.60	99.65	797.20		
1 Plumber Apprentice	51.55	412.40	79.70	637.60		
1 Welder, Electric, 300 amp		106.40		117.04	3.33	3.66
32 L.H., Daily Totals		$2069.60		$3152.64	$64.67	$98.52

Crew Q-5	Hr.	Daily	Hr.	Daily	Bare Costs	Incl. O&P
1 Steamfitter	$65.55	$524.40	$101.35	$810.80	$59.00	$91.22
1 Steamfitter Apprentice	52.45	419.60	81.10	648.80		
16 L.H., Daily Totals		$944.00		$1459.60	$59.00	$91.22

Crew Q-6	Hr.	Daily	Hr.	Daily	Bare Costs	Incl. O&P
2 Steamfitters	$65.55	$1048.80	$101.35	$1621.60	$61.18	$94.60
1 Steamfitter Apprentice	52.45	419.60	81.10	648.80		
24 L.H., Daily Totals		$1468.40		$2270.40	$61.18	$94.60

Crew Q-7	Hr.	Daily	Hr.	Daily	Bare Costs	Incl. O&P
1 Steamfitter Foreman (inside)	$66.05	$528.40	$102.15	$817.20	$62.40	$96.49
2 Steamfitters	65.55	1048.80	101.35	1621.60		
1 Steamfitter Apprentice	52.45	419.60	81.10	648.80		
32 L.H., Daily Totals		$1996.80		$3087.60	$62.40	$96.49

Crew Q-8	Hr.	Daily	Hr.	Daily	Bare Costs	Incl. O&P
1 Steamfitter Foreman (inside)	$66.05	$528.40	$102.15	$817.20	$62.40	$96.49
1 Steamfitter	65.55	524.40	101.35	810.80		
1 Welder (steamfitter)	65.55	524.40	101.35	810.80		
1 Steamfitter Apprentice	52.45	419.60	81.10	648.80		
1 Welder, Electric, 300 amp		106.40		117.04	3.33	3.66
32 L.H., Daily Totals		$2103.20		$3204.64	$65.72	$100.15

Crew Q-9	Hr.	Daily	Hr.	Daily	Bare Costs	Incl. O&P
1 Sheet Metal Worker	$62.30	$498.40	$97.60	$780.80	$56.08	$87.85
1 Sheet Metal Apprentice	49.85	398.80	78.10	624.80		
16 L.H., Daily Totals		$897.20		$1405.60	$56.08	$87.85

Crew Q-10	Hr.	Daily	Hr.	Daily	Bare Costs	Incl. O&P
2 Sheet Metal Workers	$62.30	$996.80	$97.60	$1561.60	$58.15	$91.10
1 Sheet Metal Apprentice	49.85	398.80	78.10	624.80		
24 L.H., Daily Totals		$1395.60		$2186.40	$58.15	$91.10

Crew No.	Bare Costs		Incl. Subs O&P		Cost Per Labor-Hour	
Crew Q-11	**Hr.**	**Daily**	**Hr.**	**Daily**	**Bare Costs**	**Incl. O&P**
1 Sheet Metal Foreman (inside)	$62.80	$502.40	$98.40	$787.20	$59.31	$92.92
2 Sheet Metal Workers	62.30	996.80	97.60	1561.60		
1 Sheet Metal Apprentice	49.85	398.80	78.10	624.80		
32 L.H., Daily Totals		$1898.00		$2973.60	$59.31	$92.92

Crew Q-12	Hr.	Daily	Hr.	Daily	Bare Costs	Incl. O&P
1 Sprinkler Installer	$63.25	$506.00	$97.95	$783.60	$56.92	$88.17
1 Sprinkler Apprentice	50.60	404.80	78.40	627.20		
16 L.H., Daily Totals		$910.80		$1410.80	$56.92	$88.17

Crew Q-13	Hr.	Daily	Hr.	Daily	Bare Costs	Incl. O&P
1 Sprinkler Foreman (inside)	$63.75	$510.00	$98.75	$790.00	$60.21	$93.26
2 Sprinkler Installers	63.25	1012.00	97.95	1567.20		
1 Sprinkler Apprentice	50.60	404.80	78.40	627.20		
32 L.H., Daily Totals		$1926.80		$2984.40	$60.21	$93.26

Crew Q-14	Hr.	Daily	Hr.	Daily	Bare Costs	Incl. O&P
1 Asbestos Worker	$58.65	$469.20	$93.00	$744.00	$52.77	$83.67
1 Asbestos Apprentice	46.90	375.20	74.35	594.80		
16 L.H., Daily Totals		$844.40		$1338.80	$52.77	$83.67

Crew Q-15	Hr.	Daily	Hr.	Daily	Bare Costs	Incl. O&P
1 Plumber	$64.45	$515.60	$99.65	$797.20	$58.00	$89.67
1 Plumber Apprentice	51.55	412.40	79.70	637.60		
1 Welder, Electric, 300 amp		106.40		117.04	6.65	7.32
16 L.H., Daily Totals		$1034.40		$1551.84	$64.65	$96.99

Crew Q-16	Hr.	Daily	Hr.	Daily	Bare Costs	Incl. O&P
2 Plumbers	$64.45	$1031.20	$99.65	$1594.40	$60.15	$93.00
1 Plumber Apprentice	51.55	412.40	79.70	637.60		
1 Welder, Electric, 300 amp		106.40		117.04	4.43	4.88
24 L.H., Daily Totals		$1550.00		$2349.04	$64.58	$97.88

Crew Q-17	Hr.	Daily	Hr.	Daily	Bare Costs	Incl. O&P
1 Steamfitter	$65.55	$524.40	$101.35	$810.80	$59.00	$91.22
1 Steamfitter Apprentice	52.45	419.60	81.10	648.80		
1 Welder, Electric, 300 amp		106.40		117.04	6.65	7.32
16 L.H., Daily Totals		$1050.40		$1576.64	$65.65	$98.54

Crew Q-17A	Hr.	Daily	Hr.	Daily	Bare Costs	Incl. O&P
1 Steamfitter	$65.55	$524.40	$101.35	$810.80	$59.07	$91.70
1 Steamfitter Apprentice	52.45	419.60	81.10	648.80		
1 Equip. Oper. (crane)	59.20	473.60	92.65	741.20		
1 Hyd. Crane, 12 Ton		471.00		518.10		
1 Welder, Electric, 300 amp		106.40		117.04	24.06	26.46
24 L.H., Daily Totals		$1995.00		$2835.94	$83.13	$118.16

Crew Q-18	Hr.	Daily	Hr.	Daily	Bare Costs	Incl. O&P
2 Steamfitters	$65.55	$1048.80	$101.35	$1621.60	$61.18	$94.60
1 Steamfitter Apprentice	52.45	419.60	81.10	648.80		
1 Welder, Electric, 300 amp		106.40		117.04	4.43	4.88
24 L.H., Daily Totals		$1574.80		$2387.44	$65.62	$99.48

Crew Q-19	Hr.	Daily	Hr.	Daily	Bare Costs	Incl. O&P
1 Steamfitter	$65.55	$524.40	$101.35	$810.80	$59.78	$92.30
1 Steamfitter Apprentice	52.45	419.60	81.10	648.80		
1 Electrician	61.35	490.80	94.45	755.60		
24 L.H., Daily Totals		$1434.80		$2215.20	$59.78	$92.30

Crew Q-20	Hr.	Daily	Hr.	Daily	Bare Costs	Incl. O&P
1 Sheet Metal Worker	$62.30	$498.40	$97.60	$780.80	$57.13	$89.17
1 Sheet Metal Apprentice	49.85	398.80	78.10	624.80		
.5 Electrician	61.35	245.40	94.45	377.80		
20 L.H., Daily Totals		$1142.60		$1783.40	$57.13	$89.17

Crew Q-21	Hr.	Daily	Hr.	Daily	Bare Costs	Incl. O&P
2 Steamfitters	$65.55	$1048.80	$101.35	$1621.60	$61.23	$94.56
1 Steamfitter Apprentice	52.45	419.60	81.10	648.80		
1 Electrician	61.35	490.80	94.45	755.60		
32 L.H., Daily Totals		$1959.20		$3026.00	$61.23	$94.56

Crew Q-22	Hr.	Daily	Hr.	Daily	Bare Costs	Incl. O&P
1 Plumber	$64.45	$515.60	$99.65	$797.20	$58.00	$89.67
1 Plumber Apprentice	51.55	412.40	79.70	637.60		
1 Hyd. Crane, 12 Ton		471.00		518.10	29.44	32.38
16 L.H., Daily Totals		$1399.00		$1952.90	$87.44	$122.06

Crew Q-22A	Hr.	Daily	Hr.	Daily	Bare Costs	Incl. O&P
1 Plumber	$64.45	$515.60	$99.65	$797.20	$54.33	$84.86
1 Plumber Apprentice	51.55	412.40	79.70	637.60		
1 Laborer	42.10	336.80	67.45	539.60		
1 Equip. Oper. (crane)	59.20	473.60	92.65	741.20		
1 Hyd. Crane, 12 Ton		471.00		518.10	14.72	16.19
32 L.H., Daily Totals		$2209.40		$3233.70	$69.04	$101.05

Crew Q-23	Hr.	Daily	Hr.	Daily	Bare Costs	Incl. O&P
1 Plumber Foreman (outside)	$66.45	$531.60	$102.75	$822.00	$62.55	$97.07
1 Plumber	64.45	515.60	99.65	797.20		
1 Equip. Oper. (medium)	56.75	454.00	88.80	710.40		
1 Lattice Boom Crane, 20 Ton		1488.00		1636.80	62.00	68.20
24 L.H., Daily Totals		$2989.20		$3966.40	$124.55	$165.27

Crew R-1	Hr.	Daily	Hr.	Daily	Bare Costs	Incl. O&P
1 Electrician Foreman	$61.85	$494.80	$95.20	$761.60	$57.35	$88.28
3 Electricians	61.35	1472.40	94.45	2266.80		
2 Electrician Apprentices	49.10	785.60	75.55	1208.80		
48 L.H., Daily Totals		$2752.80		$4237.20	$57.35	$88.28

Crew R-1A	Hr.	Daily	Hr.	Daily	Bare Costs	Incl. O&P
1 Electrician	$61.35	$490.80	$94.45	$755.60	$55.23	$85.00
1 Electrician Apprentice	49.10	392.80	75.55	604.40		
16 L.H., Daily Totals		$883.60		$1360.00	$55.23	$85.00

Crew R-1B	Hr.	Daily	Hr.	Daily	Bare Costs	Incl. O&P
1 Electrician	$61.35	$490.80	$94.45	$755.60	$53.18	$81.85
2 Electrician Apprentices	49.10	785.60	75.55	1208.80		
24 L.H., Daily Totals		$1276.40		$1964.40	$53.18	$81.85

Crew R-1C	Hr.	Daily	Hr.	Daily	Bare Costs	Incl. O&P
2 Electricians	$61.35	$981.60	$94.45	$1511.20	$55.23	$85.00
2 Electrician Apprentices	49.10	785.60	75.55	1208.80		
1 Portable cable puller, 8000 lb.		101.60		111.76	3.17	3.49
32 L.H., Daily Totals		$1868.80		$2831.76	$58.40	$88.49

Crew R-1D	Hr.	Daily	Hr.	Daily	Bare Costs	Incl. O&P
1 Electrician	$61.35	$490.80	$94.45	$755.60	$55.23	$85.00
1 Electrician Apprentice	49.10	392.80	75.55	604.40		
1 Aerial lift		105.00		115.50	6.56	7.22
16 L.H., Daily Totals		$988.60		$1475.50	$61.79	$92.22

Crew No.	Bare Costs		Incl. Subs O&P		Cost Per Labor-Hour	
Crew R-2	**Hr.**	**Daily**	**Hr.**	**Daily**	**Bare Costs**	**Incl. O&P**
1 Electrician Foreman	$61.85	$494.80	$95.20	$761.60	$57.61	$88.90
3 Electricians	61.35	1472.40	94.45	2266.80		
2 Electrician Apprentices	49.10	785.60	75.55	1208.80		
1 Equip. Oper. (crane)	59.20	473.60	92.65	741.20		
1 S.P. Crane, 4x4, 5 Ton		378.10		415.91	6.75	7.43
56 L.H., Daily Totals		$3604.50		$5394.31	$64.37	$96.33
Crew R-3	**Hr.**	**Daily**	**Hr.**	**Daily**	**Bare Costs**	**Incl. O&P**
1 Electrician Foreman	$61.85	$494.80	$95.20	$761.60	$61.12	$94.39
1 Electrician	61.35	490.80	94.45	755.60		
.5 Equip. Oper. (crane)	59.20	236.80	92.65	370.60		
.5 S.P. Crane, 4x4, 5 Ton		189.05		207.96	9.45	10.40
20 L.H., Daily Totals		$1411.45		$2095.76	$70.57	$104.79
Crew R-4	**Hr.**	**Daily**	**Hr.**	**Daily**	**Bare Costs**	**Incl. O&P**
1 Struc. Steel Foreman (outside)	$59.65	$477.20	$97.75	$782.00	$58.79	$95.11
3 Struc. Steel Workers	57.65	1383.60	94.45	2266.80		
1 Electrician	61.35	490.80	94.45	755.60		
1 Welder, Gas Engine, 300 amp		147.00		161.70	3.67	4.04
40 L.H., Daily Totals		$2498.60		$3966.10	$62.47	$99.15
Crew R-5	**Hr.**	**Daily**	**Hr.**	**Daily**	**Bare Costs**	**Incl. O&P**
1 Electrician Foreman	$61.85	$494.80	$95.20	$761.60	$53.61	$83.79
4 Electrician Linemen	61.35	1963.20	94.45	3022.40		
2 Electrician Operators	61.35	981.60	94.45	1511.20		
4 Electrician Groundmen	39.95	1278.40	64.95	2078.40		
1 Crew Truck		166.25		182.88		
1 Flatbed Truck, 20,000 GVW		201.60		221.76		
1 Pickup Truck, 3/4 Ton		110.85		121.94		
.2 Hyd. Crane, 55 Ton		196.04		215.64		
.2 Hyd. Crane, 12 Ton		94.20		103.62		
.2 Earth Auger, Truck-Mtd.		40.10		44.11		
1 Tractor w/Winch		373.50		410.85	13.44	14.78
88 L.H., Daily Totals		$5900.54		$8674.39	$67.05	$98.57
Crew R-6	**Hr.**	**Daily**	**Hr.**	**Daily**	**Bare Costs**	**Incl. O&P**
1 Electrician Foreman	$61.85	$494.80	$95.20	$761.60	$53.61	$83.79
4 Electrician Linemen	61.35	1963.20	94.45	3022.40		
2 Electrician Operators	61.35	981.60	94.45	1511.20		
4 Electrician Groundmen	39.95	1278.40	64.95	2078.40		
1 Crew Truck		166.25		182.88		
1 Flatbed Truck, 20,000 GVW		201.60		221.76		
1 Pickup Truck, 3/4 Ton		110.85		121.94		
.2 Hyd. Crane, 55 Ton		196.04		215.64		
.2 Hyd. Crane, 12 Ton		94.20		103.62		
.2 Earth Auger, Truck-Mtd.		40.10		44.11		
1 Tractor w/Winch		373.50		410.85		
3 Cable Trailers		192.30		211.53		
.5 Tensioning Rig		55.20		60.72		
.5 Cable Pulling Rig		303.57		333.93	19.70	21.67
88 L.H., Daily Totals		$6451.61		$9280.58	$73.31	$105.46
Crew R-7	**Hr.**	**Daily**	**Hr.**	**Daily**	**Bare Costs**	**Incl. O&P**
1 Electrician Foreman	$61.85	$494.80	$95.20	$761.60	$43.60	$69.99
5 Electrician Groundmen	39.95	1598.00	64.95	2598.00		
1 Crew Truck		166.25		182.88	3.46	3.81
48 L.H., Daily Totals		$2259.05		$3542.47	$47.06	$73.80

Crew No.	Bare Costs		Incl. Subs O&P		Cost Per Labor-Hour	
Crew R-8	**Hr.**	**Daily**	**Hr.**	**Daily**	**Bare Costs**	**Incl. O&P**
1 Electrician Foreman	$61.85	$494.80	$95.20	$761.60	$54.30	$84.74
3 Electrician Linemen	61.35	1472.40	94.45	2266.80		
2 Electrician Groundmen	39.95	639.20	64.95	1039.20		
1 Pickup Truck, 3/4 Ton		110.85		121.94		
1 Crew Truck		166.25		182.88	5.77	6.35
48 L.H., Daily Totals		$2883.50		$4372.41	$60.07	$91.09
Crew R-9	**Hr.**	**Daily**	**Hr.**	**Daily**	**Bare Costs**	**Incl. O&P**
1 Electrician Foreman	$61.85	$494.80	$95.20	$761.60	$50.71	$79.79
1 Electrician Lineman	61.35	490.80	94.45	755.60		
2 Electrician Operators	61.35	981.60	94.45	1511.20		
4 Electrician Groundmen	39.95	1278.40	64.95	2078.40		
1 Pickup Truck, 3/4 Ton		110.85		121.94		
1 Crew Truck		166.25		182.88	4.33	4.76
64 L.H., Daily Totals		$3522.70		$5411.61	$55.04	$84.56
Crew R-10	**Hr.**	**Daily**	**Hr.**	**Daily**	**Bare Costs**	**Incl. O&P**
1 Electrician Foreman	$61.85	$494.80	$95.20	$761.60	$57.87	$89.66
4 Electrician Linemen	61.35	1963.20	94.45	3022.40		
1 Electrician Groundman	39.95	319.60	64.95	519.60		
1 Crew Truck		166.25		182.88		
3 Tram Cars		216.15		237.76	7.97	8.76
48 L.H., Daily Totals		$3160.00		$4724.24	$65.83	$98.42
Crew R-11	**Hr.**	**Daily**	**Hr.**	**Daily**	**Bare Costs**	**Incl. O&P**
1 Electrician Foreman	$61.85	$494.80	$95.20	$761.60	$58.36	$90.44
4 Electricians	61.35	1963.20	94.45	3022.40		
1 Equip. Oper. (crane)	59.20	473.60	92.65	741.20		
1 Common Laborer	42.10	336.80	67.45	539.60		
1 Crew Truck		166.25		182.88		
1 Hyd. Crane, 12 Ton		471.00		518.10	11.38	12.52
56 L.H., Daily Totals		$3905.65		$5765.77	$69.74	$102.96
Crew R-12	**Hr.**	**Daily**	**Hr.**	**Daily**	**Bare Costs**	**Incl. O&P**
1 Carpenter Foreman (inside)	$53.65	$429.20	$85.95	$687.60	$49.91	$79.96
4 Carpenters	53.15	1700.80	85.15	2724.80		
4 Common Laborers	42.10	1347.20	67.45	2158.40		
1 Equip. Oper. (medium)	56.75	454.00	88.80	710.40		
1 Steel Worker	57.65	461.20	94.45	755.60		
1 Dozer, 200 H.P.		1504.00		1654.40		
1 Pickup Truck, 3/4 Ton		110.85		121.94	18.35	20.19
88 L.H., Daily Totals		$6007.25		$8813.14	$68.26	$100.15
Crew R-13	**Hr.**	**Daily**	**Hr.**	**Daily**	**Bare Costs**	**Incl. O&P**
1 Electrician Foreman	$61.85	$494.80	$95.20	$761.60	$59.29	$91.58
3 Electricians	61.35	1472.40	94.45	2266.80		
.25 Equip. Oper. (crane)	59.20	118.40	92.65	185.30		
1 Equipment Oiler	50.55	404.40	79.10	632.80		
.25 Hydraulic Crane, 33 Ton		236.34		259.97	5.63	6.19
42 L.H., Daily Totals		$2726.34		$4106.47	$64.91	$97.77
Crew R-15	**Hr.**	**Daily**	**Hr.**	**Daily**	**Bare Costs**	**Incl. O&P**
1 Electrician Foreman	$61.85	$494.80	$95.20	$761.60	$60.04	$92.66
4 Electricians	61.35	1963.20	94.45	3022.40		
1 Equipment Oper. (light)	53.00	424.00	82.95	663.60		
1 Telescoping Boom Lift, to 40'		278.90		306.79	5.81	6.39
48 L.H., Daily Totals		$3160.90		$4754.39	$65.85	$99.05

Crew No.	Bare Costs		Incl. Subs O & P		Cost Per Labor-Hour	
Crew R-18	**Hr.**	**Daily**	**Hr.**	**Daily**	**Bare Costs**	**Incl. O&P**
.25 Electrician Foreman	$61.85	$123.70	$95.20	$190.40	$53.85	$82.88
1 Electrician	61.35	490.80	94.45	755.60		
2 Electrician Apprentices	49.10	785.60	75.55	1208.80		
26 L.H., Daily Totals		$1400.10		$2154.80	$53.85	$82.88
Crew R-19	**Hr.**	**Daily**	**Hr.**	**Daily**	**Bare Costs**	**Incl. O&P**
.5 Electrician Foreman	$61.85	$247.40	$95.20	$380.80	$61.45	$94.60
2 Electricians	61.35	981.60	94.45	1511.20		
20 L.H., Daily Totals		$1229.00		$1892.00	$61.45	$94.60
Crew R-21	**Hr.**	**Daily**	**Hr.**	**Daily**	**Bare Costs**	**Incl. O&P**
1 Electrician Foreman	$61.85	$494.80	$95.20	$761.60	$61.36	$94.50
3 Electricians	61.35	1472.40	94.45	2266.80		
.1 Equip. Oper. (medium)	56.75	45.40	88.80	71.04		
.1 S.P. Crane, 4x4, 25 Ton		114.40		125.84	3.49	3.84
32.8 L.H., Daily Totals		$2127.00		$3225.28	$64.85	$98.33
Crew R-22	**Hr.**	**Daily**	**Hr.**	**Daily**	**Bare Costs**	**Incl. O&P**
.66 Electrician Foreman	$61.85	$326.57	$95.20	$502.66	$56.16	$86.44
2 Electricians	61.35	981.60	94.45	1511.20		
2 Electrician Apprentices	49.10	785.60	75.55	1208.80		
37.28 L.H., Daily Totals		$2093.77		$3222.66	$56.16	$86.44
Crew R-30	**Hr.**	**Daily**	**Hr.**	**Daily**	**Bare Costs**	**Incl. O&P**
.25 Electrician Foreman (outside)	$63.35	$126.70	$97.50	$195.00	$49.66	$78.07
1 Electrician	61.35	490.80	94.45	755.60		
2 Laborers (Semi-Skilled)	42.10	673.60	67.45	1079.20		
26 L.H., Daily Totals		$1291.10		$2029.80	$49.66	$78.07
Crew R-31	**Hr.**	**Daily**	**Hr.**	**Daily**	**Bare Costs**	**Incl. O&P**
1 Electrician	$61.35	$490.80	$94.45	$755.60	$61.35	$94.45
1 Core Drill, Electric, 2.5 H.P.		62.50		68.75	7.81	8.59
8 L.H., Daily Totals		$553.30		$824.35	$69.16	$103.04

Historical Cost Indexes

The table below lists both the RSMeans® historical cost index based on Jan. 1, 1993 = 100 as well as the computed value of an index based on Jan. 1, 2020 costs. Since the Jan. 1, 2020 figure is estimated, space is left to write in the actual index figures as they become available through the quarterly *RSMeans Construction Cost Indexes*.

To compute the actual index based on Jan. 1, 2020 = 100, divide the historical cost index for a particular year by the actual Jan. 1, 2020 construction cost index. Space has been left to advance the index figures as the year progresses.

Year	Historical Cost Index Jan. 1, 1993 = 100		Current Index Based on Jan. 1, 2020 = 100		Year	Historical Cost Index Jan. 1, 1993 = 100	Current Index Based on Jan. 1, 2020 = 100		Year	Historical Cost Index Jan. 1, 1993 = 100	Current Index Based on Jan. 1, 2020 = 100	
	Est.	Actual	Est.	Actual		Actual	Est.	Actual		Actual	Est.	Actual
Oct 2020*					July 2005	151.6	63.4		July 1987	87.7	36.7	
July 2020*					2004	143.7	60.1		1986	84.2	35.2	
Apr 2020*					2003	132.0	55.2		1985	82.6	34.6	
Jan 2020*	239.1		100.0	100.0	2002	128.7	53.8		1984	82.0	34.3	
July 2019		232.2	97.1		2001	125.1	52.3		1983	80.2	33.5	
2018		222.9	93.2		2000	120.9	50.6		1982	76.1	31.8	
2017		213.6	89.3		1999	117.6	49.2		1981	70.0	29.3	
2016		207.3	86.7		1998	115.1	48.1		1980	62.9	26.3	
2015		206.2	86.2		1997	112.8	47.2		1979	57.8	24.2	
2014		204.9	85.7		1996	110.2	46.1		1978	53.5	22.4	
2013		201.2	84.1		1995	107.6	45.0		1977	49.5	20.7	
2012		194.6	81.4		1994	104.4	43.7		1976	46.9	19.6	
2011		191.2	80.0		1993	101.7	42.5		1975	44.8	18.7	
2010		183.5	76.7		1992	99.4	41.6		1974	41.4	17.3	
2009		180.1	75.3		1991	96.8	40.5		1973	37.7	15.8	
2008		180.4	75.4		1990	94.3	39.4		1972	34.8	14.6	
2007		169.4	70.8		1989	92.1	38.5		1971	32.1	13.4	
2006		162.0	67.8		1988	89.9	37.6		1970	28.7	12.0	

Adjustments to Costs

The "Historical Cost Index" can be used to convert national average building costs at a particular time to the approximate building costs for some other time.

Time Adjustment Using the Historical Cost Indexes:

$$\frac{\text{Index for Year A}}{\text{Index for Year B}} \times \text{Cost in Year B} = \text{Cost in Year A}$$

Example:

Estimate and compare construction costs for different years in the same city.

To estimate the national average construction cost of a building in 1970, knowing that it cost $900,000 in 2020:

INDEX in 1970 = 28.7

INDEX in 2020 = 239.1

$$\frac{\text{INDEX 1970}}{\text{INDEX 2020}} \times \text{Cost 2020} = \text{Cost 1970}$$

$$\frac{28.7}{239.1} \times \$900{,}000 = .120 \times \$900{,}000 = \$108{,}000$$

The construction cost of the building in 1970 was $108,000.

Note: The city cost indexes for Canada can be used to convert U.S. national averages to local costs in Canadian dollars.

Example:

To estimate and compare the cost of a building in Toronto, ON in 2020 with the known cost of $600,000 (US$) in New York, NY in 2020:

INDEX Toronto = 115.6

INDEX New York = 137.1

$$\frac{\text{INDEX Toronto}}{\text{INDEX New York}} \times \text{Cost New York} = \text{Cost Toronto}$$

$$\frac{115.6}{137.1} \times \$600{,}000 = .843 \times \$600{,}000 = \$505{,}908$$

The construction cost of the building in Toronto is $505,908 (CN$).

*Historical Cost Index updates and other resources are provided on the following website:
http://info.thegordiangroup.com/RSMeans.html

How to Use the City Cost Indexes

What you should know before you begin

RSMeans City Cost Indexes (CCI) are an extremely useful tool for when you want to compare costs from city to city and region to region.

This publication contains average construction cost indexes for 731 U.S. and Canadian cities covering over 930 three-digit zip code locations, as listed directly under each city.

Keep in mind that a City Cost Index number is a percentage ratio of a specific city's cost to the national average cost of the same item at a stated time period.

In other words, these index figures represent relative construction factors (or, if you prefer, multipliers) for material and installation costs, as well as the weighted average for Total In Place costs for each CSI MasterFormat division. Installation costs include both labor and equipment rental costs. When estimating equipment rental rates only for a specific location, use 01 54 33 EQUIPMENT RENTAL COSTS in the Reference Section.

The 30 City Average Index is the average of 30 major U.S. cities and serves as a national average.

Index figures for both material and installation are based on the 30 major city average of 100 and represent the cost relationship as of July 1, 2019. The index for each division is computed from representative material and labor quantities for that division. The weighted average for each city is a weighted total of the components listed above it. It does not include relative productivity between trades or cities.

As changes occur in local material prices, labor rates, and equipment rental rates (including fuel costs), the impact of these changes should be accurately measured by the change in the City Cost Index for each particular city (as compared to the 30 city average).

Therefore, if you know (or have estimated) building costs in one city today, you can easily convert those costs to expected building costs in another city.

In addition, by using the Historical Cost Index, you can easily convert national average building costs at a particular time to the approximate building costs for some other time. The City Cost Indexes can then be applied to calculate the costs for a particular city.

Quick calculations

Location Adjustment Using the City Cost Indexes:

$$\frac{\text{Index for City A}}{\text{Index for City B}} \times \text{Cost in City B} = \text{Cost in City A}$$

Time Adjustment for the National Average Using the Historical Cost Index:

$$\frac{\text{Index for Year A}}{\text{Index for Year B}} \times \text{Cost in Year B} = \text{Cost in Year A}$$

Adjustment from the National Average:

$$\frac{\text{Index for City A}}{100} \times \text{National Average Cost} = \text{Cost in City A}$$

Since each of the other RSMeans data sets contains many different items, any *one* item multiplied by the particular city index may give incorrect results. However, the larger the number of items compiled, the closer the results should be to actual costs for that particular city.

The City Cost Indexes for Canadian cities are calculated using Canadian material and equipment prices and labor rates in Canadian dollars. Therefore, indexes for Canadian cities can be used to convert U.S. national average prices to local costs in Canadian dollars.

How to use this section

1. Compare costs from city to city.

In using the RSMeans Indexes, remember that an index number is not a fixed number but a ratio: It's a percentage ratio of a building component's cost at any stated time to the national average cost of that same component at the same time period. Put in the form of an equation:

$$\frac{\text{Specific City Cost}}{\text{National Average Cost}} \times 100 = \text{City Index Number}$$

Therefore, when making cost comparisons between cities, do not subtract one city's index number from the index number of another city and read the result as a percentage difference. Instead, divide one city's index number by that of the other city. The resulting number may then be used as a multiplier to calculate cost differences from city to city.

The formula used to find cost differences between cities for the purpose of comparison is as follows:

$$\frac{\text{City A Index}}{\text{City B Index}} \times \text{City B Cost (Known)} = \text{City A Cost (Unknown)}$$

In addition, you can use RSMeans CCI to calculate and compare costs division by division between cities using the same basic formula. (Just be sure that you're comparing similar divisions.)

2. Compare a specific city's construction costs with the national average.

When you're studying construction location feasibility, it's advisable to compare a prospective project's cost index with an index of the national average cost.

For example, divide the weighted average index of construction costs of a specific city by that of the 30 City Average, which = 100.

$$\frac{\text{City Index}}{100} = \text{\% of National Average}$$

As a result, you get a ratio that indicates the relative cost of construction in that city in comparison with the national average.

3. Convert U.S. national average to actual costs in Canadian City.

$$\frac{\text{Index for Canadian City}}{100} \times \text{National Average Cost} = \text{Cost in Canadian City in \$ CAN}$$

4. **Adjust construction cost data based on a national average.**

When you use a source of construction cost data which is based on a national average (such as RSMeans cost data), it is necessary to adjust those costs to a specific location.

$$\frac{\text{City Index}}{100} \times \text{Cost Based on National Average Costs} = \text{City Cost (Unknown)}$$

5. **When applying the City Cost Indexes to demolition projects, use the appropriate division installation index.** For example, for removal of existing doors and windows, use the Division 8 (Openings) index.

What you might like to know about how we developed the Indexes

The information presented in the CCI is organized according to the Construction Specifications Institute (CSI) MasterFormat 2018 classification system.

To create a reliable index, RSMeans researched the building type most often constructed in the United States and Canada. Because it was concluded that no one type of building completely represented the building construction industry, nine different types of buildings were combined to create a composite model.

The exact material, labor, and equipment quantities are based on detailed analyses of these nine building types, and then each quantity is weighted in proportion to expected usage. These various material items, labor hours, and equipment rental rates are thus combined to form a composite building representing as closely as possible the actual usage of materials, labor, and equipment in the North American building construction industry.

The following structures were chosen to make up that composite model:

1. Factory, 1 story
2. Office, 2–4 stories
3. Store, Retail
4. Town Hall, 2–3 stories
5. High School, 2–3 stories
6. Hospital, 4–8 stories
7. Garage, Parking
8. Apartment, 1–3 stories
9. Hotel/Motel, 2–3 stories

For the purposes of ensuring the timeliness of the data, the components of the index for the composite model have been streamlined. They currently consist of:

- specific quantities of 66 commonly used construction materials;
- specific labor-hours for 21 building construction trades; and
- specific days of equipment rental for 6 types of construction equipment (normally used to install the 66 material items by the 21 trades.) Fuel costs and routine maintenance costs are included in the equipment cost.

Material and equipment price quotations are gathered quarterly from cities in the United States and Canada. These prices and the latest negotiated labor wage rates for 21 different building trades are used to compile the quarterly update of the City Cost Index.

The 30 major U.S. cities used to calculate the national average are:

Atlanta, GA	Memphis, TN
Baltimore, MD	Milwaukee, WI
Boston, MA	Minneapolis, MN
Buffalo, NY	Nashville, TN
Chicago, IL	New Orleans, LA
Cincinnati, OH	New York, NY
Cleveland, OH	Philadelphia, PA
Columbus, OH	Phoenix, AZ
Dallas, TX	Pittsburgh, PA
Denver, CO	St. Louis, MO
Detroit, MI	San Antonio, TX
Houston, TX	San Diego, CA
Indianapolis, IN	San Francisco, CA
Kansas City, MO	Seattle, WA
Los Angeles, CA	Washington, DC

What the CCI does not indicate

The weighted average for each city is a total of the divisional components weighted to reflect typical usage. It does not include the productivity variations between trades or cities.

In addition, the CCI does not take into consideration factors such as the following:

- managerial efficiency
- competitive conditions
- automation
- restrictive union practices
- unique local requirements
- regional variations due to specific building codes

City Cost Indexes

DIVISION		UNITED STATES 30 CITY AVERAGE			ALABAMA ANNISTON 362			BIRMINGHAM 350 - 352			BUTLER 369			DECATUR 356			DOTHAN 363		
		MAT.	INST.	TOTAL	MAT.	INST.	TOTAL	MAT.	INST.	TOTAL	MAT.	INST.	TOTAL	MAT.	INST.	TOTAL	MAT.	INST.	TOTAL
015433	**CONTRACTOR EQUIPMENT**		100.0	100.0		101.9	101.9		104.7	104.7		99.5	99.5		101.9	101.9		99.5	99.5
0241, 31 - 34	**SITE & INFRASTRUCTURE, DEMOLITION**	100.0	100.0	100.0	90.9	89.5	89.9	91.2	94.5	93.5	105.4	84.4	90.9	84.5	88.1	87.0	102.9	84.8	90.4
0310	Concrete Forming & Accessories	100.0	100.0	100.0	85.6	68.7	71.2	90.6	69.7	72.8	82.3	68.8	70.8	90.9	62.8	66.9	90.3	69.1	72.2
0320	Concrete Reinforcing	100.0	100.0	100.0	88.8	71.2	80.3	94.2	71.3	83.1	93.9	70.5	82.6	88.1	67.7	78.2	93.9	71.3	83.0
0330	Cast-in-Place Concrete	100.0	100.0	100.0	82.3	67.7	76.9	103.8	69.3	90.9	80.3	67.8	75.6	101.9	67.2	89.0	80.3	67.7	75.6
03	**CONCRETE**	100.0	100.0	100.0	89.2	70.4	81.0	92.2	71.4	83.1	90.3	70.4	81.6	92.0	67.0	81.0	89.5	70.7	81.2
04	**MASONRY**	100.0	100.0	100.0	89.2	63.1	73.2	87.1	64.3	73.1	93.7	63.1	74.9	85.2	62.4	71.2	95.1	62.9	75.3
05	**METALS**	100.0	100.0	100.0	101.8	94.8	99.7	100.5	94.2	98.7	100.7	94.8	98.9	102.9	93.1	100.0	100.7	95.4	99.1
06	**WOOD, PLASTICS & COMPOSITES**	100.0	100.0	100.0	82.0	69.9	75.3	91.0	70.0	79.4	77.1	69.9	73.1	95.7	62.1	77.2	88.1	69.9	78.1
07	**THERMAL & MOISTURE PROTECTION**	100.0	100.0	100.0	96.0	63.0	81.6	94.0	67.6	82.5	96.1	65.9	82.9	93.1	64.5	80.7	96.0	65.7	82.8
08	**OPENINGS**	100.0	100.0	100.0	94.7	69.2	88.7	102.6	69.8	94.9	94.7	69.5	88.8	106.9	64.6	97.0	94.7	69.7	88.9
0920	Plaster & Gypsum Board	100.0	100.0	100.0	81.8	69.5	73.5	88.9	69.5	75.9	79.3	69.5	72.7	92.7	61.6	71.7	88.7	69.5	75.8
0950, 0980	Ceilings & Acoustic Treatment	100.0	100.0	100.0	77.1	69.5	72.0	84.8	69.5	74.5	77.1	69.5	72.0	83.2	61.6	68.6	77.1	69.5	72.0
0960	Flooring	100.0	100.0	100.0	82.0	69.1	78.3	99.6	69.1	91.0	86.5	69.1	81.6	89.9	69.1	84.1	91.6	69.1	85.3
0970, 0990	Wall Finishes & Painting/Coating	100.0	100.0	100.0	89.1	66.7	75.9	87.0	66.7	75.0	89.1	45.9	63.5	79.4	61.0	68.5	89.1	80.4	84.0
09	**FINISHES**	100.0	100.0	100.0	80.4	68.8	74.1	91.2	69.2	79.3	83.6	66.6	74.4	85.3	63.5	73.5	86.1	70.3	77.6
COVERS	**DIVS. 10 - 14, 25, 28, 41, 43, 44, 46**	100.0	100.0	100.0	100.0	74.0	94.2	100.0	83.6	96.3	100.0	74.0	94.2	100.0	72.9	94.0	100.0	74.0	94.2
21, 22, 23	**FIRE SUPPRESSION, PLUMBING & HVAC**	100.0	100.0	100.0	101.0	51.8	81.7	100.0	66.3	86.8	98.1	67.0	85.9	100.0	64.5	86.1	98.1	65.0	85.1
26, 27, 3370	**ELECTRICAL, COMMUNICATIONS & UTIL.**	100.0	100.0	100.0	97.7	59.9	79.1	98.4	59.7	79.3	99.8	61.7	81.0	94.5	63.8	79.4	98.6	76.1	87.5
MF2018	**WEIGHTED AVERAGE**	100.0	100.0	100.0	95.9	67.7	84.0	97.4	71.8	86.6	96.1	70.6	85.3	97.1	69.2	85.3	96.2	72.7	86.3

DIVISION		ALABAMA EVERGREEN 364			GADSDEN 359			HUNTSVILLE 357 - 358			JASPER 355			MOBILE 365 - 366			MONTGOMERY 360 - 361		
		MAT.	INST.	TOTAL	MAT.	INST.	TOTAL	MAT.	INST.	TOTAL	MAT.	INST.	TOTAL	MAT.	INST.	TOTAL	MAT.	INST.	TOTAL
015433	**CONTRACTOR EQUIPMENT**		99.5	99.5		101.9	101.9		101.9	101.9		101.9	101.9		99.5	99.5		102.0	102.0
0241, 31 - 34	**SITE & INFRASTRUCTURE, DEMOLITION**	105.9	84.4	91.0	89.9	89.5	89.6	84.2	89.4	87.8	89.9	89.5	89.6	97.7	85.4	89.2	95.6	89.9	91.7
0310	Concrete Forming & Accessories	79.4	68.6	70.2	83.9	69.1	71.3	90.9	65.9	69.6	88.5	64.7	68.3	89.4	68.4	71.5	92.8	69.4	72.8
0320	Concrete Reinforcing	94.0	70.4	82.6	93.5	71.3	82.8	88.1	76.3	82.4	88.1	71.3	80.0	91.5	70.5	81.3	99.7	71.4	86.0
0330	Cast-in-Place Concrete	80.3	67.6	75.6	101.9	67.8	89.2	99.2	67.8	87.5	113.0	67.8	96.1	84.2	66.9	77.7	81.5	68.7	76.7
03	**CONCRETE**	90.7	70.2	81.7	96.5	70.7	85.1	90.8	70.1	81.7	100.1	68.7	86.3	85.5	69.9	78.7	84.1	71.1	78.4
04	**MASONRY**	93.8	62.9	74.8	83.8	63.1	71.1	86.3	63.3	72.2	81.4	63.1	70.2	92.3	61.4	73.3	89.4	63.2	73.3
05	**METALS**	100.7	94.6	98.9	100.7	95.2	99.1	102.9	96.7	101.0	100.7	95.1	99.0	102.8	94.9	100.5	101.7	94.3	99.5
06	**WOOD, PLASTICS & COMPOSITES**	73.8	69.9	71.6	86.7	69.9	77.4	95.7	65.3	79.0	93.0	64.1	77.1	86.7	69.9	77.4	90.3	70.0	79.1
07	**THERMAL & MOISTURE PROTECTION**	96.0	65.4	82.7	93.3	66.3	81.6	93.0	65.9	81.2	93.3	64.4	80.7	95.6	65.4	82.4	94.2	66.9	82.3
08	**OPENINGS**	94.7	69.3	88.8	103.4	69.7	95.6	106.6	68.5	97.7	103.4	66.6	94.8	97.3	69.5	90.8	96.2	69.8	90.0
0920	Plaster & Gypsum Board	78.3	69.5	72.4	85.1	69.5	74.6	92.7	64.8	73.9	89.5	63.6	72.1	85.6	69.5	74.8	85.9	69.5	74.9
0950, 0980	Ceilings & Acoustic Treatment	77.1	69.5	72.0	79.1	69.5	72.7	85.0	64.8	71.4	79.1	63.6	68.6	83.0	69.5	73.9	84.6	69.5	74.4
0960	Flooring	84.5	69.1	80.1	86.4	69.1	81.5	89.9	69.1	84.1	88.4	69.1	82.9	91.1	69.1	84.9	91.7	69.1	85.3
0970, 0990	Wall Finishes & Painting/Coating	89.1	45.9	63.5	79.4	57.4	66.4	79.4	63.7	70.1	79.4	66.7	71.9	92.3	45.9	64.8	91.1	66.7	76.7
09	**FINISHES**	82.9	66.5	74.1	82.9	67.8	74.7	85.7	65.9	75.0	84.0	65.4	73.9	86.2	66.2	75.4	88.3	68.9	77.8
COVERS	**DIVS. 10 - 14, 25, 28, 41, 43, 44, 46**	100.0	73.9	94.2	100.0	82.9	96.2	100.0	82.5	96.1	100.0	73.4	94.1	100.0	82.5	96.1	100.0	83.3	96.3
21, 22, 23	**FIRE SUPPRESSION, PLUMBING & HVAC**	98.1	59.8	83.0	102.3	62.3	86.6	100.0	66.9	87.0	102.3	65.7	87.9	100.0	61.4	84.8	100.0	64.3	86.0
26, 27, 3370	**ELECTRICAL, COMMUNICATIONS & UTIL.**	97.1	56.8	77.2	94.5	59.7	77.4	95.4	65.9	80.8	94.2	59.9	77.3	100.4	56.8	78.9	101.1	76.1	88.8
MF2018	**WEIGHTED AVERAGE**	95.8	68.3	84.2	97.3	70.2	85.9	97.1	71.7	86.4	97.8	69.8	86.0	96.7	68.7	84.9	96.3	73.1	86.5

DIVISION		ALABAMA PHENIX CITY 368			SELMA 367			TUSCALOOSA 354			ALASKA ANCHORAGE 995 - 996			FAIRBANKS 997			JUNEAU 998		
		MAT.	INST.	TOTAL	MAT.	INST.	TOTAL	MAT.	INST.	TOTAL	MAT.	INST.	TOTAL	MAT.	INST.	TOTAL	MAT.	INST.	TOTAL
015433	**CONTRACTOR EQUIPMENT**		99.5	99.5		99.5	99.5		101.9	101.9		110.5	110.5		112.9	112.9		110.5	110.5
0241, 31 - 34	**SITE & INFRASTRUCTURE, DEMOLITION**	109.8	85.5	93.0	102.7	85.5	90.8	84.8	89.5	88.0	118.4	121.7	120.7	117.8	124.5	122.5	133.6	121.7	125.4
0310	Concrete Forming & Accessories	85.6	67.6	70.2	83.5	69.1	71.2	90.8	69.2	72.4	120.5	116.2	116.8	124.7	115.0	116.4	125.7	116.2	117.6
0320	Concrete Reinforcing	93.9	66.9	80.9	93.9	71.3	83.0	88.1	71.3	80.0	148.4	120.7	135.0	147.2	120.7	134.4	138.9	120.7	130.1
0330	Cast-in-Place Concrete	80.3	67.9	75.7	80.3	67.8	75.6	103.4	67.9	90.2	105.1	117.5	109.7	109.7	115.6	111.9	116.1	117.5	116.6
03	**CONCRETE**	93.6	69.3	82.9	89.0	70.7	81.0	92.7	70.8	83.1	107.5	116.7	111.6	101.4	115.5	107.6	114.1	116.7	115.2
04	**MASONRY**	93.7	63.1	74.9	97.4	63.1	76.3	85.4	63.1	71.7	171.5	119.4	139.5	181.4	117.8	142.3	162.5	119.4	136.0
05	**METALS**	100.6	93.7	98.6	100.6	95.2	99.0	102.1	95.3	100.1	127.3	104.4	120.5	122.6	105.4	117.5	113.1	104.4	110.6
06	**WOOD, PLASTICS & COMPOSITES**	81.7	67.7	74.0	78.9	69.9	73.9	95.7	69.9	81.5	112.3	113.5	112.9	126.9	112.7	119.1	121.9	113.5	117.3
07	**THERMAL & MOISTURE PROTECTION**	96.5	66.1	83.2	95.9	65.9	82.8	93.1	66.3	81.4	176.9	116.9	150.8	186.2	115.8	155.6	188.4	116.9	157.3
08	**OPENINGS**	94.7	67.4	88.3	94.7	69.7	88.8	106.6	69.7	98.0	130.8	115.8	127.3	132.8	114.7	128.6	131.0	115.8	127.5
0920	Plaster & Gypsum Board	83.0	67.3	72.4	81.1	69.5	73.3	92.7	69.5	77.1	144.1	113.8	123.7	177.5	112.9	134.0	158.2	113.8	128.3
0950, 0980	Ceilings & Acoustic Treatment	77.1	67.3	70.5	77.1	69.5	72.0	85.0	69.5	74.6	131.3	113.8	119.5	123.9	112.9	116.5	136.1	113.8	121.0
0960	Flooring	88.4	69.1	83.0	87.0	69.1	81.9	89.9	69.1	84.1	118.3	119.4	118.6	116.7	119.4	117.4	125.0	119.4	123.5
0970, 0990	Wall Finishes & Painting/Coating	89.1	79.0	83.1	89.1	66.7	75.9	79.4	66.7	71.9	113.1	119.1	116.6	111.6	116.9	114.7	108.7	119.1	114.8
09	**FINISHES**	85.2	68.9	76.4	83.7	68.8	75.7	85.6	68.8	76.6	127.0	117.1	121.7	128.6	116.0	121.8	128.9	117.1	122.5
COVERS	**DIVS. 10 - 14, 25, 28, 41, 43, 44, 46**	100.0	82.7	96.1	100.0	70.4	93.4	100.0	82.9	96.2	100.0	114.1	103.2	100.0	113.6	103.0	100.0	114.1	103.2
21, 22, 23	**FIRE SUPPRESSION, PLUMBING & HVAC**	98.1	64.2	84.8	98.1	65.1	85.1	100.0	67.1	87.1	100.5	107.1	103.1	100.2	107.9	103.2	100.5	107.1	103.1
26, 27, 3370	**ELECTRICAL, COMMUNICATIONS & UTIL.**	99.2	65.4	82.6	98.3	76.1	87.3	95.0	59.7	77.6	117.2	109.4	113.3	124.2	109.4	116.9	110.2	109.4	109.8
MF2018	**WEIGHTED AVERAGE**	96.8	70.8	85.8	95.9	72.5	86.0	97.1	71.4	86.3	118.5	113.1	116.2	118.8	113.0	116.3	116.8	113.1	115.2

DIVISION		ALASKA			ARIZONA															
		KETCHIKAN			CHAMBERS			FLAGSTAFF			GLOBE			KINGMAN			MESA/TEMPE		
		999			865			860			855			864			852		
		MAT.	INST.	TOTAL	MAT.	INST.	TOTAL	MAT.	INST.	TOTAL	MAT.	INST.	TOTAL	MAT.	INST.	TOTAL	MAT.	INST.	TOTAL
015433	**CONTRACTOR EQUIPMENT**		112.9	112.9		88.2	88.2		88.2	88.2		89.3	89.3		88.2	88.2		89.3	89.3
0241, 31 - 34	**SITE & INFRASTRUCTURE, DEMOLITION**	169.1	124.6	138.4	71.5	90.1	84.4	91.2	90.1	90.4	109.0	91.0	96.5	71.4	90.1	84.3	97.8	91.0	93.1
0310	Concrete Forming & Accessories	116.3	116.2	116.2	99.2	70.3	74.6	104.6	67.4	72.8	93.5	70.4	73.8	97.4	66.1	70.7	96.5	70.5	74.3
0320	Concrete Reinforcing	110.6	120.7	115.5	105.5	76.9	91.7	105.4	76.9	91.6	110.5	76.9	94.2	105.7	76.9	91.7	111.2	76.9	94.6
0330	Cast-in-Place Concrete	221.8	117.3	182.9	89.8	67.3	81.4	89.8	67.4	81.5	83.3	67.0	77.2	89.5	67.3	81.2	84.0	67.0	77.7
03	**CONCRETE**	170.9	116.7	147.1	93.4	70.3	83.3	114.1	69.0	94.3	101.9	70.3	88.0	93.1	68.4	82.3	93.1	70.4	83.1
04	**MASONRY**	188.6	119.4	146.1	96.0	59.4	73.5	96.1	59.4	73.6	101.7	59.3	75.7	96.0	59.4	73.5	101.9	59.3	75.8
05	**METALS**	122.8	105.4	117.7	103.0	71.6	93.8	103.5	71.7	94.2	105.8	72.5	96.0	103.7	71.6	94.3	106.2	72.6	96.3
06	**WOOD, PLASTICS & COMPOSITES**	117.1	113.5	115.1	101.8	73.0	85.9	108.2	68.8	86.5	92.7	73.1	81.9	96.4	67.3	80.4	96.7	73.1	83.7
07	**THERMAL & MOISTURE PROTECTION**	191.6	116.4	158.9	99.9	69.7	86.8	101.8	69.3	87.7	99.1	68.3	85.7	99.9	69.1	86.5	99.0	68.3	85.6
08	**OPENINGS**	128.6	115.8	125.6	105.8	69.3	97.3	105.9	70.1	97.6	94.9	69.4	88.9	106.0	67.9	97.1	94.9	69.4	89.0
0920	Plaster & Gypsum Board	161.0	113.8	129.2	92.8	72.5	79.1	96.5	68.2	77.5	88.2	72.5	77.6	84.2	66.6	72.4	91.7	72.5	78.7
0950, 0980	Ceilings & Acoustic Treatment	118.4	113.8	115.3	106.8	72.5	83.6	107.6	68.2	81.1	95.5	72.5	80.0	107.6	66.6	80.0	95.5	72.5	80.0
0960	Flooring	116.4	119.4	117.2	89.7	64.5	82.6	92.0	64.5	84.3	98.6	64.5	89.0	88.4	64.5	81.6	100.4	64.5	90.3
0970, 0990	Wall Finishes & Painting/Coating	111.6	119.1	116.0	87.9	58.6	70.5	87.9	58.6	70.5	90.6	58.6	71.6	87.9	58.6	70.5	90.6	58.6	71.6
09	**FINISHES**	129.2	117.1	122.7	91.3	68.1	78.8	94.5	65.7	78.9	96.6	68.2	81.2	89.9	64.8	76.3	96.4	68.2	81.2
COVERS	**DIVS. 10 - 14, 25, 28, 41, 43, 44, 46**	100.0	114.2	103.2	100.0	81.8	95.9	100.0	81.3	95.8	100.0	82.1	96.0	100.0	81.2	95.8	100.0	82.1	96.0
21, 22, 23	**FIRE SUPPRESSION, PLUMBING & HVAC**	98.1	107.2	101.7	97.7	76.8	89.5	100.3	77.0	91.1	96.6	76.9	88.9	97.7	76.8	89.5	100.1	77.0	91.1
26, 27, 3370	**ELECTRICAL, COMMUNICATIONS & UTIL.**	124.2	109.4	116.9	102.3	66.5	84.7	101.3	60.7	81.3	97.1	62.9	80.2	102.3	62.9	82.9	94.2	62.9	78.7
MF2018	**WEIGHTED AVERAGE**	128.4	113.4	122.1	98.3	71.8	87.1	102.4	70.5	88.9	99.5	71.4	87.6	98.3	70.4	86.5	98.7	71.5	87.2

DIVISION		ARIZONA												ARKANSAS					
		PHOENIX			PRESCOTT			SHOW LOW			TUCSON			BATESVILLE			CAMDEN		
		850,853			863			859			856 - 857			725			717		
		MAT.	INST.	TOTAL	MAT.	INST.	TOTAL	MAT.	INST.	TOTAL	MAT.	INST.	TOTAL	MAT.	INST.	TOTAL	MAT.	INST.	TOTAL
015433	**CONTRACTOR EQUIPMENT**		92.5	92.5		88.2	88.2		89.3	89.3		89.3	89.3		89.7	89.7		89.7	89.7
0241, 31 - 34	**SITE & INFRASTRUCTURE, DEMOLITION**	98.3	94.3	95.6	78.4	90.1	86.5	111.4	91.0	97.3	92.9	91.0	91.6	73.0	85.7	81.8	77.2	85.8	83.1
0310	Concrete Forming & Accessories	101.2	70.8	75.3	100.8	70.3	74.8	100.6	70.5	74.9	97.1	67.5	71.8	82.5	61.4	64.5	78.6	61.9	64.4
0320	Concrete Reinforcing	109.1	76.9	93.6	105.4	76.9	91.6	111.2	76.9	94.6	92.3	76.9	84.9	84.4	69.4	77.1	91.0	69.4	80.5
0330	Cast-in-Place Concrete	83.6	67.8	77.7	89.8	67.3	81.4	83.3	67.0	77.3	86.5	67.0	79.2	72.8	74.6	73.5	76.4	74.6	75.7
03	**CONCRETE**	95.5	70.8	84.6	99.0	70.3	86.4	104.2	70.3	89.3	91.6	69.0	81.7	73.2	68.1	70.9	75.7	68.3	72.4
04	**MASONRY**	94.8	62.1	74.7	96.1	59.4	73.6	101.8	59.3	75.7	88.9	59.3	70.7	94.1	60.8	73.6	104.7	60.8	77.7
05	**METALS**	107.8	73.7	97.8	103.5	71.6	94.2	105.6	72.5	95.9	107.0	72.6	96.9	94.7	76.0	89.2	101.1	76.1	93.8
06	**WOOD, PLASTICS & COMPOSITES**	102.1	72.0	85.5	103.3	73.0	86.6	101.2	73.1	85.7	96.9	69.0	81.6	91.1	63.5	75.9	89.2	64.0	75.4
07	**THERMAL & MOISTURE PROTECTION**	100.3	69.6	86.9	100.5	69.7	87.1	99.3	68.3	85.8	99.9	67.9	86.0	102.7	61.9	84.9	97.5	61.9	82.0
08	**OPENINGS**	100.7	71.4	93.9	105.9	69.3	97.4	94.2	71.1	88.8	91.4	70.2	86.5	100.0	61.0	90.9	103.1	61.3	93.4
0920	Plaster & Gypsum Board	99.4	71.2	80.4	93.3	72.5	79.3	93.5	72.5	79.4	96.3	68.2	77.4	73.7	62.9	66.5	84.7	63.5	70.4
0950, 0980	Ceilings & Acoustic Treatment	106.5	71.2	82.7	106.0	72.5	83.4	95.5	72.5	80.0	96.4	68.2	77.4	88.1	62.9	71.1	86.3	63.5	70.9
0960	Flooring	102.4	64.5	91.7	90.6	64.5	83.2	102.2	64.5	91.6	91.8	64.8	84.3	81.8	70.9	78.7	86.1	70.9	81.9
0970, 0990	Wall Finishes & Painting/Coating	95.3	57.2	72.8	87.9	58.6	70.5	90.6	58.6	71.6	91.3	58.6	71.9	92.2	54.5	69.9	88.8	54.5	68.5
09	**FINISHES**	101.0	68.1	83.2	92.0	68.1	79.1	98.6	68.2	82.2	94.5	65.9	79.0	76.3	62.6	68.9	79.8	62.9	70.7
COVERS	**DIVS. 10 - 14, 25, 28, 41, 43, 44, 46**	100.0	83.2	96.2	100.0	81.8	95.9	100.0	82.1	96.0	100.0	81.7	95.9	100.0	66.4	92.5	100.0	66.5	92.5
21, 22, 23	**FIRE SUPPRESSION, PLUMBING & HVAC**	100.0	78.5	91.6	100.3	76.8	91.0	96.6	76.9	88.9	100.1	74.8	90.2	96.5	51.2	78.7	96.4	57.5	81.2
26, 27, 3370	**ELECTRICAL, COMMUNICATIONS & UTIL.**	99.8	60.7	80.5	101.0	62.9	82.2	94.5	62.9	78.9	96.2	60.6	78.7	95.5	60.5	78.3	95.8	58.1	77.2
MF2018	**WEIGHTED AVERAGE**	100.6	72.3	88.6	99.9	71.3	87.8	99.7	71.5	87.8	97.7	70.2	86.1	91.6	63.6	79.8	94.0	64.7	81.6

DIVISION		ARKANSAS																	
		FAYETTEVILLE			FORT SMITH			HARRISON			HOT SPRINGS			JONESBORO			LITTLE ROCK		
		727			729			726			719			724			720 - 722		
		MAT.	INST.	TOTAL	MAT.	INST.	TOTAL	MAT.	INST.	TOTAL	MAT.	INST.	TOTAL	MAT.	INST.	TOTAL	MAT.	INST.	TOTAL
015433	**CONTRACTOR EQUIPMENT**		89.7	89.7		89.7	89.7		89.7	89.7		89.7	89.7		112.7	112.7		92.4	92.4
0241, 31 - 34	**SITE & INFRASTRUCTURE, DEMOLITION**	72.5	85.7	81.6	77.7	85.5	83.1	77.7	85.7	83.3	80.2	85.7	84.0	96.7	102.4	100.7	85.3	90.3	88.8
0310	Concrete Forming & Accessories	78.4	61.8	64.3	95.7	61.6	66.7	86.5	61.8	65.4	76.3	61.8	64.0	85.7	61.9	65.4	93.1	62.4	66.9
0320	Concrete Reinforcing	84.4	62.1	73.6	85.4	66.3	76.2	84.0	69.4	76.9	89.3	69.4	79.7	81.5	65.1	73.6	90.4	69.5	80.3
0330	Cast-in-Place Concrete	72.9	74.6	73.5	83.3	75.5	80.4	80.8	74.5	78.5	78.1	74.6	76.8	79.4	75.8	78.0	80.7	76.9	79.3
03	**CONCRETE**	72.9	67.0	70.3	80.1	68.0	74.8	79.7	68.2	74.6	78.9	68.2	74.2	77.3	69.0	73.7	81.0	69.3	75.9
04	**MASONRY**	85.0	60.8	70.1	91.6	60.8	72.7	94.3	60.8	73.7	78.6	60.8	67.7	86.5	60.8	70.7	87.7	62.1	72.0
05	**METALS**	94.7	73.6	88.5	96.9	74.9	90.4	95.8	76.0	90.0	101.1	76.1	93.7	91.3	89.4	90.8	97.2	76.1	91.0
06	**WOOD, PLASTICS & COMPOSITES**	87.7	64.0	74.7	107.1	64.0	83.4	96.8	64.0	78.7	86.7	64.0	74.2	95.2	64.4	78.3	98.4	64.2	79.5
07	**THERMAL & MOISTURE PROTECTION**	103.5	61.9	85.4	103.8	61.7	85.5	103.0	61.9	85.1	97.8	61.9	82.2	108.4	61.9	88.2	98.5	63.1	83.1
08	**OPENINGS**	100.0	59.6	90.6	102.0	60.1	92.3	100.8	61.3	91.6	103.1	61.3	93.3	105.6	60.5	95.1	95.6	60.7	87.5
0920	Plaster & Gypsum Board	73.0	63.5	66.6	79.7	63.5	68.8	78.8	63.5	68.5	83.4	63.5	70.0	87.3	63.5	71.2	92.0	63.5	72.8
0950, 0980	Ceilings & Acoustic Treatment	88.1	63.5	71.5	89.8	63.5	72.0	89.8	63.5	72.0	86.3	63.5	70.9	92.1	63.5	72.8	87.3	63.5	71.2
0960	Flooring	78.9	70.9	76.7	88.3	70.9	83.4	84.1	70.9	80.4	85.1	70.9	81.1	59.3	70.9	62.6	89.2	81.0	86.9
0970, 0990	Wall Finishes & Painting/Coating	92.2	54.5	69.9	92.2	52.1	68.5	92.2	54.5	69.9	88.8	54.5	68.5	81.7	54.5	65.6	92.8	52.4	68.9
09	**FINISHES**	75.5	62.9	68.7	79.6	62.6	70.4	78.3	62.9	70.0	79.6	62.9	70.6	74.6	63.2	68.4	85.3	64.9	74.3
COVERS	**DIVS. 10 - 14, 25, 28, 41, 43, 44, 46**	100.0	66.1	92.4	100.0	77.7	95.0	100.0	68.5	93.0	100.0	66.2	92.5	100.0	67.0	92.6	100.0	78.2	95.1
21, 22, 23	**FIRE SUPPRESSION, PLUMBING & HVAC**	96.5	60.0	82.2	100.0	48.0	79.6	96.5	49.1	77.8	96.4	51.5	78.8	100.4	51.2	81.1	99.9	49.1	79.9
26, 27, 3370	**ELECTRICAL, COMMUNICATIONS & UTIL.**	90.0	53.9	72.2	93.1	57.0	75.3	94.2	57.5	76.1	97.7	65.2	81.7	99.0	62.3	81.0	100.1	66.7	83.7
MF2018	**WEIGHTED AVERAGE**	90.5	64.1	79.3	94.1	62.6	80.8	92.9	62.8	80.2	93.5	64.4	81.2	93.8	66.6	82.3	94.5	65.4	82.2

DIVISION		ARKANSAS												CALIFORNIA					
		PINE BLUFF			RUSSELLVILLE			TEXARKANA			WEST MEMPHIS			ALHAMBRA			ANAHEIM		
		716			728			718			723			917 - 918			928		
		MAT.	INST.	TOTAL	MAT.	INST.	TOTAL	MAT.	INST.	TOTAL	MAT.	INST.	TOTAL	MAT.	INST.	TOTAL	MAT.	INST.	TOTAL
015433	CONTRACTOR EQUIPMENT		89.7	89.7		89.7	89.7		90.9	90.9		112.7	112.7		95.4	95.4		99.2	99.2
0241, 31 - 34	SITE & INFRASTRUCTURE, DEMOLITION	82.3	85.6	84.6	74.2	85.7	82.2	92.9	87.8	89.3	103.1	102.8	102.9	100.8	105.3	103.9	97.7	105.5	103.1
0310	Concrete Forming & Accessories	76.1	61.9	64.0	83.1	61.7	64.8	81.4	61.5	64.5	90.6	62.2	66.4	114.4	139.5	135.8	102.9	139.7	134.3
0320	Concrete Reinforcing	90.9	69.4	80.5	84.9	69.4	77.4	90.5	69.3	80.2	81.5	63.3	72.7	105.0	133.6	118.9	95.2	133.5	113.8
0330	Cast-in-Place Concrete	78.1	74.6	76.8	76.3	74.5	75.6	85.2	74.4	81.2	83.2	75.9	80.5	83.3	127.8	99.9	86.7	131.0	103.2
03	CONCRETE	79.7	68.3	74.7	76.0	68.1	72.6	77.8	68.1	73.5	83.8	68.8	77.2	94.2	133.1	111.3	96.2	134.3	112.9
04	MASONRY	110.7	60.8	80.0	90.9	60.8	72.4	92.6	60.8	73.0	75.3	60.8	66.4	106.9	139.4	126.9	78.2	138.0	115.0
05	METALS	101.9	76.1	94.3	94.7	75.9	89.1	93.8	75.8	88.5	90.4	89.1	90.0	79.9	113.2	89.7	107.7	114.2	109.6
06	WOOD, PLASTICS & COMPOSITES	86.4	64.0	74.1	92.3	64.0	76.8	93.4	64.0	77.2	100.7	64.4	80.7	99.3	138.0	120.6	97.5	138.3	119.9
07	THERMAL & MOISTURE PROTECTION	97.9	62.0	82.3	103.7	61.9	85.5	98.5	61.9	82.6	108.9	61.9	88.4	104.2	131.8	116.2	108.9	134.9	120.2
08	OPENINGS	104.3	60.8	94.1	100.0	61.3	91.0	109.2	61.3	98.0	103.2	60.1	93.1	87.8	137.4	99.4	103.9	137.6	111.8
0920	Plaster & Gypsum Board	83.1	63.5	69.9	73.7	63.5	66.8	85.7	63.5	70.7	89.2	63.5	71.9	94.5	139.3	124.6	112.7	139.3	130.6
0950, 0980	Ceilings & Acoustic Treatment	86.3	63.5	70.9	88.1	63.5	71.5	89.6	63.5	72.0	90.2	63.5	72.2	111.0	139.3	130.1	109.3	139.3	129.5
0960	Flooring	84.9	70.9	80.9	81.3	70.9	78.4	86.9	43.4	74.7	61.6	70.9	64.2	104.8	121.1	109.4	99.2	121.1	105.3
0970, 0990	Wall Finishes & Painting/Coating	88.8	52.1	67.1	92.2	54.5	69.9	88.8	54.5	68.5	81.7	54.5	65.6	103.9	120.3	113.6	92.6	120.3	109.0
09	FINISHES	79.6	62.6	70.4	76.4	62.9	69.1	81.7	57.2	68.4	75.9	63.2	69.0	101.8	134.1	119.3	99.4	134.2	118.3
COVERS	DIVS. 10 - 14, 25, 28, 41, 43, 44, 46	100.0	77.7	95.0	100.0	66.5	92.5	100.0	66.1	92.5	100.0	67.0	92.6	100.0	119.6	104.4	100.0	120.2	104.5
21, 22, 23	FIRE SUPPRESSION, PLUMBING & HVAC	100.0	51.4	80.9	96.5	50.9	78.6	100.0	54.5	82.1	96.8	64.6	84.2	96.5	130.9	110.0	99.9	130.9	112.1
26, 27, 3370	ELECTRICAL, COMMUNICATIONS & UTIL.	95.9	57.5	77.0	93.1	57.5	75.6	97.6	57.5	77.9	100.5	65.3	83.1	121.0	129.2	125.0	91.5	113.3	102.2
MF2018	WEIGHTED AVERAGE	96.0	63.6	82.3	91.7	63.1	79.6	94.8	63.3	81.5	93.3	69.8	83.3	96.7	128.6	110.2	99.3	126.6	110.9

DIVISION		CALIFORNIA																	
		BAKERSFIELD			BERKELEY			EUREKA			FRESNO			INGLEWOOD			LONG BEACH		
		932 - 933			947			955			936 - 938			903 - 905			906 - 908		
		MAT.	INST.	TOTAL	MAT.	INST.	TOTAL	MAT.	INST.	TOTAL	MAT.	INST.	TOTAL	MAT.	INST.	TOTAL	MAT.	INST.	TOTAL
015433	CONTRACTOR EQUIPMENT		98.7	98.7		100.2	100.2		97.1	97.1		97.3	97.3		96.9	96.9		96.9	96.9
0241, 31 - 34	SITE & INFRASTRUCTURE, DEMOLITION	96.1	106.8	103.5	109.8	106.0	107.2	109.0	102.7	104.7	98.9	103.2	101.8	87.7	102.0	97.6	94.7	102.0	99.7
0310	Concrete Forming & Accessories	105.8	139.2	134.3	113.0	168.1	160.0	111.2	154.1	147.8	102.9	153.3	145.8	107.3	139.9	135.1	102.0	139.9	134.3
0320	Concrete Reinforcing	99.8	133.4	116.0	90.7	135.2	112.2	103.7	136.6	119.6	83.3	134.2	107.9	99.8	133.7	116.2	99.0	133.7	115.8
0330	Cast-in-Place Concrete	88.1	130.2	103.8	110.4	133.5	119.0	94.1	129.6	107.3	95.5	129.2	108.0	81.5	130.2	99.6	92.7	130.2	106.7
03	CONCRETE	91.7	133.7	110.2	105.1	148.2	124.0	106.8	140.8	121.7	95.5	139.9	115.0	90.9	134.2	109.9	100.5	134.2	115.3
04	MASONRY	92.1	137.5	120.0	122.7	154.9	142.5	103.3	153.4	134.1	96.9	145.0	126.5	73.1	139.5	113.9	81.7	139.5	117.2
05	METALS	102.5	112.5	105.4	109.9	116.5	111.9	107.4	116.1	109.9	102.8	115.2	106.4	88.0	114.8	95.9	87.9	114.8	95.8
06	WOOD, PLASTICS & COMPOSITES	96.9	138.3	119.7	104.3	173.4	142.3	111.2	157.3	136.6	101.2	157.3	132.1	99.0	138.4	120.7	92.5	138.4	117.8
07	THERMAL & MOISTURE PROTECTION	108.5	124.3	115.4	112.2	155.7	131.2	112.6	151.2	129.4	98.9	132.9	113.7	107.2	133.4	118.6	107.5	133.4	118.8
08	OPENINGS	92.6	134.3	102.3	91.0	161.7	107.5	103.2	143.2	112.5	94.7	145.1	106.5	87.6	137.6	99.2	87.5	137.6	99.2
0920	Plaster & Gypsum Board	94.4	139.3	124.6	113.4	175.0	154.8	117.4	158.8	145.2	90.9	158.8	136.6	100.3	139.3	126.5	96.2	139.3	125.2
0950, 0980	Ceilings & Acoustic Treatment	100.7	139.3	126.8	101.2	175.0	151.0	113.3	158.8	144.0	98.1	158.8	139.0	114.0	139.3	131.1	114.0	139.3	131.1
0960	Flooring	102.2	121.1	107.5	124.0	143.3	129.4	103.2	143.3	114.4	100.4	121.5	106.3	114.7	121.1	116.5	111.5	121.1	114.2
0970, 0990	Wall Finishes & Painting/Coating	92.5	108.6	102.0	111.6	171.3	146.9	94.1	138.6	120.4	101.3	120.7	112.8	115.1	120.3	118.2	115.1	120.3	118.2
09	FINISHES	96.4	133.0	116.2	110.1	165.4	140.0	104.6	152.0	130.3	94.6	146.3	122.6	108.2	134.3	122.4	107.2	134.3	121.9
COVERS	DIVS. 10 - 14, 25, 28, 41, 43, 44, 46	100.0	117.7	103.9	100.0	134.0	107.6	100.0	130.8	106.9	100.0	130.8	106.9	100.0	120.5	104.6	100.0	120.5	104.6
21, 22, 23	FIRE SUPPRESSION, PLUMBING & HVAC	100.1	128.8	111.4	96.7	169.6	125.3	96.4	130.6	109.9	100.1	130.1	111.9	96.1	131.0	109.8	96.1	131.0	109.8
26, 27, 3370	ELECTRICAL, COMMUNICATIONS & UTIL.	107.6	108.1	107.8	99.9	161.8	130.4	98.4	124.9	111.5	97.0	107.4	102.2	98.5	129.2	113.6	98.3	129.2	113.5
MF2018	WEIGHTED AVERAGE	98.9	124.6	109.8	103.0	151.9	123.7	102.5	134.1	115.9	98.5	129.4	111.5	93.8	128.8	108.6	95.4	128.8	109.5

DIVISION		CALIFORNIA																	
		LOS ANGELES			MARYSVILLE			MODESTO			MOJAVE			OAKLAND			OXNARD		
		900 - 902			959			953			935			946			930		
		MAT.	INST.	TOTAL	MAT.	INST.	TOTAL	MAT.	INST.	TOTAL	MAT.	INST.	TOTAL	MAT.	INST.	TOTAL	MAT.	INST.	TOTAL
015433	CONTRACTOR EQUIPMENT		103.3	103.3		97.1	97.1		97.1	97.1		97.3	97.3		100.2	100.2		96.3	96.3
0241, 31 - 34	SITE & INFRASTRUCTURE, DEMOLITION	93.7	109.2	104.4	105.6	102.6	103.5	100.6	102.6	102.0	91.9	103.4	99.8	115.2	106.0	108.8	99.2	101.6	100.8
0310	Concrete Forming & Accessories	105.4	140.1	135.0	101.6	153.6	145.9	98.1	153.7	145.5	113.6	139.3	135.5	102.3	168.2	158.4	105.5	139.8	134.7
0320	Concrete Reinforcing	100.7	133.7	116.6	103.7	134.1	118.4	107.4	134.1	120.3	100.4	133.4	116.4	92.8	136.9	114.1	98.6	133.5	115.5
0330	Cast-in-Place Concrete	84.1	131.4	101.7	105.2	129.2	114.1	94.2	129.3	107.2	82.6	130.1	100.2	104.8	133.5	115.4	96.1	130.4	108.9
03	CONCRETE	98.3	134.7	114.3	107.1	140.0	121.6	98.2	140.1	116.6	87.4	133.8	107.8	106.3	148.5	124.8	95.5	134.1	112.5
04	MASONRY	88.6	139.5	119.9	104.2	142.9	128.0	101.5	142.9	126.9	94.9	137.4	121.0	131.5	154.9	145.9	97.9	136.8	121.8
05	METALS	94.3	115.0	100.4	106.9	114.7	109.2	103.6	114.8	106.9	100.0	113.4	104.0	105.2	117.2	108.7	98.0	113.9	102.7
06	WOOD, PLASTICS & COMPOSITES	105.6	138.6	123.7	98.3	157.3	130.8	93.9	157.3	128.8	100.8	138.4	121.5	92.8	173.4	137.2	95.6	138.4	119.1
07	THERMAL & MOISTURE PROTECTION	103.4	133.9	116.7	112.0	138.9	123.7	111.6	139.5	123.7	105.6	123.7	113.5	110.8	155.7	130.4	108.6	133.3	119.3
08	OPENINGS	98.1	136.9	107.2	102.5	145.6	112.6	101.3	147.4	112.1	89.8	134.3	100.2	91.1	162.1	107.7	92.4	137.6	102.9
0920	Plaster & Gypsum Board	95.6	139.3	125.0	110.0	158.8	142.8	112.7	158.8	143.7	100.5	139.3	126.6	107.6	175.0	152.9	94.5	139.3	124.7
0950, 0980	Ceilings & Acoustic Treatment	121.4	139.3	133.5	112.5	158.8	143.7	109.3	158.8	142.6	99.4	139.3	126.3	103.8	175.0	151.8	100.8	139.3	126.8
0960	Flooring	112.5	122.2	115.3	98.8	119.5	104.6	99.2	133.0	108.7	101.3	121.1	106.8	117.0	143.3	124.4	93.8	121.1	101.5
0970, 0990	Wall Finishes & Painting/Coating	111.6	120.3	116.7	94.1	134.1	117.8	94.1	138.6	120.4	87.2	109.6	100.5	111.6	171.3	146.9	87.2	114.8	103.6
09	FINISHES	109.4	134.7	123.1	101.7	147.4	126.4	101.0	150.2	127.7	95.0	133.2	115.6	108.1	165.4	139.1	92.5	133.7	114.8
COVERS	DIVS. 10 - 14, 25, 28, 41, 43, 44, 46	100.0	121.0	104.7	100.0	130.8	106.9	100.0	130.8	106.9	100.0	117.9	104.0	100.0	134.0	107.6	100.0	120.4	104.5
21, 22, 23	FIRE SUPPRESSION, PLUMBING & HVAC	99.9	131.0	112.1	96.4	130.0	109.6	99.9	131.7	112.4	96.5	128.8	109.2	100.2	169.6	127.5	100.0	131.0	112.2
26, 27, 3370	ELECTRICAL, COMMUNICATIONS & UTIL.	96.9	131.0	113.7	95.1	120.7	107.7	97.5	109.9	103.6	95.5	108.1	101.7	99.1	161.8	130.0	101.4	117.9	109.5
MF2018	WEIGHTED AVERAGE	98.5	129.8	111.7	101.7	131.3	114.2	100.6	130.6	113.3	95.4	124.4	107.7	103.4	152.0	123.9	98.0	126.7	110.1

	DIVISION	CALIFORNIA																	
		PALM SPRINGS			PALO ALTO			PASADENA			REDDING			RICHMOND			RIVERSIDE		
		922			943			910 - 912			960			948			925		
		MAT.	INST.	TOTAL	MAT.	INST.	TOTAL	MAT.	INST.	TOTAL	MAT.	INST.	TOTAL	MAT.	INST.	TOTAL	MAT.	INST.	TOTAL
015433	CONTRACTOR EQUIPMENT		98.1	98.1		100.2	100.2		95.4	95.4		97.1	97.1		100.2	100.2		98.1	98.1
0241, 31 - 34	SITE & INFRASTRUCTURE, DEMOLITION	89.3	103.6	99.2	106.1	106.0	106.0	97.5	105.3	102.9	122.2	102.6	108.7	114.3	106.0	108.5	96.2	103.6	101.3
0310	Concrete Forming & Accessories	99.8	136.1	130.7	100.5	168.2	158.2	103.4	139.6	134.2	105.0	153.6	146.5	115.5	167.8	160.1	103.5	139.7	134.4
0320	Concrete Reinforcing	109.0	133.5	120.9	90.7	136.7	113.0	106.0	133.6	119.3	137.8	134.1	136.0	90.7	136.7	112.9	105.9	133.5	119.3
0330	Cast-in-Place Concrete	82.7	131.0	100.6	93.4	133.5	108.4	79.1	127.8	97.2	107.2	129.3	115.4	107.4	133.3	117.1	89.9	131.0	105.2
03	CONCRETE	90.5	132.7	109.0	95.3	148.5	118.7	89.9	133.1	108.9	114.0	140.1	125.4	108.0	148.2	125.7	96.4	134.3	113.1
04	MASONRY	76.5	137.7	114.1	105.1	151.5	133.6	93.3	139.4	121.6	131.9	142.9	138.7	122.5	151.5	140.3	77.5	137.7	114.5
05	METALS	108.3	114.1	110.0	102.7	116.9	106.9	80.0	113.2	89.7	103.9	114.7	107.1	102.8	116.6	106.8	107.7	114.2	109.6
06	WOOD, PLASTICS & COMPOSITES	92.1	133.5	114.9	90.4	173.4	136.1	85.4	138.0	114.4	107.8	157.3	135.0	107.5	173.4	143.8	97.5	138.3	119.9
07	THERMAL & MOISTURE PROTECTION	108.1	134.3	119.5	110.0	155.0	129.6	103.8	131.8	116.0	128.2	138.9	132.9	110.6	154.8	129.8	109.0	134.8	120.2
08	OPENINGS	99.9	134.9	108.1	91.1	160.2	107.2	87.8	137.4	99.4	115.9	145.6	122.8	91.1	160.2	107.2	102.6	137.6	110.8
0920	Plaster & Gypsum Board	107.2	134.4	125.5	106.0	175.0	152.4	88.8	139.3	122.8	112.6	158.8	143.7	114.5	175.0	155.2	111.9	139.3	130.4
0950, 0980	Ceilings & Acoustic Treatment	106.0	134.4	125.1	102.0	175.0	151.2	111.0	139.3	130.1	148.8	158.8	155.5	102.0	175.0	151.2	114.2	139.3	131.1
0960	Flooring	101.5	121.1	107.0	115.6	143.3	123.4	98.5	121.1	104.9	87.8	128.6	99.3	126.5	143.3	131.3	102.9	121.1	108.0
0970, 0990	Wall Finishes & Painting/Coating	91.1	120.3	108.4	111.6	171.3	146.9	103.9	120.3	113.6	105.2	134.1	122.3	111.6	171.3	146.9	91.1	120.3	108.4
09	FINISHES	98.0	131.4	116.1	106.5	165.4	138.4	99.0	134.1	118.0	107.2	149.0	129.8	111.6	165.4	140.7	101.0	134.2	119.0
COVERS	DIVS. 10 - 14, 25, 28, 41, 43, 44, 46	100.0	119.6	104.4	100.0	134.0	107.6	100.0	119.6	104.4	100.0	130.8	106.9	100.0	133.8	107.5	100.0	119.3	104.3
21, 22, 23	FIRE SUPPRESSION, PLUMBING & HVAC	96.4	130.9	109.9	96.7	170.6	125.7	96.5	130.9	110.0	100.3	130.1	112.0	96.7	164.0	123.1	99.9	130.9	112.1
26, 27, 3370	ELECTRICAL, COMMUNICATIONS & UTIL.	94.6	110.8	102.6	99.0	176.6	137.2	117.5	129.2	123.2	101.3	120.8	110.9	99.6	133.2	116.2	91.3	114.4	102.7
MF2018	WEIGHTED AVERAGE	97.4	125.3	109.2	99.2	153.8	122.3	94.8	128.6	109.1	107.6	131.5	117.7	102.4	146.2	120.9	99.3	126.6	110.8

	DIVISION	CALIFORNIA																	
		SACRAMENTO			SALINAS			SAN BERNARDINO			SAN DIEGO			SAN FRANCISCO			SAN JOSE		
		942, 956 - 958			939			923 - 924			919 - 921			940 - 941			951		
		MAT.	INST.	TOTAL	MAT.	INST.	TOTAL	MAT.	INST.	TOTAL	MAT.	INST.	TOTAL	MAT.	INST.	TOTAL	MAT.	INST.	TOTAL
015433	CONTRACTOR EQUIPMENT		99.0	99.0		97.3	97.3		98.1	98.1		102.2	102.2		109.8	109.8		98.7	98.7
0241, 31 - 34	SITE & INFRASTRUCTURE, DEMOLITION	95.6	111.3	106.4	112.3	103.2	106.0	76.1	103.6	95.1	105.7	107.8	107.2	116.4	113.1	114.1	131.6	97.9	108.3
0310	Concrete Forming & Accessories	100.5	156.0	147.8	108.9	156.6	149.6	107.1	139.7	134.9	103.8	129.1	125.4	106.1	168.3	159.1	103.9	167.9	158.5
0320	Concrete Reinforcing	86.0	134.4	109.4	99.1	134.6	116.3	105.9	133.5	119.3	102.6	133.4	117.5	105.6	132.1	118.4	94.7	135.0	114.2
0330	Cast-in-Place Concrete	88.6	130.2	104.1	95.0	129.6	107.8	62.1	131.0	87.8	90.5	124.5	103.2	116.9	133.3	123.0	109.4	132.4	118.0
03	CONCRETE	97.9	141.1	116.9	105.8	141.6	121.5	71.1	134.3	98.9	100.2	127.3	112.1	117.8	148.1	131.1	103.9	148.1	123.3
04	MASONRY	106.1	145.9	130.6	95.1	148.4	127.9	83.6	137.7	116.8	86.2	133.4	115.2	140.9	154.6	149.3	131.5	151.5	143.8
05	METALS	100.8	109.8	103.4	102.7	115.8	106.5	107.7	114.2	109.6	94.8	114.1	100.5	110.7	122.4	114.2	101.8	121.6	107.6
06	WOOD, PLASTICS & COMPOSITES	87.1	160.2	127.3	100.6	160.2	133.4	101.2	138.3	121.6	95.0	126.1	112.1	95.1	173.4	138.2	106.0	173.2	143.0
07	THERMAL & MOISTURE PROTECTION	122.2	142.5	131.0	106.2	145.4	123.3	107.1	134.8	119.2	108.2	121.5	114.0	116.5	155.7	133.5	108.0	155.1	128.5
08	OPENINGS	103.9	149.0	114.4	93.5	154.4	107.7	100.0	137.6	108.7	99.8	127.8	106.3	99.5	159.0	113.3	92.6	161.5	108.7
0920	Plaster & Gypsum Board	103.4	161.5	142.5	95.4	161.8	140.1	113.3	139.3	130.8	94.4	126.7	116.1	105.8	175.0	152.4	108.3	175.0	153.2
0950, 0980	Ceilings & Acoustic Treatment	102.0	161.5	142.1	99.4	161.8	141.5	109.3	139.3	129.5	121.1	126.7	124.8	111.7	175.0	154.4	109.3	175.0	153.6
0960	Flooring	115.8	128.6	119.4	95.9	143.3	109.2	105.0	121.1	109.5	108.9	121.1	112.3	117.5	143.3	124.7	92.4	143.3	106.7
0970, 0990	Wall Finishes & Painting/Coating	107.9	134.1	123.4	88.1	171.3	137.3	91.1	120.3	108.4	101.5	118.0	111.2	109.6	174.8	148.2	94.4	171.3	139.9
09	FINISHES	105.3	150.7	129.9	94.6	157.3	128.5	99.4	134.2	118.3	106.6	126.1	117.2	109.9	165.8	140.1	99.9	165.2	135.3
COVERS	DIVS. 10 - 14, 25, 28, 41, 43, 44, 46	100.0	131.7	107.1	100.0	131.2	107.0	100.0	117.7	103.9	100.0	115.7	103.5	100.0	128.4	106.3	100.0	133.4	107.5
21, 22, 23	FIRE SUPPRESSION, PLUMBING & HVAC	100.1	131.4	112.4	96.5	137.6	112.7	96.4	130.9	110.0	100.0	129.4	111.5	100.1	182.8	132.6	99.9	170.5	127.7
26, 27, 3370	ELECTRICAL, COMMUNICATIONS & UTIL.	94.8	120.8	107.6	96.5	132.1	114.0	94.6	112.6	103.5	103.8	104.3	104.0	99.2	182.8	140.4	100.7	176.6	138.1
MF2018	WEIGHTED AVERAGE	100.8	133.0	114.4	99.1	137.2	115.2	95.0	126.3	108.2	99.7	121.7	109.0	107.4	158.4	129.0	102.6	153.5	124.2

	DIVISION	CALIFORNIA																	
		SAN LUIS OBISPO			SAN MATEO			SAN RAFAEL			SANTA ANA			SANTA BARBARA			SANTA CRUZ		
		934			944			949			926 - 927			931			950		
		MAT.	INST.	TOTAL	MAT.	INST.	TOTAL	MAT.	INST.	TOTAL	MAT.	INST.	TOTAL	MAT.	INST.	TOTAL	MAT.	INST.	TOTAL
015433	CONTRACTOR EQUIPMENT		97.3	97.3		100.2	100.2		99.4	99.4		98.1	98.1		97.3	97.3		98.7	98.7
0241, 31 - 34	SITE & INFRASTRUCTURE, DEMOLITION	104.4	103.4	103.7	112.1	106.0	107.9	107.6	111.2	110.1	87.8	103.6	98.7	99.1	103.4	102.1	131.3	97.7	108.1
0310	Concrete Forming & Accessories	115.3	139.7	136.1	106.3	168.2	159.0	111.5	167.8	159.5	107.4	139.7	134.9	106.1	139.7	134.8	103.9	156.9	149.0
0320	Concrete Reinforcing	100.4	133.4	116.4	90.7	136.9	113.0	91.3	135.1	112.5	109.6	133.5	121.2	98.6	133.4	115.4	117.1	134.6	125.6
0330	Cast-in-Place Concrete	102.1	130.2	112.6	104.0	133.5	115.0	121.1	132.3	125.2	79.3	131.0	98.6	95.8	130.2	108.6	108.7	131.4	117.1
03	CONCRETE	103.3	134.0	116.8	104.6	148.5	123.9	128.2	147.4	136.6	88.0	134.3	108.3	95.4	134.0	112.4	106.4	142.6	122.3
04	MASONRY	96.5	135.9	120.7	122.2	154.6	142.1	99.1	154.6	133.2	73.4	138.0	113.1	95.2	135.9	120.2	135.3	148.6	143.4
05	METALS	100.7	113.8	104.6	102.6	117.1	106.9	103.7	112.4	106.3	107.8	114.2	109.7	98.5	113.8	103.0	109.3	120.2	112.5
06	WOOD, PLASTICS & COMPOSITES	103.0	138.4	122.4	97.9	173.4	139.5	95.0	173.2	138.0	103.0	138.3	122.4	95.6	138.4	119.1	106.0	160.4	135.9
07	THERMAL & MOISTURE PROTECTION	106.4	132.6	117.8	110.4	156.4	130.4	115.0	155.4	132.5	108.4	134.9	119.9	105.8	132.6	117.5	107.6	147.9	125.1
08	OPENINGS	91.7	134.3	101.6	91.1	162.1	107.6	101.5	159.7	115.1	99.2	137.6	108.2	93.2	137.6	103.6	93.9	154.4	108.0
0920	Plaster & Gypsum Board	101.1	139.3	126.8	111.7	175.0	154.3	113.2	175.0	154.8	114.6	139.3	131.2	94.5	139.3	124.7	116.8	161.8	147.1
0950, 0980	Ceilings & Acoustic Treatment	99.4	139.3	126.3	102.0	175.0	151.2	109.3	175.0	153.6	109.3	139.3	129.5	100.8	139.3	126.8	110.9	161.8	145.2
0960	Flooring	102.0	121.1	107.4	119.5	143.3	126.2	132.6	143.3	135.6	105.6	121.1	110.0	95.0	121.1	102.4	96.6	143.3	109.7
0970, 0990	Wall Finishes & Painting/Coating	87.2	114.8	103.6	111.6	171.3	146.9	107.8	165.5	141.9	91.1	118.0	107.0	87.2	114.8	103.6	94.5	171.3	139.9
09	FINISHES	96.3	133.7	116.5	108.9	165.4	139.5	112.0	164.6	140.5	100.9	134.0	118.8	93.0	133.7	115.0	102.6	157.4	132.3
COVERS	DIVS. 10 - 14, 25, 28, 41, 43, 44, 46	100.0	126.5	105.9	100.0	134.0	107.6	100.0	133.3	107.4	100.0	120.2	104.5	100.0	119.5	104.3	100.0	131.5	107.0
21, 22, 23	FIRE SUPPRESSION, PLUMBING & HVAC	96.5	131.0	110.0	96.7	165.7	123.8	96.7	182.8	130.5	96.4	130.9	110.0	100.0	131.0	112.2	99.9	137.9	114.8
26, 27, 3370	ELECTRICAL, COMMUNICATIONS & UTIL.	95.5	110.8	103.1	99.0	168.0	133.0	96.2	124.9	110.3	94.6	111.9	103.1	94.5	113.7	103.9	99.8	132.1	115.7
MF2018	WEIGHTED AVERAGE	98.2	125.8	109.9	101.5	152.0	122.9	104.5	149.1	123.4	97.0	126.3	109.4	97.2	126.1	109.4	104.6	137.5	118.5

	DIVISION	CALIFORNIA															COLORADO		
		SANTA ROSA			STOCKTON			SUSANVILLE			VALLEJO			VAN NUYS			ALAMOSA		
		954			952			961			945			913 - 916			811		
		MAT.	INST.	TOTAL	MAT.	INST.	TOTAL	MAT.	INST.	TOTAL	MAT.	INST.	TOTAL	MAT.	INST.	TOTAL	MAT.	INST.	TOTAL
015433	CONTRACTOR EQUIPMENT		97.6	97.6		97.1	97.1		97.1	97.1		99.4	99.4		95.4	95.4		90.2	90.2
0241, 31 - 34	SITE & INFRASTRUCTURE, DEMOLITION	101.4	102.6	102.2	100.3	102.6	101.9	129.5	102.6	110.9	96.3	111.1	106.5	115.7	105.3	108.5	141.3	84.4	102.0
0310	Concrete Forming & Accessories	100.7	167.1	157.3	101.9	155.5	147.6	106.2	153.9	146.8	102.3	166.8	157.3	109.8	139.5	135.2	103.8	66.0	71.5
0320	Concrete Reinforcing	104.6	135.1	119.3	107.4	134.1	120.3	137.8	134.1	136.0	92.5	135.0	113.0	106.0	133.6	119.3	114.0	67.9	91.7
0330	Cast-in-Place Concrete	103.3	130.8	113.5	91.7	129.2	105.7	97.5	129.3	109.3	96.4	131.4	109.5	83.4	127.8	99.9	100.1	73.1	90.1
03	CONCRETE	107.1	146.9	124.6	97.3	140.9	116.5	117.0	140.2	127.2	102.4	146.6	121.8	104.4	133.1	117.0	112.5	69.4	93.6
04	MASONRY	101.8	153.5	133.6	101.4	142.9	126.9	130.4	142.9	138.1	77.4	153.5	124.1	106.9	139.4	126.9	130.1	59.5	86.7
05	METALS	108.1	117.7	110.9	103.9	114.7	107.0	102.9	114.7	106.4	103.7	111.7	106.0	79.2	113.2	89.2	105.2	78.5	97.3
06	WOOD, PLASTICS & COMPOSITES	93.6	173.0	137.3	99.5	159.9	132.8	109.6	157.6	136.0	84.8	173.2	133.4	94.1	138.0	118.3	98.5	67.4	81.4
07	THERMAL & MOISTURE PROTECTION	108.9	154.1	128.6	112.1	138.8	123.7	130.2	138.9	134.0	113.1	154.1	130.9	105.0	131.8	116.6	111.9	67.2	92.5
08	OPENINGS	100.8	161.4	114.9	101.3	147.0	112.0	116.8	145.8	123.6	103.2	159.7	116.3	87.7	137.4	99.3	96.3	68.5	89.8
0920	Plaster & Gypsum Board	109.2	175.0	153.5	112.7	161.5	145.5	113.1	159.1	144.1	107.5	175.0	152.9	92.8	139.3	124.1	80.8	66.5	71.2
0950, 0980	Ceilings & Acoustic Treatment	109.3	175.0	153.6	117.5	161.5	147.1	139.9	159.1	152.9	111.1	175.0	154.2	107.8	139.3	129.1	102.7	66.5	78.3
0960	Flooring	102.1	136.5	111.7	99.2	133.0	108.7	88.3	128.6	99.6	126.0	143.3	130.9	101.9	121.1	107.3	107.8	70.1	97.2
0970, 0990	Wall Finishes & Painting/Coating	91.1	165.5	135.2	94.1	134.1	117.8	105.2	134.1	122.3	108.7	165.5	142.3	103.9	120.3	113.6	101.7	76.9	87.0
09	FINISHES	100.0	163.0	134.1	102.6	151.3	129.0	106.7	149.2	129.7	108.4	164.4	138.7	101.2	134.1	119.0	99.8	67.5	82.3
COVERS	DIVS. 10 - 14, 25, 28, 41, 43, 44, 46	100.0	132.4	107.2	100.0	128.0	106.2	100.0	130.9	106.9	100.0	132.9	107.3	100.0	119.6	104.4	100.0	83.0	96.2
21, 22, 23	FIRE SUPPRESSION, PLUMBING & HVAC	96.4	182.2	130.1	99.9	131.7	112.4	96.8	130.1	109.9	100.2	147.2	118.7	96.5	130.9	110.0	96.5	72.4	87.0
26, 27, 3370	ELECTRICAL, COMMUNICATIONS & UTIL.	94.9	124.9	109.7	97.5	114.3	105.8	101.7	120.8	111.1	92.4	127.5	109.7	117.5	129.2	123.2	96.3	63.2	80.0
MF2018	WEIGHTED AVERAGE	101.2	148.4	121.2	100.7	131.3	113.7	107.3	131.6	117.5	100.2	141.6	117.7	97.8	128.6	110.8	103.6	70.3	89.5

	DIVISION	COLORADO																	
		BOULDER			COLORADO SPRINGS			DENVER			DURANGO			FORT COLLINS			FORT MORGAN		
		803			808 - 809			800 - 802			813			805			807		
		MAT.	INST.	TOTAL	MAT.	INST.	TOTAL	MAT.	INST.	TOTAL	MAT.	INST.	TOTAL	MAT.	INST.	TOTAL	MAT.	INST.	TOTAL
015433	CONTRACTOR EQUIPMENT		92.2	92.2		90.2	90.2		99.6	99.6		90.2	90.2		92.2	92.2		92.2	92.2
0241, 31 - 34	SITE & INFRASTRUCTURE, DEMOLITION	97.1	91.4	93.2	99.2	86.7	90.6	104.1	102.5	103.0	134.3	84.4	99.9	109.6	91.2	96.9	99.3	90.9	93.5
0310	Concrete Forming & Accessories	107.0	77.4	81.8	97.3	63.5	68.5	106.1	66.4	72.2	109.9	65.7	72.2	104.5	65.8	71.5	107.6	65.7	71.9
0320	Concrete Reinforcing	108.7	68.1	89.0	107.9	67.9	88.6	107.9	70.1	89.6	114.0	67.9	91.7	108.8	68.0	89.1	108.9	68.0	89.2
0330	Cast-in-Place Concrete	113.8	73.9	98.9	116.8	73.4	100.7	121.8	73.6	103.9	115.1	73.1	99.5	128.5	72.5	107.6	111.6	72.5	97.1
03	CONCRETE	105.2	74.9	91.9	108.7	68.4	91.0	111.4	70.2	93.3	114.2	69.3	94.5	116.3	69.2	95.6	103.7	69.1	88.5
04	MASONRY	101.6	65.4	79.4	103.7	62.3	78.3	104.9	64.1	79.9	117.5	60.6	82.5	118.9	63.1	84.6	116.7	63.1	83.8
05	METALS	94.4	77.9	89.6	97.5	77.7	91.7	99.9	79.0	93.8	105.2	78.4	97.3	95.7	77.7	90.4	94.2	77.8	89.4
06	WOOD, PLASTICS & COMPOSITES	107.1	81.4	93.0	96.6	63.6	78.5	108.4	67.1	85.7	108.4	67.4	85.8	104.6	67.0	83.9	107.1	67.0	85.0
07	THERMAL & MOISTURE PROTECTION	108.7	73.2	93.2	109.4	69.7	92.1	107.1	71.7	91.7	111.9	67.5	92.6	109.0	70.6	92.3	108.6	69.6	91.6
08	OPENINGS	95.4	76.2	90.9	99.5	66.4	91.8	102.5	68.8	94.6	103.1	68.5	95.0	95.3	68.3	89.0	95.3	68.2	89.0
0920	Plaster & Gypsum Board	116.3	81.3	92.8	100.3	62.9	75.2	111.7	66.6	81.4	95.3	66.5	75.9	110.3	66.6	80.9	116.3	66.5	82.8
0950, 0980	Ceilings & Acoustic Treatment	91.9	81.3	84.8	100.0	62.9	75.0	104.4	66.6	78.9	102.7	66.5	78.3	91.9	66.6	74.8	91.9	66.5	74.8
0960	Flooring	111.2	76.7	101.5	102.2	69.0	92.9	108.4	76.7	99.5	113.1	69.0	100.7	107.3	76.7	98.7	111.7	76.7	101.8
0970, 0990	Wall Finishes & Painting/Coating	97.1	76.9	85.1	96.8	76.9	85.0	103.4	76.9	87.7	101.7	76.9	87.0	97.1	76.9	85.1	97.1	76.9	85.1
09	FINISHES	100.6	77.7	88.2	97.4	65.3	80.0	102.5	68.9	84.3	102.4	67.4	83.5	99.2	68.6	82.6	100.7	68.5	83.3
COVERS	DIVS. 10 - 14, 25, 28, 41, 43, 44, 46	100.0	84.5	96.5	100.0	82.2	96.0	100.0	86.3	96.9	100.0	82.9	96.2	100.0	82.1	96.0	100.0	82.1	96.0
21, 22, 23	FIRE SUPPRESSION, PLUMBING & HVAC	96.5	74.8	87.9	100.1	72.2	89.1	99.9	73.8	89.7	96.5	58.7	81.6	100.0	71.7	88.9	96.5	73.5	87.4
26, 27, 3370	ELECTRICAL, COMMUNICATIONS & UTIL.	99.6	80.6	90.3	103.2	73.1	88.4	105.0	80.4	92.9	95.8	50.3	73.4	99.6	80.6	90.2	100.0	78.1	89.2
MF2018	WEIGHTED AVERAGE	98.7	77.0	89.6	101.2	71.5	88.6	103.0	75.4	91.3	103.9	65.6	87.8	102.2	73.5	90.1	99.3	73.5	88.4

	DIVISION	COLORADO																	
		GLENWOOD SPRINGS			GOLDEN			GRAND JUNCTION			GREELEY			MONTROSE			PUEBLO		
		816			804			815			806			814			810		
		MAT.	INST.	TOTAL	MAT.	INST.	TOTAL	MAT.	INST.	TOTAL	MAT.	INST.	TOTAL	MAT.	INST.	TOTAL	MAT.	INST.	TOTAL
015433	CONTRACTOR EQUIPMENT		93.3	93.3		92.2	92.2		93.3	93.3		92.2	92.2		91.7	91.7		90.2	90.2
0241, 31 - 34	SITE & INFRASTRUCTURE, DEMOLITION	150.9	91.9	110.2	110.4	91.4	97.3	133.7	92.1	105.0	95.9	91.2	92.7	144.0	87.9	105.3	125.2	84.5	97.1
0310	Concrete Forming & Accessories	100.6	65.5	70.7	99.7	66.7	71.5	108.7	66.6	72.8	102.4	65.8	71.2	100.1	65.5	70.6	106.1	63.7	70.0
0320	Concrete Reinforcing	112.8	68.0	91.2	108.9	68.0	89.2	113.2	67.8	91.2	108.7	68.0	89.0	112.7	67.8	91.0	109.0	67.9	89.1
0330	Cast-in-Place Concrete	100.1	72.1	89.6	111.7	73.9	97.6	110.8	73.5	96.9	107.3	72.5	94.3	100.1	72.6	89.8	99.4	73.7	89.8
03	CONCRETE	118.0	68.9	96.4	114.1	70.0	94.7	110.7	69.9	92.8	100.3	69.1	86.6	108.5	69.0	91.1	100.9	68.7	86.7
04	MASONRY	103.1	63.0	78.5	119.7	65.4	86.3	137.8	61.9	91.2	111.9	62.7	81.7	110.0	59.6	79.0	99.5	60.3	75.4
05	METALS	104.8	78.2	97.0	94.3	77.8	89.5	106.5	78.2	98.2	95.7	77.7	90.4	103.9	78.1	96.3	108.3	78.7	99.6
06	WOOD, PLASTICS & COMPOSITES	93.6	67.1	79.0	98.9	67.0	81.3	106.1	67.2	84.7	101.7	67.0	82.6	94.6	67.2	79.6	101.2	63.9	80.7
07	THERMAL & MOISTURE PROTECTION	111.8	68.2	92.8	109.8	71.6	93.2	110.9	69.3	92.8	108.2	70.4	91.8	112.0	67.2	92.5	110.4	67.8	91.8
08	OPENINGS	102.1	68.3	94.2	95.4	68.2	89.0	102.8	68.4	94.7	95.3	68.3	89.0	103.2	68.4	95.1	98.0	66.6	90.7
0920	Plaster & Gypsum Board	121.2	66.5	84.4	107.5	66.5	79.9	135.4	66.6	89.1	108.7	66.6	80.4	79.9	66.5	70.9	84.7	62.9	70.1
0950, 0980	Ceilings & Acoustic Treatment	101.9	66.5	78.0	91.9	66.5	74.8	101.9	66.6	78.1	91.9	66.6	74.8	102.7	66.5	78.3	110.0	62.9	78.3
0960	Flooring	106.9	76.7	98.4	104.9	76.7	97.0	112.4	70.1	100.5	106.2	73.1	96.9	110.4	73.1	100.0	109.1	73.1	99.0
0970, 0990	Wall Finishes & Painting/Coating	101.7	76.9	87.0	97.1	76.9	85.1	101.7	76.9	87.0	97.1	76.9	85.1	101.7	76.9	87.0	101.7	76.9	87.0
09	FINISHES	105.2	68.7	85.4	98.9	69.1	82.8	106.8	67.9	85.8	97.9	67.9	81.6	100.5	68.0	82.9	100.2	66.3	81.9
COVERS	DIVS. 10 - 14, 25, 28, 41, 43, 44, 46	100.0	82.3	96.1	100.0	82.9	96.2	100.0	83.1	96.2	100.0	82.1	96.0	100.0	82.6	96.1	100.0	82.9	96.2
21, 22, 23	FIRE SUPPRESSION, PLUMBING & HVAC	96.5	58.7	81.6	96.5	74.8	87.9	100.0	74.8	90.1	100.0	71.7	88.9	96.5	58.7	81.6	100.0	72.2	89.1
26, 27, 3370	ELECTRICAL, COMMUNICATIONS & UTIL.	93.4	52.3	73.1	100.0	80.6	90.4	95.5	54.6	75.3	99.6	80.6	90.2	95.5	50.3	73.2	96.4	63.3	80.0
MF2018	WEIGHTED AVERAGE	103.9	66.8	88.2	100.9	74.7	89.8	105.7	70.5	90.8	99.3	73.4	88.4	102.6	65.8	87.1	101.8	70.0	88.3

DIVISION		COLORADO			CONNECTICUT															
		SALIDA			BRIDGEPORT			BRISTOL			HARTFORD			MERIDEN			NEW BRITAIN		
		812			066			060			061			064			060		
		MAT.	INST.	TOTAL	MAT.	INST.	TOTAL	MAT.	INST.	TOTAL	MAT.	INST.	TOTAL	MAT.	INST.	TOTAL	MAT.	INST.	TOTAL
015433	CONTRACTOR EQUIPMENT		91.7	91.7		95.7	95.7		95.7	95.7		99.0	99.0		96.1	96.1		95.7	95.7
0241, 31 - 34	SITE & INFRASTRUCTURE, DEMOLITION	134.0	88.2	102.4	106.1	97.8	100.3	105.1	97.7	100.0	101.6	102.4	102.2	102.8	98.5	99.8	105.3	97.7	100.1
0310	Concrete Forming & Accessories	108.6	65.9	72.2	103.4	116.7	114.7	103.4	116.6	114.6	103.5	116.8	114.8	103.1	116.6	114.6	103.9	116.6	114.7
0320	Concrete Reinforcing	112.4	67.9	90.9	116.3	144.5	130.0	116.3	144.5	130.0	111.6	144.6	127.5	116.3	144.5	130.0	116.3	144.5	130.0
0330	Cast-in-Place Concrete	114.7	72.8	99.1	108.5	127.5	115.6	101.6	127.5	111.2	106.5	128.5	114.7	97.8	127.5	108.8	103.3	127.5	112.3
03	CONCRETE	109.1	69.3	91.6	104.1	124.5	113.0	100.9	124.4	111.2	101.1	124.8	111.5	99.1	124.4	110.2	101.7	124.4	111.7
04	MASONRY	138.6	60.6	90.7	109.4	130.9	122.6	100.8	130.9	119.3	100.5	131.0	119.2	100.4	130.9	119.2	102.5	130.9	120.0
05	METALS	103.6	78.6	96.2	98.1	116.6	103.5	98.1	116.5	103.5	103.1	116.0	106.9	95.4	116.5	101.6	94.5	116.5	100.9
06	WOOD, PLASTICS & COMPOSITES	102.4	67.2	83.0	105.0	114.2	110.1	105.0	114.2	110.1	96.2	114.4	106.2	105.0	114.2	110.1	105.0	114.2	110.1
07	THERMAL & MOISTURE PROTECTION	110.8	67.5	92.0	99.7	123.5	110.0	99.8	120.6	108.9	104.4	121.1	111.7	99.8	120.6	108.9	99.8	120.7	108.9
08	OPENINGS	96.3	68.4	89.8	96.4	120.6	102.0	96.4	120.6	102.0	98.1	120.7	103.4	98.5	120.6	103.7	96.4	120.6	102.0
0920	Plaster & Gypsum Board	80.2	66.5	71.0	115.7	114.3	114.8	115.7	114.3	114.8	102.5	114.3	110.5	117.3	114.3	115.3	115.7	114.3	114.8
0950, 0980	Ceilings & Acoustic Treatment	102.7	66.5	78.3	102.0	114.3	110.3	102.0	114.3	110.3	98.8	114.3	109.3	107.5	114.3	112.1	102.0	114.3	110.3
0960	Flooring	115.9	69.9	103.0	92.6	127.0	102.3	92.6	124.5	101.6	96.9	127.0	105.4	92.6	124.5	101.6	92.6	124.5	101.6
0970, 0990	Wall Finishes & Painting/Coating	101.7	76.9	87.0	92.0	125.5	111.8	92.0	129.1	113.9	98.1	129.1	116.4	92.0	129.1	113.9	92.0	129.1	113.9
09	FINISHES	101.1	67.4	82.8	94.7	118.9	107.8	94.7	118.8	107.8	96.2	119.4	108.8	96.0	118.8	108.4	94.7	118.8	107.8
COVERS	DIVS. 10 - 14, 25, 28, 41, 43, 44, 46	100.0	82.7	96.1	100.0	113.9	103.1	100.0	113.9	103.1	100.0	114.4	103.2	100.0	114.0	103.1	100.0	113.9	103.1
21, 22, 23	FIRE SUPPRESSION, PLUMBING & HVAC	96.5	73.6	87.5	100.1	118.6	107.4	100.1	118.6	107.4	100.0	118.6	107.3	96.6	118.6	105.2	100.1	118.6	107.4
26, 27, 3370	ELECTRICAL, COMMUNICATIONS & UTIL.	95.7	63.2	79.7	93.0	105.0	98.9	93.0	103.7	98.3	92.7	109.4	100.9	92.9	105.8	99.3	93.0	103.7	98.3
MF2018	WEIGHTED AVERAGE	103.1	70.9	89.5	99.3	117.0	106.8	98.5	116.7	106.2	99.6	118.0	107.3	97.2	117.0	105.6	98.1	116.7	106.0

DIVISION		CONNECTICUT																	
		NEW HAVEN			NEW LONDON			NORWALK			STAMFORD			WATERBURY			WILLIMANTIC		
		065			063			068			069			067			062		
		MAT.	INST.	TOTAL	MAT.	INST.	TOTAL	MAT.	INST.	TOTAL	MAT.	INST.	TOTAL	MAT.	INST.	TOTAL	MAT.	INST.	TOTAL
015433	CONTRACTOR EQUIPMENT		96.1	96.1		96.1	96.1		95.7	95.7		95.7	95.7		95.7	95.7		95.7	95.7
0241, 31 - 34	SITE & INFRASTRUCTURE, DEMOLITION	105.3	98.4	100.5	97.2	98.1	97.8	105.8	97.6	100.2	106.5	97.7	100.4	105.8	97.8	100.2	105.8	97.7	100.2
0310	Concrete Forming & Accessories	103.2	116.5	114.6	103.1	116.5	114.5	103.4	116.6	114.7	103.4	116.9	114.9	103.4	116.6	114.7	103.4	116.4	114.5
0320	Concrete Reinforcing	116.3	144.5	130.0	91.2	144.5	117.0	116.3	144.5	130.0	116.3	144.5	130.0	116.3	144.5	130.0	116.3	144.5	130.0
0330	Cast-in-Place Concrete	105.1	125.5	112.7	89.5	125.5	102.9	106.7	126.7	114.2	108.5	126.8	115.3	108.5	127.5	115.6	101.3	125.5	110.3
03	CONCRETE	116.3	123.7	119.6	89.2	123.7	104.4	103.3	124.1	112.4	104.1	124.3	113.0	104.1	124.4	113.0	100.8	123.6	110.8
04	MASONRY	101.1	130.9	119.4	99.3	130.9	118.7	100.6	130.1	118.7	101.4	130.1	119.0	101.4	130.9	119.5	100.6	130.9	119.3
05	METALS	94.7	116.4	101.1	94.4	116.4	100.9	98.1	116.5	103.5	98.1	116.8	103.6	98.1	116.5	103.5	97.9	116.3	103.3
06	WOOD, PLASTICS & COMPOSITES	105.0	114.2	110.1	105.0	114.2	110.1	105.0	114.2	110.1	105.0	114.2	110.1	105.0	114.2	110.1	105.0	114.2	110.1
07	THERMAL & MOISTURE PROTECTION	99.9	120.8	109.0	99.7	120.3	108.7	99.9	123.2	110.0	99.8	123.2	110.0	99.8	121.1	109.1	100.0	120.3	108.8
08	OPENINGS	96.4	120.6	102.0	98.7	120.6	103.8	96.4	120.6	102.0	96.4	120.6	102.0	96.4	120.6	102.0	98.8	120.6	103.9
0920	Plaster & Gypsum Board	115.7	114.3	114.8	115.7	114.3	114.8	115.7	114.3	114.8	115.7	114.3	114.8	115.7	114.3	114.8	115.7	114.3	114.8
0950, 0980	Ceilings & Acoustic Treatment	102.0	114.3	110.3	100.2	114.3	109.7	102.0	114.3	110.3	102.0	114.3	110.3	102.0	114.3	110.3	100.2	114.3	109.7
0960	Flooring	92.6	127.0	102.3	92.6	127.0	102.3	92.6	124.5	101.6	92.6	127.0	102.3	92.6	127.0	102.3	92.6	121.1	100.6
0970, 0990	Wall Finishes & Painting/Coating	92.0	125.5	111.8	92.0	129.1	113.9	92.0	125.5	111.8	92.0	125.5	111.8	92.0	129.1	113.9	92.0	129.1	113.9
09	FINISHES	94.7	118.9	107.8	93.9	119.3	107.6	94.7	118.5	107.6	94.8	118.9	107.8	94.6	119.3	107.9	94.5	118.3	107.3
COVERS	DIVS. 10 - 14, 25, 28, 41, 43, 44, 46	100.0	114.0	103.1	100.0	114.0	103.1	100.0	113.9	103.1	100.0	114.1	103.2	100.0	114.0	103.1	100.0	114.0	103.1
21, 22, 23	FIRE SUPPRESSION, PLUMBING & HVAC	100.1	118.6	107.4	96.6	118.6	105.2	100.1	118.6	107.4	100.1	118.6	107.4	100.1	118.6	107.4	100.1	118.4	107.3
26, 27, 3370	ELECTRICAL, COMMUNICATIONS & UTIL.	92.9	106.0	99.4	90.5	107.9	99.1	93.0	104.8	98.8	93.0	150.1	121.1	92.6	107.1	99.7	93.0	108.8	100.8
MF2018	WEIGHTED AVERAGE	99.9	117.0	107.1	95.2	117.2	104.5	98.8	116.7	106.4	99.0	123.2	109.2	98.9	117.2	106.6	98.7	117.1	106.5

DIVISION		D.C.			DELAWARE									FLORIDA					
		WASHINGTON			DOVER			NEWARK			WILMINGTON			DAYTONA BEACH			FORT LAUDERDALE		
		200 - 205			199			197			198			321			333		
		MAT.	INST.	TOTAL	MAT.	INST.	TOTAL	MAT.	INST.	TOTAL	MAT.	INST.	TOTAL	MAT.	INST.	TOTAL	MAT.	INST.	TOTAL
015433	CONTRACTOR EQUIPMENT		105.3	105.3		118.2	118.2		119.0	119.0		118.4	118.4		99.5	99.5		92.8	92.8
0241, 31 - 34	SITE & INFRASTRUCTURE, DEMOLITION	103.0	97.8	99.4	106.0	107.5	107.0	104.5	108.7	107.4	102.2	107.8	106.1	117.6	85.0	95.1	95.0	73.2	80.0
0310	Concrete Forming & Accessories	97.7	73.4	77.0	98.0	99.4	99.2	97.0	99.2	98.9	95.9	99.4	98.9	97.1	61.6	66.8	93.4	59.4	64.4
0320	Concrete Reinforcing	103.5	86.8	95.4	101.4	115.1	108.0	97.2	115.1	105.9	101.0	115.1	107.8	96.0	61.0	79.1	94.6	59.9	77.8
0330	Cast-in-Place Concrete	103.6	79.0	94.5	108.1	107.7	107.9	91.2	106.7	97.0	102.6	107.7	104.5	91.8	66.0	82.2	96.3	62.3	83.6
03	CONCRETE	106.0	78.6	94.0	100.4	106.0	102.9	93.4	105.7	98.8	97.7	106.0	101.4	90.5	64.8	79.2	93.4	62.3	79.7
04	MASONRY	97.6	87.8	91.6	99.0	98.5	98.7	101.0	98.5	99.5	94.4	98.5	96.9	86.4	60.9	70.7	87.6	54.4	67.2
05	METALS	103.9	95.6	101.5	102.9	121.9	108.5	104.7	123.1	110.1	103.1	121.9	108.6	101.9	88.9	98.1	96.9	88.2	94.4
06	WOOD, PLASTICS & COMPOSITES	95.4	72.2	82.6	92.3	97.6	95.2	91.0	97.4	94.5	85.7	97.6	92.2	93.2	60.1	75.0	78.3	60.6	68.6
07	THERMAL & MOISTURE PROTECTION	104.3	87.1	96.8	105.2	110.4	107.4	109.2	109.8	109.4	105.1	110.4	107.4	101.4	65.4	85.7	106.3	60.4	86.3
08	OPENINGS	100.7	74.6	94.6	91.2	107.4	95.0	91.2	107.3	94.9	88.6	107.4	93.0	92.9	59.4	85.1	94.2	59.5	86.1
0920	Plaster & Gypsum Board	104.3	71.4	82.2	96.0	97.4	96.9	96.7	97.4	97.2	97.3	97.4	97.4	92.7	59.5	70.4	107.2	60.0	75.5
0950, 0980	Ceilings & Acoustic Treatment	110.9	71.4	84.3	100.2	97.4	98.3	96.0	97.4	96.9	90.8	97.4	95.2	78.3	59.5	65.6	84.8	60.0	68.1
0960	Flooring	95.4	75.5	89.8	96.9	105.4	99.3	91.3	105.4	95.3	96.0	105.4	98.7	99.5	62.7	89.2	97.1	62.7	87.5
0970, 0990	Wall Finishes & Painting/Coating	103.7	73.1	85.6	92.6	119.1	108.3	88.2	119.1	106.4	95.3	119.1	109.4	103.8	61.4	78.7	96.5	58.3	73.9
09	FINISHES	98.5	72.8	84.6	95.9	101.6	99.0	89.7	101.5	96.1	94.2	101.6	98.2	91.5	61.1	75.1	91.3	59.4	74.1
COVERS	DIVS. 10 - 14, 25, 28, 41, 43, 44, 46	100.0	94.5	98.8	100.0	103.7	100.8	100.0	103.4	100.8	100.0	103.7	100.8	100.0	81.4	95.9	100.0	80.7	95.7
21, 22, 23	FIRE SUPPRESSION, PLUMBING & HVAC	100.0	88.4	95.4	100.0	119.4	107.6	100.3	119.3	107.7	100.1	119.4	107.7	99.9	75.6	90.4	100.0	66.7	86.9
26, 27, 3370	ELECTRICAL, COMMUNICATIONS & UTIL.	97.7	100.0	98.8	95.1	109.7	102.3	96.9	109.7	103.2	95.1	109.7	102.3	96.7	58.9	78.1	94.7	67.4	81.3
MF2018	WEIGHTED AVERAGE	101.1	87.5	95.4	99.0	109.6	103.5	98.3	109.7	103.1	97.9	109.7	102.9	97.2	69.5	85.5	96.1	66.4	83.6

	DIVISION	FLORIDA																	
		FORT MYERS			GAINESVILLE			JACKSONVILLE			LAKELAND			MELBOURNE			MIAMI		
		339, 341			326, 344			320, 322			338			329			330 - 332, 340		
		MAT.	INST.	TOTAL	MAT.	INST.	TOTAL	MAT.	INST.	TOTAL	MAT.	INST.	TOTAL	MAT.	INST.	TOTAL	MAT.	INST.	TOTAL
015433	CONTRACTOR EQUIPMENT		99.5	99.5		99.5	99.5		99.5	99.5		99.5	99.5		99.5	99.5		94.9	94.9
0241, 31 - 34	SITE & INFRASTRUCTURE, DEMOLITION	106.7	85.2	91.9	127.1	84.8	97.9	117.6	85.2	95.2	108.7	85.2	92.5	125.6	85.1	97.7	96.5	78.9	84.3
0310	Concrete Forming & Accessories	89.6	63.8	67.6	92.5	57.1	62.4	96.9	60.2	65.6	86.4	63.8	67.1	93.7	61.7	66.4	99.3	60.2	66.0
0320	Concrete Reinforcing	95.7	76.1	86.2	101.8	61.5	82.3	96.0	61.3	79.2	98.0	76.4	87.6	97.1	65.6	81.9	101.6	60.0	81.5
0330	Cast-in-Place Concrete	100.4	66.0	87.6	105.4	65.7	90.6	92.7	65.8	82.7	102.7	66.1	89.1	110.7	66.0	94.1	93.1	63.9	82.2
03	CONCRETE	94.0	68.3	82.7	101.7	62.7	84.6	90.9	64.2	79.2	95.8	68.5	83.8	102.1	65.6	86.1	91.7	63.2	79.2
04	MASONRY	82.5	58.8	67.9	99.3	60.8	75.7	86.1	58.7	69.3	96.8	61.0	74.8	84.2	60.9	69.9	88.0	55.5	68.0
05	METALS	99.0	93.6	97.4	100.7	88.4	97.1	100.4	88.6	96.9	98.9	94.7	97.7	110.6	90.6	104.7	97.1	87.6	94.3
06	WOOD, PLASTICS & COMPOSITES	75.4	63.2	68.7	87.2	54.6	69.3	93.2	58.6	74.2	71.0	63.2	66.7	88.9	60.1	73.0	92.9	60.7	75.2
07	THERMAL & MOISTURE PROTECTION	106.1	63.2	87.4	101.8	62.5	84.7	101.7	62.4	84.6	106.0	63.8	87.6	101.9	64.1	85.5	106.1	61.4	86.6
08	OPENINGS	95.5	64.6	88.3	92.5	56.5	84.1	92.9	58.8	84.9	95.5	64.7	88.3	92.2	60.5	84.8	96.6	59.5	88.0
0920	Plaster & Gypsum Board	103.3	62.7	76.0	89.6	53.8	65.6	92.7	58.0	69.3	100.2	62.7	75.0	89.6	59.5	69.4	96.2	60.0	71.9
0950, 0980	Ceilings & Acoustic Treatment	79.7	62.7	68.2	72.9	53.8	60.1	78.3	58.0	64.6	79.7	62.7	68.2	77.5	59.5	65.4	85.6	60.0	68.3
0960	Flooring	94.1	77.2	89.4	97.0	62.7	87.4	99.5	62.7	89.2	92.2	62.7	83.9	97.2	62.7	87.5	98.6	62.7	88.5
0970, 0990	Wall Finishes & Painting/Coating	101.4	63.2	78.8	103.8	63.2	79.8	103.8	63.2	79.8	101.4	63.2	78.8	103.8	80.9	90.2	100.2	58.3	75.4
09	FINISHES	91.1	66.1	77.6	90.4	58.1	72.9	91.6	60.4	74.7	90.2	63.1	75.6	91.0	63.2	76.0	90.9	59.8	74.1
COVERS	DIVS. 10 - 14, 25, 28, 41, 43, 44, 46	100.0	80.0	95.5	100.0	80.8	95.7	100.0	78.0	95.1	100.0	80.0	95.5	100.0	81.4	95.9	100.0	81.5	95.9
21, 22, 23	FIRE SUPPRESSION, PLUMBING & HVAC	98.1	59.3	82.9	98.6	63.1	84.7	99.9	63.1	85.5	98.1	59.9	83.1	99.9	74.2	89.8	100.0	63.8	85.8
26, 27, 3370	ELECTRICAL, COMMUNICATIONS & UTIL.	96.6	60.0	78.6	97.0	58.6	78.0	96.4	61.9	79.4	95.0	58.5	77.0	97.7	62.7	80.5	98.4	77.9	88.3
MF2018	WEIGHTED AVERAGE	96.5	67.7	84.3	98.8	65.8	84.8	96.9	66.6	84.1	97.1	67.6	84.6	100.1	70.3	87.5	96.7	68.0	84.6

	DIVISION	FLORIDA																	
		ORLANDO			PANAMA CITY			PENSACOLA			SARASOTA			ST. PETERSBURG			TALLAHASSEE		
		327 - 328, 347			324			325			342			337			323		
		MAT.	INST.	TOTAL	MAT.	INST.	TOTAL	MAT.	INST.	TOTAL	MAT.	INST.	TOTAL	MAT.	INST.	TOTAL	MAT.	INST.	TOTAL
015433	CONTRACTOR EQUIPMENT		102.0	102.0		99.5	99.5		99.5	99.5		99.5	99.5		99.5	99.5		102.0	102.0
0241, 31 - 34	SITE & INFRASTRUCTURE, DEMOLITION	115.8	89.6	97.7	131.4	85.2	99.5	131.3	85.0	99.3	119.0	85.2	95.7	110.4	85.0	92.9	110.2	89.8	96.1
0310	Concrete Forming & Accessories	101.8	61.5	67.4	96.2	66.2	70.6	94.2	61.6	66.4	93.7	63.7	68.1	92.8	60.8	65.5	99.0	62.2	67.6
0320	Concrete Reinforcing	104.9	65.6	85.9	100.3	67.4	84.4	102.8	67.4	85.7	96.6	76.2	86.7	98.0	76.0	87.4	98.4	61.5	80.5
0330	Cast-in-Place Concrete	111.6	66.7	94.9	97.4	65.9	85.7	120.4	65.3	99.9	105.1	66.1	90.6	103.8	65.8	89.7	92.3	66.7	82.8
03	CONCRETE	100.8	65.7	85.4	100.0	67.9	85.9	109.7	65.6	90.3	95.3	68.4	83.5	97.3	66.9	83.9	90.0	65.3	79.2
04	MASONRY	94.8	60.9	74.0	90.6	60.3	72.0	109.1	59.9	78.9	88.7	61.0	71.6	132.4	58.7	87.1	85.2	60.3	69.9
05	METALS	96.7	89.1	94.5	101.5	90.9	98.4	102.6	90.8	99.1	102.5	93.9	100.0	99.8	93.5	98.0	101.7	87.7	97.6
06	WOOD, PLASTICS & COMPOSITES	91.7	60.2	74.4	92.0	66.6	78.0	90.3	61.1	74.2	92.5	63.2	76.4	79.5	59.3	68.4	95.6	61.2	76.7
07	THERMAL & MOISTURE PROTECTION	109.2	66.0	90.4	102.0	64.1	85.5	101.9	63.2	85.1	100.1	63.8	84.3	106.2	62.2	87.1	96.7	64.0	82.4
08	OPENINGS	97.8	60.6	89.1	91.0	64.5	84.9	91.0	61.4	84.1	97.8	64.2	90.0	94.3	62.5	86.9	97.9	60.2	89.1
0920	Plaster & Gypsum Board	93.3	59.5	70.6	91.9	66.1	74.6	99.6	60.5	73.3	97.2	62.7	74.0	105.6	58.7	74.0	100.0	60.5	73.4
0950, 0980	Ceilings & Acoustic Treatment	89.7	59.5	69.3	77.5	66.1	69.8	77.5	60.5	66.0	83.9	62.7	69.6	81.6	58.7	66.1	86.2	60.5	68.9
0960	Flooring	95.7	62.7	86.4	99.1	72.5	91.6	95.1	62.7	86.0	103.5	54.9	89.9	96.1	60.8	86.2	97.6	61.7	87.5
0970, 0990	Wall Finishes & Painting/Coating	95.6	60.3	74.7	103.8	63.2	79.8	103.8	63.2	79.8	99.0	63.2	77.8	101.4	63.2	78.8	101.6	63.2	78.9
09	FINISHES	93.2	61.1	75.8	92.6	66.8	78.6	92.1	61.6	75.6	96.0	61.8	77.5	92.6	60.5	75.2	94.4	61.8	76.7
COVERS	DIVS. 10 - 14, 25, 28, 41, 43, 44, 46	100.0	81.7	95.9	100.0	80.6	95.7	100.0	79.7	95.5	100.0	80.0	95.5	100.0	79.6	95.4	100.0	81.5	95.9
21, 22, 23	FIRE SUPPRESSION, PLUMBING & HVAC	100.0	56.4	82.9	99.9	63.2	85.5	99.9	62.7	85.3	99.9	58.5	83.6	100.0	59.9	84.2	100.0	65.5	86.5
26, 27, 3370	ELECTRICAL, COMMUNICATIONS & UTIL.	98.4	63.1	81.0	95.5	58.6	77.3	99.0	50.7	75.2	96.9	58.5	78.0	95.0	60.6	78.0	103.7	58.6	81.4
MF2018	WEIGHTED AVERAGE	99.0	66.6	85.3	98.6	68.3	85.8	101.2	65.8	86.2	98.8	67.0	85.4	99.7	66.8	85.8	98.2	67.7	85.3

	DIVISION	FLORIDA						GEORGIA											
		TAMPA			WEST PALM BEACH			ALBANY			ATHENS			ATLANTA			AUGUSTA		
		335 - 336, 346			334, 349			317, 398			306			300 - 303, 399			308 - 309		
		MAT.	INST.	TOTAL	MAT.	INST.	TOTAL	MAT.	INST.	TOTAL	MAT.	INST.	TOTAL	MAT.	INST.	TOTAL	MAT.	INST.	TOTAL
015433	CONTRACTOR EQUIPMENT		99.5	99.5		92.8	92.8		93.7	93.7		92.6	92.6		96.6	96.6		92.6	92.6
0241, 31 - 34	SITE & INFRASTRUCTURE, DEMOLITION	110.9	87.0	94.4	91.6	73.2	78.9	107.3	76.7	86.1	102.8	90.8	94.5	99.7	96.0	97.1	95.9	91.9	93.2
0310	Concrete Forming & Accessories	95.4	64.1	68.7	96.7	59.1	64.7	90.6	66.5	70.1	90.0	43.6	50.4	94.3	72.3	75.5	90.9	73.0	75.6
0320	Concrete Reinforcing	94.6	76.5	85.8	97.3	57.5	78.1	96.5	71.3	84.3	94.5	63.5	79.5	93.9	71.3	83.0	94.9	70.7	83.2
0330	Cast-in-Place Concrete	101.5	66.2	88.3	91.5	62.2	80.6	94.6	69.1	85.1	109.7	69.5	94.7	113.1	71.1	97.5	103.6	70.0	91.1
03	CONCRETE	95.9	68.7	83.9	90.1	61.8	77.7	86.9	70.1	79.5	99.6	57.4	81.1	101.8	72.4	88.9	92.5	72.2	83.6
04	MASONRY	88.3	61.0	71.5	87.1	52.2	65.7	92.5	68.7	77.9	75.0	77.0	76.2	87.8	68.9	76.2	88.1	68.8	76.2
05	METALS	99.0	95.0	97.8	95.9	87.0	93.3	106.2	97.4	103.6	95.3	79.1	90.6	96.2	83.5	92.5	95.0	82.6	91.4
06	WOOD, PLASTICS & COMPOSITES	83.3	63.2	72.2	83.1	60.6	70.7	80.1	66.4	72.6	92.5	35.8	61.3	97.8	74.0	84.7	93.8	75.3	83.6
07	THERMAL & MOISTURE PROTECTION	106.5	63.8	87.9	106.0	60.6	86.2	100.6	68.0	86.4	95.4	68.3	83.6	96.8	73.8	86.8	95.1	71.4	84.8
08	OPENINGS	95.5	64.7	88.3	93.8	59.0	85.7	85.7	68.5	81.7	91.3	50.2	81.7	99.4	73.5	93.3	91.4	74.0	87.3
0920	Plaster & Gypsum Board	108.2	62.7	77.6	111.9	60.0	77.0	101.1	66.0	77.5	93.0	34.4	53.6	95.0	73.6	80.6	94.0	75.0	81.2
0950, 0980	Ceilings & Acoustic Treatment	84.8	62.7	69.9	79.7	60.0	66.4	77.5	66.0	69.8	96.6	34.4	54.7	89.7	73.6	78.8	97.5	75.0	82.4
0960	Flooring	97.1	62.7	87.5	98.9	54.9	86.6	99.4	70.8	91.4	96.2	81.9	92.2	99.1	70.8	91.1	96.4	70.8	89.2
0970, 0990	Wall Finishes & Painting/Coating	101.4	63.2	78.8	96.5	58.3	73.9	88.7	94.0	91.8	92.4	94.0	93.3	96.1	97.3	96.8	92.4	90.2	91.1
09	FINISHES	93.9	63.1	77.3	91.2	57.8	73.2	91.7	69.7	79.8	95.0	53.9	72.7	95.0	74.5	83.9	94.7	74.5	83.7
COVERS	DIVS. 10 - 14, 25, 28, 41, 43, 44, 46	100.0	80.0	95.6	100.0	80.7	95.7	100.0	85.9	96.8	100.0	82.7	96.1	100.0	87.2	97.2	100.0	87.0	97.1
21, 22, 23	FIRE SUPPRESSION, PLUMBING & HVAC	100.0	59.9	84.3	98.1	58.5	82.6	100.0	68.9	87.8	96.6	66.3	84.7	100.0	69.4	88.0	100.1	69.0	87.9
26, 27, 3370	ELECTRICAL, COMMUNICATIONS & UTIL.	94.7	63.1	79.1	95.7	67.4	81.8	95.5	61.9	79.0	97.8	63.9	81.1	97.6	71.5	84.7	98.3	67.3	83.1
MF2018	WEIGHTED AVERAGE	97.6	68.4	85.3	95.1	64.1	82.0	96.5	71.9	86.1	95.6	66.9	83.5	98.2	75.0	88.4	96.0	73.9	86.6

DIVISION		GEORGIA																	
		COLUMBUS			DALTON			GAINESVILLE			MACON			SAVANNAH			STATESBORO		
		318 - 319			307			305			310 - 312			313 - 314			304		
		MAT.	INST.	TOTAL	MAT.	INST.	TOTAL	MAT.	INST.	TOTAL	MAT.	INST.	TOTAL	MAT.	INST.	TOTAL	MAT.	INST.	TOTAL
015433	CONTRACTOR EQUIPMENT		93.7	93.7		107.6	107.6		92.6	92.6		103.2	103.2		96.7	96.7		95.5	95.5
0241, 31 - 34	SITE & INFRASTRUCTURE, DEMOLITION	107.2	76.7	86.2	102.6	95.8	97.9	102.5	90.7	94.3	108.6	91.0	96.4	107.0	82.0	89.7	103.7	78.0	85.9
0310	Concrete Forming & Accessories	90.5	66.6	70.1	83.2	64.5	67.3	93.3	40.7	48.5	90.3	66.6	70.1	93.4	72.0	75.2	78.5	52.2	56.1
0320	Concrete Reinforcing	96.6	71.3	84.4	94.1	60.9	78.0	94.4	63.4	79.4	97.8	71.3	85.0	104.1	70.7	88.0	93.7	71.2	82.8
0330	Cast-in-Place Concrete	94.2	68.8	84.8	106.6	68.0	92.2	115.3	68.7	98.0	93.0	68.3	83.8	98.3	69.5	87.6	109.5	68.6	94.3
03	CONCRETE	86.8	70.0	79.4	98.9	67.0	84.9	101.4	55.8	81.4	86.3	69.8	79.1	88.5	72.5	81.5	99.3	63.4	83.5
04	MASONRY	92.5	68.7	77.9	75.8	75.5	75.6	83.2	76.2	78.9	104.8	68.9	82.7	87.4	68.8	75.9	77.8	77.0	77.3
05	METALS	105.9	97.6	103.4	96.4	94.0	95.7	94.6	78.5	89.9	100.9	97.6	100.0	102.1	96.0	100.3	100.0	98.9	99.7
06	WOOD, PLASTICS & COMPOSITES	80.1	66.4	72.6	75.9	65.2	70.0	96.1	33.2	61.5	85.9	66.5	75.2	87.4	73.9	80.0	69.7	47.2	57.3
07	THERMAL & MOISTURE PROTECTION	100.5	68.9	86.8	97.3	70.1	85.5	95.4	67.7	83.3	99.0	70.6	86.6	98.5	70.4	86.3	95.9	67.1	83.4
08	OPENINGS	85.7	69.3	81.9	92.8	65.8	86.5	91.3	48.8	81.4	85.5	69.3	81.7	95.3	73.2	90.2	93.7	58.5	85.5
0920	Plaster & Gypsum Board	101.1	66.0	77.5	81.9	64.7	70.3	94.9	31.8	52.4	104.7	66.0	78.7	103.3	73.6	83.3	83.6	46.2	58.4
0950, 0980	Ceilings & Acoustic Treatment	77.5	66.0	69.8	110.2	64.7	79.5	96.6	31.8	52.9	72.8	66.0	68.3	87.3	73.6	78.0	105.8	46.2	65.6
0960	Flooring	99.4	70.8	91.4	97.0	81.9	92.8	97.8	81.9	93.3	77.9	70.8	75.9	95.6	70.8	88.6	114.6	81.9	105.4
0970, 0990	Wall Finishes & Painting/Coating	88.7	87.4	87.9	82.8	70.1	75.3	92.4	94.0	93.3	90.7	94.0	92.6	88.0	86.1	86.9	90.8	70.1	78.5
09	FINISHES	91.6	68.9	79.3	104.5	68.2	84.9	95.5	52.1	72.0	80.5	69.7	74.7	92.9	73.2	82.3	107.9	57.9	80.9
COVERS	DIVS. 10 - 14, 25, 28, 41, 43, 44, 46	100.0	85.9	96.9	100.0	85.3	96.7	100.0	35.5	85.6	100.0	86.0	96.9	100.0	87.0	97.1	100.0	83.8	96.4
21, 22, 23	FIRE SUPPRESSION, PLUMBING & HVAC	100.1	66.8	87.0	96.7	60.8	82.6	96.6	65.6	84.4	100.1	68.2	87.6	100.1	64.8	86.3	97.2	67.3	85.5
26, 27, 3370	ELECTRICAL, COMMUNICATIONS & UTIL.	95.7	67.7	81.9	106.9	62.2	84.9	97.8	70.8	84.5	94.2	62.7	78.6	99.1	63.7	81.7	98.3	63.4	81.1
MF2018	WEIGHTED AVERAGE	96.5	72.2	86.2	97.5	71.4	86.5	96.2	65.6	83.2	95.2	73.1	85.8	97.3	72.7	86.9	97.8	69.7	85.9

DIVISION		GEORGIA						HAWAII									IDAHO		
		VALDOSTA			WAYCROSS			HILO			HONOLULU			STATES & POSS., GUAM			BOISE		
		316			315			967			968			969			836 - 837		
		MAT.	INST.	TOTAL	MAT.	INST.	TOTAL	MAT.	INST.	TOTAL	MAT.	INST.	TOTAL	MAT.	INST.	TOTAL	MAT.	INST.	TOTAL
015433	CONTRACTOR EQUIPMENT		93.7	93.7		93.7	93.7		97.8	97.8		99.3	99.3		160.5	160.5		94.6	94.6
0241, 31 - 34	SITE & INFRASTRUCTURE, DEMOLITION	116.9	76.7	89.1	113.7	77.0	88.3	142.5	103.1	115.3	150.0	106.6	120.0	186.3	98.5	125.7	86.8	92.0	90.4
0310	Concrete Forming & Accessories	81.7	41.6	47.5	83.5	64.0	66.8	108.2	123.8	121.5	120.2	123.7	123.2	110.5	51.5	60.2	100.2	81.1	83.9
0320	Concrete Reinforcing	98.7	59.5	79.8	98.7	59.8	79.9	143.4	127.0	135.5	166.8	127.0	147.6	253.7	27.4	144.2	107.6	79.5	94.0
0330	Cast-in-Place Concrete	92.6	68.7	83.7	104.7	68.6	91.2	180.3	123.3	159.1	145.0	123.5	137.0	156.3	95.7	133.7	90.6	96.6	92.8
03	CONCRETE	91.2	56.6	76.0	94.5	66.5	82.2	144.7	123.2	135.3	137.9	123.1	131.4	147.9	63.9	111.0	97.6	86.4	92.7
04	MASONRY	97.8	77.1	85.1	98.7	77.0	85.4	148.6	122.1	132.3	135.5	122.2	127.3	206.7	34.8	101.0	126.3	86.2	101.7
05	METALS	105.3	93.0	101.7	104.4	88.9	99.8	110.8	108.1	110.0	124.1	107.1	119.1	143.1	76.0	123.4	109.6	82.7	101.7
06	WOOD, PLASTICS & COMPOSITES	68.7	33.0	49.0	70.2	63.4	66.5	111.4	124.3	118.5	132.8	124.2	128.1	123.4	53.9	85.1	93.5	79.9	86.0
07	THERMAL & MOISTURE PROTECTION	100.8	66.6	86.0	100.6	69.1	86.9	128.7	119.3	124.6	146.3	120.0	134.9	150.7	58.2	110.4	101.4	86.6	94.9
08	OPENINGS	82.5	47.6	74.4	82.8	61.8	77.9	114.1	123.2	116.2	128.1	123.2	126.9	118.4	43.8	101.0	96.9	75.9	92.0
0920	Plaster & Gypsum Board	93.8	31.6	52.0	93.8	62.9	73.0	115.2	124.8	121.7	160.8	124.8	136.6	236.1	42.5	105.8	92.3	79.4	83.6
0950, 0980	Ceilings & Acoustic Treatment	75.9	31.6	46.0	74.0	62.9	66.5	136.6	124.8	128.6	144.8	124.8	131.3	257.2	42.5	112.4	102.8	79.4	87.0
0960	Flooring	93.0	83.5	90.3	94.3	81.9	90.8	103.8	139.3	113.8	121.3	139.3	126.4	121.2	39.5	98.2	92.9	90.7	92.3
0970, 0990	Wall Finishes & Painting/Coating	88.7	94.0	91.8	88.7	70.1	77.7	99.2	143.8	125.6	110.1	143.8	130.1	105.2	31.5	61.6	92.6	39.5	61.2
09	FINISHES	89.4	52.5	69.4	89.0	67.5	77.4	110.8	129.1	120.7	125.5	129.1	127.4	185.1	48.0	110.9	92.7	78.2	84.9
COVERS	DIVS. 10 - 14, 25, 28, 41, 43, 44, 46	100.0	82.2	96.0	100.0	85.5	96.8	100.0	113.7	103.1	100.0	113.6	103.0	100.0	66.0	92.4	100.0	88.6	97.4
21, 22, 23	FIRE SUPPRESSION, PLUMBING & HVAC	100.1	68.5	87.7	97.9	63.6	84.4	100.3	112.4	105.0	100.4	112.3	105.1	102.6	33.5	75.5	100.1	74.2	89.9
26, 27, 3370	ELECTRICAL, COMMUNICATIONS & UTIL.	94.1	57.8	76.2	97.7	63.4	80.8	108.1	124.6	116.2	109.8	124.5	117.1	158.1	35.9	97.9	97.2	68.2	82.9
MF2018	WEIGHTED AVERAGE	96.7	66.2	83.8	96.8	70.0	85.5	114.8	118.4	116.3	119.1	118.5	118.9	137.1	51.4	100.9	101.0	79.9	92.1

DIVISION		IDAHO															ILLINOIS		
		COEUR D'ALENE			IDAHO FALLS			LEWISTON			POCATELLO			TWIN FALLS			BLOOMINGTON		
		838			834			835			832			833			617		
		MAT.	INST.	TOTAL	MAT.	INST.	TOTAL	MAT.	INST.	TOTAL	MAT.	INST.	TOTAL	MAT.	INST.	TOTAL	MAT.	INST.	TOTAL
015433	CONTRACTOR EQUIPMENT		89.7	89.7		94.6	94.6		89.7	89.7		94.6	94.6		94.6	94.6		101.9	101.9
0241, 31 - 34	SITE & INFRASTRUCTURE, DEMOLITION	85.5	86.5	86.2	85.1	92.1	89.9	92.2	87.2	88.7	88.3	92.0	90.9	95.4	91.8	92.9	94.9	96.0	95.7
0310	Concrete Forming & Accessories	108.7	81.4	85.4	94.1	78.3	80.6	113.9	81.1	85.9	100.4	80.5	83.4	101.6	76.5	80.2	84.0	116.0	111.2
0320	Concrete Reinforcing	115.7	96.2	106.3	109.6	77.9	94.3	115.7	96.2	106.3	108.0	79.4	94.2	110.0	77.8	94.4	93.0	102.4	97.5
0330	Cast-in-Place Concrete	98.0	85.2	93.2	86.3	83.4	85.2	101.8	82.0	94.5	93.1	96.4	94.3	95.6	82.3	90.6	98.0	113.7	103.9
03	CONCRETE	103.9	85.3	95.7	89.6	80.2	85.5	107.5	84.1	97.2	96.7	86.0	92.0	104.6	79.0	93.4	91.9	113.6	101.4
04	MASONRY	128.0	87.3	103.0	121.5	87.2	100.4	128.5	84.7	101.6	124.0	86.2	100.8	126.7	82.8	99.7	110.2	121.7	117.3
05	METALS	103.0	88.1	98.6	118.1	81.6	107.4	102.4	88.2	98.3	118.2	82.0	107.6	118.2	81.4	107.4	95.2	123.1	103.4
06	WOOD, PLASTICS & COMPOSITES	96.7	79.8	87.4	87.2	76.3	81.2	102.6	79.8	90.1	93.5	79.9	86.0	94.6	76.3	84.5	81.7	113.0	99.0
07	THERMAL & MOISTURE PROTECTION	158.9	84.3	126.4	100.4	75.3	89.5	159.2	83.1	126.1	101.0	75.0	89.7	101.8	80.5	92.5	95.4	112.1	102.6
08	OPENINGS	113.8	75.9	105.0	99.8	71.3	93.2	106.7	79.6	100.4	97.6	68.8	90.9	100.6	62.4	91.7	91.3	118.7	97.7
0920	Plaster & Gypsum Board	167.0	79.4	108.1	80.3	75.7	77.2	168.9	79.4	108.7	81.9	79.4	80.2	84.0	75.7	78.4	89.1	113.5	105.5
0950, 0980	Ceilings & Acoustic Treatment	136.3	79.4	98.0	104.3	75.7	85.0	136.3	79.4	98.0	110.0	79.4	89.4	106.8	75.7	85.8	83.9	113.5	103.9
0960	Flooring	130.9	77.0	115.8	92.7	77.0	88.3	134.3	77.0	118.2	96.2	77.0	90.8	97.5	77.0	91.7	86.0	120.8	95.8
0970, 0990	Wall Finishes & Painting/Coating	111.2	73.8	89.1	92.6	39.5	61.2	111.2	71.3	87.6	92.5	39.5	61.1	92.6	39.5	61.2	87.6	134.8	115.5
09	FINISHES	158.3	79.5	115.7	90.9	74.0	81.8	159.7	78.9	116.0	93.7	75.9	84.1	94.5	72.9	82.8	86.1	119.0	103.9
COVERS	DIVS. 10 - 14, 25, 28, 41, 43, 44, 46	100.0	93.0	98.4	100.0	87.4	97.2	100.0	92.6	98.3	100.0	88.5	97.4	100.0	85.9	96.9	100.0	104.5	101.0
21, 22, 23	FIRE SUPPRESSION, PLUMBING & HVAC	99.4	85.5	94.0	101.1	81.8	93.5	100.9	85.1	94.6	100.0	73.6	89.6	100.0	72.2	89.1	96.5	105.4	100.0
26, 27, 3370	ELECTRICAL, COMMUNICATIONS & UTIL.	88.7	81.5	85.2	88.3	70.9	79.7	86.9	78.2	82.6	94.4	66.0	80.4	89.6	70.9	80.4	95.9	90.3	93.1
MF2018	WEIGHTED AVERAGE	108.1	84.4	98.1	100.5	79.9	91.8	108.2	83.5	97.8	102.0	78.4	92.0	103.2	76.8	92.1	95.0	109.5	101.1

	DIVISION	ILLINOIS																	
		CARBONDALE			CENTRALIA			CHAMPAIGN			CHICAGO			DECATUR			EAST ST. LOUIS		
		629			628			618 - 619			606 - 608			625			620 - 622		
		MAT.	INST.	TOTAL	MAT.	INST.	TOTAL	MAT.	INST.	TOTAL	MAT.	INST.	TOTAL	MAT.	INST.	TOTAL	MAT.	INST.	TOTAL
015433	CONTRACTOR EQUIPMENT		110.3	110.3		110.3	110.3		102.8	102.8		100.3	100.3		102.8	102.8		110.3	110.3
0241, 31 - 34	SITE & INFRASTRUCTURE, DEMOLITION	100.9	96.8	98.1	101.3	98.6	99.5	104.0	97.2	99.3	105.4	104.4	104.7	94.8	97.0	96.3	103.5	97.7	99.5
0310	Concrete Forming & Accessories	90.0	108.9	106.1	91.6	114.2	110.9	90.3	116.7	112.8	97.4	159.7	150.5	92.1	117.9	114.1	87.5	114.2	110.2
0320	Concrete Reinforcing	82.2	103.4	92.5	82.2	104.0	92.8	93.0	101.6	97.2	103.8	153.3	127.7	80.6	99.8	89.9	82.2	104.0	92.7
0330	Cast-in-Place Concrete	91.5	102.7	95.7	92.0	119.6	102.3	113.7	110.8	112.6	126.0	155.6	137.0	99.9	114.6	105.4	93.6	118.4	102.8
03	CONCRETE	79.9	107.3	91.9	80.4	115.8	96.0	104.1	112.7	107.9	108.0	156.5	129.3	90.3	114.3	100.8	81.4	115.4	96.3
04	MASONRY	75.0	111.5	97.4	75.0	122.9	104.5	133.1	125.4	128.4	105.3	163.4	141.0	70.9	122.8	102.8	75.3	123.0	104.6
05	METALS	100.9	130.4	109.6	100.9	133.5	110.5	95.2	120.3	102.6	95.9	146.8	110.8	104.9	120.7	109.5	102.1	133.5	111.3
06	WOOD, PLASTICS & COMPOSITES	87.7	105.8	97.7	90.2	111.2	101.7	88.5	114.0	102.5	100.0	158.4	132.1	88.7	116.8	104.2	84.8	111.2	99.3
07	THERMAL & MOISTURE PROTECTION	90.6	101.3	95.3	90.7	111.5	99.8	96.2	115.6	104.6	95.2	150.7	119.4	95.9	113.3	103.5	90.7	110.7	99.4
08	OPENINGS	87.3	116.3	94.1	87.3	119.2	94.8	91.8	116.9	97.7	100.9	169.3	116.9	98.2	118.2	102.9	87.4	118.4	94.6
0920	Plaster & Gypsum Board	93.7	106.2	102.1	95.0	111.6	106.2	91.0	114.5	106.8	101.3	160.0	140.8	96.5	117.4	110.6	92.2	111.6	105.3
0950, 0980	Ceilings & Acoustic Treatment	80.6	106.2	97.8	80.6	111.6	101.5	83.9	114.5	104.5	100.4	160.0	140.6	87.1	117.4	107.5	80.6	111.6	101.5
0960	Flooring	113.7	116.6	114.5	114.6	116.6	115.1	89.2	116.6	96.9	91.3	163.8	111.7	101.8	117.1	106.1	112.6	116.6	113.7
0970, 0990	Wall Finishes & Painting/Coating	103.1	105.0	104.2	103.1	111.6	108.2	87.6	114.9	103.7	96.7	170.3	140.2	95.5	117.8	108.7	103.1	111.6	108.2
09	FINISHES	92.1	110.7	102.1	92.5	114.4	104.4	88.1	116.9	103.7	95.8	162.4	131.8	92.2	118.3	106.3	91.6	114.4	103.9
COVERS	DIVS. 10 - 14, 25, 28, 41, 43, 44, 46	100.0	106.9	101.5	100.0	105.8	101.3	100.0	109.2	102.0	100.0	129.3	106.5	100.0	108.6	101.9	100.0	108.3	101.9
21, 22, 23	FIRE SUPPRESSION, PLUMBING & HVAC	96.3	106.8	100.4	96.3	97.8	96.9	96.5	107.0	100.6	99.9	137.4	114.6	99.9	98.8	99.5	99.9	99.2	99.6
26, 27, 3370	ELECTRICAL, COMMUNICATIONS & UTIL.	91.9	107.8	99.8	93.1	106.3	99.6	99.0	95.4	97.2	97.9	136.2	116.8	95.3	103.7	99.5	92.7	99.7	96.2
MF2018	WEIGHTED AVERAGE	92.5	109.6	99.7	92.8	111.2	100.5	98.5	110.5	103.6	100.1	145.6	119.3	96.6	110.1	102.3	93.8	110.4	100.9

	DIVISION	ILLINOIS																	
		EFFINGHAM			GALESBURG			JOLIET			KANKAKEE			LA SALLE			NORTH SUBURBAN		
		624			614			604			609			613			600 - 603		
		MAT.	INST.	TOTAL	MAT.	INST.	TOTAL	MAT.	INST.	TOTAL	MAT.	INST.	TOTAL	MAT.	INST.	TOTAL	MAT.	INST.	TOTAL
015433	CONTRACTOR EQUIPMENT		102.8	102.8		101.9	101.9		93.7	93.7		93.7	93.7		101.9	101.9		93.7	93.7
0241, 31 - 34	SITE & INFRASTRUCTURE, DEMOLITION	99.3	95.9	97.0	97.3	95.8	96.3	101.2	97.2	98.5	94.7	97.1	96.4	96.7	96.8	96.8	100.5	97.9	98.7
0310	Concrete Forming & Accessories	96.5	114.6	111.9	90.2	117.0	113.0	96.8	156.7	147.8	90.5	145.4	137.3	103.7	125.3	122.1	96.1	158.5	149.3
0320	Concrete Reinforcing	83.3	93.8	88.4	92.5	105.6	98.8	103.8	136.6	119.7	104.6	135.1	119.3	92.7	132.9	112.1	103.8	149.1	125.7
0330	Cast-in-Place Concrete	99.6	109.9	103.4	100.9	109.2	104.0	114.6	143.9	125.5	106.8	135.8	117.6	100.8	121.8	108.6	114.6	152.9	128.9
03	CONCRETE	91.0	109.9	99.3	94.8	113.0	102.8	101.0	147.9	121.6	95.0	139.7	114.7	95.6	126.1	109.0	101.0	154.1	124.3
04	MASONRY	79.1	115.3	101.3	110.4	122.4	117.8	104.5	156.7	136.6	100.6	147.0	129.1	110.4	129.7	122.3	101.2	158.8	136.6
05	METALS	101.9	114.9	105.7	95.2	123.6	103.5	94.1	134.7	106.0	94.1	133.4	105.6	95.3	142.0	109.0	95.1	141.6	108.8
06	WOOD, PLASTICS & COMPOSITES	91.0	114.0	103.6	88.3	115.4	103.2	96.6	157.7	130.2	90.0	144.7	120.1	103.8	123.2	114.5	95.4	158.2	130.0
07	THERMAL & MOISTURE PROTECTION	95.4	109.1	101.4	95.6	110.0	101.9	99.5	144.2	119.0	98.6	139.5	116.4	95.8	122.5	107.4	99.9	147.6	120.7
08	OPENINGS	93.3	114.5	98.2	91.3	115.9	97.0	99.3	161.7	113.9	92.4	156.2	107.3	91.3	133.3	101.1	99.4	167.9	115.4
0920	Plaster & Gypsum Board	96.3	114.5	108.5	91.0	115.9	107.8	93.7	159.5	138.0	92.5	146.1	128.6	98.3	124.0	115.6	97.2	160.0	139.5
0950, 0980	Ceilings & Acoustic Treatment	80.6	114.5	103.4	83.9	115.9	105.5	100.9	159.5	140.4	100.9	146.1	131.4	83.9	124.0	110.9	100.9	160.0	140.8
0960	Flooring	103.0	116.6	106.8	89.1	120.8	98.0	89.4	149.8	106.4	86.3	150.7	104.4	95.7	124.8	103.9	89.9	159.5	109.4
0970, 0990	Wall Finishes & Painting/Coating	95.5	109.3	103.7	87.6	98.5	94.0	89.0	163.0	132.8	89.0	135.7	116.6	87.6	133.0	114.5	91.0	163.9	134.1
09	FINISHES	91.3	115.1	104.2	87.4	116.3	103.1	91.6	157.4	127.2	90.0	145.8	120.2	90.4	124.7	109.0	92.2	160.1	129.0
COVERS	DIVS. 10 - 14, 25, 28, 41, 43, 44, 46	100.0	104.5	101.0	100.0	104.4	101.0	100.0	122.6	105.0	100.0	120.6	104.6	100.0	105.6	101.2	100.0	124.1	105.4
21, 22, 23	FIRE SUPPRESSION, PLUMBING & HVAC	96.4	104.5	99.6	96.5	105.7	100.1	100.0	130.0	111.8	96.5	129.6	109.5	96.5	128.4	109.0	99.9	136.2	114.1
26, 27, 3370	ELECTRICAL, COMMUNICATIONS & UTIL.	93.5	107.8	100.5	96.7	86.5	91.7	96.9	133.3	114.8	92.4	137.0	114.4	94.0	137.0	115.2	96.7	135.5	115.9
MF2018	WEIGHTED AVERAGE	95.1	109.1	101.0	95.7	108.5	101.1	98.3	138.8	115.4	95.0	134.9	111.9	95.9	127.0	109.0	98.3	142.9	117.1

	DIVISION	ILLINOIS																	
		PEORIA			QUINCY			ROCK ISLAND			ROCKFORD			SOUTH SUBURBAN			SPRINGFIELD		
		615 - 616			623			612			610 - 611			605			626 - 627		
		MAT.	INST.	TOTAL	MAT.	INST.	TOTAL	MAT.	INST.	TOTAL	MAT.	INST.	TOTAL	MAT.	INST.	TOTAL	MAT.	INST.	TOTAL
015433	CONTRACTOR EQUIPMENT		101.9	101.9		102.8	102.8		101.9	101.9		101.9	101.9		93.7	93.7		105.1	105.1
0241, 31 - 34	SITE & INFRASTRUCTURE, DEMOLITION	97.9	96.0	96.6	98.1	96.4	96.9	95.5	94.8	95.0	97.4	96.8	97.0	100.5	97.8	98.6	99.8	100.5	100.3
0310	Concrete Forming & Accessories	93.0	115.9	112.5	94.3	113.6	110.8	91.6	99.1	98.0	97.2	127.7	123.2	96.1	158.5	149.3	93.2	115.6	112.3
0320	Concrete Reinforcing	90.1	104.7	97.2	83.0	86.2	84.5	92.5	99.0	95.6	85.3	134.1	108.9	103.8	149.0	125.7	83.2	99.8	91.2
0330	Cast-in-Place Concrete	97.9	115.6	104.5	99.8	106.6	102.3	98.7	98.4	98.6	100.2	126.6	110.0	114.6	152.8	128.8	94.9	109.2	100.2
03	CONCRETE	92.0	114.6	102.0	90.7	107.1	97.9	92.7	100.0	95.9	92.7	129.0	108.7	101.0	154.0	124.3	88.5	111.2	98.5
04	MASONRY	109.7	121.5	116.9	98.9	113.2	107.7	110.2	100.2	104.1	85.6	142.3	120.4	101.2	158.7	136.6	80.7	122.2	106.2
05	METALS	97.9	124.2	105.6	101.9	112.6	105.1	95.2	118.5	102.1	97.9	141.5	110.7	95.1	141.4	108.7	102.5	119.1	107.3
06	WOOD, PLASTICS & COMPOSITES	96.4	113.2	105.6	88.6	114.0	102.6	90.0	97.5	94.1	96.4	123.6	111.3	95.4	158.2	130.0	90.0	114.1	103.3
07	THERMAL & MOISTURE PROTECTION	96.3	112.4	103.3	95.3	106.8	100.3	95.5	98.4	96.8	98.8	129.0	111.9	99.9	147.6	120.7	98.3	114.4	105.3
08	OPENINGS	96.6	119.5	102.0	94.0	112.1	98.2	91.3	103.3	94.1	96.6	136.5	105.9	99.4	167.9	115.4	99.2	116.4	103.2
0920	Plaster & Gypsum Board	95.5	113.7	107.7	95.0	114.5	108.1	91.0	97.6	95.5	95.5	124.4	114.9	97.2	160.0	139.5	99.9	114.5	109.7
0950, 0980	Ceilings & Acoustic Treatment	89.6	113.7	105.8	80.6	114.5	103.4	83.9	97.6	93.2	89.6	124.4	113.1	100.9	160.0	140.8	91.3	114.5	107.0
0960	Flooring	92.6	119.5	100.2	101.8	113.6	105.1	90.2	94.6	91.4	92.6	124.9	101.7	89.9	159.5	109.4	104.8	117.1	108.3
0970, 0990	Wall Finishes & Painting/Coating	87.6	134.8	115.5	95.5	111.5	105.0	87.6	95.5	92.3	87.6	142.7	120.2	91.0	163.9	134.1	98.0	111.5	106.0
09	FINISHES	90.2	118.7	105.6	90.7	114.2	103.4	87.7	97.6	93.0	90.2	128.5	110.9	92.2	160.1	129.0	98.0	115.9	107.7
COVERS	DIVS. 10 - 14, 25, 28, 41, 43, 44, 46	100.0	104.4	101.0	100.0	106.9	101.5	100.0	97.5	99.4	100.0	114.2	103.2	100.0	124.1	105.4	100.0	108.3	101.9
21, 22, 23	FIRE SUPPRESSION, PLUMBING & HVAC	100.0	100.2	100.0	96.4	102.9	98.9	96.5	97.5	96.9	100.1	116.1	106.4	99.9	136.2	114.1	99.9	103.7	101.4
26, 27, 3370	ELECTRICAL, COMMUNICATIONS & UTIL.	97.7	91.5	94.6	91.4	82.2	86.9	89.3	93.1	91.2	98.0	125.6	111.6	96.7	135.5	115.9	98.1	88.7	93.5
MF2018	WEIGHTED AVERAGE	97.6	108.8	102.3	95.7	104.2	99.3	94.6	99.5	96.7	96.6	125.5	108.8	98.3	142.8	117.1	97.5	108.3	102.1

	DIVISION	INDIANA																	
		ANDERSON			BLOOMINGTON			COLUMBUS			EVANSVILLE			FORT WAYNE			GARY		
		460			474			472			476 - 477			467 - 468			463 - 464		
		MAT.	INST.	TOTAL	MAT.	INST.	TOTAL	MAT.	INST.	TOTAL	MAT.	INST.	TOTAL	MAT.	INST.	TOTAL	MAT.	INST.	TOTAL
015433	CONTRACTOR EQUIPMENT		94.5	94.5		81.5	81.5		81.5	81.5		110.3	110.3		94.5	94.5		94.5	94.5
0241, 31 - 34	SITE & INFRASTRUCTURE, DEMOLITION	98.7	89.0	92.0	86.5	87.7	87.3	83.2	87.4	86.1	91.9	115.1	107.9	99.8	88.8	92.2	99.3	92.8	94.8
0310	Concrete Forming & Accessories	94.9	77.4	80.0	100.8	82.1	84.9	95.0	80.2	82.4	94.3	78.6	80.9	92.9	73.7	76.5	95.0	111.1	108.7
0320	Concrete Reinforcing	104.9	83.0	94.3	90.4	86.9	88.7	90.8	86.9	88.9	99.0	81.0	90.3	104.9	77.9	91.9	104.9	115.8	110.2
0330	Cast-in-Place Concrete	106.3	74.9	94.6	101.3	75.7	91.7	100.8	76.5	91.8	96.8	83.2	91.7	113.0	75.2	98.9	111.1	112.2	111.5
03	CONCRETE	95.7	78.1	88.0	99.2	80.3	90.9	98.5	79.6	90.2	99.5	80.8	91.3	98.6	75.7	88.6	97.9	112.1	104.1
04	MASONRY	87.4	74.0	79.2	88.9	74.4	80.0	88.8	74.3	79.9	84.3	77.3	80.0	90.7	71.4	78.8	88.8	110.3	102.0
05	METALS	97.7	88.2	94.9	99.1	75.7	92.2	99.1	74.7	91.9	91.9	83.2	89.3	97.7	86.3	94.3	97.7	107.1	100.5
06	WOOD, PLASTICS & COMPOSITES	93.8	77.6	84.9	109.5	82.9	94.9	104.0	80.5	91.1	90.8	77.7	83.6	93.7	73.8	82.7	91.3	109.1	101.1
07	THERMAL & MOISTURE PROTECTION	109.2	75.0	94.4	95.8	77.9	88.0	95.2	78.7	88.0	100.0	81.7	92.1	108.9	76.8	94.9	107.7	106.2	107.0
08	OPENINGS	93.2	75.8	89.2	97.5	80.0	93.4	93.9	78.6	90.3	91.8	76.0	88.1	93.2	71.2	88.1	93.2	114.6	98.2
0920	Plaster & Gypsum Board	106.4	77.3	86.8	98.6	83.2	88.2	95.4	80.7	85.5	94.2	76.7	82.4	105.8	73.4	84.0	99.5	109.7	106.4
0950, 0980	Ceilings & Acoustic Treatment	89.8	77.3	81.4	77.9	83.2	81.5	77.9	80.7	79.8	81.6	76.7	78.3	89.8	73.4	78.7	89.8	109.7	103.2
0960	Flooring	93.5	75.0	88.3	99.6	83.2	95.0	94.4	83.2	91.2	94.1	71.4	87.7	93.5	71.2	87.2	93.5	110.5	98.3
0970, 0990	Wall Finishes & Painting/Coating	92.8	65.8	76.9	84.4	81.1	82.5	84.4	81.1	82.5	90.3	84.4	86.8	92.8	70.2	79.5	92.8	120.7	109.3
09	FINISHES	91.3	75.7	82.9	90.6	82.4	86.2	88.6	81.0	84.5	89.3	77.7	83.0	91.1	72.9	81.3	90.3	111.8	102.0
COVERS	DIVS. 10 - 14, 25, 28, 41, 43, 44, 46	100.0	87.4	97.2	100.0	88.3	97.4	100.0	88.0	97.3	100.0	91.2	98.0	100.0	87.2	97.1	100.0	107.9	101.8
21, 22, 23	FIRE SUPPRESSION, PLUMBING & HVAC	100.0	77.0	91.0	99.7	79.6	91.8	96.2	79.1	89.5	99.9	78.6	91.5	100.0	72.2	89.1	100.0	106.2	102.4
26, 27, 3370	ELECTRICAL, COMMUNICATIONS & UTIL.	87.7	83.2	85.5	99.9	86.6	93.3	99.1	86.7	93.0	95.7	81.4	88.7	88.4	74.9	81.7	99.5	110.1	104.7
MF2018	WEIGHTED AVERAGE	96.0	79.8	89.1	97.8	81.1	90.7	96.1	80.5	89.5	95.5	82.7	90.1	96.6	76.3	88.0	97.5	108.1	102.0

	DIVISION	INDIANA																	
		INDIANAPOLIS			KOKOMO			LAFAYETTE			LAWRENCEBURG			MUNCIE			NEW ALBANY		
		461 - 462			469			479			470			473			471		
		MAT.	INST.	TOTAL	MAT.	INST.	TOTAL	MAT.	INST.	TOTAL	MAT.	INST.	TOTAL	MAT.	INST.	TOTAL	MAT.	INST.	TOTAL
015433	CONTRACTOR EQUIPMENT		86.1	86.1		94.5	94.5		81.5	81.5		100.6	100.6		92.8	92.8		90.6	90.6
0241, 31 - 34	SITE & INFRASTRUCTURE, DEMOLITION	99.8	91.3	93.9	94.9	89.0	90.8	83.9	87.4	86.3	81.2	102.2	95.7	86.6	87.9	87.5	78.5	89.8	86.3
0310	Concrete Forming & Accessories	100.4	85.2	87.4	98.0	78.2	81.1	92.5	78.6	80.7	91.5	78.4	80.3	92.3	76.9	79.2	90.1	76.9	78.9
0320	Concrete Reinforcing	106.1	87.2	97.0	94.9	87.2	91.1	90.4	82.9	86.8	89.7	79.8	84.9	100.0	82.9	91.7	91.0	82.6	86.9
0330	Cast-in-Place Concrete	101.2	84.8	95.1	105.2	81.8	96.5	101.4	77.6	92.5	94.8	74.9	87.4	106.4	73.9	94.3	97.9	73.1	88.7
03	CONCRETE	99.0	84.8	92.8	92.5	81.6	87.7	98.7	78.7	89.9	91.8	78.0	85.7	97.4	77.6	88.7	97.3	76.8	88.3
04	MASONRY	89.8	78.7	83.0	87.1	77.2	81.0	94.0	74.0	81.7	73.8	74.2	74.1	90.5	74.1	80.4	80.2	71.1	74.6
05	METALS	94.7	76.0	89.2	94.2	89.8	92.9	97.4	73.7	90.5	93.9	86.2	91.7	100.8	88.0	97.0	96.0	82.2	91.9
06	WOOD, PLASTICS & COMPOSITES	100.1	86.0	92.4	96.8	77.4	86.1	101.2	79.2	89.1	89.1	78.3	83.2	102.6	77.1	88.6	91.2	77.6	83.7
07	THERMAL & MOISTURE PROTECTION	98.7	82.0	91.4	108.1	77.5	94.8	95.2	77.2	87.4	100.8	76.9	90.4	98.2	76.4	88.7	87.6	72.3	80.9
08	OPENINGS	103.9	81.9	98.8	88.4	76.9	85.7	92.4	76.7	88.8	93.6	75.0	89.3	91.0	75.6	87.4	91.4	76.3	87.9
0920	Plaster & Gypsum Board	96.8	85.8	89.4	111.0	77.1	88.2	93.3	79.4	83.9	71.7	78.4	76.2	94.2	77.3	82.8	92.0	77.5	82.2
0950, 0980	Ceilings & Acoustic Treatment	92.7	85.8	88.0	90.6	77.1	81.5	73.9	79.4	77.6	84.0	78.4	80.2	77.9	77.3	77.5	81.6	77.5	78.8
0960	Flooring	97.2	83.2	93.2	97.3	88.4	94.8	93.3	79.2	89.3	68.9	83.2	72.9	93.7	75.0	88.5	91.6	54.0	81.1
0970, 0990	Wall Finishes & Painting/Coating	96.5	81.1	87.4	92.8	68.8	78.6	84.4	81.4	82.6	85.2	73.0	78.0	84.4	65.8	73.4	90.3	66.3	76.1
09	FINISHES	94.9	84.6	89.4	93.1	79.1	85.5	87.2	79.2	82.8	78.8	78.9	78.8	87.9	75.4	81.1	88.5	71.4	79.2
COVERS	DIVS. 10 - 14, 25, 28, 41, 43, 44, 46	100.0	93.7	98.6	100.0	88.6	97.5	100.0	86.6	97.0	100.0	88.0	97.3	100.0	86.3	97.0	100.0	87.7	97.3
21, 22, 23	FIRE SUPPRESSION, PLUMBING & HVAC	100.1	79.9	92.2	96.5	79.3	89.7	96.2	76.2	88.3	97.0	75.9	88.7	99.7	76.9	90.7	96.4	78.6	89.4
26, 27, 3370	ELECTRICAL, COMMUNICATIONS & UTIL.	101.9	86.7	94.4	92.0	78.3	85.3	98.6	78.6	88.7	93.6	74.1	84.0	91.4	74.0	82.9	94.3	74.6	84.6
MF2018	WEIGHTED AVERAGE	98.7	83.2	92.1	94.2	81.1	88.6	95.8	78.1	88.3	92.5	79.6	87.0	96.2	78.2	88.6	93.9	77.3	86.9

	DIVISION	INDIANA									IOWA								
		SOUTH BEND			TERRE HAUTE			WASHINGTON			BURLINGTON			CARROLL			CEDAR RAPIDS		
		465 - 466			478			475			526			514			522 - 524		
		MAT.	INST.	TOTAL	MAT.	INST.	TOTAL	MAT.	INST.	TOTAL	MAT.	INST.	TOTAL	MAT.	INST.	TOTAL	MAT.	INST.	TOTAL
015433	CONTRACTOR EQUIPMENT		108.2	108.2		110.3	110.3		110.3	110.3		98.8	98.8		98.8	98.8		95.7	95.7
0241, 31 - 34	SITE & INFRASTRUCTURE, DEMOLITION	97.7	94.2	95.3	93.4	115.4	108.6	93.2	115.7	108.7	98.0	91.8	93.7	87.1	92.6	90.9	99.9	91.4	94.0
0310	Concrete Forming & Accessories	95.1	76.0	78.9	95.0	77.1	79.7	96.0	80.5	82.8	96.5	94.2	94.5	84.1	83.9	83.9	102.5	83.9	86.6
0320	Concrete Reinforcing	104.1	83.6	94.2	99.0	83.0	91.3	91.6	82.6	87.2	94.7	97.8	96.2	95.4	85.8	90.8	95.3	80.2	88.0
0330	Cast-in-Place Concrete	108.2	78.4	97.1	93.7	78.3	88.0	102.0	85.0	95.6	107.3	54.8	87.8	107.3	82.2	97.9	107.6	83.4	98.6
03	CONCRETE	97.7	79.7	89.8	102.5	78.8	92.1	108.3	82.6	97.0	95.8	81.5	89.6	94.7	84.3	90.1	95.9	83.8	90.6
04	MASONRY	91.2	74.0	80.6	92.0	73.3	80.5	84.5	79.6	81.5	99.5	71.9	82.5	101.1	72.4	83.5	105.0	80.0	89.6
05	METALS	101.5	102.4	101.7	92.6	84.4	90.2	87.1	84.7	86.4	87.9	99.1	91.2	87.9	95.4	90.1	90.3	93.1	91.2
06	WOOD, PLASTICS & COMPOSITES	96.6	75.2	84.8	93.0	76.7	84.0	93.2	79.2	85.5	91.6	98.1	95.2	77.8	88.4	83.6	98.9	83.6	90.5
07	THERMAL & MOISTURE PROTECTION	103.2	78.7	92.5	100.1	79.1	91.0	100.0	83.5	92.8	104.4	77.1	92.5	104.7	78.8	93.4	105.5	80.1	94.4
08	OPENINGS	93.1	75.5	89.0	92.3	75.3	88.3	89.3	77.1	86.4	94.5	95.8	94.8	98.8	84.2	95.4	99.3	81.2	95.1
0920	Plaster & Gypsum Board	97.7	74.7	82.2	94.2	75.7	81.8	94.0	78.3	83.4	105.0	98.3	100.5	100.9	88.3	92.4	110.3	83.6	92.3
0950, 0980	Ceilings & Acoustic Treatment	94.2	74.7	81.1	81.6	75.7	77.6	76.7	78.3	77.8	99.0	98.3	98.5	99.0	88.3	91.8	101.4	83.6	89.4
0960	Flooring	90.7	86.2	89.5	94.1	75.7	88.9	95.1	83.2	91.7	93.7	69.1	86.8	88.0	79.5	85.6	107.9	84.3	101.3
0970, 0990	Wall Finishes & Painting/Coating	94.3	82.5	87.3	90.3	79.8	84.1	90.3	84.4	86.8	92.8	84.6	87.9	92.8	84.6	87.9	94.7	70.8	80.6
09	FINISHES	92.5	78.3	84.8	89.3	76.9	82.6	88.9	81.3	84.7	93.9	89.7	91.6	90.1	83.6	86.6	99.3	82.7	90.3
COVERS	DIVS. 10 - 14, 25, 28, 41, 43, 44, 46	100.0	88.7	97.5	100.0	88.7	97.5	100.0	91.8	98.2	100.0	93.4	98.5	100.0	89.1	97.6	100.0	93.4	98.5
21, 22, 23	FIRE SUPPRESSION, PLUMBING & HVAC	99.9	75.0	90.1	99.9	76.5	90.7	96.4	81.0	90.4	96.5	83.8	91.5	96.5	77.8	89.2	100.0	81.2	92.6
26, 27, 3370	ELECTRICAL, COMMUNICATIONS & UTIL.	101.3	83.9	92.7	94.1	83.5	88.8	94.5	83.6	89.2	101.1	72.2	86.9	101.8	76.9	89.5	98.8	80.1	89.5
MF2018	WEIGHTED AVERAGE	98.4	81.8	91.4	96.2	81.7	90.1	94.6	84.8	90.4	95.7	84.3	90.9	95.5	82.4	89.9	98.0	83.7	92.0

	DIVISION	IOWA																	
		COUNCIL BLUFFS			CRESTON			DAVENPORT			DECORAH			DES MOINES			DUBUQUE		
		515			508			527 - 528			521			500 - 503,509			520		
		MAT.	INST.	TOTAL	MAT.	INST.	TOTAL	MAT.	INST.	TOTAL	MAT.	INST.	TOTAL	MAT.	INST.	TOTAL	MAT.	INST.	TOTAL
015433	**CONTRACTOR EQUIPMENT**		95.0	95.0		98.8	98.8		98.8	98.8		98.8	98.8		102.4	102.4		94.5	94.5
0241, 31 - 34	**SITE & INFRASTRUCTURE, DEMOLITION**	104.5	88.5	93.4	92.9	93.6	93.4	98.4	94.6	95.8	96.6	91.6	93.2	98.3	99.3	99.0	97.7	88.7	91.5
0310	Concrete Forming & Accessories	83.5	73.6	75.0	79.2	86.7	85.6	102.0	95.9	96.8	94.0	71.8	75.1	95.6	89.5	90.4	84.9	81.5	82.0
0320	Concrete Reinforcing	97.3	79.8	88.8	96.3	85.9	91.3	95.3	99.0	97.1	94.7	85.2	90.1	101.0	101.6	101.3	94.0	79.9	87.2
0330	Cast-in-Place Concrete	111.8	79.1	99.6	114.6	85.7	103.8	103.6	95.2	100.5	104.4	76.0	93.8	97.6	92.1	95.6	105.3	83.0	97.0
03	**CONCRETE**	98.1	77.6	89.1	97.7	86.8	92.9	94.0	96.7	95.2	93.8	76.6	86.3	91.0	92.8	91.8	92.7	82.6	88.3
04	**MASONRY**	106.5	76.2	87.8	99.7	79.9	87.5	102.0	92.2	96.0	120.3	69.2	88.9	87.7	89.0	88.5	105.9	68.8	83.1
05	**METALS**	95.2	92.8	94.5	88.4	96.0	90.6	90.3	105.5	94.8	88.0	94.2	89.9	94.1	96.8	94.9	88.9	92.4	89.9
06	**WOOD, PLASTICS & COMPOSITES**	76.5	72.1	74.1	69.7	88.4	80.0	98.9	95.3	96.9	88.5	70.9	78.8	87.6	88.5	88.1	78.3	82.2	80.4
07	**THERMAL & MOISTURE PROTECTION**	104.8	74.8	91.8	106.2	82.2	95.8	104.9	91.6	99.1	104.6	69.9	89.5	98.5	87.6	93.8	105.1	76.5	92.6
08	**OPENINGS**	98.4	76.8	93.3	108.6	85.9	103.3	99.3	96.7	98.7	97.4	77.9	92.9	100.8	89.6	98.2	98.4	82.4	94.6
0920	Plaster & Gypsum Board	100.9	71.8	81.3	95.8	88.3	90.8	110.3	95.4	100.2	103.7	70.3	81.2	92.8	88.3	89.8	100.9	82.1	88.3
0950, 0980	Ceilings & Acoustic Treatment	99.0	71.8	80.6	90.2	88.3	88.9	101.4	95.4	97.3	99.0	70.3	79.6	93.4	88.3	90.0	99.0	82.1	87.6
0960	Flooring	87.0	84.3	86.2	81.3	69.1	77.8	96.1	89.4	94.2	93.3	69.1	86.5	95.2	94.5	95.0	99.1	69.1	90.7
0970, 0990	Wall Finishes & Painting/Coating	89.6	59.8	72.0	83.6	84.6	84.2	92.8	90.5	91.4	92.8	84.6	87.9	91.6	84.6	87.5	93.9	78.4	84.8
09	**FINISHES**	91.0	73.7	81.7	85.1	83.4	84.2	95.7	94.1	94.8	93.5	72.4	82.1	93.6	90.0	91.6	94.8	78.9	86.2
COVERS	**DIVS. 10 - 14, 25, 28, 41, 43, 44, 46**	100.0	91.4	98.1	100.0	91.6	98.1	100.0	96.9	99.3	100.0	88.6	97.5	100.0	95.6	99.0	100.0	92.3	98.3
21, 22, 23	**FIRE SUPPRESSION, PLUMBING & HVAC**	100.0	72.8	89.4	96.3	79.8	89.8	100.0	93.9	97.6	96.5	74.5	87.9	99.8	86.8	94.7	100.0	75.6	90.4
26, 27, 3370	**ELECTRICAL, COMMUNICATIONS & UTIL.**	103.9	82.7	93.4	93.5	76.9	85.3	96.9	87.6	92.3	98.8	47.2	73.4	105.1	84.3	94.8	102.5	76.9	89.9
MF2018	**WEIGHTED AVERAGE**	98.9	79.2	90.6	95.7	84.2	90.8	97.1	94.5	96.0	96.4	73.7	86.8	97.2	90.2	94.3	97.2	79.9	89.9

	DIVISION	IOWA																	
		FORT DODGE			MASON CITY			OTTUMWA			SHENANDOAH			SIBLEY			SIOUX CITY		
		505			504			525			516			512			510 - 511		
		MAT.	INST.	TOTAL	MAT.	INST.	TOTAL	MAT.	INST.	TOTAL	MAT.	INST.	TOTAL	MAT.	INST.	TOTAL	MAT.	INST.	TOTAL
015433	**CONTRACTOR EQUIPMENT**		98.8	98.8		98.8	98.8		94.5	94.5		95.0	95.0		98.8	98.8		98.8	98.8
0241, 31 - 34	**SITE & INFRASTRUCTURE, DEMOLITION**	101.4	90.5	93.9	101.5	91.5	94.6	98.0	86.7	90.2	102.6	88.6	92.9	107.7	91.6	96.6	109.6	93.3	98.3
0310	Concrete Forming & Accessories	79.8	78.0	78.3	83.6	71.4	73.2	91.9	87.1	87.8	85.1	77.2	78.3	85.7	37.8	44.9	102.5	75.5	79.4
0320	Concrete Reinforcing	96.3	85.2	90.9	96.2	85.2	90.9	94.7	98.0	96.3	97.3	85.8	91.7	97.3	85.1	91.4	95.3	100.8	98.0
0330	Cast-in-Place Concrete	107.6	41.9	83.1	107.6	71.2	94.0	108.0	63.9	91.6	108.1	82.6	98.6	105.9	55.5	87.2	106.6	89.3	100.2
03	**CONCRETE**	93.1	67.6	81.9	93.3	74.8	85.2	95.4	81.5	89.3	95.6	81.5	89.4	94.7	54.1	76.8	95.3	85.6	91.0
04	**MASONRY**	98.6	51.9	69.9	111.6	68.0	84.8	102.3	55.2	73.3	106.0	74.5	86.6	124.6	52.2	80.1	99.2	70.0	81.2
05	**METALS**	88.5	94.0	90.1	88.5	94.2	90.2	87.9	99.7	91.3	94.2	95.6	94.6	88.1	93.3	89.7	90.3	101.3	93.6
06	**WOOD, PLASTICS & COMPOSITES**	70.1	88.4	80.2	74.0	70.9	72.3	85.7	97.9	92.4	78.3	78.5	78.4	79.1	34.2	54.4	98.9	74.4	85.4
07	**THERMAL & MOISTURE PROTECTION**	105.5	65.2	88.0	105.0	70.5	90.0	105.3	70.3	90.1	104.0	73.3	90.7	104.3	55.2	82.9	104.9	73.4	91.2
08	**OPENINGS**	102.5	75.9	96.3	94.7	77.9	90.8	98.8	92.8	97.4	90.2	76.4	87.0	95.6	46.0	84.0	99.3	81.6	95.1
0920	Plaster & Gypsum Board	95.8	88.3	90.8	95.8	70.3	78.6	101.5	98.3	99.3	100.9	78.3	85.7	100.9	32.5	54.9	110.3	73.8	85.8
0950, 0980	Ceilings & Acoustic Treatment	90.2	88.3	88.9	90.2	70.3	76.8	99.0	98.3	98.5	99.0	78.3	85.0	99.0	32.5	54.2	101.4	73.8	82.8
0960	Flooring	82.6	69.1	78.8	84.6	69.1	80.2	102.2	69.1	92.9	87.7	73.9	83.8	88.9	69.1	83.3	96.1	73.3	89.7
0970, 0990	Wall Finishes & Painting/Coating	83.6	81.3	82.2	83.6	84.6	84.2	93.9	84.6	88.4	89.6	84.6	86.6	92.8	83.6	87.3	92.8	69.1	78.8
09	**FINISHES**	87.0	78.0	82.1	87.6	72.1	79.2	95.9	84.8	89.9	91.1	77.1	83.5	93.3	46.4	67.9	97.2	73.8	84.5
COVERS	**DIVS. 10 - 14, 25, 28, 41, 43, 44, 46**	100.0	84.8	96.6	100.0	88.2	97.4	100.0	87.3	97.2	100.0	88.2	97.4	100.0	78.9	95.3	100.0	91.4	98.1
21, 22, 23	**FIRE SUPPRESSION, PLUMBING & HVAC**	96.3	71.4	86.6	96.3	77.7	89.0	96.5	73.1	87.3	96.5	83.9	91.6	96.5	71.5	86.7	100.0	81.2	92.6
26, 27, 3370	**ELECTRICAL, COMMUNICATIONS & UTIL.**	99.8	69.4	84.8	98.9	47.2	73.4	100.9	71.0	86.2	98.8	78.4	88.7	98.8	47.2	73.4	98.8	74.0	86.6
MF2018	**WEIGHTED AVERAGE**	95.5	73.7	86.3	95.3	74.0	86.3	96.3	78.7	88.9	96.1	81.9	90.1	96.8	62.4	82.3	97.7	81.5	90.9

	DIVISION	IOWA						KANSAS											
		SPENCER			WATERLOO			BELLEVILLE			COLBY			DODGE CITY			EMPORIA		
		513			506 - 507			669			677			678			668		
		MAT.	INST.	TOTAL	MAT.	INST.	TOTAL	MAT.	INST.	TOTAL	MAT.	INST.	TOTAL	MAT.	INST.	TOTAL	MAT.	INST.	TOTAL
015433	**CONTRACTOR EQUIPMENT**		98.8	98.8		98.8	98.8		103.7	103.7		103.7	103.7		103.7	103.7		101.9	101.9
0241, 31 - 34	**SITE & INFRASTRUCTURE, DEMOLITION**	107.8	90.4	95.7	106.9	92.7	97.1	111.3	92.0	98.0	106.7	92.7	97.0	109.4	91.7	97.2	103.1	89.7	93.8
0310	Concrete Forming & Accessories	91.8	37.5	45.5	94.4	68.7	72.5	94.8	54.3	60.3	97.5	61.2	66.6	91.1	61.1	65.5	86.2	70.0	72.4
0320	Concrete Reinforcing	97.3	85.1	91.4	96.9	80.3	88.9	98.9	102.6	100.7	97.1	102.6	99.8	94.7	102.3	98.4	97.5	102.8	100.1
0330	Cast-in-Place Concrete	105.9	65.5	90.9	115.2	84.0	103.6	121.0	83.5	107.1	119.5	87.2	107.5	121.6	86.9	108.7	117.0	87.3	106.0
03	**CONCRETE**	95.0	57.4	78.5	99.0	77.2	89.4	110.5	74.3	94.6	108.3	78.7	95.3	109.7	78.5	96.0	103.1	82.8	94.2
04	**MASONRY**	124.6	52.2	80.1	99.4	75.1	84.5	86.4	58.5	69.3	100.5	64.6	78.4	110.7	59.6	79.3	91.8	66.0	76.0
05	**METALS**	88.1	93.0	89.5	90.7	93.0	91.4	95.6	99.2	96.7	92.8	100.0	94.9	94.2	98.6	95.5	95.3	100.2	96.8
06	**WOOD, PLASTICS & COMPOSITES**	85.5	34.2	57.3	87.2	64.4	74.6	93.3	51.2	70.1	96.9	57.4	75.2	88.9	57.4	71.5	84.3	68.8	75.8
07	**THERMAL & MOISTURE PROTECTION**	105.3	55.8	83.8	105.3	77.2	93.0	91.4	62.1	78.7	97.8	65.4	83.7	97.8	66.6	84.2	89.5	75.4	83.3
08	**OPENINGS**	106.7	46.0	92.5	95.1	72.9	90.0	94.1	62.5	86.7	97.9	65.9	90.4	97.8	65.9	90.4	92.1	74.4	88.0
0920	Plaster & Gypsum Board	101.5	32.5	55.1	103.8	63.6	76.7	88.7	50.0	62.6	96.8	56.3	69.6	90.5	56.3	67.5	85.5	68.1	73.8
0950, 0980	Ceilings & Acoustic Treatment	99.0	32.5	54.2	91.8	63.6	72.8	76.5	50.0	58.6	77.3	56.3	63.2	77.3	56.3	63.2	76.5	68.1	70.8
0960	Flooring	91.6	69.1	85.3	89.6	78.9	86.6	92.9	68.4	86.0	87.6	68.4	82.2	83.8	68.4	79.4	88.3	68.4	82.7
0970, 0990	Wall Finishes & Painting/Coating	92.8	57.2	71.7	83.6	81.3	82.2	89.7	57.9	70.9	96.1	57.9	73.5	96.1	57.9	73.5	89.7	57.9	70.9
09	**FINISHES**	94.2	42.3	66.1	90.7	70.9	80.0	86.1	55.7	69.6	84.7	60.8	71.8	82.8	60.8	70.9	83.3	67.6	74.8
COVERS	**DIVS. 10 - 14, 25, 28, 41, 43, 44, 46**	100.0	78.9	95.3	100.0	91.0	98.0	100.0	83.3	96.3	100.0	86.1	96.9	100.0	86.0	96.9	100.0	85.9	96.9
21, 22, 23	**FIRE SUPPRESSION, PLUMBING & HVAC**	96.5	71.5	86.7	99.9	80.1	92.1	96.4	70.3	86.1	96.5	71.1	86.5	100.0	71.1	88.7	96.4	73.5	87.4
26, 27, 3370	**ELECTRICAL, COMMUNICATIONS & UTIL.**	100.4	47.2	74.2	95.4	61.6	78.7	104.1	62.7	83.7	95.5	67.9	81.9	92.9	67.9	80.6	101.4	66.9	84.4
MF2018	**WEIGHTED AVERAGE**	98.2	62.3	83.0	96.8	77.4	88.6	97.9	70.7	86.4	97.3	74.0	87.4	98.6	73.3	87.9	96.1	76.5	87.8

	DIVISION	KANSAS																	
		FORT SCOTT			HAYS			HUTCHINSON			INDEPENDENCE			KANSAS CITY			LIBERAL		
		667			676			675			673			660 - 662			679		
		MAT.	INST.	TOTAL	MAT.	INST.	TOTAL	MAT.	INST.	TOTAL	MAT.	INST.	TOTAL	MAT.	INST.	TOTAL	MAT.	INST.	TOTAL
015433	CONTRACTOR EQUIPMENT		102.8	102.8		103.7	103.7		103.7	103.7		103.7	103.7		100.4	100.4		103.7	103.7
0241, 31 - 34	SITE & INFRASTRUCTURE, DEMOLITION	99.9	89.7	92.8	111.4	92.2	98.2	90.6	92.7	92.0	110.2	92.7	98.1	94.3	90.2	91.5	111.3	92.3	98.2
0310	Concrete Forming & Accessories	102.8	82.9	85.9	95.2	58.9	64.3	86.0	56.6	61.0	105.8	67.8	73.4	99.4	98.4	98.6	91.5	58.8	63.7
0320	Concrete Reinforcing	96.9	100.8	98.8	94.7	102.6	98.5	94.7	102.6	98.5	94.1	100.8	97.3	94.0	105.3	99.4	96.1	102.3	99.1
0330	Cast-in-Place Concrete	108.5	82.4	98.8	94.2	83.8	90.3	87.3	86.9	87.2	122.1	87.1	109.1	92.9	97.4	94.6	94.2	83.5	90.3
03	CONCRETE	98.4	86.6	93.2	99.8	76.5	89.6	83.0	76.5	80.2	110.8	81.3	97.8	90.7	99.7	94.6	101.7	76.3	90.5
04	MASONRY	92.7	56.1	70.2	109.5	58.6	78.2	100.2	64.5	78.3	97.8	64.5	77.3	93.5	98.1	96.3	108.2	53.6	74.6
05	METALS	95.3	99.0	96.4	92.3	99.8	94.5	92.1	99.0	94.1	92.0	98.8	94.0	103.1	106.2	104.0	92.6	98.7	94.4
06	WOOD, PLASTICS & COMPOSITES	103.8	90.5	96.5	93.9	57.4	73.8	84.0	51.4	66.0	107.3	65.9	84.5	99.6	98.5	99.0	89.6	57.4	71.9
07	THERMAL & MOISTURE PROTECTION	90.5	72.8	82.8	98.1	62.8	82.8	96.6	64.3	82.5	97.8	74.9	87.9	90.3	99.3	94.2	98.3	60.9	82.0
08	OPENINGS	92.1	85.9	90.6	97.8	65.9	90.4	97.8	62.6	89.6	95.9	70.2	89.9	93.2	97.4	94.2	97.9	65.9	90.4
0920	Plaster & Gypsum Board	91.2	90.4	90.7	94.3	56.3	68.7	89.2	50.2	63.0	105.0	65.1	78.1	84.8	98.6	94.1	91.4	56.3	67.8
0950, 0980	Ceilings & Acoustic Treatment	76.5	90.4	85.9	77.3	56.3	63.2	77.3	50.2	59.0	77.3	65.1	69.0	76.5	98.6	91.4	77.3	56.3	63.2
0960	Flooring	102.6	66.9	92.5	86.4	68.4	81.3	80.9	68.4	77.3	92.0	66.9	84.9	82.4	95.3	86.0	84.1	68.4	79.6
0970, 0990	Wall Finishes & Painting/Coating	91.3	77.5	83.1	96.1	57.9	73.5	96.1	57.9	73.5	96.1	57.9	73.5	97.1	102.7	100.4	96.1	57.9	73.5
09	FINISHES	88.5	80.7	84.3	84.5	59.3	70.9	80.3	57.3	67.9	87.3	66.1	75.8	83.6	98.3	91.5	83.7	59.3	70.5
COVERS	DIVS. 10 - 14, 25, 28, 41, 43, 44, 46	100.0	86.8	97.1	100.0	84.0	96.4	100.0	85.4	96.7	100.0	87.0	97.1	100.0	94.5	98.8	100.0	84.0	96.4
21, 22, 23	FIRE SUPPRESSION, PLUMBING & HVAC	96.4	67.4	85.0	96.5	67.4	85.1	96.5	71.1	86.5	96.5	70.4	86.3	99.9	99.1	99.6	96.5	69.3	85.8
26, 27, 3370	ELECTRICAL, COMMUNICATIONS & UTIL.	100.6	67.9	84.5	94.6	67.9	81.4	90.5	61.0	75.9	92.3	72.8	82.7	106.2	98.1	102.2	92.9	67.9	80.6
MF2018	WEIGHTED AVERAGE	96.0	77.1	88.0	96.5	72.0	86.1	92.5	71.9	83.8	97.2	76.0	88.2	97.3	98.6	97.8	96.5	71.7	86.0

	DIVISION	KANSAS									KENTUCKY								
		SALINA			TOPEKA			WICHITA			ASHLAND			BOWLING GREEN			CAMPTON		
		674			664 - 666			670 - 672			411 - 412			421 - 422			413 - 414		
		MAT.	INST.	TOTAL	MAT.	INST.	TOTAL	MAT.	INST.	TOTAL	MAT.	INST.	TOTAL	MAT.	INST.	TOTAL	MAT.	INST.	TOTAL
015433	CONTRACTOR EQUIPMENT		103.7	103.7		105.0	105.0		106.7	106.7		97.4	97.4		90.6	90.6		96.8	96.8
0241, 31 - 34	SITE & INFRASTRUCTURE, DEMOLITION	99.8	92.2	94.6	98.4	94.4	95.6	96.8	98.5	97.9	114.0	78.8	89.7	78.7	89.4	86.1	87.5	90.7	89.7
0310	Concrete Forming & Accessories	87.8	61.4	65.3	96.7	66.1	70.7	92.9	57.0	62.3	87.2	89.6	89.3	86.5	78.8	79.9	89.3	80.9	82.2
0320	Concrete Reinforcing	94.1	100.5	97.2	93.5	103.8	98.5	92.4	100.4	96.3	92.5	94.2	93.3	89.7	76.8	83.5	90.6	93.0	91.7
0330	Cast-in-Place Concrete	105.6	83.8	97.5	97.6	86.4	93.4	98.7	79.5	91.5	89.1	94.6	91.1	88.4	69.7	81.4	98.6	68.1	87.2
03	CONCRETE	96.8	77.3	88.2	92.3	80.8	87.3	92.4	73.7	84.2	93.1	93.3	93.2	91.6	75.7	84.6	95.0	78.8	87.9
04	MASONRY	125.4	54.9	82.1	87.7	64.3	73.3	97.5	54.1	70.8	90.7	92.0	91.5	92.7	69.2	78.3	89.7	54.7	68.2
05	METALS	94.0	98.9	95.5	99.5	100.0	99.6	96.2	97.5	96.6	95.1	107.7	98.8	96.7	82.8	92.6	96.0	89.8	94.2
06	WOOD, PLASTICS & COMPOSITES	85.4	60.7	71.8	99.9	65.2	80.8	98.7	54.4	74.3	73.4	87.7	81.3	85.7	80.6	82.9	84.2	88.3	86.5
07	THERMAL & MOISTURE PROTECTION	97.2	62.1	82.0	94.0	74.7	85.6	96.2	60.8	80.8	91.2	88.3	90.0	87.6	77.9	83.4	100.2	67.8	86.1
08	OPENINGS	97.8	67.3	90.7	103.5	74.1	96.7	101.3	63.8	92.5	90.6	89.0	90.2	91.4	77.1	88.1	92.6	87.4	91.4
0920	Plaster & Gypsum Board	89.2	59.8	69.4	99.5	64.2	75.7	95.7	53.1	67.0	59.3	87.6	78.4	87.6	80.6	82.9	87.6	87.6	87.6
0950, 0980	Ceilings & Acoustic Treatment	77.3	59.8	65.5	87.2	64.2	71.7	88.0	53.1	64.5	77.0	87.6	84.2	81.6	80.6	80.9	81.6	87.6	85.7
0960	Flooring	82.2	68.4	78.3	98.2	68.4	89.8	94.3	68.4	87.0	73.8	81.3	75.9	89.5	61.9	81.8	91.6	64.3	83.9
0970, 0990	Wall Finishes & Painting/Coating	96.1	57.9	73.5	92.8	67.7	78.0	94.9	57.9	73.0	91.6	90.4	90.9	90.3	67.6	76.9	90.3	53.9	68.8
09	FINISHES	81.5	61.3	70.6	93.9	65.7	78.6	90.3	57.7	72.7	76.1	88.1	82.6	87.1	74.6	80.3	87.9	75.9	81.4
COVERS	DIVS. 10 - 14, 25, 28, 41, 43, 44, 46	100.0	85.3	96.7	100.0	80.3	95.6	100.0	85.1	96.7	100.0	88.2	97.4	100.0	84.4	96.5	100.0	47.8	88.4
21, 22, 23	FIRE SUPPRESSION, PLUMBING & HVAC	100.0	70.5	88.5	100.0	73.2	89.5	99.8	71.0	88.5	96.2	85.0	91.8	99.9	77.0	90.9	96.4	76.4	88.6
26, 27, 3370	ELECTRICAL, COMMUNICATIONS & UTIL.	92.7	73.8	83.4	105.1	71.9	88.7	95.9	73.8	85.0	91.9	88.3	90.1	94.6	74.7	84.8	92.1	88.2	90.2
MF2018	WEIGHTED AVERAGE	97.2	73.5	87.2	98.6	76.5	89.3	97.0	72.6	86.7	93.2	89.7	91.7	94.6	77.2	87.3	94.4	78.0	87.5

	DIVISION	KENTUCKY																	
		CORBIN			COVINGTON			ELIZABETHTOWN			FRANKFORT			HAZARD			HENDERSON		
		407 - 409			410			427			406			417 - 418			424		
		MAT.	INST.	TOTAL	MAT.	INST.	TOTAL	MAT.	INST.	TOTAL	MAT.	INST.	TOTAL	MAT.	INST.	TOTAL	MAT.	INST.	TOTAL
015433	CONTRACTOR EQUIPMENT		96.8	96.8		100.6	100.6		90.6	90.6		100.1	100.1		96.8	96.8		110.3	110.3
0241, 31 - 34	SITE & INFRASTRUCTURE, DEMOLITION	91.6	91.2	91.3	82.7	101.9	95.9	72.9	89.3	84.3	90.0	96.8	94.7	85.3	91.8	89.7	81.7	114.6	104.5
0310	Concrete Forming & Accessories	85.0	76.0	77.3	84.9	70.8	72.8	81.3	72.1	73.5	102.0	78.5	82.0	85.8	81.8	82.4	92.3	77.1	79.3
0320	Concrete Reinforcing	89.6	92.3	90.9	89.3	78.6	84.1	90.2	83.7	87.0	94.0	84.0	89.2	91.0	92.5	91.7	89.8	84.4	87.2
0330	Cast-in-Place Concrete	91.0	72.5	84.1	94.3	79.6	88.9	79.9	67.6	75.3	91.1	77.0	85.8	94.8	69.6	85.4	78.1	84.5	80.5
03	CONCRETE	84.2	78.0	81.5	93.4	76.1	85.8	83.5	73.1	79.0	87.1	79.3	83.6	91.8	79.7	86.5	88.7	81.3	85.5
04	MASONRY	81.9	60.0	68.4	104.0	72.0	84.3	77.4	61.8	67.8	78.6	74.2	75.9	88.6	56.8	69.0	96.0	78.7	85.4
05	METALS	90.9	89.3	90.5	93.9	88.1	92.2	95.9	85.3	92.7	93.1	86.4	91.1	96.0	89.8	94.2	86.8	87.0	86.9
06	WOOD, PLASTICS & COMPOSITES	72.1	77.6	75.1	81.8	68.3	74.3	80.6	73.9	76.9	101.5	77.7	88.4	80.9	88.3	85.0	88.4	75.8	81.5
07	THERMAL & MOISTURE PROTECTION	103.8	69.5	88.9	101.1	72.1	88.5	87.0	68.5	79.0	101.8	75.7	90.5	100.1	69.5	86.8	99.3	82.8	92.1
08	OPENINGS	87.3	70.5	83.4	94.5	71.8	89.2	91.4	73.0	87.1	98.9	78.4	94.1	93.0	87.2	91.6	89.6	78.5	87.0
0920	Plaster & Gypsum Board	88.4	76.6	80.5	68.2	68.0	68.1	86.3	73.7	77.8	91.6	76.6	81.5	86.3	87.6	87.2	90.9	74.8	80.0
0950, 0980	Ceilings & Acoustic Treatment	74.7	76.6	76.0	84.0	68.0	73.2	81.6	73.7	76.2	90.5	76.6	81.1	81.6	87.6	85.7	76.7	74.8	75.4
0960	Flooring	88.4	64.3	81.6	66.3	83.5	71.1	86.9	73.8	83.2	98.9	78.6	93.2	89.8	64.3	82.6	93.2	76.0	88.4
0970, 0990	Wall Finishes & Painting/Coating	87.2	62.0	72.3	85.2	71.1	76.9	90.3	68.1	77.2	94.2	91.7	92.8	90.3	53.9	68.8	90.3	86.7	88.2
09	FINISHES	83.0	72.5	77.3	77.8	72.7	75.0	85.8	71.6	78.1	93.2	80.2	86.2	87.0	76.4	81.3	87.1	77.9	82.1
COVERS	DIVS. 10 - 14, 25, 28, 41, 43, 44, 46	100.0	89.8	97.7	100.0	85.9	96.8	100.0	70.0	93.3	100.0	58.5	90.8	100.0	48.5	88.5	100.0	55.8	90.1
21, 22, 23	FIRE SUPPRESSION, PLUMBING & HVAC	96.5	74.2	87.7	97.0	76.0	88.8	96.6	77.9	89.3	99.9	81.0	92.5	96.4	77.5	89.0	96.6	79.3	89.8
26, 27, 3370	ELECTRICAL, COMMUNICATIONS & UTIL.	90.0	88.2	89.1	95.6	72.4	84.2	91.9	77.9	85.0	100.6	77.9	89.4	92.1	88.2	90.2	94.0	77.9	86.1
MF2018	WEIGHTED AVERAGE	90.8	78.0	85.4	94.3	77.8	87.3	91.3	75.6	84.7	95.4	80.3	89.0	93.8	78.7	87.4	92.2	81.9	87.8

	DIVISION	KENTUCKY																	
		LEXINGTON			LOUISVILLE			OWENSBORO			PADUCAH			PIKEVILLE			SOMERSET		
		403 - 405			400 - 402			423			420			415 - 416			425 - 426		
		MAT.	INST.	TOTAL	MAT.	INST.	TOTAL	MAT.	INST.	TOTAL	MAT.	INST.	TOTAL	MAT.	INST.	TOTAL	MAT.	INST.	TOTAL
015433	**CONTRACTOR EQUIPMENT**		96.8	96.8		93.7	93.7		110.3	110.3		110.3	110.3		97.4	97.4		96.8	96.8
0241, 31 - 34	**SITE & INFRASTRUCTURE, DEMOLITION**	93.8	92.8	93.1	87.8	94.4	92.3	91.8	115.3	108.0	84.4	114.8	105.4	125.2	78.0	92.6	77.9	91.2	87.1
0310	Concrete Forming & Accessories	96.5	73.7	77.1	93.5	79.1	81.2	90.8	76.8	78.9	88.8	81.2	82.3	96.0	85.6	87.2	86.8	76.6	78.1
0320	Concrete Reinforcing	98.4	84.0	91.4	94.8	84.2	89.6	89.8	77.3	83.8	90.4	82.5	86.6	93.0	94.1	93.5	90.2	92.5	91.3
0330	Cast-in-Place Concrete	93.1	84.4	89.9	89.3	70.7	82.4	91.1	83.9	88.4	83.2	81.1	82.4	97.9	89.4	94.8	78.1	87.9	81.7
03	**CONCRETE**	86.9	79.7	83.8	85.8	77.4	82.1	100.7	79.8	91.5	93.3	81.7	88.2	106.8	89.7	99.3	78.6	83.7	80.8
04	**MASONRY**	80.6	71.1	74.8	78.7	72.0	74.6	88.8	78.6	82.5	91.5	79.2	83.9	88.5	83.2	85.3	83.3	63.2	71.0
05	**METALS**	93.3	86.7	91.3	95.5	85.5	92.5	88.3	84.7	87.3	85.3	87.5	86.0	95.0	107.5	98.7	95.9	90.0	94.2
06	**WOOD, PLASTICS & COMPOSITES**	87.3	71.6	78.6	88.8	80.8	84.4	86.3	75.8	80.5	84.0	81.4	82.6	83.3	87.7	85.7	81.5	77.6	79.3
07	**THERMAL & MOISTURE PROTECTION**	104.1	73.9	90.9	101.0	75.7	90.0	100.1	76.6	89.9	99.4	82.6	92.1	92.1	80.5	87.0	99.3	71.5	87.2
08	**OPENINGS**	87.5	74.6	84.5	88.8	76.8	86.0	89.6	76.7	86.6	89.0	81.1	87.1	91.2	85.7	89.9	92.1	77.1	88.6
0920	Plaster & Gypsum Board	97.7	70.4	79.4	93.7	80.6	84.9	89.3	74.8	79.5	88.6	80.6	83.2	63.1	87.6	79.6	86.3	76.6	79.8
0950, 0980	Ceilings & Acoustic Treatment	77.9	70.4	72.9	82.8	80.6	81.3	76.7	74.8	75.4	76.7	80.6	79.3	77.0	87.6	84.2	81.6	76.6	78.2
0960	Flooring	93.4	64.3	85.2	95.0	62.9	86.0	92.6	61.9	84.0	91.5	76.0	87.2	78.1	64.3	74.2	90.1	64.3	82.8
0970, 0990	Wall Finishes & Painting/Coating	87.2	79.7	82.8	93.9	68.1	78.7	90.3	86.7	88.2	90.3	73.1	80.2	91.6	91.7	91.7	90.3	68.1	77.2
09	**FINISHES**	86.4	72.0	78.6	90.8	74.9	82.2	87.3	74.7	80.5	86.5	79.4	82.7	78.7	82.6	80.8	86.4	73.3	79.3
COVERS	**DIVS. 10 - 14, 25, 28, 41, 43, 44, 46**	100.0	92.1	98.2	100.0	90.5	97.9	100.0	93.0	98.4	100.0	91.1	98.0	100.0	48.2	88.4	100.0	89.2	97.6
21, 22, 23	**FIRE SUPPRESSION, PLUMBING & HVAC**	100.0	76.4	90.8	99.9	80.0	92.1	99.9	77.1	91.0	96.6	78.2	89.4	96.2	82.3	90.7	96.6	74.0	87.7
26, 27, 3370	**ELECTRICAL, COMMUNICATIONS & UTIL.**	92.5	74.7	83.7	96.4	77.9	87.3	94.1	73.7	84.0	96.3	77.3	86.9	94.8	88.3	91.6	92.5	88.2	90.4
MF2018	**WEIGHTED AVERAGE**	93.1	78.1	86.7	93.9	79.6	87.8	94.7	81.0	88.9	92.4	83.2	88.6	95.8	85.4	91.4	91.6	79.5	86.5

	DIVISION	LOUISIANA																	
		ALEXANDRIA			BATON ROUGE			HAMMOND			LAFAYETTE			LAKE CHARLES			MONROE		
		713 - 714			707 - 708			704			705			706			712		
		MAT.	INST.	TOTAL	MAT.	INST.	TOTAL	MAT.	INST.	TOTAL	MAT.	INST.	TOTAL	MAT.	INST.	TOTAL	MAT.	INST.	TOTAL
015433	**CONTRACTOR EQUIPMENT**		90.9	90.9		91.9	91.9		89.3	89.3		89.3	89.3		88.8	88.8		90.9	90.9
0241, 31 - 34	**SITE & INFRASTRUCTURE, DEMOLITION**	98.7	87.7	91.1	102.1	91.2	94.6	101.2	85.7	90.5	102.5	87.9	92.4	103.2	87.1	92.0	98.7	87.6	91.1
0310	Concrete Forming & Accessories	77.8	60.0	62.6	99.0	71.8	75.8	79.0	54.9	58.5	95.6	67.8	71.9	96.4	67.8	72.0	77.4	59.4	62.1
0320	Concrete Reinforcing	92.2	54.3	73.9	93.3	54.3	74.5	92.1	54.3	73.8	93.4	54.3	74.5	93.4	54.3	74.5	91.2	54.3	73.3
0330	Cast-in-Place Concrete	89.0	66.2	80.5	88.8	73.1	82.9	87.6	62.6	78.3	87.1	66.1	79.3	91.7	66.1	82.2	89.0	64.1	79.7
03	**CONCRETE**	82.8	62.0	73.6	85.5	69.8	78.6	85.8	58.3	73.7	86.8	65.4	77.4	89.0	65.4	78.6	82.6	61.0	73.1
04	**MASONRY**	109.3	64.3	81.6	91.7	66.7	76.3	95.3	60.0	73.6	95.3	66.6	77.7	94.7	66.6	77.5	103.9	63.1	78.8
05	**METALS**	92.6	70.9	86.2	96.9	74.0	90.1	88.4	69.3	82.8	87.7	69.6	82.4	87.7	69.6	82.4	92.6	70.8	86.2
06	**WOOD, PLASTICS & COMPOSITES**	89.2	58.8	72.5	103.0	73.2	86.6	83.0	54.3	67.2	103.7	68.0	84.1	101.9	68.0	83.3	88.6	58.8	72.2
07	**THERMAL & MOISTURE PROTECTION**	99.0	65.6	84.5	97.7	69.2	85.3	96.7	62.9	82.0	97.3	67.4	84.3	97.1	67.2	84.1	99.0	64.8	84.1
08	**OPENINGS**	110.8	58.6	98.6	97.9	70.4	91.5	96.2	54.9	86.6	99.8	62.2	91.0	99.8	62.2	91.0	110.7	57.1	98.2
0920	Plaster & Gypsum Board	83.6	58.1	66.4	101.0	72.7	82.0	100.0	53.4	68.7	108.6	67.6	81.0	108.6	67.6	81.0	83.3	58.1	66.3
0950, 0980	Ceilings & Acoustic Treatment	87.1	58.1	67.6	92.9	72.7	79.3	96.9	53.4	67.6	94.4	67.6	76.3	95.3	67.6	76.6	87.1	58.1	67.6
0960	Flooring	85.1	68.2	80.4	93.0	68.2	86.0	89.9	68.2	83.8	98.2	68.2	89.8	98.2	68.2	89.8	84.7	68.2	80.1
0970, 0990	Wall Finishes & Painting/Coating	88.8	59.7	71.6	94.7	59.7	74.0	97.7	59.9	75.3	97.7	59.7	75.2	97.7	59.7	75.2	88.8	59.7	71.6
09	**FINISHES**	80.9	61.2	70.2	92.2	70.3	80.3	90.4	57.5	72.6	93.5	67.2	79.3	93.7	67.2	79.4	80.7	60.9	70.0
COVERS	**DIVS. 10 - 14, 25, 28, 41, 43, 44, 46**	100.0	79.1	95.3	100.0	84.2	96.5	100.0	80.7	95.7	100.0	83.3	96.3	100.0	83.3	96.3	100.0	78.9	95.3
21, 22, 23	**FIRE SUPPRESSION, PLUMBING & HVAC**	100.0	63.2	85.6	99.9	64.1	85.9	96.7	61.0	82.7	100.2	64.2	86.0	100.2	64.5	86.2	100.0	61.9	85.0
26, 27, 3370	**ELECTRICAL, COMMUNICATIONS & UTIL.**	94.3	61.4	78.1	101.7	58.0	80.2	97.0	69.6	83.5	98.1	63.8	81.2	97.7	65.9	82.0	95.9	57.3	76.9
MF2018	**WEIGHTED AVERAGE**	95.9	65.6	83.1	96.6	69.3	85.1	93.7	64.4	81.3	95.4	67.9	83.8	95.6	68.2	84.1	95.8	64.4	82.5

	DIVISION	LOUISIANA									MAINE								
		NEW ORLEANS			SHREVEPORT			THIBODAUX			AUGUSTA			BANGOR			BATH		
		700 - 701			710 - 711			703			043			044			045		
		MAT.	INST.	TOTAL	MAT.	INST.	TOTAL	MAT.	INST.	TOTAL	MAT.	INST.	TOTAL	MAT.	INST.	TOTAL	MAT.	INST.	TOTAL
015433	**CONTRACTOR EQUIPMENT**		88.6	88.6		93.9	93.9		89.3	89.3		98.4	98.4		95.7	95.7		95.7	95.7
0241, 31 - 34	**SITE & INFRASTRUCTURE, DEMOLITION**	104.3	93.9	97.1	101.1	92.3	95.0	103.6	87.6	92.6	88.2	98.0	95.0	90.6	95.5	94.0	88.3	94.2	92.4
0310	Concrete Forming & Accessories	95.7	68.9	72.8	92.9	60.3	65.1	90.2	64.5	68.2	98.6	78.6	81.5	93.0	76.5	79.0	89.1	78.4	80.0
0320	Concrete Reinforcing	92.9	54.4	74.3	92.5	54.3	74.0	92.1	54.3	73.8	102.6	81.1	92.2	94.1	81.1	87.8	93.1	81.1	87.3
0330	Cast-in-Place Concrete	90.7	69.7	82.9	92.0	65.2	82.0	94.1	64.3	83.0	89.8	115.3	99.3	70.9	114.1	87.0	70.9	114.3	87.1
03	**CONCRETE**	94.4	66.6	82.2	86.2	61.7	75.4	90.6	63.2	78.5	95.5	92.3	94.1	88.5	91.0	89.6	88.6	91.9	90.1
04	**MASONRY**	99.8	62.6	77.0	94.5	63.4	75.4	119.9	62.8	84.8	101.0	94.1	96.8	115.8	93.7	102.2	122.5	94.1	105.0
05	**METALS**	99.1	62.6	88.4	96.7	70.2	88.9	88.4	69.5	82.8	105.4	92.3	101.6	93.7	93.0	93.5	92.1	92.9	92.3
06	**WOOD, PLASTICS & COMPOSITES**	100.3	71.1	84.2	98.0	59.7	76.9	90.5	65.5	76.8	93.5	75.6	83.6	90.7	73.0	80.9	85.1	75.4	79.8
07	**THERMAL & MOISTURE PROTECTION**	95.0	68.0	83.3	97.4	65.6	83.6	96.9	65.2	83.1	111.0	100.6	106.5	109.0	100.0	105.1	108.9	100.1	105.1
08	**OPENINGS**	98.8	64.2	90.7	104.7	56.6	93.5	100.7	56.9	90.5	102.7	78.5	97.1	96.4	77.1	91.9	96.4	78.4	92.2
0920	Plaster & Gypsum Board	96.5	70.5	79.0	91.5	58.8	69.5	101.6	65.0	77.0	109.7	74.5	86.0	116.8	71.9	86.6	112.0	74.5	86.8
0950, 0980	Ceilings & Acoustic Treatment	97.5	70.5	79.3	92.8	58.8	69.9	96.9	65.0	75.4	107.1	74.5	85.1	89.6	71.9	77.7	88.6	74.5	79.1
0960	Flooring	103.6	68.2	93.7	91.2	68.2	84.8	95.8	68.2	88.0	89.9	110.6	95.7	82.5	110.6	90.4	80.9	110.6	89.2
0970, 0990	Wall Finishes & Painting/Coating	103.6	60.6	78.2	87.7	59.7	71.1	98.9	59.9	75.8	96.0	88.9	91.8	90.6	88.9	89.6	90.6	88.9	89.6
09	**FINISHES**	96.6	68.2	81.2	87.7	61.5	73.5	92.6	64.8	77.6	95.8	84.7	89.8	89.7	83.0	86.1	88.3	84.6	86.3
COVERS	**DIVS. 10 - 14, 25, 28, 41, 43, 44, 46**	100.0	86.5	97.0	100.0	82.6	96.1	100.0	81.8	95.9	100.0	98.0	99.6	100.0	99.6	99.9	100.0	100.9	100.2
21, 22, 23	**FIRE SUPPRESSION, PLUMBING & HVAC**	100.1	62.5	85.4	99.9	62.8	85.3	96.7	62.5	83.2	100.0	74.3	89.9	100.3	74.1	90.0	96.7	74.3	87.9
26, 27, 3370	**ELECTRICAL, COMMUNICATIONS & UTIL.**	102.1	70.6	86.6	101.8	65.5	83.9	95.7	69.6	82.8	101.6	76.8	89.3	100.0	69.4	84.9	98.2	76.8	87.6
MF2018	**WEIGHTED AVERAGE**	99.0	68.5	86.1	97.1	66.3	84.1	96.0	67.0	83.8	100.4	85.6	94.2	97.1	83.9	91.5	95.9	85.4	91.4

	DIVISION	MAINE																	
		HOULTON			KITTERY			LEWISTON			MACHIAS			PORTLAND			ROCKLAND		
		047			039			042			046			040 - 041			048		
		MAT.	INST.	TOTAL	MAT.	INST.	TOTAL	MAT.	INST.	TOTAL	MAT.	INST.	TOTAL	MAT.	INST.	TOTAL	MAT.	INST.	TOTAL
015433	CONTRACTOR EQUIPMENT		95.7	95.7		95.7	95.7		95.7	95.7		95.7	95.7		99.2	99.2		95.7	95.7
0241, 31 - 34	SITE & INFRASTRUCTURE, DEMOLITION	90.2	94.2	93.0	79.5	94.3	89.7	88.2	95.5	93.3	89.5	94.2	92.8	87.4	100.6	96.5	86.0	94.2	91.7
0310	Concrete Forming & Accessories	96.7	78.4	81.1	88.8	78.7	80.2	98.2	76.6	79.8	93.8	78.4	80.7	99.8	76.8	80.2	94.9	78.4	80.9
0320	Concrete Reinforcing	94.1	81.1	87.8	89.7	81.1	85.5	115.6	81.1	98.9	94.1	81.1	87.8	106.5	81.1	94.2	94.1	81.1	87.8
0330	Cast-in-Place Concrete	70.9	113.2	86.7	70.7	114.4	87.0	72.4	114.1	87.9	70.9	114.3	87.1	85.8	115.2	96.8	72.4	114.3	88.0
03	CONCRETE	89.5	91.5	90.4	82.1	92.1	86.5	88.1	91.0	89.4	89.0	91.9	90.3	91.5	91.4	91.5	86.0	91.9	88.6
04	MASONRY	97.9	94.1	95.6	106.5	94.1	98.9	98.5	93.7	95.5	97.9	94.1	95.6	104.7	93.7	98.0	92.5	94.1	93.5
05	METALS	92.3	92.8	92.4	86.8	93.1	88.6	97.1	93.0	95.9	92.3	92.8	92.5	103.0	92.5	99.9	92.2	92.9	92.4
06	WOOD, PLASTICS & COMPOSITES	94.5	75.4	84.0	88.2	75.4	81.2	96.3	73.0	83.5	91.5	75.4	82.7	95.8	73.1	83.3	92.4	75.4	83.1
07	THERMAL & MOISTURE PROTECTION	109.1	100.1	105.1	107.8	100.1	104.4	108.7	100.0	104.9	109.0	100.1	105.1	113.6	100.6	107.9	108.6	100.1	104.9
08	OPENINGS	96.5	78.4	92.3	96.1	82.0	92.8	99.3	77.1	94.1	96.5	78.4	92.3	96.8	77.2	92.2	96.4	78.4	92.2
0920	Plaster & Gypsum Board	119.0	74.5	89.0	106.1	74.5	84.8	121.8	71.9	88.2	117.4	74.5	88.5	108.0	71.9	83.7	117.4	74.5	88.5
0950, 0980	Ceilings & Acoustic Treatment	88.6	74.5	79.1	101.8	74.5	83.4	99.3	71.9	80.8	88.6	74.5	79.1	107.4	71.9	83.5	88.6	74.5	79.1
0960	Flooring	83.7	110.6	91.2	86.6	110.6	93.4	85.2	110.6	92.3	82.9	110.6	90.7	89.6	110.6	95.5	83.3	110.6	91.0
0970, 0990	Wall Finishes & Painting/Coating	90.6	88.9	89.6	81.8	102.1	93.8	90.6	88.9	89.6	90.6	88.9	89.6	94.6	88.9	91.2	90.6	88.9	89.6
09	FINISHES	90.2	84.6	87.2	91.6	86.0	88.6	92.6	83.0	87.4	89.7	84.6	86.9	96.2	83.2	89.1	89.4	84.6	86.8
COVERS	DIVS. 10 - 14, 25, 28, 41, 43, 44, 46	100.0	97.6	99.5	100.0	100.9	100.2	100.0	99.7	99.9	100.0	97.6	99.5	100.0	100.1	100.0	100.0	100.9	100.2
21, 22, 23	FIRE SUPPRESSION, PLUMBING & HVAC	96.7	74.3	87.9	96.7	80.4	90.3	100.3	74.1	90.0	96.7	74.3	87.9	100.0	74.1	89.9	96.7	74.3	87.9
26, 27, 3370	ELECTRICAL, COMMUNICATIONS & UTIL.	101.9	76.8	89.5	90.3	76.8	83.6	101.9	76.7	89.5	101.9	76.8	89.5	105.0	76.7	91.0	101.8	76.8	89.5
MF2018	WEIGHTED AVERAGE	95.6	85.2	91.2	92.6	87.0	90.2	97.5	84.9	92.2	95.4	85.3	91.1	99.6	85.4	93.6	94.6	85.4	90.7

	DIVISION	MAINE			MARYLAND														
		WATERVILLE			ANNAPOLIS			BALTIMORE			COLLEGE PARK			CUMBERLAND			EASTON		
		049			214			210 - 212			207 - 208			215			216		
		MAT.	INST.	TOTAL	MAT.	INST.	TOTAL	MAT.	INST.	TOTAL	MAT.	INST.	TOTAL	MAT.	INST.	TOTAL	MAT.	INST.	TOTAL
015433	CONTRACTOR EQUIPMENT		95.7	95.7		104.5	104.5		104.1	104.1		107.8	107.8		102.4	102.4		102.4	102.4
0241, 31 - 34	SITE & INFRASTRUCTURE, DEMOLITION	90.0	94.2	92.9	100.0	93.5	95.5	100.1	97.7	98.4	100.1	93.6	95.6	92.0	90.3	90.8	98.8	87.7	91.1
0310	Concrete Forming & Accessories	88.6	78.4	79.9	101.0	78.1	81.5	98.4	76.2	79.5	85.1	76.2	77.5	92.3	81.9	83.5	90.4	73.6	76.1
0320	Concrete Reinforcing	94.1	81.1	87.8	108.6	94.9	102.0	101.2	87.3	94.5	103.5	94.0	98.9	93.7	87.4	90.6	92.9	87.1	90.1
0330	Cast-in-Place Concrete	70.9	114.3	87.1	114.1	79.4	101.2	113.8	77.1	100.1	109.4	76.4	97.2	98.0	82.9	92.4	108.8	64.8	92.4
03	CONCRETE	90.2	91.9	90.9	101.4	82.7	93.2	108.2	79.4	95.5	105.2	80.9	94.5	91.4	84.5	88.4	99.7	74.2	88.5
04	MASONRY	109.8	94.1	100.2	94.8	76.6	83.6	104.7	73.1	85.2	111.1	74.6	88.7	99.3	85.4	90.8	113.5	58.6	79.7
05	METALS	92.3	92.8	92.4	105.5	105.2	105.4	103.8	96.6	101.7	90.2	108.3	95.5	100.7	104.2	101.8	101.0	99.7	100.6
06	WOOD, PLASTICS & COMPOSITES	84.6	75.4	79.5	97.1	77.2	86.1	102.5	76.7	88.3	77.3	75.7	76.4	86.7	81.0	83.6	84.5	80.8	82.5
07	THERMAL & MOISTURE PROTECTION	109.1	100.1	105.2	103.5	83.4	94.8	102.7	80.7	93.2	105.3	81.7	95.0	100.9	81.9	92.6	101.0	73.3	89.0
08	OPENINGS	96.5	78.4	92.3	102.3	83.0	97.8	100.5	80.7	95.9	92.7	82.4	90.3	97.6	84.6	94.6	96.0	83.0	92.9
0920	Plaster & Gypsum Board	112.0	74.5	86.8	94.4	76.8	82.5	100.3	76.2	84.0	95.3	75.2	81.8	99.9	80.8	87.0	99.9	80.6	86.9
0950, 0980	Ceilings & Acoustic Treatment	88.6	74.5	79.1	89.1	76.8	80.8	100.5	76.2	84.1	115.1	75.2	88.2	99.0	80.8	86.7	99.0	80.6	86.6
0960	Flooring	80.6	110.6	89.0	90.7	79.5	87.6	92.8	75.7	88.0	87.9	78.4	85.2	85.8	93.1	87.9	85.0	75.7	82.4
0970, 0990	Wall Finishes & Painting/Coating	90.6	88.9	89.6	99.2	72.7	83.5	101.3	75.3	85.9	104.8	72.5	85.7	98.2	84.7	90.2	98.2	72.5	83.0
09	FINISHES	88.4	84.6	86.4	88.5	77.3	82.4	97.0	75.7	85.5	95.6	75.6	84.8	93.3	84.2	88.4	93.5	74.4	83.2
COVERS	DIVS. 10 - 14, 25, 28, 41, 43, 44, 46	100.0	97.5	99.4	100.0	90.3	97.8	100.0	86.8	97.1	100.0	86.7	97.0	100.0	91.3	98.1	100.0	82.9	96.2
21, 22, 23	FIRE SUPPRESSION, PLUMBING & HVAC	96.7	74.3	87.9	100.0	87.0	94.9	100.1	82.0	93.0	96.5	87.4	93.0	96.4	72.0	86.8	96.4	72.8	87.1
26, 27, 3370	ELECTRICAL, COMMUNICATIONS & UTIL.	101.9	76.8	89.5	101.6	88.9	95.3	99.6	86.9	93.3	97.3	101.9	99.5	98.2	80.8	89.6	97.7	62.3	80.3
MF2018	WEIGHTED AVERAGE	96.0	85.2	91.4	100.4	86.3	94.4	101.8	83.2	93.9	97.3	87.7	93.2	96.9	83.8	91.4	98.6	74.8	88.5

	DIVISION	MARYLAND															MASSACHUSETTS		
		ELKTON			HAGERSTOWN			SALISBURY			SILVER SPRING			WALDORF			BOSTON		
		219			217			218			209			206			020 - 022, 024		
		MAT.	INST.	TOTAL	MAT.	INST.	TOTAL	MAT.	INST.	TOTAL	MAT.	INST.	TOTAL	MAT.	INST.	TOTAL	MAT.	INST.	TOTAL
015433	CONTRACTOR EQUIPMENT		102.4	102.4		102.4	102.4		102.4	102.4		99.7	99.7		99.7	99.7		103.0	103.0
0241, 31 - 34	SITE & INFRASTRUCTURE, DEMOLITION	86.3	89.0	88.1	90.6	90.4	90.4	98.8	87.6	91.1	88.9	85.7	86.7	95.1	85.7	88.6	92.7	102.1	99.2
0310	Concrete Forming & Accessories	96.2	92.3	92.9	91.5	78.8	80.7	104.5	50.2	58.2	93.0	75.5	78.1	99.7	75.4	79.0	105.1	139.7	134.6
0320	Concrete Reinforcing	92.9	117.3	104.7	93.7	87.4	90.6	92.9	65.2	79.5	102.3	93.9	98.3	103.0	93.9	98.6	117.7	156.9	136.7
0330	Cast-in-Place Concrete	88.1	72.2	82.2	93.3	82.9	89.5	108.8	62.9	91.7	112.1	76.8	98.9	125.6	76.7	107.4	99.2	141.8	115.1
03	CONCRETE	84.0	90.6	86.9	87.7	83.1	85.7	100.5	59.2	82.4	102.8	80.5	93.0	113.3	80.5	98.9	105.4	142.7	121.8
04	MASONRY	98.6	67.0	79.2	105.1	85.4	93.0	112.9	55.2	77.4	110.6	75.0	88.7	94.7	75.0	82.6	112.0	146.3	133.1
05	METALS	101.0	114.0	104.8	100.9	104.3	101.9	101.0	90.9	98.1	94.8	104.3	97.6	94.8	104.3	97.5	102.8	136.1	112.6
06	WOOD, PLASTICS & COMPOSITES	91.9	101.3	97.0	85.8	76.7	80.8	102.6	51.4	74.4	84.0	75.0	79.0	90.8	75.0	82.1	103.8	140.2	123.8
07	THERMAL & MOISTURE PROTECTION	100.5	79.4	91.3	101.2	83.0	93.3	101.3	68.5	87.0	108.4	86.5	98.9	109.0	86.5	99.2	107.2	137.1	120.2
08	OPENINGS	96.0	102.1	97.4	95.9	81.2	92.5	96.2	60.8	87.9	84.9	82.1	84.2	85.5	81.5	84.5	99.9	147.3	111.0
0920	Plaster & Gypsum Board	102.8	101.7	102.0	99.9	76.4	84.1	109.4	50.4	69.7	101.9	75.2	83.9	104.8	75.2	84.8	109.0	141.0	130.6
0950, 0980	Ceilings & Acoustic Treatment	99.0	101.7	100.8	101.7	76.4	84.6	99.0	50.4	66.2	123.2	75.2	90.8	123.2	75.2	90.8	100.5	141.0	127.8
0960	Flooring	87.2	75.7	84.0	85.4	93.1	87.6	90.7	75.7	86.5	93.8	78.4	89.5	97.2	78.4	92.0	90.5	163.5	111.0
0970, 0990	Wall Finishes & Painting/Coating	98.2	72.5	83.0	98.2	72.5	83.0	98.2	72.5	83.0	112.7	72.5	88.9	112.7	72.5	88.9	94.3	160.4	133.4
09	FINISHES	93.7	88.5	90.9	93.5	80.4	86.4	96.4	56.2	74.6	95.8	75.0	84.5	97.5	75.1	85.4	98.7	147.0	124.8
COVERS	DIVS. 10 - 14, 25, 28, 41, 43, 44, 46	100.0	56.7	90.4	100.0	90.8	98.0	100.0	76.5	94.8	100.0	85.2	96.7	100.0	83.1	96.2	100.0	119.1	104.3
21, 22, 23	FIRE SUPPRESSION, PLUMBING & HVAC	96.4	78.7	89.5	99.9	83.1	93.3	96.4	70.7	86.3	96.5	87.9	93.1	96.5	87.8	93.1	100.1	127.4	110.8
26, 27, 3370	ELECTRICAL, COMMUNICATIONS & UTIL.	99.4	86.9	93.2	98.0	80.8	89.5	96.6	60.1	78.6	94.6	101.9	98.2	92.2	101.9	97.0	102.6	129.5	115.9
MF2018	WEIGHTED AVERAGE	95.9	86.2	91.8	97.4	85.3	92.3	99.0	66.9	85.4	96.5	86.7	92.4	97.3	86.6	92.8	101.9	134.0	115.5

	DIVISION	MASSACHUSETTS																	
		BROCKTON			BUZZARDS BAY			FALL RIVER			FITCHBURG			FRAMINGHAM			GREENFIELD		
		023			025			027			014			017			013		
		MAT.	INST.	TOTAL	MAT.	INST.	TOTAL	MAT.	INST.	TOTAL	MAT.	INST.	TOTAL	MAT.	INST.	TOTAL	MAT.	INST.	TOTAL
015433	CONTRACTOR EQUIPMENT		98.1	98.1		98.1	98.1		98.9	98.9		95.7	95.7		97.4	97.4		95.7	95.7
0241, 31 - 34	SITE & INFRASTRUCTURE, DEMOLITION	91.1	98.0	95.8	81.2	98.1	92.9	90.3	98.0	95.6	82.7	97.8	93.1	79.5	97.8	92.1	86.4	97.3	93.9
0310	Concrete Forming & Accessories	100.7	121.1	118.1	98.5	120.6	117.4	100.7	120.8	117.8	94.0	118.5	114.9	101.4	121.3	118.4	92.4	117.7	114.0
0320	Concrete Reinforcing	109.7	142.7	125.7	88.0	122.3	104.6	109.7	122.3	115.8	87.9	142.5	114.3	87.9	142.8	114.4	91.5	126.5	108.4
0330	Cast-in-Place Concrete	88.1	136.6	106.2	73.2	136.4	96.7	85.2	136.9	104.4	78.4	136.5	100.0	78.4	136.7	100.1	80.5	122.2	96.0
03	CONCRETE	95.1	129.8	110.3	80.7	126.0	100.6	93.8	126.2	108.0	79.5	128.3	101.0	82.3	130.0	103.2	83.0	120.3	99.4
04	MASONRY	105.4	135.1	123.6	97.9	135.1	120.8	105.9	135.1	123.8	98.2	131.9	118.9	104.7	136.1	124.0	102.7	119.1	112.8
05	METALS	98.8	125.6	106.7	93.6	117.0	100.5	98.8	117.3	104.3	95.3	122.1	103.2	95.3	125.8	104.3	97.8	113.8	102.5
06	WOOD, PLASTICS & COMPOSITES	98.1	119.7	109.9	94.7	119.7	108.4	98.1	119.9	110.1	94.5	116.5	106.6	101.0	119.4	111.2	92.5	119.7	107.4
07	THERMAL & MOISTURE PROTECTION	103.2	128.4	114.2	102.0	125.7	112.3	103.0	124.9	112.5	103.2	122.1	111.4	103.3	128.3	114.2	103.3	109.9	106.2
08	OPENINGS	97.4	128.4	104.6	93.4	118.9	99.3	97.4	118.3	102.2	99.2	126.7	105.6	89.8	128.3	98.7	99.4	120.3	104.3
0920	Plaster & Gypsum Board	92.3	119.9	110.9	87.3	119.9	109.3	92.3	119.9	110.9	111.3	116.7	114.9	114.4	119.9	118.1	112.6	119.9	117.5
0950, 0980	Ceilings & Acoustic Treatment	104.3	119.9	114.9	86.2	119.9	108.9	104.3	119.9	114.9	93.9	116.7	109.3	93.9	119.9	111.5	103.7	119.9	114.6
0960	Flooring	85.0	162.0	106.6	82.6	162.0	104.9	83.8	162.0	105.8	86.7	162.0	107.8	88.2	162.0	108.9	85.9	135.0	99.7
0970, 0990	Wall Finishes & Painting/Coating	85.0	139.2	117.1	85.0	139.2	117.1	85.0	139.2	117.1	86.2	139.2	117.6	87.1	139.2	117.9	86.2	109.3	99.9
09	FINISHES	90.8	130.8	112.5	85.2	130.8	109.9	90.5	131.0	112.4	89.5	128.9	110.9	90.2	130.6	112.1	91.7	120.9	107.5
COVERS	DIVS. 10 - 14, 25, 28, 41, 43, 44, 46	100.0	109.7	102.2	100.0	109.7	102.2	100.0	110.2	102.3	100.0	104.2	100.9	100.0	109.3	102.1	100.0	102.9	100.7
21, 22, 23	FIRE SUPPRESSION, PLUMBING & HVAC	100.4	103.9	101.7	96.8	103.5	99.4	100.4	103.5	101.6	97.0	104.5	99.9	97.0	121.1	106.4	97.0	98.5	97.6
26, 27, 3370	ELECTRICAL, COMMUNICATIONS & UTIL.	101.3	97.7	99.5	98.5	100.1	99.3	101.2	97.7	99.4	100.6	104.1	102.3	97.4	124.4	110.7	100.5	96.1	98.4
MF2018	WEIGHTED AVERAGE	98.5	116.7	106.2	93.2	115.2	102.5	98.3	114.9	105.3	94.6	116.2	103.7	94.0	124.2	106.8	95.9	109.0	101.4

	DIVISION	MASSACHUSETTS																	
		HYANNIS			LAWRENCE			LOWELL			NEW BEDFORD			PITTSFIELD			SPRINGFIELD		
		026			019			018			027			012			010 - 011		
		MAT.	INST.	TOTAL	MAT.	INST.	TOTAL	MAT.	INST.	TOTAL	MAT.	INST.	TOTAL	MAT.	INST.	TOTAL	MAT.	INST.	TOTAL
015433	CONTRACTOR EQUIPMENT		98.1	98.1		98.1	98.1		95.7	95.7		98.9	98.9		95.7	95.7		95.7	95.7
0241, 31 - 34	SITE & INFRASTRUCTURE, DEMOLITION	87.4	98.1	94.8	91.6	98.0	96.0	90.6	98.1	95.8	88.7	98.0	95.1	91.6	96.7	95.2	91.1	97.1	95.2
0310	Concrete Forming & Accessories	92.7	120.6	116.5	102.5	121.6	118.8	99.0	122.6	119.1	100.7	120.8	117.9	99.0	102.9	102.3	99.3	117.7	115.0
0320	Concrete Reinforcing	88.0	122.3	104.6	108.8	148.3	127.9	109.7	148.1	128.2	109.7	122.4	115.8	90.8	114.1	102.1	109.7	126.5	117.8
0330	Cast-in-Place Concrete	80.4	136.4	101.3	90.7	136.8	107.9	82.4	138.5	103.3	75.1	136.9	98.1	90.0	113.4	98.7	85.8	122.2	99.4
03	CONCRETE	86.7	126.0	104.0	96.2	131.1	111.5	88.0	131.9	107.3	89.1	126.3	105.4	89.0	108.3	97.5	89.5	120.2	103.0
04	MASONRY	104.3	135.1	123.2	110.1	135.9	125.9	97.3	134.9	120.4	104.2	135.7	123.6	98.0	111.5	106.3	97.6	119.1	110.8
05	METALS	95.1	117.0	101.5	98.2	128.4	107.1	98.2	125.0	106.1	98.8	117.4	104.3	98.0	108.2	101.0	100.9	113.7	104.6
06	WOOD, PLASTICS & COMPOSITES	88.0	119.7	105.4	101.5	119.7	111.5	100.8	119.7	111.2	98.1	119.9	110.1	100.8	102.5	101.8	100.8	119.7	111.2
07	THERMAL & MOISTURE PROTECTION	102.5	125.7	112.6	104.0	128.6	114.7	103.7	128.0	114.3	102.9	125.1	112.6	103.8	104.3	104.0	103.7	109.9	106.4
08	OPENINGS	93.9	118.9	99.8	93.4	129.9	101.9	100.4	129.9	107.3	97.4	123.0	103.3	100.4	107.6	102.1	100.4	120.3	105.1
0920	Plaster & Gypsum Board	83.2	119.9	107.9	117.3	119.9	119.1	117.3	119.9	119.1	92.3	119.9	110.9	117.3	102.3	107.2	117.3	119.9	119.1
0950, 0980	Ceilings & Acoustic Treatment	96.2	119.9	112.2	105.5	119.9	115.2	105.5	119.9	115.2	104.3	119.9	114.9	105.5	102.3	103.4	105.5	119.9	115.2
0960	Flooring	80.0	162.0	103.0	88.8	162.0	109.4	88.8	162.0	109.4	83.8	162.0	105.8	89.1	129.8	100.6	88.2	135.0	101.3
0970, 0990	Wall Finishes & Painting/Coating	85.0	139.2	117.1	86.3	139.2	117.6	86.2	139.2	117.6	85.0	139.2	117.1	86.2	109.3	99.9	87.0	109.3	100.2
09	FINISHES	86.3	130.8	110.4	93.8	130.8	113.8	93.7	131.6	114.2	90.4	131.0	112.3	93.8	108.6	101.8	93.6	120.9	108.4
COVERS	DIVS. 10 - 14, 25, 28, 41, 43, 44, 46	100.0	109.7	102.2	100.0	109.8	102.2	100.0	110.8	102.4	100.0	110.2	102.3	100.0	99.5	99.9	100.0	102.9	100.7
21, 22, 23	FIRE SUPPRESSION, PLUMBING & HVAC	100.4	104.2	101.9	100.1	119.7	107.8	100.1	121.2	108.4	100.4	103.4	101.5	100.1	94.2	97.8	100.1	98.5	99.5
26, 27, 3370	ELECTRICAL, COMMUNICATIONS & UTIL.	98.9	97.7	98.3	99.7	122.9	111.1	100.1	119.3	109.6	102.1	97.7	99.9	100.1	96.1	98.2	100.1	93.0	96.6
MF2018	WEIGHTED AVERAGE	95.7	115.0	103.8	98.4	124.2	109.3	97.5	123.8	108.6	97.7	115.2	105.1	97.7	102.6	99.7	98.2	108.5	102.5

	DIVISION	MASSACHUSETTS			MICHIGAN														
		WORCESTER			ANN ARBOR			BATTLE CREEK			BAY CITY			DEARBORN			DETROIT		
		015 - 016			481			490			487			481			482		
		MAT.	INST.	TOTAL	MAT.	INST.	TOTAL	MAT.	INST.	TOTAL	MAT.	INST.	TOTAL	MAT.	INST.	TOTAL	MAT.	INST.	TOTAL
015433	CONTRACTOR EQUIPMENT		95.7	95.7		109.1	109.1		96.1	96.1		109.1	109.1		109.1	109.1		97.8	97.8
0241, 31 - 34	SITE & INFRASTRUCTURE, DEMOLITION	91.0	97.9	95.7	81.0	92.2	88.7	93.6	81.7	85.3	72.5	91.7	85.8	80.8	92.3	88.7	98.2	101.2	100.2
0310	Concrete Forming & Accessories	99.6	118.5	115.7	96.3	104.1	103.0	96.0	77.3	80.0	96.4	81.5	83.7	96.2	104.7	103.5	100.4	105.3	104.6
0320	Concrete Reinforcing	109.7	155.5	131.9	101.2	103.5	102.3	98.5	80.9	90.0	101.2	102.7	101.9	101.2	103.6	102.3	101.4	104.8	103.0
0330	Cast-in-Place Concrete	85.3	136.5	104.4	88.3	96.4	91.3	86.1	91.0	87.9	84.5	83.5	84.1	86.4	97.2	90.4	104.5	101.3	103.3
03	CONCRETE	89.3	130.6	107.5	89.8	102.2	95.2	85.7	82.6	84.3	88.0	87.3	87.7	88.9	102.8	95.0	103.1	103.2	103.2
04	MASONRY	97.2	131.9	118.5	99.6	96.4	97.6	99.1	78.4	86.4	99.2	77.7	86.0	99.5	98.2	98.7	102.6	99.8	100.9
05	METALS	100.9	127.4	108.7	102.7	115.7	106.5	103.7	83.7	97.9	103.3	114.0	106.4	102.8	115.8	106.6	104.3	94.3	101.3
06	WOOD, PLASTICS & COMPOSITES	101.2	116.5	109.6	89.2	106.6	98.8	88.4	75.6	81.3	89.2	81.9	85.2	89.2	106.6	98.8	97.1	106.7	102.4
07	THERMAL & MOISTURE PROTECTION	103.8	122.1	111.8	105.3	98.2	102.2	96.1	79.0	88.7	102.7	81.0	93.2	103.6	100.7	102.3	102.6	102.9	102.7
08	OPENINGS	100.4	130.2	107.4	94.2	100.8	95.7	86.5	74.1	83.6	94.2	83.5	91.7	94.2	101.1	95.8	97.2	101.9	98.3
0920	Plaster & Gypsum Board	117.3	116.7	116.9	107.4	106.5	106.8	91.7	71.7	78.2	107.4	81.1	89.7	107.4	106.5	106.8	104.1	106.5	105.7
0950, 0980	Ceilings & Acoustic Treatment	105.5	116.7	113.0	84.8	106.5	99.4	78.7	71.7	74.0	85.7	81.1	82.6	84.8	106.5	99.4	96.9	106.5	103.4
0960	Flooring	88.8	154.9	107.4	91.2	107.2	95.7	91.5	70.4	85.6	91.2	77.0	87.2	90.6	101.6	93.6	95.7	106.7	98.8
0970, 0990	Wall Finishes & Painting/Coating	86.2	139.2	117.6	82.7	95.5	90.3	82.3	76.3	78.7	82.7	79.1	80.6	82.7	93.8	89.3	93.1	100.1	97.3
09	FINISHES	93.7	127.5	112.0	90.8	104.3	98.1	84.7	75.2	79.6	90.6	79.9	84.8	90.6	103.5	97.6	97.7	105.4	101.9
COVERS	DIVS. 10 - 14, 25, 28, 41, 43, 44, 46	100.0	104.2	100.9	100.0	93.7	98.6	100.0	93.7	98.6	100.0	88.0	97.3	100.0	94.1	98.7	100.0	102.4	100.5
21, 22, 23	FIRE SUPPRESSION, PLUMBING & HVAC	100.1	104.7	101.9	100.0	91.3	96.6	100.1	82.6	93.2	100.0	78.6	91.6	100.0	100.0	100.0	99.9	102.7	101.0
26, 27, 3370	ELECTRICAL, COMMUNICATIONS & UTIL.	100.1	104.1	102.1	97.3	103.0	100.1	96.6	77.7	87.3	96.4	82.5	89.5	97.3	96.1	96.7	99.9	103.5	101.6
MF2018	WEIGHTED AVERAGE	98.1	117.0	106.1	97.1	99.8	98.2	95.5	80.4	89.1	96.6	85.3	91.8	96.9	100.9	98.6	100.7	102.0	101.3

	DIVISION	MICHIGAN																	
		FLINT			GAYLORD			GRAND RAPIDS			IRON MOUNTAIN			JACKSON			KALAMAZOO		
		484 - 485			497			493, 495			498 - 499			492			491		
		MAT.	INST.	TOTAL	MAT.	INST.	TOTAL	MAT.	INST.	TOTAL	MAT.	INST.	TOTAL	MAT.	INST.	TOTAL	MAT.	INST.	TOTAL
015433	CONTRACTOR EQUIPMENT		109.1	109.1		103.6	103.6		98.3	98.3		91.1	91.1		103.6	103.6		96.1	96.1
0241, 31 - 34	SITE & INFRASTRUCTURE, DEMOLITION	70.2	91.8	85.1	87.4	79.6	82.0	92.9	85.9	88.1	96.4	88.5	90.9	111.0	81.2	90.4	93.9	81.6	85.4
0310	Concrete Forming & Accessories	99.1	81.8	84.4	95.0	73.9	77.0	95.6	75.7	78.6	86.6	78.6	79.8	91.6	82.3	83.7	96.0	77.1	79.9
0320	Concrete Reinforcing	101.2	103.1	102.1	91.9	93.0	92.4	103.4	80.7	92.4	91.6	85.0	88.4	89.3	103.1	96.0	98.5	76.2	87.8
0330	Cast-in-Place Concrete	88.9	85.8	87.8	85.8	78.4	83.1	90.4	89.0	89.9	101.7	67.6	89.0	85.7	89.8	87.2	87.8	91.0	89.0
03	CONCRETE	90.2	88.4	89.4	82.6	80.5	81.6	89.7	81.1	85.9	90.9	76.6	84.6	77.8	89.8	83.1	88.9	81.7	85.7
04	MASONRY	99.7	85.9	91.2	109.7	72.7	87.0	92.0	74.1	81.0	95.4	79.7	85.7	89.2	86.9	87.8	97.6	78.4	85.8
05	METALS	102.8	114.6	106.3	105.5	110.8	107.0	100.8	83.2	95.6	104.8	91.1	100.8	105.7	112.7	107.8	103.7	81.9	97.3
06	WOOD, PLASTICS & COMPOSITES	92.6	80.7	86.1	81.9	73.4	77.2	92.2	73.9	82.2	78.2	78.9	78.6	80.8	79.8	80.2	88.4	75.6	81.3
07	THERMAL & MOISTURE PROTECTION	103.0	84.5	95.0	94.7	74.7	86.0	98.0	71.8	86.6	98.4	74.8	88.1	94.0	88.3	91.5	96.1	79.0	88.7
08	OPENINGS	94.2	82.6	91.5	86.0	78.7	84.3	101.8	73.8	95.2	92.5	69.8	87.2	85.2	85.5	85.3	86.5	73.5	83.4
0920	Plaster & Gypsum Board	109.0	79.8	89.4	91.3	71.8	78.2	99.8	70.2	79.8	51.3	79.0	69.9	90.1	78.3	82.1	91.7	71.7	78.2
0950, 0980	Ceilings & Acoustic Treatment	84.8	79.8	81.5	77.1	71.8	73.5	91.1	70.2	77.0	76.0	79.0	78.0	77.1	78.3	77.9	78.7	71.7	74.0
0960	Flooring	91.2	86.0	89.7	84.9	85.6	85.1	95.7	76.1	90.2	102.0	91.0	98.9	83.4	78.7	82.1	91.5	70.4	85.6
0970, 0990	Wall Finishes & Painting/Coating	82.7	78.3	80.1	78.7	81.6	80.4	89.3	76.9	82.0	94.8	69.6	79.9	78.7	93.8	87.7	82.3	76.3	78.7
09	FINISHES	90.1	81.6	85.5	84.1	76.2	79.8	91.4	75.2	82.6	84.6	80.0	82.1	85.2	81.9	83.4	84.7	75.2	79.6
COVERS	DIVS. 10 - 14, 25, 28, 41, 43, 44, 46	100.0	88.8	97.5	100.0	77.9	95.1	100.0	93.0	98.4	100.0	85.9	96.8	100.0	91.6	98.1	100.0	93.7	98.6
21, 22, 23	FIRE SUPPRESSION, PLUMBING & HVAC	100.0	83.3	93.4	96.8	79.0	89.8	100.0	79.6	92.0	96.7	83.9	91.7	96.8	85.7	92.4	100.1	78.2	91.5
26, 27, 3370	ELECTRICAL, COMMUNICATIONS & UTIL.	97.3	89.1	93.2	94.5	76.3	85.5	102.2	82.9	92.7	100.8	80.7	90.9	98.3	103.0	100.6	96.4	74.1	85.4
MF2018	WEIGHTED AVERAGE	96.8	88.6	93.3	94.5	80.6	88.6	97.9	79.9	90.3	96.5	81.7	90.2	94.0	90.7	92.6	95.8	78.6	88.6

	DIVISION	MICHIGAN															MINNESOTA		
		LANSING			MUSKEGON			ROYAL OAK			SAGINAW			TRAVERSE CITY			BEMIDJI		
		488 - 489			494			480, 483			486			496			566		
		MAT.	INST.	TOTAL	MAT.	INST.	TOTAL	MAT.	INST.	TOTAL	MAT.	INST.	TOTAL	MAT.	INST.	TOTAL	MAT.	INST.	TOTAL
015433	CONTRACTOR EQUIPMENT		111.0	111.0		96.1	96.1		89.8	89.8		109.1	109.1		91.1	91.1		97.7	97.7
0241, 31 - 34	SITE & INFRASTRUCTURE, DEMOLITION	90.7	96.5	94.7	91.5	81.7	84.8	84.8	92.6	90.2	73.5	91.7	86.1	81.8	87.5	85.7	95.1	95.1	95.1
0310	Concrete Forming & Accessories	93.8	74.9	77.7	96.3	79.4	81.9	92.3	105.2	103.2	96.3	79.7	82.1	86.6	72.4	74.5	88.0	82.8	83.6
0320	Concrete Reinforcing	104.8	102.9	103.8	99.3	80.7	90.3	91.7	104.8	98.0	101.2	102.7	101.9	93.1	80.2	86.9	97.4	101.8	99.5
0330	Cast-in-Place Concrete	97.9	85.1	93.1	85.8	90.5	87.5	77.3	97.5	84.8	87.3	83.4	85.8	79.3	76.7	78.3	101.8	104.5	102.8
03	CONCRETE	91.4	84.9	88.5	84.2	83.3	83.8	77.1	101.6	87.9	89.3	86.5	88.1	74.8	76.1	75.4	91.0	95.0	92.8
04	MASONRY	94.5	84.6	88.4	96.3	76.2	83.9	93.2	99.5	97.1	101.2	77.7	86.7	93.5	74.8	82.0	98.9	102.4	101.0
05	METALS	101.6	112.7	104.9	101.5	83.1	96.1	106.3	92.2	102.2	102.8	113.8	106.0	104.8	88.9	100.1	91.2	117.9	99.0
06	WOOD, PLASTICS & COMPOSITES	87.0	71.2	78.3	85.4	78.0	81.3	85.0	106.6	96.9	85.9	79.7	82.5	78.2	71.9	74.8	68.8	77.0	73.3
07	THERMAL & MOISTURE PROTECTION	102.0	83.8	94.1	95.0	73.5	85.7	101.2	100.9	101.1	103.9	80.9	93.9	97.3	74.2	87.3	106.3	90.7	99.5
08	OPENINGS	102.7	77.2	96.8	85.8	76.5	83.6	94.0	101.3	95.7	92.4	82.2	90.0	92.5	65.4	86.2	99.7	101.3	100.1
0920	Plaster & Gypsum Board	100.1	70.2	79.9	71.8	74.2	73.4	104.8	106.5	106.0	107.4	78.8	88.2	51.3	71.8	65.1	104.3	76.8	85.8
0950, 0980	Ceilings & Acoustic Treatment	83.9	70.2	74.6	78.7	74.2	75.6	84.1	106.5	99.2	84.8	78.8	80.8	76.0	71.8	73.2	128.8	76.8	93.7
0960	Flooring	97.2	78.7	92.0	90.2	72.3	85.1	88.2	106.7	93.4	91.2	77.0	87.2	102.0	85.6	97.4	90.7	91.4	90.9
0970, 0990	Wall Finishes & Painting/Coating	91.1	76.9	82.7	80.5	79.7	80.0	84.2	93.8	89.9	82.7	79.1	80.6	94.8	39.4	62.1	89.6	105.8	99.2
09	FINISHES	90.2	74.6	81.7	81.1	77.1	78.9	89.7	104.3	97.6	90.5	78.6	84.0	83.5	71.0	76.7	98.2	86.0	91.6
COVERS	DIVS. 10 - 14, 25, 28, 41, 43, 44, 46	100.0	88.8	97.5	100.0	94.5	98.8	100.0	100.0	100.0	100.0	87.7	97.3	100.0	84.4	96.5	100.0	95.6	99.0
21, 22, 23	FIRE SUPPRESSION, PLUMBING & HVAC	99.9	83.7	93.5	99.9	83.4	93.4	96.7	101.4	98.5	100.0	78.3	91.5	96.7	78.9	89.7	96.7	84.8	92.0
26, 27, 3370	ELECTRICAL, COMMUNICATIONS & UTIL.	99.1	87.6	93.4	96.8	73.0	85.1	99.4	103.1	101.3	95.4	84.5	90.0	96.2	76.2	86.4	105.0	100.7	102.8
MF2018	WEIGHTED AVERAGE	98.0	86.8	93.3	94.3	80.0	88.3	95.1	100.3	97.3	96.5	85.1	91.7	93.4	77.7	86.8	96.7	95.4	96.1

	DIVISION	MINNESOTA																	
		BRAINERD			DETROIT LAKES			DULUTH			MANKATO			MINNEAPOLIS			ROCHESTER		
		564			565			556 - 558			560			553 - 555			559		
		MAT.	INST.	TOTAL	MAT.	INST.	TOTAL	MAT.	INST.	TOTAL	MAT.	INST.	TOTAL	MAT.	INST.	TOTAL	MAT.	INST.	TOTAL
015433	CONTRACTOR EQUIPMENT		100.2	100.2		97.7	97.7		102.9	102.9		100.2	100.2		108.4	108.4		100.7	100.7
0241, 31 - 34	SITE & INFRASTRUCTURE, DEMOLITION	96.0	99.9	98.7	93.3	95.4	94.7	98.2	101.7	100.6	92.6	99.4	97.3	95.0	108.8	104.5	96.0	97.6	97.1
0310	Concrete Forming & Accessories	88.7	83.6	84.4	84.9	82.8	83.1	98.4	94.6	95.2	96.9	96.6	96.6	99.3	114.0	111.8	99.2	94.8	95.4
0320	Concrete Reinforcing	96.2	102.2	99.1	97.4	101.8	99.6	102.6	102.0	102.3	96.0	109.2	102.4	96.1	109.4	102.5	98.9	108.8	103.7
0330	Cast-in-Place Concrete	110.7	109.1	110.1	98.9	107.5	102.1	94.3	97.7	95.6	101.9	99.5	101.0	95.6	115.3	102.9	94.8	95.3	95.0
03	CONCRETE	94.8	97.1	95.8	88.6	96.1	91.9	95.3	98.1	96.5	90.5	100.8	95.0	97.6	114.3	104.9	91.4	98.6	94.6
04	MASONRY	122.5	116.3	118.7	122.6	109.8	114.8	95.3	105.7	101.7	111.3	110.8	111.0	114.5	119.5	117.6	102.9	103.5	103.3
05	METALS	92.2	118.3	99.9	91.1	117.5	98.9	100.2	119.1	105.8	92.1	121.6	100.7	98.7	124.5	106.3	99.3	123.1	106.3
06	WOOD, PLASTICS & COMPOSITES	84.9	74.2	79.1	65.8	74.4	70.6	93.9	91.9	92.8	94.5	95.4	95.0	96.1	110.8	104.2	95.2	92.9	93.9
07	THERMAL & MOISTURE PROTECTION	104.7	102.4	103.7	106.1	99.4	103.2	97.9	102.2	99.8	105.1	93.4	100.1	101.2	115.9	107.6	105.3	90.0	98.6
08	OPENINGS	86.4	99.8	89.5	99.6	99.9	99.7	103.3	104.6	103.6	90.9	112.2	95.9	99.3	121.5	104.5	98.6	112.5	101.8
0920	Plaster & Gypsum Board	91.5	74.1	79.8	103.7	74.1	83.8	94.5	92.1	92.9	95.6	95.9	95.8	101.1	111.2	107.9	103.2	93.2	96.5
0950, 0980	Ceilings & Acoustic Treatment	58.8	74.1	69.1	128.8	74.1	91.9	89.2	92.1	91.1	58.8	95.9	83.8	103.1	111.2	108.6	93.9	93.2	93.4
0960	Flooring	89.6	85.6	88.4	89.5	85.6	88.4	93.2	123.9	101.8	91.4	81.8	88.7	100.6	114.6	104.5	92.3	81.8	89.3
0970, 0990	Wall Finishes & Painting/Coating	84.0	105.8	96.9	89.6	83.4	85.9	86.0	104.4	96.9	95.0	103.7	100.2	101.3	123.8	114.6	85.3	98.2	92.9
09	FINISHES	82.1	85.2	83.8	97.6	82.5	89.4	89.4	100.2	95.3	83.7	94.9	89.7	98.9	115.0	107.6	90.0	93.1	91.7
COVERS	DIVS. 10 - 14, 25, 28, 41, 43, 44, 46	100.0	97.3	99.4	100.0	97.1	99.4	100.0	94.6	98.8	100.0	97.5	99.4	100.0	105.8	101.3	100.0	101.8	100.4
21, 22, 23	FIRE SUPPRESSION, PLUMBING & HVAC	96.0	88.6	93.1	96.7	87.5	93.1	99.8	94.5	97.7	96.0	84.6	91.5	100.0	110.6	104.2	100.0	91.6	96.7
26, 27, 3370	ELECTRICAL, COMMUNICATIONS & UTIL.	102.7	101.7	102.2	104.8	71.1	88.2	100.3	101.7	101.0	109.1	93.1	101.2	107.2	110.3	108.7	103.0	93.1	98.1
MF2018	WEIGHTED AVERAGE	95.6	98.6	96.8	97.4	92.4	95.3	98.6	101.3	99.7	95.7	98.5	96.9	100.7	114.1	106.3	98.3	98.7	98.5

	DIVISION	MINNESOTA															MISSISSIPPI		
		SAINT PAUL			ST. CLOUD			THIEF RIVER FALLS			WILLMAR			WINDOM			BILOXI		
		550 - 551			563			567			562			561			395		
		MAT.	INST.	TOTAL	MAT.	INST.	TOTAL	MAT.	INST.	TOTAL	MAT.	INST.	TOTAL	MAT.	INST.	TOTAL	MAT.	INST.	TOTAL
015433	CONTRACTOR EQUIPMENT		102.9	102.9		100.2	100.2		97.7	97.7		100.2	100.2		100.2	100.2		100.1	100.1
0241, 31 - 34	SITE & INFRASTRUCTURE, DEMOLITION	96.5	102.5	100.7	91.4	100.8	97.9	94.0	94.9	94.7	90.6	99.7	96.9	84.5	98.9	94.4	105.2	85.9	91.9
0310	Concrete Forming & Accessories	98.6	117.0	114.3	86.0	113.5	109.5	88.8	81.9	82.9	85.8	86.8	86.7	90.1	82.2	83.4	92.8	64.1	68.4
0320	Concrete Reinforcing	105.4	109.9	107.6	96.2	109.4	102.6	97.8	101.6	99.6	95.8	109.2	102.3	95.8	108.1	101.8	93.4	48.8	71.8
0330	Cast-in-Place Concrete	99.9	116.3	106.0	97.7	115.0	104.1	100.9	81.3	93.6	99.2	79.5	91.9	85.8	83.3	84.8	114.2	66.3	96.4
03	CONCRETE	95.3	116.1	104.4	86.5	114.0	98.5	89.8	86.5	88.3	86.3	89.5	87.7	77.6	88.5	82.4	94.4	64.0	81.0
04	MASONRY	105.3	127.4	118.9	107.3	121.8	116.2	98.9	102.3	101.0	112.1	114.5	113.6	122.0	87.9	101.0	90.4	62.8	73.4
05	METALS	100.1	125.0	107.4	92.9	123.1	101.8	91.3	116.7	98.7	92.0	121.5	100.7	91.9	119.5	100.0	90.3	84.5	88.6
06	WOOD, PLASTICS & COMPOSITES	95.0	114.2	105.6	82.1	110.3	97.6	70.0	77.0	73.8	81.8	79.9	80.8	86.3	79.9	82.8	93.1	64.5	77.3
07	THERMAL & MOISTURE PROTECTION	102.0	120.5	110.1	104.9	111.7	107.9	107.1	88.1	98.8	104.7	101.0	103.1	104.6	83.0	95.2	100.0	63.1	83.9
08	OPENINGS	98.2	124.3	104.3	91.4	122.1	98.5	99.7	101.3	100.1	88.4	102.8	91.7	91.9	102.8	94.4	96.3	55.0	86.7
0920	Plaster & Gypsum Board	94.5	115.0	108.3	91.5	111.2	104.7	104.3	76.8	85.8	91.5	79.9	83.7	91.5	79.9	83.7	103.8	64.0	77.0
0950, 0980	Ceilings & Acoustic Treatment	91.6	115.0	107.4	58.8	111.2	94.1	128.8	76.8	93.7	58.8	79.9	73.0	58.8	79.9	73.0	86.4	64.0	71.3
0960	Flooring	92.4	122.8	101.0	86.4	120.9	96.1	90.4	85.6	89.0	87.8	85.6	87.2	90.3	85.6	89.0	94.0	65.0	85.8
0970, 0990	Wall Finishes & Painting/Coating	92.1	129.1	114.0	95.0	123.8	112.0	89.6	83.4	85.9	89.6	83.4	85.9	89.6	103.7	98.0	86.0	46.9	62.8
09	FINISHES	91.1	119.1	106.3	81.5	115.9	100.1	98.1	82.6	89.7	81.6	85.3	83.6	81.8	84.8	83.4	89.4	62.8	75.0
COVERS	DIVS. 10 - 14, 25, 28, 41, 43, 44, 46	100.0	108.7	101.9	100.0	103.2	100.7	100.0	95.5	99.0	100.0	97.3	99.4	100.0	93.7	98.6	100.0	71.9	93.7
21, 22, 23	FIRE SUPPRESSION, PLUMBING & HVAC	99.9	116.7	106.5	99.5	110.5	103.8	96.7	84.4	91.9	96.0	101.6	98.2	96.0	81.0	90.1	100.0	55.3	82.4
26, 27, 3370	ELECTRICAL, COMMUNICATIONS & UTIL.	102.9	115.1	108.9	102.7	115.1	108.8	102.1	71.0	86.8	102.7	83.7	93.4	109.1	93.1	101.2	101.0	53.7	77.7
MF2018	WEIGHTED AVERAGE	99.0	117.6	106.9	95.1	114.2	103.2	96.3	89.3	93.3	94.0	98.0	95.7	94.3	91.2	93.0	96.2	63.9	82.6

	DIVISION	MISSISSIPPI																	
		CLARKSDALE			COLUMBUS			GREENVILLE			GREENWOOD			JACKSON			LAUREL		
		386			397			387			389			390 - 392			394		
		MAT.	INST.	TOTAL	MAT.	INST.	TOTAL	MAT.	INST.	TOTAL	MAT.	INST.	TOTAL	MAT.	INST.	TOTAL	MAT.	INST.	TOTAL
015433	CONTRACTOR EQUIPMENT		100.1	100.1		100.1	100.1		100.1	100.1		100.1	100.1		102.6	102.6		100.1	100.1
0241, 31 - 34	SITE & INFRASTRUCTURE, DEMOLITION	101.4	84.4	89.7	103.9	85.2	91.0	107.5	86.1	92.7	104.5	84.1	90.4	101.1	90.6	93.9	109.6	84.5	92.2
0310	Concrete Forming & Accessories	83.8	43.9	49.8	81.9	46.3	51.6	80.7	62.7	65.4	92.5	44.3	51.4	92.0	64.6	68.6	82.0	59.5	62.8
0320	Concrete Reinforcing	104.2	66.5	85.9	99.9	66.6	83.8	104.7	66.7	86.3	104.2	66.5	86.0	104.0	51.7	78.7	100.6	32.4	67.6
0330	Cast-in-Place Concrete	104.1	58.8	87.2	116.5	61.1	95.9	107.2	65.7	91.8	112.0	58.9	92.2	99.6	66.5	87.3	113.9	60.1	93.9
03	CONCRETE	93.8	55.1	76.8	95.8	57.1	78.8	99.3	66.2	84.8	100.1	55.4	80.4	89.2	64.7	78.4	97.9	56.9	79.9
04	MASONRY	89.1	50.8	65.6	115.4	53.6	77.4	132.2	61.4	88.7	89.8	50.7	65.8	95.7	61.5	74.7	111.5	49.4	73.4
05	METALS	92.5	87.9	91.1	87.4	90.4	88.3	93.6	91.2	92.8	92.5	87.8	91.1	96.9	84.6	93.3	87.5	76.0	84.1
06	WOOD, PLASTICS & COMPOSITES	79.4	44.4	60.1	78.8	45.2	60.3	76.2	62.8	68.8	91.8	44.4	65.7	96.3	65.5	79.4	79.9	64.5	71.4
07	THERMAL & MOISTURE PROTECTION	97.7	52.0	77.8	100.0	55.4	80.6	98.2	62.1	82.5	98.1	54.2	79.0	98.3	62.9	82.9	100.2	56.5	81.2
08	OPENINGS	95.1	48.2	84.1	95.9	48.6	84.9	94.8	58.1	86.2	95.1	48.2	84.1	99.5	56.0	89.4	93.2	51.8	83.5
0920	Plaster & Gypsum Board	89.9	43.3	58.6	94.3	44.2	60.6	89.6	62.3	71.2	100.6	43.3	62.1	90.7	64.9	73.3	94.3	64.0	73.9
0950, 0980	Ceilings & Acoustic Treatment	82.3	43.3	56.0	81.5	44.2	56.3	85.0	62.3	69.7	82.3	43.3	56.0	89.7	64.9	73.0	81.5	64.0	69.7
0960	Flooring	97.8	45.0	83.0	87.9	65.0	81.4	96.2	65.0	87.4	103.5	65.0	92.7	92.1	65.0	84.5	86.6	65.0	80.5
0970, 0990	Wall Finishes & Painting/Coating	94.6	46.9	66.4	86.0	46.9	62.8	94.6	56.7	72.1	94.6	46.9	66.4	89.9	56.7	70.3	86.0	46.9	62.8
09	FINISHES	90.2	43.9	65.1	85.4	49.1	65.8	90.8	62.5	75.5	93.7	48.0	68.9	89.9	64.1	75.9	85.6	60.5	72.0
COVERS	DIVS. 10 - 14, 25, 28, 41, 43, 44, 46	100.0	48.0	88.4	100.0	49.0	88.6	100.0	71.3	93.6	100.0	48.0	88.4	100.0	71.9	93.7	100.0	34.2	85.3
21, 22, 23	FIRE SUPPRESSION, PLUMBING & HVAC	98.4	51.3	79.9	98.0	53.1	80.4	100.0	57.4	83.3	98.4	51.7	80.1	100.0	59.1	83.9	98.1	47.0	78.0
26, 27, 3370	ELECTRICAL, COMMUNICATIONS & UTIL.	96.8	41.6	69.6	98.6	55.1	77.2	96.8	55.9	76.6	96.8	39.0	68.3	102.7	55.9	79.6	100.1	57.9	79.3
MF2018	WEIGHTED AVERAGE	95.3	55.2	78.4	95.9	59.1	80.4	98.7	65.5	84.7	96.6	55.5	79.2	97.3	65.6	83.9	96.1	57.8	79.9

	DIVISION	MISSISSIPPI									MISSOURI								
		MCCOMB			MERIDIAN			TUPELO			BOWLING GREEN			CAPE GIRARDEAU			CHILLICOTHE		
		396			393			388			633			637			646		
		MAT.	INST.	TOTAL	MAT.	INST.	TOTAL	MAT.	INST.	TOTAL	MAT.	INST.	TOTAL	MAT.	INST.	TOTAL	MAT.	INST.	TOTAL
015433	CONTRACTOR EQUIPMENT		100.1	100.1		100.1	100.1		100.1	100.1		107.1	107.1		107.1	107.1		101.5	101.5
0241, 31 - 34	SITE & INFRASTRUCTURE, DEMOLITION	96.3	84.3	88.0	100.7	86.1	90.7	98.9	84.3	88.8	87.6	90.0	89.3	89.3	89.9	89.7	101.2	88.4	92.3
0310	Concrete Forming & Accessories	81.9	45.5	50.9	79.4	62.8	65.2	81.2	45.9	51.1	96.0	94.0	94.3	88.9	82.0	83.0	86.2	95.7	94.3
0320	Concrete Reinforcing	101.2	34.3	68.8	99.9	51.6	76.6	101.9	66.5	84.8	93.4	96.5	94.9	94.6	79.6	87.3	100.4	102.6	101.5
0330	Cast-in-Place Concrete	101.1	58.4	85.2	108.2	66.0	92.5	104.1	67.9	90.6	90.7	95.5	92.5	89.8	86.6	88.6	91.7	86.0	89.6
03	CONCRETE	85.9	50.2	70.2	90.1	63.8	78.6	93.4	59.2	78.4	91.2	96.4	93.5	90.4	85.0	88.0	95.0	94.3	94.7
04	MASONRY	116.7	50.3	75.9	90.0	62.1	72.9	121.7	53.3	79.6	109.6	97.9	102.4	105.9	80.5	90.3	99.2	92.2	94.9
05	METALS	87.6	75.6	84.1	88.5	85.6	87.7	92.4	87.9	91.1	91.8	117.7	99.4	92.9	109.4	97.8	85.1	110.6	92.5
06	WOOD, PLASTICS & COMPOSITES	78.8	46.8	61.2	76.2	62.8	68.8	76.7	45.2	59.4	96.7	94.4	95.4	89.2	80.0	84.1	94.0	97.0	95.7
07	THERMAL & MOISTURE PROTECTION	99.4	54.2	79.8	99.6	62.4	83.4	97.6	53.9	78.6	96.4	99.2	97.6	95.9	85.3	91.3	92.7	92.7	92.7
08	OPENINGS	96.0	41.3	83.2	95.6	54.8	86.1	95.0	51.4	84.9	98.7	98.2	98.5	98.6	76.1	93.4	86.0	96.2	88.3
0920	Plaster & Gypsum Board	94.3	45.8	61.7	94.3	62.3	72.8	89.6	44.2	59.0	101.7	94.6	96.9	101.1	79.8	86.8	98.2	96.9	97.3
0950, 0980	Ceilings & Acoustic Treatment	81.5	45.8	57.4	82.4	62.3	68.8	82.3	44.2	56.6	88.9	94.6	92.8	88.9	79.8	82.8	87.4	96.9	93.8
0960	Flooring	87.9	65.0	81.4	86.5	65.0	80.5	96.5	65.0	87.6	95.7	96.9	96.0	92.4	84.5	90.2	94.9	99.7	96.2
0970, 0990	Wall Finishes & Painting/Coating	86.0	46.9	62.8	86.0	46.9	62.8	94.6	45.2	65.4	95.4	106.3	101.8	95.4	67.1	78.6	94.3	103.2	99.6
09	FINISHES	84.8	49.2	65.5	84.8	61.6	72.3	89.7	48.9	67.6	98.2	95.7	96.8	97.0	79.8	87.7	97.1	97.6	97.4
COVERS	DIVS. 10 - 14, 25, 28, 41, 43, 44, 46	100.0	50.9	89.1	100.0	71.5	93.6	100.0	49.0	88.6	100.0	81.4	95.8	100.0	94.3	98.7	100.0	81.3	95.8
21, 22, 23	FIRE SUPPRESSION, PLUMBING & HVAC	98.0	50.7	79.5	100.0	57.6	83.3	98.5	52.7	80.5	96.5	99.5	97.7	100.0	98.7	99.5	96.6	99.9	97.9
26, 27, 3370	ELECTRICAL, COMMUNICATIONS & UTIL.	97.3	56.0	76.9	100.1	55.6	78.2	96.5	55.0	76.1	99.2	76.8	88.2	99.2	98.5	98.9	94.9	75.8	85.5
MF2018	WEIGHTED AVERAGE	94.4	55.8	78.0	94.6	64.5	81.9	96.6	59.0	80.7	96.4	95.5	96.0	97.0	91.2	94.5	93.7	93.8	93.7

DIVISION		MISSOURI																	
		COLUMBIA			FLAT RIVER			HANNIBAL			HARRISONVILLE			JEFFERSON CITY			JOPLIN		
		652			636			634			647			650 - 651			648		
		MAT.	INST.	TOTAL	MAT.	INST.	TOTAL	MAT.	INST.	TOTAL	MAT.	INST.	TOTAL	MAT.	INST.	TOTAL	MAT.	INST.	TOTAL
015433	CONTRACTOR EQUIPMENT		110.3	110.3		107.1	107.1		107.1	107.1		101.5	101.5		113.1	113.1		104.8	104.8
0241, 31 - 34	SITE & INFRASTRUCTURE, DEMOLITION	94.2	93.8	93.9	90.2	89.7	89.8	85.4	89.8	88.4	92.9	89.5	90.6	94.4	98.7	97.4	102.0	92.6	95.5
0310	Concrete Forming & Accessories	84.4	81.0	81.5	102.4	88.3	90.4	94.3	82.5	84.3	83.4	99.0	96.7	96.3	78.6	81.2	97.4	74.2	77.7
0320	Concrete Reinforcing	85.6	93.3	89.3	94.6	102.9	98.6	92.9	96.5	94.6	100.0	110.9	105.3	92.4	93.3	92.9	103.7	84.3	94.3
0330	Cast-in-Place Concrete	88.6	84.8	87.2	93.7	90.6	92.5	85.9	93.5	88.7	93.9	99.7	96.0	93.9	84.1	90.2	99.3	75.5	90.4
03	CONCRETE	80.8	86.3	83.2	94.1	93.3	93.8	87.6	90.5	88.9	91.1	102.0	95.9	87.5	84.8	86.3	93.9	77.7	86.7
04	MASONRY	136.9	86.4	105.9	106.5	76.4	88.0	101.6	94.6	97.3	94.0	99.4	97.3	96.0	86.5	90.1	92.4	80.2	84.9
05	METALS	96.3	115.4	101.9	91.7	119.2	99.8	91.8	117.2	99.3	85.4	115.1	94.1	95.8	114.0	101.1	87.9	98.8	91.1
06	WOOD, PLASTICS & COMPOSITES	80.3	78.6	79.3	105.2	89.6	96.6	94.9	81.1	87.3	90.4	98.7	95.0	97.1	75.6	85.2	106.1	73.5	88.2
07	THERMAL & MOISTURE PROTECTION	90.6	86.1	88.7	96.6	91.0	94.2	96.2	93.4	95.0	91.8	100.8	95.7	97.5	86.1	92.5	91.9	81.2	87.3
08	OPENINGS	94.4	81.7	91.4	98.6	97.4	98.4	98.7	83.7	95.2	85.7	103.2	89.8	94.5	80.0	91.1	86.9	76.3	84.5
0920	Plaster & Gypsum Board	84.7	78.2	80.3	107.7	89.7	95.6	101.4	81.0	87.7	94.1	98.6	97.1	96.2	74.9	81.9	105.4	72.7	83.4
0950, 0980	Ceilings & Acoustic Treatment	88.5	78.2	81.5	88.9	89.7	89.4	88.9	81.0	83.6	87.4	98.6	95.0	93.3	74.9	80.9	88.3	72.7	77.8
0960	Flooring	90.2	98.6	92.5	98.9	84.5	94.9	95.0	96.9	95.5	90.3	100.8	93.2	97.6	71.5	90.3	120.1	72.2	106.6
0970, 0990	Wall Finishes & Painting/Coating	93.1	81.9	86.4	95.4	71.7	81.4	95.4	94.1	94.6	98.6	107.8	104.0	91.2	81.9	85.7	93.9	75.9	83.2
09	FINISHES	84.0	83.5	83.7	100.0	85.1	92.0	97.7	85.3	91.0	94.8	100.2	97.7	91.5	77.1	83.7	103.8	74.4	87.9
COVERS	DIVS. 10 - 14, 25, 28, 41, 43, 44, 46	100.0	93.6	98.6	100.0	94.0	98.7	100.0	78.8	95.3	100.0	82.7	96.2	100.0	93.6	98.6	100.0	79.5	95.4
21, 22, 23	FIRE SUPPRESSION, PLUMBING & HVAC	100.0	96.9	98.8	96.5	96.8	96.6	96.5	97.7	97.0	96.6	100.8	98.2	100.0	96.9	98.8	100.1	71.1	88.7
26, 27, 3370	ELECTRICAL, COMMUNICATIONS & UTIL.	97.8	81.4	89.7	103.8	98.5	101.2	98.0	76.8	87.5	101.5	101.2	101.4	103.2	81.4	92.4	92.9	66.1	79.7
MF2018	WEIGHTED AVERAGE	96.1	90.7	93.8	97.3	94.2	96.0	95.3	91.6	93.7	93.2	100.8	96.4	96.4	89.9	93.7	95.1	77.7	87.7

DIVISION		MISSOURI																	
		KANSAS CITY			KIRKSVILLE			POPLAR BLUFF			ROLLA			SEDALIA			SIKESTON		
		640 - 641			635			639			654 - 655			653			638		
		MAT.	INST.	TOTAL	MAT.	INST.	TOTAL	MAT.	INST.	TOTAL	MAT.	INST.	TOTAL	MAT.	INST.	TOTAL	MAT.	INST.	TOTAL
015433	CONTRACTOR EQUIPMENT		104.4	104.4		97.4	97.4		99.6	99.6		110.3	110.3		100.4	100.4		99.6	99.6
0241, 31 - 34	SITE & INFRASTRUCTURE, DEMOLITION	94.6	98.6	97.3	88.8	84.9	86.1	76.1	88.5	84.7	92.9	94.1	93.8	92.1	89.3	90.2	79.6	89.1	86.2
0310	Concrete Forming & Accessories	97.2	103.5	102.6	86.9	78.4	79.6	87.4	79.3	80.5	91.6	94.0	93.6	89.6	78.2	79.9	88.2	79.4	80.7
0320	Concrete Reinforcing	98.5	111.1	104.6	93.8	83.3	88.7	96.7	76.0	86.7	86.0	93.4	89.6	84.8	110.3	97.1	96.0	76.0	86.4
0330	Cast-in-Place Concrete	97.3	102.6	99.3	93.6	82.2	89.4	72.0	83.5	76.3	90.6	94.7	92.1	94.7	81.3	89.7	77.0	83.5	79.4
03	CONCRETE	93.8	104.9	98.7	106.9	81.8	95.9	81.5	81.4	81.5	82.5	95.6	88.2	95.7	86.1	91.5	85.2	81.5	83.6
04	MASONRY	99.6	103.3	101.9	113.2	84.8	95.8	104.5	74.1	85.8	110.8	85.3	95.2	117.1	81.6	95.3	104.2	74.1	85.7
05	METALS	94.9	112.8	100.2	91.5	101.0	94.3	92.1	97.8	93.7	95.7	115.9	101.6	94.5	112.8	99.9	92.4	97.9	94.0
06	WOOD, PLASTICS & COMPOSITES	100.1	103.8	102.2	82.2	76.5	79.1	81.3	80.0	80.6	88.0	95.6	92.2	81.6	76.7	78.9	82.9	80.0	81.3
07	THERMAL & MOISTURE PROTECTION	91.8	105.2	97.7	102.2	90.8	97.3	100.5	83.0	92.9	90.9	92.9	91.8	96.1	87.8	92.5	100.7	82.1	92.6
08	OPENINGS	96.5	106.0	98.7	104.1	78.8	98.2	105.0	75.2	98.1	94.4	91.1	93.6	99.4	87.0	96.5	105.0	75.2	98.1
0920	Plaster & Gypsum Board	100.9	104.0	103.0	96.3	76.2	82.8	96.6	79.8	85.3	86.9	95.6	92.8	80.5	76.2	77.6	98.5	79.8	85.9
0950, 0980	Ceilings & Acoustic Treatment	90.5	104.0	99.6	87.3	76.2	79.8	88.9	79.8	82.8	88.5	95.6	93.3	88.5	76.2	80.2	88.9	79.8	82.8
0960	Flooring	95.9	101.8	97.6	73.8	96.4	80.2	87.3	84.5	86.5	93.8	96.4	94.5	72.5	72.9	72.6	87.8	84.5	86.9
0970, 0990	Wall Finishes & Painting/Coating	97.2	107.8	103.5	91.3	78.3	83.6	90.6	67.1	76.7	93.1	90.2	91.4	93.1	103.2	99.1	90.6	67.1	76.7
09	FINISHES	98.1	103.6	101.1	97.4	81.1	88.6	96.9	78.4	86.9	85.4	94.0	90.1	82.9	79.1	80.8	97.5	78.7	87.3
COVERS	DIVS. 10 - 14, 25, 28, 41, 43, 44, 46	100.0	98.4	99.7	100.0	78.3	95.2	100.0	92.2	98.3	100.0	96.7	99.3	100.0	87.9	97.3	100.0	92.2	98.3
21, 22, 23	FIRE SUPPRESSION, PLUMBING & HVAC	100.0	103.4	101.4	96.6	97.1	96.8	96.6	94.9	95.9	96.5	98.6	97.3	96.4	94.9	95.8	96.6	95.0	95.9
26, 27, 3370	ELECTRICAL, COMMUNICATIONS & UTIL.	103.3	101.2	102.2	98.2	76.8	87.6	98.4	98.4	98.4	96.4	79.4	88.0	97.2	101.2	99.2	97.5	98.4	97.9
MF2018	WEIGHTED AVERAGE	97.9	103.8	100.4	98.9	86.5	93.7	95.1	87.7	92.0	94.1	94.3	94.2	96.4	91.5	94.3	95.7	87.8	92.3

DIVISION		MISSOURI									MONTANA								
		SPRINGFIELD			ST. JOSEPH			ST. LOUIS			BILLINGS			BUTTE			GREAT FALLS		
		656 - 658			644 - 645			630 - 631			590 - 591			597			594		
		MAT.	INST.	TOTAL	MAT.	INST.	TOTAL	MAT.	INST.	TOTAL	MAT.	INST.	TOTAL	MAT.	INST.	TOTAL	MAT.	INST.	TOTAL
015433	CONTRACTOR EQUIPMENT		102.8	102.8		101.5	101.5		108.6	108.6		97.9	97.9		97.7	97.7		97.7	97.7
0241, 31 - 34	SITE & INFRASTRUCTURE, DEMOLITION	94.4	91.4	92.4	96.7	87.4	90.3	94.4	97.8	96.7	91.8	93.5	93.0	97.3	93.3	94.5	101.0	93.3	95.7
0310	Concrete Forming & Accessories	97.7	75.4	78.7	96.3	89.1	90.2	99.5	102.5	102.1	99.3	68.3	72.9	86.5	68.2	70.9	99.2	68.2	72.8
0320	Concrete Reinforcing	82.5	92.9	87.5	97.3	110.4	103.6	86.5	105.2	95.5	98.0	81.8	90.2	106.3	81.7	94.4	98.0	81.8	90.2
0330	Cast-in-Place Concrete	96.3	73.7	87.9	92.2	95.5	93.4	101.3	102.5	101.7	118.7	72.7	101.6	130.6	72.5	109.0	138.1	72.5	113.7
03	CONCRETE	92.1	79.0	86.3	90.3	96.0	92.8	97.1	104.0	100.1	98.3	73.1	87.2	101.9	73.0	89.2	106.9	73.0	92.0
04	MASONRY	89.1	79.9	83.5	95.3	88.7	91.2	91.1	107.1	100.9	125.8	83.8	100.0	120.9	82.7	97.4	125.4	83.8	99.8
05	METALS	100.7	101.7	101.0	91.5	113.8	98.0	97.4	119.0	103.8	107.0	89.6	101.9	101.2	89.4	97.7	104.4	89.5	100.0
06	WOOD, PLASTICS & COMPOSITES	89.4	75.0	81.5	106.1	88.7	96.5	97.6	101.0	99.5	90.1	64.6	76.1	76.8	64.6	70.1	91.4	64.6	76.6
07	THERMAL & MOISTURE PROTECTION	94.6	75.5	86.3	92.2	88.7	90.7	94.3	105.5	99.2	108.3	73.7	93.2	107.8	73.3	92.8	108.5	73.7	93.4
08	OPENINGS	101.7	84.3	97.6	89.9	96.7	91.5	100.3	105.8	101.6	98.6	65.9	91.0	96.9	65.9	89.7	99.8	65.9	91.9
0920	Plaster & Gypsum Board	87.8	74.4	78.8	106.9	88.3	94.4	106.5	101.5	103.1	118.3	64.0	81.8	118.8	64.0	81.9	128.7	64.0	85.2
0950, 0980	Ceilings & Acoustic Treatment	88.5	74.4	79.0	94.8	88.3	90.4	90.5	101.5	97.9	96.6	64.0	74.6	104.1	64.0	77.0	105.7	64.0	77.6
0960	Flooring	92.8	71.5	86.8	99.4	99.0	99.3	98.0	99.1	98.3	88.1	80.8	86.0	85.4	90.9	87.0	91.9	80.8	88.8
0970, 0990	Wall Finishes & Painting/Coating	87.6	99.4	94.6	94.3	107.8	102.3	96.4	106.3	102.2	93.8	91.0	92.2	92.3	67.8	77.8	92.3	91.0	91.5
09	FINISHES	87.6	77.6	82.2	100.3	92.8	96.3	100.5	101.9	101.3	92.5	72.3	81.6	93.0	71.8	81.5	96.8	72.3	83.5
COVERS	DIVS. 10 - 14, 25, 28, 41, 43, 44, 46	100.0	90.5	97.9	100.0	94.0	98.7	100.0	102.8	100.6	100.0	91.6	98.1	100.0	91.6	98.1	100.0	91.6	98.1
21, 22, 23	FIRE SUPPRESSION, PLUMBING & HVAC	100.0	69.4	88.0	100.1	87.5	95.2	100.0	105.5	102.2	100.0	76.4	90.7	100.0	71.2	88.7	100.0	71.2	88.7
26, 27, 3370	ELECTRICAL, COMMUNICATIONS & UTIL.	101.3	68.8	85.3	101.5	75.8	88.9	101.8	98.5	100.2	102.3	75.2	88.9	109.2	71.3	90.5	101.7	70.8	86.5
MF2018	WEIGHTED AVERAGE	97.6	79.0	89.7	96.1	90.9	93.9	98.8	104.5	101.2	101.6	78.4	91.8	101.5	76.5	90.9	102.9	76.6	91.8

	DIVISION	MONTANA																	
		HAVRE			HELENA			KALISPELL			MILES CITY			MISSOULA			WOLF POINT		
		595			596			599			593			598			592		
		MAT.	INST.	TOTAL	MAT.	INST.	TOTAL	MAT.	INST.	TOTAL	MAT.	INST.	TOTAL	MAT.	INST.	TOTAL	MAT.	INST.	TOTAL
015433	CONTRACTOR EQUIPMENT		97.7	97.7		99.9	99.9		97.7	97.7		97.7	97.7		97.7	97.7		97.7	97.7
0241, 31 - 34	SITE & INFRASTRUCTURE, DEMOLITION	104.2	93.1	96.6	91.1	96.8	95.0	87.5	93.3	91.5	93.5	93.2	93.3	80.5	93.3	89.3	110.2	93.2	98.4
0310	Concrete Forming & Accessories	79.6	67.2	69.0	100.2	68.3	73.0	89.7	68.3	71.4	97.7	67.3	71.8	89.7	68.4	71.5	90.4	67.3	70.7
0320	Concrete Reinforcing	107.1	78.9	93.5	113.0	81.8	97.9	109.1	85.8	97.8	106.7	78.9	93.3	108.1	85.8	97.3	108.2	78.2	93.7
0330	Cast-in-Place Concrete	140.7	71.2	114.9	102.1	73.3	91.4	113.4	72.5	98.2	124.2	71.3	104.5	96.2	72.6	87.4	139.1	70.2	113.5
03	CONCRETE	109.8	71.6	93.0	94.7	73.3	85.3	91.7	73.7	83.8	98.8	71.7	86.9	80.5	73.8	77.6	113.2	71.2	94.8
04	MASONRY	121.9	80.6	96.5	114.2	82.7	94.8	119.6	82.7	96.9	127.6	80.6	98.7	144.5	82.7	106.5	128.9	80.6	99.2
05	METALS	97.3	88.2	94.6	103.2	88.5	98.9	97.2	90.8	95.3	96.5	88.4	94.1	97.7	91.0	95.7	96.6	88.1	94.1
06	WOOD, PLASTICS & COMPOSITES	68.1	64.6	66.2	93.8	64.7	77.8	80.2	64.6	71.6	88.4	64.6	75.3	80.2	64.6	71.6	79.5	64.6	71.3
07	THERMAL & MOISTURE PROTECTION	108.3	66.2	90.0	103.1	73.9	90.4	107.4	73.6	92.7	107.8	67.7	90.3	106.9	75.6	93.3	108.9	67.5	90.9
08	OPENINGS	97.4	65.2	89.9	96.3	65.9	89.2	97.4	66.8	90.3	96.9	65.2	89.5	96.9	66.8	89.9	96.9	65.1	89.5
0920	Plaster & Gypsum Board	114.1	64.0	80.4	113.5	64.0	80.2	118.8	64.0	81.9	127.8	64.0	84.9	118.8	64.0	81.9	121.8	64.0	82.9
0950, 0980	Ceilings & Acoustic Treatment	104.1	64.0	77.0	107.9	64.0	78.3	104.1	64.0	77.0	101.6	64.0	76.3	104.1	64.0	77.0	101.6	64.0	76.3
0960	Flooring	83.0	90.9	85.3	95.9	90.9	94.5	87.1	90.9	88.2	91.8	90.9	91.6	87.1	90.9	88.2	88.4	90.9	89.1
0970, 0990	Wall Finishes & Painting/Coating	92.3	67.8	77.8	97.6	67.8	79.9	92.3	91.0	91.5	92.3	67.8	77.8	92.3	91.0	91.5	92.3	67.8	77.8
09	FINISHES	92.3	71.3	80.9	100.1	71.8	84.8	92.9	74.3	82.8	95.4	71.3	82.3	92.3	74.3	82.6	95.0	71.3	82.1
COVERS	DIVS. 10 - 14, 25, 28, 41, 43, 44, 46	100.0	88.8	97.5	100.0	91.9	98.2	100.0	89.5	97.7	100.0	88.9	97.5	100.0	92.4	98.3	100.0	88.9	97.5
21, 22, 23	FIRE SUPPRESSION, PLUMBING & HVAC	96.5	68.9	85.6	100.0	71.2	88.7	96.5	69.4	85.8	96.5	75.2	88.1	100.0	71.2	88.7	96.5	75.2	88.1
26, 27, 3370	ELECTRICAL, COMMUNICATIONS & UTIL.	101.7	69.2	85.7	109.0	70.8	90.2	106.0	68.1	87.3	101.7	74.4	88.3	107.0	70.2	88.9	101.7	74.4	88.3
MF2018	WEIGHTED AVERAGE	100.4	74.8	89.6	100.9	76.7	90.7	98.1	76.2	88.9	99.2	77.0	89.8	98.6	77.1	89.5	101.4	76.9	91.1

	DIVISION	NEBRASKA																	
		ALLIANCE			COLUMBUS			GRAND ISLAND			HASTINGS			LINCOLN			MCCOOK		
		693			686			688			689			683 - 685			690		
		MAT.	INST.	TOTAL	MAT.	INST.	TOTAL	MAT.	INST.	TOTAL	MAT.	INST.	TOTAL	MAT.	INST.	TOTAL	MAT.	INST.	TOTAL
015433	CONTRACTOR EQUIPMENT		95.3	95.3		101.9	101.9		101.9	101.9		101.9	101.9		105.0	105.0		101.9	101.9
0241, 31 - 34	SITE & INFRASTRUCTURE, DEMOLITION	98.2	96.2	96.8	101.8	90.6	94.1	106.8	90.7	95.7	105.4	90.6	95.2	94.9	95.7	95.4	101.5	90.6	94.0
0310	Concrete Forming & Accessories	87.0	55.3	60.0	96.5	75.0	78.2	96.1	68.8	72.9	99.3	72.5	76.5	94.4	75.5	78.3	92.3	55.9	61.2
0320	Concrete Reinforcing	107.4	87.4	97.7	100.4	85.9	93.4	99.8	76.1	88.4	99.8	76.1	88.4	97.6	76.4	87.3	100.1	76.3	88.6
0330	Cast-in-Place Concrete	108.1	81.8	98.3	107.3	81.8	97.8	113.6	77.7	100.2	113.6	74.0	98.9	87.2	83.1	85.6	116.9	73.9	100.9
03	CONCRETE	114.3	70.9	95.2	100.2	80.4	91.5	104.9	74.5	91.5	105.1	74.9	91.8	87.5	79.4	83.9	103.7	67.4	87.7
04	MASONRY	106.0	77.4	88.4	111.1	77.4	90.4	104.3	73.8	85.6	112.8	74.0	88.9	93.6	78.2	84.1	101.9	77.4	86.8
05	METALS	99.1	85.2	95.0	91.4	98.0	93.3	93.2	93.7	93.3	94.0	93.7	93.9	95.0	93.3	94.5	93.9	93.8	93.9
06	WOOD, PLASTICS & COMPOSITES	82.6	48.7	64.0	94.3	74.9	83.6	93.4	66.3	78.5	97.2	72.1	83.4	99.0	75.0	85.8	90.7	49.7	68.1
07	THERMAL & MOISTURE PROTECTION	100.7	75.8	89.9	101.6	80.4	92.4	101.8	78.3	91.5	101.8	77.6	91.3	100.1	81.1	91.8	96.0	76.1	87.3
08	OPENINGS	90.8	58.1	83.2	91.3	73.7	87.2	91.3	66.8	85.6	91.3	69.8	86.3	105.4	68.8	96.9	91.2	55.8	83.0
0920	Plaster & Gypsum Board	79.8	47.5	58.1	93.4	74.3	80.6	92.5	65.5	74.3	94.4	71.4	78.9	102.7	74.3	83.6	91.0	48.4	62.4
0950, 0980	Ceilings & Acoustic Treatment	88.6	47.5	60.9	83.4	74.3	77.3	83.4	65.5	71.3	83.4	71.4	75.3	98.3	74.3	82.1	83.9	48.4	60.0
0960	Flooring	91.4	87.3	90.2	84.0	87.3	84.9	83.7	78.9	82.4	85.0	81.3	84.0	96.3	86.6	93.6	90.0	87.3	89.2
0970, 0990	Wall Finishes & Painting/Coating	155.8	51.7	94.2	74.9	60.1	66.1	74.9	62.5	67.6	74.9	60.1	66.1	94.4	79.3	85.5	87.6	45.0	62.4
09	FINISHES	90.8	59.2	73.7	84.9	75.5	79.8	84.9	69.3	76.5	85.6	72.7	78.6	95.6	77.9	86.0	88.4	59.1	72.5
COVERS	DIVS. 10 - 14, 25, 28, 41, 43, 44, 46	100.0	83.9	96.4	100.0	83.5	96.3	100.0	89.7	97.7	100.0	83.1	96.2	100.0	91.0	98.0	100.0	84.1	96.5
21, 22, 23	FIRE SUPPRESSION, PLUMBING & HVAC	96.7	75.0	88.2	96.6	75.3	88.2	100.1	79.5	92.0	96.6	74.6	87.9	100.0	79.6	92.0	96.5	75.2	88.1
26, 27, 3370	ELECTRICAL, COMMUNICATIONS & UTIL.	93.6	65.6	79.8	97.3	81.4	89.5	95.9	65.7	81.0	95.3	79.4	87.5	109.2	65.7	87.8	95.6	65.7	80.9
MF2018	WEIGHTED AVERAGE	98.7	73.3	87.9	96.0	80.7	89.6	97.4	76.8	88.7	97.1	78.2	89.1	98.3	79.7	90.5	96.3	73.1	86.5

	DIVISION	NEBRASKA												NEVADA					
		NORFOLK			NORTH PLATTE			OMAHA			VALENTINE			CARSON CITY			ELKO		
		687			691			680 - 681			692			897			898		
		MAT.	INST.	TOTAL	MAT.	INST.	TOTAL	MAT.	INST.	TOTAL	MAT.	INST.	TOTAL	MAT.	INST.	TOTAL	MAT.	INST.	TOTAL
015433	CONTRACTOR EQUIPMENT		91.7	91.7		101.9	101.9		94.2	94.2		95.0	95.0		97.9	97.9		94.6	94.6
0241, 31 - 34	SITE & INFRASTRUCTURE, DEMOLITION	84.1	89.7	88.0	102.9	90.4	94.3	89.1	93.9	92.4	86.3	95.2	92.4	83.3	97.4	93.0	67.8	91.3	84.0
0310	Concrete Forming & Accessories	83.2	74.1	75.5	94.8	75.2	78.1	93.1	76.7	79.1	83.5	53.1	57.6	106.1	76.3	80.7	111.4	96.3	98.6
0320	Concrete Reinforcing	100.5	66.5	84.1	99.5	76.2	88.3	100.8	76.5	89.0	100.1	66.3	83.7	115.8	116.1	116.0	124.5	114.5	119.7
0330	Cast-in-Place Concrete	107.9	71.3	94.3	116.9	60.6	96.0	91.4	78.6	86.7	103.1	54.0	84.8	98.0	82.2	92.1	95.0	72.4	86.6
03	CONCRETE	98.8	72.4	87.2	103.8	71.4	89.6	89.8	77.8	84.5	101.2	56.9	81.7	101.0	85.3	94.1	98.5	90.8	95.1
04	MASONRY	117.1	77.3	92.7	90.4	76.9	82.1	93.0	80.0	85.0	101.4	76.8	86.3	118.2	68.8	87.8	125.0	67.7	89.8
05	METALS	94.8	80.4	90.6	93.1	93.2	93.2	95.0	83.9	91.7	105.1	79.4	97.5	111.4	94.1	106.3	115.0	93.6	108.7
06	WOOD, PLASTICS & COMPOSITES	78.4	74.4	76.2	92.7	76.9	84.0	93.4	76.6	84.2	77.5	47.4	61.0	93.6	74.4	83.0	104.7	102.3	103.4
07	THERMAL & MOISTURE PROTECTION	101.2	78.3	91.2	95.9	78.4	88.3	96.7	80.8	89.8	96.5	73.9	86.7	118.1	79.1	101.2	114.2	73.6	96.5
08	OPENINGS	92.7	68.7	87.1	90.6	72.1	86.3	99.8	76.8	94.4	92.7	53.4	83.6	101.1	77.6	95.6	102.5	92.6	100.2
0920	Plaster & Gypsum Board	93.2	74.3	80.5	91.0	76.4	81.2	102.9	76.4	85.1	92.5	46.6	61.6	100.8	73.7	82.6	106.9	102.5	103.9
0950, 0980	Ceilings & Acoustic Treatment	97.6	74.3	81.9	83.9	76.4	78.8	101.0	76.4	84.4	100.4	46.6	64.1	96.0	73.7	80.9	94.5	102.5	99.9
0960	Flooring	104.1	87.3	99.4	91.1	78.9	87.7	100.6	87.3	96.9	116.8	87.3	108.5	99.5	67.1	90.4	103.0	67.1	92.9
0970, 0990	Wall Finishes & Painting/Coating	132.5	60.1	89.7	87.6	58.9	70.6	104.2	61.0	78.6	155.3	61.0	99.5	95.8	79.3	86.1	94.2	79.3	85.4
09	FINISHES	101.8	75.2	87.4	88.7	74.8	81.2	100.4	76.9	87.7	109.1	59.3	82.1	96.5	74.0	84.3	94.0	90.4	92.0
COVERS	DIVS. 10 - 14, 25, 28, 41, 43, 44, 46	100.0	85.1	96.7	100.0	61.1	91.3	100.0	90.2	97.8	100.0	56.1	90.2	100.0	100.2	100.0	100.0	91.0	98.0
21, 22, 23	FIRE SUPPRESSION, PLUMBING & HVAC	96.3	74.3	87.7	100.0	73.7	89.6	100.0	80.3	92.3	96.1	73.5	87.2	100.0	77.0	91.0	98.3	76.9	89.9
26, 27, 3370	ELECTRICAL, COMMUNICATIONS & UTIL.	96.2	81.4	88.9	93.9	65.7	80.0	103.9	83.7	94.0	91.4	65.7	78.7	102.2	88.3	95.3	98.8	88.3	93.6
MF2018	WEIGHTED AVERAGE	97.4	77.5	89.0	96.3	75.6	87.6	97.6	81.5	90.8	98.6	69.2	86.2	102.9	82.5	94.3	102.3	85.2	95.0

	DIVISION	NEVADA									NEW HAMPSHIRE								
		ELY			LAS VEGAS			RENO			CHARLESTON			CLAREMONT			CONCORD		
		893			889 - 891			894 - 895			036			037			032 - 033		
		MAT.	INST.	TOTAL	MAT.	INST.	TOTAL	MAT.	INST.	TOTAL	MAT.	INST.	TOTAL	MAT.	INST.	TOTAL	MAT.	INST.	TOTAL
015433	CONTRACTOR EQUIPMENT		94.6	94.6		94.6	94.6		94.6	94.6		95.7	95.7		95.7	95.7		98.4	98.4
0241, 31 - 34	SITE & INFRASTRUCTURE, DEMOLITION	73.4	92.5	86.6	76.6	95.5	89.7	73.4	92.6	86.7	80.6	96.1	91.3	74.6	96.1	89.5	88.7	101.2	97.3
0310	Concrete Forming & Accessories	104.3	101.6	102.0	105.2	105.6	105.5	100.6	76.3	79.9	86.6	83.1	83.6	92.1	83.2	84.5	97.2	94.8	95.1
0320	Concrete Reinforcing	123.1	114.7	119.0	113.5	125.9	119.5	116.4	124.2	120.1	89.7	88.6	89.1	89.7	88.6	89.1	100.9	88.7	95.0
0330	Cast-in-Place Concrete	102.0	95.2	99.4	98.7	108.2	102.2	107.9	81.1	97.9	84.7	116.1	96.4	77.8	116.1	92.1	100.9	117.8	107.2
03	CONCRETE	107.5	101.2	104.7	102.4	109.5	105.5	106.7	86.3	97.8	90.4	95.8	92.7	82.8	95.8	88.5	96.8	101.6	98.9
04	MASONRY	130.5	74.8	96.3	117.4	106.2	110.5	124.2	68.7	90.1	89.8	99.5	95.8	89.9	99.5	95.8	97.8	101.4	100.0
05	METALS	115.0	96.5	109.5	124.1	103.9	118.2	116.7	97.4	111.1	94.2	92.1	93.5	94.2	92.1	93.6	101.3	91.8	98.5
06	WOOD, PLASTICS & COMPOSITES	94.8	104.4	100.1	92.4	102.3	97.8	88.4	74.3	80.6	86.6	79.4	82.6	92.4	79.4	85.3	93.3	93.6	93.5
07	THERMAL & MOISTURE PROTECTION	114.7	92.2	104.9	129.4	101.9	117.4	114.2	78.3	98.6	107.3	106.5	107.0	107.1	106.5	106.8	111.3	109.5	110.5
08	OPENINGS	102.4	93.8	100.4	101.4	109.6	103.3	100.2	79.2	95.3	96.1	80.6	92.5	97.2	80.6	93.3	97.0	88.5	95.1
0920	Plaster & Gypsum Board	101.2	104.6	103.5	93.9	102.5	99.7	89.0	73.7	78.7	106.1	78.5	87.5	107.0	78.5	87.9	108.7	93.1	98.2
0950, 0980	Ceilings & Acoustic Treatment	94.5	104.6	101.3	101.0	102.5	102.0	97.8	73.7	81.5	101.8	78.5	86.1	101.8	78.5	86.1	106.1	93.1	97.4
0960	Flooring	100.5	67.1	91.1	92.0	104.1	95.4	97.3	67.1	88.8	85.1	113.1	92.9	87.6	113.1	94.7	96.4	113.1	101.1
0970, 0990	Wall Finishes & Painting/Coating	94.2	112.0	104.7	96.8	119.4	110.2	94.2	79.3	85.4	81.8	90.0	86.6	81.8	90.0	86.6	95.9	90.0	92.4
09	FINISHES	92.9	97.0	95.1	90.8	106.4	99.3	90.8	73.9	81.7	90.2	88.5	89.3	90.6	88.5	89.5	95.9	97.4	96.7
COVERS	DIVS. 10 - 14, 25, 28, 41, 43, 44, 46	100.0	70.7	93.5	100.0	105.1	101.1	100.0	99.8	100.0	100.0	88.5	97.4	100.0	88.5	97.4	100.0	109.5	102.1
21, 22, 23	FIRE SUPPRESSION, PLUMBING & HVAC	98.3	96.2	97.4	100.1	103.0	101.3	100.0	77.1	91.0	96.7	82.2	91.0	96.7	82.2	91.0	100.0	87.2	95.0
26, 27, 3370	ELECTRICAL, COMMUNICATIONS & UTIL.	99.0	93.0	96.0	103.5	107.8	105.6	99.4	88.3	93.9	91.4	75.2	83.4	91.4	75.2	83.4	90.2	75.2	82.8
MF2018	WEIGHTED AVERAGE	103.7	93.2	99.2	105.0	105.1	105.0	103.6	82.6	94.7	94.1	88.4	91.7	93.1	88.4	91.1	98.0	93.2	96.0

	DIVISION	NEW HAMPSHIRE															NEW JERSEY		
		KEENE			LITTLETON			MANCHESTER			NASHUA			PORTSMOUTH			ATLANTIC CITY		
		034			035			031			030			038			082, 084		
		MAT.	INST.	TOTAL	MAT.	INST.	TOTAL	MAT.	INST.	TOTAL	MAT.	INST.	TOTAL	MAT.	INST.	TOTAL	MAT.	INST.	TOTAL
015433	CONTRACTOR EQUIPMENT		95.7	95.7		95.7	95.7		99.0	99.0		95.7	95.7		95.7	95.7		93.3	93.3
0241, 31 - 34	SITE & INFRASTRUCTURE, DEMOLITION	88.0	96.2	93.6	74.7	95.2	88.8	87.8	101.3	97.1	89.5	96.3	94.2	83.3	97.0	92.7	88.2	99.7	96.1
0310	Concrete Forming & Accessories	90.7	83.4	84.5	102.4	77.3	81.0	97.5	95.1	95.5	98.5	94.9	95.4	87.9	94.1	93.2	114.8	142.0	138.0
0320	Concrete Reinforcing	89.7	88.6	89.2	90.4	88.6	89.5	108.5	88.7	98.9	112.0	88.7	100.7	89.7	88.7	89.2	81.1	138.0	108.6
0330	Cast-in-Place Concrete	85.1	116.2	96.7	76.3	107.5	87.9	99.2	118.3	106.3	80.5	117.4	94.3	76.3	117.2	91.5	78.5	137.1	100.3
03	CONCRETE	89.9	95.9	92.6	82.2	90.1	85.7	97.0	101.9	99.2	90.0	101.6	95.1	82.4	101.2	90.7	87.7	138.1	109.8
04	MASONRY	92.5	99.5	96.8	100.9	84.1	90.6	94.4	101.4	98.7	94.4	101.4	98.7	90.2	99.8	96.1	104.1	140.0	126.1
05	METALS	94.8	92.5	94.2	94.9	92.2	94.1	102.0	92.2	99.1	100.5	92.7	98.2	96.4	94.3	95.8	98.8	115.0	103.6
06	WOOD, PLASTICS & COMPOSITES	90.9	79.4	84.6	101.9	79.4	89.5	93.6	93.7	93.7	100.2	93.6	96.6	87.8	93.6	91.0	122.8	143.7	134.3
07	THERMAL & MOISTURE PROTECTION	107.9	106.4	107.2	107.2	99.5	103.8	112.5	109.5	111.2	108.3	108.9	108.6	107.8	108.1	107.9	102.1	134.0	116.0
08	OPENINGS	94.8	84.1	92.3	98.1	80.6	94.0	95.7	93.3	95.2	99.0	91.3	97.2	99.6	81.9	95.5	96.4	139.2	106.4
0920	Plaster & Gypsum Board	106.4	78.5	87.6	120.6	78.5	92.3	99.7	93.1	95.3	115.6	93.1	100.5	106.1	93.1	97.4	114.9	144.6	134.9
0950, 0980	Ceilings & Acoustic Treatment	101.8	78.5	86.1	101.8	78.5	86.1	101.2	93.1	95.8	114.1	93.1	100.0	102.7	93.1	96.3	96.4	144.6	128.9
0960	Flooring	87.2	113.1	94.5	97.9	113.1	102.2	93.8	113.1	99.2	91.0	113.1	97.2	85.2	113.1	93.0	97.5	162.5	115.8
0970, 0990	Wall Finishes & Painting/Coating	81.8	103.4	94.6	81.8	90.0	86.6	95.3	119.0	109.3	81.8	102.2	93.9	81.8	102.2	93.9	81.9	141.8	117.4
09	FINISHES	92.1	89.9	90.9	95.5	84.6	89.6	94.8	100.6	97.9	96.8	98.7	97.8	91.1	98.2	95.0	93.5	147.2	122.6
COVERS	DIVS. 10 - 14, 25, 28, 41, 43, 44, 46	100.0	94.7	98.8	100.0	95.6	99.0	100.0	109.7	102.2	100.0	109.4	102.1	100.0	108.7	101.9	100.0	114.9	103.3
21, 22, 23	FIRE SUPPRESSION, PLUMBING & HVAC	96.7	82.3	91.0	96.7	74.1	87.8	100.1	87.3	95.1	100.2	87.3	95.1	100.2	86.3	94.7	99.7	133.1	112.8
26, 27, 3370	ELECTRICAL, COMMUNICATIONS & UTIL.	91.4	75.2	83.4	92.2	49.1	71.0	93.9	79.3	86.7	93.2	79.3	86.4	91.7	75.3	83.6	94.2	143.4	118.5
MF2018	WEIGHTED AVERAGE	94.5	89.0	92.2	94.4	80.2	88.4	98.2	94.5	96.6	97.5	93.7	95.9	94.9	92.4	93.8	96.8	133.2	112.2

	DIVISION	NEW JERSEY																	
		CAMDEN			DOVER			ELIZABETH			HACKENSACK			JERSEY CITY			LONG BRANCH		
		081			078			072			076			073			077		
		MAT.	INST.	TOTAL	MAT.	INST.	TOTAL	MAT.	INST.	TOTAL	MAT.	INST.	TOTAL	MAT.	INST.	TOTAL	MAT.	INST.	TOTAL
015433	CONTRACTOR EQUIPMENT		93.3	93.3		95.7	95.7		95.7	95.7		95.7	95.7		93.3	93.3		92.9	92.9
0241, 31 - 34	SITE & INFRASTRUCTURE, DEMOLITION	89.3	99.3	96.2	103.8	101.4	102.2	108.9	101.4	103.7	104.8	101.4	102.5	95.2	101.4	99.5	100.0	100.5	100.4
0310	Concrete Forming & Accessories	105.0	137.3	132.5	96.4	148.3	140.6	108.0	148.5	142.5	96.4	148.3	140.7	100.6	148.4	141.3	101.1	139.0	133.4
0320	Concrete Reinforcing	106.5	128.5	117.1	75.3	156.5	114.6	75.3	156.5	114.6	75.3	156.5	114.6	97.7	156.5	126.2	75.3	156.1	114.4
0330	Cast-in-Place Concrete	76.1	133.8	97.6	81.3	131.5	100.0	69.8	140.3	96.1	79.4	140.3	102.1	63.4	131.6	88.8	70.5	133.8	94.0
03	CONCRETE	88.2	133.2	108.0	86.1	142.3	110.8	83.3	145.5	110.6	84.5	145.4	111.2	81.0	142.2	107.9	85.1	138.6	108.6
04	MASONRY	94.2	135.0	119.3	91.2	145.6	124.7	107.2	145.6	130.8	95.3	145.6	126.2	85.3	145.6	122.4	99.6	135.5	121.6
05	METALS	104.7	111.0	106.5	96.5	125.1	104.9	98.0	125.3	106.0	96.5	125.1	104.9	102.4	122.2	108.2	96.6	121.0	103.7
06	WOOD, PLASTICS & COMPOSITES	110.1	137.9	125.4	94.9	148.3	124.3	109.7	148.3	131.0	94.9	148.3	124.3	95.8	148.3	124.7	96.9	139.3	120.2
07	THERMAL & MOISTURE PROTECTION	102.0	133.0	115.5	103.7	142.8	120.7	104.0	144.1	121.4	103.5	136.7	117.9	103.2	142.8	120.4	103.4	129.0	114.5
08	OPENINGS	98.4	132.0	106.3	102.0	145.1	112.1	100.4	145.1	110.8	99.8	145.1	110.4	98.5	145.1	109.4	94.6	138.5	104.8
0920	Plaster & Gypsum Board	110.9	138.6	129.6	111.5	149.4	137.0	119.0	149.4	139.5	111.5	149.4	137.0	114.6	149.4	138.0	113.7	140.2	131.5
0950, 0980	Ceilings & Acoustic Treatment	106.3	138.6	128.1	92.3	149.4	130.8	94.1	149.4	131.4	92.3	149.4	130.8	102.2	149.4	134.0	92.3	140.2	124.6
0960	Flooring	93.7	162.5	113.0	83.7	188.8	113.2	88.7	188.8	116.8	83.7	188.8	113.2	84.7	188.8	114.0	84.9	176.9	110.8
0970, 0990	Wall Finishes & Painting/Coating	81.9	141.8	117.4	83.8	148.6	122.1	83.8	148.6	122.1	83.8	148.6	122.1	83.9	148.6	122.2	83.9	141.8	118.2
09	FINISHES	93.7	143.7	120.8	90.0	156.2	125.8	93.4	156.2	127.4	89.9	156.2	125.8	92.4	156.6	127.1	91.0	147.2	121.4
COVERS	DIVS. 10 - 14, 25, 28, 41, 43, 44, 46	100.0	114.3	103.2	100.0	131.7	107.1	100.0	131.7	107.1	100.0	131.7	107.1	100.0	131.7	107.1	100.0	114.5	103.2
21, 22, 23	FIRE SUPPRESSION, PLUMBING & HVAC	100.0	126.3	110.3	99.7	135.7	113.8	100.0	137.1	114.6	99.7	135.7	113.8	100.0	135.7	114.0	99.7	131.1	112.1
26, 27, 3370	ELECTRICAL, COMMUNICATIONS & UTIL.	98.4	132.1	115.0	96.0	138.5	116.9	96.6	138.5	117.2	96.0	142.3	118.8	100.5	142.3	121.1	95.7	129.9	112.6
MF2018	WEIGHTED AVERAGE	98.0	127.8	110.6	96.4	137.5	113.7	97.5	138.3	114.7	96.1	138.3	114.0	96.5	137.8	114.0	95.8	130.9	110.7

DIVISION		NEW JERSEY																	
		NEW BRUNSWICK			NEWARK			PATERSON			POINT PLEASANT			SUMMIT			TRENTON		
		088 - 089			070 - 071			074 - 075			087			079			085 - 086		
		MAT.	INST.	TOTAL	MAT.	INST.	TOTAL	MAT.	INST.	TOTAL	MAT.	INST.	TOTAL	MAT.	INST.	TOTAL	MAT.	INST.	TOTAL
015433	CONTRACTOR EQUIPMENT		92.9	92.9		98.3	98.3		95.7	95.7		92.9	92.9		95.7	95.7		97.0	97.0
0241, 31 - 34	SITE & INFRASTRUCTURE, DEMOLITION	101.1	100.9	100.9	110.0	106.2	107.4	106.6	101.4	103.0	102.6	100.5	101.2	106.3	101.4	102.9	88.6	105.4	100.2
0310	Concrete Forming & Accessories	108.5	148.0	142.2	100.7	148.7	141.6	98.3	148.3	140.9	102.4	139.1	133.6	99.0	148.5	141.2	103.5	138.8	133.6
0320	Concrete Reinforcing	82.1	156.1	117.9	96.2	156.6	125.4	97.7	156.5	126.2	82.1	156.1	117.9	75.3	156.5	114.6	107.7	116.2	111.8
0330	Cast-in-Place Concrete	97.0	136.9	111.8	89.8	140.9	108.8	80.8	140.2	102.9	97.0	135.8	111.5	67.5	140.3	94.6	95.6	134.4	110.0
03	CONCRETE	104.2	143.8	121.6	93.1	145.7	116.2	88.8	145.3	113.7	103.9	139.4	119.5	80.6	145.5	109.1	97.4	132.1	112.6
04	MASONRY	102.5	140.6	125.9	94.9	145.7	126.1	91.7	145.6	124.9	91.4	135.5	118.5	93.8	145.6	125.6	95.4	135.5	120.0
05	METALS	98.9	121.4	105.5	104.2	124.7	110.3	97.5	125.1	105.6	98.9	121.0	105.4	96.5	125.3	104.9	104.3	108.8	105.6
06	WOOD, PLASTICS & COMPOSITES	115.7	148.3	133.6	97.8	148.4	125.6	97.5	148.3	125.5	107.3	139.3	124.9	98.7	148.3	126.0	101.6	139.4	122.4
07	THERMAL & MOISTURE PROTECTION	102.5	138.2	118.0	105.2	144.7	122.4	103.8	136.7	118.1	102.5	131.3	115.0	104.1	144.1	121.5	102.9	135.2	117.0
08	OPENINGS	91.6	145.1	104.1	100.5	145.1	110.9	105.0	145.1	114.3	93.4	141.0	104.5	106.1	145.1	115.2	96.9	130.0	104.6
0920	Plaster & Gypsum Board	113.0	149.4	137.5	106.2	149.4	135.3	114.6	149.4	138.0	107.9	140.2	129.6	113.7	149.4	137.7	101.6	140.2	127.5
0950, 0980	Ceilings & Acoustic Treatment	96.4	149.4	132.2	103.0	149.4	134.3	102.2	149.4	134.0	96.4	140.2	125.9	92.3	149.4	130.8	101.2	140.2	127.5
0960	Flooring	95.1	188.8	121.4	96.7	188.8	122.6	84.7	188.8	114.0	92.5	162.5	112.2	84.9	188.8	114.1	100.1	176.9	121.7
0970, 0990	Wall Finishes & Painting/Coating	81.9	148.6	121.4	94.1	148.6	126.3	83.8	148.6	122.1	81.9	141.8	117.4	83.8	148.6	122.1	94.9	141.8	122.7
09	FINISHES	93.8	156.2	127.6	97.2	156.7	129.4	92.5	156.2	127.0	92.3	144.7	120.7	91.0	156.2	126.3	97.0	147.3	124.2
COVERS	DIVS. 10 - 14, 25, 28, 41, 43, 44, 46	100.0	131.5	107.0	100.0	131.8	107.1	100.0	131.7	107.1	100.0	111.6	102.6	100.0	131.7	107.1	100.0	114.7	103.3
21, 22, 23	FIRE SUPPRESSION, PLUMBING & HVAC	99.7	133.8	113.1	100.0	139.2	115.4	100.0	135.5	114.0	99.7	131.1	112.0	99.8	137.1	114.4	100.1	130.8	112.1
26, 27, 3370	ELECTRICAL, COMMUNICATIONS & UTIL.	94.8	133.7	114.0	105.0	142.3	123.4	100.5	138.5	119.2	94.2	129.9	111.8	96.6	138.5	117.2	102.6	128.7	115.4
MF2018	WEIGHTED AVERAGE	98.7	135.6	114.3	100.3	139.7	117.0	98.0	137.7	114.8	98.1	130.8	111.9	96.4	138.3	114.1	99.7	128.9	112.0

DIVISION		NEW JERSEY			NEW MEXICO														
		VINELAND			ALBUQUERQUE			CARRIZOZO			CLOVIS			FARMINGTON			GALLUP		
		080, 083			870 - 872			883			881			874			873		
		MAT.	INST.	TOTAL	MAT.	INST.	TOTAL	MAT.	INST.	TOTAL	MAT.	INST.	TOTAL	MAT.	INST.	TOTAL	MAT.	INST.	TOTAL
015433	CONTRACTOR EQUIPMENT		93.3	93.3		107.5	107.5		107.5	107.5		107.5	107.5		107.5	107.5		107.5	107.5
0241, 31 - 34	SITE & INFRASTRUCTURE, DEMOLITION	92.5	99.4	97.3	91.1	97.5	95.5	109.7	97.5	101.3	96.6	97.5	97.2	97.3	97.5	97.5	107.2	97.5	100.5
0310	Concrete Forming & Accessories	99.2	137.6	131.9	101.0	66.2	71.3	98.6	66.2	71.0	98.6	66.1	70.9	101.1	66.2	71.4	101.1	66.2	71.4
0320	Concrete Reinforcing	81.1	131.2	105.3	99.9	71.5	86.1	116.0	71.5	94.5	117.3	71.4	95.1	109.2	71.5	90.9	104.5	71.5	88.5
0330	Cast-in-Place Concrete	84.6	133.9	103.0	90.5	70.1	82.9	94.8	70.1	85.6	94.7	70.1	85.5	91.3	70.1	83.4	86.0	70.1	80.1
03	CONCRETE	92.2	133.8	110.5	92.9	69.8	82.7	116.7	69.8	96.1	104.3	69.7	89.1	96.4	69.8	84.7	103.3	69.8	88.6
04	MASONRY	92.6	135.5	118.9	106.0	60.5	78.1	108.5	60.5	79.0	108.5	60.5	79.0	113.8	60.5	81.1	101.6	60.5	76.4
05	METALS	98.8	112.6	102.8	108.1	90.6	102.9	106.6	90.6	101.9	106.3	90.5	101.6	105.7	90.6	101.3	104.8	90.6	100.7
06	WOOD, PLASTICS & COMPOSITES	103.7	137.9	122.5	102.1	67.2	82.9	93.5	67.2	79.0	93.5	67.2	79.0	102.2	67.2	82.9	102.2	67.2	82.9
07	THERMAL & MOISTURE PROTECTION	101.9	132.2	115.1	101.0	72.8	88.7	106.3	72.8	91.7	105.0	72.8	91.0	101.3	72.8	88.9	102.5	72.8	89.6
08	OPENINGS	92.9	134.9	102.7	98.5	67.0	91.1	96.5	67.0	89.6	96.6	67.0	89.7	100.8	67.0	92.9	100.8	67.0	93.0
0920	Plaster & Gypsum Board	106.4	138.6	128.1	113.0	66.1	81.5	80.2	66.1	70.7	80.2	66.1	70.7	99.4	66.1	77.0	99.4	66.1	77.0
0950, 0980	Ceilings & Acoustic Treatment	96.4	138.6	124.9	99.8	66.1	77.1	102.7	66.1	78.1	102.7	66.1	78.1	98.6	66.1	76.7	98.6	66.1	76.7
0960	Flooring	91.7	162.5	111.6	88.9	66.8	82.7	97.2	66.8	88.7	97.2	66.8	88.7	90.4	66.8	83.8	90.4	66.8	83.8
0970, 0990	Wall Finishes & Painting/Coating	81.9	141.8	117.4	97.8	52.6	71.1	92.6	52.6	68.9	92.6	52.6	68.9	92.1	52.6	68.7	92.1	52.6	68.7
09	FINISHES	91.1	143.9	119.6	92.3	64.7	77.4	94.2	64.7	78.3	92.9	64.7	77.7	90.7	64.7	76.6	92.1	64.7	77.3
COVERS	DIVS. 10 - 14, 25, 28, 41, 43, 44, 46	100.0	114.5	103.2	100.0	85.1	96.7	100.0	85.1	96.7	100.0	85.1	96.7	100.0	85.1	96.7	100.0	85.1	96.7
21, 22, 23	FIRE SUPPRESSION, PLUMBING & HVAC	99.7	126.6	110.3	100.3	69.0	88.0	97.9	69.0	86.5	97.9	68.6	86.4	100.2	69.0	88.0	98.1	69.0	86.6
26, 27, 3370	ELECTRICAL, COMMUNICATIONS & UTIL.	94.2	143.3	118.5	87.9	69.5	78.8	90.1	69.5	80.0	87.9	69.5	78.8	85.9	69.5	77.8	85.3	69.5	77.5
MF2018	WEIGHTED AVERAGE	96.3	129.8	110.5	98.6	72.5	87.6	101.6	72.5	89.3	99.3	72.4	87.9	99.0	72.5	87.8	99.0	72.5	87.8

DIVISION		NEW MEXICO																	
		LAS CRUCES			LAS VEGAS			ROSWELL			SANTA FE			SOCORRO			TRUTH/CONSEQUENCES		
		880			877			882			875			878			879		
		MAT.	INST.	TOTAL	MAT.	INST.	TOTAL	MAT.	INST.	TOTAL	MAT.	INST.	TOTAL	MAT.	INST.	TOTAL	MAT.	INST.	TOTAL
015433	CONTRACTOR EQUIPMENT		83.5	83.5		107.5	107.5		107.5	107.5		110.4	110.4		107.5	107.5		83.5	83.5
0241, 31 - 34	SITE & INFRASTRUCTURE, DEMOLITION	97.8	77.5	83.8	96.2	97.5	97.1	99.0	97.5	98.0	99.1	102.8	101.6	92.8	97.5	96.1	112.5	77.5	88.3
0310	Concrete Forming & Accessories	95.4	65.2	69.7	101.1	66.2	71.4	98.6	66.2	71.0	100.0	66.3	71.2	101.1	66.2	71.4	98.8	65.1	70.1
0320	Concrete Reinforcing	113.1	71.3	92.9	106.3	71.5	89.5	117.3	71.5	95.1	99.3	71.5	85.9	108.4	71.5	90.5	101.7	71.3	87.0
0330	Cast-in-Place Concrete	89.5	63.1	79.7	88.8	70.1	81.9	94.7	70.1	85.6	94.3	70.9	85.6	87.0	70.1	80.7	95.4	63.1	83.4
03	CONCRETE	82.1	66.5	75.3	94.0	69.8	83.4	105.3	69.8	89.7	92.9	70.0	82.9	92.9	69.8	82.8	86.1	66.4	77.4
04	MASONRY	104.0	60.2	77.1	101.9	60.5	76.5	119.5	60.5	83.3	95.0	60.6	73.9	101.8	60.5	76.4	98.7	60.2	75.0
05	METALS	105.0	83.0	98.5	104.5	90.6	100.4	107.5	90.6	102.6	101.6	89.6	98.1	104.8	90.6	100.7	104.4	83.0	98.1
06	WOOD, PLASTICS & COMPOSITES	82.9	66.2	73.7	102.2	67.2	82.9	93.5	67.2	79.0	99.0	67.2	81.5	102.2	67.2	82.9	93.5	66.2	78.5
07	THERMAL & MOISTURE PROTECTION	92.2	68.2	81.7	100.8	72.8	88.6	105.2	72.8	91.1	103.5	73.7	90.5	100.8	72.8	88.6	90.1	68.2	80.5
08	OPENINGS	92.0	66.4	86.1	97.3	67.0	90.3	96.4	67.0	89.6	98.8	67.0	91.4	97.2	67.0	90.1	90.6	66.4	85.0
0920	Plaster & Gypsum Board	78.5	66.1	70.2	99.4	66.1	77.0	80.2	66.1	70.7	114.1	66.1	81.8	99.4	66.1	77.0	100.6	66.1	77.4
0950, 0980	Ceilings & Acoustic Treatment	87.3	66.1	73.0	98.6	66.1	76.7	102.7	66.1	78.1	95.2	66.1	75.6	98.6	66.1	76.7	86.2	66.1	72.7
0960	Flooring	128.1	66.8	110.9	90.4	66.8	83.8	97.2	66.8	88.7	100.9	66.8	91.3	90.4	66.8	83.8	119.2	66.8	104.5
0970, 0990	Wall Finishes & Painting/Coating	81.8	52.6	64.5	92.1	52.6	68.7	92.6	52.6	68.9	99.8	52.6	71.9	92.1	52.6	68.7	84.5	52.6	65.6
09	FINISHES	103.0	64.0	81.9	90.5	64.7	76.6	93.0	64.7	77.7	98.2	64.8	80.1	90.4	64.7	76.5	102.4	64.0	81.6
COVERS	DIVS. 10 - 14, 25, 28, 41, 43, 44, 46	100.0	82.7	96.1	100.0	85.1	96.7	100.0	85.1	96.7	100.0	85.2	96.7	100.0	85.1	96.7	100.0	82.6	96.1
21, 22, 23	FIRE SUPPRESSION, PLUMBING & HVAC	100.4	68.7	88.0	98.1	69.0	86.6	100.0	69.0	87.8	100.3	69.0	88.0	98.1	69.0	86.6	98.0	68.7	86.5
26, 27, 3370	ELECTRICAL, COMMUNICATIONS & UTIL.	90.1	83.8	87.0	87.4	69.5	78.6	89.2	69.5	79.5	100.3	69.5	85.2	85.7	69.5	77.7	89.5	69.5	79.6
MF2018	WEIGHTED AVERAGE	96.9	71.4	86.1	97.2	72.5	86.8	100.8	72.5	88.9	99.0	72.9	88.0	96.8	72.5	86.6	96.6	69.4	85.1

DIVISION		NEW MEXICO			NEW YORK															
		TUCUMCARI			ALBANY			BINGHAMTON			BRONX			BROOKLYN			BUFFALO			
		884			120 - 122			137 - 139			104			112			140 - 142			
		MAT.	INST.	TOTAL	MAT.	INST.	TOTAL	MAT.	INST.	TOTAL	MAT.	INST.	TOTAL	MAT.	INST.	TOTAL	MAT.	INST.	TOTAL	
015433	**CONTRACTOR EQUIPMENT**		107.5	107.5		115.6	115.6		117.6	117.6		104.4	104.4		109.4	109.4		100.5	100.5	
0241, 31 - 34	**SITE & INFRASTRUCTURE, DEMOLITION**	96.2	97.5	97.1	80.0	105.1	97.4	94.3	89.3	90.8	97.3	109.7	105.9	118.1	120.6	119.8	97.3	102.3	100.7	
0310	Concrete Forming & Accessories	98.6	66.1	70.9	97.3	107.3	105.8	99.4	93.4	94.3	95.7	189.1	175.3	105.1	189.1	176.7	101.3	117.1	114.7	
0320	Concrete Reinforcing	115.0	71.4	93.9	101.3	115.2	108.0	95.8	106.3	100.9	94.8	182.0	136.9	97.2	242.5	167.5	98.9	116.4	107.3	
0330	Cast-in-Place Concrete	94.7	70.1	85.5	81.0	116.4	94.2	106.3	106.9	106.5	85.6	172.0	117.8	108.3	170.5	131.5	109.1	123.5	114.4	
03	**CONCRETE**	103.6	69.7	88.7	87.7	112.8	98.7	94.3	102.6	97.9	88.0	181.1	128.9	106.4	189.9	143.1	104.9	118.7	111.0	
04	**MASONRY**	119.9	60.5	83.4	88.0	117.9	106.4	104.3	104.9	104.7	88.8	188.9	150.3	114.8	188.8	160.3	112.6	122.2	118.5	
05	**METALS**	106.3	90.5	101.6	102.2	125.0	108.9	95.0	133.6	106.3	86.4	174.6	112.3	103.1	175.0	124.2	97.0	107.5	100.1	
06	**WOOD, PLASTICS & COMPOSITES**	93.5	67.2	79.0	94.6	104.0	99.8	101.7	90.2	95.4	97.2	188.6	147.5	103.7	188.3	150.3	98.6	115.8	108.1	
07	**THERMAL & MOISTURE PROTECTION**	105.0	72.8	91.0	105.4	110.4	107.6	109.2	94.8	102.9	102.5	168.8	131.3	109.3	168.2	134.9	102.5	111.7	106.5	
08	**OPENINGS**	96.4	67.0	89.5	96.4	103.8	98.1	90.4	93.7	91.2	92.3	197.7	116.9	87.8	197.5	113.4	99.8	111.4	102.5	
0920	Plaster & Gypsum Board	80.2	66.1	70.7	97.8	103.9	101.9	109.2	89.7	96.1	102.8	191.0	162.1	103.3	191.0	162.3	105.8	116.1	112.7	
0950, 0980	Ceilings & Acoustic Treatment	102.7	66.1	78.1	97.1	103.9	101.7	96.6	89.7	92.0	89.1	191.0	157.8	91.8	191.0	158.7	104.2	116.1	112.2	
0960	Flooring	97.2	66.8	88.7	90.9	110.3	96.4	103.1	102.4	102.9	99.9	182.5	123.1	110.2	182.5	130.5	96.3	116.6	102.0	
0970, 0990	Wall Finishes & Painting/Coating	92.6	52.6	68.9	95.9	104.4	100.9	89.3	106.3	99.4	106.2	167.4	142.4	116.0	167.4	146.4	96.9	117.8	109.3	
09	**FINISHES**	92.8	64.7	77.6	90.6	107.1	99.5	93.6	95.9	94.8	98.0	186.2	145.7	107.1	186.0	149.8	102.0	117.6	110.5	
COVERS	**DIVS. 10 - 14, 25, 28, 41, 43, 44, 46**	100.0	85.1	96.7	100.0	103.8	100.9	100.0	96.6	99.2	100.0	141.5	109.3	100.0	140.8	109.1	100.0	106.5	101.4	
21, 22, 23	**FIRE SUPPRESSION, PLUMBING & HVAC**	97.9	68.6	86.4	100.1	110.3	104.1	100.6	97.0	99.2	100.3	177.9	130.7	99.8	177.9	130.5	100.0	102.4	100.9	
26, 27, 3370	**ELECTRICAL, COMMUNICATIONS & UTIL.**	90.1	69.5	80.0	100.0	108.4	104.1	98.7	101.0	99.8	90.6	184.5	136.9	98.5	184.5	140.9	99.5	105.3	102.4	
MF2018	**WEIGHTED AVERAGE**	99.9	72.4	88.3	96.8	111.2	102.8	97.3	101.5	99.1	93.8	175.3	128.3	101.9	177.3	133.8	100.8	110.3	104.8	

DIVISION		NEW YORK																	
		ELMIRA			FAR ROCKAWAY			FLUSHING			GLENS FALLS			HICKSVILLE			JAMAICA		
		148 - 149			116			113			128			115, 117, 118			114		
		MAT.	INST.	TOTAL	MAT.	INST.	TOTAL	MAT.	INST.	TOTAL	MAT.	INST.	TOTAL	MAT.	INST.	TOTAL	MAT.	INST.	TOTAL
015433	**CONTRACTOR EQUIPMENT**		120.0	120.0		109.4	109.4		109.4	109.4		112.6	112.6		109.4	109.4		109.4	109.4
0241, 31 - 34	**SITE & INFRASTRUCTURE, DEMOLITION**	97.4	89.6	92.0	121.2	120.6	120.8	121.3	120.6	120.8	71.4	99.6	90.9	111.1	119.1	116.6	115.5	120.6	119.0
0310	Concrete Forming & Accessories	85.0	96.3	94.6	92.2	189.1	174.8	95.9	189.1	175.3	81.5	99.9	97.2	88.8	157.2	147.1	95.9	189.1	175.3
0320	Concrete Reinforcing	98.6	105.3	101.8	97.2	242.5	167.5	98.9	242.5	168.4	96.9	112.7	104.5	97.2	174.4	134.6	97.2	242.5	167.5
0330	Cast-in-Place Concrete	98.8	106.4	101.6	117.2	170.5	137.1	117.2	170.5	137.1	78.1	111.6	90.6	99.5	161.5	122.6	108.3	170.5	131.5
03	**CONCRETE**	90.9	103.7	96.5	112.8	189.9	146.7	113.2	189.9	146.9	81.2	107.4	92.8	98.2	161.8	126.2	105.8	189.9	142.8
04	**MASONRY**	101.4	106.7	104.6	118.6	188.8	161.8	112.8	188.8	159.5	93.4	112.6	105.2	109.0	174.4	149.2	117.2	188.8	161.2
05	**METALS**	95.5	135.8	107.3	103.2	175.0	124.3	103.2	175.0	124.3	95.5	124.5	104.0	104.7	172.4	124.6	103.2	175.0	124.3
06	**WOOD, PLASTICS & COMPOSITES**	84.1	94.4	89.8	87.1	188.3	142.8	91.7	188.3	144.9	82.7	96.1	90.1	83.7	154.0	122.4	91.7	188.3	144.9
07	**THERMAL & MOISTURE PROTECTION**	108.9	95.2	102.9	109.2	168.3	134.9	109.3	168.3	134.9	98.6	107.1	102.3	108.9	157.8	130.2	109.1	168.3	134.8
08	**OPENINGS**	96.7	95.7	96.4	86.6	197.5	112.5	86.6	197.5	112.5	90.0	99.1	92.1	87.0	178.6	108.3	86.6	197.5	112.5
0920	Plaster & Gypsum Board	99.5	94.2	96.0	91.7	191.0	158.5	94.2	191.0	159.3	91.0	96.0	94.4	91.4	155.6	134.6	94.2	191.0	159.3
0950, 0980	Ceilings & Acoustic Treatment	106.0	94.2	98.1	81.2	191.0	155.2	81.2	191.0	155.2	86.4	96.0	92.9	80.3	155.6	131.1	81.2	191.0	155.2
0960	Flooring	87.6	102.4	91.8	105.0	182.5	126.8	106.5	182.5	127.9	82.3	107.7	89.5	104.0	179.7	125.2	106.5	182.5	127.9
0970, 0990	Wall Finishes & Painting/Coating	97.2	93.3	94.9	116.0	167.4	146.4	116.0	167.4	146.4	89.3	104.4	98.2	116.0	167.4	146.4	116.0	167.4	146.4
09	**FINISHES**	94.2	96.9	95.7	102.4	186.0	147.6	103.2	186.0	148.0	84.0	101.1	93.2	101.0	161.6	133.8	102.7	186.0	147.8
COVERS	**DIVS. 10 - 14, 25, 28, 41, 43, 44, 46**	100.0	97.6	99.5	100.0	140.8	109.1	100.0	140.8	109.1	100.0	96.9	99.3	100.0	132.1	107.1	100.0	140.8	109.1
21, 22, 23	**FIRE SUPPRESSION, PLUMBING & HVAC**	96.6	95.5	96.2	96.2	177.9	128.3	96.2	177.9	128.3	96.7	109.3	101.6	99.8	161.0	123.8	96.2	177.9	128.3
26, 27, 3370	**ELECTRICAL, COMMUNICATIONS & UTIL.**	96.4	104.1	100.2	105.3	184.5	144.4	105.3	184.5	144.4	94.3	105.0	99.6	98.0	141.0	119.2	97.0	184.5	140.1
MF2018	**WEIGHTED AVERAGE**	96.3	102.5	98.9	102.2	177.3	133.9	102.1	177.3	133.9	92.0	107.3	98.5	99.9	157.1	124.1	100.3	177.3	132.8

DIVISION		NEW YORK																	
		JAMESTOWN			KINGSTON			LONG ISLAND CITY			MONTICELLO			MOUNT VERNON			NEW ROCHELLE		
		147			124			111			127			105			108		
		MAT.	INST.	TOTAL	MAT.	INST.	TOTAL	MAT.	INST.	TOTAL	MAT.	INST.	TOTAL	MAT.	INST.	TOTAL	MAT.	INST.	TOTAL
015433	**CONTRACTOR EQUIPMENT**		90.6	90.6		109.4	109.4		109.4	109.4		109.4	109.4		104.4	104.4		104.4	104.4
0241, 31 - 34	**SITE & INFRASTRUCTURE, DEMOLITION**	98.8	89.9	92.6	139.6	115.9	123.2	119.2	120.6	120.2	134.9	115.8	121.7	102.7	105.4	104.5	102.2	105.3	104.3
0310	Concrete Forming & Accessories	85.1	90.1	89.3	82.7	130.7	123.7	100.0	189.1	176.0	89.3	130.8	124.7	86.7	138.8	131.1	100.5	135.3	130.1
0320	Concrete Reinforcing	98.9	111.2	104.8	97.3	160.4	127.8	97.2	242.5	167.5	96.5	160.4	127.4	93.5	180.8	135.7	93.6	180.6	135.7
0330	Cast-in-Place Concrete	102.4	105.2	103.4	105.5	146.3	120.7	111.8	170.5	133.7	98.7	146.3	116.5	95.6	150.0	115.8	95.6	149.7	115.7
03	**CONCRETE**	93.9	99.0	96.2	102.4	140.8	119.3	109.0	189.9	144.5	97.3	140.8	116.4	97.2	150.4	120.6	96.4	148.7	119.4
04	**MASONRY**	110.2	104.6	106.8	108.1	155.4	137.2	111.7	188.8	159.1	100.8	155.4	134.3	94.6	156.4	132.6	94.6	156.4	132.6
05	**METALS**	93.0	101.9	95.6	104.0	135.9	113.4	103.1	175.0	124.2	104.0	136.0	113.4	86.2	169.8	110.7	86.4	169.0	110.7
06	**WOOD, PLASTICS & COMPOSITES**	82.9	86.2	84.7	83.9	124.0	106.0	97.8	188.3	147.6	90.4	124.0	108.9	87.6	132.4	112.3	104.5	128.7	117.8
07	**THERMAL & MOISTURE PROTECTION**	108.5	95.4	102.8	121.5	145.0	131.7	109.2	168.3	134.9	121.1	145.0	131.5	103.4	147.3	122.5	103.5	144.7	121.4
08	**OPENINGS**	96.5	93.0	95.7	92.1	142.5	103.9	86.6	197.5	112.5	88.0	142.4	100.7	92.3	168.4	110.1	92.4	166.3	109.6
0920	Plaster & Gypsum Board	89.2	85.8	86.9	91.0	124.9	113.8	99.3	191.0	161.0	91.7	124.8	114.0	98.3	133.2	121.8	111.0	129.3	123.3
0950, 0980	Ceilings & Acoustic Treatment	102.7	85.8	91.3	77.5	124.9	109.4	81.2	191.0	155.2	77.5	124.8	109.4	87.5	133.2	118.3	87.5	129.3	115.7
0960	Flooring	90.3	102.4	93.7	98.6	160.5	116.0	108.2	182.5	129.1	101.0	160.5	117.8	91.2	179.7	116.1	99.2	164.0	117.4
0970, 0990	Wall Finishes & Painting/Coating	98.8	100.6	99.9	116.6	130.9	125.0	116.0	167.4	146.4	116.6	123.4	120.6	104.6	167.4	141.8	104.6	167.4	141.8
09	**FINISHES**	93.2	92.5	92.9	97.9	135.4	118.2	104.1	186.0	148.4	98.3	134.6	117.9	95.0	148.0	123.7	98.9	142.6	122.5
COVERS	**DIVS. 10 - 14, 25, 28, 41, 43, 44, 46**	100.0	96.8	99.3	100.0	121.2	104.7	100.0	140.8	109.1	100.0	121.3	104.7	100.0	129.2	106.5	100.0	113.0	102.9
21, 22, 23	**FIRE SUPPRESSION, PLUMBING & HVAC**	96.5	93.6	95.4	96.6	134.6	111.5	99.8	177.9	130.5	96.6	138.9	113.2	96.8	145.9	116.1	96.8	145.6	116.0
26, 27, 3370	**ELECTRICAL, COMMUNICATIONS & UTIL.**	95.4	94.9	95.1	95.7	116.3	105.9	97.5	184.5	140.4	95.7	116.3	105.9	88.8	170.6	129.1	88.8	141.7	114.9
MF2018	**WEIGHTED AVERAGE**	96.4	96.0	96.3	100.6	133.8	114.6	101.6	177.3	133.6	99.1	134.6	114.1	94.1	150.6	118.0	94.4	144.8	115.7

		NEW YORK																	
DIVISION		NEW YORK			NIAGARA FALLS			PLATTSBURGH			POUGHKEEPSIE			QUEENS			RIVERHEAD		
		100 - 102			143			129			125 - 126			110			119		
		MAT.	INST.	TOTAL	MAT.	INST.	TOTAL	MAT.	INST.	TOTAL	MAT.	INST.	TOTAL	MAT.	INST.	TOTAL	MAT.	INST.	TOTAL
015433	CONTRACTOR EQUIPMENT		104.6	104.6		90.6	90.6		94.5	94.5		109.4	109.4		109.4	109.4		109.4	109.4
0241, 31 - 34	SITE & INFRASTRUCTURE, DEMOLITION	106.6	111.9	110.3	100.9	90.8	94.0	109.3	96.7	100.6	135.9	114.8	121.3	114.3	120.6	118.7	112.1	118.6	116.6
0310	Concrete Forming & Accessories	104.3	188.9	176.4	85.0	113.1	108.9	86.8	91.4	90.7	82.7	167.6	155.1	89.0	189.1	174.3	93.2	156.2	146.9
0320	Concrete Reinforcing	100.3	178.3	138.0	97.5	112.0	104.5	101.4	112.0	106.5	97.3	160.0	127.6	98.9	242.5	168.4	99.1	215.6	155.5
0330	Cast-in-Place Concrete	98.9	172.2	126.2	106.0	124.1	112.7	95.5	100.6	97.4	102.2	135.4	114.5	102.9	170.5	128.1	101.2	159.7	123.0
03	CONCRETE	101.6	180.3	136.2	96.2	116.1	105.0	94.3	98.0	95.9	99.7	153.6	123.4	101.5	189.9	140.3	98.7	166.5	128.5
04	MASONRY	102.1	188.8	155.4	117.5	125.1	122.2	88.3	97.4	93.9	100.6	139.5	124.5	106.0	188.8	156.9	114.3	173.3	150.6
05	METALS	97.5	171.0	119.1	95.6	102.4	97.6	99.8	99.5	99.7	104.0	135.6	113.3	103.1	175.0	124.2	105.1	160.6	121.4
06	WOOD, PLASTICS & COMPOSITES	102.0	188.6	149.7	82.8	108.2	96.8	88.8	88.2	88.5	83.9	179.3	136.4	83.8	188.3	141.3	88.7	154.0	124.6
07	THERMAL & MOISTURE PROTECTION	105.2	170.0	133.4	108.6	110.0	109.2	116.0	97.2	107.8	121.4	144.1	131.3	108.8	168.3	134.7	110.0	154.8	129.5
08	OPENINGS	96.0	197.5	119.7	96.5	104.7	98.4	97.4	94.7	96.8	92.1	170.5	110.4	86.6	197.5	112.5	87.0	172.1	106.8
0920	Plaster & Gypsum Board	108.6	191.0	164.0	89.2	108.4	102.1	108.2	87.4	94.2	91.0	181.7	152.0	91.4	191.0	158.4	92.5	155.6	135.0
0950, 0980	Ceilings & Acoustic Treatment	104.5	191.0	162.8	102.7	108.4	106.6	107.3	87.4	93.9	77.5	181.7	147.7	81.2	191.0	155.2	81.1	155.6	131.4
0960	Flooring	101.3	182.5	124.1	90.3	112.1	96.4	104.2	105.1	104.4	98.6	158.6	115.5	104.0	182.5	126.0	105.0	166.9	122.4
0970, 0990	Wall Finishes & Painting/Coating	102.7	167.4	141.0	98.8	111.5	106.3	111.2	95.1	101.7	116.6	123.4	120.6	116.0	167.4	146.4	116.0	167.4	146.4
09	FINISHES	102.9	186.2	148.0	93.3	112.9	103.9	95.0	93.5	94.2	97.7	164.2	133.7	101.3	186.0	147.1	101.5	158.6	132.4
COVERS	DIVS. 10 - 14, 25, 28, 41, 43, 44, 46	100.0	141.5	109.3	100.0	102.7	100.6	100.0	92.7	98.4	100.0	116.5	103.7	100.0	140.8	109.1	100.0	120.1	104.5
21, 22, 23	FIRE SUPPRESSION, PLUMBING & HVAC	100.2	178.3	130.9	96.5	105.1	99.9	96.6	100.3	98.1	96.6	120.3	105.9	99.8	177.9	130.5	100.0	155.9	121.9
26, 27, 3370	ELECTRICAL, COMMUNICATIONS & UTIL.	96.8	184.5	140.0	94.1	98.9	96.4	92.1	91.7	91.9	95.7	121.2	108.3	98.0	184.5	140.6	99.5	134.9	116.9
MF2018	WEIGHTED AVERAGE	99.7	175.2	131.6	97.4	107.5	101.6	97.0	96.6	96.8	99.8	136.8	115.4	100.0	177.3	132.6	100.6	153.5	122.9

		NEW YORK																	
DIVISION		ROCHESTER			SCHENECTADY			STATEN ISLAND			SUFFERN			SYRACUSE			UTICA		
		144 - 146			123			103			109			130 - 132			133 - 135		
		MAT.	INST.	TOTAL	MAT.	INST.	TOTAL	MAT.	INST.	TOTAL	MAT.	INST.	TOTAL	MAT.	INST.	TOTAL	MAT.	INST.	TOTAL
015433	CONTRACTOR EQUIPMENT		117.6	117.6		112.6	112.6		104.4	104.4		104.4	104.4		112.6	112.6		112.6	112.6
0241, 31 - 34	SITE & INFRASTRUCTURE, DEMOLITION	88.8	105.1	100.0	81.4	100.1	94.3	107.0	109.7	108.9	99.4	103.7	102.4	93.0	98.5	96.8	71.9	98.1	90.0
0310	Concrete Forming & Accessories	105.8	100.3	101.2	96.4	107.1	105.5	86.1	189.3	174.1	94.4	140.8	133.9	98.5	89.1	90.5	99.5	86.3	88.2
0320	Concrete Reinforcing	101.4	105.0	103.1	95.9	115.1	105.2	94.8	216.0	153.4	93.6	152.0	121.9	96.8	106.1	101.3	96.8	100.3	98.5
0330	Cast-in-Place Concrete	97.8	103.3	99.9	94.4	115.4	102.2	95.6	172.1	124.1	92.7	138.9	109.9	98.7	102.3	100.0	90.3	101.0	94.3
03	CONCRETE	99.5	103.3	101.2	94.9	112.4	102.6	99.1	186.4	137.5	93.8	141.3	114.7	97.6	98.1	97.8	95.8	95.4	95.6
04	MASONRY	93.6	103.4	99.7	90.7	117.9	107.4	100.9	188.9	154.9	94.5	142.2	123.8	97.0	101.0	99.5	88.4	99.4	95.2
05	METALS	104.6	119.3	108.9	99.7	126.1	107.4	84.6	174.8	111.1	84.6	131.9	98.5	98.6	119.8	104.9	96.7	117.5	102.9
06	WOOD, PLASTICS & COMPOSITES	108.4	100.2	103.9	100.1	103.8	102.1	86.2	188.6	142.6	96.9	142.5	122.0	98.2	86.0	91.5	98.2	82.7	89.7
07	THERMAL & MOISTURE PROTECTION	116.4	99.8	109.2	100.1	109.8	104.3	102.9	168.8	131.5	103.4	140.9	119.7	104.0	94.1	99.7	92.7	94.0	93.3
08	OPENINGS	100.7	99.3	100.3	95.3	103.7	97.3	92.3	197.7	116.9	92.4	150.7	106.0	92.4	89.5	91.8	95.2	86.3	93.1
0920	Plaster & Gypsum Board	103.8	100.1	101.3	100.3	103.9	102.7	98.4	191.0	160.7	102.1	143.5	130.0	96.8	85.6	89.3	96.8	82.2	87.0
0950, 0980	Ceilings & Acoustic Treatment	100.4	100.1	100.2	92.8	103.9	100.3	89.1	191.0	157.8	87.5	143.5	125.2	96.6	85.6	89.2	96.6	82.2	86.9
0960	Flooring	90.4	105.0	94.5	89.6	110.3	95.4	95.1	182.5	119.6	95.0	171.6	116.5	92.0	92.3	92.1	90.0	92.4	90.7
0970, 0990	Wall Finishes & Painting/Coating	97.7	98.4	98.1	89.3	104.4	98.2	106.2	167.4	142.4	104.6	126.9	117.8	91.9	98.7	95.9	85.6	98.7	93.4
09	FINISHES	95.7	101.4	98.8	89.2	107.0	98.9	96.7	186.2	145.2	96.3	145.8	123.1	92.4	90.0	91.1	90.8	87.9	89.2
COVERS	DIVS. 10 - 14, 25, 28, 41, 43, 44, 46	100.0	98.9	99.7	100.0	103.5	100.8	100.0	141.5	109.3	100.0	114.7	103.3	100.0	95.1	98.9	100.0	89.8	97.7
21, 22, 23	FIRE SUPPRESSION, PLUMBING & HVAC	99.9	88.1	95.3	100.2	106.2	102.6	100.3	177.9	130.8	96.8	123.7	107.4	100.3	94.4	98.0	100.3	92.5	97.2
26, 27, 3370	ELECTRICAL, COMMUNICATIONS & UTIL.	99.5	89.9	94.8	98.7	108.4	103.5	90.6	184.5	136.9	94.8	114.6	104.6	98.6	101.0	99.8	96.6	101.0	98.8
MF2018	WEIGHTED AVERAGE	100.3	99.2	99.8	97.0	109.9	102.4	95.6	176.1	129.6	94.1	130.2	109.4	97.8	98.3	98.0	96.0	96.6	96.2

		NEW YORK									NORTH CAROLINA								
DIVISION		WATERTOWN			WHITE PLAINS			YONKERS			ASHEVILLE			CHARLOTTE			DURHAM		
		136			106			107			287 - 288			281 - 282			277		
		MAT.	INST.	TOTAL	MAT.	INST.	TOTAL	MAT.	INST.	TOTAL	MAT.	INST.	TOTAL	MAT.	INST.	TOTAL	MAT.	INST.	TOTAL
015433	CONTRACTOR EQUIPMENT		112.6	112.6		104.4	104.4		104.4	104.4		99.0	99.0		100.4	100.4		104.1	104.1
0241, 31 - 34	SITE & INFRASTRUCTURE, DEMOLITION	79.5	98.6	92.7	97.1	105.4	102.8	104.5	105.4	105.1	96.4	77.8	83.5	98.6	81.9	87.1	98.5	86.3	90.1
0310	Concrete Forming & Accessories	85.3	93.0	91.9	99.2	149.3	141.9	99.4	149.1	141.8	90.5	61.4	65.7	96.6	61.5	66.6	96.2	61.4	66.5
0320	Concrete Reinforcing	97.5	106.1	101.6	93.6	180.8	135.8	97.3	180.8	137.7	95.4	66.0	81.2	99.6	66.8	83.7	104.4	66.8	86.2
0330	Cast-in-Place Concrete	105.1	104.7	105.0	84.9	150.0	109.1	95.0	150.0	115.4	105.9	71.5	93.1	108.0	71.6	94.4	106.3	71.2	93.2
03	CONCRETE	108.8	100.7	105.3	87.6	155.2	117.3	96.6	155.1	122.3	91.1	67.4	80.7	91.2	67.5	80.8	95.2	67.4	83.0
04	MASONRY	89.7	104.8	99.0	93.9	156.4	132.3	97.1	156.4	133.6	84.9	64.4	72.3	88.6	64.4	73.8	84.3	64.4	72.0
05	METALS	96.8	119.8	103.6	86.1	169.8	110.7	94.1	169.9	116.4	101.2	89.9	97.9	102.1	88.9	98.2	117.9	90.2	109.7
06	WOOD, PLASTICS & COMPOSITES	80.3	89.9	85.6	102.4	146.7	126.8	102.3	146.3	126.5	87.6	59.0	71.9	89.9	59.0	72.9	90.6	59.0	73.2
07	THERMAL & MOISTURE PROTECTION	93.0	97.2	94.8	103.2	148.7	123.0	103.5	149.3	123.4	99.1	63.8	83.8	93.7	64.5	81.0	102.8	63.8	85.8
08	OPENINGS	95.2	93.8	94.8	92.4	176.2	112.0	96.0	176.5	114.8	89.0	59.8	82.2	99.0	60.0	89.9	95.6	60.0	87.3
0920	Plaster & Gypsum Board	87.6	89.6	89.0	105.3	147.8	133.9	108.7	147.4	134.7	102.2	57.8	72.3	98.1	57.8	71.0	89.9	57.8	68.3
0950, 0980	Ceilings & Acoustic Treatment	96.6	89.6	91.9	87.5	147.8	128.2	103.7	147.4	133.2	79.4	57.8	64.9	83.5	57.8	66.2	82.8	57.8	66.0
0960	Flooring	83.1	92.4	85.7	97.5	179.7	120.6	96.9	182.5	121.0	94.9	66.8	87.0	95.3	66.8	87.3	102.0	66.8	92.1
0970, 0990	Wall Finishes & Painting/Coating	85.6	93.0	90.0	104.6	167.4	141.8	104.6	167.4	141.8	102.5	57.1	75.7	95.9	57.1	72.9	104.4	57.1	76.4
09	FINISHES	88.2	92.4	90.5	97.1	156.4	129.2	101.0	156.7	131.1	88.8	61.6	74.1	88.3	61.6	73.9	90.4	61.6	74.8
COVERS	DIVS. 10 - 14, 25, 28, 41, 43, 44, 46	100.0	96.4	99.2	100.0	127.5	106.1	100.0	130.7	106.8	100.0	82.5	96.1	100.0	82.5	96.1	100.0	82.5	96.1
21, 22, 23	FIRE SUPPRESSION, PLUMBING & HVAC	100.3	87.8	95.4	100.4	145.9	118.3	100.4	146.0	118.3	100.4	61.3	85.0	99.9	63.1	85.5	100.5	61.3	85.1
26, 27, 3370	ELECTRICAL, COMMUNICATIONS & UTIL.	98.6	90.2	94.4	88.8	170.6	129.1	94.9	170.6	132.2	100.8	57.8	79.6	100.0	60.3	80.4	96.5	57.4	77.2
MF2018	WEIGHTED AVERAGE	97.7	96.8	97.4	93.8	152.8	118.8	97.9	153.0	121.2	96.4	66.6	83.8	97.4	67.6	84.8	100.1	67.2	86.2

DIVISION		NORTH CAROLINA																	
		ELIZABETH CITY			FAYETTEVILLE			GASTONIA			GREENSBORO			HICKORY			KINSTON		
		279			283			280			270, 272 - 274			286			285		
		MAT.	INST.	TOTAL	MAT.	INST.	TOTAL	MAT.	INST.	TOTAL	MAT.	INST.	TOTAL	MAT.	INST.	TOTAL	MAT.	INST.	TOTAL
015433	CONTRACTOR EQUIPMENT		108.0	108.0		104.1	104.1		99.0	99.0		104.1	104.1		104.1	104.1		104.1	104.1
0241, 31 - 34	SITE & INFRASTRUCTURE, DEMOLITION	102.7	87.9	92.4	95.6	86.1	89.0	96.1	78.1	83.7	98.3	86.4	90.1	95.1	85.1	88.2	93.9	84.9	87.7
0310	Concrete Forming & Accessories	82.6	63.6	66.4	90.1	60.1	64.5	96.4	61.4	66.6	96.0	61.3	66.5	87.1	61.2	65.0	83.6	59.9	63.4
0320	Concrete Reinforcing	102.3	69.8	86.6	99.1	66.8	83.4	95.8	66.8	81.7	103.2	66.8	85.6	95.4	66.7	81.5	94.9	66.7	81.3
0330	Cast-in-Place Concrete	106.5	71.7	93.5	111.1	69.3	95.5	103.6	71.2	91.5	105.5	71.2	92.7	105.9	71.2	93.0	102.2	69.2	90.0
03	CONCRETE	95.4	69.1	83.8	93.1	66.2	81.3	89.8	67.5	80.0	94.7	67.4	82.7	90.9	67.3	80.5	88.1	66.1	78.4
04	MASONRY	95.4	61.0	74.2	88.3	61.0	71.5	89.4	64.4	74.0	81.8	64.4	71.1	74.6	64.4	68.3	81.2	61.0	68.8
05	METALS	103.5	92.3	100.2	122.2	90.2	112.8	101.8	90.2	98.4	110.0	90.2	104.1	101.3	90.0	98.0	100.0	90.0	97.1
06	WOOD, PLASTICS & COMPOSITES	75.1	63.4	68.7	86.5	59.0	71.4	95.4	59.0	75.4	90.2	59.0	73.1	82.4	59.0	69.6	79.3	59.0	68.2
07	THERMAL & MOISTURE PROTECTION	102.0	62.5	84.8	98.6	62.3	82.8	99.3	63.8	83.9	102.6	63.8	85.7	99.5	63.8	84.0	99.3	62.3	83.2
08	OPENINGS	92.9	63.1	86.0	89.1	60.0	82.3	92.3	60.0	84.8	95.6	60.0	87.3	89.0	60.0	82.3	89.1	60.0	82.3
0920	Plaster & Gypsum Board	84.4	61.7	69.1	105.8	57.8	73.5	107.9	57.8	74.2	91.1	57.8	68.7	102.2	57.8	72.3	102.2	57.8	72.3
0950, 0980	Ceilings & Acoustic Treatment	82.8	61.7	68.5	81.9	57.8	65.7	83.5	57.8	66.2	82.8	57.8	66.0	79.4	57.8	64.9	83.5	57.8	66.2
0960	Flooring	94.1	66.8	86.4	95.1	66.8	87.1	98.0	66.8	89.2	102.0	66.8	92.1	94.8	66.8	86.9	92.2	66.8	85.0
0970, 0990	Wall Finishes & Painting/Coating	104.4	57.1	76.4	102.5	57.1	75.7	102.5	57.1	75.7	104.4	57.1	76.4	102.5	57.1	75.7	102.5	57.1	75.7
09	FINISHES	87.7	63.4	74.6	89.7	60.8	74.0	91.1	61.6	75.2	90.6	61.6	74.9	88.9	61.6	74.1	88.7	60.8	73.6
COVERS	DIVS. 10 - 14, 25, 28, 41, 43, 44, 46	100.0	85.4	96.8	100.0	81.4	95.8	100.0	82.5	96.1	100.0	79.8	95.5	100.0	82.5	96.1	100.0	81.3	95.8
21, 22, 23	FIRE SUPPRESSION, PLUMBING & HVAC	96.9	58.6	81.9	100.2	59.5	84.2	100.4	60.1	84.6	100.4	61.3	85.0	96.9	60.2	82.5	96.9	58.3	81.7
26, 27, 3370	ELECTRICAL, COMMUNICATIONS & UTIL.	96.2	64.6	80.6	100.6	57.4	79.3	100.3	60.4	80.6	95.7	57.8	77.0	98.6	60.4	79.8	98.4	55.8	77.4
MF2018	WEIGHTED AVERAGE	96.9	68.3	84.8	100.2	66.1	85.8	97.0	66.8	84.2	98.5	67.2	85.3	94.7	67.3	83.1	94.4	65.5	82.2

DIVISION		NORTH CAROLINA															NORTH DAKOTA		
		MURPHY			RALEIGH			ROCKY MOUNT			WILMINGTON			WINSTON-SALEM			BISMARCK		
		289			275 - 276			278			284			271			585		
		MAT.	INST.	TOTAL	MAT.	INST.	TOTAL	MAT.	INST.	TOTAL	MAT.	INST.	TOTAL	MAT.	INST.	TOTAL	MAT.	INST.	TOTAL
015433	CONTRACTOR EQUIPMENT		99.0	99.0		106.5	106.5		104.1	104.1		99.0	99.0		104.1	104.1		99.9	99.9
0241, 31 - 34	SITE & INFRASTRUCTURE, DEMOLITION	97.4	76.3	82.8	98.8	90.8	93.3	100.7	86.3	90.8	97.4	77.6	83.7	98.7	86.4	90.2	99.8	97.8	98.4
0310	Concrete Forming & Accessories	97.0	59.8	65.3	96.9	61.0	66.3	88.5	60.9	64.9	91.8	60.1	64.8	97.9	61.4	66.8	110.6	76.6	81.6
0320	Concrete Reinforcing	94.9	64.3	80.1	105.4	66.8	86.7	102.3	66.8	85.1	96.1	66.8	81.9	103.2	66.8	85.6	92.6	98.3	95.3
0330	Cast-in-Place Concrete	109.6	69.2	94.6	109.3	71.3	95.1	104.2	70.5	91.6	105.5	69.3	92.0	108.1	71.2	94.4	106.3	86.6	98.9
03	CONCRETE	93.9	65.6	81.4	93.9	67.2	82.2	95.9	66.9	83.2	91.1	66.2	80.1	96.0	67.4	83.5	93.3	84.4	89.4
04	MASONRY	77.3	61.0	67.3	80.5	63.1	69.8	74.4	63.1	67.4	75.1	61.0	66.4	82.0	64.4	71.2	105.3	83.4	91.8
05	METALS	99.0	89.1	96.1	102.2	89.1	98.3	102.6	90.2	99.0	100.7	90.2	97.6	107.0	90.2	102.1	95.6	94.2	95.2
06	WOOD, PLASTICS & COMPOSITES	96.1	58.9	75.6	90.9	59.2	73.4	82.0	59.0	69.4	89.2	59.0	72.6	90.2	59.0	73.1	106.6	72.7	88.0
07	THERMAL & MOISTURE PROTECTION	99.3	62.3	83.2	97.2	63.8	82.7	102.5	63.2	85.4	99.1	62.3	83.1	102.6	63.8	85.7	109.3	86.0	99.2
08	OPENINGS	89.0	59.4	82.1	97.6	60.1	88.8	92.2	60.0	84.7	89.1	60.0	82.3	95.6	60.0	87.3	103.7	81.9	98.6
0920	Plaster & Gypsum Board	106.9	57.7	73.8	86.5	57.8	67.2	85.8	57.8	67.0	103.7	57.8	72.8	91.1	57.8	68.7	102.4	72.2	82.1
0950, 0980	Ceilings & Acoustic Treatment	79.4	57.7	64.8	83.5	57.8	66.2	80.4	57.8	65.2	81.9	57.8	65.7	82.8	57.8	66.0	108.5	72.2	84.1
0960	Flooring	98.3	66.8	89.4	96.8	66.8	88.4	97.6	66.8	88.9	95.5	66.8	87.5	102.0	66.8	92.1	86.9	53.7	77.6
0970, 0990	Wall Finishes & Painting/Coating	102.5	57.1	75.7	97.8	57.1	73.7	104.4	57.1	76.4	102.5	57.1	75.7	104.4	57.1	76.4	90.9	59.8	72.5
09	FINISHES	90.6	60.7	74.4	90.0	61.4	74.5	88.6	61.3	73.8	89.6	60.8	74.0	90.6	61.6	74.9	96.1	71.1	82.5
COVERS	DIVS. 10 - 14, 25, 28, 41, 43, 44, 46	100.0	81.3	95.8	100.0	82.3	96.1	100.0	82.1	96.0	100.0	81.4	95.8	100.0	82.5	96.1	100.0	94.8	98.8
21, 22, 23	FIRE SUPPRESSION, PLUMBING & HVAC	96.9	58.3	81.7	100.0	60.6	84.5	96.9	59.5	82.2	100.4	59.5	84.3	100.4	61.3	85.0	99.9	77.0	90.9
26, 27, 3370	ELECTRICAL, COMMUNICATIONS & UTIL.	101.7	55.8	79.1	99.1	56.3	78.0	98.0	57.4	78.0	101.1	55.8	78.8	95.7	57.8	77.0	98.7	73.4	86.3
MF2018	WEIGHTED AVERAGE	95.5	64.7	82.5	97.4	67.0	84.5	96.0	66.6	83.6	96.0	65.2	83.0	98.2	67.3	85.2	98.9	81.6	91.6

DIVISION		NORTH DAKOTA																	
		DEVILS LAKE			DICKINSON			FARGO			GRAND FORKS			JAMESTOWN			MINOT		
		583			586			580 - 581			582			584			587		
		MAT.	INST.	TOTAL	MAT.	INST.	TOTAL	MAT.	INST.	TOTAL	MAT.	INST.	TOTAL	MAT.	INST.	TOTAL	MAT.	INST.	TOTAL
015433	CONTRACTOR EQUIPMENT		97.7	97.7		97.7	97.7		99.9	99.9		97.7	97.7		97.7	97.7		97.7	97.7
0241, 31 - 34	SITE & INFRASTRUCTURE, DEMOLITION	107.5	93.7	98.0	115.7	93.7	100.5	101.3	97.6	98.8	111.5	93.7	99.2	106.5	93.7	97.7	108.9	94.1	98.7
0310	Concrete Forming & Accessories	107.0	71.6	76.8	95.5	71.5	75.1	99.4	72.1	76.1	99.7	71.6	75.7	97.1	71.6	75.4	95.1	71.8	75.2
0320	Concrete Reinforcing	94.8	98.0	96.4	95.7	98.2	96.9	95.7	98.7	97.2	93.4	98.0	95.6	95.4	98.7	97.0	96.6	98.1	97.4
0330	Cast-in-Place Concrete	126.2	81.6	109.6	114.1	81.5	102.0	102.5	85.6	96.2	114.1	81.5	102.0	124.6	81.6	108.6	114.1	81.8	102.1
03	CONCRETE	104.4	80.4	93.8	103.5	80.4	93.3	98.3	82.1	91.2	100.9	80.3	91.8	102.9	80.5	93.1	99.4	80.6	91.1
04	MASONRY	118.0	79.2	94.2	120.1	81.4	96.4	102.7	90.0	94.9	112.1	79.2	91.9	130.9	89.9	105.7	110.1	80.7	92.1
05	METALS	94.7	94.1	94.5	94.6	94.0	94.4	100.2	94.8	98.6	94.6	93.8	94.4	94.6	94.9	94.7	94.9	94.8	94.9
06	WOOD, PLASTICS & COMPOSITES	99.0	67.5	81.7	85.6	67.5	75.7	95.9	67.6	80.4	90.4	67.5	77.8	87.7	67.5	76.6	85.3	67.5	75.5
07	THERMAL & MOISTURE PROTECTION	107.1	82.3	96.3	107.7	83.7	97.2	104.1	87.4	96.8	107.3	83.0	96.7	106.9	85.7	97.7	107.0	83.9	97.0
08	OPENINGS	100.9	79.0	95.8	100.9	79.0	95.8	101.0	79.1	95.9	99.5	79.0	94.7	100.9	79.0	95.8	99.7	79.0	94.8
0920	Plaster & Gypsum Board	118.2	67.0	83.8	109.1	67.0	80.8	101.1	67.0	78.2	110.3	67.0	81.2	110.0	67.0	81.1	109.1	67.0	80.8
0950, 0980	Ceilings & Acoustic Treatment	105.6	67.0	79.6	105.6	67.0	79.6	95.4	67.0	76.3	105.6	67.0	79.6	105.6	67.0	79.6	105.6	67.0	79.6
0960	Flooring	93.8	53.7	82.5	87.4	53.7	77.9	101.2	53.7	87.9	89.2	53.7	79.2	88.0	53.7	78.4	87.1	53.7	77.7
0970, 0990	Wall Finishes & Painting/Coating	86.6	56.4	68.7	86.6	56.4	68.7	93.1	67.4	77.9	86.6	66.0	74.4	86.6	56.4	68.7	86.6	57.5	69.4
09	FINISHES	95.2	66.8	79.8	93.0	66.8	78.8	96.2	68.1	81.0	93.2	67.8	79.5	92.3	66.8	78.5	92.1	66.9	78.5
COVERS	DIVS. 10 - 14, 25, 28, 41, 43, 44, 46	100.0	87.2	97.2	100.0	87.2	97.2	100.0	93.6	98.6	100.0	87.2	97.2	100.0	87.2	97.2	100.0	93.3	98.5
21, 22, 23	FIRE SUPPRESSION, PLUMBING & HVAC	96.6	78.3	89.4	96.6	73.3	87.4	100.0	74.1	89.8	100.1	72.3	89.2	96.6	72.3	87.0	100.1	72.1	89.1
26, 27, 3370	ELECTRICAL, COMMUNICATIONS & UTIL.	94.3	68.3	81.5	101.9	67.9	85.1	98.5	69.4	84.2	97.3	68.3	83.0	94.3	68.3	81.5	100.1	72.6	86.6
MF2018	WEIGHTED AVERAGE	99.2	78.8	90.6	99.9	77.9	90.6	99.7	80.3	91.5	99.4	77.6	90.2	99.2	78.8	90.6	99.3	78.6	90.5

City Cost Indexes

DIVISION		NORTH DAKOTA			OHIO														
		WILLISTON			AKRON			ATHENS			CANTON			CHILLICOTHE			CINCINNATI		
		588			442 - 443			457			446 - 447			456			451 - 452		
		MAT.	INST.	TOTAL	MAT.	INST.	TOTAL	MAT.	INST.	TOTAL	MAT.	INST.	TOTAL	MAT.	INST.	TOTAL	MAT.	INST.	TOTAL
015433	**CONTRACTOR EQUIPMENT**		97.7	97.7		88.7	88.7		84.9	84.9		88.7	88.7		95.5	95.5		95.9	95.9
0241, 31 - 34	**SITE & INFRASTRUCTURE, DEMOLITION**	109.2	91.5	97.0	96.2	94.1	94.8	107.3	85.3	92.1	96.3	93.9	94.7	93.8	94.9	94.6	90.0	98.2	95.7
0310	Concrete Forming & Accessories	101.7	71.3	75.8	102.8	82.8	85.8	95.3	78.9	81.4	102.8	74.3	78.5	98.0	81.5	84.0	101.5	80.2	83.3
0320	Concrete Reinforcing	97.5	98.1	97.8	94.7	91.0	92.9	86.9	89.2	88.0	94.7	75.0	85.2	84.0	88.8	86.3	89.1	77.8	83.6
0330	Cast-in-Place Concrete	114.1	81.4	101.9	102.6	89.0	97.6	111.2	95.8	105.5	103.6	87.2	97.5	100.9	92.4	97.8	96.4	77.5	89.3
03	**CONCRETE**	100.8	80.2	91.7	99.8	85.9	93.7	101.0	86.3	94.6	100.3	78.7	90.8	95.6	86.9	91.8	94.7	79.0	87.8
04	**MASONRY**	105.1	80.7	90.1	90.9	89.4	90.0	73.6	97.2	88.1	91.5	80.7	84.9	80.2	90.5	86.5	83.0	79.8	81.0
05	**METALS**	94.8	93.6	94.4	97.5	80.2	92.4	97.1	80.5	92.3	97.5	73.7	90.5	89.4	89.8	89.5	91.5	81.7	88.6
06	**WOOD, PLASTICS & COMPOSITES**	92.1	67.5	78.6	105.9	81.1	92.2	87.4	74.0	80.0	106.2	71.8	87.3	100.3	78.0	88.0	102.3	80.1	90.1
07	**THERMAL & MOISTURE PROTECTION**	107.3	83.5	96.9	104.9	91.1	98.9	102.9	91.5	98.0	106.0	87.2	97.8	104.7	88.4	97.6	102.7	80.9	93.2
08	**OPENINGS**	101.0	79.0	95.8	107.5	82.7	101.7	96.2	75.8	91.5	101.3	69.8	93.9	88.1	78.0	85.7	98.3	76.4	93.2
0920	Plaster & Gypsum Board	110.3	67.0	81.2	98.3	80.5	86.3	93.1	73.2	79.7	99.2	71.0	80.2	96.6	77.8	83.9	96.1	79.9	85.2
0950, 0980	Ceilings & Acoustic Treatment	105.6	67.0	79.6	90.6	80.5	83.8	105.9	73.2	83.8	90.6	71.0	77.4	98.7	77.8	84.6	93.6	79.9	84.3
0960	Flooring	90.1	53.7	79.9	91.7	83.3	89.3	120.7	75.3	108.0	91.8	74.1	86.8	97.5	75.3	91.3	98.7	78.5	93.0
0970, 0990	Wall Finishes & Painting/Coating	86.6	56.4	68.7	98.3	90.9	93.9	100.1	89.1	93.6	98.3	73.9	83.9	97.4	89.1	92.5	97.0	71.7	82.0
09	**FINISHES**	93.4	66.8	79.0	95.0	83.6	88.8	99.4	78.7	88.2	95.2	73.4	83.4	96.5	80.4	87.8	95.9	79.1	86.8
COVERS	**DIVS. 10 - 14, 25, 28, 41, 43, 44, 46**	100.0	87.2	97.1	100.0	90.3	97.8	100.0	87.9	97.3	100.0	88.4	97.4	100.0	86.8	97.1	100.0	88.4	97.4
21, 22, 23	**FIRE SUPPRESSION, PLUMBING & HVAC**	96.6	73.2	87.4	100.0	87.8	95.2	96.5	81.2	90.5	100.0	78.7	91.6	97.0	92.4	95.2	99.9	76.9	90.9
26, 27, 3370	**ELECTRICAL, COMMUNICATIONS & UTIL.**	97.7	68.7	83.4	98.6	81.8	90.4	97.9	89.7	93.9	97.9	85.0	91.5	97.2	89.7	93.5	96.3	74.2	85.4
MF2018	**WEIGHTED AVERAGE**	98.4	77.7	89.6	99.4	86.0	93.8	97.1	84.8	91.9	98.9	79.9	90.9	94.3	88.5	91.8	96.1	80.0	89.3

DIVISION		OHIO																	
		CLEVELAND			COLUMBUS			DAYTON			HAMILTON			LIMA			LORAIN		
		441			430 - 432			453 - 454			450			458			440		
		MAT.	INST.	TOTAL	MAT.	INST.	TOTAL	MAT.	INST.	TOTAL	MAT.	INST.	TOTAL	MAT.	INST.	TOTAL	MAT.	INST.	TOTAL
015433	**CONTRACTOR EQUIPMENT**		92.0	92.0		92.3	92.3		88.9	88.9		95.5	95.5		88.3	88.3		88.7	88.7
0241, 31 - 34	**SITE & INFRASTRUCTURE, DEMOLITION**	94.9	96.2	95.8	101.8	92.3	95.3	90.0	94.2	92.9	90.0	94.5	93.1	101.1	84.9	89.9	95.5	94.3	94.7
0310	Concrete Forming & Accessories	102.3	90.2	92.0	98.7	78.1	81.1	99.9	77.2	80.6	99.9	77.4	80.7	95.3	76.3	79.1	102.8	74.6	78.7
0320	Concrete Reinforcing	95.2	91.4	93.4	99.9	78.7	89.7	89.1	79.1	84.2	89.1	77.2	83.3	86.9	79.3	83.2	94.7	91.3	93.1
0330	Cast-in-Place Concrete	99.5	97.2	98.6	102.1	80.9	94.2	86.5	80.6	84.3	92.6	80.9	88.3	102.1	89.5	97.4	97.7	91.4	95.4
03	**CONCRETE**	100.1	92.4	96.7	98.6	79.2	90.0	87.0	78.6	83.3	89.9	79.1	85.1	93.9	81.5	88.5	97.5	83.1	91.2
04	**MASONRY**	96.9	97.9	97.5	91.4	85.2	87.6	79.1	76.4	77.4	79.5	81.2	80.5	98.8	78.1	86.1	87.8	93.1	91.0
05	**METALS**	99.1	83.3	94.5	98.1	79.3	92.5	90.9	76.8	86.7	90.9	85.6	89.4	97.2	80.1	92.2	98.1	81.6	93.3
06	**WOOD, PLASTICS & COMPOSITES**	99.8	88.5	93.6	98.5	77.7	87.0	104.2	77.1	89.3	103.0	77.1	88.7	87.3	75.1	80.6	105.9	70.1	86.2
07	**THERMAL & MOISTURE PROTECTION**	102.0	96.9	99.8	93.6	83.8	89.3	108.8	79.5	96.1	104.8	79.9	94.0	102.4	83.7	94.3	105.9	91.0	99.4
08	**OPENINGS**	100.3	86.4	97.0	98.3	74.0	92.7	95.0	74.6	90.3	92.8	74.8	88.6	96.3	73.7	91.0	101.3	76.6	95.5
0920	Plaster & Gypsum Board	97.9	88.1	91.3	94.1	77.0	82.6	98.1	76.9	83.9	98.1	76.9	83.9	93.1	74.3	80.5	98.3	69.2	78.7
0950, 0980	Ceilings & Acoustic Treatment	86.3	88.1	87.5	93.5	77.0	82.4	99.6	76.9	84.3	98.7	76.9	84.0	105.9	74.3	84.6	90.6	69.2	76.2
0960	Flooring	93.1	89.8	92.2	95.1	75.3	89.5	101.2	72.4	93.1	98.5	78.5	92.9	119.8	77.2	107.8	91.8	89.8	91.2
0970, 0990	Wall Finishes & Painting/Coating	101.2	90.9	95.1	99.5	79.6	87.7	97.4	70.7	81.6	97.4	71.3	82.0	100.2	76.5	86.2	98.3	90.9	93.9
09	**FINISHES**	95.3	90.1	92.5	95.0	77.5	85.5	97.7	75.4	85.6	96.7	76.9	86.0	98.7	76.0	86.4	95.0	78.0	85.8
COVERS	**DIVS. 10 - 14, 25, 28, 41, 43, 44, 46**	100.0	96.4	99.2	100.0	88.2	97.4	100.0	85.1	96.7	100.0	85.3	96.7	100.0	84.4	96.5	100.0	90.9	98.0
21, 22, 23	**FIRE SUPPRESSION, PLUMBING & HVAC**	100.0	89.9	96.0	100.0	84.1	93.7	100.7	81.2	93.0	100.5	74.9	90.5	96.5	91.7	94.6	100.0	88.4	95.4
26, 27, 3370	**ELECTRICAL, COMMUNICATIONS & UTIL.**	98.3	92.7	95.5	99.9	80.2	90.2	94.9	75.7	85.4	95.2	75.5	85.5	98.2	75.7	87.1	98.0	78.2	88.2
MF2018	**WEIGHTED AVERAGE**	99.1	91.6	95.9	98.4	82.0	91.4	94.8	79.2	88.2	94.8	79.5	88.3	97.2	81.7	90.7	98.4	84.7	92.6

DIVISION		OHIO																	
		MANSFIELD			MARION			SPRINGFIELD			STEUBENVILLE			TOLEDO			YOUNGSTOWN		
		448 - 449			433			455			439			434 - 436			444 - 445		
		MAT.	INST.	TOTAL	MAT.	INST.	TOTAL	MAT.	INST.	TOTAL	MAT.	INST.	TOTAL	MAT.	INST.	TOTAL	MAT.	INST.	TOTAL
015433	**CONTRACTOR EQUIPMENT**		88.7	88.7		88.7	88.7		88.9	88.9		92.4	92.4		92.4	92.4		88.7	88.7
0241, 31 - 34	**SITE & INFRASTRUCTURE, DEMOLITION**	91.7	94.0	93.3	95.7	90.6	92.2	90.3	94.2	93.0	141.5	98.2	111.6	99.8	90.9	93.6	96.1	93.9	94.6
0310	Concrete Forming & Accessories	92.1	73.5	76.3	96.6	79.4	81.9	99.9	77.1	80.5	97.6	79.0	81.7	100.3	84.9	87.1	102.8	76.6	80.5
0320	Concrete Reinforcing	86.2	79.0	82.7	92.1	79.0	85.8	89.1	79.1	84.2	89.7	94.9	92.2	99.9	82.9	91.7	94.7	85.2	90.1
0330	Cast-in-Place Concrete	95.1	87.8	92.3	88.5	88.1	88.3	88.8	80.3	85.6	96.1	89.3	93.6	96.9	90.2	94.4	101.7	87.2	96.3
03	**CONCRETE**	92.1	79.3	86.4	85.4	82.2	84.0	88.1	78.5	83.9	89.4	84.9	87.4	93.2	86.4	90.2	99.4	81.5	91.5
04	**MASONRY**	89.9	88.9	89.3	92.7	89.9	90.9	79.3	75.9	77.2	80.7	91.1	87.1	98.9	91.8	94.5	91.1	86.6	88.4
05	**METALS**	98.4	76.0	91.8	97.1	78.9	91.7	90.9	76.8	86.7	93.6	81.8	90.1	97.8	85.3	94.2	97.5	77.9	91.8
06	**WOOD, PLASTICS & COMPOSITES**	92.8	70.1	80.3	92.8	77.5	84.4	105.6	77.1	89.9	88.1	76.4	81.6	96.9	83.9	89.8	105.9	74.6	88.7
07	**THERMAL & MOISTURE PROTECTION**	104.1	88.4	97.3	90.5	88.8	89.8	108.7	79.2	95.9	102.0	87.4	95.7	91.0	91.2	91.1	106.1	88.2	98.3
08	**OPENINGS**	102.3	71.0	95.0	91.2	75.1	87.5	93.2	74.6	88.9	91.3	78.4	88.3	93.8	81.3	90.9	101.3	75.2	95.2
0920	Plaster & Gypsum Board	90.9	69.2	76.3	95.3	77.0	83.0	98.1	76.9	83.9	93.5	75.4	81.3	97.2	83.6	88.0	98.3	73.9	81.9
0950, 0980	Ceilings & Acoustic Treatment	91.4	69.2	76.4	99.5	77.0	84.3	99.6	76.9	84.3	96.7	75.4	82.3	99.5	83.6	88.8	90.6	73.9	79.3
0960	Flooring	86.6	92.7	88.3	94.4	92.7	93.9	101.2	72.4	93.1	122.5	92.7	114.1	94.8	94.4	94.7	91.8	90.0	91.3
0970, 0990	Wall Finishes & Painting/Coating	98.3	77.8	86.2	104.0	77.8	88.5	97.4	70.7	81.6	117.8	87.1	99.6	104.0	87.0	94.0	98.3	80.0	87.5
09	**FINISHES**	92.4	76.6	83.8	96.3	81.2	88.1	97.7	75.3	85.6	114.5	81.6	96.7	97.0	86.6	91.4	95.1	78.8	86.3
COVERS	**DIVS. 10 - 14, 25, 28, 41, 43, 44, 46**	100.0	88.3	97.4	100.0	85.9	96.8	100.0	84.9	96.6	100.0	86.2	96.9	100.0	91.3	98.1	100.0	88.5	97.4
21, 22, 23	**FIRE SUPPRESSION, PLUMBING & HVAC**	96.5	87.5	92.9	96.5	93.1	95.1	100.7	80.9	92.9	96.9	91.3	94.7	100.0	93.6	97.5	100.0	83.9	93.7
26, 27, 3370	**ELECTRICAL, COMMUNICATIONS & UTIL.**	95.7	90.8	93.3	94.0	90.8	92.4	94.9	80.2	87.7	88.7	106.4	97.4	100.0	103.4	101.7	98.0	74.3	86.3
MF2018	**WEIGHTED AVERAGE**	96.4	84.2	91.3	94.3	86.7	91.1	94.8	79.7	88.4	96.1	90.0	93.5	97.6	91.2	94.9	98.7	81.9	91.6

DIVISION		OHIO			OKLAHOMA														
		ZANESVILLE			ARDMORE			CLINTON			DURANT			ENID			GUYMON		
		437 - 438			734			736			747			737			739		
		MAT.	INST.	TOTAL	MAT.	INST.	TOTAL	MAT.	INST.	TOTAL	MAT.	INST.	TOTAL	MAT.	INST.	TOTAL	MAT.	INST.	TOTAL
015433	CONTRACTOR EQUIPMENT		88.7	88.7		80.7	80.7		79.9	79.9		79.9	79.9		79.9	79.9		79.9	79.9
0241, 31 - 34	SITE & INFRASTRUCTURE, DEMOLITION	98.9	90.5	93.1	96.9	92.6	94.0	98.1	91.2	93.3	93.4	88.6	90.1	100.0	91.2	93.9	102.4	90.5	94.2
0310	Concrete Forming & Accessories	93.7	78.3	80.6	88.0	55.4	60.2	86.8	55.4	60.1	82.9	55.0	59.1	89.9	55.6	60.7	92.6	55.1	60.7
0320	Concrete Reinforcing	91.6	92.6	92.1	79.5	67.7	73.8	80.0	67.7	74.0	88.6	63.6	76.5	79.5	67.7	73.8	80.0	63.3	71.9
0330	Cast-in-Place Concrete	93.2	86.5	90.7	96.6	70.5	86.9	93.4	70.5	84.9	88.6	70.3	81.8	93.4	70.6	84.9	93.4	70.0	84.7
03	CONCRETE	89.2	83.5	86.7	86.7	62.8	76.2	86.2	62.8	75.9	84.1	61.8	74.3	86.6	62.9	76.2	89.4	61.7	77.2
04	MASONRY	90.7	86.6	88.2	95.2	57.0	71.8	119.3	57.0	81.0	88.6	62.1	72.3	101.3	57.0	74.1	97.5	55.4	71.7
05	METALS	98.5	84.2	94.3	97.3	62.2	87.0	97.4	62.2	87.0	92.7	60.6	83.2	98.8	62.4	88.1	97.8	59.3	86.5
06	WOOD, PLASTICS & COMPOSITES	88.2	77.5	82.3	97.4	54.4	73.7	96.5	54.4	73.4	87.8	54.4	69.5	99.7	54.4	74.8	102.9	54.4	76.2
07	THERMAL & MOISTURE PROTECTION	90.6	84.5	88.0	100.7	64.8	85.1	100.8	64.8	85.1	96.4	64.6	82.6	100.9	64.8	85.2	101.3	62.2	84.3
08	OPENINGS	91.2	79.0	88.4	103.7	55.0	92.3	103.7	55.0	92.3	96.0	54.0	86.2	104.9	55.9	93.5	103.8	54.0	92.2
0920	Plaster & Gypsum Board	91.2	77.0	81.7	90.5	53.7	65.8	90.2	53.7	65.7	78.5	53.7	61.8	91.1	53.7	66.0	91.1	53.7	66.0
0950, 0980	Ceilings & Acoustic Treatment	99.5	77.0	84.3	89.1	53.7	65.3	89.1	53.7	65.3	83.0	53.7	63.3	89.1	53.7	65.3	89.1	53.7	65.3
0960	Flooring	92.6	75.3	87.7	82.7	57.3	75.6	81.7	57.3	74.9	89.7	51.7	79.0	83.3	72.9	80.4	84.8	57.3	77.1
0970, 0990	Wall Finishes & Painting/Coating	104.0	89.1	95.2	86.4	44.8	61.8	86.4	44.8	61.8	92.8	44.8	64.4	86.4	44.8	61.8	86.4	42.2	60.3
09	FINISHES	95.4	78.2	86.1	82.7	53.5	66.9	82.5	53.5	66.8	83.2	52.4	66.5	83.2	56.7	68.8	84.0	54.5	68.0
COVERS	DIVS. 10 - 14, 25, 28, 41, 43, 44, 46	100.0	83.3	96.3	100.0	79.4	95.4	100.0	79.4	95.4	100.0	79.3	95.4	100.0	79.4	95.4	100.0	79.4	95.4
21, 22, 23	FIRE SUPPRESSION, PLUMBING & HVAC	96.5	89.8	93.9	96.5	67.5	85.1	96.5	67.5	85.1	96.6	67.4	85.2	100.0	67.5	87.2	96.5	63.6	83.6
26, 27, 3370	ELECTRICAL, COMMUNICATIONS & UTIL.	94.1	89.7	92.0	95.0	71.0	83.2	95.9	71.0	83.7	97.3	69.2	83.4	95.9	71.0	83.7	97.6	65.6	81.8
MF2018	WEIGHTED AVERAGE	94.9	85.7	91.0	95.2	65.6	82.7	96.4	65.5	83.3	93.1	65.1	81.3	97.0	66.0	83.9	96.3	63.3	82.4

DIVISION		OKLAHOMA																	
		LAWTON			MCALESTER			MIAMI			MUSKOGEE			OKLAHOMA CITY			PONCA CITY		
		735			745			743			744			730 - 731			746		
		MAT.	INST.	TOTAL	MAT.	INST.	TOTAL	MAT.	INST.	TOTAL	MAT.	INST.	TOTAL	MAT.	INST.	TOTAL	MAT.	INST.	TOTAL
015433	CONTRACTOR EQUIPMENT		80.7	80.7		79.9	79.9		90.9	90.9		90.9	90.9		85.6	85.6		79.9	79.9
0241, 31 - 34	SITE & INFRASTRUCTURE, DEMOLITION	96.2	92.7	93.7	86.8	90.5	89.4	88.1	88.1	88.1	88.7	87.9	88.1	94.2	98.0	96.8	93.9	90.9	91.8
0310	Concrete Forming & Accessories	92.5	55.6	61.1	81.1	41.4	47.3	93.2	55.2	60.8	97.3	54.8	61.1	92.8	63.8	68.0	89.0	55.4	60.3
0320	Concrete Reinforcing	79.7	67.7	73.9	88.3	63.3	76.2	86.9	67.7	77.6	87.7	62.8	75.7	87.9	67.8	78.2	87.7	67.7	78.0
0330	Cast-in-Place Concrete	90.3	70.6	83.0	77.7	70.0	74.8	81.4	71.4	77.7	82.4	71.1	78.2	90.3	73.4	84.0	91.0	70.5	83.4
03	CONCRETE	83.0	62.9	74.2	75.3	55.4	66.6	79.7	64.0	72.8	81.2	62.8	73.1	86.1	67.6	78.0	86.1	62.7	75.8
04	MASONRY	97.4	57.0	72.6	106.1	56.9	75.9	91.3	57.1	70.3	108.1	49.2	71.9	99.8	57.4	73.8	84.1	57.0	67.5
05	METALS	102.5	62.3	90.7	92.6	59.4	82.9	92.6	77.0	88.0	94.1	73.9	88.2	95.7	63.7	86.3	92.6	62.1	83.6
06	WOOD, PLASTICS & COMPOSITES	102.0	54.4	75.8	85.5	36.2	58.4	100.0	54.6	75.0	104.3	54.6	76.9	92.5	65.1	77.4	95.6	54.4	72.9
07	THERMAL & MOISTURE PROTECTION	100.7	64.8	85.1	96.0	61.2	80.9	96.5	64.1	82.4	96.8	61.1	81.2	92.1	66.9	81.1	96.6	65.0	82.9
08	OPENINGS	106.6	55.9	94.8	96.0	43.9	83.9	96.0	55.0	86.5	97.2	54.0	87.1	100.0	61.8	91.1	96.0	55.0	86.4
0920	Plaster & Gypsum Board	92.8	53.7	66.5	77.6	35.0	48.9	84.2	53.7	63.7	86.4	53.7	64.4	94.7	64.6	74.4	82.9	53.7	63.3
0950, 0980	Ceilings & Acoustic Treatment	96.4	53.7	67.6	83.0	35.0	50.6	83.0	53.7	63.3	92.7	53.7	66.4	87.4	64.6	72.0	83.0	53.7	63.3
0960	Flooring	85.0	72.9	81.6	88.6	57.3	79.8	95.7	51.7	83.3	98.2	33.2	79.9	86.4	72.9	82.6	92.8	57.3	82.8
0970, 0990	Wall Finishes & Painting/Coating	86.4	44.8	61.8	92.8	42.2	62.9	92.8	42.2	62.9	92.8	42.2	62.9	89.5	44.8	63.1	92.8	44.8	64.4
09	FINISHES	84.9	56.7	69.6	82.2	42.4	60.7	85.1	52.2	67.3	88.1	48.8	66.9	86.3	63.1	73.7	84.9	54.4	68.4
COVERS	DIVS. 10 - 14, 25, 28, 41, 43, 44, 46	100.0	79.4	95.4	100.0	77.4	95.0	100.0	76.3	94.7	100.0	76.3	94.7	100.0	80.9	95.8	100.0	79.4	95.4
21, 22, 23	FIRE SUPPRESSION, PLUMBING & HVAC	100.0	67.5	87.2	96.6	63.5	83.6	96.6	63.6	83.7	100.1	61.3	84.9	99.9	67.8	87.3	96.6	63.6	83.7
26, 27, 3370	ELECTRICAL, COMMUNICATIONS & UTIL.	97.6	69.2	83.6	95.8	67.0	81.6	97.1	67.0	82.3	95.4	67.2	81.5	103.4	71.1	87.5	95.4	65.6	80.7
MF2018	WEIGHTED AVERAGE	97.3	65.8	84.0	92.4	60.5	78.9	92.7	65.1	81.1	95.0	62.9	81.4	96.4	68.7	84.7	93.1	64.0	80.8

DIVISION		OKLAHOMA												OREGON					
		POTEAU			SHAWNEE			TULSA			WOODWARD			BEND			EUGENE		
		749			748			740 - 741			738			977			974		
		MAT.	INST.	TOTAL	MAT.	INST.	TOTAL	MAT.	INST.	TOTAL	MAT.	INST.	TOTAL	MAT.	INST.	TOTAL	MAT.	INST.	TOTAL
015433	CONTRACTOR EQUIPMENT		89.7	89.7		79.9	79.9		90.9	90.9		79.9	79.9		97.3	97.3		97.3	97.3
0241, 31 - 34	SITE & INFRASTRUCTURE, DEMOLITION	74.8	83.8	81.0	96.9	90.9	92.7	94.8	87.2	89.5	98.5	91.2	93.5	105.9	98.5	100.8	96.6	98.5	97.9
0310	Concrete Forming & Accessories	87.0	54.9	59.7	82.8	55.3	59.4	97.4	56.0	62.1	86.9	43.7	50.1	107.5	96.6	98.2	104.2	96.6	97.7
0320	Concrete Reinforcing	88.7	67.7	78.5	87.7	65.4	76.9	88.0	67.7	78.2	79.5	67.7	73.8	91.0	112.8	101.5	94.9	112.8	103.5
0330	Cast-in-Place Concrete	81.4	71.3	77.7	93.9	70.5	85.2	89.7	73.2	83.6	93.4	70.5	84.9	117.6	100.6	111.3	113.9	100.6	108.9
03	CONCRETE	81.9	63.8	73.9	87.7	62.3	76.6	86.2	65.0	76.9	86.5	57.5	73.7	110.5	100.3	106.0	101.3	100.3	100.9
04	MASONRY	91.6	57.1	70.4	107.5	57.0	76.5	92.2	57.1	70.6	90.8	57.0	70.0	105.1	100.6	102.3	102.1	100.6	101.1
05	METALS	92.6	76.7	87.9	92.5	61.3	83.3	97.0	77.1	91.2	97.4	62.1	87.0	108.0	96.0	104.5	108.7	95.9	105.0
06	WOOD, PLASTICS & COMPOSITES	92.4	54.6	71.6	87.7	54.4	69.4	103.5	55.6	77.2	96.6	38.7	64.8	99.6	96.1	97.7	95.4	96.1	95.8
07	THERMAL & MOISTURE PROTECTION	96.6	64.1	82.5	96.6	63.9	82.4	96.8	64.3	82.7	100.9	63.3	84.5	118.3	100.2	110.4	117.4	102.7	111.0
08	OPENINGS	96.0	55.0	86.5	96.0	54.4	86.3	98.9	56.2	88.9	103.7	46.3	90.3	96.9	100.0	97.6	97.2	100.0	97.8
0920	Plaster & Gypsum Board	81.3	53.7	62.8	78.5	53.7	61.8	86.4	54.8	65.1	90.2	37.6	54.8	122.7	95.9	104.6	120.8	95.9	104.0
0950, 0980	Ceilings & Acoustic Treatment	83.0	53.7	63.3	83.0	53.7	63.3	92.7	54.8	67.1	89.1	37.6	54.4	85.7	95.9	92.6	86.6	95.9	92.9
0960	Flooring	92.2	51.7	80.8	89.7	57.3	80.6	97.0	61.6	87.1	81.7	54.6	74.1	105.8	105.7	105.7	104.2	105.7	104.6
0970, 0990	Wall Finishes & Painting/Coating	92.8	44.8	64.4	92.8	42.2	62.9	92.8	51.3	68.2	86.4	44.8	61.8	98.6	77.3	86.0	98.6	69.6	81.4
09	FINISHES	82.9	52.5	66.4	83.5	53.2	67.1	88.0	55.8	70.5	82.6	43.7	61.5	100.1	95.9	97.8	98.6	95.0	96.7
COVERS	DIVS. 10 - 14, 25, 28, 41, 43, 44, 46	100.0	79.6	95.5	100.0	79.4	95.4	100.0	76.4	94.7	100.0	77.7	95.0	100.0	102.0	100.4	100.0	102.0	100.4
21, 22, 23	FIRE SUPPRESSION, PLUMBING & HVAC	96.6	63.6	83.7	96.6	67.5	85.2	100.1	63.7	85.8	96.5	67.5	85.1	96.6	105.2	100.0	100.1	98.6	99.5
26, 27, 3370	ELECTRICAL, COMMUNICATIONS & UTIL.	95.5	67.0	81.5	97.4	71.0	84.4	97.3	67.0	82.4	97.4	71.0	84.4	102.0	99.8	100.9	100.7	99.8	100.2
MF2018	WEIGHTED AVERAGE	92.3	64.9	80.7	94.5	65.2	82.2	95.9	65.8	83.2	95.3	62.9	81.6	102.6	100.2	101.6	101.7	98.7	100.5

	DIVISION	OREGON																	
		KLAMATH FALLS			MEDFORD			PENDLETON			PORTLAND			SALEM			VALE		
		976			975			978			970 - 972			973			979		
		MAT.	INST.	TOTAL	MAT.	INST.	TOTAL	MAT.	INST.	TOTAL	MAT.	INST.	TOTAL	MAT.	INST.	TOTAL	MAT.	INST.	TOTAL
015433	CONTRACTOR EQUIPMENT		97.3	97.3		97.3	97.3		94.9	94.9		97.3	97.3		99.9	99.9		94.9	94.9
0241, 31 - 34	SITE & INFRASTRUCTURE, DEMOLITION	109.8	98.4	102.0	104.0	98.4	100.2	103.0	91.9	95.4	99.1	98.5	98.7	92.6	102.1	99.2	90.6	91.8	91.5
0310	Concrete Forming & Accessories	100.7	96.3	97.0	99.7	96.3	96.8	101.2	96.6	97.3	105.3	96.8	98.1	105.8	96.8	98.1	107.5	95.3	97.1
0320	Concrete Reinforcing	91.0	112.7	101.5	92.6	112.7	102.3	90.3	112.8	101.2	95.6	112.8	103.9	102.7	112.8	107.6	88.1	112.5	99.9
0330	Cast-in-Place Concrete	117.6	97.0	110.0	117.6	100.4	111.2	118.4	97.5	110.7	117.0	100.6	110.9	107.6	101.7	105.4	93.2	97.8	94.9
03	CONCRETE	113.5	98.9	107.1	107.8	100.1	104.4	93.7	99.3	96.2	103.0	100.4	101.8	98.7	100.7	99.6	77.8	98.8	87.0
04	MASONRY	119.1	100.6	107.7	99.0	100.6	100.0	109.9	100.6	104.2	104.0	100.6	101.9	108.2	100.6	103.5	107.7	100.6	103.4
05	METALS	108.0	95.6	104.4	108.3	95.6	104.6	116.3	96.4	110.4	110.1	96.1	106.0	117.1	95.4	110.8	116.2	95.1	110.0
06	WOOD, PLASTICS & COMPOSITES	90.4	96.1	93.5	89.3	96.1	93.0	92.3	96.2	94.5	96.4	96.1	96.3	90.3	96.3	93.6	100.9	96.2	98.3
07	THERMAL & MOISTURE PROTECTION	118.5	96.8	109.1	118.1	96.1	108.5	110.7	95.4	104.1	117.3	100.2	109.9	114.1	100.9	108.3	110.1	91.4	102.0
08	OPENINGS	96.9	100.0	97.6	99.6	100.0	99.7	93.2	100.0	94.8	95.1	100.0	96.3	102.6	100.0	102.0	93.2	89.0	92.2
0920	Plaster & Gypsum Board	117.1	95.9	102.8	116.5	95.9	102.6	102.4	95.9	98.0	120.3	95.9	103.9	115.4	95.9	102.3	109.0	95.9	100.2
0950, 0980	Ceilings & Acoustic Treatment	93.0	95.9	95.0	98.5	95.9	96.7	61.9	95.9	84.8	88.5	95.9	93.5	96.0	95.9	95.9	61.9	95.9	84.8
0960	Flooring	102.8	105.7	103.6	102.2	105.7	103.2	70.8	105.7	80.6	101.9	105.7	102.9	107.4	105.7	106.9	72.9	105.7	82.1
0970, 0990	Wall Finishes & Painting/Coating	98.6	65.2	78.8	98.6	65.2	78.8	88.3	77.3	81.8	98.4	77.3	85.9	97.3	75.2	84.2	88.3	77.3	81.8
09	FINISHES	100.3	94.6	97.2	100.4	94.6	97.3	70.4	96.0	84.2	98.2	95.9	97.0	99.7	95.8	97.6	71.0	96.0	84.5
COVERS	DIVS. 10 - 14, 25, 28, 41, 43, 44, 46	100.0	101.9	100.4	100.0	101.9	100.4	100.0	96.5	99.2	100.0	102.1	100.5	100.0	102.4	100.5	100.0	102.2	100.5
21, 22, 23	FIRE SUPPRESSION, PLUMBING & HVAC	96.6	105.2	100.0	100.1	105.2	102.1	98.7	112.0	104.0	100.1	111.5	104.6	100.1	106.9	102.8	98.7	71.1	87.9
26, 27, 3370	ELECTRICAL, COMMUNICATIONS & UTIL.	100.7	82.6	91.8	104.1	82.6	93.5	92.8	96.8	94.8	100.9	108.7	104.7	109.3	99.8	104.6	92.8	68.1	80.6
MF2018	WEIGHTED AVERAGE	103.5	97.3	100.9	103.3	97.4	100.8	98.6	100.3	99.3	102.1	102.8	102.4	104.4	100.8	102.9	96.2	87.0	92.3

	DIVISION	PENNSYLVANIA																	
		ALLENTOWN			ALTOONA			BEDFORD			BRADFORD			BUTLER			CHAMBERSBURG		
		181			166			155			167			160			172		
		MAT.	INST.	TOTAL	MAT.	INST.	TOTAL	MAT.	INST.	TOTAL	MAT.	INST.	TOTAL	MAT.	INST.	TOTAL	MAT.	INST.	TOTAL
015433	CONTRACTOR EQUIPMENT		112.6	112.6		112.6	112.6		110.7	110.7		112.6	112.6		112.6	112.6		111.8	111.8
0241, 31 - 34	SITE & INFRASTRUCTURE, DEMOLITION	91.7	97.2	95.5	95.1	97.1	96.5	103.6	94.8	97.6	90.5	96.0	94.3	86.3	97.8	94.2	86.3	95.3	92.5
0310	Concrete Forming & Accessories	97.9	108.2	106.7	83.7	82.6	82.8	82.0	81.2	81.3	85.8	96.5	94.9	85.1	94.1	92.8	88.0	76.8	78.5
0320	Concrete Reinforcing	96.8	113.3	104.8	93.8	105.7	99.6	93.0	105.7	99.1	95.8	105.8	100.6	94.4	119.3	106.5	94.5	112.2	103.1
0330	Cast-in-Place Concrete	89.4	99.9	93.3	99.7	86.9	94.9	109.1	85.9	100.5	95.2	90.6	93.5	88.0	96.2	91.0	93.0	93.8	93.3
03	CONCRETE	92.3	107.0	98.8	87.5	89.5	88.4	102.2	88.5	96.2	94.2	97.0	95.5	79.5	100.3	88.7	97.1	90.5	94.2
04	MASONRY	92.1	93.1	92.7	95.7	83.4	88.1	107.1	80.2	90.6	93.1	82.0	86.3	97.6	93.0	94.8	93.8	79.1	84.8
05	METALS	98.9	120.5	105.3	92.9	114.8	99.3	102.2	113.4	105.5	96.7	113.2	101.6	92.6	121.7	101.2	98.3	118.8	104.3
06	WOOD, PLASTICS & COMPOSITES	97.6	111.5	105.3	76.7	81.8	79.5	79.4	81.7	80.7	82.7	101.2	92.9	78.1	94.1	86.9	84.3	75.6	79.5
07	THERMAL & MOISTURE PROTECTION	104.0	107.8	105.7	102.7	90.0	97.2	98.3	87.4	93.6	103.8	88.5	97.2	102.2	94.4	98.8	96.1	83.1	90.4
08	OPENINGS	92.4	107.3	95.9	86.3	85.6	86.1	93.4	85.6	91.6	92.4	95.3	93.1	86.2	99.3	89.3	89.2	81.4	87.4
0920	Plaster & Gypsum Board	94.9	111.8	106.3	86.3	81.2	82.9	97.5	81.2	86.6	87.1	101.2	96.6	86.3	93.9	91.4	108.9	74.9	86.0
0950, 0980	Ceilings & Acoustic Treatment	88.5	111.8	104.2	90.9	81.2	84.4	101.1	81.2	87.7	91.0	101.2	97.9	91.8	93.9	93.2	96.0	74.9	81.8
0960	Flooring	92.0	95.1	92.8	85.3	98.9	89.2	95.0	103.7	97.5	86.2	103.7	91.1	86.3	105.3	91.6	90.5	78.0	87.0
0970, 0990	Wall Finishes & Painting/Coating	91.9	101.8	97.8	87.2	106.2	98.4	97.6	105.7	102.4	91.9	105.7	100.0	87.2	106.2	98.4	92.0	100.7	97.1
09	FINISHES	90.6	105.8	98.8	88.2	87.5	87.8	100.6	87.4	93.5	88.4	99.4	94.3	88.1	97.2	93.0	90.8	78.8	84.3
COVERS	DIVS. 10 - 14, 25, 28, 41, 43, 44, 46	100.0	99.6	99.9	100.0	94.5	98.8	100.0	93.2	98.5	100.0	96.1	99.1	100.0	97.3	99.4	100.0	92.7	98.4
21, 22, 23	FIRE SUPPRESSION, PLUMBING & HVAC	100.3	114.5	105.9	99.8	83.6	93.4	96.6	84.4	91.8	96.8	90.6	94.4	96.3	95.4	96.0	96.7	89.7	94.0
26, 27, 3370	ELECTRICAL, COMMUNICATIONS & UTIL.	98.0	95.8	96.9	88.7	107.7	98.0	94.2	107.7	100.9	92.1	107.7	99.8	89.2	107.7	98.3	92.4	85.8	89.2
MF2018	WEIGHTED AVERAGE	96.7	105.8	100.6	93.3	92.8	93.1	98.8	92.1	96.0	94.9	97.1	95.8	91.3	100.6	95.2	95.1	89.4	92.7

	DIVISION	PENNSYLVANIA																	
		DOYLESTOWN			DUBOIS			ERIE			GREENSBURG			HARRISBURG			HAZLETON		
		189			158			164 - 165			156			170 - 171			182		
		MAT.	INST.	TOTAL	MAT.	INST.	TOTAL	MAT.	INST.	TOTAL	MAT.	INST.	TOTAL	MAT.	INST.	TOTAL	MAT.	INST.	TOTAL
015433	CONTRACTOR EQUIPMENT		91.5	91.5		110.7	110.7		112.6	112.6		110.7	110.7		114.7	114.7		112.6	112.6
0241, 31 - 34	SITE & INFRASTRUCTURE, DEMOLITION	104.8	85.9	91.7	108.4	95.2	99.3	92.1	97.4	95.7	99.6	96.7	97.6	86.6	100.3	96.1	85.0	97.1	93.3
0310	Concrete Forming & Accessories	82.9	126.8	120.3	81.6	83.7	83.4	97.2	85.8	87.5	87.8	88.6	88.5	100.5	85.7	87.9	80.6	88.8	87.6
0320	Concrete Reinforcing	93.6	150.9	121.3	92.3	119.4	105.4	95.8	107.3	101.4	92.3	119.0	105.2	103.4	111.0	107.1	94.0	113.1	103.2
0330	Cast-in-Place Concrete	84.5	129.7	101.3	105.2	94.1	101.1	98.0	88.7	94.5	101.3	95.7	99.2	91.7	96.9	93.6	84.5	95.6	88.6
03	CONCRETE	88.0	131.2	107.0	104.4	94.7	100.1	86.4	91.9	88.8	97.2	97.5	97.4	91.9	95.3	93.4	85.0	96.7	90.2
04	MASONRY	95.4	134.9	119.7	107.8	93.4	98.9	85.0	86.8	86.1	117.5	89.1	100.0	88.8	83.9	85.8	104.7	89.5	95.4
05	METALS	96.5	123.9	104.5	102.2	118.8	107.1	93.1	115.6	99.7	102.1	119.7	107.3	105.1	117.4	108.7	98.7	119.7	104.8
06	WOOD, PLASTICS & COMPOSITES	78.5	125.8	104.5	78.3	81.7	80.2	94.0	84.4	88.7	85.3	87.1	86.3	101.8	85.3	92.7	77.1	87.0	82.6
07	THERMAL & MOISTURE PROTECTION	101.1	131.7	114.4	98.7	92.9	96.1	103.2	88.6	96.8	98.2	92.2	95.6	98.9	100.4	99.6	103.3	100.7	102.2
08	OPENINGS	94.8	133.2	103.7	93.4	88.7	92.3	86.4	88.0	86.8	93.4	95.5	93.9	100.9	86.5	97.5	93.0	93.3	93.0
0920	Plaster & Gypsum Board	85.4	126.5	113.0	96.3	81.2	86.2	94.9	83.9	87.5	98.3	86.8	90.5	113.8	84.7	94.2	85.8	86.6	86.4
0950, 0980	Ceilings & Acoustic Treatment	87.7	126.5	113.8	101.1	81.2	87.7	88.5	83.9	85.4	100.3	86.8	91.2	104.0	84.7	91.0	89.4	86.6	87.5
0960	Flooring	76.2	139.3	93.9	94.8	103.7	97.3	92.2	98.9	94.1	98.4	78.0	92.7	94.7	90.4	93.5	83.5	89.3	85.1
0970, 0990	Wall Finishes & Painting/Coating	91.5	143.7	122.4	97.6	106.2	102.7	98.2	92.3	94.7	97.6	106.2	102.7	95.8	84.8	89.3	91.9	104.1	99.1
09	FINISHES	81.9	130.5	108.2	100.9	88.8	94.4	91.6	88.4	89.9	101.3	88.1	94.1	96.0	85.8	90.5	86.6	90.2	88.6
COVERS	DIVS. 10 - 14, 25, 28, 41, 43, 44, 46	100.0	113.0	102.9	100.0	95.2	98.9	100.0	96.0	99.1	100.0	96.4	99.2	100.0	95.4	99.0	100.0	95.5	99.0
21, 22, 23	FIRE SUPPRESSION, PLUMBING & HVAC	96.3	132.6	110.6	96.6	87.5	93.0	99.8	93.6	97.4	96.6	87.3	92.9	100.1	91.7	96.8	96.8	95.5	96.3
26, 27, 3370	ELECTRICAL, COMMUNICATIONS & UTIL.	91.6	132.4	111.7	94.8	107.7	101.1	90.3	93.1	91.7	94.8	107.7	101.2	99.7	85.8	92.9	92.9	90.0	91.5
MF2018	WEIGHTED AVERAGE	94.0	127.1	108.0	99.3	96.0	97.9	93.2	93.9	93.5	98.7	96.3	97.7	98.7	93.0	96.3	94.4	95.9	95.0

DIVISION		PENNSYLVANIA																	
		INDIANA			JOHNSTOWN			KITTANNING			LANCASTER			LEHIGH VALLEY			MONTROSE		
		157			159			162			175 - 176			180			188		
		MAT.	INST.	TOTAL	MAT.	INST.	TOTAL	MAT.	INST.	TOTAL	MAT.	INST.	TOTAL	MAT.	INST.	TOTAL	MAT.	INST.	TOTAL
015433	CONTRACTOR EQUIPMENT		110.7	110.7		110.7	110.7		112.6	112.6		111.8	111.8		112.6	112.6		112.6	112.6
0241, 31 - 34	SITE & INFRASTRUCTURE, DEMOLITION	97.8	95.8	96.4	104.1	96.0	98.5	88.8	97.4	94.7	78.4	95.6	90.3	88.9	97.5	94.8	87.5	97.2	94.2
0310	Concrete Forming & Accessories	82.6	92.9	91.4	81.6	82.5	82.4	85.1	89.7	89.0	90.0	86.4	86.9	91.7	110.3	107.6	81.5	89.3	88.2
0320	Concrete Reinforcing	91.7	119.4	105.1	93.0	118.9	105.5	94.4	119.2	106.4	94.1	111.0	102.3	94.0	108.0	100.8	98.4	116.2	107.0
0330	Cast-in-Place Concrete	99.3	94.4	97.5	110.1	86.7	101.4	91.5	94.7	92.7	79.1	96.7	85.7	91.5	99.2	94.3	89.7	93.0	90.9
03	CONCRETE	94.7	99.1	96.6	103.2	91.6	98.1	82.1	97.7	88.9	85.2	95.6	89.8	91.4	106.9	98.2	90.1	96.6	93.0
04	MASONRY	103.8	93.5	97.5	104.7	83.7	91.8	100.1	88.8	93.1	99.6	86.7	91.7	92.1	96.5	94.8	92.0	91.5	91.7
05	METALS	102.3	119.9	107.4	102.2	117.9	106.8	92.7	120.6	100.9	98.3	118.7	104.3	98.6	119.0	104.6	96.8	120.4	103.7
06	WOOD, PLASTICS & COMPOSITES	80.1	94.0	87.7	78.3	81.7	80.2	78.1	89.3	84.3	87.4	85.1	86.2	89.1	112.6	102.0	77.8	87.0	82.9
07	THERMAL & MOISTURE PROTECTION	98.1	94.2	96.4	98.4	88.5	94.1	102.3	92.4	98.0	95.4	101.0	97.8	103.7	111.9	107.3	103.3	91.0	97.9
08	OPENINGS	93.4	95.4	93.9	93.4	88.7	92.3	86.3	96.7	88.7	89.2	86.4	88.5	92.9	106.6	96.1	89.8	91.9	90.3
0920	Plaster & Gypsum Board	97.8	93.9	95.2	96.1	81.2	86.1	86.3	89.0	88.1	111.8	84.7	93.6	88.0	112.9	104.7	86.2	86.6	86.5
0950, 0980	Ceilings & Acoustic Treatment	101.1	93.9	96.2	100.3	81.2	87.5	91.8	89.0	89.9	96.0	84.7	88.4	89.4	112.9	105.2	91.0	86.6	88.1
0960	Flooring	95.6	103.7	97.8	94.8	78.0	90.1	86.3	103.7	91.2	91.5	95.7	92.7	88.9	94.1	90.4	84.1	103.7	89.6
0970, 0990	Wall Finishes & Painting/Coating	97.6	106.2	102.7	97.6	106.2	102.7	87.2	106.2	98.4	92.0	86.6	88.8	91.9	101.0	97.3	91.9	104.1	99.1
09	FINISHES	100.4	96.1	98.1	100.3	83.9	91.4	88.3	93.4	91.0	90.9	87.3	88.9	88.8	106.5	98.4	87.4	92.7	90.3
COVERS	DIVS. 10 - 14, 25, 28, 41, 43, 44, 46	100.0	96.5	99.2	100.0	94.4	98.8	100.0	96.2	99.1	100.0	95.7	99.0	100.0	102.5	100.6	100.0	96.2	99.2
21, 22, 23	FIRE SUPPRESSION, PLUMBING & HVAC	96.6	85.4	92.2	96.6	78.9	89.6	96.3	90.7	94.1	96.7	92.7	95.1	96.8	117.6	104.9	96.8	94.8	96.0
26, 27, 3370	ELECTRICAL, COMMUNICATIONS & UTIL.	94.8	107.7	101.2	94.8	107.7	101.1	88.7	107.7	98.1	93.7	93.3	93.5	92.9	129.4	110.9	92.1	95.2	93.6
MF2018	WEIGHTED AVERAGE	97.6	97.7	97.7	98.9	91.9	95.9	91.8	97.9	94.4	93.8	94.5	94.1	94.9	111.6	102.0	93.8	96.8	95.1

DIVISION		PENNSYLVANIA																	
		NEW CASTLE			NORRISTOWN			OIL CITY			PHILADELPHIA			PITTSBURGH			POTTSVILLE		
		161			194			163			190 - 191			150 - 152			179		
		MAT.	INST.	TOTAL	MAT.	INST.	TOTAL	MAT.	INST.	TOTAL	MAT.	INST.	TOTAL	MAT.	INST.	TOTAL	MAT.	INST.	TOTAL
015433	CONTRACTOR EQUIPMENT		112.6	112.6		97.5	97.5		112.6	112.6		99.8	99.8		99.7	99.7		111.8	111.8
0241, 31 - 34	SITE & INFRASTRUCTURE, DEMOLITION	86.7	97.7	94.3	97.3	95.1	95.8	85.2	96.3	92.9	99.8	101.1	100.7	103.7	96.1	98.5	81.3	95.8	91.3
0310	Concrete Forming & Accessories	85.1	93.6	92.3	83.7	125.3	119.1	85.1	92.7	91.6	99.8	142.4	136.2	97.5	97.5	97.5	82.0	87.7	86.8
0320	Concrete Reinforcing	93.3	100.2	96.7	95.9	150.8	122.5	94.4	94.9	94.7	100.4	143.1	121.0	93.5	122.9	107.7	93.4	114.1	103.4
0330	Cast-in-Place Concrete	88.8	94.3	90.8	86.1	127.4	101.5	86.3	94.4	89.3	89.8	135.6	106.8	108.6	101.2	105.8	84.3	96.4	88.8
03	CONCRETE	79.9	96.2	87.0	88.0	129.7	106.3	78.3	95.0	85.6	99.4	139.1	116.8	103.8	103.1	103.5	88.7	96.7	92.2
04	MASONRY	97.3	92.6	94.4	108.3	131.2	122.4	96.5	88.7	91.7	100.2	139.7	124.5	100.4	101.4	101.0	93.3	87.6	89.8
05	METALS	92.7	114.3	99.0	100.3	124.0	107.3	92.7	113.2	98.7	103.8	125.3	110.1	103.7	107.4	104.8	98.6	120.4	105.0
06	WOOD, PLASTICS & COMPOSITES	78.1	94.1	86.9	76.2	125.7	103.4	78.1	94.1	86.9	99.1	143.5	123.6	99.6	96.7	98.0	77.2	85.1	81.6
07	THERMAL & MOISTURE PROTECTION	102.2	92.0	97.8	108.7	130.1	118.0	102.1	90.7	97.1	103.4	139.7	119.2	98.6	99.3	98.9	95.6	99.6	97.3
08	OPENINGS	86.3	91.4	87.5	86.5	133.1	97.4	86.3	94.9	88.3	97.6	144.9	108.6	96.8	102.4	98.1	89.2	93.0	90.1
0920	Plaster & Gypsum Board	86.3	93.9	91.4	84.5	126.5	112.7	86.3	93.9	91.4	103.0	144.6	131.0	97.5	96.4	96.7	106.4	84.7	91.8
0950, 0980	Ceilings & Acoustic Treatment	91.8	93.9	93.2	91.1	126.5	114.9	91.8	93.9	93.2	104.2	144.6	131.5	95.2	96.4	96.0	96.0	84.7	88.4
0960	Flooring	86.3	105.3	91.6	86.7	139.3	101.5	86.3	103.7	91.2	97.8	156.2	114.2	103.1	107.3	104.3	87.5	103.7	92.1
0970, 0990	Wall Finishes & Painting/Coating	87.2	106.2	98.4	89.4	143.7	121.5	87.2	106.2	98.4	97.2	152.6	130.0	100.7	111.3	107.0	92.0	104.1	99.1
09	FINISHES	88.2	96.9	92.9	85.3	129.5	109.2	88.0	96.1	92.4	98.3	146.5	124.4	102.4	100.3	101.2	89.2	91.3	90.3
COVERS	DIVS. 10 - 14, 25, 28, 41, 43, 44, 46	100.0	97.3	99.4	100.0	109.1	102.0	100.0	96.7	99.3	100.0	119.2	104.3	100.0	103.0	100.7	100.0	97.5	99.4
21, 22, 23	FIRE SUPPRESSION, PLUMBING & HVAC	96.3	96.8	96.5	96.6	130.6	110.0	96.3	94.2	95.5	100.1	141.2	116.3	100.0	99.8	99.9	96.7	95.8	96.3
26, 27, 3370	ELECTRICAL, COMMUNICATIONS & UTIL.	89.2	98.2	93.7	91.9	142.0	116.6	90.9	107.7	99.2	98.8	159.0	128.4	97.1	111.3	104.1	92.0	91.3	91.6
MF2018	WEIGHTED AVERAGE	91.4	97.8	94.1	94.7	128.0	108.8	91.3	97.8	94.0	100.2	138.7	116.5	100.7	102.7	101.5	93.7	96.1	94.7

DIVISION		PENNSYLVANIA																	
		READING			SCRANTON			STATE COLLEGE			STROUDSBURG			SUNBURY			UNIONTOWN		
		195 - 196			184 - 185			168			183			178			154		
		MAT.	INST.	TOTAL	MAT.	INST.	TOTAL	MAT.	INST.	TOTAL	MAT.	INST.	TOTAL	MAT.	INST.	TOTAL	MAT.	INST.	TOTAL
015433	CONTRACTOR EQUIPMENT		119.0	119.0		112.6	112.6		111.8	111.8		112.6	112.6		112.6	112.6		110.7	110.7
0241, 31 - 34	SITE & INFRASTRUCTURE, DEMOLITION	101.8	107.7	105.9	92.2	97.2	95.7	82.7	95.7	91.7	86.6	97.3	94.0	92.6	96.7	95.4	98.3	96.5	97.1
0310	Concrete Forming & Accessories	98.8	86.2	88.0	98.0	86.6	88.3	84.3	82.8	83.0	86.5	90.5	89.9	93.1	84.8	86.0	76.5	93.9	91.4
0320	Concrete Reinforcing	97.2	148.6	122.1	96.8	116.3	106.2	95.1	105.8	100.3	97.1	116.4	106.4	96.0	112.3	103.9	92.3	119.5	105.5
0330	Cast-in-Place Concrete	77.2	97.0	84.6	93.4	93.2	93.3	90.0	87.0	88.9	88.0	94.5	90.4	92.2	94.9	93.2	99.3	95.6	97.9
03	CONCRETE	87.1	102.1	93.7	94.1	95.4	94.7	94.6	89.7	92.5	88.8	97.7	92.7	92.0	94.5	93.1	94.4	100.0	96.9
04	MASONRY	97.5	91.5	93.8	92.5	93.8	93.3	97.8	83.9	89.3	90.3	96.3	94.0	93.6	81.0	85.8	119.3	95.4	104.6
05	METALS	100.6	133.0	110.1	100.9	120.5	106.6	96.6	115.1	102.0	98.7	120.9	105.2	98.3	119.4	104.5	102.0	120.2	107.3
06	WOOD, PLASTICS & COMPOSITES	93.7	82.4	87.5	97.6	83.3	89.8	84.5	81.8	83.0	83.5	87.0	85.5	85.4	85.1	85.2	72.8	94.0	84.4
07	THERMAL & MOISTURE PROTECTION	109.0	103.3	106.6	103.9	91.3	98.4	102.9	97.6	100.6	103.5	91.3	98.2	96.8	92.7	95.0	98.0	95.1	96.7
08	OPENINGS	90.9	99.4	92.9	92.4	89.8	91.8	89.6	85.6	88.7	93.0	94.5	93.3	89.3	89.9	89.4	93.3	99.3	94.7
0920	Plaster & Gypsum Board	94.1	81.9	85.9	96.8	82.8	87.4	87.8	81.2	83.4	86.7	86.6	86.6	105.9	84.7	91.6	94.2	93.9	94.0
0950, 0980	Ceilings & Acoustic Treatment	83.7	81.9	82.5	96.6	82.8	87.3	88.5	81.2	83.6	87.7	86.6	87.0	93.6	84.7	87.6	100.3	93.9	96.0
0960	Flooring	91.3	97.9	93.2	92.0	92.9	92.2	89.4	97.9	91.8	86.8	94.1	88.8	88.3	103.7	92.6	92.0	103.7	95.3
0970, 0990	Wall Finishes & Painting/Coating	88.2	101.8	96.3	91.9	108.3	101.6	91.9	106.2	100.4	91.9	104.1	99.1	92.0	104.1	99.1	97.6	110.0	104.9
09	FINISHES	86.8	88.4	87.7	92.4	89.3	90.7	88.0	87.3	87.6	87.5	91.8	89.8	90.2	89.6	89.9	98.7	97.0	97.8
COVERS	DIVS. 10 - 14, 25, 28, 41, 43, 44, 46	100.0	95.8	99.1	100.0	94.6	98.8	100.0	94.6	98.8	100.0	97.2	99.4	100.0	93.1	98.5	100.0	97.2	99.4
21, 22, 23	FIRE SUPPRESSION, PLUMBING & HVAC	100.3	106.8	102.8	100.3	96.0	98.6	96.8	91.4	94.7	96.8	97.9	97.2	96.7	87.4	93.0	96.6	91.7	94.6
26, 27, 3370	ELECTRICAL, COMMUNICATIONS & UTIL.	97.8	91.3	94.6	98.0	95.2	96.6	91.3	107.7	99.4	92.9	141.8	117.0	92.4	91.9	92.1	91.8	107.7	99.7
MF2018	WEIGHTED AVERAGE	96.5	101.5	98.6	97.4	96.5	97.1	94.5	94.6	94.5	94.3	104.7	98.7	94.6	92.7	93.8	97.8	99.8	98.6

	DIVISION	PENNSYLVANIA																	
		WASHINGTON			WELLSBORO			WESTCHESTER			WILKES-BARRE			WILLIAMSPORT			YORK		
		153			169			193			186 - 187			177			173 - 174		
		MAT.	INST.	TOTAL	MAT.	INST.	TOTAL	MAT.	INST.	TOTAL	MAT.	INST.	TOTAL	MAT.	INST.	TOTAL	MAT.	INST.	TOTAL
015433	CONTRACTOR EQUIPMENT		110.7	110.7		112.6	112.6		97.5	97.5		112.6	112.6		112.6	112.6		111.8	111.8
0241, 31 - 34	SITE & INFRASTRUCTURE, DEMOLITION	98.3	96.8	97.3	94.0	96.6	95.8	103.4	96.1	98.4	84.5	97.2	93.3	83.7	96.8	92.8	82.4	95.6	91.5
0310	Concrete Forming & Accessories	82.8	94.0	92.4	85.2	84.2	84.3	90.0	126.6	121.2	89.1	85.3	85.8	90.1	85.6	86.3	85.1	86.3	86.1
0320	Concrete Reinforcing	92.3	119.5	105.5	95.1	116.1	105.3	95.0	150.8	122.0	95.8	116.3	105.7	95.3	110.9	102.8	96.0	111.0	103.3
0330	Cast-in-Place Concrete	99.3	95.9	98.0	94.4	88.2	92.1	95.4	129.5	108.1	84.5	93.0	87.6	78.0	90.0	82.5	84.7	96.7	89.2
03	CONCRETE	94.8	100.1	97.2	96.9	92.5	95.0	95.8	131.0	111.3	85.8	94.7	89.7	80.2	92.9	85.8	89.8	95.6	92.4
04	MASONRY	103.6	95.5	98.6	98.7	81.4	88.1	102.8	134.9	122.5	105.0	93.5	97.9	85.9	86.1	86.0	94.5	86.7	89.7
05	METALS	101.9	120.6	107.4	96.6	119.7	103.4	100.3	124.0	107.3	96.8	120.4	103.7	98.3	118.1	104.1	100.0	118.6	105.5
06	WOOD, PLASTICS & COMPOSITES	80.2	94.0	87.8	82.1	84.4	83.4	82.8	125.7	106.4	85.9	81.7	83.6	81.9	85.1	83.7	80.7	85.1	83.1
07	THERMAL & MOISTURE PROTECTION	98.1	95.1	96.8	104.1	86.0	96.2	109.1	127.0	116.9	103.3	90.9	97.9	96.1	88.3	92.7	95.6	101.0	98.0
08	OPENINGS	93.3	99.3	94.7	92.4	90.0	91.8	86.5	133.1	97.4	89.8	88.9	89.6	89.3	89.6	89.4	89.2	86.4	88.5
0920	Plaster & Gypsum Board	97.7	93.9	95.1	86.5	83.9	84.8	85.2	126.5	112.9	87.4	81.2	83.2	106.4	84.7	91.8	106.7	84.7	91.9
0950, 0980	Ceilings & Acoustic Treatment	100.3	93.9	96.0	88.5	83.9	85.4	91.1	126.5	114.9	91.0	81.2	84.4	96.0	84.7	88.4	95.1	84.7	88.1
0960	Flooring	95.7	103.7	97.9	85.8	103.7	90.9	89.6	139.3	103.6	87.7	92.9	89.1	87.2	86.8	87.1	88.9	95.7	90.8
0970, 0990	Wall Finishes & Painting/Coating	97.6	110.0	104.9	91.9	104.1	99.1	89.4	143.7	121.5	91.9	101.8	97.8	92.0	101.8	97.8	92.0	84.8	87.7
09	FINISHES	100.3	97.0	98.5	88.1	89.6	88.9	86.7	130.5	110.4	88.4	87.6	88.0	89.7	86.7	88.0	89.5	87.1	88.2
COVERS	DIVS. 10 - 14, 25, 28, 41, 43, 44, 46	100.0	97.2	99.4	100.0	96.2	99.2	100.0	115.3	103.4	100.0	94.3	98.7	100.0	95.5	99.0	100.0	95.6	99.0
21, 22, 23	FIRE SUPPRESSION, PLUMBING & HVAC	96.6	93.2	95.2	96.8	87.4	93.1	96.6	132.6	110.7	96.8	96.6	96.7	96.7	92.2	94.9	100.2	92.7	97.3
26, 27, 3370	ELECTRICAL, COMMUNICATIONS & UTIL.	94.2	107.7	100.9	92.1	95.2	93.6	91.7	132.4	111.8	92.9	90.0	91.5	92.8	83.4	88.2	93.7	82.9	88.4
MF2018	WEIGHTED AVERAGE	97.5	100.2	98.6	95.5	92.9	94.4	95.8	127.9	109.3	94.0	95.5	94.7	92.5	92.3	92.4	95.2	93.0	94.3

	DIVISION	PUERTO RICO			RHODE ISLAND						SOUTH CAROLINA								
		SAN JUAN			NEWPORT			PROVIDENCE			AIKEN			BEAUFORT			CHARLESTON		
		009			028			029			298			299			294		
		MAT.	INST.	TOTAL	MAT.	INST.	TOTAL	MAT.	INST.	TOTAL	MAT.	INST.	TOTAL	MAT.	INST.	TOTAL	MAT.	INST.	TOTAL
015433	CONTRACTOR EQUIPMENT		87.3	87.3		97.9	97.9		100.1	100.1		103.7	103.7		103.7	103.7		103.7	103.7
0241, 31 - 34	SITE & INFRASTRUCTURE, DEMOLITION	81.4	87.9	85.9	86.3	98.0	94.4	89.4	102.2	98.3	128.3	85.2	98.5	123.4	84.7	96.7	108.2	85.2	92.3
0310	Concrete Forming & Accessories	105.9	21.6	34.0	100.7	124.7	121.1	99.9	124.7	121.0	94.0	64.6	68.9	93.1	38.9	46.9	92.3	64.9	69.0
0320	Concrete Reinforcing	114.9	18.7	68.4	109.7	127.1	118.1	103.7	127.1	115.0	94.3	60.6	78.0	93.4	66.7	80.5	93.2	67.3	80.7
0330	Cast-in-Place Concrete	70.9	32.7	56.6	71.7	118.5	89.1	87.5	119.0	99.2	92.9	66.9	83.2	92.9	66.8	83.2	109.1	67.2	93.5
03	CONCRETE	89.8	26.1	61.8	87.5	122.4	102.9	91.8	122.5	105.3	101.0	66.4	85.8	98.4	55.7	79.7	94.9	67.8	83.0
04	MASONRY	110.0	24.2	57.3	99.2	123.1	113.8	103.4	123.1	115.5	79.9	64.8	70.6	94.9	64.8	76.4	96.4	67.5	78.6
05	METALS	116.8	40.6	94.4	98.8	116.7	104.1	103.2	115.9	106.9	102.6	89.1	98.7	102.7	90.7	99.2	104.6	92.1	100.9
06	WOOD, PLASTICS & COMPOSITES	145.4	20.5	76.6	98.0	124.0	112.3	98.4	124.0	112.5	89.7	66.8	77.1	88.4	32.5	57.6	87.2	66.8	76.0
07	THERMAL & MOISTURE PROTECTION	134.1	27.7	87.8	102.6	123.0	111.5	108.3	123.5	114.9	98.1	64.2	83.4	97.8	56.2	79.7	96.8	66.1	83.5
08	OPENINGS	96.8	19.1	78.7	97.4	124.4	103.7	99.0	124.4	104.9	94.0	62.5	86.7	94.1	44.9	82.6	97.6	65.1	90.0
0920	Plaster & Gypsum Board	76.4	25.2	48.0	91.4	124.3	113.6	104.9	124.3	118.0	83.4	65.8	71.6	87.0	30.5	49.0	88.1	65.8	73.1
0950, 0980	Ceilings & Acoustic Treatment	158.8	19.3	64.7	90.3	124.3	113.2	104.1	124.3	117.7	78.8	65.8	70.1	82.1	30.5	47.3	82.1	65.8	71.1
0960	Flooring	117.8	29.7	93.1	83.8	127.3	96.0	86.4	127.3	97.9	97.5	89.8	95.3	99.0	75.8	92.5	98.6	81.9	93.9
0970, 0990	Wall Finishes & Painting/Coating	90.2	24.6	51.4	85.0	118.9	105.0	90.8	118.9	107.4	95.2	66.0	77.9	95.2	56.9	72.5	95.2	70.0	80.3
09	FINISHES	106.0	23.2	61.2	87.7	125.1	107.9	93.7	125.1	110.7	89.0	69.7	78.5	90.1	46.0	66.2	88.2	68.9	77.8
COVERS	DIVS. 10 - 14, 25, 28, 41, 43, 44, 46	100.0	26.6	83.6	100.0	110.2	102.3	100.0	110.3	102.3	100.0	70.3	93.4	100.0	80.1	95.6	100.0	70.3	93.4
21, 22, 23	FIRE SUPPRESSION, PLUMBING & HVAC	88.5	18.8	61.1	100.4	114.8	106.0	100.2	114.8	105.9	96.9	55.2	80.5	96.9	55.6	80.7	100.4	58.4	83.9
26, 27, 3370	ELECTRICAL, COMMUNICATIONS & UTIL.	84.4	25.0	55.2	102.1	95.5	98.9	102.3	95.5	99.0	96.2	61.9	79.3	99.7	30.9	65.8	98.1	58.9	78.8
MF2018	WEIGHTED AVERAGE	98.5	29.8	69.4	97.0	114.7	104.5	99.3	115.0	105.9	97.6	67.1	84.7	98.3	57.5	81.1	98.7	68.2	85.8

	DIVISION	SOUTH CAROLINA															SOUTH DAKOTA		
		COLUMBIA			FLORENCE			GREENVILLE			ROCK HILL			SPARTANBURG			ABERDEEN		
		290 - 292			295			296			297			293			574		
		MAT.	INST.	TOTAL	MAT.	INST.	TOTAL	MAT.	INST.	TOTAL	MAT.	INST.	TOTAL	MAT.	INST.	TOTAL	MAT.	INST.	TOTAL
015433	CONTRACTOR EQUIPMENT		105.6	105.6		103.7	103.7		103.7	103.7		103.7	103.7		103.7	103.7		97.7	97.7
0241, 31 - 34	SITE & INFRASTRUCTURE, DEMOLITION	105.9	89.5	94.6	117.6	84.9	95.1	112.6	85.4	93.8	110.0	84.3	92.2	112.4	85.4	93.8	99.2	93.4	95.2
0310	Concrete Forming & Accessories	93.2	64.9	69.1	81.1	64.8	67.2	91.8	64.9	68.8	90.1	64.0	67.9	94.4	64.9	69.2	98.9	75.7	79.1
0320	Concrete Reinforcing	98.3	67.3	83.3	92.8	67.2	80.4	92.8	64.4	79.1	93.5	65.8	80.1	92.8	67.3	80.4	101.0	71.4	86.7
0330	Cast-in-Place Concrete	110.3	67.5	94.4	92.9	67.0	83.2	92.9	67.2	83.3	92.8	66.4	83.0	92.9	67.2	83.3	109.5	78.8	98.1
03	CONCRETE	93.2	67.8	82.1	93.3	67.6	82.0	92.0	67.3	81.2	90.1	66.8	79.9	92.2	67.8	81.5	98.7	76.7	89.1
04	MASONRY	89.9	67.5	76.1	79.9	67.4	72.2	77.9	67.5	71.5	101.6	64.0	78.5	79.9	67.5	72.3	109.4	73.3	87.2
05	METALS	101.6	91.0	98.5	103.5	91.4	99.9	103.5	91.7	100.0	102.7	90.5	99.1	103.5	92.1	100.1	93.8	82.1	90.4
06	WOOD, PLASTICS & COMPOSITES	88.6	66.9	76.7	74.8	66.8	70.4	86.8	66.8	75.8	85.4	66.8	75.2	90.6	66.8	77.5	97.6	74.2	84.7
07	THERMAL & MOISTURE PROTECTION	93.3	66.3	81.5	97.2	66.1	83.7	97.1	66.1	83.6	96.9	59.2	80.5	97.1	66.1	83.6	103.1	77.2	91.8
08	OPENINGS	99.7	65.1	91.7	94.1	65.1	87.3	94.0	64.8	87.2	94.1	63.6	87.0	94.1	65.1	87.3	97.2	62.2	89.0
0920	Plaster & Gypsum Board	88.7	65.8	73.3	78.2	65.8	69.9	82.1	65.8	71.2	81.8	65.8	71.1	84.7	65.8	72.0	110.1	73.9	85.7
0950, 0980	Ceilings & Acoustic Treatment	86.2	65.8	72.5	79.6	65.8	70.3	78.8	65.8	70.1	78.8	65.8	70.1	78.8	65.8	70.1	98.8	73.9	82.0
0960	Flooring	90.7	80.4	87.8	89.9	80.4	87.2	96.3	80.4	91.8	95.2	75.8	89.7	97.6	80.4	92.8	91.1	43.8	77.8
0970, 0990	Wall Finishes & Painting/Coating	93.9	70.0	79.7	95.2	70.0	80.3	95.2	70.0	80.3	95.2	66.0	77.9	95.2	70.0	80.3	89.8	33.4	56.4
09	FINISHES	86.5	68.7	76.8	85.0	68.7	76.2	86.9	68.7	77.0	86.3	67.1	75.9	87.6	68.7	77.4	92.4	65.9	78.1
COVERS	DIVS. 10 - 14, 25, 28, 41, 43, 44, 46	100.0	70.5	93.4	100.0	70.3	93.4	100.0	70.3	93.4	100.0	70.0	93.3	100.0	70.3	93.4	100.0	84.8	96.6
21, 22, 23	FIRE SUPPRESSION, PLUMBING & HVAC	100.0	58.3	83.6	100.4	58.2	83.9	100.4	58.2	83.9	96.9	54.8	80.4	100.4	58.2	83.9	100.0	54.2	82.0
26, 27, 3370	ELECTRICAL, COMMUNICATIONS & UTIL.	98.5	62.8	80.9	96.1	62.8	79.7	98.2	59.8	79.3	98.2	59.9	79.3	98.2	59.8	79.3	101.1	65.3	83.4
MF2018	WEIGHTED AVERAGE	97.6	68.9	85.5	96.9	68.6	84.9	97.0	68.2	84.8	96.7	66.4	83.9	97.2	68.3	85.0	98.6	70.1	86.5

	DIVISION	SOUTH DAKOTA																	
		MITCHELL			MOBRIDGE			PIERRE			RAPID CITY			SIOUX FALLS			WATERTOWN		
		573			576			575			577			570 - 571			572		
		MAT.	INST.	TOTAL	MAT.	INST.	TOTAL	MAT.	INST.	TOTAL	MAT.	INST.	TOTAL	MAT.	INST.	TOTAL	MAT.	INST.	TOTAL
015433	CONTRACTOR EQUIPMENT		97.7	97.7		97.7	97.7		99.9	99.9		97.7	97.7		100.9	100.9		97.7	97.7
0241, 31 - 34	SITE & INFRASTRUCTURE, DEMOLITION	95.8	93.2	94.0	95.7	93.2	94.0	97.2	96.3	96.6	97.5	92.8	94.3	92.5	98.7	96.8	95.6	93.2	94.0
0310	Concrete Forming & Accessories	97.9	44.2	52.1	88.2	44.6	51.0	99.7	45.6	53.6	106.9	56.6	64.0	103.3	78.1	81.8	84.6	74.5	76.0
0320	Concrete Reinforcing	100.4	70.5	86.0	103.0	70.5	87.3	97.8	100.2	99.0	94.4	100.4	97.3	106.6	100.6	103.7	97.6	70.5	84.5
0330	Cast-in-Place Concrete	106.4	52.3	86.3	106.4	77.2	95.6	103.1	77.8	93.7	105.6	77.2	95.0	88.3	78.8	84.7	106.4	77.2	95.5
03	CONCRETE	96.5	53.1	77.4	96.2	62.0	81.2	91.8	67.4	81.1	95.9	72.3	85.5	88.6	82.7	86.0	95.3	75.5	86.6
04	MASONRY	97.8	75.5	84.1	106.5	72.8	85.8	106.5	72.6	85.7	106.0	74.0	86.3	91.1	75.6	81.5	132.0	73.7	96.2
05	METALS	92.9	81.6	89.6	92.9	82.0	89.7	96.3	89.5	94.3	95.5	91.2	94.2	95.3	92.6	94.5	92.9	81.9	89.6
06	WOOD, PLASTICS & COMPOSITES	96.6	33.7	62.0	84.4	33.4	56.4	104.2	34.0	65.6	102.2	48.0	72.3	101.6	76.5	87.8	80.5	74.2	77.0
07	THERMAL & MOISTURE PROTECTION	102.8	72.9	89.8	102.9	75.1	90.8	104.1	72.7	90.4	103.6	78.5	92.7	103.2	83.3	94.6	102.6	76.7	91.3
08	OPENINGS	96.3	39.6	83.1	98.9	39.0	84.9	99.8	58.4	90.2	101.0	66.1	92.9	104.2	81.8	99.0	96.3	61.5	88.2
0920	Plaster & Gypsum Board	109.0	32.2	57.3	102.2	32.0	54.9	108.1	32.5	57.2	109.8	46.9	67.5	103.6	76.1	85.1	100.5	73.9	82.6
0950, 0980	Ceilings & Acoustic Treatment	95.5	32.2	52.8	98.8	32.0	53.7	98.6	32.5	54.0	100.4	46.9	64.3	97.7	76.1	83.1	95.5	73.9	80.9
0960	Flooring	90.7	43.8	77.6	86.5	46.4	75.3	98.2	31.2	79.4	90.5	72.7	85.5	93.9	75.0	88.6	85.2	43.8	73.6
0970, 0990	Wall Finishes & Painting/Coating	89.8	36.6	58.3	89.8	37.5	58.8	97.1	95.6	96.2	89.8	95.6	93.2	100.3	95.6	97.5	89.8	33.4	56.4
09	FINISHES	91.4	41.1	64.2	89.9	41.6	63.7	98.4	45.2	69.6	92.5	61.8	75.9	96.4	79.1	87.1	88.6	64.7	75.7
COVERS	DIVS. 10 - 14, 25, 28, 41, 43, 44, 46	100.0	82.0	96.0	100.0	81.9	96.0	100.0	82.3	96.1	100.0	83.5	96.3	100.0	87.0	97.1	100.0	85.8	96.8
21, 22, 23	FIRE SUPPRESSION, PLUMBING & HVAC	96.5	49.0	77.8	96.5	69.0	85.7	100.0	76.8	90.9	100.0	76.8	90.9	99.9	70.2	88.2	96.5	52.1	79.0
26, 27, 3370	ELECTRICAL, COMMUNICATIONS & UTIL.	99.4	63.4	81.7	101.1	39.8	70.9	103.9	47.7	76.2	97.6	47.7	73.0	102.6	63.4	83.3	98.6	63.4	81.2
MF2018	WEIGHTED AVERAGE	96.3	60.9	81.3	96.9	62.9	82.5	99.0	68.5	86.1	98.3	72.0	87.2	97.7	78.4	89.5	97.3	69.0	85.4

	DIVISION	TENNESSEE																	
		CHATTANOOGA			COLUMBIA			COOKEVILLE			JACKSON			JOHNSON CITY			KNOXVILLE		
		373 - 374			384			385			383			376			377 - 379		
		MAT.	INST.	TOTAL	MAT.	INST.	TOTAL	MAT.	INST.	TOTAL	MAT.	INST.	TOTAL	MAT.	INST.	TOTAL	MAT.	INST.	TOTAL
015433	CONTRACTOR EQUIPMENT		104.7	104.7		99.5	99.5		99.5	99.5		105.5	105.5		98.8	98.8		98.8	98.8
0241, 31 - 34	SITE & INFRASTRUCTURE, DEMOLITION	106.6	93.8	97.8	91.5	84.6	86.7	97.3	81.6	86.5	100.9	94.0	96.2	113.6	81.1	91.2	92.6	83.8	86.5
0310	Concrete Forming & Accessories	95.8	57.3	63.0	80.9	62.4	65.1	81.1	31.9	39.2	87.5	41.0	47.8	82.7	57.5	61.2	94.3	61.7	66.5
0320	Concrete Reinforcing	101.6	65.1	83.9	91.4	65.0	78.6	91.4	64.6	78.4	91.3	67.2	79.7	102.2	61.9	82.7	101.6	61.8	82.3
0330	Cast-in-Place Concrete	97.5	64.1	85.1	92.8	63.8	82.0	105.2	56.7	87.2	102.7	66.0	89.0	78.4	58.0	70.8	91.6	64.0	81.3
03	CONCRETE	93.1	62.8	79.8	92.4	65.1	80.4	102.4	48.7	78.8	93.2	56.5	77.0	102.7	60.2	84.0	90.7	64.2	79.0
04	MASONRY	98.6	56.3	72.6	111.4	55.5	77.0	106.7	39.4	65.3	111.7	44.4	70.4	111.0	43.2	69.3	76.3	50.7	60.6
05	METALS	93.0	88.1	91.6	93.5	88.9	92.1	93.5	87.6	91.8	96.0	89.5	94.1	90.3	86.4	89.2	93.4	86.5	91.4
06	WOOD, PLASTICS & COMPOSITES	106.5	56.3	78.9	67.8	63.4	65.4	68.0	29.5	46.8	82.9	39.7	59.1	78.7	62.6	69.8	93.1	62.6	76.4
07	THERMAL & MOISTURE PROTECTION	99.2	61.7	82.9	95.6	62.9	81.4	96.1	52.6	77.2	98.0	55.5	79.5	95.3	55.6	78.0	92.8	60.8	78.9
08	OPENINGS	102.0	57.5	91.6	91.0	53.3	82.2	91.1	35.5	78.1	97.7	44.3	85.2	98.3	59.5	89.3	95.7	55.2	86.2
0920	Plaster & Gypsum Board	80.0	55.6	63.6	87.4	62.9	70.9	87.4	28.0	47.4	90.0	38.5	55.3	101.0	62.1	74.8	109.1	62.1	77.5
0950, 0980	Ceilings & Acoustic Treatment	97.9	55.6	69.4	75.8	62.9	67.1	75.8	28.0	43.6	83.9	38.5	53.3	93.5	62.1	72.3	94.3	62.1	72.6
0960	Flooring	98.3	55.2	86.2	81.7	53.9	73.9	81.7	50.0	72.8	81.0	54.2	73.5	93.3	43.9	79.4	98.4	49.7	84.7
0970, 0990	Wall Finishes & Painting/Coating	98.0	59.5	75.2	85.0	56.5	68.1	85.0	56.5	68.1	87.0	56.5	68.9	94.9	58.7	73.5	94.9	57.6	72.8
09	FINISHES	94.4	56.6	73.9	86.2	60.1	72.1	86.8	36.2	59.4	86.1	43.4	63.0	98.5	55.3	75.1	91.4	58.9	73.8
COVERS	DIVS. 10 - 14, 25, 28, 41, 43, 44, 46	100.0	67.8	92.8	100.0	68.0	92.9	100.0	35.9	85.7	100.0	62.0	91.5	100.0	75.8	94.6	100.0	79.3	95.4
21, 22, 23	FIRE SUPPRESSION, PLUMBING & HVAC	100.1	59.3	84.1	97.8	73.6	88.3	97.8	67.1	85.8	100.1	61.4	84.9	100.0	55.5	82.5	100.0	60.5	84.5
26, 27, 3370	ELECTRICAL, COMMUNICATIONS & UTIL.	100.8	82.9	92.0	94.0	50.6	72.6	95.5	61.3	78.6	100.5	65.7	83.3	91.5	40.8	66.5	97.0	54.9	76.2
MF2018	WEIGHTED AVERAGE	98.0	68.1	85.3	94.8	66.5	82.8	96.2	56.9	79.6	97.6	61.4	82.3	98.1	58.6	81.4	94.8	63.7	81.6

	DIVISION	TENNESSEE									TEXAS								
		MCKENZIE			MEMPHIS			NASHVILLE			ABILENE			AMARILLO			AUSTIN		
		382			375, 380 - 381			370 - 372			795 - 796			790 - 791			786 - 787		
		MAT.	INST.	TOTAL	MAT.	INST.	TOTAL	MAT.	INST.	TOTAL	MAT.	INST.	TOTAL	MAT.	INST.	TOTAL	MAT.	INST.	TOTAL
015433	CONTRACTOR EQUIPMENT		99.5	99.5		102.0	102.0		105.5	105.5		90.9	90.9		93.9	93.9		92.5	92.5
0241, 31 - 34	SITE & INFRASTRUCTURE, DEMOLITION	96.9	81.8	86.5	91.1	95.4	94.0	102.8	98.1	99.6	91.9	87.4	88.8	91.3	91.3	91.3	95.5	90.1	91.7
0310	Concrete Forming & Accessories	88.4	34.1	42.1	94.6	64.7	69.1	95.9	68.1	72.2	94.0	59.2	64.3	95.6	51.7	58.2	95.3	53.0	59.3
0320	Concrete Reinforcing	91.5	65.5	78.9	110.9	66.0	89.2	107.1	66.8	87.6	88.5	51.7	70.7	95.3	49.6	73.2	86.8	47.2	67.6
0330	Cast-in-Place Concrete	102.9	57.9	86.2	96.5	76.7	89.1	88.2	69.0	81.1	84.6	64.9	77.3	83.2	65.7	76.7	94.0	65.9	83.5
03	CONCRETE	101.0	50.2	78.7	96.5	70.2	84.9	92.9	69.4	82.6	84.5	60.7	74.1	87.1	57.2	74.0	89.4	57.4	75.4
04	MASONRY	110.5	45.1	70.3	92.6	56.3	70.3	87.7	55.9	68.2	99.9	60.6	75.8	97.6	60.2	74.6	91.7	60.7	72.7
05	METALS	93.5	88.0	91.9	87.2	82.5	85.8	101.6	85.1	96.7	107.1	70.4	96.3	101.9	68.7	92.1	100.7	67.4	90.9
06	WOOD, PLASTICS & COMPOSITES	76.6	31.7	51.9	98.9	65.4	80.4	100.1	70.1	83.6	99.2	60.8	78.1	102.6	51.0	74.2	92.9	52.6	70.7
07	THERMAL & MOISTURE PROTECTION	96.1	53.6	77.6	93.0	66.4	81.4	96.9	65.0	83.0	98.3	63.4	83.1	100.1	62.3	83.6	95.1	63.8	81.5
08	OPENINGS	91.1	37.2	78.5	99.5	61.3	90.6	101.0	67.5	93.2	101.8	56.7	91.2	103.2	50.8	91.0	101.1	49.6	89.1
0920	Plaster & Gypsum Board	90.9	30.3	50.1	90.9	64.5	73.1	95.9	69.3	78.0	88.2	60.1	69.3	100.7	50.0	66.6	88.9	51.6	63.8
0950, 0980	Ceilings & Acoustic Treatment	75.8	30.3	45.1	90.8	64.5	73.0	94.1	69.3	77.4	99.2	60.1	72.9	100.7	50.0	66.5	93.3	51.6	65.2
0960	Flooring	84.5	53.9	75.9	97.8	55.2	85.9	97.1	57.5	86.0	93.2	67.6	86.0	93.7	63.8	85.3	94.1	63.8	85.6
0970, 0990	Wall Finishes & Painting/Coating	85.0	44.2	60.8	93.2	59.3	73.1	100.7	68.6	81.7	92.8	51.7	68.5	91.0	51.7	67.7	95.7	44.2	65.2
09	FINISHES	88.0	37.5	60.7	95.9	61.9	77.5	97.8	66.1	80.6	86.1	60.1	72.0	92.6	53.5	71.4	91.1	53.6	70.8
COVERS	DIVS. 10 - 14, 25, 28, 41, 43, 44, 46	100.0	24.0	83.1	100.0	80.5	95.7	100.0	81.1	95.8	100.0	78.9	95.3	100.0	72.7	93.9	100.0	76.5	94.8
21, 22, 23	FIRE SUPPRESSION, PLUMBING & HVAC	97.8	58.6	82.4	100.1	70.3	88.4	99.9	75.1	90.2	100.1	52.1	81.3	99.9	51.1	80.7	99.9	58.7	83.8
26, 27, 3370	ELECTRICAL, COMMUNICATIONS & UTIL.	95.3	55.4	75.6	102.9	63.6	83.5	93.5	63.2	78.6	97.9	54.0	76.3	100.9	58.9	80.2	97.5	57.2	77.6
MF2018	WEIGHTED AVERAGE	96.3	55.1	78.9	96.6	69.8	85.3	98.0	71.8	86.9	97.8	61.4	82.4	98.2	60.0	82.1	97.2	61.4	82.1

	DIVISION	TEXAS																	
		BEAUMONT			BROWNWOOD			BRYAN			CHILDRESS			CORPUS CHRISTI			DALLAS		
		776 - 777			768			778			792			783 - 784			752 - 753		
		MAT.	INST.	TOTAL	MAT.	INST.	TOTAL	MAT.	INST.	TOTAL	MAT.	INST.	TOTAL	MAT.	INST.	TOTAL	MAT.	INST.	TOTAL
015433	CONTRACTOR EQUIPMENT		95.3	95.3		90.9	90.9		95.3	95.3		90.9	90.9		100.6	100.6		106.9	106.9
0241, 31 - 34	SITE & INFRASTRUCTURE, DEMOLITION	90.6	91.8	91.4	101.3	87.3	91.7	81.5	91.9	88.7	101.8	85.8	90.8	144.2	82.8	101.8	106.0	97.6	100.2
0310	Concrete Forming & Accessories	102.3	54.6	61.7	97.6	58.7	64.4	82.1	58.2	61.7	92.6	58.6	63.6	98.4	52.2	59.0	95.0	62.3	67.1
0320	Concrete Reinforcing	85.7	62.0	74.2	81.9	51.1	67.0	87.9	51.2	70.2	88.7	51.0	70.5	79.8	50.9	65.8	91.4	52.3	72.5
0330	Cast-in-Place Concrete	90.7	65.2	81.2	96.8	58.6	82.6	72.6	59.6	67.8	86.7	58.6	76.3	114.1	67.1	96.6	92.6	71.4	84.7
03	CONCRETE	91.0	60.7	77.7	90.3	58.2	76.2	75.8	58.5	68.2	93.4	58.2	77.9	95.2	59.1	79.4	93.3	65.3	81.0
04	MASONRY	102.9	61.7	77.6	122.8	60.6	84.6	143.3	60.6	92.5	103.5	60.2	76.9	81.4	60.7	68.7	99.5	60.2	75.3
05	METALS	99.4	76.2	92.6	102.6	69.8	92.9	99.5	72.1	91.4	104.4	69.6	94.2	97.3	84.4	93.5	101.7	82.6	96.1
06	WOOD, PLASTICS & COMPOSITES	113.6	54.1	80.9	105.8	60.8	81.0	78.6	59.3	68.0	98.6	60.8	77.8	115.6	51.5	80.3	100.8	63.6	80.3
07	THERMAL & MOISTURE PROTECTION	92.8	64.3	80.4	92.0	61.9	78.9	85.2	63.8	75.9	99.0	60.4	82.2	97.7	62.2	82.2	87.3	67.2	78.6
08	OPENINGS	95.0	55.0	85.6	100.4	56.6	90.2	95.7	56.0	86.4	96.1	56.6	86.9	101.3	50.0	89.4	96.5	59.5	87.9
0920	Plaster & Gypsum Board	101.0	53.2	68.9	82.6	60.1	67.5	88.2	58.7	68.3	87.6	60.1	69.1	96.5	50.5	65.5	92.7	62.5	72.4
0950, 0980	Ceilings & Acoustic Treatment	105.3	53.2	70.2	86.6	60.1	68.7	96.7	58.7	71.1	96.8	60.1	72.1	98.8	50.5	66.2	102.4	62.5	75.5
0960	Flooring	116.9	74.6	105.0	75.7	54.6	69.8	86.7	67.0	81.1	91.6	54.6	81.2	107.4	67.7	96.3	94.2	69.4	87.2
0970, 0990	Wall Finishes & Painting/Coating	94.9	53.9	70.6	92.8	51.7	68.4	92.2	57.4	71.6	92.8	51.7	68.5	110.1	44.2	71.1	96.1	54.8	71.6
09	FINISHES	98.0	57.8	76.2	77.6	57.4	66.7	85.4	59.4	71.3	86.2	57.4	70.6	101.9	53.8	75.9	97.9	62.7	78.8
COVERS	DIVS. 10 - 14, 25, 28, 41, 43, 44, 46	100.0	77.8	95.0	100.0	73.4	94.1	100.0	77.0	94.9	100.0	73.4	94.1	100.0	77.8	95.0	100.0	82.5	96.1
21, 22, 23	FIRE SUPPRESSION, PLUMBING & HVAC	100.0	63.0	85.5	96.5	49.5	78.0	96.5	62.4	83.1	96.6	51.7	79.0	100.0	48.7	79.9	99.9	62.3	85.1
26, 27, 3370	ELECTRICAL, COMMUNICATIONS & UTIL.	100.1	63.4	82.0	98.6	46.1	72.7	97.8	63.4	80.8	97.9	55.8	77.1	94.0	63.4	78.9	95.2	63.2	79.4
MF2018	WEIGHTED AVERAGE	97.9	65.5	84.2	97.4	58.7	81.1	95.2	64.9	82.4	97.6	60.3	81.8	98.9	61.4	83.1	98.2	68.0	85.4

	DIVISION	TEXAS																	
		DEL RIO			DENTON			EASTLAND			EL PASO			FORT WORTH			GALVESTON		
		788			762			764			798 - 799, 885			760 - 761			775		
		MAT.	INST.	TOTAL	MAT.	INST.	TOTAL	MAT.	INST.	TOTAL	MAT.	INST.	TOTAL	MAT.	INST.	TOTAL	MAT.	INST.	TOTAL
015433	CONTRACTOR EQUIPMENT		89.9	89.9		100.2	100.2		90.9	90.9		93.9	93.9		93.9	93.9		107.9	107.9
0241, 31 - 34	SITE & INFRASTRUCTURE, DEMOLITION	122.2	85.5	96.9	101.4	79.7	86.4	104.0	85.5	91.2	89.7	92.7	91.8	97.4	92.7	94.2	108.4	90.6	96.1
0310	Concrete Forming & Accessories	95.1	51.0	57.5	105.0	58.8	65.7	98.4	58.6	64.4	95.6	56.7	62.5	98.2	59.5	65.2	90.9	55.8	61.0
0320	Concrete Reinforcing	80.4	47.6	64.5	83.2	51.1	67.7	82.1	51.1	67.1	94.2	51.1	73.4	88.1	50.9	70.1	87.5	59.7	74.0
0330	Cast-in-Place Concrete	122.9	58.4	98.9	74.9	59.7	69.2	102.3	58.5	86.0	73.9	68.0	71.7	89.3	65.9	80.6	96.4	60.6	83.1
03	CONCRETE	115.8	54.0	88.6	70.6	59.7	65.8	94.8	58.1	78.7	82.7	60.5	72.9	86.1	61.0	75.1	92.1	60.3	78.1
04	MASONRY	95.4	60.5	74.0	131.0	59.3	86.9	92.7	60.6	73.0	91.6	62.7	73.8	92.5	59.3	72.1	99.0	60.7	75.5
05	METALS	97.2	67.2	88.4	102.1	84.9	97.1	102.4	69.6	92.7	102.1	69.4	92.5	104.3	69.5	94.1	101.2	91.4	98.3
06	WOOD, PLASTICS & COMPOSITES	96.1	50.4	71.0	117.8	60.9	86.5	113.0	60.8	84.3	90.2	57.1	72.0	99.6	60.9	78.3	95.3	55.9	73.6
07	THERMAL & MOISTURE PROTECTION	94.7	62.0	80.5	90.0	62.4	78.0	92.4	61.9	79.1	91.9	64.4	80.0	89.7	64.4	78.7	84.6	64.4	75.8
08	OPENINGS	96.6	47.8	85.2	118.7	56.2	104.2	70.6	56.6	67.4	96.3	52.8	86.2	101.2	56.6	90.8	99.7	56.0	89.5
0920	Plaster & Gypsum Board	93.3	49.5	63.8	87.3	60.1	69.0	82.6	60.1	67.5	97.3	56.2	69.6	89.8	60.1	69.8	94.8	55.0	68.0
0950, 0980	Ceilings & Acoustic Treatment	95.3	49.5	64.4	90.7	60.1	70.1	86.6	60.1	68.7	94.2	56.2	68.6	92.7	60.1	70.7	100.8	55.0	69.9
0960	Flooring	91.9	54.6	81.4	71.9	54.6	67.1	95.9	54.6	84.3	97.3	69.7	89.5	94.5	64.5	86.0	102.4	67.0	92.5
0970, 0990	Wall Finishes & Painting/Coating	97.8	42.6	65.1	103.3	48.9	71.1	94.2	51.7	69.0	98.6	49.6	69.6	101.2	51.8	72.0	104.6	53.7	74.5
09	FINISHES	94.5	50.3	70.6	76.3	57.2	66.0	84.1	57.4	69.6	92.3	58.1	73.8	89.6	59.5	73.3	95.4	56.9	74.6
COVERS	DIVS. 10 - 14, 25, 28, 41, 43, 44, 46	100.0	73.6	94.1	100.0	79.2	95.4	100.0	71.7	93.7	100.0	76.6	94.8	100.0	79.2	95.4	100.0	78.3	95.2
21, 22, 23	FIRE SUPPRESSION, PLUMBING & HVAC	96.5	59.4	81.9	96.5	55.8	80.5	96.5	49.5	78.0	99.8	65.3	86.3	99.9	55.4	82.4	96.5	62.5	83.2
26, 27, 3370	ELECTRICAL, COMMUNICATIONS & UTIL.	95.9	59.8	78.1	101.0	58.7	80.1	98.5	58.6	78.8	97.5	50.5	74.3	100.2	58.6	79.7	99.6	63.5	81.8
MF2018	WEIGHTED AVERAGE	99.7	60.4	83.1	97.3	62.9	82.7	94.2	60.3	79.8	95.9	63.6	82.3	97.6	62.9	82.9	97.7	66.5	84.5

	DIVISION	TEXAS																	
		GIDDINGS			GREENVILLE			HOUSTON			HUNTSVILLE			LAREDO			LONGVIEW		
		789			754			770 - 772			773			780			756		
		MAT.	INST.	TOTAL	MAT.	INST.	TOTAL	MAT.	INST.	TOTAL	MAT.	INST.	TOTAL	MAT.	INST.	TOTAL	MAT.	INST.	TOTAL
015433	CONTRACTOR EQUIPMENT		89.9	89.9		100.4	100.4		103.3	103.3		95.3	95.3		89.9	89.9		91.8	91.8
0241, 31 - 34	SITE & INFRASTRUCTURE, DEMOLITION	106.8	85.8	92.3	96.4	82.4	86.7	107.7	95.5	99.3	96.8	91.6	93.2	100.2	86.2	90.5	93.7	88.6	90.2
0310	Concrete Forming & Accessories	92.6	51.1	57.3	86.9	57.1	61.5	96.0	58.9	64.4	89.2	54.1	59.2	95.1	52.1	58.5	82.9	58.8	62.3
0320	Concrete Reinforcing	80.9	48.4	65.2	91.9	51.1	72.2	87.3	52.2	70.3	88.2	51.1	70.3	80.4	50.9	66.1	90.9	50.7	71.5
0330	Cast-in-Place Concrete	104.1	58.7	87.2	92.1	59.6	80.0	89.6	68.9	81.9	99.9	59.5	84.9	87.8	65.0	79.3	107.1	58.5	89.0
03	CONCRETE	92.6	54.3	75.8	86.7	58.7	74.4	91.4	62.5	78.7	98.7	56.6	80.2	87.3	57.4	74.2	102.7	58.1	83.1
04	MASONRY	102.2	60.6	76.7	158.6	59.3	97.5	98.0	63.1	76.6	141.3	60.6	91.7	88.9	60.6	71.5	154.2	59.2	95.8
05	METALS	96.7	68.2	88.4	99.2	83.2	94.5	103.9	77.3	96.1	99.4	71.8	91.3	99.6	69.4	90.7	91.8	68.4	84.9
06	WOOD, PLASTICS & COMPOSITES	95.3	50.4	70.6	87.2	58.4	71.4	100.6	59.2	77.8	88.1	54.1	69.4	96.1	51.4	71.5	81.2	60.8	70.0
07	THERMAL & MOISTURE PROTECTION	95.0	62.7	81.0	87.3	61.5	76.1	86.3	68.5	78.5	86.5	63.2	76.4	93.8	63.0	80.4	89.1	60.9	76.8
08	OPENINGS	95.8	48.7	84.8	92.2	55.3	83.6	106.6	56.4	94.9	95.7	52.5	85.6	96.7	49.8	85.8	83.4	56.2	77.0
0920	Plaster & Gypsum Board	92.4	49.5	63.5	82.8	57.5	65.8	94.0	58.0	69.8	92.3	53.2	66.0	94.1	50.5	64.7	81.7	60.1	67.2
0950, 0980	Ceilings & Acoustic Treatment	95.3	49.5	64.4	98.1	57.5	70.7	97.9	58.0	71.0	96.7	53.2	67.4	98.6	50.5	66.1	93.2	60.1	70.9
0960	Flooring	91.7	54.6	81.2	87.7	54.6	78.4	103.5	71.2	94.4	91.3	54.6	81.0	91.8	63.8	83.9	93.7	54.6	82.7
0970, 0990	Wall Finishes & Painting/Coating	97.8	44.2	66.1	90.4	51.7	67.5	102.1	58.6	76.3	92.2	57.4	71.6	97.8	44.2	66.1	82.1	48.9	62.4
09	FINISHES	93.1	50.5	70.0	92.3	56.1	72.7	102.7	60.8	80.1	88.3	54.1	69.8	93.4	52.9	71.5	98.1	57.2	76.0
COVERS	DIVS. 10 - 14, 25, 28, 41, 43, 44, 46	100.0	70.3	93.4	100.0	76.5	94.8	100.0	83.1	96.2	100.0	69.8	93.3	100.0	75.7	94.6	100.0	79.0	95.3
21, 22, 23	FIRE SUPPRESSION, PLUMBING & HVAC	96.5	61.4	82.7	96.5	58.0	81.3	100.0	65.3	86.3	96.5	62.4	83.1	100.0	58.9	83.8	96.4	57.5	81.1
26, 27, 3370	ELECTRICAL, COMMUNICATIONS & UTIL.	93.1	55.6	74.6	92.1	58.7	75.6	101.8	67.8	85.1	97.8	59.1	78.7	96.0	57.1	76.8	92.6	50.9	72.0
MF2018	WEIGHTED AVERAGE	96.1	60.4	81.0	97.4	62.9	82.8	100.4	68.2	86.8	98.7	62.8	83.6	96.4	61.1	81.5	97.5	61.0	82.1

DIVISION		TEXAS																	
		LUBBOCK			LUFKIN			MCALLEN			MCKINNEY			MIDLAND			ODESSA		
		793 - 794			759			785			750			797			797		
		MAT.	INST.	TOTAL	MAT.	INST.	TOTAL	MAT.	INST.	TOTAL	MAT.	INST.	TOTAL	MAT.	INST.	TOTAL	MAT.	INST.	TOTAL
015433	CONTRACTOR EQUIPMENT		102.9	102.9		91.8	91.8		100.7	100.7		100.4	100.4		102.9	102.9		90.9	90.9
0241, 31 - 34	SITE & INFRASTRUCTURE, DEMOLITION	115.0	86.7	95.5	88.6	90.0	89.5	149.0	82.9	103.3	92.8	82.4	85.6	117.7	85.8	95.6	92.1	88.1	89.3
0310	Concrete Forming & Accessories	92.9	53.0	58.9	85.9	54.1	58.8	99.1	51.2	58.3	86.0	57.1	61.4	96.4	59.1	64.6	93.9	59.2	64.3
0320	Concrete Reinforcing	89.7	51.8	71.4	92.6	66.7	80.1	80.0	50.9	65.9	91.9	51.1	72.2	90.7	51.7	71.8	88.5	51.6	70.7
0330	Cast-in-Place Concrete	84.7	68.2	78.6	95.8	58.3	81.8	124.0	59.7	100.1	86.3	59.6	76.4	90.0	65.9	81.1	84.6	64.9	77.3
03	CONCRETE	83.1	60.1	73.0	95.1	58.6	79.0	102.8	56.2	82.3	82.1	58.8	71.9	87.1	62.1	76.1	84.5	60.7	74.1
04	MASONRY	99.2	61.0	75.7	117.3	60.5	82.4	95.1	60.7	74.0	170.8	59.3	102.3	116.2	60.2	81.8	99.9	60.2	75.5
05	METALS	110.9	85.8	103.5	99.4	73.3	91.7	97.1	84.2	93.3	99.2	83.3	94.5	109.2	85.4	102.2	106.4	70.4	95.8
06	WOOD, PLASTICS & COMPOSITES	98.6	52.1	73.0	88.6	54.2	69.6	114.3	50.5	79.2	86.1	58.4	70.9	103.0	60.9	79.8	99.2	60.8	78.1
07	THERMAL & MOISTURE PROTECTION	88.2	63.5	77.5	88.8	60.0	76.3	97.6	62.9	82.5	87.1	61.5	76.0	88.5	63.3	77.5	98.3	62.7	82.8
08	OPENINGS	108.6	52.4	95.5	63.3	56.6	61.7	100.4	48.6	88.3	92.2	55.3	83.6	110.0	56.7	97.6	101.8	56.7	91.2
0920	Plaster & Gypsum Board	88.4	51.1	63.3	80.6	53.2	62.2	97.5	49.5	65.2	82.8	57.5	65.8	90.0	60.1	69.9	88.2	60.1	69.3
0950, 0980	Ceilings & Acoustic Treatment	100.0	51.1	67.0	88.3	53.2	64.7	98.6	49.5	65.5	98.1	57.5	70.7	97.6	60.1	72.3	99.2	60.1	72.9
0960	Flooring	87.5	69.0	82.3	126.2	54.6	106.1	106.9	72.8	97.3	87.2	54.6	78.1	88.6	63.8	81.7	93.2	63.8	84.9
0970, 0990	Wall Finishes & Painting/Coating	103.4	53.7	74.0	82.1	51.7	64.1	110.1	42.6	70.2	90.4	51.7	67.5	103.4	51.7	72.8	92.8	51.7	68.5
09	FINISHES	88.5	55.5	70.6	106.5	53.9	78.0	102.4	54.1	76.3	91.9	56.1	72.5	89.0	59.4	72.9	86.1	59.3	71.6
COVERS	DIVS. 10 - 14, 25, 28, 41, 43, 44, 46	100.0	78.3	95.2	100.0	70.1	93.3	100.0	75.8	94.6	100.0	79.1	95.3	100.0	73.8	94.1	100.0	73.5	94.1
21, 22, 23	FIRE SUPPRESSION, PLUMBING & HVAC	99.7	53.1	81.4	96.4	58.1	81.4	96.5	48.7	77.7	96.5	60.2	82.2	96.1	47.4	77.0	100.1	52.5	81.4
26, 27, 3370	ELECTRICAL, COMMUNICATIONS & UTIL.	96.7	60.1	78.6	93.9	61.4	77.8	93.8	31.6	63.1	92.2	58.7	75.6	96.7	58.9	78.1	98.0	58.9	78.7
MF2018	WEIGHTED AVERAGE	99.2	62.9	83.9	94.7	62.6	81.1	99.7	56.5	81.4	97.2	63.5	82.9	99.7	62.2	83.8	97.7	61.8	82.6

DIVISION		TEXAS																	
		PALESTINE			SAN ANGELO			SAN ANTONIO			TEMPLE			TEXARKANA			TYLER		
		758			769			781 - 782			765			755			757		
		MAT.	INST.	TOTAL	MAT.	INST.	TOTAL	MAT.	INST.	TOTAL	MAT.	INST.	TOTAL	MAT.	INST.	TOTAL	MAT.	INST.	TOTAL
015433	CONTRACTOR EQUIPMENT		91.8	91.8		90.9	90.9		94.1	94.1		90.9	90.9		91.8	91.8		91.8	91.8
0241, 31 - 34	SITE & INFRASTRUCTURE, DEMOLITION	94.2	88.8	90.5	97.8	87.7	90.8	101.5	93.7	96.1	86.7	87.6	87.3	83.4	91.1	88.7	92.9	88.7	90.0
0310	Concrete Forming & Accessories	77.5	58.8	61.5	97.9	51.4	58.3	94.7	52.6	58.8	101.2	51.2	58.5	93.0	59.4	64.3	87.6	58.9	63.1
0320	Concrete Reinforcing	90.2	51.0	71.3	81.8	51.6	67.2	89.6	47.7	69.3	82.0	50.8	66.9	90.1	50.8	71.1	90.9	50.8	71.5
0330	Cast-in-Place Concrete	87.5	58.5	76.7	91.4	64.9	81.5	85.7	68.3	79.2	74.9	58.6	68.8	88.2	64.9	79.5	105.2	58.6	87.8
03	CONCRETE	97.1	58.1	80.0	86.1	57.2	73.4	88.0	58.0	74.8	73.4	54.7	65.2	88.0	60.6	75.9	102.3	58.2	82.9
04	MASONRY	111.9	59.2	79.5	119.4	60.6	83.3	89.7	60.8	71.9	130.1	60.6	87.4	175.3	59.2	103.9	164.4	59.2	99.7
05	METALS	99.1	68.5	90.1	102.7	70.3	93.2	102.0	65.4	91.3	102.4	69.2	92.7	91.7	69.0	85.0	99.0	68.5	90.1
06	WOOD, PLASTICS & COMPOSITES	79.3	60.8	69.1	106.1	50.4	75.5	98.1	51.7	72.5	115.7	50.4	79.8	93.7	60.8	75.6	90.3	60.8	74.1
07	THERMAL & MOISTURE PROTECTION	89.4	61.8	77.4	91.8	62.6	79.1	86.8	65.6	77.5	91.3	61.7	78.4	88.6	63.0	77.5	89.2	61.8	77.3
08	OPENINGS	63.3	56.6	61.7	100.5	51.0	88.9	102.1	49.5	89.8	67.3	50.8	63.4	83.3	56.5	77.1	63.2	56.5	61.7
0920	Plaster & Gypsum Board	79.0	60.1	66.3	82.6	49.5	60.3	93.3	50.5	64.5	82.6	49.5	60.3	86.8	60.1	68.8	80.6	60.1	66.8
0950, 0980	Ceilings & Acoustic Treatment	88.3	60.1	69.3	86.6	49.5	61.6	95.2	50.5	65.1	86.6	49.5	61.6	93.2	60.1	70.9	88.3	60.1	69.3
0960	Flooring	116.4	54.6	99.0	75.8	54.6	69.9	101.2	67.7	91.8	97.5	54.6	85.4	102.7	63.8	91.7	128.7	54.6	107.8
0970, 0990	Wall Finishes & Painting/Coating	82.1	51.7	64.1	92.8	51.7	68.4	99.4	44.2	66.7	94.2	44.2	64.6	82.1	51.7	64.1	82.1	51.7	64.1
09	FINISHES	104.0	57.5	78.8	77.4	51.3	63.3	104.2	54.0	77.0	83.3	50.5	65.5	100.9	59.3	78.4	107.5	57.5	80.4
COVERS	DIVS. 10 - 14, 25, 28, 41, 43, 44, 46	100.0	73.5	94.1	100.0	76.0	94.7	100.0	78.0	95.1	100.0	72.3	93.8	100.0	79.1	95.3	100.0	79.0	95.3
21, 22, 23	FIRE SUPPRESSION, PLUMBING & HVAC	96.4	56.7	80.8	96.5	52.2	79.1	100.0	59.8	84.2	96.5	53.2	79.5	96.4	55.5	80.3	96.4	58.2	81.4
26, 27, 3370	ELECTRICAL, COMMUNICATIONS & UTIL.	90.1	48.4	69.5	102.8	54.6	79.1	97.5	59.8	78.9	99.8	52.0	76.2	93.7	54.5	74.4	92.6	53.7	73.4
MF2018	WEIGHTED AVERAGE	94.2	60.4	79.9	97.1	59.4	81.2	98.3	62.3	83.1	92.5	58.5	78.2	96.7	62.0	82.1	97.8	61.6	82.5

DIVISION		TEXAS															UTAH		
		VICTORIA			WACO			WAXAHACHIE			WHARTON			WICHITA FALLS			LOGAN		
		779			766 - 767			751			774			763			843		
		MAT.	INST.	TOTAL	MAT.	INST.	TOTAL	MAT.	INST.	TOTAL	MAT.	INST.	TOTAL	MAT.	INST.	TOTAL	MAT.	INST.	TOTAL
015433	CONTRACTOR EQUIPMENT		106.6	106.6		90.9	90.9		100.4	100.4		107.9	107.9		90.9	90.9		93.9	93.9
0241, 31 - 34	SITE & INFRASTRUCTURE, DEMOLITION	113.4	88.0	95.8	95.1	88.0	90.2	94.7	82.5	86.3	118.7	90.0	98.9	95.8	88.1	90.5	99.8	89.6	92.7
0310	Concrete Forming & Accessories	90.6	51.3	57.1	99.8	59.2	65.2	86.0	59.0	63.0	85.7	53.1	57.9	99.8	59.2	65.2	103.5	67.2	72.6
0320	Concrete Reinforcing	84.0	50.9	68.0	81.6	47.2	65.0	91.9	51.1	72.2	87.4	50.9	69.7	81.6	51.7	67.1	108.0	84.4	96.6
0330	Cast-in-Place Concrete	108.3	60.5	90.5	81.0	64.9	75.0	91.1	59.8	79.5	111.3	59.5	92.0	86.5	65.0	78.5	86.4	74.3	81.9
03	CONCRETE	99.7	56.7	80.8	79.6	60.0	71.0	85.8	59.7	74.3	103.8	57.1	83.3	82.2	60.8	72.8	105.4	73.2	91.2
04	MASONRY	116.9	60.7	82.4	90.9	60.6	72.3	159.2	59.3	97.8	100.3	60.6	75.9	91.4	60.1	72.2	109.5	60.5	79.4
05	METALS	99.4	87.8	96.0	104.8	68.4	94.1	99.2	83.6	94.6	101.2	87.4	97.1	104.7	70.4	94.6	109.5	82.5	101.5
06	WOOD, PLASTICS & COMPOSITES	98.4	50.5	72.1	113.9	60.8	84.7	86.1	61.0	72.3	88.4	52.7	68.8	113.9	60.8	84.7	85.9	67.1	75.5
07	THERMAL & MOISTURE PROTECTION	88.0	61.1	76.3	92.2	63.7	79.8	87.2	61.6	76.1	85.0	63.7	75.7	92.2	62.9	79.5	103.0	68.8	88.1
08	OPENINGS	99.5	51.1	88.2	78.4	55.6	73.1	92.2	56.7	83.9	99.7	52.3	88.6	78.4	56.7	73.4	92.2	66.1	86.1
0920	Plaster & Gypsum Board	90.9	49.5	63.0	82.7	60.1	67.5	83.2	60.1	67.7	90.4	51.7	64.4	82.7	60.1	67.5	80.6	66.2	70.9
0950, 0980	Ceilings & Acoustic Treatment	101.6	49.5	66.5	87.4	60.1	69.0	99.7	60.1	73.0	100.8	51.7	67.7	87.4	60.1	69.0	104.3	66.2	78.7
0960	Flooring	100.7	54.6	87.7	96.8	63.8	87.5	87.2	54.6	78.1	99.5	66.6	90.2	97.4	71.9	90.2	97.2	60.5	86.9
0970, 0990	Wall Finishes & Painting/Coating	104.8	57.4	76.8	94.2	51.7	69.0	90.4	51.7	67.5	104.6	55.7	75.7	96.1	51.7	69.8	92.6	60.7	73.7
09	FINISHES	92.4	52.0	70.6	83.7	59.3	70.5	92.5	57.6	73.6	94.7	55.2	73.3	84.1	60.9	71.5	93.9	64.9	78.2
COVERS	DIVS. 10 - 14, 25, 28, 41, 43, 44, 46	100.0	69.7	93.2	100.0	78.9	95.3	100.0	79.4	95.4	100.0	70.0	93.3	100.0	73.5	94.1	100.0	84.4	96.5
21, 22, 23	FIRE SUPPRESSION, PLUMBING & HVAC	96.5	62.3	83.1	100.0	58.7	83.8	96.5	58.0	81.4	96.5	61.9	82.9	100.0	52.1	81.2	100.0	69.2	87.9
26, 27, 3370	ELECTRICAL, COMMUNICATIONS & UTIL.	104.9	54.0	79.8	103.0	54.1	78.9	92.1	58.7	75.6	103.9	59.2	81.9	105.0	56.0	80.8	95.2	71.3	83.4
MF2018	WEIGHTED AVERAGE	99.7	62.9	84.2	94.4	62.4	80.9	97.2	63.5	83.0	99.8	64.3	84.8	95.0	61.6	80.9	100.9	71.7	88.5

	DIVISION	UTAH												VERMONT					
		OGDEN			PRICE			PROVO			SALT LAKE CITY			BELLOWS FALLS			BENNINGTON		
		842, 844			845			846 - 847			840 - 841			051			052		
		MAT.	INST.	TOTAL	MAT.	INST.	TOTAL	MAT.	INST.	TOTAL	MAT.	INST.	TOTAL	MAT.	INST.	TOTAL	MAT.	INST.	TOTAL
015433	CONTRACTOR EQUIPMENT		93.9	93.9		92.9	92.9		92.9	92.9		93.8	93.8		95.7	95.7		95.7	95.7
0241, 31 - 34	SITE & INFRASTRUCTURE, DEMOLITION	87.3	89.6	88.9	97.1	87.4	90.4	95.9	88.0	90.5	87.0	89.5	88.7	86.8	96.1	93.3	86.2	96.1	93.1
0310	Concrete Forming & Accessories	103.5	67.2	72.6	106.0	64.2	70.4	105.3	67.2	72.9	105.8	67.2	72.9	96.5	101.7	100.9	94.0	101.6	100.4
0320	Concrete Reinforcing	107.6	84.4	96.4	115.8	84.0	100.4	116.8	84.4	101.1	110.0	84.4	97.6	85.1	86.3	85.7	85.1	86.3	85.6
0330	Cast-in-Place Concrete	87.7	74.3	82.7	86.5	71.6	80.9	86.5	74.3	81.9	95.8	74.3	87.8	89.6	115.9	99.4	89.6	115.8	99.4
03	CONCRETE	94.4	73.2	85.1	106.6	70.8	90.8	104.9	73.2	91.0	114.2	73.2	96.2	88.7	103.7	95.3	88.5	103.6	95.2
04	MASONRY	104.0	60.5	77.3	115.1	65.1	84.4	115.3	60.5	81.6	118.1	60.5	82.7	96.3	86.2	90.1	105.1	86.2	93.5
05	METALS	110.0	82.5	101.9	106.5	82.5	99.5	107.5	82.5	100.1	113.7	82.5	104.5	95.2	91.1	94.0	95.1	90.9	93.9
06	WOOD, PLASTICS & COMPOSITES	85.9	67.1	75.5	89.2	64.0	75.3	87.5	67.1	76.3	87.8	67.1	76.4	102.8	106.8	105.0	100.1	106.8	103.8
07	THERMAL & MOISTURE PROTECTION	101.8	68.8	87.4	104.8	68.3	88.9	104.9	68.8	89.2	110.0	68.8	92.1	100.9	89.1	95.8	100.9	89.1	95.8
08	OPENINGS	92.2	66.1	86.1	96.1	76.3	91.5	96.1	66.1	89.1	94.1	66.1	87.5	100.1	97.5	99.5	100.1	97.5	99.5
0920	Plaster & Gypsum Board	80.6	66.2	70.9	83.4	63.1	69.8	81.2	66.2	71.1	90.9	66.2	74.3	108.9	106.7	107.4	107.9	106.7	107.1
0950, 0980	Ceilings & Acoustic Treatment	104.3	66.2	78.7	104.3	63.1	76.5	104.3	66.2	78.7	97.1	66.2	76.3	97.2	106.7	103.6	97.2	106.7	103.6
0960	Flooring	95.2	60.5	85.4	98.4	56.1	86.5	98.2	60.5	87.6	99.0	60.5	88.1	92.0	98.9	94.0	91.2	98.9	93.4
0970, 0990	Wall Finishes & Painting/Coating	92.6	60.7	73.7	92.6	60.7	73.7	92.6	60.7	73.7	95.7	60.7	75.0	86.4	101.5	95.3	86.4	101.5	95.3
09	FINISHES	92.0	64.9	77.3	95.0	61.7	77.0	94.5	64.9	78.5	93.7	64.9	78.1	90.6	101.8	96.7	90.3	101.8	96.5
COVERS	DIVS. 10 - 14, 25, 28, 41, 43, 44, 46	100.0	84.4	96.5	100.0	81.6	95.9	100.0	84.4	96.5	100.0	84.4	96.5	100.0	91.8	98.2	100.0	91.7	98.2
21, 22, 23	FIRE SUPPRESSION, PLUMBING & HVAC	100.0	69.2	87.9	98.1	63.2	84.4	100.0	69.2	87.9	100.1	69.2	88.0	96.7	87.6	93.1	96.7	87.6	93.1
26, 27, 3370	ELECTRICAL, COMMUNICATIONS & UTIL.	95.5	71.3	83.6	100.2	68.8	84.7	95.7	68.6	82.3	98.1	71.3	84.9	108.3	79.5	94.1	108.3	54.5	81.8
MF2018	WEIGHTED AVERAGE	98.8	71.7	87.4	101.4	70.0	88.1	101.2	71.2	88.5	103.5	71.7	90.1	96.7	92.1	94.8	97.0	88.6	93.4

	DIVISION	VERMONT																	
		BRATTLEBORO			BURLINGTON			GUILDHALL			MONTPELIER			RUTLAND			ST. JOHNSBURY		
		053			054			059			056			057			058		
		MAT.	INST.	TOTAL	MAT.	INST.	TOTAL	MAT.	INST.	TOTAL	MAT.	INST.	TOTAL	MAT.	INST.	TOTAL	MAT.	INST.	TOTAL
015433	CONTRACTOR EQUIPMENT		95.7	95.7		99.0	99.0		95.7	95.7		99.0	99.0		95.7	95.7		95.7	95.7
0241, 31 - 34	SITE & INFRASTRUCTURE, DEMOLITION	87.8	96.1	93.5	92.6	101.0	98.4	85.8	91.9	90.0	89.7	100.8	97.3	90.4	96.1	94.3	85.9	95.2	92.3
0310	Concrete Forming & Accessories	96.7	101.7	101.0	99.4	80.2	83.0	94.6	95.1	95.1	97.3	101.3	100.7	97.0	80.0	82.5	93.5	95.6	95.2
0320	Concrete Reinforcing	84.1	86.3	85.2	107.3	86.2	97.1	85.8	86.1	86.0	101.6	86.2	94.2	106.4	86.1	96.6	84.1	86.2	85.1
0330	Cast-in-Place Concrete	92.5	115.9	101.2	109.6	116.7	112.2	86.9	107.2	94.4	104.5	116.7	109.0	87.8	115.6	98.2	86.9	107.3	94.5
03	CONCRETE	90.7	103.7	96.4	101.7	94.2	98.4	86.2	97.7	91.3	98.5	103.7	100.8	91.4	93.8	92.5	85.9	97.9	91.2
04	MASONRY	104.6	86.2	93.3	103.9	86.4	93.1	104.8	71.5	84.4	101.4	86.4	92.2	87.2	86.3	86.7	131.1	71.5	94.5
05	METALS	95.1	91.1	93.9	102.8	89.8	99.0	95.2	90.0	93.7	101.2	89.9	97.9	101.1	90.3	97.9	95.2	90.4	93.8
06	WOOD, PLASTICS & COMPOSITES	103.1	106.8	105.1	99.2	78.5	87.8	99.0	106.8	103.3	96.0	106.9	102.0	103.2	78.3	89.5	94.1	106.8	101.0
07	THERMAL & MOISTURE PROTECTION	101.0	89.1	95.8	107.4	86.6	98.3	100.7	82.4	92.8	109.4	89.7	100.8	101.1	86.1	94.6	100.6	82.4	92.7
08	OPENINGS	100.1	97.8	99.6	102.6	78.4	96.9	100.1	94.0	98.7	103.1	94.1	101.0	103.2	78.3	97.4	100.1	94.0	98.7
0920	Plaster & Gypsum Board	108.9	106.7	107.4	109.9	77.4	88.1	114.5	106.7	109.2	110.4	106.7	107.9	109.4	77.4	87.9	115.8	106.7	109.6
0950, 0980	Ceilings & Acoustic Treatment	97.2	106.7	103.6	103.9	77.4	86.0	97.2	106.7	103.6	104.9	106.7	106.1	103.2	77.4	85.8	97.2	106.7	103.6
0960	Flooring	92.2	98.9	94.1	101.8	98.9	101.0	95.0	98.9	96.1	98.5	98.9	98.6	92.0	98.9	94.0	98.3	98.9	98.5
0970, 0990	Wall Finishes & Painting/Coating	86.4	101.5	95.3	98.0	88.3	92.3	86.4	88.3	87.5	97.8	88.3	92.2	86.4	88.3	87.5	86.4	88.3	87.5
09	FINISHES	90.8	101.8	96.8	99.1	83.8	90.8	92.2	96.7	94.6	97.9	100.6	99.4	92.0	83.7	87.5	93.3	96.7	95.2
COVERS	DIVS. 10 - 14, 25, 28, 41, 43, 44, 46	100.0	91.8	98.2	100.0	89.1	97.6	100.0	86.7	97.0	100.0	92.2	98.3	100.0	88.6	97.5	100.0	86.8	97.0
21, 22, 23	FIRE SUPPRESSION, PLUMBING & HVAC	96.7	87.6	93.1	100.0	68.3	87.6	96.7	60.5	82.5	96.5	68.3	85.4	100.2	68.2	87.7	96.7	60.5	82.5
26, 27, 3370	ELECTRICAL, COMMUNICATIONS & UTIL.	108.3	79.5	94.1	107.6	53.6	81.0	108.3	54.5	81.8	108.1	53.6	81.3	108.4	53.6	81.4	108.3	54.5	81.8
MF2018	WEIGHTED AVERAGE	97.4	92.1	95.2	101.9	79.8	92.5	96.8	79.0	89.3	100.2	84.4	93.5	98.9	79.4	90.7	98.1	79.3	90.2

	DIVISION	VERMONT			VIRGINIA														
		WHITE RIVER JCT.			ALEXANDRIA			ARLINGTON			BRISTOL			CHARLOTTESVILLE			CULPEPER		
		050			223			222			242			229			227		
		MAT.	INST.	TOTAL	MAT.	INST.	TOTAL	MAT.	INST.	TOTAL	MAT.	INST.	TOTAL	MAT.	INST.	TOTAL	MAT.	INST.	TOTAL
015433	CONTRACTOR EQUIPMENT		95.7	95.7		105.3	105.3		104.1	104.1		104.1	104.1		108.1	108.1		104.1	104.1
0241, 31 - 34	SITE & INFRASTRUCTURE, DEMOLITION	90.3	95.2	93.7	113.7	89.0	96.7	123.9	87.0	98.5	108.2	86.0	92.8	112.7	88.3	95.9	111.4	86.5	94.2
0310	Concrete Forming & Accessories	91.4	96.1	95.4	91.5	71.9	74.8	90.5	71.6	74.4	86.6	64.1	67.5	85.1	61.7	65.2	82.4	71.0	72.7
0320	Concrete Reinforcing	85.1	86.2	85.6	86.6	85.0	85.9	97.8	85.1	91.6	97.7	83.7	91.0	97.1	73.0	85.4	97.8	84.9	91.6
0330	Cast-in-Place Concrete	92.5	108.1	98.3	108.1	77.7	96.8	105.2	77.2	94.8	104.8	44.5	82.3	109.0	75.5	96.5	107.7	74.4	95.3
03	CONCRETE	92.6	98.4	95.2	99.9	77.7	90.1	104.5	77.4	92.6	100.8	62.4	83.9	101.3	70.2	87.7	98.4	76.0	88.6
04	MASONRY	117.2	72.9	90.0	89.8	73.7	79.9	102.3	72.6	84.0	92.3	45.4	63.5	116.9	55.3	79.0	105.1	70.8	84.1
05	METALS	95.2	90.4	93.8	106.0	100.8	104.5	104.6	101.8	103.8	103.4	97.8	101.8	103.7	95.8	101.4	103.7	99.6	102.5
06	WOOD, PLASTICS & COMPOSITES	97.1	106.8	102.4	91.0	70.1	79.5	87.5	70.1	78.0	80.2	69.8	74.5	78.7	59.8	68.3	77.4	70.1	73.4
07	THERMAL & MOISTURE PROTECTION	101.1	83.0	93.2	102.5	81.1	93.2	104.6	80.5	94.1	103.9	58.8	84.3	103.5	69.7	88.8	103.7	78.9	92.9
08	OPENINGS	100.1	94.0	98.7	95.4	74.3	90.5	93.7	74.3	89.2	96.4	67.9	89.7	94.8	65.7	88.0	95.1	73.8	90.1
0920	Plaster & Gypsum Board	106.6	106.7	106.7	98.0	69.2	78.6	94.8	69.2	77.6	91.1	68.9	76.2	91.1	57.9	68.8	91.3	69.2	76.4
0950, 0980	Ceilings & Acoustic Treatment	97.2	106.7	103.6	92.4	69.2	76.8	90.8	69.2	76.2	90.0	68.9	75.8	90.0	57.9	68.4	90.8	69.2	76.2
0960	Flooring	90.2	98.9	92.6	97.1	77.4	91.5	95.6	75.7	90.0	92.3	55.7	82.0	90.8	55.7	80.9	90.8	74.4	86.2
0970, 0990	Wall Finishes & Painting/Coating	86.4	88.3	87.5	115.8	76.0	92.3	115.8	73.1	90.6	102.1	55.3	74.4	102.1	58.6	76.3	115.8	73.1	90.6
09	FINISHES	90.1	97.1	93.9	93.8	72.6	82.3	93.8	71.6	81.8	90.6	62.3	75.3	90.1	59.4	73.5	90.8	71.2	80.2
COVERS	DIVS. 10 - 14, 25, 28, 41, 43, 44, 46	100.0	87.2	97.1	100.0	86.6	97.0	100.0	84.4	96.5	100.0	75.1	94.5	100.0	83.2	96.2	100.0	83.6	96.3
21, 22, 23	FIRE SUPPRESSION, PLUMBING & HVAC	96.7	61.2	82.8	100.4	86.2	94.8	100.4	85.6	94.6	96.8	46.0	76.9	96.8	69.1	86.0	96.8	85.3	92.3
26, 27, 3370	ELECTRICAL, COMMUNICATIONS & UTIL.	108.3	54.3	81.7	96.6	100.2	98.4	94.4	100.2	97.2	96.2	36.3	66.7	96.2	70.4	83.5	98.5	96.9	97.7
MF2018	WEIGHTED AVERAGE	98.2	79.8	90.4	99.6	84.8	93.3	100.4	84.2	93.6	98.2	59.3	81.8	99.4	71.0	87.4	98.7	83.0	92.1

	DIVISION	VIRGINIA																	
		FAIRFAX			FARMVILLE			FREDERICKSBURG			GRUNDY			HARRISONBURG			LYNCHBURG		
		220 - 221			239			224 - 225			246			228			245		
		MAT.	INST.	TOTAL	MAT.	INST.	TOTAL	MAT.	INST.	TOTAL	MAT.	INST.	TOTAL	MAT.	INST.	TOTAL	MAT.	INST.	TOTAL
015433	CONTRACTOR EQUIPMENT		104.1	104.1		108.1	108.1		104.1	104.1		104.1	104.1		104.1	104.1		104.1	104.1
0241, 31 - 34	SITE & INFRASTRUCTURE, DEMOLITION	122.4	86.7	97.8	108.5	87.3	93.9	110.9	86.4	93.9	106.0	85.3	91.7	119.3	85.0	95.6	106.8	86.7	92.9
0310	Concrete Forming & Accessories	85.1	63.8	67.0	98.9	60.3	66.0	85.1	65.4	68.3	89.6	38.2	45.8	81.4	56.6	60.2	86.6	70.5	72.9
0320	Concrete Reinforcing	97.8	90.3	94.2	92.9	67.9	80.8	98.5	80.1	89.6	96.4	44.1	71.1	97.8	57.3	78.2	97.1	84.5	91.0
0330	Cast-in-Place Concrete	105.2	78.5	95.2	106.8	83.9	98.3	106.8	74.2	94.7	104.8	50.9	84.7	105.2	52.6	85.6	104.8	74.3	93.4
03	CONCRETE	104.2	75.2	91.5	101.8	71.5	88.5	98.1	72.7	86.9	99.3	45.8	75.8	102.0	57.1	82.3	99.2	75.8	88.9
04	MASONRY	102.2	73.6	84.6	101.8	50.0	69.9	104.0	66.7	81.1	94.9	55.4	70.6	101.4	57.9	74.6	109.3	63.9	81.4
05	METALS	103.8	102.5	103.4	101.3	90.8	98.2	103.8	98.0	102.1	103.4	75.2	95.1	103.7	87.2	98.9	103.6	99.8	102.5
06	WOOD, PLASTICS & COMPOSITES	80.2	59.2	68.6	93.0	59.8	74.7	80.2	64.6	71.6	83.1	30.6	54.2	76.3	59.2	66.9	80.2	71.5	75.4
07	THERMAL & MOISTURE PROTECTION	104.4	79.9	93.7	105.2	69.4	89.6	103.7	76.3	91.8	103.9	57.3	83.6	104.1	69.6	89.1	103.7	70.4	89.2
08	OPENINGS	93.7	69.6	88.0	94.3	54.9	85.1	94.8	68.5	88.7	96.4	30.1	80.9	95.0	57.0	86.2	95.0	68.8	88.9
0920	Plaster & Gypsum Board	91.3	57.9	68.9	103.1	57.9	72.7	91.3	63.6	72.7	91.1	28.5	49.0	91.1	57.9	68.8	91.1	70.7	77.4
0950, 0980	Ceilings & Acoustic Treatment	90.8	57.9	68.6	86.5	57.9	67.2	90.8	63.6	72.4	90.0	28.5	48.6	90.0	57.9	68.4	90.0	70.7	77.0
0960	Flooring	92.5	76.7	88.1	95.6	50.5	82.9	92.5	72.4	86.9	93.5	28.2	75.2	90.6	74.4	86.1	92.3	65.4	84.7
0970, 0990	Wall Finishes & Painting/Coating	115.8	73.6	90.8	99.6	57.2	74.5	115.8	57.0	81.0	102.1	31.2	60.2	115.8	48.3	75.9	102.1	55.3	74.4
09	FINISHES	92.4	65.7	77.9	90.6	57.7	72.8	91.3	65.0	77.1	90.7	33.9	60.0	91.3	58.9	73.7	90.4	67.8	78.2
COVERS	DIVS. 10 - 14, 25, 28, 41, 43, 44, 46	100.0	84.2	96.5	100.0	77.1	94.9	100.0	78.3	95.2	100.0	73.6	94.1	100.0	62.8	91.7	100.0	78.8	95.3
21, 22, 23	FIRE SUPPRESSION, PLUMBING & HVAC	96.8	88.1	93.4	96.9	52.0	79.3	96.8	79.9	90.2	96.8	65.7	84.6	96.8	65.4	84.5	96.8	69.5	86.1
26, 27, 3370	ELECTRICAL, COMMUNICATIONS & UTIL.	97.3	101.1	99.1	91.2	68.3	79.9	94.5	91.8	93.2	96.2	68.3	82.5	96.4	86.4	91.5	97.2	67.8	82.7
MF2018	WEIGHTED AVERAGE	99.5	83.6	92.8	97.9	65.3	84.1	98.3	78.7	90.0	98.1	58.8	81.5	99.0	68.8	86.3	98.7	73.8	88.2

	DIVISION	VIRGINIA																	
		NEWPORT NEWS			NORFOLK			PETERSBURG			PORTSMOUTH			PULASKI			RICHMOND		
		236			233 - 235			238			237			243			230 - 232		
		MAT.	INST.	TOTAL	MAT.	INST.	TOTAL	MAT.	INST.	TOTAL	MAT.	INST.	TOTAL	MAT.	INST.	TOTAL	MAT.	INST.	TOTAL
015433	CONTRACTOR EQUIPMENT		108.2	108.2		110.2	110.2		108.1	108.1		108.0	108.0		104.1	104.1		109.6	109.6
0241, 31 - 34	SITE & INFRASTRUCTURE, DEMOLITION	107.6	88.4	94.3	106.9	93.8	97.9	111.1	88.4	95.4	106.2	88.2	93.8	105.4	85.7	91.8	103.4	92.8	96.1
0310	Concrete Forming & Accessories	98.1	61.1	66.6	103.6	61.2	67.5	91.8	61.8	66.2	87.8	61.3	65.2	89.6	40.4	47.7	94.2	61.9	66.7
0320	Concrete Reinforcing	92.8	66.2	79.9	101.2	66.2	84.3	92.4	73.6	83.3	92.4	66.2	79.7	96.4	83.8	90.3	102.1	73.6	88.3
0330	Cast-in-Place Concrete	103.8	64.5	89.2	111.0	76.1	98.0	110.2	76.6	97.7	102.8	65.6	89.0	104.8	82.4	96.5	95.3	76.6	88.4
03	CONCRETE	98.9	65.0	84.0	102.1	69.0	87.6	104.5	70.8	89.7	97.7	65.5	83.6	99.3	64.7	84.1	94.5	70.7	84.0
04	MASONRY	97.3	55.0	71.3	97.1	55.0	71.3	110.2	64.6	82.2	103.0	55.6	73.9	87.9	54.3	67.3	93.3	64.7	75.7
05	METALS	103.8	93.1	100.6	105.3	91.8	101.3	101.4	96.7	100.0	102.7	93.1	99.9	103.5	96.0	101.3	105.9	95.3	102.8
06	WOOD, PLASTICS & COMPOSITES	91.8	59.8	74.2	99.3	59.7	77.5	83.2	59.5	70.2	79.6	59.8	68.7	83.1	32.0	55.0	93.7	59.7	75.0
07	THERMAL & MOISTURE PROTECTION	105.1	67.7	88.9	104.4	70.0	89.5	105.1	72.3	90.9	105.1	68.0	89.0	103.9	66.6	87.6	101.0	72.9	88.8
08	OPENINGS	94.6	60.1	86.6	95.2	60.1	87.0	94.0	65.6	87.4	94.7	56.0	85.7	96.4	42.9	83.9	101.7	65.7	93.3
0920	Plaster & Gypsum Board	104.3	57.9	73.1	98.5	57.9	71.2	97.8	57.7	70.8	98.6	57.9	71.2	91.1	30.0	50.0	99.5	57.9	71.5
0950, 0980	Ceilings & Acoustic Treatment	91.4	57.9	68.8	88.8	57.9	68.0	88.2	57.7	67.6	91.4	57.9	68.8	90.0	30.0	49.6	89.5	57.9	68.2
0960	Flooring	95.6	51.1	83.1	96.2	51.1	83.5	91.7	71.7	86.1	88.9	51.7	78.4	93.5	55.7	82.9	94.0	71.7	87.8
0970, 0990	Wall Finishes & Painting/Coating	99.6	59.7	76.0	101.2	59.7	76.6	99.6	58.6	75.3	99.6	59.7	76.0	102.1	48.3	70.3	95.4	58.6	73.6
09	FINISHES	91.5	58.4	73.6	90.5	58.3	73.1	89.3	62.5	74.8	88.8	58.6	72.5	90.7	40.9	63.8	91.9	62.5	76.0
COVERS	DIVS. 10 - 14, 25, 28, 41, 43, 44, 46	100.0	83.4	96.3	100.0	83.2	96.3	100.0	83.1	96.2	100.0	83.6	96.3	100.0	73.2	94.0	100.0	83.0	96.2
21, 22, 23	FIRE SUPPRESSION, PLUMBING & HVAC	100.4	66.8	87.2	100.1	66.8	87.0	96.9	69.1	86.0	100.4	67.1	87.4	96.8	66.0	84.7	100.0	69.1	87.9
26, 27, 3370	ELECTRICAL, COMMUNICATIONS & UTIL.	93.6	64.2	79.1	98.0	60.7	79.6	93.8	70.5	82.3	92.1	60.7	76.7	96.2	83.4	89.9	96.8	70.4	83.8
MF2018	WEIGHTED AVERAGE	98.8	68.3	85.9	99.9	68.7	86.7	98.7	72.6	87.7	98.3	67.8	85.4	97.8	67.1	84.8	99.2	72.8	88.1

	DIVISION	VIRGINIA									WASHINGTON								
		ROANOKE			STAUNTON			WINCHESTER			CLARKSTON			EVERETT			OLYMPIA		
		240 - 241			244			226			994			982			985		
		MAT.	INST.	TOTAL	MAT.	INST.	TOTAL	MAT.	INST.	TOTAL	MAT.	INST.	TOTAL	MAT.	INST.	TOTAL	MAT.	INST.	TOTAL
015433	CONTRACTOR EQUIPMENT		104.1	104.1		108.1	108.1		104.1	104.1		90.2	90.2		99.3	99.3		101.6	101.6
0241, 31 - 34	SITE & INFRASTRUCTURE, DEMOLITION	106.0	86.8	92.7	109.3	87.3	94.1	118.1	86.4	96.2	98.1	86.1	89.8	92.1	105.6	101.4	92.8	108.2	103.5
0310	Concrete Forming & Accessories	95.7	71.2	74.8	89.3	61.4	65.5	83.6	66.4	68.9	107.0	64.2	70.5	111.6	104.6	105.7	101.6	104.3	103.9
0320	Concrete Reinforcing	97.5	84.6	91.2	97.1	83.8	90.7	97.1	80.2	88.9	108.4	95.9	102.3	109.2	114.5	111.8	119.5	114.4	117.0
0330	Cast-in-Place Concrete	119.0	86.2	106.8	109.0	85.5	100.2	105.2	63.9	89.8	82.3	82.9	82.5	102.6	109.4	105.1	95.3	109.8	100.7
03	CONCRETE	103.5	80.2	93.3	100.7	75.2	89.5	101.5	69.5	87.4	87.6	76.6	82.8	95.1	107.4	100.5	93.7	107.2	99.7
04	MASONRY	96.3	65.9	77.6	105.0	55.3	74.5	98.7	69.8	80.9	97.8	91.7	94.0	121.0	101.8	109.2	115.3	101.7	107.0
05	METALS	105.8	100.2	104.2	103.6	96.0	101.4	103.8	97.9	102.1	92.8	87.3	91.2	116.8	98.4	111.4	117.8	96.1	111.4
06	WOOD, PLASTICS & COMPOSITES	91.9	71.5	80.7	83.1	59.8	70.3	78.7	64.8	71.1	104.4	58.5	79.2	107.5	104.7	105.9	97.5	104.8	101.5
07	THERMAL & MOISTURE PROTECTION	103.6	76.0	91.6	103.4	67.5	87.8	104.2	75.9	91.9	159.7	82.9	126.3	112.4	105.9	109.6	108.3	103.0	106.0
08	OPENINGS	95.4	68.8	89.2	95.0	58.5	86.5	96.5	68.3	89.9	116.4	65.0	104.4	106.0	107.3	106.3	109.6	104.8	108.5
0920	Plaster & Gypsum Board	98.0	70.7	79.6	91.1	57.9	68.8	91.3	63.8	72.8	154.4	57.3	89.1	112.7	104.9	107.5	106.5	104.9	105.4
0950, 0980	Ceilings & Acoustic Treatment	92.4	70.7	77.8	90.0	57.9	68.4	90.8	63.8	72.6	103.8	57.3	72.4	101.2	104.9	103.7	98.7	104.9	102.9
0960	Flooring	97.1	67.3	88.7	93.1	33.1	76.2	92.0	74.4	87.1	85.4	77.5	83.2	109.9	96.0	106.0	97.6	96.0	97.1
0970, 0990	Wall Finishes & Painting/Coating	102.1	55.3	74.4	102.1	30.7	59.8	115.8	77.6	93.2	84.4	72.1	77.2	97.6	95.2	96.2	99.6	93.1	95.8
09	FINISHES	92.7	68.5	79.6	90.6	52.6	70.0	91.8	68.3	79.1	108.2	65.8	85.2	105.4	101.6	103.4	94.9	101.5	98.5
COVERS	DIVS. 10 - 14, 25, 28, 41, 43, 44, 46	100.0	79.2	95.4	100.0	78.2	95.1	100.0	82.7	96.1	100.0	89.3	97.6	100.0	103.7	100.8	100.0	99.9	100.0
21, 22, 23	FIRE SUPPRESSION, PLUMBING & HVAC	100.4	67.9	87.6	96.8	59.3	82.1	96.8	81.4	90.8	96.7	80.6	90.4	100.0	105.2	102.0	99.9	100.6	100.2
26, 27, 3370	ELECTRICAL, COMMUNICATIONS & UTIL.	96.2	60.8	78.8	95.2	68.4	82.0	94.9	91.8	93.4	89.4	95.6	92.5	107.0	104.1	105.6	105.4	97.7	101.6
MF2018	WEIGHTED AVERAGE	100.1	73.6	88.9	98.6	67.9	85.6	98.9	79.5	90.7	99.1	81.9	91.8	105.1	104.0	104.6	104.0	101.8	103.1

	DIVISION	WASHINGTON																	
		RICHLAND			SEATTLE			SPOKANE			TACOMA			VANCOUVER			WENATCHEE		
		993			980 - 981, 987			990 - 992			983 - 984			986			988		
		MAT.	INST.	TOTAL	MAT.	INST.	TOTAL	MAT.	INST.	TOTAL	MAT.	INST.	TOTAL	MAT.	INST.	TOTAL	MAT.	INST.	TOTAL
015433	CONTRACTOR EQUIPMENT		90.2	90.2		102.4	102.4		90.2	90.2		99.3	99.3		95.8	95.8		99.3	99.3
0241, 31 - 34	SITE & INFRASTRUCTURE, DEMOLITION	100.3	86.3	90.6	97.8	108.9	105.5	99.7	86.3	90.4	95.2	105.1	102.1	105.2	93.1	96.8	104.0	102.8	103.2
0310	Concrete Forming & Accessories	107.1	80.0	84.0	107.4	109.5	109.2	112.0	79.5	84.3	102.3	104.1	103.9	102.7	95.7	96.7	104.0	76.5	80.5
0320	Concrete Reinforcing	103.9	95.4	99.8	109.2	117.3	113.1	104.6	95.3	100.1	107.8	114.3	110.9	108.8	114.8	111.7	108.8	95.9	102.5
0330	Cast-in-Place Concrete	82.5	84.4	83.2	105.9	113.9	108.9	85.7	84.2	85.2	105.5	108.8	106.7	117.8	99.9	111.2	107.7	90.8	101.4
03	CONCRETE	87.2	84.2	85.9	102.8	111.8	106.7	89.2	83.9	86.8	97.0	106.9	101.4	107.2	100.2	104.1	105.3	84.9	96.3
04	MASONRY	98.8	83.7	89.5	123.5	107.2	113.5	99.4	83.7	89.7	117.6	100.2	106.9	118.8	105.5	110.6	120.9	94.1	104.4
05	METALS	93.2	87.1	91.4	116.1	101.9	111.9	95.4	86.6	92.8	118.7	96.5	112.2	116.0	99.3	111.1	116.0	87.3	107.6
06	WOOD, PLASTICS & COMPOSITES	104.6	77.9	89.9	103.4	108.7	106.3	113.5	77.9	93.9	97.1	104.7	101.2	89.5	94.9	92.5	98.7	73.3	84.8
07	THERMAL & MOISTURE PROTECTION	161.1	84.4	127.7	111.6	110.8	111.3	157.4	84.7	125.8	112.1	102.1	107.8	111.8	101.6	107.4	111.6	88.3	101.5
08	OPENINGS	114.4	74.7	105.2	106.1	110.7	107.2	115.0	74.7	105.6	106.7	104.7	106.2	102.8	101.5	102.5	106.2	73.0	98.5
0920	Plaster & Gypsum Board	154.4	77.2	102.5	111.2	109.1	109.8	144.2	77.2	99.1	110.6	104.9	106.8	109.0	95.2	99.7	113.9	72.7	86.2
0950, 0980	Ceilings & Acoustic Treatment	109.2	77.2	87.6	106.6	109.1	108.3	105.6	77.2	86.4	104.5	104.9	104.8	102.4	95.2	97.5	97.8	72.7	80.9
0960	Flooring	85.7	77.5	83.4	109.5	103.4	107.8	84.8	77.5	82.7	102.2	96.0	100.5	107.4	107.1	107.3	105.3	77.5	97.5
0970, 0990	Wall Finishes & Painting/Coating	84.4	75.1	78.9	107.1	95.8	100.4	84.6	76.4	79.8	97.6	93.1	95.0	100.3	79.9	88.2	97.6	70.9	81.8
09	FINISHES	109.5	78.1	92.5	108.9	106.4	107.6	107.3	78.2	91.6	103.6	101.4	102.4	101.3	95.6	98.3	104.7	74.9	88.5
COVERS	DIVS. 10 - 14, 25, 28, 41, 43, 44, 46	100.0	92.2	98.3	100.0	105.2	101.2	100.0	92.2	98.3	100.0	99.6	99.9	100.0	101.5	100.3	100.0	91.2	98.0
21, 22, 23	FIRE SUPPRESSION, PLUMBING & HVAC	100.3	108.6	103.6	99.9	123.0	109.0	100.3	83.7	93.8	100.0	100.6	100.2	100.0	110.7	104.2	96.5	92.0	94.7
26, 27, 3370	ELECTRICAL, COMMUNICATIONS & UTIL.	87.2	92.8	90.0	106.2	118.9	112.5	85.4	77.1	81.3	106.8	97.7	102.3	113.1	107.3	110.2	107.7	94.5	101.2
MF2018	WEIGHTED AVERAGE	99.8	90.0	95.6	106.3	112.5	108.9	100.1	82.4	92.6	105.4	101.3	103.7	106.6	102.8	105.0	105.6	88.7	98.5

	DIVISION	WASHINGTON			WEST VIRGINIA														
		YAKIMA			BECKLEY			BLUEFIELD			BUCKHANNON			CHARLESTON			CLARKSBURG		
		989			258 - 259			247 - 248			262			250 - 253			263 - 264		
		MAT.	INST.	TOTAL	MAT.	INST.	TOTAL	MAT.	INST.	TOTAL	MAT.	INST.	TOTAL	MAT.	INST.	TOTAL	MAT.	INST.	TOTAL
015433	CONTRACTOR EQUIPMENT		99.3	99.3		104.1	104.1		104.1	104.1		104.1	104.1		106.8	106.8		104.1	104.1
0241, 31 - 34	SITE & INFRASTRUCTURE, DEMOLITION	97.6	103.8	101.9	100.8	87.4	91.6	99.9	87.4	91.3	106.0	87.4	93.1	100.4	93.0	95.3	106.8	87.4	93.4
0310	Concrete Forming & Accessories	102.9	99.2	99.8	85.0	87.1	86.8	86.6	87.1	87.0	86.1	83.7	84.0	96.0	88.6	89.7	83.6	83.8	83.7
0320	Concrete Reinforcing	108.3	95.4	102.1	98.0	87.7	93.1	96.0	82.5	89.5	96.7	82.4	89.8	106.1	87.8	97.2	96.7	95.3	96.0
0330	Cast-in-Place Concrete	112.7	83.0	101.7	99.2	92.5	96.7	102.5	92.5	98.8	102.1	92.0	98.4	95.9	96.6	96.2	111.9	87.6	102.9
03	CONCRETE	101.8	92.4	97.7	91.8	90.2	91.1	95.1	89.3	92.6	98.2	87.5	93.5	90.7	92.2	91.4	102.2	88.2	96.1
04	MASONRY	110.3	81.0	92.3	91.0	89.0	89.8	89.5	89.0	89.2	100.6	85.8	91.5	88.4	90.7	89.8	104.1	85.8	92.9
05	METALS	116.7	87.3	108.0	98.2	103.3	99.7	103.6	101.5	103.0	103.8	101.6	103.2	96.8	102.5	98.5	103.8	106.0	104.5
06	WOOD, PLASTICS & COMPOSITES	97.4	104.7	101.4	80.2	87.4	84.2	82.1	87.4	85.0	81.4	83.0	82.3	89.3	88.2	88.7	78.0	83.0	80.8
07	THERMAL & MOISTURE PROTECTION	112.3	86.1	100.9	105.0	86.5	96.9	103.6	86.5	96.1	104.0	86.6	96.4	101.2	88.1	95.5	103.8	86.3	96.2
08	OPENINGS	106.2	100.5	104.9	94.5	83.0	91.8	96.9	81.8	93.4	96.9	79.4	92.8	94.9	83.4	92.2	96.9	82.1	93.4
0920	Plaster & Gypsum Board	110.4	104.9	106.7	93.5	87.0	89.1	90.2	87.0	88.0	90.8	82.5	85.2	98.7	87.6	91.2	88.6	82.5	84.5
0950, 0980	Ceilings & Acoustic Treatment	99.6	104.9	103.2	78.1	87.0	84.1	87.5	87.0	87.2	90.0	82.5	84.9	89.6	87.6	88.2	90.0	82.5	84.9
0960	Flooring	103.2	77.5	96.0	91.0	98.9	93.2	90.1	98.9	92.6	89.8	95.2	91.3	97.1	98.9	97.6	88.8	95.2	90.6
0970, 0990	Wall Finishes & Painting/Coating	97.6	76.4	85.1	89.3	92.1	91.0	102.1	89.8	94.8	102.1	88.0	93.7	88.6	91.9	90.5	102.1	88.0	93.7
09	FINISHES	103.1	93.3	97.8	86.3	90.2	88.4	88.7	89.9	89.4	89.7	86.1	87.8	92.8	91.0	91.9	89.0	86.1	87.4
COVERS	DIVS. 10 - 14, 25, 28, 41, 43, 44, 46	100.0	96.1	99.1	100.0	90.3	97.8	100.0	90.3	97.8	100.0	90.6	97.9	100.0	92.5	98.3	100.0	90.6	97.9
21, 22, 23	FIRE SUPPRESSION, PLUMBING & HVAC	100.0	107.4	102.9	97.1	80.3	90.5	96.8	84.0	91.8	96.8	89.7	94.0	100.1	81.4	92.8	96.8	89.8	94.1
26, 27, 3370	ELECTRICAL, COMMUNICATIONS & UTIL.	109.9	92.9	101.5	92.9	84.1	88.6	95.2	84.1	89.8	96.6	89.3	93.0	98.0	87.1	92.6	96.6	89.3	93.0
MF2018	WEIGHTED AVERAGE	105.6	95.4	101.3	95.2	87.7	92.0	97.0	88.1	93.2	98.3	88.8	94.3	96.5	89.4	93.5	98.9	89.5	94.9

	DIVISION	WEST VIRGINIA																	
		GASSAWAY			HUNTINGTON			LEWISBURG			MARTINSBURG			MORGANTOWN			PARKERSBURG		
		266			255 - 257			249			254			265			261		
		MAT.	INST.	TOTAL	MAT.	INST.	TOTAL	MAT.	INST.	TOTAL	MAT.	INST.	TOTAL	MAT.	INST.	TOTAL	MAT.	INST.	TOTAL
015433	CONTRACTOR EQUIPMENT		104.1	104.1		104.1	104.1		104.1	104.1		104.1	104.1		104.1	104.1		104.1	104.1
0241, 31 - 34	SITE & INFRASTRUCTURE, DEMOLITION	103.5	87.4	92.4	105.4	88.5	93.7	115.5	87.4	96.1	104.7	88.0	93.2	100.9	88.2	92.1	109.3	88.4	94.9
0310	Concrete Forming & Accessories	85.5	86.9	86.7	96.0	89.3	90.3	83.9	86.6	86.2	85.0	79.5	80.3	83.9	83.8	83.9	88.0	86.0	86.3
0320	Concrete Reinforcing	96.7	91.6	94.2	99.5	93.1	96.4	96.7	82.4	89.8	98.0	95.1	96.6	96.7	95.3	96.0	96.0	82.3	89.4
0330	Cast-in-Place Concrete	106.9	94.0	102.1	108.2	94.0	102.9	102.5	92.3	98.7	104.0	88.3	98.1	102.1	92.0	98.4	104.4	89.0	98.7
03	CONCRETE	98.6	91.3	95.4	96.8	92.6	95.0	105.1	89.0	98.0	95.3	86.5	91.4	94.9	89.8	92.7	100.3	87.5	94.7
04	MASONRY	105.0	87.9	94.5	91.7	93.4	92.7	94.3	89.0	91.1	93.1	83.9	87.5	122.6	85.8	100.0	80.1	85.2	83.2
05	METALS	103.8	104.8	104.1	100.6	105.7	102.1	103.7	101.0	102.9	98.6	104.4	100.3	103.9	106.0	104.5	104.5	101.4	103.6
06	WOOD, PLASTICS & COMPOSITES	80.4	87.4	84.3	90.8	88.7	89.6	78.3	87.4	83.3	80.2	78.6	79.3	78.3	83.0	80.9	81.5	85.6	83.8
07	THERMAL & MOISTURE PROTECTION	103.6	87.9	96.8	105.2	88.4	97.9	104.6	86.4	96.7	105.3	80.8	94.6	103.7	86.6	96.2	103.9	86.7	96.4
08	OPENINGS	95.2	83.9	92.6	93.7	84.8	91.7	96.9	81.8	93.4	96.2	73.8	91.0	98.0	82.1	94.3	95.8	80.4	92.2
0920	Plaster & Gypsum Board	89.9	87.0	87.9	99.5	88.3	91.9	88.6	87.0	87.5	93.5	77.9	83.0	88.6	82.5	84.5	91.1	85.1	87.1
0950, 0980	Ceilings & Acoustic Treatment	90.0	87.0	88.0	78.1	88.3	85.0	90.0	87.0	88.0	78.1	77.9	78.0	90.0	82.5	84.9	90.0	85.1	86.7
0960	Flooring	89.6	98.9	92.2	98.4	103.3	99.7	88.9	98.9	91.7	91.0	98.9	93.2	88.9	95.2	90.7	92.6	91.9	92.4
0970, 0990	Wall Finishes & Painting/Coating	102.1	91.9	96.0	89.3	89.8	89.6	102.1	64.5	79.8	89.3	83.7	86.0	102.1	88.0	93.7	102.1	83.7	91.2
09	FINISHES	89.2	89.9	89.6	89.3	92.0	90.8	90.1	87.2	88.5	86.6	83.0	84.6	88.6	86.1	87.3	90.6	86.8	88.6
COVERS	DIVS. 10 - 14, 25, 28, 41, 43, 44, 46	100.0	91.1	98.0	100.0	92.3	98.3	100.0	65.4	92.3	100.0	81.4	95.8	100.0	83.5	96.3	100.0	91.3	98.1
21, 22, 23	FIRE SUPPRESSION, PLUMBING & HVAC	96.8	83.4	91.6	100.6	90.6	96.7	96.8	80.2	90.3	97.1	80.9	90.8	96.8	89.8	94.1	100.3	88.3	95.6
26, 27, 3370	ELECTRICAL, COMMUNICATIONS & UTIL.	96.6	87.1	91.9	96.6	89.6	93.1	92.8	84.1	88.5	98.6	76.9	87.9	96.7	89.3	93.1	96.7	88.3	92.5
MF2018	WEIGHTED AVERAGE	98.3	89.0	94.3	97.8	92.2	95.4	98.7	86.1	93.4	96.7	84.1	91.4	98.8	89.6	94.9	98.6	88.6	94.4

	DIVISION	WEST VIRGINIA									WISCONSIN								
		PETERSBURG			ROMNEY			WHEELING			BELOIT			EAU CLAIRE			GREEN BAY		
		268			267			260			535			547			541 - 543		
		MAT.	INST.	TOTAL	MAT.	INST.	TOTAL	MAT.	INST.	TOTAL	MAT.	INST.	TOTAL	MAT.	INST.	TOTAL	MAT.	INST.	TOTAL
015433	CONTRACTOR EQUIPMENT		104.1	104.1		104.1	104.1		104.1	104.1		99.0	99.0		99.9	99.9		97.7	97.7
0241, 31 - 34	SITE & INFRASTRUCTURE, DEMOLITION	100.2	88.3	92.0	103.0	88.3	92.9	109.9	88.2	94.9	97.1	101.7	100.3	96.0	99.0	98.1	99.5	95.3	96.6
0310	Concrete Forming & Accessories	87.0	86.4	86.5	83.3	86.8	86.3	89.7	83.8	84.7	99.8	96.2	96.7	98.5	94.7	95.3	108.2	97.5	99.1
0320	Concrete Reinforcing	96.0	91.4	93.8	96.7	95.4	96.1	95.5	95.3	95.4	95.0	137.1	115.4	94.2	114.2	103.9	92.3	109.4	100.5
0330	Cast-in-Place Concrete	102.1	91.8	98.3	106.9	83.5	98.2	104.4	92.0	99.8	105.8	100.2	103.7	99.9	94.8	98.0	103.3	99.9	102.1
03	CONCRETE	95.0	90.2	92.9	98.4	88.2	93.9	100.3	89.8	95.7	96.7	104.7	100.2	93.1	98.4	95.4	96.2	100.6	98.2
04	MASONRY	96.1	85.8	89.8	94.7	87.9	90.5	103.4	85.7	92.5	99.3	100.3	99.9	91.4	96.1	94.3	122.7	97.8	107.4
05	METALS	103.9	104.1	104.0	104.0	105.9	104.5	104.7	106.1	105.1	98.6	112.0	102.5	93.8	105.8	97.3	96.4	104.2	98.7
06	WOOD, PLASTICS & COMPOSITES	82.3	87.4	85.1	77.5	87.4	82.9	83.1	83.0	83.1	96.9	94.1	95.4	100.6	94.2	97.1	107.8	97.2	101.9
07	THERMAL & MOISTURE PROTECTION	103.7	82.5	94.5	103.8	83.8	95.1	104.1	86.7	96.6	105.4	93.4	100.2	104.0	94.6	99.9	106.1	97.4	102.3
08	OPENINGS	98.1	83.6	94.7	98.0	84.5	94.8	96.6	82.1	93.2	98.1	108.3	100.5	103.1	99.3	102.2	99.1	101.2	99.6
0920	Plaster & Gypsum Board	90.8	87.0	88.2	88.3	87.0	87.4	91.1	82.5	85.3	91.6	94.5	93.5	105.5	94.5	98.1	103.3	97.5	99.4
0950, 0980	Ceilings & Acoustic Treatment	90.0	87.0	88.0	90.0	87.0	88.0	90.0	82.5	84.9	84.4	94.5	91.2	90.9	94.5	93.3	83.6	97.5	93.0
0960	Flooring	90.6	95.2	91.9	88.7	98.9	91.6	93.5	95.2	94.0	94.8	118.5	101.4	80.5	106.5	87.8	96.0	114.8	101.3
0970, 0990	Wall Finishes & Painting/Coating	102.1	88.0	93.7	102.1	88.0	93.7	102.1	88.0	93.7	95.0	103.6	100.1	83.3	76.4	79.2	92.0	80.0	84.9
09	FINISHES	89.4	88.7	89.0	88.7	89.5	89.1	90.9	86.1	88.3	92.3	100.7	96.9	87.3	94.9	91.4	91.4	99.1	95.6
COVERS	DIVS. 10 - 14, 25, 28, 41, 43, 44, 46	100.0	55.0	90.0	100.0	91.0	98.0	100.0	84.4	96.5	100.0	98.6	99.7	100.0	94.1	98.7	100.0	95.7	99.0
21, 22, 23	FIRE SUPPRESSION, PLUMBING & HVAC	96.8	83.7	91.7	96.8	83.6	91.6	100.4	89.8	96.2	99.8	95.5	98.1	100.0	88.2	95.4	100.2	83.4	93.6
26, 27, 3370	ELECTRICAL, COMMUNICATIONS & UTIL.	99.7	76.9	88.4	99.0	76.9	88.1	94.1	89.3	91.7	100.6	84.3	92.5	104.3	84.4	94.5	99.0	81.2	90.2
MF2018	WEIGHTED AVERAGE	98.0	85.9	92.9	98.3	87.2	93.6	99.6	89.6	95.4	98.6	98.9	98.7	97.5	94.1	96.1	99.4	93.4	96.9

	DIVISION	WISCONSIN																	
		KENOSHA			LA CROSSE			LANCASTER			MADISON			MILWAUKEE			NEW RICHMOND		
		531			546			538			537			530, 532			540		
		MAT.	INST.	TOTAL	MAT.	INST.	TOTAL	MAT.	INST.	TOTAL	MAT.	INST.	TOTAL	MAT.	INST.	TOTAL	MAT.	INST.	TOTAL
015433	CONTRACTOR EQUIPMENT		97.1	97.1		99.9	99.9		99.0	99.0		101.4	101.4		90.2	90.2		100.2	100.2
0241, 31 - 34	SITE & INFRASTRUCTURE, DEMOLITION	103.0	98.4	99.9	89.8	99.0	96.1	96.0	101.1	99.5	94.7	105.3	102.0	93.6	96.0	95.2	93.3	99.6	97.7
0310	Concrete Forming & Accessories	108.7	106.1	106.5	85.2	94.5	93.2	99.1	95.2	95.7	103.8	95.8	97.0	102.9	113.9	112.3	93.6	92.5	92.6
0320	Concrete Reinforcing	94.8	110.5	102.4	93.9	106.2	99.8	96.3	106.2	101.1	97.8	106.4	101.9	99.9	115.1	107.2	91.3	114.1	102.3
0330	Cast-in-Place Concrete	115.2	102.7	110.5	89.8	96.5	92.3	105.2	99.7	103.2	101.4	100.3	101.0	93.9	112.1	100.7	104.0	102.7	103.5
03	CONCRETE	101.5	105.5	103.3	84.6	97.5	90.3	96.4	98.8	97.4	96.3	99.2	97.6	94.5	112.5	102.4	90.8	100.1	94.9
04	MASONRY	97.1	108.0	103.8	90.5	96.1	93.9	99.4	100.1	99.9	97.4	99.1	98.5	102.3	117.2	111.4	117.9	98.4	105.9
05	METALS	99.6	103.3	100.7	93.7	102.8	96.4	96.0	100.3	97.3	102.8	100.4	102.1	97.5	97.5	97.5	94.1	105.4	97.4
06	WOOD, PLASTICS & COMPOSITES	103.9	105.7	104.9	85.2	94.2	90.2	96.1	94.1	95.0	99.4	94.3	96.6	101.2	112.3	107.3	89.5	90.2	89.9
07	THERMAL & MOISTURE PROTECTION	105.5	104.0	104.9	103.4	93.5	99.1	105.2	93.7	100.2	106.3	100.5	103.7	104.8	112.7	108.2	104.9	100.2	102.9
08	OPENINGS	92.5	105.4	95.5	103.1	92.0	100.5	94.0	91.9	93.5	101.6	98.9	100.9	103.0	111.5	105.0	88.6	90.3	89.0
0920	Plaster & Gypsum Board	80.2	106.4	97.8	100.9	94.5	96.6	90.9	94.5	93.3	101.7	94.5	96.8	96.4	112.6	107.3	91.7	90.5	90.9
0950, 0980	Ceilings & Acoustic Treatment	84.4	106.4	99.2	90.0	94.5	93.0	81.1	94.5	90.1	86.5	94.5	91.9	93.1	112.6	106.3	57.1	90.5	79.6
0960	Flooring	112.8	114.8	113.4	74.7	111.6	85.1	94.4	111.0	99.0	94.3	111.0	99.0	101.7	118.6	106.4	89.7	114.5	96.7
0970, 0990	Wall Finishes & Painting/Coating	106.4	118.7	113.7	83.3	81.7	82.4	95.0	97.7	96.6	94.9	103.6	100.0	104.5	124.3	116.3	95.0	81.7	87.1
09	FINISHES	97.1	109.4	103.7	84.4	96.5	90.9	91.4	98.4	95.2	93.7	99.3	96.7	99.5	116.1	108.5	82.5	95.2	89.4
COVERS	DIVS. 10 - 14, 25, 28, 41, 43, 44, 46	100.0	97.3	99.4	100.0	92.8	98.4	100.0	84.2	96.5	100.0	96.8	99.3	100.0	105.2	101.2	100.0	95.1	98.9
21, 22, 23	FIRE SUPPRESSION, PLUMBING & HVAC	100.0	95.9	98.4	100.0	88.1	95.3	96.3	89.2	93.5	99.8	95.3	98.0	99.8	107.0	102.6	96.0	89.0	93.2
26, 27, 3370	ELECTRICAL, COMMUNICATIONS & UTIL.	100.9	97.1	99.0	104.6	84.4	94.7	100.3	84.9	92.7	101.5	94.0	97.8	100.2	102.4	101.3	102.4	84.9	93.7
MF2018	WEIGHTED AVERAGE	99.4	102.0	100.5	95.9	93.5	94.9	96.7	94.3	95.7	99.7	98.1	99.0	99.2	107.9	102.9	95.4	94.6	95.0

	DIVISION	WISCONSIN																	
		OSHKOSH			PORTAGE			RACINE			RHINELANDER			SUPERIOR			WAUSAU		
		549			539			534			545			548			544		
		MAT.	INST.	TOTAL	MAT.	INST.	TOTAL	MAT.	INST.	TOTAL	MAT.	INST.	TOTAL	MAT.	INST.	TOTAL	MAT.	INST.	TOTAL
015433	CONTRACTOR EQUIPMENT		97.7	97.7		99.0	99.0		99.0	99.0		97.7	97.7		100.2	100.2		97.7	97.7
0241, 31 - 34	SITE & INFRASTRUCTURE, DEMOLITION	91.0	95.3	94.0	86.9	100.9	96.6	96.9	102.1	100.5	103.2	95.3	97.7	90.1	99.3	96.5	86.9	95.8	93.0
0310	Concrete Forming & Accessories	90.7	94.7	94.1	91.3	95.7	95.1	100.1	106.1	105.2	88.3	94.6	93.7	91.6	88.6	89.0	90.1	97.4	96.3
0320	Concrete Reinforcing	92.4	109.3	100.6	96.4	106.4	101.2	95.0	110.5	102.5	92.6	106.4	99.3	91.3	110.2	100.4	92.6	106.2	99.2
0330	Cast-in-Place Concrete	95.8	98.0	96.6	90.3	98.3	93.3	103.8	102.5	103.3	108.6	97.9	104.6	98.0	98.3	98.1	89.3	92.4	90.5
03	CONCRETE	86.2	98.6	91.7	84.2	98.7	90.6	95.8	105.4	100.0	98.0	98.1	98.1	85.5	96.2	90.2	81.4	97.4	88.4
04	MASONRY	104.7	97.8	100.4	98.2	100.2	99.4	99.3	108.0	104.6	121.3	97.6	106.8	117.1	100.4	106.8	104.2	97.6	100.1
05	METALS	94.3	103.5	97.0	96.7	104.3	98.9	100.2	103.3	101.1	94.2	102.5	96.6	95.1	104.6	97.9	94.0	102.9	96.6
06	WOOD, PLASTICS & COMPOSITES	86.7	94.2	90.9	85.8	94.1	90.4	97.2	105.7	101.9	84.2	94.2	89.7	87.9	85.8	86.7	86.2	97.2	92.2
07	THERMAL & MOISTURE PROTECTION	105.0	83.3	95.6	104.5	100.1	102.6	105.5	102.7	104.3	106.0	83.1	96.0	104.5	95.9	100.8	104.8	95.1	100.6
08	OPENINGS	95.6	96.2	95.7	94.1	99.7	95.4	98.1	105.4	99.8	95.6	91.9	94.7	88.0	91.9	88.9	95.8	97.1	96.1
0920	Plaster & Gypsum Board	90.7	94.5	93.2	83.8	94.5	91.0	91.6	106.4	101.5	90.7	94.5	93.2	92.1	86.0	88.0	90.7	97.5	95.3
0950, 0980	Ceilings & Acoustic Treatment	83.6	94.5	90.9	83.6	94.5	90.9	84.4	106.4	99.2	83.6	94.5	90.9	58.8	86.0	77.1	83.6	97.5	93.0
0960	Flooring	87.8	114.8	95.4	90.5	114.5	97.3	94.8	114.8	100.4	87.2	114.5	94.9	90.8	123.1	99.9	87.7	114.5	95.2
0970, 0990	Wall Finishes & Painting/Coating	89.6	104.3	98.3	95.0	97.7	96.6	95.0	115.9	107.4	89.6	80.0	83.9	84.0	105.8	96.9	89.6	83.2	85.8
09	FINISHES	86.5	100.0	93.8	89.2	99.4	94.7	92.3	109.1	101.4	87.4	97.3	92.7	82.1	96.4	89.8	86.2	99.4	93.3
COVERS	DIVS. 10 - 14, 25, 28, 41, 43, 44, 46	100.0	85.6	96.8	100.0	85.6	96.8	100.0	97.3	99.4	100.0	85.9	96.8	100.0	93.9	98.6	100.0	95.7	99.0
21, 22, 23	FIRE SUPPRESSION, PLUMBING & HVAC	96.7	83.1	91.3	96.3	95.8	96.1	99.8	96.0	98.3	96.7	88.6	93.5	96.0	91.1	94.1	96.7	88.8	93.6
26, 27, 3370	ELECTRICAL, COMMUNICATIONS & UTIL.	103.0	79.1	91.2	104.0	94.0	99.1	100.4	98.0	99.2	102.3	78.1	90.4	107.2	98.5	102.9	104.1	78.1	91.3
MF2018	WEIGHTED AVERAGE	95.4	91.9	93.9	95.2	98.0	96.4	98.7	102.3	100.2	97.9	92.2	95.5	95.1	96.5	95.7	94.7	93.4	94.2

	DIVISION	WYOMING																	
		CASPER			CHEYENNE			NEWCASTLE			RAWLINS			RIVERTON			ROCK SPRINGS		
		826			820			827			823			825			829 - 831		
		MAT.	INST.	TOTAL	MAT.	INST.	TOTAL	MAT.	INST.	TOTAL	MAT.	INST.	TOTAL	MAT.	INST.	TOTAL	MAT.	INST.	TOTAL
015433	CONTRACTOR EQUIPMENT		97.3	97.3		94.6	94.6		94.6	94.6		94.6	94.6		94.6	94.6		94.6	94.6
0241, 31 - 34	SITE & INFRASTRUCTURE, DEMOLITION	97.8	94.7	95.7	91.6	90.1	90.6	83.6	89.3	87.6	97.5	89.3	91.9	91.1	89.3	89.8	87.4	89.5	88.9
0310	Concrete Forming & Accessories	101.3	53.6	60.6	102.5	64.6	70.2	93.7	70.5	74.0	97.6	70.8	74.7	92.7	60.3	65.0	99.7	64.7	69.9
0320	Concrete Reinforcing	111.1	83.4	97.7	106.4	83.4	95.3	114.5	83.2	99.3	114.2	83.2	99.2	115.2	83.2	99.7	115.2	81.6	99.0
0330	Cast-in-Place Concrete	103.8	77.8	94.1	97.9	77.4	90.3	99.0	72.8	89.2	99.0	72.9	89.3	99.0	72.8	89.2	99.0	75.0	90.0
03	CONCRETE	103.3	67.9	87.7	100.9	72.8	88.6	101.1	73.8	89.1	116.0	74.0	97.5	110.1	69.2	92.1	101.6	71.7	88.4
04	MASONRY	102.0	64.6	79.0	105.3	65.2	80.7	102.2	59.3	75.8	102.2	59.3	75.8	102.2	59.3	75.8	164.5	61.8	101.4
05	METALS	102.8	80.1	96.1	105.3	80.8	98.1	101.4	79.9	95.1	101.5	80.1	95.2	101.6	79.9	95.2	102.3	79.5	95.6
06	WOOD, PLASTICS & COMPOSITES	95.2	47.8	69.1	91.8	62.6	75.7	82.4	72.7	77.1	86.2	72.7	78.8	81.4	58.7	68.9	91.3	62.6	75.5
07	THERMAL & MOISTURE PROTECTION	112.7	65.9	92.3	107.2	68.0	90.2	108.8	63.9	89.3	110.3	74.9	94.9	109.7	67.7	91.4	108.9	69.5	91.8
08	OPENINGS	109.3	59.4	97.7	107.2	67.7	98.0	111.4	72.3	102.3	111.1	72.3	102.0	111.3	64.6	100.4	111.8	66.5	101.2
0920	Plaster & Gypsum Board	101.7	46.2	64.4	90.6	61.7	71.2	87.6	72.0	77.1	87.9	72.0	77.2	87.6	57.7	67.5	100.2	61.7	74.3
0950, 0980	Ceilings & Acoustic Treatment	105.3	46.2	65.5	98.5	61.7	73.7	100.3	72.0	81.2	100.3	72.0	81.2	100.3	57.7	71.6	100.3	61.7	74.3
0960	Flooring	102.2	73.8	94.2	100.3	73.8	92.8	94.2	44.5	80.2	97.0	44.5	82.3	93.6	60.3	84.2	99.3	59.9	88.3
0970, 0990	Wall Finishes & Painting/Coating	95.7	58.7	73.8	97.9	58.7	74.7	94.6	60.5	74.4	94.6	60.5	74.4	94.6	60.5	74.4	94.6	60.5	74.4
09	FINISHES	99.6	56.3	76.2	96.3	65.1	79.4	91.4	64.6	76.9	93.6	64.6	77.9	92.0	59.5	74.4	94.6	62.7	77.3
COVERS	DIVS. 10 - 14, 25, 28, 41, 43, 44, 46	100.0	92.5	98.3	100.0	87.4	97.2	100.0	91.8	98.2	100.0	91.8	98.2	100.0	77.3	94.9	100.0	86.9	97.1
21, 22, 23	FIRE SUPPRESSION, PLUMBING & HVAC	100.0	70.5	88.4	100.1	73.3	89.6	98.2	71.8	87.8	98.2	71.9	87.8	98.2	71.8	87.8	100.0	73.9	89.8
26, 27, 3370	ELECTRICAL, COMMUNICATIONS & UTIL.	94.1	60.3	77.5	95.2	69.0	82.3	94.0	62.3	78.4	94.0	62.3	78.4	94.0	59.1	76.8	92.5	66.0	79.4
MF2018	WEIGHTED AVERAGE	101.5	68.9	87.7	101.1	72.7	89.1	99.6	71.1	87.5	102.0	71.5	89.1	101.0	68.6	87.3	103.4	71.4	89.9

	DIVISION	WYOMING												CANADA					
		SHERIDAN			WHEATLAND			WORLAND			YELLOWSTONE NAT'L PA			BARRIE, ONTARIO			BATHURST, NEW BRUNSWICK		
		828			822			824			821								
		MAT.	INST.	TOTAL	MAT.	INST.	TOTAL	MAT.	INST.	TOTAL	MAT.	INST.	TOTAL	MAT.	INST.	TOTAL	MAT.	INST.	TOTAL
015433	CONTRACTOR EQUIPMENT		94.6	94.6		94.6	94.6		94.6	94.6		94.6	94.6		100.1	100.1		99.7	99.7
0241, 31 - 34	SITE & INFRASTRUCTURE, DEMOLITION	91.5	90.1	90.5	88.1	89.3	88.9	85.5	89.3	88.1	85.6	89.3	88.1	117.0	94.7	101.6	105.1	90.6	95.1
0310	Concrete Forming & Accessories	100.5	61.6	67.4	95.4	48.4	55.3	95.5	61.1	66.1	95.5	61.1	66.2	125.6	82.3	88.7	106.1	57.8	64.9
0320	Concrete Reinforcing	115.2	83.2	99.7	114.5	82.6	99.1	115.2	83.2	99.7	117.2	82.3	100.3	175.2	86.9	132.5	140.5	57.5	100.4
0330	Cast-in-Place Concrete	102.3	77.4	93.1	103.3	72.7	91.9	99.0	72.7	89.2	99.0	72.8	89.2	157.7	82.1	129.6	114.0	56.2	92.5
03	CONCRETE	109.9	71.4	93.0	106.3	63.7	87.6	101.3	69.5	87.3	101.6	69.4	87.4	138.9	83.5	114.5	110.3	58.3	87.4
04	MASONRY	102.5	62.0	77.6	102.6	51.2	71.0	102.2	59.3	75.8	102.2	59.3	75.8	165.6	89.3	118.7	159.8	57.4	96.9
05	METALS	105.1	80.5	97.8	101.4	79.1	94.8	101.6	79.8	95.2	102.2	79.1	95.4	111.1	91.5	105.3	114.1	73.6	102.2
06	WOOD, PLASTICS & COMPOSITES	93.3	58.7	74.3	84.2	43.0	61.5	84.3	60.0	70.9	84.3	60.0	70.9	116.5	80.8	96.9	97.5	57.9	75.7
07	THERMAL & MOISTURE PROTECTION	110.1	66.7	91.2	109.1	58.1	86.9	108.9	64.8	89.7	108.3	64.8	89.4	115.3	86.1	102.6	111.8	58.4	88.6
08	OPENINGS	112.0	64.6	101.0	110.0	55.9	97.4	111.6	65.3	100.8	104.5	64.8	95.2	90.5	80.9	88.3	84.0	51.2	76.3
0920	Plaster & Gypsum Board	111.2	57.7	75.2	87.6	41.5	56.6	87.6	58.9	68.3	87.8	58.9	68.4	153.5	80.3	104.3	124.2	56.7	78.8
0950, 0980	Ceilings & Acoustic Treatment	103.9	57.7	72.7	100.3	41.5	60.6	100.3	58.9	72.4	101.1	58.9	72.7	94.3	80.3	84.9	116.8	56.7	76.3
0960	Flooring	97.9	60.3	87.4	95.6	43.8	81.0	95.6	44.5	81.3	95.6	44.5	81.3	119.3	87.3	110.3	99.6	41.5	83.3
0970, 0990	Wall Finishes & Painting/Coating	96.8	58.7	74.2	94.6	58.4	73.2	94.6	60.5	74.4	94.6	60.5	74.4	105.6	83.8	92.7	111.3	47.9	73.8
09	FINISHES	99.2	60.4	78.2	92.1	46.7	67.5	91.8	57.1	73.0	92.0	57.2	73.2	111.3	83.5	96.2	107.7	53.8	78.5
COVERS	DIVS. 10 - 14, 25, 28, 41, 43, 44, 46	100.0	93.4	98.5	100.0	81.4	95.9	100.0	76.1	94.7	100.0	76.2	94.7	139.2	65.0	122.7	131.1	57.9	114.8
21, 22, 23	FIRE SUPPRESSION, PLUMBING & HVAC	98.2	73.3	88.4	98.2	71.8	87.8	98.2	71.8	87.8	98.2	71.8	87.8	103.7	93.5	99.7	103.8	64.8	88.5
26, 27, 3370	ELECTRICAL, COMMUNICATIONS & UTIL.	96.1	59.1	77.9	94.0	76.6	85.4	94.0	76.6	85.4	93.1	86.1	89.7	117.0	82.9	100.2	113.5	55.9	85.1
MF2018	WEIGHTED AVERAGE	102.5	70.2	88.8	100.3	67.1	86.2	99.7	70.7	87.5	99.0	71.9	87.6	116.5	87.1	104.1	110.7	62.3	90.2

	DIVISION	CANADA																	
		BRANDON, MANITOBA			BRANTFORD, ONTARIO			BRIDGEWATER, NOVA SCOTIA			CALGARY, ALBERTA			CAP-DE-LA-MADELEINE, QUEBEC			CHARLESBOURG, QUEBEC		
		MAT.	INST.	TOTAL	MAT.	INST.	TOTAL	MAT.	INST.	TOTAL	MAT.	INST.	TOTAL	MAT.	INST.	TOTAL	MAT.	INST.	TOTAL
015433	CONTRACTOR EQUIPMENT		101.5	101.5		99.7	99.7		99.4	99.4		127.1	127.1		100.2	100.2		100.2	100.2
0241, 31 - 34	SITE & INFRASTRUCTURE, DEMOLITION	126.4	92.9	103.3	116.9	94.8	101.6	101.1	92.3	95.0	125.5	118.4	120.6	96.9	93.7	94.7	96.9	93.7	94.7
0310	Concrete Forming & Accessories	145.4	66.1	77.8	127.2	89.1	94.7	99.7	67.9	72.6	125.0	94.9	99.3	132.9	79.2	87.1	132.9	79.2	87.1
0320	Concrete Reinforcing	170.2	54.5	114.3	163.7	85.6	125.9	139.3	47.6	95.0	136.4	82.0	110.1	139.3	72.4	107.0	139.3	72.4	107.0
0330	Cast-in-Place Concrete	109.4	70.2	94.8	130.7	101.8	119.9	134.3	66.9	109.2	136.4	104.3	124.4	105.5	87.5	98.8	105.5	87.5	98.8
03	CONCRETE	118.1	66.6	95.5	121.8	93.2	109.3	120.1	65.1	95.9	126.3	96.8	113.3	107.5	81.5	96.1	107.5	81.5	96.1
04	MASONRY	212.8	60.7	119.3	166.1	93.4	121.4	161.8	65.6	102.6	212.4	89.8	137.1	162.4	76.3	109.5	162.4	76.3	109.5
05	METALS	127.2	78.2	112.8	112.3	92.0	106.3	111.6	76.0	101.1	131.0	102.4	122.6	110.7	85.0	103.1	110.7	85.0	103.1
06	WOOD, PLASTICS & COMPOSITES	150.5	66.8	104.4	120.3	87.7	102.4	89.6	67.4	77.4	102.6	94.4	98.1	131.6	79.0	102.7	131.6	79.0	102.7
07	THERMAL & MOISTURE PROTECTION	128.4	68.7	102.4	121.6	91.4	108.5	116.1	67.6	95.0	132.0	98.0	117.2	114.8	84.1	101.5	114.8	84.1	101.5
08	OPENINGS	100.3	59.9	90.8	87.7	86.6	87.5	82.4	61.3	77.5	82.3	83.7	82.6	89.1	72.3	85.2	89.1	72.3	85.2
0920	Plaster & Gypsum Board	115.4	65.6	81.9	115.8	87.4	96.7	124.3	66.5	85.4	126.4	93.3	104.1	147.6	78.4	101.0	147.6	78.4	101.0
0950, 0980	Ceilings & Acoustic Treatment	123.8	65.6	84.6	103.5	87.4	92.7	103.5	66.5	78.6	149.2	93.3	111.5	103.5	78.4	86.6	103.5	78.4	86.6
0960	Flooring	131.2	61.9	111.7	113.6	87.3	106.2	95.7	58.9	85.4	118.1	84.8	108.7	113.6	86.0	105.8	113.6	86.0	105.8
0970, 0990	Wall Finishes & Painting/Coating	117.9	54.1	80.1	110.3	92.0	99.5	110.3	59.3	80.1	113.7	106.4	109.4	110.3	83.6	94.5	110.3	83.6	94.5
09	FINISHES	122.9	64.3	91.2	108.7	89.3	98.2	103.8	65.5	83.1	123.6	94.8	108.0	112.2	81.2	95.4	112.2	81.2	95.4
COVERS	DIVS. 10 - 14, 25, 28, 41, 43, 44, 46	131.1	60.1	115.3	131.1	66.9	116.8	131.1	60.8	115.5	131.1	92.7	122.6	131.1	75.6	118.8	131.1	75.6	118.8
21, 22, 23	FIRE SUPPRESSION, PLUMBING & HVAC	104.0	78.9	94.1	103.8	96.3	100.9	103.8	79.5	94.3	105.1	90.2	99.2	104.2	84.2	96.4	104.2	84.2	96.4
26, 27, 3370	ELECTRICAL, COMMUNICATIONS & UTIL.	113.9	63.1	88.8	111.5	82.5	97.2	115.8	58.7	87.6	107.1	93.9	100.6	110.2	65.2	88.0	110.2	65.2	88.0
MF2018	WEIGHTED AVERAGE	120.6	70.5	99.4	113.1	90.7	103.6	111.3	70.3	94.0	119.7	95.6	109.5	110.6	80.0	97.7	110.6	80.0	97.7

CANADA

	DIVISION	CHARLOTTETOWN, PRINCE EDWARD ISLAND			CHICOUTIMI, QUEBEC			CORNER BROOK, NEWFOUNDLAND			CORNWALL, ONTARIO			DALHOUSIE, NEW BRUNSWICK			DARTMOUTH, NOVA SCOTIA		
		MAT.	INST.	TOTAL	MAT.	INST.	TOTAL	MAT.	INST.	TOTAL	MAT.	INST.	TOTAL	MAT.	INST.	TOTAL	MAT.	INST.	TOTAL
015433	**CONTRACTOR EQUIPMENT**		116.8	116.8		100.3	100.3		100.0	100.0		99.7	99.7		99.7	99.7		98.9	98.9
0241, 31 - 34	**SITE & INFRASTRUCTURE, DEMOLITION**	134.6	101.5	111.7	102.0	93.6	96.2	130.4	90.5	102.8	115.1	94.3	100.7	101.0	90.6	93.8	117.9	92.0	100.0
0310	Concrete Forming & Accessories	123.7	53.1	63.5	134.9	89.8	96.4	121.9	75.6	82.4	125.0	82.4	88.7	105.9	58.0	65.1	111.8	67.8	74.3
0320	Concrete Reinforcing	156.7	46.6	103.5	106.0	94.5	100.4	156.0	48.8	104.2	163.7	85.3	125.8	142.5	57.6	101.4	163.2	47.6	107.3
0330	Cast-in-Place Concrete	152.1	57.6	116.9	107.4	94.7	102.7	127.3	63.6	103.6	117.5	92.8	108.3	110.5	56.3	90.3	122.9	66.8	102.0
03	**CONCRETE**	136.8	55.4	101.0	101.6	92.4	97.6	149.1	67.5	113.3	115.7	87.0	103.1	113.1	58.4	89.0	131.9	65.0	102.5
04	**MASONRY**	187.9	54.9	106.2	161.9	89.8	117.6	208.6	74.3	126.1	165.0	85.6	116.2	163.3	57.4	98.3	222.6	65.6	126.1
05	**METALS**	136.4	79.6	119.8	113.9	91.9	107.4	127.7	75.2	112.2	112.1	90.7	105.8	105.6	73.7	96.2	128.1	75.7	112.7
06	**WOOD, PLASTICS & COMPOSITES**	104.3	52.5	75.8	131.8	90.3	108.9	128.1	81.4	102.4	118.5	81.6	98.2	95.8	57.9	74.9	115.9	67.4	89.2
07	**THERMAL & MOISTURE PROTECTION**	137.8	57.4	102.8	112.3	96.4	105.4	133.0	65.9	103.8	121.4	86.2	106.1	121.0	58.4	93.8	130.1	67.6	102.9
08	**OPENINGS**	85.7	45.8	76.4	87.8	76.3	85.2	106.3	66.6	97.0	89.1	80.6	87.1	85.5	51.2	77.5	91.3	61.3	84.3
0920	Plaster & Gypsum Board	126.3	50.5	75.3	143.9	89.9	107.6	151.2	80.9	103.9	174.9	81.1	111.8	129.5	56.7	80.5	146.6	66.5	92.7
0950, 0980	Ceilings & Acoustic Treatment	128.9	50.5	76.1	116.0	89.9	98.4	124.6	80.9	95.1	107.6	81.1	89.8	106.6	56.7	72.9	131.1	66.5	87.5
0960	Flooring	115.1	55.3	98.3	115.7	86.0	107.3	112.8	49.5	95.0	113.6	86.0	105.8	101.9	63.4	91.1	107.6	58.9	93.9
0970, 0990	Wall Finishes & Painting/Coating	117.3	39.9	71.5	111.3	104.1	107.1	117.8	56.5	81.5	110.3	85.7	95.7	113.8	47.9	74.8	117.8	59.3	83.2
09	**FINISHES**	119.8	52.3	83.3	114.6	91.6	102.1	122.6	70.2	94.2	117.3	83.6	99.0	108.4	58.2	81.2	120.3	65.5	90.7
COVERS	**DIVS. 10 - 14, 25, 28, 41, 43, 44, 46**	131.1	58.4	114.9	131.1	81.5	120.1	131.1	60.6	115.4	131.1	64.6	116.3	131.1	57.8	114.8	131.1	60.8	115.5
21, 22, 23	**FIRE SUPPRESSION, PLUMBING & HVAC**	104.1	59.0	86.4	103.8	81.0	94.9	104.0	65.9	89.0	104.2	94.3	100.3	103.9	64.8	88.5	104.0	79.5	94.4
26, 27, 3370	**ELECTRICAL, COMMUNICATIONS & UTIL.**	110.0	47.5	79.2	109.9	84.1	97.2	110.9	51.5	81.6	111.6	83.4	97.7	113.9	52.8	83.8	115.0	58.7	87.2
MF2018	**WEIGHTED AVERAGE**	121.3	60.2	95.5	110.4	87.6	100.7	124.7	68.4	100.9	113.1	87.3	102.2	110.2	62.5	90.0	121.5	70.2	99.8

CANADA

	DIVISION	EDMONTON, ALBERTA			FORT MCMURRAY, ALBERTA			FREDERICTON, NEW BRUNSWICK			GATINEAU, QUEBEC			GRANBY, QUEBEC			HALIFAX, NOVA SCOTIA		
		MAT.	INST.	TOTAL	MAT.	INST.	TOTAL	MAT.	INST.	TOTAL	MAT.	INST.	TOTAL	MAT.	INST.	TOTAL	MAT.	INST.	TOTAL
015433	**CONTRACTOR EQUIPMENT**		129.1	129.1		101.9	101.9		113.9	113.9		100.2	100.2		100.2	100.2		115.6	115.6
0241, 31 - 34	**SITE & INFRASTRUCTURE, DEMOLITION**	125.1	121.8	122.8	121.7	95.4	103.5	116.0	100.8	105.5	96.7	93.7	94.6	97.2	93.7	94.8	102.8	103.6	103.4
0310	Concrete Forming & Accessories	128.3	94.9	99.8	125.0	87.5	93.1	126.9	58.6	68.7	132.9	79.1	87.0	132.9	79.0	87.0	124.1	87.7	93.0
0320	Concrete Reinforcing	136.0	82.0	109.9	151.5	81.9	117.8	141.0	57.8	100.8	147.4	72.4	111.1	147.4	72.4	111.1	156.3	77.9	118.3
0330	Cast-in-Place Concrete	145.8	104.3	130.4	173.5	98.2	145.5	112.7	57.5	92.1	104.0	87.5	97.9	107.5	87.4	100.0	99.4	85.1	94.0
03	**CONCRETE**	130.8	96.8	115.9	139.8	90.6	118.2	116.9	59.7	91.8	107.9	81.4	96.2	109.5	81.4	97.1	111.9	86.1	100.6
04	**MASONRY**	212.7	89.8	137.2	206.1	86.3	132.5	189.1	58.8	109.0	162.3	76.3	109.5	162.6	76.3	109.6	185.4	86.4	124.6
05	**METALS**	134.6	102.4	125.1	139.8	90.9	125.5	136.6	83.3	121.0	110.7	84.8	103.1	110.9	84.8	103.2	136.3	99.2	125.4
06	**WOOD, PLASTICS & COMPOSITES**	101.6	94.4	97.6	114.5	86.7	99.2	106.5	58.2	79.9	131.6	79.0	102.7	131.6	79.0	102.7	103.7	87.8	95.0
07	**THERMAL & MOISTURE PROTECTION**	143.5	98.0	123.7	129.8	92.6	113.6	137.4	59.4	103.5	114.8	84.1	101.5	114.8	82.6	100.8	137.4	88.7	116.2
08	**OPENINGS**	80.9	83.7	81.5	89.1	79.5	86.9	88.0	50.3	79.2	89.1	67.9	84.2	89.1	67.9	84.2	88.9	79.2	86.7
0920	Plaster & Gypsum Board	129.8	93.3	105.2	116.6	86.1	96.1	130.3	56.7	80.8	114.4	78.4	90.2	114.4	78.4	90.2	117.8	87.1	97.1
0950, 0980	Ceilings & Acoustic Treatment	144.5	93.3	110.0	112.5	86.1	94.7	137.2	56.7	82.9	103.5	78.4	86.6	103.5	78.4	86.6	139.8	87.1	104.2
0960	Flooring	121.8	84.8	111.4	113.6	84.8	105.5	115.6	66.2	101.7	113.6	86.0	105.8	113.6	86.0	105.8	109.5	81.5	101.6
0970, 0990	Wall Finishes & Painting/Coating	115.6	107.4	110.8	110.4	88.3	97.3	118.8	61.1	84.6	110.3	83.6	94.5	110.3	83.6	94.5	118.4	90.9	102.1
09	**FINISHES**	123.6	94.9	108.0	111.8	87.4	98.6	125.1	60.4	90.1	107.8	81.2	93.4	107.8	81.2	93.4	117.7	87.5	101.3
COVERS	**DIVS. 10 - 14, 25, 28, 41, 43, 44, 46**	131.1	93.6	122.8	131.1	90.1	122.0	131.1	58.5	115.0	131.1	75.6	118.8	131.1	75.6	118.8	131.1	68.9	117.3
21, 22, 23	**FIRE SUPPRESSION, PLUMBING & HVAC**	104.9	90.4	99.2	104.2	93.2	99.9	104.2	73.7	92.2	104.2	84.2	96.3	103.8	84.2	96.1	104.9	82.3	96.0
26, 27, 3370	**ELECTRICAL, COMMUNICATIONS & UTIL.**	115.0	93.9	104.6	105.4	77.5	91.7	112.2	69.7	91.2	110.2	65.2	88.0	110.9	65.2	88.4	113.3	90.6	102.1
MF2018	**WEIGHTED AVERAGE**	121.8	95.9	110.9	121.9	88.4	107.7	119.3	69.0	98.0	110.3	79.8	97.4	110.5	79.7	97.5	117.9	88.0	105.2

CANADA

	DIVISION	HAMILTON, ONTARIO			HULL, QUEBEC			JOLIETTE, QUEBEC			KAMLOOPS, BRITISH COLUMBIA			KINGSTON, ONTARIO			KITCHENER, ONTARIO		
		MAT.	INST.	TOTAL	MAT.	INST.	TOTAL	MAT.	INST.	TOTAL	MAT.	INST.	TOTAL	MAT.	INST.	TOTAL	MAT.	INST.	TOTAL
015433	**CONTRACTOR EQUIPMENT**		115.5	115.5		100.2	100.2		100.2	100.2		103.3	103.3		101.9	101.9		101.6	101.6
0241, 31 - 34	**SITE & INFRASTRUCTURE, DEMOLITION**	105.8	107.0	106.6	96.7	93.7	94.6	97.3	93.7	94.8	120.1	97.1	104.3	115.1	98.0	103.3	94.2	99.8	98.0
0310	Concrete Forming & Accessories	130.0	94.7	99.9	132.9	79.1	87.0	132.9	79.2	87.1	124.3	82.3	88.5	125.1	82.5	88.7	117.3	88.0	92.3
0320	Concrete Reinforcing	136.9	103.3	120.7	147.4	72.4	111.1	139.3	72.4	107.0	109.3	75.8	93.1	163.7	85.3	125.8	96.4	103.1	99.6
0330	Cast-in-Place Concrete	107.6	102.1	105.6	104.0	87.5	97.9	108.4	87.5	100.6	93.8	91.8	93.1	117.5	92.8	108.3	111.7	97.2	106.3
03	**CONCRETE**	110.2	99.3	105.4	107.9	81.4	96.2	108.8	81.5	96.8	117.2	85.0	103.0	117.5	87.0	104.1	97.1	94.1	95.8
04	**MASONRY**	173.2	100.5	128.5	162.3	76.3	109.5	162.7	76.3	109.6	169.4	84.0	116.9	171.9	85.7	118.9	142.2	98.6	115.4
05	**METALS**	134.7	106.4	126.4	110.9	84.8	103.2	110.9	85.0	103.3	112.8	87.1	105.3	113.5	90.7	106.8	124.1	98.0	116.5
06	**WOOD, PLASTICS & COMPOSITES**	111.4	94.0	101.8	131.6	79.0	102.7	131.6	79.0	102.7	102.2	80.7	90.3	118.5	81.7	98.3	110.3	85.9	96.9
07	**THERMAL & MOISTURE PROTECTION**	134.0	101.5	119.8	114.8	84.1	101.5	114.8	84.1	101.5	131.2	81.7	109.7	121.4	87.3	106.6	116.4	98.4	108.5
08	**OPENINGS**	85.3	93.2	87.1	89.1	67.9	84.2	89.1	72.3	85.2	86.1	78.5	84.3	89.1	80.3	87.1	80.0	87.1	81.7
0920	Plaster & Gypsum Board	131.0	93.1	105.5	114.4	78.4	90.2	147.6	78.4	101.0	101.1	79.8	86.8	177.6	81.2	112.8	107.2	85.6	92.6
0950, 0980	Ceilings & Acoustic Treatment	139.9	93.1	108.4	103.5	78.4	86.6	103.5	78.4	86.6	103.5	79.8	87.5	119.0	81.2	93.5	107.9	85.6	92.9
0960	Flooring	115.3	95.0	109.6	113.6	86.0	105.8	113.6	86.0	105.8	112.5	49.2	94.7	113.6	86.0	105.8	101.3	95.1	99.6
0970, 0990	Wall Finishes & Painting/Coating	113.0	104.1	107.8	110.3	83.6	94.5	110.3	83.6	94.5	110.3	76.4	90.3	110.3	79.2	91.9	104.4	93.9	98.2
09	**FINISHES**	121.0	95.6	107.3	107.8	81.2	93.4	112.2	81.2	95.4	108.3	75.8	90.7	119.9	82.9	99.9	102.1	89.4	95.2
COVERS	**DIVS. 10 - 14, 25, 28, 41, 43, 44, 46**	131.1	92.5	122.5	131.1	75.6	118.8	131.1	75.6	118.8	131.1	84.4	120.7	131.1	64.6	116.3	131.1	89.6	121.9
21, 22, 23	**FIRE SUPPRESSION, PLUMBING & HVAC**	105.2	92.1	100.0	103.8	84.2	96.1	103.8	84.2	96.1	103.8	87.5	97.4	104.2	94.4	100.4	104.2	91.3	99.1
26, 27, 3370	**ELECTRICAL, COMMUNICATIONS & UTIL.**	108.5	103.6	106.1	112.3	65.2	89.1	110.9	65.2	88.4	114.3	74.2	94.5	111.6	82.1	97.1	109.6	99.8	104.8
MF2018	**WEIGHTED AVERAGE**	116.3	98.8	108.9	110.5	79.8	97.5	110.8	80.0	97.8	113.0	83.5	100.5	114.1	87.4	102.8	108.6	94.6	102.7

	DIVISION	CANADA																	
		LAVAL, QUEBEC			LETHBRIDGE, ALBERTA			LLOYDMINSTER, ALBERTA			LONDON, ONTARIO			MEDICINE HAT, ALBERTA			MONCTON, NEW BRUNSWICK		
		MAT.	INST.	TOTAL	MAT.	INST.	TOTAL	MAT.	INST.	TOTAL	MAT.	INST.	TOTAL	MAT.	INST.	TOTAL	MAT.	INST.	TOTAL
015433	**CONTRACTOR EQUIPMENT**		100.2	100.2		101.9	101.9		101.9	101.9		116.7	116.7		101.9	101.9		99.7	99.7
0241, 31 - 34	**SITE & INFRASTRUCTURE, DEMOLITION**	97.2	93.9	94.9	114.5	96.0	101.7	114.5	95.4	101.3	103.5	106.8	105.8	113.2	95.5	101.0	104.5	92.5	96.2
0310	Concrete Forming & Accessories	133.2	81.1	88.8	126.4	87.6	93.3	124.6	78.3	85.1	129.1	89.1	95.0	126.4	78.2	85.3	106.1	66.1	72.0
0320	Concrete Reinforcing	147.4	74.4	112.1	151.5	81.9	117.8	151.5	81.8	117.8	126.7	102.1	114.8	151.5	81.8	117.8	140.5	69.8	106.3
0330	Cast-in-Place Concrete	107.5	89.4	100.7	130.1	98.2	118.2	120.7	94.6	111.0	120.2	100.8	113.0	120.7	94.6	111.0	109.8	68.0	94.3
03	**CONCRETE**	109.5	83.4	98.0	120.0	90.6	107.1	115.5	85.2	102.2	114.6	96.2	106.5	115.7	85.2	102.3	108.4	68.5	90.9
04	**MASONRY**	162.6	78.4	110.8	180.1	86.3	122.4	161.6	80.1	111.6	183.8	99.5	132.0	161.6	80.1	111.6	159.4	74.6	107.3
05	**METALS**	110.7	85.9	103.5	134.1	91.0	121.5	112.8	90.8	106.4	134.7	106.8	126.5	113.0	90.8	106.5	114.1	86.8	106.1
06	**WOOD, PLASTICS & COMPOSITES**	131.7	81.1	103.9	118.0	86.7	100.8	114.5	77.4	94.1	114.8	86.8	99.4	118.0	77.4	95.7	97.5	65.0	79.6
07	**THERMAL & MOISTURE PROTECTION**	115.4	86.0	102.6	127.1	92.6	112.1	123.5	88.2	108.1	129.9	99.0	116.5	130.5	88.2	112.1	116.6	72.4	97.4
08	**OPENINGS**	89.1	69.8	84.6	89.1	79.5	86.9	89.1	74.4	85.7	81.6	87.9	83.1	89.1	74.4	85.7	84.0	61.9	78.8
0920	Plaster & Gypsum Board	114.7	80.5	91.7	106.4	86.1	92.7	102.1	76.6	84.9	134.4	85.8	101.7	104.3	76.6	85.6	124.2	64.1	83.7
0950, 0980	Ceilings & Acoustic Treatment	103.5	80.5	88.0	112.5	86.1	94.7	103.5	76.6	85.3	138.5	85.8	102.9	103.5	76.6	85.3	116.8	64.1	81.2
0960	Flooring	113.6	88.4	106.5	113.6	84.8	105.5	113.6	84.8	105.5	109.9	95.1	105.7	113.6	84.8	105.5	99.6	65.6	90.1
0970, 0990	Wall Finishes & Painting/Coating	110.3	85.9	95.8	110.2	96.4	102.1	110.4	75.1	89.5	113.5	100.5	105.8	110.2	75.1	89.4	111.3	83.6	94.9
09	**FINISHES**	107.8	83.3	94.5	109.6	88.3	98.0	107.4	78.9	92.0	121.5	91.0	105.0	107.5	78.9	92.0	107.7	67.5	86.0
COVERS	**DIVS. 10 - 14, 25, 28, 41, 43, 44, 46**	131.1	77.5	119.2	131.1	89.2	121.8	131.1	87.1	121.3	131.1	91.6	122.3	131.1	86.1	121.1	131.1	61.0	115.5
21, 22, 23	**FIRE SUPPRESSION, PLUMBING & HVAC**	104.3	86.5	97.3	104.1	90.1	98.6	104.2	90.0	98.6	105.1	89.2	98.8	103.8	86.8	97.2	103.8	73.7	92.0
26, 27, 3370	**ELECTRICAL, COMMUNICATIONS & UTIL.**	110.9	67.0	89.3	107.1	77.5	92.5	104.6	77.5	91.2	103.5	102.0	102.8	104.6	77.5	91.2	117.3	88.5	103.1
MF2018	**WEIGHTED AVERAGE**	110.7	81.6	98.4	117.0	87.9	104.7	111.5	84.7	100.2	116.3	96.5	108.0	111.7	84.0	100.0	110.9	76.0	96.2

	DIVISION	CANADA																	
		MONTREAL, QUEBEC			MOOSE JAW, SASKATCHEWAN			NEW GLASGOW, NOVA SCOTIA			NEWCASTLE, NEW BRUNSWICK			NORTH BAY, ONTARIO			OSHAWA, ONTARIO		
		MAT.	INST.	TOTAL	MAT.	INST.	TOTAL	MAT.	INST.	TOTAL	MAT.	INST.	TOTAL	MAT.	INST.	TOTAL	MAT.	INST.	TOTAL
015433	**CONTRACTOR EQUIPMENT**		117.7	117.7		98.3	98.3		98.9	98.9		99.7	99.7		99.3	99.3		101.6	101.6
0241, 31 - 34	**SITE & INFRASTRUCTURE, DEMOLITION**	113.0	105.5	107.8	114.4	90.0	97.5	111.9	92.0	98.1	105.1	90.6	95.1	126.8	93.6	103.9	104.6	99.6	101.2
0310	Concrete Forming & Accessories	130.9	90.4	96.4	109.6	54.6	62.7	111.8	67.8	74.3	106.1	58.0	65.1	146.2	80.0	89.8	122.8	91.3	96.0
0320	Concrete Reinforcing	126.8	94.6	111.3	106.9	61.4	84.9	156.0	47.6	103.6	140.5	57.6	100.4	184.7	84.8	136.4	152.7	103.7	129.0
0330	Cast-in-Place Concrete	124.4	97.4	114.3	117.1	63.7	97.2	122.9	66.8	102.0	114.0	56.3	92.6	118.4	80.1	104.1	129.2	106.4	120.7
03	**CONCRETE**	116.7	94.3	106.8	101.4	60.0	83.2	131.0	65.0	102.0	110.3	58.4	87.5	132.7	81.4	110.2	118.3	98.8	109.8
04	**MASONRY**	178.8	89.8	124.1	160.3	56.4	96.4	208.2	65.6	120.5	159.8	57.4	96.9	214.4	81.9	133.0	145.1	101.9	118.6
05	**METALS**	144.3	102.2	131.9	109.6	75.0	99.4	125.6	75.7	111.0	114.1	73.8	102.3	126.5	90.2	115.8	114.2	98.6	109.6
06	**WOOD, PLASTICS & COMPOSITES**	111.3	90.8	100.0	99.4	53.2	74.0	115.9	67.4	89.2	97.5	57.9	75.7	153.5	80.2	113.2	117.6	89.4	102.1
07	**THERMAL & MOISTURE PROTECTION**	125.9	97.2	113.4	114.1	60.1	90.6	130.1	67.6	102.9	116.6	58.4	91.3	136.4	82.7	113.0	117.3	104.1	111.6
08	**OPENINGS**	87.4	78.7	85.4	85.3	51.4	77.4	91.3	61.3	84.3	84.0	51.2	76.3	98.6	78.2	93.8	85.0	90.2	86.2
0920	Plaster & Gypsum Board	126.2	89.9	101.8	98.2	51.9	67.1	145.0	66.5	92.2	124.2	56.7	78.8	140.6	79.8	99.7	110.7	89.2	96.2
0950, 0980	Ceilings & Acoustic Treatment	146.1	89.9	108.2	103.5	51.9	68.7	123.8	66.5	85.2	116.8	56.7	76.3	123.8	79.8	94.1	103.9	89.2	94.0
0960	Flooring	112.4	90.5	106.2	103.6	53.7	89.6	107.6	58.9	93.9	99.6	63.4	89.5	131.2	86.0	118.5	104.2	97.6	102.4
0970, 0990	Wall Finishes & Painting/Coating	115.0	104.1	108.5	110.3	61.0	81.1	117.8	59.3	83.2	111.3	47.9	73.8	117.8	85.0	98.4	104.4	108.1	106.6
09	**FINISHES**	121.6	92.8	106.0	103.7	54.8	77.2	118.7	65.5	89.9	107.7	58.2	80.9	125.9	81.9	102.1	103.2	93.7	98.1
COVERS	**DIVS. 10 - 14, 25, 28, 41, 43, 44, 46**	131.1	82.8	120.4	131.1	57.5	114.7	131.1	60.8	115.5	131.1	57.9	114.8	131.1	63.4	116.0	131.1	90.4	122.0
21, 22, 23	**FIRE SUPPRESSION, PLUMBING & HVAC**	105.1	81.2	95.7	104.2	70.3	90.9	104.0	79.5	94.4	103.8	64.8	88.5	104.0	92.4	99.4	104.2	91.9	99.3
26, 27, 3370	**ELECTRICAL, COMMUNICATIONS & UTIL.**	107.9	84.1	96.2	113.4	56.3	85.3	111.3	58.7	85.4	112.9	55.9	84.8	112.6	83.3	98.1	110.6	103.8	107.3
MF2018	**WEIGHTED AVERAGE**	119.1	90.1	106.9	109.0	63.9	89.9	119.6	70.2	98.7	110.7	62.9	90.5	122.5	85.2	106.8	110.8	97.2	105.0

	DIVISION	CANADA																	
		OTTAWA, ONTARIO			OWEN SOUND, ONTARIO			PETERBOROUGH, ONTARIO			PORTAGE LA PRAIRIE, MANITOBA			PRINCE ALBERT, SASKATCHEWAN			PRINCE GEORGE, BRITISH COLUMBIA		
		MAT.	INST.	TOTAL	MAT.	INST.	TOTAL	MAT.	INST.	TOTAL	MAT.	INST.	TOTAL	MAT.	INST.	TOTAL	MAT.	INST.	TOTAL
015433	**CONTRACTOR EQUIPMENT**		117.8	117.8		100.1	100.1		99.7	99.7		101.9	101.9		98.3	98.3		103.3	103.3
0241, 31 - 34	**SITE & INFRASTRUCTURE, DEMOLITION**	107.0	107.2	107.1	117.0	94.6	101.5	116.9	94.2	101.3	115.4	93.2	100.1	110.2	90.1	96.3	123.4	97.1	105.3
0310	Concrete Forming & Accessories	124.8	88.4	93.7	125.6	78.7	85.6	127.2	81.0	87.8	126.6	65.7	74.7	109.6	54.4	62.5	115.1	77.5	83.0
0320	Concrete Reinforcing	137.0	102.0	120.1	175.2	86.9	132.5	163.7	85.3	125.8	151.5	54.5	104.6	111.7	61.4	87.3	109.3	75.8	93.1
0330	Cast-in-Place Concrete	122.3	103.0	115.1	157.7	76.6	127.5	130.7	81.8	112.5	120.7	69.8	101.8	106.2	63.6	90.3	117.6	91.8	108.0
03	**CONCRETE**	116.6	96.7	107.9	138.9	79.9	113.0	121.8	82.5	104.6	109.5	66.2	90.5	96.9	59.9	80.7	127.5	82.8	107.9
04	**MASONRY**	173.2	99.4	127.8	165.6	87.1	117.3	166.1	88.0	118.1	164.8	59.7	100.2	159.5	56.4	96.1	171.4	84.0	117.7
05	**METALS**	135.5	108.9	127.7	111.1	91.3	105.3	112.3	90.8	106.0	113.0	78.4	102.8	109.7	74.8	99.4	112.8	87.2	105.3
06	**WOOD, PLASTICS & COMPOSITES**	108.4	85.7	95.9	116.5	77.3	94.9	120.3	79.0	97.6	118.0	66.8	89.8	99.4	53.2	74.0	102.2	74.1	86.7
07	**THERMAL & MOISTURE PROTECTION**	137.3	99.7	121.0	115.3	83.5	101.5	121.6	88.0	107.0	114.6	68.2	94.4	114.0	59.0	90.1	124.5	81.0	105.6
08	**OPENINGS**	90.3	87.5	89.7	90.5	77.6	87.5	87.7	79.9	85.9	89.1	59.9	82.3	84.3	51.4	76.6	86.1	74.9	83.5
0920	Plaster & Gypsum Board	129.1	84.6	99.1	153.5	76.7	101.8	115.8	78.5	90.7	104.0	65.6	78.2	98.2	51.9	67.1	101.1	73.0	82.2
0950, 0980	Ceilings & Acoustic Treatment	142.3	84.6	103.4	94.3	76.7	82.4	103.5	78.5	86.7	103.5	65.6	78.0	103.5	51.9	68.7	103.5	73.0	83.0
0960	Flooring	108.3	90.6	103.4	119.3	87.3	110.3	113.6	86.0	105.8	113.6	61.9	99.1	103.6	53.7	89.6	109.0	67.4	97.3
0970, 0990	Wall Finishes & Painting/Coating	113.5	95.9	103.1	105.6	83.8	92.7	110.3	87.1	96.6	110.4	54.1	77.1	110.3	52.0	75.8	110.3	76.4	90.3
09	**FINISHES**	122.3	89.0	104.3	111.3	80.8	94.8	108.7	82.6	94.5	107.4	64.1	84.0	103.7	53.8	76.7	107.2	75.0	89.8
COVERS	**DIVS. 10 - 14, 25, 28, 41, 43, 44, 46**	131.1	89.5	121.9	139.2	63.9	122.4	131.1	64.8	116.3	131.1	59.8	115.2	131.1	57.5	114.7	131.1	83.7	120.6
21, 22, 23	**FIRE SUPPRESSION, PLUMBING & HVAC**	105.1	90.7	99.4	103.7	92.4	99.3	103.8	95.8	100.7	103.8	78.4	93.8	104.2	63.4	88.2	103.8	87.5	97.4
26, 27, 3370	**ELECTRICAL, COMMUNICATIONS & UTIL.**	104.4	101.9	103.2	118.5	82.1	100.6	111.5	82.9	97.4	113.0	54.4	84.2	113.4	56.3	85.3	111.5	74.2	93.1
MF2018	**WEIGHTED AVERAGE**	117.5	96.8	108.7	116.7	85.4	103.4	113.1	87.1	102.1	111.6	69.1	93.6	108.2	62.2	88.8	114.0	82.8	100.8

DIVISION		CANADA																	
		QUEBEC CITY, QUEBEC			RED DEER, ALBERTA			REGINA, SASKATCHEWAN			RIMOUSKI, QUEBEC			ROUYN-NORANDA, QUEBEC			SAINT HYACINTHE, QUEBEC		
		MAT.	INST.	TOTAL	MAT.	INST.	TOTAL	MAT.	INST.	TOTAL	MAT.	INST.	TOTAL	MAT.	INST.	TOTAL	MAT.	INST.	TOTAL
015433	CONTRACTOR EQUIPMENT		117.4	117.4		101.9	101.9		128.3	128.3		100.2	100.2		100.2	100.2		100.2	100.2
0241, 31 - 34	SITE & INFRASTRUCTURE, DEMOLITION	114.1	104.7	107.6	113.2	95.5	101.0	126.4	119.6	121.7	97.0	93.6	94.6	96.7	93.7	94.6	97.2	93.7	94.8
0310	Concrete Forming & Accessories	127.4	90.5	95.9	139.7	78.2	87.3	131.0	91.5	97.4	132.9	89.8	96.2	132.9	79.1	87.0	132.9	79.1	87.0
0320	Concrete Reinforcing	124.2	94.6	109.9	151.5	81.8	117.8	123.8	88.8	106.9	105.1	94.5	100.0	147.4	72.4	111.1	147.4	72.4	111.1
0330	Cast-in-Place Concrete	131.6	97.0	118.7	120.7	94.6	111.0	153.8	97.7	132.9	109.4	94.7	103.9	104.0	87.5	97.9	107.5	87.5	100.0
03	CONCRETE	119.4	94.3	108.4	116.5	85.2	102.7	131.3	94.0	114.9	105.0	92.4	99.5	107.9	81.4	96.2	109.5	81.4	97.1
04	MASONRY	173.6	89.8	122.1	161.6	80.1	111.6	205.2	86.7	132.4	162.1	89.8	117.6	162.3	76.3	109.5	162.6	76.3	109.6
05	METALS	138.0	103.3	127.8	113.0	90.8	106.5	139.1	101.4	128.0	110.4	91.8	104.9	110.9	84.8	103.2	110.9	84.8	103.2
06	WOOD, PLASTICS & COMPOSITES	114.2	90.7	101.3	118.0	77.4	95.7	108.0	92.2	99.3	131.6	90.3	108.9	131.6	79.0	102.7	131.6	79.0	102.7
07	THERMAL & MOISTURE PROTECTION	125.1	97.2	112.9	141.4	88.2	118.3	143.8	86.6	118.9	114.8	96.4	106.8	114.8	84.1	101.5	115.2	84.1	101.7
08	OPENINGS	88.4	86.0	87.8	89.1	74.4	85.7	87.9	79.8	86.1	88.7	76.3	85.8	89.1	67.9	84.2	89.1	67.9	84.2
0920	Plaster & Gypsum Board	128.5	89.9	102.6	104.3	76.6	85.6	140.8	91.2	107.4	147.4	89.9	108.7	114.2	78.4	90.1	114.2	78.4	90.1
0950, 0980	Ceilings & Acoustic Treatment	131.7	89.9	103.5	103.5	76.6	85.3	154.3	91.2	111.7	102.7	89.9	94.1	102.7	78.4	86.3	102.7	78.4	86.3
0960	Flooring	108.9	90.5	103.8	116.0	84.8	107.2	123.6	95.1	115.6	114.8	86.0	106.7	113.6	86.0	105.8	113.6	86.0	105.8
0970, 0990	Wall Finishes & Painting/Coating	120.8	104.1	111.0	110.2	75.1	89.4	118.0	89.0	100.9	113.4	104.1	107.9	110.3	83.6	94.5	110.3	83.6	94.5
09	FINISHES	117.4	92.7	104.0	108.2	78.9	92.4	130.9	92.9	110.3	112.6	91.6	101.2	107.6	81.2	93.3	107.6	81.2	93.3
COVERS	DIVS. 10 - 14, 25, 28, 41, 43, 44, 46	131.1	82.5	120.3	131.1	86.1	121.1	131.1	70.9	117.7	131.1	81.5	120.1	131.1	75.6	118.8	131.1	75.6	118.8
21, 22, 23	FIRE SUPPRESSION, PLUMBING & HVAC	105.1	81.2	95.7	103.8	86.8	97.2	104.8	86.9	97.8	103.8	81.0	94.9	103.8	84.2	96.1	100.5	84.2	94.1
26, 27, 3370	ELECTRICAL, COMMUNICATIONS & UTIL.	114.7	84.1	99.6	104.6	77.5	91.2	110.8	91.5	101.3	110.9	84.1	97.7	110.9	65.2	88.4	111.5	65.2	88.7
MF2018	WEIGHTED AVERAGE	118.7	90.4	106.7	112.1	84.0	100.2	123.1	92.4	110.2	110.2	87.6	100.6	110.3	79.8	97.4	109.8	79.8	97.1

DIVISION		CANADA																	
		SAINT JOHN, NEW BRUNSWICK			SARNIA, ONTARIO			SASKATOON, SASKATCHEWAN			SAULT STE MARIE, ONTARIO			SHERBROOKE, QUEBEC			SOREL, QUEBEC		
		MAT.	INST.	TOTAL	MAT.	INST.	TOTAL	MAT.	INST.	TOTAL	MAT.	INST.	TOTAL	MAT.	INST.	TOTAL	MAT.	INST.	TOTAL
015433	CONTRACTOR EQUIPMENT		99.7	99.7		99.7	99.7		98.4	98.4		99.7	99.7		100.2	100.2		100.2	100.2
0241, 31 - 34	SITE & INFRASTRUCTURE, DEMOLITION	105.2	92.5	96.4	115.5	94.3	100.9	112.7	92.8	98.9	105.6	93.9	97.5	97.2	93.7	94.8	97.3	93.7	94.8
0310	Concrete Forming & Accessories	126.2	63.4	72.7	125.9	87.6	93.3	109.7	90.7	93.5	115.1	86.2	90.5	132.9	79.1	87.0	132.9	79.2	87.1
0320	Concrete Reinforcing	140.5	69.8	106.3	116.5	86.7	102.1	114.2	88.7	101.9	105.1	85.4	95.6	147.4	72.4	111.1	139.3	72.4	107.0
0330	Cast-in-Place Concrete	112.4	68.0	95.9	120.6	94.2	110.8	113.5	93.1	105.9	108.3	80.5	98.0	107.5	87.5	100.0	108.4	87.5	100.6
03	CONCRETE	110.9	67.3	91.7	111.1	90.0	101.8	101.3	91.2	96.9	97.4	84.7	91.8	109.5	81.4	97.1	108.8	81.5	96.8
04	MASONRY	180.8	74.4	115.4	177.6	90.5	124.1	172.1	86.6	119.5	162.9	90.1	118.2	162.6	76.3	109.6	162.7	76.3	109.6
05	METALS	114.0	86.8	106.0	112.3	91.2	106.1	106.5	88.6	101.2	111.5	95.2	106.7	110.7	84.8	103.1	110.9	85.0	103.3
06	WOOD, PLASTICS & COMPOSITES	120.7	61.4	88.0	119.4	86.7	101.4	96.0	91.3	93.4	106.5	87.9	96.3	131.6	79.0	102.7	131.6	79.0	102.7
07	THERMAL & MOISTURE PROTECTION	116.9	72.2	97.5	121.7	91.5	108.6	117.0	84.7	103.0	120.4	87.3	106.0	114.8	84.1	101.5	114.8	84.1	101.5
08	OPENINGS	83.9	58.9	78.0	90.5	83.5	88.8	85.3	79.4	83.9	82.4	86.1	83.2	89.1	67.9	84.2	89.1	72.3	85.2
0920	Plaster & Gypsum Board	137.9	60.3	85.7	143.2	86.4	105.0	115.6	91.2	99.1	107.1	87.6	94.0	114.2	78.4	90.1	147.4	78.4	101.0
0950, 0980	Ceilings & Acoustic Treatment	121.7	60.3	80.3	109.2	86.4	93.9	124.5	91.2	102.0	103.5	87.6	92.8	102.7	78.4	86.3	102.7	78.4	86.3
0960	Flooring	111.4	65.6	98.5	113.6	94.1	108.1	106.5	95.1	103.3	106.5	92.1	102.4	113.6	86.0	105.8	113.6	86.0	105.8
0970, 0990	Wall Finishes & Painting/Coating	111.3	83.6	94.9	110.3	98.5	103.3	113.8	89.0	99.2	110.3	91.3	99.1	110.3	83.6	94.5	110.3	83.6	94.5
09	FINISHES	114.0	65.4	87.7	113.5	90.2	100.9	112.1	92.3	101.4	104.7	88.2	95.8	107.6	81.2	93.3	112.0	81.2	95.3
COVERS	DIVS. 10 - 14, 25, 28, 41, 43, 44, 46	131.1	61.0	115.5	131.1	66.1	116.6	131.1	69.3	117.4	131.1	87.9	121.5	131.1	75.6	118.8	131.1	75.6	118.8
21, 22, 23	FIRE SUPPRESSION, PLUMBING & HVAC	103.8	75.1	92.6	103.8	101.5	102.9	104.4	86.7	97.5	103.8	90.5	98.6	104.2	84.2	96.3	103.8	84.2	96.1
26, 27, 3370	ELECTRICAL, COMMUNICATIONS & UTIL.	119.9	88.5	104.4	114.7	85.0	100.0	114.6	91.4	103.2	113.2	83.3	98.5	110.9	65.2	88.4	110.9	65.2	88.4
MF2018	WEIGHTED AVERAGE	113.2	75.7	97.3	113.2	91.3	103.9	109.9	88.5	100.9	108.6	88.7	100.2	110.6	79.8	97.6	110.8	80.0	97.8

DIVISION		CANADA																	
		ST. CATHARINES, ONTARIO			ST JEROME, QUEBEC			ST. JOHN'S, NEWFOUNDLAND			SUDBURY, ONTARIO			SUMMERSIDE, PRINCE EDWARD ISLAND			SYDNEY, NOVA SCOTIA		
		MAT.	INST.	TOTAL	MAT.	INST.	TOTAL	MAT.	INST.	TOTAL	MAT.	INST.	TOTAL	MAT.	INST.	TOTAL	MAT.	INST.	TOTAL
015433	CONTRACTOR EQUIPMENT		99.5	99.5		100.2	100.2		123.4	123.4		99.5	99.5		98.8	98.8		98.9	98.9
0241, 31 - 34	SITE & INFRASTRUCTURE, DEMOLITION	94.2	96.1	95.6	96.7	93.7	94.6	115.4	111.8	112.9	94.4	95.7	95.3	121.8	89.4	99.5	108.0	92.0	96.9
0310	Concrete Forming & Accessories	115.3	93.8	97.0	132.9	79.1	87.0	125.3	85.2	91.1	111.0	88.9	92.1	112.0	52.5	61.3	111.8	67.8	74.3
0320	Concrete Reinforcing	97.2	103.2	100.1	147.4	72.4	111.1	172.1	82.4	128.7	98.0	101.4	99.7	154.0	46.5	102.0	156.0	47.6	103.6
0330	Cast-in-Place Concrete	106.7	98.7	103.7	104.0	87.5	97.9	137.1	98.5	122.8	107.6	95.3	103.0	116.7	54.7	93.6	94.8	66.8	84.4
03	CONCRETE	94.8	97.2	95.8	107.9	81.4	96.2	132.7	90.3	114.1	95.0	93.5	94.3	139.1	53.4	101.5	118.1	65.0	94.8
04	MASONRY	141.8	100.8	116.6	162.3	76.3	109.5	200.1	89.9	132.4	141.9	94.9	113.0	207.4	54.8	113.7	205.8	65.6	119.6
05	METALS	114.2	97.9	109.4	110.9	84.8	103.2	137.9	99.1	126.5	113.6	97.1	108.8	125.6	69.2	109.1	125.6	75.7	111.0
06	WOOD, PLASTICS & COMPOSITES	107.8	93.2	99.8	131.6	79.0	102.7	106.9	83.4	94.0	103.5	87.9	94.9	116.4	51.9	80.9	115.9	67.4	89.2
07	THERMAL & MOISTURE PROTECTION	116.4	102.0	110.1	114.8	84.1	101.5	140.2	94.7	120.4	115.7	96.3	107.3	129.4	57.3	98.0	130.1	67.6	102.9
08	OPENINGS	79.5	91.4	82.3	89.1	67.9	84.2	85.4	74.3	82.8	80.2	86.7	81.7	102.5	45.4	89.2	91.3	61.3	84.3
0920	Plaster & Gypsum Board	100.8	93.1	95.6	114.2	78.4	90.1	150.0	82.1	104.3	100.0	87.6	91.7	145.9	50.5	81.8	145.0	66.5	92.2
0950, 0980	Ceilings & Acoustic Treatment	103.9	93.1	96.6	102.7	78.4	86.3	143.7	82.1	102.2	98.2	87.6	91.0	123.8	50.5	74.4	123.8	66.5	85.2
0960	Flooring	100.2	91.9	97.8	113.6	86.0	105.8	116.8	52.2	98.7	98.3	92.1	96.6	107.6	55.3	92.9	107.6	58.9	93.9
0970, 0990	Wall Finishes & Painting/Coating	104.4	104.1	104.2	110.3	83.6	94.5	120.8	98.6	107.7	104.4	95.3	99.0	117.8	39.9	71.7	117.8	59.3	83.2
09	FINISHES	100.1	94.6	97.1	107.6	81.2	93.3	129.2	80.4	102.8	98.3	89.9	93.7	119.8	51.9	83.1	118.7	65.5	89.9
COVERS	DIVS. 10 - 14, 25, 28, 41, 43, 44, 46	131.1	69.7	117.4	131.1	75.6	118.8	131.1	68.6	117.2	131.1	89.6	121.9	131.1	57.0	114.6	131.1	60.8	115.5
21, 22, 23	FIRE SUPPRESSION, PLUMBING & HVAC	104.2	91.1	99.0	103.8	84.2	96.1	104.8	83.0	96.2	103.6	89.4	98.1	104.0	58.9	86.2	104.0	79.5	94.4
26, 27, 3370	ELECTRICAL, COMMUNICATIONS & UTIL.	111.3	101.8	106.6	111.6	65.2	88.7	115.6	81.4	98.7	109.5	102.1	105.9	110.2	47.4	79.2	111.3	58.7	85.4
MF2018	WEIGHTED AVERAGE	106.6	95.6	101.9	110.4	79.8	97.4	122.6	87.4	107.7	106.1	93.6	100.8	121.9	57.9	94.9	117.8	70.2	97.7

DIVISION		CANADA																	
		THUNDER BAY, ONTARIO			TIMMINS, ONTARIO			TORONTO, ONTARIO			TROIS RIVIERES, QUEBEC			TRURO, NOVA SCOTIA			VANCOUVER, BRITISH COLUMBIA		
		MAT.	INST.	TOTAL	MAT.	INST.	TOTAL	MAT.	INST.	TOTAL	MAT.	INST.	TOTAL	MAT.	INST.	TOTAL	MAT.	INST.	TOTAL
015433	**CONTRACTOR EQUIPMENT**		99.5	99.5		99.7	99.7		117.5	117.5		99.7	99.7		99.4	99.4		135.3	135.3
0241, 31 - 34	**SITE & INFRASTRUCTURE, DEMOLITION**	98.8	96.0	96.9	116.9	93.9	101.0	108.4	107.7	107.9	108.2	93.4	97.9	101.3	92.3	95.1	117.0	122.8	121.0
0310	Concrete Forming & Accessories	122.8	92.5	97.0	127.2	80.0	87.0	128.7	102.5	106.3	153.7	79.1	90.1	99.7	67.9	72.6	131.0	91.6	97.4
0320	Concrete Reinforcing	86.9	102.6	94.5	163.7	84.8	125.6	136.9	103.9	120.9	156.0	72.4	115.6	139.3	47.6	95.0	132.5	95.2	114.5
0330	Cast-in-Place Concrete	117.5	98.0	110.2	130.7	80.2	111.9	108.4	113.7	110.4	98.2	87.4	94.2	135.7	66.9	110.1	123.5	90.6	111.3
03	**CONCRETE**	101.9	96.2	99.4	121.8	81.5	104.1	110.5	107.1	109.0	120.3	81.4	103.2	120.8	65.1	96.3	119.7	93.0	108.0
04	**MASONRY**	142.4	100.5	116.7	166.1	81.9	114.4	173.3	108.7	133.6	209.8	76.3	127.8	161.9	65.6	102.7	165.3	85.6	116.3
05	**METALS**	114.2	97.0	109.1	112.3	90.4	105.9	135.3	109.4	127.7	124.9	84.7	113.1	111.6	76.0	101.1	133.5	109.8	126.6
06	**WOOD, PLASTICS & COMPOSITES**	117.6	91.3	103.1	120.3	80.3	98.3	114.8	100.5	107.0	168.2	79.0	119.1	89.6	67.4	77.4	110.3	91.0	99.7
07	**THERMAL & MOISTURE PROTECTION**	116.6	99.3	109.1	121.6	82.7	104.7	135.6	109.4	124.2	128.6	84.1	109.2	116.1	67.6	95.0	141.7	88.6	118.6
08	**OPENINGS**	78.9	89.8	81.4	87.7	78.2	85.5	83.9	98.9	87.4	100.3	72.3	93.7	82.4	61.3	77.5	84.5	88.7	85.5
0920	Plaster & Gypsum Board	128.0	91.2	103.2	115.8	79.8	91.6	129.1	99.9	109.5	174.3	78.4	109.8	124.3	66.5	85.4	127.2	89.7	102.0
0950, 0980	Ceilings & Acoustic Treatment	98.2	91.2	93.4	103.5	79.8	87.5	139.9	99.9	112.9	122.2	78.4	92.6	103.5	66.5	78.6	151.0	89.7	109.7
0960	Flooring	104.2	98.5	102.6	113.6	86.0	105.8	109.1	100.8	106.8	131.2	86.0	118.5	95.7	58.9	85.4	124.7	88.5	114.6
0970, 0990	Wall Finishes & Painting/Coating	104.4	96.2	99.5	110.3	85.0	95.3	109.8	108.1	108.8	117.8	83.6	97.6	110.3	59.3	80.1	113.7	100.3	105.8
09	**FINISHES**	104.1	93.9	98.6	108.7	81.9	94.2	117.8	102.8	109.7	129.3	81.1	103.2	103.8	65.5	83.1	128.0	92.1	108.6
COVERS	**DIVS. 10 - 14, 25, 28, 41, 43, 44, 46**	131.1	69.7	117.4	131.1	63.4	116.0	131.1	95.4	123.2	131.1	75.6	118.8	131.1	60.8	115.5	131.1	90.8	122.1
21, 22, 23	**FIRE SUPPRESSION, PLUMBING & HVAC**	104.2	91.2	99.1	103.8	92.4	99.3	105.1	99.6	103.0	104.0	84.2	96.2	103.8	79.5	94.3	105.2	87.4	98.2
26, 27, 3370	**ELECTRICAL, COMMUNICATIONS & UTIL.**	109.6	101.2	105.4	113.2	83.3	98.5	106.8	103.8	105.3	111.6	65.2	88.8	110.5	58.7	85.0	108.7	78.6	93.9
MF2018	**WEIGHTED AVERAGE**	107.8	95.0	102.4	113.3	85.3	101.4	116.0	104.2	111.0	120.2	79.9	103.2	110.8	70.3	93.7	117.9	92.4	107.2

DIVISION		CANADA																	
		VICTORIA, BRITISH COLUMBIA			WHITEHORSE, YUKON			WINDSOR, ONTARIO			WINNIPEG, MANITOBA			YARMOUTH, NOVA SCOTIA			YELLOWKNIFE, NWT		
		MAT.	INST.	TOTAL	MAT.	INST.	TOTAL	MAT.	INST.	TOTAL	MAT.	INST.	TOTAL	MAT.	INST.	TOTAL	MAT.	INST.	TOTAL
015433	**CONTRACTOR EQUIPMENT**		106.3	106.3		132.8	132.8		99.5	99.5		125.5	125.5		98.9	98.9		130.7	130.7
0241, 31 - 34	**SITE & INFRASTRUCTURE, DEMOLITION**	124.3	101.3	108.4	135.7	119.6	124.6	90.8	96.0	94.4	113.7	115.3	114.8	111.7	92.0	98.1	146.3	123.6	130.6
0310	Concrete Forming & Accessories	114.9	88.9	92.7	135.0	56.7	68.2	122.8	90.6	95.3	130.4	65.4	75.0	111.8	67.8	74.3	137.3	75.8	84.8
0320	Concrete Reinforcing	111.7	95.0	103.6	167.7	63.3	117.2	95.2	101.9	98.5	127.3	61.0	95.3	156.0	47.6	103.6	138.6	65.8	103.4
0330	Cast-in-Place Concrete	117.6	87.0	106.2	152.4	71.2	122.2	109.3	99.7	105.7	143.6	72.0	116.9	121.6	66.8	101.2	174.9	87.6	142.4
03	**CONCRETE**	129.4	89.5	111.9	144.0	64.8	109.1	96.2	95.9	96.0	127.0	68.4	101.3	130.4	65.0	101.7	150.7	79.5	119.4
04	**MASONRY**	175.0	85.5	120.0	246.5	58.1	130.7	141.9	100.1	116.3	194.6	65.3	115.2	208.1	65.6	120.5	234.9	68.9	132.9
05	**METALS**	107.7	92.2	103.2	145.1	90.5	129.1	114.2	97.3	109.2	141.6	88.0	125.8	125.6	75.7	111.0	140.5	92.7	126.4
06	**WOOD, PLASTICS & COMPOSITES**	100.0	88.3	93.6	121.3	55.3	85.0	117.6	89.0	101.9	107.7	66.0	84.8	115.9	67.4	89.2	130.1	77.1	100.9
07	**THERMAL & MOISTURE PROTECTION**	126.9	86.3	109.2	145.8	63.3	109.9	116.4	99.1	108.9	136.1	70.3	107.5	130.1	67.6	102.9	143.7	78.8	115.5
08	**OPENINGS**	86.8	82.9	85.9	99.5	53.0	88.6	78.7	88.9	81.1	87.5	59.7	81.0	91.3	61.3	84.3	93.9	65.5	87.3
0920	Plaster & Gypsum Board	109.1	87.7	94.7	178.3	53.0	94.0	111.4	88.8	96.2	133.0	64.2	86.7	145.0	66.5	92.2	192.2	75.5	113.7
0950, 0980	Ceilings & Acoustic Treatment	105.8	87.7	93.6	169.3	53.0	90.9	98.2	88.8	91.8	145.3	64.2	90.6	123.8	66.5	85.2	169.0	75.5	105.9
0960	Flooring	112.0	67.4	99.5	122.4	55.2	103.5	104.2	95.9	101.9	115.9	68.5	102.6	107.6	58.9	93.9	121.9	82.9	111.0
0970, 0990	Wall Finishes & Painting/Coating	113.8	100.3	105.8	120.7	53.1	80.7	104.4	97.1	100.1	113.9	53.5	78.1	117.8	59.3	83.2	123.3	75.9	95.2
09	**FINISHES**	110.7	86.8	97.8	143.7	55.8	96.1	101.6	92.1	96.5	126.9	65.0	93.4	118.7	65.5	89.9	147.5	76.6	109.1
COVERS	**DIVS. 10 - 14, 25, 28, 41, 43, 44, 46**	131.1	65.8	116.6	131.1	60.6	115.4	131.1	69.2	117.3	131.1	63.8	116.1	131.1	60.8	115.5	131.1	63.9	116.1
21, 22, 23	**FIRE SUPPRESSION, PLUMBING & HVAC**	103.9	88.7	97.9	104.3	71.6	91.5	104.2	90.7	98.9	105.2	64.6	89.2	104.0	79.5	94.4	104.8	88.5	98.4
26, 27, 3370	**ELECTRICAL, COMMUNICATIONS & UTIL.**	112.8	82.1	97.7	130.5	57.1	94.3	114.5	100.4	107.6	113.1	63.4	88.6	111.3	58.7	85.4	125.8	77.4	102.0
MF2018	**WEIGHTED AVERAGE**	114.1	87.6	102.9	132.2	69.2	105.6	107.1	94.4	101.8	122.0	71.2	100.5	119.5	70.2	98.7	131.4	83.2	111.0

Location Factors - Commercial

Costs shown in RSMeans cost data publications are based on national averages for materials and installation. To adjust these costs to a specific location, simply multiply the base cost by the factor and divide by 100 for that city. The data is arranged alphabetically by state and postal zip code numbers. For a city not listed, use the factor for a nearby city with similar economic characteristics.

STATE/ZIP	CITY	MAT.	INST.	TOTAL
ALABAMA				
350-352	Birmingham	97.4	71.8	86.6
354	Tuscaloosa	97.1	71.4	86.3
355	Jasper	97.8	69.8	86.0
356	Decatur	97.1	69.2	85.3
357-358	Huntsville	97.1	71.7	86.4
359	Gadsden	97.3	70.2	85.9
360-361	Montgomery	96.3	73.1	86.5
362	Anniston	95.9	67.7	84.0
363	Dothan	96.2	72.7	86.3
364	Evergreen	95.8	68.3	84.2
365-366	Mobile	96.7	68.7	84.9
367	Selma	95.9	72.5	86.0
368	Phenix City	96.8	70.8	85.8
369	Butler	96.1	70.6	85.3
ALASKA				
995-996	Anchorage	118.5	113.1	116.2
997	Fairbanks	118.8	113.0	116.3
998	Juneau	116.8	113.1	115.2
999	Ketchikan	128.4	113.4	122.1
ARIZONA				
850,853	Phoenix	100.6	72.3	88.6
851,852	Mesa/Tempe	98.7	71.5	87.2
855	Globe	99.5	71.4	87.6
856-857	Tucson	97.7	70.2	86.1
859	Show Low	99.7	71.5	87.8
860	Flagstaff	102.4	70.5	88.9
863	Prescott	99.9	71.3	87.8
864	Kingman	98.3	70.4	86.5
865	Chambers	98.3	71.8	87.1
ARKANSAS				
716	Pine Bluff	96.0	63.6	82.3
717	Camden	94.0	64.7	81.6
718	Texarkana	94.8	63.3	81.5
719	Hot Springs	93.5	64.4	81.2
720-722	Little Rock	94.5	65.4	82.2
723	West Memphis	93.3	69.8	83.3
724	Jonesboro	93.8	66.6	82.3
725	Batesville	91.6	63.6	79.8
726	Harrison	92.9	62.8	80.2
727	Fayetteville	90.5	64.1	79.3
728	Russellville	91.7	63.1	79.6
729	Fort Smith	94.1	62.6	80.8
CALIFORNIA				
900-902	Los Angeles	98.5	129.8	111.7
903-905	Inglewood	93.8	128.8	108.6
906-908	Long Beach	95.4	128.8	109.5
910-912	Pasadena	94.8	128.6	109.1
913-916	Van Nuys	97.8	128.6	110.8
917-918	Alhambra	96.7	128.6	110.2
919-921	San Diego	99.7	121.7	109.0
922	Palm Springs	97.4	125.3	109.2
923-924	San Bernardino	95.0	126.3	108.2
925	Riverside	99.3	126.6	110.8
926-927	Santa Ana	97.0	126.3	109.4
928	Anaheim	99.3	126.6	110.9
930	Oxnard	98.0	126.7	110.1
931	Santa Barbara	97.2	126.1	109.4
932-933	Bakersfield	98.9	124.6	109.8
934	San Luis Obispo	98.2	125.8	109.9
935	Mojave	95.4	124.4	107.7
936-938	Fresno	98.5	129.4	111.5
939	Salinas	99.1	137.2	115.2
940-941	San Francisco	107.4	158.4	129.0
942,956-958	Sacramento	100.8	133.0	114.4
943	Palo Alto	99.2	153.8	122.3
944	San Mateo	101.5	152.0	122.9
945	Vallejo	100.2	141.6	117.7
946	Oakland	103.4	152.0	123.9
947	Berkeley	103.0	151.9	123.7
948	Richmond	102.4	146.2	120.9
949	San Rafael	104.5	149.1	123.4
950	Santa Cruz	104.6	137.5	118.5

STATE/ZIP	CITY	MAT.	INST.	TOTAL
CALIFORNIA (CONT'D)				
951	San Jose	102.6	153.5	124.2
952	Stockton	100.7	131.3	113.7
953	Modesto	100.6	130.6	113.3
954	Santa Rosa	101.2	148.4	121.2
955	Eureka	102.5	134.1	115.9
959	Marysville	101.7	131.3	114.2
960	Redding	107.6	131.5	117.7
961	Susanville	107.3	131.6	117.5
COLORADO				
800-802	Denver	103.0	75.4	91.3
803	Boulder	98.7	77.0	89.6
804	Golden	100.9	74.7	89.8
805	Fort Collins	102.2	73.5	90.1
806	Greeley	99.3	73.4	88.4
807	Fort Morgan	99.3	73.5	88.4
808-809	Colorado Springs	101.2	71.5	88.6
810	Pueblo	101.8	70.0	88.3
811	Alamosa	103.6	70.3	89.5
812	Salida	103.1	70.9	89.5
813	Durango	103.9	65.6	87.8
814	Montrose	102.6	65.8	87.1
815	Grand Junction	105.7	70.5	90.8
816	Glenwood Springs	103.9	66.8	88.2
CONNECTICUT				
060	New Britain	98.1	116.7	106.0
061	Hartford	99.6	118.0	107.3
062	Willimantic	98.7	117.1	106.5
063	New London	95.2	117.2	104.5
064	Meriden	97.2	117.0	105.6
065	New Haven	99.9	117.0	107.1
066	Bridgeport	99.3	117.0	106.8
067	Waterbury	98.9	117.2	106.6
068	Norwalk	98.8	116.7	106.4
069	Stamford	99.0	123.2	109.2
D.C.				
200-205	Washington	101.1	87.5	95.4
DELAWARE				
197	Newark	98.3	109.7	103.1
198	Wilmington	97.9	109.7	102.9
199	Dover	99.0	109.6	103.5
FLORIDA				
320,322	Jacksonville	96.9	66.6	84.1
321	Daytona Beach	97.2	69.5	85.5
323	Tallahassee	98.2	67.7	85.3
324	Panama City	98.6	68.3	85.8
325	Pensacola	101.2	65.8	86.2
326,344	Gainesville	98.8	65.8	84.8
327-328,347	Orlando	99.0	66.6	85.3
329	Melbourne	100.1	70.3	87.5
330-332,340	Miami	96.7	68.0	84.6
333	Fort Lauderdale	96.1	66.4	83.6
334,349	West Palm Beach	95.1	64.1	82.0
335-336,346	Tampa	97.6	68.4	85.3
337	St. Petersburg	99.7	66.8	85.8
338	Lakeland	97.1	67.6	84.6
339,341	Fort Myers	96.5	67.7	84.3
342	Sarasota	98.8	67.0	85.4
GEORGIA				
300-303,399	Atlanta	98.2	75.0	88.4
304	Statesboro	97.8	69.7	85.9
305	Gainesville	96.2	65.6	83.2
306	Athens	95.6	66.9	83.5
307	Dalton	97.5	71.4	86.5
308-309	Augusta	96.0	73.9	86.6
310-312	Macon	95.2	73.1	85.8
313-314	Savannah	97.3	72.7	86.9
315	Waycross	96.8	70.0	85.5
316	Valdosta	96.7	66.2	83.8
317,398	Albany	96.5	71.9	86.1
318-319	Columbus	96.5	72.2	86.2

STATE/ZIP	CITY	MAT.	INST.	TOTAL
HAWAII				
967	Hilo	114.8	118.4	116.3
968	Honolulu	119.1	118.5	118.9
STATES & POSS.				
969	Guam	137.1	51.4	100.9
IDAHO				
832	Pocatello	102.0	78.4	92.0
833	Twin Falls	103.2	76.8	92.1
834	Idaho Falls	100.5	79.9	91.8
835	Lewiston	108.2	83.5	97.8
836-837	Boise	101.0	79.9	92.1
838	Coeur d'Alene	108.1	84.4	98.1
ILLINOIS				
600-603	North Suburban	98.3	142.9	117.1
604	Joliet	98.3	138.8	115.4
605	South Suburban	98.3	142.8	117.1
606-608	Chicago	100.1	145.6	119.3
609	Kankakee	95.0	134.9	111.9
610-611	Rockford	96.6	125.5	108.8
612	Rock Island	94.6	99.5	96.7
613	La Salle	95.9	127.0	109.0
614	Galesburg	95.7	108.5	101.1
615-616	Peoria	97.6	108.8	102.3
617	Bloomington	95.0	109.5	101.1
618-619	Champaign	98.5	110.5	103.6
620-622	East St. Louis	93.8	110.4	100.9
623	Quincy	95.7	104.2	99.3
624	Effingham	95.1	109.1	101.0
625	Decatur	96.6	110.1	102.3
626-627	Springfield	97.5	108.3	102.1
628	Centralia	92.8	111.2	100.5
629	Carbondale	92.5	109.6	99.7
INDIANA				
460	Anderson	96.0	79.8	89.1
461-462	Indianapolis	98.7	83.2	92.1
463-464	Gary	97.5	108.1	102.0
465-466	South Bend	98.4	81.8	91.4
467-468	Fort Wayne	96.6	76.3	88.0
469	Kokomo	94.2	81.1	88.6
470	Lawrenceburg	92.5	79.6	87.0
471	New Albany	93.9	77.3	86.9
472	Columbus	96.1	80.5	89.5
473	Muncie	96.2	78.2	88.6
474	Bloomington	97.8	81.1	90.7
475	Washington	94.6	84.8	90.4
476-477	Evansville	95.5	82.7	90.1
478	Terre Haute	96.2	81.7	90.1
479	Lafayette	95.8	78.1	88.3
IOWA				
500-503,509	Des Moines	97.2	90.2	94.3
504	Mason City	95.3	74.0	86.3
505	Fort Dodge	95.5	73.7	86.3
506-507	Waterloo	96.8	77.4	88.6
508	Creston	95.7	84.2	90.8
510-511	Sioux City	97.7	81.5	90.9
512	Sibley	96.8	62.4	82.3
513	Spencer	98.2	62.3	83.0
514	Carroll	95.5	82.4	89.9
515	Council Bluffs	98.9	79.2	90.6
516	Shenandoah	96.1	81.9	90.1
520	Dubuque	97.2	79.9	89.9
521	Decorah	96.4	73.7	86.8
522-524	Cedar Rapids	98.0	83.7	92.0
525	Ottumwa	96.3	78.7	88.9
526	Burlington	95.7	84.3	90.9
527-528	Davenport	97.1	94.5	96.0
KANSAS				
660-662	Kansas City	97.3	98.6	97.8
664-666	Topeka	98.6	76.5	89.3
667	Fort Scott	96.0	77.1	88.0
668	Emporia	96.1	76.5	87.8
669	Belleville	97.9	70.7	86.4
670-672	Wichita	97.0	72.6	86.7
673	Independence	97.2	76.0	88.2
674	Salina	97.2	73.5	87.2
675	Hutchinson	92.5	71.9	83.8
676	Hays	96.5	72.0	86.1
677	Colby	97.3	74.0	87.4

STATE/ZIP	CITY	MAT.	INST.	TOTAL
KANSAS (CONT'D)				
678	Dodge City	98.6	73.3	87.9
679	Liberal	96.5	71.7	86.0
KENTUCKY				
400-402	Louisville	93.9	79.6	87.8
403-405	Lexington	93.1	78.1	86.7
406	Frankfort	95.4	80.3	89.0
407-409	Corbin	90.8	78.0	85.4
410	Covington	94.3	77.8	87.3
411-412	Ashland	93.2	89.7	91.7
413-414	Campton	94.4	78.0	87.5
415-416	Pikeville	95.8	85.4	91.4
417-418	Hazard	93.8	78.7	87.4
420	Paducah	92.4	83.2	88.6
421-422	Bowling Green	94.6	77.2	87.3
423	Owensboro	94.7	81.0	88.9
424	Henderson	92.2	81.9	87.8
425-426	Somerset	91.6	79.5	86.5
427	Elizabethtown	91.3	75.6	84.7
LOUISIANA				
700-701	New Orleans	99.0	68.5	86.1
703	Thibodaux	96.0	67.0	83.8
704	Hammond	93.7	64.4	81.3
705	Lafayette	95.4	67.9	83.8
706	Lake Charles	95.6	68.2	84.1
707-708	Baton Rouge	96.6	69.3	85.1
710-711	Shreveport	97.1	66.3	84.1
712	Monroe	95.8	64.4	82.5
713-714	Alexandria	95.9	65.6	83.1
MAINE				
039	Kittery	92.6	87.0	90.2
040-041	Portland	99.6	85.4	93.6
042	Lewiston	97.5	84.9	92.2
043	Augusta	100.4	85.6	94.2
044	Bangor	97.1	83.9	91.5
045	Bath	95.9	85.4	91.4
046	Machias	95.4	85.3	91.1
047	Houlton	95.6	85.2	91.2
048	Rockland	94.6	85.4	90.7
049	Waterville	96.0	85.2	91.4
MARYLAND				
206	Waldorf	97.3	86.6	92.8
207-208	College Park	97.3	87.7	93.2
209	Silver Spring	96.5	86.7	92.4
210-212	Baltimore	101.8	83.2	93.9
214	Annapolis	100.4	86.3	94.4
215	Cumberland	96.9	83.8	91.4
216	Easton	98.6	74.8	88.5
217	Hagerstown	97.4	85.3	92.3
218	Salisbury	99.0	66.9	85.4
219	Elkton	95.9	86.2	91.8
MASSACHUSETTS				
010-011	Springfield	98.2	108.5	102.5
012	Pittsfield	97.7	102.6	99.7
013	Greenfield	95.9	109.0	101.4
014	Fitchburg	94.6	116.2	103.7
015-016	Worcester	98.1	117.0	106.1
017	Framingham	94.0	124.2	106.8
018	Lowell	97.5	123.8	108.6
019	Lawrence	98.4	124.2	109.3
020-022, 024	Boston	101.9	134.0	115.5
023	Brockton	98.5	116.7	106.2
025	Buzzards Bay	93.2	115.2	102.5
026	Hyannis	95.7	115.0	103.8
027	New Bedford	97.7	115.2	105.1
MICHIGAN				
480,483	Royal Oak	95.1	100.3	97.3
481	Ann Arbor	97.1	99.8	98.2
482	Detroit	100.7	102.0	101.3
484-485	Flint	96.8	88.6	93.3
486	Saginaw	96.5	85.1	91.7
487	Bay City	96.6	85.3	91.8
488-489	Lansing	98.0	86.8	93.3
490	Battle Creek	95.5	80.4	89.1
491	Kalamazoo	95.8	78.6	88.6
492	Jackson	94.0	90.7	92.6
493,495	Grand Rapids	97.9	79.9	90.3
494	Muskegon	94.3	80.0	88.3

STATE/ZIP	CITY	MAT.	INST.	TOTAL
MICHIGAN (CONT'D)				
496	Traverse City	93.4	77.7	86.8
497	Gaylord	94.5	80.6	88.6
498-499	Iron Mountain	96.5	81.7	90.2
MINNESOTA				
550-551	Saint Paul	99.0	117.6	106.9
553-555	Minneapolis	100.7	114.1	106.3
556-558	Duluth	98.6	101.3	99.7
559	Rochester	98.3	98.7	98.5
560	Mankato	95.7	98.5	96.9
561	Windom	94.3	91.2	93.0
562	Willmar	94.0	98.0	95.7
563	St. Cloud	95.1	114.2	103.2
564	Brainerd	95.6	98.6	96.8
565	Detroit Lakes	97.4	92.4	95.3
566	Bemidji	96.7	95.4	96.1
567	Thief River Falls	96.3	89.3	93.3
MISSISSIPPI				
386	Clarksdale	95.3	55.2	78.4
387	Greenville	98.7	65.5	84.7
388	Tupelo	96.6	59.0	80.7
389	Greenwood	96.6	55.5	79.2
390-392	Jackson	97.3	65.6	83.9
393	Meridian	94.6	64.5	81.9
394	Laurel	96.1	57.8	79.9
395	Biloxi	96.2	63.9	82.6
396	McComb	94.4	55.8	78.0
397	Columbus	95.9	59.1	80.4
MISSOURI				
630-631	St. Louis	98.8	104.5	101.2
633	Bowling Green	96.4	95.5	96.0
634	Hannibal	95.3	91.6	93.7
635	Kirksville	98.9	86.5	93.7
636	Flat River	97.3	94.2	96.0
637	Cape Girardeau	97.0	91.2	94.5
638	Sikeston	95.7	87.8	92.3
639	Poplar Bluff	95.1	87.7	92.0
640-641	Kansas City	97.9	103.8	100.4
644-645	St. Joseph	96.1	90.9	93.9
646	Chillicothe	93.7	93.8	93.7
647	Harrisonville	93.2	100.8	96.4
648	Joplin	95.1	77.7	87.7
650-651	Jefferson City	96.4	89.9	93.7
652	Columbia	96.1	90.7	93.8
653	Sedalia	96.4	91.5	94.3
654-655	Rolla	94.1	94.3	94.2
656-658	Springfield	97.6	79.0	89.7
MONTANA				
590-591	Billings	101.6	78.4	91.8
592	Wolf Point	101.4	76.9	91.1
593	Miles City	99.2	77.0	89.8
594	Great Falls	102.9	76.6	91.8
595	Havre	100.4	74.8	89.6
596	Helena	100.9	76.7	90.7
597	Butte	101.5	76.5	90.9
598	Missoula	98.6	77.1	89.5
599	Kalispell	98.1	76.2	88.9
NEBRASKA				
680-681	Omaha	97.6	81.5	90.8
683-685	Lincoln	98.3	79.7	90.5
686	Columbus	96.0	80.7	89.6
687	Norfolk	97.4	77.5	89.0
688	Grand Island	97.4	76.8	88.7
689	Hastings	97.1	78.2	89.1
690	McCook	96.3	73.1	86.5
691	North Platte	96.3	75.6	87.6
692	Valentine	98.6	69.2	86.2
693	Alliance	98.7	73.3	87.9
NEVADA				
889-891	Las Vegas	105.0	105.1	105.0
893	Ely	103.7	93.2	99.2
894-895	Reno	103.6	82.6	94.7
897	Carson City	102.9	82.5	94.3
898	Elko	102.3	85.2	95.0
NEW HAMPSHIRE				
030	Nashua	97.5	93.7	95.9
031	Manchester	98.2	94.5	96.6

STATE/ZIP	CITY	MAT.	INST.	TOTAL
NEW HAMPSHIRE (CONT'D)				
032-033	Concord	98.0	93.2	96.0
034	Keene	94.5	89.0	92.2
035	Littleton	94.4	80.2	88.4
036	Charleston	94.1	88.4	91.7
037	Claremont	93.1	88.4	91.1
038	Portsmouth	94.9	92.4	93.8
NEW JERSEY				
070-071	Newark	100.3	139.7	117.0
072	Elizabeth	97.5	138.3	114.7
073	Jersey City	96.5	137.8	114.0
074-075	Paterson	98.0	137.7	114.8
076	Hackensack	96.1	138.3	114.0
077	Long Branch	95.8	130.9	110.7
078	Dover	96.4	137.5	113.7
079	Summit	96.4	138.3	114.1
080,083	Vineland	96.3	129.8	110.5
081	Camden	98.0	127.8	110.6
082,084	Atlantic City	96.8	133.2	112.2
085-086	Trenton	99.7	128.9	112.0
087	Point Pleasant	98.1	130.8	111.9
088-089	New Brunswick	98.7	135.6	114.3
NEW MEXICO				
870-872	Albuquerque	98.6	72.5	87.6
873	Gallup	99.0	72.5	87.8
874	Farmington	99.0	72.5	87.8
875	Santa Fe	99.0	72.9	88.0
877	Las Vegas	97.2	72.5	86.8
878	Socorro	96.8	72.5	86.6
879	Truth/Consequences	96.6	69.4	85.1
880	Las Cruces	96.9	71.4	86.1
881	Clovis	99.3	72.4	87.9
882	Roswell	100.8	72.5	88.9
883	Carrizozo	101.6	72.5	89.3
884	Tucumcari	99.9	72.4	88.3
NEW YORK				
100-102	New York	99.7	175.2	131.6
103	Staten Island	95.6	176.1	129.6
104	Bronx	93.8	175.3	128.3
105	Mount Vernon	94.1	150.6	118.0
106	White Plains	93.8	152.8	118.8
107	Yonkers	97.9	153.0	121.2
108	New Rochelle	94.4	144.8	115.7
109	Suffern	94.1	130.2	109.4
110	Queens	100.0	177.3	132.6
111	Long Island City	101.6	177.3	133.6
112	Brooklyn	101.9	177.3	133.8
113	Flushing	102.1	177.3	133.9
114	Jamaica	100.3	177.3	132.8
115,117,118	Hicksville	99.9	157.1	124.1
116	Far Rockaway	102.2	177.3	133.9
119	Riverhead	100.6	153.5	122.9
120-122	Albany	96.8	111.2	102.8
123	Schenectady	97.0	109.9	102.4
124	Kingston	100.6	133.8	114.6
125-126	Poughkeepsie	99.8	136.8	115.4
127	Monticello	99.1	134.6	114.1
128	Glens Falls	92.0	107.3	98.5
129	Plattsburgh	97.0	96.6	96.8
130-132	Syracuse	97.8	98.3	98.0
133-135	Utica	96.0	96.6	96.2
136	Watertown	97.7	96.8	97.4
137-139	Binghamton	97.3	101.5	99.1
140-142	Buffalo	100.8	110.3	104.8
143	Niagara Falls	97.4	107.5	101.6
144-146	Rochester	100.3	99.2	99.8
147	Jamestown	96.4	96.0	96.3
148-149	Elmira	96.3	102.5	98.9
NORTH CAROLINA				
270,272-274	Greensboro	98.5	67.2	85.3
271	Winston-Salem	98.2	67.3	85.2
275-276	Raleigh	97.4	67.0	84.5
277	Durham	100.1	67.2	86.2
278	Rocky Mount	96.0	66.6	83.6
279	Elizabeth City	96.9	68.3	84.8
280	Gastonia	97.0	66.8	84.2
281-282	Charlotte	97.4	67.6	84.8
283	Fayetteville	100.2	66.1	85.8
284	Wilmington	96.0	65.2	83.0
285	Kinston	94.4	65.5	82.2

Location Factors - Commercial

STATE/ZIP	CITY	MAT.	INST.	TOTAL
NORTH CAROLINA (CONT'D)				
286	Hickory	94.7	67.3	83.1
287-288	Asheville	96.4	66.6	83.8
289	Murphy	95.5	64.7	82.5
NORTH DAKOTA				
580-581	Fargo	99.7	80.3	91.5
582	Grand Forks	99.4	77.6	90.2
583	Devils Lake	99.2	78.8	90.6
584	Jamestown	99.2	78.8	90.6
585	Bismarck	98.9	81.6	91.6
586	Dickinson	99.9	77.9	90.6
587	Minot	99.3	78.6	90.5
588	Williston	98.4	77.7	89.6
OHIO				
430-432	Columbus	98.4	82.0	91.4
433	Marion	94.3	86.7	91.1
434-436	Toledo	97.6	91.2	94.9
437-438	Zanesville	94.9	85.7	91.0
439	Steubenville	96.1	90.0	93.5
440	Lorain	98.4	84.7	92.6
441	Cleveland	99.1	91.6	95.9
442-443	Akron	99.4	86.0	93.8
444-445	Youngstown	98.7	81.9	91.6
446-447	Canton	98.9	79.9	90.9
448-449	Mansfield	96.4	84.2	91.3
450	Hamilton	94.8	79.5	88.3
451-452	Cincinnati	96.1	80.0	89.3
453-454	Dayton	94.8	79.2	88.2
455	Springfield	94.8	79.7	88.4
456	Chillicothe	94.3	88.5	91.8
457	Athens	97.1	84.8	91.9
458	Lima	97.2	81.7	90.7
OKLAHOMA				
730-731	Oklahoma City	96.4	68.7	84.7
734	Ardmore	95.2	65.6	82.7
735	Lawton	97.3	65.8	84.0
736	Clinton	96.4	65.5	83.3
737	Enid	97.0	66.0	83.9
738	Woodward	95.3	62.9	81.6
739	Guymon	96.3	63.3	82.4
740-741	Tulsa	95.9	65.8	83.2
743	Miami	92.7	65.1	81.1
744	Muskogee	95.0	62.9	81.4
745	McAlester	92.4	60.5	78.9
746	Ponca City	93.1	64.0	80.8
747	Durant	93.1	65.1	81.3
748	Shawnee	94.5	65.2	82.2
749	Poteau	92.3	64.9	80.7
OREGON				
970-972	Portland	102.1	102.8	102.4
973	Salem	104.4	100.8	102.9
974	Eugene	101.7	98.7	100.5
975	Medford	103.3	97.4	100.8
976	Klamath Falls	103.5	97.3	100.9
977	Bend	102.6	100.2	101.6
978	Pendleton	98.6	100.3	99.3
979	Vale	96.2	87.0	92.3
PENNSYLVANIA				
150-152	Pittsburgh	100.7	102.7	101.5
153	Washington	97.5	100.2	98.6
154	Uniontown	97.8	99.8	98.6
155	Bedford	98.8	92.1	96.0
156	Greensburg	98.7	96.3	97.7
157	Indiana	97.6	97.7	97.7
158	Dubois	99.3	96.0	97.9
159	Johnstown	98.9	91.9	95.9
160	Butler	91.3	100.6	95.2
161	New Castle	91.4	97.8	94.1
162	Kittanning	91.8	97.9	94.4
163	Oil City	91.3	97.8	94.0
164-165	Erie	93.2	93.9	93.5
166	Altoona	93.3	92.8	93.1
167	Bradford	94.9	97.1	95.8
168	State College	94.5	94.6	94.5
169	Wellsboro	95.5	92.9	94.4
170-171	Harrisburg	98.7	93.0	96.3
172	Chambersburg	95.1	89.4	92.7
173-174	York	95.2	93.0	94.3
175-176	Lancaster	93.8	94.5	94.1

STATE/ZIP	CITY	MAT.	INST.	TOTAL
PENNSYLVANIA (CONT'D)				
177	Williamsport	92.5	92.3	92.4
178	Sunbury	94.6	92.7	93.8
179	Pottsville	93.7	96.1	94.7
180	Lehigh Valley	94.9	111.6	102.0
181	Allentown	96.7	105.8	100.6
182	Hazleton	94.4	95.9	95.0
183	Stroudsburg	94.3	104.7	98.7
184-185	Scranton	97.4	96.5	97.1
186-187	Wilkes-Barre	94.0	95.5	94.7
188	Montrose	93.8	96.8	95.1
189	Doylestown	94.0	127.1	108.0
190-191	Philadelphia	100.2	138.7	116.5
193	Westchester	95.8	127.9	109.3
194	Norristown	94.7	128.0	108.8
195-196	Reading	96.5	101.5	98.6
PUERTO RICO				
009	San Juan	98.5	29.8	69.4
RHODE ISLAND				
028	Newport	97.0	114.7	104.5
029	Providence	99.3	115.0	105.9
SOUTH CAROLINA				
290-292	Columbia	97.6	68.9	85.5
293	Spartanburg	97.2	68.3	85.0
294	Charleston	98.7	68.2	85.8
295	Florence	96.9	68.6	84.9
296	Greenville	97.0	68.2	84.8
297	Rock Hill	96.7	66.4	83.9
298	Aiken	97.6	67.1	84.7
299	Beaufort	98.3	57.5	81.1
SOUTH DAKOTA				
570-571	Sioux Falls	97.7	78.4	89.5
572	Watertown	97.3	69.0	85.4
573	Mitchell	96.3	60.9	81.3
574	Aberdeen	98.6	70.1	86.5
575	Pierre	99.0	68.5	86.1
576	Mobridge	96.9	62.9	82.5
577	Rapid City	98.3	72.0	87.2
TENNESSEE				
370-372	Nashville	98.0	71.8	86.9
373-374	Chattanooga	98.0	68.1	85.3
375,380-381	Memphis	96.6	69.8	85.3
376	Johnson City	98.1	58.6	81.4
377-379	Knoxville	94.8	63.7	81.6
382	McKenzie	96.3	55.1	78.9
383	Jackson	97.6	61.4	82.3
384	Columbia	94.8	66.5	82.8
385	Cookeville	96.2	56.9	79.6
TEXAS				
750	McKinney	97.2	63.5	82.9
751	Waxahachie	97.2	63.5	83.0
752-753	Dallas	98.2	68.0	85.4
754	Greenville	97.4	62.9	82.8
755	Texarkana	96.7	62.0	82.1
756	Longview	97.5	61.0	82.1
757	Tyler	97.8	61.6	82.5
758	Palestine	94.2	60.4	79.9
759	Lufkin	94.7	62.6	81.1
760-761	Fort Worth	97.6	62.9	82.9
762	Denton	97.3	62.9	82.7
763	Wichita Falls	95.0	61.6	80.9
764	Eastland	94.2	60.3	79.8
765	Temple	92.5	58.5	78.2
766-767	Waco	94.4	62.4	80.9
768	Brownwood	97.4	58.7	81.1
769	San Angelo	97.1	59.4	81.2
770-772	Houston	100.4	68.2	86.8
773	Huntsville	98.7	62.8	83.6
774	Wharton	99.8	64.3	84.8
775	Galveston	97.7	66.5	84.5
776-777	Beaumont	97.9	65.5	84.2
778	Bryan	95.2	64.9	82.4
779	Victoria	99.7	62.9	84.2
780	Laredo	96.4	61.1	81.5
781-782	San Antonio	98.3	62.3	83.1
783-784	Corpus Christi	98.9	61.4	83.1
785	McAllen	99.7	56.5	81.4
786-787	Austin	97.2	61.4	82.1

Location Factors - Commercial

STATE/ZIP	CITY	MAT.	INST.	TOTAL
TEXAS (CONT'D)				
788	Del Rio	99.7	60.4	83.1
789	Giddings	96.1	60.4	81.0
790-791	Amarillo	98.2	60.0	82.1
792	Childress	97.6	60.3	81.8
793-794	Lubbock	99.2	62.9	83.9
795-796	Abilene	97.8	61.4	82.4
797	Midland	99.7	62.2	83.8
798-799,885	El Paso	95.9	63.6	82.3
UTAH				
840-841	Salt Lake City	103.5	71.7	90.1
842,844	Ogden	98.8	71.7	87.4
843	Logan	100.9	71.7	88.5
845	Price	101.4	70.0	88.1
846-847	Provo	101.2	71.2	88.5
VERMONT				
050	White River Jct.	98.2	79.8	90.4
051	Bellows Falls	96.7	92.1	94.8
052	Bennington	97.0	88.6	93.4
053	Brattleboro	97.4	92.1	95.2
054	Burlington	101.9	79.8	92.5
056	Montpelier	100.2	84.4	93.5
057	Rutland	98.9	79.4	90.7
058	St. Johnsbury	98.1	79.3	90.2
059	Guildhall	96.8	79.0	89.3
VIRGINIA				
220-221	Fairfax	99.5	83.6	92.8
222	Arlington	100.4	84.2	93.6
223	Alexandria	99.6	84.8	93.3
224-225	Fredericksburg	98.3	78.7	90.0
226	Winchester	98.9	79.5	90.7
227	Culpeper	98.7	83.0	92.1
228	Harrisonburg	99.0	68.8	86.3
229	Charlottesville	99.4	71.0	87.4
230-232	Richmond	99.2	72.8	88.1
233-235	Norfolk	99.9	68.7	86.7
236	Newport News	98.8	68.3	85.9
237	Portsmouth	98.3	67.8	85.4
238	Petersburg	98.7	72.6	87.7
239	Farmville	97.9	65.3	84.1
240-241	Roanoke	100.1	73.6	88.9
242	Bristol	98.2	59.3	81.8
243	Pulaski	97.8	67.1	84.8
244	Staunton	98.6	67.9	85.6
245	Lynchburg	98.7	73.8	88.2
246	Grundy	98.1	58.8	81.5
WASHINGTON				
980-981,987	Seattle	106.3	112.5	108.9
982	Everett	105.1	104.0	104.6
983-984	Tacoma	105.4	101.3	103.7
985	Olympia	104.0	101.8	103.1
986	Vancouver	106.6	102.8	105.0
988	Wenatchee	105.6	88.7	98.5
989	Yakima	105.6	95.4	101.3
990-992	Spokane	100.1	82.4	92.6
993	Richland	99.8	90.0	95.6
994	Clarkston	99.1	81.9	91.8
WEST VIRGINIA				
247-248	Bluefield	97.0	88.1	93.2
249	Lewisburg	98.7	86.1	93.4
250-253	Charleston	96.5	89.4	93.5
254	Martinsburg	96.7	84.1	91.4
255-257	Huntington	97.8	92.2	95.4
258-259	Beckley	95.2	87.7	92.0
260	Wheeling	99.6	89.6	95.4
261	Parkersburg	98.6	88.6	94.4
262	Buckhannon	98.3	88.8	94.3
263-264	Clarksburg	98.9	89.5	94.9
265	Morgantown	98.8	89.6	94.9
266	Gassaway	98.3	89.0	94.3
267	Romney	98.3	87.2	93.6
268	Petersburg	98.0	85.9	92.9
WISCONSIN				
530,532	Milwaukee	99.2	107.9	102.9
531	Kenosha	99.4	102.0	100.5
534	Racine	98.7	102.3	100.2
535	Beloit	98.6	98.9	98.7
537	Madison	99.7	98.1	99.0

STATE/ZIP	CITY	MAT.	INST.	TOTAL
WISCONSIN (CONT'D)				
538	Lancaster	96.7	94.3	95.7
539	Portage	95.2	98.0	96.4
540	New Richmond	95.4	94.6	95.0
541-543	Green Bay	99.4	93.4	96.9
544	Wausau	94.7	93.4	94.2
545	Rhinelander	97.9	92.2	95.5
546	La Crosse	95.9	93.5	94.9
547	Eau Claire	97.5	94.1	96.1
548	Superior	95.1	96.5	95.7
549	Oshkosh	95.4	91.9	93.9
WYOMING				
820	Cheyenne	101.1	72.7	89.1
821	Yellowstone Nat'l Park	99.0	71.9	87.6
822	Wheatland	100.3	67.1	86.2
823	Rawlins	102.0	71.5	89.1
824	Worland	99.7	70.7	87.5
825	Riverton	101.0	68.6	87.3
826	Casper	101.5	68.9	87.7
827	Newcastle	99.6	71.1	87.5
828	Sheridan	102.5	70.2	88.8
829-831	Rock Springs	103.4	71.4	89.9
CANADIAN FACTORS (reflect Canadian currency)				
ALBERTA				
	Calgary	119.7	95.6	109.5
	Edmonton	121.8	95.9	110.9
	Fort McMurray	121.9	88.4	107.7
	Lethbridge	117.0	87.9	104.7
	Lloydminster	111.5	84.7	100.2
	Medicine Hat	111.7	84.0	100.0
	Red Deer	112.1	84.0	100.2
BRITISH COLUMBIA				
	Kamloops	113.0	83.5	100.5
	Prince George	114.0	82.8	100.8
	Vancouver	117.9	92.4	107.2
	Victoria	114.1	87.6	102.9
MANITOBA				
	Brandon	120.6	70.5	99.4
	Portage la Prairie	111.6	69.1	93.6
	Winnipeg	122.0	71.2	100.5
NEW BRUNSWICK				
	Bathurst	110.7	62.3	90.2
	Dalhousie	110.2	62.5	90.0
	Fredericton	119.3	69.0	98.0
	Moncton	110.9	76.0	96.2
	Newcastle	110.7	62.9	90.5
	Saint John	113.2	75.7	97.3
NEWFOUNDLAND				
	Corner Brook	124.7	68.4	100.9
	St. John's	122.6	87.4	107.7
NORTHWEST TERRITORIES				
	Yellowknife	131.4	83.2	111.0
NOVA SCOTIA				
	Bridgewater	111.3	70.3	94.0
	Dartmouth	121.5	70.2	99.8
	Halifax	117.9	88.0	105.2
	New Glasgow	119.6	70.2	98.7
	Sydney	117.8	70.2	97.7
	Truro	110.8	70.3	93.7
	Yarmouth	119.5	70.2	98.7
ONTARIO				
	Barrie	116.5	87.1	104.1
	Brantford	113.1	90.7	103.6
	Cornwall	113.1	87.3	102.2
	Hamilton	116.3	98.8	108.9
	Kingston	114.1	87.4	102.8
	Kitchener	108.6	94.6	102.7
	London	116.3	96.5	108.0
	North Bay	122.5	85.2	106.8
	Oshawa	110.8	97.2	105.0
	Ottawa	117.5	96.8	108.7
	Owen Sound	116.7	85.4	103.4
	Peterborough	113.1	87.1	102.1
	Sarnia	113.2	91.3	103.9

Location Factors - Commercial

STATE/ZIP	CITY	MAT.	INST.	TOTAL
ONTARIO (CONT'D)				
	Sault Ste. Marie	108.6	88.7	100.2
	St. Catharines	106.6	95.6	101.9
	Sudbury	106.1	93.6	100.8
	Thunder Bay	107.8	95.0	102.4
	Timmins	113.3	85.3	101.4
	Toronto	116.0	104.2	111.0
	Windsor	107.1	94.4	101.8
PRINCE EDWARD ISLAND				
	Charlottetown	121.3	60.2	95.5
	Summerside	121.9	57.9	94.9
QUEBEC				
	Cap-de-la-Madeleine	110.6	80.0	97.7
	Charlesbourg	110.6	80.0	97.7
	Chicoutimi	110.4	87.6	100.7
	Gatineau	110.3	79.8	97.4
	Granby	110.5	79.7	97.5
	Hull	110.5	79.8	97.5
	Joliette	110.8	80.0	97.8
	Laval	110.7	81.6	98.4
	Montreal	119.1	90.1	106.9
	Quebec City	118.7	90.4	106.7
	Rimouski	110.2	87.6	100.6
	Rouyn-Noranda	110.3	79.8	97.4
	Saint-Hyacinthe	109.8	79.8	97.1
	Sherbrooke	110.6	79.8	97.6
	Sorel	110.8	80.0	97.8
	Saint-Jerome	110.4	79.8	97.4
	Trois-Rivieres	120.2	79.9	103.2
SASKATCHEWAN				
	Moose Jaw	109.0	63.9	89.9
	Prince Albert	108.2	62.2	88.8
	Regina	123.1	92.4	110.2
	Saskatoon	109.9	88.5	100.9
YUKON				
	Whitehorse	132.2	69.2	105.6

R011105-05 Tips for Accurate Estimating

1. Use pre-printed or columnar forms for orderly sequence of dimensions and locations and for recording telephone quotations.
2. Use only the front side of each paper or form except for certain pre-printed summary forms.
3. Be consistent in listing dimensions: For example, length x width x height. This helps in rechecking to ensure that, the total length of partitions is appropriate for the building area.
4. Use printed (rather than measured) dimensions where given.
5. Add up multiple printed dimensions for a single entry where possible.
6. Measure all other dimensions carefully.
7. Use each set of dimensions to calculate multiple related quantities.
8. Convert foot and inch measurements to decimal feet when listing. Memorize decimal equivalents to .01 parts of a foot (1/8″ equals approximately .01′).
9. Do not "round off" quantities until the final summary.
10. Mark drawings with different colors as items are taken off.
11. Keep similar items together, different items separate.
12. Identify location and drawing numbers to aid in future checking for completeness.
13. Measure or list everything on the drawings or mentioned in the specifications.
14. It may be necessary to list items not called for to make the job complete.
15. Be alert for: Notes on plans such as N.T.S. (not to scale); changes in scale throughout the drawings; reduced size drawings; discrepancies between the specifications and the drawings.
16. Develop a consistent pattern of performing an estimate. For example:
    a. Start the quantity takeoff at the lower floor and move to the next higher floor.
    b. Proceed from the main section of the building to the wings.
    c. Proceed from south to north or vice versa, clockwise or counterclockwise.
    d. Take off floor plan quantities first, elevations next, then detail drawings.
17. List all gross dimensions that can be either used again for different quantities, or used as a rough check of other quantities for verification (exterior perimeter, gross floor area, individual floor areas, etc.).
18. Utilize design symmetry or repetition (repetitive floors, repetitive wings, symmetrical design around a center line, similar room layouts, etc.). Note: Extreme caution is needed here so as not to omit or duplicate an area.
19. Do not convert units until the final total is obtained. For instance, when estimating concrete work, keep all units to the nearest cubic foot, then summarize and convert to cubic yards.
20. When figuring alternatives, it is best to total all items involved in the basic system, then total all items involved in the alternates. Therefore you work with positive numbers in all cases. When adds and deducts are used, it is often confusing whether to add or subtract a portion of an item; especially on a complicated or involved alternate.

R011105-50 Metric Conversion Factors

Description: This table is primarily for converting customary U.S. units in the left hand column to SI metric units in the right hand column. In addition, conversion factors for some commonly encountered Canadian and non-SI metric units are included.

	If You Know		Multiply By		To Find
Length	Inches	x	25.4[a]	=	Millimeters
	Feet	x	0.3048[a]	=	Meters
	Yards	x	0.9144[a]	=	Meters
	Miles (statute)	x	1.609	=	Kilometers
Area	Square inches	x	645.2	=	Square millimeters
	Square feet	x	0.0929	=	Square meters
	Square yards	x	0.8361	=	Square meters
Volume (Capacity)	Cubic inches	x	16,387	=	Cubic millimeters
	Cubic feet	x	0.02832	=	Cubic meters
	Cubic yards	x	0.7646	=	Cubic meters
	Gallons (U.S. liquids)[b]	x	0.003785	=	Cubic meters[c]
	Gallons (Canadian liquid)[b]	x	0.004546	=	Cubic meters[c]
	Ounces (U.S. liquid)[b]	x	29.57	=	Milliliters[c, d]
	Quarts (U.S. liquid)[b]	x	0.9464	=	Liters[c, d]
	Gallons (U.S. liquid)[b]	x	3.785	=	Liters[c, d]
Force	Kilograms force[d]	x	9.807	=	Newtons
	Pounds force	x	4.448	=	Newtons
	Pounds force	x	0.4536	=	Kilograms force[d]
	Kips	x	4448	=	Newtons
	Kips	x	453.6	=	Kilograms force[d]
Pressure, Stress, Strength (Force per unit area)	Kilograms force per square centimeter[d]	x	0.09807	=	Megapascals
	Pounds force per square inch (psi)	x	0.006895	=	Megapascals
	Kips per square inch	x	6.895	=	Megapascals
	Pounds force per square inch (psi)	x	0.07031	=	Kilograms force per square centimeter[d]
	Pounds force per square foot	x	47.88	=	Pascals
	Pounds force per square foot	x	4.882	=	Kilograms force per square meter[d]
Flow	Cubic feet per minute	x	0.4719	=	Liters per second
	Gallons per minute	x	0.0631	=	Liters per second
	Gallons per hour	x	1.05	=	Milliliters per second
Bending Moment Or Torque	Inch-pounds force	x	0.01152	=	Meter-kilograms force[d]
	Inch-pounds force	x	0.1130	=	Newton-meters
	Foot-pounds force	x	0.1383	=	Meter-kilograms force[d]
	Foot-pounds force	x	1.356	=	Newton-meters
	Meter-kilograms force[d]	x	9.807	=	Newton-meters
Mass	Ounces (avoirdupois)	x	28.35	=	Grams
	Pounds (avoirdupois)	x	0.4536	=	Kilograms
	Tons (metric)	x	1000	=	Kilograms
	Tons, short (2000 pounds)	x	907.2	=	Kilograms
	Tons, short (2000 pounds)	x	0.9072	=	Megagrams[e]
Mass per Unit Volume	Pounds mass per cubic foot	x	16.02	=	Kilograms per cubic meter
	Pounds mass per cubic yard	x	0.5933	=	Kilograms per cubic meter
	Pounds mass per gallon (U.S. liquid)[b]	x	119.8	=	Kilograms per cubic meter
	Pounds mass per gallon (Canadian liquid)[b]	x	99.78	=	Kilograms per cubic meter
Temperature	Degrees Fahrenheit	(F-32)/1.8		=	Degrees Celsius
	Degrees Fahrenheit	(F+459.67)/1.8		=	Degrees Kelvin
	Degrees Celsius	C+273.15		=	Degrees Kelvin

[a]The factor given is exact
[b]One U.S. gallon = 0.8327 Canadian gallon
[c]1 liter = 1000 milliliters = 1000 cubic centimeters
1 cubic decimeter = 0.001 cubic meter
[d]Metric but not SI unit
[e]Called "tonne" in England and "metric ton" in other metric countries

R011105-60 Weights and Measures

Measures of Length
1 Mile = 1760 Yards = 5280 Feet
1 Yard = 3 Feet = 36 inches
1 Foot = 12 Inches
1 Mil = 0.001 Inch
1 Fathom = 2 Yards = 6 Feet
1 Rod = 5.5 Yards = 16.5 Feet
1 Hand = 4 Inches
1 Span = 9 Inches
1 Micro-inch = One Millionth Inch or 0.000001 Inch
1 Micron = One Millionth Meter + 0.00003937 Inch

Surveyor's Measure
1 Mile = 8 Furlongs = 80 Chains
1 Furlong = 10 Chains = 220 Yards
1 Chain = 4 Rods = 22 Yards = 66 Feet = 100 Links
1 Link = 7.92 Inches

Square Measure
1 Square Mile = 640 Acres = 6400 Square Chains
1 Acre = 10 Square Chains = 4840 Square Yards = 43,560 Sq. Ft.
1 Square Chain = 16 Square Rods = 484 Square Yards = 4356 Sq. Ft.
1 Square Rod = 30.25 Square Yards = 272.25 Square Feet = 625 Square Lines
1 Square Yard = 9 Square Feet
1 Square Foot = 144 Square Inches
An Acre equals a Square 208.7 Feet per Side

Cubic Measure
1 Cubic Yard = 27 Cubic Feet
1 Cubic Foot = 1728 Cubic Inches
1 Cord of Wood = 4 x 4 x 8 Feet = 128 Cubic Feet
1 Perch of Masonry = 16½ x 1½ x 1 Foot = 24.75 Cubic Feet

Avoirdupois or Commercial Weight
1 Gross or Long Ton = 2240 Pounds
1 Net or Short Ton = 2000 Pounds
1 Pound = 16 Ounces = 7000 Grains
1 Ounce = 16 Drachms = 437.5 Grains
1 Stone = 14 Pounds

Power
1 British Thermal Unit per Hour = 0.2931 Watts
1 Ton (Refrigeration) = 3.517 Kilowatts
1 Horsepower (Boiler) = 9.81 Kilowatts
1 Horsepower (550 ft-lb/s) = 0.746 Kilowatts

Shipping Measure
For Measuring Internal Capacity of a Vessel:
1 Register Ton = 100 Cubic Feet

For Measurement of Cargo:
Approximately 40 Cubic Feet of Merchandise is considered a Shipping Ton, unless that bulk would weigh more than 2000 Pounds, in which case Freight Charge may be based upon weight.

40 Cubic Feet = 32.143 U.S. Bushels = 31.16 Imp. Bushels

Liquid Measure
1 Imperial Gallon = 1.2009 U.S. Gallon = 277.42 Cu. In.
1 Cubic Foot = 7.48 U.S. Gallons

R011110-10 Architectural Fees

Tabulated below are typical percentage fees by project size, for good professional architectural service. Fees may vary from those listed depending upon degree of design difficulty and economic conditions in any particular area.

Rates can be interpolated horizontally and vertically. Various portions of the same project requiring different rates should be adjusted proportionately. For alterations, add 50% to the fee for the first $500,000 of project cost and add 25% to the fee for project cost over $500,000.

Architectural fees tabulated below include Structural, Mechanical, and Electrical Engineering Fees. They do not include the fees for special consultants such as kitchen planning, security, acoustical, interior design, etc

Civil Engineering fees are included in the Architectural fee for project sites requiring minimal design such as city sites. However, separate Civil Engineering fees must be added when utility connections require design, drainage calculations are needed, stepped foundations are required, or provisions are required to protect adjacent wetlands.

	Total Project Size in Thousands of Dollars						
Building Types	100	250	500	1,000	5,000	10,000	50,000
Factories, garages, warehouses, repetitive housing	9.0%	8.0%	7.0%	6.2%	5.3%	4.9%	4.5%
Apartments, banks, schools, libraries, offices, municipal buildings	12.2	12.3	9.2	8.0	7.0	6.6	6.2
Churches, hospitals, homes, laboratories, museums, research	15.0	13.6	12.7	11.9	9.5	8.8	8.0
Memorials, monumental work, decorative furnishings	—	16.0	14.5	13.1	10.0	9.0	8.3

R011110-30 Engineering Fees

Typical **Structural Engineering Fees** based on type of construction and total project size. These fees are included in Architectural Fees.

Type of Construction	Total Project Size (in thousands of dollars)			
	$500	$500-$1,000	$1,000-$5,000	Over $5000
Industrial buildings, factories & warehouses	Technical payroll times 2.0 to 2.5	1.60%	1.25%	1.00%
Hotels, apartments, offices, dormitories, hospitals, public buildings, food stores		2.00%	1.70%	1.20%
Museums, banks, churches and cathedrals		2.00%	1.75%	1.25%
Thin shells, prestressed concrete, earthquake resistive		2.00%	1.75%	1.50%
Parking ramps, auditoriums, stadiums, convention halls, hangars & boiler houses		2.50%	2.00%	1.75%
Special buildings, major alterations, underpinning & future expansion	↓	Add to above 0.5%	Add to above 0.5%	Add to above 0.5%

For complex reinforced concrete or unusually complicated structures, add 20% to 50%.

Typical **Mechanical and Electrical Engineering Fees** are based on the size of the subcontract. The fee structure for both is shown below. These fees are included in Architectural Fees.

Type of Construction	Subcontract Size							
	$25,000	$50,000	$100,000	$225,000	$350,000	$500,000	$750,000	$1,000,000
Simple structures	6.4%	5.7%	4.8%	4.5%	4.4%	4.3%	4.2%	4.1%
Intermediate structures	8.0	7.3	6.5	5.6	5.1	5.0	4.9	4.8
Complex structures	10.1	9.0	9.0	8.0	7.5	7.5	7.0	7.0

For renovations, add 15% to 25% to applicable fee.

General Requirements R0121 Allowances

R012153-10 Repair and Remodeling

Cost figures are based on new construction utilizing the most cost-effective combination of labor, equipment, and material with the work scheduled in proper sequence to allow the various trades to accomplish their work in an efficient manner.

The costs for repair and remodeling work must be modified due to the following factors that may be present in any given repair and remodeling project:

1. Equipment usage curtailment due to the physical limitations of the project, with only hand-operated equipment being used.
2. Increased requirement for shoring and bracing to hold up the building while structural changes are being made and to allow for temporary storage of construction materials on above-grade floors.
3. Material handling becomes more costly due to having to move within the confines of an enclosed building. For multi-story construction, low capacity elevators and stairwells may be the only access to the upper floors.
4. Large amount of cutting and patching and attempting to match the existing construction is required. It is often more economical to remove entire walls rather than create many new door and window openings. This sort of trade-off has to be carefully analyzed.
5. Cost of protection of completed work is increased since the usual sequence of construction usually cannot be accomplished.
6. Economies of scale usually associated with new construction may not be present. If small quantities of components must be custom fabricated due to job requirements, unit costs will naturally increase. Also, if only small work areas are available at a given time, job scheduling between trades becomes difficult and subcontractor quotations may reflect the excessive start-up and shut-down phases of the job.
7. Work may have to be done on other than normal shifts and may have to be done around an existing production facility which has to stay in production during the course of the repair and remodeling.
8. Dust and noise protection of adjoining non-construction areas can involve substantial special protection and alter usual construction methods.
9. Job may be delayed due to unexpected conditions discovered during demolition or removal. These delays ultimately increase construction costs.
10. Piping and ductwork runs may not be as simple as for new construction. Wiring may have to be snaked through walls and floors.
11. Matching "existing construction" may be impossible because materials may no longer be manufactured. Substitutions may be expensive.
12. Weather protection of existing structure requires additional temporary structures to protect building at openings.
13. On small projects, because of local conditions, it may be necessary to pay a tradesman for a minimum of four hours for a task that is completed in one hour.

All of the above areas can contribute to increased costs for a repair and remodeling project. Each of the above factors should be considered in the planning, bidding and construction stage in order to minimize the increased costs associated with repair and remodeling jobs.

R012153-60 Security Factors

Contractors entering, working in, and exiting secure facilities often lose productive time during a normal workday. The recommended allowances in this section are intended to provide for the loss of productivity by increasing labor costs. Note that different costs are associated with searches upon entry only and searches upon entry and exit. Time spent in a queue is unpredictable and not part of these allowances. Contractors should plan ahead for this situation.

Security checkpoints are designed to reflect the level of security required to gain access or egress. An extreme example is when contractors, along with any materials, tools, equipment, and vehicles, must be physically searched and have all materials, tools, equipment, and vehicles inventoried and documented prior to both entry and exit.

Physical searches without going through the documentation process represent the next level and take up less time.

Electronic searches—passing through a detector or x-ray machine with no documentation of materials, tools, equipment, and vehicles—take less time than physical searches.

Visual searches of materials, tools, equipment, and vehicles represent the next level of security.

Finally, access by means of an ID card or displayed sticker takes the least amount of time.

Another consideration is if the searches described above are performed each and every day, or if they are performed only on the first day with access granted by ID card or displayed sticker for the remainder of the project. The figures for this situation have been calculated to represent the initial check-in as described and subsequent entry by ID card or displayed sticker for up to 20 days on site. For the situation described above, where the time period is beyond 20 days, the impact on labor cost is negligible.

There are situations where tradespeople must be accompanied by an escort and observed during the work day. The loss of freedom of movement will slow down productivity for the tradesperson. Costs for the observer have not been included. Those costs are normally born by the owner.

R012157-20 Construction Time Requirements

The table below lists the construction durations for various building types along with their respective project sizes and project values. Design time runs 25% to 40% of construction time.

Building Type	Size S.F.	Project Value	Construction Duration
Industrial/Warehouse	100,000	$8,000,000	14 months
	500,000	$32,000,000	19 months
	1,000,000	$75,000,000	21 months
Offices/Retail	50,000	$7,000,000	15 months
	250,000	$28,000,000	23 months
	500,000	$58,000,000	34 months
Institutional/Hospitals/Laboratory	200,000	$45,000,000	31 months
	500,000	$110,000,000	52 months
	750,000	$160,000,000	55 months
	1,000,000	$210,000,000	60 months

R012909-80 Sales Tax by State

State sales tax on materials is tabulated below (5 states have no sales tax). Many states allow local jurisdictions, such as a county or city, to levy additional sales tax.

Some projects may be sales tax exempt, particularly those constructed with public funds.

State	Tax (%)	State	Tax (%)	State	Tax (%)	State	Tax (%)
Alabama	4	Illinois	6.25	Montana	0	Rhode Island	7
Alaska	0	Indiana	7	Nebraska	5.5	South Carolina	6
Arizona	5.6	Iowa	6	Nevada	6.85	South Dakota	4.5
Arkansas	6.5	Kansas	6.5	New Hampshire	0	Tennessee	7
California	7.25	Kentucky	6	New Jersey	6.625	Texas	6.25
Colorado	2.9	Louisiana	4.45	New Mexico	5.125	Utah	4.85
Connecticut	6.35	Maine	5.5	New York	4	Vermont	6
Delaware	0	Maryland	6	North Carolina	4.75	Virginia	4.3
District of Columbia	6	Massachusetts	6.25	North Dakota	5	Washington	6.5
Florida	6	Michigan	6	Ohio	5.75	West Virginia	6
Georgia	4	Minnesota	6.875	Oklahoma	4.5	Wisconsin	5
Hawaii	4	Mississippi	7	Oregon	0	Wyoming	4
Idaho	6	Missouri	4.225	Pennsylvania	6	Average	5.06 %

Sales Tax by Province (Canada)

GST - a value-added tax, which the government imposes on most goods and services provided in or imported into Canada. PST - a retail sales tax, which five of the provinces impose on the prices of most goods and some services. QST - a value-added tax, similar to the federal GST, which Quebec imposes. HST - Three provinces have combined their retail sales taxes with the federal GST into one harmonized tax.

Province	PST (%)	QST(%)	GST(%)	HST (%)
Alberta	0	0	5	0
British Columbia	7	0	5	0
Manitoba	7	0	5	0
New Brunswick	0	0	0	15
Newfoundland	0	0	0	15
Northwest Territories	0	0	5	0
Nova Scotia	0	0	0	15
Ontario	0	0	0	13
Prince Edward Island	0	0	0	15
Quebec	9.975	0	5	14.975
Saskatchewan	6	0	5	0
Yukon	0	0	5	0

R012909-85 Unemployment Taxes and Social Security Taxes

State unemployment tax rates vary not only from state to state, but also with the experience rating of the contractor. The federal unemployment tax rate is 6.0% of the first $7,000 of wages. This is reduced by a credit of up to 5.4% for timely payment to the state. The minimum federal unemployment tax is 0.6% after all credits.

Social security (FICA) for 2020 is estimated at time of publication to be 7.65% of wages up to $137,100.

R012909-86 Unemployment Tax by State

Information is from the U.S. Department of Labor, state unemployment tax rates.

State	Tax (%)	State	Tax (%)	State	Tax (%)	State	Tax (%)
Alabama	6.80	Illinois	6.93	Montana	6.30	Rhode Island	9.49
Alaska	5.4	Indiana	7.4	Nebraska	5.4	South Carolina	5.46
Arizona	12.76	Iowa	7.5	Nevada	5.4	South Dakota	9.35
Arkansas	14.3	Kansas	7.6	New Hampshire	7.5	Tennessee	10.0
California	6.2	Kentucky	9.3	New Jersey	5.8	Texas	6.5
Colorado	8.15	Louisiana	6.2	New Mexico	5.4	Utah	7.1
Connecticut	6.8	Maine	5.46	New York	9.1	Vermont	7.7
Delaware	8.20	Maryland	7.50	North Carolina	5.76	Virginia	6.21
District of Columbia	7	Massachusetts	12.65	North Dakota	10.74	Washington	7.73
Florida	5.4	Michigan	10.3	Ohio	9	West Virginia	8.5
Georgia	5.4	Minnesota	9.1	Oklahoma	5.5	Wisconsin	12.0
Hawaii	5.6	Mississippi	5.6	Oregon	5.4	Wyoming	8.8
Idaho	5.4	Missouri	8.37	Pennsylvania	11.03	Median	7.40 %

R012909-90 Overtime

One way to improve the completion date of a project or eliminate negative float from a schedule is to compress activity duration times. This can be achieved by increasing the crew size or working overtime with the proposed crew.

To determine the costs of working overtime to compress activity duration times, consider the following examples. Below is an overtime efficiency and cost chart based on a five, six, or seven day week with an eight through twelve hour day. Payroll percentage increases for time and one half and double times are shown for the various working days.

Days per Week	Hours per Day	Production Efficiency					Payroll Cost Factors	
		1st Week	2nd Week	3rd Week	4th Week	Average 4 Weeks	@ 1-1/2 Times	@ 2 Times
5	8	100%	100%	100%	100%	100%	1.000	1.000
	9	100	100	95	90	96	1.056	1.111
	10	100	95	90	85	93	1.100	1.200
	11	95	90	75	65	81	1.136	1.273
	12	90	85	70	60	76	1.167	1.333
6	8	100	100	95	90	96	1.083	1.167
	9	100	95	90	85	93	1.130	1.259
	10	95	90	85	80	88	1.167	1.333
	11	95	85	70	65	79	1.197	1.394
	12	90	80	65	60	74	1.222	1.444
7	8	100	95	85	75	89	1.143	1.286
	9	95	90	80	70	84	1.183	1.365
	10	90	85	75	65	79	1.214	1.429
	11	85	80	65	60	73	1.240	1.481
	12	85	75	60	55	69	1.262	1.524

R013113-40 Builder's Risk Insurance

Builder's risk insurance is insurance on a building during construction. Premiums are paid by the owner or the contractor. Blasting, collapse and underground insurance would raise total insurance costs.

R013113-50 General Contractor's Overhead

There are two distinct types of overhead on a construction project: Project overhead and main office overhead. Project overhead includes those costs at a construction site not directly associated with the installation of construction materials. Examples of project overhead costs include the following:

1. Superintendent
2. Construction office and storage trailers
3. Temporary sanitary facilities
4. Temporary utilities
5. Security fencing
6. Photographs
7. Cleanup
8. Performance and payment bonds

The above project overhead items are also referred to as general requirements and therefore are estimated in Division 1. Division 1 is the first division listed in the CSI MasterFormat but it is usually the last division estimated. The sum of the costs in Divisions 1 through 49 is referred to as the sum of the direct costs.

All construction projects also include indirect costs. The primary components of indirect costs are the contractor's main office overhead and profit. The amount of the main office overhead expense varies depending on the following:

1. Owner's compensation
2. Project managers' and estimators' wages
3. Clerical support wages
4. Office rent and utilities
5. Corporate legal and accounting costs
6. Advertising
7. Automobile expenses
8. Association dues
9. Travel and entertainment expenses

These costs are usually calculated as a percentage of annual sales volume. This percentage can range from 35% for a small contractor doing less than $500,000 to 5% for a large contractor with sales in excess of $100 million.

R013113-55 Installing Contractor's Overhead

Installing contractors (subcontractors) also incur costs for general requirements and main office overhead.

Included within the total incl. overhead and profit costs is a percent mark-up for overhead that includes:

1. Compensation and benefits for office staff and project managers
2. Office rent, utilities, business equipment, and maintenance
3. Corporate legal and accounting costs
4. Advertising
5. Vehicle expenses (for office staff and project managers)
6. Association dues
7. Travel, entertainment
8. Insurance
9. Small tools and equipment

R013113-60 Workers' Compensation Insurance Rates by Trade

The table below tabulates the national averages for workers' compensation insurance rates by trade and type of building. The average "Insurance Rate" is multiplied by the "% of Building Cost" for each trade. This produces the "Workers' Compensation" cost by % of total labor cost, to be added for each trade by building type to determine the weighted average workers' compensation rate for the building types analyzed.

Trade	Insurance Rate (% Labor Cost) Range			Insurance Rate Average	% of Building Cost Office Bldgs.	% of Building Cost Schools & Apts.	% of Building Cost Mfg.	Workers' Compensation Office Bldgs.	Workers' Compensation Schools & Apts.	Workers' Compensation Mfg.
Excavation, Grading, etc.	2.3 %	to	16.9%	9.6%	4.8%	4.9%	4.5%	0.46%	0.47%	0.43%
Piles & Foundations	3.3	to	28.0	15.7	7.1	5.2	8.7	1.11	0.82	1.37
Concrete	3.1	to	25.8	14.4	5.0	14.8	3.7	0.72	2.13	0.53
Masonry	3.3	to	52.1	27.7	6.9	7.5	1.9	1.91	2.08	0.53
Structural Steel	4.4	to	31.8	18.1	10.7	3.9	17.6	1.94	0.71	3.19
Miscellaneous & Ornamental Metals	2.9	to	21.4	12.2	2.8	4.0	3.6	0.34	0.49	0.44
Carpentry & Millwork	3.4	to	29.1	16.2	3.7	4.0	0.5	0.60	0.65	0.08
Metal or Composition Siding	4.8	to	124.6	64.7	2.3	0.3	4.3	1.49	0.19	2.78
Roofing	4.8	to	110.4	57.6	2.3	2.6	3.1	1.32	1.50	1.79
Doors & Hardware	3.1	to	29.1	16.1	0.9	1.4	0.4	0.14	0.23	0.06
Sash & Glazing	3.9	to	21.6	12.8	3.5	4.0	1.0	0.45	0.51	0.13
Lath & Plaster	2.7	to	30.1	16.4	3.3	6.9	0.8	0.54	1.13	0.13
Tile, Marble & Floors	2.0	to	19.0	10.5	2.6	3.0	0.5	0.27	0.32	0.05
Acoustical Ceilings	1.7	to	29.7	15.7	2.4	0.2	0.3	0.38	0.03	0.05
Painting	3.3	to	37.4	20.3	1.5	1.6	1.6	0.30	0.32	0.32
Interior Partitions	3.4	to	29.1	16.2	3.9	4.3	4.4	0.63	0.70	0.71
Miscellaneous Items	1.9	to	97.7	10.3	5.2	3.7	9.7	0.54	0.38	1.00
Elevators	1.3	to	9.0	5.1	2.1	1.1	2.2	0.11	0.06	0.11
Sprinklers	1.8	to	14.6	8.2	0.5	—	2.0	0.04	—	0.16
Plumbing	1.4	to	13.3	7.4	4.9	7.2	5.2	0.36	0.53	0.38
Heat., Vent., Air Conditioning	2.8	to	15.8	9.3	13.5	11.0	12.9	1.26	1.02	1.20
Electrical	1.7	to	10.7	6.2	10.1	8.4	11.1	0.63	0.52	0.69
Total	1.3 %	to	124.6%	—	100.0%	100.0%	100.0%	15.54%	14.79%	16.13%
				Overall Weighted Average	15.49%					

Workers' Compensation Insurance Rates by States

The table below lists the weighted average Workers' Compensation base rate for each state with a factor comparing this with the national average of 9.9%.

State	Weighted Average	Factor	State	Weighted Average	Factor	State	Weighted Average	Factor
Alabama	12.4%	126	Kentucky	10.5%	108	North Dakota	6.4%	65
Alaska	8.6	88	Louisiana	17.5	178	Ohio	5.5	56
Arizona	7.7	79	Maine	8.0	81	Oklahoma	7.4	76
Arkansas	5.1	52	Maryland	9.7	99	Oregon	7.5	76
California	19.6	200	Massachusetts	8.8	90	Pennsylvania	17.2	175
Colorado	6.2	63	Michigan	6.3	64	Rhode Island	10.0	102
Connecticut	12.4	126	Minnesota	13.4	137	South Carolina	16.2	165
Delaware	11.8	120	Mississippi	9.2	94	South Dakota	7.6	77
District of Columbia	7.7	78	Missouri	12.0	122	Tennessee	6.0	61
Florida	8.4	86	Montana	7.1	73	Texas	4.9	50
Georgia	30.0	307	Nebraska	11.7	120	Utah	5.7	58
Hawaii	8.4	86	Nevada	7.6	77	Vermont	8.9	91
Idaho	7.8	79	New Hampshire	8.8	90	Virginia	6.7	69
Illinois	17.5	179	New Jersey	14.2	145	Washington	7.3	75
Indiana	3.1	32	New Mexico	13.0	132	West Virginia	4.2	42
Iowa	9.9	101	New York	15.6	159	Wisconsin	11.2	115
Kansas	5.6	57	North Carolina	11.6	119	Wyoming	4.8	49
			Weighted Average for U.S. is	9.9% of payroll = 100%				

The weighted average skilled worker rate for 35 trades is 9.8%. For bidding purposes, apply the full value of Workers' Compensation directly to total labor costs, or if labor is 38%, materials 42% and overhead and profit 20% of total cost, carry 38/80 x 9.8% = 4.66% of cost (before overhead and profit) into overhead. Rates vary not only from state to state but also with the experience rating of the contractor.

Rates are the most current available at the time of publication.

R013113-80 Performance Bond

This table shows the cost of a Performance Bond for a construction job scheduled to be completed in 12 months. Add 1% of the premium cost per month for jobs requiring more than 12 months to complete. The rates are "standard" rates offered to contractors that the bonding company considers financially sound and capable of doing the work. Preferred rates are offered by some bonding companies based upon financial strength of the contractor. Actual rates vary from contractor to contractor and from bonding company to bonding company. Contractors should prequalify through a bonding agency before submitting a bid on a contract that requires a bond.

Contract Amount	Building Construction Class B Projects	Highways & Bridges	
		Class A New Construction	Class A-1 Highway Resurfacing
First $ 100,000 bid	$25.00 per M	$15.00 per M	$9.40 per M
Next 400,000 bid	$ 2,500 plus $15.00 per M	$ 1,500 plus $10.00 per M	$ 940 plus $7.20 per M
Next 2,000,000 bid	8,500 plus 10.00 per M	5,500 plus 7.00 per M	3,820 plus 5.00 per M
Next 2,500,000 bid	28,500 plus 7.50 per M	19,500 plus 5.50 per M	15,820 plus 4.50 per M
Next 2,500,000 bid	47,250 plus 7.00 per M	33,250 plus 5.00 per M	28,320 plus 4.50 per M
Over 7,500,000 bid	64,750 plus 6.00 per M	45,750 plus 4.50 per M	39,570 plus 4.00 per M

R015423-10 Steel Tubular Scaffolding

On new construction, tubular scaffolding is efficient up to 60′ high or five stories. Above this it is usually better to use a hung scaffolding if construction permits. Swing scaffolding operations may interfere with tenants. In this case, the tubular is more practical at all heights.

In repairing or cleaning the front of an existing building the cost of tubular scaffolding per S.F. of building front increases as the height increases above the first tier. The first tier cost is relatively high due to leveling and alignment.

The minimum efficient crew for erecting and dismantling is three workers. They can set up and remove 18 frame sections per day up to 5 stories high. For 6 to 12 stories high, a crew of four is most efficient. Use two or more on top and two on the bottom for handing up or hoisting. They can also set up and remove 18 frame sections per day. At 7' horizontal spacing, this will run about 800 S.F. per day of erecting and dismantling. Time for placing and removing planks must be added to the above. A crew of three can place and remove 72 planks per day up to 5 stories. For over 5 stories, a crew of four can place and remove 80 planks per day.

The table below shows the number of pieces required to erect tubular steel scaffolding for 1000 S.F. of building frontage. This area is made up of a scaffolding system that is 12 frames (11 bays) long by 2 frames high.

For jobs under twenty-five frames, add 50% to rental cost. Rental rates will be lower for jobs over three months duration. Large quantities for long periods can reduce rental rates by 20%.

Description of Component	Number of Pieces for 1000 S.F. of Building Front	Unit
5′ Wide Standard Frame, 6′-4″ High	24	Ea.
Leveling Jack & Plate	24	↓
Cross Brace	44	↓
Side Arm Bracket, 21″	12	↓
Guardrail Post	12	↓
Guardrail, 7′ section	22	↓
Stairway Section	2	↓
Stairway Starter Bar	1	↓
Stairway Inside Handrail	2	↓
Stairway Outside Handrail	2	↓
Walk-Thru Frame Guardrail	2	↓

Scaffolding is often used as falsework over 15′ high during construction of cast-in-place concrete beams and slabs. Two foot wide scaffolding is generally used for heavy beam construction. The span between frames depends upon the load to be carried with a maximum span of 5′.

Heavy duty shoring frames with a capacity of 10,000#/leg can be spaced up to 10′ O.C. depending upon form support design and loading.

Scaffolding used as horizontal shoring requires less than half the material required with conventional shoring.

On new construction, erection is done by carpenters.

Rolling towers supporting horizontal shores can reduce labor and speed the job. For maintenance work, catwalks with spans up to 70′ can be supported by the rolling towers.

R015423-20 Pump Staging

Pump staging is generally not available for rent. The table below shows the number of pieces required to erect pump staging for 2400 S.F. of building frontage. This area is made up of a pump jack system that is 3 poles (2 bays) wide by 2 poles high.

Item	Number of Pieces for 2400 S.F. of Building Front	Unit
Aluminum pole section, 24′ long	6	Ea.
Aluminum splice joint, 6′ long	3	↓
Aluminum foldable brace	3	↓
Aluminum pump jack	3	↓
Aluminum support for workbench/back safety rail	3	↓
Aluminum scaffold plank/workbench, 14″ wide x 24′ long	4	↓
Safety net, 22′ long	2	↓
Aluminum plank end safety rail	2	↓

The cost in place for this 2400 S.F. will depend on how many uses are realized during the life of the equipment.

R015433-10 Contractor Equipment

Rental Rates shown elsewhere in the data set pertain to late model high quality machines in excellent working condition, rented from equipment dealers. Rental rates from contractors may be substantially lower than the rental rates from equipment dealers depending upon economic conditions; for older, less productive machines, reduce rates by a maximum of 15%. Any overtime must be added to the base rates. For shift work, rates are lower. Usual rule of thumb is 150% of one shift rate for two shifts; 200% for three shifts.

For periods of less than one week, operated equipment is usually more economical to rent than renting bare equipment and hiring an operator.

Costs to move equipment to a job site (mobilization) or from a job site (demobilization) are not included in rental rates, nor in any Equipment costs on any Unit Price line items or crew listings. These costs can be found elsewhere. If a piece of equipment is already at a job site, it is not appropriate to utilize mob/demob costs in an estimate again.

Rental rates vary throughout the country with larger cities generally having lower rates. Lease plans for new equipment are available for periods in excess of six months with a percentage of payments applying toward purchase.

Rental rates can also be treated as reimbursement costs for contractor-owned equipment. Owned equipment costs include depreciation, loan payments, interest, taxes, insurance, storage, and major repairs.

Monthly rental rates vary from 2% to 5% of the cost of the equipment depending on the anticipated life of the equipment and its wearing parts. Weekly rates are about 1/3 the monthly rates and daily rental rates are about 1/3 the weekly rates.

The hourly operating costs for each piece of equipment include costs to the user such as fuel, oil, lubrication, normal expendables for the equipment, and a percentage of the mechanic's wages chargeable to maintenance. The hourly operating costs listed do not include the operator's wages.

The daily cost for equipment used in the standard crews is figured by dividing the weekly rate by five, then adding eight times the hourly operating cost to give the total daily equipment cost, not including the operator. This figure is in the right hand column of the Equipment listings under Equipment Cost/Day.

Pile Driving rates shown for the pile hammer and extractor do not include leads, cranes, boilers or compressors. Vibratory pile driving requires an added field specialist during set-up and pile driving operation for the electric model. The hydraulic model requires a field specialist for set-up only. Up to 125 reuses of sheet piling are possible using vibratory drivers. For normal conditions, crane capacity for hammer type and size is as follows.

Crane Capacity	Hammer Type and Size		
	Air or Steam	Diesel	Vibratory
25 ton	to 8,750 ft.-lb.		70 H.P.
40 ton	15,000 ft.-lb.	to 32,000 ft.-lb.	170 H.P.
60 ton	25,000 ft.-lb.		300 H.P.
100 ton		112,000 ft.-lb.	

Cranes should be specified for the job by size, building and site characteristics, availability, performance characteristics, and duration of time required.

Backhoes & Shovels rent for about the same as equivalent size cranes but maintenance and operating expenses are higher. The crane operator's rate must be adjusted for high boom heights. Average adjustments: for 150′ boom add 2% per hour; over 185′, add 4% per hour; over 210′, add 6% per hour; over 250′, add 8% per hour and over 295′, add 12% per hour.

Tower Cranes of the climbing or static type have jibs from 50′ to 200′ and capacities at maximum reach range from 4,000 to 14,000 pounds. Lifting capacities increase up to maximum load as the hook radius decreases.

Typical rental rates, based on purchase price, are about 2% to 3% per month.

Erection and dismantling run between 500 and 2000 labor hours. Climbing operation takes 10 labor hours per 20′ climb. Crane dead time is about 5 hours per 40′ climb. If crane is bolted to side of the building add cost of ties and extra mast sections. Climbing cranes have from 80′ to 180′ of mast while static cranes have 80′ to 800′ of mast.

Truck Cranes can be converted to tower cranes by using tower attachments. Mast heights over 400′ have been used.

A single 100′ high material **Hoist and Tower** can be erected and dismantled in about 400 labor hours; a double 100′ high hoist and tower in about 600 labor hours. Erection times for additional heights are 3 and 4 labor hours per vertical foot respectively up to 150′, and 4 to 5 labor hours per vertical foot over 150′ high. A 40′ high portable Buck hoist takes about 160 labor hours to erect and dismantle. Additional heights take 2 labor hours per vertical foot to 80′ and 3 labor hours per vertical foot for the next 100′. Most material hoists do not meet local code requirements for carrying personnel.

A 150′ high **Personnel Hoist** requires about 500 to 800 labor hours to erect and dismantle. Budget erection time at 5 labor hours per vertical foot for all trades. Local code requirements or labor scarcity requiring overtime can add up to 50% to any of the above erection costs.

Earthmoving Equipment: The selection of earthmoving equipment depends upon the type and quantity of material, moisture content, haul distance, haul road, time available, and equipment available. Short haul cut and fill operations may require dozers only, while another operation may require excavators, a fleet of trucks, and spreading and compaction equipment. Stockpiled material and granular material are easily excavated with front end loaders. Scrapers are most economically used with hauls between 300′ and 1-1/2 miles if adequate haul roads can be maintained. Shovels are often used for blasted rock and any material where a vertical face of 8′ or more can be excavated. Special conditions may dictate the use of draglines, clamshells, or backhoes. Spreading and compaction equipment must be matched to the soil characteristics, the compaction required and the rate the fill is being supplied.

R015433-15 Heavy Lifting

Hydraulic Climbing Jacks

The use of hydraulic heavy lift systems is an alternative to conventional type crane equipment. The lifting, lowering, pushing, or pulling mechanism is a hydraulic climbing jack moving on a square steel jackrod from 1-5/8″ to 4″ square, or a steel cable. The jackrod or cable can be vertical or horizontal, stationary or movable, depending on the individual application. When the jackrod is stationary, the climbing jack will climb the rod and push or pull the load along with itself. When the climbing jack is stationary, the jackrod is movable with the load attached to the end and the climbing jack will lift or lower the jackrod with the attached load. The heavy lift system is normally operated by a single control lever located at the hydraulic pump.

The system is flexible in that one or more climbing jacks can be applied wherever a load support point is required, and the rate of lift synchronized.

Economic benefits have been demonstrated on projects such as: erection of ground assembled roofs and floors, complete bridge spans, girders and trusses, towers, chimney liners and steel vessels, storage tanks, and heavy machinery. Other uses are raising and lowering offshore work platforms, caissons, tunnel sections and pipelines.

R015436-50 Mobilization

Costs to move rented construction equipment to a job site from an equipment dealer's or contractor's yard (mobilization) or off the job site (demobilization) are not included in the rental or operating rates, nor in the equipment cost on a unit price line or in a crew listing. These costs can be found consolidated in the Mobilization section of the data and elsewhere in particular site work sections. If a piece of equipment is already on the job site, it is not appropriate to include mob/demob costs in a new estimate that requires use of that equipment. The following table identifies approximate sizes of rented construction equipment that would be hauled on a towed trailer. Because this listing is not all-encompassing, the user can infer as to what size trailer might be required for a piece of equipment not listed.

3-ton Trailer	20-ton Trailer	40-ton Trailer	50-ton Trailer
20 H.P. Excavator	110 H.P. Excavator	200 H.P. Excavator	270 H.P. Excavator
50 H.P. Skid Steer	165 H.P. Dozer	300 H.P. Dozer	Small Crawler Crane
35 H.P. Roller	150 H.P. Roller	400 H.P. Scraper	500 H.P. Scraper
40 H.P. Trencher	Backhoe	450 H.P. Art. Dump Truck	500 H.P. Art. Dump Truck

R019313-10 Facility Maintenance - Frequency Table

The following table lists "average" frequency data for selected facility maintenance activities. The frequencies given are for a normal standard of maintenance under average conditions.

Activity	Average Frequency	Notes
Acoustical tile, cleaning Heavy smoking Non-smoking	 2-3 years 10 years	
Carpet, cleaning Heavy traffic Light traffic	 Weekly Every 6 weeks	Frequency depends on the type of carpet and the occupancy in the building.
Carpet, vacuuming Heavy traffic Light traffic Offices	 Daily Twice weekly Weekly	6-8 passes by machine in key areas.
Corridor and lobby, policing Main Secondary	 4 times daily Daily	Includes picking up loose trash, removing cigarette butts from in and around jardinieres and/or sand urns. Frequency depends on occupancy in the building.
Elevator, cleaning Passenger Freight	 Daily Weekly	The frequency of elevator cleaning is dependent upon the amount of traffic, size of the elevator car and the occupancy in the building.
Elevator lobby, cleaning	Daily	Includes sweeping, mopping and rinsing.
Escalator, cleaning	Daily	Frequency given is for a normal standard of cleaning under average conditions. Approximately 40% of escalator treads are exposed when escalator is stopped.
Floors, mopping Main corridors Secondary	 Daily Weekly	Frequency given is for a normal standard of cleaning under average conditions.
Floors, sweeping	Daily	Frequency given is for a normal standard of cleaning under average conditions.
Floors, waxing and polishing Office area Open area	 Every 9 weeks Every 9 weeks	Frequency given is for a normal standard of cleaning under average conditions.
Flower beds Fall clean-up Fertilize Mulch Police-up Weed With mulch No mulch	 Every year 2 times per year Every year 30 times per year 15 times per year 25 times per year	Frequency depends on geographic location.
Lawn, mowing Fertilize Sweep Weed control Edge-trim Walks Shrub	30 times per year 2 times per year 3 times per year 2 times per year 30 times per year 10 times per year	Frequency depends on geographic location.
Light fixtures Dusting Washing	 Every month Every 3 months	Frequency depends on occupancy in the building.
Shrub Areas Fertilize Mulch Police-up Prune Weed	 Every year 2 times per year 30 times per year 5 times per year 10 times per year	Frequency depends on geographic location.
Stairway Sweeping and dusting Mopping or scrubbing	 Daily Weekly	Frequency is dependent upon weather conditions, type of building, type of occupancy, and amount of traffic.
Toilet, cleaning	Daily	Includes toilet cleaning, collecting waste, and cleaning wash basins, urinals, water closets, partitions, walls and floors.
Trash collection	Daily	
Trees Fertilize Prune Pest control Spray Systemic	 Every year 2 times per year 3 times per year Every year	Frequency depends on geographic location.
Urn and Jardiniere cleaning	Daily	Includes the removal of refuse and debris, cleaning and polishing.
Walks, sweeping	30 times per year	
Walls, periodic cleaning	Every 6 months	Damp wipe, spot removal.
Windows, washing	Every month	Frequency given is for a normal standard of cleaning under average conditions.

R019313-20 Facility Maintenance Labor-Hours

This section lists minimum and maximum cleaning times per unit.
For more information, see *Means Facilities Maintenance Standards.*

	Unit	
	Minimum	Maximum
Blast Cleaning		
White-metal	100 S.F./Hr	
Near-white	175 S.F./Hr.	
Commercial	370 S.F./Hr.	
Brush-off	870 S.F./Hr.	
Paint Application		
Brushing	125 S.F./Hr.	
Rolling	125 S.F./Hr	
Spraying	500 S.F./Hr.	
Plaster Cleaning		
Wall dusting	2 sec./S.F.	3 sec./S.F.
Vacuuming	4 sec./S.F.	5 sec./S.F.
Spot washing	125 S.F./Hr.	175 S.F./Hr.
Thorough cleaning	275 S.F./Hr.	
Plaster Repair		
Gypsum and lime repair	5 S.Y./Hr.	10 S.Y./Hr.
Ceramic Tile Repair		
General	7 S.F./Hr.	10 S.F./Hr.
Adhesive tile setting	9S.F./Hr	12 S.F./Hr.
Pointing tile joints	10 S.F./Hr.	15 S.F./Hr.
Floor Cleaning		
Manual sweeping	10 min./1000 S.F.	25 min./1000 S.F.
Dust mopping	5 min./1000 S.F.	20 min./1000 S.F.
Buffing	15 min./1000 S.F.	40 min./1000 S.F.
Spray buffing	20 min./1000 S.F.	50 min./1000 S.F.
Damp mopping	15 min./1000 S.F.	30 min./1000 S.F.
Wet mopping	30 min./1000 S.F.	50 min./1000 S.F.
Scrubbing	50 min./1000 S.F.	140 min./1000 S.F.
Scrubbing using electric floor machine		
General	15 min./1000 S.F.	30 min./1000 S.F.
Stripping	100 min./1000 S.F.	200 min./1000 S.F.
Waxing and Buffing using power maching		
Rewaxing	15 min./1000 S.F.	30 min./1000 S.F.
Stripping and rewaxing (two coats)	100 min./1000 S.F.	300 min./1000 S.F.
Waxing and buffing (one coat)	30 min./1000 S.F.	70 min./1000 S.F.
Carpets		
Dry vacuuming	15 min./1000 S.F.	40 min./1000 S.F.
Wet vacuuming	30 min./1000 S.F.	50 min./1000 S.F.
Carpet mopping	20 min./1000 S.F.	40 min./1000 S.F.
Shampooing	175 min./1000 S.F.	250 min./1000 S.F.
Resilient Floor Repair		
Grinding	50 S.F./Hr.	80 S.F./Hr.
Floor Replacement		
Removal (by hand)		
Tiles	100 S.F./Hr.	130 S.F./Hr.
Sheet	120 S.F./Hr	160 S.F./Hr.
Hardwood	40 S.F./Hr.	60 S.F./Hr.
Replacement		
Ceramic	10 S.F./Hr.	20 S.F./Hr.
Resilient	40 S.F./Hr.	70 S.F./Hr.
Hardwood	25 S.F./Hr.	35 S.F./Hr.
Add for related items:		
Replace wood subfloor	80 S.F./Hr.	100 S.F./Hr.
Replace underlayment	75 S.F./Hr.	90 S.F./Hr.
Replace floor moulding	10 S.F./Hr.	30 S.F./Hr.

Unit	Minimum	Maximum
Seamless Floor Repair	5 S.F./Hr.	15 S.F./Hr.
Wood Floors		
Sanding	40 S.F./Hr.	60 S.F./Hr.
Sealing	200 S.F/Hr.	300 S.F./Hr.
Waxing*		
Wood Floor Repair		
Loose boards or tiles	50 S.F./Hr.	250S.F./Hr.
Wood strip floor replacement	30 S.F./Hr.	60 S.F./Hr.
Window Washing	300 S.F./Hr.	450 S.F./Hr.
Venetian Blinds, cleaning	15 min./set of blinds	

This section lists average cleaning times per each, or square foot	Average
Floor Operations	
Sweeping	
Halls and Corridors	15 min./1000 S.F.
General Rooms	30 min./100 S.F.
Dust Mop (unobstructed)	10 min./1000 S.F.
Dust Mop (obstructed)	15 min./1000 S.F.
Damp Mop (unobstructed)	20 min./1000 S.F.
Damp Mop (obstructed)	40 min./1000 S.F.
Wet Mop and Rinse	100 min./1000 S.F.
Hand Scrubbing 12″ brush	300 min./1000 S.F.
Deck Scrubbing	100 min./1000 S.F.
Machine Scrubbing	
12″ diameter	50 min./1000 S.F.
14″ diameter	40 min./1000 S.F.
16″ diameter	35 min./1000 S.F.
18″ diameter	31 min./1000 S.F.
19″ diameter	28 min./1000 S.F.
21″ diameter	25 min./1000 S.F.
23″ diameter	23 min./1000 S.F.
24″ diameter	20 min./1000 S.F.
32″ diameter	18 min./1000 S.F.
36″ diameter	15 min./1000 S.F.
Automatic Scrub Machine (24″)	5 min./1000 S.F.
Vacuum (unobstructed)	20 min./1000 S.F.
Vacuum (obstructed)	30 min./1000 S.F.
Waxing	30 min./1000 S.F.
Machine Polish (19″ machine)	15 min./1000 S.F.
Rectangular Machine (48″ plate)	5 min./1000 S.F.
Buff with steel wool	20 min./1000 S.F.
Strip and Rewax	150 min./1000 S.F.
Dry Strip and Rewax	120 min./1000 S.F.
Spray Buffing (unobstructed)	30 min./1000 S.F.
Spray Buffing (obstructed)	45 min./1000 S.F.
Carpeting	
Vacuuming (unobstructed)	20 min./1000 S.F.
Vacuuming (obstructed)	30 min./1000 S.F.
Spot Vacuuming	15 min./1000 S.F.
Shampoo (dry foam)	60 min. 1000 S.F.
Pile Lift	30 min 1000 S.F.
Lockers	.20 min.
Radiators	.30 min.
Tables (medium)	.50 min.
Telephones	.15 min.
Towel dispensers	.12 min.

*See waxing and buffing using electric power machine.

R019313-20 Facility Maintenance Labor-Hours (cont.)

	Average
Towel Disposal Cans	.40 min.
Typewriter and Stand	.50 min
Wash Basin (office)	.60 min
Waste Basin (office)	.50 min.
Window Sill	.20 min.
Venetian Blinds, std. size	3.50 min.
Washrooms	
Cleaning commode	4 min.
Door (spot wash both sides)	1 min.
Mirrors	1 min.
Sanitary napkin dispenser	.50 min.
Urinals	3 min.
Wash basin-soap dispenser	3 min.
General cleaning	120 min./1000 S.F.
Wall Washing	
Painted walls (manual)	240 min./1000 S.F.
Painted walls (machine)	150 min./1000 S.F.
Marble walls (manual)	90 min./1000 S.F.
Ceiling Washing	
Ceiling washing (manual)	300 min./1000 S.F.
Ceiling washing (machine)	180 min./1000 S.F.
Window Washing	
Single pane	125 min./1000 S.F.
Multi-pane	170 min./1000 S.F.
Frosted single pane	190 min./1000 S.F.
Opaque glass	50 min./1000 S.F.
Plate glass	35 min./1000 S.F.
Office partitions (glass)	110 min./1000 S.F.
Dusting Lamps and Lighting Fixtures	
Wall fluorescent fixtures	.15 min.
Desk fluorescent lamp	.30 min.
Table lamp and shade	.60 min.
Floor lamp and shade	.60 min.
Washing Fluorescent Light Fixtures	
Ceiling fixtures (egg crate) 4′ ea.	9 min.
Ceiling fixtures (egg crate) 8′ ea.	12 min.
Dusting	
Air conditioners	.30 min.
Ash trays (desk)	.25 min.
Book cases (3-tier set)	.30 min.
Chairs	.30 min.
Cigarette stands	.40 min.

	Average
Couch	.25 min.
Desks	.80 min.
Desk Trays	.15 min.
File cabinets (4 drawer)	.40 min.
Fabric Upholstery Cleaning	
Whisk or vacuum armless chair	.50 min.
Armchair	1 min.
Couch	2 min.
Shampooing armless chair	4 min.
Armchair	7 min.
Couch	20 min.
Stairway Cleaning	
Sweep and dust, 1 flight, 15 steps	6 min.
Damp mop, 1 flight, 15 steps	5 min.
Scrubbing (hand)	20 min.
Carpentry	
Repair door surface closer	1-1/2 Hrs.
Repair concealed door closer	4 Hrs.
Repair door damage at shop	4 Hrs.
Repair and replace screens	1 Hr.
Repair broken glass	1-1/2 Hrs.
Repair and replace ceiling tile	2 Hrs.
Repair small drywall damage	4 Hrs.
Repair and replace sash balance	4 Hrs.
Change lock on door	1-1/2 Hrs.
Painting	
Repaint 20′ x 15′ room, 1 coat	34 Hrs.
Repaint small bathroom, 1 coat	11-1/2 Hrs.
Repaint exterior window	2-1/2 Hrs.
Plumbing and Steamfitting	
Clear stopped water closet	3 Hrs.
Clear stopped basin	1-1/2 Hrs.
Replace leaking radiator valve	6 Hrs.
Clear external sewer stoppage	9 Hrs.
Install replacement valve, faucet, or trap	10 Hrs.
Electrical	
Replace fluorescent lamp ballast	2 Hrs.
Replace fractional HP motor	10 Hrs.
Replace blown fuse or reset Circuit breaker	1-1/2 Hrs.
Repair exterior light damage	4 Hrs.
Control circuit problems	5 Hrs.

R019313-30 A Review of Major Building Materials–Advantages and Disadvantages

Material	Advantages	Disadvantages
Aluminum	Lightweight High resistance to corrosion Low electrical resistance Good conductor	Softness Limited strength for structural uses Low stiffness High rate of thermal expansion Low rate of fire resistance Relatively high cost
Concrete	High compressive resistance Durable Resistance to moisture, rot, insects, fire, and wear Watertight (depending on water-cement ratio)	Workability Lack of tensile strength Lack of resistance to many types of chemical exposure, such as salt Hard to remove stains
Copper	High resistance to corrosion Good electrical conductivity Workable Forms its own surface protection Thermal contraction and expansion not high	Costly Strength varies with treatment and mechanical working
Epoxy	Can have varied properties depending on composition Liquid use very applicable Controllable Excellent strength properties Small creep during curing Hard Tough Resistant to abrasion Resistant to corrosion, salts, acids, petroleum products, solvents, and other chemicals Adhesion to surfaces good	Color Form
Glass	Considerably strong Non-corrosive	Brittle, subject to shattering under shock Transmits energy at a rapid rate Large sizes expensive
Lead	Good resistance to corrosion	Heavy High coefficient of thermal expansion Difficult to hold in place
Masonry	Available in small units Appearance (available in many textures, sizes, and colors) Good insulator	Stains hard to remove Porous (absorption rate high) Shrinkage of mortar Thermal and expansion cracking Others — see concrete
Paper	Readily available Low in cost Used in conjunction with other materials	Susceptible to water damage Susceptible to rot Relatively weak Highly combustible
Plastics — general	Applicable to many uses Relatively low in cost Workable into many shapes High strength Lightweight Non-corrosive Others — see specific plastic entry	Lack of resistance to fire Low stiffness High rate of thermal expansion Low thermal conductivity Some cases of chemical or physical instability with time Non-salvageable Others — see specific plastic entry

R019313-30 A Review of Major Building Materials—Advantages and Disadvantages (cont.)

Material	Advantages	Disadvantages
Plastics — specific		
Acrylics	Transparent Hard Weather resistant Shatter resistant	Easily scratched
Alkyds	Water resistant Touch Good adhesive properties	
Melamines	Hard Durable Abrasive resistant Chemical and heat resistant	
Polyamides (nylon)	Hard Tough Wear resistant	Costly
Polyesters	Weather and chemical resistant Stiff Hard	
Polyethylene	Flexible Tough Translucent Low cost	Easily scratched
Polystyrene	Hard Clear Water and chemical resistant Low cost	Brittle
Vinyls	Tough Wear resistant Stain resistant	
Plywood	Stronger than standard lumber Durable High resistance to impact and load Not as affected by moisture changes as standard wood Workability See Wood for others	Dimensional stability not as good as that of standard structural lumber Poor quality glues are possible Thermal expansion and contraction more of a problem
Steel	Strong Most resistant to aging Most reliable in quality, non-combustible, non-rotting, dimensionally stable with time and moisture change Resistant to staining	Costly Resultant loss of strength when exposed to intense heat, heat, rapid heat gain and loss Corrosive when exposed to moisture and air or other corrosive conditions
Tin	Extremely workable Very resistant to corrosion	Heavy Loss of strength when exposed to heat
Wood	Readily available Relatively low in cost Simple to work with Available in many shapes and forms Good insulating properties	Paintability changes with moisture content Combustible Susceptible to rot and insect infestation Soft and easily damaged (porous) Dimensional changes due to changes in temperature and moisture Strength changes with changing moisture content
Zinc	Fairly high resistance to corrosion Forms its own protective surface Workable	Brittle

R024119-10 Demolition Defined

Whole Building Demolition - Demolition of the whole building with no concern for any particular building element, component, or material type being demolished. This type of demolition is accomplished with large pieces of construction equipment that break up the structure, load it into trucks, and haul it to a disposal site, but disposal or dump fees are not included. Demolition of below-grade foundation elements, such as footings, foundation walls, grade beams, slabs on grade, etc., is not included. Certain mechanical equipment containing flammable liquids or ozone-depleting refrigerants, electric lighting elements, communication equipment components, and other building elements may contain hazardous waste, and must be removed, either selectively or carefully, as hazardous waste before the building can be demolished.

Foundation Demolition - Demolition of below-grade foundation footings, foundation walls, grade beams, and slabs on grade. This type of demolition is accomplished by hand or pneumatic hand tools, and does not include saw cutting, or handling, loading, hauling, or disposal of the debris.

Gutting - Removal of building interior finishes and electrical/mechanical systems down to the load-bearing and sub-floor elements of the rough building frame, with no concern for any particular building element, component, or material type being demolished. This type of demolition is accomplished by hand or pneumatic hand tools, and includes loading into trucks, but not hauling, disposal or dump fees, scaffolding, or shoring. Certain mechanical equipment containing flammable liquids or ozone-depleting refrigerants, electric lighting elements, communication equipment components, and other building elements may contain hazardous waste, and must be removed, either selectively or carefully, as hazardous waste, before the building is gutted.

Selective Demolition - Demolition of a selected building element, component, or finish, with some concern for surrounding or adjacent elements, components, or finishes (see the first Subdivision(s) at the beginning of appropriate Divisions). This type of demolition is accomplished by hand or pneumatic hand tools, and does not include handling, loading, storing, hauling, or disposal of the debris, scaffolding, or shoring. "Gutting" methods may be used in order to save time, but damage that is caused to surrounding or adjacent elements, components, or finishes may have to be repaired at a later time.

Careful Removal - Removal of a piece of service equipment, building element or component, or material type, with great concern for both the removed item and surrounding or adjacent elements, components or finishes. The purpose of careful removal may be to protect the removed item for later re-use, preserve a higher salvage value of the removed item, or replace an item while taking care to protect surrounding or adjacent elements, components, connections, or finishes from cosmetic and/or structural damage. An approximation of the time required to perform this type of removal is 1/3 to 1/2 the time it would take to install a new item of like kind (see Reference Number R220105-10). This type of removal is accomplished by hand or pneumatic hand tools, and does not include loading, hauling, or storing the removed item, scaffolding, shoring, or lifting equipment.

Cutout Demolition - Demolition of a small quantity of floor, wall, roof, or other assembly, with concern for the appearance and structural integrity of the surrounding materials. This type of demolition is accomplished by hand or pneumatic hand tools, and does not include saw cutting, handling, loading, hauling, or disposal of debris, scaffolding, or shoring.

Rubbish Handling - Work activities that involve handling, loading or hauling of debris. Generally, the cost of rubbish handling must be added to the cost of all types of demolition, with the exception of whole building demolition.

Minor Site Demolition - Demolition of site elements outside the footprint of a building. This type of demolition is accomplished by hand or pneumatic hand tools, or with larger pieces of construction equipment, and may include loading a removed item onto a truck (check the Crew for equipment used). It does not include saw cutting, hauling, or disposal of debris, and, sometimes, handling or loading.

R024119-20 Dumpsters

Dumpster rental costs on construction sites are presented in two ways.

The cost per week rental includes the delivery of the dumpster, its pulling or emptying once per week, and its final removal. The assumption is made that the dumpster contractor could choose to empty a dumpster by simply bringing in an empty unit and removing the full one. These costs also include the disposal of the materials in the dumpster.

The Alternate Pricing can be used when actual planned conditions are not approximated by the weekly numbers. For example, these lines can be used when a dumpster is needed for 4 weeks and will need to be emptied 2 or 3 times per week. Conversely the Alternate Pricing lines can be used when a dumpster will be rented for several weeks or months but needs to be emptied only a few times over this period.

R024119-30 Rubbish Handling Chutes

To correctly estimate the cost of rubbish handling chute systems, the individual components must be priced separately. First choose the size of the system; a 30-inch diameter chute is quite common, but the sizes range from 18 to 36 inches in diameter. The 30-inch chute comes in a standard weight and two thinner weights. The thinner weight chutes are sometimes chosen for cost savings, but they are more easily damaged.

There are several types of major chute pieces that make up the chute system. The first component to consider is the top chute section (top intake hopper) where the material is dropped into the chute at the highest point. After determining the top chute, the intermediate chute pieces called the regular chute sections are priced. Next, the number of chute control door sections (intermediate intake hoppers) must be determined. In the more complex systems, a chute control door section is provided at each floor level. The last major component to consider is bolt down frames; these are usually provided at every other floor level.

There are a number of accessories to consider for safe operation and control. There are covers for the top chute and the chute door sections. The top chute can have a trough that allows for better loading of the chute. For the safest operation, a chute warning light system can be added that will warn the other chute intake locations not to load while another is being used. There are dust control devices that spray a water mist to keep down the dust as the debris is loaded into a Dumpster. There are special breakaway cords that are used to prevent damage to the chute if the dumpster is removed without disconnecting from the chute. There are chute liners that can be installed to protect the chute structure from physical damage from rough abrasive materials. Warning signs can be posted at each floor level that is provided with a chute control door section.

In summary, a complete rubbish handling chute system will include one top section, several intermediate regular sections, several intermediate control door (intake hopper) sections and bolt down frames at every other floor level starting with the top floor. If so desired, the system can also include covers and a light warning system for a safer operation. The bottom of the chute should always be above the Dumpster and should be tied off with a breakaway cord to the Dumpster.

Existing Conditions — R0265 Underground Storage Tank Removal

R026510-20 Underground Storage Tank Removal

Underground Storage Tank Removal can be divided into two categories: Non-Leaking and Leaking. Prior to removing an underground storage tank, tests should be made, with the proper authorities present, to determine whether a tank has been leaking or the surrounding soil has been contaminated.

To safely remove Liquid Underground Storage Tanks:

1. Excavate to the top of the tank.
2. Disconnect all piping.
3. Open all tank vents and access ports.
4. Remove all liquids and/or sludge.
5. Purge the tank with an inert gas.
6. Provide access to the inside of the tank and clean out the interior using proper personal protective equipment (PPE).
7. Excavate soil surrounding the tank using proper PPE for on-site personnel.
8. Pull and properly dispose of the tank.
9. Clean up the site of all contaminated material.
10. Install new tanks or close the excavation.

Existing Conditions — R0282 Asbestos Remediation

R028213-20 Asbestos Removal Process

Asbestos removal is accomplished by a specialty contractor who understands the federal and state regulations regarding the handling and disposal of the material. The process of asbestos removal is divided into many individual steps. An accurate estimate can be calculated only after all the steps have been priced.

The steps are generally as follows:

1. Obtain an asbestos abatement plan from an industrial hygienist.
2. Monitor the air quality in and around the removal area and along the path of travel between the removal area and transport area. This establishes the background contamination.
3. Construct a two part decontamination chamber at entrance to removal area.
4. Install a HEPA filter to create a negative pressure in the removal area.
5. Install wall, floor, and ceiling protection as required by the plan, usually 2 layers of fireproof 6 mil polyethylene.
6. Industrial hygienist visually inspects work area to verify compliance with plan.
7. Provide temporary supports for conduit and piping affected by the removal process.
8. Proceed with asbestos removal and bagging process. Monitor air quality as described in Step #2. Discontinue operations when contaminate levels exceed applicable standards.
9. Document the legal disposal of materials in accordance with EPA standards.
10. Thoroughly clean removal area including all ledges, crevices, and surfaces.
11. Post abatement inspection by industrial hygienist to verify plan compliance.
12. Provide a certificate from a licensed industrial hygienist attesting that contaminate levels are within acceptable standards before returning area to regular use.

R028319-60 Lead Paint Remediation Methods

Lead paint remediation can be accomplished by the following methods:

1. Abrasive blast
2. Chemical stripping
3. Power tool cleaning with vacuum collection system
4. Encapsulation
5. Remove and replace
6. Enclosure

Each of these methods has strengths and weaknesses depending on the specific circumstances of the project. The following is an overview of each method.

1. **Abrasive blasting** is usually accomplished with sand or recyclable metallic blast. Before work can begin, the area must be contained to ensure the blast material with lead does not escape to the atmosphere. The use of vacuum blast greatly reduces the containment requirements. Lead abatement equipment that may be associated with this work includes a negative air machine. In addition, it is necessary to have an industrial hygienist monitor the project on a continual basis. When the work is complete, the spent blast sand with lead must be disposed of as a hazardous material. If metallic shot was used, the lead is separated from the shot and disposed of as hazardous material. Worker protection includes disposable clothing and respiratory protection.
2. **Chemical stripping** requires strong chemicals to be applied to the surface to remove the lead paint. Before the work can begin, the area under/adjacent to the work area must be covered to catch the chemical and removed lead. After the chemical is applied to the painted surface it is usually covered with paper. The chemical is left in place for the specified period, then the paper with lead paint is pulled or scraped off. The process may require several chemical applications. The paper with chemicals and lead paint adhered to it, plus the containment and loose scrapings collected by a HEPA (High Efficiency Particulate Air Filter) vac, must be disposed of as hazardous material. The chemical stripping process usually requires a neutralizing agent and several wash downs after the paint is removed. Worker protection includes a neoprene or other compatible protective clothing and respiratory protection with face shield. An industrial hygienist is required intermittently during the process.
3. **Power tool cleaning** is accomplished using shrouded needle blasting guns. The shrouding with different end configurations is held up against the surface to be cleaned. The area is blasted with hardened needles and the shroud captures the lead with a HEPA vac and deposits it in a holding tank. An industrial hygienist monitors the project. Protective clothing and a respirator are required until air samples prove otherwise. When the work is complete the lead must be disposed of as hazardous material.
4. **Encapsulation** is a method that leaves the well bonded lead paint in place after the peeling paint has been removed. Before the work can begin, the area under/adjacent to the work must be covered to catch the scrapings. The scraped surface is then washed with a detergent and rinsed. The prepared surface is covered with approximately 10 mils of paint. A reinforcing fabric can also be embedded in the paint covering. The scraped paint and containment must be disposed of as hazardous material. Workers must wear protective clothing and respirators.
5. **Removing and replacing** are effective ways to remove lead paint from windows, gypsum walls and concrete masonry surfaces. The painted materials are removed and new materials are installed. Workers should wear a respirator and tyvek suit. The demolished materials must be disposed of as hazardous waste if it fails the TCLP (Toxicity Characteristic Leachate Process) test.
6. **Enclosure** is the process that permanently seals lead painted materials in place. This process has many applications such as covering lead painted drywall with new drywall, covering exterior construction with tyvek paper then residing, or covering lead painted structural members with aluminum or plastic. The seams on all enclosing materials must be securely sealed. An industrial hygienist monitors the project, and protective clothing and a respirator are required until air samples prove otherwise.

All the processes require clearance monitoring and wipe testing as required by the hygienist.

R031113-60 Formwork Labor-Hours

Item	Hours Required				Total Hours	Multiple Use		
	Unit	Fabricate	Erect & Strip	Clean & Move	1 Use	2 Use	3 Use	4 Use
Beam and Girder, interior beams, 12″ wide	100 S.F.	6.4	8.3	1.3	16.0	13.3	12.4	12.0
Hung from steel beams		5.8	7.7	1.3	14.8	12.4	11.6	11.2
Beam sides only, 36″ high		5.8	7.2	1.3	14.3	11.9	11.1	10.7
Beam bottoms only, 24″ wide		6.6	13.0	1.3	20.9	18.1	17.2	16.7
Box out for openings		9.9	10.0	1.1	21.0	16.6	15.1	14.3
Buttress forms, to 8′ high		6.0	6.5	1.2	13.7	11.2	10.4	10.0
Centering, steel, 3/4″ rib lath			1.0		1.0			
3/8″ rib lath or slab form			0.9		0.9			
Chamfer strip or keyway	100 L.F.		1.5		1.5	1.5	1.5	1.5
Columns, fiber tube 8″ diameter			20.6		20.6			
12″			21.3		21.3			
16″			22.9		22.9			
20″			23.7		23.7			
24″			24.6		24.6			
30″			25.6		25.6			
Columns, round steel, 12″ diameter			22.0		22.0	22.0	22.0	22.0
16″			25.6		25.6	25.6	25.6	25.6
20″			30.5		30.5	30.5	30.5	30.5
24″			37.7		37.7	37.7	37.7	37.7
Columns, plywood 8″ x 8″	100 S.F.	7.0	11.0	1.2	19.2	16.2	15.2	14.7
12″ x 12″		6.0	10.5	1.2	17.7	15.2	14.4	14.0
16″ x 16″		5.9	10.0	1.2	17.1	14.7	13.8	13.4
24″ x 24″		5.8	9.8	1.2	16.8	14.4	13.6	13.2
Columns, steel framed plywood 8″ x 8″			10.0	1.0	11.0	11.0	11.0	11.0
12″ x 12″			9.3	1.0	10.3	10.3	10.3	10.3
16″ x 16″			8.5	1.0	9.5	9.5	9.5	9.5
24″ x 24″			7.8	1.0	8.8	8.8	8.8	8.8
Drop head forms, plywood		9.0	12.5	1.5	23.0	19.0	17.7	17.0
Coping forms		8.5	15.0	1.5	25.0	21.3	20.0	19.4
Culvert, box			14.5	4.3	18.8	18.8	18.8	18.8
Curb forms, 6″ to 12″ high, on grade		5.0	8.5	1.2	14.7	12.7	12.1	11.7
On elevated slabs		6.0	10.8	1.2	18.0	15.5	14.7	14.3
Edge forms to 6″ high, on grade	100 L.F.	2.0	3.5	0.6	6.1	5.6	5.4	5.3
7″ to 12″ high	100 S.F.	2.5	5.0	1.0	8.5	7.8	7.5	7.4
Equipment foundations		10.0	18.0	2.0	30.0	25.5	24.0	23.3
Flat slabs, including drops		3.5	6.0	1.2	10.7	9.5	9.0	8.8
Hung from steel		3.0	5.5	1.2	9.7	8.7	8.4	8.2
Closed deck for domes		3.0	5.8	1.2	10.0	9.0	8.7	8.5
Open deck for pans		2.2	5.3	1.0	8.5	7.9	7.7	7.6
Footings, continuous, 12″ high		3.5	3.5	1.5	8.5	7.3	6.8	6.6
Spread, 12″ high		4.7	4.2	1.6	10.5	8.7	8.0	7.7
Pile caps, square or rectangular		4.5	5.0	1.5	11.0	9.3	8.7	8.4
Grade beams, 24″ deep		2.5	5.3	1.2	9.0	8.3	8.0	7.9
Lintel or Sill forms		8.0	17.0	2.0	27.0	23.5	22.3	21.8
Spandrel beams, 12″ wide		9.0	11.2	1.3	21.5	17.5	16.2	15.5
Stairs			25.0	4.0	29.0	29.0	29.0	29.0
Trench forms in floor		4.5	14.0	1.5	20.0	18.3	17.7	17.4
Walls, Plywood, at grade, to 8′ high		5.0	6.5	1.5	13.0	11.0	9.7	9.5
8′ to 16′		7.5	8.0	1.5	17.0	13.8	12.7	12.1
16′ to 20′		9.0	10.0	1.5	20.5	16.5	15.2	14.5
Foundation walls, to 8′ high		4.5	6.5	1.0	12.0	10.3	9.7	9.4
8′ to 16′ high		5.5	7.5	1.0	14.0	11.8	11.0	10.6
Retaining wall to 12′ high, battered		6.0	8.5	1.5	16.0	13.5	12.7	12.3
Radial walls to 12′ high, smooth		8.0	9.5	2.0	19.5	16.0	14.8	14.3
2′ chords		7.0	8.0	1.5	16.5	13.5	12.5	12.0
Prefabricated modular, to 8′ high		—	4.3	1.0	5.3	5.3	5.3	5.3
Steel, to 8′ high		—	6.8	1.2	8.0	8.0	8.0	8.0
8′ to 16′ high		—	9.1	1.5	10.6	10.3	10.2	10.2
Steel framed plywood to 8′ high		—	6.8	1.2	8.0	7.5	7.3	7.2
8′ to 16′ high		—	9.3	1.2	10.5	9.5	9.2	9.0

R032110-10 Reinforcing Steel Weights and Measures

Bar Designation No.**	Nominal Weight Lb./Ft.	U.S. Customary Units			SI Units			
		Nominal Dimensions*			Nominal Dimensions*			
		Diameter in.	Cross Sectional Area, in.^{2}	Perimeter in.	Nominal Weight kg/m	Diameter mm	Cross Sectional Area, cm^2	Perimeter mm
3	.376	.375	.11	1.178	.560	9.52	.71	29.9
4	.668	.500	.20	1.571	.994	12.70	1.29	39.9
5	1.043	.625	.31	1.963	1.552	15.88	2.00	49.9
6	1.502	.750	.44	2.356	2.235	19.05	2.84	59.8
7	2.044	.875	.60	2.749	3.042	22.22	3.87	69.8
8	2.670	1.000	.79	3.142	3.973	25.40	5.10	79.8
9	3.400	1.128	1.00	3.544	5.059	28.65	6.45	90.0
10	4.303	1.270	1.27	3.990	6.403	32.26	8.19	101.4
11	5.313	1.410	1.56	4.430	7.906	35.81	10.06	112.5
14	7.650	1.693	2.25	5.320	11.384	43.00	14.52	135.1
18	13.600	2.257	4.00	7.090	20.238	57.33	25.81	180.1

* The nominal dimensions of a deformed bar are equivalent to those of a plain round bar having the same weight per foot as the deformed bar.
** Bar numbers are based on the number of eighths of an inch included in the nominal diameter of the bars.

R032110-80 Shop-Fabricated Reinforcing Steel

The material prices for reinforcing, shown in the unit cost sections of the data set, are for 50 tons or more of shop-fabricated reinforcing steel and include:

1. Mill base price of reinforcing steel
2. Mill grade/size/length extras
3. Mill delivery to the fabrication shop
4. Shop storage and handling
5. Shop drafting/detailing
6. Shop shearing and bending
7. Shop listing
8. Shop delivery to the job site

Both material and installation costs can be considerably higher for small jobs consisting primarily of smaller bars, while material costs may be slightly lower for larger jobs.

R033053-50 Industrial Chimneys

Foundation requirements in C.Y. of concrete for various sized chimneys.

Size Chimney	2 Ton Soil	3 Ton Soil	Size Chimney	2 Ton Soil	3 Ton Soil	Size Chimney	2 Ton Soil	3 Ton Soil
75′ x 3′-0″	13 C.Y.	11 C.Y.	160′ x 6′-6″	86 C.Y.	76 C.Y.	300′ x 10′-0″	325 C.Y.	245 C.Y.
85′ x 5′-6″	19	16	175′ x 7′-0″	108	95	350′ x 12′-0″	422	320
100′ x 5′-0″	24	20	200′ x 6′-0″	125	105	400′ x 14′-0″	520	400
125′ x 5′-6″	43	36	250′ x 8′-0″	230	175	500′ x 18′-0″	725	575

R033105-20 Materials for One C.Y. of Concrete

This is an approximate method of figuring quantities of cement, sand and coarse aggregate for a field mix with waste allowance included.

With crushed gravel as coarse aggregate, to determine barrels of cement required, divide 10 by total mix; that is, for 1:2:4 mix, 10 divided by 7 = 1-3/7 barrels.

If the coarse aggregate is crushed stone, use 10-1/2 instead of 10 as given for gravel.

To determine tons of sand required, multiply barrels of cement by parts of sand and then by 0.2; that is, for the 1:2:4 mix, as above, 1-3/7 x 2 x .2 = .57 tons.

Tons of crushed gravel are in the same ratio to tons of sand as parts in the mix, or 4/2 x .57 = 1.14 tons.

1 bag cement = 94#	1 C.Y. sand or crushed gravel = 2700#	1 C.Y. crushed stone = 2575#
4 bags = 1 barrel	1 ton sand or crushed gravel = 20 C.F.	1 ton crushed stone = 21 C.F.

Average carload of cement is 692 bags; of sand or gravel is 56 tons.

Do not stack stored cement over 10 bags high.

R033105-70 Placing Ready-Mixed Concrete

For ground pours allow for 5% waste when figuring quantities.

Prices in the front of the data set assume normal deliveries. If deliveries are made before 8 A.M. or after 5 P.M. or on Saturday afternoons add 30%. Negotiated discounts for large volumes are not included in prices in front of the data set.

For the lower floors without truck access, concrete may be wheeled in rubber-tired buggies, conveyer handled, crane handled or pumped. Pumping is economical if there is top steel. Conveyers are more efficient for thick slabs.

At higher floors the rubber-tired buggies may be hoisted by a hoisting tower and wheeled to the location. Placement by a conveyer is limited to three floors and is best for high-volume pours. Pumped concrete is best when the building has no crane access. Concrete may be pumped directly as high as thirty-six stories using special pumping techniques. Normal maximum height is about fifteen stories.

The best pumping aggregate is screened and graded bank gravel rather than crushed stone.

Pumping downward is more difficult than pumping upward. The horizontal distance from pump to pour may increase preparation time prior to pour. Placing by cranes, either mobile, climbing or tower types, continues as the most efficient method for high-rise concrete buildings.

R033105-80 Slab on Grade

General: Ground slabs are classified on the basis of use. Thickness is generally controlled by the heaviest concentrated load supported. If load area is greater than 80 sq. in., soil bearing may be important. The base granular fill must be a uniformly compacted material of limited capillarity, such as gravel or crushed rock. Concrete is placed on this surface of the vapor barrier on top of the base.

Ground slabs are either single or two course floors. Single course floors are widely used. Two course floors have a subsequent wear resistant topping.

Reinforcement is provided to maintain tightly closed cracks.

Control joints limit crack locations and provide for differential horizontal movement only. Isolation joints allow both horizontal and vertical differential movement.

Use of Table: Determine the appropriate type of slab (A, B, C, or D) by considering the type of use or amount of abrasive wear of traffic type.

Determine thickness by maximum allowable wheel load or uniform load, opposite 1st column, thickness. Increase the controlling thickness if details require, and select either plain or reinforced slab thickness and type.

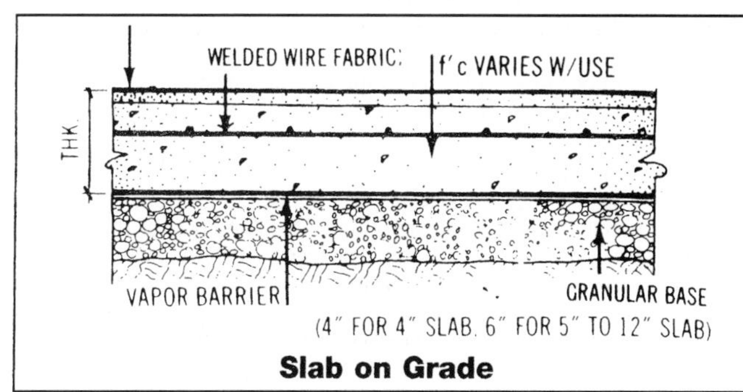

Slab on Grade

Thickness and Loading Assumptions by Type of Use

		A	B	C	D	◄ Slab I.D.
SLAB THICKNESS (IN.)	TYPE	Non	Light	Normal	Heavy	◄ Industrial
		Little	Light	Moderate	Severe	◄ Abrasion
		Foot Only	Pneumatic Wheels	Solid Rubber Wheels	Steel Tires	◄ Type of Traffic
		Load* (K)	Load* (K)	Load* (K)	Load* (K)	Max. Uniform Load to Slab ▼ (PSF)
4"	Reinf. Plain	4K				100
5"	Reinf. Plain	6K	4K			200
6"	Reinf. Plain		8K	6K	6K	500 to 800
7"	Reinf. Plain			9K	8K	1,500
8"	Reinf. Plain				11K	* Max. Wheel Load in Kips (incl. impact)
10"	Reinf. Plain				14K	
12"	Reinf. Plain					
DESIGN ASSUMPTIONS	Concrete, Chuted	f'c = 3.5 KSI	4 KSI	4.5 KSI	Slab @ 3.5 KSI	ASSUMPTIONS BY SLAB TYPE
	Toppings			1" Integral	1" Bonded	
	Finish	Steel Trowel	Steel Trowel	Steel Trowel	Screed & Steel Trowel	
	Compacted Granular Base	4" deep for 4" slab thickness 6" deep for 5" slab thickness & greater				ASSUMPTIONS FOR ALL SLAB TYPES
	Vapor Barrier	6 mil polyethylene				
	Forms & Joints	Allowances included				
	Rein-forcement	WWF as required ≥ 60,000 psi				

Concrete — R0335 Concrete Finishing

R033543-10 Polished Concrete Floors

A polished concrete floor has a glossy mirror-like appearance and is created by grinding the concrete floor with finer and finer diamond grits, similar to sanding wood, until the desired level of reflective clarity and sheen are achieved. The technical term for this type of polished concrete is bonded abrasive polished concrete. The basic piece of equipment used in the polishing process is a walk-behind planetary grinder for working large floor areas. This grinder drives diamond-impregnated abrasive discs, which progress from coarse- to fine-grit discs.

The process begins with the use of very coarse diamond segments or discs bonded in a metallic matrix. These segments are coarse enough to allow the removal of pits, blemishes, stains, and light coatings from the floor surface in preparation for final smoothing. The condition of the original concrete surface will dictate the grit coarseness of the initial grinding step which will generally end up being a three- to four-step process using ever finer grits. The purpose of this initial grinding step is to remove surface coatings and blemishes and to cut down into the cream for very fine aggregate exposure, or deeper into the fine aggregate layer just below the cream layer, or even deeper into the coarse aggregate layer. These initial grinding steps will progress up to the 100/120 grit. If wet grinding is done, a waste slurry is produced that must be removed between grit changes and disposed of properly. If dry grinding is done, a high performance vacuum will pick up the dust during grinding and collect it in bags which must be disposed of properly.

The process continues with honing the floor in a series of steps that progresses from 100-grit to 400-grit diamond abrasive discs embedded in a plastic or resin matrix. At some point during, or just prior to, the honing step, one or two coats of stain or dye can be sprayed onto the surface to give color to the concrete, and two coats of densifier/hardener must be applied to the floor surface and allowed to dry. This sprayed-on densifier/hardener will penetrate about 1/8″ into the concrete to make the surface harder, denser and more abrasion-resistant.

The process ends with polishing the floor surface in a series of steps that progresses from resin-impregnated 800-grit (medium polish) to 1500-grit (high polish) to 3000-grit (very high polish), depending on the desired level of reflective clarity and sheen.

The Concrete Polishing Association of America (CPAA) has defined the flooring options available when processing concrete to a desired finish. The first category is aggregate exposure, the grinding of a concrete surface with bonded abrasives, in as many abrasive grits necessary, to achieve one of the following classes:

A. Cream – very little surface cut depth; little aggregate exposure
B. Fine aggregate (salt and pepper) – surface cut depth of 1/16″; fine aggregate exposure with little or no medium aggregate exposure at random locations
C. Medium aggregate – surface cut depth of 1/8″; medium aggregate exposure with little or no large aggregate exposure at random locations
D. Large aggregate – surface cut depth of 1/4″; large aggregate exposure with little or no fine aggregate exposure at random locations

The second CPAA defined category is reflective clarity and sheen, the polishing of a concrete surface with the minimum number of bonded abrasives as indicated to achieve one of the following levels:

1. Ground – flat appearance with none to very slight diffused reflection; none to very low reflective sheen; using a minimum total of 4 grit levels up to 100-grit
2. Honed – matte appearance with or without slight diffused reflection; low to medium reflective sheen; using a minimum total of 5 grit levels up to 400-grit
3. Semi-polished – objects being reflected are not quite sharp and crisp but can be easily identified; medium to high reflective sheen; using a minimum total of 6 grit levels up to 800-grit
4. Highly-polished – objects being reflected are sharp and crisp as would be seen in a mirror-like reflection; high to highest reflective sheen; using a minimum total of up to 8 grit levels up to 1500-grit or 3000-grit

The CPAA defines reflective clarity as the degree of sharpness and crispness of the reflection of overhead objects when viewed 5′ above and perpendicular to the floor surface. Reflective sheen is the degree of gloss reflected from a surface when viewed at least 20′ from and at an angle to the floor surface. These terms are relatively subjective. The final outcome depends on the internal makeup and surface condition of the original concrete floor, the experience of the floor polishing crew, and the expectations of the owner. Before the grinding, honing, and polishing work commences on the main floor area, it might be beneficial to do a mock-up panel in the same floor but in an out of the way place to demonstrate the sequence of steps with increasingly fine abrasive grits and to demonstrate the final reflective clarity and reflective sheen. This mock-up panel will be within the area of, and part of, the final work.

Masonry — R0405 Common Work Results for Masonry

R040519-50 Masonry Reinforcing

Horizontal joint reinforcing helps prevent wall cracks where wall movement may occur and in many locations is required by code. Horizontal joint reinforcing is generally not considered to be structural reinforcing and an unreinforced wall may still contain joint reinforcing.

Reinforcing strips come in 10′ and 12′ lengths and in truss and ladder shapes, with and without drips. Field labor runs between 2.7 to 5.3 hours per 1000 L.F. for wall thicknesses up to 12″.

The wire meets ASTM A82 for cold drawn steel wire and the typical size is 9 ga. sides and ties with 3/16″ diameter also available. Typical finish is mill galvanized with zinc coating at .10 oz. per S.F. Class I (.40 oz. per S.F.) and Class III (.80 oz. per S.F.) are also available, as is hot dipped galvanizing at 1.50 oz. per S.F.

R042110-20 Common and Face Brick

Common building brick manufactured according to ASTM C62 and facing brick manufactured according to ASTM C216 are the two standard bricks available for general building use.

Building brick is made in three grades: SW, where high resistance to damage caused by cyclic freezing is required; MW, where moderate resistance to cyclic freezing is needed; and NW, where little resistance to cyclic freezing is needed. Facing brick is made in only the two grades SW and MW. Additionally, facing brick is available in three types: FBS, for general use; FBX, for general use where a higher degree of precision and lower permissible variation in size than FBS is needed; and FBA, for general use to produce characteristic architectural effects resulting from non-uniformity in size and texture of the units.

In figuring the material cost of brickwork, an allowance of 25% mortar waste and 3% brick breakage was included. If bricks are delivered palletized with 280 to 300 per pallet, or packaged, allow only 1-1/2% for breakage. Packaged or palletized delivery is practical when a job is big enough to have a crane or other equipment available to handle a package of brick. This is so on all industrial work but not always true on small commercial buildings.

The use of buff and gray face is increasing, and there is a continuing trend to the Norman, Roman, Jumbo and SCR brick.

Common red clay brick for backup is not used that often. Concrete block is the most usual backup material with occasional use of sand lime or cement brick. Building brick is commonly used in solid walls for strength and as a fire stop.

Brick panels built on the ground and then crane erected to the upper floors have proven to be economical. This allows the work to be done under cover and without scaffolding.

R042110-50 Brick, Block & Mortar Quantities

Running Bond					
Number of Brick per S.F. of Wall - Single Wythe with 3/8″ Joints				C.F. of Mortar per M Bricks, Waste Included	
Type Brick	Nominal Size (incl. mortar) L H W	Modular Coursing	Number of Brick per S.F.	3/8″ Joint	1/2″ Joint
Standard	8 x 2-2/3 x 4	3C=8″	6.75	8.1	10.3
Economy	8 x 4 x 4	1C=4″	4.50	9.1	11.6
Engineer	8 x 3-1/5 x 4	5C=16″	5.63	8.5	10.8
Fire	9 x 2-1/2 x 4-1/2	2C=5″	6.40	550 # Fireclay	—
Jumbo	12 x 4 x 6 or 8	1C=4″	3.00	22.5	29.2
Norman	12 x 2-2/3 x 4	3C=8″	4.50	11.2	14.3
Norwegian	12 x 3-1/5 x 4	5C=16″	3.75	11.7	14.9
Roman	12 x 2 x 4	2C=4″	6.00	10.7	13.7
SCR	12 x 2-2/3 x 6	3C=8″	4.50	21.8	28.0
Utility	12 x 4 x 4	1C=4″	3.00	12.3	15.7

For Other Bonds Standard Size Add to S.F. Quantities in Table to Left		
Bond Type	Description	Factor
Common	full header every fifth course	+20%
	full header every sixth course	+16.7%
English	full header every second course	+50%
Flemish	alternate headers every course	+33.3%
	every sixth course	+5.6%
Header = W x H exposed		+100%
Rowlock = H x W exposed		+100%
Rowlock stretcher = L x W exposed		+33.3%
Soldier = H x L exposed		—
Sailor = W x L exposed		-33.3%

Concrete Blocks Nominal Size	Approximate Weight per S.F.		Blocks per 100 S.F.	Mortar per M block, waste included	
	Standard	Lightweight		Partitions	Back up
2″ x 8″ x 16″	20 PSF	15 PSF	113	27 C.F.	36 C.F.
4″	30	20		41	51
6″	42	30		56	66
8″	55	38		72	82
10″	70	47		87	97
12″	85	55		102	112

Brick & Mortar Quantities
©Brick Industry Association. 2009 Feb. Technical Notes on Brick Construction 10:
Dimensioning and Estimating Brick Masonry. Reston (VA): BIA. Table 1 Modular Brick Sizes and Table 4 Quantity Estimates for Brick Masonry.

Masonry R0422 Concrete Unit Masonry

R042210-20 Concrete Block

The material cost of special block such as corner, jamb and head block can be figured at the same price as ordinary block of equal size. Labor on specials is about the same as equal-sized regular block.

Bond beams and 16″ high lintel blocks are more expensive than regular units of equal size. Lintel blocks are 8″ long and either 8″ or 16″ high.

Use of a motorized mortar spreader box will speed construction of continuous walls.

Hollow non-load-bearing units are made according to ASTM C129 and hollow load-bearing units according to ASTM C90.

R050516-30 Coating Structural Steel

On field-welded jobs, the shop-applied primer coat is necessarily omitted. All painting must be done in the field and usually consists of red oxide rust inhibitive paint or an aluminum paint. The table below shows paint coverage and daily production for field painting.

See Division 05 05 13.50 for hot-dipped galvanizing and Division 09 97 13.23 for field-applied cold galvanizing and other paints and protective coatings.

See Division 05 01 10.51 for steel surface preparation treatments such as wire brushing, pressure washing and sand blasting.

Type Construction	Surface Area per Ton	Coat	One Gallon Covers		In 8 Hrs. Person Covers		Average per Ton Spray	
			Brush	Spray	Brush	Spray	Gallons	Labor-hours
Light Structural	300 S.F. to 500 S.F.	1st	500 S.F.	455 S.F.	640 S.F.	2000 S.F.	0.9 gals.	1.6 L.H.
		2nd	450	410	800	2400	1.0	1.3
		3rd	450	410	960	3200	1.0	1.0
Medium	150 S.F. to 300 S.F.	All	400	365	1600	3200	0.6	0.6
Heavy Structural	50 S.F. to 150 S.F.	1st	400	365	1920	4000	0.2	0.2
		2nd	400	365	2000	4000	0.2	0.2
		3rd	400	365	2000	4000	0.2	0.2
Weighted Average	225 S.F.	All	400	365	1350	3000	0.6	0.6

R050521-20 Welded Structural Steel

Usual weight reductions with welded design run 10% to 20% compared with bolted or riveted connections. This amounts to about the same total cost compared with bolted structures since field welding is more expensive than bolts. For normal spans of 18′ to 24′ figure 6 to 7 connections per ton.

Trusses — For welded trusses add 4% to weight of main members for connections. Up to 15% less steel can be expected in a welded truss compared to one that is shop bolted. Cost of erection is the same whether shop bolted or welded.

General — Typical electrodes for structural steel welding are E6010, E6011, E60T and E70T. Typical buildings vary between 2# to 8# of weld rod per ton of steel. Buildings utilizing continuous design require about three times as much welding as conventional welded structures. In estimating field erection by welding, it is best to use the average linear feet of weld per ton to arrive at the welding cost per ton. The type, size and position of the weld will have a direct bearing on the cost per linear foot. A typical field welder will deposit 1.8# to 2# of weld rod per hour manually. Using semiautomatic methods can increase production by as much as 50% to 75%.

R050523-10 High Strength Bolts

Common bolts (A307) are usually used in secondary connections (see Division 05 05 23.10).

High strength bolts (A325 and A490) are usually specified for primary connections such as column splices, beam and girder connections to columns, column bracing, connections for supports of operating equipment or of other live loads which produce impact or reversal of stress, and in structures carrying cranes of over 5-ton capacity.

Allow 20 field bolts per ton of steel for a 6 story office building, apartment house or light industrial building. For 6 to 12 stories allow 25 bolts per ton, and above 12 stories, 30 bolts per ton. On power stations, 20 to 25 bolts per ton are needed.

Metals | R0512 Structural Steel Framing

R051223-10 Structural Steel

The bare material prices for structural steel, shown in the unit cost sections of the data set, are for 100 tons of shop-fabricated structural steel and include:

1. Mill base price of structural steel
2. Mill scrap/grade/size/length extras
3. Mill delivery to a metals service center (warehouse)
4. Service center storage and handling
5. Service center delivery to a fabrication shop
6. Shop storage and handling
7. Shop drafting/detailing
8. Shop fabrication
9. Shop coat of primer paint
10. Shop listing
11. Shop delivery to the job site

In unit cost sections of the data set that contain items for field fabrication of steel components, the bare material cost of steel includes:

1. Mill base price of structural steel
2. Mill scrap/grade/size/length extras
3. Mill delivery to a metals service center (warehouse)
4. Service center storage and handling
5. Service center delivery to the job site

R051223-15 Structural Steel Estimating for Repair and Remodeling Projects

The correct approach to estimating structural steel is dependent upon the amount of steel required for the particular project. If the project is a sizable addition to a building, the data can be used directly from the cost data set. This is not the case however if the project requires a small amount of steel, for instance to reinforce existing roof or floor structural systems. To better understand this, please refer to the unit price line for a W16x31 beam with bolted connections. Assume your project requires the reinforcement of the structural members of the roof system of a 3 story building required for the installation a new piece of HVAC equipment. The project will need 4 pieces of W16x31 that are 30′ long. After pricing this 120 L.F. job using the unit prices for a W16x31, an analysis will reveal that the price is wholly inadequate.

The first problem is apparent if you examine the amount of steel that can be installed per day. The unit price line indicates 900 linear feet per day can be installed by a 5 man crew with an 90 ton crane. This productivity is correct for new construction but certainly is not for repair and remodeling work. Installation of new structural steel considers that each member is installed from the foundation to the roof of the structure in a planned and systematic manner with a crane having unrestricted access to all parts of the project. Additionally each connection is planned and detailed with full field access for fit-up and final bolting. The erection is planned and progresses such that interferences and conflicts with other structural members are minimized, if not completely eliminated. All of these assumptions are clearly not the case with a repair and remodeling job, and a significant decrease in the stated productivity will be observed.

A crane will certainly be needed to lift the members into the general area of the project but in most cases will not be able to place the beams into their final position. An opening in the existing roof may not be large enough to permit the beams to pass through and it may be necessary to bring them into the building through existing windows or doors. Moving the beams to the actual area where they will be installed may involve hand labor and the use of dollies. Finally, hoists and/or jacks may be needed for final positioning.

The connection of new members to existing can often be accomplished by field bolting with accurate field measurements and good planning but in many cases access to both sides of existing members is not possible and field welding becomes the only alternative. In addition to the cost of the actual welding, protection of existing finishes, systems and structure and fire protection must be considered.

New beams can never be installed tight to the existing decks or floors which they must support and the use of shims and tack welding becomes necessary. Additionally, further planning and cost is involved in assuring that existing loads are minimized during the installation and shimming process.

It is apparent that installation of structural steel as part of a repair and remodeling project involves more than simply installing the members and estimating the cost in the same manner as new construction. The best procedure for estimating the total cost is adequate planning and coordination of each process and activity that will be needed. Unit costs for the materials, labor and equipment can then be attached to each needed activity and a final, complete price can be determined.

R051223-20 Steel Estimating Quantities

One estimate on erection is that a crane can handle 35 to 60 pieces per day. Say the average is 45. With usual sizes of beams, girders, and columns, this would amount to about 20 tons per day. The type of connection greatly affects the speed of erection. Moment connections for continuous design slow down production and increase erection costs.

Short open web bar joists can be set at the rate of 75 to 80 per day, with 50 per day being the average for setting long span joists.

After main members are calculated, add the following for usual allowances: base plates 2% to 3%; column splices 4% to 5%; and miscellaneous details 4% to 5%, for a total of 10% to 13% in addition to main members.

The ratio of column to beam tonnage varies depending on type of steels used, typical spans, story heights and live loads.

It is more economical to keep the column size constant and to vary the strength of the column by using high strength steels. This also saves floor space. Buildings have recently gone as high as ten stories with 8″ high strength columns. For light columns under W8X31 lb. sections, concrete filled steel columns are economical.

High strength steels may be used in columns and beams to save floor space and to meet head room requirements. High strength steels in some sizes sometimes require long lead times.

Round, square and rectangular columns, both plain and concrete filled, are readily available and save floor area, but are higher in cost per pound than rolled columns. For high unbraced columns, tube columns may be less expensive.

Below are average minimum figures for the weights of the structural steel frame for different types of buildings using A36 steel, rolled shapes and simple joints. For economy in domes, rise to span ratio = .13. Open web joist framing systems will reduce weights by 10% to 40%. Composite design can reduce steel weight by up to 25% but additional concrete floor slab thickness may be required. Continuous design can reduce the weights up to 20%. There are many building codes with different live load requirements and different structural requirements, such as hurricane and earthquake loadings, which can alter the figures.

Structural Steel Weights per S.F. of Floor Area

Type of Building	No. of Stories	Avg. Spans	L.L. #/S.F.	Lbs. Per S.F.	Type of Building	No. of Stories	Avg. Spans	L.L. #/S.F.	Lbs. Per S.F.
Steel Frame Mfg.	1	20′x20′	40	8	Apartments	2-8	20′x20′	40	8
		30′x30′		13		9-25			14
		40′x40′		18	Office	to 10	Various	80	10
Parking garage	4	Various	80	8.5		20			18
Domes (Schwedler)*	1	200′	30	10		30			26
		300′		15		over 50			35

R051223-25 Common Structural Steel Specifications

ASTM A992 (formerly A36, then A572 Grade 50) is the all-purpose carbon grade steel widely used in building and bridge construction.

The other high-strength steels listed below may each have certain advantages over ASTM A992 structural carbon steel, depending on the application. They have proven to be economical choices where, due to lighter members, the reduction of dead load and the associated savings in shipping cost can be significant.

ASTM A588 atmospheric weathering, high-strength, low-alloy steels can be used in the bare (uncoated) condition, where exposure to normal atmosphere causes a tightly adherant oxide to form on the surface, protecting the steel from further oxidation. ASTM A242 corrosion-resistant, high-strength, low-alloy steels have enhanced atmospheric corrosion resistance of at least two times that of carbon structural steels with copper, or four times that of carbon structural steels without copper. The reduction or elimination of maintenance resulting from the use of these steels often offsets their higher initial cost.

Steel Type	ASTM Designation	Minimum Yield Stress in KSI	Shapes Available
Carbon	A36	36	All structural shape groups, and plates & bars up through 8″ thick
	A529	50	Structural shape group 1, and plates & bars up through 2″ thick
High-Strength Low-Alloy Quenched & Self-Tempered	A913	50	All structural shape groups
		60	
		65	
		70	
High-Strength Low-Alloy Columbium-Vanadium	A572	42	All structural shape groups, and plates & bars up through 6″ thick
		50	All structural shape groups, and plates & bars up through 4″ thick
		55	Structural shape groups 1 & 2, and plates & bars up through 2″ thick
		60	Structural shape groups 1 & 2, and plates & bars up through 1-1/4″ thick
		65	Structural shape group 1, and plates & bars up through 1-1/4″ thick
High-Strength Low-Alloy Columbium-Vanadium	A992	50	All structural shape groups
Weathering High-Strength Low-Alloy	A242	42	Structural shape groups 4 & 5, and plates & bars over 1-1/2″ up through 4″ thick
		46	Structural shape group 3, and plates & bars over 3/4″ up through 1-1/2″ thick
		50	Structural shape groups 1 & 2, and plates & bars up through 3/4″ thick
Weathering High-Strength Low-Alloy	A588	42	Plates & bars over 5″ up through 8″ thick
		46	Plates & bars over 4″ up through 5″ thick
		50	All structural shape groups, and plates & bars up through 4″ thick
Quenched and Tempered Low-Alloy	A852	70	Plates & bars up through 4″ thick
Quenched and Tempered Alloy	A514	90	Plates & bars over 2-1/2″ up through 6″ thick
		100	Plates & bars up through 2-1/2″ thick

R051223-30 High Strength Steels

The mill price of high strength steels may be higher than A992 carbon steel, but their proper use can achieve overall savings through total reduced weights. For columns with L/r over 100, A992 steel is best; under 100, high strength steels are economical. For heavy columns, high strength steels are economical when cover plates are eliminated. There is no economy using high strength steels for clip angles or supports or for beams where deflection governs. Thinner members are more economical than thick.

The per ton erection and fabricating costs of the high strength steels will be higher than for A992 since the same number of pieces, but less weight, will be installed.

R051223-35 Common Steel Sections

The upper portion of this table shows the name, shape, common designation, and basic characteristics of commonly used steel sections. The lower portion explains how to read the designations used for the above illustrated common sections.

Shape & Designation	Name & Characteristics	Shape & Designation	Name & Characteristics
W	W Shape Parallel flange surfaces	MC	Miscellaneous Channel Infrequently rolled by some producers
S	American Standard Beam (I Beam) Sloped inner flange	L	Angle Equal or unequal legs, constant thickness
M	Miscellaneous Beams Cannot be classified as W, HP or S; infrequently rolled by some producers	T	Structural Tee Cut from W, M or S on center of web
C	American Standard Channel Sloped inner flange	HP	Bearing Pile Parallel flanges and equal flange and web thickness

Common drawing designations follow:

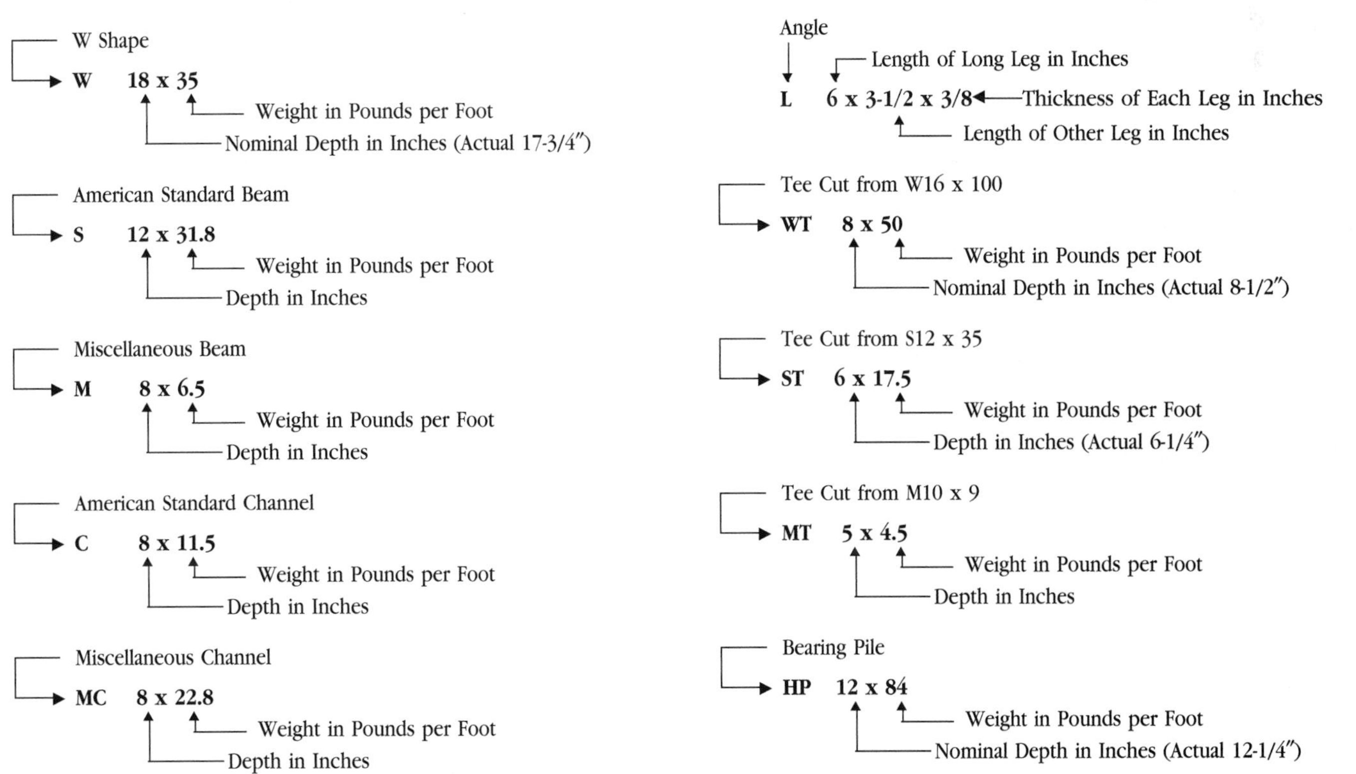

R051223-45 Installation Time for Structural Steel Building Components

The following tables show the expected average installation times for various structural steel shapes. Table A presents installation times for columns, Table B for beams, Table C for light framing and bolts, and Table D for structural steel for various project types.

Table A

Description	Labor-Hours	Unit
Columns		
Steel, Concrete Filled		
3-1/2″ Diameter	.933	Ea.
6-5/8″ Diameter	1.120	Ea.
Steel Pipe		
3″ Diameter	.933	Ea.
8″ Diameter	1.120	Ea.
12″ Diameter	1.244	Ea.
Structural Tubing		
4″ x 4″	.966	Ea.
8″ x 8″	1.120	Ea.
12″ x 8″	1.167	Ea.
W Shape 2 Tier		
W8 x 31	.052	L.F.
W8 x 67	.057	L.F.
W10 x 45	.054	L.F.
W10 x 112	.058	L.F.
W12 x 50	.054	L.F.
W12 x 190	.061	L.F.
W14 x 74	.057	L.F.
W14 x 176	.061	L.F.

Table B

Description	Labor-Hours	Unit	Labor-Hours	Unit
Beams, W Shape				
W6 x 9	.949	Ea.	.093	L.F.
W10 x 22	1.037	Ea.	.085	L.F.
W12 x 26	1.037	Ea.	.064	L.F.
W14 x 34	1.333	Ea.	.069	L.F.
W16 x 31	1.333	Ea.	.062	L.F.
W18 x 50	2.162	Ea.	.088	L.F.
W21 x 62	2.222	Ea.	.077	L.F.
W24 x 76	2.353	Ea.	.072	L.F.
W27 x 94	2.581	Ea.	.067	L.F.
W30 x 108	2.857	Ea.	.067	L.F.
W33 x 130	3.200	Ea.	.071	L.F.
W36 x 300	3.810	Ea.	.077	L.F.

Table C

Description	Labor-Hours	Unit
Light Framing		
Angles 4″ and Larger	.055	lbs.
Less than 4″	.091	lbs.
Channels 8″ and Larger	.048	lbs.
Less than 8″	.072	lbs.
Cross Bracing Angles	.055	lbs.
Rods	.034	lbs.
Hanging Lintels	.069	lbs.
High Strength Bolts in Place		
3/4″ Bolts	.070	Ea.
7/8″ Bolts	.076	Ea.

Table D

Description	Labor-Hours	Unit	Labor-Hours	Unit
Apartments, Nursing Homes, etc.				
1-2 Stories	4.211	Piece	7.767	Ton
3-6 Stories	4.444	Piece	7.921	Ton
7-15 Stories	4.923	Piece	9.014	Ton
Over 15 Stories	5.333	Piece	9.209	Ton
Offices, Hospitals, etc.				
1-2 Stories	4.211	Piece	7.767	Ton
3-6 Stories	4.741	Piece	8.889	Ton
7-15 Stories	4.923	Piece	9.014	Ton
Over 15 Stories	5.120	Piece	9.209	Ton
Industrial Buildings				
1 Story	3.478	Piece	6.202	Ton

R051223-50 Subpurlins

Bulb tee subpurlins are structural members designed to support and reinforce a variety of roof deck systems such as precast cement fiber roof deck tiles, monolithic roof deck systems, and gypsum or lightweight concrete over formboard. Other uses include interstitial service ceiling systems, wall panel systems, and joist anchoring in bond beams. See the Unit Price section for pricing on a square foot basis at 32-5/8″ O.C. Maximum span is based on a 3-span condition with a total allowable vertical load of 40 psf.

R051223-80 Dimensions and Weights of Sheet Steel

Gauge No.	Approximate Thickness				Weight		
	Inches (in fractions)	Inches (in decimal parts)		Millimeters			
	Wrought Iron	Wrought Iron	Steel	Steel	per S.F. in Ounces	per S.F. in Lbs.	per Square Meter in Kg.
0000000	1/2″	.5	.4782	12.146	320	20.000	97.650
000000	15/32″	.46875	.4484	11.389	300	18.750	91.550
00000	7/16″	.4375	.4185	10.630	280	17.500	85.440
0000	13/32″	.40625	.3886	9.870	260	16.250	79.330
000	3/8″	.375	.3587	9.111	240	15.000	73.240
00	11/32″	.34375	.3288	8.352	220	13.750	67.130
0	5/16″	.3125	.2989	7.592	200	12.500	61.030
1	9/32″	.28125	.2690	6.833	180	11.250	54.930
2	17/64″	.265625	.2541	6.454	170	10.625	51.880
3	1/4″	.25	.2391	6.073	160	10.000	48.820
4	15/64″	.234375	.2242	5.695	150	9.375	45.770
5	7/32″	.21875	.2092	5.314	140	8.750	42.720
6	13/64″	.203125	.1943	4.935	130	8.125	39.670
7	3/16″	.1875	.1793	4.554	120	7.500	36.320
8	11/64″	.171875	.1644	4.176	110	6.875	33.570
9	5/32″	.15625	.1495	3.797	100	6.250	30.520
10	9/64″	.140625	.1345	3.416	90	5.625	27.460
11	1/8″	.125	.1196	3.038	80	5.000	24.410
12	7/64″	.109375	.1046	2.657	70	4.375	21.360
13	3/32″	.09375	.0897	2.278	60	3.750	18.310
14	5/64″	.078125	.0747	1.897	50	3.125	15.260
15	9/128″	.0713125	.0673	1.709	45	2.813	13.730
16	1/16″	.0625	.0598	1.519	40	2.500	12.210
17	9/160″	.05625	.0538	1.367	36	2.250	10.990
18	1/20″	.05	.0478	1.214	32	2.000	9.765
19	7/160″	.04375	.0418	1.062	28	1.750	8.544
20	3/80″	.0375	.0359	.912	24	1.500	7.324
21	11/320″	.034375	.0329	.836	22	1.375	6.713
22	1/32″	.03125	.0299	.759	20	1.250	6.103
23	9/320″	.028125	.0269	.683	18	1.125	5.490
24	1/40″	.025	.0239	.607	16	1.000	4.882
25	7/320″	.021875	.0209	.531	14	.875	4.272
26	3/160″	.01875	.0179	.455	12	.750	3.662
27	11/640″	.0171875	.0164	.417	11	.688	3.357
28	1/64″	.015625	.0149	.378	10	.625	3.052

R053100-10 Decking Descriptions

General - All Deck Products

A steel deck is made by cold forming structural grade sheet steel into a repeating pattern of parallel ribs. The strength and stiffness of the panels are the result of the ribs and the material properties of the steel. The deck lengths can be varied to suit job conditions, but because of shipping considerations, are usually less than 40 feet. Standard deck width varies with the product used but full sheets are usually 12″, 18″, 24″, 30″, or 36″. The deck is typically furnished in a standard width with the ends cut square. Any cutting for width, such as at openings or for angular fit, is done at the job site.

The deck is typically attached to the building frame with arc puddle welds, self-drilling screws, or powder or pneumatically driven pins. Sheet-to-sheet fastening is done with screws, button punching (crimping), or welds.

Composite Floor Deck

After installation and adequate fastening, a floor deck serves several purposes. It (a) acts as a working platform, (b) stabilizes the frame, (c) serves as a concrete form for the slab, and (d) reinforces the slab to carry the design loads applied during the life of the building. Composite decks are distinguished by the presence of shear connector devices as part of the deck. These devices are designed to mechanically lock the concrete and deck together so that the concrete and the deck work together to carry subsequent floor loads. These shear connector devices can be rolled-in embossments, lugs, holes, or wires welded to the panels. The deck profile can also be used to interlock concrete and steel.

Composite deck finishes are either galvanized (zinc coated) or phosphatized/painted. Galvanized deck has a zinc coating on both the top and bottom surfaces. The phosphatized/painted deck has a bare (phosphatized) top surface that will come into contact with the concrete. This bare top surface can be expected to develop rust before the concrete is placed. The bottom side of the deck has a primer coat of paint.

A composite floor deck is normally installed so the panel ends do not overlap on the supporting beams. Shear lugs or panel profile shapes often prevent a tight metal to metal fit if the panel ends overlap; the air gap caused by overlapping will prevent proper fusion with the structural steel supports when the panel end laps are shear stud welded.

Adequate end bearing of the deck must be obtained as shown on the drawings. If bearing is actually less in the field than shown on the drawings, further investigation is required.

Roof Deck

A roof deck is not designed to act compositely with other materials. A roof deck acts alone in transferring horizontal and vertical loads into the building frame. Roof deck rib openings are usually narrower than floor deck rib openings. This provides adequate support of the rigid thermal insulation board.

A roof deck is typically installed to endlap approximately 2″ over supports. However, it can be butted (or lapped more than 2″) to solve field fit problems. Since designers frequently use the installed deck system as part of the horizontal bracing system (the deck as a diaphragm), any fastening substitution or change should be approved by the designer. Continuous perimeter support of the deck is necessary to limit edge deflection in the finished roof and may be required for diaphragm shear transfer.

Standard roof deck finishes are galvanized or primer painted. The standard factory applied paint for roof decks is a primer paint and is not intended to weather for extended periods of time. Field painting or touching up of abrasions and deterioration of the primer coat or other protective finishes is the responsibility of the contractor.

Cellular Deck

A cellular deck is made by attaching a bottom steel sheet to a roof deck or composite floor deck panel. A cellular deck can be used in the same manner as a floor deck. Electrical, telephone, and data wires are easily run through the chase created between the deck panel and the bottom sheet.

When used as part of the electrical distribution system, the cellular deck must be installed so that the ribs line up and create a smooth cell transition at abutting ends. The joint that occurs at butting cell ends must be taped or otherwise sealed to prevent wet concrete from seeping into the cell. Cell interiors must be free of welding burrs, or other sharp intrusions, to prevent damage to wires.

When used as a roof deck, the bottom flat plate is usually left exposed to view. Care must be maintained during erection to keep good alignment and prevent damage.

A cellular deck is sometimes used with the flat plate on the top side to provide a flat working surface. Installation of the deck for this purpose requires special methods for attachment to the frame because the flat plate, now on the top, can prevent direct access to the deck material that is bearing on the structural steel. It may be advisable to treat the flat top surface to prevent slipping.

A cellular deck is always furnished galvanized or painted over galvanized.

Form Deck

A form deck can be any floor or roof deck product used as a concrete form. Connections to the frame are by the same methods used to anchor floor and roof decks. Welding washers are recommended when welding a deck that is less than 20 gauge thickness.

A form deck is furnished galvanized, prime painted, or uncoated. A galvanized deck must be used for those roof deck systems where a form deck is used to carry a lightweight insulating concrete fill.

Wood, Plastics & Comp. — R0611 Wood Framing

R061110-30 Lumber Product Material Prices

The price of forest products fluctuates widely from location to location and from season to season depending upon economic conditions. The bare material prices in the unit cost sections of the data set show the National Average material prices in effect Jan. 1 of this data year. It must be noted that lumber prices in general may change significantly during the year.

Availability of certain items depends upon geographic location and must be checked prior to firm-price bidding.

Wood, Plastics & Comp. — R0616 Sheathing

R061636-20 Plywood

There are two types of plywood used in construction: interior, which is moisture-resistant but not waterproofed, and exterior, which is waterproofed.

The grade of the exterior surface of the plywood sheets is designated by the first letter: A, for smooth surface with patches allowed; B, for solid surface with patches and plugs allowed; C, which may be surface plugged or may have knot holes up to 1″ wide; and D, which is used only for interior type plywood and may have knot holes up to 2-1/2″ wide. "Structural Grade" is specifically designed for engineered applications such as box beams. All CC & DD grades have roof and floor spans marked on them.

Underlayment-grade plywood runs from 1/4″ to 1-1/4″ thick. Thicknesses 5/8″ and over have optional tongue and groove joints which eliminate the need for blocking the edges. Underlayment 19/32″ and over may be referred to as Sturd-i-Floor.

The price of plywood can fluctuate widely due to geographic and economic conditions.

Typical uses for various plywood grades are as follows:

AA-AD Interior — cupboards, shelving, paneling, furniture

BB Plyform — concrete form plywood

CDX — wall and roof sheathing

Structural — box beams, girders, stressed skin panels

AA-AC Exterior — fences, signs, siding, soffits, etc.

Underlayment — base for resilient floor coverings

Overlaid HDO — high density for concrete forms & highway signs

Overlaid MDO — medium density for painting, siding, soffits & signs

303 Siding — exterior siding, textured, striated, embossed, etc.

Thermal & Moist. Protec. — R0752 Modified Bituminous Membrane Roofing

R075213-30 Modified Bitumen Roofing

The cost of modified bitumen roofing is highly dependent on the type of installation that is planned. Installation is based on the type of modifier used in the bitumen. The two most popular modifiers are atactic polypropylene (APP) and styrene butadiene styrene (SBS). The modifiers are added to heated bitumen during the manufacturing process to change its characteristics. A polyethylene, polyester or fiberglass reinforcing sheet is then sandwiched between layers of this bitumen. When completed, the result is a pre-assembled, built-up roof that has increased elasticity and weatherability. Some manufacturers include a surfacing material such as ceramic or mineral granules, metal particles or sand.

The preferred method of adhering SBS-modified bitumen roofing to the substrate is with hot-mopped asphalt (much the same as built-up roofing). This installation method requires a tar kettle/pot to heat the asphalt, as well as the labor, tools and equipment necessary to distribute and spread the hot asphalt.

The alternative method for applying APP and SBS modified bitumen is as follows. A skilled installer uses a torch to melt a small pool of bitumen off the membrane. This pool must form across the entire roll for proper adhesion. The installer must unroll the roofing at a pace slow enough to melt the bitumen, but fast enough to prevent damage to the rest of the membrane.

Modified bitumen roofing provides the advantages of both built-up and single-ply roofing. Labor costs are reduced over those of built-up roofing because only a single ply is necessary. The elasticity of single-ply roofing is attained with the reinforcing sheet and polymer modifiers. Modifieds have some self-healing characteristics and because of their multi-layer construction, they offer the reliability and safety of built-up roofing.

R078413-30 Firestopping

Firestopping is the sealing of structural, mechanical, electrical, and other penetrations through fire-rated assemblies. The basic components of firestop systems are safing insulation and firestop sealant on both sides of wall penetrations and the top side of floor penetrations.

Pipe penetrations are assumed to be through concrete, grout, or joint compound and can be sleeved or unsleeved. Costs for the penetrations and sleeves are not included. An annular space of 1″ is assumed. Escutcheons are not included.

A metallic pipe is assumed to be copper, aluminum, cast iron, or similar metallic material. An insulated metallic pipe is assumed to be covered with a thermal insulating jacket of varying thickness and materials.

A non-metallic pipe is assumed to be PVC, CPVC, FR Polypropylene, or similar plastic piping material. Intumescent firestop sealant or wrap strips are included. Collars on both sides of wall penetrations and a sheet metal plate on the underside of floor penetrations are included.

Ductwork is assumed to be sheet metal, stainless steel or similar metallic material. Duct penetrations are assumed to be through concrete, grout or joint compound. Costs for penetrations and sleeves are not included. An annular space of 1/2″ is assumed.

Multi-trade openings include costs for sheet metal forms, firestop mortar, wrap strips, collars, and sealants as necessary.

Structural penetrations joints are assumed to be 1/2″ or less. CMU walls are assumed to be within 1-1/2″ of the metal deck. Drywall walls are assumed to be tight to the underside of metal decking.

Metal panel, glass, or curtain wall systems include a spandrel area of 5′ filled with mineral wool foil-faced insulation. Fasteners and stiffeners are included.

Openings — R0813 Metal Doors

R081313-20 Steel Door Selection Guide

Standard steel doors are classified into four levels, as recommended by the Steel Door Institute in the chart below. Each of the four levels offers a range of construction models and designs to meet architectural requirements for preference and appearance, including full flush, seamless, and stile & rail. Recommended minimum gauge requirements are also included.

For complete standard steel door construction specifications and available sizes, refer to the Steel Door Institute Technical Data Series, ANSI A250.8-98 (SDI-100), and ANSI A250.4-94 Test Procedure and Acceptance Criteria for Physical Endurance of Steel Door and Hardware Reinforcements.

Level		Model	Construction	For Full Flush or Seamless		
				Min. Gauge	Thickness (in)	Thickness (mm)
I	Standard Duty	1 2	Full Flush Seamless	20	0.032	0.8
II	Heavy Duty	1 2	Full Flush Seamless	18	0.042	1.0
III	Extra Heavy Duty	1 2 3	Full Flush Seamless *Stile & Rail	16	0.053	1.3
IV	Maximum Duty	1 2	Full Flush Seamless	14	0.067	1.6

*Stiles & rails are 16 gauge; flush panels, when specified, are 18 gauge

R087110-10 Hardware Finishes

This table describes hardware finishes used throughout the industry. It also shows the base metal and the respective symbols in the three predominate systems of identification. Many of these are used in pricing descriptions in Division Eight.

US″	BMHA*	CDN	Base	Description
US P	600	CP	Steel	Primed for Painting
US 1B	601	C1B	Steel	Bright Black Japanned
US 2C	602	C2C	Steel	Zinc Plated
US 2G	603	C2G	Steel	Zinc Plated
US 3	605	C3	Brass	Bright Brass, Clear Coated
US 4	606	C4	Brass	Satin Brass, Clear Coated
US 5	609	C5	Brass	Satin Brass, Blackened, Satin Relieved, Clear Coated
US 7	610	C7	Brass	Satin Brass, Blackened, Bright Relieved, Clear Coated
US 9	611	C9	Bronze	Bright Bronze, Clear Coated
US 10	612	C10	Bronze	Satin Bronze, Clear Coated
US 10A	641	C10A	Steel	Antiqued Bronze, Oiled and Lacquered
US 10B	613	C10B	Bronze	Antiqued Bronze, Oiled
US 11	616	C11	Bronze	Satin Bronze, Blackened, Satin Relieved, Clear Coated
US 14	618	C14	Brass/Bronze	Bright Nickel Plated, Clear Coated
US 15	619	C15	Brass/Bronze	Satin Nickel, Clear Coated
US 15A	620	C15A	Brass/Bronze	Satin Nickel Plated, Blackened, Satin Relieved, Clear Coated
US 17A	621	C17A	Brass/Bronze	Nickel Plated, Blackened, Relieved, Clear Coated
US 19	622	C19	Brass/Bronze	Flat Black Coated
US 20	623	C20	Brass/Bronze	Statuary Bronze, Light
US 20A	624	C20A	Brass/Bronze	Statuary Bronze, Dark
US 26	625	C26	Brass/Bronze	Bright Chromium
US 26D	626	C26D	Brass/Bronze	Satin Chromium
US 20	627	C27	Aluminum	Satin Aluminum Clear
US 28	628	C28	Aluminum	Anodized Dull Aluminum
US 32	629	C32	Stainless Steel	Bright Stainless Steel
US 32D	630	C32D	Stainless Steel	Stainless Steel
US 3	632	C3	Steel	Bright Brass Plated, Clear Coated
US 4	633	C4	Steel	Satin Brass, Clear Coated
US 7	636	C7	Steel	Satin Brass Plated, Blackened, Bright Relieved, Clear Coated
US 9	637	C9	Steel	Bright Bronze Plated, Clear Coated
US 5	638	C5	Steel	Satin Brass Plated, Blackened, Bright Relieved, Clear Coated
US 10	639	C10	Steel	Satin Bronze Plated, Clear Coated
US 10B	640	C10B	Steel	Antique Bronze, Oiled
US 10A	641	C10A	Steel	Antiqued Bronze, Oiled and Lacquered
US 11	643	C11	Steel	Satin Bronze Plated, Blackened, Bright Relieved, Clear Coated
US 14	645	C14	Steel	Bright Nickel Plated, Clear Coated
US 15	646	C15	Steel	Satin Nickel Plated, Clear Coated
US 15A	647	C15A	Steel	Nickel Plated, Blackened, Bright Relieved, Clear Coated
US 17A	648	C17A	Steel	Nickel Plated, Blackened, Relieved, Clear Coated
US 20	649	C20	Steel	Statuary Bronze, Light
US 20A	650	C20A	Steel	Statuary Bronze, Dark
US 26	651	C26	Steel	Bright Chromium Plated
US 26D	652	C26D	Steel	Satin Chromium Plated

* - BMHA Builders Hardware Manufacturing Association
″ - US Equivalent
^ - Canadian Equivalent
Japanning is imitating Asian lacquer work

R092910-10 Levels of Gypsum Board Finish

In the past, contract documents often used phrases such as "industry standard" and "workmanlike finish" to specify the expected quality of gypsum board wall and ceiling installations. The vagueness of these descriptions led to unacceptable work and disputes.

In order to resolve this problem, four major trade associations concerned with the manufacture, erection, finish and decoration of gypsum board wall and ceiling systems developed an industry-wide *Recommended Levels of Gypsum Board Finish.*

The finish of gypsum board walls and ceilings for specific final decoration is dependent on a number of factors. A primary consideration is the location of the surface and the degree of decorative treatment desired. Painted and unpainted surfaces in warehouses and other areas where appearance is normally not critical may simply require the taping of wallboard joints and 'spotting' of fastener heads. Blemish-free, smooth, monolithic surfaces often intended for painted and decorated walls and ceilings in habitated structures, ranging from single-family dwellings through monumental buildings, require additional finishing prior to the application of the final decoration.

Other factors to be considered in determining the level of finish of the gypsum board surface are (1) the type of angle of surface illumination (both natural and artificial lighting), and (2) the paint and method of application or the type and finish of wallcovering specified as the final decoration. Critical lighting conditions, gloss paints, and thin wall coverings require a higher level of gypsum board finish than heavily textured surfaces which are subsequently painted or surfaces which are to be decorated with heavy grade wall coverings.

The following descriptions were developed by the Association of the Wall and Ceiling Industries-International (AWCI), Ceiling & Interior Systems Construction Association (CISCA), Gypsum Association (GA), and Painting and Decorating Contractors of America (PDCA) as a guide.

Level 0: Used in temporary construction or wherever the final decoration has not been determined. Unfinished. No taping, finishing or corner beads are required. Also could be used where non-predecorated panels will be used in demountable-type partitions that are to be painted as a final finish.

Level 1: Frequently used in plenum areas above ceilings, in attics, in areas where the assembly would generally be concealed, or in building service corridors and other areas not normally open to public view. Some degree of sound and smoke control is provided; in some geographic areas, this level is referred to as "fire-taping," although this level of finish does not typically meet fire-resistant assembly requirements. Where a fire resistance rating is required for the gypsum board assembly, details of construction should be in accordance with reports of fire tests of assemblies that have met the requirements of the fire rating acceptable.

All joints and interior angles shall have tape embedded in joint compound. Accessories are optional at specifier discretion in corridors and other areas with pedestrian traffic. Tape and fastener heads need not be covered with joint compound. Surface shall be free of excess joint compound. Tool marks and ridges are acceptable.

Level 2: It may be specified for standard gypsum board surfaces in garages, warehouse storage, or other similar areas where surface appearance is not of primary importance.

All joints and interior angles shall have tape embedded in joint compound and shall be immediately wiped with a joint knife or trowel, leaving a thin coating of joint compound over all joints and interior angles. Fastener heads and accessories shall be covered with a coat of joint compound. Surface shall be free of excess joint compound. Tool marks and ridges are acceptable.

Level 3: Typically used in areas receiving heavy texture (spray or hand applied) finishes before final painting, or where commercial-grade (heavy duty) wall coverings are to be applied as the final decoration. This level of finish should not be used where smooth painted surfaces or where lighter weight wall coverings are specified. The prepared surface shall be coated with a drywall primer prior to the application of final finishes.

All joints and interior angles shall have tape embedded in joint compound and shall be immediately wiped with a joint knife or trowel, leaving a thin coating of joint compound over all joints and interior angles. One additional coat of joint compound shall be applied over all joints and interior angles. Fastener heads and accessories shall be covered with two separate coats of joint compound. All joint compounds shall be smooth and free of tool marks and ridges. The prepared surface shall be covered with a drywall primer prior to the application of the final decoration.

Level 4: This level should be used where residential grade (light duty) wall coverings, flat paints, or light textures are to be applied. The prepared surface shall be coated with a drywall primer prior to the application of final finishes. Release agents for wall coverings are specifically formulated to minimize damage if coverings are subsequently removed.

The weight, texture, and sheen level of the wall covering material selected should be taken into consideration when specifying wall coverings over this level of drywall treatment. Joints and fasteners must be sufficiently concealed if the wall covering material is lightweight, contains limited pattern, has a glossy finish, or has any combination of these features. In critical lighting areas, flat paints applied over light textures tend to reduce joint photographing. Gloss, semi-gloss, and enamel paints are not recommended over this level of finish.

All joints and interior angles shall have tape embedded in joint compound and shall be immediately wiped with a joint knife or trowel, leaving a thin coating of joint compound over all joints and interior angles. In addition, two separate coats of joint compound shall be applied over all flat joints and one separate coat of joint compound applied over interior angles. Fastener heads and accessories shall be covered with three separate coats of joint compound. All joint compounds shall be smooth and free of tool marks and ridges. The prepared surface shall be covered with a drywall primer like Sheetrock first coat prior to the application of the final decoration.

Level 5: The highest quality finish is the most effective method to provide a uniform surface and minimize the possibility of joint photographing and of fasteners showing through the final decoration. This level of finish is required where gloss, semi-gloss, or enamel is specified; when flat joints are specified over an untextured surface; or where critical lighting conditions occur. The prepared surface shall be coated with a drywall primer prior to the application of the final decoration.

All joints and interior angles shall have tape embedded in joint compound and be immediately wiped with a joint knife or trowel, leaving a thin coating of joint compound over all joints and interior angles. Two separate coats of joint compound shall be applied over all flat joints and one separate coat of joint compound applied over interior angles. Fastener heads and accessories shall be covered with three separate coats of joint compound.

A thin skim coat of joint compound shall be trowel applied to the entire surface. Excess compound is immediately troweled off, leaving a film or skim coating of compound completely covering the paper. As an alternative to a skim coat, a material manufactured especially for this purpose may be applied such as Sheetrock Tuff-Hide primer surfacer. The surface must be smooth and free of tool marks and ridges. The prepared surface shall be covered with a drywall primer prior to the application of the final decoration.

R099100-10 Painting Estimating Techniques

Proper estimating methodology is needed to obtain an accurate painting estimate. There is no known reliable shortcut or square foot method. The following steps should be followed:

- List all surfaces to be painted, with an accurate quantity (area) of each. Items having similar surface condition, finish, application method and accessibility may be grouped together.
- List all the tasks required for each surface to be painted, including surface preparation, masking, and protection of adjacent surfaces. Surface preparation may include minor repairs, washing, sanding and puttying.
- Select the proper Means line for each task. Review and consider all adjustments to labor and materials for type of paint and location of work. Apply the height adjustment carefully. For instance, when applying the adjustment for work over 8' high to a wall that is 12' high, apply the adjustment only to the area between 8' and 12' high, and not to the entire wall.

When applying more than one percent (%) adjustment, apply each to the base cost of the data, rather than applying one percentage adjustment on top of the other.

When estimating the cost of painting walls and ceilings remember to add the brushwork for all cut-ins at inside corners and around windows and doors as a LF measure. One linear foot of cut-in with a brush equals one square foot of painting.

All items for spray painting include the labor for roll-back.

Deduct for openings greater than 100 SF or openings that extend from floor to ceiling and are greater than 5' wide. Do not deduct small openings.

The cost of brushes, rollers, ladders and spray equipment are considered part of a painting contractor's overhead, and should not be added to the estimate. The cost of rented equipment such as scaffolding and swing staging should be added to the estimate.

R099100-20 Painting

Item	Coat	One Gallon Covers			In 8 Hours a Laborer Covers			Labor-Hours per 100 S.F.		
		Brush	Roller	Spray	Brush	Roller	Spray	Brush	Roller	Spray
Paint wood siding	prime	250 S.F.	225 S.F.	290 S.F.	1150 S.F.	1300 S.F.	2275 S.F.	.695	.615	.351
	others	270	250	290	1300	1625	2600	.615	.492	.307
Paint exterior trim	prime	400	—	—	650	—	—	1.230	—	—
	1st	475	—	—	800	—	—	1.000	—	—
	2nd	520	—	—	975	—	—	.820	—	—
Paint shingle siding	prime	270	255	300	650	975	1950	1.230	.820	.410
	others	360	340	380	800	1150	2275	1.000	.695	.351
Stain shingle siding	1st	180	170	200	750	1125	2250	1.068	.711	.355
	2nd	270	250	290	900	1325	2600	.888	.603	.307
Paint brick masonry	prime	180	135	160	750	800	1800	1.066	1.000	.444
	1st	270	225	290	815	975	2275	.981	.820	.351
	2nd	340	305	360	815	1150	2925	.981	.695	.273
Paint interior plaster or drywall	prime	400	380	495	1150	2000	3250	.695	.400	.246
	others	450	425	495	1300	2300	4000	.615	.347	.200
Paint interior doors and windows	prime	400	—	—	650	—	—	1.230	—	—
	1st	425	—	—	800	—	—	1.000	—	—
	2nd	450	—	—	975	—	—	.820	—	—

R133419-10 Pre-Engineered Steel Buildings

These buildings are manufactured by many companies and normally erected by franchised dealers throughout the U.S. The four basic types are: Rigid Frames, Truss type, Post and Beam and the Sloped Beam type. The most popular roof slope is a low pitch of 1″ in 12″. The minimum economical area of these buildings is about 3000 S.F. of floor area. Bay sizes are usually 20′ to 24′ but can go as high as 30′ with heavier girts and purlins. Eave heights are usually 12′ to 24′ with 18′ to 20′ most typical.

Material prices shown in the Unit Price section are bare costs for the building shell only and do not include floors, foundations, anchor bolts, interior finishes or utilities. Costs assume at least three bays of 24′ each, a 1″ in 12″ roof slope, and they are based on a 30 psf roof load and a 20 psf wind load and no unusual requirements. Wind load is a function of wind speed, building height, and terrain characteristics; this should be determined by a registered structural engineer. Costs include the structural frame, 26 ga. non-insulated colored corrugated or ribbed roofing and siding panels, fasteners, closures, trim and flashing but no allowance for insulation, doors, windows, skylights, gutters or downspouts. Very large projects would generally cost less for materials than the prices shown. For roof panel substitutions and wall panel substitutions, see appropriate Unit Price sections.

Conditions at the site, weather, shape and size of the building, and labor availability will affect the erection cost of the building.

R142000-10 Freight Elevators

Capacities run from 2,000 lbs. to over 100,000 lbs. with 3,000 lbs. to 10,000 lbs. most common. Travel speeds are generally lower and the control is less intricate than on passenger elevators. Costs in the Unit Price sections are for hydraulic and geared elevators.

R142000-20 Elevator Selective Costs See R142000-40 for cost development.

	Passenger		Freight		Hospital	
A. Base Unit	Hydraulic	Electric	Hydraulic	Electric	Hydraulic	Electric
Capacity	1,500 lb.	2,000 lb.	2,000 lb.	4,000 lb.	4,000 lb.	4,000 lb.
Speed	100 F.P.M.	200 F.P.M.	100 F.P.M.	200 F.P.M.	100 F.P.M.	200 F.P.M.
#Stops/Travel Ft.	2/12	4/40	2/20	4/40	2/20	4/40
Push Button Oper.	Yes	Yes	Yes	Yes	Yes	Yes
Telephone Box & Wire	"	"	"	"	"	"
Emergency Lighting	"	"	No	No	"	"
Cab	Plastic Lam. Walls	Plastic Lam. Walls	Painted Steel	Painted Steel	Plastic Lam. Walls	Plastic Lam. Walls
Cove Lighting	Yes	Yes	No	No	Yes	Yes
Floor	V.C.T.	V.C.T.	Wood w/Safety Treads	Wood w/Safety Treads	V.C.T.	V.C.T.
Doors, & Speedside Slide	Yes	Yes	Yes	Yes	Yes	Yes
Gates, Manual	No	No	No	No	No	No
Signals, Lighted Buttons	Car and Hall	Car and Hall	Car and Hall	Car and Hall	Car and Hall	Car and Hall
O.H. Geared Machine	N.A.	Yes	N.A.	Yes	N.A.	Yes
Variable Voltage Contr.	"	"	N.A.	"	"	"
Emergency Alarm	Yes	"	Yes	"	Yes	"
Class "A" Loading	N.A.	N.A.	"	"	N.A.	N.A.

R142000-30 Passenger Elevators

Electric elevators are used generally but hydraulic elevators can be used for lifts up to 70′ and where large capacities are required. Hydraulic speeds are limited to 200 F.P.M. but cars are self leveling at the stops. On low rises, hydraulic installation runs about 15% less than standard electric types but on higher rises this installation cost advantage is reduced. Maintenance of hydraulic elevators is about the same as the electric type but the underground portion is not included in the maintenance contract.

In electric elevators there are several control systems available, the choice of which will be based upon elevator use, size, speed and cost criteria. The two types of drives are geared for low speeds and gearless for 450 F.P.M. and over.

The tables above illustrate typical installed costs of the various types of elevators available.

R142000-40 Elevator Cost Development

To price a new car or truck from the factory, you must start with the manufacturer's basic model, then add or exchange optional equipment and features. The same is true for pricing elevators.

Example:

Requirement: One-passenger elevator, five-story hydraulic, 2,500 lb. capacity, 12′ floor to floor, speed 150 F.P.M., emergency power switching and maintenance contract.

Description	Adjustment
A. Base Elevator: Hydraulic Passenger, 1500 lb. Capacity, 100 fpm, 2 Stops, Standard Finish	1 Ea.
B. Capacity Adjustment (2,500 lb.)	1 Ea.
C. Excess Travel Adjustment: 48′ Total Travel (4 x 12′) minus 12′ Base Unit Travel =	36 V.L.F.
D. Stops Adjustment: 5 Total Stops minus 2 Stops (Base Unit) =	3 Stops
E. Speed Adjustment (150 F.P.M.)	1 Ea.
F. Options:	
1. Intercom Service	1 Ea.
2. Emergency Power Switching, Automatic	1 Ea.
3. Stainless Steel Entrance Doors	5 Ea.
4. Maintenance Contract (12 Months)	1 Ea.
5. Position Indicator for main floor level (none indicated in Base Unit)	1 Ea.

Conveying Equipment | R1432 Moving Walks

R143210-20 Moving Ramps and Walks

These are a specialized form of conveyor 3′ to 6′ wide with capacities of 3,600 to 18,000 persons per hour. Maximum speed is 140 F.P.M. and normal incline is 0° to 15°.

Local codes will determine the maximum angle. Outdoor units would require additional weather protection.

Fire Suppression | R2112 Fire-Suppression Standpipes

R211226-10 Standpipe Systems

The basis for standpipe system design is National Fire Protection Association NFPA 14. However, the authority having jurisdiction should be consulted for special conditions, local requirements and approval.

Standpipe systems, properly designed and maintained, are an effective and valuable time saving aid for extinguishing fires, especially in the upper stories of tall buildings, the interior of large commercial or industrial malls, or other areas where construction features or access make the laying of temporary hose lines time consuming and/or hazardous. Standpipes are frequently installed with automatic sprinkler systems for maximum protection.

There are three general classes of service for standpipe systems:
Class I for use by fire departments and personnel with special training for heavy streams (2-1/2″ hose connections).
Class II for use by building occupants until the arrival of the fire department (1-1/2″ hose connector with hose).
Class III for use by either fire departments and trained personnel or by the building occupants (both 2-1/2″ and 1-1/2″ hose connections or one 2-1/2″ hose valve with an easily removable 2-1/2″ by 1-1/2″ adapter).

Standpipe systems are also classified by the way water is supplied to the system. The four basic types are:
Type 1: Wet standpipe system having supply valve open and water pressure maintained at all times.
Type 2: Standpipe system so arranged through the use of approved devices as to admit water to the system automatically by opening a hose valve.
Type 3: Standpipe system arranged to admit water to the system through manual operation of approved remote control devices located at each hose station.
Type 4: Dry standpipe having no permanent water supply.

R211226-20 NFPA 14 Basic Standpipe Design

Class	Design-Use	Pipe Size Minimums	Water Supply Minimums
Class I	2-1/2″ hose connection on each floor All areas within 150′ of an exit in every exit stairway Fire Department Trained Personnel	Height to 100′, 4″ dia. Heights above 100′, 6″ dia. (275′ max. except with pressure regulators 400′ max.)	For each standpipe riser 500 GPM flow For common supply pipe allow 500 GPM for first standpipe plus 250 GPM for each additional standpipe (2500 GPM max. total) 30 min. duration 65 PSI at 500 GPM
Class II	1-1/2″ hose connection with hose on each floor All areas within 130′ of hose connection measured along path of hose travel Occupant personnel	Height to 50′, 2″ dia. Height above 50′, 2-1/2″ dia.	For each standpipe riser 100 GPM flow For multiple riser common supply pipe 100 GPM 300 min. duration, 65 PSI at 100 GPM
Class III	Both of above. Class I valved connections will meet Class III with additional 2-1/2″ by 1-1/2″ adapter and 1-1/2″ hose.	Same as Class I	Same as Class I

*Note: Where 2 or more standpipes are installed in the same building or section of building they shall be interconnected at the bottom.

Combined Systems

Combined systems are systems where the risers supply both automatic sprinklers and 2-1/2″ hose connection outlets for fire department use. In such a system the sprinkler spacing pattern shall be in accordance with NFPA 13 while the risers and supply piping will be sized in accordance with NFPA 14. When the building is completely sprinklered the risers may be sized by hydraulic calculation. The minimum size riser for buildings not completely sprinklered is 6″.

The minimum water supply of a completely sprinklered, light hazard, high-rise occupancy building will be 500 GPM while the supply required for other types of completely sprinklered high-rise buildings is 1000 GPM.

General System Requirements

1. Approved valves will be provided at the riser for controlling branch lines to hose outlets.
2. A hose valve will be provided at each outlet for attachment of hose.
3. Where pressure at any standpipe outlet exceeds 100 PSI a pressure reducer must be installed to limit the pressure to 100 PSI. Note that the pressure head due to gravity in 100′ of riser is 43.4 PSI. This must be overcome by city pressure, fire pumps, or gravity tanks to provide adequate pressure at the top of the riser.
4. Each hose valve on a wet system having a linen hose shall have an automatic drip connection to prevent valve leakage from entering the hose.
5. Each riser will have a valve to isolate it from the rest of the system.
6. One or more fire department connections as an auxiliary supply shall be provided for each Class I or Class III standpipe system. In buildings having two or more zones, a connection will be provided for each zone.
7. There will be no shutoff valve in the fire department connection, but a check valve will be located in the line before it joins the system.
8. All hose connections street side will be identified on a cast plate or fitting as to purpose.

R211313-10 Sprinkler Systems (Automatic)

Sprinkler systems may be classified by type as follows:

1. **Wet Pipe System.** A system employing automatic sprinklers attached to a piping system containing water and connected to a water supply so that water discharges immediately from sprinklers opened by a fire.
2. **Dry Pipe System.** A system employing automatic sprinklers attached to a piping system containing air under pressure, the release of which, as from the opening of sprinklers, permits the water pressure to open a valve known as a "dry pipe valve." The water then flows into the piping system and out the opened sprinklers.
3. **Pre-Action System.** A system employing automatic sprinklers attached to a piping system containing air that may or may not be under pressure, with a supplemental heat responsive system of generally more sensitive characteristics than the automatic sprinklers themselves, installed in the same areas as the sprinklers; actuation of the heat responsive system, as from a fire, opens a valve which permits water to flow into the sprinkler piping system and be discharged from any sprinklers which may be open.
4. **Deluge System.** A system employing open sprinklers attached to a piping system connected to a water supply through a valve which is opened by the operation of a heat responsive system installed in the same areas as the sprinklers. When this valve opens, water flows into the piping system and discharges from all sprinklers attached thereto.
5. **Combined Dry Pipe and Pre-Action Sprinkler System.** A system employing automatic sprinklers attached to a piping system containing air under pressure with a supplemental heat responsive system of generally more sensitive characteristics than the automatic sprinklers themselves, installed in the same areas as the sprinklers; operation of the heat responsive system, as from a fire, actuates tripping devices which open dry pipe valves simultaneously and without loss of air pressure in the system. Operation of the heat responsive system also opens approved air exhaust valves at the end of the feed main which facilitates the filling of the system with water which usually precedes the opening of sprinklers. The heat responsive system also serves as an automatic fire alarm system.
6. **Limited Water Supply System.** A system employing automatic sprinklers and conforming to these standards but supplied by a pressure tank of limited capacity.
7. **Chemical Systems.** Systems using halon, carbon dioxide, dry chemical or high expansion foam as selected for special requirements. Agent may extinguish flames by chemically inhibiting flame propagation, suffocate flames by excluding oxygen, interrupting chemical action of oxygen uniting with fuel or sealing and cooling the combustion center.
8. **Firecycle System.** Firecycle is a fixed fire protection sprinkler system utilizing water as its extinguishing agent. It is a time delayed, recycling, preaction type which automatically shuts the water off when heat is reduced below the detector operating temperature and turns the water back on when that temperature is exceeded. The system senses a fire condition through a closed circuit electrical detector system which controls water flow to the fire automatically. Batteries supply up to 90 hour emergency power supply for system operation. The piping system is dry (until water is required) and is monitored with pressurized air. Should any leak in the system piping occur, an alarm will sound, but water will not enter the system until heat is sensed by a firecycle detector.

Area coverage sprinkler systems may be laid out and fed from the supply in any one of several patterns as shown below. It is desirable, if possible, to utilize a central feed and achieve a shorter flow path from the riser to the furthest sprinkler. This permits use of the smallest sizes of pipe possible with resulting savings.

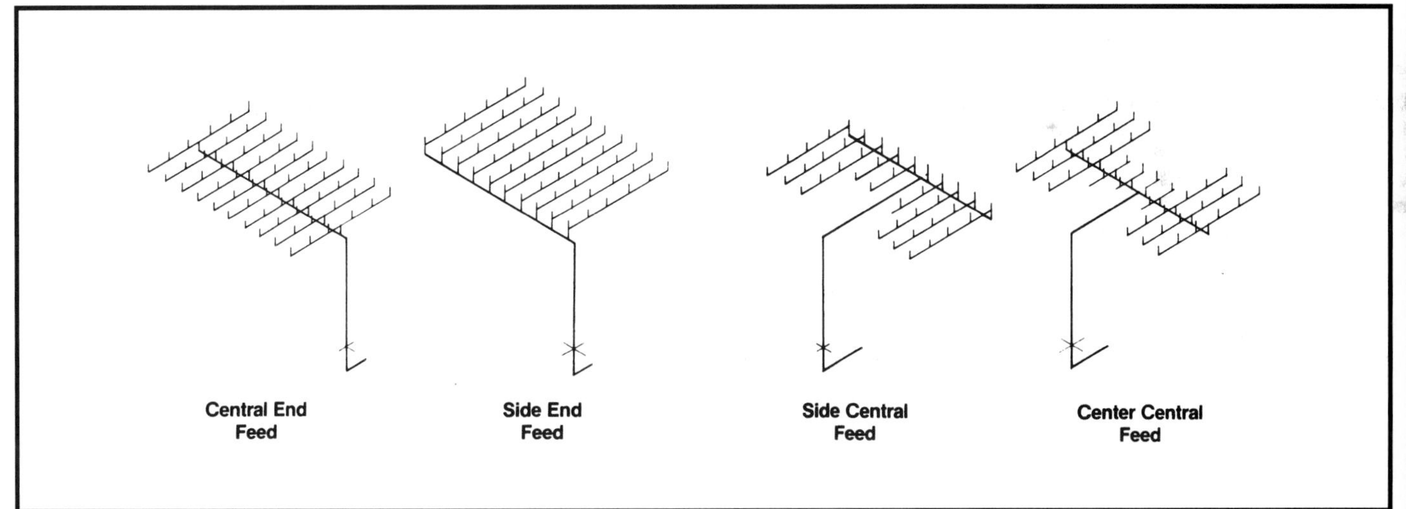

R211313-20 System Classification

System Classification

Rules for installation of sprinkler systems vary depending on the classification of occupancy falling into one of three categories as follows:

Light Hazard Occupancy

The protection area allotted per sprinkler should not exceed 225 S.F., with the maximum distance between lines and sprinklers on lines being 15′. The sprinklers do not need to be staggered. Branch lines should not exceed eight sprinklers on either side of a cross main. Each large area requiring more than 100 sprinklers and without a sub-dividing partition should be supplied by feed mains or risers sized for ordinary hazard occupancy.
Maximum system area = 52,000 S.F.

Included in this group are:

Churches
Clubs
Educational
Hospitals
Institutional
Libraries
  (except large stack rooms)
Museums
Nursing Homes
Offices
Residential
Restaurants
Theaters and Auditoriums
  (except stages and prosceniums)
Unused Attics

Ordinary Hazard Occupancy

The protection area allotted per sprinkler shall not exceed 130 S.F. of noncombustible ceiling and 130 S.F. of combustible ceiling. The maximum allowable distance between sprinkler lines and sprinklers on lines is 15′. Sprinklers shall be staggered if the distance between heads exceeds 12′. Branch lines should not exceed eight sprinklers on either side of a cross main.
Maximum system area = 52,000 S.F.

Included in this group are:

Group 1
Automotive Parking and Showrooms
Bakeries
Beverage Manufacturing
Canneries
Dairy Products Manufacturing/Processing
Electronic Plans
Glass and Glass Products Manufacturing
Laundries
Restaurant Service Areas

Group 2
Cereal Mills
Chemical Plants—Ordinary
Confectionery Products
Distilleries
Dry Cleaners
Feed Mills
Horse Stables
Leather Goods Manufacturing
Libraries—Large Stack Room Areas
Machine Shops
Metal Working
Mercantile
Paper and Pulp Mills
Paper Process Plants
Piers and Wharves
Post Offices
Printing and Publishing
Repair Garages
Stages
Textile Manufacturing
Tire Manufacturing
Tobacco Products Manufacturing
Wood Machining
Wood Product Assembly

Extra Hazard Occupancy

The protection area allotted per sprinkler shall not exceed 100 S.F. of noncombustible ceiling and 100 S.F. of combustible ceiling. The maximum allowable distance between lines and between sprinklers on lines is 12′. Sprinklers on alternate lines shall be staggered if the distance between sprinklers on lines exceeds 8′. Branch lines should not exceed six sprinklers on either side of a cross main.
Maximum system area:
  Design by pipe schedule = 25,000 S.F.
  Design by hydraulic calculation = 40,000 S.F.

Included in this group are:

Group 1
Aircraft hangars
Combustible Hydraulic Fluid Use Area
Die Casting
Metal Extruding
Plywood/Particle Board Manufacturing
Printing (inks with flash points < 100 degrees F)
Rubber Reclaiming, Compounding, Drying, Milling, Vulcanizing
Saw Mills
Textile Picking, Opening, Blending, Garnetting, Carding, Combing of Cotton, Synthetics, Wood Shoddy, or Burlap
Upholstering with Plastic Foams

Group 2
Asphalt Saturating
Flammable Liquids Spraying
Flow Coating
Manufactured/Modular Home Building Assemblies (where finished enclosure is present and has combustible interiors)
Open Oil Quenching
Plastics Processing
Solvent Cleaning
Varnish and Paint Dipping

R211313-40 Adjustment for Sprinkler/Standpipe Installations

Quality/Complexity Multiplier (For all installations)

Economy installation, add	0 to 5%
Good quality, medium complexity, add	5 to 15%
Above average quality and complexity, add	15 to 25%

R220102-20 Labor Adjustment Factors

Labor Adjustment Factors are provided for Divisions 21, 22, and 23 to assist the mechanical estimator account for the various complexities and special conditions of any particular project. While a single percentage has been entered on each line of Division 22 01 02.20, it should be understood that these are just suggested midpoints of ranges of values commonly used by mechanical estimators. They may be increased or decreased depending on the severity of the special conditions.

The group for "existing occupied buildings" has been the subject of requests for explanation. Actually there are two stages to this group: buildings that are existing and "finished" but unoccupied, and those that also are occupied. Buildings that are "finished" may result in higher labor costs due to the workers having to be more careful not to damage finished walls, ceilings, floors, etc. and may necessitate special protective coverings and barriers. Also corridor bends and doorways may not accommodate long pieces of pipe or larger pieces of equipment. Work above an already hung ceiling can be very time consuming. The addition of occupants may force the work to be done on premium time (nights and/or weekends), eliminate the possible use of some preferred tools such as pneumatic drivers, powder charged drivers, etc. The estimator should evaluate the access to the work area and just how the work is going to be accomplished to arrive at an increase in labor costs over "normal" new construction productivity.

R220105-10 Demolition (Selective vs. Removal for Replacement)

Demolition can be divided into two basic categories.

One type of demolition involves the removal of material with no concern for its replacement. The labor-hours to estimate this work are found under "Selective Demolition" in the Fire Protection, Plumbing and HVAC Divisions. It is selective in that individual items or all the material installed as a system or trade grouping such as plumbing or heating systems are removed. This may be accomplished by the easiest way possible, such as sawing, torch cutting, or sledge hammering as well as simple unbolting.

The second type of demolition is the removal of some items for repair or replacement. This removal may involve careful draining, opening of unions, disconnecting and tagging electrical connections, capping pipes/ducts to prevent entry of debris or leakage of the material contained as well as transporting the item away from its in-place location to a truck/dumpster. An approximation of the time required to accomplish this type of demolition is to use half of the time indicated as necessary to install a new unit. For example: installation of a new pump might be listed as requiring 6 labor-hours so if we had to estimate the removal of the old pump we would allow an additional 3 hours for a total of 9 hours. That is, the complete replacement of a defective pump with a new pump would be estimated to take 9 labor-hours.

R220523-90 Valve Selection Considerations

INTRODUCTION: In any piping application, valve performance is critical. Valves should be selected to give the best performance at the lowest cost.

The following is a list of performance characteristics generally expected of valves.

1. Stopping flow or starting it.
2. Throttling flow (Modulation).
3. Flow direction changing.
4. Checking backflow (Permitting flow in only one direction).
5. Relieving or regulating pressure.

In order to properly select the right valve, some facts must be determined.

A. What liquid or gas will flow through the valve?

B. Does the fluid contain suspended particles?

C. Does the fluid remain in liquid form at all times?

D. Which metals does fluid corrode?

E. What are the pressure and temperature limits? (As temperature and pressure rise, so will the price of the valve.)

F. Is there constant line pressure?

G. Is the valve merely an on-off valve?

H. Will checking of backflow be required?

I. Will the valve operate frequently or infrequently?

Valves are classified by design type into such classifications as Gate, Globe, Angle, Check, Ball, Butterfly and Plug. They are also classified by end connection, stem, pressure restrictions and material such as bronze, cast iron, etc. Each valve has a specific use. A quality valve used correctly will provide a lifetime of trouble-free service, but a high quality valve installed in the wrong service may require frequent attention.

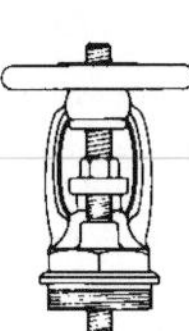

STEM TYPES
(O.S. & Y)—Rising Stem-Outside Screw and Yoke

Offers a visual indication of whether the valve is open or closed. Recommended where high temperatures, corrosives, and solids in the line might cause damage to inside-valve stem threads. The stem threads are engaged by the yoke bushing so the stem rises through the hand wheel as it is turned.

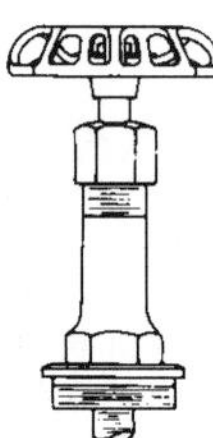

(R.S.)—Rising Stem-Inside Screw

Adequate clearance for operation must be provided because both the hand wheel and the stem rise.
The valve wedge position is indicated by the position of the stem and hand wheel.

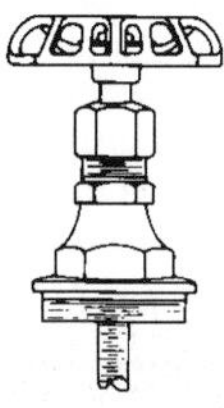

(N.R.S.)—Non-Rising Stem-Inside Screw

A minimum clearance is required for operating this type of valve. Excessive wear or damage to stem threads inside the valve may be caused by heat, corrosion, and solids. Because the hand wheel and stem do not rise, wedge position cannot be visually determined.

VALVE TYPES
Gate Valves

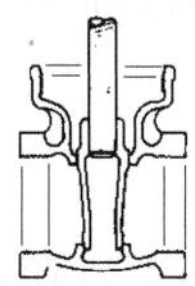

Provide full flow, minute pressure drop, minimum turbulence and minimum fluid trapped in the line.
They are normally used where operation is infrequent.

Globe Valves

Globe valves are designed for throttling and/or frequent operation with positive shut-off. Particular attention must be paid to the several types of seating materials available to avoid unnecessary wear. The seats must be compatible with the fluid in service and may be composition or metal. The configuration of the globe valve opening causes turbulence which results in increased resistance. Most bronze globe valves are rising stem-inside screw, but they are also available on O.S. & Y.

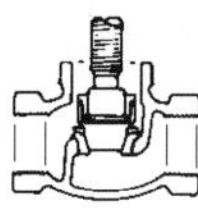

Angle Valves

The fundamental difference between the angle valve and the globe valve is the fluid flow through the angle valve. It makes a 90° turn and offers less resistance to flow than the globe valve while replacing an elbow. An angle valve thus reduces the number of joints and installation time.

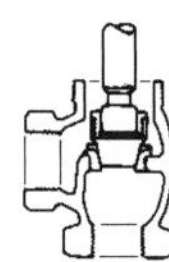

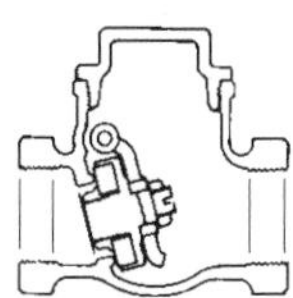

Check Valves

Check valves are designed to prevent backflow by automatically seating when the direction of fluid is reversed.
Swing check valves are generally installed with gate valves, as they provide comparable full flow. Usually recommended for lines where flow velocities are low and should not be used on lines with pulsating flow. Recommended for horizontal installation or in vertical lines only where flow is upward.

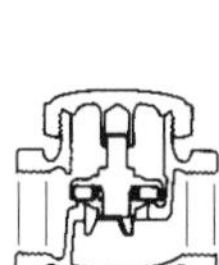

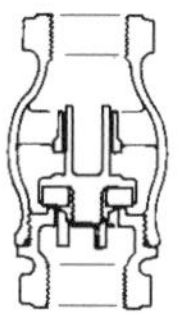

Lift Check Valves

These are commonly used with globe and angle valves since they have similar diaphragm seating arrangements and are recommended for preventing backflow of steam, air, gas and water, and on vapor lines with high flow velocities. For horizontal lines, horizontal lift checks should be used and vertical lift checks for vertical lines.

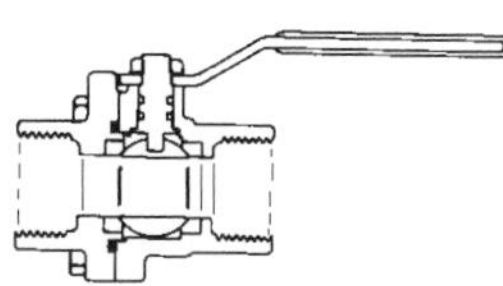

Ball Valves

Ball valves are light and easily installed, yet because of modern elastomeric seats, provide tight closure. Flow is controlled by rotating up to 90° a drilled ball which fits tightly against resilient seals. This ball seats with flow in either direction, and valve handle indicates the degree of opening. Recommended for frequent operation readily adaptable to automation, ideal for installation where space is limited.

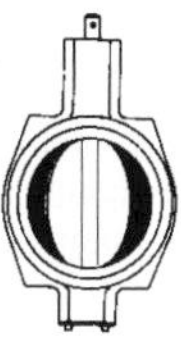

Butterfly Valves

Butterfly valves provide bubble-tight closure with excellent throttling characteristics. They can be used for full-open, closed and for throttling applications.

The butterfly valve consists of a disc within the valve body which is controlled by a shaft. In its closed position, the valve disc seals against a resilient seat. The disc position throughout the full 90° rotation is visually indicated by the position of the operator.

A butterfly valve is only a fraction of the weight of a gate valve and requires no gaskets between flanges in most cases. Recommended for frequent operation and adaptable to automation where space is limited.

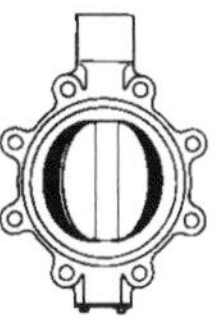

Wafer and lug type bodies, when installed between two pipe flanges, can be easily removed from the line. The pressure of the bolted flanges holds the valve in place.
Locating lugs makes installation easier.

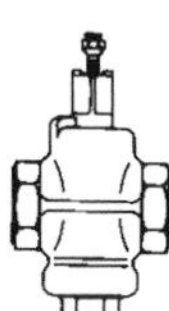

Plug Valves

Lubricated plug valves, because of the wide range of services to which they are adapted, may be classified as all purpose valves. They can be safely used at all pressure and vacuums, and at all temperatures up to the limits of available lubricants. They are the most satisfactory valves for the handling of gritty suspensions and many other destructive, erosive, corrosive and chemical solutions.

R221113-40 Plumbing Approximations for Quick Estimating

Water Control

Water Meter; Backflow Preventer, Shock Absorbers; Vacuum Breakers; Mixer. .... 10 to 15% of Fixtures

Pipe And Fittings .... 30 to 60% of Fixtures

Note: Lower percentage for compact buildings or larger buildings with plumbing in one area.
Larger percentage for large buildings with plumbing spread out.
In extreme cases pipe may be more than 100% of fixtures.
Percentages **do not** include special purpose or process piping.

Plumbing Labor

1 & 2 Story Residential .... Rough-in Labor = 80% of Materials
Apartment Buildings .... Rough-in Labor = 90 to 100% of Materials
Labor for handling and placing fixtures is approximately 25 to 30% of fixtures

Quality/Complexity Multiplier (for all installations)

Economy installation, add.......... 0 to 5%
Good quality, medium complexity, add .... 5 to 15%
Above average quality and complexity, add .... 15 to 25%

R221113-50 Pipe Material Considerations

1. Malleable fittings should be used for gas service.
2. Malleable fittings are used where there are stresses/strains due to expansion and vibration.
3. Cast fittings may be broken as an aid to disassembling heating lines frozen by long use, temperature and minerals.
4. A cast iron pipe is extensively used for underground and submerged service.
5. Type M (light wall) copper tubing is available in hard temper only and is used for nonpressure and less severe applications than K and L.
6. Type L (medium wall) copper tubing, available hard or soft for interior service.
7. Type K (heavy wall) copper tubing, available in hard or soft temper for use where conditions are severe. For underground and interior service.
8. Hard drawn tubing requires fewer hangers or supports but should not be bent. Silver brazed fittings are recommended, but soft solder is normally used.
9. Type DMV (very light wall) copper tubing designed for drainage, waste and vent plus other non-critical pressure services.

Domestic/Imported Pipe and Fittings Costs

The prices shown in this publication for steel/cast iron pipe and steel, cast iron, and malleable iron fittings are based on domestic production sold at the normal trade discounts. The above listed items of foreign manufacture may be available at prices 1/3 to 1/2 of those shown. Some imported items after minor machining or finishing operations are being sold as domestic to further complicate the system.

Caution: Most pipe prices in this data set also include a coupling and pipe hangers, which, for the larger sizes, can add significantly to the per-foot cost and should be taken into account when comparing "book cost" with the quoted supplier's cost.

R221113-70 Piping to 10′ High

When taking off pipe, it is important to identify the different material types and joining procedures, as well as distances between supports and components required for proper support.

During the takeoff, measure through all fittings. Do not subtract the lengths of the fittings, valves, or strainers, etc. This added length plus the final rounding of the totals will compensate for nipples and waste.

When rounding off totals always increase the actual amount to correspond with the manufacturer's shipping lengths.

A. Both red brass and yellow brass pipes are normally furnished in 12′ lengths, plain end. The Unit Price section includes in the linear foot costs two field threads and one coupling per 10′ length. A carbon steel clevis type hanger assembly every 10′ is also prorated into the linear foot costs, including both material and labor.

B. Cast iron soil pipe is furnished in either 5′ or 10′ lengths. For pricing purposes, the Unit Price section features 10′ lengths with a joint and a carbon steel clevis hanger assembly every 5′ prorated into the per foot costs of both material and labor.

Three methods of joining are considered: lead and oakum poured joints or push-on gasket type joints for the bell and spigot pipe and a joint clamp for the no-hub soil pipe. The labor and material costs for each of these individual joining procedures are also prorated into the linear costs per foot.

C. Copper tubing covers types K, L, M, and DWV which are furnished in 20′ lengths. Means pricing data is based on a tubing cut each length and a coupling and two soft soldered joints every 10′. A carbon steel, clevis type hanger assembly every 10′ is also prorated into the per foot costs. The prices for refrigeration tubing are for materials only. Labor for full lengths may be based on the type L labor but short cut measures in tight areas can increase the installation labor-hours from 20 to 40%.

D. Corrosion-resistant piping does not lend itself to one particular standard of hanging or support assembly due to its diversity of application and placement. The several varieties of corrosion-resistant piping do not include any material or labor costs for hanger assemblies (See the Unit Price section for appropriate selection).

E. Glass pipe is furnished in standard lengths either 5′ or 10′ long, beaded on one end. Special orders for diverse lengths beaded on both ends are also available. For pricing purposes, RSMeans features 10′ lengths with a coupling and a carbon steel band hanger assembly every 10′ prorated into the per foot linear costs.

Glass pipe is also available with conical ends and standard lengths ranging from 6″ through 3′ in 6″ increments, then up to 10′ in 12″ increments. Special lengths can be customized for particular installation requirements.

For pricing purposes, Means has based the labor and material pricing on 10′ lengths. Included in these costs per linear foot are the prorated costs for a flanged assembly every 10′ consisting of two flanges, a gasket, two insertable seals, and the required number of bolts and nuts. A carbon steel band hanger assembly based on 10′ center lines has also been prorated into the costs per foot for labor and materials.

F. Plastic pipe of several compositions and joining methods are considered. Fiberglass reinforced pipe (FRP) is priced, based on 10′ lengths (20′ lengths are also available), with coupling and epoxy joints every 10′. FRP is furnished in both "General Service" and "High Strength." A carbon steel clevis hanger assembly, 3 for every 10′, is built into the prorated labor and material costs on a per foot basis.

The PVC and CPVC pipe schedules 40, 80 and 120 plus SDR ratings are all based on 20′ lengths with a coupling installed every 10′, as well as a carbon steel clevis hanger assembly every 3′. The PVC and ABS type DWV piping is based on 10′ lengths with solvent weld couplings every 10′, and with carbon steel clevis hanger assemblies, 3 for every 10′. The rest of the plastic piping in this section is based on flexible 100′ coils and does not include any coupling or supports.

This section ends with PVC drain and sewer piping based on 10′ lengths with bell and spigot ends and 0-ring type, push-on joints.

G. Stainless steel piping includes both weld end and threaded piping, both in the type 304 and 316 specification and in the following schedules, 5, 10, 40, 80, and 160. Although this piping is usually furnished in 20′ lengths, this cost grouping has a joint (either heli-arc butt-welded or threads and coupling) every 10′. A carbon steel clevis type hanger assembly is also included at 10′ intervals and prorated into the linear foot costs.

H. Carbon steel pipe includes both black and galvanized. This section encompasses schedules 40 (standard) and 80 (extra heavy).

Several common methods of joining steel pipe such as thread and coupled, butt welded, and flanged (150 lb. weld neck flanges) are also included.

For estimating purposes, it is assumed that the piping is purchased in 20′ lengths and that a compatible joint is made up every 10′. These joints are prorated into the labor and material costs per linear foot. The following hanger and support assemblies every 10′ are also included: carbon steel clevis for the T & C pipe, and single rod roll type for both the welded and flanged piping. All of these hangers are oversized to accommodate pipe insulation 3/4″ thick through 5″ pipe size and 1-1/2″ thick from 6″ through 12″ pipe size.

I. Grooved joint steel pipe is priced both black and galvanized, in schedules 10, 40, and 80, furnished in 20′ lengths. This section describes two joining methods: cut groove and roll groove. The schedule 10 piping is roll-grooved, while the heavier schedules are cut-grooved. The labor and material costs are prorated into per linear foot prices, including a coupled joint every 10′, as well as a carbon steel clevis hanger assembly.

Notes:

The pipe hanger assemblies mentioned in the preceding paragraphs include the described hanger; appropriately sized steel, box-type insert and nut; plus 18″ of threaded hanger rod.

C clamps are used when the pipe is to be supported from steel shapes rather than anchored in the slab. C clamps are slightly less costly than inserts. However, to save time in estimating, it is advisable to use the given line number cost, rather than substituting a C clamp for the insert.

Add to piping labor for elevated installation:

10′ to 14.5′ high	10%	30′ to 34.5′ high	40%
15′ to 19.5′ high	20%	35′ to 39.5′ high	50%
20′ to 24.5′ high	25%	40′ and higher	55%
25′ to 29.5′ high	35%		

When using the percentage adds for elevated piping installations as shown above, bear in mind that the given heights are for the pipe supports, even though the insert, anchor, or clamp may be several feet higher than the pipe itself.

An allowance has been included in the piping installation time for testing and minor tightening of leaking joints, fittings, stuffing boxes, packing glands, etc. For extraordinary test requirements such as x-rays, prolonged pressure or demonstration tests, a percentage of the piping labor, based on the estimator's experience, must be added to the labor total. A testing service specializing in weld x-rays should be consulted for pricing if it is an estimate requirement. Equipment installation time includes start-up with associated adjustments.

R224000-10 Water Consumption Rates

Fixture Type	Water Supply Fixture Unit Value		
	Hot Water	Cold Water	Combined
Bathtub	1.0	1.0	1.4
Clothes Washer	1.0	1.0	1.4
Dishwasher	1.4	0.0	1.4
Kitchen Sink	1.0	1.0	1.4
Laundry Tub	1.0	1.0	1.4
Lavatory	0.5	0.5	0.7
Shower Stall	1.0	1.0	1.4
Water Closet (tank type)	0.0	2.2	2.2
Full Bath Group			
w/bathtub or shower stall	1.5	2.7	3.6
Half Bath Group			
w/W.C. and Lavatory	0.5	2.5	2.6
Kitchen Group			
w/Dishwasher and Sink	1.9	1.0	2.5
Laundry Group			
w/Clothes Washer and Laundry Tub	1.8	1.8	2.5
Hose Bibb (sillcock)	0.0	2.5	2.5

Notes:
Typically, WSFU = 1GPM

Supply loads in the building water-distribution system shall be determined by total load on the pipe being sized, in terms of water supply fixture units (WSFU) and gallons per minute (GPM) flow rates. For fixtures not listed, choose a WSFU value of a fixture with similar flow characteristics. Water Fixture Supply Units determined the required water supply to fixtures and their service systems. Fixture units are equal to one (1) cubic foot of water drained in a 1-1/4″ pipe per minute. It is not a flow rate unit but a design factor.

R224000-20 Fixture Demands in Gallons per Fixture per Hour

The table below is based on a 140°F final temperature except for dishwashers in public places where 180°F water is mandatory.

Supply Systems for Flush Tanks			Supply Systems for Flushometer Valves		
Load	Demand		Load	Demand	
WSFU	GPM	CU. FT.	WSFU	GPM	CU. FT.
1.0	3.0	0.041040	-	-	-
2.0	5.0	0.068400	-	-	-
3.0	6.5	0.868920	-	-	-
4.0	8.0	1.069440	-	-	-
5.0	9.4	1.256592	5.0	15	2.0052
6.0	10.7	1.430376	6.0	17.4	2.326032
7.0	11.8	1.577424	7.0	19.8	2.646364
8.0	12.8	1.711104	8.0	22.2	2.967696
9.0	13.7	1.831416	9.0	24.6	3.288528
10.0	14.6	1.951728	10.0	27	3.60936
11.0	15.4	2.058672	11.0	27.8	3.716304
12.0	16.0	2.138880	12.0	28.6	3.823248
13.0	16.5	2.205720	13.0	29.4	3.930192
14.0	17.0	2.272560	14.0	30.2	4.037136
15.0	17.5	2.339400	15.0	31	4.14408
16.0	18.0	2.906240	16.0	31.8	4.241024
17.0	18.4	2.459712	17.0	32.6	4.357968
18.0	18.8	2.513184	18.0	33.4	4.464912
19.0	19.2	2.566656	19.0	34.2	4.571856
20.0	19.6	2.620218	20.0	35.0	4.678800
25.0	21.5	2.874120	25.0	38.0	5.079840
30.0	23.3	3.114744	30.0	42.0	5.611356
35.0	24.9	3.328632	35.0	44.0	5.881920
40.0	26.3	3.515784	40.0	46.0	6.149280
45.0	27.7	3.702936	45.0	48.0	6.416640
50.0	29.1	3.890088	50.0	50.0	6.684000

Notes:
When designing a plumbing system that utilizes fixtures other than, or in addition to, water closets, use the data provided in the Supply Systems for Flush Tanks section of the above table.

To obtain the probable maximum demand, multiply the total demands for the fixtures (gal./fixture/hour) by the demand factor. The heater should have a heating capacity in gallons per hour equal to this maximum. The storage tank should have a capacity in gallons equal to the probable maximum demand multiplied by the storage capacity factor.

R224000-30 Minimum Plumbing Fixture Requirements

Classification	Occupancy	Description	Water Closet Male	Water Closet Female	Lavatories Male	Lavatories Female	Bathtubs/Showers	Drinking Fountains	Other
Assembly	A-1	Theaters and other buildings for the performing arts and motion pictures	1:125	1:65	1:200			1:500	1 Service Sink
	A-2	Nightclubs, bars, taverns, dance halls	1:40		1:75			1:500	1 Service Sink
		Restaurants, banquet halls, food courts	1:75		1:200			1:500	1 Service Sink
	A-3	Auditorium w/o permanent seating, art galleries, exhibition halls, museums, lecture halls, libraries, arcades & gymnasiums	1:125	1:65	1:200			1:500	1 Service Sink
		Passenger terminals and transportation facilities	1:500		1:750			1:1000	1 Service Sink
		Places of worship and other religious services	1:150	1:75	1:200			1:1000	1 Service Sink
	A-4	Indoor sporting events and activities, coliseums, arenas, skating rinks, pools, and tennis courts	1:75 for the first 1500, then 1:120 for the remainder	1:40 for the first 1520, then 1:60 for the remainder	1:200	1:150		1:1000	1 Service Sink
	A-5	Outdoor sporting events and activities, stadiums, amusement parks, bleachers, grandstands	1:75 for the first 1500, then 1:120 for the remainder	1:40 for the first 1520, then 1:60 for the remainder	1:200	1:150		1:1000	1 Service Sink
Business	B	Buildings for the transaction of business, professional services, other services involving merchandise, office buildings, banks, light industrial	1:25 for the first 50, then 1:50 for the remainder		1:40 for the first 80, then 1:80 for the remainder			1:100	1 Service Sink
Educational	E	Educational facilities	1:50		1:50			1:100	1 Service Sink
Factory and industrial	F-1 and F-2	Structures in which occupants are engaged in work fabricating, assembly or processing of products or materials	1:100		1:100		See *International Plumbing Code*	1:400	1 Service Sink
Institutional	I-1	Residential care	1:10		1:10		1:8	1:100	1 Service Sink
	I-2	Hospitals, ambulatory nursing home care recipient	1 per room		1 per room		1:15	1:100	1 Service Sink
		Employees, other than residential care	1:25		1:35			1:100	
		Visitors, other than residential care	1:75		1:100			1:500	
	I-3	Prisons	1 per cell		1 per cell		1:15	1:100	1 Service Sink
		Reformatories, detention and correction centers	1:15		1:15		1:15	1:100	1 Service Sink
		Employees	1:25		1:35			1:100	
	I-4	Adult and child day care	1:15		1:15		1	1:100	1 Service Sink
Mercantile	M	Retail stores, service stations, shops, salesrooms, markets and shopping centers	1:500		1:750			1:1000	1 Service Sink
Residential	R-1	Hotels, motels, boarding houses (transient)	1 per sleeping unit		1 per sleeping unit		1 per sleeping unit		1 Service Sink
	R-2	Dormitories, fraternities, sororities and boarding houses (not transient)	1:10		1:10		1:8	1:100	1 Service Sink
		Apartment house	1 per dwelling unit		1 per dwelling unit		1 per dwelling unit		1 Kitchen sink per dwelling; 1 clothes washer connection per 20 dwellings
	R-3	1 and 2 Family dwellings	1 per dwelling unit		1:10		1 per dwelling unit		1 Kitchen sink per dwelling; 1 clothes washer connection per dwelling
	R-3	Congregate living facilities w/<16 people	1:10		1:10		1:8	1:100	1 Service Sink
	R-4	Congregate living facilities w/>16 people	1:10		1:10		1:8	1:100	1 Service Sink
Storage	S-1 and S-2	Structures for the storage of goods, warehouses, storehouses and freight depots, low and moderate hazard	1:100		1:100		See *International Plumbing Code*	1;1000	1 Service Sink

Table 2902.1

Excerpted from the 2012 *International Building Code*, Copyright 2011. Washington, D.C.: International Code Council. Reproduced with permission.

R230500-10 Subcontractors

On the unit cost pages of the RSMeans Cost Data sets, the last column is entitled "Total Incl. O&P". This is normally the cost of the installing contractor. In the HVAC Division, this is the cost of the mechanical contractor. If the particular work being estimated is to be performed by a sub to the mechanical contractor, the mechanical's profit and handling charge (usually 10%) is added to the total of the last column.

R233100-40 Steel Sheet Metal Calculator (Weight in Lb./Ft. of Length)

Gauge	26	24	22	20	18	16
Wt.-Lb./S.F.	.906	1.156	1.406	1.656	2.156	2.656
SMACNA Max. Dimension – Long Side		30″	54″	84″	85″ Up	
Sum-2 sides						
2	.3	.40	.50	.60	.80	.90
3	.5	.65	.80	.90	1.1	1.4
4	.7	.85	1.0	1.2	1.5	1.8
5	.8	1.1	1.3	1.5	1.9	2.3
6	1.0	1.3	1.5	1.7	2.3	2.7
7	1.2	1.5	1.8	2.0	2.7	3.2
8	1.3	1.7	2.0	2.3	3.0	3.6
9	1.5	1.9	2.3	2.6	3.4	4.1
10	1.7	2.2	2.5	2.9	3.8	4.5
11	1.8	2.4	2.8	3.2	4.2	5.0
12	2.0	2.6	3.0	3.5	4.6	5.4
13	2.2	2.8	3.3	3.8	4.9	5.9
14	2.3	3.0	3.5	4.1	5.3	6.3
15	2.5	3.2	3.8	4.4	5.7	6.8
16	2.7	3.4	4.0	4.6	6.1	7.2
17	2.8	3.7	4.3	4.9	6.5	7.7
18	3.0	3.9	4.5	5.2	6.8	8.1
19	3.2	4.1	4.8	5.5	7.2	8.6
20	3.3	4.3	5.0	5.8	7.6	9.0
21	3.5	4.5	5.3	6.1	8.0	9.5
22	3.7	4.7	5.5	6.4	8.4	9.9
23	3.8	5.0	5.8	6.7	8.7	10.4
24	4.0	5.2	6.0	7.0	9.1	10.8
25	4.2	5.4	6.3	7.3	9.5	11.3
26	4.3	5.6	6.5	7.5	9.9	11.7
27	4.5	5.8	6.8	7.8	10.3	12.2
28	4.7	6.0	7.0	8.1	10.6	12.6
29	4.8	6.2	7.3	8.4	11.0	13.1
30	5.0	6.5	7.5	8.7	11.4	13.5
31	5.2	6.7	7.8	9.0	11.8	14.0
32	5.3	6.9	8.0	9.3	12.2	14.4
33	5.5	7.1	8.3	9.6	12.5	14.9
34	5.7	7.3	8.5	9.9	12.9	15.3
35	5.8	7.5	8.8	10.2	13.3	15.8
36	6.0	7.8	9.0	10.4	13.7	16.2
37	6.2	8.0	9.3	10.7	14.1	16.7
38	6.3	8.2	9.5	11.0	14.4	17.1
39	6.5	8.4	9.8	11.3	14.8	17.6
40	6.7	8.6	10.0	11.6	15.2	18.0
41	6.8	8.8	10.3	11.9	15.6	18.5
42	7.0	9.0	10.5	12.2	16.0	18.9
43	7.2	9.2	10.8	12.5	16.3	19.4
44	7.3	9.5	11.0	12.8	16.7	19.8
45	7.5	9.7	11.3	13.1	17.1	20.3
46	7.7	9.9	11.5	13.3	17.5	20.7
47	7.8	10.1	11.8	13.6	17.9	21.2
48	8.0	10.3	12.0	13.9	18.2	21.6
49	8.2	10.5	12.3	14.2	18.6	22.1
50	8.3	10.7	12.5	14.5	19.0	22.5
51	8.5	11.0	12.8	14.8	19.4	23.0
52	8.7	11.2	13.0	15.1	19.8	23.4
53	8.8	11.4	13.3	15.4	20.1	23.9
54	9.0	11.6	13.5	15.7	20.5	24.3
55	9.2	11.8	13.8	16.0	20.9	24.8

Gauge	26	24	22	20	18	16
Wt.-Lb./S.F.	.906	1.156	1.406	1.656	2.156	2.656
SMACNA Max. Dimension – Long Side		30″	54″	84″	85″ Up	
Sum-2 Sides						
56	9.3	12.0	14.0	16.2	21.3	25.2
57	9.5	12.3	14.3	16.5	21.7	25.7
58	9.7	12.5	14.5	16.8	22.0	26.1
59	9.8	12.7	14.8	17.1	22.4	26.6
60	10.0	12.9	15.0	17.4	22.8	27.0
61	10.2	13.1	15.3	17.7	23.2	27.5
62	10.3	13.3	15.5	18.0	23.6	27.9
63	10.5	13.5	15.8	18.3	24.0	28.4
64	10.7	13.7	16.0	18.6	24.3	28.8
65	10.8	13.9	16.3	18.9	24.7	29.3
66	11.0	14.1	16.5	19.1	25.1	29.7
67	11.2	14.3	16.8	19.4	25.5	30.2
68	11.3	14.6	17.0	19.7	25.8	30.6
69	11.5	14.8	17.3	20.0	26.2	31.1
70	11.7	15.0	17.5	20.3	26.6	31.5
71	11.8	15.2	17.8	20.6	27.0	32.0
72	12.0	15.4	18.0	20.9	27.4	32.4
73	12.2	15.6	18.3	21.2	27.7	32.9
74	12.3	15.8	18.5	21.5	28.1	33.3
75	12.5	16.1	18.8	21.8	28.5	33.8
76	12.7	16.3	19.0	22.0	28.9	34.2
77	12.8	16.5	19.3	22.3	29.3	34.7
78	13.0	16.7	19.5	22.6	29.6	35.1
79	13.2	16.9	19.8	22.9	30.0	35.6
80	13.3	17.1	20.0	23.2	30.4	36.0
81	13.5	17.3	20.3	23.5	30.8	36.5
82	13.7	17.5	20.5	23.8	31.2	36.9
83	13.8	17.8	20.8	24.1	31.5	37.4
84	14.0	18.0	21.0	24.4	31.9	37.8
85	14.2	18.2	21.3	24.7	32.3	38.3
86	14.3	18.4	21.5	24.9	32.7	38.7
87	14.5	18.6	21.8	25.2	33.1	39.2
88	14.7	18.8	22.0	25.5	33.4	39.6
89	14.8	19.0	22.3	25.8	33.8	40.1
90	15.0	19.3	22.5	26.1	34.2	40.5
91	15.2	19.5	22.8	26.4	34.6	41.0
92	15.3	19.7	23.0	26.7	35.0	41.4
93	15.5	19.9	23.3	27.0	35.3	41.9
94	15.7	20.1	23.5	27.3	35.7	42.3
95	15.8	20.3	23.8	27.6	36.1	42.8
96	16.0	20.5	24.0	27.8	36.5	43.2
97	16.2	20.8	24.3	28.1	36.9	43.7
98	16.3	21.0	24.5	28.4	37.2	44.1
99	16.5	21.2	24.8	28.7	37.6	44.6
100	16.7	21.4	25.0	29.0	38.0	45.0
101	16.8	21.6	25.3	29.3	38.4	45.5
102	17.0	21.8	25.5	29.6	38.8	45.9
103	17.2	22.0	25.8	29.9	39.1	46.4
104	17.3	22.3	26.0	30.2	39.5	46.8
105	17.5	22.5	26.3	30.5	39.9	47.3
106	17.7	22.7	26.5	30.7	40.3	47.7
107	17.8	22.9	26.8	31.0	40.7	48.2
108	18.0	23.1	27.0	31.3	41.0	48.6
109	18.2	23.3	27.3	31.6	41.4	49.1
110	18.3	23.5	27.5	31.9	41.8	49.5

Example: If duct is 34″ x 20″ x 15′ long, 34″ is greater than 30″ maximum for 24 ga. so it must be 22 ga. 34″ + 20″ = 54″. Going across from 54″ find 13.5 lb. per foot. 13.5 x 15′ = 202.5 lbs. For S.F. of surface area 202.5 ÷ 1.406 = 144 S.F.

Note: Figures include an allowance for scrap.
*__Do Not__ use unless engineer specified. 26GA is very light and sometimes used for toilet exhaust. 16GA is heavy plate and mostly specified for plenums and hoods.

R233700-60 Diffuser Evaluation

CFM = V x An x K where V = outlet velocity in feet per minute, An = neck area in square feet and K = diffuser delivery factor. An undersized diffuser for a desired CFM will produce a high velocity and noise level. When air moves past people at a velocity in excess of 25 FPM, an annoying draft is felt. An oversized diffuser will result in low velocity with poor mixing. Consideration must be given to avoid vertical stratification or horizontal areas of stagnation.

R235000-10 Heating Systems

Heating Systems

The basic function of a heating system is to bring an enclosed volume up to a desired temperature and then maintain that temperature within a reasonable range. To accomplish this, the selected system must have sufficient capacity to offset transmission losses resulting from the temperature difference on the interior and exterior of the enclosing walls in addition to losses due to cold air infiltration through cracks, crevices and around doors and windows. The amount of heat to be furnished is dependent upon the building size, construction, temperature difference, air leakage, use, shape, orientation and exposure. Air circulation is also an important consideration. Circulation will prevent stratification which could result in heat losses through uneven temperatures at various levels. For example, the most efficient use of unit heaters can usually be achieved by circulating the space volume through the total number of units once every 20 minutes or 3 times an hour. This general rule must, of course, be adapted for special cases such as large buildings with low ratios of heat transmitting surface to cubical volume. The type of occupancy of a building will have considerable bearing on the number of heat transmitting units and the location selected. It is axiomatic, however, that the basis of any successful heating system is to provide the maximum amount of heat at the points of maximum heat loss such as exposed walls, windows, and doors. Large roof areas, wind direction, and wide doorways create problems of excessive heat loss and require special consideration and treatment.

Heat Transmission

Heat transfer is an important parameter to consider during selection of the exterior wall style, material and window area. A high rate of transfer will permit greater heat loss during the wintertime with the resultant increase in heating energy costs and a greater rate of heat gain in the summer with proportionally greater cooling cost. Several terms are used to describe various aspects of heat transfer. However, for general estimating purposes this data set lists U values for systems of construction materials. U is the "overall heat transfer coefficient." It is defined as the heat flow per hour through one square foot when the temperature difference in the air on either side of the structure wall, roof, ceiling or floor is one degree Fahrenheit. The structural segment may be a single homogeneous material or a composite.

Total heat transfer is found using the following equation:

$Q = AU(T_2 - T_1)$ where

- Q = Heat flow, BTU per hour
- A = Area, square feet
- U = Overall heat transfer coefficient
- $(T_2 - T_1)$ = Difference in temperature of air on each side of the construction component. (Also abbreviated TD)

Note that heat can flow through all surfaces of any building and this flow is in addition to heat gain or loss due to ventilation, infiltration and generation (appliances, machinery, people).

R235000-20 Heating Approximations for Quick Estimating

Oil Piping & Boiler Room Piping

Small System	20 to 30% of Boiler
Complex System with Pumps, Headers, Etc.	80 to 110% of Boiler

Breeching With Insulation:

Small	10 to 15% of Boiler
Large	15 to 25% of Boiler
Coils:	15 to 30% of Containing Unit
Balancing (Independent)	1/2% of H.V.A.C. Estimate

Quality/Complexity Adjustment: For all heating installations add these adjustments to the estimate to more closely allow for the equipment and conditions of the particular job under consideration.

Economy installation, add	0 to 5% of System
Good quality, medium complexity, add	5 to 15% of System
Above average quality and complexity, add	15 to 25% of System

R235000-30 The Basics of a Heating System

The function of a heating system is to achieve and maintain a desired temperature in a room or building by replacing the amount of heat being dissipated. There are four kinds of heating systems: hot-water, steam, warm-air and electric resistance. Each has certain essential and similar elements with the exception of electric resistance heating.

The basic elements of a heating system are:

A. A **combustion chamber** in which fuel is burned and heat transferred to a conveying medium.

B. The **"fluid"** used for conveying the heat (water, steam or air).

C. **Conductors** or pipes for transporting the fluid to specific desired locations.

D. A means of disseminating the heat, sometimes called **terminal units.**

A. The **combustion chamber** in a furnace heats air which is then distributed. This is called a warm-air system.

The combustion chamber in a boiler heats water which is either distributed as hot water or steam and this is termed a hydronic system.

The maximum allowable working pressures are limited by the ASME "Code for Heating Boilers" to 15 PSI for steam and 160 PSI for hot water heating boilers, with a maximum temperature limitation of 250° F. Hot water boilers are generally rated for a working pressure of 30 PSI. High pressure boilers are governed by the ASME "Code for Power Boilers" which is used almost universally for boilers operating over 15 PSIG. High pressure boilers used for a combination of heating/process loads are usually designed for 150 PSIG.

Boiler ratings are usually indicated as either gross or net output. The gross load is equal to the net load plus a piping and pickup allowance. When this allowance cannot be determined, divide the gross output rating by 1.25 for a value equal to or greater than the net heat loss requirement of the building.

B. Of the three **fluids** used, steam carries the greatest amount of heat per unit volume. This is due to the fact that it gives up its latent heat of vaporization at a temperature considerably above room temperature. Another advantage is that the pressure to produce a positive circulation is readily available. Piping conducts the steam to terminal units and returns condensate to the boiler.

The **steam system** is well adapted to large buildings because of its positive circulation, its comparatively economical installation and its ability to deliver large quantities of heat. Nearly all large office buildings, stores, hotels, and industrial buildings are so heated, in addition to many residences.

Hot water, when used as the heat carrying fluid, gives up a portion of its sensible heat and then returns to the boiler or heating apparatus for reheating. As the heat conveyed by each pound of water is about one-fiftieth of the heat conveyed by a pound of steam, it is necessary to circulate about fifty times as much water as steam by weight (although only one-thirtieth as much by volume). The hot water system is usually, although not necessarily, designed to operate at temperatures below that of the ordinary steam system and so the amount of heat transfer surface must be correspondingly greater. A temperature of 190° F to 200° F is normally the maximum. Circulation in small buildings may depend on the difference in density between hot water and the cool water returning to the boiler; circulating pumps are normally used to maintain a desired rate of flow. Pumps permit a greater degree of flexibility and better control.

In **warm-air** furnace systems, cool air is taken from one or more points in the building, passed over the combustion chamber and flue gas passages and then distributed through a duct system. A disadvantage of this system is that the ducts take up much more building volume than steam or hot water pipes. Advantages of this system are the relative ease with which humidification can be accomplished by the evaporation of water as the air circulates through the heater, and the lack of need for expensive disseminating units as the warm air simply becomes part of the interior atmosphere of the building.

C. Conductors (pipes and ducts) have been lightly treated in the discussion of conveying fluids. For more detailed information such as sizing and distribution methods, please refer to technical publications such as the American Society of Heating, Refrigerating and Air-Conditioning Engineers' "Handbook of Fundamentals."

D. Terminal units come in an almost infinite variety of sizes and styles, but the basic principles of operation are very limited. As previously mentioned, warm-air systems require only a simple register or diffuser to mix heated air with that present in the room. Special application items such as radiant coils and infrared heaters are available to meet particular conditions but are not usually considered for general heating needs. Most heating is accomplished by having air flow over coils or pipes containing the heat transporting medium (steam, hot-water, electricity). These units, while varied, may be separated into two general types, (1) radiator/convectors and (2) unit heaters.

Radiator/convectors may be cast, fin-tube or pipe assemblies. They may be direct, indirect, exposed, concealed or mounted within a cabinet enclosure, upright or baseboard style. These units are often collectively referred to as "radiators" or "radiation" although none gives off heat either entirely by radiation or by convection but rather a combination of both. The air flows over the units as a gravity "current." It is necessary to have one or more heat-emitting units in each room. The most efficient placement is low along an outside wall or under a window to counteract the cold coming into the room and achieve an even distribution.

In contrast to radiator/convectors which operate most effectively against the walls of smaller rooms, **unit heaters** utilize a fan to move air over heating coils and are very effective in locations of relatively large volume. Unit heaters, while usually suspended overhead, may be floor mounted. They also may take in fresh outside air for ventilation. The heat distributed by unit heaters may be from a remote source and conveyed by a fluid or it may be from the combustion of fuel in each individual heater. In the latter case the only piping required would be for fuel, however, a vent for the products of combustion would be necessary.

The following list gives many of the advantages of unit heaters for applications other than office or residential:

a. Large capacity so smaller number of units are required,
b. Piping system simplified, **c.** Space saved where they are located overhead out of the way, **d.** Rapid heating directed where needed with effective wide distribution, **e.** Difference between floor and ceiling temperature reduced, **f.** Circulation of air obtained, and ventilation with introduction of fresh air possible, **g.** Heat output flexible and easily controlled.

R235616-60 Solar Heating (Space and Hot Water)

Collectors should face as close to due South as possible, but variations of up to 20 degrees on either side of true South are acceptable. Local climate and collector type may influence the choice between east or west deviations. Obviously they should be located so they are not shaded from the sun's rays. Incline collectors at a slope of latitude minus 5 degrees for domestic hot water and latitude plus 15 degrees for space heating.

Flat plate collectors consist of a number of components as follows: Insulation to reduce heat loss through the bottom and sides of the collector. The enclosure which contains all the components in this assembly is usually weatherproof and prevents dust, wind and water from coming in contact with the absorber plate. The cover plate usually consists of one or more layers of a variety of glass or plastic and reduces the reradiation by creating an air space which traps the heat between the cover and the absorber plates.

The absorber plate must have a good thermal bond with the fluid passages. The absorber plate is usually metallic and treated with a surface coating which improves absorptivity. Black or dark paints or selective coatings are used for this purpose, and the design of this passage and plate combination helps determine a solar system's effectiveness.

Heat transfer fluid passage tubes are attached above and below or integral with an absorber plate for the purpose of transferring thermal energy from the absorber plate to a heat transfer medium. The heat exchanger is a device for transferring thermal energy from one fluid to another.

Piping and storage tanks should be well insulated to minimize heat losses.

Size domestic water heating storage tanks to hold 20 gallons of water per user, minimum, plus 10 gallons per dishwasher or washing machine. For domestic water heating an optimum collector size is approximately 3/4 square foot of area per gallon of water storage. For space heating of residences and small commercial applications the collector is commonly sized between 30% and 50% of the internal floor area. For space heating of large commercial applications, collector areas less than 30% of the internal floor area can still provide significant heat reductions.

A supplementary heat source is recommended for Northern states for December through February.

The solar energy transmission per square foot of collector surface varies greatly with the material used. Initial cost, heat transmittance and useful life are obviously interrelated.

R236000-10 Air Conditioning

General: The purpose of air conditioning is to control the environment of a space so that comfort is provided for the occupants and/or conditions are suitable for the processes or equipment contained therein. The several items which should be evaluated to define system objectives are:

- Temperature Control
- Humidity Control
- Cleanliness
- Odor, smoke and fumes
- Ventilation

Efforts to control the above parameters must also include consideration of the degree or tolerance of variation, the noise level introduced, the velocity of air motion and the energy requirements to accomplish the desired results.

The variation in **temperature** and **humidity** is a function of the sensor and the controller. The controller reacts to a signal from the sensor and produces the appropriate suitable response in either the terminal unit, the conductor of the transporting medium (air, steam, chilled water, etc.), or the source (boiler, evaporating coils, etc.).

The **noise level** is a by-product of the energy supplied to moving components of the system. Those items which usually contribute the most noise are pumps, blowers, fans, compressors and diffusers. The level of noise can be partially controlled through use of vibration pads, isolators, proper sizing, shields, baffles and sound absorbing liners.

Some **air motion** is necessary to prevent stagnation and stratification. The maximum acceptable velocity varies with the degree of heating or cooling which is taking place. Most people feel air moving past them at velocities in excess of 25 FPM as an annoying draft. However, velocities up to 45 FPM may be acceptable in certain cases. Ventilation, expressed as air changes per hour and percentage of fresh air, is usually an item regulated by local codes.

Selection of the system to be used for a particular application is usually a trade-off. In some cases the building size, style, or room available for mechanical use limits the range of possibilities. Prime factors influencing the decision are first cost and total life (operating, maintenance and replacement costs). The accuracy with which each parameter is determined will be an important measure of the reliability of the decision and subsequent satisfactory operation of the installed system.

Heat delivery may be desired from an air conditioning system. Heating capability usually is added as follows: A gas fired burner or hot water/steam/electric coils may be added to the air handling unit directly and heat all air equally. For limited or localized heat requirements the water/steam/electric coils may be inserted into the duct branch supplying the cold areas. Gas fired duct furnaces are also available.

Note: When water or steam coils are used the cost of the piping and boiler must also be added. For a rough estimate use the cost per square foot of the appropriate sized hydronic system with unit heaters. This will provide a cost for the boiler and piping, and the unit heaters of the system would equate to the approximate cost of the heating coils.

R236000-20 Air Conditioning Requirements

BTU's per hour per S.F. of floor area and S.F. per ton of air conditioning.

Type of Building	BTU/Hr per S.F.	S.F. per Ton	Type of Building	BTU/Hr per S.F.	S.F. per Ton	Type of Building	BTU/Hr per S.F.	S.F. per Ton
Apartments, Individual	26	450	Dormitory, Rooms	40	300	Libraries	50	240
Corridors	22	550	Corridors	30	400	Low Rise Office, Exterior	38	320
Auditoriums & Theaters	40	300/18*	Dress Shops	43	280	Interior	33	360
Banks	50	240	Drug Stores	80	150	Medical Centers	28	425
Barber Shops	48	250	Factories	40	300	Motels	28	425
Bars & Taverns	133	90	High Rise Office—Ext. Rms.	46	263	Office (small suite)	43	280
Beauty Parlors	66	180	Interior Rooms	37	325	Post Office, Individual Office	42	285
Bowling Alleys	68	175	Hospitals, Core	43	280	Central Area	46	260
Churches	36	330/20*	Perimeter	46	260	Residences	20	600
Cocktail Lounges	68	175	Hotel, Guest Rooms	44	275	Restaurants	60	200
Computer Rooms	141	85	Corridors	30	400	Schools & Colleges	46	260
Dental Offices	52	230	Public Spaces	55	220	Shoe Stores	55	220
Dept. Stores, Basement	34	350	Industrial Plants, Offices	38	320	Shop'g. Ctrs., Supermarkets	34	350
Main Floor	40	300	General Offices	34	350	Retail Stores	48	250
Upper Floor	30	400	Plant Areas	40	300	Specialty	60	200

*Persons per ton
12,000 BTU = 1 ton of air conditioning

R236000-30 Psychrometric Table

Dewpoint or Saturation Temperature (F)

Relative humidity (%)															
100	32	35	40	45	50	55	60	65	70	75	80	85	90	95	100
90	30	33	37	42	47	52	57	62	67	72	77	82	87	92	97
80	27	30	34	39	44	49	54	58	64	68	73	78	83	88	93
70	24	27	31	36	40	45	50	55	60	64	69	74	79	84	88
60	20	24	28	32	36	41	46	51	55	60	65	69	74	79	83
50	16	20	24	28	33	36	41	46	50	55	60	64	69	73	78
40	12	15	18	23	27	31	35	40	45	49	53	58	62	67	71
30	8	10	14	18	21	25	29	33	37	42	46	50	54	59	62
20	6	7	8	9	13	16	20	24	28	31	35	40	43	48	52
10	4	4	5	5	6	8	9	10	13	17	20	24	27	30	34
	32	**35**	**40**	**45**	**50**	**55**	**60**	**65**	**70**	**75**	**80**	**85**	**90**	**95**	**100**

Dry bulb temperature (F)

This table shows the relationship between RELATIVE HUMIDITY, DRY BULB TEMPERATURE AND DEWPOINT.

As an example, assume that the thermometer in a room reads 75° F, and we know that the relative humidity is 50%. The chart shows the dewpoint temperature to be 55° F. That is, any surface colder than 55° F will "sweat" or collect condensing moisture. This surface could be the outside of an uninsulated chilled water pipe in the summertime, or the inside surface of a wall or deck in the wintertime. After determining the extreme ambient parameters, the table at the left is useful in determining which surfaces need insulation or vapor barrier protection.

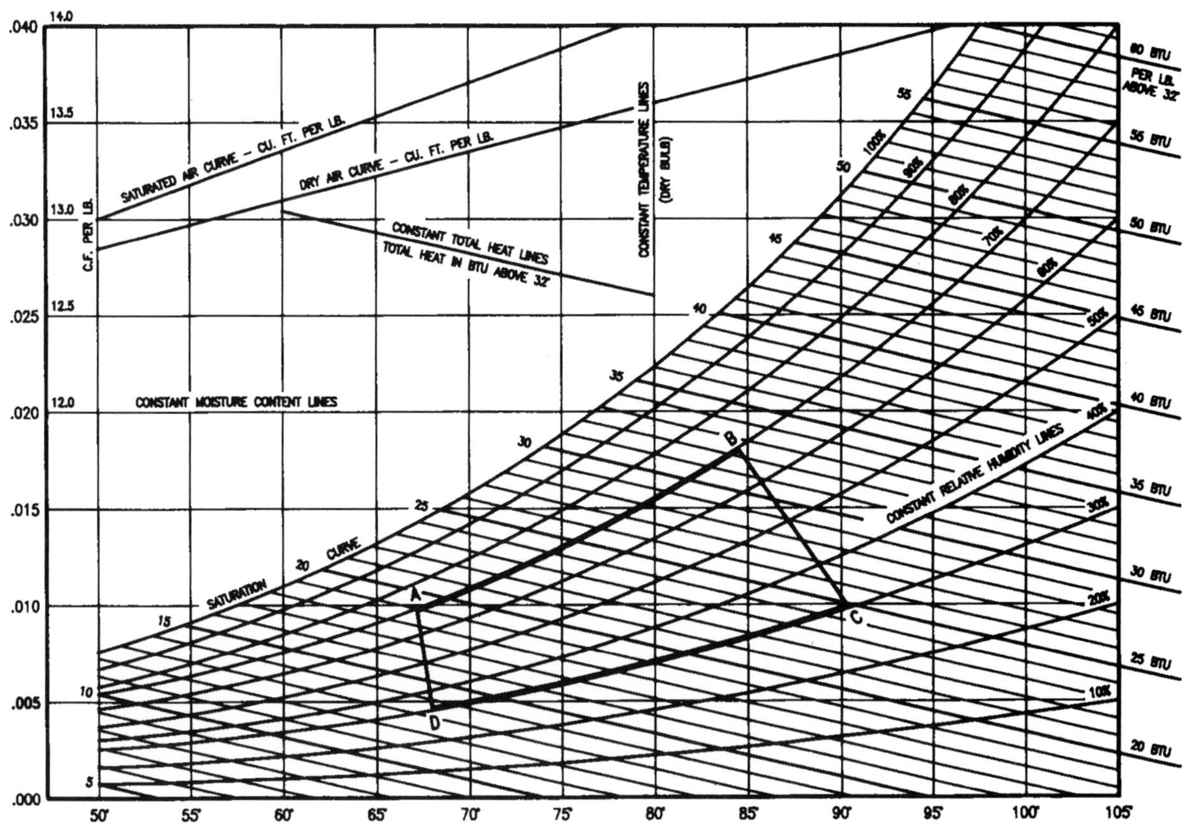

TEMPERATURE DEGREES FAHRENHEIT
TOTAL PRESSURE = 14.696 LB. PER SQ. IN. ABS.

Psychrometric chart showing different variables based on one pound of dry air. Space marked A B C D is temperature-humidity range which is most comfortable for most people.

R236000-90 Quality/Complexity Adjustment for Air Conditioning Systems

Economy installation, add	0 to 5%
Good quality, medium complexity, add	5 to 15%
Above average quality and complexity, add	15 to 25%

Add the above adjustments to the estimate to more closely allow for the equipment and conditions of the particular job under consideration.

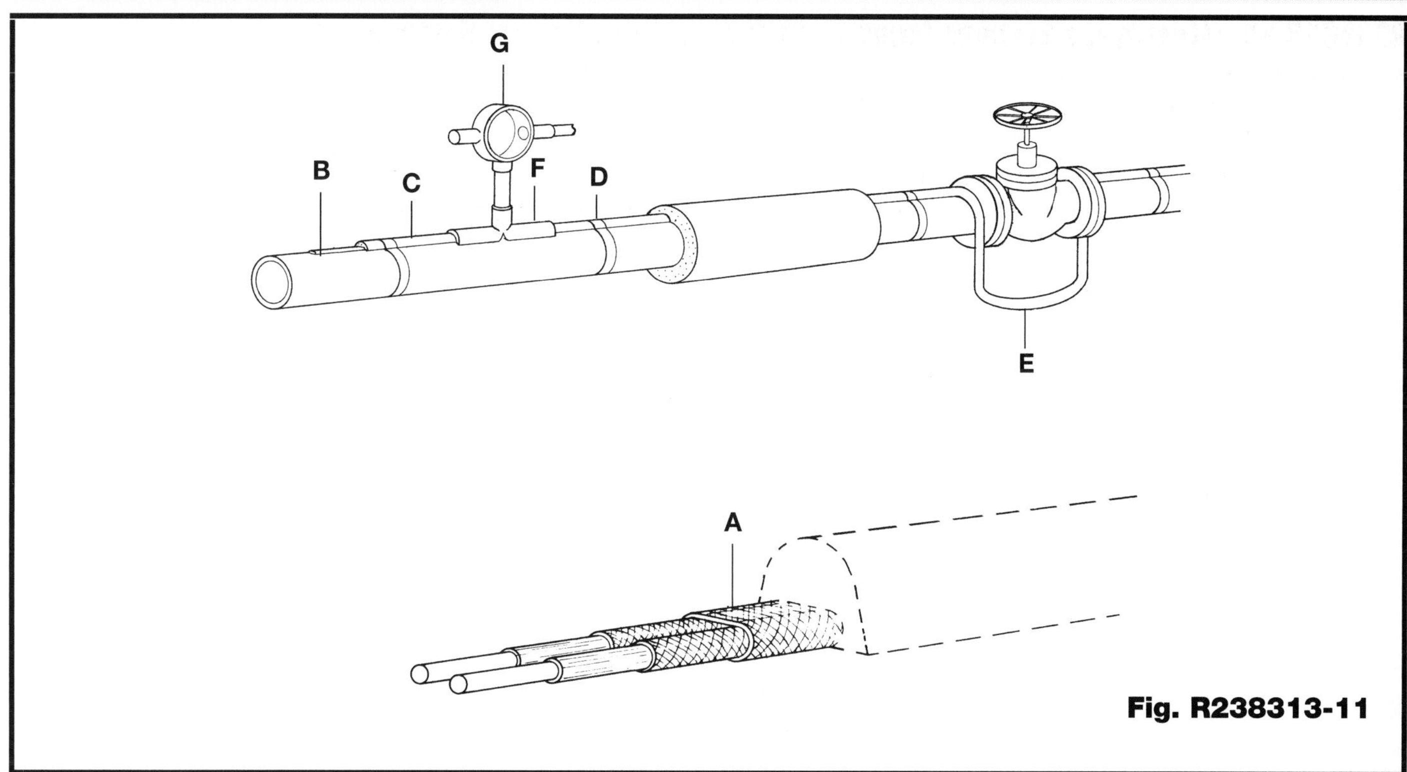

Fig. R238313-11

R238313-10 Heat Trace Systems

Before you can determine the cost of a HEAT TRACE installation the method of attachment must be established. There are four (4) common methods:

1. Cable is simply attached to the pipe with polyester tape every 12′.
2. Cable is attached with a continuous cover of 2″ wide aluminum tape.
3. Cable is attached with factory extruded heat transfer cement and covered with metallic raceway with clips every 10′.
4. Cable is attached between layers of pipe insulation using either clips or polyester tape.

Example: Components for method 3 must include:

A. Heat trace cable by voltage and watts per linear foot.
B. Heat transfer cement, 1 gallon per 60 linear feet of cover.
C. Metallic raceway by size and type.
D. Raceway clips by size of pipe.

When taking off linear foot lengths of cable add the following for each valve in the system. (E)

In all of the above methods each component of the system must be priced individually.

SCREWED OR WELDED VALVE:

Size		Length
1/2″	=	6″
3/4″	=	9″
1″	=	1′ -0″
1-1/2″	=	1′ -6″
2″	=	2′
2-1/2″	=	2′ -6″
3″	=	2′ -6″
4″	=	4′ -0″
6″	=	7′ -0″
8″	=	9′ -6″
10″	=	12′ -6″
12″	=	15′ -0″
14″	=	18′ -0″
16″	=	21′ -6″
18″	=	25′ -6″
20″	=	28′ -6″
24″	=	34′ -0″
30″	=	40′ -0″

FLANGED VALVE:

Size		Length
1/2″	=	1′ -0″
3/4″	=	1′ -6″
1″	=	2′ -0″
1-1/2″	=	2′ -6″
2″	=	2′ -6″
2-1/2″	=	3′ -0″
3″	=	3′ -6″
4″	=	4′ -0″
6″	=	8′ -0″
8″	=	11′ -0″
10″	=	14′ -0″
12″	=	16′ -6″
14″	=	19′ -6″
16″	=	23′ -0″
18″	=	27′ -0″
20″	=	30′ -0″
24″	=	36′ -0″
30″	=	42′ -0″

BUTTERFLY VALVE:

Size		Length
1/2″	=	0′
3/4″	=	0′
1″	=	1′ -0″
1-1/2″	=	1′ -6″
2″	=	2′ -0″
2-1/2″	=	2′ -6″
3″	=	2′ -6″
4″	=	3′ -0″
6″	=	3′ -6″
8″	=	4′ -0″
10″	=	4′ -0″
12″	=	5′ -0″
14″	=	5′ -6″
16″	=	6′ -0″
18″	=	6′ -6″
20″	=	7′ -0″
24″	=	8′ -0″
30″	=	10′ -0″

R238313-10 Heat Trace Systems (cont.)

Add the following quantities of heat transfer cement to linear foot totals for each valve:

Nominal Valve Size	Gallons of Cement per Valve
1/2″	0.14
3/4″	0.21
1″	0.29
1-1/2″	0.36
2″	0.43
2-1/2″	0.70
3″	0.71
4″	1.00
6″	1.43
8″	1.48
10″	1.50
12″	1.60
14″	1.75
16″	2.00
18″	2.25
20″	2.50
24″	3.00
30″	3.75

The following must be added to the list of components to accurately price HEAT TRACE systems:

1. Expediter fitting and clamp fasteners (F)
2. Junction box and nipple connected to expediter fitting (G)
3. Field installed terminal blocks within junction box
4. Ground lugs
5. Piping from power source to expediter fitting
6. Controls
7. Thermostats
8. Branch wiring
9. Cable splices
10. End of cable terminations
11. Branch piping fittings and boxes

Deduct the following percentages from labor if cable lengths in the same area exceed:

150′ to 250′	10%	351′ to 500′	20%
251′ to 350′	15%	Over 500′	25%

Add the following percentages to labor for elevated installations:

15′ to 20′ high	10%	31′ to 35′ high	40%
21′ to 25′ high	20%	36′ to 40′ high	50%
26′ to 30′ high	30%	Over 40′ high	60%

R238313-20 Spiral-Wrapped Heat Trace Cable (Pitch Table)

In order to increase the amount of heat, occasionally a heat trace cable is wrapped in a spiral fashion around a pipe, increasing the number of feet of heater cable per linear foot of pipe.

Engineers first determine the heat loss per foot of pipe (based on the insulating material, its thickness, and the temperature differential across it). A ratio is then calculated by the formula:

$$\text{Feet of Heat Trace per Foot of Pipe} = \frac{\text{Watts/Foot of Heat Loss}}{\text{Watts/Foot of the Cable}}$$

The linear distance between wraps (pitch) is then taken from a chart or table. Generally, the pitch is listed on a drawing leaving the estimator to calculate the total length of heat tape required. An approximation may be taken from this table.

Feet of Heat Trace Per Foot of Pipe																
	Nominal Pipe Size in Inches															
Pitch In Inches	1	1¼	1½	2	2½	3	4	6	8	10	12	14	16	18	20	24
3.5	1.80															
4	1.65															
5	1.46	1.60	1.80													
6	1.34	1.45	1.55	1.75												
7	1.25	1.35	1.43	1.57	1.75											
8	1.20	1.28	1.34	1.45	1.60	1.80										
9	1.16	1.23	1.28	1.37	1.51	1.68										
10	1.13	1.19	1.24	1.32	1.44	1.57	1.82									
15	1.06	1.08	1.10	1.15	1.21	1.29	1.42	1.78								
20	1.04	1.05	1.06	1.08	1.13	1.17	1.25	1.49	1.73							
25		1.04	1.04	1.06	1.08	1.11	1.17	1.33	1.51	1.72						
30				1.04	1.05	1.07	1.12	1.24	1.37	1.54	1.70	1.80				
35					1.06	1.06	1.09	1.17	1.28	1.42	1.54	1.64	1.78			
40						1.05	1.07	1.14	1.22	1.33	1.44	1.52	1.64	1.75		
50							1.05	1.09	1.15	1.22	1.29	1.35	1.44	1.53	1.64	1.83
60								1.06	1.11	1.16	1.21	1.25	1.31	1.39	1.46	1.62
70								1.05	1.08	1.12	1.17	1.19	1.24	1.30	1.35	1.47
80									1.06	1.09	1.13	1.15	1.19	1.24	1.28	1.38
90									1.04	1.06	1.10	1.13	1.16	1.19	1.23	1.32
100										1.05	1.08	1.10	1.13	1.15	1.19	1.23

Note: Common practice would normally limit the lower end of the table to 5% of additional heat. Above 80% an engineer would likely opt for two (2) parallel cables.

R260105-30 Electrical Demolition (Removal for Replacement)

The purpose of this reference number is to provide a guide to users for electrical "removal for replacement" by applying the rule of thumb: 1/3 of new installation time (typical range from 20% to 50%) for removal. Remember to use reasonable judgment when applying the suggested percentage factor. For example:

Contractors have been requested to remove an existing fluorescent lighting fixture and replace it with a new fixture utilizing energy saver lamps and electronic ballast.

In order to fully understand the extent of the project, contractors should visit the job site and estimate the time to perform the renovation work in accordance with applicable national, state and local regulation and codes.

The contractor may need to add extra labor-hours to his estimate if he discovers unknown concealed conditions such as: contaminated asbestos ceiling, broken acoustical ceiling tile and the need to repair, patch and touch-up paint all the damaged or disturbed areas, tasks normally assigned to general contractors. In addition, the owner could request that the contractors salvage the materials removed and turn over the materials to the owner or dispose of the materials to a reclamation station. The normal removal item is 0.5 labor-hour for a lighting fixture and 1.5 labor-hours for new installation time. Revise the estimate times from 2 labor-hours of work up to a minimum of 4 labor-hours work for just fluorescent lighting fixture.

For removal of large concentrations of lighting fixtures in the same area, apply an "economy of scale" to reduce estimating labor-hours.

R260519-80 Undercarpet Systems

Takeoff Procedure for Power Systems: List components for each fitting type, tap, splice, and bend on your quantity takeoff sheet. Each component must be priced separately. Start at the power supply transition fittings and survey each circuit for the components needed. List the quantities of each component under a specific circuit number. Use the floor plan layout scale to get cable footage.

Reading across the list, combine the totals of each component in each circuit and list the total quantity in the last column. Calculate approximately 5% for scrap for items such as cable, top shield, tape, and spray adhesive. Also provide for final variations that may occur on-site.

Suggested guidelines are:
1. Equal amounts of cable and top shield should be priced.
2. For each roll of cable, price a set of cable splices.
3. For every 1 ft. of cable, price 2-1/2 ft. of hold-down tape.
4. For every 3 rolls of hold-down tape, price 1 can of spray adhesive.

Adjust final figures wherever possible to accommodate standard packaging of the product. This information is available from the distributor.

Each transition fitting requires:
1. 1 base
2. 1 cover
3. 1 transition block

Each floor fitting requires:
1. 1 frame/base kit
2. 1 transition block
3. 2 covers (duplex/blank)

Each tap requires:
1. 1 tap connector for each conductor
2. 1 pair insulating patches
3. 2 top shield connectors

Each splice requires:
1. 1 splice connector for each conductor
2. 1 pair insulating patches
3. 3 top shield connectors

Each cable bend requires:
1. 2 top shield connectors

Each cable dead end (outside of transition block) requires:
1. 1 pair insulating patches

Labor does not include:
1. Patching or leveling uneven floors
2. Filling in holes or removing projections from concrete slabs.
3. Sealing porous floors
4. Sweeping and vacuuming floors
5. Removal of existing carpeting
6. Carpet square cut-outs
7. Installation of carpet squares

Takeoff Procedures for Telephone Systems: After reviewing floor plans identify each transition. Number or letter each cable run from that fitting.

Start at the transition fitting and survey each circuit for the components needed. List the cable type, terminations, cable length, and floor fitting type under the specific circuit number. Use the floor plan layout scale to get the cable footage. Add some extra length (next higher increment of 5 feet) to preconnectorized cable.

Transition fittings require:
1. 1 base plate
2. 1 cover
3. 1 transition block

Floor fittings require:
1. 1 frame/base kit
2. 2 covers
3. Modular jacks

Reading across the list, combine the list of components in each circuit and list the total quantity in the last column. Calculate the necessary scrap factors for such items as tape, bottom shield and spray adhesive. Also provide for final variations that may occur on-site.

Adjust final figures whenever possible to accommodate standard packaging. Check that items such as transition fittings, floor boxes, and floor fittings that are to utilize both power and telephone have been priced as combination fittings, so as to avoid duplication.

Make sure to include marking of floors and drilling of fasteners if fittings specified are not the adhesive type.

Labor does not include:
1. Conduit or raceways before transition of floor boxes
2. Telephone cable before transition boxes
3. Terminations before transition boxes
4. Floor preparation as described in power section

Be sure to include all cable folds when pricing labor.

Takeoff Procedure for Data Systems: Start at the transition fittings and take off quantities in the same manner as the telephone system, keeping in mind that data cable does not require top or bottom shields.

The data cable is simply cross-taped on the cable run to the floor fitting.

Data cable can be purchased in either bulk form in which case coaxial connector material and labor must be priced, or in preconnectorized cut lengths.

Data cable cannot be folded and must be notched at 1 inch intervals. A count of all turns must be added to the labor portion of the estimate. (Note: Some manufacturers have prenotched cable.)

Notching required:
1. 90 degree turn requires 8 notches per side
2. 180 degree turn requires 16 notches per side

Floor boxes, transition boxes, and fittings are the same as described in the power and telephone procedures.

Since undercarpet systems require special hand tools, be sure to include this cost in proportion to the number of crews involved in the installation.

Job Conditions: Productivity is based on new construction in an unobstructed area. Staging area is assumed to be within 200′ of work being performed.

R260519-90 Wire

Wire quantities are taken off by either measuring each cable run or by extending the conduit and raceway quantities times the number of conductors in the raceway. Ten percent should be added for waste and tie-ins. Keep in mind that the unit of measure of wire is C.L.F. not L.F. as in raceways so the formula would read:

$$\frac{\text{(L.F. Raceway x No. of Conductors) x 1.10}}{100} = \text{C.L.F.}$$

Price per C.L.F. of wire includes:
1. Setting up wire coils or spools on racks
2. Attaching wire to pull in means
3. Measuring and cutting wire
4. Pulling wire into a raceway
5. Identifying and tagging

Price does not include:
1. Connections to breakers, panelboards, or equipment
2. Splices

Job Conditions: Productivity is based on new construction to a height of 15′ using rolling staging in an unobstructed area. Material staging is assumed to be within 100′ of work being performed.

Economy of Scale: If more than three wires at a time are being pulled, deduct the following percentages from the labor of that grouping:

4-5 wires	25%
6-10 wires	30%
11-15 wires	35%
over 15 wires	40%

If a wire pull is less than 100′ in length and is interrupted several times by boxes, lighting outlets, etc., it may be necessary to add the following lengths to each wire being pulled:

Junction box to junction box	2 L.F.
Lighting panel to junction box	6 L.F.
Distribution panel to sub panel	8 L.F.
Switchboard to distribution panel	12 L.F.
Switchboard to motor control center	20 L.F.
Switchboard to cable tray	40 L.F.

Measure of Drops and Riser: It is important when taking off wire quantities to include the wire for drops to electrical equipment. If heights of electrical equipment are not clearly stated, use the following guide:

	Bottom A.F.F.	Top A.F.F.	Inside Cabinet
Safety switch to 100A	5′	6′	2′
Safety switch 400 to 600A	4′	6′	3′
100A panel 12 to 30 circuit	4′	6′	3′
42 circuit panel	3′	6′	4′
Switch box	3′	3′6″	1′
Switchgear	0′	8′	8′
Motor control centers	0′	8′	8′
Transformers - wall mount	4′	8′	2′
Transformers - floor mount	0′	12′	4′

R260519-92 Minimum Copper and Aluminum Wire Size Allowed for Various Types of Insulation

Minimum Wire Sizes									
	Copper		Aluminum			Copper		Aluminum	
Amperes	THW THWN or XHHW	THHN XHHW *	THW XHHW	THHN XHHW *	Amperes	THW THWN or XHHW	THHN XHHW *	THW XHHW	THHN XHHW *
15A	#14	#14	#12	#12	195	3/0	2/0	250kcmil	4/0
20	#12	#12	#10	#10	200	3/0	3/0	250kcmil	4/0
25	#10	#10	#10	#10	205	4/0	3/0	250kcmil	4/0
30	#10	#10	# 8	# 8	225	4/0	3/0	300kcmil	250kcmil
40	# 8	# 8	# 8	# 8	230	4/0	4/0	300kcmil	250kcmil
45	# 8	# 8	# 6	# 8	250	250kcmil	4/0	350kcmil	300kcmil
50	# 8	# 8	# 6	# 6	255	250kcmil	4/0	400kcmil	300kcmil
55	# 6	# 8	# 4	# 6	260	300kcmil	4/0	400kcmil	350kcmil
60	# 6	# 6	# 4	# 6	270	300kcmil	250kcmil	400kcmil	350kcmil
65	# 6	# 6	# 4	# 4	280	300kcmil	250kcmil	500kcmil	350kcmil
75	# 4	# 6	# 3	# 4	285	300kcmil	250kcmil	500kcmil	400kcmil
85	# 4	# 4	# 2	# 3	290	350kcmil	250kcmil	500kcmil	400kcmil
90	# 3	# 4	# 2	# 2	305	350kcmil	300kcmil	500kcmil	400kcmil
95	# 3	# 4	# 1	# 2	310	350kcmil	300kcmil	500kcmil	500kcmil
100	# 3	# 3	# 1	# 2	320	400kcmil	300kcmil	600kcmil	500kcmil
110	# 2	# 3	1/0	# 1	335	400kcmil	350kcmil	600kcmil	500kcmil
115	# 2	# 2	1/0	# 1	340	500kcmil	350kcmil	600kcmil	500kcmil
120	# 1	# 2	1/0	1/0	350	500kcmil	350kcmil	700kcmil	500kcmil
130	# 1	# 2	2/0	1/0	375	500kcmil	400kcmil	700kcmil	600kcmil
135	1/0	# 1	2/0	1/0	380	500kcmil	400kcmil	750kcmil	600kcmil
150	1/0	# 1	3/0	2/0	385	600kcmil	500kcmil	750kcmil	600kcmil
155	2/0	1/0	3/0	3/0	420	600kcmil	500kcmil		700kcmil
170	2/0	1/0	4/0	3/0	430		500kcmil		750kcmil
175	2/0	2/0	4/0	3/0	435		600kcmil		750kcmil
180	3/0	2/0	4/0	4/0	475		600kcmil		

*Dry Locations Only

Notes:

1. Size #14 to 4/0 is in AWG units (American Wire Gauge).
2. Size 250 to 750 is in kcmil units (Thousand Circular Mils).
3. Use next higher ampere value if exact value is not listed in table.
4. For loads that operate continuously increase ampere value by 25% to obtain proper wire size.
5. Table R260519-92 has been written for estimating only. It is based on an ambient temperature of 30°C (86° F). For ambient temperature other than 30°C (86° F), ampacity correction factors will be applied.

R260533-20 Conduit To 15′ High

List conduit by quantity, size, and type. Do not deduct for lengths occupied by fittings, since this will be allowance for scrap. Example:

A. Aluminum — size
B. Rigid galvanized — size
C. Steel intermediate (IMC) — size
D. Rigid steel, plastic-coated 20 Mil. — size
E. Rigid steel, plastic-coated 40 Mil. — size
F. Electric metallic tubing (EMT) — size
G. PVC Schedule 40 — size

Types (A) through (E) listed above contain the following per 100 L.F.:

1. (11) Threaded couplings
2. (11) Beam-type hangers
3. (2) Factory sweeps
4. (2) Fiber bushings
5. (4) Locknuts
6. (2) Field threaded pipe terminations
7. (2) Removal of concentric knockouts

Type (F) contains per 100 L.F.:

1. (11) Set screw couplings
2. (11) Beam clamps
3. (2) Field bends on 1/2″ and 3/4″ diameter
4. (2) Factory sweeps for 1″ and above
5. (2) Set screw steel connectors
6. (2) Removal of concentric knockouts

Type (G) contains per 100 L.F.:

1. (11) Field cemented couplings
2. (34) Beam clamps
3. (2) Factory sweeps
4. (2) Adapters
5. (2) Locknuts
6. (2) Removal of concentric knockouts

Labor-hours for all conduit to 15′ include:

1. Unloading by hand
2. Hauling by hand to an area up to 200′ from loading dock
3. Setup of rolling staging
4. Installation of conduit and fittings as described in conduit models (A) through (G)

Not included in the material and labor are:

1. Staging rental or purchase
2. Structural modifications
3. Wire
4. Junction boxes
5. Fittings in excess of those described in conduit models (A) through (G)
6. Painting of conduit

Fittings

Only those fittings listed above are included in the linear foot totals, although they should be listed separately from conduit lengths, without prices, to ensure proper quantities for material procurement.

If the fittings required exceed the quantities included in the model conduit runs, then material and labor costs must be added to the difference. If actual needs per 100 L.F. of conduit are: (2) sweeps, (4) LBs and (1) field bend, then, (4) LBs and (1) field bend must be priced additionally.

R260533-21 Hangers

It is sometimes desirable to substitute an alternate style of hanger if the support being used is not the type described in the conduit models.

One approach is the substitution method:

1. Find the cost of the type hanger described in the conduit model.
2. Calculate the cost of the desired type hanger (it may be necessary to calculate individual components such as drilling, expansion shields, etc.).
3. Calculate the cost difference (delta) between the two types of hangers.
4. Multiply the cost delta by the number of hangers in the model.
5. Divide the total delta cost for hangers in the model by the length of the model to find the delta cost per L.F. for that model.
6. Modify the given unit costs per L.F. for the model by the delta cost per L.F.

Another approach to hanger configurations would be to start with the conduit only and add all the supports and any other items as separate lines. This procedure is most useful if the project involves racking many runs of conduit on a single hanger, for instance, a trapeze type hanger.

Example: Five (5) 2″ RGS conduits, 50 L.F. each, are to be run on trapeze hangers from one pull box to another. The run includes one 90° bend.

1. List the hangers' components to create an assembly cost for each 2′-wide trapeze.
2. List the components for the 50′ conduit run, noting that 6 trapeze supports will be required.

Job Conditions: Productivities are based on new construction to 15′ high, using scaffolding in an unobstructed area. Material storage is assumed to be within 100′ of work being performed.

Add to labor for elevated installations:

15′ to 20′ High–10%	30′ to 35 High–30%
20′ to 25′ High–20%	35′ to 40′ High–35%
25′ to 30′ High–25%	Over 40′ High–40%

Add these percentages to the L.F. labor cost, but not to fittings. Add these percentages only to quantities exceeding the different height levels, rather than the total conduit quantities.

Linear foot price for labor does not include penetrations in walls or floors and must be added to the estimate.

R260533-22 Conductors in Conduit

The table below lists the maximum number of conductors for various sized conduit using THW, TW or THWN insulations.

Copper Wire Size	Conduit Size																										
	½″			¾″			1″			1-¼″			1-½″			2″			2-½″			3″		3-½″		4″	
	TW	THW	THWN	TW	THW	THWN	TW	THW	THWN	TW	THW	THWN	TW	THW	THWN	TW	THW	THWN	TW	THW	THWN	THW	THWN	THW	THWN	THW	THWN
#14	9	6	13	15	10	24	25	16	39	44	29	69	60	40	94	99	65	154	142	93		143		192			
#12	7	4	10	12	8	18	19	13	29	35	24	51	47	32	70	78	53	114	111	76	164	117		157			
#10	5	4	6	9	6	11	15	11	18	26	19	32	36	26	44	60	43	73	85	61	104	95	160	127		163	
#8	2	1	3	4	3	5	7	5	9	12	10	16	17	13	22	28	22	36	40	32	51	49	79	66	106	85	136
#6		1	1		2	4		4	6		7	11		10	15		16	26		23	37	36	57	48	76	62	98
#4		1	1		1	2		3	4		5	7		7	9		12	16		17	22	27	35	36	47	47	60
#3		1	1		1	1		2	3		4	6		6	8		10	13		15	19	23	29	31	39	40	51
#2		1	1		1	1		2	3		4	5		5	7		9	11		13	16	20	25	27	33	34	43
#1					1	1		1	1		3	3		4	5		6	8		9	12	14	18	19	25	25	32
1/0					1	1		1	1		2	3		3	4		5	7		8	10	12	15	16	21	21	27
2/0					1	1		1	1		1	2		3	3		5	6		7	8	10	13	14	17	18	22
3/0					1	1		1	1		1	1		2	3		4	5		6	7	9	11	12	14	15	18
4/0						1		1	1		1	1		1	2		3	4		5	6	7	9	10	12	13	15
250 kcmil								1	1		1	1		1	1		2	3		4	4	6	7	8	10	10	12
300								1	1		1	1		1	1		2	3		3	4	5	6	7	8	9	11
350									1		1	1		1	1		1	2		3	3	4	5	6	7	8	9
400											1	1		1	1		1	1		2	3	4	5	5	6	7	8
500											1	1		1	1		1	1		1	2	3	4	4	5	6	7
600												1		1	1		1	1		1	1	3	3	4	4	5	5
700														1	1		1	1		1	1	2	3	3	4	4	5
750 ▼														1	1		1	1		1	1	2	2	3	3	4	4

R260533-30 Labor for Couplings and Fittings

The labor included in the unit price lines for couplings is for their installation separate from fitting. However, the labor included in the unit price lines for fittings covers their complete installation, which may include the installation of two couplings. If couplings are required to complete a fitting installation, the material only for two couplings needs to be added to the appropriate unit price lines.

Electrical | R2624 Switchboards & Panelboards

R262416-50 Load Centers and Panelboards

When pricing load centers list panels by size and type. List breakers in a separate column of the "Quantity Sheet," and define by phase and ampere rating.

Material and labor prices include breakers; for example: a 100A, 3-wire, 102/240V, 18 circuit panel w/ main breaker, as described in the unit cost section, contains 18 single pole 20A breakers.

If you do not choose to include a full panel of single pole breakers, use the following method to adjust material and labor costs.

Example: In an 18 circuit panel only 16 single pole breakers are desired, requiring that the cost of 2 breakers be subtracted from the panel cost.

1. Go to the appropriate unit cost section of the data set to find the unit prices of the given circuit breaker type.
2. Modify those costs as follows:
   Bare material price x 0.50
   Bare labor cost x 0.60.
3. Multiply those modified bare costs by 2 (breakers in this example).
4. Subtract the modified costs for the 2 breakers from the given cost of the panel.

Labor-hours for load center installation includes:

1. Unloading, uncrating, and handling enclosures 200′ from unloading area
2. Measuring and marking
3. Drilling four (4) lead anchor type fasteners using a hammer drill
4. Mounting and leveling panel to a height of 6′
5. Preparation and termination of feeder cable to lugs or main breaker
6. Branch circuit identification
7. Lacing using tie wraps
8. Testing and load balancing
9. Marking panel directory

Not included in the material and labor are:

1. Modifications to enclosure
2. Structural supports
3. Additional lugs
4. Plywood backboards
5. Painting or lettering

Note: Knockouts are included in the price of terminating pipe runs and need not be added to the load center costs.

Job Conditions: Productivity is based on new construction to a height of 6′, in an unobstructed area. Material staging area is assumed to be within 100′ of work being performed.

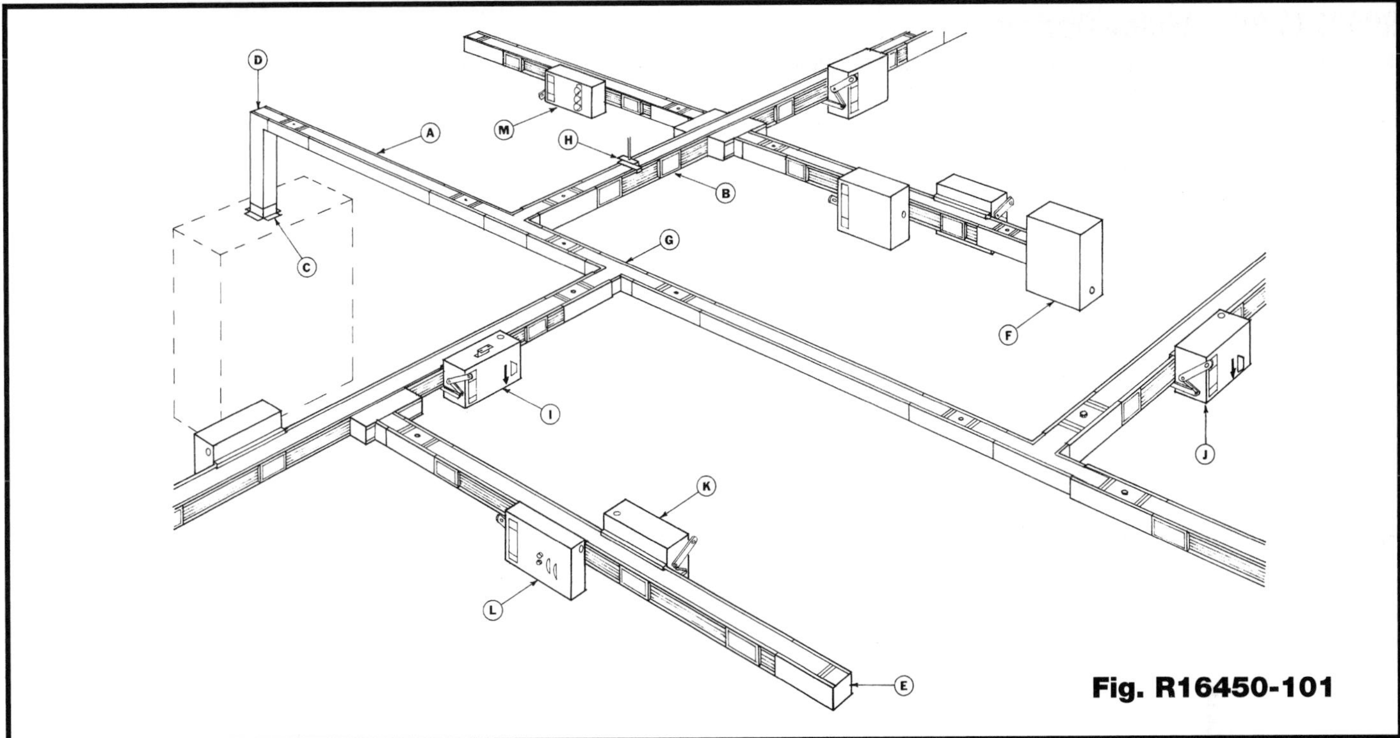

Fig. R16450-101

R262513-10 Aluminum Bus Duct

When taking off bus duct identify the system as either:

1. Aluminum
2. Copper

List straight lengths by type and size

A. Feeder — 800 A
B. Plug in — 800 A

Do not measure through fittings as you would on conduit, since there is no allowance for scrap in bus duct systems.

If upon taking off linear foot quantities of bus duct you find your quantities are not divisible evenly by 10 ft., then the remainder must be priced as a special item and quoted from the manufactuer. Do not use the bare material cost per L.F. for these special items. You can, however, safely use the bare labor cost per L.F. for the entire length.

Identify fittings by type and ampere rating.

Example:

C. Switchboard stub 800 A

D. Elbows 800 A

E. End box 800 A

F. Cable tap box 800 A

G. Tee Fittings 800 A

H. Hangers

Plug-in Units – List separately plug-in units and identify by type and ampere rating

I. Plug-in switches 600 Volt 3 phase 60 A

J. Plug-in molded case C.B. 60 A

K. Combination starter FVNR NEMA 1

L. Combination contactor & fused switch NEMA 1

M. Combination fusible switch & lighting control 60 A

Labor-hours for feeder and plug-in sections include:

1. Unloading and uncrating
2. Hauling up to 200 ft. from loading dock
3. Measuring and marking
4. Setup of rolling staging
5. Installing hangers
6. Hanging and bolting sections
7. Aligning and leveling
8. Testing

Labor-hours do not include:

1. Modifications to existing structure for hanger supports
2. Threaded rod in excess of 2 ft.
3. Welding
4. Penetrations through walls
5. Staging rental

Deduct the following percentages from labor only:

150 ft. to 250 ft. — 10%
251 ft. to 350 ft. — 15%
351 ft. to 500 ft. — 20%
Over 500 ft. — 25%

Deduct percentage only if runs are contained in the same area.

Example: If the job entails running 100 ft. in 5 different locations do not deduct 20%, but if the duct is being run in 1 area and the quantity is 500 ft. then you would deduct 20%.

Deduct only from straight lengths, not fittings or plug-in units.

R262726-90 Wiring Devices

Wiring devices should be priced on a separate takeoff form which includes boxes, covers, conduit and wire.

Labor-hours for devices include:

1. Stripping of wire
2. Attaching wire to device using terminators on the device itself, lugs, set screws etc.
3. Mounting of device in box

Labor-hours do not include:

1. Conduit
2. Wire
3. Boxes
4. Plates

Economy of Scale – for large concentrations of devices in the same area deduct the following percentages from labor-hours:

1	to	10	0%
11	to	25	20%
26	to	50	25%
51	to	100	30%
	over	100	35%

R262726-90 Wiring Devices (cont.)

NEMA No.	15 R	20 R	30 R	50 R	60 R
1 125V 2 Pole, 2 Wire	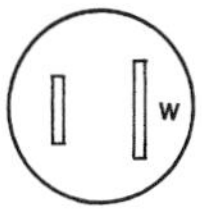				
2 250V 2 Pole, 2 Wire		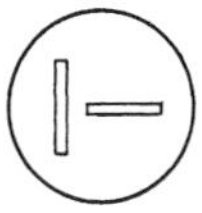			
5 125V 2 Pole, 3 Wire	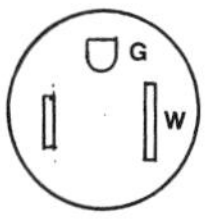	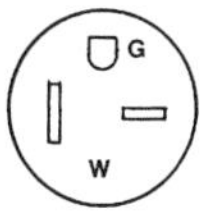		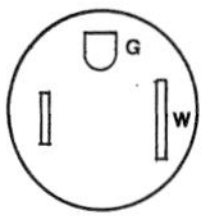	
6 250V 2 Pole, 3 Wire	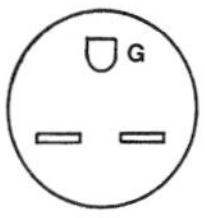		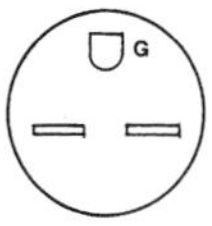	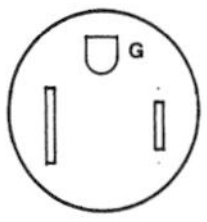	
7 277V, AC 2 Pole, 3 Wire	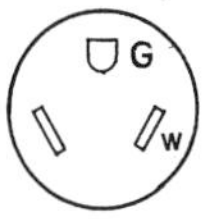			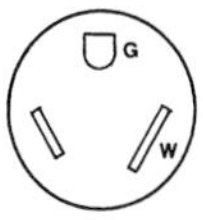	
10 125/250V 3 Pole, 3 Wire		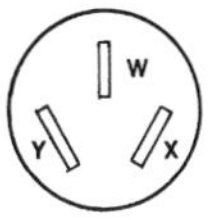	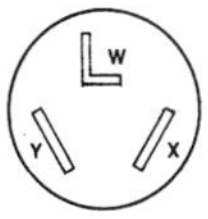	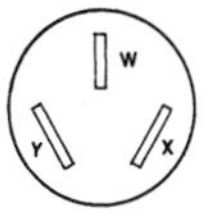	
11 3 Phase 250V 3 Pole, 3 Wire	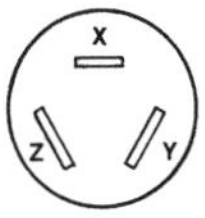	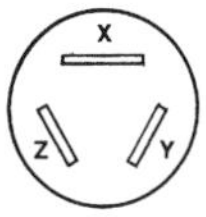	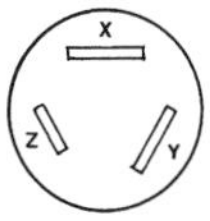	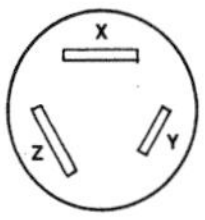	
14 125/250V 3 Pole, 4 Wire	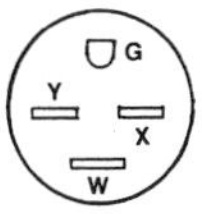	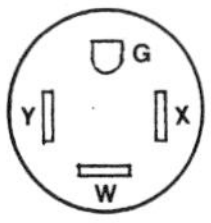	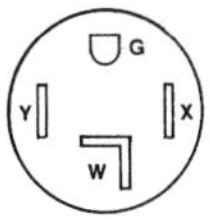	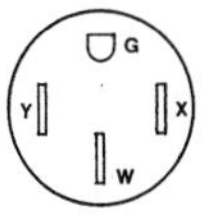	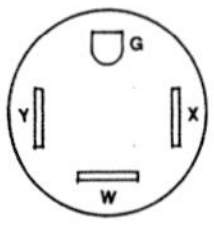
15 3 Phase 250V 3 Pole, 4 Wire	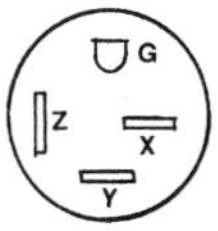		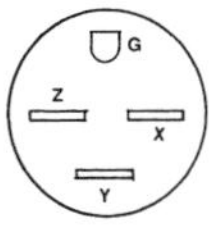	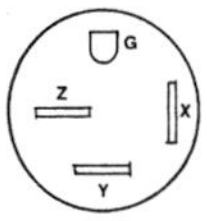	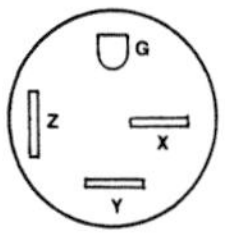
18 3 Phase 208Y/120V 4 Pole, 4 Wire	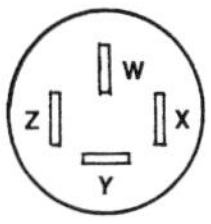	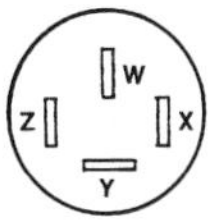	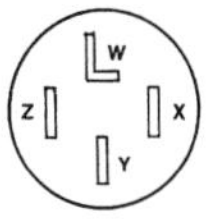	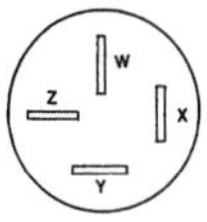	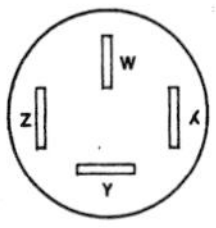

R262726-90 Wiring Devices (cont.)

NEMA No.	15 R	20 R	30 R	NEMA No.	15 R	20 R	30 R
L 1 125V 2 Pole, 2 Wire				**L 13** 3 Phase 600V 3 Pole, 3 Wire			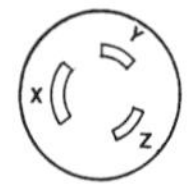
L 2 250V 2 Pole, 2 Wire		15A		**L 14** 125/250V 3 Pole, 4 Wire			
L 5 125 V 2 Pole, 3 Wire				**L 15** 3 Phase 250 V 3 Pole, 4 Wire			
L 6 250 V 2 Pole, 3 Wire				**L 16** 3 Phase 480V 3 Pole, 4 Wire			
L 7 227 V, AC 2 Pole, 3 Wire				**L 17** 3 Phase 600 V 3 Pole, 4 Wire			
L 8 480 V 2 Pole, 3 Wire				**L 18** 3 Phase 208Y/120V 4 Pole, 4 Wire			
L 9 600 V 2 Pole, 3 Wire				**L 19** 3 Phase 480Y/277V 4 Pole, 4 Wire			
L 10 125 /250V 3 Pole, 3 Wire				**L 20** 3 Phase 600Y/347V 4 Pole, 4 Wire		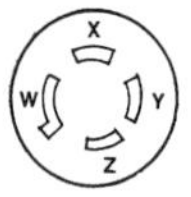	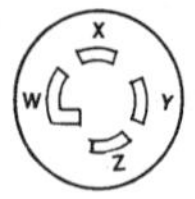
L 11 3 Phase 250 V 3 Pole, 3 Wire				**L 21** 3 Phase 208Y/120V 4 Pole, 5 Wire			
L 12 3 Phase 480 V 3 Pole, 3 Wire				**L 22** 3 Phase 480Y/277V 4 Pole, 5 Wire			
				L 23 3 Phase 600Y/347V 4 Pole, 5 Wire		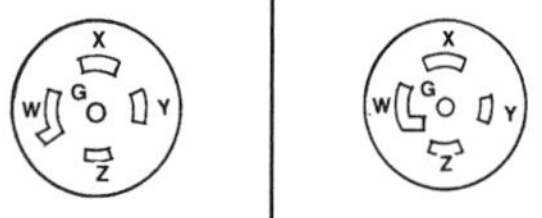	

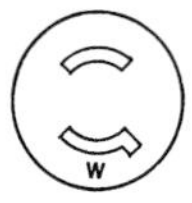

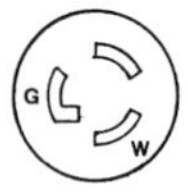

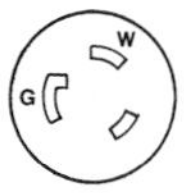

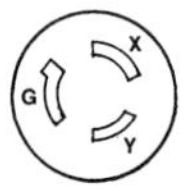

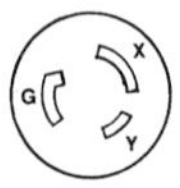

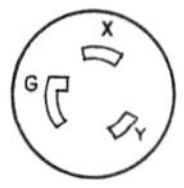

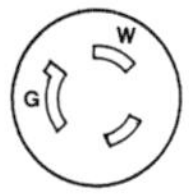

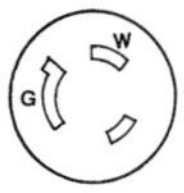

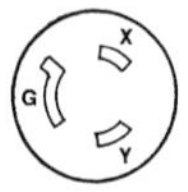

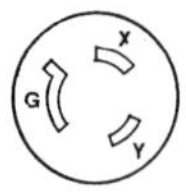

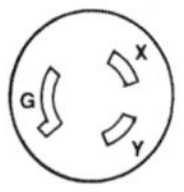

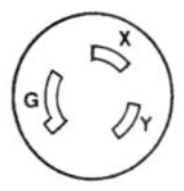

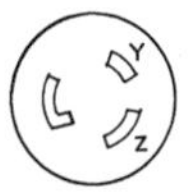

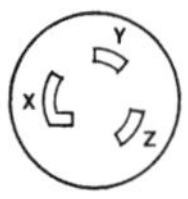

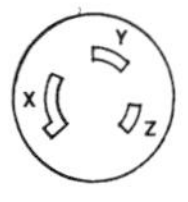

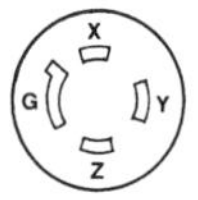

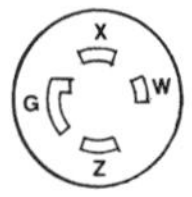

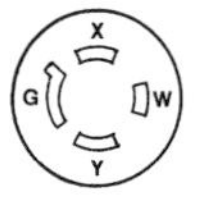

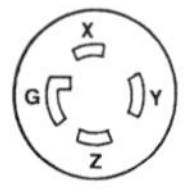

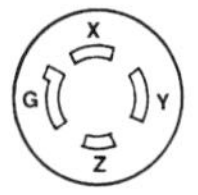

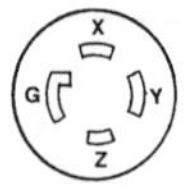

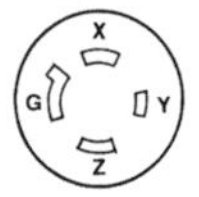

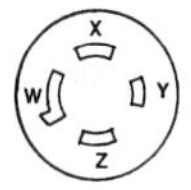

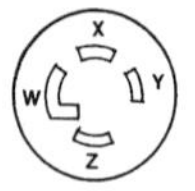

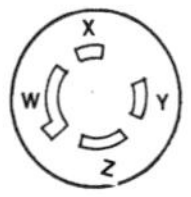

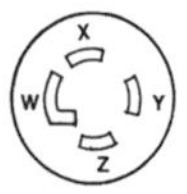

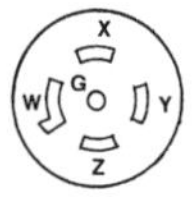

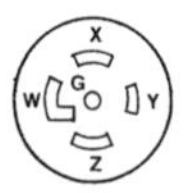

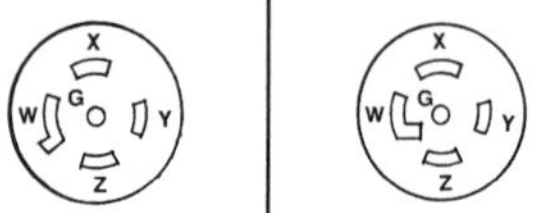

R265113-40 Interior Lighting Fixtures

When taking off interior lighting fixtures, it is advisable to set up your quantity work sheet to conform to the lighting schedule as it appears on the print. Include the alpha-numeric code plus the symbol on your work sheet.

Take off a particular section or floor of the building and count each type of fixture before going on to another type. It would also be advantageous to include on the same work sheet the pipe, wire, fittings and circuit number associated with each type of lighting fixture. This will help you identify the costs associated with any particular lighting system and in turn make material purchases more specific as to when and how much to order under the classification of lighting.

By taking off lighting first you can get a complete "walk through" of the job. This will become helpful when doing other phases of the project.

Materials for a recessed fixture include:

1. Fixture
2. Lamps
3. 6′ of jack chain
4. (2) S hooks
5. (2) Wire nuts

Labor for interior recessed fixtures include:

1. Unloading by hand
2. Hauling by hand to an area up to 200′ from loading dock
3. Uncrating
4. Layout
5. Installing fixture
6. Attaching jack chain & S hooks
7. Connecting circuit power
8. Reassembling fixture
9. Installing lamps
10. Testing

Material for surface mounted fixtures includes:

1. Fixture
2. Lamps
3. Either (4) lead type anchors, (4) toggle bolts, or (4) ceiling grid clips
4. (2) Wire nuts

Material for pendent mounted fixtures includes:

1. Fixture
2. Lamps
3. (2) Wire nuts
4. Rigid pendents as required by type of fixtures
5. Canopies as required by type of fixture

Labor-hours include the following for both surface and pendent fixtures:

1. Unloading by hand
2. Hauling by hand to an area up to 200′ from loading dock
3. Uncrating
4. Layout and marking
5. Drilling (4) holes for either lead anchors or toggle bolts using a hammer drill
6. Installing fixture
7. Leveling fixture
8. Connecting circuit power
9. Installing lamps
10. Testing

Labor for surface or pendent fixtures does not include:

1. Conduit
2. Boxes or covers
3. Connectors
4. Fixture whips
5. Special support
6. Switching
7. Wire

Economy of Scale: For large concentrations of lighting fixtures in the same area deduct the following percentages from labor:

25 to 50 fixtures	15%
51 to 75 fixtures	20%
76 to 100 fixtures	25%
101 and over	30%

Job Conditions: Productivity is based on new construction in an unobstructed first floor location, using rolling staging to 15′ high.

Material staging is assumed to be within 100′ of work being performed.

Add the following percentages to labor for elevated installations:

15′ to 20′ high	10%
21′ to 25′ high	20%
26′ to 30′ high	30%
31′ to 35′ high	40%
36′ to 40′ high	50%
41′ and over	60%

Communications R2713 Communications Backbone Cabling

R271323-40 Fiber Optics

Fiber optic systems use optical fiber such as plastic, glass, or fused silica, a transparent material, to transmit radiant power (i.e. light) for control, communication, and signaling applications. The types of fiber optic cables can be nonconductive, conductive, or composite. The composite cables contain fiber optics and current-carrying electrical conductors. The configuration for one of the fiber optic systems is as follows:

The transceiver module acts as transmitting and receiving equipment in a common house, which converts electrical energy to light energy or vice versa.

Pricing the fiber optic system is not an easy task. The performance of the whole system will affect the cost significantly. New specialized tools and techniques decrease the installing cost tremendously. In the fiber optic section of Means Electrical Cost Data, a benchmark for labor-hours and material costs is set up so that users can adjust their costs according to unique project conditions.

Units for Measure: Fiber optic cable is measured in hundred linear feet (C.L.F.) or industry units of measure - meter (m) or kilometer (km). The connectors are counted as units (ea.)

Material Units: Generally, the material costs include only the cable. All the accessories shall be priced separately.

Labor Units: The following procedures are generally included for the installation of fiber optic cables:

- Receiving
- Material handling
- Setting up pulling equipment
- Measuring and cutting cable
- Pulling cable

These additional items are listed and extended: Terminations

Takeoff Procedure: Cable should be taken off by type, size, number of fibers, and number of terminations required. List the lengths of each type of cable on the takeoff sheets. Total and add 10% for waste. Transfer the figures to a cost analysis sheet and extend.

R271513-75 High Performance Cable

There are several categories used to describe high performance cable. The following information includes a description of categories CAT 3, 5, 5e, 6, and 7, and details classifications of frequency and specific standards. The category standards have evolved under the sponsorship of organizations such as the Telecommunication Industry Association (TIA), the Electronic Industries Alliance (EIA), the American National Standards Institute (ANSI), the International Organization for Standardization (ISO), and the International Electrotechnical Commission (IEC), all of which have catered to the increasing complexities of modern network technology. For network cabling, users must comply with national or international standards. A breakdown of these categories is as follows:

Category 3: Designed to handle frequencies up to 16 MHz.

Category 5: (TIA/EIA 568A) Designed to handle frequencies up to 100 MHz.

Category 5e: Additional transmission performance to exceed Category 5.

Category 6 (draft): Development by TIA and other international groups to handle frequencies of 250 MHz.

Category 7 (draft): Under development to handle a frequency range from 1 to 600 MHz.

Exterior Improvements R3292 Turf & Grasses

R329219-50 Seeding

The type of grass is determined by light, shade and moisture content of soil plus intended use. Fertilizer should be disked 4″ before seeding. For steep slopes disk five tons of mulch and lay two tons of hay or straw on surface per acre after seeding. Surface mulch can be staked, lightly disked or tar emulsion sprayed. Material for mulch can be wood chips, peat moss, partially rotted hay or straw, wood fibers, and sprayed emulsions. Hemp seed blankets with fertilizer are also available. For spring seeding, watering is necessary. Late fall seeding may have to be reseeded in the spring. Hydraulic seeding, power mulching, and aerial seeding can be used on large areas.

Utilities R3311 Water Utility Distribution Piping

R331113-80 Piping Designations

There are several systems currently in use to describe pipe and fittings. The following paragraphs will help to identify and clarify classifications of piping systems used for water distribution.

Piping may be classified by schedule. Piping schedules include 5S, 10S, 10, 20, 30, Standard, 40, 60, Extra Strong, 80, 100, 120, 140, 160, and Double Extra Strong. These schedules are dependent upon the pipe wall thickness. The wall thickness of a particular schedule may vary with pipe size.

Ductile iron pipe for water distribution is classified by Pressure Classes such as Class 150, 200, 250, 300, and 350. These classes are actually the rated water working pressure of the pipe in pounds per square inch (psi). The pipe in these pressure classes is designed to withstand the rated water working pressure plus a surge allowance of 100 psi.

The American Water Works Association (AWWA) provides standards for various types of **plastic pipe**. C-900 is the specification for polyvinyl chloride (PVC) piping used for water distribution in sizes ranging from 4″ through 12″. C-901 is the specification for polyethylene (PE) pressure pipe, tubing and fittings used for water distribution in sizes ranging from 1/2″ through 3″. C-905 is the specification for PVC piping sizes 14″ and greater.

PVC pressure-rated pipe is identified using the standard dimensional ratio (SDR) method. This method is defined by the American Society for Testing and Materials (ASTM) Standard D 2241. This pipe is available in SDR numbers 64, 41, 32.5, 26, 21, 17, and 13.5. A pipe with an SDR of 64 will have the thinnest wall while a pipe with an SDR of 13.5 will have the thickest wall. When the pressure rating (PR) of a pipe is given in psi, it is based on a line supplying water at 73 degrees F.

The National Sanitation Foundation (NSF) seal of approval is applied to products that can be used with potable water. These products have been tested to ANSI/NSF Standard 14.

Valves and strainers are classified by American National Standards Institute (ANSI) Classes. These Classes are 125, 150, 200, 250, 300, 400, 600, 900, 1500 and 2500. Within each class there is an operating pressure range dependent upon temperature. Design parameters should be compared to the appropriate material dependent, pressure-temperature rating chart for accurate valve selection.

R337119-30 Concrete for Conduit Encasement

Table below lists C.Y. of concrete for 100 L.F. of trench. Conduits separation center to center should meet 7.5″ (N.E.C.).

Number of Conduits	1	2	3	4	6	8	9	Number of Conduits
Trench Dimension	11.5″ x 11.5″	11.5″ x 19″	11.5″ x 27″	19″ x 19″	19″ x 27″	19″ x 38″	27″ x 27″	Trench Dimension
Conduit Diameter 2.0″	3.29	5.39	7.64	8.83	12.51	17.66	17.72	Conduit Diameter 2.0″
2.5″	3.23	5.29	7.49	8.62	12.19	17.23	17.25	2.5″
3.0″	3.15	5.13	7.24	8.29	11.71	16.59	16.52	3.0″
3.5″	3.08	4.97	7.02	7.99	11.26	15.98	15.84	3.5″
4.0″	2.99	4.80	6.76	7.65	10.74	15.30	15.07	4.0″
5.0″	2.78	4.37	6.11	6.78	9.44	13.57	13.12	5.0″
6.0″	2.52	3.84	5.33	5.74	7.87	11.48	10.77	6.0″

R347216-10 Single Track R.R. Siding

The costs for a single track RR siding in the Unit Price section include the components shown in the table below.

Description of Component	Qty. per L.F. of Track	Unit
Ballast, 1-1/2″ crushed stone	.667	C.Y.
6″ x 8″ x 8′-6″ Treated timber ties, 22″ O.C.	.545	Ea.
Tie plates, 2 per tie	1.091	Ea.
Track rail	2.000	L.F.
Spikes, 6″, 4 per tie	2.182	Ea.
Splice bars w/ bolts, lock washers & nuts, @ 33′ O.C.	.061	Pair
Crew B-14 @ 57 L.F./Day	.018	Day

R347216-20 Single Track, Steel Ties, Concrete Bed

The costs for a R.R. siding with steel ties and a concrete bed in the Unit Price section include the components shown in the table below.

Description of Component	Qty. per L.F. of Track	Unit
Concrete bed, 9′ wide, 10″ thick	.278	C.Y.
Ties, W6x16 x 6′-6″ long, @ 30″ O.C.	.400	Ea.
Tie plates, 4 per tie	1.600	Ea.
Track rail	2.000	L.F.
Tie plate bolts, 1″, 8 per tie	3.200	Ea.
Splice bars w/bolts, lock washers & nuts, @ 33′ O.C.	.061	Pair
Crew B-14 @ 22 L.F./Day	.045	Day

Change Orders

Change Order Considerations

A change order is a written document usually prepared by the design professional and signed by the owner, the architect/engineer, and the contractor. A change order states the agreement of the parties to: an addition, deletion, or revision in the work; an adjustment in the contract sum, if any; or an adjustment in the contract time, if any. Change orders, or "extras", in the construction process occur after execution of the construction contract and impact architects/ engineers, contractors, and owners.

Change orders that are properly recognized and managed can ensure orderly, professional, and profitable progress for everyone involved in the project. There are many causes for change orders and change order requests. In all cases, change orders or change order requests should be addressed promptly and in a precise and prescribed manner. The following paragraphs include information regarding change order pricing and procedures.

The Causes of Change Orders

Reasons for issuing change orders include:

- Unforeseen field conditions that require a change in the work
- Correction of design discrepancies, errors, or omissions in the contract documents
- Owner-requested changes, either by design criteria, scope of work, or project objectives
- Completion date changes for reasons unrelated to the construction process
- Changes in building code interpretations, or other public authority requirements that require a change in the work
- Changes in availability of existing or new materials and products

Procedures

Properly written contract documents must include the correct change order procedures for all parties—owners, design professionals, and contractors—to follow in order to avoid costly delays and litigation.

Being "in the right" is not always a sufficient or acceptable defense. The contract provisions requiring notification and documentation must be adhered to within a defined or reasonable time frame.

The appropriate method of handling change orders is by a written proposal and acceptance by all parties involved. Prior to starting work on a project, all parties should identify their authorized agents who may sign and accept change orders, as well as any limits placed on their authority.

Time may be a critical factor when the need for a change arises. For such cases, the contractor might be directed to proceed on a "time and materials" basis, rather than wait for all paperwork to be processed—a delay that could impede progress. In this situation, the contractor must still follow the prescribed change order procedures including, but not limited to, notification and documentation.

Lack of documentation can be very costly, especially if legal judgments are to be made, and if certain field personnel are no longer available. For time and material change orders, the contractor should keep accurate daily records of all labor and material allocated to the change.

Owners or awarding authorities who do considerable and continual building construction (such as the federal government) realize the inevitability of change orders for numerous reasons, both predictable and unpredictable. As a result, the federal government, the American Institute of Architects (AIA), the Engineers Joint Contract Documents Committee (EJCDC), and other contractor, legal, and technical organizations have developed standards and procedures to be followed by all parties to achieve contract continuance and timely completion, while being financially fair to all concerned.

Pricing Change Orders

When pricing change orders, regardless of their cause, the most significant factor is when the change occurs. The need for a change may be perceived in the field or requested by the architect/engineer *before* any of the actual installation has begun, or may evolve or appear *during* construction when the item of work in question is partially installed. In the latter cases, the original sequence of construction is disrupted, along with all contiguous and supporting systems. Change orders cause the greatest impact when they occur *after* the installation has been completed and must be uncovered, or even replaced. Post-completion changes may be caused by necessary design changes, product failure, or changes in the owner's requirements that are not discovered until the building or the systems begin to function.

Specified procedures of notification and record keeping must be adhered to and enforced regardless of the stage of construction: *before, during,* or *after* installation. Some bidding documents anticipate change orders by requiring that unit prices including overhead and profit percentages—for additional as well as deductible changes—be listed. Generally these unit prices do not fully take into account the ripple effect, or impact on other trades, and should be used for general guidance only.

When pricing change orders, it is important to classify the time frame in which the change occurs. There are two basic time frames for change orders: *pre-installation change orders,* which occur before the start of construction, and *post-installation change orders,* which involve reworking after the original installation. Change orders that occur between these stages may be priced according to the extent of work completed using a combination of techniques developed for pricing *pre-* and *post-installation* changes.

Factors To Consider When Pricing Change Orders

As an estimator begins to prepare a change order, the following questions should be reviewed to determine their impact on the final price.

General

- *Is the change order work* pre-installation *or* post-installation?

  Change order work costs vary according to how much of the installation has been completed. Once workers have the project scoped in their minds, even though they have not started, it can be difficult to refocus. Consequently they may spend more than the normal amount of time understanding the change. Also, modifications to work in place, such as trimming or refitting, usually take more time than was initially estimated. The greater the amount of work in place, the more reluctant workers are to change it. Psychologically they may resent the change and as a result the rework takes longer than normal. Post-installation change order estimates must include demolition of existing work as required to accomplish the change. If the work is performed at a later time, additional obstacles, such as building finishes, may be present which must be protected. Regardless of whether the change occurs pre-installation or post-installation, attempt to isolate the identifiable factors and price them separately. For example, add shipping costs that may be required pre-installation or any demolition required post-installation. Then analyze the potential impact on productivity of psychological and/or learning curve factors and adjust the output rates accordingly. One approach is to break down the typical workday into segments and quantify the impact on each segment.

Change Order Installation Efficiency

The labor-hours expressed (for new construction) are based on average installation time, using an efficiency level. For change order situations, adjustments to this efficiency level should reflect the daily labor-hour allocation for that particular occurrence.

- *Will the change substantially delay the original completion date?*

  A significant change in the project may cause the original completion date to be extended. The extended schedule may subject the contractor to new wage rates dictated by relevant labor contracts. Project supervision and other project overhead must also be extended beyond the original completion date. The schedule extension may also put installation into a new weather season. For example, underground piping scheduled for October installation was delayed until January. As a result, frost penetrated the trench area, thereby changing the degree of difficulty of the task. Changes and delays may have a ripple effect throughout the project. This effect must be analyzed and negotiated with the owner.

- *What is the net effect of a deduct change order?*

  In most cases, change orders resulting in a deduction or credit reflect only bare costs. The contractor may retain the overhead and profit based on the original bid.

Materials

- *Will you have to pay more or less for the new material, required by the change order, than you paid for the original purchase?*

  The same material prices or discounts will usually apply to materials purchased for change orders as new construction. In some instances, however, the contractor may forfeit the advantages of competitive pricing for change orders. Consider the following example:

  A contractor purchased over $20,000 worth of fan coil units for an installation and obtained the maximum discount. Some time later it was determined the project required an additional matching unit. The contractor has to purchase this unit from the original supplier to ensure a match. The supplier at this time may not discount the unit because of the small quantity, and he is no longer in a competitive situation. The impact of quantity on purchase can add between 0% and 25% to material prices and/or subcontractor quotes.

- *If materials have been ordered or delivered to the job site, will they be subject to a cancellation charge or restocking fee?*

  Check with the supplier to determine if ordered materials are subject to a cancellation charge. Delivered materials not used as a result of a change order may be subject to a restocking fee if returned to the supplier. Common restocking charges run between 20% and 40%. Also, delivery charges to return the goods to the supplier must be added.

Labor

- *How efficient is the existing crew at the actual installation?*

  Is the same crew that performed the initial work going to do the change order? Possibly the change consists of the installation of a unit identical to one already installed; therefore, the change should take less time. Be sure to consider this potential productivity increase and modify the productivity rates accordingly.

- *If the crew size is increased, what impact will that have on supervision requirements?*

  Under most bargaining agreements or management practices, there is a point at which a working foreman is replaced by a nonworking foreman. This replacement increases project overhead by adding a nonproductive worker. If additional workers are added to accelerate the project or to perform changes while maintaining the schedule, be sure to add additional supervision time if warranted. Calculate the hours involved and the additional cost directly if possible.

- *What are the other impacts of increased crew size?*

  The larger the crew, the greater the potential for productivity to decrease. Some of the factors that cause this productivity loss are: overcrowding (producing restrictive conditions in the working space) and possibly a shortage of any special tools and equipment required. Such factors affect not only the crew working on the elements directly involved in the change order, but other crews whose movements may also be hampered. As the crew increases, check its basic composition for changes by the addition or deletion of apprentices or nonworking foreman, and quantify the potential effects of equipment shortages or other logistical factors.

- *As new crews, unfamiliar with the project, are brought onto the site, how long will it take them to become oriented to the project requirements?*

  The orientation time for a new crew to become 100% effective varies with the site and type of project. Orientation is easiest at a new construction site and most difficult at existing, very restrictive renovation sites. The type of work also affects orientation time. When all elements of the work are exposed, such as concrete or masonry work, orientation is decreased. When the work is concealed or less visible, such as existing electrical systems, orientation takes longer. Usually orientation can be accomplished in one day or less. Costs for added orientation should be itemized and added to the total estimated cost.

- *How much actual production can be gained by working overtime?*

  Short term overtime can be used effectively to accomplish more work in a day. However, as overtime is scheduled to run beyond several weeks, studies have shown marked decreases in output. The following chart shows the effect of long term overtime on worker efficiency. If the anticipated change requires extended overtime to keep the job on schedule, these factors can be used as a guide to predict the impact on time and cost. Add project overhead, particularly supervision, that may also be incurred.

Days per Week	Hours per Day	Production Efficiency					Payroll Cost Factors	
		1st Week	2nd Week	3rd Week	4th Week	Average 4 Weeks	@ 1-1/2 Times	@ 2 Times
5	8	100%	100%	100%	100%	100%	100%	100%
	9	100	100	95	90	96.25	105.6	111.1
	10	100	95	90	85	92.50	110.0	120.0
	11	95	90	75	65	81.25	113.6	127.3
	12	90	85	70	60	76.25	116.7	133.3
6	8	100	100	95	90	96.25	108.3	116.7
	9	100	95	90	85	92.50	113.0	125.9
	10	95	90	85	80	87.50	116.7	133.3
	11	95	85	70	65	78.75	119.7	139.4
	12	90	80	65	60	73.75	122.2	144.4
7	8	100	95	85	75	88.75	114.3	128.6
	9	95	90	80	70	83.75	118.3	136.5
	10	90	85	75	65	78.75	121.4	142.9
	11	85	80	65	60	72.50	124.0	148.1
	12	85	75	60	55	68.75	126.2	152.4

Effects of Overtime

Caution: Under many labor agreements, Sundays and holidays are paid at a higher premium than the normal overtime rate.

The use of long-term overtime is counterproductive on almost any construction job; that is, the longer the period of overtime, the lower the actual production rate. Numerous studies have been conducted, and while they have resulted in slightly different numbers, all reach the same conclusion. The figure above tabulates the effects of overtime work on efficiency.

As illustrated, there can be a difference between the *actual* payroll cost per hour and the *effective* cost per hour for overtime work. This is due to the reduced production efficiency with the increase in weekly hours beyond 40. This difference between actual and effective cost results from overtime work over a prolonged period. Short-term overtime work does not result in as great a reduction in efficiency and, in such cases, effective cost may not vary significantly from the actual payroll cost. As the total hours per week are increased on a regular basis, more time is lost due to fatigue, lowered morale, and an increased accident rate.

As an example, assume a project where workers are working 6 days a week, 10 hours per day. From the figure above (based on productivity studies), the average effective productive hours over a 4-week period are:

$$0.875 \times 60 = 52.5$$

Depending upon the locale and day of week, overtime hours may be paid at time and a half or double time. For time and a half, the overall (average) *actual* payroll cost (including regular and overtime hours) is determined as follows:

$$\frac{40 \text{ reg. hrs.} + (20 \text{ overtime hrs.} \times 1.5)}{60 \text{ hrs.}} = 1.167$$

Based on 60 hours, the payroll cost per hour will be 116.7% of the normal rate at 40 hours per week. However, because the effective production (efficiency) for 60 hours is reduced to the equivalent of 52.5 hours, the effective cost of overtime is calculated as follows:

For time and a half:

$$\frac{40 \text{ reg. hrs.} + (20 \text{ overtime hrs.} \times 1.5)}{52.5 \text{ hrs.}} = 1.33$$

The installed cost will be 133% of the normal rate (for labor).

Thus, when figuring overtime, the actual cost per unit of work will be higher than the apparent overtime payroll dollar increase, due to the reduced productivity of the longer work week. These efficiency calculations are true only for those cost factors determined by hours worked. Costs that are applied weekly or monthly, such as equipment rentals, will not be similarly affected.

Equipment

- *What equipment is required to complete the change order?*

  Change orders may require extending the rental period of equipment already on the job site, or the addition of special equipment brought in to accomplish the change work. In either case, the additional rental charges and operator labor charges must be added.

Summary

The preceding considerations and others you deem appropriate should be analyzed and applied to a change order estimate. The impact of each should be quantified and listed on the estimate to form an audit trail.

Change orders that are properly identified, documented, and managed help to ensure the orderly, professional, and profitable progress of the work. They also minimize potential claims or disputes at the end of the project.

Back by customer demand!

You asked and we listened. For customer convenience and estimating ease, we have made the 2020 Project Costs available for download at **RSMeans.com/2020books**. You will also find sample estimates, an RSMeans data overview video, and a book registration form to receive quarterly data updates throughout 2020.

Estimating Tips

- The cost figures available in the download were derived from hundreds of projects contained in the RSMeans database of completed construction projects. They include the contractor's overhead and profit. The figures have been adjusted to January of the current year.
- These projects were located throughout the U.S. and reflect a tremendous variation in square foot (S.F.) costs. This is due to differences, not only in labor and material costs, but also in individual owners' requirements. For instance, a bank in a large city would have different features than one in a rural area. This is true of all the different types of buildings analyzed. Therefore, caution should be exercised when using these Project Costs. For example, for courthouses, costs in the database are local courthouse costs and will not apply to the larger, more elaborate federal courthouses.
- None of the figures "go with" any others. All individual cost items were computed and tabulated separately. Thus, the sum of the median figures for plumbing, HVAC, and electrical will not normally total up to the total mechanical and electrical costs arrived at by separate analysis and tabulation of the projects.
- Each building was analyzed as to total and component costs and percentages. The figures were arranged in ascending order with the results tabulated as shown. The 1/4 column shows that 25% of the projects had lower costs and 75% had higher. The 3/4 column shows that 75% of the projects had lower costs and 25% had higher. The median column shows that 50% of the projects had lower costs and 50% had higher.
- Project Costs are useful in the conceptual stage when no details are available. As soon as details become available in the project design, the square foot approach should be discontinued and the project should be priced as to its particular components. When more precision is required, or for estimating the replacement cost of specific buildings, the current edition of *Square Foot Costs with RSMeans data* should be used.
- In using the figures in this section, it is recommended that the median column be used for preliminary figures if no additional information is available. The median figures, when multiplied by the total city construction cost index figures (see City Cost Indexes) and then multiplied by the project size modifier at the end of this section, should present a fairly accurate base figure, which would then have to be adjusted in view of the estimator's experience, local economic conditions, code requirements, and the owner's particular requirements. There is no need to factor in the percentage figures, as these should remain constant from city to city.
- The editors of this data would greatly appreciate receiving cost figures on one or more of your recent projects, which would then be included in the averages for next year. All cost figures received will be kept confidential, except that they will be averaged with other similar projects to arrive at square foot cost figures for next year.

See the website above for details and the discount available for submitting one or more of your projects.

	50 17 00 | Project Costs		UNIT	UNIT COSTS			% OF TOTAL			
				1/4	MEDIAN	3/4	1/4	MEDIAN	3/4	
01	0000	**Auto Sales with Repair**	S.F.							01
	0100	Architectural		106	119	128	58%	64%	67%	
	0200	Plumbing		8.90	9.30	12.40	4.84%	5.20%	6.80%	
	0300	Mechanical		11.90	15.95	17.60	6.40%	8.70%	10.15%	
	0400	Electrical		18.30	22.50	28.50	9.05%	11.70%	15.90%	
	0500	Total Project Costs		178	186	191				
02	0000	**Banking Institutions**	S.F.							02
	0100	Architectural		160	197	239	59%	65%	69%	
	0200	Plumbing		6.45	9	12.50	2.12%	3.39%	4.19%	
	0300	Mechanical		12.80	17.70	21	4.41%	5.10%	10.75%	
	0400	Electrical		31.50	38	58	10.45%	13.05%	15.90%	
	0500	Total Project Costs		266	299	370				
03	0000	**Court House**	S.F.							03
	0100	Architectural		84.50	166	166	54.50%	58.50%	58.50%	
	0200	Plumbing		3.19	3.19	3.19	2.07%	2.07%	2.07%	
	0300	Mechanical		19.95	19.95	19.95	12.95%	12.95%	12.95%	
	0400	Electrical		25.50	25.50	25.50	16.60%	16.60%	16.60%	
	0500	Total Project Costs		154	284	284				
04	0000	**Data Centers**	S.F.							04
	0100	Architectural		191	191	191	68%	68%	68%	
	0200	Plumbing		10.45	10.45	10.45	3.71%	3.71%	3.71%	
	0300	Mechanical		26.50	26.50	26.50	9.45%	9.45%	9.45%	
	0400	Electrical		25.50	25.50	25.50	9%	9%	9%	
	0500	Total Project Costs		281	281	281				
05	0000	**Detention Centers**	S.F.							05
	0100	Architectural		177	187	198	52%	53%	60.50%	
	0200	Plumbing		18.65	22.50	27.50	5.15%	7.10%	7.25%	
	0300	Mechanical		23.50	34	40.50	7.55%	9.50%	13.80%	
	0400	Electrical		39	46	60	10.90%	14.85%	17.95%	
	0500	Total Project Costs		299	315	370				
06	0000	**Fire Stations**	S.F.							06
	0100	Architectural		100	129	187	49%	54.50%	61.50%	
	0200	Plumbing		10.75	13.95	16.95	4.67%	5.60%	6.30%	
	0300	Mechanical		14.95	21	29.50	6.10%	8.45%	10.20%	
	0400	Electrical		23	29.50	35.50	10.75%	12.75%	14.95%	
	0500	Total Project Costs		210	238	320				
07	0000	**Gymnasium**	S.F.							07
	0100	Architectural		88.50	116	116	57%	64.50%	64.50%	
	0200	Plumbing		2.17	7.10	7.10	1.58%	3.48%	3.48%	
	0300	Mechanical		3.32	30	30	2.42%	14.65%	14.65%	
	0400	Electrical		10.90	21	21	7.95%	10.35%	10.35%	
	0500	Total Project Costs		137	204	204				
08	0000	**Hospitals**	S.F.							08
	0100	Architectural		107	176	191	43%	47.50%	48%	
	0200	Plumbing		7.85	15.05	32.50	6%	7.45%	7.65%	
	0300	Mechanical		52	58.50	76.50	14.20%	17.95%	23.50%	
	0400	Electrical		23.50	47.50	61.50	10.95%	13.75%	16.85%	
	0500	Total Project Costs		250	375	405				
09	0000	**Industrial Buildings**	S.F.							09
	0100	Architectural		45	71.50	232	46%	54%	56.50%	
	0200	Plumbing		1.74	6.55	13.25	2%	3.06%	6.30%	
	0300	Mechanical		4.81	9.15	43.50	4.77%	5.55%	14.80%	
	0400	Electrical		7.35	8.40	70	7.85%	13.55%	16.20%	
	0500	Total Project Costs		80	104	430				
10	0000	**Medical Clinics & Offices**	S.F.							10
	0100	Architectural		89.50	121	161	48.50%	55.50%	62.50%	
	0200	Plumbing		8.90	13.15	21	4.47%	6.60%	8.65%	
	0300	Mechanical		13.95	22.50	44	7.80%	10.95%	16.10%	
	0400	Electrical		20	27	37.50	9.70%	11.65%	14.05%	
	0500	Total Project Costs		166	217	292				

	50 17 00 | Project Costs		UNIT	UNIT COSTS 1/4	UNIT COSTS MEDIAN	UNIT COSTS 3/4	% OF TOTAL 1/4	% OF TOTAL MEDIAN	% OF TOTAL 3/4	
11	0000	**Mixed Use**	S.F.							11
	0100	Architectural		91.50	129	211	48.50%	57%	62.50%	
	0200	Plumbing		6.15	9.35	11.80	3.23%	3.44%	4.18%	
	0300	Mechanical		16.40	24.50	47.50	6.10%	13.75%	18.85%	
	0400	Electrical		16	26.50	41.50	8.30%	11.40%	14%	
	0500	Total Project Costs	↓	189	216	345				
12	0000	**Multi-Family Housing**	S.F.							12
	0100	Architectural		76	114	170	56.50%	62%	66.50%	
	0200	Plumbing		6.55	11.55	15.05	5.30%	6.85%	8%	
	0300	Mechanical		7	9.45	27	4.92%	6.90%	10.40%	
	0400	Electrical		9.90	15.45	19.85	6.20%	7.90%	10.25%	
	0500	Total Project Costs	↓	111	208	271				
13	0000	**Nursing Home & Assisted Living**	S.F.							13
	0100	Architectural		71.50	94	118	51.50%	55.50%	63.50%	
	0200	Plumbing		7.75	11.60	12.80	6.25%	7.40%	8.80%	
	0300	Mechanical		6.35	9.35	18.30	4.04%	6.70%	9.55%	
	0400	Electrical		10.50	16.55	23	7%	10.75%	13.10%	
	0500	Total Project Costs	↓	122	160	192				
14	0000	**Office Buildings**	S.F.							14
	0100	Architectural		94.50	129	177	56%	61%	69%	
	0200	Plumbing		5.10	8	15.05	2.65%	3.56%	5.85%	
	0300	Mechanical		11	17	26	5.60%	8.20%	11.10%	
	0400	Electrical		12.75	21.50	34	7.75%	10%	12.70%	
	0500	Total Project Costs	↓	161	200	282				
15	0000	**Parking Garage**	S.F.							15
	0100	Architectural		31.50	38.50	40.50	70%	79%	88%	
	0200	Plumbing		1.04	1.09	2.04	2.05%	2.70%	2.83%	
	0300	Mechanical		.81	1.24	4.72	2.11%	3.62%	3.81%	
	0400	Electrical		2.78	3.05	6.35	5.30%	6.35%	7.95%	
	0500	Total Project Costs	↓	38.50	47	51				
16	0000	**Parking Garage/Mixed Use**	S.F.							16
	0100	Architectural		103	112	114	61%	62%	65.50%	
	0200	Plumbing		3.30	4.32	6.60	2.47%	2.72%	3.66%	
	0300	Mechanical		14.10	15.85	23	7.80%	13.10%	13.60%	
	0400	Electrical		14.75	21	22	8.20%	12.65%	18.15%	
	0500	Total Project Costs	↓	168	175	181				
17	0000	**Police Stations**	S.F.							17
	0100	Architectural		116	130	164	49%	56.50%	61%	
	0200	Plumbing		15.30	18.40	18.50	5.05%	5.55%	9.05%	
	0300	Mechanical		34.50	48.50	50.50	13%	14.55%	16.55%	
	0400	Electrical		26	28.50	30.50	9.15%	12.10%	14%	
	0500	Total Project Costs	↓	217	267	305				
18	0000	**Police/Fire**	S.F.							18
	0100	Architectural		113	113	345	55.50%	66%	68%	
	0200	Plumbing		9.05	9.35	34.50	5.45%	5.50%	5.55%	
	0300	Mechanical		13.85	22	79	8.35%	12.70%	12.80%	
	0400	Electrical		15.75	20	90.50	9.50%	11.75%	14.55%	
	0500	Total Project Costs	↓	166	171	625				
19	0000	**Public Assembly Buildings**	S.F.							19
	0100	Architectural		115	159	238	57.50%	61.50%	66%	
	0200	Plumbing		6.10	8.90	13.40	2.60%	3.36%	4.79%	
	0300	Mechanical		12.85	23	35.50	6.55%	8.95%	12.45%	
	0400	Electrical		19	26	41	8.60%	10.75%	13%	
	0500	Total Project Costs	↓	185	253	375				
20	0000	**Recreational**	S.F.							20
	0100	Architectural		106	168	236	53.50%	60%	66%	
	0200	Plumbing		7.80	14.75	21.50	3.08%	5%	6.85%	
	0300	Mechanical		13.15	20	31.50	5.15%	6.95%	11.70%	
	0400	Electrical		15.20	26.50	39.50	7.35%	8.95%	10.75%	
	0500	Total Project Costs	↓	168	281	445				

50 17 00 | Project Costs

		50 17 00 | Project Costs	UNIT	UNIT COSTS 1/4	UNIT COSTS MEDIAN	UNIT COSTS 3/4	% OF TOTAL 1/4	% OF TOTAL MEDIAN	% OF TOTAL 3/4	
21	0000	**Restaurants**	S.F.							21
	0100	Architectural		126	193	248	59%	60%	63.50%	
	0200	Plumbing		13.80	31.50	40	7.35%	7.75%	8.95%	
	0300	Mechanical		14.85	17.55	37	6.50%	8.15%	11.15%	
	0400	Electrical		14.80	24	48.50	7.10%	10.30%	11.60%	
	0500	Total Project Costs	↓	208	305	415				
22	0000	**Retail**	S.F.							22
	0100	Architectural		56	86	127	60%	62%	64.50%	
	0200	Plumbing		7	9.25	12.10	5.05%	6.70%	9%	
	0300	Mechanical		6.55	9.25	17.05	5.70%	6.20%	10.20%	
	0400	Electrical		10.45	18.95	31.50	8.05%	11.25%	12.45%	
	0500	Total Project Costs	↓	85.50	114	186				
23	0000	**Schools**	S.F.							23
	0100	Architectural		97	125	164	52.50%	56%	61%	
	0200	Plumbing		7.75	10.60	16.05	3.85%	4.82%	7.25%	
	0300	Mechanical		18.90	26.50	38.50	9.50%	12.35%	15.10%	
	0400	Electrical		17.85	25	32.50	9.45%	11.45%	13.30%	
	0500	Total Project Costs	↓	169	227	300				
24	0000	**University, College & Private School Classroom & Admin Buildings**	S.F.							24
	0100	Architectural		124	153	192	50.50%	55%	59.50%	
	0200	Plumbing		7.05	10.95	15.45	2.74%	4.30%	6.35%	
	0300	Mechanical		26.50	38.50	46	10.10%	12.15%	14.70%	
	0400	Electrical		19.95	28	34	7.65%	9.50%	11.55%	
	0500	Total Project Costs	↓	205	284	375				
25	0000	**University, College & Private School Dormitories**	S.F.							25
	0100	Architectural		81	142	151	54.50%	65%	68.50%	
	0200	Plumbing		10.70	15.10	22.50	6.45%	7.30%	9.15%	
	0300	Mechanical		4.79	20.50	32.50	4.13%	9%	12.05%	
	0400	Electrical		5.70	19.75	30	4.75%	7.35%	12.30%	
	0500	Total Project Costs	↓	119	227	268				
26	0000	**University, College & Private School Science, Eng. & Lab Buildings**	S.F.							26
	0100	Architectural		143	164	193	50.50%	56.50%	58%	
	0200	Plumbing		9.60	15.15	26.50	3.29%	3.95%	8.40%	
	0300	Mechanical		43.50	68.50	70	12.75%	19.40%	23.50%	
	0400	Electrical		29	33.50	38.50	9%	11.55%	13.15%	
	0500	Total Project Costs	↓	290	315	365				
27	0000	**University, College & Private School Student Union Buildings**	S.F.							27
	0100	Architectural		110	289	289	54.50%	54.50%	59.50%	
	0200	Plumbing		16.65	16.65	25	3.13%	4.27%	11.45%	
	0300	Mechanical		31.50	51	51	9.60%	9.60%	14.55%	
	0400	Electrical		27.50	48	48	9.05%	12.80%	13.15%	
	0500	Total Project Costs	↓	217	530	530				
28	0000	**Warehouses**	S.F.							28
	0100	Architectural		47	73.50	174	61.50%	67.50%	72%	
	0200	Plumbing		2.45	5.25	10.10	2.82%	3.72%	5%	
	0300	Mechanical		2.91	16.55	26	4.56%	8.20%	10.70%	
	0400	Electrical		5.30	19.85	33	7.50%	10.10%	18.30%	
	0500	Total Project Costs	↓	70.50	125	243				

Square Foot Project Size Modifier

One factor that affects the S.F. cost of a particular building is the size. In general, for buildings built to the same specifications in the same locality, the larger building will have the lower S.F. cost. This is due mainly to the decreasing contribution of the exterior walls plus the economy of scale usually achievable in larger buildings. The Area Conversion Scale shown below will give a factor to convert costs for the typical size building to an adjusted cost for the particular project.

The Square Foot Base Size lists the median costs, most typical project size in our accumulated data, and the range in size of the projects.

The Size Factor for your project is determined by dividing your project area in S.F. by the typical project size for the particular Building Type. With this factor, enter the Area Conversion Scale at the appropriate Size Factor and determine the appropriate Cost Multiplier for your building size.

Example: Determine the cost per S.F. for a 152,600 S.F. Multi-family housing.

$$\frac{\text{Proposed building area} = 152{,}600 \text{ S.F.}}{\text{Typical size from below} = 49{,}900 \text{ S.F.}} = 2.00$$

Enter Area Conversion Scale at 2.0, intersect curve, read horizontally the appropriate cost multiplier of .94. Size adjusted cost becomes .94 x $208.00 = $195.52 based on national average costs.

Note: For Size Factors less than .50, the Cost Multiplier is 1.1
For Size Factors greater than 3.5, the Cost Multiplier is .90

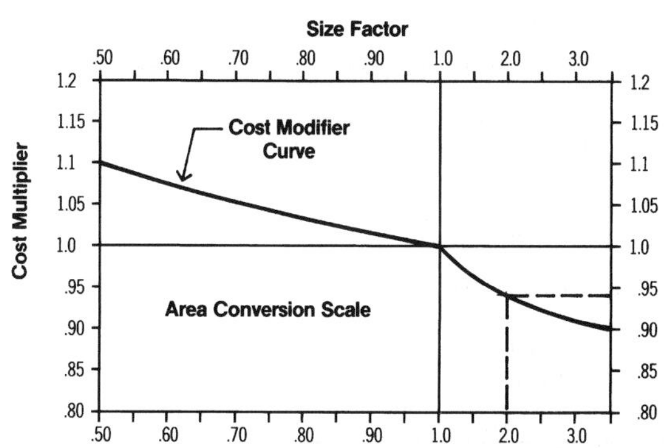

System	Median Cost (Total Project Costs)	Typical Size Gross S.F. (Median of Projects)	Typical Range (Low – High) (Projects)
Auto Sales with Repair	$186.00	24,900	4,700 – 29,300
Banking Institutions	299.00	9,300	3,300 – 38,100
Court House	284.00	47,600	24,700 – 70,500
Data Centers	281.00	14,400	14,400 – 14,400
Detention Centers	315.00	37,800	12,300 – 183,300
Fire Stations	238.00	13,100	6,300 – 49,600
Gymnasium	204.00	52,400	22,800 – 82,000
Hospitals	375.00	87,100	22,400 – 410,300
Industrial Buildings	104.00	22,100	5,100 – 200,600
Medical Clinics & Offices	217.00	20,900	2,300 – 327,000
Mixed Use	216.00	28,500	7,200 – 188,900
Multi-Family Housing	208.00	49,900	2,500 – 1,161,500
Nursing Home & Assisted Living	160.00	38,200	1,500 – 242,600
Office Buildings	200.00	20,500	1,100 – 930,000
Parking Garage	47.00	151,800	99,900 – 287,000
Parking Garage/Mixed Use	175.00	254,200	5,300 – 318,000
Police Stations	267.00	28,500	15,400 – 88,600
Police/Fire	171.00	44,300	8,600 – 50,300
Public Assembly Buildings	253.00	21,000	2,200 – 235,300
Recreational	281.00	28,800	1,000 – 223,800
Restaurants	305.00	6,000	4,000 – 42,000
Retail	114.00	28,700	5,200 – 84,300
Schools	227.00	70,600	1,300 – 410,800
University, College & Private School Classroom & Admin Buildings	284.00	48,300	9,400 – 196,200
University, College & Private School Dormitories	227.00	28,900	1,500 – 126,900
University, College & Private School Science, Eng. & Lab Buildings	315.00	39,800	5,300 – 117,600
University, College & Private School Student Union Buildings	530.00	48,700	42,100 – 50,000
Warehouses	125.00	10,400	600 – 303,800

Abbreviations

A	Area Square Feet; Ampere
AAFES	Army and Air Force Exchange Service
ABS	Acrylonitrile Butadiene Stryrene; Asbestos Bonded Steel
A.C., AC	Alternating Current; Air-Conditioning; Asbestos Cement; Plywood Grade A & C
ACI	American Concrete Institute
ACR	Air Conditioning Refrigeration
ADA	Americans with Disabilities Act
AD	Plywood, Grade A & D
Addit.	Additional
Adh.	Adhesive
Adj.	Adjustable
af	Audio-frequency
AFFF	Aqueous Film Forming Foam
AFUE	Annual Fuel Utilization Efficiency
AGA	American Gas Association
Agg.	Aggregate
A.H., Ah	Ampere Hours
A hr.	Ampere-hour
A.H.U., AHU	Air Handling Unit
A.I.A.	American Institute of Architects
AIC	Ampere Interrupting Capacity
Allow.	Allowance
alt., alt	Alternate
Alum.	Aluminum
a.m.	Ante Meridiem
Amp.	Ampere
Anod.	Anodized
ANSI	American National Standards Institute
APA	American Plywood Association
Approx.	Approximate
Apt.	Apartment
Asb.	Asbestos
A.S.B.C.	American Standard Building Code
Asbe.	Asbestos Worker
ASCE	American Society of Civil Engineers
A.S.H.R.A.E.	American Society of Heating, Refrig. & AC Engineers
ASME	American Society of Mechanical Engineers
ASTM	American Society for Testing and Materials
Attchmt.	Attachment
Avg., Ave.	Average
AWG	American Wire Gauge
AWWA	American Water Works Assoc.
Bbl.	Barrel
B&B, BB	Grade B and Better; Balled & Burlapped
B&S	Bell and Spigot
B.&W.	Black and White
b.c.c.	Body-centered Cubic
B.C.Y.	Bank Cubic Yards
BE	Bevel End
B.F.	Board Feet
Bg. cem.	Bag of Cement
BHP	Boiler Horsepower; Brake Horsepower
B.I.	Black Iron
bidir.	bidirectional
Bit., Bitum.	Bituminous
Bit., Conc.	Bituminous Concrete
Bk.	Backed
Bkrs.	Breakers
Bldg., bldg	Building
Blk.	Block
Bm.	Beam
Boil.	Boilermaker
bpm	Blows per Minute
BR	Bedroom
Brg., brng.	Bearing
Brhe.	Bricklayer Helper
Bric.	Bricklayer
Brk., brk	Brick
brkt	Bracket
Brs.	Brass
Brz.	Bronze
Bsn.	Basin
Btr.	Better
BTU	British Thermal Unit
BTUH	BTU per Hour
Bu.	Bushels
BUR	Built-up Roofing
BX	Interlocked Armored Cable
°C	Degree Centigrade
c	Conductivity, Copper Sweat
C	Hundred; Centigrade
C/C	Center to Center, Cedar on Cedar
C-C	Center to Center
Cab	Cabinet
Cair.	Air Tool Laborer
Cal.	Caliper
Calc	Calculated
Cap.	Capacity
Carp.	Carpenter
C.B.	Circuit Breaker
C.C.A.	Chromate Copper Arsenate
C.C.F.	Hundred Cubic Feet
cd	Candela
cd/sf	Candela per Square Foot
CD	Grade of Plywood Face & Back
CDX	Plywood, Grade C & D, exterior glue
Cefi.	Cement Finisher
Cem.	Cement
CF	Hundred Feet
C.F.	Cubic Feet
CFM	Cubic Feet per Minute
CFRP	Carbon Fiber Reinforced Plastic
c.g.	Center of Gravity
CHW	Chilled Water; Commercial Hot Water
C.I., CI	Cast Iron
C.I.P., CIP	Cast in Place
Circ.	Circuit
C.L.	Carload Lot
CL	Chain Link
Clab.	Common Laborer
Clam	Common Maintenance Laborer
C.L.F.	Hundred Linear Feet
CLF	Current Limiting Fuse
CLP	Cross Linked Polyethylene
cm	Centimeter
CMP	Corr. Metal Pipe
CMU	Concrete Masonry Unit
CN	Change Notice
Col.	Column
CO_2	Carbon Dioxide
Comb.	Combination
comm.	Commercial, Communication
Compr.	Compressor
Conc.	Concrete
Cont., cont	Continuous; Continued, Container
Corkbd.	Cork Board
Corr.	Corrugated
Cos	Cosine
Cot	Cotangent
Cov.	Cover
C/P	Cedar on Paneling
CPA	Control Point Adjustment
Cplg.	Coupling
CPM	Critical Path Method
CPVC	Chlorinated Polyvinyl Chloride
C.Pr.	Hundred Pair
CRC	Cold Rolled Channel
Creos.	Creosote
Crpt.	Carpet & Linoleum Layer
CRT	Cathode-ray Tube
CS	Carbon Steel, Constant Shear Bar Joist
Csc	Cosecant
C.S.F.	Hundred Square Feet
CSI	Construction Specifications Institute
CT	Current Transformer
CTS	Copper Tube Size
Cu	Copper, Cubic
Cu. Ft.	Cubic Foot
cw	Continuous Wave
C.W.	Cool White; Cold Water
Cwt.	100 Pounds
C.W.X.	Cool White Deluxe
C.Y.	Cubic Yard (27 cubic feet)
C.Y./Hr.	Cubic Yard per Hour
Cyl.	Cylinder
d	Penny (nail size)
D	Deep; Depth; Discharge
Dis., Disch.	Discharge
Db	Decibel
Dbl.	Double
DC	Direct Current
DDC	Direct Digital Control
Demob.	Demobilization
d.f.t.	Dry Film Thickness
d.f.u.	Drainage Fixture Units
D.H.	Double Hung
DHW	Domestic Hot Water
DI	Ductile Iron
Diag.	Diagonal
Diam., Dia	Diameter
Distrib.	Distribution
Div.	Division
Dk.	Deck
D.L.	Dead Load; Diesel
DLH	Deep Long Span Bar Joist
dlx	Deluxe
Do.	Ditto
DOP	Dioctyl Phthalate Penetration Test (Air Filters)
Dp., dp	Depth
D.P.S.T.	Double Pole, Single Throw
Dr.	Drive
DR	Dimension Ratio
Drink.	Drinking
D.S.	Double Strength
D.S.A.	Double Strength A Grade
D.S.B.	Double Strength B Grade
Dty.	Duty
DWV	Drain Waste Vent
DX	Deluxe White, Direct Expansion
dyn	Dyne
e	Eccentricity
E	Equipment Only; East; Emissivity
Ea.	Each
EB	Encased Burial
Econ.	Economy
E.C.Y	Embankment Cubic Yards
EDP	Electronic Data Processing
EIFS	Exterior Insulation Finish System
E.D.R.	Equiv. Direct Radiation
Eq.	Equation
EL	Elevation
Elec.	Electrician; Electrical
Elev.	Elevator; Elevating
EMT	Electrical Metallic Conduit; Thin Wall Conduit
Eng.	Engine, Engineered
EPDM	Ethylene Propylene Diene Monomer
EPS	Expanded Polystyrene
Eqhv.	Equip. Oper., Heavy
Eqlt.	Equip. Oper., Light
Eqmd.	Equip. Oper., Medium
Eqmm.	Equip. Oper., Master Mechanic
Eqol.	Equip. Oper., Oilers
Equip.	Equipment
ERW	Electric Resistance Welded

Abbreviations

E.S.	Energy Saver
Est.	Estimated
esu	Electrostatic Units
E.W.	Each Way
EWT	Entering Water Temperature
Excav.	Excavation
excl	Excluding
Exp., exp	Expansion, Exposure
Ext., ext	Exterior; Extension
Extru.	Extrusion
f.	Fiber Stress
F	Fahrenheit; Female; Fill
Fab., fab	Fabricated; Fabric
FBGS	Fiberglass
F.C.	Footcandles
f.c.c.	Face-centered Cubic
f'c.	Compressive Stress in Concrete; Extreme Compressive Stress
F.E.	Front End
FEP	Fluorinated Ethylene Propylene (Teflon)
F.G.	Flat Grain
F.H.A.	Federal Housing Administration
Fig.	Figure
Fin.	Finished
FIPS	Female Iron Pipe Size
Fixt.	Fixture
FJP	Finger jointed and primed
Fl. Oz.	Fluid Ounces
Flr.	Floor
Flrs.	Floors
FM	Frequency Modulation; Factory Mutual
Fmg.	Framing
FM/UL	Factory Mutual/Underwriters Labs
Fdn.	Foundation
FNPT	Female National Pipe Thread
Fori.	Foreman, Inside
Foro.	Foreman, Outside
Fount.	Fountain
fpm	Feet per Minute
FPT	Female Pipe Thread
Fr	Frame
F.R.	Fire Rating
FRK	Foil Reinforced Kraft
FSK	Foil/Scrim/Kraft
FRP	Fiberglass Reinforced Plastic
FS	Forged Steel
FSC	Cast Body; Cast Switch Box
Ft., ft	Foot; Feet
Ftng.	Fitting
Ftg.	Footing
Ft lb.	Foot Pound
Furn.	Furniture
FVNR	Full Voltage Non-Reversing
FVR	Full Voltage Reversing
FXM	Female by Male
Fy.	Minimum Yield Stress of Steel
g	Gram
G	Gauss
Ga.	Gauge
Gal., gal.	Gallon
Galv., galv	Galvanized
GC/MS	Gas Chromatograph/Mass Spectrometer
Gen.	General
GFI	Ground Fault Interrupter
GFRC	Glass Fiber Reinforced Concrete
Glaz.	Glazier
GPD	Gallons per Day
gpf	Gallon per Flush
GPH	Gallons per Hour
gpm, GPM	Gallons per Minute
GR	Grade
Gran.	Granular
Grnd.	Ground
GVW	Gross Vehicle Weight
GWB	Gypsum Wall Board
H	High Henry
HC	High Capacity
H.D., HD	Heavy Duty; High Density
H.D.O.	High Density Overlaid
HDPE	High Density Polyethylene Plastic
Hdr.	Header
Hdwe.	Hardware
H.I.D., HID	High Intensity Discharge
Help.	Helper Average
HEPA	High Efficiency Particulate Air Filter
Hg	Mercury
HIC	High Interrupting Capacity
HM	Hollow Metal
HMWPE	High Molecular Weight Polyethylene
HO	High Output
Horiz.	Horizontal
H.P., HP	Horsepower; High Pressure
H.P.F.	High Power Factor
Hr.	Hour
Hrs./Day	Hours per Day
HSC	High Short Circuit
Ht.	Height
Htg.	Heating
Htrs.	Heaters
HVAC	Heating, Ventilation & Air-Conditioning
Hvy.	Heavy
HW	Hot Water
Hyd.; Hydr.	Hydraulic
Hz	Hertz (cycles)
I.	Moment of Inertia
IBC	International Building Code
I.C.	Interrupting Capacity
ID	Inside Diameter
I.D.	Inside Dimension; Identification
I.F.	Inside Frosted
I.M.C.	Intermediate Metal Conduit
In.	Inch
Incan.	Incandescent
Incl.	Included; Including
Int.	Interior
Inst.	Installation
Insul., insul	Insulation/Insulated
I.P.	Iron Pipe
I.P.S., IPS	Iron Pipe Size
IPT	Iron Pipe Threaded
I.W.	Indirect Waste
J	Joule
J.I.C.	Joint Industrial Council
K	Thousand; Thousand Pounds; Heavy Wall Copper Tubing, Kelvin
K.A.H.	Thousand Amp. Hours
kcmil	Thousand Circular Mils
KD	Knock Down
K.D.A.T.	Kiln Dried After Treatment
kg	Kilogram
kG	Kilogauss
kgf	Kilogram Force
kHz	Kilohertz
Kip	1000 Pounds
KJ	Kilojoule
K.L.	Effective Length Factor
K.L.F.	Kips per Linear Foot
Km	Kilometer
KO	Knock Out
K.S.F.	Kips per Square Foot
K.S.I.	Kips per Square Inch
kV	Kilovolt
kVA	Kilovolt Ampere
kVAR	Kilovar (Reactance)
KW	Kilowatt
KWh	Kilowatt-hour
L	Labor Only; Length; Long; Medium Wall Copper Tubing
Lab.	Labor
lat	Latitude
Lath.	Lather
Lav.	Lavatory
lb.; #	Pound
L.B., LB	Load Bearing; L Conduit Body
L. & E.	Labor & Equipment
lb./hr.	Pounds per Hour
lb./L.F.	Pounds per Linear Foot
lbf/sq.in.	Pound-force per Square Inch
L.C.L.	Less than Carload Lot
L.C.Y.	Loose Cubic Yard
Ld.	Load
LE	Lead Equivalent
LED	Light Emitting Diode
L.F.	Linear Foot
L.F. Hdr	Linear Feet of Header
L.F. Nose	Linear Foot of Stair Nosing
L.F. Rsr	Linear Foot of Stair Riser
Lg.	Long; Length; Large
L & H	Light and Heat
LH	Long Span Bar Joist
L.H.	Labor Hours
L.L., LL	Live Load
L.L.D.	Lamp Lumen Depreciation
lm	Lumen
lm/sf	Lumen per Square Foot
lm/W	Lumen per Watt
LOA	Length Over All
log	Logarithm
L-O-L	Lateralolet
long.	Longitude
L.P., LP	Liquefied Petroleum; Low Pressure
L.P.F.	Low Power Factor
LR	Long Radius
L.S.	Lump Sum
Lt.	Light
Lt. Ga.	Light Gauge
L.T.L.	Less than Truckload Lot
Lt. Wt.	Lightweight
L.V.	Low Voltage
M	Thousand; Material; Male; Light Wall Copper Tubing
M^2CA	Meters Squared Contact Area
m/hr.; M.H.	Man-hour
mA	Milliampere
Mach.	Machine
Mag. Str.	Magnetic Starter
Maint.	Maintenance
Marb.	Marble Setter
Mat; Mat'l.	Material
Max.	Maximum
MBF	Thousand Board Feet
MBH	Thousand BTU's per hr.
MC	Metal Clad Cable
MCC	Motor Control Center
M.C.F.	Thousand Cubic Feet
MCFM	Thousand Cubic Feet per Minute
M.C.M.	Thousand Circular Mils
MCP	Motor Circuit Protector
MD	Medium Duty
MDF	Medium-density fibreboard
M.D.O.	Medium Density Overlaid
Med.	Medium
MF	Thousand Feet
M.F.B.M.	Thousand Feet Board Measure
Mfg.	Manufacturing
Mfrs.	Manufacturers
mg	Milligram
MGD	Million Gallons per Day
MGPH	Million Gallons per Hour
MH, M.H.	Manhole; Metal Halide; Man-Hour
MHz	Megahertz
Mi.	Mile
MI	Malleable Iron; Mineral Insulated
MIPS	Male Iron Pipe Size
mj	Mechanical Joint
m	Meter
mm	Millimeter
Mill.	Millwright
Min., min.	Minimum, Minute

Abbreviations

Misc.	Miscellaneous
ml	Milliliter, Mainline
M.L.F.	Thousand Linear Feet
Mo.	Month
Mobil.	Mobilization
Mog.	Mogul Base
MPH	Miles per Hour
MPT	Male Pipe Thread
MRGWB	Moisture Resistant Gypsum Wallboard
MRT	Mile Round Trip
ms	Millisecond
M.S.F.	Thousand Square Feet
Mstz.	Mosaic & Terrazzo Worker
M.S.Y.	Thousand Square Yards
Mtd., mtd., mtd	Mounted
Mthe.	Mosaic & Terrazzo Helper
Mtng.	Mounting
Mult.	Multi; Multiply
MUTCD	Manual on Uniform Traffic Control Devices
M.V.A.	Million Volt Amperes
M.V.A.R.	Million Volt Amperes Reactance
MV	Megavolt
MW	Megawatt
MXM	Male by Male
MYD	Thousand Yards
N	Natural; North
nA	Nanoampere
NA	Not Available; Not Applicable
N.B.C.	National Building Code
NC	Normally Closed
NEMA	National Electrical Manufacturers Assoc.
NEHB	Bolted Circuit Breaker to 600V.
NFPA	National Fire Protection Association
NLB	Non-Load-Bearing
NM	Non-Metallic Cable
nm	Nanometer
No.	Number
NO	Normally Open
N.O.C.	Not Otherwise Classified
Nose.	Nosing
NPT	National Pipe Thread
NQOD	Combination Plug-on/Bolt on Circuit Breaker to 240V.
N.R.C., NRC	Noise Reduction Coefficient/ Nuclear Regulator Commission
N.R.S.	Non Rising Stem
ns	Nanosecond
NTP	Notice to Proceed
nW	Nanowatt
OB	Opposing Blade
OC	On Center
OD	Outside Diameter
O.D.	Outside Dimension
ODS	Overhead Distribution System
O.G.	Ogee
O.H.	Overhead
O&P	Overhead and Profit
Oper.	Operator
Opng.	Opening
Orna.	Ornamental
OSB	Oriented Strand Board
OS&Y	Outside Screw and Yoke
OSHA	Occupational Safety and Health Act
Ovhd.	Overhead
OWG	Oil, Water or Gas
Oz.	Ounce
P.	Pole; Applied Load; Projection
p.	Page
Pape.	Paperhanger
P.A.P.R.	Powered Air Purifying Respirator
PAR	Parabolic Reflector
P.B., PB	Push Button
Pc., Pcs.	Piece, Pieces
P.C.	Portland Cement; Power Connector
P.C.F.	Pounds per Cubic Foot
PCM	Phase Contrast Microscopy
PDCA	Painting and Decorating Contractors of America
P.E., PE	Professional Engineer; Porcelain Enamel; Polyethylene; Plain End
P.E.C.I.	Porcelain Enamel on Cast Iron
Perf.	Perforated
PEX	Cross Linked Polyethylene
Ph.	Phase
P.I.	Pressure Injected
Pile.	Pile Driver
Pkg.	Package
Pl.	Plate
Plah.	Plasterer Helper
Plas.	Plasterer
plf	Pounds Per Linear Foot
Pluh.	Plumber Helper
Plum.	Plumber
Ply.	Plywood
p.m.	Post Meridiem
Pntd.	Painted
Pord.	Painter, Ordinary
pp	Pages
PP, PPL	Polypropylene
P.P.M.	Parts per Million
Pr.	Pair
P.E.S.B.	Pre-engineered Steel Building
Prefab.	Prefabricated
Prefin.	Prefinished
Prop.	Propelled
PSF, psf	Pounds per Square Foot
PSI, psi	Pounds per Square Inch
PSIG	Pounds per Square Inch Gauge
PSP	Plastic Sewer Pipe
Pspr.	Painter, Spray
Psst.	Painter, Structural Steel
P.T.	Potential Transformer
P. & T.	Pressure & Temperature
Ptd.	Painted
Ptns.	Partitions
Pu	Ultimate Load
PVC	Polyvinyl Chloride
Pvmt.	Pavement
PRV	Pressure Relief Valve
Pwr.	Power
Q	Quantity Heat Flow
Qt.	Quart
Quan., Qty.	Quantity
Q.C.	Quick Coupling
r	Radius of Gyration
R	Resistance
R.C.P.	Reinforced Concrete Pipe
Rect.	Rectangle
recpt.	Receptacle
Reg.	Regular
Reinf.	Reinforced
Req'd.	Required
Res.	Resistant
Resi.	Residential
RF	Radio Frequency
RFID	Radio-frequency Identification
Rgh.	Rough
RGS	Rigid Galvanized Steel
RHW	Rubber, Heat & Water Resistant; Residential Hot Water
rms	Root Mean Square
Rnd.	Round
Rodm.	Rodman
Rofc.	Roofer, Composition
Rofp.	Roofer, Precast
Rohe.	Roofer Helpers (Composition)
Rots.	Roofer, Tile & Slate
R.O.W.	Right of Way
RPM	Revolutions per Minute
R.S.	Rapid Start
Rsr	Riser
RT	Round Trip
S.	Suction; Single Entrance; South
SBS	Styrene Butadiere Styrene
SC	Screw Cover
SCFM	Standard Cubic Feet per Minute
Scaf.	Scaffold
Sch., Sched.	Schedule
S.C.R.	Modular Brick
S.D.	Sound Deadening
SDR	Standard Dimension Ratio
S.E.	Surfaced Edge
Sel.	Select
SER, SEU	Service Entrance Cable
S.F.	Square Foot
S.F.C.A.	Square Foot Contact Area
S.F. Flr.	Square Foot of Floor
S.F.G.	Square Foot of Ground
S.F. Hor.	Square Foot Horizontal
SFR	Square Feet of Radiation
S.F. Shlf.	Square Foot of Shelf
S4S	Surface 4 Sides
Shee.	Sheet Metal Worker
Sin.	Sine
Skwk.	Skilled Worker
SL	Saran Lined
S.L.	Slimline
Sldr.	Solder
SLH	Super Long Span Bar Joist
S.N.	Solid Neutral
SO	Stranded with oil resistant inside insulation
S-O-L	Socketolet
sp	Standpipe
S.P.	Static Pressure; Single Pole; Self-Propelled
Spri.	Sprinkler Installer
spwg	Static Pressure Water Gauge
S.P.D.T.	Single Pole, Double Throw
SPF	Spruce Pine Fir; Sprayed Polyurethane Foam
S.P.S.T.	Single Pole, Single Throw
SPT	Standard Pipe Thread
Sq.	Square; 100 Square Feet
Sq. Hd.	Square Head
Sq. In.	Square Inch
S.S.	Single Strength; Stainless Steel
S.S.B.	Single Strength B Grade
sst, ss	Stainless Steel
Sswk.	Structural Steel Worker
Sswl.	Structural Steel Welder
St.; Stl.	Steel
STC	Sound Transmission Coefficient
Std.	Standard
Stg.	Staging
STK	Select Tight Knot
STP	Standard Temperature & Pressure
Stpi.	Steamfitter, Pipefitter
Str.	Strength; Starter; Straight
Strd.	Stranded
Struct.	Structural
Sty.	Story
Subj.	Subject
Subs.	Subcontractors
Surf.	Surface
Sw.	Switch
Swbd.	Switchboard
S.Y.	Square Yard
Syn.	Synthetic
S.Y.P.	Southern Yellow Pine
Sys.	System
t.	Thickness
T	Temperature; Ton
Tan	Tangent
T.C.	Terra Cotta
T & C	Threaded and Coupled
T.D.	Temperature Difference
TDD	Telecommunications Device for the Deaf
T.E.M.	Transmission Electron Microscopy
temp	Temperature, Tempered, Temporary
TFFN	Nylon Jacketed Wire

TFE	Tetrafluoroethylene (Teflon)
T. & G.	Tongue & Groove; Tar & Gravel
Th., Thk.	Thick
Thn.	Thin
Thrded	Threaded
Tilf.	Tile Layer, Floor
Tilh.	Tile Layer, Helper
THHN	Nylon Jacketed Wire
THW.	Insulated Strand Wire
THWN	Nylon Jacketed Wire
T.L., TL	Truckload
T.M.	Track Mounted
Tot.	Total
T-O-L	Threadolet
tmpd	Tempered
TPO	Thermoplastic Polyolefin
T.S.	Trigger Start
Tr.	Trade
Transf.	Transformer
Trhv.	Truck Driver, Heavy
Trlr	Trailer
Trlt.	Truck Driver, Light
TTY	Teletypewriter
TV	Television
T.W.	Thermoplastic Water Resistant Wire
UCI	Uniform Construction Index
UF	Underground Feeder
UGND	Underground Feeder
UHF	Ultra High Frequency
U.I.	United Inch
U.L., UL	Underwriters Laboratory
Uld.	Unloading
Unfin.	Unfinished
UPS	Uninterruptible Power Supply
URD	Underground Residential Distribution
US	United States
USGBC	U.S. Green Building Council
USP	United States Primed
UTMCD	Uniform Traffic Manual For Control Devices
UTP	Unshielded Twisted Pair
V	Volt
VA	Volt Amperes
VAT	Vinyl Asbestos Tile
V.C.T.	Vinyl Composition Tile
VAV	Variable Air Volume
VC	Veneer Core
VDC	Volts Direct Current
Vent.	Ventilation
Vert.	Vertical
V.F.	Vinyl Faced
V.G.	Vertical Grain
VHF	Very High Frequency
VHO	Very High Output
Vib.	Vibrating
VLF	Vertical Linear Foot
VOC	Volatile Organic Compound
Vol.	Volume
VRP	Vinyl Reinforced Polyester
W	Wire; Watt; Wide; West
w/	With
W.C., WC	Water Column; Water Closet
W.F.	Wide Flange
W.G.	Water Gauge
Wldg.	Welding
W. Mile	Wire Mile
W-O-L	Weldolet
W.R.	Water Resistant
Wrck.	Wrecker
WSFU	Water Supply Fixture Unit
W.S.P.	Water, Steam, Petroleum
WT., Wt.	Weight
WWF	Welded Wire Fabric
XFER	Transfer
XFMR	Transformer
XHD	Extra Heavy Duty
XHHW	Cross-Linked Polyethylene Wire
XLPE	Insulation
XLP	Cross-linked Polyethylene
Xport	Transport
Y	Wye
yd	Yard
yr	Year
Δ	Delta
%	Percent
~	Approximately
Ø	Phase; diameter
@	At
#	Pound; Number
<	Less Than
>	Greater Than
Z	Zone

A

B

C

E

F

Index

H

I

J

K

L

M

N

O

P

Q

S

T

U

V

W

X

Y

Z

Notes

Notes

Notes

Notes

Notes

Notes

Notes

Division Notes

	CREW	DAILY OUTPUT	LABOR-HOURS	UNIT	BARE COSTS MAT.	LABOR	EQUIP.	TOTAL	TOTAL INCL O&P

Division Notes

	CREW	DAILY OUTPUT	LABOR-HOURS	UNIT	BARE COSTS MAT.	LABOR	EQUIP.	TOTAL	TOTAL INCL O&P

Division Notes

	CREW	DAILY OUTPUT	LABOR-HOURS	UNIT	BARE COSTS MAT.	LABOR	EQUIP.	TOTAL	TOTAL INCL O&P

Division Notes

I	CREW	DAILY OUTPUT	LABOR-HOURS	UNIT	BARE COSTS				TOTAL INCL O&P
					MAT.	LABOR	EQUIP.	TOTAL	

Division Notes

		CREW	DAILY OUTPUT	LABOR-HOURS	UNIT	BARE COSTS MAT.	LABOR	EQUIP.	TOTAL	TOTAL INCL O&P

Division Notes

	CREW	DAILY OUTPUT	LABOR-HOURS	UNIT	BARE COSTS MAT.	BARE COSTS LABOR	BARE COSTS EQUIP.	BARE COSTS TOTAL	TOTAL INCL O&P

Division Notes

	CREW	DAILY OUTPUT	LABOR-HOURS	UNIT	BARE COSTS MAT.	LABOR	EQUIP.	TOTAL	TOTAL INCL O&P

Cost Data Selection Guide

The following table provides definitive information on the content of each cost data publication. The number of lines of data provided in each unit price or assemblies division, as well as the number of crews, is listed for each data set. The presence of other elements such as reference tables, square foot models, equipment rental costs, historical cost indexes, and city cost indexes, is also indicated. You can use the table to help select the RSMeans data set that has the quantity and type of information you most need in your work.

Unit Cost Divisions	Building Construction	Mechanical	Electrical	Commercial Renovation	Square Foot	Site Work Landsc.	Green Building	Interior	Concrete Masonry	Open Shop	Heavy Construction	Light Commercial	Facilities Construction	Plumbing	Residential
1	609	444	465	564	0	533	198	365	495	608	550	310	1078	450	217
2	754	278	87	710	0	970	181	397	219	753	737	479	1197	285	274
3	1745	341	232	1265	0	1537	1043	355	2274	1745	1930	538	2028	317	445
4	960	22	0	920	0	724	180	613	1158	928	614	532	1175	0	446
5	1890	158	155	1094	0	853	1788	1107	729	1890	1026	980	1907	204	746
6	2462	18	18	2121	0	110	589	1544	281	2458	123	2151	2135	22	2671
7	1593	215	128	1633	0	580	761	532	523	1590	26	1326	1693	227	1046
8	2140	80	3	2733	0	255	1138	1813	105	2142	0	2328	2966	0	1552
9	2125	86	45	1943	0	313	464	2216	424	2062	15	1779	2379	54	1544
10	1088	17	10	684	0	232	32	898	136	1088	34	588	1179	237	224
11	1096	199	166	540	0	135	56	924	29	1063	0	230	1116	162	108
12	539	0	2	297	0	219	147	1546	14	506	0	272	1565	23	216
13	740	149	157	252	0	365	124	250	77	716	266	109	756	115	103
14	273	36	0	223	0	0	0	257	0	273	0	12	293	16	6
21	127	0	41	37	0	0	0	293	0	127	0	121	665	685	259
22	1165	7543	160	1226	0	2010	1061	849	20	1154	2109	875	7505	9400	719
23	1170	6906	546	940	0	157	865	775	38	1153	98	887	5143	1919	486
25	0	0	14	14	0	0	0	0	0	0	0	0	0	0	0
26	1513	491	10465	1293	0	860	646	1159	55	1439	649	1361	10246	399	636
27	95	0	448	102	0	0	0	71	0	95	39	67	389	0	56
28	143	79	223	124	0	0	28	97	0	127	0	70	209	57	41
31	1511	733	610	807	0	3263	286	7	1216	1456	3280	607	1568	660	616
32	896	49	8	937	0	4523	408	417	361	867	1941	486	1800	140	533
33	1255	1088	565	260	0	3078	33	0	241	532	3213	135	1726	2101	161
34	107	0	47	4	0	190	0	0	31	62	221	0	136	0	0
35	18	0	0	0	0	327	0	0	0	18	442	0	84	0	0
41	63	0	0	34	0	8	0	22	0	62	31	0	69	14	0
44	75	79	0	0	0	0	0	0	0	0	0	0	75	75	0
46	23	16	0	0	0	274	261	0	0	23	264	0	33	33	0
48	8	0	36	2	0	0	21	0	0	8	15	8	21	0	8
Totals	26183	19027	14631	20759	0	21516	10310	16507	8426	24945	17623	16251	51136	17595	13113

Assem Div	Building Construction	Mechanical	Elec-trical	Commercial Renovation	Square Foot	Site Work Landscape	Assemblies	Green Building	Interior	Concrete Masonry	Heavy Construction	Light Commercial	Facilities Construction	Plumbing	Asm Div	Residential
A		15	0	188	164	577	598	0	0	536	571	154	24	0	1	378
B		0	0	848	2554	0	5661	56	329	1976	368	2094	174	0	2	211
C		0	0	647	954	0	1334	0	1641	146	0	844	251	0	3	591
D		1057	941	712	1858	72	2538	330	824	0	0	1345	1104	1088	4	851
E		0	0	85	261	0	301	0	5	0	0	258	5	0	5	391
F		0	0	0	114	0	143	0	0	0	0	114	0	0	6	357
G		527	447	318	312	3378	792	0	0	535	1349	205	293	677	7	307
															8	760
															9	80
															10	0
															11	0
															12	0
Totals		1599	1388	2798	6217	4027	11367	386	2799	3193	2288	5014	1851	1765		3926

Reference Section	Building Construction Costs	Mechanical	Electrical	Commercial Renovation	Square Foot	Site Work Landscape	Assem.	Green Building	Interior	Concrete Masonry	Open Shop	Heavy Construction	Light Commercial	Facilities Construction	Plumbing	Resi.
Reference Tables	yes	yes	yes	yes	no	yes	yes	yes	yes	yes	yes	yes	yes	yes	yes	yes
Models					111			25					50			28
Crews	582	582	582	561		582		582	582	582	560	582	560	561	582	560
Equipment Rental Costs	yes	yes	yes	yes		yes		yes	yes	yes	yes	yes	yes	yes	yes	yes
Historical Cost Indexes	yes	yes	yes	yes	yes	yes	yes	yes	yes	yes	yes	yes	yes	yes	yes	no
City Cost Indexes	yes	yes	yes	yes	yes	yes	yes	yes	yes	yes	yes	yes	yes	yes	yes	yes

2020 Seminar Schedule

📞 877.620.6245

Note: call for exact dates, locations, and details as some cities are subject to change.

Location	Dates	Location	Dates
Seattle, WA	January and August	San Francisco, CA	June
Dallas/Ft. Worth, TX	January	Bethesda, MD	June
Austin, TX	February	Dallas, TX	September
Jacksonville, FL	February	Raleigh, NC	October
Anchorage, AK	March and September	Baltimore, MD	November
Las Vegas, NV	March	Orlando, FL	November
Washington, D.C.	April and September	San Diego, CA	December
Charleston, SC	April	San Antonio, TX	December
Toronto	May		
Denver, CO	May		

Gordian also offers a suite of online RSMeans data self-paced offerings.
Check our website at RSMeans.com/products/training.aspx for more information.

Facilities Construction Estimating

In this two-day course, professionals working in facilities management can get help with their daily challenges to establish budgets for all phases of a project.

Some of what you'll learn:

- Determining the full scope of a project
- Understanding of Means data and what is included in prices
- Identifying appropriate factors to be included in your estimate
- Creative solutions to estimating issues
- Organizing estimates for presentation and discussion
- Special estimating techniques for repair/remodel and maintenance projects
- Appropriate use of contingency, city cost indexes, and reference notes
- Techniques to get to the correct estimate quickly

Who should attend: facility managers, engineers, contractors, facility tradespeople, planners, and project managers.

Mechanical & Electrical Estimating

This two-day course teaches attendees how to prepare more accurate and complete mechanical/electrical estimates, avoid the pitfalls of omission and double-counting, and understand the composition and rationale within the RSMeans mechanical/electrical database.

Some of what you'll learn:

- The unique way mechanical and electrical systems are interrelated
- M&E estimates—conceptual, planning, budgeting, and bidding stages
- Order of magnitude, square foot, assemblies, and unit price estimating
- Comparative cost analysis of equipment and design alternatives

Who should attend: architects, engineers, facilities managers, mechanical and electrical contractors, and others who need a highly reliable method for developing, understanding, and evaluating mechanical and electrical contracts.

Construction Cost Estimating: Concepts and Practice

This one or two day introductory course to improve estimating skills and effectiveness starts with the details of interpreting bid documents and ends with the summary of the estimate and bid submission.

Some of what you'll learn:

- Using the plans and specifications to create estimates
- The takeoff process—deriving all tasks with correct quantities
- Developing pricing using various sources; how subcontractor pricing fits in
- Summarizing the estimate to arrive at the final number
- Formulas for area and cubic measure, adding waste and adjusting productivity to specific projects
- Evaluating subcontractors' proposals and prices
- Adding insurance and bonds
- Understanding how labor costs are calculated
- Submitting bids and proposals

Who should attend: project managers, architects, engineers, owners' representatives, contractors, and anyone who's responsible for budgeting or estimating construction projects.

Assessing Scope of Work for Facilities Construction Estimating

This two-day practical training program addresses the vital importance of understanding the scope of projects in order to produce accurate cost estimates for facilities repair and remodeling.

Some of what you'll learn:

- Discussions of site visits, plans/specs, record drawings of facilities, and site-specific lists
- Review of CSI divisions, including means, methods, materials, and the challenges of scoping each topic
- Exercises in scope identification and scope writing for accurate estimating of projects
- Hands-on exercises that require scope, take-off, and pricing

Who should attend: corporate and government estimators, planners, facility managers, and others who need to produce accurate project estimates.

Maintenance & Repair Estimating for Facilities

This two-day course teaches attendees how to plan, budget, and estimate the cost of ongoing and preventive maintenance and repair for existing buildings and grounds.

Some of what you'll learn:

- The most financially favorable maintenance, repair, and replacement scheduling and estimating
- Auditing and value engineering facilities
- Preventive planning and facilities upgrading
- Determining both in-house and contract-out service costs
- Annual, asset-protecting M&R plan

Who should attend: facility managers, maintenance supervisors, buildings and grounds superintendents, plant managers, planners, estimators, and others involved in facilities planning and budgeting.

Practical Project Management for Construction Professionals

In this two-day course, acquire the essential knowledge and develop the skills to effectively and efficiently execute the day-to-day responsibilities of the construction project manager.

Some of what you'll learn:

- General conditions of the construction contract
- Contract modifications: change orders and construction change directives
- Negotiations with subcontractors and vendors
- Effective writing: notification and communications
- Dispute resolution: claims and liens

Who should attend: architects, engineers, owners' representatives, and project managers.

Life Cycle Cost Estimating for Facilities Asset Managers

Life Cycle Cost Estimating will take the attendee through choosing the correct RSMeans database to use and then correctly applying RSMeans data to their specific life cycle application. Conceptual estimating through RSMeans' new building models, conceptual estimating of major existing building projects through RSMeans' renovation models, pricing specific renovation elements, estimating repair, replacement and preventive maintenance costs today and forward up to 30 years will be covered.

Some of what you'll learn:

- Cost implications of managing assets
- Planning projects and initial & life cycle costs
- How to use RSMeans data online

Who should attend: facilities owners and managers and anyone involved in the financial side of the decision making process in the planning, design, procurement, and operation of facilities real assets.

Please bring a laptop with ability to access the internet.

Building Systems and the Construction Process

This one-day course was written to assist novices and those outside the industry in obtaining a solid understanding of the construction process - from both a building systems and construction administration approach.

Some of what you'll learn:

- Various systems used and how components come together to create a building
- Start with foundation and end with the physical systems of the structure such as HVAC and Electrical
- Focus on the process from start of design through project closeout

This training session requires you to bring a laptop computer to class.

Who should attend: building professionals or novices to help make the crossover to the construction industry; suited for anyone responsible for providing high level oversight on construction projects.

Training for our Online Estimating Solution

Construction estimating is vital to the decision-making process at each state of every project. Our online solution works the way you do. It's systematic, flexible and intuitive. In this one-day class you will see how you can estimate any phase of any project faster and better.

Some of what you'll learn:

- Customizing our online estimating solution
- Making the most of RSMeans "Circle Reference" numbers
- How to integrate your cost data
- Generating reports, exporting estimates to MS Excel, sharing, collaborating and more

Also offered as a self-paced or on-site training program!

Training for our CD Estimating Solution

This one-day course helps users become more familiar with the functionality of the CD. Each menu, icon, screen, and function found in the program is explained in depth. Time is devoted to hands-on estimating exercises.

Some of what you'll learn:

- Searching the database using all navigation methods
- Exporting RSMeans data to your preferred spreadsheet format
- Viewing crews, assembly components, and much more
- Automatically regionalizing the database

This training session requires you to bring a laptop computer to class.

When you register for this course you will receive an outline for your laptop requirements.

Also offered as a self-paced or on-site training program!

Site Work Estimating with RSMeans data

This one-day program focuses directly on site work costs. Accurately scoping, quantifying, and pricing site preparation, underground utility work, and improvements to exterior site elements are often the most difficult estimating tasks on any project. Some of what you'll learn:

- Evaluation of site work and understanding site scope including: site clearing, grading, excavation, disposal and trucking of materials, backfill and compaction, underground utilities, paving, sidewalks, and seeding & planting.
- Unit price site work estimates—Correct use of RSMeans site work cost data to develop a cost estimate.
- Using and modifying assemblies—Save valuable time when estimating site work activities using custom assemblies.

Who should attend: Engineers, contractors, estimators, project managers, owner's representatives, and others who are concerned with the proper preparation and/or evaluation of site work estimates.

Please bring a laptop with ability to access the internet.

Facilities Estimating Using the CD

This two-day class combines hands-on skill-building with best estimating practices and real-life problems. You will learn key concepts, tips, pointers, and guidelines to save time and avoid cost oversights and errors.

Some of what you'll learn:

- Estimating process concepts
- Customizing and adapting RSMeans cost data
- Establishing scope of work to account for all known variables
- Budget estimating: when, why, and how
- Site visits: what to look for and what you can't afford to overlook
- How to estimate repair and remodeling variables

This training session requires you to bring a laptop computer to class.

Who should attend: facility managers, architects, engineers, contractors, facility tradespeople, planners, project managers, and anyone involved with JOC, SABRE, or IDIQ.

Registration Information

Register early to save up to $100!!!

Register 45+ days before the date of a class and save $50 off each class. This savings cannot be combined with any other promotion or discounting of the regular price of classes!

How to register

By Phone
Register by phone at 877.620.6245

Online
Register online at
RSMeans.com/products/seminars.aspx

Note: Purchase Orders or Credits Cards are required to register.

Two-day seminar registration fee - $1,200*.

One-Day Construction Cost Estimating or Building Systems and the Construction Process - $765*.

Government pricing

All federal government employees save off the regular seminar price. Other promotional discounts cannot be combined with the government discount. Call 781.422.5115 for government pricing.

CANCELLATION POLICY:

If you are unable to attend a seminar, substitutions may be made at any time before the session starts by notifying the seminar registrar at 781.422.5115 or your sales representative.
If you cancel twenty-one (21) days or more prior to the seminar, there will be no penalty and your registration fees will be refunded. These cancellations must be received by the seminar registrar or your sales representative and will be confirmed to be eligible for cancellation.
If you cancel fewer than twenty-one (21) days prior to the seminar, you will forfeit the registration fee.
In the unfortunate event of an RSMeans cancellation, RSMeans will work with you to reschedule your attendance in the same seminar at a later date or will fully refund your registration fee. RSMeans cannot be responsible for any non-refundable travel expenses incurred by you or another as a result of your registration, attendance at, or cancellation of an RSMeans seminar.
Any on-demand training modules are not eligible for cancellation, substitution, transfer, return or refund.

AACE approved courses

Many seminars described and offered here have been approved for 14 hours (1.4 recertification credits) of credit by the AACE International Certification Board toward meeting the continuing education requirements for recertification as a Certified Cost Engineer/Certified Cost Consultant.

AIA Continuing Education

We are registered with the AIA Continuing Education System (AIA/CES) and are committed to developing quality learning activities in accordance with the CES criteria. Many seminars meet the AIA/CES criteria for Quality Level 2. AIA members may receive 14 learning units (LUs) for each two-day RSMeans course.

Daily course schedule

The first day of each seminar session begins at 8:30 a.m. and ends at 4:30 p.m. The second day begins at 8:00 a.m. and ends at 4:00 p.m. Participants are urged to bring a hand-held calculator since many actual problems will be worked out in each session.

Continental breakfast

Your registration includes the cost of a continental breakfast and a morning and afternoon refreshment break. These informal segments allow you to discuss topics of mutual interest with other seminar attendees. (You are free to make your own lunch and dinner arrangements.)

Hotel/transportation arrangements

We arrange to hold a block of rooms at most host hotels. To take advantage of special group rates when making your reservation, be sure to mention that you are attending the RSMeans Institute data seminar. You are, of course, free to stay at the lodging place of your choice. (Hotel reservations and transportation arrangements should be made directly by seminar attendees.)

Important

Class sizes are limited, so please register as soon as possible.

***Note: Pricing subject to change.**